INDEX and DIRECTORY of INDUSTRY STANDARDS

11th Edition
1994

Volume III

INTERNATIONAL AND NON-U.S. NATIONAL STANDARDS

Subject Index
(A-L)

Compiled by the
Index Development Staff
Information Handling Services

GLOBAL PROFESSIONAL PUBLICATIONS
Global Engineering Documents

DENVER   WASHINGTON   ST. LOUIS   SANTA ANA
MIAMI   PARIS   HONG KONG

INDEX AND DIRECTORY OF INDUSTRY STANDARDS

Copyright © 1994 by Information Handling Services Inc.

All rights reserved. Except as permitted under the United States copyright Act of 1976, no part of this publication may be reproduced or distributed in any form or by any means, or stored in a data base or retrieval system, without the prior written permission of the publisher.

Printed and Bound in the United Sates of America.

ISBN 1-57053-000-9

Library of Congress Catalog Card Number: 81-20150

Library of Congress Cataloging in Publication Data Main Entry under title:

Index and directory of industry standards.

    Includes index.
    1. Standardization—Directories.
I. Information Handling Services.
T59.2.U615   1994   016.602'18   81-20150
ISBN 1-57053-000-9

Published by

GLOBAL PROFESSIONAL PUBLICATIONS
Global Engineering Documents
15 Inverness Way East
Englewood, Colorado 80112

CONTENTS

VOLUME I

Acknowledgments	v
Introduction	vii
Organization Acronyms	ix
U.S. STANDARDS Subject Index	3

VOLUME II

Introduction	v
ANSI Acronyms	ii
U.S. STANDARDS Contents to Organization/Numeric Listing	ix
U.S. STANDARDS Organization/Numeric Listing	3
ANSI Number Concordance	947
Organization Directory	979

VOLUME III

Introduction	v
Organization Acronyms	vii
INTERNATIONAL AND NON-U.S. NATIONAL STANDARDS Subject Index (A-L)	3

VOLUME IV

Introduction	v
Organization Acronyms	vii
INTERNATIONAL AND NON-U.S. NATIONAL STANDARDS Subject Index (M-Z)	1573

VOLUME V

Introduction	v
INTERNATIONAL AND NON-U.S. NATIONAL STANDARDS Contents to Organization/Numeric Listings	vii
INTERNATIONAL AND NON-U.S. NATIONAL STANDARDS Organization/Numeric Listings	3
Organization Directory	1463

INTRODUCTION

The *Index and Directory of Industry Standards* provides access to standards in four major sections: Subject Index, Numeric Index, ANSI Number Concordance, and Organization Directory. The first two volumes contain information on U.S. Standards and Volumes III, IV, and V cover International and Non-U.S. National Standards.

This *Index* is the first product of its kind to provide access to the primary international standards as well as to the majority of U.S., Canadian, Japanese, German (English translations), British and British Defence standards and the joint European standards. Subject access to these standards has been very limited; until now there has been no comprehensive subject approach. The National Institute for Standards and Technology (formerly the National Bureau of Standards) publishes a computer-produced title keyword I*ndex of U.S. Voluntary Engineering Standards*. Only words from document titles are used as access points and no international documents are included.

Some of the standards-producing bodies have prepared subject indexes of their own documents. These vary in level of subject indexing from the detailed annual volume prepared by ASTM to much less ambitious publications.

DEFINITION

For purposes of this compilation, the term "standard" is used for documents that have been developed and approved in accordance with established standardization procedures and applies collectively to regulatory standards, test methods, and voluntary industry standards. Also included are specifications, which contain a definitive technical procedure for the identification, measurement, or evaluation of one or more qualities, characteristics, or properties of material, product, system or service along with the procedures for determining whether each of the requirements is satisfied. Handbooks, codes and manuals, which through usage, have become de facto standards, are included where appropriate.

There are over 400 organizations outside the government which prepare standards in the U.S., whereas most other countries have one centralized agency which is principally responsible for generating national standards, e.g., the Canadian Standards Association (CSA) and the British Standards Institution (BSI). In addition, there are international groups which issue standards that are applicable across national boundaries, e.g., the International Organization for Standardization (ISO) and the International Electrotechnical Commission (IEC).

The creation of a single European market is generating the development of "harmonized" European standards, published by CEN/CENELEC, the joint European standards institute. These standards are included in the International volumes. Deutsches Institut fur Normung e.V. (DIN) is one of the largest standards producing bodies in the world, and many of their standards have not been translated into English. Those that do have official English translations including VDE are included in the International volumes with an English title.

A few of the major groups produce most of the documents; the remainder issue very few each. Often organizations work together to prepare standards and many adopt standards originated by others.

Adherence to a particular standard is often mandated by law, but a major reason for use of standards is economic. Standard materials, components, and products are less expensive; savings in engineering time and materials are realized, as is high quality.

INDEXING METHODOLOGY

Standards currently in force from 423 organizations are indexed by subject using a unique and extensive thesaurus of technical terms developed specifically for this *Index*. Nomenclature was derived from many published technical thesauri, dictionaries, encyclopedias, handbooks, and other reference tools, as well as the indexed documents themselves. The terminology was designed to provide uniform and consistent treatment and retrieval of a wide range of topics treated in industry standards and engineering. Vocabulary may not, therefore, reflect the very specialized usage of a particular discipline.

Documents are indexed according to subject as determined by an analysis of the text of the standard.

Documents are indexed at a level of specificity appropriate to the content of the data base. That is, if there is a very large number of standards on a topic, that topic is subdivided, where possible, to facilitate retrieval by keeping to a manageable number the entries a user has to scan to locate the desired standard.

The very broad subject for a particular standard may not be indexed. For example, not all SAE automotive standards will be found under a general motor vehicle heading. For such a broad approach to locating standards information, see the Organization/Numeric Listings.

COVERAGE

With a few exceptions, the two Subject Indexes cover the 145,430 documents from 423 standards-generating bodies as indicated in the Organization/Numeric Listings. Some documents which are issued by one body and merely adopted by another may be identified in the Subject Index with only the original issuing body. For example, standards approved but not issued by ANSI are identified by the originating organization. Similarly, an organization may issue the same material in different forms (book, chapters and pamphlets); in this case only one form is generally subject indexed.

INDEX ORGANIZATION

Due to the difference in coverage and usage, U.S. Standards are included in one Subject Index, Volume I, and International and Non-U.S. National Standards in Volumes III and IV.

Subject terms to which documents apply are displayed in bold type and listed in alphabetical order. Under a term, document citations are sorted first by organization acronym (see list on page ix for Volume I, page vii for Volumes III and IV).

Under a term, there may be a note indicating the scope of the word or phrase. It may also suggest other areas of the Index where related documents may be found.

Two kinds of cross-references indicate other specific indexing terms. When another form of a term or a synonym has been used, there will be a *Use* reference to the preferred term. *See Also* references direct the user to broader, narrower, and related terms within the Index.

When there would be many entries under a term, it may be subdivided into smaller categories. For example, documents dealing with the general topic of "Aircraft" will be indexed under that term; documents covering specific types or aspects of "Aircraft," such as "Agricultural" will be listed under that subheading. The subheadings are displayed in bold type preceded by a dash.

CONTENT OF CITATION

1. Standards body which published the document. A list of acronyms used is on page ix for Volume I, page vii for Volumes III and IV.
2. Document number assigned by the issuing body. It may be alphanumeric or numeric and is generally followed by the year of latest issue. Sometimes there is no document number, and in some cases a date may be used instead.
3. Title as it appears on the document. Sometimes this information is inverted so the topic of the standard is listed first, followed by "Method for" or "Spec for."
4. Date or other identification of latest revision or reapproval.
5. Number of pages in the document.

SCOPE NOTE → **Aircraft** — **MAIN HEADING**

Scope Note: For specific types of aircraft equipment, see the subheading Aircraft under specific types of equipment *Use For:* Aeroplanes; Airplanes; Fixed Wing Aircraft *See Also:* Aerodynamics; Aeronautical Communications; Aeronautical Meteorology; Aeronautics; Aerospace Vehicles; Air Cargo Transportation; Air Navigation; Air Traffic Control; Air Transportation; Aircraft; Accidents and Incidents; Aircraft Doors; Aircraft Engines; Aircraft Equipment; Aircraft Fuel Systems; Aircraft Fuels; Aircraft Gasoline; Aircraft Instruments; Aircraft Interception; Aircraft Lighting; Aircraft Noise; Aircraft Offenses and Criminal Acts; Aircraft Propellers; Aircraft Refueling Equipment; Aircraft Safety; Aircraft Surveillance Radar; Airport Lighting; Airport Runways; Airport Surveillance Radar; Airports; Airspeed Indicators; Avionics; Balloons; Flight Crews; Flight Dynamics; Flight Envelopes; Flight Operations; Flight Testing; Gliders; Helicopters; Military Aircraft; Pilot Tubes; Pressurized Cabins; Rotary Wing Aircraft; Seaplanes; Supersonic Transports; Vehicles; Vertical Speed Indicators

— **SYNONYMOUS TERM**

— **CROSS REFERENCES**

— **Acceleration Testing** — **SUBHEADING**

1. → ISO 2669-78. Environmental Tests for Aircraft Equipment-Part 3.2: Steady State Acceleration First Edition. 6 pp.

2. → — **Airworthiness**

ICAO 9051. Airworthiness Technical Manual Second Edition-1987; (Corrigendum 1 06/01/87) (Amendment 10 11/14/88). 265 pp ← 4.

— **Glossaries**

DIN 9020 Pt 2-83. Aerospace; Mass Breakdown for Aircraft Heavier Than Air; Main Mass Groups and Terms of Mass; Definitions (Oct). 6 pp.

JIS W 0106-84. Glossary of Terms for Aircraft ← 5.
General.

3. → JIS W 0108-76. Glossary of Terms for Aircraft (Structure).

INTERNATIONAL AND NON-U.S. NATIONAL STANDARDS ORGANIZATION ACRONYMS

Following is a list of acronyms used to identify organizations' documents in the Subject Index.

AECMA
Association Europeene des Constructeurs De Materiel Aerospatial

BSI
British Standards Institution

CAA
Civil Aviation Authority

CCIR
International Radio Consultative Committee

CCITT
International Telegraph & Telephone Consultative Committee

CECC
CENELEC Electronic Components Committee

CEN
European Committee for Standardization

CENELEC
European Committee for Electrotechnical Standardization

CEPT
Conference Europeenne des Administrations des Postes et des Telecommunications

CGSB
Canadian General Standards Board

CGSB
Canadian General Standards Board French Edition

CNS
Chinese National Standards

CPPA
Canadian Pulp & Paper Association

CSA
Canadian Standards Association

CSA
Canadian Standards Association French Edition

DIN ENGL
Din Deutsches Institut Fur Normung E.V.

DIN VDE
Verband Deutscher Elektrotechniker

EC
European Council/Commission Legislative Documents

ECMA
European Computer Manufacturers Association

ETSI
European Telecommunications Standards Institute

EURO
Includes Docs From More T Han One Body. Please Cal L Global Or Ti To Order.

EUROCAE
European Organization for Civil Aviation Electronics

ICAO
International Civil Aviation Organization

IEC
International Electrotechnical Commission

IECQ
Iec Quality Assessment System for Electronic Components

ISO
International Organization for Standardization

JIS
Japanese Industrial Standards

JTC1
Joint Technical Committee

MOD UK
British Defence Standards

NATO
North Atlantic Treaty Organization

OSI
Open Systems Interconnection-International Standards Organization

SAA
Standards Association of Australia

SBAC
Society of British Aerospace Companies

SNZ
Standards New Zealand

INTERNATIONAL AND NON-U.S. NATIONAL STANDARDS

Subject Index
(A-L)

INTERNATIONAL AND NON-U.S. NATIONAL STANDARDS
SUBJECT INDEX

Abbreviations

A/D Converters
Use: Analog to Digital Converters

A Programming Language (APL)
Use: APL (A Programming Language)

A Rivets
Use: Blind Rivets

AAIC
Use: Accounting Authority Identification Codes

AAL (OSI)
Use: ATM Adaptation Layer (OSI)

AAS
Use: Atomic Absorption Spectrophotometry

Abacus
Use For: Soroban See Also: Calculators
JIS S 6048-78. SOROBAN.

Abalone Mushrooms
Use: Mushrooms

Abattoirs
Use: Meat Processing Facilities

Abbe Refractometers
See Also: Measuring Instruments; Refractometers
—Packaging—Export
CNS Z5091-81. Packaging and Packing of Abbe Refractometer for Export (Jun)(7543).

Abbreviated Test Language for Avionic Systems
Use: ATLAS

Abbreviations
Use For: Acronyms; Initials See Also: Symbols
—Aeronautical
ICAO 8400. Procedures for Air Navigation Services ICAO Abbreviations and Codes Fourth Edition—1989; (Incorporating Amendments 1 to 20) (Amendment 21 7/1/93). 133 pp.
—Aircraft
SBAC RS 1 (V). Standard Abbreviations.
—Atomic Time Scales
CCIR RECMN 536-78. Time-Scale Notations—Section 7B—Specifications for the Standard-Frequency and Time-Signal Services. 2 pp.
—Bibliographic References
ISO 832-75. Documentation—Bibliographical References—Abbreviations of Typical Words First Edition. 43 pp.
—CCITT No. 7 Signaling Systems
CCITT FASCICLE VI.7-88. Specifications of Signalling System No. 7 Recommendations Q.700—Q.716. 593 pp.
CCITT FASCICLE VI.8-88. Specifications of Signalling System No. 7 Recommendations Q.721—Q.766. 475 pp.
—Construction
SAA HB24-92. Symbols and Abbreviations for Building and Construction. 111 pp.
—Coordinated Universal Time
CCIR RECMN 536-78. Time-Scale Notations—Section 7B—Specifications for the Standard-Frequency and Time-Signal Services. 2 pp.
—Designations
DIN ENGL 1353 Pt 1-71. Abbreviations of Designations; Elementary Abbreviations (Apr). 3 pp.
—Directory Services
CCITT RECMN F.500-92. International Public Directory Services (Study Group I) 39 pp. 39 pp.
CCITT RECMN F.500-89. International Public Directory Services—Message Handling Services—Operations and Definition of Service (Study Group I) 32 pp. 32 pp.
—Documents
NATO STANAG 1059 ED 5 AMD 0-88. National Distinguishing Letters for Use by NATO Forces. 5 pp.
—Fasteners
DIN ENGL 918-79. Fasteners; Terminology; Spelling of Terms; Abbreviations (Sept). 4 pp.
—Full Scale Layout—Aircraft
SBAC RS 196 (V). Full Scale, Layout Reproduction, Recommended Terms and Abbreviations.

Abbreviations *(Cont.)*
—International Atomic Time
CCIR RECMN 536-78. Time-Scale Notations—Section 7B—Specifications for the Standard-Frequency and Time-Signal Services. 2 pp.
—Interpersonal Messaging Services
CCITT RECMN F.420-92. Message Handling Services: Public Interpersonal Messaging Service (Study Group I) 16 pp. 16 pp.
CCITT RECMN F.420-89. Message Handling Services: the Public Interpersonal Messaging Service—Message Handling and Directory Services—Operations and Definition of Service (Study Group I) 15 pp. 15 pp.
CCITT RECMN F.421-89. Message Handling Services: Intercommunication Between the IPM Service and the Telex Service—Message Handling and Directory Services—Operations and Definition of Service (Study Group I) 12 pp (Same as Recmn F.85). 12 pp.
CCITT RECMN F.422-89. Message Handling Services: Intercommunication Between the IPM Service and the Teletex Service—Message Handling and Directory Services—Operations and Definition of Service (Study Group I) 6 pp. 6 pp.
—Jigs
DIN ENGL 6300-70. Jigs and Fixtures for Form-Modifying Production Processes; Denominations and Their Abbreviations (June). 2 pp.
—Marine
JIS F 0061-84. Abbreviations in English for Marine Equipment.
JIS F 7007-79. Abbreviation Used in Name Plate for Marine Valve.
—Message Handling Systems
CCITT RECMN F.400-92. Message Handling Services: Message Handling System and Service Overview (Study Group I) 82 pp (Same as Recmn X.400). 82 pp.
CCITT RECMN F.400-89. Message Handling System and Service Overview—Message Handling and Directory Services—Operations and Definition of Service (Study Group I) 73 pp. 73 pp.
CCITT RECMN F.401-92. Message Handling Services: Naming and Addressing for Public Message Handling Services (Study Group I) 21 pp. 21 pp.
CCITT RECMN F.401-89. Message Handling Services: Naming and Addressing for Public Message Handling Services—Message Handling and Directory Services—Operations and Definition of Service (Study Group I) 11 pp. 11 pp.
CCITT RECMN F.415-89. Message Handling Services: Intercommunication with Public Physical Delivery Services—Message Handling and Directory Services—Operations and Definition of Service (Study Group I) 15 pp (Erratum in Recmn F.410). 15 pp.
CCITT RECMN X.400-93. Message Handling Services: Message Handling System and Service Overview (Study Group VII) 82 pp (Same as Recmn F.400). 82 pp.
—Message Transfer Systems
CCITT RECMN F.410-92. Message Handling Services: Public Message Transfer Service (Study Group I) 11 pp. 11 pp.
CCITT RECMN F.410-89. Message Handling Services: Message Transfer Service—Message Handling and Directory Services—Operations and Definition of Service (Study Group I) 10 pp. 10 pp.
—Military Documents
NATO STANAG 2066 ED 4 AMD 6-79. Layout for Military Correspondence. 9 pp.
NATO STANAG 2066 ED 5 AMD 0-90. Layout for Military Correspondence. 9 pp.
—Periodical Titles
JIS X 0801-89. Abbreviation of Titles of Periodicals for Information Interchange.
—Plastics—Compounding Ingredients
BSI BS 4589-70. (WITHDRAWN) 1970 Abbreviations for Rubber and Plastics Compounding Materials (Superseded by BS 3502: Part 3: 1993). 11 pp.
—Rubber—Compounding Ingredients
BSI BS 4589-70. (WITHDRAWN) 1970 Abbreviations for Rubber and Plastics Compounding Materials (Superseded by BS 3502: Part 3: 1993). 11 pp.
ISO 6472-86. Rubber Compounding Ingredients—Abbreviations First Edition. 4 pp.
—Scientific/Engineering
CSA Z85-1983. ABBR Abbreviations for Scientific and Engineering Terms. 280 pp.
—Semifinished Products
DIN ENGL 1353 Pt 2-71. Abbreviations of Denominations for Half-Finished Products (Apr). 4 pp.

Abbreviations *(Cont.)*
—Ships
MOD UK NES 707: Part 2-88. Symbols and Abbreviations Part 2: Guide to Abbreviations and System Code Letters Issue 2 (08.88). 111 pp.
—Standards Preparation
EURO STANDARDIZAT ION-REF ST 1-88. Glossary. 3 pp.
SNZ NZMP 8201-86. Abbreviations Encountered in Standards Work. 30 pp.
—Steel Wire
DIN ENGL 1653-79. Surface Condition of Commercial Steel Wires; Denominations and Abbreviations Thereof (Jan). 4 pp.
—Telecommunication Equipment
CCIR RECMN 666-1-90. Abbreviations and Initials Used in Telecommunications—Section C—Other Means of Expression. 17 pp.
—Telecommunication Services
CCIR RECMN 666-1-90. Abbreviations and Initials Used in Telecommunications—Section C—Other Means of Expression. 17 pp.
—Telecommunication Services—Books
CCITT RECMN F.92-89. Service Codes—Telegraph and Mobile Services Operations and Quality of Service (Study Group I) 3 pp. 3 pp.
—Telecommunication Systems
CCIR RECMN 666-1-90. Abbreviations and Initials Used in Telecommunications—Section C—Other Means of Expression. 17 pp.
CCITT FASCICLE I.3-88. Terms and Definitions Abbreviations and Acronyms Recommendations on Means of Expression (Series B), General Telecommunications Statistics (Series C). 901 pp.
CCITT RECMN B.19-89. Abbreviations and Initials Used in Telecommunications—Terms and Definitions Abbreviations and Acronyms Recommendations on Means of Expression (Series B) General Telecommunications Statistics (Series C) 3 pp. 3 pp.
—Telegraphy
CCITT RECMN F.1-89. Operational Provisions for the International Public Telegram Service—Telegraph and Mobile Services Operations and Quality of Service (Study Group I) 54 pp. 54 pp.
—Teletex Communications
CCITT RECMN F.422-89. Message Handling Services: Intercommunication Between the IPM Service and the Teletex Service—Message Handling and Directory Services—Operations and Definition of Service (Study Group I) 6 pp. 6 pp.
CCITT RECMN T.64-89. Conformance Testing Procedures for the Teletex Recommendations—Conformance Testing Procedures for the Teletex Recommendations (Study Group VIII) 121 pp. 121 pp.
—Telex Communications
CCITT RECMN F.421-89. Message Handling Services. Intercommunication Between the IPM Service and the Telex Service—Message Handling and Directory Services—Operations and Definition of Service (Study Group I) 12 pp (Same as Recmn F.85). 12 pp.
—Time Scales
CCIR RECMN 536-78. Time-Scale Notations—Section 7B—Specifications for the Standard-Frequency and Time-Signal Services. 2 pp.
—Titles
BSI BS 4148-85. 1985 Abbreviation of Title Words and Titles of Publications. 8 pp.
ISO 4-84. Documentation—Rules for the Abbreviation of Title Words and Titles of Publications Second Edition. 7 pp.
—Units of Measurement—Quantity—Data Communication
CCITT RECMN B.14-89. Terms and Abbreviations for Information Quantities in Telecommunications—Terms and Definitions Abbreviations and Acronyms Recommendations on Means of Expression (Series B) General Telecommunications Statistics (Series C) 2 pp. 2 pp.
—Units of Measurement—Quantity—Data Transmission
CCITT RECMN B.14-89. Terms and Abbreviations for Information Quantities in Telecommunications—Terms and Definitions Abbreviations and Acronyms Recommendations on Means of Expression (Series B) General Telecommunications Statistics (Series C) 2 pp. 2 pp.

INTERNATIONAL AND NON-U.S. NATIONAL STANDARDS
SUBJECT INDEX

Abbreviations

Abbreviations (Cont.)
—Universal Time
CCIR RECMN 536-78. Time-Scale Notations—Section 7B—Specifications for the Standard-Frequency and Time-Signal Services. 2 pp.

—Voice Messaging Services
CCITT RECMN F.440-92. Message Handling Services: Voice Messaging Service (Study Group I) 34 pp. 34 pp.

Abdominal Supports
Use: Orthopedic Supports

Abel Testers
See Also: Flash Point; Testing Equipment

—Test Tubes
MOD UK DSTAN 66-33-84. Test Tube MK11 (Apparatus, Abel Heat Test) Issue 1. 6 pp.

Abrasion
Use: Abrasion Testing

Abrasion Resistance
Use: Abrasion Testing

Abrasion Resistance Testing
Use: Abrasion Testing

Abrasion Testers
BSI BS 6161: Part 9-87. 1987 Methods of Test for Anodic Oxidation Coatings on Aluminum and Its Alloys Part 9: Measurement of Wear Properties with an Abrasive Wheel Wear Test Apparatus. 8 pp.

BSI BS 6161: Part 10-87. 1987 Methods of Test for Anodic Oxidation Coatings on Aluminum and Its Alloys Part 10: Measurement of of Mean Specific Abrasion Resistance with an Abrasive Jet Test Apparatus. 12 pp.

BSI BS AU 148: Part 4-69. 1969 Methods of Test for Motor Vehicle Paints Part 4: Resistance to Abrasion. 3 pp.

CNS K6729-82. Method of Test for Abrasion Resistance of Coating of Paint, Varnish, Lacquer and Related Products with the Air Blast Abrasion Tester (Sep)(9405).

ISO 4649-85. Rubber—Determination of Abrasion Resistance Using a Rotating Cylindrical Drum Device First Edition. 11 pp.

ISO 8252-87. Anodized Aluminium and Aluminium Alloys—Measurement of Mean Specific Abrasion Resistance of Anodic Oxidation Coatings with an Abrasive Jet Test Apparatus First Edition. 12 pp.

—Aggregates
CNS A3059-85. Method of Test for Resistance to Abrasion of Large Size Coarse Aggregate by Use of the Los Angeles Machine (Greater Than 19 mm) (Jun)(3408). 2 pp.

JIS A 1121-89. Method of Test for Abrasion of Coarse Aggregates by Use of the Los Angeles Machine. 8 pp.

—Anodic Coatings
BSI BS 6161: Part 9-87. 1987 Methods of Test for Anodic Oxidation Coatings on Aluminum and Its Alloys Part 9: Measurement of Wear Properties with an Abrasive Wheel Wear Test Apparatus. 8 pp.

BSI BS 6161: Part 10-87. 1987 Methods of Test for Anodic Oxidation Coatings on Aluminum and Its Alloys Part 10: Measurement of of Mean Specific Abrasion Resistance with an Abrasive Jet Test Apparatus. 12 pp.

ISO 8252-87. Anodized Aluminium and Aluminium Alloys—Measurement of Mean Specific Abrasion Resistance of Anodic Oxidation Coatings with an Abrasive Jet Test Apparatus First Edition. 12 pp.

—Rubber
ISO 4649-85. Rubber—Determination of Abrasion Resistance Using a Rotating Cylindrical Drum Device First Edition. 11 pp.

—Vitreous Enamels
ISO 6370 Pt 1-91. Vitreous and Porcelain Enamels—Determination of the Resistance to Abrasion—Part 1: Abrasion Testing Apparatus First Edition. 6 pp.

Abrasion Testing
Use For: Rub Resistance *See Also:* Abrasion Testers; Scrub Resistance; Surface Strength; Wear Testing

DIN ENGL 52108-88. Testing the Abrasive Wear of Inorganic Non-Metallic Materials Using the Bohme Disk Abrader (Aug). 4 pp.

—Aggregates
BSI BS 812: Part 113-90. 1990 Testing Aggregates Part 113: Methods for Determination of Aggregate Abrasion Value. 10 pp.

Abrasion Testing (Cont.)
—Aggregates (Cont.)
BSI BS 812: Part 113-01. 1990 Amd 1 Testing Aggregates Part 113: Method for Determination of Aggregate Abrasion Value (AAV) (AMD 6986) December 24, 1991. 11 pp.

CNS A3009-86. Method of Test for Resistance to Abrasion of Small Size Coarse Aggregate by Use of the Los Angeles Machine (Aug)(490).

CNS A3059-85. Method of Test for Resistance to Abrasion of Large Size Coarse Aggregate by Use of the Los Angeles Machine (Greater Than 19 mm) (Jun)(3408). 2 pp.

JIS A 1121-89. Method of Test for Abrasion of Coarse Aggregates by Use of the Los Angeles Machine. 8 pp.

SNZ NZS 4407: Part 3.12-91. Methods of Sampling and Testing Road Aggregates Part 3: Methods of Testing Road Aggregates—Laboratory Tests Test 3.12: The Abrasion Resistance of Aggregate by Use of the Los Angeles Machine. 4 pp.

—Anodic Coatings
BSI BS 6161: Part 9-87. 1987 Methods of Test for Anodic Oxidation Coatings on Aluminum and Its Alloys Part 9: Measurement of Wear Properties with an Abrasive Wheel Wear Test Apparatus. 8 pp.

BSI BS 6161: Part 10-87. 1987 Methods of Test for Anodic Oxidation Coatings on Aluminum and Its Alloys Part 10: Measurement of Mean Specific Abrasion Resistance with an Abrasive Jet Test Apparatus. 12 pp.

BSI BS 6161: Part 18-91. 1991 Methods of Test for Anodic Oxidation Coatings on Aluminium and Its Alloys Part 18: Determination of Surface Abrasion Resistance. 8 pp.

CNS H2064-82. Methods of Test for Abrasion Resistance of Anodic Oxidation Coatings on Aluminium and Aluminium Alloys (Jan)(8411). 5 pp.

ISO 8252-87. Anodized Aluminium and Aluminium Alloys—Measurement of Mean Specific Abrasion Resistance of Anodic Oxidation Coatings with an Abrasive Jet Test Apparatus First Edition. 12 pp.

JIS H 8682-88. Test Methods for Abrasion Resistance of Anodic Oxidation Coatings on Aluminium and Aluminium Alloys. 18 pp.

—Anodic Coatings—Aluminum
SAA AS 2039.5.1-78. Methods for Testing Anodic Oxidation Coatings on Aluminium and Aluminium Alloys—Part 5.1: Abrasion Resistance Tests—Jet Test for Abrasion Resistance of Anodic Oxidation Coatings Reconfirmed 1990. 4 pp.

SAA AS 2039.5.2-78. Methods for Testing Anodic Oxidation Coatings on Aluminium and Aluminium Alloys—Part 5.2: Abrasion Resistance Tests—ASTM Test for Abrasion Resistance of Anodic Oxidation Coatings Reconfirmed 1990. 6 pp.

—Athletic Fields—Artifical Surfaces
BSI BS 7044: Sec 2.3-90. 1990 Artificial Sports Surfaces Part 2: Methods of Test Section 2.3: Methods of Determination of Durability. 12 pp.

—Building Stones
CNS A3227-85. Method of Test for Abrasion Resistance of Stone Subjected to Foot Traffic (Aug)(11320).

—Cables (Electric)
CEN PREN 3475 (Part 503)-92. Cables, Electrical, Aircraft Use Test Methods Part 503—Scrape Abrasion. 3 pp.

—Carpets
CNS L3120-81. Method of Test for Abrasion Resistance of Carpets (Jun)(7495).

JIS L 1023-92. Testing Methods for Several Characteristics of Textile Floor Coverings. 26 pp.

—Ceramic Tiles
BSI BS 6431: Part 14-83. 1983 Ceramic Floor and Wall Tiles Part 14: Method for Determination of Resistance to Deep Abrasion. Unglazed Tiles. 10 pp.

BSI BS 6431: Part 14-01. 1983 Amd 1 Ceramic Floor and Wall Tiles Part 14: Method for Determination of Resistance to Deep Abrasion. Unglazed Tiles (AMD 7103) July 15, 1992. 13 pp.

BSI BS 6431: Part 20-84. 1984 Ceramic Floor and Wall Tiles Part 20: Method of Determination of Resistance to Surface Abrasion. Glazed Tiles. 10 pp.

BSI BS 6431: Part 20-01. 1984 Amd 1 Ceramic Floor and Wall Tiles Part 20: Method for Determination of Resistance to Surface Abrasion. Glazed Tiles (AMD 7109) July 15, 1992. 13 pp.

CEN EN 102-82. Ceramic Tiles: Determination of Resistance to Deep Abrasion; Unglazed Tiles. 8 pp.

CEN EN 102-91. Ceramic Tiles—Determination of Resistance to Deep Abrasion—Unglazed Tiles. 7 pp.

CEN EN 154-84. Ceramic Tiles: Determination of Resistance to Surface Abrasion; Glazed Tiles. 8 pp.

CEN EN 154-91. Ceramic Tiles—Determination of Resistance to Surface Abrasion—Glazed Tiles. 7 pp.

Abrasion Testing (Cont.)
—Coal
BSI BS 1016: Part 19-80. (WITHDRAWN) 1980 Methods for Analysis and Testing of Coal and Coke Part 19: Determination of the Index of Abrasion of Coal (Superseded by BS 1016: Part 111: 1990). 7 pp.

BSI BS 1016: Part 111-93. 1993 Methods for Analysis and Testing of Coal and Coke Part 111: Determination of Abrasion Index of Coal (W). 12 pp.

BSI BS 1016: Part 111-90. 1990 Methods for the Analysis and Testing of Coal and Coke Part 111: Determination of Abrasion Index of Coal. 11 pp.

SAA AS 1038.19-89. Methods for the Analysis and Testing of Coal and Coke—Part 19: Determination of the Abrasion Index of Higher Rank Coal. 6 pp.

—Coated Fabrics—Rubber
BSI BS 3424: Part 24-90. 1990 Testing Coated Fabrics Part 24: Methods 27A and 27B. Determinaton of Abrasion Resistance. 8 pp.

ISO 5470-80. Rubber or Plastics Coated Fabrics—Determination of Abrasion Resistance First Edition. 5 pp.

ISO 5981-82. Rubber or Plastics Coated Fabrics—Determination of Flex Abrasion First Edition. 6 pp.

—Coatings
JIS A 1436-91. Test Methods for Movement Capability of Coatings and Sheets Fully Adhered on Substrate. 15 pp.

—Construction Materials
CNS A3202-84. Method of Abrasion Test for Building Materials and Building Construction Parts (Falling Sand Method) (Mar)(10784).

CNS A3203-84. Method of Abrasion Test for Building Materials and Building Construction Parts (Abrasive-Paper Method) (Mar)(10785).

JIS A 1452-72. Method of Abrasion Test for Building Materials and Part of Building Construction (Falling Sand Method).

JIS A 1453-73. Method of Abrasion Test for Building Materials and Part of Building Construction (Abrasive-Paper Method).

—Detergents
CGSB 2-GP-11M METH 15.1-83. Methods of Testing and Analysis of Soaps and Detergents Abrasion Test. 1 p.

—Elastomers
DIN ENGL 53516-87. Testing of Rubber and Elastomers; Determination of Abrasion Resistance (June). 6 pp.

—Fabrics
BSI BS 5690-91. 1991 Abrasion Resistance of Fabrics. 13 pp.

SAA AS 2001.2.25-90. Methods of Test for Textiles—Part 2: Physical Tests—Part 2.25: Determination of Flat Abrasion Resistance of Textile Fabrics (Martindale Abrasion Method). 6 pp.

SAA AS 2001.2.26-90. Methods of Test for Textiles—Part 2: Physical Tests—Part 2.26: Determination of Abrasion Resistance of Textile Fabrics (Flexing and Abrasion Method). 6 pp.

SAA AS 2001.2.27-90. Methods of Test for Textiles—Part 2: Physical Tests—Part 2.27: Determination of Abrasion Resistance of Textile Fabrics (Inflated Diaphragm Method). 5 pp.

SAA AS 2001.2.28-92. Methods of Test for Textiles—Part 2: Physical Tests—Part 2.28: Determination of Abrasion Resistance of Textile Fabrics (Rotary Platform, Double-Head Method). 8 pp.

—Firebricks
BSI BS 1902: Sec 4.6-85. 1985 Methods for Testing Refractory Materials Part 4: Properties Measured Under an Applied Stress: Section 4.6: Determination of Resistance to Abrasion at Ambient Temperature (Abradability Index at Ambient Temperature) (Method 1902-406). 10 pp.

BSI BS 1902: Sec 4.7-85. (OBSOLESCENT) 1985 Methods for Testing Refractory Materials Part 4: Properties Measured Under an Applied Stress Section 4.7: Determination of Resistance to Abrasion at Elevated Temperature up to 1400 Degrees C (Abradability Index at Elevated Temperatures). 9 pp.

—Floor Coverings
CEN PREN 668-92. Rubber Floor Coverings—Determination of Abrasion Resistance Using a Rotating Cylindrical Drum Device and Non-Rotating Sample Holder. 14 pp.

—Floor Tiles—Asbestos
CNS K6312-77. Method of Test for Polyvinyl Chloride Asbestos Tiles for Flooring (Dec)(3309). 4 pp.

INTERNATIONAL AND NON-U.S. NATIONAL STANDARDS
SUBJECT INDEX

Abrasion Testing (Cont.)
—Floors
CNS A3160-82. Method of Abrasion Test for Building Material and Part of Building Construction (Method of Abrasion Test for Flooring Materials Method with Rotating Disk Fitted Friction and Impact) (Jun)(8912).
JIS A 1451-70. Method of Abrasion Test for Building Materials and Part of Building Construction (Method of Abrasion Test for Flooring Materials Method with Rotating Disk Fitted Friction and Impact).

—Footwear—Paperboard
BSI BS 5131: Sec 4.12-90. 1990 Methods of Test for Footwear and Footwear Materials Part 4: Other Components Section 4.12: Fastness of Fibreboard Finishes to Rubbing in the Presence of Water and Perspiration. 10 pp.

—Glass
CNS Z9069-86. Method of Test for Abrasion in Optical Glasses (Dec)(9940).

—Gold Coatings (Made From Gold)
ISO 3160 Pt 3-93. Watch Cases and Accessories—Gold Alloy Coverings—Part 3: Abrasion Resistance Tests of a Type of Coating on Standard Guages First Edition. 7 pp.

—Hoses—Plastic
BSI BS 5173: Sec 103.9-86. 1986 Rubber and Plastics Hoses and Hose Assemblies Part 103: Physical Tests Section 103.9: Determination of Abrasion Resistance of the Outer Cover. 6 pp.
ISO 4649-85. Rubber—Determination of Abrasion Resistance Using a Rotating Cylindrical Drum Device First Edition. 11 pp.
ISO 7662-88. Rubber and Plastics Hoses—Determination of Abrasion of Lining First Edition. 6 pp.

—Hoses—Rubber
BSI BS 5173: Sec 103.9-86. 1986 Rubber and Plastics Hoses and Hose Assemblies Part 103: Physical Tests Section 103.9: Determination of Abrasion Resistance of the Outer Cover. 6 pp.
ISO 4649-85. Rubber—Determination of Abrasion Resistance Using a Rotating Cylindrical Drum Device First Edition. 11 pp.
ISO 7662-88. Rubber and Plastics Hoses—Determination of Abrasion of Lining First Edition. 6 pp.

—Insulated Cables
DIN VDE 0472 Pt 605-85. Testing of Cables, Insulated Wires and Flexible Cords; Abrasion (Jan) (Partially Supersedes 0472/09.71 and VDE 0472 d/12.77). 8 pp.

—Laces (Footwear)—Shoes
BSI BS 5131: Sec 3.6-91. 1991 Methods of Test for Footwear and Footwear Materials Part 3: Uppers, Textiles and Threads Section 3.6: Abrasion Resistance of Shoe Laces. 6 pp.

—Lacquers
CNS K6729-82. Method of Test for Abrasion Resistance of Coating of Paint, Varnish, Lacquer and Related Products with the Air Blast Abrasion Tester (Sep)(9405).

—Leather
BSI BS 3144: Add 1-81. 1981 Methods of Sampling and Physical Testing of Leather Addendum 1: Determination of Resistance to Bending and Abrasion of Heavy Leather. 12 pp.

—Leather—Polyurethane
CNS K6771-83. Method of Test for Polyurethane Leather (May)(10270). 3 pp.

—Molybdenum Disulfide
MOD UK DSTAN 05-50: Part 37-91. Methods for Testing Fuels, Lubricants and Associated Products Part 37: The Assessment of the Abrasiveness of Molybdenum Disulphide Powders Issue 1. 12 pp.

—Nonmetals
DIN ENGL 52108-88. Testing the Abrasive Wear of Inorganic Non-Metallic Materials Using the Bohme Disk Abrader (Aug). 4 pp.

—Oxide Coatings
BSI BS 6161: Part 9-87. 1987 Methods of Test for Anodic Oxidation Coatings on Aluminum and Its Alloys Part 9: Measurement of Wear Properties with an Abrasive Wheel Wear Test Apparatus. 8 pp.
BSI BS 6161: Part 10-87. 1987 Methods of Test for Anodic Oxidation Coatings on Aluminum and Its Alloys Part 10: Measurement of Mean Specific Abrasion Resistance with an Abrasive Jet Test Apparatus. 12 pp.

Abrasion Testing (Cont.)
—Oxide Coatings (Cont.)
BSI BS 6161: Part 18-91. 1991 Methods of Test for Anodic Oxidation Coatings on Aluminium and Its Alloys Part 18: Determination of Surface Abrasion Resistance. 8 pp.
CNS H2064-82. Methods of Test for Abrasion Resistance of Anodic Oxidation Coatings on Aluminium and Aluminium Alloys (Jan)(8411). 5 pp.
ISO 8252-87. Anodized Aluminium and Aluminium Alloys—Measurement of Mean Specific Abrasion Resistance of Anodic Oxidation Coatings with an Abrasive Jet Test Apparatus First Edition. 12 pp.
JIS H 8682-88. Test Methods for Abrasion Resistance of Anodic Oxidation Coatings on Aluminium and Aluminium Alloys. 18 pp.

—Paints
CGSB 1-GP-71 METH 104.1-74. Methods of Testing Paints and Pigments Abrasion Resistance. 1 p.
CNS K6143-87. Method of Test for Traffic Paints (May)(1334). 19 pp.
CNS K6729-82. Method of Test for Abrasion Resistance of Coating of Paint, Varnish, Lacquer and Related Products with the Air Blast Abrasion Tester (Sep)(9405).

—Paints—Automotive
CNS K6143-87. Method of Test for Traffic Paints (May)(1334).

—Paints—Enamel
BSI BS 1344: Part 4-68. (OBSOLESCENT) 1968 Methods of Testing Vitreous Enamel Finishes Part 4: Resistance to Abrasion. 8 pp.
ISO 6370 Pt 2-91. Vitreous and Porcelain Enamels—Determination of the Resistance to Abrasion—Part 2: Loss in Mass After Sub-Surface Abrasion First Edition. 10 pp.

—Paperboard
CNS P3045-81. Method of Test for Abrasion Resistance of Paperboard (Jan) (6951). 2 pp.
JIS P 8136-76. Testing Method of Abrasion Resistance of Paperboard (R 1984). 6 pp.

—Pigments
CGSB 1-GP-71 METH 104.1-74. Methods of Testing Paints and Pigments Abrasion Resistance. 1 p.

—Plastics
BSI BS 2782:Pt3: METH 370-90. 1990 Methods of Testing Plastics Part 3: Mechanical Properties Method 370: Determination of Resistance to Wear by Abrasive Wheels. 11 pp.
BSI BS 5173: Sec 103.9-86. 1986 Rubber and Plastics Hoses and Hose Assemblies Part 103: Physical Tests Section 103.9: Determination of Abrasion Resistance of the Outer Cover. 6 pp.
ISO 7662-88. Rubber and Plastics Hoses—Determination of Abrasion of Lining First Edition. 6 pp.
ISO 9352-89. Plastics—Determination of Resistance to Wear by Abrasive Wheels First Edition. 8 pp.
JIS K 7204-77. Testing Method for Abrasion Resistance of Plastics by Abrasive Wheels (R 1980). 11 pp.
JIS K 7205-77. Testing Method for Abrasion Resistance of Plastics by Abrasive (R 1980). 10 pp.

—Plywood
CNS O2042-82. Method of Test for Abrasion of Special Plywood (Feb)(8540). 2 pp.

—Polishes
CGSB 25-GP-1M METH 39.1-84. Methods of Sampling and Testing Waxes and Polishes Abrasive Properties. 1 p.

—Polyurethane Resins—Athletic Fields
CNS K6591-86. Method of Test for Polyurethane (PU) Athletic Installation Material (May)(6483). 6 pp.

—Printing
BSI BS 3110-59. 1959 Methods for Measuring the Rub Resistance of Print. 12 pp.

—Protective Clothing
CEN PREN 530-91. Abrasion Resistance of Protective Clothing Material. 9 pp.

—Refractory Materials
SAA AS 1774.23.1-92. Refractories and Refractory Materials—Physical Test Methods—Part 23: Abradability Index—Part 23.1: Oblique Method (Supersedes AS 1774.23—1987). 6 pp.

—Rubber
BSI BS 903: Part A9-88. 1988 Methods of Testing Vulcanized Rubber Part A9: Determination of Abrasion Resistance. 20 pp.

Abrasion Testing (Cont.)
—Rubber (Cont.)
BSI BS 5173: Sec 103.9-86. 1986 Rubber and Plastics Hoses and Hose Assemblies Part 103: Physical Tests Section 103.9: Determination of Abrasion Resistance of the Outer Cover. 6 pp.
CNS K6047-56. Testing Standard for Abrasion Resistance of Rubber Products (May)(734).
DIN ENGL 53516-87. Testing of Rubber and Elastomers; Determination of Abrasion Resistance (June). 6 pp.
ISO 4649-85. Rubber—Determination of Abrasion Resistance Using a Rotating Cylindrical Drum Device First Edition. 11 pp.
ISO 6945-91. Rubber Hoses—Determination of Abrasion Resistance of the Outer Cover Second Edition. 7 pp.
ISO 7662-88. Rubber and Plastics Hoses—Determination of Abrasion of Lining First Edition. 6 pp.
MOD UK TS 10227. Rubber Sheet, Standard Abrasive.

—Soaps
CGSB 2-GP-11M METH 15.1-83. Methods of Testing and Analysis of Soaps and Detergents Abrasion Test. 1 p.

—Stoneware
CNS R3062-70. Method of Test for Abrasive Resistance of Chemical Stoneware (Dec) (3189).

—Varnishes
CNS K6729-82. Method of Test for Abrasion Resistance of Coating of Paint, Varnish, Lacquer and Related Products with the Air Blast Abrasion Tester (Sep)(9405).

—Vulcanized Rubber
SAA AS 1683.21-82. Methods of Test for Elastomers—Part 21: Rubber—Vulcanized—Determination of Abrasion Resistance Using a Rotating Cylindrical Device. 7 pp.

—Waveguides
CNS C6312-88. Method of Test for Fiber Optic Devices (FOTP-66 Measuring Relative Abrasion Resistance of Optical Waveguide Coating and Buffers) (Jul)(12359).

—Waxes
CGSB 25-GP-1M METH 39.1-84. Methods of Sampling and Testing Waxes and Polishes Abrasive Properties. 1 p.

—Wood
CNS O2009-81. Method of Test for Abrasion of Wood (Mar)(458). 3 pp.
JIS Z 2141-78. Method of Abrasion Test for Wood.
JIS Z 2141-60. Method of Abrasion Test for Wood. 5 pp.

Abrasive Belts
See Also: Abrasives; Belts (Machinery)
BSI BS 7007: Part 3-88. 1988 Abrasive Products Part 3: Dimensional Tolerances on Non-Standard Coated Abrasive Products. 4 pp.
ISO 1929-93. Abrasive Belts—Dimensions, Tolerances and Designation Second Edition. 6 pp.
ISO 2976-73. Abrasive Belts—Selection of Width/Length Combinations First Edition. 3 pp.
ISO 8366-87. Coated Abrasive Products—Dimensional Tolerances on Non-Standard Conversions First Edition. 4 pp.
JIS R 6254-76. Endless Abrasive Belts. 8 pp.

—Jointed
CNS R2092-84. Abrasive Belts (Nov) (4207).
JIS R 6256-76. Abrasive Belts. 14 pp.

Abrasive Blast Cleaning
Use: Blast Cleaning

Abrasive Blasting
—Breathing Apparatus
CEN PREN 271-89. Respiratory Protective Devices: Compressed Air Line Breathing Apparatus for Use in Abrasive Blasting Operations; Requirements, Testing, Marking. 8 pp.

—Metals
SAA AS 1627.4-89. Metal Finishing—Preparation and Pretreatment of Surfaces—Part 4: Abrasive Glass Cleaning (This is a Joint Standard with SANZ NZS 1627). 17 pp.
SNZ NZS/AS 1627. 4-89. Metal Finishing-Preparation and Pretreatment of Surfaces Part 4: Abrasive Blast Cleaning (This is a Joint Standard with SAA AS 1627.4). 17 pp.

—Steel—Ships
MOD UK NES 755-01. Abrasive Blasting of Steel Issue 2 (08.87); Amendment 3. 27 pp.

INDUSTRY STANDARDS

INTERNATIONAL AND NON-U.S. NATIONAL STANDARDS
SUBJECT INDEX

Abrasive

Abrasive Blasting *(Cont.)*
—Steel—Submarines
MOD UK NES 755-01. Abrasive Blasting of Steel Issue 2 (08.87); Amendment 3. 27 pp.

—Titanium—Aerospace
AECMA PREN2497-84. Dry Abrasive Blasting of Titanium and Titanium Alloys. 4 pp.
BSI BS EN 2497-90. 1990 Dry Abrasive Blasting of Titanium and Titanium Alloys. 7 pp.
CEN EN 2497-89. Dry Abrasive Blasting of Titanium and Titanium Alloys. 4 pp.

—Titanium Alloys—Aerospace
AECMA PREN2497-84. Dry Abrasive Blasting of Titanium and Titanium Alloys. 4 pp.
BSI BS EN 2497-90. 1990 Dry Abrasive Blasting of Titanium and Titanium Alloys. 7 pp.
CEN EN 2497-89. Dry Abrasive Blasting of Titanium and Titanium Alloys. 4 pp.

Abrasive Cloth
See Also: Abrasives; Cleaning Cloths; Fabrics
BSI BS 871-81. 1981 Abrasive Papers and Cloths. 12 pp.
CNS R2039-84. Abrasive Cloths (Jan)(1076). 4 pp.
CNS R3038-80. Method of Test for Abrasive Cloth (Oct) (1077).
JIS R 6251-76. Abrasive Cloths. 10 pp.
MOD UK DSTAN 53-4-78. Abrasive Cloths and Papers Issue 3. 8 pp.

—Glossaries/Symbols
JIS R 6004-90. Glossary of Terms and Marks Used in Abrasives, Grinding Wheels and Coated Abrasives. 20 pp.

—Metal Working
CNS R2039-84. Abrasive Cloths (Jan)(1076). 4 pp.

—Woodworking
CNS R2039-84. Abrasive Cloths (Jan)(1076). 4 pp.

Abrasive Disks
See Also: Abrasives
BSI BS 7007: Part 3-88. 1988 Abrasive Products Part 3: Dimensional Tolerances on Non-Standard Coated Abrasive Products. 4 pp.
CNS R2091-85. Abrasive Discs (May) (4206).
ISO 3017-81. Abrasive Discs—Designation, Dimensions and Tolerances—Selection of Disc Outside Diameter/Centre Hole Diameter Combinations Second Edition. 4 pp.
ISO 8366-87. Coated Abrasive Products—Dimensional Tolerances on Non-Standard Conversions First Edition. 4 pp.
JIS R 6255-76. Abrasive Discs. 7 pp.
SAA AS B172.5-70. Bonded Abrasive Products —Part 5: Threaded Insert Discs. 15 pp.

—Inserted Nut—Aluminum Oxide
CNS R2093-87. Inserted Nut Abrasive Discs and Inserted Nut Ring Wheels (Oct)(4208).
JIS R 6216-86. Inserted Nut Abrasive Discs and Inserted Nut Ring Wheels. 12 pp.

—Inserted Nut—Silicon Carbide
CNS R2093-87. Inserted Nut Abrasive Discs and Inserted Nut Ring Wheels (Oct)(4208).
JIS R 6216-86. Inserted Nut Abrasive Discs and Inserted Nut Ring Wheels. 12 pp.

Abrasive Flap Wheels
Use: Grinding Wheels

Abrasive Grains
See Also: Abrasives; Grit

—Bonded—Grain Size Analysis
ISO 8486-86. Bonded Abrasives—Grain Size Analysis—Designation and Determination of Grain Size Distribution of Macrogrits F4 to F220 First Edition. 5 pp.

—Grain Size Analysis
CNS R2082-84. Abrasive Grain Sizes (Nov)(3787).
CNS R3098-86. Methods of Test for Abrasive Grain Size (Jan)(7530). 10 pp.
JIS R 6001-87. Abrasive Grain Sizes. 8 pp.
JIS R 6002-87. Testing Methods for Abrasive Grain Size. 17 pp.
JIS R 6011-91. Testing Method for Grain Size of Coated Abrasive Macrogrits (P12—P220). 9 pp.

—Sampling
BSI BS 7425: Part 3-93. 1993 Abrasive Grains Part 3: Method for Sampling and Splitting (ISO 9138: 1993) (F). 8 pp.
ISO 9138-93. Abrasive Grains—Sampling and Splitting First Edition. 5 pp.
JIS R 6003-73. Methods for Sampling of Abrasive Grains. 7 pp.

Abrasive Grains *(Cont.)*
—Sieves
BSI BS 7425: Part 4-93. 1993 Abrasive Grains Part 4: Specification for Test-Sieving Machines (ISO 9284: 1992) (F). 10 pp.
ISO 9284-92. Abrasive Grains—Test-Sieving Machines First Edition. 7 pp.

Abrasive Papers
Use For: Sandpaper *See Also:* Abrasives; Papers
BSI BS 871-81. 1981 Abrasive Papers and Cloths. 12 pp.
CNS R2038-84. Abrasive Papers (Jan)(1074). 4 pp.
CNS R3037-81. Method of Test for Abrasive Papers (May) (1075).
JIS R 6252-76. Abrasive Papers. 10 pp.
MOD UK DSTAN 53-4-78. Abrasive Cloths and Papers Issue 3. 8 pp.

—Glossaries/Symbols
JIS R 6004-90. Glossary of Terms and Marks Used in Abrasives, Grinding Wheels and Coated Abrasives. 20 pp.

—Metal Working
CNS R2038-84. Abrasive Papers (Jan)(1074). 4 pp.

—Symbols
JIS R 6004-90. Glossary of Terms and Marks Used in Abrasives, Grinding Wheels and Coated Abrasives. 20 pp.

—Waterproof
CNS R2037-80. Water Proof Abrasive Papers (Oct)(1072). 3 pp.
CNS R3036-80. Method of Test for Waterproof Abrasive Papers (Oct)(1073).
JIS R 6253-76. Waterproof Abrasive Papers. 10 pp.

—Woodworking
CNS R2038-84. Abrasive Papers (Jan)(1074). 4 pp.

Abrasive Rolls
See Also: Abrasives
ISO 3366-75. Coated Abrasives—General Purpose Rolls—Any Backing—Designation and Dimensions First Edition. 3 pp.
ISO 3367-75. Coated Abrasives—Rolls for Widths of 50 mm and Greater—Any Backing—Designation and Dimensions First Edition. 3 pp.
ISO 3368-75. Coated Abrasives—Cloth Rolls up to and Including 40 mm Width—Designation and Dimensions First Edition. 3 pp.

Abrasive Sheets
See Also: Abrasives; Emery Paper; Glass Paper
ISO 2235-93. Abrasive Sheets—Dimensions, Tolerances and Designation Third Edition. 5 pp.

Abrasive Sleeves
See Also: Abrasives; Sleeves (Fittings)
BSI BS 7007: Part 1-88. 1988 Abrasive Products Part 1: Dimensions and Designation of Truncated Cone Abrasive Sleeves. 3 pp.
CNS R2190-87. Cylindrical Abrasive Sleeves (Nov)(12167).
ISO 2421-72. Cylindrical Abrasive Sleeves—Designation—Dimensions—Tolerances First Edition. 4 pp.
ISO 2422-86. Truncated Cone Abrasive Sleeves—Dimensions and Designation Second Edition. 3 pp.
JIS R 6257-80. Cylindrical Abrasive Sleeves.

—Cone
ISO 2422-86. Truncated Cone Abrasive Sleeves—Dimensions and Designation Second Edition. 3 pp.

—Cylindrical
ISO 2421-72. Cylindrical Abrasive Sleeves—Designation—Dimensions—Tolerances First Edition. 4 pp.
JIS R 6257-80. Cylindrical Abrasive Sleeves.

Abrasive Testing
Use: Abrasion Testing

Abrasive Wheels
Use: Grinding Wheels

Abrasives
Scope Note: For additional listings, use a more specific term *Use For:* Artificial Abrasives
See Also: Abrasive Belts; Abrasive Cloth; Abrasive Disks; Abrasive Grains; Abrasive Papers; Abrasive Rolls; Abrasive Sheets; Abrasive Sleeves; Aluminous Abrasives; Aluminum Oxide; Ceramics; Cleaning Agents; Coated Abrasives; Emery Paper; Glass Paper; Grinding Wheels; Grit; Metals; Paint Removers; Polishes; Silicon Carbide
CNS R2083-87. Artificial Abrasives (Nov)(3788).
JIS R 6111-87. Artificial Abrasives. 7 pp.
SAA AS 1863-76. Coated Abrasives (Technical Products). 12 pp.

Abrasives *(Cont.)*
—Bulk Density
BSI BS 7425: Part 1-91. 1991 Abrasive Grains Part 1: Method for the Determination of the Bulk Density of Macrograins. 8 pp.
CNS R3101-81. Method of Test for Bulk Density of Artificial Abrasives (Aug)(7841).
ISO 9136-89. Abrasive Macrograins—Determination of Bulk Density First Edition. 6 pp.
JIS R 6126-70. Testing Method for Bulk Density of Artificial Abrasives. 5 pp.

—Capillarity
BSI BS 7425: Part 2-91. 1991 Abrasive Grains Part 2: Method for Determining the Capillarity (ISO 9137: 1990). 8 pp.
CNS R3102-81. Method of Test for Capillarity of Artificial Abrasives (Aug)(7842).
ISO 9137-90. Abrasive Grains—Determination of Capillarity First Edition. 6 pp.
JIS R 6127-69. Testing Method for Capillarity of Artificial Abrasives. 5 pp.

—Density
CNS R3099-81. Method of Test for Specific Gravity of Artificial Abrasives (Jun)(7531). 2 pp.
JIS R 6125-76. Testing Method for Specific Gravity of Artificial Abrasives (R 1979). 6 pp.

—Designations
BSI BS 4481: Part 1-89. 1989 Bonded Abrasive Products Part 1: General Features of Abrasive Wheels, Segments, Bricks and Sticks. 16 pp.
ISO 525-86. Bonded Abrasive Products—General—Designation, Marking, Range of Outside Diameters and Tolerances Second Edition. 15 pp.
SAA AS B172.1-67. Bonded Abrasive Products —Part 1: Designation and Marking. 6 pp.
SNZ NZS/ISO 525-86. Bonded Abrasive Products—General—Designation, Marking, Range of Outside Diameters and Tolerances. 13 pp.

—Glossaries
JIS R 6004-90. Glossary of Terms and Marks Used in Abrasives, Grinding Wheels and Coated Abrasives. 20 pp.

—Grain Size Analysis
SAA AS 1487-75. Abrasive Grain Size. 30 pp.

—Identification Systems
BSI BS 4481: Part 1-89. 1989 Bonded Abrasive Products Part 1: General Features of Abrasive Wheels, Segments, Bricks and Sticks. 16 pp.
ISO 525-86. Bonded Abrasive Products—General—Designation, Marking, Range of Outside Diameters and Tolerances Second Edition. 15 pp.
SAA AS B172.1-67. Bonded Abrasive Products —Part 1: Designation and Marking. 6 pp.
SNZ NZS/ISO 525-86. Bonded Abrasive Products—General—Designation, Marking, Range of Outside Diameters and Tolerances. 13 pp.

—Magnetic Materials Content
CNS R3106-82. Method of Test for Magnetic Matter in Artificial Abrasives (Sep)(9436).
JIS R 6121-77. Testing Method for Magnetic Matter in Artificial Abrasives.
JIS R 6121-61. Testing Method for Magnetic Matter in Artificial Abrasives. 3 pp.

—Metallic—Wear Testing—Centrifugal Blasting
DIN ENGL 50315-88. Testing the Wear and Effectiveness of Metallic Abrasives by Centrifugal Blasting (Oct). 10 pp.

—PH
CNS R3103-81. Method of Test for pH of Artificial Abrasives (Aug)(7843).
JIS R 6129-76. Testing Method for pH of Artificial Abrasives.

—Sampling
CNS R3097-86. Method for Sampling of Abrasive Grains (Jan) (7529).

—Symbols
JIS R 6004-90. Glossary of Terms and Marks Used in Abrasives, Grinding Wheels and Coated Abrasives. 20 pp.

—Toughness—Ball Mill Grindability Testing
CNS R3107-82. Method of Test for Toughness of Artificial Abrasives (Ball Mill Method) (Sep)(9437).
JIS R 6128-87. Ball Mill Test for Toughness of Artificial Abrasives. 7 pp.

—Turbine Components
MOD UK TS 10126. Grain, Abrasive, Soft 12/30 Mesh (Withdrawn).

INTERNATIONAL AND NON-U.S. NATIONAL STANDARDS
SUBJECT INDEX

ABS Resins
Scope Note: For additional listings, see also specific products made from ABS resins
Use For: Acrylonitrile Butadiene Styrene Resins
See Also: Plastics; Styrene Butadiene Resins; Thermoplastic Resins

CNS K3072-85. Rigid Acrylonitrile-Butadiene-Styrene (ABS) Plastics (Jun)(11285). 2 pp.
CNS K6810-85. Method of Test for Rigid Acrylonitrile-Butadiene-Styrene (ABS) Plastics (Jun)(11286). 2 pp.

—Comprehensive Testing
JIS K 6874-86. Testing Methods for ABS Resins. 7 pp.

—Extrusion Materials
BSI BS 4935-89. 1989 Method of Specifying Acrylonitrile-Butadiene--Styrene (ABS) Moulding and Extrusion Materials. 10 pp.
ISO 2580 Pt 1-90. Plastics—Acrylonitrile/Butadiene/Styrene (ABS) Moulding and Extrusion Materials—Part 1: Designation Second Edition. 8 pp.
ISO 2580 Pt 2-82. Plastics—Acrylonitrile/Butadiene/Styrene (ABS) Moulding and Extrusion Materials—Part 2: Determination of Properties First Edition. 7 pp.

—Molding Materials
BSI BS 4935-89. 1989 Method of Specifying Acrylonitrile-Butadiene--Styrene (ABS) Moulding and Extrusion Materials. 10 pp.
ISO 2580 Pt 1-90. Plastics—Acrylonitrile/Butadiene/Styrene (ABS) Moulding and Extrusion Materials—Part 1: Designation Second Edition. 8 pp.
ISO 2580 Pt 2-82. Plastics—Acrylonitrile/Butadiene/Styrene (ABS) Moulding and Extrusion Materials—Part 2: Determination of Properties First Edition. 7 pp.

—Molding Materials—Test Specimens
DIN ENGL 16772 Pt 2-88. Plastics Moulding Materials; Acrylonitrile /Butadiene/Styrene (ABS) Moulding Materials; Preparation of Specimens and Determination of Their Properties (Dec). 4 pp.

—Sheets
CNS K3074-85. Acrylonitrile-Butadiene-Styrene (ABS) Sheets (Aug)(11337). 2 pp.
CNS K6812-85. Method of Test for Acrylonitrile-Butadiene-Styrene Sheets (Aug)(11338). 4 pp.
JIS K 6873-75. Acrylonitrile-Butadiene-Styrene (ABS) Sheets. 7 pp.
MOD UK TS 10027C. ABS Sheet Type 'P'.

Absolute Density
Use: Density

Absolute Humidity
Use For: Vapor Concentration; Vapor Density; Vapour Density *See Also:* Density; Humidity

—Radio Wave Propagation
CCIR RECMN 836-92. Surface Water Vapour Density. 3 pp.

Absolute Pressure Gages
Use For: Absolute Pressure Transmitters
See Also: Barometers; Differential Pressure Gages; Gage Pressure Transmitters; Pressure Gages; Pressure Transmitters

BSI BS 6447-84. 1984 Amd 1 Absolute and Gauge Pressure Transmitters with Electrical Outputs. 13 pp.

Absolute Pressure Transmitters
Use: Absolute Pressure Gages

Absolute Viscosity
Use: Viscosity

Absorbance
Use: Absorptance

Absorbent Materials
Use: Absorbents

Absorbent Papers
Scope Note: For additional listings, use a more specific term *Use For:* Bibulous Papers
See Also: Absorbents; Blotting Papers; Filter Paper; Papers; Sanitary Papers; Toilet Papers
MOD UK TS 409A. Paper, Absorbent.

—Water Absorption
CNS P3019-87. Method of Test for Water Absorbtivity of Bibulous Paper (Capillary Method) (Feb)(2645). 2 pp.
CNS P3087-87. Method of Test for Water Absorbency of Bibulous Papers—Absorption Rate Method (Sep)(12106).
CPPA F.4-92. Absorption of Water and Ink by Bibulous and Blotting Paper. 2 pp.

Absorbents
Use For: Absorbent Materials *See Also:* Absorbent Papers; Acoustics; Attenuators; Fuller's Earth; Insulating Board

—Oil
CGSB CAN/CGSB-25.14-M89. Absorbent Material for Oils, Liquid Fuels and Water. 8 pp.

—Water
CGSB CAN/CGSB-25.14-M89. Absorbent Material for Oils, Liquid Fuels and Water. 8 pp.

Absorbers (Equipment)
See Also: Air Conditioning Equipment; Clarifiers; Cooling Systems; Dehumidifiers; Separators (Mechanical); Shock Absorbers

—Elastomers—Solar Water Heaters
BSI BS 7431-91. 1991 Assessing Solar Water Heaters—Elastomeric Materials for Absorbers, Connecting Pipes and Fittings (ISO 9808: 1990). 8 pp.
ISO 9808-90. Solar Water Heaters—Elastomeric Materials for Absorbers, Connecting Pipes and Fittings—Method of Assessment First Edition. 6 pp.

Absorptance
Use For: Absorbance *See Also:* Density; Optical Properties; Reflectance; Surface Properties; Translucence; Transmissivity; Transmittance

—Caprolactam—Spectrometry
ISO 7059-82. Caprolactam for Industrial Use—Determination of Absorbance at a Wavelength of 290 nm First Edition. 3 pp.

Absorptiometric Analysis
See Also: Absorption Number; Chemical Analysis
CNS K0012-80. General Rules for Absorptiometric Analysis (Sep)(6494).
JIS K 0115-92. General Rules for Molecular Absorptiometric Analysis. 20 pp.

—Aluminum
BSI BS 1728: Part 5-53. 1953 Amd 1 Methods for the Analysis of Aluminium and Aluminium Alloys Part 5: Determination of Copper (Absorptiometric Method). 8 pp.
BSI BS 1728: Part 8-57. 1957 Methods for the Analysis of Aluminium and Aluminium Alloys Part 8: Determination of Iron (Absorptiometric 1:10-Phenanthroline Method). 8 pp.
BSI BS 1728: Part 10-57. 1957 Methods for the Analysis of Aluminium and Aluminium Alloys: Part 10: Determination of Manganese (Absorptiometric Method). 8 pp.
BSI BS 1728: Part 12-61. 1961 Amd 1 Methods for the Analysis of Aluminium and Aluminium Alloys: Part 12: Determination of Silicon (Absorptiometric Molybdenum Blue Method). 9 pp.

—Aluminum Alloys
BSI BS 1728: Part 5-53. 1953 Amd 1 Methods for the Analysis of Aluminium and Aluminium Alloys Part 5: Determination of Copper (Absorptiometric Method). 8 pp.
BSI BS 1728: Part 8-57. 1957 Methods for the Analysis of Aluminium and Aluminium Alloys Part 8: Determination of Iron (Absorptiometric 1:10-Phenanthroline Method). 8 pp.
BSI BS 1728: Part 10-57. 1957 Methods for the Analysis of Aluminium and Aluminium Alloys: Part 10: Determination of Manganese (Absorptiometric Method). 8 pp.
BSI BS 1728: Part 12-61. 1961 Amd 1 Methods for the Analysis of Aluminium and Aluminium Alloys: Part 12: Determination of Silicon (Absorptiometric Molybdenum Blue Method). 9 pp.

—Coal
CNS M3149-84. Method for Determination of Phosphorus of Coal and Coke (Molybdenum Blue Absorptiometric Method) (Mar)(10830).

—Coke
CNS M3149-84. Method for Determination of Phosphorus of Coal and Coke (Molybdenum Blue Absorptiometric Method) (Mar)(10830).

—Copper
JIS H 1061-89. Methods for Determination of Silicon in Copper and Copper Alloys. 10 pp.

—Copper Alloys
JIS H 1060-89. Methods for Determination of Cobalt in Copper and Copper Alloys. 12 pp.
JIS H 1061-89. Methods for Determination of Silicon in Copper and Copper Alloys. 10 pp.
JIS H 1063-89. Methods for Determination of Beryllium in Copper Alloys. 12 pp.
MOD UK M 9201/59. Determination of Phosphorous in Copper Alloys: Absorptiometric Method (No Information).

Absorptiometric Analysis (Cont.)

—Dolomite
CNS M3105-82. Method for Determining Iron Oxide in Dolomite (O-Phenanthroline Absorptiometric Method) (Sep)(9423).
CNS M3109-82. Method for Determining Phosphorus Pentoxide in Dolomite (MIBK Absorptiometric Method) (Sep)(9427).

—Feldspars
CNS M3082-82. Method for Determination of Silicon Dioxide in Feldspar (Dehydration Gravimetric and Absorptiometric Method) (Jan)(8437).
CNS M3083-82. Method for Determination of Silicon Dioxide in Feldspar (Flocculation Gravimetric and Absorptiometric Method) (Jan)(8438).
CNS M3086-82. Method for Determination of Iron Oxide in Feldspar (O-Phenanthroline Absorptiometric Method) (Jan)(8441).
CNS M3087-82. Method for Determination of Titanium Oxide in Feldspar (Diantipyrylmethane Absorptiometric Method) (Jan)(8442).
CNS M3090-82. Method for Determination of Sodium Oxide in Feldspar (Atomic Absorptiometric Method) (Jan)(8445).
CNS M3093-82. Method for Determination of Potassium Oxide in Feldspar (Atomic Absorptiometric Method) (Jan)(8448).

—Flue Gases
JIS K 0083-76. Method for Determination of Vanadium in Stack Gas. 12 pp.

—Insecticides—Aerosol
MOD UK M 5003/66. Examination of Insecticide Concentrate Space Spray and Insecticide Space Spray.

—Iron Ores
CNS M3016-80. Method for Determination of Cobalt in Iron Ores (Mar)(5378).
CNS M3079-81. Method for Determination of Copper in Iron Ores (BCOD Absorptiometric Method) (Oct)(8047).
CNS M3126-83. Method for Determining Titanium Dioxide in Iron Ores (Diantipyrylmethane Absorptiometric Method) (Apr)(10181).
CNS M3131-83. Method for Determining Tin in Iron Ores (Phenylfluorone Absorptiometric Method) (Apr)(10186).
CNS M3134-83. Method for Determining Zinc in Iron Ores (Zincon Absorptiometric Method) (Apr)(10189).
CNS M3136-83. Method for Determining Lead in Iron Ores (Dithizone Absorptiometric Method) (Apr)(10191).
JIS M 8210-83. Methods for Determination of Cobalt in Iron Ores. 11 pp.
JIS M 8218-83. Methods for Determination of Copper in Iron Ores. 14 pp.
JIS M 8223-83. Methods for Determination of Nickel in Iron Ores. 16 pp.
JIS M 8227-83. Methods for Determination of Tin in Iron Ores. 15 pp.
JIS M 8228-83. Methods for Determination of Zinc in Iron Ores.
JIS M 8228-71. Methods for Determination of Zinc in Iron Ores. 11 pp.
JIS M 8229-83. Methods for Determination of Lead in Iron Ores. 15 pp.
JIS M 8230-83. Methods for Determination of Bismuth in Iron Ores. 14 pp.

—Manganese Ores
CNS M3066-81. Method for Determination of Iron in Manganese Ores (o-Phenanthroline Absorptiometric Method) (Aug)(7831).
JIS M 8234-92. Manganese Ores—Determination of Iron Content. 16 pp.
JIS M 8242-89. Methods for Determination of Copper in Manganese Ores. 11 pp.

—Nickel Castings
JIS H 1276-88. Methods for Determination of Silicon in Nickel and Nickel Alloy Castings. 8 pp.
JIS H 1278-88. Methods for Determination of Phosphorus in Nickel and Nickel Alloy Castings. 10 pp.
JIS H 1281-88. Method for Determination of Vanadium in Nickel Alloy Castings. 7 pp.

—Ores
CNS M3050-81. Method for Determination of Nickel in Ores (Dimethylglyoxime Absorptiometric Method) (Jun)(7512).
CNS M3057-81. Method for Determination of Cobalt in Ores (Absorptiometric Method) (Jun)(7519).
CNS M3076-81. Method for Determination of Copper in Ores (Cupric Ammonia Absorptiometric Method) (Oct)(8044).

—Silicon
CNS G2112-82. Method of Chemical Analysis for Phosphorus in Metallic Silicon (Molybdate Blue Absorptimetric Method) (Jan)(8286).

INDUSTRY STANDARDS

INTERNATIONAL AND NON-U.S. NATIONAL STANDARDS
SUBJECT INDEX

Absorptiometric

Absorptiometric Analysis (Cont.)
—Silicon (Cont.)
  CNS G2136-82. Method of Chemical Analysis for Ferrite in Metallic Silicon (Sulfosalicylic Acid Absorptimetric Method) (Oct)(9495).
  CNS G2139-82. Method of Chemical Analysis for Aluminum in Metallic Silicon (Chrome Azurol S Absorptimetric Method) (Oct)(9498).
  CNS G2142-82. Method of Chemical Analysis for Calcium in Metallic Silicon (GHA Absorptimetric Method) (Oct)(9501).
—Water
  JIS K 0555-90. Testing Methods for Determination of Silica in Highly Purified Water. 9 pp.
—Zinc
  JIS H 1109-89. Methods for Determination of Iron in Zinc Metal. 10 pp.
—Zirconium
  JIS H 1661-86. Methods for Determination of Aluminium in Zirconium and Zirconium Alloys. 9 pp.
—Zirconium Alloys
  JIS H 1661-86. Methods for Determination of Aluminium in Zirconium and Zirconium Alloys. 9 pp.

Absorption
Scope Note: For absorption of specific products or materials, see the subheading Absorption under the specific material or product See: Absorptiometric Analysis; Absorption Number; Ink Absorption; Light Absorption; Oil Absorption; Water Absorption

Absorption Number
Use For: Absorption Value
See Also: Absorptiometric Analysis
—Carbon Black
  BSI BS 5293: Part 17-93. 1993 Sampling and Testing Carbon Black for Use in the Rubber Industry Part 17: Method for Determination of Dibutyl Phthalate Absorption Number Using an Absorptometer (ISO 4656-1: 1992) (V). 15 pp.
  BSI BS 5293: Part 17-88. 1988 Sampling and Testing Carbon Black for Use in the Rubber Industry Part 17: Method for Determination of Dibutylphthalate Absorption Number Using an Absorptometer. 11 pp.
  BSI BS 5293: Part 18-92. 1992 Sampling and Testing Carbon Black for Use in the Rubber Industry Part 18: Method for Determination of Dibutyl Phthalate Absorption Number Using Plastograph or Plasticorder (ISO 4656-2: 1991). 11 pp.
  BSI BS 5293: Part 19-91. 1991 Sampling and Testing Carbon Black for Use in the Rubber Industry Part 19: Preparation of Samples for Determination of Dibutyl Phthalate Absorption Number (Compressed Sample) (ISO 6894: 1991). 10 pp.
  DIN ENGL 53601-78. Testing of Carbon Blacks; Determination of the Dibutylphthalate Absorption of Carbon Blacks (Dec). 3 pp.
  ISO 4656 Pt 1-92. Rubber Compounding Ingredients—Carbon Black—Determination of Dibutyl Phthalate Absorption Number—Part 1: Method Using Absorptometer Third Edition. 12 pp.
  ISO 4656 Pt 2-91. Rubber Compounding Ingredients—Carbon Black—Determination of Dibutyl Phthalate Absorption Number—Part 2: Method Using Plastograph or Plasticorder Second Edition. 9 pp.
  ISO 6894-91. Rubber Compounding Ingredients—Carbon Black—Preparation of Samples for Determination of Dibutylphthalate Absorption Number (Compressed Sample) Second Edition. 7 pp.

Absorption Spectrophotometry
Scope Note: Use a more specific term See: Atomic Absorption Spectrophotometry; Infrared Spectrophotometry; Molecular Absorption Spectrophotometry; Spectrophotometry; Ultraviolet Spectrophotometry

Absorption Value
Use: Absorption Number

Absorptivity
See Also: Density; Opacity; Optical Properties
—Drying Agents
  BSI BS 3482: Part 6-91. 1991 Methods of Test for Desiccants Part 6: Determination of Adsorptive Capacity. 7 pp.
—Reagents
  CNS K7623-82. Chemical Reagent (Absorptivity Determination) (Jun)(8995).

Abstract Sheets
See Also: Abstracts; Periodicals

Abstract Sheets (Cont.)
—Periodicals
  ISO 5122-79. Documentation—Abstract Sheets in Serial Publications First Edition. 7 pp.

Abstract Syntax Notation One
Use: Open Systems Interconnection—Abstract Syntax Notation One

Abstracts
Use For: Summaries See Also: Abstract Sheets
  ISO 214-76. Documentation—Abstracts for Publications and Documentation First Edition. 13 pp.
—Computer Programs
  CNS C5147-83. Computer Program Abstracts (Feb)(9980).
—NES (Naval Engineering Standards)
  MOD UK NES 2: Part 2-91. Index of Standards for Ships and Weapons Part 2: Abstracts of Naval Engineering Standards NES 0 to 99 (Administrative/Procedural) Issue 3 (09.91). 12 pp.
  MOD UK NES 2: Part 3-91. Index of Standards for Ships and Weapons Part 3: Abstracts of Naval Engineering Standards NES 100 to 299 (Constructive) Issue 4 (09.91). 17 pp.
  MOD UK NES 2: Part 4-91. Index of Standards for Ships and Weapons Part 4: Abstracts of Naval Engineering Standards NES 300 to 499 (Mechanical) Issue 4 (09.91). 16 pp.
  MOD UK NES 2: Part 5-91. Index of Standards for Ships and Weapons Part 5: Abstracts of Naval Engineering Standards NES 500 to 699 (Electrical) Issue 4 (09.91). 23 pp.
  MOD UK NES 2: Part 6-91. Index of Standards for Ships and Weapons Part 6: Abstracts of Naval Engineering Standards NES 700 to 899 (General/Common) Issue 4 (09.91). 37 pp.
  MOD UK NES 2: Part 9-91. Index of Standards for Ships and Weapons Part 9: Abstracts of Naval Engineering Standards NES 1000 to 1099 (Weapons) Issue 3 (09.91). 11 pp.
  MOD UK NES 2: Part 10-91. Index of Standards for Ships and Weapons Part 10: Abstracts of Naval Engineering Standards NES 2000 to 3999 Issue 1 (09.91). 13 pp.

AC Circuit Breakers
Use: Circuit Breakers

AC Generators
Use For: Alternating Current Generators; Synchronous Generators See Also: Generators
  CNS C1007-88. Scheduled Information to Be Given on Enquiring and Ordering of Electric Machines (Medium and Small Size A. C. Synchronous Generators) (Jan)(104).
  CNS C4080-84. Alternating Current Generator (Small Size) (Dec)(2901). 9 pp.
  SAA AS 1359.40-77. Rotating Electrical Machines—General Requirements—Part 40: Characteristics of Synchronous Generators. 7 pp.
—Aircraft—Constant Frequency
  BSI 2G 124: Part 1-66. 1966 Amd 1 A.C. Generators for Aircraft Part 1: Constant Frequency. 7 pp.
—Aircraft—Variable Frequency
  BSI 2G 124: Part 2-66. 1966 A.C. Generators for Aircraft Part 2: Variable Frequency. 5 pp.
—Design—Interoperability
  NATO STANAG 4135 ED 1 AMD 0-77. Electrical Characteristics of Rotating Alternating Current Generating Sets. 14 pp.
  NATO STANAG 4135 ED 1 AMD 3-77. Electrical Characteristics of Rotating Alternating. 14 pp.
—Generator Sets—Internal Combustion Piston Engines
  ISO 8528 Pt 3-93. Reciprocating Internal Combustion Engine Driven Alternating Current Generating Sets—Part 3: Alternating Current Generators for Generating Sets First Edition. 16 pp.
—Safety
  CSA CAN/CSA-C22. 2 NO 100-92. Motors and Generators; (Gen Instr 1). 74 pp.
  CSA 1169 Bull. Electrical Bulletin 1169 June 27, 1978 to C22.2 NO 100. 2 pp.
—Ships
  CNS F5107-86. A.C. Generators in Ships (Dec)(11790).
  MOD UK NES 342-01. Trials Oil Driven Generators On-Board Procedures Issue 2 (12.81); Amendment 1. 28 pp.
  MOD UK NES 630-88. Requirements for Main 60 Hz Generators Issue 3 (11.88). 24 pp.

AC Generators (Cont.)
—Submarines
  MOD UK NES 630-88. Requirements for Main 60 Hz Generators Issue 3 (11.88). 24 pp.

AC Motors
See Also: Induction Motors; Motors; Single Phase Motors; Squirrel Cage Motors; Universal Motors
—Electric Vehicles
  IEC 349 Pt 2-93. Electric Traction—Rotating Electrical Machines for Rail and Road Vehicles—Part 2: Electronic Convertor-Fed Alternating Current Motors First Edition. 62 pp.
—Motor Protectors
  CSA C22.2 NO 77-1988. Motors with Inherent Overheating Protection; (Gen Instr 1). 23 pp.
  CSA 1169 Bull. Electrical Bulletin 1169 June 27, 1978 to C22.2 NO 77. 2 pp.
—Railroad Traction Equipment
  IEC 349 Pt 2-93. Electric Traction—Rotating Electrical Machines for Rail and Road Vehicles—Part 2: Electronic Convertor-Fed Alternating Current Motors First Edition. 62 pp.
  JIS E 6102-90. AC Traction Motors of Railway Rolling Stock—Test Methods.
—Safety
  CSA CAN/CSA-C22. 2 NO 100-92. Motors and Generators; (Gen Instr 1). 74 pp.
  CSA 1169 Bull. Electrical Bulletin 1169 June 27, 1978 to C22.2 NO 100. 2 pp.
—Ships—Controllers
  CNS F5119-87. Starters and Controllers for A.C. Motor in Ships (Jun)(11997).
—Ships—Starters
  CNS F5119-87. Starters and Controllers for A.C. Motor in Ships (Jun)(11997).
—Thermal Protection
  SAA AS 1023.2-89. Low Voltage Switchgear and Controlgear—Protection of Electric Motors—Part 2: Current Sensing Protection Devices for a.c. Motors. 40 pp.
  SAA AS 1023.3-73. Low Voltage Switchgear and Controlgear—Protection of Electric Motors—Part 3: Inherent Overheat Protectors Reconfirmed 1984. 16 pp.
  SNZ NZS/AS 1023. 2-89. Low Voltage Switchgear and Controlgear—Protection of Electric Motors Part 2: Current-Sensing Protection Devices for a.c. Motors (This is a Joint Standard with SAA AS 1023.2). 40 pp.

AC Power Supplies
See Also: Power Supplies
—Stabilized
  IEC 686-80. Stabilized Power Supplies, A.C. Output First Edition. 68 pp.
—Three Phase—Cables (Electric)—Ships
  NATO STANAG 4143 ED 1 AMD 4-76. Ship/Shore Connection Terminals for 3-Phase AC Power. 6 pp.

AC Power Transmission
Use: Power Transmission—AC

ac Watt Hour Meters
Use: Watt Hour Meters

Acaroid Resins
See Also: Resins
—Pyrotechnics
  MOD UK DSTAN 13-123-91. Acariod Resin Issue 1. 8 pp.
  MOD UK DSTAN 13-134-93. Composition SR 239 Issue 1. 12 pp.

Accelerated Aging Testing
Use: Aging Testing

Accelerated Corrosion Testing
See Also: Accelerated Testing
—Atmospheric Corrosion Testing
  IEC 355-71. Appraisal of the Problems of Accelerated Testing for Atmospheric Corrosion First Edition. 21 pp.

Accelerated Testing
See Also: Accelerated Corrosion Testing; Acid Resistance Testing; Corrodkote Testing; Corrosion Testing; Immersion Testing
—Aging—Terminal Blocks
  CSA C22.2 NO 158-1987. Terminal Blocks; (Gen Instr 1 Thru 4). 32 pp.

INDEX and DIRECTORY of

INTERNATIONAL AND NON-U.S. NATIONAL STANDARDS
SUBJECT INDEX
Accelerators

Accelerated Testing (Cont.)

—Aluminum Coatings (On Aluminum)
CGSB 1-GP-71 METH 122.8-79. Methods of Testing Paints and Pigments Accelerated Weathering Test for One Coat of Air-Drying Material on Steel or Aluminum Panels. 1 p.

CGSB 1-GP-71 METH 122.10-79. Methods of Testing Paints and Pigments Accelerated Weathering Test for Aircraft Finishes on Primed Anodized Aluminum Panels. 1 p.

—Anodic Coatings
BSI BS 6161: Part 7-84. 1984 Methods of Test for Anodic Oxidation Coatings on Aluminium and Its Alloys Part 7: Accelerated Determination of Light Fastness of Coloured Anodic Oxidation Coatings Using Artificial Light. 4 pp.

ISO 2135-84. Anodizing of Aluminium and Its Alloys—Accelerated Test of Light Fastness of Coloured Anodic Oxide Coatings Using Artificial Light Second Edition. 4 pp.

—Coatings
CGSB 1-GP-71 METH 122.2-74. Methods of Testing Paints and Pigments Accelerated Weathering General Procedure. 2 pp.

—Coatings—Test Panels
CGSB 1-GP-71 METH 122.1-74. Methods of Testing Paints and Pigments Accelerated Weathering Selection, Exposure and Examination of Panels. 1 p.

—Electrical Components
BSI BS 2011: Part 2.1GA-84. (WITHDRAWN) 1984 Amd 1 Basic Environmental Testing Procedures Part 2.1: Tests Part 2.1GA: Acceleration, Steady State (Superseded by BS EN 60068-2-7: 1993). 12 pp.

BSI BS 2011: Sec 4.1-83. (WITHDRAWN) 1983 Basic Environmental Testing Procedures Part 4: Miscellaneous Section 4.1: Specification for Mounting of Components, Equipment and Other Articles for Dynamic Tests (Renumbered as BS EN 60068-2-47: 1993). 15 pp.

BSI BS EN 60068-2-7-93. 1993 Environmental Testing Part 2: Tests Test Ga and Guidance: Acceleration, Steady State (IEC 68-2-7: 1983) (G). 16 pp.

BSI BS EN 60068-2-47-93. 1993 Environmental Testing Part 2: Tests Mounting of Components, Equipment and Other Articles for Dynamic Tests Including Shock (Ea), Bump (Eb), Vibration (Fc and Fd) and Steady-State Acceleration (Ga) and Guidance (IEC 68-2-47: 1982) (G). 23 pp.

CENELEC HD 323.2.7 S2-87. Basic Environmental Testing Procedures—Part 2: Tests. Test Ga and Guidance: Acceleration, Steady State. 3 pp.

CENELEC HD 323.2.47 S1-88. Basic Environmental Testing Procedures—Part 2: Tests. Mounting of Components. Equipment and Other Articles for Dynamic Tests Including Shock (Es). Bump (Eb). Vibration (Fc and Fd) and Steady-State Acceleration (Ga) and Guidance. 3 pp.

CENELEC EN 60068-2-7-93. Basic Environmental Testing Procedures Part 2: Tests Test Ga and Guidance: Acceleration, Steady State (IEC 68-2-7: 1983 + A1: 1986). 11 pp.

CENELEC EN 60068-2-47-93. Basic Environmental Testing Procedures Part 2: Tests Mounting of Components, Equipment and Other Articles for Dynamic Tests Including Shock (Ea), Bump (Eb), Vibration (Fc and Fd) and Steady-State Acceleration (Ga) and Guidance (IEC 68-2-47: 1982). 16 pp.

IEC 68 Pt 2-7-83. Basic Environmental Testing Procedures Part 2: Tests—Test Ga and Guidance: Acceleration, Steady State Second Edition; (Amendment 1-1986) (CENELEC EN 60068-2-7: 1993). 32 pp.

IEC 68 Pt 2-47-82. Basic Environmental Testing Procedures Part 2: Tests Mounting of Components, Equipment and Other Articles for Dynamic Tests Including Shock (Ea), Bump (Eb), Vibration (Fc and Fd) and Steady-State Acceleration (Ga) and Guidance First Edition (CENE EN 60068-2-47:93). 29 pp.

SAA AS 1099.2.7-90. Basic Environmental Testing Procedures for Electrotechnology—Part 2: Tests—Part 2.7: Test Ga—Acceleration, Steady State (IEC 68-2-7). 8 pp.

SAA AS 1099.3.7-81. Basic Environmental Testing Procedures for Electrotechnology—Part 3: Background Information—Part 3.7: Section 7—Appraisal of the Problems of Accelerated Testing for Atmospheric Corrosion. 4 pp.

SNZ IEC 68: Part 2-7-83. Basic Environmental Testing Procedures Part 2.7: Test Ga and Guidance: Acceleration, Steady State. 23 pp.

SNZ IEC 68: Part 2-47-82. Basic Environmental Testing Procedures Part 2-47: Mounting of Components and Other Articles for Dynamic Tests Including Shock (Ea), Bump (Eb), Vibration (Fc and Fd) and Steady-State Acceleration (Ga) and Guidance. 25 pp.

Accelerated Testing (Cont.)

—Electromechanical Equipment
BSI BS 5772: Part 4-79. 1979 Basic Testing Procedures and Measuring Methods for Electromechancial Components for Electronic Equipment Part 4: Dynamic Stress Tests. 8 pp.

IEC 512 Pt 4-76. Electromechanical Components for Electronic Equipment; Basic Testing Procedures and Measuring Methods Part 4: Dynamic Stress Tests First Edition. 17 pp.

—Masonry Coatings
CGSB 1-GP-71 METH 122.13-79. Methods of Testing Paints and Pigments Accelerated Weathering Test for Masonry Coatings. 1 p.

—Oxide Coatings
BSI BS 6161: Part 7-84. 1984 Methods of Test for Anodic Oxidation Coatings on Aluminium and Its Alloys Part 7: Accelerated Determination of Light Fastness of Coloured Anodic Oxidation Coatings Using Artificial Light. 4 pp.

ISO 2135-84. Anodizing of Aluminium and Its Alloys—Accelerated Test of Light Fastness of Coloured Anodic Oxide Coatings Using Artificial Light Second Edition. 4 pp.

—Paints
CGSB 1-GP-71 METH 30.3-74. Methods of Testing Paints and Pigments Accelerated Stability at 60 Degrees C (140 Degrees F). 1 p.

CGSB 1-GP-71 METH 30.4-80. Methods of Testing Paints and Pigments Accelerated Stability at 45 Degrees C. 1 p.

CGSB 1-GP-71 METH 30.6-80. Methods of Testing Paints and Pigments Accelerated Stability Heat Stability. 1 p.

CGSB 1-GP-71 METH 122.4-79. Methods of Testing Paints and Pigments Accelerated Weathering Tests for House Paints on Wooden Panels. 1 p.

—Primers
CGSB 1-GP-71 METH 122.11-79. Methods of Testing Paints and Pigments Accelerated Weathering Test for Wood Primer on Sealed Wooden Panels. 1 p.

CGSB 1-GP-71 METH 122.12-79. Methods of Testing Paints and Pigments Accelerated Weathering Test for Wood Primer on Sealed and Topcoated Wooden Panels. 1 p.

—Sealing Materials
CGSB CAN2-19.0-M77METH17.1-78. Methods of Testing Putty, Caulking and Sealing Compounds Accelerated Weathering General Procedure. 2 pp.

—Steel Coatings
CGSB 1-GP-71 METH 122.6-79. Methods of Testing Paints and Pigments Accelerated Weathering Test for Two Coats of Air-Drying Material on Steel Panels. 1 p.

CGSB 1-GP-71 METH 122.7-80. Methods of Testing Paints and Pigments Accelerated Weathering Test for Two Coats of Baking Material on Steel Panels. 1 p.

CGSB 1-GP-71 METH 122.8-79. Methods of Testing Paints and Pigments Accelerated Weathering Test for One Coat of Air-Drying Material on Steel or Aluminum Panels. 1 p.

CGSB 1-GP-71 METH 122.9-79. Methods of Testing Paints and Pigments Accelerated Weathering Test for One Coat Material on Primed Steel Panels. 1 p.

—Wood Coatings
CGSB 1-GP-71 METH 122.19-79. Methods of Testing Paints and Pigments Accelerated Weathering Test for Two Coats of Pigmented Stain on Wood Panels. 1 p.

—Wood Coatings—Test Panels
CGSB 1-GP-71 METH 97.1-74. Methods of Testing Paints and Pigments Preparation of Wooden Panels General Conditions. 2 pp.

Accelerated Weathering Testing
Use: Environmental Testing

Accelerated Yellowing
Use: Yellowing

Acceleration Testing
See Also: Accelerometers; Collision Research; Deceleration Testing; Mechanical Shock; Vibration

—Aircraft
BSI 3G 100:Pt 2: SUBSEC 3.6-72. 1972 Amd 1 General Requirements for Equipment for Use in Aircraft Part 2: All Equipment Section 3: Environmental Conditions Subsection 3.6: Acceleration Requirements. 7 pp.

ISO 2669-78. Environmental Tests for Aircraft Equipment—Part 3.2: Steady State Acceleration First Edition. 6 pp.

Acceleration Testing (Cont.)

—Automotive
CNS D3016-89. Method of Acceleration Test for Automobiles (Apr)(2735).

CNS D3099-89. Accelerate Pump Test Code of Carburetors for Automobiles (Jan)(8256).

JIS D 1014-82. Acceleration Test Method of Automobiles. 6 pp.

—Carburetors—Automotive
CNS D3093-89. Acceleration Test Code of Carburetors for Automobiles (Jan)(8250).

—Connectors
BSI BS EN 2591-D1-92. 1992 Elements of Electrical and Optical Connection—Test Methods Part D1: Acceleration, Steady State. 9 pp.

CEN PREN 2591 (Part D1)-91. Elements of Electrical and Optical Connection Test Methods Part D1—Acceleration, Steady State. 4 pp.

CEN EN 2591-D1-92. Aerospace Series—Elements of Electrical and Optical Connection—Test Methods—Part D1: Acceleration, Steady State. 4 pp.

CNS C6184-88. Method of Test for Low Frequency (Below 3 MHz) Electrical Connectors (TP-1 Acceleration) (Apr)(9363).

—Fiber Optics
CNS C6219-87. Method of Test for Fiber Optic Devices (FOTP-18 Acceleration Test) (Sep)(10874).

—Ground Effect Machines
CAA Chapter B4-3 11.72. Collision Accelerations. 1 p.

—Motorcycles
CNS D3031-88. Method of Acceleration Test for Motor Cycles (May)(3107). 5 pp.

JIS D 1035-82. Method of Acceleration Test for Motor Cycles (R 1987). 8 pp.

—Semiconductor Devices
CNS C6050-88. Environmental Testing Methods and Endurance Testing Methods for Discrete Semiconductor Devices (Test of Constant Acceleration) (Jun)(5075).

—Test Specimens—Mounting
CNS C6337-90. Basic Environmental Testing Procedures (Part 2: Tests, Mounting of Components, Equipment and Articles for Dynamic Tests Including Shock (Ea), Bump (Eb), Vibration (Fc and Fd) and Steady—State Acceleration (Ga) and Guidance (May)(12716).

—Wheelchairs—Electric
CSA CAN/CSA-Z323.4.6-M89. Wheelchairs—Determination of Maximum Speed, Acceleration, and Retardation of Electric Wheelchairs (ISO 7176-6:1988); (Gen Instr 1). 16 pp.

ISO 7176 Pt 6-88. Wheelchairs—Part 6: Determination of Maximum Speed, Acceleration and Retardation of Electric Wheelchairs First Edition. 4 pp.

Accelerators (Materials)
See Also: Admixtures; Retarders (Materials)
DIN ENGL 16945-89. Testing of Resins, Hardeners and Accelerators, and Catalyzed Resins (Mar). 16 pp.

—1,3-Diphenylguanidine
CNS K4036-78. Rubber Vulcanization Accelerator DPG (1, 3-Diphenyl Guanidine) (Oct)(4630).

—Epoxy Resins
ISO 4597 Pt 1-83. Plastics—Hardeners and Accelerators for Epoxide Resins—Part 1: Designation First Edition. 4 pp.

—Rubber—Vulcanization—Organic
CNS K4034-78. Rubber Vulcanization Accelerator MBT (2-Mercapto-Benzothiazolyl) (Oct)(4628).

CNS K4035-78. Rubber Vulcanization Accelerator MBTS (Benzothiozolyl Disulfide or Dibenzothiazyl Disulfide) (Oct)(4629).

CNS K4036-78. Rubber Vulcanization Accelerator DPG (1, 3-Diphenyl Guanidine) (Oct)(4630).

CNS K4037-78. Rubber Vulcanization Accelerator ZnMDC (Zinc Dimethyl Dithiocarbamate) (Oct)(4631).

CNS K4038-78. Rubber Vulcanization Accelerator TMTD (Tetramethyl Thiuram Disulfide) (Oct)(4632).

CNS K4039-78. Rubber Vulcanization Accelerator ZnEPDC (Zinc Ethylphenyl Dithiocarbamate) (Oct)(4633).

CNS K4040-78. Rubber Vulcanization Accelerator NaDBC (Sodium Di-N-Butyl Dithiocarbamate) (Oct)(4634).

CNS K4041-78. Rubber Vulcanization Accelerator TMTM (Tetramethyl Thiuram Monosulfide) (Oct)(4635).

CNS K4042-78. Rubber Vulcanization Accelerator DOTG (Dio-Toyl Guanidine) (Oct)(4636).

CNS K4049-79. Rubber Vulcanization Accelertor CBS (N-Cyclo-Hexyl-2-Benzothiazyl Sulfenamide) (Oct)(5005).

INDUSTRY STANDARDS

Accelerators (Materials) (Cont.)
—Vulcanization—Organic
JIS K 6202-79. Organic Vulcanization Accelerators for Rubber.

Accelerators (Mechanical)
See Also: Automotive Equipment; Electron Accelerators; Speed Controls

—Dry Pipe Valves
ISO 6182 Pt 4-93. Fire Protection—Automatic Sprinkler Systems—Part 4: Requirements and Test Methods for Quick-Opening Devices First Edition. 11 pp.

Accelerometers
See Also: Acceleration Testing; Rate of Climb Indicators; Servomechanisms; Speedometers; Vibration Meters

—Aircraft
MOD UK DSTAN 66-37-86. Indicating Accelerometers Issue 1. 8 pp.
NATO STANAG 3330 ED 3 AMD 5-77. Indicating Accelerometers. 8 pp.
NATO STANAG 3330 ED 3 AMD 6-77. Indicating Accelerometers. 8 pp.

—Aircraft—Inertial Navigation Systems
NATO STANAG 4218 ED 1 AMD 2-84. Accelerometer Test Procedure for an Aircraft Inertial Navigation System. 31 pp.

—Gliders
CAA JAR-22 SUBPART F. Equipment (Joint Airworthiness Requirements). 4 pp.
CAA JAR-22 SUBPART G. Operating Limitations and Information (Joint Airworthiness Requirements). 13 pp.

—Mechanical Shock
BSI BS 7129-89. 1989 Recommendations for Mechanical Mounting of Accelerometers for Measuring Mechanical Vibration and Shock. 11 pp.
ISO 5348-87. Mechanical Vibration and Shock—Mechanical Mounting of Accelerometers First Edition. 10 pp.
MOD UK DSTAN 05-4-68. Methods for Specifying and Calibrating Shock and Vibration Instruments Issue 1. 4 pp.

—Mounting
BSI BS 7129-89. 1989 Recommendations for Mechanical Mounting of Accelerometers for Measuring Mechanical Vibration and Shock. 11 pp.
ISO 5348-87. Mechanical Vibration and Shock—Mechanical Mounting of Accelerometers First Edition. 10 pp.

—Vibration
BSI BS 7129-89. 1989 Recommendations for Mechanical Mounting of Accelerometers for Measuring Mechanical Vibration and Shock. 11 pp.
ISO 5348-87. Mechanical Vibration and Shock—Mechanical Mounting of Accelerometers First Edition. 10 pp.

Accent Lighting
Use: Display Lighting

Acceptance Sampling
Use: Sampling

Access Codes
See Also: Codes

—Telex Communications
CCITT RECMN E.216-89. Selection Procedures for the INMARSAT Mobile-Satellite Telephone and ISDN Services—Telephone Network and ISDN—Operation, Numbering, Routing and Mobile Service (Study Group II) 12 pp. 12 pp.
CCITT RECMN F.60-89. Operational Provisions for the International Telex Service—Telegraph and Mobile Services Operations and Quality of Service (Study Group I) 17 pp. 17 pp.
CCITT RECMN F.68-89. Establishment of the Automatic Intercontinental Telex Network—Telegraph and Mobile Services Operations and Quality of Service (Study Group I) 6 pp. 6 pp.
CCITT RECMN F.126-89. Selection Procedures for the INMARSAT Mobile-Satellite Telex Service—Telegraph and Mobile Services Operations and Quality of Service (Study Group I) 11 pp. 11 pp.

—Telex Communications—Tables (Data)
CCITT RECMN E.216-89. Selection Procedures for the INMARSAT Mobile-Satellite Telephone and ISDN Services—Telephone Network and ISDN—Operation, Numbering, Routing and Mobile Service (Study Group II) 12 pp. 12 pp.

Access Codes (Cont.)
—Telex Communications—Tables (Data) (Cont.)
CCITT RECMN E.252-89. Mode of Application of the Flat-Rate Price Procedure Set Forthe in Recommendations D.67 and D.150 for Remuneration of Facilities Made Available to the Administrations of Other Countries—Telephone Network and ISDN—Operation, Numbering, Routing and Mobile. 1 p.
CCITT RECMN F.126-89. Selection Procedures for the INMARSAT Mobile-Satellite Telex Service—Telegraph and Mobile Services Operations and Quality of Service (Study Group I) 11 pp. 11 pp.

Access Control Equipment
Use: Access Control Systems

Access Control Systems
Use For: Access Control Equipment; Entry Control Systems; Proximity Access Control Systems; Proximity Systems *See Also:* Alarm Systems; Control Systems Equipment; Gates (Barriers); Security Systems; Traffic Barriers

—MAC/Packet System
BSI BS EN 50094-93. 1993 Access Control System for the MAC/Packet Family: EURCRYPT (S). 200 pp.
CENELEC EN 50094-92. Access Control System for the MAC/Packet Family: EUROCRYPT. 197 pp.

—Open Systems Interconnection
IEC DIS 10164 Pt 9-93. Information Technology—Open Systems Interconnection—Systems Management: Objects and Attributes for Access Control; (CCITT RECMN X.741) ***CD-ROM ONLY***. 97 pp.
ISO DIS 10164 Pt 9-93. Information Technology—Open Systems Interconnection—Systems Management: Objects and Attributes for Access Control; (CCITT RECMN X.741) ***CD-ROM ONLY***. 97 pp.
JTC1 DIS10164 Pt 9-93. Information Technology—Open Systems Interconnection—Systems Management: Objects and Attributes for Access Control; (CCITT RECMN X.741) ***CD-ROM ONLY***. 97 pp.

Access Covers
Use: Access Doors

Access Doors
Use For: Access Panels *See Also:* Access Fittings; Doors

—Aircraft—Assembly
SBAC AS 1933 ISSUE 4 (I). Assembly of Access Door. 1 p.

—Aircraft—Frames
SBAC AS 1934 ISSUE 4 (I). Frame for Access Door. 1 p.

—Aircraft—Symbols
BSI M 43-73. 1973 Amd 1 Methods of Zoning Aircraft and Referencing Access Doors and Panels. 9 pp.
ISO 2529-73. Aircraft—Zones, Access Doors and Panels—Referencing System First Edition. 8 pp.

—Buildings
SAA AS 1428.1-93. Design for Access and Mobility—Part 1: General Requirements for Access—Buildings (in Professional Packages 17, 20, 21, 30, 55, 61, 62, 63, 64, 65, 66, 67, 68, 69) Amdt 1 October 1993. 59 pp.
SAA AS 1428.1 Supp 1-90. Design for Access and Mobility—Part 1: General Requirements for Access—Buildings—Supplement 1: Commentary (Supplement to AS 1428.1 —1988). 18 pp.

—Earthmoving Equipment
BSI BS 6112-83. (WITHDRAWN) 1983 Minimum Access Dimensions for Earth-Moving Machinery (Superseded by BS 6912: part 5: 1992). 6 pp.
BSI BS 6112-01. (WITHDRAWN) 1983 Amd 1 Minimum Access Dimensions for Earth-Moving Machinery (AMD 6877) May 1, 1992 (Superseded by BS 6912: Part 5: 1992). 11 pp.
BSI BS 6912: Part 5-92. 1992 Safety of Earth-Moving Machinery Part 5: Recommendations for Minimum Access Dimensions (ISO 2860: 1992). 9 pp.
CEN EN 22860-85. Earth-Moving Machinery Minimum Access Dimensions. 6 pp.
CNS A4014-88. Earth-Moving Machinery-Minimum Access Dimensions (Dec)(9945).
CNS A4015-88. Earth-Moving Machinery Access Systems (Dec)(9946).
CSA CAN/CSA-M2860-M91. Minimum Access Dimensions for Servicing Machines (EMM,FM) (ISO 2860-1983). 15 pp.
ISO 2860-92. Earth-Moving Machinery—Minimum Access Dimensions Fourth Edition. 7 pp.

Access Doors (Cont.)
—Earthmoving Equipment (Cont.)
ISO 2867-89. Earth-Moving Machinery—Access Systems Fourth Edition. 10 pp.
SAA AS 3868-91. Earth-Moving Machinery—Design Guide for Access Systems. 10 pp.

—Ergonomics—Body Dimensions
DIN ENGL 33402 Pt 4-86. Human Body Dimensions; Principles of Dimensioning Passages and Accesses (Oct). 6 pp.

Access Fittings
See Also: Access Doors

—Chimneys
BSI BS 3572-86. 1986 Amd 1 Access Fittings for Chimneys and Other High Structures in Concrete or Brickwork. 20 pp.
BSI BS 3678-86. 1986 Access Hooks for Chimneys and Other High Structures. 8 pp.

—Construction Equipment
JIS A 8302-86. Construction Machinery—Access Systems. 10 pp.

—Earthmoving Equipment
CNS A4015-88. Earth-Moving Machinery Access Systems (Dec)(9946).
ISO 2867-89. Earth-Moving Machinery—Access Systems Fourth Edition. 10 pp.

—Ships
MOD UK NES 127-81. Access Fittings and Equipment Issue 1 (10.81). 38 pp.
MOD UK NES 149-83. Access Policy in Surface Ships Issue 1 (07.83). 19 pp.

Access Lines (Telecommunications)
See Also: Telephone Circuits; Telephone Lines

—Telephone Circuits—Maintenance
CCITT RECMN O.11-89. Maintenance Access Lines—Specifications for Measuring Equipment (Study Group IV) 6 pp. 6 pp.

Access Networks
Use For: AN *See Also:* Telecommunication Networks

—Interfaces—V5.1—Local Exchanges
CENELEC PRETS 300 324-1-93. Signalling Protocols and Switching (SPS); V5.1 Interface Specification for the Support of Access Network. 270 pp.
ETSI PRETS 300 324-1-93. Signalling Protocols and Switching (SPS); V5.1 Interface Specification for the Support of Access Network. 270 pp.

Access Panels
Use: Access Doors

Access Systems
Use: Access Fittings

Access Units, Messaging
Use: Message Access Units

Accident Analysis
Use: Accident Investigations

Accident Investigations
Use For: Accident Analysis *See Also:* Aircraft Accidents; Collision Research; Incident Investigations

—Aircraft
CAA Section 8 CAP 393. Investigation of Accidents. 35 pp.
ICAO Annex 13. Aircraft Accident Investigation Seventh Edition—May 1988; (Supplement 3/4/89). 62 pp.
ICAO 6920. Manual of Aircraft Accident Investigation Fourth Edition—1970; (Amendment 10 10/17/86). 705 pp.
NATO STANAG 3531 ED 5 AMD 3-84. Investigation of Aircraft Missile Accidents/Incidents. 15 pp.
NATO STANAG 3531 ED 6 AMD 0-91. Safety Investigation and Reporting of Accidents/Incidents Involving Military Aircraft and/or Missiles. 11 pp.
NATO STANAG 3531 ED 6 AMD 1-91. Safety Investigation and Reporting of Accidents/Incidents Involving Military Aircraft and/or Missiles. 12 pp.
NATO STANAG 3531 ED 6 AMD 2-91. Safety Investigation and Reporting of Accidents/Incidents Involving Military Aircraft and/or Missiles. 13 pp.

—Aircraft—Human Factors
ICAO Circular 240. Human Factors Digest No. 7 Investigation of Human Factors in Accidents and Incidents—1993. 61 pp.

INTERNATIONAL AND NON-U.S. NATIONAL STANDARDS
SUBJECT INDEX

Accident Investigations *(Cont.)*
—**Missiles**
NATO STANAG 3531 ED 5 AMD 3-84. Investigation of Aircraft Missile Accidents/Incidents. 15 pp.
NATO STANAG 3531 ED 6 AMD 0-91. Safety Investigation and Reporting of Accidents/Incidents Involving Military Aircraft and/or Missiles. 11 pp.
NATO STANAG 3531 ED 6 AMD 1-91. Safety Investigation and Reporting of Accidents/Incidents Involving Military Aircraft and/or Missiles. 12 pp.
NATO STANAG 3531 ED 6 AMD 2-91. Safety Investigation and Reporting of Accidents/Incidents Involving Military Aircraft and/or Missiles. 13 pp.

Accident Prevention
Scope Note: For additional listings, use a more specific term *See Also:* Aircraft Accidents; Alarm Systems; Collision Avoidance; Firesafety; Gas Detectors; Occupational Safety and Health; Reflective Materials; Warning Systems

—**Children**
ISO Guide 50-87. Child Safety and Standards—General Guidelines First Edition. 16 pp.

—**Classification**
BSI BS 1000: (614)-90. 1990 Universal Decimal Classification (UDC). English Full Edition (614): Public Health and Hygiene. Accident Prevention. 26 pp.
SNZ NZS/BS 1000 (614)-69. Universal Decimal Classification Public Health. Accident Prevention. 20 pp.

—**Colors**
BSI BS 5378: Part 1-80. 1980 Amd 1 Safety Signs and Colours Part 1: Colour and Design. 12 pp.
BSI BS 5378: Part 2-80. 1980 Amd 2 Safety Signs and Colours Part 2: Colormetric and Photometric Properties of Materials. 11 pp.
BSI BS 5378: Part 3-82. 1982 Safety Signs and Colours Part 3: Specification for Additional Signs to Those Given in BS 5378: Part 1. 15 pp.
ISO 3864-84. Safety Colours and Safety Signs First Edition. 16 pp.
MOD UK NES 784-01. Requirements for Safety Signs and Colours Issue 3 (11.90); Amendment 1. 31 pp.

—**Floor Coverings**
SNZ NZS 5841: Part 1-88. Code of Practice for the Reduction of Slip Hazards Part 1: Guidelines for the Selection, Installation, Care and Maintenance of Flooring and Other Surfaces. 12 pp.

—**Lighting—Colored**
CNS Z1027-87. General Rules of Coloured Light for Safety (Mar)(9331).
JIS Z 9104-87. General Rules of Coloured Light for Safety. 9 pp.
JIS Z 9106-90. Fluorescent Safety Colours—General Rules for Application.

—**Signs**
SAA AS 1319-83. Safety Signs for the Occupational Environment Amdt 1 May 1988 (This is a Joint Standard with SANZ NZS 1319). 27 pp.
SAA AS 1319/E-79. Symbolic Safety Signs 1319/E/English 1319/E/Arabic 1319/E/Greek 1319/E/Italian 1319/E/Serbo-Croatian 1319/E/Spanish 1319/E/Turkish.
SNZ NZS/AS 1319-83. Safety Signs for the Occupational Environment Amend: 1 (This is a Joint Standard with SAA AS 1319). 27 pp.

—**Signs—Safety**
BSI BS 5378: Part 1-80. 1980 Amd 1 Safety Signs and Colours Part 1: Colour and Design. 12 pp.
BSI BS 5378: Part 2-80. 1980 Amd 2 Safety Signs and Colours Part 2: Colormetric and Photometric Properties of Materials. 11 pp.
BSI BS 5378: Part 3-82. 1982 Safety Signs and Colours Part 3: Specification for Additional Signs to Those Given in BS 5378: Part 1. 15 pp.
BSI BS AU 47-65. 1965 Amd 1 Advance Warning Triangle to Indicate Temporary Obstructions. 9 pp.
CNS Z1026-87. Safety Sign (Mar)(9330).
CNS Z1028-87. Safety Sign Boards (Mar)(9332).
CNS Z8014-78. Testing Standard for Reflective Safety Signs (Mar)(4344). 5 pp.
CNS Z8044-87. Method of Test for Safety Sign Boards (Mar)(9334).
CNS Z8045-87. Method of Test for Fluorescent Safety Signs (Mar)(9651).
CSA CAN3-Z321-77. Signs and Symbols for the Occupational Environment; (Erratum May 1978). 37 pp.
ISO 3864-84. Safety Colours and Safety Signs First Edition. 16 pp.
JIS Z 9100-87. Phosphorescent Safety Sign Boards. 19 pp.
JIS Z 9103-86. Safety Signs. 21 pp.
JIS Z 9105-84. Retroreflective Safety Signs. 25 pp.
JIS Z 9107-86. Safety Sign Boards. 16 pp.

Accident Prevention *(Cont.)*
—**Signs—Safety** *(Cont.)*
JIS Z 9108-90. Fluorescent Safety Signs. 13 pp.
JIS Z 9115-79. Self-Luminous Safety Signs (R 1984). 10 pp.
MOD UK NES 784-01. Requirements for Safety Signs and Colours Issue 3 (11.90); Amendment 1. 31 pp.

Accommodation Ladders
Use: Ladders

Accordions
See Also: Musical Instruments
JIS S 8504-84. Accordions.

Accounting
See Also: Accounting Authority Identification Codes; Banking; Currencies; Expenses; Financial Disclosure; Inventory Control Systems; Record Layouts; Reverse Charging; Transferred Account Services

—**Administration Management Domain**
CCITT RECMN D.36-91. General Accounting Principles Applicable to Message Handling Services (Study Group III) 14 pp. 14 pp.

—**Aeronautical Mobile Radio Services**
CCITT RECMN D.94-92. Charging, Billing and Accounting Principles for International Aeronautical Mobil Service, and International Aeronautical Mobile-Satellite Service (Study Group III) 7 pp. 7 pp.

—**Aeronautical Mobile Satellite Communications**
CCITT RECMN D.94-92. Charging, Billing and Accounting Principles for International Aeronautical Mobil Service, and International Aeronautical Mobile-Satellite Service (Study Group III) 7 pp. 7 pp.

—**Automated Telephone Credit Cards**
CCITT RECMN E.118-89. Automated International Telephone Credit Card System—Telephone Network and ISDN—Operation, Numbering, Routing and Mobile Service (Study Group II) 7 pp. 7 pp.

—**Automatic Services (Telephone)**
CCITT RECMN D.140-92. Accounting Rate Principles for International Telephone Services (Study Group III) 6 pp. 6 pp.

—**Automatic Services—Call Duration**
CCITT RECMN E.260-89. Basic Technical Problems Concerning the Measurement and Recording of Call Durations—Telephone Network and ISDN—Operation, Numbering, Routing and Mobile Service (Study Group II) 4 pp. 4 pp.

—**Automatic Services—Europe**
CCITT RECMN D.390R-89. Accounting System in the International Automatic Telephone Service—General Tariff Principles—Charging and Accounting in International Telecommunications Services (Study Group III) 2 pp. 2 pp.

—**Automatic Services—Mediterranean**
CCITT RECMN D.390R-89. Accounting System in the International Automatic Telephone Service—General Tariff Principles—Charging and Accounting in International Telecommunications Services (Study Group III) 2 pp. 2 pp.

—**Bearer Services**
CEPT T/PGT 41 E-91. Charging and Accounting Principles for the International 2 Mbit/s Switched Bearer Service. 1 p.

—**Bureaufax Communications**
CCITT RECMN D.70 (REV 1)-92. General Tariff Principles for the International Public Facsimile Service Between Public Bureaux (Bureaufax Service) (Study Group III) 6 pp. 6 pp.
CCITT RECMN D.70-89. General Tariff Principles for the International Public Facsimile Service Between Public Bureaux (Bureaufax Service)—General Tariff Principles—Charging and Accounting in International Telecommunications Services (Study Group III) 3 pp. 3 pp.

—**Cellular Mobile Radio Services**
CCITT RECMN D.93-89. Charging and Accounting in the International Land Mobile Telephone Service (Provided Via Cellular Radio Systems)—General Tariff Principles—Charging and Accounting in International Telecommunications Services (Study Group III) 11 pp. 11 pp.

Accounting *(Cont.)*
—**Cellular Mobile Radio Services—Routing**
CCITT RECMN D.93-89. Charging and Accounting in the International Land Mobile Telephone Service (Provided Via Cellular Radio Systems)—General Tariff Principles—Charging and Accounting in International Telecommunications Services (Study Group III) 11 pp. 11 pp.

—**Circuit Mode Bearer Services—Integrated Services Digital Networks**
CCITT RECMN D.220-91. Charging and Accounting Principles to Be Applied to International Circuit-Mode Demand Bearer Services Provided over the Integrated Services Digital Network (ISDN) (Study Group III) 5 pp. 5 pp.
CCITT RECMN D.220-89. Charging and Accounting Principles to Be Applied to International Circuit Mode Demand Bearer Services Provided over the Integrated Services Digital Network (ISDN)—General Tariff Principles—Charging and Accounting of Intl Telecom. Serv. (Study Group III) 2 pp. 2 pp.
CEPT T/PGT 33 E-90. Charging and Accounting Principles Applicable to Circuit Mode on Demand Bearer Services over the International ISDN. 1 p.

—**Circuit Switched Public Data Communication**
CCITT RECMN D.20-89. Special Tariff Principles for the International Circuit-Switched Public Data Communication Services—General Tariff Principles—Charging and Accounting in International Telecommunications Services (Study Group III) 4 pp. 4 pp.

—**Conference Calling**
CCITT RECMN D.110 (REV 1)-92. Charging and Accounting for Conference Calls (Study Group III) 4 pp. 4 pp.
CCITT RECMN D.110-89. Charging and Accounting for Conference Calls—General Tariff Principles—Charging and Accounting in International Telecommunications Services (Study Group III) 2 pp. 2 pp.

—**Country Direct Telephone Services**
CCITT RECMN D.116-92. Charging and Accounting Principles Relating to the Home Country Direct Telephone Service (Study Group III) 4 pp. 4 pp.

—**Data Terminal Equipment—Access Methods**
CEPT T/PGT 35 E-90. Accounting Principles for the Use of X.32 Access Method. 1 p.

—**Data Transmission—Telecommunication Administrations**
CCITT RECMN D.160-89. Mode of Application of the Flat-Rate Price Procedure Set Forth in Recommendation D.67 and Recommendation D.150 for Remuneration of Facilities Made Available to the Administrations of Other Countries—General Tariff Principles—Charging and Accounting. 5 pp.
CCITT RECMN E.252-89. Mode of Application of the Flat-Rate Price Procedure Set Forthe in Recommendations D.67 and D.150 for Remuneration of Facilities Made Available to the Administrations of Other Countries—Telephone Network and ISDN—Operation, Numbering, Routing and Mobile. 1 p.

—**Fast Select Services—Packet Switched Public Data Networks**
CCITT RECMN D.21-89. Special Tariff Principles for Short Transaction Transmissions on the International Packet Switched Public Data Networks Using the Fast Select Facility with Restriction—General Tariff Principles—Charging and Accounting in International Telecommunications. 1 p.

—**FAX Communications**
CCITT RECMN D.71 (REV 1)-92. General Tariff Principles for the Public Facsimile Service Between Subscriber Stations (Telefax Service) (Study Group III) 5 pp. 5 pp.
CCITT RECMN D.71-89. General Tariff Principles for the Public Facsimile Between Subscriber Stations (Telefax Service)—General Tariff Principles—Charging and Accounting in International Telecommunications Services (Study Group III) 2 pp. 2 pp.

—**FAX Communications—Directories**
CCITT RECMN F.180-89. General Operational Provisions for the International Public Facsimile Service Between Subscribers' Stations (Telefax)—Telematic, Data Transmission and Teleconference Services—Operations and Quality of Service (Study Group I) 5 pp. 5 pp.

INTERNATIONAL AND NON-U.S. NATIONAL STANDARDS
SUBJECT INDEX

Accounting (Cont.)

—Freephone Services

CCITT RECMN D.115-89. Tariff Principles and Accounting for the International Freephone Service (IFS)—General Tariff Principles—Charging and Accounting in International Telecommunications Services (Study Group III) 2 pp. 2 pp.

—Groups—Telecommunication Administrations

CCITT RECMN D.160-89. Mode of Application of the Flat-Rate Price Procedure Set Forth in Recommendation D.67 and Recommendation D.150 for Remuneration of Facilities Made Available to the Administrations of Other Countries—General Tariff Principles—Charging and Accounting. 5 pp.

CCITT RECMN E.252-89. Mode of Application of the Flat-Rate Price Procedure Set Forthe in Recommendations D.67 and D.150 for Remuneration of Facilities Made Available to the Administrations of Other Countries—Telephone Network and ISDN—Operation, Numbering, Routing and Mobile. 1 p.

—Groups—Telephone Relations—Africa

CCITT RECMN D.600R-89. Determination of Accounting Rate Shares and Collection Charges in Telephone Relations Between Countries in Africa—General Tariff Principles—Charging and Accounting in International Telecommunications Services (Study Group III) 11 pp. 11 pp.

—Integrated Services Digital Networks

CCITT RECMN D.232 (REV 1)-92. Specific Tariff and Accounting Principles Applicable to ISDN Supplementary Services (Study Group III) 10 pp. 10 pp.

CCITT RECMN D.232-91. Specific Tariff and Accounting Principles Applicable to ISDN Supplementary Services (Study Group III) 9 pp. 9 pp.

CCITT RECMN D.250-91. General Charging and Accounting Principles for Non-Voice Services Provided by Interworking Between the ISDN and Existing Public Data Networks (Study Group III) 5 pp. 5 pp.

CCITT RECMN D.250-89. General Charging and Accounting Principles for Non-Voice Services Provided by Interworking Between the ISDN and Existing Public Data Networks—General Tariff Prin.—Charging and Accounting in Intl Telecommunications Services (Study Group III) 1 pp. 1 p.

CCITT RECMN D.260-91. Charging and Accounting Capabilities to Be Applied on the ISDN (Study Group III) 6 pp. 6 pp.

—International Private Leased Circuits

CCITT RECMN D.1-91. General Principles for the Lease of International (Continental and Intercontinental) Private Telecommunication Circuits and Networks (Study Group III) 11 pp. 11 pp.

CCITT RECMN D.1-89. General Principles for the Lease of International (Continental and Intercontinental) Private Telecommunication Circuits—General Tariff Principles—Charging and Accounting in International Telecommunications. 8 pp.

—International Private Leased Circuits—Sound-Programs

CCITT RECMN D.4 (REV 1)-92. Special Conditions for the Lease of International (Continental and Intercontinental) Sound-and Television-Programme Circuits for Private Service (Study Group III) 8 pp. 8 pp.

CCITT RECMN D.4-89. Special Conditions for the Lease of International (Continental and Intercontinental) Sound-and Television-Programme Circuits for Private Service—General Tariff Principles—Charging and Accounting in International. 5 pp.

—International Private Leased Circuits—Television Transmission

CCITT RECMN D.4 (REV 1)-92. Special Conditions for the Lease of International (Continental and Intercontinental) Sound-and Television-Programme Circuits for Private Service (Study Group III) 8 pp. 8 pp.

CCITT RECMN D.4-89. Special Conditions for the Lease of International (Continental and Intercontinental) Sound-and Television-Programme Circuits for Private Service—General Tariff Principles—Charging and Accounting in International. 5 pp.

—Interpersonal Messaging Services

CCITT RECMN D.35-89. General Charging and Accounting Principles in the International Public Interpersonal Messaging (IPM) Service—General Tariff Principles—Charging and Accounting in International Telecommunications Services (Study Group III) 1 pp. 1 p.

Accounting (Cont.)

—Land Mobile Satellite Communications

CCITT RECMN D.95-92. Charging, Billing, Accounting and Refunds in the Data Messaging Land/Maritime Mobile-Satellite Service (Study Group III) 11 pp. 11 pp.

—Maritime Mobile Satellite Communications

CCITT RECMN D.95-92. Charging, Billing, Accounting and Refunds in the Data Messaging Land/Maritime Mobile-Satellite Service (Study Group III) 11 pp. 11 pp.

—Maritime Mobile Services

CCITT FASCICLE II.1-88. General Tariff Principles—Charging and Accounting in International Telecommunications Services Series D Recommendations. 371 pp.

CCITT RECMN D.90-92. Charging, Accounting and Refunds in the Maritime Mobile Service (Study Group III) 23 pp. 23 pp.

CCITT RECMN D.90-89. Charging, Accounting and Refunds in the Maritime Mobile Service—General Tariff Principles—Charging and Accounting in International Telecommunications Services (Study Group III) 19 pp. 19 pp.

—Maritime Mobile Services—Codes

CCITT RECMN D.91-91. Transmission in Encoded Form of Maritime Telecommunications Accounting Information (Study Group III) 13 pp. 13 pp.

CCITT RECMN D.91-89. Transmission in Encoded Form of Maritime Telecommunications Accounting Information—General Tariff Principles—Charging and Accounting in International Telecommunications Services (Study Group III) 8 pp. 8 pp.

—Maritime Mobile Services—Encoded Data Transmission

CCITT RECMN D.91-91. Transmission in Encoded Form of Maritime Telecommunications Accounting Information (Study Group III) 13 pp. 13 pp.

CCITT RECMN D.91-89. Transmission in Encoded Form of Maritime Telecommunications Accounting Information—General Tariff Principles—Charging and Accounting in International Telecommunications Services (Study Group III) 8 pp. 8 pp.

—Maritime Mobile Services—Radiotelegraphy

CCITT RECMN D.90-92. Charging, Accounting and Refunds in the Maritime Mobile Service (Study Group III) 23 pp. 23 pp.

CCITT RECMN D.90-89. Charging, Accounting and Refunds in the Maritime Mobile Service—General Tariff Principles—Charging and Accounting in International Telecommunications Services (Study Group III) 19 pp. 19 pp.

CCITT RECMN E.200-89. Operational Provisions for the Maritime Mobile Service—Telephone Network and ISDN—Operation, Numbering, Routing and Mobile Service (Study Group II) 21 pp (Same as Recmn F.110). 21 pp.

—Maritime Mobile Services—Radiotelephony

CCITT RECMN D.90-92. Charging, Accounting and Refunds in the Maritime Mobile Service (Study Group III) 23 pp. 23 pp.

CCITT RECMN D.90-89. Charging, Accounting and Refunds in the Maritime Mobile Service—General Tariff Principles—Charging and Accounting in International Telecommunications Services (Study Group III) 19 pp. 19 pp.

—Maritime Mobile Services—Radioteletype Communications

CCITT RECMN D.90-92. Charging, Accounting and Refunds in the Maritime Mobile Service (Study Group III) 23 pp. 23 pp.

CCITT RECMN D.90-89. Charging, Accounting and Refunds in the Maritime Mobile Service—General Tariff Principles—Charging and Accounting in International Telecommunications Services (Study Group III) 19 pp. 19 pp.

—Maritime Mobile Services—Record Layouts

CCITT RECMN D.91-91. Transmission in Encoded Form of Maritime Telecommunications Accounting Information (Study Group III) 13 pp. 13 pp.

CCITT RECMN D.91-89. Transmission in Encoded Form of Maritime Telecommunications Accounting Information—General Tariff Principles—Charging and Accounting in International Telecommunications Services (Study Group III) 8 pp. 8 pp.

—Message Handling Systems

CCITT RECMN D.36-91. General Accounting Principles Applicable to Message Handling Services (Study Group III) 14 pp. 14 pp.

Accounting (Cont.)

—Message Handling Systems (Cont.)

CCITT RECMN D.95-92. Charging, Billing, Accounting and Refunds in the Data Messaging Land/Maritime Mobile-Satellite Service (Study Group III) 11 pp. 11 pp.

CCITT RECMN F.415-89. Message Handling Services: Intercommunication with Public Physical Delivery Services—Message Handling and Directory Services—Operations and Definition of Service (Study Group I) 15 pp (Erratum in Recmn F.410). 15 pp.

—Open Systems Interconnection—Metering

IEC DIS 10164 Pt 10-93. Information Technology—Open Systems Interconnection—Systems Management—Part 10: Accounting Metering Function ***CD-ROM ONLY***. 39 pp.

ISO DIS 10164 Pt 10-92. Information Technology—Open Systems Interconnection—Systems Management—Part 10: Accounting Metering Function ***CD-ROM ONLY***. 39 pp.

JTC1 DIS10164 Pt 10-92. Information Technology—Open Systems Interconnection—Systems Management—Part 10: Accounting Metering Function ***CD-ROM ONLY***. 39 pp.

—Packet Switched Public Data Communication—Telephone Relations

CCITT RECMN D.13-89. Guiding Principles to Govern the Apportionment of Accounting Rates in International Packet-Switched Public Data Communication Relations—General Tariff Principles—Charging and Accounting in International Telecommunications. 2 pp.

—Phototelegraphy

CCITT RECMN D.80-89. Accounting and Refunds for Phototelegrams—General Tariff Principles—Charging and Accounting in International Telecommunications Services (Study Group III) 2 pp. 2 pp.

CCITT RECMN D.81-89. Accounting and Refunds for Private Phototelegraph Calls—General Tariff Principles—Charging and Accounting in International Telecommunications Services (Study Group III) 2 pp. 2 pp.

—Phototelegraphy—Telecommunication Administrations

CCITT RECMN D.170-89. Monthly Telephone and Telex Accounts—General Tariff Principles—Charging and Accounting in International Telecommunications Services (Study Group III) 9 pp (Same as Recmn E.270). 9 pp.

CCITT RECMN E.270-89. Monthly Telephone and Telex Accounts—Telephone Network and ISDN—Operation, Numbering, Routing and Mobile Service (Study Group II)1 pp (Same as Recmn D.170). 1 p.

—Public Data Communication

CCITT RECMN D.10-91. General Tariff Principles for International Public Data Communication Services (Study Group III) 6 pp. 6 pp.

CCITT RECMN D.10-89. General Tariff Principles for International Public Data Communication Services—General Tariff Principles—Charging and Accounting in International Telecommunications Services (Study Group III) 3 pp. 3 pp.

—Public Data Communication—Telephone Relations

CCITT RECMN D.13-89. Guiding Principles to Govern the Apportionment of Accounting Rates in International Packet-Switched Public Data Communication Relations—General Tariff Principles—Charging and Accounting in International Telecommunications. 2 pp.

—Public Data Networks

CCITT RECMN D.250-91. General Charging and Accounting Principles for Non-Voice Services Provided by Interworking Between the ISDN and Existing Public Data Networks (Study Group III) 5 pp. 5 pp.

CCITT RECMN D.250-89. General Charging and Accounting Principles for Non-Voice Services Provided by Interworking Between the ISDN and Existing Public Data Networks—General Tariff Prin.—Charging and Accounting in Intl Telecommunications Services (Study Group III) 1 pp. 1 p.

—Public Land Mobile Networks—Routing

CCITT RECMN D.93-89. Charging and Accounting in the International Land Mobile Telephone Service (Provided Via Cellular Radio Systems)—General Tariff Principles—Charging and Accounting in International Telecommunications Services (Study Group III) 11 pp. 11 pp.

INTERNATIONAL AND NON-U.S. NATIONAL STANDARDS
SUBJECT INDEX

Accounting

Accounting (Cont.)

—Radiotelephony
  CCITT FASCICLE II.1-88. General Tariff Principles—Charging and Accounting in International Telecommunications Services Series D Recommendations. 371 pp.
  CCITT RECMN D.151-89. Old System for Accounting in International Telephony—General Tariff Principles—Charging and Accounting in International Telecommunications Services (Study Group III) 2 pp (Same as Recmn E.251). 2 pp.
  CCITT RECMN E.251-89. Old System for Accounting In International Telephony—Telephone Network and ISDN—Operation, Numbering, Routing and Mobile Serivce (Study Group II) 1 pp (Same as Recmn D.151). 1 p.

—Radiotelephony—Public Land Mobile Networks
  CCITT RECMN D.93-89. Charging and Accounting in the International Land Mobile Telephone Service (Provided Via Cellular Radio Systems)—General Tariff Principles—Charging and Accounting in International Telecommunications Services (Study Group III) 11 pp. 11 pp.

—Reverse Charging—Encoded Data Transmission
  CCITT RECMN D.176 (REV 1)-92. Transmission in Encoded Form of Telephone Reversed Charge Billing and Accounting Information (Study Group III) 9 pp (Same as Recmn E.276). 9 pp.
  CCITT RECMN D.176-89. Transmission in Encoded Form of Telephone Reversed Charge Billing and Accounting Information—General Tariff Principles—Charging and Accounting in International Telecommunications Services (Study Group III) 6 pp. 6 pp.
  CCITT RECMN E.276-89. Transmission in Encoded Form of Telephone Reversed Charge Billing and Accounting Information—Telephone Network and ISDN—Operation, Numbering, Routing and Mobile Service (Study Group II) 1 pp (Same as Recmn D.176). 1 p.

—Reverse Charging—Packet Switched Networks
  CEPT T/CAC E12 E-92. Tariff and Accounting Principles of the International Reverse Charging Facility in PSPDNs. 1 p.

—Reverse Charging—Record Layouts
  CCITT RECMN D.176 (REV 1)-92. Transmission in Encoded Form of Telephone Reversed Charge Billing and Accounting Information (Study Group III) 9 pp (Same as Recmn E.276). 9 pp.
  CCITT RECMN D.176-89. Transmission in Encoded Form of Telephone Reversed Charge Billing and Accounting Information—General Tariff Principles—Charging and Accounting in International Telecommunications Services (Study Group III) 6 pp. 6 pp.
  CCITT RECMN E.276-89. Transmission in Encoded Form of Telephone Reversed Charge Billing and Accounting Information—Telephone Network and ISDN—Operation, Numbering, Routing and Mobile Service (Study Group II) 1 pp (Same as Recmn D.176). 1 p.

—Satellite Communications—Regional
  CCITT FASCICLE II.1-88. General Tariff Principles—Charging and Accounting in International Telecommunications Services Series D Recommendations. 371 pp.

—Satellite Communications—Telegraph Circuits—Europe
  CCITT RECMN D.301R-89. Determination of Accounting Rate Shares and Collection Charges in Telex Relations Between Countries in Europe and the Mediterranean Basin—General Tariff Principles—Charging and Accounting in Intl Telecom. Serv. (Study Group III) 12 pp. 12 pp.

—Satellite Communications—Telegraph Circuits—Mediterranean
  CCITT RECMN D.301R-89. Determination of Accounting Rate Shares and Collection Charges in Telex Relations Between Countries in Europe and the Mediterranean Basin—General Tariff Principles—Charging and Accounting in Intl Telecom. Serv. (Study Group III) 12 pp. 12 pp.

—Satellite Communications—Telephone Circuits—Europe
  CCITT RECMN D.300R-92. Determination of Accounting Rate Shares in Telephone Relations Between Countries in Europe and the Mediterranean Basin (Study Group III) 12 pp. 12 pp.
  CCITT RECMN D.300R-91. Determination of Accounting Rate Shares in Telephone Relations Between Countries in Europe and the Mediterranean Basin (Study Group III) 16 pp. 16 pp.

Accounting (Cont.)

—Satellite Communications—Telephone Circuits—Europe (Cont.)
  CCITT RECMN D.300R-89. Determination of Accounting Rate Shares and Collection Charges in Telephone Relations Between Countries in Europe and the Mediterranean Basin—General Tariff Principles—Charging and Accounting in Intl Telecom. Serv. (Study Group III) 13 pp. 13 pp.

—Satellite Communications—Telephone Circuits—Mediterranean
  CCITT RECMN D.300R-92. Determination of Accounting Rate Shares in Telephone Relations Between Countries in Europe and the Mediterranean Basin (Study Group III) 12 pp. 12 pp.
  CCITT RECMN D.300R-91. Determination of Accounting Rate Shares in Telephone Relations Between Countries in Europe and the Mediterranean Basin (Study Group III) 16 pp. 16 pp.
  CCITT RECMN D.300R-89. Determination of Accounting Rate Shares and Collection Charges in Telephone Relations Between Countries in Europe and the Mediterranean Basin—General Tariff Principles—Charging and Accounting in Intl Telecom. Serv. (Study Group III) 13 pp. 13 pp.

—Satellite Communications—Telephone Services—Africa
  CCITT RECMN D.600R-89. Determination of Accounting Rate Shares and Collection Charges in Telephone Relations Between Countries in Africa—General Tariff Principles—Charging and Accounting in International Telecommunications Services (Study Group III) 11 pp. 11 pp.

—Satellite Communications—Television Transmission
  CCITT RECMN D.180-89. Occasional Provision of Circuits for International Sound-and Television-Programme Transmissions—General Tariff Principles—Charging and Accounting in International Telecommunications Services (Study Group III) 13 pp. 13 pp.

—Satellite Communications—Television Transmission—Europe
  CCITT RECMN D.303R-89. Determination of Accounting Rate Shares and Collection Charges Applicable by Countries in Europe and the Mediterranean Basin to the Occasional Provision of Circuits for Sound-and Television-Programme Transmissions—General Tariff Principles—Charging and Accounting. 14 pp.

—Satellite Communications—Television Transmission—Mediterranean
  CCITT RECMN D.303R-89. Determination of Accounting Rate Shares and Collection Charges Applicable by Countries in Europe and the Mediterranean Basin to the Occasional Provision of Circuits for Sound-and Television-Programme Transmissions—General Tariff Principles—Charging and Accounting. 14 pp.

—Satellite Communications—Telex Communications—Africa
  CCITT RECMN D.601R-89. Determination of Accounting Rate Shares and Collection Charges in Telex Relations Between Countries in Africa—General Tariff Principles—Charging and Accounting in International Telecommunications Services (Study Group III) 10 pp. 10 pp.

—Satellite Communications—Telex Communications—Latin America
  CCITT RECMN D.401R-89. Accounting Rates Applicable to Telex Relations Between Countries in Latin America—General Tariff Principles—Charging and Accounting in International Telecommunications Services (Study Group III) 2 pp. 2 pp.

—Satellite Transmissions—Television Transmission
  CCITT RECMN D.185-89. General Tariff and Accounting Principles for International One-Way Point-to-Multipoint Satellite Services—General Tariff Principles—Charging and Accounting in Intl Telecommunications Services (Study Group III) 4 pp. 4 pp.

—Signal Transfer Points—CCITT No. 7 Signaling Systems
  CCITT RECMN D.211-89. International Accounting for the Use of the Signal Transfer Point (STP) in CCITT Signalling System No. 7—General Tariff Principles—Charging and Accounting in International Telecommunications Services (Study Group III) 1 pp. 1 p.

Accounting (Cont.)

—Sound-Program Circuits
  CCITT RECMN D.180-89. Occasional Provision of Circuits for International Sound-and Television-Programme Transmissions—General Tariff Principles—Charging and Accounting in International Telecommunications Services (Study Group III) 13 pp. 13 pp.

—Sound-Program Circuits—Europe
  CCITT RECMN D.303R-89. Determination of Accounting Rate Shares and Collection Charges Applicable by Countries in Europe and the Mediterranean Basin to the Occasional Provision of Circuits for Sound-and Television-Programme Transmissions—General Tariff Principles—Charging and Accounting. 14 pp.

—Sound-Program Circuits—Mediterranean
  CCITT RECMN D.303R-89. Determination of Accounting Rate Shares and Collection Charges Applicable by Countries in Europe and the Mediterranean Basin to the Occasional Provision of Circuits for Sound-and Television-Programme Transmissions—General Tariff Principles—Charging and Accounting. 14 pp.

—Sound-Program Circuits—Telecommunication Administrations
  CCITT RECMN D.170-89. Monthly Telephone and Telex Accounts—General Tariff Principles—Charging and Accounting in International Telecommunications Services (Study Group III) 9 pp (Same as Recmn E.270). 9 pp.
  CCITT RECMN E.270-89. Monthly Telephone and Telex Accounts—Telephone Network and ISDN—Operation, Numbering, Routing and Mobile Service (Study Group II)1 pp (Same as Recmn D.170). 1 p.

—Supergroups—Regional
  CCITT FASCICLE II.1-88. General Tariff Principles—Charging and Accounting in International Telecommunications Services Series D Recommendations. 371 pp.

—Supergroups—Telecommunication Administrations
  CCITT RECMN D.160-89. Mode of Application of the Flat-Rate Price Procedure Set Forth in Recommendation D.67 and Recommendation D.150 for Remuneration of Facilities Made Available to the Administrations of Other Countries—General Tariff Principles—Charging and Accounting. 5 pp.
  CCITT RECMN E.252-89. Mode of Application of the Flat-Rate Price Procedure Set Forthe in Recommendations D.67 and D.150 for Remuneration of Facilities Made Available to the Administrations of Other Countries—Telephone Network and ISDN—Operation, Numbering, Routing and Mobile. 1 p.

—Supergroups—Telephone Relations—Africa
  CCITT RECMN D.600R-89. Determination of Accounting Rate Shares and Collection Charges in Telephone Relations Between Countries in Africa—General Tariff Principles—Charging and Accounting in International Telecommunications Services (Study Group III) 11 pp. 11 pp.

—Supplementary Services—ISDN
  CEPT T/PGT 34 E-91. Charging and Accounting Principles for Specific Supplementary Services Provided over the ISDN. 1 p.

—Telecommunication Circuits—Television Transmission
  CCITT RECMN D.180-89. Occasional Provision of Circuits for International Sound-and Television-Programme Transmissions—General Tariff Principles—Charging and Accounting in International Telecommunications Services (Study Group III) 13 pp. 13 pp.

—Telecommunication Circuits—Television Transmission—Europe
  CCITT RECMN D.303R-89. Determination of Accounting Rate Shares and Collection Charges Applicable by Countries in Europe and the Mediterranean Basin to the Occasional Provision of Circuits for Sound-and Television-Programme Transmissions—General Tariff Principles—Charging and Accounting. 14 pp.

—Telecommunication Circuits—Television Transmission—Mediterranean
  CCITT RECMN D.303R-89. Determination of Accounting Rate Shares and Collection Charges Applicable by Countries in Europe and the Mediterranean Basin to the Occasional Provision of Circuits for Sound-and Television-Programme Transmissions—General Tariff Principles—Charging and Accounting. 14 pp.

INDUSTRY STANDARDS

INTERNATIONAL AND NON-U.S. NATIONAL STANDARDS
SUBJECT INDEX

Accounting

Accounting (Cont.)

—Telecommunication Services

CCITT FASCICLE II.1-88. General Tariff Principles—Charging and Accounting in International Telecommunications Services Series D Recommendations. 371 pp.

CCITT RECMN D.172-89. Accounting for Calls Circulated over International Routes for Which Accounting Rates Have Not Been Established—General Tariff Principles—Charging and Accounting in International Telecommunications Services (Study Group III) 1 pp. 1 p.

CCITT RECMN D.192 (REV 1)-92. Principles for Charging and Accounting of Service Telecommunications (Study Group III) 5 pp. 5 pp.

CCITT RECMN D.192-91. Principles for Charging and Accounting of Service Telecommunications (Study Group III) 4 pp. 4 pp.

—Telecommunication Services—Balance of Accounts—Clearing

CCITT RECMN D.196-92. Clearing of International Telecommunication Balances of Accounts (Study Group III) 4 pp. 4 pp.

—Telecommunication Services—Change of Address

CCITT RECMN D.197-91. Notification of Change of Address(es) for Accounting and Settlement Purposes (Study Group III) 6 pp. 6 pp.

—Telecommunication Services—Integrated Services Digital Networks

CCITT RECMN D.210-89. General Charging and Accounting Principles for International Telecommunication Services Provided over the Integrated Services Digital Network (ISDN)—General Tariff Principles—Charging and Accounting in Intl Telecom. Serv. (Study Group III) 4 pp. 4 pp.

CCITT RECMN D.230-89. General Charging and Accounting Principles for Supplementary Services Associated with International Telecommunication Services Provided over the Integrated Services Digital Network (ISDN)—General Tariff Principles—Charging and Accounting in. 2 pp.

CEPT T/CAC E39 E-92. General Tariff and Accounting Principles for Packet Mode Services Provided over the ISDN. 2 pp.

—Telecommunication Services—Preferential Rates—Africa

CCITT RECMN D.606R-89. Preferential Rates in Telecommunication Relations Between Countries in Africa—General Tariff Principles—Charging and Accounting in International Telecommunications Services (Study Group III) 1 pp. 1 p.

—Telecommunication Services—Special Drawing Rights (IMF)

CCITT RECMN D.195-89. Settlement of International Telecommunication Balances of Accounts—General Tariff Principles—Charging and Accounting in International Telecommunications Services (Study Group III) 2 pp. 2 pp.

—Telecommunication Services—Telephone Relations—Europe

CCITT RECMN D.300R-92. Determination of Accounting Rate Shares in Telephone Relations Between Countries in Europe and the Mediterranean Basin (Study Group III) 12 pp. 12 pp.

CCITT RECMN D.300R-91. Determination of Accounting Rate Shares in Telephone Relations Between Countries in Europe and the Mediterranean Basin (Study Group III) 16 pp. 16 pp.

CCITT RECMN D.300R-89. Determination of Accounting Rate Shares and Collection Charges in Telephone Relations Between Countries in Europe and the Mediterranean Basin—General Tariff Principles—Charging and Accounting in Intl Telecom. Serv. (Study Group III) 13 pp. 13 pp.

—Telecommunication Services—Telephone Relations—Mediterranean

CCITT RECMN D.300R-92. Determination of Accounting Rate Shares in Telephone Relations Between Countries in Europe and the Mediterranean Basin (Study Group III) 12 pp. 12 pp.

CCITT RECMN D.300R-91. Determination of Accounting Rate Shares in Telephone Relations Between Countries in Europe and the Mediterranean Basin (Study Group III) 16 pp. 16 pp.

CCITT RECMN D.300R-89. Determination of Accounting Rate Shares and Collection Charges in Telephone Relations Between Countries in Europe and the Mediterranean Basin—General Tariff Principles—Charging and Accounting in Intl Telecom. Serv. (Study Group III) 13 pp. 13 pp.

Accounting (Cont.)

—Telecommunication Systems—Glossaries

CCITT RECMN D.000-89. Terms and Definitions for the Series D Recommendations—General Tariff Principles—Charging and Accounting in International Telecommunications Services (Study Group III) 5 pp. 5 pp.

—Telegraph Circuits—Telecommunication Administrations

CCITT RECMN D.160-89. Mode of Application of the Flat-Rate Price Procedure Set Forth in Recommendation D.67 and Recommendation D.150 for Remuneration of Facilities Made Available to the Administrations of Other Countries—General Tariff Principles—Charging and Accounting. 5 pp.

CCITT RECMN E.252-89. Mode of Application of the Flat-Rate Price Procedure Set Forthe in Recommendations D.67 and D.150 for Remuneration of Facilities Made Available to the Administrations of Other Countries—Telephone Network and ISDN—Operation, Numbering, Routing and Mobile. 1 p.

—Telegraphy

CCITT RECMN D.40 (REV 1)-92. General Tariff Principles Applicable to Telegrams Exchanged in the International Public Telegram Service (Study Group III) 7 pp. 7 pp.

CCITT RECMN D.40-89. General Tariff Principles Applicable to Telegrams Exchanged in the International Public Telegram Service—General Tariff Principles—Charging and Accounting in International Telecommunications Services (Study Group III) 4 pp. 4 pp.

CCITT RECMN D.41-89. Introduction of Accounting Rates by Zones in the International Public Telegram Service—General Tariff Principles—Charging and Accounting in International Telecommunications Services (Study Group III) 14 pp. 14 pp.

CCITT RECMN D.42-89. Accounting in the International Public Telegram Service—General Tariff Principles—Charging and Accounting in International Telecommunications Services (Study Group III) 3 pp. 3 pp.

CCITT RECMN D.95-92. Charging, Billing, Accounting and Refunds in the Data Messaging Land/Maritime Mobile-Satellite Service (Study Group III) 11 pp. 11 pp.

—Telegraphy—Encoded Data Transmission

CCITT RECMN D.190-89. Transmission in Encoded Form of Monthly International Accounting Information—General Tariff Principles—Charging and Accounting in International Telecommunications Services (Study Group III) 6 pp (Same as Recmn E.275). 6 pp.

CCITT RECMN E.275-89. Transmission in Encoded Form of Monthly International Accounting Information—Telephone Network and ISDN—Operation, Numbering, Routing and Mobile Service (Study Group II) 1 pp (Same as Recmn D.190). 1 p.

—Telegraphy—Europe

CCITT RECMN D.302R-89. Determination of Accounting Rate Shares and Collection Charges for the International Public Telegram Service Applicable to Telegrams Exchanged Between Countries in Europe and the Mediterranean Basin—General Tariff Principles—Charging and Accounting in. 4 pp.

—Telegraphy—Mediterranean

CCITT RECMN D.302R-89. Determination of Accounting Rate Shares and Collection Charges for the International Public Telegram Service Applicable to Telegrams Exchanged Between Countries in Europe and the Mediterranean Basin—General Tariff Principles—Charging and Accounting in. 4 pp.

—Telegraphy—Record Layouts

CCITT RECMN D.190-89. Transmission in Encoded Form of Monthly International Accounting Information—General Tariff Principles—Charging and Accounting in International Telecommunications Services (Study Group III) 6 pp (Same as Recmn E.275). 6 pp.

CCITT RECMN E.275-89. Transmission in Encoded Form of Monthly International Accounting Information—Telephone Network and ISDN—Operation, Numbering, Routing and Mobile Service (Study Group II) 1 pp (Same as Recmn D.190). 1 p.

—Telegraphy—Tables (Data)

CCITT RECMN D.41-89. Introduction of Accounting Rates by Zones in the International Public Telegram Service—General Tariff Principles—Charging and Accounting in International Telecommunications Services (Study Group III) 14 pp. 14 pp.

Accounting (Cont.)

—Telegraphy—Telecommunication Administrations

CCITT RECMN D.160-89. Mode of Application of the Flat-Rate Price Procedure Set Forth in Recommendation D.67 and Recommendation D.150 for Remuneration of Facilities Made Available to the Administrations of Other Countries—General Tariff Principles—Charging and Accounting. 5 pp.

CCITT RECMN E.252-89. Mode of Application of the Flat-Rate Price Procedure Set Forthe in Recommendations D.67 and D.150 for Remuneration of Facilities Made Available to the Administrations of Other Countries—Telephone Network and ISDN—Operation, Numbering, Routing and Mobile. 1 p.

—Teleinformatic Services—PSPDN—Europe

CCITT RECMN D.306R-91. Remuneration of Public Packet-Switched Data Transmission Networks Between the Countries of Europe and the Mediterranean Basin (Study Group III) 5 pp. 5 pp.

CCITT RECMN D.306R-89. Remuneration of Public Packet-Switched Data Transmission Networks Between the Countries of Europe and the Mediterranean Basin—General Tariff Principles—Charging and Accounting in International Telecom. Serv. (Study Group III) 1 pp. 1 p.

—Teleinformatic Services—PSPDN—Mediterranean

CCITT RECMN D.306R-91. Remuneration of Public Packet-Switched Data Transmission Networks Between the Countries of Europe and the Mediterranean Basin (Study Group III) 5 pp. 5 pp.

CCITT RECMN D.306R-89. Remuneration of Public Packet-Switched Data Transmission Networks Between the Countries of Europe and the Mediterranean Basin—General Tariff Principles—Charging and Accounting in International Telecom. Serv. (Study Group III) 1 pp. 1 p.

—Teleinformatic Services—Public Data Networks

CCITT RECMN D.15-89. General Charging and Accounting Principles for Non-Voice Services Provided by Interworking Between Public Data Networks—General Tariff Principles—Charging and Accounting in International Telecommunications Services (Study Group III) 2 pp. 2 pp.

—Telemessage Services

CCITT RECMN D.45 (REV 1)-92. Charging and Accounting Principles for the International Telemessage Service (Study Group III) 5 pp. 5 pp.

CCITT RECMN D.45-89. Charging and Accounting Principles for the International Telemessage Service—General Tariff Principles—Charging and Accounting in International Telecommunications Services (Study Group III) 2 pp. 2 pp.

—Telephone Circuits—Regional

CCITT FASCICLE II.1-88. General Tariff Principles—Charging and Accounting in International Telecommunications Services Series D Recommendations. 371 pp.

—Telephone Circuits—Telecommunication Administrations

CCITT RECMN D.160-89. Mode of Application of the Flat-Rate Price Procedure Set Forth in Recommendation D.67 and Recommendation D.150 for Remuneration of Facilities Made Available to the Administrations of Other Countries—General Tariff Principles—Charging and Accounting. 5 pp.

CCITT RECMN E.252-89. Mode of Application of the Flat-Rate Price Procedure Set Forthe in Recommendations D.67 and D.150 for Remuneration of Facilities Made Available to the Administrations of Other Countries—Telephone Network and ISDN—Operation, Numbering, Routing and Mobile. 1 p.

—Telephone Circuits—Telephone Relations—Africa

CCITT RECMN D.600R-89. Determination of Accounting Rate Shares and Collection Charges in Telephone Relations Between Countries in Africa—General Tariff Principles—Charging and Accounting in International Telecommunications Services (Study Group III) 11 pp. 11 pp.

—Telephone Credit Cards—Numbering Systems

CCITT RECMN E.116-89. International Telephone Credit Cards for Use in a Non-Automated Environment—Telephone Network and ISDN—Operation, Numbering, Routing and Mobile Service (Study Group II) 3 pp. 3 pp.

INDEX and DIRECTORY of

INTERNATIONAL AND NON-U.S. NATIONAL STANDARDS
SUBJECT INDEX

Accounting

Accounting *(Cont.)*

—**Telephone Directories—Telecommunication Administrations**
CCITT RECMN D.170-89. Monthly Telephone and Telex Accounts—General Tariff Principles—Charging and Accounting in International Telecommunications Services (Study Group III) 9 pp (Same as Recmn E.270). 9 pp.
CCITT RECMN E.114-89. Supply of Lists of Subscribers (Directories and Other Means)—Telephone Network and ISDN—Operation, Numbering, Routing and Mobile Service (Study Group II) 1 pp. 1 p.
CCITT RECMN E.270-89. Monthly Telephone and Telex Accounts—Telephone Network and ISDN—Operation, Numbering, Routing and Mobile Service (Study Group II)1 pp (Same as Recmn D.170). 1 p.

—**Telephone Services**
CCITT RECMN D.140-92. Accounting Rate Principles for International Telephone Services (Study Group III) 6 pp. 6 pp.
CCITT RECMN D.150-92. New System for Accounting in International Telephony (Study Group III) 14 pp. 14 pp.
CCITT RECMN D.150-89. New System for Accounting in International Telephony—General Tariff Principles—Charging and Accounting in International Telecommunications Services (Study Group III) 12 pp. 12 pp.
CCITT RECMN D.151-89. Old System for Accounting in International Telephony—General Tariff Principles—Charging and Accounting in International Telecommunications Services (Study Group III) 2 pp (Same as Recmn E.251). 2 pp.
CCITT RECMN E.250-89. New System for Accounting in International Telephony—Telephone Network and ISDN—Operation, Numbering, Routing and Mobile Serivce (Study Group II) 1 pp (Same as Recmn D.150). 1 p.
CCITT RECMN E.251-89. Old System for Accounting In International Telephony—Telephone Network and ISDN—Operation, Numbering, Routing and Mobile Serivce (Study Group II) 1 pp (Same as Recmn D.151). 1 p.

—**Telephone Services—Alternative Routing**
CCITT RECMN D.155-92. Guiding Principles Governing the Apportionment of Accounting Rates in Intercontinental Telephone Relations (Study Group III) 4 pp. 4 pp.
CCITT RECMN D.155-89. Guiding Principles Governing the Apportionment of Accounting Rates in Intercontinental Telephone Relations—General Tariff Principles—Charging and Accounting in Intl Telecommunications Services (Study Group III) 2 pp. 2 pp.

—**Telephone Services—Call Duration**
CCITT RECMN E.260-89. Basic Technical Problems Concerning the Measurement and Recording of Call Durations—Telephone Network and ISDN—Operation, Numbering, Routing and Mobile Service (Study Group II) 4 pp. 4 pp.

—**Telephone Services—Encoded Data Transmission**
CCITT RECMN D.190-89. Transmission in Encoded Form of Monthly International Accounting Information—General Tariff Principles—Charging and Accounting in International Telecommunications Services (Study Group III) 6 pp (Same as Recmn E.275). 6 pp.
CCITT RECMN E.275-89. Transmission in Encoded Form of Monthly International Accounting Information—Telephone Network and ISDN—Operation, Numbering, Routing and Mobile Service (Study Group II) 1 pp (Same as Recmn D.190). 1 p.

—**Telephone Services—Europe**
CCITT FASCICLE II.1-88. General Tariff Principles—Charging and Accounting in International Telecommunications Services Series D Recommendations. 371 pp.

—**Telephone Services—Integrated Services Digital Networks**
CCITT RECMN D.251-89. General Charging and Accounting Principles for the Basic Telephone Service Provided over the ISDN or by Interconnection Between the ISDN and the Public Switched Telephone Network—General Tariff Principles—Charging and Accounting in International. 2 pp.

—**Telephone Services—Mediterranean**
CCITT FASCICLE II.1-88. General Tariff Principles—Charging and Accounting in International Telecommunications Services Series D Recommendations. 371 pp.

Accounting *(Cont.)*

—**Telephone Services—Record Layouts**
CCITT RECMN D.190-89. Transmission in Encoded Form of Monthly International Accounting Information—General Tariff Principles—Charging and Accounting in International Telecommunications Services (Study Group III) 6 pp (Same as Recmn E.275). 6 pp.
CCITT RECMN E.275-89. Transmission in Encoded Form of Monthly International Accounting Information—Telephone Network and ISDN—Operation, Numbering, Routing and Mobile Service (Study Group II) 1 pp (Same as Recmn D.190). 1 p.

—**Telephone Services—Regional**
CCITT FASCICLE II.1-88. General Tariff Principles—Charging and Accounting in International Telecommunications Services Series D Recommendations. 371 pp.

—**Telephone Services—Semiautomatic Demand Operating**
CCITT RECMN D.178-89. Monthly Accounts for Semi-Automatic Telephone Calls (Ordinary and Urgent Calls, with or Without Special Facilities)—General Tariff Principles—Charging and Accounting in International Telecommunications Services (Study Group III) 1 pp. 1 p.

—**Telephone Services—Telecommunication Administrations**
CCITT RECMN D.160-89. Mode of Application of the Flat-Rate Price Procedure Set Forth in Recommendation D.67 and Recommendation D.150 for Remuneration of Facilities Made Available to the Administrations of Other Countries—General Tariff Principles—Charging and Accounting. 5 pp.
CCITT RECMN D.170-89. Monthly Telephone and Telex Accounts—General Tariff Principles—Charging and Accounting in International Telecommunications Services (Study Group III) 9 pp (Same as Recmn E.270). 9 pp.
CCITT RECMN E.252-89. Mode of Application of the Flat-Rate Price Procedure Set Forthe in Recommendations D.67 and D.150 for Remuneration of Facilities Made Available to the Administrations of Other Countries—Telephone Network and ISDN—Operation, Numbering, Routing and Mobile. 1 p.
CCITT RECMN E.270-89. Monthly Telephone and Telex Accounts—Telephone Network and ISDN—Operation, Numbering, Routing and Mobile Service (Study Group II)1 pp (Same as Recmn D.170). 1 p.

—**Telephone Services—Telephone Relations**
CCITT RECMN D.150-92. New System for Accounting in International Telephony (Study Group III) 14 pp. 14 pp.
CCITT RECMN D.150-89. New System for Accounting in International Telephony—General Tariff Principles—Charging and Accounting in International Telecommunications Services (Study Group III) 12 pp. 12 pp.
CCITT RECMN D.155-92. Guiding Principles Governing the Apportionment of Accounting Rates in Intercontinental Telephone Relations (Study Group III) 4 pp. 4 pp.
CCITT RECMN D.155-89. Guiding Principles Governing the Apportionment of Accounting Rates in Intercontinental Telephone Relations—General Tariff Principles—Charging and Accounting in Intl Telecommunications Services (Study Group III) 2 pp. 2 pp.

—**Telephone Services—Telephone Relations—Africa**
CCITT RECMN D.600R-89. Determination of Accounting Rate Shares and Collection Charges in Telephone Relations Between Countries in Africa—General Tariff Principles—Charging and Accounting in International Telecommunications Services (Study Group III) 11 pp. 11 pp.

—**Telephone Services—Telephone Relations—Asia**
CCITT RECMN D.500R-89. Accounting Rates Applicable to Telephone Relations Between Countries in Asia and Oceania—General Tariff Principles—Charging and Accounting in International Telecommunications Services (Study Group III) 1 pp. 1 p.

—**Telephone Services—Telephone Relations—Europe**
CCITT RECMN D.305R-89. Remuneration for Facilities Used for the Switched-Transit Handling of Intercontinental Telephone Traffic in a Country in Europe or the Mediterranean Basin—General Tariff Principles—Charging and Accounting in International. 1 p.

Accounting *(Cont.)*

—**Telephone Services—Telephone Relations—Latin America**
CCITT RECMN D.400R-89. Accounting Rates Applicable in Telephone Relations Between Countries in Latin America—General Tariff Principles—Charging and Accounting in International Telecommunications Services (Study Group III) 2 pp. 2 pp.

—**Telephone Services—Telephone Relations—Mediterranean**
CCITT RECMN D.305R-89. Remuneration for Facilities Used for the Switched-Transit Handling of Intercontinental Telephone Traffic in a Country in Europe or the Mediterranean Basin—General Tariff Principles—Charging and Accounting in International. 1 p.

—**Telephone Services—Telephone Relations—Oceania**
CCITT RECMN D.500R-89. Accounting Rates Applicable to Telephone Relations Between Countries in Asia and Oceania—General Tariff Principles—Charging and Accounting in International Telecommunications Services (Study Group III) 1 pp. 1 p.

—**Telephone Services—Traffic Units**
CCITT RECMN D.150-92. New System for Accounting in International Telephony (Study Group III) 14 pp. 14 pp.
CCITT RECMN D.150-89. New System for Accounting in International Telephony—General Tariff Principles—Charging and Accounting in International Telecommunications Services (Study Group III) 12 pp. 12 pp.
CCITT RECMN E.250-89. New System for Accounting in International Telephony—Telephone Network and ISDN—Operation, Numbering, Routing and Mobile Serivce (Study Group II) 1 pp (Same as Recmn D.150). 1 p.

—**Teleservices**
CCITT RECMN D.240-91. Charging and Accounting Principles for Teleservices Supported by the ISDN (Study Group III) 5 pp. 5 pp.

—**Teletex Communications**
CCITT RECMN D.50-89. Tariff and International Accounting Principles for the International Teletex Service—Charging and Accounting in International Telecommunications Services (Study Group III) 4 pp. 4 pp.

—**Television Transmission—Telecommunication Administrations**
CCITT RECMN D.170-89. Monthly Telephone and Telex Accounts—General Tariff Principles—Charging and Accounting in International Telecommunications Services (Study Group III) 9 pp (Same as Recmn E.270). 9 pp.
CCITT RECMN E.270-89. Monthly Telephone and Telex Accounts—Telephone Network and ISDN—Operation, Numbering, Routing and Mobile Service (Study Group II)1 pp (Same as Recmn D.170). 1 p.

—**Telex Communications**
CCITT RECMN D.60-91. Guiding Principles to Govern the Apportionment of Accounting Rates in Intercontinental Telex Relations (Study Group III) 5 pp. 5 pp.
CCITT RECMN D.60-89. Guiding Principles to Govern the Apportionment of Accounting Rates in Intercontinental Telex Relations—General Tariff Principles—Charging and Accounting in International Telecommunications Services (Study Group III) 2 pp. 2 pp.
CCITT RECMN D.67-91. Charging and Accounting in the International Telex Service (Study Group III) 7 pp. 7 pp.
CCITT RECMN D.67-89. Charging and Accounting in the International Telex Service—General Tariff Principles—Charging and Accounting in International Telecommunications Services (Study Group III) 4 pp. 4 pp.
CCITT RECMN D.95-92. Charging, Billing, Accounting and Refunds in the Data Messaging Land/Maritime Mobile-Satellite Service (Study Group III) 11 pp. 11 pp.

—**Telex Communications—Encoded Data Transmission**
CCITT RECMN D.190-89. Transmission in Encoded Form of Monthly International Accounting Information—General Tariff Principles—Charging and Accounting in International Telecommunications Services (Study Group III) 6 pp (Same as Recmn E.275). 6 pp.

INDUSTRY STANDARDS

Accounting (Cont.)

—Telex Communications—Encoded Data Transmission (Cont.)
CCITT RECMN E.275-89. Transmission in Encoded Form of Monthly International Accounting Information—Telephone Network and ISDN—Operation, Numbering, Routing and Mobile Service (Study Group II) 1 pp (Same as Recmn D.190). 1 p.

—Telex Communications—Europe
CCITT FASCICLE II.1-88. General Tariff Principles—Charging and Accounting in International Telecommunications Services Series D Recommendations. 371 pp.

—Telex Communications—Mediterranean
CCITT FASCICLE II.1-88. General Tariff Principles—Charging and Accounting in International Telecommunications Services Series D Recommendations. 371 pp.

—Telex Communications—Multiple Address Messages—Store and Forward
CCITT RECMN D.65-89. General Charging and Accounting Principles in the International Telex Service for Multi-Address Messages Via Store-and-Forward Units—General Tariff Principles—Charging and Accounting in Intl Telecommunications Services (Study Group III) 2 pp. 2 pp.

—Telex Communications—Record Layouts
CCITT RECMN D.190-89. Transmission in Encoded Form of Monthly International Accounting Information—General Tariff Principles—Charging and Accounting in International Telecommunications Services (Study Group III) 6 pp (Same as Recmn E.275). 6 pp.
CCITT RECMN E.275-89. Transmission in Encoded Form of Monthly International Accounting Information—Telephone Network and ISDN—Operation, Numbering, Routing and Mobile Service (Study Group II) 1 pp (Same as Recmn D.190). 1 p.

—Telex Communications—Regional
CCITT FASCICLE II.1-88. General Tariff Principles—Charging and Accounting in International Telecommunications Services Series D Recommendations. 371 pp.

—Telex Communications—Telecommunication Administrations
CCITT RECMN D.160-89. Mode of Application of the Flat-Rate Price Procedure Set Forth in Recommendation D.67 and Recommendation D.150 for Remuneration of Facilities Made Available to the Administrations of Other Countries—General Tariff Principles—Charging and Accounting. 5 pp.
CCITT RECMN D.170-89. Monthly Telephone and Telex Accounts—General Tariff Principles—Charging and Accounting in International Telecommunications Services (Study Group III) 9 pp (Same as Recmn E.270). 9 pp.
CCITT RECMN E.252-89. Mode of Application of the Flat-Rate Price Procedure Set Forthe in Recommendations D.67 and D.150 for Remuneration of Facilities Made Available to the Administrations of Other Countries—Telephone Network and ISDN—Operation, Numbering, Routing and Mobile. 1 p.
CCITT RECMN E.270-89. Monthly Telephone and Telex Accounts—Telephone Network and ISDN—Operation, Numbering, Routing and Mobile Service (Study Group II)1 pp (Same as Recmn D.170). 1 p.

—Telex Communications—Telephone Relations
CCITT RECMN D.60-91. Guiding Principles to Govern the Apportionment of Accounting Rates in Intercontinental Telex Relations (Study Group III) 5 pp. 5 pp.
CCITT RECMN D.60-89. Guiding Principles to Govern the Apportionment of Accounting Rates in Intercontinental Telex Relations—General Tariff Principles—Charging and Accounting in International Telecommunications Services (Study Group III) 2 pp. 2 pp.

—Telex Communications—Telephone Relations—Africa
CCITT RECMN D.601R-89. Determination of Accounting Rate Shares and Collection Charges in Telex Relations Between Countries in Africa—General Tariff Principles—Charging and Accounting in International Telecommunications Services (Study Group III) 10 pp. 10 pp.

—Telex Communications—Telephone Relations—Asia
CCITT RECMN D.501R-89. Accounting Rates Applicable to Telex Relations Between Countries in Asia and Oceania—General Tariff Principles—Charging and Accounting in International Telecommunications Services (Study Group III) 1 pp. 1 p.

—Telex Communications—Telephone Relations—Europe
CCITT RECMN D.301R-89. Determination of Accounting Rate Shares and Collection Charges in Telex Relations Between Countries in Europe and the Mediterranean Basin—General Tariff Principles—Charging and Accounting in Intl Telecom. Serv. (Study Group III) 12 pp. 12 pp.

—Telex Communications—Telephone Relations—Latin America
CCITT RECMN D.401R-89. Accounting Rates Applicable to Telex Relations Between Countries in Latin America—General Tariff Principles—Charging and Accounting in International Telecommunications Services (Study Group III) 2 pp. 2 pp.

—Telex Communications—Telephone Relations—Mediterranean
CCITT RECMN D.301R-89. Determination of Accounting Rate Shares and Collection Charges in Telex Relations Between Countries in Europe and the Mediterranean Basin—General Tariff Principles—Charging and Accounting in Intl Telecom. Serv. (Study Group III) 12 pp. 12 pp.

—Telex Communications—Telephone Relations—Oceania
CCITT RECMN D.501R-89. Accounting Rates Applicable to Telex Relations Between Countries in Asia and Oceania—General Tariff Principles—Charging and Accounting in International Telecommunications Services (Study Group III) 1 pp. 1 p.

—Telex Communications—Traffic Units
CCITT RECMN D.67-91. Charging and Accounting in the International Telex Service (Study Group III) 7 pp. 7 pp.
CCITT RECMN D.67-89. Charging and Accounting in the International Telex Service—General Tariff Principles—Charging and Accounting in International Telecommunications Services (Study Group III) 4 pp. 4 pp.

—Transferred Account Services—Telegraphy
CCITT RECMN F.13-89. Operational Provisions for Participation in the Transferred Account Telegraph and Telematic Service—Telegraph and Mobile Services Operations and Quality of Service (Study Group I) 6 pp (Renumbered from F.41). 6 pp.
CCITT RECMN F.41-91. Interworking Between the Telemessage Service and the International Public Telegram Service (Study Group I) 5 pp (See Recmn F.13) (Renumbered from Recmn F.51). 5 pp.

—Transferred Account Services—Teleinformatic Services
CCITT RECMN F.13-89. Operational Provisions for Participation in the Transferred Account Telegraph and Telematic Service—Telegraph and Mobile Services Operations and Quality of Service (Study Group I) 6 pp (Renumbered from F.41). 6 pp.
CCITT RECMN F.41-91. Interworking Between the Telemessage Service and the International Public Telegram Service (Study Group I) 5 pp (See Recmn F.13) (Renumbered from Recmn F.51). 5 pp.

—Universal Personal Telecommunication Services
CENELEC ETR 055-3-92. Universal Personal Telecommunication (UPT); the Service Concept Part 3: Service Aspects of Charging, Billing and Accounting. 15 pp.
CENELEC ETR 065-92. Universal Personal Telecommunication (UPT); Requirements on Charging, Billing and Accounting. 9 pp.
ETSI ETR 055-3-92. Universal Personal Telecommunication (UPT); the Service Concept Part 3: Service Aspects of Charging, Billing and Accounting. 15 pp.
ETSI ETR 065-92. Universal Personal Telecommunication (UPT); Requirements on Charging, Billing and Accounting. 9 pp.

—User to User Information Services—ISDN
CCITT RECMN D.231-89. Charging and Accounting Principles Relating to the User-To-User Information (UUI) Supplementary Service—General Tariff Principles—Charging and Accounting in International Telecommunications Services (Study Group III) 2 pp. 2 pp.

—Videoconferencing Services
CCITT RECMN D.188-92. General Charging and Accounting Principles Applicable to an International Videoconferencing Service (Study Group III) 5 pp. 5 pp.
CEPT T/PGT 29 E-91. General Tariff and Accounting Principles of the European Multi-Point Videoconferencing Service. 2 pp.

—Videotex Communications
CCITT RECMN D.79-91. Charging and Accounting Principles for the International Videotex Service (Study Group III) 8 pp. 8 pp.

Accounting Authority Identification Codes
Use For: AAIC *See Also:* Accounting; Codes; Identification Systems

—Maritime Mobile Services—Radiotelegraphy
CCITT RECMN F.110-89. Operational Provisions for the Maritime Mobile Service—Telegraph and Mobile Services Operations and Quality of Service (Study Group I) 21 pp (Same as Recmn E.200). 21 pp.

—Telecommunication Administrations
CCITT RECMN D.90-92. Charging, Accounting and Refunds in the Maritime Mobile Service (Study Group III) 23 pp. 23 pp.
CCITT RECMN D.90-89. Charging, Accounting and Refunds in the Maritime Mobile Service—General Tariff Principles—Charging and Accounting in International Telecommunications Services (Study Group III) 19 pp. 19 pp.

Accounting Machines
Use: Calculators

Accreditation (Laboratories)
Use: Laboratory Accreditation

Accumulators
See Also: Concentrators; Hydraulic Accumulators; Hydropneumatic Accumulators

—Hazardous Materials—Disposal
EC 91/157/EEC-91. Council Directive on Batteries and Accumulators Containing Certain Dangerous Substances. 4 pp.

—Refrigeration Equipment
CSA C22.2 NO 140.3-M1987. Refrigerant-Containing Components for Use in Electrical Equipment (R 1993). 19 pp.

Accumulators (Batteries)
Use: Storage Batteries

Accumulators, Electrical
Use: Storage Batteries

Accumulators, Hydraulic
Use: Hydraulic Accumulators

Accumulators, Hydropneumatic
Use: Hydropneumatic Accumulators

Accuracy Testing
Use For: Precision Testing *See Also:* Reproducibility

—Aircraft—Altimeters
EUROCAE ED-26 03.79. MPS for Airborne Altitude Measurement and Coding Systems. 57 pp.

—Aircraft—Area Navigation Computing Instruments
EUROCAE ED-28 07.82. MPS for Airborne Area Navigation Computing Equipment Based on VOR and DME as Sensors. 31 pp.
EUROCAE ED-40 06.84. MPS for Airborne Computing Equipment for Area Navigation System Using Two DME as Sensors. 33 pp.

—Aircraft—Area Navigation Systems
EUROCAE ED-27 10.79. MOPR for Airborne Area Navigation Systems, Based on VOR and DME as Sensors. 63 pp.
EUROCAE ED-39 06.84. MOPR for Airborne Area Navigation Systems, Based on Two DME as Sensors. 67 pp.

INTERNATIONAL AND NON-U.S. NATIONAL STANDARDS
SUBJECT INDEX

Accuracy

Accuracy Testing (Cont.)

—Aircraft—Area Navigation Systems (Cont.)
EUROCAE ED-58 06.88. MOPS for Area Navigation Equipment Using Multi-Sensor Inputs. 356 pp.

—Aircraft—Distance Measuring Equipment
EUROCAE ED-57 12.86. MPS for Distance Measuring Equipment (DME/N and DME/P). 96 pp.

—Aircraft—Flowmeters
EUROCAE ED-42 10.83. MPS for a Fuel Flowmeter to Aircraft Standards. 24 pp.

—Aircraft—Fuel Gages
EUROCAE ED-41 12.83. MPS for Airborne Fuel Quantity Gauging Systems. 30 pp.

—Aircraft—Fuel Tanks—Float Gages
EUROCAE ED-41 12.83. MPS for Airborne Fuel Quantity Gauging Systems. 30 pp.

—Aircraft—Glide Path Receiving Instruments
EUROCAE ED-47A 01.88. MPS for Airborne ILS Receiving Equipment (Glide Path). 223 pp.

—Aircraft—Low Range Radio Altimeters
EUROCAE ED-30 03.80. MPS for Airborne Low Range Radio (Radar) Altimeter Equipment. 22 pp.

—Aircraft—VOR Receiving Equipment
EUROCAE ED-22B 01.88. MPS for Airborne VOR Receiving Equipment. 195 pp.

—Ball Screws
BSI BS 6101: Part 2-91. 1991 Machine Tool Ball Screws Part 2: Specification for Accuracy, Including Geometrical Tests. 19 pp.

—Band Saws
CNS B7154-86. Test Code for Accuracy of Feed Carriages of Band Saw Mills (Feb)(6305).
JIS B 6519-90. Test Methods for Performance and Accuracy of Band Scroll Saws. 11 pp.

—Boring Machines
CNS B7007-85. Test Code for Accuracy of Table Type Horizontal Boring Machines (Jan)(100).
CNS B7008-85. Test Code for Accuracy of Floor Type Horizontal Boring Machines (Jan)(101).
CNS B7098-82. Accuracy Inspection of Vertical Boring and Turning Mills (Aug)(4350).
JIS B 6334-86. Test Code for Performance and Accuracy of Numerically Controlled Horizontal Boring Machines (Table Type). 36 pp.

—Centers (Machine Tool Components)
ISO 2891-77. Modular Units for Machine Tool Construction—Centre Bases and Columns Second Edition. 5 pp.

—Construction
BSI BS 5606-90. 1990 Accuracy in Building. 57 pp.

—Distance Measuring Equipment
BSI BS 7334: Part 8-92. 1992 Measuring Instruments for Building Construction Part 8: Methods for Determining Accuracy in Use of Electronic Distance-Measuring Instruments up to 150 m (ISO 8322-8: 1992). 26 pp.
ISO 8322 Pt 8-92. Building Construction—Measuring Instruments—Procedures for Determining Accuracy in Use—Part 8: Electronic Distance-Measuring Instruments up to 150 m First Edition. 24 pp.

—Drilling Machines
CNS B7002-82. Accuracy Inspection Upright Drilling Machines (Aug)(95). 8 pp.
CNS B7003-86. Test Code for Performance and Accuracy of Radial Drilling Machines (Feb)(96). 8 pp.
CNS B7120-85. Test Code for Accuracy of Bench Drilling Machines (Jan)(4963).
CNS B7121-82. Accuracy Inspection of Drilling and Milling Combined Machines (Dec)(4964).
CNS B7215-86. Test Code for Accuracy of Numerically Controlled Turret and Single Spindle Drilling Machines with Vertical Spindle (Aug)(9476).
JIS B 6332-86. Test Code for Performance and Accuracy of Numerically Controlled Turret and Single Spindle Drilling Machines with Vertical Spindle. 22 pp.

—Gears
BSI BS 4185: Part 11-83. 1983 Machine Tool Components Part 11: Recommendations for Accuracy Grades of Gears. 4 pp.
CNS B1241-84. Accuracy of Spur Gear and Helical Gear (Aug)(7187).
ISO 1328-75. Parallel Involute Gears—ISO System of Accuracy First Edition. 38 pp.

Accuracy Testing (Cont.)

—Grinders
BSI BS 4656: Part 23-79. 1979 Accuracy of Machine Tools and Methods of Test Part 23: Surface Grinding Machines with Two Columns-Slideway Grinding Machines. 17 pp.
BSI BS 4656: Part 24-93. 1993 Accuracy of Machine Tools and Methods of Test Part 24: Specification for Cylindrical External Centreless Grinding Machines (ISO 3875: 1990). 20 pp.
BSI BS 4656: Part 24-80. 1980 Accuracy of Machine Tools and Methods of Test Part 24: Cylindrical External Centreless Grinding Machines. 12 pp.
CNS B7088-81. Accuracy Inspection of Internal Cylindrical Grinding Machines (Oct)(4261). 4 pp.
CNS B7100-78. Accuracy Inspection of Centerless Grinding Machines (Jul)(4441). 7 pp.
ISO 3875-90. Acceptance Conditions for External Cylindrical Centreless Grinding Machines—Testing of the Accuracy Second Edition. 18 pp.
JIS B 6211-86. Test Code for Performance and Accuracy of Internal Cylindrical Grinding Machines. 24 pp.
JIS B 6212-86. Test Code for Performance and Accuracy of External Cylindrical and Universal Grinding Machines. 28 pp.
JIS B 6220-90. External Cylindrical Centreless Grinding Machines—Test Code for Performance and Accuracy. 14 pp.

—Hacksaws
CNS B7115-82. Accuracy Inspection of Hack Sawing Machines (Dec)(4895). 3 pp.

—Helical Gears
BSI BS 436: Part 1-67. 1967 Amd 1 Spur and Helical Gears Part 1: Basic Rack Form, Pitches and Accuracy (Diametral Pitch Series). 31 pp.
BSI BS 436: Part 2-70. 1970 Amd 1 Spur and Helical Gears Part 2: Basic Rack Form, Modules and Accuracy (1 to 50 Metric Module). 33 pp.
JIS B 1702-76. Accuracy for Spur and Helical Gears. 36 pp.

—Hobbing Machines
CNS B7099-82. Accuracy Inspection of Hobbing Machines (Dec)(4351).
ISO 6545-92. Acceptance Conditions for Gear Hobbing Machines—Testing of the Accuracy First Edition. 33 pp.

—Jointers
CNS B7112-79. Accuracy Inspection Method of Jointer (Apr)(4810). 2 pp.

—Laser Instruments (Distance Measurement)
BSI BS 7334: Part 6-92. 1992 Measuring Instruments for Building Construction Part 6: Methods for Determining Accuracy in Use of Laser Instruments (ISO 8322-6: 1991). 19 pp.
ISO 8322 Pt 6-91. Building Construction—Measuring Instruments—Procedures for Determining Accuracy in Use—Part 6: Laser Instruments First Edition. 17 pp.

—Lathes
BSI BS 4361:Pt2: Sec 13-74. 1974 Woodworking Machines Part 2: Accuracy Tests Section 13: Turning Lathes. 6 pp.
BSI BS 4656: Part 31-82. 1982 Amd 1 Accuracy of Machine Tools and Methods of Test Part 31: Capstan, Turret and Automatic Lathes, Singles Spindle Type Greater Than 25 mm Diameter Capacity. 23 pp.
BSI BS 4656: Part 35-87. 1987 Accuracy of Machine Tools and Methods of Test Part 35: Capstan, Turret and Automatic Lathes, Single Spindle Type Less Than 25 mm Diameter Capacity. 27 pp.
ISO 6155 Pt 1-81. Acceptance Conditions for Horizontal Spindle Capstan, Turret and Single Spindle Automatic Lathes—Testing of the Accuracy—Part 1: Machinable Bar Diameters Greater Than 25 mm First Edition. 24 pp.
ISO 6155 Pt 2-86. Acceptance Conditions for Horizontal Spindle Capstan, Turret and Single Spindle Automatic Lathes—Testing of the Accuracy—Part 2: Machinable Bar Diameters 25 mm or Less and Chuck Diameter up to 160 mm First Edition. 25 pp.

—Machine Tools
BSI BS 3800-64. (WITHDRAWN) 1964 Amd 1 Methods for Testing the Accuracy of Machine Tools (Superseded by BS 3800: Part 1: 1990). 41 pp.
BSI BS 3800: Part 1-90. 1990 Amd 1 General Tests for Machine Tools Part 1: Code of Practice for Testing Geometric Accuracy of Machines Operating Under No Load or Finishing Conditions (AMD 6616) December 21, 1990. 81 pp.
BSI BS 3800: Part 2-91. 1991 General Tests for Machine Tools Part 2: Statistical Methods for Determination of Accuracy and Repeatability of Machine Tools. 30 pp.

Accuracy Testing (Cont.)

—Machine Tools (Cont.)
BSI BS 3800: Part 3-90. 1990 General Tests for Machine Tools Part 3: Method of Testing Performance of Machines Operating Under Loaded Conditions in Respect of Thermal Distortion. 13 pp.
BSI BS 4656: Part 16-85. (WITHDRAWN) 1985 Amd 1 Accuracy of Machine Tools and Methods of Test Part 16: Determin. of Accuracy and Repeatability of Positioning of Numeric. Controlled Machine Tools (AMD 5599) September 30 (Superseded by BS 3800: Part 2: 1991). 12 pp.
CNS B7082-86. Test Code Accuracy of Machine Tools (Nov)(4168). 24 pp.
CNS B7226-86. Test Code for Accuracy of Machining Centers (Aug)(9785). 25 pp.
ISO 230 Pt 2-88. Acceptance Code for Machine Tools—Part 2: Determination of Accuracy and Repeatability of Positioning of Numerically Controlled Machine Tools First Edition. 10 pp.
JIS B 6201-90. Machine Tools—General Test Code for Performance and Accuracy. 108 pp.
JIS B 6338-85. Test Code for Performance and Accuracy of Machining Centers (Vertical Type). 38 pp.

—Measuring Instruments
BSI BS 7334: Part 1-90. 1990 Measuring Instruments for Building Construction Part 1: Methods for Determining Accuracy in Use: Theory. 10 pp.
BSI BS 7334: Part 2-90. 1990 Measuring Instruments for Building Construction Part 2: Methods for Determining Accuracy in Use: Measuring Tapes. 11 pp.
BSI BS 7334: Part 3-90. 1990 Measuring Instruments for Building Construction Part 3: Methods for Determining Accuracy in Use: Optical Levelling Instruments. 15 pp.
ISO 8322 Pt 1-89. Building Construction—Measuring Instruments—Procedures for Determining Accuracy in Use—Part 1: Theory First Edition. 8 pp.
ISO 8322 Pt 2-89. Building Construction—Measuring Instruments—Procedures for Determining Accuracy in Use—Part 2: Measuring Tapes First Edition. 8 pp.
ISO 8322 Pt 3-89. Building Construction—Measuring Instruments—Procedures for Determining Accuracy in Use—Part 3: Optical Levelling Instruments First Edition. 12 pp.
ISO 9373-89. Cranes and Related Equipment—Accuracy Requirements for Measuring Parameters During Testing First Edition. 4 pp.

—Measuring Tapes
BSI BS 7334: Part 7-92. 1992 Measuring Instruments for Building Construction Part 7: Methods for Determining Accuracy in Use of Instruments When Used for Setting Out (ISO 8322-7: 1991). 16 pp.
ISO 8322 Pt 7-91. Building Construction—Measuring Instruments—Procedures for Determining Accuracy in Use—Part 7: Instruments When Used for Setting out First Edition. 15 pp.

—Milling Machines
BSI BS 4656: Part 30-92. 1992 Accuracy of Machine Tools and Methods of Test Part 30: Specification for Machining Centres and Computer Numerically Controlled Milling Machines, Horizontal and Vertical Spindle Types (Supersedes BS 4656: Parts 20 & 27: 1985) (F). 137 pp.
BSI BS 4656: Part 36-87. 1987 Accuracy of Machine Tools and Methods of Test Part 36: Plano-Milling Machines, Portal Type. 38 pp.
BSI BS 4656: Part 37-89. 1989 Accuracy of Machine Tools and Methods of Test Part 37: Plano-Milling Machines, Gantry Type. 36 pp.
CNS B7086-82. Accuracy Inspection of Knee Type Horizontal Milling Machines and Universal Milling Machines (Oct)(4259). 9 pp.
CNS B7087-82. Accuracy Inspection of Knee Vertical Milling Machine (Oct)(4260). 7 pp.
CNS B7098-82. Accuracy Inspection of Vertical Boring and Turning Mills (Aug)(4350).
CNS B7121-82. Accuracy Inspection of Drilling and Milling Combined Machines (Dec)(4964).
CNS B7181-86. Test Code for Accuracy of Numerically Controlled Knee Type Vertical Milling Machine (Feb)(8352).
CNS B7203-82. Accuracy Inspection of Knee Type Vertical-Horizontal Milling Machines (Oct)(9460). 11 pp.
CNS B7208-82. Test Code for Accuracy of Bed Type Vertical Milling Machine (Oct)(9465). 9 pp.
ISO 8636 Pt 1-87. Acceptance Conditions for Plano-Milling Machines—Testing of the Accuracy—Part 1: Portal-Type Machines First Edition. 36 pp.
ISO 8636 Pt 2-88. Acceptance Conditions for Plano-Milling Machines—Testing of the Accuracy—Part 2: Gantry-Type Machines First Edition. 35 pp.
JIS B 6228-84. Test Code for Performance and Accuracy of Plano-Milling Machines (Double Column Type). 35 pp.

INDUSTRY STANDARDS

Accuracy Testing (Cont.)

—Milling Machines (Cont.)
JIS B 6333-86. Test Code for Performance and Accuracy of Numerically Controlled Knee Type Vertical Milling Machines. 40 pp.

—Molding Machines
CNS B7142-86. Test Code for Accuracy of Three-Side Planers and Moulders (Feb)(5663).
CNS B7143-86. Test Code for Accuracy of Four-Side Planers and Moulders (Feb)(5664).
JIS B 6518-90. Test Methods for Performance and Accuracy of Moulding Machine. 16 pp.

—Optical Plummets
BSI BS 7334: Part 5-92. 1992 Measuring Instruments for Building Construction Part 5: Methods for Determining Accuracy in Use of Optical Plumbing Instruments (ISO 8322-5: 1991). 16 pp.
ISO 8322 Pt 5-91. Building Construction—Measuring Instruments—Procedures for Determining Accuracy in Use—Part 5: Optical Plumbing Instruments First Edition. 14 pp.

—Planers (Tools)
CNS B7004-86. Test Code for Performance and Accuracy of Double Housing Type Planing Machines (Feb)(97). 15 pp.
CNS B7113-79. Accuracy Inspection Method of Single Surface Planers (Apr)(4811).
CNS B7114-79. Accuracy Test of Double Surface Planers (Apr)(4812). 2 pp.
CNS B7142-86. Test Code for Accuracy of Three-Side Planers and Moulders (Feb)(5663).
CNS B7143-86. Test Code for Accuracy of Four-Side Planers and Moulders (Feb)(5664).
JIS B 6502-90. Test Methods for Performance and Accuracy of Wood Planers. 13 pp.

—Plotters
BSI BS 6808: Part 2-87. 1987 Coordinate Measuring Machines Part 2: Methods for Verifying Performance. 10 pp.
BSI BS 7172-89. 1989 Guide to Assessment of Position, Size and Departure from Nominal Form of Geometric Features. 19 pp.
JIS B 7440-87. Test Code for Accuracy of Coordinate Measuring Machines. 29 pp.

—Presses
CNS B7045-81. Accuracy Test Standard for Crank Press (Tentative) (Nov)(3157).

—Routers
BSI BS 4361:Pt2: Sec 12-74. 1974 Woodworking Machines Part 2: Accuracy Tests Section 12: Routing Machines. 6 pp.

—Routers (Tools)—Numerical Control
JIS B 6572-92. Numerically Controlled Routing Machines—Test Methods for Performance and Accuracy. 23 pp.

—Scroll Saws
JIS B 6519-90. Test Methods for Performance and Accuracy of Band Scroll Saws. 11 pp.

—Shaping Machines
CNS B7005-82. Accuracy Inspection of Shaping Machines (Aug)(98). 4 pp.

—Shears
CNS B7119-82. Accuracy Inspection of Shearing Machines (Dec)(4962).

—Slotting Machines
CNS B7006-82. Accuracy Inspection of Slotting Machines (Aug)(99). 4 pp.
JIS B 6206-56. Accuracy Inspection of Slotting Machines (R 1971). 6 pp.

—Spur Gears
BSI BS 436: Part 1-67. 1967 Amd 1 Spur and Helical Gears Part 1: Basic Rack Form, Pitches and Accuracy (Diametral Pitch Series). 31 pp.
BSI BS 436: Part 2-70. 1970 Amd 1 Spur and Helical Gears Part 2: Basic Rack Form, Modules and Accuracy (1 to 50 Metric Module). 33 pp.
JIS B 1702-76. Accuracy for Spur and Helical Gears. 36 pp.

—Tenoning Cutters
CNS B7242-83. Test Code for Accuracy of Tenoners (May)(10222).

—Theodolites
BSI BS 7334: Part 4-92. 1992 Measuring Instruments for Building Construction Part 4: Methods for Determining Accuracy in Use of Theodolites (ISO 8322-4: 1991). 19 pp.
BSI BS 7334: Part 7-92. 1992 Measuring Instruments for Building Construction Part 7: Methods for Determining Accuracy in Use of Instruments When Used for Setting Out (ISO 8322-7: 1991). 16 pp.

Accuracy Testing (Cont.)

—Theodolites (Cont.)
ISO 8322 Pt 4-91. Building Construction—Measuring Instruments—Procedures for Determining Accuracy in Use—Part 4: Theodolites First Edition. 17 pp.
ISO 8322 Pt 7-91. Building Construction—Measuring Instruments—Procedures for Determining Accuracy in Use—Part 7: Instruments When Used for Setting out First Edition. 15 pp.

—Turning Tools
BSI BS 4656: Part 28-88. 1988 Amd 1 Accuracy of Machine Tools and Methods of Test Part 28: Specification for Numerically Controlled Turning Machines up to and Including 1500 mm Turning Diameter (AMD 6665) March 28, 1991. 23 pp.

—Woodworking Equipment—Numerical Control
CNS B7111-84. General Rules for Accuracy Inspection of Woodworking Machine (Jul)(4809).
JIS B 6501-75. Test Code for Performance and Accuracy of Wood Working Machinery (R 1978). 29 pp.

Acetaldehyde

Use For: Acetic Aldehyde; Ethanal; Ethyl Aldehyde
See Also: Acetaldehyde Content Analysis
CNS K1176-75. Acetaldehyde for Industry Use (May)(3778). 1 p.
CNS K6380-75. Method of Test for Acetaldehyde (for Industrial Use) (May)(3779).
CNS K7575-81. Chemical Reagent (Acetaldehyde) (Aug)(7808).
JIS K 8030-80. Acetaldehyde.

—Carbonyl Compound Content—Volumetric Analysis
ISO 2885-73. Acetaldehyde for Industrial Use—Determination of Total Content of Carbonyl Compounds—Volumetric Method First Edition. 4 pp.

—Density
ISO 2513-74. Acetaldehyde for Industrial Use—Determination of Density at 15 Degrees Celsius First Edition. 4 pp.

—Iron Content—Photometry
ISO 2886-73. Acetaldehyde for Industrial Use—Determination of Iron Content—2,2'-Bipyridyl Photometric Method First Edition. 5 pp.

—Karl Fischer Method
ISO 2514-74. Acetaldehyde for Industrial Use—Determination of Water Content—Karl Fischer Method First Edition. 4 pp.

Acetaldehyde Content Analysis

See Also: Acetaldehyde

—Acetic Acid—Volumetric Analysis
ISO 753 Pt 4-81. Acetic Acid for Industrial Use—Methods of Test—Part 4: Determination of Acetaldehyde Monomer Content—Titrimetric Method First Edition. 4 pp.
ISO 753 Pt 5-81. Acetic Acid for Industrial Use—Methods of Test—Part 5: Determination of Total Acetaldehyde Content—Titrimetric Method First Edition. 5 pp.

Acetamides

See Also: Acetanilides
CNS K7576-81. Chemical Reagent (Acetamide) (Aug)(7809).
JIS K 8029-79. Acetamide.

Acetanilides

See Also: Acetamides; Acetoacetanilides
CNS K2179-87. Acetanilide (Dec)(12195). 2 pp.
JIS K 4111-77. Acetanilide.

Acetate Fabrics

Scope Note: For additional listings, see also specific products made from acetate fabrics
See Also: Acetate Fibers; Fabrics

—Linings—Satin
CGSB CAN/CGSB-4.89-M90. Acetate Satin Cloth Lining, 155 g/m2; (Corrigendum—Aug 1990). 10 pp.

—Linings—Twill
CGSB CAN/CGSB-4.90-M90. Acetate Twill Cloth Lining, 135 g/m2; (Corrigendum—Aug 1990). 10 pp.

Acetate Fibers

See Also: Acetate Fabrics; Cellulose Fibers; Fibers; Triacetate Fibers

Acetate Fibers (Cont.)

—Binary Mixtures—Quantitative Analysis
CGSB CAN/CGSB-4.2 NO.14.7-M88. Textile Test Methods Quantitative Analysis of Fibre Mixtures—Binary Mixtures Containing Acetate Fibres—Acetone Method. 12 pp.
CGSB CAN/CGSB-4.2 NO.14.11-M89. Textile Test Methods Quantitative Analysis of Fibre Mixtures—Binary Mixtures Containing Secondary and Triacetate Fibres—70% Acetone Method. 11 pp.
CGSB CAN/CGSB-4.2 NO.14.16-M88. Textile Test Methods Quantitative Analysis of Fibre Mixtures—Binary Mixtures Containing Acetate and Polyvinyl Chloride Fibres—Glacial Acetic Acid Method. 11 pp.

Acetates

Scope Note: For additional listings, use a more specific term *See Also:* Aluminum Acetate; Amyl Acetate; Barium Acetate; n-Butyl Acetate; sec-Butyl Acetate; tert-Butyl Acetate; Cadmium Acetate; Calcium Acetate; Cellulose Acetate; Cobaltous Acetate; Cupric Acetate; Esters; Ethyl Acetate; Isobutyl Acetate; Isopropyl Acetate; Lead Acetate Trihydrate; Lead Subacetate; Magnesium Acetate; Mercuric Acetate; Mercurous Acetate; Methyl Acetate; 3-Methylbutyl Acetate; Nickel Acetate; Potassium Acetate; n-Propyl Acetate; Silver Acetate; Sodium Acetates; Uranyl Acetate; Vinyl Acetate; Zinc Acetate
CNS K6234-78. Method of Test for Solvent Acetate of Industrial Use (Mar)(2618).
ISO 1386-83. Solvent Acetates for Industrial Use—Methods of Test First Edition. 6 pp.

Acetic Acid

Use For: Glacial Acetic Acid *See Also:* Acetic Acid Content Analysis; Fatty Acids; Hazardous Materials; Vinegars
BSI BS 576: Part 1-88. 1988 Acetic Acid for Industrial Use Part 1: Acetic Acid. 4 pp.
BSI BS 576: Part 2-88. 1988 Acetic Acid For Industrial Use Part 2: Methods of Test for Acetic Acid. 17 pp.
CNS K1052-80. Glacial Acetic Acid (Jul)(1340). 1 p.
CNS K6144-60. Method of Test for Glacial Acetic Acid (for Industrial Use) (Dec)(1341)(R 1971).
CNS K7020-61. Chemical Reagent (Glacial Acetic Acid) (Dec)(1520).
CNS K7021-61. Chemical Reagent (Glacial Acetic Acid Conformed to Dischromate Test) (Dec)(1521).
CNS K7022-61. Chemical Reagent (Acetic Acid 36 Percent) (Dec)(1522).
ISO 753 Pt 1-81. Acetic Acid for Industrial Use—Methods of Test—Part 1: General First Edition. 5 pp.
JIS K 1351-66. Acetic Acid. 12 pp.
JIS K 8355-88. Acetic Acid.

—Acetaldehyde Content—Volumetric Analysis
ISO 753 Pt 4-81. Acetic Acid for Industrial Use—Methods of Test—Part 4: Determination of Acetaldehyde Monomer Content—Titrimetric Method First Edition. 4 pp.
ISO 753 Pt 5-81. Acetic Acid for Industrial Use—Methods of Test—Part 5: Determination of Total Acetaldehyde Content—Titrimetric Method First Edition. 5 pp.

—Chloride Content—Visual Limit Testing
ISO 753 Pt 8-81. Acetic Acid for Industrial Use—Methods of Test—Part 8: Visual Limit Test for Inorganic Chlorides First Edition. 4 pp.

—Dichromate Number
ISO 753 Pt 7-81. Acetic Acid for Industrial Use—Methods of Test—Part 7: Determination of Dichromate Index First Edition. 4 pp.

—Formic Acid Content—Iodometry
ISO 753 Pt 3-83. Acetic Acid for Industrial Use—Methods of Test—Part 3: Determination of Formic Acid Content—Iodometric Method First Edition. 6 pp.

—Heavy Metals Content—Visual Limit Testing
ISO 753 Pt 10-81. Acetic Acid for Industrial Use—Methods of Test—Part 10: Visual Limit Test for Heavy Metals (Including Iron) First Edition. 4 pp.

—Permanganate Number
ISO 753 Pt 6-81. Acetic Acid for Industrial Use—Methods of Test—Part 6: Determination of Permanganate Index First Edition. 4 pp.

—Photographic Chemicals
BSI BS 5569-78. 1978 Photographic Grade Glacial Acetic Acid. 7 pp.
CNS Z9013-80. Glacial Acetic Acid (Photographic Grade) (Jan)(5154).

INTERNATIONAL AND NON-U.S. NATIONAL STANDARDS
SUBJECT INDEX

Acid

Acetic Acid (Cont.)
—Photographic Chemicals (Cont.)
ISO 3298-76. Photographic Grade Glacial Acetic Acid—Specification First Edition. 5 pp.
JIS K 7723-86. Photographic Grade Glacial Acetic Acid.

—Sulfate Content—Visual Limit Testing
ISO 753 Pt 9-81. Acetic Acid for Industrial Use—Methods of Test—Part 9: Visual Limit Test for Inorganic Sulphates First Edition. 4 pp.

—Volumetric Analysis
ISO 753 Pt 2-81. Acetic Acid for Industrial Use—Methods of Test—Part 2: Determination of Acetic Acid Content—Titrimetric Method First Edition. 4 pp.

Acetic Acid Butyl Ester
Use: n-Butyl Acetate

Acetic Acid Content Analysis
See Also: Acetic Acid

—Cellulose Acetate
DIN ENGL 53730-69. Testing of Plastics; Determination of the Acetic Acid Yield of Non-Plasticized Cellulose Acetate Thermoplastics (Oct). 2 pp.
ISO 1597-75. Plastics—Unplasticized Cellulose Acetate—Determination of Acetic Acid Yield First Edition. 5 pp.

Acetic Acid Salt Spray Testing
Use: Salt Spray Testing

Acetic Aldehyde
Use: Acetaldehyde

Acetic Anhydride
BSI BS 2068-70. 1970 Acetic Anhydride. 12 pp.
BSI BS 2068-01. 1970 Amd 1 Acetic Anhydride (AMD 7627) May 15, 1993 (W). 13 pp.
CNS K1256-81. Acetic Anhydride (Mar)(7156).
CNS K7023-61. Chemical Reagent (Acetic Anhydride) (Dec)(1523).
ISO 754-82. Acetic Anhydride for Industrial Use—Methods of Test First Edition. 13 pp.
JIS K 1352-56. Acetic Anhydride. 5 pp.
JIS K 8886-79. Acetic Anhydride.

—Bromine Number
ISO 761-77. Acetic Anhydride and Butan-1-ol for Industrial Use—Determination of Bromine Number First Edition. 4 pp.

—Highway Transportation—Emergency Procedures
SAA AS 1678.8.0. 005-93. Emergency Procedure Guides—Transport—Part 8.0.005: Acetic Anhydride (in Professional Package 37A).

—Safety—Information Cards
SAA AS 2508.8.00 1-91. Safe Storage and Handling Information Cards for Hazardous Materials—Part 8.001: Acetic Anhydride (in Professional Package 38). 2 pp.

Acetic Ester
Use: Ethyl Acetate

Acetoacetanilides
See Also: Acetanilides
CNS K2208-88. Acetoacetanilides (Apr)(12276).
JIS K 4173-81. Acetoacetanilides (Acetoacetanilide 2-Methylaceto-acetanilide 2',4'-Dimethylaceto-acetanilide.2'-Chloroacetoacetanilide).

Acetone
Use For: Dimethylketone; 2-Propanone
See Also: Acetylacetone; Diacetone Alcohol; Hazardous Materials; Ketones
BSI BS 509: Part 1-87. 1987 Acetone for Industrial Use Part 1: Acetone. 4 pp.
BSI BS 509: Part 2-84. 1984 Amd 1 Acetone for Industrial Use Part 2: Methods of Test. 10 pp.
CGSB CAN/CGSB-15.50-92. Technical Grade Acetone. 7 pp.
CNS K1023-84. Acetone, Industrial Grade (Sep)(197). 1 p.
CNS K6154-84. Method of Test for Acetone (Sep)(1419).
CNS K7024-61. Chemical Reagent (Acetone) (Dimethyl Ketone) (Dec)(1524).
ISO 757 Pt 1-82. Acetone for Industrial Use—Methods of Test—Part 1: General First Edition. 4 pp.
JIS K 1503-59. Acetone.
JIS K 8034-91. Acetone.

Acetone (Cont.)
—Acidity—Volumetric Analysis
ISO 757 Pt 2-82. Acetone for Industrial Use—Methods of Test—Part 2: Determination of Acidity to Phenolphthalein—Titrimetric Method First Edition. 4 pp.

—Agulhon's Reagents
ISO 757 Pt 5-82. Acetone for Industrial Use—Methods of Test—Part 5: Control Test with Agulhon's Reagent First Edition. 4 pp.

—Chemical Analysis
CNS K6362-73. Chemical Analysis of Acetone Extracted from Natural Rubber (May)(3571).

—Miscibility
ISO 757 Pt 3-82. Acetone for Industrial Use—Methods of Test—Part 3: Test for Miscibility with Water First Edition. 4 pp.

—Permanganate Number
ISO 757 Pt 4-83. Acetone for Industrial Use—Methods of Test—Part 4: Determination of Permanganate Time First Edition. 4 pp.

—Plastic Pipes—PVC—Unplasticized
SAA AS 1462.12-84. Methods of Test for Unplasticized PVC (UPVC) Pipes and Fittings—Part 12: Method for Determining the Effect of Immersion in Anhydrous Acetone on UPVC Pipes. 2 pp.

Acetone Content Analysis
—Ethylene—Gas Chromatography
ISO 8174-86. Ethylene and Propylene for Industrial Use—Determination of Acetone, Acetonitrile, Propan-2-ol and Methanol—Gas Chromatographic Method First Edition. 8 pp.

—Methanols
CNS K6263-80. Method of Test for Methyl Alcohol (Methanol) (Feb)(2790). 7 pp.

—Propylene—Gas Chromatography
ISO 8174-86. Ethylene and Propylene for Industrial Use—Determination of Acetone, Acetonitrile, Propan-2-ol and Methanol—Gas Chromatographic Method First Edition. 8 pp.

Acetonitrile
See Also: Acetonitrile Content Analysis
CNS K7025-61. Chemical Reagent (Acetonitril) (Methyl Cyanide) (Dec)(1525).
JIS K 8032-91. Acetonitrile.

Acetonitrile Content Analysis
See Also: Acetonitrile

—Ethylene—Gas Chromatography
ISO 8174-86. Ethylene and Propylene for Industrial Use—Determination of Acetone, Acetonitrile, Propan-2-ol and Methanol—Gas Chromatographic Method First Edition. 8 pp.

—Propylene—Gas Chromatography
ISO 8174-86. Ethylene and Propylene for Industrial Use—Determination of Acetone, Acetonitrile, Propan-2-ol and Methanol—Gas Chromatographic Method First Edition. 8 pp.

Acetophenones
See Also: Ketones
CNS K7588-88. Chemical Reagent (Acetophenone) (Oct)(8026).
JIS K 8033-88. Acetophenone.

Acetyl Chloride
CNS K7026-61. Chemical Reagent (Acetyl Chloride) (Dec)(1526).

Acetyl Value
—Fats
BSI BS 684: Sec 2.9-77. 1977 Methods of Analysis of Fats and Fatty Oils Part 2: Other Methods Section 2.9: Determination of the Hydroxyl Value and the Acetyl Value. 4 pp.

Acetylacetone
Use For: Diacetylmethane; 2,4-Pentanedione
See Also: Acetone
CNS K7666-83. Chemical Reagent (Acetylacetone) (Mar)(10105).
JIS K 8027-88. 2,4pentanedione (Acetylacetone).

Acetylation
See Also: Esterification
—Essential Oils
CNS K6611-80. Estimation of Free Alcohols Content by Determination of Ester Value After Acetylation of Essential Oil (Dec)(6677).

Acetylation (Cont.)
—Essential Oils (Cont.)
CNS K6614-80. Estimation of Free Alcohols Content by Determination of Ester Value After Acetylation of Essential Oils (Containing Tertiary Alcohols) (Dec)(6680).
CNS K6615-80. Estimation of Primary and Secondary Free Alcohols Content of Essential Oil by Acetylation in Pyridine (Dec)(6681).
ISO 1241-80. Essential Oils—Determination of Ester Value After Acetylation and Evaluation of Free Alcohols and Total Alcohols Content First Edition. 5 pp.
ISO 3793-76. Essential Oils—Estimation of Primary and Secondary Free Alcohols Content by Acetylation in Pyridine First Edition. 4 pp.
ISO 3794-76. Essential Oils (Containing Tertiary Alcohols)—Estimation of Free Alcohols Content by Determination of Ester Value After Acetylation First Edition. 4 pp.

Acetylcellulose
Use: Cellulose Acetate

Acetylene
Use For: Dissolved Acetylene *See Also:* Acetylene Black; Hazardous Materials; Hydrocarbons
CNS K1072-63. Acetylene for Industrial Use (Nov)(2157)(R 1973).
CNS K6167-63. Method of Test for Acetylene (for Industrial Use) (Nov)(2158)(R 1968).
JIS K 1902-80. Dissolved Acetylene. 8 pp.

—Cylinders
BSI BS 6061-81. 1981 Amd 1 Transportable Acetylene Containers (AMD 6322) January 31, 1991. 11 pp.
BSI BS 6071-81. 1981 Periodic Inspection and Maintenance of Transportable Gas Containers for Dissolved Acetylene. 12 pp.
ISO 3807-77. Dissolved Acetylene Cylinders—Basic Requirements First Edition. 9 pp.
SAA AS B189-62. Welded Capsule Type Steel for the Storage and Transport of Compressed Acetylene Dissolved in Acetone Amdt 1 February 1963 Amdt 2 March 1965 Andt 3 December 1968. 16 pp.
SAA AS CB4. Gas Cylinders Code—Interpretations to SAA Gas Cylinders Code No 6 (April 1973) Welded Three-Piece Type Cylinders for Acetylene Amdt 1 January 1976 See AS 2527—1982 Annex 2. 21 pp.

—Cylinders—Steel
JIS B 8234-88. Refillable Welded Steel Cylinders for Dissolved Acetylene. 50 pp.

—Cylinders—Valves
JIS B 8244-89. Valves for Dissolved Acetylene Cylinder. 8 pp.

Acetylene Black
See Also: Acetylene

—Dry Cells
CNS K1143-72. Acetylene Black (Jan)(3139). 1 p.
CNS K6286-70. Method of Test for Acetylene Black (Sep)(3140).
JIS K 1469-84. Acetylene Black.
JIS K 1469-66. Acetylene Black. 9 pp.

Acetylene Content Analysis
—Ethylene
CNS K6397-76. Method of Test for Trace Acetylene and Carbon Dioxide in Ethylene (Dec)(4054).

Acetylene Cylinders
Use: Gas Cylinders—Acetylene

Acid Anhydrides
Scope Note: See listings under Anhydrides

Acid Capacity
Use: Acidity

Acid Content Analysis
Use: Acidity

Acid Copper Chromate
Use: Cupric Chromate

Acid Immersion Testing
Use: Immersion Testing—Acid

Acid Number
Use: Saponification Number

Acid Pickling
Use: Chemical Cleaning

Acid Resistance Testing
See Also: Accelerated Testing; Corrosion Testing; Immersion Testing; Salt Spray Testing

INDUSTRY STANDARDS

Acid Resistance Testing (Cont.)

—Chemical Porcelain
CNS R3092-85. Method of Test for Acid-Proof Porcelain for Chemical Industry (May) (6516).

JIS R 1503-91. Testing Methods for Acid Proof Porcelain of Chemical Industry. 6 pp.

—Citric Acid—Vitreous Enamels
BSI BS 1344: Part 2-75. 1975 Methods of Testing Vitreous Enamel Finishes Part 2: Resistance to Citric Acid at Room Temperature. 8 pp.

BSI BS 1344: Part 8-84. 1984 Methods of Testing Vitreous Enamel Finishes Part 8: Resistance to Boiling Citric Acid Solution. 6 pp.

CNS R3096-80. Method of Test for Resistance of Porcelain Enameled Utensils to Boiling Acid (Dec)(6852). 5 pp.

ISO 2722-73. Vitreous and Porcelain Enamels—Determination of Resistance to Citric Acid at Room Temperature First Edition. 5 pp.

ISO 2742-83. Vitreous and Porcelain Enamels—Determination of Resistance to Boiling Citric Acid Second Edition. 4 pp.

—Fabrics
ISO 105 Pt E05-89. Textiles—Tests for Colour Fastness—Part E05: Colour Fastness to Spotting: Acid Third Edition. 4 pp.

—Gaskets
CNS O2014-73. Method of Test for Cork Binder Discs (May)(2303). 1 p.

—Glass
ISO 8424-87. Raw Optical Glass—Testing of the Resistance to Attack by Aqueous Acidic Solutions at 25 Degrees C and Classification First Edition. 8 pp.

ISO 9689-90. Raw Optical Glass—Resistance to Attack by Aqueous Alkaline Phosphate-Containing Detergent Solutions at 50 Degrees Celsius—Testing and Classification First Edition. 11 pp.

—Glass Coatings (On Glass)
CGSB 1-GP-71 METH 105.2-79. Methods of Testing Paints and Pigments Acid Resistance Enamel Films on Glass Panels. 1 p.

—Glazes
CNS R3089-80. Method of Test for Resistance of Glaze to Acid and Alkali for Porcelain Laboratory Apparatus (Apr)(5478).

—Hydrochloric Acid—Glass
BSI BS 3473: Part 5-87. 1987 Chemical Resistance of Glass Used in the Production of Laboratory Glassware Part 5: Method for Determination of Resistance of Glass to Attack by 6 mol/L Hydrochloric Acid at 100 Degrees Celcius. 10 pp.

ISO 1776-85. Glass—Resistance to Attack by Hydrochloric Acid at 100 Degrees Celsius—Flame Emission or Flame Atomic Absorption Spectrometric Method First Edition. 7 pp.

—Hydrochloric Acid—Vitreous Enamels
BSI BS 1344: Part 10-87. 1987 Methods of Testing Vitreous Enamel Finishes Part 10: Determination of Resistance to Condensing Hydrochloric Acid Vapour. 6 pp.

ISO 2743-86. Vitreous and Porcelain Enamels—Determination of Resistance to Condensing Hydrochloric Acid Vapour Second Edition. 4 pp.

—Hydrofluoric Acid—Steels
CNS G2055-86. Method of Nitro-Hydrofluoric Acid Test for Stainless Steels (Jul)(4765).

JIS G 0574-80. Method of Nitric-Hydrofluoric Acid Test for Stainless Steels. 6 pp.

—Nitric Acid—Steels
CNS G2054-86. Method of Sixty Five Percent Nitric Acid Test for Stainless Steels (Jul)(4764).

CNS G2055-86. Method of Nitro-Hydrofluoric Acid Test for Stainless Steels (Jul)(4765).

JIS G 0573-80. Method of 65 Per Cent Nitric Acid Test for Stainless Steels. 6 pp.

JIS G 0574-80. Method of Nitric-Hydrofluoric Acid Test for Stainless Steels. 6 pp.

—Oxalic Acid—Steels
CNS G2170-83. Method of Ten Percent Oxalic Acid Etch Test for Stainless Steels (Apr)(10170).

—Paints
CGSB 1-GP-71 METH 105.1-80. Methods of Testing Paints and Pigments Acid Resistance Paint Films on Steel Panels. 1 p.

—Plywood
CNS O2048-82. Method of Test for Acid, Alkali and Thinner Resistance of Special Plywood (Apr)(8735). 1 p.

Acid Resistance Testing (Cont.)

—Slate Roofing
DIN ENGL 52206-75. Testing of Roofing Slates; Acid Resistance Test (Mar). 1 p.

—Stoneware
CNS R3064-76. Method of Test for Acid Resistance of Chemical Stoneware (Jun)(3191).

—Sulfuric Acid—Plastic Pipes
CNS K6596-80. Unplasticized Polyvinyl Chloride (PVC) Pipes Effect of Sulphuric Acid Test Method (Sep)(6489).

ISO 3473-77. Unplasticized Polyvinyl Chloride (PVC) Pipes—Effect of Sulphuric Acid—Requirement and Test Method First Edition. 4 pp.

—Sulfuric Acid—Refractory Materials
BSI BS 1902: Sec 3.12-89. 1989 Methods for Testing Refractory Materials Part 3: General and Textural Properties Section 3.12: Determination of Resistance to Sulphuric Acid (Powders) (Method 1902-312). 4 pp.

CEN PREN 993-16-93. Dense Shaped Refractory Products—Methods of Test—Part 16: Determination of Resistance to Sulphuric Acid. 6 pp.

ISO 8890-88. Dense Shaped Refractory Products—Determination of Resistance to Sulfuric Acid First Edition. 4 pp.

ISO 10080-90. Refractory Products—Classification of Dense, Shaped Acid-Resisting Products First Edition. 4 pp.

—Sulfuric Acid—Steels
CNS G2052-86. Method of Five Percent Sulfuric Acid Test for Stainless Steels (Jul)(4762).

CNS G2053-86. Method of Ferric Sulfate-Sulfuric Acid Test for Stainless Steels (Jul)(4763).

CNS G2056-86. Method of Copper Sulfate-Sulfuric Acid Test for Stainless Steels (Jul)(4766).

JIS G 0572-84. Method of Ferric Sulfate-Sulfuric Acid Test for Stainless Steels. 6 pp.

JIS G 0575-80. Method of Copper Sulfate-Sulfuric Acid Test for Stainless Steels. 7 pp.

JIS G 0591-80. Method of 5 Per Cent Sulfuric Acid Test for Stainless Steels. 5 pp.

—Sulfuric Acid—Vitreous Enamels
BSI BS 1344: Part 3-88. 1988 Methods of Testing Vitreous Enamel Finishes Part 3: Determination of Resistance to Sulphuric Acid at Room Temperature. 7 pp.

BSI BS 1344: Part 3-67. 1967 Methods of Testing Vitreous Enamel Finishes Part 3: Resistance to Products of Combustion Containing Sulphur Compounds. 10 pp.

ISO 8290-87. Vitreous and Porcelain Enamels—Determination of Resistance to Sulfuric Acid at Room Temperature First Edition. 6 pp.

Acid Resistance Testing Equipment

—Enamels
BSI BS 1344: Part 14-84. 1984 Methods of Testing Vitreous Enamel Finishes Part 14: Apparatus for Testing with Acid and Neutral Liquids and Their Vapours. 9 pp.

ISO 2733-83. Vitreous and Porcelain Enamels—Apparatus for Testing with Acid and Neutral Liquids and Their Vapours Second Edition. 7 pp.

Acid Value
Use: Saponification Number

2S-Acid
CNS K2189-87. 2S-Acid (Monosodium Salt) (Dec)(12205). 2 pp.

Acidimetric Analysis
Use: Acidimetry

Acidimetry
See Also: Volumetric Analysis

—Dairy Products
CNS N6057-72. Method of Test for Milk and Milk Products (Titration of Acidity) (Oct)(3441). 1 p.

—Milk
CNS N6057-72. Method of Test for Milk and Milk Products (Titration of Acidity) (Oct)(3441). 1 p.

—Soaps
ISO 6387-83. Surface Active Agents—Determination of the Power to Disperse Calcium Soap—Acidimetric Method (Modified Schonfeldt Method) First Edition. 5 pp.

—Starches
ISO 5379-83. Starches and Derived Products—Determination of Sulfur Dioxide Content—Acidimetric Method and Nephelometric Method First Edition. 8 pp.

Acidimetry (Cont.)

—Water
SAA AS 3550.3-92. Waters—Part 3: Determination of Alkalinity—Acidimetric Titration Method (Supersedes AS 2449—1981). 4 pp.

Acidity
Use For: Acid Capacity; Acid Content Analysis; Free Acidity; Nonvolatile Acid Content
See Also: Alkalinity; Amino Acid Content Analysis; pH

—Acetone—Volumetric Analysis
ISO 757 Pt 2-82. Acetone for Industrial Use—Methods of Test—Part 2: Determination of Acidity to Phenolphthalein—Titrimetric Method First Edition. 4 pp.

—Adipates—Volumetric Analysis
ISO 2525-74. Adipate Esters for Industrial Use—Determination of Acidity to Phenolphthalein—Volumetric Method First Edition. 4 pp.

—Alcohols—Volumetric Analysis
ISO 1843 Pt II-77. Higher Alcohols for Industrial Use—Methods of Test—Part II: Determination of Acidity to Phenolphthalein—Titrimetric Method First Edition. 4 pp.

—Aluminum Sulfate
CPPA J.6-86. Analysis of Alum Reprinted—1988. 5 pp.

MOD UK M 9508/67. Examination of Aluminium Sulphate.

—Ammonium Nitrate—Volumetric Analysis
ISO 2364-72. Ammonium Nitrate for Industrial Use—Determination of Free Acidity—Volumetric Method First Edition. 3 pp.

—Ammonium Sulphate—Volumetric Analysis
ISO 2993-74. Ammonium Sulphate for Industrial Use—Determination of Free Acidity—Titrimetric Method First Edition. 4 pp.

—Amyl Acetate
BSI BS 552-70. 1970 Amyl Acetate. 12 pp.

BSI BS 552-01. 1970 Amd 1 Amyl Acetate (AMD 7218) September 15, 1992. 14 pp.

—Anhydrous Hydrogen Fluoride—Volumetric Analysis
BSI BS 5365: Part 2-76. (WITHDRAWN) 1976 Methods of Sampling and Test for Anhydrous Hydrogen Fluoride for Industrial Use Part 2: Determination of Non-Volatile Acid Content: Titrimetric Method. 4 pp.

ISO 3138-74. Anhydrous Hydrogen Fluoride for Industrial Use—Determination of Non-Volatile Acid Content—Titrimetric Method First Edition. 4 pp.

—Aromatic Hydrocarbons
CNS K6257-74. Method of Test for Acidity and Alkalinity of Aromatic Hydrocarbons (Oct)(2758).

—Beeswax
MOD UK M 2012/71. Examination of Beeswax GS and Beeswax, Lead Free.

MOD UK M 9517/65. Examination of Beeswax and Beeswax, Lead Free (LAG 931A, No Information).

—Beverages
CNS N6091-91. Method of Test for Fruit and Vegetable Juices and Drinks (General Rules) (Jun)(3736). 1 p.

CNS N6176-82. Method of Test for Beverage: Determination of Acidity (Sep)(9434). 2 pp.

—Butanols—Volumetric Analysis
ISO 755 Pt 2-81. Butan-1-ol for Industrial Use—Methods of Test—Part 2: Determination of Acidity—Titrimetric Method First Edition. 4 pp.

ISO 2887-73. secButyl Alcohol, Methyl Ethyl Ketone, isoButyl Methyl Ketone, isoAmyl Ethyl Ketone, Diacetone Alcohol, and Hexylene Glycol for Industrial Use—Determination of Acidity to Phenolphthalein—Volumetric Method First Edition. 4 pp.

—Butyl Acetate
BSI BS 551-90. 1990 Butyl Acetate for Industrial Use. 8 pp.

—Cables (Electric)—Combustion Gases
BSI BS 6425: Part 2-93. 1993 Test on Gases Evolved During the Combustion of Materials from Cables Part 2: Determination of Degree of Acidity (Corrosivity) of Gases by Measuring pH and Conductivity (E). 19 pp.

INTERNATIONAL AND NON-U.S. NATIONAL STANDARDS
SUBJECT INDEX

Acidity

Acidity (Cont.)

—Cables (Electric)—Combustion Gases (Cont.)
CENELEC HD 602 S1-92. Test on Gases Evolved During Combustion of Electric Cables Part 2: Determination of Degree of Acidity of Gases Evolved During the Combustion of Materials Taken from Electric Cables by Measuring pH and Conductivity. 4 pp.
CENELEC HD 602 S1-92. Test on Gases Evolved During Combustion of Materials from Cables Determination of Degree of Acidity (Corrosivity) of Gases by Measuring pH and Conductivity. 4 pp.
IEC 754 Pt 2-91. Test on Gases Evolved During Combustion of Electric Cables Part 2: Determination of Degree of Acidity of Gases Evolved During the Combustion of Materials Taken from Electric Cables by Measuring pH and Conductivity First Edition. 33 pp.

—Camphor
MOD UK M 9555/72. Examination of Camphor for Celluloid (Superseded by CS 2914).

—Canning Liquids—Volumetric Analysis
CNS N6019-87. Method of Test for Canned Food: Determination of Packing Liquid (Feb)(975). 3 pp.

—Caseins
BSI BS 6248: Part 5-82. 1982 Caseins and Caseinates Part 5: Method for Determination of Free Acidity of Caseins (Reference Method). 4 pp.
SAA AS 2300.10.6-91. Methods of Chemical and Physical Testing for the Dairying Industry—Part 10: Caseins, Caseinates and Coprecipitates—Part 10.6: Determination of Free Acidity of Caseins (Supersedes AS N60—1970 (in Part)). 3 pp.

—Cellulose Acetate—Volumetric Analysis
BSI BS 2782:Pt4: METH 459A-91. 1991 Methods of Testing Plastics Part 4: Chemical Properties Method 459A: Determination of Free Acidity in Cellulose Acetate (ISO 1061: 1990). 6 pp.
DIN ENGL 53729-68. Testing of Plastics; Determination of Free Acid in Non-Plasticized Cellulose Acetate (Nov). 2 pp.
ISO 1061-90. Plastics—Unplasticized Cellulose Acetate—Determination of Free Acidity Second Edition. 5 pp.

—Cellulose Nitrate
CNS K6185-87. Method of Test for Nitrocellulose for Industrial Use (Jun)(2358). 7 pp.

—Chlorofluorocarbons—Volumetric Analysis
BSI BS 5598: Part 6-79. 1979 Methods of Sampling and Test for Halogenated Hydrocarbons Part 6: Determination of Acidity of Fluorochlorinated Hydrocarbons. 6 pp.
ISO 3363-76. Fluorochlorinated Hydrocarbons for Industrial Use—Determination of Acidity—Titrimetric Method First Edition. 5 pp.

—Chloroform
BSI BS 4774-72. 1972 Methods of Test for Chloroform for Industrial Use. 12 pp.

—Cleaning Agents
CGSB 31-GP-0A METH 51.1-57. Methods of Testing Corrosion-Prevention Materials and Processes Free Alkali or Free Acid. 1 p.

—Cosmetics
CNS S2073-82. Method of Hygienic Test for Cosmetics Value, Acidity and Alkalinity (Jun)(9036).

—Creosote Oil
CNS K6074-57. Method of Test for Acidity of Creosote Oil (Jul)(917)(R 1971).

—Dairy Products—Volumetric Analysis
BSI BS 1741: Sec 10.1-89. 1989 Methods for Chemical Analysis of Liquid Milk and Cream Part 10: Determination of Titratable Acidity Section 10.1: Method for Liquid Milk. 4 pp.
BSI BS 1743: Sec 7.1-82. 1982 Analysis of Dried Milk and Dried Milk Products Part 7: Determination of the Titratable Acidity of Dried Milk Section 7.1: Reference Method. 4 pp.
BSI BS 1743: Sec 7.2-82. 1982 Analysis of Dried Milk and Dried Milk Products Part 7: of the Titratable Acidity of Dried Milk Section 7.2: Routine Method. 4 pp.
CNS N6195-88. Method of Test for Milk and Milk Products: Determination of Titrable Acidity (Sep)(11026). 1 p.
ISO 5547-78. Caseins—Determination of Free Acidity (Reference Method) First Edition. 4 pp.
ISO 6091-80. Dried Milk—Determination of Titratable Acidity (Reference Method) First Edition. 4 pp.
ISO 6092-80. Dried Milk—Determination of Titratable Acidity (Routine Method) First Edition; (Erratum—Sept 1981). 5 pp.

Acidity (Cont.)

—Detergents
CGSB 2-GP-11M METH 6.1-83. Methods of Testing and Analysis of Soaps and Detergents Free Alkali or Free Acid. 1 p.

—Diacetone Alcohol
BSI BS 549-70. 1970 Diacetone Alcohol. 12 pp.

—Diacetone Alcohol—Volumetric Analysis
ISO 2887-73. secButyl Alcohol, Methyl Ethyl Ketone, isoButyl Methyl Ketone, isoAmyl Ethyl Ketone, Diacetone Alcohol, and Hexylene Glycol for Industrial Use—Determination of Acidity to Phenolphthalein—Volumetric Method First Edition. 4 pp.

—Eggs—Frozen
CNS N6050-80. Method of Test for Frozen Eggs (Jul)(2923). 3 pp.

—Ethanols
BSI BS 6392: Part 1-83. 1983 Amd 1 Testing of Ethanol for Industrial Use Part 1: Method for Detection of Alkalinity or Determination of Acidity to Phenolphthalein. 5 pp.
ISO 1388 Pt 2-81. Ethanol for Industrial Use—Methods of Test—Part 2: Detection of Alkalinity or Determination of Acidity to Phenolphthalein First Edition. 4 pp.

—Ethyl Acetate
BSI BS 553-90. 1990 Ethyl Acetate for Industrial Use. 8 pp.

—Ethyl Ether
BSI BS 579-57. 1957 Diethyl Ether (Technical). 12 pp.

—Extenders
BSI BS 3483: Part C3-82. 1982 Methods for Testing Pigments in Paints Part C3: Determination of Acidity or Alkalinity of the Aqueous Extract. 4 pp.
ISO 787 Pt 4-81. General Methods of Test for Pigments and Extenders—Part 4: Determination of Acidity or Alkalinity of the Aqueous Extract First Edition. 5 pp.

—Fats
BSI BS 684: Sec 2.10-88. 1988 Amd 1 Methods of Analysis of Fats and Fatty Oils Part 2: Other Methods Section 2.10: Determination of Acidity (AMD 6050) May 31, 1990. 7 pp.
BSI BS 4289: Part 5-89. 1989 Methods for the Analysis of Oilseeds Part 5: Determination of Acidity of Fat. 4 pp.
CNS N6082-82. Methods of Test for Edible Oils and Fats (Determination of Acid Value) (Jan)(3647). 1 p.
ISO 660-83. Animal and Vegetable Fats and Oils—Determination of Acid Value and of Acidity First Edition. 6 pp.

—Fish Meal
CNS N4015-78. Method of Test for Fish Meal and Paste (Mar)(2245). 2 pp.

—Fish Paste
CNS N4015-78. Method of Test for Fish Meal and Paste (Mar)(2245). 2 pp.

—Formaldehyde
BSI BS 2942-57. (WITHDRAWN) 1957 Amd 2 Formaldehyde Solution. 12 pp.

—Formaldehyde—Volumetric Analysis
ISO 2225-72. Formaldehyde Solutions for Industrial Use—Determination of Acidity First Edition. 4 pp.

—Formic Acid
BSI BS 4341-68. 1968 Methods of Test for Formic Acid. 13 pp.

—Formic Acid—Potentiometric Analysis
ISO 731 Pt III-77. Formic Acid for Industrial Use—Methods of Test—Part III: Determination of Content of Other Acids—Potentiometric Method First Edition. 4 pp.

—Formic Acid—Volumetric Analysis
ISO 731 Pt II-77. Formic Acid for Industrial Use—Methods of Test—Part II: Determination of Total Acidity—Titrimetric Method First Edition. 4 pp.
ISO 731 Pt VII-77. Formic Acid for Industrial Use—Methods of Test—Part VII: Determination of Low Contents of Other Volatile Acids—Titrimetric Method After Distillation First Edition. 6 pp.

—Fruit Juices
CNS N6091-91. Method of Test for Fruit and Vegetable Juices and Drinks (General Rules) (Jun)(3736). 1 p.
CNS N6216-89. Method of Test for Fruit and Vegetable Juices and Drinks Determination of Acidity (Jul)(12570).

Acidity (Cont.)

—Fruits—Potentiometric Analysis
CNS N6167-91. Method of Test for Fruit and Vegetable Products-Determination of Titratable Acidity (Jun)(8626). 3 pp.
ISO 750-81. Fruit and Vegetable Products—Determination of Titratable Acidity First Edition. 5 pp.

—Furfurals—Volumetric Analysis
ISO 2888-73. Furfural for Industrial Use—Determination of Acidity to Phenolphthalein—Volumetric Method First Edition. 4 pp.

—Glycerol—Volumetric Analysis
BSI BS 5711: Part 5-79. 1979 Methods of Sampling and Test for Glycerol. 2 pp.
ISO 1615-76. Glycerines for Industrial Use—Determination of Alkalinity or Acidity—Titrimetric Method First Edition. 4 pp.

—Halogenated Hydrocarbons—Volumetric Analysis
BSI BS 5598: Part 4-79. 1979 Methods of Sampling and Test for Halogenated Hydrocarbons Part 4: Determination of Acidity, Titrimetric Method. 4 pp.
CNS K6561-80. Method of Test for Acidity of Liquid Halogenated Hydrocarbons (Titrimetric Method) (Aug)(6206).
ISO 1393-77. Liquid Halogenated Hydrocarbons for Industrial Use—Determination of Acidity—Titrimetric Method First Edition. 4 pp.

—Hardeners
JIS K 7239-91. Determination of Free Acid in Acid Anhydride-Based Hardeners for Epoxide Resins. 8 pp.

—Hexylene Glycol—Volumetric Analysis
ISO 2887-73. secButyl Alcohol, Methyl Ethyl Ketone, isoButyl Methyl Ketone, isoAmyl Ethyl Ketone, Diacetone Alcohol, and Hexylene Glycol for Industrial Use—Determination of Acidity to Phenolphthalein—Volumetric Method First Edition. 4 pp.

—Honey
CNS N6027-74. Method of Test for Honey (Oct)(1344). 8 pp.

—Hydrocarbons
CNS K6518-80. Method of Test for Acidity of Distillation Residues or Hydrocarbon Liquid (Jul)(5835).

—Hydrochloric Acid—Volumetric Analysis
ISO 904-76. Hydrochloric Acid for Industrial Use—Determination of Total Acidity—Titrimetric Method First Edition. 4 pp.

—Hydrofluoric Acid—Volumetric Analysis
BSI BS 5366-76. (WITHDRAWN) 1976 Amd 1 Methods of Sampling and Test for Aqueous Hydrofluoric Acid for Industrial Use (ISO 3139: 1976) (AMD 3784) October 30, 1981. 8 pp.

—Hydrogen Peroxide—Volumetric Analysis
BSI BS 7546: Part 3-92. 1992 Hydrogen Peroxide for Industrial Use Part 3: Method for Determination of Apparent Acidity (W). 6 pp.

—Ice Cream
CNS N6143-86. Method of Test for Ice Cream (Jan)(6509). 2 pp.

—Industrial Water
CPPA H.4P(F)-67. Determination of Acidity in Process Waters. 2 pp.

—Isopropyl Acetate
BSI BS 1834-68. 1968 Isopropyl Acetate. 11 pp.
BSI BS 1834-01. 1968 Amd 1 Isopropyl Acetate (AMD 7219) September 15, 1992. 13 pp.

—Isopropyl Alcohol—Volumetric Analysis
ISO 756 Pt 2-81. Propan-2-ol for Industrial Use—Methods of Test—Part 2: Determination of Acidity—Titrimetric Method First Edition. 4 pp.

—Ketones—Volumetric Analysis
ISO 2887-73. secButyl Alcohol, Methyl Ethyl Ketone, isoButyl Methyl Ketone, isoAmyl Ethyl Ketone, Diacetone Alcohol, and Hexylene Glycol for Industrial Use—Determination of Acidity to Phenolphthalein—Volumetric Method First Edition. 4 pp.

—Lactose Monohydrate
MOD UK M 821/71. Examination of Lactose.

—Maleic Anhydride—Potentiometric Analysis
CNS K6567-80. Determination of Free Acidity of Maleic Anhydride for Industrial Use (Potentiometric Method) (Aug)(6212).

INDUSTRY STANDARDS

INTERNATIONAL AND NON-U.S. NATIONAL STANDARDS
SUBJECT INDEX

Acidity

Acidity (Cont.)

—Maleic Anhydride—Potentiometric Analysis (Cont.)
ISO 1390 Pt III-77. Maleic Anhydride for Industrial Use—Methods of Test—Part III: Determination of Free Acidity—Potentiometric Method First Edition. 4 pp.

—Methanols
CNS K6263-80. Method of Test for Methyl Alcohol (Methanol) (Feb)(2790). 7 pp.

—Milk—Volumetric Analysis
CNS N6195-88. Method of Test for Milk and Milk Products: Determination of Titrable Acidity (Sep)(11026). 1 p.

—Nitric Acid—Volumetric Analysis
ISO 1980-77. Nitric Acid for Industrial Use—Determination of Total Acidity—Titrimetric Method First Edition. 4 pp.

—Oilseeds
BSI BS 4289: Part 5-89. 1989 Methods for the Analysis of Oilseeds Part 5: Determination of Acidity of Fat. 4 pp.
ISO 729-88. Oilseeds—Determination of Acidity of Oils Second Edition. 6 pp.

—Organic Coatings—Insoluble Matter
CNS K6804-19-89. Method of Test for Organic Coating (Chemical Analysis) — Acidity Test of Water Extract of Solvent Insolubles (Jan)(10880-19).

—Paint Removers
CGSB 31-GP-0A METH 51.4-62. Methods of Testing Corrosion-Prevention Materials and Processes Alkalinity and Acidity of Paint Remover. 1 p.

—Paints
CGSB 1-GP-71 METH 53.5-80. Methods of Testing Paints and Pigments Alkalinity and Acidity. 1 p.

—Paraffin Wax
CNS K6035-75. Method of Test for Paraffin Wax (Mar)(632). 2 pp.
MOD UK M 2009/67. Examination of Paraffin Wax.

—Paraformaldehyde
BSI BS 2941-57. (WITHDRAWN) 1957 Amd 1 Paraformaldehyde. 14 pp.

—Petroleum Products
BSI BS 2000: Part 1-93. 1993 Methods of Test for Petroleum and Its Products Part 1: Determination of Acidity of (W). 4 pp.
BSI BS 2000: Part 1-82. 1982 Amd 1 Methods of Test for Petroleum and Its Products Part 1: Acidity of Petroleum Products and Lubricants (Neutralization Value) (AMD 6372) June 28, 1991. 7 pp.
BSI BS 2000: Part 182-93. 1993 Methods of Test for Petroleum and Its Products Part 182: Determination of Inorganic Acidity of Petroleum Products—Colour Indicator Titration Method (W). 3 pp.
BSI BS 2000: Part 182-91. 1991 Petroleum and Its Products Part 182: Acidity (Inorganic) of Petroleum Products (Colour Indicator Titration Method) (Identical with IP 182/82(88)). 4 pp.
JIS K 2252-80. Testing Method for Reaction of Petroleum Products (R 1985). 5 pp.

—Phenol Red—Volumetric Analysis
CNS K6553-80. Determination of Acidity of Tritolyl Phosphate for Industrial Use (Volumetric Method) (Aug)(6198).
ISO 2521-74. Tritolyl Phosphate for Industrial Use—Determination of Acidity to Phenol Red—Volumetric Method First Edition. 4 pp.

—Phosphoric Acids
CNS K6235-72. Method of Test for Phosphoric Acid (Industrial Use) (Jun)(2620). 2 pp.

—Photographic Chemicals
ISO 10349 Pt 7-92. Photography—Photographic-Grade Chemicals—Test Methods—Part 7: Determination of Alkalinity or Acidity First Edition. 6 pp.

—Phthalate Esters—Volumetric Analysis
ISO 1385 Pt IV-77. Phthalate Esters for Industrial Use—Methods of Test—Part IV: Determination of Acidity to Phenolphthalein—Titrimetric Method First Edition. 4 pp.

—Phthalic Anhydride—Potentiometric Analysis
ISO 1389 Pt V-77. Phthalic Anhydride for Industrial Use—Methods of Test—Part V: Determination of Free Acidity—Potentiometric Method First Edition. 4 pp.

Acidity (Cont.)

—Pigments
BSI BS 3483: Part C3-82. 1982 Methods for Testing Pigments in Paints Part C3: Determination of Acidity or Alkalinity of the Aqueous Extract. 4 pp.
CGSB 1-GP-71 METH 53.5-80. Methods of Testing Paints and Pigments Alkalinity and Acidity. 1 p.
ISO 787 Pt 4-81. General Methods of Test for Pigments and Extenders—Part 4: Determination of Acidity or Alkalinity of the Aqueous Extract First Edition. 5 pp.

—Polishes
CGSB 25-GP-1M METH 14.1-84. Methods of Sampling and Testing Waxes and Polishes Acidity. 1 p.
CGSB 25-GP-1M METH 20.1-84. Methods of Sampling and Testing Waxes and Polishes Free Alkali and Free Acid. 1 p.

—Polyvinyl Acetate
JIS K 6725-77. Testing Method for Polyvinyl Acetate (R 1980). 7 pp.

—Potassium Sulfate—Volumetric Analysis
ISO 2489-73. Potassium Sulphate for Industrial Use—Determination of Acidity to Methyl Orange First Edition. 3 pp.

—Potentiometric Analysis
CNS J2035-82. Free Acid by Lodate Precipitation-Potentiometric Titration Method (Jan)(8414).

—Radioactive Isotopes
CNS J2015-81. Method for Analysis of Radioisotopes Free Acid (May)(7392).

—Reagents
CNS K7615-82. Chemical Reagent (Acidity and Alkalinity Determination) (Jun) (8987).

—Red Phosphorus
CNS K6195-65. Method of Test for Red Phosphorus (Apr)(2452) (R 1971). 3 pp.

—Residues
CNS K6518-80. Method of Test for Acidity of Distillation Residues or Hydrocarbon Liquid (Jul)(5835).

—Shellac
CGSB 1-GP-71 METH 56.3-80. Methods of Testing Paints and Pigments Acidity of Shellac. 1 p.

—Soaps
CGSB 2-GP-11M METH 6.1-83. Methods of Testing and Analysis of Soaps and Detergents Free Alkali or Free Acid. 1 p.

—Sodium Hexafluorosilicate—Volumetric Analysis
BSI BS 5705: Part 1-79. (WITHDRAWN) 1979 Sodium Hexafluorosilicate for Industrial Use Part 1: Determination of Free Acidity and Total Hexafluorosilicate Content. 4 pp.
ISO 4281-77. Sodium Hexafluorosilicate for Industrial Use—Determination of Free Acidity and Total Hexafluorosilicate Content—Titrimetric Method First Edition. 4 pp.

—Sodium Nitrate
MOD UK M 869/91. Examination of Sodium Nitrate, Grade 1.

—Sodium Sulfate
CNS K6085-74. Methods of Test for Sodium Sulfate of Industrial Grade (Oct)(1004). 5 pp.
ISO 3240-75. Sodium Sulphate for Industrial Use—Determination of Acidity or Alkalinity First Edition. 5 pp.

—Soldering Fluxes
MOD UK M 9524/66. Examination of Flux Soldering Solution (Withdrawn).
MOD UK M 9542/70. Examination of Flux, Soldering (Zinc Chloride Solution).

—Solvents
CGSB 31-GP-0A METH 51.3-57. Methods of Testing Corrosion-Prevention Materials and Processes Free Alkali or Free Acid. 1 p.
CNS K6543-80. Method of Test for Acidity in Volatile Solvents and Chemical Intermediates Used in Paint, Varnish, Lacquer and Related Products (Aug)(6188).

—Starches
DIN ENGL 10389-85. Determination of pH Value and Acidity of Starch and Starch Products (Aug). 2 pp.

—Sulfur—Volumetric Analysis
ISO 3704-76. Sulphur for Industrial Use—Determination of Acidity—Titrimetric Method First Edition. 4 pp.

Acidity (Cont.)

—Sulfuric Acid—Volumetric Analysis
ISO 910-77. Sulphuric Acid and Oleum for Industrial Use—Determination of Total Acidity, and Calculation of Free Sulphur Trioxide Content of Oleum—Titrimetric Method First Edition. 5 pp.

—Surfactants—Volumetric Analysis
BSI BS 6829: Sec 1.2-88. 1988 Analysis of Surface Active Agents (Raw Materials) Part 1: General Methods Section 1.2: Method for Determination of Free Alkalinity or Free Acidity. 4 pp.
BSI BS 6829: Sec 1.2-01. 1988 Amd 1 Analysis of Surface Active Agents (Raw Materials) Part 1: General Methods Section 1.2: Method for Determination of Free Alkalinity or Free Acidity (AMD 7848) September 15, 1993 (W). 5 pp.
ISO 4314-77. Surface Active Agents—Determination of Free Alkalinity or Free Acidity—Titrimetric Method First Edition. 4 pp.

—Triphenyl Phosphate
BSI BS 1998-70. 1970 Triphenyl Phosphate. 14 pp.

—Urea
CNS K6158-75. Method of Test for Urea Technical Grade (Jan)(1426). 2 pp.

—Vegetable Juices
CNS N6091-91. Method of Test for Fruit and Vegetable Juices and Drinks (General Rules) (Jun)(3736). 1 p.
CNS N6216-89. Method of Test for Fruit and Vegetable Juices and Drinks Determination of Acidity (Jul)(12570).

—Vegetable Oils
CNS N6082-82. Methods of Test for Edible Oils and Fats (Determination of Acid Value) (Jan)(3647). 1 p.

—Vegetables—Potentiometric Analysis
CNS N6167-91. Method of Test for Fruit and Vegetable Products-Determination of Titratable Acidity (Jun)(8626). 3 pp.
ISO 750-81. Fruit and Vegetable Products—Determination of Titratable Acidity First Edition. 5 pp.

—Vinegars
CNS N6025-83. Method of Test for Vinegar (Jul)(1071). 1 p.

—Volumetric Analysis
CNS J2036-82. Free Acid by Titration in an Oxalate Solution (Jan)(8415).

—Waste Water
CNS K9007-81. Method of Test for Acidity in Polluted and Waste Water (Apr)(3795).

—Water Pollution
CNS K9007-81. Method of Test for Acidity in Polluted and Waste Water (Apr)(3795).

—Water—Potentiometric Analysis
BSI BS 2690: Part 109-84. 1984 Water Used in Industry Part 109: Alkalinity, Acidity, pH Value and Carbon Dioxide. 8 pp.

—Water—Volumetric Analysis
DIN ENGL 38409 Pt 7-79. German Standard Methods for the Examination of Water, Waste Water and Sludge; Summary Indices of Actions and Substances (Group H); Determination of Acid and Base Capacity (H7) (May). 5 pp.

—Waxes
CGSB 25-GP-1M METH 14.1-84. Methods of Sampling and Testing Waxes and Polishes Acidity. 1 p.
CGSB 25-GP-1M METH 20.1-84. Methods of Sampling and Testing Waxes and Polishes Free Alkali and Free Acid. 1 p.

—Wines
EC EEC/506/92-92. Commission Regulation on Transitional Measures Regarding the Total Acidity Content of Wines Produced in Spain and Released to the Spanish Market for 1992. 2 pp.

—Wool Fibers
BSI BS 6981-88. 1988 Determination of Acid Content of Wool. 7 pp.
CGSB CAN/CGSB-4.2 NO.73.5-M91. Textile Test Methods Wool—Determination of Acid Content (ISO 3073:1975). 7 pp.
ISO 3073-75. Wool—Determination of Acid Content First Edition. 4 pp.

Acids

Scope Note: Use a more specific term *See:* Acetic Acid; 2S-Acid; Acidity; 5-Amino-2-Chlorotoluene-4-sulfonic Acid; 1-Amino-2-naphthol-4-sulfonic Acid; 2-Amino-5-naphthol-7-sulfonic Acid; o-Aminobenzoic

INTERNATIONAL AND NON-U.S. NATIONAL STANDARDS
SUBJECT INDEX

Acoustic

Acids *(Cont.)*
See: *(Cont.)*
Acid; Amsonic Acid; Aqua Regia; Arsenic Acid; Ascorbic Acid; Asparagine Monohydrate; Aspartic Acid; Benzenearsonic Acid; Boric Acid; C Acid; alpha-Camphoric Acid; n-Capric Acid; n-Caproic Acid; Carbonyl J Acid; Chloric Acid; Chlorine Sulfonic Acid; Chloroacetic Acid; Chlorosulfonic Acid; Chromotropic Acid; Citric Acid; o-Cresotic Acid; Cresylic Acids; L-Cystine; EDTA; Erucic Acid; Fatty Acids; Fluoboric Acid; Formic Acid; Gallic Acid; Gamma Acid; Gluconic Acid; L-Glutamic Acid; L-Glutamine; Glycine; Hydriodic Acid; Hydrobromic Acid; Hydrochloric Acid; Hydroiodic Acid; 2-Hydroxy-1-(2-hydroxy-4-sulfo-1-naphthylazo)-3-naphthoic Acid; Hydroxyproline; Hypophosphorus Acid; Iodic Acid; Iodoxyquinsulfonic Acid; L-Isoleucine; Lactic Acid; L-Leucine; Maleic Acid; Metanilic Acid; L-Methionine; Molybdic Acid, 85%; 2-Naphthol-3,6-disulfonic Acid; 2-Naphthol-6,8-disulfonic Acid; 1-Naphthylamine-8-sulfonic Acid; 2-Naphthylamine Sulfonic Acid; 2-Naphthylamine-1-Sulfonic Acid; Nitric Acid; Oleic Acid; Oxalic Acid; Palmitic Acid; Perchloric Acid; Periodic Acid; Phenyl J Acid; L-Phenylalanine; Phenylhydrazine-p-sulfonic Acid; Phosphomolybdic Acid; Phosphoric Acid; Phosphoric Acid, Meta; Phosphorous Acid; Phosphotungstic Acid; Picric Acid; Picrolonic Acid; Platinic Chloride; Polybasic Organic Acids; L-Proline; Quinaldic Acid; Ricinoleic Acid; Rubeanic Acid; Selenious Acid; L-Serine; Silicotungstic Acid; Stearic Acid; Succinic Acid; Sulfanilic Acid; Sulfosalicylic Acid; Sulfosalicylic Acid Dihydrate; Sulfuric Acid; Sulfurous Acid; Tannic Acid; Tar Acids; Tartaric Acid; Thioglycolic Acid; Threonine; Trichloroacetic Acid See: L-Tyrosine; Uric Acid; L-Valine

Acorn Nuts
Use: Cap Nuts

Acoustic Absorption
Use For: Sound Absorption See Also: Acoustics; Atmospheric Attenuation; Noise Reduction; Sound Attenuation

—Acoustical Insulation
JIS A 1405-63. Methods of Test for Sound Absorption of Acoustical Materials by the Tube Method (R 1988). 16 pp.

—Outdoor
ISO 9613 Pt 1-93. Acoustics—Attenuation of Sound During Propagation Outdoors—Part 1: Calculation of the Absorption of Sound by the Atmosphere First Edition. 30 pp.

—Reverberation Chambers
BSI BS 3638-87. 1987 Method for Measurement of Sound Absorption in a Reverberation Room (Renumbered as BS EN 20354: 1993). 14 pp.
BSI BS EN 20354-93. 1993 Acoustics—Measurment of Sound Absorption in a Reverberation Room (ISO 354: 1985) (R). 20 pp.
CEN EN 20354-93. Acoustics—Measurement of Sound Absorption in a Reverberation Room (ISO 354: 1985). 10 pp.
CNS A3165-86. Method for Measurement of Sound Absorption Coefficients in a Reverberation Room (Jan)(9056).
ISO 354-85. Acoustics—Measurement of Sound Absorption in a Reverberation Room First Edition (CEN EN 20354: 1993). 11 pp.
JIS A 1409-77. Method of Test for Sound Absorption Coefficients in a Reverberation Room.
SAA AS 1045-88. Acoustics—Measurement of Sound Absorption in a Reverberation Room. 14 pp.

Acoustic Couplers
See Also: Computer Equipment; Couplers; Modems; Telephones

—Data Transmission
CCITT RECMN V.15-89. Use of Acoustic Coupling for Data Transmission—Data Communication over the Telephone Network (Study Group XVII) 2 pp. 2 pp.

Acoustic Emission Examinations
Use: Acoustic Emissions

Acoustic Emissions
Use For: Stress Ware Emissions See Also: Acoustic Measuring Instruments; Cracking (Fracturing); Nondestructive Testing; Ultrasonic Testing

—Pressure Pipes
JIS Z 2342-91. Methods for Acoustic Emission Testing of Pressure Vessels During Pressure Test. 8 pp.

Acoustic Emissions *(Cont.)*
—Pressure Vessels
JIS Z 2342-91. Methods for Acoustic Emission Testing of Pressure Vessels During Pressure Test. 8 pp.

Acoustic Insulation
Use: Acoustical Insulation

Acoustic Intrusion Detectors
Use: Audio Detectors

Acoustic Materials
Use: Acoustical Materials

Acoustic Measurement
Use: Acoustics

Acoustic Measuring Equipment
Use: Acoustic Measuring Instruments

Acoustic Measuring Instruments
Use For: Acoustic Measuring Equipment; Acoustoelectric Measuring Instruments; Electroacoustic Measuring Instruments
See Also: Acoustic Emissions; Acoustics; Artificial Ears; Artificial Mouths; Audiometers; Circuit Noise Meters; Hydrophones; Measuring Instruments; Modulated Noise Reference Units; Noise; Noise Meters; Octave Band Analyzers; Sound Level Meters

—Aircraft Noise
BSI BS 5647-79. 1979 Electro-Acoustical Measuring Equipment for Aircraft Noise Certification. 16 pp.
IEC 561-76. Electro-Acoustical Measuring Equipment for Aircraft Noise Certification First Edition. 26 pp.

—Aural Impedance/Admittance
BSI BS EN 61027-93. 1993 Instruments for the Measurement of Aural Acoustic Impedance/Admittance (IEC 1027: 1991). 27 pp.
CENELEC EN 61027-93. Instruments for the Measurement of Aural Acoustic Impedance/Admittance (IEC 1027:1991). 5 pp.
IEC 1027-91. Instruments for the Measurement of Aural Acoustic Impedance/Admittance First Edition; (CENELEC EN 61027:1993). 45 pp.

—Burglar Alarms
BSI BS 4737: Sec 3.6-78. 1978 Amd 1 Intruder Alarm Systems in Buildings Part 3: Components Section 3.6: Acoustic Detectors. 4 pp.

Acoustic Power
Use: Sound Power

Acoustic Resonators
Scope Note: Use a more specific term See: Tuning Forks

Acoustic Sealants
Use: Acoustical Sealants

Acoustic Signals
Use For: Audio Signals; Sound Signals
See Also: Acoustics; Annunciators; Bells; Buzzers; Digital Audio Coding; Signal Devices; Sound-Program Signals; Sound Transmission
CSA C22.2 NO 205-M1983. Signal Equipment (R 1992); (Gen Instr 1). 32 pp.

—Automotive
ISO 512-79. Road Vehicles—Sound Signalling Devices—Technical Specifications Second Edition. 5 pp.
ISO 6969-81. Road Vehicles—Sound Signalling Devices—Tests After Mounting on Vehicle First Edition. 4 pp.

—Digital—Modulation—Sound Broadcasting
CCIR QUESTION 89/10-90. Methods of Modulation for the Emission of Digital Sound Signals in Broadcasting—Questions Concerning Study Group 10—Broadcasting Service (Sound). 1 p.

—Digital—Multiplexing—Sound Broadcasting
CCIR QUESTION 88/10-90. Multiplexing Methods for the Emission of Digital Sound Signals in Broadcasting—Questions Concerning Study Group 10—Broadcasting Service (Sound). 1 p.

—Digital—Synchronization—Sampling
CCIR Report 1071-86. Sampling Frequency Conversion and Synchronization of Digital Sound Signals—Section 10C—Audio-Frequency Characteristics of Sound-Broadcasting Signals. 1 p.

Acoustic Signals *(Cont.)*
—Digital Transmission
CCIR Report 1236-90. Digital Transmission of Medium-and High-Quality Sound Signals at Low Bit Rates—Section CMTT C—Transmission Standards and Performance Objectives for Sound-Programme Channels. 8 pp.
CCIR QUESTION 51-2/10-90. Standards for Digital Techniques for Sound in Broadcasting—Questions Concerning Study Group 10—Broadcasting Service (Sound). 1 p.

—Encoding—Sound Broadcasting
CCIR RECMN 646-86. Source Encoding for Digital Sound Signals in Broadcasting Studios—Section 10C—Audio-Frequency Characteristics of Sound-Broadcasting Signals. 1 p.
CCIR RECMN 646-1-92. Source Encoding for Digital Sound Signals in Broadcasting Studios—Section 10C—Audio-Frequency Characteristics of Sound-Broadcasting Signals. 1 p.
CCIR QUESTION 87/10-90. Digital Coding in Broadcasting for Emission of Sound Signals—Questions Concerning Study Group 10—Broadcasting Service (Sound). 1 p.

—Multiplexing—Broadcasting Satellite Services
CCIR Report 954-2-90. Multiplexing Methods for the Emission of Several Digital Audio Signals and Also Data Signals in Broadcasting—Section 10/11B—Systems. 8 pp.

—Pulse Code Modulation—Broadcasting Satellite Services
CCIR RECMN 651-86. Digital PCM Coding for the Emission of High-Quality Sound Signals in Satellite Broadcasting (15 kHz Nominal Bandwidth)—Section 10/11B—Systems. 2 pp.

—Secondary Distribution
CCIR QUESTION 25/CMTT-90. Standards for Secondary Distribution of Sound and Television Signals—Questions Concerning the CMTT CCIR/CCITT Joint Study Group for Television and Sound Transmission. 1 p.
CCIR QUESTION 77/CMTT-90. Signal Characteristics and Coding Methods Sound and Television Signals—Questions Concerning the CMTT CCIR/CCITT Joint Study Group for Television and Sound Transmission. 1 p.

—Secondary Distribution—Coding
CCIR QUESTION 77/CMTT-90. Signal Characteristics and Coding Methods Sound and Television Signals—Questions Concerning the CMTT CCIR/CCITT Joint Study Group for Television and Sound Transmission. 1 p.

—Ships
CNS F4013-85. Application Standard of Audio-Signalings and Pilot Lamps in Shipboard (Jan)(11185).
IEC 92 Pt 203-85. Electrical Installations in Ships Part 203: System Design—Acoustic and Optical Signals First Edition. 21 pp.
JIS F 0413-87. Application of Audio-Signalings and Pilot Lamps in Shipboard.
JIS F 8078-87. Electrical Installations in Ships Part 203 System Design—Acoustic and Optical Signals (IEC 92-203-1985).

—Stereophonic—Monophonic—Compatibility
CCIR QUESTION 99/10-90. Compatibility of a Monophonic Signal Obtained from a Stereophonic Source—Questions Concerning Study Group 10—Broadcasting Service (Sound). 1 p.

—Studio Quality—Digital Transmission—Communication Channels
CCIR RECMN 724-90. Transmission of Digital Studio Quality Sound Signals over H1 Channels —Section CMTT C—Transmission Standards and Performance Objectives for Sound-Programme Channels. 9 pp.

—Video Signals—Multiplexing
CCIR QUESTION 21-2/CMTT-86. Standards for Transmission of Signals with Multiplexing of Video, Sound and Other Types of Signal—Questions Concerning the CMTT CCIR/CCITT Joint Study Group for Television and Sound Transmission. 1 p.

Acoustic Testing
Use For: Listening Tests See Also: Acoustics; Physical Testing; Ultrasonic Testing
IEC 543-76. Informative Guide for Subjective Listening Tests First Edition. 24 pp.

—Backup Alarms—Earthmoving Equipment
BSI BS 6912: Part 3-90. 1990 Safety of Earth Moving Machinery Part 3: Sound Test Method for Machine-Mounted Forward and Reverse Warning Alarm. 12 pp.

INTERNATIONAL AND NON-U.S. NATIONAL STANDARDS
SUBJECT INDEX

Acoustic

Acoustic Testing *(Cont.)*

—Backup Alarms—Earthmoving Equipment *(Cont.)*
ISO 9533-89. Earth-Moving Machinery—Machine-Mounted Forward and Reverse Audible Warning Alarm—Sound Test Method First Edition. 8 pp.

—Digital Circuit Multiplication Equipment—Telephone Transmission
CCITT RECMN P.84-89. Subjective Listening Test Method for Evaluating Digital Circuit Multiplication and Packetized Voice Systems—Telephone Transmission Quality (Study Group XII) 27 pp. 27 pp.

—Hearing Aids
IEC 959-90. Provisional Head and Torso Simulator for Acoustic Measurements on Air Conduction Hearing Aids First Edition. 60 pp.

—Horns—Motorcycle
CNS D3036-88. Method of Sound Level Test of the Horn for Motorcycles (May)(3113). 2 pp.
JIS D 1041-87. Method of Acoustic Test of Horns for Motor Cycles.

—Packetized Voice Systems—Telephone Transmission
CCITT RECMN P.84-89. Subjective Listening Test Method for Evaluating Digital Circuit Multiplication and Packetized Voice Systems—Telephone Transmission Quality (Study Group XII) 27 pp. 27 pp.

—Radio Receivers
BSI BS 5877-80. 1980 Guide for Subjective Listening Tests for Sound Radio Receivers. 10 pp.

—Speakers
BSI BS 6840: Part 13-87. 1987 Sound System Equipment Part 13: Guide for Listening Tests on Loudspeakers. 37 pp.
IEC 268 Pt 13-85. Sound System Equipment Part 13: Listening Tests on Loudspeakers First Edition. 75 pp.

Acoustic Testing Rooms
Scope Note: Use a more specific term
See: Acoustics; Anechoic Chambers; Reverberation Chambers; Test Chambers

Acoustic Tiles
Use: Acoustical Tiles

Acoustic-To-Digital Transmission Systems
Use: Voice/Data Systems

Acoustic Transducers
Use: Electroacoustic Transducers

Acoustic Vibration
Use: Elastic Waves

Acoustic Waves
Use: Elastic Waves

Acoustical Hoods (Telephones)
See Also: Telephone Booths; Telephones

—Telephone Transmission—Quality of Service
CCITT RECMN P.32-89. Evaluation of the Efficiency of Telephone Booths and Acousting Hoods—Telephone Transmission Quality (Study Group XII) 7 pp. 7 pp.

Acoustical Insulation
Use For: Acoustic Insulation; Noise Damping Materials; Sound Absorbers; Sound Control Insulation; Sound Insulation; Soundproofing
See Also: Acoustical Materials; Acoustical Panels; Acoustical Tiles; Acoustics; Anechoic Chambers; Blanket Insulation; Ceilings; Fiberglass Insulation; Insulating Board; Mineral Wool Insulation; Noise; Noise Reduction; Panels; Sound Transmission
DIN ENGL 18165 Pt 2-87. Fibre Insulating Building Materials; Impact Sound Insulating Materials (Mar). 9 pp.

—Airborne—Buildings
CEN PREN 20140-10-91. Acoustics—Measurement of Sound Insulation in Buildings and of Building Elements—Part 10: Laboratory Measurement of Airborne Sound Insulation of Small Building Elements. 16 pp.
CEN EN 20140-10-92. Acoustics—Measurement of Sound Insulation in Buildings and of Building Elements—Part 10: Laboratory Measurement of Airborne Sound Insulation of Small Building Elements. 9 pp.

Acoustical Insulation *(Cont.)*

—Airborne—Buildings *(Cont.)*
DIN ENGL EN 20140 Pt 10-92. Acoustics; Measurement of Sound Insulation in Buildings and of Building Elements; Laboratory Measurement of Airborne Sound Insulation of Small Building Elements (ISO 140-10:1991) (Sept). 10 pp.

—Board—Polystyrene
BSI BS 3837: Part 1-86. 1986 Expanded Polystyrene Boards Part 1: Boards Manufactured from Expandable Beads. 9 pp.

—Board—Rock Wool
CNS A2177-84. Dressed Rockwool Boards for Acoustic Use (Sep)(10994).
CNS A3213-84. Method of Test for Dressed Rockwool Boards for Acoustic Use (Sep)(10995).
JIS A 6307-80. Dressed Rockwool Boards for Acoustic Use. 15 pp.

—Buildings
BSI BS 2750: Part 1-80. 1980 Measurement of Sound Insulation in Buildings and of Building Elements Part 1: Recommendations for Laboratories. 6 pp.
BSI BS 2750: Part 2-93. 1993 Acoustics—Measurements of Sound Insulation in Buildings and of Building Elements Part 2: Determination, Verification and Application of Precision Data (F). 21 pp.
BSI BS 2750: Part 2-80. 1980 Measurement of Sound Insulation in Buildings and of Building Elements Part 2: Statement of Precision Requirements. 10 pp.
BSI BS 2750: Part 3-80. 1980 Measurement of Sound Insulation in Buildings and of Building Elements Part 3: Laboratory Measurements of Airborne Sound Insulation of Building Elements. 8 pp.
BSI BS 2750: Part 4-80. 1980 Measurement of Sound Insulation in Buildings and of Building Elements Part 4: Field Measurements of Airborne Sound Insulation Between Rooms. 8 pp.
BSI BS 8233-87. 1987 Amd 1 Sound Insulation and Noise Reduction for Buildings. 59 pp.
BSI BS EN 20140-10-92. 1992 Acoustics—Measurement of Sound Insulation in Buildings and of Building Elements Part 10: Laboratory Measurement of Airborne Sound Insulation of Small Building Elements (ISO 140-10: 1991) (F). 14 pp.
BSI BS EN 20140-2-93. 1993 Acoustics—Measurement 4f Sound Insulation in Buildings and of Building Elements Part 2: Determination, Verification and Application of Precision Data (ISO 140-2: 1991) (Supersedes BS 2750: Part 2: 1980). 21 pp.
CEN EN 20140-2-93. Acoustics—Measurement of Sound Insulation in Buildings and of Building Elements—Part 2: Determination, Verification and Application of Precision Data (ISO 140-2: 1991). 15 pp.
CEN PREN 20140-3-92. Methods of Measurement of Sound Insulation in Buildings and of Building Elements: Part 3: Laboratory Measurements of Airborne Sound Insulation of Building Elements. 23 pp.
CNS A1031-87. Classification of Air-Borne and Impact Sound Insulation for Buildings (Mar)(8465).
DIN ENGL 4109-89. Sound Insulation in Buildings; Requirements and Testing (Nov). 25 pp.
DIN ENGL 52210 Pt 3-87. Testing of Acoustics in Buildings; Airborne and Impact Sound Insulation; Laboratory Measurement of Airborne Sound Insulation of Building Elements and Field Measurements Between Rooms (Feb). 12 pp.
DIN ENGL 52210 Pt 5-85. Testing in Building Acoustics; Airborne and Impact Sound Insulation; Field Measurements of Airborne Sound Insulation of Exterior Building Elements (July). 10 pp.
DIN ENGL 52210 Pt 6-89. Testing of Acoustics in Buildings; Airborne Impact and Sound Insulation; Measurement of Level Difference (May). 4 pp.
DIN ENGL 52210 Pt 7-89. Testing of Acoustics in Buildings; Airborne Impact and Sound Insulation; Measurement of Lateral Sound Reduction Index (May). 4 pp.
ISO 140 Pt I-78. Acoustics—Measurement of Sound Insulation in Buildings and of Building Elements—Part I: Requirements for Laboratories First Edition. 4 pp.
ISO 140 Pt 2-91. Acoustics—Measurement of Sound Insulation in Buildings and of Building Elements—Part 2: Determination, Verification and Application of Precision Data Second Edition; (CEN EN 20140-2: 1993). 19 pp.
ISO 140 Pt 3-78. Acoustics—Measurement of Sound Insulation in Buildings and of Building Elements—Part 3: Laboratory Measurements of Airborne Sound Insulation of Building Elements First Edition; (Amendment 1-1990). 15 pp.
ISO 140 Pt IV-78. Acoustics—Measurement of Sound Insulation in Buildings and of Building Elements—Part IV: Field Measurements of Airborne Sound Insulation Between Rooms First Edition. 7 pp.

Acoustical Insulation *(Cont.)*

—Buildings *(Cont.)*
ISO 140 Pt 10-91. Acoustics—Measurement of Sound Insulation in Buildings and of Building Elements—Part 10: Laboratory Measurement of Airborne Sound Insulation of Small Building Elements First Edition. 11 pp.
JIS A 1419-92. Classification of Air-Borne and Impact Sound Insulation for Buildings. 7 pp.
SNZ NZS/ISO 140: Part 1-90. Acoustics Part 1: Requirements for Laboratories. 4 pp.
SNZ NZS/ISO 140: Part 2-78. Acoustics Part 2: Statement of Precision Requirements. 5 pp.
SNZ NZS/ISO 140: Part 3-78. Acoustics Part 3: Laboratory Measurements of Airborne Sound Insulation. 5 pp.
SNZ NZS/ISO 140: Part 4-78. Acoustics Part 4: Field Measurements of Airborne Sound Insulation Between Rooms. 5 pp.

—Buildings—Ratings
BSI BS 5821: Part 1-84. 1984 Method for Rating Sound Insulation in Buildings and of Building Elements Part 1: Rating the Airborne Sound Insulation in Buildings and of Interior Building Elements. 6 pp.
BSI BS 5821: Part 2-84. 1984 Method for Rating Sound Insulation in Buildings and of Building Elements Part 2: Rating the Impact Sound Insulation. 11 pp.
BSI BS 5821: Part 3-84. 1984 Method for Rating Sound Insulation in Buildings and of Building Elements Part 3: Rating the Airborne Sound Insulation of Facade Elements and Facades. 6 pp.
ISO 717 Pt 1-82. Acoustics—Rating of Sound Insulation in Buildings and of Building Elements—Part 1: Insulation in Buildings and of Interior Building Elements First Edition. 5 pp.
ISO 717 Pt 2-82. Acoustics—Rating of Sound Insulation in Buildings and of Building Elements—Part 2: Impact Sound Insulation First Edition. 9 pp.
ISO 717 Pt 3-82. Acoustics—Rating of Sound Insulation in Buildings and of Building Elements—Part 3: Airborne Sound Insulation of Facade Elements and Facades First Edition. 5 pp.

—Ceilings
BSI BS 2750: Part 9-87. 1987 Measurement of Sound Insulation in Buildings and of Building Elements Part 9: Method for Labor-tory Measurement of Room-To-Room Airborne Sound Insulation of a Suspended Ceiling with a Plenum Above it. 10 pp.
ISO 140 Pt 9-85. Acoustics—Measurements of Sound Insulation in Buildings and of Building Elements—Part 9: Laboratory Measurement of Room-to-Room Airborne Sound Insulation of a Suspended Ceiling with a Plenum Above It First Edition. 8 pp.

—Cellular Plastics
DIN ENGL 18164 Pt 2-91. Rigid Cellular Plastics Insulating Building Materials; Polystyrene Foam Impact Sound Insulating Materials (Mar). 9 pp.

—Doors
JIS A 1520-88. Method for Field Measurements of Sound Insulation of Windows and Doors. 15 pp.

—Facades
BSI BS 2750: Part 5-80. 1980 Measurement of Sound Insulation in Buildings and of Building Elements Part 5: Field Measurements of Airborne Sound Insulation of Facade Elements and Facades. 12 pp.
ISO 140 Pt V-78. Acoustics—Measurement of Sound Insulation in Buildings and of Building Elements—Part V: Field Measurements of Airborne Sound Insulation of Facade Elements and Facades First Edition. 10 pp.
SNZ NZS/ISO 140: Part 5-78. Acoustics Part 5: Field Measurements of Airborne Sound Insulation of Facade Elements and Facades. 8 pp.

—Fiberboard
CNS A2038-86. Organic Fiber Sound Proof Boards (Dec)(2777). 1 p.
CNS A3058-86. Method of Test for Organic Fiber Soundproof Boards (Dec)(2778). 3 pp.
CNS O1029-83. Insulation Fiberboard for Acoustic Use (Jul)(10468). 8 pp.
JIS A 6304-72. Soft Fibreboards for Acoustic Use.

—Floating Floors—Dynamic Stiffness
BSI BS EN 29052-1-92. 1992 Acoustics—Method for the Determination of Dynamic Stiffness Part 1: Materials Used Under Floating Floors in Dwellings (ISO 9052-1: 1989). 11 pp.
CEN EN 29052-1-92. Acoustics—Determination of Dynamic Stiffness—Part 1: Materials Used Under Floating Floors in Dwellings. 6 pp.
DIN ENGL 52214 (S)-84. Testing of Acoustics in Buildings; Determination of the Dynamic Stiffness of Insulating Layers for Floating Screeds (Dec) (Superseded by DIN EN 29052 Part 1). 3 pp.

INTERNATIONAL AND NON-U.S. NATIONAL STANDARDS
SUBJECT INDEX
Acoustics

Acoustical Insulation (Cont.)
—**Floating Floors—Dynamic Stiffness** (Cont.)
DIN ENGL EN 29052 Pt 1-92. Acoustics; Determination of Dynamic Stiffness; Materials Used Under Floating Screed in Dwellings (ISO 9052-1: 1989) (Aug) (Supersedes DIN 52214. December 1984 Edition). 8 pp.
ISO 9052 Pt 1-89. Acoustics—Determination of Dynamic Stiffness—Part 1: Materials Used Under Floating Floors in Dwellings First Edition. 7 pp.

—**Floating Floors—Glass Wool**
CNS A2167-83. Glass Wool Isolating Material for Floating Floors (Nov)(10638).
JIS A 6322-79. Glass Wool Isolating Material for Floating Floors (R 1983). 12 pp.

—**Floating Floors—Rock Wool**
CNS A2166-83. Rock Wool Isolating Material for Floating Floors (Nov)(10637).
JIS A 6321-79. Rock Wool Isolating Material for Floating Floors.

—**Floor Coverings**
BSI BS 2750: Part 8-80. 1980 Measurement of Sound Insulation in Buildings and of Building Elements Part 8: Laboratory Measurements of the Reduction of Transmitted Impact Noise by Floor Coverings on a Standard Floor. 11 pp.
ISO 140 Pt VIII-78. Acoustics—Measurement of Sound Insulation in Buildings and of Building Elements—Part VIII: Laboratory Measurements of the Reduction of Transmitted Impact Noise by Floor Coverings on a Standard Floor First Edition. 9 pp.
SNZ NZS/ISO 140: Part 8-78. Acoustics Part 8: Laboratory Measurements of the Reduction of Transmitted Impact Noise by Floor Coverings on a Standard Floor. 7 pp.

—**Floors**
BSI BS 2750: Part 6-80. 1980 Measurement of Sound Insulation in Buildings and of Building Elements Part 6: Laboratory Measurements of Impact Sound Insulation of Floors. 10 pp.
BSI BS 2750: Part 7-80. 1980 Measurement of Sound Insulation in Buildings and of Building Elements Part 7: Field Measurements of Impact Sound Insulation of Floors. 8 pp.
ISO 140 Pt VI-78. Acoustics—Measurement of Sound Insulation in Buildings and of Building Elements—Part VI: Laboratory Measurements of Impact Sound Insulation of Floors First Edition. 7 pp.
ISO 140 Pt VII-78. Acoustics—Measurement of Sound Insulation in Buildings and of Building Elements—Part VII: Field Measurements of Impact Sound Insulation of Floors First Edition. 7 pp.
JIS A 1418-78. Method for Field Measurement of Floor Impact Sound Level (R 1983). 11 pp.
SNZ NZS/ISO 140: Part 6-78. Acoustics Part 6: Laboratory Measurements of Impact Sound Insulation of Floors. 5 pp.
SNZ NZS/ISO 140: Part 7-78. Acoustics Part 7: Field Measurements of Impact Sound Insulation of Floors. 5 pp.

—**Glass Wool**
CNS A2140-82. Glass Wool Acoustic Materials (Jul)(9057).
CNS A3166-82. Method of Test for Glass Wool Acoustic Materials (Jul)(9058).
JIS A 6306-91. Glass Wool Acoustic Materials. 20 pp.

—**Machine Rooms**
CEN PREN 31546-2-92. Acoustics—Determination of Sound Insulation Performances of Enclosures: Part 2: Measurements of In-Situ Sound Insulation Performance of Enclosures (for Acceptance/Verification Purposes). 26 pp.

—**Machine Rooms—Laboratory Testing**
CEN PREN 31546-1-92. Acoustics—Determination of Sound Insulation Performances of Enclosures: Part 1: Measurements in Small Enclosures Under Laboratory Conditions. 26 pp.

—**Naval Ships**
MOD UK NES 703-01. Thermal and Acoustic Insulation of Hull and Machinery Issue 1 (01.81); Corrigendum. 118 pp.
MOD UK NES 703-81. Thermal and Acoustic Insulation of Hull and Machinery Issue 1 (01.81) (Amendment 1 Incorporated). 120 pp.

—**Perlite**
CNS A2096-81. Perlite Powder (Mar)(6992). 4 pp.

—**Plasterboards—Perforated**
CNS A2070-90. Perforated Gypsum Boards for Acoustic Use (Oct)(4965).
CNS A3081-90. Method of Test for Perforated Gypsum Boards for Acoustic Use (Oct)(4966).
JIS A 6301-84. Perforated Gypsum Boards for Acoustic Use. 17 pp.

Acoustical Insulation (Cont.)
—**Plastics—Vibration Damping**
ISO 6721-83. Plastics—Determination of Damping Properties and Complex Modulus by Bending Vibration First Edition. 7 pp.

—**Rock Wool**
CNS A2143-82. Rock Wool Acoustic Materials (Dec)(9659).
CNS A3173-82. Method of Test for Rock Wool Acoustic Materials (Dec)(9660).
JIS A 6303-86. Rock Wool Acoustic Materials.
JIS A 6303-80. Rock Wool Acoustic Materials. 20 pp.

—**Sheets—Asbestos Cement**
CNS A2155-83. Perforated Asbestos Cement Flat Sheets for Acoustic Use (May) (10211).
JIS A 6302-75. Perforated Asbestos Cement Flat Sheets for Acoustic Use. 18 pp.

—**Skylights**
BSI CP 153: Part 3-72. 1972 Windows and Rooflights Part 3: Sound Insulation. 22 pp.

—**Submarines**
MOD UK NES 703-01. Thermal and Acoustic Insulation of Hull and Machinery Issue 1 (01.81); Corrigendum. 118 pp.
MOD UK NES 703-81. Thermal and Acoustic Insulation of Hull and Machinery Issue 1 (01.81) (Amendment 1 Incorporated). 120 pp.

—**Windows**
BSI CP 153: Part 3-72. 1972 Windows and Rooflights Part 3: Sound Insulation. 22 pp.
CNS A3196-85. Method of Test for Sound Insulation Windows (Sep)(10486).
JIS A 1520-88. Method for Field Measurements of Sound Insulation of Windows and Doors. 15 pp.

Acoustical Materials
Use For: Acoustic Materials *See Also:* Acoustical Insulation; Acoustical Panels; Acoustical Sealants; Acoustical Tiles
CGSB CAN/CGSB-92.1-M89. Sound Absorptive Prefabricated Acoustical Units. 13 pp.
CGSB CAN/CGSB-92.2-M90. Trowel or Spray Applied Acoustical Material. 11 pp.
JIS A 1405-63. Methods of Test for Sound Absorption of Acoustical Materials by the Tube Method (R 1988). 16 pp.

—**Air Flow**
BSI BS EN 29053-93. 1993 Acoustics—Materials for Acoustical Applications—Determination of Airflow Resistance (ISO 9053: 1991). 16 pp.
CEN EN 29053-93. Acoustics—Materials for Acoustical Applications—Determination of Airflow Resistance (ISO 9053: 1991). 11 pp.
ISO 9053-91. Acoustics—Materials for Acoustical Applications—Determination of Airflow Resistance First Edition; (CEN EN 29053: 1993). 13 pp.

Acoustical Measurement
Use: Acoustics

Acoustical Panels
See Also: Acoustical Insulation; Acoustical Materials; Acoustics; Panels

—**Perforated—Aluminum**
CNS A2144-83. Perforated Aluminium Panel for Acoustic Use (Feb)(9957).
JIS A 6305-78. Perforated Aluminium Panel for Acoustic Use.

—**Sound Attenuation**
ISO 10053-91. Acoustics—Measurement of Office Screen Sound Attenuation Under Specific Laboratory Conditions First Edition. 11 pp.

Acoustical Sealants
Use For: Acoustic Sealants *See Also:* Acoustical Materials; Sealants

—**Interior**
CGSB CAN/CGSB-19.21-M87. Sealing and Bedding Compound, Acoustical. 8 pp.

Acoustical Tiles
Use For: Acoustic Tiles *See Also:* Acoustical Insulation; Acoustical Materials; Acoustics; Tiles

—**Asbestos Cement**
CGSB CAN/CGSB-34.19-M89. Tiles or Panels, Asbestos-Cement, Acoustical. 10 pp.

—**Cork**
ISO 2509-89. Sound-Absorbing Expanded Pure Agglomerated Cork in Tiles Second Edition. 4 pp.
ISO 2510-89. Sound-Reducing Composition Cork in Tiles Second Edition. 4 pp.

Acoustics
Scope Note: Methods of measurement and testing in the field of physical acoustics; excludes acoustical noise pertaining to safety, tolerance, and comfort
See Also: Absorbents; Acoustic Absorption; Acoustic Measuring Instruments; Acoustic Signals; Acoustic Testing; Acoustical Insulation; Acoustical Panels; Acoustical Tiles; Anechoic Chambers; Architecture; Bioacoustics; Electroacoustics; Harmonics; Loudness Ratings; Noise; Noise Reduction; Physics; Pitch (Frequency); Pressure Measurement; Resonance Frequency; Sound Attenuation; Sound Power; Sound Pressure; Sound Pressure Levels; Sound Transmission; Speech; Ultrasonic Frequencies; Vibration; Vibration Damping; Vibration Meters
BSI BS 3383-88. 1988 Normal Equal-Loudness Contours for Pure Tones and Normal Threshold of Hearing Under Free-Field Listening Conditions. 12 pp.
BSI BS 3593-63. 1963 Recommendation on Preferred Frequencies for Acoustical Measurements. 8 pp.
CNS Z7196-7-85. Quantities and Units of Acoustics (Oct)(11296-7).
CSA Z107.52-M1983. Recommended Practice for the Prediction of Sound Pressure Levels in Large Rooms Containing Sound Sources; (Gen Instr 1). 23 pp.
ISO 226-87. Acoustics—Normal Equal-Loudness Level Contours First Edition; (Supersedes 454). 11 pp.
ISO 1683-83. Acoustics—Preferred Reference Quantities for Acoustic Levels First Edition. 4 pp.
JIS Z 8731-83. Methods for Measurement and Description of A-Weighted Sound Pressure Level. 13 pp.
SAA AS 1055.1-89. Acoustics—Description and Measurement of Environment Noise—Part 1: General Procedures. 11 pp.
SAA AS 1055.2-89. Acoustics—Description and Measurement of Environment Noise—Part 2: Application to Specific Situations. 5 pp.
SAA AS 1935-76. Method for Measurement of Normal Incidence Sound Absorption Coefficient and Specific Normal Acoustic Impedance of Acoustic Materials by the Tube Method Reconfirmed 1986. 24 pp.
SNZ NZS/ISO 226-87. Acoustics. Normal Equal-Loudness Level Contours. 8 pp.
SNZ NZS 6801-91. Measurement of Sound—Methods of Measuring Noise. 28 pp.

—**Auditoriums**
BSI BS 5363-76. 1976 Measurement of Reverberation Time in Auditoria. 6 pp.
ISO 3382-75. Acoustics—Measurement of Reverberation Time in Auditoria First Edition. 5 pp.

—**Buildings—Ergonomics**
BSI BS 7643: Part 3-93. 1993 Building Construction—Expression of Users' Requirements Part 3: Acoustical Requirements (ISO 6242-3: 1992). 12 pp.
ISO 6242 Pt 3-92. Building Construction—Expression of Users' Requirements—Part 3: Acoustical Requirements First Edition. 9 pp.

—**Buildings—Partitions**
SAA AS 1191-85. Acoustics—Method for Laboratory Measurement of Airborne Sound Transmission Loss of Building Partitions Amdt 1 January 1987. 16 pp.

—**Classification**
BSI BS 1000: Am. (681.8/9)-72. 1972 Amd 1 Universal Decimal Classification (UDC). English Full Edition (681.8/.9): Technical Acoustics. Musical Instruments. Engraving and Sculpting Machines. 26 pp.
SAA AS 1469-83. Acoustics—Methods for the Determination of Noise Rating Numbers. 9 pp.
SNZ NZS/BS 1000 (681.8/.9)-72. Universal Decimal Classification Technical Acoustics. Musical Instruments. Engraving and Sculpting Machines Amend: 1. 24 pp.

—**Control Rooms—Sound Broadcasting**
CCIR QUESTION 81/10-90. Determination of the Acoustical Properties of Control Rooms and High-Quality Listening Rooms in Broadcasting—Questions Concerning Study Group 10—Broadcasting Service (Sound). 1 p.

—**Glossaries**
BSI BS 4727:Pt3: Group 08-85. 1985 Electrotechnical, Power, Telecommunication, Electronics, Lighting and Colour Terms Part 3: Terms Particular to Telecommunications and Electronics Group 08: Acoustics and Electroacoustics Terminology. 36 pp.
CSA Z107.0-1984. Definitions of Common Acoustical Terms Used in CSA Standards; (Gen Instr 1). 13 pp.
JIS Z 8106-88. Glossary of Acoustical Terms (General). 39 pp.
SAA AS 1633-85. Acoustics—Glossary of Terms and Related Symbols. 46 pp.

INDUSTRY STANDARDS

INTERNATIONAL AND NON-U.S. NATIONAL STANDARDS
SUBJECT INDEX

Acoustics (Cont.)
—Glossaries (Cont.)
SAA AS 1852.801-88. International Electrotechnical Vocabulary—Part 801: Acoustics and Electro-Acoustics. 116 pp.

—Land Use Planning
SAA AS 1055.3-89. Acoustics—Description and Measurement of Environment Noise—Part 3: Acquisition of Data Pertinent to Land Use. 5 pp.

—Listening Rooms—Sound Broadcasting
CCIR QUESTION 81/10-90. Determination of the Acoustical Properties of Control Rooms and High-Quality Listening Rooms in Broadcasting—Questions Concerning Study Group 10—Broadcasting Service (Sound). 1 p.

—Preferred Frequencies
ISO 266-75. Acoustics—Preferred Frequencies for Measurements First Edition. 4 pp.
SNZ NZS/ISO 266-75. Acoustics. Preferred Frequencies for Measurements. 2 pp.

—Sound Broadcasting
CCIR QUESTION 50-1/10-90. Audio-Frequency Characteristics of Broadcasting Signals—Questions Concerning Study Group 10—Broadcasting Service (Sound). 1 p.

—Symbols
BSI BS 5775: Part 7-93. 1993 Quantities, Units and Symbols Part 7: Acoustics (ISO 31-7: 1992) (G). 19 pp.
BSI BS 5775: Part 7-79. 1979 Amd 1 Specification for Quantities, Units and Symbols Part 7: Acoustics (AMD 5852) July 31, 1989. 17 pp.
ISO 31 Pt 7-92. Quantities and Units—Part 7: Acoustics Second Edition. 18 pp.
SAA AS 1633-85. Acoustics—Glossary of Terms and Related Symbols. 46 pp.

—Television Receivers
SAA AS 1173.2-86. Recommended Methods of Measurement on Receivers for Television Broadcast Transmissions—Part 2: Electrical and Acoustic Measurements at Audio-Frequencies. 22 pp.

—Units of Measurement
BSI BS 5775: Part 7-93. 1993 Quantities, Units and Symbols Part 7: Acoustics (ISO 31-7: 1992) (G). 19 pp.
BSI BS 5775: Part 7-79. 1979 Amd 1 Specification for Quantities, Units and Symbols Part 7: Acoustics (AMD 5852) July 31, 1989. 17 pp.
ISO 31 Pt 7-92. Quantities and Units—Part 7: Acoustics Second Edition. 18 pp.

Acoustoelectric Measuring Instruments
Use: Acoustic Measuring Instruments

Acquisition
Scope Note: Use a more specific term See: Contract Pricing; Data Acquisition Systems; Document Acquisition; Procurement

Acrolein Content Analysis
—Exhaust Gases
JIS K 0089-83. Methods for Determination of Acrolein in Exhaust Gas. 12 pp.

Acronyms
Use: Abbreviations

Acrylate Coatings
Use: Acrylic Paints

Acrylate Esters
—Methyl Ethyl Hydroquinone Content
CNS K6609-80. Method of Test for Methyl Ethyl of Hydroquinone in Colorless Monomeric Acrylate Esters (Dec)(6671).

—Purity—Gas Chromatography
CNS K6610-80. Purity of Acrylate Esters by Gas Chromatography (Dec)(6672).

Acrylate Resin Coatings
Use: Acrylic Paints

Acrylates
See Also: Acrylate Esters; n-Butyl Acrylate; Ethyl Acrylate; 2-Ethylhexyl Acrylate

—Highway Transportation—Emergency Procedures
SAA AS 1678.3.0. 023-80. Emergency Procedure Guide—Transport—Part 3.0.023: Acrylates (Methyl, Ethyl, and Butyl).

Acrylic Coatings
Use: Acrylic Paints

Acrylic Coatings (Made From Acrylic)
Use: Acrylic Paints

Acrylic Emulsion Sealants
Use: Acrylic Sealants

Acrylic Enamel Paints
See Also: Acrylic Paints; Enamel Paints
CNS K2084-87. Acrylic Resin Baking Enamel (Jul)(4933). 2 pp.
CNS K6875-86. Method of Test for Acrylic Resin Baking Enamel (Nov)(11605).
JIS K 5654-92. Acrylic Resin Enamel. 13 pp.

Acrylic Fibers
See Also: Fibers; Modacrylic Fibers; Synthetic Fibers

—Binary Mixtures—Quantitative Analysis
CGSB CAN/CGSB-4.2 NO.14.5-M91. Textile Test Methods Quantitative Analysis of Fibre Mixtures—Binary Mixtures Containing Acrylic or Modacrylic Fibres—Butyrolactone Method. 9 pp.
CGSB CAN/CGSB-4.2 NO.14.6-M91. Textile Test Methods Quantitative Analysis of Fibre Mixtures—Binary Mixtures Containing Acrylic and Modacrylic Fibres—Cyclohexanone Method. 9 pp.
CGSB CAN/CGSB-4.2 NO.14.7-M85. Textile Test Methods Quantitative Analysis of Fibre Mixtures—Binary Mixtures Containing Acetate Fibres—Acetone Method. 12 pp.
CGSB CAN/CGSB-4.2 NO.14.8-M88. Textile Test Methods Quantitative Analysis of Fibre Mixtures—Binary Mixtures Containing Triacetate Fibres—Dichloromethane Method. 12 pp.
CGSB CAN/CGSB-4.2 NO.14.14-M89. Textile Test Methods Quantitative Analysis of Fibre Mixtures—Binary Mixtures Containing Polyvinyl Chloride Fibres—Carbon Disulphide/Acetone Method. 11 pp.
CGSB CAN/CGSB-4.2 NO.14.15-M89. Textile Test Methods Quantitative Analysis of Fibre Mixtures—Binary Mixtures Containing Polypropylene Fibres—Xylene Method. 10 pp.
EC 87/185/EEC-87. Commission Recommendation on Quantitative Methods of Analysis for the Identification of Acrylic and Modacrylic Fibres, Chlorofibres and Trivinyl Fibres. 6 pp.

Acrylic Glass
Use: Polymethyl Methacrylate

Acrylic Lacquers
See Also: Acrylic Nitrocellulose Lacquers; Lacquers

—Copper
MOD UK DSTAN 80-24-73. Lacquer, Acrylic (Incralac) Issue 1. 7 pp.
MOD UK DSTAN 80-24-92. Lacquer, Acrylic, Type QX Types: Brushing Spraying Issue 2. 11 pp.

Acrylic Nitrocellulose Lacquers
See Also: Acrylic Lacquers; Nitrocellulose Lacquers

—Aircraft
CGSB CAN/CGSB-1.159-92. Acrylic Cellulose Nitrate Gloss Lacquer. 11 pp.
CGSB 1-GP-160A-72. Lacquer: Acrylic Cellulose Nitrate, Camouflage, Standard for. 10 pp.

—Thinners
CGSB CAN/CGSB-1.197-92. Thinner for Epoxy Coatings. 7 pp.

Acrylic Paints
Use For: Acrylate Coatings; Acrylate Resin Coatings; Acrylic Coatings; Acrylic Coatings (Made From Acrylic) See Also: Acrylic Enamel Paints; Paints

—Aircraft—Flat—Exterior
MOD UK DTD-5599A-01. Selectively Strippable Acrylic Finishing Schemes—Matt and Glossy—for Use on Aircraft; Amendment 1. 11 pp.

—Aircraft—Gloss—Exterior
MOD UK DTD-5599A-01. Selectively Strippable Acrylic Finishing Schemes—Matt and Glossy—for Use on Aircraft; Amendment 1. 11 pp.

—Emulsions—Interior
MOD UK TS 10205B. Paint, Emulsion, Matt, for Interior Use.
MOD UK TS 10207B. Paint, Emulsion, Vinyl Silk, for Interior Use.
MOD UK TS 10221B. Paint, Emulsion, Vinyl Silk, Mould-Resistant for Interior Use.
MOD UK TS 10225B. Paint, Emulsion, Matt Mould-Resistant, for Interior Use.

Acrylic Paints (Cont.)
—Roofs—Waterproof
CNS A2131-88. Roof Coating for Waterproofing (Acrylic Resin Type) (Jan)(8643).

—Thinners
MOD UK DSTAN 80-38-85. Thinners for: Paint Epoxy Two-Pack, Cellulose Nitrate Paints, Dopes and Lacquers Issue 2. 8 pp.
MOD UK DSTAN 80-99-01. Thinners for Paint Finishing Acrylic, Spraying Quality Issue 1; Amendment 1. 7 pp.

Acrylic Plastics
Use: Acrylic Resins

Acrylic Resins
Scope Note: Use a more specific term
Use For: Acrylic Plastics See Also: ABS Resins; Acrylic Styrene Acrylonitrile Resins; Acrylic Styrene Resins; Latex; Polyacrylamides; Polyacrylonitrile; Polymethyl Methacrylate; Resins; Thermoplastic Resins; Thermosetting Resins

—Sheets—Fabrication—Aerospace
MOD UK DTD-925D-69. Fabrication of Acrylic Panels and Shapings. 3 pp.

—Sheets—Glazing—Aircraft
MOD UK DTD-5592A-81. Acrylic Sheets for Aircraft Glazing. 12 pp.
MOD UK DTD-5634-82. Stretched or Pressed Acrylic Sheet of Superior Environmental Resistance, for Aircraft Glazing. 15 pp.

Acrylic Sealants
See Also: Sealants

—Joints
CGSB 19-GP-5M-76. Sealing Compound, One Component, Acrylic Base, Solvent Curing, Standard for (R 1984). 8 pp.
CGSB CAN/CGSB-19.17-M90. One-Component, Acrylic Emulsion Base Sealing Compound. 9 pp.
CNS A2209-87. Joint Treatment Materials for Gypsum Boards (Jun)(11990).
CNS A3262-87. Method of Test for Joint Treatment Materials for Gypsum Boards (Jun)(11991).
JIS A 6914-85. Joint Treatment Materials for Gypsum Boards. 9 pp.

Acrylic Styrene Acrylonitrile Resins
Scope Note: See specific products made from acrylic styrene acrylonitrile resin See Also: Acrylic Resins; Styrene Resins

—Molding Materials—Test Specimens
DIN ENGL 16777 Pt 2-88. Plastics Moulding Materials; Impact Resistant Acrylonitrile/Styrene (ASA, AES, ACS) Moulding Materials (Except for Butadiene-Modified Moulding Materials) Preparation of Specimens and Determination of Their Properties (Dec). 4 pp.

Acrylic Styrene Resins
See Also: Resins
JIS K 6875-87. Testing Methods for AS Resins. 7 pp.

Acrylic Varnishes
See Also: Varnishes
CNS K2164-87. Acrylic Resin Varnish (Oct)(12143). 2 pp.
CNS K6926-87. Method of Test for Acrylic Resin Varnish (Oct)(12144). 4 pp.
JIS K 5653-92. Acrylic Resin Varnish. 12 pp.

Acrylonitrile Butadiene Rubber
Use: Nitrile Rubber

Acrylonitrile Butadiene Styrene Resins
Use: ABS Resins

Acrylonitrile Content Analysis
—Elastomers—Kjeldahl Method
DIN ENGL 53621 Pt 3-76. Testing of Rubber and Elastomers; Quantitative Determination of Polymers; Determination of the Acrylonitrile Content and Calculation of the Acrylonitrile-Butadiene Copolymer Content (July). 3 pp.

—Nitrile Latex
BSI BS 6057: Sec 3.22-90. 1990 Rubber Latices Part 3: Methods of Test Section 3.22: Determination of Residual Acrylonitrile Content in Nitrile Latex. 6 pp.
ISO 3899-88. Rubber—Nitrile Latex—Determination of Residual Acrylonitrile Content Second Edition. 4 pp.
ISO 3900-76. Rubber—Nitrile Latex—Determination of Bound Acrylonitrile Content First Edition. 4 pp.

Acrylonitrile Content Analysis (Cont.)
—Nitrile Rubber
MOD UK NES 2005-89. Specification for Vulcanized Butadiene—Acrylonitrile Rubber Compounds—High Acrylonitrile Content Issue 2 (06.89). 12 pp.

—Rubber—Kjeldahl Method
DIN ENGL 53621 Pt 3-76. Testing of Rubber and Elastomers; Quantitative Determination of Polymers; Determination of the Acrylonitrile Content and Calculation of the Acrylonitrile-Butadiene Copolymer Content (July). 3 pp.

—Styrene Acrylonitrile Resins—Gas Chromatography
ISO 4581-87. Plastics—Styrene/Acrylonitrile Copolymers—Determination of Residual Acrylonitrile Monomer Content—Gas Chromatography Method First Edition. 8 pp.

Acrylonitrile Copolymers
See Also: Modacrylic Fibers; Polyacrylonitrile; Rubber

—Explosives
MOD UK TS 10285. Carboxy Terminated Acrylonitrile/Butadiene Reactive Liquid Polymer Type 2, QX.

—Flares
MOD UK TS 10251. Carboxy Terminated Acrylonitrile/Butadiene Reactive Liquid Polymer Type 1.

Acrylonitrile Styrene Acrylester Resins
Use: Acrylic Styrene Acrylonitrile Resins

Acrylonitrile Styrene Copolymers
—Extrusion Materials
ISO 6402 Pt 1-90. Plastics—Impact-Resistant Acrylonitrile/ Styrene Moulding and Extrusion Materials (ASA, AES, ACS), Excluding Butadiene-Modified Materials—Part 1: Designation First Edition. 8 pp.

—Molding Materials
ISO 6402 Pt 1-90. Plastics—Impact-Resistant Acrylonitrile/ Styrene Moulding and Extrusion Materials (ASA, AES, ACS), Excluding Butadiene-Modified Materials—Part 1: Designation First Edition. 8 pp.

Acrylonitriles
See Also: Plastics; Resins

—Comprehensive Testing
JIS K 6759-77. Testing Methods for Acrylonitrile (R 1980). 13 pp.

—Food Contact
SAA AS 2070.4-92. Plastics Materials for Food Contact Use—Part 4: Acrylonitrile Plastics Materials Amdt 1 August 1993. 4 pp.

—Highway Transportation—Emergency Procedures
SAA AS 1678.3.0. 022-89. Emergency Procedure Guide—Transport—Part 3.0.022: Acrylonitrile (Inhibited) (Vinyl Cyanide).

Activated Alumina
Use For: Aluminum Oxide, Brockmann
See Also: Aluminum Oxide
CNS K7036-61. Chemical Reagent (Aluminum Oxide) (for Chomatography) (Dec)(1536).

—Drying Agents
MOD UK TS 10311. Desiccant, Activated Alumina, Pellets (Withdrawn).

Activated Carbon
Use For: Activated Charcoal *See Also:* Carbon; Charcoal
CNS K7133-64. Chemical Reagent (Carbon, Decolorizing) (Activated Charcoal) (Jul)(1633).
JIS K 1474-91. Test Methods for Activated Carbon. 50 pp.
MOD UK TS 647. Charcoal, Activated Nutshell, Powdered (Dormant).
MOD UK TS 10250B. Charcoal, Activated, Coal-Based. Types: 1100, 1120 and 1140.

—Air Filters
CGSB CAN/CGSB-115.16-M82. Activated Carbon for Odor Removal from Ventilating Systems. 9 pp.

—Air Filters—NBC
MOD UK DSTAN 68-133-90. Carbon, Activated, Impregnated, ASC/TEDA Type 1101 Issue 1. 24 pp.

Activated Carbon (Cont.)
—Air Filters—NBC (Cont.)
MOD UK DSTAN 68-133-92. Carbon, Activated, Impregnated, ASC/TEDA, Type 1101 Issue 2. 20 pp.
MOD UK DSTAN 68-135-90. Charcoal, Activated, Type 1121 Issue 1. 14 pp.
MOD UK DSTAN 68-135-01. Charcoal, Activated, Type 1121 Issue 1; Amendment 1. 15 pp.
MOD UK DSTAN 68-139-01. Charcoal, Activated, Type 1142 Issue 1; Corrigendum. 15 pp.

—Ash Content
CNS K6478-80. Method of Test for Total Ash Content of Activated Carbon (May)(5581).

—Bulk Density
CNS K6477-80. Method of Test for Apparent Density of Activation Carbon (May)(5580).

—Granular
MOD UK TS 10318A. Charcoal, Activated, Nutshell, Granular.

—Moisture Content
CNS K6479-80. Method of Test for Moisture in Activated Carbon (May)(5582).

—Particle Size Distribution
CNS K6480-80. Method of Test for Particle Size Distribution of Granular Activated Carbon (May)(5583).

—Powder
CNS K1041-65. Activated Carbon Powder for Industrial Use (Aug)(697)(R 1971).
CNS K6037-65. Method of Test for Activated Carbon (Powder, for Industrial Use) (Aug)(698)(R 1971).
MOD UK DSTAN 68-36-93. Charcoal Activated Nutshell, Powdered Issue 1. 16 pp.

—Water Treatment
DIN ENGL 19603-69. Activated Charcoal for Water Treatment; Technical Conditions of Delivery (May). 4 pp.

Activated Charcoal
Use: Activated Carbon

Activated Clay
—Drying Agent
BSI BS 7529-91. 1991 Desiccant Activated Clay (V). 7 pp.
MOD UK TS 487B. Desiccant, Activated Clay (Withdrawn).

—Packaging
MOD UK DSTAN 81-68-01. Bags, Desiccant, Silica Gel and Bags, Desiccant, Activated Clay Issue 1; Amendment 2. 15 pp.

Activated Sludge
Use: Sludge (Sewage)

Activated Sludge Process
Use: Sludge (Sewage)

Active Detectors
Use For: Active Sensors *See Also:* Passive Detectors; Sensors

—Earth Exploration Satellites—Frequency Bands
CCIR RECMN 516-78. Frequency Bands for Active Sensors Used on Earth Exploration and Meteorological Satellites—Section 2F—Earth Exploration Satellites. 1 p.

—Meteorological Satellites—Frequency Bands
CCIR RECMN 516-78. Frequency Bands for Active Sensors Used on Earth Exploration and Meteorological Satellites—Section 2F—Earth Exploration Satellites. 1 p.

—Space Communications—Frequency Bands
CCIR RECMN 577-2-90. Preferred Frequency Bands for Active Sensing Measurements—Section 2F—Earth Exploration Satellites. 1 p.

Active Sensors
Use: Active Detectors

Active Speech Level
See Also: Speech Transmission Systems

—Measurement—Telephone Transmission
CCITT Volume V. Telephone Transmission Quality Series P Recommendations (Study Group XII). 438 pp.

Active Speech Level (Cont.)
—Measurement—Telephone Transmission (Cont.)
CCITT RECMN P.56-89. Objective Measurement of Active Speech Level—Telephone Transmission Quality (Study Group XII) 12 pp. 12 pp.

Actuating Levers
Use: Lever Switches

Actuators
Use For: Electric Actuators *See Also:* Automatic Control Equipment; Control Systems Equipment; Linear Actuators; Regulators; Rotary Actuators; Servomotors; Servovalves; Solenoids; Starters; Valve Actuators
BSI BS 6013-80. 1980 Standard Direction of Movement for Actuators Which Control the Operation of Electrical Apparatus. 10 pp.
CENELEC HD 331-76. Standard Direction of Movement for Actuators Which Control the Operation of Electrical Appartus. 2 pp.

—Design—Ships
MOD UK NES 375-84. Valve Design and Manufacture Issue 1 (10.84). 54 pp.

—Design—Submarines
MOD UK NES 375-84. Valve Design and Manufacture Issue 1 (10.84). 54 pp.

—Electric Switches
CNS C6092-88. Method of Test for Electromechanical Switches (Actuator/Mounting Bushing Resistance) (Dec)(6155).

—Ergonomics—Machinery
CEN PREN 894-1-92. Safety of Machinery—Ergonomics Requirements for the Design of Displays and Control Actuators—Part 1: Human Interactions with Displays and Control Actuators. 19 pp.
CEN PREN 894-3-92. Safety of Machinery—Ergonomics Requirements for the Design of Displays and Control Actuators—Part 3: Control Actuators. 37 pp.
IEC 447-93. Man-Machine Interface (MMI)—Actuating Principles Second Edition. 41 pp.

—Internal Combustion Piston Engines—Glossaries
BSI BS 7016: Part 3-88. 1988 Components and Systems of Reciprocating Internal Combustion Engines Part 3: Glossary of Terms of Valves, Camshaft Drive and Actuating Mechanisms. 19 pp.
ISO 7967 Pt 3-87. Reciprocating Internal Combustion Engines—Vocabulary of Components and Systems—Part 3: Valves, Camshaft Drive and Actuating Mechanisms First Edition. 19 pp.

ADA
Scope Note: Programming language named for Augusta Ada Byron, born 1815, died 1852, who wrote coded instruction for Charles Babbage's mechanical calculators and is considered the first computer programmer *See Also:* Programming Languages
BSI BS 7145-90. 1990 Programming Language Ada. 4 pp.
CEN EN 28 652-89. Programming Languages Ada. 339 pp.
ISO 8652-87. Programming Languages—ADA First Edition; (Corrected and Reprinted -1987). 4 pp.
JIS X 3009-91. Programming Language Ada (ISO 8652:1988).
JTC1 8652-87. Programming Languages—ADA First Edition; (Corrected and Reprinted -1987). 4 pp.
OSI ISO 8652-87. Programming Languages—Ada. 341 pp.
SAA AS 4007-92. Ada as a Program Design Language (ANSI/IEEE 990—1987) (in Professional Package 26A). 5 pp.

—Elementary Functions
IEC DIS 11430-91. Information Technology—Generic Package of Elementary Functions for ADA ***CD-ROM ONLY***. 39 pp.
ISO DIS 11430-91. Information Technology—Generic Package of Elementary Functions for ADA ***CD-ROM ONLY***. 39 pp.

—Language Binding—Graphical Kernel System
BSI BS 7040: Part 3-89. (WITHDRAWN) 1989 Computer Graphics: Graphical Kernel System (GKS) Language Bindings Part 3: GKS Language Binding for Ada (Renumbered as BS EN 28651-3: 1992). 188 pp.
BSI BS EN 28651-3-92. 1992 Information Processing Systems—Computer Graphics—Graphical Kernel System (GKS) Language Bindings Part 3: Ada (ISO 8651-3: 1988) (AMD 7539) December 15, 1992. 194 pp.

INTERNATIONAL AND NON-U.S. NATIONAL STANDARDS
SUBJECT INDEX

ADA (Cont.)

—Language Binding—Graphical Kernel System (Cont.)
CEN EN 28651-3-92. Information Processing Systems—Computer Graphics—Graphical Kernel System (GKS) Language Bindings Part 3 Ada; (ISO 8651-3: 1988). 186 pp.
ISO 8651 Pt 3-88. Information Processing Systems—Computer Graphics—Graphical Kernel System (GKS) Language Bindings—Part 3: ADA First Edition; (CEN EN 28651-3:1992). 188 pp.
JTC1 8651 Pt 3-88. Information Processing Systems—Computer Graphics—Graphical Kernel System (GKS) Language Bindings—Part 3: ADA First Edition. 188 pp.
OSI ISO 8651-3-88. Information Processing Systems—Computer Graphics—Graphical Kernal System (GKS) Language Bindings Part 3: Ada. 188 pp.
OSI ISO DP 8651-3.2-87. Computer Graphics—GKS—Language Bindings—Part 3: Ada. 185 pp.
SNZ NZS/ISO 8651. 3-88. Computer Graphics—Graphical Kernel System (GKS) Language Bindings Part 3: Ada. 184 pp.

—Language Binding—Portable Common Tool Environment
ECMA ECMA 162-93. Portable Common Tool Environment (PCTE) ADA Programming Language Binding Second Edition. 153 pp.
IEC DIS 13719 Pt 3-93. Information Technology—Portable Common Tool Environment (PCTE)—Part 3: Ada Programming Language Binding ***CD-ROM ONLY***. 156 pp.
ISO DIS 13719 Pt 3-93. Information Technology—Portable Common Tool Environment (PCTE)—Part 3: Ada Programming Language Binding ***CD-ROM ONLY***. 156 pp.
JTC1 DIS 13719 Pt 3-93. Information Technology—Portable Common Tool Environment (PCTE)—Part 3: Ada Programming Language Binding ***CD-ROM ONLY***. 156 pp.

—Language Binding—Programmers' Hierarchical Interactive Graphics
BSI BS ISO/IEC 9593/3-90. 1990 Information Technology—Computer Graphics—Programmers Hierarchical Interactive Graphics System (PHIGS) Language Bindings 9593/3: Ada. 294 pp.
IEC 9593 Pt 3-90. Information Technology—Computer Graphics—Programmer's Hierarchical Interactive Graphics System (PHIGS) Language Bindings—Part 3: Ada First Edition; (Technical Corrigendum 1—1993) (ANSI 9593.3-1990) (FIPS PUB 153). 292 pp.
IEC 9593 Pt 3 Draft AMD 1. Information Technology—Computer Graphics—Programmer's Hierarchical Interactive Graphics System (PHIGS) Language Bindings—Part 3: Ada Amendment 1: Incorporation of PHIGS Plus; (1992) ***CD-ROM ONLY***. 339 pp.
ISO 9593 Pt 3-90. Information Technology—Computer Graphics—Programmer's Hierarchical Interactive Graphics System (PHIGS) Language Bindings—Part 3: Ada First Edition; (Technical Corrigendum 1—1993) (ANSI 9593.3-1990) (FIPS PUB 153). 292 pp.
ISO 9593 Pt 3 Draft AMD 1. Information Technology—Computer Graphics—Programmer's Hierarchical Interactive Graphics System (PHIGS) Language Bindings—Part 3: Ada Amendment 1: Incorporation of PHIGS Plus; (1992) ***CD-ROM ONLY***. 339 pp.
JTC1 9593 Pt 3-90. Information Technology—Computer Graphics—Programmer's Hierarchical Interactive Graphics System (PHIGS) Language Bindings—Part 3: Ada First Edition; (Technical Corrigendum 1—1993) (ANSI 9593.3-1990) (FIPS PUB 153). 292 pp.
JTC1 9593 Pt3 Draft AMD 1. Information Technology—Computer Graphics—Programmer's Hierarchical Interactive Graphics System (PHIGS) Language Bindings—Part 3: Ada Amendment 1: Incorporation of PHIGS Plus; (1992) ***CD-ROM ONLY***. 339 pp.
OSI ISO/IEC 9593-3-90. Information Technology—Computer Graphics—Programmer's Hierarchical Interactive Graphics System (PHIGS) Language Bindings—Part 3: Ada. 291 pp.
SAA AS 3794.3-91. Computer Graphics—Programmer's Hierarchical Interactive Graphics System (PHIGS) Language Bindings Part 3: Ada (ISO/IEC 9593-3) (In Professional Packages 26A, 26D). 285 pp.
SNZ NZS/ISO-IEC 9593.3-90. Computer Graphics—Programmer's Hierarchical Interactive Graphics System (PHIGS) Language Bindings Part 3: Ada. 285 pp.

—Real Time Software—Avionics
NATO STANAG 3912 ED 1 AMD 1-85. Standardization of a Real-Time High Order Computer Programming Language for Avionics Systems Applications. 8 pp.

ADA (Cont.)

—Real Time Software—Avionics (Cont.)
NATO STANAG 3912 ED 1 AMD 2-85. Standardization of a Real-Time High Order Computer Programming Language for Avionics Systems Applications. 8 pp.

—System Software—Interfaces
IEC DIS 11729-93. Information Technology—Programming Languages, Their Environments and System Software Interfaces—Generic Package of Primitive Functions for Ada ***CD-ROM ONLY***. 24 pp.
ISO DIS 11729-93. Information Technology—Programming Languages, Their Environments and System Software Interfaces—Generic Package of Primitive Functions for Ada ***CD-ROM ONLY***. 24 pp.
JTC1 DIS 11729-93. Information Technology—Programming Languages, Their Environments and System Software Interfaces—Generic Package of Primitive Functions for Ada ***CD-ROPM ONLY***. 24 pp.

Adapters (Electric)
Use For: Electric Adapters See Also: Adapters (Fittings); Coaxial Adapters; Electric Plug Adapters; Waveguide Adapters
CNS C4351-82. Adapter (Jul)(9078).

—Cables—Mining Equipment
BSI BS 3454-84. 1984 1.9/3.3 kV, 300 A Bolted Flameproof Cable Couplers and Adaptors (Including 380/660 V and 640/1100 V, 300 A Adaptors), Primarily for Use in Mines. 25 pp.
BSI BS 3905-84. 1984 3.8/6.6 kV, 300 A Bolted Flameproof Cable Couplers and Adapters, Primarily for Use in Mines. 24 pp.
BSI BS 7019-89. 1989 Fully Screened 400 A Bolted Flameproof Cable Couplers and Adaptors for Use on Systems up to and Including 6.6 kV, Primarily for Use in Mines. 26 pp.

—Cords
CNS C4412-5-85. Adapter Cord Sets (Feb)(10917-5).

—Data Terminal Equipment
CEPT T/GSI 04-04-87. Adaptateurs De Terminaux. 1 p.
CEPT T/GSI 04-04 E-87. Terminal Adaptors. 1 p.

—Electric Conduits
BSI BS 731: Part 1-52. 1952 Amd 3 Flexible Steel Conduit for Cable Protection and Flexible Steel Tubing to Enclose Flexible Drives Part 1: Flexible Steel Conduit and Adaptors for the Protection of Electric Cable. 16 pp.
BSI DD 200-91. 1991 Pliable Non-Rigid Conduit and Fittings for Direct Burial Underground (Type DB). 13 pp.

—Electric Outlets
BSI BS 1363: Part 3-89. 1989 Amd 1 General Purpose Fuse Links for Domestic and Socket-Outlets and Boxes Part 3: Adaptors (AMD 6223) May 31, 1989. 57 pp.
DIN ENGL 1808-64. Adaptor Sockets for Cotter Retention (June). 2 pp.
DIN VDE 0620 A7 (D)-89. Plugs and Socket Outlets up to 389 V 25 A: Amendment 7: Socket-Adapters and Appliances with Integrated Plugs (June) (Supersedes Draft DIN VDE 0620 A2/07.86). 8 pp.

—Electric Outlets—Aircraft
SBAC AS 2482 ISSUE 2. Outlet Adapter. 1 p.

—Electric Outlets—Inserts—Aircraft
SBAC AS 2483 ISSUE 2. Insert for Outlet Adaptor. 1 p.

Adapters (Fittings)
See Also: Adapters (Electric); Connectors; Flanged Adapters; Hose Adapters; Integrated Circuit Socket Adapters; Pipe Adapters; Sleeves (Fittings); Tube Adapters

—Air Cargo Handling Equipment—Compatibility
NATO STANAG 3616 ED 1 AMD 4-72. Responsibility for the Design and Provision of Adaptors Necessary for the Compatibility of Air Cargo Loading, Securing, Unloading or Dropping Systems in Fixed Wing Aircraft. 5 pp.
NATO STANAG 3616 ED 1 AMD 5-72. Responsibility for the Design and Provision of Adaptors Necessary for the Compatibility of Air Cargo Loading, Securing, Unloading or Dropping Systems in Fixed Wing Aircraft. 6 pp.

—Aircraft
MOD UK DSTAN 47-17-78. Adaptors and Seals for Pipe Connections to Aircraft Pitot/Static Instruments and Associated Equipment Issue 1. 7 pp.

Adapters (Fittings) (Cont.)

—Aircraft—Drop Tanks
SBAC AS 6606 ISSUE 1. Adaptor (Drop Tanks).

—Aircraft—Forgings
SBAC AS 2954 (V). Flanged Adaptor Forging. 2 pp.
SBAC AS 2978 (V). Flanged Adaptor Forging. 2 pp.

—Aircraft—Stop Valves—Drop Tanks
SBAC AS 4897 ISSUE 3. Adaptor (Drop Tanks). 1 p.

—Antifriction Bearings
JIS B 1552-93. Rolling Bearings—Adapterassemblies and Adapter Sleeves. 24 pp.

—Bits (Tools)—Ratchet Screwdrivers
CNS B3327-80. Assembly Tools for Screws and Nuts Hexagon Insert Bits for Use with Spiral Ratchet Drivers (Jan)(5106).

—Cap Lifters—Aircraft
SBAC AS 6371 ISSUE 2. Adaptor—Cap Lifter.

—Compasses (Drawing)
CNS Z7157-83. Indian Ink Writing and Drawing Instruments, Adapters for Compasses for Ink Writing Pen for Precision Drawing Instruments (P) (Jan)(9933).
ISO 9176-88. Tubular Technical Pens—Adaptor for Compasses First Edition. 7 pp.

—Drainage Connections—Aircraft
SBAC AS 6290 ISSUE 1. Adaptor—1 1/4 Inch. 1 p.
SBAC AS 6314 ISSUE 1. Adaptor. 1 p.

—Drills
CNS B3078-52. Adapter, Countersink Drill and Reamer, with Square Head (Sep)(252)(R 1970).

—Flanged—Aircraft
SBAC AS 55460-469 ISSUE 1. 2 Bolt Flanged Adaptor—Straight Spigot—Without Integral Filler—for Metric Tube, "O" Ring Groove & Fasteners.
SBAC AS 55470-479 ISSUE 1. 2 Bolt Flanged Adaptor—90 Degree Elbow—Without Integral Filler—for Metric Tube, "O" Ring Groove & Fasteners.
SBAC AS 55480-489 (V). 2 Bolt Flanged Adaptor—Straight Housing—Without Integral Filler—for Metric Tube, "O" Ring Groove & Fasteners.
SBAC AS 61970-979 ISSUE 1. 2 Bolt Flanged Adaptor—Straight Spigot—with Integral Filler—for Metric Tube, "O" Ring Groove & Fasteners.
SBAC AS 61980-989 ISSUE 1. 2 Bolt Flanged Adaptor—90 Degree Elbow—with Integral Filler—for Metric Tube, "O" Ring Groove & Fasteners.
SBAC AS 61990-999 (V). 2 Bolt Flanged Adaptor—Straight Housing—with Integral Filler—for Metric Tube, "O" Ring Groove & Fasteners.

—Flanged—Aircraft—O Rings
SBAC AS 55460-469 ISSUE 1. 2 Bolt Flanged Adaptor—Straight Spigot—Without Integral Filler—for Metric Tube, "O" Ring Groove & Fasteners.
SBAC AS 55470-479 ISSUE 1. 2 Bolt Flanged Adaptor—90 Degree Elbow—Without Integral Filler—for Metric Tube, "O" Ring Groove & Fasteners.
SBAC AS 55480-489 (V). 2 Bolt Flanged Adaptor—Straight Housing—Without Integral Filler—for Metric Tube, "O" Ring Groove & Fasteners.
SBAC AS 61970-979 ISSUE 1. 2 Bolt Flanged Adaptor—Straight Spigot—with Integral Filler—for Metric Tube, "O" Ring Groove & Fasteners.
SBAC AS 61980-989 ISSUE 1. 2 Bolt Flanged Adaptor—90 Degree Elbow—with Integral Filler—for Metric Tube, "O" Ring Groove & Fasteners.
SBAC AS 61990-999 (V). 2 Bolt Flanged Adaptor—Straight Housing—with Integral Filler—for Metric Tube, "O" Ring Groove & Fasteners.

—Gas Cylinder Valves
MOD UK DSTAN 81-44-01. Adaptors for Gas Cylinder Valves and Regulating Equipment Issue 1; Amendment 1 (Withdrawn). 43 pp.

—Hoses—Aircraft
SBAC AS 6374 ISSUE 2. Adaptor—Hose.

—Hoses—Aircraft—Assembly
SBAC AS 2583 ISSUE 5. Assembly of Valve Lifter and Adaptor. 1 p.

—Hydraulic Equipment
BSI BS 5200-86. 1986 Dimensions of Hydraulic Connectors and Adaptors. 19 pp.

—Hydraulic Hose Assemblies
JIS B 8363-88. End Fittings and Adapters for Hydraulic Hose Assemblies. 25 pp.

—Laboratory Ware
CNS R2143-86. Glass Adapters for Chemical Analysis (Mar)(7319).

INTERNATIONAL AND NON-U.S. NATIONAL STANDARDS
SUBJECT INDEX

Adapters (Fittings) *(Cont.)*

—**Morse Taper Reamers**
CNS B3077-52. Adapter, Countersink Drill and Reamer, with Morse Taper Shank (Sep)(251)(R 1970).

—**Morse Taper Shanks**
DIN ENGL 6327 Pt 1-85. Adjustable Adaptors for Tapered Tools; Short Types (June). 10 pp.

—**Morse Taper Shanks—Drills**
CNS B3077-52. Adapter, Countersink Drill and Reamer, with Morse Taper Shank (Sep)(251)(R 1970).

—**Nipples (Fittings)**
BSI BS 1486: Part 1-59. (OBSOLESCENT) 1959 Amd 1 Lubricating Nipples Part 1: Lubricating Nipples and Adaptors for Use on Machinery and Vehicles. 34 pp.
MOD UK DSTAN 47-7: Addendum. Lubricating Nipples, and Adaptors, Lubricating Nipple Issue 2. 10 pp.

—**Reamers**
CNS B3078-52. Adapter, Countersink Drill and Reamer, with Square Head (Sep)(252)(R 1970).

—**Screwdriver Bits—Ratchet Screwdrivers**
DIN ENGL 7433-78. Adapters for Hexagon Insert Bits for Use with Spiral Ratchet Drivers (Aug). 2 pp.

—**Socket Wrenches**
CNS B3320-80. Assembly Tools for Screws and Nuts Socket Wrench (Jan)(5099).

—**Taps (Threading Tools)**
CNS B3097-53. Tap Adaptor (Jul)(312) (R 1970).

—**Valve Lifters—Aircraft**
SBAC AS 6373 ISSUE 2. Assembly—Valve Lifter Adaptor.

—**Welding Electrodes—Resistance Spot Welding**
BSI BS EN 25183-1-92. 1992 Resistance Spot Welding—Electrode Adaptors, Male Taper 1:10 Part 1: Conical Fixing, Taper 1:10 (ISO 5183-1: 1988). 9 pp.
BSI BS EN 25183-2-92. 1992 Resistance Spot Welding—Electrode Adaptors, Male Taper 1:10 Part 2: Parallel Shank Fixing for End-Thrust Electrodes. 10 pp.
CEN EN 25183-1-91. Resistance Spot Welding—Electrode Adaptors, Male Taper 1: 10—Part 1: Conical Fixing, Taper 1: 10. 3 pp.
CEN EN 25183-2-91. Resistance Spot Welding—Electrode Adaptors, Male Taper 1: 10—Part 2: Parallel Shank Fixing for End-Thrust Electrodes. 3 pp.
DIN ENGL EN 25183 Pt 1-92. Resistance Spot Welding; Electrode Adaptors, Male Taper 1: 10; Conical Fixing, Taper 1: 10 (ISO 5183: 1988) (Feb). 5 pp.
DIN ENGL EN 25183 Pt 2-92. Resistance Spot Welding; Electrode Adaptors, Male Taper 1: 10; Parallel Shank Fixing for End-Thrust Electrodes (ISO 5183-2: 1988) (Feb) (Supersedes DIN 44750 Part 7, September 1976 Edition). 6 pp.
ISO 5183 Pt 1-88. Resistance Spot Welding—Electrode Adaptors, Male Taper 1: 10—Part 1: Conical Fixing, Taper 1: 10 First Edition. 5 pp.
ISO 5183 Pt 2-88. Resistance Spot Welding—Electrode Adaptors, Male Taper 1: 10—Part 2: Parallel Shank Fixing for End-Thrust Electrodes First Edition. 6 pp.

—**Welding Electrodes—Resistance Welding**
BSI BS 4215: Part 1-87. (WITHDRAWN) 1987 Resistance Spot Welding Electrodes, Electrode Holders and Ancillary Equipment Part 1: Resistance Spot Welding Electrode Adaptors (Superseded by BS 4215: Parts 9 & 10: 1990). 4 pp.
BSI BS 4215: Part 10-90. (WITHDRAWN) 1990 Resistance Spot Welding Electrodes, Electrode Holders and Ancillary Equipment Part 10: Electrode Adaptors, Male Taper 1: 10, Parallel Shank Fixing for End Thrust Electrodes (Superseded by BS EN 25183-2: 1992). 8 pp.

—**Welding Electrodes—Spot Welding**
BSI BS 4215: Part 1-87. (WITHDRAWN) 1987 Resistance Spot Welding Electrodes, Electrode Holders and Ancillary Equipment Part 1: Resistance Spot Welding Electrode Adaptors (Superseded by BS 4215: Parts 9 & 10: 1990). 4 pp.
BSI BS 4215: Part 6-87. 1987 Resistance Spot Welding Electrodes, Electrode Holders and Ancillary Equipment Part 6: Electrode Adaptors, Female Taper 1:10. 4 pp.

Adapters (Fittings) *(Cont.)*

—**Welding Electrodes—Spot Welding** *(Cont.)*
BSI BS 4215: Part 6-01. 1987 Amd 1 Resistance Spot Welding Electrodes, Electrode Holders and Ancillary Equipment Part 6: Specification for Electrode Adaptors, Female Taper 1:10 (AMD 7130) July 15, 1992. 5 pp.
BSI BS 4215: Part 9-90. (WITHDRAWN) 1990 Resistance Spot Welding Electrodes, Electrode Holders and Ancillary Equipment Part 9: Electrode Adaptors, Male Taper 1: 10, Conical Fixing 1: 10 (Superseded by BS EN 25183-1: 1992). 7 pp.
ISO 5829-84. Resistance Spot Welding—Electrode Adaptors, Female Taper 1: 10 First Edition. 4 pp.

Adaptive Antenna Arrays
See Also: Antennas
CCIR RECMN 856-92. Use of Interference Cancellers, Screens and Adaptive Antennas. 1 p.

—**Interference Cancellation Systems**
CCIR Report 1128-90. Adaptive Antenna Arrays for Interference Cancellation—Section 3Ab—Antennas Characteristics. 5 pp.

—**Mobile Radio Equipment—Co-Channel Interference**
CCIR Report 973-86. Adaptive Antennas for Land Mobile Applications in Reducing Co-Channel Interference—Section 1D—Spectrum Utilization and Applications. 5 pp.

Adaptive Delta Pulse Code Modulation
Use For: ADPCM *See Also:* Pulse Code Modulation

—**DCME**
CCITT RECMN G.723-89. Extensions of Recommendation G.721 Adaptive Differential Pulse Code Modulation to 24 and 40 kbit/s for Digital Circuit Multiplication Equipment Application—General Aspects of Digital Transmission Systems; Terminal Equipments (Stdy Grp XV/XVIII)18 pp. 18 pp.

—**DCME—Digital Speech Interpolation**
CCITT RECMN G.763-91. Digital Circuit Multiplication Equipment Using 32 kBit/s ADPCM and Digital Speech Interpolation (Study Group XV) 368 pp. 368 pp.
CCITT RECMN G.763-89. Digital Circuit Multiplication Equipment Using 32 kBit/s ADPCM and Digital Speech Interpolation—General Aspects of Digital Transmission Systems; Terminal Equipments (Study Groups XV and XVIII) 37 pp. 37 pp.

—**Encoder/Decoders—16 kbit/s**
CCITT RECMN G.726-90. 40, 32, 24, 16 kbit/s Adaptive Differential Pulse Code Modulation (ADPCM) (Study Group XV) 60 pp (Replaces Recmn G.721 and G.723). 60 pp.

—**Encoder/Decoders—2-Bit**
CCITT RECMN G.727-90. 5-. 4-, 3-amd 2-Bits Sample Embedded Adaptive Differential Pulse Code Modulation (ADPCM) (Study Group XV) 58 pp. 58 pp.

—**Encoder/Decoders—24 kbit/s**
CCITT RECMN G.726-90. 40, 32, 24, 16 kbit/s Adaptive Differential Pulse Code Modulation (ADPCM) (Study Group XV) 60 pp (Replaces Recmn G.721 and G.723). 60 pp.

—**Encoder/Decoders—3-Bit**
CCITT RECMN G.727-90. 5-. 4-, 3-amd 2-Bits Sample Embedded Adaptive Differential Pulse Code Modulation (ADPCM) (Study Group XV) 58 pp. 58 pp.

—**Encoder/Decoders—32 kbit/s**
CCITT RECMN G.721-89. 32 kbit/s Adaptive Differential Pulse Code Modulation (ADPCM)—General Aspects of Digital Transmission Systems; Terminal Equipments (Study Groups XV and XVIII) 38 pp (Replaced by Recmn G.726). 38 pp.
CCITT RECMN G.726-90. 40, 32, 24, 16 kbit/s Adaptive Differential Pulse Code Modulation (ADPCM) (Study Group XV) 60 pp (Replaces Recmn G.721 and G.723). 60 pp.

—**Encoder/Decoders—4-Bit**
CCITT RECMN G.727-90. 5-. 4-, 3-amd 2-Bits Sample Embedded Adaptive Differential Pulse Code Modulation (ADPCM) (Study Group XV) 58 pp. 58 pp.

—**Encoder/Decoders—40 kbit/s**
CCITT RECMN G.726-90. 40, 32, 24, 16 kbit/s Adaptive Differential Pulse Code Modulation (ADPCM) (Study Group XV) 60 pp (Replaces Recmn G.721 and G.723). 60 pp.

Adaptive Delta Pulse Code Modulation *(Cont.)*

—**Encoder/Decoders—5-Bit**
CCITT RECMN G.727-90. 5-. 4-, 3-amd 2-Bits Sample Embedded Adaptive Differential Pulse Code Modulation (ADPCM) (Study Group XV) 58 pp. 58 pp.

—**Encoder/Decoders—64 kbit/s**
CCITT RECMN G.722-89. 7 kHz Audio-Coding Within 64 kBit/s—General Aspects of Digital Transmission Systems; Terminal Equipments (Study Groups XV and XVIII) 73 pp. 73 pp.

—**Quantizing Distortion**
CCITT RECMN G.113-89. Transmission Impairments—General Characteristics of International Telephone Connections and Circuits (Study Groups XII and XV) 22 pp. 22 pp.

—**Voice Band Data Transmission**
CCITT RECMN G.113-89. Transmission Impairments—General Characteristics of International Telephone Connections and Circuits (Study Groups XII and XV) 22 pp. 22 pp.

—**Voice Band Data Transmission—Bit Rate Encoding**
CCITT RECMN G.724-89. Characteristics of a 48-Channel Low Bit Rate Encoding Primary Multiplex Operating at 1544 kbit/s—General Aspects of Digital Transmission Systems; Terminal Equipments (Study Groups XV and XVIII) 6 pp. 6 pp.

Adaptive Differential Pulse Code Modulation
Use: Adaptive Delta Pulse Code Modulation

Adaptive Equalizers
See Also: Telecommunication Equipment; Transmission Systems

—**Digital Microwave Radios**
IEC 835 Pt 2-8-93. Methods of Measurement for Equipment Used in Digital Microwave Radio Transmissions Systems Part 2: Measurements on Terrestrial Radio-Relay Systems Section 8: Adaptive Equalizer First Edition. 44 pp.

Adaptors
Use: Adapters (Fittings)

Add On Conference
See Also: Conference Calling; Telephone Services

—**Integrated Services Digital Networks**
CENELEC PRETS 300 183-91. Integrated Services Digital Network (ISDN); Conference Call Add-on (CONF) Supplementary Service Service Description (T/NA1 (89)25). 17 pp.
CENELEC PRETS 300 183-92. Integrated Services Digital Network (ISDN); Conference Call, Add-On (CONF) Supplementary Service Service Description. 20 pp.
CENELEC ETS 300 183-92. Integrated Services Digital Network (ISDN); Conference Call, Add-on (CONF) Supplementary Service Service Description. 20 pp.
CENELEC PRETS 300 184-91. Integrated Services Digital Network (ISDN); Conference Call Add-on (CONF) Supplementary Service Functional Capabilities and Information Flows (T/S 22-12). 41 pp.
CENELEC PRETS 300 184-92. Integrated Services Digital Network (ISDN); Conference Call, Add-on (CONF) Supplementary Service Functional Capabilities and Information Flows. 40 pp.
CENELEC ETS 300 184-93. Integrated Services Digital Network (ISDN); Conference Call, Add-on (CONF) Supplementary Service Functional Capabilities and Information Flows. 40 pp.
CENELEC PRETS 300 185-91. Integrated Services Digital Network (ISDN); Conference Call Add-on (CONF) Supplementary Service Digital Subscriber Signalling System No. One (DSS1) Protocol (T/S 46-33J1). 38 pp.
CENELEC PRETS 300 185-92. Integrated Services Digital Network (ISDN); Conference Call, Add-on (CONF) Supplementary Service Digital Subscriber Signalling System No. One (DSS1) Protocol. 44 pp.
CENELEC ETS 300 185-93. Integrated Services Digital Network (ISDN); Conference Call, Add-on (CONF) Supplementary Service Digital Subscriber Signalling System No. One (DSS1) Protocol. 44 pp.
ETSI ETS 300 183-92. Integrated Services Digital Network (ISDN); Conference Call, Add-on (CONF) Supplementary Service Service Description. 20 pp.
ETSI PRETS 300 183-91. Integrated Services Digital Network (ISDN); Conference Call Add-on (CONF) Supplementary Service Service Description (T/NA1 (89)25). 17 pp.

Add On Conference (Cont.)
—Integrated Services Digital Networks (Cont.)
ETSI ETS 300 184-93. Integrated Services Digital Network (ISDN); Conference Call, Add-on (CONF) Supplementary Service Functional Capabilities and Information Flows. 40 pp.
ETSI PRETS 300 184-92. Integrated Services Digital Network (ISDN); Conference Call, Add-on (CONF) Supplementary Service Functional Capabilities and Information Flows. 40 pp.
ETSI PRETS 300 184-91. Integrated Services Digital Network (ISDN); Conference Call Add-on (CONF) Supplementary Service Functional Capabilities and Information Flows (T/S 22-12). 41 pp.
ETSI ETS 300 185-93. Integrated Services Digital Network (ISDN); Conference Call, Add-on (CONF) Supplementary Service Digital Subscriber Signalling System No. One (DSS1) Protocol. 44 pp.
ETSI PRETS 300 185-92. Integrated Services Digital Network (ISDN); Conference Call, Add-on (CONF) Supplementary Service Digital Subscriber Signalling System No. One (DSS1) Protocol. 44 pp.
ETSI PRETS 300 185-91. Integrated Services Digital Network (ISDN); Conference Call Add-on (CONF) Supplementary Service Digital Subscriber Signalling System No. One (DSS1) Protocol (T/S 46-33J1). 38 pp.

Adding Machines
See Also: Calculators; Office Machines
—Glossaries
BSI BS 5478: Part 4-85. 1985 Calculators and Adding Machines Part 4: Glossary of Terms for Calculators. 18 pp.
—Keyboards
BSI BS 5478: Part 1-77. 1977 Calculators and Adding Mahchines Part 1:Numeric Section of Ten-Key Keyboards for Calculators and Adding Machines. 3 pp.
BSI BS 5478: Part 2-77. 1977 Calculators and Adding Machines Part 2: Layout of Function Keys on Adding Machines. 4 pp.
CNS C5119-81. 10-Key Keyboard for Adding and Calculating Machines (Jul)(7649).
CNS C5195-85. Calculating Machines—Numeric Section of Ten-Key Keyboards (Mar)(11219).
CNS C5196-85. Calculating Machines—Keytop and Printed or Displayed Symbols (Mar)(11220).
ISO 1092-74. Adding Machines and Calculating Machines—Numeric Section of Ten-Key Keyboards First Edition. 3 pp.
ISO 3792-76. Adding Machines—Layout of Function Keyboard First Edition. 4 pp.
JIS B 9517-90. Keyboards Arrangement for Calculating Machines with Ten-Key. 6 pp.
JTC1 1092-74. Adding Machines and Calculating Machines—Numeric Section of Ten-Key Keyboards First Edition. 3 pp.
JTC1 3792-76. Adding Machines—Layout of Function Keyboard First Edition. 4 pp.
OSI ISO 1092-74. Adding Machines and Calculating Machines—Numeric Sectionb of Ten-Key Keyboards. 3 pp.
OSI ISO 3792-76. Adding Machines—Layout of Function Keyboard. 4 pp.
—Papers
CGSB CAN/CGSB-53.24-M88. Tape, Paper, Computing Machine. 8 pp.
—Symbols
BSI BS 5478: Part 1-77. 1977 Calculators and Adding Mahchines Part 1:Numeric Section of Ten-Key Keyboards for Calculators and Adding Machines. 3 pp.
BSI BS 5478: Part 3-82. 1982 Calculators and Adding Machines Part 3: Specification for Keytop and Printed or Displayed Symbols. 4 pp.
ISO 1092-74. Adding Machines and Calculating Machines—Numeric Section of Ten-Key Keyboards First Edition. 3 pp.
JIS B 9516-90. Keytop and Printed or Displayed Symbols for Adding Machines. 5 pp.
JTC1 1092-74. Adding Machines and Calculating Machines—Numeric Section of Ten-Key Keyboards First Edition. 3 pp.
OSI ISO 1092-74. Adding Machines and Calculating Machines—Numeric Sectionb of Ten-Key Keyboards. 3 pp.
SAA AS 1412.1-76. Adding and Calculating Machines—Part 1: Basic Informative Symbols. 4 pp.

Additional Signaling
See Also: Telephone Services
CEPT T/CS 20-26-81. Caracteristiques Complement De Signalisation. 2 pp.
CEPT T/CS 20-26 E-81. Additional Signalling. 2 pp.

Additions, Cement
Use: Cement Additives

Additives
Scope Note: For additional listings, use a more specific term *See Also:* Admixtures; Antifreezes; Antiicing Additives; Antioxidants; Antistatic Agents; Cement Additives; Coatings; Corrosion Inhibitors; Diluents; Dopes; Extenders; Fire Retardants; Food Additives; Furnish; Gelling Agents; Lubricants; Lubricating Oil Additives; Master Alloys; Naphtha; Paint Thinners; Paints; Pigments; Plasticizers; Pour Point Depressants; Solvents; Stabilizers (Agents)
—Animal Feed
EC 87/153/EEC-87. Council Directive Fixing Guidelines for the Assessment of Additives in Animal Nutrition. 10 pp.
EC 92/113/EEC-92. Commission Directive Amending Council Directive 70/524/EEC Concerning Additives in Feedingstuffs. 2 pp.
—Petroleum—Visual Inspection
CNS K6267-68. Method of Test for Appearance of Petroleum Products and Additives (Oct)(2914)(R 1973).

Address Codes
See Also: Called Line Identification; Calling Line Identification; Codes; Destination Indicators (Telegraphy); End of Address; Location Registers; Mobile Global Titles; Originator/Recipient Addresses (Message Handling); Signaling Connection Control Part; Subaddressing; Telex Destination Codes
—FAX Communications
CCITT RECMN F.1-89. Operational Provisions for the International Public Telegram Service—Telegraph and Mobile Services Operations and Quality of Service (Study Group I) 54 pp. 54 pp.
—FAX Communications—Store and Forward Mode
CCITT RECMN F.162-89. Operational Requirements of an International Store-and-Forward Facsimile Switching Service (Comfax)—Telematic, Data Transmission and Teleconference Services —Operations and Quality of Service—(Study Group I) 7 pp. 7 pp.
—Magnetic Tapes
BSI BS 6288: Part 10-89. 1989 Amd 1 Magnetic Tape Sound Recording and Reproducing Systems Part 10: Specification for Time and Address Codes (IEC 94-10: 1988) (AMD 6405) August 30, 1991. 11 pp.
IEC 94 Pt 10-88. Magnetic Tape Sound Recording and Reproducing Systems Part 10: Time and Address Codes First Edition. 19 pp.
IEC 94 Pt 11-88. Magnetic Tape Sound Recording and Reproducing Systems Part 11: Address Code for Compact Cassettes First Edition. 20 pp.
—Maritime Mobile Services—Radiotelegraphy
CCITT RECMN F.110-89. Operational Provisions for the Maritime Mobile Service—Telegraph and Mobile Services Operations and Quality of Service (Study Group I) 21 pp (Same as Recmn E.200). 21 pp.
—Maritime Mobile Services—SELCAL Services
CCIR RECMN 493-4-90. Digital Selective-Calling System for Use in the Maritime Mobile Service—Section 8B—Maritime Mobile Service; Telegraphy and Related Subjects. 26 pp.
CCIR RECMN 493-5-92. Digital Selective-Calling System for Use in the Maritime Mobile Service—Section 8B—Maritime Mobile Service; Telegraphy and Related Subjects. 24 pp.
—Phototelegraphy
CCITT RECMN F.80-91. Basic Requirements for Interworking Relations Between the International Telex Service and Other Services (Study Group I) 6 pp. 6 pp.
—Routing—Register Signaling—CCITT R1 Signaling Systems
CCITT RECMN Q.324-89. Analysis of Address Information for Routing—Specifications of Signalling Systems R1 and R2 (Study Group XI) 1 pp. 1 p.
—Telegraph Repeaters
CCITT RECMN F.31-89. Telegram Retransmission System—Telegraph and Mobile Services Operations and Quality of Service (Study Group I) 12 pp. 12 pp.
—Telegraphy
CCITT RECMN F.1-89. Operational Provisions for the International Public Telegram Service—Telegraph and Mobile Services Operations and Quality of Service (Study Group I) 54 pp. 54 pp.

Address Codes (Cont.)
—Teletex Communications
CCITT RECMN F.1-89. Operational Provisions for the International Public Telegram Service—Telegraph and Mobile Services Operations and Quality of Service (Study Group I) 54 pp. 54 pp.
—Telex Communications
CCITT RECMN F.1-89. Operational Provisions for the International Public Telegram Service—Telegraph and Mobile Services Operations and Quality of Service (Study Group I) 54 pp. 54 pp.
—Telex Communications—Store and Forward Units
CCITT RECMN F.72-89. International Telex Store and Forward—General Principles and Operational Aspects—Telegraph and Mobile Services Operations and Quality of Service (Study Group I) 18 pp. 18 pp.

Address Masters
Use: Address Plates

Address Plates
See Also: Office Machines
BSI BS 5560-78. 1978 Office Machines. Line and Character Capacity of Address Masters. 2 pp.
ISO 3883-77. Office Machines—Line and Character Capacity of Address Masters First Edition. 3 pp.
JTC1 3883-77. Office Machines—Line and Character Capacity of Address Masters First Edition. 3 pp.
OSI ISO 3883-77. Office Machines—Line and Character Capacity of Address Masters. 3 pp.

Addressing Machines
See Also: Office Machines; Postal Addressing
ISO 3883-77. Office Machines—Line and Character Capacity of Address Masters First Edition. 3 pp.
JTC1 3883-77. Office Machines—Line and Character Capacity of Address Masters First Edition. 3 pp.
OSI ISO 3883-77. Office Machines—Line and Character Capacity of Address Masters. 3 pp.
—Glossaries
BSI BS 6191: Part 3-82. 1982 Mail Processing Machines Part 3: Glossary of Terms for Addressing Machines. 14 pp.
ISO 5138 Sec 03-81. Office Machines—Vocabulary—Section 03: Addressing Machines First Edition. 31 pp.
JTC1 5138 Sec 03-81. Office Machines—Vocabulary—Section 03: Addressing Machines First Edition. 31 pp.
OSI ISO 5138-3. Office Machines—Vocabulary—Section 03: Addressing Machines. 31 pp.
—Ribbons—Cotton
CGSB CAN/CGSB-53.97-M88. Cotton Ribbon for Address Plate Machine. 8 pp.

Addressing Systems
See Also: Telecommunication Equipment; Telecommunication Systems
—Connectionless Broadband Data Service
CENELEC PRETS 300 217-3-91. Network Aspects (NA); Connectionless Broadband Data Service (CBDS) Part 3: Definition of Supplementary Services (DE/NA-53201-3). 9 pp.
CENELEC PRETS 300 217-3-92. Network Aspects (NA); Connectionless Broadband Data Service (CBDS) Part 3: Definition of Supplementary Services. 8 pp.
CENELEC ETS 300 217-3-92. Network Aspects (NA); Connectionless Broadband Data Service (CBDS) Part 3: Definition of Supplementary Services. 8 pp.
CENELEC PRETS 300 217-4-91. Network Aspects (NA); Connectionless Broadband Data Service (CBDS) Part 4: Address Screening Supplementary Services (DE/NA-53201-4). 11 pp.
CENELEC PRETS 300 217-4-92. Network Aspects (NA); Connectionless Broadband Data Service (CBDS) Part 4: Address Screening Supplementary Service. 10 pp.
CENELEC ETS 300 217-4-92. Network Aspects (NA); Connectionless Broadband Data Service (CBDS) Part 4: Address Screening Supplementary Service. 10 pp.
ETSI ETS 300 217-3-92. Network Aspects (NA); Connectionless Broadband Data Service (CBDS) Part 3: Definition of Supplementary Services. 8 pp.
ETSI PRETS 300 217-3-92. Network Aspects (NA); Connectionless Broadband Data Service (CBDS) Part 3: Definition of Supplementary Services. 8 pp.
ETSI PRETS 300 217-3-91. Network Aspects (NA); Connectionless Broadband Data Service (CBDS) Part 3: Definition of Supplementary Services (DE/NA-53201-3). 9 pp.
ETSI ETS 300 217-4-92. Network Aspects (NA); Connectionless Broadband Data Service (CBDS) Part 4: Address Screening Supplementary Service. 10 pp.

INTERNATIONAL AND NON-U.S. NATIONAL STANDARDS
SUBJECT INDEX

Adhesion

Addressing Systems (Cont.)
—**Connectionless Broadband Data Service** (Cont.)

ETSI PRETS 300 217-4-92. Network Aspects (NA); Connectionless Broadband Data Service (CBDS) Part 4: Address Screening Supplementary Service. 10 pp.

ETSI PRETS 300 217-4-91. Network Aspects (NA); Connectionless Broadband Data Service (CBDS) Part 4: Address Screening Supplementary Services (DE/NA-53201-4). 11 pp.

—**European Digital Cordless Telecommunications**

CENELEC PRETS 300 175-6-91. Radio Equipment and Systems Digital European Cordless Telecommunications Common Interface Part 6: Identities and Addressing (DE/RES 3001-6). 40 pp.

CENELEC PRETS 300 175-6-92. Radio Equipment and Systems (RES); Digital European Cordless Telecommunications (DECT) Common Interface Part 6: Identities and Addressing. 130 pp.

CENELEC ETS 300 175-6-92. Radio Equipment and Systems (RES); Digital European Cordless Telecommunications (DECT) Common Interface Part 6: Identities and Addressing. 40 pp.

CENELEC PRI-ETS 300 176-91. Radio Equipment and Systems Digital European Cordless Telecommunications Approval Test Specification (DI/RES 3002). 109 pp.

CENELEC PRI-ETS 300 176-92. Radio Equipment and Systems (RES); Digital European Cordless Telecommunications (DECT) Approval Test Specification. 117 pp.

CENELEC I-ETS 300 176-92. Radio Equipment and Systems (RES); Digital European Cordless Telecommunications (DECT) Approval Test Specification. 118 pp.

ETSI ETS 300 175-6-92. Radio Equipment and Systems (RES); Digital European Cordless Telecommunications (DECT) Common Interface Part 6: Identities and Addressing. 40 pp.

ETSI PRETS 300 175-6-92. Radio Equipment and Systems (RES); Digital European Cordless Telecommunications (DECT) Common Interface Part 6: Identities and Addressing. 130 pp.

ETSI PRETS 300 175-6-91. Radio Equipment and Systems Digital European Cordless Telecommunications Common Interface Part 6: Identities and Addressing (DE/RES 3001-6). 40 pp.

ETSI I-ETS 300 176-92. Radio Equipment and Systems (RES); Digital European Cordless Telecommunications (DECT) Approval Test Specification. 118 pp.

ETSI PRI-ETS 300 176-92. Radio Equipment and Systems (RES); Digital European Cordless Telecommunications (DECT) Approval Test Specification. 117 pp.

ETSI PRI-ETS 300 176-91. Radio Equipment and Systems Digital European Cordless Telecommunications Approval Test Specification (DI/RES 3002). 109 pp.

—**Integrated Services Digital Networks**

BSI BS ISO 7498/3-89. 1989 Information Processing Systems—Open Systems Interconnection—Basic Reference Model 7498/3: Naming and Addressing. 28 pp.

CCITT RECMN I.330-89. ISDN Numbering and Addressing Principles—Integrated Services Digital Network (ISDN)—Overall Network Aspects and Functions, ISDN User-Network Interfaces (Study Group XVIII) 8 pp. 8 pp.

CENELEC ETR 006-90. Network Aspects; Numbering and Addressing for the Memorandum of Understanding (MOU) on Integrated Services Digital Network (ISDN) (Priorities 1 and 2). 15 pp.

CEPT T/GSI 03-07-85. Numerotation Et Adressage. 1 p.

CEPT T/GSI 03-07 E-85. Numbering and Addressing. 1 p.

CEPT T/N 21-02 E-88. Numbering and Addressing for Gap Phase II. 5 pp.

ETSI ETR 006-90. Network Aspects; Numbering and Addressing for the Memorandum of Understanding (MOU) on Integrated Services Digital Network (ISDN) (Priorities 1 and 2). 15 pp.

—**ISDN—Network Layer—Numbering Plans**

CCITT RECMN I.334-89. Principles Relating ISDN Numbers/Subaddresses to the OSI Reference Model Network Layer Addresses—Integrated Services Digital Network (ISDN)—Overall Network Aspects and Functions, ISDN User-Network Interfaces (Study Group XVIII) 5 pp. 5 pp.

—**ISDN—Network Layer—Subaddressing**

CCITT RECMN I.334-89. Principles Relating ISDN Numbers/Subaddresses to the OSI Reference Model Network Layer Addresses—Integrated Services Digital Network (ISDN)—Overall Network Aspects and Functions, ISDN User-Network Interfaces (Study Group XVIII) 5 pp. 5 pp.

Addressing Systems (Cont.)
—**Open Systems Interconnection**

BSI BS 7306-01. 1990 Amd 1 Procedures for the Operation of the UK Scheme for the Allocation of ISO DCC Format OSI NSAP Addresses (Including the Operation of the UK Registration Authority) (AMD 7652) May 15, 1993 (S). 17 pp.

BSI BS ISO/IEC TR 10730-93. 1993 Information Technology—Open Systems Interconnection—Tutorial on Naming and Addressing (S). 43 pp.

CCITT RECMN X.650-92. Open Systems Interconnection (OSI)—Reference Model for Naming and Addressing (Study Group VII) 32 pp. 32 pp.

CSA CAN/CSA-Z243. 100.3-92. Information Processing Systems—Open Systems Interconnection—Basic Reference Model—Part 3: Naming and Addressing (ISO 7498-3:1989); (Gen Instr 1). 38 pp.

ECMA ECMA-TR 20-84. Layer 4 to 1 Addressing. 25 pp.

IEC 8348 Draft AMD 5. Information Processing Systems—Data Communications—Network Service Definition Amendment 5: Group Network Addressing; (1993) (CCITT RECMN X.213) ***CD-ROM ONLY*** 9 pp.

IEC TR10730-93. Information Technology—Open Systems Interconnection—Tutorial on Naming and Addressing First Edition. 40 pp.

ISO 7498 Pt 3-89. Information Processing Systems—Open Systems Interconnection—Basic Reference Model—Part 3: Naming and Addressing First Edition. 26 pp.

ISO 8348 Draft AMD 5. Information Processing Systems—Data Communications—Network Service Definition Amendment 5: Group Network Addressing; (1993) (CCITT RECMN X.213) ***CD-ROM ONLY*** 9 pp.

ISO TR10730-93. Information Technology—Open Systems Interconnection—Tutorial on Naming and Addressing First Edition. 40 pp.

JIS X 5005-90. Information Processing Systems—Open Systems Interconnection—Basic Reference Model—Part 3:Naming and Addressing (ISO 7498/3:1989).

JTC1 7498 Pt 3-89. Information Processing Systems—Open Systems Interconnection—Basic Reference Model—Part 3: Naming and Addressing First Edition. 26 pp.

JTC1 8348 Draft AMD 5. Information Processing Systems—Data Communications—Network Service Definition Amendment 5: Group Network Addressing; (1993) (CCITT RECMN X.213) ***CD-ROM ONLY*** 9 pp.

JTC1 TR 10730-93. Information Technology—Open Systems Interconnection—Tutorial on Naming and Addressing First Edition. 40 pp.

OSI ISO/IEC DIS 7498-3-88. Information Processing Systems—Open Systems Interconnection—Basic Reference Model Part 3: Naming and Addressing.

OSI ISO 8348 DAD 2-87. Information Processing Systems—Data Communications—Network Service Definition Addendum 2: Network Layer Addressing.

SAA MP59-91. Naming and Addressing in the Australian OSI Environment. 9 pp.

SNZ NZS/ISO 7498. 3-88. Information Processing Systems—Open Systems Interconnection—Basic Reference Model Part 3: Naming and Addressing. 21 pp.

—**Open Systems Interconnection—NATO Reference Model**

NATO STANAG 4250 Pt3 ED1 AMD0-00. (Draft) NATO Reference Model for Open System Interconnection Part 3: Addressing. 6 pp.

—**Packet Switched Networks—ISDN**

CENELEC ETR 020-91. Network Aspects Numbering and Addressing for X.31 Services. 17 pp.

ETSI ETR 020-91. Network Aspects Numbering and Addressing for X.31 Services. 17 pp.

—**PISN—Network Layer—Numbering Plans**

IEC DIS 11571-92. Information Technology—Telecommunications and Information Exchange Between Systems—Numbering and Sub-Addressing in Private Integrated Services Networks ***CD-ROM ONLY***. 23 pp.

ISO DIS 11571-92. Information Technology—Telecommunications and Information Exchange Between Systems—Numbering and Sub-Addressing in Private Integrated Services Networks ***CD-ROM ONLY***. 23 pp.

JTC1 DIS11571-92. Information Technology—Telecommunications and Information Exchange Between Systems—Numbering and Sub-Addressing in Private Integrated Services Networks ***CD-ROM ONLY***. 23 pp.

—**PISN—Network Layer—Subaddressings**

IEC DIS 11571-92. Information Technology—Telecommunications and Information Exchange Between Systems—Numbering and Sub-Addressing in Private Integrated Services Networks ***CD-ROM ONLY***. 23 pp.

Addressing Systems (Cont.)
—**PISN—Network Layer—Subaddressings** (Cont.)

ISO DIS 11571-92. Information Technology—Telecommunications and Information Exchange Between Systems—Numbering and Sub-Addressing in Private Integrated Services Networks ***CD-ROM ONLY***. 23 pp.

JTC1 DIS11571-92. Information Technology—Telecommunications and Information Exchange Between Systems—Numbering and Sub-Addressing in Private Integrated Services Networks ***CD-ROM ONLY***. 23 pp.

—**Private Switched Networks**

CENELEC PRETS 300 189-91. Addressing in Private Telecommunications Networks (Standard ECMA-155. June 1991). 31 pp.

CENELEC PRETS 300 189-92. Private Telecommunications Network (PTN); Addressing. 30 pp.

CENELEC ETS 300 189-92. Private Telecommunications Network (PTN); Addressing. 30 pp.

ETSI ETS 300 189-92. Private Telecommunication Network (PTN); Addressing. 30 pp.

ETSI PRETS 300 189-92. Private Telecommunications Network (PTN); Addressing. 30 pp.

ETSI PRETS 300 189-91. Addressing in Private Telecommunications Networks (Standard ECMA-155. June 1991). 31 pp.

—**Universal Personal Telecommunication Services**

CENELEC ETR 055-9-92. Universal Personal Telecommunication (UPT); the Service Concept Part 9: Service Requirements on UPT Numbering, Addressing and Identification. 9 pp.

ETSI ETR 055-9-92. Universal Personal Telecommunication (UPT); the Service Concept Part 9: Service Requirements on UPT Numbering, Addressing and Identification. 9 pp.

ADF
Use: Automatic Direction Finders

ADF Receivers
Use: Automatic Direction Finders

Adhesion
See Also: Adhesion Testing; Adhesive Bonding; Adhesive Strength; Adhesives

—**Belt Conveyors**

SAA AS 1334.7-82. Methods of Testing Conveyor and Elevator Belting—Part 7: Determination of Ply Adhesion of Conveyor Belting. 2 pp.

—**Designations**

ISO 10365-92. Adhesives—Designation of Main Failure Patterns First Edition. 6 pp.

—**Glossaries**

CNS K3036-80. Standard Definitions of Terms Relating to Adhesive (Mar)(5345).

JIS K 6800-85. Glossary of Terms Used in Adhesives and Adhesion.

—**Paperboard**

CNS P3047-81. Method of Test for Ply Adhesion of Paperboard (Apr)(7297).

JIS P 8139-78. Testing Method for Ply Adhesion of Paperboard (R 1989). 8 pp.

Adhesion Testing
Use For: Scale Adhesion Test; Static Adhesion Testing; Tackiness *See Also:* Adhesion; Fusion (Melting); Physical Testing; Shear Testing; Tape Testing

—**Adhesive Tapes**

CPPA D.21H-88. Wet Adhesiveness of Gummed Paper Tape. 2 pp.

—**Anaerobic Adhesives to Metal**

BSI BS 5350: Part G3-87. 1987 Methods of Test for Adhesives Group G: Tests for Anaerobic Adhesives Part G3: Determination of Ability of Anaerobic Adhesives to Set on Metal Surfaces. 4 pp.

—**Anodic Coatings**

AECMA PREN3002-90. Chromic Acid Anodizing Testing of Adhesives. 8 pp.

—**Aperture Cards to Protection Sheets**

BSI BS 6247-82. 1982 Adhesion of Aperture-Card Protection Sheets. 8 pp.

ISO 6343-81. Micrographics—Unitized Microfilm Carrier (Aperture Card)—Determination of Adhesion of Protection Sheet to Aperture Adhesive First Edition. 6 pp.

INDUSTRY STANDARDS

INTERNATIONAL AND NON-U.S. NATIONAL STANDARDS
SUBJECT INDEX

Adhesion

Adhesion Testing (Cont.)

—**Autoclave Tape to Tray Wrap Material**
BSI BS 6988-89. 1989 Method for Determination of the Adhesion of Autoclave Tape to Tray Wrap Material. 4 pp.

—**Belt Conveyors**
ISO 252-88. Conveyor Belts—Ply Adhesion Between Constitutive Elements—Test Methods and Requirements Second Edition. 10 pp.
ISO 8094-84. Steel Cord Conveyor Belts—Adhesion Strength Test of the Cover to the Core Layer First Edition. 4 pp.

—**Bituminous Coatings**
DIN ENGL 1996 Pt 19-84. Testing of Asphalt; Determination of Extensibility and Adhesion Using a Rabe Joint Model (May). 6 pp.
DIN ENGL 52006 Pt 1-80. Testing of Bituminous Binders; Effect of Water on Binder Coatings; Bitumen Emulsion Binder Coating (Dec). 2 pp.
DIN ENGL 52006 Pt 2-80. Testing of Bituminous Binders; Effect of Water on Binder Coatings; Rapid Curing Cutback or Cold Tar Binder Coating (Dec). 2 pp.
DIN ENGL 52006 Pt 3-85. Bitumen and Coal Tar Pitch; Effect of Water on Binder Coatings; Testing of Fluxed Bitumen Binder Coatings (Sept). 2 pp.
MOD UK M 3004/67. Protective PX-2 and Composition Rust Preventative Type B.

—**Brake Linings to Ferroalloys**
BSI BS AU 180: Part 6-82. 1982 Brake Linings Part 6: Method for Assessment of Seizure to Ferrous Mating Surface Due to Corrosion. 4 pp.
ISO 6315-80. Road Vehicles—Brake Linings—Seizure to Ferrous Mating Surface Due to Corrosion—Test Procedure First Edition. 4 pp.

—**Brake Linings to Ferroalloys—Automotive**
CNS D3178-87. Method of Test of Seizure to Ferrous Mating Surface Due to Corrosion for Brake Linings and Pads of Automobiles (Dec)(12174).
JIS D 4414-86. Test Procedure of Seizure to Ferrous Mating Surface Due to Corrosion for Brake Linings and Pads of Automobiles. 8 pp.

—**Brake Pads to Ferroalloys—Automotive**
CNS D3178-87. Method of Test of Seizure to Ferrous Mating Surface Due to Corrosion for Brake Linings and Pads of Automobiles (Dec)(12174).
JIS D 4414-86. Test Procedure of Seizure to Ferrous Mating Surface Due to Corrosion for Brake Linings and Pads of Automobiles. 8 pp.

—**Cable Insulation**
CNS C6284-87. Method of Test for Fiber Optic Devices (FOTP-84 Jacket Self-Adhesion (Blocking) Test for Fiber Optic Cable) (Apr)(11906).
DIN VDE 0472 Pt 606-83. Testing of Cables, Wires and Flexible Cords. Wipe Resistance and Absence of Sticking (Aug). 8 pp.

—**Cable Insulation—Aircraft**
CEN PREN 3475 (Part 701)-92. Cables, Electrical, Aircraft Use Test Methods Part 701—Strippability and Adherence of Insulation to the Conductor. 3 pp.

—**Coatings**
CGSB 1-GP-71 METH 135.1-74. Methods of Testing Paints and Pigments Adhesion Knife Test. 1 p.
CGSB 1-GP-71 METH 135.2-78. Methods of Testing Paints and Pigments Adhesion Toughness and Adhesion. 1 p.
CGSB 1-GP-71 METH 135.8-74. Methods of Testing Paints and Pigments Adhesion Cross-Cut Adhesion Test. 2 pp.

—**Coatings to Furniture**
CNS S2135-86. Method of Test for Adhesion of Furniture Coatings (Aug)(11684).

—**Coatings to Steel**
CGSB 1-GP-71 METH 151-80. Methods of Testing Paints and Pigments Resistance to Cathodic Protection. 6 pp.

—**Copper**
BSI BS 5909-80. 1980 Scale Adhesion Test for Oxygen-Free Copper. 2 pp.
ISO 4746-77. Oxygen-Free Copper—Scale Adhesion Test First Edition. 3 pp.

—**Corrosion Inhibitors**
CGSB 31-GP-0A METH 7.3-57. Methods of Testing Corrosion-Prevention Materials and Processes Paint Adherence (Tape Test); (Amended September 1974) (Re-Edited March 1982). 1 p.

—**Corrosion Inhibitors to Metals**
CGSB 31-GP-0A METH 7.1-57. Methods of Testing Corrosion-Prevention Materials and Processes Low Temperature Adhesion. 2 pp.

Adhesion Testing (Cont.)

—**Corrugated Board**
CPPA D.10U-77. Ply Separation of Combined Container Board. 1 p.

—**Corrugated Fiberboard**
JIS Z 0402-88. Test Method for Adhesion of Corrugated Fibreboard. 7 pp.

—**Dusting Powders**
MOD UK DSTAN 68-119-92. Powder, Dusting for Rubber, Types: 1 and 2 Issue 1. 15 pp.

—**Elastomers**
DIN ENGL 53539-79. Testing of Elastomers; Evaluation of Tear Propagation, Adhesion and Peel Tests (Sept). 2 pp.

—**Electrodeposited Coatings**
BSI BS 5411: Part 10-81. 1981 Methods of Test for Metallic and Related Coatings Part 10: Review of Methods Available for Testing Adhesion of Electro-Deposited and Chemically Deposited Metallic Coatings on Metallic Substrates. 10 pp.
ISO 2819-80. Metallic Coatings on Metallic Substrates—Electrodeposited and Chemically Deposited Coatings—Review of Methods Available for Testing Adhesion Second Edition. 8 pp.

—**Enamels to Aluminum**
DIN ENGL 51173-86. Testing of Vitreous and Porcelain Enamels; Determination of the Adhesion of Enamels on Aluminium Under the Action of Electrolytic Solution (Spall Test) (Oct). 2 pp.

—**Fiberboard**
CPPA D.10U-77. Ply Separation of Combined Container Board. 1 p.

—**Finishes to Aluminum**
CGSB 1-GP-71 METH 135.6-78. Methods of Testing Paints and Pigments Adhesion Adhesion of Finishes to Primed Anodized Aluminum. 1 p.

—**Finishing Compounds to Leather**
CNS K6818-85. Method of Test for Adhesion of Finish to Leather (Dec)(11455).
JIS K 6555-79. Testing Method for Adhesion of Finish to Leathers.

—**Floor Polishes**
CGSB 25-GP-1M METH 24.1-89. Methods of Sampling and Testing Waxes and Polishes Tackiness; (Amendment 2 Feb 1989). 2 pp.

—**Floor Waxes**
CGSB 25-GP-1M METH 24.1-89. Methods of Sampling and Testing Waxes and Polishes Tackiness; (Amendment 2 Feb 1989). 2 pp.

—**Glazing Compounds to Aluminum**
CGSB CAN2-19.0-M77METH 9.3-78. Methods of Testing Putty, Caulking and Sealing Compounds Oil Bleeding and Spotting of Face Glazing Compounds. 1 p.

—**Gold Coatings (Made From Gold)**
BSI BS 6670: Part 5-86. 1986 Methods of Test for Electroplated Gold and Gold Alloy Coatings Part 5: Adhesion Tests. 4 pp.
ISO 3160 Pt 2-92. Watch Cases and Accessories—Gold Alloy Coverings—Part 2: Determination of Fineness, Thickness, Corrosion Resistance and Adhesion Second Edition. 12 pp.
ISO 4524 Pt 5-85. Metallic Coatings—Test Methods for Electrodeposited Gold and Gold Alloy Coatings—Part 5: Adhesion Tests First Edition. 3 pp.

—**Hoses—Plastic**
BSI BS 5173: Sec 103.1-92. (WITHDRAWN) 1992 Rubber and Plastics Hoses and Hose Assemblies Part 103: Physical Tests Section 103.1: Determination of Adhesion Between Components (ISO 8033: 1991) (NOW KNOWN AS BS EN 28033: 1993). 13 pp.
BSI BS 5173: Sec 103.1-86. 1986 Rubber and Plastics Hoses and Hose Assemblies Part 103: Physical Tests Section 103.1: Determination of Adhesion Between Components (Renumbered as BS EN 28033: 1993). 11 pp.
BSI BS EN 28033-93. 1993 Amd 1 Rubber and Plastics Hose —Determination of Adhesion Between Components (ISO 8033: 1991) (AMD 7563) August 15, 1993 (E). 16 pp.
CEN EN 28033-93. Rubber and Plastics Hose—Determination of Adhesion Between Components (ISO 8033: 1991). 12 pp.
ISO 8033-91. Rubber and Plastics Hose—Determination of Adhesion Between Components Second Edition (CEN EN 28033: 1993). 12 pp.
SAA AS 1180.4A-72. Methods of Test for Hose Made from Elastomeric Materials—Part 4A: Ply Adhesion—Dead Weight Method Reconfirmed 1988.

Adhesion Testing (Cont.)

—**Hoses—Plastic (Cont.)**
SAA AS 1180.4B-72. Methods of Test for Hose Made from Elastomeric Materials—Part 4B: Ply Adhesion—Autographic Method Corrig. Reconfirmed 1988.

—**Hoses—Rubber**
BSI BS 5173: Sec 103.1-92. (WITHDRAWN) 1992 Rubber and Plastics Hoses and Hose Assemblies Part 103: Physical Tests Section 103.1: Determination of Adhesion Between Components (ISO 8033: 1991) (NOW KNOWN AS BS EN 28033: 1993). 13 pp.
BSI BS 5173: Sec 103.1-86. 1986 Rubber and Plastics Hoses and Hose Assemblies Part 103: Physical Tests Section 103.1: Determination of Adhesion Between Components (Renumbered as BS EN 28033: 1993). 11 pp.
BSI BS EN 28033-93. 1993 Amd 1 Rubber and Plastics Hose —Determination of Adhesion Between Components (ISO 8033: 1991) (AMD 7563) August 15, 1993 (E). 16 pp.
CEN EN 28033-93. Rubber and Plastics Hose—Determination of Adhesion Between Components (ISO 8033: 1991). 12 pp.
ISO 8033-91. Rubber and Plastics Hose—Determination of Adhesion Between Components Second Edition (CEN EN 28033: 1993). 12 pp.
SAA AS 1180.4A-72. Methods of Test for Hose Made from Elastomeric Materials—Part 4A: Ply Adhesion—Dead Weight Method Reconfirmed 1988.
SAA AS 1180.4B-72. Methods of Test for Hose Made from Elastomeric Materials—Part 4B: Ply Adhesion—Autographic Method Corrig. Reconfirmed 1988.

—**Interlaminar Insulation—Magnetic Materials**
BSI BS 6404: Part 12-93. 1993 Magnetic Materials Part 12: Guide to Methods of Assessment of Temperature Capability of Interlaminar Insulation Coatings (IEC 404-12: 1992) (V). 16 pp.
IEC 404 Pt 12-92. Magnetic Materials Part 12: Guide to Methods of Assessment of Temperature Capability of Interlaminar Insulation Coatings First Edition. 28 pp.

—**Joint Sealants**
BSI BS 3712: Part 4-85. 1985 Building and Construction Sealants Part 4: Methods of Test for Adhesion in Peel, Tensile Extension and Recovery and Loss of Mass After Heat Ageing. 13 pp.
BSI BS EN 29046-91. 1991 Building Construction—Sealants—Determination of Adhesion/Cohesion Properties at Constant Temperature (ISO 9046: 1987). 10 pp.
CEN EN 29 046-90. Building Construction—Jointing Products—Determination of Adhesion Properties at Constant Temperatures. 9 pp.
DIN ENGL 52455 Pt 1-87. Testing of Building Construction Sealants; Determination of Adhesion/Cohesion After Conditioning in Standard Atmosphere, Water, or at Elevated Temperature (Apr). 4 pp.
DIN ENGL 52455 Pt 4-87. Testing of Building Construction Sealants; Determination of Adhesion/Cohesion When Subjected to an Extension/Compression Cycle at Alternating Temperature (Apr). 3 pp.
DIN ENGL EN 29046-91. Sealants in Building Construction; Determination of Adhesion/Cohesion Properties at Constant Temperature (ISO 9046: 1987) (May). 6 pp.
ISO 9046-87. Building Construction—Sealants—Determination of Adhesion/Cohesion Properties at Constant Temperature First Edition. 6 pp.
ISO 9047-89. Building Construction—Sealants—Determination of Adhesion/Cohesion Properties at Variable Temperatures First Edition. 6 pp.
ISO 10590-91. Building Construction—Sealants—Determination of Adhesion/Cohesion Properties at Maintained Extension After Immersion in Water First Edition. 6 pp.
ISO 10591-91. Building Construction—Sealants—Determination of Adhesion/Cohesion Properties After Immersion in Water First Edition. 6 pp.

—**Joint Sealants to Window Glazing**
DIN ENGL 52455 Pt 3-74. Testing of Materials for Joint and Glazing Seals in Building Construction; Adhesion and Extension Test; Exposure to Light (Sept). 2 pp.

—**Lacquers**
CNS K6386-76. Method of Test for Lacquer in Surface of Leather Shoes (Mar) (3922). 2 pp.
SAA AS 1580.408. 2-93. Paints and Related Materials—Methods of Test—Part 408.2: Adhesion—Knife Test (in Professional Package 39). 3 pp.
SAA AS 1580.408. 4-93. Paints and Related Materials—Methods of Test—Part 408.4: Adhesion (Cross-Cut) (in Professional Package 39A). 5 pp.

INTERNATIONAL AND NON-U.S. NATIONAL STANDARDS
SUBJECT INDEX
Adhesion

Adhesion Testing *(Cont.)*

—**Lacquers** *(Cont.)*
SNZ NZS/AS 1580. 408.2-84. Methods of Test for Paints and Related Materials Part 408.2: Adhesion—Knife Test (This is a Joint Standard with SAA AS 1580.408.2). 4 pp.
SNZ NZS/AS 1580. 408.4-80. Methods of Test for Paints and Related Materials Part 408.4: Adhesion (Cross-Cut) (This is a Joint Standard with SAA AS 1580.408.4). 4 pp.

—**Laminated Fabrics**
DIN ENGL 53530-81. Testing of Organic Materials; Separation Test on Fabric Plies Bonded Together (Feb). 4 pp.

—**Magnetic Tapes**
CNS C5051-80. Recommended Test Method Layer-to-Layer Adhesion of Magnetic Tape (Jul)(5761).

—**Metal Coatings (Made From Metal)**
BSI BS 5411: Part 10-81. 1981 Methods of Test for Metallic and Related Coatings Part 10: Review of Methods Available for Testing Adhesion of Electro-Deposited and Chemically Deposited Metallic Coatings on Metallic Substrates. 10 pp.
JIS H 8504-90. Methods of Adhesion Test for Metallic Coatings. 16 pp.

—**Metal Coatings (Made From Metal)—Aerospace**
BSI BS EN 2828-93. 1993 Adhesion Test for Metallic Coatings by Burnishing (S). 8 pp.
BSI BS EN 2830-93. 1993 Adhesion Test for Metallic Coatings by Shearing Action (S). 8 pp.
CEN PREN 2828-90. Adhesion Test for Metallic Coatings by Burnishing. 2 pp.
CEN EN 2828-93. Aerospace Series—Adhesion Test for Metallic Coatings by Burnishing. 3 pp.
CEN PREN 2830-90. Adhesion Test for Metallic Coatings by Shearing Action. 2 pp.
CEN EN 2830-93. Aerospace Series—Adhesion Test for Metallic Coatings by Shearing Action. 3 pp.

—**Metal Coatings to Cans**
CNS Z6069-83. Method of Test for Coating Materials of Metal Cans for Foods—Adhesive Test for Lacquered Film (Sep)(10585). 2 pp.

—**Paints**
BSI BS 3900: Part E10-79. (WITHDRAWN) 1979 Methods of Test for Paints Group E: Mechanical Tests on Paint Films Part E10: Pull-off Test for Adhesion (Superseded by BS EN 24624: 1993). 9 pp.
BSI BS EN 24624-93. 1993 Paints and Varnishes—Pull-off Test (ISO 4624: 1978) (Supersedes BS 3900: Part E10: 1979). 14 pp.
CEN EN 24624-92. Paints and Varnishes—Pull-off Test; (ISO 4624: 1978). 9 pp.
DIN ENGL EN 24624-92. Paints and Varnishes; Pull-Off Test for Adhesion (ISO 4624:1978) (Sep) (Supersedes DIN ISO 4624, June 1980 Edition). 11 pp.
ISO 4624-78. Paints and Varnishes—Pull-off Test for Adhesion First Edition; (CEN EN 24624:1992). 9 pp.
SAA AS 1580.408. 2-93. Paints and Related Materials—Methods of Test—Part 408.2: Adhesion—Knife Test (in Professional Package 39). 3 pp.
SAA AS 1580.408. 4-93. Paints and Related Materials—Methods of Test—Part 408.4: Adhesion (Cross-Cut) (in Professional Package 39A). 5 pp.
SNZ NZS/AS 1580. 408.1-84. Methods of Test for Paints and Related Materials Part 408.1: Adhesion—Paint Inspection Gauge (This is a Joint Standard with SAA AS 1580.408.1). 4 pp.
SNZ NZS/AS 1580. 408.2-84. Methods of Test for Paints and Related Materials Part 408.2: Adhesion—Knife Test (This is a Joint Standard with SAA AS 1580.408.2). 4 pp.
SNZ NZS/AS 1580. 408.4-80. Methods of Test for Paints and Related Materials Part 408.4: Adhesion (Cross-Cut) (This is a Joint Standard with SAA AS 1580.408.4). 4 pp.

—**Paints—Automotive**
BSI BS AU 148: Part 3-69. 1969 Methods of Test for Motor Vehicle Paints Part 3: Flexibility and Adhesion. 3 pp.

—**Paints to Conversion Coatings**
CGSB 31-GP-0A METH 7.2-57. Methods of Testing Corrosion-Prevention Materials and Processes Adhesion of Paint to Chemical Films; (Amended June 1974) (Re-Edited March 1982). 1 p.

—**Paper Laminates**
JIS Z 0219-57. Testing Method for Tackiness Resistance of Laminate Paper for Wrappings.

Adhesion Testing *(Cont.)*

—**Plastic Sheets**
BSI BS 2782:Pt8: METH 826A-92. 1992 Methods of Testing Plastics Part 8: Other Properties Method 826A: Determination of Adhesion of Print on Plastics Sheet. 7 pp.

—**Plastics to Fabrics**
ISO 2411-91. Rubber-or Plastics-Coated Fabrics—Determination of Coating Adhesion Second Edition. 10 pp.
SAA AS 1441.5-73. Methods of Test for Coated Fabrics—Part 5: Method for Determination of Coating and Ply Adhesion.

—**Plating—Shielded Conductor Cables—Aircraft**
CEN PREN 3475 (Part 507)-92. Cables, Electrical, Aircraft Use Test Methods Part 507—Adherence of Plating. 2 pp.

—**Plating—Stranded Conductors—Aircraft**
CEN PREN 3475 (Part 507)-92. Cables, Electrical, Aircraft Use Test Methods Part 507—Adherence of Plating. 2 pp.

—**Primers to Aluminum**
CGSB 1-GP-71 METH 135.5-78. Methods of Testing Paints and Pigments Adhesion Adhesion of Primer to Pretreated Anodized Aluminum. 1 p.
CGSB 1-GP-71 METH 135.7-78. Methods of Testing Paints and Pigments Adhesion Adhesion of Topcoated Primers; (Amended September 1980) (Re-Edited September 1982). 1 p.

—**Rubber to Fabrics**
ISO 2411-91. Rubber-or Plastics-Coated Fabrics—Determination of Coating Adhesion Second Edition. 10 pp.
ISO 4637-79. Rubber-Coated Fabrics—Determination of Rubber-to-Fabric Adhesion—Direct Tension Method First Edition. 5 pp.

—**Rubber to Rigid Materials**
DIN ENGL 53531 Pt 1-90. Determination of the Adhesion of Rubber to Rigid Materials by the One-Plate Method (Oct). 3 pp.
DIN ENGL 53531 Pt 2-90. Determination of the Adhesion of Rubber to Rigid Materials Using Conical Ended Cylinders (Aug). 3 pp.

—**Sandwich Structures**
DIN ENGL 53295-82. Testing of Sandwiches; Peel Test by Means of a Drum (Feb). 3 pp.

—**Sealants**
CGSB CAN2-19.0-M77METH 2.1-78. Methods of Testing Putty, Caulking and Sealing Compounds Tackiness of Sealing Compounds. 1 p.
CGSB CAN2-19.0-M77METH14.5-78. Methods of Testing Putty, Caulking and Sealing Compounds Adhesion in Extension Cycling. 1 p.
ISO 11431-93. Building Construction—Sealants—Determination of Adhesion/Cohesion Properties After Exposure to Artificial Light Through Glass First Edition. 7 pp.

—**Semiconductor Devices**
CNS C6043-88. Environmental Testing Methods and Endurance Testing Methods for Discrete Semiconductor Devices (Solderability Testing for Adhesion) (Jun)(5068).

—**Silver Coatings (Made From Silver)**
BSI BS 6669: Part 2-86. 1986 Methods of Test for Electroplated Silver and Silver Alloy Coatings Part 2: Adhesion Tests. 7 pp.
ISO 4522 Pt 2-85. Metallic Coatings—Test Methods for Electrodeposited Silver and Silver Alloy Coatings—Part 2: Adhesion Tests First Edition. 5 pp.

—**Soil to Agricultural Equipment**
DIN ENGL 19683 Pt 7-73. Methods of Soil Analysis for Water Management for Agricultural Purposes; Physical Laboratory Tests; Determination of Adhesion (Apr). 2 pp.

—**Soldering Fluxes—Residues**
ISO 9455 Pt 14-91. Soft Soldering Fluxes—Test Methods—Part 14: Assessment of Tackiness of Flux Residues First Edition. 7 pp.

—**Soles (Footwear)**
BSI BS 5131: Sec 5.1-90. 1990 Methods of Test for Footwear and Footwear Materials Part 5: Testing of Complete Footwear Section 5.1: Adhesion of Stuck-on and Moulded-on Soles. 6 pp.

Adhesion Testing *(Cont.)*

—**Test Specimens—Footwear**
BSI BS 5131: SUB SEC 1.1.3-76. (WITHDRAWN) 1976 Amd 1 Methods of Test for Footwear and Footwear Materials Part 1: Adhesives Section 1.1: Resistance of Adhesive Joints to Heat and to Peeling Subsection 1.1.3: Preparation of Test Assemblies for Adhesion Tests. 6 pp.
BSI BS 5131: Sec 1.6-79. 1979 Methods of Test for Footwear and Footwear Materials Part 1: Adhesives Section 1.6: Recommended Environmental Storage Conditions for Adhesive Joints Prior to Heat Resistance or Peeling Tests. 2 pp.

—**Tin Coatings (Made From Tin)**
ISO 2093-86. Electroplated Coatings of Tin—Specification and Test Methods Second Edition. 10 pp.

—**Varnishes**
BSI BS 3900: Part E10-79. (WITHDRAWN) 1979 Methods of Test for Paints Group E: Mechanical Tests on Paint Films Part E10: Pull-off Test for Adhesion (Superseded by BS EN 24624: 1993). 9 pp.
BSI BS EN 24624-93. 1993 Paints and Varnishes—Pull-off Test (ISO 4624: 1978) (Supersedes BS 3900: Part E10: 1979). 14 pp.
CEN EN 24624-92. Paints and Varnishes—Pull-off Test; (ISO 4624: 1978). 9 pp.
DIN ENGL EN 24624-92. Paints and Varnishes; Pull-Off Test for Adhesion (ISO 4624:1978) (Sep) (Supersedes DIN ISO 4624, June 1980 Edition). 11 pp.
ISO 4624-78. Paints and Varnishes—Pull-off Test for Adhesion First Edition; (CEN EN 24624:1992). 9 pp.
SAA AS 1580.408. 2-93. Paints and Related Materials—Methods of Test—Part 408.2: Adhesion—Knife Test (in Professional Package 39). 3 pp.
SAA AS 1580.408. 4-93. Paints and Related Materials—Methods of Test—Part 408.4: Adhesion (Cross-Cut) (in Professional Package 39A). 5 pp.
SNZ NZS/AS 1580. 408.2-84. Methods of Test for Paints and Related Materials Part 408.2: Adhesion—Knife Test (This is a Joint Standard with SAA AS 1580.408.2). 4 pp.
SNZ NZS/AS 1580. 408.4-80. Methods of Test for Paints and Related Materials Part 408.4: Adhesion (Cross-Cut) (This is a Joint Standard with SAA AS 1580.408.4). 4 pp.

—**Vinyl Coatings to Steel**
CGSB 1-GP-71 METH 135.4-78. Methods of Testing Paints and Pigments Adhesion Adhesion of Vinyl Coatings. 1 p.

—**Vinyl Primers to Steel**
CGSB 1-GP-71 METH 135.3-78. Methods of Testing Paints and Pigments Adhesion Adhesion of Vinyl Primer. 1 p.

—**Vulcanized Rubber to Fabrics**
BSI BS 903: Part A12-75. 1975 Methods of Testing Vulcanized Rubber: Part A12: Determination of the Adhesion Strength of Vulcanized Rubbers to Fabrics (Ply Adhesion). 8 pp.
BSI BS 903: Part A27-86. 1986 Methods of Testing Vulcanized Rubber Part A27: Determination of Rubber to Fabric Adhesion: Direct Tension Method. 7 pp.
ISO 36-93. Rubber, Vulcanized or Thermoplastic—Determination of Adhesion to Textile Fabric Second Edition. 6 pp.

—**Vulcanized Rubber to Metals**
BSI BS 903: Part A21-89. 1989 Methods of Testing Vulcanized Rubber Part A21: Determination of Rubber-To-Metal Bond Strength. 10 pp.
BSI BS 903: Part A37-87. 1987 Methods of Testing Vulcanized Rubber Part A37: Determination of Adhesion to and Corrosion of Metals. 10 pp.
CNS K6349-85. Method of Test for Adhesion of Vulcanized Rubber to Metal (Nov)(3558).
ISO 813-86. Rubber, Vulcanized—Determination of Adhesion to Metal—One-Plate Method Second Edition. 5 pp.
ISO 814-86. Rubber, Vulcanized—Determination of Adhesion to Metal—Two-Plate Method Second Edition. 6 pp.
SAA AS 1683.14-76. Methods of Test for Elastomers—Part 14: Adhesion of Vulcanized Rubber to Metal (Superseded (in Part) by AS 1683.14.1—1992 and AS 1683.14.2—1992).
SAA AS 1683.14.1-92. Methods of Test for Elastomers—Part 14: Rubber, Vulcanized—Determination of Adhesion to Metal—Part 14.1: One-Plate Method (ISO 813:1986) (Supersedes AS 1683.14—1976 (in Part)). 3 pp.
SAA AS 1683.14.2-92. Methods of Test for Elastomers—Part 14: Rubber, Vulcanized—Determination of Adhesion to Metal—Part 14.2: Two-Plate Method (ISO 814:1986) (Supersedes AS 1683.14—1976 (in Part)). 4 pp.

INDUSTRY STANDARDS

INTERNATIONAL AND NON-U.S. NATIONAL STANDARDS
SUBJECT INDEX
Adhesion

Adhesion Testing (Cont.)

—Vulcanized Rubber to Metals—Avoidance
ISO 6505-84. Rubber, Vulcanized—Determination of Adhesion to, and Corrosion of, Metals First Edition. 6 pp.

—Vulcanized Rubber to Rigid Materials
BSI BS 903: Part A40-88. 1988 Methods of Testing Vulcanized Rubber Part A40: Determination of Adhesion to Rigid Materials Using Conical Shaped Parts. 7 pp.
ISO 5600-86. Rubber—Determination of Adhesion to Rigid Materials Using Conical Shaped Parts Second Edition. 5 pp.

—Vulcanized Rubber to Tire Cords
BSI BS 903: Part A48-84. 1984 Methods of Testing Vulcanized Rubber Part A48: Determination of Static Adhesion to Textile Cord (H-Pull Test). 12 pp.
ISO 4647-82. Rubber, Vulcanized—Determination of Static Adhesion to Textile Cord—H-Pull Test First Edition. 10 pp.

—Vulcanized Rubber to Wire Rope
BSI BS 903: Part A56-89. 1989 Methods of Testing Vulcanized Rubber Part A56: Determination of Adhesion to Wire Cord (ISO 5603: 1986) (V). 16 pp.
ISO 5603-86. Rubber, Vulcanized—Determination of Adhesion to Wire Cord First Edition. 14 pp.

—Zinc Coatings (Made From Zinc)
DIN ENGL 50978-85. Testing of Metallic Coatings; Adherence of Hot-Dip Zinc Coatings (Oct). 3 pp.

—Zinc Coatings to Steel Wire
BSI BS 443-82. 1982 Amd 1 Testing Zinc Coatings on Steel Wire and for Quality Requirements (AMD 6158) December 22, 1989. 16 pp.

Adhesive Bonding
See Also: Adhesion; Bonding Agents
ISO 9653-91. Adhesives—Test Method for Shear Impact Strength of Adhesive Bonds First Edition. 8 pp.

—Flexible Materials—Peel Testing
ISO 11339-93. Adhesives—180 Degrees Peel Test for Flexable-to-Flexable Bonded Assemblies (T-Peel Test) First Edition. 6 pp.

—Metals
ISO 4588-89. Adhesives—Preparation of Metal Surfaces for Adhesive Bonding First Edition. 8 pp.

—Metals—Cleavage Strength—Impact Testing
ISO 11343-93. Adhesives—Determination of Dynamic Resistance to Cleavage of High Strength Adhesive Bonds Under Impact Conditions—Wedge Impact Method First Edition. 9 pp.

—Rubber—Aircraft
SBAC TS 65 ISSUE 1. Recommendations on the Adhesive Bonding of Rubbers to Other Substrates.

—Shear Strength
ISO 11003 Pt 1-93. Adhesives—Determination of Shear Behaviour of Structural Bonds—Part 1: Torsion Test Method Using Butt-Bonded Hollow Cylinders First Edition. 10 pp.
ISO 11003 Pt 2-93. Adhesives—Determination of Shear Behaviour of Structural Bonds—Part 2: Thick-Adherend Tensile-Test Method First Edition. 15 pp.

Adhesive Bonding Strength
Use: Adhesive Strength

Adhesive Joints
Use: Bonded Joints

Adhesive Papers
Use For: Adhesive Sheets See Also: Adhesive Tapes; Papers
CNS Z6086-87. Method of Test for Pressure Sensitive Tapes and Sheets (Mar)(11888).
JIS Z 0237-91. Testing Methods of Pressure Sensitive Adhesive Tapes and Sheets. 27 pp.

—Glossaries
JIS Z 0109-81. Glossary of Terms Used in Pressure Sensitive Adhesive Tapes and Sheets.

—Printing
CNS P2016-84. Mechanical Printing Paper (Oct)(1461). 2 pp.
CNS Z5113-82. Pressure Sensitive Adhesive Paper for Printing (Sep)(9449).
CNS Z6054-82. Method of Test for Pressure Sensitive Adhesive Paper for Printing (Sep)(9450).
JIS Z 1538-91. Pressure Sensitive Adhesive Papers for Printing. 10 pp.

Adhesive Sheets
Use: Adhesive Papers

Adhesive Strength
Use For: Adhesive Bonding Strength
See Also: Adhesion; Adhesives; Bonding Strength; Shear Strength; Shear Testing; Surface Properties; Tensile Testing
CNS K6447-80. Method of Measurement for Strength Development of Adhesive Bonds (Jan)(5137).
CNS K6498-80. Method of Test for Bonding Strength of Adhesives (General Rules) (May)(5604).
JIS K 6848-87. General Rules for Testing Methods of Bonding Strength of Adhesives. 14 pp.

—Adhesive Tapes
SAA AS 1635.3.1-74. Methods of Testing Pressure Sensitive Adhesive Tape—Part 3.1: Adhesion Strength. 2 pp.
SAA AS 1635.3.5-74. Methods of Testing Pressure Sensitive Adhesive Tape—Part 3.5: Adhesion Strength After Water Immersion. 1 p.
SAA AS 1635.3.6-74. Methods of Testing Pressure Sensitive Adhesive Tape—Part 3.6: Adhesion Strength to Own Backing. 1 p.
SAA AS 1635.10.1-74. Methods of Testing Pressure Sensitive Adhesive Tape—Part 10.1: Holding Power to Standard Panel. 1 p.

—Belt Conveyors
BSI BS 490: Sec 10.4-83. 1983 Conveyor and Elevator Belting Part 10: Testing for Physical Properties Section 10.4: Method for Determination of Adhesion Strength of Rubber and Plastics Belting of Textile Construction. 4 pp.
SNZ NZS/BS 490: Pt10:Sec10.4-83. Conveyor and Elevator Belting Part 10: Testing and Physical Properties Section 10.4: Method for Determination of Adhesion Strength of Rubber and Plastics Belting of Textile Construction. 2 pp.

—Cables (Electric)
DIN VDE 0472 Pt 618-83. Testing of Cables, Wires and Flexible Cords; Adhesive Strength (Jan). 6 pp.

—Ceiling Board
JIS A 1612-84. Testing Methods for Bonding Strength of Ceiling Boards Adhesives. 11 pp.

—Chemical Resistance
CNS K6503-80. Method of Test for Resistance of Adhesive Bonds to Chemical Substances (May)(5609).
JIS K 6858-74. Testing Methods for Resistance of Adhesive Bonds to Chemical Substances (R 1983). 8 pp.

—Cleavage Strength
BSI BS 5350: Part C1-86. 1986 Methods of Test for Adhesives Group C: Adhesively Bonded Joints: Mechanical Tests Part C1: Determination of Cleavage Strength of Adhesive Bonds. 6 pp.
CNS K6507-80. Methods of Test for Cleavage Strength of Adhesives (Jul)(5811).
JIS K 6853-77. Testing Method for Cleavage Strength of Adhesives.
JIS K 6853-73. Testing Method for Cleavage Strength of Adhesives. 5 pp.

—Coated Fabrics
BSI BS 903: Part A47-82. 1982 Methods of Testing Vulcanized Rubber Part A47: Analysis of Multi-Peak Traces Obtained in Determinations of Tear Strength and Adhesion Strength. 7 pp.
BSI BS 3424: Part 7-82. 1982 Testing Coated Fabrics Part 7: Method 9. Methods for Determination of Coating Adhesion Strength. 6 pp.
ISO 6133-81. Rubber and Plastics—Analysis of Multi-Peak Traces Obtained in Determinations of Tear Strength and Adhesion Strength First Edition. 5 pp.

—Communication Cables
DIN VDE 0472 Pt 618-83. Testing of Cables, Wires and Flexible Cords; Adhesive Strength (Jan). 6 pp.

—Cords (Electric)
DIN VDE 0472 Pt 618-83. Testing of Cables, Wires and Flexible Cords; Adhesive Strength (Jan). 6 pp.

—Creep Rupture Strength
BSI BS 5350: Part C7-90. 1990 Methods of Test for Adhesives Group C: Adhesively Bonded Joints: Mechanical Tests Part C7: Determination of Creep and Resistance to Sustained Application of Force. 4 pp.
JIS K 6859-80. Testing Method for Creep Rupture of Adhesive Bonds. 11 pp.

—Designations
ISO 10365-92. Adhesives—Designation of Main Failure Patterns First Edition. 6 pp.

Adhesive Strength (Cont.)

—Environmental Testing
JIS K 6860-74. General Recommended Practices for Atmospheric Exposure of Adhesive Bonds (R 1989). 8 pp.

—Fatigue (Materials)—Shear Testing
ISO 9664-93. Adhesives—Test Methods for Fatigue Properties of Structural Adhesives in Tensile Shear First Edition. 13 pp.

—Flexural Strength
JIS K 6856-77. Testing Methods for Flexural Strength of Adhesives.
JIS K 6856-73. Testing Methods for Flexural Strength of Adhesives. 7 pp.

—Gummed Tapes
JIS Z 0218-75. Testing Methods for Adhesive Strength of Gummed Tape.

—Heat Resistance—Footwear
BSI BS 5131: Sec 1.1-91. 1991 Methods of Test for Footwear and Footwear Materials Part 1: Adhesives Section 1.1: Resistance of Adhesive Joints to Heat (Creep Test). 9 pp.
BSI BS 5131: SUB SEC 1.1.1-76. (WITHDRAWN) 1976 Amd 1 Methods of Test for Footwear and Footwear Materials Part 1: Adhesives Section 1.1: Resistance of Adhesive Joints to Heat and to Peeling Subsec 1.1.1: Resist. to Heat (Creep Test) (Superseded by BS 5131: Section 1.1: 1991). 5 pp.
BSI BS 5131: Sec 1.3-91. 1991 Methods of Test for Footwear and Footwear Materials Part 1: Adhesives Section 1.3: Preparation of Test Assemblies Using Adhesives (Other Than Hot Melt Adhesives) for Heat Resistance (Creep) and Peel Tests. 15 pp.
BSI BS 5131: Sec 1.7-91. 1991 Methods of Test for Footwear and Footwear Materials Part 1: Adhesives Section 1.7: Preparation of Test Assemblies Using Hot Melt Adhesives for Heat Resistance (Creep) and Peel Tests. 8 pp.

—Impact Testing
BSI BS 5350: Part C4-86. 1986 Methods of Test for Adhesives Group C: Adhesively Bonded Joints: Mechanical Tests Part C4: Determination of Impact Resistance of Adhesive Bonds. 6 pp.

—Masking Tapes
SAA AS 1635.10.2-74. Methods of Testing Pressure Sensitive Adhesive Tape—Part 10.2: Holding Power of Masking Tape Amdt 1 February 1978. 1 p.

—Mortars
CEN PREN 1015-12-93. Methods of Test for Mortar for Masonry—Part 12: Determination of Adhesion of Hardened Rendering and Plastering Mortars. 11 pp.

—Peel Strength
AECMA PREN2243-02-80. Structural Adhesives—Test Methods—Part 2—Peel Metal-Metal (C7/SC3/B1). 9 pp.
AECMA PREN2243-03-80. Structural Adhesives—Test Methods—Part 3—Peeling Test Metal-Honeycomb Core (C7/SC3/B2). 9 pp.
BSI BS 5350: Part C9-90. 1990 Methods of Test for Adhesives Group C: Adhesively Bonded Joints: Mechanical Tests Part C9: Floating Roller Peel Test. 4 pp.
BSI BS 5350: Part C10-91. (WITHDRAWN) 1991 Methods of Test for Adhesives Group C: Adhesively Bonded Joints: Mechanical Tests Part C10: 90 Degree Peel Test for a Flexible-to-Rigid Assembly (ISO 8510-1: 1990) (Renumbered as BS EN 28510-1: 1993). 9 pp.
BSI BS 5350: Part C11-91. (WITHDRAWN) 1991 Methods of Test for Adhesives Group C: Adhesively Bonded Joints: Mechanical Tests Part C11: 180 Degree Peel Test for a Flexible-to-Rigid Assembly (ISO 8510-2: 1990) (Renumbered as BS EN 28510-2: 1993). 9 pp.
BSI BS 5350: Part C12-79. 1979 Methods of Test for Adhesives Group C: Adhesively Bonded Joints: Mechanical Tests Part C12: 'T' Peel Test for a Flexible Assembly. 4 pp.
BSI BS 5350: Part C13-90. 1990 Methods of Test for Adhesives Group C: Adhesively Bonded Joints: Mechanical Tests Part 13: Climbing Drum Peel Test. 6 pp.
BSI BS 5350: Part C14-79. 1979 Methods of Test for Adhesives Group C: Adhesively Bonded Joints: Mechanical Tests Part C14: 90 Degree Peel Test for a Rigid-To-Rigid Assembly. 4 pp.
BSI BS EN 2243-2-91. 1991 Test Methods for Structural Adhesives Part 2: Peel Metal-Metal. 14 pp.
BSI BS EN 2243-3-92. 1992 Structural Adhesives. Test Methods Part 3: Peeling Test Metal-Honeycomb Core. 16 pp.
BSI BS EN 28510-1-93. 1993 Amd 1 Adhesives—Peel Test for a Flexible-Bonded-to-Rigid Test Specimen Assembly Part 1: 90 Degree Peel (ISO 8510-1: 1990) (AMD 7559) August 15, 1993 (W). 12 pp.

INTERNATIONAL AND NON-U.S. NATIONAL STANDARDS
SUBJECT INDEX
Adhesive

Adhesive Strength *(Cont.)*
—Peel Strength *(Cont.)*
BSI BS EN 28510-2-93. 1993 Amd 1 Adhesives—Peel Test for a Flexible-Bonded-to-Rigid Test Specimen Assembly Part 2: 180 Degree Peel (ISO 8510-2: 1990) (AMD 7560) August 15, 1993 (W). 17 pp.
CEN PREN 2243-2-89. Structural Adhesives Test Methods Part 2—Peel Metal-Metal. 8 pp.
CEN EN 2243-2-91. Structural Adhesives—Test Methods—Part 2: Peel Metal—Metal. 10 pp.
CEN PREN 2243-3-88. Structural Adhesives Test Methods Part 3—Peeling Test Metal—Honeycomb Core. 11 pp.
CEN EN 2243-3-91. Structural Adhesives Test Methods Part 3—Peeling Test Metal-Honeycomb Core. 12 pp.
CEN EN 28510-1-93. Adhesives—Peel Test for a Flexible-Bonded-to-Rigid Test Specimen Assembly—Part 1: 90 Degree Peel (ISO 8510-1: 1990). 7 pp.
CEN EN 28510-2-93. Adhesives—Peel Test for a Flexible-Bonded-to-Rigid Test Specimen Assembly—Part 2: 180 Degree Peel (ISO 8510-2: 1990). 7 pp.
CNS K6508-80. Methods of Test for Peel Strength of Adhesives (Jul)(5812).
ISO 4578-90. Adhesives—Determination of Peel Resistance of High-Strength Adhesive Bonds—Floating Roller Method Second Edition. 6 pp.
ISO 8510 Pt 1-90. Adhesives—Peel Test for a Flexible-Bonded-to-Rigid Test Specimen Assembly—Part 1: 90 Degree Peel First Edition (CEN EN 28510-1: 1993). 8 pp.
ISO 8510 Pt 2-90. Adhesives—Peel Test for a Flexible-Bonded-to-Rigid Test Specimen Assembly—Part 2: 180 Degree Peel First Edition (CEN EN 28510-2: 1993). 8 pp.
JIS K 6854-77. Testing Methods for Peel Strength of Adhesives (R 1986). 10 pp.

—Peel Strength—Footwear
BSI BS 5131: SUBSEC 1.1.2-76. (WITHDRAWN) 1976 Methods of Test for Footwear and Footwear Materials: Part 1: Adhesives: Section 1.1: Resistance of Adhesive Joints toHeat and to Peeling: Subsection 1.1.2: Resistance to Peeling (Superseded by BS 5131: Section 1.2: 1991). 2 pp.
BSI BS 5131: Sec 1.2-91. 1991 Methods of Test for Footwear and Footwear Materials Part 1: Adhesives Section 1.2: Resistance of Adhesive Joints to Peeling. 12 pp.
BSI BS 5131: Sec 1.7-91. 1991 Methods of Test for Footwear and Footwear Materials Part 1: Adhesives Section 1.7: Preparation of Test Assemblies Using Hot Melt Adhesives for Heat Resistance (Creep) and Peel Tests. 8 pp.

—Shear Strength
AECMA PREN2243-01-80. Structural Adhesives—Test Methods—Part 1—Single Lap Shear. 8 pp.
BSI BS 5350: Part C5-90. 1990 Methods of Test for Adhesives Group C: Adhesively Bonded Joints: Mechanical Tests Part C5: Determination of Bond Strength in Longitudinal Shear. 8 pp.
BSI BS 5350: Part C15-90. 1990 Methods of Test for Adhesives Group C: Adhesively Bonded Joints: Mechanical Tests Part C15: Determination of Bond Strength in Compressive Shear. 5 pp.
CEN PREN 2243 (Part 6)-90. Structural Adhesives Test Methods Part 6 Determination of Shear Stress and Shear Displacement. 11 pp.
CEN PREN 3793-90. Anaerobic Polymerisable Compounds Test Method Determination of Static Shear Strength. 15 pp.
CEN PREN 3794-90. Anaerobic Polymerisable Compounds Test Method Determination of Torque Strength on Threaded Fasteners. 6 pp.
CNS K6500-80. Method of Test for Strength Properties of Adhesives in Shear by Tension Loading (Jun)(5606).
CNS K6505-80. Methods of Test for Strength Properties of Adhesives in Shear by Compression Loading (Jul)(5809).
CNS K6506-80. Method of Test for Impact Shear Strength of Adhesives (Jul)(5810).
ISO 4587-79. Adhesives—Determination of Tensile Lap-Shear Strength of High-Strength Adhesive Bonds First Edition. 5 pp.
ISO 10123-90. Adhesives—Determination of Shear Strength of Anaerobic Adhesives Using Pin-and-Collar Specimens First Edition. 7 pp.
JIS K 6850-76. Testing Methods for Strength Properties of Adhesives in Shear by Tension Loading. 4 pp.
JIS K 6852-76. Testing Methods for Strength Properties of Adhesives in Shear by Compression Loading. 5 pp.
JIS K 6855-77. Testing Methods for Impact Shear Strength of Adhesives (R 1986). 8 pp.

—Shear Strength—Anaerobic Adhesives—Steels
BSI BS 5350: Part G2-87. 1987 Methods of Test for Adhesives Group G: Adhesives Part G2: Determination of Static Shear Strength of Anaerobic Adhesives. 8 pp.

Adhesive Strength *(Cont.)*
—Shear Strength—Footwear
BSI BS 5131: Sec 1.8-81. 1981 Methods of Test for Footwear and Footwear Materials Part 1: Adhesives Sectiion 1.8: Rate of Bond Strength Development in Shear of Hot Melt Adhesives for Lasting. 6 pp.

—Shear Strength—Wood
BSI BS EN 205-91. 1991 Test Methods for Wood Adhesives for Non-Structural Applications—Determination of Tensile Shear Strength of Lap Joints. 12 pp.
CEN EN 205-91. Test Methods for Wood Adhesives for Non-Structural Applications—Determination of Tensile Shear Strength of Lap Joints. 9 pp.
CNS K6504-80. Methods of Test for Strength Properties of Adhesives for Wood in Shear by Tension Loading (Jul)(5808).
DIN ENGL EN 205-91. Test Methods for Wood Adhesives for Non-Structural Applications; Determination of Tensile Shear Strength of Lap Joints (Oct). 6 pp.
ISO 6237-87. Adhesives—Wood-to-Wood Adhesive Bonds—Determination of Shear Strength by Tensile Loading First Edition. 11 pp.
ISO 6238-87. Adhesives—Wood-to-Wood Adhesive Bonds—Determination of Shear Strength by Compression Loading First Edition. 11 pp.
JIS K 6851-76. Testing Methods for Strength Properties of Adhesives for Wood in Shear by Tension Loading. 6 pp.

—Shear Strength—Wooden Structures
BSI BS EN 302-1-92. 1992 Adhesives for Load-Bearing Timber Structures: Test Methods Part 1: Determination of Bond Strength in Longitudinal Tensile Shear (Supersedes BS 1204: Part 1 & 2: 1979). 11 pp.
BSI BS EN 302-2-92. 1992 Adhesives for Load-Bearing Timber Structures: Test Methods Part 2: Determination of Resistance to Delamination (Laboratory Method) (Supersedes BS 1204: Parts 1 & 2: 1979). 10 pp.
BSI BS EN 302-2-01. 1992 Amd 1 Adhesives for Load-Bearing Timber Structures: Test Methods Part 2: Determination of Resistance to Delamination (Laboratory Method) (AMD 7719) May 15, 1993 (W). 11 pp.
CEN PREN 302 (Part 1)-89. Adhesives for Load-Bearing Timber Structures: Polycondensation Adhesives of the Phenolic and Aminoplastic Types—Test Tethods—Part 1: Determination of Bond Strength in Longitudinal Shear. 12 pp.
CEN EN 302-1-92. Adhesives for Load-Bearing Timber Structures: Test Methods—Part 1: Determination of Bond Strength in Longitudinal Tensile Shear. 11 pp.
CEN PREN 302 (Part 2)-89. Adhesives for Load-Bearing Timber Structures: Polycondensation Adhesives of the Phenolic and Aminoplastic Types—Test—Part 2: Determination of Resistance to Delamination (Laboratory Method). 10 pp.
CEN EN 302-2-92. Adhesives for Load-Bearing Timber Structures—Test Methods—Part 2: Determination of Resistance to Delamination (Laboratory Method). 9 pp.
DIN ENGL EN 302 Pt 1-92. Test Methods for Adhesives for Loadbearing Timber Structures; Determination of Bond Strength in Longitudinal Shear (Aug). 7 pp.
DIN ENGL EN 302 Pt 2-92. Test Methods for Adhesives for Loadbearing Timber Structures; Determination of Resistance to Delamination (Laboratory Method) (Aug). 5 pp.

—Tensile Testing
AECMA PREN2243-04-80. Structural Adhesives—Test Methods—Part 4—Metal-Honeycomb Core Flatwise Tensile Test (C7/SC3/C). 7 pp.
BSI BS EN 2243-4-91. 1991 Test Methods for Structural Adhesives Part 4: Metal-Honeycomb Core Flatwise Tensile Test. 13 pp.
CEN PREN 2243-4-89. Structural Adhesives Test Methods Part 4—Metal—Honeycomb Core Flatwise Tensile Test. 8 pp.
CEN EN 2243-4-91. Structural Adhesives—Test Methods—Part 4: Metal-Honeycomb Core Flatwise Tensile Test. 9 pp.
CNS K6499-80. Method of Test for Tensile Strength of Adhesives (May)(5605).
JIS K 6849-76. Testing Methods for Tensile Strength of Adhesives. 7 pp.

—Tensile Testing—Butt Joints
BSI BS 5350: Part C3-89. (WITHDRAWN) 1989 Methods of Test for Adhesives Group C: Adhesively Bonded Joints Part C3: Determination of Bond Strength in Direct Tension (Renumbered as BS EN 26922: 1993). 6 pp.
BSI BS EN 26922-93. 1993 Amd 1 Adhesives—Determination of Tensile Strength of Butt Joints (ISO 6922: 1987) (AMD 7613) August 15, 1993 (W). 9 pp.

Adhesive Strength *(Cont.)*
—Tensile Testing—Butt Joints *(Cont.)*
CEN EN 26922-93. Adhesives—Determination of Tensile Strength of Butt Joints (ISO 6922: 1987). 6 pp.
ISO 6922-87. Adhesives—Determination of Tensile Strength of Butt Joints First Edition (CEN EN 26922: 1993). 6 pp.

—Tensile Testing—Sandwich Panels
BSI BS 5350: Part C6-90. 1990 Methods of Test for Adhesives Group C: Adhesively Bonded Joints: Mechanical Tests Part 6: Determination of Bond Strength in Direct Tension in Sandwich Panels. 4 pp.

—Torsional Strength—Anaerobic Adhesives—Fasteners
BSI BS 5350: Part G1-87. 1987 Methods of Test for Adhesives Group G: Adhesives Part G1: Determination of Torque Strength of Anaerobic Adhesives on Threaded Fasteners. 4 pp.

—Vulcanized Rubber
BSI BS 903: Part A47-82. 1982 Methods of Testing Vulcanized Rubber Part A47: Analysis of Multi-Peak Traces Obtained in Determinations of Tear Strength and Adhesion Strength. 7 pp.
ISO 6133-81. Rubber and Plastics—Analysis of Multi-Peak Traces Obtained in Determinations of Tear Strength and Adhesion Strength First Edition. 5 pp.

—Wallboard
JIS A 1613-84. Testing Methods for Bonding Strength of Wall Boards Adhesives. 11 pp.

—Water Resistance
CNS K6502-80. Method of Test for Resistance of Adhesive Bonds to Water or Moisture (May)(5608).
JIS K 6857-73. Testing Methods for Resistance of Adhesive Bonds to Water or Moisture. 6 pp.

—Wood
CNS A3204-89. Method of Test for Bonding Strength of Anchoring Wooden Block Adhesives (Sep)(10842).
CNS K6178-75. Method of Test for Urea Resin Adhesives (May)(2233). 4 pp.
JIS A 1611-84. Testing Methods for Bonding Strength of Anchoring Wooden Block Adhesives.

Adhesive Tapes
Use For: Pressure Sensitive Tapes
See Also: Adhesive Papers; Electrical Insulating Tapes; Masking Tapes; Splicing Tapes
BSI BS 1133: Sec 14-85. 1985 Packaging Code Section 14: Adhesive Closing and Sealing Tapes. 10 pp.
CGSB CAN/CGSB-43.1-M91. Gummed Paper Tape. 8 pp.
CNS Z6086-87. Method of Test for Pressure Sensitive Tapes and Sheets (Mar)(11888).
JIS Z 0218-91. Testing Methods for Adhesive Strength of Gummed Tape.
JIS Z 0237-91. Testing Methods of Pressure Sensitive Adhesive Tapes and Sheets. 27 pp.
JIS Z 1511-75. Gummed Paper Tapes (for Packaging) (R 1983). 6 pp.
JIS Z 1512-75. Gummed Cloth Tapes (for Packaging) (R 1983). 6 pp.
MOD UK DSTAN 01-06: Part 1-90. Guide to the Selection of Adhesives and Adhesive Tapes for Use on Service Materiel Part 1: Introduction Issue 2. 14 pp.
MOD UK DSTAN 01-06: Part 2-90. Guide to the Selection of Adhesives and Adhesive Tapes for Use on Service Materiel Part 2: Materials Selection Guide Issue 2. 17 pp.
MOD UK DSTAN 01-06: Part 3-90. Guide to the Selection of Adhesives and Adhesive Tapes for Use on Service Materiel Part 3: Specifications Issue 2. 15 pp.
MOD UK DEF-1299-58. Tape, Kraft Paper, Gummed (Reprinted December 1965). 3 pp.
MOD UK TS 375A. Tape, Reinforced, Paper, Gummed.
MOD UK TS 447. Tape, Kraft Paper, Gummed, Non-Corrosive (Withdrawn).
MOD UK TS 10021. Paper, Gummed, for Tape Paper Undulator Gummed (Withdrawn).
SAA AS 1599-88. Pressure Sensitive Adhesive Packaging Tapes. 11 pp.
SAA AS 1602-88. Pressure Sensitive Adhesive Filmic Tapes for General Office Applications. 4 pp.
SAA AS 1635. Methods of Testing Pressure Sensitive Adhesive Tape Complete Set in Binder.

—Adhesion Testing
CPPA D.21H-88. Wet Adhesiveness of Gummed Paper Tape. 2 pp.

—Adhesive Strength
SAA AS 1635.3.1-74. Methods of Testing Pressure Sensitive Adhesive Tape—Part 3.1: Adhesion Strength. 2 pp.

Adhesive Tapes (Cont.)

—Adhesive Strength (Cont.)
SAA AS 1635.3.5-74. Methods of Testing Pressure Sensitive Adhesive Tape—Part 3.5: Adhesion Strength After Water Immersion. 1 p.
SAA AS 1635.3.6-74. Methods of Testing Pressure Sensitive Adhesive Tape—Part 3.6: Adhesion Strength to Own Backing. 1 p.
SAA AS 1635.10.1-74. Methods of Testing Pressure Sensitive Adhesive Tape—Part 10.1: Holding Power to Standard Panel. 1 p.

—Aging Testing
SAA AS 1635.4.1-74. Methods of Testing Pressure Sensitive Adhesive Tape—Part 4.1: Stability—Accelerated Ageing. 1 p.

—Aircraft—Identification Systems
SBAC RS 681 ISSUE 1. Directions for Application of P.T.F.E. Impregnated Marker Tape.

—Aluminum Foil
BSI BS 3887: Part 2-85. (WITHDRAWN) 1985 Adhesive Closing and Sealing Tapes Part 2: Aluminium Foil, Paper, Plasticized PVC, Polyester, Polyethylene, Woven Cloth and Filament Reinforced Tapes (Superseded by BS 3887: 1991). 12 pp.
BSI BS 3887-91. 1991 Specification for Pressure Sensitive Adhesive Closing and Sealing Tapes (H). 12 pp.

—Aluminum Foil—Waterproof
MOD UK DSTAN 75-5-83. Tape, Pressure-Sensitive Adhesive, Type QX (Waterproof, Aluminium Foil) Issue 1. 8 pp.

—Autoclave—Adhesion Testing
BSI BS 6988-89. 1989 Method for Determination of the Adhesion of Autoclave Tape to Tray Wrap Material. 4 pp.

—Bonding Strength
SAA AS 1635.16.1-74. Methods of Testing Pressure Sensitive Adhesive Tape—Part 16.1: Bond Strength at Elevated Temperature. 1 p.

—Breaking Load
SAA AS 1635.5.1-74. Methods of Testing Pressure Sensitive Adhesive Tape—Part 5.1: Breaking Strength. 1 p.

—Cellophane Film
CNS Z5062-78. Pressure Sensitive Adhesive Cellophane (Regenerated Cellulose Film) Tapes (Mar)(4293).
CNS Z6024-78. Method of Test for Pressure Sensitive Adhesive Cellophane (Regenerated Cellulose Film) Tapes (Mar)(4294).
JIS Z 1522-89. Pressure Sensitive Adhesive Cellophane Tapes. 7 pp.

—Cellulose Acetate Film—Transparent
CGSB 53-GP-118M-78. Tape, Pressure-Sensitive Adhesive, Mending, Cellulose Acetate, Matte Backing, Standard for. 10 pp.

—Cellulose Film
BSI BS 3887: Part 1-84. (WITHDRAWN) 1984 Amd 1 Adhesive Closing and Sealing Tapes Part 1: Polypropylene, Regenerated Cellulose and Unplasticized PVC Tapes (AMD 4963) November 29, 1985 (Superseded by BS 3887: 1991). 10 pp.
BSI BS 3887-91. 1991 Specification for Pressure Sensitive Adhesive Closing and Sealing Tapes (H). 12 pp.

—Cellulose Film—Aging Testing
SAA AS 1635.4.2-74. Methods of Testing Pressure Sensitive Adhesive Tape—Part 4.2: Stability—Accelerated Ageing of Cellulose Tape. 1 p.

—Cellulose Film—Transparent
CGSB 53-GP-20M-78. Tape, Pressure-Sensitive Adhesive, Cellulose, Transparent, Standard for. 8 pp.

—Corrosion Prevention—Waterproof
MOD UK DSTAN 80-119-88. Adhesive, Non Corrosive and Water Resistant Issue 1. 13 pp.

—Dielectric Breakdown
SAA AS 1635.15.1-74. Methods of Testing Pressure Sensitive Adhesive Tape—Part 15.1: Dielectric Breakdown and Proof Voltage. 4 pp.

—Double Sided
BSI BS 7116-90. 1990 Double Sided Pressure Sensitive Adhesive Tapes. 24 pp.
CNS Z5137-87. Pressure Sensitive Adhesive Double Coated Tapes (Mar)(11886).
CNS Z6085-87. Method of Test for Pressure Sensitive Adhesive Double Coated Tapes (Mar)(11887).
JIS Z 1528-91. Pressure Sensitive Adhesive Double Coated Tapes. 10 pp.
MOD UK TS 10006A. Tape, Pressure Sensitive Adhesive Transfer, Type QX (Dormant).

Adhesive Tapes (Cont.)

—Double Sided (Cont.)
MOD UK TS 10186. Tape Pressure Sensitive Adhesive, Double Sided (Withdrawn).

—Elongation
SAA AS 1635.6.1-74. Methods of Testing Pressure Sensitive Adhesive Tape—Part 6.1: Elongation. 1 p.

—Fabric
BSI BS 3887: Part 2-85. (WITHDRAWN) 1985 Adhesive Closing and Sealing Tapes Part 2: Aluminium Foil, Paper, Plasticized PVC, Polyester, Polyethylene, Woven Cloth and Filament Reinforced Tapes (Superseded by BS 3887: 1991). 12 pp.
BSI BS 3887-91. 1991 Specification for Pressure Sensitive Adhesive Closing and Sealing Tapes (H). 12 pp.
CGSB 43-GP-3MA-86. Tape, Adhesive, Pressure-Sensitive, Fabric Type. 21 pp.
CNS Z5059-78. Pressure Sensitive Adhesive Cloth Tapes for Packaging and Sealing (Mar)(4287).
CNS Z6021-78. Method of Test for Pressure Sensitive Cloth Tapes for Packaging and Sealing (Mar)(4288).
JIS Z 1512-75. Gummed Cloth Tapes (for Packaging) (R 1983). 6 pp.
JIS Z 1524-89. Pressure Sensitive Adhesive Cloth Tapes for Packaging. 9 pp.
MOD UK DSTAN 75-2-88. Tape, Pressure-Sensitive Adhesive (Fabric) Issue 2. 11 pp.
MOD UK DSTAN 81-25-01. Tape, Pressure-Sensitive Adhesive (Water Resistant Fabric) Issue 2; Amendment 1. 18 pp.

—Filament Reinforced
BSI BS 3887: Part 2-85. (WITHDRAWN) 1985 Adhesive Closing and Sealing Tapes Part 2: Aluminium Foil, Paper, Plasticized PVC, Polyester, Polyethylene, Woven Cloth and Filament Reinforced Tapes (Superseded by BS 3887: 1991). 12 pp.
BSI BS 3887-91. 1991 Specification for Pressure Sensitive Adhesive Closing and Sealing Tapes (H). 12 pp.
CGSB 43-GP-14MA-86. Tape, Adhesive, Pressure Sensitive, Filament Reinforced. 14 pp.
CGSB CAN/CGSB-43.28-92. Water Resistant, Filament Reinforced Gummed Paper Tape. 7 pp.

—Film
CGSB 43-GP-161M-86. Tape, Adhesive, Pressure-Sensitive, Film Type, General Packaging. 12 pp.
MOD UK DEF-1312. Tape, Pressure-Sensitive Adhesive (Printed Film) (Superseded by Def Stan 75-7).

—Film—Printing
CNS Z6045-82. Method of Test for Pressure Sensitive Adhesive Plastic Films for Printing (Mar)(8640).

—Film—Waterproof
MOD UK DSTAN 81-47-01. Tape, Pressure-Sensitive Adhesive (Heavy Duty, Waterproof Film) Issue 1; Amendment 2. 10 pp.

—Flammability Testing
SAA AS 1635.19.1-74. Methods of Testing Pressure Sensitive Adhesive Tape—Part 19.1: Flammability. 1 p.

—Glossaries
JIS Z 0109-81. Glossary of Terms Used in Pressure Sensitive Adhesive Tapes and Sheets.
SAA AS 1230-78. Glossary of Terms for Pressure Sensitive Adhesive Tapes. 12 pp.

—Identification Systems
MOD UK DSTAN 75-7-88. Tape, Pressure-Sensitive Adhesive (Identification Tape) Issue 1. 14 pp.

—Kraft Paper
CNS Z5021-72. Adhesive Tapes of Kraft Paper (Tentative) (Jun)(2389).

—Kraft Paper—Water Soluble
CNS Z7020-72. Water Soluble Kraft Paper Adhesive Tapes for Plywood Use (Tentative) (Oct)(2647).

—Length Measurement
SAA AS 1635.8.1-74. Methods of Testing Pressure Sensitive Adhesive Tape—Part 8.1: Length. 1 p.
SAA AS 1635.8.2-74. Methods of Testing Pressure Sensitive Adhesive Tape—Part 8.2: Length of Double Faced Tape. 1 p.

—Light Testing—Artificial
SAA AS 1635.21.2-85. Methods of Testing Pressure Sensitive Adhesive Tape—Part 21.2: Determination of Resistance to Accelerated Ageing by Artificial Light. 2 pp.

—Medical
CNS T2009-81. Adhesive Tape Medical (Apr)(7329).

Adhesive Tapes (Cont.)

—Mending Stability
SAA AS 1635.21.1-77. Methods of Testing Pressure Sensitive Adhesive Tape—Part 21.1: Mending Stability. 1 p.

—Penetration Resistance
SAA AS 1635.18.1-74. Methods of Testing Pressure Sensitive Adhesive Tape—Part 18.1: Resistance to Penetration at Elevated Temperatures. 1 p.

—Plastic Sheets
JIS K 6782-92. Biaxially Oriented Polypropylene Films for General Use. 8 pp.

—Polyester Film
BSI BS 3887: Part 2-85. (WITHDRAWN) 1985 Adhesive Closing and Sealing Tapes Part 2: Aluminium Foil, Paper, Plasticized PVC, Polyester, Polyethylene, Woven Cloth and Filament Reinforced Tapes (Superseded by BS 3887: 1991). 12 pp.
BSI BS 3887-91. 1991 Specification for Pressure Sensitive Adhesive Closing and Sealing Tapes (H). 12 pp.

—Polyethylene Film
BSI BS 3887: Part 2-85. (WITHDRAWN) 1985 Adhesive Closing and Sealing Tapes Part 2: Aluminium Foil, Paper, Plasticized PVC, Polyester, Polyethylene, Woven Cloth and Filament Reinforced Tapes (Superseded by BS 3887: 1991). 12 pp.
BSI BS 3887-91. 1991 Specification for Pressure Sensitive Adhesive Closing and Sealing Tapes (H). 12 pp.

—Polypropylene Film
BSI BS 3887: Part 1-84. (WITHDRAWN) 1984 Amd 1 Adhesive Closing and Sealing Tapes Part 1: Polypropylene, Regenerated Cellulose and Unplasticized PVC Tapes (AMD 4963) November 29, 1985 (Superseded by BS 3887: 1991). 10 pp.
BSI BS 3887-91. 1991 Specification for Pressure Sensitive Adhesive Closing and Sealing Tapes (H). 12 pp.
CNS Z5105-82. Pressure Sensitive Adhesive Polypropylene Tapes for Packaging and Sealing (Apr)(8749).
CNS Z6046-82. Method of Test for Pressure Sensitive Adhesive Polypropylene Tapes for Packing and Sealing (Apr)(8750).
JIS Z 1539-91. Pressure Sensitive Adhesive Polypropylene Tapes for Packaging. 8 pp.

—PVC Film
BSI BS 3887: Part 1-84. (WITHDRAWN) 1984 Amd 1 Adhesive Closing and Sealing Tapes Part 1: Polypropylene, Regenerated Cellulose and Unplasticized PVC Tapes (AMD 4963) November 29, 1985 (Superseded by BS 3887: 1991). 10 pp.
BSI BS 3887: Part 2-85. (WITHDRAWN) 1985 Adhesive Closing and Sealing Tapes Part 2: Aluminium Foil, Paper, Plasticized PVC, Polyester, Polyethylene, Woven Cloth and Filament Reinforced Tapes (Superseded by BS 3887: 1991). 12 pp.
BSI BS 3887-91. 1991 Specification for Pressure Sensitive Adhesive Closing and Sealing Tapes (H). 12 pp.
CNS Z5061-78. Pressure Sensitive Adhesive Polyvinyl Chloride for Packaging and Sealing (Mar)(4291).
CNS Z6023-78. Method of Test for Pressure Sensitive Adhesive Polyvinyl Chloride for Packaging and Sealing (Mar)(4292).
CNS Z6042-82. Method of Test for Protective Polyvinyl Chloride Tapes (Jan)(8338).
JIS Z 1525-91. Pressure Sensitive Adhesive Polyvinyl Chloride Tapes for Packaging. 8 pp.

—PVC Film—Aerospace—Waterproof
BSI 3J 10-75. (WITHDRAWN) 1975 Pressure-Sensitive Adhesive Waterproof PVC Tape (Superseded by 4J 10: 1991). 8 pp.

—PVC Film—Corrosion Prevention
JIS Z 1901-88. Pressure Sensitive Adhesive Polyvinyl Chloride Tapes for Corrosion Protection. 9 pp.

—Quickstick
SAA AS 1635.3.3-74. Methods of Testing Pressure Sensitive Adhesive Tape—Part 3.3: Quickstick. 2 pp.

—Sampling
SAA AS 1635.2.1-74. Methods of Testing Pressure Sensitive Adhesive Tape—Part 2.1: Sampling. 1 p.

—Solvent Resistance Testing
SAA AS 1635.17.1-74. Methods of Testing Pressure Sensitive Adhesive Tape—Part 17.1: Resistance to Solvent. 1 p.

—Test Specimens
SAA AS 1635.1.1-74. Methods of Testing Pressure Sensitive Adhesive Tape—Part 1.1: Preparation of Rolls and Removal of Test Specimens. 1 p.

INTERNATIONAL AND NON-U.S. NATIONAL STANDARDS
SUBJECT INDEX
Adhesives

Adhesive Tapes (Cont.)

—Thickness Measurement
SAA AS 1635.9.1-74. Methods of Testing Pressure Sensitive Adhesive Tape—Part 9.1: Thickness. 1 p.

—Traffic Control—Identification Systems
BSI PD 6518-87. 1987 Prefabricated Temporary Road Marking Materials for Use at Road Works. 9 pp.

—Transparent—Waterproof
MOD UK DSTAN 75-3-83. Tape, Pressure-Sensitive Adhesive (Waterproof, Transparent) Issue 2. 8 pp.

—Unwind Characteristics
SAA AS 1635.7.1-74. Methods of Testing Pressure Sensitive Adhesive Tape—Part 7.1: Unwind Characteristics. 2 pp.

—Vapor Transmission
SAA AS 1635.11.1-74. Methods of Testing Pressure Sensitive Adhesive Tape—Part 11.1: Water Vapour Permeability. 2 pp.

—Waterproof
CNS Z7200-86. Adhesive Tape for Nylon/Rubber Raincoat Waterproof (Hotseal Type) (Feb)(11516).
CNS Z8064-86. Method of Test for Adhesive Tape for Nylon/Rubber Raincoat Waterproof (Hotseal Type) (Feb)(11517).
MOD UK DSTAN 75-1-88. Tape, Pressure-Sensitive Adhesive (Water Resistant, Conformable) Issue 2. 13 pp.
MOD UK DSTAN 81-85-93. Tape, Pressure-Sensitive Adhesive (Water-Resistant Fabric) Low Chloride, Non-Corrosive Issue 1. 18 pp.

Adhesives

Scope Note: For additional listings, use a more specific term *See Also:* Adhesion; Adhesive Strength; Airframe Adhesives; Aminoplastic Adhesives; Anaerobic Adhesives; Anaerobic Polymers; Animal Glue; Binders (Materials); Bonding Agents; Bonding Strength; Carpet Adhesives; Casein Glue; Chloroprene Rubber Adhesives; Cohesion; Contact Adhesives; Cyanoacrylate Adhesives; Drywall Adhesives; Elastomeric Adhesives; Epoxy Adhesives; Epoxy Resins; Fasteners; Film Adhesives; Fish Glue; Glue; Joints; Marine Glue; Melamine Resin Adhesives; Metal Adhesives; Pastes; Phenolic Adhesives; Polymer Adhesives; Polystyrene Adhesives; Polyurethane Adhesives; Polyvinyl Acetate Adhesives; PVC Adhesives; Resorcinol Adhesives; Roofing Cements; Rubber Adhesives; Shellac Adhesives; Sizing Compounds; Solvent Adhesives; Starch Adhesives; Tile Adhesives; Urea Formaldehyde Resin Adhesives; Wallpaper Adhesives; Water Detecting Pastes; Wood Adhesives

BSI BS 5350: Part 0-87. 1987 Methods of Test for Adhesives Part 0: General Introduction. 4 pp.
BSI BS 5350: Part A2-90. 1990 Methods of Test for Adhesives Group A: Adherends Part A2: Guide to the Selection of Adherend Materials. 10 pp.
CNS K6440-80. Method of Test for Adhesives (General) (Jan)(5130).
JIS S 6040-86. Adhesives for General Works.
MOD UK DSTAN 01-06: Part 1-90. Guide to the Selection of Adhesives and Adhesive Tapes for Use on Service Materiel Part 1: Introduction Issue 2. 14 pp.
MOD UK DSTAN 01-06: Part 2-90. Guide to the Selection of Adhesives and Adhesive Tapes for Use on Service Materiel Part 2: Materials Selection Guide Issue 2. 17 pp.
MOD UK DSTAN 01-06: Part 3-90. Guide to the Selection of Adhesives and Adhesive Tapes for Use on Service Materiel Part 3: Specifications Issue 2. 15 pp.
MOD UK DSTAN 80-130-90. Adhesive, Transfer, Flexible, Pressure-Sensitive Issue 1. 19 pp.

—Ammunition
MOD UK DEF-71. Adhesive, Non-Corrosive and Waterproof (Superseded by Def Stan 80-119).

—Automotive
JIS K 6829-84. Testing Methods for Adhesives for Automobiles.

—Automotive—Bonding Strength
SAA AS 1937.2-77. Methods of Test for Sealers and Adhesives for Automotive Purposes—Part 2: Determination of Initial Bond Strength of Adhesives Reconfirmed 1989. 4 pp.

—Automotive—Comprehensive Testing
SAA AS 1937. Methods of Test for Sealers and Adhesives for Automotive Purposes Complete Set in Binder.

Adhesives (Cont.)

—Automotive—Drying
SAA AS 1937.3-77. Methods of Test for Sealers and Adhesives for Automotive Purposes—Part 3: Determination of Optimum Open Drying Time of Adhesives Reconfirmed 1989. 2 pp.

—Automotive—Heat Resistance
SAA AS 1937.4-77. Methods of Test for Sealers and Adhesives for Automotive Purposes—Part 4: Determination of Heat Resistance of Adhesive Bonds Reconfirmed 1989. 1 p.
SAA AS 1937.12-77. Methods of Test for Sealers and Adhesives for Automotive Purposes—Part 12: Determination of Heat Resistance of Vinyl Laminate Reconfirmed 1989. 1 p.

—Automotive—Mechanical Damping—Thick Plate Method
SAA AS 1937.10-77. Methods of Test for Sealers and Adhesives for Automotive Purposes—Part 10: Determination of Damping Coefficient by the Thick Plate Method Reconfirmed 1989. 4 pp.

—Automotive—Staining
SAA AS 1937.8-77. Methods of Test for Sealers and Adhesives for Automotive Purposes—Part 8: Determination of Staining of Automotive Paintwork by Adhesives or Sealers Reconfirmed 1989. 2 pp.

—Automotive Trim Materials—Staining
SAA AS 1937.5-77. Methods of Test for Sealers and Adhesives for Automotive Purposes—Part 5: Determination of Staining of Trim Materials by Adhesives Reconfirmed 1989. 4 pp.

—Blocking Points
CNS K6446-80. Method of Measurement for Blocking Point of Adhesives (Jan)(5136).

—Bonding Strength
ISO 9653-91. Adhesives—Test Method for Shear Impact Strength of Adhesive Bonds First Edition. 8 pp.

—Building Board—Ceilings
JIS A 5539-84. Adhesives for Ceiling Boards. 12 pp.

—Building Board—Cellular Plastics
CNS A2238-89. Adhesives for Plastic Foam Boards (Oct)(12609).
CNS A3312-89. Method of Test for Adhesives for Plastic Foam Boards (Oct)(12610).
JIS A 5547-81. Adhesives for Plastic Foam Boards. 17 pp.

—Ceilings
BSI BS 5442: Part 2-89. 1989 Classification of Adhesives for Construction Part 2: Adhesives for Use with Interior Wall and Ceiling Coverings (Excluding Decorative Flexible Materials in Roll Form). 8 pp.
BSI BS 5442: Part 2-78. (WITHDRAWN) 1978 Classification of Adhesives for Construction Part 2: Adhesives for Use with Interior Wall and Ceiling Coverings (Excluding Decorative Flexible Materials in Roll Form). 8 pp.
CNS A2237-89. Adhesives for Ceiling Boards (Sep)(12603).
CNS A3310-89. Method of Test for Adhesives for Ceiling Boards (Sep)(12604).

—Ceilings—Bonding Strength
CNS A3311-89. Method of Test for Bonding Strength of Ceiling Boards Adhesives (Sep)(12605).

—Classification
BSI BS 1000: (665)-79. 1979 Universal Decimal Classification (UDC). English Full Edition (665): Oils. Fats. Waxes. Adhesives. Gums. Resins. 34 pp.
EC 77/728/EEC-77. Council Directive on the Approximation of the Laws, Regulations and Administrative Provisions of the Member States Relating to the Classification, Packaging and Labelling of Paints, Varnishes, Printing Inks, Adhesives and Similar Products. 10 pp.
EC 81/916/EEC-81. Commission Directive Adapting to Tech. Prog. Coun. Dir. 77/728/EEC on the Approximation of the Laws, Regs. and Ad. Provs. of the Member Sts. Relating to the Class., Packaging and Labelling of Paints, Varnishes, Printing Inks, Adhesives and Similar Products. 9 pp.
EC 83/265/EEC-83. Council Directive Amending Directive 77/728/EEC on the Approximation of the Laws, Regulations and Administrative Provisions of the Member States Relating to the Classification, Packaging and Labelling of Paints, Varnishes, Printing Inks, Adhesives and Similar Products. 7 pp.
EC 86/508/EEC-86. Commn. Directive Adapting to Tech. Progress for the Second Time Council Directive 77/728/EEC on the Approximation of the Laws, Regs. and Adm. Provisions of the Member States Rel. to the CL., Packaging and Labeling of Paints, Varnishes, Printing Inks, Adhesives and Similar Products. 1 p.

Adhesives (Cont.)

—Classification (Cont.)
EC 89/451/EEC-89. Commission Directive Adapting to Tech. Prog. for the Third Time Coun. Directive 77/728/EEC on the Approximation of the Laws, Regs. and Ad. Provs. of the Member Sts. Rel. to the Class., Packaging and Labelling of Paints, Varnishes, Ptg. Inks, Adhesives and Similar Products. 2 pp.
SNZ NZS/BS 1000 (665)-79. Universal Decimal Classification Oils. Fats. Waxes. Adhesives. Gums. Resins. 36 pp.

—Comprehensive Testing
JIS K 6833-80. General Testing Methods for Adhesives. 27 pp.

—Corrugated Board—Adhesion Testing
CPPA D.10U-77. Ply Separation of Combined Container Board. 1 p.

—Density
BSI BS 5350: Part B1-78. 1978 Amd 1 Methods of Test for Adhesives Group B: Adhesives Part B1: Determination of Density (AMD 6015) July 31, 1989 (W). 8 pp.
CEN PREN 542-91. Adhesives—Determination of Density. 7 pp.
CEN PREN 543-91. Adhesives—Determination of Aparent Density of Powder and Granule Adhesives. 6 pp.
CNS K6441-80. Method of Measurement for Specific Gravity of Liquid Adhesives (Jan)(5131).

—Environmental Testing
CNS K6442-80. Method of Test for Storage Stability of Adhesives (Jan)(5132).
JIS K 6860-74. General Recommended Practices for Atmospheric Exposure of Adhesive Bonds (R 1989). 8 pp.

—Fiberboard—Adhesion Testing
CPPA D.10U-77. Ply Separation of Combined Container Board. 1 p.

—Flash Point
CEN PREN 924-92. Adhesives—Solvent-Borne and Solvent-Free Adhesives—Determination of Flashpoint. 10 pp.

—Flexible Materials—Bonding Strength
ISO 11339-93. Adhesives—180 Degrees Peel Test for Flexable-to-Flexable Bonded Assemblies (T-Peel Test) First Edition. 6 pp.

—Floor Coverings
BSI BS 5350: Part F1-83. 1983 Amd 1 Methods of Test for Adhesives Group F: Performance of Flooring Adhesives Part F1: Performance Tests for Flooring Adhesives (AMD 6699) April 30, 1991. 7 pp.
BSI BS 5442: Part 1-89. 1989 Classification of Adhesives for Construction Part 1: Classification of Adhesives for Use with Flooring Materials. 4 pp.
BSI BS 5442: Part 1-77. (WITHDRAWN) 1977 Classification of Adhesives for Construction Part 1: Adhesives for Use with Flooring Materials. 4 pp.
DIN ENGL 16864-89. Dispersion, Solvent-Borne and Multi-Part Adhesives for Elastomer Floor, Wall and Ceiling Coverings; Requirements and Testing (July). 6 pp.

—Flow Measurement
SAA AS 1937.1-91. Methods of Test for Sealers and Adhesives for Automotive Purposes—Part 1: Determination of Flow Properties in Absolute Units. 6 pp.

—Flow Measurement—Sagging Resistance
BSI BS 5350: Part B9-84. 1984 Methods of Test for Adhesives Group B: Adhesives Part B9: Determination of Resistance to Sagging (Flow After Application). 8 pp.

—Footwear
BSI BS 5131: Sec 1.4-79. 1979 Methods of Test for Footwear and Footwear Materials Part 1: Adhesives Section 1.4: Heat Activation Life of Adhesives. 3 pp.
BSI BS 5131: Sec 1.9-85. 1985 Methods of Test for Footwear and Footwear Materials Part 1: Adhesives Section 1.9: Green Strength of Adhesive Joints. 5 pp.

—Footwear—Bonding Strength
CEN PREN 522-91. Adhesives for Leather and Footwear Materials—Bond Strength—Minimum Requirements and Adhesive Classification. 17 pp.

—Fungus Resistance Testing
SAA AS 1157.10-79. Methods of Testing Materials for Resistance to Fungal Growth—Part 10: Resistance of Adhesives and Glues to Fungal Growth. 6 pp.

INDUSTRY STANDARDS

INTERNATIONAL AND NON-U.S. NATIONAL STANDARDS
SUBJECT INDEX

Adhesives

Adhesives (Cont.)

—Gelation Time
BSI BS 5350: Part B5-76. 1976 Amd 1 Methods of Test for Adhesives Part B5: Determination of Gelation Time (AMD 5359) February 27, 1987. 3 pp.

—Glossaries
BSI BS 6138-89. 1989 Terms Used in the Adhesives Industry. 28 pp.

CEN PREN 923-92. Adhesives—Terms and Definitions. 42 pp.

CNS K3036-80. Standard Definitions of Terms Relating to Adhesive (Mar)(5345).

JIS K 6800-85. Glossary of Terms Used in Adhesives and Adhesion.

SAA AS 1309-74. Glossary of Terms Used in the Adhesives Industry Reconfirmed 1988. 32 pp.

—Hot Melt
MOD UK TS 10155. Adhesive, Hot Melt No. 1, for Packaging Purposes.

—Hot Melt—Bookbinding Machinery
CGSB CAN/CGSB-21.12-M91. Hot Melt Adhesives. 7 pp.

—Hot Melt—Cracking (Fracturing)
BSI BS 5350: Part H2-82. 1982 Methods of Test for Adhesives Group H: Physical Tests on Hot Melt Adhesives Part H2: Determination of Low Temperature Flexibility or Cold Crack Temperature. 4 pp.

—Hot Melt—Heat Resistance
BSI BS 5350: Part H3-84. 1984 Methods of Test for Adhesives Group H: Physical Tests on Hot Melt Adhesives Group H3: Determination of Heat Resistance of Hot-Melt Adhesives. 4 pp.

—Hot Melt—Low Temperature Testing
BSI BS 5350: Part H2-82. 1982 Methods of Test for Adhesives Group H: Physical Tests on Hot Melt Adhesives Part H2: Determination of Low Temperature Flexibility or Cold Crack Temperature. 4 pp.

—Hot Melt—Open Time
BSI BS 5350: Part H4-84. 1984 Methods of Test for Adhesives Group H: Physical Tests on Hot Melt Adhesives Part H4: Determination of Maximun Open Time of Hot-Melt Adhesives (Oven Method). 6 pp.

—Hot Melt—Thermal Stability
BSI BS 5350: Part H1-81. 1981 Methods of Test for Adhesives Group H: Physical Tests on Hot Melt Adhesives Part H1: Determination of Heat Stability of Hot Melt Adhesives in the Applicaton of Equipment. 3 pp.

ISO 10363-92. Hot-Melt Adhesives—Determination of Thermal Stability First Edition. 6 pp.

—Hot Melt—Viscosity
JIS K 6862-84. Testing Methods for Melt Viscosity of Hot-Melt Adhesives. 14 pp.

—Identification Systems
EC 77/728/EEC-77. Council Directive on the Approximation of the Laws, Regulations and Administrative Provisions of the Member States Relating to the Classification, Packaging and Labelling of Paints, Varnishes, Printing Inks, Adhesives and Similar Products. 10 pp.

EC 81/916/EEC-81. Commission Directive Adapting to Tech. Prog. Coun. Dir. 77/728/EEC on the Approximation of the Laws, Regs. and Ad. Provs. of the Member Sts. Relating to the Class., Packaging and Labelling of Paints, Varnishes, Printing Inks, Adhesives and Similar Products. 9 pp.

EC 83/265/EEC-83. Council Directive Amending Directive 77/728/EEC on the Approximation of the Laws, Regulations and Administrative Provisions of the Member States Relating to the Classification, Packaging and Labelling of Paints, Varnishes, Printing Inks, Adhesives. 7 pp.

EC 86/508/EEC-86. Commn. Directive Adapting to Tech. Progress for the Second Time Council Directive 77/728/EEC on the Approximation of the Laws, Regs. and Adm. Provisions of the Member States Rel. to the CL., Packaging and Labeling of Paints, Varnishes, Printing Inks, Adhesives and Similar Products. 1 p.

EC 89/451/EEC-89. Commission Directive Adapting to Tech. Prog. for the Third Time Coun. Directive 77/728/EEC on the Approximation of the Laws, Regs. and Ad. Provs. of the Member Sts. Rel. to the Class., Packaging and Labelling of Paints, Varnishes, Ptg. Inks, Adhesives and Similar Products. 2 pp.

—Identification Systems—Aircraft
BSI 3J 12-91. 1991 Pressure—Sensitive Adhesive Identification Tape. 13 pp.

Adhesives (Cont.)

—Leather—Bonding Strength
CEN PREN 522-91. Adhesives for Leather and Footwear Materials—Bond Strength—Minimum Requirements and Adhesive Classification. 17 pp.

—Life (Durability)
BSI BS 5350: Part B4-93. 1993 Methods of Test for Adhesives Part B4: Determination of Pot Life (ISO 10364: 1993) (W). 6 pp.

BSI BS 5350: Part B4-76. 1976 Amd 1 Methods of Test for Adhesives Part B4: Determination of Pot Life (AMD 5358) February 27, 1987. 3 pp.

CNS K6444-80. Method of Test for Working Life of Adhesives (Jan)(5134).

ISO 10364-93. Adhesives—Determination of Working Life (Pot Life) of Multi-Component Adhesives First Edition. 6 pp.

—Measurement—Surface Tension
CEN PREN 828-92. Adhesives—Wettability—Determination by Measurement of Contact Angle and Critical Surface Tension of Solid Surface. 10 pp.

—Mineral Wool Insulation—Naval Ships
MOD UK NES 782-89. Requirements for Adhesive for Mineral Wool Marine Board Issue 2 (03.89). 14 pp.

MOD UK NES 782-92. Requirements for Adhesive for Mineral Wool Marine Board Issue 3 (08.92). 16 pp.

—Mineral Wool Insulation—Submarines
MOD UK NES 782-89. Requirements for Adhesive for Mineral Wool Marine Board Issue 2 (03.89). 14 pp.

MOD UK NES 782-92. Requirements for Adhesive for Mineral Wool Marine Board Issue 3 (08.92). 16 pp.

—Nonvolatile Matter Content
CNS K6443-80. Method of Determination for Nonvolatile Content of Adhesives (Jan)(5133).

SAA AS 1321.10-80. Methods for the Sampling and Testing of Adhesives—Part 10: Determination of Non-Volatile Matter (Solids Content) of Adhesives (R 1993). 8 pp.

—Packaging
BSI BS 1133: Sec 16-87. 1987 Packaging Code Section 16: Adhesives for Packaging. 24 pp.

BSI BS 1133: Sec 16-01. 1987 Amd 1 Packaging Code Section 16: Adhesives for Packaging (AMD 7839) August 15, 1993 (H). 25 pp.

EC 77/728/EEC-77. Council Directive on the Approximation of the Laws, Regulations and Administrative Provisions of the Member States Relating to the Classification, Packaging and Labelling of Paints, Varnishes, Printing Inks, Adhesives and Similar Products. 10 pp.

EC 81/916/EEC-81. Commission Directive Adapting to Tech. Prog. Coun. Dir. 77/728/EEC on the Approximation of the Laws, Regs. and Ad. Provs. of the Member Sts. Relating to the Class., Packaging and Labelling of Paints, Varnishes, Printing Inks, Adhesives and Similar Products. 9 pp.

EC 83/265/EEC-83. Council Directive Amending Directive 77/728/EEC on the Approximation of the Laws, Regulations and Administrative Provisions of the Member States Relating to the Classification, Packaging and Labelling of Paints, Varnishes, Printing Inks, Adhesives. 7 pp.

EC 86/508/EEC-86. Commn. Directive Adapting to Tech. Progress for the Second Time Council Directive 77/728/EEC on the Approximation of the Laws, Regs. and Adm. Provisions of the Member States Rel. to the CL., Packaging and Labeling of Paints, Varnishes, Printing Inks, Adhesives and Similar Products. 1 p.

EC 89/451/EEC-89. Commission Directive Adapting to Tech. Prog. for the Third Time Coun. Directive 77/728/EEC on the Approximation of the Laws, Regs. and Ad. Provs. of the Member Sts. Rel. to the Class., Packaging and Labelling of Paints, Varnishes, Ptg. Inks, Adhesives and Similar Products. 2 pp.

—Packaging—Glossaries
CNS Z5056-77. Terminology of Adhesives for Packaging (Oct)(4188).

—Padding
CGSB CAN/CGSB-21.13-M91. Padding Adhesives. 7 pp.

—Paper Tubes
MOD UK DEF-152-65. Adhesive for Tubes of Hard Rolled Paper for Pyrotechnic Stores. 4 pp.

Adhesives (Cont.)

—pH
SAA AS 1321.7-77. Methods for the Sampling and Testing of Adhesives—Part 7: pH Value of Aqueous Extracts of Uncured Adhesives and of Their Components, and of Cured Adhesive Films Reconfirmed 1989. 5 pp.

—Physical Testing—Statistical Analysis
BSI BS 5350: Part E1-76. 1976 Amd 1 Methods of Test For Adhesives Part E1: Guide to Statistical Analysis (AMD 6285) April 30, 1990. 5 pp.

—Plastic Pipes—PVC
DIN ENGL 16970-70. Adhesives for Bonding Pipes and Pipe System Elements of Rigid PVC; General Quality Requirements and Testings (Dec). 3 pp.

—PVC—Waterproof—Aerospace
BSI 4J 10-91. 1991 Pressure—Sensitive Adhesive Water-Resistant PVC Tape. 11 pp.

—Sampling
BSI BS 5350: Part E2-79. 1979 Methods of Test for Adhesives Group E: Miscellaneous Part E2: Guide to Sampling. 10 pp.

CEN PREN 1066-93. Adhesives—Sampling. 13 pp.

CEN PREN 1067-93. Adhesives—Examination and Preparation of Samples for Testing. 9 pp.

SAA AS 1321.1-74. Methods for the Sampling and Testing of Adhesives—Part 1: Sampling Reconfirmed 1989. 12 pp.

—Shear Strength
ISO 9653-91. Adhesives—Test Method for Shear Impact Strength of Adhesive Bonds First Edition. 8 pp.

—Softening Points
CNS K6448-80. Method of Measurement for Softening Temperature of Adhesive (Jan)(5138).

—Solids Content
BSI BS 5350: Part B2-76. 1976 Methods of Test for Adhesives Group B: Adhesives Part B2: Determination of Solids Content. 2 pp.

CEN PREN 827-92. Adhesives—Determination of Conventional Solids Content and Constant Mass Solids Content. 6 pp.

—Staining
BSI BS 5350: Part D4-76. 1976 Amd 1 Methods of Test for Adhesives Group D: Adhesively Bonded Joints: Environmental Tests Part D4: Determination of Staining Potential (AMD 5803) February 28, 1989. 6 pp.

—Storage
SAA AS 1321.8-77. Methods for the Sampling and Testing of Adhesives—Part 8: Storage Properties of Adhesives Reconfirmed 1989. 6 pp.

—Structural—Aerospace—Aging Testing
AECMA PREN2243-05-89. Structural Adhesives Test Methods for Part 5—Ageing Tests. 8 pp.

—Structural—Aerospace—Exothermicity
CEN PREN 2667 (Part 5)-91. Foaming Structural Adhesive Films Exothermicity. 5 pp.

—Structural—Aerospace—Expansion
CEN PREN 2667 (Part 3)-91. Foaming Structural Adhesive Films Test to Determine the Expansion Ratio and the Volatile Substance Ratio. 7 pp.

—Structural—Aerospace—Peel Testing
CEN PREN 2243-3-88. Structural Adhesives Test Methods Part 3—Peeling Test Metal—Honeycomb Core. 11 pp.

CEN EN 2243-3-91. Structural Adhesives Test Methods Part 3—Peeling Test Metal-Honeycomb Core. 12 pp.

—Structural—Aerospace—Shear Testing
CEN PREN 2667-01-91. Foaming Structural Adhesives Shear Strength by the Overlap Tensile—Shear Test. 12 pp.

CEN PREN 2667 (Part 2)-92. Foaming Structural Adhesives Determination of Shear Strength by a Tube Test Part 2. 10 pp.

—Structural—Aerospace—Slump Testing
CEN PREN 2667-04-91. Foaming Structural Adhesive Films Test of Vertical Slump. 6 pp.

—Structural—Fatigue (Materials)—Shear Testing
ISO 9664-93. Adhesives—Test Methods for Fatigue Properties of Structural Adhesives in Tensile Shear First Edition. 13 pp.

Adhesives (Cont.)

—Structural—Life (Durability)
ISO 10354-92. Adhesives—Characterization of Durability of Structural -Adhesive-Bonded Assemblies—Wedge Rupture Test First Edition. 7 pp.

—Structural—Shear Strength—Tensile Testing
ISO 11003 Pt 2-93. Adhesives—Determination of Shear Behaviour of Structural Bonds—Part 2: Thick-Adherend Tensile-Test Method First Edition. 15 pp.

—Structural—Shear Strength—Torsion
ISO 11003 Pt 1-93. Adhesives—Determination of Shear Behaviour of Structural Bonds—Part 1: Torsion Test Method Using Butt-Bonded Hollow Cylinders First Edition. 10 pp.

—Surface Finishing
BSI BS 5350: Part A1-76. 1976 Amd 1 Methods of Test for Adhesives Part A1: Adherend Preparation (AMD 6556) January 31, 1991. 8 pp.

—Tensile Testers
BSI BS 5214: Part 1-75. 1975 Amd 1 Testing Machines for Rubbers and Plastics Part 1: Constant Rate of Traverse Machines. 13 pp.
BSI BS 5214: Part 2-78. 1978 Amd 1 Testing Machines for Rubbers and Plastics Part 2: Constant Rate of Force Application Machines. 14 pp.

—Thermal Insulation—Polystyrene
CGSB 71-GP-24M-77. Adhesive, Flexible, for Bonding Cellular Polystyrene Insulation, Standard for; (Amendment 1 Nov 1983). 14 pp.
CSA CAN3-A451.1-M86. Polystyrene Insulation Adhesives; (Gen Instr 1). 19 pp.

—Viscosity
BSI BS 5350: Part B8-90. 1990 Methods of Test for Adhesives Part B8: Determination of Viscosity. 13 pp.
CNS K6501-80. Method of Determination for Viscosity of Adhesives (May)(5607).
SAA AS 1321.9-77. Methods for the Sampling and Testing of Adhesives—Part 9: Brookfield Viscosity of Liquid Adhesives Corrig. Reconfirmed 1989. 12 pp.

—Wall Coverings
BSI BS 3046-81. 1981 Adhesives for Hanging Flexible Wallcoverings. 14 pp.
BSI BS 5442: Part 2-89. 1989 Classification of Adhesives for Construction Part 2: Adhesives for Use with Interior Wall and Ceiling Coverings (Excluding Decorative Flexible Materials in Roll Form). 8 pp.
BSI BS 5442: Part 2-78. (WITHDRAWN) 1978 Classification of Adhesives for Construction Part 2: Adhesives for Use with Interior Wall and Ceiling Coverings (Excluding Decorative Flexible Materials in Roll Form). 8 pp.
CNS A2182-85. Adhesives for Wall Paper and Wall Coverings for Decorative Finish (Feb)(11196).
CNS A3222-85. Method of Test for Adhesives for Wall Paper and Wall Coverings for Decorative Finish (Feb)(11197).
JIS A 6922-78. Adhesives for Wall Paper and Wall Coverings for Decorative Finish (R 1983). 11 pp.

—Wallboard
CNS A2236-89. Adhesives for Wall Boards (Sep)(12600).
CNS A3308-89. Method of Test for Adhesives for Wall Boards (Sep)(12601).
JIS A 5538-84. Adhesives for Wall Boards. 15 pp.

—Wallboard—Bonding Strength
CNS A3309-89. Method of Test for Bonding Strength of Wall Boards Adhesives (Sep)(12602).
JIS A 1613-84. Testing Methods for Bonding Strength of Wall Boards Adhesives. 11 pp.

—Water Resistance Testing
SAA AS 1321.6.1-77. Methods for the Sampling and Testing of Adhesives—Part 6.1: Preferred Conditions for Testing the Water Resistance of Adhesives—Cellulosic Substrates Reconfirmed 1989. 5 pp.

—Wear Testing
SAA AS 1321.4-75. Methods for the Sampling and Testing of Adhesives—Part 4: Wearing Effect of Set Adhesives on Cutting Edges Reconfirmed 1989. 6 pp.

—Weight Measurement
CNS K6445-80. Method of Measurement for Applied Weight of Liquid Adhesives (Jan)(5135).

ADI
Use: Attitude Indicators

Adipates
See Also: Esters
ISO 2523-74. Adipate Esters for Industrial Use—List of Methods of Test First Edition. 3 pp.

—Acidity—Volumetric Analysis
ISO 2525-74. Adipate Esters for Industrial Use—Determination of Acidity to Phenolphthalein—Volumetric Method First Edition. 4 pp.

—Ash Content—Gravimetric Analysis
ISO 2526-74. Adipate Esters for Industrial Use—Determination of Ash—Gravimetric Method First Edition. 4 pp.

—Color Testing
ISO 2524-74. Adipate Esters for Industrial Use—Measurement of Colour After Heat Treatment First Edition. 4 pp.

—Ester Content—Volumetric Analysis
ISO 2527-74. Adipate Esters for Industrial Use—Determination of Ester Content—Volumetric Method First Edition. 5 pp.

Adipic Acid
CNS K2207-88. Adipic Acid HOOCCH2CH2CH2CH2COOH FW: 146.14 (Apr)(12275). 3 pp.
JIS K 4172-72. Adipic Acid.
MOD UK DSTAN 68-93-87. Adipic Acid Issue 1. 16 pp.

Adipic Esters
See Also: Esters

—Alcohol Content—Spectrometry
ISO 6795-83. Phthalic and Adipic Esters for Industrial Use—Determination of Free Alcohols Content—Spectrometric Method First Edition. 4 pp.

Adjacent Band Interference
Use: Adjacent Channel Interference

Adjacent Channel Interference
Use For: Adjacent Band Interference
See Also: Radio Frequency Interference
CCIR Report 654-3-90. Methods for Calculating Interference Power in Adjacent Bands and Channels—Section 1B—Spectrum Sharing and Planning Principles and Techniques. 13 pp.

—Broadcasting
CCIR Report 485-1-82. Contribution to the Planning of Broadcasting Services—Section 11E—Planning of Television Networks, Protection Ratios, Television Receivers and Antennas. 5 pp.

—Radio Astronomy—Radio Transmitters
CCIR RECMN 517-1-82. Protection of the Radioastronomy Service from Transmitters in Adjacent Bands—Section 2G—Radioastronomy and Radar Astronomy. 1 p.
CCIR RECMN 517-2-92. Protection of the Radioastronomy Service from Transmitters in Adjacent Bands—Section 2G—Radioastronomy and Radar Astronomy. 5 pp.

—Radio Receivers
CCIR QUESTION 56/1-77. Interference Between Services in Adjacent Bands—Questions Concerning Study Group 1 —Spectrum Management Techniques (Spectrum Engineering, Planning, Sharing, Monitoring and Utilization). 1 p.

—Radio Relay Systems
CCIR QUESTION 128/9-90. Maximum Allowable Degradations of Radio-Relay Systems Due to Energy Spread from Services in the Adjacent Bands—Questions of Study Group 9 Fixed Service. 1 p.

—Radio Transmitters—Out of Band Emissions
CCIR QUESTION 56/1-77. Interference Between Services in Adjacent Bands—Questions Concerning Study Group 1 —Spectrum Management Techniques (Spectrum Engineering, Planning, Sharing, Monitoring and Utilization). 1 p.

—Single Sideband Communication Equipment
CCIR Report 1018-1-90. Co-Channel and Adjacent-Channel Coordination Criteria for Simultaneous Use of Different Modulation Techniques in the Mobile Service—Section 8A—Land Mobile Service and Related Subjects. 6 pp.

—Television Broadcasting
CCIR QUESTION 75/11-90. Methods of Reducing Interference to the Broadcasting Service (Television) from Other Services Operating in the Same or Adjacent Bands—Questions Concerning Study Group 11—Broadcasting Service (Television). 1 p.

Adjustable Capacitors
Use: Variable Capacitors

Adjustable Wrenches
See Also: Chain Wrenches; Combination Wrenches; Hand Tools; Pipe Wrenches; Tap Wrenches; Wrenches
BSI BS 6333-83. 1983 Adjustable Wrenches. 4 pp.
CGSB 39-GP-13M-81. Wrenches, Adjustable, Open-End and Monkey, Standard for. 18 pp.
CNS B3210-84. Adjustable Wrenches (Nov)(3303).
ISO 6787-82. Assembly Tools for Screws and Nuts—Adjustable Wrenches—Width Across Flats up to 50 mm First Edition. 4 pp.

—Angle
JIS B 4604-76. Adjustable Angle Wrenches. 9 pp.

—Copper Beryllium Alloy—Nonsparking
CNS M2061-86. Non-Sparking Beryllium Copper Alloy Tools Adjustable Wrenches (Nov)(5909).

Adjusting Rings
See Also: Ring Gages; Rings; Setting Rings
CNS Z7057-78. Adjusting Ring (Mar)(4214).

Adjusting Screws
See Also: Screws

—Drawing Equipment
CNS B2380-83. Adjusting Screws for Drawing Instruments (Feb)(4945).

ADMD
Use: Administration Management Domain

Administration
Use: Management

Administration Management Domain
Use For: ADMD *See Also:* Message Handling Systems; Private Management Domain
CCITT RECMN F.400-92. Message Handling Services: Message Handling System and Service Overview (Study Group I) 82 pp (Same as Recmn X.400). 82 pp.
CCITT RECMN F.400-89. Message Handling System and Service Overview—Message Handling and Directory Services—Operations and Definition of Service (Study Group I) 73 pp. 73 pp.
CCITT RECMN X.400-93. Message Handling Services: Message Handling System and Service Overview (Study Group VII) 82 pp (Same as Recmn F.400). 82 pp.

—Accounting
CCITT RECMN D.36-91. General Accounting Principles Applicable to Message Handling Services (Study Group III) 14 pp. 14 pp.

—Interpersonal Messaging Services
CCITT RECMN F.420-92. Message Handling Services: Public Interpersonal Messaging Service (Study Group I) 16 pp. 16 pp.
CCITT RECMN F.420-89. Message Handling Services: the Public Interpersonal Messaging Service—Message Handling and Directory Services—Operations and Definition of Service (Study Group I) 15 pp. 15 pp.

—Message Transfer Agents
CCITT RECMN F.410-92. Message Handling Services: Public Message Transfer Service (Study Group I) 11 pp. 11 pp.
CCITT RECMN F.410-89. Message Handling Services: Message Transfer Service —Message Handling and Directory Services—Operations and Definition of Service (Study Group I) 10 pp. 10 pp.

—Originator/Recipient Addresses
CCITT RECMN F.401-92. Message Handling Services: Naming and Addressing for Public Message Handling Services (Study Group I) 21 pp. 21 pp.
CCITT RECMN F.401-89. Message Handling Services: Naming and Addressing for Public Message Handling Services—Message Handling and Directory Services—Operations and Definition of Service (Study Group I) 11 pp. 11 pp.

—User Agents
CCITT RECMN F.400-92. Message Handling Services: Message Handling System and Service Overview (Study Group I) 82 pp (Same as Recmn X.400). 82 pp.
CCITT RECMN F.400-89. Message Handling System and Service Overview—Message Handling and Directory Services—Operations and Definition of Service (Study Group I) 73 pp. 73 pp.
CCITT RECMN X.400-93. Message Handling Services: Message Handling System and Service Overview (Study Group VII) 82 pp (Same as Recmn F.400). 82 pp.

Admittance
Use: Electrical Impedance

INTERNATIONAL AND NON-U.S. NATIONAL STANDARDS
SUBJECT INDEX

Admixtures

See Also: Accelerators (Materials); Additives; Air Entraining Agents; Bonding Agents; Cement Additives; Concretes; Dispersants; Fly Ash; Mortars; Plasticizers; Release Agents; Retarders (Materials); Surfactants; Wetting Agents

—Concretes
BSI BS 5075: Part 1-82. 1982 Amd 2 Concrete Admixtures Part 1: Accelerating Admixtures, Retarding Admixtures, and Water Reducing Admixtures. 21 pp.
CEN PREN 104.301 (Part 1)-91. Admixtures for Concrete, Mortar and Grout Test Methods Reference Concrete and Reference Mortar for Testing. 3 pp.
CEN PREN 934-2-92. Admixtures for Concrete, Mortars and Grouts—Part 2: Concrete Admixtures—Definitions, Specifications and Conformity Criteria. 12 pp.
CNS A2170-83. Expansive Additive for Concrete (Nov)(10641).
CNS A2219-88. Chemical Admixtures for Concrete (May)(12283).
CNS A3167-82. Method of Test for Air-Entraining Admixtures for Concrete (Oct)(9208).
CNS A3272-88. Method of Test for Chemical Admixture for Concrete (May)(12284).
CSA CAN3-A266.2-M78. Chemical Admixtures for Concrete; (Amd 1 September 1978). 26 pp.
CSA CAN3-A266.4-M78. Guidelines for the Use of Admixtures in Concrete; (Amd 1-11 November 1983). 45 pp.
JIS A 6202-80. Expansive Additive for Concrete. 31 pp.
JIS A 6204-87. Chemical Admixtures for Concrete. 31 pp.
SAA AS 1478-92. Chemical Admixtures for Concrete (in Professional Packages 30, 58A) (Supersedes AS 1479—1973). 21 pp.
SAA AS 2073-77. Methods for the Testing of Expanding Admixtures for Concrete, Mortar and Grout. 28 pp.
SAA MP20.2-75. Part 2: Thickening Admixtures for Use in Concrete and Mortar. 9 pp.
SAA MP20.3-77. Part 3: Expanding Admixtures for Use in Concrete, Mortar and Grout. 12 pp.
SNZ NZS 3113-79. Specification for Chemical Admixtures for Concrete. 15 pp.

—Concretes—Chloride Content
CEN PREN 480 (Part 10)-91. Admixtures for Concrete, Mortar and Grout—Test Methods—Part 10: Determination of the Chloride Content. 7 pp.

—Concretes—Density
CEN PREN 480 (Part 7)-91. Admixtures for Concrete, Mortar and Grout—Test Methods—Part 7: Determination of the Density of Liquid Admixtures. 4 pp.

—Concretes—Dry Matter Content Analysis
CEN PREN 480 (Part 8)-91. Admixtures for Concrete, Mortar and Grout—Test Methods—Part 8: Determination of the Conventional Dry Material Content. 5 pp.

—Concretes—Infrared Analysis
CEN PREN 480 (Part 6)-91. Admixtures for Concrete, Mortar and Grout—Test Methods—Part 6: Infrared Analysis. 5 pp.

—Concretes—PH
CEN PREN 480 (Part 9)-91. Admixtures for Concrete, Mortar and Grout—Test Methods—Part 9: Determination of the Ph Value. 6 pp.

—Concretes—Sampling
SAA AS 2072-77. Methods for the Sampling of Expanding Admixtures for Concrete, Mortar and Grout. 4 pp.

—Glossaries
BSI BS 6100: Sec 6.4-86. 1986 Glossary of Building and Civil Engineering Terms Part 6: Concrete and Plaster Section 6.4: Admixtures. 5 pp.
BSI BS 6100: Sec 6.4-01. 1986 Amd 1 Building and Civil Engineering Terms Part 6: Concrete and Plaster Section 6.4: Admixtures (AMD 7266) August 15, 1992. 6 pp.

—Grout
CEN PREN 104.301 (Part 1)-91. Admixtures for Concrete, Mortar and Grout Test Methods Reference Concrete and Reference Mortar for Testing. 3 pp.
SAA AS 2073-77. Methods for the Testing of Expanding Admixtures for Concrete, Mortar and Grout. 28 pp.
SAA MP20.3-77. Part 3: Expanding Admixtures for Use in Concrete, Mortar and Grout. 12 pp.

Admixtures (Cont.)

—Grout—Sampling
SAA AS 2072-77. Methods for the Sampling of Expanding Admixtures for Concrete, Mortar and Grout. 4 pp.

—Mortars
CEN PREN 104.301 (Part 1)-91. Admixtures for Concrete, Mortar and Grout Test Methods Reference Concrete and Reference Mortar for Testing. 3 pp.
SAA AS 2073-77. Methods for the Testing of Expanding Admixtures for Concrete, Mortar and Grout. 28 pp.
SAA MP20.2-75. Part 2: Thickening Admixtures for Use in Concrete and Mortar. 9 pp.
SAA MP20.3-77. Part 3: Expanding Admixtures for Use in Concrete, Mortar and Grout. 12 pp.

—Mortars—Chloride Content
CEN PREN 480 (Part 10)-91. Admixtures for Concrete, Mortar and Grout—Test Methods—Part 10: Determination of the Chloride Content. 7 pp.

—Mortars—Density
CEN PREN 480 (Part 7)-91. Admixtures for Concrete, Mortar and Grout—Test Methods—Part 7: Determination of the Density of Liquid Admixtures. 4 pp.

—Mortars—Dry Matter Content Analysis
CEN PREN 480 (Part 8)-91. Admixtures for Concrete, Mortar and Grout—Test Methods—Part 8: Determination of the Conventional Dry Material Content. 5 pp.

—Mortars—Infrared Analysis
CEN PREN 480 (Part 6)-91. Admixtures for Concrete, Mortar and Grout—Test Methods—Part 6: Infrared Analysis. 5 pp.

—Mortars—PH
CEN PREN 480 (Part 9)-91. Admixtures for Concrete, Mortar and Grout—Test Methods—Part 9: Determination of the Ph Value. 6 pp.

—Mortars—Sampling
SAA AS 2072-77. Methods for the Sampling of Expanding Admixtures for Concrete, Mortar and Grout. 4 pp.

Adonitol
Use For: Ribitol
JIS K 9058-88. Adonitol.

ADPCM
Use: Adaptive Delta Pulse Code Modulation

ADREP
Use: Aircraft Accidents

Adsorption
See Also: Gas Adsorption; Heat of Combustion; Water Treatment

—Aluminum Oxide
BSI BS 4140: Part 23-87. 1987 Methods of Test for Aluminium Oxide Part 23: Determination of Specific Surface Area by Nitrogen Absorption (Single-Point Method). 11 pp.
ISO 2961-74. Aluminium Oxide Primarily Used for the Production of Aluminium—Determination of an Adsorption Index First Edition. 6 pp.
ISO 8008-86. Aluminium Oxide Primarily Used for the Production of Aluminium—Determination of Specific Surface Area by Nitrogen Adsorption—Single-Point Method First Edition. 11 pp.
SAA AS 2879.4-91. Alumina Determination of Specific Surface Area by Nitrogen Adsorption. 2 pp.

—Carbon Black
BSI BS 5293: Part 12-88. 1988 Sampling and Testing Carbon Black for Use in the Rubber Industry Part 12: Methods for Determination of Surface Area by Surfactant Absorption. 15 pp.
ISO 4652-81. Rubber Compounding Ingredients—Carbon Black—Determination of Specific Surface Area—Nitrogen Adsorption Methods First Edition. 17 pp.
ISO 6810-85. Rubber Compounding Ingredients—Carbon Black—Determination of Surface Area—Surfactant Adsorption Methods First Edition. 13 pp.

—Waste Water
CPPA H.6P-91. Determination of Adsorbable Organic Halogens (AOX) in Waters and Wastewaters. 7 pp.
DIN ENGL 38409 Pt 14-85. German Standard Methods for the Examination of Water, Waste Water and Sludge; Summary Indices of Actions and Substances (Group H); Determination of Adsorbable Organically Bonded Halogens (AOX) (H 14) (Mar). 6 pp.

Adsorption (Cont.)

—Water
CPPA H.6P-91. Determination of Adsorbable Organic Halogens (AOX) in Waters and Wastewaters. 7 pp.
DIN ENGL 38409 Pt 14-85. German Standard Methods for the Examination of Water, Waste Water and Sludge; Summary Indices of Actions and Substances (Group H); Determination of Adsorbable Organically Bonded Halogens (AOX) (H 14) (Mar). 6 pp.
ISO 9562-89. Water Quality—Determination of Adsorbable Organic Halogens (AOX) First Edition. 11 pp.

Advance Warning Triangles
Use: Warning Triangles

Advanced Testing Methods (OSI)
Use For: ATM (Advanced Testing Methods)
See Also: Open Systems Interconnection

—Conformance
CENELEC ETR 040-92. Advanced Testing Methods (ATM); Profile Test Specifications and Conformance Test Reports. 33 pp.
ETSI ETR 040-92. Advanced Testing Methods (ATM); Profile Test Specifications and Conformance Test Reports. 33 pp.

—Profile Test Specifications
CENELEC ETR 040-92. Advanced Testing Methods (ATM); Profile Test Specifications and Conformance Test Reports. 33 pp.
ETSI ETR 040-92. Advanced Testing Methods (ATM); Profile Test Specifications and Conformance Test Reports. 33 pp.

Advice of Charge Services
See Also: Tariffs (Telecommunications); Telephone Services

—Integrated Services Digital Networks
CCITT RECMN I.256-89. Charging Supplementary Services—Integrated Services Digital Network (ISDN)—General Structure and Service Capabilities (Study Group XVIII) 16 pp. 16 pp.
CCITT RECMN I.256. 2-88. Advice of Charge—Integrated Services Digital Network (ISDN)—General Structure and Service Capabilities (Study Group XVIII) 15 pp. 15 pp.
CENELEC PRETS 300 178-91. Integrated Services Digital Network (ISDN); Advice of Charge: Charging Information at Call Set-up Time (AOC-S) Supplementary Service Service Description (T/NA1(89)13). 18 pp.
CENELEC PRETS 300 178-92. Integrated Services Digital Network (ISDN); Advice of Charge: Charging Information at Call Set-up Time (AOC-S) Supplementary Service Service Description. 18 pp.
CENELEC ETS 300 178-92. Integrated Services Digital Network (ISDN); Advice of Charge: Charging Information at Call Set-up Time (AOC-S) Supplementary Service Service Description. 18 pp.
CENELEC PRETS 300 179-91. Integrated Services Digital Network (ISDN); Advice of Charge: Charging Information During the Call (AOC-D) Supplementary Service Service Description (T/NA1(89)14). 16 pp.
CENELEC PRETS 300 179-92. Integrated Services Digital Network (ISDN); Advice of Charge: Charging Information During the Call (AOC-D) Supplementary Service Service Description. 16 pp.
CENELEC ETS 300 179-92. Integrated Services Digital Network (ISDN); Advice of Charge: Charging Information During the Call (AOC-D) Supplementary Service Service Description. 16 pp.
CENELEC PRETS 300 180-91. Integrated Services Digital Network (ISDN); Advice of Charge: Charging Information at the End of the Call (AOC-E) Supplementary Service Service Description (T/NA1(89)15). 16 pp.
CENELEC PRETS 300 180-92. Integrated Services Digital Network (ISDN); Advice of Charge: Charging Information at the End of the Call (AOC-E) Supplementary Service Service Description. 16 pp.
CENELEC ETS 300 180-92. Integrated Services Digital Network (ISDN); Advice of Charge: Charging Information at the End of the Call (AOC-E) Supplementary Service Service Description. 16 pp.
CENELEC PRETS 300 181-91. Integrated Services Digital Network (ISDN); Advice of Charge (AOC) Supplementary Service Functional Capabilities and Information Flows (T/S 22-04). 46 pp.
CENELEC PRETS 300 181-92. Integrated Services Digital Network (ISDN); Advice of Charge (AOC) Supplementary Service Functional Capabilities and Information Flows. 46 pp.
CENELEC ETS 300 181-93. Integrated Services Digital Network (ISDN); Advice of Charge (AOC) Supplementary Service Functional Capabilities and Information Flows. 46 pp.

INTERNATIONAL AND NON-U.S. NATIONAL STANDARDS
SUBJECT INDEX
Aerial

Advice of Charge Services *(Cont.)*
—Integrated Services Digital Networks *(Cont.)*
ETSI PRETS 300 178-91. Integrated Services Digital Network (ISDN); Advice of Charge: Charging Information at Call Set-up Time (AOC-S) Supplementary Service Service Description (T/NA1(89)13). 18 pp.
ETSI ETS 300 179-92. Integrated Services Digital Network (ISDN); Advice of Charge: Charging Information During the Call (AOC-D) Supplementary Service Service Description. 16 pp.
ETSI PRETS 300 179-91. Integrated Services Digital Network (ISDN); Advice of Charge: Charging Information During the Call (AOC-D) Supplementary Service Service Description (T/NA1(89)14). 16 pp.
ETSI ETS 300 180-92. Integrated Services Digital Network (ISDN); Advice of Charge: Charging Information at the End of the Call (AOC-E) Supplementary Service Service Description. 16 pp.
ETSI PRETS 300 180-91. Integrated Services Digital Network (ISDN); Advice of Charge: Charging Information at the End of the Call (AOC-E) Supplementary Service Service Description (T/NA1(89)15). 16 pp.
ETSI ETS 300 181-93. Integrated Services Digital Network (ISDN); Advice of Charge (AOC) Supplementary Service Functional Capabilities and Information Flows. 46 pp.
ETSI PRETS 300 181-92. Integrated Services Digital Network (ISDN); Advice of Charge (AOC) Supplementary Service Functional Capabilities and Information Flows. 46 pp.
ETSI PRETS 300 181-91. Integrated Services Digital Network (ISDN); Advice of Charge (AOC) Supplementary Service Functional Capabilities and Information Flows (T/S 22-04). 46 pp.

—Integrated Services Digital Networks—DSS1
CENELEC PRETS 300 182-91. Integrated Services Digital Network (ISDN); Advice of Charge (AOC) Supplementary Service Digital Subscriber Signalling System No. One (DSS1) Protocol (T/S 46-33K). 37 pp.
CENELEC PRETS 300 182-92. Integrated Services Digital Network (ISDN); Advice of Charge (AOC) Supplementary Service Digital Subscriber Signalling System No. One (DSS1) Protocol. 41 pp.
CENELEC PRETS 300 182-92. Integrated Services Digital Network (ISDN); Advice of Charge (AOC) Supplementary Service Digital Subscriber Signalling System No. One (DSS1) Protocol. 41 pp.
CENELEC ETS 300 182-93. Integrated Services Digital Network (ISDN); Advice of Charge (AOC) Supplementary Service Digital Subscriber Signalling System No. One (DSS1) Protocol. 41 pp.
CEPT T/S 22-04 E-88. Advice of Charge Service. 24 pp.
ETSI ETS 300 178-92. Integrated Services Digital Network (ISDN); Advice of Charge: Charging Information at Call Set-up Time (AOC-S) Supplementary Service Service Description. 18 pp.
ETSI ETS 300 182-93. Integrated Services Digital Network (ISDN); Advice of Charge (AOC) Supplementary Service Digital Subscriber Signalling System No. One (DSS1) Protocol. 41 pp.
ETSI PRETS 300 182-92. Integrated Services Digital Network (ISDN); Advice of Charge (AOC) Supplementary Service Digital Subscriber Signalling System No. One (DSS1) Protocol. 41 pp.
ETSI PRETS 300 182-91. Integrated Services Digital Network (ISDN); Advice of Charge (AOC) Supplementary Service Digital Subscriber Signalling System No. One (DSS1) Protocol (T/S 46-33K). 37 pp.

—Integrated Services Digital Networks—Functions/Information Flows
CCITT RECMN Q.86-89. Charging Supplementary Services—General Recommendations on Telephone Switching and Signalling—Functions and Information Flows for Services in the ISDN—Supplements (Study Group XI) 22 pp. 22 pp.

Adzes
See Also: Cutting Tools; Tools
—Handles
CGSB 39-GP-10M-79. Handle, Hickory, Striking Tool (Metric), Standard for; (Amendment 2 Jan 1984). 15 pp.

AECMA
Use For: Association Europeene des Constructeurs de Materiel Aerospatial ; European Association of Aerospace Manufacturers *See Also:* European Standards; Standards
—Certification Management Committee
EURO AECMA-CERT-91. AECMA-CERT Information Brochure Procedures. 55 pp.

Aerated Concrete Blocks
Use: Concrete Blocks

Aerated Concretes
See Also: Cellular Concretes; Concretes
—Autoclaved
CEN PREN 680-92. Determination of the Drying Shrinkage of Autoclaved Aerated Concrete. 5 pp.
CEN PREN 991-92. Determination of the Dimensions of Prefabricated Reinforced Components Made of Autoclaved Aerated Concrete or Lightweight Aggregate Concrete with Open Structure. 4 pp.

—Autoclaved—Compression Testing
CEN PREN 679-92. Determination of the Compressive Strength of Autoclaved Aerated Concrete. 5 pp.

—Autoclaved—Density
CEN PREN 678-92. Determination of the Dry Density of Autoclaved Aerated Concrete. 5 pp.

—Autoclaved—Shrinkage
CEN PREN 680-92. Determination of the Drying Shrinkage of Autoclaved Aerated Concrete. 5 pp.
CEN PREN 991-92. Determination of the Dimensions of Prefabricated Reinforced Components Made of Autoclaved Aerated Concrete or Lightweight Aggregate Concrete with Open Structure. 4 pp.

—Panels—Autoclaved
DIN ENGL 4166-86. Autoclaved Aerated Concrete Slabs and Panels (Dec). 3 pp.
JIS A 5416-85. Autoclaved Lightweight Aerated Concrete Panels. 20 pp.

—Slabs—Autoclaved
DIN ENGL 4166-86. Autoclaved Aerated Concrete Slabs and Panels (Dec). 3 pp.

Aeration Cabinets
Use: Aerator Cabinets

Aerator Cabinets
See Also: Medical Equipment
—Ethylene Oxide
SAA AS 1862-76. Aeration Cabinets (for Use with Ethylene Oxide Sterilizers). 12 pp.

Aerial Cableways
Use: Aerial Ropeways

Aerial Camera Films
Use: Aerial Films

Aerial Delivery
See Also: Air Transportation; Transportation
ICAO 9408. Manual on Aerial Work First Edition—1984. 171 pp.

—Cargo Nets
MOD UK DSTAN 16-9: Part 1-91. Nets and Slings, Cargo, Helicopter Part 1: Nets, Cargo, Aerial Delivery Issue 2. 11 pp.

—Information Interchange
NATO STANAG 3428 ED 4 AMD 2-88. Exchange of Information on Aerial Delivery Systems. 10 pp.

—Paper—Honeycomb Cores
NATO STANAG 3778 ED 1 AMD 3-80. Performance Criteria for Honeycomb Paper Used as Energy Dissipating Material. 6 pp.
NATO STANAG 3778 ED 1 AMD 4-80. Performance Criteria for Honeycomb Paper Used as Energy Dissipating Material. 6 pp.

Aerial Equipment
Scope Note: Use a more specific term *See:* Aerial Ropeways; Materials Handling Equipment

Aerial Films
Use For: Aerial Camera Films *See Also:* Aerial Photography; Photographic Films

— Black and White
ISO 7829-86. Photography—Black-and-White Aerial Camera Films—Determination of ISO Speed and Average Gradient First Edition. 11 pp.

— Film Cores
MOD UK DSTAN 67-5-01. Spools and Cores, Photographic Film for Airborne Cameras Issue 1; Amendment 1. 10 pp.

— Glossaries
MOD UK DSTAN 67-7-01. Nomenclature for Rolled Air Film Issue 1; Amendment 1. 11 pp.

Aerial Films *(Cont.)*
— Spools
MOD UK DSTAN 67-5-01. Spools and Cores, Photographic Film for Airborne Cameras Issue 1; Amendment 1. 10 pp.

— 16mm—Winding
MOD UK DSTAN 67-2-74. Dimensions, Tolerances, and Winding Characteristics of 16mm and 35mm Rolled Film for Airborne Cameras Issue 4. 5 pp.

— 16mm—Winding—Cameras
NATO STANAG 3179 ED 6 AMD 0-90. 16 mm and 35 mm Roll Film for Aerial Cameras. 6 pp.

— 35mm—Winding
MOD UK DSTAN 67-2-74. Dimensions, Tolerances, and Winding Characteristics of 16mm and 35mm Rolled Film for Airborne Cameras Issue 4. 5 pp.

— 35mm—Winding—Cameras
NATO STANAG 3179 ED 6 AMD 0-90. 16 mm and 35 mm Roll Film for Aerial Cameras. 6 pp.

— 70mm
MOD UK DSTAN 67-4-01. Film, Photographic, Aerial Issue 2; Amendment 1. 7 pp.

— 70mm—Cameras
NATO STANAG 3178 ED 5 AMD 1-85. Rolled Air Film and Air Framing Camera Standard Image Format Sizes. 9 pp.
NATO STANAG 3178 ED 5 AMD 2-85. Rolled Air Film and Air Framing Camera Standard Image Format Sizes. 7 pp.

—Mapping—Titling
NATO STANAG 3189 ED 6 AMD 0-90. Titling of Air Reconnaissance, Air Survey and Air Mapping Imagery. 13 pp.
NATO STANAG 3189 ED 6 AMD 1-90. Titling of Air Reconnaissance, Air Survey and Air Mapping Imagery. 13 pp.

—Reconnaissance—Numbering
NATO STANAG 3599 ED 2 AMD 0-89. Edge Numbering of Films Used for Air Reconnaissance. 5 pp.

—Reconnaissance—Titling
NATO STANAG 3189 ED 6 AMD 0-90. Titling of Air Reconnaissance, Air Survey and Air Mapping Imagery. 13 pp.
NATO STANAG 3189 ED 6 AMD 1-90. Titling of Air Reconnaissance, Air Survey and Air Mapping Imagery. 13 pp.

—Spools
NATO STANAG 3419 ED 3 AMD 0-91. Film Spools for Aerial Cameras. 7 pp.

—Surveys—Titling
NATO STANAG 3189 ED 6 AMD 0-90. Titling of Air Reconnaissance, Air Survey and Air Mapping Imagery. 13 pp.
NATO STANAG 3189 ED 6 AMD 1-90. Titling of Air Reconnaissance, Air Survey and Air Mapping Imagery. 13 pp.

—126mm
MOD UK DSTAN 67-4-01. Film, Photographic, Aerial Issue 2; Amendment 1. 7 pp.

—126mm—Cameras
NATO STANAG 3178 ED 5 AMD 1-85. Rolled Air Film and Air Framing Camera Standard Image Format Sizes. 9 pp.
NATO STANAG 3178 ED 5 AMD 2-85. Rolled Air Film and Air Framing Camera Standard Image Format Sizes. 7 pp.

—240mm
MOD UK DSTAN 67-4-01. Film, Photographic, Aerial Issue 2; Amendment 1. 7 pp.

—240mm—Cameras
NATO STANAG 3178 ED 5 AMD 1-85. Rolled Air Film and Air Framing Camera Standard Image Format Sizes. 9 pp.
NATO STANAG 3178 ED 5 AMD 2-85. Rolled Air Film and Air Framing Camera Standard Image Format Sizes. 7 pp.

Aerial Photography
See Also: Aerial Films
—Mapping—Vertical
NATO STANAG 2216 ED 3 AMD 8-70. Vertical Aerial Cartographic Photography. 20 pp.

—Target Pattern
NATO STANAG 3704 ED 1 AMD 3-74. Ground Resolution Targets for Aerial Photography. 7 pp.

INTERNATIONAL AND NON-U.S. NATIONAL STANDARDS
SUBJECT INDEX

Aerial

Aerial Positions
Use: Antenna Positions

Aerial Reconnaissance
Use For: Air Reconnaissance *See Also:* Photographic Reconnaissance; Reconnaissance

—**Films—Numbering**
NATO STANAG 3599 ED 2 AMD 0-89. Edge Numbering of Films Used for Air Reconnaissance. 5 pp.

—**Films—Titling**
NATO STANAG 3189 ED 6 AMD 0-90. Titling of Air Reconnaissance, Air Survey and Air Mapping Imagery. 13 pp.
NATO STANAG 3189 ED 6 AMD 1-90. Titling of Air Reconnaissance, Air Survey and Air Mapping Imagery. 13 pp.

—**Imagery—Cassette Tape Recorders**
NATO STANAG 7024 ED 1 AMD 0-93. Imagery Air Reconnaissance Tape Recorder Standard. 77 pp.

—**Imagery Data**
NATO STANAG 7023 ED 1 AMD 0-93. Air Reconnaissance Imagery Data Architecture. 190 pp.

—**Imagery Interpretation**
NATO STANAG 3884 ED 1 AMD 2-81. Air Imagery Interpretation Annotation. 11 pp.
NATO STANAG 3884 ED 1 AMD 3-81. Air Imagery Interpretation Annotation. 12 pp.

—**Interdepartmental Procurement**
NATO STANAG 3781 ED 1 AMD 1-85. Reconnaissance Cross-Servicing. 11 pp.

—**Military Intelligence—Glossaries**
NATO STANAG 3483 ED 3 AMD 0-90. Air Reconnaissance Intelligence Reporting Nomenclature—ATP-26(A). 4 pp.
NATO STANAG 3483 ED 3 AMD 1-90. Air Reconnaissance Intelligence Reporting Nomenclature—ATP-26(A). 5 pp.

—**Military Intelligence—Reporting**
NATO STANAG 3596 ED 4 AMD 5-80. Air Reconnaissance Requesting and Target Reporting Guide. 41 pp.
NATO STANAG 3596 ED 4 AMD 6-80. Air Reconnaissance Requesting and Target Reporting Guide. 42 pp.

—**Reporting—Military Intelligence**
NATO STANAG 3377 ED 5 AMD 6-80. Air Reconnaissance Intelligence Report Forms. 18 pp.

—**Reporting—Technical Manuals**
NATO STANAG 3920 ED 2 AMD 0-90. Handbook for Air Reconnaissance Tasking and Reporting-ATP-47. 5 pp.

—**Request/Task Forms**
NATO STANAG 3277 ED 6 AMD 5-78. Air Reconnaissance Request/Task Form. 10 pp.

Aerial Resupply
Use: Air Drop Operations

Aerial Rods
IEC 223-66. Dimensions of Aerial Rods and Slabs of Ferromagnetic Oxides First Edition; (Supplement A-1972) (Supplement B-1977). 32 pp.

—**Magnetic Measurement**
IEC 492-74. Measuring Methods for Aerial Rods First Edition. 12 pp.

Aerial Ropeways
Use For: Aerial Cableways
CNS A1025-80. Aerial Ropeways (Jan)(5085).

—**Buckets**
CNS M2083-80. Bucket for Rope Ways (Dec)(6844).

—**Glossaries**
BSI BS 3810: Part 7-73. 1973 Glossary of Terms Used in Materials Handling Part 7: Terms Used in Connection with Aerial Ropeways and Cableways. 22 pp.

—**Safety**
CSA CAN/CSA-Z98-M91. Passenger Ropeways; (Gen Instr 1) (Supplement 1 1992). 130 pp.

Aerial Rudders
Use: Rudders—Aerial

Aerial Slabs
IEC 223-66. Dimensions of Aerial Rods and Slabs of Ferromagnetic Oxides First Edition; (Supplement A-1972) (Supplement B-1977). 32 pp.

Aerial Surveys
—**Films—Titling**
NATO STANAG 3189 ED 6 AMD 0-90. Titling of Air Reconnaissance, Air Survey and Air Mapping Imagery. 13 pp.
NATO STANAG 3189 ED 6 AMD 1-90. Titling of Air Reconnaissance, Air Survey and Air Mapping Imagery. 13 pp.

Aerials
Use: Antennas

Aero Commander 200 Aircraft
See Also: Aircraft

—**Antenna Positions**
CAA. Aero Commander 200 (Approved Aerial Positions). 1 p.

Aero Commander 500 Aircraft
See Also: Aircraft

—**Antenna Positions**
CAA. Aero Commander 500/520/560 (Approved Aerial Positions). 1 p.

Aero Commander 520 Aircraft
See Also: Aircraft

—**Antenna Positions**
CAA. Aero Commander 500/520/560 (Approved Aerial Positions). 1 p.

Aero Commander 560 Aircraft
See Also: Aircraft

—**Antenna Positions**
CAA. Aero Commander 500/520/560 (Approved Aerial Positions). 1 p.

Aero Commander 680 Aircraft
See Also: Aircraft

—**Antenna Positions**
CAA. Aero Commander 680, 680E and 680F (Approved Aerial Positions). 1 p.

Aero Commander 680E Aircraft
See Also: Aircraft

—**Antenna Positions**
CAA. Aero Commander 680, 680E and 680F (Approved Aerial Positions). 1 p.

Aero Commander 680F Aircraft
See Also: Aircraft

—**Antenna Positions**
CAA. Aero Commander 680, 680E and 680F (Approved Aerial Positions). 1 p.

Aero Commander 680T Aircraft
See Also: Aircraft

—**Antenna Positions**
CAA. Aero Commander 680T (Approved Aerial Positions). 1 p.

Aero Commander 685 Aircraft
See Also: Aircraft

—**Antenna Positions**
CAA. Aero Commander 685, 690A, 690B, 690C, 690D, 695 and 695A (Approved Aerial Positions). 1 p.

Aero Commander 690A Aircraft
See Also: Aircraft

—**Antenna Positions**
CAA. Aero Commander 685, 690A, 690B, 690C, 690D, 695 and 695A (Approved Aerial Positions). 1 p.

Aero Commander 690B Aircraft
See Also: Aircraft

—**Antenna Positions**
CAA. Aero Commander 685, 690A, 690B, 690C, 690D, 695 and 695A (Approved Aerial Positions). 1 p.

Aero Commander 690C Aircraft
See Also: Aircraft

—**Antenna Positions**
CAA. Aero Commander 685, 690A, 690B, 690C, 690D, 695 and 695A (Approved Aerial Positions). 1 p.

Aero Commander 690D Aircraft
See Also: Aircraft

Aero Commander 690D Aircraft (Cont.)
—**Antenna Positions**
CAA. Aero Commander 685, 690A, 690B, 690C, 690D, 695 and 695A (Approved Aerial Positions). 1 p.

Aero Commander 695 Aircraft
See Also: Aircraft

—**Antenna Positions**
CAA. Aero Commander 685, 690A, 690B, 690C, 690D, 695 and 695A (Approved Aerial Positions). 1 p.

Aero Commander 695A Aircraft
See Also: Aircraft

—**Antenna Positions**
CAA. Aero Commander 685, 690A, 690B, 690C, 690D, 695 and 695A (Approved Aerial Positions). 1 p.

Aero-Engines
Use: Aircraft Engines

Aero 45 Aircraft
See Also: Aircraft

—**Antenna Positions**
CAA. Aero/Super Aero 45/145 (Approved Aerial Positions). 1 p.

Aerobatic Maneuvers
Use: Aerobatics

Aerobatics
CAA NOTICE #51 ISSUE 1. Aerobatic Manoeuvres (Airworthiness Notices). 2 pp.

—**Aircraft**
CAA Chapter K2-12 10.92. Handling—Aerobatics (Light Aeroplanes). 3 pp.
CAA Chapter K2-12 App 10.92. Handling—Aerobatics (Light Aeroplanes). 2 pp.

—**Gliders**
CAA JAR-22 Appendix F. Glossary of Aerobatic Manoeuvres (Joint Airworthiness Requirements). 2 pp.

Aerobic Bacteria
See Also: Bacteria

—**Dairy Products**
BSI BS 4285: Sec 3.3-86. 1986 Microbiological Examination for Dairy Purposes Part 3: Methods for Detection and/or Enumeration of Specific Groups of Microorganisms Section 3.3: Enumeration of Aerobic Bacterial Spores. 4 pp.

—**Meat**
ISO 2293-88. Meat and Meat Products—Enumeration of Micro-Organisms—Colony Count Technique at 30 Degrees Celsius (Reference Method) Second Edition. 7 pp.

—**Water**
ISO 6222-88. Water Quality—Enumeration of Viable Micro-Organisms—Colony Count by Inoculation in or on a Nutrient Agar Culture Medium First Edition. 4 pp.

Aerobic Digestion
—**Waste Water**
BSI BS 6068: Sec 5.13-93. 1993 Water Quality—Evaluation of the Aerobic Biodegradability of Organic Compounds in an Aqueous Medium—Static Test (Zahn-Wellens Method) (ISO 9888: 1991) (N). 14 pp.
BSI BS EN 29888-93. 1993 Water Quality—Evaluation of the Aerobic Biodegradability of Organic Compounds in an Aqueous Medium Static Test (Zahn-Wellens Method) (ISO 9888: 1991) (N). 14 pp.
CEN EN 29888-93. Water Quality—Evaluation of the Aerobic Biodegradability of Organic Compounds in an Aqueous Medium—Static Test (Zahn-Wellens Method) (ISO 9888: 1991). 9 pp.
ISO 9888-91. Water Quality—Evaluation of the Aerobic Biodegradability of Organic Compounds in an Aqueous Medium—Static Test (Zahn-Wellens Method) First Edition; (CEN EN 29888:1993). 10 pp.

—**Water**
BSI BS 6068: Sec 5.6-85. 1985 Water Quality Part 5: Methods for Biological Testing Section 5.6: Evaluation in an Aqueous Medium of the 'Ultimate' Aerobic Biodegradability of Organic Compounds: Method by Analysis of Dissolved Organic Carbon (DOC). 8 pp.

INTERNATIONAL AND NON-U.S. NATIONAL STANDARDS
SUBJECT INDEX
Aeronautical

Aerobic Digestion *(Cont.)*
—Water *(Cont.)*
BSI BS 6068: Sec 5.11-93. 1993 Water Quality Evaluation in an Aqueous Medium of the 'Ultimate' Aerobic Biodegradability of Organic Compounds: Method by Determining the Oxygen Demand in a Closed Respirometer (ISO 9408: 1991) (N). 17 pp.

BSI BS 6068: Sec 5.12-93. 1993 Water Quality Evaluation in an Aqueous Medium of the 'Ultimate' Aerobic Biodegradability of Organic Compounds: Method by Analysis of Released Carbon Dioxide (ISO 9439: 1990) (N). 16 pp.

BSI BS 6068: Sec 5.13-93. 1993 Water Quality—Evaluation of the Aerobic Biodegradability of Organic Compounds in an Aqueous Medium—Static Test (Zahn-Wellens Method) (ISO 9888: 1991) (N). 14 pp.

BSI BS EN 29408-93. 1993 Water Quality Evaluation in an Aqueous Medium of the 'Ultimate' Aerobic Biodegradability of Organic Compounds: Method by Determining the Oxygen Demand in a Closed Respirometer (ISO 9408: 1991) (N). 17 pp.

BSI BS EN 29439-93. 1993 Water Quality Evaluation in an Aqueous Medium of the 'Ultimate' Aerobic Biodegradability of Organic Compounds: Method by Analysis of Released Carbon Dioxide (ISO 9439: 1990) (N). 16 pp.

BSI BS EN 29888-93. 1993 Water Quality—Evaluation of the Aerobic Biodegradability of Organic Compounds in an Aqueous Medium Static Test (Zahn-Wellens Method) (ISO 9888: 1991) (N). 14 pp.

CEN EN 29408-93. Water Quality—Evaluation in an Aqueous Medium of the 'Ultimate' Aerobic Biodegradability of Organic Compounds—Method by Determining the Oxygen Demand in a Closed Respirometer (ISO 9408: 1991). 12 pp.

CEN EN 29439-93. Water Quality—Evaluation in an Aqueous Medium of the 'Ultimate' Aerobic Biodegradability of Organic Compounds—Method by Analysis of Released Carbon Dioxide (ISO 9439: 1990). 11 pp.

CEN EN 29888-93. Water Quality—Evaluation of the Aerobic Biodegradability of Organic Compounds in an Aqueous Medium—Static Test (Zahn-Wellens Method) (ISO 9888: 1991). 9 pp.

CENELEC PREN 210634-92. Water Quality—Guidance for the Evaluation in an Aqueous Medium of the "Ultimate" Biodegradability of Poorly Soluble Organic Compounds. 14 pp.

ISO 7827-84. Water Quality—Evaluation in an Aqueous Medium of the "Ultimate" Aerobic Biodegradability of Organic Compounds—Method by Analysis of Dissolved Organic Carbon (DOC) First Edition. 7 pp.

ISO 9408-91. Water Quality—Evaluation in an Aqueous Medium of the "Ultimate" Aerobic Biodegradability of Organic Compounds—Method by Determining the Oxygen Demand in a Closed Respirometer First Edition; (Corrigendum 1-1992) (CEN EN 29408: 1993). 14 pp.

ISO 9439-90. Water Quality—Evaluation in an Aqueous Medium of the "Ultimate" Aerobic Biodegradability of Organic Compounds—Method by Analysis of Released Carbon Dioxide First Edition; (Corrigendum 1-1992) (CEN EN 29439: 1993). 13 pp.

ISO 9888-91. Water Quality—Evaluation of the Aerobic Biodegradability of Organic Compounds in an Aqueous Medium—Static Test (Zahn-Wellens Method) First Edition; (CEN EN 29888:1993). 10 pp.

Aerodromes
Use: Airports

Aerodynamic Loads
See Also: Ground Loads; Gust Loads
—Aerostats
CAA Chapter Q3-2 12.79. Design Airspeeds and Manoeuvres (Non-Rigid Airships). 1 p.
—Aircraft
CAA Chapter K3-2 App 10.69. Flight Manoeuvring Loads and Design Air Speeds (Light Aeroplanes). 2 pp.
CAA STANDARD NO. 4-3 08.69. Design Airspeeds, Gust Cases and Flight Manoeuvring Loads. 13 pp.
CAA CAP 482 SUB-Part C 03.83. Structure (Small Light Aeroplanes). 12 pp.
—Gliders
CAA JAR-22 SUBPART C. Structure (Joint Airworthiness Requirements). 17 pp.
—Rotary Wing Aircraft
CAA Chapter G3-2 01.75. Flight Loads (Rotocraft). 3 pp.
CAA CAP 524 SUB-Part C 12.86. Structure (Rotorcraft). 11 pp.

Aerodynamic Testing
Use: Aerodynamics

Aerodynamics
See Also: Aeroelasticity; Aircraft; Flight Testing; Mach Number
—Air Handling Equipment
BSI BS 4773: Sec 1.1-91. 1991 Methods for Testing and Rating Air Terminal Devices for Air Distribution Systems Part 1: Aerodynamic Testing Section 1.1: Testing Under Laboratory Conditions. 30 pp.
BSI BS 6821-88. 1988 Methods for Aerodynamic Testing of Dampers and Valves. 11 pp.
ISO 5219-84. Air Distribution and Air Diffusion—Laboratory Aerodynamic Testing and Rating of Air Terminal Devices First Edition. 30 pp.
ISO 5220-81. Air Distribution and Air Diffusion—Aerodynamic Testing and Rating of Constant and Variable Dual or Single Duct Boxes and Single Duct Units First Edition; (Addendum 1-1984). 16 pp.
ISO 7244-84. Air Distribution and Air Diffusion—Aerodynamic Testing of Dampers and Valves First Edition. 8 pp.
—Glossaries
BSI BS 185: Sec 4-69. 1969 Amd 1 Glossary of Aeronautical and Astronautical Terms Section 4: Aerodynamics (Stability and Control, Performance Fluid Motion and Model Testing). 19 pp.
—Unit Heaters
BSI BS 4773: Part 1-71. (WITHDRAWN) 1971 Methods for Testing and Rating Air Terminal Devices for Air Distribution Systems Part 1: Aerodynamic Testing (Superseded by BS 4773: Section 1.1: 1991). 53 pp.
BSI BS 4857: Part 1-72. 1972 Methods for Testing and Rating Terminal Reheat Units for Air Distribution Systems Part 1: Thermal and Aerodynamic Performance. 20 pp.
BSI BS 4954: Part 1-73. 1973 Methods for Testing and Rating Induction Units for Air Distribution Systems Part 1: Thermal and Aerodynamic Performance. 30 pp.
BSI BS 4979-86. 1986 Aerodynamic Testing of Constant and Variable Dual or Single Duct Boxes, Single Duct Units and Induction Boxes for Air Distribution Systems. 22 pp.

Aeroelasticity
See Also: Aerodynamics; Flight Dynamics
—Glossaries
BSI BS 185: Sec 3-69. 1969 Amd 1 Glossary of Aeronautical and Astronautical Terms Section 3: Structure (Detail Parts, Aeroelasticity, Fatigue). 10 pp.

Aeromedical Evacuation
See Also: Air Medical Transport Units; Military Operations; Rescue Operations; Transportation
NATO STANAG 3204 ED 4 AMD 6-73. Aeromedical Evacuation. 20 pp.

Aeronautical Charts
Use: Aeronautical Navigation Charts

Aeronautical Communication Equipment
Scope Note: Use a more specific term
See: Aeronautical Communications; Air Navigation Equipment; Avionics; Communication Equipment; Radio Equipment; Satellite Communication Equipment; Space Operation Systems

Aeronautical Communications
Scope Note: For additional listings, use a more specific term *See Also:* Aeronautical Fixed Telecommunications Network; Aeronautical Passenger Communications; Aeronautical Telecommunication Network; Aeronautics; Air Navigation; Aircraft; Fixed Satellite Services; Fixed Services (Radio Communications); Mobile Radio Services; Mobile Satellite Communications; Radio Communications; Space Communications; Space Operation Services; Space Operation Systems; Space Research Services; Space Research Systems; Telecommunication Equipment

ICAO Annex 10 Vol I. Aeronautical Telecommunications Volume I (Part I—Equipment and Systems; Part II—Radio Frequencies) Fourth Edition of Volume I—April 1985; (Incorporating Amendments 1 to 65 4/6/85) (Supplement 11/15/85). 385 pp.

ICAO Annex 10 Vol II. Aeronautical Telecommunications Volume II (Communication Procedures Including Those with PANS Status) Fourth Edition of Volume II—April 1985; (Incorporating Amendments 1 to 65 4/6/85) (Supplement 11/15/85) (Amendment 1 10/20/86 to Supplement). 122 pp.

Aeronautical Communications *(Cont.)*
ICAO 8400. Procedures for Air Navigation Services ICAO Abbreviations and Codes Fourth Edition—1989; (Incorporating Amendments 1 to 20) (Amendment 21 7/1/93). 133 pp.

Aeronautical Engineering
See Also: Aircraft
—Certification
CAA NOTICE #31 ISSUE 18. United Kingdom Aeronautical Engineering Certificates. 3 pp.
—Glossaries
BSI BS 6100: SUBSEC 2.4.3-92. 1992 Building and Civil Engineering Terms Part 2: Civil Engineering Section 2.4: Highway, Railway and Airport Engineering Subsection 2.4.3: Airport Engineering. 10 pp.

Aeronautical Fixed Telecommunications Network
Use For: AFTN *See Also:* Aeronautical Communications; Aeronautical Telecommunication Network; Aeronautics; Telecommunication Equipment
ICAO 8259. Manual on the Planning and Engineering of the Aeronautical Fixed Telecommunication Network Fifth Edition—1991. 143 pp.
—Teletypewriters
ICAO 7946. Manual of Teletypewriter Operating Practices Prepared in Consultation with the ICAO Panel of Teletypewriter Specialists Fourth Edition—1967. 66 pp.

Aeronautical Information Services
Use For: AIS
ICAO 7383. Aeronautical Information Services Provided by States 80th Edition—March 1993. 132 pp.
—Airports—Briefing
NATO STANAG 3052 ED 4 AMD 4-81. Aeronautical Briefing Facilities. 13 pp.
NATO STANAG 3052 ED 4 AMD 5-81. Aeronautical Briefing Facilities. 16 pp.
—Codes
ICAO Annex 15. Aeronautical Information Services Eighth Edition-July 1991. 50 pp.
ICAO 8126. Aeronautical Information Services Manual Fourth Edition—1991. 263 pp.
ICAO 8400. Procedures for Air Navigation Services ICAO Abbreviations and Codes Fourth Edition—1989; (Incorporating Amendments 1 to 20) (Amendment 21 7/1/93). 133 pp.
—Location Indicators
ICAO 7910. Location Indicators 71st Edition—May 1993. 174 pp.
—Navigation Charts
NATO STANAG 3412 ED 4 AMD 2-89. Aeronautical Information on Aeronautical Charts. 6 pp.
NATO STANAG 3412 ED 4 AMD 3-89. Aeronautical Information on Aeronautical Charts. 7 pp.
NATO STANAG 3412 ED 4 AMD 4-89. Aeronautical Information on Aeronautical Charts. 7 pp.
NATO STANAG 3412 ED 4 AMD 5-89. Aeronautical Information on Aeronautical Charts. 8 pp.

Aeronautical Meteorology
Use For: Aviation Meteorology
See Also: Aeronautics; Air Transportation; Aircraft
ICAO Annex 3. Meteorological Services for International Air Navigation Eleventh Edition—July 1992; (Incorporating Amendments 1 to 69 11/12/92) (Corrigendum 1/21/93). 98 pp.
ICAO 8896. Manual of Aeronautical Meteorological Practice Fourth Edition—1993. 133 pp.
ICAO 9377. Manual on Co-Ordination Between Air Traffic Services and Aeronautical Meteorological Services First Edition—1983; (Corrigendum 01/18/84). 65 pp.
—Air Traffic Control
ICAO 9377. Manual on Co-Ordination Between Air Traffic Services and Aeronautical Meteorological Services First Edition—1983; (Corrigendum 01/18/84). 65 pp.
—Glossaries
BSI BS 185: Sec 15-72. 1972 Glossary of Aeronautical and Astronautical Terms Section 15: Meteorology (General, Winds, Weather, Clouds, Synoptic Meteorology, Ice Formation in Flight, Instruments). 15 pp.

INDUSTRY STANDARDS

Aeronautical Meteorology (Cont.)

—Naval Forces

NATO STANAG 6006 ED 1 AMD 0-81. Allied Naval and Maritime Air Meteorological Procedures and Services—AWP-1. 4 pp.

Aeronautical Mobile Services

Use: Mobile Radio Services—Aeronautical

Aeronautical Navigation

Use: Air Navigation

Aeronautical Navigation Charts

See Also: Aeronautics; Air Transportation; Charts; Navigational Aids; Navigational Charts

ICAO Annex 4. Aeronautical Charts Eighth Edition—July 1985; (Incorporating Amendments 1 to 47 11/21/85) (Supplement 1/6/86) (Amendment 1 5/2/86 to Supplement) (Amendment 2 4/30/87 to Supplement) (Amendment 3 2/19/88 to Supplement) (Amendment 48 7/31/89) (Amendment 4 7/14/90 to. 154 pp.

ICAO 8697. Aeronautical Chart Manual Second Edition—1987; (Addendum 09/16/87) (Amendment 1 02/08/90). 411 pp.

—Catalogs

ICAO 7101. Aeronautical Chart Catalogue Twenty-Seventh Edition—January 1992. 292 pp.

—Data Processing

NATO STANAG 3721 ED 3 AMD 1-88. Automatic Data Processing (ADP) Master File for Land Maps and Aeronautical Charts. 23 pp.

—Edition Designation

NATO STANAG 3671 ED 2 AMD 7-72. Edition Designation System for Land Maps, Aeronautical Charts and Military Geographic Documentation. 10 pp.

NATO STANAG 3671 ED 3 AMD 0-92. Edition Designation System for Land Maps, Aeronautical Charts and Military Geographic Documentation. 10 pp.

NATO STANAG 3671 ED 3 AMD 1-92. Edition Designation System for Land Maps, Aeronautical Charts and Military Geographic Documentation. 10 pp.

NATO STANAG 3671 ED 3 AMD 2-92. Edition Designation System for Land Maps, Aeronautical Charts and Military Geographic Documentation. 11 pp.

—Elevation

NATO STANAG 3591 ED 3 AMD 0-89. Criteria for Maximum Elevation Figures for Aeronautical Charts. 7 pp.

—Evaluation

NATO STANAG 2215 ED 5 AMD 1-89. Evaluation of Land Maps, Aeronautical Charts and Digital Topographic Data. 29 pp.

—Geographic Reference Systems

NATO STANAG 3408 ED 2 AMD 5-67. Position Reference Systems for Aeronautical Charts. 11 pp.

NATO STANAG 3408 ED 2 AMD 8-67. Position Reference Systems for Aeronautical Charts. 14 pp.

—Indexes (Documentation)

NATO STANAG 3672 ED 1 AMD 5-73. Indexes to Series of Land Maps and Aeronautical Charts and Indexes to Military Geographic Information and Documentation (MGID). 10 pp.

NATO STANAG 3672 ED 1 AMD 6-73. Indexes to Series of Land Maps and Aeronautical Charts and Indexes to Military Geographic Information and Documentation (MGID). 12 pp.

NATO STANAG 3672 ED 1 AMD 7-73. Indexes to Series of Land Maps and Aeronautical Charts and Indexes to Military Geographic Information and Documentation (MGID). 13 pp.

—Information Services

NATO STANAG 3412 ED 4 AMD 2-89. Aeronautical Information on Aeronautical Charts. 6 pp.

NATO STANAG 3412 ED 4 AMD 3-89. Aeronautical Information on Aeronautical Charts. 7 pp.

NATO STANAG 3412 ED 4 AMD 4-89. Aeronautical Information on Aeronautical Charts. 7 pp.

NATO STANAG 3412 ED 4 AMD 5-89. Aeronautical Information on Aeronautical Charts. 8 pp.

—Maintenance

NATO STANAG 7016 ED 1 AMD 1-90. Maintenance of Geographic Materials. 24 pp.

NATO STANAG 7016 ED 2 AMD 0-93. Maintenance of Geographic Materials. 23 pp.

Aeronautical Navigation Charts (Cont.)

—Marginal Data

NATO STANAG 3676 ED 2 AMD 3-87. Marginal Information on Land Maps, Aeronautical Charts and Photomaps. 31 pp.

—Military Operations

NATO STANAG 3600 ED 3 AMD 4-79. Topographical Land Maps and Aeronautical Charts 1:250,000 for Joint Operations. 8 pp.

NATO STANAG 3600 ED 3 AMD 5-79. Topographical Land Maps and Aeronautical Charts 1:250,000 for Joint Operations. 9 pp.

—Projections

NATO STANAG 3409 ED 3 AMD 5-72. Projections for Aeronautical Charts. 7 pp.

—Scales

NATO STANAG 3677 ED 2 AMD 1-76. Standard Scales for Land Maps and Aeronautical Charts. 7 pp.

NATO STANAG 3677 ED 2 AMD 3-76. Standard Scales for Land Maps and Aeronautical Charts. 6 pp.

NATO STANAG 3677 ED 2 AMD 4-76. Standard Scales for Land Maps and Aeronautical Charts. 8 pp.

—Sizes

NATO STANAG 3666 ED 1 AMD 0-92. Maximum Sizes for Maps, Aeronautical Charts and other Geographic Products (Excluding Nautical Charts). 8 pp.

NATO STANAG 3666 ED 1 AMD 1-92. Maximum Sizes for Maps, Aeronautical Charts and Other Geographic Products (Excluding Nautical Charts). 11 pp.

—Symbols

NATO STANAG 3675 ED 1 AMD 11-74. Symbols on Land Maps, Aeronautical Charts and Special Naval Charts. 41 pp.

NATO STANAG 3675 ED 1 AMD 12-74. Symbols on Land Maps, Aeronautical Charts and Special Naval Charts. 42 pp.

Aeronautical Parts

Use: Aircraft Equipment

Aeronautical Passenger Communications

Use For: APC; Terrestrial Flight Telephone Systems; TFTS *See Also:* Aeronautical Communications; Aircraft Communications; Radio Communications; Telephone Systems

—Frequency Bands

CEPT T/R 42-01 E-91. Designation of Frequency Bands for the Pan-European Terrestrial Flight Telephone System (TFTS). 2 pp.

EC COM(92) 314-92. Proposal for a Council Directive on Common Frequency Bands to Be Designated for the Coordinated Introduction of the Terrestrial Flight Telecommunications System (TFTS) in the Community. 2 pp.

—Mobile Satellite Communications—Operation

CCITT RECMN F.113-92. Service Provisions for Aeronautical Passenger Communications Supported by Mobile-Satellite Systems (Study Group I) 5 pp. 5 pp.

—Mobile Satellite Communications—Quality of Service

CCITT RECMN F.113-92. Service Provisions for Aeronautical Passenger Communications Supported by Mobile-Satellite Systems (Study Group I) 5 pp. 5 pp.

—Radio Communications—Interfaces

CENELEC PRETS 300 326-1-93. Radio Equipment and Systems (RES); Terrestrial Flight Telephone System (TFTS) Part 1: Speech Services, Facilities and Requirements. 45 pp.

CENELEC PRETS 300 326-2-93. Radio Equipment and Systems (RES); Terrestrial Flight Telephone System (TFTS) Part 2: Speech Services, Radio Interface. 316 pp.

ETSI PRETS 300 326-1-93. Radio Equipment and Systems (RES); Terrestrial Flight Telephone System (TFTS) Part 1: Speech Services, Facilities and Requirements. 45 pp.

ETSI PRETS 300 326-2-93. Radio Equipment and Systems (RES); Terrestrial Flight Telephone System (TFTS) Part 2: Speech Services, Radio Interface. 316 pp.

Aeronautical Screws

Use: Aerospace Screws

Aeronautical Telecommunication Network

Use For: ATN *See Also:* Aeronautical Communications; Aeronautical Fixed Telecommunications Network; Aeronautics; Air Traffic Control; Telecommunication Equipment; Telecommunication Networks

ICAO 9578. Manual of the Aeronautical Telecommunication Network (ATN) First Edition—1991. 60 pp.

Aeronautics

Scope Note: For additional listings, use a more specific term. See also the subheading Aeronautical under specific products or materials

See Also: Aeronautical Communications; Aeronautical Fixed Telecommunications Network; Aeronautical Meteorology; Aeronautical Navigation Charts; Aeronautical Telecommunication Network; Aircraft

—Glossaries

BSI BS 185: Sec 1-69. 1969 Amd 1 Glossary of Aeronautical and Astronautical Terms Section 1: General (Including Main Types of Aircraft). 7 pp.

BSI BS 185: Sec 13-72. 1972 Glossary of Aeronautical and Astronautical Terms Section 13: Air-Traffic and Ground Services (General, Air-Traffic Control, Ground Services, Naval Terms). 12 pp.

—Indexes

BSI BS 185 Index-73. Obsolescent 1973 Glossary of Aeronautical and Astronautical Terms: Index. 38 pp.

—Symbols

JIS W 0104-60. Letter Symbols for Aeronautical Sciences.

Aeronca 15AC Aircraft

See Also: Aircraft

—Antenna Positions

CAA. Aeronca 15AC. 1 p.

Aeroplanes

Use: Aircraft

Aerosol Cans

Use: Aerosol Containers

Aerosol Containers

See Also: Containers; Pressure Vessels

EC 75/324/EEC-75. Council Directive on the Approximation of the Laws of the Member States Relating to Aerosol Dispensers. 8 pp.

—Filling—Safety

CNS Z5083-81. Safe Fill of Aerosol Containers (Mar)(7182). 1 p.

—Food

CNS Z6056-82. Method of Test for Overrun on Food Aerosol Products (Nov)(9649).

—Glass—Filling

CNS Z6040-81. Filling and Inspection of Glass Aerosol Containers (Jun)(7537). 2 pp.

—Glass—Inspection

CNS Z6040-81. Filling and Inspection of Glass Aerosol Containers (Jun)(7537). 2 pp.

—Glass—Pressure Measurement

CNS Z6032-80. Method of Test for Pressure in Glass Aerosol Bottles (Dec)(6867). 2 pp.

—Identification Systems

CNS Z5082-81. Labelling of Aerosol Cans (Mar)(7181). 1 p.

—Insecticides

BSI BS 4172-67. (WITHDRAWN) 1967 Insecticidal Efficiency of Aerosols Against Flies (Superseded by BS 4172: Parts 1 & 2: 1993). 17 pp.

BSI BS 4172: Part 1-93. 1993 Hand-Held Pressurized Aerosol Dispensers Against Houseflies Part 1: Specification for Insecticidal Efficiency (Supersedes BS 4172: 1967). 6 pp.

BSI BS 4172: Part 2-93. 1993 Hand-Held Pressurized Aerosol Dispensers Against Houseflies Part 2: Method for Determination of Insecticidal Efficiency (Supersedes BS 4172: 1967). 20 pp.

MOD UK DSTAN 68-161-93. Dispensers, Insecticide Aerosol Flying Insect Killer Issue 1. 14 pp.

MOD UK TS 10123. Dispensers, Insecticidal, Aerosol (Superseded by Def Stan 68-161).

—Inspection

CNS Z6037-81. Reporting Laboratory Results When Checking Aerosol Containers (Mar)(7179). 1 p.

—Measurement

CNS Z6059-82. Method of Measurement of A-D Dimension of Aerosol Value Dip Tubes (Dec)(9764).

Aerosol Containers (Cont.)

—Metal

BSI BS 3914-91. 1991 Specification for Non-Refillable Metal Aerosol Dispensers of 50mL to 1400 mL Capacity and up to 85mm Diameter. 12 pp.

BSI BS 3914: Part 1-74. (WITHDRAWN) 1974 Aerosol Dispensers Part 1: Non-Returnable Metal Containers up to 1400 Cubic Centimetres and 85 mm Diameter (Superseded by BS 3914: 1991). 13 pp.

CNS B4019-81. Aerosol Dispenser (Jun)(3305). 2 pp.

—Metal—Capacity Measurement

BSI BS 6966: Part 3-88. (WITHDRAWN) 1988 Light Gauge Metal Containers Part 3: Glossary of Terms for Aerosol Cans and Methods for Determination of Dimensions Capacities (Renumbered As BS EN 20090-3: 1993). 11 pp.

BSI BS EN 20090-3-93. 1993 Amd 1 Light Gauge Metal Containers. Definitions and Determination Methods for Dimensions and Capacities Part 3: Aerosol Cans (ISO 90-3: 1986) (AMD 7480) March 15, 1993 (H). 17 pp.

CEN EN 20090-3-92. Light Gauge Metal Containers—Definitions and Determination Methods for Dimensions and Capacities—Part 3: Aerosol Cans (ISO 90-3: 1986). 9 pp.

ISO 90 Pt 3-86. Light Gauge Metal Containers—Definitions and Determination Methods for Dimensions and Capacities—Part 3: Aerosol Cans First Edition; (CEN EN 20090-3: 1992). 9 pp.

—Metal—Glossaries

BSI BS 3130: Part 4-76. 1976 Glossary of Packaging Terms Part 4: Metal Containers and Aerosols. 26 pp.

—Metal—Pressure Measurement

CNS Z6033-80. Method of Test for Pressure in Metal Aerosol Containers—Can Piercing Method (Dec)(6868). 3 pp.

CNS Z6034-80. Method of Test for Pressure in Metal Aerosol Containers—Valve Measurement Method (Dec)(6869). 2 pp.

—Metal—Shipping Containers—Hazardous Materials

CGSB CAN/CGSB-43.123-M86. Containers, Metal, Aerosol, (TC-2P, TC-2Q). 7 pp.

—Plastic

BSI BS 5597-91. 1991 Non-Refillable Plastics Aerosol Dispensers up to 1000 mL Capacity. 13 pp.

—Tinplate

BSI BS 7289-90. 1990 Dimensions of Three-Piece Tinplate Aerosol Cans. 12 pp.

CNS Z5121-82. Tin-Plate Fabricated Aerosol Cans (Dec)(9763).

ISO 10154-91. Light Gauge Metal Containers—Three-Piece Necked-in Tinplate Aerosol Cans—Dimensions of the Top End First Edition. 5 pp.

Aerosol Dispensers

Use: Aerosol Containers

Aerosol Products

Use: Aerosols

Aerosols

Scope Note: For aerosols of specific materials, see the subheading Aerosols under the specific material
See Also: Air Pollution; Fumes

—Air Cleaners

JIS Z 8901-84. Dusts and Aerosols for Industrial Testing. 21 pp.

—Air Filters

JIS K 3803-90. Testing Methods for Determining Aerosol Retention of Depthfilter for Air Sterilization. 8 pp.

—Delivery Rate

CNS Z6031-80. Methods of Test for Delivery Rate of Aerosol Products (Dec)(6866). 2 pp.

—Flammability Testing

CNS Z6038-81. Test Methods for Inflammability of Aerosol Products (Mar) (7180). 4 pp.

—Mass

CNS Z6064-83. Method of Test for Delivered Mass of Aerosol Products (Mar)(10134).

—Moisture Content

CNS Z6030-80. Method of Test for Moisture in Aerosol Products (Dec)(6865). 2 pp.

—Seepage

CNS Z6062-83. Method of Test for Seepage Rate of Aerosol Products (Feb)(10036).

Aerosols (Cont.)

—Spray Patterns

CNS Z6060-83. Practice for Comparison of Spray Patterns of Aerosol Products (Jan)(9917).

—Storage

CNS Z6039-81. Storage Testing of Aerosol Products (May)(7459).

—Volatile Matter Content—Densitometry

CNS Z6053-82. Method of Test for Volatile Content of Aerosol Products (Densimetric Method) (Sep)(9448).

—Volatile Matter Content—Vacuum Distillation Method

CNS Z6052-82. Method of Test for Volatile Content of Aerosol Products (Vacuum Distillation Method) (Sep)(9447).

Aerospace Bolts

See Also: Aerospace Equipment; Aerospace Fasteners; Aerospace Nuts; Aerospace Rivets; Aerospace Screws; Bolts; Fasteners

AECMA PREN2576-87. Bolts in Heat Resisting Steel FE-PA92HT (A286) Classification 900 MPa/650 Degrees Celsius Technical Specification. 23 pp.

AECMA PREN2582-87. Bolts in Heat Resisting Nickel Base Alloy N1-P101HT (Waspaloy) Classification 1210 MPa/730 Degrees Celsius Technical Specification. 23 pp.

AECMA PREN2583-87. Bolts in Heat Resisting Nickel Base Alloy N1-P100HT (Inco 718) Classification 1275 MPa/650 Degrees Celsius Technical Specification. 23 pp.

AECMA PREN3043-90. Fasteners, Externally Threaded, in Heat Resisting Steel FE-PA92HT (A 286) Classification: 90 MPa/550 Degrees Technical Specification. 19 pp.

AECMA PREN3293-89. Bolts, T-Head Close Tolerance Medium Thread Length in Heat Resisting Nickel Base Alloy N1-P100HT (Inconel 718), Uncoated Classification: 1275 MPa/650 Degrees Celsius. 6 pp.

AECMA PREN3294-89. Bolts, T-Head, Close Tolerance in Heat Resisting Nickel Basealloy N1-P101HT (Waspaloy), Uncoated for Increased Height Nuts Classification: 1210 MPa/730 Degrees Celsius. 6 pp.

AECMA PREN3302-89. Bolts in Heat Resisting Steel FE-PM38 (FV535) Classification: 1000 MPa 1550 Degrees Celsius Technical Specification. 22 pp.

AECMA PREN3326-90. Bolts, D-Head, Close Tolerance, Medium Thread Length, in Heat Resisting Nickel Base Alloy N1-P100HT (INCO178) Classification: 1275 MPa/550 Degrees Celsius Updated. 7 pp.

BSI 2A 100-68. (WITHDRAWN) 1968 Amd 3 General Requirements for Bolts and Nuts of Strength Not Exceeding 180 000 LbF/Square Inches (125 Hbar) (Superseded by 3A 100: 1991). 24 pp.

BSI 3A 100-91. 1991 General Requirements for Bolts and Free Running Nuts of Tensile Strength not Exceeding 1249 MPa. 16 pp.

BSI 3A 100-01. 1991 Amd 1 General Requirements for Bolts and Free Running Nuts of Tensile Strength not Exceeding 1249 MPa (AMD 7661) April 15, 1993 (S). 21 pp.

BSI A 101-69. 1969 Amd 2 General Requirements for Titanium Bolts. 29 pp.

BSI A 296-86. 1986 Tolerances of Form and Position for Bolts and Screws. Metric Series. 8 pp.

BSI A 299-89. 1989 Procurement of Alloy Steel Protruding Lead Bolts with Strength Classification 250 MPa and MJ Threads. 17 pp.

BSI A 300-89. 1989 Procurement of Alloy Steel Bolts with Strength Classification 1100 MPa and MJ Threads. 18 pp.

BSI 2S 147-76. 1976 0.5 Per Cent Nickel-Chromium-Molybdenum Steel Bars for the Manufacture of Forged Bolts and Forged Nuts. 2 pp.

BSI 2S 147-01. 1976 Amd 1 0.5 Per Cent Nickel-Chromium-Molybdenum Steel Bars for the Manufacture of Forged Bolts and Forged Nuts (AMD 6836) November 29, 1991. 3 pp.

BSI 2S 149-76. 1976 1.75 Per Cent Nickel-Chromium-Molybdenum Steel Bars for the Manufacture of Forged Bolts and Forged Nuts. 2 pp.

BSI 2S 149-01. 1976 Amd 1 1.75 Per Cent Nickel-Chromium-Molybdenum Steel Bars for the Manufacture of Forged Bolts and Forged Nuts (AMD 6837) November 29, 1991. 3 pp.

BSI S 158-77. 1977 1 Per Cent Chromium-Molybdenum Steel Bars for the Manufacture of Forged Bolts and Forged Nuts. 4 pp.

CEN PREN 2650-90. Bolts, Pan Head, Slotted, Threaded to Head, in Corrosion Resisting Steel, Passivated Classification: 600 MPa (at Ambient Temperature) /425 Degrees Celsius. 7 pp.

CEN PREN 2870-90. Bolts, Bihexagonal Normal Head, Close Tolerance Normal Shank Short Thread in Titanium Alloy, Anodized, MoS2 Lubricated Classification:1100 MPa (at Ambient Temperature) /315 Degrees Celsius. 8 pp.

Aerospace Bolts (Cont.)

CEN PREN 3818-91. Bolts in Titanium Alloy TI-P63 Classification: 1100 MPa (at Ambient Temperature) Technical Specification. 21 pp.

CEN PREN 3818-93. Aerospace Series Bolts with MJ Threads, in Titanium Alloy TI-P63 Classification: 1 100 MPa (at Ambient Temperature) Technical Specification. 22 pp.

DIN ENGL LN 65009-78. Aerospace; Bolts, Steel, with 1100 N/mm2 and 1250 N/mm2 Nominal Tensile Strength, for Temperature up to 235 Degrees C, Procurement Specification (May). 22 pp.

DIN ENGL LN 65010 Pt 1-68. Specification for Screws and Bolts of Unalloyed and Low-Alloy Steel with Minimum Tensile Strength up to 900 N/mm2 (Mar). 2 pp.

DIN ENGL LN 65010 Pt 2-68. Specification for Screws and Bolts of Unalloyed and Low-Alloy Steel with Minimum Tensile Strength up to 900 N/mm2, Minimum Ultimate Tensile Loads, Proof Loads (Mar). 1 p.

ISO 5857-88. Aerospace—Alloy Steel Protruding Head Bolts with Strength Classification 1 250 MPa and MJ Threads—Procurement Specification First Edition. 19 pp.

ISO 7689-88. Aerospace—Alloy Steel Bolts with Strength Classification 1 100 MPa and MJ Threads—Procurement Specification First Edition. 18 pp.

ISO 7913-85. Aerospace—Fasteners—Tolerances of Form and Position for Bolts and Screws First Edition. 8 pp.

—Acceptance Certificates

DIN ENGL LN 65018 Pt 1-68. Manufacturer's Acceptance Certificate for Screws and Bolts (Sept). 1 p.

—Corrosion Resistant

BSI A 301-89. 1989 Procurement of Corrosion and Heat-Resisting Steel Bolts with Strength Classification 1100 MPa and MJ Threads. 18 pp.

ISO 8168-88. Aerospace—Corrosion-and Heat-Resisting Steel Bolts with Strength Classification 1 100 MPa and MJ Threads—Procurement Specification First Edition. 18 pp.

—Countersunk Head

CEN PREN 3304-92. Bolts, 100 Degree Countersunk Reduced Head, Offset Cruciform Recess, Close Tolerance Shank, Short Thread, in Titanium Alloy, MoS2 Lubricated Classification: 1100 MPa (at Ambient Temperature)/315 Degrees C. 7 pp.

CEN PREN 3305-92. Bolts, 100 Degree Countersunk Reduced Head, Offset Cruciform Recess, Close Tolerance Shank, Short Thread, in Alloy Steel, Cadmium Plated Classification: 1100 MPa (at Ambient Temperature)/235 Degrees C. 8 pp.

CEN PREN 3381-90. Bolts, 100 Degrees Countersunk Normal Head, Offset Cruciform—Ribbed Recess, Close Tolerance Shank Short Thread in Titanium Anodized, MoS Lubricated Classification: 100 MPa (at Ambient Temperature) /315 Degrees Celsius. 8 pp.

CEN PREN 6024-91. Bolts, 100 Degree Countersunk Reduced Head, Offset Cruciform Recess, Close Tolerance Shank, Short Thread, Titanium Alloy, Lubricated Classification: 1100 MPa (at Ambient Temperature) /315 Degrees C Inch Series. 9 pp.

DIN ENGL LN 29956-80. Aerospace; Bolts Countersunk Head TORQ-SET Recess; Short Thread Length; Titanium Alloy (Aug). 4 pp.

—Countersunk Head—Collars—Locking

DIN ENGL LN 65047 Pt 1-78. Aerospace; Bolts, Close Tolerance, in Titanium Alloy, with Self-Locking Collars in Aluminium Alloy; Procurement Specification; Bolts (May). 17 pp.

DIN ENGL LN 65047 Pt 2-78. Aerospace; Bolts, Close Tolerance, in Titanium Alloy, with Self-Locking Collars in Aluminium Alloy; Procurement Specification; Collars (May). 10 pp.

—Countersunk Head—Flushness Control—Gaging

BSI A 273-78. 1978 Gauging Practice for 100 Degree Countersunk Head Fasteners for Flushness Control. 7 pp.

—Countersunk Head—Recessed

AECMA PREN3381-89. Bolts, 100 Degrees Countersunk Normal Head, Offset Enciform-Ribbed Recess, Close Tolerance Shank, Short Thread, in Titanium Anodized Classification: 1100 MPa/315 Degrees Celsius. 7 pp.

AECMA PREN3543-89. Bolts, 100 Degrees Countersunk Normal Head, Offset Cruciform-Ribbed Recess, Close Tolerance Shank, Short Thread, in Corrosion Resisting Steel, Passivated Classification: 1100 MPa/425 Degrees Celsius. 8 pp.

BSI A 266-A 271-77. 1977 Amd 1 100 Degree Countersunk Head Titanium Alloy Bolts 160 000 Lbf/Square Inches (1100 MPa) with Hi-Torque Speed Drive Recesses. 11 pp.

Aerospace
INTERNATIONAL AND NON-U.S. NATIONAL STANDARDS
SUBJECT INDEX

Aerospace Bolts *(Cont.)*

—Countersunk Head—Recessed *(Cont.)*

BSI A 272-77. 1977 Hi-Torque Speed Drive Recess: Dimensions and Gauging for Countersunk Head Fasteners. 3 pp.

CEN PREN 3760-90. Bolts, 100 Degress C Countersunk Normal Head, Offset, Cruciform-Ribbed Recess, Threaded to Head, in Heat and Corrosion Resisting Steel, Passivated Classification: 1100 MPa (at Ambient Temperature)/425 Degrees C. 8 pp.

—Countersunk Head—Slotted

AECMA PREN2653-89. Bolts, 100 Degrees Countersunk Normal Head, Slotted, Threaded to Head, in Corrosion Resisting Steel, Passivated Classification: 600 MPa/425 Degrees Celsius. 7 pp.

AECMA PREN2654-89. Bolts, 100 Degrees Countersunk Normal Head, Slotted, Threaded to Head, in Brass, Tin Plated Classification: 280 MPa/ 80 Degrees Celsius. 7 pp.

CEN PREN 2653-91. Bolts, 100 Degrees Countersunk Head, Slotted, Threaded to Head, in Corrosion Resisting Steel, Passivated Classification: 600 MPa (at Ambient Temperature) /425 Degrees C. 7 pp.

CEN PREN 2654-91. Bolts, 100 Degrees Countersunk Normal Head, Slotted Threaded to Head, in Brass, Tin Plated Classification: 380 MPa (at Ambient Temperature)/80 Degrees C. 7 pp.

—Double Hexagonal

AECMA PREN3063-87. Bolts, Double Hexagon Head, Close Tolerance, Medium Thread Length, in Heat Resisting Nickel Base Alloy N1-P101HT (Waspaloy) Classification 1210 MPa/735 Degrees Celsius, Uncoated. 5 pp.

AECMA PREN3327-89. Bolts, Double Hexagon Head, Close Tolerance Medium Thread Length in Heat Resisting Nickel Base Alloy N1-P100HT (Inconel 718), Uncoated Classification: 1275 MPa/650 Degrees Celsius. 6 pp.

AECMA PREN3328-89. Bolts, Double Hexagon Head Close Tolerance Medium Thread Length in Heat Resisting Steel FE-P1738 (FV535), Uncoated Classification: 1000 MPa/550 Degrees Celsius. 6 pp.

AECMA PREN3379-89. Bolts, Double Hexagon Head, Close Tolerance in Heat Resisting Nickel Base Alloy N1-P101HT (Waspaloy), Uncoated for Increased Height Nuts Classification: 1210 MPa/730 Degrees Celsius. 6 pp.

BSI A 241-73. 1973 Amd 3 General Requirements for Steel Protruding-Head Bolts of Tensile Strength 1250 MPa (180 000 Lbf/Square Inches) or Greater. 44 pp.

CEN PREN 2874-92. Bolts, Large Bihexagonal Head, Close Tolerance Normal Shank, Medium Length Thread, in Nickel Alloy, Passivated Classification: 1 550 MPa (at Ambient Temperature) /315 Degrees C. 7 pp.

CEN PREN 3907-92. Bolts, Double Hexagon Head, Normal Shank, Long Thread, in Titanium Alloy TI-P63, MoS2 Coated Classification: 1 100 MPa (at Ambient Temperature)/350 Degrees C. 7 pp.

DIN ENGL LN 29551-81. Aerospace; Bolts, Bi-Hexagon for Temperatures up to 650 Degrees C (Dec). 4 pp.

DIN ENGL LN 29769-80. Aerospace; Bolts Bi-Hexagon; Nominal Tensile Strength of 1250 N/ mm2; Medium Thread Length (Mar). 5 pp.

DIN ENGL LN 29934-83. Aerospace; Bolts Bi-Hexagon Close Tolerance; Nominal Tensile Strength of 1550 N/mm2; Medium Thread (Oct). 5 pp.

—Double Hexagonal—Relieved Shank

AECMA PREN2925-86. Bolts with Double Hexagon Head, Relieved Shank, Long Thread, in Heat Resisting Steel FE-PA92HT (A286) Classification 900 MPa/650 Degrees Celsius. 6 pp.

AECMA PREN2926-86. Bolts with Double Hexagon Head, Relieved Shank, Long Thread, in Heat Resisting Steel FE-PA92HT (A286) Silver Plated Classification 900 MPa/650 Degrees Celsius. 6 pp.

AECMA PREN2927-86. Bolts with Double Hexagon Head, Relieved Shank, Long Thread, in Heat Resisting Nickel Base Alloy NI-P100HT (Inconel 718) Classification 1275 MPa/650 Degrees Celsius. 6 pp.

AECMA PREN2928-86. Bolts with Double Hexagon Head, Relieved Shank, Long Thread, in Heat Resisting Nickel Base Alloy NI-P100HT (Inconel 718) Silver Plated Classification 1275 MPa/650 Degrees Celsius. 6 pp.

AECMA PREN2929-86. Bolts with Double Hexagon Head, Relieved Shank, Long Thread, in Heat Resisting Nickel Base Alloy NI-P101HT (Waspaloy) Classification 1210 MPa/735 Degrees Celsius. 6 pp.

AECMA PREN2930-86. Bolts with Double Hexagon Head, Relieved Shank, Long Thread, in Heat Resisting Nickel Base Alloy NI-P101HT (Waspaloy) Silver Plated Classification 1210 MPa/735 Degrees Celsius. 6 pp.

AECMA PREN3323-90. Bolts with Double Head, Relieved Shank, Long Thread, in Heat Resisting Steel FE-PM38 (FV535) Classification: 100 MPa/550 Degrees. 7 pp.

Aerospace Bolts *(Cont.)*

—Double Hexagonal—Relieved Shank *(Cont.)*

CEN PREN 2925-92. Bolts with Double Hexagon Head, Relieved Shank, Long Thread, in Heat Resisting Steel FE-PA92HT (A286) Classification: 900 MPa /650 Degrees C. 6 pp.

CEN PREN 2926-92. Bolts with Double Hexagon Head, Relieved Shank, Long Thread, in Heat Resisting Steel FE-PA92HT (A286), Silver Plated Classification: 900 MPa /650 Degrees C. 6 pp.

CEN PREN 2927-92. Bolts with Double Hexagon Head, Relieved Shank, Long Thread, in Heat Resisting Nickel Base Alloy NI-P100HT (Inconel 718) Classification: 1 275 MPa /650 Degrees C. 6 pp.

CEN PREN 2928-92. Bolts with Double Hexagon Head, Relieved Shank, Long Thread, in Heat Resisting Nickel Base Alloy NI-P100HT (Inconel 718), Silver Plated Classification: 1 275 MPa /650 Degrees C. 6 pp.

CEN PREN 2929-92. Bolts with Double Hexagon Head, Relieved Shank, Long Thread, in Heat Resisting Nickel Base Alloy NI-P101HT (Waspaloy) Classification: 1 210 MPa /735 Degrees C. 6 pp.

CEN PREN 2930-92. Bolts with Double Hexagon Head, Relieved Shank, Long Thread, in Heat Resisting Nickel Base Alloy NI-P101HT (Waspaloy), Silver Plated Classification: 1 210 MPa /735 Degrees C. 6 pp.

CEN PREN 3686-91. Bolts with Double Hexagon Head, Relieved Shank, Long Thread, in Heat Resisting Steel, FE-PA92HT (A286), Silver Plated Classification: 1100 MPa/650 Degrees C. 7 pp.

CEN PREN 3724-92. Bolts, Double Hexagon Head, Relieved Shank, Long Thread, in Titanium Alloy TI-P63, MoS2 Coated Classification: 1 100 MPa (at Ambient Temperature). 6 pp.

CEN PREN 3832-91. Bolts, Double Hexagon Head, Relieved Shank, Long Thread, in Heat Resisting Nickel Base Alloy NI-P100HT (INCO 718), Uncoated Classification: 1550 MPa/650 Degrees C. 8 pp.

CEN PREN 3832-92. Aerospace Series Bolts, Double Hexagon Head, Relieved Shank, Long Thread, in Heat Resisting Nickel Base Alloy NI-P100HT (Inconel 718) Classification: 1 550 MPa (at Ambient Temperature)/650 Degrees C. 6 pp.

CENELEC PREN 3686-89. Bolts with Double Hexagon Lead, Relieved Shank, Long Thread, in Heat Resisting Steel FE-PA92HT (A286) Silver Plated Classification 1100 MPa/650 Degrees C. 6 pp.

—Glossaries

ISO 5843 Pt 3-87. Aerospace—List of Equivalent Terms—Part 3: Aerospace Bolts and Nuts First Edition. 25 pp.

—Heat Resistant

BSI A 301-89. 1989 Procurement of Corrosion and Heat-Resisting Steel Bolts with Strength Classification 1100 MPa and MJ Threads. 18 pp.

BSI HR 650-72. 1972 Amd 2 High Expansion Heat-Resisting Steel Bar and Wire for the Manufacture of Bolts, Studs, Set Screws and Nuts (Ni 25.5, Cr 15, Ti 2, Mn 1.5, Mo 1.25, Si 0.7, V 0.3) (Limiting Ruling Section 20 mm). 5 pp.

CEN PREN 3685-91. Bolts in Heat Resisting Steel, FE-PA92HT (A286) Classification: MPa/650 Degrees C Technical Specification. 27 pp.

CENELEC PREN 3685-89. Bolts in Heat Resisting Steel FE-PA92HT(A286) Classification: 1100 MPa/ 650 Degrees C Technical Specification. 23 pp.

ISO 8168-88. Aerospace—Corrosion-and Heat-Resisting Steel Bolts with Strength Classification 1 100 MPa and MJ Threads—Procurement Specification First Edition. 18 pp.

—Heat Resistant—Qualification Testing

CEN PREN 3685-91. Bolts in Heat Resisting Steel, FE-PA92HT (A286) Classification: MPa/650 Degrees C Technical Specification. 27 pp.

CENELEC PREN 3685-89. Bolts in Heat Resisting Steel FE-PA92HT(A286) Classification: 1100 MPa/ 650 Degrees C Technical Specification. 23 pp.

—Heat Resistant—Quality Assurance

CEN PREN 3685-91. Bolts in Heat Resisting Steel, FE-PA92HT (A286) Classification: MPa/650 Degrees C Technical Specification. 27 pp.

CENELEC PREN 3685-89. Bolts in Heat Resisting Steel FE-PA92HT(A286) Classification: 1100 MPa/ 650 Degrees C Technical Specification. 23 pp.

—Hexagonal Head

BSI A 241-73. 1973 Amd 3 General Requirements for Steel Protruding-Head Bolts of Tensile Strength 1250 MPa (180 000 Lbf/Square Inches) or Greater. 44 pp.

BSI A 254-A 259-76. 1976 Amd 2 Hexagonal Head Titanium Alloy Bolts 160 000 Lbf/ Square Inches (1100 MPa). 12 pp.

Aerospace Bolts *(Cont.)*

—Hexagonal Head *(Cont.)*

DIN ENGL LN 29930-84. Aerospace; Bolts, Hexagon, Close Tolerance with Reduced Head and Short Thread Length (Apr). 4 pp.

DIN ENGL LN 29943-81. Aerospace; Bolts, Hexagon, Close Tolerance, Short Thread, Titanium Alloy (Dec). 4 pp.

—Hexagonal Head—Close Shank

AECMA PREN2413-87. Bolts, Shouldered, Thin Hexagonal Head, Close Tolerance Shank, Short Thread, in Steel Cadmium Plated Classification 1100 MPa/235 Degrees Celsius. 7 pp.

AECMA PREN2549-87. Bolts, Hexagonal Normal Head, Close Tolerance Shank, Short Thread in Titanium, Anodised Classification 1100 MPa/315 Degrees Celsius. 7 pp.

AECMA PREN2859-89. Bolts, Hexagonal Normal Head, Close Tolerance Shank, Short Thread in Steel Cadmium Plated Classification 1100 MPa/235 Degrees Celsius. 7 pp.

AECMA PREN3052-87. Bolts, Hexagonal Normal Head, Close Tolerance Shank, Short Thread in Corrosion Resisting Steel, Passivated Classification 1100 MPa/425 Degrees Celsius. 7 pp.

BSI 3A 212-93. 1993 Cadmium-Plated Steel Bolts, 55/ 65 Tonf/Square Inches (Unified Hexagons, 3A Unified Threads and Close Tolerance Shanks) (Supersedes 2A 212: 1973) (S). 9 pp.

CEN PREN2859-87. Bolts, Hexagonal Normal Head, Close Tolerance Shank, Short Thread in Steel Cadmium Plated Classification: 1100 MPa/235 Degrees C. 9 pp.

CEN PREN 3740-92. Bolts, Shouldered, Thin Hexagonal Head, Close Tolerance Shank, Short Thread, in Titanium Alloy, MoS2 Lubricated Classificaton: 1 100 MPa (at Ambient Temperature) /315 Degrees C. 5 pp.

—Hexagonal Head—Coarse Shank

AECMA PREN 2888-88. Bolts, Hexagonal Normal Head, Coarse Tolerance Shank, Short Thread, in Corrosion Resisting Steel, Passivated Classification: 600 MPa/425 Degrees Celsius. 7 pp.

AECMA PREN2889-89. Bolts, Hexagonal Normal Head, Coarse Tolerance Shank, Short Thread, in Steel, Cadmium Plated Classification 900 MPa/235 Degrees Celsius. 2 pp.

CEN PREN2889-87. Bolts, Hexagonal Normal Head, Coarse Tolerance Shank, Short Thread, in Steel, Cadmium Plated Classification: 900 MPa/235 Degrees C. 9 pp.

—Hexagonal Head—Relieved Shank

AECMA PREN3006-87. Bolts, Hexagon Head, Relieved Shank, Long Thread, in Heat Resisting Steel FE-PA92HT (A286) Classification 900 MPa/ 650 Degrees Celsius Unplated. 7 pp.

AECMA PREN3007-87. Bolts, Hexagon Head, Relieved Shank, Long Thread, in Heat Resisting Steel FE-PA92HT (A286) Classification 900 MPa/ 650 Degrees Celsius Silver Plated. 7 pp.

AECMA PREN3008-87. Bolts, Hexagon Head, Relieved Shank, Long Thread, in Heat Resisting Nickel Base Alloy N1-P100HT (INCO 718) Classification 1275 MPa/650 Degrees Celsius Unplated. 7 pp.

AECMA PREN3009-87. Bolts, Hexagon Head, Relieved Shank, Long Thread, in Heat Resisting Nickel Base Alloy N1-P100HT (INCO 718) Classification 1275 MPa/650 Degrees Celsius Silver Plated. 7 pp.

AECMA PREN3010-87. Bolts, Hexagon Head, Relieved Shank, Long Thread, in Heat Resisting Nickel Base Alloy N1-P101HT (Waspaloy) Classification 1210 MPa/730 Degrees Celsius Unplated. 7 pp.

AECMA PREN3011-87. Bolts, Hexagon Head, Relieved Shank, Long Thread, in Heat Resisting Nickel Base Alloy N1-P101HT (Waspaloy) Classification 1210 MPa/730 Degrees Celsius Silver Plated. 7 pp.

AECMA PREN3613. Bolts, Normal Hexagon Head, Relieved Shank, Long Thread, in Heat Resisting Nickel Base Alloy NI-P100HT (Inconel 718), Silver Plated Classification: 1275 MPa/650 Degrees. 8 pp.

AECMA PREN3614-90. Bolts, Normal Hexagon Head, Relieved Shank, Long Thread, in Heat Resisting Steel FE-PA92HT (A286), Silver Plated Classification: 900 MPa/650 Degrees. 7 pp.

CEN PREN 3324-91. Bolts, Hexagon Head, Relieved Shank, Long Thread, in Heat Resisting Steel FR-PM 38 (FV 535) Classification: 1000 MPa/550 Degrees C Unplated. 7 pp.

CEN PREN 3687-91. Bolts, Normal Hexagon Head, Relieved Shank, Long Thread, in Heat Resisting Steel, FE-PA92HT (A286), Silver. 7 pp.

CEN PREN 3724-91. Bolts, Double Hexagon Head Relieved Shank, Long Thread, in Titanium Alloy TI-P63, MoS2 Coated Classification: 1100 MPa (at Ambient Temperature)/350 Degrees C. 7 pp.

INTERNATIONAL AND NON-U.S. NATIONAL STANDARDS
SUBJECT INDEX
Aerospace

Aerospace Bolts *(Cont.)*

—Hexagonal Head—Shank
ISO 3193-91. Aerospace—Bolts, Normal Hexagonal Head, Normal Shank, Short or Medium Length MJ Threads, Metallic Material, Coated or Uncoated, Strength Classes Less Than or Equal to 1 100 MPa—Dimensions First Edition; (Corrigendum 1-1992). 7 pp.

—Hexagonal Head—Shear
BSI 3A 109-85. 1985 Amd 1 Steel Bolts (Unified Hexagons and Unified Threads) (AMD 5563) July 29, 1988. 11 pp.
BSI 4A 112-92. 1992 Cadmium-Plated Steel Shear Bolts (Unified Hexagons and Unified Threads) (Supersedes 3A 112: 1985). 12 pp.
BSI 3A 112-85. (WITHDRAWN) 1985 Amd 1 Cadmium-Plated Steel Shear Bolts (Unified Hexagons and Unified Threads) (AMD 5564) September 30, 1988 (Superseded by 4A 112: 1992). 11 pp.

—Hexagonal Head—Socket—Cap
CEN PREN 3303-92. Bolts, Cap Head, Hexagon Socket, Coarse Tolerance Normal Shank, Medium Length Thread, in Alloy Steel, Cadmium Plated Classification: 1 100 MPa (at Ambient Temperature) /235 Degrees C. 6 pp.

—Hexagonal Nuts
DIN ENGL LN 29669-80. Aerospace; Hexagon Nuts for Screws and Bolts with a Nominal Tensile Strength of 1100 N/mm2 (Mar). 2 pp.

—High Expansion
BSI HR 650-72. 1972 Amd 2 High Expansion Heat-Resisting Steel Bar and Wire for the Manufacture of Bolts, Studs, Set Screws and Nuts (Ni 25.5, Cr 15, Ti 2, Mn 1.5, Mo 1.25, Si 0.7, V 0.3) (Limiting Ruling Section 20 mm). 5 pp.

—Locking Collars
DIN ENGL LN 65047 Pt 1-78. Aerospace; Bolts, Close Tolerance, in Titanium Alloy, with Self-Locking Collars in Aluminium Alloy; Procurement Specification; Bolts (May). 17 pp.
DIN ENGL LN 65047 Pt 2-78. Aerospace; Bolts, Close Tolerance, in Titanium Alloy, with Self-Locking Collars in Aluminium Alloy; Procurement Specification; Collars (May). 10 pp.

—Pan Head—Recessed
CEN PREN 3725-92. Bolts, Pan Head, 6 Lobe Recess, Long Thread, in Titanium Alloy Tl-P63, MoS2 Coated Classification: 1 100 MPa (at Ambient Temperature)/350 Degrees C. 7 pp.

—Pan Head—TORQ-SET
DIN ENGL LN 29781-80. Aerospace; Bolts, Pan Head, with TORQ-SET Recess (Aug). 3 pp.
DIN ENGL LN 29957-83. Aerospace; Bolts, Pan Head, Close Tolerance, with TORQ-SET Recess in Titanium Alloy; Reduced Thread (Oct). 5 pp.

—Phillips—Pan Head
DIN ENGL LN 9439-81. Aerospace; Bolts, Pan Head, Cross Recessed (Dec). 4 pp.

—Procurement
BSI A 274-81. 1981 Procurement of Alloy Steel Bolts, Metric, with a Minimum Tensile Strength of 1100 MPa. 20 pp.

—Quality Assurance
AECMA PREN3202-89. T-Head Bolt Traps Design Dimensions. 7 pp.
AECMA PREN3302-89. Bolts in Heat Resisting Steel FE-PM38 (FV535) Classification: 1000 MPa 1550 Degrees Celsius Technical Specification. 22 pp.

—Recessed Head
CEN PREN 3911-92. Six Lobe Recess Geometrical Definition. 4 pp.
CEN PREN 3911-93. Aerospace Series Six Lobe Recess Geometrical Definition. 3 pp.

—Recesses
DIN ENGL LN 29999 Pt 1-74. TORQ-SET Recesses; Dimensions (Aug). 2 pp.

—Rolled Threads—Runouts
BSI 2A 231-93. 1993 Rolled Threads for Bolts—Lead and Runout Requirements (ISO 3353: 1992) (S). 13 pp.
BSI A 231-82. (WITHDRAWN) 1982 Runout and Lead Threads for Rolled Threads (Superseded by BS 2A 231: 1993). 6 pp.
ISO 3353-92. Aerospace—Rolled Threads for Bolts—Lead and Runout Requirements Second Edition. 12 pp.

—Screw Threads
BSI BS 6293: Part 2-82. 1982 MJ Threads for Aerospace Construction Part 2: Dimensions for Bolts and Nuts. 9 pp.

Aerospace Bolts *(Cont.)*

—Screw Threads *(Cont.)*
DIN ENGL LN 9163 Pt 2-80. Aerospace; ISO Metric Threads; Selection of Threads and Allowances for Screws, Bolts and Nuts (Dec). 3 pp.
DIN ENGL LN 29580-80. Aerospace; ISO Metric Screws Thread; Selection of Threads and Limit Dimensions for Screws and Bolts with Nominal Tensile Strength Not Less Than 1250 N/mm2 (Dec). 4 pp.
ISO 5855 Pt 2-88. Aerospace—MJ Threads—Part 2: Limit Dimensions for Bolts and Nuts Second Edition; (Corrected and Reprinted -1989). 8 pp.

—Screw Threads—Projections
DIN ENGL LN 9378-70. Thread Ends, Projections for Metric Thread (June). 6 pp.

—Self Retaining
SBAC TS 50 ISSUE 4. Alloy Steel "JO—BOLT" Fasteners.
SBAC TS 134 ISSUE 2. Self Retaining "Impedance Type" Bolts—Metric Tensile Strength: —1275 MPa at Ambient Temperature 1065 MPa at 425 Degrees C.
SBAC TS 137 ISSUE 2. Self Retaining "Impedance Type" Bolts—Imperial Tensile Strength:—1275 MPa (185,000 lbf/in2) at Ambient Temperature 1065 MPa (154,000 lbf/in2) at 425 Degrees C.

—Self Retaining—Quality Assurance
SBAC TS 50 ISSUE 4. Alloy Steel "JO—BOLT" Fasteners.
SBAC TS 134 ISSUE 2. Self Retaining "Impedance Type" Bolts—Metric Tensile Strength: —1275 MPa at Ambient Temperature 1065 MPa at 425 Degrees C.
SBAC TS 137 ISSUE 2. Self Retaining "Impedance Type" Bolts—Imperial Tensile Strength:—1275 MPa (185,000 lbf/in2) at Ambient Temperature 1065 MPa (154,000 lbf/in2) at 425 Degrees C.

—Special Designs—Identification Systems
DIN ENGL LN 29962-76. Bolts and Accessories; Additional Procurement Data for Special Designs and Arrangement of Code Letters in Characteristic Blocks (Apr). 3 pp.

—Substitute Materials
AECMA PREN3051-87. Substitute Materials for Bolts. 3 pp.

—Tee Head
AECMA PREN2931-86. Bolts T-head, Relieved Shank, Long Thread, in Heat Resisting Steel FE-PA92HT (A286) Unplated Classification 900 MPa/650 Degrees Celsius. 5 pp.
AECMA PREN2932-86. Bolts T-Head, Relieved Shank, Long Thread, in Heat Resisting Steel FE-PA92HT (A286) Silver Plated Classification 900 MPa/650 Degrees Celsius. 5 pp.
AECMA PREN2933-86. Bolts T-Head, Relieved Shank, Long Thread, in Heat Resisting Nickel Base Alloy NI-P100HT (INCO 718) Unplated Classification 1275 MPa/650 Degrees Celsius. 5 pp.
AECMA PREN2934-86. Bolts T-Head, Relieved Shank, Long Thread, in Heat Resisting Nickel Base Alloy NI-P100HT (INCO 718) Silver Plated Classification 1275 MPa/650 Degrees Celsius. 5 pp.
AECMA PREN2935-86. Bolts T-Head, Relieved Shank, Long Thread, in Heat Resisting Nickel Base Alloy NI-P101HT (Waspaloy) Unplated Classification 1210 MPa/730 Degrees Celsius. 5 pp.
AECMA PREN2936-86. Bolts T-Head, Relieved Shank, Long Thread, in Heat Resisting Nickel Base Alloy NI-P101HT (Waspaloy) Silver Plated Classification 1210 MPa/730 Degrees Celsius. 5 pp.
AECMA PREN3293-89. Bolts, T-Head Close Tolerance Medium Thread Length in Heat Resisting Nickel Base Alloy N1-P100HT (Inconel 718), Uncoated Classification: 1275 MPa/650 Degrees Celsius. 6 pp.
AECMA PREN3294-89. Bolts, T-Head, Close Tolerance in Heat Resisting Nickel Basealloy N1-P101HT (Waspaloy), Uncoated for Increased Height Nuts Classification: 1210 MPa/730 Degrees Celsius. 6 pp.
AECMA PREN3301-89. Bolts, T-Head, Close Tolerance Medium Thread Length in Heat Resisting Steel FE-P1738 (FV535), Uncoated Classification: 1000 MPa/550 Degrees Celsius. 6 pp.
CEN PREN 3325-91. Bolts, T-head, Relieved Shank, Long Thread, in Heat Resisting Steel, FE-PM38(FV535) Classification: 1000 MPa/550 Degrees C Unplated. 7 pp.

Aerospace Clamps
See Also: Aerospace Fasteners; Clamps
AECMA PREN2900-89. Clamps Technical Specification. 11 pp.
AECMA PREN2903-89. Clamps Worm Drive Dimensions-Masses. 5 pp.

Aerospace Clamps *(Cont.)*
AECMA PREN2905-89. Rubber Cushionings for Clamps Dimensions-Masses. 5 pp.
BSI M 61-89. 1989 Envelope Dimensions of P Clamps (Loop Clamps) for Fluid Systems. 5 pp.
BSI M 62-87. 1987 Design and Qualification Testing of Clamp Blocks for Fluid System Tube Lines Having Axial Alignment. Metric Series. 11 pp.
BSI M 66-87. 1987 Dimensions of 'Q' Clamps (Centre-Mounted Clamps) for Fluid Systems. 4 pp.
ISO 7315-88. Aerospace—"P" Clamps (Loop Clamps) for Fluid Systems—Envelope Dimensions Second Edition. 6 pp.
ISO 7661-84. Aerospace—Fluid Systems—Clamp Blocks for Tube Lines Having Axial Alignment—Design Standard and Qualification Testing—Metric Series First Edition. 11 pp.
ISO 8479-86. Aerospace—"Q" Clamps (Centre-Mounted Clamps) for Fluid Systems—Dimensions First Edition. 4 pp.

—Cables (Electric)
AECMA PREN2901-89. Clamps, Loop (P Type) with Rubber Cushioning in Corrosion Resisting Steel Dimensions-Masses. 6 pp.
AECMA PREN2902-89. Clamps, Loop (P type) with Rubber Cushioning in Aluminium Alloy Dimensions-Masses. 6 pp.

—Cushions—Rubber
AECMA PREN2901-89. Clamps, Loop (P Type) with Rubber Cushioning in Corrosion Resisting Steel Dimensions-Masses. 6 pp.
AECMA PREN2902-89. Clamps, Loop (P type) with Rubber Cushioning in Aluminium Alloy Dimensions-Masses. 6 pp.

—Fluid Systems
AECMA PREN2901-89. Clamps, Loop (P Type) with Rubber Cushioning in Corrosion Resisting Steel Dimensions-Masses. 6 pp.
AECMA PREN2902-89. Clamps, Loop (P type) with Rubber Cushioning in Aluminium Alloy Dimensions-Masses. 6 pp.

—Quality Assurance
AECMA PREN2900-89. Clamps Technical Specification. 11 pp.

Aerospace Composite Materials
See Also: Aerospace Equipment; Aerospace Fabrics; Composite Materials
CEN PREN 3074-91. Drawing Annotation for Composite Laminate Structure. 16 pp.

Aerospace Electronics
Use: Avionics—Aerospace

Aerospace Engines
Use: Aircraft Engines

Aerospace Equipment
Scope Note: For additional listings, see the subheading Aerospace under specific types of equipment *Use For:* Aerospace Systems
See Also: Aerospace Bolts; Aerospace Composite Materials; Aerospace Fabrics; Aerospace Fasteners; Aerospace Materials; Aerospace Nuts; Aerospace Vehicles; Aircraft Equipment; Airframe Bearings; Missiles; Propulsion Systems
EURO EC/88/20420/E,F-89. Panorama of EC Industry—Aerospace Industry. 32 pp.
EURO 89/EEPSC/206 55-89. Equipment Technology—Identification of Changes. 11 pp.
EURO AECMA/23396/92-92. European Aeronautical Industry Towards the 21st Century. 34 pp.

—Components
MOD UK DTD-941-69. Surface Coating of Parts by Use of Detonation, Flame and Plasma Spraying Processes. 32 pp.

—Contracts—Inspection
AECMA PREN2078-82. Manufacturing Schedule—Inspection Schedule and Inspection Report—General Definitions (C5/34A). 4 pp.

—Contracts—Manufacturing
AECMA PREN2078-82. Manufacturing Schedule—Inspection Schedule and Inspection Report—General Definitions (C5/34A). 4 pp.

—Design—Interchangeability
EURO CTI/CO/79/88 22-79. Designing for Interchangeability in the Aerospace Industry. 13 pp.

—Electrical Bonding
JIS W 2009-78. Bonding, Electrical and Lightning Protection, for Aerospace Systems.

—Electromagnetic Compatibility
JIS W 7005-84. Electromagnetic Compatibility Requirements for Aerospace Systems.

Aerospace Equipment *(Cont.)*

—Environmental Testing
JIS W 0801-83. Environmental Test Methods for Aerospace Equipment.

—Glossaries
ISO 5843 Pt 4-90. Aerospace—List of Equivalent Terms—Part 4: Flight Dynamics Second Edition. 35 pp.
ISO 5843 Pt 9-88. Aerospace—List of Equivalent Terms—Part 9: Aircraft First Edition. 19 pp.

—Life Cycle Costs
EURO PSSWG/92/23335-92. AECMA Product Support Services Working Group Life Cycle Cost (LCC) an Investigative Survey of LCC Models Used by Companies in the European Aerospace Industry. 58 pp.

—Liquid Penetrant Testing
BSI M 39-72. 1972 Method for Penetrant Inspection of Aerospace Materials and Components. 12 pp.
JIS W 0904-90. Inspection Process of Liquid Penetrant for Aerospace Use.

—Magnetic Particle Testing
JIS W 4031-82. Inspection Process of Magnetic Particle for Aerospace Use.

—Nondestructive Testing
BSI M 38-71. 1971 Guide to Compilation of Instructions and Reports for the In-Service and Non-Destructive Testing of Aerospace Products. 3 pp.

—Pressure—Symbols
SBAC RS 717 ISSUE 1. Practical Expression of Aerospace System, Pressures.

—Proprietary Materials/Processes
MOD UK DTD-900Z Addendum:. Approval Procedure for Proprietary Materials and Processes (Superseded by DTD-900AA).

—Quality Assurance
AECMA PREN2000-89. Quality Assurance EN Aerospace Products Approval of the Quality System of Manufacturers. 6 pp.
AECMA PREN3042-89. Quality Assurance EN Aerospace Products Qualification Procedure. 7 pp.
BSI BS EN 2000-92. 1992 Quality Assurance—EN Aerospace Products—Approval of the Quality System of Manufacturers. 11 pp.
BSI BS EN 3042-92. 1992 Quality Assurance—EN Aerospace Products—Qualification Procedure. 12 pp.
CEN PREN2000-85. Quality Assurance Requirements for the Manufacture and Procurement of EN Aerospace Standard Products. 10 pp.
CEN EN 2000-91. Quality Assurance EN Aerospace Products Approval of the Quality System of Manufacturers. 7 pp.
CEN EN 3042-91. Quality Assurance EN Aerospace Products Qualification Procedure. 8 pp.

—Radiography
BSI M 34-70. 1970 Method of Preparation and Use of Radiographic Techniques. 7 pp.

—Targets
BSI F 133-88. 1988 Warp Knitted Nylon Fabric for Aerospace Purposes. 8 pp.
MOD UK DTD-5635-81. Banner Target and Target Towing Cable Materials. 7 pp.

—Test Specimens—Forged
CEN PREN 2957-91. Method of Preparation of Forged Samples. 8 pp.

—Testing Personnel—Certification
JIS W 0905-84. Aerospace Nondestructive Inspection Personnel Qualification and Certification.

—Warranties
EURO EEPSC/91—23108-91. AECMA—EEPSC Equipment Warranties Current Practices and Trends. 19 pp.

Aerospace Fabrics
See Also: Aerospace Composite Materials; Aerospace Equipment; Fabrics

—Canvas—Cotton/Polyester
MOD UK DSTAN 83-80-91. Polyester/Cotton Core-Spun Canvas Issue 1. 14 pp.

—Coated
BSI F 127-83. (WITHDRAWN) 1983 Amd 1 Nylon Fabrics Suitable for Coating with Natural or Synthetic Elastomers (Superseded by BS 2F 127: 1991). 6 pp.
BSI 2F 127-91. 1991 Nylon Fabrics Suitable for Coating with Natural or Synthetic Elastomers for Aerospace Purposes. 8 pp.

Aerospace Fabrics *(Cont.)*

—Coated *(Cont.)*
BSI 2F 127-01. 1991 Amd 1 Nylon Fabrics Suitable for Coating with Natural or Synthetic Elastomers for Aerospace Purposes (AMD 7967) October 15, 1993 (S). 9 pp.
BSI F 131-86. 1986 Amd 1 Chlorosulphonated Polyethylene Rubber Coated Nylon Fabric (300 G/M2) for Aerospace Purposes (AMD 5795) December 30, 1988. 5 pp.
BSI F 136-89. 1989 Pigmented Chlorosulphonated Polyethylene Rubber-Coated Nylon Fabric (190 g/m Squared) for Aerospace Purposes. 4 pp.
BSI F 139-89. 1989 Aluminized Polychloroprene Rubber-Coated Nylon Fabric (350 g/m Squared) for Aerospace Purposes. 4 pp.

—Comprehensive Testing
BSI 4F 100-82. 1982 Amd 1 Inspection and Testing of Textiles. 37 pp.
BSI 5F 100-91. 1991 Inspection and Testing of Textiles for Aerospace Puposes. 59 pp.

—Inspection
BSI 4F 100-82. 1982 Amd 1 Inspection and Testing of Textiles. 37 pp.

—Nylon
BSI 2F 123-84. (WITHDRAWN) 1984 Woven Nylon Narrow Fabrics for Aerospace Purposes (Nominally 235 DTEX Yarns) (Superseded by 3F 123: 1990). 5 pp.
BSI 3F 123-90. 1990 Specification for Woven Nylon Narrow Fabrics for Aerospace Purposes (Norminally 235 Dtex Yarns). 4 pp.
BSI 3F 124-91. 1991 Woven Nylon 6.6 Narrow Fabrics for Aerospace Purposes (Nominally 940 dtex Yarns). 11 pp.
BSI F 127-83. (WITHDRAWN) 1983 Amd 1 Nylon Fabrics Suitable for Coating with Natural or Synthetic Elastomers (Superseded by BS 2F 127: 1991). 6 pp.
BSI 2F 127-91. 1991 Nylon Fabrics Suitable for Coating with Natural or Synthetic Elastomers for Aerospace Purposes. 8 pp.
BSI 2F 127-01. 1991 Amd 1 Nylon Fabrics Suitable for Coating with Natural or Synthetic Elastomers for Aerospace Purposes (AMD 7967) October 15, 1993 (S). 9 pp.
BSI F 131-86. 1986 Amd 1 Chlorosulphonated Polyethylene Rubber Coated Nylon Fabric (300 G/M2) for Aerospace Purposes (AMD 5795) December 30, 1988. 5 pp.

—Nylon—Construction
BSI 2F 124-85. (WITHDRAWN) 1985 Amd 1 Woven Nylon Narrow Fabrics for Aerospace Purposes (Nominally 940 Dtex Yarns) (AMD 5162) April 30, 1986 (Superseded by 3F 124: 1991). 9 pp.

—Nylon—Hypalon (R) Coated
BSI F 136-89. 1989 Pigmented Chlorosulphonated Polyethylene Rubber-Coated Nylon Fabric (190 g/m Squared) for Aerospace Purposes. 4 pp.

—Nylon—Identification Systems
BSI 2F 124-85. (WITHDRAWN) 1985 Amd 1 Woven Nylon Narrow Fabrics for Aerospace Purposes (Nominally 940 Dtex Yarns) (AMD 5162) April 30, 1986 (Superseded by 3F 124: 1991). 9 pp.

—Nylon—Neoprene Coated
BSI F 139-89. 1989 Aluminized Polychloroprene Rubber-Coated Nylon Fabric (350 g/m Squared) for Aerospace Purposes. 4 pp.

—Polyester
BSI 2F 129-91. 1991 Woven Polyester Narrow Fabrics for Aerospace Purposes. 14 pp.

—Resins—Sampling
AECMA PREN2375-89. Resin Preimpregnated Materials Production Batch Sampling Procedure. 3 pp.
CEN EN 2375-92. Resin Preimpregnated Materials—Production Batch Sampling Procedure. 5 pp.

—Wool Felts
MOD UK DSTAN 83-19: Part 3-01. Cloth, Compressed Felts, Wool, and Blended Wool Part 3: For Aerospace Purposes Issue 1; Amendment 1. 10 pp.

Aerospace Fasteners
Scope Note: For additional listings, use a more specific term *See Also:* Aerospace Bolts; Aerospace Clamps; Aerospace Equipment; Aerospace Nuts; Aerospace Rivets; Aerospace Screws; Aerospace Washers; Aircraft Fasteners; Fasteners; Threaded Fasteners
BSI S 159-81. 1981 Amd 1 12% Chromium-Nickel-Molybdenum-Vanadium Heat-Resisting Steel Bars for the Manufacture of Fasteners (1100-1300 MPa; Limiting Ruling Section 50 mm). 4 pp.

Aerospace Fasteners *(Cont.)*
DIN ENGL 29895-84. Aerospace; Alternative Materials for Fasteners (Mar). 12 pp.

—Electroplating
CEN PREN 2786-90. Electrolytic Silver Plating of Fasteners. 4 pp.
CEN PREN 2786-93. Aerospace Series Electrolytic Silver Plating of Fasteners. 4 pp.

—Heat Resistant
BSI 2HR 601-75. (WITHDRAWN) 1975 Nickel-Chromium-Titanium Aluminium Heat-Resisting Alloy Bar and Wire for the Manufacture of Fasteners (Maximum Diameter or Minor Sectional Dimensions 25 mm) (Nickel Base Cr 19.5, Ti 2.2, Al 1.4) (Superseded by 3HR 601: 1989). 3 pp.
BSI 3HR 601-89. 1989 Amd 1 Ni-Cr-Ti-Al Heat-Resisting Alloy Bar Wire Manufacture of Fasteners (Max Diameter or Minor Sectional Dimension 30 mm) (Nickel Base Cr 19.5-Ti 2.2-Al 1.4) (AMD 6529) May 31, 1990. 6 pp.

—Identification Systems
AECMA PREN2424-88. Identification Marking of Standard Fasteners. 5 pp.
SBAC RS 732 (V). Manufacturer Identification, Monogram for Aerospace Fasteners.

—Locking Thread Inserts
CEN PREN 2942-90. Insert, Screw Thread, Helical Coil, Self-Locking, in Heat Resisting Alloy, Nickel Base Alloy (INCO X750-EN 3018) Silver Plated. 7 pp.
CEN PREN 2943-90. Wire Thread Inserts, Prevailing Torque (Self-Locking) Technical Specification. 18 pp.
CEN PREN 2944-90. Insert, Screw Thread, Helical Coil, Self-Locking in Corrosion Resisting Alloy (Z10CN18-09-EN2947) Unplated. 7 pp.
CEN PREN 2945-90. Assembly with Self-Locking Thread Inserts. 7 pp.
CEN PREN 3831-91. Inserts, Thick Wall, Self-Locking, in Heat Resisting Steel FE-PM61 (17-4PH), MoS2 Lubricated Classification: 1100 MPa (at Ambient Temperature)/350 Degrees C. 5 pp.
CEN PREN 3899-91. Inserts, Thick Wall, Self-Locking, in Heat Resisting Steel FE-PM61 (17-4PH), MoS2 Lubricated Classification: 1100 MPa/350 Degrees C Technical Specification. 28 pp.
SBAC AS 62130-139 ISSUE 1. Insert Salvage—Metric, Thinwall—Short, Screw Locking (INCO 718).
SBAC AS 62140-149 ISSUE 1. Insert Salvage—Inch Thinwall—Short, Screw Locking (INCO 718).
SBAC AS 62150-159 ISSUE 1. Insert Salvage—Metric, Thinwall—Long Screw Locking (INCO 718).
SBAC AS 62160-169 ISSUE 1. Insert Salvage—Inch Thinwall—Long Screw Locking (INCO 718).

—Quick Release
CEN PREN 3776-90. Pins, Quick Release, Two Balls, Single Acting, Single Locking, "L" Type Handle, in Corrosion Resisting Steel. 8 pp.
CEN PREN 3777-90. Pins, Quick Release, Single and Double Acting Technical Specification. 18 pp.
CEN PREN 3778-90. Pins, Quick Release, Two Balls, Double Acting, Single Locking, "L" Type Handle, in Corrosion Resisting Steel. 8 pp.

Aerospace Ground Equipment
Use: Ground Support Equipment

Aerospace Materials
Scope Note: For materials used in aerospace applications, see the subheading Aerospace under the specific material *See Also:* Aerospace Equipment
MOD UK DTD-900AA-88. Approval Procedure for Proprietary Materials and Processes. 22 pp.

—Radiography
BSI M 34-70. 1970 Method of Preparation and Use of Radiographic Techniques. 7 pp.
JIS W 0903-84. Radiographic Inspection for Aerospace Quality Materials.

Aerospace Medicine
Use For: Aviation Medicine
ICAO 8984. Manual of Civil Aviation Medicine Second Edition—1985. 269 pp.

—Glossaries
BSI BS 185: Sec 17-73. 1973 Glossary of Aeronautical and Astronautical Terms Section 17: Aerospace Medicine (General, Disorders and Illusions, Systems and Equipment). 5 pp.

—Training—Flight Crews
NATO STANAG 3114 ED 6 AMD 5-86. Aeromedical Training of Flight Personnel. 16 pp.
NATO STANAG 3114 ED 6 AMD 6-86. Aeromedical Training of Flight Personnel. 19 pp.
NATO STANAG 3114 ED 6 AMD 7-86. Aeromedical Training of Flight Personnel. 19 pp.

INTERNATIONAL AND NON-U.S. NATIONAL STANDARDS
SUBJECT INDEX
Aerospace

Aerospace Medicine *(Cont.)*

—Training—Flight Crews—NBC Equipment

NATO STANAG 3497 ED 1 AMD 2-88. Aeromedical Training of Aircrew NBC Equipment and Procedures. 12 pp.
NATO STANAG 3497 ED 1 AMD 3-88. Aeromedical Training of Aircrew in Aircrew NBC Equipment and Procedures. 13 pp.

Aerospace Nuts

See Also: Aerospace Bolts; Aerospace Equipment; Aerospace Fasteners; Fasteners; Nuts (Fasteners)
BSI 2A 100-68. (WITHDRAWN) 1968 Amd 3 General Requirements for Bolts and Nuts of Strength Not Exceeding 180 000 LbF/Square Inches (125 Hbar) (Superseded by 3A 100: 1991). 24 pp.
BSI 3A 100-91. 1991 General Requirements for Bolts and Free Running Nuts of Tensile Strength not Exceeding 1249 MPa. 16 pp.
BSI 3A 100-01. 1991 Amd 1 General Requirements for Bolts and Free Running Nuts of Tensile Strength not Exceeding 1249 MPa (AMD 7661) April 15, 1993 (S). 21 pp.
DIN ENGL LN 65015-70. Nuts from Steel for Screws with Minimum Tensile Strength of 900 N/mm2 and 1100 N/mm2; Technical Specification (Dec). 13 pp.

—Acceptance Certificates

DIN ENGL LN 65018 Pt 2-68. Manufacturer's Acceptance Certificate for Nuts and Self-Locking Nuts (Sept). 1 p.

—Anchor

DIN ENGL LN 29638 Pt 1-69. Press Nuts; Self-Locking for Temperatures up to 260 Degrees C (Apr). 1 p.
DIN ENGL LN 29639 Pt 1-69. Press Nuts; Self-Locking, Floating for Temperatures up to 260 Degrees C (Apr). 1 p.
DIN ENGL LN 29679-77. Nuts Anchor; Self-Locking; Floating; Double Lug for Temperatures up to 235 Degrees C (July). 3 pp.
DIN ENGL LN 29791-80. Aerospace; Nuts, Anchor, Right Angle; Self-Locking, Floating for Temperatures up to 235 Degrees C (Aug). 3 pp.
DIN ENGL LN 29951-74. Nuts, Anchor; Self-Locking, Pressure Tight, Double Lug for Temperatures from-55 to +100 Degrees C (Dec). 2 pp.
DIN ENGL LN 29954-80. Aerospace; Nuts, Anchor; Self-Locking, Pressure Tight, Double Lug for Temperatures from-55 to +230 Degrees C (Mar). 3 pp.
DIN ENGL LN 29982-78. Aerospace; Nuts Anchor; Self-Locking; Deep Counterbore; Double Lug for Temperatures up to 235 Degrees C (Mar). 3 pp.

—Anchor—Locking

CEN PREN 3653-92. Nuts, Anchor, Self-Locking, Floating, Self-Aligning, One Lug, in Steel, Cadmium Plated, MoS2 Lubricated Classificaton: 900 MPa (at Ambient Temperature) /235 Degrees C. 5 pp.
DIN ENGL LN 29671-77. Nuts Anchor; Self-Locking; Double Lug for Temperatures up to 235 Degrees C (May). 3 pp.

—Anchor—Locking—Counterbore

CEN PREN 2862-90. Nuts Anchor, Self-Locking, Fixed, 90 Degrees Corner, with Counterbore, in Alloy Steel, Cadmium Plated, MO2 Lubricated Classification: 1100 MPa (at Ambient Temperature) /235 Degrees Celsius. 6 pp.
CEN PREN 2862-93. Aerospace Series Nuts, Anchor, Self-Locking, Fixed, 90 Degree Corner, with Counterbore, in Alloy Steel, Cadmium Plated, MoS2 Lubricated Classification: 1 100 MPa (at Ambient Temperature)/235 Degrees C. 5 pp.
CEN PREN 2862-93. CORRIGENDUM Aerospace Series Nuts, Anchor, Self-Locking, Fixed, 90 Degree Corner, with Counterbore, in Alloy Steel, Cadmium Plated, MoS2 Lubricated Classification: 1 100 MPa (at Ambient Temperature) /235 Degrees C. 1 p.
CEN PREN 2863-90. Nuts, Anchor Self-Locking, Fixed, 90 Degrees Corner, with Counterbore, in Heat Resisting Steel, Passivated, MOS2 Lubricated Classification: 1100 MPa (at Ambient Temperature) /315 Degrees Celsius. 6 pp.
CEN PREN 2863-93. Aerospace Series Nuts, Anchor, Self-Locking, Fixed, 90 Degree Corner, with Counterbore, in Heat Resisting Steel, MoS2 Lubricated Classification: 1 100 MPa (at Ambient Temperature)/315 Degrees C. 5 pp.
CEN PREN 2863-93. CORRIGENDUM Aerospace Series Nuts, Anchor, Self-Locking, Fixed, 90 Degree Corner, with Counterbore, in Heat Resisting Steel, MoS2 Lubricated Classification: 1 100 MPa (at Ambient Temperature) /315 Degrees C. 1 p.
CEN PREN 2865-93. Aerospace Series Nuts, Anchor, Self-Locking, Floating, Two Lug, with Counterbore, in Heat Resisting Steel, MoS2 Lubricated Classification: 1 100 MPa (at Ambient Temperature)/315 Degrees C. 5 pp.

Aerospace Nuts *(Cont.)*

—Anchor—Locking—Counterbore *(Cont.)*

CEN PREN 2865-93. CORRIGENDUM Aerospace Series Nuts, Anchor, Self-Locking, Floating, Two Lug, with Counterbore, in Heat Resisting Steel, MoS2 Lubricated Classification: 1 100 MPa (at Ambient Temperature) /315 Degrees C. 1 p.
CEN PREN 2866-92. Nuts, Anchor, Self-Locking, Floating, One Lug, with Counterbore, in Steel, Cadmium Plated, MoS2 Lubricated Classification: 1 110 MPa (at Ambient Temperature) /235 Degrees C. 5 pp.
CEN PREN 2867-92. Nuts, Anchor, Self-Locking, Floating, One Lug, with Counterbore, in Heat Resisting Steel, MoS2 Lubricated Classification: 1 110 MPa (at Ambient Temperature) /315 Degrees C. 5 pp.
CEN PREN 3435-93. Aerospace Series Nuts, Anchor, Self-Locking, Floating, Two Lug, Reduced Series, with Counterbore, in Heat Resisting Steel, MoS2 Lubricated Classification: 1 100 MPa (at Ambient Temperature)/315 Degrees C. 5 pp.
CEN PREN 3435-93. CORRIGENDUM Aerospace Series Nuts, Anchor, Self-Locking, Floating, Two Lug, Reduced Series, with Counterbore, in Heat Resisting Steel, MoS2 Lubricated Classification: 1 100 MPa (at Ambient Temperature) /315 Degrees C. 1 p.
CEN PREN 3537-90. Nuts, Anchor, Self-Locking, Fixed, Two Lug, with Counterbore, in Heat Resisting Steel, Passivated, MoS2 Lubricated Classification:1100 MPa (at Ambient Temperature) /315 Degrees Celsius. 6 pp.
CEN PREN 3537-93. Aerospace Series Nuts, Anchor, Self-Locking, Fixed, Two Lug, with Counterbore, in Heat Resisting Steel, MoS2 Lubricated Classification: 1 100 MPa (at Ambient Temperature)/315 Degrees C. 5 pp.
CEN PREN 3537-93. CORRIGENDUM Aerospace Series Nuts, Anchor, Self-Locking, Fixed, Two Lug, with Counterbore, in Heat Resisting Steel, MoS2 Lubricated Classification: 1 100 MPa (at Ambient Temperature) /315 Degrees C. 1 p.
CEN PREN 3538-90. Nuts, Anchor, Self-Locking, Fixed, Two Lug, Reduced Series with Counterbore, in Heat Resisting Steel Passivated, MoS 2 Lubricated Classification: 100 MPa (at Ambient Temperature) /315 Degrees Celsius. 6 pp.
CEN PREN 3538-93. Aerospace Series Nuts, Anchor, Self-Locking, Fixed, Two Lug, Reduced Series, with Counterbore, in Heat Resisting Steel, MoS2 Lubricated Classification: 1 100 MPa (at Ambient Temperature)/315 Degrees C. 5 pp.
CEN PREN 3538-93. CORRIGENDUM Aerospace Series Nuts, Anchor, Self-Locking, Fixed, Two Lug, Reduced Series, with Counterbore, in Heat Resisting Steel, MoS2 Lubricated Classification: 1 100 MPa (at Ambient Temperature) /315 Degrees C. 1 p.
CEN PREN 3539-90. Nuts Anchor, Self-Locking, One Lug, Fixed, with Counterbore, in Heat Resisting Steel, MoS2 Lubricated Classification: 1100 MPa (at Ambient Temperature) /315 Degrees Celsius. 5 pp.
CEN PREN 3539-93. Aerospace Series Nuts, Anchor, Self-Locking, One Lug, Fixed, with Counterbore, in Heat Resisting Steel, MoS2 Lubricated Classification: 1 100 MPa (at Ambient Temperature) /315 Degrees C. 5 pp.
CEN PREN 3712-91. Nuts, Anchor, Self-Locking, One Lug Fixed, Reduced Series, with Counterbore, in Steel, Cadmium Plated, MoS2 Lubricated Classification: 1100 MPa (at Ambient Temperature)/ 235 Degrees C. 6 pp.
CEN PREN 3714-93. Aerospace Series Nuts, Anchor, Self-Locking, Floating, Two Lug, with Counterbore, in Heat Resisting Steel, Silver Plated Classification: 1 100 MPa (at Ambient Temperature)/425 Degrees C. 5 pp.
CEN PREN 3714-93. CORRIGENDUM Aerospace Series Nuts, Anchor, Self-Locking, Floating, Two Lug, with Counterbore, in Heat Resisting Steel, Silver Plated Classification: 1 100 MPa (at Ambient Temperature) /425 Degrees C. 1 p.
CEN PREN 3750-90. Nuts, Anchor, Self-Locking, Fixed 90 Degrees Corner, Reduced Series with Counterbore, in Heat Resisting Steel, Passivated, MoS2 Lubricated/Classification:100 MPa (at Ambient Temperature) /315 Degrees Celsius. 6 pp.
CEN PREN 3750-93. Aerospace Series Nuts, Anchor, Self-Locking, Fixed, 90 Degree Corner, Reduced Series, with Counterbore, in Heat Resisting Steel, MoS2 Lubricated Classification: 1 100 MPa (at Ambient Temperature)/315 Degrees C. 5 pp.
CEN PREN 3750-93. CORRIGENDUM Aerospace Series Nuts, Anchor, Self-Locking, Fixed, 90 Degree Corner, Reduced Series, with Counterbore, in Heat Resisting Steel, MoS2 Lubricated Classification: 1 100 MPa (at Ambient Temperature) /315 Degrees C. 1 p.
CEN PREN 3751-90. Nuts, Anchor, Self-Locking, Fixed, Closed Corner, Reduced Series, with Counterbore, in Heat Resisting Steel, Passivated, MoS2 Lubricated Classification: 1100 MPa (at Ambient Temperature)/315 Degrees Celsius. 6 pp.

Aerospace Nuts *(Cont.)*

—Anchor—Locking—Counterbore *(Cont.)*

CEN PREN 3751-93. Aerospace Series Nuts, Anchor, Self-Locking, Fixed, Closed Corner, Reduced Series, with Counterbore, in Heat Resisting Steel, MoS2 Lubricated Classification: 1 100 MPa (at Ambient Temperature)/315 Degrees C. 5 pp.
CEN PREN 3751-93. CORRIGENDUM Aerospace Series Nuts, Anchor, Self-Locking, Fixed, Closed Corner, Reduced Series, with Counterbore, in Heat Resisting Steel, MoS2 Lubricated Classification: 1 100 MPa (at Ambient Temperature) /315 Degrees C. 1 p.
CEN PREN 3753-90. Nuts, Anchor, Self-Locking, Fixed, Closed Corner, with Counterbore, in Alloy Steel, Cadmium Plated, MoS2 Lubricated Classification:1100 MPa (at Ambient Temperature) /235 Degrees Celsius. 6 pp.
CEN PREN 3753-93. Aerospace Series Nuts, Anchor, Self-Locking, Fixed, 60 Degree Corner, with Counterbore, in Alloy Steel, Cadmium Plated, MoS2 Lubricated Classification: 1 100 MPa (at Ambient Temperature)/235 Degrees C. 5 pp.
CEN PREN 3753-93. CORRIGENDUM Aerospace Series Nuts, Anchor, Self-Locking, Fixed, 60 Degree Corner, with Counterbore, in Alloy Steel, Cadmium Plated, MoS2 Lubricated Classification: 1 100 MPa (at Ambient Temperature) /235 Degrees C. 1 p.
CEN PREN 3754-90. Nuts, Anchor, Self-Locking, Fixed, Closed Corner, with Counterbore, in Heat Resisting Steel, Passivated, MoS2 Lubricated Classification: 100 MPa (at Ambient Temperature) /315 Degrees Celsius. 6 pp.
CEN PREN 3754-93. Aerospace Series Nuts, Anchor, Self-Locking, Fixed, 60 Degree Corner, with Counterbore, in Heat Resisting Steel, MoS2 Lubricated Classification: 1 100 MPa (at Ambient Temperature)/315 Degrees C. 5 pp.
CEN PREN 3754-93. CORRIGENDUM Aerospace Series Nuts, Anchor, Self-Locking, Fixed, 60 Degree Corner, with Counterbore, in Heat Resisting Steel, MoS2 Lubricated Classification: 1 100 MPa (at Ambient Temperature) /315 Degrees C. 1 p.
CEN PREN 3757-93. Aerospace Series Nuts, Anchor, Self-Locking, Floating, Self-Aligning, Two Lug, in Heat Resisting Steel, MoS2 Lubricated Classification: 900 MPa (at Ambient Temperature) /315 Degrees C. 5 pp.
CEN PREN 3757-93. CORRIGENDUM Aerospace Series Nuts, Anchor, Self-Locking, Floating, Self-Aligning, Two Lug, in Heat Resisting Steel, MoS2 Lubricated Classification: 900 MPa (at Ambient Temperature) /315 Degrees C. 1 p.
CEN PREN 3768-90. Nuts, Anchor, Self-Locking, One Lug, Fixed, Reduced Series, with Counterbore, in Heat Resisting Steel, MoS2 Lubricated Classification: 1100 MPa (at Ambient Temperature) /315 Degrees Celsius. 5 pp.
CEN PREN 3768-93. Aerospace Series Nuts, Anchor, Self-Locking, One Lug, Fixed, Reduced Series, with Counterbore, in Heat Resisting Steel, MoS2 Lubricated Classification: 1 100 MPa (at Ambient Temperature) /315 Degrees C. 5 pp.
CEN PREN 3834-91. Nuts, Anchor, Self-Locking, Floating, Two Lug, Incremental Counterbore, in Heat Resisting Steel, MoS2 Lubricated Classification: 900 MPa (at Ambient Temperature)/315 Degrees C. 7 pp.
CEN PREN 3834-93. Aerospace Series Nuts, Anchor, Self-Locking, Floating, Two Lug, Incremental Counterbore, in Heat Resisting Steel, MoS2 Lubricated Classification: 900 MPa (at Ambient Temperature) /315 Degrees C. 5 pp.
CEN PREN 4084-93. Aerospace Series Nuts, Anchor, Self-Locking, Fixed, Two Lug, with Counterbore, in Alloy Steel, Cadmium Plated, MoS2 Lubricated Classification: 1 100 MPa (at Ambient Temperature) /235 Degrees C. 5 pp.
DIN ENGL LN 29982-78. Aerospace; Nuts Anchor; Self-Locking; Deep Counterbore; Double Lug for Temperatures up to 235 Degrees C (Mar). 3 pp.
DIN ENGL LN 29983-77. Nuts Anchor; Self-Locking; Deep Counterbore; Double Lug Reduced for Temperatures up to 235 Degrees C (July). 3 pp.
DIN ENGL LN 29984-77. Nuts Anchor; Self-Locking; Deep Counterbore; Single Lug for Temperatures up to 235 Degrees C (July). 3 pp.
DIN ENGL LN 29985-77. Nuts Anchor; Self-Locking; Floating; Deep Counterbore; Double Lug for Temperatures up to 235 Degrees C (Dec). 3 pp.
DIN ENGL LN 29986-77. Anchor Nuts; Self-Locking, Floating, Deep Counterbore, Double Lug, Miniature, for Temperatures up to 235 Degrees C (May). 3 pp.
DIN ENGL LN 29987-77. Nuts Anchor; Self-Locking; Floating; Deep Counterbore; Single Lug for Temperatures up to 235 Degrees C (Dec). 3 pp.
DIN ENGL LN 29988-77. Nuts Anchor; Self-Locking; Deep Counterbore; Corner Lug for Temperatures up to 235 Degrees C (July). 3 pp.
DIN ENGL LN 29989-77. Nuts Anchor; Self-Locking; Deep Counterbore; Corner Lug Reduced for Temperatures up to 235 Degrees C (July). 3 pp.

INDUSTRY STANDARDS
49

Aerospace Nuts *(Cont.)*

—Anchor—Locking—Counterbore *(Cont.)*

DIN ENGL LN 29990-77. Anchor Nuts; Self-Locking, Deep Counterbore, Double Lug, for Temperatures up to 315 Degrees C and up to 425 Degrees C (July). 3 pp.

DIN ENGL LN 29991-77. Anchor Nuts; Self-Locking, Deep Counterbore, Double Lug, Miniature for Temperatures up to 315 Degrees C and up to 425 Degrees C (July). 3 pp.

DIN ENGL LN 29992-77. Nuts, Anchor; Self-Locking, Deep Counterbore, Single Lug for Temperatures up to 315 Degrees C and up to 425 Degrees C (July). 3 pp.

DIN ENGL LN 29996-77. Nuts Anchor; Self-Locking; Deep Counterbore; Single Lug Reduced for Temperatures up to 315 Degrees C and up to 425 Degrees C (May). 3 pp.

ISO 3168-86. Aerospace—Self-Locking, Fixed, Single-Lug Anchor Nuts with Counterbore, Classification 1 100 MPa/235 Degrees Celsius First Edition. 5 pp.

ISO 3191-85. Aerospace—Self-Locking, Fixed, Single-Lug Anchor Nuts, Reduced Series, with Counterbore, Strength Classification 1 100 MPa and Maximum Operating Temperature 235 Degrees Celsius First Edition. 5 pp.

ISO 3209-89. Aerospace—Nuts, Anchor, Self-Locking, Floating, Two-Lug, with Counter-bore, with MJ Threads, Coated or Uncoated, Classification ! 100 MPa/235 Degrees Celsius, 1 100 MPa/315 Degrees Celsius or 1 100 MPa/425 Degrees Celsius—Dimensions First Edition. 6 pp.

ISO 3221-89. Aerospace—Nuts, Anchor, Self-Locking, Fixed, 90 Degree Corner, with Counterbore, with MJ Threads, Coated or Uncoated, Classification 1 100 MPa/235 Degrees Celsius, 1 100 MPa/315 Degrees Celsius or 1 100 MPa/425 Degrees Celsius—Dimensions First Edition. 5 pp.

ISO 3222-89. Aerospace—Nuts, Anchor, Self-Locking, Fixed, Closed Corner, Reduced Series, with Counter-bore, with MJ Threads, Coated or Uncoated, Classification 1 100 MPa/235 Degrees C, 1 100 MPa/315 Degrees C or 1 100 MPa/425 Degrees C—Dimensions First Edition. 5 pp.

ISO 3223-89. Aerospace—Nuts, Anchor, Self-Locking, Fixed, Two-Lug, with Counter-bore, with MJ Threads, Coated or Uncoated, Classification 1 100 MPa/235 Degrees Celsius, 1 100 MPa/315 Degrees Celsius or 1 100 MPa/425 Degrees Celsius—Dimensions First Edition. 6 pp.

ISO 3224-85. Aerospace—Self-Locking, Floating, Single-Lug Anchor Nuts, with Counterbore, Classification 1 100 MPa/235 Degrees Celsius First Edition. 5 pp.

ISO 3225-85. Aerospace—Self-Locking, Fixed, Two Lug Anchor Nuts, Reduced Series, with Counterbore, Strength Classification 1 100 MPa and Maximum Operating Temperature 235 Degrees Celsius First Edition. 5 pp.

ISO 7332-83. Metric Fasteners for Aerospace Construction—Nuts, Anchor, Self-Locking, Floating, Two Lug, Reduced Series, with Counterbore—Strength Classification 1 100 MPa—Maximum Operating Temperature 235 Degrees Celsius First Edition. 5 pp.

ISO 8940-88. Aerospace—Nuts, Anchor, Self-Locking, Sealing, Floating, Two-Lug, with Counterbore, Classifications 900 MPa/ 120 Degrees Celsius, 900 MPa/175 Degrees Celsius and 900 MPa/235 Degrees Celsius Dimensions First Edition. 4 pp.

ISO 9156-89. Aerospace—Nuts, Anchor, Self-Locking, Fixed, 90 Degree Corner, Reduced Series, with Counter-bore, with MJ Threads, Coated or Uncoated, Classification 1 100 MPa/235 Degrees Celsius, 1 100 MPa/315 Degrees Celsius or 1 100 MPa/425 Degrees Celsius—Dimensions First Edition. 5 pp.

—Anchor—Plate

AECMA PREN3019-89. Self-Locking Plate Nuts Floating, Two-Lug in Heat Resisting Steel FE-PA2HT (A286), Unplated Classification: 1100 MPa/ 650 Degrees Celsius. 5 pp.

DIN ENGL LN 29910-83. Aerospace; Nuts, Plate; Self-Locking, Floating, Double Lug, Reduced, for Temperatures up to 235 Degrees C (Oct). 3 pp.

—Anchor—Plate—Locking

DIN ENGL LN 29680-77. Aerospace; Nuts, Plate; Self-Locking, Floating, Self-Aligning, Double Lug for Temperatures up to 235 Degrees C (July). 3 pp.

—Castle

BSI A 242-A 245-74. 1974 Hexagonal Castle Nuts (of Class 3B UNJ Thread). 5 pp.

BSI A 246-A 249-74. 1974 Slotted Hexagonal Thick Nuts (of Class 3B UNJ Thread). 5 pp.

CEN PREN 2868-91. Nuts, Hexagonal, Slotted/ Castellated, Normal Height, Normal Across Flats, in Heat Resisting Steel, Silver Plated Classification: 1100 MPa (at Ambient Temperature)/650 Degrees C. 6 pp.

Aerospace Nuts *(Cont.)*

—Castle *(Cont.)*

CEN PREN 2868-93. Aerospace Series Nuts, Hexagonal, Slotted/Castellated, Normal Height, Normal Across Flats, in Heat Resisting Steel, Silver Plated Classification: 1 100 MPa (at Ambient Temperature) /650 Degrees C. 5 pp.

CEN PREN 2869-91. Nuts, Hexagonal, Slotted/ Castellated, Normal Height, Normal Across Flats, in Heat Resisting Steel, Passivated Classification: 1100 MPa (at Ambient Temperature) /650 Degrees C. 6 pp.

CEN PREN 2869-93. Aerospace Series Nuts, Hexagonal, Slotted/Castellated, Normal Height, Normal Across Flats, in Heat Resisting Steel, Passivated Classification: 1 100 MPa (at Ambient Temperature) /650 Degrees C. 5 pp.

DIN ENGL LN 9345-80. Aerospace; Nuts, Castle (Aug). 3 pp.

—Channel—Locking—Counterbore

DIN ENGL LN 29993-74. Gang Channel Nuts; Deep Counterbore, Self-Locking, Floating for Temperatures up to 120 Degrees C (Oct). 2 pp.

—Clip—Locking

AECMA PREN3240-89. Nuts, Self-Locking, Clip in Heat Resisting Steel FE-PA92HT (A286), Uncoated Classification: 1100 MPa/425 Degrees Celsius. 4 pp.

AECMA PREN3241-89. Nuts, Self-Locking, Clip in Heat Resisting Steel FE-PA92HT (A286), Silvercoated, Classification: 1100 MPa/425 Degrees Celsius. 4 pp.

CEN PREN 3240-92. Nuts, Self-Locking, Clip, in Heat Resisting Steel FE-PA92HT (A286), Uncoated Classification: 1 100 MPa (at Ambient Temperature) /425 Degrees C. 4 pp.

CEN PREN 3241-92. Nuts, Self-Locking, Clip, in Heat Resisting Steel FE-PA92HT (A286), Silver Coated Classification: 1 100 MPa (at Ambient Temperature) /425 Degrees C. 4 pp.

—Double Hexagonal

CEN PREN 3771-90. Nuts, Bi-Hexagonal, Plain with Wire Locking Holes, in Heat Resisting Steel, Passivated, MoS2 Lubricated Classification: 1550 MPa (at Ambient Temperature) /315 Degrees Celsius. 5 pp.

CEN PREN 3771-93. Aerospace Series Nuts, Bi-Hexagonal, Plain, in Heat Resisting Steel, MoS2 Lubricated Classification: 1 100 MPa (at Ambient Temperature) /315 Degrees C. 5 pp.

ISO 4095-78. Fasteners for Aerospace Construction—Bi-Hexagonal Wrenching Configuration First Edition. 4 pp.

—Double Hexagonal—Locking

AECMA PREN2906-86. Nuts, Self Locking, Bi Hexagonal, in Heat Resisting Steel FE-PA92HT (A286) Unplated Classification 1100 MPa/650 Degrees. 4 pp.

AECMA PREN2907-86. Nuts, Self Locking, Bi Hexagonal, in Heat Resisting Steel FE-PA92HT (A286) Silver Plated Classification 1100 MPa/650 Degrees Celsius. 4 pp.

AECMA PREN2908-86. Nuts, Self Locking, Bi Hexagonal, Deep Counterbored, in Heat Resisting Steel FE-PA92HT (A286) Unplated Classification 1100 MPa/650 Degrees Celsius. 4 pp.

AECMA PREN2909-86. Nuts, Self Locking, Bi Hexagonal, Deep Counterbored, in Heat Resisting Steel FE-PA92HT (A286) Silver Plated Classification 1100 MPa/650 Degrees Celsius. 4 pp.

AECMA PREN3012-88. Nuts, Self-Locking, Bihexagonal in Heat Resisting Nickel Base Alloy NI-P101HT (Waspaloy) Unplated Classification: 1210 MPa/730 Degrees Celsius. 5 pp.

AECMA PREN3013-88. Nuts, Self-Locking, Bihexagonal in Heat Resisting Nickel Base Alloy NI-P101HT (Waspaloy) Completely Silver Plated: Classification: 1210 MPa/730 Degrees Celsius. 5 pp.

AECMA PREN3239-88. Nuts, Self-Locking, Bihexagonal in Heat Resisting Nickel Base Alloy NI-P101HT (Waspaloy) Silver Plated Thread Classification: 1210 MPa/730 Degrees Celsius. 4 pp.

AECMA PREN3637-90. Nuts, Self-Locking, Bi-Hexagonal (Double Reduced), in Heat Resisting Nickel Base Alloy N1-P101HT (Waspaloy), Silver Plated Classification: 1210 MPa/730 Degrees Celsius. 5 pp.

CEN PREN 3713-91. Nuts, Anchor, Self-Locking, Bihexagonal, in Steel, Cadmium Plated, MoS2 Lubricated Classification: 1550 MPa (at Ambient Temperature)/235 Degrees C. 6 pp.

CEN PREN 3720-91. Nuts, Bihexagonal, Self-Locking, in Heat Resisting Steel FE-PA92HT (A286), MoS2 Coated Classification:1100 (at Ambient Temperature) /425 Degrees C. 5 pp.

CEN PREN 4011-93. Aerospace Series Nuts, Bihexagonal, Self-Locking, in Heat Resisting Nickel Base Alloy NI-P100HT (Inconel 718), Silver Plated Classification: 1 550 MPa (at Ambient Temperature) /600 Degrees C. 4 pp.

Aerospace Nuts *(Cont.)*

—Double Hexagonal—Locking *(Cont.)*

CEN PREN 4012-93. Aerospace Series Nuts, Bihexagonal, Self-Locking, in Heat Resisting Nickel Base Alloy NI-P100HT (Inconel 718), MoS2 Coated Classification: 1 550 MPa (at Ambient Temperature) /425 Degrees C. 4 pp.

DIN ENGL LN 29528-80. Aerospace; Nut Bihexagonal with Flange; Selflocking for Temperatures up to 235 Degrees C for Bolts and Screws with Nominal Tensile Strength up to 1800 N/mm2 (Nov). 2 pp.

DIN ENGL LN 29942-85. Aerospace; Nuts, Bi-Hexagon Flange, Deep Counterbore, Self-Locking for Temperatures up to 235 Degrees C for Bolts and Screws with Nominal Tensile Strength of 1550 and 1800 N/mm2 (Feb). 3 pp.

ISO 9199-87. Aerospace—Self-Locking Bihexagonal Nuts, Classificatons 1 100 MPa/650 Degrees Celsius, 1 250 MPa/760 Degrees Celsius, 1 550 MPa/235 Degrees Celsius and 1 550 MPa/650 Degrees Celsius—Dimensions First Edition. 4 pp.

—Double Hexagonal—Locking—Counterbore

AECMA PREN2908-86. Nuts, Self Locking, Bi Hexagonal, Deep Counterbored, in Heat Resisting Steel FE-PA92HT (A286) Unplated Classification 1100 MPa/650 Degrees Celsius. 4 pp.

AECMA PREN2909-86. Nuts, Self Locking, Bi Hexagonal, Deep Counterbored, in Heat Resisting Steel FE-PA92HT (A286) Silver Plated Classification 1100 MPa/650 Degrees Celsius. 4 pp.

BSI 2L 87-71. 1971 Hexagonal Bars for Nuts, Couplings and Hollow Machined Parts of Al-Cu-Mg-Si-Mn Alloy (Solution and Precipita-tion Treated) (Free from Peripheral and Asymmetric Coarse Grain) (MW 14 mm, MW 36mm Across Flats) (Cu 4.4, Mg 0.5, Si 0.7, Mn 0.8). 12 pp.

BSI 2S 147-76. 1976 0.5 Per Cent Nickel-Chromium-Molybdenum Steel Bars for the Manufacture of Forged Bolts and Forged Nuts. 2 pp.

BSI 2S 147-01. 1976 Amd 1 0.5 Per Cent Nickel-Chromium-Molybdenum Steel Bars for the Manufacture of Forged Bolts and Forged Nuts (AMD 6836) November 29, 1991. 3 pp.

BSI 2S 149-76. 1976 1.75 Per Cent Nickel-Chromium-Molybdenum Steel Bars for the Manufacture of Forged Bolts and Forged Nuts. 2 pp.

BSI 2S 149-01. 1976 Amd 1 1.75 Per Cent Nickel-Chromium-Molybdenum Steel Bars for the Manufacture of Forged Bolts and Forged Nuts (AMD 6837) November 29, 1991. 3 pp.

BSI S 158-77. 1977 1 Per Cent Chromium-Molybdenum Steel Bars for the Manufacture of Forged Bolts and Forged Nuts. 4 pp.

CEN PREN 3721-91. Nuts, Bihexagonal, Self-Locking, Deep Counterbored, in Heat Resisting Steel FE-PA92HT (A286), MoS Coated Classification:1100 MpA (at Ambient Temperature)/Degrees C. 5 pp.

CEN PREN 3772-90. Nuts, Bi-Hexagonal, with Flange, Deep Counterbore, Self Locking, in Heat Resisting Steel, Passivated Classification: 1100 MPa (at Ambient Temperature) /425 Degrees Celsius. 5 pp.

CEN PREN 3843-93. Aerospace Series Nuts, Bi-Hexagonal, Self-Locking, with Counterbore, in Heat Reisisting Steel, Passivated Classification: 1 100 MPa (at Ambient Temperature) /650 Degrees C. 5 pp.

ISO 8788-87. Aerospace—Fasteners—Tolerances of Form and Position for Nuts First Edition. 12 pp.

—Glossaries

ISO 5843 Pt 3-87. Aerospace—List of Equivalent Terms—Part 3: Aerospace Bolts and Nuts First Edition. 25 pp.

—Grommet

CEN PREN 3660-003-91. Part 003—Detail Specification Sheet Grommet Nut, Style A. 11 pp.

CEN PREN 3660-003-93. Aerospace Series Cable Outlet Accessories for Circular and Rectangular Electrical and Optical Connectors Part 003—Grommet Nut, Type A, 003 Product Standard. 7 pp.

—Heat Resistant

BSI HR 650-72. 1972 Amd 2 High Expansion Heat-Resisting Steel Bar and Wire for the Manufacture of Bolts, Studs, Set Screws and Nuts (Ni 25.5, Cr 15, Ti 2, Mn 1.5, Mo 1.25, Si 0.7, V 0.3) (Limiting Ruling Section 20 mm). 5 pp.

—Hexagonal

BSI A 237-A 240-71. 1971 Amd 1 Plain Hexagon Nuts (of Class 3B UNJ Thread). 5 pp.

BSI A 237-A 240-02. 1971 Amd 2 Plain Hexagonal Nut (of Class 3B UNJ Thread) (AMD 7770) August 15, 1993 (S). 6 pp.

CEN PREN 2895-91. Nuts, Hexagonal, Plain, Normal Height, Normal Across Flats, in Corrosion Resisting Steel, Passivated Classification: 900 MPa (at Ambient Temperature)/425 Degrees C. 6 pp.

INTERNATIONAL AND NON-U.S. NATIONAL STANDARDS
SUBJECT INDEX
Aerospace

Aerospace Nuts (Cont.)
—Hexagonal (Cont.)
CEN PREN 2895-93. Aerospace Series Nuts, Hexagonal, Plain, Normal Height, Normal Across Flats, in Corrosion Resisting Steel, Passivated Classification: 900 MPa (at Ambient Temperature) /425 Degrees C. 4 pp.

CEN PREN 2922-92. Nuts, Hexagon, Plain, Reduced Height, Reduced Across Flats, in Heat Resisting Steel, Passivated, Left Hand Thread Classification: 600 MPa (at Ambient Temperature) /650 Degrees C. 5 pp.

CEN PREN 2923-92. Nuts, Hexagon, Plain, Reduced Height, Reduced Across Flats, in Heat Resisting Steel, Silver Plated Classification: 600 MPa (at Ambient Temperature) /425 Degrees C. 5 pp.

CEN PREN 2924-92. Nuts, Hexagon, Plain, Reduced Height, Reduced Across Flats, in Heat Resisting Steel, Silver Plated, Left Hand Thread Classification: 600 MPa (at Ambient Temperature) /425 Degrees C. 5 pp.

CEN PREN 3226-90. Nuts, Hexagon, Plain, Normal Height, Normal Across Flats, in Steel, Cadmium Plated Classification: 1100 MPa (at Ambient Temperature) 235 Degrees Celsius. 7 pp.

CEN PREN 3226-92. Nuts, Hexagon, Plain, Normal Height, Normal Across Flats, in Steel, Cadmium Plated Classification: 1 100 MPa (at Ambient Temperature) /235 Degrees C. 5 pp.

CEN PREN 3227-90. Nuts, Hexagon, Plain, Normal Heights, Normal Across Flats, in Steel, Cadmium Plated, Left Hand Thread, Classification: 1100 MPa (at Ambient Temperature) 1235 Degrees Celsius. 7 pp.

CEN PREN 3227-92. Nuts, Hexagon, Plain, Normal Height, Normal Across Flats, in Steel, Cadmium Plated, Left Hand Thread Classification: 1 100 MPa (at Ambient Temperature) /235 Degrees C. 5 pp.

CEN PREN 3228-90. Nuts, Hexagon, Plain Reduced Height, Normal Across Flats, in Steel, Cadmium Plated Classification: 900 MPa (at Ambient Temperature) 1235 Degrees Celsius. 7 pp.

CEN PREN 3228-92. Nuts, Hexagon, Plain, Reduced Height, Normal Across Flats, in Steel, Cadmium Plated Classification: 900 MPa (at Ambient Temperature) /235 Degrees C. 5 pp.

CEN PREN 3229-90. Nuts, Hexagon, Plain, Reduced Height, Normal Across Flats in Steel, Cadmium Plated, Left Hand Thread, Classification: 900 MPa (at Ambient Temperature) /235 Degrees Celsius. 7 pp.

CEN PREN 3229-92. Nuts, Hexagon, Plain, Reduced Height, Normal Across Flats, in Steel, Cadmium Plated, Left Hand Thread Classification: 900 MPa (at Ambient Temperature) /235 Degrees C. 5 pp.

CEN PREN 3742-90. Nuts, Hexagonal, Slotted/ Catellated, Thin, Reduced Across Flats, in Heat Resisting Steel, Passivated Classification: 600 MPa (at Ambient Temperature) /650 Degrees Celsius. 5 pp.

CEN PREN 3742-92. Nuts, Hexagonal, Slotted/ Castellated, Reduced Height, Reduced Across Flats, in Heat Resisting Steel, Passivated Classification: 600 MPa (at Ambient Temperature)/650 Degrees C. 5 pp.

DIN ENGL LN 9343-85. Aerospace; Nuts, Hexagon (Feb). 3 pp.

DIN ENGL LN 29669-80. Aerospace; Hexagon Nuts for Screws and Bolts with a Nominal Tensile Strength of 1100 N/mm2 (Mar). 2 pp.

—Hexagonal—Locking
AECMA PREN3033-89. Nuts, Self-Locking, Hexagonal with Captive Washer, in Heat Resisting Steel FE-PA92HT (A286), Uncoated Classification: 1100 MPa/425 Degrees Celsius. 5 pp.

AECMA PREN3034-89. Nuts, Self-Locking, Hexagonal with Captive Washer, in Heat Resisting Steel FE-PA92HT (A286), Silver Coated Classification: 1100 MPA/425 Degrees Celsius. 5 pp.

AECMA PREN3196-87. Nuts, Self-Locking, Hexagonal, in Heat Resisting Steel FE-PA92HT (A286), Silver Coated Classification 1100 MPa/425 Degrees Celsius. 4 pp.

AECMA PREN3377-89. Nuts, Self-Locking, Hexagonal in Heat Resisting Steel FE-PA92HT (A286), Uncoated Classification: 1100 MPa/425 Degrees Celsius. 5 pp.

BSI A 275-A 280-81. 1981 Amd 1 Hexagon Self-Locking Nuts with Non-Metallic Locking Inserts. Metric Series. 10 pp.

CEN PREN 3723-91. Nuts, Hexagonal,Self-Locking, in Heat Resisting Steel FE-PA92HT (A286), MoS2 Coated Classification: 1100 MPa (at Ambient Temperature)/425 Degrees C. 5 pp.

CEN PREN 3763-90. Nuts, Hexagonal, Self-Locking, Ball Seat, in Heat Resisting Steel, Passivated, MOS2 Lubricated Classification: 900MPa (at Ambient Temperature) /315 Degrees C. 7 pp.

CEN PREN 3763-93. Aerospace Series Nuts, Hexagonal, Self-Locking, Ball Seat, in Heat Resisting Steel, MoS2 Lubricated Classification: 900 MPa (at Ambient Temperature) /315 Degrees C. 6 pp.

—Hexagonal—Locking (Cont.)
CEN PREN 3763-93. CORRIGENDUM Aerospace Series Nuts, Hexagonal, Self-Locking, Ball Seat, in Heat Resisting Steel, MoS2 Lubricated Classification: 900 MPa (at Ambient Temperature) /315 Degrees C. 1 p.

—Hexagonal—Locking—Counterbore
CEN PREN2882-90. Nuts, Hexagonal, Self-Locking, with Counterbore and Captive Washer, in Steel, Cadmium Plated, MoS2 Lubricated Classification: 1100MPa (at Ambient Temperature) /235 Degrees Celsius. 9 pp.

ISO 8538-86. Aerospace—Self-Locking Hexagon Nuts with Counterbore and Captive Washer, Classification 1 100 MPa/235 Degrees Celsius First Edition. 5 pp.

SBAC TS 135 ISSUE 1. All Metal Self-Locking Nuts (MJ Threads) Reduced Hexagon—Deep Counterbored Tensile Strength:—1100 MPa at Ambient Temperature.

—Hexagonal—Locking—Counterbore—Washer Head
CEN PREN2882-90. Nuts, Hexagonal, Self-Locking, with Counterbore and Captive Washer, in Steel, Cadmium Plated, MoS2 Lubricated Classification: 1100MPa (at Ambient Temperature) /235 Degrees Celsius. 9 pp.

—Hexagonal—Locking—Flanged
CEN PREN 3536-91. Nuts, Hexagon, Self-Locking, in Heat Resisting Steel, MoS2 Lubricated Classification: 1100MPa (at Ambient Temperature)/315 Degrees C. 5 pp.

CEN PREN 3626-91. Nuts, Hexagon, Self-Locking, in Steel, Cadmium Plated, MoS2 Lubricated Classification: 1100 MPa (at Ambient Temperature)/235 Degrees C. 5 pp.

DIN ENGL LN 9161-77. Nuts Hexagon Flanged; Self-Locking for Temperatures up to 315 Degrees Celsius and up to 425 Degrees Celsius (May). 3 pp.

DIN ENGL LN 9338-85. Aerospace; Nuts Hexagon Flanged; Self-Locking for Temperatures up to 235 Degrees C (Feb). 3 pp.

—Hexagonal—Locking—Self Aligning
DIN ENGL LN 29795-80. Aerospace; Hexagon Nuts; Self-Aligning; Self-Locking for Temperatures up to 235 Degrees C (Sept). 3 pp.

—Hexagonal—Locking—Thin
BSI A 281-A 286-81. 1981 Hexagon (Thin) Self-Locking Nuts with Non-Metallic Locking Inserts. Metric Series. 8 pp.

—Hexagonal—Locking—Washer Head
AECMA PREN3033-89. Nuts, Self-Locking, Hexagonal with Captive Washer, in Heat Resisting Steel FE-PA92HT (A286), Uncoated Classification: 1100 MPa/425 Degrees Celsius. 5 pp.

—Hexagonal—Thin
BSI A 233-A 236-71. 1971 Amd 1 Plain Hexagonal Thin Nuts (of Class 3B UNJ Thread). 5 pp.

BSI A 233-A 236-02. 1971 Amd 2 Plain Hexagonal Thin Nut (of Class 3B UNJ Thread) (AMD 7769) August 15, 1993 (S). 6 pp.

CEN PREN 2876-91. Nuts, Hexagon, Plain, Thin Normal Across Flats, in Aluminium Alloy, Anodized Classification: 450 MPa (at Ambient Temperature) /120 Degrees C. 8 pp.

DIN ENGL LN 9342-84. Aerospace; Nuts, Hexagon, Thin, Steel (Feb). 3 pp.

—Hexagonal—Union
SBAC AS 62350 ISSUE 1. Nut Union (Cres). 2 pp.
SBAC AS 62351 ISSUE 1. Nut Union-Male (Cres). 2 pp.

—Hexagonal—Washer Head—Locking
DIN ENGL LN 29790-76. Nuts Hexagon with Captive Washer; Self-Locking for Temperatures up to 235 Degrees C (Oct). 3 pp.

—High Expansion
BSI HR 650-72. 1972 Amd 2 High Expansion Heat-Resisting Steel Bar and Wire for the Manufacture of Bolts, Studs, Set Screws and Nuts (Ni 25.5, Cr 15, Ti 2, Mn 1.5, Mo 1.25, Si 0.7, V 0.3) (Limiting Ruling Section 20 mm). 5 pp.

—Insert—Locking
AECMA PREN3297. Inserts, Thin Wall, Self-Locking in Heat Resisting Nickel Base Alloy N1-P100HT (Inconel 718) Classification: 1275 MPa/550 Degrees Celsius. 21 pp.

—Locking
BSI A 293-83. 1983 Amd 1 Procurement of Self-Locking Nuts with Non-Metallic Locking Element. Metric Series (AMD 5767) April 30, 1990. 20 pp.

BSI A 295-85. 1985 Methods of Test for Self-Locking Nuts with Maximum Operating Temperature Less than or Equal to 425 Degrees C. 13 pp.

—Locking (Cont.)
BSI A 298-88. 1988 Methods of Test for Self-Locking Nuts with Maximum Operating Temprature Greater than 425 Degrees C. 14 pp.

BSI A 302-92. 1992 Self-Locking Nuts with Maximum Operating Temperature Less Than or Equal to 425 Degrees C—Procurement Specification (ISO 5858: 1991). 23 pp.

CEN PREN 3726-91. Nuts, Self-Locking, Clip, in Heat Resisting Steel FE-PA92HT (A286), MoS2 Coated Classification: 1100 MPa (at Ambient Temperature)/425 Degrees C. 6 pp.

CEN PREN 3726-93. Aerospace Series Nuts, Self-Locking, Clip, in Heat Resisting Steel FE-PA92HT (A286), MoS2 Coated Classification: 1 100 MPa (at Ambient Temperature) /425 Degrees C. 4 pp.

CEN PREN 3752-91. Nuts, Self-Locking, in Heat Resisting Steel FE-PA92HT (A286), MoS2 Coated Classification: 1100 MPa/425 Degrees C Technical Specification. 24 pp.

DIN ENGL LN 65016-70. Self-Locking Nuts for Temperatures up to 425 Degrees C; Procurement Specification (Dec). 10 pp.

DIN ENGL LN 65100-75. Self-Locking Nuts for Temperature Class 650 Degrees C; Procurement Specification (Jan). 10 pp.

ISO 5858-91. Aerospace—Self-Locking Nuts with Maximum Operating Temperature Less Than or Equal to 425 Degrees Celsius—Procurement Specification First Edition. 23 pp.

ISO 7481-84. Aerospace—Fasteners—Self-Locking Nuts with Maximum Operating Temperature Less Than or Equal to 425 Degrees Celsius—Test Methods First Edition. 14 pp.

ISO 8641-87. Aerospace—Self-Locking Nuts with Maximum Operating Temperature Greater Than 425 Degrees Celsius—Procurement Specification First Edition. 19 pp.

ISO 8642-86. Aerospace—Self-Locking Nuts with Maximum Operating Temperature Greater Than 425 Degrees Celsius—Test Methods First Edition. 14 pp.

MOD UK DSTAN 53-89-01. Requirements for the Substitutability of Unified Self-Locking Nuts (Stiffnuts) Issue 1; Amendment 2. 89 pp.

—Lockwire Holes—Identification Systems
DIN ENGL LN 29962-76. Bolts and Accessories; Additional Procurement Data for Special Designs and Arrangement of Code Letters in Characteristic Blocks (Apr). 3 pp.

—Plate—Locking
AECMA PREN3019-89. Self-Locking Plate Nuts Floating, Two-Lug in Heat Resisting Steel FE-PA2HT (A286), Unplated Classification: 1100 MPa/ 650 Degrees Celsius. 5 pp.

AECMA PREN3020-89. Self Locking Plate Nuts Floating, Two-Lug in Heat Resisting Steel FE-PA2HT (A286), Silver-Coated Classification: 1100 MPa/650 Degrees Celsius. 5 pp.

CEN PREN 3019-92. Self-Locking Plate Nuts, Floating, Two-Lug, in Heat Resisting Steel FE-PA92HT (A286) Classification: 1 100 MPa (at Ambient Temperature) /650 Degrees C. 4 pp.

CEN PREN 3020-92. Self-Locking Plate Nuts, Floating, Two-Lug, in Heat Resisting Steel FE-PA92HT (A286), Silver Plated Classification: 1 100 MPa (at Ambient Temperature) /650 Degrees C. 4 pp.

DIN ENGL LN 29680-77. Aerospace; Nuts, Plate; Self-Locking, Floating, Self-Aligning, Double Lug for Temperatures up to 235 Degrees C (July). 3 pp.

DIN ENGL LN 29910-83. Aerospace; Nuts, Plate; Self-Locking, Floating, Double Lug, Reduced, for Temperatures up to 235 Degrees C (Oct). 3 pp.

—Screw Threads
BSI BS 6293: Part 2-82. 1982 MJ Threads for Aerospace Construction Part 2: Dimensions for Bolts and Nuts. 9 pp.

ISO 5855 Pt 2-88. Aerospace—MJ Threads—Part 2: Limit Dimensions for Bolts and Nuts Second Edition; (Corrected and Reprinted -1989). 8 pp.

—Screw Threads—Identification Systems
DIN ENGL LN 29962-76. Bolts and Accessories; Additional Procurement Data for Special Designs and Arrangement of Code Letters in Characteristic Blocks (Apr). 3 pp.

—Screw Threads—Selection
DIN ENGL LN 9163 Pt 2-80. Aerospace; ISO Metric Threads; Selection of Threads and Allowances for Screws, Bolts and Nuts (Dec). 3 pp.

—Shaft—Locking
CEN PREN 3295-92. Shaft-Nuts, Self-Locking, in FE-PA92HT (A286), Silver Plated. 6 pp.

—Shank—Locking
AECMA PREN3014-87. Self-Locking Serrated Shank Nuts in Heat Resisting Steel FE-PA92HT Classification 1100 MPa/650 Degrees Celsius. 4 pp.

INDUSTRY STANDARDS

INTERNATIONAL AND NON-U.S. NATIONAL STANDARDS
SUBJECT INDEX
Aerospace

Aerospace Nuts *(Cont.)*
—Shank—Locking *(Cont.)*
AECMA PREN3015-87. Self-Locking Serrated Shank Nuts in Heat Resisting Steel FE-PA92HT Silver Plated Classification 1100 MPa/650 Degrees Celsius. 4 pp.
CEN PREN 3722-91. Shank Nuts, Self-Locking, in Heat Resisting Steel FE-PA92HT (A286), MoS2 Coated Classification: 1100 MPa (at Ambient Temperature)/425 Degrees C. 5 pp.
CEN PREN 4013-93. Aerospace Series Shank Nuts, Self-Locking, in Heat Resisting Nickel Base Alloy NI-P100HT (Inconel 718), Silver Plated Classification: 1 550 MPa (at Ambient Temperature) /600 Degrees C. 4 pp.

—Slotted
BSI A 242-A 245-74. 1974 Hexagonal Castle Nuts (of Class 3B UNJ Thread). 5 pp.
BSI A 246-A 249-74. 1974 Slotted Hexagonal Thick Nuts (of Class 3B UNJ Thread). 5 pp.
CEN PREN 2868-91. Nuts, Hexagonal, Slotted/Castellated, Normal Height, Normal Across Flats, in Heat Resisting Steel, Silver Plated Classification: 1100 MPa (at Ambient Temperature)/650 Degrees C. 6 pp.
CEN PREN 2868-93. Aerospace Series Nuts, Hexagonal, Slotted/Castellated, Normal Height, Normal Across Flats, in Heat Resisting Steel, Silver Plated Classification: 1 100 MPa (at Ambient Temperature) /650 Degrees C. 5 pp.
CEN PREN 2869-91. Nuts, Hexagonal, Slotted/Castellated, Normal Height, Normal Across Flats, in Heat Resisting Steel, Passivated Classification: 1100 MPa (at Ambient Temperature) /650 Degrees C. 6 pp.
CEN PREN 2869-93. Aerospace Series Nuts, Hexagonal, Slotted/Castellated, Normal Height, Normal Across Flats, in Heat Resisting Steel, Passivated Classification: 1 100 MPa (at Ambient Temperature) /650 Degrees C. 5 pp.
DIN ENGL LN 9345-80. Aerospace; Nuts, Castle (Aug). 3 pp.

—Spline
ISO 7403-83. Fasteners for Aerospace Construction—Spline Drive Wrenching Configuration—Metric Series First Edition. 5 pp.

—Union
CEN PREN 3851-91. Nuts, for Spherical Tube Coupling, in Titanium Alloy TI-P63 Use Temperature 350 Degrees C. 5 pp.

Aerospace Rivets
See Also: Aerospace Bolts; Aerospace Fasteners; Fasteners; Rivets
AECMA PREN2345-82. Aluminium and Aluminium Alloy Rivets—Technical Specification (C3/D44). 10 pp.
AECMA PREN2898-88. Corrosion and Heat Resisting Steel Rivets Technical Specification. 8 pp.
AECMA PREN2941-88. Nickel Alloy Rivets Technical Specification. 8 pp.
AECMA PREN3659-89. Titanium/Columbium Solid Rivets Technical Specification. 8 pp.
BSI 2SP 157-163-73. 1973 Amd 2 Solid Rivets with Universal Head Made from L 86 (SP 157 and SP 163), L 58 (SP 160 and SP 161), L 37 (SP 162) and DTD 204 (SP 158 and SP 159) Materials. 9 pp.
CENELEC PREN 3627-88. Titanium Solid Rivets Technical Specification. 11 pp.
DIN ENGL LN 65020-70. Rivet Wires and Rivets of Aluminium and Wrought Aluminium Alloys; Technical Specification (Jan). 7 pp.
ISO 9417-90. Aerospace—Metric Solid Rivets—Test Methods First Edition. 6 pp.
MOD UK DTD-5036-01. Low Carbon Chromium-Nickel Corrosion-Resisting Steel Wire, Rivets and Split Pins (Weldable); Amendment 1. 5 pp.

—Cold Forged
BSI 7L 37-89. 1989 Amd 1 Wire for Solid Cold-Forged Rivets of Aluminium-Copper-Magnesium-Silicon-Manganese Alloy (AMD 6625) December 21, 1990. 5 pp.

—Countersunk Head
AECMA PREN2550-85. Rivets, Solid, Countersunk Head in Aluminium EN2114. 6 pp.
AECMA PREN3394-90. Rivets, Solid, 100 Degrees Normal Countersunk Head, in Aluminium Alloy 2117, Metric Series. 6 pp.
AECMA PREN3643-90. Rivets, Solid, 100 Degrees Normal Countersunk Head, in Titanium T1-P02, Anodized. 6 pp.
BSI BS EN 2550-93. 1993 Rivets, Solid, 100 Degree Normal Countersunk Head, in Aluminium 1050A, Inch Based Series. 11 pp.
BSI 2SP 68-71-73. 1973 Amd 2 100 Degree Countersunk Precision Head Aluminium and Aluminium Alloy Rivets for Aircraft. 15 pp.

Aerospace Rivets *(Cont.)*
—Countersunk Head *(Cont.)*
BSI 2SP 142-143-73. 1973 Solid Rivets with 100 Degree Countersunk Truncated Radiused Head Made from L 86 (Sp 142) and L 37 (Sp 143) Materials. 8 pp.
CEN PREN2550-89. Rivets, Solid, 100 Degrees Normal Countersunk Head, in Aluminium EN 2114. 6 pp.
CEN EN 2550-92. Aerospace Series—Rivets, Solid, 100 Degrees Normal Countersunk Head, in Aluminium 1050A, Inch Based Series. 6 pp.
CEN EN 2556-92. Aerospace Series—Rivets, Solid, 100 Degrees Normal Countersunk Head with Dome, inAluminium Alloy 5056A, Anodized or Chromated, Inch Based Series. 6 pp.
CEN PREN 3137-91. Rivets, Solid, 100 Degree Normal Countersunk Head with Dome, in Corrosion Resisting Steel FE-PA11, Passivated. 6 pp.
CEN PREN 3138-91. Rivets, Solid, 100 Degree Normal Countersunk Head with Dome, in Corrosion Resisting Steel FE-PA92HT, Passivated. 6 pp.
CEN PREN 3139-91. Rivets, Solid, 100 Degree Normal Countersunk Head, in Corrosion Resisting Steel FE-PA11, Passivated. 6 pp.
CEN PREN 3140-91. Rivets, Solid, 100 Degree Normal Countersunk Head, in Corrosion Resisting Steel FE-PA92HT, Passivated. 6 pp.
CEN PREN 3399-90. Rivets Solid, 100 Degrees Normal Countersunk Head, in Aluminum Alloy 5056A, Metric Series. 6 pp.
CEN PREN 3399-91. Rivets, Solid, 100 Degree Normal Countersunk Head, in Aluminium Alloy 5056A, Metric Series. 6 pp.
CEN PREN 3400-91. Rivets, Solid, 100 Degree Normal Countersunk Head, in Aluminium Alloy 5056A, Anodized or Chromated, Metric Series. 6 pp.
CEN PREN 3401-92. Rivets, Solid, 100 Degrees Normal Countersunk Head, in Aluminium Alloy 7050, Metric Series. 6 pp.
CEN PREN 3402-92. Rivets, Solid, 100 Degrees Normal Countersunk Head, in Aluminium Alloy 7050, Anodized or Chromated, Metric Series. 6 pp.
CEN PREN 3406-91. Rivets, Solid, 100 Degree Normal Countersunk Head with Dome, in Aluminium Alloy 2117, Anodized or Chromated, Metric Series. 6 pp.
CEN PREN 3407-91. Rivets, Solid, 100 Degree Normal Countersunk Head, in Aluminium alloy 2117, Anodized or Chromated, Metric Series. 6 pp.
CEN PREN 3408-90. Rivets, Solid, 100 Degree Normal Countersunk Head, in Aluminum Alloy 2017A, Metric Series. 6 pp.
CEN PREN 3408-91. Rivets, Solid, 100 Degree Normal Countersunk Head, in Aluminium Alloy 2017A, Metric Series. 6 pp.
CEN PREN 3410-93. Aerospace Series Rivets, Solid, 100 Degrees Normal Countersunk Head, in Aluminium 1050A, Metric Series. 5 pp.
CEN PREN 3418-90. Rivets, Solid 100 Degrees Countersunk Head in Nickel Base Alloy NI-P11, Cadmium Plated, Metric Series. 6 pp.
CEN PREN 3418-91. Rivets, Solid, 100 Degree Normal Countersunk Head, in Nickel Base Alloy NI-P11, Cadmium Plated, Metric Series. 6 pp.
CEN PREN 3419-90. Rivets, Solid 100 Degrees Normal Countersunk Head in Nickel Base Alloy NI-P11, Metric Series. 6 pp.
CEN PREN 3419-91. Rivets, Solid, 100 Degree Normal Countersunk Head, in Nickel Base Alloy NI-P11, Metric Series. 6 pp.
CEN PREN 3739-91. Rivets, Solid, 100 Degree Normal Countersunk Head, in Titanium Alloy Ti 45,5 Cb, Metric Series. 6 pp.

—Countersunk Head—Domed
AECMA PREN2551-87. Rivets, Solid, Countersunk Head with Dome in Aluminium Alloy EN 2627. 6 pp.
AECMA PREN2552-87. Rivets, Solid, Countersunk Head with Dome in Aluminium Alloy EN 2627 Anodised. 6 pp.
AECMA PREN2553-85. Rivets, Solid, Countersunk Head with Dome in Aluminium Alloy EN2116. 6 pp.
AECMA PREN2555-85. Rivets, Solid, Countersunk Head with Dome in Aluminium Alloy EN2117. 6 pp.
AECMA PREN2556-85. Rivets, Solid, Countersunk Head with Dome in Aluminium Alloy EN2117—Anodised. 6 pp.
AECMA PREN3137-89. Rivets, Solid, 100 Degrees Normal Countersunk Head, in Corrosion Resisting Steel EN 2470, Passivated. 5 pp.
AECMA PREN3393-90. Rivets, Solid, 100 Degrees Normal Countersunk Head with Dome, in Aluminium Alloy 2117, Metric Series. 6 pp.
AECMA PREN3642-90. Rivets, Solid, 100 Degrees Normal Countersunk Head with Dome, in Titanium T1-P02, Anodized. 6 pp.
BSI BS EN 2551-93. 1993 Rivets, Solid, 100 Degree Normal Countersunk Head with Dome, in Aluminium Alloy 2117, Inch Based Series. 11 pp.

Aerospace Rivets *(Cont.)*
—Countersunk Head—Domed *(Cont.)*
BSI BS EN 2552-93. 1993 Rivets, Solid, 100 Degree Normal Countersunk Head with Dome, in Aluminium Alloy 2117, Anodized or Chromated, Inch Based Series. 11 pp.
BSI BS EN 2553-93. 1993 Rivets, Solid, 100 Degree Normal Countersunk Head with Dome, in Aluminium Alloy 2017A, Inch Based Series. 12 pp.
BSI BS EN 2555-93. 1993 Rivets, Solid, 100 Degree Normal Countersunk Head with Dome, in Aluminium Alloy 5056A, Inch Based Series. 11 pp.
BSI BS EN 2556-93. 1993 Rivets, Solid, 100 Degree Normal Countersunk Head with Dome, in Aluminium Alloy 5056A, Anodized or Chromated, Inch Based Series. 11 pp.
CEN EN 2551-92. Aerospace Series—Rivets, Solid, 100 Degrees Normal Countersunk Head with Dome, in Aluminium Alloy 2117, Inch Based Series. 6 pp.
CEN EN 2552-92. Aerospace Series—Rivets, Solid, 100 Degrees Normal Countersunk Head with Dome, in Aluminium Alloy 2117, Anodized or Chromated, Inch Based Series. 6 pp.
CEN PREN2553-89. Rivets, Solid, 100 Degrees Normal Countersunk Head with Dome, in Aluminium Alloy EN 2116. 6 pp.
CEN EN 2553-92. Aerospace Series—Rivets, Solid, 100 Degrees Normal Countersunk Head with Dome, in Aluminium Alloy 2017A, Inch Based Series. 7 pp.
CEN PREN2555-89. Rivets, Solid, 100 Degrees Countersunk Head with Dome, in Aluminum Alloy EN 2116. 6 pp.
CEN EN 2555-92. Aerospace Series—Rivets, Solid, 100 Degrees Normal Countersunk Head with Dome, in Aluminium Alloy 5056A, Inch Based Series. 6 pp.
CEN PREN2556-89. Rivets, Solid, 100 Degrees Normal Countersunk Head with Dome, in Aluminum Alloy EN 2117. 6 pp.
CEN PREN 3137-91. Rivets, Solid, 100 Degree Normal Countersunk Head with Dome, in Corrosion Resisting Steel FE-PA11, Passivated. 6 pp.
CEN PREN 3138-91. Rivets, Solid, 100 Degree Normal Countersunk Head with Dome, in Corrosion Resisting Steel FE-PA92HT, Passivated. 6 pp.
CEN PREN 3370-92. Rivets, Solid, 100 Degree Normal Countersunk Head with Dome, in Aluminium Alloy 7050, Anodized or Chromated, Inch Based Series. 7 pp.
CEN PREN 3395-90. Rivets, Solid, 100 Degrees Normal Countersunk Head with Dome, in Aluminum Alloy 2017A, Metric Series. 6 pp.
CEN PREN 3395-91. Rivets, Solid, 100 Degrees Normal Countersunk Head with Dome, in Aluminium Alloy 2017A, Metric Series. 6 pp.
CEN PREN 3396-90. Rivets, Solid, 100 Degrees Normal Countersunk Head with Dome, in Aluminum Alloy 5056A, Metric Series. 6 pp.
CEN PREN 3396-91. Rivets, Solid, 100 Degrees Normal Countersunk Head with Dome, in Aluminium Alloy 5056A, Metric Series. 6 pp.
CEN PREN 3405-92. Rivets, Solid, 100 Degrees Normal Countersunk Head with Dome, in Aluminium Alloy 7050, Anodized or Chromated, Metric Series. 6 pp.
CEN PREN 3412-90. Rivets, Solid, 100 Degrees Countersunk Head with Dome, in Aluminum Alloy 5056A, Anodized or Chromated, Metric Series. 6 pp.
CEN PREN 3412-91. Rivets, Solid, 100 Degree Normal Countersunk Head with Dome, in Aluminium Alloy 5056A, Anodized or Chromated, Metric Series. 6 pp.
CEN PREN 3413-92. Rivets, Solid, 100 Degree Normal Countersunk Head with Dome, in Aluminium Alloy 7050, Metric Series. 6 pp.
CEN PREN 3738-91. Rivets, Solid, 100 Degree Normal Countersunk Head with Dome, in Titanium Alloy Ti 45,5 Cb, Metric Series. 6 pp.

—Countersunk Head—Flushness Control—Gaging
BSI A 273-78. 1978 Gauging Practice for 100 Degree Countersunk Head Fasteners for Flushness Control. 7 pp.

—Countersunk Head—Torque Recessed
BSI A 272-77. 1977 Hi-Torque Speed Drive Recess: Dimensions and Gauging for Countersunk Head Fasteners. 3 pp.

—Engineering Drawings
AECMA PREN2544-83. Representation of Rivets on Drawings for Aerospace Equipment. 7 pp.
BSI BS EN 2544-90. 1990 Representation of Rivets on Drawings for Aerospace Equipment. 10 pp.
CEN EN 2544-87. Representation of Rivets on Drawings for Aerospace Equipment. 7 pp.

—Glossaries
ISO 5843 Pt 2-90. Aerospace—List of Equivalent Terms—Part 2: Aerospace Rivets Second Edition. 27 pp.

INTERNATIONAL AND NON-U.S. NATIONAL STANDARDS
SUBJECT INDEX
Aerospace

Aerospace Rivets *(Cont.)*
—Heat Resistant
BSI 2HR 504-73. 1973 Amd 2 Nickel-Chromium-Titanium Heat-Resisting Alloy Bar, and Wire for Rivets, and Rivets (Nickel Base, Cr 19.5, Ti 0.4). 4 pp.

—Hole Size
AECMA PREN2309-79. Hole Sizes for Solid Rivets (C3/D55). 3 pp.
BSI BS EN 2309-90. 1990 Hole Sizes for Solid Rivets. 6 pp.
CEN EN 2309-89. Hole Sizes for Solid Rivets. 3 pp.

—Oval Head
BSI 2SP 83-85-73. 1973 Mushroom Head Aluminium Alloy Rivets for Aircraft. 10 pp.

—Universal Head
AECMA PREN2143-83. Rivets, Solid, Universal Head, in Aluminium, EN 2114. 6 pp.
AECMA PREN2144-83. Rivets, Solid, Universal Head in Aluminium Alloy, EN 2115. 6 pp.
AECMA PREN2145-83. Rivets, Solid, Universal Head, in Aluminium Alloy, EN 2115. 6 pp.
AECMA PREN2146-83. Rivets, Solid, Universal Head, in Aluminium Alloy, EN 2116. 6 pp.
AECMA PREN2148-83. Rivets, Solid, Universal Head, in Aluminium Alloy, EN 2117. 6 pp.
AECMA PREN2149-83. Rivets, Solid, Universal Head, in Aluminium Alloy, EN 2117 Anodised. 6 pp.
AECMA PREN3135-90. Rivets, Solid, Universal Head, in Corrison Resisting Steel FE-PA11, Passivated. 5 pp.
AECMA PREN3136-90. Rivets, Solid, Universal Head in Corrosion Resisting Steel FE-PA92HT, Passivated. 5 pp.
AECMA PREN3141-90. Rivets, Solid, Universal Head, in Nickel Base Alloy NI-P11, Cadmium Plated. 5 pp.
AECMA PREN3142-90. Rivets, Solid, Universal Head, in Nickel Alloy N1-P11. 5 pp.
AECMA PREN3392-90. Rivets, Solid, Universal Head in Aluminium Alloy 2117, Anodized or Chromated, Metric Series. 6 pp.
AECMA PREN3644-90. Rivets, Solid, Universal Head, in Titanium T1 P02, Anodized. 6 pp.
BSI BS EN 2143-93. 1993 Rivets, Solid, Universal Head, in Aluminium 1050A, Inch Based Series. 11 pp.
BSI BS EN 2146-93. 1993 Rivets, Solid, Universal Head, in Aluminium Alloy 2017A, Inch Based Series. 12 pp.
BSI BS EN 2148-93. 1993 Rivets, Solid, Universal Head, in Aluminium Alloy 5056A, Inch Based Series. 11 pp.
BSI BS EN 2149-93. 1993 Rivets, Solid, Universal Head, in Aluminium Alloy 5056A, Anodized or Chromated, Inch Based Series. 11 pp.
CEN PREN 2143-89. Rivets, Solid, Universal Head, in Aluminium EN 2114. 6 pp.
CEN EN 2143-92. Aerospace Series—Rivets, Solid, Head, in Aluminium 1050A, Inch Based Series. 6 pp.
CEN PREN 2144-93. Aerospace Series Rivets, Solid, Universal Head, in Aluminium Alloy 2117, Inch Based Series. 6 pp.
CEN PREN 2145-93. Aerospace Series Rivets, Solid, Universal Head, in Aluminium Alloy 2117, Anodized or Chromated, Inch Based Series. 7 pp.
CEN PREN2146-89. Rivets, Solid, Universal Head, in Aluminium Alloy EN 2116. 6 pp.
CEN EN 2146-92. Aerospace Series—Rivets, Solid, Universal Head, in Aluminium Alloy 2017A, Inch Based Series. 7 pp.
CEN PREN2148-89. Rivets, Solid, Universal Head, in Aluminium Alloy EN 2117. 6 pp.
CEN EN 2148-92. Aerospace Series—Rivets, Solid, Universal Head in Aluminium Alloy 5056A, Inch Based Series. 6 pp.
CEN PREN2149-89. Rivets, Solid, Universal Head, in Aluminium Alloy EN 2117 Anodized. 6 pp.
CEN EN 2149-92. Aerospace Series—Rivets, Solid, Universal Head, in Aluminium Alloy 5056A, Anodized or Chromated Inch Based Series. 6 pp.
CEN PREN 3369-92. Rivets, Solid, Universal Head, in Aluminium Alloy 7050, Anodized or Chromated, Inch Based Series. 6 pp.
CEN PREN 3391-91. Rivets, Solid, Universal Head, in Aluminium Alloy 2117, Metric Series. 4 pp.
CEN PREN 3397-92. Rivets, Solid, Universal Head, in Aluminium Alloy 7050, Metric Series. 6 pp.
CEN PREN 3398-92. Rivets, Solid, Universal Head, in Aluminium Alloy 7050, Anodized or Chromated, Metric Series. 6 pp.
CEN PREN 3403-90. Rivets, Solid, Universal Head, in Aluminium Alloy 5056A, Metric Series. 6 pp.
CEN PREN 3403-91. Rivets, Solid, Universal Head in Aluminium Alloy 5056A, Metric Series. 6 pp.
CEN PREN 3404-90. Rivets, Solid, Universal Head, in Aluminium Alloy 5056A, Anodized or Chromated Metric Series. 6 pp.

Aerospace Rivets *(Cont.)*
—Universal Head *(Cont.)*
CEN PREN 3404-91. Rivets, Solid, Universal Head, in Aluminium Alloy 5056A, Anodized or Chromated, Metric Series. 6 pp.
CEN PREN 3409-90. Universal Head in Aluminium 105A Metric Series. 6 pp.
CEN PREN 3409-93. Aerospace Series Rivets, Solid, Universal Head, in Aluminium 1050A, Metric Series. 5 pp.
CEN PREN 3410-90. Rivets, Solid, Universal Head, in Nickel Base Alloy NI-P11, Cadmium Plated, Metric Series. 6 pp.
CEN PREN 3411-90. Rivets, Solid, Universal Head, in Aluminium Alloy 2017A, Metric Series. 6 pp.
CEN PREN 3411-91. Rivets, Solid, Universal Head, in Aluminium Alloy 2017A, Metric Series. 6 pp.
CEN PREN 3416-91. Rivets, Solid, Universal Head, in Nickel Base Alloy NI-P11, Cadmium Plated, Metric Series. 6 pp.
CEN PREN 3417-90. Rivets, Solid, Universal Head, in Nickel Base Alloy NI-P11, Metric Series. 6 pp.
CEN PREN 3417-91. Rivets, Solid, Universal Head, in Nickel Base Alloy NI-P11, Metric Series. 6 pp.
CEN PREN 3424-91. Rivets, Solid, Universal Head, in Titanium Alloy Ti 45,5 Cb, Metric Series. 6 pp.

Aerospace Screws
Use For: Aeronautical Screws *See Also:* Aerospace Bolts; Aerospace Fasteners; Fasteners; Screws
BSI A 296-86. 1986 Tolerances of Form and Position for Bolts and Screws. Metric Series. 8 pp.
DIN ENGL LN 65010 Pt 1-68. Specification for Screws and Bolts of Unalloyed and Low-Alloy Steel with Minimum Tensile Strength up to 900 N/mm2 (Mar). 22 pp.
DIN ENGL LN 65010 Pt 2-68. Specification for Screws and Bolts of Unalloyed and Low-Alloy Steel with Minimum Tensile Strength up to 900 N/mm2, Minimum Ultimate Tensile Loads, Proof Loads (Mar). 1 p.
ISO 7913-85. Aerospace—Fasteners—Tolerances of Form and Position for Bolts and Screws First Edition. 6 pp.
SBAC AS 2426 ISSUE 2. Clamping Screw—2 BA Tank Straps. 1 p.
SBAC AS 6686 ISSUE 1. Clamping Screw-No.10 UNF Tank Straps.
SBAC TS 27 ISSUE 2. Studs, Bolts and Screws in BS HR601 (Nimonic) Material. 1000 MPa (145,000 lbf/in2) Tensile at Room Temperature (for Which Forging is Not Mandatory and Qualification Not a Requirement).
SBAC TS 108 ISSUE 3. Studs, Bolts and Screws in BS.S159 Material 1100 MPa (159,000 lbf/in2) Tensile, 450 Degrees C (for Which Forging is Not Mandatory and Qualification Not a Requirement).
SBAC TS 109 ISSUE 3. S.B.A.C. Manufacturing Specification for Studs, Bolts and Screws in BSHR 650 Material 900 MPa (130,000 lbf/in2) Tensile, 500 Degrees C (for Which Forging is Not Mandatory and Qualification Not a Requirement).
SBAC TS 121 ISSUE 2. Studs, Bolts and Screws in DTD 5639 (WASPLOY) Material 1210 MPa (175,000 Lbf/in2) Tensile at Room Temperature (For Which Forging) is Not Mandatory and Qualification Not a Requirement.

—Acceptance Certificates
DIN ENGL LN 65018 Pt 1-68. Manufacturer's Acceptance Certificate for Screws and Bolts (Sept). 1 p.

—Classification
AECMA PREN3043-90. Fasteners, Externally Threaded, in Heat Resisting Steel FE-PA92HT (A 286) Classification: 90 MPa/550 Degrees Technical Specification. 19 pp.

—Countersunk Head
BSI A 260-A 265-76. 1976 Amd 2 Specification for Metric Screws. 8 pp.
CEN PREN 3782-91. Holes for 100 Degrees Countersunk Head Screws Design Standard. 7 pp.
DIN ENGL LN 29940-80. Aerospace; Countersunk Head Screws, with TORQ-SET Recess; Non-Magnetizable (Mar). 3 pp.
DIN ENGL LN 65019-77. Bolts, Countersunk Head with TORQ-SET Recess (Dec). 3 pp.

—Countersunk Head—Machine
BSI 2A 204-63. 1963 Amd 2 2A 206, 2A 208, A 217-225 Machine Screws and Nuts (Unified Threads) Below 1/4 in for Aeronautical Purposes. 12 pp.

—Countersunk Head—Phillips
DIN ENGL LN 9438-83. Aerospace; Screws, Countersunk Head with Cross Recess (Nov). 4 pp.
ISO 5856-91. Aerospace—Screws, 100 Degree Normal Counter-sunk Head, Internal off-set Cruciform Ribbed Drive, Normal Shank, Short or Medium Length MJ Threads, Metallic Material, Coated or Uncoated, Strength Classes Less Than or Equal to 1 100 MPa—Dimensions First Edition. 6 pp.

Aerospace Screws *(Cont.)*
—Countersunk Head—Phillips—Nonmagnetic
DIN ENGL LN 9136-81. Aerospace; Bolts, Countersunk Head, Cross-Recessed, Non Magnetic (Dec). 4 pp.

—Countersunk Head—Recessed
AECMA PREN3381-89. Bolts, 100 Degrees Countersunk Normal Head, Offset Enciform-Ribbed Recess, Close Tolerance Shank, Short Thread, in Titanium Anodized Classification: 1100 MPa/315 Degrees Celsius. 7 pp.
AECMA PREN3543-89. Bolts, 100 Degrees Countersunk Normal Head, Offset Cruciform-Ribbed Recess, Close Tolerance Shank, Short Thread, in Corrosion Resisting Steel, Passivated Classification: 1100 MPa/425 Degrees Celsius. 8 pp.
BSI A 266-A 271-77. 1977 Amd 1 100 Degree Countersunk Head Titanium Alloy Bolts 160 000 Lbf/Square Inches (1100 MPa) with Hi-Torque Speed Drive Recesses. 11 pp.
BSI A 272-77. 1977 Hi-Torque Speed Drive Recess; Dimensions and Gauging for Countersunk Head Fasteners. 3 pp.
CEN PREN 3306-91. Screws, 100 Degree Countersunk Normal Head, Offset Cruciform-Ribbed Recess, Threaded to Head, in Titanium Alloy, MoS2 Lubricated Classification: 1100 MPa (at Ambient Temperature)/315 Degrees C. 6 pp.
CEN PREN 3760-90. Bolts, 100 Degress C Countersunk Normal Head, Offset, Cruciform-Ribbed Recess, Threaded to Head, in Heat and Corrosion Resisting Steel, Passivated Classification: 1100 MPa (at Ambient Temperature)/425 Degrees C. 8 pp.
CEN PREN 3760-92. Screws, 100 Degree Countersunk Normal Head, Offset Cruciform Recess, Threaded to Head, in Heat and Corrosion Resisting Steel, Passivated Classification: 1 100 MPa (at Ambient Temperature) /425 Degrees C. 5 pp.
CEN PREN 6024-92. Screws, 100 Degree Countersunk Reduced Normal Head, Offset Cruciform Recess, Close Tolerance Shank, Short Thread, in Titanium Alloy, Anodized, MoS2 Lubricated, Classification: 1100 MPa (at Ambient Temperature) /315 Degrees C Inch Series. 8 pp.

—Countersunk Head—Slotted
AECMA PREN2652-87. Screws, 100 Degrees Countersunk Normal Head, Slotted, Fully Threaded, in Steel Cadmium Plated Classification 900 MPa/235 Degrees C.

—Countersunk Head—TORQ-SET
DIN ENGL LN 29787-74. Screws Countersunk Head TORQ-SET Recess; Fully Threaded to Head; Titanium Alloy (Dec). 3 pp.
DIN ENGL LN 29794-80. Aerospace; Screws, Countersunk Head with TORQ-SET Recess, Nearly Threaded to the Head (Mar). 4 pp.

—Double Hexagonal Head
ISO 4095-78. Fasteners for Aerospace Construction—Bi-Hexagonal Wrenching Configuration First Edition. 4 pp.

—Hexagonal Head
AECMA PREN2887-88. Screws, Hexagonal Normal Head, Fully Threaded, in Corrosion Resisting Steel, Passivated Classification: 600 MPa/425 Degrees Celsius. 8 pp.
AECMA PREN2937-86. Screws, Hexagon Head, Thread to Head, in Heat Resisting Steel FE-PA92HT (A286) Unplated Classification 900 MPa/650 Degrees Celsius. 5 pp.
AECMA PREN2938-86. Screws, Hexagon Head, Thread to Head, in Heat Resisting Steel FE-PA92HT (A286) Silver Plated Classification 900 MPa/650 Degrees Celsius. 5 pp.
AECMA PREN3112-87. Screws, Hexagonal Normal Head, Fully Threaded in Steel, Cadmium Plated Classification 900 MPa/235 Degrees Celsius. 7 pp.
AECMA PREN3113-87. Screws, Hexagonal Normal Head, Fully Threaded in Titanium Alloy Anodized Classification 900 MPa/315 Degrees Celsius. 7 pp.
CEN PREN 3308-91. Screws, Normal Hexagonal Head, Threaded to Head, in Titanium Alloy, MoS2 Lubricated Classification: 1100MPa (at Ambient Temperature)/315 Degrees C. 7 pp.
DIN ENGL LN 9038-81. Aerospace; Screws, Hexagon Head, Fully Threaded (Dec). 4 pp.

—Hexagonal Nuts
DIN ENGL LN 29669-80. Aerospace; Hexagon Nuts for Screws and Bolts with a Nominal Tensile Strength of 1100 N/mm2 (Mar). 2 pp.

—Pan Head—Phillips
AECMA PREN3636-90. Screws, Reduced Pan Head, Offset Cruciform Recess, Relieved Shank, Long Thread, in Heat Resisting Steel FE-PA92HT (A286), Silver Plated Classification: 900 MPa/650 Degrees. 6 pp.

INDUSTRY STANDARDS

Aerospace Screws (Cont.)

—Pan Head—Phillips (Cont.)
CEN PREN 3307-92. Screws, Pan Head, Offset Cruciform Recess, Threaded to Head, in Titanium Alloy, MoS2 Lubricated Classification: 1100 MPa (at Ambient Temperature)/315 Degrees C. 6 pp.
ISO 3202-91. Aerospace—Screws, Pan Head, Internal Offset Cruciform Ribbed Drive, Threaded to Head, MJ Threads, Metallic Material, Coated or Uncoated. Strength Classes Less Than or Equal to 1 100 MPa—Dimensions First Edition. 6 pp.

—Pan Head—TORQ-SET
DIN ENGL LN 29785-81. Aerospace; Screws, Pan Head with TORQ-SET Recess, Fully Threaded (Dec). 4 pp.
DIN ENGL LN 29958-83. Aerospace; Screws, Pan Head, with TORQ-SET Recess in Titanium Alloy; Fully Threaded (Oct). 4 pp.
DIN ENGL LN 65022-80. Aerospace; Pan Head Screws with TORQ-SET Recess; Threaded to Head (Mar). 4 pp.

—Phillips
BSI A 297-87. 1987 Internal Drive, Offset Cruciform Recess (Torq-Set) for Rotary Fastening Devices. Metric Series. 12 pp.
ISO 7994-85. Aerospace—Internal Drive, Offset Cruciform Recess (Torq-Set) for Rotary Fastening Devices—Metric Series First Edition. 12 pp.

—Phillips—Screwdrivers
BSI A 297-87. 1987 Internal Drive, Offset Cruciform Recess (Torq-Set) for Rotary Fastening Devices. Metric Series. 12 pp.
ISO 7994-85. Aerospace—Internal Drive, Offset Cruciform Recess (Torq-Set) for Rotary Fastening Devices—Metric Series First Edition. 12 pp.

—Round Head—Phillips—Nonmagnetic
DIN ENGL LN 9139-83. Aerospace; Screws, Round Head, with Cross Recess; Non-Magnetic (June). 3 pp.

—Socket Head
ISO 7403-83. Fasteners for Aerospace Construction—Spline Drive Wrenching Configuration—Metric Series First Edition. 5 pp.

—TORQ-SET—Recesses
DIN ENGL LN 29999 Pt 1-74. TORQ-SET Recesses; Dimensions (Aug). 2 pp.

Aerospace Systems
Use: Aerospace Equipment

Aerospace Threaded Fasteners
Use: Aerospace Fasteners

Aerospace Vehicles
See Also: Aerospace Equipment; Aircraft

—Components—Identification Systems
DIN ENGL LN 9051 Suppl. 1-72. Identification Marking of Controlled Components in Manufacturing Aircraft and Space Vehicles; Examples (Aug). 2 pp.

—Electric Wiring
JIS W 2010-89. Wiring for Aerospace Vehicle.

Aerospace Washers
See Also: Aerospace Fasteners; Fasteners; Washers (Fasteners)
AECMA PREN2948-86. Washers, Tab; Heat Resisting Steel. 5 pp.
AECMA PREN2949-86. Washer, Bent Tab; Heat Resisting Steel. 4 pp.
AECMA PREN3017-90. Washers Facing/Packing Heat Resisting Steel. 5 pp.
BSI 2SP 148-150-79. 1979 Unified Aluminium Alloy Washers. 3 pp.
BSI 2SP 151-153-79. 1979 Unified Alloy Steel Washers. 3 pp.
BSI 2SP 154-156-79. 1979 Unified Corrosion-Resistant Steel Washers. 4 pp.
CEN PREN 3696-91. Washer in Heat Resisting Steel. 7 pp.
CEN PREN 3822-91. Washers, in Heat Resisting Steel, Passivated. 5 pp.
CEN PREN 3822-93. Aerospace Series Washers, Flat, Thick, in Heat Resisting Steel, Passivated. 3 pp.
CEN PREN 3835-93. Aerospace Series Washers, 100 Degrees Dimpled, in Heat Resisting Steel, Passivated. 4 pp.
CEN PREN 3902-92. Washers for Rivet Assemblies, in Aluminium Clad or Unclad, Anodized, Metric Series. 5 pp.
CEN PREN 3903-92. Washers, Laminated, in Corrosion Resisting Steel FE-PA13. 5 pp.
DIN ENGL LN 9025-80. Aerospace; Washers (Apr). 4 pp.
DIN ENGL LN 29790-76. Nuts Hexagon with Captive Washer; Self-Locking for Temperatures up to 235 Degrees C (Oct). 3 pp.
DIN ENGL LN 29905-80. Aerospace; Washers; Self-Aligning (Apr). 2 pp.

—Bevel
DIN ENGL LN 9016-80. Aerospace; Washers; Bevelled (Apr). 3 pp.

—Concave
AECMA PREN2648-87. Washers, Concave. 4 pp.
CEN PREN 3764-90. Washers, Spherical, Concave, in Geat Resisting Steel Passivated. 5 pp.
CEN PREN 3764-93. Aerospace Series Washers, Concave, in Heat Resisting Steel, Passivated. 4 pp.
CEN PREN3764-93. CORRIGENDUM Aerospace Series Washers, Concave, in Heat Resisting Steel, Passivated. 3 pp.

—Control Rods
AECMA PREN2328-80. Washers Tab for Flight Control Rods—Dimensions (C1/23-3). 5 pp.
BSI BS EN 2328-90. 1990 Washers, Tab for Flight Control Rods. Dimensions. 7 pp.
CEN EN 2328-88. Washers, Tab for Flight Control Rods Dimensions. 7 pp.

—Counterbore
CEN PREN 2414-93. Aerospace Series Washers, Chamfered, with Counterbore, in Alloy Steel, Cadmium Plated. 4 pp.
CEN PREN 3821-91. Washers, with Cylindrical Counterbore, in Heat Resisting Steel, Passivated. 6 pp.
CEN PREN 3821-93. Aerospace Series Washers, Chamfered, with Counterbore, in Heat Resisting Steel, Passivated. 4 pp.
DIN ENGL LN 9028-80. Aerospace; Counterbored Washers (Apr). 3 pp.

—Countersunk
AECMA PREN2998-86. Washers Countersunk; Heat Resisting Steel. 4 pp.
AECMA PREN3016-90. Washers Countersunk, Load Spreading Heat Resisting Steel. 5 pp.
CEN PREN 3835-91. Washers 100 Degrees Countersunk, in Heat Resisting Steel, Passivated. 6 pp.

—Lock
AECMA PREN3432-89. Lockwashers Steel, Cadmium Plated. 6 pp.
AECMA PREN3433-89. Lockwashers Heat Resisting Steel, Passivated. 6 pp.
CEN PREN 3904-92. Washers, Bent Tab, in Aluminium Alloy 2024, Clad, Anodized. 5 pp.

—Lock—Control Rods
AECMA PREN2586-86. Washers, Lock for Flight Control Rods Dimensions. 3 pp.
BSI BS EN 2327-89. 1989 Washers, Lock with Radial Serrations in Alloy Steel. Dimensions. 8 pp.
CEN EN 2327-87. Washers, Lock with Radial Serrations in Alloy Steel Dimensions. 5 pp.
CEN EN 2586-91. Washers, Lock for Flight Control Rods—Dimensions. 5 pp.

—Lock—Corrosion Resistant Steel—Control Rods
BSI BS EN 2546-90. 1990 Washers, Lock with Radial Serrations in Corrosion Resisting Steel. Dimensions. 8 pp.
CEN EN 2546-88. Washers, Lock with Radial Serrations in Corrosions Resisting Steel Dimensions. 5 pp.

—Plain
AECMA PREN2122-82. Washers, Flat, Aluminium Alloy. 5 pp.
AECMA PREN2138-82. Washers, Flat, Steel. 5 pp.
AECMA PREN2139-82. Washers, Flat, Heat Resisting Steel. 5 pp.
AECMA PREN2913-86. Washers, Flat, Large Diameter Steel. 3 pp.
AECMA PREN2914-86. Washers, Flat, Large Diameter Heat Resisting Steel. 4 pp.
BSI BS EN 2122-90. 1990 Washers, Flat Aluminium Alloy. 8 pp.
BSI BS EN 2138-93. 1993 Washers, Flat, in Steel, Cadmium Plated. 10 pp.
BSI BS EN 2138-90. 1990 Washers, Flat Steel. 8 pp.
BSI BS EN 2139-90. 1990 Washers, Flat Heat Resisting Steel. 8 pp.
CEN EN 2122-89. Washers, Flat Aluminium Alloy. 5 pp.
CEN PREN 2122-93. Aerospace Series Washers, Flat, in Aluminium Alloy, Anodized or Chromated. 7 pp.
CEN EN 2138-89. Washers, Flat Steel. 5 pp.
CEN EN 2138-93. Aerospace Series—Washers, Flat, in Steel, Cadmium Plated. 5 pp.
CEN EN 2139-89. Washers, Flat Heat Resisting Steel. 5 pp.
CEN PREN2912-86. Washers, Flat, Large Diameter Aluminium Alloy. 5 pp.

—Rivet Assemblies
CEN PREN 3902-92. Washers for Rivet Assemblies, in Aluminium Clad or Unclad, Anodized, Metric Series. 5 pp.

Aerospatiale AS 332 Aircraft
See Also: Aircraft

—Foreign Airworthiness Directives
CAA. Aerospatiale AS 332 Series Helcopters (Foreign Airworthiness Directives). 12 pp.
CAA. Procaer F-15 Series (Foreign Airworthiness Directives). 1 p.

Aerospatiale AS 332C Helicopters
See Also: Helicopters

—Certification
CAA. Aerospatiale AS332C and L. 8 pp.

Aerospatiale AS 332L Helicopters
See Also: Helicopters

—Antenna Positions
CAA. Aerospatiale AS332L (Approved Aerial Positions). 1 p.

—Certification
CAA. Aerospatiale AS332C and L. 8 pp.

Aerospatiale AS 332L Supa Puma Aircraft
See Also: Aircraft

—Data Sheets
CAA FR16 ISSUE 1. Aerospatiale AS 332L Supa Puma. 3 pp.

Aerospatiale AS 350 Aircraft
See Also: Aircraft

—Foreign Airworthiness Directives
CAA. Aerospatiale AS 350 Series Helicopters (Foreign Airworthiness Directives). 11 pp.

Aerospatiale AS 350B Aircraft
See Also: Aircraft

—Antenna Positions
CAA. Aerospatiale AS350B, AS355, AS355F1 and AS355N (Approved Aerial Positions). 1 p.

Aerospatiale AS 350B Helicopters
See Also: Helicopters

—Certification
CAA. Aerospatiale AS350B, BA, B1 and B2. 4 pp.

—Data Sheets
CAA FR8 ISSUE 7. Aerospatiale AS350B Ecureuil. 2 pp.

Aerospatiale AS 350B1 Helicopters
See Also: Helicopters

—Certification
CAA. Aerospatiale AS350B, BA, B1 and B2. 4 pp.

Aerospatiale AS 350B2 Helicopters
See Also: Helicopters

—Certification
CAA. Aerospatiale AS350B, BA, B1 and B2. 4 pp.

Aerospatiale AS 355 Aircraft
See Also: Aircraft

—Antenna Positions
CAA. Aerospatiale AS350B, AS355, AS355F1 and AS355N (Approved Aerial Positions). 1 p.

—Foreign Airworthiness Directives
CAA. Aerospatiale AS 355 Series Helicopters (Foreign Airworthiness Directives). 9 pp.

Aerospatiale AS 355F II Helicopters
See Also: Helicopters

—Data Sheets
CAA FR15 ISSUE 2. Aerospatiale AS 355F Ecureuil II. 2 pp.

Aerospatiale AS 365N2 Helicopters
See Also: Helicopters

—Certification
CAA. Aerospatiale SA365N, N1 and AS365N2. 7 pp.

INTERNATIONAL AND NON-U.S. NATIONAL STANDARDS
SUBJECT INDEX
Aerostats

Aerospatiale ATR 42 Aircraft
See Also: Aircraft
—Foreign Airworthiness Directives
CAA. Aerospatiale ATR 42 Series Aircraft. 6 pp.

Aerospatiale SA 315B Lama Aircraft
See Also: Aircraft
—Data Sheets
CAA FR5 ISSUE 3. Aerospatiale SE3160 Alouette III, SA315B Lama, SA316B Alouette III, SA316C Alouette III, SA391B Alouette III. 3 pp.
—Foreign Airworthiness Directives
CAA. Aerospatiale SA 315B Lama Series Helcopters (Foreign Airworthiness Directives). 7 pp.

Aerospatiale SA 316B Alouette III Aircraft
See Also: Aircraft
—Data Sheets
CAA FR5 ISSUE 3. Aerospatiale SE3160 Alouette III, SA315B Lama, SA316B Alouette III, SA316C Alouette III, SA391B Alouette III. 3 pp.

Aerospatiale SA 316C Alouette III Aircraft
See Also: Aircraft
—Data Sheets
CAA FR5 ISSUE 3. Aerospatiale SE3160 Alouette III, SA315B Lama, SA316B Alouette III, SA316C Alouette III, SA391B Alouette III. 3 pp.

Aerospatiale SA 318B/C Alouette Astazou Aircraft
See Also: Aircraft
—Foreign Airworthiness Directives
CAA. Aerospatiale AS 332 Series Helcopters (Foreign Airworthiness Directives). 12 pp.
CAA. Aerospatiale SE 3130 and SE 313B Alouette II and Aerospatiale SA 3180 and SA 318 B/C Alouette Astazou Series Helicopters. 10 pp.

Aerospatiale SA 3180 Aircraft
See Also: Aircraft
—Foreign Airworthiness Directives
CAA. Aerospatiale AS 332 Series Helcopters (Foreign Airworthiness Directives). 12 pp.

Aerospatiale SA 3180 Alouette Astazou Aircraft
See Also: Aircraft
—Foreign Airworthiness Directives
CAA. Aerospatiale SE 3130 and SE 313B Alouette II and Aerospatiale SA 3180 and SA 318 B/C Alouette Astazou Series Helicopters. 10 pp.

Aerospatiale SA 319 Alouette III Aircraft
See Also: Aircraft
—Foreign Airworthiness Directives
CAA. Aerospatiale SE 316 and SA 319 Alouette III Series Helicopters. 8 pp.

Aerospatiale SA 330 Puma Helicopters
See Also: Helicopters
—Accidents
CAA. Aerospatiale SA330ffi332 Puma Series (World Helicopter Accident Summary). 1 p.
—Certification
CAA. Aerospatiale SA330 Puma, SA330G and SA330J. 5 pp.
—Foreign Airworthiness Directives
CAA. Aerospatiale SA 330 Puma Series Helicopters (Foreign Airworthiness Directives). 10 pp.

Aerospatiale SA 330G Puma Helicopters
See Also: Helicopters
—Certification
CAA. Aerospatiale SA330 Puma, SA330G and SA330J. 5 pp.
—Data Sheets
CAA FR6 ISSUE 2. Aerospatiale Puma SA330G and SA330J. 3 pp.

Aerospatiale SA 330J Puma Helicopters
See Also: Helicopters
—Certification
CAA. Aerospatiale SA330 Puma, SA330G and SA330J. 5 pp.
—Data Sheets
CAA FR6 ISSUE 2. Aerospatiale Puma SA330G and SA330J. 3 pp.

Aerospatiale SA 332 Puma Helicopters
See Also: Helicopters
—Accidents
CAA. Aerospatiale SA330ffi332 Puma Series (World Helicopter Accident Summary). 1 p.

Aerospatiale SA 341G Gazelle Aircraft
See Also: Aircraft
—Data Sheets
CAA FR4 ISSUE 1. Aerospatiale,Gazelle SA341G and Gazelle SA341G Series 1. 3 pp.
—Foreign Airworthiness Directives
CAA. Aerospatiale SA 341 Series Helicopters (Foreign Airworthiness Directives). 6 pp.

Aerospatiale SA 365 Helicopters
See Also: Helicopters
—Certification
CAA. Aerospatiale AS355F, F1 and F2. 16 pp.

Aerospatiale SA 365C Helicopters
See Also: Helicopters
—Antenna Positions
CAA. Aerospatiale SA365C (Approved Aerial Positions). 1 p.
—Certification
CAA. Aerospatiale AS355F, F1 and F2. 16 pp.
CAA. Aerospatiale SA365C C1, C2 and C3. 12 pp.
—Data Sheets
CAA FR9 ISSUE 6. Aerospatiale SA365C, SA365C1, SA365C2 and SA365N Dauphin. 4 pp.
—Foreign Airworthiness Directives
CAA. Aerospatiale SA 365C Series Helicopters (Foreign Airworthiness Directives). 6 pp.

Aerospatiale SA 365C1 Helicopters
See Also: Helicopters
—Certification
CAA. Aerospatiale AS355F, F1 and F2. 16 pp.
CAA. Aerospatiale SA365C C1, C2 and C3. 12 pp.
—Data Sheets
CAA FR9 ISSUE 6. Aerospatiale SA365C, SA365C1, SA365C2 and SA365N Dauphin. 4 pp.

Aerospatiale SA 365C2 Helicopters
See Also: Helicopters
—Certification
CAA. Aerospatiale AS355F, F1 and F2. 16 pp.
—Data Sheets
CAA FR9 ISSUE 6. Aerospatiale SA365C, SA365C1, SA365C2 and SA365N Dauphin. 4 pp.

Aerospatiale SA 365C3 Helicopters
See Also: Helicopters
—Certification
CAA. Aerospatiale SA365C C1, C2 and C3. 12 pp.

Aerospatiale SA 365N Dauphin Aircraft
See Also: Aircraft
—Data Sheets
CAA FR9 ISSUE 6. Aerospatiale SA365C, SA365C1, SA365C2 and SA365N Dauphin. 4 pp.

Aerospatiale SA 365N Helicopters
See Also: Helicopters
—Antenna Positions
CAA. Aerospatiale SA365N (Approved Aerial Positions). 1 p.
—Certification
CAA. Aerospatiale SA365N, N1 and AS365N2. 7 pp.

Aerospatiale SA 365N Helicopters (Cont.)
—Foreign Airworthiness Directives
CAA. Aerospatiale SA 365N Series Helicopters (Foreign Airworthiness Directives). 5 pp.

Aerospatiale SA 365N1 Helicopters
See Also: Helicopters
—Certification
CAA. Aerospatiale SA365N, N1 and AS365N2. 7 pp.

Aerospatiale SA 391B Alouette III Aircraft
See Also: Aircraft
—Data Sheets
CAA FR5 ISSUE 3. Aerospatiale SE3160 Alouette III, SA315B Lama, SA316B Alouette III, SA316C Alouette III, SA391B Alouette III. 3 pp.

Aerospatiale SE 313B Alouette II Aircraft
See Also: Aircraft
—Foreign Airworthiness Directives
CAA. Aerospatiale SE 3130 and SE 313B Alouette II and Aerospatiale SA 3180 and SA 318 B/C Alouette Astazou Series Helicopters. 10 pp.

Aerospatiale SE 313B Alouette III Aircraft
See Also: Aircraft
—Foreign Airworthiness Directives
CAA. Aerospatiale AS 332 Series Helcopters (Foreign Airworthiness Directives). 12 pp.

Aerospatiale SE 3130 Aircraft
See Also: Aircraft
—Foreign Airworthiness Directives
CAA. Aerospatiale AS 332 Series Helcopters (Foreign Airworthiness Directives). 12 pp.
CAA. Aerospatiale SE 3130 and SE 313B Alouette II and Aerospatiale SA 3180 and SA 318 B/C Alouette Astazou Series Helicopters. 10 pp.

Aerospatiale SE 316 Alouette III Aircraft
See Also: Aircraft
—Foreign Airworthiness Directives
CAA. Aerospatiale SE 316 and SA 319 Alouette III Series Helicopters. 8 pp.

Aerospatiale SE 3160 Alouette III Aircraft
See Also: Aircraft
—Data Sheets
CAA FR5 ISSUE 3. Aerospatiale SE3160 Alouette III, SA315B Lama, SA316B Alouette III, SA316C Alouette III, SA391B Alouette III. 3 pp.

Aerosport 2 Aircraft
See Also: Aircraft
—Antenna Positions
CAA. Aerosport 2. 1 p.

Aerostar 601P Aircraft
See Also: Aircraft
—Antenna Positions
CAA. Aerostar 601P (Approved Aerial Positions). 1 p.

Aerostats
Use For: Lighter-Than-Air Aircraft; Non-Rigid Aircraft See Also: Aircraft
CAA. Contents 12.79 (Non-Rigid Airships). 2 pp.
CAA. Foreword 12.79 (Non-Rigid Airships). 6 pp.
CAA Chapter Q1-1 12.79. General and Definitions—General (Non-Rigid Airships). 3 pp.
—Air Supply Systems
CAA Chapter Q6-2 12.79. Gas and Air Supply Systems (Non-Rigid Airships). 2 pp.
—Aircraft Lighting
CAA Chapter Q6-7 12.79. External Lights (Non-Rigid Airships). 5 pp.
—Automatic Pilots
CAA Chapter Q6-4 12.79. Automatic Pilots (Non-Rigid Airships). 1 p.

INDUSTRY STANDARDS

INTERNATIONAL AND NON-U.S. NATIONAL STANDARDS
SUBJECT INDEX

Aerostats *(Cont.)*
—Ballast Systems
CAA Chapter Q6-3 12.79. Ballast Systems (Non-Rigid Airships). 1 p.
—Cabins—Flight Crew—Design
CAA Chapter Q4-2 12.79. Flight Crew Accommodation Design (Non-Rigid Airships). 3 pp.
—Cabins—Safety
CAA Chapter Q4-3 App 12.79. Compartment Design and Safety Provisions (Non-Rigid Airships). 2 pp.
—Compasses—Installation
CAA Chapter Q6-1 12.79. Equipment Installations—General (Non-Rigid Airships). 7 pp.
—Control Systems Equipment—Deflation
CAA Chapter Q4-9 12.79. Envelope Design (Non-Rigid Airships). 2 pp.
—Control Systems Equipment—Engines
CAA Chapter Q5-7 12.79. Controls (Non-Rigid Airships). 2 pp.
—Cooling Systems
CAA Chapter Q5-4 12.79. Cooling Systems (Non-Rigid Airships). 3 pp.
—Design Speeds
CAA 864 (Q). Section Q Non-Rigid Airships—Airship Design Speeds. 1 p.
CAA Chapter Q3-2 12.79. Design Airspeeds and Manoeuvres (Non-Rigid Airships). 1 p.
—Electrical Bonding
CAA Chapter Q4-6 12.79. Electrical Bonding and Lightning Discharge Protection (Non-Rigid Airships). 5 pp.
CAA Chapter Q4-6 App 12.79. Primary Conductors (Non-Rigid Airships). 1 p.
—Electrical Equipment
CAA 858 (Q). Section Q Non-Rigid Airships Electrical Supply, Systems and Equipment. 45 pp.
—Electrical Installations
CAA 858 (Q). Section Q Non-Rigid Airships Electrical Supply, Systems and Equipment. 45 pp.
—Engines
CAA 790 (Q). Section Q Non-Rigid Airships Engines Propellers and Transmissions (Amending Blue Paper 772) (Blue Papers). 4 pp.
—Engines—Air Intakes
CAA Chapter Q5-5 12.79. Engine Air Intake and Ice Protection Systems (Non-Rigid Airships). 2 pp.
—Engines—Fire Protection
CAA Chapter Q5-8 12.79. Fire Precautions (Non-Rigid Airships). 3 pp.
—Engines—Ice Formation
CAA Chapter Q5-5 12.79. Engine Air Intake and Ice Protection Systems (Non-Rigid Airships). 2 pp.
—Engines—Lubrication
CAA Chapter Q5-3 12.79. Oil Systems (Non-Rigid Airships). 3 pp.
CAA Chapter Q5-3 App 12.79. Oil Systems (Non-Rigid Airships). 1 p.
—Equipment—Installation
CAA Chapter Q6-1 12.79. Equipment Installations—General (Non-Rigid Airships). 7 pp.
—Exhaust Systems
CAA Chapter Q5-6 12.79. Exhaust Systems (Non-Rigid Airships). 2 pp.
—Fabrication
CAA Chapter Q4-1 12.79. Design and Construction—General (Non-Rigid Airships). 3 pp.
—Fabrication—Tear Resistant
CAA 862 (Q). Section Q Non-Rigid Airships—Airship Fabric Tear Resistance. 2 pp.
—Fire Protection
CAA Chapter Q5-8 App 12.79. Fire Precautions (Non-Rigid Airships). 1 p.
—Flight Dynamics
CAA 819 (Q). Section Q Non-Rigid Airships Performance Requirements (Blue Papers). 11 pp.
CAA Chapter Q2-2 12.79. Performance—General (Non-Rigid Airships). 4 pp.
CAA Chapter Q2-5 12.79. Handling—General (Non-Rigid Airships). 2 pp.
CAA Chapter Q2-6 12.79. Controllability (Non-Rigid Airships). 3 pp.
CAA Chapter Q2-7 12.79. Handling—Ability to Trim (Non-Rigid Airships). 1 p.
CAA Chapter Q2-8 12.79. Stability (Non-Rigid Airships). 1 p.

Aerostats *(Cont.)*
—Flight Operation—Signs
CAA Chapter Q7-3 12.79. Markings and Placards (Non-Rigid Airships). 4 pp.
CAA Chapter Q7-5 App #4 12.79. Flight Manuals—Supplements (Non-Rigid Airships). 2 pp.
—Flight Operations
CAA 819 (Q). Section Q Non-Rigid Airships Performance Requirements (Blue Papers). 11 pp.
CAA 843 (Q). Section Q Non-Rigid Airships Performance—Take off Distance Factors. 1 p.
CAA Chapter Q2-1 12.79. Flight—General (Non-Rigid Airships). 1 p.
CAA Chapter Q2-2 App 02.85. Performance—General (Non-Rigid Airships). 22 pp.
CAA Chapter Q2-3 12.79. Performance—Take-Off and Landing (Non-Rigid Airships). 3 pp.
CAA Chapter Q2-3 App 02.85. Performance—Take-Off and Landing (Non-Rigid Airships). 1 p.
CAA Chapter Q2-4 12.79. Performance—Climb and Level Flight (Non-Rigid Airships). 2 pp.
CAA Chapter Q2-4 App 02.85. Performance—Climb (Non-Rigid Airships). 1 p.
CAA Chapter Q7-1 12.79. Operating Limitations and Information—General (Non-Rigid Airships). 1 p.
CAA Chapter Q7-2 12.79. Operating Information (Non-Rigid Airships). 4 pp.
—Flight Operations—Procedures
CAA 819 (Q). Section Q Non-Rigid Airships Performance Requirements (Blue Papers). 11 pp.
CAA Chapter Q2-2 12.79. Performance—General (Non-Rigid Airships). 4 pp.
CAA Chapter Q7-5 12.79. Flight Manuals (Non-Rigid Airships). 6 pp.
CAA Chapter Q7-5 App #1 12.79. Flight Manuals—General (Non-Rigid Airships). 2 pp.
CAA Chapter Q7-5 App #2 12.79. Flight Manuals—Limitations (Non-Rigid Airships). 1 p.
CAA Chapter Q7-5 App #3 12.79. Flight Manuals—Procedures (Non-Rigid Airships). 1 p.
—Fuel Tanks
CAA Chapter Q5-2 App 12.79. Fuel Systems (Non-Rigid Airships). 1 p.
—Gas Supply Systems
CAA Chapter Q6-2 12.79. Gas and Air Supply Systems (Non-Rigid Airships). 2 pp.
—Glossaries
BSI BS 185: Sec 7-69. 1969 Glossary of Aeronautical and Astronautical Terms Section 7: Lighter-Than-Air Aircraft (Aerostats) (Hull, Envelope, Mooring and Handling). 9 pp.
CAA 819 (Q). Section Q Non-Rigid Airships Performance Requirements (Blue Papers). 11 pp.
CAA Chapter Q1-2 App 12.79. Definitions Climatic Conditions (Non-Rigid Airships). 3 pp.
CAA Chapter Q2-2 12.79. Performance—General (Non-Rigid Airships). 4 pp.
—Heaters
CAA Chapter Q6-5 12.79. Combustion Heater Systems (Non-Rigid Airships). 1 p.
—Ice Formation
CAA Chapter Q4-7 12.79. Flight in Precipitation Conditions (Non-Rigid Airships). 1 p.
—Instruments—Installation
CAA Chapter Q6-1 12.79. Equipment Installations—General (Non-Rigid Airships). 7 pp.
—Landing Gear
CAA Chapter Q4-5 12.79. Landing Gear Design (Non-Rigid Airships). 1 p.
—Landings
CAA Chapter Q2-6 12.79. Controllability (Non-Rigid Airships). 3 pp.
—Landings—Crash—Safety
CAA Chapter Q3-8 12.79. Crashworthiness Conditions (Non-Rigid Airships). 2 pp.
—Life Rafts
CAA Chapter Q6-6 12.79. Life Rafts (Non-Rigid Airships). 1 p.
—Lighting—Installation
CAA Chapter Q6-1 12.79. Equipment Installations—General (Non-Rigid Airships). 7 pp.
—Lightning Protection
CAA Chapter Q4-6 12.79. Electrical Bonding and Lightning Discharge Protection (Non-Rigid Airships). 5 pp.
CAA Chapter Q4-6 App 12.79. Primary Conductors (Non-Rigid Airships). 1 p.
—Loads (Forces)
CAA Chapter Q3-1 12.79. Structures—General (Non-Rigid Airships). 3 pp.

Aerostats *(Cont.)*
—Loads (Forces) *(Cont.)*
CAA Chapter Q3-2 12.79. Design Airspeeds and Manoeuvres (Non-Rigid Airships). 1 p.
CAA Chapter Q3-3 12.79. Gust Loads (Non-Rigid Airships). 2 pp.
CAA Chapter Q3-4 12.79. Engine and Propeller Loads (Non-Rigid Airships). 2 pp.
CAA Chapter Q3-5 12.79. Ground Loads (Non-Rigid Airships). 3 pp.
—Moorings—Loads (Forces)
CAA Chapter Q3-13 12.79. Mooring and Groundhandling. 2 pp.
—Propellers
CAA 790 (Q). Section Q Non-Rigid Airships Engines Propellers and Transmissions (Amending Blue Paper 772) (Blue Papers). 4 pp.
—Propellers—Clearance
CAA Chapter Q5-1 12.79. Power Plant Installations—General (Non-Rigid Airships). 5 pp.
—Safety Belts—Design
CAA Chapter Q4-4 12.79. Seats, Safety Belts and Harnesses (Non-Rigid Airships). 3 pp.
—Seats—Design
CAA Chapter Q4-4 12.79. Seats, Safety Belts and Harnesses (Non-Rigid Airships). 3 pp.
CAA Chapter Q4-4 App 12.79. Seats, Safety Belts and Harnesses (Non-Rigid Airships). 1 p.
—Structural Design
CAA Chapter Q3-1 12.79. Structures—General (Non-Rigid Airships). 3 pp.
CAA Chapter Q4-1 12.79. Design and Construction—General (Non-Rigid Airships). 3 pp.
—Structural Design—Aerodynamic Loads
CAA Chapter Q4-9 12.79. Envelope Design (Non-Rigid Airships). 2 pp.
—Transmissions (Power Systems)
CAA 790 (Q). Section Q Non-Rigid Airships Engines Propellers and Transmissions (Amending Blue Paper 772) (Blue Papers). 4 pp.

AESL Aircraft
See Also: Aircraft
—Foreign Airworthiness Directives
CAA. AESL and Glos Air Airtourer Series (Foreign Airworthiness Directives). 14 pp.

AFC
Use: Automatic Frequency Control

Aflatoxin Content Analysis
—Animal Feed
BSI BS 5766: Part 7-88. 1988 Method for Analysis of Animal Feeding Stuffs Part 7: Determination of Aflatoxin B Less Than of 12 Greater Than 1. 12 pp.
CNS N4070-81. Methods for Determining Aflatoxins in Feedstuff (Feb)(7066). 6 pp.
ISO 6651-87. Animal Feeding Stuffs—Determination of Aflatoxin B1 Content Second Edition. 11 pp.
—Shortening
CNS N6111-86. Method of Test for Shortening (Nov)(4990). 2 pp.
—Vegetable Fats
CNS N6097-79. Method of Test for Edible Fats and Oils (Determination of Aflatoxin) (Apr)(4090). 3 pp.
—Vegetable Oils
CNS N6097-79. Method of Test for Edible Fats and Oils (Determination of Aflatoxin) (Apr)(4090). 3 pp.

African Characters
See Also: Graphic Characters
—Coded Character Sets—Bibliographic References
ISO 6438-83. Documentation—African Coded Character Set for Bibliographic Information Interchange First Edition. 8 pp.
OSI ISO 6438-83. Documentation—African Coded Character Set for Bibliographic Information Interchange. 8 pp.

African Swine Fever
Use: Hog Cholera

AFS
Use: Atomic Fluorescence Spectrometry

INTERNATIONAL AND NON-U.S. NATIONAL STANDARDS
SUBJECT INDEX
Aggregates

Aftertack
—Coatings
CGSB 1-GP-71 METH 5.10-74. Methods of Testing Paints and Pigments Drying Time After-Tack. 1 p.

AFTN
Use: Aeronautical Fixed Telecommunications Network

Agar
See Also: Algae; Culture Media
CNS K7027-61. Chemical Reagent (Agar)(Agar-agar) (Dec)(1527).
CNS N5159-79. Agar-Agar (Oct)(3367). 2 pp.
JIS K 8263-61. Agar.

—Ash Content
CNS N6113-79. Method of Test for Agar-Agar (Oct)(5006). 2 pp.

—Gel Strength
CNS N6113-79. Method of Test for Agar-Agar (Oct)(5006). 2 pp.

—Insoluble Matter Content
CNS N6113-79. Method of Test for Agar-Agar (Oct)(5006). 2 pp.

—Moisture Content
CNS N6113-79. Method of Test for Agar-Agar (Oct)(5006). 2 pp.

—Protein Content
CNS N6113-79. Method of Test for Agar-Agar (Oct)(5006). 2 pp.

—Sensory Analysis
CNS N6113-79. Method of Test for Agar-Agar (Oct)(5006). 2 pp.

—Starch Content
CNS N6113-79. Method of Test for Agar-Agar (Oct)(5006). 2 pp.

Agar-Agar
Use: Agar

Agaric Mushrooms
Use: Mushrooms

AGC
Use: Automatic Gain Control

AGC Services
Use: Audiographic Conferencing Services

Ageing Testing
Use: Aging Testing

Agglomerates
—Iron Ores
BSI BS 6212-87. 1987 Method for Determination of Tumbler Strength of Iron Ores. 9 pp.
ISO 3271-85. Iron Ores—Determination of Tumbler Strength Second Edition. 8 pp.

Aggregate Coatings
See Also: Aggregates; Coatings

—Ceilings
JIS A 6917-83. Lightweight Aggregate Coating Materials. 16 pp.

—Walls
JIS A 6917-83. Lightweight Aggregate Coating Materials. 16 pp.

Aggregate Dryers
See Also: Dryers
BSI BS 2096-54. 1954 Method of Testing Oil-Fired Rotary Dryers for Use in Asphalt and Coated Macadam Plant. 12 pp.

Aggregates
See Also: Aggregate Coatings; Alkali Aggregate Reactions; Clinkers; Concretes; Crushed Stone; Dolomite; Gravel; Limestone; Macadam Aggregates; Mineral Aggregates; Rocks; Rubble; Sands; Slags

—Abrasion—Los Angeles Machine
CNS A3059-85. Method of Test for Resistance to Abrasion of Large Size Coarse Aggregate by Use of the Los Angeles Machine (Greater Than 19 mm) (Jun)(3408). 2 pp.
JIS A 1121-89. Method of Test for Abrasion of Coarse Aggregates by Use of the Los Angeles Machine. 8 pp.

—Abrasion Testing
BSI BS 812: Part 113-90. 1990 Testing Aggregates Part 113: Methods for Determination of Aggregate Abrasion Value. 10 pp.

—Abrasion Testing *(Cont.)*
BSI BS 812: Part 113-01. 1990 Amd 1 Testing Aggregates Part 113: Method for Determination of Aggregate Abrasion Value (AAV) (AMD 6986) December 24, 1991. 11 pp.
CNS A3009-86. Method of Test for Resistance to Abrasion of Small Size Coarse Aggregate by Use of the Los Angeles Machine (Aug)(490).
SNZ NZS 4407: Part 3.12-91. Methods of Sampling and Testing Road Aggregates Part 3: Methods of Testing Road Aggregates—Laboratory Tests Test 3.12: The Abrasion Resistance of Aggregate by Use of the Los Angeles Machine. 4 pp.

—Binders
BSI BS 598: Part 108-90. 1990 Sampling and Examination of Bituminous Mixtures for Roads and Other Paved Areas Part 108: Methods for Determination of the Condition of the Binder on Coated Chippings and for Measurement of the Rate of Spread of Coated Chippings. 8 pp.

—Bituminous Coatings
CNS A3287-88. Method of Test for Determining Degree of Particle Coating of Bituminous—Aggregate Mixtures (Aug)(12389).
CNS A3289-88. Method of Test for Effect of Water on Bituminous—Coated Aggregate Quick Field (Aug)(12391).
CNS A3292-88. Method of Test for Coating and Stripping of Bituminous—Aggregate Mixtures (Aug)(12394).

—Bituminous Coatings—Stripping
CNS A3292-88. Method of Test for Coating and Stripping of Bituminous—Aggregate Mixtures (Aug)(12394).

—Breaking Load—Magnesium Sulfate
CNS A3031-85. Method of Test for Soundness of Aggregates by Use of Sodium Sulfate or Magnesium Sulfate (Jun)(1167).

—Breaking Load—Sodium Sulfate
CNS A3031-85. Method of Test for Soundness of Aggregates by Use of Sodium Sulfate or Magnesium Sulfate (Jun)(1167).
JIS A 1122-89. Method of Test for Soundness of Aggregates by Use of Sodium Sulfate. 11 pp.

—Broken Faces Content
SNZ NZS 4407: Part 3.14-91. Methods of Sampling and Testing Road Aggregates Part 3: Methods of Testing Road Aggregates—Laboratory Tests Test 3.14: The Broken Faces Content of Aggregate. 2 pp.

—Bulk Density
BSI BS 812: Part 2-75. 1975 Amd 1 Testing Aggregates Part 2: Physical Properties. 24 pp.

—California Bearing Ratio
SNZ NZS 4407: Part 3.15-91. Methods of Sampling and Testing Road Aggregates Part 3: Methods of Testing Road Aggregates—Laboratory Tests Test 3.15: The California Bearing Ratio (CBR). 10 pp.

—Cement Mortars—Impurities Content
CNS A3028-83. Method of Test for Organic Impurities in Fine Aggregate (Jan)(1164).
JIS A 1105-76. Method of Test for Organic Impurities in Fine Aggregate. 5 pp.

—Chloride Ion Content
BSI BS 812: Part 117-88. 1988 Testing Aggregates Part 117: Method for Determination of Water-Soluble Chloride Salts. 11 pp.

—Clay Content
CNS A3035-86. Method of Test for Clay Lumps and Friable Particles in Aggregate (Aug)(1171).
JIS A 1137-89. Method of Test for Clay Contained in Aggregates. 6 pp.
SNZ NZS 4407: Part 3.5-91. Methods of Sampling and Testing Road Aggregates Part 3: Methods of Testing Road Aggregates—Laboratory Tests Test 3.5: The Clay Index. 6 pp.

—Cleanness Value
SNZ NZS 4407: Part 3.9-91. Methods of Sampling and Testing Road Aggregates Part 3: Methods of Testing Road Aggregates—Laboratory Tests Test 3.9: The Cleanness Value of Coarse Aggregate. 4 pp.

—Cohesion
SAA AS 1141.52-87. Methods of Sampling and Testing Aggregates—Part 52: Unconfined Cohesion of Compacted Pavement Materials. 6 pp.

—Compactibility
BSI BS 5835: Part 1-80. 1980 Amd 3 Recommendations for Testing of Aggregates Part 1: Compactibility Test for Graded Aggregates. 21 pp.

—Comprehensive Testing
SAA AS 1141. Methods for Sampling and Testing Aggregates (Complete Set in Binder).
SAA AS 1141-74. Methods for Sampling and Testing Aggregates—Includes Parts 1,4,5,6, 12,13,14,16,21,24, 31 to 36,38,39, and 61. 52 pp.

—Compression Testing
SAA AS 1141.51-85. Methods for Sampling and Testing Aggregates—Part 51: Unconfined Compressive Strength of Compacted Bound Materials. 4 pp.
SNZ NZS 4407: Part 3.10-91. Methods of Sampling and Testing Road Aggregates Part 3: Methods of Testing Road Aggregates—Laboratory Tests Test 3.10: The Crushing Resistance of Coarse Aggregate Under a Specific Load. 4 pp.

—Compression Testing—Aggregate Crushing Value
BSI BS 812: Part 110-90. 1990 Testing Aggregates Part 110: Method for Determination of Aggregate Crushing Value (ACV). 12 pp.

—Compression Testing—Ten Per Cent Fines Value
BSI BS 812: Part 111-90. 1990 Testing Aggregates Part 111: Methods for Determination of Ten per Cent Fines Value (TFV). 13 pp.

—Concretes
BSI BS 877: Part 2-73. (WITHDRAWN) 1973 Amd 1 Foamed or Expanded Blast-Furnace Slag Light Weight Aggregate for Concrete Part 2: Metric Units (Superseded by BS 3797: 1990). 9 pp.
BSI BS 882-92. 1992 Aggregates from Natural Sources for Concrete. 12 pp.
BSI BS 882-83. 1983 Amd 1 Aggregates from Natural Sources for Concrete. 11 pp.
BSI BS 3797-90. 1990 Amd 1 Lightweight Aggregates for Masonry Units and Structural Concrete (AMD 6796) August 30, 1991. 12 pp.
BSI BS 3797: Part 2-76. (WITHDRAWN) 1976 Amd 1 Lightweight Aggregates for Concrete Part 2: Metric Units (Superseded by BS 3797: 1990). 8 pp.
BSI BS 4550: Part 4-78. 1978 Methods of Testing Cement Part 4: Standard Coarse Aggregate for Concrete Cubes. 2 pp.
CNS A2029-86. Concrete Aggregates (Aug)(1240).
CNS A2046-74. Lightweight Aggregate for Structural Concrete (Jan)(3691). 5 pp.
CNS A2202-87. Air-Cooled Iron-Blast-Furnace Slag Coarse Aggregate for Concrete (Feb)(11824).
DIN ENGL 4226 Pt 1-83. Aggregates for Concrete; Aggregates of Dense Structure (Heavy Aggregates); Terminology, Designation and Requirements (Apr). 8 pp.
DIN ENGL 4226 Pt 2-83. Aggregates for Concrete; Aggregates of Porous Structure (Lightweight Aggregates); Terminology, Designation and Requirements (Apr). 7 pp.
JIS A 5002-78. Light Weight Aggregates for Structural Concrete. 12 pp.
SNZ NZS 3111-86. Methods of Test for Water and Aggregate for Concrete Amend: 1, 1988. 50 pp.
SNZ NZS 3121-86. Specification for Water and Aggregate for Concrete. 8 pp.

—Concretes—Alkali Reactions
JIS A 1804-92. Mehtods of Test for Production Control of Concrete—Method of Rapid Test for Identification of the Alkali Reactivity of Aggregates. 9 pp.

—Concretes—Ashes
BSI BS 1165-85. (WITHDRAWN) 1985 Clinker and Furnace Bottom Ash Aggregates for Concrete (R) (Superseded by BS 3797: 1990). 4 pp.

—Concretes—Carbon Content
BSI BS 3681: Part 2-73. (WITHDRAWN) 1973 Methods for the Sampling and Testing of Lightweight Aggregates for Concrete Part 2: Metric Units (Superseded by BS 3797: 1990). 19 pp.

—Concretes—Chloride Content
SAA AS 1012.20-92. Methods of Testing Concrete—Part 20: Determination of Chloride and Sulfate in Hardened Concrete and Concrete Aggregates (in Professional Packages 30, 58). 4 pp.

—Concretes—Comprehensive Testing
DIN ENGL 4226 Pt 3-83. Aggregates for Concrete; Testing of Heavy and Lightweight Aggregates (Apr). 8 pp.

—Concretes—Density
BSI BS 3681: Part 2-73. (WITHDRAWN) 1973 Methods for the Sampling and Testing of Lightweight Aggregates for Concrete Part 2: Metric Units (Superseded by BS 3797: 1990). 19 pp.

INDUSTRY STANDARDS

INTERNATIONAL AND NON-U.S. NATIONAL STANDARDS
SUBJECT INDEX

Aggregates

Aggregates (Cont.)

—Concretes—Density (Cont.)
CEN PREN 992-92. Determination of the Dry Density of Lightweight Aggregate Concrete with Open Structure. 3 pp.
ISO 6782-82. Aggregates for Concrete—Determination of Bulk Density First Edition. 4 pp.
JIS A 1134-89. Methods of Test for Bulk Specific Gravity and Absorption of Light Weight Fine Aggregates for Structural Concrete. 9 pp.
JIS A 1135-89. Methods of Test for Bulk Specific Gravity and Absorption of Light Weight Coarse Aggregates for Structural Concrete. 6 pp.

—Concretes—Impurities Content
CNS A3028-83. Method of Test for Organic Impurities in Fine Aggregate (Jan)(1164).
JIS A 1105-76. Method of Test for Organic Impurities in Fine Aggregate. 5 pp.

—Concretes—Inspection
DIN ENGL 4226 Pt 4-83. Aggregates for Concrete; Inspection (Apr). 5 pp.

—Concretes—Mass
CNS A3027-86. Method of Test for Unit Weight and Voids in Aggregate (Aug)(1163).
JIS A 1104-76. Method of Test for Unit Weight of Aggregate and Solid Content in Aggregate. 7 pp.
JIS A 1119-89. Method of Test for Variability of Constituents in Freshly Mixed Concrete. 9 pp.

—Concretes—Moisture Content
CNS A3008-84. Method of Test for Surface Moisture in Fine Aggregate (Apr)(489). 2 pp.
JIS A 1111-76. Method of Test for Surface Moisture in Fine Aggregate. 7 pp.
JIS A 1802-89. Methods of Test for Production Control of Concrete (Method of Test for Surface Moisture in Fine Aggregate by Centrifugal Force). 7 pp.
JIS A 1803-91. Method of Test for Production Control of Concrete—Method of Test for Surface Moisture in Coarse Aggregate. 6 pp.

—Concretes—Particle Density
ISO 6783-82. Coarse Aggregates for Concrete—Determination of Particle Density and Water Absorption—Hydrostatic Balance Method First Edition. 4 pp.
ISO 7033-87. Fine and Coarse Aggregates for Concrete—Determination of the Particle Mass-per-Volume and Water Absorption—Pyknometer Method First Edition. 6 pp.

—Concretes—Sampling
BSI BS 3681: Part 2-73. (WITHDRAWN) 1973 Methods for the Sampling and Testing of Lightweight Aggregates for Concrete Part 2: Metric Units (Superseded by BS 3797: 1990). 19 pp.

—Concretes—Sand Content
JIS A 1801-89. Methods of Test for Production Control of Concrete (Method of Test for Sand Equivalent Value of Fine Aggregates for Concrete). 7 pp.

—Concretes—Shrinkage
BSI BS 812: Part 120-89. 1989 Testing Aggregates Part 120: Method for Testing and Classifying Drying Shrinkage of Aggregates in Concrete. 8 pp.

—Concretes—Sieve Analysis
BSI BS 3681: Part 2-73. (WITHDRAWN) 1973 Methods for the Sampling and Testing of Lightweight Aggregates for Concrete Part 2: Metric Units (Superseded by BS 3797: 1990). 19 pp.
CNS A3005-59. Method of Sieve Analysis for Coarse and Fine Aggregates (Oct)(486) (R 1970).
CNS A3010-84. Method of Test for Substance in Aggregates Finer Than 75 Micrometer CNS 386 Test Sieve (Dec)(491). 2 pp.
ISO 6274-82. Concrete—Sieve Analysis of Aggregates First Edition. 4 pp.
JIS A 1102-89. Method of Test for Sieve Analysis of Aggregates. 6 pp.
JIS A 1103-89. Method of Test for Amount of Material Passing Standard Sieve 75 Micrometer in Aggregates. 6 pp.

—Concretes—Slag
CNS A2204-87. Granulated Blast Furnace Slag Fine Aggregate for Concrete (Apr)(11890).
CNS A3254-87. Method of Test for Air-Cooled Iron-Blast-Furnace Slag Coarse Aggregate for Concrete (Feb)(11825).
CNS A3257-87. Method of Test for Granulated Blast Furnace Slag Fine Aggregate for Concrete (Apr)(11891).
JIS A 5011-77. Air-Cooled Iron-Blast-Furnace Slag Aggregate for Concrete. 16 pp.
JIS A 5012-81. Granulated Blast Furnace Slag Fine Aggregate for Concrete. 10 pp.

Aggregates (Cont.)

—Concretes—Solids Content
CNS A3027-86. Method of Test for Unit Weight and Voids in Aggregate (Aug)(1163).
JIS A 1104-76. Method of Test for Unit Weight of Aggregate and Solid Content in Aggregate. 7 pp.

—Concretes—Sulfate Content
BSI BS 812: Part 118-88. 1988 Testing Aggregates Part 118: Method for Determination of Sulphate Content. 15 pp.
BSI BS 3681: Part 2-73. (WITHDRAWN) 1973 Methods for the Sampling and Testing of Lightweight Aggregates for Concrete Part 2: Metric Units (Superseded by BS 3797: 1990). 19 pp.
SAA AS 1012.20-92. Methods of Testing Concrete—Part 20: Determination of Chloride and Sulfate in Hardened Concrete and Concrete Aggregates (in Professional Packages 30, 58). 4 pp.

—Concretes—Volatile Matter Content
BSI BS 3681: Part 2-73. (WITHDRAWN) 1973 Methods for the Sampling and Testing of Lightweight Aggregates for Concrete Part 2: Metric Units (Superseded by BS 3797: 1990). 19 pp.

—Cone Penetration Limit
SNZ NZS 4407: Part 3.2-91. Methods of Sampling and Testing Road Aggregates Part 3: Methods of Testing Road Aggregates—Laboratory Tests Test 3.2: The Cone Penetration Limit. 5 pp.

—Degradation
SAA AS 1141.25-81. Methods for Sampling and Testing Aggregates—Part 25: Degradation Factor—Source Rock. 6 pp.

—Density
BSI BS 812: Part 2-75. 1975 Amd 1 Testing Aggregates Part 2: Physical Properties. 24 pp.
CNS A3006-86. Method of Test for Specific Gravity and Absorption of Fine Aggregate (Aug)(487). 6 pp.
CNS A3007-72. Method of Test for Specific Gravity and Absorption of Coarse Aggregate (Jun)(488).
JIS A 1109-76. Method of Test for Specific Gravity and Absorption of Fine Aggregate. 7 pp.
JIS A 1110-89. Method of Test for Specific Gravity and Absorption of Coarse Aggregates. 6 pp.
SNZ NZS 4407: Part 3.7.1-91. Methods of Sampling and Testing Road Aggregates Part 3: Methods of Testing Road Aggregates—Laboratory Tests Test 3: The Solid Density of Aggregate Particles. Test 3.7.1: Pycnometer Method for Particles Passing the 19 mm Test Sieve. 3 pp.
SNZ NZS 4407: Part 3.7.2-91. Methods of Sampling and Testing Road Aggregates Part 3: Methods of Testing Road Aggregates—Laboratory Tests Test 3.7: The Solid Density of Aggregate Particles. Test 3.7.2: Immersion in Water Method for Coarse Aggregate. 4 pp.
SNZ NZS 4407: Part 4.1.1-91. Methods of Sampling and Testing Road Aggregates Part 4: Methods of Testing Road Aggregates—Field Tests Test 4.1: The Density of Compacted Aggregate Test 4.1.1: Sand Replacement Method. 7 pp.
SNZ NZS 4407: Pt 4.2.1-91. Methods of Sampling and Testing Road Aggregates Part 4: Methods of Testing Road Aggregates—Field Tests Test 4.2: Theield Water Content and Field Dry Density of Compacted Aggregate Test 4.2.1: Method Using a Nuclear Surface Moisture-Density Gauge-. 3 pp.
SNZ NZS 4407: Part 4.2.2-91. Methods of Sampling and Testing Road Aggregates Part 4: Methods of Testing Road Aggregates—Field Tests Test 4.2: The Field Water Content and Field Dry Density of Compacted Aggregate Test 4.2.2: Method Using a Nuclear Surface Moisture-Density Gauge—Backscatter Mode. 3 pp.
SNZ NZS 4407: Part 4.2.3-91. Methods of Sampling and Testing Road Aggregates Part 4: Methods of Testing Road Aggregates—Field Tests Test 4.2: The Field Water Content and Field Dry Density of Compacted Aggregate Test 4.2.3: Calibration of Nuclear Surface Moisture-Density Gauge for Field Use. 3 pp.
SNZ NZS 4407: Part 4.2.4-91. Methods of Sampling and Testing Road Aggregates Part 4: Methods of Testing Road Aggregates—Field Tests Test 4.2: The Field Water Content and Field Dry Density of Compacted Aggregate Test 4.2.4: Calibration of Nuclear Surface Moisture-Density Gauge Using Standard Blocks. 6 pp.

—Elongation
BSI BS 812: Part 105.2-90. 1990 Testing Aggregates Part 105.2: Elongation Index of Coarse Aggregate. 9 pp.

—Flakiness
BSI BS 812: Sec 105.1-89. 1989 Testing Aggregates Part 105: Methods for Determination of Particle Shape Section 105.1: Flakiness Index. 8 pp.
CEN PREN 933-6-92. Tests for Geometrical Properties of Aggregates —Part 6: Determination of Particle Shape—Flakiness Index. 9 pp.

Aggregates (Cont.)

—Flakiness (Cont.)
SAA AS 1141.15-88. Methods for Sampling and Testing Aggregates—Part 15: Flakiness Index. 3 pp.

—Friction
SAA AS 1141.42-84. Methods for Sampling and Testing Aggregates—Part 42: Pendulum Friction Test (PAFV). 7 pp.

—Frost Heave Testing
BSI BS 812: Part 124-89. 1989 Testing Aggregates Part 124: Method for Determination of Frost-Leave. 26 pp.

—Glossaries
BSI BS 6100: Sec 6.3-84. 1984 Glossary of Building and Civil Engineering Terms Part 6: Concrete and Plaster Section 6.3: Aggregates. 6 pp.
BSI BS 6100: Sec 6.3-01. 1984 Amd 1 Glossary of Building and Civil Engineering Terms Part 6: Concrete and Plaster Section 6.3: Aggregates (AMD 7265) August 15, 1992. 7 pp.

—Grouting
CNS A2036-70. Aggregate of Masonry Concrete for Grouting (Jan)(2466). 3 pp.

—Impact Testing
BSI BS 812: Part 112-90. 1990 Testing Aggregates Part 112: Methods for Determination of Aggregate Impact Value (AIV). 12 pp.

—Iron Unsoundness
SAA AS 1141.37-82. Methods for Sampling and Testing Aggregates—Part 37: Iron Unsoundness. 1 p.

—Los Angeles Value
SAA AS 1141.23-80. Methods for Sampling and Testing Aggregates—Part 23: Los Angeles Value. 4 pp.

—Mechanical Properties
BSI BS 812: Part 3-75. (WITHDRAWN) 1975 Amd 5 Testing Aggregates Part 3: Mechanical Properties (AMD 4845) February 28, 1985 (Superseded by BS 812: Parts 110, 111, 113: 1990 & 114: 1989). 28 pp.

—Moisture Content
BSI BS 812: Part 2-75. 1975 Amd 1 Testing Aggregates Part 2: Physical Properties. 24 pp.
BSI BS 812: Part 109-90. 1990 Testing Aggregates Part 109: Methods for Determination of Moisture Content. 11 pp.

—Moisture Content—Drying
BSI BS 812: Part 109-90. 1990 Testing Aggregates Part 109: Methods for Determination of Moisture Content. 11 pp.
CNS A3225-85. Method of Test for Total Moisture Content of Aggregate by Drying (Jul)(11298).
JIS A 1125-76. Method of Test for Total Moisture Content and Surface Moisture of Aggregate by Drying (R 1980). 7 pp.

—Mortars
CNS A2039-86. Aggregate for Masonry Mortar (Dec)(3001). 3 pp.
CNS A3029-65. Method of Test for Measuring Mortar-Making Properties of Fine Aggregates (Apr)(1165).
CSA A82.56-1950. Aggregate for Masonry Mortar (R 1971); (Rev 1 June 1967). 7 pp.
CSA A82.56M-1976. Aggregate for Masonry Mortar. 6 pp.

—Mortars—Compression Testing
CNS A3220-84. Method of Test for Effect of Organic Impurities in Fine Aggregate on Strength of Mortar (Dec)(11153). 4 pp.

—Particle Content
CNS A3035-86. Method of Test for Clay Lumps and Friable Particles in Aggregate (Aug)(1171).
CNS A3037-59. Method of Test for Soft Particles in Coarse Aggregates (Sep)(1173)(R 1970).

—Particle Density
SAA AS 1141.7-80. Methods for Sampling and Testing Aggregates—Part 7: Apparent Particle Density of Filler. 2 pp.

—Particle Size Analysis
SAA AS 1141.11-80. Methods for Sampling and Testing Aggregates—Part 11: Particle Size Distribution by Dry Sieving. 2 pp.
SAA AS 1141.19-80. Methods for Sampling and Testing Aggregates—Part 19: Fine Particle Size Distribution in Road Materials by Sieving and Decantation. 6 pp.
SAA AS 1141.20-82. Methods for Sampling and Testing Aggregates—Part 20: Average Least Dimension of Aggregate by Direct Measurement. 3 pp.

INTERNATIONAL AND NON-U.S. NATIONAL STANDARDS
SUBJECT INDEX

Aggregates (Cont.)
—Particle Size Analysis (Cont.)
SNZ NZS 4407: Part 3.13-91. Methods of Sampling and Testing Road Aggregates Part 3: Methods of Testing Road Aggregates—Laboratory Tests Test 3.13: The Size and Shape of Aggregate Particles. 5 pp.

—Particle Size Analysis—Scratch Tester
JIS A 1126-89. Method of Test for Soft Particles in Coarse Aggregates by Use of Scratch Tester. 7 pp.

—Particle Size Distribution
BSI BS 812: Part 1-75. 1975 Amd 4 Testing Aggregates Part 1: Methods for Determination of Particle Size and Shape (AMD 6587) April 30, 1991. 14 pp.

BSI BS 812: Sec 103.1-85. 1985 Amd 1 Testing Aggregates Part 103: Methods for Determination of Particle Size Distribution Sec 103.1: Sieve Tests (AMD 6003) June 30, 1989. 11 pp.

BSI BS 812: Sec 103.2-89. 1989 Testing Aggregates Part 103: Methods for Determination of Particle Size Distribution Section 103.2: Sedimentation Test. 10 pp.

CEN PREN 933-1-92. Tests for Geometrical Properties of Aggregates —Part 1: Determination of Particle Size Distribution—Granulometric Analysis (Sieving Method). 12 pp.

SNZ NZS 4407: Part 3.6-91. Methods of Sampling and Testing Road Aggregates Part 3: Methods of Testing Road Aggregates—Laboratory Tests Test 3.6: The Sand Equivalent. 7 pp.

SNZ NZS 4407: Part 3.8.1-91. Methods of Sampling and Testing Road Aggregates Part 3: Methods of Testing Road Aggregates—Laboratory Tests Test 3.8: The Particle-Size Distribution Test 3.8.1: Preferred Method by Wet Sieving. 5 pp.

SNZ NZS 4407: Part 3.8.2-91. Methods of Sampling and Testing Road Aggregates Part 3: Methods of Testing Road Aggregates—Laboratory Tests Test 3.8: The Particle-Size Distribution Test 3.8.2: Subsidiary Method by Dry Sieving. 5 pp.

—Paving Materials
BSI BS 63: Part 1-87. 1987 Road Aggregates Part 1: Single-Sized Aggregate for General Purposes. 4 pp.

—Paving Materials—Grading
SAA AS 2891.3.1-91. Methods of Sampling and Testing Asphalt—Part 3: Bitumen Content and Aggregate Grading—Part 3.1: Reflux Method. 4 pp.

SAA AS 2891.3.2-91. Methods of Sampling and Testing Asphalt—Part 3: Bitumen Content and Aggregate Grading—Part 3.2: Centrifugal Extraction. 4 pp.

SAA AS 2891.3.3-91. Methods of Sampling and Testing Asphalt—Part 3: Bitumen Content and Aggregate Grading—Part 3.3: Pressure Filter Method. 4 pp.

SAA AS 2891.4-91. Methods of Sampling and Testing Asphalt—Part 4: Tar Content and Aggregated Grading. 4 pp.

—Petrography
CEN PREN 932-3-93. Test for General Properties of Aggregates —Part 3: Methods for Description and Petrography—Simplified Procedure. 12 pp.

—Plaster
CSA A82.57-M1977. Inorganic Aggregates for Use in Interior Plaster (R 1992); (Erratum January 1985). 11 pp.

—Plastic Properties
SNZ NZS 4407: Part 3.3-91. Methods of Sampling and Testing Road Aggregates Part 3: Methods of Testing Road Aggregates—Laboratory Tests Test 3.3: The Plastic Limit. 5 pp.

SNZ NZS 4407: Part 3.4-91. Methods of Sampling and Testing Road Aggregates Part 3: Methods of Testing Road Aggregates—Laboratory Tests Test 3.4: The Plasticity Index. 2 pp.

—Polishing
BSI BS 812: Part 114-89. 1989 Testing Aggregates Part 114: Method for Determination of the Polished-Stone Value. 12 pp.

SAA AS 1141.40-84. Methods for Sampling and Testing Aggregates—Part 40: Laboratory Polishing of Aggregate Using the Vertical Road-Wheel Machine. 10 pp.

SAA AS 1141.41-84. Methods for Sampling and Testing Aggregates—Part 41: Laboratory Polishing of Aggregate Using the Horizontal Bed Machine. 6 pp.

—Refractory Materials
BSI BS 6043: Sec 4.2-91. 1991 Methods of Sampling and Test for Carbonaceous Materials Used in Aluminium Manufacture Part 4: Cold Ramming Pastes Section 4.2: Determination of Effective Binder Content and Aggregate Content by Extraction with Quinoline; Determination. 9 pp.

Aggregates (Cont.)
—Refractory Materials—Particle Size Distribution
BSI BS 6043: Sec 4.2-91. 1991 Methods of Sampling and Test for Carbonaceous Materials Used in Aluminium Manufacture Part 4: Cold Ramming Pastes Section 4.2: Determination of Effective Binder Content and Aggregate Content by Extraction with Quinoline; Determination. 9 pp.

BSI BS 6043: Sec 4.3-91. 1991 Methods of Sampling and Test for Carbonaceous Materials Used in Aluminium Manufacture Part 4: Cold Ramming Pastes Section 4.3: Determination of Aggregate Size Distribution After Extraction by Dichloromethane. 8 pp.

—Sampling
BSI BS 812: Part 101-84. 1984 Testing Aggregates Part 101: Guide to Sampling and Test Procedures. 8 pp.

BSI BS 812: Part 102-89. 1989 Testing Aggregates Part 102: Methods of Sampling. 12 pp.

BSI BS 812: Part 102-84. (WITHDRAWN) 1984 Testing Aggregates Part 102: Methods for Sampling. 11 pp.

CEN PREN 932-1-92. Tests for General Properties of Aggregates —Part 1: Methods for Sampling. 28 pp.

CNS A3209-84. Method for Reducing Field Samples of Aggregate to Testing Size (Sep) (10989).

CNS A3210-84. Method of Test for Lightweight Pieces in Aggregate (Sep)(10990).

SAA AS 1141. Methods for Sampling and Testing Aggregates (Complete Set in Binder).

SAA AS 1141-74. Methods for Sampling and Testing Aggregates—Includes Parts 1,4,5,6, 12,13,14,16,21,24, 31 to 36,38,39, and 61. 52 pp.

SAA AS 1141.3-86. Methods for Sampling and Testing Aggregates—Part 3: Sampling of Aggregates and Rock. 10 pp.

SNZ NZS 4407: Part 1-91. Methods of Sampling and Testing Road Aggregates Part 1: Preliminary and General. 10 pp.

SNZ NZS 4407: Part 2-91. Methods of Sampling and Testing Road Aggregates Part 2: Methods of Sampling Road Aggregates. 6 pp.

—Screeds
DIN ENGL 1100-89. Hard Aggregate for Cement-Bound Floor Screed (Oct). 2 pp.

—Shell Content
BSI BS 812: Part 1-75. 1975 Amd 4 Testing Aggregates Part 1: Methods for Determination of Particle Size and Shape (AMD 6587) April 30, 1991. 14 pp.

BSI BS 812: Part 106-85. 1985 Testing Aggregates Part 106: Method for Determination of Shell Content in Coarse Aggregate. 4 pp.

—Slag—Chemical Analysis
CNS A3255-87. Method of Chemical Analysis for Air-Cooled Iron-Blast-Furnace Slag Aggregate (Feb)(11826).

—Soluble Matter Content
BSI BS 812: Part 119-85. 1985 Testing Aggregates Part 119: Method for Determination of Acid-Soluble Material in Fine Aggregate. 8 pp.

SAA AS 1141.8-80. Methods for Sampling and Testing Aggregates—Part 8: Water Soluble Fraction of Filler. 2 pp.

—Spreading Rate
BSI BS 598: Part 108-90. 1990 Sampling and Examination of Bituminous Mixtures for Roads and Other Paved Areas Part 108: Methods for Determination of the Condition of the Binder on Coated Chippings and for Measurement of the Rate of Spread of Coated Chippings. 8 pp.

—Stripping Resistance
SAA AS 1141.50-83. Methods for Sampling and Testing Aggregates—Part 50: Resistance to Stripping of Cover Aggregates from Binders. 2 pp.

—Sulfate Content
BSI BS 812: Part 118-88. 1988 Testing Aggregates Part 118: Method for Determination of Sulphate Content. 15 pp.

—Test Specimens
BSI BS 812: Part 121-89. 1989 Testing Aggregates Part 121: Method for Determination of Soundness. 10 pp.

SNZ NZS 4407: Part 1-91. Methods of Sampling and Testing Road Aggregates Part 1: Preliminary and General. 10 pp.

—Testing Equipment
SAA AS 1141.2-88. Methods for Sampling and Testing Aggregates—Part 2: Basic Testing Equipment. 16 pp.

Aggregates (Cont.)
—Testing Equipment—Calibration
BSI BS 812: Part 100-90. 1990 Testing Aggregates Part 100: General Requirements for Apparatus and Calibration. 15 pp.

—Testing Sieves—Plates
CEN PREN 933-2-93. Tests for Geometrical Properties of Aggregates —Part 2: Determination of Particle Size Distribution—Test Sieves, Nominal Size of Apertures. 5 pp.

—Testing Sieves—Wire Cloth
CEN PREN 933-2-93. Tests for Geometrical Properties of Aggregates —Part 2: Determination of Particle Size Distribution—Test Sieves, Nominal Size of Apertures. 5 pp.

—Unsound Particles
SAA AS 1141.30-83. Methods for Sampling and Testing Aggregates—Part 30: Unsound Particles (Clay Lumps, Friable and Weathered Aggregate). 2 pp.

—Voids
BSI BS 812: Part 2-75. 1975 Amd 1 Testing Aggregates Part 2: Physical Properties. 24 pp.

SAA AS 1141.17-80. Methods for Sampling and Testing Aggregates—Part 17: Voids in Dry Compacted Filler. 4 pp.

—Water Absorption
BSI BS 812: Part 2-75. 1975 Amd 1 Testing Aggregates Part 2: Physical Properties. 24 pp.

CNS A3006-86. Method of Test for Specific Gravity and Absorption of Fine Aggregate (Aug)(487). 6 pp.

CNS A3007-72. Method of Test for Specific Gravity and Absorption of Coarse Aggregate (Jun)(488).

JIS A 1109-76. Method of Test for Specific Gravity and Absorption of Fine Aggregate. 7 pp.

JIS A 1110-89. Method of Test for Specific Gravity and Absorption of Coarse Aggregate. 6 pp.

—Water Content
SNZ NZS 4407: Part 3.1-91. Methods of Sampling and Testing Road Aggregates Part 3: Methods of Testing Road Aggregates—Laboratory Tests Test 3.1: The Water Content of Aggregate. 3 pp.

SNZ NZS 4407: Pt 4.2.1-91. Methods of Sampling and Testing Road Aggregates Part 4: Methods of Testing Road Aggregates—Field Tests Test 4.2: Theield Water Content and Field Dry Density of Compacted Aggregate Test 4.2.1: Method Using a Nuclear Surface Moisture-Density Gauge-. 3 pp.

SNZ NZS 4407: Part 4.2.2-91. Methods of Sampling and Testing Road Aggregates Part 4: Methods of Testing Road Aggregates—Field Tests Test 4.2: The Field Water Content and Field Dry Density of Compacted Aggregate Test 4.2.2: Method Using a Nuclear Surface Moisture-Density Gauge—Backscatter Mode. 3 pp.

SNZ NZS 4407: Part 4.2.3-91. Methods of Sampling and Testing Road Aggregates Part 4: Methods of Testing Road Aggregates—Field Tests Test 4.2: The Field Water Content and Field Dry Density of Compacted Aggregate Test 4.2.3: Calibration of Nuclear Surface Moisture-Density Gauge for Field Use. 3 pp.

SNZ NZS 4407: Part 4.2.4-91. Methods of Sampling and Testing Road Aggregates Part 4: Methods of Testing Road Aggregates—Field Tests Test 4.2: The Field Water Content and Field Dry Density of Compacted Aggregate Test 4.2.4: Calibration of Nuclear Surface Moisture-Density Gauge Using Standard Blocks. 6 pp.

—Wear Testing
CEN PREN 1097-1-93. Tests for Mechanical and Physical Properties of Aggregates—Part 1: Determination of the Resistance to Wear (Micro-Deval). 9 pp.

—Weather Resistance
SNZ NZS 4407: Part 3.11-91. Methods of Sampling and Testing Road Aggregates Part 3: Methods of Testing Road Aggregates—Laboratory Tests Test 3.11: The Weathering Quality Index of Coarse Aggregate. 4 pp.

—Wet/Dry Strength
SAA AS 1141.22-80. Methods for Sampling and Testing Aggregates—Part 22: Wet/Dry Strength Variation. 8 pp.

Aging Persons
Use For: Elderly Persons; Old People
See Also: Handicapped Persons

—Anthropometry
BSI BS 4467-91. 1991 Guide to Dimensions in Designing for Elderly People. 20 pp.

—Ergonokics
BSI BS 4467-91. 1991 Guide to Dimensions in Designing for Elderly People. 20 pp.

Aging Testing

Use For: Accelerated Aging Testing
See Also: Materials Testing; Physical Testing; Strain Hardening; Yellowing

—Adhesive Tapes

SAA AS 1635.4.1-74. Methods of Testing Pressure Sensitive Adhesive Tape—Part 4.1: Stability—Accelerated Ageing. 1 p.

SAA AS 1635.4.2-74. Methods of Testing Pressure Sensitive Adhesive Tape—Part 4.2: Stability—Accelerated Ageing of Cellulose Tape. 1 p.

—Adhesives—Aerospace

AECMA PREN2243-05-89. Structural Adhesives Test Methods for Part 5—Ageing Tests. 8 pp.

—Bonded Joints

BSI BS EN 2243-5-92. 1992 Test Methods for Structural Adhesives Part 5: Ageing Tests. 13 pp.

BSI BS EN 29142-93. 1993 Adhesives—Guide to the Selection of Standard Laboratory Ageing Conditions for Testing Bonded Joints (ISO 9142: 1990) (W). 28 pp.

CEN EN 2243-5-92. Structural Adhesives—Test Methods—Part 5: Ageing Tests. 8 pp.

CEN EN 29142-93. Adhesives—Guide to the Selection of Standard Laboratory Ageing Conditions for Testing Bonded Joints (ISO 9142: 1990). 22 pp.

ISO 9142-90. Adhesives—Guide to the Selection of Standard Laboratory Ageing Conditions for Testing Bonded Joints First Edition (CEN EN 29142: 1993). 24 pp.

—Cables (Electric)

DIN VDE 0472 Pt 303-90. Testing of Cables, Wires and Flexible Cords; Ageing Procedures (May). 15 pp.

—Cables (Electric)—Aircraft

CEN PREN 3475 (Part 401)-92. Cables, Electrical, Aircraft Use Test Methods Part 401—Accelerated Ageing. 3 pp.

CEN PREN 3475 (Part 409)-92. Cables, Electrical, Aircraft Use Test Methods Part 409—Air-Excluded Ageing. 2 pp.

—Cellular Materials

BSI BS 4443: Part 4-76. 1976 Methods of Test for Flexible Cellular Materials Part 4: Method 10. Determination of Solvent Swelling Method 11. Humidity Ageing at an Elevated Temperature Method 12. Heat Ageing. 7 pp.

DIN ENGL 53578-88. Testing the Ageing Behaviour of Latex and Polyurethane Foams (Dec). 3 pp.

ISO 2440-83. Polymeric Materials, Cellular Flexible—Accelerated Ageing Tests Second Edition. 4 pp.

—Coated Fabrics

BSI BS 3424: Part 12-90. 1990 Testing Coated Fabrics Part 12: Methods 14A, 14B, 14C, 14D and 14E. Accelerated Ageing Tests. 6 pp.

ISO 1419-77. Fabrics Coated with Rubber or Plastics—Accelerated Ageing and Simulated Service Tests First Edition. 4 pp.

—Connectors

CNS C6133-87. Method of Test for Low Frequency (Below 3 MHz) Electrical Connectors (TP-24 Maintenance Aging Test) (Oct)(7663).

—Contraceptive Diaphragms

ISO 8009 Pt 6-85. Reusable Rubber Contraceptive Diaphragms—Part 6: Determination of Deterioration After Accelerated Ageing First Edition. 3 pp.

—Conveyor Belts—Covers

SAA AS 1334.6-85. Methods of Testing Conveyor and Elevator Belting—Part 6: Determination of Resistance of Covers to Ageing. 2 pp.

—Cords (Electric)

DIN VDE 0472 Pt 303-90. Testing of Cables, Wires and Flexible Cords; Ageing Procedures (May). 15 pp.

—Elastomers

DIN ENGL 53508-77. Testing of Elastomers; Accelerated Ageing (July). 4 pp.

DIN ENGL 53509 Pt 2-77. Testing of Rubber and Elastomers; Accelerated Test of Ageing in Elastomers by Exposure to Ozone; Determination of Ozone Concentration (Reference Method) (Mar). 3 pp.

—Electrical Insulation

BSI BS 5691: Part 1-79. 1979 Determination of Thermal Endurance Properties of Electrical Insulating Materials Part 1: General Procedures for the Determination of Thermal Endurance Properties, Temperature Indices and Thermal Endurance Profiles. 19 pp.

Aging Testing (Cont.)

—Electrical Insulation (Cont.)

BSI BS 5691: Part 2-79. 1979 Amd 1 Determination of Thermal Endurance Properties of Electrical Insulating Materials Part 2: List of Materials and Available Tests. 11 pp.

CENELEC HD 611.1 S1-92. Guide for the Determination of Thermal Endurance Properties of Electrical Insulating Materials Part 1: General Guidelines for Ageing Procedures and Evaluation of Test Results. 6 pp.

IEC 216 Pt 1-90. Guide for the Determination of Thermal Endurance Properties of Electrical Insulating Materials Part 1: General Guidelines for Ageing Procedures and Evaluation of Test Results Fourth Edition. 55 pp.

IEC 610-78. Principal Aspects of Functional Evaluation of Electrical Insulation Systems: Aging Mechanisms and Diagnostic Procedures First Edition. 23 pp.

IEC 727 Pt 1-82. Evaluation of Electrical Endurance of Electrical Insulation Systems Part 1: General Considerations and Evaluation Procedures Based on Normal Distributions First Edition. 34 pp.

IEC 941-88. Mechanical Endurance Functional Tests for Electrical Insulation Systems First Edition. 24 pp.

—Fiber Optic Equipment

CNS C6222-84. Method of Test for Fiber Optic Devices (Maintenance Aging) (May)(10877).

CNS C6330-89. Method of Test for Fiber Optic Devices (FOTP-101 Accelerated Oxygen Aging) (Jul)(12561).

—Footwear

BSI BS 5131: Sec 5.3-90. 1990 Methods of Test for Footwear and Footwear Materials Part 5: Testing of Complete Footwear Section 5.3: Resistance of Complete Footwear to Heat. 5 pp.

—Fuel Oils

ISO 10307 Pt 2-93. Petroleum Products—Total Sediment in Residual Fuel Oils—Part 2: Determination Using Standard Procedures for Ageing First Edition. 9 pp.

—Glossaries

DIN ENGL 50035 Pt 1-72. Terms and Definitions Used in Connection with the Ageing of Materials; Basic Terms and Definitions (Mar). 2 pp.

—Hoses

SAA AS 1180.3-72. Methods of Test for Hose Made from Elastomeric Materials—Part 3: Accelerated Ageing Reconfirmed 1988.

—Hydraulic Fluids

DIN ENGL 51554 Pt 3-78. Testing of Mineral Oils; Test of Susceptibility to Ageing According to Baader; Testing at 95 Deg. C (Sept). 2 pp.

DIN ENGL 51587-74. Testing of Lubricants; Determination of the Ageing Behaviour of Steam Turbine Oils and Hydraulic Oils Containing Additives (Aug). 6 pp.

—Lubricant Additives

DIN ENGL 51587-74. Testing of Lubricants; Determination of the Ageing Behaviour of Steam Turbine Oils and Hydraulic Oils Containing Additives (Aug). 6 pp.

—Oils

BSI BS EN 61065-93. 1993 Method for Evaluating the Low Temperature Flow Properties of Mineral Insulating Oils After Aging (IEC 1065: 1991). 14 pp.

CENELEC EN 61065-93. Method for Evaluating the Low Temperature Flow Properties of Mineral Insulating Oils After Ageing (IEC 1065:1991). 5 pp.

DIN ENGL 51352 Pt 1-85. Testing of Lubricants; Determination of Ageing Characteristics of Lubricating Oils; Increase in Conradson Carbon Residue After Ageing by Passing Air Through the Lubricating Oil (Aug). 4 pp.

DIN ENGL 51352 Pt 2-85. Testing of Lubricants; Determination of Ageing Characteristics of Lubricating Oils; Conradson Carbon Residue After Ageing by Passing Air Through the Lubricating Oil in the Presence of Iron (III) Oxide (Aug). 3 pp.

DIN ENGL 51554 Pt 1-78. Testing of Mineral Oils; Test of Susceptibility to Ageing According to Baader; Purpose, Sampling, Ageing (Sept). 2 pp.

DIN ENGL 51554 Pt 2-78. Testing of Mineral Oils; Test of Susceptibility to Ageing According to Baader; Testing at 110 Deg. C (Sept). 3 pp.

DIN ENGL 51554 Pt 3-78. Testing of Mineral Oils; Test of Susceptibility to Ageing According to Baader; Testing at 95 Deg. C (Sept). 2 pp.

DIN ENGL 51586-77. Testing of Lubricants; Determination of the Ageing Properties of Lubricating Oils for High Pressure Load (Aug). 4 pp.

Aging Testing (Cont.)

—Oils (Cont.)

DIN ENGL 51587-74. Testing of Lubricants; Determination of the Ageing Behaviour of Steam Turbine Oils and Hydraulic Oils Containing Additives (Aug). 6 pp.

IEC 1065-91. Method for Evaluating the Low Temperature Flow Properties of Mineral Insulating Oils After Ageing First Edition; (CENELEC EN 61065:1993). 19 pp.

—Paints

DIN ENGL 53231-91. Artificial Weathering and Ageing of Paint Coatings by Exposure to Filtered Xenon Arc Radiation (Apr). 9 pp.

—Paperboard

BSI BS 6388: Part 1-91. 1991 Accelerated Ageing of Paper and Board Part 1: Method for Dry Heat Treatment at 105 Degrees Celsius (ISO 5630-1: 1991). 8 pp.

BSI BS 6388: Part 2-87. 1987 Accelerated Ageing of Paper and Board Part 2: Method for Dry Heat Treatment at 120 Degrees C or 150 Degrees C. 7 pp.

BSI BS 6388: Part 3-87. 1987 Accelerated Ageing of Paper and Board Part 3: Method for Moist Heat Treatment at 80 Degrees Celsius and 65 Percent Relative Humidity. 8 pp.

ISO 5630 Pt 1-91. Paper and Board—Accelerated Ageing—Part 1: Dry Heat Treatment at 105 Degrees Celsius Second Edition. 7 pp.

ISO 5630 Pt 3-86. Paper and Board—Accelerated Ageing—Part 3: Moist Heat Treatment at 80 Degrees Celsius and 65% Relative Humidity First Edition. 7 pp.

ISO 5630 Pt 4-86. Paper and Board—Accelerated Ageing—Part 4: Dry Heat Treatment at 120 or 150 Degrees Celsius First Edition. 5 pp.

—Papers

BSI BS 6388: Part 1-91. 1991 Accelerated Ageing of Paper and Board Part 1: Method for Dry Heat Treatment at 105 Degrees Celsius (ISO 5630-1: 1991). 8 pp.

BSI BS 6388: Part 2-87. 1987 Accelerated Ageing of Paper and Board Part 2: Method for Dry Heat Treatment at 120 Degrees C or 150 Degrees C. 7 pp.

BSI BS 6388: Part 3-87. 1987 Accelerated Ageing of Paper and Board Part 3: Method for Moist Heat Treatment at 80 Degrees Celsius and 65 Percent Relative Humidity. 8 pp.

ISO 5630 Pt 1-91. Paper and Board—Accelerated Ageing—Part 1: Dry Heat Treatment at 105 Degrees Celsius Second Edition. 7 pp.

ISO 5630 Pt 3-86. Paper and Board—Accelerated Ageing—Part 3: Moist Heat Treatment at 80 Degrees Celsius and 65% Relative Humidity First Edition. 7 pp.

ISO 5630 Pt 4-86. Paper and Board—Accelerated Ageing—Part 4: Dry Heat Treatment at 120 or 150 Degrees Celsius First Edition. 5 pp.

—Plastic Pipes

CNS K6139-60. Method of Test of Soft Polyvinyl Chloride Pipes (Apr)(1297)(R 1968). 3 pp.

—Plastic Sheets

CNS K6307-71. Method of Test for Polymethylmethacrylate Corrugated Sheet (Aging Test) (Apr)(3278). 2 pp.

—Polyurethane Resins—Athletic Fields

CNS K6591-86. Method of Test for Polyurethane (PU) Athletic Installation Material (May)(6483). 6 pp.

—Pulps

CPPA E.4P-88. Accelerated Aging Test for Brightness Reversion of Pulp. 2 pp.

—Sealants—Consistency

SAA AS 1937.7-77. Methods of Test for Sealers and Adhesives for Automotive Purposes—Part 7: Determination by Pressure Extrudiometer of the Consistency of Sealers After Accelerated Ageing Reconfirmed 1989. 2 pp.

—Sealants—Embrittlement

SAA AS 1937.11-77. Methods of Test for Sealers and Adhesives for Automotive Purposes—Part 11: Determination of Characteristics of Medium to Heavy Bodied Sealers Resistant to Embrittlement from Ageing Reconfirmed 1989. 5 pp.

—Shunt Power Capacitors

IEC 831 Pt 2-88. Shunt Power Capacitors of the Self-Healing Type for a.c. Systems Having a Rated Voltage up to and Including 660 V Part 2: Ageing Test, Self-Healing Test and Destruction Test First Edition; (Amendment 1-1991) (Amendment 2-1993). 37 pp.

INTERNATIONAL AND NON-U.S. NATIONAL STANDARDS
SUBJECT INDEX
Agricultural

Aging Testing *(Cont.)*
—**Shunt Power Capacitors** *(Cont.)*
IEC 931 Pt 2-92. Shunt Power Capacitors of the Non-Self-Healing Type for a.c. Systems Having a Rated Voltage up to and Including 1 000 V Part 2: Ageing Test and Destruction Test First Edition. 20 pp.

—**Statistical Analysis**
IEC 493 Pt 1-74. Guide for the Statistical Analysis of Ageing Test Data Part 1: Methods Based on Mean Values of Normally Distributed Test Results First Edition. 41 pp.

—**Storage Tanks**
CEN PREN 978-92. Underground Tanks of Glass-Reinforced Plastics (GRP)—Determination of Creep Factor and Ageing Factor. 10 pp.

—**Terminal Blocks**
CSA C22.2 NO 158-1987. Terminal Blocks; (Gen Instr 1 Thru 4). 32 pp.

—**Thermal—Cable Insulation**
BSI BS 6469: Sec 1.2-92. 1992 Insulation and Sheathing Materials of Electric Cables Part 1: Methods of Test for General Application Section 1.2: Thermal Ageing Methods. 23 pp.
BSI BS 6469: Sec 4.2-92. 1992 Insulating and Sheathing Materials of Electric Cables Part 4: Methods of Test Specific to Polyethylene and Polypropylene Compounds Section 4.2: Elongation at Break After Pre-Conditioning—Wrapping Test After Pre-Conditioning-Wrapping. 24 pp.
CENELEC HD 505.4.2 S1-92. Common Test Methods for Insulating and Sheathing Materials of Electric Cables Part 4: Methods Specific to Polyethylene and Polypropylene Compounds Section Two—Elongation at Break After Pre-Conditioning—Wrapping Test After Thermal Ageing in Air—Measurement of Mass. 6 pp.
IEC 811 Pt 1-2-85. Common Test Methods for Insulating and Sheathing Materials of Electric Cables Part 1: Methods for General Application Section Two—Thermal Ageing Methods First Edition; (Corrigendum—May 1986) (Amendment 1-1989). 41 pp.
IEC 811 Pt 4-2-90. Common Test Methods for Insulating and Sheathing Materials of Electric Cables Part 4: Methods Specific to Polyethylene and Polypropylene Compounds Section Two—Elongation at Break After Pre-Conditioning—Wrapping Test After Pre-Conditioning—Wrapping Test After. 38 pp.

—**Thermal—Electrical Insulation**
DIN VDE 0472 Pt 617-85. Testing of Cables, Wires and Flexible Cords; Mechanical Behaviour of Polyethylene After Thermal Ageing (Nov). 9 pp.

—**Thermal—Plastic Sheets—Thermoplastic Resins**
CNS K6967-89. General Rules for Tests for Thermal Ageing Properties of Thermoplastics in the Form of Sheet by Means of Ovens (Nov)(12627).
JIS K 7212-77. General Rules for Tests for Thermal Ageing Properties of Thermoplastics in the Form of Sheet by Means of Ovens (R 1980). 12 pp.

—**Thermal—Sealants**
CGSB CAN2-19.0-M77METH 8.2-78. Methods of Testing Putty, Caulking and Sealing Compounds Shore "A" Hardness After Heat Aging. 1 p.
CGSB CAN2-19.0-M77METH14.1-78. Methods of Testing Putty, Caulking and Sealing Compounds Tensile Extension After Heat Aging and Water Immersion. 4 pp.

—**Varnishes**
DIN ENGL 53231-91. Artificial Weathering and Ageing of Paint Coatings by Exposure to Filtered Xenon Arc Radiation (Apr). 9 pp.

—**Vulcanized Rubber**
BSI BS 903: Part A19-86. 1986 Methods of Testimg Vulcanized Rubber Part A19: Heat Resistance and Accelerated Aging Tests. 8 pp.
BSI BS 903: Part A52-86. 1986 Methods of Testing Vulcanized Rubber Part A52: Determination of Aging Characteristics by Measurement of Stress at a Given Elongation. 8 pp.
CNS K6347-85. Method of Test for Accelerated Aging of Vulcanized Rubber (Nov)(3556).
ISO 188-82. Rubber, Vulcanized—Accelerated Aging or Heat-Resistance Tests Second Edition. 6 pp.
ISO 6914-85. Rubber, Vulcanized—Determination of Ageing Characteristics by Measurement of Stress at a Given Elongation First Edition. 6 pp.

—**Wood Preservatives**
BSI BS 5761: Part 1-89. 1989 Wood Preservatives. Accelerated Ageing of Treated Wood Prior to Biological Testing Part 1: Evaporative Ageing Procedure. 8 pp.

Aging Testing *(Cont.)*
—**Wood Preservatives** *(Cont.)*
BSI BS 5761: Part 2-90. 1990 Wood Preservatives. Accelerated Ageing of Treated Wood Prior to Biological Testing Part 2: Leaching Procedure. 8 pp.
CEN EN 73-78. Wood Preservatives: Accelerated Ageing of Treateed Wood Prior to Biological Testing; Evaporative Ageing Procedure. 6 pp.
CEN EN 84-79. Wood Preservatives: Accelerated Ageing of Treated Wood Prior to Biological Testing; Leaching Procedure. 8 pp.
CEN EN 84-89. Wood Preservatives: Accelerated Ageing of Treated Wood Prior to Biological Testing; Leaching Procedure. 7 pp.
DIN ENGL EN 73-90. Wood Preservatives; Accelerated Ageing of Treated Wood Prior to Biological Testing; Evaporative Ageing Procedure (Apr). 5 pp.

Agitator Vessels
Use: Containers—Agitator

Agitators
Use For: Anchor Agitators; Impeller Agitators
See Also: Centrifuges; Classifiers; Contactors; Laundry Equipment; Separators (Mechanical)

—**Rings**
DIN ENGL 28084-82. Ring Supports; Dimensions, Maximum Operating Mass (Sept). 10 pp.
DIN ENGL 28084 Suppl. 1-82. Ring Supports; Examples of Calculation (Sept). 4 pp.

Agricultural Aircraft
See Also: Aircraft; Transport Aircraft

—**Loads (Forces)**
CAA NOTICE #90 ISSUE 1. Maximum Total Weight Authorised for Agricultural Operations and Other Aerial Applications (Airworthiness Notices). 3 pp.

Agricultural Buildings
Use For: Agricultural Facilities *See Also:* Meat Processing Facilities; Silos
BSI BS 5502: Part 0-92. 1992 Buildings and Structures for Agriculture Part 0: Introduction. 16 pp.

—**Building Codes**
SNZ NZS 1900: Chapter 11.2-85. Model Building Bylaw Chapter 11.2: Special Structures. Division 11.2 Farm Buildings Amend: 1, 1992. 20 pp.

—**Colors**
BSI BS 4903-79. (WITHDRAWN) 1979 External Colours for Farm Buildings (Superseded by BS 5502: Part 20: 1990). 10 pp.

—**Construction**
BSI BS 5502: Part 22-87. 1987 Design of Buildings and Structures for Agriculture Part 22: Code of Practice for Design, Construction and Loading. 28 pp.
BSI BS 5502: Part 42-90. 1990 Design of Buildings and Structures for Agriculture Part 42: Code of Practice for Design and Construction of Pig Buildings. 14 pp.

—**Construction—Chemical Products**
BSI BS 5502: Part 81-89. 1989 Design of Buildings and Structures for Agriculture Part 81: Code of Practice for Design and Construction of Chemical Stores (R). 12 pp.

—**Construction—Floors**
BSI BS 5502: Part 51-91. 1991 Buildings and Structures for Agriculture Part 51: Code of Practice for Design and Construction of Slatted, Perforated and Mesh Floors for Livestock. 14 pp.
BSI BS 5502: Part 51-01. 1991 Amd 1 Buildings and Structures for Agriculture Part 51: Code of Practice for Design and Construction of Slatted, Perforated and Mesh Floors for Livestock (AMD 7666) April 15, 1993 (R). 15 pp.

—**Construction—Storage**
BSI BS 5502: Part 72-92. 1992 Buildings and Structures for Agriculture Part 72: Code of Practice for Design and Construction of Controlled Environment Stores for Vegetables, Fruit and Flowers. 19 pp.

—**Construction—Vegetables**
BSI BS 5502: Part 60-92. 1992 Buildings and Structures for Agriculture Part 60: Code of Practice for Design and Construction of Buildings for Mushrooms. 11 pp.

—**Construction—Ventilation Equipment**
BSI BS 5502: Part 70-92. 1992 Buildings and Structures for Agriculture Part 70: Code of Practice for Design and Construction of Ventilated on Floor Stores for Combinable Crops. 15 pp.

Agricultural Buildings *(Cont.)*
—**Construction—Ventilation Equipment** *(Cont.)*
BSI BS 5502: Part 71-92. 1992 Buildings and Structures for Agriculture Part 71: Code of Practice for Design and Construction of Ventilated Stores for Potatoes and Onions. 14 pp.

—**Design**
BSI BS 5502: Sec 1.0-86. (WITHDRAWN) 1986 Design of Buildings and Structures for Agriculture Part 1: General Considerations Section 1.0: Introduction (Superseded by BS 5502: Part 0: 1992). 7 pp.
BSI BS 5502: Sec 1.1-86. 1986 Design of Buildings and Structures for Agriculture Part 1: General Considerations Section 1.1: Materials. 8 pp.
BSI BS 5502: Sec 1.5-78. (WITHDRAWN) 1978 Amd 1 Design of Buildings and Structures for Agriculture Part 1: General Considerations Section 1.5: Services (Superseded by BS 5502: Part 25: 1991). 17 pp.
BSI BS 5502: Sec 2.1-81. (WITHDRAWN) 1981 Design of Buildings and Structures for Agriculture Part 2: General Considerations Section 2.1: General Design and Planning (Superseded by BS 5502: Part 20: 1990). 3 pp.
BSI BS 5502: Sec 2.2-81. 1981 Amd 1 Design of Buildings and Structures for Agriculture Part 2: General Considerations Section 2.2: Livestock Buildings. 19 pp.
BSI BS 5502: Sec 2.5-81. 1981 Design of Buildings and Structures for Agriculture Part 2: General Considerations Section 2.5: Service Buildings. 6 pp.
BSI BS 5502: Sec 3.6-86. (WITHDRAWN) 1986 Design of Buildings and Structures for Agriculture Part 3: Appendices: Legislation, Technical Data and References Section 3.6: Reference Data: Space (Superseded by BS 5502: Part 80: 1990). 6 pp.
BSI BS 5502: Sec 3.12-86. (WITHDRAWN) 1986 Design of Buildings and Structures for Agriculture Part 3: Appendices: Legislation, Technical Data and References Section 3.12: Sources of Information (Superseded by BS 5502: Part 11: 1990). 6 pp.
BSI BS 5502: Part 11-90. 1990 Design of Buildings and Structures for Agriculture Part 11: Guide to Regulations and Sources of Information. 11 pp.
BSI BS 5502: Part 20-90. 1990 Design of Buildings and Structures for Agriculture Part 20: Code of Practice for General Design Considerations. 20 pp.
BSI BS 5502: Part 22-87. 1987 Design of Buildings and Structures for Agriculture Part 22: Code of Practice for Design, Construction and Loading. 28 pp.
BSI BS 5502: Part 25-91. 1991 Buildings and Structures for Agriculture Part 25: Code of Practice for Design and Installation of Services and Facilities. 28 pp.
BSI BS 5502: Part 42-90. 1990 Design of Buildings and Structures for Agriculture Part 42: Code of Practice for Design and Construction of Pig Buildings. 14 pp.
BSI BS 5502: Part 49-90. 1990 Design of Buildings and Structures for Agriculture Part 49: Code of Practice for Design and Construction of Milking Premises. 11 pp.
BSI BS 5502: Part 80-90. 1990 Design of Buildings and Structures for Agriculture Part 80: Code of Practice for Design and Construction of Workshops, Maintenance and Inspection Facilities (R). 15 pp.
BSI BS 5502: Part 82-90. 1990 Design of Buildings and Structures for Agriculture Part 82: Code of Practice for Design of Amenity Buildings. 8 pp.

—**Design—Animal Welfare**
BSI BS 5502: Sec 1.6-86. (WITHDRAWN) 1986 Design of Buildings and Structures for Agriculture Part 1: General Considerations Section 1.6: Human and Animal Welfare (Superseded by BS 5502: Part 20: 1990 and BS 5502: Part 80:1990). 7 pp.
BSI BS 5502: Sec 3.5-78. (WITHDRAWN) 1978 Amd 1 Design of Buildings and Structures for Agriculture Part 3: Appendices: Legislation, Technical Data and References Section 3.5: Reference Data: Environment (Superseded by BS 5502: Parts 20, 80 and 82: 1990). 4 pp.

—**Design—Cattle**
BSI BS 5502: Part 40-90. 1990 Design of Buildings and Structures for Agriculture Part 40: Code of Practice for Design and Construction of Cattle Buildings. 18 pp.
BSI BS 5502: Part 40-01. 1990 Amd 1 Buildings and Structures for Agriculture Part 40: Code of Practice for Design and Construction of Cattle Buildings (AMD 7665) June 15, 1993 (R). 20 pp.

—**Design—Chemical Products**
BSI BS 5502: Part 81-89. 1989 Design of Buildings and Structures for Agriculture Part 81: Code of Practice for Design and Construction of Chemical Stores (R). 12 pp.

INDUSTRY STANDARDS

Agricultural Buildings (Cont.)

—Design—Chitting Potatoes
BSI BS 5502: Part 66-92. 1992 Buildings and Structures for Agriculture Part 66: Code of Practice for Design and Construction of Chitting Houses. 10 pp.

—Design—Colors
BSI BS 5502: Sec 3.11-78. (WITHDRAWN) 1978 Design of Buildings and Structures for Agriculture Part 3: Appendices: Legislation, Technical Data and References Section 3.11: External Colours for Farm Buildings (Superseded by BS 5502: Part 20: 1990). 2 pp.

—Design—Construction Materials
BSI BS 5502: Part 21-90. 1990 Design of Buildings and Structures for Agriculture Part 21: Code of Practice for Selection and Use of Construction Materials. 23 pp.
BSI BS 5502: Part 21-01. 1990 Amd 1 Buildings and Structures for Agriculture Part 21: Code of Practice for Selection and Use of Construction Materials (AMD 7676) June 15, 1993 (R). 28 pp.

—Design—Crop Storage
BSI BS 5502: Sec 2.3-81. (WITHDRAWN) 1981 Design of Buildings and Structures for Agriculture Part 2: General Considerations Section 2.3: Storage, Conditioning and Processing Buildings (Superseded by BS 5502: Parts 60, 65, 66 and 72: 1992). 15 pp.
BSI BS 5502: Part 75-93. 1993 Buildings and Structures for Agriculture Part 75: Code of Practice for the Design and Construction of Forage Stores (R). 13 pp.

—Design—Energy Conservation
BSI BS 5502: Sec 1.4-86. 1986 Design of Buildings and Structures for Agriculture Part 1: General Considerations Section 1.4: Energy. 4 pp.

—Design—Ergonomics
BSI BS 5502: Sec 1.6-86. (WITHDRAWN) 1986 Design of Buildings and Structures for Agriculture Part 1: General Considerations Section 1.6: Human and Animal Welfare (Superseded by BS 5502: Part 20: 1990 and BS 5502: Part 80:1990). 7 pp.
BSI BS 5502: Sec 3.5-78. (WITHDRAWN) 1978 Amd 1 Design of Buildings and Structures for Agriculture Part 3: Appendices: Legislation, Technical Data and References Section 3.5: Reference Data: Environment (Superseded by BS 5502: Parts 20, 80 and 82: 1990). 4 pp.

—Design—Fire Alarms and Detection Systems
BSI BS 5502: Part 52-91. 1991 Buildings and Structures for Agriculture Part 52: Code of Practice for Design of Alarm Systems and Emergency Ventilation for Livestock Housing. 14 pp.

—Design—Fire Protection
BSI BS 5502: Sec 1.3-86. (WITHDRAWN) 1986 Design of Buildings and Structures for Agriculture Part 1: General Considerations Section 1.3: Fire Protection (R) (Superseded by BS 5588: Part 1: 1990). 6 pp.
BSI BS 5502: Part 23-90. 1990 Design of Buildings and Structures for Agriculture Part 23: Code of Practice for Fire Precautions. 16 pp.

—Design—Floors
BSI BS 5502: Part 51-91. 1991 Buildings and Structures for Agriculture Part 51: Code of Practice for Design and Construction of Slatted, Perforated and Mesh Floors for Livestock. 14 pp.
BSI BS 5502: Part 51-01. 1991 Amd 1 Buildings and Structures for Agriculture Part 51: Code of Practice for Design and Construction of Slatted, Perforated and Mesh Floors for Livestock (AMD 7666) April 15, 1993 (R). 15 pp.

—Design—Horticultural Processing
BSI BS 5502: Part 65-92. 1992 Buildings and Structures for Agriculture Part 65: Code of Practice for Design and Construction of Crop Processing Buildings. 11 pp.

—Design—Legislation
BSI BS 5502: Sec 3.1-86. (WITHDRAWN) 1986 Design of Buildings and Structures for Agriculture Part 3: Appendices: Legislation, Technical Data and References Section 3.1: Legislation (Superseded by BS 5502: Part 11: 1990). 8 pp.

—Design—Modular Construction
BSI BS 5502: Sec 3.8-78. (WITHDRAWN) 1978 Amd 1 Design of Buildings and Structures for Agriculture Part 3: Appendices: Legislation, Technical Data and References Section 3.8: Dimensional Co-Ordination (AMD 5003) April 30, 1986 (Superseded by BS 5502:. 5 pp.

Agricultural Buildings (Cont.)

—Design—Noise Reduction
BSI BS 5502: Part 32-90. 1990 Design of Buildings and Structures for Agriculture Part 32: Guide to Noise Attenuation. 12 pp.

—Design—Pest Control
BSI BS 5502: Sec 1.7-78. (WITHDRAWN) 1978 Design of Buildings and Structures for Agriculture Part 1: General Considerations Section 1.7: Infestation (Superseded by BS 5502: Part 30: 1992). 2 pp.

—Design—Poultry
BSI BS 5502: Part 43-90. 1990 Design of Buildings and Structures for Agriculture Part 43: Code of Practice for Design and Construction of Poultry Buildings. 13 pp.

—Design—Sheep
BSI BS 5502: Part 41-90. 1990 Design of Buildings and Structures for Agriculture Part 41: Code of Practice for Design and Construction of Sheep Buildings and Pens. 12 pp.

—Design—Sheeting
BSI BS 5502: Sec 3.2-78. 1978 Amd 1 Design of Buildings and Structures for Agriculture Part 3: Appendices: Legislation, Technical Data and References Section 3.2: Reference Data: Roof and Wall Sheeting and Durability of Materials. 10 pp.

—Design—Storage
BSI BS 5502: Part 72-92. 1992 Buildings and Structures for Agriculture Part 72: Code of Practice for Design and Construction of Controlled Environment Stores for Vegetables, Fruit and Flowers. 19 pp.

—Design—Storage Silos
BSI BS 5502: Part 74-91. 1991 Buildings and Structures for Agriculture Part 74: Code of Practice for Design and Construction of Bins and Silos for Combinable Crops. 15 pp.

—Design—Vegetables
BSI BS 5502: Part 60-92. 1992 Buildings and Structures for Agriculture Part 60: Code of Practice for Design and Construction of Buildings for Mushrooms. 11 pp.

—Design—Ventilation Equipment
BSI BS 5502: Part 52-91. 1991 Buildings and Structures for Agriculture Part 52: Code of Practice for Design of Alarm Systems and Emergency Ventilation for Livestock Housing. 14 pp.
BSI BS 5502: Part 70-92. 1992 Buildings and Structures for Agriculture Part 70: Code of Practice for Design and Construction of Ventilated on Floor Stores for Combinable Crops. 15 pp.
BSI BS 5502: Part 71-92. 1992 Buildings and Structures for Agriculture Part 71: Code of Practice for Design and Construction of Ventilated Stores for Potatoes and Onions. 14 pp.

—Drainage Systems
BSI BS 5502: Part 25-91. 1991 Buildings and Structures for Agriculture Part 25: Code of Practice for Design and Installation of Services and Facilities. 28 pp.

—Electrical Installations
BSI BS 5502: Sec 1.5-78. (WITHDRAWN) 1978 Amd 1 Design of Buildings and Structures for Agriculture Part 1: General Considerations Section 1.5: Services (Superseded by BS 5502: Part 25: 1991). 17 pp.
BSI BS 5502: Part 25-91. 1991 Buildings and Structures for Agriculture Part 25: Code of Practice for Design and Installation of Services and Facilities. 28 pp.
CENELEC HD 384.7.705 S1-91. Electrical Installations of Buildings Part 7: Requirements for Special Installations or Locations Section 705—Electrical Installations of Agricultural and Horticultural Premises. 5 pp.
IEC 364 Pt 7-705-84. Electrical Installations of Buildings Part 7: Requirements for Special Installations or Locations Section 705—Electrical Installations of Agricultural and Horticultural Premises First Edition. 13 pp.

—Gas Supply Installations
BSI BS 5502: Part 25-91. 1991 Buildings and Structures for Agriculture Part 25: Code of Practice for Design and Installation of Services and Facilities. 28 pp.

—Ladders
BSI BS 4211-87. 1987 Ladders for Permanent Access to Chimneys, Other High Structures, Silos and Bins. 16 pp.
BSI BS 4211-01. 1987 Amd 1 Ladders for Permanent Access to Chimneys, Other High Structures, Silos and Bins (AMD 7064) June 15, 1992 (R). 18 pp.

Agricultural Buildings (Cont.)

—Loads (Forces)
BSI BS 5502: Part 22-87. 1987 Design of Buildings and Structures for Agriculture Part 22: Code of Practice for Design, Construction and Loading. 28 pp.

—Odors
BSI BS 5502: Part 33-91. 1991 Buildings and Structures for Agriculture Part 33: Guide to the Control of Odour Pollution. 13 pp.

—Pest Control
BSI BS 5502: Part 30-92. 1992 Buildings and Structures for Agriculture Part 30: Code of Practice for Control of Infestation. 10 pp.

—Water Supply Installations
BSI BS 5502: Part 25-91. 1991 Buildings and Structures for Agriculture Part 25: Code of Practice for Design and Installation of Services and Facilities. 28 pp.

Agricultural Chemicals

See Also: Fertilizers

—Protective
EC COM(93) 351 FINAL SYN465-93. Commission Proposal for a Council Directive Concerning the Placing of Biocidal Products on the Market. 38 pp.
EC 91/414/EEC-91. Council Directive Concerning the Placing of Plant Protection Products on the Market. 33 pp.
EC 93/71/EEC-93. Commission Directive Amending Council Directive 91/414/EEC Concerning the Placing of Plant Protection Products on the Market. 10 pp.

Agricultural Equipment

See Also: Agricultural Spraying Equipment; Agriculture; Balers; Binders (Agricultural); Brooders; Combines; Crop Dryers; Cultivators; Fertilizing Equipment; Harvesting Equipment; Haymakers; Hoes; Incubators; Industrial Equipment; Irrigation Equipment (Agricultural); Livestock Equipment; Machinery; Manure Spreaders; Mowers; Pesticide Distribution Equipment; Planting Equipment; Plows; Reapers; Rice Milling Equipment; Self Propelled Machinery; Silage Cutters; Slurry Storage Systems; Sowing Equipment; Spreaders; Threshing Equipment; Tines; Tractors; Transplanters; Viticultural Equipment; Weeding Equipment; Yokes

—Aircraft
CAA Chapter K4-9 04.72. Agricultural Equipment (Light Aeroplanes). 3 pp.

—Bolts
CNS B2355-82. Countersunk Double-Rib Bolts and Counter Sinkings for Agricultural Equipments (Oct)(4698).
ISO 5713-90. Equipment for Working the Soil—Fixing Bolts for Soil Working Elements First Edition. 8 pp.

—Brakes
BSI BS 4639-87. 1987 Brakes and Braking Systems for Towed Agricultural Vehicles. 8 pp.
BSI BS 6384-83. 1983 Determination of Braking Performance of Agricultural and Forestry Vehicles. 18 pp.
ISO 5697-82. Agricultural and Forestry Vehicles—Determination of Braking Performance First Edition. 18 pp.

—Classification
ISO 3339 Pt 0-86. Tractors and Machinery for Agriculture and Forestry—Classification and Terminology—Part 0: Classification System and Classification Second Edition. 33 pp.

—Connectors
BSI BS 6777: Part 1-86. (WITHDRAWN) 1986 Electrical Connections for Agricultural Machinery Part 1: Double-Pole Connectors (Superseded by BS EN 24165: 1992). 3 pp.

—Countersinks
CNS B2355-82. Countersunk Double-Rib Bolts and Counter Sinkings for Agricultural Equipments (Oct)(4698).

—Diesel Engines
CNS B4007-66. Water-Cooled Diesel Engines for Land Use (Small Size) (Dec)(2695). 5 pp.
CNS B4011-69. Air-Cooled Diesel Engines for Land Use (Small Size) (Jul)(3002). 6 pp.
JIS B 8012-77. Small Size Water Cooled Diesel Engines for Land Use (R 1982). 9 pp.
JIS B 8016-77. Small Size Forced Air-Cooled Diesel Engines for Land Use (R 1982). 10 pp.
JIS B 8018-89. Test Method of Performance of Small Size Diesel Engines for Land Use. 25 pp.

INTERNATIONAL AND NON-U.S. NATIONAL STANDARDS
SUBJECT INDEX
Agricultural

Agricultural Equipment (Cont.)

—Drawbars—Jacks (Lifts)
BSI BS 6792-86. 1986 Agricultural Trailer and Trailed Machinery Drawbar Jacks. 4 pp.

—Ducts—Conveyors
BSI BS 4286: Part 2-70. 1970 Steel Ducting for Grain and Fodder Conveying Part 2: Metric Units. 11 pp.

—Engines—Comprehensive Testing
BSI BS 4539-70. (WITHDRAWN) 1970 Methods of Testing Small Spark Ignition Engines for Agricultural Use. 33 pp.

—Engines—Power Ratings
BSI BS 6507-84. (WITHDRAWN) 1984 Bench Testing Engines for Agricultural Tractors and Machines for Net Power. 19 pp.
ISO 2288-89. Agricultural Tractors and Machines—Engine Test Code (Bench Test)—Net Power Second Edition. 20 pp.

—Feed Chains
CNS B4032-82. Feed Chains for Agricultural Machinery (Jan)(6007).
JIS B 9204-81. Feed Chains for Agricultural Machinery (R 1987). 11 pp.

—Filter Elements
DIN ENGL 71459-87. Air Filter Elements for Commercial Vehicles; Dimensions (May). 6 pp.

—Fuel Oils
BSI BS 2869: Part 2-88. 1988 Amd 1 Fuel Oils for Non-Marine Use Part 2: Specification for Fuel Oil for Agricultural and Industrial Engines and Burners (Classes A2,C1, C2,D,E,F,G and H) (AMD 6505) February 28, 1991 (Supersedes BS 2869: 1993). 16 pp.

—Gasoline Engines
CNS B4012-69. Air-Cooled Gasoline Engines for Land Use (Small Size) (Jul)(3003). 6 pp.

—Glossaries
BSI BS 2468-63. 1963 Amd 1 Glossary of Terms Relating to Agricultural Machinery and Implements (AMD 4395) October 21, 1983. 96 pp.

—Grooved Pulleys
BSI BS 3733-74. 1974 Endless V-Belt Drivers for Agricultural Purposes. 37 pp.
ISO 3410-89. Agricultural Machinery—Endless Variable-Speed V-Belts and Groove Sections of Corresponding Pulleys Second Edition. 8 pp.
ISO 5289-92. Agricultural Machinery—Endless Hexagonal Belts and Groove Sections of Corresponding Pulleys Second Edition. 8 pp.

—Hand Signals
BSI BS 6736-86. 1986 Code of Practice for Hand Signalling for Use in Agricultural Operations. 7 pp.

—Hand Tools—Handles
CGSB 39-GP-14M-79. Handles, Tool, Agricultural, Standard for. 9 pp.

—Hitches
ISO 11374-93. Agricultural Tractors and Implements—Four-Point Rigid Hitch—Specifications First Edition. 8 pp.

—Hydraulic Couplings
BSI BS 4742: Part 2-93. 1993 Hydraulic Equipment for Agricultural Machinery Part 2: Specification for Quick-Action Hydraulic Couplers for General Purposes (ISO 5675: 1992) (E). 8 pp.
BSI BS 4742: Part 2-83. 1983 Hydraulic Equipment for Agricultural Machinery Part 2: Specification for Hydraulic Couplers for General Purposes. 6 pp.
BSI BS 4742: Part 4-85. 1985 Hydraulic Equipment for Agricultural Machinery Part 4: Hydraulic Couplings on Braking Systems for Trailers and Trailed Equipment. 4 pp.
BSI BS 4742: Part 7-89. 1989 Hydraulic Equipment for Agricultural Machinery Part 7: Port Connections for Agricultural Trailed Vehicle Hydraulic Brake Couplings. 4 pp.
CSA CAN/CSA-M5675-92. Agricultural Tractors and Machinery—General Purpose Quick-Action Hydraulic Couplers (ISO 5675:1992); (Gen Instr 1). 19 pp.
ISO 5675-92. Agricultural Tractors and Machinery—General Purpose Quick-Action Hydraulic Couplers Second Edition. 7 pp.

—Hydraulic Equipment
BSI BS 6142-81. 1981 Presentation of Technical Data on Hydraulic Systems of Agricultural Tractors and Machinery. 2 pp.

—Internal Combustion Engines
CNS B7036-66. Testing Standard for Small Internal Combustion Engines of Land Use (Dec)(2696) (R 1973).

Agricultural Equipment (Cont.)

—Internal Combustion Engines (Cont.)
CNS B7037-66. Method of Test for Small Internal Combustion Engines of Land Use (Dec)(2697)(R 1973).

—Manual Controls
BSI BS 3904-65. (WITHDRAWN) 1965 Recommendations for the Location and Direction of Motion of Operator's Controls for Self-Propelled Agricultural Machines. 8 pp.
BSI BS 4746-71. (WITHDRAWN) 1971 Recommendations for the Location and Direction of Movement of Controls of Pedestrian-Operated Tractors, Agricultural and Horticultural Machines. 8 pp.
BSI BS 6735-87. 1987 Reach Volumes for the Location of Controls on Agricultural Tractors and Machinery. 63 pp.
ISO 3789 Pt 1-82. Tractors, Machinery for Agriculture and Forestry, Powered Lawn and Garden Equipment—Location and Method of Operation of Operator Controls—Part 1: Common Controls First Edition. 6 pp.
ISO 3789 Pt 2-82. Tractors, Machinery for Agriculture and Forestry, Powered Lawn and Garden Equipment—Location and Method of Operation of Operator Controls—Part 2: Controls for Agricultural Tractors and Machinery First Edition. 9 pp.
SAA AS 1246-72. Location and Direction of Motion of Operator's Controls for Agricultural Tractors and Self-Propelled Agricultural Machines Reconfirmed 1985. 9 pp.

—Manual Controls—Symbols
BSI BS 4964: Part 1-93. 1993 Symbols for Control Markings and Displays on Tractors and Machinery for Agricultural and Forestry, and on Powered Lawn and Garden Equipment Part 1: Specification for Common Symbols (ISO 3767-1: 1991) (E). 37 pp.
BSI BS 4964: Part 1-81. 1981 Symbols for Control Markings and Displays on Tractors and Machinery for Agricultural and Forestry, and on Powered Lawn and Garden Equipment Part 1: Common Symbols. 24 pp.
BSI BS 4964: Part 2-93. 1993 Symbols for Control Markings and Displays on Tractors and Machinery for Agricul & Forestry, and on Powered Lawn and Garden Equipment Part 2: Specification for Symbols for Agricultural Tractors and Machinery (ISO 3767-2: 1991) (E). 23 pp.
BSI BS 4964: Part 2-81. 1981 Symbols for Control Markings and Displays on Tractors and Machinery for Agricultural and Forestry, and on Powered Lawn and Garden Equipment Part 2: Agricultural Tractors and Machines. 8 pp.
CNS B4029-80. Agricultural Tractors and Machines Operator Controls (Aug)(6004).
ISO 3767 Pt 1-91. Tractors, Machinery for Agriculture and Forestry, Powered Lawn and Garden Equipment—Symbols for Operator Controls and Other Displays—Part 1: Common Symbols Second Edition; (Incorporating Addendum 1). 33 pp.
ISO 3767 Pt 2-91. Tractors, Machinery for Agriculture and Forestry, Powered Lawn and Garden Equipment—Symbols for Operator Controls and Other Displays—Part 2: Symbols for Agricultural Tractors and Machinery Second Edition; (Incorporating Addendum 1). 19 pp.
JIS B 9126-91. Agricultural Tractors and Machines—Symbols for Operator Controls. 14 pp.
SAA AS 1064-87. Agricultural and Light Industrial Equipment—Operator Controls—Symbols. 7 pp.
SNZ NZS 5104: Part 1-86. Tractors, Machinery for Agriculture, Forestry, Powered Lawn and Garden Equipment Part 1: Common Symbols Addendum 1; Amend: A, 1986; 1A, 1986. 5 pp.
SNZ NZS 5104: Part 2-86. Tractors, Machinery for Agriculture, Forestry, Powered Lawn and Garden Equipment Part 2: Symbols for Agricultural Tractors and Machinery Addendum 1; Amend: A, 1986; 1A, 1986. 3 pp.

—Nets—Shading
CNS K3095-90. Polyethylene Shading Net for Agriculture Use (May)(12719). 2 pp.
CNS K6970-90. Method of Test for Polyethylene Shading Net for Agriculture Use (May)(12720). 3 pp.

—Noise
CSA CAN/CSA-Z107.31-M86. Test Procedures for the Measurement of Sound Levels from Agricultural Machines; (Gen Instr 1). 28 pp.
ISO 5131-82. Acoustics—Tractors and Machinery for Agriculture and Forestry—Measurement of Noise at the Operator's Position—Survey Method Amendment 1: Annex D—Forestry Forwarders and Skidders First Edition; (Amendment 1-1992). 15 pp.

Agricultural Equipment (Cont.)

—Noise (Cont.)
ISO 7216-92. Acoustics—Agricultural and Forestry Wheeled Tractors and Self-Propelled Machines—Measurement of Noise Emitted When in Motion First Edition. 7 pp.

—Nozzles
SAA AS N66-68. Methods of Testing Nozzles Used for Agricultural Purposes. 6 pp.

—Power Takeoffs
BSI BS 3417: Part 3-86. 1986 Power Take-Off Shaft and Guards for Tractors and Machinery for Agriculture and Forestry Part 3: Supplementary Requirements to Part 1 'Specification for Power Take-Off Drive Shafts' and to Part 2 'Methods for Testing Guards' (Sup's BS 3417: 1974). 4 pp.

—Power Takeoffs—Drive Shafts
BSI BS 3417: Part 1-84. 1984 1974 Power Take-Off Shafts and Guards for Tractors and Machinery for Agriculture and Forestry Part 1: Power Take-Off Drive Shafts. 8 pp.
BSI BS 3417: Part 2-84. 1984 1974 Power Take-Off Shafts and Guards for Tractors and Machinery for Agriculture and Forestry Part 2: Methods for Testing Guards. 8 pp.
CEN PREN 1152-93. Tractors and Machinery for Agriculture and Forestry—Guards for Power Take-Off (PTO) Drive Shafts—Wear and Strength Tests. 12 pp.
ISO 5673-80. Agricultural Tractors—Power Take-off Drive Shafts for Machines and Implements First Edition. 7 pp.
ISO 5674 Pt 1-92. Tractors and Machinery for Agriculture and Forestry—Guards for Power Take-off (PTO) Drive-Shafts—Part 1: Strength Test First Edition. 9 pp.

—Power Takeoffs—Drive Shafts—Wear Testing
CEN PREN 1152-93. Tractors and Machinery for Agriculture and Forestry—Guards for Power Take-Off (PTO) Drive Shafts—Wear and Strength Tests. 12 pp.
ISO 5674 Pt 2-92. Tractors and Machinery for Agriculture and Forestry—Guards for Power Take-off (PTO) Drive-Shafts—Part 2: Wear Test First Edition. 8 pp.

—Roll Over Protective Structures
CSA B352-M1980. Rollover Protective Structures (ROPS) for Agricultural, Construction, Earthmoving, Forestry, Industrial, and Mining Machines; (Gen Instr 1). 76 pp.
SNZ NZS 5101-78. Specification for Safety Frames and Safety Cabs for Attachment to Agricultural Wheeled Tractors. 8 pp.

—Roller Chains
BSI BS 2947-85. 1985 Steel Roller Chains, Types S and C, Attachments and Chain Wheels for Agricultural and Similar Machinery. 22 pp.
DIN ENGL 8189 Pt 1-87. Roller Chains for Agricultural Machinery; Chains and Connecting Links (Mar). 4 pp.
DIN ENGL 8189 Pt 2-87. Roller Chains for Agricultural Machinery; Links with Attachment Plates and Driving Plates (Mar). 4 pp.

—Safety
CEN PREN 474-2-93. Earth-Moving Machinery—Safety—Part 2: Requirements for Tractor-Dozers. 7 pp.
CEN PREN 707-92. Safety Requirements for Agricultural and Forestry Machinery—Slurry Tankers. 12 pp.
CEN PREN 708-92. Safety Requirements for Agricultural and Forestry Machinery—Soil Working Machines with Powered Tools. 11 pp.
CEN PREN 709-92. Safety Requirements for Agricultural and Forestry Machinery—Pedestrian Controlled Tractors with Mounted Rotary Cultivators, Motor Hoes, Motor Hoes with Drive Wheel(s). 27 pp.
DIN ENGL 11000-80. Agricultural Machines and Tractors; Technical Safety Requirements (Aug). 4 pp.
DIN ENGL 11001 Pt 1-85. Tractors and Machinery for Agriculture; Agricultural Tractors; Special Safety Requirements and Testing (May). 4 pp.
ISO 4254 Pt 1-89. Tractors and Machinery for Agriculture and Forestry—Technical Means for Ensuring Safety—Part 1: General Second Edition. 13 pp.
ISO 4254 Pt 3-92. Tractors and Machinery for Agriculture and Forestry—Technical Means for Ensuring Safety—Part 3: Tractors First Edition. 8 pp.
JIS B 9220-82. General Code of Safety for Agricultural Machinery (R 1988). 8 pp.

Agricultural Equipment (Cont.)

—Seats

ISO 3462-80. Tractors and Machinery for Agriculture and Forestry—Seat Reference Point—Method of Determination First Edition; (Amendment 1-1992). 8 pp.

—Sprockets—Roller Chains

DIN ENGL 8199-87. Toothing of Chain Wheels for Roller Chains for Agricultural Machinery; Profile Dimensions (Mar). 3 pp.

—Technical Manuals

BSI BS 5401-90. 1990 Guide to Information Content and Presentation of Operators' Manuals Provided for Tractors and Machinery for Agriculture and Forestry. 18 pp.

BSI BS 5401-82. (WITHDRAWN) 1982 Guide for Presentation of Operator Manuals and Technical Publications Dealing with Tractors and Machinery for Agriculture and Forestry. 18 pp.

ISO 3600-81. Tractors and Machinery for Agriculture and Forestry—Operator Manuals and Technical Publications—Presentation Second Edition. 5 pp.

—Tire Tubes

CGSB 20-GP-6E-81. Tubes, Inner, Vehicle and Mobile Ground Equipment, Standard for. 26 pp.

—Tire Valves

BSI BS AU 50: Pt 3: Sec 5-83. 1983 Tyres and Wheels Part 3: Valves Section 5: Dimensions for ISO Core Chamber No. 1 for Tyre Inflation Valves. 4 pp.

BSI BS AU 50: Pt 3: Sec 6-82. 1982 Tyres and Wheels Part 3: Valves Section 6: Dimensions for ISO Core Chamber No. 2 for Tyre Inflation Valves. 4 pp.

ISO 6762-82. Tyre Valves—ISO Core Chamber No. 2 Second Edition. 5 pp.

ISO 7442-82. Tyre Valves—ISO Core Chamber No. 1 First Edition. 6 pp.

—Tires

BSI BS 3486: Part 2-70. 1970 Wheels for Agricultural Machinery, Implements, and Trailers Part 2: Tyre and Rim Sizes. 8 pp.

BSI BS AU 50:Pt 1:SUBSEC4.1A-93. 1993 Tyres and Wheels Part 1: Tyres Section 4: Agricultural Tractor and Machine Tyres Subsection 4.1a: Specification for Designations and Dimensions for Ply Rating Marked Series Tyres (ISO 4251-1: 1992) (E). 13 pp.

BSI BS AU 50:Pt 1:SUBSEC 4.1-89. 1989 Tyres and Wheels Part 1: Tyres Section 4: Agricultural Tractor and Machine Tyres Subsection 4.1: Designations and Dimensions for Existing (Inch) Series Tyres. 13 pp.

BSI BS AU 50:Pt 1:SUBSEC4.2A-93. 1993 Tyres and Wheels Part 1: Tyres Section 4: Agricultural Tractor and Machine Tyres Subsection 4.2a: Specification for Load Ratings for Ply Rating Marked Series Tyres (ISO 4251-2: 1992) (E). 16 pp.

BSI BS AU 50:Pt 1:SUBSEC 4.2-89. 1989 Tyres and Wheels Part 1: Tyres Section 4: Agricultural Tractor and Machine Tyres Subsection 4.2: Load Ratings for Existing (Inch) Series Tyres. 12 pp.

BSI BS AU 50:Pt 1:SUBSEC 4.3-89. (WITHDRAWN) 1989 Tyres and Wheels Part 1: Tyres Section 4: Agricultural Tractor and Machine Tyres Subsection 4.3: Classification Codes and Nomenclature for Existing (Inch) Series Tyres (Superseded by BS AU 50: Subsection 4.3a: 1992). 5 pp.

BSI BS AU 50:Pt 1:SUBSEC4.3A-92. 1992 Tyres and Wheels Part 1: Tyres Section 4: Agricultural Tractor and Machine Tyres Subsection 4.3a: Specification for Classification Codes and Nomenclature for Ply Rating Marked Series Tyres (ISO 4251-4: 1992) (Sup. BS AU 50: Pt 1: Subsec. 4.3: 1989) (E). 6 pp.

BSI BS AU 50:Pt 1:SUBSEC 4.4-89. (WITHDRAWN) 1989 Tyres and Wheels Part 1: Tyres Section 4: Agricultural Tractor and Machine Tyres Subsection 4.4: Existing (Inch) Series Diagonal Tyres for Log Skidders (Superseded by BS AU 50: Subsection 4.4a: 1992). 7 pp.

BSI BS AU 50:Pt 1:SUBSEC4.4A-92. 1992 Tyres and Wheels Part 1: Tyres Section 4: Agricultural Tractor and Machine Tyres Subsection 4.4a: Specification for Ply Rating Marked Series Tyres for Log Skidders (ISO 4251-5: 1992) (Superseding BS AU 50: Part 1: Subsection 4.4: 1989) (E). 7 pp.

BSI BS AU 50:Pt 1:SUBSEC 4.5-92. 1992 Tyres and Wheels Part 1: Tyres Section 4: Agricultural Tractor and Machine Tyres Subsection 4.5: Specification for Designations, Dimensions and Marking for Metric Series Tyres (ISO 7867-1: 1992) (E). 11 pp.

CGSB 20-GP-8-80. Tires, Pneumatic, Agricultural, Standard for. 17 pp.

CNS K4032-80. Dimensions of Tires for Agricultural Machines and Implements (Feb)(3675). 12 pp.

CNS K4046-89. Tires for Agricultural Implements and Machineries (Oct)(4880). 4 pp.

Agricultural Equipment (Cont.)

—Tires (Cont.)

ISO 4251 Pt 1-92. Tyres (Ply Rating Marked Series) and Rims for Agricultural Tractors and Machines—Part 1: Tyre Designation and Dimensions Fourth Edition. 12 pp.

ISO 4251 Pt 2-92. Tyres (Ply Rating Marked Series) and Rims for Agricultural Tractors and Machines—Part 2: Tyre Load Ratings Fourth Edition. 14 pp.

ISO 4251 Pt 4-92. Tyres (Ply Rating Marked Series) and Rims for Agricultural Tractors and Machines—Part 4: Tyre Classification and Nomenclature Second Edition. 5 pp.

ISO 4251 Pt 5-92. Tyres (Ply Rating Marked Series) and Rims for Agricultural Tractors and Machines—Part 5: Log Skidder Tyres Second Edition. 6 pp.

ISO 7867 Pt 1-92. Tyres and Rims (Metric Series) for Agricultural Tractors and Machines—Part 1: Tyre Designation, Dimensions and Marking Second Edition. 10 pp.

JIS B 9202-82. Dimensions and Load Ratings of Tires for Agricultural Machines and Implements. 22 pp.

JIS B 9205-87. Tires for Agricultural Implements and Machineries. 7 pp.

SAA AS D22-71. Dimensions of Agricultural Tyres, Rims and Hubs Corrig.. 8 pp.

—Towing Attachments—Connectors

CSA CAN/CSA-M663-92. Seven-Pin Electrical Connector and Cable for Agricultural Towing/Towed Equipment; (Gen Instr 1). 21 pp.

—Tracks

BSI BS 6508-84. 1984 Track Widths for Agricultural Tractors and Machinery. 3 pp.

ISO 4004-83. Agricultural Tractors and Machinery—Track Widths First Edition. 3 pp.

—V Belts

BSI BS 3733-74. 1974 Endless V-Belt Drivers for Agricultural Purposes. 37 pp.

ISO 3410-89. Agricultural Machinery—Endless Variable-Speed V-Belts and Groove Sections of Corresponding Pulleys Second Edition. 8 pp.

ISO 5289-92. Agricultural Machinery—Endless Hexagonal Belts and Groove Sections of Corresponding Pulleys Second Edition. 8 pp.

—Vibration

BSI BS 6794-86. 1986 Reporting Measured Vibration Data for Land Vehicles. 14 pp.

ISO 8002-86. Mechanical Vibrations—Land Vehicles—Method for Reporting Measured Data First Edition. 12 pp.

—Wheels

BSI BS 3486: Part 1-62. 1962 Amd 1 Wheels for Agricultural Machinery, Implements, and Trailers Part 1: Wheel and Hub Centre Dimensions. 20 pp.

BSI BS 3486: Part 2-70. 1970 Wheels for Agricultural Machinery, Implements, and Trailers Part 2: Tyre and Rim Sizes. 8 pp.

BSI BS 3486: Part 3-70. 1970 Wheels for Agricultural Machinery, Implements, and Trailers Part 3: Nuts. 8 pp.

BSI BS AU 50:Pt 1:SUBSEC 4.4-89. (WITHDRAWN) 1989 Tyres and Wheels Part 1: Tyres Section 4: Agricultural Tractor and Machine Tyres Subsection 4.4: Existing (Inch) Series Diagonal Tyres for Log Skidders (Superseded by BS AU 50: Subsection 4.4a: 1992). 7 pp.

BSI BS AU 50:Pt 1:SUBSEC4.4A-92. 1992 Tyres and Wheels Part 1: Tyres Section 4: Agricultural Tractor and Machine Tyres Subsection 4.4a: Specification for Ply Rating Marked Series Tyres for Log Skidders (ISO 4251-5: 1992) (Superseding BS AU 50: Part 1: Subsection 4.4: 1989) (E). 7 pp.

BSI BS AU 50:Pt 1:SUBSEC 4.5-92. 1992 Tyres and Wheels Part 1: Tyres Section 4: Agricultural Tractor and Machine Tyres Subsection 4.5: Specification for Designations, Dimensions and Marking for Metric Series Tyres (ISO 7867-1: 1992) (E). 11 pp.

BSI BS AU 50:Pt4: SUBSEC 4.1-91. 1991 Tyres and Wheels Part 4: Rim Profiles and Dimensions Section 4: Agricultural Tractor and Machine Rims Subsection 4.1: Specification for Existing (Inch) Series Rims (ISO 4251-3: 1985). 16 pp.

ISO 4251 Pt 3-85. Tyres and Rims (Existing Series) for Agricultural Tractors and Machines—Part 3: Rims Second Edition. 13 pp.

ISO 4251 Pt 5-92. Tyres (Ply Rating Marked Series) and Rims for Agricultural Tractors and Machines—Part 5: Log Skidder Tyres Second Edition. 6 pp.

ISO 7867 Pt 1-92. Tyres and Rims (Metric Series) for Agricultural Tractors and Machines—Part 1: Tyre Designation, Dimensions and Marking Second Edition. 10 pp.

ISO 8016-85. Machinery for Agriculture—Wheels with Integral Hub First Edition. 5 pp.

JIS B 9203-71. Contours of Rims for Agricultural Machines and Implements. 12 pp.

SAA AS D22-71. Dimensions of Agricultural Tyres, Rims and Hubs Corrig.. 8 pp.

Agricultural Facilities

Use: Agricultural Buildings

Agricultural Products

Scope Note: For additional listings, use a more specific term *Use For:* Farm Crops; Farm Produce *See Also:* Agriculture; Biogas; Carob; Cereals; Coba; Dairy Products; Farms; Food; Forage Crops; Fruits; Meat; Nuts (Food); Oilseeds; Seeds; Sugarcane; Tobacco; Vegetable Oils; Vegetables

EC EEC/330/92-92. Council Regulation on Urgent Action for the Supply of Agricultural Products in Particular to the People of Moscow and St Petersburg. 2 pp.

—Classification

BSI BS 1000: (633/635)-85. 1985 Universal Decimal Classification (UDC). English Full Edition (633/635): Specific Crops. Horticulture. 32 pp.

SNZ NZS/BS 1000 (633/635)-85. Universal Decimal Classification Specific Crops. Horticulture. 32 pp.

—Organic

EC EEC/2092/91-91. Council Regulation on Organic Production of Agricultural Products and Indications Referring Thereto on Agricultural Products and Foodstuffs. 16 pp.

Agricultural Spraying Equipment

See Also: Agricultural Equipment; Sprayers

BSI BS 6356: Part 5-92. 1992 Spraying Equipment for Crop Protection Part 5: Specification for Tank Nominal Volume and Filling Hole Diameter (ISO 9357: 1990). 5 pp.

BSI BS 6550-85. 1985 Recommendation for Working Widths of Equipment for Sowing, Planting, Distributing Fertilizers and Spraying. 3 pp.

ISO 5682 Pt 2-86. Equipment for Crop Protection—Spraying Equipment—Part 2: Test Methods for Agricultural Sprayers First Edition. 7 pp.

ISO 6720-89. Agricultural Machinery—Equipment for Sowing, Planting, Distributing Fertilizers and Spraying—Recommended Working Widths Second Edition. 4 pp.

JIS B 9222-89. Standard Form of Specification for Air-Blast Sprayers. 10 pp.

—Ammonia—Safety

ISO 4254 Pt 2-86. Tractors and Machinery for Agriculture and Forestry—Technical Means for Providing Safety—Part 2: Anhydrous Ammonia Applicators First Edition. 6 pp.

—Glossaries

BSI BS 6355-83. 1983 Equipment for Crop Protection. 23 pp.

ISO 5681-92. Equipment for Crop Protection—Vocabulary Second Edition. 19 pp.

—Hoses

JIS K 6339-82. Spray Hoses for Agricultural Use. 7 pp.

—Hoses—Rubber

CNS K4054-80. Sprayer Rubber Hoses for Agricultural Purpose (Sep)(6491).

ISO 1401-87. Rubber Hoses for Agricultural Spraying First Edition. 4 pp.

—Layouts

BSI BS 6356: Part 7-93. 1993 Spraying Equipment for Crop Protection Part 7: Guide for Typical Data Sheet Layout (E). 20 pp.

ISO 10627 Pt 1-92. Agricultural Sprayers—Data Sheet—Part 1: Typical Layout First Edition. 18 pp.

—Nozzles

BSI BS 6356: Part 1-83. 1983 Spraying Equipment for Crop Protection Part 1: Method of Test for Sprayer Nozzles. 17 pp.

BSI BS 6356: Part 2-85. 1985 Spraying Equipment for Crop Protection Part 2: Connecting Dimensions for Nozzles and Manometers. 4 pp.

BSI BS 6356: Part 4-89. 1989 Spraying Equipment for Crop Protection Part 4: Performance Limits for Field Spraying Nozzles. 4 pp.

BSI BS 6356: Part 6-92. 1992 Spraying Equipment for Crop Protection Part 6: Specification for Connecting Dimensions for Nozzles with Bayonet Fixing (ISO 10626: 1991). 6 pp.

ISO 5682 Pt 1-81. Equipment for Crop Protection—Spraying Equipment—Part 1: Test Methods of Sprayer Nozzles First Edition. 16 pp.

ISO 8169-84. Equipment for Crop Protection—Sprayers—Connecting Dimensions for Nozzles and Manometers First Edition. 5 pp.

ISO 10626-91. Equipment for Crop Protection—Sprayers—Connecting Dimensions for Nozzles with Bayonet Fixing First Edition. 4 pp.

—Nozzles—Antidrip Devices

ISO 6686-81. Equipment for Crop Protection—Anti-Drip Devices—Determination of Reduction of Nozzle Flow Rate First Edition. 6 pp.

INTERNATIONAL AND NON-U.S. NATIONAL STANDARDS
SUBJECT INDEX
Air

Agricultural Spraying Equipment *(Cont.)*
—Safety
CEN PREN 907-92. Safety Requirements for Agricultural and Forestry Machinery—Sprayers and Liquid Fertilizer Distributors. 14 pp.

—Screw Threads
BSI BS 6356: Part 3-85. 1985 Spraying Equipment for Crop Protection Part 3: Dimensions of Connecting Threading. 3 pp.
ISO 4102-84. Equipment for Crop Protection—Sprayers—Connection Threading First Edition. 3 pp.

—Spraying Oils
SAA AS 1888-76. Agricultural Spraying Oils (Emulsifiable) Corrig. May 1977 Corrig. June 1978. 20 pp.

—Tanks (Containers)
BSI BS 6356: Part 5-92. 1992 Spraying Equipment for Crop Protection Part 5: Specification for Tank Nominal Volume and Filling Hole Diameter (ISO 9357: 1990). 5 pp.

—Tanks (Containers)—Capacities
ISO 9357-90. Equipment for Crop Protection—Agricultural Sprayers—Tank Nominal Volume and Filling Hole Diameter First Edition. 4 pp.

—Tanks (Containers)—Filler Necks
ISO 9357-90. Equipment for Crop Protection—Agricultural Sprayers—Tank Nominal Volume and Filling Hole Diameter First Edition. 4 pp.

—Tanks (Containers)—Level Indicators
ISO 9357-90. Equipment for Crop Protection—Agricultural Sprayers—Tank Nominal Volume and Filling Hole Diameter First Edition. 4 pp.

Agriculture
See Also: Agricultural Equipment; Agricultural Products

—Classification
BSI BS 1000: (63/632)-81. 1981 Universal Decimal Classification (UDC). English Full Edition (63/632): Agriculture in General. Forestry. Plant Injuries, Diseases and Pests. Plant Protection. 58 pp.
BSI BS 1000: (633/635)-85. 1985 Universal Decimal Classification (UDC). English Full Edition (633/635): Specific Crops. Horticulture. 32 pp.
BSI BS 1000: (636/639)-81. 1981 Universal Decimal Classification (UDC). English Full Edition (636/639): Animal Breeding. Animal Produce. Hunting. Fishing. 56 pp.
SNZ NZS/BS 1000 (63/632)-81. Universal Decimal Classification Agriculture in General. Forestry. Plant Injuries, Diseases and Pests. Plant Protection. 60 pp.
SNZ NZS/BS 1000 (633/635)-85. Universal Decimal Classification Specific Crops. Horticulture. 32 pp.
SNZ NZS/BS 1000 (636/639)-81. Universal Decimal Classification Animal Breeding. Animal Produce. Hunting. Fishing. 56 pp.

—European Communities
EURO 1990-1. Community's Agricultural Policy on the Threshold of the 1990's. 11 pp.

Agronomy
Scope Note: Use a more specific term *See:* Nursery Stock

Aguillettes
Use: Fourrageres

Agulhon's Reagent
—Acetone
ISO 757 Pt 5-82. Acetone for Industrial Use—Methods of Test—Part 5: Control Test with Agulhon's Reagent First Edition. 4 pp.

Agusta A 1 Helicopters
See Also: Helicopters

—Certification
CAA. Agusta A.109A AII and C Models. 7 pp.

Agusta A 109 Helicopters
See Also: Helicopters

—Foreign Airworthiness Directives
CAA. Agusta A109 and A109A Helicopters (Foreign Airworthiness Directives). 11 pp.

Agusta A 109A Helicopters
See Also: Helicopters

—Antenna Positions
CAA. Agusta A109A and 109C (Approved Aerial Positions). 1 p.

Agusta A 109A Helicopters *(Cont.)*
—Certification
CAA. Agusta A.109A AII and C Models. 7 pp.

—Data Sheets
CAA FR11 ISSUE 1. Agusta A109A. 2 pp.

—Foreign Airworthiness Directives
CAA. Agusta A109 and A109A Helicopters (Foreign Airworthiness Directives). 11 pp.

Agusta A 109C Helicopters
See Also: Helicopters

—Antenna Positions
CAA. Agusta A109A and 109C (Approved Aerial Positions). 1 p.

Agusta Bell 47 Helicopters
See Also: Helicopters

—Foreign Airworthiness Directives
CAA. Agusta Bell 47 Series (Foreign Airworthiness Directives). 17 pp.

Agusta Bell 206 Helicopters
See Also: Helicopters

—Foreign Airworthiness Directives
CAA. Agusta Bell 206 Series Helicopters (Foreign Airworthiness Directives). 33 pp.

Agusta Bell 212 Helicopters
See Also: Helicopters

—Accidents
CAA. Bell/Augusta Bell 212 Series (World Helicopter Accident Summary). 3 pp.

Ailerons
See Also: Control Surfaces

—Gliders
CAA JAR-22 SUBPART C. Structure (Joint Airworthiness Requirements). 17 pp.

Air
See Also: Air Monitoring; Compressed Air; Gases

—Aircraft—Replenishment—Connectors
NATO STANAG 3806 ED 1 AMD 5-79. Aircraft Gaseous Air/Nitrogen Systems Replenishment Connectors. 6 pp.

—Aromatic Hydrocarbon Content—Gas Chromatography
BSI BS 6069: Sec 3.4-91. 1991 Characterization of Air Quality Part 3: Workplace Atmospheres Section 3.4: Method for the Determination of Vaporous Aromatic Hydrocarbons by Charcoal Tube/Solvent Desorption/ Gas Chromatography (ISO 9487: 1991). 18 pp.
ISO 9487-91. Workplace Air—Determination of Vaporous Aromatic Hydrocarbons—Charcoal Tube/Solvent Desorption/ Gas Chromatographic Method First Edition. 15 pp.

—Carbon Monoxide Content
CSA Z223.21-M1978. Method for the Measurement of Carbon Monoxide. 27 pp.
ISO 8760-90. Work-Place Air—Determination of Mass Concentration of Carbon Monoxide—Method Using Detector Tubes for Short-Term Sampling with Direct Indication First Edition. 10 pp.
SAA AS 3580.7.1-92. Methods for Sampling and Analysis of Ambient Air—Part 7: Determination of Carbon Monoxide—Part 7.1: Direct-Reading Instrumental Method. 4 pp.

—Halogenated Hydrocarbon Content—Gas Chromatography
BSI BS 6069: Sec 3.3-91. 1991 Characterization of Air Quality Part 3: Workplace Atmospheres Section 3.3: Method for the Determination of Vaporous Chlorinated Hydrocarbons by Charcoal Tube/Solvent Desorption/ Gas Chromatography (ISO 9486: 1991). 18 pp.
ISO 9486-91. Workplace Air—Determination of Vaporous Chlorinated Hydrocarbons—Charcoal Tube/Solvent Desorption/ Gas Chromatographic Method First Edition. 15 pp.

—Lead Content—Atomic Absorption Spectrometry
BSI BS 6069: Sec 3.2-91. 1991 Characterization of Air Quality Part 3: Workplace Atmospheres Section 3.2: Method for the Determination of Particulate Lead and Lead Compounds by Flame Atomic Absorption Spectrometry (ISO 8518: 1990). 15 pp.
ISO 8518-90. Workplace Air—Determination of Particulate Lead and Lead Compounds—Flame Atomic Absorption Spectrometric Method First Edition. 13 pp.

Air *(Cont.)*
—Microorganisms Content
SAA AS 1095.4. Microbiological Methods for the Dairy Industry—Part 4: Methods for the Examination of Water and Air (Subset Including the Series 1095.4.1.1—1095.4.2).
SAA AS 1095.4.2-80. Microbiological Methods for the Dairy Industry —Part 4: Methods for the Examination of Water and Air—Part 4.2: Microbiological Examination of Air. 4 pp.

—Ozone Content—Chemiluminescence
BSI BS 1747: Part 12-93. 1993 Methods for Measurement of Air Pollution Part 12: Determination of Mass the Concentration of Ozone in Ambient Air: Chemiluminescence Method (ISO 10313: 1993) (N). 14 pp.
CSA Z223.23-M1981. Method for the Measurement of Ozone in Air; (Gen Instr 1). 17 pp.
ISO 10313-93. Ambient Air—Determination of the Mass Concentration of Ozone—Chemiluminescence Method First Edition. 12 pp.

Air Ambulances
Use: Air Medical Transport Units

Air Attack
See Also: Warfare

—Warning Systems
NATO STANAG 2047 ED 6 AMD 2-85. Emergency Alarms of Hazard or Attack (NBC and Air Attack Only). 11 pp.

Air Batteries
See Also: Air Cells; Batteries
CNS C3087-80. Method of Test for Air Dry Batteries (Aug)(6031).
CNS C4204-80. Air Dry Batteries (Aug)(6030).

Air Brakes
Use For: Pneumatic Brakes *See Also:* Automotive Equipment; Brake Pipes; Brakes; Disk Brakes; Drum Brakes; Master Cylinders; Pneumatic Equipment

—Adjusters
CNS D2163-86. Air Brake Slack Adjusters (Oct)(9848).

—Aircraft—Control Knobs
SBAC AS 6384 ISSUE 2. Knob (Airbrake Control).

—Automotive
CNS D3109-86. Method of Test for Air Brakes of Automobiles (Oct)(8566).
CNS D3110-86. Air Brake Performance Requirements for Automobiles (Oct)(8567).
CNS D3156-83. Air Servo and Vacuum Servo Brake Performance Requirements (Nov)(10652).
CNS D3157-83. Air Servo and Vacuum Servo Brake Test Procedure (Nov)(10653).
ISO 1728-80. Road Vehicles—Pneumatic Braking Connections Between Motor Vehicles and Towed Vehicles—Interchangeability Second Edition. 8 pp.
ISO 6597-91. Road Vehicles—Hydraulic Braking Systems—Measurement of Braking Performance Second Edition. 23 pp.
ISO 7375 Pt 1-86. Road Vehicles—Coiled Pipe Assemblies for Pneumatic Braking Connection Between Motor Vehicles and Towed Vehicles—Part 1: Dimensions Second Edition. 4 pp.
JIS D 2606-80. Air Brake Hoses for Automobiles (R 1985). 12 pp.
JIS D 2620-86. Identification of Connections on Air Brake Units for Automobiles. 7 pp.

—Automotive—Connectors—Identification Systems
JIS D 2620-86. Identification of Connections on Air Brake Units for Automobiles. 7 pp.

—Automotive—Couplings
MOD UK DSTAN 25-1-93. Vehicle Braking Systems, Hose Assemblies and Air Brake Coupling Adaptors, Brake, Intervehicular Issue 2. 10 pp.

—Automotive—Hose Assemblies
MOD UK DSTAN 25-1-93. Vehicle Braking Systems, Hose Assemblies and Air Brake Coupling Adaptors, Brake, Intervehicular Issue 2. 10 pp.
MOD UK DSTAN 25-1: Part 2-75. Vehicle Braking Systems Part 2: Hose Assemblies Brake Intervehicular Issue 1 (Superseded by Def Stan 25-1). 4 pp.
SNZ NZS/SAEJ1402-85. Automotive Air Brake Hose and Hose Assemblies. 4 pp.

—Automotive—Hoses
CNS D2162-86. Air Brake Hoses for Automobiles (Oct)(9847).
CNS D2172-86. Nylon Tubes for Air Brake Piping of Automobiles (Oct)(10332).

INDUSTRY STANDARDS

INTERNATIONAL AND NON-U.S. NATIONAL STANDARDS
SUBJECT INDEX

Air Brakes (Cont.)
—Automotive—Hoses (Cont.)
CNS D2174-86. Vacuum Brake Hose for Automobiles (Oct)(10709).
JIS D 2606-80. Air Brake Hoses for Automobiles (R 1985). 12 pp.
SNZ NZS/SAEJ1402-85. Automotive Air Brake Hose and Hose Assemblies. 4 pp.

—Automotive—Hoses—Nylon
CNS D2172-86. Nylon Tubes for Air Brake Piping of Automobiles (Oct)(10332).

—Automotive—Pipe Fittings
ISO 4039-77. Road Vehicles—Pneumatic Braking Systems—Pipes, Tapped Holes and Male Fittings First Edition. 3 pp.

—Automotive—Pipe Fittings—Identification Systems
ISO 6786-80. Road Vehicles—Air Braking Systems—Identification of Connections on Units First Edition. 3 pp.

—Automotive—Pipes
ISO 4039-77. Road Vehicles—Pneumatic Braking Systems—Pipes, Tapped Holes and Male Fittings First Edition. 3 pp.

—Automotive—Plastic Pipes
ISO 7628 Pt 1-85. Road Vehicles—Thermoplastic Tubing for Use in Air Braking Systems—Part 1: Dimensions and Marking First Edition. 3 pp.

—Automotive—Plastic Pipes—Identification Systems
ISO 7628 Pt 1-85. Road Vehicles—Thermoplastic Tubing for Use in Air Braking Systems—Part 1: Dimensions and Marking First Edition. 3 pp.

—Automotive—Plastic Pipes—Installation
ISO TR7628 Pt 2-86. Road Vehicles—Thermoplastics Tubing for Use in Air Braking Systems—Part 2: Installation on the Vehicle and Test Methods First Edition. 11 pp.

—Automotive—Pressure Taps
BSI BS AU 163A-85. 1985 Pressure Test Connection for Compressed-Air Pneumatic Braking Equipment for Road Vehicles. 4 pp.
ISO 3583-84. Road Vehicles—Pressure Test Connection for Compressed-Air Pneumatic Braking Equipment Third Edition. 4 pp.

—Automotive—Pressure Vessels
BSI BS EN 286: Part 2-92. 1992 Simple Unfired Pressure Vessels Designed to Contain Air or Nitrogen Part 2: Specification for Pressure Vessels for Air Braking and Auxiliary Systems for Motor Vehicles and Their Trailers (Q). 48 pp.
BSI BS EN 286: Part 2-01. 1992 AMD 1 Simple Unfired Pressure Vessels Designed to Contain Air or Nitrogen Part 2: Specification for Pressure Vessels for Air Braking and Auxiliary Systems for Motor Vehicles and Their Trailers (AMD 7570) December 15, 1992. 49 pp.
CEN PREN 286 (Part 2)-89. Simple Unifired Pressure Vessels for Air Braking and Auxillary Systems for Motor-Vehicles Trailers. 60 pp.
CEN EN 286-2-92. Simple Unfired Pressure Vessels Designed to Contain Air or Nitrogen: Part 2: Pressure Vessels for Air Braking and Auxiliary Systems for Motor Vehicles and Their Trailers. 42 pp.
CEN EN 286-2-92. Simple Unfired Pressure Vessels Designed to Contain Air or Nitrogen—Part 2: Pressure Vessels for Air Braking and Auxiliary Systems for Motor Vehicles and Their Trailers. 67 pp.
CEN EN 286-2 AC-92. CORRIGENDUM Simple Unfired Pressure Vessels Designated to Contain Air or Nitrogen—Part 2: Pressure Vessels for Air Braking and Auxiliary Systems for Motor Vehicles and Their Trailers. 2 pp.

—Automotive—Tapped Holes
ISO 4039-77. Road Vehicles—Pneumatic Braking Systems—Pipes, Tapped Holes and Male Fittings First Edition. 3 pp.

—Hose Couplings—Prime Movers
SAA AS D8-71. Hose Couplings for Use with Vacuum and Air-Pressure Braking Systems on Prime Movers,Trailers and Semitrailers. 22 pp.

—Hose Couplings—Semitrailers
SAA AS D8-71. Hose Couplings for Use with Vacuum and Air-Pressure Braking Systems on Prime Movers,Trailers and Semitrailers. 22 pp.

—Hose Couplings—Trailers
SAA AS D8-71. Hose Couplings for Use with Vacuum and Air-Pressure Braking Systems on Prime Movers,Trailers and Semitrailers. 22 pp.

Air Brakes (Cont.)
—Hoses
DIN ENGL 74310 Pt 1-76. Air Braking Systems; Hoses; Dimensions; Material; Marking (Aug). 2 pp.
DIN ENGL 74310 Pt 2-76. Air Braking Systems; Hoses; Requirements; Tests (Aug). 3 pp.

—Hoses—Rubber
BSI BS AU 110-65. 1965 Amd 1 Air Pressure Rubber Brake Hose. 5 pp.

—Military Vehicles
MOD UK DSTAN 25-19-01. Military Vehicle Two (2) —Line Air Pressure Braking Systems Issue 1; Amendment 1. 54 pp.

—Overflow Valves
DIN ENGL 74279-77. Compressed-Air Braking Systems; Overflow Valves (Sept). 2 pp.

—Rolling Stock—Couplings
BSI BS 3710-64. (OBSOLESCENT) 1964 Brake Hose Couplings for Locomotives and Rolling Stock. 29 pp.

—Rolling Stock—Hoses
CNS E1013-82. Air Brake Hose for Railway Rolling Stock (Dec)(9694).
JIS E 4305-59. Air Brake Hose for Rolling Stock.

—Towed/Towing Vehicles
ISO 1728-80. Road Vehicles—Pneumatic Braking Connections Between Motor Vehicles and Towed Vehicles—Interchangeability Second Edition. 8 pp.
ISO 7375 Pt 1-86. Road Vehicles—Coiled Pipe Assemblies for Pneumatic Braking Connection Between Motor Vehicles and Towed Vehicles—Part 1: Dimensions Second Edition. 4 pp.
MOD UK DSTAN 25-1: Part 1-75. Vehicle Braking Systems Part 1: Air Pressure Braking Systems for Towing Vehicles, Trailers and Other Towed Equipment Issue 1 (Withdrawn). 4 pp.

—Towed/Towing Vehicles—Couplings
BSI BS AU 138A-80. 1980 Dimensions of Contact Type Couplings for Air Pressure Braking Systems on Trailers and Semi-Trailers and Their Towing Vehicles, and the Arrangement of These Couplings on Articulated and Drawbar Combinations. 10 pp.

Air Break Switches
Use: Air Switches

Air Bricks
See Also: Bricks; Ventilation Equipment
BSI BS 493-70. 1970 Amd 2 Airbricks and Gratings for Wall Ventilation. 20 pp.

Air Capacitors
Use For: Air Condensers; Air Dielectric Capacitors; Air Dielectric Condensers See Also: Capacitors; Variable Capacitors
IEC 334 Pt 1-70. Air Dielectric Rotary Variable Capacitors Part 1: General Requirements for Tests and Measuring Methods First Edition; (Supplement A-1974). 59 pp.
MOD UK DEF-5141-61. Capacitors, Variable, Air Dielectric (Reprinted March 1970, Incorporating Amendments Nos. 1, 2 and 3). 26 pp.

—Precision
MOD UK DEF-5143-60. Capacitors, Variable, Precision, Air Dielectric (Reprinted April 1965, Incorporating Amendment Nos. 1 & 2). 19 pp.

—Radio Receivers
CNS C7003-77. Variable Air Capacitors for Radio Receivers (Jul)(1263).

—Trimmer—Rectangular—Preferred Products List
CECC CECC MUAHAG Vol 1 IS 4-90. Preferred Products List; Capacitors (En, Fr, Ge). 64 pp.

—Trimmer—Tubular—Microwave—Preferred Products List
CECC CECC MUAHAG Vol 1 IS 4-90. Preferred Products List; Capacitors (En, Fr, Ge). 64 pp.

—Trimmer—Tubular Miniature—Microwave—Preferred Products List
CECC CECC MUAHAG Vol 1 IS 4-90. Preferred Products List; Capacitors (En, Fr, Ge). 64 pp.

—Tuning
MOD UK DSTAN 59-70: 90/001-01. Capacitors, Variable, Air Dielectric Tuning Type A Issue 1; Amendment 1. 15 pp.

Air Cargo Containers
Use: Shipping Containers—Air Cargo

Air Cargo Handling Equipment
Use For: Aircraft Loading Equipment See Also: Air Cargo Pallets; Air Cargo Transportation; Materials Handling Equipment; Shipping
ISO 6966-82. Aircraft—Basic Requirements for Aircraft Loading Equipment First Edition. 7 pp.

—Adapters (Fittings)—Compatibility
NATO STANAG 3616 ED 1 AMD 5-72. Responsibility for the Design and Provision of Adaptors Necessary for the Compatibility of Air Cargo Loading, Securing, Unloading or Dropping Systems in Fixed Wing Aircraft. 6 pp.

—Design—Adapters (Fittings)
NATO STANAG 3616 ED 1 AMD 4-72. Responsibility for the Design and Provision of Adaptors Necessary for the Compatibility of Air Cargo Loading, Securing, Unloading or Dropping Systems in Fixed Wing Aircraft. 5 pp.

—Transport Aircraft—Information Interchange
NATO STANAG 3543 ED 3 AMD 3-86. Air Transport Cargo/Passenger Handling Systems—Request for Information. 14 pp.
NATO STANAG 3543 ED 3 AMD 4-86. Air Transport Cargo/Passenger Handling Systems—Request for Information. 17 pp.

—Unit Loads
BSI M 68-92. 1992 Air Cargo Unit Load Devices for Transportation of Horses (ISO 9469: 1991). 10 pp.
ISO 9469-91. Air Cargo Equipment—Unit Load Devices for Transportation of Horses First Edition. 9 pp.

Air Cargo Pallets
See Also: Air Cargo Handling Equipment; Pallets
BSI M 69-93. 1993 Air Cargo Equipment—Air and Air/Land Cargo Pallets—Specification and Testing (ISO 4117: 1993) (S). 29 pp.
ISO 4117-93. Air and Air/Land Cargo Pallets—Specification and Testing Second Edition. 28 pp.
ISO 4171-93. Air Cargo Equipment—Interline Pallets Second Edition. 11 pp.
NATO STANAG 3774 ED 3 AMD 0-87. Control Procedures for Pallets and Associated Restraint Equipment Used in Combined Air Transport Operations. 5 pp.
NATO STANAG 3774 ED 3 AMD 3-87. Control Procedures for Pallets and Associated Restraint Equipment Used in Combined Air Transport Operations. 10 pp.

—Nets
ISO 4115-87. Air Cargo Equipment—Air/Land Pallet Nets Second Edition. 10 pp.
ISO 4170-87. Air Cargo Equipment—Interline Pallet Nets Second Edition. 8 pp.

—Unit Loads
ISO 8097-93. Aircraft—Minimum Airworthiness Requirements and Test Conditions for Certified Air Cargo Unit Load Devices Second Edition. 5 pp.
NATO STANAG 3467 ED 1 AMD 7-74. Characteristics of Air Transport (Airlanded) Pallets for Carriage Internally. 9 pp.

—Unit Loads—Tie Downs—Fabrics—Environmental Testing
ISO TR8647-90. Environmental Degradation of Textiles Used in Air Cargo Restraint Equipment First Edition. 21 pp.

Air Cargo Transportation
See Also: Air Cargo Handling Equipment; Air Transportation; Aircraft; Materials Handling Equipment; Shipping

—Automotive—Shipping Containers
ISO 8268-87. Air Cargo Equipment—Automobile Transport Devices—Basic Requirements First Edition. 15 pp.

—Cargo Claims—Tracing
NATO STANAG 3740 ED 2 AMD 3-86. Procedures for Tracing Air Transported Baggage and Missing Air Transported Cargo. 18 pp.
NATO STANAG 3740 ED 2 AMD 4-86. Procedures for Tracing Air Transported Baggage and Missing Air Transported Cargo. 19 pp.
NATO STANAG 3740 ED 3 AMD 0-93. Procedures for Tracing Air Transported Baggage and Missing Air Transported Cargo. 24 pp.

—Cargo Nets
ISO 4115-87. Air Cargo Equipment—Air/Land Pallet Nets Second Edition. 10 pp.
ISO 4170-87. Air Cargo Equipment—Interline Pallet Nets Second Edition. 8 pp.
ISO 8097-93. Aircraft—Minimum Airworthiness Requirements and Test Conditions for Certified Air Cargo Unit Load Devices Second Edition. 5 pp.

INTERNATIONAL AND NON-U.S. NATIONAL STANDARDS
SUBJECT INDEX

Air Cargo Transportation (Cont.)
—Cargo Nets (Cont.)
MOD UK DSTAN 16-9-69. Nets and Slings, Cargo, Helicopter (Section on Slings is Superseded) Issue 1. 5 pp.
MOD UK DSTAN 16-9: Addendum. Nets and Slings, Helicopter Issue 1. 3 pp.

—Floors—Fittings
MOD UK DSTAN 16-23-82. Floor Fittings for Cargo Transport Aircraft Apertures for Tie-Down Points Issue 2. 6 pp.

—Ground Vehicles—Unit Loads
ISO 7715-85. Air Cargo Equipment—Ground Handling and Transport Systems for Unit Load Devices—Minimum Requirements First Edition. 10 pp.
ISO 7716-85. Air Cargo Equipment—Unit Load Devices Transport Vehicle (UTV)—Functional Requirements First Edition. 9 pp.

—Hazardous Materials
CAA Section 9 CAP 393. Air Navigation (Dangerous Goods) Regulations 1985. 6 pp.
ICAO Annex 18. Safe Transport of Dangerous Goods by Air Second Edition—July 1989; (Supplement 8/1/90). 40 pp.
ICAO 9284 Supplement. Technical Instructions for the Safe Transport of Dangerous Goods by Air—1986 Edition. 117 pp.
ICAO 9375 Book 1. Dangerous Goods Training Programme Book 1 Shippers and Packers Second Edition—October 1985. 102 pp.
ICAO 9375 Book 2. Dangerous Goods Training Programme Book 2 Cargo Agents Second Edition—October 1985. 91 pp.
ICAO 9375 Book 3. Dangerous Goods Training Programme Book 3 Operator's Cargo Acceptance Staff Second Edition—October 1985. 89 pp.
ICAO 9375 Book 4. Dangerous Goods Training Programme Book 4 Load Planners and Flight Crew Second Edition—January 1986. 36 pp.
ICAO 9375 Book 5. Dangerous Goods Training Programme Book 5 Passenger Handling Staff and Flight Attendants Second Edition—January 1986. 19 pp.
ICAO 9375 Book 6. Dangerous Goods Training Programme Book 6 Loading and Warehouse Personnel Second Edition—January 1986. 35 pp.
NATO STANAG 3854 ED 2 AMD 2-88. Policies and Procedures Governing the Air Transportation of Dangerous Cargo. 29 pp.
NATO STANAG 3854 ED 2 AMD 3-88. Policies and Procedures Governing the Air Transportation of Dangerous Cargo. 31 pp.
NATO STANAG 3854 ED 2 AMD 4-88. Policies and Procedures Governing the Air Transportation of Dangerous Cargo. 32 pp.

—Pallets
BSI M 69-93. 1993 Air Cargo Equipment—Air and Air/Land Cargo Pallets—Specification and Testing (ISO 4117: 1993) (S). 29 pp.
ISO 4117-93. Air and Air/Land Cargo Pallets—Specification and Testing Second Edition. 28 pp.
ISO 4171-93. Air Cargo Equipment—Interline Pallets Second Edition. 11 pp.
ISO 8097-93. Aircraft—Minimum Airworthiness Requirements and Test Conditions for Certified Air Cargo Unit Load Devices Second Edition. 5 pp.
NATO STANAG 3774 ED 3 AMD 0-87. Control Procedures for Pallets and Associated Restraint Equipment Used in Combined Air Transport Operations. 5 pp.
NATO STANAG 3774 ED 3 AMD 3-87. Control Procedures for Pallets and Associated Restraint Equipment Used in Combined Air Transport Operations. 10 pp.

—Rails
BSI 2C 8-88. 1988 Rail and Stud Configuration for Aircraft Passenger Equipment and Cargo Restraint. 5 pp.
ISO 7166-85. Aircraft—Rail and Stud Configuration for Passenger Equipment and Cargo Restraint First Edition. 5 pp.
ISO 9788-90. Air Cargo Equipment—Cast Components of Double Stud Fitting Assembly with a Load Capacity of 22 250 N (5 000 lbf), for Aircraft Cargo Restraint First Edition. 8 pp.

—Refrigerated
NATO STANAG 3775 ED 3 AMD 2-87. Air Movement and in-Transit Storage of Cargo Requiring Controlled Temperature Conditions. 15 pp.
NATO STANAG 3775 ED 3 AMD 3-87. Air Movement and In-Transit Storage of Cargo Requiring Controlled Temperature Conditions. 16 pp.

Air Cargo Transportation (Cont.)
—Shipping Containers
BSI M 70-93. 1993 Air Cargo Equipment—Base-Restrained Certified Containers Exclusively for the Lower Deck of High-Capacity Aircraft (ISO 6517: 1992) (S). 17 pp.
CAA NOTICE #92 ISSUE 1. (This Notice Gives Details of a Mandatory Action) Cargo Containment (Airworthiness Notices). 3 pp.
ISO 4118-80. Non-Certified Lower Deck Containers for Air Transport—Specification and Testing First Edition. 12 pp.
ISO 4128-85. Aircraft—Air Mode Modular Containers First Edition. 19 pp.
ISO 6517-92. Air Cargo Equipment—Base-Restrained Certified Containers Exclusively for the Lower Deck of High-Capacity Aircraft Second Edition. 17 pp.

—Thermal Containers
ISO 8058-85. Air Cargo Equipment—Air Mode Insulated Containers—Thermal Efficiency Requirements First Edition. 9 pp.

—Tie Downs
NATO STANAG 3548 ED 2 AMD 6-78. Tie-Down Fittings on Air Transport and Airdropped Equipment and Cargo Carried Internally by Fixed Wing Aircraft. 9 pp.
NATO STANAG 3548 ED 2 AMD 7-78. Tie-Down Fittings on Air Transported and Airdropped Equipment and Cargo Carried Internally by Fixed Wing Aircraft. 10 pp.

—Unit Loads
ISO 4116-86. Air Cargo Equipment—Ground Equipment Requirements for Compatibility with Aircraft Unit Load Devices Second Edition. 6 pp.
ISO 6965-82. Aircraft—Self-Propelled Gantry for Lifting Air Cargo Containers and Outside Cargoes—Functional Requirements First Edition. 7 pp.
ISO 6966-82. Aircraft—Basic Requirements for Aircraft Loading Equipment First Edition. 7 pp.
ISO 6967-83. Aircraft—Wide Body Aircraft Main Deck Container/Pallet Loader—Functional Requirements First Edition. 5 pp.
ISO 6968-83. Aircraft—Wide Body Aircraft Lower Deck Container/Pallet Loader—Functional Requirements First Edition. 5 pp.
ISO 7715-85. Air Cargo Equipment—Ground Handling and Transport Systems for Unit Load Devices—Minimum Requirements First Edition. 10 pp.
ISO 8097-93. Aircraft—Minimum Airworthiness Requirements and Test Conditions for Certified Air Cargo Unit Load Devices Second Edition. 5 pp.

—Unit Loads—Symbols
BSI M 67-88. 1988 Symbols for the Pictorial Representation of Aircraft Cargo Handling Systems for Unit Load Devices (ULDs). 7 pp.
ISO 9031-87. Air Cargo Equipment—Handling Systems for Unit Load Devices (ULDs)—Symbols for Pictorial Representation First Edition. 8 pp.

—Unit Loads—Tie Downs—Fabrics—Environmental Testing
ISO TR8647-90. Environmental Degradation of Textiles Used in Air Cargo Restraint Equipment First Edition. 21 pp.

Air Cells
See Also: Air Batteries; Cells (Electric)
CNS C4004-82. Air Depolarized Primary Cell (or Air Cell) (Dec)(419).

Air Circuit Breakers
See Also: Circuit Breakers
CNS C1106-85. Minimum Breaking Distance for Air Break Switches and Air Circuit Breakers (Under or Equal to AC 600V) (Jun)(9356).
JIS C 0921-84. Minimum Breaking Distance for Air Break Switches and Air Circuit Breakers (Under or Equal to AC 600 V). 5 pp.

— 41-100 kV
CSA CAN/CSA-C22. 2 NO 31-M89. Switchgear Assemblies; (Gen Instr 1 Thru 3). 54 pp.
CSA 1169 Bull. Electrical Bulletin 1169 June 27, 1978 to C22.2 NO 31. 2 pp.

Air Cleaners
Use For: Air Cleaning Equipment See Also: Air Filters; Dust Collectors; Electrostatic Precipitators; Exhaust Systems; Gas Cleaning Equipment; Heating, Ventilating and Air Conditioning Equipment; Ventilation Equipment
BSI BS 5295: Part 1-89. 1989 Environmental Cleanliness in Enclosed Spaces Part 1: Specification for Clean Rooms and Clean Air Devices (N). 15 pp.
CNS C3121-81. Method of Test of Air Cleaners (Jul)(7620).
CNS C4323-81. Air Cleaners (Jul)(7619).

Air Cleaners (Cont.)
JIS C 9615-76. Air Cleaners. 20 pp.

—Aerosols
JIS Z 8901-84. Dusts and Aerosols for Industrial Testing. 21 pp.

—Automotive
CNS D3138-82. Method of Test of Air Cleaners for Automobiles (Sep)(9372).

—Clean Rooms
BSI BS 5295: Part 1-89. 1989 Environmental Cleanliness in Enclosed Spaces Part 1: Specification for Clean Rooms and Clean Air Devices (N). 15 pp.
BSI BS 5295: Part 1-01. 1989 Amd 1 Environmental Cleanliness in Enclosed Spaces Part 1: Clean Rooms and Clean Air Devices (AMD 6602) December 21, 1990. 17 pp.
BSI BS 5295: Part 2-89. 1989 Environmental Cleanliness in Enclosed Spaces Part 2: Method for Specifying the Design, Construction and Commissioning of Clean Rooms and Clean Air Devices (N). 14 pp.
BSI BS 5295: Part 2-01. 1989 Amd 1 Environmental Cleanliness in Enclosed Spaces Part 2: Method for Specifying the Design, Construction and Commissioning of Clean Rooms and Clean Air Devices (AMD 6603) December 21, 1990. 15 pp.
BSI BS 5295: Part 3-89. 1989 Environmental Cleanliness in Enclosed Spaces Part 3: Guide to Operational Procedures and Disciplines Applicable to Clean Rooms and Clean Air Devices. 8 pp.

—Dust
JIS Z 8901-84. Dusts and Aerosols for Industrial Testing. 21 pp.

—Glossaries
BSI BS 6202-82. 1982 Cleaning Equipment for Air or Other Gases. 10 pp.
ISO 3649-80. Cleaning Equipment for Air or Other Gases—Vocabulary First Edition. 10 pp.

—Infrared Detectors
MOD UK DSTAN 58-96-92. Pure Air Systems for Detector Cooling Applications Issue 2. 17 pp.
MOD UK DSTAN 81-91-93. High Pressure Pure Air Equipment for Detector Cooling Applications Issue 1. 34 pp.

—Nuclear Power Plants
CSA CAN3-N288 .3.2-M85. High Efficiency Air-Cleaning Assemblies for Normal Operation of Nuclear Facilities (R 1992). 38 pp.

—Safety
DIN VDE 0700 Pt 259-88. Safety of Household and Similar Electrical Appliances; Air Cleaning Appliances (July). 16 pp.
IEC 335 Pt 2-65-93. Safety of Household and Similar Electrical Appliances Part 2: Particular Requirements for Air-Cleaning Appliances First Edition. 29 pp.

—Ships
MOD UK NES 316-90. Requirements for Compressed Air Purification Equipment Issue 2 (07.90). 23 pp.

—Thermal Imaging Systems
MOD UK DSTAN 58-96-87. Pure Air Systems for Thermal Imaging Equipment Issue 1. 16 pp.

Air Cleaning Equipment
Use: Air Cleaners

Air Compressors
See Also: Compressors (Pressure Equipment); Reciprocating Compressors; Rotary Compressors
CNS B7044-69. Method of Test for Air Compressors (Jul)(3038).

—Acceptance Testing
BSI BS 1571: Part 2-75. 1975 Amd 3 Testing of Positive Displacement Compressors and Exhausters Part 2: Methods for Simplified Acceptance Testing for Air Compressors and Exhausters (AMD 5576) August 28, 1987. 22 pp.
BSI BS 1571: Part 2-04. 1975 Amd 4 Testing of Positive Displacement Compressors and Exhausters Part 2: Methods for Simplified Acceptance Testing for Air Compressors and Exhausters (AMD 6879) February 28, 1992. 23 pp.
SNZ NZS/BS 1571: Part 2-75. Testing of Positive Displacement Compressors and Exhausters Part 2: Simplified Acceptance Tests for Air Compressors and Exhausters Amend: 3; 4, 1992. 24 pp.

—Automotive—Commercial
ISO 8719-86. Commercial Vehicles and Buses—Four-Hole Flanges for Gear-Driven Air Compressors First Edition. 5 pp.

INDUSTRY STANDARDS

INTERNATIONAL AND NON-U.S. NATIONAL STANDARDS
SUBJECT INDEX

Air Compressors (Cont.)

—Automotive—Commercial—Flanges
BSI BS AU 211-87. 1987 Dimensions of Four-Hole Flanges for Gear-Driven Air Compressors for Commercial Vehicles and Buses. 5 pp.

—Automotive—Mounting
ISO 7652-83. Road Vehicles—Base-Mounted Air Compressors, Single Cylinder V-Belt Drive—Mounting Dimensions First Edition; (Amendment 1-1992). 9 pp.

—Buses (Vehicles)—Flanges
BSI BS AU 211-87. 1987 Dimensions of Four-Hole Flanges for Gear-Driven Air Compressors for Commercial Vehicles and Buses. 5 pp.
ISO 8719-86. Commercial Vehicles and Buses—Four-Hole Flanges for Gear-Driven Air Compressors First Edition. 5 pp.

—Construction Equipment
MOD UK DEF-1017-A-65. Air Compressors for Construction Purposes. 22 pp.

—Electric
CSA CAN/CSA-C22. 2 NO 68-92. Motor-Operated Appliances (Household and Commercial); (Gen Instr 1 Thru 2). 115 pp.
CSA 1169 Bull. Electrical Bulletin 1169 June 27, 1978 to C22.2 NO 68. 2 pp.

—Lubricants
BSI BS 6413:Pt3: Sec 3.1-88. 1988 Lubricants, Industrial Oils and Related Products (Class L) Part 3: Classification for Family D (Compressors) Section 3.1: Air Compressors. 7 pp.
BSI BS 6413:Pt3: Sec 3.2-89. 1989 Lubricants, Industrial Oils and Related Products (Class L) Part 3: Classification for Family D (Compressors) Section 3.2: Gas and Refrigeration Compressors. 7 pp.
CNS K5049-88. Compressor Oil (Jan)(2974). 2 pp.
DIN ENGL 51506-85. Lubricants; VB and VC Lubricating Oils with and Without Additives and VDL Lubricating Oils; Classification and Requirements (Sept). 3 pp.
ISO 6743 Pt 3A-87. Lubricants, Industrial Oils and Related Products (Class L)—Classification—Part 3A: Family D (Compressors) First Edition. 7 pp.
ISO 6743 Pt 3B-88. Lubricants, Industrial Oils and Related Products (Class L)—Classification—Part 3B: Family D (Gas and Refrigeration Compressors) First Edition. 6 pp.
MOD UK DSTAN 91-42-78. Lubricating Oil, Petroleum: Compressor, Light Joint Service Designation: OM-58 Lubricating Oil, Petroleum: Compressor, Medium Joint Service Designation: OM-160 Issue 1. 6 pp.

—Maintenance
MOD UK DL-16. Air Compressor Plant for Servicing and Workshop Purposes.

—Recreational Vehicles—Electrical Codes
CSA 1020B Bull. Electrical Bulletin 1020B January 28, 1977 to C22.2 NO 68. 3 pp.

—Rolling Stock
CNS E2008-87. Method of Test for Air Compressors for Railway Rolling Stock (Jan)(11813).
JIS E 5002-91. Air Compressors for Railway Rolling Stock—Test Methods.

—Safety
BSI BS 6244-82. 1982 Stationary Air Compressors. 24 pp.
ISO 5388-81. Stationary Air Compressors—Safety Rules and Code of Practice First Edition. 24 pp.

—Ships—Design
MOD UK NES 315-87. Air Compressors Issue 2 (07.87). 56 pp.

—Ships—Installation
MOD UK NES 315-87. Air Compressors Issue 2 (07.87). 56 pp.

—Submarines—Design
MOD UK NES 315-87. Air Compressors Issue 2 (07.87). 56 pp.

—Submarines—Installation
MOD UK NES 315-87. Air Compressors Issue 2 (07.87). 56 pp.

Air Condensers
Use: Air Capacitors

Air Conditioners
Use For: Central Air Conditioners *See Also:* Air Conditioning Equipment; Air Coolers; Cooling Equipment; Cooling Systems; Environmental Engineering; Heat Pumps; Heating, Ventilating and Air Conditioning Equipment; Limit Switches; Refrigeration Equipment; Unit Air Conditioners

Air Conditioners (Cont.)

CSA CAN/CSA-C22. 2 NO 236-M90. Heating and Cooling Equipment; (Gen Instr 1) (ANSI/UL 1995). 136 pp.
CSA CAN/CSA-C746-93. Performance Standard for Rating Large Air Conditioners and Heat Pumps; (Gen Instr 1). 31 pp.

—Air Cooled
CEN PREN 814-2-92. Air Cooled Air Conditioners—Part 2: Testing and Requirements for Marking. 10 pp.

—Air Cooled—Identification Systems
CEN PREN 814-2-92. Air Cooled Air Conditioners—Part 2: Testing and Requirements for Marking. 10 pp.

—Air Cooled—Room—Built In
CSA C22.2 NO 117-1970. Room Air Conditioners (R 1992); (Rev 1-3 February 1974). 29 pp.
CSA 1169 Bull. Electrical Bulletin 1169 June 27, 1978 to C22.2 NO 117. 2 pp.

—Air Cooled—Room—Console
CSA C22.2 NO 117-1970. Room Air Conditioners (R 1992); (Rev 1-3 February 1974). 29 pp.
CSA 1169 Bull. Electrical Bulletin 1169 June 27, 1978 to C22.2 NO 117. 2 pp.

—Air Cooled—Room—Window
CSA C22.2 NO 117-1970. Room Air Conditioners (R 1992); (Rev 1-3 February 1974). 29 pp.
CSA 1169 Bull. Electrical Bulletin 1169 June 27, 1978 to C22.2 NO 117. 2 pp.

—Air Filters
SAA AS 1324-73. Air Filters for Use in Air Conditioning and General Ventilation. 8 pp.

—Air Filters—Glossaries
SAA AS 1323-73. Glossary of Terms for Air Filters for Use in Air Conditioning and General Ventilation Corrig.. 8 pp.

—Air Quantity
CNS A3119-89. Method of Measurement of Air Quantity for Ventilation and Air Conditioning System (Jan)(6874).
JIS A 1431-74. Method of Measurement of Air Quantity for Ventilation and Air Conditioning System. 5 pp.

—Aircraft
CAA STANDARD NO. 7-1&AP 07.69. Air Conditioning Systems and Cosmic Radiation. 13 pp.

—Aircraft—Air Quality
NATO STANAG 3610 ED 1 AMD 3-73. Characteristics of Conditioned Breathable Air Supplied to Aircraft on the Ground. 10 pp.

—Aircraft—Connectors
NATO STANAG 3208 ED 4 AMD 4-82. Air Conditioning Connections. 9 pp.
NATO STANAG 3208 ED 5 AMD 0-92. Air Conditioning Connections. 9 pp.
NATO STANAG 3208 ED 5 AMD 1-92. Air Conditioning Connections. 10 pp.

—Aircraft—Glossaries
BSI BS 185: Sec 10-70. 1970 Glossary of Aeronautical and Astronautical Terms Section 10: Auxiliary Services (Hydraulic, Pneumatic, Electrical, Air Conditioning and Refuelling). 7 pp.

—Aircraft—Ground Support Equipment—Connectors
BSI C 10-58. 1958 Amd 1 Coupling Dimensions for Aircraft Ground Air-Conditioning Connections. 4 pp.
ISO 1034-73. Aircraft—Ground Air-Conditioning Connections First Edition; (Erratum—March 1974). 5 pp.
NATO STANAG 3917 ED 2 AMD 2-84. Air Conditioning Connections, Ground Half Connectors. 9 pp.
NATO STANAG 3917 ED 2 AMD 3-84. Air Conditioning Connections, Ground Half Connectors. 10 pp.

—Aircraft—Subsystems—Air Cycle
JIS W 3103-84. Air Conditioning Subsystems, Air-Cycle, Aircraft, General Specification for.

—Buildings
BSI BS 5720-79. 1979 Code of Practice for Mechanical Ventilation and Air Conditioning in Buildings. 82 pp.
JIS A 4412-76. Air-Conditioning Unit for Dwellings (R 1983). 32 pp.

Air Conditioners (Cont.)

—Buildings (Cont.)
SAA AS 1668.2 Supp 1-91. Use of Mechanical Ventilation and Air-Con-ditioning in Buildings—Part 2: Mechanical Ven-tilation for Acceptable Indoor-Air Quality—Commentary (Supplement to AS 1668.2—1991) (Supersedes SAA MP47.C2—1980) (in Professional Packages 20, 21, 44, 50). 24 pp.

—Buildings—Hot Water Supply
JIS A 1713-76. Method of Test for Hot-Water Supply of Air-Conditioning Unit for Dwellings.

—Cabinets (Electrical)
CEN PREN 814-5-92. Air Conditioners—Part 5: Requirements for Control Cabinet Air Conditioners. 8 pp.

—Condensing Units
CSA CAN/CSA-C746-93. Performance Standard for Rating Large Air Conditioners and Heat Pumps; (Gen Instr 1). 31 pp.

—Ducts
CSA B228.1-1968. Pipes, Ducts, and Fittings for Residential Type Air Conditioning Systems. 21 pp.

—Electric
CEN PREN 814-4-92. Air Conditioners—Part 4: Requirements for Comfort Cooling Air Conditioners, Single Duct and Spot Air Conditioners, Close Control Air Conditioners. 9 pp.

—Electric Rolling Stock
JIS E 6602-89. Cooling Unit Apparatus for Electric Cars.

—Electrical Codes
CSA 732 Bull. Electrical Bulletin 732 August 29, 1968 to C22.2 NO 117. 2 pp.
CSA 732A Bull. Electrical Bulletin 732A March 28, 1969 to C22.2 NO 117. 1 p.

—Fittings
CSA B228.1-1968. Pipes, Ducts, and Fittings for Residential Type Air Conditioning Systems. 21 pp.

—Glossaries
CEN PREN 814-1-92. Air Conditioners—Part 1: Terms, Definitions and Designations. 6 pp.
DIN ENGL 1946 Pt 1-88. Heating, Ventilation and Air Conditioning; Terminology and Graphical Symbols (VDI Code of Practice) (Oct). 38 pp.

—Installation—Construction Contracts
DIN ENGL 18379-90. Tendering and Performance Stipulations in Contracts for Construction Works (VOB); Part C: General Technical Specifications in Contracts for Construction Works (ATV); Installation of Air Conditioning Systems (July) (This. 14 pp.

—Medical Equipment
MOD UK DSTAN 00-30-82. Air-Conditioning for Medical Purposes Issue 1. 5 pp.

—Pipes
CSA B228.1-1968. Pipes, Ducts, and Fittings for Residential Type Air Conditioning Systems. 21 pp.

—Refrigerant Compressors—Motor
CSA CAN/CSA-C22. 2NO140.2-M91. Hermetic Refrigerant Motor-Compressors Third Edition; (Gen Instr 1) (UL 984). 80 pp.

—Room
BSI BS 2852: Part 1-82. 1982 Testing for Rating of Room Air-Conditioners Part 1: Cooling Performance. 16 pp.
CNS B7048-88. Room Air Conditioners (Nov)(3615). 28 pp.
CSA CAN/CSA-C368.1-M90. Performance Standard for Room Air Conditioners; (Gen Instr 1). 25 pp.
ISO R859-68. Testing and Rating Room Air Conditioners First Edition. 28 pp.
JIS C 9612-89. Room Air Conditioners. 120 pp.
MOD UK DEF-1405-62. Air-Conditioner, Refrigerated. 10 pp.
SAA AS 1861.1-88. Air-Conditioning Units—Methods of Assessing and Rating Performance—Part 1: Refrigerated Room Air Conditioners. 21 pp.
SAA AS 3179-89. Approval and Test Specifications—Small Self-Contained Refrigerated Air Conditioners Amdt 1 May 1990 (Superseded by AS/NZS 3179:1993).
SNZ NZS/AS 3179-93. Approval and Test Specification—Refrigerated Room Air-Conditioners. 9 pp.

—Room—Electrical Codes
CSA 836A Bull. Electrical Bulletin 836A October 20, 1971 to C22.2 NO 117. 2 pp.

Air Conditioners (Cont.)

—Room—Safety
DIN ENGL 1946 Pt 2-83. Air Conditioning; Health Requirements (VDI Ventilation Rules) (Jan). 9 pp.

—Safety
CENELEC PREN 60 335-2-402-89. Safety of Household and Similar Electrical Appliances Part 2: Particular Requirements for Electric Heat Pumps, Air Conditioners and Dehumidifiers for Commercial and Light Industrial Applications. 37 pp.

—Self Propelled Machinery
ISO TR8953-87. Tractors and Self-Propelled Machines for Agriculture and Forestry—Test Method for Performance of Air-Conditioning System First Edition. 6 pp.

—Ships—Design
BSI BS MA 103-86. 1986 Design Conditions and Basis of Calculations for Air-Conditioning and Ventilation of Accomodation Spaces in Ships. 11 pp.
ISO 7547-85. Air-Conditioning and Ventilation of Accommodation Spaces on Board Ships—Design Conditions and Basis of Calculations First Edition. 10 pp.
ISO 8862-87. Air-Conditioning and Ventilation of Machinery Control-Rooms on Board Ships—Design Conditions and Basis of Calculations First Edition. 4 pp.
ISO 8864-87. Air-Conditioning and Ventilation of Wheelhouse on Board Ships—Design Conditions and Basis of Calculations First Edition. 4 pp.
ISO 9099-87. Air-Conditioning and Ventilation of Dry Provision Rooms on Board Ships—Design Conditions and Basis of Calculations First Edition. 4 pp.
MOD UK NES 102: Volume 1-83. Requirements for Air Conditioning and Ventilation Design Issue 1 (08.83). 278 pp.
MOD UK NES 102: Volume 2-83. Requirements for Air Conditioning and Ventilation Design Issue 1 (08.83). 231 pp.

—Ships—Ducts—Thermal Insulation
JIS F 0302-77. Standard Practice for Thermal Insulation Work for Small Ships' Airconditioning Ducts.

—Submarines—Design
MOD UK NES 102: Volume 1-83. Requirements for Air Conditioning and Ventilation Design Issue 1 (08.83). 278 pp.
MOD UK NES 102: Volume 2-83. Requirements for Air Conditioning and Ventilation Design Issue 1 (08.83). 231 pp.

—Symbols
DIN ENGL 1946 Pt 1-88. Heating, Ventilation and Air Conditioning; Terminology and Graphical Symbols (VDI Code of Practice) (Oct). 38 pp.

—Symbols—Engineering Drawings
ISO 4067 Pt 1-84. Technical Drawings—Installations—Part 1: Graphical Symbols for Plumbing, Heating, Ventilation and Ducting First Edition. 12 pp.

—Tractors
ISO TR8953-87. Tractors and Self-Propelled Machines for Agriculture and Forestry—Test Method for Performance of Air-Conditioning System First Edition. 6 pp.

—Water Cooled
CEN PREN 814-3-92. Water Cooled Air Conditioners—Part 3: Testing and Requirements for Marking. 10 pp.

—Water Cooled—Identification Systems
CEN PREN 814-3-92. Water Cooled Air Conditioners—Part 3: Testing and Requirements for Marking. 10 pp.

Air Conditioning Equipment

Scope Note: For additional listings, use a more specific term *Use For:* Air Conditioning Systems
See Also: Absorbers (Equipment); Air Conditioners; Air Coolers; Air Handling Equipment; Air Terminal Devices; Blowers (Ventilation); Compressor/Evaporators; Compressors (Pressure Equipment); Condensing Units; Coolers; Cooling Coils; Cooling Systems; Cooling Towers; Dehumidifiers; Evaporative Coolers; Exhaust Systems; Expansion Valves; Fans; Heat Pumps; Heating, Ventilating and Air Conditioning Equipment; Humidifiers
SAA HB40-92. Australian Refrigeration and Air Conditioning Code of Good Practice. 22 pp.

—Automotive
CNS D3079-81. Method of Test for Automobile Air Conditioners (Sep)(7897).
JIS D 1618-86. Testing Method for Automobile Air Conditioners.

—Chillers
CSA C743-93. Performance Standard for Rating Packaged Water Chillers; (Gen Instr 1). 38 pp.

—Hygiene
SNZ NZS 4302-87. Control of Hygiene in Air and Water Systems in Buildings Amend: 1, 1991. 30 pp.

—Naval Ships—Installation
MOD UK NES 103-01. Air Conditioning and Ventilation Installation in Ships Not to Full RN Requirements Issue 2 (01.82); Amendment 1 (Withdrawn). 35 pp.

—Refrigerants—Electrical Equipment
CSA C22.2 NO 140.3-M1987. Refrigerant-Containing Components for Use in Electrical Equipment (R 1993). 19 pp.

—Rolling Stock
JIS E 4015-89. Measuring Methods for Air Conditioning and Herting Temperature of Railway Rolling Stock.

—Ships
SAA AS 1921-76. Air Conditioning and Ventilation in Ships' Cabins and Living Spaces of Merchant Ships Amdt 1 February 1984. 32 pp.
SAA AS 1925-76. Air Conditioning and Ventilation in Ships—Machinery Control Rooms of Merchant Ships Amdt 1 February 1984 Reconfirmed 1984. 16 pp.

—Temperature Controllers
CSA C22.2 NO 24-93. Temperature-Indicating and-Regulating Equipment; (Gen Instr 1). 88 pp.

—Temperature Indicators
CSA C22.2 NO 24-93. Temperature-Indicating and-Regulating Equipment; (Gen Instr 1). 88 pp.

—Thermostats
CSA C22.2 NO 24-93. Temperature-Indicating and-Regulating Equipment; (Gen Instr 1). 88 pp.

—Volt Ampere Limits
CSA 551 Bull. Electrical Bulletin 551 July 19, 1962 to C22.2 NO 117. 1 p.

Air Conditioning Systems
Use: Air Conditioning Equipment

Air Content Analysis

—Concretes
SAA AS 1012.4-83. Methods of Testing Concrete—Part 4: Methods for the Determination of Air Content of Freshly Mixed Concrete. 27 pp.

—Mortars
CEN PREN 1015-7-93. Methods of Test for Mortar for Masonry—Part 7: Determination of Air Content of Fresh Mortar. 10 pp.

Air Cooled Transformers
Use: Dry Type Transformers

Air Coolers

See Also: Air Conditioners; Coolers
BSI DD ENV 328-93. 1993 Heat Exchangers—Test Procedures for Establishing Peformance of Unit Air Coolers for Refrigeration (F). 31 pp.
CEN PRENV 328-90. Heat Exchangers; Test Procedures for Establishing Performance of Forced Convection Air Coolers. 25 pp.
CEN ENV 328-92. Heat Exchangers—Test Procedures for Establishing Performance of Unit Air Coolers for Refrigeration. 26 pp.

—Residential Requirements
CSA CAN/CSA-F280-M90. Determining the Required Capacity of Residential Space Heating and Cooling Appliances; (Gen Instr 1). 102 pp.

Air Cooling

See Also: Coolants; Cooling Systems; Ventilation Equipment

—Rotating Machines
BSI BS 4999: Part 106-87. 1987 General Requirements for Rotating Electrical Machines Part 106: Classification of Methods of Cooling. 14 pp.
CENELEC HD 53.6-74. Rotating Electrical Machines Part 6: Methods of Cooling Rotating Machinery. 2 pp.
CENELEC HD 53.6 S1-88. Rotating Electrical Machines Part 6: Methods of Cooling Rotating Machinery. 3 pp.
IEC 34 Pt 6-91. Rotating Electrical Machines Part 6: Methods of Cooling (IC Code) Second Edition. 43 pp.

Air Cushion Vehicles

Use For: Ground Effect Machines; Hovercraft
See Also: Aircraft; Hoverplatforms
CAA. Contents 01.91 (British Hovercraft Safety Requirements). 6 pp.
CAA. Preface 01.91 (British Hovercraft Safety Requirements). 7 pp.
CAA. Foreword 01.91 (British Hovercraft Safety Requirements). 3 pp.
CAA. Index 01.91 (British Hovercraft Safety Requirements). 15 pp.
CAA 57.
CAA Chapter B6-1 01.91. Systems General. 3 pp.

—Accidents—Acceleration Testing
CAA Chapter B4-3 11.72. Collision Accelerations. 1 p.

—Air Handling Equipment—Contamination
CAA 42.
CAA Chapter B6-6 01.91. Heating and Ventilation Systems. 1 p.

—Anchors (Marine)
CAA Chapter B5-9 04.79. Anchoring, Towing and Berthing. 4 pp.
CAA Chapter B5-9 App 04.79. Anchoring, Towing and Berthing. 2 pp.

—Approved
CAA Chapter A2-2 App 01.74. Approval of Organisations for Design of Hovercraft or Items Other Than Hoverplatforms. 1 p.

—Approved Authorities
CAA. Introductory Note 04.79. 3 pp.
CAA. Introductory Note 4.79.

—Authorities—Design
CAA Chapter A2-2 App 01.74. Approval of Organisations for Design of Hovercraft or Items Other Than Hoverplatforms. 1 p.

—Balancing
CAA Chapter A5-1 App 08.81. Weight and Balance Data. 2 pp.
CAA Chapter A5-1 07.89. General. 2 pp.

—Berths
CAA Chapter B5-9 04.79. Anchoring, Towing and Berthing. 4 pp.

—Bilge Pumps
CAA 44.
CAA Chapter B6-12 01.91. Bilge Pumping and Drainage. 1 p.

—Buoyancy
CAA Chapter B5-8 08.83. Buoyancy and Displacement Mode. 3 pp.

—Cabins—Flight Crew—Design
CAA Chapter B5-2 08.83. Operating Crew Compartments. 3 pp.

—Cabins—Passenger—Design
CAA Chapter B5-3 11.72. Passenger Compartments. 1 p.
CAA Chapter B5-3 App 11.72. Passenger Compartments. 1 p.

—Center of Gravity
CAA Chapter A5-1 App 08.81. Weight and Balance Data. 2 pp.

—Certification
CAA 53.
CAA Chapter A3-1 01.74. General. 2 pp.
CAA Chapter A3-1 App 01.74. General. 1 p.
CAA Chapter B1-1 01.91. General. 2 pp.
CAA Chapter B2-2 06.90. Type Certification or Validation. 4 pp.
CAA Chapter B2-4. Type Certification or Validation of a Variant. 1 p.

—Compartments—Baggage—Design
CAA Chapter B5-4 11.72. Cargo and Baggage Compartments. 1 p.

—Compartments—Cargo—Design
CAA Chapter B5-4 11.72. Cargo and Baggage Compartments. 1 p.

—Compasses
CAA 52.
CAA Chapter B6-13 01.91. Compasses. 3 pp.

—Construction
CAA 29.
CAA Chapter B8 01.91. Construction. 3 pp.

—Construction—Certification
CAA Chapter A2-3 04.79. Flight Testing for Type Certification. 1 p.
CAA Chapter A2-3 App 01.74. Flight Testing for Type Certification. 1 p.

INTERNATIONAL AND NON-U.S. NATIONAL STANDARDS
SUBJECT INDEX

Air Cushion Vehicles *(Cont.)*
—Control Systems Equipment
 CAA 58.
 CAA Chapter B6-3 01.91. Control Systems. 3 pp.
—Data Sheets
 CAA 53.
 CAA Chapter A3-3 01.91. Type Certificates for Hovercraft. 4 pp.
 CAA Chapter A3-3 App 01.91. Type Certificates for Hovercraft. 3 pp.
 CAA Chapter A3-4 01.74. Type Certificates for Hovercraft Items. 5 pp.
 CAA Chapter A3-4 App 01.91. Type Certificates for Hovercraft Items. 1 p.
—Design
 CAA Chapter B5-1 08.83. General. 1 p.
—Design—Certification
 CAA Chapter A2-2 01.74. Approval of Organisations for Design of Hovercraft or Items Other Than Hoverplatforms. 2 pp.
—Doors
 CAA Chapter B5-6 01.91. Exits. 4 pp.
 CAA Chapter B5-6 App 11.72. Exits. 3 pp.
—Drive Shafts
 CAA Chapter B7-4 04.79. Transmission Items. 1 p.
—Electrical Installations
 CAA Chapter B6-7 01.91. Electrical Systems. 8 pp.
—Emergency Exits
 CAA Chapter B5-6 01.91. Exits. 4 pp.
 CAA Chapter B5-6 App 11.72. Exits. 3 pp.
—Emergency Lighting
 CAA Chapter B5-6 01.91. Exits. 4 pp.
 CAA Chapter B6-7 App 08.83. Electrical Systems. 1 p.
—Emergency Power Supplies
 CAA Chapter B6-7 App 08.83. Electrical Systems. 1 p.
—Engines
 CAA 41.
 CAA Chapter B6-2 01.91. Power Plant Installations. 3 pp.
—Engines—Comprehensive Testing
 CAA Chapter B7-3 04.79. Engine Testing. 3 pp.
—Engines—Design
 CAA Chapter B7-2 04.79. Engine Design and Construction. 3 pp.
—Experimental Certificates
 CAA Chapter A3-2 04.79. Experimental Certs. 3 pp.
 CAA Chapter A3-2 App 08.81. Experimental Certs. 2 pp.
—Export—Certification
 CAA Chapter A3-6 08.83. Certificates of Export Hovercraft. 1 p.
 CAA Chapter A3-6 App 01.91. Certificates for Export Hovercraft. 2 pp.
—Fire Alarm and Extinguishing Equipment
 CAA Chapter B6-10 08.83. Fire Extinguishing Systems. 2 pp.
 CAA Chapter B6-10 App 08.83. Fire Detection and Extinguishing Systems. 2 pp.
—Fire Fighting Equipment
 CAA Chapter B6-10 08.83. Fire Extinguishing Systems. 2 pp.
 CAA Chapter B6-10 App 08.83. Fire Detection and Extinguishing Systems. 2 pp.
—Fire Protection
 CAA Chapter B5-7 01.91. Fire Precautions. 6 pp.
—Flexible Structures
 CAA Chapter B5-10 01.91. Flexible Structures. 2 pp.
 CAA Chapter B5-10 App 08.83. Flexible Structures. 2 pp.
—Flight Operations
 CAA 53.
 CAA Chapter B3-1 01.91. Determination of Handling and Performance. 5 pp.
 CAA Chapter B3-1 App 08.81. Determination of Handling and Performance. 2 pp.
 CAA Chapter B3-2 08.81. Scheduling of Handling and Performance Procedure. 2 pp.
 CAA Chapter B3-3 08.83. Operating Limitations and Approved Information. 2 pp.
 CAA Chapter B3-3 App 01.91. Operating Limitations and Approved Information. 4 pp.
—Flight Testing
 CAA Chapter A5-2 App 08.81. Vehicle Trails. 2 pp.

Air Cushion Vehicles *(Cont.)*
—Flight Testing—Certification
 CAA Chapter A5-2 08.81. Vehicle Trails. 3 pp.
 CAA Chapter B2-3 09.91. Flight Testing for Type Certification or Validation. 2 pp.
—Forgings
 CAA Chapter B8-5 11.72. Forgings. 3 pp.
—Fuel Systems
 CAA Chapter B6-5 01.91. Fuel Systems. 4 pp.
—Gas Turbine Engines—Comprehensive Testing
 CAA Chapter B7-3 App 04.79. Engine Testing—Type Tests for Gas Turbine Shaft Power Engines. 7 pp.
—Gear Boxes
 CAA Chapter B7-4 04.79. Transmission Items. 1 p.
—Glossaries
 CAA 45.
 CAA Chapter A1-2 09.91. Definitions. 8 pp.
—Harnesses
 CAA Chapter B5-5 11.72. Seats, Safety Belts and Harnesses. 2 pp.
—Hulls—Design
 CAA Chapter B5-1 App 08.83. Hull Hydrodynamic Design (Amphibious Craft). 1 p.
—Hydraulic Systems
 CAA 56.
 CAA Chapter B6-4 01.91. Hydraulic Systems. 3 pp.
—Identification Systems
 CAA Chapter B2-1 08.81. General. 2 pp.
 CAA Chapter B2-1 App 08.81. General. 1 p.
 CAA Chapter B2-6 01.91. Marking and Placards. 3 pp.
 CAA Chapter B2-6 App 08.81. Markings and Placards. 1 p.
—Impact Shock—Water
 CAA Chapter B4-2 App 01.91. Loading Cases. 3 pp.
—Instruments
 CAA 54.
 CAA Chapter B6-8 01.91. Instruments. 2 pp.
 CAA Chapter B6-8 App 01.91. Instruments. 1 p.
—Lighting
 CAA Chapter B6-7 01.91. Electrical Systems. 8 pp.
—Loads (Forces)
 CAA Chapter B4-1 01.91. General. 2 pp.
 CAA Chapter B4-2 08.83. Loading Cases. 3 pp.
 CAA Chapter B4-4 11.72. Fluctuating Loads. 1 p.
—Machinery
 CAA Chapter B7-1 08.83. General. 3 pp.
 CAA Chapter B7-1 App 08.83. General Reliability. 2 pp.
—Maintenance
 CAA Chapter B2-3 08.81. Type Servicing Schedule. 2 pp.
 CAA Chapter B2-3 App 8.81. Type Servicing Schedule. 1 p.
—Maintenance—Certification
 CAA Chapter A2-4 04.79. Appoval of Organisations for Maintenance of Hovercraft or Items Other Than Hoverplatforms. 4 pp.
 CAA Chapter A2-4 App 01.74. Appoval of Organisations for Maintenance of Hovercraft or Items Other Than Hoverplatforms. 1 p.
 CAA Chapter A4-1 08.81. Maintenance Procedure. 5 pp.
 CAA Chapter A4-1 App 08.81. Maintenance Procedure. 2 pp.
—Mechanical Drives
 CAA Chapter B6-11 04.79. Transmission Systems. 3 pp.
—Modification
 CAA Chapter A4-2 08.83. Modification Procedures. 5 pp.
 CAA Chapter A4-2 App 08.83. Modification Procedure. 3 pp.
 CAA Chapter B2-5 07.89. Approval of Modifications. 4 pp.
—Pneumatic Systems
 CAA 56.
 CAA Chapter B6-4 01.91. Hydraulic Systems. 3 pp.
—Propellers
 CAA Chapter B7-5 04.79. Fans and Air Propellers. 3 pp.
—Safety
 CAA. Explanatory Note. 1 p.

Air Cushion Vehicles *(Cont.)*
—Safety *(Cont.)*
 CAA 34.
 CAA Chapter B1-2 01.91. Safety Assessment of Systems. 2 pp.
—Safety Belts
 CAA Chapter B5-5 11.72. Seats, Safety Belts and Harnesses. 2 pp.
—Safety—Certification
 CAA Chapter A3-5 04.79. Safety Certificates. 3 pp.
—Seats
 CAA Chapter B5-5 11.72. Seats, Safety Belts and Harnesses. 2 pp.
 CAA Chapter B5-5 App 11.72. Strength of Seats. 2 pp.
—Signs
 CAA Chapter B2-1 08.81. General. 2 pp.
 CAA Chapter B2-1 App 08.81. General. 1 p.
 CAA Chapter B2-6 01.91. Marking and Placards. 3 pp.
 CAA Chapter B2-6 App 08.81. Markings and Placards. 1 p.
—Stability Testing—Water
 CAA Chapter B5-8 08.83. Buoyancy and Displacement Mode. 3 pp.
—Structural Design
 CAA Chapter B4-1 01.91. General. 2 pp.
—Technical Manuals
 CAA Chapter B2-1 08.81. General. 2 pp.
 CAA Chapter B2-1 App 08.81. General. 1 p.
 CAA Chapter B2-2 08.81. Craft Type Operating Manual. 3 pp.
 CAA Chapter B2-2 App 08.81. Craft Type Operating Manual. 3 pp.
 CAA Chapter B2-3 App 8.81. Type Servicing Schedule. 1 p.
 CAA Chapter B2-4 08.81. Type Maintenance Manual. 1 p.
 CAA Chapter B2-5 08.81. Service Bulletins. 2 pp.
—Towing
 CAA Chapter B5-9 04.79. Anchoring, Towing and Berthing. 4 pp.
—Turbine Engines—Fatigue (Materials)
 CAA Chapter B7-3 App 04.79. Engine Testing—Type Tests for Gas Turbine Shaft Power Engines. 7 pp.
—Turbine Engines—Life (Durability)
 CAA Chapter B7-2 App 04.79. Engine Design and Construction. 3 pp.
—Vibration
 CAA Chapter B4-1 01.91. General. 2 pp.
 CAA Chapter B4-1 App 11.72. General. 1 p.
—Weight Measurement
 CAA Chapter A5-1 App 08.81. Weight and Balance Data. 2 pp.

Air Dielectric Capacitors
Use: Air Capacitors

Air Dielectric Condensers
Use: Air Capacitors

Air Diffusion
See Also: Air Handling Equipment; Louvers
—Glossaries
 ISO 3258-76. Air Distribution and Air Diffusion— Vocabulary First Edition. 36 pp.

Air Distribution Equipment
Use: Air Handling Equipment

Air Drop Operations
Use For: Aerial Resupply *See Also:* Air Transportation; Airborne Operations; Drop Zones; Extraction Zones; Transportation
 ICAO 9408. Manual on Aerial Work First Edition— 1984. 171 pp.
—Materiel—Color Coding
 NATO STANAG 3427 ED 4 AMD 1-77. Colours for the Identification of Airdropped Supplies. 6 pp.
 NATO STANAG 3427 ED 4 AMD 2-77. Colours for the Identification of Airdropped Supplies. 7 pp.
—Procedures—Equipment/Personnel
 NATO STANAG 3464 ED 3 AMD 2-79. Procedures for Aircrew and Airlifted Personnel in Tactical Air Transport Operations. 16 pp.
 NATO STANAG 3464 ED 3 AMD 5-79. Procedures for Aircrew and Airlifted Personnel in Tactical Air Transport Operations. 13 pp.

INTERNATIONAL AND NON-U.S. NATIONAL STANDARDS
SUBJECT INDEX

Air

Air Drop Operations *(Cont.)*
—Procedures—Equipment/Personnel *(Cont.)*
NATO STANAG 3922 ED 2 AMD 1-89. Airdrop Systems for Personnel and Supply/Equipment—ATP-46(A). 4 pp.

—Tie Downs
NATO STANAG 3548 ED 2 AMD 6-78. Tie-Down Fittings on Air Transport and Airdropped Equipment and Cargo Carried Internally by Fixed Wing Aircraft. 9 pp.

NATO STANAG 3548 ED 2 AMD 7-78. Tie-Down Fittings on Air Transported and Airdropped Equipment and Cargo Carried Internally by Fixed Wing Aircraft. 10 pp.

Air Dryers
Use: Dryers

Air Drying
See Also: Drying

—Coatings
CGSB 1-GP-71 METH 103.1-79. Methods of Testing Paints and Pigments Drying Conditions Air Drying. 1 p.

—Coatings—Low Temperature
CGSB 1-GP-71 METH 103.3-79. Methods of Testing Paints and Pigments Drying Conditions Low-Temperature Conditioning in Air. 1 p.

Air Ejectors
See Also: Materials Handling Equipment; Vacuum Pumps

—Steam Jet
JIS B 8111-76. Testing Methods for Steam Jet Air Ejectors.

Air Entraining Admixtures
Use: Air Entraining Agents

Air Entraining Agents
Use For: Air Entraining Admixtures
See Also: Admixtures; Surfactants

—Concretes
BSI BS 5075: Part 2-82. 1982 Amd 2 Concrete Admixtures Part 2: Air-Entraining Admixtures. 12 pp.
CNS A2043-86. Air-Entraining Admixtures for Concrete (Dec)(3091). 5 pp.
CNS A3167-82. Method of Test for Air-Entraining Admixtures for Concrete (Oct)(9208).
CSA CAN3-A266.1-M78. Air-Entraining Admixtures for Concrete; (Amd 1 September 1978) (Amd 2 November 1981). 23 pp.
JIS A 6204-87. Chemical Admixtures for Concrete. 31 pp.

—Mortars
BSI BS 4887: Part 1-86. 1986 Mortar Admixtures Part 1: Air-Entraining (Plasticizing) Admixtures. 11 pp.

—Portland Cements
CNS R2077-85. Air-Entraining Additions for Use in the Manufacture of Air-Entraining Portland Cement (May)(3589).

Air Exhaust Equipment
Use: Exhaust Systems

Air Filters
See Also: Air Cleaners; Dust Collectors; Oxygen Supply Equipment; Protective Masks; Separators (Mechanical); Ventilation Equipment

CNS B4047-81. Air Filter Units for Ventilation (Sep)(7869).
ISO 5782 Pt 1-90. Pneumatic Fluid Power—Compressed Air Filters—Part 1: Main Characteristics to Be Included in Commercial Literature and Specific Requirements First Edition. 9 pp.
JIS B 9908-91. Air Filter Units for Ventilation. 23 pp.
JIS K 0901-91. Form, Size and Performance Testing Methods of Filtration Media for Collecting Airborne Particulate Matters. 16 pp.
SAA AS 1132. Methods of Test for Air Filters for Use in Air Conditioning and General Ventilation (Bound Together AS 1132 Series).
SAA AS 1132.3-73. Methods of Test for Air Filters for Use in Air Conditioning and General Ventilation—Part 3: Determination of Oil Carry-Over.
SAA AS 1132.8-73. Methods of Test for Air Filters for Use in Air Conditioning and General Ventilation—Part 8: Determination of Effectiveness of Renewing Mechanism of Filters Incorporating Restoration to Original Condition.

Air Filters *(Cont.)*
—Activated Carbon
CGSB CAN/CGSB-115.16-M82. Activated Carbon for Odor Removal from Ventilating Systems. 9 pp.

—Aerosols
JIS K 3803-90. Testing Methods for Determining Aerosol Retention of Depthfilter for Air Sterilization. 8 pp.

—Air Conditioners
SAA AS 1324-73. Air Filters for Use in Air Conditioning and General Ventilation. 8 pp.

—Air Conditioners—Glossaries
SAA AS 1323-73. Glossary of Terms for Air Filters for Use in Air Conditioning and General Ventilation Corrig.. 8 pp.

—Arrestance Efficiency
SAA AS 1132.4-73. Methods of Test for Air Filters for Use in Air Conditioning and General Ventilation—Part 4: Determination of Arrestance Efficiency, Average Arrestance Efficiency, Dust-Holding Capacity Per Unit of Effective Face Area for Test Dust No 4.
SAA AS 1132.5-73. Methods of Test for Air Filters for Use in Air Conditioning and General Ventilation—Part 5: Detn. of Arrestance Eff., Avg. Arrestance Eff., Dust-Holding Capcy., and Dust-Holding Capcy. Per Unit of Effective Face Area for Test Dusts Nos 1, 2, and 3.

—Atmospheric Dust Spot Efficiency
BSI BS 6540: Part 1-85. 1985 Air Filters Used in Air Conditioning and General Ventilation Part 1: Methods of Test for Atmospheric Dust Spot Efficiency and Synthetic Dust Weight Arrestance (Supersedes BS 2831: 1971). 32 pp.

—Automotive
ISO 8027-84. Road Vehicles—Air Filter Elements for Passenger Cars—Types P and R—Dimensions First Edition. 6 pp.
JIS D 1612-89. Test Methods of Air Cleaners for Automobiles. 48 pp.

—Automotive—Connections
ISO 7312-84. Road Vehicles—Air Filter Connections—Types A and B First Edition. 5 pp.

—Bag
CGSB CAN/CGSB-115.11-M85. Filters, Air, High Efficiency, Disposable, Bag Type. 8 pp.
CGSB CAN/CGSB-115.12-M85. Filters, Air, Medium Efficiency, Disposable, Bag Type. 8 pp.

—Breathing Apparatus
BSI BS EN 141-91. 1991 Gas Filters and Combined Filters Used in Respiratory Protective Equipment. 13 pp.
BSI BS EN 143-91. 1991 Particle Filters Used in Respiratory Protective Equipment (N). 30 pp.
BSI BS EN 371-92. 1992 AX Gas Filters and Combined Filters Against Low Boiling Organic Compounds Used in Respiratory Protective Equipment. 11 pp.
BSI BS EN 372-92. 1992 SX Gas Filters and Combined Filters Against Specific Named Compounds Used in Respiratory Protective Equipment. 11 pp.
CEN EN 141-90. Respiratory Protective Devices—Gas Filters and Combined Filters—Requirements, Testing, Marking. 13 pp.
CEN EN 143-90. Respiratory Protective Devices—Particle Filters—Requirements, Testing, Marking. 34 pp.
CEN PREN 371-90. Respiratory Protective Devices: AX Gas Filters and Combined Filters Against Low Boiling Organic Compounds: Requirements, Testing, Marking. 10 pp.
CEN EN 371-92. Respiratory Protective Devices—AX Gas Filters and Combined Filters Against Low Boiling Organic Compounds—Requirements, Testing, Marking. 14 pp.
CEN PREN 372-90. Respiratory Protective Devices: SX Gas Filters and Combined Filters Against Specific Named Compounds; Requirements, Testing, Marking. 11 pp.
CEN EN 372-92. Respiratory Protective Devices—SX Gas Filters and Combined Filters Against Specific Named Compounds—Requirements, Testing, Marking. 14 pp.

—Breathing Apparatus—Identification Systems
BSI BS EN 141-91. 1991 Gas Filters and Combined Filters Used in Respiratory Protective Equipment. 13 pp.
BSI BS EN 143-91. 1991 Particle Filters Used in Respiratory Protective Equipment (N). 30 pp.
BSI BS EN 371-92. 1992 AX Gas Filters and Combined Filters Against Low Boiling Organic Compounds Used in Respiratory Protective Equipment. 11 pp.

Air Filters *(Cont.)*
—Breathing Apparatus—Identification Systems *(Cont.)*
BSI BS EN 372-92. 1992 SX Gas Filters and Combined Filters Against Specific Named Compounds Used in Respiratory Protective Equipment. 11 pp.
CEN EN 141-90. Respiratory Protective Devices—Gas Filters and Combined Filters—Requirements, Testing, Marking. 13 pp.
CEN EN 143-90. Respiratory Protective Devices—Particle Filters—Requirements, Testing, Marking. 34 pp.
CEN PREN 371-90. Respiratory Protective Devices: AX Gas Filters and Combined Filters Against Low Boiling Organic Compounds: Requirements, Testing, Marking. 10 pp.
CEN EN 371-92. Respiratory Protective Devices—AX Gas Filters and Combined Filters Against Low Boiling Organic Compounds—Requirements, Testing, Marking. 14 pp.
CEN PREN 372-90. Respiratory Protective Devices: SX Gas Filters and Combined Filters Against Specific Named Compounds; Requirements, Testing, Marking. 11 pp.
CEN EN 372-92. Respiratory Protective Devices—SX Gas Filters and Combined Filters Against Specific Named Compounds—Requirements, Testing, Marking. 14 pp.

—Cartridges
CGSB CAN/CGSB-115.14-M91. High Efficiency Cartridge Type Supported Air Filters for the Removal of Particulate Matter from Ventilating Systems. 7 pp.

—Comprehensive Testing
SAA AS 1132.1-73. Methods of Test for Air Filters for Use in Air Conditioning and General Ventilation—Part 1: Apparatus, Materials, Reagents and Test Samples Prerequisite and Common to Methods Constituting an Evaluation.

—Compressors (Pressure Equipment)
BSI BS 1701-70. (WITHDRAWN) 1970 Air Filters for Air Supply to Internal Combustion Engines and Compressors Other Than for Aircraft (Superseded by BS 7226: 1989). 40 pp.
BSI BS 2806-56. (OBSOLESCENT) 1956 Limiting Dimensions of Air Filters for Internal Combustion Engines and Compressors, Other Than for Aircraft. 17 pp.
BSI BS 7226-89. 1989 Methods of Test for Performance of Inlet Air Cleaning Equipment for Internal Combustion Engines and Compressors. 28 pp.
DIN ENGL 71459-87. Air Filter Elements for Commercial Vehicles; Dimensions (May). 6 pp.
ISO 5011-88. Inlet Air Cleaning Equipment for Internal Combustion Engines and Compressors—Performance Testing First Edition. 28 pp.

—Dust Holding Capacity
SAA AS 1132.4-73. Methods of Test for Air Filters for Use in Air Conditioning and General Ventilation—Part 4: Determination of Arrestance Efficiency, Average Arrestance Efficiency, Dust-Holding Capacity Per Unit of Effective Face Area for Test Dust No 4.
SAA AS 1132.5-73. Methods of Test for Air Filters for Use in Air Conditioning and General Ventilation—Part 5: Detn. of Arrestance Eff., Avg. Arrestance Eff., Dust-Holding Capcy., and Dust-Holding Capcy. Per Unit of Effective Face Area for Test Dusts Nos 1, 2, and 3.

—Dynamic Arrestance Efficiency
SAA AS 1132.6-73. Methods of Test for Air Filters for Use in Air Conditioning and General Ventilation—Part 6: Detn. of Dynamic Arrestance Eff. (Mtg. Res.), Dynamic Arrestance Eff. (Steady State), and Dynamic Dust -Holding Capy. of Self-Renewable (Cl. D), Fresh Med. (Kind I) Filters.
SAA AS 1132.7-73. Methods of Test for Air Filters for Use in Air Conditioning and General Ventilation—Part 7: Detn. of Dynamic Arrestance Eff. (Mtg. Res.) and Dynamic Arrestance Eff. (Steady State) of Self Renewable (Class D), Reconditioned Medium (Kind II) Filters.

—Dynamic Dust Holding Capacity
SAA AS 1132.6-73. Methods of Test for Air Filters for Use in Air Conditioning and General Ventilation—Part 6: Detn. of Dynamic Arrestance Eff. (Mtg. Res.), Dynamic Arrestance Eff. (Steady State), and Dynamic Dust -Holding Capy. of Self-Renewable (Cl. D), Fresh Med. (Kind I) Filters.
SAA AS 1132.7-73. Methods of Test for Air Filters for Use in Air Conditioning and General Ventilation—Part 7: Detn. of Dynamic Arrestance Eff. (Mtg. Res.) and Dynamic Arrestance Eff. (Steady State) of Self Renewable (Class D), Reconditioned Medium (Kind II) Filters.

INDUSTRY STANDARDS

Air Filters (Cont.)

—Fabrics
JIS Z 8908-89. Filter Fabrics for Dust Collection. 12 pp.

—HEPA—Biological Hazards
JIS K 3801-89. Penetration Test Method of HEPA Filters for Microbiological Filtration. 10 pp.

—HEPA—Clean Rooms
SAA AS 1807.6-89. Cleanrooms, Workstations and Safety Cabinets—Methods of Test—Part 6: Determination of Integrity of Terminally Mounted HEPA Filter Installations. 2 pp.
SAA AS 1807.7-89. Cleanrooms, Workstations and Safety Cabinets—Methods of Test—Part 7: Determination of Integrity of HEPA Filter Installations Not Terminally Mounted. 2 pp.

—HEPA—Radiation Protection
JIS Z 4812-75. HEPA Filters for Radioactive Aerosols.

—HEPA—Workstations—Clean
SAA AS 1807.6-89. Cleanrooms, Workstations and Safety Cabinets—Methods of Test—Part 6: Determination of Integrity of Terminally Mounted HEPA Filter Installations. 2 pp.
SAA AS 1807.7-89. Cleanrooms, Workstations and Safety Cabinets—Methods of Test—Part 7: Determination of Integrity of HEPA Filter Installations Not Terminally Mounted. 2 pp.

—Internal Combustion Engines
BSI BS 1701-70. (WITHDRAWN) 1970 Air Filters for Air Supply to Internal Combustion Engines and Compressors Other Than for Aircraft (Superseded by BS 7226: 1989). 40 pp.
BSI BS 2806-56. (OBSOLESCENT) 1956 Limiting Dimensions of Air Filters for Internal Combustion Engines and Compressors, Other Than for Aircraft. 17 pp.
BSI BS 7226-89. 1989 Methods of Test for Performance of Inlet Air Cleaning Equipment for Internal Combustion Engines and Compressors. 28 pp.
DIN ENGL 24189-86. Testing of Air Cleaners for Internal Combustion Engines and Compressors; Test Methods (Jan). 10 pp.
ISO 5011-88. Inlet Air Cleaning Equipment for Internal Combustion Engines and Compressors—Performance Testing First Edition. 28 pp.

—Internal Combustion Engines—Flammability Testing
BSI BS 7525-92. 1992 Flammability of Air Cleaner Elements for Internal Combustion Engines. 13 pp.

—Leakage
SAA AS 1132.9-73. Methods of Test for Air Filters for Use in Air Conditioning and General Ventilation—Part 9: Determination of Local Area Flaws and Pinhole Leaks.

—Lubricating Oils
JIS K 2243-83. Air Filter Oil. 12 pp.

—NBC—Activated Carbon
MOD UK DSTAN 68-133-90. Carbon, Activated, Impregnated, ASC/TEDA Type 1101 Issue 1. 24 pp.
MOD UK DSTAN 68-133-92. Carbon, Activated, Impregnated, ASC/TEDA, Type 1101 Issue 2. 20 pp.
MOD UK DSTAN 68-135-90. Charcoal, Activated, Type 1121 Issue 1. 14 pp.
MOD UK DSTAN 68-135-01. Charcoal, Activated, Type 1121 Issue 1; Amendment 1. 15 pp.
MOD UK DSTAN 68-139-01. Charcoal, Activated, Type 1142 Issue 1; Corrigendum. 15 pp.

—Panels
CGSB CAN/CGSB-115.10-M90. Disposable Air Filters for the Removal of Particulate Matter from Ventilating Systems. 9 pp.
CGSB CAN/CGSB-115.18-M85. Filter, Air, Extended Area Panel Type, Medium Efficiency. 9 pp.

—Particulate Removal
BSI BS EN 779-93. 1993 Particulate Air Filters for General Ventilation—Requirements, Testing, Marking (N). 28 pp.

—Pneumatic Equipment
JIS B 8371-81. Filters for Pneumatic Use. 15 pp.
MOD UK NES 316-90. Requirements for Compressed Air Purification Equipment Issue 2 (07.90). 23 pp.

—Resistance
SAA AS 1132.2-73. Methods of Test for Air Filters for Use in Air Conditioning and General Ventilation—Part 2: Determination of Initial Resistance.

—Respirators—NBC—Flight Crews
NATO STANAG 3501 ED 2 AMD 1-90. Performance of Portable Filter—Blowers for Aircrew NBC Respirators. 9 pp.
NATO STANAG 3501 ED 2 AMD 2-90. Performance of Portable Filter—Blowers for Aircrew NBC Respirators. 12 pp.

—Respirators—Sodium Chloride Particulate Testing
BSI BS 4400-69. 1969 Amd 1 Sodium Chloride Particulate Test for Respirator Filters. 48 pp.

—Rigid Type
CGSB CAN/CGSB-115.15-M91. High Efficiency Rigid Type Air Filters for the Removal of Particulate Matter from Ventilating Systems. 7 pp.

—Rolls
CGSB CAN/CGSB-115.13-M85. Filter Media, Automatic Roll. 10 pp.

—Sodium Flame Testing
BSI BS 3928-69. 1969 Method for Sodium Flame Test for Air Filters (Other Than for Air Supply to I.C. Engines and Compressors). 57 pp.

—Spacers
MOD UK TS 10147. Base Material for Spacer Card.
MOD UK TS 10148. Spacer Card.

—Submicron Particles
DIN ENGL 24184-90. Type Testing of High Efficiency Particulate Air Filters Using Paraffin Oil Mist as Test Aerosol (Dec). 4 pp.

—Synthetic Dust Weight Arrestance
BSI BS 6540: Part 1-85. 1985 Air Filters Used in Air Conditioning and General Ventilation Part 1: Methods of Test for Atmospheric Dust Spot Efficiency and Synthetic Dust Weight Arrestance (Supersedes BS 2831: 1971). 32 pp.

—Tractors
BSI BS 6347: Part 8-93. 1993 Performance Assessment of Agricultural Tractors Part 8: Method of Testing Engine Air Cleaners (ISO 789-8: 1991) (E). 12 pp.
ISO 789 Pt 8-91. Agricultural Tractors—Test Procedures—Part 8: Engine Air Cleaner First Edition. 11 pp.

—Trucks
ISO 7750 Pt 1-92. Commercial Vehicles—Dimensions of Air Filter Elements—Part 1: Types A and B Second Edition. 7 pp.
ISO 7750 Pt 2-84. Road Vehicles—Air Filter Elements for Commercial Vehicles—Dimensions—Part 2: Types C and D First Edition. 7 pp.

—Ventilation Equipment
SAA AS 1324-73. Air Filters for Use in Air Conditioning and General Ventilation. 8 pp.

—Ventilation Equipment—Glossaries
SAA AS 1323-73. Glossary of Terms for Air Filters for Use in Air Conditioning and General Ventilation Corrig.. 8 pp.

Air Flow
See Also: Flow Measurement; Gas Flow

—Acoustical Materials
BSI BS EN 29053-93. 1993 Acoustics—Materials for Acoustical Applications—Determination of Airflow Resistance (ISO 9053: 1991). 16 pp.
CEN EN 29053-93. Acoustics—Materials for Acoustical Applications—Determination of Airflow Resistance (ISO 9053: 1991). 11 pp.
ISO 9053-91. Acoustics—Materials for Acoustical Applications—Determination of Airflow Resistance First Edition; (CEN EN 29053: 1993). 13 pp.

—Cellular Plastics
BSI BS 4443: Part 6-91. 1991 Methods of Test for Flexible Cellular Materials Part 6: Method 14. Preparation of Water Extract Method 15. Determination of Tear Strength Method 16. Determination of Air Flow Value. 14 pp.
ISO 7231-84. Polymeric Materials, Cellular Flexible—Method of Assessment of Air Flow Value at Constant Pressure-Drop First Edition. 6 pp.

—Cereals—Pressure Measurement
ISO 4174-80. Cereals and Pulses—Measurement of Unit Pressure Losses Due to Single-Dimension Air Flow Through a Batch of Grain First Edition. 13 pp.

—Clean Rooms
SAA AS 1807.3-89. Cleanrooms, Workstations and Safety Cabinets—Methods of Test—Part 3: Determination of Air Velocity and Uniformity of Air Velocity in Laminar Flow Cleanrooms. 2 pp.
SAA AS 1807.4-89. Cleanrooms, Workstations and Safety Cabinets—Methods of Test—Part 4: Determination of Performance of Laminar Flow Cleanrooms Under Loaded Filter Conditions. 2 pp.
SAA AS 1807.11-89. Cleanrooms, Workstations and Safety Cabinets—Methods of Test—Part 11: Determination of Airflow Parallelism in Laminar Flow Cleanrooms. 1 p.

—Containment Cabinets
SAA AS 1807.1-89. Cleanrooms, Workstations and Safety Cabinets—Methods of Test—Part 1: Determination of Air Velocity and Uniformity of Air Velocity in Clean Workstations and Laminar Flow Safety Cabinets. 1 p.
SAA AS 1807.2-89. Cleanrooms, Workstations and Safety Cabinets—Methods of Test—Part 2: Determination of Performance of Clean Workstations and Laminar Flow Safety Cabinets Under Loaded Filter Conditions. 2 pp.
SAA AS 1807.21-89. Cleanrooms, Workstations and Safety Cabinets—Methods of Test—Part 21: Determination of Inward Air Velocity of Class 1 Biological Safety Cabinets. 2 pp.

—Fans
ISO 9097-91. Small Craft—Electric Fans First Edition. 6 pp.

—Legumes—Pressure Measurement
ISO 4174-80. Cereals and Pulses—Measurement of Unit Pressure Losses Due to Single-Dimension Air Flow Through a Batch of Grain First Edition. 13 pp.

—Workstations—Clean Rooms
SAA AS 1807.1-89. Cleanrooms, Workstations and Safety Cabinets—Methods of Test—Part 1: Determination of Air Velocity and Uniformity of Air Velocity in Clean Workstations and Laminar Flow Safety Cabinets. 1 p.
SAA AS 1807.2-89. Cleanrooms, Workstations and Safety Cabinets—Methods of Test—Part 2: Determination of Performance of Clean Workstations and Laminar Flow Safety Cabinets Under Loaded Filter Conditions. 2 pp.

Air Flow Angle Transmitters
See Also: Aircraft; Aircraft Instruments; Transmitters
BSI G 234-84. 1984 Self-Aligning Air Flow Transmitters for Use on Aircraft. 5 pp.

Air Flow Equipment
Use: Ventilation Equipment

Air Force Operations
See Also: Airborne Operations; Military Operations

—Air Movements—Forms (Paper)
NATO STANAG 3345 ED 5 AMD 2-88. Data/Forms for Planning Air Movements. 13 pp.
NATO STANAG 3345 ED 5 AMD 3-88. Data/Forms for Planning Air Movements. 12 pp.

—Airports
NATO STANAG 3739 ED 3 AMD 2-86. Combined Air Terminal Operations. 12 pp.
NATO STANAG 3739 ED 3 AMD 4-86. Combined Air Terminal Operations. 7 pp.

Air Furnaces
See Also: Furnaces
JIS A 4003-77. Warm Air Furnaces.

Air Gages
CNS B6046-81. Flow Type Air Gauges (Jun)(7547).
JIS B 7535-82. Flow Type Air Gauges. 15 pp.

Air Gauging
Use: Pneumatic Length Measurement

Air Gratings
Use: Registers (Air Circulation)

Air Grinders
Use: Pneumatic Grinders

Air Guns
Use For: Air Pistols; Air Rifles
See Also: Pneumatic Equipment

—Bores
JIS S 7101-68. Dimensions of Bore and Chamber for Small Bore Rifles and Air Guns.

—Chambers
JIS S 7101-68. Dimensions of Bore and Chamber for Small Bore Rifles and Air Guns.

INTERNATIONAL AND NON-U.S. NATIONAL STANDARDS
SUBJECT INDEX
Air

Air Handling Equipment
See Also: Air Diffusion; Air Terminal Devices; Blowers (Ventilation); Building Services; Convectors; Dampers; Heating, Ventilating and Air Conditioning Equipment; Heating Equipment; Ventilation Equipment

—Aerodynamics
BSI BS 4979-86. 1986 Aerodynamic Testing of Constant and Variable Dual or Single Duct Boxes, Single Duct Units and Induction Boxes for Air Distribution Systems. 22 pp.
ISO 5220-81. Air Distribution and Air Diffusion—Aerodynamic Testing and Rating of Constant and Variable Dual or Single Duct Boxes and Single Duct Units First Edition; (Addendum 1-1984). 16 pp.

—Ceilings—Compatibility
SAA AS 2946-91. Suspended Ceilings, Recessed Luminaires and Air Diffusers—Interface Requirements for Physical Compatability (In Professional Package 50). 22 pp.

—Engineering Drawings—Symbols
ISO 4067 Pt 1-84. Technical Drawings—Installations—Part 1: Graphical Symbols for Plumbing, Heating, Ventilation and Ducting First Edition. 12 pp.

—Fire Protection—Maintenance
SAA AS 1851.6-83. Maintenance of Fire Protection Equipment—Part 6: Management Procedures for Maintaining the Fire Precaution Features of Air-Handling Systems. 16 pp.
SAA AS 1851.6C-83. Maintenance of Fire Protection Equipment—Part 6C: Air-Handling Systems—Fire Precaution Features—Typical Maintenance Schedule 630 X 430mm Enlargement of Table C1 of AS 1851, Part 6..

—Glossaries
ISO 3258-76. Air Distribution and Air Diffusion—Vocabulary First Edition. 36 pp.

—Ground Effect Machines—Contamination
CAA 42.
CAA Chapter B6-6 01.91. Heating and Ventilation Systems. 1 p.

—Induction Units—Aerodynamics
BSI BS 4954: Part 1-73. 1973 Methods for Testing and Rating Induction Units for Air Distribution Systems Part 1: Thermal and Aerodynamic Performance. 30 pp.

—Induction Units—Noise
BSI BS 4954: Part 2-78. 1978 Methods for Testing and Rating Induction Units for Air Distribution Systems Part 2: Acoustic Testing and Rating. 16 pp.

—Induction Units—Thermal Measurement
BSI BS 4954: Part 1-73. 1973 Methods for Testing and Rating Induction Units for Air Distribution Systems Part 1: Thermal and Aerodynamic Performance. 30 pp.

—Lighting
CENELEC EN 60 598-2-19-89. Luminaires Part 2: Particular Requirements Section Nineteen-Air-Handling Luminaires (Safety Requirements). 7 pp.
IEC 598 Pt 2-19-81. Luminaires Part 2: Particular Requirements Section Nineteen—Air-Handling Luminaires (Safety Requirements) First Edition; (Amendment 1-1987). 26 pp.

—Lighting—Safety
BSI BS 4533: Sec 102.19-90. 1990 Luminaires Part 102: Particular Requirements Section 102.19: Specification for Air-Handling Luminaires (Safety Requirements). 19 pp.

—Microorganism Control
SAA HB32-92. Control of Microbial Growth in Air-Handling and Water Systems in Buildings (in Professional Packages 20, 21, 44, 50B, 62, 63, 64, 65, 66, 67, 68). 54 pp.

—Noise
DIN ENGL 45635 Pt 56-86. Measurement of Noise Emitted by Machines; Airborne Noise Emission; Enveloping Surface Method and In-Duct Method; Fan-Assisted Warm Air Generators, Fan-Assisted Air Heaters and Fan Units of Air Handling Devices (Oct). 6 pp.

—Silencers
BSI BS 4718-71. 1971 Methods of Test for Silencers for Air Distribution Systems. 37 pp.

—Sound Power
CEN EN 25135-91. Acoustics—Determination of Sound Power Levels of Noise from Air Terminal Devices, High/Low Velocity/Pressure Assemblies, Dampers and Valves by Measurement in a Reverberation Room. 3 pp.

Air Handling Equipment (Cont.)
—Sound Power (Cont.)
DIN ENGL EN 25135-91. Determination of Sound Power Levels of Noise from Air Terminal Devices, High/Low Velocity/Pressure Assemblies, Dampers and Valves by Measurement in a Reverberation Room (ISO 5135: 1984) (Nov). 15 pp.
ISO 5135-84. Acoustics—Determination of Sound Power Levels of Noise from Air Terminal Devices, High/Low Velocity/Pressure Assemblies, Dampers and Valves by Measurement in a Reverberation Room First Edition. 13 pp.

Air Heaters
Use: Air Preheaters

Air Hose Assemblies
See Also: Air Hoses

—Brakes—Automotive
SNZ NZS/SAEJ1402-85. Automotive Air Brake Hose and Hose Assemblies. 4 pp.

Air Hoses
See Also: Air Hose Assemblies; Hoses; Pneumatic Hoses
JIS K 6336-77. Air Pump Hose.

—Aircraft—Compressed Air—Interdepartmental Procurement
NATO STANAG 3054 ED 4 AMD 4-73. Characteristics of Compressed Air for Technical Purposes, Supply Pressure and Hoses. 6 pp.
NATO STANAG 3054 ED 5 AMD 0-89. Characteristics of Compressed Air for Technical Purposes, Supply Pressure and Hoses. 6 pp.
NATO STANAG 3054 ED 5 AMD 1-89. Characteristics of Compressed Air for Technical Purposes, Supply Pressure and Hoses. 6 pp.

—Aircraft—Nitrogen
NATO STANAG 3624 ED 2 AMD 2-89. Characteristics of Oil Free Compressed Nitrogen, Supply Pressure and Hoses. 7 pp.

—Brakes—Automotive
SNZ NZS/SAEJ1402-85. Automotive Air Brake Hose and Hose Assemblies. 4 pp.

—Brakes—Automotive—Rubber
BSI BS AU 110-65. 1965 Amd 1 Air Pressure Rubber Brake Hose. 5 pp.
CNS D2162-86. Air Brake Hoses for Automobiles (Oct)(9847).
CNS D2172-86. Nylon Tubes for Air Brake Piping of Automobiles (Oct)(10332).
CNS D2174-86. Vacuum Brake Hose for Automobiles (Oct)(10709).
DIN ENGL 74310 Pt 1-76. Air Braking Systems; Hoses; Dimensions; Material; Marking (Aug). 2 pp.
DIN ENGL 74310 Pt 2-76. Air Braking Systems; Hoses; Requirements; Tests (Aug). 3 pp.
JIS D 2606-80. Air Brake Hoses for Automobiles (R 1985). 12 pp.
JIS D 2607-80. Vacuum Brake Hoses for Automobiles (R 1985). 11 pp.
SNZ NZS/BSAU 110-65. Specification for Air Pressure Rubber Brake Hose Amend: 1. 5 pp.

—Brakes—Rolling Stock
CNS E1013-82. Air Brake Hose for Railway Rolling Stock (Dec)(9694).
JIS E 4305-59. Air Brake Hose for Rolling Stock.

—Diving
CNS K4069-82. Diving Hoses (Nov)(9618).

—Mining
SAA AS 2660-91. Hose and Hose Assemblies —Air/Water—for Underground Coal Mines Amdt 1 May 1992 (in Professional Package 38). 1 p.

—Rubber—Diving
JIS K 6345-82. Rubber Hoses for Diving.

Air Intakes
See Also: Supersonic Inlets

—Aircraft Engines
CAA Chapter G5-5 06.76. Air Intake Systems (Rotorcraft). 2 pp.
CAA Chapter K5-5 04.74. Engine Air Intake and Ice Protection Systems (Light Aeroplanes). 3 pp.
CAA Chapter Q5-5 12.79. Engine Air Intake and Ice Protection Systems (Non-Rigid Airships). 2 pp.
CAA CAP 524 SUB-Part E 12.86. Power-Plant (Rotorcraft). 27 pp.

Air Kerma Rate Meters
Use: Kerma Rate Meters

Air Leakage
See Also: Leakage

Air Leakage (Cont.)
—Buildings
CGSB CAN/CGSB-149.10-M86. Determination of the Airtightness of Building Envelopes by the Fan Depressurization Method. 39 pp.

—Connectors
CEN PREN 2591 (Part C12)-92. Elements of Electrical and Optical Connection Test Methods Part C12—Air Leakage. 5 pp.

Air Lifts (Materials Handling)
Use: Gas Lifts

Air Mattresses
See Also: Bedding; Mattresses

—Medical—Alternating Pressure
BSI BS 7068-89. 1989 Alternating Pressure Air Mattresses. 16 pp.

Air Medical Transport Units
Use For: Air Ambulances *See Also:* Aeromedical Evacuation; Air Transportation; Aircraft; Ambulances; Helicopters; Transport Aircraft; Vehicles (Transportation)

—Military Operations
NATO STANAG 2087 ED 4 AMD 1-83. Medical Employment of Air Transport in the Forward Area. 8 pp.
NATO STANAG 2087 ED 4 AMD 3-83. Medical Employment of Air Transport in the Forward Area. 9 pp.

Air Monitoring
Scope Note: Includes pollution measurements in the ambient air; for monitoring of specific toxins, see also specific substances *See Also:* Air; Air Pollution; Ambient Temperature; Exhaust Gases; Flue Gases
ISO TR4227-89. Planning of Ambient Air Quality Monitoring First Edition. 20 pp.
ISO 6879-83. Air Quality—Performance Characteristics and Related Concepts for Air Quality Measuring Methods First Edition. 6 pp.
ISO 7168-85. Air Quality—Presentation of Ambient Air Quality Data in Alphanumerical Form First Edition. 10 pp.

—Carbon Monoxide Analyzers
JIS B 7951-86. Continuous Analyzer for Carbon Monoxide in Ambient Air. 14 pp.

—Chlorine Analyzers
JIS B 7955-79. Continuous Analyzers for Chlorine in Ambient Air.

—Classification
ISO TR4227-89. Planning of Ambient Air Quality Monitoring First Edition. 20 pp.

—Dust
BSI BS 1747: Part 5-72. 1972 Methods for the Measurement of Air Pollution Part 5: Directional Dust Gauges. 21 pp.
CNS K9013-76. Method of Determination for Dust-Fall in Atmosphere (Mar)(3916).
CNS K9066-81. General Rule of Measuring Methods for Airborne Dust Concentration in Environmental Atmosphere (Apr)(7252).
JIS Z 8813-79. General Rules of Measuring Methods for Airborne Dust Concentration in Environmental Atmosphere.
JIS Z 8813-69. General Rules of Measuring Methods for Airborne Dust Concentration in Environmental Atmosphere. 7 pp.

—Exhaust Gas Analyzers
JIS K 0055-86. General Rules for Calibration Method of Gas Analyzer. 10 pp.

—Fluorine Analyzers
JIS B 7958-80. Continuous Analyzers for Fluorine Compounds in Ambient Air.

—Gases—Workplace
CEN PREN 838-92. Workplace Atmospheres—Requirements and Test Methods for Diffusive Samplers for the Determination of Gases and Vapours. 40 pp.

—Glossaries
BSI BS 6069: Part 2-83. 1983 Characterization of Air Quality Part 2: Glossary. 7 pp.
ISO 4225-80. Air Quality—General Aspects—Vocabulary First Edition. 10 pp.
ISO 6879-83. Air Quality—Performance Characteristics and Related Concepts for Air Quality Measuring Methods First Edition. 6 pp.

—Hydrocarbon Analyzers
JIS B 7956-79. Continuous Analyzers for Hydrocarbons in Ambient Air.

INDUSTRY STANDARDS

Air Monitoring (Cont.)

—Oxidizer Analyzers
JIS B 7957-92. Continuous Analyzers for Oxidants in Ambient Air. 29 pp.

—Ozone
ECMA ECMA-TR 56-91. Information Technology Equipment Recommended Measuring Method for Ozone Emission. 17 pp.

—Particulates
BSI BS 1747: Part 1-69. 1969 Methods for the Measurement of Air Pollution Part 1: Deposit Gauges. 17 pp.
ISO TR7708-83. Air Quality—Particle Size Fraction Definitions for Health-Related Sampling First Edition. 13 pp.
JIS B 7954-88. Automatic Monitors for Suspended Particulate Matter in Ambient Air. 21 pp.

—Sampling
ISO 9359-89. Air Quality—Stratified Sampling Method for Assessment of Ambient Air Quality First Edition. 15 pp.

—Smoke
BSI BS 1747: Part 2-69. 1969 Methods for the Measurement of Air Pollution Part 2: Determination of Concentration of Suspended Matter. 17 pp.

—Sulfur Dioxide Analyzers
BSI BS 1747: Part 3-69. 1969 Methods for the Measurement of Air Pollution Part 3: Determination of Sulphur Dioxide. 13 pp.
BSI BS 1747: Part 6-83. 1983 Methods for the Measurement of Air Pollution Part 6: Sampling Equipment Used for the Determination of Gaseous Sulphur Compounds in Ambient Air. 8 pp.
BSI BS 1747: Part 7-83. 1983 Methods for the Measurement of Air Pollution Part 7: Determination of Mass Concentration of Sulphur Dioxide in Ambient Air. Thorin Spectrophotometric Method. 11 pp.
ISO 4219-79. Air Quality—Determination of Gaseous Sulphur Compounds in Ambient Air—Sampling Equipment First Edition. 6 pp.
JIS B 7952-77. Continuous Analyzers for Sulfur Dioxide in Ambient Air. 25 pp.

—Units of Measurement
BSI BS 6069: Part 1-81. 1981 Characterization of Air Quality Part 1: Units of Measurement. 4 pp.
ISO 4226-80. Air Quality—General Aspects—Units of Measurement First Edition. 4 pp.

—Vapors—Workplace
CEN PREN 838-92. Workplace Atmospheres—Requirements and Test Methods for Diffusive Samplers for the Determination of Gases and Vapours. 40 pp.

Air Motors
Use: Pneumatic Motors

Air Moving Products
Use: Ventilation Equipment

Air Navigation
Use For: Aeronautical Navigation; Aircraft Navigation; International Air Navigation
See Also: Aeronautical Communications; Air Navigation Facilities; Air Navigation Plans; Air Transportation; Aircraft; Flight Control Systems; Flight Levels; Global Positioning Systems; Hyperbolic Navigation Systems; Inertial Navigation Systems; Instrument Flight; Instrument Landing Systems; Microwave Landing Systems; Navigational Aids; Omega Navigation Systems; Radio Navigation Equipment; Vertical Separation

—Accuracy Determination
NATO STANAG 4278 ED 2 AMD 0-86. Method of Expressing Navigation Accuracies. 13 pp.
NATO STANAG 4278 ED 3 AMD 0-89. (DRAFT) Method of Expressing Navigation Accuracies. 14 pp.

—Aeronautical Information Services
CAA Part 10 CAP 168. Aeronautical Information. 6 pp.
ICAO 8126. Aeronautical Information Services Manual Fourth Edition—1991. 263 pp.

—Aeronautical Meteorology
ICAO Annex 3. Meteorological Services for International Air Navigation Eleventh Edition—July 1992; (Incorporating Amendments 1 to 69 11/12/92) (Corrigendum 1/21/93). 98 pp.

—All-Weather Operations
ICAO 9365. Manual of All-Weather Operations Second Edition—1991. 69 pp.

Air Navigation (Cont.)

—Anticollision Lights
NATO STANAG 3153 ED 6 AMD 5-73. Aircraft Navigation and Anti-Collision Lights. 11 pp.

—Civil Aviation Authority—Orders and Regulations
CAA. Contents 11.91. 1 p.
CAA. Foreword 11.89. 1 p.
CAA. Check List for Pages 04.93. 2 pp.
CAA Section 3 CAP 393. References in the Air Navigation (General) Regulations 1981 to the Air Navigation Order. 34 pp.

—Facilitation
ICAO Annex 9. Facilitation Ninth Edition—July 1990; (Incorporating Amendments 1 to 14 and Corrigenda 1) (Supplement 7/1/91) (Corrigendum 11/9/92 to Supplement) (Amendment 15 7/26/93). 150 pp.
ICAO 7891. Aims of ICAO in the Field of Facilitation Second Edition—1965. 26 pp.

—Helicopters—Military Communication
NATO STANAG 2863 ED 5 AMD 2-88. Navigational and Communicational Capabilities for Helicopters in Multinational Land Operations. 26 pp.
NATO STANAG 2863 ED 5 AMD 4-88. Navigational and Communicational Capabilities for Helicopters in Multinational Land Operations. 28 pp.

—ICAO Special Committee on Future Air Navigation Systems
ICAO 9503. Special Committee on Future Air Navigation Systems Third Meeting Montreal, 3-21 November 1986 (Report); (Corrigendum/Supplement 1 07/31/87) (Corrigendum/Addendum 1 08/10/87). 190 pp.
ICAO 9524. Special Committee on Future Air Navigation Systems Fourth Meeting Montreal, 2-20 May 1988; (Supplement 1). 232 pp.

—Plans
ICAO 9573. Manual of Area Navigation (RNAV) Operations First Edition—1991. 32 pp.

—Procedures
ICAO 7030. Regional Supplementary Procedures Fourth Edition—1987; (Incorporating Amendments 1 to 168) (Amendment 169 8/21/87) (Amendment 170 3/25/88) (Amendment 171 9/22/88) (Amendment 172 5/15/89) (Amendment 173 10/23/89) (Amendment 174 4/2/90) (Amendment 175. 175 pp.
ICAO 9573. Manual of Area Navigation (RNAV) Operations First Edition—1991. 32 pp.

—Radio
ICAO 8071 Vol I. Manual on Testing of Radio Navigation Aids Volume 1 Third Edition—1972. 142 pp.
ICAO 8071 Vol II. Manual on Testing of Radio Navigation Aids Volume II ILS (Instrument Landing System) Third Edition—1972. 146 pp.

—Radio—Glossaries
BSI BS 185: Sec 14-72. 1972 Glossary of Aeronautical and Astronautical Terms Section 14: Radiocommunication and Radiolocation (Services and Stations, Radio-Navigational Aids, Radio Beacons, Equipment, Miscellaneous). 4 pp.

—Search and Rescue
ICAO Annex 12. Search and Rescue Sixth Edition—March 1975 (Incorporating Amendments 1-11); (Supplement 2/15/84) (Amendment 1 5/15/86 to Supplement) (Amendment 14 7/30/90) (Amendment 2 3/21/91) to Supplement) (Amendment 15 7/26/93). 59 pp.
ICAO 7333 Part 1. Search and Rescue Manual Part 1 the Search and Rescue Organization Third Edition—1970. 77 pp.
ICAO 7333 Part 2. Search and Rescue Manual Part 2 Search and Rescue Procedures Third Edition—1970. 126 pp.

Air Navigation Equipment
Use For: Aircraft Navigation Equipment
See Also: Aircraft Equipment; Aircraft Instruments; Altimeters; Flight Control Systems; Global Positioning Systems; Gyrocompasses; Localizers; Navigational Aids; Radio Navigation Equipment; Space Operation Systems

—Manual Controls
NATO STANAG 3258 ED 5 AMD 0-91. Position of Pilot Operated Navigation and Radio Controls. 5 pp.
NATO STANAG 3258 ED 6 AMD 0-92. Position of Pilot Operated Navigation and Radio Controls. 5 pp.

Air Navigation Facilities
See Also: Air Navigation; Air Transportation; Airports

—Charges
ICAO 9082. Statements by the Council to Contracting States on Charges for Airports and Air Navigation Services Fourth Edition—1992. 27 pp.

—Economics
ICAO 9161. Manual on Route Air Navigation Facility Economics Second Edition—1986. 67 pp.

—Tariffs
ICAO 7100. Manual of Airport and Air Navigation Facility Tariffs—1990. 438 pp.

Air Navigation Plans
See Also: Air Navigation; Air Transportation
ICAO 7474. Air Navigation Plan—Africa-Indian Ocean Region Twenty-Sixth Edition—1989; (Amendment 1 4/30/90) (Amendment 2 1/31/91) (Amendment 3 1/10/92). 393 pp.
ICAO 7754. Air Navigation Plan—European Region Twenty-Third Edition—1985; (Amd 1 9/27/85) (Amd 2 1/24/86) (Corr 3/21/86) (Amd 3 4/10/86) (Amd 4 7/23/86) (Amd 5 11/26/86) (Amd 6 2/13/87) (Amd 7 5/6/87) (Amd 8 7/29/87) (Addendum 8/14/87 to Amd 8) (Amd 9 10/21/87) (Amd 10 2/17/88). 1772 pp.
ICAO 8700. Air Navigation Plan—Middle East and Asia Regions Fifteenth Edition—1989; (Amendment 1 8/31/90) (Amendment 2 5/6/91) (Amendment 3 11/29/91) (Amendment 4 8/7/92) (Amendment 5 1/5/93). 436 pp.
ICAO 8733. Air Navigation Plan—Caribbean and South American Regions Fourteenth Edition—1991; (Amendment 1 02/07/92). 342 pp.
ICAO 8755. Air Navigation Plan—North Atlantic, North American and Pacific Regions Thirteenth Edition October—1990; (Corrigendum 2/13/91) (Amendment 1 10/15/91) (Amendment 2 7/15/92) (Amendment 3 1/5/93). 367 pp.

Air Operations
Use: Airborne Operations

Air Permeability
Use: Gas Permeability

Air Pistols
Use: Air Guns

Air Pollution
Use For: Air Quality *See Also:* Aerosols; Air Monitoring; Black Smoke Index; Catalytic Converters; Exhaust Emission Control Equipment; Exhaust Gases; Exhaust Systems; Fly Ash; Fumes; Industrial Wastes; Odors; Oxidizers; Particulates; Pollution Control Equipment; Smoke; Symbols; Waste Disposal
EURO 1991 Aug. Pollution Knows No Frontiers. 9 pp.

—Aircraft Engines
ICAO Annex 16 Vol II. Environmental Protection Volume II—Aircraft Engine Emissions Second Edition—July 1993 (Incorporating Amendments 1-2). 64 pp.
ICAO 9259. Committee on Aircraft Engine Emissions First Meeting Montreal, 12-22 June 1978; (Supplement 1 12/15/78). 195 pp.
ICAO 9304. Committee on Aircraft Engine Emissions Second Meeting Montreal, 14-29 May 1980; (Supplement 1 06/27/80). 154 pp.
ICAO 9499. Committee on Aviation Environmental Protection First Meeting Montreal, 9-20 June 1986. 230 pp.

—Aluminum Coatings (On Aluminum)—Anodic Oxide
ISO TR11728-93. Anodized Aluminium and Aluminium Alloys—Accelerated Test of Weather Fastness of Coloured Anodic Oxide Coatings Using Cyclic Artificial Light and Pollution Gas First Edition. 9 pp.

—Buildings—Human Factors Engineering
BSI BS 7643: Part 2-93. 1993 Building Construction—Expression of Users' Requirements Part 2: Air Purity Requirements (ISO 6242-2: 1992). 9 pp.
ISO 6242 Pt 2-92. Building Construction—Expression of Users' Requirements—Part 2: Air Purity Requirements First Edition. 8 pp.

—Electrical Insulators
BSI BS EN 60507-93. 1993 Artificial Pollution Tests on High-Voltage Insulators to be Used on a.c. Systems (IEC 507: 1991) (Q). 38 pp.
CENELEC EN 60507-93. Artificial Pollution Tests on High-Voltage Insulators to Be Used on A.C. Systems (IEC 507: 1991). 33 pp.

INTERNATIONAL AND NON-U.S. NATIONAL STANDARDS
SUBJECT INDEX

Air Pollution (Cont.)
—**Electrical Insulators** (Cont.)
IEC 507-91. Artificial Pollution Tests on High-Voltage Insulators to Be Used on a.c. Systems Second Edition (CENELEC EN 60507:1993). 59 pp.
IEC 815-86. Guide for the Selection of Insulators in Respect of Polluted Conditions First Edition. 42 pp.

—**Infrared Analyzers**
BSI BS 5849-80. 1980 Method of Expression of Performance of Air Quality Infra-Red Analysers. 20 pp.
IEC 528-75. Expression of Performance of Air Quality Infra-Red Analyzers First Edition. 37 pp.

—**Inorganic Fiber Content—Phase Contrast Microscopy**
ISO 8672-93. Air Quality—Determination of the Number Concentration of Airborne Inorganic Fibres by Phase Contrast Optical Microscopy—Membrane Filter Method First Edition. 30 pp.

—**Machinery—Test Methods**
CEN PREN 1093-1-93. Safety of Machinery—Evaluation of the Emission of Airborne Hazardous Substances—Part 1: Selection of Test Methods. 13 pp.

—**Mopeds**
ISO 6855-83. Road Vehicles—Measurement Methods for Gaseous Pollutants Emitted by Mopeds Equipped with a Controlled Ignition Engine Second Edition. 15 pp.
ISO TR6970-81. Road Vehicles—Pollution Tests for Motorcycles and Mopeds—Bench (Chassis Dynamometer) First Edition. 10 pp.

—**Motor Vehicles**
CNS D3166-87. Method of Test for Positive Crankcase Ventilation System for Gasoline Engine Automobiles (Sep)(11496). 7 pp.
CNS D3173-87. Method of Test for Evaporative Emission Control System for Gasoline Engine Automobiles (Sep)(11534). 10 pp.
EC COM(87) 706-88. Proposal for a Council Directive Amending Directive 70/220/EEC on the Approximation of the Laws of Member States Relating to Measures to be Taken Against air Pollution by Gases from the Engines of Motor Vehicles—European Emission Standard for Cars Below 1.4 Litres. 13 pp.
EC COM(88) 675-88. Amendment to the Proposal for a Council Directive amending Directive 70/220/EEC on the Approximation of the Laws of the Member States Relating to Measures to be Taken Against Air Pollution by Gases from the Engines of Motor Vehicles (European Emission). 4 pp.
EC COM(92) 572-93. Proposal for a Council Directive Relating to Measures to Be Taken Against Air Pollution by Emissions from Motor Vehicles and Amending Directive 70/220/EEC. 10 pp.
EC COM(93) 277-93. Re-Examined Proposal for a Council Directive Amending Directive 70/220/EEC on the Approximation of the Laws of the Member States Relating to Measures to Be Taken Against Air Pollution by Emmissions from Motor Vehicles. 14 pp.
EC 88/76/EEC-87. Council Directive Amending Directive 70/220/EEC on the Approximation of the Laws of the Members States Relating to Measures to be Taken Against Air Pollution by Gases from the Engines of Motor Vehicles. 32 pp.
EC 88/436/EEC-88. Council Directive Amending Directive 70/220/EEC on the Approximation of the Laws of the Member States Relating to Measures to be Taken Against Air Pollution by Gases from Engines of Motor Vehicles (Restriction of Particulate Pollutant). 18 pp.
EC 89/458/EEC-89. Council Directive Amending with Regard to European Emission Standards for Cars Below, 1,4 Litres, Directive 70/220/EEC on the Approximation of the Laws of the Member States Relating to Measures to Be Taken Against Air Pollution by Emissions from Motor. 3 pp.
EC 91/441/EEC-91. Council Directive Amending Directive 70/220/EEC on the Approximation of the Laws of the Member States Relating to Measures to be Taken Against Air Pollution by Emissions from Motor Vehicles. 106 pp.

—**Motor Vehicles—Glossaries**
JIS D 0108-85. Glossary of Terms Relating to Control of Air Pollutions from Automobile Emissions. 53 pp.

—**Motorcycles**
ISO 6460-81. Road Vehicles—Measurement Method of Gaseous Pollutants Emitted by Motorcycles Equipped with a Controlled Ignition Engine First Edition. 16 pp.
ISO TR6970-81. Road Vehicles—Pollution Tests for Motorcycles and Mopeds—Bench (Chassis Dynamometer) First Edition. 10 pp.

Air Pollution (Cont.)
—**Surge Arresters**
IEC 99 Pt 3-90. Surge Arresters Part 3: Artificial Pollution Testing of Surge Arresters First Edition. 26 pp.

—**Welding**
DIN ENGL 32507-84. Sampling for the Purpose of Determining the Concentration of Air Pollutants Within the Operator's Breathing Zone During Welding and Allied Processes (May). 6 pp.

Air Preheaters
Use For: Air Heaters *See Also:* Heating Equipment
CEN PREN 621-91. Requirements for Non-Domestic Gas-Fired Forced Convection Air Heaters for Space Heating. 92 pp.
CSA C22.2 NO 92-1971. Dehumidifiers (R 1992); (Rev 1-3 April 1973). 32 pp.
CSA 1169 Bull. Electrical Bulletin 1169 June 27, 1978 to C22.2 NO 92. 2 pp.

—**Deposits—Sampling**
BSI BS 2455: Part 2-83. 1983 Methods of Sampling and Examining Deposits from Boilers and Associated Industrial Plant Part 2: Methods for Sampling and Examining Free-Side Deposits. 41 pp.
BSI BS 2455: Part 2-01. 1983 Amd 1 Sampling and Examining Deposits from Boilers and Associated Industrial Plant Part 2: Methods for Sampling and Examining Fire-Side Deposits (AMD 7751) October 15, 1993 (Q). 42 pp.

—**Flues**
BSI BS 5440: Part 1-90. 1990 Code of Practice for Flues and Air Supply for Gas Appliances of Rated Input Not Exceeding 60 kW (1st and 2nd Family Gases) Part 1: Installation of Flues. 44 pp.
BSI BS 5440: Part 1-78. (WITHDRAWN) 1978 Amd 1 Code of Practice for Flues and Air Supply for Gas Appliances of Rated Input Not Exceeding 60 kW (1st and 2nd Family Gases) Part 1: Flues. 42 pp.

—**Gas Fired—Safety**
CEN PREN 1020-93. Requirements for Non-Domestic Gas-Fired Forced Convection Air Heaters for Space Heating Incorporating a Fan to Assist Transportation of Combustion Air and/or Combustion Products. 137 pp.

—**Household**
CEN PREN 778-92. Requirements for Domestic Gas-Fired Forced Convection Air Heaters for Space Heating. 99 pp.

—**Noise**
DIN ENGL 45635 Pt 56-86. Measurement of Noise Emitted by Machines; Airborne Noise Emission; Enveloping Surface Method and In-Duct Method; Fan-Assisted Warm Air Generators, Fan-Assisted Air Heaters and Fan Units of Air Handling Devices (Oct). 6 pp.

—**Portable—Heating Elements—Open Coil**
CSA C22.2 NO 46-M1988. Electric Air-Heaters. 76 pp.

—**Safety**
BSI BS 5990-90. 1990 Amd 1 Direct Gas-Fired Forced Convection Air Heaters with Rated Heat Inputs up to 2 MW for Ind. and Commercial Space Heating Safety and Performance Requirements (Excl. Electrical Reqmts.) (2nd Family Gases) (AMD 6649) October 31, 1990. 26 pp.

—**Thermal Efficiency**
BSI BS 6332: Part 6-90. 1990 Thermal Performance of Domestic Gas Appliances Part 6: Thermal Performance of Combined Appliances: Fanned-Circulatopm Ducted-Air Heater/Circulator. 17 pp.

Air Pumps
See Also: Pumps

—**Automotive**
CNS D3057-89. Method of Test for Air Pumps for Automobiles (Jul)(5777).

—**Hoses**
JIS K 6336-77. Air Pump Hose.

—**Hoses—Rubber**
CNS K4011-86. Rubber Hoses for Air Pump (Oct)(812).

—**Hydrobaths—Household—Portable**
CSA 1268 Bull. Electrical Bulletin 1268 May 16, 1980 to C22.2 NO 108. 5 pp.

—**Tubes—Sorbent—Workplace Atmospheres**
CEN PREN 1076-93. Workplace Atmospheres—Pumped Sorbent Tubes for the Determination of Gases and Vapours—Requirements and Test Methods. 41 pp.

Air Quality
Use: Air Pollution

Air Quality Infrared Analyzers
Use: Infrared Analyzers

Air Quality Measurement
Use: Air Monitoring

Air Reconnaissance
Use: Aerial Reconnaissance

Air Registers
Use: Registers (Air Circulation)

Air Reservoirs
Use: Pneumatic Reservoirs

Air Resistance
—**Paper Products**
CPPA D.14-77. Air Resistance of Paper (Gurley Method). 2 pp.

—**Papers**
CPPA D.14-77. Air Resistance of Paper (Gurley Method). 2 pp.

Air Rifles
Use: Air Guns

Air Shutters
Use: Air Valves

Air Springs
See Also: Springs (Elastic)

—**Bellows—Rubber**
CNS D3080-81. Testing Method of Rubber Bellows for Air Springs (Oct)(7977).
JIS D 4101-77. Testing Method of Rubber Bellows for Air Springs.

Air Standardization Coordinating Committee
Use: ASCC

Air Supported Structures
BSI BS 6661-86. 1986 Design, Construction and Maintenance of Single-Skin Air Supported Structures. 20 pp.
CSA CAN3-S367-M81. Air-Supported Structures; (Gen Instr 1 Thru 2). 53 pp.

Air Surveillance
—**Message Formats—Military Communications—Interoperability**
NATO STANAG 4312 Pt1 ED1 AMD0-90. Interoperability of Army Short Range Air Defence Surveillance, Command and Control Systems Part I: Information Exchange Requirements (IERs) for Developing Common Character Oriented Message Text Formats (MFTs) for Use in Immediate Future. 16 pp.

—**Military Communications—Interoperability**
NATO STANAG 4312 Pt2 ED1 AMD0-00. Interoperability of Army Short Range Air Defence Surveillance, Command and Control Systems Part II: Interface Requirements and Bit-Oriented Messages. 60 pp.
NATO STANAG 4312 ED 2 AMD 0-89. (DRAFT) Interoperability of Army Short Range Air Defence Surveillance, Command and Control Systems Part II: Common Interface Requirements and Bit-Oriented Messages. 12 pp.

Air Switches
Use For: Air Break Switches *See Also:* Busbars; Circuit Breakers; Electric Switches; Knife Switches; Switches
BSI BS 5419-77. (WITHDRAWN) 1977 Amd 1 Air-Break Switches, Air-Break Disconnectors, Air-Break Switch Disconnectors and Fuse Combination Units for Voltages up to and Including 1000 V a.c. and 1200 V d.c. (Superseded by BS EN 60947-3: 1992). 51 pp.
CNS C1106-85. Minimum Breaking Distance for Air Break Switches and Air Circuit Breakers (Under or Equal to AC 600V) (Jun)(9356).
JIS C 0921-84. Minimum Breaking Distance for Air Break Switches and Air Circuit Breakers (Under or Equal to AC 600 V). 5 pp.
SAA AS 1775-84. Low Voltage Switchgear and Controlgear—Air-Break Switches, Isolators and Fuse-Combination Units (up to and Including 1 000 V a.c. and 1 200 V d.c.) Amdt 1 January 1985. 38 pp.
SAA AS 3133-89. Approval and Test Specification—Air Break Switches Amdt 1 October 1990 (This is a Joint Standard with SANZ NZS 3133). 9 pp.

Air Switches (Cont.)
SNZ NZS/AS 3133-89. Approval and Test Specification—Air Break Switches Amend: 1, 1990 (This is a Joint Standard with SAA AS 3133). 9 pp.

—**Enclosed**
CSA CAN/CSA-C22. 2 NO 4-M89. Enclosed Switches; (Gen Instr 1 Thru 3). 78 pp.
CSA 655 Bull. Electrical Bulletin 655 October 12, 1966 to C22.2 NO 4. 4 pp.

—**Low Voltage**
CENELEC HD 422-82. Low-Voltage Air-Break Switches, Air-Break Disconnectors, Air-Break Switch-Disconnectors and Fuse-Combination Units. 7 pp.

—**Miniature**
BSI BS 3871: Part 1-65. 1965 Amd 4 Miniature and Moulded Case Circuit-Breakers Part 1: Miniature Air-Break Circuit-Breakers for a.c. Circuits (AMD 6431) March 30, 1990. 58 pp.
SNZ NZS 2205: Part 1-68. Specification for Miniature and Moulded Case Circuit-Breakers Part 1: Miniature Air-Break Circuit-Breakers for Alternating Current Circuits Amend: A, 1968. 52 pp.

Air Technical Publications
Use: Technical Writing—Air Technical Publications

Air Terminal Devices
See Also: Air Handling Equipment; Registers (Air Circulation); Ventilation Equipment

—**Aerodynamics**
BSI BS 4773: Part 1-71. (WITHDRAWN) 1971 Methods for Testing and Rating Air Terminal Devices for Air Distribution Systems Part 1: Aerodynamic Testing (Superseded by BS 4773: Section 1.1: 1991). 53 pp.
BSI BS 4773: Sec 1.1-91. 1991 Methods for Testing and Rating Air Terminal Devices for Air Distribution Systems Part 1: Aerodynamic Testing Section 1.1: Testing Under Laboratory Conditions. 30 pp.
ISO 5219-84. Air Distribution and Air Diffusion—Laboratory Aerodynamic Testing and Rating of Air Terminal Devices First Edition. 30 pp.

—**Engineering Drawings**
ISO 6412 Pt 3-93. Technical Drawings—Simplified Representation of Pipelines—Part 3: Terminal Features of Ventilation and Drainage Systems First Edition. 8 pp.

—**Noise**
BSI BS 4773: Part 2-89. 1989 Methods for Testing and Rating Air Terminal Devices for Air Distribution Systems Part 2: Acoustic Testing. 16 pp.
BSI BS 4773: Part 2-01. 1989 Amd 1 Methods for Testing and Rating Air Terminal Devices for Air Distribution Systems Part 2: Acoustic Testing (ISO 5135: 1984) (AMD 6971) April 1, 1992. 22 pp.
BSI BS 4773: Part 2-76. 1976 Methods for Testing and Rating Air Terminal Devices for Air Distribution Systems Part 2: Acoustic Testing. 16 pp.
BSI BS 4857: Part 2-78. 1978 Methods for Testing and Rating Terminal Reheat Units for Air Distribution Systems Part 2: Acoustic Testing and Rating. 18 pp.

—**Sound Power**
CEN EN 25135-91. Acoustics—Determination of Sound Power Levels of Noise from Air Terminal Devices, High/Low Velocity/Pressure Assemblies, Dampers and Valves by Measurement in a Reverberation Room. 3 pp.
DIN ENGL EN 25135-91. Determination of Sound Power Levels of Noise from Air Terminal Devices, High/Low Velocity/Pressure Assemblies, Dampers and Valves by Measurement in a Reverberation Room (ISO 5135: 1984) (Nov). 15 pp.
ISO 5135-84. Acoustics—Determination of Sound Power Levels of Noise from Air Terminal Devices, High/Low Velocity/Pressure Assemblies, Dampers and Valves by Measurement in a Reverberation Room First Edition. 13 pp.

Air Terminals (Buildings)
Use: Airports

Air Traffic Control
See Also: Aeronautical Telecommunication Network; Air Traffic Control Equipment; Air Transportation; Aircraft; Aircraft Communications; Aircraft Landing Approaches; Aircraft Landings; Airports; Collision Avoidance; Flight Paths; Instrument Landing Systems; Takeoff

Air Traffic Control (Cont.)
—**Aeronautical Meteorology**
ICAO 9377. Manual on Co-Ordination Between Air Traffic Services and Aeronautical Meteorological Services First Edition—1983; (Corrigendum 01/18/84). 65 pp.

—**Airspace—Combat Zones—Allied Tactical Publications**
NATO STANAG 3805 ED 2 AMD 0-80. Doctrine and Procedures for Airspace Control in the Combat Zone. 5 pp.
NATO STANAG 3805 ED 3 AMD 0-89. Doctrine and Procedures for Airspace Control in the Combat Zone—ATP-40. 5 pp.

—**Automatic Dependent Surveillance**
ICAO Circular 226. Automatic Dependent Surveillance—1990. 30 pp.

—**Ergonomics**
ICAO Circular 241. Human Factors Digest No. 8 Human Factors in Air Traffic Control—1993. 48 pp.

—**Flight Information—Pilot**
CAA Part 10 CAP 168. Aeronautical Information. 6 pp.
ICAO Circular 211. Aerodrome Flight Information Service (AFIS)—1988. 15 pp.

—**Glossaries**
BSI BS 185: Sec 13-72. 1972 Glossary of Aeronautical and Astronautical Terms Section 13: Air-Traffic and Ground Services (General, Air-Traffic Control, Ground Services, Naval Terms). 12 pp.
NATO STANAG 3993 ED 1 AMD 1-88. Air Control Terms and Definitions. 10 pp.

—**Licenses—Charges**
CAA Section 6 CAP 393. CAA Charges. 46 pp.

—**Minimum Operating Strips**
NATO STANAG 7025 ED 1 AMD 0-92. Air Traffic Management and Control of Minimum Operating Strips (MOS) Operations. 7 pp.
NATO STANAG 7025 ED 1 AMD 1-92. Air Traffic Management and Control of Minimum Operating Strips (MOS) Operations. 6 pp.

—**Navigational Aids—Failure Procedures**
NATO STANAG 3530 ED 4 AMD 3-84. Radio and/or Navigational Aid Failure Procedures for Operational Air Traffic (OAT) Flights. 10 pp.
NATO STANAG 3530 ED 4 AMD 4-84. Radio and/or Navigational Aid Failure Procedures for Operational Air Traffic (OAT) Flights. 11 pp.

—**Radio Communications**
EUROCAE ED-67 04.91. MOPS for Devices That Prevent Unintentional or Continuous Transmissions. 49 pp.

—**Radio Communications—Failure Procedures**
NATO STANAG 3530 ED 4 AMD 3-84. Radio and/or Navigational Aid Failure Procedures for Operational Air Traffic (OAT) Flights. 10 pp.
NATO STANAG 3530 ED 4 AMD 4-84. Radio and/or Navigational Aid Failure Procedures for Operational Air Traffic (OAT) Flights. 11 pp.

—**Rules**
CAA. Contents 11.91. 1 p.
CAA. Foreword 11.89. 1 p.
CAA Section 1 CAP 393. Air Navigation Order 1989. 188 pp.
CAA Section 2 CAP 393. Rules of the Air Regulations 1991. 51 pp.
CAA Section 3 CAP 393. References in the Air Navigation (General) Regulations 1981 to the Air Navigation Order. 34 pp.

—**Rules—Flying Displays—Air Shows**
NATO STANAG 3533 ED 4 AMD 3-85. Safety Rules for Flying and Static Displays. 16 pp.
NATO STANAG 3533 ED 5 AMD 0-92. Safety Rules for Flying and Static Displays. 15 pp.
NATO STANAG 3533 ED 5 AMD 1-92. Flying and Static Displays. 17 pp.

—**Services**
ICAO Annex 11. Air Traffic Services Air Traffic Control Service Flight Information Service Alerting Service Ninth Edition—July 1990 (Incorporating Amendments 1-33); (Corrigendum 2/13/91) (Amendment 34 7/26/93). 67 pp.
ICAO 4444. Procedures for Air Navigation Services Rules of the Air and Air Traffic Services Twelfth Edition—1985. (Amendment 1 11/20/86) (Amendment 2 10/22/87) (Amendment 3 11/14/91) (Amendment 4 11/11/93). 298 pp.

Air Traffic Control (Cont.)
—**Services (Cont.)**
ICAO 9426. Air Traffic Services Planning Manual—First (Provisional) Edition—1984; (Amendment 1 07/15/85) (Amendment 2 09/11/86) (Amendment 3 11/03/88). 413 pp.

—**Signal Lights**
NATO STANAG 3758 ED 2 AMD 0-86. Signals Used by Air Traffic Service Units for Control of Pedestrian and Vehicular Traffic in the Manoeuvring Area of Airfields. 8 pp.
NATO STANAG 3758 ED 2 AMD 2-86. Signals Used by Air Traffic Service Units for Control of Pedestrian and Vehicular Traffic in the Manoeuvring Area of Airfields. 6 pp.

—**Surface Movement**
ICAO 9476. Manual of Surface Movement Guidance and Control Systems (SMGCS) First Edition—1986; (Amendment 1 05/06/87) (Amendment 2 12/01/87). 89 pp.

—**Visual Signals**
NATO STANAG 3379 ED 7 AMD 1-90. In-Flight Visual Signals. 11 pp.
NATO STANAG 3379 ED 7 AMD 2-90. In-Flight Visual Signals. 12 pp.

Air Traffic Control Equipment
See Also: Air Traffic Control; Collision Avoidance

—**Aircraft Type Designators—Symbols**
ICAO 8643. Aircraft Type Designators 23rd Edition—February 1993. 58 pp.

—**Microwave Landing Systems**
ICAO Circular 165. Microwave Landing System (MLS) Advisory Circular Issue No. 1—1981. 38 pp.

—**Transponders**
CAA Chapter R4-6 App #8 04.74. Tests for Air Traffic Control Transponder Systems (Radio). 1 p.

Air Traffic Controllers
See Also: Air Transportation; Aviation Personnel

—**Aircraft Accidents—Reporting**
CAA Section 8 CAP 382. Reporting by Air Traffic Controllers (Civil Air Publications: Mandatory Occurrence Reporting Scheme). 1 p.

—**Training Manuals**
ICAO 7192 Part D-2. Training Manual Part D-2 Air Traffic Controller First Edition—1977 (OUT OF PRINT). 89 pp.

Air Traffic Forecasting
See Also: Air Transportation
ICAO Circular 236. Investment Requirements for Aircraft Fleets and for Airport and Route Facility Infrastructure to the Year 2010—1992. 39 pp.
ICAO Circular 237. Outlook for Air Transport to the Year 2001—1992. 54 pp.
ICAO 8991. Manual on Air Traffic Forecasting Second Edition—1985. 51 pp.

Air Transportation
Scope Note: For additional listings, use a more specific term *Use For:* Civil Aviation
See Also: Aerial Delivery; Aeronautical Meteorology; Aeronautical Navigation Charts; Air Cargo Transportation; Air Drop Operations; Air Medical Transport Units; Air Navigation; Air Navigation Facilities; Air Navigation Plans; Air Traffic Control; Air Traffic Controllers; Air Traffic Forecasting; Aircraft; Aircraft Accidents; Aircraft Safety; Airports; Aviation Personnel; Civil Engineering; Flight Control Systems; Shipping; Tactical Air Transport Operations; Transportation
EURO JLJ/FMR/23179-92. AECMA Views on European Air Transport Policy. 8 pp.
ICAO Circular 200. Economic Situation of Air Transport Review and Outlook—1986. 77 pp.
ICAO Circular 201. International Air Passenger and Freight Transport Asia and Pacific—1986 (Supersedes Circular 160). 177 pp.
ICAO Circular 221. International Air Passenger and Freight Transport Middle East—1989. 97 pp.
ICAO Circular 222. Economic Situation of Air Transport Review and Outlook 1978 to the Year 2000—1989. 80 pp.

—**Competition Laws**
ICAO Circular 215. Guidance Material on the Avoidance or Resolution of Conflicts over the Application of Competition Laws to International Air Transport—1989. 19 pp.

INTERNATIONAL AND NON-U.S. NATIONAL STANDARDS
SUBJECT INDEX

Air Transportation (Cont.)

—Competition Rules
EC EEC/3975/87-87. Council Regulation Laying Down the Procedure for the Application of the Rules on Competition to Undertakings in the Air Transport Sector. 8 pp.

—Computer Reservation Systems
ICAO Circular 214. Guidance Material on the Regulation of Computer Reservation Systems—1988. 17 pp.

—Environmental Protection
ICAO 9592. Committee on Aviation Environmental Protection Second Meeting Montreal, 2-13 December 1991. 191 pp.

—Facilitation
ICAO Circular 152. Selection of ICAO Facilitation B-Type Recommendations—1980. 44 pp.

—Fares
EC 87/601/EEC-87. Council Directive on Fares for Scheduled Air Services Between Member States. 7 pp.
EC EEC/2342/90-90. Council Regulation on Fares for Scheduled Air Services. 7 pp.
ICAO Circular 198. Survey of International Air Transport Fares and Rates September 1985. 119 pp.
ICAO Circular 199. Regional Differences in Fares, Rates and Costs for International Air Transport 1984. 53 pp.
ICAO Circular 204. Survey of International Air Transport Fares and Rates—September 1986 (OUT OF PRINT). 120 pp.
ICAO Circular 206. Regional Differences in Fares, Rates and Costs for International Air Transport—1985. 55 pp.
ICAO Circular 208. Survey of International Air Transport Fares and Rates September 1987—1988. 145 pp.
ICAO Circular 210. Regional Differences in Fares, Rates and Costs for International Air Transport 1986—1988. 52 pp.
ICAO Circular 219. Survey of International Air Transport Fares and Rates September—1988. 131 pp.
ICAO Circular 220. Regional Differences in Fares, Rates and Costs for International Air Transport—1987. 49 pp.
ICAO Circular 224. Survey of International Air Transport Fares and Rates September—1989. 124 pp.
ICAO Circular 228. Regional Differences in Fares, Rates and Costs for International Air Transport—1988. 49 pp.
ICAO Circular 231. Survey of International Air Transport Fares and Rates September—1990. 113 pp.
ICAO Circular 239. Survey of International Air Transport Fares and Rates—September 1991. 113 pp.
ICAO 9587. Policy and Guidance Material on the Regulation of International Air Transport First Edition—1992 (Replacing Doc 9440). 98 pp.

—International—Fares
ICAO Circular 231. Survey of International Air Transport Fares and Rates September—1990. 113 pp.
ICAO Circular 235. Regional Differences in Fares, Rates and Costs for International Air Transport—1989. 49 pp.

—International—Rates
ICAO Circular 231. Survey of International Air Transport Fares and Rates September—1990. 113 pp.
ICAO Circular 235. Regional Differences in Fares, Rates and Costs for International Air Transport—1989. 49 pp.

—Licenses—Acceptance
EC 91/670/EEC-91. Council Directive on Mutual Acceptance of Personnel Licences for the Exercise of Functions in Civil Aviation. 6 pp.

—Measurement Units
BSI A 22-47. 1947 Units of Measurement in Civil Aviation. 1 p.
ICAO Annex 5. Units of Measurement to Be Used in Air and Ground Operations Fourth Edition—July 1979; (Amendment 14 07/30/84) (Supplement 5/1/85) (Amendment 1 12/16/85) (Supplement 5/1/88) (Amendment 1 8/1/90). 65 pp.

—Military—Baggage Tags
NATO STANAG 3770 ED 2 AMD 2-86. Air Transport Baggage Tags. 9 pp.
NATO STANAG 3770 ED 2 AMD 3-86. Air Transport Baggage Tags. 10 pp.

—Military Operations
NATO STANAG 3998 ED 1 AMD 0-89. NATO Air Transport Policies and Procedures-ATP-53. 5 pp.

—Organizations—Designations
ICAO 8585. Designators for Aircraft Operating Agencies, Aeronautical Authorities and Services 87th Edition—May 1993. 185 pp.

—Passenger—Capacity Sharing
EC EEC/2343/90-90. Council Regulation on Access for Air Carriers to Scheduled Intra-Community Air Service Routes and on the Sharing of Passenger Capacity Between Air Carriers on Scheduled Air Services Between Member States. 7 pp.
EC 87/602/87-87. Council Decision on the Sharing of Passenger Capacity Between Air Carriers on Scheduled Air Services Between Member States and on Access for Air Carriers to Scheduled Air-Service Routes Between Member States. 7 pp.

—Passenger—Cost Estimates
ICAO Circular 236. Investment Requirements for Aircraft Fleets and for Airport and Route Facility Infrastructure to the Year 2010—1992. 39 pp.

—Passenger—Service Contracts
MOD UK DEFCON 112AE-91. Basic Set of Conditions of Contract for Air Transport 8/91. 6 pp.

—Passenger—Space
CAA NOTICE #64 ISSUE 1. Minimum Space for Seated Passengers. 5 pp.

—Publications
CAA LEAFLET 13-1 07.90. Cap Listing. 8 pp.

—Rates
ICAO Circular 198. Survey of International Air Transport Fares and Rates September 1985. 119 pp.
ICAO Circular 199. Regional Differences in Fares, Rates and Costs for International Air Transport 1984. 53 pp.
ICAO Circular 204. Survey of International Air Transport Fares and Rates—September 1986 (OUT OF PRINT). 120 pp.
ICAO Circular 206. Regional Differences in Fares, Rates and Costs for International Air Transport—1985. 55 pp.
ICAO Circular 208. Survey of International Air Transport Fares and Rates September 1987—1988. 145 pp.
ICAO Circular 210. Regional Differences in Fares, Rates and Costs for International Air Transport 1986—1988. 52 pp.
ICAO Circular 219. Survey of International Air Transport Fares and Rates September—1988. 131 pp.
ICAO Circular 220. Regional Differences in Fares, Rates and Costs for International Air Transport—1987. 49 pp.
ICAO Circular 224. Survey of International Air Transport Fares and Rates September—1989. 124 pp.
ICAO Circular 228. Regional Differences in Fares, Rates and Costs for International Air Transport—1988. 49 pp.
ICAO Circular 231. Survey of International Air Transport Fares and Rates September—1990. 113 pp.
ICAO Circular 239. Survey of International Air Transport Fares and Rates—September 1991. 113 pp.
ICAO 9587. Policy and Guidance Material on the Regulation of International Air Transport First Edition—1992 (Replacing Doc 9440). 98 pp.

—Rules
ICAO 4444. Procedures for Air Navigation Services Rules of the Air and Air Traffic Services Twelfth Edition—1985; (Amendment 1 11/20/86) (Amendment 2 10/22/87) (Amendment 3 11/14/91) (Amendment 4 11/11/93). 298 pp.
ICAO 7632. Protocol to Amend the Convention for the Unification of Certain Rules Relating to International Carriage by Air Signed at Warsaw on 12 October 1929 as Signed at Hague on 28 September 1955. 16 pp.
ICAO 8181. Convention, Supplementary to the Warsaw Convention, for the Unification of Certain Rules Relating to International Carriage by Air Performed by a Person Other Than the Contracting Carrier, Signed at Guadalajara on 18 September 1961. 9 pp.
ICAO 8932. Protocol to Amend the Convention for the Unification of Certain Rules Relating to International Carriage by Air Signed at Warsaw on 12 October 1929 as Amended by the Protocol Done at the Hague on 28 September 1955 Process-Verbal 05/31/83. 31 pp.
ICAO 9145. Additional Protocol No. 1 to Amend the Convention for the Unification of Certain Rules Relating to International Carriage by Air Signed at Warsaw on 12 October 1929 Signed at Montreal on 25 September 1975. 16 pp.
ICAO 9146. Additional Protocol No. 2 to Amend the Convention for the Unification of Certain Rules Relating to International Carriage by Air Signed at Warsaw on 12 October 1929 as Amended by the Protocol Done at the Hague on 28 September 1955 Signed at Montreal on 25 September 1975. 16 pp.
ICAO 9147. Additional Protocol No. 3 to Amend the Convention for the Unification of Certain Rules Relating to International Carriage by Air Signed at Warsaw on 12 October 1929 as Amended by the Protocols Done at the Hague on 28 September 1955 and at Guatemala City on 8 March 1971. 18 pp.
ICAO 9148. Montreal Protocol No. 4 to Amend the Convention for the Unification of Certain Rules Relating to International Carriage by Air Signed at Warsaw on 12 October 1929 as Amended by the Protocol Done at the Hague on 28 September 1955 Signed at Montreal on 25 September 1975. 31 pp.
ICAO 9587. Policy and Guidance Material on the Regulation of International Air Transport First Edition—1992 (Replacing Doc 9440). 98 pp.

—Statistical Analysis
ICAO 9060. Manual on the ICAO Statistics Programme Third Edition—1984. 210 pp.

—Statistics
ICAO 9180. ICAO Statistical Yearbook Civil Aviation Statistics of the World Sixteenth Edition—1991. 174 pp.

—Tariffs
ICAO 9364. Manual on the Establishment of International Air Carrier Tariffs 1983. 126 pp.
ICAO 9538. Models of Bilateral Tariff Clauses—1989 (Supersedes 9228). 24 pp.
ICAO 9587. Policy and Guidance Material on the Regulation of International Air Transport First Edition—1992 (Replacing Doc 9440). 98 pp.

—Taxes
ICAO 8632. ICAO's Policies on Taxation in the Field of International Air Transport—1966; (Supplement 09/01/82) (Amendment 3 05/01/83) (Amendment 4 02/01/84) (Amendment 5 01/01/85) (Amendment 6 05/30/85) (Amendment 7 09/12/86) (Amendment 8 10/14/87) (Amendment 9 08/18/88). 88 pp.

Air Valves
Use For: Air Shutters *See Also:* Pneumatic Valves; Tire Valves; Valves

—Aerostats
CAA Chapter Q6-2 12.79. Gas and Air Supply Systems (Non-Rigid Airships). 2 pp.

—Aircraft
SBAC AGS 2096/A ISSUE 5. 1/4" B.S.P.F. (Air and Mineral Base Hydraulic Fluid) Charging Valve.
SBAC AGS 2096/B (V). 1/4" B.S.P.F. (Air and Mineral Base Hydraulic Fluid) Charging Valve.
SBAC AGS 3034/A ISSUE 5. Valve.
SBAC AGS 3034/B (V). Valve.

—Aircraft—Filler Caps
SBAC AS 3167 ISSUE 2. Assembly of Filler Cap—Air Valve Type. 1 p.
SBAC AS 3168 ISSUE 1. Air Valve for Filler Cap.AS.1964. 1 p.

—Aircraft—Manual Controls
SBAC AS 1910 ISSUE 6. Knob (Air Shutter Control). 1 p.
SBAC AS 6388 ISSUE 2. Knob (Air Shutter Control).

—Compressed Air—Aircraft
ISO 1023-74. Aircraft—High Pressure Air Charging Valves First Edition. 4 pp.

—Compressed Air—Angle—Marine
CNS F3147-81. Marine Forged Steel Angle Valves for Compressed Air (Nov)(8111).
JIS F 7337-88. Forged Steel Angle Air Valves for Marine Use.
JIS F 7475-88. Cast Steel Angle Air Valves for Marine Use.

—Compressed Air—Globe—Marine
CNS F3146-87. Marine Forged Steel Globe Valves for Compressed Air (Dec)(8110).
CNS F3148-87. Marine Cast Steel Globe Valves for Compressed Air (Dec)(8112).
JIS F 7336-88. Forged Steel Globe Air Valves for Marine Use.
JIS F 7340-88. Cast Steel Globe Air Valves for Marine Use.

—Compressed Air—Pressure Control—Aircraft—Couplings
BSI C 9-70. 1970 Coupling Dimensions for High Pressure Air Charging Valves for Aircraft. 2 pp.

INDUSTRY STANDARDS

Air

Air Valves (Cont.)
—Compressed Air—Safety
BSI BS 6759: Part 2-84. 1984 Amd 1 Safety Valves Part 2: Specification for Safety Valves for Compressed Air or Inert Gases (AMD 5494) March 31, 1987. 25 pp.

—Waterworks
JIS B 2063-87. Air Vent Valves for Water Works. 30 pp.

Air Vessels
See Also: Pipelines
CNS B7238-83. Method of Test for Air Vessel for Filling Use (Total Pressure Ratio Below 30) and for Small Size Displacement Compressor (Shaft Horsepower Below 5.5KW) (May)(10214).

Air Weapons Ranges
See Also: Military Facilities; Ranges (Facilities); Weapons

—Control Installations—Identification Systems
NATO STANAG 3783 ED 2 AMD 4-78. Air Weapons Ranges—Identification of Control Installations and Spectator Sites During Daylight Operations. 5 pp.
NATO STANAG 3783 ED 2 AMD 6-78. Air Weapons Ranges—Identification of Control Installations and Spectator Sites During Daylight Operations. 6 pp.

—Spectator Sites—Identification Systems
NATO STANAG 3783 ED 2 AMD 4-78. Air Weapons Ranges—Identification of Control Installations and Spectator Sites During Daylight Operations. 5 pp.
NATO STANAG 3783 ED 2 AMD 6-78. Air Weapons Ranges—Identification of Control Installations and Spectator Sites During Daylight Operations. 6 pp.

Airborne Area Navigation Computing Equipment
Use: Area Navigation Computing Instruments

Airborne Area Navigation Systems
Use: Area Navigation Systems

Airborne Computers
Use: Avionic Computers

Airborne Electronics
Use: Avionics

Airborne Equipment
Use: Aircraft Equipment

Airborne Operations
Use For: Air Operations *See Also:* Air Drop Operations; Air Force Operations; Airmobile Operations; Forward Air Controllers; Helicopters; Military Operations; Tactical Air Operations; Tactical Air Support; Tactical Air Transport Operations

—Counter
NATO STANAG 3880 ED 2 AMD 0-90. Counter Air Operations—ATP-42(A). 4 pp.

—Electronic Warfare
NATO STANAG 3873 ED 3 AMD 0-90. Electronic Warfare (EW) in Air Operations—ATP-44(A). 5 pp.

—Information Interchange
NATO STANAG 3572 ED 1 AMD 8-77. Exchange of Information on Tactical Air Transport Operations. 8 pp.

—Information Interchange—Offensive
NATO STANAG 3982 ED 1 AMD 2-85. Messages Supplement to ATP-27—Offensive Air Support Operations (Withdrawn 93-02). 10 pp.

—Safety—Signal Lights—Emergency Planning
NATO STANAG 3465 ED 4 AMD 10-73. Safety, Emergency and Signalling Procedures for Military Air Movement—Fixed Wing Aircraft. 15 pp.

—Warfare—Technical Manuals
NATO STANAG 1167 ED 10 AMD 1-92. NATO Above Water Warfare Manual—ATP-31. 6 pp.

Airborne Particulates
Use: Particulates—Airborne

Airborne Radar
—Aircraft—Environmental Testing
EUROCAE ED-38 06.83. MPS for Airborne Weather, Ground Mapping and Assisted Approach Pulse Radars. 82 pp.

—Assisted Approach
EUROCAE ED-38 06.83. MPS for Airborne Weather, Ground Mapping and Assisted Approach Pulse Radars. 82 pp.

—Design
EUROCAE ED-38 06.83. MPS for Airborne Weather, Ground Mapping and Assisted Approach Pulse Radars. 82 pp.

—Electrical Properties
EUROCAE WG9/1-71 06.72. MPS for Airborne Secondary Surveillance Radar Transponder Apparatus (Amendment No. 1 Incorporated). 29 pp.

—Environmental Testing
EUROCAE WG9/1-71 06.72. MPS for Airborne Secondary Surveillance Radar Transponder Apparatus (Amendment No. 1 Incorporated). 29 pp.

—Weather
EUROCAE ED-38 06.83. MPS for Airborne Weather, Ground Mapping and Assisted Approach Pulse Radars. 82 pp.

Airbus A300 Aircraft
See Also: Aircraft

—Accidents
CAA. Airbus A300 (World Airline Accident Summary). 1 p.

—Antenna Positions
CAA. Airbus A300 (Approved Aerial Positions). 1 p.

—Foreign Airworthiness Directives
CAA. Airbus Industrie A300 and A310 Series Aircraft (Foreign Airworthiness Directives). 21 pp.

Airbus A300 B4-203 Aircraft
See Also: Aircraft

—Certification
CAA. Airbus Industrie A300 B4-203. 5 pp.

—Data Sheets
CAA FA24 ISSUE 1. Airbus A300 B4-203. 3 pp.

Airbus A300 B4-600 Aircraft
See Also: Aircraft

—Certification
CAA. Airbus A300 B4-600/-600R. 18 pp.

Airbus A300 B4-600R Aircraft
See Also: Aircraft

—Certification
CAA. Airbus A300 B4-600/-600R. 18 pp.

Airbus A310 Aircraft
See Also: Aircraft

—Accidents
CAA. Airbus A310 (World Airline Accident Summary). 1 p.

—Antenna Positions
CAA. Airbus A310 (Approved Aerial Positions). 1 p.

—Foreign Airworthiness Directives
CAA. Airbus Industrie A300 and A310 Series Aircraft (Foreign Airworthiness Directives). 21 pp.

Airbus A310-203 Aircraft
See Also: Aircraft

—Certification
CAA. Airbus Industrie A310-203. 11 pp.

—Data Sheets
CAA FA31 ISSUE 1. Airbus Industrie A310-203. 3 pp.

Airbus A320 Aircraft
See Also: Aircraft

—Accidents
CAA. Airbus A320 (World Airline Accident Summary). 1 p.

—Antenna Positions
CAA. Airbus A320. 1 p.

—Foreign Airworthiness Directives
CAA. Airbus Industrie A320 Series Aircraft (Foreign Airworthiness Directives). 6 pp.

AIRCOM
Use For: Airways Communications System
See Also: Mobile Satellite Communications; Radio Communications; Satellite Communications
CCIR Report 1173-90. Technical and Operational Considerations for Aeronautical Mobile-Satellite Communications—Section 8F—Frequencies, Orbits and Systems. 40 pp.

Aircraft
Scope Note: For specific types of aircraft equipment, see the subheading Aircraft under specific types of equipment *Use For:* Airplanes; Fixed Wing Aircraft
See Also: Aero Commander 200 Aircraft; Aero Commander 500 Aircraft; Aero Commander 520 Aircraft; Aero Commander 560 Aircraft; Aero Commander 680 Aircraft; Aero Commander 680E Aircraft; Aero Commander 680F Aircraft; Aero Commander 680T Aircraft; Aero Commander 685 Aircraft; Aero Commander 690A Aircraft; Aero Commander 690B Aircraft; Aero Commander 690C Aircraft; Aero Commander 690D Aircraft; Aero Commander 695 Aircraft; Aero Commander 695A Aircraft; Aero 45 Aircraft; Aerodynamics; Aeronautical Engineering; Aeronautical Meteorology; Aeronautics; Aeronca 15AC Aircraft; Aerospace Vehicles; Aerospatiale AS 332 Aircraft; Aerospatiale AS 332L Supa Puma Aircraft; Aerospatiale AS 350 Aircraft; Aerospatiale AS 350B Aircraft; Aerospatiale AS 355 Aircraft; Aerospatiale ATR 42 Aircraft; Aerospatiale SA 315B Lama Aircraft; Aerospatiale SA 316B Alouette III Aircraft; Aerospatiale SA 316C Alouette III Aircraft; Aerospatiale SA 318B/C Alouette Astazou Aircraft; Aerospatiale SA 3180 Aircraft; Aerospatiale SA 3180 Alouette Astazou Aircraft; Aerospatiale SA 319 Alouette III Aircraft; Aerospatiale SA 341G Gazelle Aircraft; Aerospatiale SA 365N Dauphin Aircraft; Aerospatiale SA 391B Alouette III Aircraft; Aerospatiale SE 313B Alouette II Aircraft; Aerospatiale SE 313B Alouette III Aircraft; Aerospatiale SE 3130 Alouette III Aircraft; Aerospatiale SE 316 Alouette III Aircraft; Aerospatiale SE 3160 Alouette III Aircraft; Aerosport 2 Aircraft; Aerostar 601P Aircraft; Aerostats; AESL Aircraft; Agricultural Aircraft; Air Cargo Transportation; Air Cushion Vehicles; Air Flow Angle Transmitters; Air Medical Transport Units; Air Navigation; Air Traffic Control; Air Transportation; Airbus A300 Aircraft; Airbus A300 B4-203 Aircraft; Airbus A300 B4-600 Aircraft; Airbus A300 B4-600R Aircraft; Airbus A310 Aircraft; Airbus A310-203 Aircraft; Airbus A320 Aircraft; Aircraft Accidents; Aircraft Cabins; Aircraft Communications; Aircraft Compartments; Aircraft Doors; Aircraft Engines; Aircraft Equipment; Aircraft Fuel Systems; Aircraft Fuels; Aircraft Gasoline; Aircraft Instruments; Aircraft Interception; Aircraft Landings; Aircraft Lighting; Aircraft Marshallers; Aircraft Noise; Aircraft Propellers; Aircraft Refueling Equipment; Aircraft Safety; Airport Lighting; Airport Runways; Airports; Airships; Airspeed Ambassador Aircraft; Airspeed Indicators; Airtourer T6-24 Aircraft; Airtourer 100 Aircraft; Airtourer 150 Aircraft; Alon A-2A Aircraft; Alon Aircoupe Aircraft; Anglo Polish Sailplanes SZD45A Ogar; Antenna Positions; Aries Aircraft; ARV Aviation ARV 1 Aircraft; ARV Aviation Super 2 Aircraft; ATL 98 Aircraft; ATR 42 Aircraft
See Also: Auster Aircraft; Auster Beagle A 61 Aircraft; Auster 3 Aircraft; Auster 4 Aircraft; Auster 5 Aircraft; Aviation Traders ATL 98 Carvair Aircraft; Avionics; Avions Mudry/CAARP CAP 10B Aircraft; Avions Mudry/CAARP CAP 20 L/S 200 Aircraft; Avro Aircraft; Avro Anson Aircraft; Avro Lancaster Aircraft; Avro Lancastrian Aircraft; Avro Tudor Aircraft; Avro XIX Aircraft; Avro York Aircraft; Avro 19 Aircraft; AW 650 Argosy Aircraft; Ayres S2R Aircraft; BAC/SNIAS Concorde Aircraft; BAC Super VC 10 Aircraft; BAC 1-11 Aircraft; BAe ATP Aircraft; BAe Jetstream Series Aircraft; BAe Jetstream 31 Aircraft; BAe 125 Aircraft; BAe 146 Aircraft; BAe 748 Aircraft; Balloons; Beagle A 109 Aircraft; Beagle-Auster A 61 Aircraft; Beagle-Auster A109 Aircraft; Beagle B. (Bulldog) Aircraft; Beagle B 121 Aircraft; Beagle B 125 Aircraft; Beagle B 206 Aircraft; Beagle B 206Y Aircraft; Beagle 121 Aircraft; Beagle 206 Aircraft; Bede BD 4-150 Aircraft; Beech; Beech; Beech; Beech; Beech; Beech; Beech; Beech; Beech; Beech; Beech; Beech; Beech; Beech 100 Aircraft; Beech 200 Aircraft; Beech 200 Super King Air Aircraft; Beech 300 Aircraft; Beech 300LW Aircraft; Beech A 36 Aircraft; Beech A 100 King Air Aircraft; Beech B 60 Aircraft; Beech B 95A Aircraft; Beech B 200 Aircraft; Beech B 200 Super King Air Aircraft; Beech D 17S Aircraft; Beech D 55 Aircraft; Beech D 95A Aircraft; Beech E 18 Aircraft; Beech F 33A Aircraft; Beech F 90 Super King Air Aircraft; Beech

INTERNATIONAL AND NON-U.S. NATIONAL STANDARDS
SUBJECT INDEX
Aircraft

Aircraft *(Cont.)*

See Also: (Cont.)
H 18 Aircraft; Beech Super King Air 200 Aircraft; Beech Super King Air 300 Aircraft; Beechjet 400 Aircraft; Behez Aircraft; Belfast SC 5 Aircraft; Bellanca (Champion) 7 Aircraft; Bellanca (Champion) 8 Aircraft; Bellanca 17-30A Aircraft; Berths; Blanik Gliders; BN 2A B and T Islander Aircraft; BN 2A MK III Trislander Aircraft; Boeing A75 N1 Aircraft; Boeing Kawasaki 107 Aircraft; Boeing Vertol 44 Aircraft; Boeing Vertol 107 Aircraft; Boeing Vertol 234 Aircraft; Boeing Vertol 234 Chinook Aircraft; Boeing Vertol 234LR Aircraft ***See Also:*** Boeing 377 Stratocruiser Aircraft; Boeing 707 Aircraft; Boeing 720 Aircraft; Boeing 720B Aircraft; Boeing 727 Aircraft; Boeing 737 Aircraft; Boeing 747 Aircraft; Boeing 747-400 Aircraft; Boeing 757 Aircraft; Boeing 767 Aircraft; Boeing 767-300ER Aircraft; Bolkow BO 105 Aircraft; Bolkow BO 208 Aircraft; Bolkow BO 209 Aircraft; Bolkow 107 Aircraft; Bolkow 207 Aircraft; Bolkow 208 Junior Aircraft; Bolkow 209 Aircraft; Brantly B2 Aircraft; Brantly 305 Aircraft; Brasov IS 28M24 Aircraft; Bristol Britannia Aircraft; Bristol 170 Mark 31 Aircraft; Bristol 170 MK 21 Aircraft; Bristol 170 MK 32 Aircraft; Bristol 170 Wayfarer Aircraft; Bristol 171 Aircraft; Britannia 100 Aircraft; Britannia 300 Aircraft; British Aerospace ATP Aircraft; British Aerospace Bulldog Series 100/ 120 Aircraft; British Aerospace C1 Chipmunk Aircraft; British Aerospace DH 82 Tiger Moth Aircraft; British Aerospace HS 125 Aircraft; British Aerospace PLC Jetstream Aircraft; British Aerospace Scottish Division Bulldog Series Aircraft; British Aircraft Corporation/Aerospatiale Concorde Aircraft; British Aircraft Corporation VC 10 Aircraft; Britten-Norman Trislander BN 2A Mark III Aircraft; Britten-Norman Trislander BN 2A Mark III-1 Aircraft; Britten-Norman Trislander BN 2A Mark III-2 Aircraft; Brooklands A07 Optica Aircraft; Canadair C 4 Aircraft; Canadair Challenger Aircraft; Canadair CL 44 Aircraft; Canadair CL 600 Series Aircraft; Canadair CL 600-2B16 Challenger 601-3A Aircraft; Canadair DC 4-M-2 Aircraft; Candair CL-44 Aircraft; CAP 10 Aircraft; CAP 20 Aircraft; CAP 231 Aircraft; CASA C-212 Aircraft; Catalina Aircraft; Cessna F 406 Caravan II Aircraft; Cessna S 550 Aircraft; Cessna Series Aircraft; Cessna T 303 Aircraft; Cessna 120 Aircraft; Cessna 140 Aircraft; Cessna 150 Aircraft; Cessna 152 Aircraft; Cessna 170 Aircraft; Cessna 172 Aircraft; Cessna 175 Aircraft; Cessna 177 Aircraft; Cessna 180 Aircraft; Cessna 182 Aircraft; Cessna 185 Aircraft; Cessna 205 Aircraft; Cessna 206 Aircraft; Cessna 207 Aircraft; Cessna 208 Caravan 1; Cessna 210 Aircraft; Cessna 300 Aircraft; Cessna 310 Aircraft; Cessna 320 Aircraft; Cessna 335 Aircraft; Cessna 336 Aircraft; Cessna 337 Aircraft; Cessna 340 Aircraft; Cessna 400 Aircraft; Cessna 401 Aircraft; Cessna 402 Aircraft; Cessna 402A Aircraft; Cessna 402B Aircraft; Cessna 402C Aircraft; Cessna 404 Aircraft; Cessna 406 Aircraft; Cessna 411 Aircraft; Cessna 414 Aircraft; Cessna 414A Chancellor Aircraft; Cessna 421 Aircraft; Cessna 421B Aircraft; Cessna 425 Aircraft; Cessna 441 Aircraft; Cessna 500 Aircraft; Cessna 500 Citation I Aircraft; Cessna 500 Citation II Aircraft
See Also: Cessna 550 Aircraft; Cessna 560 Citation V Aircraft; Cessna 600 Citation I Aircraft; Cessna 600 Citation II Aircraft; Cessna 600 Citation III Aircraft; Cessna 650 Aircraft; Cessna 650 Citation III Aircraft; Chipmunk Aircraft; Christen Pitts S-1 Aircraft; Christen Pitts S-2 Aircraft; CIT/A1 Aircraft; Comet 4 Aircraft; Comet 4B Aircraft; Comet 4C Aircraft; Concorde Aircraft; Convair CV240 Aircraft; Convair CV340 Aircraft; Convair CV440 Aircraft; Convair CV580 Aircraft; Convair CV640 Aircraft; Convair CV880 Aircraft; Convair CV990 Aircraft; Convair PBY 5 Catalina Aircraft; Crash Locator Beacons; Curtiss C 46 Commando Aircraft; Dakota Aircraft; Dassault Mystere (Fan Jet Falcon) Aircraft; Dassault Mystere-Falcon 900 Aircraft; De Havilland Canada DHC-8-102 Aircraft; De Havilland Canada 2 Beaver Aircraft; De Havilland Canada 2 Turbo Beaver Aircraft; De Havilland Canada 6 Twin Otter Aircraft; De Havilland Canada 7 Aircraft; De Havilland Canada 8 Aircraft; DeHavilland DHC-2 Beaver MK1 Aircraft; DeHavilland DHC-4 Aircraft; DeHavilland DHC-5 Aircraft; DeHavilland DHC-8 Aircraft; DeHavilland DHC-8 Dash 8 Series 300; Desford Trainer Aircraft; DH 104 Riley Dove Aircraft; DH 112/Venom MK1 Aircraft; DH 80A Puss Moth Aircraft; DH 82A Tiger Moth Aircraft; DH 85 Leopard Moth Aircraft; DH 87B Aircraft; DH 89A Rapide Aircraft; DH 90 Dragonfly Aircraft; DH 90A Dragonfly Aircraft; Dornier DO 27 Series Aircraft; Dornier DO 28 Aircraft; Dornier 228 Aircraft; Dornier 228-100 Aircraft; Dornier 228-200 Aircraft; Douglas C 47

Aircraft *(Cont.)*

See Also: (Cont.)
Dakota Aircraft; Douglas C 54 Aircraft; Douglas Dakota Aircraft; Douglas DC 2 Aircraft; Douglas DC 3 Aircraft; Douglas DC 4 Aircraft; Douglas DC 6 Aircraft; Douglas DC 7 Aircraft; Douglas DC 9 Series 10 Aircraft; Douglas DC 9 Series 30 Aircraft; Edgley EA 7 Optica Aircraft; Embraer Bandeirante EMB-P1 Aircraft; Embraer Bandeirante EMB-P2 Aircraft; Embraer Bandeirante EMB-110 Aircraft; Embraer Bandeirante EMB-110K1 Aircraft; Embraer Brasilia EMB-120 Aircraft; Embraer Brasilia EMB-120RT Aircraft; Embraer EMB-110 Aircraft; Embraer EMB-110 Bandeirante Aircraft; Embraer EMB-120 Aircraft; Embraer EMB-120 Brasilia Aircraft; Embraer EMB-121 Aircraft; Embraer Xingu EMB-121 Aircraft; Embraer 110 P1 Aircraft; Embraer 110 P2 Aircraft; Embraer 121 Aircraft; Emeraude (Piel) Aircraft; Emeraude (Rousseau) Aircraft; Emeraude (Scintex) Aircraft; Enstrom F 28A Aircraft; Enstrom 280 Aircraft; EP 9 Aircraft; Ercoupe 415 Forney FIA Aircraft; Extra 230 Aircraft; Extra 300 Aircraft; Fairchild Argus Aircraft; Fairchild F-27 Aircraft; Fairchild-Hiller FH-227 Aircraft; Fairchild SA227-AC Metro III Aircraft; Fairchild/Swearingen Merlin III B Aircraft; Fairchild 24R-46A Argus III Aircraft; Falco F8L Aircraft; Fan Jet Falcon 20 Series D Aircraft; Fan Jet Falcon 20 Series E Aircraft; Fan Jet Falcon 20 Series F Aircraft ***See Also:*** Flight Control Systems; Flight Crews; Flight Decks; Flight Dynamics; Flight Envelopes; Flight Testing; Fokker F27 Aircraft; Fokker F28 Aircraft; Fokker F28 2000 Aircraft; Fokker F28 4000 Aircraft; Forney F1A Aircraft; Fournier RF 4 Aircraft; Fournier RF 6 Aircraft; Fuji FA 200 Aircraft; Fuselages; G.A.L. Cygnet Aircraft; Gardan GY 80 Aircraft; Gemini Aircraft; Gliders; Glos Air Airtourer Aircraft; Glos Airtourer Super 150 Aircraft; Glos Airtourer T 3 Aircraft; Glos Airtourer 115 Aircraft; Glos Airtourer 150 Aircraft; Gloster Meteor NFII Aircraft; Grob G 109B Aircraft; Grob G 115 Aircraft; Grumman American AA 1 Aircraft; Grumman American AA 5 Aircraft; Grumman American GA 7 Aircraft; Grumman American General AG-5B Aircraft; Grumman G 159 Gulfstream 1 Aircraft; Grumman G 164A Aircraft; Grumman G 1159 Gulfstream 2 Aircraft; Grumman G 1159 Gulfstream 3 Aircraft; Grumman G 1159A Gulfstream 3 Aircraft; Grumman Gulfstream II Aircraft; Gulfstream Aerospace G-159 Aircraft; Gulfstream Aerospace GA-7 Aircraft; Gulfstream Aerospace Gulfstream III Aircraft; Gulfstream Aerospace Gulfstream IV Aircraft; Gulfstream Aerospace 112 Aircraft; Gulfstream Aerospace 114 Aircraft; Gulfstream Aerospace 685 Aircraft; Gulfstream Aerospace 690 Aircraft; Gulfstream Aerospace 695 Aircraft; Gulfstream Aircraft; Gulfstream American AA 1 Aircraft; Gulfstream American G 159 Aircraft; Gulfstream American GA 7 Aircraft; Gyroflug SC 01 Aircraft; Gyroflug SC 01B 160 Aircraft; Gyroplanes; Handley Page Halifax Aircraft; Handley Page Halton Aircraft; Handley Page Herald Aircraft; Handley Page Hermes Aircraft; Handley Page Marathon Aircraft; Harbin Y-12 II Aircraft; Harvard AT 6D Aircraft; Hawker Siddeley Argosy Aircraft; Hawker Siddeley Comet 1 Aircraft; Hawker Siddeley/De Havilland 89A Rapide Aircraft; Hawker Siddeley/De Havilland 104 Dove Aircraft; Hawker Siddeley/De Havilland 114 Heron Aircraft; Hawker Siddeley/De Havilland 121 Trident Aircraft; Hawker Siddeley DH.106 Comet 4 Aircraft; Hawker Siddeley 114 Heron Aircraft; Hawker Siddeley 121 Trident Aircraft; Hawker Siddeley 125 Aircraft; Hawker Siddeley 748 Aircraft; Helicopters; Helio H-295 Aircraft; Helio H-395 Aircraft; Herald (Dart) 100 Aircraft; Herald (Dart) 200 Aircraft; Heron Aircraft; Hiller FH 1100 Aircraft; Hiller UH 12A Aircraft; Hiller UH 12C Aircraft; Hiller UH 12E Aircraft; Hiller UH 12E4 Aircraft; HP 137 Jetstream Mark 1 Aircraft; HS 125-1000 Aircraft; HS 125-200 Aircraft; HS 125-400 Aircraft; HS 125-600 Aircraft; HS 125-700 Aircraft; HS 125-800 Aircraft; HS 748 Aircraft; Hunting President 2 Aircraft; Hunting President 2A Aircraft; ICA Brasov Motor Glider Aircraft; Israeli Aircraft Industries IAI 101 Arava Aircraft; Israeli Aircraft Industries IAI 102 Arava Aircraft; Israeli Aircraft Industries IAI 201 Arava Aircraft; Jodel (CEA) 253 Aircraft; Jodel (CEA) 315 Aircraft; Jodel (CEA) 340 Aircraft ***See Also:*** Jodel (CEA) 380 Aircraft; Jodel D 117 Aircraft; Jodel D 120 Aircraft; Jodel D 140 Aircraft; Jodel D 150 Aircraft; Jodel DR 100A Aircraft; Jodel DR 1050 Aircraft; Jodel DR 1051 Aircraft; Jodel DR 200 Aircraft; Jodel DR 221 Aircraft; Jodel Series Aircraft; Junkers JV52/3M; L 40 Meta Sokol Aircraft; Laboratories; Lake LA 4 Aircraft; Latecoere 631 Aircraft; Learjet Series Aircraft; Learjet 25B Aircraft; Learjet 35A Aircraft;

Aircraft *(Cont.)*

See Also: (Cont.)
Learjet 36A Aircraft; Learjet 55 Aircraft; LET L 410AG Turbolet Aircraft; Let Z 37 Aircraft; Linnet Aircraft; Lockheed Constellation Aircraft; Lockheed Hudson Aircraft; Lockheed L; Lockheed L 100 Hercules Aircraft; Lockheed L 188 Electra Aircraft; Lockheed L 382 Hercules Aircraft; Lockheed L 382B Hercules Aircraft; Lockheed L 1011 Aircraft; Lockheed L 1011-385-1 Aircraft; Lockheed L 1011-385-3 Aircraft; Lockheed Super Constellation Aircraft; Lockheed Tri-Star L1011 Aircraft; Martin 202 Aircraft; Martin 204 Aircraft; Martin 404 Aircraft; Maule M5 235C Aircraft; Maule M6 235 Aircraft; Maule Series Aircraft; MBB BK 117 B-1 Aircraft; McDonnell Douglas DC 10 Aircraft; McDonnell Douglas DC 4 Aircraft; McDonnell Douglas DC 6B Aircraft; McDonnell Douglas DC 7B Aircraft; McDonnell Douglas DC 7C Aircraft; McDonnell Douglas DC 8 Aircraft; McDonnell Douglas DC 9 Aircraft; McDonnell Douglas MD 80 Series Aircraft; Messerschmitt-Bolkow-Blohm GmbH BK 117 Aircraft; Messerschmitt-Bolkow-Blohm GmbH BO 105 Aircraft; Meta/Sokol L 40 Aircraft; Microlight Aircraft; Miles M 14 (Hawk) Aircraft; Miles M 14A Hawk Trainer 111 (Magister) Aircraft; Miles M 38 (Messenger) Aircraft; Miles M 65 Gemini Aircraft; Mitsubishi MU 300 Aircraft; Mooney Aircraft; Mooney M 20 Aircraft; Morane Saulnier MS 760 Aircraft; Morane Saulnier MS 880B Aircraft; Morane Saulnier MS 885 Aircraft; Morane Saulnier MS 890 Aircraft; Morane Saulnier MS 892 Aircraft; Morane Saulnier MS 893 Aircraft; Morane Saulnier MS 894 Rallye 100ST Aircraft; Morane Saulnier MS 894 Rallye 150ST Aircraft; Morane Saulnier Rallye Varients Aircraft; Morane Saulnier Rallye 100 Aircraft; Morane Saulnier Rallye 110 Aircraft; Morane Saulnier Rallye 150 Aircraft; Morane Saulnier Rallye 180 Aircraft; Morane Saulnier Rallye 235 Aircraft; Morane Saulnier 760 Aircraft; Morane Saulnier 880 Aircraft; Morane Saulnier 890 Aircraft; Morava L 200 Aircraft; Morava L 200A Aircraft; Moravan Zlin Z50L Series Aircraft; NAMC YS-11 Aircraft; Nash Petrel Aircraft; NDN Aircraft NDN-1; NDN Aircraft NDN-1T Firecracker; Noorduyn/Canadian Car/ Foundry Harvard Series Aircraft; Nord 1002 Aircraft; Nord 1101 Aircraft; Nord 262 Aircraft; Norman NAC 6 Fieldmaster Aircraft; North American Harvard Aircraft; Optica Industries Ltd Aircraft; Partenavia P 64B Aircraft; Percival Prince Aircraft; Percival Proctor Aircraft; Piaggio FW P149D Aircraft; Piaggio P 166 Aircraft; Piel CP 301A Aircraft ***See Also:*** Pierre Robin CEA DR 400 Aircraft; Pierre Robin HR 100/200 Aircraft; Pierre Robin HR 100/210 Aircraft; Pierre Robin HR 200/ 100 Aircraft; Pierre Robin R 1180 (T) Aircraft; Pierre Robin R 2100A Aircraft; Pierre Robin R 2112 Aircraft; Pierre Robin R 2160 Aircraft; Pierre Robin R 3000-120 Aircraft; Pierre Robin Series Aircraft; PIK 16C Vasama Sailplanes; Pilatus Britten-Norman Islander BN 2 Aircraft; Pilatus Britten-Norman Islander BN 2A-8 Aircraft; Pilatus Britten-Norman Islander BN 2A Aircraft; Pilatus Britten-Norman Islander BN 2A-20 Aircraft; Pilatus Britten-Norman Islander BN 2A-21 Aircraft; Pilatus Britten-Norman Islander BN 2A-26 Aircraft; Pilatus Britten-Norman Islander BN 2A-27 Aircraft; Pilatus Britten-Norman Islander BN 2B Aircraft; Pilatus Britten-Norman Islander BN 2B-20 Aircraft; Pilatus Britten-Norman Islander BN 2B-21 Aircraft; Pilatus Britten-Norman Islander BN 2B-26 Aircraft; Pilatus Britten-Norman Islander BN 2B-27 Aircraft; Pilatus Britten-Norman Islander BN 2T-2 Aircraft; Pilatus Britten-Norman Islander BN 2T Aircraft; Pilatus Britten-Norman Trilander BN 2A MK III Aircraft; Pilatus PC 6 Aircraft; Piper Aircraft; Piper J 3C-65 Aircraft; Piper P 31T1 Aircraft; Piper P 31T3 Aircraft; Piper PA 12 Aircraft; Piper PA 18 Aircraft; Piper PA 19 Aircraft; Piper PA 20 Aircraft; Piper PA 22 Aircraft; Piper PA 23 Aircraft; Piper PA 24 Aircraft; Piper PA 25/150 Aircraft; Piper PA 25/235 Aircraft; Piper PA 28 Aircraft; Piper PA 28-201T Aircraft; Piper PA 28-236 Aircraft; Piper PA 28R 200-2 Aircraft; Piper PA 30 Aircraft; Piper PA 31 Aircraft; Piper PA 31-325 Aircraft; Piper PA 31-350 Aircraft; Piper PA 31B Aircraft; Piper PA 31P Aircraft; Piper PA 32 Aircraft; Piper PA 32-30IT Aircraft; Piper PA 32R Aircraft; Piper PA 32RT Aircraft; Piper PA 34 Aircraft; Piper PA 36-375 Aircraft; Piper PA 38 Aircraft; Piper PA 39 Aircraft; Piper PA 42-720 Aircraft; Piper PA 42 Aircraft; Piper PA 42-1000 Aircraft; Piper PA 44 Aircraft; Piper PA 46-31OP Aircraft; Piper PA 60-601 P Aircraft; Pitot Tubes; Pitts S1 Aircraft; Pitts S2 Aircraft; Prentice Aircraft; Pressurized Cabins; Prince Aircraft; Procaer F-15

INDUSTRY STANDARDS
79

Aircraft

Aircraft (Cont.)
See Also: (Cont.)
Aircraft; Proctor Aircraft; Propulsion Systems; Provost Aircraft; PZL-104 Wilga Aircraft; Rate of Climb Indicators; Reims-Cessna Aircraft; Robinson R22 Helicopters; Rockwell (North American) Sabre Aircraft; Rockwell (North American) Sabreliner Aircraft; Rockwell Commander 112 Aircraft; Rockwell Commander 114 Aircraft; Rockwell Commander 690A Turbo Aircraft; Rockwell International 690C Aircraft; Rockwell International 690D Aircraft; Rockwell International 695 Aircraft; Rockwell International 695A Aircraft; Rockwell 1121 Jet Commander/Westwind Aircraft; Rockwell 1123 Jet Commander/Westwind Aircraft; Rockwell 1124 Jet Commander/Westwind Aircraft; Rockwell 500S Aircraft; Rollason Druine Condor Aircraft; Rollason D62B Condor Aircraft; Russian-Built Aircraft; S E 161 Languedoc Aircraft; S 51 Aircraft; S 55 Aircraft; SA 330J Puma Aircraft; SAAB-Fairchild SF 340 Aircraft See Also: SAAB Scandia Aircraft; SAAB 91A Aircraft; SAAB 91C Safir Aircraft; Saro Skeeter Aircraft; Saunders ST 27 Aircraft; Saunders ST 27-100 Aircraft; Scheibe SF25E Aircraft; Schweizer 300C Aircraft; Scintex Aircraft; Scottish Aviation/British Aerospace Jetstream Aircraft; Scottish Aviation Twin Pioneer Aircraft; Seaplanes; Shorts D 330 Aircraft; Shorts S 25 Sunderland, Sandringham/Hythe Aircraft; Shorts S 45 Solent Aircraft; Shorts SC 5 Aircraft; Shorts SC 5 Belfast Aircraft; Shorts SC 7 Aircraft; Shorts SD 3 Aircraft; Shorts SD 3 Sherpa Aircraft; Shorts SD 330 Aircraft; Shorts SD 360 Aircraft; Shorts Skyvan Aircraft; Siai Marchetti F 260 Aircraft; Siai Marchetti S 205 Aircraft; Slingsby Gliders; Slingsby T53B Aircraft; Slingsby T59D Aircraft; Slingsby T61 Aircraft; Slingsby T65A Vega Aircraft; Slingsby T67 Aircraft; Slingsby T67A, B, C and M Aircraft; Socata ST 10 Aircraft; Socata TB 10 Aircraft; Socata TB 20 Aircraft; Socata TB 9 Aircraft; Spitfire Aircraft; Sportavia-Putzer RF4 Motor Gliders; Sportavia-Putzer RF5 Motor Gliders; Stampe SV 4 Aircraft; Stampe SV 4C Aircraft; Stinson V77 Reliant Aircraft; Stinson 108-3 Aircraft; Sud Aviation SE 210 Caravelle Aircraft; Sud SA 315 Aircraft; Super Aero 145 Aircraft; Super Aero 45 Aircraft; Super VC 10 Aircraft; Supersonic Transports; Swearingen/Fairchild SA226 Merlin Aircraft; Swearingen/Fairchild SA227 Merlin Aircraft; Swearingen/Fairchild SA227 Metro Aircraft; Swearingen/Fairchild SA26 Merlin Aircraft; Swearingen/Fairchild SA26 Metro Aircraft; Swearingen SA 226T Aircraft; Taylor Craft Plus C and D Aircraft; Thurston TSC 1A Aircraft; Thurston TSC 1A1 Aircraft; Tipsy Nipper Aircraft; Trago Mills SAH 1 Aircraft; Trident 1 Aircraft; Trident 2E Aircraft; Trident 3B Aircraft; Trim Controls; Twin Pioneer Aircraft; Type Certificates; Valentin Taifun 17E Aircraft; Vanguard 953 Aircraft; Varga 2150A Aircraft; VC 10 Aircraft; Vega Gull Aircraft; Vehicles (Transportation); Vertical Takeoff Aircraft; Vickers Vanguard Aircraft; Vickers Vanguard 951 Aircraft; Vickers Vanguard 952 Aircraft; Vickers Vanguard 953 Aircraft; Vickers Viking Aircraft; Vickers Viscount Aircraft; Vickers Viscount 700 Aircraft; Vickers Viscount 800 Aircraft; Vickers Viscount 810 Aircraft; Victa Airtourer 100 Aircraft; Victa Airtourer 115 Aircraft; Viscount 700 Aircraft; Viscount 800 Aircraft; Wassmer WA 40 Aircraft; Wassmer WA 41 Aircraft; Wassmer WA 51 Aircraft; Wassmer WA 51A Aircraft; Wassmer WA 54 Aircraft; Wessex 60 Aircraft; Westland Bell 47 Aircraft; Westland Gazelle SA 341-G Aircraft; Westland S 51 Series 2 Aircraft; Westland S 55 Series 1 Aircraft; Westland S 55 Series 2 Aircraft See Also: Westland W 30 Aircraft; Westland Wessex 60 Aircraft; Wilga 80 Aircraft; Wittman W8 Tailwind Aircraft; Zlin Z 226 Aircraft; Zlin Z 326 Aircraft; Zlin Z 526 Aircraft
CAA. Explanatory Note 10.92 (Light Aeroplanes). 4 pp.
CAA. Contents 10.92 (Light Aeroplanes). 4 pp.
CAA. Foreword 10.92 (Light Aeroplanes). 4 pp.
CAA. Index 10.92 (Light Aeroplanes). 20 pp.
CAA Chapter K1-1 04.72. General and Definitions—General (Light Aeroplanes). 3 pp.
CAA. Explanatory Note 03.83 (Small Light Aeroplanes). 2 pp.
CAA. Contents 03.83 (Small Light Aeroplanes). 3 pp.
CAA CAP 482 SUB-Part A 03.83. General (Small Light Aeroplanes). 2 pp.
CAA. Contents 12.87 (Civil Air Publications: Light Aeroplanes). 18 pp.
CAA. Foreword 12.87 (Civil Air Publications: Light Aeroplanes). 2 pp.
CAA. Preambles 12.87 (Civil Air Publications: Light Aeroplanes). 1 p.
CAA CAP 531 Section 1. Regulations (Civil Air Publications: Light Aeroplanes). 14 pp.

Aircraft (Cont.)
CAA CAP 531 Section 2. Advisory Circulars—British (Civil Air Publications: Light Aeroplanes). 6 pp.
CAA. Contents (Joint Airworthiness Requirements). 18 pp.
CAA. Foreword (Joint Airworthiness Requirements). 1 p.
CAA. Check List of Pages (Joint Airworthiness Requirements). 10 pp.
CAA. Preambles (Joint Airworthiness Requirements). 40 pp.
CAA. Foreword. 1 p.
CAA. Check List of Pages. 1 p.
CAA. Preambles. 1 p.
CAA JAR-VLA Section 1. Requirements. 100 pp.
JIS W 0101-77. Dimensional Units for Aircraft.
MOD UK DSTAN 05-122-86. Military Listing of Particular Civil-Owned Military Type Aircraft Issue 1. 11 pp.

—Acceleration Testing
ISO 2669-78. Environmental Tests for Aircraft Equipment—Part 3.2: Steady State Acceleration First Edition. 6 pp.

—Access Doors—Symbols
BSI M 43-73. 1973 Amd 1 Methods of Zoning Aircraft and Referencing Access Doors and Panels. 9 pp.
ISO 2529-73. Aircraft—Zones, Access Doors and Panels—Referencing System First Edition. 8 pp.

—Accidents
CAA. Miscellaneous Aircraft (World Airline Accident Summary). 1 p.
CAA. Unknown Aircraft (World Airline Accident Summary). 1 p.
CAA. Miscellaneous Aircraft. 3 pp.
CAA. Unknown Aircraft. 1 p.
CAA. Unknown (World Helicopter Accident Summary). 1 p.

—Aerial Recovery
NATO STANAG 2970 ED 1 AMD 0-90. Aerial Recovery Equipment and Techniques for Helicopters. 8 pp.
NATO STANAG 2970 ED 2 AMD 0-92. Aerial Recovery Equipment and Techniques for Helicopters. 9 pp.

—Airworthiness
CAA NOTICE #74 ISSUE 3. Airworthiness Concessions in Respect of Foreign Built Aeroplanes (Appendices Included). 2 pp.
CAA. Explanatory Note (Provisional Airworthiness Requirements for Civil Powered Lift Aircraft). 7 pp.
CAA. Contents (Provisional Airworthiness Requirements for Civil Powered Lift Aircraft). 3 pp.
CAA. Foreword (Provisional Airworthiness Requirements for Civil Powered Lift Aircraft). 12 pp.
CAA Chapter P1-1. General (Provisional Airworthiness Requirements for Civil Powered Lift Aircraft). 2 pp.
CAA. Preface 04.92. 1 p.
CAA. Cross Reference Index. 4 pp.
CAA. General Contents 09.91. 2 pp.
CAA. General Foreword 06.90. 4 pp.
CAA. Foreword. 3 pp.
CAA. List of Effective Pages 07.89. 5 pp.
CAA Chapter A1-2 09.91. Categories of Aircraft. 2 pp.
CAA. Preface 09.91. 3 pp.
CAA. Foreword. 3 pp.
CAA. List of Effective Pages 07.89. 4 pp.
CAA. Contents 06.92. 7 pp.
CAA. Foreword. 1 p.
CAA. List of Effective Pages 07.90. 12 pp.
ICAO Annex 8. Airworthiness of Aircraft Eighth Edition—July 1988. 52 pp.
ICAO Circular 95. Continuing Airworthiness of Aircraft in Service Sixth Edition—1985. 186 pp.
ICAO 9051. Airworthiness Technical Manual Second Edition—1987. (Corrigendum 1 06/01/87) (Amendment 10 11/14/88). 255 pp.
ICAO 9388. Manual of Model Regulations for National Control of Flight Operations and Continuing Airworthiness of Aircraft Second Edition—1987. 108 pp.
ICAO 9389. Manual of Procedures for an Airworthiness Organization First Edition—1983. 182 pp.

—Airworthiness—Acceptance Testing
CAA Chapter A2-5 06.90. Approval of Modifications. 5 pp.

—Airworthiness—Certification
CAA Chapter A1-3. Certification of Collaborative Projects. 2 pp.

—Airworthiness—Confidential Information
CAA NOTICE #66 ISSUE 2. Aircraft Insurance (Airworthiness Notices). 1 p.

Aircraft (Cont.)
—Airworthiness—Specification
CAA Chapter A1-4 09.91. Specifications. 2 pp.

—Airworthiness—Type Certificates
CAA Chapter A2-2 06.90. Type Certification. 2 pp.
CAA Chapter A2-4. Type Certification of a Variant. 3 pp.

—Airworthiness—Type Certificates—Flight Testing
CAA Chapter A2-3 09.91. Flight Testing for Type Certification. 4 pp.

—Availability
EURO PSC/85/15859-84. A Discussion Paper Regarding the Availability Concept in Aircraft Applications. 27 pp.

—Balancing
CAA Chapter A5-4 09.91. Weight and Balance of Aircraft. 2 pp.
CAA Chapter A6-4 09.91. Weight and Balance of Aircraft. 2 pp.
CAA Chapter A6-4 App 1 07.89. Weight and Balance of Aircraft—Fleet Mean Weight and Fleet Mean Centre-of-Gravity. 2 pp.
CAA Chapter B5-4 07.89. Weight and Balance of Aircraft. 2 pp.
CAA Chapter B6-4 07.89. Weight and Balance of Aircraft. 2 pp.
CAA Chapter B7-10 07.89. Weight and Balance Report. 2 pp.
CAA LEAFLET 1-4 07.90. Weight and Balance of Aircraft. 27 pp.

—Banner Towing
CAA 806 (K). Section K Light Aeroplanes Banner Towing. 13 pp.
CAA Chapter K4-11 10.92. Banner Towing Installations. 2 pp.
CAA Chapter K4-11 App 10.92. Banner Towing Installations. 1 p.

—Break-In Points—Emergency Evacuation
CAA Spec. NO. 7 ISSUE 2.

—Camouflage
NATO STANAG 3687 ED 2 AMD 1-85. Camouflage of Aircraft. 5 pp.
NATO STANAG 3687 ED 3 AMD 1-90. Camouflage of Aircraft. 5 pp.
NATO STANAG 3687 ED 3 AMD 2-90. Camouflage of Aircraft. 6 pp.

—Categories
CAA Part 1 CAP 396. Navigation Order (Civil Air Publications: Air Navigation—Order and Regulations). 6 pp.

—Center of Gravity
CAA Chapter A6-4 App 1 07.89. Weight and Balance of Aircraft—Fleet Mean Weight and Fleet Mean Centre-of-Gravity. 2 pp.
CAA Chapter B6-4 App 1 7.89. Weight and Balance of Aircraft—Fleet Mean Weight and Fleet Mean Centre-Of-Gravity. 2 pp.
CAA Chapter B7-1 0 App1 07.89. Weight and Centre-Of-Gravity Schedules for Aircraft Exceeding 2740 kg. 3 pp.
CAA Chapter B7-1 0 App2 07.89. Weight and Centre-Of-Gravity and Loading and Distribution Schedules—Aircraft Not Exceeding 2730 kg. 3 pp.

—Center of Gravity—Scheduling
CAA Chapter A7-10 App 1. Weight and Centre-of-Gravity Schedules for Aircraft Exceeding 2730 kg. 3 pp.
CAA Chapter A7-1 0 App2 09.91. Weight and Centre-of-Gravity and Loading and Distribution Schedules—Aircraft Not Exceeding 2730 kg. 5 pp.

—Certificate of Airworthiness
CAA Part 2 CAP 396. Obtaining Certification (Civil Air Publications: Air Navigation—Order and Regulations). 20 pp.
CAA NOTICE #32 ISSUE 2. Overhauls, Modifications, Repairs and Replacements to Aircraft Not Exceeding 2730 kg with Certificates of Airworthiness in the Special Category (Airworthiness Notices). 2 pp.
CAA. Contents—Section A 07.91. 5 pp.
CAA Chapter B1-2 07.89. Categories of Aircraft. 3 pp.
CAA Chapter B3-2 07.89. Issue of Certificate of Airworthiness. 3 pp.
CAA Chapter B3-4 07.89. Renewal of Certificate of Airworthiness. 5 pp.
CAA Chapter B3-6 09.91. Certificates of Airworthiness for Export from the UK. 2 pp.
CAA Chapter B3-8 09.91. 'A' Conditions. 1 p.
CAA Chapter B3-9 06.90. 'B' Conditions. 2 pp.
CAA Supp TO Sec B 09.91. Sub-Section A8-Approvals. 110 pp.

INTERNATIONAL AND NON-U.S. NATIONAL STANDARDS
SUBJECT INDEX
Aircraft

Aircraft *(Cont.)*

—**Certificate of Airworthiness—Fees**
CAA NOTICE #25 ISSUE 19. Charges for the CAA's Airworthiness and Noise Certification (Airworthiness Notices). 13 pp.

—**Certificate of Airworthiness—Inspection**
CAA Chapter B6-7 07.89. Certification of Inspections, Overhauls, Modifications, Repairs and Replacements. 5 pp.

—**Certificate of Airworthiness—Maintenance/Repair**
CAA Chapter B6-7 07.89. Certification of Inspections, Overhauls, Modifications, Repairs and Replacements. 5 pp.

—**Certificate of Airworthiness—Modification**
CAA Chapter B6-7 07.89. Certification of Inspections, Overhauls, Modifications, Repairs and Replacements. 5 pp.

—**Certificate of Airworthiness—Radio**
CAA Chapter B3-11 07.89. Radio Installations. 3 pp.

—**Certification**
CAA. Foreword (Civil Air Publications: Registration, Certification, and Maintenance of Aircraft). 1 p.
CAA. Contents (Civil Air Publications: Registration, Certification, and Maintenance of Aircraft). 1 p.
CAA Part 1 CAP 396. Navigation Order (Civil Air Publications: Air Navigation—Order and Regulations). 6 pp.
CAA NOTICE #65 ISSUE 1. CAA Use of Confidential Information (Airworthiness Notices). 3 pp.

—**Circuit Diagrams—Computer Aided Design**
EURO CTI/79/9343-79. Recommendation for the Automation of Wiring Diagrams. 50 pp.

—**Cleaning Agents**
MOD UK TS 10306. Liquid Cleaner for Use on Aircraft Transparencies.

—**Cleanliness**
CAA LEAFLET 2-6 07.90. Cleanliness of Aircraft. 7 pp.

—**Coated Fabrics**
CAA NOTICE #20 ISSUE 6. Cotton, Linen and Synthetic Fabric Covered Aircraft (Airworthiness Notices). 3 pp.

—**Cockpit Voice Recorders**
CAA LEAFLET 14-14 06.91. Cockpit Voice Recorder System Fairchild A100 Series. 2 pp.
CAA LEAFLET 14-16 06.92. Cockpit Voice Recorder System Sundstrand AV557 Series. 2 pp.

—**Components—Interchangeability**
EURO CTI/74/3282-74. Interchangeability and Mastering Policy in Airframe Manufacturing. 68 pp.

—**Construction—Mass**
DIN ENGL 9020 Pt 1-83. Aerospace; Mass Breakdown for Aircraft Heavier Than Air; Main Mass Groups and Terms of Mass; Survey (Oct). 3 pp.
DIN ENGL 9020 Pt 2-83. Aerospace; Mass Breakdown for Aircraft Heavier Than Air; Main Mass Groups and Terms of Mass; Definitions (Oct). 6 pp.
DIN ENGL 9020 Pt 5-92. Aerospace Series; Mass Breakdown for Aircraft Heavier Than Air; Dimensional and Construction Data (Feb). 8 pp.

—**Construction—Mass—Forms (Paper)**
DIN ENGL 9020 Pt 3-83. Aerospace; Mass Breakdown for Aircraft Heavier Than Air; Group Mass Statement (Oct). 16 pp.
DIN ENGL 9020 Pt 3 Suppl. 1-83. Aerospace; Mass Breakdown for Aircraft Heavier Than Air; Group Mass Statement; Condensed Version (Oct). 4 pp.
DIN ENGL 9020 Pt 4-83. Aerospace; Mass Breakdown for Aircraft Heavier Than Air; Detail Mass Statement (Oct). 77 pp.

—**Contamination—Carbon Monoxide**
CAA NOTICE #40 ISSUE 1. Carbon Monoxide Contamination in Aircraft (Airworthiness Notices). 2 pp.

—**Control Systems Equipment**
NATO STANAG 3220 ED 4 AMD 6-74. Location, Actuation and Shape of Airframe Controls for Fixed Wing Aircraft. 11 pp.

—**Control Systems Equipment—Identification Systems**
CAA CAP 482 SUB-Part G 03.83. Operating Limitations and Information (Small Light Aeroplanes). 6 pp.

Aircraft *(Cont.)*

—**Corrosion**
CAA NOTICE #73 ISSUE 2. Corrosion of Aircraft Structures (Airworthiness Notices). 2 pp.
CAA Chapter K4-1 App 04.74. Protection Against Corrosion and Other Effects of the Presence of Fluids (Light Aeroplanes). 3 pp.
CAA Chapter Q4-1 App 12.79. Protection Against Corrosion and Other Effects of the Presence of Fluids (Non-Rigid Airships). 4 pp.

—**Costs**
EURO PSC/87/18535-87. Life Cycle Cost Perspectives—Military Aircraft Applications. 48 pp.

—**Costs—Glossaries**
EURO ECO/CTI/3691-77. Definition of an Aircraft Cost Terminology for Use in Aircraft Pricing. 28 pp.
EURO CTI/86/16946 /D,E,F-85. Definition of an AECMA-Cost-Weight and AECMA-Cost-Hours for Aircraft Manufacture. 25 pp.

—**Costs—Tooling**
EURO CTI/83/12301-82. Tooling Costs Investigation—Final Report. 112 pp.

—**Cross Servicing**
NATO STANAG 7028 ED 1 AMD 1-91. Identical Aircraft for Cross-Servicing. 9 pp.
NATO STANAG 7028 ED 1 AMD 2-91. Identical Aircraft for Cross-Servicing. 10 pp.

—**Data Sheets**
CAA. Contents Issue 104 (Type Certificate Data Sheets). 3 pp.
CAA. Foreword (Type Certificate Data Sheets). 3 pp.

—**Decals**
CGSB 62-GP-3M-80. Prefabricated Markings, Exterior, for Aircraft, Standard for. 18 pp.
CGSB 62-GP-8M-80. Prefabricated Opaque Markings, General Purpose, for Use on Exterior and Interior Surfaces, Standard for. 12 pp.
CGSB 62-GP-9M-80. Prefabricated Markings, Positionable, Exterior, for Aircraft, Ground Equipment and Facilities, Standard for. 17 pp.

—**Deicing Equipment**
CAA. Contents (Civil Air Publications: Ground De-Icing of Aircraft). 1 p.
CAA. Foreword (Civil Air Publications: Ground De-Icing of Aircraft). 1 p.
CAA CAP 512 Part 1. Maintenance Aspects (Civil Air Publications: Ground De-Icing of Aircraft). 1 p.
CAA CAP 512 Part 2. Operational Aspects (Civil Air Publications: Ground De-Icing of Aircraft). 7 pp.
CAA CAP 512 Part 3. Common Practices or Suggested Practices for Safe Cold Weather Operations (Civil Air Publications: Ground De-Icing of Aircraft). 3 pp.
CAA CAP 512 Appendix. General Information Relating to Ground and Flight Operations in Conditions Conductive to Aircraft Icing (Civil Air Publications: Ground De-Icing of Aircraft). 6 pp.

—**Design**
CAA Chapter P4-1. General (Provisional Airworthiness Requirements for Civil Powered Lift Aircraft). 4 pp.
CAA CAP 482 SUB-Part D 03.83. Design and Construction (Small Light Aeroplanes). 11 pp.
MOD UK DSTAN 00-970: Pt 3:Vol 1-01. Structural Strength and Design for Operation on Specified Surfaces Issue 1; Amendment 10. 226 pp.
MOD UK DSTAN 00-970: Pt 3:Vol 1-02. Structural Strength and Design for Operation on Specified Surfaces Issue 1; Amendment 11. 207 pp.
MOD UK DSTAN 00-970: Pt 3:Vol 1-03. Structural Strength and Design for Operation on Specified Surfaces Issue 1; Amendment 11. 207 pp.
MOD UK DSTAN 05-123: Part 3-01. Control of Designs and Design Records Issue 1; Amendment 55. 146 pp.
MOD UK DSTAN 05-123: Part 3-02. Control of Designs and Design Records Issue 1; Amendment 55. 156 pp.
MOD UK DSTAN 05-123: Part 3-03. Control of Designs and Design Records Issue 1; Amendment 58. 155 pp.
MOD UK DSTAN 05-123: Part 3-04. Control of Designs and Design Records Issue 1; Amendment 59. 152 pp.

—**Design—Control Systems**
EURO CTI/82/12440 /D,E,F-83. Basic Outlines for Design Change Control Procedures. 24 pp.

—**Design Loads**
CAA Chapter P3-2. Flight Manoeuvering Loads and Design Airspeeds (Provisional Airworthiness Requirements for Civil Powered Lift Aircraft). 13 pp.

Aircraft *(Cont.)*

—**Design Speeds**
CAA Chapter P3-2. Flight Manoeuvering Loads and Design Airspeeds (Provisional Airworthiness Requirements for Civil Powered Lift Aircraft). 13 pp.

—**Designations—Codes**
NATO STANAG 3236 ED 8 AMD 0-91. Designation System for Aircraft and Designation of Role Codes of Aircraft Codes of Aircraft. 10 pp.

—**Designations—Warsaw Pact—Codes**
NATO STANAG 3236 ED 7 AMD 0-88. Designation System for Warsaw Pact Aircraft and Guided Missiles and Designation of Role Codes of Aircraft. 10 pp.

—**Elastic Waves**
ISO 2671-82. Environmental Tests for Aircraft Equipment—Part 3.4: Acoustic Vibration First Edition. 11 pp.

—**Electric Wiring**
AECMA PREN2283-88. Testing of Aircraft Wiring. 9 pp.

—**Electrical Bonding**
NATO STANAG 3659 ED 1 AMD 6-75. Bonding and in-Flight Lightning Protection for Aircraft. 17 pp.
NATO STANAG 3659 ED 1 AMD 7-75. Bonding and In-Flight Lightning Protection for Aircraft. 19 pp.

—**Environmental Testing**
BSI 3G 100:Pt 2: SUBSEC 3.0-72. 1972 Amd 1 General Requirements for Equipment for Use in Aircraft Part 2: All Equipment Section 3: Environmental Conditions Subsection 3.0: Standard Test Requirements. 4 pp.
BSI 3G 100:Pt 2: SUBSEC 3.2-70. 1970 Amd 2 General Requirements for Equipment for Use in Aircraft Part 2: All Equipment Section 3: Environmental Conditions Subsection 3.2: Temperature-Pressure Requirements. 15 pp.
BSI 3G 100:Pt 2: SUBSEC 3.8-72. 1972 Amd 1 General Requirements for Equipment for Use in Aircraft Part 2: All Equipment Section 3: Environmental Conditions Subsection 3.8: Salt Mist. 3 pp.
BSI 2G 229-87. (WITHDRAWN) 1987 Environmental Conditions and Test Procedures for Airborne Equipment (Superseded by BS 36 229: 1993). 4 pp.
BSI 3G 229-93. 1993 Environmental Conditions and Test Procedures for Airborne Equipment (ISO 7137: 1992) (S). 6 pp.
EUROCAE ED-14C/RTCA DO160C 12.89. Environmental Conditions and Test Procedures for Airborne Equipment. 480 pp.
ISO 7137-92. Aircraft—Environmental Conditions and Test Procedures for Airborne Equipment Third Edition. 8 pp.

—**Environments**
BSI 2G 229-87. (WITHDRAWN) 1987 Environmental Conditions and Test Procedures for Airborne Equipment (Superseded by BS 36 229: 1993). 4 pp.
BSI 3G 229-93. 1993 Environmental Conditions and Test Procedures for Airborne Equipment (ISO 7137: 1992) (S). 6 pp.
EUROCAE ED-14C/RTCA DO160C 12.89. Environmental Conditions and Test Procedures for Airborne Equipment. 480 pp.
ISO 7137-92. Aircraft—Environmental Conditions and Test Procedures for Airborne Equipment Third Edition. 8 pp.

—**Environments—Glossaries**
JIS W 0110-87. Glossary of Terms for Aircraft Environmental Control.

—**Evacuation Equipment**
CAA LEAFLET 14-18. Air Cruisers Company Evacuation Systems. 1 p.

—**Fabrication**
CAA Chapter P4-1. General (Provisional Airworthiness Requirements for Civil Powered Lift Aircraft). 4 pp.

—**Fatigue (Materials)**
CAA 799 (K). Semi-Aerobatic Aeroplanes (Blue Papers). 25 pp.
CAA Chapter P3-9. Flutter Preventions and Structural Stiffness of Aerodynamic Surfaces (Provisional Airworthiness Requirements for Civil Powered Lift Aircraft). 3 pp.

—**Fire Zones**
ISO 2685-92. Aircraft—Environmental Conditions and Test Procedures for Airborne Equipment—Resistance to Fire in Designated Fire Zones First Edition. 33 pp.

INDUSTRY STANDARDS

INTERNATIONAL AND NON-U.S. NATIONAL STANDARDS
SUBJECT INDEX

Aircraft

Aircraft (Cont.)
—**First Aid Equipment**
NATO STANAG 3746 ED 1 AMD 4-75. Medical First Aid Equipment in Aircraft. 9 pp.

—**Flameout Procedures**
NATO STANAG 3297 ED 3 AMD 5-74. Flame-Out Procedures. 15 pp.
NATO STANAG 3297 ED 4 AMD 0-92. Flame-Out Procedures. 7 pp.
NATO STANAG 3297 ED 4 AMD 1-92. Flame-Out Procedures. 9 pp.
NATO STANAG 3297 ED 4 AMD 2-92. Flame-Out Procedures. 10 pp.

—**Flight Dynamics**
CAA 711 (K). Section K Light Aeroplanes Spinning. 6 pp.

—**Flight Recorders**
CAA LEAFLET 14-7 06.91. Plessey/Davall Date Recorder Sustems. 3 pp.
CAA LEAFLET 14-10 06.91. Plessey Flight Data Recorder System Type PV1584 Series. 3 pp.
CAA LEAFLET 14-13 06.91. Universal/Sundstrand Flight Data Recorder Model 980-4100 Series. 3 pp.
CAA Spec. NO. 10 A ISSUE 3.

—**Flight Restrictions**
CAA Section 4 CAP 393. Restriction of Flying. 11 pp.
CAA Chapter P7-1. General (Provisional Airworthiness Requirements for Civil Powered Lift Aircraft). 11 pp.

—**Flight Testing**
CAA 781 (K). Section K Light Aeroplanes Handling Requirements—Miscellaneous Amendments. 8 pp.
CAA Chapter B3-3 09.91. Flight Testing for Issue of Certifcate of Airworthiness or a Permit to Fly. 4 pp.
CAA Chapter B3-5 07.89. Flight Testing for Renewal of Certificates of Airworthiness or a Permit to Fly. 6 pp.
CAA Chapter B3-7 09.91. Issue and Renewal of Permits to Fly. 4 pp.
CAA Chapter B3-7 App 1 06.90. Evidenance to Substantiate Applications. 1 p.
CAA Chapter B3-7 App 2 07.89. Flight Release Certificate. 1 p.
CAA Chapter B6-8 07.89. Flight Testing After Modification or Repair. 1 p.

—**Fluids**
MOD UK EL 2491-74. Fluids, Aviation, Thrust Augmentation Issue 3. 8 pp.

—**Flutter**
CAA 820 (K). Section K Light Aeroplanes Flutter Fail Safe Criteria (Blue Papers). 11 pp.
CAA Chapter K3-9 10.92. Flutter Prevention and Structural Stiffness (Light Aeroplanes). 4 pp.
CAA Chapter K3-9 App 10.69. Flutter Prevention and Structural Stiffness (Light Aeroplanes). 6 pp.

—**Glossaries**
BSI BS 185: Sec 2-69. 1969 Amd 1 Glossary of Aeronautical and Astronautical Terms Section 2: Motion of Aircraft (In Flight and on the Earth's Surface). 6 pp.
BSI BS 185: Sec 3-69. 1969 Amd 1 Glossary of Aeronautical and Astronautical Terms Section 3: Structure (Detail Parts, Aeroelasticity, Fatigue). 10 pp.
BSI BS 185: Sec 5-69. 1969 Amd 3 Glossary of Aeronautical and Astronautical Terms Section 5: Heavier-Than-Air Aircraft (Aerodynes) (Shape, and Disposition of Surfaces, Component Parts, Instruments, Loadings, Weights and Rotorcraft). 23 pp.
CAA Chapter K1-2 App 09.66. Definitions Climatic Conditions (Light Aeroplanes). 1 p.
CAA Chapter P1-2. Definitions (Provisional Airworthiness Requirements for Civil Powered Lift Aircraft). 16 pp.
DIN ENGL 9020 Pt 2-83. Aerospace; Mass Breakdown for Aircraft Heavier Than Air; Main Mass Groups and Terms of Mass; Definitions (Oct). 6 pp.
JIS W 0106-84. Glossary of Terms for Aircraft General.
JIS W 0108-76. Glossary of Terms for Aircraft (Structure).

—**Greases**
MOD UK DSTAN 91-12-72. Grease, Aircraft: General Purpose XG-271, NATO G-382 Issue 1. 9 pp.

—**Ground Handling**
CAA Chapter P3-13. Ground Handling (Provisional Airworthiness Requirements for Civil Powered Lift Aircraft). 4 pp.

Aircraft (Cont.)
—**Ground Loads**
CAA Chapter P3-5. Ground Loads (Provisional Airworthiness Requirements for Civil Powered Lift Aircraft). 18 pp.

—**Gust Loads**
CAA Chapter P3-3. Gust Loads (Provisional Airworthiness Requirements for Civil Powered Lift Aircraft). 11 pp.

—**Harnesses (Safety)**
CAA LEAFLET 14-15 06.92. Maintenance and Inspection of Crew Harnesses and Passenger Seat Belts (Metal to Metal Attachment). 1 p.

—**Ice Formation**
CAA Chapter K4-7 04.72. Flight in Ice Forming Conditions (Light Aeroplanes). 2 pp.
CAA CAP 512 Appendix. General Information Relating to Ground and Flight Operations in Conditions Conductive to Aircraft Icing (Civil Air Publications: Ground De-Icing of Aircraft). 6 pp.

—**Identification Systems**
CAA Chapter K7-3 10.92. Markings and Placards (Light Aeroplanes). 4 pp.
SBAC AGS 4714 (V). Call up of Glass Fibre Tape System Identification Markers.

—**Installations**
MOD UK DSTAN 00-970: Pt 7:Vol 1-01. Design and Airworthiness Requirements for Service Aircraft Volume 1—Aeroplanes Book 3 (Chapters 711 ON and Parts 8 to 10) Part 7: Installations Issue 1; Amendment 10. 555 pp.
MOD UK DSTAN 00-970: Pt 7:Vol 1-02. Design and Airworthiness Requirements for Service Aircraft Volume 1—Aeroplanes Book 3 (Chapters 711 ON and Parts 8 to 10) Part 7: Installations Issue 1; Amendment 11. 774 pp.
MOD UK DSTAN 00-970: Pt 7:Vol 1-03. Design and Airworthiness Requirements for Service Aircraft Volume 1—Aeroplanes Book 3 (Chapters 711 ON and Parts 8 to 10) Part 7: Installations Issue 1; Amendment 12. 773 pp.

—**Insurance**
CAA NOTICE #66 ISSUE 2. Aircraft Insurance (Airworthiness Notices). 1 p.

—**Lightning Protection**
CAA Chapter K4-6 03.67. Electrical Bonding and Lightning Discharge Protection (Light Aeroplanes). 5 pp.
CAA Chapter K4-6 App 03.67. Electrical Bonding and Lightning Discharge Protection (Light Aeroplanes). 2 pp.
CAA Chapter P4-6. Electrical Bonding and Lightning Discharge Protection (Provisional Airworthiness Requirements for Civil Powered Lift Aircraft). 18 pp.
NATO STANAG 3659 ED 1 AMD 6-75. Bonding and in-Flight Lightning Protection for Aircraft. 17 pp.
NATO STANAG 3659 ED 1 AMD 7-75. Bonding and In-Flight Lightning Protection for Aircraft. 19 pp.

—**Loading—Scheduling**
CAA Chapter A7-1 0 App2 09.91. Weight and Centre-of-Gravity and Loading and Distribution Schedules—Aircraft Not Exceeding 2730 kg. 5 pp.

—**Loads (Forces)**
CAA NOTICE #90 ISSUE 1. Maximum Total Weight Authorised for Agricultural Operations and Other Aerial Applications (Airworthiness Notices). 3 pp.
CAA Chapter K3-1 10.92. Structures—General (Light Aeroplanes). 3 pp.
CAA Chapter K3-2 App 10.69. Flight Manoeuvring Loads and Design Air Speeds (Light Aeroplanes). 2 pp.
CAA Chapter K3-3 10.69. Gust Loads (Light Aeroplanes). 4 pp.
CAA Chapter K3-5 04.72. Ground Loads (Light Aeroplanes). 8 pp.
CAA Chapter K3-13 10.69. Ground Handling. 2 pp.
CAA CAP 482 SUB-Part C 03.83. Structure (Small Light Aeroplanes). 12 pp.

—**Logistics Support**
EURO PSC/88/20178-89. Integrated Logistic Support in Aircraft Application. 110 pp.

—**Logs (Records)**
CAA Chapter B7-8 07.89. Technical Logs. 3 pp.
CAA LEAFLET 1-5 07.90. Aircraft, Engine and Propeller Log Books. 8 pp.

—**Maintainability**
EURO PSC/83/12418-83. Supply of Basic Maintainablity and Reliability Data. 11 pp.

Aircraft (Cont.)
—**Maintenance**
CAA. Foreword (Civil Air Publications: Registration, Certification, and Maintenance of Aircraft). 1 p.
CAA. Contents (Civil Air Publications: Registration, Certification, and Maintenance of Aircraft). 1 p.
CAA Part 1 CAP 396. Navigation Order (Civil Air Publications: Air Navigation—Order and Regulations). 6 pp.
CAA Part 3 CAP 396. Maintaining Airworthiness (Civil Air Publications: Air Navigation—Order and Regulations). 15 pp.
CAA NOTICE #63 ISSUE 9. Certification and Maintenance of Aircraft Not Exceeding 2730 kg (Airworthiness Notices). 4 pp.
CAA Chapter A6-2 07.89. Maintenance of Aircraft. 8 pp.
CAA Chapter A6-2 App 1 07.89. Condition Monitored Maintenance Programmes. 7 pp.
CAA Chapter A7-5 07.89. Maintenance Schedules. 3 pp.
CAA Chapter B5-2 07.89. Maintenance Review Board (MRB). 5 pp.
CAA Chapter B6-2 09.91. Maintenance of Aircraft. 8 pp.
CAA Chapter B6-2 App 1 07.89. Maintenance Programmes. 6 pp.
CAA Chapter B7-5 07.89. Maintenance Schedules. 3 pp.
CAA LEAFLET 1-6 07.90. Maintenance of Aircraft Not Exceeding 2730 kg MTWA, Including the Star Inspection. 22 pp.
CAA LEAFLET 1-7 12.90. Condition Monitored Maintenance. 29 pp.
CAA LEAFLET 1-8 12.90. Storage Conditions for Aeronautical Supplies. 15 pp.
CAA LEAFLET 14-11 06.91. Specific Maintenance Schedules for Aircraft Below 2730 kg. 3 pp.
CAA. Contents 03.92. 2 pp.
CAA. Introductory Note 03.92. 1 p.
CAA. Checklist of Pages. 2 pp.
CAA. Amendments to Document 03.92. 1 p.
CAA Chapter 8 03.92. Procedures for the Continued JAR-145 Approval by the JAA National Aviation Authority. 1 p.
CAA Chapter 9 03.92. Procedures for the Revocation of JAR-145 Approval by the JAA National Aviation Authority. 1 p.
CAA Chapter 10 03.92. Procedures for JAR-145 Applications from Organisations Located Outside the JAA States. 1 p.
CAA Chapter 11 03.92. Provisions for the Provision of Temporary Guidance Material by the JAA Maintenance Division Pending Possible NPA Action. 1 p.
CAA Chapter 14 03.92. Role of the JAA Maintenance Division. 1 p.
CAA Appendix 2 03.92. Notified JAA-NAA Approval Reference Designators for the JAA Form Three Approval Certificate. 1 p.

—**Maintenance—Certification**
CAA LEAFLET 1-10 06.92. Aeronautical Maintenance Certificate. 2 pp.

—**Maintenance—Classifications**
CAA LEAFLET 14-8 06.91. Airworthiness Classifications of Items in Approved Maintenance Schedules. 1 p.

—**Maintenance—Condition Monitored**
CAA. Foreword (Civil Air Publications: Conditioned Monitored Maintenance). 1 p.
CAA. Contents (Civil Air Publications: Conditioned Monitored Maintenance). 1 p.
CAA Section 1 CAP 418. Introduction (Civil Air Publications: Conditioned Monitored Maintenance). 1 p.
CAA Section 2 CAP 418. Primary Maintenance (Civil Air Publications: Conditioned Monitored Maintenance). 1 p.
CAA Section 3 CAP 418. Conditioned Monitored Maintenance (Civil Air Publications: Conditioned Monitored Maintenance). 11 pp.
CAA Section 4 CAP 418. Programme Document (Civil Air Publications: Conditioned Monitored Maintenance). 2 pp.
CAA Section 5 CAP 418. Conditioned Monitored Maintenance and the Airworthiness Authority (Civil Air Publications: Conditioned Monitored Maintenance). 1 p.
CAA Appendix A CAP 418. Short Introduction to the Basic Principles of Maintenance Steering Group Logic Analysis (Civil Air Publications: Conditioned Monitored Maintenance). 4 pp.
CAA Appendix B CAP 418. Typical Organisation and Data Flow Chart (Civil Air Publications: Conditioned Monitored Maintenance). 1 p.
CAA Apprndix C CAP 418. Alert Level Calculations (Civil Air Publications: Conditioned Monitored Maintenance). 5 pp.
CAA Appendix D CAP 418. Typical Data Displays (Civil Air Publications: Conditioned Monitored Maintenance). 7 pp.

INDEX and DIRECTORY of

INTERNATIONAL AND NON-U.S. NATIONAL STANDARDS
SUBJECT INDEX

Aircraft

Aircraft (Cont.)

—Maintenance—Condition Monitored (Cont.)
CAA Appendix E CAP 418. Defined Terms and Abbreviations (Civil Air Publications: Conditioned Monitored Maintenance). 2 pp.

—Maintenance Contracts
MOD UK DSTAN 00-970: Pt 8:Vol 1-01. Maintenance Issue 1; Amendment 10. 90 pp.
MOD UK DSTAN 00-970: Pt 8:Vol 1-02. Maintenance Issue 1; Amendment 9. 47 pp.
MOD UK DSTAN 00-970: Pt 8:Vol 1-03. Maintenance Issue 1; Amendment 12. 40 pp.
MOD UK DEFCON 112AG-83. Conditions of Contract for Aircraft Maintenance and Airfield Services 9/83. 6 pp.

—Maintenance—Cost Estimates
EURO PSC/88/20184-89. Discussion Paper on the Subject "Prediction of Maintenance Costs". 25 pp.

—Maintenance—Documentation
CAA CAP 482 SUB-Part G 03.83. Operating Limitations and Information (Small Light Aeroplanes). 6 pp.
CAA Appendix 1 03.92. Example Documents. 18 pp.

—Maintenance—Licensing
CAA LEAFLET 1-11 06.92. Introduction to the Licensing of Aircraft Maintenance Engineers. 7 pp.

—Maintenance Personnel
CAA Chapter 12 03.92. Procedures for the Conduct of Visits by the Maintenance Standardisation Team (MAST) to the JAA National Aviation Authority and Follow up Action by the JAA Maintenance Division. 1 p.
CAA Chapter 15 03.92. Role of the JAA Maintenance Committee, the JAA Maintenance Board and the Joint JAA Maintenance Board. 2 pp.

—Maintenance—Symbols
SBAC AGS 4672 (V). Servicing Point, Emblem—Self Adhesive.
SBAC AGS 4673 (V). Maintenance Point, Emblem—Self Adhesive.
SBAC AGS 4674 (V). Ground Handling Point, Emblem—Self Adhesive.

—Mandatory Aircraft Modifications and Inspections Summaries
CAA Chapter B5-6 07.89. Mandatory Modifications and Inspections; Procedure for Classifications. 3 pp.
CAA Chapter B6-6 07.89. Mandatory Modifications and Inspections; Procedure for Implementation. 4 pp.

—Manuals—Crew
CAA Chapter A7-3 07.89. Crew Manuals. 5 pp.
CAA Chapter B7-3 07.89. Crew Manuals. 5 pp.

—Manuals—Flight
CAA Chapter B7-2 09.91. Flight Manuals.

—Manuals—Maintenance
CAA Chapter B5-3 06.90. Maintenance, Overhaul and Repair Manuals. 4 pp.

—Manuals—Maintenance/Overhaul/Repair
CAA Chapter A5-3 07.89. Maintenance, Overhaul an Repair Manuals. 5 pp.
CAA Chapter B7-4 7.89. Maintenance, Overhaul and Repair Manuals.

—Manuals—Product Support Systems
EURO PSC/PSSWG/91—22505-90. AECMA—PSSWG Product Support Services for Civil Operators in the 1990's. 24 pp.
EURO PSSWG-90. Product Support Services for Civil Operators in the 1990's. 23 pp.

—Manufacturing
CAA LEAFLET 1-9 06.91. Concessions During Manufacture. 2 pp.

—Metal—Inspection
CAA LEAFLET 6-3 07.90. Inspection of Metal Aircraft After Abnormal Occurrences. 8 pp.

—Military Support
NATO STANAG 3113 ED 5 AMD 5-82. Provision of Support to Visiting Personnel, Aircraft and Vehicles. 18 pp.
NATO STANAG 3113 ED 6 AMD 0-93. Provision of Support to Visiting Personnel, Aircraft and Vehicles. 12 pp.

—Modification—Records and Reports
CAA Chapter B7-9 07.89. Modification Record Book. 2 pp.

Aircraft (Cont.)

—Nationality Symbols
ICAO Annex 7. Aircraft Nationality and Registration Marks Fourth Edition—July 1981; (Supplement 2 2/4/85) (Amendment 2 12/17/85) (Amendment 3 5/1/88) (Supplement 1 8/1/88) (Amendment 1 8/1/90) (Amendment 4 8/1/90). 50 pp.

—Nitrogen
ISO 2435-73. Nitrogen for Use in Aircraft First Edition. 3 pp.
MOD UK DSTAN 68-95-90. Nitrogen, Compressed, High Purity Oil—Free: for All Applications Issue 2. 8 pp.
MOD UK DSTAN 68-95-01. Nitrogen Gas, Compressed, High Purity Oil-Free: for All Applications Issue 2; Amendment 2. 10 pp.

—Noise—Building Sites
SAA AS 2021-85. Acoustics—Aircraft Noise Intrusion—Building Siting and Construction. 42 pp.

—Nondestructive Testing
CAA NOTICE #94 ISSUE 6. Non-Destructive Testing of Aircraft, Engines and Components (Airworthiness Notices). 8 pp.

—Painting
CAA NOTICE #38 ISSUE 1. Painting of Aircraft. 5 pp.

—Permit to Fly
CAA. Contents—Section A 07.91. 5 pp.

—Pressurized Cabins
CAA. Pressurised Aircraft Under 12,500 lb. AUW. 5 pp.

—Procurement
MOD UK DSTAN 05-123-01. Technical Procedures for the Procurement of Aircraft, Weapon and Electronic Systems Issue 1; Amendment 55. 20 pp.
MOD UK DSTAN 05-123-02. Technical Procedures for the Procurement of Aircraft, Weapon and Electronic Systems Issue 1; Amendment 57. 23 pp.
MOD UK DSTAN 05-123-03. Technical Procedures for the Procurement of Aircraft, Weapon and Electronic Systems Issue 1; Amendment 58. 25 pp.
MOD UK DSTAN 05-123-04. Technical Procedures for the Procurement of Aircraft, Weapon and Electronic Systems Issue 1; Amendment 59. 25 pp.

—Radiotelephony
CCIR Report 1051-1-90. Public Mobile Telephone Service with Aircraft—Section 8K—Aeronautical Mobile Service (Terrestrial). 18 pp.
CCIR QUESTION 74-1/8-90. Public Mobile Telephone Service with Aircraft—Questions Concerning Study Group 8—Mobile, Radiodetermination, Amateur and Related Satellite Services. 1 p.

—Registration
CAA. Foreword (Civil Air Publications: Registration, Certification, and Maintenance of Aircraft). 1 p.
CAA. Contents (Civil Air Publications: Registration, Certification, and Maintenance of Aircraft). 1 p.
CAA Part 1 CAP 396. Navigation Order (Civil Air Publications: Air Navigation—Order and Regulations). 6 pp.
CAA NOTICE #72 ISSUE 1. Registrations of Aircraft and Continuing Airworthiness (Airworthiness Notices). 1 p.

—Registration Symbols
CAA Part 1 CAP 396. Navigation Order (Civil Air Publications: Air Navigation—Order and Regulations). 6 pp.
ICAO Annex 7. Aircraft Nationality and Registration Marks Fourth Edition—July 1981; (Supplement 2 2/4/85) (Amendment 2 12/17/85) (Amendment 3 5/1/88) (Supplement 1 8/1/88) (Amendment 1 8/1/90) (Amendment 4 8/1/90). 50 pp.

—Reliability
EURO PSC/83/12418-83. Supply of Basic Maintainablity and Reliability Data. 11 pp.

—Reliability—Glossaries
ISO 5843 Pt 8-88. Aerospace—List of Equivalent Terms—Part 8: Aircraft Reliability First Edition. 18 pp.

—Replenishment—Methanol/Water
NATO STANAG 3416 ED 4 AMD 4-83. Methanol/Water and Demineralized Water Replenishment. 6 pp.
NATO STANAG 3416 ED 4 AMD 5-83. Methanol/Water and Demineralized Water Replenishment. 7 pp.

—Replenishment—Water
NATO STANAG 3416 ED 4 AMD 4-83. Methanol/Water and Demineralized Water Replenishment. 6 pp.

Aircraft (Cont.)

—Replenishment—Water (Cont.)
NATO STANAG 3416 ED 4 AMD 5-83. Methanol/Water and Demineralized Water Replenishment. 7 pp.

—Reporting
CAA LEAFLET 1-3 07.90. Occurance Reporting and the Engineer's Role. 7 pp.

—Safety Belts
CAA LEAFLET 14-15 06.92. Maintenance and Inspection of Crew Harnesses and Passenger Seat Belts (Metal to Metal Attachment). 1 p.

—Safety—Symbols
SBAC AGS 4675 (V). Safety/Hazard, Emblem—Self Adhesive.

—Sales Financing—China (People's Republic)
EURO AFS/HS/4650/435-79. China-Sales Financing of Civil Aircraft. 10 pp.

—Search and Rescue—Medical Equipment
NATO STANAG 3744 ED 2 AMD 3-88. Minimum Requirements of Medical Equipment in Search and Rescue (SAR) Aircraft. 8 pp.

—Sonar Beacons
CAA LEAFLET 14-17. N15F210B Underwater Locating Beacon. 3 pp.

—Statistics
ICAO 9180. ICAO Statistical Yearbook Civil Aviation Statistics of the World Sixteenth Edition—1991. 174 pp.

—Structural Design
CAA NOTICE #15 ISSUE 3. U.K. Certification of Foreign Aircraft of MIWA Not Exceeding 5700 kg (Airworthiness Notices). 4 pp.
CAA Chapter K4-1 04.74. Design and Construction—General (Light Aeroplanes). 3 pp.
CAA Chapter P3-1. General (Provisional Airworthiness Requirements for Civil Powered Lift Aircraft). 21 pp.
CAA CAP 482 SUB-Part C 03.83. Structure (Small Light Aeroplanes). 12 pp.
CAA CAP 482 SUB-Part D 03.83. Design and Construction (Small Light Aeroplanes). 11 pp.

—Structural Timber
CAA LEAFLET 2-3 07.90. Timber Conversion—Spruce. 4 pp.

—Suspension Systems—Lugs (Fasteners)
MOD UK DSTAN 13-8-78. Dimensional Requirements for Airborne Stores, Associated Suspension and Release Systems, and Electrical Control Connections Issue 4 (Withdrawn). 12 pp.
MOD UK DSTAN 13-45-01. Lug, Suspension, Airborne Store for Aircraft Twin Lug Suspension and Release Systems Issue 3; Amendment 2 (Withdrawn). 15 pp.
MOD UK DSTAN 13-46-69. Lug, Suspension, Airborne Store, for Aircraft 'Single Lug' Suspension and Release Systems Issue 1 (Withdrawn). 6 pp.

—Symbols
SBAC AGS 4676 (V). Armament Handling Point, Emblem—Self Adhesive.

—Symbols—Identification Systems
MOD UK DSTAN 05-18-01. Symbol Markings of Servicing and Safety/Hazard Points on Aircraft, Ground Support Equipment and Guided Weapons Systems Issue 4; Amendment 2. 20 pp.

—Technical Manuals—Maintenance
CAA Chapter 13 03.92. Procedure for Resolution of Conflicts About JAA National Aviation Authority Maintenance Standards. 1 p.

—Test Equipment—Software
CAA Chapter A7-4 App 1 09.91. Automatic Test Equipment Software. 3 pp.
CAA Chapter B7-4 App 1 07.89. Automatic Test Equipment Software. 3 pp.

—Tie Downs—Shipborne
MOD UK DSTAN 17-4-86. Picketing (Tie-Down) Fittings for Shipborne Aircraft (Including Helicopters) Issue 3. 7 pp.
NATO STANAG 1095 ED 5 AMD 0-86. Tie-Down Fittings on Shipboard Aircraft. 7 pp.
NATO STANAG 1095 ED 5 AMD 1-86. Tie-Down Fittings on Shipborne Aircraft. 8 pp.

—Title (Ownership)
ICAO 7620. Convention on the International Recognition of Rights in Aircraft Signed at Geneva, on 19 June 1948. 14 pp.

INDUSTRY STANDARDS

INTERNATIONAL AND NON-U.S. NATIONAL STANDARDS
SUBJECT INDEX

Aircraft

Aircraft (Cont.)

—**Tooling—Glossaries**
EURO CTI/78/6728-78. Tooling Definitions. 42 pp.

—**Turbulence**
CAA LEAFLET 6-6 07.90. Effect of Disturbed Airflow on Aeroplane Behaviour. 8 pp.

—**Type Certificates**
CAA NOTICE #15 ISSUE 3. U.K. Certification of Foreign Aircraft of MIWA Not Exceeding 5700 kg (Airworthiness Notices). 4 pp.

—**Type Certificates—Data Sheets**
CAA NOTICE #43 ISSUE 2. Aircraft Type Certificates Data Sheets (Airworthiness Notices). 2 pp.

—**Type Designators—Symbols**
ICAO 8643. Aircraft Type Designators 23rd Edition—February 1993. 58 pp.

—**Weight Distribution**
CAA LEAFLET 1-4 07.90. Weight and Balance of Aircraft. 27 pp.
JIS W 0602-84. Weight Distribution of Aircraft.

—**Weight Distribution—Scheduling**
CAA Chapter A7-1 0 App2 09.91. Weight and Centre-of-Gravity and Loading and Distribution Schedules—Aircraft Not Exceeding 2730 kg. 5 pp.

—**Weight—Glossaries**
EURO CTI/81/11760-80. Report on Aircraft Weight Terms and Definitions. 122 pp.

—**Weight Measurement**
CAA Chapter A5-4 09.91. Weight and Balance of Aircraft. 2 pp.
CAA Chapter A6-4 09.91. Weight and Balance of Aircraft. 2 pp.
CAA Chapter A6-4 App 1 07.89. Weight and Balance of Aircraft—Fleet Mean Weight and Fleet Mean Centre-of-Gravity. 2 pp.
CAA Chapter B5-4 07.89. Weight and Balance of Aircraft. 2 pp.
CAA Chapter B6-4 07.89. Weight and Balance of Aircraft. 2 pp.
CAA Chapter B6-4 App 1 7.89. Weight and Balance of Aircraft—Fleet Mean Weight and Fleet Mean Centre-Of-Gravity.
CAA Chapter B7-10 07.89. Weight and Balance Report. 2 pp.
CAA Chapter B7-1 0 App1 07.89. Weight and Centre-Of-Gravity Schedules for Aircraft Exceeding 2740 kg. 3 pp.
CAA Chapter B7-1 0 App2 07.89. Weight and Centre-Of-Gravity and Loading and Distribution Schedules—Aircraft Not Exceeding 2730 kg. 3 pp.
CAA LEAFLET 1-4 07.90. Weight and Balance of Aircraft. 27 pp.

—**Weight Measurement—Scheduling**
CAA Chapter A7-10 App 1. Weight and Centre-of-Gravity Schedules for Aircraft Exceeding 2730 kg. 3 pp.

—**Welding**
JIS W 0901-85. Aerospace Fusion Welders Qualification.

—**Wind Shear**
ICAO Circular 186. Wind Shear—1987. 192 pp.

—**Wooden Structures—Deterioration**
CAA NOTICE #50 ISSUE 3. Deterioration of Wooden Aircraft Structures (Airworthiness Notices). 2 pp.

Aircraft Accessories
Use: Aircraft Equipment

Aircraft Accidents
Scope Note: Includes documents on aircraft accidents resulting in serious or fatal injury, aircraft damage, or structural failure as well as incidents such as engine failure, fire, decompression, etc.
Use For: ADREP; Aircraft Incidents
See Also: Accident Investigations; Accident Prevention; Air Transportation; Aircraft; Aircraft Landings; Aircraft Safety; Collision Research; Crash Locator Beacons; Hijacking; Near Misses (Aircraft)
CAA. Contents 04.69 (World Airline Accident Summary). 3 pp.
CAA. Acknowledgements 04.88 (World Airline Accident Summary). 1 p.
CAA. Foreword 04.88 (World Airline Accident Summary). 3 pp.
CAA. Check List of Pages (World Airline Accident Summary). 5 pp.
CAA. Contents. 3 pp.
CAA. Acknowledgements. 2 pp.
CAA. Foreword 04.92. 6 pp.
CAA. Checklist of Pages. 4 pp.

Aircraft Accidents (Cont.)
CAA. Supplement No. 90 (World Airline Accident Summary). 1 p.
CAA. Contents (World Helicopter Accident Summary). 1 p.
CAA. Acknowledgements (World Helicopter Accident Summary). 1 p.
CAA. Foreword 04.92 (World Helicopter Accident Summary). 1 p.
CAA. Check List of Pages 07.92 (World Helicopter Accident Summary). 1 p.
CAA. Supplement No. 48 (World Helicopter Accident Summary). 1 p.
CAA LEAFLET #11 01.93. Lead Airlines and Lead Authorities to Deal with Emergencies on Specific Aeroplane Types (Information Leaflets). 3 pp.
EURO CE/RC/77/700 8/0/F & E/RE-77. Product Liability (catastrophe scheme). 30 pp.

—**By Airplane Type**
CAA Chapter B4-3 11.72. Collision Accelerations. 1 p.
CAA. British Aerospace Jetstream Series Aircraft. 21 pp.
CAA. Airspeed Ambassador. 1 p.
CAA. Avro Lancaster/Lancastrian (World Airline Accident Summary). 1 p.
CAA. Avro Tudor. 1 p.
CAA. Avro York (World Airline Accident Summary). 1 p.
CAA. British Aircraft Corporation BAC 1-11 (World Airline Accident Summary). 1 p.
CAA. British Aircraft Corporation VC10 Series (World Airline Accident Summary). 1 p.
CAA. Boeing 307 Stratoline (World Airline Accident Summary). 1 p.
CAA. British 377 Stratocruiser (World Airline Accident Summary). 1 p.
CAA. Boeing 707/720 Series (World Airline Accident Summary). 3 pp.
CAA. Boeing 727 Series (World Airline Accident Summary). 1 p.
CAA. Boeing 737 (World Airline Accident Summary). 1 p.
CAA. Boeing 747 (World Airline Accident Summary). 1 p.
CAA. Bristol Britannia (World Airline Accident Summary). 1 p.
CAA. Bristol 170 Wayfarer/Freighter (World Airline Accident Summary). 1 p.
CAA. Canadair C4 and DC4-M-2 (World Airline Accident Summary). 1 p.
CAA. Canadair CL 44 Series (World Airline Accident Summary). 1 p.
CAA. Convair PBY5 Catalina. 1 p.
CAA. Convair CV240, 340 and 440 (World Airline Accident Summary). 3 pp.
CAA. Convair CV580 and 640 Series (World Airline Accident Summary). 1 p.
CAA. Convair CV880 and CV990 Series (World Airline Accident Summary). 1 p.
CAA. Curtiss C46 Commando (World Airline Accident Summary). 6 pp.
CAA. Dassault Mystere/Fan Jet Falcon (World Airline Accident Summary). 1 p.
CAA. DHC-6 Twin Otter (World Airline Accident Summary). 1 p.
CAA. Douglas DC-3/C47 Dakota (World Airline Accident Summary). 12 pp.
CAA. Douglas C54/DC4 (World Airline Accident Summary). 4 pp.
CAA. Douglas DC6 Series (World Airline Accident Summary). 3 pp.
CAA. Douglas DC-7 Series (World Airline Accident Summary). 1 p.
CAA. Embraer EMB—110 Bandeirante (World Airline Accident Summary). 1 p.
CAA. Fairchild Hiller FH-227 (World Airline Accident Summary). 1 p.
CAA. Fokker/Fairchild F-27 Series (World Airline Accident Summary). 2 pp.
CAA. Fokker F-28 Series (World Airline Accident Summary). 1 p.
CAA. Gates Learjet (World Airline Accident Summary). 1 p.
CAA. Handley Page Halifax/Halton (World Airline Accident Summary). 1 p.
CAA. Handley Page Herald (World Airline Accident Summary). 1 p.
CAA. Handley Page Hermes (World Airline Accident Summary). 1 p.
CAA. Handley Page Marathon (World Airline Accident Summary). 1 p.
CAA. Hawker Siddeley Argosy (World Airline Accident Summary). 1 p.
CAA. Hawker Siddeley Comet 1 (World Airline Accident Summary). 1 p.
CAA. Hawker Siddeley DH.106 Comet 4 Series (World Airline Accident Summary). 1 p.
CAA. Hawker Siddeley 114 Heron/Riley Heron (World Airline Accident Summary). 1 p.
CAA. Hawker Siddeley 121 Trident (World Airline Accident Summary). 1 p.
CAA. Hawker Siddeley/125 Series (World Airline Accident Summary). 1 p.

Aircraft Accidents (Cont.)
—**By Airplane Type (Cont.)**
CAA. Hawker Siddeley/748 Series (World Airline Accident Summary). 1 p.
CAA. Israeli Aircraft Industries/North American Rockwell Jet Commander 1121 and Series (World Airline Accident Summary). 2 pp.
CAA. Junkers JU52/3M (World Airline Accident Summary). 1 p.
CAA. Latecoere 631 (World Airline Accident Summary). 1 p.
CAA. Lockheed Constellation and Super Constellation (World Airline Accident Summary). 3 pp.
CAA. Lockheed L188 Electra (World Airline Accident Summary). 1 p.
CAA. Lockheed L100 and L382B Hercules (World Airline Accident Summary). 1 p.
CAA. Lockheed Hudson (World Airline Accident Summary). 1 p.
CAA. Lockheed L18 Lodestar (World Airline Accident Summary). 1 p.
CAA. Lockheed L-1011 Tristar (World Airline Accident Summary). 1 p.
CAA. Martin 202/404 (World Airline Accident Summary). 1 p.
CAA. MBB HFB-320 Hansa (World Airline Accident Summary). 1 p.
CAA. McDonnell Douglas DC-8 Series (World Airline Accident Summary). 1 p.
CAA. McDonnell Douglas DC-9/Series (World Airline Accident Summary). 1 p.
CAA. McDonnell Douglas DC-10 Series (World Airline Accident Summary). 2 pp.
CAA. Miscellaneous Aircraft (World Airline Accident Summary). 1 p.
CAA. NAMC YS-11 Series (World Airline Accident Summary). 1 p.
CAA. Nord 262 Series (World Airline Accident Summary). 1 p.
CAA. Rockwell (North American) Sabreliner and Sabre (World Airline Accident Summary). 1 p.
CAA. Russian-Built Aircraft (World Airline Accident Summary). 1 p.
CAA. Saab Scania (World Airline Accident Summary). 1 p.
CAA. Scottish Aviation Twin Pioneer. 1 p.
CAA. S.E.161 Languedoc (World Airline Accident Summary). 1 p.
CAA. Short S25 Sunderland, Sandringham/Hythe (World Airline Accident Summary). 1 p.
CAA. Short S45 Solent (World Airline Accident Summary). 1 p.
CAA. Short Skyvan (World Airline Accident Summary). 1 p.
CAA. Sud Aviation SE-210 Caravelle Series (World Airline Accident Summary). 1 p.
CAA. Unknown Aircraft (World Airline Accident Summary). 1 p.
CAA. Vickers Supermarine Stranraer (World Airline Accident Summary). 1 p.
CAA. Vickers Vanguard (World Airline Accident Summary). 1 p.
CAA. Vickers Viking (World Airline Accident Summary). 1 p.
CAA. Vickers Viscount (World Airline Accident Summary). 1 p.
CAA. Airbus A300 (World Airline Accident Summary). 1 p.
CAA. Airbus A310 (World Airline Accident Summary). 1 p.
CAA. Airbus A320 (World Airline Accident Summary). 1 p.
CAA. ATR 42 and 72. 1 p.
CAA. Beachcraft 200 and 300 Super Kingair. 1 p.
CAA. Boeing 707/720 Series. 2 pp.
CAA. Boeing 727. 4 pp.
CAA. Boeing 737. 2 pp.
CAA. Boeing 747. 2 pp.
CAA. Boeing 757 (World Airline Accident Summary). 1 p.
CAA. Boeing 767 (World Airline Accident Summary). 1 p.
CAA. Bristol Britannia. 1 p.
CAA. Bristol 170 Wayfarer/Freighter. 1 p.
CAA. British Aerospace 146. 1 p.
CAA. British Aircraft Corporation/Aerospatiale Concorde (World Airline Accident Summary). 1 p.
CAA. British Aircraft Corporation BAC 1-11. 1 p.
CAA. Canadair Challenger. 1 p.
CAA. Canadair CL 44 Series. 1 p.
CAA. CASA 212. 1 p.
CAA. Cessna 500 and 600 Series Citation I, II and III (World Airline Accident Summary). 1 p.
CAA. Convair PBY5 Catalina. 1 p.
CAA. Convair CV240, 340 and 440. 1 p.
CAA. Convair CV580 and 640 Series. 1 p.
CAA. Convair CV880 and CV990 Series. 1 p.
CAA. Curtis C46 Commando. 1 p.
CAA. Dassult Mystere/Fan Jet Falcon 10, 100, 20, 200, 50 and 900. 1 p.
CAA. DHC-4 (World Airline Accident Summary). 1 p.
CAA. DHC-5 (World Airline Accident Summary). 1 p.

INTERNATIONAL AND NON-U.S. NATIONAL STANDARDS
SUBJECT INDEX
Aircraft

Aircraft Accidents *(Cont.)*
—By Airplane Type *(Cont.)*
CAA. DHC-6 Twin Otter. 3 pp.
CAA. DHC-7 (World Airline Accident Summary). 1 p.
CAA. DHC-8 (World Airline Accident Summary). 1 p.
CAA. Dornier 228 (World Airline Accident Summary). 1 p.
CAA. Douglas DC-3/C47 Dakota. 3 pp.
CAA. Douglas DC-4/C54. 1 p.
CAA. Douglas DC-6 Series. 1 p.
CAA. Douglas DC-7 Series. 1 p.
CAA. Embraer EMB-110 Bandeirante. 1 p.
CAA. Embraer EMB-120 Brasitia (World Airline Accident Summary). 1 p.
CAA. Fairchild Hiller FH-227. 1 p.
CAA. Fokker/Fairchild F-27 Series. 2 pp.
CAA. Fokker F-28 Series. 1 p.
CAA. Gates Learjet. 2 pp.
CAA. Handley Page Herald. 1 p.
CAA. Hawker Siddely Argosy. 1 p.
CAA. Hawker Siddely 114 Heron/Riley Heron. 1 p.
CAA. Hawker Siddely 121 Trident. 1 p.
CAA. Hawker Siddely/BAe 125 Series. 1 p.
CAA. Hawker Siddely/BAe 748 Series. 1 p.
CAA. 1AI, 101, and 201 Arava (World Airline Accident Summary). 1 p.
CAA. 1AI/North American Rockwell 1121, 1123, and 1124 Jet Commander/Westwind (World Airline Accident Summary). 1 p.
CAA. Lockheed Constellation and Super Constellation. 1 p.
CAA. Lockheed L188 Electra. 1 p.
CAA. Lockheed L100 and L382-Hercules. 1 p.
CAA. Lockheed L18 Lodestar. 1 p.
CAA. Lockheed L-1011 Tristar. 1 p.
CAA. Martin 202/204/404. 1 p.
CAA. MBB HFB-320 Hansa. 1 p.
CAA. McDonnell Douglas DC-8 Series. 1 p.
CAA. McDonnell Douglas DC-9/MD80 Series. 2 pp.
CAA. McDonnell Douglas DC-10 Series. 1 p.
CAA. Miscellaneous Aircraft. 3 pp.
CAA. NAMC Y5-11 Series. 1 p.
CAA. Nord 262 Series. 1 p.
CAA. Rockwell (North American) Sabreliner and Sabre. 1 p.
CAA. Russian-Built Aircraft. 3 pp.
CAA. Scottish Aviation/British Aerospace Jetstream (World Airline Accident Summary). 1 p.
CAA. Scottish Aviation Twin Pioneer. 1 p.
CAA. Shorts Skyvan. 1 p.
CAA. Shorts SD3-30 and 3-60. 1 p.
CAA. Sud Aviation SE-210 Caravelle Series. 1 p.
CAA. Swearingen/Fairchild SA26, 226 & 227 Merlin and Metro (World Airline Accident Summary). 2 pp.
CAA. Unknown Aircraft. 1 p.
CAA. Vickers Viscount. 1 p.
CAA. Lockheed L18 Lodestar. 1 p.

—By Cause
CAA Section 7 CAP 382. Reporting of Airmiss and Birdstrike Occurences (Civil Air Publications: Mandatory Occurrence Reporting Scheme). 1 p.
CAA Section 9 CAP 382. Reporting of Radio Faults (Civil Air Publications: Mandatory Occurrence Reporting Scheme). 1 p.
CAA NOTICE #23 ISSUE 1. Fuel Additives—Health Hazards (Airworthiness Notices). 1 p.
CAA NOTICE #33 ISSUE 3. Unprotected Starter Circuits in Aircraft Not Exceeding 12500 lb (Airworthiness Notices). 2 pp.
CAA NOTICE #40 ISSUE 1. Carbon Monoxide Contamination in Aircraft (Airworthiness Notices). 2 pp.
CAA NOTICE #41 ISSUE 8. Maintenance of Cockpit and Cabin Combustion Heaters and Their Associated Exhaust Systems (Airworthiness Notices). 2 pp.
CAA NOTICE #42 ISSUE 1. Internal Emergency Lighting Systems (Airworthiness Notices). 2 pp.
CAA NOTICE #53 ISSUE 1. Vertical Speed Indicators on Imported Aircraft (Airworthiness Notices). 2 pp.
CAA NOTICE #57 ISSUE 1. Toilet Flush Motors (Airworthiness Notices). 2 pp.
CAA NOTICE #60 ISSUE 1. Cabin and Toilet Fire Protection (Airworthiness Notices). 3 pp.
CAA NOTICE #70 ISSUE 2. Tyre Bursts in Flight—Inflation Media (Airworthiness Notices). 2 pp.
CAA NOTICE #81 ISSUE 1. Emergency Power Supply for Electrically Operated Gyroscopic Bank and Pitch Indicators (Artificial Horizons) (Airworthiness Notices). 5 pp.
CAA NOTICE #82 ISSUE 1. Electrical Generation Systems Aircraft Not Exceeding 5700 kg Maximum Authorised Weight (Airworthiness Notices). 4 pp.
CAA. Aircraft Shot at by Ground Fire or Shot or Forced down by Fighter Aircraft (World Airline Accident Summary). 1 p.
CAA. Airframe Failure (Excluding Sabotage) (World Airline Accident Summary). 1 p.

Aircraft Accidents *(Cont.)*
—By Cause *(Cont.)*
CAA. Aquaplaning/Hydroplaning (World Airline Accident Summary). 1 p.
CAA. Bird Strike/Ingestion (World Airline Accident Summary). 1 p.
CAA. Cargo Breaking Loose (World Airline Accident Summary). 1 p.
CAA. Collision with High Ground (World Airline Accident Summary). 2 pp.
CAA. Collision with Water (World Airline Accident Summary). 2 pp.
CAA. Crew Incapacitation (World Airline Accident Summary). 1 p.
CAA. Crew Shot (World Airline Accident Summary). 1 p.
CAA. Door, Windows Opening/Failing in Flight (World Airline Accident Summary). 1 p.
CAA. Electrical System Failure/Malfunction (World Airline Accident Summary). 1 p.
CAA. Failure of All Power Units (World Airline Accident Summary). 1 p.
CAA. Flying Control System Malfunction (World Airline Accident Summary). 1 p.
CAA. Fuel Contamination (World Airline Accident Summary). 1 p.
CAA. Fuel Exhaustion, Starvation, Mismanagement (World Airline Accident Summary). 1 p.
CAA. Hail Damage (World Airline Accident Summary). 1 p.
CAA. Ice/Snow Accretion—Airframe/Engine (World Airline Accident Summary). 1 p.
CAA. Inflight Accidents Due to the Carriage or Hiding of Bombs on Aircraft (World Airline Accident Summary). 1 p.
CAA. Inflight Fire/Smoke (Excluding Sabotage) (World Airline Accident Summary). 1 p.
CAA. Instruments—Incorrectly Set—Misread—Failure—Malfunction—Design (World Airline Accident Summary). 1 p.
CAA. Lightning Strike (World Airline Accident Summary). 1 p.
CAA. Major Power Plant Disruption/Loss of Propeller in Flight (World Airline Accident Summary). 3 pp.
CAA. Mid-Air Collisions (Both Aircraft Airborne) (World Airline Accident Summary). 1 p.
CAA. Third Party Accidents (World Airline Accident Summary). 1 p.
CAA. Tyre Burst After Retraction (World Airline Accident Summary). 1 p.
CAA. Aircraft Shot at by Ground Fire or Shot or Forced down by Fighter Aircraft. 1 p.
CAA. Airframe Failure (Excluding Sabotage). 1 p.
CAA. Aquaplaning/Hydroplaning. 1 p.
CAA. Bird Strike/Ingestion. 1 p.
CAA. Cargo Breaking Loose. 1 p.
CAA. Collision with High Ground. 3 pp.
CAA. Collision with Water (Excluding Flying Boats and Amphibians). 2 pp.
CAA. Crew Incapacitation. 1 p.
CAA. Crew Shot. 1 p.
CAA. Door, Window Opening/Failing in Flight. 1 p.
CAA. Electrical System Failure/Malfunction. 1 p.
CAA. Failure of All Power Units. 2 pp.
CAA. Flying Control System Malfunction. 1 p.
CAA. Fuel Contamination. 1 p.
CAA. Fuel Exhaustion, Starvation, Mismanagement. 1 p.
CAA. Hail Damage. 1 p.
CAA. Ice/Snow/Accretion Airframe/Engine. 1 p.
CAA. Inflight Accidents Due to the Carriage or Hiding of Bombs on Aircraft. 1 p.
CAA. Inflight Fire/Smoke (Excluding Sabotage). 2 pp.
CAA. Instruments Incorrectly Set-Misread-Failure-Malfunction-Design. 1 p.
CAA. Lightning Strike. 1 p.
CAA. Major Power Plant Disruption/Loss of Propeller in Flight. 2 pp.
CAA. Mid-Air Collisions (Both Aircraft Airborne). 1 p.
CAA. Over Running/Veering off Runway. 7 pp.
CAA. Third Party Accidents. 2 pp.
CAA. Tyre Burst After Retraction. 1 p.
CAA 1990 Part B.
CAA Part B 1989. (World Helicopter Accident Summary). 4 pp.
CAA Part B 1990.

—By Helicopter Type
CAA. Aerospatiale SA330ffi332 Puma Series (World Helicopter Accident Summary). 1 p.
CAA. Bell 205 Series (World Helicopter Accident Summary). 1 p.
CAA. Bell/Augusta Bell 212 Series (World Helicopter Accident Summary). 3 pp.
CAA. Bell 214 Series (World Helicopter Accident Summary). 1 p.
CAA. Bell 412 Series (World Helicopter Accident Summary). 1 p.
CAA. Boeing Vertol 44 Series (World Helicopter Accident Summary). 1 p.
CAA. Boeing Vertol/Kawasaki 107 Series (World Helicopter Accident Summary). 1 p.

Aircraft Accidents *(Cont.)*
—By Helicopter Type *(Cont.)*
CAA. Boeing Vertol 234 Chinook Series (World Helicopter Accident Summary). 1 p.
CAA. Russian Built Helicopters (World Helicopter Accident Summary). 2 pp.
CAA. Sikorsky S58 Series (World Helicopter Accident Summary). 3 pp.
CAA. Sikorsky S61 Series (World Helicopter Accident Summary). 2 pp.
CAA. Sikorsky S64 Series (World Helicopter Accident Summary). 1 p.
CAA. Sikorsky S67 Series (World Helicopter Accident Summary). 1 p.
CAA. Sikorsky S70 Series. 1 p.
CAA. Sikorsky S76 Series (World Helicopter Accident Summary). 1 p.
CAA. Unknown (World Helicopter Accident Summary). 1 p.
CAA. Wessex 60 Series (World Helicopter Accident Summary). 1 p.
CAA. Westland W30 Series (World Helicopter Accident Summary). 1 p.

—By Year
CAA. Douglas DC-2. 1 p.
CAA 1946 Part C. (World Airline Accident Summary). 21 pp.
CAA 1947 Part C. (World Airline Accident Summary). 28 pp.
CAA 1948 Part C. (World Airline Accident Summary). 32 pp.
CAA 1949 Part C. (World Airline Accident Summary). 26 pp.
CAA 1950 Part C. (World Airline Accident Summary). 23 pp.
CAA 1951 Part C. (World Airline Accident Summary). 26 pp.
CAA 1952 Part C. (World Airline Accident Summary). 21 pp.
CAA 1953 Part C. (World Airline Accident Summary). 25 pp.
CAA 1954 Part C. (World Airline Accident Summary). 27 pp.
CAA 1955 Part C. (World Airline Accident Summary). 25 pp.
CAA 1956 Part C. (World Airline Accident Summary). 25 pp.
CAA 1957 Part C. (World Airline Accident Summary). 30 pp.
CAA 1958 Part C. (World Airline Accident Summary). 24 pp.
CAA 1959 Part C. (World Airline Accident Summary). 30 pp.
CAA 1960 Part C. (World Airline Accident Summary). 34 pp.
CAA 1961 Part C. (World Airline Accident Summary). 34 pp.
CAA 1962 Part C. (World Airline Accident Summary). 37 pp.
CAA 1963 Part C. (World Airline Accident Summary). 27 pp.
CAA 1964 Part C. (World Airline Accident Summary). 35 pp.
CAA 1965 Part C. (World Airline Accident Summary). 35 pp.
CAA 1966 Part C. (World Airline Accident Summary). 38 pp.
CAA 1967 Part C. (World Airline Accident Summary). 29 pp.
CAA 1968 Part C. (World Airline Accident Summary). 28 pp.
CAA 1969 Part C. (World Airline Accident Summary). 29 pp.
CAA 1970 Part C. (World Airline Accident Summary). 33 pp.
CAA 1971 Part C. (World Airline Accident Summary). 29 pp.
CAA 1972 Part C. (World Airline Accident Summary). 40 pp.
CAA 1973 Part C. (World Airline Accident Summary). 42 pp.
CAA 1974 Part C. (World Airline Accident Summary). 30 pp.
CAA 1975 Part C. (World Airline Accident Summary). 34 pp.
CAA 1976 Part B. (World Airline Accident Summary). 37 pp.
CAA 1977 Part B. (World Airline Accident Summary). 49 pp.
CAA 1978 Part B. (World Airline Accident Summary). 41 pp.
CAA 1979 Part B. (World Airline Accident Summary). 42 pp.
CAA 1980 Part B. (World Airline Accident Summary). 36 pp.
CAA 1981 Part B. (World Airline Accident Summary). 30 pp.
CAA 1982 Part B. (World Airline Accident Summary). 32 pp.
CAA 1983 Part B. (World Airline Accident Summary). 9 pp.
CAA 1984 Part B. (World Airline Accident Summary). 3 pp.

INDUSTRY STANDARDS

Aircraft Accidents (Cont.)
—By Year (Cont.)
CAA 1985 Part B. (World Airline Accident Summary). 31 pp.
CAA 1986 Part B. (World Airline Accident Summary). 35 pp.
CAA 1987 Part B. (World Airline Accident Summary). 7 pp.
CAA 1988 Part B. (World Airline Accident Summary). 13 pp.
CAA 1989 Part B. (World Airline Accident Summary). 23 pp.
CAA 1991 Part B.
CAA 1992 Part B.
CAA Part B 1956. (World Helicopter Accident Summary). 1 p.
CAA Part B 1957. (World Helicopter Accident Summary). 1 p.
CAA Part B 1958. (World Helicopter Accident Summary). 1 p.
CAA Part B 1959. (World Helicopter Accident Summary). 1 p.
CAA Part B 1960. (World Helicopter Accident Summary). 1 p.
CAA Part B 1961. (World Helicopter Accident Summary). 1 p.
CAA Part B 1963. (World Helicopter Accident Summary). 1 p.
CAA Part B 1964. (World Helicopter Accident Summary). 2 pp.
CAA Part B 1965. (World Helicopter Accident Summary). 1 p.
CAA Part B 1966. (World Helicopter Accident Summary). 1 p.
CAA Part B 1967. (World Helicopter Accident Summary). 2 pp.
CAA Part B 1968. (World Helicopter Accident Summary). 2 pp.
CAA Part B 1969. (World Helicopter Accident Summary). 2 pp.
CAA Part B 1970. (World Helicopter Accident Summary). 2 pp.
CAA Part B 1971. (World Helicopter Accident Summary). 1 p.
CAA Part B 1972. (World Helicopter Accident Summary). 2 pp.
CAA Part B 1973. (World Helicopter Accident Summary). 2 pp.
CAA Part B 1974. (World Helicopter Accident Summary). 4 pp.
CAA Part B 1975. (World Helicopter Accident Summary). 3 pp.
CAA Part B 1976. (World Helicopter Accident Summary). 4 pp.
CAA Part B 1977. (World Helicopter Accident Summary). 4 pp.
CAA Part B 1978. (World Helicopter Accident Summary). 4 pp.
CAA Part B 1979. (World Helicopter Accident Summary). 3 pp.
CAA Part B 1980. (World Helicopter Accident Summary). 4 pp.
CAA Part B 1981. (World Helicopter Accident Summary). 4 pp.
CAA Part B 1982. (World Helicopter Accident Summary). 4 pp.
CAA Part B 1983. (World Helicopter Accident Summary). 6 pp.
CAA Part B 1984. (World Helicopter Accident Summary). 4 pp.
CAA Part B 1985. (World Helicopter Accident Summary). 5 pp.
CAA Part B 1986. (World Helicopter Accident Summary). 4 pp.
CAA Part B 1987. (World Helicopter Accident Summary). 4 pp.
CAA Part B 1988. (World Helicopter Accident Summary). 6 pp.
CAA Part B 1989. (World Helicopter Accident Summary). 4 pp.
CAA Part B 1991.
CAA Part B 1992.
ICAO Circular 196. Aircraft Accident Digest—1983 (No. 30). 161 pp.
ICAO Circular 197. Accident/Incident Reporting (ADREP) Annual Statistics—1983. 53 pp.
ICAO Circular 202. Aircraft Accident Digest 1984 (No. 31). 136 pp.
ICAO Circular 203. Accident/Incident Reporting (ADREP) Annual Statistics—1984. 33 pp.
ICAO Circular 209. Accident/Incident Reporting (ADREP) Annual Statistics—1985. 31 pp.
ICAO Circular 223. Accident/Incident Reporting (ADREP) Annual Statistics—1986. 23 pp.
ICAO Circular 230. Accident/Incident Reporting (ADREP) Annual Statistics—1987. 20 pp.
ICAO 9156. Accident/Incident Reporting Manual (ADREP Manual) Second Edition—1987. 162 pp.

—Compensation
ICAO 7364. Convention on Damage Caused by Foreign Aircraft to Third Parties on the Surface Signed at Rome, on 7 October 1952. 23 pp.
ICAO 9257. Protocol to Amend the Convention on Damage Caused by Foreign Aircraft to Third Parties on the Surface Signed at Rome on 7 October 1952 Signed at Montreal on 23 September 1978. 31 pp.

—Emergency Planning—Hazardous Materials
ICAO 9481. Emergency Response Guidance for Aircraft Incidents Involving Dangerous Goods 1993-1994 Edition. 78 pp.

—Ergonomics
ICAO Circular 216. Human Factors Digest No. 1 Fundamental Human Factors Concepts—1989. 36 pp.
ICAO Circular 217. Human Factors Digest No. 2 Flight Crew Training: Cockpit Resource Management (CRM) and Line-Oriented Flight Training (LOFT)—1989. 67 pp.
ICAO Circular 227. Human Factors Digest No. 3 Training of Operational Personnel in Human Factors—1991. 58 pp.
ICAO Circular 229. Human Factors Digest No. 4 Proceedings of the ICAO Human Factors Seminar—1990. 628 pp.
ICAO Circular 240. Human Factors Digest No. 7 Investigation of Human Factors in Accidents and Incidents—1993. 61 pp.

—Information Interchange
NATO STANAG 3101 ED 8 AMD 1-89. Exchange of Accident/Incident Information Concerning Aircraft and Missiles. 35 pp.
NATO STANAG 3101 ED 8 AMD 2-89. Exchange of Accident/Incident Information Concerning Aircraft and Missiles. 36 pp.
NATO STANAG 3101 ED 9 AMD 0-93. Exchange of Accident/Incident Information Concerning Aircraft and Missiles. 36 pp.

—Investigation
CAA Section 8 CAP 393. Investigation of Accidents. 35 pp.
ICAO Annex 13. Aircraft Accident Investigation Seventh Edition—May 1988; (Supplement 3/4/89). 62 pp.
ICAO 6920. Manual of Aircraft Accident Investigation Fourth Edition—1970; (Amendment 10 10/17/86). 705 pp.
NATO STANAG 3531 ED 5 AMD 3-84. Investigation of Aircraft Missile Accidents/Incidents. 15 pp.
NATO STANAG 3531 ED 6 AMD 0-91. Safety Investigation and Reporting of Accidents/Incidents Involving Military Aircraft and/or Missiles. 11 pp.
NATO STANAG 3531 ED 6 AMD 1-91. Safety Investigation and Reporting of Accidents/Incidents Involving Military Aircraft and/or Missiles. 12 pp.
NATO STANAG 3531 ED 6 AMD 2-91. Safety Investigation and Reporting of Accidents/Incidents Involving Military Aircraft and/or Missiles. 13 pp.

—Investigation—Medical Personnel
NATO STANAG 3318 ED 4 AMD 5-81. Aeromedical Aspects of Aircraft Accident/Incident Investigation. 28 pp.

—Prevention
ICAO 9422. Accident Prevention Manual First Edition—1984. 147 pp.

—Prevention—International Cooperation
NATO STANAG 3102 ED 4 AMD 0-90. Flight Safety in Common Ground/Air Space. 5 pp.
NATO STANAG 3102 ED 4 AMD 1-90. Flight Safety Co-Operation in Common Ground/Air Space. 6 pp.

—Reporting
CAA. Contents (Civil Air Publications: Mandatory Occurrence Reporting Scheme). 2 pp.
CAA. Foreword (Civil Air Publications: Mandatory Occurrence Reporting Scheme). 1 p.
CAA. Statement by the Chairman of the Authority (Civil Air Publications: Mandatory Occurrence Reporting Scheme). 2 pp.
CAA Section 1 CAP 382. Objectives of the Scheme (Civil Air Publications: Mandatory Occurrence Reporting Scheme). 1 p.
CAA Section 2 CAP 382. Division of Responsibilities (Civil Air Publications: Mandatory Occurrence Reporting Scheme). 1 p.
CAA Section 3 CAP 382. Relationship with Accidents Investigating Branch, Department of Transport (Civil Air Publications: Mandatory Occurrence Reporting Scheme). 1 p.
CAA Section 4 CAP 382. Legislation (Civil Air Publications: Mandatory Occurrence Reporting Scheme). 1 p.
CAA Section 5 CAP 382. Applicability (Civil Air Publications: Mandatory Occurrence Reporting Scheme). 1 p.
CAA Section 6 CAP 382. Reporting Procedure (Civil Air Publications: Mandatory Occurrence Reporting Scheme). 4 pp.
CAA Section 7 CAP 382. Reporting of Airmiss and Birdstrike Occurences (Civil Air Publications: Mandatory Occurrence Reporting Scheme). 1 p.
CAA Section 9 CAP 382. Reporting of Radio Faults (Civil Air Publications: Mandatory Occurrence Reporting Scheme). 1 p.
CAA Section 10 CAP 382. Processing of Occurence Reports and Publication of Occurence Information (Civil Air Publications: Mandatory Occurrence Reporting Scheme). 4 pp.
CAA Appendix A CAP 382. Advice on the Completion of the CAA Occurence Report Form (Civil Air Publications: Mandatory Occurrence Reporting Scheme). 6 pp.
CAA Appendix B CAP 382. Occurences Required to Be Reported (Civil Air Publications: Mandatory Occurrence Reporting Scheme). 1 p.
CAA Appendix C CAP 382. Occurence Publications (Civil Air Publications: Mandatory Occurrence Reporting Scheme). 3 pp.
CAA LEAFLET 13-3 07.90. Mandatory Occurrence Reporting Scheme. 47 pp.

—Reporting—Air Traffic Controllers
CAA Section 8 CAP 382. Reporting by Air Traffic Controllers (Civil Air Publications: Mandatory Occurrence Reporting Scheme). 1 p.

—Spare Parts
CAA NOTICE #97 ISSUE 1. Return to Service of Aircraft Items Recovered from Aircraft Involved in Accidents/Incidents (Airworthiness Notices). 3 pp.

—Statistical Analysis
ICAO Circular 233. Accident/Incident Reporting (ADREP) Annual Statistics—1988. 21 pp.

Aircraft Armament Systems
Use: Weapons Systems—Aircraft

Aircraft Arresting Systems
Use For: Arresting Systems, Aircraft
See Also: Aircraft Equipment; Landing Gear
NATO STANAG 3697 ED 3 AMD 1-85. Aircraft Arresting Systems. 14 pp.
NATO STANAG 3697 ED 3 AMD 3-85. Aircraft Arresting Systems. 10 pp.
NATO STANAG 3697 ED 3 AMD 4-85. Aircraft Arresting Systems. 11 pp.

Aircraft Baggage Compartments
Use: Aircraft Compartments—Baggage

Aircraft Batteries
See Also: Aircraft Equipment; Batteries
CAA 706 (K). Section K Light Aeroplanes Electrical Supply, Systems and Equipment (Blue Papers). 54 pp.
CAA Chapter J2-1 09.66. Supply and Distribution. 5 pp.

—Battery Chargers—Aircraft
CAA LEAFLET 9-2 07.90. Charging Rooms for Aircraft Batteries. 5 pp.

—Bolts
SBAC AS 416 ISSUE 4(I). Bolt—Detail—Accumulator Attachment. 1 p.
SBAC AS 6202 ISSUE 2. Bolt Assembly—Attachment of Lead Acid Accumulators. 1 p.
SBAC AS 6203 ISSUE 1. Bolt Accumulator Attachment. 1 p.

—Capacity
CAA NOTICE #82 ISSUE 1. Electrical Generation Systems Aircraft Not Exceeding 5700 kg Maximum Authorised Weight (Airworthiness Notices). 4 pp.

—Connectors
IEC 952 Pt 3-93. Aircraft Batteries Part 3: External Electrical Connectors First Edition. 15 pp.

—Ground Recharging
NATO STANAG 3454 ED 4 AMD 0-90. Ground Recharging of Aircraft Main Batteries. 14 pp.
NATO STANAG 3454 ED 4 AMD 1-90. Ground Recharging of Aircraft Main Batteries. 13 pp.
NATO STANAG 3454 ED 4 AMD 2-90. Ground Recharging of Aircraft Main Batteries. 14 pp.

—Installation
CAA Chapter J2-1 App #2 09.66. Generator Switchgear and Battery Installation. 3 pp.

INTERNATIONAL AND NON-U.S. NATIONAL STANDARDS
SUBJECT INDEX

Aircraft

Aircraft Batteries (Cont.)

—Interchangeability
NATO STANAG 3514 ED 3 AMD 7-84. Aircraft Main Batteries. 33 pp.

—Lithium
BSI G 239-87. (WITHDRAWN) 1987 Primary Active Lithium Batteries for Use in Aircraft (Superseded by BS 2G 239: 1992). 16 pp.
BSI 2G 239-92. Primary Active Lithium Batteries for Use in Aircraft. 22 pp.

—Storage
CAA CAP 482 SUB-Part F 03.83. Equipment (Small Light Aeroplanes). 3 pp.

—Storage—Assembly
SBAC AS 4906 ISSUE 1. Accumulator Through 24V 40AH. 1 p.

—Storage—Troughs—Absorbent Pads
SBAC AS 5347 ISSUE 1. Absorbent, Pads—Accumulator Trough 40 AH. 1 p.
SBAC AS 5348 ISSUE 2(I). Absorbent Pads—Bottom Accumulator Trough 15A.H. and 25A.H. 25A.H.. 1 p.
SBAC AS 5350 ISSUE 1. Absorbent Pads—Bottom Accumulator Trough 40 A.H.. 1 p.

—Storage—Troughs—Assembly
SBAC AS 5339 ISSUE 1. Assembly of Accumulator Trough 12V 15AH. 1 p.
SBAC AS 5353 ISSUE 1. Assembly of Accumulator Trough 12 V. 25 A.H.. 1 p.
SBAC AS 5354 ISSUE 1. Assembly of Accumulator Trough 12 V. 40 A.H.. 1 p.

—Storage—Troughs—Boundary Rings
SBAC AS 4908 ISSUE 2. Boundary, Ring for Accumulator Trough. 1 p.
SBAC AS 5342 ISSUE 1. Boundary, Ring for Plastic Trough 25 AH. 1 p.
SBAC AS 5343 ISSUE 1. Boundary, Ring for Plastic Trough 40 AH. 1 p.

—Vented
CENELEC EN 60952-2-93. Aircraft Batteries Part 2: Design and Construction Requirements (IEC 952-2:1991). 5 pp.
IEC 952 Pt 1-88. Aircraft Batteries Part 1: General Test Requirements and Performance Levels First Edition. 30 pp.
IEC 952 Pt 2-91. Aircraft Batteries Part 2: Design and Construction Requirements First Edition; (CENELEC EN 60952-2: 1993). 43 pp.

Aircraft Bearings
Use: Airframe Bearings

Aircraft Bolts
See Also: Aircraft Fasteners; Bolts; Fasteners
CAA LEAFLET 3-3 07.90. Bolts and Screws of British Manufacture. 11 pp.
MOD UK DTD-487-41. Nickel-Copper-Aluminium Alloy Cold-Headed Bolts (Not Exceeding 5/8 in Diameter) (Reprinted March 1963). 4 pp.
SBAC AS 4754 ISSUE 2. Bolts—Weldable, (B.A., B.S.F.). 1 p.
SBAC AS 6547 ISSUE 2. Bolt.
SBAC AS 6615 ISSUE 2. Special Bolt.
SBAC AS 6737 ISSUE 4. Bolt, Weldable (Unified).
SBAC AS 23900-999 ISSUE 3. Bolt, Externally Relieved Body—Double Hexagon Extended Washer Head, BS.S.159, 190-32 UNJF-3A.B.S.4084. 1 p.
SBAC AS 50200-299 ISSUE 2. Bolt, Hexagon Head, P.D. Shank INCO 718 0.1900-32 UNJF-3A. BS4084.
SBAC AS 50300-399 ISSUE 2. Bolt, Hexagon Head, P.D. Shank INCO 718 0.2500-28 UNJF-3A. BS4084.
SBAC AS 50400-499 ISSUE 2. Bolt, Hexagon Head, P.D. Shank INCO 718 0.3125-24 UNJF-3A. BS4084.
SBAC AS 50500-599 ISSUE 2. Bolt, Hexagon Head, P.D. Shank INCO 718 0.3750-24 UNJF-3A. BS4084.
SBAC TS 21 ISSUE 7. Bolts 900 MPa (130,000 lbf/in2) Tensile, 650 Degrees C.
SBAC TS 22 ISSUE 7. Bolts 1100 MPa (159,000 lbf/in2) Tensile, 450 Degrees C.
SBAC TS 23 ISSUE 6. Bolts 1000 MPa (145,000 lbf/in2) Tensile, 650 Degrees C.
SBAC TS 24 ISSUE 1. Bolts Tensile Strength: 1275 MPa (185,000 lbf/in2), at Ambient Temperature 1000 MPa (145,000 lbf/in2), 650 Degrees C.
SBAC TS 25 ISSUE 4. Bolts 1210 MPa (175,000 lbf/in2) Tensile at Ambient Temperature.
SBAC TS 27 ISSUE 2. Studs, Bolts and Screws in BS HR601 (Nimonic) Material. 1000 MPa (145,000 lbf/in2) Tensile at Room Temperature (for Which Forging is Not Mandatory and Qualification Not a Requirement).

Aircraft Bolts (Cont.)
SBAC TS 108 ISSUE 3. Studs, Bolts and Screws in BS.S159 Material 1100 MPa (159,000 lbf/in2) Tensile, 450 Degrees C (for Which Forging is Not Mandatory and Qualification Not a Requirement).
SBAC TS 109 ISSUE 3. S.B.A.C. Manufacturing Specification for Studs, Bolts and Screws in BSHR 650 Material 900 MPa (130,000 lbf/in2) Tensile, 500 Degrees C (for Which Forging is Not Mandatory and Qualification Not a Requirement).
SBAC TS 121 ISSUE 2. Studs, Bolts and Screws in DTD 5639 (WASPLOY) Material 1210 MPa (175,000 Lbf/in2) Tensile at Room Temperature (For Which Forging is Not Mandatory and Qualification Not a Requirement.
SBAC TS 145 ISSUE 1. Bolts EN 2493 (FV 535) Class 1000 MPa (145,000 lbf/in2).
SBAC TS 148 ISSUE 1. Studs, Bolts and Screws in DTD5638 (INCO 718) Material 1275 MPa (185,000 lbf/in2) Tensile at Room Temperature (for Which Forging is Not Mandatory and Qualification Not a Requirement).

—Anchor Nut Assemblies
SBAC AS 4572 (V). Typical Assembly of Retained Bolts-Double Anchor Nuts. 2 pp.

—Banjo—Adapters
SBAC AS 6727 (V). Single End Cone Banjo Body (3/16 Inch to 1/2 Inch) for Use with Banjo Bolts AS 6721-6.
SBAC AS 6728 (V). Double End Cone Banjo Body (3/16 Inch to 1/2 Inch) for Use with Banjo Bolts AS 6721-6.
SBAC AS 6729 (V). Double R.A. Cone Banjo Body (3/16 Inch to 1/2 Inch) for Use with Banjo Bolts AS 6721-6.
SBAC AS 6758 (V). Pipe Couplings (Unified Adaptor) Steel Banjo Bolt with Hole for Bleeder Screw.
SBAC AS 6759 (V). Pipe Couplings (Unified Adaptor) Aluminium Alloy Banjo Bolt with Hole for Bleeder Screw.

—Banjo—Pipe Couplings
SBAC AS 6721 (V). Pipe Couplings (Unified Adaptor) Aluminium Alloy Banjo Bolt.
SBAC AS 6722 (V). Pipe Couplings (Unified Adaptor) Banjo Bolt Steel.
SBAC AS 6723 (V). Pipe Couplings (Unified Adaptor) Aluminium Alloy Banjo Bolt with Cone Head Connection.
SBAC AS 6724 (V). Pipe Couplings (Unified Adaptor) Steel Banjo Bolt with Cone Head Connection.
SBAC AS 6725 (V). Pipe Coupling (Unified Adaptor) Aluminium Alloy Banjo Bolt with Union Head Connection.
SBAC AS 6726 (V). Pipe Couplings (Unified Adaptor) Steel, Banjo Bolt with Union Head Connection.

—Banjo—Pitot Static Tubes
SBAC AS 8680 ISSUE 1. Banjo Bolt, Pitot, Pitot Static System.
SBAC AS 8682 ISSUE 1. Banjo Bolt, Pitot, Pitot Static System.
SBAC AS 8683 ISSUE 1. Banjo Bolt, Static, Pitot Static System.

—Banjo—Union
SBAC AS 6730 (V). Single End Union Banjo Body (3/16 Inch to 1 Inch) for Use with Banjo Bolts AS 6721-6.
SBAC AS 6731 (V). Double End Union Banjo Body (3/16 Inch to 1 Inch) for Use with Banjo Bolts AS 6721-6.
SBAC AS 6732 (V). Double R.A. Union Banjo Body (3/16 Inch to 1/2 Inch) for Use with Banjo Bolts AS 6721-6.

—Batteries
SBAC AS 6202 ISSUE 2. Bolt Assembly—Attachment of Lead Acid Accumulators. 1 p.
SBAC AS 6203 ISSUE 1. Bolt Accumulator Attachment. 1 p.

—Chains
SBAC AS 46792 ISSUE 1. Nut and Bolt Assembly (for Chains).

—Chains—Assembly
SBAC AS 174 ISSUE 10. Assembly of Chain End Bolt. 1 p.
SBAC AS 175 ISSUE 10. Assembly of Chain End Bolt (Roller Omitted). 1 p.

—Control Columns
SBAC AS 6252 ISSUE 2. Adjustable Stop, Control Column. 1 p.
SBAC AS 6255 ISSUE 2. Bolt 5/16 Inch Diameter (Unified) Control Column. 1 p.

—Countersunk Head
BSI 2A 173-62. 1962 Amd 6 100 Degrees Countersunk Head Steel Bolts (Unified Treads) for Aircraft. 19 pp.

Aircraft Bolts (Cont.)

—Countersunk Head (Cont.)
BSI 2A 175-62. 1962 Amd 2 100 Degree Countersunk Head Aluminium Alloy Bolts (Unified Threads) for Aircraft. 15 pp.
SBAC AS 4563 ISSUE 7. Bolt, 90 Degree Countersunk Head Aluminium Alloy. 1 p.
SBAC AS 4599-4600 ISSUE 3(I). 120 Degree Countersunk Head Bolt. 2 pp.
SBAC AS 6208 ISSUE 2(I). 120 Degree Countersunk Head, Bolt (6 UNC). 1 p.
SBAC AGS 3818 (V). 'Jo-Bolt' Fastners 100 Degree Countersunk Head.

—Countersunk Head—Corrosion Resistant
BSI 2A 174-62. 1962 Amd 2 100 Degree Countersunk Head Corrosion-Resisting Steel Bolts (Unified Threads) for Aircraft. 15 pp.

—Countersunk Head—Phillips
SBAC AS 2929 ISSUE 2(I). Bolt M.S.—"Phillip's" 120 Degree Countersunk Head. 1 p.
SBAC AS 3294 ISSUE 7. Bolt—Phillip's 90 Degree Countersunk Head. 1 p.
SBAC AS 3297 ISSUE 7. Bolt—Phillip's 120 Degree Countersunk Head. 1 p.

—Countersunk Head—Phillips—Flat
SBAC AS 2926 ISSUE 2(I). Bolt M.S.—"Phillip's" 90 Degree Countersunk Head. 1 p.

—Countersunk Head—Recessed
AECMA PREN3035-87. Bolts, 100 Degrees Countersunk Head, Torq-Setrm Recess, Close Tolerance Shank, Short Thread, in Steel Cadmium Plated Classification 1100 MPa/235 Degrees Celsius. 7 pp.
BSI A 211-61. 1961 Amd 4 100 Degree Countersunk Head Steel Bolts (Unified Threads and Cruciform Recesses) for Aircraft. 19 pp.
BSI 2A 230-74. 1974 Amd 2 100 Degree Countersunk Head Steel Bolts (160 000 Lbf/Square Inches (1100 MPa) with Hi-Torque Speed Drive Recesses) (AMD 6151) May 31, 1990. 4 pp.
BSI A 232-68. 1968 Amd 4 100 Degree Countersunk Head Steel Bolts (160 000 Lbf/Square Inches (110 hbar) with Torq-Set Recesses). 21 pp.

—Countersunk Head—Slotted—Flat—High Strength
SBAC AS 1242 ISSUE 18. Bolt, 90 Degree Countersunk Head H.T.S.. 1 p.

—Dee Head
SBAC AS 29000-099 ISSUE 3. Bolt, Externally Relieved Body Dee Head, DTD5066, 10(.190)-32 UNJF-3A BS4084.
SBAC AS 29200-299 ISSUE 3. Bolt, Externally Relieved Body—Dee Head, BS 5159, 3125-24 UNJF-3A BS4084.
SBAC AS 29300-399 ISSUE 3. Bolt, Externally Relieved Body—Dee Head, BS 5159, 375-24 UNJF-3A BS4084.
SBAC AS 29400-499 ISSUE 3. Bolt, Externally Relieved Body—Dee Head, BS5159, 4375-20 UNJF-3A BS4084.
SBAC AS 29500-599 ISSUE 3. Bolt, Externally Relieved Body—Dee Head, BS5159, 500-20 UNJF-3A BS4084.

—Doors
SBAC AS 5382 ISSUE 1. Pivot Bolt. 1 p.

—Doors—Locks
SBAC AS 6531 ISSUE 1. Pivot Bolt.
SBAC AS 6532 ISSUE 1. Stop, Bolt.

—Double Hexagonal
BSI 3A 228-79. 1979 Steel Double Hexagon Bolts 1250 MPa (180 000 Lbf/Square Inches) for Aircraft. 9 pp.
MOD UK DTD-5182-65. Bolt, Double Hexagon, External Wrenching—180,000 lbf/in2. 2 pp.
SBAC AS 57100-199 ISSUE 1. Bolt, Close Tolerance—Double Hexagon Extended Washer head, Waspaloy 0.1900-32 UNJF—3A MOD BS 4084.

—Double Hexagonal—Machine
SBAC AS 43200-299 (V). Bolt, Machine, Double Hexagon Extended Washer Head, Drilled, BS S 159, Silver Coated, 10(.190)-32 UNJF-3A. BS4084.

—Double Hexagonal—Machine—Shank
SBAC AS 48400-499 ISSUE 2. Bolt, Machine, Double Hexagon Extended Washer Head, P.D. Shank, INCO 718 BS 4084.
SBAC AS 48500-599 ISSUE 2. Bolt, Machine, Double Hexagon Extended Washer Head, P.D. Shank, INCO 718 BS 4084.
SBAC AS 48600-699 ISSUE 2. Bolt, Machine, Double Hexagon Extended Washer Head, P.D. Shank, INCO 718 BS 4084.
SBAC AS 48700-799 ISSUE 2. Bolt, Machine, Double Hexagon Extended Washer Head, P.D. Shank, INCO 718 BS 4084.

INDUSTRY STANDARDS

INTERNATIONAL AND NON-U.S. NATIONAL STANDARDS
SUBJECT INDEX

Aircraft

Aircraft Bolts *(Cont.)*

—Double Hexagonal—Machine—Shank *(Cont.)*
SBAC AS 48800-899 ISSUE 2. Bolt, Machine, Double Hexagon Extended Washer Head, P.D. Shank, Waspaloy UNJF-3A. BS 4084.
SBAC AS 48900-999 ISSUE 2. Bolt, Machine, Double Hexagon Extended Washer Head, P.D. Shank, Waspaloy 0.3125-24 UNJF-3A. BS4084.
SBAC AS 49000-099 ISSUE 2. Bolt, Machine, Double Hexagon Extended Washer Head, P.D. Shank, Waspaloy 0.3750-24 UNJF-3A. BS4084.
SBAC AS 62900-999 ISSUE 1. Bolt, Machine, Double Hexagon Extended Washer Head, P.D. Shank, Waspaloy 0.1900—32UNJF-3A. BS 4084.

—Double Hexagonal—Washers (Fasteners)
SBAC AS 21300-399 ISSUE 4. Bolt, Machine—Double Hexagonal Extended Washer Head, D.T.D.5026 8(.164) -36 UNJF-3A. B.S.4084.
SBAC AS 22400-499 ISSUE 4. Bolt, Close Tolerance—Double Hexagonal Extended Washer Head, BS.5159. 190-32 UNJF-3A.MOD.B.S. 4084. 1 p.

—Electrical Grounding
SBAC RS 682 ISSUE 1. Standard Servicing Earth Bolt (Ground Contact).

—Hexagonal Head
BSI 2A 60-05. 1962 Amd 5 Cadmium-Plated Steel Hexagonal-Headed Shear Bolts (B.S.F. Threads) for Aircraft (AMD 7660) April 15, 1993 (S). 10 pp.
BSI 3A 111-62. (WITHDRAWN) 1962 Amd 4 Cadmium-Plated Steel Bolts (Unified Hexagons and Unified Threads) with Close Tolerance Shanks for Aircraft. 11 pp.
BSI 4A 111-93. 1993 Cadmium-Plated Steel Bolts (Unified Hexagons and Unified Threads) with Close Tolerance Shanks (S). 9 pp.
BSI 3A 169-62. 1962 Amd 1 Aluminium Alloy Bolts (Unified Hexagons and Unified Threads) for Aircraft. 9 pp.
BSI A 226-65. 1965 Amd 2 Steel Hexagon Head Bolts (Short Thread: Class 3A). 7 pp.
BSI 2A 229-72. 1972 Amd 3 Hexagonal Head Steel Bolts 160 000 Lbf/Square Inches (1100 MPa) (AMD 6142) December 21, 1990. 16 pp.
SBAC AS 2504 ISSUE 10. Bolts, Hexagon, Close Tolerance. 1 p.
SBAC AS 4569 ISSUE 7. Bolts, 2 BA Hexagon, Close Tolerance. 1 p.
SBAC AS 52000-599 ISSUE 1. Bolt—Hexagon—Self Retaining—Impedance Type—Metric Series.
SBAC AGS 3817 (V). 'Jo-Bolt' Fastners Hexagon Head.

—Hexagonal Head—Blanking Caps
SBAC AS 2942 ISSUE 3. Bolt. 1 p.

—Hexagonal Head—Control Column
SBAC AS 2292 ISSUE 6. Bolt 3/8 Inch Dia-Control Column. 1 p.
SBAC AS 2293 ISSUE 6. Bolt 5/16 Inch Dia-Control Column. 1 p.

—Hexagonal Head—Corrosion Resistant
BSI 3A 104-62. 1962 Amd 1 Corrosion-Resisting Steel Bolts (Unified Hexagons and Unified Threads) for Aircraft. 12 pp.

—Hexagonal Head—Machine
SBAC AS 36500-599 (V). Screw, Machine, Hexagon Head, Washer Face, HR650 4(.112)-40 UNJC-3A BS4084.
SBAC AS 36600-699 (V). Screw, Machine, Hexagon Head, Washer Face, HR650 6(.138)-32 UNJF-3A BS4084.
SBAC AS 36700-799 (V). Screw, Machine, Hexagon Head, Washer Face, HR650 8(.164)-32 UNJC-3A BS4084.
SBAC AS 44702 ISSUE 2. Bolt, Machine—Hexagon Head—.190-32—(Balance Weight).

—Hexagonal Head—Rings
SBAC AS 2006-2020 ISSUE 9. Ring, Bolt. 1 p.
SBAC AS 2021-2025 ISSUE 5. Bolt for Inner Ring. 1 p.

—Hexagonal Head—Rudder Bars
SBAC AS 2196 ISSUE 6. Special Bolt (Rudder Bar). 1 p.
SBAC AS 6332 ISSUE 1. Special Bolt (Rudder Bar).

—Hexagonal Head—Self Retaining
SBAC AS 53761-543 60 ISSUE 1. Bolt—Hexagon—Self Retaining—Impedance Type—Inch Series.

—Hexagonal Head—Self Retaining—Hole Size
SBAC RS 747 ISSUE 1. Hole Sizes for Self Retaining "Impedance Type" Bolts. Inch Series.
SBAC RS 748 ISSUE 1. Hole Sizes for Self Retaining "Impedance Type" Bolts. Metric Series.

Aircraft Bolts *(Cont.)*

—Hexagonal Head—Shank
SBAC AS 49800-899 ISSUE 1. Bolt, Hexagon Head, P.D. Shank BS. HR650. 190-32 UNJF-3A. BS4084.
SBAC AS 49900-999 ISSUE 1. Bolt, Hexagon Head, P.D. Shank BS.HR650. 250-28 UNJF-3A. BS4084.
SBAC AS 50000-099 ISSUE 1. Bolt, Hexagon Head, P.D. Shank BS.HR650 3125-24 UNJF-3A. BS4084.
SBAC AS 50100-199 ISSUE 1. Bolt, Hexagon Head, P.D. Shank BS.HR650 UNJF-3A. BS4084.
SBAC AS 50200-299 ISSUE 2. Bolt, Hexagon Head, P.D. Shank INCO 718 0.1900-32 UNJF-3A. BS4084.
SBAC AS 50300-399 ISSUE 2. Bolt, Hexagon Head, P.D. Shank INCO 718 0.2500-28 UNJF-3A. BS4084.
SBAC AS 50400-499 ISSUE 2. Bolt, Hexagon Head, P.D. Shank INCO 718 0.3125-24 UNJF-3A. BS4084.
SBAC AS 50500-599 ISSUE 2. Bolt, Hexagon Head, P.D. Shank INCO 718 0.3750-24 UNJF-3A. BS4084.
SBAC AS 56300-399 ISSUE 1. Bolt, Hexagon Head, PD Shank, INCO 718, 0.4375—20 UNJF—3A. BS4084.
SBAC AS 56400-99 ISSUE 1. Bolt, Hexagon Head, PD Shank, INCO 718, 0.5000—20 UNJF—3A. BS4084.
SBAC AS 56600-699 ISSUE 1. Bolt, Hexagon Head, P.D. Shank, Waspaloy, 0.2500—28 UNJF—3A. BS4084.
SBAC AS 56700-799 ISSUE 1. Bolt, Hexagon Head, P.D. Shank, Waspaloy, 0.3125—24 UNJF—3A. BS4084.
SBAC AS 56800-899 ISSUE 1. Bolt, Hexagon Head, P.D. Shank, Waspaloy, 0.3750—24 UNJF—3A. BS4084.
SBAC AS 56900-999 ISSUE 1. Bolt, Hexagon Head, P.D. Shank, Waspaloy, 0.4375—20 UNJF—3A. BS4084.
SBAC AS 57000-099 ISSUE 1. Bolt, Hexagon Head, P.D. Shank, Waspaloy, 0.5000—20 UNJF—3A. BS 4084.
SBAC AS 58900-999 ISSUE 1. Bolt, Hexagon Head, P.D. Shank, FV535, 0.1900-32 UNJF-3A. BS4084.
SBAC AS 59000-099 ISSUE 1. Bolt, Hexagon Head, P.D. Shank, FV535, 0.2500—28 UNJF—3A. BS4084.
SBAC AS 59100-199 ISSUE 1. Bolt, Hexagon Head, P.D. Shank, FV535, 0.3125-24 UNJF-3A. BS4084.
SBAC AS 59200-299 ISSUE 1. Bolt, Hexagon Head, P.D. Shank, FV535, 0.3750-24 UNJF-3A. BS4084.
SBAC AS 59300-399 ISSUE 1. Bolt, Hexagon Head, P.D. Shank, FV535, 0.4375-20 UNJF-3A. BS4084.
SBAC AS 59400-499 ISSUE 1. Bolt, Hexagon Head, P.D. Shank, FV535, 0.5000-20 UNJF-3A. BS4084.

—Hexagonal Head—Steel
BSI 3A 102-62. 1962 Amd 4 Steel Bolts (Unified Hexagons and Unified Threads) for Aircraft (AMD 5388) August 31, 1988. 14 pp.

—Hexagonal Head—Washers (Fasteners)
BSI SP 107-112-54. 1954 Amd 2 Tab Washers for Unified Hexagons for Aircraft. 4 pp.
BSI SP 122-125-58. 1958 Washers for Unified Hexagons for Aircraft (Primarily for Packing Purposes). 5 pp.
BSI SP 126-127-59. 1959 Washers for Unified Hexagons for Aircraft (Primarily for Facing Purposes). 4 pp.

—Jo-Bolts
SBAC TS 50 ISSUE 4. Alloy Steel "JO—BOLT" Fasteners.

—Locking Wire Holes
SBAC RS 704 (V). Recommended Size and Location of Locking Wire Holes in Bolts Screws and Nuts.

—Nut Assemblies—Chains
SBAC AS 168 ISSUE 4. Nut and Bolt Assembly for Chains. 1 p.

—Oval Head
BSI 3A 113-62. 1962 Amd 4 Steel Mushroom Head Bolts (Unified Threads) for Aircraft. 12 pp.
BSI 3A 114-62. 1962 Amd 2 Corrosion-Resisting Steel Mushroom Head Bolts (Unified Threads) for Aircraft. 10 pp.
BSI 3A 170-62. 1962 Amd 1 Aluminium Alloy Mushroom Head Bolts (Unified Threads) for Aircraft. 7 pp.
SBAC AS 4566 ISSUE 6. Bolt Mushroom Head Aluminium Alloy. 1 p.

—Oval Head—Countersunk
SBAC AS 4564 ISSUE 6. Bolt Raised 90 Degree Countersunk Head Aluminium Alloy. 1 p.

—Oval Head—Phillips
SBAC AS 2925 ISSUE 2(I). Bolt M.S.—"Phillip's" Mushroom Head. 1 p.
SBAC AS 4598 ISSUE 6. Bolt—Phillips Mushroom Head. 1 p.

Aircraft Bolts *(Cont.)*

—Oval Head—Phillips—Countersunk
SBAC AS 2927 ISSUE 2(I). Bolt M.S.—"Phillip's" Raised 90 Degree Countersunk Head. 1 p.
SBAC AS 2928 ISSUE 2(I). Bolt M.S.—"Phillip's" Raised 120 Degree Countersunk Head. 1 p.
SBAC AS 3295 ISSUE 7. Bolt—"Phillip's" Raised 90 Degree Countersunk Head. 1 p.
SBAC AS 3296 ISSUE 7. Bolt—"Phillip's" Raised 120 Degree Countersunk Head. 1 p.

—Oval Head—Slotted
SBAC AS 2923 ISSUE 6. Bolt, Mushroom Head Stainless Steel. 1 p.

—Oval Head—Slotted—Countersunk
SBAC AS 2921 ISSUE 6. Bolt, Raised 90 Degree Countersunk Head Stainless Steel. 1 p.

—Oval Head—Slotted—High Strength
SBAC AS 1248 ISSUE 16. Bolt Mushroom Head (H.T.S.). 1 p.

—Oval Head—Square Neck
DIN ENGL 603-81. Mushroom Head Square Neck Bolts (Oct). 5 pp.

—Pan Head
BSI 3A 116-62. 1962 Amd 4 Steel Pan Head Bolts (Unified Threads) for Aircraft (AMD 5386) June 30, 1988. 10 pp.
BSI 3A 171-62. 1962 Amd 1 Aluminium Alloy Pan Head Bolts (Unified Threads) for Aircraft. 6 pp.
BSI A 227-65. 1965 Amd 1 Steel Pan Head Bolts (Short Thread: Class 3A). 6 pp.

—Pedal Pivot—Rudder Bars
SBAC AS 6337-6338 ISSUE 1. Pedal Pivot Bolt (Rudder Bar).

—Pilot Seat—Gears
SBAC AS 6234 ISSUE 3(I). Bearing Bolt—Pilot's Seat Raising Gear. 1 p.

—Pivot—Rudder Bars
SBAC AS 2202-2203 ISSUE 5. Pivot Bolts (Rudder Bar). 1 p.
SBAC AS 2208-2209 ISSUE 5. Pedal Pivot Bolt (Rudder Bar). 1 p.
SBAC AS 6335-6336 ISSUE 1. Pivot Bolt (Rudder Bar).

—Recessed Head—Countersunk
AECMA PREN3035-87. Bolts, 100 Degrees Countersunk Head, Torq-Setrm Recess, Close Tolerance Shank, Short Thread, Active Cadmium Plated Classification 1100 MPa/235 Degrees Celsius. 7 pp.
BSI A 211-61. 1961 Amd 4 100 Degree Countersunk Head Steel Bolts (Unified Threads and Cruciform Recesses) for Aircraft. 19 pp.
BSI 2A 230-74. 1974 Amd 2 100 Degree Countersunk Head Steel Bolts (160 000 Lbf/Square Inches (1100 MPa) with Hi-Torque Speed Drive Recesses) (AMD 6151) May 31, 1990. 22 pp.
BSI A 232-68. 1968 Amd 4 100 Degree Countersunk Head Steel Bolts (160 000 Lbf/Square Inches (110 hbar) with Torq-Set Recesses). 21 pp.

—Rings
SBAC AS 6647-6661 ISSUE 1. Ring, (Bolt).
SBAC AS 6662-6666 ISSUE 1. Bolt for Inner Ring.

—Round Head—Phillips
SBAC AS 2924 ISSUE 2(I). Bolt M.S.—"Phillip's" Round Head. 1 p.
SBAC AS 4597 ISSUE 6. Bolt—Phillips Round Head. 1 p.

—Rudder Bars
SBAC AS 6343 ISSUE 1. Taper Bolt Rudder Bar.
SBAC AS 6344 ISSUE 1. Eye Bolt Rudder Bar.

—Screw Threads
SBAC RS 723 ISSUE 1. Selected Aerospace Diameter-Pitch Combinations of Metric, 60 Degree Screw Threads for Nuts and Bolts.

—Stops
SBAC AS 5379 ISSUE 1. Stop Bolt. 1 p.

—Taper—Rudder Bars
SBAC AS 3399 ISSUE 4. Taper Bolt—Rudder Bar. 1 p.

—Tee Head
SBAC TS 200 ISSUE 5. Bolt, Tee Head, Close Tol Shank FV535 0.4375—20 UNJF—3A. MOD BS4084.

—Tee Head—Shank
SBAC AS 49100-199 ISSUE 2. Bolt, Tee Head, P.D. Shank, INCO 718 0.1900-32 UNJF-3A BS 4084.
SBAC AS 49200-299 ISSUE 2. Bolt, Tee Head, P.D. Shank, INCO 718 0.2500-28 UNJF-3A BS 4084.

INTERNATIONAL AND NON-U.S. NATIONAL STANDARDS
SUBJECT INDEX
Aircraft

Aircraft Bolts *(Cont.)*
—Tee Head—Shank *(Cont.)*
SBAC AS 49300-399 ISSUE 2. Bolt, Tee Head, P.D. Shank, INCO 718 0.3125-24 UNJF-3A BS 4084.
SBAC AS 49400-499 ISSUE 3. Bolt, Tee head, P.D. Shank, INCO 718 0.3750-24 UNJF-3A BS 4084.
SBAC AS 49500-599 ISSUE 3. Bolt, Tee Head, P.D. Shank, Waspaloy 0.2500-28 UNJF-3A BS 4084.
SBAC AS 49600-699 ISSUE 3. Bolt, Tee Head, P.D. Shank, WASPALOY 0.3125-24 UNJF-3A BS4084.
SBAC AS 49700-799 ISSUE 3. Bolt,Tee Head, P.D. Shank, Waspaloy. 03750-24 UNJF-3A BS4084.
SBAC AS 57900-999 ISSUE 1. Bolt, Tee Head, Close Tolerance Shank WASPALOY 0.3125-24 UNJF-3A.MOD. BS.4084.

—Washers (Fasteners)
BSI SP 10-11-49. 1949 Amd 1 Washers for Aircraft (Primarily for Facing Purposes). 2 pp.
BSI SP 122-125-58. 1958 Washers for Unified Hexagons for Aircraft (Primarily for Packing Purposes). 5 pp.
BSI SP 126-127-59. 1959 Washers for Unified Hexagons for Aircraft (Primarily for Facing Purposes). 4 pp.

—Wing—Cradles
SBAC AS 1895 ISSUE 3. Bolt, Wing—Oxygen Cradle. 1 p.
SBAC AS 6246 ISSUE 2. Bolt Wing—Oxygen Cradle. 1 p.

—Wing—Shanks
SBAC AS 1896 ISSUE 3. Shank for Wing Bolt. 1 p.
SBAC AS 6247 ISSUE 1. Shank for Wing Bolt. 1 p.

Aircraft Brakes
See Also: Aircraft Equipment; Brakes
CAA Chapter K4-5 04.74. Landing Gear Design (Light Aeroplanes). 4 pp.
CAA LEAFLET 5-8 07.90. Wheels and Brakes. 19 pp.
CAA Spec. NO. 17 ISSUE 1.

—Control Wheels
SBAC AS 3048 ISSUE 1. Pilot's Control Wheel Brake Assembly (Spade Type). 1 p.
SBAC AS 3049 ISSUE 1. Pilot's Control Wheel Brake Assembly (Horn Type). 1 p.
SBAC AS 6580 ISSUE 2. Brake off Stop Pilot's Control Wheel.
SBAC AS 6581 ISSUE 2. Brake on Stop Pilot's Control Wheel.

—Control Wheels—Assembly
SBAC AS 6570 ISSUE 1. Pilot's Control Wheel, Brake Assembly—Spade Type.
SBAC AS 6571 ISSUE 1. Pilot's Control Wheel, Brake Assembly (Horn Type).

—Control Wheels—Locking Rings
SBAC AS 6586 ISSUE 1. Locking, Ring for Brake Stop for Pilot's Control Wheel.

—Rotary Wing Aircraft—Design
CAA Chapter G4-5 12.80. Landing Gear Design (Rotocraft). 3 pp.

—Screws—Adjusting
SBAC AS 6584-6585 ISSUE 1. Adjusting Brake Screw Sto

—Supersonic Transports
CAA STANDARD NO. 5-3&AP 01.70. Tyres—Wheels—Brakes. 24 pp.

Aircraft Cabins
See Also: Aircraft; Aircraft Compartments; Aircraft Equipment; Cockpits; Pressurized Cabins

—Aerostats—Safety
CAA Chapter Q4-3 App 12.79. Compartment Design and Safety Provisions (Non-Rigid Airships). 2 pp.

—Design
CAA Chapter K4-3 App 04.74. Compartment Design and Safety Provisions (Light Aeroplanes). 1 p.

—Doors
CAA Chapter K4-3 04.74. Compartment Design and Safety Provisions (Light Aeroplanes). 3 pp.

—Emergency Exits
CAA Chapter K4-3 04.74. Compartment Design and Safety Provisions (Light Aeroplanes). 3 pp.

—Fire Protection
CAA Chapter K4-3 04.74. Compartment Design and Safety Provisions (Light Aeroplanes). 3 pp.

—Fire Testing
CAA Spec. NO. 8 FIRST Draft.

Aircraft Cabins *(Cont.)*
—Flammability Testing
CAA NOTICE #61 ISSUE 2. Improved Flammability Test Standards for Cabin Interior Materials (Airworthiness Notices). 4 pp.

—Flight Crew—Aerostats—Design
CAA Chapter Q4-2 12.79. Flight Crew Accommodation Design (Non-Rigid Airships). 3 pp.

—Flight Crew—Aerostats—Fire Protection
CAA Chapter Q4-3 12.79. Compartment Design and Safety Provisions (Non-Rigid Airships). 5 pp.

—Flight Crew—Aerostats—Ventilation
CAA Chapter Q4-3 12.79. Compartment Design and Safety Provisions (Non-Rigid Airships). 5 pp.

—Flight Crew—Design
CAA 673 (K). Section K Light Aeroplanes Pilot Intercommunication in Light Aeroplanes (Blue Papers). 1 p.
CAA Chapter K4-2 10.92. Flight Crew Accommodation Design (Light Aeroplanes). 3 pp.
CAA Chapter P4-2. Flight Crew Compartment Design (Provisional Airworthiness Requirements for Civil Powered Lift Aircraft). 16 pp.
CAA Chapter P4-3. Compartment Design and Safety Provisions (Provisional Airworthiness Requirements for Civil Powered Lift Aircraft). 12 pp.
NATO STANAG 3639 ED 3 AMD 5-81. Aircrew Station Dimensional Design Factors. 9 pp.
NATO STANAG 3639 ED 4 AMD 0-92. Aircrew Station Dimensional Design Factors. 9 pp.
NATO STANAG 3639 ED 4 AMD 1-92. Aircrew Station Dimensional Design Factors. 10 pp.

—Flight Crew—Design—Color
NATO STANAG 3701 ED 2 AMD 4-78. Aircraft Interior Colour Schemes. 14 pp.
NATO STANAG 3701 ED 3 AMD 0-92. Aircraft Interior Colour Schemes. 6 pp.
NATO STANAG 3701 ED 3 AMD 1-92. Aircraft Interior Colour Schemes. 7 pp.

—Flight Crew—Design—Display Devices—Glossaries
NATO STANAG 3869 ED 1 AMD 4-80. Aircrew Station Control Panels. 26 pp.
NATO STANAG 3871 ED 1 AMD 2-85. NATO Glossary of Displays and Aircrew Stations (AI) Specialist Terminology and Abbreviations. 26 pp.
NATO STANAG 3871 ED 1 AMD 3-85. NATO Glossary of Displays and Aircrew Stations (AI) Specialist Terminology and Abbreviations. 38 pp.

—Flight Crew—Design—Night Vision Goggles
NATO STANAG 3800 ED 1 AMD 3-85. Night Vision Goggles Lighting Compatibility Design Criteria (English Version). 12 pp.

—Flight Crew—Display Devices
NATO STANAG 3705 ED 2 AMD 1-81. Principles of Presentation of Information in Aircrew Stations. 11 pp.
NATO STANAG 3705 ED 2 AMD 3-81. Principles of Presentation of Information in Aircrew Stations. 10 pp.

—Flight Crew—Display Devices—Antireflection Coatings
NATO STANAG 3643 ED 2 AMD 7-73. Coating, Reflection Reducing for Glass Elements Used in Aircrew Station Displays. 6 pp.
NATO STANAG 3643 ED 2 AMD 9-73. Coating, Reflection Reducing for Glass Elements Used in Aircrew Station Displays. 6 pp.

—Flight Crew—Display Devices—Ergonomics
NATO STANAG 3705 ED 3 AMD 0-92. Human Engineering Design Criteria for Controls and Displays in Aircrew Stations. 21 pp.
NATO STANAG 3705 ED 3 AMD 1-92. Human Engineering Design Criteria for Controls and Displays in Aircrew Stations. 22 pp.

—Flight Crew—Display Devices—Glossaries
NATO STANAG 3647 ED 3 AMD 1-86. Nomenclature in Aircrew Stations. 26 pp.
NATO STANAG 3647 ED 3 AMD 2-86. Nomenclature in Aircrew Stations. 22 pp.

—Flight Crew—Display Devices—Symbols
NATO STANAG 3329 ED 6 AMD 3-84. Numerials and Letters in Aircrew Stations. 14 pp.
NATO STANAG 3329 ED 6 AMD 4-84. Numerals and Letters in Aircrew Stations. 13 pp.

—Flight Crew—External Vision
NATO STANAG 3622 ED 3 AMD 6-73. External Vision from Aircrew Stations. 14 pp.

Aircraft Cabins *(Cont.)*
—Flight Crew—Lighting
NATO STANAG 3224 ED 5 AMD 0-86. Aircrew Station Lighting. 20 pp.
NATO STANAG 3224 ED 5 AMD 1-86. Aircrew Station Lighting. 18 pp.
NATO STANAG 3224 ED 5 AMD 2-86. Aircrew Station Lighting. 19 pp.

—Flight Crew—Manual Controls
NATO STANAG 3217 ED 5 AMD 5-81. Operation of Controls and Switches at Aircrew Stations. 9 pp.

—Flight Crew—Manual Controls—Human Factors Engineering
NATO STANAG 3705 ED 3 AMD 0-92. Human Engineering Design Criteria for Controls and Displays in Aircrew Stations. 21 pp.
NATO STANAG 3705 ED 3 AMD 1-92. Human Engineering Design Criteria for Controls and Displays in Aircrew Stations. 22 pp.

—Flight Crew—Rotary Wing Aircraft—Design
CAA Chapter G4-2 12.80. Flight Crew Compartment Design (Rotocraft). 3 pp.

—Flight Crew—Rotary Wing Aircraft—Temperature
CAA Chapter G4-2 App 12.80. Flight-Crew Compartment Design (Rotocraft). 1 p.

—Flight Crew—Safety
CAA Chapter P4-3. Compartment Design and Safety Provisions (Provisional Airworthiness Requirements for Civil Powered Lift Aircraft). 12 pp.

—Flight Crew—Safety—Emergency Lighting
NATO STANAG 3870 ED 1 AMD 2-86. Emergency Escape/Evacuation Lighting. 10 pp.

—Flight Crew—Supersonic Transports—Design
CAA STANDARD NO. 5-1&AP 05.73. Crew Compartment Design and Flight Deck View. 29 pp.

—Flight Crew—Switches
NATO STANAG 3217 ED 5 AMD 5-81. Operation of Controls and Switches at Aircrew Stations. 9 pp.

—Flight Crew—Warning Systems
CAA Chapter P6-1. Flight Information (Provisional Airworthiness Requirements for Civil Powered Lift Aircraft). 5 pp.
NATO STANAG 3370 ED 5 AMD 0-91. Aircrew Station Warning, Cautionary and Advisory Signals. 12 pp.
NATO STANAG 3370 ED 5 AMD 1-91. Aircrew Station Warning, Cautionary and Advisory Signals. 13 pp.

—Ground Effect Machines—Air Handling Equipment
CAA 42.
CAA Chapter B6-6 01.91. Heating and Ventilation Systems. 1 p.

—Ground Effect Machines—Flight Crew—Design
CAA Chapter B5-2 08.83. Operating Crew Compartments. 3 pp.

—Ground Effect Machines—Passenger—Design
CAA Chapter B5-3 11.72. Passenger Compartments. 1 p.
CAA Chapter B5-3 App 11.72. Passenger Compartments. 1 p.

—Passenger—Aerostats—Fire Protection
CAA Chapter Q4-3 12.79. Compartment Design and Safety Provisions (Non-Rigid Airships). 5 pp.

—Passenger—Aerostats—Ventilation
CAA Chapter Q4-3 12.79. Compartment Design and Safety Provisions (Non-Rigid Airships). 5 pp.

—Passenger—Breathing Apparatus
EUROCAE ED-65 04.91. MOPS for Passenger Protective Breathing Equipment. 62 pp.

—Passenger—Design
CAA Chapter P4-3. Compartment Design and Safety Provisions (Provisional Airworthiness Requirements for Civil Powered Lift Aircraft). 12 pp.

—Passenger—Safety
CAA Chapter P4-3. Compartment Design and Safety Provisions (Provisional Airworthiness Requirements for Civil Powered Lift Aircraft). 12 pp.

INDUSTRY STANDARDS

Aircraft Cabins (Cont.)

—Passenger—Safety—Emergency Lighting
NATO STANAG 3870 ED 1 AMD 2-86. Emergency Escape/Evacuation Lighting. 10 pp.

—Pressure Measurement—Connections
JIS W 3104-91. Aircraft—Ground Pressure Testing Connections for Pressure Cabins (ISO 11:1976).
NATO STANAG 3315 ED 7 AMD 4-83. Aircraft Cabin Pressurizing Test Connections. 18 pp.
NATO STANAG 3315 ED 7 AMD 5-83. Aircraft Cabin Pressurizing Test Connections. 18 pp.

—Rotary Wing Aircraft—Design
CAA CAP 524 SUB-Part D 12.86. Design and Construction (Rotorcraft). 23 pp.

—Rotary Wing Aircraft—Flammability Testing
CAA 786 (G). Section G Rotorcraft Compartment Fire Precautions Chapter G4-3 (Blue Papers). 15 pp.
CAA Chapter G4-2 App 12.80. Flight-Crew Compartment Design (Rotocraft). 1 p.

—Safety
CAA Chapter K4-3 04.74. Compartment Design and Safety Provisions (Light Aeroplanes). 3 pp.
CAA Chapter K4-3 App 04.74. Compartment Design and Safety Provisions (Light Aeroplanes). 1 p.

—Supersonic Transports—Design
CAA STANDARD NO. 5-2&AP 05.73. Compartment Design and Safety Provisions. 22 pp.

—Supersonic Transports—Safety
CAA STANDARD NO. 5-2&AP 05.73. Compartment Design and Safety Provisions. 22 pp.

—Ventilation
CAA Chapter K4-3 04.74. Compartment Design and Safety Provisions (Light Aeroplanes). 3 pp.

Aircraft Cargo Compartments
Use: Aircraft Compartments—Cargo

Aircraft Certification
Scope Note: See the subheading Certification under the specific type of aircraft

Aircraft Communications
Scope Note: For additional listings, use a more specific term *See Also:* Aeronautical Communications; Aeronautical Passenger Communications; Air Navigation; Air Traffic Control; Aircraft; Radio Communications; Space Communications; Telecommunication Equipment; Telecommunication Systems

—Glossaries
NATO STANAG 3817 ED 4 AMD 3-89. Standard R/T Phraseology to Be Used for Aircraft. 84 pp.
NATO STANAG 3817 ED 4 AMD 4-89. Standard R/T Phraseology to Be Used for Aircraft. 85 pp.

—Public Data Networks
CEPT T/SF 49-87. Aspects Services Et Facilities, D'un Service De Correspondance Publique Pour L'Aviation Civile. 3 pp.
CEPT T/SF 49 E-87. Services and Facilities Aspects of a System for Public Correspondence for Civil Aviation. 3 pp.

Aircraft Compartments
Scope Note: Includes areas in aircraft for baggage, cargo, engines *See Also:* Aircraft; Aircraft Cabins; Aircraft Equipment; Cargo Holds; Cockpits

—Aerostats—Baggage—Design
CAA Chapter Q4-3 12.79. Compartment Design and Safety Provisions (Non-Rigid Airships). 5 pp.

—Aerostats—Baggage—Fire Protection
CAA Chapter Q4-3 12.79. Compartment Design and Safety Provisions (Non-Rigid Airships). 5 pp.

—Aerostats—Cargo—Design
CAA Chapter Q4-3 12.79. Compartment Design and Safety Provisions (Non-Rigid Airships). 5 pp.

—Aerostats—Cargo—Fire Protection
CAA Chapter Q4-3 12.79. Compartment Design and Safety Provisions (Non-Rigid Airships). 5 pp.

—Baggage—Design
CAA Chapter P4-3. Compartment Design and Safety Provisions (Provisional Airworthiness Requirements for Civil Powered Lift Aircraft). 12 pp.

—Cargo—Design
CAA Chapter P4-3. Compartment Design and Safety Provisions (Provisional Airworthiness Requirements for Civil Powered Lift Aircraft). 12 pp.

Aircraft Compartments (Cont.)

—Cargo—Design (Cont.)
NATO STANAG 3980 ED 1 AMD 0-88. (DRAFT) Aircraft Cargo Compartment Design Criteria for Fixed Wing Aircraft. 7 pp.

—Cargo—Tie Downs
NATO STANAG 3400 ED 3 AMD 2-83. Restraint of Cargo in Fixed Wing Aircraft. 8 pp.
NATO STANAG 3400 ED 3 AMD 3-83. Restraint of Cargo in Fixed Wing Aircraft. 8 pp.
NATO STANAG 3400 ED 3 AMD 4-83. Restraint of Cargo in Fixed Wing Aircraft. 9 pp.

—Design
CAA Chapter K4-3 04.74. Compartment Design and Safety Provisions (Light Aeroplanes). 3 pp.
CAA Chapter K4-3 App 04.74. Compartment Design and Safety Provisions (Light Aeroplanes). 1 p.

—Fire Protection
CAA Chapter K4-3 04.74. Compartment Design and Safety Provisions (Light Aeroplanes). 3 pp.
CAA Chapter Q4-3 12.79. Compartment Design and Safety Provisions (Non-Rigid Airships). 5 pp.
ISO 2685-92. Aircraft—Environmental Conditions and Test Procedures for Airborne Equipment—Resistance to Fire in Designated Fire Zones First Edition. 33 pp.

—Fire Testing
CAA Spec. NO. 8 FIRST Draft.
ISO 2685-92. Aircraft—Environmental Conditions and Test Procedures for Airborne Equipment—Resistance to Fire in Designated Fire Zones First Edition. 33 pp.

—Ground Effect Machines—Air Handling Equipment
CAA 42.
CAA Chapter B6-6 01.91. Heating and Ventilation Systems. 1 p.

—Ground Effect Machines—Baggage—Design
CAA Chapter B5-4 11.72. Cargo and Baggage Compartments. 1 p.

—Ground Effect Machines—Cargo—Design
CAA Chapter B5-4 11.72. Cargo and Baggage Compartments. 1 p.

—Rotary Wing Aircraft—Design
CAA CAP 524 SUB-Part D 12.86. Design and Construction (Rotorcraft). 23 pp.

—Rotary Wing Aircraft—Flammability Testing
CAA 786 (G). Section G Rotorcraft Compartment Fire Precautions Chapter G4-3 (Blue Papers). 15 pp.
CAA Chapter G4-3 App #2. Flame Resistance Testing for Rotorcraft Interior Materials (Rotocraft). 7 pp.

—Safety
CAA Chapter K4-3 04.74. Compartment Design and Safety Provisions (Light Aeroplanes). 3 pp.
CAA Chapter K4-3 App 04.74. Compartment Design and Safety Provisions (Light Aeroplanes). 1 p.

Aircraft Components
Use: Aircraft Equipment

Aircraft Control
Use: Flight Dynamics

Aircraft Cross Servicing
See Also: Interdepartmental Procurement; Services

—Ground Crews—Training
NATO STANAG 3812 ED 5 AMD 1-90. Responsibilities for Aircraft Cross-Servicing Ground Crew Training. 14 pp.
NATO STANAG 3812 ED 5 AMD 2-90. Responsibilities for Aircraft Cross-Servicing Ground Crew Training. 14 pp.
NATO STANAG 3812 ED 6 AMD 0-93. Responsibilities for Aircraft Cross-Servicing Ground Crew Training. 18 pp.

Aircraft Design
Scope Note: See the subheading Design or Structural Design under the specific type of aircraft

Aircraft Doors
Use For: Exits (Aircraft) *See Also:* Aircraft; Aircraft Equipment; Doors

—Aluminum Alloy—Covers
SBAC AS 2056-2060 ISSUE 4. Cover, Inspection Door. 1 p.

Aircraft Doors (Cont.)

—Cabins
CAA Chapter K4-3 04.74. Compartment Design and Safety Provisions (Light Aeroplanes). 3 pp.

—Ground Effect Machines
CAA Chapter B5-6 01.91. Exits. 4 pp.
CAA Chapter B5-6 App 11.72. Exits. 3 pp.

—Locks (Security)
SBAC AS 2542 ISSUE 5. Main Body Left Hand. 1 p.
SBAC AS 2543 ISSUE 5. Main Body Right Hand. 1 p.
SBAC AS 6400-6519 (V). Assembly of Flush Fitting Door Lock.
SBAC AS 6520 ISSUE 1. Main Body—Left Hand.
SBAC AS 6521 ISSUE 1. Main Body—Right Hand.

—Locks (Security)—Assembly
SBAC AS 2840-2899 (V). Assembly of Flush Fitting Door Lock. 2 pp.

—Passenger Loading Systems—Interfaces
ISO 7718-84. Aircraft—Connection of Passenger Loading Bridge or Transfer Vehicle—Interface Requirements in the Vicinity of Main Deck Passenger Doors First Edition. 4 pp.

Aircraft Electronics
Use: Avionics

Aircraft Engines
Use For: Power Plant Installations (Aircraft)
See Also: Aircraft; Aircraft Equipment; Gas Turbine Engines; Helicopter Engines; Internal Combustion Engines; Internal Combustion Piston Engines; Jet Propulsion; Ramjet Engines; Spark Ignition Engines; Turbine Engines; Turbofan Engines; Turbojet Engines; Turboprop Engines

CAA 798 (C). Section C Engines and Propellers Miscellaneous Amendments to Section C (Blue Papers). 5 pp.
CAA Chapter P5-1. General (Provisional Airworthiness Requirements for Civil Powered Lift Aircraft). 16 pp.
CAA CAP 482 SUB-Part E 03.83. Power-Plant (Small Light Aeroplanes). 5 pp.
CAA CAP 482 SUB-Part F 03.83. Equipment (Small Light Aeroplanes). 3 pp.
CAA. Contents (Joint Airworthiness Requirements). 5 pp.
CAA. Foreword (Joint Airworthiness Requirements). 1 p.
CAA. Check List of Pages (Joint Airworthiness Requirements). 3 pp.
CAA JAR-E Section 1. Requirements (Joint Airworthiness Requirements). 60 pp.
CAA JAR-E Section 2. Acceptable Means of Compliance and Interpretations (Joint Airworthiness Requirements). 54 pp.
CAA JAR-E Section 4. Approval of Engines and Associated Equipment. 5 pp.
CAA JAR-E Appendix. Cross-Reference Tables. 58 pp.

—Aerostats
CAA 790 (Q). Section Q Non-Rigid Airships Engines Propellers and Transmissions (Amending Blue Paper 772) (Blue Papers). 4 pp.
CAA Chapter Q5-9. Engines, Transmissions and Propellers (Non-Rigid Airships). 3 pp.

—Aerostats—Air Intakes
CAA Chapter Q5-5 12.79. Engine Air Intake and Ice Protection Systems (Non-Rigid Airships). 2 pp.

—Aerostats—Control Systems Equipment
CAA Chapter Q5-7 12.79. Controls (Non-Rigid Airships). 2 pp.

—Aerostats—Fire Protection
CAA Chapter Q5-8 12.79. Fire Precautions (Non-Rigid Airships). 3 pp.

—Aerostats—Ice Formation
CAA Chapter Q5-5 12.79. Engine Air Intake and Ice Protection Systems (Non-Rigid Airships). 2 pp.

—Aerostats—Loads (Forces)
CAA Chapter Q3-4 12.79. Engine and Propeller Loads (Non-Rigid Airships). 2 pp.

—Aerostats—Lubrication
CAA Chapter Q5-3 12.79. Oil Systems (Non-Rigid Airships). 3 pp.
CAA Chapter Q5-3 App 12.79. Oil Systems (Non-Rigid Airships). 1 p.

—Air Intakes
CAA Chapter P5-5. Air Intake Systems (Provisional Airworthiness Requirements for Civil Powered Lift Aircraft). 4 pp.

INTERNATIONAL AND NON-U.S. NATIONAL STANDARDS
SUBJECT INDEX
Aircraft

Aircraft Engines (Cont.)
—**Air Pollution**
ICAO Annex 16 Vol II. Environmental Protection Volume II—Aircraft Engine Emissions Second Edition—July 1993 (Incorporating Amendments 1-2). 64 pp.
ICAO 9259. Committee on Aircraft Engine Emissions First Meeting Montreal, 12-22 June 1978; (Supplement 1 12/15/78). 195 pp.
ICAO 9304. Committee on Aircraft Engine Emissions Second Meeting Montreal, 14-29 May 1980; (Supplement 1 06/27/80). 154 pp.
ICAO 9499. Committee on Aviation Environmental Protection First Meeting Montreal, 9-20 June 1986. 230 pp.

—**Airworthiness Notices**
CAA NOTICE #9 ISSUE 2. Piston Engines Supplied in Kit Form (Airworthiness Notices). 2 p.

—**Allison**
CAA. Allison Series Engines. 2 p.

—**Alvis Leonides**
CAA. Alvis Leonides Engines. 1 p.

—**Avco Lycoming**
CAA. Avco Lycoming Series Engines. 5 pp.

—**Bombardier-Rotax**
CAA. Bombardier—Rotax Series Engines. 2 pp.

—**British Aerospace Dynamics Group**
CAA. British Aerospace Dynamics Group. 13 pp.

—**Certificate of Airworthiness**
CAA NOTICE #16 ISSUE 9. Aircraft Engines, Propellers and Related Equipment Obtained from Sources Not Under the Airworthiness Control of the CAA (Airworthiness Notices). 6 pp.
CAA Chapter B4-2 06.90. Type Certification or Validation of Engines.

—**CFM56-3**
CAA. CFM International CFM56-3 Series Engines. 1 p.

—**CFM56-5**
CAA. CFM International CFM56-5 Series Engines. 1 p.

—**Chip Detectors—Housings**
MOD UK DSTAN 47-3-82. Metal Chip Detectors (Housings) Issue 4. 6 pp.

—**Connectors—Lubrication**
BSI C 19-86. 1986 Aircraft Pressure Replenishment Connection for Engine Oil (Inch Dimensions). 4 pp.
ISO 451-76. Aircraft—Pressure Re-Oiling Connection First Edition; (Erratum—Feb 1980). 5 pp.

—**Connectors—Starters—Air**
NATO STANAG 3372 ED 5 AMD 2-88. Low Pressure Air and Associated Electrical Connections for Aircraft Engine Starting. 10 pp.
NATO STANAG 3372 ED 5 AMD 3-88. Low Pressure Air and Associated Electrical Connections for Aircraft Engine Starting. 11 pp.

—**Connectors—Starting—Low Pressure**
MOD UK DSTAN 16-10-01. Low Pressure Air Starting of Aircraft Engines Issue 3; Amendment 1. 10 pp.
NATO STANAG 3947 ED 1 AMD 5-83. Ground Half Air and Electrical Connectors for Low Pressure Air Starting of Aircraft Engines. 8 pp.
NATO STANAG 3947 ED 1 AMD 6-83. Ground Half Air and Electrical Connectors for Low Pressure Air Starting of Aircraft Engines. 9 pp.

—**Control Systems Equipment**
CAA Chapter K5-7 04.74. Controls (Light Aeroplanes). 2 pp.
CAA Chapter P5-7. Control Systems (Provisional Airworthiness Requirements for Civil Powered Lift Aircraft). 6 pp.
CAA CAP 482 SUB-Part E 03.83. Power-Plant (Small Light Aeroplanes). 5 pp.

—**Design**
CAA Chapter K5-1 10.92. Power Plant Installations—General (Light Aeroplanes). 6 pp.

—**Design—Quality Assurance—Procurement**
MOD UK AVP 115: Part 1. Technical Procecdure for Engine Design (Superseded by Def Stan 05-124). 101 pp.

—**Electronic Control Equipment**
CAA JAR-E Section 3. Advisory Material Joint (Joint Airworthiness Requirements). 23 pp.

—**Emissions**
CAA Chapter M4-2 05.86. Test Conditions (Emissions Certification). 2 pp.

Aircraft Engines (Cont.)
—**Emissions** (Cont.)
CAA Chapter M5-1 12.86. Modifications (Emissions Certification). 1 p.

—**Emissions—Certification**
CAA Chapter M2-1 05.86. Fuel Venting Certification (Emissions Certification). 1 p.
CAA Chapter M2-2 05.86. Engine Emissions Certification (Emissions Certification). 1 p.

—**Engineering Drawings—Quality Assurance—Procurement**
MOD UK AVP 115: Part 3-01. Requirements and Definitions for Drawings, Drawing Introduction Sheet and Design Changes for Equipment Under the Design Control of DG/ENG (PE); Amendment 24 (Superseded by Def Stan 05-124). 89 pp.

—**Exhaust Gases—Smoke Content**
CAA Chapter M4-1 05.86. General (Emissions Certification). 1 p.
CAA Chapter M4-3 05.86. Smoke Emission Evaluation (Emissions Certification). 5 pp.
CAA Chapter M4-6 05.86. Test Schedules—Subsonic Engines (Emissions Certification). 1 p.
CAA Chapter M4-7 05.86. Test Schedules—Supersonic Engines (Emissions Certification). 1 p.
CAA Chapter M4-8 05.86. Calculation of Smoke Numbers (Emissions Certification). 1 p.

—**Exhaust Systems**
CAA Chapter K5-6 04.74. Exhaust Systems (Light Aeroplanes). 2 pp.
CAA Chapter P5-6. Compressed Air and Exhaust Systems (Provisional Airworthiness Requirements for Civil Powered Lift Aircraft). 2 pp.

—**Fire Extinguishers**
MOD UK DSTAN 42-25-01. Extinguisher, Fire, Vaporizing Liquid (Bromochlorodifluoro-methane (BCF)) 12 Kilogram Capacity, Skid Type-Metric Issue 1; Amendment 1. 13 pp.

—**Fire Fighting**
CAA Chapter G5-8 06.76. Fire Precautions (Rotocraft). 8 pp.

—**Fire Protection**
CAA Chapter G5-8 06.76. Fire Precautions (Rotocraft). 8 pp.
CAA Chapter K5-8 04.74. Fire Precautions (Light Aeroplanes). 3 pp.
CAA Chapter P5-8. Fire Precautions (Provisional Airworthiness Requirements for Civil Powered Lift Aircraft). 15 pp.

—**Foreign Airworthiness Directives**
CAA. Allison Series Engines. 2 pp.
CAA. Avco Lycoming Series Engines. 5 pp.
CAA. CFM International CFM56-3 Series Engines. 1 p.
CAA. Garrett Series Engines. 1 p.
CAA. General Electric Series Engines. 1 p.
CAA. Pratt and Whitney Engines. 10 pp.
CAA. Teledyne Continental Motors Series Engines. 1 p.
CAA. Textron Lycoming Series Engines. 4 pp.
CAA. Bombardier—Rotax Series Engines. 2 pp.
CAA. CFM International CFM56-5 Series Engines. 1 p.
CAA. Hirth F10 Engines (Foreign Airworthiness Directives). 1 p.
CAA. Limbach Engines (Foreign Airworthiness Directives). 1 p.
CAA. M462-RF Engines (Foreign Airworthiness Directives). 2 pp.
CAA. Microturbo APU Saphir I and II (Foreign Airworthiness Directives). 1 p.
CAA. Potez 4E20 Series Engines (Foreign Airworthiness Directives). 2 pp.
CAA. Pratt and Whitney Canada JT15D Series Engines (Foreign Airworthiness Directives). 3 pp.
CAA. Pratt and Whitney of Canada PT6 Series Engines (Foreign Airworthiness Directives). 5 pp.
CAA. Pratt and Whitney Canada PW100 Series Engines (Foreign Airworthiness Directives). 6 pp.
CAA. Pratt and Whitney of Canada ST6L Series Engines (Foreign Airworthiness Directives). 2 pp.
CAA. Renault Engines (Foreign Airworthiness Directives). 1 p.
CAA. Turbomeca Series Engines (Foreign Airworthiness Directives). 6 pp.
CAA. Walter M4-111 Engines (Foreign Airworthiness Directives). 2 pp.
CAA. Walter Minor 6-111 Engines (Foreign Airworthiness Directives). 2 pp.

—**Garrett**
CAA. Garrett Series Engines. 1 p.

—**Gas Turbine—Air Intakes—Design**
CAA Chapter K5-5 04.74. Engine Air Intake and Ice Protection Systems (Light Aeroplanes). 3 pp.

Aircraft Engines (Cont.)
—**Gas Turbine—Decarbonizing**
MOD UK TS 10127. 2-Butoxyethanol, Technical.

—**Gas Turbine—Flight Testing**
JIS W 4012-87. Tests, Ground and Flight, Aircraft Gas Turbine Propulsion System Installation.

—**Gas Turbine—Glossaries**
BSI BS 185: Sec 8-70. 1970 Glossary of Aeronautical and Astronautical Terms Section 8: Power Plant (Piston Engines, Gas Turbines and Jet Propulsion). 20 pp.

—**Gas Turbine—Ice Formation**
CAA Chapter K5-5 04.74. Engine Air Intake and Ice Protection Systems (Light Aeroplanes). 3 pp.

—**Gas Turbine—Lubricating Oils**
MOD UK DERD 2490-01. Lubricating Oil, Aircraft Turbine Engine, Petroleum Oil OM-11 NATO Code No. 0-135 Issue 2; Amendment 1. 23 pp.
MOD UK DERD 2493-01. Lubricating Oil: Aircraft Turbine Engines: Synthetic Type (D.Eng.R.D.2487) Reclaimed for Bench Test and Flight Usage Issue 3; Amendment 2. 9 pp.
MOD UK DERD 2497. Lubricating Oil, Aircraft Turbine Engine, Synthetic Issue 3.

—**Gas Turbine—Overhauling**
MOD UK DSTAN 28-8-86. General Requirements for the Overhaul of Gas Turbine Aero Engines Engine Change Units, Engine Modules and Jet Pipes Issue 1. 30 pp.
MOD UK DSTAN 28-8-93. General Requirements for the Overhaul of Aero Engines, Engine Modules, Jet Pipes and Engine Accessories Including Propellers Issue 2. 31 pp.
MOD UK DSTAN 28-8-01. General Requirements for the Overhaul of Aero Engines, Engine Modules, Jet Pipes and Engine Accessories Including Propellers Issue 2; Amendment 1. 33 pp.

—**Gas Turbine—Preservation**
MOD UK DERD 2028-73. Preservation of Gas Turbine Aero-Engines and Modules During Storage and Transit Issue 5. 5 pp.

—**Gas Turbine—Rotary Wing**
CAA 774 (G). Section G Rotorcraft Installational Assumptions Involved in Engine Certification (Blue Papers). 1 p.
CAA 775 (K). Section K Light Aeroplanes Installational Assumptions Involved in Engine Certification (Blue Papers). 1 p.
CAA Chapter G5-1 06.76. Power-Plant Installations—General (Rotorcraft). 6 pp.
CAA Chapter G5-1 App 06.76. Power-Plant Installations—General (Rotorcraft). 2 pp.

—**Gas Turbine—Supersonic Transports—Fire Protection**
CAA STANDARD NO. 6-3 03.76. Power Plant Installations: Fire Precautions. 11 pp.

—**Gas Turbine—Temperature Indicators**
NATO STANAG 3742 ED 1 AMD 6-76. Circular Dial Engine Temperature Indicators. 10 pp.
NATO STANAG 3742 ED 1 AMD 7-76. Circular Dial Engine Temperature Indicators. 11 pp.

—**Gaskets**
JIS W 1102-56. Gasket Sheet for Aircraft Engine.

—**General Electric**
CAA. General Electric Series Engines. 1 p.

—**Glossaries**
BSI BS 185: Sec 8-70. 1970 Glossary of Aeronautical and Astronautical Terms Section 8: Power Plant (Piston Engines, Gas Turbines and Jet Propulsion). 20 pp.
CAA 795 (C). Section C Engines and Propellers Propeller Requirements (Blue Papers). 22 pp.
CAA 798 (C). Section C Engines and Propellers Miscellaneous Amendments to Section C (Blue Papers). 5 pp.
JIS W 0109-77. Glossary of Terms for Aircraft (Engine).

—**Ground Effect Machines**
CAA 41.
CAA Chapter B6-2 01.91. Power Plant Installations. 3 pp.

—**Ground Effect Machines—Comprehensive Testing**
CAA Chapter B7-3 04.79. Engine Testing. 3 pp.
CAA Chapter B7-3 App 04.79. Engine Testing—Type Tests for Gas Turbine Shaft Power Engines. 7 pp.

—**Ground Effect Machines—Design**
CAA Chapter B7-2 04.79. Engine Design and Construction. 3 pp.

INDUSTRY STANDARDS

Aircraft Engines (Cont.)

—Ground Effect Machines—Turbine—Fatigue (Materials)
CAA Chapter B7-3 App 04.79. Engine Testing—Type Tests for Gas Turbine Shaft Power Engines. 7 pp.

—Ground Effect Machines—Turbine—Life (Durability)
CAA Chapter B7-2 App 04.79. Engine Design and Construction. 3 pp.

—Hirth F10
CAA. Hirth F10 Engines (Foreign Airworthiness Directives). 1 p.

—Hose Fittings
BSI C 17-73. 1973 Coupling Dimensions for Low Pressure Air Connections for Engine Starting. 4 pp.

—Hose Fittings—Engine Coolants
BSI 2C 4-73. 1973 Amd 2 Coupling Dimensions for Aero-Engine Refrigerant Pressure Replenishment Connections. 9 pp.

—ICAO Committee on Aircraft Engine Emissions
ICAO 9259. Committee on Aircraft Engine Emissions First Meeting Montreal, 12-22 June 1978; (Supplement 1 12/15/78). 195 pp.
ICAO 9304. Committee on Aircraft Engine Emissions Second Meeting Montreal, 14-29 May 1980; (Supplement 1 06/27/80). 154 pp.

—ICAO Committee on Aviation Environmental Protection
ICAO 9499. Committee on Aviation Environmental Protection First Meeting Montreal, 9-20 June 1986. 230 pp.

—Insulated Cables—Fire Resistant
BSI G 241-88. 1988 Fire Proof Electric Cables for Engine Fire Zone and Airframe Use. 18 pp.

—Jet Propulsion—Glossaries
BSI BS 185: Sec 8-70. 1970 Glossary of Aeronautical and Astronautical Terms Section 8: Power Plant (Piston Engines, Gas Turbines and Jet Propulsion). 20 pp.

—Limbach
CAA. Limbach Engines (Foreign Airworthiness Directives). 1 p.

—Limitations
CAA Chapter K7-2 10.92. Operating Information (Light Aeroplanes). 4 pp.

—Loads (Forces)
CAA Chapter K3-4 10.92. Engine and Propeller Loads (Light Aeroplanes). 5 pp.
CAA Chapter P3-4. Engine, Fan, Rotor, Propeller and Transmission System Loads (Provisional Airworthiness Requirements for Civil Powered Lift Aircraft). 10 pp.

—Logs (Reports)
CAA LEAFLET 1-5 07.90. Aircraft, Engine and Propeller Log Books. 8 pp.

—Lubricants—Spectrography
NATO STANAG 7017 ED 1 AMD 0-88. (DRAFT) Spectrographic Analysis of Aircraft Engine Lubricants. 3 pp.

—Lubrication
CAA Chapter K5-3 04.74. Oil Systems (Light Aeroplanes). 3 pp.
CAA Chapter K5-3 App 03.67. Oil Systems (Light Aeroplanes). 1 p.
CAA Chapter P5-3. Oil Systems (Provisional Airworthiness Requirements for Civil Powered Lift Aircraft). 5 pp.
CAA CAP 482 SUB-Part E 03.83. Power-Plant (Small Light Aeroplanes). 5 pp.

—Mandatory Aircraft Modifications and Inspections Summaries
CAA. Introduction. 1 p.
CAA. Alvis Leonides Engines. 1 p.
CAA. British Aerospace Dynamics Group. 13 pp.
CAA. Rolls-Royce Avon Engines. 1 p.
CAA. Rolls-Royce (Bristol) Proteus 750/760 Series Engines. 1 p.
CAA. Rolls-Royce (Bristol Siddeley) Centaurus 661 Engines. 2 pp.
CAA. Rolls-Royce (Bristol Siddeley) Cheetah Engines. 1 p.
CAA. Rolls-Royce Continental Engines. 1 p.
CAA. Rolls-Royce Conway Engines. 2 pp.
CAA. Rolls-Royce Dart Engines. 5 pp.
CAA. Rolls Royce Gem Series Engines. 1 p.
CAA. Rolls-Royce Gnome Engines. 2 pp.
CAA. Rolls-Royce M45H-501 Engines. 2 pp.

Aircraft Engines (Cont.)

—Mandatory Aircraft Modifications and Inspections Summaries (Cont.)
CAA. Rolls-Royce/SNECMA Olympus 593 Engines. 6 pp.
CAA. Rolls-Royce RB211 Engines. 15 pp.
CAA. Rolls-Royce Small Engines. 6 pp.
CAA. Rolls-Royce Spey Engines. 5 pp.
CAA. Rolls-Royce Tay Engines. 1 p.
CAA. Rolls-Royce Tyne Engines. 2 pp.
CAA. Rolls-Royce Viper Engines. 3 pp.

—Microturbo APU Saphir
CAA. Microturbo APU Saphir I and II (Foreign Airworthiness Directives). 1 p.

—Motor Oils
MOD UK DSTAN 91-40-81. Corrosion Preventive Oil, Aircraft Engine: Piston, Metallic NATO Code No: C-615 Joint Service Designation: PX-27 Issue 2. 10 pp.

—Motor Oils—Corrosion Inhibitors
MOD UK DTD-791C-72. Corrosion Preventive Oil, Aircraft Piston Engine: Static Preservation, Upper Cylinder NATO Code Number: C-613 Joint Service Designation: PX-13. 6 pp.
MOD UK DERD 2493-01. Lubricating Oil: Aircraft Turbine Engines: Synthetic Type (D.Eng.R.D.2487) Reclaimed for Bench Test and Flight Usage Issue 3; Amendment 3.
MOD UK DERD 2497. Lubricating Oil, Aircraft Turbine Engine, Synthetic Issue 3.

—Motor Oils—Pressure Replenishment
NATO STANAG 3765 ED 1 AMD 3-79. Pressure Replenishment of Aircraft Engine Oil. 5 pp.

—M462-RF
CAA. M462-RF Engines (Foreign Airworthiness Directives). 2 pp.

—Nacelles—Apertures—Fire Extinguishers
BSI C 6-66. 1966 Aircraft Engine Nacelle Fire Extinguisher Doors. 1 p.
ISO 1021-80. Aircraft—Engine Nacelle Fire Extinguisher Apertures and Doors First Edition. 3 pp.

—Nondestructive Testing
CAA NOTICE #94 ISSUE 6. Non-Destructive Testing of Aircraft, Engines and Components (Airworthiness Notices). 8 pp.

—Numbering
BSI 2M 41-77. 1977 Methods of Numbering Propulsion Units and Components and Describing Their Direction of Rotation. 6 pp.
ISO 482-77. Aircraft—Propulsion Units and Components—Methods of Numbering and Describing Direction of Rotation First Edition. 6 pp.
NATO STANAG 3593 ED 3 AMD 9-73. Numbering of Engines and Their Associated Controls and Displays in Aircraft. 8 pp.

—Piston—Air Intakes—Design
CAA Chapter K5-5 04.74. Engine Air Intake and Ice Protection Systems (Light Aeroplanes). 3 pp.

—Piston—Glossaries
BSI BS 185: Sec 8-70. 1970 Glossary of Aeronautical and Astronautical Terms Section 8: Power Plant (Piston Engines, Gas Turbines and Jet Propulsion). 20 pp.

—Piston—Ice Formation
CAA Chapter K5-5 04.74. Engine Air Intake and Ice Protection Systems (Light Aeroplanes). 3 pp.

—Piston—Kits
CAA NOTICE #9 ISSUE 2. Piston Engines Supplied in Kit Form (Airworthiness Notices). 2 pp.

—Piston—Overhaul
CAA NOTICE #35 ISSUE 17. Light Aircraft Piston Engine Overhaul Periods (Airworthiness Notices). 7 pp.

—Piston—Rotary Wing
CAA 774 (G). Section G Rotorcraft Installational Assumptions Involved in Engine Certification (Blue Papers). 1 p.
CAA 775 (K). Section K Light Aeroplanes Installational Assumptions Involved in Engine Certification (Blue Papers). 1 p.
CAA Chapter G5-1 06.76. Power-Plant Installations—General (Rotocraft). 6 pp.
CAA Chapter G5-1 App 06.76. Power-Plant Installations—General (Rotocraft). 2 pp.

—Piston—Rotary Wing—Fire Testing
CAA Chapter G5-8 06.76. Fire Precautions (Rotocraft). 8 pp.

Aircraft Engines (Cont.)

—Piston—Spark Plugs
JIS W 4501-77. Spark Plugs for Aircraft Piston Engines (R 1983). 13 pp.

—Piston—Spark Plugs—Terminals
JIS W 4502-77. Spark Plug Terminals for Aircraft Piston Engines.

—Pneumatic Systems
CAA Chapter P5-6. Compressed Air and Exhaust Systems (Provisional Airworthiness Requirements for Civil Powered Lift Aircraft). 2 pp.

—Potez
CAA. Potez 4E20 Series Engines (Foreign Airworthiness Directives). 2 pp.

—Pratt and Whitney
CAA. Pratt and Whitney Engines. 10 pp.

—Pratt and Whitney of Canada
CAA. Pratt and Whitney Canada JT15D Series Engines (Foreign Airworthiness Directives). 3 pp.
CAA. Pratt and Whitney of Canada PT6 Series Engines (Foreign Airworthiness Directives). 5 pp.
CAA. Pratt and Whitney of Canada PW100 Series Engines (Foreign Airworthiness Directives). 6 pp.
CAA. Pratt and Whitney of Canada ST6L Series Engines (Foreign Airworthiness Directives). 2 pp.

—Pressure Reduction
JIS W 2017-84. Environmental Control, Environmental Protection, and Engine Bleed Air Systems, Aircraft, General Specification for.

—Procurement
MOD UK DSTAN 05-124-92. Procedures for the Procurement of Aircraft Engines and Their Accessories Issue 1. 160 pp.
MOD UK DSTAN 05-124-01. Procedures for the Procurement of Aircraft Engines and Their Accessories Issue 1; Amendment 1. 161 pp.

—Procurement—Modification
MOD UK AVP 115: Part 3-01. Requirements and Definitions for Drawings, Drawing Introduction Sheet and Design Changes for Equipment Under the Design Control of DG/ENG (PE); Amendment 24 (Superseded by Def Stan 05-124). 89 pp.

—Quality Assurance—Procurement
MOD UK AVP 115. Director General (Engines) Manual of Technical Procedures and Management Practices (Superseded by Def Stan 05-124). 11 pp.
MOD UK AVP 115: Part 0-01. Approval Procedure; Amendment 25 (Superseded by Def Stan 05-124). 25 pp.

—Quality Assurance—Procurement—Management
MOD UK AVP 115. Director General (Engines) Manual of Technical Procedures and Management Practices (Superseded by Def Stan 05-124). 11 pp.
MOD UK AVP 115: Part 2-01. Management Control Procedures; Amendment 26 (Superseded by Def Stan 05-124). 59 pp.
MOD UK AVP 115: Part 5. Notes for the Guidance of Engine Project Officers (Superseded by Def Stan 05-124). 13 pp.

—Quick Release Fasteners—Accessories
BSI M 32-69. 1969 Amd 3 Dimensions for Aircraft Accessory Drives and Mounting Pads. 20 pp.
ISO 1971-75. Aircraft—Accessory Drives and Mounting Pads First Edition. 19 pp.

—Reciprocating—Fuel Air Mixture—Manual Controls
SBAC AS 3199 ISSUE 3. Knob (Mixture Control). 1 p.

—Reciprocating—Lubricating Oils
MOD UK DSTAN 91-40-81. Corrosion Preventive Oil, Aircraft Engine: Piston, Metallic NATO Code No: C-615 Joint Service Designation: PX-27 Issue 2. 10 pp.
MOD UK DTD-791C-72. Corrosion Preventive Oil, Aircraft Piston Engine: Static Preservation, Upper Cylinder NATO Code Number: C-613 Joint Service Designation: PX-13. 6 pp.

—Reciprocating—Spark Plugs—Preservation
MOD UK DERD 2027-01. Preservation of Piston Type Aero-Engines During Storage and Transit Issue 2; Amendment 1. 5 pp.

—Renault
CAA. Renault Engines (Foreign Airworthiness Directives). 1 p.

—Rolls Royce
CAA. Rolls-Royce Avon Engines. 1 p.
CAA. Rolls-Royce (Bristol) Proteus 750/760 Series Engines. 1 p.

INTERNATIONAL AND NON-U.S. NATIONAL STANDARDS
SUBJECT INDEX
Aircraft

Aircraft Engines (Cont.)

—Rolls Royce (Cont.)
CAA. Rolls-Royce (Bristol Siddeley) Centaurus 661 Engines. 2 pp.
CAA. Rolls-Royce (Bristol Siddeley) Cheetah Engines. 1 p.
CAA. Rolls-Royce Continental Engines. 1 p.
CAA. Rolls-Royce Conway Engines. 2 pp.
CAA. Rolls-Royce Dart Engines. 5 pp.
CAA. Rolls Royce Gem Series Engines. 1 p.
CAA. Rolls-Royce Gnome Engines. 2 pp.
CAA. Rolls-Royce M45H-501 Engines. 2 pp.
CAA. Rolls-Royce/SNECMA Olympus 593 Engines. 6 pp.
CAA. Rolls-Royce RB211 Engines. 15 pp.
CAA. Rolls-Royce Small Engines. 6 pp.
CAA. Rolls-Royce Spey Engines. 5 pp.
CAA. Rolls-Royce Tay Engines. 1 p.
CAA. Rolls-Royce Tyne Engines. 2 pp.
CAA. Rolls-Royce Viper Engines. 3 pp.

—Rotary Wing
CAA Chapter G4-1 App #1 01.75. Operation of Essential Services Following Power-Unit Failure (Rotocraft). 1 p.
CAA Chapter G7-2 06.66. Test Conditions—Piston and Turbine Engine Installations (Rotocraft). 3 pp.
CAA CAP 524 SUB-Part E 12.86. Power-Plant (Rotorcraft). 27 pp.
CAA CAP 524 SUB-Part G 12.86. Operating Limitations and Information (Rotorcraft). 13 pp.
CAA CAP524 SUB-Part E 12.86. (Rotorcraft). 19 pp.

—Rotary Wing—Air Intakes
CAA Chapter G5-5 06.76. Air Intake Systems (Rotocraft). 2 pp.
CAA CAP 524 SUB-Part E 12.86. Power-Plant (Rotorcraft). 27 pp.

—Rotary Wing—Control Systems and Equipment
CAA CAP 524 SUB-Part E 12.86. Power-Plant (Rotorcraft). 27 pp.

—Rotary Wing—Design
CAA 778 (G). Section G Rotorcraft Sub-Section G4—Design and Construction Chapter G4-9—Rotor and Transmission Systems (Blue Papers). 14 pp.
CAA Chapter G4-9 10.85. Rotor and Transmission Systems (Rotocraft). 10 pp.
CAA Chapter G4-9 App 10.85. Rotor and Transmission Systems (Rotocraft). 4 pp.

—Rotary Wing—Fire Protection
CAA CAP 524 SUB-Part E 12.86. Power-Plant (Rotorcraft). 27 pp.

—Rotary Wing—Loads (Forces)
CAA Chapter G3-4 06.66. Engine and Transmission Systems (Rotocraft). 1 p.

—Rotary Wing—Lubrication
CAA CAP 524 SUB-Part E 12.86. Power-Plant (Rotorcraft). 27 pp.

—Rotary Wing—Oil Systems
CAA Chapter G5-3 06.75. Oil Systems (Rotocraft). 5 pp.

—Rotary Wing—Safety
CAA 778 (G). Section G Rotorcraft Sub-Section G4—Design and Construction Chapter G4-9—Rotor and Transmission Systems (Blue Papers). 14 pp.
CAA Chapter G4-9 10.85. Rotor and Transmission Systems (Rotocraft). 10 pp.
CAA Chapter G4-9 App 10.85. Rotor and Transmission Systems (Rotocraft). 4 pp.

—Spark Plugs
MOD UK DERD 2021-01. Approved Sparking Plugs for Aero Engines in Current Use Issue 10; Amendment 1. 7 pp.

—Starters
NATO STANAG 3368 ED 2 AMD 7-68. Internal Aircraft Engine Starting Systems. 9 pp.
NATO STANAG 3368 ED 2 AMD 8-68. Internal Aircraft Engine Starting Systems. 10 pp.

—Storage
CAA LEAFLET 7-4 12.90. Storage Procedures. 5 pp.

—Supersonic Transports—Acceptance Testing
CAA STANDARD NO. 6-9 & App. Engine Unit—Acceptance Tests for Series, Overhauled or Repaired Engine Units and Modules. 5 pp.

—Supersonic Transports—Construction
CAA STANDARD NO. 6-6&AP 05.73. Engine Unit—Design and Construction. 23 pp.

—Supersonic Transports—Design
CAA STANDARD NO. 6-6&AP 05.73. Engine Unit—Design and Construction. 23 pp.

—Supersonic Transports—Glossaries
CAA STANDARD NO. 6-5&AP 05.78. Engine Unit—Applicability and Definitions. 10 pp.

—Supersonic Transports—Installation
CAA STANDARD NO. 6-0&AP 03.76. Power Plant Installation. 15 pp.

—Supersonic Transports—Loads (Forces)
CAA STANDARD NO. 4-6 08.69. Engine Mounting Loads. 3 pp.

—Supersonic Transports—Test Conditions
CAA STANDARD NO. 6-7 03.76. Engine Unit—Test Conditions. 6 pp.

—Supersonic Transports—Type Testing
CAA STANDARD NO. 6-8&AP 03.76. Engine Unit—Type Tests. 43 pp.

—Teledyne Continental
CAA. Teledyne Continental Motors Series Engines. 1 p.

—Temperature Testing
CAA Chapter P5-4. Temperature Suitability (Provisional Airworthiness Requirements for Civil Powered Lift Aircraft). 5 pp.

—Textron Lycoming
CAA. Textron Lycoming Series Engines. 4 pp.

—Threads
DIN ENGL LN 9163 Pt 5-80. Aerospace; ISO Metric Threads; Selection of Threads and Allowances for Engines (Dec). 3 pp.

—Thrust Augmentation—Refrigerant Replenishment Equipment
MOD UK DSTAN 17-2-89. Replenishment Equipment for Aero Engine Thrust Augmentation (Refrigerant Injection) Systems Issue 3. 8 pp.

—Torque
CAA Chapter G3-4 06.66. Engine and Transmission Systems (Rotocraft). 1 p.
CAA Chapter K3-4 10.92. Engine and Propeller Loads (Light Aeroplanes). 5 pp.

—Turbofan
JIS W 4601-85. General Specification of Turbojet and Turbofan Engines for Aircraft.

—Turbojet
JIS W 4601-85. General Specification of Turbojet and Turbofan Engines for Aircraft.

—Turbojet—Comprehensive Testing
CAA 792 (C). Section C Engines and Propellers Compressor Fan and Turbine Engine Shafts (Refer To Blue Paper No. C831 Amending C 792) (Blue Papers). 3 pp.
CAA 798 (C). Section C Engines and Propellers Miscellaneous Amendments to Section C (Blue Papers). 5 pp.
CAA 831 (C). Section C Engines and Propellers Amendment to Blue Paper C792—Compressor Fan and Turbine Engine Shafts (Blue Papers). 2 pp.

—Turbojet—Design
CAA 791 (C). Section C Engines and Propellers Oil/Fuel Filters with or Without by-Passes (Blue Papers). 3 pp.
CAA 798 (C). Section C Engines and Propellers Miscellaneous Amendments to Section C (Blue Papers). 5 pp.

—Turbomeca
CAA. Turbomeca Series Engines (Foreign Airworthiness Directives). 6 pp.

—Turboprop
JIS W 4606-85. Turboshaft and Turboprop Engines for Aircraft, General Specification of.

—Turboprop—Comprehensive Testing
CAA 792 (C). Section C Engines and Propellers Compressor Fan and Turbine Engine Shafts (Refer To Blue Paper No. C831 Amending C 792) (Blue Papers). 3 pp.
CAA 798 (C). Section C Engines and Propellers Miscellaneous Amendments to Section C (Blue Papers). 5 pp.
CAA 831 (C). Section C Engines and Propellers Amendment to Blue Paper C792—Compressor Fan and Turbine Engine Shafts (Blue Papers). 2 pp.

—Turboprop—Design
CAA 791 (C). Section C Engines and Propellers Oil/Fuel Filters with or Without by-Passes (Blue Papers). 3 pp.
CAA 798 (C). Section C Engines and Propellers Miscellaneous Amendments to Section C (Blue Papers). 5 pp.

—Turboshaft
JIS W 4606-85. Turboshaft and Turboprop Engines for Aircraft, General Specification of.

—Type Certificates
CAA Chapter A4-2 07.89. Type Certification of Engines and Associated Equipment. 10 pp.
CAA Chapter A4-2 App 1 09.91. Engines and Associated Equipment. 2 pp.

—Walter
CAA. Walter M4-111 Engines (Foreign Airworthiness Directives). 2 pp.
CAA. Walter Minor 6-111 Engines (Foreign Airworthiness Directives). 2 pp.

Aircraft Equipment

Scope Note: For specific types of aircraft equipment, see the subheading Aircraft under specific types of equipment *Use For:* Aeronautical Parts; Airborne Equipment; Aircraft Accessories; Aircraft Components; Aircraft Parts *See Also:* Aerospace Equipment; Air Navigation Equipment; Aircraft; Aircraft Arresting Systems; Aircraft Batteries; Aircraft Brakes; Aircraft Cabins; Aircraft Compartments; Aircraft Doors; Aircraft Engines; Aircraft Instruments; Aircraft Lighting; Aircraft Propellers; Aircraft Refueling Equipment; Aircraft Stores; Airframe Bearings; Airframes; Arrester Hooks; Avionics; Cockpit Voice Recorders; Cockpits; Control Wheels; Evacuation Equipment; Flight Control Systems; Flight Recorders; Glide Slope Guidance Systems; Landing Gear; Localizers; Pressurized Cabins; Propulsion Systems; Suspension Equipment (Aircraft); Thrust Vector Controls

BSI 3G 100: Part 0-80. 1980 Amd 1 General Requirements for Equipment for Use in Aircraft Part 0: Introduction (AMD 5366) April 30, 1987. 5 pp.
CAA Chapter K6-1 10.92. Equipment Installations—General (Light Aeroplanes). 8 pp.
MOD UK DSTAN 05-123: Part 0-01. General Information Issue 1; Amendment 55. 27 pp.
MOD UK DSTAN 05-123: Part 0-02. General Information Issue 1; Amendment 57. 40 pp.
MOD UK DSTAN 05-123: Part 0-03. General Information Issue 1; Amendment 57. 30 pp.
MOD UK DSTAN 05-123: Part 0-04. General Information Issue 1; Amendment 57. 30 pp.
SBAC RS 700 (V). General Requirements for A.S. and A.G.S. Parts.

—Acceleration Testing
BSI 3G 100:Pt 2: SUBSEC 3.6-72. 1972 Amd 1 General Requirements for Equipment for Use in Aircraft Part 2: All Equipment Section 3: Environmental Conditions Subsection 3.6: Acceleration Requirements. 7 pp.

—Bogus
CAA NOTICE #19 ISSUE 10. Problem of Bogus Parts (Airworthiness Notices). 5 pp.

—Certificate of Airworthiness
CAA NOTICE #16 ISSUE 9. Aircraft Engines, Propellers and Related Equipment Obtained from Sources Not Under the Airworthiness Control of the CAA (Airworthiness Notices). 6 pp.
CAA NOTICE #17 ISSUE 2. Acceptance of Aircraft Components (Appendices Included). 12 pp.
CAA NOTICE #18 ISSUE 6. Acceptance Standards for the Maintenance Overhaul and Repair of Second-Hand Imported Aircraft for Which a UK C of A is Sought (Airworthiness Notices). 2 pp.
CAA Chapter B5-7 09.91. Master Minimum Equipment List. 4 pp.
CAA Chapter B6-5 09.91. Minimum Equipment Lists. 1 p.
CAA Chapter B7-6 09.91. Minimum Equipment Lists. 1 p.

—Clean Rooms
CAA LEAFLET 2-2 07.90. Clean Rooms. 19 pp.

—Components—Identification Systems
DIN ENGL LN 9051 Suppl. 1-72. Identification Marking of Controlled Components in Manufacturing Aircraft and Space Vehicles; Examples (Aug). 2 pp.

—Construction
BSI 3G 100:Pt 2: Sec 1-73. 1973 General Requirements for Equipment for Use in Aircraft Part 2: All Equipment Section 1: Construction. 5 pp.

—Demand Order Repair Contracts
MOD UK DEFCON 112AP-87. Conditions of Contract Demand Order/Repair Airframes and Equipment 5/87. 5 pp.

INDUSTRY STANDARDS

INTERNATIONAL AND NON-U.S. NATIONAL STANDARDS
SUBJECT INDEX

Aircraft

Aircraft Equipment (Cont.)

—**Design**
CAA Chapter K4-1 04.74. Design and Construction—General (Light Aeroplanes). 3 pp.
CAA Chapter P4-1. General (Provisional Airworthiness Requirements for Civil Powered Lift Aircraft). 4 pp.
CAA Chapter Q4-1 12.79. Design and Construction—General (Non-Rigid Airships). 3 pp.
CAA Chapter B4-8 09.91. Design Approval of Aircraft Equipment and Accessories. 8 pp.
MOD UK DSTAN 05-123: Part 3-01. Control of Designs and Design Records Issue 1; Amendment 55. 146 pp.
MOD UK DSTAN 05-123: Part 3-02. Control of Designs and Design Records Issue 1; Amendment 55. 156 pp.
MOD UK DSTAN 05-123: Part 3-03. Control of Designs and Design Records Issue 1; Amendment 58. 155 pp.
MOD UK DSTAN 05-123: Part 3-04. Control of Designs and Design Records Issue 1; Amendment 59. 152 pp.

—**Design—Airworthiness**
CAA Chapter A4-8. Design Approval of Aircraft Equipment and Accessories. 8 pp.

—**Differential Pressure**
BSI 3G 100:Pt 2: SUBSEC 3.4-72. 1972 General Requirements for Equipment for Use in Aircraft Part 2: All Equipment Section 3: Environmental Conditions Subsection 3.4: Differential Pressure Requirements. 6 pp.

—**Documentation**
CAA LEAFLET 11-1 07.90. Materials, Parts or Appliances Approved to USA Technical Standards Orders (TSO's) for Part 21 Sub-Part 0. 3 pp.

—**Elastic Waves**
BSI 3G 100:Pt 2: SUBSEC 3.14-73. 1973 General Requirements for Equipment for Use in Aircraft Part 2: All Equipment Section 3: Environmental Conditions Subsection 3.14: Acoustic Vibration. 10 pp.

—**Electromagnetic Compatibility**
NATO STANAG 3516 ED 2 AMD 4-77. Electromagnatic Compatibility for Aircraft Electrical and Electronic Equipment. 47 pp.
NATO STANAG 3614 ED 3 AMD 0-89. Electromagnetic Compatibility (EMC) of Aircraft Systems. 9 pp.

—**Electromagnetic Compatibility—Design**
MOD UK AVP 118: Chapter 8-81. Equipment Design for Electromagnetic Compatibility. 34 pp.

—**Electromagnetic Compatibility—Installation**
MOD UK AVP 118: Chapter 7-76. Installation of Equipment and Systems. 56 pp.

—**Electromagnetic Interference**
MOD UK AVP 118-76. Guide to Electromagnetic Compatibility in Aircraft Systems. 5 pp.
MOD UK AVP 118: Chapter 1-76. Nature of Electromagnetic Interference in Aircraft. 4 pp.
MOD UK AVP 118: Chapter 3-76. Types of Interference. 3 pp.
MOD UK AVP 118: Chapter 4-76. Propagation of Interference. 4 pp.
MOD UK AVP 118: Chapter 5-76. Effects of Interference. 5 pp.
MOD UK AVP 118: Chapter 6-76. Control of Interference. 19 pp.
NATO STANAG 3516 ED 3 AMD 0-93. Electromagnetic Interference and Test Methods for Aircraft Electrical and Electronic Equipment. 51 pp.

—**Electromagnetic Interference—Measurement**
MOD UK AVP 118: Chapter 2-76. Methods of Measurement. 9 pp.

—**Environmental Conditions—Fluid Contamination**
BSI 3G 100:Pt 2: SUBSEC 3.12-91. 1991 General Requirements for Equipment for Use on Aircraft Part 2: All Equipment Section 3: Environmental Conditions Subsection 3.12: Fluid Contamination. 11 pp.

—**Environmental Conditions—Glossaries**
ISO 5843 Pt 5-90. Aerospace—List of Equivalent Terms—Part 5: Environmental and Operating Conditions for Aircraft Equipment First Edition. 39 pp.

—**Environmental Testing**
BSI 2G 229-87. (WITHDRAWN) 1987 Environmental Conditions and Test Procedures for Airborne Equipment (Superseded by BS 36 229: 1993). 4 pp.

Aircraft Equipment (Cont.)

—**Environmental Testing (Cont.)**
BSI 3G 229-93. 1993 Environmental Conditions and Test Procedures for Airborne Equipment (ISO 7137: 1992) (S). 6 pp.
EUROCAE ED-14C/RTCA DO160C 12.89. Environmental Conditions and Test Procedures for Airborne Equipment. 480 pp.
ISO 7137-92. Aircraft—Environmental Conditions and Test Procedures for Airborne Equipment Third Edition. 8 pp.
JIS W 7002-81. Environmental Conditions and Test Procedures for Airborne Equipment and Instruments.
MOD UK DTD-1085C-61. Climatic Testing of Airborne Instruments and Equipment. 3 pp.
MOD UK DTD-1085C-01. Climatic Testing of Airborne Instruments and Equipment; Amendment 1. 4 pp.
NATO STANAG 3518 ED 4 AMD 0-90. Environmental Test Methods (Withdrawn 93-06). 5 pp.

—**Explosion Proof**
BSI 3G 100:Pt 2: SUBSEC 3.5-72. 1972 General Requirements for Equipment for Use in Aircraft Part 2: All Equipment Section 3: Environmental Conditions Subsection 3.5: Explosion-Proofness. 12 pp.

—**Fabrication**
CAA Chapter K4-1 04.74. Design and Construction—General (Light Aeroplanes). 3 pp.
CAA Chapter P4-1. General (Provisional Airworthiness Requirements for Civil Powered Lift Aircraft). 4 pp.
CAA Chapter Q4-1 12.79. Design and Construction—General (Non-Rigid Airships). 3 pp.

—**Fatigue (Materials)**
CAA NOTICE #62 ISSUE 3. Fatigue Lives (Airworthiness Notices). 2 pp.

—**Fire Fighting**
CAA Chapter G5-8 06.76. Fire Precautions (Rotocraft). 8 pp.

—**Fire Protection**
CAA Chapter G5-8 06.76. Fire Precautions (Rotocraft). 8 pp.
CAA Chapter K5-8 04.74. Fire Precautions (Light Aeroplanes). 3 pp.
CAA Chapter K5-8 App 04.74. Fire Precautions (Light Aeroplanes). 1 p.

—**Fire Testing**
BSI 3G 100:Pt 2: SUBSEC 3.13-73. 1973 General Requirements for Equipment for Use in Aircraft Part 2: All Equipment Section 3: Environmental Conditions Subsection 3.13: Resistance to Fire in Designated Fire Zones. 20 pp.
ISO 2685-92. Aircraft—Environmental Conditions and Test Procedures for Airborne Equipment—Resistance to Fire in Designated Fire Zones First Edition. 33 pp.

—**Fire Zones**
ISO 2685-92. Aircraft—Environmental Conditions and Test Procedures for Airborne Equipment—Resistance to Fire in Designated Fire Zones First Edition. 33 pp.

—**Foreign Airworthiness Directives**
CAA. Instruments (Foreign Airworthiness Directives). 2 pp.
CAA. Aircraft Instruments (Foreign Airworthiness Directives). 1 p.
CAA. Aircraft Equipment (Foreign Airworthiness Directives). 21 pp.
CAA. Aircraft—General (Foreign Airworthiness Directives). 1 p.

—**Frames—Inspection—Wood**
SBAC AGS 582 ISSUE 6. Frames Inspection (Woods Type) 6.625 Inches X 6.625 Inches.
SBAC AGS 583 ISSUE 5. Frames Inspection (Woods Type) 4 Inches X 4 Inches.

—**Glossaries**
ISO 5843 Pt 5-90. Aerospace—List of Equivalent Terms—Part 5: Environmental and Operating Conditions for Aircraft Equipment First Edition. 39 pp.
ISO 5843 Pt 10-88. Aerospace—List of Equivalent Terms—Part 10: Aircraft Structure First Edition. 30 pp.

—**Humidity**
BSI 3G 100:Pt 2: SUBSEC 3.7-72. 1972 General Requirements for Equipment for Use in Aircraft Part 2: All Equipment Section 3: Environmental Conditions Subsection 3.7: Tropical Exposure. 2 pp.

Aircraft Equipment (Cont.)

—**Ice Formation**
BSI 3G 100:Pt 2: SUBSEC 3.9-72. 1972 General Requirements for Equipment for Use in Aircraft Part 2: All Equipment Section 3: Environmental Conditions Subsection 3.9: Ice Formation. 3 pp.
BSI 3G 100:Pt 2: SUBSEC 3.10-74. 1974 General Requirements for Equipment for Use in Aircraft Part 2: All Equipment Section 3: Environmental Conditions Subsection 3.10: Impact Icing. 12 pp.

—**Identification Systems**
BSI 3G 100: Part 1-73. 1973 General Requirements for Equipment for Use in Aircraft Part 1: Identification and Declarations. 8 pp.
CAA LEAFLET 3-1 07.90. Identification Marking Processes for Aircraft Parts. 11 pp.
SBAC RS 610 (V). Identification Markings of Parts and Assemblies.

—**Information Services**
MOD UK DSTAN 05-123: Part 4-01. Supply of Technical Information Issue 1; Amendment 55. 175 pp.
MOD UK DSTAN 05-123: Part 4-02. Supply of Technical Information Issue 1; Amendment 57. 188 pp.
MOD UK DSTAN 05-123: Part 4-03. Supply of Technical Information Issue 1; Amendment 58. 179 pp.
MOD UK DSTAN 05-123: Part 4-04. Supply of Technical Information Issue 1; Amendment 59. 181 pp.

—**Inspection**
MOD UK DSTAN 05-61: Part 9-87. Quality Assurance Procedural Requirements Part 9: Special Inspection Requirements for Aircraft Systems Issue 2. 7 pp.

—**Inspection Stamps**
SBAC TS 131 ISSUE 1. Guidelines for the Registration and Use of Final Inspection Stamps.

—**Life (Durability)**
CAA NOTICE #62 ISSUE 3. Fatigue Lives (Airworthiness Notices). 2 pp.

—**Liquid Penetrant Testing**
MOD UK DTD-929-57. Penetrant Methods of Flaw Detection (Reprinted December 1965). 1 p.

—**Magnetic Influence**
BSI 3G 100:Pt 2: Sec 2-72. 1972 Amd 1 General Requirements for Equipment for Use in Aircraft Part 2: All Equipment Section 2: Magnetic Influence. 3 pp.

—**Mandatory Aircraft Modifications and Inspections Summary**
CAA. Instruments and Equipment. 47 pp.

—**Master Minimum Equipment Lists**
CAA Chapter A5-7 06.90. Master Minimum Equipment Lists. 5 pp.
CAA Chapter A7-6 07.89. Master Minimum Equipment Lists and Minimum Equipment Lists. 4 pp.

—**Minimum Equipment Lists**
CAA. Contents. 1 p.
CAA. Foreword. 1 p.
CAA. Introduction. 1 p.
CAA. Appendix. 5 pp.
CAA Chapter A6-5 07.89. Minimum Equipment Lists. 2 pp.
CAA Chapter A7-6 07.89. Master Minimum Equipment Lists and Minimum Equipment Lists. 4 pp.

—**Mold Growth Testing**
BSI 3G 100:Pt 2: SUBSEC 3.3-72. 1972 Amd 1 General Requirements for Equipment for Use in Aircraft Part 2: All Equipment Section 3: Environmental Conditions Subsection 3.3: Mould Growth. 2 pp.

—**Nondestructive Testing**
CAA NOTICE #94 ISSUE 6. Non-Destructive Testing of Aircraft, Engines and Components (Airworthiness Notices). 8 pp.

—**Product Support—Manufacturers'**
EURO PSCOP/86/175 53-86. Recommended Code of Practice for Commercial Aviation Product Support by Equipment Manufacturers. 40 pp.

—**Production Methods**
MOD UK DSTAN 05-123: Part 5-01. Production Procedures Issue 1; Amendment 55. 46 pp.
MOD UK DSTAN 05-123: Part 5-02. Production Procedures Issue 1; Amendment 55. 46 pp.
MOD UK DSTAN 05-123: Part 5-03. Production Procedures Issue 1; Amendment 58. 48 pp.

INTERNATIONAL AND NON-U.S. NATIONAL STANDARDS
SUBJECT INDEX
Aircraft

Aircraft Equipment (Cont.)
—Production Methods (Cont.)
MOD UK DSTAN 05-123: Part 5-04. Production Procedures Issue 1; Amendment 59. 50 pp.

—Production Methods—Identification Systems
DIN ENGL LN 9051-72. Identification Marking of Controlled Components in Manufacturing Aircraft and Space Vehicles (Aug). 5 pp.

—Qualification Testing
SBAC TS 201 ISSUE 1. Method Change Approval for Qualified Parts.

—Quality Assurance
MOD UK DSTAN 05-123: Part 1-01. Approval Procedure and Responsibilities Issue 1; Amendment 53. 39 pp.
MOD UK DSTAN 05-123: Part 1-02. Approval Procedure and Responsibilities Issue 1; Amendment 53. 49 pp.
MOD UK DSTAN 05-123: Part 1-03. Approval Procedure and Responsibilities Issue 1; Amendment 53. 39 pp.
MOD UK DSTAN 05-123: Part 1-04. Approval Procedure and Responsibilities Issue 1; Amendment 53. 39 pp.

—Quality Assurance—Contractors'
MOD UK DSTAN 05-61: Part 4-01. Quality Assurance Procedural Requirements Part 4: Contractors Working Parties for Aeronautical Equipments Issue 1; Amendment 2. 5 pp.
MOD UK DSTAN 05-61: Part 4-91. Quality Assurance Procedural Requirements Part 4: Contractors Working Parties for Aeronautical Stores Issue 2. 10 pp.

—Radiography
CGSB CAN/CGSB-48.16-92. Radiographic Inspection of Aircraft Structures. 24 pp.

—Reclaimed
CAA NOTICE #97 ISSUE 1. Return to Service of Aircraft Items Recovered from Aircraft Involved in Accidents/Incidents (Airworthiness Notices). 3 pp.

—Research and Development
MOD UK DSTAN 05-123: Part 2-01. Development Procedures Issue 1; Amendment 55. 143 pp.
MOD UK DSTAN 05-123: Part 2-02. Development Procedures Issue 1; Amendment 57. 158 pp.
MOD UK DSTAN 05-123: Part 2-03. Development Procedures Issue 1; Amendment 57. 148 pp.
MOD UK DSTAN 05-123: Part 2-04. Development Procedures Issue 1; Amendment 57. 148 pp.

—Rotary Wing Aircraft
CAA 235 (G). Section G Rotorcraft Miscellaneous Amendments Derived from Section D, "Aeroplanes" (Blue Papers). 6 pp.
CAA 815 (G). Section G Rotorcraft Radio Altimeters (Blue Papers). 1 p.
CAA Chapter G6-1 08.82. Equipment Installations—General (Rotorcraft). 11 pp.
CAA Chapter G6-1 App 12.80. Flight Path Computing Systems (Rotorcraft). 1 p.

—Tables (Data)—Pressure/Altitude
MOD UK DSTAN 16-20-01. Standard Pressure/Altitude Relationship Tables for Aircraft Equipment (Specification) Issue 1; Amendment 1. 25 pp.

—Targets
MOD UK DTD-778A-59. Proofed Fabric for Targets. 2 pp.

—Temperature Change Testing
BSI 3G 100:Pt 2: SUBSEC 3.15-78. 1978 General Requirements for Equipment for Use in Aircraft Part 2: All Equipment Section 3: Environmental Conditions Subsection 3.15: Change of Temperature. 8 pp.

—Testing—Identification Systems
DIN ENGL LN 9051-72. Identification Marking of Controlled Components in Manufacturing Aircraft and Space Vehicles (Aug). 5 pp.

—Water Resistance Testing
BSI 3G 100:Pt 2: SUBSEC 3.11-73. 1973 General Requirements for Equipment for Use in Aircraft Part 2: All Equipment Section 3: Environmental Conditions Subsection 3.11: Waterproofness. 7 pp.

Aircraft Exits
Use: Aircraft Doors

Aircraft Fasteners
See Also: Aerospace Fasteners; Aircraft Bolts; Aircraft Nuts; Aircraft Rivets; Aircraft Screws; Aircraft Washers; Fasteners

Aircraft Fasteners (Cont.)
—Cables (Electric)
SBAC AS 2506 ISSUE 4. Flexible Strip for Cable Fastening. 1 p.
SBAC AS 4528 ISSUE 3. Flexible. Strip for Cable Fastening. 1 p.

—Connectors
SBAC TS 368 ISSUE 1. Test for Electrical Connectors Robustness of Protective Cover Attachment.

—Double Hexagonal Head
SBAC RS 683 ISSUE 1. Minimum Effective Wrench Pad Heights for Double Hexagon, Fasteners.

—Latches—Harness Release Gears
SBAC AS 2829 ISSUE 4. Catch for Harness Release Gear. 1 p.

—Locking—Brake—Spindles
SBAC AS 2950 ISSUE 2. Spindle for Brake Lock. 1 p.
SBAC AS 6602 ISSUE 2. Spindle for Brake Lock. 1 p.

—Locking—Wire
SBAC AS 4561 ISSUE 1. Wire Locking Tabs. 1 p.

—Lockplates
SBAC AGS 3421 ISSUE 3. Locking Plate.

—Lockplates—Lever Switches
SBAC AS 3132 ISSUE 2. Locking Plate—Switch Lever. 1 p.
SBAC AS 3306 ISSUE 2. Locking, Plate—Switch Lever. 1 p.

—Quick Release
SBAC AS 2434-2438 ISSUE 9. Quick-Release Pins (Ringed). 1 p.
SBAC AS 2439-2443 (V). Assembly of Quick-Release Pin. 2 pp.
SBAC AS 2444-2445 (V). Assembly of Quick-Release Pin (Ringed). 2 pp.
SBAC AS 2451-2452 ISSUE 4. Pin (Quick-Release, Ringed). 1 p.
SBAC AS 46773 (V). Pins, Quick Release, (Ball Type) Non C.R. Steel 'T' Handle Single Acting.
SBAC AS 46774 (V). Pins, Quick Release, (Ball Type)Non C.R. Steel, Ring Handle, Single Acting.
SBAC AS 46775 (V). Pins, Quick Release, (Ball Type) Non C.R. Steel 'L' Handle Single Acting.
SBAC AS 46776 (V). Pins, Quick Release, (Ball Type) C.R. Steel 'T' Handle Single Acting.
SBAC AS 46777 (V). Pins, Quick Release, (Ball Type) C.R. Steel Ring Handle Single Acting.
SBAC AS 46778 (V). Pins, Quick Release, (Ball Type) C.R. Steel 'L' Handle Single Acting.
SBAC AS 46779 (V). Pins, Quick Release, (Ball Type) Alloy Steel 'T' Handle Double Acting.
SBAC AS 46780 (V). Pin, Quick Release, (Ball Type) Alloy Steel Ring Handle, Double Acting.
SBAC AS 46781 (V). Pin, Quick Release, (Ball Type) Alloy Steel 'L' Handle, Double Acting.
SBAC TS 115 ISSUE 1. SBAC Specification for the Manufacture, Testing and Procurement of: Quick Release Pins (Ball Type) Metric.

—Quick Release—Retaining Bars
SBAC AS 2453-2457 ISSUE 3. Retaining Bar for Quick Release Pin. 1 p.
SBAC AS 2458-2459 ISSUE 3. Retaining Bar for Quick Release Pin (Ringed). 1 p.

—Quick Release—Springs (Elastic)
SBAC AS 2460-2466 ISSUE 2. Springs for Quick-Release Pins. 1 p.

—Quick Release—Washers
SBAC AS 2467 ISSUE 4. Spring Washer for Quick Release Pin. 1 p.

—Tags
SBAC AS 2961 ISSUE 3. Tab. 1 p.
SBAC AS 41089-090 ISSUE 2. Tag Attachment.

—Tags—Chains
SBAC AS 2570 ISSUE 3. Chain Tag. 1 p.

—Tags—Chains—Drain Valves
SBAC AS 3338 ISSUE 2. Chain Tag, Washer—1/4 Inch Bore Drain Valve Drain Valve. 1 p.

—Tags—Pushbutton Switches
SBAC AS 3159 ISSUE 2. Connector, Tag Push Button. 1 p.
SBAC AS 3160 ISSUE 2. Connector Tag for Push Button. 1 p.
SBAC AS 3200 ISSUE 2. Connector, Tag Push Button. 1 p.
SBAC AS 3201 ISSUE 2. Connector Tag for Push Button. 1 p.

Aircraft Fasteners (Cont.)
—Taper Pins
BSI SP 28-29-51. 1951 Amd 3 Solid Taper Pins for Aeronautical Purposes. 5 pp.
BSI SP 31-32-51. 1951 Amd 2 Split Taper Pins for Aeronautical Purposes. 5 pp.
BSI SP 65-55. 1955 Amd 1 Aluminium Alloy Taper Pins for Aeronautical Purposes. 3 pp.

Aircraft Flight Crew Compartments
Use: Aircraft Cabins—Flight Crew

Aircraft Fuel Systems
See Also: Aircraft; Aircraft Fuels; Aircraft Instruments; Aircraft Refueling Equipment; Fuel Systems; Fuel Tanks
CAA Chapter P5-2. Fuel Systems (Provisional Airworthiness Requirements for Civil Powered Lift Aircraft). 16 pp.
CAA CAP 482 SUB-Part E 03.83. Power-Plant (Small Light Aeroplanes). 5 pp.

—Antistatic—Design
NATO STANAG 3991 ED 2 AMD 1-90. Design Criteria to Minimize Generation of Static Electricity Within Aircraft Fuel Systems. 8 pp.

—Components
JIS W 3201-83. Aircraft Fuel System Components, General Specification for.

—Construction—Landing Areas
NATO STANAG 3784 ED 2 AMD 2-87. Technical Guidance for the Design and Construction of POL Installations on NATO Airfields. 14 pp.
NATO STANAG 3784 ED 3 AMD 0-92. Technical Guidance for the Design and Construction of Aviation and Ground Fuel Installations on NATO Airfields. 17 pp.

—Contamination
CAA NOTICE #21 ISSUE 3. Microbiological Contamination of Fuel Tanks of Turbine Engined Aircraft (Airworthiness Notices). 3 pp.
MOD UK DERD 2153-78. Fuel and Control Systems for Gas Turbine and Ramjet Engines: Rig Tests for Assessing Sensitivity to Abrasion and Blockage by Fuel Borne Solids Issue 4. 13 pp.

—Defuelling
NATO STANAG 3334 ED 3 AMD 1-86. Defuelling of Aircraft. 5 pp.
NATO STANAG 3334 ED 3 AMD 3-86. Defuelling of Aircraft. 5 pp.
NATO STANAG 3334 ED 3 AMD 4-86. Defuelling of Aircraft. 6 pp.
NATO STANAG 3334 ED 3 AMD 5-86. Defuelling of Aircraft. 7 pp.
NATO STANAG 3681 ED 1 AMD 6-74. Criteria for Pressure Fuelling of Aircraft. 6 pp.
NATO STANAG 3681 ED 2 AMD 0-93. Criteria for Pressure Fuelling/Defuelling of Aircraft. 7 pp.

—Design
CAA Chapter K5-2 04.74. Fuel Systems (Light Aeroplanes). 8 pp.
CAA Chapter K5-2 App 04.74. Fuel Systems (Light Aeroplanes). 1 p.

—Design—Landing Areas
NATO STANAG 3784 ED 2 AMD 2-87. Technical Guidance for the Design and Construction of POL Installations on NATO Airfields. 14 pp.
NATO STANAG 3784 ED 3 AMD 0-92. Technical Guidance for the Design and Construction of Aviation and Ground Fuel Installations on NATO Airfields. 17 pp.

—Ground Effect Machines
CAA Chapter B6-5 01.91. Fuel Systems. 4 pp.

—Inspection—Shipborne
MOD UK NES 367-91. Requirements for Trials and Inspections of Aviation Fuel Systems Issue 3 (10.91). 12 pp.

—Installation
CAA Chapter K5-2 04.74. Fuel Systems (Light Aeroplanes). 8 pp.

—Knobs
SBAC RS 434 ISSUE 2. Knob for High Pressure Fuel Control, (Incorporating Re-Light Switch).

—Maintenance
NATO STANAG 3609 ED 2 AMD 2-88. Standards for Maintenance of Fixed Aviation Fuel Receipt, Storage and Dispensing Systems. 34 pp.
NATO STANAG 3609 ED 2 AMD 3-88. Standards for Maintenance of Fixed Aviation Fuel Receipt, Storage and Dispensing Systems. 35 pp.

INDUSTRY STANDARDS

Aircraft Fuel Systems (Cont.)

—Orifices—Gravity Filling
BSI 2C 13-88. 1988 Sizes of Aircraft Gravity Filling Orifices and Associated Replenishment Nozzles (Metric Series). 4 pp.
ISO 102-76. Aircraft—Gravity Filling Orifices First Edition. 3 pp.

—Rotary Wing Aircraft
CAA Chapter G5-2 08.82. Fuel Systems (Rotocraft). 11 pp.
CAA Chapter G5-2 App 08.82. Fuel Systems (Rotocraft). 1 p.
CAA CAP 524 SUB-Part E 12.86. Power-Plant (Rotorcraft). 27 pp.
CAA CAP524 SUB-Part E 12.86. (Rotorcraft). 19 pp.
NATO STANAG 2946 ED 1 AMD 2-83. Forward Area Refuelling Equipment. 8 pp.
NATO STANAG 2946 ED 1 AMD 4-83. Forward Area Refuelling Equipment. 8 pp.
NATO STANAG 2947 ED 1 AMD 3-83. Technical Criteria for a Closed Circuit Refuelling System. 11 pp.

—Supersonic Transports
CAA STANDARD NO. 6-1 07.69. Fuel Systems. 17 pp.

Aircraft Fuels
Use For: Aviation Fuels *See Also:* Aircraft; Aircraft Fuel Systems; Aircraft Gasoline; Jet Engine Fuels
JIS K 2209-91. Aviation Turbine Fuels. 9 pp.
JIS K 2276-89. Testing Methods for Aviation Fuels. 174 pp.
SBAC TS 98 ISSUE 2. Tables of Aerospace Fuels, Corrosion Preventatives, and Miscellaneous Products.

—Additives
EUROCAE ED-42 10.83. MPS for a Fuel Flowmeter to Aircraft Standards. 24 pp.

—Antiicing Additives—Safety
CAA NOTICE #23 ISSUE 1. Fuel Additives—Health Hazards (Airworthiness Notices). 2 pp.

—Antistatic Agents—Electrical Resistivity
CNS K6965-89. Method of Test for Electrical Conductivity of Aviation and Distillate Fuels Containing a Static Dissipator Additive (Oct)(12616).
ISO 6297-83. Petroleum Products—Aviation and Distillate Fuels Containing a Static Dissipator Additive—Determination of Electrical Conductivity First Edition. 6 pp.

—Contaminants
EUROCAE ED-42 10.83. MPS for a Fuel Flowmeter to Aircraft Standards. 24 pp.

—Corrosion Inhibitors
MOD UK TS 10067E. Fluid AL-38 (Superseded by Def Stan 68-150).
MOD UK DERD 2461: APL. Approved Products List of Aircraft Materials to Specification DERD 2461 Issue 1.
SBAC TS 98 ISSUE 2. Tables of Aerospace Fuels, Corrosion Preventatives, and Miscellaneous Products.

—Emissions
CAA Chapter M4-2 App 05.86. Specification for Fuel to Be Used in Aircraft Turbine Engine Emmission Testing (Emissions Certification). 1 p.

—Freezing Points
ISO 3013-74. Aviation Fuels—Determination of Freezing Point First Edition. 6 pp.

—Gum Testing—Jet Evaporation Method
BSI BS 4348-76. 1976 Determination of Existent Gum in Fuels by Jet Evaporation. 12 pp.
CEN EN 5-74. Determination of Existent Gum in Fuels by Jet Evaporation. 10 pp.
CNS K6321-72. Method of Test for Existing Gum Content in Jet and Motor Fuels (Oct)(3382).
ISO 6246-81. Petroleum Products—Motor Gasoline and Aviation Fuels—Determination of Existent Gum—Jet Evaporation Method First Edition. 10 pp.

—Heat of Combustion
ISO 3648-76. Aviation Fuels—Estimation of Net Heat of Combustion First Edition. 8 pp.

—Hose Couplings—Replenishment at Sea
NATO STANAG 1357 ED 1 AMD 0-89. NATO Standard F-44 Hose Coupling. 8 pp.
NATO STANAG 1357 ED 1 AMD 1-92. NATO Standard F-44 Hose Coupling. 9 pp.

—Moisture Content—Solubility
CNS K6544-80. Method of Test for Water Reaction of Aviation Fuels (Aug)(6189).

—Octane Number
BSI BS 2637-91. 1991 Amd 1 Determination of Knock Characteristics of Motor and Aviation-Type Fuels (Motor Method). 7 pp.

—Octane Number (Cont.)
ISO 5163-90. Motor and Aviation-Type Fuels—Determination of Knock Characteristics—Motor Method Second Edition. 6 pp.

—Oxidation Resistance
BSI BS 2000: Part 138-93. 1993 Methods of Test for Petroleum and Its Products Part 138: Determination of Oxidatioin Stability of Aviation Fuel—Potential Residue Method (W). 7 pp.
BSI BS 2000: Part 138-83. 1983 Amd 1 Petroleum and Its Products Part 138: Oxidation Stability of Aviation Fuels (Potential Residue Method). 9 pp.
CNS K6964-89. Method of Test for Oxidation Stability of Aviation Fuels (Potential Residue Method) (Oct)(12615).

—Separators/Filters
NATO STANAG 3967 ED 1 AMD 0-93. Design and Performance Requirements for Aviation Fuel Filter Separator Vessels and Coalescer and Separator Elements. 33 pp.

—Separators/Filters—Design
NATO STANAG 3967 ED 1 AMD 0-93. Design and Performance Requirements for Aviation Fuel Filter Separator Vessels and Coalescer and Separator Elements. 33 pp.

—Separators/Filters—Differential Pressure Gages
NATO STANAG 3583 ED 2 AMD 3-74. Standards of Accuracy for Differential Pressure Gauges for Aviation Fuel Filters and Filter Separators. 8 pp.
NATO STANAG 3583 ED 3 AMD 0-93. Standards of Accuracy for Differential Pressure Gauges for Aviation Fuel Filters and Filter Separator Vessels. 7 pp.
NATO STANAG 3583 ED 3 AMD 1-93. Standards of Accuracy for Differential Pressure Gauges for Aviation Fuel Filters and Filter Separator Vessels. 8 pp.

—Smoke Point
ISO 3014-81. Aviation Turbine Fuels—Determination of Smoke Point Second Edition. 8 pp.

—Thiol Content—Amperometric Titration
BSI BS 2000: Part 342-93. 1993 Methods of Test for Petroleum and Its Products Part 342: Gasoline, Kerosine and Distillate Fuels—Determination of Mercaptan Sulfur—Potentiometric Method (ISO 3012: 1991) (W). 7 pp.
CNS K6545-80. Method of Test for Mercaptan Sulfur in Aviation Turbine Fuels (Amperometric Method) (Aug)(6190).
ISO 3012-91. Gasoline, Kerosene and Distillate Fuels—Determination of Mercaptan Sulfur—Potentiometric Method Second Edition. 8 pp.

—Thiol Content—Colorimetry
CNS K6369-73. Method of Test for Mercaptan Sulfur in Aviation Fuels (Color Indicator Method) (Nov)(3478).

—Thiol Content—Potentiometric Analysis
BSI BS 2000: Part 342-93. 1993 Methods of Test for Petroleum and Its Products Part 342: Gasoline, Kerosine and Distillate Fuels—Determination of Mercaptan Sulfur—Potentiometric Method (ISO 3012: 1991) (W). 7 pp.
CNS K6546-80. Method of Test for Mercaptan Sulfur in Aviation Turbine Fuels (Potentiometric Method) (Aug)(6191).
CNS K6547-80. Method of Test for Mercaptan Sulfur in Gasoline, Kerosene, Aviation Turbine and Distillate Fuels (Potentiometric Method) (Aug)(6192).
ISO 3012-91. Gasoline, Kerosene and Distillate Fuels—Determination of Mercaptan Sulfur—Potentiometric Method Second Edition. 8 pp.

—Water Content—Solubility
ISO 6250-82. Aviation Fuels—Determination of Water Reaction First Edition. 4 pp.

—Water Reaction
DIN ENGL 51415-81. Testing of Liquid Fuels; Testing of the Behaviour of Aviation Fuels in the Presence of Water (July). 3 pp.

Aircraft Gasoline
Use For: Aviation Gasoline *See Also:* Aircraft; Aircraft Fuels; Gasoline; Jet Engine Fuels
CAA NOTICE #98 ISSUE 1. Use of Motor Gasoline (Mogas) in Certain Light Aircraft (Airworthiness Notices). 11 pp.
CAA NOTICE #98A ISSUE 1. Use of Filling Station Forecourt Motor Gasoline (Mogas) in Certain Light Aircraft. 4 pp.
CGSB CAN/CGSB-3.25-M89. Gasoline, Aviation (Grades 80, 100 and 100LL); (Amendment 1 August 1990). 13 pp.

Aircraft Gasoline (Cont.)
CNS K5029-83. Aviation Gasoline (Dec)(2259). 3 pp.
JIS K 2206-91. Aviation Gasoline. 10 pp.

—Distillation Methods
BSI BS 7392-90. 1990 Method for Determination of Distillation Characteristics of Petroleum Products. 26 pp.
ISO 3405-88. Petroleum Products—Determination of Distillation Characteristics Second Edition. 24 pp.

—Gum Testing
JIS K 2261-92. Petroleum Products—Motor Gasoline and Aviation Fuels—Determination of Existent Gum—Jet Evaporation Method. 20 pp.

—Lead Content—Volumetric Analysis
BSI BS 5290-76. 1976 Determination of Lead Content of Gasoline Volumetric Chromate Method (Supersedes BS 2878: 1976). 11 pp.
CEN EN 13-74. Determination of Lead Content of Gasoline: Volumetric Chromate Method. 9 pp.

—Piston Engines
MOD UK DERD 2485-83. Aviation Gasolines AVGAS 80 AVGAS 100LL F-18 AVGAS 100 AVGAS 115 Issue 9. 10 pp.

—Vapor Pressure
DIN ENGL 51754-83. Testing of Liquid Fuels; Determination of Vapour Pressure; Reid Method (Sept). 4 pp.

Aircraft Geometry
Use: Flight Dynamics

Aircraft Glazing
Use: Aircraft Glazing Materials

Aircraft Glazing Materials
See Also: Glazing Materials
AECMA PREN2155-80. Test Methods of Transparent Materials for Aircraft Glazing (C7/SC2/01). 4 pp.
AECMA PREN2314-84. Shock Resistant Acrylic Sheets for Aircraft Glazing (C7/SC2/08). 13 pp.

—Acrylic—Heat Resistant
AECMA PREN2342-82. Heat and Crazing Resistant Acrylic Sheets for Aircraft Glazing—Technical Specification. 12 pp.

—Acrylic Sheets
MOD UK DTD-5592A-81. Acrylic Sheets for Aircraft Glazing. 12 pp.
MOD UK DTD-5634-82. Stretched or Pressed Acrylic Sheet of Superior Environmental Resistance, for Aircraft Glazing. 15 pp.

—Cracking (Fracturing)
AECMA PREN2155-19-81. Test Methods for Transparent Materials for Aircraft Glazing—Part 19—Determination of Craze Resistance. 5 pp.
AECMA PREN2155-21-81. Test Methods for Transparent Materials for Aircraft Glazing—Part 21—Determination of Resistance to Crack Propagation Less Than Factor Greater Than. 7 pp.
BSI BS EN 2155-21-89. 1989 Test Methods for Transparent Materials for Aircraft Glazing Part 21: Determination of Resistance to Crack Propagation (K Factor). 11 pp.
CEN EN 2155 (Part 21)-89. Test Methods for Transparent Materials for Aircraft Glazing Part 21: Determination of Resistance to Crack Propagation (K Factor). 8 pp.

—Deflection Temperature
AECMA PREN2155-13-89. Test Methods for Transparent Materials for Aircraft Glazing Part 13—Determination of Temperature at Deflection Under Load. 8 pp.
BSI BS EN 2155-13-93. 1993 Test Methods for Transparent Materials for Aircraft Glazing Part 13: Determination of Temperature at Deflection Under Load (S). 13 pp.
CEN EN 2155-13-93. Aerospace Series Test Methods for Transparent Materials for Aircraft Glazing Part 13: Determination of Temperature at Deflection Under Load. 8 pp.
CEN PREN 2155-14-91. Test Methods for Transparent Materials for Aircraft Glazing Part 14—Determination of the 1/10 Vicat Softening Temperature. 5 pp.

—Density
AECMA PREN2155-01-84. Test Methods for Transparent Materials for Aircraft Glazing Part 1—Determination of the Density and Relative Density. 6 pp.

—Haze
AECMA PREN2155-09-81. Test Methods for Transparent Materials for Aircraft Glazing—Part 9—Determination of Haze (C7/SC2/01). 6 pp.
BSI BS EN 2155: Part 9-89. 1989 Determination of Haze. 10 pp.

INTERNATIONAL AND NON-U.S. NATIONAL STANDARDS
SUBJECT INDEX
Aircraft

Aircraft Glazing Materials *(Cont.)*

—Haze *(Cont.)*
CEN EN 2155 (Part 9)-89. Test Methods for Transparent Materials for Aircraft Glazing Part 9: Determination of Haze. 7 pp.

—Light Transmission
AECMA PREN2155-05-84. Test Methods for Transparent Materials for Aircraft Glazing Part 5—Determination of Visible Light Transmission. 7 pp.
BSI BS EN 2155-5-89. 1989 Test Methods for Transparent Materials for Aircraft Glazing Part 5: Determination of Visible Light Transmission. 11 pp.
CEN EN 2155 (Part 5)-89. Test Methods for Transparent Materials for Aircraft Glazing Part 5: Determination of Visible Light Transmission. 8 pp.

—Optical Properties
AECMA PREN2155-06-80. Test Methods for Transparent Materials for Aircraft Glazing—Part 6—Determination of Optical Defects. 4 pp.
AECMA PREN2155-07-82. Test Methods for Transparent Materials for Aircraft Glazing—Part 7—Determination of Optical Deviation. 4 pp.
AECMA PREN2155-08-80. Test Methods for Transparent Materials for Aircraft Glazing—Part 8—Determination of Optical Distortion. 4 pp.
BSI BS EN 2155-8-89. 1989 Determination of Optical Distortion. 8 pp.
CEN EN 2155 (Part 8)-89. Test Method for Transparent Materials for Aircraft Glazing Part 8: Determination of Optical Distortion. 5 pp.

—Polymethyl Methacrylate
AECMA PREN2263-79. Polymethyl Methacrylate Less Than as Cast Greater Than Sheets for Aircraft Glazing—Technical Specification (C7/SC2/05). 11 pp.

—Refractive Index
BSI BS EN 2155-3-93. 1993 Test Methods for Transparent Materials for Aircraft Glazing Part 3: Determination of Refractive Index (S). 9 pp.
CEN PREN 2155-3-91. Test Methods for Transparent Materials for Aircraft Glazing Part 3—Determination of Refractive Index. 4 pp.

—Security—Bullet Resistant
MOD UK DTD-870A-57. Bullet-Resistant Safety Glass, High Light Transmission. 3 pp.

—Thermal Expansion
AECMA PREN2155-12-81. Test Methods for Transparent Materials for Aircraft Glazing—Part 12—Determination of Linear Thermal Expansion. 6 pp.

—Thermal Stability
AECMA PREN2155-15-81. Test Methods for Transparent Materials for Aircraft Glazing—Part 15—Determination of Thermal Stability. 4 pp.

—Ultraviolet Radiation
AECMA PREN2155-04-80. Test Methods for Transparent Materials for Aircraft Glazing—Part 4—Determination of Ultra-Violet Radiation Transmission at 290 to 330 NM (C7/SC2/01). 3 pp.

—Vicat Softening Temperature
BSI BS EN 2155-14-93. 1993 Test Methods for Transparent Materials for Aircraft Glazing Part 14: Determination of the 1/10 Vicat Softening Temperature (S). 11 pp.
CEN EN 2155-14-93. Aerospace Series—Test Methods for Transparent Materials for Aircraft Glazing—Part 14: Determination of the 1/10 Vicat Softening Temperature. 6 pp.

—Vinyl Acetal Resin Sheets
MOD UK DTD-5637-85. Polyvinylbutyral for Aircraft Transparencies. 10 pp.

—Water Absorption
BSI BS EN 2155-2-93. 1993 Test Methods for Transparent Materials for Aircraft Glazing Part 2: Determination of Water Absorption (S). 9 pp.
CEN PREN 2155-2-91. Test Methods for Transparent Materials for Aircraft Glazing Part 2—Determination of Water Absorption. 4 pp.
CEN EN 2155-2-93. Aerospace Series Test Methods for Transparent Materials for Aircraft Glazing Part 2: Determination of Water Absorption. 4 pp.

Aircraft Incidents
Use: Aircraft Accidents

Aircraft Inspection
Scope Note: See the subheading Inspection under the specific type of aircraft

Aircraft Instruments
See Also: Air Flow Angle Transmitters; Air Navigation Equipment; Aircraft; Aircraft Equipment; Aircraft Fuel Systems; Airspeed Indicators;

Aircraft Instruments *(Cont.)*
See Also: (Cont.)
Altimeters; Artificial Horizons; Attitude Indicators; Automatic Direction Finders; Automatic Pilots; Ground Proximity Warning Systems; Gyrocompasses; Gyrohorizons; Head Up Displays; Horizontal Indicators; Indicating Instruments; Instrument Panels; Machmeters; Multifunction Displays; Navigational Aids; Pressure Altimeters; Rate of Climb Indicators; Static Tubes; Turn and Bank Indicators

CAA NOTICE #54 ISSUE 1. Instruments with Unusual Presentations (Airworthiness Notices). 2 pp.
CAA Chapter P6-1. Flight Information (Provisional Airworthiness Requirements for Civil Powered Lift Aircraft). 5 pp.
CAA CAP 482 SUB-Part F 03.83. Equipment (Small Light Aeroplanes). 3 pp.
MOD UK DSTAN 66-34-86. General Requirements for Aircraft Instruments and Displays Machmeters Issue 1. 16 pp.

—Accuracy Testing
EUROCAE ED-26 03.79. MPS for Airborne Altitude Measurement and Coding Systems. 57 pp.

—Adapters (Fittings)
MOD UK DSTAN 47-17-78. Adaptors and Seals for Pipe Connections to Aircraft Pitot/Static Instruments and Associated Equipment Issue 1. 7 pp.

—Aerostats—Installation
CAA Chapter Q6-1 12.79. Equipment Installations—General (Non-Rigid Airships). 7 pp.

—Altitude
EUROCAE ED-26 03.79. MPS for Airborne Altitude Measurement and Coding Systems. 57 pp.

—Cases
MOD UK DSTAN 66-26: Part 1-77. General Requirements for Aircraft Instruments and Displays Part 1: Aircraft Instrument Cases Issue 1. 16 pp.

—Color Coding
MOD UK DSTAN 66-26: Part 7-86. General Requirements for Aircraft Instruments and Displays Part 7: Colours and Markings Used to Denote Operating Ranges of Instruments Issue 1. 11 pp.
NATO STANAG 3436 ED 4 AMD 3-85. Colours and Markings Used to Denote Operating Ranges of Aircraft Instruments. 15 pp.
NATO STANAG 3436 ED 4 AMD 4-85. Colours and Markings Used to Denote Operating Ranges of Aircraft Instruments. 16 pp.

—Color Coding—Emergency
NATO STANAG 3341 ED 3 AMD 6-72. Emergency Control Colour Schemes (Withdrawn 93-06). 6 pp.

—Design
EUROCAE ED-26 03.79. MPS for Airborne Altitude Measurement and Coding Systems. 57 pp.

—Display Devices—Engine Location
NATO STANAG 3359 ED 5 AMD 3-85. Location and Arrangement of Engine Displays in Aircraft. 6 pp.

—Display Devices—Flight Data
NATO STANAG 3216 ED 5 AMD 3-84. Layout of Flight Data in Pilots' Displays. 7 pp.
NATO STANAG 3648 ED 2 AMD 5-75. Electronically and/or Optically Generated Aircraft Displays for Fixed Wing Aircraft. 11 pp.
NATO STANAG 3648 ED 2 AMD 6-75. Electronically and/or Optically Generated Aircraft Displays for Fixed Wing Aircraft. 11 pp.

—Electromechanical
BSI 3G 101-87. 1987 General Requirements for Mechanical and Electromechanical Aircraft Indicators. 22 pp.
ISO 268-80. Aircraft—Mechanical and Electromechanical Indicators—General Requirements First Edition. 7 pp.
JIS W 6011-91. Aircraft—Mechanical and Electromechanical Indicators—General Requirements (ISO 268:1980).

—Environmental Conditions
BSI 2G 100: Part 2-62. 1962 Amd 8 General Requirements for Electrical Equipment and Indicating Instruments for Aircraft Part 2: Environmental and Operating Conditions. 14 pp.

—Environmental Testing
EUROCAE ED-26 03.79. MPS for Airborne Altitude Measurement and Coding Systems. 57 pp.

—Failure (Quality Control)
EUROCAE ED-26 03.79. MPS for Airborne Altitude Measurement and Coding Systems. 57 pp.

Aircraft Instruments *(Cont.)*

—Flow Meters—Fuel
EUROCAE ED-42 10.83. MPS for a Fuel Flowmeter to Aircraft Standards. 24 pp.

—Flow Rate—Fuels
NATO STANAG 3872 ED 1 AMD 5-79. Rate of Fuel Flow Indicators (Amd 5). 9 pp.

—Glass—Antireflection Coatings
BSI G 211-71. 1971 Amd 1 Reflection-Reducing Coating of Instrument Windows and Lighting Wedges. 6 pp.
MOD UK DSTAN 66-26: Part 3-79. General Requirements for Aircraft Instruments and Displays Part 3: Reflection-Reducing Coatings Used on Glass Elements in Aircrew Station Displays (Specification) Issue 1. 7 pp.

—Glass—Covers
MOD UK DSTAN 66-26: Part 4-79. General Requirements for Aircraft Instruments and Displays Part 4: Glass, Aircraft Instrument, Lighting Wedge and Cover (Specification) Issue 1. 9 pp.

—Gliders
CAA JAR-22 SUBPART D. Design and Construction (Joint Airworthiness Requirements). 15 pp.

—Holes
SBAC AGS 3090 ISSUE 3. 4375-20 UNJF-3B Tapped Hole for Instruments (Pressure).
SBAC AGS 3091 ISSUE 3. 500-20 UNJF-3B Tapped Hole for Instruments (Static).

—Hose Assemblies
MOD UK DSTAN 47-21-79. Flexible Hose Assemblies for Pitot and Static Systems in Aircraft (Metric) Issue 1 (Withdrawn). 4 pp.

—Identification Systems
MOD UK DSTAN 66-26: Part 7-86. General Requirements for Aircraft Instruments and Displays Part 7: Colours and Markings Used to Denote Operating Ranges of Instruments Issue 1. 11 pp.

—Indicator Lights
BSI 2G 191-88. 1988 Lighting for Aircraft Indicators Using Integral Filament Lamps. 4 pp.

—Low Voltage
BSI G 133-51. 1951 Amd 1 Electro-Magnetic Indicators for Extra Low Voltage Aircraft Systems. 4 pp.

—Magnetic
ISO 5065 Pt 1-86. Aircraft—Magnetic Indicators—Part 1: Characteristics First Edition. 9 pp.

—Magnetic—Comprehensive Testing
ISO 5065 Pt 2-86. Aircraft—Magnetic Indicators—Part 2: Tests First Edition. 9 pp.

—Mandatory Aircraft Modifications and Inspections Summary
CAA. Instruments and Equipment. 47 pp.

—Mechanical
BSI 3G 101-87. 1987 General Requirements for Mechanical and Electromechanical Aircraft Indicators. 22 pp.
ISO 268-80. Aircraft—Mechanical and Electromechanical Indicators—General Requirements First Edition. 7 pp.
JIS W 6011-91. Aircraft—Mechanical and Electromechanical Indicators—General Requirements (ISO 268:1980).
MOD UK DSTAN 66-26: Part 5-79. General Requirements for Aircraft Instruments and Displays Part 5: Principles of Presentation of Information (Specification) Issue 1. 7 pp.

—Mineral Oils
MOD UK DTD-417B-64. Lubricating Oil, Aircraft Controls: Anti-Freezing Joint Services Designation OM-150 NATO Code Number 0-140. 4 pp.

—Mounting Clamps
NATO STANAG 3492 ED 5 AMD 5-83. Clamps, Mounting (Imperial and Metric) for Aircraft Instruments. 19 pp.

—Numbering
NATO STANAG 3593 ED 3 AMD 9-73. Numbering of Engines and Their Associated Controls and Displays in Aircraft. 8 pp.

—Operating Conditions
BSI 2G 100: Part 2-62. 1962 Amd 8 General Requirements for Electrical Equipment and Indicating Instruments for Aircraft Part 2: Environmental and Operating Conditions. 14 pp.

INDUSTRY STANDARDS

INTERNATIONAL AND NON-U.S. NATIONAL STANDARDS
SUBJECT INDEX

Aircraft

Aircraft Instruments (Cont.)
—Quality Assurance
EUROCAE ED-26 03.79. MPS for Airborne Altitude Measurement and Coding Systems. 57 pp.

—Rotary Wing Aircraft
CAA CAP 524 SUB-Part F 12.86. Equipment (Rotorcraft). 17 pp.
CAA CAP 524 SUB-Part G 12.86. Operating Limitations and Information (Rotorcraft). 13 pp.

—Signs
CAA CAP 482 SUB-Part G 03.83. Operating Limitations and Information (Small Light Aeroplanes). 6 pp.

—Terminals
SBAC AGS 1759 (V). Terminals Instrument (Single).
SBAC AGS 1760 (V). Terminals, Instrument (Double).

—Tube Connectors
NATO STANAG 3323 ED 5 AMD 5-83. Pipe (Tube) Connectors to Aircraft Instruments (Imperial and Metric). 26 pp.
NATO STANAG 3323 ED 5 AMD 6-83. Pipe (Tube) Connections to Aircraft Instruments (Imperial and Metric). 27 pp.

—Washers—Screws
BSI SP 22-27-50. 1950 Amd 1 B.A. Washers for Aircraft Purposes (Primarily for Instrument Use). 4 pp.

Aircraft Interception
See Also: Aircraft
ICAO 9433. Manual Concerning Interception of Civil Aircraft (Consolidation of Current ICAO Provisions and Special Recommendations) Second Edition—1990. 66 pp.

Aircraft Landing Approaches
See Also: Air Traffic Control; Aircraft Landing Areas; Aircraft Landings; Instrument Approach Procedures

—Single Frequency
NATO STANAG 3642 ED 2 AMD 5-72. Single Frequency Approaches. 7 pp.
NATO STANAG 3642 ED 3 AMD 0-91. Single Frequency Approaches. 6 pp.
NATO STANAG 3642 ED 3 AMD 1-92. Single Frequency Approaches. 7 pp.
NATO STANAG 3642 ED 3 AMD 2-91. Single Frequency Approaches. 8 pp.

Aircraft Landing Areas
Use For: Airfields; Landing Fields; Landing Zones
See Also: Aircraft Landing Approaches; Aircraft Landings; Aircraft Marshallers; Airport Runways; Airport Taxiways; Airports; Flight Decks

—Diagrams
NATO STANAG 3970 ED 1 AMD 0-87. (DRAFT) Production Specifications for Flight Information Publications (Terminal High/Low Instrument Approach Procedures and Instrument Departure Procedures)—APATC-2. 3 pp.
NATO STANAG 3970 ED 1 AMD 0-92. Content and Format of Flight Information Publication (FLIP) Terminal High/Low Instrument Approach Procedures, Instrument Departure Procedures, and Aerodrome Diagrams/Layouts. 14 pp.
NATO STANAG 3970 ED 1 AMD 1-92. Content and Format of Flight Information Publication (FLIP) Terminal High/Low Instrument Approach Procedures, Instrument Departure Procedures, and Aerodrome Diagrams/Layouts. 15 pp.

—Fuel Systems
NATO STANAG 3784 ED 2 AMD 2-87. Technical Guidance for the Design and Construction of POL Installations on NATO Airfields. 14 pp.
NATO STANAG 3784 ED 3 AMD 0-92. Technical Guidance for the Design and Construction of Aviation and Ground Fuel Installations on NATO Airfields. 17 pp.

—Identification Systems—Camouflage
NATO STANAG 3111 ED 1 AMD 2-86. Airfield Marking Tone Down. 16 pp.
NATO STANAG 3111 ED 1 AMD 3-86. Airfield Marking Tone Down. 17 pp.

—Inertial Navigation Systems
NATO STANAG 7010 ED 1 AMD 1-89. Positions for INS Settings on Airfields. 9 pp.

—Marking
NATO STANAG 3685 ED 2 AMD 2-88. Airfield Portable Marking. 11 pp.
NATO STANAG 3685 ED 2 AMD 3-88. Airfield Portable Marking. 12 pp.

Aircraft Landing Areas (Cont.)
—Markings—Color Coding
NATO STANAG 3711 ED 1 AMD 6-76. Airfield Marking and Lighting Colour Standards. 16 pp.
NATO STANAG 3711 ED 2 AMD 0-92. Airfield Marking and Lighting Colour Standards. 16 pp.
NATO STANAG 3711 ED 2 AMD 1-92. Airfield Marking and Lighting Colour Standards. 16 pp.

—Military—Restricted Areas
NATO STANAG 2952 ED 1 AMD 3-87. Procedures for Providing Restricted Areas for NATO Military Aircraft While Using Military Airfields of Other NATO Nations. 9 pp.

—Transport Aircraft—Identification Systems
NATO STANAG 3601 ED 3 AMD 3-86. Criteria for Selection and Marking of Landing Zones for Fixed Wing Transport Aircraft. 27 pp.
NATO STANAG 3601 ED 3 AMD 4-86. Criteria for Selection and Marking of Landing Zones for Fixed Wing Transport Aircraft. 26 pp.

Aircraft Landings
Use For: Emergency Alighting See Also: Air Traffic Control; Aircraft; Aircraft Accidents; Aircraft Landing Approaches; Aircraft Landing Areas; Flight Control Systems; Flight Maneuvers; Instrument Landing Systems; Landing Gear; Takeoff
CAA Chapter P2-5. Performance—Landing (Provisional Airworthiness Requirements for Civil Powered Lift Aircraft). 6 pp.

—Aerostats
CAA Chapter Q2-6 12.79. Controllability (Non-Rigid Airships). 3 pp.

—Aerostats—Crash—Safety
CAA Chapter Q3-8 12.79. Crashworthiness Conditions (Non-Rigid Airships). 2 pp.

—Crash—Fire Fighting—Evaluation Guides
NATO STANAG 3929 ED 1 AMD 3-88. Evaluation Guide for NATO Crash/Fire/Rescue Services. 13 pp.
NATO STANAG 3929 ED 1 AMD 4-88. Evaluation Guide for NATO Crash/Fire/Rescue Services. 12 pp.

—Crash—Rescue Operations—Evaluation Guides
NATO STANAG 3929 ED 1 AMD 3-88. Evaluation Guide for NATO Crash/Fire/Rescue Services. 13 pp.
NATO STANAG 3929 ED 1 AMD 4-88. Evaluation Guide for NATO Crash/Fire/Rescue Services. 12 pp.

—Ditching—Safety
CAA Chapter P3-8. Emergency Alighting Conditions (Provisional Airworthiness Requirements for Civil Powered Lift Aircraft). 4 pp.

—Restricted Visibility
CAA Chapter P6-12. Airworthiness Requirements for Landing in Restricted Visability (Provisional Airworthiness Requirements for Civil Powered Lift Aircraft). 10 pp.

—Rotary Wing Aircraft—Crash
CAA 235 (G). Section G Rotorcraft Miscellaneous Amendments Derived from Section D, "Aeroplanes" (Blue Papers). 6 pp.
CAA 805 (G). Section G Rotocraft Crash Landing—Protection of Occupants. 3 pp.
CAA Chapter G3-8 11.75. Crash Landing Conditions (Rotocraft). 1 p.
CAA Chapter G3-8 App 02.63. Crash Landing Conditions (Rotocraft). 1 p.

—Rotary Wing Aircraft—Water
CAA 779 (G). Section G Rotorcraft Emergency Alighting on Water (Blue Papers). 4 pp.
CAA Chapter G4-10 01.75. Emergency Alighting on Water (Rotocraft). 2 pp.
CAA Chapter G4-10 App 01.75. Emergency Alighting on Water (Rotocraft). 1 p.

—Safety
CAA Chapter K3-8 04.72. Emergency Alighting Conditions (Light Aeroplanes). 2 pp.
CAA Chapter P3-8. Emergency Alighting Conditions (Provisional Airworthiness Requirements for Civil Powered Lift Aircraft). 4 pp.

—Supersonic Transports
CAA STANDARD NO. 2-4 & App. Landing. 8 pp.

—Supersonic Transports—Crash
CAA STANDARD NO. 5-5&AP 07.69. Crashworthiness and Ditching. 12 pp.

Aircraft Landings (Cont.)
—Supersonic Transports—Ditching
CAA STANDARD NO. 5-5&AP 07.69. Crashworthiness and Ditching. 12 pp.

—Supersonic Transports—Low Visibility
CAA STANDARD NO. 1-2 & App. Landing in Low Visibility Conditions. 10 pp.

Aircraft Lighting
See Also: Aircraft; Aircraft Equipment; Aircraft Safety; Anticollision Lights; Landing Lights; Lighting; Navigation Lights
CNS C4175-84. Miniature Lamps for Aircraft (Jun)(5420). 17 pp.
JIS C 7522-74. Miniature Lamps for Aircraft.

—Aerostats
CAA Chapter Q6-1 12.79. Equipment Installations—General (Non-Rigid Airships). 7 pp.

—Aircraft Cabins—Flight Crew
NATO STANAG 3224 ED 5 AMD 0-86. Aircrew Station Lighting. 20 pp.
NATO STANAG 3224 ED 5 AMD 1-86. Aircrew Station Lighting. 18 pp.
NATO STANAG 3224 ED 5 AMD 2-86. Aircrew Station Lighting. 19 pp.

—Controllers—Cockpit
BSI G 217-75. 1975 General Requirements for Cockpit Lighting Controllers. 10 pp.

—Emergency
CAA NOTICE #42 ISSUE 1. Internal Emergency Lighting Systems (Airworthiness Notices). 2 pp.
CAA NOTICE #56 ISSUE 4. Floor Proximity Emergency Escape Path Marking (Airworthiness Notices). 6 pp.

—Glossaries
JIS W 0107-87. Glossary of Terms for Aircraft Electrical and Lighting Systems.

—Ground Effect Machines
CAA Chapter B6-7 01.91. Electrical Systems. 8 pp.

—Incandescent
IEC 434-73. Aircraft Electrical Filament Lamps First Edition; (Amendment 1-1981) (Amendment 2-1984). 64 pp.

—Incandescent—Indicating Instruments
BSI 2G 191-88. 1988 Lighting for Aircraft Indicators Using Integral Filament Lamps. 4 pp.

—Incandescent—Lampholders
AECMA PREN2241-90. Lamp Base Dimensions. 16 pp.

—Lamps—Miniature
JIS C 7522-74. Miniature Lamps for Aircraft.

—Rotary Wing Aircraft
CAA CAP 524 SUB-Part F 12.86. Equipment (Rotorcraft). 17 pp.

—Supersonic Transports
CAA STANDARD NO. 7-7. External Lights. 8 pp.

Aircraft Loading Equipment
Use: Air Cargo Handling Equipment

Aircraft Maintenance Engineers
Use: Maintenance Personnel—Aircraft

Aircraft Marshallers
See Also: Aircraft; Aircraft Landing Areas; Flight Decks; Military Personnel

—Hand Signals
NATO STANAG 3117 ED 5 AMD 0-85. Aircraft Marshalling Signals. 35 pp.
NATO STANAG 3117 ED 5 AMD 3-85. Aircraft Marshalling Signals. 36 pp.

—Protective Clothing
NATO STANAG 3117 ED 5 AMD 0-85. Aircraft Marshalling Signals. 35 pp.
NATO STANAG 3117 ED 5 AMD 3-85. Aircraft Marshalling Signals. 36 pp.

Aircraft Materials
Scope Note: For materials used in aircraft applications, see the subheading Aircraft under the specific material or product

Aircraft Navigation
Use: Air Navigation

Aircraft Navigation Equipment
Use: Air Navigation Equipment

INTERNATIONAL AND NON-U.S. NATIONAL STANDARDS
SUBJECT INDEX

Aircraft

Aircraft Noise

Use For: Jet Aircraft Noise *See Also:* Aircraft; ICAO Sonic Boom Committee; Noise; Sonic Boom

BSI BS 5727-79. 1979 Method for Describing Aircraft Noise Heard on the Ground. 28 pp.
CAA 738 (K). Section K Light Aeroplanes Amendments to Performance Requirements to Achieve Consistency with Section N—Noise (Blue Papers). 2 pp.
CAA 810 (N). Section N Amendments from ICAO Annex 16 (Not Including Helicopters) and to Include Microlight Aeroplanes (Blue Papers). 36 pp.
CAA 835 (N). Section N Noise Amendments from Can 7 Changes to ICAO Annex 16 (Blue Papers). 47 pp.
CAA. Contents 08.90. 3 pp.
CAA. Foreword 08.90. 4 pp.
CAA. Explanatory Note 08.90. 1 p.
CAA Chapter N3-1 01.88. Maximum Permissible Noise Levels—Maximum Permissible Noise Levels for Subsonic Jet Aeroplanes (Noise). 3 pp.
CAA Chapter N3-1 App 08.90. Maximum Permissible Noise Levels for Subsonic Jet Aeroplanes (Noise). 2 pp.
CAA Chapter N3-3 08.90. Maximum Permissible Noise Levels for Propeller Driven Aeroplanes Whose Prototype Has a Maximum Take-Off Weight Not Exceeding 9000 kg (Noise). 3 pp.
CAA Chapter N3-3 App 08.90. Maximum Permissible Noise Levels for Propeller-Driven Aeroplanes Whose Prototype Has a Maximum Take-Off Weight Exceeding 9000 kg (Noise). 4 pp.
CAA Chapter N3-4 08.90. Maximum Permissible Noise Levels for Propeller-Driven Aeroplanes Whose Prototype Has a Maximum Take-Off Weight Not Exceeding 9000 kg Application for Certificate of Airworthiness for the Prototype Accepted Before 1 January 1988. 2 pp.
CAA Chapter N3-5 08.90. Maximum Permissible Noise Levels for Propeller-Driven Aeroplanes Having a Maximum Take-Off Weight Not Exceeding 9000 kg—Application for Certificate of Airworthiness for the Prototype or Prototype (Modified) Accepted on or After 1 January 1988. 1 p.
ICAO Annex 16 Vol I. Environmental Protection Volume I—Aircraft Noise Third Edition—July 1993 (Incorporating Amendments 1-4). 133 pp.
ICAO Circular 157. Assessment of Technological Progress Made in Reduction of Noise from Subsonic and Supersonic Jet Aeroplanes—1981. 115 pp.
ICAO 9499. Committee on Aviation Environmental Protection First Meeting Montreal, 9-20 June 1986. 230 pp.
ISO 3891-78. Acoustics—Procedure for Describing Aircraft Noise Heard on the Ground First Edition. 27 pp.

—**Acoustic Measuring Instruments**
BSI BS 5647-79. 1979 Electro-Acoustical Measuring Equipment for Aircraft Noise Certification. 16 pp.
IEC 561-76. Electro-Acoustical Measuring Equipment for Aircraft Noise Certification First Edition. 26 pp.

—**Aircraft Modifications**
CAA 810 (N). Section N Amendments from ICAO Annex 16 (Not Including Helicopters) and to Include Microlight Aeroplanes (Blue Papers). 36 pp.
CAA Chapter N2-1 08.90. Modifications Which May Effect Noise of an Aircraft (Noise). 2 pp.

—**Certification**
BSI BS 5647-79. 1979 Electro-Acoustical Measuring Equipment for Aircraft Noise Certification. 16 pp.
CAA 810 (N). Section N Amendments from ICAO Annex 16 (Not Including Helicopters) and to Include Microlight Aeroplanes (Blue Papers). 36 pp.
CAA Section 7 CAP 393. Air Navigation (Noise Certification) Order 1990. 36 pp.
CAA Chapter N1-1 08.90. Noise Type Certificates and Noise Certificates for Aircraft Designed and Constructed in the United Kingdom (Noise). 3 pp.
CAA Chapter N1-2 08.90. Noise Type Certificates and Noise Certificates for Aircraft Designed and Constructed Outside the U.K. (Noise). 2 pp.
IEC 561-76. Electro-Acoustical Measuring Equipment for Aircraft Noise Certification First Edition. 26 pp.

—**Certification—Technical Manuals**
ICAO 9501. Environmental Technical Manual on the Use of Procedures in the Noise Certification of Aircraft First Edition—1988. 42 pp.

—**Economics**
ICAO Circular 218. Economic Implications of Future Noise Restrictions on Subsonic Jet Aircraft—1989 (Corrigendum 08/01/89). 63 pp.

—**Flight Testing**
CAA 810 (N). Section N Amendments from ICAO Annex 16 (Not Including Helicopters) and to Include Microlight Aeroplanes (Blue Papers). 36 pp.
CAA Chapter N4-1 08.90. Flight Testing of Aircraft—General (Noise). 4 pp.

Aircraft Noise (Cont.)

—**Flight Testing (Cont.)**
CAA Chapter N4-2 08.90. Flight Testing of Jet Aeroplanes and of Propeller-Driven Aeroplanes Whose Prototype Has a Maximum Take-Off Weight Exceeding 9000 kg (Noise). 4 pp.
CAA Chapter N4-3 08.90. Flight Testing of Propeller-Driven Aeroplanes Having a Maximum Take-Off Weight Not Exceeding 5700 kg Flight Testing of Propeller-Driven Aeroplanes Whose Prototype Has a Maximum Take-Off Weight Not Exceeding 9000 kg—Application for. 2 pp.
CAA Chapter N4-4 08.90. Flight Testing of Propeller-Driven Aeroplanes Whose Prototype Has a Maximum Take-Off Weight Not Exceeding 9000 kg Application for Certificate of Airworthiness for Prototype or Prototype (Modified) Accepted on or After 1 January 1988. 3 pp.
CAA Chapter N4-5 08.90. Flight Testing of Helicopters (Noise). 7 pp.

—**Flight Testing—Indicating Instruments**
CAA 810 (N). Section N Amendments from ICAO Annex 16 (Not Including Helicopters) and to Include Microlight Aeroplanes (Blue Papers). 36 pp.
CAA Chapter N4-6 01.88. Instrumentation and Technique for Flight Path Measurement (Noise). 1 p.

—**Frequency Weighting**
BSI BS 5721-79. 1979 Frequency Weighting for the Measurement of Aircraft Noise (D-Weighting). 4 pp.
IEC 537-76. Frequency Weighting for the Measurement of Aircraft Noise (D-Weighting) First Edition. 9 pp.

—**Helicopters**
CAA Chapter N3-6 08.90. Maximum Permissible Noise Levels for Microlight Aeroplanes (Noise). 1 p.
CAA Chapter N3-7 08.90. Maximum Permissible Noise Levels for Helicopters (Noise). 2 pp.
CAA Chapter N3-7 App 08.90. Maximum Permissible Noise Levels for Helicopters (Noise). 1 p.

—**ICAO Committee on Aircraft Noise**
ICAO 8993. Committee on Aircraft Noise Second Meeting Montreal 15-26 November 1971; (Supplement 02/28/72). 66 pp.
ICAO 9063. Committee on Aircraft Noise Third Meeting Montreal 5-23 March 1973; (Supplement 06/15/73). 112 pp.
ICAO 9133. Committee on Aircraft Noise Fourth Meeting Montreal, 27 January—14 February 1975; (Supplement 1 03/02/75) (Supplement 2 06/02/75). 175 pp.
ICAO 9197. Committee on Aircraft Noise Fifth Meeting Montreal, 15-30 November 1976; (Supplement 04/04/77). 115 pp.
ICAO 9286. Committee on Aircraft Noise Sixth Meeting Montreal, 23 May-7 June 1979; (Supplement 11/14/79). 214 pp.
ICAO 9419. Committee on Aircraft Noise Seventh Meeting Montreal, 2-13 May 1983; (Corrigendum) (Supplement 12/05/83). 152 pp.

—**Internal**
ISO 5129-87. Acoustics—Measurement of Noise Inside Aircraft Second Edition. 7 pp.

—**Measurement**
CAA Chapter N4-1 08.90. Flight Testing of Aircraft—General (Noise). 4 pp.
CAA Chapter N4-7 08.90. Instrumentation and Technique for Measurement of Effective Perceived Noise Level in EPDdB (Noise). 8 pp.
CAA Chapter N4-7 App #1 08.90. Calculation of Effective Perceived Noise Level from Measured Noise Data (Noise). 1 p.
CAA Chapter N4-7 App #3 08.90. Guidance Material for Removing the Effects of Ambient Noise Levels from Aircraft Noise Data (Noise). 2 pp.
CAA Chapter N4-8 App 08.90. Calculation of Sound Exposure Level Lax from Measured Noise Data (Noise). 1 p.
CAA Chapter N4-9 08.90. Analysis and Adjustment of Flight Test Results for Jet and Large Propeller-Driven Aeroplanes (Noise). 3 pp.
CAA Chapter N4-9 App 08.90. Analysis and Adjustment of Flight Test Results for Jet and Large Propeller-Driven Aeroplanes (Noise). 14 pp.
CAA Chapter N4-10 08.90. Adjustment of Flight Test Results for Small Propeller-Driven Aeroplanes for Which an Application for Certificate of Airworthiness for Prototype Was Accepted Before 1 January 1988. 2 pp.
CAA Chapter N4-11 08.90. Adjustment of Flight Test Results for Small Propeller-Driven Aeroplanes for Which an Application for Certificate of Airworthiness for Prototype Was Accepted on or After 1 January 1988. 3 pp.
CAA Chapter N4-12 08.90. Adjustment of Flight Test Results for Microlight Aeroplanes. 1 p.
CAA Chapter N4-13 08.90. Analysis and Adjustment of Flight Test Results for Helicopters. 3 pp.

Aircraft Noise (Cont.)

—**Measurement (Cont.)**
CAA Chapter N4-13 App 08.90. Analysis and Adjustment of Flight Test Results for Helicopters. 4 pp.

—**Monitoring Systems**
DIN ENGL 45643 Pt 1-84. Measurement and Assessment of Aircraft Noise; Quantities and Parameters (Oct) (This Standard, Together with DIN 45643 Parts 2 and 3, October 1984 Editions, Supersedes DIN 45643, August 1974 Edition). 4 pp.
DIN ENGL 45643 Pt 2-84. Measurement and Assessment of Aircraft Noise; Aircraft Noise Monitoring Systems Within the Meaning of Article 19a of the Luftverkehrsgesetz (Civil Aviation Law) (Oct) (This Standard, Together with DIN 45643 Pts. 1 & 3, 10/84 Eds., Supsds.45643 8/74 Ed.). 5 pp.

—**Rating Levels**
DIN ENGL 45643 Pt 3-84. Measurement and Assessment of Aircraft Noise; Determination of Rating Level of Aircraft Noise Exposure (Oct) (This Standard, Together with DIN 45643 Parts 1 and 2, October 1984 Editions, Supersedes DIN 45643, August 1974 Edition). 3 pp.

—**Sound Pressure**
CAA Chapter N4-8 08.90. Instrumentation and Technique for Measurement of Weighted Overall Sound Pressure Level in dB(A) (Noise). 3 pp.

—**Supersonic Aircraft**
CAA 810 (N). Section N Amendments from ICAO Annex 16 (Not Including Helicopters) and to Include Microlight Aeroplanes (Blue Papers). 36 pp.
CAA Chapter N3-2 08.90. Maximum Permissible Noise Levels for Supersonic Aeroplanes (Noise). 1 p.

Aircraft Nuts

See Also: Aircraft Fasteners; Fasteners; Nuts (Fasteners)

CAA LEAFLET 3-4 07.90. Nuts of British Manufacture. 14 pp.
SBAC AS 3003 ISSUE 2. Ring Nut—.625 Inch Diameter. 1 p.
SBAC AS 3004 ISSUE 2. Ring Nut. 750 Inch Diameter. 1 p.
SBAC AS 3005 ISSUE 2. Ring Nut. 875 Inch Diameter. 1 p.
SBAC AS 3006 ISSUE 2. Ring Nut 1.000 Inch Diameter. 1 p.
SBAC AS 3007 ISSUE 2. Ring Nut—1.125 Inch Diameter. 1 p.
SBAC AS 3008 ISSUE 2. Ring Nut—1.250 Inch Diameter. 1 p.
SBAC AS 3009 ISSUE 2. Ring Nut—1.375 Inch Diameter. 1 p.
SBAC AS 3010 ISSUE 3. Ring Nut—1.500 Inch Diameter. 1 p.
SBAC AS 3011 ISSUE 2. Ring Nut 1.625 Inch Diameter. 1 p.
SBAC AS 3012 ISSUE 2. Ring Nut—1.750 Inch Diameter. 1 p.
SBAC AS 3013 ISSUE 2. Ring Nut—1.875 Inch Diameter. 1 p.
SBAC AS 3014 ISSUE 2. Ring Nut—2.000 Inch Diameter. 1 p.
SBAC AS 3015 ISSUE 2. Ring Nut—2.125 Inch Diameter. 1 p.
SBAC AS 3016 ISSUE 2. Ring Nut—2.250 Inch Diameter. 1 p.
SBAC AS 3017 ISSUE 2. Ring Nut—2.375 Inch Diameter. 1 p.
SBAC AS 3018 ISSUE 2. Ring Nut—2.500 Inch Diameter. 1 p.
SBAC AS 3019 ISSUE 2. Ring Nut—2.625 Inch Diameter. 1 p.
SBAC AS 3020 ISSUE 2. Ring Nut—2.750 Inch Diameter. 1 p.
SBAC AS 3021 ISSUE 2. Ring Nut—2.875 Inch Diameter. 1 p.
SBAC AS 3022 ISSUE 2. Ring Nut—3.000 Inch Diameter. 1 p.
SBAC AS 3023 ISSUE 2. Ring Nut—3.125 Inch Diameter. 1 p.
SBAC AS 3024 ISSUE 2. Ring Nut—3.250 Inch Diameter. 1 p.
SBAC AS 3025 ISSUE 2. Ring Nut—3.375 Inch Diameter. 1 p.
SBAC AS 3026 ISSUE 2. Ring Nut—3.500 Inch Diameter. 1 p.
SBAC AS 3027 ISSUE 2. Ring Nut—3.625 Inch Diameter. 1 p.
SBAC AS 3028 ISSUE 2. Ring Nut—3.750 Inch Diameter. 1 p.
SBAC AS 3029 ISSUE 2. Ring Nut—3.875 Inch Diameter. 1 p.
SBAC AS 3030 ISSUE 2. Ring Nut—4.000 Inch Diameter. 1 p.
SBAC AS 6608 ISSUE 1. Brass Nut, 4 U.N.C..
SBAC AS 6609-6610 ISSUE 1. Hand Nut.
SBAC AS 28001 ISSUE 2. Nut, 6 Point, Self Aligning.

INTERNATIONAL AND NON-U.S. NATIONAL STANDARDS
SUBJECT INDEX
Aircraft

Aircraft Nuts (Cont.)

—Anchor
BSI 2A122-2A124-81. 1981 Clinch Nuts (Unified Threads) for Aircraft. 15 pp.
SBAC TS 120 ISSUE 2. Technical Specifications for AGS 2000 Series Stiffnuts (Including Clinch Nuts) (BA and BSF Threads) for Aircraft.

—Anchor—Blind
SBAC AGS 3037 ISSUE 1. Anchor, Blind, Avlok 4 BA and No.6. U.N.C. Blind Anchor Nuts (Steel).
SBAC AGS 3038 ISSUE 1. Avlok 2 BA, No.10 UNF/No.8 U.N.C. Blind Anchor (Steel).
SBAC AGS 3039 ISSUE 1. Avlok 1/4 Inch BSF and 1/4 Inch U.N.F. Blind Anchor (Steel).
SBAC AGS 3040 ISSUE 1. Avlok 5/16 Inch BSF and 5/16 Inch U.N.F. Blind Anchor (Steel).

—Anchor—Bolt Assemblies
SBAC AS 4572 (V). Typical Assembly of Retained Bolts-Double Anchor Nuts. 2 pp.

—Anchor—Counterbore
AECMA PREN2264-82. Nuts, Anchor, Self Locking Two Lug, Floating Incremental Counterbore Classification 1100 MPa/235 Degrees Celsius. 6 pp.
AECMA PREN2752-86. Nuts, Anchor, Self Locking Fixed, Two Lug, Reduced Series, with Counterbore Classification 1100 MPa/235 Degrees Celsius. 4 pp.
AECMA PREN2753-87. Nuts, Anchor, Self Locking One Lug, with Counterbore Classification 1100 MPa/235 Degrees Celsius. 6 pp.
AECMA PREN2754-87. Nuts, Anchor, Self Locking Two Lug, with Counterbore Classification 1100 MPa/235 Degrees Celsius. 6 pp.
AECMA PREN2855-89. Nuts, Anchor, Self-Locking Fixed, 90 Degrees Corner, Reduced Series, with Counterbore Classification: 1100 MPa/235 Degrees Celsius. 6 pp.
AECMA PREN2856-89. Nuts, Anchor, Self-Locking, Fixed, Closed Corner, Reduced Series, with Counterbore Classification: 1100 MPa/235 Degrees Celsius. 6 pp.
BSI BS EN 2752-90. 1990 Nuts, Anchor, Self Locking Fixed, Two Lug, Reduced Series with Counterbore. Classification: 1100 MPa/235 Degrees C. 8 pp.
CEN EN 2752-89. Nuts, Anchor, Self Locking Fixed, Two Lug, Reduced Series, with Counterbore Classification: 1100 MPa/235 Degrees C. 5 pp.

—Anchor—Locking
SBAC AS 8602 ISSUE 4. Self-Locking Nuts, Double Lug Anchor, Steel, Cadmium Plated, 125,000 lbf/sq.in. 250 Degrees C.
SBAC AS 8603 ISSUE 4. Self-Locking Nuts, Miniature Double Lug Anchor Steel, Cadmium Plated, 125,000 lbf/sq. in. 250 Degrees C.
SBAC AS 8604 ISSUE 4. Self-Locking Nuts, Single Long Lug Anchor, Steel, Cadmium Plated, 125,000 lbf/sq.in. 250 Degrees C.
SBAC AS 8605 ISSUE 4. Self-Locking Nuts, Miniature, Single Long Lug Anchor, Steel, Cadmium Plated, 125,000 lbf/sq.in. 250 Degrees C.
SBAC AS 8606 ISSUE 4. Self-Locking Nuts, Miniature, Single Short Lug Anchor, Steel, Cadmium Plated, 125,000 lbf/sq.in. 250 Degrees C.
SBAC AS 8607 ISSUE 4. Self-Locking Nuts, Corner Anchor Steel, Cadmium Plated, 125,000 lbf/sq. in. 250 Degrees C.
SBAC AS 8608 ISSUE 4. Self-Locking Nuts, Miniature Corner Anchor, Steel, Cadmium Plated, 125,000 lbf/sq.in. 250 Degrees C.
SBAC AS 8609 ISSUE 4. Self-Locking Nuts, Double Lug Floating Anchor, Steel, Cadmium Plated, 125,000 lbf/sq.in. 250 Degrees C.
SBAC AS 8611 ISSUE 4. Self-Locking Nuts, Single Long Lug Floating Anchor Steel, Cadmium Plated, 125,000 lbf/sq.in. 250 Degrees C.
SBAC AS 8625 ISSUE 2. Self-Locking Nuts, Double Lug Anchor, CR Steel, Silver Plated, lbf/in.2 450 Degrees C.
SBAC AS 8626 ISSUE 2. Self-Locking Nuts, Miniature, Double Lug Anchor, CR Steel, Silver Plated, 125,000 lbf/in.2 450 Degrees C.
SBAC AS 8627 ISSUE 2. Self-Locking Nuts,Single Long Lug Anchor, CR Steel, Silver Plated, 125,000 lbf/in.2 450 Degrees C.
SBAC AS 8628 ISSUE 2. Self-Locking Nuts, Miniature, Single, Long Lug Anchor, CR Steel, Silver Plated, 125,000 lbf/in.2 450 Degrees C.
SBAC AS 8629 ISSUE 2. Self-Locking Nuts, Miniature, Single, Short Lug Anchor, CR Steel, Silver Plated, 125,000 lbf/in.2 450 Degrees C.
SBAC AS 8630 ISSUE 2. Self-Locking Nuts, Corner Anchor, CR Steel, Silver Plated 125,000 lbf/in.2 450 Degrees C.
SBAC AS 8631 ISSUE 2. Self-Locking Nuts, Miniature, Corner Anchor, CR Steel, Silver Plated, 125,000 lbf/in.2 450 Degrees C.
SBAC AS 8632 ISSUE 2. Self-Locking Nuts, Double Lug Anchor, CR Steel, Silver Plated, 125,000 lbf/in.2.
SBAC AS 8633 ISSUE 3(I). Self-Locking Nuts, Miniature, Double Lug Floating Anchor C.R. Steel, Silver-Plated, 125,000 lbf/in2 450 Degrees C.
SBAC AS 8634 ISSUE 2. Self-Locking Nuts, Single Long Lug Floating Anchor, CR Steel, Silver Plated, 125,000 lbf/in.2 450 Degrees C.
SBAC AS 8652 ISSUE 4. Self-Locking Nuts, Double Lug Anchor, CR Steel, 125,000 lbf/in.2 250 Degrees C.
SBAC AS 8653 ISSUE 4. Self-Locking Nuts, Miniature, Double Lug Anchor, CR Steel, 125,000 lbf/in.2 250 Degrees C.
SBAC AS 8654 ISSUE 4. Self-Locking Nuts, Single, Long Lug Anchor, CR Steel, 125,000 lbf/in.2 250 Degrees C.
SBAC AS 8655 ISSUE 4. Self-Locking Nuts, Miniature, Single Long Lug Anchor, CR Steel, 125, 000 lbf/in.2 250 Degrees C.
SBAC AS 8656 ISSUE 4. Self-Locking Nuts, Miniature, Single Short Lug Anchor, CR Steel, 125, 000 lbf/in.2 250 Degrees C.
SBAC AS 8657 ISSUE 4. Self-Locking Nuts, Corner Anchor, CR Steel, 125,000 lbf/in 250 Degrees C.
SBAC AS 8658 ISSUE 4. Self-Locking Nuts, Miniature, Corner Anchor, CR Steel, 125,000 lbf/in.2 250 Degrees C.
SBAC AS 8659 ISSUE 4. Self-Locking Nuts, Double Lug Floating Anchor, CR Steel, 125,000 lbf/in.2 250 Degrees C.
SBAC AS 8660 ISSUE 5(I). Self-Locking Nuts, Miniature, Double Lug Floating Anchor CR Steel, 125,000 lbf/in.2 250 Degrees C.
SBAC AS 8661 ISSUE 4. Self-Locking Nuts, Single, Long Lug Floating Anchor, CR Steel, 125,000 lbf/in.2 250 Degrees C.
SBAC AS 28003 ISSUE 2. Nut, Self-Locking, Non C.R. Steel, Long Lug Anchor, Cadmium Coated.

—Anchor—Locking—Countersunk
SBAC AGS 2009 ISSUE 14(I). Stiffnuts—Double Anchor—Countersunk.
SBAC AGS 2014 ISSUE 14(I). Stiffnuts—Floating Anchor—Countersunk.
SBAC AGS 2020 ISSUE 13(I). Stiffnuts—Single Anchor—Countersunk.

—Anchor—Locking—Thick
SBAC AGS 2007 ISSUE 14(I). Stiffnuts—Double Anchor—Thick.
SBAC AGS 2012 ISSUE 14(I). Stiffnuts—Floating Anchor—Thick.
SBAC AGS 2018 ISSUE 12(I). Stiffnuts—Single Anchor—Thick.

—Anchor—Locking—Thin
SBAC AGS 2008 ISSUE 14(I). Stiffnuts—Double Anchor—Thin.
SBAC AGS 2013 ISSUE 15(I). Stiffnuts—Floating Anchor—Thin.
SBAC AGS 2019 ISSUE 12(I). Stiffnuts—Single Anchor—Thin.

—Anchor—Packing Plates
SBAC AS 6753 ISSUE 1. Packing Plate for Steel Retaining Rings—Double Anchor Nuts.
SBAC AS 6754 ISSUE 1. Packing Plate for Nylon Retaining Rings—Double Anchor Nuts.
SBAC AS 6755 ISSUE 1. Packing Plate for Steel Retaining Rings—Single Anchor Nuts.
SBAC AS 6756 ISSUE 1. Packing Plate for Nylon Retaining Rings—Single Anchor Nuts.
SBAC AS 8431 ISSUE 1. Packing Plate—Plain for Use with B.S. A.125-201-Double Anchor Stiffnuts.
SBAC AS 8432 ISSUE 1. Packing Plate—Plain for Use with B.S. A.125-201-Single Anchor Stiffnuts.

—Anchor—Self Aligning
SBAC AS 28008 ISSUE 2. Nut, Anchor, Double Lug, Floating, Self-Aligning, Cadmium Coated.

—Blind—Cages
SBAC AGS 3041 ISSUE 1. Cage for Blind Nuts.
SBAC AGS 3042 ISSUE 1. Cage for Blind Nuts.
SBAC AGS 3043 (V). Cage for Blind Nuts.

—Cages
SBAC AGS 3044 ISSUE 1. Nut for AGS 3041 Cage.
SBAC AGS 3045 ISSUE 1. Nut for AGS 3042 Cage.
SBAC AGS 3046 ISSUE 1. Nut for AGS 3043 Cage.
SBAC AGS 3047 ISSUE 1. Nut for AGS 3043 Cage.

—Chains
SBAC AS 46792 ISSUE 1. Nut and Bolt Assembly (for Chains).

—Countersunk—Anchor—Locking
SBAC AGS 2009 ISSUE 14(I). Stiffnuts—Double Anchor—Countersunk.
SBAC AGS 2014 ISSUE 14(I). Stiffnuts—Floating Anchor—Countersunk.
SBAC AGS 2020 ISSUE 13(I). Stiffnuts—Single Anchor—Countersunk.

—Countersunk—Hexagonal—Locking
SBAC AGS 2003 ISSUE 14(I). Stiffnuts—Hexagon—Countersunk.

—Countersunk—Locking
SBAC AGS 2017 ISSUE 14(I). Stiffnuts—Stripnut—Countersunk.

—Double Hexagonal—Locking
SBAC AS 20620-639 ISSUE 8. Nut, Self-Locking, Extended Washer, Double Hexagon.
SBAC AS 42910-915 ISSUE 2. Nut, Self-Locking—Bi-Hexagon Waspaloy.

—Drain Valves
SBAC AGS 3033 ISSUE 1. Nut for Drain Cock.

—Fittings
SBAC AGS 3048 ISSUE 1. Fitting Data for Avlock Blind Nuts.

—Grounding Plates
SBAC AS 3116 ISSUE 1. Earthing Plate (For D.A. Nut 4 Or 2 B.A.). 1 p.

—Hexagonal
AECMA PREN2370-83. Nuts, Hexagon Steel, Cadmium Plated Classification 1100 MPa/235 Degrees Celsius. 6 pp.
AECMA PREN2372-83. Nuts, Hexagon, Thin Steel, Cadmium Plated Classification 1100 MPa/235 Degrees Celsius. 6 pp.
AECMA PREN2852-87. Nuts, Hexagonal, Plain, Normal Height, Normal Across Flats, Heat Resisting Steel Passivated Classification 1100 MPa/650 Degrees Celsius. 6 pp.
AECMA PREN2921-87. Nuts, Hexagon, Thin, Reduced Across Flats, Heat Resisting Steel, Passivated Classification 900 MPa/650 Degrees Celsius. 5 pp.
BSI 3A 107-62. 1962 Amd 1 Aluminium Alloy Nuts (Unified Hexagons and Unified Threads) (Ordinary and Slotted) for Aircraft. 5 pp.
BSI 2A 110-62. 1962 Amd 4 Steel Nuts (Unified Hexagons and Unified Threads) for Shear Bolts for Aircraft. 8 pp.
BSI 2A 210-62. 1962 Brass Nuts (Unified Hexagons and Unified Threads) for Aircraft. 4 pp.
ISO 8279-85. Aerospace—Plain Hexagon Nuts with Strength Classification 1 100 MPa and Maximum Operating Temperature 235 Degrees Celsius First Edition. 5 pp.

—Hexagonal—Bolt Assemblies—Chains
SBAC AS 168 ISSUE 4. Nut and Bolt Assembly for Chains. 1 p.

—Hexagonal—Corrosion Resistant
BSI 2A 24-62. 1962 Amd 2 Corrosion-Resisting Steel Hexagon Nuts (B.A. and B.S.F. Threads) (Ordinary, Thin, Slotted and Castle) for Aircraft (AMD 6066) May 31, 1990. 8 pp.
BSI 3A 105-62. 1962 Corrosion-Resisting Steel Nuts (Unified Hexagons and Unified Threads) (Ordinary, Thin, Slotted and Castle) for Aircraft. 6 pp.

—Hexagonal—Left Hand Thread
AECMA PREN2371-83. Nuts, Hexagon Steel, Cadmium Plated Left Hand Thread Classification 1100 MPa/235 Degrees Celsius. 6 pp.
AECMA PREN2373-83. Nuts, Hexagon, Thin Steel, Cadmium Plated Left Hand Thread Classification 1100 MPa/235 Degrees Celsius. 6 pp.

—Hexagonal—Locking
AECMA PREN2647-87. Nuts, Hexagon, Self Locking Ball Seat Classification 900 MPa/235 Degrees Celsius. 6 pp.
ISO 7995-88. Aerospace—Nuts, Hexagonal, Self-Locking, with MJ Threads, Coated or Uncoated, Classification 1 100 MPa/235 Degrees Celsius, 1 100 MPa/315 Degrees Celsius or 1 100 MPa/425 Degrees Celsius—Dimensions Second Edition. 5 pp.

—Hexagonal—Locking—Captive Washers
SBAC AS 27830-832 (V). Nut, Self-Locking, Six-Point, Captive Washer.
SBAC AS 28000 ISSUE 2. Nut, Self-Locking, Non CR Steel, Captive Washer, Cadmium Coated.

—Hexagonal—Locking—Counterbore
ISO 8538-86. Aerospace—Self-Locking Hexagon Nuts with Counterbore and Captive Washer, Classification 1 100 MPa/235 Degrees Celsius First Edition. 5 pp.
SBAC AS 52600-699 ISSUE 1. Nut—Reduced—Hexagon—Self Locking—Deep Counterbored—Metric Series.
SBAC AS 54361-54460 (V). Nut—Reduced Hexagon—Self Locking—Deep Counterbored—Inch Series.

—Hexagonal—Locking—Flanged
SBAC AS 8600 ISSUE 4. Self-Locking Nuts, Hexagon, Flange Type, Steel, Cadmium Plated, 160,000LBS/SQ.IN. 250 Degrees Celsius.

INTERNATIONAL AND NON-U.S. NATIONAL STANDARDS
SUBJECT INDEX

Aircraft

Aircraft Nuts (Cont.)
—Hexagonal—Locking—Flanged (Cont.)
SBAC AS 8623 ISSUE 2. Self-Locking Nuts, Hexagon, Flange Type, C.R. Steel, Silver Plated, 125,000 lbf/in2. 450 Degrees C.
SBAC AS 8650 ISSUE 4. Self-Locking Nuts, Hexagon, Flange Type, CR Steel, 125,000 lbf/in.2 250 Degrees C.

—Hexagonal—Locking—Thick
SBAC AGS 2001 ISSUE 14(I). Stiffnuts—Hexagon—Thick.

—Hexagonal—Locking—Thin
SBAC AGS 2002 ISSUE 14(I). Stiffnuts—Hexagon—Thin.

—Hexagonal—Pipe Fittings
SBAC AS 43099 ISSUE 1. Nut—Thin (Aluminium Alloy).

—Hexagonal—Pipe Unions
SBAC AS 43042 ISSUE 1. Elbow Assembly—90 Degree, Nipple End, Female Nut Union (Steel).
SBAC AS 43046 ISSUE 1. Tee Assembly—Nipple Branch, Female Nut Union (Steel).
SBAC AS 43048 ISSUE 1. Tee Assembly—Nipple End, Female Nut Union (Steel).
SBAC AS 43064 ISSUE 1. Elbow Assembly—135 Degree, Nipple End, Female Nut Union (Steel).
SBAC AS 43191 ISSUE 1. Reducer-Adaptor Assembly —Nipple End, Male Nut Union (Steel).
SBAC AS 43192 ISSUE 1. Reducer-Adaptor, Nipple End, Male Nut Union (Steel).
SBAC AS 43194 ISSUE 1. Nut Union, Cone Blank (Steel).
SBAC AS 46733 ISSUE 1. Union Adaptor Assembly—Nipple End Male Nut Union (Steel).
SBAC AS 46734 ISSUE 1. Union—Adaptor, Nipple End, Male Nut Union (Steel).
SBAC AS 46735 ISSUE 1. Tee Assembly—Nipple Branch, Male Nut Union (Steel).
SBAC AS 46737 ISSUE 1. Tee Assembly—Nipple End, Male Nut Union (Steel).
SBAC AS 46739 ISSUE 1. Elbow Assembly, 90 Degree—Nipple End Male Nut Union (Steel).
SBAC AS 46741 ISSUE 1. Elbow Assembly, 135 Degree Nipple End Male Nut Union (Steel).

—Hexagonal—Rudder Bars
SBAC AS 2197 ISSUE 5. Special Nut Rudder Bar. 1 p.
SBAC AS 6333 ISSUE 1. Special, Nut (Rudder Bar).
SBAC AS 6342 ISSUE 1. Cap, Nut (Rudder Bar).

—Hexagonal—Slotted
AECMA PREN3230-89. Nuts, Hexagon, Slotted/Castellated, Thin Normal Across Flats, in Steel, Cadmium Plated Classification: 900 MPa/235 Degrees Celsius. 6 pp.

—Hexagonal—Stiffnuts
SBAC AS 8624 ISSUE 2(I). Stiffnuts—Hexagon, (Standard Hexagon).
SBAC AS 8651 ISSUE 2(I). Stiffnuts—Hexagon, (Standard Hexagon).

—Hexagonal—Thin
BSI 3A 103-62. 1962 Amd 3 Steel Nuts (Unified Hexagons and Unified Threads) (Ordinary, Thin, Slotted and Castle) for Aircraft. 9 pp.
BSI 2A 210-62. 1962 Brass Nuts (Unified Hexagons and Unified Threads) for Aircraft. 4 pp.

—Hexagonal—Thin—Corrosion Resistant
BSI 2A 24-62. 1962 Amd 2 Corrosion-Resisting Steel Hexagon Nuts (B.A. and B.S.F. Threads) (Ordinary, Thin, Slotted and Castle) for Aircraft (AMD 6066) May 31, 1990. 8 pp.
BSI 3A 105-62. 1962 Corrosion-Resisting Steel Nuts (Unified Hexagons and Unified Threads) (Ordinary, Thin, Slotted and Castle) for Aircraft. 6 pp.

—Hexagonal—Union
SBAC AS 15700-711 (V). Nut, Union—Tube and Hose Coupling.
SBAC AS 43021 ISSUE 1. Nut Union (Steel).
SBAC AS 43025 ISSUE 1. Nut Union—Male (Steel).
SBAC AS 43042 ISSUE 1. Elbow Assembly—90 Degree, Nipple End, Female Nut Union (Steel).
SBAC AS 43044 ISSUE 1. Nut Union—THRUST WIRE—FEMALE (Steel).
SBAC AS 43046 ISSUE 1. Tee Assembly—Nipple Branch, Female Nut Union (Steel).
SBAC AS 43048 ISSUE 1. Tee Assembly—Nipple End, Female Nut Union (Steel).
SBAC AS 43064 ISSUE 1. Elbow Assembly—135 Degree, Nipple End, Female Nut Union (Steel).
SBAC AS 43091 ISSUE 1. Nut Union—Female (Aluminium Alloy).
SBAC AS 43095 ISSUE 1. Nut Union—Male (Aluminium Alloy).
SBAC AS 43191 ISSUE 1. Reducer-Adaptor Assembly—Nipple End, Male Nut Union (Steel).

Aircraft Nuts (Cont.)
—Hexagonal—Union (Cont.)
SBAC AS 43194 ISSUE 1. Nut Union, Cone Blank (Steel).
SBAC AS 44441 ISSUE 2. Union Nut 3/8 Inch O/D Pipe.
SBAC AS 44444 ISSUE 1. Assembly Union Nut—3/8 Inch Diameter Pipe.
SBAC AS 46733 ISSUE 1. Union Adaptor Assembly—Nipple End Male Nut Union (Steel).
SBAC AS 46734 ISSUE 1. Union—Adaptor, Nipple End, Male Nut Union (Steel).
SBAC AS 46735 ISSUE 1. Tee Assembly—Nipple Branch, Male Nut Union (Steel).
SBAC AS 46737 ISSUE 1. Tee Assembly—Nipple End, Male Nut Union (Steel).
SBAC AS 46739 ISSUE 1. Elbow Assembly, 90 Degree—Nipple End Male Nut Union (Steel).
SBAC AS 46741 ISSUE 1. Elbow Assembly, 135 Degree Nipple End Male Nut Union (Steel).

—Hexagonal—Union—Thrust Wires
SBAC AS 46730 ISSUE 1. Nut Union—Thrust Wire—Male (Steel).
SBAC AS 46731 ISSUE 1. Nut Union—Thrust Wire, Male 5,0 mm Size (Steel).

—Identification Systems
SBAC RS 440 (V). Identification of B.A. and B.S.F. Nuts and Bolts.

—Locking
BSI 3A 125-132-81. 1981 3A 136-143, 3A 147-168, 3A 180-181, 3A 186-187, 3A 192-193, 3A 200-201, 2A 213-216 (Unified Threads) for Aircraft. 30 pp.
SBAC AGS 1710 ISSUE 5. Locknut L.T. Union.

—Locking—Shank—Flange Restrained
SBAC AS 54470 ISSUE 1. Nut, Self Locking, Shank, Flange Restrained 0.1900—32 UNJF—3B BS 4084, Based on AS44750 with 1g, 2g, 3g and 6g of Additional Mass.
SBAC AS 54471 ISSUE 2. Nut, Self Locking, Shank, Flange Restrained, 0.2500—28 UNJF—3B BS4084, Based on AS44751 with 2g, 5g and 12g of Additional Mass.
SBAC AS 54472 ISSUE 2. Nut, Self Locking, Shank, Flange Restrained, 0.3125—24 UNJF—3B BS4084, Based on AS44752 with 2g, 5g and 14g of Additional Mass.
SBAC AS 54473 ISSUE 2. Nut, Self Locking, Shank, Flange Restrained, 0.3750—24 UNJF—3B BS4084, Based on AS44753 with 3g, 5g and 15g of Additional Mass.
SBAC AS 55200 ISSUE 1. Nut, Self Locking, Shank, Flange Restrained, 0.1900-32 UNJF-3B BS4084, Based on AS27852 with 1g, 2g and 3g of Additional Mass.
SBAC AS 55201 ISSUE 1. Nut, Self Locking, Shank, Flange Restrained, 0.2500-28 UNJF-3B BS4084, Based on AS27857 with 2g, 4g and 6g of Additional Mass.
SBAC AS 55203 ISSUE 1. Nut, Self Locking, Shank, Flange Restrained, 0.3750-24 UNJF-3B BS4084, Based on AS27883 with 3g, 6g and 10g of Additional Mass.

—Locking—Union
SBAC AS 52700-709 ISSUE 1. Nut, Union, Fraction, Locked (Steel).
SBAC AS 52730-739 ISSUE 3. Nut, Union, Friction, Locked (Titanium).

—Locking—Union—Fuel Filters
SBAC AS 3272 ISSUE 2. Union, Locknut—Filter Fuel Type 'B'. 1 p.

—Lockwire Holes
SBAC RS 704 (V). Recommended Size and Location of Locking Wire Holes in Bolts Screws and Nuts.

—Pipe Fittings
SBAC AGS 1148 ISSUE 10. Thin, Nuts 3/16 Inch to 1 1/2 Inch Pipe Couplings (Aluminium Alloy).

—Plate—Locking
SBAC AS 46720-722 (V). Nut, Self Locking, Plate.
SBAC TS 8 ISSUE 6. All Metal Self-Locking Nuts (UNJ Threads) 1100 MPa (160,000 lbf/in2) Minimum Tensile Strength at Room Temperature.
SBAC TS 26 ISSUE 2. All Metal Self-Locking Nuts (UNJ Threads) 1210 MPa (175,000 Lbf/in2) Minimum Tensile Strength at Ambient Temperature.
SBAC TS 128 ISSUE 3. All Metal Self-Locking Nuts (UNJ Threads) 1100 MPa (160,000 lbf/in2) Minimum Tensile Strength at Ambient Temperature.
SBAC TS 149 ISSUE 1. All Metal Self-Locking Nuts (UNJ Threads) 1550 MPa (225,000 lbf/in2) Minimum Tensile Strength at Ambient Temperature.

—Release—Drain Valves
SBAC AS 3303 ISSUE 1. Release Nut. 1 p.

Aircraft Nuts (Cont.)
—Screw Threads
SBAC RS 723 ISSUE 1. Selected Aerospace Diameter-Pitch Combinations of Metric, 60 Degree Screw Threads for Nuts and Bolts.

—Stirrup—Fuel Filters
SBAC AS 3242 ISSUE 1. Stirrup Nut Filter Fuel Type 'A'. 1 p.
SBAC AS 3270 ISSUE 1. Stirrup Nut—Filter Fuel Type 'B'. 1 p.

—Swivel—Fluid Power Equipment
AECMA PREN3250-89. Pipe Coupling 8 Degrees in Titanium Alloy Elbow, 90 Degrees Swivel Nut. 5 pp.

—Swivel—Fluid Power Equipment—Welded
AECMA PREN3252-89. Pipe Coupling 8 Degrees 30' in Titanium Alloy Elbow 90 Degrees Swivel Nut, Welded. 6 pp.

—Union—Washers
SBAC AS 44443 ISSUE 1. Washer—for 3/8 Inch Diameter Pipe Union Nut.

—Washers
SBAC AS 2568 ISSUE 2. Washer Jointing (Nut). 1 p.
SBAC AS 28000 ISSUE 2. Nut, Self-Locking, Non CR Steel, Captive Washer, Cadmium Coated.
SBAC AS 28002 ISSUE 1. Washer—for Use with AS 28001 Nuts.

—Wing
SBAC AS 2160-2162 ISSUE 3. Hand Nut. 1 p.
SBAC AS 2163-2164 ISSUE 4. Hand Nut. 1 p.
SBAC AS 6241 ISSUE 2. Hand Nut. 1 p.
SBAC AGS 3413 ISSUE 2. Wing Nuts.

—Wing—Accumulator Trays
SBAC AS 6219 ISSUE 2(I). Wing Nut—Accumulator Tray. 1 p.

—Wing—Cradles
SBAC AS 6240 ISSUE 2(I). Wing Nut—Oxygen Cradle. 1 p.

Aircraft Offenses and Criminal Acts
Use: Hijacking—Aircraft

Aircraft Operations
Use: Flight Control Systems

Aircraft Parts
Use: Aircraft Equipment

Aircraft Performance (Dynamics)
Use: Flight Dynamics

Aircraft Personnel
Use: Flight Crews

Aircraft Propellers
Use For: Airscrews *See Also:* Aircraft; Aircraft Equipment
CAA E-2(C). Section C Engines and Propellers Type Substantiation (Blue Papers). 1 p.
CAA 795 (C). Section C Engines and Propellers Propeller Requirements (Blue Papers). 22 pp.
CAA 798 (C). Section C Engines and Propellers Miscellaneous Amendments to Section C (Blue Papers). 5 pp.
CAA. Contents (Joint Airworthiness Requirements). 1 p.
CAA. Foreword (Joint Airworthiness Requirements). 1 p.
CAA. Check List of Pages (Joint Airworthiness Requirements). 2 pp.
CAA. Preambles (Joint Airworthiness Requirements). 2 pp.
CAA Chapter 3 03.92. JAA Maintenance Policy (Airworthiness Notices). 1 p.

—Aerostats
CAA 790 (Q). Section Q Non-Rigid Airships Engines Propellers and Transmissions (Amending Blue Paper 772) (Blue Papers). 4 pp.
CAA Chapter Q5-9. Engines, Transmissions and Propellers (Non-Rigid Airships). 3 pp.

—Aerostats—Loads (Forces)
CAA Chapter Q3-4 12.79. Engine and Propeller Loads (Non-Rigid Airships). 2 pp.

—Avia
CAA. Avia Propellers (Foreign Airworthiness Directives). 2 pp.

—Blades—Maintenance
CAA NOTICE #55 ISSUE 2. Routine Maintenance Propeller Blades (Airworthiness Notices). 2 pp.

—British Aerospace Dynamics Group
CAA. British Aerospace Dynamics Group. 13 pp.

INDUSTRY STANDARDS

Aircraft Propellers (Cont.)

—Certificate of Airworthiness
CAA JAR-P Section 3. Approval of Propellers and Associated Equipment (Joint Airworthiness Requirements). 4 pp.

—Clearance
CAA Chapter K5-1 10.92. Power Plant Installations—General (Light Aeroplanes). 6 pp.
CAA CAP 482 SUB-Part E 03.83. Power-Plant (Small Light Aeroplanes). 5 pp.

—Comprehensive Testing
CAA 795 (C). Section C Engines and Propellers Propeller Requirements (Blue Papers). 22 pp.

—Control Knobs
SBAC AS 6391 ISSUE 2. Knob (Propeller Control).

—Control Systems Equipment
CAA Chapter K5-7 04.74. Controls (Light Aeroplanes). 2 pp.

—Covers—Cellulose Nitrate—Sheets
MOD UK DTD-808-49. Cellulose Nitrate Sheet for Covering Wooden Propeller Blades (Reprinted November 1952). 2 pp.

—Design
CAA 795 (C). Section C Engines and Propellers Propeller Requirements (Blue Papers). 22 pp.
CAA Chapter P4-9. Rotor, Fan, Propeller and Transmission Systems (Provisional Airworthiness Requirements for Civil Powered Lift Aircraft). 3 pp.
CAA CAP 482 SUB-Part J 03.83. Propellers (Small Light Aeroplanes). 1 p.
CAA JAR-P Section 1. Relationship of JAR-P to the Basic Code and Complementary Technical Conditions (Joint Airworthiness Requirements). 10 pp.

—Direction of Movement
BSI 2M 41-77. 1977 Methods of Numbering Propulsion Units and Components and Describing Their Direction of Rotation. 6 pp.
ISO 482-77. Aircraft—Propulsion Units and Components—Methods of Numbering and Describing Direction of Rotation First Edition. 6 pp.

—Dowty Rotol
CAA. Dowty Rotol Propellers and Accessory Drive Equipment. 29 pp.

—Electronic Control Equipment
CAA JAR-E Section 3. Advisory Material Joint (Joint Airworthiness Requirements). 23 pp.

—Evra
CAA. Evra Propellers (Foreign Airworthiness Directives). 1 p.

—Fairey Reed
CAA. Fairey Reed Fixed Pitch Metal Propellers. 1 p.

—Foreign Airworthiness Directives
CAA. Hartzell Propellers (Foreign Airworthiness Directives). 1 p.
CAA. McCauley Propellers (Foreign Airworthiness Directives). 1 p.
CAA. Avia Propellers (Foreign Airworthiness Directives). 2 pp.
CAA. Evra Propellers (Foreign Airworthiness Directives). 1 p.
CAA. Hoffman Propellers (Foreign Airworthiness Directives). 1 p.
CAA. Legere Propellers (Foreign Airworthiness Directives). 1 p.
CAA. MT Propellers. 1 p.
CAA. Ratier Propellers (Foreign Airworthiness Directives). 2 pp.

—Gliders
CAA JAR-22 SUBPART E. Power Plant (Joint Airworthiness Requirements). 10 pp.
CAA JAR-22 SUBPART H. Engines (Joint Airworthiness Requirements). 3 pp.
CAA JAR-22 SUBPART J. Propellers (Joint Airworthiness Requirements). 2 pp.

—Glossaries
BSI BS 185: Sec 9-70. 1970 Glossary of Aeronautical and Astronautical Terms Section 9: Propellers. 4 pp.
CAA 795 (C). Section C Engines and Propellers Propeller Requirements (Blue Papers). 22 pp.
CAA 798 (C). Section C Engines and Propellers Miscellaneous Amendments to Section C (Blue Papers). 5 pp.

—Ground Effect Machines
CAA Chapter B7-5 04.79. Fans and Air Propellers. 3 pp.

—Hartzell
CAA. Hartzell Propellers (Foreign Airworthiness Directives). 1 p.

Aircraft Propellers (Cont.)

—Hoffman
CAA. Hoffman Propellers (Foreign Airworthiness Directives). 1 p.

—Inspection
CAA NOTICE #75 ISSUE 9. Overhaul and Inspection Requirements for Variable Pitch Propellers (Airworthiness Notices). 4 pp.

—Legere
CAA. Legere Propellers (Foreign Airworthiness Directives). 1 p.

—Light Metal Alloy—Forgings
MOD UK DTD-150A-40. Light Alloy Airscrew Forgings (Detachable Blades) (Reprinted November 1959, Incorporating Amendments Nos. 1 & 2). 4 pp.

—Loads (Forces)
CAA Chapter K3-4 10.92. Engine and Propeller Loads (Light Aeroplanes). 5 pp.
CAA Chapter P3-4. Engine, Fan, Rotor, Propeller and Transmission System Loads (Provisional Airworthiness Requirements for Civil Powered Lift Aircraft). 10 pp.

—Logs (Reports)
CAA LEAFLET 1-5 07.90. Aircraft, Engine and Propeller Log Books. 8 pp.

—Mandatory Aircraft Modifications and Inspections Summaries
CAA. Introduction. 1 p.
CAA. British Aerospace Dynamics Group. 13 pp.
CAA. Dowty Rotol Propellers and Accessory Drive Equipment. 29 pp.
CAA. Fairey Reed Fixed Pitch Metal Propellers. 1 p.
CAA. Permali-Horden Richmond Propellers. 1 p.

—Manual Controls
SBAC AS 1914 ISSUE 6. Knob (Propeller Control). 1 p.

—McCauley
CAA. McCauley Propellers (Foreign Airworthiness Directives). 1 p.

—MT
CAA. MT Propellers. 1 p.

—Overhauling
CAA NOTICE #75 ISSUE 9. Overhaul and Inspection Requirements for Variable Pitch Propellers (Airworthiness Notices). 4 pp.

—Permali-Horden Richmond
CAA. Permali-Horden Richmond Propellers. 1 p.

—Ratier
CAA. Ratier Propellers (Foreign Airworthiness Directives). 2 pp.

—Rotary Wing Aircraft
CAA Chapter G5-1 App 06.76. Power-Plant Installations—General (Rotocraft). 2 pp.

—Type Certificates
CAA Chapter A4-4 09.91. Type Certification of Propellers. 6 pp.

Aircraft Refueling Equipment
See Also: Aircraft; Aircraft Equipment; Aircraft Fuel Systems; Ground Support Equipment
NATO STANAG 3681 ED 1 AMD 6-74. Criteria for Pressure Fuelling of Aircraft. 6 pp.
NATO STANAG 3681 ED 2 AMD 0-93. Criteria for Pressure Fuelling/Defuelling of Aircraft. 7 pp.

—Connectors
BSI 3C 14-82. (WITHDRAWN) 1982 Amd 2 Coupling Requirements for Aircraft Pressure Fuelling Connection (AMD 5666) September 30, 1987 (Superseded by BS 4C 14: 1991). 7 pp.
BSI 4C 14-91. 1991 Aircraft Pressure Refuelling Connectors (ISO 45: 1990). 8 pp.
JIS W 3202-90. Aircraft-Pressure Refueling Connections (ISO 43:1976).
NATO STANAG 3681 ED 1 AMD 6-74. Criteria for Pressure Fuelling of Aircraft. 6 pp.
NATO STANAG 3681 ED 2 AMD 0-93. Criteria for Pressure Fuelling/Defuelling of Aircraft. 7 pp.

—Fuel Nozzles
BSI 2C 13-88. 1988 Sizes of Aircraft Gravity Filling Orifices and Associated Replenishment Nozzles (Metric Series). 4 pp.

—Glossaries
BSI BS 185: Sec 10-70. 1970 Glossary of Aeronautical and Astronautical Terms Section 10: Auxiliary Services (Hydraulic, Pneumatic, Electrical, Air Conditioning and Refuelling). 7 pp.

Aircraft Refueling Equipment (Cont.)

—Hazardous Locations
NATO STANAG 7013 ED 1 AMD 0-90. Aircraft Fuelling Hazard Zones. 7 pp.
NATO STANAG 7013 ED 1 AMD 1-90. Aircraft Fuelling Hazard Zones. 8 pp.
NATO STANAG 7013 ED 1 AMD 2-92. Aircraft Fuelling Hazard Zones. 9 pp.
NATO STANAG 7013 ED 1 AMD 3-90. Aircraft Fuelling Hazard Zones. 10 pp.

—Helicopters
MOD UK DSTAN 17-11-88. Forward Area Refuelling Equipment for Helicopters Issue 1. 8 pp.

—Hoses
BSI BS 3158-85. 1985 Rubber Hoses and Hose Assemblies for Aircraft Ground Fuelling and Defuelling. 19 pp.
ISO 1825-75. Rubber Hoses for Aircraft Ground Fuelling Without Static Conducting Wire First Edition. 4 pp.

—In Flight
NATO STANAG 3971 ED 1 AMD 0-90. Air-to-Air Refuelling—ATP-56. 6 pp.

—In Flight—Couplings
NATO STANAG 3447 ED 3 AMD 2-90. Aerial Refuelling Equipment Dimensional and Functional Characteristics. 10 pp.

—In Flight—Fuel Nozzles
MOD UK DSTAN 16-13-89. Reception Couplings and Nozzles for In-Flight Refuelling of Aircraft Issue 3. 9 pp.
NATO STANAG 3447 ED 3 AMD 2-90. Aerial Refuelling Equipment Dimensional and Functional Characteristics. 10 pp.

—Maintenance
NATO STANAG 3609 ED 2 AMD 2-88. Standards for Maintenance of Fixed Aviation Fuel Receipt, Storage and Dispensing Systems. 34 pp.
NATO STANAG 3609 ED 2 AMD 3-88. Standards for Maintenance of Fixed Aviation Fuel Receipt, Storage and Dispensing Systems. 35 pp.

—Shipborne
MOD UK NES 319-91. Requirements for Aviation Fuel Systems in HM Ships Issue 3 (11.91). 98 pp.

—Surges—Pressure Measurement
ISO 4153-81. Aircraft—Pressure Fuel Dispensing System—Test Procedure and Limit Value for Shut-off Surge Pressure First Edition. 6 pp.

—Tank Trucks—Fuel Nozzles—Electrical Grounding
BSI G 175-59. 1959 Amd 1 Aircraft Fuel Nozzle Grounding Plugs and Sockets. 2 pp.
ISO 46-73. Aircraft—Fuel Nozzle Grounding Plugs and Sockets First Edition. 3 pp.

Aircraft Rivets
See Also: Aircraft Fasteners; Fasteners; Rivets

—Blanking Caps
SBAC AS 2976 ISSUE 4. Chain Rivet. 1 p.

—Blind
SBAC TS 38 ISSUE 2. Rivet, Blind, Self-Plugging, Flush Break, Mechanically Locked Stem, Corrosion-Resisting Steel, Plain and Cadmium Plated AS 46788 to AS 46791.
SBAC TS 39 ISSUE 2. Rivet, Blind, Self-Plugging, Flush Break, Mechanically Locked Stem, Aluminum Alloys AS 46784 to AS 46787.

—Countersunk Head
BSI SP 86-59. 1959 100 Degree Countersunk Head Steel Rivets for Aircraft. 5 pp.
BSI 2SP 87-88-73. 1973 100 Degree Countersunk Head High Nickel-Copper Alloy Rivets for Aircraft. 10 pp.
SBAC AS 462 ISSUE 5(I). Rivet, Countersunk 90 Degree. 1 p.
SBAC AS 463 ISSUE 1. Rivet, Countersunk 120 Degree. 1 p.
SBAC AS 465 ISSUE 5(I). Rivet, Countersunk 120 Degree. 1 p.
SBAC AS 467 ISSUE 5. Rivet, Csk, 90 Degree. 1 p.
SBAC AS 2918 ISSUE 3. Close Tolerance Rivet (90 Degree Countersunk Head). 1 p.
SBAC AS 2919 ISSUE 4. Close Tolerance Rivet (120 Degree Countersunk Head). 1 p.
SBAC AS 16400-599 ISSUE 2. Rivet, Solid, 100 Degree Countersunk Head, DTD 5036, (Corrosion and Heat Resistant Steel).
SBAC AS 16600-799 ISSUE 2. Rivet, Solid, 100 Degree Countersunk Head, BS.HR 504, (Heat Resistant Alloy).

INTERNATIONAL AND NON-U.S. NATIONAL STANDARDS
SUBJECT INDEX
Aircraft

Aircraft Rivets *(Cont.)*

—**Countersunk Head** *(Cont.)*
SBAC AGS 2041 ISSUE 5. Rivets, Chobert, Countersunk Head, 120 Degree (Steel) D.T.D. 720 Countersunk Head.
SBAC AGS 2044 ISSUE 4. Rivets, Chobert, Countersunk Head, 120 Degree, Aluminium Alloy (L.37) Countersunk Head.
SBAC AGS 2046 ISSUE 6. Rivets, Chobert, 120 Degree, Aluminium Alloy Countersunk Head.
SBAC AGS 2049 ISSUE 6(I). Rivets, Tucker, Aluminium Alloy 'Pop' C'sk, Head.
SBAC AGS 2054 ISSUE 4(I). Rivets, Tucker Aluminium Alloy Cup Countersunk Head.
SBAC AGS 2056 ISSUE 3. Rivet Tucker Monel Cup, Countersunk Head.
SBAC AGS 2060 ISSUE 5. Rivets, Tucker, Monel 'Pop', Countersunk Head Unplated.
SBAC AGS 2066 ISSUE 7. Avdel Self Sealing 100 Degree Countersunk Rivets, (Aluminium Alloy).
SBAC AGS 2067 ISSUE 6. Chobert, 100 Degree Countersunk, Rivet Steel.
SBAC AGS 2068 ISSUE 5. Chobert, 100 Degree Countersunk Rivets, (Aluminium Alloy).
SBAC AGS 2069 ISSUE 4(I). Chobert 100 Degree Countersunk Rivets.
SBAC AGS 2070 ISSUE 4. Rivets, Tucker 100 Degree Monel 'Pop', Countersunk Head, (Plated).
SBAC AGS 2071 ISSUE 4. Rivets, Tucker, 100 Degree Monel 'Pop', Countersunk Head, (Unplated).
SBAC AGS 2073 ISSUE 2(I). Rivets, Tucker Aluminium Alloy 'Pop' Countersunk Head.
SBAC AGS 4716 (V)(I). Rivet Self Sealing 120 Degree Countersunk Avdel, Aluminium Alloy.
SBAC AGS 4718 (V). Rivet, Tucker 'Pop', Countersunk Head, Aluminium Alloy.

—**Countersunk Head—Blind**
SBAC AS 46786-787 (V). Rivet, Blind, Self-Plugging, Flush Break, Mechanically Locked Spindle, Aluminium Alloy, 100 Degree CSK Head.
SBAC AS 46790 ISSUE 2. Rivet, Blind, Self-Plugging, Flush Break, Mechanically Locked Stem, Corrosion-Resisting Steel, 100 Degree CSK Head, Cadmium Plated.
SBAC AS 46791 ISSUE 1. Rivet, Blind, Self-Plugging, Flush Break, Mechanically Locked Stem, Corrosion-Resisting Steel, 100 Degree CSK Head, Unplated.
SBAC AS 63250-255 ISSUE 1. Rivet Blind, Self-Plugging, Flush Break, Visible Mechanically Locked Stem, Aluminium Alloys 100 Degree Countersunk Head.
SBAC AS 63270-271 ISSUE 1. Rivet Blind, Self-Plugging, Flush Break, Visible Mechanically Locked Stem, Corrosion Resisting Steel 100 Degree Countersunk Head.
SBAC AGS 3921 ISSUE 2. Rivet Blind, 100 Degree Countersunk Head, Self Plugging, Interference Lock (Avdel), C.R. Steel.

—**Domed Head**
SBAC AGS 2048 ISSUE 6(I). Rivets, Tucker, Aluminium Alloy 'Pop', Domed Head.
SBAC AGS 2050 ISSUE 6. Rivets, Tucker, Monel, 'Pop', Domed Head.
SBAC AGS 2053 ISSUE 4(I). Rivets, Tucker Aluminium Alloy Cup Domed Head.
SBAC AGS 2055 ISSUE 3. Rivets, Tucker, Monel Cup Domed Head.
SBAC AGS 2059 ISSUE 5. Rivets, Tucker, Monel, 'Pop', Domed Head (Unplated).
SBAC AGS 2074 ISSUE 2(I). Rivets, Tucker Aluminium Alloy 'Pop' Domed Head.
SBAC AGS 4717 (V). Rivet, Tucker 'Pop', Aluminium Alloy Domed Head.

—**Flat Head**
SBAC AS 469 ISSUE 3. Rivet, Flat Head. 1 p.

—**Mandrels**
SBAC AGS 2038 ISSUE 3. Rivets Chobert, Mandrels.

—**Oval Head**
BSI 2SP 83-85-73. 1973 Mushroom Head Aluminium Alloy Rivets for Aircraft. 10 pp.

—**Rudder Bars**
SBAC AS 2232 ISSUE 2. Drum Stop Rivet. 1 p.

—**Sealing Pins**
SBAC AGS 2042 ISSUE 7. Rivets, Chobert, Steel, Pins, Sealing.
SBAC AGS 2047 ISSUE 8. Rivets, Chobert, Aluminium Alloy Pins, Sealing.

—**Snap Head**
SBAC AGS 2043 ISSUE 5. Rivets, Chobert, Snap Head, Aluminium Alloy (L.37) Snap Head.
SBAC AGS 2045 ISSUE 6. Rivets, Chobert, Aluminium Alloy Snap Head.
SBAC AGS 2065 ISSUE 7. Avdel Self Sealing Snap Head Rivets (Aluminium Alloy).

—**Stop Valves—Connecting Rods**
SBAC AS 4895 ISSUE 1. Rivet. 1 p.

Aircraft Rivets *(Cont.)*

—**Stop Valves—Floats**
SBAC AS 4885 ISSUE 1. Rivet—Float. 1 p.
SBAC AS 4888 ISSUE 1. Rivet—Float Strap to Bracket. 1 p.

—**Tails**
SBAC RS 713 ISSUE 1. Rivet Nails.

—**Universal Head**
SBAC AS 16000-199 ISSUE 2. Rivet, Solid, Universal Head, DTD5036, (Corrosion and Heat Resistant).
SBAC AS 16200-399 ISSUE 2. Rivet, Solid, Universal Head, BS.HR504, (Heat Resistant Alloy).

—**Universal Head—Blind**
SBAC AS 46788 ISSUE 2. Rivet, Blind, Self-Plugging, Flush Break, Mechanically Locked Stem, Corrosion-Resisting Steel, Universal Head Cadmium Plated.
SBAC AS 46789 ISSUE 2. Rivet, Blind, Self-Plugging, Flush Break, Mechanically Locked Stem, Corrosion-Resisting Steel, Universal Head, Unplated.
SBAC AS 63240-245 ISSUE 1. Rivet Blind, Self-Plugging, Flush Break, Visible Mechanically Locked Stem, Aluminium Alloys Universal Head.
SBAC AS 63260-261 ISSUE 1. Rivet Blind, Self-Plugging, Flush Break, Visible Mechanically Locked Stem, Corrosion Resisting Steel Universal Head.
SBAC AS 4719 (V). Rivet, Blind, Self Plugging, Flush Break, Mech. Locked Spindle (Avdel MBC), Alloy, Head.

Aircraft Safety
Use For: Flight Safety *See Also:* Air Transportation; Aircraft; Aircraft Accidents; Aircraft Lighting; Avionics; Burst Disks; Crash Locator Beacons; Emergency Exits; Flight Envelopes; Hijacking; Near Misses (Aircraft); Skid Resistance; Wind Shear
CAA Chapter P6-1. Flight Information (Provisional Airworthiness Requirements for Civil Powered Lift Aircraft). 5 pp.

—**Aerostats**
CAA Chapter Q4-3 App 12.79. Compartment Design and Safety Provisions (Non-Rigid Airships). 2 pp.

—**Air Shows**
NATO STANAG 3533 ED 4 AMD 3-85. Safety Rules for Flying and Static Displays. 16 pp.
NATO STANAG 3533 ED 5 AMD 0-92. Safety Rules for Flying and Static Displays. 15 pp.
NATO STANAG 3533 ED 5 AMD 1-92. Flying and Static Displays. 17 pp.

—**Airworthiness**
CAA STANDARD NO. 1-1 & App. Airworthiness Objectives and System Analysis. 6 pp.
CAA. General Contents. 1 p.
CAA. Contents (Joint Airworthiness Requirements). 1 p.
CAA. Foreword (Joint Airworthiness Requirements). 1 p.
CAA. Check List of Pages (Joint Airworthiness Requirements). 2 pp.
CAA. Preambles (Joint Airworthiness Requirements). 2 pp.
ICAO Annex 8. Airworthiness of Aircraft Eighth Edition—July 1988. 52 pp.
ICAO Circular 95. Continuing Airworthiness of Aircraft in Service Sixth Edition—1985. 186 pp.
ICAO 9051. Airworthiness Technical Manual Second Edition—1987; (Corrigendum 1 06/01/87) (Amendment 10 11/14/88). 255 pp.
ICAO 9388. Manual of Model Regulations for National Control of Flight Operations and Continuing Airworthiness of Aircraft Second Edition—1987. 108 pp.
ICAO 9389. Manual of Procedures for an Airworthiness Organization First Edition—1983. 182 pp.
MOD UK DSTAN 00-970: Volume 1-01. Design and Airworthiness Requirements for Service Aircraft Volume 1—Aeroplanes Book 1 (Parts 0 to 3) Issue 1; Amendment 10. 100 pp.
MOD UK DSTAN 00-970: Volume 1-02. Design and Airworthiness Requirements for Service Aircraft Volume 1—Aeroplanes Book 1 (Parts 0 to 3) Issue 1; Amendment 11. 47 pp.
MOD UK DSTAN 00-970: Volume 1-03. Design and Airworthiness Requirements for Service Aircraft Volume 1—Aeroplanes Book 1 (Parts 0 to 3) Issue 1; Amendment 12. 37 pp.
MOD UK DSTAN 00-970: Pt 0:Vol 1-01. List of Important Changes Issue 1; Amendment 10. 51 pp.
MOD UK DSTAN 00-970: Pt 0:Vol 1-02. List of Important Changes Issue 1; Amendment 11. 15 pp.
MOD UK DSTAN 00-970: Pt 0:Vol 1-03. List of Important Changes Issue 1; Amendment 12. 16 pp.
MOD UK DSTAN 00-970: Pt 1:Vol 1-01. General and Operational Requirements Issue 1; Amendment 10. 342 pp.
MOD UK DSTAN 00-970: Pt 1:Vol 1-02. General and Operational Requirements Issue 1; Amendment 11. 293 pp.

Aircraft Safety *(Cont.)*

—**Airworthiness** *(Cont.)*
MOD UK DSTAN 00-970: Pt 1:Vol 1-03. General and Operational Requirements Issue 1; Amendment 12. 302 pp.
MOD UK DSTAN 00-970: Pt 2:Vol 1-01. Structural Strength and Design for Flight Issue 1; Amendment 10. 272 pp.
MOD UK DSTAN 00-970: Pt 2:Vol 1-02. Structural Strength and Design for Flight Issue 1; Amendment 10. 230 pp.
MOD UK DSTAN 00-970: Pt 2:Vol 1-03. Structural Strength and Design for Flight Issue 1; Amendment 12. 236 pp.
MOD UK DSTAN 00-970: Pt 4:Vol 1-01. Design and Airworthiness Requirements for Service Aircraft Volume 1—Aeroplanes Book 2 (Parts 4 to 6 and Chapters 700 to 710) Part 4: Detail Design and Strength of Materials Issue 1; Amendment 10. 502 pp.
MOD UK DSTAN 00-970: Pt 4:Vol 1-02. Design and Airworthiness Requirements for Service Aircraft Volume 1—Aeroplanes Book 2 (Parts 4 to 6 and Chapters 700 to 710) Part 4: Detail Design and Strength of Materials Issue 1; Amendment 11. 214 pp.
MOD UK DSTAN 00-970: Pt 4:Vol 1-03. Design and Airworthiness Requirements for Service Aircraft Volume 1—Aeroplanes Book 2 (Parts 4 to 6 and Chapters 700 to 710) Part 4: Detail Design and Strength of Materials Issue 1; Amendment 12. 211 pp.
MOD UK DSTAN 00-970: Pt 0:Vol 2-01. List of Important Changes Issue 1; Amendment 8. 22 pp.
MOD UK DSTAN 00-970: Pt 0:Vol 2-02. List of Important Changes Issue 1; Amendment 8. 9 pp.
MOD UK DSTAN 00-970: Pt 0:Vol 2-03. List of Important Changes Issue 1; Amendment 9. 10 pp.
MOD UK DSTAN 00-970: Pt 1:Vol 2-01. General and Operational Requirements Issue 1; Amendment 8. 231 pp.
MOD UK DSTAN 00-970: Pt 1:Vol 2-02. General and Operational Requirements Issue 1; Amendment 8. 199 pp.
MOD UK DSTAN 00-970: Pt 1:Vol 2-03. General and Operational Requirements Issue 1; Amendment 9. 202 pp.
MOD UK DSTAN 00-970: Pt 2:Vol 2-01. Structural Strength and Design for Flight Issue 1; Amendment 8. 76 pp.
MOD UK DSTAN 00-970: Pt 2:Vol 2-02. Structural Strength and Design for Flight Issue 1; Amendment 8. 62 pp.
MOD UK DSTAN 00-970: Pt 2:Vol 2-03. Structural Strength and Design for Flight Issue 1; Amendment 9. 100 pp.
MOD UK DSTAN 00-970: Pt 3:Vol 2-03. Structural Strength and Design for Operation on Specified Surfaces Issue 1; Amendment 8. 135 pp.
MOD UK DSTAN 00-970: Pt 4:Vol 2-01. Design and Airworthiness Requirements for Service Aircraft Volume 2—Rotorcraft Book 2 (Parts 4 to 6) Part 4: Detail Design and Strength of Materials Issue 1; Amendment 8. 207 pp.
MOD UK DSTAN 00-970: Pt 4:Vol 2-02. Design and Airworthiness Requirements for Service Aircraft Volume 2—Rotorcraft Book 2 (Parts 4 to 6) Part 4: Detail Design and Strength of Materials Issue 1; Amendment 8. 214 pp.
MOD UK DSTAN 00-970: Pt 4:Vol 2-03. Design and Airworthiness Requirements for Service Aircraft Volume 2—Rotorcraft Book 2 (Parts 4 to 6) Part 4: Detail Design and Strength of Materials Issue 1; Amendment 9. 226 pp.

—**Airworthiness—Environmental Testing**
CAA JAR-1 Section 2. Interpretations (Joint Airworthiness Requirements). 1 p.

—**Airworthiness—Glossaries**
CAA JAR-1 Section 1. Definitions and Abbreviations (Joint Airworthiness Requirements). 25 pp.

—**Bird Strikes**
CAA Part 5 CAP 168. Bird Strike Hazard. 2 pp.
CAA Section 7 CAP 382. Reporting of Airmiss and Birdstrike Occurences (Civil Air Publications: Mandatory Occurrence Reporting Scheme). 1 p.
ICAO 9332. Manual on the ICAO Bird Strike Information System (IBIS) Third Edition—1989. 72 pp.
NATO STANAG 3879 ED 2 AMD 4-85. Birdstrike Risk/Warning Procedures (Europe). 15 pp.
NATO STANAG 3879 ED 2 AMD 5-85. Birdstrike Risk/Warning Procedures (Europe). 16 pp.

—**Collision Avoidance**
ICAO Circular 174. Secondary Surveillance Radar Mode S Advisory Circular—1983. 42 pp.
ICAO Circular 213. Pilot Skills to Make "Look-Out" More Effective in Visual Collision Avoidance—1989. 16 pp.

INDUSTRY STANDARDS

Aircraft

Aircraft Safety (Cont.)

—Collision Avoidance (Cont.)
ICAO 9274. Manual on the Use of the Collision Risk Model (CRM) for ILS Operations First Edition—1980. 307 pp.

—Ergonomics
ICAO Circular 216. Human Factors Digest No. 1 Fundamental Human Factors Concepts—1989. 36 pp.
ICAO Circular 217. Human Factors Digest No. 2 Flight Crew Training: Cockpit Resource Management (CRM) and Line-Oriented Flight Training (LOFT)—1989. 67 pp.
ICAO Circular 227. Human Factors Digest No. 3 Training of Operational Personnel in Human Factors—1991. 58 pp.
ICAO Circular 229. Human Factors Digest No. 4 Proceedings of the ICAO Human Factors Seminar—1990. 628 pp.
ICAO Circular 240. Human Factors Digest No. 7 Investigation of Human Factors in Accidents and Incidents—1993. 61 pp.
ICAO Circular 241. Human Factors Digest No. 8 Human Factors in Air Traffic Control—1993. 48 pp.

—Flight Restrictions—Nuclear Facilities
CAA Section 4 CAP 393. Restriction of Flying. 11 pp.

—Flight Time Limitations
ICAO Circular 52. Flight Crew Fatigue and Flight Time Limitations Sixth Edition—1984. 314 pp.

—International Cooperation
NATO STANAG 3102 ED 4 AMD 0-90. Flight Safety in Common Ground/Air Space. 5 pp.
NATO STANAG 3102 ED 4 AMD 1-90. Flight Safety Co-Operation in Common Ground/Air Space. 6 pp.

—Landings
CAA 235 (G). Section G Rotorcraft Miscellaneous Amendments Derived from Section D, "Aeroplanes" (Blue Papers). 6 pp.
CAA 805 (G). Section G Rotocraft Crash Landing—Protection of Occupants. 3 pp.
CAA Chapter G3-8 11.75. Crash Landing Conditions (Rotocraft). 1 p.
CAA Chapter G3-8 App 02.63. Crash Landing Conditions (Rotocraft). 1 p.
CAA Chapter K3-8 04.72. Emergency Alighting Conditions (Light Aeroplanes). 1 p.
CAA Chapter Q3-8 12.79. Crashworthiness Conditions (Non-Rigid Airships). 2 pp.

—Lockwire
NATO STANAG 3752 ED 2 AMD 3-89. Witness (Breaking) Wire for Aircraft Emergency Controls and Equipment. 8 pp.
NATO STANAG 3752 ED 2 AMD 4-89. Witness (Breaking) Wire for Aircraft Emergency Controls and Equipment. 9 pp.

—Military Activities
ICAO 9554. Manual Concerning Safety Measures Relating to Military Activities Potentially Hazardous to Civil Aircraft Operations First Edition—1990. 21 pp.

—Obstacles—Airport Air Space
CAA Part 4 CAP 168. Assessment and Treatment of Obstacles. 31 pp.

—Occurrence Reports
CAA NOTICE #86 ISSUE 3. Communication on Safety Matters (Airworthiness Notices). 2 pp.

—Statistical Analysis
ICAO Circular 233. Accident/Incident Reporting (ADREP) Annual Statistics—1988. 21 pp.

—Symbols—Hazard Points
BSI M 44-74. 1974 Identification of Aircraft Servicing, Maintenance, Ground Handling and Safety-Hazard Points. 10 pp.
ISO 1950-74. Aircraft—Identification of Servicing, Maintenance, Ground Handling and Safety/Hazard Points First Edition. 9 pp.
NATO STANAG 3109 ED 5 AMD 4-86. Symbol Marking of Aircraft Servicing and Safety/Hazard Points. 21 pp.

—Warning Systems—Streamers
NATO STANAG 3714 ED 2 AMD 0-86. Streamers, Warning. 8 pp.
NATO STANAG 3714 ED 2 AMD 2-86. Streamers, Warning. 7 pp.

Aircraft Screws

See Also: Aircraft Fasteners; Fasteners; Screws
CAA LEAFLET 3-3 07.90. Bolts and Screws of British Manufacture. 11 pp.

Aircraft Screws (Cont.)

SBAC AS 6535 ISSUE 1. Special Mushroom Head, Screw.
SBAC TS 148 ISSUE 1. Studs, Bolts and Screws in DTD5638 (INCO 718) Material 1275 MPa (185,000 lbf/in2) Tensile at Room Temperature (for Which Forging is Not Mandatory and Qualification Not a Requirement).

—Control Wheels
SBAC AS 3070 ISSUE 1. Switch Contact Screw. 1 p.
SBAC AS 6593 ISSUE 1. Switch Contact, Screw for Pilot's Control Wheel.

—Countersunk Head—Phillips
SBAC AS 2993 ISSUE 4. Screw-"Phillip's" 90 Degree Countersunk Head. 1 p.
SBAC AS 2996 ISSUE 3. Screw—"Phillip's" 120 Degree Countersunk Head. 1 p.

—Countersunk Head—Phillips—Flat
SBAC AS 54700-799 ISSUE 2. Bolt, Close Tolerance—Double Hexagon Extended Washer Head, INCO 718, 0.3125—24 UNJF—3A MOD BS4084.
SBAC AS 54800-899 ISSUE 2. Bolt, Close Tolerance—Double Hexagon Extended Washer Head, INCO 718, 03750—24 UNJF—3A MOD BS4084.
SBAC AS 54900-999 ISSUE 2. Bolt, Close Tolerance—Double Hexagon Extended Washer Head, INCO 718, 04375—20 UNJF—3A MOD BS4084.

—Countersunk Head—Phillips—Oval
SBAC AS 2994 ISSUE 3. Screw-"Phillip's" Raised 90 Countersunk Head. 1 p.
SBAC AS 2995 ISSUE 3. Screw-"Phillip's" Raised 120 Degree Countersunk Head. 1 p.

—Countersunk Head—Recessed—Wood
BSI SP 128-133-65. 1965 Wood Screws for Aircraft. 5 pp.

—Electric Terminals
BSI G 203-67. 1967 Unified Screws with Captive Facing and Locking Washers. 4 pp.

—Filler Caps
SBAC AS 2407 ISSUE 3. Screw (Filler Cap). 1 p.
SBAC AS 6350 ISSUE 1. Pivot, Screw (Filler Cap).
SBAC AS 6351 ISSUE 1. Screw (Filler Cap).

—Locking
SBAC AS 6553 ISSUE 1. Locking, Screw.
SBAC AS 52790-799 ISSUE 1. Insert—Inch, Thinwall—Long Screw Locking (INCO 718).

—Lockwire Holes
SBAC RS 704 (V). Recommended Size and Location of Locking Wire Holes in Bolts Screws and Nuts.

—Oval Head
SBAC AS 2554 ISSUE 2. Special Mushroom Head Screw. 1 p.

—Oval Head—Doors—Locks
SBAC AS 6535 ISSUE 1. Special Mushroom Head, Screw.

—Oval Head—Phillips
SBAC AS 2925 ISSUE 2(I). Bolt M.S.-"Phillip's" Mushroom Head. 1 p.
SBAC AS 2992 ISSUE 4. Screw-"Phillip's" Mushroom Head. 1 p.
SBAC AS 4598 ISSUE 6. Bolt—Phillips Mushroom Head. 1 p.

—Round Head—Phillips
SBAC AS 2991 ISSUE 4. Screw—"Phillip's" Round Head. 1 p.

—Stirrups
SBAC AS 3328 ISSUE 1. Stirrup Screw. 1 p.
SBAC AS 5433 ISSUE 1. Stirrup Screw. 1 p.

—Stops
SBAC AS 3127-3128 ISSUE 1. Adjusting Brake Screw Stop. 1 p.

—Washers
BSI SP 22-27-50. 1950 Amd 1 B.A. Washers for Aircraft Purposes (Primarily for Instrument Use). 4 pp.

—Washers—Instruments
BSI SP 22-27-50. 1950 Amd 1 B.A. Washers for Aircraft Purposes (Primarily for Instrument Use). 4 pp.

Aircraft Stability
Use: Flight Dynamics

Aircraft Stores
See Also: Aircraft Equipment; Weapons Systems

Aircraft Stores (Cont.)

—Arming Devices
NATO STANAG 3605 ED 2 AMD 1-86. Compatibility of Mechanical Fuzing Systems and Arming Devices for Expendable Aircraft Stores for Fixed Wing Aircraft. 12 pp.
NATO STANAG 3605 ED 3 AMD 0-92. Compatibility of Arming Units and Expendable Aircraft Stores. 9 pp.

—Certification
NATO STANAG 3859 ED 2 AMD 3-84. Standardized Data List for Interoperability Studies and Certification of Aircraft Stores. 14 pp.
NATO STANAG 3859 ED 3 AMD 0-92. Standardized Data List for Interoperability Studies and Certification of Aircraft Stores. 10 pp.

—Connectors—Location
NATO STANAG 3558 ED 3 AMD 0-86. Locations for Aircraft Electrical Control Connections for Aircraft Stores. 2 pp.
NATO STANAG 3558 ED 3 AMD 1-86. Locations for Aircraft Electrical Control Connections for Aircraft Stores. 9 pp.
NATO STANAG 3558 ED 3 AMD 4-86. Locations for Aircraft Electrical Control Connections for Aircraft Stores. 9 pp.
NATO STANAG 3558 ED 3 AMD 5-86. Locations for Aircraft Electrical Control Connections for Aircraft Stores. 10 pp.

—Design
NATO STANAG 3441 ED 4 AMD 4-84. Design of Aircraft Stores. 16 pp.
NATO STANAG 3441 ED 5 AMD 0-92. Design of Aircraft Stores. 8 pp.

—Ejector Racks
NATO STANAG 3575 ED 3 AMD 1-86. Aircraft Stores Ejector Racks. 8 pp.
NATO STANAG 3575 ED 4 AMD 0-92. Aircraft Stores Ejector Racks. 10 pp.

—Environmental Testing
NATO STANAG 3557 ED 1 AMD 1-00. Standard NATO Safety Tests for Airborne Pyrotechnic Stores. 16 pp.

—Fit—Compatibility
NATO STANAG 3899 ED 1 AMD 4-84. Ground Fit and Compatibility Criteria for Aircraft Stores. 20 pp.
NATO STANAG 3899 ED 2 AMD 0-93. Ground Fit and Compatibility Criteria for Aircraft Stores. 14 pp.

—Fuses (Ordnance)
NATO STANAG 3605 ED 2 AMD 1-86. Compatibility of Mechanical Fuzing Systems and Arming Devices for Expendable Aircraft Stores for Fixed Wing Aircraft. 12 pp.

—Interconnection Systems
NATO STANAG 3837 ED 2 AMD 2-86. Aircraft Stores Electrical Interconnection System. 7 pp.

—Interoperability
NATO STANAG 3791 ED 2 AMD 1-89. Interoperability of NATO Aircraft Stores—AOP-11. 4 pp.
NATO STANAG 3791 ED 3 AMD 0-91. Interoperability of NATO Aircraft and Stores—AOP-11(D). 4 pp.
NATO STANAG 3859 ED 2 AMD 3-84. Standardized Data List for Interoperability Studies and Certification of Aircraft Stores. 14 pp.
NATO STANAG 3859 ED 3 AMD 0-92. Standardized Data List for Interoperability Studies and Certification of Aircraft Stores. 10 pp.

—Suspension Equipment
NATO STANAG 3635 ED 2 AMD 2-86. Suspension Units with Gravity Release Capability for Aircraft Stores. 7 pp.
NATO STANAG 3635 ED 3 AMD 0-92. Aircraft Stores Suspension Units with Gravity Release Capability. 8 pp.
NATO STANAG 3635 ED 3 AMD 1-92. Aircraft Stores Suspension Units with Gravity Release Capability. 10 pp.

—Technical Manuals
NATO STANAG 3898 ED 3 AMD 1-91. Aircraft Stores Interface Manual—AOP-12. 4 pp.

—Undercarriages—Lugs
NATO STANAG 3726 ED 3 AMD 4-83. Bail (Portal) Lugs for the Suspension of Aircraft Stores. 8 pp.
NATO STANAG 3726 ED 4 AMD 0-92. Bail (Portal) Lugs for the Suspension of Aircraft Stores. 9 pp.

Aircraft Structures
Use: Airframes

Aircraft Surveillance Radar
Use: Surveillance Radar—Aircraft

Aircraft Valves
See Also: Valves
SBAC AS 2955 ISSUE 3. Valve Retainer. 1 p.
SBAC AS 2982 ISSUE 2. Valve Retainer. 1 p.
SBAC AS 6271 (V). Valve Body. 1 p.
SBAC AS 6272 (V). Valve. 1 p.

—Air
CAA Chapter Q6-2 12.79. Gas and Air Supply Systems (Non-Rigid Airships). 2 pp.
SBAC AGS 2096/A ISSUE 5. 1/4" B.S.P.F. (Air and Mineral Base Hydraulic Fluid) Charging Valve.
SBAC AGS 3034/A ISSUE 5. Valve.
SBAC AGS 3034/B (V). Valve.

—Air—Manual Controls
SBAC AS 1910 ISSUE 6. Knob (Air Shutter Control). 1 p.
SBAC AS 6388 ISSUE 2. Knob (Air Shutter Control).

—Caps (Lids)
SBAC AS 6364 ISSUE 4. Cap—Valve.

—Check
SBAC AS 3382 ISSUE 3. 3/8 Inch Non-Return, Valve (Hydraulic). 1 p.
SBAC AS 3388 ISSUE 3. 1/2 Inch Non Return, Valve—Hydraulic. 1 p.

—Check—Aluminum Alloy
SBAC AS 3377 ISSUE 1. Body—Non-Return Valve. 1 p.
SBAC AS 3383 ISSUE 1. Body. 1 p.
SBAC AS 3389 ISSUE 1. Body. 1 p.

—Check—Aluminum Alloy—Stoppers
SBAC AS 3378 ISSUE 1. Plug. 1 p.
SBAC AS 3384 ISSUE 1. Plug. 1 p.
SBAC AS 3390 ISSUE 1. Plug. 1 p.

—Check—Guards (Protective)—Pipe Threads
SBAC AGS 3415 ISSUE 1. Non Return Valve, Safety Guard for Use with 1/4 Inch B.S.P.F. Thread.

—Check—Sealing Rings
SBAC AS 3381 ISSUE 1. Sealing, Ring. 1 p.
SBAC AS 3387 ISSUE 1. Sealing, Ring. 1 p.
SBAC AS 3393 ISSUE 1. Sealing, Ring. 1 p.

—Check—Springs (Elastic)
SBAC AS 3380 ISSUE 1. Spring. 1 p.
SBAC AS 3386 ISSUE 1. Spring. 1 p.
SBAC AS 3392 ISSUE 1. Spring. 1 p.

—Check—Steel
SBAC AS 3379 ISSUE 1. Valve. 1 p.
SBAC AS 3385 ISSUE 1. Valve. 1 p.
SBAC AS 3391 ISSUE 1. Valve. 1 p.

—Compressed Air
ISO 1023-74. Aircraft—High Pressure Air Charging Valves First Edition. 4 pp.

—Compressed Air—Couplings
BSI C 9-70. 1970 Coupling Dimensions for High Pressure Air Charging Valves for Aircraft. 2 pp.

—Drain
SBAC AS 2564 (V). Body. 1 p.
SBAC AS 2574 (V). Body. 1 p.
SBAC AS 3300 ISSUE 2. Drain, Valve Assembly. 1 p.
SBAC AS 3301 ISSUE 1. Valve. 1 p.
SBAC AS 3334 ISSUE 2. Valve 1/4 Inch Bore Drain Valve. 1 p.
SBAC AS 4585 ISSUE 1. Flush Water/Sediment Drain Valve. 1 p.
SBAC AS 4586 ISSUE 1. Body Assembly. 1 p.
SBAC AS 4587 ISSUE 1. Valve. 1 p.
SBAC AS 4760 ISSUE 4. 1/4 Inch Bore Sediment Drain Valve. 1 p.
SBAC AS 4761 ISSUE 1. Body (1/4 Inch Drain Valve). 1 p.
SBAC AS 6297 ISSUE 3. Valve. 1 p.
SBAC AS 6361 ISSUE 4(I). Body—Drain Valve.
SBAC AS 6370 ISSUE 3. Drainage Connection Drain Valve.
SBAC AS 6617 ISSUE 1. Flush Water/Sediment Drain Valve.
SBAC AS 6619 ISSUE 1. Valve.
SBAC AS 6700 ISSUE 1. Assembly—Drain Valve.
SBAC AS 6701 ISSUE 2. Body—Drain Valve.
SBAC AS 8433 ISSUE 1. 1/4 Inch Bore Sediment Drain Valve, Unified.
SBAC AS 8434 ISSUE 1. Body 1/4 Bore Drain Valve.
SBAC AGS 3031 ISSUE 1. Body for Drain Cock.

Aircraft Valves *(Cont.)*

—Drain—Assembly
SBAC AS 2560 ISSUE 5. Drain Valve Assembly 2 Inch. 1 p.
SBAC AS 2573 ISSUE 6. Drain Valve Assembly 1 1/4 Inch. 1 p.
SBAC AS 6270 ISSUE 2. Assembly of Flush Drain Valve—1 1/4 Inch Bore. 1 p.
SBAC AS 6295 ISSUE 1. Assembly of Flush Drain Valve 2 Inch Bore. 1 p.
SBAC AS 6360 ISSUE 3(I). Assembly—Drain Valve.
SBAC AS 6618 ISSUE 1. Body Assembly.

—Drain—Cap Seals
SBAC AS 6277 ISSUE 3. Cap, Seal. 1 p.
SBAC AS 6301 ISSUE 3. Cap, Seal. 1 p.

—Drain—Caps (Lids)
SBAC AS 2565 ISSUE 1. Screw Cap. 1 p.
SBAC AS 2572 ISSUE 3. Cap Assembly. 1 p.
SBAC AS 2575 ISSUE 4. Screw Cap. 1 p.
SBAC AS 4762 ISSUE 1. Cap 1/4 Inch Bore Drain Valve. 1 p.
SBAC AS 6274 ISSUE 3. Valve Cap. 1 p.
SBAC AS 6299 ISSUE 3. Valve Cap. 2 pp.
SBAC AS 8435 ISSUE 1. Cap 1/4 Bore Drain Valve.

—Drain—Chain Clips
SBAC AS 3339 ISSUE 2. Chain Clip 1/4 Bore Drain Hole. 1 p.
SBAC AS 8437 ISSUE 1. Chain, Clip 1/4 Inch Bore Drain Valve.

—Drain—Connections
SBAC AS 6289 ISSUE 1. Tank Drainage Connection Drain Valve AS 6270, 1 1/4 Inch. 1 p.
SBAC AS 6313 ISSUE 1. Tank Drainage Connection Drain Valve AS 6295, 2 Inches. 1 p.

—Drain—Guides
SBAC AS 2562 ISSUE 4. Valve Guide. 1 p.
SBAC AS 6273 ISSUE 2. Valve Guide. 1 p.
SBAC AS 6298 ISSUE 2. Valve Guide. 1 p.

—Drain—Guides—Bushings
SBAC AS 2563 ISSUE 2. Bush for Valve Guide. 1 p.

—Drain—Helical Springs
SBAC AS 6276 ISSUE 1. Cap, Spring. 1 p.
SBAC AS 6300 ISSUE 1. Valve Spring. 1 p.
SBAC AS 6366 ISSUE 3. Spring—Cap—Oil Drain Valve.

—Drain—Nuts (Fasteners)
SBAC AS 3302 ISSUE 1. Union, Nut. 1 p.
SBAC AS 3303 ISSUE 1. Release Nut. 1 p.
SBAC AGS 3033 ISSUE 1. Nut for Drain Cock.

—Drain—Oil
SBAC AS 6362 ISSUE 4. Valve—Oil Drain Valve.

—Drain—Pitot Tubes
SBAC AS 40665 ISSUE 1. Cock, Poppet, Drain Assembly (Pitot).
SBAC AS 40669 ISSUE 1. Cock, Poppet, Drain Assembly (Pitot).
SBAC AS 40673 ISSUE 1. Cock, Poppet, Drain Assembly Pitot.

—Drain—Plugs
SBAC AGS 3032 ISSUE 1. Plug for Drain Cock.

—Drain—Rings
SBAC AS 6279-6286 (V). Tapped, Ring for Drain Valve. 2 pp.
SBAC AS 6287 (V). Tapped Ring for Drain Valve. 2 pp.
SBAC AS 6303-6310 (V). Tapped, Ring for Drain Valve AS 6295. 2 pp.

—Drain—Rings—Stampings
SBAC AS 6288 ISSUE 2. Stamping Tapped Ring for Drain Valve AS6270. 1 p.
SBAC AS 6312 ISSUE 2. Stamping Tapped Ring for Drain Valve AS 6295. 1 p.

—Drain—Seals
SBAC AS 4588 ISSUE 2. Seal. 1 p.
SBAC AS 6278 ISSUE 3. Valve, Seal. 1 p.
SBAC AS 6302 ISSUE 3. Valve Seal. 1 p.
SBAC AS 8436 ISSUE 1. Seal, 1/4 Inch Bore Drain Valve.

—Drain—Spindles
SBAC AS 3332 ISSUE 3. Valve, Spindle 1/4 Inch Bore Drain Valve. 1 p.
SBAC AS 3333 ISSUE 2. Insert 1/4 Inch Bore Drain Valve. 1 p.

—Drain—Springs
SBAC AS 4524 ISSUE 2. Spring—1/4 Inch Bore Drain Valve. 1 p.
SBAC AS 6275 ISSUE 1. Valve Spring. 1 p.

Aircraft Valves *(Cont.)*

—Drain—Static Tubes
SBAC AS 40667 ISSUE 1. Cock, Poppet, Drain Assembly (Static).
SBAC AS 40671 ISSUE 1. Cock, Poppet, Drain Assembly (Static).
SBAC AS 40675 ISSUE 1. Cock, Poppet, Drain Assembly (Static).

—Drain—Stops
SBAC AS 3304 ISSUE 1. Stop for Drain Valve. 1 p.

—Drain—Tanks (Containers)
SBAC AS 2580 ISSUE 5. Tank Drainage Connection—Drain Valve 2 Inch. 1 p.
SBAC AS 2584 ISSUE 5. Tank Drainage Connection—Drain Valve 1 1/4 Inch. 1 p.

—Drain—Washers (Fasteners)
SBAC AS 2567 ISSUE 3. Joint, Washer. 1 p.
SBAC AS 3337 ISSUE 2. Washer 1/4 Inch Bore Drain Valve. 1 p.

—Drain—Washers—Chain Tags
SBAC AS 3338 ISSUE 2. Chain Tag, Washer—1/4 Inch Bore Drain Valve Drain Valve. 1 p.

—Filler Caps
SBAC AS 3167 ISSUE 2. Assembly of Filler Cap—Air Valve Type. 1 p.
SBAC AS 3168 ISSUE 1. Air Valve for Filler Cap.AS.1964. 1 p.

—Fuel—Control Knobs
SBAC AS 6395 ISSUE 1. Fuel Cock, Knob (Low Pressure).

—Fuel—Couplings
SBAC AS 2958 ISSUE 4. Valve. 1 p.
SBAC AS 2979 ISSUE 3. Valve. 1 p.
SBAC AS 2980 ISSUE 3. Spring. 1 p.

—Gas
CAA Chapter Q6-2 12.79. Gas and Air Supply Systems (Non-Rigid Airships). 2 pp.

—Guides
SBAC AS 2563 ISSUE 2. Bush for Valve Guide. 1 p.

—Guides—Aluminum Alloys
SBAC AS 6273 ISSUE 2. Valve Guide. 1 p.
SBAC AS 6298 ISSUE 2. Valve Guide. 1 p.
SBAC AS 6363 ISSUE 4. Guide—Valve.

—Helical Springs
SBAC AS 6365 ISSUE 3. Spring Valve.

—Hydraulic
SBAC AGS 2096/A ISSUE 5. 1/4" B.S.P.F. (Air and Mineral Base Hydraulic Fluid) Charging Valve.
SBAC AGS 3034/A ISSUE 5. Valve.
SBAC AGS 3034/B (V). Valve.

—Hydraulic—Check
SBAC AS 3382 ISSUE 3. 3/8 Inch Non-Return, Valve (Hydraulic). 1 p.
SBAC AS 3388 ISSUE 3. 1/2 Inch Non Return, Valve—Hydraulic. 1 p.

—Lifters
SBAC AS 2582 ISSUE 3. Valve Lifter. 1 p.
SBAC AS 2586 ISSUE 3. Valve Lifter. 1 p.

—Lifters—Adapters
SBAC AS 6375 ISSUE 2. Segment—Valve Lifter.

—Lifters—Adapters—Assembly
SBAC AS 2583 ISSUE 5. Assembly of Valve Lifter and Adaptor. 1 p.
SBAC AS 2587 ISSUE 5. Assembly of Valve Lifter and Adaptor. 1 p.
SBAC AS 6373 ISSUE 2. Assembly—Valve Lifter Adaptor.

—Pipe Threads—Safety
SBAC AGS 3415 ISSUE 1. Non Return Valve, Safety Guard for Use with 1/4 Inch B.S.P.F. Thread.

—Retainers
SBAC AS 2955 ISSUE 3. Valve Retainer. 1 p.
SBAC AS 2982 ISSUE 2. Valve Retainer. 1 p.
SBAC AS 6548 ISSUE 1. Valve Retainer.
SBAC AS 6556 ISSUE 1. Valve Retainer.

—Sealing Rings
SBAC AS 2910 ISSUE 2. Valve Seal. 1 p.

—Seals
SBAC AS 6367 ISSUE 3. Seal—Valve.

—Thermal Relief
SBAC AS 4551 ISSUE 2. Valve Body. 1 p.

—Thermal Relief—Caps (Lids)
SBAC AS 4552 ISSUE 3. Valve Cap. 1 p.

INTERNATIONAL AND NON-U.S. NATIONAL STANDARDS
SUBJECT INDEX
Aircraft

Aircraft Valves *(Cont.)*

—**Thermal Relief—Gaskets**
SBAC AS 3357 ISSUE 2. Gasket. 1 p.

—**Thermal Relief—Plungers**
SBAC AS 4553 ISSUE 2. Plunger. 1 p.

—**Thermal Relief—Shims**
SBAC AS 4555 ISSUE 1. Shim. 1 p.

—**Thermal Relief—Sleeves**
SBAC AS 3353 ISSUE 3. Sleeve. 1 p.

—**Thermal Relief—Springs (Elastic)**
SBAC AS 4554 ISSUE 4. Spring for Valve. 1 p.

—**Thermal Relief—Springs (Elastic)—Holders**
SBAC AS 3356 ISSUE 3. Spring, Retainer. 1 p.

Aircraft Washers
See Also: Aircraft Fasteners; Fasteners; Washers (Fasteners)
SBAC AS 410 ISSUE 5(I). Washer, Joint. 1 p.
SBAC AS 1963 ISSUE 5. Washer. 1 p.
SBAC AS 2026-2040 ISSUE 4. Washer, Joint (Outer). 1 p.
SBAC AS 2041-2055 ISSUE 5. Washer, Joint (Inner). 1 p.
SBAC AS 2242 ISSUE 4. Tab Washer. 1 p.
SBAC AS 2516 ISSUE 3. Sealing Washer. 1 p.
SBAC AS 2555 ISSUE 3. Chamfered Washer. 1 p.
SBAC AS 8596 ISSUE 1. Washer.
SBAC AS 22900-999 ISSUE 2. Bolt, Close Tolerance—Double Hexagon Extended Washer Head, BSS.159. 190-32 UNJF-3A.MOD.B.S. 4084. 1 p.
SBAC AS 44409-412 ISSUE 2. Washer, Spring Tension, Metric.
SBAC AS 44691-694 ISSUE 2. Washer, Double Key—Unified Hexagon.
SBAC AGS 567 ISSUE 9. Jointing, Washers Thin (Hard Fibre).
SBAC AGS 568 ISSUE 5. Jointing Washers (Aluminium).
SBAC AGS 2072 ISSUE 1. Washer 1/4 Inch.
SBAC AS 3409 ISSUE 1. Washer.
SBAC AGS 3820-69 ISSUE 1. Washer, Seal, Flat.

—**Bolts**
BSI SP 10-11-49. 1949 Amd 1 Washers for Aircraft (Primarily for Facing Purposes). 2 pp.
BSI SP 122-125-58. 1958 Washers for Unified Hexagons for Aircraft (Primarily for Packing Purposes). 5 pp.
BSI SP 126-127-59. 1959 Washers for Unified Hexagons for Aircraft (Primarily for Facing Purposes). 4 pp.

—**Caps (Lids)**
SBAC AS 2576 ISSUE 4. Jointing Washer (Screw Cap). 1 p.
SBAC AS 3112 ISSUE 1. Sealing Washer for Cap. 1 p.

—**Chain Tags—Drain Valves**
SBAC AS 3338 ISSUE 2. Chain Tag, Washer—1/4 Inch Bore Drain Valve Drain Valve. 1 p.

—**Control Columns**
SBAC AS 860 ISSUE 6. Combined Dust Washer and Bush Control Column. 1 p.
SBAC AS 861 ISSUE 4. Distance Piece—Control Column. 1 p.
SBAC AS 5362 ISSUE 2. Washer—Front Spigot—Control Column. 1 p.

—**Drain Plugs**
SBAC AS 2422 ISSUE 3. Washer—Drain Plug. 1 p.

—**Drain Valves**
SBAC AS 2567 ISSUE 3. Joint, Washer. 1 p.
SBAC AS 3337 ISSUE 2. Washer 1/4 Inch Bore Drain Valve. 1 p.

—**Filler Caps**
SBAC AS 2404 ISSUE 4. Washer (Filler Cap). 1 p.
SBAC AS 2405 ISSUE 4. Washer—Seating (Filler Cap). 1 p.
SBAC AS 2517 ISSUE 2. Washer for 2 Inch Filler Cap. 1 p.
SBAC AS 3031-3036 ISSUE 1. Packing Washer (Flush Filler Cap). 1 p.

—**Fuel Filters**
SBAC AS 3224 ISSUE 1. Jointing, Washer Filter Fuel Type 'A' Type 'A'. 1 p.
SBAC AS 3254 ISSUE 1. Jointing, Washer Filter Fuel Type 'B'. 1 p.

—**Lock—Brake Locks**
SBAC AS 2951 ISSUE 2. Lock Washer Brake Lock Adjustment. 1 p.

—**Lock—Cup**
SBAC AS 8690-8699 ISSUE 3(I). Washer, Cup, Lock.

Aircraft Washers *(Cont.)*

—**Lock—Terminals**
BSI G 203-67. 1967 Unified Screws with Captive Facing and Locking Washers. 4 pp.

—**Lock Tooth—External—Countersunk—Steel**
SBAC AGS 2036 ISSUE 8. Lock/Washers Shakeproof Countersunk, External Teeth, Steel.

—**Lock Tooth—External—Steel**
SBAC AGS 2034 ISSUE 9. Lockwashers Shakeproof Flat, External Teeth, Steel.

—**Lock Tooth—Internal—Bronze**
SBAC AGS 2037 ISSUE 6(I). Lockwashers Shakeproof Flat, Internal Teeth, Phos. Bronze.

—**Lock Tooth—Internal—Steel**
SBAC AGS 2035 ISSUE 9. Lockwasher Shakeproof Flat, Internal Teeth, Steel.

—**Nuts**
SBAC AS 2568 ISSUE 2. Washer Jointing (Nut). 1 p.
SBAC AS 28000 ISSUE 2. Nut, Self-Locking, Non CR Steel, Captive Washer, Cadmium Coated.
SBAC AS 28002 ISSUE 1. Washer—for Use with AS 28001 Nuts.

—**Packaging**
BSI SP 13-16-49. 1949 Amd 3 Washers for Aircraft (Primarily for Packing Purposes). 3 pp.

—**Pins**
SBAC AS 2467 ISSUE 4. Spring Washer for Quick Release Pin. 1 p.

—**Pipe Couplings**
SBAC AGS 1139 ISSUE 10. Pipe Couplings, Jointing Washers (Copper).

—**Pipe Fittings**
SBAC AS 43098 ISSUE 2. Washer Plate (Aluminium Alloy).
SBAC AS 43103 ISSUE 2. Washer Plate Triangular (Aluminium Alloy).
SBAC AS 43105 ISSUE 2. Washer Plate (Aluminium Alloy).
SBAC AGS 1138 ISSUE 8. Pipe Couplings Jointing Washers (Aluminium).

—**Plain**
SBAC AS 46712 ISSUE 1. Washer, Flat, Brass, Metric. 1 p.
SBAC AGS 164 ISSUE 14. Washers Round Jointing (Hard Fibre).

—**Plain—Precision**
SBAC AS 4562 ISSUE 2. Plain, Washers (Precision). 1 p.

—**Plugs**
SBAC AS 6717 ISSUE 2. Round Jointing Washers, Fibre to Suit Unified Plugs.

—**Rudder Bars**
SBAC AS 3398 ISSUE 2. Washer for Rudder Bar. 1 p.

—**Saddle**
SBAC AS 1903 ISSUE 4. Saddle Washers (Aluminium Alloy). 1 p.
SBAC AS 1904 ISSUE 3. Saddle Washers (Aluminium Alloy). 1 p.
SBAC AS 1905 ISSUE 2. Saddle Washers (Steel). 1 p.
SBAC AS 1906 ISSUE 2. Saddle Washers (Steel). 1 p.

—**Screws**
BSI SP 22-27-50. 1950 Amd 1 B.A. Washers for Aircraft Purposes (Primarily for Instrument Use). 4 pp.

—**Screws—Instruments**
BSI SP 22-27-50. 1950 Amd 1 B.A. Washers for Aircraft Purposes (Primarily for Instrument Use). 4 pp.

—**Signal Pistols**
SBAC AS 4635 ISSUE 1. Washer. 1 p.

—**Stop Valves—Connecting Rods**
SBAC AS 4894 ISSUE 1. Washer. 1 p.

—**Tab**
BSI SP 41-46-51. 1951 Amd 2 Tab Washers for Aircraft. 4 pp.
BSI SP 107-112-54. 1954 Amd 2 Tab Washers for Unified Hexagons for Aircraft. 4 pp.
SBAC AS 3274 ISSUE 2. Tab, Washer Filter Fuel Type B. 1 p.
SBAC AS 4904 ISSUE 3. Tab Washer—Float Bracket. 1 p.

Aircraft Washers *(Cont.)*

—**Tab** *(Cont.)*
SBAC AS 4905 ISSUE 3. Tab Washer 1/2 Inch Gas. 1 p.

—**Taper**
SBAC AS 2114 ISSUE 1. Washer, Taper. 1 p.

—**Terminals**
BSI G 203-67. 1967 Unified Screws with Captive Facing and Locking Washers. 4 pp.

—**Thermometers**
SBAC AGS 839 ISSUE 3. Washer Thermometer (Copper).

—**Timber**
BSI SP 18-19-49. 1949 Amd 2 Washers for Aircraft (Primarily for Use with Timber). 4 pp.

—**Trim Controls**
SBAC AS 892 ISSUE 5. Washer—Trim Control. 1 p.

—**Union Nuts**
SBAC AS 44443 ISSUE 1. Washer—for 3/8 Inch Diameter Pipe Union Nut.

—**Weights**
SBAC AS 54461-544 64 ISSUE 1. Balance Weights, Washer Type to Suit Metric Fasteners.

Aircrews
Use: Flight Crews

Airfield Lighting
Use: Airport Lighting

Airfield Taxiways
Use: Airport Taxiways

Airfields
Use: Aircraft Landing Areas

Airframe Adhesives
See Also: Adhesives

—**Heat Stable**
MOD UK DTD-5577-65. Heat Stable Structural Adhesives. 11 pp.

Airframe Bearings *See Also:* Aerospace Equipment; Aircraft Equipment; Antifriction Bearings; Bearings
MOD UK DSTAN 31-4: Part 2A-01. Bearings, Ball, Roller and Needle Roller, for Aerospace Use Part 2A: Rolling Bearings for Inch Based Airframe Applications (Excluding Tapered Roller Bearings) Issue 1; Amendment 3. 15 pp.
MOD UK DSTAN 31-6-01. Requirements for Alternative Airframe Rolling Bearings for Use on Sepecat Jaguar Aircraft Issue 1; Amendment 1. 13 pp.
MOD UK DTD-244-34. White Metal Bearings (Reprinted August 1060). 1 p.

—**Ball—Loads (Forces)**
AECMA PREN3045-87. Bearings-Airframe Rolling Rigid Single Row Ball Bearings in Steel Diameter Series 0 and 2 Reduced Clearance Category Dimensions and Loads. 5 pp.
AECMA PREN3046-87. Bearings-Airframe Rolling Rigid Single Row Ball Bearings in Steel Cadmium Plated Diameter Series 0 and 2 Reduced Clearance Category Dimensions and Loads. 5 pp.
AECMA PREN3047-87. Bearings-Airframe Rolling Rigid Single Row Ball Bearings in Corrosion Resisting Steel Diameter Series 0 and 2 Reduced Clearance Category Dimensions and Loads. 5 pp.
AECMA PREN3056-87. Bearings-Airframe Rolling Rigid Two Row Ball Bearings in Steel Dimensions and Loads. 5 pp.
AECMA PREN3057-87. Bearings-Airframe Rolling Rigid Two Row Ball Bearings in Steel Cadmium Plated Dimensions and Loads. 5 pp.
AECMA PREN3058-87. Bearings-Airframe Rolling Rigid Two Row Ball Bearings in Corrosion Resisting Steel Dimensions and Loads. 5 pp.
AECMA PREN3281-87. Bearings-Airframe Rolling Rigid, Single Row Ball Bearings in Steel Diameter Series 8 and 9 Dimensions, Torques, Clearances and Loads. 7 pp.
AECMA PREN3282-87. Bearings-Airframe Rolling Rigid, Single Row Ball Bearings in Steel, Cadmium Plated Diameter Series 8 and 9 Dimensions, Torques, Clearances and Loads. 6 pp.
AECMA PREN3283-87. Bearings-Airframe Rolling Rigid, Single Row Ball Bearings in Corrosion Resisting Steel Diameter Series 8 and 9 Dimensions, Torques, Clearances and Loads. 6 pp.

INTERNATIONAL AND NON-U.S. NATIONAL STANDARDS
SUBJECT INDEX
Airport

Airframe Bearings *(Cont.)*
—Ball—Loads (Forces) *(Cont.)*
AECMA PREN3284-87. Bearings-Airframe Rolling Rigid, Single Row Ball Bearings in Steel Diameter Series 0 and 2 Normal Clearance Category Dimensions, Torques, and Loads. 6 pp.

AECMA PREN3285-87. Bearings-Airframe Rolling Rigid, Single Row Ball Bearings in Steel, Cadmium Plated Diameter Series 0 and 2 Normal Clearance Category Dimensions, Torques, and Loads. 6 pp.

AECMA PREN3286-87. Bearings-Airframe Rolling Rigid, Single Row Ball Bearings in Corrosion Resisting Steel Diameter Series 0 and 2 Normal Clearance Category Dimensions, Torques, and Loads. 6 pp.

BSI BS EN 2009-92. 1992 Bearings—Airframe Rolling Rigid, Single Row Ball Bearings in Steel Diameter Series 8 and 9 Dimensions and Loads. 10 pp.

BSI BS EN 2011-89. 1989 Bearings—Airframe Rolling Rigid Single Row Ball Bearings in Corrosion Resisting Steel Diameter Series 8 and 9. Dimensions and Loads. 8 pp.

BSI BS EN 2012-89. 1989 Bearings—Airframe Rolling Rigid, Single Row Ball Bearings in Steel Diameter Series 0 and 2. Dimensions and Loads. 8 pp.

BSI BS EN 2014-89. 1989 Bearings—Airframe Rolling Rigid Single Row Ball Bearings in Corrosion Resisting Steel Diameter Series 0 and 2. Dimensions and Loads. 8 pp.

CEN PREN2009-82. Bearings Airframe Rolling Rigid Single Row Ball Bearings in Steel—Diameter Series 8 and 9—Dimensions and Loads (C1/04).

CEN EN2009-84. Bearings-Airframe Rolling Rigid, Single Roll Ball Bearings in Steel Diameter Series 8 and 9 Dimensions and Loads. 4 pp.

CEN PREN2010-82. Bearings Airframe Rolling Rigid Single Row Ball Bearings in Steel Cadmium Plated—Diameters Series 8 and 9—Dimensions and Loads (C1/04). 5 pp.

CEN PREN2011-82. Bearings Airframe Rolling Rigid Single Row Ball Bearings in Corrosion Resisting Steel—Diameter Series 8 and 9—Dimensions and Loads (C1/04).

CEN EN 2011-84. Bearings-Airframe Rolling Rigid Single Row Ball Bearings in Corrosion Resisting Steel Diameter Series 8 and 9 Dimensions and Loads. 5 pp.

CEN PREN2012-82. Bearings Airframe Rolling Rigid Single Row Ball Bearings in Steel—Diameter Series 0 and 2—Dimensions and Loads (C1/05).

CEN EN 2012-84. Bearings-Airframe Rolling Rigid, Single Row Ball Bearings in Steel Diameter Series 0 and 2 Dimensions and Loads. 5 pp.

CEN PREN2013-82. Bearings Airframe Rolling Rigid Single Row Ball Bearings in Steel Cadmium Plated—Diameter Series 0 and 2—Dimensions and Loads (C1/05). 5 pp.

CEN PREN2014-82. Bearings Airframe Rolling Rigid Single Row Ball Bearings in Corrosion Resisting Steel—Diameter Series 0 and 2—Dimensions and Loads (C1/05).

CEN EN 2014-84. Bearings—Airframe Rolling Rigid Single Row Ball Bearings in Corrosion Resisting Steel Diametes Series 0 and 2 Dimensions and Loads Aerospace Series. 7 pp.

CEN PREN 3059-91. Bearings-Airframe Rolling Rigid Single Row Ball Bearings in Steel with Flanged Alignment Bush Dimensions and Loads. 7 pp.

CEN PREN 3060-91. Bearings-Airframe Rolling Rigid Single Row Ball Bearings in Steel Cadmium Plated with Flanged Alignment Bush Dimensions and Loads. 7 pp.

CEN PREN 3061-91. Bearings-Airframe Rolling Rigid Single Row Ball Bearings in Corrosion Resisting Steel with Flanged Alignment Bush Dimensions and Loads. 7 pp.

—Ball—Self Aligning—Loads (Forces)
AECMA PREN3287-87. Bearings-Airframe Rolling, Double Row Self-Aligning Ball Bearings in Steel Diameter Series 2 Dimensions, Torques, Clearances and Loads. 6 pp.

AECMA PREN3288-87. Bearings-Airframe Rolling Double Row, Self-Aligning Ball Bearings in Steel, Cadmium Plated Diameter Series 2 Dimensions, Torques, Clearances and Loads. 6 pp.

AECMA PREN3289-87. Bearings-Airframe Rolling Double Row, Self-Aligning Ball Bearings in Corrosion Resisting Steel Diameter Series 2 Dimensions, Torques, Clearances and Loads. 6 pp.

BSI BS EN 2015-89. 1989 Bearings—Airframe Rolling Double Row, Self Aligning Ball Bearings in Steel Diameter Series 2. Dimensions and Loads. 8 pp.

BSI BS EN 2017-89. 1989 Bearings—Airframe Rolling Double Row Self Aligning Ball Bearings in Corrosion Resisting Steel Diameter Series 2. Dimensions and Loads. 8 pp.

CEN PREN2015-82. Bearing—Airframe Rolling Double Row Self Aligning Ball Bearings in Steel—Diameter Series—Dimensions and Loads (C1/06).

CEN EN 2015-84. Bearings-Airframe Rolling Double Row, Self Aligning Ball Bearings in Steel Diameter Series 2 Dimensions and Loads. 5 pp.

Airframe Bearings *(Cont.)*
—Ball—Self Aligning—Loads (Forces) *(Cont.)*
CEN PREN2016-82. Bearings—Airframe Rolling Double Row Self Aligning Ball Bearings in Steel Cadmium Plated—Diameter Series 2—Dimensions and Loads (C1/06). 5 pp.

CEN PREN2017-82. Bearings—Airframe Rolling Double Row Self Aligning Ball Bearings in Corrosion Resisting Steel—Diameter Series 2—Dimensions and Loads (C1/06).

CEN EN 2017-84. Bearings-Airframe Rolling Double Row Self Aligning Ball Bearings in Corrosion Resisting Steel Diameter Series 2 Dimensions and Loads. 5 pp.

CEN PREN2018-82. Bearings—Airframe Rolling Single Row Self Aligning Roller Bearings in Steel—Diameter Series 3 and 4—Dimensions and Loads (C1/07).

CEN EN 2018-84. Bearings-Airframe Rolling Single Row, Self Aligning Roller Bearings in Steel Diameter Series 3 and 4 Dimensions and Loads. 5 pp.

CEN PREN2019-82. Bearings—Airframe Rolling Single Row Self Aligning Roller Bearings in Steel Cadmium Plated—Diameter Series 3 and 4—Dimensions and Loads (C1/07). 5 pp.

—Control Wheels—Bushings
SBAC AS 3121 ISSUE 2. Bearing Bush for Pilot's Control Wheel. 1 p.

—Cylindrical
SBAC AGS 3808 (V). Bearings, Airframe, Self Lubricating Plain Cylindrical Aluminium Alloy Shell.

SBAC AGS 3809 (V). Bearings, Airframe, Self Lubricating Plain Cylindrical Steel Shell.

—Cylindrical—Flanged
SBAC AGS 3810 (V). Bearings, Airframe, Self Lubricating Flanged Cylindrical Aluminium Alloy Shell.

SBAC AGS 3811 (V). Bearings, Airframe, Self Lubricating Flanged Cylindrical Steel Shell.

—Filter Heads
SBAC AS 3312 ISSUE 3. Bearing for Filter Head. 1 p.

—Rods
SBAC AGS 3050-69 (V). Rose Self Aligning Spherical Bearing, Rod Ends Unified Female Series.

SBAC AGS 3070-89 (V). Rose Self Aligning Spherical Bearing, Rod Ends Unified Male Series.

—Roller
AECMA PREN2063-75. Airframe Rolling Bearings—Technical Specification (C1/03). 14 pp.

—Roller—Loads (Forces)
BSI BS EN 2063-92. 1992 Airframe Rolling Bearings—Technical Specification. 22 pp.

CEN PREN 2063-91. Airframe Rolling Bearings—Technical Specification. 18 pp.

CEN EN 2063-92. Airframe Rolling Bearings—Technical Specification. 18 pp.

—Roller—Self Aligning—Loads (Forces)
BSI BS EN 2018-89. 1989 Bearings—Airframe Rolling Single Row, Self Aligning Roller Bearings in Steel Diameter Series 3 and 4. Dimensions and Loads. 8 pp.

BSI BS EN 2020-89. 1989 Bearings—Airframe Rolling Single Row Self Aligning Roller Bearings in Corrosion Resisting Steel Diameter Series 3 and 4. Dimensions and Loads. 7 pp.

CEN PREN2020-82. Bearings—Airframe Rolling Single Row Self Aligning Roller Bearings in Corrosion Resisting Steel—Diameter Series 3 and 4—Dimensions and Loads (C1/07).

CEN EN 2020-84. Bearings-Airframe Rolling Single Row Self Aligning Roller Bearings in Corrosion Resisting Steel Diameter Series 3 and 4 Dimensions and Loads. 5 pp.

—Self Aligning
AECMA PREN3280-89. Bearings, Airframe Rolling Rigid or Self-Aligning Technical Specification. 23 pp.

SBAC AGS 3812 (V)(I). Bearing, Self Aligning, Grooved Outer Ring, P.T.F.E. Lined, Wide.

SBAC AGS 3813 (V)(I). Bearing, Self Aligning, Grooved Outer Ring P.T.F.E. Lined, Narrow.

SBAC AGS 3814 (V)(I). Bearing, Self Aligning, Plain Outer Ring, P.T.F.E. Lined Wide.

SBAC AGS 3815 (V)(I). Bearing, Self Aligning, Plain Outer Ring, P.T.F.E. Lined, Narrow.

—Static Load Ratings
BSI 3SP 89-83. 1983 Characteristics, Boundary Dimensions, Tolerances and Static Load Ratings of Airframe Bearings. 20 pp.

Airframe Bearings *(Cont.)*
—Static Load Ratings *(Cont.)*
ISO 1002-83. Rolling Bearings—Airframe Bearings—Characteristics, Boundary Dimensions, Tolerances, Static Load Ratings First Edition. 21 pp.

Airframes
See Also: Aircraft Equipment; Landing Gear
JIS W 0620-87. Aircraft Structural Integrity Program. Airplane Requirements.

—Demand Order Repair Contracts
MOD UK DEFCON 112AP-87. Conditions of Contract Demand Order/Repair Airframes and Equipment 5/87. 5 pp.

—Eddy Current Testing
CAA LEAFLET 4-8 07.90. Eddy Current Methods. 13 pp.

—Insulated Cables
BSI 2G 231-90. 1990 Conductors for General-Purpose Aircraft Electrical Cables and Aerospace Applications. 8 pp.

BSI 2G 232-87. 1987 Electrical Cables for General Airframe or Equipment Interconnect Use (135 Degrees C), Wrapped Insulation. 18 pp.

BSI 2G 233-87. (WITHDRAWN) 1987 Electric Cables for General Airframe or Equipment Interconnect Use (135 Degrees C), Extruded Insulation (Superseded by 3G 233: 1990). 20 pp.

BSI 3G 233-90. 1990 Electric Cables for General Airframe or Equipment Interconnect Use (135 Degrees Celsius), Extruded Insulation. 28 pp.

BSI G 235-87. 1987 Electric Cables for General Airframe or Equipment Interconnect Use (150 Degrees Celsius) Wrapped Insulation with Silver Plated Conductors. 20 pp.

BSI G 236-87. 1987 Electric Cables for General Airframe or Equipment Interconnect Use (200 Degrees Celsius), Wrapped Insulation with Nickel Plated Conductors. 18 pp.

BSI G 237-87. 1987 Electric Cables for General Airframe or Equipment Interconnect Use (200 Degrees Celsius), Extruded Insulation with Nickel Plated Conductors. 19 pp.

BSI G 238-87. 1987 Electric Cables for General Airframe or Equipment Interconnect Use (260 Degrees Celsius), Wrapped Insulation with Nickel Plated Conductors. 18 pp.

—Metal—Aircraft—Repair
CAA LEAFLET 6-4 07.90. Repair of Metal Airframes. 12 pp.

—Radiation Detection and Measurement
CAA LEAFLET 4-6 07.90. Radiological Examination of Aircraft Structures. 13 pp.

—Radiography
CGSB CAN/CGSB-48.16-92. Radiographic Inspection of Aircraft Structures. 24 pp.

Airman
—Aircraft
SBAC RS 186 (V). Standard Airman.

Airmiss
Use: Near Misses (Aircraft)

Airmobile Operations
See Also: Airborne Operations; Military Operations
—Helicopters
NATO STANAG 2904 ED 1 AMD 3-79. Airmobile Operations (ATP-41). 4 pp.

Airplanes
Use: Aircraft

Airport Lighting
Use For: Airfield Lighting; Aviation Lighting
See Also: Aircraft; Airports; Lighting; Runway Lighting; Taxiway Lighting.
BSI BS 3224: Part 1-88. 1988 Lighting Fittings for Civil Land Aerodromes Part 1: Lighting Units for Use in Precision Approach Path Indicator Systems (Supersedes BS 3224: Sections B1 1970, D1: 1960, D2, D6, & D7: 1961). 10 pp.

BSI BS 3224: Part 4-88. 1988 Lighting Fittings for Civil Land Aerodromes Part 4: Elevated Lighting Units (Supersedes BS 3224: Section B1: 1970 & Section D1: 1960). 8 pp.

BSI BS 3224: Part 5-88. 1988 Lighting Fittings for Civil Land Aerodromes Part 5: Inset Lighting Units (Supersedes BS 3224: Section B2: 1970, D3 & D8 1962: D4: 1963, D5: 1964). 8 pp.

BSI BS 3224: Part 6-88. 1988 Lighting Fittings for Civil Land Aerodromes Part 6: Obstacle Lighting Units. 8 pp.

INDUSTRY STANDARDS

Airport Lighting (Cont.)

ICAO 9157 Part 4. Aerodrome Design Manual Part 4 Visual Aids Second Edition—1983; (Amendment 1 09/23/85) (Amendment 2 06/30/86) (Corrigendum 1 03/26/87) (Amendment 3 11/10/87). 199 pp.
JIS W 8302-76. Lamps for Airport and Airway Lighting.
NATO STANAG 3316 ED 8 AMD 0-91. Airfield Lighting. 55 pp.
NATO STANAG 3316 ED 8 AMD 1-91. Airfield Lighting. 56 pp.
NATO STANAG 3316 ED 8 AMD 2-91. Airfield Lighting. 61 pp.

—Cables (Electric)

CSA C22.2 NO 179-M1987. Airport Series Lighting Cables (R 1993); (Gen Instr 1). 21 pp.

—Color Coding

NATO STANAG 3711 ED 1 AMD 6-76. Airfield Marking and Lighting Colour Standards. 16 pp.
NATO STANAG 3711 ED 2 AMD 0-92. Airfield Marking and Lighting Colour Standards. 15 pp.
NATO STANAG 3711 ED 2 AMD 1-92. Airfield Marking and Lighting Colour Standards. 16 pp.

—Floodlights

NATO STANAG 3892 ED 2 AMD 2-86. Movement Area Floodlighting. 8 pp.
NATO STANAG 3892 ED 3 AMD 0-93. Movement Area Floodlighting. 7 pp.

—Lamps

JIS W 8302-76. Lamps for Airport and Airway Lighting.

—Portable

NATO STANAG 3534 ED 3 AMD 0-87. Airfield Portable Lighting. 16 pp.
NATO STANAG 3534 ED 4 AMD 1-88. Airfield Portable Lighting. 15 pp.
NATO STANAG 3534 ED 4 AMD 2-88. Airfield Portable Lighting. 16 pp.
NATO STANAG 3534 ED 4 AMD 3-88. Airfield Portable Lighting. 17 pp.

—Series Isolation Transformers

CSA C22.2 NO 180-M1983. Series Isolating Transformers for Airport Lighting (R 1993); (Gen Instr 1 Thru 2). 14 pp.

—Surface Marking—Colors

JIS W 8301-87. Aeronautical Ground Light and Surface Marking Colours.

Airport Runways

See Also: Aircraft; Aircraft Landing Areas; Airport Taxiways; Airports; Minimum Operating Strips; Pavements; Runway Lighting
CAA Part 3 CAP 168. Physical Characteristics. 43 pp.
ICAO Circular 120. Methodology for the Derivation of Separation Minima Applied to the Spacing Between Parallel Tracks in ATS Route Structures Second Edition—1976. 273 pp.
ICAO Circular 207. Simultaneous Operations on Parallel or Near-Parallel Instrument Runways (Soir)—1988. 41 pp.
ICAO 9137 Part 2. Airport Services Manual Part 2 Pavement Surface Conditions Second Edition—1984; (Corrigendum 1 03/13/85). 99 pp.
ICAO 9157 Part 1. Aerodrome Design Manual Part 1 Runways Second Edition—1984; (Corrigendum 1 10/21/88) (Corrigendum 2 07/31/90) (Amendment 1 01/23/91). 70 pp.
ICAO 9328. Manual of Runway Visual Range Observing and Reporting Practices First Edition—1981; (Corrigendum 1 02/08/82) (Amendment 1 02/13/84). 77 pp.

—Braking Conditions

NATO STANAG 3634 ED 2 AMD 0-90. Runway Breaking Conditions. 8 pp.
NATO STANAG 3634 ED 3 AMD 0-92. Runway Friction and Braking Conditions. 5 pp.
NATO STANAG 3634 ED 3 AMD 1-92. Runaway Friction and Braking Conditions. 7 pp.

—Deicers

MOD UK TS 10228A. Ice Control Agents for Aircraft Runways.

—Friction

NATO STANAG 3634 ED 2 AMD 0-90. Runway Breaking Conditions. 8 pp.
NATO STANAG 3634 ED 3 AMD 0-92. Runway Friction and Braking Conditions. 5 pp.
NATO STANAG 3634 ED 3 AMD 1-92. Runaway Friction and Braking Conditions. 7 pp.
NATO STANAG 3811 ED 1 AMD 1-86. Runway Friction Standards. 11 pp.

—Local Air Space—Obstacles

CAA Part 4 CAP 168. Assessment and Treatment of Obstacles. 31 pp.

Airport Runways (Cont.)

—Marking

NATO STANAG 3158 ED 4 AMD 1-89. Day Marking of Airfield Runways and Taxiways. 20 pp.
NATO STANAG 3158 ED 4 AMD 3-89. Day Marking of Airfield Runways and Taxiways. 20 pp.
NATO STANAG 3158 ED 4 AMD 4-89. Day Marking of Airfield Runways and Taxiways. 20 pp.
NATO STANAG 3158 ED 4 AMD 5-89. Day Marking of Airfield Runways and Taxiways. 21 pp.
NATO STANAG 3685 ED 2 AMD 2-88. Airfield Portable Marking. 11 pp.
NATO STANAG 3685 ED 2 AMD 3-88. Airfield Portable Marking. 12 pp.

—Marking—Obstacles

NATO STANAG 3346 ED 4 AMD 5-85. Marking and Lighting of Airfield Obstructions. 19 pp.
NATO STANAG 3346 ED 5 AMD 0-92. Marking and Lighting of Airfield Obstructions. 13 pp.
NATO STANAG 3346 ED 5 AMD 1-92. Marking and Lighting of Airfield Obstructions. 14 pp.

—Visual Ground Aids

CAA Part 7 CAP 168. Signals, Signs and Markings. 24 pp.

Airport Surveillance Radar

Use: Surveillance Radar—Airports

Airport Taxiways

Use For: Airfield Taxiways; Taxiways
See Also: Aircraft Landing Areas; Airport Runways; Airports; Pavements

—Marking

NATO STANAG 3158 ED 4 AMD 1-89. Day Marking of Airfield Runways and Taxiways. 20 pp.
NATO STANAG 3158 ED 4 AMD 3-89. Day Marking of Airfield Runways and Taxiways. 20 pp.
NATO STANAG 3158 ED 4 AMD 4-89. Day Marking of Airfield Runways and Taxiways. 20 pp.
NATO STANAG 3158 ED 4 AMD 5-89. Day Marking of Airfield Runways and Taxiways. 21 pp.

Airports

Use For: Aerodromes; Air Terminals (Buildings)
See Also: Air Navigation Facilities; Air Traffic Control; Air Transportation; Aircraft; Aircraft Landing Areas; Airport Lighting; Airport Runways; Airport Taxiways; Heliports; Passenger Loading Systems
CAA Part 3 CAP 168. Physical Characteristics. 43 pp.

—Aeronautical Information Services

NATO STANAG 3052 ED 4 AMD 4-81. Aeronautical Briefing Facilities. 13 pp.
NATO STANAG 3052 ED 4 AMD 5-81. Aeronautical Briefing Facilities. 16 pp.

—Air Force Operations

NATO STANAG 3739 ED 3 AMD 2-86. Combined Air Terminal Operations. 12 pp.
NATO STANAG 3739 ED 3 AMD 4-86. Combined Air Terminal Operations. 7 pp.

—All-Weather Operations

ICAO 9365. Manual of All-Weather Operations Second Edition—1991. 69 pp.

—Aprons

ICAO 9157 Part 2. Aerodrome Design Manual Part 2 Taxiways, Aprons and Holding Bays Second Edition—1983; (Corrigendum 1 01/09/84). 144 pp.

—Baggage Claims—Tracing

NATO STANAG 3740 ED 2 AMD 3-86. Procedures for Tracing Air Transported Baggage and Missing Air Transported Cargo. 18 pp.
NATO STANAG 3740 ED 2 AMD 4-86. Procedures for Tracing Air Transported Baggage and Missing Air Transported Cargo. 19 pp.
NATO STANAG 3740 ED 3 AMD 0-93. Procedures for Tracing Air Transported Baggage and Missing Air Transported Cargo. 24 pp.

—Bird Control

ICAO 9137 Part 1. Airport Services Manual Part 1 Rescue and Fire Fighting Third Edition—1990. 223 pp.

—Cargo Claims—Tracing

NATO STANAG 3740 ED 2 AMD 3-86. Procedures for Tracing Air Transported Baggage and Missing Air Transported Cargo. 18 pp.
NATO STANAG 3740 ED 2 AMD 4-86. Procedures for Tracing Air Transported Baggage and Missing Air Transported Cargo. 19 pp.
NATO STANAG 3740 ED 3 AMD 0-93. Procedures for Tracing Air Transported Baggage and Missing Air Transported Cargo. 24 pp.

Airports (Cont.)

—Charges

ICAO 9082. Statements by the Council to Contracting States on Charges for Airports and Air Navigation Services Fourth Edition—1992. 27 pp.
ICAO 9562. Airport Economics Manual First Edition—1991; (Supplement 3/31/93). 120 pp.

—Cost Estimates

ICAO Circular 236. Investment Requirements for Aircraft Fleets and for Airport and Route Facility Infrastructure to the Year 2010—1992. 39 pp.

—Departure—Radiotelephony—Peacetime

NATO STANAG 7012 ED 2 AMD 0-91. Minimum Radiotelephony (R/T) Aerodrome Departure Procedures. 6 pp.

—Departure—Radiotelephony—Times of Tension and War

NATO STANAG 7012 ED 2 AMD 0-91. Minimum Radiotelephony (R/T) Aerodrome Departure Procedures. 6 pp.

—Design

ICAO Annex 14 Vol I. Aerodromes Volume I Aerodrome Design and Operations First Edition—July 1990; (Corrigendum 1 5/31/91) (Supplement 6/15/91). 225 pp.
ICAO 9157 Part 3. Aerodrome Design Manual Part 3 Pavements Second Edition—1983; (Amendment 1 10/25/85) (Amendment 2 08/31/89). 355 pp.
ICAO 9157 Part 5. Aerodrome Design Manual Part 5 Electrical Systems First Edition—1983. 98 pp.

—Electrical Systems

ICAO 9157 Part 5. Aerodrome Design Manual Part 5 Electrical Systems First Edition—1983. 98 pp.

—Emergency Planning

CAA Part 8 CAP 168. Rescue and Fire Fighting Services. 29 pp.
CAA Part 9 CAP 168. Medical Services. 4 pp.
ICAO 9137 Part 7. Airport Services Manual Part 7 Airport Emergency Planning First Edition—1980. 76 pp.

—Fire Extinguishing Agents

CAA Part 8 CAP 168. Rescue and Fire Fighting Services. 29 pp.

—Fire Fighters—Training

CAA Part 8 CAP 168. Rescue and Fire Fighting Services. 29 pp.

—Fire Fighting

ICAO 9137 Part 1. Airport Services Manual Part 1 Rescue and Fire Fighting Third Edition—1990. 223 pp.
NATO STANAG 3712 ED 2 AMD 4-83. Airport Rescue and Fire Fighting Services—Identification Categories. 6 pp.
NATO STANAG 3712 ED 2 AMD 5-83. Airport Rescue and Fire Fighting Services—Identification Categories. 7 pp.

—Flight Information

ICAO 9249. Dynamic Flight-Related Public Information Displays—1978. 21 pp.

—Flight Information—Pilot

CAA Part 10 CAP 168. Aeronautical Information. 6 pp.
ICAO Circular 211. Aerodrome Flight Information Service (AFIS)—1988. 15 pp.

—Glossaries

CAA. Glossary of Terms. 4 pp.

—Holding Bays

ICAO 9157 Part 2. Aerodrome Design Manual Part 2 Taxiways, Aprons and Holding Bays Second Edition—1983; (Corrigendum 1 01/09/84). 144 pp.

—Land Use Planning

SNZ NZS 6805-92. Airport Noise Management and Land Use Planning. 28 pp.

—Licensing

CAA. Check List of Pages. 2 pp.
CAA. Contents. 4 pp.
CAA. Foreword. 2 pp.
CAA. Index. 11 pp.
CAA Part 1 CAP 168. Licensing Process. 19 pp.

—Local Air Space—Obstacles

CAA Part 4 CAP 168. Assessment and Treatment of Obstacles. 31 pp.

—Medical Equipment

CAA Part 9 CAP 168. Medical Services. 4 pp.

—Microwave Landing Systems

ICAO Circular 165. Microwave Landing System (MLS) Advisory Circular Issue No. 1—1981. 38 pp.

INTERNATIONAL AND NON-U.S. NATIONAL STANDARDS
SUBJECT INDEX
Airworthiness

Airports *(Cont.)*
—Noise
ICAO Circular 205. Recommended Method for Computing Noise Contours Around Airports—1988 (Supersedes Circular 116). 37 pp.
SNZ NZS 6805-92. Airport Noise Management and Land Use Planning. 28 pp.

—Operations
ICAO Annex 14 Vol I. Aerodromes Volume I Aerodrome Design and Operations First Edition—July 1990; (Corrigendum 1 5/31/91) (Supplement 6/15/91). 225 pp.

—Operations—Technical Manuals
CAA Part 2 CAP 168. Aerodrome Manual. 17 pp.

—Passports
ICAO 9303 Part 1. Machine Readable Travel Documents Part 1 Machine Readable Passports Third Edition—1992. 49 pp.

—Pavements
ICAO 9157 Part 3. Aerodrome Design Manual Part 3 Pavements Second Edition—1983; (Amendment 1 10/25/85) (Amendment 2 08/31/89). 355 pp.

—Pavements—Static Loads
CNS A3290-88. Method of Test for Nonrepetitive Static Plate Load of Soils and Flexible Pavement Components, for Use in Evaluation and Design of Airport and Highway Pavements (Aug)(12392).
CNS A3291-88. Method of Test for Repetitive Static Plate Load of Soils and Flexible Pavement Components, for Use in Evaluation and Design of Airport and Highway Pavements (Aug)(12393).

—Planning
ICAO 9184 Part 1. Airport Planning Manual Part 1. Master Planning Second Edition—1987; (Corrigendum 2 07/31/90). 86 pp.
ICAO 9184 Part 2. Airport Planning Manual Part 2 Land Use and Environmental Control Second Edition—1985; (Corrigendum 05/14/85). 46 pp.
ICAO 9184 Part 3. Airport Planning Manual Part 3 Guidelines for Consultant/Construction Services First Edition—1983. 106 pp.

—Public Information Displays
ICAO 9249. Dynamic Flight-Related Public Information Displays—1978. 21 pp.

—Rescue Equipment
CAA Part 8 CAP 168. Rescue and Fire Fighting Services. 29 pp.

—Rescue Systems
ICAO 9137 Part 1. Airport Services Manual Part 1 Rescue and Fire Fighting Third Edition—1990. 223 pp.

—Service Contracts
MOD UK DEFCON 112AG-83. Conditions of Contract for Aircraft Maintenance and Airfield Services 9/83. 6 pp.

—Services
CAA Part 8 CAP 168. Rescue and Fire Fighting Services. 29 pp.
CAA Part 9 CAP 168. Medical Services. 4 pp.
ICAO 9137 Part 1. Airport Services Manual Part 1 Rescue and Fire Fighting Third Edition—1990. 223 pp.
ICAO 9137 Part 2. Airport Services Manual Part 2 Pavement Surface Conditions Second Edition—1984; (Corrigendum 1 03/13/85). 99 pp.
ICAO 9137 Part 3. Airport Services Manual Part 3 Bird Control and Reduction Third Edition—1991. 26 pp.
ICAO 9137 Part 5. Airport Services Manual Part 5 Removal of Disabled Aircraft Second Edition—1986; (Corrigendum 07/24/89) (Corrigendum 3 05/29/90). 39 pp.
ICAO 9137 Part 6. Airport Services Manual Part 6 Control of Obstacles Second Edition—1983. 72 pp.
ICAO 9137 Part 7. Airport Services Manual Part 7 Airport Emergency Planning First Edition—1980. 76 pp.
ICAO 9137 Part 8. Airport Services Manual Part 8 Airport Operational Services First Edition—1983. 62 pp.
ICAO 9137 Part 9. Airport Services Manual Part 9 Airport Maintenance Practices First Edition—1984. 55 pp.

—Short Takeoff and Landing
ICAO 9150. Stolport Manual First Edition—1976. 68 pp.

—Soils—Static Loads
CNS A3290-88. Method of Test for Nonrepetitive Static Plate Load of Soils and Flexible Pavement Components, for Use in Evaluation and Design of Airport and Highway Pavements (Aug)(12392).

Airports *(Cont.)*
—Soils—Static Loads *(Cont.)*
CNS A3291-88. Method of Test for Repetitive Static Plate Load of Soils and Flexible Pavement Components, for Use in Evaluation and Design of Airport and Highway Pavements (Aug)(12393).

—Surface Movement
ICAO 9476. Manual of Surface Movement Guidance and Control Systems (SMGCS) First Edition—1986; (Amendment 1 05/06/87) (Amendment 2 12/01/87). 89 pp.

—Symbols—International
ICAO 9430. International Signs to Provide Guidance to Persons at Airports—1984. 49 pp.

—Tariffs
ICAO 7100. Manual of Airport and Air Navigation Facility Tariffs—1990. 438 pp.

—Taxiways
CAA Part 3 CAP 168. Physical Characteristics. 43 pp.
ICAO 9157 Part 2. Aerodrome Design Manual Part 2 Taxiways, Aprons and Holding Bays Second Edition—1983; (Corrigendum 1 01/09/84). 144 pp.

—Visual Ground Aids
CAA Part 7 CAP 168. Signals, Signs and Markings. 24 pp.
ICAO 9157 Part 4. Aerodrome Design Manual Part 4 Visual Aids Second Edition—1983; (Amendment 1 09/23/85) (Amendment 2 06/30/86) (Corrigendum 1 03/26/87) (Amendment 3 11/10/87). 199 pp.

Airscrews
Use: Aircraft Propellers

Airships
See Also: Aircraft; SKS 500 Airship; SKS 600 Airship; Skyship 500 Airship; Skyship 600 Airships; Thunder and Colt GA 42 Airships

—Data Sheets
CAA BAS1 ISSUE 1. Airship Industries. 2 pp.
CAA BAS2 ISSUE 1. Airship Industries. 3 pp.

Airslide Conveyors
See Also: Conveyors
ISO 2326-72. Continuous Mechanical Handling Equipment for Loose Bulk Materials—Aeroslides First Edition. 3 pp.

Airspace
See Also: Boundaries; Flight Levels; Flight Maneuvers; Flight Paths; Vertical Separation

—Air Traffic Control—Combat Zones—Allied Tactical Publications
NATO STANAG 3805 ED 2 AMD 0-80. Doctrine and Procedures for Airspace Control in the Combat Zone. 5 pp.
NATO STANAG 3805 ED 3 AMD 0-89. Doctrine and Procedures for Airspace Control in the Combat Zone—ATP-40. 5 pp.

Airspeed
See Also: Flight Maneuvers; Flight Paths; Mach Number; Speed

—Tables (Data)
BSI 2G 199-84. 1984 Amd 1 Tables Relating to Altitudes, Airspeed and Mach Numbers for Use in Aeronautical Instrument Design and Calibration. 41 pp.

Airspeed Ambassador Aircraft
See Also: Aircraft

—Accidents
CAA. Airspeed Ambassador. 1 p.

Airspeed Indicators
See Also: Aircraft; Aircraft Instruments; Machmeters; Measuring Instruments; Navigational Aids; Pitot Tubes
BSI 2G 117-75. 1975 Pressure-Driven Airspeed Indicators. 3 pp.
MOD UK DSTAN 66-36-86. Airspeed Indicators Imperial and Metric Issue 1. 22 pp.
MOD UK DSTAN 66-42-87. General Requirements for Aircraft Instruments and Displays Combined Speed Indicators Issue 1. 12 pp.
NATO STANAG 3451 ED 2 AMD 3-84. Combined Speed Indicator. 16 pp.
NATO STANAG 3451 ED 2 AMD 4-84. Combined Speed Indicator. 17 pp.
NATO STANAG 3636 ED 3 AMD 4-81. Airspeed Indicators (Imperial and Metric). 24 pp.
NATO STANAG 3636 ED 3 AMD 5-81. Airspeed Indicators (Imperial and Metric). 25 pp.

Airspeed Indicators *(Cont.)*
—Differential Pressure
JIS W 0202-79. Table of Differential Pressure for Airspeed Indicators.

—Foreign Airworthiness Directives
CAA. Aircraft Instruments (Foreign Airworthiness Directives). 1 p.

—Gliders
CAA JAR-22 Appendix F. Glossary of Aerobatic Manoeuvres (Joint Airworthiness Requirements). 2 pp.

Airtourer T6-24 Aircraft
See Also: Aircraft

—Antenna Positions
CAA. Airtourer T6-24, 100 and 150 (Approved Aerial Positions). 1 p.

Airtourer 100 Aircraft
See Also: Aircraft

—Antenna Positions
CAA. Airtourer T6-24, 100 and 150 (Approved Aerial Positions). 1 p.

Airtourer 150 Aircraft
See Also: Aircraft

—Antenna Positions
CAA. Airtourer T6-24, 100 and 150 (Approved Aerial Positions). 1 p.

Airways
—Navigational Lights
JIS W 8302-76. Lamps for Airport and Airway Lighting.

Airways Communications System
Use: AIRCOM

Airworthiness Directives, Foreign
Use: Foreign Airworthiness Directives

Airworthiness Leaflets
See Also: Joint Airworthiness Requirements Administration
CAA LEAFLET 1-1 07.90. Approval of Organisations. 3 pp.
CAA LEAFLET 1-2 07.90. Legislation and Requirements. 7 pp.
CAA LEAFLET 1-3 07.90. Occurance Reporting and the Engineer's Role. 7 pp.
CAA LEAFLET 1-4 07.90. Weight and Balance of Aircraft. 27 pp.
CAA LEAFLET 1-5 07.90. Aircraft, Engine and Propeller Log Books. 8 pp.
CAA LEAFLET 1-6 07.90. Maintenance of Aircraft Not Exceeding 2730 kg MTWA, Including the Star Inspection. 22 pp.
CAA LEAFLET 1-7 12.90. Condition Monitored Maintenance. 29 pp.
CAA LEAFLET 1-8 12.90. Storage Conditions for Aeronautical Supplies. 15 pp.
CAA LEAFLET 1-9 06.91. Concessions During Manufacture. 2 pp.
CAA LEAFLET 1-10 06.92. Aeronautical Maintenance Certificate. 2 pp.
CAA LEAFLET 1-11 06.92. Introduction to the Licensing of Aircraft Maintenance Engineers. 7 pp.
CAA LEAFLET 1-12 06.92. Airworthiness Publications—General Information. 14 pp.
CAA LEAFLET 2-1 07.90. Engineering Drawings. 17 pp.
CAA LEAFLET 2-2 07.90. Clean Rooms. 19 pp.
CAA LEAFLET 2-3 07.90. Timber Conversion—Spruce. 4 pp.
CAA LEAFLET 2-4 07.90. Synthetic Resin Adhesives. 7 pp.
CAA LEAFLET 2-6 07.90. Cleanliness of Aircraft. 7 pp.
CAA LEAFLET 2-7 07.90. Paint Finishing of Metal Aircraft. 10 pp.
CAA LEAFLET 2-8 07.90. Fabric Covering. 15 pp.
CAA LEAFLET 2-9 07.90. Doping. 10 pp.
CAA LEAFLET 2-10 07.90. Thread Inserts. 8 pp.
CAA LEAFLET 2-11 07.90. Torque Loading. 4 pp.
CAA LEAFLET 2-12 07.90. Cable—Splicing and Swaging. 10 pp.
CAA LEAFLET 2-13 07.90. Control Systems. 13 pp.
CAA LEAFLET 3-1 07.90. Identification Marking Processes for Aircraft Parts. 11 pp.
CAA LEAFLET 3-2 07.90. Identification Markings on Metallic Materials. 7 pp.
CAA LEAFLET 3-3 07.90. Bolts and Screws of British Manufacture. 11 pp.
CAA LEAFLET 3-4 07.90. Nuts of British Manufacture. 14 pp.
CAA LEAFLET 3-5 07.90. Standard Fasteners of American Manufacture. 26 pp.

INDUSTRY STANDARDS

INTERNATIONAL AND NON-U.S. NATIONAL STANDARDS
SUBJECT INDEX

Airworthiness

Airworthiness Leaflets (Cont.)

CAA LEAFLET 4-1 07.90. Oil and Chalk Processes. 4 pp.
CAA LEAFLET 4-2 07.90. Penetrant Dye Processes. 6 pp.
CAA LEAFLET 4-3 07.90. Fluorescent Penetrant Processes. 6 pp.
CAA LEAFLET 4-4 07.90. Performance Testing of Penetrant Testing Materials. 3 pp.
CAA LEAFLET 4-5 07.90. Ultrasonic Flaw Detection and Thickness Measurement. 12 pp.
CAA LEAFLET 4-6 07.90. Radiological Examination of Aircraft Structures. 13 pp.
CAA LEAFLET 4-7 07.90. Magnetic Flaw Detection. 11 pp.
CAA LEAFLET 4-8 07.90. Eddy Current Methods. 13 pp.
CAA LEAFLET 5-2 07.90. Lifejackets. 7 pp.
CAA LEAFLET 5-3 07.90. Carbon Monoxide Contamination. 4 pp.
CAA LEAFLET 5-4 07.90. Control Chains, Chain Wheels and Pulleys. 8 pp.
CAA LEAFLET 5-5 07.90. Hose and Hose Assemblies. 17 pp.
CAA LEAFLET 5-6 07.90. Installation and Maintenance of Rigid Pipes. 8 pp.
CAA LEAFLET 5-7 07.90. Tyres. 20 pp.
CAA LEAFLET 5-8 07.90. Wheels and Brakes. 19 pp.
CAA LEAFLET 5-9 07.90. Oxygen Systems. 17 pp.
CAA LEAFLET 6-1 07.90. Inspection of Wooden Structures. 8 pp.
CAA LEAFLET 6-3 07.90. Inspection of Metal Aircraft After Abnormal Occurrences. 8 pp.
CAA LEAFLET 6-4 07.90. Repair of Metal Airframes. 12 pp.
CAA LEAFLET 6-6 07.90. Effect of Disturbed Airflow on Aeroplane Behaviour. 8 pp.
CAA LEAFLET 6-7 07.90. Assembly and Maintenance of Critical Bolted Joints. 7 pp.
CAA LEAFLET 6-8 07.90. Glass Windscreen Assemblies. 7 pp.
CAA LEAFLET 7-1 12.90. Piston Engine Overhaul—Test Requirements for Overhauled Engines. 14 pp.
CAA LEAFLET 7-2 07.90. Piston Engine Overhaul—Dynamometer Testing of Overhauled Engines. 8 pp.
CAA LEAFLET 7-3 12.90. Piston Engine Overhaul—Fan Testing of Overhauled Engines. 12 pp.
CAA LEAFLET 7-4 12.90. Storage Procedures. 5 pp.
CAA LEAFLET 7-5 12.90. Piston Engine Overhaul—Correcting Engine Test Results. 10 pp.
CAA LEAFLET 7-6 07.90. Piston Engines—Operation Beyond Recommended Overhaul Periods. 4 pp.
CAA LEAFLET 8-1 07.90. Compass Base Surveying. 8 pp.
CAA LEAFLET 9-1 07.90. Bonding and Circuit Testing. 12 pp.
CAA LEAFLET 9-2 07.90. Charging Rooms for Aircraft Batteries. 5 pp.
CAA LEAFLET 9-3 07.90. Cables—Installation and Maintenance. 13 pp.
CAA LEAFLET 9-4 07.90. Antistatic Protection. 8 pp.
CAA LEAFLET 10-1 07.90. Aircraft Handling. 15 pp.
CAA LEAFLET 11-1 07.90. Materials, Parts or Appliances Approved to USA Technical Standards Orders (TSO's) for Part 21 Sub-Part O. 3 pp.
CAA LEAFLET 11-2 07.90. Information and Advice on the Procedures for the Issue of a Permit to Fly for Amateur Constructed Aeroplanes. 3 pp.
CAA LEAFLET 11-3 07.90. CAA Flight Testing Policy. 2 pp.
CAA LEAFLET 11-4 12.90. UK Certification of Imported Aircraft Not Exceeding 2730 kg MTWA Which Are Eligible for the Issue of Airworthiness. 6 pp.
CAA LEAFLET 11-5 12.90. Aircraft Electrical Cables. 19 pp.
CAA LEAFLET 11-6 12.90. Use of 'B' Conditions Overseas. 4 pp.
CAA LEAFLET 11-7 06.92. Authorisation of Personnel to Issue Certificates of Release to Service (BCAR Chapter A8-13). 3 pp.
CAA LEAFLET 11-8 06.92. Approval-Certification Authorisations Under Limited and Simple Procedures (BCAR Chapter A8-13). 6 pp.
CAA LEAFLET 11-9 06.92. CAA Recognition of Aircraft Maintenance Engineer Type Training Courses for License Endorsement. 3 pp.
CAA LEAFLET 11-10 06.92. Electrical Generation Systems—Bus-Bar Low Voltage Warning Single Engined Aircraft with a UK Certificate of Airworthiness. 5 pp.
CAA LEAFLET 11-11 06.92. Steep Approaches. 3 pp.
CAA LEAFLET 11-12 06.92. Installation of High Intensity Strobe Lights (HISL) on Helicopters. 6 pp.
CAA LEAFLET 11-13 06.92. DH 82 Tiger Moth Anti-Spin Strakes. 2 pp.
CAA LEAFLET 11-14. Design Requirements and Quality Assurance Associated with Bearing Assemblies. 2 pp.
CAA LEAFLET 11-15 07.90. CAA Approval of Aircraft Maintenance Engineer Type Training Arrangements Within A8-13 and A8-18. 1 p.

Airworthiness Leaflets (Cont.)

CAA LEAFLET 12-1 07.90. Aerospace Specifications. 4 pp.
CAA LEAFLET 13-1 07.90. Cap Listing. 8 pp.
CAA LEAFLET 13-2 07.90. Nationality and Registration Marks. 10 pp.
CAA LEAFLET 13-3 07.90. Mandatory Occurrence Reporting Scheme. 47 pp.
CAA LEAFLET 13-4 07.90. Condition Monitored Maintenance—An Explanatory Handbook. 36 pp.
CAA LEAFLET 13-5 12.90. Light Aircraft Maintenance. 41 pp.
CAA LEAFLET 13-6. Ground De-Icing of Aircraft. 18 pp.
CAA LEAFLET 14-0 06.91. Standard Maintenance Practices—Introduction. 2 pp.
CAA LEAFLET 14-1 06.91. Midas Flight Data Recording System Type CMM/3RB/2. 3 pp.
CAA LEAFLET 14-2 06.91. Sadas Flight Data Recording Equipment. 3 pp.
CAA LEAFLET 14-3 06.91. Plessey/Davall Types PV710 and PV710A Data Recorders. 3 pp.
CAA LEAFLET 14-4 06.91. Sunstrand Flight Data Recorder Model FB542. 3 pp.
CAA LEAFLET 14-6 06.91. EFDAS Flgiht Data Recording Equipment AO and AO1 Systems. 3 pp.
CAA LEAFLET 14-7 06.91. Plessey/Davall Date Recorder Sustems. 3 pp.
CAA LEAFLET 14-8 06.91. Airworthiness Classifications of Items in Approved Maintenance Schedules. 1 p.
CAA LEAFLET 14-10 06.91. Plessey Flight Data Recorder System Type PV1584 Series. 3 pp.
CAA LEAFLET 14-11 06.91. Specific Maintenance Schedules for Aircraft Below 2730 kg. 3 pp.
CAA LEAFLET 14-13 06.91. Universal/Sundstrand Flight Data Recorder Model 980-4100 Series. 3 pp.
CAA LEAFLET 14-14 06.91. Cockpit Voice Recorder System Fairchild A100 Series. 2 pp.
CAA LEAFLET 14-15 06.92. Maintenance and Inspection of Crew Harnesses and Passenger Seat Belts (Metal to Metal Attachment). 1 p.
CAA LEAFLET 14-16 06.92. Cockpit Voice Recorder System Sundstrand AV557 Series. 2 pp.
CAA LEAFLET 14-17. N15F210B Underwater Locating Beacon. 3 pp.
CAA LEAFLET 14-18. Air Cruisers Company Evacuation Systems. 1 p.
CAA INFORMATION LEAFLETS.
CAA LEAFLET #1. Structure, Committees and Groups (Information Leaflets). 32 pp.
CAA LEAFLET #2 01.93. Bodies Directly Involved or Interested in Airworthiness in Europe (Information Leaflets). 4 pp.
CAA LEAFLET #3 01.93. JAR Directory (Information Leaflets). 1 p.
CAA LEAFLET #4 01.93. Review of Amendments (Information Leaflets). 10 pp.
CAA LEAFLET #5 01.93. (Information Leaflets). 3 pp.
CAA LEAFLET #6 01.93. Adoption of JAR's (Information Leaflets). 7 pp.
CAA LEAFLET #7 01.93. JAR-25/FAR Part 25. Comparison Showing Where JAR-25 is Less Severe Than FAR Part 25 (Information Leaflets). 1 p.
CAA LEAFLET #8 01.93. JAA Secretariat File References (Information Leaflets). 1 p.
CAA LEAFLET #9 01.93. Complimentary Copies of JAR Publications (Information Leaflets). 1 p.
CAA LEAFLET #10 01.93. Discussion Papers (Information Leaflets). 1 p.
CAA LEAFLET #11 01.93. Lead Airlines and Lead Authorities to Deal with Emergencies on Specific Aeroplane Types (Information Leaflets). 3 pp.
CAA LEAFLET #12 01.93. Arrangements Concerning the Development and the Acceptance of Joint Airworthiness Requirements (Information Leaflets). 28 pp.
CAA LEAFLET #12A 01.93. Memorandum of Understanding Between Certain European Civil Aviation Authorities on Future Airworthiness Procedures. 21 pp.
CAA LEAFLET #13 01.93. Contacts with FAA (Information Leaflets). 4 pp.
CAA LEAFLET #14 01.93. Type Certification to a JAR Code of an Aeroplane Originating from a Non-JAR Country. 6 pp.
CAA LEAFLET #15 01.93. Type Certification Basis. 10 pp.
CAA LEAFLET #16 01.93. Special Conditions. 3 pp.
CAA LEAFLET #17 01.93. Reserved. 1 p.
CAA LEAFLET #18 01.93. Procedure for Establishing the Joint Type Certification Basis for Derivative Large Aeroplanes. 22 pp.

Airworthiness Notices

See Also: Joint Airworthiness Requirements Administration

BSI 3G 100: Part 0-80. 1980 Amd 1 General Requirements for Equipment for Use in Aircraft Part 0: Introduction (AMD 5366) April 30, 1987. 5 pp.
CAA. Contents Issue 110. 7 pp.
CAA. Information Sheet No. 1 08.89. 1 p.

Airworthiness Notices (Cont.)

CAA. Information Sheet No. 3 03.93. 20 pp.
CAA NOTICE #1 ISSUE 11. Foreword. 25 pp.
CAA NOTICE #2 ISSUE 1. Certificates of Airworthiness—Grant of Supplementary Approval Rating to Recommend C of A Renewal. 5 pp.
CAA NOTICE #3 ISSUE 9. Type Rated Licensed Aircraft Maintenance Engineers. 20 pp.
CAA NOTICE #4 ISSUE 14. Propellers Approved for Use on Civil Aircraft Constructed in the United Kingdom. 2 pp.
CAA NOTICE #5 ISSUE 1. Tyre Wear Limitations. 11 pp.
CAA NOTICE #6 ISSUE 37. Airworthiness Publications—General Information (Appendices Included). 19 pp.
CAA NOTICE #7 ISSUE 47. Airworthiness Publications—Prices, Publication Dates and Issue Numbers. 12 pp.
CAA NOTICE #8 ISSUE 1. Cessna 300 and 400 Series Aircraft—Fuel Icing. 1 p.
CAA NOTICE #9 ISSUE 2. Piston Engines Supplied in Kit Form (Airworthiness Notices). 2 pp.
CAA NOTICE #10 ISSUE 13. Aircraft Maintenance Engineers' Licences—Type Ratings (Airworthiness Notices). 9 pp.
CAA NOTICE #12 ISSUE 39. Experience from Incidents (Appendices Included). 4 pp.
CAA NOTICE #13 ISSUE 1. Design Organisation Approvals. 4 pp.
CAA NOTICE #14 ISSUE 3. Approved Maintenance Organisations (JAR 145)—Implementation Procedure. 6 pp.
CAA NOTICE #15 ISSUE 3. U.K. Certification of Foreign Aircraft of MIWA Not Exceeding 5700 kg (Airworthiness Notices). 4 pp.
CAA NOTICE #16 ISSUE 9. Aircraft Engines, Propellers and Related Equipment Obtained from Sources Not Under the Airworthiness Control of the CAA (Airworthiness Notices). 6 pp.
CAA NOTICE #17 ISSUE 2. Acceptance of Aircraft Components (Appendices Included). 12 pp.
CAA NOTICE #18 ISSUE 6. Acceptance Standards for the Maintenance Overhaul and Repair of Second-Hand Imported Aircraft for Which a UK C of A is Sought (Airworthiness Notices). 2 pp.
CAA NOTICE #19 ISSUE 10. Problem of Bogus Parts (Airworthiness Notices). 5 pp.
CAA NOTICE #20 ISSUE 6. Cotton, Linen and Synthetic Fabric Covered Aircraft (Airworthiness Notices). 3 pp.
CAA NOTICE #21 ISSUE 3. Microbiological Contamination of Fuel Tanks of Turbine Engined Aircraft (Airworthiness Notices). 3 pp.
CAA NOTICE #22 ISSUE 6. Overseas Airworthiness Authorities (Airworthiness Notices). 2 pp.
CAA NOTICE #23 ISSUE 1. Fuel Additives—Health Hazards (Airworthiness Notices). 2 pp.
CAA NOTICE #24 ISSUE 31. UK Airworthiness Course (Airworthiness Notices). 6 pp.
CAA NOTICE #25 ISSUE 19. Charges for the CAA's Airworthiness and Noise Certification (Airworthiness Notices). 13 pp.
CAA NOTICE #26 ISSUE 1. Information for Continued Airworthiness of UK Manufactured Aeroplanes. 5 pp.
CAA NOTICE #27 ISSUE 2. Helicopter Emergency Escape Facilities. 4 pp.
CAA NOTICE #32 ISSUE 2. Overhauls, Modifications, Repairs and Replacements to Aircraft Not Exceeding 2730 kg with Certificates of Airworthiness in the Special Category (Airworthiness Notices). 2 pp.
CAA NOTICE #33 ISSUE 3. Unprotected Starter Circuits in Aircraft Not Exceeding 12500 lb (Airworthiness Notices). 2 pp.
CAA NOTICE #34 ISSUE 6. Civil Aviation Authority—Bureau Veritas Agreement (Airworthiness Notices). 4 pp.
CAA NOTICE #34A ISSUE 1. Civil Aviation Authority—Bureau Veritas Arrangement (Airworthiness Notices). 4 pp.
CAA NOTICE #35 ISSUE 17. Light Aircraft Piston Engine Overhaul Periods (Airworthiness Notices). 7 pp.
CAA NOTICE #36 ISSUE 10. Mandatory Modifications and Inspections (Airworthiness Notices). 6 pp.
CAA NOTICE #38 ISSUE 1. Painting of Aircraft. 5 pp.
CAA NOTICE #39 ISSUE 4. Selection and Procurement of Electronic Components (Airworthiness Notices). 4 pp.
CAA NOTICE #40 ISSUE 1. Carbon Monoxide Contamination in Aircraft (Airworthiness Notices). 2 pp.
CAA NOTICE #41 ISSUE 8. Maintenance of Cockpit and Cabin Combustion Heaters and Their Associated Exhaust Systems (Airworthiness Notices). 2 pp.
CAA NOTICE #42 ISSUE 1. Internal Emergency Lighting Systems (Airworthiness Notices). 2 pp.
CAA NOTICE #43 ISSUE 2. Aircraft Type Certificates Data Sheets (Airworthiness Notices). 2 pp.

INDEX and DIRECTORY of

INTERNATIONAL AND NON-U.S. NATIONAL STANDARDS
SUBJECT INDEX
Alarm

Airworthiness Notices *(Cont.)*
CAA NOTICE #44 ISSUE 5. Gas Turbine Engine Parts Subject to Retirement or Ultimate (Scrap) Lives (Airworthiness Notices). 3 pp.
CAA NOTICE #45 ISSUE 1. Software Management (Airworthiness Notices). 4 pp.
CAA NOTICE #45A ISSUE 1. Software Management and Certification Guidelines (Appendices Included) (Airworthiness Notices). 7 pp.
CAA NOTICE #46 ISSUE 7. Aircraft Maintenance Engineers Licensing—BCAR Section L Issue 7 Ratings (Airworthiness Notices). 2 pp.
CAA NOTICE #48 ISSUE 1. Airworthiness Flight Tests (Airworthiness Notices). 1 p.
CAA NOTICE #49 ISSUE 1. Eligibility of Organisations for Design and Manufacturing Approvals. 5 pp.
CAA NOTICE #50 ISSUE 3. Deterioration of Wooden Aircraft Structures (Airworthiness Notices). 2 pp.
CAA NOTICE #51 ISSUE 1. Aerobatic Maneouvres (Airworthiness Notices). 2 pp.
CAA NOTICE #52 ISSUE 1. Flight in UK Airspace of Foreign-Registered Home-Built Aircraft (Airworthiness Notices). 4 pp.
CAA NOTICE #53 ISSUE 1. Vertical Speed Indicators on Imported Aircraft (Airworthiness Notices). 2 pp.
CAA NOTICE #54 ISSUE 1. Instruments with Unusual Presentations (Airworthiness Notices). 2 pp.
CAA NOTICE #55 ISSUE 2. Routine Maintenance Propeller Blades (Airworthiness Notices). 2 pp.
CAA NOTICE #56 ISSUE 4. Floor Proximity Emergency Escape Path Marking (Airworthiness Notices). 6 pp.
CAA NOTICE #57 ISSUE 1. Toilet Flush Motors (Airworthiness Notices). 2 pp.
CAA NOTICE #58 ISSUE 4. Flame Resistant Furnishing Materials (Airworthiness Notices). 2 pp.
CAA NOTICE #59 ISSUE 3. (This Notice Gives Details of a Mandatory Action) Aircraft Seats—Resistance to Fire (Airworthiness Notices). 3 pp.
CAA NOTICE #60 ISSUE 1. Cabin and Toilet Fire Protection (Airworthiness Notices). 3 pp.
CAA NOTICE #61 ISSUE 2. Improved Flammability Test Standards for Cabin Interior Materials (Airworthiness Notices). 4 pp.
CAA NOTICE #62 ISSUE 3. Fatigue Lives (Airworthiness Notices). 2 pp.
CAA NOTICE #63 ISSUE 9. Certification and Maintenance of Aircraft Not Exceeding 2730 kg (Airworthiness Notices). 4 pp.
CAA NOTICE #64 ISSUE 1. Minimum Space for Seated Passengers. 5 pp.
CAA NOTICE #65 ISSUE 1. CAA Use of Confidential Information (Airworthiness Notices). 3 pp.
CAA NOTICE #66 ISSUE 2. Aircraft Insurance (Airworthiness Notices). 1 p.
CAA NOTICE #67 ISSUE 1. Portable Oxygen Equipment Pressure Relief. 2 pp.
CAA NOTICE #68 ISSUE 1. ATC Transponders and Traffic Alert and Collision Avoidance Systems (TCAS) Ground Testing. 4 pp.
CAA NOTICE #69 ISSUE 1. CAA Approval of Test Houses Holding—Names Accreditation (Airworthiness Notices). 3 pp.
CAA NOTICE #70 ISSUE 2. Tyre Bursts in Flight—Inflation Media (Airworthiness Notices). 2 pp.
CAA NOTICE #71 ISSUE 2. Filament Wound Fibre Glass High Pressure Vessels (Airworthiness Notices). 2 pp.
CAA NOTICE #72 ISSUE 1. Registrations of Aircraft and Continuing Airworthiness (Airworthiness Notices). 1 p.
CAA NOTICE #74 ISSUE 3. Airworthiness Concessions in Respect of Foreign Built Aeroplanes (Appendices Included). 2 pp.
CAA NOTICE #75 ISSUE 9. Overhaul and Inspection Requirements for Variable Pitch Propellers (Airworthiness Notices). 4 pp.
CAA NOTICE #76 ISSUE 3. (This Notice Gives Details of Mandatory Action) Electrical Supplies for Aircraft Radio Systems (Airworthiness Notices). 3 pp.
CAA NOTICE #77 ISSUE 1. Counter/Pointer Altimeters (Airworthiness Notices). 2 pp.
CAA NOTICE #80 ISSUE 2. Class C and D Cargo or Baggage Compartment-Fire Containment Capability (Airworthiness Notices). 3 pp.
CAA NOTICE #81 ISSUE 1. Emergency Power Supply for Electrically Operated Gyroscopic Bank and Pitch Indicators (Artificial Horizons) (Airworthiness Notices). 5 pp.
CAA NOTICE #82 ISSUE 1. Electrical Generation Systems Aircraft Not Exceeding 5700 kg Maximum Authorised Weight (Airworthiness Notices). 4 pp.
CAA NOTICE #83 ISSUE 3. Fire Precautions—Aircraft Toilets (Airworthiness Notices). 4 pp.
CAA NOTICE #85 ISSUE 2. (This Notice Gives Details of a Mandatory Modification) Automatic Direction Finding Equipment on Turbine-Engined Aeroplanes and Helicopters (Airworthiness Notices). 2 pp.

Airworthiness Notices *(Cont.)*
CAA NOTICE #86 ISSUE 3. Communication on Safety Matters (Airworthiness Notices). 2 pp.
CAA NOTICE #87 ISSUE 1. (This Notice Gives Details of a Mandatory Action) Failure of Mechanical Products Inc. Circuit Breakers (Airworthiness Notices). 2 pp.
CAA NOTICE #88 ISSUE 2. (This Notice Gives Details of a Mandatory Action) Electrical Generation Systems—Single-Engined Aircraft (Airworthiness Notices). 3 pp.
CAA NOTICE #90 ISSUE 1. Maximum Total Weight Authorised for Agricultural Operations and Other Aerial Applications (Airworthiness Notices). 3 pp.
CAA NOTICE #91 ISSUE 2. (This Notice Gives Details of a Mandatory Action) Communications Transmitters in the VHF Radio Telephony Band 118—136 MHz (Airworthiness Notices). 2 pp.
CAA NOTICE #92 ISSUE 1. (This Notice Gives Details of a Mandatory Action) Cargo Containment (Airworthiness Notices). 3 pp.
CAA NOTICE #93 ISSUE 3. (This Notice Gives Details of a Mandatory Action) Tyres and Wheels Fitted to Aircraft Certificated in the Transport Category (Airworthiness Notices). 3 pp.
CAA NOTICE #94 ISSUE 6. Non-Destructive Testing of Aircraft, Engines and Components (Airworthiness Notices). 8 pp.
CAA NOTICE #95 ISSUE 1. Use of High Intensity Ultra-Violet Lamps in Fluorescent Penetrant and Magnetic Particle Inspections (Airworthiness Notices). 3 pp.
CAA NOTICE #97 ISSUE 1. Return to Service of Aircraft Items Recovered from Aircraft Involved in Accidents/Incidents (Airworthiness Notices). 3 pp.
CAA NOTICE #98 ISSUE 11. Use of Motor Gasoline (Mogas) in Certain Light Aircraft (Airworthiness Notices). 11 pp.
CAA NOTICE #98A ISSUE 1. Use of Filling Station Forecourt Motor Gasoline (Mogas) in Certain Light Aircraft. 4 pp.
CAA NOTICE #99 ISSUE 1. Galley Equipment (Airworthiness Notices). 4 pp.
CAA Chapter 1 03.92. Basis for This Document (Airworthiness Notices). 1 p.
CAA Chapter 2 03.92. Definitions and Abbreviations. 3 pp.
CAA Chapter 3 03.92. JAA Maintenance Policy (Airworthiness Notices). 1 p.
CAA Chapter 4 03.92. JAA Implementation Policy for JAR-145 (Airworthiness Notices). 2 pp.
CAA Chapter 5 03.92. JAA Common Procedures Policy for Maintenance Approval and Result and Mutual Recognition Policy (Airworthiness Notices). 2 pp.
CAA Chapter 6 03.92. Procedures for the Grant of JAR-145 Approval by the JAA National Aviation Authority (Airworthiness Notices). 4 pp.
CAA Chapter 7 03.92. Procedures for the Renewal of the JAR-145 Approval by the JAA National Aviation Authority (Airworthiness Notices). 1 p.

—**Maintenance Personnel**
CAA NOTICE #28 ISSUE 3. Application for Civil Aviation Authority Supervision During the Maintenance, Overhaul, Repair or Modification of a Foreign Registered Aircraft in the United Kingdom. 2 pp.
CAA NOTICE #46 ISSUE 7. Aircraft Maintenance Engineers Licensing—BCAR Section L Issue 7 Ratings (Airworthiness Notices). 2 pp.
CAA Chapter 2 03.92. Definitions and Abbreviations. 3 pp.

—**Safety Regulation Group**
CAA NOTICE #29 ISSUE 9. Airworthiness Division Head Officer (Appendices Included) (Airworthiness Notices). 8 pp.

AIS
Use: Aeronautical Information Services

Aisles
Use: Walkways

Alabaster
See Also: Gypsum

—**Chemical Analysis**
JIS M 8855-91. Methods for Chemical Analysis of Roseki.

Alanines
See Also: L-Serine
JIS K 9101-76. L-Alanine.

Alarm Cables
Use For: Fire Alarm Cables *See Also:* Alarm Systems; Cables (Electric); Electric Conductors; Fire Alarm and Extinguishing Equipment
CSA C22.2 NO 208-M1986. Fire Alarm and Signal Cable (R 1992); (Gen Instr 1 Thru 5). 28 pp.

Alarm Cables *(Cont.)*
—**Multiconductor**
CSA C22.2 NO 208-M1986. Fire Alarm and Signal Cable (R 1992); (Gen Instr 1 Thru 5). 28 pp.

—**PVC Insulated**
BSI BS 4737: Sec 3.30-86. 1986 Intruder Alarm Systems in Buildings Part 3: Components Section 3.30: PVC Insulated Cables for Interconnecting Wiring. 7 pp.

Alarm Call Services
See Also: Telephone Services

—**Call Handling**
CEPT T/CS 21-12-84. Traitement D'Appel Pour Les Services Du Reveil. 2 pp.
CEPT T/CS 21-12 E-84. Call Handling for Alarm Call Services. 2 pp.

Alarm Monitoring Systems
Use: Alarm Systems

Alarm Signals
See Also: Alarm Systems; Fire Alarm Signals; Signal Devices

—**Power Failure**
CSA C22.2 NO 205-M1983. Signal Equipment (R 1992); (Gen Instr 1). 32 pp.

—**Ships**
CNS F4012-85. Alarms and Indications for Marine Use (Jan)(11184).
JIS F 0412-90. Alarming and Indicating System of Equipment Used for Engine Department in Ships.

Alarm Systems
Use For: Electronic Alarms *See Also:* Access Control Systems; Accident Prevention; Alarm Cables; Alarm Signals; Backup Alarms; Burglar Alarms; Fire Alarm and Extinguishing Equipment; Floor Mat Detectors; Foil on Glass Intruder Systems; Gas Alarms; Monitors; Passive Infrared Detectors; Photoelectric Alarm Systems; Radiation Monitors; Security Systems; Vibration Detection Systems; Warning Systems
BSI BS 2740-69. 1969 Amd 1 Simple Smoke Alarms and Alarm Metering Devices. 16 pp.
BSI BS 7524-91. 1991 Code of Practice for Documentation for Social Alarm Systems. 10 pp.
CENELEC PREN 50134-1-2-93. Alarm Systems—Social Alarm Systems Part 1-2: Application Guidelines. 13 pp.
DIN VDE 0830 Pt 11 (D)-88. Alarm Systems: General Requirements; Identical with IEC 79 Central Office 14 (June). 21 pp.
IEC 839 Pt 1-1-88. Alarm Systems Part 1: General Requirements Section One—General First Edition. 28 pp.
IEC 839 Pt 1-4-89. Alarm Systems Part 1: General Requirements Section Four—Code of Practice First Edition. 32 pp.

—**Annunciators**
CENELEC PREN 50136-4-93. Alarm Transmission System Part 4: Annunciation Equipment. 31 pp.

—**Automotive**
BSI BS 6803: Part 1-93. 1993 Vehicle Security Alarm Systems Part 1: Specification for Installed Systems (Supersedes BS 6803: Part 2: 1986). 11 pp.
BSI BS 6803: Part 1-87. 1987 Vehicle Security Alarm Systems Part 1: Systems Installed as Original Vehicle Equipment. 8 pp.
BSI BS 6803: Part 2-86. (WITHDRAWN) 1986 Vehicle Security Alarm Systems Part 2: Code of Practice for Systems Installed After Vehicle Marking (Superseded by BS 6803: Part 1: 1993). 8 pp.
BSI BS 6803: Part 3-90. 1990 Vehicle Security Alarm Systems Part 3: Code of Practice for the Protection of Vehicles and Goods in Transit. 12 pp.
BSI BS AU 209: Part 2-88. 1988 Vehicle Security Part 2: Security Systems Against Theft of In-Car Equipment for Entertainment and Communication Purposes. 4 pp.

—**Boats (Marine)—Panels**
JIS F 0415-89. Arrangement of Main Engine Control Stand in Wheelhouse and Local Alarm Panel for Small Shi

—**Chemical Defense—Silicone Rubber**
MOD UK DSTAN 93-69-93. Silicone Rubbers Grades 35, 55 and 65 IRHD Issue 1. 12 pp.

—**Elevators (Lifts)**
CEN PREN 627-92. Specification for Data Logging, Alarm Reporting and Remote Monitoring of Lifts, Escalators and Passenger Conveyors. 10 pp.

INDUSTRY STANDARDS

INTERNATIONAL AND NON-U.S. NATIONAL STANDARDS
SUBJECT INDEX

Alarm Systems (Cont.)

—Environmental Testing
DIN VDE 0830 Pt 13 (D)-87. Alarm Systems; General Requirements; Environmental Testing for Alarm Systems (Aug). 69 pp.
IEC 839 Pt 1-3-88. Alarm Systems Part 1: General Requirements Section Three—Environmental Testing First Edition. 90 pp.

—Escalators
CEN PREN 627-92. Specification for Data Logging, Alarm Reporting and Remote Monitoring of Lifts, Escalators and Passenger Conveyors. 10 pp.

—Gas Pressure
CSA C22.2 NO 205-M1983. Signal Equipment (R 1992); (Gen Instr 1). 32 pp.

—Medical
CEN PREN 475-91. Medical Devices—Electrically—Generated Alarm Signals. 15 pp.
JIS T 1031-91. General Rules for Alarms of Medical Equipment. 11 pp.

—Naval Ships
MOD UK NES 599-88. Policy Requirements and Design Guidance for Alarm and Warning Systems Issue 3 (11.88). 19 pp.
MOD UK NES 599-92. Policy Requirements and Design Guidance for Alarm and Warning Systems Issue 4 (11.92). 19 pp.

—Oxygen Supply Equipment
CNS T2014-82. Oxygen Indicator and Oxygen Alarm (Jan)(8457).
CNS T4003-82. Method of Test for Oxygen Indicator and Oxygen Alarm (Jan)(8458).
JIS T 8201-90. Oxygen Indicator/Alarm. 10 pp.

—Passenger Conveyors
CEN PREN 627-92. Specification for Data Logging, Alarm Reporting and Remote Monitoring of Lifts, Escalators and Passenger Conveyors. 10 pp.

—Personal
BSI BS 6800-86. 1986 Home and Personal Security Devices. 8 pp.
BSI BS 6804-86. 1986 Social Alarm Systems. 15 pp.

—Power Supplies
DIN VDE 0830 Pt 121 (D)-87. Alarm Systems; Power Units; Requirements, Test Methods and Performance Criteria; Identical with IEC 79(CO) 5 and IEC 79(CO)10 (June). 16 pp.
IEC 839 Pt 1-2-87. Alarm Systems Part 1: General Requirements Section Two—Power Units, Test Methods and Performance Criteria First Edition. 20 pp.

—Public Switched Telephone Networks
BSI BS 7606-92. 1992 Social Alarm System Apparatus for Connection to Public Switched Telephone Networks. 8 pp.

—Radiotelephony—Maritime Mobile Services
CCIR RECMN 219-1-66. Alarm Signal for Use on the Maritime Radiotelephony Distress Frequency of 2182 kHz—Section 8C—Maritime Mobile Service; Telephony and Related Subjects. 2 pp.

—Restoration Switching Control Equipment
CCITT RECMN G.180-89. Characteristics of N + M Type Direct Transmission Restoration Systems for Use on Digital Sections, Links or Equipment—General Characteristics of International Telephone Connections and Circuits (Study Groups XII and XV) 11 pp. 11 pp.

—Ships—Compasses
BSI BS MA 2: Part 8-85. 1985 Magnetic Compasses and Binnacles Part 8: Transmitting Systems and Off-Course Alarms for Class A Magnetic Compasses. 4 pp.

—Ships—Panels
CNS F5123-87. Extension Alarm Panels (Nov)(12157).
JIS F 8530-84. Extension Alarm Panels.

—Signaling Systems—Connections
IEC 839 Pt 5-1-91. Alarm Systems Part 5: Requirements for Alarm Transmission Systems Section 1: General Requirements for Systems First Edition. 34 pp.

—Signaling Systems—Faults
CCITT RECMN Q.117-89. Alarms for Technical Staff and Arrangements in Case of Faults—General Recommendations on Telephone Switching and Signalling—Functions and Information Flows for Services in the ISDN—Supplements (Study Group XI) 1 pp. 1 p.

—Speech Transmission Systems—Public Switched Telephone Networks
IEC 839 Pt 5-6-91. Alarm Systems Part 5: Requirements for Alarm Transmission Systems Section 6: Requirements for Voice Communicator Systems Using the Public Switched Telephone Network First Edition. 18 pp.

—Submarines
MOD UK NES 599-88. Policy Requirements and Design Guidance for Alarm and Warning Systems Issue 3 (11.88). 19 pp.
MOD UK NES 599-92. Policy Requirements and Design Guidance for Alarm and Warning Systems Issue 4 (11.92). 19 pp.

—Supersonic Transports
CAA STANDARD NO. 3-2&AP 07.69. Limitations and Alarms. 6 pp.

—Symbols
CNS C1143-4-89. Graphical Symbols for Use on Electrical and Electronic Equipment (Alarm, Signal, Filter) (Mar)(12491-4).
CNS C5097-81. Special Symbols for Danger Alarm Systems (Mar)(7017).

—Telephone Circuits
CCITT RECMN Q.33-89. Protection Against the Effects of Faulty Transmission on Groups of Circuits—General Recommendations on Telephone Switching and Signalling—Functions and Information Flows for Services in the ISDN —Supplements (Study Group XI) 17 pp. 17 pp.

—Telephone Systems—HF
CCITT RECMN G.343-89. 4 MHz Systems on Standardized 1.2/4.4 mm Coaxial Cable Pairs—International Analogue Carrier Systems (Study Group XV) 4 pp. 4 pp.
CCITT RECMN G.344-89. 6 MHz Systems on Standardized 1.2/4.4 mm Coaxial Cable Pairs—International Analogue Carrier Systems (Study Group XV) 4 pp. 4 pp.

—Telephone Systems—HF/VHF
CCITT RECMN G.333-89. 60 MHz Systems on Standardized 2.6/9.5 mm Coaxial Cable Pairs—International Analogue Carrier Systems (Study Group XV) 10 pp. 10 pp.

—Telephone Systems—MF
CCITT RECMN G.341-89. 1.3 MHz Systems on Standardized 1.2/4.4 mm Coaxial Cable Pairs—International Analogue Carrier Systems (Study Group XV) 5 pp. 5 pp.

—Tone Signaling Protocol
BSI BS 7369-91. 1991 Multi-Frequency Tone Signalling Protocol for Social Alarm Systems. 17 pp.

—Transmission Systems
IEC 839 Pt 5-1-91. Alarm Systems Part 5: Requirements for Alarm Transmission Systems Section 1: General Requirements for Systems First Edition. 34 pp.
IEC 839 Pt 5-2-91. Alarm Systems Part 5: Requirements for Alarm Transmission Systems Section 2: General Requirements for Equipment First Edition. 28 pp.

—Transmission Systems—Dedicated Service
IEC 839 Pt 5-4-91. Alarm Systems Part 5: Requirements for Alarm Transmission Systems Section 4: Alarm Transmission Systems Using Dedicated Alarm Transmission Paths First Edition. 16 pp.

—Transmission Systems—Maintenance
CCITT RECMN M.32-89. Principles for Using Alarm Information for Maintenance of International Transmission Systems and Equipment—General Maintenance Principles—Maintenance of International Transmission Systems and Telephone Circuits (Study Group IV) 4 pp. 4 pp.

—Transmission Systems—Public Switched Telephone Networks
IEC 839 Pt 5-5-91. Alarm Systems Part 5: Requirements for Alarm Transmission Systems Section 5: Requirements for Digital Communicator Systems Using the Public Switched Telephone Network First Edition. 20 pp.

—Transmultiplexers
CCITT FASCICLE III.4-89. General Aspects of Digital Transmission Systems Terminal Equipments—Recommendations G.700-G.795. 621 pp.

Alarm Valves
Use For: Wet Alarm Valves

Alarm Valves (Cont.)
See Also: Fire Alarm Signals; Fire Sprinkler Equipment; Flow Control Valves; Valves
ISO 6182 Pt 2-93. Fire Protection—Automatic Sprinkler Systems—Part 2: Requirements and Test Methods for Wet Alarm Valves, Retard Chambers and Water Motor Alarms First Edition. 20 pp.

Alarms
Use: Alarm Systems

Albumin, Egg
Use: Egg Albumin

Albumin, Serum
Use: Serum Albumin

Alcohol Content Analysis
Use For: Free Alcohols Content Analysis; Higher Alcohol Content Analysis *See Also:* Alcohols
CNS K7620-82. Chemical Reagent (Fatty Oil, Fatty Acid and Higher Alcohol Determination) (Jun)(8992).

—Adipic Esters—Spectrometry
ISO 6795-83. Phthalic and Adipic Esters for Industrial Use—Determination of Free Alcohols Content—Spectrometric Method First Edition. 4 pp.

—Citronella Oils
CNS K6063-65. Method of Test for Citronella Oil (Aug)(817)(R 1971). 3 pp.

—Detergents—Gas Chromatography
BSI BS 3762: Sec 3.24-88. 1988 Analysis of Formulated Detergents Part 3: Quantitative Test Methods Section 3.24: Methods for Determination of Low Molecular Mass Alcohols Content. 6 pp.

—Essential Oils
ISO 4096-78. Essential Oils (Containing Tertiary Alcohols)—Evaluation of Free Alcohols Content by Determination of Ester Value After Cold Formylation First Edition. 4 pp.

—Essential Oils—Acetylation
CNS K6611-80. Estimation of Free Alcohols Content by Determination of Ester Value After Acetylation of Essential Oil (Dec)(6677).
CNS K6614-80. Estimation of Free Alcohols Content by Determination of Ester Value After Acetylation of Essential Oils (Containing Tertiary Alcohols) (Dec)(6680).
CNS K6615-80. Estimation of Primary and Secondary Free Alcohols Content of Essential Oil by Acetylation in Pyridine (Dec)(6681).
ISO 1241-80. Essential Oils—Determination of Ester Value After Acetylation and Evaluation of Free Alcohols and Total Alcohols Content First Edition. 5 pp.
ISO 3793-76. Essential Oils—Estimation of Primary and Secondary Free Alcohols Content by Acetylation in Pyridine First Edition. 4 pp.
ISO 3794-76. Essential Oils (Containing Tertiary Alcohols)—Estimation of Free Alcohols Content by Determination of Ester Value After Acetylation First Edition. 4 pp.

—Gasoline—Gas Chromatography
DIN ENGL 51413 Pt 1-84. Analysis of Liquid Petroleum Products by Gas Chromatography; Determination of Alcohol Content (Nov). 3 pp.

—Ketones—Volumetric Analysis
ISO 2501-74. Methyl Ethyl Ketone, isoButyl Methyl Ketone and isoAmyl Ethyl Ketone for Industrial Use—Determination of Alcoholic Impurities—Volumetric Method First Edition. 5 pp.

—Phthalic Esters—Spectrometry
ISO 6795-83. Phthalic and Adipic Esters for Industrial Use—Determination of Free Alcohols Content—Spectrometric Method First Edition. 4 pp.

—Vegetables—Perserved
CNS N6126-80. Method of Test for Preserved Vegetable—Determination of Alcohol (Aug)(6246). 1 pp.

Alcohol Fuels
See Also: Diesel Fuels; Gasohol; Liquid Fuels

—Octane Number
DIN ENGL 51756 Pt 7-86. Testing of Gasolines; Determination of Knock Characteristics (Octane Number) of Alcohols and Alcohol/Fuel Mixtures Using the CFR Engine (Feb). 4 pp.

Alcohol Resistance Testing

—Coatings
CGSB 1-GP-71 METH 146.1-78. Methods of Testing Paints and Pigments Alcohol Resistance. 1 p.

INTERNATIONAL AND NON-U.S. NATIONAL STANDARDS
SUBJECT INDEX

Alcoholic Beverages
Use For: Spiritous Beverages *See Also:* Beer; Beverages; Rum; Wines
- EC COM(82) 328-82. Proposal for a Council Regulation (EEC) laying down general rules on the definition, description and presentation of spiritous beverages and of vermouths and other wines of fresh grapes flavoured with plants or other aromatic substances. 35 pp.
- EC COM(86) 159-86. Amended Proposals for Council Regulations (EEC)—Laying Down General Rules on the Definition, Description and Presentation of Spirituous Beverages—Vermouths and Other Wines of Fresh Grapes Flavoured with Plants or Other Aromatic Substances. 36 pp.
- EC EEC/1576/89-89. Council Regulation Laying Down General Rules on the Definition, Description and Presentation of Spirit Drinks. 17 pp.
- EC EEC/351/92-92. Commission Regulation Correcting the German Version of Regulation (EEC) No 3664/91 Laying down Transitional Measures for Aromatized Wines, Aromatized Wine-Based Drinks and Aromatized Wine-Product Cocktails. 1 p.

—Advertising
- EC COM(82) 626-82. Proposal for a Council Directive amending Directive 79/112/EEC on the approximation of the laws of the Member States relating to the labelling, presentation and advertising of foodstuffs for sale to the ultimate consumer. 6 pp.
- EC 86/197/EEC-86. Council Directive Amending Directive 79/112/EEC on the Approximation of the Laws of the Member States Relating to the Labelling, Presentation and Advertising of Foodstuffs for Sale to the Ultimate Consumer. 2 pp.

—Alcohol Content—Identification Systems
- EC COM(82) 626-82. Proposal for a Council Directive amending Directive 79/112/EEC on the approximation of the laws of the Member States relating to the labelling, presentation and advertising of foodstuffs for sale to the ultimate consumer. 6 pp.
- EC 86/197/EEC-86. Council Directive Amending Directive 79/112/EEC on the Approximation of the Laws of the Member States Relating to the Labelling, Presentation and Advertising of Foodstuffs for Sale to the Ultimate Consumer. 2 pp.

—Excise Taxes
- EC COM(85) 150-85. Proposal for a Council Directive Laying Down Certain Rules on Indirect Taxes Which Affect the Consumption of Alcoholic Drinks. 10 pp.
- EC COM(85) 151-85. Proposal for a Council Directive Concerning the Harmonization of Excise Duties on Fortified Wine and Similar Products. 24 pp.
- EC COM(87) 328-87. Proposal for a Council Directive Concerning Approximation of the Rates of Excise Duty on Alcoholic Beverages and on the Alcohol Contained in Other Products. 18 pp.

—Tasting Glasses
- ISO 5494-78. Sensory Analysis—Apparatus—Tasting Glass for Liquid Products First Edition. 5 pp.

Alcoholometers
See Also: Alcoholometry; Measuring Instruments
- EC 76/765/EEC-76. Council Directive on the Approximation of the Laws of the Member States Relating to Alcoholometers and Alcohol Hydrometers. 6 pp.
- EC 82/624/EEC-82. Commission Directive Adapting to Technical Progress Council Directive 76/765/EEC on the Approximation of the Laws of the Member States Relating to Alcoholometers and Alcohol Hydrometers. 2 pp.
- ISO 4801-79. Glass Alcoholometers and Alcohol Hydrometers Not Incorporating a Thermometer First Edition; (Erratum—April 1980). 10 pp.

—Thermometers
- ISO 4805-82. Laboratory Glassware—Thermo-Alcoholometers and Alcohol-Thermohydrometers First Edition. 9 pp.
- ISO 6152-82. Thermometers for Use with Alcoholometers and Alcohol Hydrometers First Edition. 6 pp.

Alcoholometry
See Also: Alcoholometers

—Tables (Data)
- EC 76/766/EEC-76. Council Directive on the Approximation of the Laws of the Member States Relating to Alcohol Tables. 4 pp.

Alcohols
Use For: Higher Alcohols *See Also:* Alcohol Content Analysis; Amyl Alcohols; Benzyl Alcohol; n-Butyl Alcohol; sec-Butyl Alcohol; Cyclohexanols; Diacetone Alcohol; 1,2-Dihydroxybenzene-3,5-disulfonic Acid Disodium Salt; Ethanol; 2-Ethylhexyl Alcohol; Glycerol; Inositol; Isobutyl Alcohol; Isopropyl Alcohol; Lead-2,4-Dihydroxybenzoate; Mannitol; Methanols; Methylamyl Alcohol; n-Octyl Alcohol; Pentaerythritol; Phenol; n-Propyl Alcohol; Solvents; Sorbitol; Stearyl Alcohol
- BSI BS 4583-91. 1991 Methods of Test for Higher Alcohols for Industrial Use. 13 pp.
- CNS K6009-75. Method of Test for Alcohol (Apr)(115).
- JIS K 1505-54. Technical Alcohol.

—Acidity—Volumetric Analysis
- ISO 1843 Pt II-77. Higher Alcohols for Industrial Use—Methods of Test—Part II: Determination of Acidity to Phenolphthalein—Titrimetric Method First Edition. 4 pp.

—Ash Content
- ISO 1843 Pt VI-77. Higher Alcohols for Industrial Use—Methods of Test—Part VI: Determination of Ash First Edition. 3 pp.

—Ash Content—Gravimetric Analysis
- CNS K6413-78. Determination of Ash of Higher Alcohols for Industrial Use (Gravimetric Method) (Mar)(4282).

—Industrial
- CNS K1059-75. Alcohol for Industrial Use (Apr)(1397).
- ISO 1843 Pt I-77. Higher Alcohols for Industrial Use—Methods of Test—Part I: General First Edition. 4 pp.

—Industrial—Bromine Number—Volumetric Analysis
- ISO 1843 Pt IV-77. Higher Alcohols for Industrial Use—Methods of Test—Part IV: Determination of Bromine Number—Titrimetric Method in the Presence of Mercury (II) Chloride First Edition. 4 pp.

—Industrial—Carbonyl Compound Content—Potentiometric Analysis
- ISO 1843 Pt III-77. Higher Alcohols for Industrial Use—Methods of Test—Part III: Determination of Carbonyl Compounds Content—Potentiometric Method First Edition. 4 pp.

—Industrial—Color Testing
- CNS K6414-78. Test for Colour with Sulphuric Acid of Higher Alcohols for Industrial Use (Mar)(4283).
- ISO 1843 Pt 8-82. Higher Alcohols for Industrial Use—Methods of Test—Part 8: Sulphuric Acid Colour Test First Edition. 4 pp.

—Industrial—Distillation Methods
- ISO 1843 Pt 7-82. Higher Alcohols for Industrial Use—Methods of Test—Part 7: Determination of Distillation Yield First Edition. 4 pp.

—Industrial—Volumetric Analysis
- CNS K6412-78. Determination of Total Alcohols Content of Higher Alcohols for Industrial Use (Volumetric Method) (Mar)(4281).
- ISO 1843 Pt V-77. Higher Alcohols for Industrial Use—Methods of Test—Part V: Determination of Total Alcohols Content—Titrimetric Method First Edition. 4 pp.

Aldehyde Content Analysis

—Ethanols—Colorimetric Analysis
- BSI BS 6392: Part 4-83. 1983 Amd 1 Testing of Ethanol for Industrial Use Part 4: Method for Determination of Aldehydes Content. 5 pp.
- ISO 1388 Pt 5-81. Ethanol for Industrial Use—Methods of Test—Part 5: Determination of Aldehydes Content—Visual Colorimetric Method First Edition. 5 pp.

—Formaldehyde
- BSI BS 2942-57. (WITHDRAWN) 1957 Amd 2 Formaldehyde Solution. 12 pp.

—Paraformaldehyde
- BSI BS 2941-57. (WITHDRAWN) 1957 Amd 1 Paraformaldehyde. 14 pp.

—Paraformaldehyde—Volumetric Analysis
- ISO 1391 Pt 5-81. Paraformaldehyde for Industrial Use—Methods of Test—Determination of Aldehyde Content—Titrimetric Method First Edition. 4 pp.

Aldehydes
Scope Note: For additional listings, use a more specific term *See Also:* Acetaldehyde; Aldehyde Content Analysis; Benzaldehyde; p-Dimethylaminobenzaldehyde; Formaldehyde; Furfural; o-Nitrobenzaldehyde; Paraformaldehyde; Vanillin

—Purity
- CNS K6476-80. Method of Test for Purity of Aldehydes and Ketones (Apr)(5459).

Algae
Use For: Chlorella; Chlorophyta; Gelidium; Gracilaria; Green Algae; Kelp; Seaweed
See Also: Agar; Plants (Botany)
- CNS N1056-79. Gelidium and Gracilaria (Oct)(2378). 2 pp.
- CNS N5134-85. Edible Chlorella (Aug)(4202). 2 pp.

—Ash Content
- CNS N6106-78. Method of Test for Edible Chlorella (Oct)(4597). 9 pp.

—Cadmium Content
- CNS N6106-78. Method of Test for Edible Chlorella (Oct)(4597). 9 pp.

—Chlorophyll Content
- CNS N6106-78. Method of Test for Edible Chlorella (Oct)(4597). 9 pp.

—Dried—Condiments
- CNS N5190-81. Flavored Condiments (Oct)(8055). 3 pp.

—Growth Factor
- CNS N6106-78. Method of Test for Edible Chlorella (Oct)(4597). 9 pp.

—Lead Content
- CNS N6106-78. Method of Test for Edible Chlorella (Oct)(4597). 9 pp.

—Mercury Content
- CNS N6106-78. Method of Test for Edible Chlorella (Oct)(4597). 9 pp.

—Microbiological Analysis
- CNS N6106-78. Method of Test for Edible Chlorella (Oct)(4597). 9 pp.

—Moisture Content
- CNS N4016-79. Method of Test for Celidium and Gracilaria (Oct)(2416). 1 p.
- CNS N6106-78. Method of Test for Edible Chlorella (Oct)(4597). 9 pp.

—Paints—Environmental Testing
- SAA AS 1580.481. 1.13-92. Paints and Related Materials—Methods of Test—Pt. 481: Coatings—Pt. 481.1: Assmt. of Indiv. Defects of Exposed Films—Pt. 481. 1.13: Exposedto Weathering—Degree to Fungal or Algal Growth Amdt 1 January/February 1993 (in Professional Packages 30, 39). 5 pp.
- SNZ NZS/AS 1580. 481.1.13-92. Methods of Test for Paints and Related Materials Part 481.1.13: Coatings—Exposed to Weathering—Degree of Fungal or Algal Growth (This is a Joint Standard with SAA AS 1580.481.13). 3 pp.

—Protein Content
- CNS N6106-78. Method of Test for Edible Chlorella (Oct)(4597). 9 pp.

—Sampling
- CNS N4016-79. Method of Test for Celidium and Gracilaria (Oct)(2416). 1 p.

—Waste Water—Toxicity
- DIN ENGL 38412 Pt 33-91. German Standard Methods for the Examination of Water, Waste Water and Sludge; Bio-Assays (Group L); Determining the Tolerance of Green Algae to the Toxicity of Waste Water (Scenedesmus Chlorophyll Fluorescence Test) by Way of a Dilution Series (L 33) (Mar). 5 pp.

—Water—Toxicity
- BSI BS 6068: Sec 5.10-90. 1990 Water Quality Part 5: Methods for Biological Testing Section 5.10: Fresh Water Algal Growth Inhibition Test with Scenedesmus Subspicatus and Selenastrum Capricornutum. 11 pp.
- BSI BS 6068: Sec 5.10-01. 1990 Amd 1 Water Quality—Fresh Water Algal Growth Inhibition Test with Scenedesmus Subspicatus and Selenastrum Capricornutum (ISO 8692: 1989) (AMD 7430) May 15, 1993 (N). 17 pp.
- BSI BS EN 28692-93. 1993 Amd 1 Water Quality—Fresh Water Algal Growth Inhibition Test with Scenedesmus Subspicatus and Selenastrum Capricornutum (ISO 8692: 1989) (AMD 7430) May 15, 1993 (N). 17 pp.
- CEN EN 28692-93. Water Quality—Fresh Water Algal Growth Inhibition Test with Scenedesmus Subspicatus and Selenastrum Capricornutum (ISO 8692: 1989). 10 pp.

Algae (Cont.)
—Water—Toxicity (Cont.)
DIN ENGL 38412 Pt 9-89. German Standard Methods for the Examination of Water, Waste Water and Sludge; Bio-Assays (Group L); Determination of the Inhibitory Effect of Water Constituents on the Growth of Scenedesmus Green Algae (L 9) (May). 7 pp.
ISO 8692-89. Water Quality—Fresh Water Algal Growth Inhibition Test with Scenedesmus Subspicatus and Selenastrum Capricornutum First Edition (CENE EN 28692: 1993). 8 pp.

Alginates
See Also: Esters
—Dental Materials
BSI BS 4269: Part 2-91. 1991 Dental Elastic Impression Material Part 2: Specification for Alginate Impression Material (ISO 1563: 1990). 16 pp.
BSI BS 4269: Part 2-01. 1991 Amd 1 Dental Elastic Impression Material Part 2: Specification for Alginate Impression Material (ISO 1563: 1990) (AMD 7037) May 1, 1992. 21 pp.
CEN EN 21563-91. Alginate Dental Impression Material. 3 pp.
CNS T3058-86. Alginate Dental Impression Material (Aug)(11686).
CSA CAN/CSA-Z349.38-93. Dental Alginate Impression Material (ISO 1563-1990); (Gen Instr 1). 39 pp.
ISO 1563-90. Dental Alginate Impression Material Second Edition. 15 pp.
JIS T 6505-89. Dental Alginate Impression Material. 11 pp.

ALGOL
Use For: ALGOL 60 *See Also:* Programming Languages
OSI ISO 1538-84. Programming Languages—ALGOL 60. 20 pp.
—Symbols
OSI ISO TR 1672-77. Hardware Representation of ALGOL Basic Symbols in the ISO 7-Bit Coded Character Set for Information Processing Interchange. 7 pp.

ALGOL 60
Use: ALGOL

Algorithms
See Also: Computer Programming; Computer Programs
—Banking—Data Encryption
BSI BS ISO 10126-2-91. 1991 Banking—Procedures for Message Encipherment (Wholesale)—Part 2: DEA Algorithm (G). 10 pp.
ISO 10126 Pt 2-91. Banking—Procedures for Message Encipherment (Wholesale)—Part 2: DEA Algorithm First Edition. 7 pp.
JTC1 10126 Pt 2-91. Banking—Procedures for Message Encipherment (Wholesale)—Part 2: DEA Algorithm First Edition. 7 pp.
—Banking—Message Authentication
BSI BS 7101-89. (WITHDRAWN) 1989 Procedures for Protecting Authentic Wholesale Messages Between Financial Institutions (Superseded by BS ISO 8730: 1990). 11 pp.
BSI BS 7102: Part 1-89. 1989 Recommendations for Algorithms for Use in Banking Message Authentication Part 1: Data Encryption Algorithm (DEA). 6 pp.
BSI BS 7102: Part 2-89. (WITHDRAWN) 1989 Recommendations for Algorithms for Use in Banking Message Authentication Part 2: Message Authenticator Algorithm (MAA) (Superseded by BS ISO 8731-2: 1992). 12 pp.
BSI BS ISO 8730-90. 1990 Banking—Requirements for Message Authentication (Wholesale). 37 pp.
BSI BS ISO 9807-91. 1991 Banking and Related Financial Services—Requirements for Message Authentication (Retail) (G). 19 pp.
ISO 8730-90. Banking—Requirements for Message Authentication (Wholesale) Second Edition. 31 pp.
ISO 8731 Pt 1-87. Banking—Approved Algorithms for Message Authentication—Part 1: DEA First Edition. 4 pp.
ISO 8731 Pt 2-92. Banking—Approved Algorithms for Message Authentication—Part 2: Message Authenticator Algorithm Second Edition. 23 pp.
ISO 9807-91. Banking and Related Financial Services—Requirements for Message Authentication (Retail) First Edition. 16 pp.
JTC1 8730-90. Banking—Requirements for Message Authentication (Wholesale) Second Edition. 31 pp.
JTC1 8731 Pt 1-87. Banking—Approved Algorithms for Message Authentication—Part 1: DEA First Edition. 4 pp.
JTC1 8731 Pt 2-92. Banking—Approved Algorithms for Message Authentication—Part 2: Message Authenticator Algorithm Second Edition. 23 pp.
JTC1 9807-91. Banking and Related Financial Services—Requirements for Message Authentication (Retail) First Edition. 16 pp.
OSI ISO 8730-91. Banking—Requirements for Message Authentication (Wholesale). 31 pp.
OSI ISO 8730-86. Banking—Requirements for Message Authentication (Wholesale). 9 pp.
OSI ISO 8730-86. Banking—Requirements for Message Authentication (Wholesale). 8 pp.
OSI ISO DIS 8731-2-86. Banking—Approved Algorithm for Message Authentication—Part 2: Message Authenticator Algorithm. 21 pp.
OSI ISO/DIS 9807/DAM-90. Banking and Related Financial Services—Requirements for Message Authentication (Retail). 5 pp.
OSI ISO DIS 9807-89. Banking and Related Financial Services—Requirements for Message Authentication (Retail). 18 pp.
OSI ISO DP 9807-87. Banking—for Message Authentication (Retail). 14 pp.

—Banking—Personal Identification Numbers
BSI BS ISO 9564-2-91. 1991 Banking—Personal Identification Number Management and Security—Part 2: Approved Algorithm(s) for PIN Encipherment (G). 7 pp.
ISO 9564 Pt 2-91. Banking—Personal Identification Number Management and Security—Part 2: Approved Algorithm(s) for PIN Encipherment First Edition. 5 pp.
JTC1 9564 Pt 2-91. Banking—Personal Identification Number Management and Security—Part 2: Approved Algorithm(s) for PIN Encipherment First Edition. 5 pp.

—Block Cipher—Electronic Funds Transfer
SAA AS 2805.5.2-92. Electronic Funds Transfer—Requirements for Interfaces—Part 5: Ciphers—Part 5.2: Modes of Operation for an n-Bit Block Cipher Algorithm (Supersedes AS 2805.5—1985 (in Part)). 12 pp.

—Block Cipher—Message Authentication
BSI BS ISO/IEC 9797-90. 1990 Data Cryptographic Techniques—Data Integrity Mechanism Using a Cryptographic Check Function Employing a Block Cipher Algorithm. 9 pp.
IEC 9797-89. Data Cryptographic Techniques—Data Integrity Mechanism Using a Cryptographic Check Function Employing a Block Cipher Algorithm First Edition. 7 pp.
IEC DIS 9797-93. Information Technology—Security Techniques—Data Integrity Mechanism Using a Cryptographic Check Function Employing a Block Cipher Algorithm (Revision of First Edition (ISO/IEC 9797: 1989)) ***CD-ROM ONLY***. 11 pp.
ISO 9797-89. Data Cryptographic Techniques—Data Integrity Mechanism Using a Cryptographic Check Function Employing a Block Cipher Algorithm First Edition. 7 pp.
ISO DIS 9797-93. Information Technology—Security Techniques—Data Integrity Mechanism Using a Cryptographic Check Function Employing a Block Cipher Algorithm (Revision of First Edition (ISO/IEC 9797: 1989)) ***CD-ROM ONLY***. 11 pp.
JTC1 9797-89. Data Cryptographic Techniques—Data Integrity Mechanism Using a Cryptographic Check Function Employing a Block Cipher Algorithm First Edition. 7 pp.
JTC1 DIS9797-93. Information Technology—Security Techniques—Data Integrity Mechanism Using a Cryptographic Check Function Employing a Block Cipher Algorithm (Revision of First Edition (ISO/IEC 9797: 1989)) ***CD-ROM ONLY***. 11 pp.
OSI ISO IEC 9797-89. Data Cryptographic Techniques—Data Integrity Mechanism Using a Coyptographic Check Function Employing a Block Cipher Algorithm. 7 pp.
OSI ISO IEC DIS 9797-88. Data Cryptographic Techniques—Data Integrity Mechanism Using a Cryptographic Check Function Employing an N-Bit Algorithm with Truncation. 6 pp.
OSI ISO DP 9797-87. Data Integrity Mechanism Using a Cryptographic Check-Function Employing an N-Bit Alogrithm with Truncation. 6 pp.

—Data Compression
BSI BS ISO/IEC 11558-92. 1992 Information Technology—Data Compression for Information Interchange—Adaptive Coding with Embedded Dictionary—DCLZ Algorithm. 18 pp.
ECMA ECMA 151-91. Data Compression for Information Interchange Adaptive Coding with Embedded Dictionary DCLZ Algorithm. 14 pp.
ECMA ECMA 159-91. Data Compression for Information Interchange Binary Arithmetic Coding Algorithm. 21 pp.
IEC 11558-92. Information Technology—Data Compression for Information Interchange—Adaptive Coding with Embedded Dictionary—DCLZ Algorithm First Edition. 15 pp.
IEC DIS 11576-92. Information Technology—Procedure for the Registration of Algorithms for the Lossless Compression of Data ***CD-ROM ONLY***. 7 pp.
IEC DIS 12042-92. Information Technology—Data Compression for Information Interchange—Binary Arithmetic Coding Algorithm ***CD-ROM ONLY***. 25 pp.
ISO 11558-92. Information Technology—Data Compression for Information Interchange—Adaptive Coding with Embedded Dictionary—DCLZ Algorithm First Edition. 15 pp.
ISO DIS 11576-92. Information Technology—Procedure for the Registration of Algorithms for the Lossless Compression of Data ***CD-ROM ONLY***. 7 pp.
ISO DIS 12042-92. Information Technology—Data Compression for Information Interchange—Binary Arithmetic Coding Algorithm ***CD-ROM ONLY***. 25 pp.
JTC1 DIS11576-92. Information Technology—Procedure for the Registration of Algorithms for the Lossless Compression of Data ***CD-ROM ONLY***. 7 pp.

—Data Encipherment—Electronic Funds Transfer
SAA AS 2805.5.1-92. Electronic Funds Transfer—Requirements for Interfaces—Part 5: Ciphers—Part 5.1: Data Encipherment Algorithm 1 (DEA 1) (Supersedes AS 2805.5—1985 (in Part)). 10 pp.
SAA AS 2805.5.3-92. Electronic Funds Transfer—Requirements for Interfaces—Part 5: Ciphers—Part 5.3: Data Encipherment Algorithm 2 (DEA 2). 5 pp.

—Data Encryption
CNS C5164-83. Data Encryption Algorithm (Sep)(10545).

—Data Encryption—Register Entry Format
IEC 9979-91. Data Cryptographic Techniques—Procedures for the Registration of Cryptographic Algorithms First Edition. 8 pp.
ISO 9979-91. Data Cryptographic Techniques—Procedures for the Registration of Cryptographic Algorithms First Edition. 8 pp.
JTC1 9979-91. Data Cryptographic Techniques—Procedures for the Registration of Cryptographic Algorithms First Edition. 8 pp.

—Data Encryption—Registration Procedures
IEC 9979-91. Data Cryptographic Techniques—Procedures for the Registration of Cryptographic Algorithms First Edition. 8 pp.
ISO 9979-91. Data Cryptographic Techniques—Procedures for the Registration of Cryptographic Algorithms First Edition. 8 pp.
JTC1 9979-91. Data Cryptographic Techniques—Procedures for the Registration of Cryptographic Algorithms First Edition. 8 pp.

—Frequency Assignment—Radio Spectra
CCIR QUESTION 66/1-90. Methods and Algorithms for Frequency Planning—Questions Concerning Study Group 1—Spectrum Management Techniques (Spectrum Engineering, Planning, Sharing, Monitoring and Utilization). 1 p.

—Message Authentication—Registration Procedures
OSI ISO IEC DIS 9979-2-90. Data Cryptographic Techniques—Procedures for the Registration of Cryptographic Algorithms. 9 pp.
OSI ISO IEC DIS 9979-2-89. Data Cryptographic Techniques—Procedures for the Registration of Cryptographic Algorithms. 8 pp.

—n-bit Block Cipher
BSI BS ISO/IEC 10116-91. 1991 Information Technology—Modes of Operation for an n-Bit Block Cipher Algorithm. 17 pp.
IEC 10116-91. Information Technology—Modes of Operation for an n-Bit Block Cipher Algorithm First Edition. 14 pp.
ISO 10116-91. Information Technology—Modes of Operation for an n-Bit Block Cipher Algorithm First Edition. 14 pp.
OSI ISO IEC DIS 10116-89. Information Processing—Modes of Operation for an F2n F1-Bit Block Cipher Algorithm. 12 pp.

—n-bit Block Cipher—Hash Functions
IEC DIS 10118 Pt 2-93. Information Technology—Security Techniques—Hash-Functions—Part 2: Hash-Functions Using an n-Bit Cipher Algorithm ***CD-ROM ONLY***. 10 pp.

INTERNATIONAL AND NON-U.S. NATIONAL STANDARDS
SUBJECT INDEX

Algorithms (Cont.)
—n-bit Block Cipher—Hash Functions (Cont.)
ISO DIS 10118 Pt 2-93. Information Technology—Security Techniques—Hash-Functions—Part 2: Hash-Functions Using an n-Bit Cipher Algorithm ***CD-ROM ONLY***. 10 pp.

JTC1 DIS10118 Pt 2-93. Information Technology—Security Techniques—Hash-Functions—Part 2: Hash-Functions Using an n-Bit Cipher Algorithm ***CD-ROM ONLY***. 10 pp.

—Public Key—Message Authentication
OSI ISO DP 9799-87. Information Processing—Data Cryptographic Techniques—Peer Entity Authentication Mechanism Using a Public-Key Algorithm with a Two-Way Handshake. 7 pp.

—Secret Key—Message Authentication
OSI ISO DP 9798-87. Information Processing—Data Cryptographic Techniques—Peer Entity Authentication Mechanisms Using an N-Bit Secret-Key Algorithm. 14 pp.

—Time Scales
CCIR Report 579-4-90. Time Scale Algorithm and Associated Averaging Problems—Section 7D—Characterization of Sources and Time Scales Formation. 5 pp.

CCIR QUESTION 108/7-90. Time Scale Algorithms and Statistical Problems—Questions Concerning Study Group 7—Science Services. 1 p.

—64-bit Block Cipher
BSI BS 7111-91. 1991 Modes of Operation for a 64-Bit Block Cipher Algorithm. 13 pp.

CNS C5228-89. Information Processing Modes of Operation for a 64-Bit Block Cipher Algorithm (Dec)(12645).

ISO 8372-87. Information Processing—Modes of Operation for a 64-Bit Block Cipher Algorithm First Edition. 9 pp.

JIS X 5052-90. Modes of Operation for a 64-Bit Block Cipher Algorithem (ISO 8372:1987).

JTC1 8372-87. Information Processing—Modes of Operation for a 64-Bit Block Cipher Algorithm First Edition. 9 pp.

OSI ISO 8372-87. Information Processing—Modes of Operation for a 64-Bit Block Cipher Algorithm. 9 pp.

Alidades
See Also: Surveying Instruments
CNS B6043-81. Alidades (May)(7341).

Aligners
Use: Alignment Equipment

Aligning Equipment
Use: Alignment Equipment

Aligning Systems
Use: Alignment Equipment

Alignment Equipment
Use For: Mechanical Alignment Equipment; Misalignment Equipment

—Motor Operated—Automotive
CSA CAN/CSA-C22. 2 NO 68-92. Motor-Operated Appliances (Household and Commercial); (Gen Instr 1 Thru 2). 115 pp.

CSA 1169 Bull. Electrical Bulletin 1169 June 27, 1978 to C22.2 NO 68. 2 pp.

Alimentary Pasta
Use: Pasta

Aliphatic Hydrocarbons
Scope Note: For additional listings, use a more specific term See Also: Acetylene; Aromatic Hydrocarbons; Butadienes; Butanes; Butylene; Ethylene; Halogenated Hydrocarbons; Heptane; Hydrocarbons; Methane; Octane; Olefins; Petroleum Gases; Propane; Propylene

—Metal Cleaners
MOD UK TS 10320. Hydrocarbon Solvent for Cleaning and Degreasing Applications.

Alizarin Red
Use For: Sodium Alizarinsulfonate
See Also: Pigments
CNS K7442-78. Chemical Reagent (Sodium Alizarinsulfonate) (Alizarin Reds) (Nov)(1941).
JIS K 8057-72. Alizarin Red S (Sodium Alizarinsulfonate).

Alizarin Yellow GG
Use: Metachrome Yellow

Alizarin Yellow R
See Also: Pigments
CNS K7578-81. Chemical Reagent (Alizarin Yellow R) (Aug)(7813).
JIS K 8055-72. Alizarin Yellow R.

Alkali Aggregate Reactions
See Also: Aggregates; Alkali Resistance Testing; Cements; Concretes

—Concretes
JIS A 1804-92. Mehtods of Test for Production Control of Concrete—Method of Rapid Test for Identification of the Alkali Reactivity of Aggregates. 9 pp.

Alkali Content Analysis
See Also: Alkalinity

—Bleach Liquors
CPPA J.22P-86. Analysis of Chlorine Solutions, Hypochlorite Bleach Liquors, and Spent Bleach Liquors. 3 pp.

—Carbonated Beverage Bottles
CNS S2111-83. Method of Test for Lead, Arsenic and Alkali Content Extracted from Carbonated Beverage Bottles (Oct)(10633).

—Cementitious Materials
SAA AS 3583.12-91. Methods of Test for Supplementery Cementitous Materials for Use with Portland Cement—Part 12: Determination of Available Alkali (In Professional Packages 30,58). 4 pp.

—Cements—Volumetric Analysis
BSI BS EN 196-21-92. 1992 Methods of Testing Cement Part 21: Determination of the Chloride, Carbon Dioxide and Alkali Content of Cement. 22 pp.

CEN EN 196 (Part 21)-89. Methods of Testing Cement: Determination of the Chloride, Carbon Dioxide and Alkali Content of Cement. 28 pp.

—Cosmetics
EC 83/514/EEC-83. Third Commission Directive on the Approximation of the Laws of the Member States Relating to Methods of Analysis Necessary for Checking the Composition of Cosmetic Products. 38 pp.

—Fiberglass Reinforced Plastics
CNS K6662-81. Method of Test for Fiber Glass Products (May)(7397). 9 pp.

—Polishes
CGSB 25-GP-1M METH 20.1-84. Methods of Sampling and Testing Waxes and Polishes Free Alkali and Free Acid. 1 p.

—Scouring Powders
CGSB 2-GP-11M METH 6.2-83. Methods of Testing and Analysis of Soaps and Detergents Free Alkali (Scouring Compounds). 1 p.

—Slag Cements—Volumetric Analysis
BSI BS EN 196-21-92. 1992 Methods of Testing Cement Part 21: Determination of the Chloride, Carbon Dioxide and Alkali Content of Cement. 22 pp.

CEN EN 196 (Part 21)-89. Methods of Testing Cement: Determination of the Chloride, Carbon Dioxide and Alkali Content of Cement. 28 pp.

—Soaps
CGSB 2-GP-11M METH 6.1-83. Methods of Testing and Analysis of Soaps and Detergents Free Alkali or Free Acid. 1 p.

ISO 456-73. Surface Active Agents—Analysis of Soaps—Determination of Free Caustic Alkali First Edition. 5 pp.

—Soaps—Volumetric Analysis
BSI BS 1715: Sec 2.1-89. 1989 Analysis of Soaps Part 2: Quantitative Test Methods Section 2.1: Method for Determination of Total Alkali Content and Total Fatty Matter Content (ISO 685-1975). 7 pp.

BSI BS 1715: Sec 2.2-89. 1989 Analysis of Soaps Part 2: Quantitative Test Methods Section 2.2: Method for Determination of Total Free Alkali Content (ISO 684-1974). 4 pp.

BSI BS 1715: Sec 2.3-89. 1989 Analysis of Soaps Part 2: Quantitative Test Methods Section 2.3: Methods for Determination of Free Caustic Alkali Content (ISO 456: 1973). 7 pp.

ISO 684-74. Analysis of Soaps—Determination of Total Free Alkali First Edition. 4 pp.

ISO 685-75. Analysis of Soaps—Determination of Total Alkali Content and Total Fatty Matter Content First Edition. 6 pp.

MOD UK M 9532/67. Examination of Soap, Saddle, Glycerine (No Information) (Withdrawn).

Alkali Content Analysis (Cont.)
—Sodium Hydroxide
CNS K6039-63. Method of Test for Sodium Hydroxide (for Industrial Use) (Jun)(726)(R 1968). 2 pp.

—Sodium Sulfates
CNS K6085-74. Methods of Test for Sodium Sulfate of Industrial Grade (Oct)(1004). 5 pp.

—Spent Liquors
CPPA J.22P-86. Analysis of Chlorine Solutions, Hypochlorite Bleach Liquors, and Spent Bleach Liquors. 3 pp.

—Waxes
CGSB 25-GP-1M METH 20.1-84. Methods of Sampling and Testing Waxes and Polishes Free Alkali and Free Acid. 1 p.

—Wool Fabrics
SAA AS 2001.3.2-77. Methods of Test for Textiles—Part 3: Chemical Tests—Part 3.2: Determination of Alkali Content of Wool Reconfirmed 1986. 4 pp.

—Wool Fibers
CGSB CAN/CGSB-4.2 NO.73.4-M91. Textile Test Methods Wool—Determination of Alkali Content (ISO 2916:1975). 8 pp.
ISO 2916-75. Wool—Determination of Alkali Content First Edition. 5 pp.

Alkali Glass
Use For: Soda Lime Glass See Also: Glass; Glassware

—Structural Design
CGSB CAN/CGSB-12.20-M89. Structural Design of Glass for Buildings. 51 pp.

Alkali Resistance Testing
See Also: Alkali Aggregate Reactions; Corrosion Testing

—Enamels
BSI BS 1344: Part 6-71. 1971 Amd 1 Methods of Testing Vitreous Enamel Finishes Part 6: Resistance to Alkali. 10 pp.

—Enamels—Testing Equipment
BSI BS 1344: Part 15-84. 1984 Methods of Testing Vitreous Enamel Finishes Part 15: Apparatus for Testing with Alkaline Liquids. 7 pp.

ISO 2734-83. Vitreous and Porcelain Enamels—Apparatus for Testing with Alkaline Liquids Second Edition. 6 pp.

—Fabrics
ISO 105 Pt E06-89. Textiles—Tests for Colour Fastness—Part E06: Colour Fastness to Spotting: Alkali Third Edition. 4 pp.

—Glass
BSI BS 3473: Part 1-91. 1991 Chemical Resistance of Glass Used in the Production of Laboratory Glassware Part 1: Method for Determination of Resistance of Glass to Attack by a Boiling Aqueous Solution of Mixed Alkali. 11 pp.

CNS R3104-82. Standard Test Method for Glass Apparatus Chemical Analysis (May)(8861). 4 pp.

ISO 695-91. Glass—Resistance to Attack by a Boiling Aqueous Solution of Mixed Alkali—Method of Test and Classification Third Edition. 8 pp.

—Glass Coatings (On Glass)
CGSB 1-GP-71 METH 106.1-79. Methods of Testing Paints and Pigments Alkali Resistance Air-Dried Films. 1 p.

CGSB 1-GP-71 METH 106.2-79. Methods of Testing Paints and Pigments Alkali Resistance Baked Films. 1 p.

—Glazes
CNS R3089-80. Method of Test for Resistance of Glaze to Acid and Alkali for Porcelain Laboratory Apparatus (Apr)(5478).

—Lacquers
SAA AS 1580.460. 2-82. Paints and Related Materials—Methods of Test—Part 460.2: Resistance to Alkaline Conditions (Immersion Test) (R 1992). 1 p.

SNZ NZS/AS 1580. 460.2-82. Methods of Test for Paints and Related Materials Part 460.2: Resistance to Alkaline Conditions (Immersion Test) (This is a Joint Standard with SAA AS 1580.460.2). 1 p.

—Paints
CGSB 1-GP-71 METH 106.3-78. Methods of Testing Paints and Pigments Alkali Resistance Emulsion Paints. 1 p.

SAA AS 1580.460. 2-82. Paints and Related Materials—Methods of Test—Part 460.2: Resistance to Alkaline Conditions (Immersion Test) (R 1992). 1 p.

Alkali Resistance Testing (Cont.)
—Paints (Cont.)
SNZ NZS/AS 1580. 460.2-82. Methods of Test for Paints and Related Materials Part 460.2: Resistance to Alkaline Conditions (Immersion Test) (This is a Joint Standard with SAA AS 1580.460.2). 1 p.

—Plywood
CNS O2048-82. Method of Test for Acid, Alkali and Thinner Resistance of Special Plywood (Apr)(8735). 1 p.

—Prints
ISO 2838-74. Prints and Printing Inks—Assessment of Resistance to Alkalis First Edition. 4 pp.

—Varnishes
SAA AS 1580.460. 2-82. Paints and Related Materials—Methods of Test—Part 460.2: Resistance to Alkaline Conditions (Immersion Test) (R 1992). 1 p.
SNZ NZS/AS 1580. 460.2-82. Methods of Test for Paints and Related Materials Part 460.2: Resistance to Alkaline Conditions (Immersion Test) (This is a Joint Standard with SAA AS 1580.460.2). 1 p.

Alkali Silica Aggregate Reactions
Use: Alkali Aggregate Reactions

Alkaline Batteries
See Also: Batteries; Leclanche Batteries; Manganese Dioxide Batteries; Nickel Cadmium Batteries; Primary Batteries; Storage Batteries
CNS C3088-87. Method of Test for Alkaline Primary Cells and Batteries (Sep)(6033).
CNS C4205-87. Alkaline Primary Cells and Batteries (Sep)(6032).
JIS C 8511-85. Alkaline Primary Cells and Batteries. 19 pp.
JIS C 8706-89. Stationary Nickel Cadmium Alkaline Batteries. 26 pp.
MOD UK DSTAN 61-3: Pt 1: Supp 8-73. Battery, Dry (Leclanche), 3 V, No 1 Issue 2. 4 pp.
MOD UK DSTAN 61-3: Part 2-01. Batteries Primary Part 2: Batteries Dry and Batteries Water Activated Issue 2; Amendment 4. 41 pp.

—Portable
MOD UK DSTAN 61-9: Part 2-01. Batteries Secondary Part 2: Batteries Secondary—Portable Lead Acid and Alkaline Types Issue 2; Amendment 4 (Superseded by Def Stan 61-9: Part 1). 51 pp.

—Preferred Products List
CECC CECC MUAHAG Vol 15 IS 1-91. Preferred Products List; Batteries (En). 20 pp.

—Safety
BSI BS 6132-83. 1983 Safe Operation of Alkaline Secondary Cells and Batteries. 13 pp.

Alkaline Cells
See Also: Nickel Cadmium Cells
CNS C3088-87. Method of Test for Alkaline Primary Cells and Batteries (Sep)(6033).
CNS C4205-87. Alkaline Primary Cells and Batteries (Sep)(6032).
JIS C 8511-85. Alkaline Primary Cells and Batteries. 19 pp.

—Potassium Hydroxides
BSI BS 5634-78. 1978 Potassium Hydroxide. 24 pp.

—Safety
BSI BS 6132-83. 1983 Safe Operation of Alkaline Secondary Cells and Batteries. 13 pp.

Alkaline Cleaners
CGSB CAN/CGSB-31.201-M90. High Pressure (Steam) Cleaning Compound. 10 pp.
CGSB CAN/CGSB-31.206-M91. Alkali Cleaning Compound. 7 pp.
CGSB 31-GP-0A METH 38.3-62. Methods of Testing Corrosion-Prevention Materials and Processes Cleaning Efficiency of Alkaline Cleaning Compounds (Oil and Asphalt Removal). 2 pp.

—Alkalinity
CGSB 31-GP-0A METH 57.1-57. Methods of Testing Corrosion-Prevention Materials and Processes Total Alkalinity. 1 p.
MOD UK M 9522/66. Examination of Degreasing Compound, Alkaline, Type A (Superseded by Def Stan 68-160).

—Insoluble Matter Content
MOD UK M 9522/66. Examination of Degreasing Compound, Alkaline, Type A (Superseded by Def Stan 68-160).

Alkaline Cleaners (Cont.)
—Metals—Surface Finishing
SAA AS 1627.1-89. Metal Finishing—Preparation and Pretreatment of Surfaces—Part 1: Cleaning Using Liquid Solvents and Alkaline Solutions (This is a Joint Standard with SANZ NZS 1627). 6 pp.
SNZ NZS/AS 1627. 1-89. Metal Finishing-Preparation and Pretreatment of Surfaces Part 1: Cleaning Using Liquid Solvents and Alkaline Solutions (This is a Joint Standard with SAA AS 1627.1). 6 pp.

—PH
CGSB 31-GP-0A METH 33.2-57. Methods of Testing Corrosion-Prevention Materials and Processes Determination of pH. 1 p.

—Sodium Hydroxide Content
MOD UK M 9522/66. Examination of Degreasing Compound, Alkaline, Type A (Superseded by Def Stan 68-160).

—Sodium Metasilicate Content
MOD UK M 9522/66. Examination of Degreasing Compound, Alkaline, Type A (Superseded by Def Stan 68-160).

—Stability Testing
CGSB 31-GP-0A METH 9.5-62. Methods of Testing Corrosion-Prevention Materials and Processes Solution Stability of Alkali Cleaners. 1 p.

—Surface Tension
CGSB 31-GP-0A METH 41.1-57. Methods of Testing Corrosion-Prevention Materials and Processes Determination of Surface Tension; (Re-Edited April 1982). 1 p.

Alkaline Earth Metal Content Analysis
—Phosphoric Acids
CNS K6235-72. Method of Test for Phosphoric Acid (Industrial Use) (Jun)(2620). 2 pp.

Alkaline Earth Metals
Scope Note: Use a more specific term *See:* Barium; Metals

Alkaline Earth Oxides
Scope Note: Use a more specific term *See:* Barium Oxides; Beryllium Oxides; Calcium Oxide

Alkalinity
Use For: Basicity; Causticity; Free Alkalinity
See Also: Acidity; Alkali Content Analysis; Buffers (Chemical); pH
CNS K7615-82. Chemical Reagent (Acidity and Alkalinity Determination) (Jun) (8987).

—Alkaline Cleaners
CGSB 31-GP-0A METH 57.1-57. Methods of Testing Corrosion-Prevention Materials and Processes Total Alkalinity. 1 p.
MOD UK M 9522/66. Examination of Degreasing Compound, Alkaline, Type A (Superseded by Def Stan 68-160).

—Ammonia Liquor
MOD UK M 2006/60. Examination of Ammonia Liquor.

—Ammonium Bicarbonate—Volumetric Analysis
ISO 2516-73. Ammonium Hydrogen Carbonate for Industrial Use (Including Foodstuffs)—Determination of Total Alkalinity—Volumetric Method First Edition. 3 pp.

—Antifreezes
CNS K6481-80. Method of Test for Reserve Alkalinity of Engine Antifreezes Antirusts and Coolants (May)(5584).

—Aromatic Hydrocarbons
CNS K6257-74. Method of Test for Acidity and Alkalinity of Aromatic Hydrocarbons (Oct)(2758).

—Bleach Liquors
CPPA J.22P-86. Analysis of Chlorine Solutions, Hypochlorite Bleach Liquors, and Spent Bleach Liquors. 3 pp.

—Boiler Water Treatment
MOD UK M 9501/59. Determination of Solution Boiler Treatment (No Information) (Withdrawn).

—Calcium Silicide
MOD UK M 805/84. Examination of Calcium Silicide.

—Cleaning Agents
CGSB 31-GP-0A METH 51.1-57. Methods of Testing Corrosion-Prevention Materials and Processes Free Alkali or Free Acid. 1 p.

Alkalinity (Cont.)
—Corrosion Inhibitors
CNS K6481-80. Method of Test for Reserve Alkalinity of Engine Antifreezes Antirusts and Coolants (May)(5584).

—Cosmetics
CNS S2073-82. Method of Hygienic Test for Cosmetics Value, Acidity and Alkalinity (Jun)(9036).

—Degreasers (Cleaning Agents)
MOD UK M 9522/66. Examination of Degreasing Compound, Alkaline, Type A (Superseded by Def Stan 68-160).

—Engine Coolants
CNS K6481-80. Method of Test for Reserve Alkalinity of Engine Antifreezes Antirusts and Coolants (May)(5584).

—Ethanols
BSI BS 6392: Part 1-83. 1983 Amd 1 Testing of Ethanol for Industrial Use Part 1: Method for Detection of Alkalinity or Determination of Acidity to Phenolphthalein. 5 pp.
ISO 1388 Pt 2-81. Ethanol for Industrial Use—Methods of Test—Part 2: Detection of Alkalinity or Determination of Acidity to Phenolphthalein First Edition. 4 pp.

—Extenders
BSI BS 3483: Part C3-82. 1982 Methods for Testing Pigments in Paints Part C3: Determination of Acidity or Alkalinity of the Aqueous Extract. 4 pp.
ISO 787 Pt 4-81. General Methods of Test for Pigments and Extenders—Part 4: Determination of Acidity or Alkalinity of the Aqueous Extract First Edition. 5 pp.

—Fats—Volumetric Analysis
BSI BS 684: Sec 2.5-89. 1989 Amd 1 Methods of Analysis of Fats and Fatty Oils Part 2: Other Methods Section 2.5: Determination of Alkalinity (AMD 6550) November 30, 1990. 5 pp.

—Fruits—Ash Analysis
ISO 5520-81. Fruits, Vegetables and Derived Products—Determination of Alkalinity of Total Ash and of Water-Soluble Ash First Edition. 5 pp.

—Glass Powder
MOD UK M 809/91. Examination of Glass Powder.

—Glycerol—Volumetric Analysis
BSI BS 5711: Part 5-79. 1979 Methods of Sampling and Test for Glycerol. 2 pp.
ISO 1615-76. Glycerines for Industrial Use—Determination of Alkalinity or Acidity—Titrimetric Method First Edition. 4 pp.

—Green Liquors
CPPA G.11U-77. Alkalinity of Sulphate Green and White Liquor. 2 pp.

—Hair Care Products
EC 80/1335/EEC-80. First Commission Directive on the Approximation of the Laws of the Member States Relating to Methods of Analysis Neccessary for Checking the Composition of Cosmetic Products. 20 pp.

—Industrial Water
CPPA H.4P(D)-67. Determination of Alkalinity in Process Waters. 2 pp.

—Lactose Monohydrate
MOD UK M 821/71. Examination of Lactose.

—Latex
BSI BS 6057: Sec 3.3-90. 1990 Rubber Latices Part 3: Methods of Test Section 3.3: Determination of Alkalinity of Natural Rubber Latex Concentrate. 6 pp.
ISO 125-90. Natural Rubber Latex Concentrate—Determination of Alkalinity Fourth Edition. 4 pp.

—Magnesium Stearate
CNS K6473-80. Method of Test for Magnesium Stearate (Apr)(5454). 4 pp.

—Paint Removers
CGSB 31-GP-0A METH 51.4-62. Methods of Testing Corrosion-Prevention Materials and Processes Alkalinity and Acidity of Paint Remover. 1 p.

—Paints
CGSB 1-GP-71 METH 53.5-80. Methods of Testing Paints and Pigments Alkalinity and Acidity. 1 p.

—Paraffin Wax
MOD UK M 2009/67. Examination of Paraffin Wax.

—Perchloroethylene
BSI BS 1593-63. (WITHDRAWN) 1963 Amd 1 Perchloroethylene (Tetrachloroethylene). 13 pp.

INTERNATIONAL AND NON-U.S. NATIONAL STANDARDS
SUBJECT INDEX

Alkalinity (Cont.)
—Perchloroethylene (Cont.)
SAA AS K105-66. Perchloroethylene (Tetrachloroethylene) Being BS 1593:1963, Endorsed Without Amendment. 12 pp.

—Petroleum Products
JIS K 2252-80. Testing Method for Reaction of Petroleum Products (R 1985). 5 pp.

—Photographic Chemicals
ISO 10349 Pt 7-92. Photography—Photographic-Grade Chemicals—Test Methods —Part 7: Determination of Alkalinity or Acidity First Edition. 6 pp.

—Pigments
BSI BS 3483: Part C3-82. 1982 Methods for Testing Pigments in Paints Part C3: Determination of Acidity or Alkalinity of the Aqueous Extract. 4 pp.
CGSB 1-GP-71 METH 53.5-80. Methods of Testing Paints and Pigments Alkalinity and Acidity. 1 p.
ISO 787 Pt 4-81. General Methods of Test for Pigments and Extenders—Part 4: Determination of Acidity or Alkalinity of the Aqueous Extract First Edition. 5 pp.

—Potassium Hydroxides
BSI BS 5634-78. 1978 Potassium Hydroxide. 24 pp.

—Potassium Silicates—Volumetric Analysis
BSI BS 6092: Part 5-81. 1981 Sampling and Test for Sodium and Potassium Silicates for Industrial Use Part 5: Determination of Total Alkali Content. 4 pp.
ISO 1692-76. Sodium and Potassium Silicates for Industrial Use—Determination of Total Alkalinity—Titrimetric Method First Edition. 4 pp.

—Scouring Powders
MOD UK M 9504/59. Determination of Powder, Cleaning.

—Sodium Azide
MOD UK M 807/91. Examination of Sodium Azide.

—Sodium Carbonates
CPPA J.11-90. Analysis of Soda Ash. 3 pp.

—Sodium Carbonates—Volumetric Analysis
BSI BS 6070: Part 1-81. 1981 Methods of Sampling and Test for Sodium Carbonate for Industrial Use Part 1: Determination of Total Soluble Alkali Content. 4 pp.
ISO 740-76. Sodium Carbonate for Industrial Use—Determination of Total Soluble Alkalinity—Titrimetric Method First Edition. 4 pp.

—Sodium Chloride—Potentiometric Analysis
BSI BS 7319: Part 10-90. 1990 Analysis of Sodium Chloride for Industrial Use Part 10: Method for Determination of pH and Total Alkalinity. 6 pp.

—Sodium Nitrate
MOD UK M 869/91. Examination of Sodium Nitrate, Grade 1.

—Sodium Silicates—Volumetric Analysis
BSI BS 6092: Part 5-81. 1981 Sampling and Test for Sodium and Potassium Silicates for Industrial Use Part 5: Determination of Total Alkali Content. 4 pp.
ISO 1692-76. Sodium and Potassium Silicates for Industrial Use—Determination of Total Alkalinity—Titrimetric Method First Edition. 4 pp.

—Sodium Sulfates
ISO 3240-75. Sodium Sulphate for Industrial Use—Determination of Acidity or Alkalinity First Edition. 5 pp.

—Spent Liquors
CPPA J.22P-86. Analysis of Chlorine Solutions, Hypochlorite Bleach Liquors, and Spent Bleach Liquors. 3 pp.

—Surfactants—Volumetric Analysis
BSI BS 6829: Sec 1.2-88. 1988 Analysis of Surface Active Agents (Raw Materials) Part 1: General Methods Section 1.2: Method for Determination of Free Alkalinity or Free Acidity. 4 pp.
BSI BS 6829: Sec 1.2-01. 1988 Amd 1 Analysis of Surface Active Agents (Raw Materials) Part 1: General Methods Section 1.2: Method for Determination of Free Alkalinity or Free Acidity (AMD 7848) September 15, 1993 (W). 5 pp.
BSI BS 6829: Sec 1.3-88. 1988 Analysis of Surface Active Agents (Raw Materials) Part 1: General Methods Section 1.3: Method for Determination of Alkalinity. 4 pp.
ISO 4314-77. Surface Active Agents—Determination of Free Alkalinity or Free Acidity—Titrimetric Method First Edition. 4 pp.
ISO 4315-77. Surface Active Agents—Determination of Alkalinity—Titrimetric Method First Edition. 4 pp.

Alkalinity (Cont.)
—Urea—Volumetric Analysis
ISO 1593-77. Urea for Industrial Use—Determination of Alkalinity—Titrimetric Method First Edition. 4 pp.

—Vegetables—Ash Analysis
ISO 5520-81. Fruits, Vegetables and Derived Products—Determination of Alkalinity of Total Ash and of Water-Soluble Ash First Edition. 5 pp.

—Water—Acidimetry
SAA AS 3550.3-92. Waters—Part 3: Determination of Alkalinity—Acidimetric Titration Method (Supersedes AS 2449—1981). 4 pp.

—Water—Potentiometric Analysis
BSI BS 2690: Part 109-84. 1984 Water Used in Industry Part 109: Alkalinity, Acidity, pH Value and Carbon Dioxide. 8 pp.

—White Liquors
CPPA G.11U-77. Alkalinity of Sulphate Green and White Liquor. 2 pp.

Alkaloid Content Analysis
—Cigarette Filters—Spectrometry
BSI BS 5202: Part 9-92. 1992 Methods for Chemical Analysis of Tobacco and Tobacco Products Part 9: Determination of Alkaloid Retention by Filters of Cigarettes (Spectrometric Method) (ISO 3401: 1991) (W). 15 pp.
BSI BS 5202: Part 9-77. 1977 Methods for Chemical Analysis of Tobacco and Tobacco Products Part 9: Determination of Alkaloid Retention by Filters of Cigarettes. 10 pp.

—Cigarette Filters—Spectrophotometry
BSI BS 5202: Part 9-77. 1977 Methods for Chemical Analysis of Tobacco and Tobacco Products Part 9: Determination of Alkaloid Retention by Filters of Cigarettes. 10 pp.
CNS N4051-80. Methods of Test for Tobacco and Tobacco Products—Determination of Alkaloid Retention by Filters of Cigarettes (Jul)(5920). 7 pp.
ISO 3401-91. Cigarettes—Determination of Alkaloid Retention by the Filters—Spectrometric Method Second Edition. 12 pp.

—Cigarette Smoke—Spectrophotometry
BSI BS 5202: Part 7-90. 1990 Methods for Chemical Analysis of Tobacco and Tobacco Products Part 7: Determination of Alkaloids in Smoke Condensate of Cigarettes (Spectrophotometric Method). 8 pp.
CNS N4050-85. Method of Test for Tobacco and Tobacco Products-Determination of Alkaloids in Cigarette Smoke Condensates-Spectrophotometric Method (Dec)(5919). 3 pp.
ISO 3400-89. Cigarettes—Determination of Alkaloids in Smoke Condensates—Spectrometric Method Second Edition. 6 pp.

—Tobacco—Spectrophotometry
BSI BS 5202: Part 4-92. 1992 Methods for Chemical Analysis of Tobacco and Tobacco Products Part 4: Determination of Alkaloid Content (Spectrometric Method) (ISO 2881: 1992). 10 pp.
BSI BS 5202: Part 4-76. 1976 Methods for Chemical Analysis of Tobacco and Tobacco Products Part 4: Determination of Alkaloids in Tobacco: Spectrophotometric Method. 4 pp.
CNS N4049-85. Method of Test for Tobacco and Tobacco Products-Determination of Alkaloids in Tobacco (Spectrophotometric Method) (Dec)(5918). 3 pp.
ISO 2881-92. Tobacco and Tobacco Products—Determination of Alkaloid Content—Spectrophotometric Method Third Edition. 7 pp.

Alkane Content Analysis
CNS K6260-67. Method of Test for Paraffins in Aromatic Hydrocarbons (Mar)(2761)(R 1973).

—Gasoline—Gas Chromatography
CGSB CAN/CGSB-3.0 NO.14.3-M91. Methods of Testing Petroleum and Associated Products Standard Test Method for the Identification of Hydrocarbon Components in Automotive Gasoline Using Gas Chromatography. 32 pp.

—Waxed Paperboard
CPPA G.15-81. Paraffin in Paper and Paperboard. 1 p.

—Waxed Papers
CPPA G.15-81. Paraffin in Paper and Paperboard. 1 p.

Alkane Sulfonate Content Analysis
—Surfactants
BSI BS 6829: Sec 2.1-88. 1988 Analysis of Surface Active Agents (Raw Materials) Part 2: Alkane Sulphonates Section 2.1: Method for Determination of Total Alkane Sulphonates Content. 4 pp.
BSI BS 6829: Sec 2.2-89. 1989 Analysis of Surface Active Agents (Raw Materials) Part 2: Alkane Sulphonates Section 2.2: Method for Determination of Mean Relative Molecular Mass and Content of Alkane Monosulphonates. 8 pp.
BSI BS 6829: Sec 2.2-01. 1989 Amd 1 Analysis of Surface Active Agents (Raw Materials) Part 2: Alkane Sulphonates Section 2.2: Method for Determination of Mean Relative Molecular Mass and Content of Alkane Monosulphonates (ISO 6845: 1989) (AMD 7846) August 15, 1993 (W). 9 pp.
BSI BS 6829: Sec 2.3-89. 1989 Analysis of Surface Active Agents (Raw Materials) Part 2: Alkane Sulphonates Section 2.3: Method for Determination of Alkane Monosulphonates Content. 8 pp.
ISO 6122-78. Surface Active Agents—Technical Alkane Sulphonates—Determination of Total Alkane Sulphonates Content First Edition. 4 pp.
ISO 6845-89. Surface Active Agents—Technical Alkane Sulfonates—Determination of the Mean Relative Molecular Mass of the Alkane Monosulfonates and the Alkane Monosulfonate Content Second Edition. 7 pp.

—Surfactants—Volumetric Analysis
ISO 6121-88. Surface Active Agents—Technical Alkane Sulfonates—Determination of Alkane Monosulfonates Content by Direct Two-Phase Titration Second Edition. 7 pp.

Alkanolamides Content Analysis
—Detergents
BSI BS 3762: Sec 3.8-86. 1986 Analysis of Formulated Detergents Part 3: Quantitative Test Methods Section 3.8: Method for Determination of Alkanalamides Content. 4 pp.

Alkyd Base Primers
Use: Alkyd Primers

Alkyd Enamel Coatings
Use: Alkyd Enamel Paints

Alkyd Enamel Paints
Use For: Alkyd Enamel Coatings; Alkyd Resin Enamel Paints *See Also:* Alkyd Paints; Aminoalkyd Enamel Paints; Enamel Paints; Vinyl Alkyd Enamel Paints
CNS K2029-87. Alkyd Enamel (May)(1157).
CNS K6106-87. Method of Test for Alkyd Enamel (May)(1158). 6 pp.
CNS K6837-87. Method of Test for Alkyd Resin Baking Enamel (May)(11550).

—Automotive
CGSB CAN/CGSB-1.88-92. Gloss Alkyd Enamel, Air Drying and Baking. 9 pp.
MOD UK DSTAN 80-50-01. Paint System, Defence Equipment, High Gloss, Alkyd Paint, Priming, Defence Equipment, Zinc Phosphate/Zinc Chrome, Paint, Undercoat, Defence Equipment Paint, Finishing, Defence Equipment, High Gloss, Alkyd Types: Brushing Spraying. 26 pp.

—Exterior
CGSB CAN/CGSB-1.28-M89. Alkyd, Exterior House Paint. 10 pp.

—Flat—Interior
CGSB CAN/CGSB-1.118-M89. Interior Alkyd, Flat Finish. 11 pp.

—Formulas—Rapid Dry—Flat
CGSB 1-GP-71 METH 150.1-74. Methods of Testing Paints and Pigments Standard Formulas for Test Materials Formula for Quick-Drying Styrenated Alkyd Enamel, Flat Green 503-301; (Re-Edited September 1982). 1 p.

—Gloss—Exterior
MOD UK DSTAN 80-48-01. Paint, Finishing, General Service, Gloss, Stoving, Types: Brushing Spraying Electrostatic Spraying Dipping Issue 1; Amendment 2. 11 pp.

—Gloss—Interior
CGSB CAN/CGSB-1.60-M89. Interior Alkyd Gloss Enamel. 10 pp.

—Marine—Interior/Exterior
CGSB 1-GP-61MA-85. Enamel, Alkyd, Marine, Exterior and Interior. 13 pp.

—Marine—Interior—Fire Retardant
CGSB 1-GP-52M-77. Enamel, Alkyd, Marine, Interior, Fire Retardant, Standard for; (Amendment 2 September 1984 Incorporates Amendment 1). 12 pp.

INTERNATIONAL AND NON-U.S. NATIONAL STANDARDS
SUBJECT INDEX

Alkyd

Alkyd Enamel Paints (Cont.)
—Metal—Gloss
  CGSB CAN/CGSB-1.88-92. Gloss Alkyd Enamel, Air Drying and Baking. 9 pp.
—Metal—Gloss—Exterior
  CGSB CAN/CGSB-1.59-M89. Alkyd, Exterior Gloss Enamel. 11 pp.
—Metal—Heat Resistant—Interior/Exterior
  CGSB CAN/CGSB-1.143-M90. Heat Resistant Aluminum Enamel, Silicone Alkyd. 10 pp.
—Metal—Interior/Exterior
  CGSB CAN/CGSB-1.141-M89. Aluminum Alkyd Baking Enamel. 10 pp.
  CNS K2028-86. Alkyd Resin Baking Enamel (Apr)(1112). 2 pp.
—Metal—Semigloss—Interior/Exterior
  CGSB CAN/CGSB-1.104-M91. Semigloss Alkyd Air Drying and Baking Enamel. 9 pp.
—Semigloss—Interior
  CGSB CAN/CGSB-1.57-M90. Alkyd, Interior, Semigloss Enamel. 10 pp.
—Traffic
  CGSB 1-GP-74M-79. Standard for: Paint, Traffic, Alkyd; (Amendment 1 May 1981) (QPL Aug 1986). 14 pp.
  CGSB CAN/CGSB-1.206-M89. Hot Applied Alkyd Traffic Paint. 10 pp.
—Wood—Glare Resistant—Flat
  CGSB CAN/CGSB-1.135-M91. Flat Alkyd Enamel for Equipment; (Corrigendum—November 1992). 9 pp.
—Wood—Gloss—Exterior
  CGSB CAN/CGSB-1.59-M89. Alkyd, Exterior Gloss Enamel. 11 pp.
—Wood—Primers
  CGSB 1-GP-71 METH 130.5-79. Methods of Testing Paints and Pigments Behavior Towards Topcoats Wood Primers. 1 p.

Alkyd Modified Cellulose Nitrate Primers
Use: Nitrocellulose Primers

Alkyd Paints
See Also: Alkyd Enamel Paints; Paints
—Ammunition
  MOD UK DSTAN 80-25-01. Paint, Finishing, Ammunition, Stoving Paint, Priming, Ammunition, Stoving Scheme A: System of Primer and Finish Scheme B: One-Coat Finish Types: Brushing Spraying Electrostatic Spraying Dipping Issue 2; Amendment 3. 15 pp.
  MOD UK DSTAN 80-27-01. Paint, Finishing, Ammunition Paint, Priming, Ammunition Scheme A: System of Primer and Finish Scheme B: One-Coat Finish Types: Brushing Spraying Electrostatic Spraying Dipping Issue 2; Amendment 1. 13 pp.
—Automotive—Heat Resistant—Gloss
  MOD UK DSTAN 80-50-01. Paint System, Defence Equipment, High Gloss, Alkyd Paint, Priming, Defence Equipment, Zinc Phosphate/Zinc Chrome, Paint, Undercoat, Defence Equipment Paint, Finishing, Defence Equipment, High Gloss, Alkyd Types: Brushing Spraying. 26 pp.
—Metal—Flat
  MOD UK DSTAN 80-41-91. Paint System, Defence Equipment, IRR, Matt Paint, Undercoat, Defence Equipment Paint, Finishing, Defence Equipment, IRR Matt Types: Brushing Spraying Dipping Issue 2. 23 pp.
—Walls—Scrub Resistance—Flat—Interior
  CGSB 1-GP-71 METH 125.4-79. Methods of Testing Paints and Pigments Scrubbability Interior Flat Wall Finishes, Emulsion and Alkyd Types. 1 p.
—Wood—Flat
  MOD UK DSTAN 80-41-91. Paint System, Defence Equipment, IRR, Matt Paint, Undercoat, Defence Equipment Paint, Finishing, Defence Equipment, IRR Matt Types: Brushing Spraying Dipping Issue 2. 23 pp.

Alkyd Primers
Use For: Alkyd Base Primers
  CGSB CAN/CGSB-1.166-M90. Basic Lead Silicochromate Primer, Oil Alkyd Type. 17 pp.
—Aluminum
  CGSB CAN/CGSB-1.81-M90. Air Drying and Baking Alkyd Primer for Vehicles and Equipment; (Corrigendum—Aug 1990). 11 pp.

Alkyd Primers (Cont.)
—Aluminum (Cont.)
  CNS K2201-88. Alkyd Resin Aluminum Tri-Polyphosphate Anti-Corrosive Primer (Apr)(12266). 2 pp.
—Aluminum—Tripolyphosphate Content
  CNS K6952-88. Method of Test for Alkyd Resin Aluminum Tri-Polyphosphate Anticorrosive Primer (Apr)(12267). 3 pp.
—Automotive
  CGSB CAN/CGSB-1.81-M90. Air Drying and Baking Alkyd Primer for Vehicles and Equipment; (Corrigendum—Aug 1990). 11 pp.
—Steel
  CGSB CAN/CGSB-1.40-M89. Primer, Structural Steel Oil Alkyd Type. 10 pp.
  CGSB CAN/CGSB-1.81-M90. Air Drying and Baking Alkyd Primer for Vehicles and Equipment; (Corrigendum—Aug 1990). 11 pp.
  CGSB CAN/CGSB-1.140-M89. Oil-Alkyd Type Red Lead, Iron Oxide Primer. 12 pp.
—Wood—Environmental Testing
  CGSB 1-GP-71 METH 143.9-78. Methods of Testing Paints and Pigments Long-Term Outdoor Performance Test for Alkyd Primer on Wood. 1 p.
—Wood—Exterior
  CGSB CAN/CGSB-1.84-M91. Alkyd Primer for Hardwood. 5 pp.
  CGSB 1-GP-189M-78. Standard for: Primer, Alkyd, Wood, Exterior; (Amendment 1 Aug 1984) (QPL Aug 1986). 12 pp.

Alkyd Resin Enamel Paints
Use: Alkyd Enamel Paints

Alkyd Resins
See Also: Polyester Resins; Resins; Thermosetting Resins
  CNS K2099-80. Alkyd Resin (Medium Oil Type) (Feb)(5213). 1 p.
  CNS K2100-80. Alkyd Resin (Long Oil Type) (Feb)(5214). 1 p.
  CNS K2101-80. Alkyd Resin (Short Oil Type) (Feb)(5215). 1 p.
—Molding Materials—Viscosity
  BSI BS 2782:Pt7: METH 720B-79. 1979 Methods of Testing Plastics Part 7: Rheological Properties Method 720B: Cup Flow of Phenolic and Alkyd Moulding Materials. 3 pp.
—Paints—Binders
  ISO 6744-84. Binders for Paints and Varnishes—Alkyd Resins—General Methods of Test First Edition. 9 pp.
—Varnishes—Binders
  ISO 6744-84. Binders for Paints and Varnishes—Alkyd Resins—General Methods of Test First Edition. 9 pp.

Alkyd Stains
—Wood—Interior/Exterior
  CGSB CAN/CGSB-1.145-M90. Solvent-Based Pigmented Stain. 8 pp.

Alkylbenzene Sulfonates Content Analysis
—Detergents
  BSI BS 3762: Sec 3.10-89. 1989 Analysis of Formulated Detergents Part 3: Quantitative Test Methods Section 3.10: Methods for Determination of Short-Chain Alkylbenzenesulphonates Content. 6 pp.

Alkylphenyl Polyoxyethylene Sulfates
Use: Polyethoxylated Alkylphenol Sulfates

All-Or-Nothing Relays
See Also: Protective Relays; Relays
  BSI BS 142: Sec 2.1-82. 1982 Electrical Protection Relays Part 2: Requirements for the Principal Families of Protection Relays Section 2.1: All-or-Nothing Electrical Protection Relays. 8 pp.
  BSI BS 5992: Part 2-80. 1980 Electrical Relays Part 2: All-or-Nothing Electrical Relays. 36 pp.
  BSI BS 5992: Part 6-84. 1984 Electrical Relays Part 6: Basic Modules for the Dimensions of General Purpose All-or-Nothing Relays. 10 pp.
  CECC CECC 16 000 ISSUE 1-86. Generic Specification: Electromechanical All-or-Nothing Relays: Generic Data and Methods of Test for Time Delay Relays (En, Fr, Ge). 72 pp.
  CECC CECC 16 100 ISSUE 1-79. Sectional Specification: Electromechanical All-or-Nothing Relays (En, Fr, Ge). 56 pp.

All-Or-Nothing Relays (Cont.)
  CECC CECC 16 101 ISSUE 1-80. Blank Detail Specification: Electromechanical All-or-Nothing Relays (En, Fr, Ge). 45 pp.
  CECC CECC 16 200 ISSUE 1-90. Sectional Specification: Electromechanical All-or-Nothing Relays (En, Fr, Ge) AMD 1 (En, Fr, Ge). 75 pp.
  CECC EN 116 500-92. Sectional Specification: Electromechanical All-or-Nothing TELECOM Relays of Assessed Quality. 17 pp.
  CECC EN 116 501-92. Blank Detail Specification: Electromechanical All-or-Nothing TELECOM Relays of Assessed Quality. 23 pp.
  DIN VDE 0435 Pt 201 A1-90. Electrical Relays; All-or-Nothing Relays; Amendment 1 (May). 4 pp.
  IEC 255 Pt 1-00-75. All-or-Nothing Electrical Relays First Edition. 70 pp.
  IEC 255 Pt 7-91. Electrical Relays Part 7: Test and Measurement Procedures for Electromechanical All-or-Nothing Relays Second Edition (IECQ QC 160000). 105 pp.
  IEC 255 Pt 18-82. Electrical Relays Part 18: Dimensions for General Purpose All-or-Nothing Relays First Edition. 24 pp.
  IEC 255 Pt 19-83. Electrical Relays Part 19: Sectional Specification: Electromechanical All-or-Nothing Relays of Assessed Quality First Edition (IECQ QC 160100). 40 pp.
  IEC 255 Pt 19-1-83. Electrical Relays Part 19: Blank Detail Specification: Electromechanical All-or-Nothing Relays of Assessed Quality Test Schedules 1, 2 and 3 First Edition (IECQ QC 160101). 80 pp.
  IECQ QC 160000-91. Electrical Relays Part 7: Test and Measurement Procedures for Electromechanical All-or-Nothing Relays (IEC 255-7 ED 2). 106 pp.
  IECQ QC 160100-83. Electrical Relays Part 19: Sectional Specification: Electromechanical All-or-Nothing Relays of Assessed Quality (IEC 255-19 ED 1). 40 pp.
  IECQ QC 160101-83. Electrical Relays Part 19: Blank Detail Specification: Electromechanical All-or-Nothing Relays of Assessed Quality Test Schedules 1, 2 and 3 (IEC 255-19-1 ED 1). 81 pp.

—0-49 V DC—Crystal Can—Single Shot—Severe Environments—PPL
  CECC CECC MUAHAG Vol 5 IS 3-88. Preferred Products List; Relays (En, Fr, Ge). 45 pp.

—0-49 V DC—Four Pole—Hermetically Sealed
  CECC CECC 16 101-012 ISSUE 3-85. All or Nothing Relays; Type: One Cubic Inch—4 PDT—10 Amp—Relay (En). 20 pp.
  CECC CECC 16 101-021 ISSUE 1. BS CECC 16 101-021; Electromechanical All or Nothing Relays; Half Crystal Can Relay Hermetically Sealed, 2 Pole Changeover Contacts Light Duty with Low Level Capability; Solderable Straight or Hooked Terminations (En). 19 pp.

—0-49 V DC—Hermetically Sealed—Single Shot
  CECC CECC 16 207-005 ISSUE 1. CEI CECC 16 207-XXA Issue 1; All-Or-Nothing Relays; Monostable Not Polarized—2 Change over 1A Base and with Suppression Device. 20 pp.
  CECC CECC 16 207-006 ISSUE 1. CEI CECC 16 207-XXB Is 1; All-Or-Nothing Relays; Monostable Not Polarized—2 Change over 1A Base and with Suppression De-vice (Sensitive Relay); CA 0, 10 mA Max—CA 2, 1 A Resistive; Hermeti-cally Sealed—Terminals 2,54 mm—Grid Pattern—Solderable Straight Terminations (En). 20 pp.
  CECC CECC 16 207-007 ISSUE 1. CEI CECC 16 207-007 Issue 1; All-Or-Nothing Relays; Monostable Not Polarized—2 Change over 1A Base and with Suppression Device; CA 0, 10 mA Max—CA 2, 1 A Resistive; Hermeti-cally Sealed—Terminals 2,54 mm—Grid Pattern—Solderable Straight Terminations (En). 20 pp.
  CECC CECC 16 207-008 ISSUE 1. CEI CECC 16 207-008 Is 1; All-Or-Nothing Relays; Monostable Not Polarized—2 Change over 1A Base and with Suppression De-vice (Sensitive Relay); CA 0, 10 mA Max—CA 2, 1 A Resistive; Hermeti-cally Sealed—Terminals 2,54 mm—Grid Pattern—Solderable Straight Terminations (En). 20 pp.

—0-49 V DC—Polarized—Hermetically Sealed—Single Shot
  CECC CECC 16 101-001 ISSUE 1-87. BS CECC 16 101-001; Electromechanical All-Or-Nothing Relay; Polarized Monostable Hermetically Sealed with Internal ARC Barriers; Hooked or Straight Termination (En). 23 pp.
  CECC CECC 16 200-801 ISSUE 1-90. Detailed Specification: Electromechanical All-or-Nothing Relays; Polarized Monostable Hermetically Sealed General Purpose Relays (En). 30 pp.

All-Or-Nothing Relays (Cont.)

— 0-49 V DC—Single Shot—Severe Environments—PPL
CECC CECC MUAHAG Vol 5 IS 3-88. Preferred Products List; Relays (En, Fr, Ge). 45 pp.

— 0-49 V DC—Subminiature—Single Shot—Severe Environments—PPL
CECC CECC MUAHAG Vol 5 IS 3-88. Preferred Products List; Relays (En, Fr, Ge). 45 pp.

— 0-49 V DC—Telecommunication Equipment—Printed Circuit Mount
CECC CECC 16 501-001 ISSUE 1-92. Electromechanical All-Or-Nothing Telecom Relays with 20 x 15 mm Base, 4 Change-Over Contacts; Monostable, Non-Polarized 4 Change-Over Contacts (En). 20 pp.
CECC CECC 16 502-001 ISSUE 1-92. Electromechanical All-Or-Nothing Telecom Relays with 20 x 10 mm Base, 2 Change-Over Contacts; Monostable, Non-Polarized, 2 Change-Over Contacts (En). 22 pp.

— 0-49 V DC—Two Pole—Hermetically Sealed
CECC CECC 16 101-003 ISSUE 3-82. CNR CEI CECC 16 101-003; Two Pole Changeover Contact Arrangement (2 Form C); Light Duty with Low Level Switching Capability; Solderable Straight Terminations; Test Schedule 3 with Additions (En). 20 pp.
CECC CECC 16 101-004 ISSUE 3-82. CNR CEI CECC 16 101-004; Two Pole Changeover Contact Arrangement (2 Form C); Light Duty with Low Level Switching Capability; Solderable Straight Terminations; Test Schedule 3 with Additions (En). 20 pp.
CECC CECC 16 101-005 ISSUE 3-82. CNR CEI CECC 16 101-005; Two Pole Changeover Contact Arrangement (2 Form C); Light Duty with Low Level Switching Capability; Solderable Straight Terminations; Test Schedule 3 with Additions (En). 17 pp.
CECC CECC 16 101-006 ISSUE 3-82. CNR CEI CECC 16 101-006; Two Pole Changeover Contact Arrangement (2 Form C); Light Duty with Low Level Switching Capability; Solderable Straight Terminations; Test Schedule 3 with Additions (En). 17 pp.
CECC CECC 16 101-011 ISSUE 3-85. All or Nothing Relays; Type: Half Cubic Inch—2 PDT—10 Amp—Relay (En). 20 pp.

— 0-49 V DC—Two Pole—Hermetically Sealed—Crystal Can
CECC CECC 16 101-007 ISSUE 3-82. CNR CEI CECC 16 101-007; Two Pole Changeover Contact Arrangement (2 Form C); Light Duty with Low Level Switching Capability; Solderable Straight or Hooked Terminations; Test Schedule 3 with Additions (En). 23 pp.
CECC CECC 16 101-008 ISSUE 4-88. CEI CECC 16 101-008; Two Pole Changeover Contact Arrangement (2 Form C); Light Duty with Low Level Switching Capability; Solderable Straight or Hooked Terminations; Test Schedule 3 with Additions (En). 23 pp.

— 50-599 V DC—Crystal Can—Single Shot—Severe Environments—PPL
CECC CECC MUAHAG Vol 5 IS 3-88. Preferred Products List; Relays (En, Fr, Ge). 45 pp.

— 50-599 V DC—Polarized—Hermetically Sealed
CECC CECC 16 200-801 ISSUE 1-90. Detailed Specification: Electromechanical All-Or-Nothing Relays; Polarized Monostable Hermetically Sealed General Purpose Relays (En). 30 pp.

— 50-599 V DC—Polarized—Hermetically Sealed—Single Shot
CECC CECC 16 101-001 ISSUE 1-87. BS CECC 16 101-001; Electromechanical All-Or-Nothing Relay; Polarized Monostable Hermetically Sealed with Internal ARC Barriers; Hooked or Straight Termination (En). 23 pp.

— 50-599 V DC—Single Shot—Severe Environments—PPL
CECC CECC MUAHAG Vol 5 IS 3-88. Preferred Products List; Relays (En, Fr, Ge). 45 pp.

— 50-599 V DC—Two Pole—Hermetically Sealed—Crystal Can
CECC CECC 16 101-008 ISSUE 4-88. CEI CECC 16 101-008; Two Pole Changeover Contact Arrangement (2 Form C); Light Duty with Low Level Switching Capability; Solderable Straight or Hooked Terminations; Test Schedule 3 with Additions (En). 23 pp.

—Bistable—0-49 V DC—Crystal Can—Severe Environments—PPL
CECC CECC MUAHAG Vol 5 IS 3-88. Preferred Products List; Relays (En, Fr, Ge). 45 pp.

—Bistable—0-49 V DC—Severe Environments—Preferred Products List
CECC CECC MUAHAG Vol 5 IS 3-88. Preferred Products List; Relays (En, Fr, Ge). 45 pp.

—Bistable—50-599 V DC—Crystal Can—Severe Environments—PPL
CECC CECC MUAHAG Vol 5 IS 3-88. Preferred Products List; Relays (En, Fr, Ge). 45 pp.

—Bistable—50-599 V DC—Severe Environments—Preferred Products List
CECC CECC MUAHAG Vol 5 IS 3-88. Preferred Products List; Relays (En, Fr, Ge). 45 pp.

—Coaxial—0-49 V DC—Crystal Can—Severe Environments—PPL
CECC CECC MUAHAG Vol 5 IS 3-88. Preferred Products List; Relays (En, Fr, Ge). 45 pp.

—Dual In Line—Equipment
CECC EN 116 502-92. Blank Detail Specification: Electromechanical All-or-Nothing TELECOM Relays of Assessed Quality, Dual-in-Line, with 20 x 10 mm Base. 22 pp.
CECC EN 116 503-92. Blank Detail Specification: Electromechanical All-or-Nothing TELECOM Relays of Assessed Quality, Dual-in-Line, with 14 x 9 mm Base. 28 pp.

—Dual In Line—Printed Circuit Mount
CECC EN 116 502-92. Blank Detail Specification: Electromechanical All-or-Nothing TELECOM Relays of Assessed Quality, Dual-in-Line, with 20 x 10 mm Base. 22 pp.
CECC EN 116 503-92. Blank Detail Specification: Electromechanical All-or-Nothing TELECOM Relays of Assessed Quality, Dual-in-Line, with 14 x 9 mm Base. 28 pp.

—High Voltage—Severe Environments—Preferred Products List
CECC CECC MUAHAG Vol 5 IS 3-88. Preferred Products List; Relays (En, Fr, Ge). 45 pp.

—Monostable—0-49 V DC
CECC CECC 16 303-001 ISSUE 1-93. VG 95 302 Teil 400; Electromechanical All-Or-Nothing Heavy Load Relays (En). 28 pp.
CECC CECC 16 303-002 ISSUE 1-93. VG 95 302 Teil 401; Electromechanical All-Or-Nothing Heavy Load Relays (En). 29 pp.
CECC CECC 16 303-003 ISSUE 1-93. VG 95 302 Teil 402; Electromechanical All-Or-Nothing Heavy Load Relays (En). 28 pp.
CECC CECC 16 303-004 ISSUE 1-93. VG 95 302 Teil 403; Electromechanical All-Or-Nothing Heavy Load Relays (En). 29 pp.

—Packaging
MOD UK DSTAN 81-33-01. Packaging of Relays, Electrical Issue 2; Amendment 1. 16 pp.

—Quality Assurance
BSI BS 9151-72. 1972 Amd 2 Rules for the Preparation of Detail Specifications for All-Or-Nothing Electromechanical Relays of Assessed Quality. 19 pp.
BSI BS 9153-82. 1982 Amd 1 All-Or-Nothing Electromechanical Relays of Assessed Quality Primarily Intended for Telecommunication Applications: Full and Basic Assessment Levels. 30 pp.
BSI BS QC 160000 Part 1-87. 1987 Electromechanical All-Or-Nothing Relays. Electrical Relays: Generic Specification Part 1: Test and Measurement Procedures. 50 pp.
BSI BS QC 160000 Part 2-87. 1987 Electromechanical All-Or-Nothing Relays. Electrical Relays: Generic Specification Part 2: Quality Assessment Procedures. 19 pp.
BSI BS QC 160100-87. 1987 Electromechanical All-Or-Nothing Relays. Electrical Relays: Sectional Specification. 20 pp.
BSI BS QC 160101-87. 1987 Electromechanical All-Or-Nothing Relays Test Schedules 1, 2, and 3: Blank Detail Specification. 40 pp.
BSI BS CECC 16000: Pt 1-92. 1992 Generic Specification: Electromechanical All-Or-Nothing Relays Part 1: General. 86 pp.
BSI BS CECC 16000-80. 1980 Amd 1 Electromechanical All-Or-Nothing Relays: Generic Specifications. 70 pp.
BSI BS CECC 16100-80. 1980 Electromechanical All-Or-Nothing Relays: Sectional Specification. 21 pp.
BSI BS CECC 16101-80. 1980 Electromechanical All-Or-Nothing Relays: Blank Detail Specification: Test Schedule 3. 17 pp.
BSI BS CECC 16200-92. 1992 Sectional Specification: Electromechanical All-or-Nothing Relays. 28 pp.
BSI BS CECC 17000-92. 1992 Generic Specification: Solid State All-or-Nothing Relays. Generic Data and Methods of Test. 63 pp.
CENELEC EN 116 500-92. Sectional Specification: Electromechanical All-or-Nothing TELECOM Relays of Assessed Quality. 17 pp.
CENELEC EN 116 501-92. Blank Detail Specification: Electromechanical All-or-Nothing TELECOM Relays of Assessed Quality. 23 pp.
CENELEC EN 116 502-92. Blank Detail Specification: Electromechanical All-or-Nothing TELECOM Relays of Assessed Quality, Dual-in-Line, with 20 x 10 mm Base. 22 pp.
CENELEC EN 116 503-92. Blank Detail Specification: Electromechanical All-or-Nothing TELECOM Relays of Assessed Quality, Dual-in-Line, with 14 x 9 mm Base. 28 pp.
IEC 255 Pt 10-79. Electrical Relays Part 10: Application of the IEC Quality Assessment System for Electronic Components to All-or-Nothing Relays First Edition. 38 pp.
IECQ QC 160101-83. Electrical Relays Part 19: Blank Detail Specification: Electromechanical All-or-Nothing Relays of Assessed Quality Test Schedules 1, 2 and 3 (IEC 255-19-1 ED 1). 81 pp.

—Reed—Severe Environments—Preferred Products List
CECC CECC MUAHAG Vol 5 IS 3-88. Preferred Products List; Relays (En, Fr, Ge). 45 pp.

—Sensitive—0-49 V DC—Crystal Can—Single Shot—PPL
CECC CECC MUAHAG Vol 5 IS 3-88. Preferred Products List; Relays (En, Fr, Ge). 45 pp.

—Sensitive—0-49 V DC—Single Shot—Severe Environments—PPL
CECC CECC MUAHAG Vol 5 IS 3-88. Preferred Products List; Relays (En, Fr, Ge). 45 pp.

—Sensitive—50-599 V DC—Crystal Can—Single Shot—PPL
CECC CECC MUAHAG Vol 5 IS 3-88. Preferred Products List; Relays (En, Fr, Ge). 45 pp.

—Solid State—Severe Environments—Preferred Products List
CECC CECC MUAHAG Vol 5 IS 3-88. Preferred Products List; Relays (En, Fr, Ge). 45 pp.

—Telecommunication Equipment
CECC EN 116 500-92. Sectional Specification: Electromechanical All-or-Nothing TELECOM Relays of Assessed Quality. 17 pp.
CECC EN 116 501-92. Blank Detail Specification: Electromechanical All-or-Nothing TELECOM Relays of Assessed Quality. 23 pp.

—Time Delay—Environments—Preferred Products List
CECC CECC MUAHAG Vol 5 IS 3-88. Preferred Products List; Relays (En, Fr, Ge). 45 pp.

—Time Delay—Quality Assurance
BSI BS CECC 16000: Pt 2-86. 1986 Electromechanical All-or-Nothing Relays: Generic Specifications Part 2: Generic Data and Methods of Test for Time Delay Relays. 27 pp.

All Weather Operations
CAA. Contents (Joint Airworthiness Requirements). 1 p.
CAA. Foreword (Joint Airworthiness Requirements). 2 pp.
CAA. Check List of Pages (Joint Airworthiness Requirements). 2 pp.
CAA. Preambles (Joint Airworthiness Requirements). 1 p.

Allen Wrenches
See Also: Hexagonal Head Screws; Wrenches
ISO 2936-83. Assembly Tools for Screws and Nuts—Hexagon Socket Screw Keys—Metric Series Third Edition; (Corrected and Reprinted -1984). 5 pp.

Allied Quality Assurance Publications
Use: Quality Assurance—Allied Publications; Technical Writing—Allied Quality Assurance Publications

Allison Engines
Use: Aircraft Engines—Allison

Allocation of Frequencies
Use: Frequency Assignment

Allophane Content Analysis
—Soil Analysis
SNZ NZS 4402: Pt 3: TEST 3.4-86. Methods of Testing Soils for Civil Engineering Purposes Part 3: Soil Chemical Tests. Test 3.4: Detection of Presence of Allophane in Soils. 1 p.

Alloy Powder
Use: Metal Powders

Alloy Steel Castings
See Also: Alloy Steels; High Strength Steel Castings; Steel Castings

—Heat Treated
ISO 9477-92. High Strength Cast Steels for General Engineering and Structural Purposes First Edition. 6 pp.

Alloy Steels
Scope Note: For additional listings, see also specific products made from alloy steels *Use For:* Steel Alloys *See Also:* Alloy Steel Castings; Alloys; Aluminum Chromium Molybdenum Steels; Austenitic Stainless Steels; Boron Steels; Carbon Manganese Steels; Carbon Steels; Chromium Molybdenum Steels; Chromium Molybdenum Vanadium Steels; Chromium Steels; Corrosion Resistant Steels; Electrical Steels; Ferritic Stainless Steels; Ferroalloys; Free Machining Steels; Heat Resistant Steels; High Alloy Steels; High Strength Steels; Low Alloy Steels; Magnetic Materials; Manganese Steels; Maraging Steels; Martensitic Stainless Steels; Nickel Chromium Molybdenum Steels; Nickel Chromium Steels; Nickel Steels; Silicon Steels; Stainless Steels; Steels; Structural Steels; Tool Steels; Tungsten Steels
SAA AS 1444-86. Wrought Alloy Steels—Standard and Hardenability (H) Series. 43 pp.
SAA AS 1449-80. Wrought Alloy Steels—Stainless and Heat-Resisting Steel Plate, Sheet and Strip. 19 pp.

—Aerospace—Baumann Testing
AECMA PREN2003-13-86. Test Methods for Steel Products—Part 13—Macrographic Examination by Sulphur Print (Baumann Method). 1 p.

—Bars
ISO 4954-93. Steels for Cold Heading and Cold Extruding Second Edition. 42 pp.
MOD UK DSTAN 01-8-85. Steel Specifications (Group System) Issue 1. 32 pp.

—Bars—Bolts
CNS G3105-78. Alloy Steel Bars for Special Application Bolting Materials (Jul)(4443). 11 pp.
JIS G 4108-88. Alloy Steel Bars for Special Application Bolting Materials. 14 pp.

—Bars—Cold Finished
SAA AS 1444-86. Wrought Alloy Steels—Standard and Hardenability (H) Series. 43 pp.

—Bars—Hot Rolled
SAA AS 1444-86. Wrought Alloy Steels—Standard and Hardenability (H) Series. 43 pp.

—Bars—Hot Rolled—Round
ISO 9443-91. Heat-Treatable and Alloy Steels—Surface Quality Classes for Hot-Rolled Round Bars and Wire Rods —Technical Delivery Conditions First Edition. 9 pp.

—Baumann Testing
BSI BS 6285-82. 1982 Macrographic Examination of Steel by Sulphur Print (Baumann Method). 4 pp.
ISO 4968-79. Steel—Macrographic Examination by Sulphur Print (Baumann Method) First Edition. 5 pp.

—Billets
SAA AS 1444-86. Wrought Alloy Steels—Standard and Hardenability (H) Series. 43 pp.

—Blooms
SAA AS 1444-86. Wrought Alloy Steels—Standard and Hardenability (H) Series. 43 pp.

—Bolts
CNS G3210-83. Alloy Steel Bolting Materials for High Temperature Service (Jul)(10439).
JIS G 4107-88. Alloy Steel Bolting Materials for High Temperature Services. 10 pp.

—Calcium Content—Atomic Absorption Spectrometry
CEN EN 10 177-89. Chemical Analysis of Ferrous Materials Determination of Calcium in Steels Flame Atomic Absorption Spectrometric Method. 6 pp.

Alloy Steels (Cont.)
—Calcium Content—Atomic Absorption Spectrometry (Cont.)
DIN ENGL EN 10177-90. Chemical Analysis of Ferrous Materials; Determination of Calcium in Steels; Flame Atomic Absorption Spectrometric Method (Apr). 7 pp.

—Chromium Content—Atomic Absorption Spectrometry
ISO 10138-91. Steel and Iron—Determination of Chromium Content—Flame Atomic Absorption Spectrometric Method First Edition. 12 pp.

—Classification
ISO 4948 Pt 1-82. Steels—Classification—Part 1: Classification of Steels into Unalloyed and Alloy Steels Based on Chemical Composition First Edition. 4 pp.
ISO 4948 Pt 2-81. Steels—Classification—Part 2: Classification of Unalloyed and Alloy Steels According to Main Quality Classes and Main Property or Application Characteristics First Edition. 9 pp.

—Cold Worked—Bars
CNS G2133-82. Method of Test for Cold Finished Carbon and Alloy Steel Bars (Aug)(9277).
CNS G3194-85. Cold Finished Carbon and Alloy Steel Bars (Oct) (9276). 5 pp.
JIS G 3123-87. Cold Finished Carbon and Alloy Steel Bars. 11 pp.
SAA AS 1313-89. Steel Tendons for Prestressed Concrete—Cold-Worked High-Tensile Alloy Steel Bars for Prestressed Concrete. 5 pp.

—Elongation
BSI BS 3894: Part 1-65. 1965 Conversion for Elongation Values for Steel Part 1: Carbon and Low Alloy Steels. 25 pp.

—Forgings
BSI BS 4670-71. 1971 Alloy Steel Forgings. 46 pp.
SAA AS 1444-86. Wrought Alloy Steels—Standard and Hardenability (H) Series. 43 pp.

—Forgings—Aircraft
CAA Chapter K3-12 10.69. Forgings. 2 pp.
CAA Chapter Q3-12 12.79. Forgings. 2 pp.

—Forgings—Pressure Vessels
CNS G2225-84. Method of Test for Quenched and Tempered Alloy Steel Forgings for Pressure Vessels (Oct)(11065).
CNS G2226-84. Method of Test for Alloy Steel Forgings for Pressure Vessels for High Temperature Service (Oct)(11066).
CNS G2227-84. Method of Test for Carbon and Alloy Steel Forgings for Pressure Vessels for Low Temperature Service (Oct)(11068).
CNS G3173-84. Quenched and Tempered Alloy Steel Forgings for Pressure Vessels (Oct)(8700).
CNS G3174-84. Alloy Steel Forgings for Pressure Vessels for High-Temperature Service (Oct)(8701).
CNS G3222-84. Carbon and Alloy Steel Forgings for Pressure Vessels for Low Temperature Service (Oct)(11067).

—Physical Properties—Marine
MOD UK DSTAN 01-2-69. Guide to Engineering Alloys Used in Naval Service: Data Sheets (Withdrawn). 247 pp.
MOD UK DSTAN 01-2: Part 1-91. Guide to Engineering Alloys Used in Navy Service: Data Sheets Part 1: Ferrous Alloys (F) Issue 2. 111 pp.

—Plates
BSI BS 1501: Part 2-88. 1988 Steels for Fired and Unfired Pressure Vessels. Plates Part 2: Alloy Steels. 24 pp.
BSI BS 1501: Part 2-01. 1988 Amd 1 Steels for Pressure Purposes Part 2: Specification for Alloy Steels (AMD 6823) October 31, 1991. 29 pp.
BSI BS 1501: Part 2-02. 1988 Amd 2 Steels for Pressure Purposes Part 2: Specification for Alloy Steels: Plates (AMD 7025) June 15, 1992 (Q). 29 pp.
BSI BS 1501: Part 2-03. 1988 Amd 3 Steels for Pressure Purposes Part 2: Specification for Alloy Steels: Plates. (AMD 7313) Feburay 15, 1993 (Q). 29 pp.
SNZ NZS/BS 1501: Part 2-88. Steels for Pressure Purposes: Plates Part 2: Specification for Alloy Steels Amend: 1, 1991. 24 pp.

—Plates—Hot Rolled
BSI BS EN 10051-92. 1992 Continuously Hot-Rolled Uncoated Plate, Sheet and Strip of Non-Alloy and Alloy Steels—Tolerances on Dimensions and Shape (V). 20 pp.
CEN PREN 10 051-89. Continuously Hot Rolled Uncoated Sheet and Strip of Non-Alloyed Steel: Tolerance on Dimensions and Shape. 12 pp.

Alloy Steels (Cont.)
—Plates—Hot Rolled (Cont.)
CEN EN 10051-91. Continuously Hot-Rolled Uncoated Plate, Sheet and Strip of Non-Alloy and Alloy Steels—Tolerances on Dimensions and Shape. 24 pp.
DIN ENGL EN 10051-92. Continously Hot Rolled Uncoated Unalloyed and Alloy Steel Plate, Sheet and Strip; Dimensional and Geometrical Tolerances (Feb) (Supersedes Part of DIN 1016, June 1987 Edition). 18 pp.

—Sheets—Hot Rolled
BSI BS EN 10051-92. 1992 Continuously Hot-Rolled Uncoated Plate, Sheet and Strip of Non-Alloy and Alloy Steels—Tolerances on Dimensions and Shape (V). 20 pp.
CEN PREN 10 051-89. Continuously Hot Rolled Uncoated Sheet and Strip of Non-Alloyed Steel: Tolerance on Dimensions and Shape. 12 pp.
CEN EN 10051-91. Continuously Hot-Rolled Uncoated Plate, Sheet and Strip of Non-Alloy and Alloy Steels—Tolerances on Dimensions and Shape. 24 pp.
DIN ENGL EN 10051-92. Continously Hot Rolled Uncoated Unalloyed and Alloy Steel Plate, Sheet and Strip; Dimensional and Geometrical Tolerances (Feb) (Supersedes Part of DIN 1016, June 1987 Edition). 18 pp.

—Slabs
SAA AS 1444-86. Wrought Alloy Steels—Standard and Hardenability (H) Series. 43 pp.

—Stress Corrosion Cracking
DIN ENGL 50915 Pt 1-85. Testing the Unalloyed and Low Alloy Steels Resistance to Intergranular Stress Corrosion Cracking; Unwelded Products (Dec). 3 pp.

—Strips—Cold Rolled—Magnetic
BSI BS 6404: Sec 8.2-86. 1986 Magnetic Materials Part 8: Individual Materials Section 8.2: Cold-Rolled, Magnetic, Non-Oriented, Alloyed Steel Strip Delivered inthe Semi-Processed State. 11 pp.
IEC 404 Pt 8-2-85. Magnetic Materials Part 8: Specifications for Individual Materials Section Two—Specification for Cold-Rolled Magnetic Alloyed Steel Strip Delivered in the Semi-Processed State First Edition. 27 pp.

—Strips—Hot Rolled
BSI BS EN 10051-92. 1992 Continuously Hot-Rolled Uncoated Plate, Sheet and Strip of Non-Alloy and Alloy Steels—Tolerances on Dimensions and Shape (V). 20 pp.
CEN PREN 10 051-89. Continuously Hot Rolled Uncoated Sheet and Strip of Non-Alloyed Steel: Tolerance on Dimensions and Shape. 12 pp.
CEN EN 10051-91. Continuously Hot-Rolled Uncoated Plate, Sheet and Strip of Non-Alloy and Alloy Steels—Tolerances on Dimensions and Shape. 24 pp.
DIN ENGL EN 10051-92. Continously Hot Rolled Uncoated Unalloyed and Alloy Steel Plate, Sheet and Strip; Dimensional and Geometrical Tolerances (Feb) (Supersedes Part of DIN 1016, June 1987 Edition). 18 pp.

—Titanium Content—Atomic Absorption Spectrometry
CEN PREN 10211-91. Chemical Analysis of Ferrous Materials—Determination of Titanium in Steel and Iron—Flame Atomic Absorption Spectrometric Method. 16 pp.

—Tubes
CNS G2059-88. Method of Test for Alloy Steel Tubes for Machine Purpose (Mar)(5801). 2 pp.
CNS G3118-85. Alloy Steel Tubes for Machine Purposes (Mar)(5800). 5 pp.

—Valve Steels—Hot Worked—Bars
ISO 683 Pt 15-92. Heat-Treatable Steels, Alloy Steels and Free-Cutting Steels—Part 15: Valve Steels for Internal Combustion Engines Second Edition. 16 pp.

—Weld Metal
SAA AS 1553.1-83. Covered Electrodes for Welding—Part 1: Low Carbon Steel Electrodes for Manual Metal-Arc Welding of Carbon and Carbon-Manganese Steels. 26 pp.
SAA AS 1553.2-87. Covered Electrodes for Welding—Part 2: Low and Intermediate Alloy Steel Electrodes for Manual Metal-Arc Welding of Carbon Steels and Low and Intermediate Alloy Steels. 29 pp.

—Welding—Edge Preparation
DIN ENGL 8551 Pt 1-76. Edge Preparation for Welding; Edge Forms on Steel; Gas Welding, Manual Arc Welding and Gas-Shielded Arc Welding (June). 5 pp.
DIN ENGL 8551 Pt 4-76. Edge Preparation for Welding; Edge Forms on Steel; Submerged Arc Welding (Nov). 7 pp.

INTERNATIONAL AND NON-U.S. NATIONAL STANDARDS
SUBJECT INDEX
Alphanumeric

Alloy Steels *(Cont.)*
—Wire Rods—Hot Rolled
ISO 9443-91. Heat-Treatable and Alloy Steels—Surface Quality Classes for Hot-Rolled Round Bars and Wire Rods —Technical Delivery Conditions First Edition. 9 pp.

Alloy Wire
See Also: Alloys; Wire

—Electrical Resistance—Temperature Coefficient
BSI BS 3467-62. 1962 Method of Test for Temperature Coefficient of Resistance of Alloy Wire for Precision Resistors. 13 pp.

Alloys
Scope Note: For additional listings, use a more specific term *See Also:* Alloy Steels; Alloy Wire; Aluminum Alloys; Aluminum Silicon Bronze Alloys; Aluminum Zinc Alloys; Bearing Alloys; Beryllium Copper Alloys; Brazing Alloys; Cobalt Alloys; Copper Alloys; Ferroalloys; Glassy Alloys; Gold Alloys; Gunmetal; Hard Metals; Heat Resistant Alloys; Lead Alloys; Light Metal Alloys; Magnesium Alloys; Master Alloys; Metals; Nickel Alloys; Nickel Iron Alloys; Nonferrous Alloys; Pewter; Phosphor Copper; Platinum Alloys; Powder Metallurgy; Resistance Alloys; Shape Memory Alloys; Silver Alloys; Silver Indium Cadmium Alloys; Tin Alloys; Titanium Alloys; Titanium Palladium Alloys; White Metal Bearing Alloys; Zinc Alloys

—Atmospheric Corrosion Testing
BSI BS 6917-87. 1987 Method for Corrosion Testing in Artificial Atmospheres: General Principles. 7 pp.
BSI BS 7561-92. 1992 Method for Corrosion Tests in Artificial Atmospheres at Very Low Concentrations of Polluting Gas(es) (ISO 10062: 1991) (V). 14 pp.
ISO 7384-86. Corrosion Tests in Artificial Atmosphere—General Requirements First Edition. 6 pp.
ISO 8565-92. Metals and Alloys—Atmospheric Corrosion Testing—General Requirements for Field Tests First Edition. 12 pp.
ISO 10062-91. Corrosion Tests in Artificial Atmosphere at Very Low Concentrations of Polluting Gas(es) First Edition. 12 pp.

—Atmospheric Corrosion Testing—Atmosphere Classification
ISO 9223-92. Corrosion of Metals and Alloys—Corrosivity of Atmospheres—Classification First Edition. 18 pp.
ISO 9225-92. Corrosion of Metals and Alloys—Corrosivity of Atmospheres—Measurement of Pollution First Edition. 14 pp.
ISO 9226-92. Corrosion of Metals and Alloys—Corrosivity of Atmospheres—Determination of Corrosion Rate of Standard Specimens for the Evaluation of Corrosivity First Edition. 8 pp.

—Classification
BSI BS 1000: (669)-86. 1986 Universal Decimal Classification (UDC). English Full Edition (669): Metallurgy. 51 pp.
SNZ NZS/BS 1000 (669)-86. Universal Decimal Classification Metallurgy. 52 pp.

—Corrosion Testing
BSI BS 6980: Part 1-88. 1988 Stress Corrosion Testing Part 1: Guide to Testing Procedures. 15 pp.
BSI BS 6980: Part 7-90. 1990 Stress Corrosion Testing Part 7: Method for Slow Strain Rate Testing (ISO 7539-7: 1989). 10 pp.
ISO 7539 Pt 1-87. Corrosion of Metals and Alloys—Stress Corrosion Testing—Part 1: General Guidance on Testing Procedures First Edition. 14 pp.
ISO 7539 Pt 7-89. Corrosion of Metals and Alloys—Stress Corrosion Testing—Part 7: Slow Strain Rate Testing First Edition. 7 pp.

—Forging—Aerospace—Ultrasonic Testing
BSI M 36-70. 1970 Amd 1 Ultrasonic Testing of Special Forgings by an Immersion Technique Using Flat-Bottomed Holes as a Reference Standard. 21 pp.

—Low Melting—Chemical Analysis
CNS G2005-53. Method of Chemical Test for Low Melting Alloys (Jul)(294)(R 1973).

—Test Specimens—Corrosion Testing
BSI BS 6980: Part 2-90. 1990 Stress Corrosion Testing Part 2: Method for the Preparation and Use of Bent-Beam Specimens (ISO 7539-2: 1989). 12 pp.
BSI BS 6980: Part 3-90. 1990 Stress Corrosion Testing Part 3: Method for the Preparation and Use of U-Bend Specimens (ISO 7539-3: 1989). 8 pp.
BSI BS 6980: Part 4-90. 1990 Stress Corrosion Testing Part 4: Method for the Preparation and Use of Uniaxially Loaded Tension Specimens (ISO 7539-4: 1989). 9 pp.

Alloys *(Cont.)*
—Test Specimens—Corrosion Testing *(Cont.)*
BSI BS 6980: Part 5-90. 1990 Stress Corrosion Testing Part 5: Method for the Preparation and Use of C-Ring Specimens (ISO 7539-5: 1989). 15 pp.
BSI BS 6980: Part 6-90. 1990 Stress Corrosion Testing Part 6: Method for the Preparation and Use of Pre-Cracked Specimens (ISO 7539-6: 1989). 36 pp.
ISO 7539 Pt 2-89. Corrosion of Metals and Alloys—Stress Corrosion Testing—Part 2: Preparation and Use of Bent-Beam Specimens First Edition. 10 pp.
ISO 7539 Pt 3-89. Corrosion of Metals and Alloys—Stress Corrosion Testing—Part 3: Preparation and Use of U-Bend Specimens First Edition. 7 pp.
ISO 7539 Pt 4-89. Corrosion of Metals and Alloys—Stress Corrosion Testing—Part 4: Preparation and Use of Uniaxially Loaded Tension Specimens First Edition. 7 pp.
ISO 7539 Pt 5-89. Corrosion of Metals and Alloys—Stress Corrosion Testing—Part 5: Preparation and Use of C-Ring Specimens First Edition. 13 pp.
ISO 7539 Pt 6-89. Corrosion of Metals and Alloys—Stress Corrosion Testing—Part 6: Preparation and Use of Pre-Cracked Specimens First Edition. 34 pp.

Allspice
See Also: Spices
CNS N5194-82. Spices and Condiments Pimenta (Allspice) Whole and Ground Specification (Apr)(8733). 3 pp.
ISO 973-80. Spices and Condiments—Pimento (Allspice), Whole or Ground—Specification First Edition. 6 pp.

Allspice Oil
Use: Pimenta Oil

Allyl Alcohols
Use For: 2-Propen-1-ol
JIS K 8062-80. 2-Propen-1-ol.

Allyl Isothiocyanate Content Analysis
—Mustard Seeds
CNS N4060-80. Determination of Allyl Isothiocyanate in Mustard Seed (Oct) (6604). 3 pp.

Almost Differential Quasi-Ternary Code
See Also: Line Signaling
CCITT FASCICLE III.5-89. Digital Networks, Digital Sections and Digital Line Systems—Recommendations G.801—G.961. 292 pp.

Alon A-2A Aircraft
See Also: Aircraft
—Antenna Positions
CAA. Alon Aircoupe and Alon A-2A. 1 p.

Alon Aircoupe Aircraft
See Also: Aircraft
—Antenna Positions
CAA. Alon Aircoupe and Alon A-2A. 1 p.
CAA. Ercoupe 415 (Approved Aerial Positions). 1 p.

Alpha-Beta Contamination Meters
Use: Radiation Meters

Alpha-Beta Contamination Monitors
Use: Radiation Monitors

Alpha Contamination Meters
Use: Radiation Meters

Alpha Contamination Monitors
Use: Radiation Monitors

Alpha Irradiation
See Also: Irradiation
—Radioactive Wastes
ISO 6962-82. Standard Method for Testing the Long Term Alpha Irradiation Stability of Solidified High-Level Radioactive Waste Forms First Edition. 10 pp.

Alpha Particle Spectrometry
See Also: Spectrometry
—Uranium Hexafluoride
CNS J2082-83. Method for Determining Uranium-232 in UF6 Alpha Spectrometric Method (Jan)(9881).
—Water
CNS J2031-82. Practice for Alpha Spectrometry of Water (Jan)(8299).

Alpha Particles
See Also: Radiation Detection and Measurement
—Radiation Detection and Measurement—Surface Contamination
ISO 7503 Pt 1-88. Evaluation of Surface Contamination—Part 1: Beta-Emitters (Maximum Beta Energy Greater Than 0.15 MeV) and Alpha-Emitters First Edition. 15 pp.

—Water
BSI BS 6068: Sec 3.1-93. 1993 Water Quality Part 3: Radiological Methods Section 3.1: Measurement of Gross Alpha Activity in Non-Saline Water: Thick Source Method (ISO 9696: 1992). 15 pp.
CNS J2032-82. Method of Test for Alpha Particle Radioactivity of Water (Jan)(8300).
ISO 9696-92. Water Quality—Measurement of Gross Alpha Activity in Non-Saline Water—Thick Source Method First Edition. 14 pp.

Alpha Radiation Monitors
See Also: Radiation Measuring Instruments; Radiation Monitors
—Calibration
ISO 8769-88. Reference Sources for the Calibration of Surface Contamination Monitors—Beta-Emitters (Maximum Beta Energy Greater Than 0,15 MeV) and Alpha-Emitters First Edition. 10 pp.
JIS Z 4334-92. Reference Sources for the Calibration of Surface Contamination Monitors. 10 pp.

—Personnel Exposure
IEC 1098-92. Installed Personnel Surface Contamination Monitoring Assemblies for Alpha and Beta Emitters First Edition. 77 pp.

Alphabets
Use: Character Sets

Alphanumeric Character Sets
See Also: Character Sets; Information Interchange; Numeric Character Sets
—Dot Matrix Printers
ECMA ECMA 42-73. Alphanumeric Character Set for 7 X 9 Matrix Printers. 14 pp.
ECMA ECMA 51-77. Implementation of the Numeric OCR-A Font with 9 X 9 Matrix Printers. 7 pp.

—Keyboards
IEC DIS 9995 Pt 2-91. Information Technology—Keyboard Layouts for Text and Office Systems—Part 2: Alphanumeric Section ***CD-ROM ONLY***. 20 pp.
IEC DIS 9995 Pt 3-91. Information Technology—Keyboard Layouts for Text and Office Systems—Part 3: Common Secondary Layout of the Alphanumeric Zone of the Alphanumeric Section ***CD-ROM ONLY***. 9 pp.
ISO DIS 9995 Pt 2-91. Information Technology—Keyboard Layouts for Text and Office Systems—Part 2: Alphanumeric Section ***CD-ROM ONLY***. 20 pp.
ISO DIS 9995 Pt 3-91. Information Technology—Keyboard Layouts for Text and Office Systems—Part 3: Common Secondary Layout of the Alphanumeric Zone of the Alphanumeric Section ***CD-ROM ONLY***. 9 pp.

—Operators' Directory Assistance
CCITT RECMN E.115-91. Computerized Information Service for Telephone Subscriber Numbers in Foreign Countries (Directory Assistance), Reserved for Operators (Study Group I) 19 pp. 19 pp.
CCITT RECMN E.115-89. Computerized Information Service for Telephone Subscriber Numbers in Foreign Countries (Directory Assistance), Reserved for Operators—Telephone Network and ISDN—Operation, Numbering, Routing and Mobile Service (Study Group II) 11 pp. 11 pp.

—Optical Character Recognition
JIS X 9001-76. Alphanumeric Character Sets for Optical Recognition.
JIS X 9006-79. Handprinted Numerals for Optical Character Recognition.
JIS X 9007-81. Handprinted Alphabets for Optical Character Recognition.

—Optical Character Recognition—HIRAGANA
JIS X 9009-91. Handprinted HIRAGANA Characters for Optical Character Recognition. 26 pp.

—Optical Character Recognition—KATAKANA
JIS X 9003-80. KATAKANA Character Set for Optical Recognition (R 1985). 45 pp.
JIS X 9005-79. Handprinted KATAKANA Characters for Optical Character Recognition.

INDUSTRY STANDARDS

INTERNATIONAL AND NON-U.S. NATIONAL STANDARDS SUBJECT INDEX

Alphanumeric

Alphanumeric Character Sets *(Cont.)*

—**Optical Character Recognition—OCR-A**
BSI BS 5464: Part 1-77. 1977 Optical Character Recognition Part 1: Character Set OCR-A. Shapes and Dimensions of the Printed Image. 29 pp.
CNS C5248-90. Alphanumeric Character Sets for Optical Recognition—Part 1: Character Set OCR-A—Shapes and Dimensions of the Printed Image (Dec)(12822).
ECMA ECMA 8-77. Nominal Character Dimensions of the Numeric OCR-A Font. 9 pp.
ECMA ECMA 51-77. Implementation of the Numeric OCR-A Font with 9 X 9 Matrix Printers. 7 pp.
ISO 1073 Pt I-76. Alphanumeric Character Sets for Optical Recognition—Part 1: Character Set OCR-A—Shapes and Dimensions of the Printed Image First Edition; (Amendment Slip-1978). 29 pp.
JTC1 1073 Pt I-76. Alphanumeric Character Sets for Optical Recognition—Part 1: Character Set OCR-A—Shapes and Dimensions of the Printed Image First Edition; (Amendment Slip-1978). 29 pp.
OSI ISO 1073-1-76. Alphanumeric Character Sets for Optical Recognition—Part 1: Character Set OCR-A—Shapes and Dimensions of the Printed Images. 29 pp.

—**Optical Character Recognition—OCR-B**
CNS C5223-88. Alphanumeric Character Sets for Optical Recognition—Part 2: Character Set OCR-B-Shapes and Dimensions of the Printed (Jul)(12365).
ECMA ECMA 11-76. Alphanumeric Character Set OCR-B for Optical Recognition. 38 pp.
ECMA ECMA 30-76. OCR-B Subsets for Numeric Applications. 4 pp.
ISO 1073 Pt II-76. Alphanumeric Character Sets for Optical Recognition—Part II: Character Set OCR-B—Shapes and Dimensions of the Printed Image First Edition; (Amendment Slip-1978). 45 pp.
JTC1 1073 Pt II-76. Alphanumeric Character Sets for Optical Recognition—Part II: Character Set OCR-B—Shapes and Dimensions of the Printed Image First Edition; (Amendment Slip-1978). 45 pp.
OSI ISO 1073-2-76. Alphanumeric Character Sets for Optical Recognition—Part 2: Character Set OCR-B—Shapes and Dimensions of the Printed Images. 45 pp.
SAA AS 1436-83. Alphanumeric Character Set OCR-B for Optical Recognition. 56 pp.

—**Sorting**
CSA Z243.4.1-1992P. Canadian Alphanumeric Ordering Standard for Character Sets of CSA Standard CAN/CSA-Z243.4; (Gen Instr 1). 44 pp.

Alpine Ski Bindings
Use: Ski Bindings—Alpine

Alpine Skis
Use: Skis—Alpine

Alternating Current

—**Corrosion**
CCITT RECMN L.8-89. Corrosion Caused by Alternating Current—Construction, Installation and Protection of Cable and Other Elements of Outside Plant (Study Group VI) 1 pp. 1 p.

Alternating Current Disconnectors
Use: Disconnecting Switches

Alternating Current Generators
Use: AC Generators

Alternating Current Motors
Use: AC Motors

Alternating Current Power Transmission
Use: Power Transmission—AC

Alternative Routing
See Also: Call Handling; Call Redirection; Network Management Systems; Rerouting; Routing (Telecommunications); Telephone Services

—**Automatic**
CCITT RECMN E.170-92. Traffic Routing (Study Group II) 10 pp. 10 pp.
CCITT RECMN Q.12-89. Overflow—Alternative Routing—Rerouting—Automatic Repeat Attempt—General Recommendations on Telephone Switching and Signalling—Functions and Information Flows for Services in the ISDN —Supplements (Study Group XI) 1 pp. 1 p.

—**Circuit Groups—Expenses**
CCITT RECMN E.522-89. Number of Circuits in a High-Usage Group—Telephone Network and ISDN—Quality of Service, Network Management and Traffic Engineering (Study Group II) 10 pp. 10 pp.

Alternative Routing *(Cont.)*

—**Circuit Groups—Traffic Units**
CCITT RECMN E.522-89. Number of Circuits in a High-Usage Group—Telephone Network and ISDN—Quality of Service, Network Management and Traffic Engineering (Study Group II) 10 pp. 10 pp.

—**Network Management Systems**
CCITT RECMN E.412-89. Network Management Controls—Telephone Network and ISDN—Quality of Service, Network Management and Traffic Engineering (Study Group II) 9 pp. 9 pp.

—**Overflow Traffic—Grade of Service**
CCITT RECMN E.525 (REV 1)-92. Designing Networks to Control Grade of Service (Study Group II) 10 pp. 10 pp.

—**Public Data Transmission**
CCITT RECMN F.600-89. Service and Operational Principles for Public Data Transmission Services—Telematic, Data Transmission and Teleconference Services—Operations and Quality of Service (Study Group I) 4 pp. 4 pp.

—**Telecommunication Networks**
CCITT FASCICLE II.3-88. Telephone Network and ISDN Quality of Service, Network Management and Traffic Engineering. Recommendations E.401-E.880. 368 pp.

—**Telecommunication Services—Grade of Service**
CCITT RECMN E.525-89. Service Protection Methods—Telephone Network and ISDN—Quality of Service, Network Management and Traffic Engineering (Study Group II) 5 pp. 5 pp.

—**Telecommunication Services—Overflow Traffic**
CCITT RECMN E.525-89. Service Protection Methods—Telephone Network and ISDN—Quality of Service, Network Management and Traffic Engineering (Study Group II) 5 pp. 5 pp.

—**Telephone Services—Accounting**
CCITT RECMN D.155-92. Guiding Principles Governing the Apportionment of Accounting Rates in Intercontinental Telephone Relations (Study Group III) 4 pp. 4 pp.
CCITT RECMN D.155-89. Guiding Principles Governing the Apportionment of Accounting Rates in Intercontinental Telephone Relations—General Tariff Principles—Charging and Accounting in Intl Telecommunications Services (Study Group III) 2 pp. 2 pp.

—**Telephone Services—Traffic Units**
CCITT RECMN D.150-92. New System for Accounting in International Telephony (Study Group III) 14 pp. 14 pp.
CCITT RECMN D.150-89. New System for Accounting in International Telephony—General Tariff Principles—Charging and Accounting in International Telecommunications Services (Study Group III) 12 pp. 12 pp.

—**Telex Communications**
CCITT RECMN F.68-89. Establishment of the Automatic Intercontinental Telex Network—Telegraph and Mobile Services Operations and Quality of Service (Study Group I) 6 pp. 6 pp.

Alternators
See Also: Generators

—**Automotive**
BSI BS AU 232-89. 1989 Methods of Test and General Requirements for Road Vehicle Alternators with Regulators. 6 pp.
CNS D3042-87. Mounting Dimensions of Alternators for Automobiles (Oct)(5319).
CNS D3062-87. Method of Test for Alternators of Automobiles (Oct)(6169).
ISO 8854-88. Road Vehicles—Alternators with Regulators—Test Methods and General Requirements First Edition. 7 pp.
JIS D 1615-89. Inspection of Alternators for Automobiles.
JIS D 5205-76. Mounting Dimensions of Alternators for Automobiles (R 1984). 7 pp.

—**Buses (Vehicles)**
BSI BS AU 250-93. 1993 Mounting Dimensions of Alternators, Types 1, 2 and 3 for Commercial Vehicles and Buses (ISO 7651: 1991) (E). 9 pp.
ISO 7651-91. Commercial Vehicles and Buses—Mounting Dimensions for Alternators of Types 1, 2 and 3 First Edition. 8 pp.

Alternators *(Cont.)*

—**Buses (Vehicles)—Cylindrical Shaft Ends**
BSI BS AU 227-88. 1988 Dimensions of Cylindrical Shaft Ends and Hubs for Alterations for Commercial Vehicles and Buses. 4 pp.
ISO 7467-87. Commercial Vehicles and Buses—Cylindrical Shaft Ends and Hubs for Alternators First Edition; (Corrigendum 1-1990). 7 pp.

—**Buses (Vehicles)—Hubs**
BSI BS AU 227-88. 1988 Dimensions of Cylindrical Shaft Ends and Hubs for Alterations for Commercial Vehicles and Buses. 4 pp.
ISO 7467-87. Commercial Vehicles and Buses—Cylindrical Shaft Ends and Hubs for Alternators First Edition; (Corrigendum 1-1990). 7 pp.

—**Internal Combustion Piston Engines**
BSI BS AU 250-93. 1993 Mounting Dimensions of Alternators, Types 1, 2 and 3 for Commercial Vehicles and Buses (ISO 7651: 1991) (E). 9 pp.
ISO 7651-91. Commercial Vehicles and Buses—Mounting Dimensions for Alternators of Types 1, 2 and 3 First Edition. 8 pp.

—**Trucks**
BSI BS AU 250-93. 1993 Mounting Dimensions of Alternators, Types 1, 2 and 3 for Commercial Vehicles and Buses (ISO 7651: 1991) (E). 9 pp.
ISO 7651-91. Commercial Vehicles and Buses—Mounting Dimensions for Alternators of Types 1, 2 and 3 First Edition. 8 pp.

—**Trucks—Cylindrical Shaft Ends**
BSI BS AU 227-88. 1988 Dimensions of Cylindrical Shaft Ends and Hubs for Alterations for Commercial Vehicles and Buses. 4 pp.
ISO 7467-87. Commercial Vehicles and Buses—Cylindrical Shaft Ends and Hubs for Alternators First Edition; (Corrigendum 1-1990). 7 pp.

—**Trucks—Hubs**
BSI BS AU 227-88. 1988 Dimensions of Cylindrical Shaft Ends and Hubs for Alterations for Commercial Vehicles and Buses. 4 pp.
ISO 7467-87. Commercial Vehicles and Buses—Cylindrical Shaft Ends and Hubs for Alternators First Edition; (Corrigendum 1-1990). 7 pp.

Altimeters
See Also: Air Navigation Equipment; Aircraft Instruments; Altitude; Barometers; Distance Measuring Equipment; Navigational Aids; Pressure Altimeters; Radar Altimeters; Radio Altimeters
BSI 2G 115-56. 1956 Amd 1 Sensitive Altimeters for Aircraft. 7 pp.
BSI G 190-64. 1964 Cabin Altimeters for Aircraft. 4 pp.

—**Counter/Pointer**
CAA NOTICE #77 ISSUE 1. Counter/Pointer Altimeters (Airworthiness Notices). 2 pp.

—**Gliders**
CAA JAR-22 SUBPART F. Equipment (Joint Airworthiness Requirements). 4 pp.

Altimetry
Use: Altitude Measurement

Altitude
See Also: Altimeters; Geostationary Orbits; Height; Orbits; Polar Orbits; Position (Location)

—**Geographic Representation**
BSI BS 5249: Part 3-83. 1983 Representation of Elements of Data in Interchanges Using Data Processing Systems Part 3: Representation of Latitude, Longitude and Altitude for Geographical Point Locations. 4 pp.
ISO 6709-83. Standard Representation of Latitude, Longitude and Altitude for Geographic Point Locations First Edition. 5 pp.
JTC1 6709-83. Standard Representation of Latitude, Longitude and Altitude for Geographic Point Locations First Edition. 5 pp.
OSI ISO 6709-83. Standard Representation of Latitude, Longitude and Altitude for Geographic Part Locations. 5 pp.

—**Tables (Data)—Aircraft**
BSI 2G 199-84. 1984 Amd 1 Tables Relating to Altitudes, Airspeed and Mach Numbers for Use in Aeronautical Instrument Design and Calibration. 41 pp.

Altitude Immersion Testing
Use: Immersion Testing

Altitude Measurement
Use For: Altimetry *See Also:* Measurement; Navigation

INTERNATIONAL AND NON-U.S. NATIONAL STANDARDS
SUBJECT INDEX

Aluminum

Altitude Measurement *(Cont.)*
CCIR Report 827-82. Radar Spectrum Utilization—Section 1D—Spectrum Utilization and Applications. 9 pp.

Altitude Testing
—Carburetors
CNS D3091-88. Method of Altitude Test of Carburetors for Automobiles (Dec)(8248).

Alum Potassium
Use: Aluminum Potassium Sulfate

Alumina
Use: Aluminum Oxide

Alumina Cements
See Also: Cements
BSI BS 915: Part 2-72. 1972 Amd 3 High Alumina Cement Part 2: Metric Units. 8 pp.

—Refractories
CNS R2070-85. Alumine Cement for Refractories (Nov)(3324).
JIS R 2511-83. Alumina Cements for Refractories. 6 pp.

Alumina Content Analysis
Use: Aluminum Oxide Content Analysis

Alumina Refractories
Use: Aluminous Refractories

Aluminate Cements
Use For: Aluminous Cements **See Also:** Cements

—Refractory Materials—Chemical Analysis
JIS R 2522-85. Chemical Analysis of Aluminous Cement for Refractories.

—Refractory Materials—Physical Testing
JIS R 2521-85. Physical Testing Method of Aluminous Cement for Refractories.

Aluminium
Use: Aluminum

Aluminizing
Use: Aluminum Coatings (Made From Aluminum)

Aluminon
CNS K7030-61. Chemical Reagent (Aluminium) (Ammonium Aurintricarboxylate) (Dec)(1530).
JIS K 8011-78. Aurintricarboxylic Acid Ammonium Salt, Aluminon.

Aluminosilicate Refractories
See Also: Aluminous Refractories; Refractory Materials
BSI BS 7225: Sec 1.2-89. 1989 Classification of Refractories Part 1: Shaped Refractory Products Section 1.2: Alumina, Silica and Alumino—Silicate Products. 4 pp.

—Chemical Analysis
BSI BS 1902: Sec 2.1-88. 1988 Methods for Testing Refractory Materials Part 2: Chemical Analysis (Wet Methods) Section 2.1: Chemical Analysis of Refractories Containing Silica and/or Alumina. 15 pp.

—Chemical Analysis—X-Ray Fluorescence Spectrometry
BSI BS 1902: Sec 9.1-87. 1987 Methods for Testing Refractory Materials Part 9: Chemical Analysis by Instrumental Methods Section 9.1: Analysis of Alumino-Silicate Refractories by X-Ray Fluorescence. 16 pp.

—Classification
BSI BS 7225: Sec 2.1-90. 1990 Classification of Refractories Part 2: Unshaped Refractory Products Section 2.1: Alumina, Silica and Alumino-Silicate Products. 10 pp.

—Coking
BSI BS 6886-87. 1987 Alumino-Silicate Refractories for Use in Coke Ovens. 12 pp.

—Refractoriness
BSI BS 1902: Sec 5.2-83. 1983 Amd 1 Methods for Testing Refractory Materials Part 5: Refractory and Thermal Properties Section 5.2: Pyrometric Cone Equivalent (Refractories) (Method 1902-502). 10 pp.

Aluminous Abrasives
See Also: Abrasives
MOD UK TS 10099. Grain, Abrasive, Fused Aluminium Oxide.

Aluminous Abrasives *(Cont.)*
—Chemical Analysis
CNS R3108-82. Method for Chemical Analysis of Aluminous Abrasives (Sep)(9438).
JIS R 6123-87. Method for Chemical Analysis of Aluminous Abrasives. 38 pp.

—Sinterability
CNS R3100-81. Method of Sintering Test for Fused Aluminous Abrasives (Aug)(7840).
JIS R 6122-61. Method of Sintering Test for Fused Aluminous Abrasives. 3 pp.

Aluminous Cements
Use: Aluminate Cements

Aluminous Refractories
Use For: High Alumina Refractories
See Also: Refractory Materials
BSI BS 7225: Sec 1.2-89. 1989 Classification of Refractories Part 1: Shaped Refractory Products Section 1.2: Alumina, Silica and Alumino—Silicate Products. 4 pp.

—Castable
JIS R 2541-76. High Alumina and Fire Clay Castable Refractories (R 1986). 6 pp.

—Castable—Classification
CNS R1015-76. Standard Classification of Fireclay and High-Alumina Plastics and Ramming Mixes (Jun)(3971).

—Chemical Analysis
BSI BS 1902: Sec 2.1-88. 1988 Methods for Testing Refractory Materials Part 2: Chemical Analysis (Wet Methods) Section 2.1: Chemical Analysis of Refractories Containing Silica and/or Alumina. 15 pp.
CEN PREN 955-2-92. Chemical Analysis of Refractory Products—Part 2: Products Containing Silica and/or Alumina (Wet Method). 24 pp.

—Classification
BSI BS 7225: Sec 2.1-90. 1990 Classification of Refractories Part 2: Unshaped Refractory Products Section 2.1: Alumina, Silica and Alumino-Silicate Products. 10 pp.

—Plastic
CGSB CAN/CGSB-10.7-93. Fireclay and Alumina Refractory Plastic. 7 pp.
JIS R 2561-76. High Alumina and Fire Clay Plastic Refractories (R 1986). 6 pp.

—Plastic—Classification
CNS R1015-76. Standard Classification of Fireclay and High-Alumina Plastics and Ramming Mixes (Jun)(3971).

—Plastic—Compression Testing
CNS R3153-87. Method of Test for Crushing Strength and Modulus of Rupture of High Alumina and Fireclay Plastic Refractories (May) (11973).
JIS R 2575-92. Testing Method for Crushing Strength and Modulus of Rupture of High Alumina and Fireclay Plastic Refractories. 10 pp.

—Plastic—Modulus of Rupture
CNS R3153-87. Method of Test for Crushing Strength and Modulus of Rupture of High Alumina and Fireclay Plastic Refractories (May) (11973).
JIS R 2575-92. Testing Method for Crushing Strength and Modulus of Rupture of High Alumina and Fireclay Plastic Refractories. 10 pp.

—Plastic—Packaging
JIS R 2571-85. Sampling Method for High Aluminous Plastic Refractories and Fireclay Plastic Refractories. 4 pp.

—Plastic—Rate of Linear Change
JIS R 2577-81. Testing Method for the Rate of Linear Change of High Alumina and Fire Clay Plastic Refractories.

—Plastic—Sampling
JIS R 2571-85. Sampling Method for High Aluminous Plastic Refractories and Fireclay Plastic Refractories. 4 pp.

—Plastic—Water Content
JIS R 2572-85. Testing Method for Water Content of High Aluminous Plastic Refractories and Fireclay Plastic Refractories. 5 pp.

—Refractoriness
BSI BS 1902: Sec 5.2-83. 1983 Amd 1 Methods for Testing Refractory Materials Part 5: Refractory and Thermal Properties Section 5.2: Pyrometric Cone Equivalent (Refractories) (Method 1902-502). 10 pp.

Aluminum
Scope Note: For additional listings, see also specific products made from aluminum **Use For:** Aluminium
See Also: Aluminum Coatings (Made From Aluminum); Aluminum Coatings (On Aluminum); Aluminum Content Analysis; Aluminum Paints (Made From Aluminum); Aluminum Products; Cryolite; Foil Papers; Metals; Nonferrous Metals
JIS K 8069-61. Aluminum.

—Aerospace
BSI 4L 100-93. 1993 Procedure for Inspection, Testing and Acceptance of Wrought Aluminium and Aluminium Alloys (S). 81 pp.
CEN PREN 2500-2-91. Instructions for the Drafting and Use of Metallic Material Standards Part 2—Specific Requirements for Aluminium, Aluminium Alloys and Magnesium Alloys. 15 pp.
CEN PREN 2500-2-92. Instructions for the Drafting and Use of Metallic Material Standards Part 2—Specific Requirements for Aluminium, Aluminium Alloys and Magnesium Alloys. 15 pp.

—Aerospace—Tempering—Designations
CEN PREN 3350-91. Aerospace Series Aluminium, Aluminium Alloys Temper Designations. 21 pp.
CEN PREN 3350-91. Aluminium, Aluminium Alloys Temper Designations. 18 pp.

—Aircraft
SBAC RS 116 (V). S.B.A.C. Protective Treatment Chart.

—Aircraft—Welded
JIS W 0901-85. Aerospace Fusion Welders Qualification.

—Angles—Extruded
DIN ENGL 1771-81. Aluminium and Wrought Aluminium Alloy; Extruded Angles; Dimensions; Static Values (Sept). 6 pp.

—Anodized—Test Panels
CGSB CAN2-19.0-M77METH12.3-78. Methods of Testing Putty, Caulking and Sealing Compounds Preparation of Anodized Aluminum Panels. 1 p.

—Arc Welding
BSI BS EN 288: Part 4-92. 1992 Specification and Approval of Welding Procedures for Metallic Materials Part 4: Welding Procedure Tests for the Arc Welding of Aluminium and its Alloys (Supersedes BS 4870: Part 2: 1982). 32 pp.
CEN PREN 288-4-91. Specification and Approval of Welding Procedures for Metallic Materials—Part 4: Welding Procedure Tests for the Arc Welding of Aluminium and its Alloys. 37 pp.
CEN EN 288-4-92. Specification and Approval of Welding Procedures for Metallic Materials—Part 4: Welding Procedure Tests for the Arc Welding of Aluminium and Its Alloys. 27 pp.
CEN EN 288-4-92. CORRIGENDUM Specification and Approval of Welding Procedures for Metallic Materials—Part 4: Welding Procedure Tests for the Arc Welding of Aluminium and its Alloys. 2 pp.
CEN EN 288-4-92. Specification and Approval of Welding Procedures for Metallic Materials—Part 4: Welding Procedure Tests for the Arc Welding of Aluminium and Its Alloys. 28 pp.
DIN ENGL EN 288 Pt 4-92. Specification and Approval of Procedures for Welding Metallic Materials; Welding Procedure Tests for the Arc Welding of Aluminium and Its Alloys (Oct). 30 pp.

—Atmospheric Corrosion Testing
JIS H 0521-68. Testing Method for Atmospheric Corrosion of Aluminium and Aluminium Alloys. 12 pp.

—Bars
BSI BS 1474-87. 1987 Wrought Aluminium and Aluminium Alloys for General Engineering Purposes; Bars, Extruded Round Tubes and Sections. 28 pp.
BSI BS 2898-70. 1970 Amd 1 Wrought Aluminium and Aluminium Alloys for Electrical Purposes. Bars, Extruded Round Tube and Sections. 32 pp.
CNS H3048-84. Aluminium and Aluminium Alloy Rods, Bars and Wires (Sep)(3667).
DIN ENGL 40501 Pt 3-85. Aluminium for Electrical Purposes; E-Al and E-AlMgSiO,5 Bars and Sections; Technical Delivery Conditions (June). 4 pp.
JIS H 4040-88. Aluminium and Aluminium Alloy Rods, Bars and Wires. 48 pp.
SAA AS 1865-86. Aluminium and Aluminium Alloys—Drawn Wire, Rod, Bar and Strip (This is a Joint Standard with SANZ NZS 1865:1986). 11 pp.
SNZ NZS/AS 1865-86. Aluminium and Aluminium Alloys—Drawn Wire, Rod, Bar and Strip (This is a Joint Standard with SAA AS 1865). 11 pp.

—Bars—Aerospace
BSI 5L 34-85. 1985 Forging Stock, Bars, Extruded Sections and Forgings of 99% Aluminium. 4 pp.

INDUSTRY STANDARDS

Aluminum (Cont.)

—Bars—Chemical Analysis
DIN ENGL 1712 Pt 3-76. Aluminium; Half-Finished Products (Dec). 3 pp.

—Bars—Drawn
CEN PREN 754-1-92. Aluminium and Aluminium Alloys—Wrought Products—Cold Drawn Rod/Bar and Tube—Part 1: Technical Conditions for Inspection and Delivery. 16 pp.

DIN ENGL 1747 Pt 2-83. Wrought Aluminium and Aluminium Alloy Rod and Bar; Technical Delivery Conditions (Feb). 4 pp.

ISO 6363 Pt 1-88. Wrought Aluminium and Aluminium Alloy Cold-Drawn Rods/Bars and Tubes—Part 1: Technical Conditions for Inspection and Delivery First Edition; (Supersedes 5191). 6 pp.

—Bars—Drawn—Aerospace
AECMA PREN2070-03-84. Aluminium and Aluminium Alloy Wrought Products—Part 3—Bar and Section (CS/44). 13 pp.

BSI BS EN 2070: Part 3-91. 1991 Aluminium and Aluminium Alloy Wrought Products. Technical Specification Part 3: Bar and Section. 18 pp.

CEN EN 2070-3-89. Aerospace Series Aluminium and Aluminium Alloy Wrought Products Technical Specification Part 3: Bar and Section. 13 pp.

—Bars—Drawn—Hexagonal
CEN PREN 754-6-92. Wrought Aluminium and Aluminium Alloys—Cold Drawn Rod/Bar and Tube—Part 6: Drawn Hexagonal Bars, Dimensions and Form Tolerances. 7 pp.

DIN ENGL 1797-86. Wrought Aluminium and Aluminium Alloy Drawn Hexagonal Bars with Sharp Edges; Dimensions, Dimensional Deviations and Form Tolerances (Sept). 4 pp.

ISO 6363 Pt 5-92. Wrought Aluminium and Aluminium Alloy Cold-Drawn Rods/Bars and Tubes—Part 5: Drawn Square and Hexagonal Bars—Tolerances on Form and Dimensions First Edition. 5 pp.

—Bars—Drawn—Mechanical Properties
CEN PREN 754-2-92. Aluminium and Aluminium Alloys—Wrought Products—Cold Drawn Rod/Bar and Tube—Part 2: Mechanical Properties. 40 pp.

DIN ENGL 1747 Pt 1-83. Wrought Aluminium and Aluminium Alloy Rod and Bar; Properties (Feb). 4 pp.

—Bars—Drawn—Rectangular
CEN PREN 754-5-92. Wrought Aluminium and Aluminium Alloys—Cold Drawn Rod/Bar and Tube—Part 5: Drawn Rectangular Bars, Dimensions and Form Tolerances. 9 pp.

DIN ENGL 1769-86. Wrought Aluminium and Aluminium Alloy Drawn Rectangular Bars with Square Edges; Dimensions, Dimensional Deviations and Form Tolerances, Static Values (Sept). 10 pp.

DIN ENGL 46433-59. Rectangular Wires and Rectangular Bars; Drawn, with Radiused Edges; Dimensions (Nov). 8 pp.

DIN ENGL 46433 Extr. 1-59. Rectangular Wires and Rectangular Bars; Drawn, with Radiused Edges; Dimensions; Selected Sizes for Electrical Machines and Switchgear (Nov). 8 pp.

DIN ENGL 46433 Extr. 3-59. Rectangular Bars; Drawn, with Radiused Edges; Dimensions; Selected Sizes for Switchgears (Nov). 2 pp.

ISO 6363 Pt 4-91. Wrought Aluminium and Aluminium Alloy Cold-Drawn Rods/Bars and Tubes—Part 4: Drawn Rectangular Bars—Tolerances on Form and Dimensions First Edition. 7 pp.

—Bars—Drawn—Round
CEN PREN 754-3-92. Wrought Aluminium and Aluminium Alloys—Cold Drawn Rod/Bar and Tube—Part 3: Drawn Round Bars, Dimensions and Form Tolerances. 7 pp.

DIN ENGL 1798-86. Wrought Aluminium and Aluminium Alloy Drawn Round Bars; Dimensions, Dimensional Deviations and Form Tolerances (Sept). 4 pp.

ISO 5193-81. Wrought Aluminium and Aluminium Alloys—Drawn Round Bars—Tolerances on Shape and Dimensions (Symmetric Plus and Minus Tolerances on Diameter) First Edition. 3 pp.

ISO 7274-81. Wrought Aluminium and Aluminium Alloys—Drawn Round Bars—Tolerances on Shape and Dimensions (All Minus Tolerances on Diameter) First Edition. 3 pp.

—Bars—Drawn—Square
CEN PREN 754-4-92. Wrought Aluminium and Aluminium Alloys—Cold Drawn Rod/Bar and Tube—Part 4: Drawn Square Bars—Dimension and Form Tolerances. 7 pp.

DIN ENGL 1796-86. Wrought Aluminium and Aluminium Alloy Drawn Square Bars with Sharp Edges; Dimensions, Dimensional Deviations and Form Tolerances (Sept). 4 pp.

ISO 6363 Pt 5-92. Wrought Aluminium and Aluminium Alloy Cold-Drawn Rods/Bars and Tubes—Part 5: Drawn Square and Hexagonal Bars—Tolerances on Form and Dimensions First Edition. 5 pp.

—Bars—Extruded
CEN PREN 755-1-92. Aluminium and Aluminium Alloys—Wrought Products—Extruded Rod/Bar, Tube and Profile—Part 1: Technical Conditions for Inspection and Delivery. 18 pp.

DIN ENGL 1747 Pt 2-83. Wrought Aluminium and Aluminium Alloy Rod and Bar; Technical Delivery Conditions (Feb). 4 pp.

ISO 6362 Pt 1-86. Wrought Aluminium and Aluminium Alloys Extruded Rods/Bars, Tubes and Profiles—Part 1: Technical Conditions for Inspection and Delivery First Edition; (Supersedes 5191). 5 pp.

ISO 6362 Pt 4-88. Wrought Aluminium and Aluminium Alloy Extruded Rods/Bars, Tubes and Profiles—Part 4: Extruded Profiles—Tolerances on Shape and Dimensions First Edition. 11 pp.

SAA AS 1866-86. Aluminium and Aluminium Alloys—Extruded Rod, Bar, Solid and Hollow Shapes (This is a Joint Standard with SANZ NZS 1866:1986). 45 pp.

SNZ NZS/AS 1866-86. Aluminium and Aluminium Alloys—Extruded Rod, Bar, Solid and Hollow Shapes (This is a Joint Standard with SAA AS 1866). 45 pp.

—Bars—Extruded—Aerospace
AECMA PREN2070-03-84. Aluminium and Aluminium Alloy Wrought Products—Part 3—Bar and Section (CS/44). 13 pp.

BSI BS EN 2070: Part 3-91. 1991 Aluminium and Aluminium Alloy Wrought Products. Technical Specification Part 3: Bar and Section. 18 pp.

CEN EN 2070-3-89. Aerospace Series Aluminium and Aluminium Alloy Wrought Products Technical Specification Part 3: Bar and Section. 13 pp.

—Bars—Extruded—Hexagonal
CEN PREN 755-6-92. Wrought Aluminium and Aluminium Alloys—Extruded Rod/Bar, Tube and Profile—Part 6: Extruded Hexagonal Bars—Dimension and Form Tolerances. 7 pp.

DIN ENGL 59701-86. Wrought Aluminium and Aluminium Alloy Extruded Hexagonal Bars; Dimensions, Dimensional Deviations and Form Tolerances (Aug). 6 pp.

ISO 6362 Pt 5-91. Wrought Aluminium and Aluminium Alloy Extruded Rods/Bars, Tubes and Profiles—Part 5: Extruded Round, Square and Hexagonal Bars—Tolerances on Form and Dimensions First Edition; (Replaces 7273). 7 pp.

—Bars—Extruded—Mechanical Properties
CEN PREN 755-2-92. Aluminium and Aluminium Alloys—Wrought Products—Extruded Rod/Bar, Tube and Profile—Part 2: Mechanical Properties. 56 pp.

DIN ENGL 1747 Pt 1-83. Wrought Aluminium and Aluminium Alloy Rod and Bar; Properties (Feb). 4 pp.

—Bars—Extruded—Rectangular
CEN PREN 755-5-92. Wrought Aluminium and Aluminium Alloys—Extruded Rod/Bar, Tube and Profile—Part 5: Extruded Rectangular Bars, Dimensions and Form Tolerances. 10 pp.

DIN ENGL 1770-87. Wrought Aluminium and Aluminium Alloy Extruded Rectangular Bars; Dimensions, Dimensional Deviations and Form Tolerances, Static Values (Jan). 9 pp.

ISO 6362 Pt 3-90. Wrought Aluminium and Aluminium Alloy Extruded Rods/Bars, Tubes and Profiles—Part 3: Extruded Rectangular Bars—Tolerances on Dimensions First Edition. 8 pp.

—Bars—Extruded—Rectangular—Aerospace
AECMA PREN2341-80. Aluminium and Aluminium Alloy Square and Rectangular Extruded Bars—Dimensions (C6/A03-04). 9 pp.

—Bars—Extruded—Round
CEN PREN 755-3-92. Wrought Aluminium and Aluminium Alloys—Extruded Rod/Bar, Tube and Profile—Part 3: Extruded Round Bars, Dimensions and Form Tolerances. 5 pp.

DIN ENGL 1799-86. Wrought Aluminium and Aluminium Alloy Extruded Round Bars; Dimensions, Dimensional Deviations and Form Tolerances (Sept). 4 pp.

ISO 6362 Pt 5-91. Wrought Aluminium and Aluminium Alloy Extruded Rods/Bars, Tubes and Profiles—Part 5: Extruded Round, Square and Hexagonal Bars—Tolerances on Form and Dimensions First Edition; (Replaces 7273). 7 pp.

—Bars—Extruded—Square
CEN PREN 755-4-92. Wrought Aluminium and Aluminium Alloys—Extruded Rod/Bar, Tube and Profile—Part 4: Extruded Square Bars—Dimension and Form Tolerances. 8 pp.

DIN ENGL 59700-86. Wrought Aluminium and Aluminium Alloy Extruded Square Bars; Dimensions, Dimensional Deviations and Form Tolerances (Aug). 6 pp.

ISO 6362 Pt 5-91. Wrought Aluminium and Aluminium Alloy Extruded Rods/Bars, Tubes and Profiles—Part 5: Extruded Round, Square and Hexagonal Bars—Tolerances on Form and Dimensions First Edition; (Replaces 7273). 7 pp.

—Bars—Extruded—Square—Aerospace
AECMA PREN2341-80. Aluminium and Aluminium Alloy Square and Rectangular Extruded Bars—Dimensions (C6/A03-04). 9 pp.

—Bars—Rivets
DIN ENGL 59675-57. High Grade Aluminium and Aluminium Wrought Alloy; Wire and Bars for Rivets; Drawn (May). 2 pp.

—Bars—Rolled—Aerospace
AECMA PREN2070-03-84. Aluminium and Aluminium Alloy Wrought Products—Part 3—Bar and Section (CS/44). 13 pp.

BSI BS EN 2070: Part 3-91. 1991 Aluminium and Aluminium Alloy Wrought Products. Technical Specification Part 3: Bar and Section. 18 pp.

CEN EN 2070-3-89. Aerospace Series Aluminium and Aluminium Alloy Wrought Products Technical Specification Part 3: Bar and Section. 13 pp.

—Boron Content
JIS H 1365-73. Method for Determination of Boron in Aluminium Alloy.

—Channels—Extruded
DIN ENGL 9713-81. Aluminium and Wrought Aluminium Alloy Extruded Channel Sections; Dimensions; Static Values (Sept). 4 pp.

—Chemical Analysis
CNS H2026-84. Method of Chemical Analysis of Aluminium and Aluminium Alloy (May)(2069).

JIS H 1351-72. General Rules for Chemical Analysis of Aluminium and Aluminium Alloy.

—Chromium Content—Atomic Absorption Spectrometry
ISO 4193-81. Aluminium and Aluminium Alloys—Determination of Chromium Content—Flame Atomic Absorption Spectrometric Method First Edition. 5 pp.

—Chromium Content—Photometry
BSI BS 1728: Part 16-68. 1968 Methods for the Analysis of Aluminium and Aluminium Alloys: Part 16: Chromium (Photometric Method). 10 pp.

—Chromium Content—Spectrophotometry
ISO 3978-76. Aluminium and Aluminium Alloys—Determination of Chromium—Spectrophotometric Method Using Diphenylcarbazide, After Extraction First Edition. 6 pp.

—Chromium Content—Volumetric Analysis
BSI BS 1728: Part 17-68. 1968 Methods for the Analysis of Aluminium and Aluminium Alloys: Part 17: Chromium (Volumetric Method). 8 pp.

—Circle—Cold Rolled
CEN PREN 851-92. Aluminium and Aluminium Alloys—Wrought Products—Circle and Circle Stock for Culinary Utensil Applications. 11 pp.

—Circle—Hot Rolled
CEN PREN 851-92. Aluminium and Aluminium Alloys—Wrought Products—Circle and Circle Stock for Culinary Utensil Applications. 11 pp.

—Copper Content
BSI BS 1728: Part 1-51. 1951 Amd 1 Methods for the Analysis of Aluminium and Aluminium Alloys Part 1: Determination of Copper. 9 pp.

JIS H 1354-72. Methods for Determination of Copper in Aluminium and Aluminium Alloy.

—Copper Content—Absorptiometric Analysis
BSI BS 1728: Part 5-53. 1953 Amd 1 Methods for the Analysis of Aluminium and Aluminium Alloys Part 5: Determination of Copper (Absorptiometric Method). 8 pp.

—Copper Content—Atomic Absorption Spectrophotometry
BSI BS 1728: Part 23-76. 1976 Methods for the Analysis of Aluminium and Aluminium Alloys: Part 23: Copper (Atomic Absorption Method). 4 pp.

INTERNATIONAL AND NON-U.S. NATIONAL STANDARDS
SUBJECT INDEX
Aluminum

Aluminum *(Cont.)*

—**Copper Content—Atomic Absorption Spectrophotometry** *(Cont.)*
ISO 3980-77. Aluminium and Aluminium Alloys—Determination of Copper—Atomic Absorption Spectrophotometric Method First Edition. 5 pp.

—**Copper Content—Electrolytic Analysis**
ISO 796-73. Aluminium Alloys—Determination of Copper—Electrolytic Method First Edition. 7 pp.

—**Copper Content—Photometry**
ISO 795-76. Aluminium and Aluminium Alloys—Determination of Copper Content—Oxalyldihydrazide Photometric Method First Edition. 6 pp.

—**Design Loads**
CSA CAN3-S157-M83. Strength Design in Aluminum; (Erratum March 1984) (Erratums April 1988). 75 pp.

—**Extruded**
CNS H2031-82. Method of Test for Aluminium Shapes Extruded (Dec)(2258).
CNS H3028-65. Silver Solders (Sep) (2474).
JIS H 4100-88. Aluminium and Aluminium Alloy Extruded Shapes. 28 pp.

—**Extrusions—Aerospace—Ultrasonic Testing**
CEN PREN2004-02-75. Test Methods for Aluminium and Aluminium Alloy Products—Part 2—Ultrasonic Testing of Plates Forgings and Extrusions (C5/21A). 7 pp.

—**Extrusions—Test Specimens**
CGSB CAN2-19.0-M77METH12.2-78. Methods of Testing Putty, Caulking and Sealing Compounds Preparation of Plain Aluminum Panels and Extrusions. 1 p.

—**Forgings**
BSI BS 1472-72. 1972 Amd 2 Wrought Aluminium and Aluminium Alloys for General Engineering Purposes; Forging Stock. 36 pp.
BSI BS 4300/11-69. 1969 Amd 1 Specification (Supplementary Series) for Wrought Aluminium and Aluminium Alloys for General Engineering Purposes /11: 5454 Forging Stock and Forgings. 30 pp.
CEN PREN 603-1-91. Wrought Forging Stock in Aluminium and Aluminium Alloys—Part 1: Technical Conditions for Inspection and Delivery. 12 pp.
CEN PREN 605-1-91. Wrought Aluminium and Aluminium Alloy Forgings—Part 1: Technical Conditions for Inspection and Delivery. 11 pp.
CNS H3114-82. Aluminium and Aluminium Alloy Forgings (Oct)(9612).
DIN ENGL 1749 Pt 2-73. Aluminium Drop Forgings; (Highest Grade Aluminium, High Grade Aluminium and Wrought Aluminium Alloys); Technical Conditions of Delivery (Nov). 3 pp.
DIN ENGL 1749 Pt 3-74. Aluminium Drop Forgings; (Highest Grade Aluminium, High Grade Aluminium and Wrought Aluminium Alloys); Design Principles (Jan). 7 pp.
DIN ENGL 1749 Pt 4-74. Aluminium Drop Forgings; (Highest Grade Aluminium, High Grade Aluminium and Wrought Aluminium Alloys); Permissible Variations (Jan). 5 pp.
JIS H 4140-88. Aluminium and Aluminium Alloy Forgings. 26 pp.

—**Forgings—Aerospace**
AECMA PREN2070-07-87. Aluminium and Aluminium Alloy Wrought Products—Technical Specification—Part 7—Wrought Forging Stock. 6 pp.
BSI BS EN 2070: Part 7-91. 1991 Aluminium and Aluminium Alloy Wrought Products Technical Specification Part 7: Wrought Forging Stock. 9 pp.
BSI 5L 34-85. 1985 Forging Stock, Bars, Extruded Sections and Forgings of 99% Aluminium. 4 pp.
CEN EN 2070-7-89. Aerospace Series Aluminium and Aluminium Alloy Wrought Products Technical Specification Part 7: Wrought Forging Stock. 6 pp.

—**Forgings—Aerospace—Macrographic Inspection**
CEN PREN 2715-90. Macrographic Examination of Aluminium and Aluminium Alloy Forging Stock, Forgings and Wrought Products. 4 pp.

—**Forgings—Aerospace—Ultrasonic Testing**
CEN PREN2004-02-75. Test Methods for Aluminium and Aluminium Alloy Products—Part 2—Ultrasonic Testing of Plates Forgings and Extrusions (C5/21A). 7 pp.

—**Forgings—Mechanical Properties**
CEN PREN 586-2-91. Forgings in Aluminium and Aluminium Alloys—Part 2: Mechanical Properties and Additional Property Requirements. 12 pp.

Aluminum *(Cont.)*

—**Forgings—Mechanical Properties** *(Cont.)*
CEN PREN 603-2-91. Wrought Forging Stock in Aluminium and Aluminium Alloys—Part 2: Mechanical Properties. 10 pp.
DIN ENGL 1749 Pt 1-76. Drop Forgings of Aluminium and Wrought Aluminium Alloys; Strength Properties (Dec). 2 pp.

—**Fusion Welding—Company Certification**
CSA W47.2-M1987. Certification of Companies for Fusion Welding of Aluminum. 75 pp.

—**Hydrogen Peroxide**
MOD UK DEF-60-58. Selection and Treatment of Aluminium Base Materials for Use with Concentrated Hydrogen Peroxide (H.T.P.) (Reprinted August, 1973) (Withdrawn). 6 pp.

—**I Beams—Extruded**
DIN ENGL 9712-69. Aluminium and Magnesium Beams; Extruded; Dimensions; Static Values (Aug). 3 pp.

—**Ingots**
BSI BS 1490-88. 1988 Amd 1 Aluminium and Aluminium Alloy Ingots and Castings for General Engineering Purposes (V) (AMD 6309) March 30, 1990. 27 pp.
BSI BS 1490-02. 1988 Amd 2 Aluminium and Aluminium Alloy Ingots and Castings for General Engineering Purposes (AMD 6605) December 21, 1990. 29 pp.
CEN PREN 576-91. Aluminium and Aluminium Alloys—Unalloyed Aluminium Ingots. 12 pp.
JIS H 2102-88. Virgin Aluminium Ingots.
SAA AS 1874-88. Aluminium and Aluminium Alloys—Ingots and Castings. 14 pp.
SNZ NZS/AS 1874-88. Aluminium and Aluminium Alloys—Ingots and Castings (This is a Joint Standard with SAA AS 1874). 14 pp.

—**Ingots—Chemical Analysis**
DIN ENGL 1712 Pt 1-76. Aluminium; Ingots (Dec). 2 pp.

—**Ingots—Classification**
ISO R115-68. Classification and Composition of Unalloyed Aluminium Ingots for Remelting Second Edition. 4 pp.

—**Ingots—Extrusion**
CEN PREN 486-91. Aluminium and Aluminium Alloys—Extrusions Ingots Cast in a Form Suitable for Extruding. 11 pp.

—**Ingots—Refined**
JIS H 2111-68. Refined Aluminium Ingots (R 1971). 4 pp.

—**Ingots—Rolling**
CEN PREN 487-91. Aluminium and Aluminium Alloys—Rolling Ingots Obtained by Semi-Continuous Casting. 17 pp.

—**Ingots—Secondary**
CNS H3110-82. Secondary Aluminium Ingot (Sep)(9396).
JIS H 2103-65. Secondary Aluminium Ingots (R 1971). 4 pp.

—**Ingots—Spectrochemical Analysis**
JIS H 1303-76. Method for Emission Spectrochemical Analysis of Aluminium Ingot.

—**Ingots—Virgin**
CNS H3003-84. Virgin Aluminum Ingots (Nov)(8).
JIS H 2102-68. Virgin Aluminium Ingots (R 1971). 5 pp.

—**Ingots—Virgin—Electrical**
JIS H 2110-68. Virgin Aluminium Ingots for Electrical Purposes (R 1971). 4 pp.

—**Iron Content**
JIS H 1353-72. Methods for Determination of Iron in Aluminium and Aluminium Alloy.

—**Iron Content—Absorptiometric Analysis**
BSI BS 1728: Part 8-57. 1957 Methods for the Analysis of Aluminium and Aluminium Alloys Part 8: Determination of Iron (Absorptiometric 1:10-Phenanthroline Method). 8 pp.

—**Iron Content—Photometry**
ISO 793-73. Aluminium and Aluminium Alloys—Determination of Iron—Orthophenanthroline Photometric Method First Edition. 6 pp.

—**Iron Content—Volumetric Analysis**
BSI BS 1728: Part 6-55. 1955 Methods for the Analysis of Aluminium and Aluminium Alloys Part 6: Determination of Iron (Volumetric: Titanous Chloride Method). 8 pp.

Aluminum *(Cont.)*

—**Lead Content—Atomic Absorption Spectrometry**
ISO 4192-81. Aluminium and Aluminium Alloys—Determination of Lead Content—Flame Atomic Absorption Spectrometric Method First Edition. 5 pp.

—**Lead Content—Atomic Absorption Spectrophotometry**
BSI BS 1728: Part 20-71. 1971 Methods for the Analysis of Aluminium and Aluminium Alloys: Part 20: Lead (Atomic Absorption Method). 9 pp.

—**Liquid Metals**
CEN PREN 577-91. Aluminium and Aluminium Alloys—Liquid Metal. 8 pp.

—**Magnesium Content—Atomic Absorption Spectrophotometry**
BSI BS 1728: Part 19-71. 1971 Methods for the Analysis of Aluminium and Aluminium Alloys: Part 19: Magnesium (Atomic Absorption Method). 9 pp.
ISO 3256-77. Aluminium and Aluminium Alloys—Determination of Magnesium—Atomic Absorption Spectrophotometric Method First Edition. 6 pp.

—**Magnesium Content—Complexometric Titrations**
ISO 2297-73. Chemical Analysis of Aluminium and Its Alloys—Complexometric Determination of Magnesium First Edition. 7 pp.

—**Magnesium Content—Volumetric Analysis**
BSI BS 1728: Part 22-72. 1972 Methods for the Analysis of Aluminium and Aluminium Alloys: Part 22: Magnesium (Volumetric CDTA Method). 13 pp.

—**Manganese Content**
JIS H 1355-72. Methods for Determination of Manganese in Aluminium and Aluminium Alloy.

—**Manganese Content—Absorptiometric Analysis**
BSI BS 1728: Part 10-57. 1957 Methods for the Analysis of Aluminium and Aluminium Alloys Part 10: Determination of Manganese (Absorptiometric Method). 8 pp.

—**Manganese Content—Photometry**
ISO 886-73. Aluminium and Aluminium Alloys—Determination of Manganese—Photometric Method (Manganese Content Between 0,005 and 1,5 %) First Edition. 7 pp.

—**Manganese Content—Volumetric Analysis**
BSI BS 1728: Part 9-57. 1957 Methods for the Analysis of Aluminium and Aluminium Alloys Part 9: Determination of Manganese (Volumetric: Arsenite/Nitrite Method). 8 pp.

—**Master—Chemical Analysis**
CEN PREN 575-91. Aluminium and Aluminium Alloys—Master Alloys. 14 pp.

—**Neutron Flux Density—Radioactivation Analysis**
CNS J2055-82. Method for Determining Fast-Neutron Flux by Radioactivation of Aluminium (Jun)(8981).

—**Nickel Content—Atomic Absorption Spectrophotometry**
BSI BS 1728: Part 24-76. 1976 Methods for the Analysis of Aluminium and Aluminium Alloys: Part 24: Nickel (Atomic Absorption Method). 4 pp.
ISO 3981-77. Aluminium and Aluminium Alloys—Determination of Nickel—Atomic Absorption Spectrophotometric Method First Edition. 5 pp.

—**Nickel Content—Gravimetric Analysis**
BSI BS 1728: Part 14-65. 1965 Methods for the Analysis of Aluminium and Aluminium Alloys: Part 14: Determination of Nickel (Gravimetric Method). 8 pp.

—**Nickel Content—Photometry**
BSI BS 1728: Part 15-66. 1966 Methods for the Analysis of Aluminium and Aluminium Alloys: Part 15: Nickel (Photometric Method). 8 pp.

—**Nickel Content—Spectrophotometry**
ISO 3979-77. Aluminium and Aluminium Alloys—Determination of Nickel—Spectrophotometric Method Using Dimethylglyoxime First Edition. 7 pp.

—**Pigments**
BSI BS 388-72. 1972 Aluminium Pigments. 25 pp.
CGSB CAN/CGSB-1.22-M90. Aluminum Pigment Powder for Paint; (Amendment 1 October 1991). 11 pp.
CGSB CAN/CGSB-1.24-M89. Aluminum Pigment Paste for Paint. 10 pp.

INDUSTRY STANDARDS

Aluminum (Cont.)

—Pigments (Cont.)
ISO 1247-74. Aluminium Pigments for Paints First Edition; (Amendment 1-1982). 15 pp.

—Plates
BSI BS 1470-87. 1987 Amd 1 Wrought Aluminium and Aluminium Alloys for General Engineering Purposes; Plate, Sheet and Strip (AMD 6032) May 31, 1989. 24 pp.

CEN PREN 485 (Part 1)-91. Wrought Aluminium and Aluminium Alloy Sheets, Strips and Plates—Part 1: Technical Conditions for Inspection and Delivery. 16 pp.

CNS H2029-87. Method of Test for Aluminium and Aluminium Alloy Sheets and Plate (Jun)(2254).

CNS H3025-87. Aluminium and Aluminium Alloy Sheets and Plates (Feb)(2253).

DIN ENGL 40501 Pt 1-85. Aluminium for Electrical Purposes; E-Al Plate, Sheet and Strip; Technical Delivery Conditions (June). 3 pp.

JIS H 4000-88. Aluminium and Aluminium Alloy Sheets and Plates, Strips and Coiled Sheets. 145 pp.

SAA AS 1734-86. Aluminium and Aluminium Alloys—Flat Sheet, Coiled Sheet and Plate (This is a Joint Standard with SANZ NZS 1734:1986). 20 pp.

SNZ NZS/AS 1734-86. Aluminium and Aluminium Alloys—Flat Sheet, Coiled Sheet and Plate (This is a Joint Standard with SAA AS 1734). 20 pp.

—Plates—Aerospace
AECMA PREN2070-02-84. Aluminium and Aluminium Alloy Wrought Products—Part 2—Sheet, Strip Formed Profiles and Plate (CS/44). 14 pp.

BSI BS EN 2070: Part 2-91. 1991 Aluminium and Aluminium Alloy Wrought Products. Technical Specification Part 2: Sheet, Strip Formed Profiles and Plate. 19 pp.

CEN EN 2070-2-89. Aerospace Series Aluminium and Aluminium Alloy Wrought Products Technical Specification Part 2: Sheet, Strip Formed Profiles and Plate. 14 pp.

—Plates—Aerospace—Ultrasonic Testing
CEN PREN2004-02-75. Test Methods for Aluminium and Aluminium Alloy Products—Part 2—Ultrasonic Testing of Plates Forgings and Extrusions (C5/21A). 7 pp.

—Plates—Aircraft
EURO CTI/80/10834-80. Study on the Supply of Aircraft Quality Thick Aluminium. 8 pp.

—Plates—Chemical Machining
MOD UK DTD-932-58. Chemical Contouring of Aluminium and Aluminium Alloy Sheet and Plate (Reprinted January 1965). 2 pp.

—Plates—Mechanical Properties
CEN PREN 485-2-91. Wrought Aluminium and Aluminium Alloy Sheets, Strips and Plates—Part 2 Mechanical Properties. 45 pp.

—Plates—Rolled
CEN PREN 485 (Part 3)-91. Wrought Aluminium and Aluminium Alloy Sheets, Strips and Plates—Part 3: Hot-Rolled Sheets, Strips and Plates—Tolerances on Shape and Dimensions. 11 pp.

CEN PREN 485 (Part 4)-91. Wrought Aluminium and Aluminium Alloy Sheets, Strips and Plates—Part 4: Cold-Rolled Sheets, Strips and Plates—Tolerances on Shape and Dimensions. 14 pp.

CEN PREN 683-1-92. Aluminium and Aluminium Alloys—Wrought Products, Finstock—Part 1: Technical Conditions for Inspection and Delivery. 10 pp.

CEN PREN 941-92. Aluminium and Aluminium Alloys—Wrought Products—Circle and Circle Stock for General Applications. 11 pp.

DIN ENGL 1745 Pt 2-83. Wrought Aluminium and Aluminium Alloy Plate, Sheet and Strip Greater Than 0,35 mm in Thickness; Technical Delivery Conditions (Feb). 4 pp.

DIN ENGL 1783-81. Strips, Plates and Sheets of Aluminium and Wrought Aluminium Alloys with Thicknesses over 0,35 mm, Cold-Rolled; Dimensions (Apr). 9 pp.

DIN ENGL 59600-81. Strips, Plates and Sheets of Aluminium and Wrought Aluminium Alloys; Hot Rolled; Dimensions (Apr). 6 pp.

ISO 6361 Pt 1-86. Wrought Aluminium and Aluminium Alloys Sheets, Strips and Plates—Part 1: Technical Conditions for Inspection and Delivery First Edition; (Supersedes 5191). 5 pp.

ISO 6361 Pt 4-88. Wrought Aluminium and Aluminium Alloy Sheets, Strips and Plates—Part 4: Sheets and Plates—Tolerances on Shape and Dimensions First Edition; (Supersedes TR7735). 12 pp.

—Plates—Rolled—Mechanical Properties
CEN PREN 683-2-92. Aluminium and Aluminium Alloys—Wrought Products, Finstock—Part 2: Mechanical Properties. 9 pp.

DIN ENGL 1745 Pt 1-83. Wrought Aluminium and Aluminium Alloy Plate, Sheet and Strip Greater Than 0,35 mm in Thickness; Properties (Feb). 6 pp.

ISO 6361 Pt 2-90. Wrought Aluminium and Aluminium Alloy Sheets, Strips and Plates—Part 2: Mechanical Properties Second Edition; (Supersedes TR2136). 17 pp.

—Profiles—Aerospace
AECMA PREN2070-02-84. Aluminium and Aluminium Alloy Wrought Products—Part 2—Sheet, Strip Formed Profiles and Plate (CS/44). 14 pp.

BSI BS EN 2070: Part 2-91. 1991 Aluminium and Aluminium Alloy Wrought Products. Technical Specification Part 2: Sheet, Strip Formed Profiles and Plate. 19 pp.

CEN EN 2070-2-89. Aerospace Series Aluminium and Aluminium Alloy Wrought Products Technical Specification Part 2: Sheet, Strip Formed Profiles and Plate. 14 pp.

—Profiles—Extruded
CEN PREN 755-1-92. Aluminium and Aluminium Alloys—Wrought Products—Extruded Rod/Bar, Tube and Profile—Part 1: Technical Conditions for Inspection and Delivery. 18 pp.

DIN ENGL 1748 Pt 4-81. Aluminium and Wrought Aluminium Alloy; Extruded Profiles; Permissible Deviations (Nov). 10 pp.

ISO 6362 Pt 1-86. Wrought Aluminium and Aluminium Alloys Extruded Rods/Bars, Tubes and Profiles—Part 1: Technical Conditions for Inspection and Delivery First Edition; (Supersedes 5191). 5 pp.

ISO 6362 Pt 4-88. Wrought Aluminium and Aluminium Alloy Extruded Rods/Bars, Tubes and Profiles—Part 4: Extruded Profiles—Tolerances on Shape and Dimensions First Edition. 11 pp.

—Profiles—Extruded—Mechanical Properties
CEN PREN 755-2-92. Aluminium and Aluminium Alloys—Wrought Products—Extruded Rod/Bar, Tube and Profile—Part 2: Mechanical Properties. 56 pp.

ISO 6362 Pt 2-90. Wrought Aluminium and Aluminium Alloy Extruded Rods/Bars, Tubes and Profiles—Part 2: Mechanical Properties Second Edition. 15 pp.

—Rivets
BSI BS 1473-72. 1972 Amd 2 Wrought Aluminium and Aluminium Alloys for General Engineering Purposes; Rivet, Bolt and Screw Stock. 25 pp.

—Rivets—Bars—Drawn
DIN ENGL 59675-57. High Grade Aluminium and Aluminium Wrought Alloy; Wire and Bars for Rivets; Drawn (May). 2 pp.

—Rods
CNS H3048-84. Aluminium and Aluminium Alloy Rods, Bars and Wires (Sep)(3667).

CNS H3131-84. Aluminium and Aluminium Alloy Rivet Wires and Rods (Sep)(11016).

JIS H 4040-88. Aluminium and Aluminium Alloy Rods, Bars and Wires. 48 pp.

SAA AS 1865-86. Aluminium and Aluminium Alloys—Drawn Wire, Rod, Bar and Strip (This is a Joint Standard with SANZ NZS 1865:1986). 11 pp.

SNZ NZS/AS 1865-86. Aluminium and Aluminium Alloys—Drawn Wire, Rod, Bar and Strip (This is a Joint Standard with SAA AS 1865). 11 pp.

—Rods—Drawn
DIN ENGL 1747 Pt 1-83. Wrought Aluminium and Aluminium Alloy Rod and Bar; Properties (Feb). 4 pp.

ISO 6363 Pt 1-88. Wrought Aluminium and Aluminium Alloy Cold-Drawn Rods/Bars and Tubes—Part 1: Technical Conditions for Inspection and Delivery First Edition; (Supersedes 5191). 6 pp.

—Rods—Drawn—Mechanical Properties
DIN ENGL 1747 Pt 1-83. Wrought Aluminium and Aluminium Alloy Rod and Bar; Properties (Feb). 4 pp.

—Rods—Extruded
DIN ENGL 1747 Pt 1-83. Wrought Aluminium and Aluminium Alloy Rod and Bar; Properties (Feb). 4 pp.

ISO 6362 Pt 1-86. Wrought Aluminium and Aluminium Alloys Extruded Rods/Bars, Tubes and Profiles—Part 1: Technical Conditions for Inspection and Delivery First Edition; (Supersedes 5191). 5 pp.

ISO 6362 Pt 4-88. Wrought Aluminium and Aluminium Alloy Extruded Rods/Bars, Tubes and Profiles—Part 4: Extruded Profiles—Tolerances on Shape and Dimensions First Edition. 11 pp.

SAA AS 1866-86. Aluminium and Aluminium Alloys—Extruded Rod, Bar, Solid and Hollow Shapes (This is a Joint Standard with SANZ NZS 1866:1986). 45 pp.

SNZ NZS/AS 1866-86. Aluminium and Aluminium Alloys—Extruded Rod, Bar, Solid and Hollow Shapes (This is a Joint Standard with SAA AS 1866). 45 pp.

—Rods—Extruded—Mechanical Properties
DIN ENGL 1747 Pt 1-83. Wrought Aluminium and Aluminium Alloy Rod and Bar; Properties (Feb). 4 pp.

ISO 6362 Pt 2-90. Wrought Aluminium and Aluminium Alloy Extruded Rods/Bars, Tubes and Profiles—Part 2: Mechanical Properties Second Edition. 15 pp.

—Scrap
CNS H3111-82. Method of Classification for Aluminium and Aluminium Alloy Scrap (Sep)(9397).

JIS H 2119-84. Classification Standard of Aluminium and Aluminium Alloy Scraps. 12 pp.

—Sections
DIN ENGL 1748 Pt 2-83. Wrought Aluminium and Aluminium Alloy Extruded Sections; Technical Delivery Conditions (Feb). 4 pp.

DIN ENGL 40501 Pt 3-85. Aluminium for Electrical Purposes; E-Al and E-AlMgSiO,5 Bars and Sections; Technical Delivery Conditions (June). 4 pp.

—Sections—Drawn—Aerospace
AECMA PREN2070-03-84. Aluminium and Aluminium Alloy Wrought Products—Part 3—Bar and Section (CS/44). 13 pp.

BSI BS EN 2070: Part 3-91. 1991 Aluminium and Aluminium Alloy Wrought Products. Technical Specification Part 3: Bar and Section. 18 pp.

CEN EN 2070-3-89. Aerospace Series Aluminium and Aluminium Alloy Wrought Products Technical Specification Part 3: Bar and Section. 13 pp.

—Sections—Extruded
BSI BS 1474-87. 1987 Wrought Aluminium and Aluminium Alloys for General Engineering Purposes; Bars, Extruded Round Tubes and Sections. 28 pp.

BSI BS 2898-70. 1970 Amd 1 Wrought Aluminium and Aluminium Alloys for Electrical Purposes. Bars, Extruded Round Tube and Sections. 32 pp.

DIN ENGL 1748 Pt 3-68. Aluminium and Extruded Sections; (Highest Grade Aluminium, High Grade Aluminium and Wrought Aluminium Alloys); Design (Dec). 6 pp.

—Sections—Extruded—Aerospace
AECMA PREN2066-78. Extruded Sections in Aluminium and Aluminium Alloys—General Tolerances (C6/A04-00). 10 pp.

AECMA PREN2070-03-84. Aluminium and Aluminium Alloy Wrought Products—Part 3—Bar and Section (CS/44). 13 pp.

BSI BS EN 2070: Part 3-91. 1991 Aluminium and Aluminium Alloy Wrought Products. Technical Specification Part 3: Bar and Section. 18 pp.

BSI 5L 34-85. 1985 Forging Stock, Bars, Extruded Sections and Forgings of 99% Aluminium. 4 pp.

CEN EN 2070-3-89. Aerospace Series Aluminium and Aluminium Alloy Wrought Products Technical Specification Part 3: Bar and Section. 13 pp.

—Sections—Extruded—Mechanical Properties
DIN ENGL 1748 Pt 1-83. Wrought Aluminium and Aluminium Extruded Sections; Properties (Feb). 3 pp.

—Sections—Rolled—Aerospace
AECMA PREN2070-03-84. Aluminium and Aluminium Alloy Wrought Products—Part 3—Bar and Section (CS/44). 13 pp.

BSI BS EN 2070: Part 3-91. 1991 Aluminium and Aluminium Alloy Wrought Products. Technical Specification Part 3: Bar and Section. 18 pp.

CEN EN 2070-3-89. Aerospace Series Aluminium and Aluminium Alloy Wrought Products Technical Specification Part 3: Bar and Section. 13 pp.

—Semifinished Products—Anodized
DIN ENGL 17611-85. Anodized Wrought Aluminium and Aluminium Alloy Semi-Finished Products with Coating Thicknesses of at Least 10 Glm; Technical Delivery Conditions (June). 5 pp.

—Semifinished Products—Chemical Analysis
DIN ENGL 1712 Pt 3-76. Aluminium; Half-Finished Products (Dec). 3 pp.

—Shapes—Extruded
CNS H3027-82. Aluminium Shapes Extruded (Dec)(2257).

INTERNATIONAL AND NON-U.S. NATIONAL STANDARDS
SUBJECT INDEX

Aluminum

Aluminum (Cont.)
—Shapes—Extruded (Cont.)
SAA AS 1866-86. Aluminium and Aluminium Alloys—Extruded Rod, Bar, Solid and Hollow Shapes (This is a Joint Standard with SANZ NZS 1866:1986). 45 pp.
SNZ NZS/AS 1866-86. Aluminium and Aluminium Alloys—Extruded Rod, Bar, Solid and Hollow Shapes (This is a Joint Standard with SAA AS 1866). 45 pp.

—Sheets
BSI BS 1470-87. 1987 Amd 1 Wrought Aluminium and Aluminium Alloys for General Engineering Purposes; Plate, Sheet and Strip (AMD 6032) May 31, 1989. 24 pp.
BSI BS 4868-72. 1972 Profiled Aluminium Sheet for Building. 8 pp.
CEN PREN 485 (Part 1)-91. Wrought Aluminium and Aluminium Alloy Sheets, Strips and Plates—Part 1: Technical Conditions for Inspection and Delivery. 16 pp.
CNS H2029-87. Method of Test for Aluminium and Aluminium Alloy Sheets and Plate (Jun)(2254).
CNS H3025-87. Aluminium and Aluminium Alloy Sheets and Plates (Feb)(2253).
CNS H3031-82. Corrugated Aluminium Sheets (Nov)(2609).
DIN ENGL 40501 Pt 1-85. Aluminium for Electrical Purposes; E-Al Plate, Sheet and Strip; Technical Delivery Conditions (June). 3 pp.
JIS H 4000-88. Aluminium and Aluminium Alloy Sheets and Plates, Strips and Coiled Sheets. 145 pp.
JIS S 2010-86. Aluminium Sheet Wares.
SAA AS 1734-86. Aluminium and Aluminium Alloys—Flat Sheet, Coiled Sheet and Plate (This is a Joint Standard with SANZ NZS 1734:1986). 20 pp.
SNZ NZS/AS 1734-86. Aluminium and Aluminium Alloys—Flat Sheet, Coiled Sheet and Plate (This is a Joint Standard with SAA AS 1734). 20 pp.

—Sheets—Aerospace
AECMA PREN2070-02-84. Aluminium and Aluminium Alloy Wrought Products—Part 2—Sheet, Strip Formed Profiles and Plate (CS/44). 14 pp.
AECMA PREN2071-89. Sheets in Aluminium and Aluminium Alloys—Thickness A Less Than or Equal to 6 mm—Dimensions (C6/A01). 8 pp.
BSI BS EN 2070: Part 2-91. 1991 Aluminium and Aluminium Alloy Wrought Products. Technical Specification Part 2: Sheet, Strip Formed Profiles and Plate. 19 pp.
BSI 6L 16-85. 1985 Sheet and trip of 99% Aluminium (Temper Designation H14 or H24). 4 pp.
BSI 6L 17-85. 1985 Sheet and Strip of 99% Aluminium (Temper Designation—0). 4 pp.
CEN EN 2070-2-89. Aerospace Series Aluminium and Aluminium Alloy Wrought Products Technical Specification Part 2: Sheet, Strip Formed Profiles and Plate. 14 pp.
CEN PREN2071-80. Sheets in Aluminium and Aluminium Alloys Thickness a Less Than or Equal to 6 mm Dimensions. 7 pp.
CEN PREN 2071-89. Sheets Aluminium and Aluminium Alloys. 9 pp.
CEN PREN 2072-93. Aerospace Series—Aluminium AL-P1050A H14—Sheet and Strip 0,4 mm Less Than or Equal to a Less Than or Equal to 6 mm. 5 pp.
CEN PREN 3996-93. Aerospace Series Aluminium AL-P1100-H14 Sheet and Strip 0,3 mm Less Than or Equal to a Less Than or Equal to 6 mm. 4 pp.

—Sheets—Chemical Analysis
BSI BS 4300/16-84. 1984 Specification (Supplementary Series) for Wrought Aluminium and Aluminium Alloys for General Engineering Purposes /16: Specification for Sheet and Strip 8011 Material. 11 pp.

—Sheets—Chemical Machining
MOD UK DTD-932-58. Chemical Contouring of Aluminium and Aluminium Alloy Sheet and Plate (Reprinted January 1965). 2 pp.

—Sheets—Designations
BSI BS 4300/16-84. 1984 Specification (Supplementary Series) for Wrought Aluminium and Aluminium Alloys for General Engineering Purposes /16: Specification for Sheet and Strip 8011 Material. 11 pp.

—Sheets—Mechanical Properties
CEN PREN 485-2-91. Wrought Aluminium and Aluminium Alloy Sheets, Strips and Plates—Part 2 Mechanical Properties. 45 pp.

—Sheets—Rolled
CEN PREN 485 (Part 3)-91. Wrought Aluminium and Aluminium Alloy Sheets, Strips and Plates—Part 3: Hot-Rolled Sheets, Strips and Plates—Tolerances on Shape and Dimensions. 11 pp.

Aluminum (Cont.)
—Sheets—Rolled (Cont.)
CEN PREN 485 (Part 4)-91. Wrought Aluminium and Aluminium Alloy Sheets, Strips and Plates—Part 4: Cold-Rolled Sheets, Strips and Plates—Tolerances on Shape and Dimensions. 14 pp.
CEN PREN 941-92. Aluminium and Aluminium Alloys—Wrought Products—Circle and Circle Stock for General Applications. 11 pp.
DIN ENGL 1745 Pt 2-83. Wrought Aluminium and Aluminium Alloy Plate, Sheet and Strip Greater Than 0,35 mm in Thickness; Technical Delivery Conditions (Feb). 4 pp.
DIN ENGL 1783-81. Strips, Plates and Sheets of Aluminium and Wrought Aluminium Alloys with Thicknesses over 0,35 mm, Cold-Rolled; Dimensions (Apr). 9 pp.
DIN ENGL 1784-81. Strips, Sheets and Sizes of Aluminium and Wrought Aluminium Alloys with Thicknesses from 0,021 to 0,35 mm, Cold-Rolled; Dimensions (Apr). 7 pp.
DIN ENGL 59600-81. Strips, Plates and Sheets of Aluminium and Wrought Aluminium Alloys; Hot Rolled; Dimensions (Apr). 6 pp.
ISO 6361 Pt 1-86. Wrought Aluminium and Aluminium Alloys Sheets, Strips and Plates—Part 1: Technical Conditions for Inspection and Delivery First Edition; (Supersedes 5191). 5 pp.
ISO 6361 Pt 4-88. Wrought Aluminium and Aluminium Alloys, Sheets, Strips and Plates—Part 4: Sheets and Plates—Tolerances on Shape and Dimensions First Edition; (Supersedes TR7735). 12 pp.

—Sheets—Rolled—Aerospace
AECMA PREN2072-85. Aluminium 1050A-H14 Sheet and Strip 0.4 Less Than or Equal to a Less Than or Equal to 6 mm. 3 pp.

—Sheets—Rolled—Mechanical Properties
CEN PREN 683-2-92. Aluminium and Aluminium Alloys—Wrought Products, Finstock—Part 2: Mechanical Properties. 9 pp.
DIN ENGL 1745 Pt 1-83. Wrought Aluminium and Aluminium Alloy Plate, Sheet and Strip Greater Than 0,35 mm in Thickness; Properties (Feb). 6 pp.
DIN ENGL 1788-83. Wrought Aluminium and Aluminium Alloy Sheet and Strip with Thicknesses Between 0,021 and 0,350 mm; Properties (Feb). 4 pp.
ISO 6361 Pt 2-90. Wrought Aluminium and Aluminium Alloy Sheets and Plates—Part 2: Mechanical Properties Second Edition; (Supersedes TR2136). 17 pp.

—Sheets—Test Specimens
CGSB CAN2-19.0-M77METH12.2-78. Methods of Testing Putty, Caulking and Sealing Compounds Preparation of Plain Aluminum Panels and Extrusions. 1 p.

—Silicon Content
BSI BS 1728: Part 11-60. 1960 Methods for the Analysis of Aluminium and Aluminium Alloys: Part 11: Determination of Silicon (Perchloric Acid Method). 9 pp.
JIS H 1352-72. Methods for Determination of Silicon in Aluminium and Aluminium Alloy.

—Silicon Content—Absorptiometric Analysis
BSI BS 1728: Part 12-61. 1961 Amd 1 Methods for the Analysis of Aluminium and Aluminium Alloys: Part 12: Determination of Silicon (Absorptiometric Molybdenum Blue Method). 9 pp.

—Silicon Content—Gravimetric Analysis
ISO 797-73. Aluminium and Aluminium Alloys—Determination of Silicon—Gravimetric Method First Edition. 7 pp.

—Silicon Content—Spectrophotometry
ISO 808-73. Aluminium and Aluminium Alloys—Determination of Silicon—Spectrophotometric Method with the Reduced Silicomolybdic Complex First Edition. 6 pp.

—Slugs—Extruded
CEN PREN 570-91. Wrought Aluminium and Aluminium Alloys—Slugs for Impact Extrusion. 14 pp.
DIN ENGL 59604-87. Wrought Aluminium and Aluminium Alloy Slugs for Impact Extrusion (Jan) (Supersedes DIN 59604 Part 1, July 1973 Edition, and DIN 59604 Part 2, November 1970 Edition). 5 pp.

—Solders
JIS Z 3281-88. Solders for Aluminium and Aluminium Alloys. 8 pp.

—Spectrochemical Analysis
JIS H 1305-76. Method for Photoelectric Emission Spectrochemical Analysis of Aluminium and Aluminium Alloy.

Aluminum (Cont.)
—Spectrochemical Analysis (Cont.)
JIS H 1306-92. Methods for Atomic Absorption Spectrometric Analysis of Aluminium and Aluminium Alloys. 28 pp.

—Strips
BSI BS 1470-87. 1987 Amd 1 Wrought Aluminium and Aluminium Alloys for General Engineering Purposes; Plate, Sheet and Strip (AMD 6032) May 31, 1989. 24 pp.
CEN PREN 485 (Part 1)-91. Wrought Aluminium and Aluminium Alloy Sheets, Strips and Plates—Part 1: Technical Conditions for Inspection and Delivery. 16 pp.
DIN ENGL 40501 Pt 1-85. Aluminium for Electrical Purposes; E-Al Plate, Sheet and Strip; Technical Delivery Conditions (June). 3 pp.
JIS H 4000-88. Aluminium and Aluminium Alloy Sheets and Plates, Strips and Coiled Sheets. 145 pp.
SAA AS 1865-86. Aluminium and Aluminium Alloys—Drawn Wire, Rod, Bar and Strip (This is a Joint Standard with SANZ NZS 1865:1986). 11 pp.
SNZ NZS/AS 1865-86. Aluminium and Aluminium Alloys—Drawn Wire, Rod, Bar and Strip (This is a Joint Standard with SAA AS 1865). 11 pp.

—Strips—Aerospace
AECMA PREN2070-02-84. Aluminium and Aluminium Alloy Wrought Products—Part 2—Sheet, Strip Formed Profiles and Plate (CS/44). 14 pp.
AECMA PREN2599-86. Strip in Aluminium and Aluminium Alloys 0.3 Less Than or Equal to a Less Than or Equal to 3.2 mm Dimensions. 4 pp.
BSI BS EN 2070: Part 2-91. 1991 Aluminium and Aluminium Alloy Wrought Products. Technical Specification Part 2: Sheet, Strip Formed Profiles and Plate. 19 pp.
BSI 6L 16-85. 1985 Sheet and trip of 99% Aluminium (Temper Designation H14 or H24). 4 pp.
BSI 6L 17-85. 1985 Sheet and Strip of 99% Aluminium (Temper Designation—0). 4 pp.
CEN EN 2070-2-89. Aerospace Series Aluminium and Aluminium Alloy Wrought Products Technical Specification Part 2: Sheet, Strip Formed Profiles and Plate. 14 pp.
CEN PREN 2072-93. Aerospace Series—Aluminium AL-P1050A H14—Sheet and Strip 0,4 mm Less Than or Equal to a Less Than or Equal to 6 mm. 5 pp.
CEN PREN 3996-93. Aerospace Series Aluminium AL-P1100-H14 Sheet and Strip 0,3 mm Less Than or Equal to a Less Than or Equal to 6 mm. 4 pp.

—Strips—Chemical Analysis
BSI BS 4300/16-84. 1984 Specification (Supplementary Series) for Wrought Aluminium and Aluminium Alloys for General Engineering Purposes /16: Specification for Sheet and Strip 8011 Material. 11 pp.

—Strips—Designations
BSI BS 4300/16-84. 1984 Specification (Supplementary Series) for Wrought Aluminium and Aluminium Alloys for General Engineering Purposes /16: Specification for Sheet and Strip 8011 Material. 11 pp.

—Strips—Drawn
BSI BS 2897-70. 1970 Amd 1 Wrought Aluminium for Electrical Purposes. Strip with Drawn or Rolled Edges. 16 pp.

—Strips—Mechanical Properties
CEN PREN 485-2-91. Wrought Aluminium and Aluminium Alloy Sheets, Strips and Plates—Part 2 Mechanical Properties. 45 pp.

—Strips—Rolled
BSI BS 2897-70. 1970 Amd 1 Wrought Aluminium for Electrical Purposes. Strip with Drawn or Rolled Edges. 16 pp.
CEN PREN 485 (Part 3)-91. Wrought Aluminium and Aluminium Alloy Sheets, Strips and Plates—Part 3: Hot-Rolled Sheets, Strips and Plates—Tolerances on Shape and Dimensions. 11 pp.
CEN PREN 485 (Part 4)-91. Wrought Aluminium and Aluminium Alloy Sheets, Strips and Plates—Part 4: Cold-Rolled Sheets, Strips and Plates—Tolerances on Shape and Dimensions. 14 pp.
CEN PREN 941-92. Aluminium and Aluminium Alloys—Wrought Products—Circle and Circle Stock for General Applications. 11 pp.
DIN ENGL 1745 Pt 2-83. Wrought Aluminium and Aluminium Alloy Plate, Sheet and Strip Greater Than 0,35 mm in Thickness; Technical Delivery Conditions (Feb). 4 pp.
DIN ENGL 1783-81. Strips, Plates and Sheets of Aluminium and Wrought Aluminium Alloys with Thicknesses over 0,35 mm, Cold-Rolled; Dimensions (Apr). 9 pp.

INDUSTRY STANDARDS

INTERNATIONAL AND NON-U.S. NATIONAL STANDARDS
SUBJECT INDEX

Aluminum

Aluminum (Cont.)

—Strips—Rolled (Cont.)
- DIN ENGL 1784-81. Strips, Sheets and Sizes of Aluminium and Wrought Aluminium Alloys with Thicknesses from 0,021 to 0,35 mm, Cold-Rolled; Dimensions (Apr). 7 pp.
- DIN ENGL 59600-81. Strips, Plates and Sheets of Aluminium and Wrought Aluminium Alloys; Hot Rolled; Dimensions (Apr). 6 pp.
- ISO 6361 Pt 1-86. Wrought Aluminium and Aluminium Alloys Sheets, Strips and Plates—Part 1: Technical Conditions for Inspection and Delivery First Edition; (Supersedes 5191). 5 pp.
- ISO 6361 Pt 3-85. Wrought Aluminium and Aluminium Alloy Sheets, Strips and Plates—Part 3: Strips—Tolerances on Shape and Dimensions First Edition. 5 pp.

—Strips—Rolled—Aerospace
- AECMA PREN2072-85. Aluminium 1050A-H14 Sheet and Strip 0.4 Less Than or Equal to A Less Than or Equal to 6 mm. 3 pp.

—Strips—Rolled—Mechanical Properties
- CEN PREN 683-2-92. Aluminium and Aluminium Alloys—Wrought Products, Finstock—Part 2: Mechanical Properties. 9 pp.
- DIN ENGL 1745 Pt 1-83. Wrought Aluminium and Aluminium Alloy Plate, Sheet and Strip Greater Than 0,35 mm in Thickness; Properties (Feb). 6 pp.
- DIN ENGL 1788-83. Wrought Aluminium and Aluminium Alloy Sheet and Strip with Thicknesses Between 0,021 and 0,350 mm; Properties (Feb). 4 pp.
- ISO 6361 Pt 2-90. Wrought Aluminium and Aluminium Alloy Sheets, Strips and Plates—Part 2: Mechanical Properties Second Edition; (Supersedes TR2136). 17 pp.

—Surface Finishing
- CNS H1008-86. Glossary of Terms Used in the Surface Treatment of Aluminum (Feb)(8506).
- CNS H3133-87. Aluminium and Aluminium Alloy Surface Chromate Treatment (Aug)(11111).
- JIS H 0201-87. Glossary of Terms Used in the Surface Treatment of Aluminium. 46 pp.

—Surface Finishing—Acids
- MOD UK DSTAN 68-109-89. Chromium Trioxide (Chromic Acid) Issue 1. 12 pp.
- MOD UK CS 3119. Chromic Acid (Superseded by Def Stan 68-109).

—Surface Finishing—Adhesive Bonding
- ISO 4588-89. Adhesives—Preparation of Metal Surfaces for Adhesive Bonding First Edition. 8 pp.

—T Beams—Extruded
- DIN ENGL 9714-81. Aluminium and Wrought Aluminium Alloy Extruded Tee Sections; Dimensions; Static Values (Sept). 5 pp.

—Tempering
- CNS H3021-87. Alloy and Temper Designation Systems for Aluminium and Aluminium Alloys (May)(2068).
- JIS H 0001-88. Temper Designation for Aluminium and Aluminium Alloys. 12 pp.

—Tempering—Designations
- ISO 2107-83. Aluminium, Magnesium and Their Alloys—Temper Designations First Edition. 5 pp.

—Titanium Content
- JIS H 1359-72. Methods for Determination of Titanium in Aluminium and Aluminium Alloy.

—Titanium Content—Photometry
- ISO 6827-81. Aluminium and Aluminium Alloys—Determination of Titanium Content—Diantipyrylmethane Photometric Method First Edition. 5 pp.

—Titanium Content—Spectrophotometry
- BSI BS 1728: Part 25-80. 1980 Methods for the Analysis of Aluminium and Aluminium Alloys: Part 25: Titanium (Spectrophotometric Chromotropic Acid Method). 9 pp.
- ISO 1118-78. Aluminium and Aluminium Alloys—Determination of Titanium—Spectrophotometric Chromotropic Acid Method First Edition. 8 pp.

—Tubes
- CNS H3019-85. Aluminium and Aluminium Alloy Tubes (Apr)(1308).
- DIN ENGL 40501 Pt 2-85. Aluminium for Electrical Purposes; E-Al and E-AlMgSiO,5 Tubes; Technical Delivery Conditions (June). 3 pp.
- MOD UK DSTAN 47-15-76. Aluminium and Aluminium Alloy Tubing for General Purpose Applications Metric Units Issue 1. 8 pp.

Aluminum (Cont.)

—Tubes—Aerospace
- AECMA PREN2073-77. Aluminium 1050A-H14 (AL-P99, 5-H14)—Tubes for Structures D 5-100 mm—A 0.4-2 mm (C5/40). 3 pp.

—Tubes—Circular—Aerospace
- AECMA PREN2257-79. Circular Structural Tubes in Aluminium and Aluminium Alloys—Dimensions (C6/A06-02). 5 pp.

—Tubes—Drawn
- BSI BS 1471-72. 1972 Amd 2 Wrought Aluminium and Aluminium Alloys for General Engineering Purposes; Drawn Tube. 25 pp.
- CEN PREN 754-1-92. Aluminium and Aluminium Alloys—Wrought Products—Cold Drawn Rod/Bar and Tube—Part 1: Technical Conditions for Inspection and Delivery. 16 pp.
- DIN ENGL 1746 Pt 2-83. Wrought Aluminium and Aluminium Alloy Tubes; Technical Delivery Conditions (Feb). 3 pp.
- ISO 6363 Pt 1-88. Wrought Aluminium and Aluminium Alloy Cold-Drawn Rods/Bars and Tubes—Part 1: Technical Conditions for Inspection and Delivery First Edition; (Supersedes 5191). 6 pp.
- SAA AS 1867-86. Aluminium and Aluminium Alloys—Drawn Tubes (This is a Joint Standard with SANZ NZS 1867:1986). 13 pp.
- SNZ NZS/AS 1867-86. Aluminium and Aluminium Alloys—Drawn Tubes (This is a Joint Standard with SAA AS 1867). 13 pp.

—Tubes—Drawn—Aerospace
- AECMA PREN2070-04-87. Aluminium and Aluminium Alloy Wrought Products—Technical Specification—Part 4—Tube for Structures. 9 pp.
- AECMA PREN2070-05-87. Aluminium and Aluminium Alloy Wrought Products—Technical Specification—Part 5—Tube Used Under Pressure. 12 pp.
- BSI BS EN 2070: Part 4-91. 1991 Aluminium and Aluminium Alloy Wrought Products Technical Specification Part 4: Tube for Structures. 12 pp.
- BSI BS EN 2070: Part 5-91. 1991 Aluminium and Aluminium Alloy Wrought Products Technical Specification Part 5: Tube Used Under Pressure. 15 pp.
- BSI 4L 54-86. 1986 Amd 1 Tube of 99% Aluminium (Cold Drawn: Seamless: Tested Hydraulically) (Not Exceeding 12 mm Wall Thickness) (AMD 5994) August 31, 1988. 5 pp.
- CEN EN 2070-4-89. Aerospace Series Aluminium and Aluminium Alloy Wrought Products Technical Specification Part 4: Tube for Structures. 9 pp.
- CEN EN 2070-5-89. Aerospace Series Aluminium and Aluminium Alloy Wrought Products Technical Specification Part 5: Tube Used Under Pressure. 12 pp.

—Tubes—Drawn—Mechanical Properties
- CEN PREN 754-2-92. Aluminium and Aluminium Alloys—Wrought Products—Cold Drawn Rod/Bar and Tube—Part 2: Mechanical Properties. 40 pp.
- DIN ENGL 1746 Pt 1-87. Wrought Aluminium and Aluminium Alloy Tubes; Properties (Jan). 4 pp.
- ISO TR2778-77. Wrought Aluminium and Aluminium Alloys—Drawn Tubes—Mechanical Properties First Edition. 6 pp.

—Tubes—Extruded
- BSI BS 1474-87. 1987 Wrought Aluminium and Aluminium Alloys for General Engineering Purposes; Bars, Extruded Round Tubes and Sections. 28 pp.
- BSI BS 2898-70. 1970 Amd 1 Wrought Aluminium and Aluminium Alloys for Electrical Purposes. Bars, Extruded Round Tube and Sections. 32 pp.
- CEN PREN 755-1-92. Aluminium and Aluminium Alloys—Wrought Products—Extruded Rod/Bar, Tube and Profile—Part 1: Technical Conditions for Inspection and Delivery. 18 pp.
- DIN ENGL 1746 Pt 2-83. Wrought Aluminium and Aluminium Alloy Tubes; Technical Delivery Conditions (Feb). 3 pp.
- DIN ENGL 9107-87. Wrought Aluminium and Aluminium Alloy Seamless Extruded Round Tubes; Dimensions, Dimensional Tolerances and Form Tolerances (Feb). 6 pp.
- ISO 6362 Pt 1-86. Wrought Aluminium and Aluminium Alloys Extruded Rods/Bars, Tubes and Profiles—Part 1: Technical Conditions for Inspection and Delivery First Edition; (Supersedes 5191). 5 pp.
- ISO 6362 Pt 4-88. Wrought Aluminium and Aluminium Alloy Extruded Rods/Bars, Tubes and Profiles—Part 4: Extruded Profiles—Tolerances on Shape and Dimensions First Edition. 11 pp.

—Tubes—Extruded—Aerospace
- AECMA PREN2070-04-87. Aluminium and Aluminium Alloy Wrought Products—Technical Specification—Part 4—Tube for Structures. 9 pp.

Aluminum (Cont.)

—Tubes—Extruded—Aerospace (Cont.)
- AECMA PREN2070-05-87. Aluminium and Aluminium Alloy Wrought Products—Technical Specification—Part 5—Tube Used Under Pressure. 12 pp.
- BSI BS EN 2070: Part 4-91. 1991 Aluminium and Aluminium Alloy Wrought Products Technical Specification Part 4: Tube for Structures. 12 pp.
- BSI BS EN 2070: Part 5-91. 1991 Aluminium and Aluminium Alloy Wrought Products Technical Specification Part 5: Tube Used Under Pressure. 15 pp.
- CEN EN 2070-4-89. Aerospace Series Aluminium and Aluminium Alloy Wrought Products Technical Specification Part 4: Tube for Structures. 9 pp.
- CEN EN 2070-5-89. Aerospace Series Aluminium and Aluminium Alloy Wrought Products Technical Specification Part 5: Tube Used Under Pressure. 12 pp.

—Tubes—Extruded—Mechanical Properties
- CEN PREN 755-2-92. Aluminium and Aluminium Alloys—Wrought Products—Extruded Rod/Bar, Tube and Profile—Part 2: Mechanical Properties. 56 pp.
- DIN ENGL 1746 Pt 1-87. Wrought Aluminium and Aluminium Alloy Tubes; Properties (Jan). 4 pp.
- ISO 6362 Pt 2-90. Wrought Aluminium and Aluminium Alloy Extruded Rods/Bars, Tubes and Profiles—Part 2: Mechanical Properties Second Edition. 15 pp.

—Tubes—Seamless—Drawn
- DIN ENGL 1795-87. Wrought Aluminium and Aluminium Alloy Seamless Drawn Round Tubes; Dimensions, Dimensional Tolerances and Form Tolerances (Feb). 8 pp.

—Tubes—Seamless—Drawn—Aerospace
- BSI 2L 116-85. 1985 Tube of 99% Aluminium (Cold Drawn: Seamless: Not Tested Hydraulically) (Not Exceeding 12 mm Wall Thickness). 4 pp.

—Vanadium Content
- JIS H 1362-72. Method for Determination of Vanadium in Aluminium.

—Welding
- SAA AS 1665-92. Welding of Aluminium Structures (in Professional Package 36). 25 pp.

—Wire Bars—Chemical Analysis
- DIN ENGL 1712 Pt 3-76. Aluminium; Half-Finished Products (Dec). 3 pp.

—Wrought
- CEN PREN 546-1-91. Wrought Aluminium and Aluminium Alloy Foil—Part 1: Technical Conditions for Inspection and Delivery. 10 pp.
- CEN PREN 546-3-92. Aluminium and Aluminium Alloys—Wrought Products, Foil—Part 3: Dimensional Tolerances. 7 pp.
- CEN PREN 683-1-92. Aluminium and Aluminium Alloys—Wrought Products, Finstock—Part 1: Technical Conditions for Inspection and Delivery. 10 pp.
- CEN PREN 683-3-92. Aluminium and Aluminium Alloys—Wrought Products, Finstock—Part 3: Dimensional Tolerances. 9 pp.
- ISO 209 Pt 2-89. Wrought Aluminium and Aluminium Alloys—Chemical Composition and Forms of Products—Part 2: Forms of Products First Edition. 5 pp.

—Wrought—Aerospace
- AECMA PREN2070-01-84. Aluminium and Aluminium Alloy Wrought Products—Part 1—General Requirements (CS/44). 10 pp.
- BSI BS EN 2070: Part 1-91. 1991 Aluminium and Aluminium Alloy Wrought Products. Technical Specification Part 1: General Requirements. 15 pp.
- BSI BS EN 2070: Part 1-01. 1991 Amd 1 Aluminium and Aluminium Alloy Wrought Products. Technical Specification Part 1: General Requirements (AMD 7705) October 15, 1993 (S). 22 pp.
- BSI 4L 100-01. 1993 Amd 1 Procedure for Inspection, Testing and Acceptance of Wrought Aluminium and Aluminium Alloys (AMD 7884) September 15, 1993 (S). 82 pp.
- BSI 3L 100-71. (WITHDRAWN) 1971 Amd 4 Procedure for Inspection and Testing of Wrought Aluminium and Aluminium Alloys. 44 pp.
- CEN EN 2070-1-89. Aerospace Series Aluminium and Aluminium Alloy Wrought Products Technical Specification Part 1: General Requirements. 10 pp.
- CEN EN 2070 Part 1-89. Aerospace Series Aluminium and Aluminium Alloy Wrought Products Technical Specification Part 1: General Requirements. 12 pp.
- ISO 8591 Pt 1-89. Aerospace—Wrought Aluminium and Aluminium Alloys—Inspection, Testing and Supply Requirements—Part 1: General Requirements First Edition. 7 pp.

INTERNATIONAL AND NON-U.S. NATIONAL STANDARDS
SUBJECT INDEX
Aluminum

Aluminum (Cont.)

—Wrought—Aerospace—Macrographic Inspection
CEN PREN 2715-90. Macrographic Examination of Aluminium and Aluminium Alloy Forging Stock, Forgings and Wrought Products. 4 pp.

—Wrought—Bend Testing
ISO 2142-81. Wrought Aluminium, Magnesium and Their Alloys—Selection of Specimens and Test Pieces for Mechanical Testing First Edition. 5 pp.

—Wrought—Chemical Analysis
CEN PREN 573-3-91. Wrought Aluminium and Aluminium Alloys—Chemical Composition and Forms of Products—Part 3: Chemical Composition. 13 pp.
ISO 209 Pt 1-89. Wrought Aluminium and Aluminium Alloys—Chemical Composition and Forms of Products—Part 1: Chemical Composition First Edition; (Supersedes 2779 and 3335). 9 pp.

—Wrought—Designations
CEN PREN 573-1-91. Wrought Aluminium and Aluminium Alloys—Chemical Composition and Forms of Products—Part 1: Numerical Designation System. 8 pp.
CEN PREN 573-2-91. Wrought Aluminium and Aluminium Alloys—Chemical Composition and Forms of Products—Part 2: Chemical Symbol Based Designation System. 9 pp.

—Wrought—Food Preparation Equipment—Chemical Analysis
CEN PREN 602-91. Wrought Aluminium and Aluminium Alloys—Chemical Composition of the Metal Used for the Production of Materials and Articles Intended to Be in Contact with Food. 7 pp.

—Wrought—Glossaries
BSI BS 3660-76. (WITHDRAWN) 1976 Glossary of Terms Used in the Wrought Aluminium Industry (Superseded by BS EN 23134-1, EN 23134-2, EN 23134-3, and EN 23134-4: 1991). 28 pp.

—Wrought—Mechanical Properties
CEN PREN 546-2-92. Aluminium and Aluminium Alloys—Wrought Products, Foil—Part 2: Mechanical Properties. 7 pp.

—Wrought—Shapes
CEN PREN 573-4-91. Wrought Aluminium and Aluminium Alloys—Chemical Composition and Forms of Products—Part 4: Forms of Products. 13 pp.

—Wrought—Structural Shapes
BSI BS 8118: Part 2-91. 1991 Structural Use of Aluminium Part 2: Specification for Materials, Workmanship and Protection. 28 pp.

—Wrought—Structural Shapes—Design
BSI BS 8118: Part 1-91. 1991 Structural Use of Aluminium Part 1: Code of Practice for Design. 160 pp.

—Wrought—Tempering—Designations
BSI BS EN 515-93. 1993 Aluminium and Aluminium Alloys—Wrought Products—Temper Designations (V). 21 pp.
CEN PREN 515-91. Wrought Aluminium and Aluminium Alloys—Temper Designations. 20 pp.

—Wrought—Tensile Testing
ISO 2142-81. Wrought Aluminium, Magnesium and Their Alloys—Selection of Specimens and Test Pieces for Mechanical Testing First Edition. 5 pp.

—Zinc Content
BSI BS 1728: Part 18-70. 1970 Methods for the Analysis of Aluminium and Aluminium Alloys: Part 18: Determination of Zinc (Ion-Exchange-Volumetric EDTA or Polargraphic Method). 12 pp.
BSI BS 1728: Part 21-73. 1973 Methods for the Analysis of Aluminium and Aluminium Alloys: Part 21: Zinc (Atomic Absorption Method). 8 pp.
JIS H 1356-72. Methods for Determination of Zinc in Aluminium and Aluminium Alloy.

—Zinc Content—Atomic Absorption Spectrometry
ISO 5194-81. Aluminium and Aluminium Alloys—Determination of Zinc Content—Flame Atomic Absorption Spectrometric Method First Edition. 6 pp.

Aluminum Acetate
Use For: Aluminum Acetate Basic
See Also: Acetates; Esters
CNS K7562-81. Chemical Reagent (Aluminum Acetate, Basic) (Jul)(7720).

Aluminum Acetate Basic
Use: Aluminum Acetate

Aluminum Alkyl Halides
—Highway Transportation—Emergency Procedures
SAA AS 1678.4.2. 001-93. Emergency Procedure Guides—Transport—Part 4.2.001: Aluminium Alkyls and Aluminium Alkyl Halides (in Professional Package 37A).

Aluminum Alkyls
—Highway Transportation—Emergency Procedures
SAA AS 1678.4.2. 001-93. Emergency Procedure Guides—Transport—Part 4.2.001: Aluminium Alkyls and Aluminium Alkyl Halides (in Professional Package 37A).

Aluminum Alloy Castings
Use: Aluminum Castings

Aluminum Alloy Wire
Use: Aluminum Wire

Aluminum Alloys
Scope Note: For additional listings, see also specific products made from aluminum alloys.
See Also: Alloys; Aluminum Bronzes; Aluminum Foil; Aluminum Magnesium Alloys; Aluminum Magnesium Silicon Alloys; Aluminum Manganese Alloys; Aluminum Silicon Alloys; Aluminum Silicon Bronze Alloys; Aluminum Zinc Alloys; Aluminum Zinc Magnesium Alloys; Brazing Alloys; Nonferrous Alloys
CNS H3129-84. Aluminium Alloy (Aug)(10982).

—Aerospace
BSI 4L 100-93. 1993 Procedure for Inspection, Testing and Acceptance of Wrought Aluminium and Aluminium Alloys (S). 81 pp.
CEN PREN 2500-2-91. Instructions for the Drafting and Use of Metallic Material Standards Part 2—Specific Requirements for Aluminium, Aluminium Alloys and Magnesium Alloys. 15 pp.
CEN PREN 2500-2-92. Instructions for the Drafting and Use of Metallic Material Standards Part 2—Specific Requirements for Aluminium, Aluminium Alloys and Magnesium Alloys. 15 pp.

—Aerospace—Surface Finishing
CEN PREN 3837-91. Paints and Varnishes Nature and Method for Preparation of Surface of Test Specimens in Aluminium Alloys. 15 pp.

—Aerospace—Tempering—Designations
CEN PREN 3350-91. Aerospace Series Aluminium, Aluminium Alloys Temper Designations. 21 pp.
CEN PREN 3350-91. Aluminium, Aluminium Alloys Temper Designations. 18 pp.

—Aircraft
SBAC RS 116 (V). S.B.A.C. Protective Treatment Chart.

—Aircraft—Welded
JIS W 2016-85. Fusion Welding Process of Aerospace Aluminium Alloys.

—Angles—Aerospace
DIN ENGL LN 9411-86. Aerospace; Wrought Aluminium Alloy Folded Profiles, Angle; Dimensions, Static Values, Masses (Dec). 5 pp.
DIN ENGL LN 9412-86. Aerospace; Wrought Aluminium Alloy Folded Profiles, Angle with Internally Lipped Flanges; Dimensions, Static Values, Masses (Dec). 5 pp.

—Angles—Extruded
DIN ENGL 1771-81. Aluminium and Wrought Aluminium Alloy; Extruded Angles; Dimensions; Static Values (Sept). 6 pp.

—Annealed—Bars—Drawn—Aerospace
AECMA PREN2699-88. Aluminium Alloy (5086) Annealed and Straightened (H111) Drawn Bar 6 Less Than or Equal to D Less Than or Equal to 50 mm. 4 pp.

—Arc Welding
BSI BS EN 288: Part 4-92. 1992 Specification and Approval of Welding Procedures for Metallic Materials Part 4: Welding Procedure Tests for the Arc Welding of Aluminium and its Alloys (Supersedes BS 4870: Part 2: 1982). 32 pp.
CEN PREN 288-4-91. Specification and Approval of Welding Procedures for Metallic Materials—Part 4: Welding Procedure Tests for the Arc Welding of Aluminium and its Alloys. 37 pp.

Aluminum Alloys (Cont.)

—Arc Welding (Cont.)
CEN EN 288-4-92. Specification and Approval of Welding Procedures for Metallic Materials—Part 4: Welding Procedure Tests for the Arc Welding of Aluminium and Its Alloys. 27 pp.
CEN EN 288-4-92. CORRIGENDUM Specification and Approval of Welding Procedures for Metallic Materials—Part 4: Welding Procedure Tests for the Arc Welding of Aluminium and its Alloys. 2 pp.
CEN EN 288-4-92. Specification and Approval of Welding Procedures for Metallic Materials—Part 4: Welding Procedure Tests for the Arc Welding of Aluminium and Its Alloys. 28 pp.
DIN ENGL EN 288 Pt 4-92. Specification and Approval of Procedures for Welding Metallic Materials; Welding Procedure Tests for the Arc Welding of Aluminium and Its Alloys (Oct). 30 pp.

—Atmospheric Corrosion Testing
JIS H 0521-68. Testing Method for Atmospheric Corrosion of Aluminium and Aluminium Alloys. 12 pp.

—Bars
BSI BS 1474-87. 1987 Wrought Aluminium and Aluminium Alloys for General Engineering Purposes; Bars, Extruded Round Tubes and Sections. 28 pp.
BSI BS 2898-70. 1970 Amd 1 Wrought Aluminium and Aluminium Alloys for Electrical Purposes. Bars, Extruded Round Tube and Sections. 32 pp.
BSI BS 4300/5-73. 1973 Amd 1 Specification (Supplementary Series) for Wrought Aluminium and Aluminium Alloys for General Engineering Purposes /5: 2011 Free-Cutting Bar and Wire-Alloy. 16 pp.
BSI BS 4300/12-69. 1969 Amd 1 Specification (Supplementary Series) for Wrought Aluminium and Aluminium Alloys for General Engineering Purposes /12: 5454 Bar, Extruded Round Tube and Sections. 36 pp.
BSI BS 4300/15-73. 1973 Amd 1 Specification (Supplementary Series) for Wrought Aluminium and Aluminium Alloys for General Engineering Purposes /15: 7020 Bar, Extruded Round Tube and Sections. 31 pp.
CNS H3048-84. Aluminium and Aluminium Alloy Rods, Bars and Wires (Sep)(3667).
JIS H 4040-88. Aluminium and Aluminium Alloy Rods, Bars and Wires. 48 pp.
SAA AS 1865-86. Aluminium and Aluminium Alloys—Drawn Wire, Rod, Bar and Strip (This is a Joint Standard with SANZ NZS 1865:1986). 11 pp.
SNZ NZS/AS 1865-86. Aluminium and Aluminium Alloys—Drawn Wire, Rod, Bar and Strip (This is a Joint Standard with SAA AS 1865). 11 pp.

—Bars—Drawn
CEN PREN 754-1-92. Aluminium and Aluminium Alloys—Wrought Products—Cold Drawn Rod/Bar and Tube—Part 1: Technical Conditions for Inspection and Delivery. 16 pp.
DIN ENGL 1747 Pt 2-83. Wrought Aluminium and Aluminium Alloy Rod and Bar; Technical Delivery Conditions (Feb). 4 pp.
ISO 6363 Pt 1-88. Wrought Aluminium and Aluminium Alloy Cold-Drawn Rods/Bars and Tubes—Part 1: Technical Conditions for Inspection and Delivery First Edition; (Supersedes 5191). 6 pp.

—Bars—Drawn—Aerospace
AECMA PREN2070-03-84. Aluminium and Aluminium Alloy Wrought Products—Part 3—Bar and Section (CS/44). 13 pp.
AECMA PREN2128-85. Aluminium Alloy 7075-T7351 Drawn Bar a Less Than or Equal to 75 mm. 3 pp.
AECMA PREN2317-85. Aluminium Alloy 7075-T73 Drawn Bar a Less Than or Equal to 75 mm. 3 pp.
AECMA PREN2319-85. Aluminium Alloy 2024-T3510 Drawn Bar a Less Than or Equal to 75 mm. 3 pp.
AECMA PREN2320-85. Aluminium Alloy 2024-T4 Drawn Bar a Less Than or Equal to 75 mm. 3 pp.
AECMA PREN2704-85. Aluminium Alloy 2024-T3511 Drawn Bar a Less Than or Equal to 75 mm. 3 pp.
BSI BS EN 2070: Part 3-91. 1991 Aluminium and Aluminium Alloy Wrought Products. Technical Specification Part 3: Bar and Section. 18 pp.
BSI BS EN 2128-92. 1992 Aluminium Alloy AL-P7075-T7351 Drawn Bars 6 mm Than or Equal to a or D Less Than or Equal to 75 mm. 10 pp.
CEN EN 2070-3-89. Aerospace Series Aluminium and Aluminium Alloy Wrought Products Technical Specification Part 3: Bar and Section. 18 pp.
CEN PREN 2128-91. Aluminium Alloy AL-P7075-T7351 Drawn Bars 6 mm Less Than or Equal to a or D Less Than or Equal to 75 mm. 6 pp.
CEN EN 2128-92. Aluminium Alloy AL-P7075-T7351 Drawn Bars 6 mm Less Than or Equal to a or D Less Than or Equal to 75 mm. 5 pp.

INDUSTRY STANDARDS

Aluminum

INTERNATIONAL AND NON-U.S. NATIONAL STANDARDS
SUBJECT INDEX

Aluminum Alloys (Cont.)

—Bars—Drawn—Hexagonal

CEN PREN 754-6-92. Wrought Aluminium and Aluminium Alloys—Cold Drawn Rod/Bar and Tube—Part 6: Drawn Hexagonal Bars, Dimensions and Form Tolerances. 7 pp.

DIN ENGL 1797-86. Wrought Aluminium and Aluminium Alloy Drawn Hexagonal Bars with Sharp Edges; Dimensions, Dimensional Deviations and Form Tolerances (Sept). 4 pp.

ISO 6363 Pt 5-92. Wrought Aluminium and Aluminium Alloy Cold-Drawn Rods/Bars and Tubes—Part 5: Drawn Square and Hexagonal Bars—Tolerances on Form and Dimensions First Edition. 5 pp.

—Bars—Drawn—Hexagonal—Aerospace

AECMA PREN2046-76. Hexagonal Aluminium Alloy Bars Drawn—Dimensions—Tolerance H11 (C6/A03-05). 4 pp.

—Bars—Drawn—Mechanical Properties

CEN PREN 754-2-92. Aluminium and Aluminium Alloys—Wrought Products—Cold Drawn Rod/Bar and Tube—Part 2: Mechanical Properties. 40 pp.

DIN ENGL 1747 Pt 1-83. Wrought Aluminium and Aluminium Alloy Rod and Bar; Properties (Feb). 4 pp.

—Bars—Drawn—Rectangular

CEN PREN 754-5-92. Wrought Aluminium and Aluminium Alloys—Cold Drawn Rod/Bar and Tube—Part 5: Drawn Rectangular Bars, Dimensions and Form Tolerances. 9 pp.

DIN ENGL 1769-86. Wrought Aluminium and Aluminium Alloy Drawn Rectangular Bars with Square Edges; Dimensions, Dimensional Deviations and Form Tolerances, Static Values (Sept). 10 pp.

ISO 6363 Pt 4-91. Wrought Aluminium and Aluminium Alloy Cold-Drawn Rods/Bars and Tubes—Part 4: Drawn Rectangular Bars—Tolerances on Form and Dimensions First Edition. 7 pp.

—Bars—Drawn—Round

CEN PREN 754-3-92. Wrought Aluminium and Aluminium Alloys—Cold Drawn Rod/Bar and Tube—Part 3: Drawn Round Bars, Dimensions and Form Tolerances. 7 pp.

DIN ENGL 1798-86. Wrought Aluminium and Aluminium Alloy Drawn Round Bars; Dimensions, Dimensional Deviations and Form Tolerances (Sept). 4 pp.

ISO 5193-81. Wrought Aluminium and Aluminium Alloys—Drawn Round Bars—Tolerances on Shape and Dimensions (Symmetric Plus and Minus Tolerances on Diameter) First Edition. 3 pp.

ISO 7274-81. Wrought Aluminium and Aluminium Alloys—Drawn Round Bars—Tolerances on Shape and Dimensions (All Minus Tolerances on Diameter) First Edition. 3 pp.

—Bars—Drawn—Round—Aerospace

AECMA PREN2044-76. Round Aluminium Alloy Bars Drawn—Dimensions—Tolerance H11 (C6/A03-01). 4 pp.

CEN PREN 2044-89. Round Bars, Drawn Tolerance in 11 Aluminium Alloy. 5 pp.

—Bars—Drawn—Square

CEN PREN 754-4-92. Wrought Aluminium and Aluminium Alloys—Cold Drawn Rod/Bar and Tube—Part 4: Drawn Square Bars—Dimension and Form Tolerances. 7 pp.

DIN ENGL 1796-86. Wrought Aluminium and Aluminium Alloy Drawn Square Bars with Sharp Edges; Dimensions, Dimensional Deviations and Form Tolerances (Sept). 4 pp.

ISO 6363 Pt 5-92. Wrought Aluminium and Aluminium Alloy Cold-Drawn Rods/Bars and Tubes—Part 5: Drawn Square and Hexagonal Bars—Tolerances on Form and Dimensions First Edition. 5 pp.

—Bars—Drawn—Square—Aerospace

AECMA PREN2045-76. Square Aluminium Alloy Bars Drawn—Dimensions—Tolerance H11 (C6/A03-03). 4 pp.

—Bars—Extruded

BSI BS 4300/4-73. 1973 Amd 1 Specification (Supplementary Series) for Wrought Aluminium and Aluminium Alloys for General Engineering Purposes /4: 6463 Solid Extruded Bars and Sections Suitable for for Bright Trim/Reflector Applications. 27 pp.

CEN PREN 755-1-92. Aluminium and Aluminium Alloys—Wrought Products—Extruded Rod/Bar, Tube and Profile—Part 1: Technical Conditions for Inspection and Delivery. 18 pp.

DIN ENGL 1747 Pt 2-83. Wrought Aluminium and Aluminium Alloy Rod and Bar; Technical Delivery Conditions (Feb). 4 pp.

ISO 6362 Pt 1-86. Wrought Aluminium and Aluminium Alloys Extruded Rods/Bars, Tubes and Profiles—Part 1: Technical Conditions for Inspection and Delivery First Edition; (Supersedes 5191). 5 pp.

Aluminum Alloys (Cont.)

—Bars—Extruded (Cont.)

ISO 6362 Pt 4-88. Wrought Aluminium and Aluminium Alloy Extruded Rods/Bars, Tubes and Profiles—Part 4: Extruded Profiles—Tolerances on Shape and Dimensions First Edition. 11 pp.

SAA AS 1866-86. Aluminium and Aluminium Alloys—Extruded Rod, Bar, Solid and Hollow Shapes (This is a Joint Standard with SANZ NZS 1866:1986). 45 pp.

SNZ NZS/AS 1866-86. Aluminium and Aluminium Alloys—Extruded Rod, Bar, Solid and Hollow Shapes (This is a Joint Standard with SAA AS 1866). 45 pp.

—Bars—Extruded—Aerospace

AECMA PREN2070-03-84. Aluminium and Aluminium Alloy Wrought Products—Part 3—Bar and Section (CS/44). 13 pp.

AECMA PREN2100-80. Aluminium Alloy 2014A-T451—Bars and Sections A Less Than or Equal to 200 mm (C5/40). 3 pp.

AECMA PREN2127-80. Aluminium Alloy 7075-T7351—Bars and Sections A Less Than or Equal to 100 mm (C5/40). 3 pp.

AECMA PREN2315-80. Aluminium Alloy 7075-T73510 or T73511—Bars and Sections A Less Than or Equal to 100 mm (C5/40). 3 pp.

AECMA PREN2316-80. Aluminium Alloy 7075-T73—Bars and Sections—A Less Than or Equal to 100 mm (C5/40). 3 pp.

AECMA PREN2318-80. Aluminium Alloy 2024-T3510 or T3511—Bars and Sections—A 1.2-150 mm (C5/40). 3 pp.

AECMA PREN2321-80. Aluminium Alloy 2024-T3—Bars and Sections—A Less Than or Equal to 150 mm (C5/40). 3 pp.

AECMA PREN2323-80. Aluminium Alloy 2014A-T651—Bars and Sections—A Less Than or Equal to 200 mm (C5/40). 3 pp.

AECMA PREN2324-80. Aluminium Alloy 2014A-T6—Bars and Sections—A Less Than or Equal to 150 mm (C5/40). 3 pp.

AECMA PREN2325-80. Aluminium Alloy 2014A-T6—Bars A Less Than or Equal to 100 mm (C5/40). 3 pp.

AECMA PREN2326-80. Aluminium Alloy 6082-T6—Bars and Sections—A Less Than or Equal to 200 mm (C5/40). 3 pp.

AECMA PREN2384-83. Aluminium Alloy 2014A-T6511 Bars and Sections A Less Than or Equal to 150 mm. 3 pp.

AECMA PREN2385-82. Aluminum Alloy 7009-T73651—Bars and Sections—A Less Than or Equal to 125 mm (C5/40). 3 pp.

AECMA PREN2394-83. Aluminium Alloy 7075-T6511 Bars and Sections A Less Than or Equal to 125 mm.

AECMA PREN2630-85. Aluminium Alloy 7009-T736511 Bar and Section 1.2 Less Than or Equal to (a or D) Less Than or Equal to 125 mm with Peripheral Coarse Grain Control. 3 pp.

AECMA PREN2631-85. Aluminium Alloy 7075-T6511 Bar and Section 1.2 Less Than or Equal to (a or D) Less Than or Equal to 125 mm with Peripheral Coarse Grain Control. 3 pp.

AECMA PREN2632-85. Aluminium Alloy 7075-T73511 Bar and Section 1.2 Less Than or Equal to (a or D) Less Than or Equal to 100 mm with Peripheral Coarse Grain Control. 3 pp.

AECMA PREN2633-85. Aluminium Alloy 2024-T3511 Bar and Section 1.2 Less Than or Equal to (a or D) Less Than or Equal to 150 mm with Peripheral Coarse Grain Control. 3 pp.

AECMA PREN2634-85. Aluminium Alloy 2014-T4511 Bar and Section 1.2 Less Than or Equal to (a or D) Less Than or Equal to 200 mm with Peripheral Coarse Grain Control. 3 pp.

AECMA PREN2635-85. Aluminium Alloy 2014A-T6511 Bar and Section 1.2 Less Than or Equal to (a or D) Less Than or Equal to 150 mm with Peripheral Coarse Grain Control. 3 pp.

AECMA PREN2636-85. Aluminium Alloy 6082-T6 Bar and Section 1.2 Less Than or Equal to (a or D) Less Than or Equal to 200 mm with Peripheral Coarse Grain Control. 3 pp.

AECMA PREN2637-86. Aluminium Alloy 7075-T73 Extruded Bar and Section 1.2 Less Than or Equal to (a or D) Less Than or Equal to 100 mm with Coarse Peripheral Grain Control. 3 pp.

AECMA PREN2638-86. Aluminium Alloy 2024-T3 Extruded Bar and Section 1.2 Less Than or Equal to (a or D) Less Than or Equal to 150 mm with Coarse Peripheral Grain Control. 3 pp.

AECMA PREN2639-86. Aluminium Alloy 2014A-T6 Extruded Bar and Section 1.2 Less Than or Equal to (a or D) Less Than or Equal to 150 mm with Coarse Peripheral Grain Control. 3 pp.

AECMA PREN2640-85. Aluminium Alloy 2017A-T4 Extruded Bar and Section 1.2 Less Than or Equal to (a or D) Less Than or Equal to 200 mm with Peripheral Coarse Grain Control. 3 pp.

Aluminum Alloys (Cont.)

—Bars—Extruded—Aerospace (Cont.)

AECMA PREN2655-85. Aluminium Alloy 2017A-T42 Extruded Bar and Section 1.2 Less Than or Equal to (a or D) Less Than or Equal to 150 mm with Peripheral Coarse Grain Control. 3 pp.

AECMA PREN2702-86. Aluminium Alloy 6061-T6 Extruded Bar and Section 1.2 Less Than or Equal to (a or D) Less Than or Equal to 150 mm. 3 pp.

AECMA PREN2706-85. Aluminium Alloy 7009-T736510 Bar and Section 1.2 Less Than or Equal to (a or D) Less Than or Equal to 125 mm with Peripheral Coarse Grain Control. 3 pp.

AECMA PREN2707-85. Aluminium Alloy 7075-T6510 Bar and Section 1.2 Less Than or Equal to (a or D) Less Than or Equal to 125 mm with Peripheral Coarse Grain Control. 3 pp.

AECMA PREN2708-85. Aluminium Alloy 7075-T73510 Bar and Section 1.2 Less Than or Equal to (a or D) Less Than or Equal to 100 mm with Peripheral Coarse Grain Control. 3 pp.

AECMA PREN2709-85. Aluminium Alloy 2024-T3510 Bar and Section 1.2 Less Than or Equal to (a or D) Less Than or Equal to 200 mm. 3 pp.

AECMA PREN2710-85. Aluminium Alloy 2014A-T4510 Bar and Section 1.2 Less Than or Equal to (a or D) Less Than or Equal to 200 mm with Peripheral Coarse Grain Control. 3 pp.

AECMA PREN2711-85. Aluminium Alloy 2014A-T6510 Bar and Section 1.2 Less Than or Equal to (a or D) Less Than or Equal to 150 mm with Peripheral Coarse Grain Control. 3 pp.

BSI BS EN 2070: Part 3-91. 1991 Aluminium and Aluminium Alloy Wrought Products. Technical Specification Part 3: Bar and Section. 18 pp.

BSI BS EN 2100-92. 1992 Aluminium Alloy AL-P2014A-T4511 Extruded Bars and Sections a or D Less Than or Equal to 200 mm (Supersedes BS L 102: 1971). 10 pp.

BSI BS EN 2127-92. 1992 Aluminium Alloy AL-P7075-T73511 Extruded Bars and Sections a or D Less Than or Equal to 100 mm (Supersedes BS L 160: 1976). 10 pp.

BSI BS EN 2318-92. 1992 Aluminium Alloy AL-P2024-T3511 Extruded Bars and Sections 1,2 mm Less Than or Equal to a or D Less Than or Equal to 150 mm (S). 10 pp.

BSI BS EN 2326-92. 1992 Aluminium Alloy AL-P6082-T6 Extruded Bars and Sections a or D Less Than or Equal to 200 mm (Supersedes BS L 111: 1971). 10 pp.

BSI BS EN 2384-92. 1992 Aluminium Alloy AL-P2014A-T6511 Extruded Bars and Sections a or D Less Than or Equal to 150 mm. 10 pp.

BSI BS EN 2385-92. 1992 Aluminium Alloy AL-P7009-T74511 Extruded Bars and Sections a or D Less Than or Equal to 125 mm. 10 pp.

CEN EN 2070-3-89. Aerospace Series Aluminium and Aluminium Alloy Wrought Products Technical Specification Part 3: Bar and Section. 13 pp.

CEN PREN 2100-91. Aluminium Alloy AL-P2014A T4511 Extruded Bars and Sections a or D Less Than or Equal to 200 mm. 6 pp.

CEN EN 2100-92. Aerospace Series Aluminium Alloy AL-P2014A T4511 Extruded Bars and Sections a or D Less Than or Equal to 200 mm. 5 pp.

CEN PREN 2127-91. Aluminium Alloy AL-P7075-T73511 Extruded Bars and Sections a or D Less Than or Equal to 100 mm. 6 pp.

CEN EN 2127-92. Aluminium Alloy AL-P7075-T73511 Extruded Bars and Sections a or D Less Than or Equal to 100 mm. 5 pp.

CEN PREN 2318-91. Aluminium Alloy AL-P2024-T3511 Extruded Bars and Sections 1,2 mm Less Than or Equal to a or D Less Than or Equal to 150 mm. 6 pp.

CEN EN 2318-92. Aluminium Alloy AL-P2024-T3511 Extruded Bars and Sections 1,2 mm Less Than or Equal to a or D Less Than or Equal to 150 mm. 5 pp.

CEN PREN 2326-91. Aluminium Alloy AL-P6082-T6 Extruded Bars and Sections a or D Less Than or Equal to 200 mm. 5 pp.

CEN EN 2326-92. Aluminium Alloy AL-P6082-T6 Extruded Bars and Sections a or D Less Than or Equal to 200 mm. 5 pp.

CEN PREN 2384-91. Aluminium Alloy AL-P2014A T6511 Extruded Bars and Sections a or D Less Than or Equal to 150 mm. 6 pp.

CEN EN 2384-92. Aluminium Alloy AL-P2014A T6511 Extruded Bars and Sections a or D Less Than or Equal to 150 mm. 5 pp.

CEN PREN 2385-91. Aluminium Alloy AL-P7009-T74511 Extruded Bars and Sections a or D Less Than or Equal to 125 mm. 6 pp.

CEN EN 2385-92. Aluminium Alloy AL-P7009-T74511 Extruded Bars and Sections a or D Less Than or Equal to 125 mm. 5 pp.

CEN EN 2630-91. Aluminium Alloy AL-P7009-T74511 Extruded Bars and Sections a or D Less Than or Equal to 125 mm with Peripheral Coarse Grain Control. 5 pp.

INTERNATIONAL AND NON-U.S. NATIONAL STANDARDS
SUBJECT INDEX
Aluminum

Aluminum Alloys *(Cont.)*

—Bars—Extruded—Aerospace *(Cont.)*

CEN EN 2632-91. Aluminium Alloy AL-P7075-T73511 Extruded Bars and Sections a or D Less Than or Equal to 100 mm with Peripheral Coarse Grain Control. 5 pp.

CEN EN 2633-91. Aluminium Alloy AL-P2024-T3511 Extruded Bars and Sections 1,2 Less Than or Equal to a or D Less Than or Equal to 150 mm with Peripheral Coarse Grain Control. 5 pp.

CEN EN 2636-91. Aluminium Alloy AL-P6082-T6 Extruded Bars and Sections a or D Less Than or Equal to 200 mm with Peripheral Coarse Grain Control. 5 pp.

—Bars—Extruded—Chemical Analysis

DIN ENGL 1725 Pt 1-83. Aluminium Alloys; Wrought Alloys (Feb). 8 pp.

—Bars—Extruded—Hexagonal

CEN PREN 755-6-92. Wrought Aluminium and Aluminium Alloys—Extruded Rod/Bar, Tube and Profile—Part 6: Extruded Hexagonal Bars—Dimension and Form Tolerances. 7 pp.

DIN ENGL 59700-86. Wrought Aluminium and Aluminium Alloy Extruded Square Bars; Dimensions, Dimensional Deviations and Form Tolerances (Aug). 6 pp.

DIN ENGL 59701-86. Wrought Aluminium and Aluminium Alloy Extruded Hexagonal Bars; Dimensions, Dimensional Deviations and Form Tolerances (Aug). 6 pp.

ISO 6362 Pt 5-91. Wrought Aluminium and Aluminium Alloy Extruded Rods/Bars, Tubes and Profiles—Part 5: Extruded Round, Square and Hexagonal Bars—Tolerances on Form and Dimensions First Edition; (Replaces 7273). 7 pp.

—Bars—Extruded—Mechanical Properties

CEN PREN 755-2-92. Aluminium and Aluminium Alloys—Wrought Products—Extruded Rod/Bar, Tube and Profile—Part 2: Mechanical Properties. 56 pp.

DIN ENGL 1747 Pt 1-83. Wrought Aluminium and Aluminium Alloy Rod and Bar; Properties (Feb). 4 pp.

ISO 6362 Pt 2-90. Wrought Aluminium and Aluminium Alloy Extruded Rods/Bars, Tubes and Profiles—Part 2: Mechanical Properties Second Edition. 15 pp.

—Bars—Extruded—Rectangular

CEN PREN 755-5-92. Wrought Aluminium and Aluminium Alloys—Extruded Rod/Bar, Tube and Profile—Part 5: Extruded Rectangular Bars, Dimensions and Form Tolerances. 10 pp.

DIN ENGL 1770-87. Wrought Aluminium and Aluminium Alloy Extruded Rectangular Bars; Dimensions, Dimensional Deviations and Form Tolerances, Static Values (Jan). 9 pp.

ISO 6362 Pt 3-90. Wrought Aluminium and Aluminium Alloy Extruded Rods/Bars, Tubes and Profiles—Part 3: Extruded Rectangular Bars—Tolerances on Dimensions and Form First Edition. 8 pp.

—Bars—Extruded—Rectangular—Aerospace

AECMA PREN2341-80. Aluminium and Aluminium Alloy Square and Rectangular Extruded Bars—Dimensions (C6/A03-04). 9 pp.

—Bars—Extruded—Round

CEN PREN 755-3-92. Wrought Aluminium and Aluminium Alloys—Extruded Rod/Bar, Tube and Profile—Part 3: Extruded Round Bars, Dimensions and Form Tolerances. 5 pp.

DIN ENGL 1799-86. Wrought Aluminium and Aluminium Alloy Extruded Round Bars; Dimensions, Dimensional Deviations and Form Tolerances (Sept). 4 pp.

ISO 6362 Pt 5-91. Wrought Aluminium and Aluminium Alloy Extruded Rods/Bars, Tubes and Profiles—Part 5: Extruded Round, Square and Hexagonal Bars—Tolerances on Form and Dimensions First Edition; (Replaces 7273). 7 pp.

—Bars—Extruded—Round—Aerospace

AECMA PREN2134-89. Round Aluminium Alloy Bars—Extruded—Dimensions (C6/A03-02). 6 pp.

CEN PREN2134-76. Round Aluminium Alloy Bars—Extruded Dimensions. 5 pp.

—Bars—Extruded—Square

CEN PREN 755-4-92. Wrought Aluminium and Aluminium Alloys—Extruded Rod/Bar, Tube and Profile—Part 4: Extruded Square Bars—Dimension and Form Tolerances. 8 pp.

DIN ENGL 59700-86. Wrought Aluminium and Aluminium Alloy Extruded Square Bars; Dimensions, Dimensional Deviations and Form Tolerances (Aug). 6 pp.

Aluminum Alloys *(Cont.)*

—Bars—Extruded—Square *(Cont.)*

ISO 6362 Pt 5-91. Wrought Aluminium and Aluminium Alloy Extruded Rods/Bars, Tubes and Profiles—Part 5: Extruded Round, Square and Hexagonal Bars—Tolerances on Form and Dimensions First Edition; (Replaces 7273). 7 pp.

—Bars—Extruded—Square—Aerospace

AECMA PREN2341-80. Aluminium and Aluminium Alloy Square and Rectangular Extruded Bars—Dimensions (C6/A03-04). 9 pp.

—Bars—Forged—Aerospace

AECMA PREN2086-77. Aluminium Alloy 2618A-T851 (AL-P11-T851)—Forged Bars and Slabs A Less Than or Equal to 150 mm (C5/40). 3 pp.

AECMA PREN2256-77. Aluminium Alloy 2618A-T852 (AL-P11-T852)—Forged Bars and Slabs A Less Than or Equal to 150 mm (C5/40). 3 pp.

—Bars—Rivets

DIN ENGL 59675-57. High Grade Aluminium and Aluminium Wrought Alloy; Wire and Bars for Rivets; Drawn (May). 2 pp.

—Bars—Rolled—Aerospace

AECMA PREN2070-03-84. Aluminium and Aluminium Alloy Wrought Products—Part 3—Bar and Section (CS/44). 13 pp.

BSI BS EN 2070: Part 3-91. 1991 Aluminium and Aluminium Alloy Wrought Products. Technical Specification Part 3: Bar and Section. 18 pp.

CEN EN 2070-3-89. Aerospace Series Aluminium and Aluminium Alloy Wrought Products Technical Specification Part 3: Bar and Section. 13 pp.

—Bismuth Content

JIS H 1364-71. Methods for Determination of Bismuth and Lead in Aluminium Alloy.

—Bolts

BSI BS 1473-72. 1972 Amd 2 Wrought Aluminium and Aluminium Alloys for General Engineering Purposes; Rivet, Bolt and Screw Stock. 25 pp.

—Channels—Aerospace

DIN ENGL LN 9413-86. Aerospace; Wrought Aluminium Alloy Folded Profiles, Channel Section; Dimensions, Static Values, Masses (Dec). 5 pp.

DIN ENGL LN 9414-86. Aerospace; Wrought Aluminium Alloy Folded Profiles, Channel Section with Internally Lipped Flanges; Dimensions, Static Values, Masses (Dec). 5 pp.

—Channels—Extruded

DIN ENGL 9713-81. Aluminium and Wrought Aluminium Alloy Extruded Channel Sections; Dimensions; Static Values (Sept). 4 pp.

—Channels—Extruded—Aerospace

AECMA PREN2049-76. Channel Section Aluminium Alloy—Extrusions—Dimensions (C6/A04-03). 4 pp.

—Chemical Analysis

CNS H2026-84. Method of Chemical Analysis of Aluminium and Aluminium Alloy (May)(2069).

JIS H 1351-72. General Rules for Chemical Analysis of Aluminium and Aluminium Alloy.

—Chromium Content

JIS H 1358-72. Methods for Determination of Chromium in Aluminium Alloy.

—Chromium Content—Atomic Absorption Spectrometry

ISO 4193-81. Aluminium and Aluminium Alloys—Determination of Chromium Content—Flame Atomic Absorption Spectrometric Method First Edition. 5 pp.

—Chromium Content—Photometry

BSI BS 1728: Part 16-68. 1968 Methods for the Analysis of Aluminium and Aluminium Alloys: Part 16: Chromium (Photometric Method). 10 pp.

—Chromium Content—Spectrophotometry

ISO 3978-76. Aluminium and Aluminium Alloys—Determination of Chromium—Spectrophotometric Method Using Diphenylcarbazide, After Extraction First Edition. 6 pp.

—Chromium Content—Volumetric Analysis

BSI BS 1728: Part 17-68. 1968 Methods for the Analysis of Aluminium and Aluminium Alloys: Part 17: Chromium (Volumetric Method). 8 pp.

—Circle—Cold Rolled

CEN PREN 851-92. Aluminium and Aluminium Alloys—Wrought Products—Circle and Circle Stock for Culinary Utensil Applications. 11 pp.

Aluminum Alloys *(Cont.)*

—Circle—Hot Rolled

CEN PREN 851-92. Aluminium and Aluminium Alloys—Wrought Products—Circle and Circle Stock for Culinary Utensil Applications. 11 pp.

—Cold Worked—Heat Treated—Sheets—Aerospace

BSI L 163-78. 1978 Sheet and Strip of Aluminium-Coated Aluminium-Copper-Magnesium-Silicon-Manganese Alloy (Cu 4.4, Mg 0.5, Si 0.8, Mn 0.8). 3 pp.

—Cold Worked—Heat Treated—Strips—Aerospace

BSI L 163-78. 1978 Sheet and Strip of Aluminium-Coated Aluminium-Copper-Magnesium-Silicon-Manganese Alloy (Cu 4.4, Mg 0.5, Si 0.8, Mn 0.8). 3 pp.

—Copper Content

BSI BS 1728: Part 1-51. 1951 Amd 1 Methods for the Analysis of Aluminium and Aluminium Alloys Part 1: Determination of Copper. 9 pp.

JIS H 1354-72. Methods for Determination of Copper in Aluminium and Aluminium Alloy.

—Copper Content—Absorptiometric Analysis

BSI BS 1728: Part 5-53. 1953 Amd 1 Methods for the Analysis of Aluminium and Aluminium Alloys Part 5: Determination of Copper (Absorptiometric Method). 8 pp.

—Copper Content—Atomic Absorption Spectrophotometry

BSI BS 1728: Part 23-76. 1976 Methods for the Analysis of Aluminium and Aluminium Alloys: Part 23: Copper (Atomic Absorption Method). 4 pp.

ISO 3980-77. Aluminium and Aluminium Alloys—Determination of Copper—Atomic Absorption Spectrophotometric Method First Edition. 5 pp.

—Copper Content—Electrolytic Analysis

ISO 796-73. Aluminium Alloys—Determination of Copper—Electrolytic Method First Edition. 7 pp.

—Copper Content—Photometry

ISO 795-76. Aluminium and Aluminium Alloys—Determination of Copper Content—Oxalyldihydrazide Photometric Method First Edition. 6 pp.

—Extruded

JIS H 4100-88. Aluminium and Aluminium Alloy Extruded Shapes. 28 pp.

—Extruded—Aerospace—Ultrasonic Testing

CEN PREN2004-02-75. Test Methods for Aluminium and Aluminium Alloy Products—Part 2—Ultrasonic Testing of Plates Forgings and Extrusions (C5/21A). 7 pp.

—Extruded—Heat Treated—Aerospace

MOD UK DTD-5014A-71. Bars and Extruded Sections of Aluminium—Copper—Magnesium—Nickel—Iron Alloy (Solution Treated and Precipitation Treated) (Suitable for Use at Elevated Temperatures) (Cu 2.5, Mg 1.5, Ni 1.2, Fe 1.0). 2 pp.

—Extruded—Ultrasonic Testing

MOD UK DTD-936-66. Ultrasonic Inspection of Aluminium Alloy Extrusions and Hand Forgings. 3 pp.

—Forgings

BSI BS 1472-72. 1972 Amd 2 Wrought Aluminium and Aluminium Alloys for General Engineering Purposes; Forging Stock. 36 pp.

BSI BS 4300/11-69. 1969 Amd 1 Specification (Supplementary Series) for Wrought Aluminium and Aluminium Alloys for General Engineering Purposes /11: 5454 Forging Stock and Forgings. 30 pp.

CAA Chapter B8-5 11.72. Forgings. 3 pp.

CNS H3114-82. Aluminium and Aluminium Alloy Forgings (Oct)(9612).

DIN ENGL 1749 Pt 2-73. Aluminium Drop Forgings; (Highest Grade Aluminium, High Grade Aluminium and Wrought Aluminium Alloys); Technical Conditions of Delivery (Nov). 3 pp.

DIN ENGL 1749 Pt 3-74. Aluminium Drop Forgings; (Highest Grade Aluminium, High Grade Aluminium and Wrought Aluminium Alloys); Design Principles (Jan). 7 pp.

DIN ENGL 1749 Pt 4-74. Aluminium Drop Forgings; (Highest Grade Aluminium, High Grade Aluminium and Wrought Aluminium Alloys); Permissible Variations (Jan). 5 pp.

DIN ENGL 17606 Pt 2-73. Open Die Forgings of Wrought Aluminium Alloys; Technical Conditions of Delivery (Nov). 3 pp.

DIN ENGL 17606 Pt 3-73. Open Die Forgings of Wrought Aluminium Alloys; Design Principles (Dec). 2 pp.

INDUSTRY STANDARDS

INTERNATIONAL AND NON-U.S. NATIONAL STANDARDS
SUBJECT INDEX

Aluminum

Aluminum Alloys (Cont.)

—Forgings (Cont.)
DIN ENGL 17606 Pt 4-73. Open Die Forgings of Wrought Aluminium Alloys; Permissible Variations (Dec). 2 pp.
JIS H 4140-88. Aluminium and Aluminium Alloy Forgings. 26 pp.

—Forgings—Aerospace
AECMA PREN2082-01-86. Aluminium Alloy Forging Stock and Forgings—Technical Specification—Part 1—General Requirements. 10 pp.
AECMA PREN2082-02-86. Aluminium Alloy Forging Stock and Forgings—Technical Specification—Part 2—Forging Stock. 6 pp.
AECMA PREN2082-03-86. Aluminium Alloy Forging Stock and Forgings—Technical Specification—Part 3—Pre-Production and Production Forgings. 10 pp.
AECMA PREN2085-80. Aluminium Alloy 2618A-T6 Forgings—A Less Than or Equal to 150 mm (C5/40). 3 pp.
AECMA PREN2093-80. Aluminium Alloy 7009-T736—Hand Forgings—A Less Than or Equal to 150 mm (C5/40). 3 pp.
AECMA PREN2094-82. Aluminium Alloy 7009 T736—Die Forgings A Less Than or Equal to 150 mm (C5/40). 3 pp.
AECMA PREN2380-83. Aluminium Alloy 7075-T73 Forgings A Less Than or Equal to 125 mm. 3 pp.
AECMA PREN2381-83. Aluminium Alloy 7009-T73652 Hand Forgings A Less Than or Equal to 150 mm. 3 pp.
AECMA PREN2382-82. Aluminium Alloy 2214-T6—Forgings—A Less Than or Equal to 100 mm (C5/40). 3 pp.
AECMA PREN2383-82. Aluminium Alloy 2214-T4—Forgings—A Less Than or Equal to 100 mm (C5/40). 3 pp.
AECMA PREN2386-82. Aluminium Alloy 7075-T7352—Hand Forgings—A Less Than or Equal to 150 mm (C5/40). 3 pp.
AECMA PREN2681-84. Aluminium Alloy 7010-T736—Die Forgings a Less Than or Equal to 150 mm. 4 pp.
AECMA PREN2682-84. Aluminium Alloy 7010-T73652—Forgings—50 Less Than or Equal to a Less Than or Equal to 150 mm. 4 pp.
AECMA PREN2683-84. Aluminium Alloy 7010-T7651—Hand Forgings—80 Less Than or Equal to a Less Than or Equal to 160 mm. 4 pp.
AECMA PREN2685-84. Aluminium Alloy 7010-T7652—Forgings—80 Less Than or Equal to a Less Than or Equal to 160 mm (C5/40). 4 pp.
AECMA PREN2686-84. Aluminium Alloy 7010-T73651—Hand Forgings—50 Less Than or Equal to a Less Than or Equal to 150 mm (C5/40). 4 pp.
AECMA PREN2688-84. Aluminium Alloy 7050-T736—Die Forgings a Less Than or Equal to 150 mm (C5/40). 4 pp.
AECMA PREN2690-84. Aluminium Alloy 7050-T73652—Hand Forgings—a Less Than or Equal to 125 mm (C5/40). 4 pp.
BSI BS EN 2082: Part 1-91. 1991 Aluminium Alloy Forging Stock and Forgings. Technical Specification Part 1: General Requirements. 13 pp.
BSI BS EN 2082: Part 1-01. 1991 Amd 1 Aluminium Alloy Forging Stock and Forgings Technical Specification Part 1: General Requirements (AMD 7707) October 15, 1993 (S). 19 pp.
BSI BS EN 2082: Part 2-91. 1991 Aluminium Alloy Forging Stock and Forgings. Technical Specification Part 2: Forging Stock. 9 pp.
BSI BS EN 2082: Part 3-91. 1991 Aluminium Alloy Forging Stock and Forgings. Technical Specification Part 3: Pre-Production and Production Forgings. 13 pp.
BSI BS EN 2093-92. 1992 Aluminium Alloy AL-P7009-T74 Hand Forgings 20 mm Less Than or Equal to a Less Than or Equal to 150 mm. 10 pp.
BSI BS EN 2094-92. 1992 Aluminium Alloy AL-P7009-T74 Die Forgings 3 mm Less Than or Equal to a Less Than or Equal to 150 mm. 10 pp.
BSI BS EN 2381-92. 1992 Aluminium Alloy AL-P7009-T7452 Hand Forgings 40 mm Less Than or Equal to a Less Than or Equal to 150 mm. 10 pp.
CEN EN 2082-1-89. Aluminium Alloy Forging Stock and Forgings Technical Specification Part 1: General Requirements. 10 pp.
CEN EN 2082 Part 1-89. Aerospace Series Aluminium Alloy Forging Stock and Forgings Technical Specification Part 1: General Requirements. 12 pp.
CEN EN 2082-2-89. Aerospace Series Aluminium Alloy Forging Stock and Forgings Technical Specification Part 2: Forging Stock. 6 pp.
CEN EN 2082-3-89. Aerospace Series Aluminium Alloy Forging Stock and Forgings Technical Specification Part 3: Pre-Production and Production Forgings. 10 pp.
CEN PREN 2093-91. Aluminium Alloy AL-P7009-T74 Hand Forgings 20 mm Less Than or Equal to a Less Than or Equal to 150 mm. 6 pp.

Aluminum Alloys (Cont.)

—Forgings—Aerospace (Cont.)
CEN EN 2093-92. Aluminium Alloy AL-P7009-T74 Hand Forgings 20 mm Less Than or Equal to a Less Than or Equal to 150 mm. 5 pp.
CEN PREN 2094-91. Aluminium Alloy AL-P7009-T74 Die Forgings 3 mm Less Than or Equal to a Less Than or Equal to 150 mm. 6 pp.
CEN EN 2094-92. Aluminium Alloy AL-P7009-T74 Die Forgings 3 mm Less Than or Equal to a Less Than or Equal to 150 mm. 5 pp.
CEN PREN 2381-91. Aluminium Alloy AL-P7009-T7452 Hand Forgings 40 mm Less Than or Equal to a Less Than or Equal to 150 mm. 6 pp.
CEN EN 2381-92. Aluminium Alloy AL-P7009-T7452 Hand Forgings 40 mm Less Than or Equal to a Less Than or Equal to 150 mm. 5 pp.
MOD UK DTD-246C-71. Forging Stock and Crankcase Forgings of Aluminium—Copper—Nickel—Magnesium—Iron—Silicon Alloy (Solution Treated and Precipitation Treated) (Cu 2.0, Ni 1.0, Mg 1.0, Si 0.9, Fe 0.6). 2 pp.
MOD UK DTD-324B-71. Forging Stock and Forgings for Engine Cylinders and Pistons of Aluminium—Silicon—Magnesium—Copper—Nickel Alloy (Solution Treated and Precipitation Treated) (Si 11.5, Mg 1.1, Cu 1.0, Ni 1.0). 2 pp.
MOD UK DTD-717A-71. Forging Stock and Forgings of Aluminium-Copper-Magnesium-Nickel-Iron Alloy (Solution Treated and Precipitation Treated at 170 Degrees C) (Cu 2.5, Mg 1.5, Ni 1.2, Fe 1.0) (Suitable for Use at Elevated Temperatures). 2 pp.
MOD UK DTD-745A-71. Forging Stock and Compressor Blade Forgings of Aluminium—Copper—Magnesium—Nickel—Iron Alloy (Solution Treated and Precipitation Treated) (Cu 2.5, Mg 1.5, Ni 1.2, Fe 1.0). 2 pp.

—Forgings—Aerospace—Macrographic Inspection
CEN PREN 2715-90. Macrographic Examination of Aluminium and Aluminium Alloy Forging Stock, Forgings and Wrought Products. 4 pp.

—Forgings—Aerospace—Ultrasonic Testing
CEN PREN2004-02-75. Test Methods for Aluminium and Aluminium Alloy Products—Part 2—Ultrasonic Testing of Plates Forgings and Extrusions (C5/21A). 7 pp.

—Forgings—Aircraft
CAA Chapter K3-12 10.69. Forgings. 2 pp.
CAA Chapter P3-12. Forgings (Provisional Airworthiness Requirements for Civil Powered Lift Aircraft). 2 pp.
CAA Chapter Q3-12 12.79. Forgings. 2 pp.

—Forgings—Cast—Aerospace
AECMA PREN2485-81. Aluminium Alloy 2214-F—Cast or Extruded Forging Stocks (C5/40). 3 pp.
AECMA PREN2486-81. Aluminium Alloy 2618 A-F—Cast or Extruded Forging Stock (C5/40). 3 pp.
AECMA PREN2487-81. Aluminium Alloy 7009-F—Cast or Extruded Forging Stock (C5/40). 3 pp.
AECMA PREN2488-81. Aluminium Alloy 7075-F—Cast or Extruded Forging Stocks (C5/40). 3 pp.

—Forgings—Extruded—Aerospace
AECMA PREN2485-81. Aluminium Alloy 2214-F—Cast or Extruded Forging Stocks (C5/40). 3 pp.
AECMA PREN2486-81. Aluminium Alloy 2618 A-F—Cast or Extruded Forging Stock (C5/40). 3 pp.
AECMA PREN2487-81. Aluminium Alloy 7009-F—Cast or Extruded Forging Stock (C5/40). 3 pp.
AECMA PREN2488-81. Aluminium Alloy 7075-F—Cast or Extruded Forging Stocks (C5/40). 3 pp.

—Forgings—Mechanical Properties
CEN PREN 586-2-91. Forgings in Aluminium and Aluminium Alloys—Part 2: Mechanical Properties and Additional Property Requirements. 12 pp.
DIN ENGL 1749 Pt 1-76. Drop Forgings of Aluminium and Wrought Aluminium Alloys; Strength Properties (Dec). 2 pp.
DIN ENGL 17606 Pt 1-76. Open Die Forgings of Wrought Aluminium Alloys; Strength Properties (Dec). 2 pp.

—Forgings—Ultrasonic Testing
MOD UK DTD-936-66. Ultrasonic Inspection of Aluminium Alloy Extrusions and Hand Forgings. 3 pp.

—Forgings—Wrought
CEN PREN 603-1-91. Wrought Forging Stock in Aluminium and Aluminium Alloys—Part 1: Technical Conditions for Inspection and Delivery. 12 pp.
CEN PREN 605-1-91. Wrought Aluminium and Aluminium Alloy Forgings —Part 1: Technical Conditions for Inspection and Delivery. 11 pp.

Aluminum Alloys (Cont.)

—Forgings—Wrought—Aerospace
AECMA PREN2070-07-87. Aluminium and Aluminium Alloy Wrought Products—Technical Specification—Part 7—Wrought Forging Stock. 6 pp.
BSI BS EN 2070: Part 7-91. 1991 Aluminium and Aluminium Alloy Wrought Products Technical Specification Part 7: Wrought Forging Stock. 9 pp.
CEN EN 2070-7-89. Aerospace Series Aluminium and Aluminium Alloy Wrought Products Technical Specification Part 7: Wrought Forging Stock. 6 pp.

—Forgings—Wrought—Mechanical Properties
CEN PREN 603-2-91. Wrought Forging Stock in Aluminium and Aluminium Alloys—Part 2: Mechanical Properties. 10 pp.

—Heat Treated—Bars—Aerospace
BSI L 102-71. (OBSOLESCENT) 1971 Bars and Extruded Sections of Aluminium-Copper-Magnesium-Silicon-Manganese Alloy (Solution Treated and Aged at Room Temp.) (Not Exceeding 200 mm Diameter or Minor Sectional Dimension) (Cu 4.4, Mg 0.5, Si 0.7, Mn 0.8). 4 pp.
BSI L 111-71. (OBSOLESCENT) 1971 Bars and Extruded Sections of Aluminium-Magnesium-Manganese Alloy (Solution Treated and Precipitation Treated) (Replace by BS EN 2326: 1992). 3 pp.
BSI L 160-76. (OBSOLESCENT) 1976 AMD 1 Bars and Extruded Sections of Aluminium-Zinc-Magnesium-Copper-Chromium Alloy (Solution Treated and Artificially Aged to an Overaged Condition) (Zn 5.6, Mg 2.5, Cu 1.6, Cr 0.22). 5 pp.
BSI L 168-78. 1978 Bars and Extruded Sections of Aluminium-Copper-Magnesium-Silicon-Manganese Alloy (Cu 4.4, Mg 0.5, Si 0.8, Mn 0.8). 3 pp.
BSI L 168-01. 1978 Amd 1 Bars and Extruded Sections of Aluminium-Copper-Magnesium-Silicon-Manganese Alloy (Solution Treated and Artificially Aged) (Not Exceeding 200 mm Diameter or Minor Sectional Dimension) (Cu 4.4, Mg 0.5, Si 0.7, Mn 0.8) (AMD 7876) 8-15-93. 4 pp.
MOD UK DTD-5014A-71. Bars and Extruded Sections of Aluminium—Copper—Magnesium—Nickel—Iron Alloy (Solution Treated and Precipitation Treated) (Suitable for Use at Elevated Temperatures) (Cu 2.5, Mg 1.5, Ni 1.2, Fe 1.0). 2 pp.

—Heat Treated—Bars—Drawn—Aerospace
AECMA PREN2700-88. Aluminium Alloy (6061) Solution Treated, Water Quench and Artificially Aged (T6) Drawn Bar 6Less Than or Equal to D Less Than or Equal to 75 mm with Coarse Peripheral Grain Control. 4 pp.

—Heat Treated—Bars—Drawn—Aerospace—Mechanical Properties
AECMA PREN3342-88. Aluminium Alloy (6061) Solution Treated, Water Quenched and Aged (T4) Drawn Bar and Section 10 Less Than or Equal to (A or D) Less Than or Equal to 150 mm. 4 pp.

—Heat Treated—Bars—Extruded—Aerospace
AECMA PREN2697-88. Aluminium Alloy (2214) Solution Treated, Water Quench and Artificially Aged (T6) Extruded Bar and Section 1,2 Less Than or Equal to (a or D) Less Than or Equal to 100 mm with Coarse Peripheral Grain Control. 4 pp.
AECMA PREN2698-88. Aluminium Alloy (7075) Solution Treated, Water Quench, Controlled Stretched and Artificially Aged (T6510) Extruded Bar and Section 1, 2 Less Than or Equal to (A or D) Less Than or Equal to 100 mm. 4 pp.

—Heat Treated—Bars—Extruded—Aerospace—Mechanical Properties
AECMA PREN3337-88. Aluminium Alloy (7010) Solution Treated and Artificially Aged (T74 511) Extruded Bars and Sections (A or D) Less Than or Equal to 130 mm with Peripheral Coarse Grain Control. 4 pp.
AECMA PREN3338-88. Aluminium Alloy (7050) Solution Treated and Artificially Aged (T74511) Extruded Bars and Sections (A or D) Less Than or Equal to 130 mm with Peripheral Coarse Grain Control. 4 pp.
AECMA PREN3343-88. Aluminium Alloy (7010) Solution Treated and Artificially Aged (T76511) Extruded Bars and Sections 1 Less Than (A or D) Less Than or Equal to 130 mm with Peripheral Coarse Grain Control. 4 pp.
AECMA PREN3344-88. Aluminium Alloy (7050) Solution Treated and Artificially Aged (T76511) Extruded Bars and Sections (A or D) Less Than or Equal to 130 mm with Peripheral Coarse Grain Control. 4 pp.

Aluminum Alloys *(Cont.)*

—Heat Treated—Bars—Extruded—Aerospace—Mechanical Properties *(Cont.)*

AECMA PREN3345-88. Aluminium Alloy (7150) Solution Treated, Water Quenched Controlled Stretched and Artificially Aged (T6511) Extruded Bar and Section (A or D) Less Than or Equal to 90 mm with Peripheral Coarse Grain Control. 4 pp.

AECMA PREN3347-88. Aluminium Alloy (2024) Solution Treated, Stretched and Artificially Aged (T8511) Extruded Bars and Sections (A or D) Less Than or Equal to 150mm with Peripheral Coarse Grain Control. 4 pp.

AECMA PREN3550-88. Aluminium Alloy (2024) Solution Treated, Stretched and Artificially Aged (T8511) Extruded Bars and Sections (A or D) Less Than or Equal to 150 mm. 4 pp.

AECMA PREN3551-88. Aluminium Alloy (7150) Solution Treated, Straightened and Artificially Aged (T6511) Extruded Bars and Sections (A or D) Less Than or Equal to 90 mm. 4 pp.

AECMA PREN3553-88. Aluminium Alloy (2618A) Solution Treated, Stretched and Artificially Aged (T6511) Extruded Bars and Sections 1,2 Less Than or Equal to (a or d) Less Than or Equal to 100 mm. 4 pp.

AECMA PREN3555-88. Aluminium Alloy (7075) Solution Treated, Controlled Stretched and Artificially Aged (T79510) Extruded Bars and Sections 1,2 Less Than (a or D) Less Than or Equal to 100 mm with Peripheral Coarse Grain Control. 4 pp.

—Heat Treated—Bars—Hexagonal—Aerospace

BSI 2L 87-71. 1971 Hexagonal Bars for Nuts, Couplings and Hollow Machined Parts of Al-Cu-Mg-Si-Mn Alloy (Solution and Precipita-tion Treated) (Free from Peripheral and Asymmetric Coarse Grain) (MW 14 mm, MW 36mm Across Flats) (Cu 4.4, Mg 0.5, Si 0.7, Mn 0.8). 12 pp.

—Heat Treated—Forgings—Aerospace

BSI 2L 77-71. 1971 Forging Stock and Forgings of Aluminium-Copper-Magnesium-Silicon-Manganese Alloy (Solution Treated and Precipitation Treated) (Cu 4.4, Mg 0.5, Si 0.7, Mn 0.8). 4 pp.

BSI L 103-71. 1971 Forging Stock and Forgings of Aluminium-Copper-Magnesium-Silicon-Manganese Alloy (Solution Treated and Aged at Room Temperature) (Cu 4.4, Mg 0.5, Si 0.7, Mn 0.8). 3 pp.

BSI L 112-71. 1971 Forging Stock and Forgings Aluminium-Magnesium-Silicon-Manganese Alloy (Solution Treated and Precipitation Treated). 3 pp.

BSI L 161-76. 1976 Hand and Die Forgings of Aluminium-Zinc-Magnesium-Copper-Chromium Alloy (Solution Treated and Artificially Aged to an Overaged Condition) (Zn 5.6, Mg 2.5, Cu 1.6, Cr 0.22). 4 pp.

BSI L 162-76. 1976 Cold Compressed Hand Forgings of Aluminium-Zinc-Magnesium-Copper-Chromium Alloy (Solution Treated and Artificially Aged to an Overaged Condition) (Zn 5.6, Mg 2.5, Cu 1.6, Cr 0.22). 4 pp.

BSI L 171-90. 1990 Forging of Aluminium-Zinc-Magnesium-Manganese-Copper Alloy (Supplied As-Forged or Annealed for Subsequent Heat Treatment) (Not Exceeding 150mm Diameter or Minor Sectional Dimension) (Zn 5.7, Mg 2.7, Mn 0.5, Cu 0.5) (7014). 3 pp.

BSI L 171-01. 1990 Amd 1 Forgings of Aluminium-Zinc-Magnesium-Manganese-Copper Alloy (Supplied As-Forged or Annealed for Subsequent Heat Treatment) (Not Exceeding 150mm Diameter or Minor Sectional Dimension) (Zn 5.7, Mg 2.7, Mn 0.5, Cu 0.5) (7014). 4 pp.

MOD UK DTD-735B-71. Ingots and Castings of Aluminium-Silicon-Magnesium Alloy (Solution Treated and Precipitation Treated) (Si 5, Mg 0.5). 2 pp.

MOD UK DTD-5004A-71. Forging Stock and Forgings of Aluminium-Copper-Manganese Alloy (Solution Treated and Precipitation Treated) (Suitable for Use at Elevated Temperatures) (Cu 6.0, Mn 0.3). 2 pp.

MOD UK DTD-5084A-01. Forging Stock and Forgings of Aluminium-Copper-Magnesium-Nickel-Iron Alloy (Solution Treated and Precipitation Treated; Special Heat Treatment for Minimum Residual Internal Stress) (Suitable for Use at Elevated Temperatures) (Cu 2.5, Mg 1.5, Ni 1.2,. 3 pp.

—Heat Treated—Forgings—Aerospace—Mechanical Properties

AECMA PREN3339-88. Aluminium Alloy (7010) Solution Treated and Artificially Aged (T76) Die Forgings a Less Than or Equal to 200 mm. 5 pp.

AECMA PREN3340-88. Aluminium Alloy (7050) Solution Treated and Artificially Aged (T76) Die Forgings a Less Than or Equal to 200 mm. 5 pp.

Aluminum Alloys *(Cont.)*

—Heat Treated—Forgings—Aerospace—Mechanical Properties *(Cont.)*

AECMA PREN3554-88. Aluminium Alloy (7010) Solution Treated, Cold Compressed and Artificially Aged (T7652) Hand Forgings a Less Than or Equal to 200 mm. 5 pp.

—Heat Treated—Ingots—Aerospace

BSI 4L 35-70. 1970 Amd 1 Ingots and Castings of 'Y' Aluminium Alloy (Heat Treated) (Suitable for Pistons). 3 pp.

BSI 3L 51-70. 1970 Amd 1 Ingots and Castings of Aluminium-Silicon-Copper-Iron-Nickel-Magnesium Alloy (Precipitation Treated). 3 pp.

BSI 3L 78-70. 1970 Amd 1 Ingots and Castings of Aluminium-Silicon-Copper-Magnesium Alloy (Solution Treated and Precipitation Treated). 3 pp.

BSI 2L 99-72. 1972 Amd 2 Ingots and Castings of Aluminium-Silicon-Magnesium Alloy (Solution Treated and Precipitation Treated) (Si 7, Mg 0.3) (AMD 6543) March 28, 1991. 6 pp.

BSI L 119-75. 1975 Amd 2 Ingots and Castings of Aluminium-Copper-Nickel-Manganese-Titanium-Zirconium-Cobalt-Antimony Alloy (Solution Treated and Art. Aged) (Cu 5.0, Ni 1.5, Mn 0.25, Ti0.2, Zr 0.2, Co 0.2, Sb 0.2) (AMD 6544) December 21, 1990. 6 pp.

BSI L 154-77. 1977 Amd 2 Ingots and Castings of Aluminium-Copper-Silicon Alloy (Solution Treated and Aged at Room Temperature) (Cu 4, Si 1) (AMD 6546) April 30, 1991. 4 pp.

BSI L 155-77. 1977 Amd 2 Ingots and Castings of Aluminium-Copper-Silcon Alloy (Solution Treated and Artificially Aged) (Cu 4, Si 1) (AMD 6545) March 28, 1991. 5 pp.

BSI L 169-86. 1986 Amd 1 Ingots and Castings of Aluminium-Silicon-Magnesium Alloy (Solution Treated and Artificially Aged) (Si 7, Mg 0.6) (AMD 6547) December 21, 1990. 6 pp.

—Heat Treated—Plates—Aerospace

BSI 2L 93-71. 1971 Plate of Aluminium-Copper-Magnesium-Silicon-Manganese Alloy (Solution Treated, Controlled Stretched and Precipitation Treated) (Cu 4.4, Mg 0.5, Si 0.7, Mn 0.8). 4 pp.

BSI 2L 93-01. 1971 Amd 1 Plate of Aluminium-Copper-Magnesium-Silicon -Manganese Alloy (Solution Treated, Controlled Stretched and Precipitation Treated) (Cu 4.4, Mg 0.5, Si 0.7, Mn 0.85) (AMD 6841) November 29, 1991. 5 pp.

BSI 2L 95-71. 1971 Plate of Aluminium-Zinc-Magnesium-Copper-Chromium Alloy (Solution Treated, Controlled Stretched and Precipitation Treated) (Zn 5.8, Mg 2.5, Cu 1.6, Cr 0.15). 3 pp.

BSI 2L 95-01. 1971 Amd 1 Plate of Aluminium-Zinc-Magnesium-Copper-Chromium Alloy (Solution Treated, Controlled Stretched and Precipitation Treated) (Zn 5.8, Mg 2.5, Cu 1.6, Cr 0.15) (AMD 6842) November 29, 1991. 4 pp.

BSI 2L 97-71. 1971 Plate of Aluminium-Copper-Magnesium-Manganese Alloy (Solution Treated, Controlled Stretched and Aged at Room Temperature) (Cu 4.4, Mg 1.5, Mn 0.6). 4 pp.

BSI 2L 97-01. 1971 Amd 1 Plate of Aluminium-Copper-Magnesium-Manganese Alloy (Solution Treated, Controlled Stretched and Aged at Room Temperature) (Cu 4.4, Mg 1.5, Mn 0.6) (AMD 6843) November 29, 1991. 5 pp.

BSI L 115-71. 1971 Plate of Aluminium-Magnesium-Silicon-Manganese Alloy (Solution Treated, Controlled Stretched and Precipitation Treated). 3 pp.

BSI L 115-01. 1971 Amd 1 Plate of Aluminium-Magnesium-Silicon-Manganese Alloy (Solution Treated, Controlled Stretched and Precipitation Treated) (Not Exceeding 25 mm Thickness) (Suitable for Welding) (Mg 0.8, Si 1, Mn 0.7) (AMD 6844) Nov. 29, 1991. 4 pp.

MOD UK DTD-5010A-71. Plate of Aluminium—Copper—Magnesium—Silicon—Manganese Alloy (Solution Treated and Aged at Room Temperature) (Cu 4.4, Mg 0.5, Si 0.7, Mn 0.8). 2 pp.

MOD UK DTD-5030A-71. Aluminium-Coated Plate of Aluminium-Copper-Magnesium-Silicon-Manganese Alloy (Solution Treated and Aged at Room Temperature) (Cu 4.4, Mg 0.5, Si 0.7, Mn 0.8). 2 pp.

MOD UK DTD-5040A-71. Aluminium-Coated Plate of Aluminium—Copper—Magnesium—Silicon—Manganese Alloy (Solution Treated and Precipitation Treated) (Cu 4.4, Mg 0.5, Si 0.7, Mn 0.8). 2 pp.

MOD UK DTD-5100A-71. Aluminium-Coated Plate of Aluminium—Copper—Magnesium—Manganese Alloy (Solution Treated, Controlled Stretched and Aged at Room Temperature) (Cu 4.4, Mg 1.5, Mn 0.6). 2 pp.

—Heat Treated—Plates—Aerospace—Mechanical Properties

AECMA PREN3334-88. Aluminium Alloy (7050) Solution Treated, Controlled Stretched and Artificially Aged (T7651) Plate 6 Less Than or Equal to Less Than or Equal to 60 mm. 4 pp.

Aluminum Alloys *(Cont.)*

—Heat Treated—Plates—Aerospace—Mechanical Properties *(Cont.)*

AECMA PREN3336-88. Aluminium Alloy (7150) Solution Treated, Stretched and Artificially Aged (T651) Plate 6 Less Than a Less Than or Equal to 40 mm. 4 pp.

AECMA PREN3348-88. Aluminium Alloy (2024) Solution Treated and Artificially Aged (T62) Plate 6 Less Than or Equal to a Less Than or Equal to 50 mm. 4 pp.

—Heat Treated—Sections—Drawn—Aerospace—Mechanical Properties

AECMA PREN3342-88. Aluminium Alloy (6061) Solution Treated, Water Quenched and Aged (T4) Drawn Bar and Section 10 Less Than or Equal to (A or D) Less Than or Equal to 150 mm. 4 pp.

—Heat Treated—Sections—Extruded—Aerospace

AECMA PREN2697-88. Aluminium Alloy (2214) Solution Treated, Water Quench and Artificially Aged (T6) Extruded Bar and Section 1,2 Less Than or Equal to (a or D) Less Than or Equal to 100 mm with Coarse Peripheral Grain Control. 4 pp.

AECMA PREN2698-88. Aluminium Alloy (7075) Solution Treated, Water Quench, Controlled Stretched and Artificially Aged (T6510) Extruded Bar and Section 1, 2 Less Than or Equal to (A or D) Less Than or Equal to 100 mm. 4 pp.

AECMA PREN3337-88. Aluminium Alloy (7010) Solution Treated and Artificially Aged (T74 511) Extruded Bars and Sections (A or D) Less Than or Equal to 130 mm with Peripheral Coarse Grain Control. 4 pp.

AECMA PREN3338-88. Aluminium Alloy (7050) Solution Treated and Artificially Aged (T74511) Extruded Bars and Sections (A or D) Less Than or Equal to 130 mm with Peripheral Coarse Grain Control. 4 pp.

AECMA PREN3343-88. Aluminium Alloy (7010) Solution Treated and Artificially Aged (T76511) Extruded Bars and Sections 1 Less Than (A or D) Less Than or Equal to 130 mm with Peripheral Coarse Grain Control. 4 pp.

AECMA PREN3344-88. Aluminium Alloy (7050) Solution Treated and Artificially Aged (T76511) Extruded Bars and Sections (A or D) Less Than or Equal to 130 mm with Peripheral Coarse Grain Control. 4 pp.

AECMA PREN3345-88. Aluminium Alloy (7150) Solution Treated, Water Quenched Controlled Stretched and Artificially Aged (T6511) Extruded Bar and Section (A or D) Less Than or Equal to 90 mm with Peripheral Coarse Grain Control. 4 pp.

AECMA PREN3347-88. Aluminium Alloy (2024) Solution Treated, Stretched and Artificially Aged (T8511) Extruded Bars and Sections (A or D) Less Than or Equal to 150mm with Peripheral Coarse Grain Control. 4 pp.

AECMA PREN3550-88. Aluminium Alloy (2024) Solution Treated, Stretched and Artificially Aged (T8511) Extruded Bars and Sections (A or D) Less Than or Equal to 150 mm. 4 pp.

AECMA PREN3551-88. Aluminium Alloy (7150) Solution Treated, Straightened and Artificially Aged (T6511) Extruded Bars and Sections (A or D) Less Than or Equal to 90 mm. 4 pp.

AECMA PREN3553-88. Aluminium Alloy (2618A) Solution Treated, Stretched and Artificially Aged (T6511) Extruded Bars and Sections 1,2 Less Than or Equal to (a or d) Less Than or Equal to 100 mm. 4 pp.

AECMA PREN3555-88. Aluminium Alloy (7075) Solution Treated, Controlled Stretched and Artificially Aged (T79510) Extruded Bars and Sections 1,2 Less Than (a or D) Less Than or Equal to 100 mm with Peripheral Coarse Grain Control. 4 pp.

BSI L 102-71. (OBSOLESCENT) 1971 Bars and Extruded Sections of Aluminium-Copper-Magnesium-Silicon-Manganese Alloy (Solution Treated and Aged at Room Temp.) (Not Exceeding 200 mm Diameter or Minor Sectional Dimension) (Cu 4.4, Mg 0.5, Si 0.7, Mn 0.8). 4 pp.

BSI L 111-71. (OBSOLESCENT) 1971 Bars and Extruded Sections of Aluminium-Magnesium-Manganese Alloy (Solution Treated and Precipitation Treated) (Replace by BS EN 2326: 1992). 3 pp.

BSI L 160-76. (OBSOLESCENT) 1976 AMD 1 Bars and Extruded Sections of Aluminium-Zinc-Magnesium-Copper-Chromium Alloy (Solution Treated and Artificially Aged to an Overaged Condition) (Zn 5.6, Mg 2.5, Cu 1.6, Cr 0.22). 5 pp.

BSI L 168-78. 1978 Bars and Extruded Sections of Aluminium-Copper-Magnesium-Silicon-Manganese Alloy (Cu 4.4, Mg 0.5, Si 0.8, Mn 0.8). 3 pp.

INDUSTRY STANDARDS
133

Aluminum Alloys (Cont.)

—Heat Treated—Sections—Extruded—Aerospace (Cont.)

BSI L 168-01. 1978 Amd 1 Bars and Extruded Sections of Aluminium-Copper-Magnesium-Silicon-Manganese Alloy (Solution Treated and Artificially Aged) (Not Exceeding 200 mm Diameter or Minor Sectional Dimension) (Cu 4.4, Mg 0.5, Si 0.7, Mn 0.8) (AMD 7876) 8-15-93. 4 pp.

—Heat Treated—Sheets—Aerospace

BSI L 113-71. 1971 Sheet and Strip of Aluminium-Magnesium-Silicon-Manganese Alloy (Solution Treated and Precipitation Treated). 3 pp.

BSI L 156-78. 1978 Amd 1 Sheet and Strip of Aluminium-Copper-Magnesium-Silicon-Manganese Alloy. 4 pp.

BSI L 157-78. 1978 Sheet and Strip of Aluminium-Copper-Magnesium-Silicon-Manganese Alloy. 3 pp.

BSI L 158-78. 1978 Close Toleranced Sheet and Strip of Aluminium-Copper-Magnesium-Silicon-Manganese Alloy (Solution Treated and Aged at Room Temperature) (Cu 4.4, Mg 0.5, Si 0.8, Mn 0.8). 3 pp.

BSI L 159-78. 1978 Close Toleranced Sheet and Strip of Aluminium-Copper-Magnesium-Silicon-Manganese Alloy (Cu 4.4, Mg 0.5, Si 0.8, Mn 0.8). 3 pp.

—Heat Treated—Sheets—Aerospace—Mechanical Properties

AECMA PREN3341-88. Aluminium Alloy (6061) Solution Treated, Water Quenched and Aged (T4) Sheet and Strip 0,4 Less Than or Equal to a Less Than or Equal to 6 mm. 4 pp.

—Heat Treated—Sheets—Rolled—Aerospace

AECMA PREN2691-88. Aluminium Alloy (2017A) Solution Treated, Water Quench, Coldworked and Naturally Aged (T3) Sheet and Strip 0,4 Less Than or Equal to a Less Than or Equal to 6 mm. 4 pp.

—Heat Treated—Sheets—Rolled—Aerospace—Mechanical Properties

AECMA PREN3332-88. Aluminium Alloy (7475) Solution Treated and Artificially Aged (T762) Clad Sheet and Strip 0,8 Less Than or Equal to a Less Than or Equal to 6 mm. 4 pp.

AECMA PREN3333-88. Aluminium Alloy (7475) Solution Treated and Artificially Aged (T762) Sheet-Base 0,8 Less Than or Equal to a Less Than or Equal to 6 mm. 4 pp.

AECMA PREN3335-88. Aluminium Alloy (7475) Solution Treated and Artificially Aged (T762) Sheet for Super-Plastic Forming 0,8 Less Than or Equal to a Less Than or Equal to 6 mm. 4 pp.

AECMA PREN3552-88. Aluminium Alloy (2618A) Solution Treated, Straightened and Artificially Aged (T6) Clad Sheet and Strip 0,4 Less Than or Equal to a Less Than or Equal to 6 mm. 4 pp.

—Heat Treated—Strips—Aerospace

BSI L 113-71. 1971 Sheet and Strip of Aluminium-Magnesium-Silicon-Manganese Alloy (Solution Treated and Precipitation Treated). 3 pp.

BSI L 156-78. 1978 Amd 1 Sheet and Strip of Aluminium-Copper-Magnesium-Silicon-Manganese Alloy. 4 pp.

BSI L 157-78. 1978 Sheet and Strip of Aluminium-Copper-Magnesium-Silicon-Manganese Alloy. 3 pp.

BSI L 158-78. 1978 Close Toleranced Sheet and Strip of Aluminium-Copper-Magnesium-Silicon-Manganese Alloy (Solution Treated and Aged at Room Temperature) (Cu 4.4, Mg 0.5, Si 0.8, Mn 0.8). 3 pp.

BSI L 159-78. 1978 Close Toleranced Sheet and Strip of Aluminium-Copper-Magnesium-Silicon-Manganese Alloy (Cu 4.4, Mg 0.5, Si 0.8, Mn 0.8). 3 pp.

—Heat Treated—Strips—Aerospace—Mechanical Properties

AECMA PREN3341-88. Aluminium Alloy (6061) Solution Treated, Water Quenched and Aged (T4) Sheet and Strip 0,4 Less Than or Equal to a Less Than or Equal to 6 mm. 4 pp.

—Heat Treated—Strips—Rolled—Aerospace

AECMA PREN2691-88. Aluminium Alloy (2017A) Solution Treated, Water Quench, Coldworked and Naturally Aged (T3) Sheet and Strip 0,4 Less Than or Equal to a Less Than or Equal to 6 mm. 4 pp.

—Heat Treated—Strips—Rolled—Aerospace—Mechanical Properties

AECMA PREN3332-88. Aluminium Alloy (7475) Solution Treated and Artificially Aged (T762) Clad Sheet and Strip 0,8 Less Than or Equal to a Less Than or Equal to 6 mm. 4 pp.

AECMA PREN3333-88. Aluminium Alloy (7475) Solution Treated and Artificially Aged (T762) Sheet-Base 0,8 Less Than or Equal to a Less Than or Equal to 6 mm. 4 pp.

Aluminum Alloys (Cont.)

—Heat Treated—Strips—Rolled—Aerospace—Mechanical Properties (Cont.)

AECMA PREN3552-88. Aluminium Alloy (2618A) Solution Treated, Straightened and Artificially Aged (T6) Clad Sheet and Strip 0,4 Less Than or Equal to a Less Than or Equal to 6 mm. 4 pp.

—Heat Treated—Tubes—Aerospace

BSI 3L 63-71. 1971 Tube of Aluminium-Copper-Magnesium-Silicon-Manganese Alloy (Solution Treated and Precipitation Treated (Cu 4.4, Mg 0.5, Si 0.7, Mn 0.8). 3 pp.

BSI L 105-71. 1971 Tube of Aluminium-Copper-Magnesium-Silicon-Manganese Alloy (Solution Treated and Aged at Room Temperature) (Not Exceeding 10 mm Wall Thickness) (Cu 4.4, Mg 0.5, Si 0.7, Mn 0.8). 3 pp.

BSI L 114-71. 1971 Tube of Aluminium-Magnesium-Silicon-Manganese Alloy (Solution Treated and Precipitation Treated). 3 pp.

BSI L 117-75. 1975 Amd 1 Tube of Aluminium-Magnesium-Silicon-Copper-Chromium Alloy (Solution Treated and Artificially Aged: Not Tested Hydraulically) (Not Exceeding 10 mm Thickness) (Mg 1.0, Si 0.6, Cu 0.28, Cr 0.2). 5 pp.

BSI L 118-75. 1975 Amd 2 Tube of Aluminium-Magnesium-Silicon-Copper-Chromium Alloy (Solution Treated and Artificially Aged: Tested Hydraulically) (Not Exceeding 10 mm Thickness) Mg 1.0, Si 0.6, Cu 0.28, Cr 0.2). 7 pp.

—Heat Treated—Tubes—Drawn—Aerospace

AECMA PREN2701-88. Aluminium Alloy (2024) Solution Treated, Water Quench, Cold Worked and Naturally Aged (T3) Drawn Tube for Structures 0,6 Less Than or Equal to a Less Than or Equal to 12,5 mm. 4 pp.

—Heat Treated—Tubes—Drawn—Aerospace—Mechanical Properties

AECMA PREN3346-88. Aluminium Alloy (2014A) Solution Treated, Straightened and Naturally Aged T3 Tube for Structures 0,6 Less Than or Equal to a Less Than or Equal to 12,5 mm. 4 pp.

—Heat Treatment—Aircraft

JIS W 1103-85. Heat Treatment of Aluminum Alloys for Aircraft. 68 pp.

—Hollow Sections—Hexagonal—Drawn

DIN ENGL 59751-73. Tubes and Hollow Hexagon Sections of Wrought Aluminium Alloys for Free-Cutting Machining on Automatics; Seamless Drawn; Dimensions (Nov). 7 pp.

—Hydrogen Peroxide

MOD UK DEF-60-58. Selection and Treatment of Aluminium Base Materials for Use with Concentrated Hydrogen Peroxide (H.T.P.) (Reprinted August, 1973) (Withdrawn). 6 pp.

—I Beams—Extruded

DIN ENGL 9712-69. Aluminium and Magnesium Beams; Extruded; Dimensions; Static Values (Aug). 3 pp.

—Ingots

BSI BS 1490-88. 1988 Amd 1 Aluminium and Aluminium Alloy Ingots and Castings for General Engineering Purposes (V) (AMD 6309) March 30, 1990. 27 pp.

BSI BS 1490-02. 1988 Amd 2 Aluminium and Aluminium Alloy Ingots and Castings for General Engineering Purposes (AMD 6605) December 21, 1990. 29 pp.

SAA AS 1874-88. Aluminium and Aluminium Alloys—Ingots and Castings. 14 pp.

SNZ NZS/AS 1874-88. Aluminium and Aluminium Alloys—Ingots and Castings (This is a Joint Standard with SAA AS 1874). 14 pp.

—Ingots—Aerospace

AECMA PREN2076-01-87. Aluminium and Magnesium Alloy Ingots and Castings—Technical Specification—Part 1—General Requirements. 10 pp.

BSI BS EN 2076: Part 1-91. 1991 Aluminium and Magnesium Alloy Ingots and Castings. Technical Specification Part 1: General Requirements. 13 pp.

BSI BS EN 2076: Part 1-01. 1991 Amd 1 Aluminium and Magnesium Alloy Ingots and Castings Technical Specification Part 1: General Requirements (AMD 7706) October 15, 1993 (S). 19 pp.

BSI 3L 101-70. (WITHDRAWN) 1970 Amd 1 Procedure Inspection and Testing of Aluminium-Base and Magnesium-Base Ingots and Castings (Superseded by 4L 101: 1990). 15 pp.

BSI 4L 101-90. 1990 Procedure for Inspection, Testing and Acceptance of Aluminium-Base and Magnesium-Base Ingots and Castings. 21 pp.

CEN EN 2076-1-89. Aerospace Series Aluminium and Magnesium Alloy Ingots and Castings Technical Specification Part 1: General Requirements. 10 pp.

Aluminum Alloys (Cont.)

—Ingots—Aerospace (Cont.)

CEN EN 2076 Part 1-89. Aerospace Series Aluminium and Magnesium Alloy Ingots and Castings Technical Specification Part 1: General Requirements. 12 pp.

—Ingots—Castings—Aerospace

AECMA PREN2076-02-87. Aluminium and Magnesium Alloy Ingots and Castings—Technical Specification—Part 2—Ingots for Remelting. 4 pp.

BSI BS EN 2076: Part 2-91. 1991 Aluminium and Magnesium Alloy Ingots and Castings. Technical Specification Part 2: Ingots for Remelting. 7 pp.

CEN EN 2076-2-89. Aerospace Series Aluminium and Magnesium Alloy Ingots and Castings Technical Specification Part 2: Ingots for Remitting. 4 pp.

—Ingots—Chemical Analysis

DIN ENGL 1725 Pt 5-86. Aluminium Alloys; Casting Alloys; Ingots (Pigs); Liquid Metal; Composition (Feb). 6 pp.

—Ingots—Extrusion

CEN PREN 486-91. Aluminium and Aluminium Alloys—Extrusions Ingots Cast in a Form Suitable for Extruding. 11 pp.

—Ingots—Rolling

CEN PREN 487-91. Aluminium and Aluminium Alloys—Rolling Ingots Obtained by Semi-Continuous Casting. 17 pp.

—Iron Content

JIS H 1353-72. Methods for Determination of Iron in Aluminium and Aluminium Alloy.

—Iron Content—Absorptiometric Analysis

BSI 1728: Part 8-57. 1957 Methods for the Analysis of Aluminium and Aluminium Alloys Part 8: Determination of Iron (Absorptiometric 1:10-Phenanthroline Method). 8 pp.

—Iron Content—Photometry

ISO 793-73. Aluminium and Aluminium Alloys—Determination of Iron—Orthophenanthroline Photometric Method First Edition. 6 pp.

—Iron Content—Volumetric Analysis

BSI 1728: Part 6-55. 1955 Methods for the Analysis of Aluminium and Aluminium Alloys Part 6: Determination of Iron (Volumetric: Titanous Chloride Method). 8 pp.

—Lead Content

JIS H 1364-71. Methods for Determination of Bismuth and Lead in Aluminium Alloy.

—Lead Content—Atomic Absorption Spectrometry

BSI 1728: Part 20-71. 1971 Methods for the Analysis of Aluminium and Aluminium Alloys: Part 20: Lead (Atomic Absorption Method). 9 pp.

ISO 4192-81. Aluminium and Aluminium Alloys—Determination of Lead Content—Flame Atomic Absorption Spectrometric Method First Edition. 5 pp.

—Liquid Metals

CEN PREN 577-91. Aluminium and Aluminium Alloys—Liquid Metal. 8 pp.

—Liquid Metals—Alloying

DIN ENGL 1725 Pt 5 Suppl. 1-86. Aluminium Alloys; Casting Alloys; Ingots (Pigs); Liquid Metal; Composition; Information on Alloying Processes (Feb). 2 pp.

—Liquid Metals—Chemical Analysis

DIN ENGL 1725 Pt 5-86. Aluminium Alloys; Casting Alloys; Ingots (Pigs); Liquid Metal; Composition (Feb). 6 pp.

—Magnesium Content

JIS H 1357-72. Methods for Determination of Magnesium in Aluminium Alloy.

—Magnesium Content—Atomic Absorption Spectrophotometry

BSI 1728: Part 19-71. 1971 Methods for the Analysis of Aluminium and Aluminium Alloys: Part 19: Magnesium (Atomic Absorption Method). 9 pp.

ISO 3256-77. Aluminium and Aluminium Alloys—Determination of Magnesium—Atomic Absorption Spectrophotometric Method First Edition. 6 pp.

—Magnesium Content—Complexometric Titrations

ISO 2297-73. Chemical Analysis of Aluminium and Its Alloys—Complexometric Determination of Magnesium First Edition. 7 pp.

Aluminum Alloys (Cont.)

—Magnesium Content—Volumetric Analysis
BSI BS 1728: Part 22-72. 1972 Methods for the Analysis of Aluminium and Aluminium Alloys: Part 22: Magnesium (Volumetric CDTA Method). 13 pp.

—Manganese Content
JIS H 1355-72. Methods for Determination of Manganese in Aluminium and Aluminium Alloy.

—Manganese Content—Absorptiometric Analysis
BSI BS 1728: Part 10-57. 1957 Methods for the Analysis of Aluminium and Aluminium Alloys: Part 10: Determination of Manganese (Absorptiometric Method). 8 pp.

—Manganese Content—Photometry
ISO 886-73. Aluminium and Aluminium Alloys—Determination of Manganese—Photometric Method (Manganese Content Between 0,005 and 1,5 %) First Edition. 7 pp.

—Manganese Content—Volumetric Analysis
BSI BS 1728: Part 9-57. 1957 Methods for the Analysis of Aluminium and Aluminium Alloys Part 9: Determination of Manganese (Volumetric: Arsenite/Nitrite Method). 8 pp.

—Master—Chemical Analysis
CEN PREN 575-91. Aluminium and Aluminium Alloys—Master Alloys. 14 pp.
DIN ENGL 1725 Pt 3-73. Aluminium Alloys; Master Alloys (June). 3 pp.

—Nickel Content
JIS H 1360-72. Methods for Determination of Nickel in Aluminium Alloy.

—Nickel Content—Atomic Absorption Spectrophotometry
BSI BS 1728: Part 24-76. 1976 Methods for the Analysis of Aluminium and Aluminium Alloys: Part 24: Nickel (Atomic Absorption Method). 4 pp.
ISO 3981-77. Aluminium and Aluminium Alloys—Determination of Nickel—Atomic Absorption Spectrophotometric Method First Edition. 5 pp.

—Nickel Content—Gravimetric Analysis
BSI BS 1728: Part 14-65. 1965 Methods for the Analysis of Aluminium and Aluminium Alloys: Part 14: Determination of Nickel (Gravimetric Method). 8 pp.

—Nickel Content—Photometry
BSI BS 1728: Part 15-66. 1966 Methods for the Analysis of Aluminium and Aluminium Alloys: Part 15: Nickel (Photometric Method). 8 pp.

—Nickel Content—Spectrophotometry
ISO 3979-77. Aluminium and Aluminium Alloys—Determination of Nickel—Spectrophotometric Method Using Dimethylglyoxime First Edition. 7 pp.

—Plates
BSI BS 1470-87. 1987 Amd 1 Wrought Aluminium and Aluminium Alloys for General Engineering Purposes; Plate, Sheet and Strip (AMD 6032) May 31, 1989. 24 pp.
BSI BS 4300/14-73. 1973 Amd 1 Specification (Supplementary Series) for Wrought Aluminium and Aluminium Alloys for General Engineering Purposes /14: 7020 Plate, Sheet and Strip. 20 pp.
CEN PREN 485 (Part 1)-91. Wrought Aluminium and Aluminium Alloy Sheets, Strips and Plates—Part 1: Technical Conditions for Inspection and Delivery. 16 pp.
CNS H2029-87. Method of Test for Aluminium and Aluminium Alloy Sheets and Plate (Jun)(2254).
CNS H3025-87. Aluminium and Aluminium Alloy Sheets and Plates (Feb)(2253).
JIS H 4000-88. Aluminium and Aluminium Alloy Sheets and Plates, Strips and Coiled Sheets. 145 pp.
SAA AS 1734-86. Aluminium and Aluminium Alloys—Flat Sheet, Coiled Sheet and Plate (This is a Joint Standard with SANZ NZS 1734:1986). 20 pp.
SNZ NZS/AS 1734-86. Aluminium and Aluminium Alloys—Flat Sheet, Coiled Sheet and Plate (This is a Joint Standard with SAA AS 1734). 20 pp.

—Plates—Aerospace
AECMA PREN2070-02-84. Aluminium and Aluminium Alloy Wrought Products—Part 2—Sheet, Strip Formed Profiles and Plate (CS/44). 14 pp.
AECMA PREN2131-89. Plates in Aluminium Alloys Thickness 8 mm Less Than or Equal to A Less Than or Equal to 140 mm—Dimensions. 9 pp.
BSI BS EN 2070: Part 2-91. 1991 Aluminium and Aluminium Alloy Wrought Products. Technical Specification Part 2: Sheet, Strip Formed Profiles and Plate. 19 pp.

Aluminum Alloys (Cont.)

—Plates—Aerospace (Cont.)
BSI BS EN 2126-92. 1992 Aluminium Alloy AL-P7075-T651 Plate 6 mm Less Than a Less Than or Equal to 80 mm. 10 pp.
CEN EN 2070-2-89. Aerospace Series Aluminium and Aluminium Alloy Wrought Products Technical Specification Part 2: Sheet, Strip Formed Profiles and Plate. 14 pp.
CEN PREN 2126-91. Aluminium Alloy AL-P7075-T651 Plate 6 mm < a Less Than or Equal to 80 mm. 6 pp.
CEN EN 2126-92. Aluminium Alloy AL-P7075-T651 Plates 6 mm Less Than a Less Than or Equal to 80 mm. 5 pp.
CEN PREN6025-90. Plates Close-Tolerance Flatness Aluminium Alloy. 8 pp.
CEN PREN 6025-91. Plates Close-Tolerance Flatness Aluminium Alloys. 10 pp.
DIN ENGL LN 9074-88. Aerospace; Wrought Aluminium Alloy Sheet and Plate; Skin Quality; Rolled; Dimensions; Masses (Aug). 8 pp.

—Plates—Aerospace—Ultrasonic Testing
CEN PREN2004-02-75. Test Methods for Aluminium and Aluminium Alloy Products—Part 2—Ultrasonic Testing of Plates Forgings and Extrusions (C5/21A). 7 pp.

—Plates—Aircraft
EURO CTI/80/10834-80. Study on the Supply of Aircraft Quality Thick Aluminium. 8 pp.

—Plates—Chemical Machining
MOD UK DTD-932-58. Chemical Contouring of Aluminium and Aluminium Alloy Sheet and Plate (Reprinted January 1965). 2 pp.

—Plates—Embossed
DIN ENGL 59605-79. Embossed Plate and Sheet of Wrought Aluminium Alloys (Mar). 5 pp.

—Plates—Marine
MOD UK NES 831: Part 1-89. Requirements for Aluminium Alloy Products Held by Naval Stores Part 1: Sheet, Strip and Plate (Including Perforated Sheet) Issue 1 (03.89). 15 pp.
MOD UK NES 831: Part 1-01. Requirements for Aluminium Alloy Products Held by Naval Stores Part 1: Sheet, Strip and Plate (Including Perforated Sheet) Issue 1 (03.89); Amendment 1. 18 pp.

—Plates—Rolled
CEN PREN 485 (Part 3)-91. Wrought Aluminium and Aluminium Alloy Sheets, Strips and Plates—Part 3: Hot-Rolled Sheets, Strips and Plates—Tolerances on Shape and Dimensions. 11 pp.
CEN PREN 485 (Part 4)-91. Wrought Aluminium and Aluminium Alloy Sheets, Strips and Plates—Part 4: Cold-Rolled Sheets, Strips and Plates—Tolerances on Shape and Dimensions. 14 pp.
CEN PREN 683-1-92. Aluminium and Aluminium Alloys—Wrought Products, Finstock—Part 1: Technical Conditions for Inspection and Delivery. 10 pp.
CEN PREN 941-92. Aluminium and Aluminium Alloys—Wrought Products—Circle and Circle Stock for General Applications. 11 pp.
DIN ENGL 1745 Pt 2-83. Wrought Aluminium and Aluminium Alloy Plate, Sheet and Strip Greater Than 0,35 mm in Thickness; Technical Delivery Conditions (Feb). 4 pp.
DIN ENGL 1783-81. Strips, Plates and Sheets of Aluminium and Wrought Aluminium Alloys with Thicknesses over 0,35 mm, Cold-Rolled; Dimensions (Apr). 9 pp.
DIN ENGL 59600-81. Strips, Plates and Sheets of Aluminium and Wrought Aluminium Alloys; Hot Rolled; Dimensions (Apr). 6 pp.
ISO 6361 Pt 1-86. Wrought Aluminium and Aluminium Alloys Sheets, Strips and Plates—Part 1: Technical Conditions for Inspection and Delivery First Edition; (Supersedes 5191). 5 pp.
ISO 6361 Pt 4-88. Wrought Aluminium and Aluminium Alloys Sheets, Strips and Plates—Part 4: Sheets and Plates—Tolerances on Shape and Dimensions First Edition; (Supersedes TR7735). 12 pp.

—Plates—Rolled—Aerospace
AECMA PREN2123-80. Aluminium Alloy 2618A-T851—Plates—A Greater Than 6 mm and Less Than or Equal to 14 mm (C5/40). 3 pp.
AECMA PREN2124-85. Aluminium Alloy 2214-T651 Plate 6 Less Than a Less Than or Equal to 140 mm. 3 pp.
AECMA PREN2125-77. Aluminium Alloy AL-P16-T6151—Plates A Greater Than 6 mm and Less Than or Equal to 120 mm (c5/40). 3 pp.
AECMA PREN2126-85. Aluminium Alloy 7075T651 Plate 6 Less Than a Less Than or Equal to 80 mm. 3 pp.
AECMA PREN2395-80. Aluminium Alloy 2014-T4 or T42 Sheets and Strips. 3 pp.

Aluminum Alloys (Cont.)

—Plates—Rolled—Aerospace (Cont.)
AECMA PREN2419-85. Aluminium Alloy 2024-T351 Plate 6 Less Than a Less Than or Equal to 80 mm. 3 pp.
AECMA PREN2422-81. Aluminium Alloy 2124-T351—Plates A Greater Than 25 mm and Less Than or Equal to 120 mm (C5/40). 3 pp.
AECMA PREN2511-85. Aluminium Alloy 7075-T7351 Plate 6 Less Than a Less Than or Equal to 100 mm. 3 pp.
AECMA PREN2512-85. Aluminium Alloy 7175-T7351 Plate 6 Less Than a Less Than or Equal to 100 mm. 3 pp.
AECMA PREN2626-85. Aluminium Alloy 7475-T7351 Plate 6 Less Than a Less Than or Equal to 100 mm. 3 pp.
AECMA PREN2684-84. Aluminium Alloy 7010-T7651 Plate—6 Less Than a Less Than or Equal to 140 mm. 4 pp.
AECMA PREN2687-84. Aluminium Alloy 7010-T73651 Plate—6 Less Than a Less Than or Equal to 150 mm. 4 pp.
AECMA PREN2689-85. Aluminium Alloy 7050-T73651 Plate 6 Less Than a Less Than or Equal to 150 mm. 3 pp.
AECMA PREN2804-86. Aluminium Alloy 7075-T7651 Plate 6 Less Than A Less Than or Equal to 25 mm. 3 pp.
AECMA PREN2805-86. Aluminium Alloy 7475-T7651 Plate 6 Less Than a Less Than or Equal to 25 mm. 3 pp.

—Plates—Rolled—Mechanical Properties
CEN PREN 683-2-92. Aluminium and Aluminium Alloys—Wrought Products, Finstock—Part 2: Mechanical Properties. 9 pp.
DIN ENGL 1745 Pt 1-83. Wrought Aluminium and Aluminium Alloy Plate, Sheet and Strip Greater Than 0,35 mm in Thickness; Properties (Feb). 6 pp.
ISO 6361 Pt 2-90. Wrought Aluminium and Aluminium Alloy Sheets, Strips and Plates—Part 2: Mechanical Properties Second Edition; (Supersedes TR2136). 17 pp.

—Plates—Ultrasonic Testing
MOD UK DTD-937-66. Ultrasonic Inspection of Aluminium Alloy Plate. 3 pp.

—Plates—Wrought—Mechanical Properties
CEN PREN 485-2-91. Wrought Aluminium and Aluminium Alloy Sheets, Strips and Plates—Part 2 Mechanical Properties. 45 pp.

—Prefinished—Sheets
CGSB CAN/CGSB-93.1-M85. Sheet Aluminum Alloy, Prefinished, Residential. 11 pp.

—Profiles
CEN PREN 755-1-92. Aluminium and Aluminium Alloys—Wrought Products—Extruded Rod/Bar, Tube and Profile—Part 1: Technical Conditions for Inspection and Delivery. 18 pp.
ISO 6362 Pt 4-88. Wrought Aluminium and Aluminium Alloy Extruded Rods/Bars, Tubes and Profiles—Part 4: Extruded Profiles—Tolerances on Shape and Dimensions First Edition. 11 pp.

—Profiles—Aerospace
AECMA PREN2051-76. L-Section Aluminium Alloy Folded Profiles—Dimensions (C6/A05-01). 4 pp.
AECMA PREN2052-76. L-Section Aluminium Alloy Folded Profiles with Internally Lipped Flanges—Dimensions (C6/A05-02). 4 pp.
AECMA PREN2053-76. U-Section Aluminium Alloy Folded Profiles—Dimensions (C6/A05-03). 4 pp.
AECMA PREN2054-76. U-Section Aluminium Alloy Folded Profiles with Externally Lipped Flanges—Dimensions (C6/A05-04). 4 pp.
AECMA PREN2055-76. U-Section Aluminium Alloy Folded Profiles with Internally Lipped Flanges—Dimensions (C6/A05-05). 4 pp.
AECMA PREN2056-76. Z-Section Aluminium Alloy Folded Profiles—Dimensions (C6/A05-06). 4 pp.
AECMA PREN2057-76. Z-Section—Aluminium Alloy Folded Profiles with One Lipped Flange—Dimensions (C6/A05-07). 4 pp.
AECMA PREN2058-76. Z-Section—Aluminium Alloy Folded Profiles with Lipped Flanges—Dimensions (C6/A05-08). 4 pp.
AECMA PREN2059-76. Top Hat Section Aluminium Alloy Folded Profiles—Dimensions (C6/A05-09). 4 pp.
AECMA PREN2060-76. Bowler Hat Section Aluminium Alloy Folded Profiles—Dimensions (C6/A05-10). 4 pp.
AECMA PREN2061-76. Top Hat Section—Aluminium Alloy Folded Profiles with Lipped Flanges—Dimensions (C6/A05-11). 4 pp.
AECMA PREN2065-78. Folded Profiles Aluminium Alloys—General Tolerances (C6/A05-00). 6 pp.
AECMA PREN2070-02-84. Aluminium and Aluminium Alloy Wrought Products—Part 2—Sheet, Strip Formed Profiles and Plate (CS/44). 14 pp.

INTERNATIONAL AND NON-U.S. NATIONAL STANDARDS SUBJECT INDEX

Aluminum

Aluminum Alloys (Cont.)

—Profiles—Aerospace (Cont.)

BSI BS EN 2070: Part 2-91. 1991 Aluminium and Aluminium Alloy Wrought Products. Technical Specification Part 2: Sheet, Strip Formed Profiles and Plate. 19 pp.

CEN EN 2070-2-89. Aerospace Series Aluminium and Aluminium Alloy Wrought Products Technical Specification Part 2: Sheet, Strip Formed Profiles and Plate. 14 pp.

DIN ENGL LN 9415-86. Aerospace; Wrought Aluminium Alloy Folded Profiles, Z-Section with One Lipped Flange; Dimensions, Static Values, Masses (Dec). 6 pp.

DIN ENGL LN 9416-86. Aerospace; Wrought Aluminium Alloy Folded Profiles, LZ-Section; Dimensions, Static Values, Masses (Dec). 5 pp.

—Profiles—Extruded

DIN ENGL 1748 Pt 4-81. Aluminium and Wrought Aluminium Alloy; Extruded Profiles; Permissible Deviations (Nov). 10 pp.

—Profiles—Extruded—Mechanical Properties

CEN PREN 755-2-92. Aluminium and Aluminium Alloys—Wrought Products—Extruded Rod/Bar, Tube and Profile—Part 2: Mechanical Properties. 56 pp.

—Profiles—Mechanical Properties

ISO 6362 Pt 2-90. Wrought Aluminium and Aluminium Alloy Extruded Rods/Bars, Tubes and Profiles—Part 2: Mechanical Properties Second Edition. 15 pp.

—Profiles—Wrought

ISO 6362 Pt 1-86. Wrought Aluminium and Aluminium Alloys Extruded Rods/Bars, Tubes and Profiles—Part 1: Technical Conditions for Inspection and Delivery First Edition; (Supersedes 5191). 5 pp.

—Rods

CNS H3048-84. Aluminium and Aluminium Alloy Rods, Bars and Wires (Sep)(3667).

CNS H3070-88. Conductor Aluminium — Iron Alloy Rod (Dec)(4746).

CNS H3131-84. Aluminium and Aluminium Alloy Rivet Wires and Rods (Sep)(11016).

JIS H 4040-88. Aluminium and Aluminium Alloy Rods, Bars and Wires. 48 pp.

SAA AS 1865-86. Aluminium and Aluminium Alloys—Drawn Wire, Rod, Bar and Strip (This is a Joint Standard with SANZ NZS 1865:1986). 11 pp.

SNZ NZS/AS 1865-86. Aluminium and Aluminium Alloys—Drawn Wire, Rod, Bar and Strip (This is a Joint Standard with SAA AS 1865). 11 pp.

—Rods—Drawn

DIN ENGL 1747 Pt 2-83. Wrought Aluminium and Aluminium Alloy Rod and Bar; Technical Delivery Conditions (Feb). 4 pp.

ISO 6363 Pt 1-88. Wrought Aluminium and Aluminium Alloy Cold-Drawn Rods/Bars and Tubes—Part 1: Technical Conditions for Inspection and Delivery First Edition; (Supersedes 5191). 6 pp.

—Rods—Drawn—Mechanical Properties

DIN ENGL 1747 Pt 1-83. Wrought Aluminium and Aluminium Alloy Rod and Bar; Properties (Feb). 4 pp.

—Rods—Extruded

DIN ENGL 1747 Pt 2-83. Wrought Aluminium and Aluminium Alloy Rod and Bar; Technical Delivery Conditions (Feb). 4 pp.

ISO 6362 Pt 1-86. Wrought Aluminium and Aluminium Alloys Extruded Rods/Bars, Tubes and Profiles—Part 1: Technical Conditions for Inspection and Delivery First Edition; (Supersedes 5191). 5 pp.

ISO 6362 Pt 4-88. Wrought Aluminium and Aluminium Alloy Extruded Rods/Bars, Tubes and Profiles—Part 4: Extruded Profiles—Tolerances on Shape and Dimensions First Edition. 11 pp.

SAA AS 1866-86. Aluminium and Aluminium Alloys—Extruded Rod, Bar, Solid and Hollow Shapes (This is a Joint Standard with SANZ NZS 1866:1986). 45 pp.

SNZ NZS/AS 1866-86. Aluminium and Aluminium Alloys—Extruded Rod, Bar, Solid and Hollow Shapes (This is a Joint Standard with SAA AS 1866). 45 pp.

—Rods—Extruded—Mechanical Properties

DIN ENGL 1747 Pt 1-83. Wrought Aluminium and Aluminium Alloy Rod and Bar; Properties (Feb). 4 pp.

ISO 6362 Pt 2-90. Wrought Aluminium and Aluminium Alloy Extruded Rods/Bars, Tubes and Profiles—Part 2: Mechanical Properties Second Edition. 15 pp.

—Scrap

CNS H3111-82. Method of Classification for Aluminium and Aluminium Alloy Scrap (Sep)(9397).

Aluminum Alloys (Cont.)

—Scrap (Cont.)

JIS H 2119-84. Classification Standard of Aluminium and Aluminium Alloy Scraps. 12 pp.

—Screws

BSI BS 1473-72. 1972 Amd 2 Wrought Aluminium and Aluminium Alloys for General Engineering Purposes; Rivet, Bolt and Screw Stock. 25 pp.

—Sections—Aerospace

AECMA PREN2051-76. L-Section Aluminium Alloy Folded Profiles—Dimensions (C6/A05-01). 4 pp.

AECMA PREN2052-76. L-Section Aluminium Alloy Folded Profiles with Internally Lipped Flanges—Dimensions (C6/A05-02). 4 pp.

AECMA PREN2053-76. U-Section Aluminium Alloy Folded Profiles—Dimensions (C6/A05-03). 4 pp.

AECMA PREN2054-76. U-Section Aluminium Alloy Folded Profiles with Externally Lipped Flanges—Dimensions (C6/A05-04). 4 pp.

AECMA PREN2055-76. U-Section Aluminium Alloy Folded Profiles with Internally Lipped Flanges—Dimensions (C6/A05-05). 4 pp.

AECMA PREN2056-76. Z-Section Aluminium Alloy Folded Profiles—Dimensions (C6/A05-06). 4 pp.

AECMA PREN2057-76. Z-Section—Aluminium Alloy Folded Profiles with One Lipped Flange—Dimensions (C6/A05-07). 4 pp.

AECMA PREN2058-76. Z-Section—Aluminium Alloy Folded Profiles with Lipped Flanges—Dimensions (C6/A05-08). 4 pp.

AECMA PREN2059-76. Top Hat Section Aluminium Alloy Folded Profiles—Dimensions (C6/A05-09). 4 pp.

AECMA PREN2060-76. Bowler Hat Section Aluminium Alloy Folded Profiles—Dimensions (C6/A05-10). 4 pp.

AECMA PREN2061-76. Top Hat Section—Aluminium Alloy Folded Profiles with Lipped Flanges—Dimensions (C6/A05-11). 4 pp.

CEN PREN 2047-90. Beaded L-Section Aluminium Alloy. 10 pp.

—Sections—Drawn—Aerospace

AECMA PREN2070-03-84. Aluminium and Aluminium Alloy Wrought Products—Part 3—Bar and Section (CS/44). 13 pp.

BSI BS EN 2070: Part 3-91. 1991 Aluminium and Aluminium Alloy Wrought Products. Technical Specification Part 3: Bar and Section. 18 pp.

CEN EN 2070-3-89. Aerospace Series Aluminium and Aluminium Alloy Wrought Products Technical Specification Part 3: Bar and Section. 13 pp.

—Sections—Extruded

BSI BS 1474-87. 1987 Wrought Aluminium and Aluminium Alloys for General Engineering Purposes; Bars, Extruded Round Tubes and Sections. 28 pp.

BSI BS 2898-70. 1970 Amd 1 Wrought Aluminium and Aluminium Alloys for Electrical Purposes. Bars, Extruded Round Tube and Sections. 32 pp.

BSI BS 4300/4-73. 1973 Amd 1 Specification (Supplementary Series) for Wrought Aluminium and Aluminium Alloys for General Engineering Purposes /4: 6463 Solid Extruded Bars and Sections Suitable for for Bright Trim/Reflector Applications. 27 pp.

BSI BS 4300/12-69. 1969 Amd 1 Specification (Supplementary Series) for Wrought Aluminium and Aluminium Alloys for General Engineering Purposes /12: 5454 Bar, Extruded Round Tube and Sections. 36 pp.

BSI BS 4300/15-73. 1973 Amd 1 Specification (Supplementary Series) for Wrought Aluminium and Aluminium Alloys for General Engineering Purposes /15: 7020 Bar, Extruded Round Tube and Sections. 31 pp.

DIN ENGL 1748 Pt 2-83. Wrought Aluminium and Aluminium Alloy Extruded Sections; Technical Delivery Conditions (Feb). 4 pp.

DIN ENGL 1748 Pt 3-68. Aluminium Extruded Sections; (Highest Grade Aluminium, High Grade Aluminium and Wrought Aluminium Alloys); Design (Dec). 6 pp.

—Sections—Extruded—Aerospace

AECMA PREN2047-76. Beaded L-Section Alloy Extrusions—Dimensions (C6/A04-01). 4 pp.

AECMA PREN2048-76. L-Section Aluminium Alloy Extrusions—Dimensions (C6/A04-02). 4 pp.

AECMA PREN2050-76. T-Section Aluminium Alloy—Extrusions—Dimensions (C6/A04-04). 4 pp.

AECMA PREN2066-78. Extruded Sections in Aluminium and Aluminium Alloys—General Tolerances (C6/A04-00). 10 pp.

AECMA PREN2070-03-84. Aluminium and Aluminium Alloy Wrought Products—Part 3—Bar and Section (CS/44). 13 pp.

AECMA PREN2100-80. Aluminium Alloy 2014A-T451—Bars and Sections A Less Than or Equal to 200 mm (C5/40). 3 pp.

Aluminum Alloys (Cont.)

—Sections—Extruded—Aerospace (Cont.)

AECMA PREN2127-80. Aluminium Alloy 7075-T7351—Bars and Sections A Less Than or Equal to 100 mm (C5/40). 3 pp.

AECMA PREN2315-80. Aluminium Alloy 7075-T73510 or T73511—Bars and Sections A Less Than or Equal to 100 mm (C5/40). 3 pp.

AECMA PREN2316-80. Aluminium Alloy 7075-T73—Bars and Sections—A Less Than or Equal to 100 mm (C5/40). 3 pp.

AECMA PREN2318-80. Aluminium Alloy 2024-T3510 or T3511—Bars and Sections—A 1.2-150 mm (C5/40). 3 pp.

AECMA PREN2321-80. Aluminium Alloy 2024-T3—Bars and Sections—A Less Than or Equal to 150 mm (C5/40). 3 pp.

AECMA PREN2324-80. Aluminium Alloy 2014A-T6—Bars and Sections—A Less Than or Equal to 150 mm (C5/40). 3 pp.

AECMA PREN2326-80. Aluminium Alloy 6082-T6—Bars and Sections—A Less Than or Equal to 200 mm (C5/40). 3 pp.

AECMA PREN2384-83. Aluminium Alloy 2014A-T6511 Bars and Sections A Less Than or Equal to 150 mm. 3 pp.

AECMA PREN2385-82. Aluminium Alloy 7009-T73651—Bars and Sections—A Less Than or Equal to 125 mm (C5/40). 3 pp.

AECMA PREN2394-83. Aluminium Alloy 7075-T6511 Bars and Sections A Less Than or Equal to 125 mm.

AECMA PREN2630-85. Aluminium Alloy 7009-T736511 Bar and Section 1.2 Less Than or Equal to (a or D) Less Than or Equal to 125 mm with Peripheral Coarse Grain Control. 3 pp.

AECMA PREN2631-85. Aluminium Alloy 7075-T6511 Bar and Section 1.2 Less Than or Equal to (a or D) Less Than or Equal to 125 mm with Peripheral Coarse Grain Control. 3 pp.

AECMA PREN2632-85. Aluminium Alloy 7075-T73511 Bar and Section 1.2 Less Than or Equal to (a or D) Less Than or Equal to 100 mm with Peripheral Coarse Grain Control. 3 pp.

AECMA PREN2633-85. Aluminium Alloy 2024-T3511 Bar and Section 1.2 Less Than or Equal to (a or D) Less Than or Equal to 150 mm with Peripheral Coarse Grain Control. 3 pp.

AECMA PREN2634-85. Aluminium Alloy 2014A-T4511 Bar and Section 1.2 Less Than or Equal to (a or D) Less Than or Equal to 200 mm with Peripheral Coarse Grain Control. 3 pp.

AECMA PREN2635-85. Aluminium Alloy 2014A-T6511 Bar and Section 1.2 Less Than or Equal to (a or D) Less Than or Equal to 150 mm with Peripheral Coarse Grain Control. 3 pp.

AECMA PREN2636-85. Aluminium Alloy 6082-T6 Bar and Section 1.2 Less Than or Equal to (a or D) Less Than or Equal to 200 mm with Peripheral Coarse Grain Control. 3 pp.

AECMA PREN2637-86. Aluminium Alloy 7075-T73 Extruded Bar and Section 1.2 Less Than or Equal to (a or D) Less Than or Equal to 100 mm with Coarse Peripheral Grain Control. 3 pp.

AECMA PREN2638-86. Aluminium Alloy 2024-T3 Extruded Bar and Section 1.2 Less Than or Equal to (a or D) Less Than or Equal to 150 mm with Coarse Peripheral Grain Control. 3 pp.

AECMA PREN2639-86. Aluminium Alloy 2014A-T6 Extruded Bar and Section 1.2 Less Than or Equal to (a or D) Less Than or Equal to 150 mm with Coarse Peripheral Grain Control. 3 pp.

AECMA PREN2640-85. Aluminium Alloy 2017A-T4 Extruded Bar and Section 1.2 Less Than or Equal to (a or D) Less Than or Equal to 200 mm with Peripheral Coarse Grain Control. 3 pp.

AECMA PREN2655-85. Aluminium Alloy 2017A-T42 Extruded Bar and Section 1.2 Less Than or Equal to (a or D) Less Than or Equal to 150 mm with Peripheral Coarse Grain Control. 3 pp.

AECMA PREN2702-86. Aluminium Alloy 6061-T6 Extruded Bar and Section 1.2 Less Than or Equal to (a or D) Less Than or Equal to 150 mm. 3 pp.

AECMA PREN2706-85. Aluminium Alloy 7009-T736510 Bar and Section 1.2 Less Than or Equal to (a or D) Less Than or Equal to 125 mm with Peripheral Coarse Grain Control. 3 pp.

AECMA PREN2707-85. Aluminium Alloy 7075-T6510 Bar and Section 1.2 Less Than or Equal to (a or D) Less Than or Equal to 125 mm with Peripheral Coarse Grain Control. 3 pp.

AECMA PREN2708-85. Aluminium Alloy 7075-T73510 Bar and Section 1.2 Less Than or Equal to (a or D) Less Than or Equal to 100 mm with Peripheral Coarse Grain Control. 3 pp.

AECMA PREN2709-85. Aluminium Alloy 2024-T3510 Bar and Section 1. 3 pp.

AECMA PREN2710-85. Aluminium Alloy 2014A-T4510 Bar and Section 1.2 Less Than or Equal to (a or D) Less Than or Equal to 200 mm with Peripheral Coarse Grain Control. 3 pp.

INTERNATIONAL AND NON-U.S. NATIONAL STANDARDS
SUBJECT INDEX

Aluminum

Aluminum Alloys (Cont.)

—Sections—Extruded—Aerospace (Cont.)

AECMA PREN2711-85. Aluminium Alloy 2014A-T6510 Bar and Section 1.2 Less Than or Equal to (a or D) Less Than or Equal to 150 mm with Peripheral Coarse Grain Control. 3 pp.

AECMA PREN2806-86. Aluminium Alloy 2024-T42 Extruded Section 1.2 Less Than or Equal to a Less Than or Equal to 100 mm with Coarse Peripheral Grain Control. 3 pp.

AECMA PREN2807-86. Aluminium Alloy 7020-T6 Extruded Section 1.2 Less Than or Equal to a Less Than or Equal to 100 mm with Coarse Peripheral Grain Control. 3 pp.

BSI BS EN 2070: Part 3-91. 1991 Aluminium and Aluminium Alloy Wrought Products. Technical Specification Part 3: Bar and Section. 18 pp.

BSI BS EN 2100-92. 1992 Aluminium Alloy AL-P2014A-T4511 Extruded Bars and Sections a or D Less Than or Equal to 200 mm (Supersedes BS L 102: 1971). 10 pp.

BSI BS EN 2127-92. 1992 Aluminium Alloy AL-P7075-T73511 Extruded Bars and Sections a or D Less Than or Equal to 100 mm (Supersedes BS L 160: 1976). 10 pp.

BSI BS EN 2318-92. 1992 Aluminium Alloy AL-P2024-T3511 Extruded Bars and Sections 1,2 mm Less Than or Equal to a or D Less Than or Equal to 150 mm (S). 10 pp.

BSI BS EN 2326-92. 1992 Aluminium Alloy AL-P6082-T6 Extruded Bars and Sections a or D Less Than or Equal to 200 mm (Supersedes BS L 111: 1971). 10 pp.

BSI BS EN 2384-92. 1992 Aluminium Alloy AL-P2014A-T6511 Extruded Bars and Sections a or D Less Than or Equal to 150 mm. 10 pp.

BSI BS EN 2385-92. 1992 Aluminium Alloy AL-P7009-T74511 Extruded Bars and Sections a or D Less Than or Equal to 125 mm. 10 pp.

CEN EN 2070-3-89. Aerospace Series Aluminium and Aluminium Alloy Wrought Products Technical Specification Part 3: Bar and Section. 13 pp.

CEN PREN 2100-91. Aluminium Alloy AL-P2014A T4511 Extruded Bars and Sections a or D Less Than or Equal to 200 mm. 6 pp.

CEN EN 2100-92. Aerospace Series Aluminium Alloy AL-P2014A T4511 Extruded Bars and Sections a or D Less Than or Equal to 200 mm. 5 pp.

CEN PREN 2127-91. Aluminium Alloy AL-P7075-T73511 Extruded Bars and Sections a or D Less Than or Equal to 100 mm. 6 pp.

CEN EN 2127-92. Aluminium Alloy AL-P7075-T73511 Extruded Bars and Sections a or D Less Than or Equal to 100 mm. 5 pp.

CEN PREN 2318-91. Aluminium Alloy AL-P2024-T3511 Extruded Bars and Sections 1,2 mm Less Than or Equal to a or D Less Than or Equal to 150 mm. 6 pp.

CEN EN 2318-92. Aluminium Alloy AL-P2024-T3511 Extruded Bars and Sections 1,2 mm Less Than or Equal to a or D Less Than or Equal to 150 mm. 5 pp.

CEN PREN 2326-91. Aluminium Alloy AL-P6082-T6 Extruded Bars and Sections a or D Less Than or Equal to 200 mm. 5 pp.

CEN EN 2326-92. Aluminium Alloy AL-P6082-T6 Extruded Bars and Sections a or D Less Than or Equal to 200 mm. 5 pp.

CEN PREN 2384-91. Aluminium Alloy AL-P2014A T6511 Extruded Bars and Sections a or D Less Than or Equal to 150 mm. 6 pp.

CEN EN 2384-92. Aluminium Alloy AL-P2014A T6511 Extruded Bars and Sections a or D Less Than or Equal to 150 mm. 5 pp.

CEN PREN 2385-91. Aluminium Alloy AL-P7009-T74511 Extruded Bars and Sections a or D Less Than or Equal to 125 mm. 6 pp.

CEN EN 2385-92. Aluminium Alloy AL-P7009-T74511 Extruded Bars and Sections a or D Less Than or Equal to 125 mm. 5 pp.

CEN EN 2630-91. Aluminium Alloy AL-P7009-T74511 Extruded Bars and Sections a or D Less Than or Equal to 125 mm with Peripheral Coarse Grain Control. 5 pp.

CEN EN 2632-91. Aluminium Alloy AL-P7075-T73511 Extruded Bars and Sections a or D Less Than or Equal to 100 mm with Peripheral Coarse Grain Control. 5 pp.

CEN EN 2633-91. Aluminium Alloy AL-P2024-T3511 Extruded Bars and Sections 1,2 Less Than or Equal to a or D Less Than or Equal to 150 mm with Peripheral Coarse Grain Control. 5 pp.

CEN EN 2636-91. Aluminium Alloy AL-P6082-T6 Extruded Bars and Sections a or D Less Than or Equal to 200 mm with Peripheral Coarse Grain Control. 5 pp.

—Sections—Extruded—Marine

MOD UK NES 831: Part 2-89. Requirements for Aluminium Alloy Products Held by Naval Stores Part 2: Extruded Sections (Including Standard Extrusions and Special Extrusions) Issue 1 (10.89). 31 pp.

Aluminum Alloys (Cont.)

—Sections—Extruded—Marine (Cont.)

MOD UK NES 831: Part 2-01. Requirements for Aluminium Alloy Products Held by Naval Stores Part 2: Extruded Sections (Including Standard Extrusions and Special Extrusions) Issue 1 (10.89); Amendment 1. 32 pp.

—Sections—Extruded—Mechanical Properties

DIN ENGL 1748 Pt 1-83. Wrought Aluminium and Aluminium Extruded Sections; Properties (Feb). 3 pp.

—Sections—Rolled—Aerospace

AECMA PREN2070-03-84. Aluminium and Aluminium Alloy Wrought Products—Part 3—Bar and Section (CS/44). 13 pp.

BSI BS EN 2070: Part 3-91. 1991 Aluminium and Aluminium Alloy Wrought Products. Technical Specification Part 3: Bar and Section. 18 pp.

CEN EN 2070-3-89. Aerospace Series Aluminium and Aluminium Alloy Wrought Products Technical Specification Part 3: Bar and Section. 13 pp.

—Semifinished Products—Anodized

DIN ENGL 17611-85. Anodized Wrought Aluminium and Aluminium Alloy Semi-Finished Products with Coating Thicknesses of at Least 10 Glm; Technical Delivery Conditions (June). 5 pp.

—Shapes—Extruded

SAA AS 1866-86. Aluminium and Aluminium Alloys—Extruded Rod, Bar, Solid and Hollow Shapes (This is a Joint Standard with SANZ NZS 1866:1986). 45 pp.

SNZ NZS/AS 1866-86. Aluminium and Aluminium Alloys—Extruded Rod, Bar, Solid and Hollow Shapes (This is a Joint Standard with SAA AS 1866.). 45 pp.

—Sheets

BSI BS 1470-87. 1987 Amd 1 Wrought Aluminium and Aluminium Alloys for General Engineering Purposes; Plate, Sheet and Strip (AMD 6032) May 31, 1989. 24 pp.

BSI BS 4300/14-73. 1973 Amd 1 Specification (Supplementary Series) for Wrought Aluminium and Aluminium Alloys for General Engineering Purposes /14: 7020 Plate, Sheet and Strip. 20 pp.

CEN PREN 485 (Part 1)-91. Wrought Aluminium and Aluminium Alloy Sheets, Strips and Plates—Part 1: Technical Conditions for Inspection and Delivery. 16 pp.

CNS H2029-87. Method of Test for Aluminium and Aluminium Alloy Sheets and Plate (Jun)(2254).

CNS H3025-87. Aluminium and Aluminium Alloy Sheets and Plates (Feb)(2253).

DIN ENGL 59606-82. Wrought Aluminium and Aluminium Alloy Sheet and Strip for Cans and Sealing Caps (Nov). 6 pp.

JIS H 4000-88. Aluminium and Aluminium Alloy Sheets and Plates, Strips and Coiled Sheets. 145 pp.

SAA AS 1734-86. Aluminium and Aluminium Alloys—Flat Sheet, Coiled Sheet and Plate (This is a Joint Standard with SANZ NZS 1734:1986). 20 pp.

SNZ NZS/AS 1734-86. Aluminium and Aluminium Alloys—Flat Sheet, Coiled Sheet and Plate (This is a Joint Standard with SAA AS 1734). 20 pp.

—Sheets—Aerospace

AECMA PREN2070-02-84. Aluminium and Aluminium Alloy Wrought Products—Part 2—Sheet, Strip Formed Profiles and Plate (CS/44). 14 pp.

AECMA PREN2071-89. Sheets in Aluminium and Aluminium Alloys—Thickness A Less Than or Equal to 6 mm—Dimensions (C6/A01). 8 pp.

BSI BS EN 2070: Part 2-91. 1991 Aluminium and Aluminium Alloy Wrought Products. Technical Specification Part 2: Sheet, Strip Formed Profiles and Plate. 19 pp.

CEN PREN 2051-90. L-Section Aluminium Alloy Profiles Folded from Sheet Dimensions. 8 pp.

CEN EN 2070-2-89. Aerospace Series Aluminium and Aluminium Alloy Wrought Products Technical Specification Part 2: Sheet, Strip Formed Profiles and Plate. 14 pp.

CEN PREN2071-80. Sheets in Aluminium and Aluminium Alloys Thickness a Less Than or Equal to 6 mm Dimensions. 7 pp.

CEN PREN 2071-89. Sheets Aluminium and Aluminium Alloys. 9 pp.

CEN EN 2395-91. Aluminium Alloy AL-P2014A T4 or T42 Sheet and Strip 0,4 mm Less Than or Equal to a Less Than or Equal to 6 mm. 5 pp.

CEN PREN 3474-93. Aerospace Series Aluminium Alloy AL-P2024-T81 Sheet and Strip 0,25 mm Less Than or Equal to a Less Than or Equal to 6 mm. 4 pp.

CEN PREN 3997-93. Aerospace Series Aluminium Alloy AL-P2024-T3 Sheet and Strip 0,4 mm Less Than or Equal to a Less Than or Equal to 6 mm. 4 pp.

Aluminum Alloys (Cont.)

—Sheets—Aerospace (Cont.)

CEN PREN 3998-93. Aerospace Series Aluminium Alloy AL-P2024-T42 Sheet and Strip 0,4 mm Less Than or Equal to a Less Than or Equal to 6 mm. 4 pp.

CEN PREN 4004-93. Aerospace Series Aluminium Alloy AL-P3103-H16 Sheet and Strip 0,4 mm Less Than or Equal to a Less Than or Equal to 6 mm. 4 pp.

CEN PREN 4005-93. Aerospace Series Aluminium Alloy AL-P5052-O Sheet and Strip 0,3 mm Less Than or Equal to a Less Than or Equal to 6 mm. 4 pp.

CEN PREN 4006-93. Aerospace Series Aluminium Alloy AL-P6082-T4 or T42 Sheet and Strip 0,4 mm Less Than or Equal to a Less Than or Equal to 6 mm. 4 pp.

CEN PREN 4007-93. Aerospace Series Aluminium Alloy AL-P6082-T6 or T62 Sheet and Strip 0,4 mm Less Than or Equal to a Less Than or Equal to 6 mm. 4 pp.

CEN PREN 4100-93. Aerospace Series Aluminium Alloy AL-P2219-T62 Sheet and Strip 0,5 mm Less Than or Equal to a Less Than or Equal to 6 mm. 4 pp.

CEN PREN 4101-93. Aerospace Series Aluminium Alloy AL-P2024-T4 Sheet and Strip with Improved Stretch Forming Capability 0,4 mm Less Than or Equal to a Less Than or Equal to 6 mm. 4 pp.

DIN ENGL LN 9074-88. Aerospace; Wrought Aluminium Alloy Sheet and Plate; Skin Quality; Rolled; Dimensions; Masses (Aug). 8 pp.

—Sheets—Chemical Analysis

BSI BS 4300/16-84. 1984 Specification (Supplementary Series) for Wrought Aluminium and Aluminium Alloys for General Engineering Purposes /16: Specification for Sheet and Strip 8011 Material. 11 pp.

—Sheets—Chemical Machining

MOD UK DTD-932-58. Chemical Contouring of Aluminium and Aluminium Alloy Sheet and Plate (Reprinted January 1965). 2 pp.

—Sheets—Clad—Aerospace

CEN EN 2087-91. Aluminium Alloy AL-P2014A T6 or T62 Clad Sheet and Strip 0,4 mm Less Than or Equal to a Less Than or Equal to 6 mm. 6 pp.

CEN EN 2088-91. Aluminium Alloy AL-P2014A T4 or T42 Clad Sheet and Strip 0,4 mm Less Than or Equal to a Less Than or Equal to 6 mm. 5 pp.

CEN EN 2089-91. Aluminium Alloy AL-P2014A T6 or T62 Sheet and Strip 0,4 mm Less Than or Equal to a Less Than or Equal to 6 mm. 5 pp.

CEN PREN 2092-93. Aerospace Series—Aluminium Alloy Al-P7075 —T6 or T62 Clad Sheet and Strip 0,4 mm Less Than or Equal to a Less Than or Equal to 6 mm. 5 pp.

CEN PREN 4099-93. Aerospace Series Aluminium Alloy AL-P2219-T62 Clad Sheet and Strip 0,5 mm Less Than or Equal to a Less Than or Equal to 6 mm. 4 pp.

CEN PREN 4102-93. Aerospace Series Aluminium Alloy AL-P2219-T81 Clad Sheet and Strip 0,5 mm Less Than or Equal to a Less Than or Equal to 6 mm. 4 pp.

—Sheets—Corrugated

DIN ENGL 25512-80. Corrugated Sheet Sections for Rail Vehicles; Dimensions; Weights; Static Values (Oct). 2 pp.

—Sheets—Designations

BSI BS 4300/16-84. 1984 Specification (Supplementary Series) for Wrought Aluminium and Aluminium Alloys for General Engineering Purposes /16: Specification for Sheet and Strip 8011 Material. 11 pp.

—Sheets—Embossed

DIN ENGL 59605-79. Embossed Plate and Sheet of Wrought Aluminium Alloys (Mar). 5 pp.

—Sheets—Marine

MOD UK NES 831: Part 1-89. Requirements for Aluminium Alloy Products Held by Naval Stores Part 1: Sheet, Strip and Plate (Including Perforated Sheet) Issue 1 (03.89). 15 pp.

MOD UK NES 831: Part 1-01. Requirements for Aluminium Alloy Products Held by Naval Stores Part 1: Sheet, Strip and Plate (Including Perforated Sheet) Issue 1 (03.89); Amendment 1. 18 pp.

—Sheets—Rolled

CEN PREN 485 (Part 3)-91. Wrought Aluminium and Aluminium Alloy Sheets, Strips and Plates—Part 3: Hot-Rolled Sheets, Strips and Plates—Tolerances on Shape and Dimensions. 11 pp.

CEN PREN 485 (Part 4)-91. Wrought Aluminium and Aluminium Alloy Sheets, Strips and Plates—Part 4: Cold-Rolled Sheets, Strips and Plates—Tolerances on Shape and Dimensions. 14 pp.

INDUSTRY STANDARDS

Aluminum

INTERNATIONAL AND NON-U.S. NATIONAL STANDARDS SUBJECT INDEX

Aluminum Alloys (Cont.)

—Sheets—Rolled (Cont.)

CEN PREN 941-92. Aluminium and Aluminium Alloys—Wrought Products—Circle and Circle Stock for General Applications. 11 pp.

DIN ENGL 1745 Pt 2-83. Wrought Aluminium and Aluminium Alloy Plate, Sheet and Strip Greater Than 0,35 mm in Thickness; Technical Delivery Conditions (Feb). 4 pp.

DIN ENGL 1783-81. Strips, Plates and Sheets of Aluminium and Wrought Aluminium Alloys with Thicknesses over 0,35 mm, Cold-Rolled; Dimensions (Apr). 9 pp.

DIN ENGL 1784-81. Strips, Sheets and Sizes of Aluminium and Wrought Aluminium Alloys with Thicknesses from 0,021 to 0,35 mm, Cold-Rolled; Dimensions (Apr). 7 pp.

DIN ENGL 59600-81. Strips, Plates and Sheets of Aluminium and Wrought Aluminium Alloys; Hot Rolled; Dimensions (Apr). 6 pp.

ISO 6361 Pt 1-86. Wrought Aluminium and Aluminium Alloys Sheets, Strips and Plates—Part 1: Technical Conditions for Inspection and Delivery First Edition; (Supersedes 5191). 5 pp.

ISO 6361 Pt 4-88. Wrought Aluminium and Aluminium Alloy Sheets, Strips and Plates—Part 4: Sheets and Plates—Tolerances on Shape and Dimensions First Edition; (Supersedes TR7735). 12 pp.

—Sheets—Rolled—Aerospace

AECMA PREN2089-80. Aluminium Alloy 2014A-T6 or T62—Sheets and Strips A 1.6-6 mm (C5/40). 3 pp.

AECMA PREN2693-85. Aluminium Alloy 5086-H111 Sheet and Strip 0.4 less Than or Equal to a less Than or Equal to 6 mm. 3 pp.

AECMA PREN2694-85. Aluminium Alloy 6061-T6 Sheet and Strip 0.4 Less Than or Equal to a Less Than or Equal to 6 mm. 3 pp.

AECMA PREN2695-85. Aluminium Alloy 6081-T6 Sheet and Strip 0.3 Less Than or Equal to a Less Than or Equal to 6 mm. 3 pp.

AECMA PREN2696-85. Aluminium Alloy 7075-T6 Sheet and Strip 0.8 Less Than or Equal to a Less Than or Equal to 6 mm. 3 pp.

AECMA PREN2802-86. Aluminium Alloy 7475-T761 Sheet and Strip 0.8 Less Than or Equal to a Less Than or Equal to 6 mm. 3 pp.

CEN PREN 2693-93. Aerospace Series Aluminium Alloy AL-P5086-H111 Sheet and Strip 0,3 mm is Less Than or Equal to a Less Than or Equal to 6 mm. 5 pp.

CEN PREN 2694-93. Aerospace Series Aluminium Alloy AL-P6061-T6 or T62 Sheet and Strip 0,4 mm is Less Than or Equal to a Less Than or Equal to 6 mm. 5 pp.

CEN PREN 2695-93. Aerospace Series Aluminium Alloy AL-P6081-T6 Sheet and Strip 0,3 mm is Less Than or Equal to a Less Than or Equal to 6 mm. 5 pp.

CEN PREN 2696-93. Aerospace Series Aluminium Alloy AL-P7075-T6 or T62 Sheet and Strip 0,4 mm is Less Than or Equal to a Less Than or Equal to 6 mm. 5 pp.

—Sheets—Rolled—Mechanical Properties

CEN PREN 683-2-92. Aluminium and Aluminium Alloys—Wrought Products, Finstock—Part 2: Mechanical Properties. 9 pp.

DIN ENGL 1745 Pt 1-83. Wrought Aluminium and Aluminium Alloy Plate, Sheet and Strip Greater Than 0,35 mm in Thickness; Properties (Feb). 6 pp.

DIN ENGL 1788-83. Wrought Aluminium and Aluminium Alloy Sheet and Strip with Thicknesses Between 0,021 and 0,350 mm; Properties (Feb). 4 pp.

ISO 6361 Pt 2-90. Wrought Aluminium and Aluminium Alloy Sheets, Strips and Plates—Part 2: Mechanical Properties Second Edition; (Supersedes TR2136). 17 pp.

—Sheets—Super-Plastic Formed—Aerospace—Mechanical Properties

AECMA PREN3335-88. Aluminium Alloy (7475) Solution Treated and Artificially Aged (T762) Sheet for Super-Plastic Forming 0,8 Less Than or Equal to a Less Than or Equal to 6 mm. 4 pp.

—Sheets—Wrought—Mechanical Properties

CEN PREN 485-2-91. Wrought Aluminium and Aluminium Alloy Sheets, Strips and Plates—Part 2 Mechanical Properties. 45 pp.

—Silicon Content

BSI BS 1728: Part 11-60. 1960 Methods for the Analysis of Aluminium and Aluminium Alloys: Part 11: Determination of Silicon (Perchloric Acid Method). 9 pp.

JIS H 1352-72. Methods for Determination of Silicon in Aluminium and Aluminium Alloy.

Aluminum Alloys (Cont.)

—Silicon Content—Absorptiometric Analysis

BSI BS 1728: Part 12-61. 1961 Amd 1 Methods for the Analysis of Aluminium and Aluminium Alloys: Part 12: Determination of Silicon (Absorptiometric Molybdenum Blue Method). 9 pp.

—Silicon Content—Gravimetric Analysis

ISO 797-73. Aluminium and Aluminium Alloys—Determination of Silicon—Gravimetric Method First Edition. 7 pp.

—Silicon Content—Spectrophotometry

ISO 808-73. Aluminium and Aluminium Alloys—Determination of Silicon—Spectrophotometric Method with the Reduced Silicomolybdic Complex First Edition. 6 pp.

—Slabs—Forged—Aerospace

AECMA PREN2086-77. Aluminium Alloy 2618A-T851 (AL-P11-T851)—Forged Bars and Slabs A Less Than or Equal to 150 mm (C5/40). 3 pp.

AECMA PREN2256-77. Aluminium Alloy 2618A-T852 (AL-P11-T852)—Forged Bars and Slabs A Less Than or Equal to 150 mm (C5/40). 3 pp.

—Slugs—Extruded

DIN ENGL 59604-87. Wrought Aluminium and Aluminium Alloy Slugs for Impact Extrusion (Jan) (Supersedes DIN 59604 Part 1, July 1973 Edition, and DIN 59604 Part 2, November 1970 Edition). 5 pp.

—Slugs—Wrought—Extruded

CEN PREN 570-91. Wrought Aluminium and Aluminium Alloys—Slugs for Impact Extrusion. 14 pp.

—Solders

JIS Z 3281-88. Solders for Aluminium and Aluminium Alloys. 8 pp.

—Spectrochemical Analysis

JIS H 1305-76. Method for Photoelectric Emission Spectrochemical Analysis of Aluminium and Aluminium Alloy.

JIS H 1306-92. Methods for Atomic Absorption Spectrometric Analysis of Aluminium and Aluminium Alloys. 28 pp.

—Stress Corrosion Cracking

CNS H2065-89. Methods of Test for Stress Corrosion Cracking on Aluminium Alloys (Apr)(8412).

ISO 9591-92. Corrosion of Aluminium Alloys—Determination of Resistance to Stress Corrosion Cracking First Edition. 10 pp.

JIS H 8711-90. Test Methods for Stress Corrosion Cracking on Aluminium Alloys. 11 pp.

—Strips

BSI BS 1470-87. 1987 Amd 1 Wrought Aluminium and Aluminium Alloys for General Engineering Purposes; Plate, Sheet and Strip (AMD 6032) May 31, 1989. 24 pp.

BSI BS 4300/14-73. 1973 Amd 1 Specification (Supplementary Series) for Wrought Aluminium and Aluminium Alloys for General Engineering Purposes /14: 7020 Plate, Sheet and Strip. 20 pp.

CEN PREN 485 (Part 1)-91. Wrought Aluminium and Aluminium Alloy Sheets, Strips and Plates—Part 1: Technical Conditions for Inspection and Delivery. 16 pp.

DIN ENGL 59606-82. Wrought Aluminium and Aluminium Alloy Sheet and Strip for Cans and Sealing Caps (Nov). 6 pp.

JIS H 4000-88. Aluminium and Aluminium Alloy Sheets and Plates, Strips and Coiled Sheets. 145 pp.

SAA AS 1865-86. Aluminium and Aluminium Alloys—Drawn Wire, Rod, Bar and Strip (This is a Joint Standard with SANZ NZS 1865:1986). 11 pp.

SNZ NZS/AS 1865-86. Aluminium and Aluminium Alloys—Drawn Wire, Rod, Bar and Strip (This is a Joint Standard with SAA AS 1865). 11 pp.

—Strips—Aerospace

AECMA PREN2070-02-84. Aluminium and Aluminium Alloy Wrought Products—Part 2—Sheet, Strip Formed Profiles and Plate (CS/44). 14 pp.

AECMA PREN2599-86. Strip in Aluminium and Aluminium Alloys 0.3 Less Than or Equal to a Less Than or Equal to 3.2 mm Dimensions. 4 pp.

BSI BS EN 2070: Part 2-91. 1991 Aluminium and Aluminium Alloy Wrought Products. Technical Specification Part 2: Sheet, Strip Formed Profiles and Plate. 19 pp.

CEN EN 2070-2-89. Aerospace Series Aluminium and Aluminium Alloy Wrought Products Technical Specification Part 2: Sheet, Strip Formed Profiles and Plate. 19 pp.

CEN EN 2089-91. Aluminium Alloy AL-P2014A T6 or T62 Sheet and Strip 0,4 mm Less Than or Equal to a Less Than or Equal to 6 mm. 5 pp.

CEN EN 2395-91. Aluminium Alloy AL-P2014A T4 or T42 Sheet and Strip 0,4 mm Less Than or Equal to a Less Than or Equal to 6 mm. 5 pp.

Aluminum Alloys (Cont.)

—Strips—Aerospace (Cont.)

CEN PREN 3474-93. Aerospace Series Aluminium Alloy AL-P2024-T81 Sheet and Strip 0,25 mm Less Than or Equal to a Less Than or Equal to 6 mm. 4 pp.

CEN PREN 3997-93. Aerospace Series Aluminium Alloy AL-P2024-T3 Sheet and Strip 0,4 mm Less Than or Equal to a Less Than or Equal to 6 mm. 4 pp.

CEN PREN 3998-93. Aerospace Series Aluminium Alloy AL-P2024-T42 Sheet and Strip 0,4 mm Less Than or Equal to a Less Than or Equal to 6 mm. 4 pp.

CEN PREN 4004-93. Aerospace Series Aluminium Alloy AL-P3103-H16 Sheet and Strip 0,4 mm Less Than or Equal to a Less Than or Equal to 6 mm. 4 pp.

CEN PREN 4005-93. Aerospace Series Aluminium Alloy AL-P5052-O Sheet and Strip 0,3 mm Less Than or Equal to a Less Than or Equal to 6 mm. 4 pp.

CEN PREN 4006-93. Aerospace Series Aluminium Alloy AL-P6082-T4 or T42 Sheet and Strip 0,4 mm Less Than or Equal to a Less Than or Equal to 6 mm. 4 pp.

CEN PREN 4007-93. Aerospace Series Aluminium Alloy AL-P6082-T6 or T62 Sheet and Strip 0,4 mm Less Than or Equal to a Less Than or Equal to 6 mm. 4 pp.

CEN PREN 4100-93. Aerospace Series Aluminium Alloy AL-P2219-T62 Sheet and Strip 0,5 mm Less Than or Equal to a Less Than or Equal to 6 mm. 4 pp.

CEN PREN 4101-93. Aerospace Series Aluminium Alloy AL-P2024-T4 Sheet and Strip with Improved Stretch Forming Capability 0,4 mm Less Than or Equal to a Less Than or Equal to 6 mm. 4 pp.

—Strips—Chemical Analysis

BSI BS 4300/16-84. 1984 Specification (Supplementary Series) for Wrought Aluminium and Aluminium Alloys for General Engineering Purposes /16: Specification for Sheet and Strip 8011 Material. 11 pp.

—Strips—Clad—Aerospace

CEN EN 2087-91. Aluminium Alloy AL-P2014A T6 or T62 Clad Sheet and Strip 0,4 mm Less Than or Equal to a Less Than or Equal to 6 mm. 5 pp.

CEN EN 2088-91. Aluminium Alloy AL-P2014A T4 or T42 Clad Sheet and Strip 0,4 mm Less Than or Equal to a Less Than or Equal to 6 mm. 5 pp.

CEN PREN 2092-93. Aerospace Series—Aluminium Alloy Al-P7075 —T6 or T62 Clad Sheet and Strip 0,4 mm Less Than or Equal to a Less Than or Equal to 6 mm. 5 pp.

CEN PREN 4099-93. Aerospace Series Aluminium Alloy AL-P2219-T62 Clad Sheet and Strip 0,5 mm Less Than or Equal to a Less Than or Equal to 6 mm. 4 pp.

CEN PREN 4102-93. Aerospace Series Aluminium Alloy AL-P2219-T81 Clad Sheet and Strip 0,5 mm Less Than or Equal to a Less Than or Equal to 6 mm. 4 pp.

—Strips—Designations

BSI BS 4300/16-84. 1984 Specification (Supplementary Series) for Wrought Aluminium and Aluminium Alloys for General Engineering Purposes /16: Specification for Sheet and Strip 8011 Material. 11 pp.

—Strips—Marine

MOD UK NES 831: Part 1-89. Requirements for Aluminium Alloy Products Held by Naval Stores Part 1: Sheet, Strip and Plate (Including Perforated Sheet) Issue 1 (03.89). 15 pp.

MOD UK NES 831: Part 1-01. Requirements for Aluminium Alloy Products Held by Naval Stores Part 1: Sheet, Strip and Plate (Including Perforated Sheet) Issue 1 (03.89); Amendment 1. 18 pp.

—Strips—Rolled

CEN PREN 485 (Part 3)-91. Wrought Aluminium and Aluminium Alloy Sheets, Strips and Plates—Part 3: Hot-Rolled Sheets, Strips and Plates—Tolerances on Shape and Dimensions. 11 pp.

CEN PREN 485 (Part 4)-91. Wrought Aluminium and Aluminium Alloy Sheets, Strips and Plates—Part 4: Cold-Rolled Sheets, Strips and Plates—Tolerances on Shape and Dimensions. 14 pp.

CEN PREN 941-92. Aluminium and Aluminium Alloys—Wrought Products—Circle and Circle Stock for General Applications. 11 pp.

DIN ENGL 1745 Pt 2-83. Wrought Aluminium and Aluminium Alloy Plate, Sheet and Strip Greater Than 0,35 mm in Thickness; Technical Delivery Conditions (Feb). 4 pp.

DIN ENGL 1783-81. Strips, Plates and Sheets of Aluminium and Wrought Aluminium Alloys with Thicknesses over 0,35 mm, Cold-Rolled; Dimensions (Apr). 9 pp.

INTERNATIONAL AND NON-U.S. NATIONAL STANDARDS
SUBJECT INDEX
Aluminum

Aluminum Alloys (Cont.)

—Strips—Rolled (Cont.)

DIN ENGL 1784-81. Strips, Sheets and Sizes of Aluminum and Wrought Aluminium Alloys with Thicknesses from 0,021 to 0,35 mm, Cold-Rolled; Dimensions (Apr). 7 pp.

DIN ENGL 59600-81. Strips, Plates and Sheets of Aluminum and Wrought Aluminium Alloys; Hot Rolled; Dimensions (Apr). 6 pp.

ISO 6361 Pt 1-86. Wrought Aluminium and Aluminium Alloys Sheets, Strips and Plates—Part 1: Technical Conditions for Inspection and Delivery First Edition; (Supersedes 5191). 5 pp.

ISO 6361 Pt 3-85. Wrought Aluminium and Aluminium Alloy Sheets, Strips and Plates—Part 3: Strips—Tolerances on Shape and Dimensions First Edition. 5 pp.

—Strips—Rolled—Aerospace

AECMA PREN2089-80. Aluminium Alloy 2014A-T6 or T62—Sheets and Strips A 1.6-6 mm (C5/40). 3 pp.

AECMA PREN2693-85. Aluminium Alloy 5086-H111 Sheet and Strip 0.4 less Than or Equal to a less Than or Equal to 6 mm. 3 pp.

AECMA PREN2694-85. Aluminium Alloy 6061-T6 Sheet and Strip 0.4 Less Than or Equal to a Less Than or Equal to 6 mm. 3 pp.

AECMA PREN2695-85. Aluminium Alloy 6081-T6 Sheet and Strip 0.3 Less Than or Equal to a Less Than or Equal to 6 mm. 3 pp.

AECMA PREN2696-85. Aluminium Alloy 7075-T6 Sheet and Strip 0.8 Less Than or Equal to a Less Than or Equal to 6 mm. 3 pp.

AECMA PREN2802-86. Aluminium Alloy 7475-T761 Sheet and Strip 0.8 Less Than or Equal to a Less Than or Equal to 6 mm. 3 pp.

CEN PREN 2693-93. Aerospace Series Aluminium Alloy AL-P5086-H111 Sheet and Strip 0,3 mm is Less Than or Equal to a Less Than or Equal to 6 mm. 5 pp.

CEN PREN 2694-93. Aerospace Series Aluminium Alloy AL-P6061-T6 or T62 Sheet and Strip 0,4 mm is Less Than or Equal to a Less Than or Equal to 6 mm. 5 pp.

CEN PREN 2695-93. Aerospace Series Aluminium Alloy AL-P6081-T6 Sheet and Strip 0,3 mm is Less Than or Equal to a Less Than or Equal to 6 mm. 5 pp.

CEN PREN 2696-93. Aerospace Series Aluminium Alloy AL-P7075-T6 or T62 Sheet and Strip 0,4 mm is Less Than or Equal to a Less Than or Equal to 6 mm. 5 pp.

—Strips—Rolled—Mechanical Properties

CEN PREN 683-2-92. Aluminium and Aluminium Alloys—Wrought Products, Finstock—Part 2: Mechanical Properties. 9 pp.

DIN ENGL 1745 Pt 1-83. Wrought Aluminium and Aluminium Alloy Plate, Sheet and Strip Greater Than 0,35 mm in Thickness; Properties (Feb). 6 pp.

DIN ENGL 1788-83. Wrought Aluminium and Aluminium Alloy Sheet and Strip with Thicknesses Between 0,021 and 0,350 mm; Properties (Feb). 4 pp.

ISO 6361 Pt 2-90. Wrought Aluminium and Aluminium Alloy Sheets, Strips and Plates—Part 2: Mechanical Properties Second Edition; (Supersedes TR2136). 17 pp.

—Strips—Wrought—Mechanical Properties

CEN PREN 485-2-91. Wrought Aluminium and Aluminium Alloy Sheets, Strips and Plates—Part 2 Mechanical Properties. 45 pp.

—Surface Finishing

CNS H3133-87. Aluminium and Aluminium Alloy Surface Chromate Treatment (Aug)(11111).

—Surface Finishing—Adhesive Bonding

ISO 4588-89. Adhesives—Preparation of Metal Surfaces for Adhesive Bonding First Edition. 8 pp.

—T Beams—Extruded

DIN ENGL 9714-81. Aluminium and Wrought Aluminium Alloy Extruded Tee Sections; Dimensions; Static Values (Sept). 5 pp.

—Tempering

CNS H3021-87. Alloy and Temper Designation Systems for Aluminium and Aluminium Alloys (May)(2068).

JIS H 0001-88. Temper Designation for Aluminium and Aluminium Alloys. 12 pp.

—Tempering—Designations

ISO 2107-83. Aluminium, Magnesium and Their Alloys—Temper Designations First Edition. 5 pp.

—Tin Content

JIS H 1361-72. Methods for Determination of Tin in Aluminium Alloy.

Aluminum Alloys (Cont.)

—Titanium Content

JIS H 1359-72. Methods for Determination of Titanium in Aluminium and Aluminium Alloy.

—Titanium Content—Photometry

ISO 6827-81. Aluminium and Aluminium Alloys—Determination of Titanium Content—Diantipyrylmethane Photometric Method First Edition. 5 pp.

—Titanium Content—Spectrophotometry

BSI BS 1728: Part 25-80. 1980 Methods for the Analysis of Aluminium and Aluminium Alloys: Part 25: Titanium (Spectrophotometric Chromotropic Acid Method). 9 pp.

ISO 1118-78. Aluminium and Aluminium Alloys—Determination of Titanium—Spectrophotometric Chromotropic Acid Method First Edition. 8 pp.

—Tubes

CNS H3019-85. Aluminium and Aluminium Alloy Tubes (Apr)(1308).

MOD UK DSTAN 47-15-76. Aluminium and Aluminium Alloy Tubing for General Purpose Applications Metric Units Issue 1. 8 pp.

—Tubes—Circular—Aerospace

AECMA PREN2257-79. Circular Structural Tubes in Aluminium and Aluminium Alloys—Dimensions (C6/A06-02). 5 pp.

—Tubes—Drawn

BSI BS 1471-72. 1972 Amd 2 Wrought Aluminium and Aluminium Alloys for General Engineering Purposes; Drawn Tube. 25 pp.

BSI BS 4300/10-69. 1969 Amd 1 Specification (Supplementary Series) for Wrought Aluminium and Aluminium Alloys for General Engineerings Purposes /10: 5454 Drawn Tube. 22 pp.

CEN PREN 754-1-92. Aluminium and Aluminium Alloys—Wrought Products—Cold Drawn Rod/Bar and Tube—Part 1: Technical Conditions for Inspection and Delivery. 16 pp.

DIN ENGL 1746 Pt 2-83. Wrought Aluminium and Aluminium Alloy Tubes; Technical Delivery Conditions (Feb). 3 pp.

DIN ENGL 59751-73. Tubes and Hollow Hexagon Sections of Wrought Aluminium Alloys for Free-Cutting Machining on Automatics; Seamless Drawn; Dimensions (Nov). 7 pp.

ISO 6363 Pt 1-88. Wrought Aluminium and Aluminium Alloy Cold-Drawn Rods/Bars and Tubes—Part 1: Technical Conditions for Inspection and Delivery First Edition; (Supersedes 5191). 6 pp.

SAA AS 1867-86. Aluminium and Aluminium Alloys—Drawn Tubes (This is a Joint Standard with SANZ NZS 1867:1986). 13 pp.

SNZ NZS/AS 1867-86. Aluminium and Aluminium Alloys—Drawn Tubes (This is a Joint Standard with SAA AS 1867). 13 pp.

—Tubes—Drawn—Aerospace

AECMA PREN2070-04-87. Aluminium and Aluminium Alloy Wrought Products—Technical Specification—Part 4—Tube for Structures. 9 pp.

AECMA PREN2070-05-87. Aluminium and Aluminium Alloy Wrought Products—Technical Specification—Part 5—Tube Used Under Pressure. 12 pp.

AECMA PREN2387-81. Aluminium Alloy 2014A-T6—Tubes for Structures A 0.6-12.5 mm (C5/40). 3 pp.

AECMA PREN2388-81. Aluminium Alloy 2024-T351—Tubes for Structures A 0.6-12.5 mm (C5/40). 3 pp.

AECMA PREN2389-81. Aluminium Alloy 6082-T4—Tubes for Structures A 0.6-12.5 mm (C5/40). 3 pp.

AECMA PREN2390-81. Aluminium Alloy 6082-T6—Tubes for Structures A 6-12.5 mm (C5/40). 3 pp.

AECMA PREN2391-81. Aluminium Alloy 6061-T4—Tubes for Structures A 0.6-12.5 mm (C5/40). 3 pp.

AECMA PREN2392-81. Aluminium Alloy 6061-T6—Tubes for Structures A 0.6-12.5 mm (C5/40). 3 pp.

AECMA PREN2393-85. Aluminium Alloy 2017A-T4 Drawn Tube for Structures 0.6 Less Than or Equal to a Less Than or Equal to 12.5 mm. 3 pp.

AECMA PREN2508-82. Aluminium Alloy 5086-H111—Drawn Tubes for Structural Applications (C5/40). 3 pp.

AECMA PREN2509-82. Aluminium Alloy 2017A-T42—Drawn Tubes for Structural Applications (C5/40). 3 pp.

AECMA PREN2510-82. Aluminium Alloy 2024-T42—Drawn Tubes for Structural Applications (C5/40). 3 pp.

AECMA PREN2705-85. Aluminium Alloy 2017A-T42 Drawn Tube for Structures 0.6 Less Than or Equal to a Less Than or Equal to 12.5 mm. 3 pp.

AECMA PREN2813-86. Aluminium Alloy 6061-T6 Tube for Hydraulics 0.6 Less Than or Equal to a Less Than or Equal to 12.5 mm. 3 pp.

BSI BS EN 2070: Part 4-91. 1991 Aluminium and Aluminium Alloy Wrought Products Technical Specification Part 4: Tube for Structures. 12 pp.

Aluminum Alloys (Cont.)

—Tubes—Drawn—Aerospace (Cont.)

BSI BS EN 2070: Part 5-91. 1991 Aluminium and Aluminium Alloy Wrought Products Technical Specification Part 5: Tube Used Under Pressure. 15 pp.

CEN EN 2070-4-89. Aerospace Series Aluminium and Aluminium Alloy Wrought Products Technical Specification Part 4: Tube for Structures. 9 pp.

CEN EN 2070-5-89. Aerospace Series Aluminium and Aluminium Alloy Wrought Products Technical Specification Part 5: Tube Used Under Pressure. 12 pp.

DIN ENGL LN 9222-86. Aerospace; Internal Pressure Tubes in Wrought Aluminium Alloys, Seamless Drawn; Dimensions, Masses (Dec). 5 pp.

—Tubes—Drawn—Circular—Aerospace

AECMA PREN2258-83. Circular Tubes in Aluminium Alloys for Fluid—Dimensions. 6 pp.

—Tubes—Drawn—Mechanical Properties

CEN PREN 754-2-92. Aluminium and Aluminium Alloys—Wrought Products—Cold Drawn Rod/Bar and Tube—Part 2: Mechanical Properties. 40 pp.

DIN ENGL 1746 Pt 1-87. Wrought Aluminium and Aluminium Alloy Tubes; Properties (Jan). 4 pp.

ISO TR2778-77. Wrought Aluminium and Aluminium Alloys—Drawn Tubes—Mechanical Properties First Edition. 6 pp.

—Tubes—Extruded

BSI BS 1474-87. 1987 Wrought Aluminium and Aluminium Alloys for General Engineering Purposes; Bars, Extruded Round Tubes and Sections. 28 pp.

BSI BS 2898-70. 1970 Amd 1 Wrought Aluminium and Aluminium Alloys for Electrical Purposes. Bars, Extruded Round Tube and Sections. 32 pp.

BSI BS 4300/12-69. 1969 Amd 1 Specification (Supplementary Series) for Wrought Aluminium and Aluminium Alloys for General Engineering Purposes /12: 5454 Bar, Extruded Round Tube and Sections. 36 pp.

BSI BS 4300/15-73. 1973 Amd 1 Specification (Supplementary Series) for Wrought Aluminium and Aluminium Alloys for General Engineering Purposes /15: 7020 Bar, Extruded Round Tube and Sections. 31 pp.

CEN PREN 755-1-92. Aluminium and Aluminium Alloys—Wrought Products—Extruded Rod/Bar, Tube and Profile—Part 1: Technical Conditions for Inspection and Delivery. 18 pp.

DIN ENGL 1746 Pt 2-83. Wrought Aluminium and Aluminium Alloy Tubes; Technical Delivery Conditions (Feb). 3 pp.

ISO 6362 Pt 1-86. Wrought Aluminium and Aluminium Alloys Extruded Rods/Bars, Tubes and Profiles—Part 1: Technical Conditions for Inspection and Delivery First Edition; (Supersedes 5191). 5 pp.

ISO 6362 Pt 4-88. Wrought Aluminium and Aluminium Alloy Extruded Rods/Bars, Tubes and Profiles—Part 4: Extruded Profiles—Tolerances on Shape and Dimensions First Edition. 11 pp.

—Tubes—Extruded—Aerospace

AECMA PREN2070-04-87. Aluminium and Aluminium Alloy Wrought Products—Technical Specification—Part 4—Tube for Structures. 9 pp.

AECMA PREN2070-05-87. Aluminium and Aluminium Alloy Wrought Products—Technical Specification—Part 5—Tube Used Under Pressure. 12 pp.

BSI BS EN 2070: Part 4-91. 1991 Aluminium and Aluminium Alloy Wrought Products Technical Specification Part 4: Tube for Structures. 12 pp.

BSI BS EN 2070: Part 5-91. 1991 Aluminium and Aluminium Alloy Wrought Products Technical Specification Part 5: Tube Used Under Pressure. 15 pp.

CEN EN 2070-4-89. Aerospace Series Aluminium and Aluminium Alloy Wrought Products Technical Specification Part 4: Tube for Structures. 9 pp.

CEN EN 2070-5-89. Aerospace Series Aluminium and Aluminium Alloy Wrought Products Technical Specification Part 5: Tube Used Under Pressure. 12 pp.

—Tubes—Extruded—Circular—Aerospace

AECMA PREN2258-83. Circular Tubes in Aluminium Alloys for Fluid—Dimensions. 6 pp.

—Tubes—Extruded—Mechanical Properties

CEN PREN 755-2-92. Aluminium and Aluminium Alloys—Wrought Products—Extruded Rod/Bar, Tube and Profile—Part 2: Mechanical Properties. 56 pp.

DIN ENGL 1746 Pt 1-87. Wrought Aluminium and Aluminium Alloy Tubes; Properties (Jan). 4 pp.

ISO 6362 Pt 2-90. Wrought Aluminium and Aluminium Alloy Extruded Rods/Bars, Tubes and Profiles—Part 2: Mechanical Properties Second Edition. 15 pp.

Aluminum Alloys (Cont.)

—Tubes—Marine
MOD UK NES 831: Part 3-89. Requirements for Aluminium Alloy Products Held by Naval Stores Part 3: Tubes (Excluding Guard Rail Tubes) Issue 1 (03.89). 14 pp.

—Tubes—Seamless—Drawn—Aerospace
DIN ENGL LN 1795-86. Aerospace; Structural Tubes in Wrought Aluminium Alloys, Seamless Drawn; Dimensions, Masses (Dec). 5 pp.

—Tubes—Seamless—Drawn—Condensers (Liquefiers)
CNS H2032-81. Method of Test for Aluminium-Alloy Drawn Seamless Tubes for Condenser and Heat Exchangers Use (Jun) (2967).

CNS H3037-81. Aluminium Alloy Drawn Seamless Tubes for Condensers and Heat Exchangers (Jun)(2966).

—Tubes—Seamless—Drawn—Heat Exchangers
CNS H2032-81. Method of Test for Aluminium-Alloy Drawn Seamless Tubes for Condenser and Heat Exchangers Use (Jun) (2967).

CNS H3037-81. Aluminium Alloy Drawn Seamless Tubes for Condensers and Heat Exchangers (Jun)(2966).

—Tubes—Seamless—Extruded
DIN ENGL 9107-87. Wrought Aluminium and Aluminium Alloy Seamless Extruded Round Tubes; Dimensions, Dimensional Tolerances and Form Tolerances (Feb). 6 pp.

—Tubes—Welded
BSI BS 4300/1-67. 1967 Amd 1 Specification (Supplementary Series) for Wrought Aluminium and Aluminium Alloys Aluminium Alloy Longitudinally Welded Tube Metric Units (AMD 3381) June 30, 1980. 26 pp.

—Wedge Sections—Aerospace
DIN ENGL LN 29545-86. Aerospace; Wrought Aluminium Alloy Wedges, Dimensions, Masses (Dec). 2 pp.

—Wedge Sections—Extruded—Aerospace
DIN ENGL LN 9410-86. Aerospace; Wrought Aluminium Alloy Wedge Sections, Extruded, Dimensions, Masses (Dec). 2 pp.

—Welding—Edge Preparation
DIN ENGL 8552 Pt 1-81. Weld Preparation; Groove Forms for Aluminium and Aluminium Alloys; Gas Welding and Gas-Shielded Arc Welding (May). 5 pp.

—Welding—Quality Assurance
DIN ENGL 8563 Pt 30-85. Quality Assurance of Welding Operations; Fusion Welded Joints in Aluminium and Aluminium Alloys (Except Beam Welded Joints); Requirements, Classification (Oct). 13 pp.

—Wirebars—Chemical Analysis
DIN ENGL 1725 Pt 1-83. Aluminium Alloys; Wrought Alloys (Feb). 8 pp.

—Wrought
CEN PREN 546-1-91. Wrought Aluminium and Aluminium Alloy Foil—Part 1: Technical Conditions for Inspection and Delivery. 10 pp.

CEN PREN 546-3-92. Aluminium and Aluminium Alloys—Wrought Products, Foil—Part 3: Dimensional Tolerances. 7 pp.

CEN PREN 683-1-92. Aluminium and Aluminium Alloys—Wrought Products, Finstock—Part 1: Technical Conditions for Inspection and Delivery. 10 pp.

CEN PREN 683-3-92. Aluminium and Aluminium Alloys—Wrought Products, Finstock—Part 3: Dimensional Tolerances. 9 pp.

ISO 209 Pt 2-89. Wrought Aluminium and Aluminium Alloys—Chemical Composition and Forms of Products—Part 2: Forms of Products First Edition. 5 pp.

—Wrought—Aerospace
AECMA PREN2070-01-84. Aluminium and Aluminium Alloy Wrought Products—Part 1—General Requirements (CS/44). 10 pp.

BSI BS EN 2070: Part 1-91. 1991 Aluminium and Aluminium Alloy Wrought Products. Technical Specification Part 1: General Requirements. 15 pp.

BSI BS EN 2070: Part 1-01. 1991 Amd 1 Aluminium and Aluminium Alloy Wrought Products. Technical Specification Part 1: General Requirements (AMD 7705) October 15, 1993 (S). 22 pp.

BSI 4L 100-01. 1993 Amd 1 Procedure for Inspection, Testing and Acceptance of Wrought Aluminium and Aluminium Alloys (AMD 7884) September 15, 1993 (S). 82 pp.

BSI 3L 100-71. (WITHDRAWN) 1971 Amd 4 Procedure for Inspection and Testing of Wrought Aluminium and Aluminium Alloys. 44 pp.

CEN EN 2070-1-89. Aerospace Series Aluminium and Aluminium Alloy Wrought Products Technical Specification Part 1: General Requirements. 10 pp.

CEN EN 2070 Part 1-89. Aerospace Series Aluminium and Aluminium Alloy Wrought Products Technical Specification Part 1: General Requirements. 12 pp.

ISO 8591 Pt 1-89. Aerospace—Wrought Aluminium and Aluminium Alloys—Inspection, Testing and Supply Requirements—Part 1: General Requirements First Edition. 7 pp.

—Wrought—Aerospace—Corrosion Testing
AECMA PREN2720-90. Test Method for Metallic Materials Testing of Susceptibility to Exfoliation Corrosion in 2XXX and 7XXX Series Wrought Aluminium Alloy Products for Aerospace Constructions. 1 p.

CEN PREN 2004 (Part 4)-92. Test Methods for Aluminium and Aluminium Alloy Products Part 4—Stress Corrosion Test by Alternate Immersion for High Strength Aluminium Alloy Wrought Products. 14 pp.

CEN PREN 2716-90. Test Method for Susceptibility to Intergranular Corrosion of Wrought Products in 2XXX Series Aluminium Alloys. 4 pp.

CEN PREN 2717-90. Test Method for Susceptibility to Intergranular Corrosion of Wrought Products in 5XXX Series Aluminium Alloys with a Magnesium Content /3.5%. 7 pp.

—Wrought—Aerospace—Electrical Resistivity
BSI BS EN 2004-1-93. 1993 Test Methods for Aluminium and Aluminium Alloy Products Part 1: Determination of Electrical Conductivity of Wrought Aluminium Alloys (S). 11 pp.

CEN PREN 2004 (Part 1)-91. Test Methods for Aluminium and Aluminium Alloy Products Part 1—Determination of Electrical Conductivity of Wrought Aluminium Alloys. 7 pp.

CEN EN 2004-1-93. Aerospace Series Test Methods for Aluminium and Aluminium Alloy Products Part 1. Determination of Electrical Conductivity of Wrought Aluminium Alloys. 6 pp.

—Wrought—Aerospace—Macrographic Inspection
CEN PREN 2715-90. Macrographic Examination of Aluminium and Aluminium Alloy Forging Stock, Forgings and Wrought Products. 4 pp.

—Wrought—Bend Testing
ISO 2142-81. Wrought Aluminium, Magnesium and Their Alloys—Selection of Specimens and Test Pieces for Mechanical Testing First Edition. 5 pp.

—Wrought—Chemical Analysis
CEN PREN 573-3-91. Wrought Aluminium and Aluminium Alloys—Chemical Composition and Forms of Products—Part 3: Chemical Composition. 13 pp.

ISO 209 Pt 1-89. Wrought Aluminium and Aluminium Alloys—Chemical Composition and Forms of Products—Part 1: Chemical Composition First Edition; (Supersedes 2779 and 3335). 9 pp.

—Wrought—Designations
CEN PREN 573-1-91. Wrought Aluminium and Aluminium Alloys—Chemical Composition and Forms of Products—Part 1: Numerical Designation System. 8 pp.

CEN PREN 573-2-91. Wrought Aluminium and Aluminium Alloys—Chemical Composition and Forms of Products—Part 2: Chemical Symbol Based Designation System. 9 pp.

—Wrought—Food Preparation Equipment—Chemical Analysis
CEN PREN 602-91. Wrought Aluminium and Aluminium Alloys—Chemical Composition of the Metal Used for the Production of Materials and Articles Intended to Be in Contact with Food. 7 pp.

—Wrought—Mechanical Properties
CEN PREN 546-2-92. Aluminium and Aluminium Alloys—Wrought Products, Foil—Part 2: Mechanical Properties. 7 pp.

—Wrought—Rolling Ingots—Chemical Analysis
DIN ENGL 1725 Pt 1-83. Aluminium Alloys; Wrought Alloys (Feb). 8 pp.

—Wrought—Semifinished Products—Chemical Analysis
DIN ENGL 1725 Pt 1-83. Aluminium Alloys; Wrought Alloys (Feb). 8 pp.

—Wrought—Shapes
CEN PREN 573-4-91. Wrought Aluminium and Aluminium Alloys—Chemical Composition and Forms of Products—Part 4: Forms of Products. 13 pp.

—Wrought—Structural Shapes
BSI BS 8118: Part 2-91. 1991 Structural Use of Aluminium Part 2: Specification for Materials, Workmanship and Protection. 28 pp.

—Wrought—Structural Shapes—Design
BSI BS 8118: Part 1-91. 1991 Structural Use of Aluminium Part 1: Code of Practice for Design. 160 pp.

—Wrought—Tempering—Designations
BSI BS EN 515-93. 1993 Aluminium and Aluminium Alloys—Wrought Products—Temper Designations (V). 21 pp.

CEN PREN 515-91. Wrought Aluminium and Aluminium Alloys—Temper Designations. 20 pp.

—Wrought—Tensile Testing
ISO 2142-81. Wrought Aluminium, Magnesium and Their Alloys—Selection of Specimens and Test Pieces for Mechanical Testing First Edition. 5 pp.

—Zinc Content
JIS H 1356-72. Methods for Determination of Zinc in Aluminium and Aluminium Alloy.

—Zinc Content—Atomic Absorption Spectrometry
BSI BS 1728: Part 21-73. 1973 Methods for the Analysis of Aluminium and Aluminium Alloys: Part 21: Zinc (Atomic Absorption Method). 8 pp.

ISO 5194-81. Aluminium and Aluminium Alloys—Determination of Zinc Content—Flame Atomic Absorption Spectrometric Method First Edition. 6 pp.

—Zinc Content—Polarographic Analysis
BSI BS 1728: Part 18-70. 1970 Methods for the Analysis of Aluminium and Aluminium Alloys: Part 18: Determination of Zinc (Ion-Exchange-Volumetric EDTA or Polargraphic Method). 12 pp.

—Zinc Content—Volumetric Analysis
BSI BS 1728: Part 18-70. 1970 Methods for the Analysis of Aluminium and Aluminium Alloys: Part 18: Determination of Zinc (Ion-Exchange-Volumetric EDTA or Polargraphic Method). 12 pp.

ISO 1784-76. Aluminium Alloys—Determination of Zinc—EDTA Titrimetric Method First Edition. 5 pp.

—Zirconium Content
JIS H 1363-71. Methods for Determination of Zirconium in Aluminium Alloy.

Aluminum Ammonium Sulfate
Use For: Ammonium Alum

CNS K7032-61. Chemical Reagent (Aluminum Ammionium Sulfate) (Ammonia Alum) (Dec)(1532).

JIS K 1472-70. Ammonium Alum (Aluminium Ammonium Sulfate). 11 pp.

JIS K 8087-80. Aluminium Ammonium Sulfate 12-Water.

Aluminum/Asphalt Roof Coatings
Use: Roof Coatings—Aluminum/Asphalt

Aluminum Bars
Use: Aluminum—Bars

Aluminum Bearing Alloys
Use: Bearing Alloys—Aluminum

Aluminum Brass Pipes
See Also: Pipes

—Aircraft
MOD UK DTD-253A-36. Aluminium-Nickel-Silicon Brass Tubes (Low Pressure) (Reprinted September 1963, Incorporating Amendments Nos. 1, 2 and 3). 2 pp.

MOD UK DTD-5019-64. Aluminium-Nickel-Silicon Brass Tubes (Suitable for Pipe Lines and High Pressure Hydraulic Systems Especially Where Flaring is Required) (Reprinted March 1966). 5 pp.

Aluminum Brasses
See Also: Brasses

INTERNATIONAL AND NON-U.S. NATIONAL STANDARDS
SUBJECT INDEX
Aluminum

Aluminum Brasses (Cont.)

—**Bars—Aircraft**
MOD UK DTD-319-42. Aluminium-Nickel-Silicon Brass Bars (Reprinted November 1961). 3 pp.

—**Forgings—Marine**
MOD UK NES 749: Part 2-88. Requirements for Aluminium-Nickel-Silicon-Brass Part 2: Forgings, Forging Stock, Rod and Sections Issue 2 (11.88). 25 pp.

—**Plates—Marine**
MOD UK NES 749: Part 1-85. Requirements for Aluminium-Nickel-Silicon-Brass Part 1: Sheet, Strip and Plate Issue 1 (12.85). 21 pp.
MOD UK NES 749: Part 1-91. Requirements for Aluminium—Nickel—Silicon—Brass Part 1: Sheet, Strip and Plate Issue 2 (12.91). 19 pp.

—**Rods—Marine**
MOD UK NES 749: Part 2-88. Requirements for Aluminium-Nickel-Silicon-Brass Part 2: Forgings, Forging Stock, Rod and Sections Issue 2 (11.88). 25 pp.

—**Sections—Marine**
MOD UK NES 749: Part 2-88. Requirements for Aluminium-Nickel-Silicon-Brass Part 2: Forgings, Forging Stock, Rod and Sections Issue 2 (11.88). 25 pp.

—**Sheets—Aircraft**
MOD UK DTD-283A-38. Aluminium-Nickel-Silicon Brass Sheets (Annealed) (for Sheets Not over 24 Inches Wide) (Reprinted August 1961, Incorporating Amendment No. 1). 2 pp.

—**Sheets—Marine**
MOD UK NES 749: Part 1-85. Requirements for Aluminium-Nickel-Silicon-Brass Part 1: Sheet, Strip and Plate Issue 1 (12.85). 21 pp.
MOD UK NES 749: Part 1-91. Requirements for Aluminium—Nickel—Silicon—Brass Part 1: Sheet, Strip and Plate Issue 2 (12.91). 19 pp.

—**Strips—Marine**
MOD UK NES 749: Part 1-85. Requirements for Aluminium-Nickel-Silicon-Brass Part 1: Sheet, Strip and Plate Issue 1 (12.85). 21 pp.
MOD UK NES 749: Part 1-91. Requirements for Aluminium—Nickel—Silicon—Brass Part 1: Sheet, Strip and Plate Issue 2 (12.91). 19 pp.

—**Tubes—Marine**
MOD UK NES 749: Part 3-01. Requirements for Aluminium Nickel, Silicon, Brass Part 3: Tubes Issue 2 (05.87); Amendment 1. 20 pp.

Aluminum Bronze Castings
See Also: Aluminum Bronzes; Bronze Castings; Castings; Metal Products
CNS H3071-89. Aluminium Bronze Castings (Jan)(4819).
DIN ENGL 1714-81. Copper-Aluminium Casting Alloys; (Cast Aluminium Bronze); Castings (Nov). 6 pp.
JIS H 5114-88. Aluminium Bronze Castings. 10 pp.

—**Chemical Analysis**
DIN ENGL 1714-81. Copper-Aluminium Casting Alloys; (Cast Aluminium Bronze); Castings (Nov). 6 pp.

—**Die—Aircraft**
MOD UK DTD-412-40. Aluminium-Bronze Sand or Die Castings (Reprinted April 1963, Incorporating Amendment No. 1). 2 pp.

—**Ingots**
CNS H3063-89. Aluminium Bronze Ingots for Castings (Jan)(4335).
JIS H 2206-85. Aluminium Bronze Ingots for Castings. 6 pp.

—**Mechanical Properties**
DIN ENGL 1714-81. Copper-Aluminium Casting Alloys; (Cast Aluminium Bronze); Castings (Nov). 6 pp.
DIN ENGL 1714 Suppl. 1-81. Copper-Aluminium Casting Alloys; (Cast Aluminium Bronze); Castings; Reference Data on Mechanical and Physical Properties (Nov). 3 pp.

—**Physical Properties**
DIN ENGL 1714 Suppl. 1-81. Copper-Aluminium Casting Alloys; (Cast Aluminium Bronze); Castings; Reference Data on Mechanical and Physical Properties (Nov). 3 pp.

—**Sand—Aircraft**
MOD UK DTD-412-40. Aluminium-Bronze Sand or Die Castings (Reprinted April 1963, Incorporating Amendment No. 1). 2 pp.

Aluminum Bronzes
See Also: Aluminum Alloys; Aluminum Bronze Castings; Bronzes; Copper Alloys; Copper Aluminum Alloys; Tin Alloys; Tin Bronzes

—**Bars—Aircraft**
MOD UK DTD-164A-42. Aluminium-Nickel-Iron Bronze Bars, Forgings and Stampings (Reprinted May 1963). 4 pp.
MOD UK DTD-197A-44. Aluminium-Nickel-Iron Bronze Bars and Forgings (Reprinted November 1966). 4 pp.

—**Chemical Analysis**
CNS H2080-87. Method of Chemical Analysis for Aluminium Bronze and Special Aluminium Bronze (Oct)(12124).
JIS H 1252-77. Methods for Chemical Analysis of Aluminium Bronze and Special Aluminium Bronze.

—**Forgings—Aircraft**
MOD UK DTD-164A-42. Aluminium-Nickel-Iron Bronze Bars, Forgings and Stampings (Reprinted May 1963). 4 pp.
MOD UK DTD-197A-44. Aluminium-Nickel-Iron Bronze Bars and Forgings (Reprinted November 1966). 4 pp.

—**Semifinished Products—Chemical Analysis**
DIN ENGL 17665-83. Wrought Copper Alloys; Copper-Aluminium Alloys; (Aluminium Bronze); Composition (Dec). 6 pp.

—**Stampings—Aircraft**
MOD UK DTD-164A-42. Aluminium-Nickel-Iron Bronze Bars, Forgings and Stampings (Reprinted May 1963). 4 pp.

Aluminum Capacitors
Use For: Aluminum Electrolytic Capacitors
See Also: Aluminum Chip Capacitors; Capacitors; Electrolytic Capacitors; Fixed Capacitors
CECC CECC 30 300 ISSUE 2-88. Sectional Specification: Aluminium Electrolytic Capacitors with Solid and Non-Solid Electrolyte (En, Fr, Ge) AMD 1 (En, Fr, Ge) AMD 2 (En, Fr, Ge) AMD 3 (En, Fr, Ge) AMD 4 (En, Fr, Ge) AMD 5 (En, Fr, Ge). 137 pp.
CECC CECC 30 301 ISSUE 2-88. Blank Detail Specification: Aluminium Electrolytic Capacitors with Non-Solid Electrolyte; (Assessment Level E) (En, Fr, Ge) AMD 1 (En, Fr, Ge) AMD 2 (En, Fr, Ge). 68 pp.
CNS C7008-84. Electrolytic Capacitors for General Use (Oct)(2318). 11 pp.
CNS C7067-82. AC Mains Supply Aluminium Foil Electrolytic Capacitors (Jun)(4595).
CNS C7085-84. Electrolytic Capacitors, Liquid Electrolyte, Aluminium Foil, for Electronic Equipment (Oct)(4856). 35 pp.
DIN VDE 0560 Pt 15-63. Regulations for Capacitors; Part 15: Rules for Aluminium Electrolytic Capacitors for Rated d.c. Voltages up to 1000 V (Apr). 36 pp.
IECQ QC 300300-85. Fixed Capacitors for Use in Electronic Equipment: Part 4: Sectional Specification: Aluminium Electrolytic Capacitors with Solid and Non-Solid Electrolyte (Amendment 1-1992) (IEC 384-4-1 ED 2). 71 pp.
IECQ QC 300301-85. Fixed Capacitors for Use in Electronic Equipment: Part 4: Blank Detail Specification: Aluminium Electrolytic Capacitors with Assessment Level E (Amendment 1-1992). 38 pp.
JIS C 5141-91. Fixed Aluminium Electrolytic Capacitors with Non-Solid Electrolyte for Use in Electronic Equipment. 70 pp.
MOD UK DEF-5134-A-66. General Requirements for Capacitors, Fixed Electrolytic (Aluminium or Tantalum Anode). 13 pp.
MOD UK DEF-5134-1-61. Capacitors, Electrolytic, with Aluminum Electrodes (Reprinted July 1971, Incorporating Amendment No. 1). 25 pp.

—**Can Type**
CECC CECC 30 301-026 ISSUE 3-88. BS CECC 30 301-026; Aluminium Electrolytic Type I with Non-Solid Electrolyte Cylindrical Insulated Case; Screw Terminations; Plain Can—ALS20A/ALS20B Studded Can—ALS21A/ALS21B Style I and Style II Long Life Grade (En) AMD 1 (En). 26 pp.
CECC CECC 30 301-033 ISSUE 2-83. BS CECC 30 301-033; Aluminium Electrolytic Type 1 with Non-Solid Electrolyte Cylindrical Insulated Case; Plain Can Styles A & C; Studded Can Style B (En) AMD 1 (En). 23 pp.

—**Communication Equipment**
CNS C7009-84. Electrolytic Capacitor for Special Use (Oct)(2319). 15 pp.

Aluminum Capacitors (Cont.)

—**Cylindrical**
IECQ QC 300301/GB 0001-89. Aluminium Electrolytic Capacitors with Non-Solid Electrolyte, Cylindrical, Polar, Insulated Case, Screw Terminations, Not for Printed Board Applications. 22 pp.
IECQ QC 300301/GB 0002-89. Aluminium Electrolytic Capacitors with Non-Solid Electrolyte, Cylindrical, Polar Insulated Case, Rigid Pin and Tag Solder Terminations. 21 pp.

—**Cylindrical—Printed Circuit Mount**
CECC CECC 30 301-034 ISSUE 1-84. BS CECC 30 301-034 Issue 1; Aluminium Electrolytic Capacitor with Non-Solid Electrolyte; Cylindrical Insulated Case; Intended for Printed Circuit Application (En) AMD 1 (En). 17 pp.
CECC CECC 30 301-042 ISSUE 1-88. BS CECC 30 301-042 Issue 1; Aluminium Electrolytic Capacitor with Non-Solid Electrolyte Cylindrical Insulated Case Rigid Terminations PC Mounting Long Life Grade (En). 19 pp.
IECQ QC 300301/GB 0003-89. Aluminium Electrolytic Capacitors with Non-Solid Electrolyte, Cylindrical, Insulated Case, Rigid Terminations, PC Mounting. 19 pp.

—**Hermetically Sealed—Can Type**
CECC CECC 30 301-803 ISSUE 1-88. Detail Specification: Aluminium Electrolytic Capacitors with Non-Solid Electrolyte (En). 18 pp.

—**Hermetically Sealed—Cylindrical**
CECC CECC 30 201-026 ED 2-90. UTE C 83-112; Fixed Tantalum Capacitors with Solid Electrolyte, Porous Anode (Fr) ADD 11 (Fr). 15 pp.
CECC CECC 30 201-030. DIN 45 910 Teil 145; Polar Tantalum Electrolytic Capacitors DC 6.3 to 50 V, Long-Life-Grade, Porous Tantalum Anode with Solid Electrolyte, Rectangular Shaped Plastic Encapsulation, Radial Wire Leads (Ge). 14 pp.
CECC CECC 30 301-001. CEI CECC 30 301-001 Edition III; Aluminium Electrolytic Capacitors with Non-Solid Electrolyte (En). 15 pp.
CECC CECC 30 301-002 ISSUE 3-90. CEI CECC 30 301-002; Aluminium Electrolytic Capacitors with Non-Solid Electrolyte (En). 16 pp.
CECC CECC 30 301-021 ISSUE 3-88. BS CECC 30 301-021; Aluminium Electrolytic Capacitors with Non-Solid Electrolyte (En). 15 pp.
CECC CECC 30 301-024. BS CECC 30 301-024 Issue 1; Aluminium Electrolytic Capacitors with Non-Solid Electrolyte Long Life Grade (En) AMD 1 (En). 28 pp.
CECC CECC 30 301-027 ISSUE 1-77. NEN CECC 30 301-027 Issue 1; Aluminium Electrolytic Capacitors with Non-Solid Electrolyte and Axial Leads (En). 12 pp.
CECC CECC 30 301-030-83. SS CECC 30 301-030; Aluminium Electrolytic Capacitors with Non-Solid Electrolyte (En). 22 pp.
CECC CECC 30 301-053-89. Aluminium Electrolytic Capacitors with Non Solid Electrolyte (En). 24 pp.
CECC CECC 30 301-055. OVE-CECC 30 301-055 Edition 1; Aluminium Electrolytic Capacitors with Non-Solid Electrolyte (En). 16 pp.
CECC CECC 30 301-056. OVE-CECC 30 301-056 Edition 1; Aluminium Electrolytic Capacitors with Non-Solid Electrolyte (En). 16 pp.
CECC CECC 30 301-058 ISSUE 2-92. Aluminium Electrolytic Capacitors with Non Solid Electrolyte (En). 19 pp.
CECC CECC 30 301-803 ISSUE 1-88. Detail Specification: Aluminium Electrolytic Capacitors with Non-Solid Electrolyte (En). 18 pp.
CECC CECC 30 301-805 ISSUE 2-92. Detail Specification: Aluminium Electrolytic Capacitors with Non-Solid Electrolyte (En). 21 pp.
IECQ QC 300301/CN 0001-87. Detail Specification for Electronic Components: Fixed Aluminium Electrolytic Capacitors, Type CD11; Assessment Level E. 19 pp.
IECQ QC 300301/SE 0001-89. Detail Specification for Aluminium Electrolytic Capacitors with Non Solid Electrolyte. 24 pp.

—**Hermetically Sealed—Cylindrical—Can Type—Printed Circuit Mount**
CECC CECC 30 301-804 ISSUE 1-88. Detail Specification: Aluminium Electrolytic Capacitors with Non-Solid Electrolyte (En) AMD 1 (En, Fr, Ge). 18 pp.

—**Hermetically Sealed—Cylindrical—Printed Circuit Mount**
CECC CECC 30 301-801 ISSUE 1-88. Detail Specification: Aluminium Electrolytic Capacitors with Non-Solid Electrolyte (En) AMD 1 (En, Fr, Ge). 18 pp.
CECC CECC 30 301-802 ISSUE 1-88. Detail Specification: Aluminium Electrolytic Capacitors with Non-Solid Electrolyte (En) AMD 1 (En, Fr, Ge). 21 pp.

INDUSTRY STANDARDS

Aluminum
INTERNATIONAL AND NON-U.S. NATIONAL STANDARDS
SUBJECT INDEX

Aluminum Capacitors *(Cont.)*

—Hermetically Sealed—Cylindrical—Printed Circuit Mount *(Cont.)*

CECC CECC 30 301-805 ISSUE 2-92. Detail Specification: Aluminium Electrolytic Capacitors with Non-Solid Electrolyte (En). 21 pp.
CECC CECC 30 301-806 ISSUE 1-89. Detail Specification: Aluminium Electrolytic Capacitors with Non-Solid Electrolyte (En) AMD 1 (En, Fr, Ge). 19 pp.
CECC CECC 30 301-809 ISSUE 1-93. Detail Specification: Fixed Aluminium Electrolytic Capacitors with Non-Solid Electrolyte (En). 19 pp.
IECQ QC 300301/CN 0002-89. Detail Specification for Electronic Components: Fixed Aluminium Electrolytic Capacitors, Type CD30; Assessment Level E. 15 pp.
IECQ QC 300301/CN 0003-89. Detail Specification for Electronic Components: Fixed Aluminium Electrolytic Capacitors, Type CD110; Assessment Level E. 19 pp.
IECQ QC 300301/CN 0005-90. Detail Specification for Electronic Components Fixed Aluminium Electrolytic Capacitors, Type CD288; Assessment Level E. 16 pp.

—Hermetically Sealed—Preferred Products List

CECC CECC MUAHAG Vol 1 IS 4-90. Preferred Products List; Capacitors (En, Fr, Ge). 64 pp.

—Hermetically Sealed—Tubular

IECQ QC 300301/JP 0001-86. Detail Specification for Electronic Components: Aluminium Electrolytic Capacitors with Non-Solid Electrolyte. (Tubular, Polar, Metallic Case, Insulated, Radial Terminations). 16 pp.
IECQ QC 300301/JP 0002-90. Detail Specification for Electronic Components: Fixed Aluminium Electrolytic Capacitors with Solid Electrolyte Assessment Level(s) E. 15 pp.
IECQ QC 300301/JP 0003-91. Detail Specification for Electronic Components: Aluminium Electrolytic Capacitors with Non-Solid Electrolyte Assessment Level(s) E. 17 pp.
IECQ QC 300301/JP 0004-91. Detail Specification for Electronic Components Aluminium Electrolytic Capacitors with Non-Solid Electrolyte Assessment Level(s) E. 18 pp.
IECQ QC 300301/JP 0005-91. Detail Specification for Electronic Components Aluminium Electrolytic Capacitors with Non-Solid Electrolyte Assessment Level(s) E. 21 pp.
IECQ QC 300301/JP 0006-92. Detail Specification for Electronic Components Aluminium Electrolytic Capacitors with Non-Solid Electrolyte Assessment Level(s) E. 15 pp.
IECQ QC 300301/KR 0001-84. Detail Specification for Electronic Components Aluminium Electrolytic Capacitors with Non-Solid Electrolyte. 16 pp.
IECQ QC 300301/KR 0002-85. Detail Specification for Electronic Components Aluminium Electrolytic Capacitors with Non-Solid Electrolyte. 19 pp.
IECQ QC 300301/KR 0003-85. Detail Specification for Electronic Components Aluminium Electrolytic Capacitors with Non-Solid Electrolyte. 19 pp.
IECQ QC 300301/KR 0004-87. Detail Specification for Electronic Components Aluminium Electrolytic Capacitors with Non-Solid Electrolyte. 24 pp.
IECQ QC 300301/KR 0005-87. Detail Specification for Electronic Components Aluminium Electrolytic Capacitors with Non-Solid Electrolyte. 19 pp.
IECQ QC 300301/KR 0006-90. Detail Specification for Electronic Components Aluminium Electrolytic Capacitors with Non-Solid Electrolyte. 24 pp.

—Hermetically Sealed—Tubular—Miniature

IECQ QC 300301/US 0001-87. Capacitors for Use in Electronic Equipment: Detail Specification: Fixed Aluminium Electrolytic Capacitors with Non-Solid Electrolyte. 18 pp.
IECQ QC 300301/US 0003 ISSUE 1-90. Capacitors for Use in Electrotechnical Equipment Detail Specification: Fixed Aluminium Electrolytic Capacitors with Non-Solid Electrolyte SK Type. 18 pp.

—Hermetically Sealed—Tubular—Printed Circuit Mount

IECQ QC 300301/US 0002-87. Capacitors for Use in Electronic Equipment: Detail Specification: Fixed Aluminium Electrolytic Capacitors with Non-Solid Electrolyte—LY Type. 18 pp.

—Preferred Products List

CECC CECC MUAHAG Vol 1 IS 4-90. Preferred Products List; Capacitors (En, Fr, Ge). 64 pp.

—Printed Circuit Mount—Preferred Products List

CECC CECC MUAHAG Vol 1 IS 4-90. Preferred Products List; Capacitors (En, Fr, Ge). 64 pp.

Aluminum Capacitors *(Cont.)*

—Quality Assurance

BSI BS 9070: Sec 8-71. 1971 Amd 5 Fixed Capacitors of Assessed Quality: Generic Data and Methods of Test Section 8: Aluminium Electrolytic Capacitors. 16 pp.
BSI BS 9070: Sec 8-06. (WITHDRAWN) 1971 Amd 6 Fixed Capacitors of Assessed Quality: Generic Data and Methods of Test Section 8: Aluminium Electrolytic Capacitors (AMD 7399) March 15, 1993. 17 pp.
BSI BS 9078 N002-74. (WITHDRAWN) 1974 Amd 2 Detail Specification for Fixed Aluminium Electrolyic (Type 1) Capacitors with Non-Solid Electrolyte. Polarized Cylindrical Insulated Case, Screw Terminations at One End. Full Assessment. 14 pp.
BSI BS 9930: Part 03.0-88. 1988 Fixed Capacitors for Use in Electronic Equipment Part 03.0: Aluminium Electrolytic Capacitors with Solid or Non-Solid Electrolyte. 32 pp.
BSI BS QC 300301-93. 1993 Blank Detail Specification Fixed Capacitors for Use in Electronic Equipment. Part 4: Blank Detail Specification: Aluminium Electrolytic Capacitors with Non-Solid Electrolyte Assessment Level E. 16 pp.
BSI BS CECC 30300-92. 1992 Sectional Specification: Aluminium Electrolytic Capacitors with Solid and Non-Solid Electrolyte. 46 pp.
BSI BS CECC 30300-78. 1978 Amd 4 Aluminium Electrolytic Capacitors: Sectional Specification. 46 pp.
BSI BS CECC 30301-93. 1993 Blank Detail Specification Aluminium Electrolytic Capacitors with Non-Solid Electrolyte (Assessment Level E) (T). 23 pp.
BSI BS CECC 30301-77. 1977 Aluminium Electrolytic Capacitors with Non-Solid Electrolyte: Blank Detail Specification. 9 pp.
BSI BS CECC 30301 024-81. 1981 Amd 1 Aluminium Electrolytic Capacitors with Non-Solid Electrolyte. Long-Life Grade: Detail Specification: Full Plus Additional Assessment Level. 28 pp.
BSI BS CECC 30302-78. 1978 Aluminium Electrolytic Capacitors with Solid Electrolye: Blank Detail Specification. 9 pp.
CENELEC PREN 137 100-92. Sectional Specification: Fixed a.c. Aluminium Electrolytic Motor Starter Capacitors Qualification Approval. 30 pp.
CENELEC PREN 137 101-92. Blank Detail Specification: Fixed a.c. Aluminium Electrolytic Motor Starter Capacitors Qualification Approval. 13 pp.
IEC 384 Pt 4-85. Fixed Capacitors for Use in Electronic Equipment Part 4: Sectional Specification: Aluminium Electrolytic Capacitors with Solid and Non-Solid Electrolyte Second Edition; (Amendment 1-1992) (IECQ QC 300300). 72 pp.
IEC 384 Pt 4-1-85. Fixed Capacitors for Use in Electronic Equipment Part 4: Blank Detail Specification: Aluminium Electrolytic Capacitors with Non-Solid Electrolyte Assessment Level E First Edition; (Amendment 1-1992) (IECQ QC 300301). 38 pp.

—Reliability Assured

CNS C7191-87. Reliability Assured Aluminium Electrolytic Capacitors with Non-Solid Electrolyte (Apr)(11897).
MOD UK DSTAN 59-44: Pt 8:Sec 1-81. Capacitors, Fixed, of Assessed Quality (Listed on EPIC Database) Part 8: Capacitors, Fixed, Electrolytic-Solid Electrolyte, Aluminium Section 1: Sectional Requirements and Index to Sections Issue 1. 8 pp.
MOD UK DSTAN 59-44: Pt 8:Sec 2-81. Capacitors, Fixed, of Assessed Quality (Listed on EPIC Database) Part 8: Capacitors, Fixed, Electrolytic-Solid Electrolyte, Aluminium Section 2: List of Items Conforming to CECC 30302-001 Issue 1. 11 pp.
MOD UK DSTAN 59-44: Pt 9:Sec 1-80. Capacitors, Fixed, of Assessed Quality (Listed on EPIC Database) Part 9: Capacitors, Fixed, Electrolytic, Non-Solid Electrolyte, Aluminium Section 1: Sectional Requirements and Index to Sections Issue 1. 9 pp.
MOD UK DSTAN 59-44: Pt 9:Sec 2-80. Capacitors, Fixed, of Assessed Quality (Listed on EPIC Database) Part 9: Capacitors, Fixed, Electrolytic, Non-Solid Electrolyte, Aluminium Section 2: List of Items Conforming to BS 9078 F003 Issue 1. 11 pp.

—Solid

CECC CECC 30 302 ISSUE 1-78. Blank Detail Specification: Aluminium Electrolytic Capacitors with Solid Electrolyte (En, Fr, Ge) AMD 1 (En, Fr, Ge). 24 pp.
IECQ QC 300300-85. Fixed Capacitors for Use in Electronic Equipment: Part 4: Sectional Specification: Aluminium Electrolytic Capacitors with Solid and Non-Solid Electrolyte (Amendment 1-1992) (IEC 384-4-1 ED 2). 71 pp.

Aluminum Capacitors *(Cont.)*

—Solid *(Cont.)*

IECQ QC 300302-85. Fixed Capacitors for Use in Electronic Equipment Part 4: Blank Detail Specification: Aluminium Electrolytic Capacitors with Solid Electrolyte Assessment Level E (Amendment 1-1992) (IEC 384-4-2 ED 1). 36 pp.

—Solid—Hermetically Sealed—Cylindrical—Printed Circuit Mount

CECC CECC 30 302-001. NEN CECC 30 302-001 Edition 2; Aluminium Electrolytic Capacitors with Solid Electrolyte (En). 16 pp.
CECC CECC 30 302-001 ISSUE 2. NL CECC 30 302-001 Edition 2; Aluminium Electrolytic Capacitors with Solid Electrolyte (En). 10 pp.

—Solid—Printed Circuit Mount

CECC CECC 30 302-002 ED 1-79. NEN CECC 30 302-002; Aluminium Electrolytic Capacitors with Solid Electrolyte (En) AMD (En, Ne). 16 pp.

—Solid—Quality Assurance

IEC 384 Pt 4-85. Fixed Capacitors for Use in Electronic Equipment Part 4: Sectional Specification: Aluminium Electrolytic Capacitors with Solid and Non-Solid Electrolyte Second Edition; (Amendment 1-1992) (IECQ QC 300300). 72 pp.
IEC 384 Pt 4-2-85. Fixed Capacitors for Use in Electronic Equipment Part 4: Blank Detail Specification: Aluminium Electrolytic Capacitors with Solid Electrolyte Assessment Level E First Edition; (Amendment 1-1992) (IECQ QC 300302). 36 pp.

—Tubular—Preferred Products List

CECC CECC MUAHAG Vol 1 IS 4-90. Preferred Products List; Capacitors (En, Fr, Ge). 64 pp.

Aluminum Castings

Use For: Aluminum Alloy Castings
See Also: Castings

BSI BS 1490-88. 1988 Amd 1 Aluminium and Aluminium Alloy Ingots and Castings for General Engineering Purposes (V) (AMD 6309) March 30, 1990. 27 pp.
BSI BS 1490-02. 1988 Amd 2 Aluminium and Aluminium Alloy Ingots and Castings for General Engineering Purposes (AMD 6605) December 21, 1990. 29 pp.
CNS B1044-87. Permissible Deviations in Dimensions Without Tolerance Indication (Aluminium Alloy Castings) (Jun)(4025).
JIS B 0414-78. Permissible Deviations in Dimensions Without Tolerance Indication for Aluminium Alloy Castings. 6 pp.
JIS H 5202-92. Aluminium Alloy Castings. 19 pp.
JIS H 9151-66. Recommended Practice for Aluminium Alloy Castings (R 1972). 23 pp.
SAA AS 1874-88. Aluminium and Aluminium Alloys—Ingots and Castings. 14 pp.
SNZ NZS/AS 1874-88. Aluminium and Aluminium Alloys—Ingots and Castings (This is a Joint Standard with SAA AS 1874). 14 pp.

—Acceptance Testing

ISO 7722-85. Aluminium Alloy Castings Produced by Gravity, Sand, or Chill Casting, or by Related Processes—General Conditions for Inspection and Delivery First Edition. 10 pp.

—Aerospace

AECMA PREN2076-01-87. Aluminium and Magnesium Alloy Ingots and Castings—Technical Specification—Part 1—General Requirements. 10 pp.
AECMA PREN2076-02-87. Aluminium and Magnesium Alloy Ingots and Castings—Technical Specification—Part 2—Ingots for Remelting. 4 pp.
AECMA PREN2076-03-87. Aluminium and Magnesium Alloy Ingots and Castings—Technical Specification—Part 3—Pre-Production and Production Castings. 15 pp.
BSI BS EN 2076: Part 1-91. 1991 Aluminium and Magnesium Alloy Ingots and Castings. Technical Specification Part 1: General Requirements. 13 pp.
BSI BS EN 2076: Part 1-01. 1991 Amd 1 Aluminium and Magnesium Alloy Ingots and Castings Technical Specification Part 1: General Requirements (AMD 7706) October 15, 1993 (S). 19 pp.
BSI BS EN 2076: Part 2-91. 1991 Aluminium and Magnesium Alloy Ingots and Castings. Technical Specification Part 2: Ingots for Remelting. 7 pp.
BSI BS EN 2076: Part 3-91. 1991 Aluminium and Magnesium Alloy Ingots and Castings. Technical Specification Part 3: Pre-Production and Production Castings. 18 pp.
BSI 3L 101-70. (WITHDRAWN) 1970 Amd 1 Procedure Inspection and Testing of Aluminium-Base and Magnesium-Base Ingots and Castings (Superseded by 4L 101: 1990). 15 pp.
BSI 4L 101-90. 1990 Procedure for Inspection, Testing and Acceptance of Aluminium-Base and Magnesium-Base Ingots and Castings. 21 pp.

INTERNATIONAL AND NON-U.S. NATIONAL STANDARDS
SUBJECT INDEX
Aluminum

Aluminum Castings (Cont.)
—Aerospace (Cont.)
CEN EN 2076-1-89. Aerospace Series Aluminium and Magnesium Alloy Ingots and Castings Technical Specification Part 1: General Requirements. 10 pp.

CEN EN 2076 Part 1-89. Aerospace Series Aluminium and Magnesium Alloy Ingots and Castings Technical Specification Part 1: General Requirements. 12 pp.

CEN EN 2076-2-89. Aerospace Series Aluminium and Magnesium Alloy Ingots and Castings Technical Specification Part 2: Ingots for Remitting. 4 pp.

CEN EN 2076-3-89. Aerospace Series Aluminium and Magnesium Alloy Ingots and Castings Technical Specifications Part 3: Re—Production and Production Castings. 15 pp.

—Aircraft
CAA Chapter K3-10 10.69. Castings. 2 pp.

CAA Chapter P3-10. Castings (Provisional Airworthiness Requirements for Civil Powered Lift Aircraft). 3 pp.

CAA Chapter Q3-10 12.79. Castings. 4 pp.

—Chill
ISO 2378-72. Aluminium Alloy Chill Castings—Reference Test Bar First Edition. 4 pp.

—Chill—Aerospace
AECMA PREN2722-86. Aluminium Alloy A1-C12-T4 Chill Castings. 3 pp.

AECMA PREN2724-86. Aluminium Alloy A1-C12-T6 Chill Castings. 3 pp.

AECMA PREN2727-86. Aluminium Alloy A1-C26-T6 Chill Castings. 3 pp.

AECMA PREN2729-86. Aluminium Alloy A1-C27-T6 Chill Castings. 3 pp.

BSI L 173-91. 1991 Castings of Aluminium-Silicon-Magnesium Alloy, Chill Cast (Solution Treated and Precipitation Treated to an Overaged (T7) Condition). 6 pp.

CEN EN 3549-92. Aluminium Alloy AL-C21002 T7 Chill Castings. 4 pp.

—Die
DIN ENGL 1725 Pt 2-86. Aluminium Alloys; Casting Alloys; Sand Casting; Gravity Die Casting; Pressure Die Casting; Investment Casting (Feb). 15 pp.

JIS H 5302-90. Aluminium Alloys Die Castings. 10 pp.

—Die—Ingots
JIS H 2118-90. Aluminium-Base Alloys in Ingot for Die Castings. 7 pp.

—Filter Heads—Aircraft
SBAC AS 3311 (V). Filter Head. 2 pp.

—Food Preparation Equipment—Chemical Analysis
CEN PREN 601-91. Cast Aluminium Alloys—Chemical Composition of the Metal Used for the Production of Materials and Articles Intended to Be in Contact with Food. 7 pp.

—Forgings
CEN PREN 604-1-91. Cast Forging Stock in Aluminium and Aluminium Alloys—Part 1: Technical Conditions for Inspection and Delivery. 10 pp.

CEN PREN 604-2-91. Cast Forging Stock in Aluminium and Aluminium Alloys—Part 2: Dimensional Tolerances. 6 pp.

—Heat Treated—Aerospace
BSI 4L 35-70. 1970 Amd 1 Ingots and Castings of 'Y' Aluminium Alloy (Heat Treated) (Suitable for Pistons). 3 pp.

BSI 3L 51-70. 1970 Amd 1 Ingots and Castings of Aluminium-Silicon-Copper-Iron-Nickel-Magnesium Alloy (Precipitation Treated). 3 pp.

BSI 3L 78-70. 1970 Amd 1 Ingots and Castings of Aluminium-Silicon-Copper-Magnesium Alloy (Solution Treated and Precipitation Treated). 3 pp.

BSI 2L 99-72. 1972 Amd 2 Ingots and Castings of Aluminium-Silicon-Magnesium Alloy (Solution Treated and Precipitation Treated) (Si 7, Mg 0.3) (AMD 6543) March 28, 1991. 6 pp.

BSI L 119-75. 1975 Amd 2 Ingots and Castings of Aluminium-Copper-Nickel-Manganese-Titanium-Zirconium-Cobalt-Antimony Alloy (Solution Treated and Art. Aged) (Cu 5.0, Ni 1.5, Mn 0.25, Ti0.2, Zr 0.2, Co 0.2, Sb 0.2) (AMD 6544) December 21, 1990. 6 pp.

BSI L 154-77. 1977 Amd 2 Ingots and Castings of Aluminium-Copper-Silicon Alloy (Solution Treated and Aged at Room Temperature) (Cu 4, Si 1) (AMD 6546) April 30, 1991. 4 pp.

BSI L 155-77. 1977 Amd 2 Ingots and Castings of Aluminium-Copper-Silcon Alloy (Solution Treated and Artificially Aged) (Cu 4, Si 1) (AMD 6545) March 28, 1991. 5 pp.

Aluminum Castings (Cont.)
—Heat Treated—Aerospace (Cont.)
BSI L 169-86. 1986 Amd 1 Ingots and Castings of Aluminium-Silicon-Magnesium Alloy (Solution Treated and Artificially Aged) (Si 7, Mg 0.6) (AMD 6547) December 21, 1990. 6 pp.

MOD UK DTD-722B-71. Ingots and Castings of Aluminium-Silicon-Magnesium Alloy (Precipitation Treated) (Si 5, Mg 0.5). 2 pp.

MOD UK DTD-727B-71. Ingots and Castings of Aluminium-Silicon-Magnesium Alloy (Heat Treated) (Si 5, Mg 0.5). 2 pp.

MOD UK DTD-731B-71. Forging Stock and Forgings of Aluminium—Copper—Magnesium—Nickel—Iron Alloy (Solution Treated and Precipitation Treated at 200 Degrees C) (Suitable for Use at Elevated Temperatures) (Cu 2.5, Mg 1.5, Ni 1.2, Fe 1.0). 2 pp.

MOD UK DTD-735B-71. Ingots and Castings of Aluminium-Silicon-Magnesium Alloy (Solution Treated and Precipitation Treated) (Si 5, Mg 0.5). 2 pp.

MOD UK DTD-5008B-71. Ingots and Castings of Aluminium-Zinc-Magnesium-Chromium Alloy (Age Hardened) (Zn 5.2, Mg 0.6, Cr 0.5). 2 pp.

MOD UK DTD-5018A-71. Ingots and Castings of Aluminium-Magnesium-Zinc Alloy (Solution Treated) (Mg 7.7, Zn 1.2). 2 pp.

—Ingots
CNS H3143-87. Aluminium and Aluminium Alloy Ingots for Castings (Jun)(12000).

JIS H 2117-84. Secondary Aluminium Alloy Ingots for Castings. 7 pp.

JIS H 2118-90. Aluminium-Base Alloys in Ingot for Die Castings. 7 pp.

JIS H 2211-92. Aluminium Alloy Ingots for Castings. 8 pp.

—Investment
DIN ENGL 1725 Pt 2-86. Aluminium Alloys; Casting Alloys; Sand Casting; Gravity Die Casting; Pressure Die Casting; Investment Casting (Feb). 15 pp.

—Liquid Penetrant Testing
ISO 9916-91. Aluminium Alloy and Magnesium Alloy Castings —Liquid Penetrant Inspection First Edition. 16 pp.

—Porosity—Visual Inspection
ISO 10049-92. Aluminium Alloy Castings —Visual Method for Assessing the Porosity First Edition. 6 pp.

—Precipitation Hardened—Aerospace
MOD UK DTD-5008B-71. Ingots and Castings of Aluminium-Zinc-Magnesium-Chromium Alloy (Age Hardened) (Zn 5.2, Mg 0.6, Cr 0.5). 2 pp.

—Precision—Aerospace
CEN EN 3546-92. Aluminium Alloy AL-C42201 T6 Precision Castings. 5 pp.

CEN EN 3547-92. Aluminium Alloy AL-C42101 T6 Precision Castings. 5 pp.

—Precision—Solution Treated—Aerospace—Mechanical Properties
AECMA PREN3122-88. Aluminium Alloy AL-C27 Solution Treated and Artificially Aged (T6) High Strength Structural Precision Castings. 4 pp.

AECMA PREN3123-88. Aluminium Alloy AL-C27 Solution Treated and Artificially Aged (T6) Structural Precision Castings. 4 pp.

AECMA PREN3124-88. Aluminium Alloy AL-C26 Solution Treated and Artifically Aged (T6) High Strength Structural Precision Castings. 4 pp.

AECMA PREN3125-88. Aluminium Alloy AL-C26 Solution Treated and Artifically Aged (T6) Structural Precision Castings. 4 pp.

—Radiography
ISO 9915-92. Aluminium Alloy Castings —Radiography Testing First Edition. 21 pp.

JIS H 0522-69. Methods of Radiographic Test and Classification of Radiographs for Aluminium Castings (R 1975). 12 pp.

—Sand
DIN ENGL 1725 Pt 2-86. Aluminium Alloys; Casting Alloys; Sand Casting; Gravity Die Casting; Pressure Die Casting; Investment Casting (Feb). 15 pp.

ISO 2379-72. Aluminium Alloy Sand Castings—Reference Test Bar First Edition. 3 pp.

—Sand—Aerospace
AECMA PREN2721-86. Aluminum Alloy A1-C12-T4 Sand Castings. 3 pp.

AECMA PREN2723-86. Aluminium Alloy A1-C12-T6 Sand Castings. 3 pp.

AECMA PREN2725-86. Aluminium Alloy A1-C14-T6 Sand Castings. 3 pp.

AECMA PREN2726-86. Aluminium Alloy A1-C26-T6 Sand Castings. 3 pp.

AECMA PREN2728-86. Aluminium Alloy A1-C27-T6 Sand Castings. 3 pp.

Aluminum Castings (Cont.)
—Sand—Aerospace (Cont.)
BSI L 174-91. 1991 Castings of Aluminium-Silicon-Magnesium Alloy, Sand Cast (Solution Treated and Precipitation Treated to an Overaged (T7) Condition). 6 pp.

CEN EN 3548-92. Aluminium Alloy AL-C21002 T7 Sand Castings. 5 pp.

—Stress Corrosion Cracking
ISO 9591-92. Corrosion of Aluminium Alloys—Determination of Resistance to Stress Corrosion Cracking First Edition. 10 pp.

—Test Specimens
ISO 3522-84. Cast Aluminium Alloys—Chemical Composition and Mechanical Properties Second Edition. 6 pp.

Aluminum Chip Capacitors
See Also: Aluminum Capacitors; Chip Capacitors; Electrolytic Capacitors; Fixed Capacitors

—Quality Assurance
IEC 384 Pt 18-93. Fixed Capacitors for Use in Electronic Equipment Part 18: Sectional Specification: Fixed Aluminium Electrolytic Chip Capacitors with Solid and Non-Solid Electrolyte First Edition (IECQ QC 302300). 73 pp.

IEC 384 Pt 18-1-93. Fixed Capacitors for Use in Electronic Equipment Part 18: Blank Detail Specification: Fixed Aluminium Electrolytic Chip Capacitors with Solid Electrolyte Assessment Level E First Edition (IECQ QC 302301). 37 pp.

IEC 384 Pt 18-2-93. Fixed Capacitors for Use in Electronic Equipment Part 18: Blank Detail Specification: Fixed Aluminium Electrolytic Chip Capacitors with Non-Solid Electrolyte Assessment Level E First Edition (IECQ QC 302302). 39 pp.

Aluminum Chloride
See Also: Aluminum Chlorohydrate; Hazardous Materials

CNS K7033-61. Chemical Reagent (Aluminum Chloride, Anhydrous) (Dec)(1533).

CNS K7034-61. Chemical Reagent (Aluminum Chloride, Hydrated) (Dec)(1534).

JIS K 8114-78. Aluminium Chloride, Hexahydrate.

JIS K 8115-78. Aluminium Chloride, Anhydrous.

—Fabrics—Colorfastness Testing
CNS L3187-82. Method of Test for Colour Fastness to Carbonizing with Aluminium Chloride (Aug)(9310).

JIS L 0866-76. Testing Method for Colour Fastness to Carbonizing with Aluminium Chloride.

—Water Treatment
CEN PREN 881-92. Aluminium Chloride, Hydroxidechloride and Hydroxidechloridesulfate (Mononumeric) Used for Water Intended for Human Consumption. 14 pp.

CEN PREN 935-92. Aluminium-Iron Chloride and Aluminium Hydroxidechloride (Monomeric) Used for Water Intended for Human Consumption. 13 pp.

CNS K1278-89. Poly Aluminum Chloride for Water Works (Jun)(12537).

Aluminum Chlorohydrate
Use For: Aluminum Hydroxychloride; Poly Aluminum Hydroxychloride; Polyaluminum Hydroxychloride
See Also: Aluminum Chloride; Aluminum Chlorohydrate Silicate; Aluminum Chlorohydrate Sulfate

—Water Treatment
CEN PREN 881-92. Aluminium Chloride, Hydroxidechloride and Hydroxidechloridesulfate (Mononumeric) Used for Water Intended for Human Consumption. 14 pp.

CEN PREN 883-92. Polyaluminium Hydroxidechloride and Hydroxidechloridesulfate, Used for Water Intended for Human Consumption. 14 pp.

CEN PREN 884-92. Polyaluminium Hydroxidechloride and Hydroxidechloridesulfate, Special Grade, Used for Water Intended for Human Consumption. 14 pp.

DIN ENGL 19634-85. Aluminium Hydroxide Chlorides for Water Treatment; Technical Delivery Conditions (Mar). 4 pp.

—Waterworks
JIS K 1475-78. Poly Aluminium Chloride for Waterworks. 36 pp.

Aluminum Chlorohydrate Silicate
See Also: Aluminum Chlorohydrate; Aluminum Chlorohydrate Sulfate; Silicates

—Water Treatment
CEN PREN 885-92. Polyaluminium Hydroxidechloridesili-cate, Used for Water Intended for Human Consumption. 13 pp.

Aluminum

Aluminum Chlorohydrate Silicate (Cont.)
—**Water Treatment** (Cont.)
CEN PREN 886-92. Polyaluminium Hydroxidesulfatesilicate, Used for Water Intended for Human Consumption. 13 pp.

Aluminum Chlorohydrate Sulfate
See Also: Aluminum Chlorohydrate; Aluminum Chlorohydrate Silicate; Sulfates
—**Water Treatment**
CEN PREN 881-92. Aluminium Chloride, Hydroxidechloride and Hydroxidechloridesulfate (Mononumeric) Used for Water Intended for Human Consumption. 14 pp.
CEN PREN 883-92. Polyaluminium Hydroxidechloride and Hydroxidechloridesulfate, Used for Water Intended for Human Consumption. 14 pp.
CEN PREN 884-92. Polyaluminium Hydroxidechloride and Hydroxidechloridesulfate, Special Grade, Used for Water Intended for Human Consumption. 14 pp.

Aluminum Chromium Molybdenum Steels
Scope Note: For additional listings, see also specific products made from aluminum chromium molybdenum steel *See Also:* Alloy Steels; Chromium Molybdenum Steels; Low Alloy Steels; Steels
—**Hot Worked**
JIS G 4202-79. Aluminium Chromium Molybdenum Steels. 13 pp.
—**Structural**
CNS G3106-87. Aluminum Chromium Molybdenum Steels for Machine Structural Use (May)(4444). 9 pp.

Aluminum Coatings (Made From Aluminum)
Scope Note: Includes coatings made from aluminum alloys *Use For:* Aluminizing *See Also:* Aluminum; Aluminum Coatings (On Aluminum); Aluminum Primers (Made From Aluminum); Coatings; Metal Coatings (Made From Metal)
CSA G189-1966. Sprayed Metal Coatings for Atmospheric Corrosion Protection (R 1992). 16 pp.

—**Aerospace**
BSI 2L 88-71. 1971 Aluminium-Alloy Coated Sheet and Strip of Aluminium-Zinc-Magnesium-Copper-Chromium Alloy (Solution Treated and Precipitation Treated) (Zn 5.8, Mg 2.5, Cu 1.6, Cr 0.15). 3 pp.
BSI L 109-71. 1971 Amd 1 Aluminium-Coated Sheet and Strip of Aluminium-Copper-Magnesium-Manganese Alloy (Solution Treated and Aged at Room Temperature). 4 pp.
BSI L 110-71. 1971 Amd 1 Aluminium-Coated Sheet and Strip of Aluminium-Copper-Magnesium-Manganese Alloy (Supplied for Solution Treatment by the User). 4 pp.
BSI L 163-78. 1978 Sheet and Strip of Aluminium-Coated Aluminium-Copper-Magnesium-Silicon-Manganese Alloy (Cu 4.4, Mg 0.5, Si 0.8, Mn 0.8). 3 pp.
BSI L 164-78. 1978 Sheet and Strip of Aluminium-Coated Aluminium-Copper-Magnesium-Silicon-Manganese Alloy (Cu 4.4, Mg 0.5, Si 0.8, Mn 0.8). 4 pp.
BSI L 165-78. 1978 Amd 1 Specification for Sheet and Strip of Aluminium-Coated Aluminium-Copper-Magnesium-Silicon-Manganese Alloy (Solution Treated and Artificially Aged) (Cu 4.4, Mg 0.5, Si 0.8, Mn 0.8) (AMD 6391) July 31, 1991. 4 pp.
BSI L 166-78. 1978 Close Toleranced Sheet and Strip of Aluminium-Coated Aluminium-Copper-Magnesium-Silicon-Manganese Alloy (Cu 4.4, Mg 0.5, Si 0.8, Mn 0.8). 4 pp.
BSI L 167-78. 1978 Amd 1 Specification for Close Toleranced Sheet and Strip of Aluminium-Coated Aluminium-Copper-Magnesium-Silicon-Manganese Alloy (Solution Treated and Art. Aged) (Cu 4.4, Mg 0.5, Si 0.8, Mn 0.8) (AMD 6392) July 31, 1991. 4 pp.

—**Anodic—Primers—Adhesion Testing**
CGSB 1-GP-71 METH 135.5-78. Methods of Testing Paints and Pigments Adhesion Adhesion of Primer to Pretreated Anodized Aluminum. 1 p.
CGSB 1-GP-71 METH 135.7-78. Methods of Testing Paints and Pigments Adhesion Adhesion of Topcoated Primers; (Amended September 1980) Re-Edited September 1982). 1 p.

—**Carbon Steels—Hot Dip**
BSI BS 6536-85. 1985 Continuously Hot-Dip Aluminium/Silicon Coated Cold Reduced Carbon Steel Sheet and Strip. 16 pp.

Aluminum Coatings (Made From Aluminum) (Cont.)
—**Carbon Steels—Hot Dip** (Cont.)
ISO 5000-80. Continuous Hot-Dip Aluminium-Silicon-Coated Cold-Reduced Carbon Steel Sheet of Commercial and Drawing Qualities First Edition; DoD Adopted. 15 pp.

—**Heat Resistant—Interior**
MOD UK DSTAN 80-9-01. Paint, Aluminium, Flame Resisting Issue 1; Amendment 1. 11 pp.
MOD UK DSTAN 80-9-92. Paint, Aluminium, Flame Resisting Issue 2. 16 pp.
MOD UK DSTAN 80-11-01. Paint, Aluminium, Flame Resisting, Stoving Issue 1; Amendment 1. 12 pp.
MOD UK DSTAN 80-11-92. Paint, Aluminium, Flame Resisting, Stoving, Spraying Issue 2. 16 pp.

—**Iron—Hot Dip**
CNS H2058-82. Methods of Test for Aluminium Coating (Hot-Dipped) on Iron and Steel (Jan)(8296). 5 pp.
CNS H3099-82. Aluminum Coatings (Hot-Dipped) on Iron and Steel (Jan)(8295).
CNS H3100-82. Recommended Practice for Aluminum Coatings (Hot-Dipped) (Jan)(8297).
JIS H 8642-72. Aluminium Coating (Hot-Dipped) on Iron or Steel. 5 pp.
JIS H 8672-72. Methods of Test for Aluminium Coating (Hot-Dipped) on Iron or Steel. 12 pp.
JIS H 9126-68. Recommended Practice for Aluminium Coating (Hot-Dipped). 9 pp.

—**Iron—Sprayed**
BSI BS 2569: Part 1-64. 1964 Amd 1 Sprayed Metal Coatings Part 1: Protection of Iron and Steel by Aluminium and Zinc Against Atmospheric Corrosion. 14 pp.
CNS H2057-82. Methods of Test for Aluminium Spray Products (Jan)(8293). 6 pp.
CNS H3095-81. Aluminium Spray Coatings on Iron and Steel (Oct)(8009).
CNS H3098-82. Recommended Practice for Aluminum Spray Coatings on Iron and Steel (Jan)(8294).
CSA G189-1966. Sprayed Metal Coatings for Atmospheric Corrosion Protection (R 1992). 16 pp.
ISO 2063-91. Metallic and Other Inorganic Coatings—Thermal Spraying—Zinc, Aluminium and Their Alloys Second Edition. 14 pp.
JIS H 8301-90. Aluminium Spraying on Iron and Steel. 6 pp.
JIS H 8663-61. Testing Method for Aluminium Spray Products. 9 pp.
JIS H 9301-90. Recommended Practice for Aluminium Spray Coatings on Iron or Steel. 9 pp.

—**Metal—Vacuum Deposited**
MOD UK DSTAN 03-28: Part 1-88. Physical Vapour Deposition of Metals Part 1: Ion Vapour Deposition of Aluminium for Protection Against Corrosion Issue 1. 19 pp.

—**Steel**
SAA AS 1445-86. Hot-Dipped Zinc-Coated or Aluminium/Zinc-Coated Steel Sheet—76mm Pitch Corrugated. 4 pp.

—**Steel—Hot Dip**
CNS G2164-83. Method of Test for Hot-Dip Aluminium Coated Steel Sheets and Strips (Feb)(9999).
CNS G3206-83. Hot-Dip Aluminium Coated Steel Sheets and Strips (Feb)(9998).
CNS H2058-82. Methods of Test for Aluminium Coating (Hot-Dipped) on Iron and Steel (Jan)(8296). 5 pp.
CNS H3099-82. Aluminum Coatings (Hot-Dipped) on Iron and Steel (Jan)(8295).
CNS H3100-82. Recommended Practice for Aluminum Coatings (Hot-Dipped) (Jan)(8297).
ISO 9364-91. Continuous Hot-Dip Aluminium/Zinc-Coated Steel Sheet of Commercial, Lock-Forming and Structural Qualities First Edition. 15 pp.
JIS G 3314-77. Hot-Dip Aluminium Coated Steel Sheets (R 1980). 15 pp.
JIS H 8642-72. Aluminium Coating (Hot-Dipped) on Iron or Steel. 5 pp.
JIS H 8672-72. Methods of Test for Aluminium Coating (Hot-Dipped) on Iron or Steel. 12 pp.
JIS H 9126-68. Recommended Practice for Aluminium Coating (Hot-Dipped). 9 pp.
SAA AS 1397-93. Steel Sheet and Strip—Hot-Dipped Zinc-Coated or Aluminium/Zinc-Coated. 18 pp.

—**Steel—Sprayed**
BSI BS 2569: Part 1-64. 1964 Amd 1 Sprayed Metal Coatings Part 1: Protection of Iron and Steel by Aluminium and Zinc Against Atmospheric Corrosion. 14 pp.
CNS H2057-82. Methods of Test for Aluminium Spray Products (Jan)(8293). 6 pp.
CNS H3095-81. Aluminium Spray Coatings on Iron and Steel (Oct)(8009).

Aluminum Coatings (Made From Aluminum) (Cont.)
—**Steel—Sprayed** (Cont.)
CNS H3098-82. Recommended Practice for Aluminum Spray Coatings on Iron and Steel (Jan)(8294).
CSA G189-1966. Sprayed Metal Coatings for Atmospheric Corrosion Protection (R 1992). 16 pp.
DIN ENGL 8565-77. Protection Against Corrosion of Steel Structures by Thermal Spraying of Zinc and Aluminium; General Principles (Mar). 4 pp.
ISO 2063-91. Metallic and Other Inorganic Coatings—Thermal Spraying—Zinc, Aluminium and Their Alloys Second Edition. 14 pp.
JIS H 8301-90. Aluminium Spraying on Iron and Steel. 6 pp.
JIS H 8663-61. Testing Method for Aluminium Spray Products. 9 pp.
JIS H 9301-90. Recommended Practice for Aluminium Spray Coatings on Iron or Steel. 9 pp.

Aluminum Coatings (On Aluminum)
Scope Note: Includes coatings on aluminum alloys *See Also:* Aluminum; Aluminum Coatings (Made From Aluminum); Aluminum Primers (On Aluminum); Coatings; Metal Coatings (On Metal)

—**Accelerated Testing**
CGSB 1-GP-71 METH 122.8-79. Methods of Testing Paints and Pigments Accelerated Weathering Test for One Coat of Air-Drying Material on Steel or Aluminum Panels. 1 p.

—**Anodic**
CGSB 1-GP-71 METH 99.2-74. Methods of Testing Paints and Pigments Preparation of Aluminum Panels Chromic Acid Anodized Aluminum Panels. 1 p.

—**Anodic—Abrasion Testing**
SAA AS 2039.5.1-78. Methods for Testing Anodic Oxidation Coatings on Aluminium and Aluminium Alloys—Part 5.1: Abrasion Resistance Tests—Jet Test for Abrasion Resistance of Anodic Oxidation Coatings Reconfirmed 1990. 4 pp.
SAA AS 2039.5.2-78. Methods for Testing Anodic Oxidation Coatings on Aluminium and Aluminium Alloys—Part 5.2: Abrasion Resistance Tests—ASTM Test for Abrasion Resistance of Anodic Oxidation Coatings Reconfirmed 1990. 6 pp.

—**Anodic—Accelerated Testing**
CGSB 1-GP-71 METH 122.10-79. Methods of Testing Paints and Pigments Accelerated Weathering Test for Aircraft Finishes on Primed Anodized Aluminum Panels. 1 p.

—**Anodic—Adhesion Testing**
CGSB 1-GP-71 METH 135.6-78. Methods of Testing Paints and Pigments Adhesion Adhesion of Finishes to Primed Anodized Aluminum. 1 p.

—**Anodic—Aeronautical—Primers—Gasoline Resistance**
CGSB 1-GP-71 METH 107.4-74. Methods of Testing Paints and Pigments Gasoline Resistance Aeronautical Primer. 1 p.

—**Anodic—Aerospace**
AECMA PREN2101-87. Chromic Acid Anodizing of Aluminium and Wrought Aluminium Alloys.
BSI BS EN 2101-91. 1991 Chromic Acid Anodizing of Aluminium and Wrought Aluminium Alloys. 19 pp.
CEN PREN 2101-89. Chromic Acid Anodizing of Aluminium and Wrought Aluminium Alloys. 14 pp.
CEN EN 2101-91. Chromic Acid Anodizing of Aluminium and Wrought Aluminium Alloys. 16 pp.
CEN PREN 2284-89. Sulphuric Acid Anodizing of Aluminium and Wrought Aluminium Alloys. 9 pp.
CEN EN 2284-91. Sulphuric Acid Anodizing of Aluminium and Wrought Aluminium Alloys. 11 pp.
CEN PREN 2536-92. Aerospace Series Hard Anodizing of Aluminium Alloys. 11 pp.
ISO 8076-84. Aerospace Process—Anodic Treatment of Aluminium Alloys—Chromic Acid Process 40 V DC, Undyed Coating First Edition. 5 pp.
ISO 8077-84. Aerospace Process—Anodic Treatment of Aluminium Alloys—Chromic Acid Process 20 V DC, Undyed Coating First Edition. 5 pp.
ISO 8078-84. Aerospace Process—Anodic Treatment of Aluminium Alloys—Sulfuric Acid Process, Undyed Coating First Edition. 5 pp.
ISO 8079-84. Aerospace Process—Anodic Treatment of Aluminium Alloys—Sulfuric Acid Process, Dyed Coating First Edition. 5 pp.

—**Anodic—Aerospace—Adhesion Testing**
AECMA PREN3002-90. Chromic Acid Anodizing Testing of Adhesives. 8 pp.

—**Anodic—Aerospace—Quality Assurance**
AECMA PREN2101-87. Chromic Acid Anodizing of Aluminium and Wrought Aluminium Alloys.

INTERNATIONAL AND NON-U.S. NATIONAL STANDARDS
SUBJECT INDEX
Aluminum

Aluminum Coatings (On Aluminum) (Cont.)

—Anodic—Aerospace—Quality Assurance (Cont.)
AECMA PREN2284-87. Sulphuric Acid Anodizing of Aluminium and Wrought Aluminium Alloys.
BSI BS EN 2101-91. 1991 Chromic Acid Anodizing of Aluminium and Wrought Aluminium Alloys. 19 pp.
BSI BS EN 2284-91. 1991 Sulfuric Acid Anodizing of Aluminium and Wrought Aluminium Alloys. 14 pp.
CEN PREN 2101-89. Chromic Acid Anodizing of Aluminium and Wrought Aluminium Alloys. 14 pp.
CEN EN 2101-91. Chromic Acid Anodizing of Aluminium and Wrought Aluminium Alloys. 16 pp.
CEN PREN 2284-89. Sulphuric Acid Anodizing of Aluminium and Wrought Aluminium Alloys. 9 pp.
CEN EN 2284-91. Sulphuric Acid Anodizing of Aluminium and Wrought Aluminium Alloys. 11 pp.

—Anodic—Aircraft
JIS W 1111-83. Anodic Coatings on Aluminum and Aluminum Alloys for Aircraft.

—Anodic—Architectural
SAA AS 1231-85. Aluminium and Aluminium Alloys—Anodized Coatings for Architectural Applications (R 1993). 7 pp.
SNZ NZS/AS 1231-85. Aluminium and Aluminium Alloys—Anodized Coatings for Architectural Applications (This is a Joint Standard with SAA AS 1231). 7 pp.

—Anodic—Automotive
SAA AS 1956-76. Anodic Oxidation Coatings on Aluminium for Decorative and Automotive Applications. 16 pp.

—Anodic—Baking Enamels
CGSB 1-GP-71 METH 112.4-82. Methods of Testing Paints and Pigments Baking Resistance Enamels. 1 p.

—Anodic—Baking Lacquers
CGSB 1-GP-71 METH 112.3-82. Methods of Testing Paints and Pigments Baking Resistance Lacquers. 1 p.

—Anodic—Comprehensive Testing
SAA AS 2039. Methods for Testing Anodic Oxidation Coatings on Aluminium and Aluminium Alloys Complete Set in Binder.

—Anodic—Corrosion Testing
SAA AS 2039.3.1-78. Methods for Testing Anodic Oxidation Coatings on Aluminium and Aluminium Alloys —Part 3.1: Corrosion Tests—Neutral Salt Spray Test (NSS Test) of Anodic Oxidation Coatings Reconfirmed 1990. 6 pp.
SAA AS 2039.3.2-78. Methods for Testing Anodic Oxidation Coatings on Aluminium and Aluminium Alloys —Part 3.2: Corrosion Tests—Copper Accelerated Acetic Acid Salt Spray Test (CASS Test) of Anodic Oxidation Coatings Reconfirmed 1990. 6 pp.
SAA AS 2039.3.3-78. Methods for Testing Anodic Oxidation Coatings on Aluminium and Aluminium Alloys—Part 3.3: Corrosion Tests—Mortar Test for Clear Organic Coatings on Anodic Oxidation Coatings Reconfirmed 1990. 2 pp.

—Anodic—Decorative
SAA AS 1956-76. Anodic Oxidation Coatings on Aluminium for Decorative and Automotive Applications. 16 pp.

—Anodic—Enamels—Self Lifting Resistance
CGSB 1-GP-71 METH 133.1-79. Methods of Testing Paints and Pigments Behavior Towards Undercoats and Self-Lifting Finishes on Primed Aluminum; (Amended September 1980) (Re-Edited September 1982). 1 p.

—Anodic—Enamels—Undercoatings
CGSB 1-GP-71 METH 133.1-79. Methods of Testing Paints and Pigments Behavior Towards Undercoats and Self-Lifting Finishes on Primed Aluminum; (Amended September 1980) (Re-Edited September 1982). 1 p.

—Anodic—Environmental Testing
CGSB 1-GP-71 METH 122.10-79. Methods of Testing Paints and Pigments Accelerated Weathering Test for Aircraft Finishes on Primed Anodized Aluminum Panels. 1 p.

—Anodic—Flexibility
CGSB 1-GP-71 METH 119.15-79. Methods of Testing Paints and Pigments Flexibility Finishes on Primed Aluminum. 1 p.

Aluminum Coatings (On Aluminum) (Cont.)

—Anodic—Gasoline Resistance
CGSB 1-GP-71 METH 107.5-74. Methods of Testing Paints and Pigments Gasoline Resistance Finishes on Primed Aluminum. 1 p.

—Anodic—Insulation Testing
SAA AS 2039.7.1-78. Methods for Testing Anodic Oxidation Coatings on Aluminium and Aluminium Alloys—Part 7.1: Insulation Tests—Measurement of Breakdown Potential of Anodic Oxidation Coatings Reconfirmed 1990. 2 pp.

—Anodic—Lacquers
CGSB 1-GP-71 METH 131.1-79. Methods of Testing Paints and Pigments Behavior Towards Applied Films Lacquer on Zinc Chromate Primer. 1 p.

—Anodic—Lacquers—Oil Resistance Testing
CGSB 1-GP-71 METH 109.2-79. Methods of Testing Paints and Pigments Oil Resistance Lacquers on Primed Aluminum; (Amended September 1980) (Re-Edited September 1982). 1 p.

—Anodic—Lacquers—Self Lifting Resistance
CGSB 1-GP-71 METH 133.1-79. Methods of Testing Paints and Pigments Behavior Towards Undercoats and Self-Lifting Finishes on Primed Aluminum; (Amended September 1980) (Re-Edited September 1982). 1 p.

—Anodic—Lacquers—Solvent Resistance Testing
CGSB 1-GP-71 METH 108.3-79. Methods of Testing Paints and Pigments Solvent Resistance Resistance of Lacquers to De-Icing Fluid; (Amended September 1980) (Re-Edited September 1982). 1 p.

—Anodic—Lacquers—Undercoatings
CGSB 1-GP-71 METH 133.1-79. Methods of Testing Paints and Pigments Behavior Towards Undercoats and Self-Lifting Finishes on Primed Aluminum; (Amended September 1980) (Re-Edited September 1982). 1 p.

—Anodic—Light Fastness Testing
SAA AS 2039.6.1-78. Methods for Testing Anodic Oxidation Coatings on Aluminium and Aluminium Alloys—Part 6.1: Light Fastness Tests—Grey Scale Method for Light Fastness of Anodic Oxidation Coatings Reconfirmed 1990. 6 pp.

—Anodic—Primers—Flexibility
CGSB 1-GP-71 METH 119.12-79. Methods of Testing Paints and Pigments Flexibility Primers on Pretreated Aluminum. 1 p.
CGSB 1-GP-71 METH 119.14-79. Methods of Testing Paints and Pigments Flexibility Primers on Anodized Aluminum. 1 p.

—Anodic—Primers—Gasoline Resistance
CGSB 1-GP-71 METH 107.6-74. Methods of Testing Paints and Pigments Gasoline Resistance Primers on Pretreated Aluminum. 1 p.

—Anodic—Primers—Water Resistance Testing
CGSB 1-GP-71 METH 110.8-79. Methods of Testing Paints and Pigments Water Resistance Primers on Anodized Aluminum. 1 p.
CGSB 1-GP-71 METH 110.9-82. Methods of Testing Paints and Pigments Water Resistance Primers on Pretreated Aluminum. 1 p.

—Anodic—Primers—Zinc Chromate
CGSB 1-GP-71 METH 131.1-79. Methods of Testing Paints and Pigments Behavior Towards Applied Films Lacquer on Zinc Chromate Primer. 1 p.

—Anodic—Reflectance
SAA AS 2039.4.1-78. Methods of Testing Anodic Oxidation Coatings on Aluminium and Aluminium Alloys—Part 4.1: Reflectivity Tests—Specular Reflectivity Test of Anodic Oxidation Coatings with the Gloss Head Reconfirmed 1990. 4 pp.
SAA AS 2039.4.2-78. Methods of Testing Anodic Oxidation Coatings on Aluminium and Aluminium Alloys—Part 4.2: Reflectivity Tests—Total Reflectivity Test of Anodic Oxidation Coatings with the PRS Head Reconfirmed 1990. 4 pp.
SAA AS 2039.4.3-78. Methods of Testing Anodic Oxidation Coatings on Aluminium and Aluminium Alloys—Part 4.3: Reflectivity Tests—Image Clarity Test of Anodic Oxidation Coatings with the Gardam Grid Reconfirmed 1990. 4 pp.
SAA AS 2039.4.4-78. Methods of Testing Anodic Oxidation Coatings on Aluminium and Aluminium Alloys—Part 4.4: Reflectivity Tests—Infared Reflectivity Test of Anodic Oxidation Coatings Reconfirmed 1990. 4 pp.

Aluminum Coatings (On Aluminum) (Cont.)

—Anodic—Sealing
SAA AS 2039.2.1-78. Methods for Testing Anodic Oxidation Coatings on Aluminium and Aluminium Alloys—Part 2.1: Sealing Property Tests—Measurement of Impedance of Clear Anodic Oxidation Coatings (Aztac Method) Reconfirmed 1990. 4 pp.
SAA AS 2039.2.2-78. Methods for Testing Anodic Oxidation Coatings on Aluminium and Aluminium Alloys—Part 2.2: Sealing Property Tests—Resistance of Anodic Oxidation Coatings to Staining Reconfirmed 1990. 2 pp.
SAA AS 2039.2.3-78. Methods for Testing Anodic Oxidation Coatings on Aluminium and Aluminium Alloys—Part 2.3: Sealing Property Tests—Acetic Acid Test of Anodic Oxidation Coatings Reconfirmed 1990. 2 pp.
SAA AS 2039.2.4-78. Methods for Testing Anodic Oxidation Coatings on Aluminium and Aluminium Alloys—Part 2.4: Sealing Property Tests—Acidified Sodium Sulphite (Kape) Test of Anodic Oxidaiton Coatings Reconfirmed 1990. 2 pp.
SAA AS 2039.2.5-78. Methods for Testing Anodic Oxidation Coatings on Aluminium and Aluminium Alloys—Part 2.5: Sealing Property Tests—Phosphoric-Chromic Acid Test of Anodic Oxidation Coatings Reconfirmed 1990. 2 pp.

—Anodic—Thickness Measurement
SAA AS 2039.1.1-78. Methods for Testing Anodic Oxidation Coatings on Aluminium and Aluminium Alloys—Part 1.1: Thickness and Related Property Tests—Local Thickness of Anodic Oxidation Coatings by Micrographic Examination of Cross-Sections Reconfirmed 1990. 4 pp.
SAA AS 2039.1.2-78. Methods for Testing Anodic Oxidation Coatings on Aluminium and Aluminium Alloys—Part 1.2: Thickness and Related Property Tests—Local Thickness of Anodic Oxidation Coatings by Optical Split-Beam Microscope Reconfirmed 1990. 4 pp.
SAA AS 2039.1.3-78. Methods for Testing Anodic Oxidation Coatings on Aluminium and Aluminium Alloys—Part 1.3: Thickness and Related Property Tests—Average Thickness of Anodic Oxidation Coatings by Strip and Weigh Method Reconfirmed 1990. 4 pp.
SAA AS 2039.1.4-78. Methods for Testing Anodic Oxidation Coatings on Aluminium and Aluminium Alloys—Part 1.4: Thickness and Related Property Tests—Average Thickness of Anodic Oxidation Coatings by Eddy Current Method Reconfirmed 1990. 4 pp.
SAA AS 2039.1.5-78. Methods for Testing Anodic Oxidation Coatings on Aluminium and Aluminium Alloys—Part 1.5: Thickness and Related Property Tests—Continuity of Thin Anodic Oxidation Coatings Reconfirmed 1990. 2 pp.

—Anodic—Water Resistance Testing
CGSB 1-GP-71 METH 110.10-82. Methods of Testing Paints and Pigments Water Resistance Finishes on Primed Aluminum. 1 p.

—Chromate
DIN ENGL 50939-88. Corrosion Protection; Chromating of Aluminium; Principles and Methods of Test (Apr). 5 pp.
ISO 10546-93. Chemical Conversion Coatings—Rinsed and Non-Rinsed Chromate Conversion Coatings on Aluminium and Aluminium Alloys First Edition. 10 pp.
MOD UK DSTAN 03-18-79. Chromate Conversion Coatings (Chromate Filming Treatments) for Aluminium and Aluminium Alloys Issue 1. 12 pp.

—Chromate—Aerospace
CEN PREN2437-90. Chromate Conversion Coatings (Yellow) for Aluminium and Aluminium Alloys. 10 pp.

—Chromate—Surface Density
BSI BS 5411: Part 14-82. 1982 Method of Test for Metallic and Related Coatings Part 14: Gravimetric Method for Determination of Coating Mass per Unit Area of Conversion Coatings on Metallic Materials. 7 pp.
ISO 3892-80. Conversion Coatings on Metallic Materials—Determination of Coating Mass per Unit Area—Gravimetric Methods First Edition. 5 pp.

—Chromium—Electroplated
BSI BS 1224-70. 1970 Amd 1 Electroplated Coatings of Nickel and Chromium. 29 pp.

—Construction Contracts
DIN ENGL 18364-88. Tendering and Performance Stipulations in Contracts for Construction Works (VOB); Part C: General Technical Specifications in Contracts for Construction Works (ATV); Corrosion Protection of Steel and Aluminium Structures. 4 pp.

INDUSTRY STANDARDS

Aluminum

INTERNATIONAL AND NON-U.S. NATIONAL STANDARDS
SUBJECT INDEX

Aluminum Coatings (On Aluminum) (Cont.)

—Conversion
CGSB 31-GP-101MA-89. Chemical Conversion Films for Aluminum and Aluminum Alloys. 9 pp.

—Conversion—Aerospace
ISO 8081-85. Aerospace Process—Chemical Conversion Coating for Aluminium Alloys—General Purpose First Edition. 6 pp.

—Conversion—Aircraft
JIS W 1110-90. Chemical Conversion Coatings on Aluminum and Aluminum Alloys for Aircraft.

—Enamels—Test Specimens
DIN ENGL 51172-86. Vitreous and Porcelain Enamels for Aluminium; Production of Specimens for Testing (Oct). 1 p.

—Environmental Testing
CGSB 1-GP-71 METH 122.8-79. Methods of Testing Paints and Pigments Accelerated Weathering Test for One Coat of Air-Drying Material on Steel or Aluminum Panels. 1 p.

—Lacquers—Self Lifting Resistance
CGSB 1-GP-71 METH 132.2-79. Methods of Testing Paints and Pigments Self-Lifting Lacquers. 1 p.

—Nickel—Electroplated
BSI BS 1224-70. 1970 Amd 1 Electroplated Coatings of Nickel and Chromium. 29 pp.

DIN ENGL 50968-91. Electrodeposited Coatings of Nickel and Nickel Plus Copper (Jan). 5 pp.

—Organic
BSI BS 4842-84. 1984 Specification for Liquid Organic Coatings for Application to Aluminium Alloy Extrusions, Sheet and Preformed Sections for External Architec-tural Purposes, and for the Finish on Aluminium Alloy Extrusions, Sheet and Pref. Sect. Coated with Liq. Org. Coatings. 11 pp.

BSI BS 4842-01. 1984 Amd 1 Liquid Organic Coatings for Application to Aluminium Alloy Extrusions, Sheet and Preformed Sections for External Architectural Purposes, and for the Finish on Aluminium Alloy Extrusions, Sheet and Preformed Sections Coated with Liquid. 12 pp.

BSI BS 6496-84. 1984 Powder Organic Coatings for Application & Stov-ing to Aluminium Alloy Extrusions, Sheet & Pre-formed Sections for Ex-ternal Architectural Purposes, & for the Fin-ish on Aluminium Alloy Extrusions, Sheet & Pre-Formed Sections Coated w/Powder Org. Coatings. 12 pp.

BSI BS 6496-01. 1984 Amd 1 Powder Organic Coatings for Application and Stoving to Aluminium Alloy Extrusions, Sheet and Preformed Sections for External Architectural Purposes, and for the Finish on Aluminium Alloy Extrusions, Sheet and Preformed Sections. 13 pp.

JIS H 9502-92. Recommended Practice for Combined Coatings of Anodic Oxide and Organic Coatings on Aluminium and Aluminium Alloys. 11 pp.

—Oxide—Anodic
BSI BS 1615-87. 1987 Methods for Specifying Anodic Oxidation Coating on Aluminium and Its Alloys. 14 pp.

BSI BS 4479: Part 5-90. 1990 Design of Articles That Are to Be Coated Part 5: Recommendations for Anodic Oxidation Coatings. 13 pp.

BSI BS 5599-78. 1978 Hard Anodic Oxide Coatings on Aluminium for Engineering Purposes. 8 pp.

CNS H3101-82. Coatings Combined with Anodic Oxidation and Organic Finishing on Aluminium and Aluminium Alloys (Jan)(8405).

CNS H3105-82. Anodic Oxidation Coatings on Aluminum and Aluminum Alloys (Feb)(8507).

CNS H3107-82. Anodic Oxidation Practice by Sulphuric Acid Process on Aluminium and Aluminium Alloys (Mar)(8590). 13 pp.

CNS H3108-82. Anodic Oxidation Practice by Oxalic Acid Process on Aluminum and Aluminum Alloys (Apr)(8703).

ISO 7599-83. Anodizing of Aluminium and Its Alloys—General Specifications for Anodic Oxide Coatings on Aluminium First Edition. 9 pp.

JIS H 8601-92. Anodic Oxide Coatings on Aluminium and Aluminium Alloys. 11 pp.

JIS H 8602-92. Combined Coatings of Anodic Oxide and Organic Coatings on Aluminium and Aluminium Alloys. 20 pp.

JIS H 9500-92. Recommended Practice for Anodizing on Aluminium and Aluminium Alloys. 29 pp.

JIS H 9502-92. Recommended Practice for Combined Coatings of Anodic Oxide and Organic Coatings on Aluminium and Aluminium Alloys. 11 pp.

Aluminum Coatings (On Aluminum) (Cont.)

—Oxide—Anodic—Abrasion Testing
BSI BS 6161: Part 9-87. 1987 Methods of Test for Anodic Oxidation Coatings on Aluminum and Its Alloys Part 9: Measurement of Wear Properties with an Abrasive Wheel Wear Test Apparatus. 8 pp.

BSI BS 6161: Part 10-87. 1987 Methods of Test for Anodic Oxidation Coatings on Aluminum and Its Alloys Part 10: Measurement of of Mean Specific Abrasion Resistance with an Abrasive Jet Test Apparatus. 12 pp.

BSI BS 6161: Part 18-91. 1991 Methods of Test for Anodic Oxidation Coatings on Aluminum and Its Alloys Part 18: Determination of Surface Abrasion Resistance. 8 pp.

CNS H2064-82. Methods of Test for Abrasion Resistance of Anodic Oxidation Coatings on Aluminium and Aluminium Alloys (Jan)(8411). 5 pp.

ISO 8251-87. Anodized Aluminium and Aluminium Alloys—Measurement of Wear Resistance and Wear Index of Anodic Oxide Coatings with an Abrasive Wheel Wear Test Apparatus First Edition; (Corrected and Reprinted -1989). 8 pp.

ISO 8252-87. Anodized Aluminium and Aluminium Alloys—Measurement of Mean Specific Abrasion Resistance of Anodic Oxidation Coatings with an Abrasive Jet Test Apparatus First Edition. 12 pp.

JIS H 8682-88. Test Methods for Abrasion Resistance of Anodic Oxidation Coatings on Aluminium and Aluminium Alloys. 18 pp.

—Oxide—Anodic—Accelerated Testing
BSI BS 6161: Part 7-84. 1984 Methods of Test for Anodic Oxidation Coatings on Aluminum and Its Alloys Part 7: Accelerated Determination of Light Fastness of Coloured Anodic Oxidation Coatings Using Artificial Light. 4 pp.

ISO 2135-84. Anodizing of Aluminium and Its Alloys—Accelerated Test of Light Fastness of Coloured Anodic Oxide Coatings Using Artificial Light Second Edition. 4 pp.

—Oxide—Anodic—Buildings
BSI BS 3987-91. 1991 Anodic Oxidation Coatings on Wrought Aluminium for External Architectural Applications. 12 pp.

BSI BS 3987-74. 1974 Amd 1 Anodic Oxide Coatings on Wrought Aluminium for External Architectural Applications. 14 pp.

—Oxide—Anodic—Clarity
BSI BS 6161: Part 13-93. 1993 Methods of Test for Anodic Oxidation Coatings on Aluminium and Its Alloys Part 13: Visual Determination of Image Clarity—Chart Scale Method (ISO 10215: 1992) (F). 11 pp.

BSI BS 6161: Part 13-87. 1987 Methods of Test for Anodic Oxidation Coatings on Aluminium and Its Alloys Part 13: Image Clarity Test Using the Gardam Grid. 4 pp.

BSI BS 6161: Part 19-93. 1993 Methods of Test for Anodic Oxidation Coatings on Aluminium and Its Alloys Part 19: Determination of Image Clarity—Instrumental Method (ISO 10216: 1992) (F). 11 pp.

ISO 10215-92. Anodized Aluminium and Aluminium Alloys—Visual Determination of Image Clarity of Anodic Oxidation Coatings—Chart Scale Method First Edition. 9 pp.

ISO 10216-92. Anodized Aluminium and Aluminium Alloys—Instrumental Determination of Image Clarity of Anodic Oxidation Coatings—Instrumental Method First Edition. 9 pp.

JIS H 8686-78. Test Methods for Image Clarity of Anodic Oxidation Coatings on Aluminium and Aluminium Alloys. 11 pp.

—Oxide—Anodic—Colorfastness Testing
BSI BS 6161: Part 8-81. 1981 Methods of Test for Anodic Oxidation Coatings on Aluminium and Its Alloys Part 8: Determination of the Fastness to Ultraviolet Light to Coloured Anodic Oxide Coatings. 4 pp.

CNS H2061-82. Methods of Accelerated Test for Light Fastness of Colored Anodic Oxidation Coatings on Aluminium and Aluminium Alloys (Jan)(8408). 3 pp.

ISO 6581-80. Anodizing of Aluminium and Its Alloys—Determination of the Fastness to Ultra-Violet Light of Coloured Anodic Oxide Coatings First Edition. 4 pp.

JIS H 8685-88. Accelerated Test Methods for Lightfastness of Colored Anodic Oxidation Coatings on Aluminium and Aluminium Alloys. 11 pp.

—Oxide—Anodic—Colorimetry
ISO TR8125-84. Anodizing of Aluminium and Its Alloys—Determination of Colour and Colour Difference of Coloured Anodic Coatings First Edition. 33 pp.

Aluminum Coatings (On Aluminum) (Cont.)

—Oxide—Anodic—Continuity Testing
ISO 2085-76. Anodizing of Aluminium and Its Alloys—Check of Continuity of Thin Anodic Oxide Coatings—Copper Sulphate Test First Edition. 3 pp.

—Oxide—Anodic—Corrosion Testing
CNS H2063-82. Methods of Test for Corrosion Resistance of Anodic Oxidation Coatings on Aluminium and Aluminium Alloys (Jan)(8410). 4 pp.

—Oxide—Anodic—Cracking (Fracturing)
CNS H2062-89. Methods of Test for Resistance to Cracking by Deforming of Anodic Oxidation Coatings on Aluminium and Aluminium Alloys (Apr)(8409).

ISO 3211-77. Anodizing of Aluminium and Its Alloys—Assessment of Resistance of Anodic Oxide Coatings to Cracking by Deformation Second Edition. 4 pp.

JIS H 8684-77. Test Method for Resistance to Cracking by Deforming of Anodic Oxidation Coatings on Aluminium and Aluminium Alloys. 8 pp.

—Oxide—Anodic—Dielectric Breakdown
BSI BS 6161: Part 15-87. 1987 Methods of Test for Anodic Oxidation Coatings on Aluminium and Its Alloys Part 15: Determination of Electrical Breakdown Potential. 4 pp.

ISO 2376-72. Anodization (Anodic Oxidation) of Aluminium and Its Alloys—Insulation Check by Measurement of Breakdown Potential First Edition. 3 pp.

—Oxide—Anodic—Dyeing Testing
BSI BS 6161: Part 5-82. 1982 Methods of Test for Anodic Oxidation Coatings on Aluminium and Its Alloys Part 5: Estimation of Loss of Absorptive Power of Sealed Coatings: Dye Spot Test with Prior Acid Treatment. 7 pp.

ISO 2143-81. Anodizing of Aluminium and Its Alloys—Estimation of Loss of Absorptive Power of Anodic Oxide Coatings After Sealing—Dye Spot Test with Prior Acid Treatment First Edition. 5 pp.

—Oxide—Anodic—Electrical Impedance
DIN ENGL 50949-84. Non-Destructive Testing of Anodic Oxidation Coatings on Pure Aluminium and Aluminium Alloys by Measurement of Admittance (Feb). 3 pp.

—Oxide—Anodic—Environmental Testing
ISO TR11728-93. Anodized Aluminium and Aluminium Alloys—Accelerated Test of Weather Fastness of Coloured Anodic Oxide Coatings Using Cyclic Artificial Light and Pollution Gas First Edition. 9 pp.

—Oxide—Anodic—Gloss
BSI BS 6161: Part 12-87. 1987 Methods of Test for Anodic Oxidation Coatings on Aluminium and Its Alloys Part 12: Measurement of Specular Reflectance and Specular Gloss at Angles of 20 Degrees, 45 Degrees, 60 Degrees or 85 Degrees. 14 pp.

ISO 7668-86. Anodized Aluminium and Aluminium Alloys—Measurement of Specular Reflectance and Specular Gloss at Angles of 20 Degrees, 45 Degrees, 60 Degrees or 85 Degrees First Edition. 12 pp.

—Oxide—Anodic—Glossaries
ISO 7583-86. Anodizing of Aluminium and Its Alloys—Vocabulary First Edition. 37 pp.

—Oxide—Anodic—Light Testing
ISO TR11728-93. Anodized Aluminium and Aluminium Alloys—Accelerated Test of Weather Fastness of Coloured Anodic Oxide Coatings Using Cyclic Artificial Light and Pollution Gas First Edition. 9 pp.

—Oxide—Anodic—Pitting Testing
BSI BS 6161: Part 16-90. 1990 Methods of Test for Anodic Oxidation Coatings on Aluminium and Its Alloys Part 16: Rating System for the Evaluation of Pitting Corrosion—Chart Method. 12 pp.

ISO 8993-89. Anodized Aluminium and Aluminium Alloys—Rating System for the Evaluation of Pitting Corrosion—Chart Method First Edition. 12 pp.

—Oxide—Anodic—Rating System
BSI BS 6161: Part 17-90. 1990 Methods of Test for Anodic Oxidation Coatings on Aluminium and Its Alloys Part 17: Rating System for the Evaluation of Pitting Corrosion-Grid Method. 4 pp.

ISO 8994-89. Anodized Aluminium and Aluminium Alloys—Rating System for the Evaluation of Pitting Corrosion—Grid Method First Edition. 4 pp.

INTERNATIONAL AND NON-U.S. NATIONAL STANDARDS
SUBJECT INDEX
Aluminum

Aluminum Coatings (On Aluminum) (Cont.)

—Oxide—Anodic—Reflectance

BSI BS 6161: Part 11-85. 1985 Methods of Test for Anodic Oxidation Coatings on Aluminum and Its Alloys Part 11: Measurement of Total Reflectance Using a Photoelectric Reflectometer. 4 pp.

BSI BS 6161: Part 12-87. 1987 Methods of Test for Anodic Oxidation Coatings on Aluminium and Its Alloys Part 12: Measurement of Specular Reflectance and Specular Gloss at Angles of 20 Degrees, 45 Degrees, 60 Degrees or 85 Degrees. 14 pp.

BSI BS 6161: Part 14-87. 1987 Methods of Test for Anodic Oxidation Coatings on Aluminium and Its Alloys Part 14: Determination of Infra-Red Reflectance. 4 pp.

ISO 6719-86. Anodized Aluminium and Aluminium Alloys—Measurement of Reflectance Characteristics of Aluminium Surfaces Using Integrating-Sphere Instruments First Edition. 8 pp.

ISO 7668-86. Anodized Aluminium and Aluminium Alloys—Measurement of Specular Reflectance and Specular Gloss at Angles of 20 Degrees, 45 Degrees, 60 Degrees or 85 Degrees First Edition. 12 pp.

ISO 7759-83. Anodizing of Aluminium and Its Alloys—Measurement of Reflectivity Characteristics of Aluminium Surfaces Using Abridged Goniophotometer or Goniophotometer First Edition. 7 pp.

—Oxide—Anodic—Sealing

BSI BS 6161: Part 3-84. 1984 Methods of Test for Anodic Oxidation Coatings on Aluminium and Its Alloys Part 3: Assessment of Sealing Quality by Measurement of the Loss of Mass After Immersion in Phosphoric-Chromic Acid-Solution. 7 pp.

BSI BS 6161: Part 4-81. 1981 Methods of Test for Anodic Oxidaton Coatings on Aluminium and Its Alloys Part 4: Assessmentof Sealing Quality by Measurement of the Loss of Mass After Immersion in Acid Solution. 4 pp.

BSI BS 6161: Part 6-84. 1984 Methods of Test for Anodic Oxidation Coatings on Aluminium and Its Alloys Part 6: Assessment of Sealing Quality by Measurement of Admittance or Impedance. 7 pp.

CNS H2060-89. Methods of Test for Sealing Quality and Anodic Oxidation Coatings on Aluminum and Aluminum Alloys (Apr)(8407).

ISO 2931-83. Anodizing of Aluminium and Its Alloys—Assessment of Quality of Sealed Anodic Oxide Coatings by Measurement of Admittance or Impedance Second Edition. 5 pp.

ISO 3210-83. Anodizing of Aluminium and Its Alloys—Assessment of Quality of Sealed Anodic Oxide Coatings by Measurement of the Loss of Mass After Immersion in Phosphoric-Chromic Acid Solution Second Edition. 5 pp.

JIS H 8683-79. Test Methods for Sealing Quality of Anodic Oxidation Coatings on Aluminium and Aluminium Alloys.

JIS H 8683-73. Test Methods for Sealing Quality of Anodic Oxidation Coatings on Aluminium and Aluminium Alloys. 6 pp.

—Oxide—Anodic—Surface Density—Gravimetric Analysis

BSI BS 6161: Part 1-84. 1984 Methods of Test for Anodic Oxidation Coatings on Aluminium and Its Alloys Part 1: Determination of Mass Per Unit Area (Surface Density) of Anodic Oxidation Coatings (Gravimetric Method). 4 pp.

ISO 2106-82. Anodizing of Aluminium and Its Alloys—Determination of Mass per Unit Area (Surface Density) of Anodic Oxide Coatings—Gravimetric Method Second Edition; (Amendment 1-1983). 5 pp.

—Oxide—Anodic—Thickness Measurement

BSI BS 6161: Part 2-81. 1981 Methods of Test for Anodic Oxidation Coatings on Aluminium and Its Alloys Part 2: Determination of Thickness of Anodic Oxide Coatings: Non-Destructive Measurement by Split-Beam Microscope. 4 pp.

CNS H2059-89. Methods of Test for Thickness of Anodic Oxidation Coatings on Aluminum and Aluminum Alloys (Apr)(8406).

ISO 2128-76. Anodizing of Aluminium and Its Alloys—Determination of Thickness of Anodic Oxide Coatings—Non-Destructive Measurement by Split-beam Microscope First Edition. 4 pp.

JIS H 8680-79. Test Methods for Thickness of Anodic Oxidation Coatings on Aluminium and Aluminium Alloys (R 1984). 9 pp.

—Paints—Alkyd

MOD UK DSTAN 80-50-01. Paint System, Defence Equipment, High Gloss, Alkyd Paint, Priming, Defence Equipment, Zinc Phosphate/Zinc Chrome, Paint, Undercoat, Defence Equipment Paint, Finishing, Defence Equipment, High Gloss, Alkyd Types: Brushing Spraying. 26 pp.

—Paints—Alkyd—Matte

MOD UK DSTAN 80-41-91. Paint System, Defence Equipment, IRR, Matt Paint, Undercoat, Defence Equipment Paint, Finishing, Defence Equipment, IRR Matt Types: Brushing Spraying Dipping Issue 2. 23 pp.

—Paints—Oleoresinous—Exterior

MOD UK DSTAN 80-29-73. Paint, Finishing, General Service, Gloss, for Marine Use Type: Brushing Issue 1. 8 pp.

—Phosphate

DIN ENGL 50942-87. Phosphating of Metals; Principles, Methods of Test (May). 12 pp.

ISO 9717-90. Phosphate Conversion Coatings for Metals—Method of Specifying Requirements First Edition. 19 pp.

—Phosphate—Surface Density

BSI BS 5411: Part 14-82. 1982 Method of Test for Metallic and Related Coatings Part 14: Gravimetric Method for Determination of Coating Mass per Unit Area of Conversion Coatings on Metallic Materials. 7 pp.

ISO 3892-80. Conversion Coatings on Metallic Materials—Determination of Coating Mass per Unit Area—Gravimetric Methods First Edition. 5 pp.

—Plates—Aerospace

MOD UK DTD-5010A-71. Plate of Aluminium—Copper—Magnesium—Silicon—Manganese Alloy (Solution Treated and Aged at Room Temperature) (Cu 4.4, Mg 0.5, Si 0.7, Mn 0.8). 2 pp.

MOD UK DTD-5040A-71. Aluminium-Coated Plate of Aluminium—Copper—Magnesium—Silicon—Manganese Alloy (Solution Treated and Precipitation Treated) (Cu 4.4, Mg 0.5, Si 0.7, Mn 0.8). 2 pp.

MOD UK DTD-5100A-71. Aluminium-Coated Plate of Aluminium—Copper—Magnesium—Manganese Alloy (Solution Treated, Controlled Stretched and Aged at Room Temperature) (Cu 4.4, Mg 1.5, Mn 0.6). 2 pp.

—Primers—Zinc Chromate

CGSB 1-GP-71 METH 130.1-79. Methods of Testing Paints and Pigments Behavior Towards Topcoats Primers on Pretreated Aluminum. 1 p.

—Sheets

JIS H 4001-90. Painted Aluminium and Aluminium Alloy Sheets and Strips. 16 pp.

—Sheets—Aerospace

AECMA PREN2087-80. Aluminium Alloy 2014A-T6 or T62—Clad Sheets and Strips A 0.4-6 mm (C5/40). 3 pp.

AECMA PREN2088-80. Aluminium Alloy 2014A-T4 or T42—Clad Sheets and Strips A 0.4-6 mm (C5/40). 3 pp.

AECMA PREN2090-85. Aluminium Alloy 2024-T3 Clad Sheet and Strip 0.4 Less Than or Equal to a Less Than or Equal to 6 mm. 3 pp.

AECMA PREN2091-85. Aluminium Alloy 2024-T4 Clad Sheet and Strip 0.4 Less Than or Equal to a Less Than or Equal to 6 mm. 3 pp.

AECMA PREN2092-85. Aluminium Alloy 7075-T6 Clad Sheet and Strip 0.4 Less Than or Equal to a Less Than or Equal to 6 mm. 3 pp.

AECMA PREN2692-85. Aluminium Alloy 2017A-T3 Clad Sheet and Strip 0.4 Less Than or Equal to a Less Than or Equal to 6 mm. 3 pp.

AECMA PREN2703-85. Aluminium Alloy 2024-T42 Clad Sheet and Strip 0.4 Less Than or Equal to a Less Than or Equal to 6 mm. 3 pp.

AECMA PREN2803-86. Aluminium Alloy 7475-T761 Clad Sheet and Strip 0.8 Less Than or Equal to a Less Than or Equal to 6 mm. 3 pp.

BSI 2L 88-71. 1971 Aluminium-Alloy Coated Sheet and Strip of Aluminium-Zinc-Magnesium-Copper-Chromium Alloy (Solution Treated and Precipitation Treated) (Zn 5.8, Mg 2.5, Cu 1.6, Cr 0.15). 3 pp.

BSI L 109-71. 1971 Amd 1 Aluminium-Coated Sheet and Strip of Aluminium-Copper-Magnesium-Manganese Alloy (Solution Treated and Aged at Room Temperature). 4 pp.

BSI L 110-71. 1971 Amd 1 Aluminium-Coated Sheet and Strip of Aluminium-Copper-Magnesium-Manganese Alloy (Supplied for Solution Treatment by the User). 4 pp.

BSI L 163-78. 1978 Sheet and Strip of Aluminium-Coated Aluminium-Copper-Magnesium-Silicon-Manganese Alloy (Cu 4.4, Mg 0.5, Si 0.8, Mn 0.8). 3 pp.

BSI L 164-78. 1978 Sheet and Strip of Aluminium-Coated Aluminium-Copper-Magnesium-Silicon-Manganese Alloy (Cu 4.4, Mg 0.5, Si 0.8, Mn 0.8). 4 pp.

—Sheets—Aerospace (Cont.)

BSI L 165-78. 1978 Amd 1 Specification for Sheet and Strip of Aluminium-Coated Aluminium-Copper-Magnesium-Silicon-Manganese Alloy (Solution Treated and Artificially Aged) (Cu 4.4, Mg 0.5, Si 0.8, Mn 0.8) (AMD 6391) July 31, 1991. 4 pp.

BSI L 166-78. 1978 Close Toleranced Sheet and Strip of Aluminium-Coated Aluminium-Copper-Magnesium-Silicon-Manganese Alloy (Cu 4.4, Mg 0.5, Si 0.8, Mn 0.8). 4 pp.

BSI L 167-78. 1978 Amd 1 Specification for Close Toleranced Sheet and Strip of Aluminium-Coated Aluminium-Copper-Magnesium-Silicon-Manganese Alloy (Solution Treated and Art. Aged) (Cu 4.4, Mg 0.5, Si 0.8,Mn 0.8) (AMD 6392) July 31, 1991. 4 pp.

MOD UK DTD-5070B-71. Aluminium-Alloy-Coated Sheet and Strip of Aluminium-Copper-Magnesium-Nickel-Iron Alloy (Solution Treated and Precipitation Treated) (Suitable for Use at Elevated Temperatures) (Cu 2.5, Mg 1.5, Ni 1.2, Fe 1.0). 2 pp.

—Sprayed—Construction Contracts

DIN ENGL 18364-88. Tendering and Performance Stipulations in Contracts for Construction Works (VOB); Part C: General Technical Specifications in Contracts for Construction Works (ATV); Corrosion Protection of Steel and Aluminium Structures. 4 pp.

—Stripping

CGSB 31-GP-0A METH 55.2-62. Methods of Testing Corrosion-Prevention Materials and Processes Stripping Ability. 1 p.

—Strips

JIS H 4001-90. Painted Aluminium and Aluminium Alloy Sheets and Strips. 16 pp.

—Strips—Aerospace

AECMA PREN2087-80. Aluminium Alloy 2014A-T6 or T62—Clad Sheets and Strips A 0.4-6 mm (C5/40). 3 pp.

AECMA PREN2088-80. Aluminium Alloy 2014A-T4 or T42—Clad Sheets and Strips A 0.4-6 mm (C5/40). 3 pp.

AECMA PREN2090-85. Aluminium Alloy 2024-T3 Clad Sheet and Strip 0.4 Less Than or Equal to a Less Than or Equal to 6 mm. 3 pp.

AECMA PREN2091-85. Aluminium Alloy 2024-T4 Clad Sheet and Strip 0.4 Less Than or Equal to a Less Than or Equal to 6 mm. 3 pp.

AECMA PREN2092-85. Aluminium Alloy 7075-T6 Clad Sheet and Strip 0.4 Less Than or Equal to a Less Than or Equal to 6 mm. 3 pp.

AECMA PREN2692-85. Aluminium Alloy 2017A-T3 Clad Sheet and Strip 0.4 Less Than or Equal to a Less Than or Equal to 6 mm. 3 pp.

AECMA PREN2703-85. Aluminium Alloy 2024-T42 Clad Sheet and Strip 0.4 Less Than or Equal to a Less Than or Equal to 6 mm. 3 pp.

AECMA PREN2803-86. Aluminium Alloy 7475-T761 Clad Sheet and Strip 0.8 Less Than or Equal to a Less Than or Equal to 6 mm. 3 pp.

BSI 2L 88-71. 1971 Aluminium-Alloy Coated Sheet and Strip of Aluminium-Zinc-Magnesium-Copper-Chromium Alloy (Solution Treated and Precipitation Treated) (Zn 5.8, Mg 2.5, Cu 1.6, Cr 0.15). 3 pp.

BSI L 109-71. 1971 Amd 1 Aluminium-Coated Sheet and Strip of Aluminium-Copper-Magnesium-Manganese Alloy (Solution Treated and Aged at Room Temperature). 4 pp.

BSI L 110-71. 1971 Amd 1 Aluminium-Coated Sheet and Strip of Aluminium-Copper-Magnesium-Manganese Alloy (Supplied for Solution Treatment by the User). 4 pp.

BSI L 163-78. 1978 Sheet and Strip of Aluminium-Coated Aluminium-Copper-Magnesium-Silicon-Manganese Alloy (Cu 4.4, Mg 0.5, Si 0.8, Mn 0.8). 3 pp.

BSI L 164-78. 1978 Sheet and Strip of Aluminium-Coated Aluminium-Copper-Magnesium-Silicon-Manganese Alloy (Cu 4.4, Mg 0.5, Si 0.8, Mn 0.8). 4 pp.

BSI L 165-78. 1978 Amd 1 Specification for Sheet and Strip of Aluminium-Coated Aluminium-Copper-Magnesium-Silicon-Manganese Alloy (Solution Treated and Artificially Aged) (Cu 4.4, Mg 0.5, Si 0.8, Mn 0.8) (AMD 6391) July 31, 1991. 4 pp.

BSI L 166-78. 1978 Close Toleranced Sheet and Strip of Aluminium-Coated Aluminium-Copper-Magnesium-Silicon-Manganese Alloy (Cu 4.4, Mg 0.5, Si 0.8, Mn 0.8). 4 pp.

BSI L 167-78. 1978 Amd 1 Specification for Close Toleranced Sheet and Strip of Aluminium-Coated Aluminium-Copper-Magnesium-Silicon-Manganese Alloy (Solution Treated and Art. Aged) (Cu 4.4, Mg 0.5, Si 0.8,Mn 0.8) (AMD 6392) July 31, 1991. 4 pp.

INDUSTRY STANDARDS

Aluminum

INTERNATIONAL AND NON-U.S. NATIONAL STANDARDS SUBJECT INDEX

Aluminum Coatings (On Aluminum) (Cont.)

—Strips—Aerospace (Cont.)
MOD UK DTD-5070B-71. Aluminium-Alloy-Coated Sheet and Strip of Aluminium-Copper-Magnesium-Nickel-Iron Alloy (Solution Treated and Precipitation Treated) (Suitable for Use at Elevated Temperatures) (Cu 2.5, Mg 1.5, Ni 1.2, Fe 1.0). 2 p.

—Temporary—Aerospace
MOD UK DTD-5588-01. Strippable Temporary Protective Coating for Aluminium Alloys; Amendment 1. 4 pp.

—Test Panels
CGSB 1-GP-71 METH 99.1-82. Methods of Testing Paints and Pigments Preparation of Aluminum Panels Plain Panels. 1 p.

—Varnishes—Fire Resistant
MOD UK DSTAN 80-78-79. Paint, Finishing, Fire-Retardant, White and Tinted White Type: Brushing Issue 1. 10 pp.

—Zinc
MOD UK TS 10150A. Paint System, Defence Equipment, Matt Zinc Phosphate/Zinc Chrome.

Aluminum Conductor Wire
Use: Aluminum Conductors

Aluminum Conductors
Use For: Aluminum Conductor Wire
See Also: Aluminum Wire; Electric Conductors; Electric Wire; Power Line Hardware; Wiring Devices
BSI BS 2627-70. 1970 Amd 1 Wrought Aluminium for Electrical Purposes; Wire. 17 pp.
BSI BS 3988-70. 1970 Amd 3 Wrought Aluminium for Electrical Purposes: Solid Conductors for Insulated Cables. 23 pp.
BSI BS 6360-91. 1991 Amd 1 Conductors in Insulated Cables and Cords (AMD 6769) July 31, 1991. 20 pp.
BSI BS 6360-02. 1991 Amd 2 Conductors in Insulated Cables and Cords (AMD 7637) May 15, 1993 (F). 21 pp.
CNS C2049-87. Round Aluminium Electrical Wire, Solid (Sep)(2727).
CNS C2065-89. Half-Hard-Drawn Aluminium Wires for Electric Purpose (May)(5744).
CNS C2099-81. 600V Grade Aluminum Conductor Polyvinyl Chloride Insulated Wires (Aug)(7783).
CNS C3080-87. Method of Test for Electrical Copper and Aluminium Wires (Jul)(5745).
CNS C3118-87. Method of Test for Round Aluminum Electrical Wire, Solid (Sep)(7355).
CNS C4077-69. Fixtures for Aluminium Conducting Wires and Cables (Jul)(2831) (R 1973).
CNS H3070-88. Conductor Aluminium — Iron Alloy Rod (Dec)(4746).
CSA C49.3-1977. Aluminum Alloy 1350 Round Wire, All Tempers, for Electrical Purposes; (Amd 1-2 May 1991). 15 pp.
DIN ENGL 40501 Pt 4-73. Aluminium for Electrical Engineering; Wires of Pure Aluminium; Technical Conditions of Delivery (Aug). 3 pp.
DIN ENGL 46420-70. Round Aluminium Wires for Electrical Purposes; Drawn; Dimensions (June). 2 pp.
DIN ENGL 46425-70. Round Aluminium Wires for Electrical Purposes; Exactly Drawn; Dimensions (June). 2 pp.
DIN ENGL 46433-59. Rectangular Wires and Rectangular Bars; Drawn, with Radiused Edges; Dimensions (Nov). 8 pp.
DIN ENGL 46433 Extr. 1-59. Rectangular Wires and Rectangular Bars; Drawn, with Radiused Edges; Dimensions; Selected Sizes for Electrical Machines and Switchgear (Nov). 8 pp.
DIN ENGL 46433 Extr. 3-59. Rectangular Bars; Drawn, with Radiused Edges; Dimensions; Selected Sizes for Switchgears (Nov). 2 pp.
IEC 55 Pt 1-78. Paper-Insulated Metal-Sheathed Cables for Rated Voltages up to 18/30 kV (with Copper or Aluminium Conductors and Excluding Gas-Pressure and Oil-Filled Cables) Part 1: Tests Fourth Edition; (Amendment 1-1989). 32 pp.
IEC 55 Pt 2-81. Paper-Insulated Metal-Sheathed Cables for Rated Voltages up to 18/30 kV (with Copper or Aluminium Conductors and Excluding Gas-Pressure and Oil-Filled Cables) Part 2: General and Construction Requirements First Edition; (Amendment 1-1989). 85 pp.
IEC 121-60. Recommendation for Commercial Annealed Aluminium Electrical Conductor Wire First Edition. 5 pp.
JIS C 3002-92. Testing Methods of Electrical Copper and Aluminium Wires. 12 pp.
JIS C 3006-83. Methods of Test for Fiber or Paper Insulated Copper and Aluminum Winding Wires. 9 pp.

Aluminum Conductors (Cont.)
JIS C 3107-81. Half-Hard-Drawn Aluminium Wires for Electric Purposes. 6 pp.
JIS C 3108-78. Hard-Drawn Aluminium Wires for Electric Purposes. 7 pp.
JIS C 3372-87. 600 V Grade Aluminium Conductor Polyvinyl Chloride Insulated Wires (R 1983). 11 pp.
SNZ BS 2627-70. Specification for Wrought Aluminium for Electrical Purposes. Wire Amend: 1. 12 pp.

—Aerospace
CEN PREN 3719-91. Aluminium or Aluminium Alloy Conductors for Electrical Cables Product Standard. 7 pp.

—Automotive
CNS D2152-86. Low-Voltage Aluminum Cables for Automobiles (Nov)(9378).
CNS D3140-86. Method of Test for Low-Voltage Aluminum Cables for Automobiles (Nov)(9379).

—Communication Cables
IEC 753-82. Aluminum Electrical Conductor Wires Used in Polyolefin Insulated Telecommunication Cables First Edition. 13 pp.
IEC 762-83. Aluminium Alloy Electrical Conductor Wires Used in Polyolefin Insulated Telecommunication Cables First Edition; (Amendment 1-1985). 18 pp.

—Compression Joints
BSI BS 4579: Part 3-76. 1976 Amd 1 Performance of Mechanical and Compression Joints in Electric Cable and Wire Connectors Part 3: Mechanical and Compression Joints in Aluminium Conductors. 21 pp.

—Connectors
CSA C22.2 NO 65-93. Wire Connectors; (Gen Instr 1). 67 pp.

—Enameled
BSI BS 6811: Sec 5.1-93. 1993 Winding Wires Part 5.: Specifications for Particular Types of Enamelled Round Aluminium Windind Wires Section 5.1: Polyvinal Acetal Enamelled Round Aluminium Wire, Class 105 (IEC 317-14: 1990). 13 pp.
CNS C2095-83. Polyvinyl Formal Enameled Aluminum Wires (Feb)(7357).
CNS C3151-82. Method for Test for Enameled Aluminum Wires (Jun)(8938).
IEC 317 Pt 14-90. Specifications for Particular Types of Winding Wires Part 14: Polyvinyl Acetal Enamelled Round Aluminium Wire, Class 105 Second Edition. 19 pp.
JIS C 3003-84. Methods of Test for Enamelled Copper and Enamelled Aluminium Wires; (Erratum). 28 pp.

—Enameled—Windings
BSI BS 6811: Sec 5.0-93. 1993 Winding Wires Part 5: Specifications for Particular Types of Enamelled Round Aluminium Winding Wires Section 5.0: General Requirements (IEC 317-0-3: 1990). 20 pp.
BSI BS 6811: Sec 5.0-01. 1993 Amd 1 Winding Wires Part 5: Specifications for Particular Types of Enamelled Round Aluminium Winding Wires Section 5.0: General Requirements (IEC 317-0-3: 1990) (AMD 7905) September 15, 1993 (E). 19 pp.
BSI BS 6811: Sec 5.3-93. 1993 Winding Wires Part 5: Spec. for Particular Types of Enamelled Round Aluminium Winding Wires Sec. 5.3: Polyester or Polyesterimide Overcoated with Polyamide-imide Enamelled Round Alum. Wire, Class 200 (IEC 317-25: 1990). 14 pp.
CENELEC HD 555.0.3 S1-92. Specifications for Particular Types of Winding Wires Part 0: General Requirements Section 3: Enammelled Round Aluminium Wire. 5 pp.
CENELEC HD 555.14 S1-90. Specifications for Particular Types of Winding Wires Part 14: Polyvinyl Acetal Enamelled Round Aluminium Winding Wire, Class 105. 3 pp.
CENELEC HD 555.14 S2-92. Specifications for Particular Types of Winding Wires Part 14: Polyvinyl Acetal Enamelled Round Aluminium Wire, Class 105. 5 pp.
CENELEC HD 555.24 S1-90. Specifications for Particular Types of Winding Wires Part 24: Polyester (Imide) Enamelled Round Aluminium Winding Wire Overcoated with Polyamide, Class 180. 3 pp.
CENELEC HD 555.24 S2-92. Specifications for Particular Types of Winding Wires Part 24: Polyester or Polyesterimide Enamelled Round Aluminium Wire Overcoated with Polyamide, Class 180. 5 pp.
CENELEC EN 60182-2-90. Basic Dimensions of Winding Wires Part 2: Maximum Overall Diameters of Enamelled Round Winding Wires (Supersedes HD 42.2 S2-1979). 8 pp.

Aluminum Conductors (Cont.)

—Enameled—Windings (Cont.)
IEC 317 Pt 0-3-90. Specifications for Particular Types of Winding Wires Part 0: General Requirements Section 3: Enamelled Round Aluminium Wire First Edition; (Amendment 1-1992) (Amendment 2-1993). 57 pp.
IEC 317 Pt 15-90. Specifications for Particular Types of Winding Wires Part 15: Polyesterimide Enamelled Round Aluminium Wire, Class 180 Second Edition. 19 pp.
IEC 317 Pt 24-90. Specifications for Particular Types of Winding Wires Part 24: Polyester or Polyesterimide Enamelled Round Aluminium Wire Overcoated with Polyamide, Class 180 Second Edition. 21 pp.
IEC 317 Pt 25-90. Specifications for Particular Types of Winding Wires Part 25: Polyester or Polyesterimide Overcoated with Polyamide-Imide Enamelled Round Aluminium Wire, Class 200 Second Edition. 21 pp.
SAA AS 1194.3-84. Winding Wires—Part 3: Enamelled Round Aluminium Winding Wires. 27 pp.
SAA AS 1194.4-85. Winding Wires—Part 4: Enamelled Rectangular Aluminium Winding Wires. 24 pp.

—Fiber Insulated—Windings
JIS C 3006-83. Methods of Test for Fiber or Paper Insulated Copper and Aluminum Winding Wires. 9 pp.

—Neutral Cables
CSA C22.2 NO 129-1976. Neutral Supported Cable (R 1989); (Amd 1-6 May 1985). 25 pp.

—Overhead Power Lines
BSI BS 215: Part 1-70. 1970 Aluminium Conductors and Aluminium Conductors, Steel-Reinforced, for Overhead Power Transmission Part 1: Aluminium Stranded Conductors. 17 pp.
BSI BS 215: Part 2-70. 1970 Aluminium Conductors and Aluminium Conductors, Steel-Reinforced, for Overhead Power Transmission Part 2: Aluminium Conductors, Steel-Reinforced. 20 pp.
BSI BS 3242-70. 1970 Amd 1 Aluminium Alloy Stranded Conductors for Overhead Power Transmission. 17 pp.
BSI BS 7365-90. 1990 Hard Drawn Aluminium Wire for Overhead Line Conductors. 8 pp.
CENELEC HD 532 S1-89. Hard-Drawn Aluminium Wire for Overhead Live Conductors. 3 pp.
CNS C2060-90. Aluminium Conductors Steel Reinforced (ACSR) (Aug)(3619).
CSA CAN/CSA-C49.1-M87. Round Wire, Concentric Lay, Overhead Electrical Conductors. 39 pp.
CSA C49.2-1975. Compact Aluminum Conductors Steel Reinforced (ACSR). 9 pp.
CSA CAN3-C49.7-M85. Aluminum Round Wires for Use in Overhead Electrical Conductors (R 1992); (Gen Instr 1). 17 pp.
DIN VDE 0274-87. Crosslinked Polyethylene; Insulated Conductors for Overhead Cables; Nominal Voltage: Uo/U 0.6/1 kV (Feb). 18 pp.
IEC 104-87. Aluminium-Magnesium-Silicon Alloy Wire for Overhead Line Conductors Second Edition. 14 pp.
IEC 889-87. Hard-Drawn Aluminium Wire for Overhead Line Conductors First Edition. 12 pp.
IEC 1089-91. Round Wire Concentric Lay Overhead Electrical Stranded Conductors First Edition. 68 pp.
JIS C 3110-78. Aluminium Conductors Steel Reinforced. 10 pp.
SAA AS 1531-91. Conductors—Bare Overhead—Aluminium and Aluminium Alloy (Supersedes AS 1531.1—1974, AS 1531.2—1974 and AS 1531.3—1984). 12 pp.
SNZ BS 215: Part 1-70. Specification for Aluminium Conductors and Aluminium Conductors, Steel-Reinforced, for Over-Head Power Transmission Part 1: Aluminium Stranded Conductors. 16 pp.
SNZ BS 215: Part 2-70. Specification for Aluminium Conductors and Aluminium Conductors, Steel-Reinforced, for Over-Head Power Transmission Part 2: Aluminium Conductors, Steel Reinforced. 20 pp.

—Paper Insulated—Windings
JIS C 3006-83. Methods of Test for Fiber or Paper Insulated Copper and Aluminum Winding Wires. 9 pp.

—Polyester Enameled—Windings
BSI BS 6811: Sec 5.3-93. 1993 Winding Wires Part 5: Spec. for Particular Types of Enamelled Round Aluminium Winding Wires Sec. 5.3: Polyester or Polyesterimide Overcoated with Polyamide-imide Enamelled Round Alum. Wire, Class 200 (IEC 317-25: 1990). 14 pp.
CENELEC HD 555.25 S1-90. Specifications for Particular Types of Winding Wires Part 25: Polyester (Imide) Overcoated with Polyamide-Imide Enamelled Round Aluminium Winding Wire, Class 200. 3 pp.

INTERNATIONAL AND NON-U.S. NATIONAL STANDARDS
SUBJECT INDEX
Aluminum

Aluminum Conductors *(Cont.)*

—Polyester Enameled—Windings *(Cont.)*
CENELEC HD 555.25 S2-92. Specifications for Particular Types of Winding Wires Part 25: Polyester or Polyesterimide Overcoated with Polyamide-imide Enamelled Round Aluminium Wire, Class 200. 5 pp.

IEC 317 Pt 25-90. Specifications for Particular Types of Winding Wires Part 25: Polyester or Polyesterimide Overcoated with Polyamide-Imide Enamelled Round Aluminium Wire, Class 200 Second Edition. 21 pp.

—Polyesterimide Enameled—Windings
BSI BS 6811: Sec 5.2-93. 1993 Winding Wires Part 5: Specifications for Particular Types of Enamelled Round Aluminium Winding Wires Section 5.2: Polyesterimide Enamelled Round Aluminium Wire, Class 180 (IEC 317-15: 1990). 13 pp.

BSI BS 6811: Sec 5.3-93. 1993 Winding Wires Part 5: Spec. for Particular Types of Enamelled Round Aluminium Winding Wires Sec. 5.3: Polyester or Polyesterimide Overcoated with Polyamide-imide Enamelled Round Alum. Wire, Class 200 (IEC 317-25: 1990). 14 pp.

CENELEC HD 555.15 S1-90. Specifications for Particular Types of Winding Wires Part 15: Polyesterimide Enamelled Round Aluminium Winding Wire, class 180. 3 pp.

CENELEC HD 555.15 S2-92. Specifications for Particular Types of Winding Wires Part 15: Polyesterimide Enammelled Round Aluminium Wire, Class 180. 5 pp.

CENELEC HD 555.25 S1-90. Specifications for Particular Types of Winding Wires Part 25: Polyester (Imide) Overcoated with Polyamide-Imide Enamelled Round Aluminium Winding Wire, Class 200. 3 pp.

CENELEC HD 555.25 S2-92. Specifications for Particular Types of Winding Wires Part 25: Polyester or Polyesterimide Overcoated with Polyamide-imide Enamelled Round Aluminium Wire, Class 200. 5 pp.

IEC 317 Pt 25-90. Specifications for Particular Types of Winding Wires Part 25: Polyester or Polyesterimide Overcoated with Polyamide-Imide Enamelled Round Aluminium Wire, Class 200 Second Edition. 21 pp.

—Polyvinyl Acetal Enameled—Windings
CENELEC HD 555.14 S1-90. Specifications for Particular Types of Winding Wires Part 14: Polyvinyl Acetal Enamelled Round Aluminium Winding Wire, Class 105. 3 pp.

CENELEC HD 555.14 S2-92. Specifications for Particular Types of Winding Wires Part 14: Polyvinyl Acetal Enamelled Round Aluminium Wire, Class 105. 5 pp.

—PVC Insulated—PVC Sheathed—Drop
CNS C2104-82. Aluminum Conductor Polyvinyl Chloride Insulated Service Drop Wires (Mar)(8558).

JIS C 3371-87. Aluminum Conductor Polyvinyl Chloride Insulated Service Drop Wires. 15 pp.

—PVC Insulated—PVC Sheathed—600 V
CNS C2099-81. 600V Grade Aluminum Conductor Polyvinyl Chloride Insulated Wires (Aug)(7783).

JIS C 3372-87. 600 V Grade Aluminium Conductor Polyvinyl Chloride Insulated Wires (R 1983). 11 pp.

—Steel Reinforced
BSI BS 215: Part 2-70. 1970 Aluminium Conductors and Aluminium Conductors, Steel-Reinforced, for Overhead Power Transmission Part 2:Aluminium Conductors, Steel-Reinforced. 20 pp.

CNS C2060-90. Aluminium Conductors Steel Reinforced (ACSR) (Aug)(3619).

CNS C2098-81. ACSR Conductor Polyvinyl Chloride Insulated Out-Door Weather-Proof Wires (Aug)(7782).

CNS C2104-82. Aluminum Conductor Polyvinyl Chloride Insulated Service Drop Wires (Mar)(8558).

CSA CAN/CSA-C49.1-M87. Round Wire, Concentric Lay, Overhead Electrical Conductors. 39 pp.

CSA C49.2-1975. Compact Aluminum Conductors Steel Reinforced (ACSR). 9 pp.

DIN ENGL 48203 Pt 11-87. Wires and Stranded Conductors; Steel-Reinforced Aluminium Stranded Conductors; Technical Delivery Conditions (Dec). 6 pp.

DIN ENGL 48203 Pt 12-87. Wires and Stranded Conductors; Steel-Reinforced E-AlMgSi Stranded Conductors; Technical Delivery Conditions (Dec). 6 pp.

DIN ENGL 48204-84. Steel Reinforced Aluminium Stranded Conductors (Apr). 4 pp.

IEC 1089-91. Round Wire Concentric Lay Overhead Electrical Stranded Conductors First Edition. 68 pp.

JIS C 3110-78. Aluminium Conductors Steel Reinforced. 10 pp.

JIS C 3370-87. ACSR Conductor Polyvinyl Chloride Insulated out-Door Weather-Proof Wires. 12 pp.

Aluminum Conductors *(Cont.)*

—Steel Reinforced *(Cont.)*
JIS C 3371-87. Aluminium Conductor Polyvinyl Chloride Insulated Service Drop Wires. 15 pp.

SNZ BS 215: Part 2-70. Specification for Aluminium Conductors and Aluminium Conductors, Steel-Reinforced, for Over-Head Power Transmission Part 2: Aluminium Conductors, Steel Reinforced. 20 pp.

—Stranded
BSI BS 215: Part 1-70. 1970 Aluminium Conductors and Aluminium Conductors, Steel-Reinforced, for Overhead Power Transmission Part 1: Aluminium Stranded Conductors. 17 pp.

BSI BS 3242-70. 1970 Amd 1 Aluminium Alloy Stranded Conductors for Overhead Power Transmission. 17 pp.

BSI BS 7365-90. 1990 Hard Drawn Aluminium Wire for Overhead Line Conductors. 8 pp.

CENELEC HD 383 S2-86. Conductors of Insulated Cables Guide to the Dimensional Limits of Circular Conductors. 2 pp.

CENELEC HD 383 S2/A2-93. Conductors of Insulated Cables First Supplement: Guide to the Dimensional Limits of Circular Conductors (IEC 228:1978 + IEC 228A: 1982, Modified). 6 pp.

CNS C2050-87. Aluminum Stranded Wires for Insulated Cables (Dec)(2728). 4 pp.

DIN ENGL 48200 Pt 5-81. Wires for Stranded Conductors; Aluminium Wires (Apr). 2 pp.

DIN ENGL 48203 Pt 5-84. Aluminium Wires and Aluminium Stranded Conductors; Technical Delivery Conditions (Mar). 4 pp.

DIN ENGL 48203 Pt 6-84. E-AlMgSi Wires and E-AlMgSi Stranded Conductors; Technical Delivery Conditions (Mar). 4 pp.

DIN ENGL 48203 Pt 11-87. Wires and Stranded Conductors; Steel-Reinforced Aluminium Stranded Conductors; Technical Delivery Conditions (Dec). 6 pp.

DIN ENGL 48203 Pt 12-87. Wires and Stranded Conductors; Steel-Reinforced E-AlMgSi Stranded Conductors; Technical Delivery Conditions (Dec). 6 pp.

DIN ENGL 48204-84. Steel Reinforced Aluminium Stranded Conductors (Apr). 4 pp.

IEC 228-78. Conductors of Insulated Cables Second Edition; (Errata—Sept 1979) (Supplement A-1982) (Amendment 1-1993) (CENELEC HD 383 S2/A2: 1993). 52 pp.

IEC 889-87. Hard-Drawn Aluminium Wire for Overhead Line Conductors First Edition. 12 pp.

IEC 1089-91. Round Wire Concentric Lay Overhead Electrical Stranded Conductors First Edition. 68 pp.

SNZ BS 215: Part 1-70. Specification for Aluminium Conductors and Aluminium Conductors, Steel-Reinforced, for Over-Head Power Transmission Part 1: Aluminium Stranded Conductors. 16 pp.

Aluminum Content Analysis
See Also: Aluminum

—Aluminum Ores—Volumetric Analysis
BSI BS 6870: Sec 2.6-87. 1987 Analysis of Aluminium Ores Part 2: Chemical Methods Section 2.6: Method for Determination of Aluminium Content: EDTA Titrimetric Method. 8 pp.

ISO 6994-86. Aluminium Ores—Determination of Aluminium Content—EDTA Titrimetric Method First Edition. 8 pp.

—Aluminum Ores—X-Ray Fluorescence
SAA AS 2564-82. Aluminium Ores—Determination of Aluminium, Silicon, Iron, Titanium and Phosphorus Contents—Wavelength Dispersive X-Ray Fluorescence Spectrometric Method. 48 pp.

—Carbon Steels—Atomic Absorption Spectrometry
BSI BS 6200: SUBSEC 3.1.4-90. 1990 Sampling and Analysis of Iron, Steel and Other Ferrous Metals Part 3: Methods of Analysis Section 3.1: Determination of Aluminium Subsection 3.1.4: Non-Alloyed Steel: Flame Atomic Absorption Spectrometric Method. 13 pp.

BSI BS 6200: SUB SEC3.1.4-01. 1990 Amd 1 Sampling and Analysis of Iron, Steel and Other Ferrous Metals Part 3: Methods of Analysis Section 3.1: Determination of Aluminium Subsection 3.1.4: Non-Alloyed Steel: Flame Atomic Absorption Spectrometric Method. 18 pp.

DIN ENGL EN 29658-92. Determination of Aluminium Content of Steel by Flame Atomic Absorption Spectrometry; (ISO 9658: 1990) (July). 14 pp.

ISO 9658-90. Steel—Determination of Aluminium Content—Flame Atomic Absorption Spectrometric Method First Edition. 14 pp.

Aluminum Content Analysis *(Cont.)*

—Cast Iron—Spectrophotometry
BSI BS 6200: SUBSEC 3.1.2-91. 1991 Sampling and Analysis of Iron, Steel and Other Ferrous Metals Part 3: Methods of Analysis Section 3.1: Determination of Aluminium Subsection 3.1.2: Steel and Cast Iron: Spectrophotometric Method. 10 pp.

—Cast Iron—Volumetric Analysis
BSI BS 6200: SUBSEC 3.1.1-91. 1991 Sampling and Analysis of Iron, Steel and Other Ferrous Metals Part 3: Methods of Analysis Section 3.1: Determination of Aluminium Subsection 3.1.1: Steel, Cast Iron, Low Carbon Ferro-Chromium Metal: Volumetric Method. 11 pp.

—Chromium
CNS G2120-82. Methods of Chemical Analysis for Silicon, Phosphorus, Sulphur Ferrite and Aluminum in Chromium Metal (Jan)(8404).

—Chromium Ores—Complexometric Titrations
ISO 8889-88. Chromium Ores and Concentrates—Determination of Aluminium Content—Complexometric Method First Edition. 6 pp.

—Chromium—Volumetric Analysis
BSI BS 6200: SUBSEC 3.1.1-91. 1991 Sampling and Analysis of Iron, Steel and Other Ferrous Metals Part 3: Methods of Analysis Section 3.1: Determination of Aluminium Subsection 3.1.1: Steel, Cast Iron, Low Carbon Ferro-Chromium Metal: Volumetric Method. 11 pp.

—Copper
JIS H 1057-87. Methods for Determination of Aluminium in Copper and Copper Alloys. 9 pp.

—Copper Alloys
JIS H 1057-87. Methods for Determination of Aluminium in Copper and Copper Alloys. 9 pp.

—Copper Alloys—Photometry
BSI BS 1748: Parts 1-5-61. 1961 Methods for the Analysis of Copper Alloys Parts 1-5: Determination of Copper, Lead, Iron, Aluminium and Nickel in Copper Alloys. 23 pp.

—Copper Alloys—Volumetric Analysis
ISO 3110-75. Copper Alloys—Determination of Aluminium as Alloying Element—Volumetric Method First Edition. 5 pp.

—Cryolite
MOD UK M 874/68. Examination of Cryolite.

—Cryolite—Atomic Absorption Analysis
CNS M3118-83. Method for Determining Aluminium Content in Natural and Artificial Cryolite (Atomic Absorption Method) (Mar)(10117).

—Cryolite—Atomic Absorption Spectrophotometry
BSI BS 5050-74. 1974 Methods of Test for Cryolite. 20 pp.

ISO 2830-73. Cryolite, Natural and Artificial—Determination of Aluminium Content—Atomic Absorption Method First Edition. 5 pp.

—Cryolite—Gravimetric Analysis
BSI BS 5050-74. 1974 Methods of Test for Cryolite. 20 pp.

CNS M3117-83. Method for Determining Aluminium Content in Natural and Artificial Cryolite (Oxine (8-Hydroxyquinoline) Gravimetric Method) (Mar)(10116).

ISO 2367-72. Cryolite (Natural and Artificial)—Determination of Aluminium Content—8-Hydroxyquinoline Gravimetric Method First Edition. 5 pp.

—Drinking Water—Spectrometry
DIN ENGL 38406 Pt 9-89. German Standard Methods for the Examination of Water, Waste Water and Sludge; Cations (Group E) Determination of Aluminium by Spectrometry (E9) (Feb). 6 pp.

—Ferrochromium—Volumetric Analysis
BSI BS 6200: SUBSEC 3.1.1-91. 1991 Sampling and Analysis of Iron, Steel and Other Ferrous Metals Part 3: Methods of Analysis Section 3.1: Determination of Aluminium Subsection 3.1.1: Steel, Cast Iron, Low Carbon Ferro-Chromium Metal: Volumetric Method. 11 pp.

—Ferrosilicon—Atomic Absorption Spectrometry
BSI BS 6200: SUBSEC 3.1.5-85. 1985 Sampling and Analysis of Iron, Steel and Other Ferrous Metals Part 3: Methods of Analysis Section 3.1: Determination of Aluminium Subsection 3.1.5: Ferro-Silicon: Atomic Absorption Spectrometric Method. 6 pp.

INDUSTRY STANDARDS

Aluminum

Aluminum Content Analysis (Cont.)

—Ferrosilicon—Atomic Absorption Spectrometry (Cont.)
ISO 4139-79. Ferrosilicon—Determination of Aluminium Content—Flame Atomic Absorption Spectrometric Method First Edition. 5 pp.

—Fuel Oils—Atomic Absorption Spectrometry
DIN ENGL 51416 Pt 1-86. Testing of Liquid Fuels; Determination of the Aluminium Content of Fuel Oils; Determination by Atomic Absorption Spectrometry (AAS) After Ashing (Apr). 4 pp.

—Industrial Water
CPPA H.4P(J)-67. Analysis of Process Waters Aluminum. 2 pp.

—Iron
CNS G2242-85. Method of Determination for Aluminium in Iron and Steel (Apr)(11244).
JIS G 1224-81. Methods for Determination of Aluminium in Iron and Steel.

—Iron Ores—Atomic Absorption Spectrometry
BSI BS 7020: Sec 8.2-93. 1993 Analysis of Iron Ores Part 8: Methods for the Determination of Aluminium Content Section 8.2: Flame Atomic Absorption Spectrometric Method (ISO 4688-1: 1990) (Q). 15 pp.
ISO 4688 Pt 1-92. Iron Ores—Determination of Aluminium Content—Part 1: Flame Atomic Absorption Spectrometric Method First Edition. 12 pp.

—Iron Ores—Volumetric Analysis
BSI BS 7020: Sec 8.1-88. 1988 Analysis of Iron Ores Part 8: Methods for the Determination of Aluminium Content Section 8.1: Titrimetric Method. 12 pp.
ISO 6830-86. Iron Ores—Determination of Aluminium Content—EDTA Titrimetric Method First Edition. 10 pp.

—Iron Ores—X-Ray Fluorescence Spectrometry
ISO 9516-92. Iron Ores—Determination of Silicon, Calcium, Manganese, Aluminium, Titanium, Magnesium, Phosphorus, Sulfur and Potassium—Wavelength Dispersive X-Ray Fluorescence Spectrometric Method First Edition. 74 pp.

—Iron—Photometry
SAA AS K1.24-68. Methods for the Sampling and Analysis of Iron and Steel Method for the Determination of Aluminium in Iron and Steel (Photometric Method) Reconfirmed 1987. 7 pp.

—Kaolin
MOD UK M 2002/59. Examination of Kaolin (Withdrawn).

—Magnesium Alloys
JIS H 1332-87. Method for Determination of Aluminium in Magnesium Alloys. 8 pp.

—Magnesium Alloys—Gravimetric Analysis
BSI BS 3907: Part 1-65. 1965 Methods for the Analysis of Magnesium and Magnesium Alloys Part 1: Aluminium in Magnesium Alloys (Gravimetric Method). 8 pp.
ISO 791-73. Magnesium Alloys—Determination of Aluminium—8-Hydroxyquinoline Gravimetric Method First Edition. 5 pp.

—Magnesium Alloys—Photometry
BSI BS 3907: Part 12-71. 1971 Methods for the Analysis of Magnesium and Magnesium Alloys Part 12: Aluminium in Magnesium and Magnesium Alloys (Photometric Method). 8 pp.
ISO 3255-74. Magnesium and Magnesium Alloys—Determination of Aluminium—Chromazurol S Photometric Method First Edition. 5 pp.

—Magnesium Oxides
MOD UK M 9541/69. Examination of Magnesium Oxide and Magnesium Oxide, Special, Lead Free (Withdrawn).

—Magnesium—Photometry
BSI BS 3907: Part 12-71. 1971 Methods for the Analysis of Magnesium and Magnesium Alloys Part 12: Aluminium in Magnesium and Magnesium Alloys (Photometric Method). 8 pp.
ISO 3255-74. Magnesium and Magnesium Alloys—Determination of Aluminium—Chromazurol S Photometric Method First Edition. 5 pp.

Aluminum Content Analysis (Cont.)

—Magnets—Volumetric Analysis
BSI BS 6200: SUBSEC 3.1.6-91. 1991 Sampling and Analysis of Iron, Steel and Other Ferrous Metals Part 3: Methods of Analysis Section 3.1: Determination of Aluminium Subsection 3.1.6: Permanent Magnet Alloys: Volumetric Method. 8 pp.

—Manganese Ores—Atomic Absorption Spectrometry
ISO 5889-83. Manganese Ores and Concentrates—Determination of Aluminium, Copper, Lead and Zinc Contents—Flame Atomic Absorption Spectrometric Method First Edition. 6 pp.

—Manganese Ores—Gravimetric Analysis
ISO 4295-88. Manganese Ores and Concentrates—Determination of Aluminium Content—Photometric and Gravimetric Methods Second Edition. 8 pp.

—Manganese Ores—Photometry
ISO 4295-88. Manganese Ores and Concentrates—Determination of Aluminium Content—Photometric and Gravimetric Methods Second Edition. 8 pp.

—Nickel Alloys—Atomic Absorption Spectrometry
BSI BS 7455: Part 7-92. 1992 Analysis of Nickel Alloys by Flame Atomic Absorption Spectrometry Part 7: Method for Determination of Aluminium (ISO 7530-7: 1992). 10 pp.
ISO 7530 Pt 7-92. Nickel Alloys—Flame Atomic Absorption Spectrometric Analysis—Part 7: Determination of Aluminium Content First Edition. 7 pp.

—Nickel—Photometry
BSI BS 3727: Part 1-66. 1966 Methods for the Analysis of Nickel for Use in Electronic Tubes and Valves Part 1: Determination of Aluminium (Photometric Method). 8 pp.

—Organic Coatings—Quantitative Analysis
CNS K6804-28-85. Method of Test for Organic Coating (Chemical Analysis) — Quantative Test of Aluminium in Coating (Jan)(10880-28).

—Paints
CGSB 1-GP-71 METH 51.1-81. Methods of Testing Paints and Pigments Analysis of Metallic Portion Determination of Metallic Aluminum. 2 pp.

—Pigments
CGSB 1-GP-71 METH 51.1-81. Methods of Testing Paints and Pigments Analysis of Metallic Portion Determination of Metallic Aluminum. 2 pp.

—Potassium Cryolite
MOD UK M 828/68. Examination of Potassium Cryolite.

—Silicon—Absorptiometric Analysis
CNS G2139-82. Method of Chemical Analysis for Aluminum in Metallic Silicon (Chrome Azurol S Absorptimetric Method) (Oct)(9498).

—Silicon—Atomic Absorption Spectrometry
CNS G2140-82. Method of Chemical Analysis for Aluminum in Metallic Silicon (Atomic Absorptimetric Method) (Oct)(9499).

—Silicon—Volumetric Analysis
CNS G2138-82. Method of Chemical Analysis for Aluminum in Metallic Silicon (EDTA Titration Method) (Oct)(9497).

—Sodium Borates—Volumetric Analysis
BSI BS 5688: Part 11-79. 1979 Orthoboric Acid (Boric Acid), Diboron Trioxide (Boric Oxide), Disodium Tetraborates, Sodium Perborates and Crude Sodium Borates for Industrial Use Part 11: Determination of Total Aluminium Content of Crude Sodium Borates. 4 pp.
BSI BS 5688: Part 20-79. 1979 Orthoboric Acid (Boric Acid), Diboron Trioxide (Boric Oxide), Disodium Tetraborates, Sodium Perborates and Crude Sodium Borates for Industrial Use Part 20: Determination of Aluminium Soluble in Alkaline Medium in Crude Sodium Borates. 4 pp.
ISO 2760-75. Crude Sodium Borates for Industrial Use—Determination of Total Aluminium Content—Titrimetric Method First Edition. 4 pp.
ISO 3125-76. Crude Sodium Borates for Industrial Use—Determination of Aluminium Soluble in Alkaline Medium—EDTA Titrimetric Method First Edition. 5 pp.

—Sodium Hydroxide
MOD UK M 808/73. Examination of Sodium Hydroxide Pure (Withdrawn).

Aluminum Content Analysis (Cont.)

—Solders—Photometry
BSI BS 3338: Part 18-66. 1966 Methods for the Sampling and Analysis of Tin and Tin Alloys Part 18: Aluminium in Solders and White Metal Bearing Alloys (Photometric Method). 9 pp.

—Steels
CNS G2242-85. Method of Determination for Aluminium in Iron and Steel (Apr)(11244).
JIS G 1224-81. Methods for Determination of Aluminium in Iron and Steel.

—Steels—Photometry
SAA AS K1.24-68. Methods for the Sampling and Analysis of Iron and Steel Method for the Determination of Aluminium in Iron and Steel (Photometric Method) Reconfirmed 1987. 7 pp.

—Steels—Spectrophotometry
BSI BS 6200: SUBSEC 3.1.2-91. 1991 Sampling and Analysis of Iron, Steel and Other Ferrous Metals Part 3: Methods of Analysis Section 3.1: Determination of Aluminium Subsection 3.1.2: Steel and Cast Iron: Spectrophotometric Method. 10 pp.

—Steels—Volumetric Analysis
BSI BS 6200: SUBSEC 3.1.1-91. 1991 Sampling and Analysis of Iron, Steel and Other Ferrous Metals Part 3: Methods of Analysis Section 3.1: Determination of Aluminium Subsection 3.1.1: Steel, Cast Iron, Low Carbon Ferro-Chromium Metal: Volumetric Method. 11 pp.

—Tantalum
JIS H 1692-76. Method for Determination of Aluminium in Tantalum (R 1986). 7 pp.

—Titanium Alloys
JIS H 1622-76. Methods for Determination of Aluminium in Titanium Alloys.

—Water—Atomic Absorption Spectrometry
JIS K 0553-90. Testing Methods for Determination of Metallic Elements in Highly Purified Water. 27 pp.

—Water—Fluorometric Analysis
CNS K9042-80. Method of Test for Aluminum in Water (Fluorometric Method) (Jul)(5861).

—Water—Spectrometry
DIN ENGL 38406 Pt 9-89. German Standard Methods for the Examination of Water, Waste Water and Sludge; Cations (Group E) Determination of Aluminium by Spectrometry (E9) (Feb). 6 pp.

—Water—Spectrophotometry
BSI BS 2690: Part 106-79. 1979 Water Used in Industry Part 106: Reactive Aluminium: Spectrophotometric Method (Catechol Violet). 4 pp.

—White Metals—Photometry
BSI BS 3338: Part 18-66. 1966 Methods for the Sampling and Analysis of Tin and Tin Alloys Part 18: Aluminium in Solders and White Metal Bearing Alloys (Photometric Method). 9 pp.

—Zinc Alloys—Atomic Absorption Spectrometry
SAA AS 1329.3-73. Methods for the Analysis of Zinc and Zinc Alloys—Part 3: Aluminium in Zinc Alloys (Atomic Absorption Spectrometric Method). 6 pp.

—Zinc Alloys—Spectrophotometry
SAA AS 1329.4-78. Methods for the Analysis of Zinc and Zinc Alloys—Part 4: Aluminium in Zinc Alloys (Spectrophotometric Method). 6 pp.

—Zinc Alloys—Volumetric Analysis
BSI BS 3630: Part 6-65. 1965 Methods for the Sampling and Analysis of Zinc and Zinc Alloys Part 6: Aluminiumin Zinc Alloys (Volumetric Method). 8 pp.
ISO 1169-75. Zinc Alloys—Determination of Aluminium Content—Volumetric Method First Edition. 4 pp.
SAA AS 1329.1-86. Methods for the Analysis of Zinc and Zinc Alloys—Part 1: Determination of Aluminium Content (Titrimetric Method). 6 pp.

—Zinc—Volumetric Analysis
SAA AS 1329.1-86. Methods for the Analysis of Zinc and Zinc Alloys—Part 1: Determination of Aluminium Content (Titrimetric Method). 6 pp.

—Zirconium—Absorptiometric Analysis
JIS H 1661-86. Methods for Determination of Aluminium in Zirconium and Zirconium Alloys. 9 pp.

—Zirconium Alloys—Absorptiometric Analysis
JIS H 1661-86. Methods for Determination of Aluminium in Zirconium and Zirconium Alloys. 9 pp.

INTERNATIONAL AND NON-U.S. NATIONAL STANDARDS
SUBJECT INDEX
Aluminum

Aluminum Content Analysis *(Cont.)*

—**Zirconium Alloys—Spectrochemical Analysis**
JIS H 1661-86. Methods for Determination of Aluminium in Zirconium and Zirconium Alloys. 9 pp.

—**Zirconium—Spectrochemical Analysis**
JIS H 1661-86. Methods for Determination of Aluminium in Zirconium and Zirconium Alloys. 9 pp.

Aluminum Copper Alloys
Use: Copper Aluminum Alloys

Aluminum Electrolytic Capacitors
Use: Aluminum Capacitors

Aluminum Fluoride
CNS B3296-85. Circular Saws Blade for Woodworking Machines (Oct)(4968).
ISO 2925-73. Aluminium Fluoride for Industrial Use—Preparation and Storage of Test Samples First Edition. 3 pp.

—**Fluorine Content—Willard Winter Method**
BSI BS 4993: Part 1-74. 1974 Methods of Test for Aluminium Fluoride for Industrial Use Part 1: Determination of Fluorine Content. 13 pp.
ISO 2362-72. Aluminium Fluoride for Industrial Use—Determination of Fluorine Content—Modified Willard-Winter Method First Edition. 7 pp.

—**Iron Content—Photometry**
BSI BS 4993: Part 2-74. 1974 Methods of Test for Aluminium Fluoride for Industrial Use Part 2: Determination of Iron Content. 12 pp.
ISO 2368-72. Aluminium Fluoride for Industrial Use—Determination of Iron Content—1,10-Phenanthroline Photometric Method First Edition. 6 pp.

—**Moisture Content—Gravimetric Analysis**
BSI BS 4993: Part 6-80. 1980 Methods of Test for Aluminium Fluoride for Industrial Use Part 6: Determination of Moisture Content (Gravimetric Method). 6 pp.
CNS M3122-83. Method for Determining Moisture Content in Natural and Artificial Cryolite and Aluminium Fluoride for Industrial Use (Gravimetric Method) (Mar)(10121).
ISO 3393-76. Cryolite, Natural and Artificial, and Aluminium Fluoride for Industrial Use—Determination of Moisture Content—Gravimetric Method First Edition. 4 pp.

—**Moisture Content—Karl Fischer Method**
BSI BS 4993: Part 5-80. 1980 Methods of Test for Aluminium Fluoride for Industrial Use Part 5: Determination of Moisture Content (Karl Fischer Method). 9 pp.

—**Phosphorus Content—Atomic Absorption Spectrometry**
BSI BS 4993: Part 10-83. 1983 Methods of Test for Aluminium Fluoride for Industrial Use Parts 10 & 13: Determination for Phosphorus Content. 7 pp.
BSI BS 5050: Part 13-83. 1983 Methods of Test for Cryolite Part 13: Determination of Phosphorus Content. 7 pp.
ISO 6374-81. Cryolite, Natural and Artificial, and Aluminium Fluoride for Industrial Use—Determination of Phosphorus Content—Atomic Absorption Spectrometric Method After Extraction First Edition. 6 pp.

—**Phosphorus Content—Photometry**
CNS M3124-83. Method for Determining Phosphorus Content in Natural and Artificial Cryolite and Aluminium Fluoride for Industrial Use (Reduced Molybdophosphate Photometric Method) (Mar)(10123).
ISO 5930-79. Cryolite, Natural and Artificial, and Aluminium Fluoride for Industrial Use—Determination of Phosphorus Content—Reduced Molybdophosphate Photometric Method First Edition. 6 pp.

—**Silica Content—Spectrophotometry**
BSI BS 4993: Part 3-74. 1974 Amd 1 Methods of Test for Aluminium Fluoride for Industrial Use Part 3: Determination of Silica Content. 13 pp.
ISO 2369-72. Aluminium Fluoride for Industrial Use—Determination of Silica Content—Spectrophotometric Method Using the Reduced Silicomolybdic Complex First Edition. 6 pp.

—**Sodium Content—Atomic Emission Spectrophotometry**
BSI BS 4993: Part 7-80. 1980 Methods of Test for Aluminium Fluoride for Industrial Use Part 7: Determination of Sodium Content. 8 pp.

Aluminum Fluoride *(Cont.)*

—**Sodium Content—Atomic Emission Spectrophotometry** *(Cont.)*
ISO 4279-77. Aluminium Fluoride for Industrial Use—Determination of Sodium Content—Flame Emission Spectrophotometric Method First Edition; (Erratum—Sept 1978). 7 pp.

—**Sulfate Content—Gravimetric Analysis**
BSI BS 4993: Part 8-80. 1980 Methods of Test for Aluminium Fluoride for Industrial Use Part 8: Determination of Sulphate Content. 7 pp.
CNS M3123-83. Method for Determining Sulphate Content in Natural and Artificial Cryolite and Aluminium Fluoride for Industrial Use (Barium Sulphate Gravimetric Method) (Mar)(10122).
ISO 4280-77. Cryolite, Natural and Artificial, and Aluminium Fluoride for Industrial Use—Determination of Sulphate Content—Barium Sulphate Gravimetric Method First Edition; (Erratum—Sept 1978). 6 pp.

—**Sulfur Content—X-Ray Fluorescence Spectrometry**
BSI BS 4993: Part 9-80. 1980 Methods of Test for Aluminium Fluoride for Industrial Use Part 9: Determination of Sulphur Content (X-Ray Fluorescence Spectrometric Method). 8 pp.
CNS M3125-83. Method for Determining Sulphur Content in Natural and Artificial Cryolite and Aluminium Fluoride for Industrial Use (X-Ray Fluorescence Spectrometric Method) (Mar)(10124).
ISO 5938-79. Cryolite, Natural and Artificial, and Aluminium Fluoride for Industrial Use—Determination of Sulphur Content—X-Ray Fluorescence Spectrometric Method First Edition; (Amendment Slip-1980). 7 pp.

—**Test Specimens**
BSI BS 4993: Part 4-80. 1980 Methods of Test for Aluminium Fluoride for Industrial Use Part 4: Preparation and Storage of Test Samples. 3 pp.
CNS K6557-80. Preparation and Storage of Test Samples of Aluminium Fluoride for Industrial Use (Aug)(6202).

—**Water Content—Electrolytic Analysis**
CNS M3121-83. Method for Determining Water Content in Natural and Artificial Cryolite and Aluminium for Industrial Use (Electrometric Method) (Mar)(10120).
ISO 3392-76. Cryolite, Natural and Artificial, and Aluminium Fluoride for Industrial Use—Determination of Water Content—Electrometric Method First Edition. 7 pp.

Aluminum Foil
See Also: Aluminum Alloys; Foil Papers; Foils; Metal Products
CNS H2027-81. Method of Test for Aluminium Foil (Jun)(2250).
CNS H3023-81. Aluminium Foil (Jun) (2249).
ISO 7271-82. Aluminium and Aluminium Alloys—Foil and Thin Strip—Dimensional Tolerances First Edition. 4 pp.
JIS H 4160-85. Aluminium and Aluminium Alloy Foils. 9 pp.

—**Anodized—Photosensitive**
CGSB 62-GP-10-69. Photosensitive Anodized Aluminum Markings: Plates and Foil, Photographic, Standard for. 17 pp.

—**Caps (Lids)—Containers**
BSI BS 3313: Part 1-68. 1968 Amd 1 Aluminium Capping Foil and Strip for Dairy Product Containers Part 1: Aluminium Capping Foil for Glass Container. 10 pp.
BSI BS 3313: Part 2-68. 1968 Amd 1 Aluminium Capping Foil and Strip for Dairy Product Containers Part 2: Aluminium Cappint Foil for Skirted Closures for Plastics Containers. 14 pp.

—**Cold Rolled**
DIN ENGL 1784 Pt 3-70. Aluminium Coiled Strip; Foils; 0.007 to 0.020 mm (7 to 20 Micron m), Cold Rolled, Dimensions (June). 4 pp.

—**Food Packaging**
BSI BS 1683-87. 1987 Coated Aluminium Foil for Wrapping Processed Cheese. 8 pp.
BSI BS 5439-77. 1977 Aluminium Foil Catering Containers. 8 pp.
CGSB 43-GP-148M-84. Foil, Aluminum, Annealed. 9 pp.

—**High Purity**
JIS H 4170-91. High Purity Aluminium Foils. 7 pp.

—**Laminated**
CNS H2028-81. Method of Test for Laminated Aluminium Foil (Aug)(2252).
CNS H3024-81. Aluminium Foil, Laminated with Paper Lining (Aug) (2251).

Aluminum Foil *(Cont.)*

—**Laminated** *(Cont.)*
CNS P2037-84. Aluminum Foil Laminating Paper (Oct)(2646). 2 pp.
CNS Z7016-84. Laminated Aluminum Foils (Aug)(2521).
JIS Z 1520-90. Laminated Aluminium Foils. 6 pp.
MOD UK DSTAN 81-75-87. Barrier Material, Aluminium Foil Laminate, Flexible, Heat Sealable, Water-Vapour Resistant Issue 1. 23 pp.
MOD UK TS 429. Foil, Metal, Laminated Sheet (Heat Sealable) (Withdrawn).

—**Packaging**
BSI BS 1133: Sec 21-91. 1991 Packaging Code Section 21: Regenerated Cellulose Film, Films Made of Plastics, Aluminium Foil, Flexible Multilayer Structures and Metallized Materials (H). 36 pp.

—**Packaging—Dairy Products**
SAA AS 1813-75. Aluminium Foil/Vegetable Parchment Laminates for Wrapping Dairy Products. 8 pp.

—**Spools**
DIN ENGL 55470-72. Packaging Accessories; Rewinding Tubes for Aluminium Foils; Dimensions (July). 2 pp.

Aluminum Hydroxides
CNS K1121-72. Aluminium Hydroxide (Jun)(2623).
CNS K6237-72. Method of Test for Aluminium Hydroxide (Oct)(2624).

Aluminum Hydroxychloride
Use: Aluminum Chlorohydrate

Aluminum Inorganic Compounds
Scope Note: Use a more specific term
See: Aluminum Ammonium Sulfate; Aluminum Chloride; Aluminum Chlorohydrate; Aluminum Fluoride; Aluminum Hydroxides; Aluminum Nitrate; Aluminum Oxide; Aluminum Potassium Sulfate; Aluminum Silicate; Aluminum Sodium Sulfate; Aluminum Sulfate

Aluminum Iron Sulfate
See Also: Aluminum Sulfate; Ferric Sulfate

—**Water Treatment**
CEN PREN 887-92. Aluminium Iron-Sulfate, Used for Water Intended for Human Consumption. 15 pp.

Aluminum Magnesium Alloys
See Also: Aluminum Alloys; Aluminum Magnesium Silicon Alloys

—**Bars—Aerospace**
BSI 5L 44-85. 1985 Forging Stock, Bars, Extruded Sections and Forgings of Aluminium—2 1/4% Magnesium Alloy. 4 pp.

—**Bars—Extruded—Aircraft**
MOD UK DTD-297A-55. Aluminium—7 Per Cent Magnesium Alloy Bars, Extruded Sections and Forgings (Reprinted March 1966). 2 pp.

—**Explosives**
MOD UK DSTAN 68-157-93. Magnesium-Aluminium Alloy 50/50 Powdered, Sizes 120 and 200 Issue 1. 12 pp.

—**Forgings—Aerospace**
BSI 5L 44-85. 1985 Forging Stock, Bars, Extruded Sections and Forgings of Aluminium—2 1/4% Magnesium Alloy. 4 pp.

—**Forgings—Aircraft**
MOD UK DTD-297A-55. Aluminium—7 Per Cent Magnesium Alloy Bars, Extruded Sections and Forgings (Reprinted March 1966). 2 pp.

—**Heat Treated—Ingots**
MOD UK DTD-5018A-71. Ingots and Castings of Aluminium-Magnesium-Zinc Alloy (Solution Treated) (Mg 7.7, Zn 1.2). 2 pp.

—**Incendiary Mixtures**
MOD UK DSTAN 68-157-93. Magnesium-Aluminium Alloy 50/50 Powdered, Sizes 120 and 200 Issue 1. 12 pp.

—**Pyrotechnics**
MOD UK DSTAN 68-157-93. Magnesium-Aluminium Alloy 50/50 Powdered, Sizes 120 and 200 Issue 1. 12 pp.

—**Sections—Extruded—Aerospace**
BSI 5L 44-85. 1985 Forging Stock, Bars, Extruded Sections and Forgings of Aluminium—2 1/4% Magnesium Alloy. 4 pp.

Aluminum Magnesium Alloys *(Cont.)*

—Sections—Extruded—Aircraft

MOD UK DTD-297A-55. Aluminium—7 Per Cent Magnesium Alloy Bars, Extruded Sections and Forgings (Reprinted March 1966). 2 pp.

—Sheets—Aerospace

BSI 3L 80-85. 1985 Sheet and Strip of Aluminium-2 1/4% Magnesium Alloy (Temper Designation-0). 4 pp.

BSI 3L 81-85. 1985 Sheet and Strip of Aluminium-2 1/4% Magnesium Alloy (Temper Designation H16 or H26). 4 pp.

—Strips—Aerospace

BSI 3L 80-85. 1985 Sheet and Strip of Aluminium-2 1/4% Magnesium Alloy (Temper Designation-0). 4 pp.

BSI 3L 81-85. 1985 Sheet and Strip of Aluminium-2 1/4% Magnesium Alloy (Temper Designation H16 or H26). 4 pp.

—Tubes—Seamless—Aerospace

BSI 4L 56-86. 1986 Amd 1 Tube of Aluminium—2 1/4% Magnesium Alloy (Temper Designation—0) (Seamless: Tested Hydraulically) (Not Exceeding 12 mm Wall Thickness) (AMD 5993) August 31, 1988. 5 pp.

Aluminum Magnesium Silicon Alloys

See Also: Aluminum Alloys; Aluminum Magnesium Alloys

—Bars—Extruded—Aircraft

MOD UK DTD-372B-57. Aluminium—Magnesium—Silicon Alloy Extruded Bars and Sections (Suitable for Welding) (Reprinted May 1961). 2 pp.

—Profiles

DIN ENGL 17615 Pt 1-87. AlMgSiO,5 Precision Sections; Technical Delivery Conditions (Jan). 5 pp.
DIN ENGL 17615 Pt 3-87. AlMgSiO,5 Precision Sections; Tolerances (Jan). 7 pp.

—Profiles—Design

DIN ENGL 17615 Pt 2-87. AlMgSiO,5 Precision Sections; Design Principles (Jan). 12 pp.

—Sections—Extruded—Aircraft

MOD UK DTD-372B-57. Aluminium—Magnesium—Silicon Alloy Extruded Bars and Sections (Suitable for Welding) (Reprinted May 1961). 2 pp.

Aluminum Manganese Alloys

See Also: Aluminum Alloys

—Sheets—Aerospace

BSI 4L 59-85. 1985 Sheet and Strip of Aluminium-Manganese Alloy (Temper Designation H16 to H26). 4 pp.

BSI 4L 60-85. 1985 Sheet and Strip of Aluminium-Manganese Alloy (Temper Designation H12 or H22). 4 pp.

BSI 4L 61-85. 1985 Sheet and Strip of Aluminium-Manganese Alloy (Temper Designation-0). 4 pp.

—Strips—Aerospace

BSI 4L 59-85. 1985 Sheet and Strip of Aluminium-Manganese Alloy (Temper Designation H16 to H26). 4 pp.

BSI 4L 60-85. 1985 Sheet and Strip of Aluminium-Manganese Alloy (Temper Designation H12 or H22). 4 pp.

BSI 4L 61-85. 1985 Sheet and Strip of Aluminium-Manganese Alloy (Temper Designation-0). 4 pp.

Aluminum Nitrate

CNS K7035-61. Chemical Reagent (Aluminum Nitrate) (Dec)(1535).
JIS K 8544-78. Aluminium Nitrate, Nonahydrate.

Aluminum Ores

See Also: Bauxite; Cryolite; Ores

—Aluminum Content—Volumetric Analysis

BSI BS 6870: Sec 2.6-87. 1987 Analysis of Aluminium Ores Part 2: Chemical Methods Section 2.6: Method for Determination of Aluminium Content: EDTA Titrimetric Method. 8 pp.

ISO 6994-86. Aluminium Ores—Determination of Aluminium Content—EDTA Titrimetric Method First Edition. 8 pp.

—Aluminum Content—X-Ray Fluorescence

SAA AS 2564-82. Aluminium Ores—Determination of Aluminium, Silicon, Iron, Titanium and Phosphorus Contents—Wavelength Dispersive X-Ray Fluorescence Spectrometric Method. 48 pp.

—Iron Content—Reduction

ISO 10213-91. Aluminium Ores—Determination of Total Iron Content—Titanium Trichloride Reduction Method First Edition. 11 pp.

Aluminum Ores *(Cont.)*

—Iron Content—Volumetric Analysis

BSI BS 6870: Sec 2.5-87. 1987 Analysis of Aluminium Ores Part 2: Chemical Methods Section 2.5: Method for Determination of Iron Content: Titrimetric Method. 5 pp.

ISO 6609-85. Aluminium Ores—Determination of Iron Content—Titrimetric Method First Edition. 5 pp.

—Iron Content—X-Ray Fluorescence

SAA AS 2564-82. Aluminium Ores—Determination of Aluminium, Silicon, Iron, Titanium and Phosphorus Contents—Wavelength Dispersive X-Ray Fluorescence Spectrometric Method. 48 pp.

—Loss of Mass—Gravimetric Analysis

BSI BS 6870: Sec 2.3-87. 1987 Analysis of Aluminium Ores Part 2: Chemical Methods Section 2.3: Method for Determination of Loss of Mass at 1075 Degrees Celsius: Gravimetric Method. 4 pp.

ISO 6606-86. Aluminium Ores—Determination of Loss of Mass at 1 075 Degrees C—Gravimetric Method First Edition. 5 pp.

—Moisture Content

ISO 9033-89. Aluminium Ores—Determination of the Moisture Content of Bulk Material First Edition. 9 pp.

—Moisture Content—Gravimetric Analysis

BSI BS 6870: Sec 2.2-87. 1987 Analysis of Aluminium Ores Part 2: Chemical Methods Section 2.2: Method for Determination of Hygroscopic Moisture in Analytical Samples: Gravimetric Method. 4 pp.

ISO 8557-85. Aluminium Ores—Determination of Hygroscopic Moisture in Analytical Samples—Gravimetric Method First Edition. 4 pp.

—Phosphorus Content—Spectrophotometry

BSI BS 6870: Sec 2.8-87. 1987 Analysis of Aluminium Ores Part 2: Chemical Methods Section 2.8: Method for Determination of Phosphorus Content: Spectrophotometric Method. 8 pp.

ISO 8556-86. Aluminium Ores—Determination of Phosphorus Content—Molybdenum Blue Spectrophotometric Method First Edition. 8 pp.

—Phosphorus Content—X-Ray Fluorescence

SAA AS 2564-82. Aluminium Ores—Determination of Aluminium, Silicon, Iron, Titanium and Phosphorus Contents—Wavelength Dispersive X-Ray Fluorescence Spectrometric Method. 48 pp.

—Sampling

ISO 6140-91. Aluminium Ores—Preparation of Samples First Edition. 28 pp.

ISO 8685-92. Aluminium Ores—Sampling Procedures First Edition. 30 pp.

ISO 10226-91. Aluminium Ores—Experimental Methods for Checking the Bias of Sampling First Edition. 8 pp.

—Sampling—Heterogeneity

ISO 6138-91. Aluminium Ores—Experimental Determination of the Heterogeneity of Constitution First Edition. 6 pp.

ISO 6139-93. Aluminium Ores—Experimental Determination of the Heterogeneity of Distribution of a Lot First Edition. 10 pp.

—Silicon Content—Gravimetric Analysis

BSI BS 6870: Sec 2.4-87. 1987 Analysis of Aluminium Ores Part 2: Chemical Methods Section 2.4: Method for Determination of Silicon Content: Combined Gravimetric and Spectrophotometric Method. 7 pp.

ISO 6607-85. Aluminium Ores—Determination of Total Silicon Content—Combined Gravimetric and Spectrophotometric Method First Edition. 7 pp.

—Silicon Content—Spectrophotometry

BSI BS 6870: Sec 2.4-87. 1987 Analysis of Aluminium Ores Part 2: Chemical Methods Section 2.4: Method for Determination of Silicon Content: Combined Gravimetric and Spectrophotometric Method. 7 pp.

ISO 6607-85. Aluminium Ores—Determination of Total Silicon Content—Combined Gravimetric and Spectrophotometric Method First Edition. 7 pp.

—Silicon Content—X-Ray Fluorescence

SAA AS 2564-82. Aluminium Ores—Determination of Aluminium, Silicon, Iron, Titanium and Phosphorus Contents—Wavelength Dispersive X-Ray Fluorescence Spectrometric Method. 48 pp.

—Test Specimens

BSI BS 6870: Sec 2.1-87. 1987 Analysis of Aluminium Ores Part 2: Chemical Methods Section 2.1: Method of Preparation of Pre-Dried Test Samples. 3 pp.

ISO 8558-85. Aluminium Ores—Preparation of Pre-Dried Test Samples First Edition. 3 pp.

Aluminum Ores *(Cont.)*

—Titanium Content—Spectrophotometry

BSI BS 6870: Sec 2.7-87. 1987 Analysis of Aluminium Ores Part 2: Chemical Methods Section 2.7: Method for Determination of Titanium Content: Spectrophotometric Method. 7 pp.

ISO 6995-85. Aluminium Ores—Determination of Titanium Content—4,4'-Diantipyrylmethane Spectrophotometric Method First Edition. 7 pp.

—Titanium Content—X-Ray Fluorescence

SAA AS 2564-82. Aluminium Ores—Determination of Aluminium, Silicon, Iron, Titanium and Phosphorus Contents—Wavelength Dispersive X-Ray Fluorescence Spectrometric Method. 48 pp.

—Vanadium Content—Spectrophotometry

ISO 9208-89. Aluminium Ores—Determination of Vanadium Content—BPHA Spectrophotometric Method First Edition. 8 pp.

—X-Ray Fluorescence

BSI BS 6870: Part 3-89. 1989 Analysis of Aluminium Ores Part 3: Method for Multi-Element Analysis by Wavelength Dispersive X-Ray Fluorescence. 16 pp.

Aluminum Organic Compounds

Scope Note: Use a more specific term
See: Aluminum Stearate

Aluminum Oxide

Use For: Alumina; Corundum *See Also:* Abrasives; Activated Alumina; Metal Powders

CNS K7037-61. Chemical Reagent (Aluminum Oxide) (Ihnited Aluminum Oxide) (Dec)(1537).

JIS H 1901-77. Methods for Determination of Aluminium Oxide.

—alpha-Aluminum Oxide Content

SAA AS 2879.3-91. Alumina Determination of Alpha Alumina Content by X-Ray Diffraction. 3 pp.

—Adsorption

ISO 2961-74. Aluminium Oxide Primarily Used for the Production of Aluminium—Determination of an Adsorption Index First Edition. 6 pp.

—Angle of Repose

BSI BS 4140: Part 9-86. 1986 Methods of Test for Aluminium Oxide Part 9: Measurement of the Angle of Repose. 6 pp.

ISO 902-76. Aluminium Oxide Primarily Used for the Production of Aluminium—Measurement of the Angle of Repose First Edition. 6 pp.

—Area—Adsorption

BSI BS 4140: Part 23-87. 1987 Methods of Test for Aluminium Oxide Part 23: Determination of Specific Surface Area by Nitrogen Absorption (Single-Point Method). 11 pp.

ISO 8008-86. Aluminium Oxide Primarily Used for the Production of Aluminium—Determination of Specific Surface Area by Nitrogen Adsorption—Single-Point Method First Edition. 11 pp.

SAA AS 2879.4-91. Alumina Determination of Specific Surface Area by Nitrogen Adsorption. 2 pp.

—Boron Content—Spectrophotometry

BSI BS 4140: Part 19-86. 1986 Methods of Test for Aluminium Oxide Part 19: Determinaton of Boron Content. 7 pp.

BSI BS 4140: Part 19-01. 1986 Amd 1 Methods of Test for Aluminium Oxide Part 19: Determination of Boron Content (ISO 2865: 1973) (AMD 6896) April 1, 1992. 8 pp.

ISO 2865-73. Aluminium Oxide Primarily Used for the Production of Aluminium—Determination of Boron Content—Curcumin Spectrophotometric Method First Edition; (Corrigendum 1-1991). 8 pp.

—Calcium Content—Atomic Absorption Spectrometry

BSI BS 4140: Part 14-90. 1990 Methods of Test for Aluminium Oxide Part 14: Determination of Calcium Content (W). 6 pp.

BSI BS 4140: Part 14-01. 1990 Amd 1 Methods of Test for Aluminium Oxide Part 14: Determination of Calcium Content (AMD 6893) February 28, 1992 (W). 7 pp.

—Calcium Content—Atomic Absorption Spectrophotometry

ISO 2069-76. Aluminium Oxide Primarily Used for the Production of Aluminium—Determination of Calcium Content—Flame Atomic Absorption Method First Edition. 6 pp.

ISO 2070-81. Aluminium Oxide Primarily Used for the Production of Aluminium—Determination of Calcium Content—Naphthalhydroxamic Acid Spectrophotometric Method First Edition. 6 pp.

INTERNATIONAL AND NON-U.S. NATIONAL STANDARDS
SUBJECT INDEX
Aluminum

Aluminum Oxide (Cont.)
—Density
BSI BS 4140: Part 10-86. 1986 Methods of Test for Aluminium Oxide Part 10: Determination of Untamped Density. 4 pp.
ISO 903-76. Aluminium Oxide Primarily Used for the Production of Aluminium—Determination of Untamped Density First Edition. 4 pp.

—Density—Pycnometric Analysis
BSI BS 4140: Part 8-86. 1986 Methods of Test for Aluminium Oxide Part 8: Determination of Absolute Density Using a Pyknometer. 6 pp.
ISO 901-76. Aluminium Oxide Primarily Used for the Production of Aluminium—Determination of Absolute Density—Pyknometer Method First Edition. 6 pp.

—Drying Agents
BSI BS 2541-60. (WITHDRAWN) 1960 Amd 1 Activated Alumina for Use as Desiccant (Superseded by Various Parts of the 1991 Edition of BS 3482). 13 pp.

—Fluorine Content—Spectrophotometry
BSI BS 4140: Part 17-86. 1986 Methods of Test for Aluminium Oxide Part 17: Determination of Fluorine Content. 5 pp.
ISO 2828-73. Aluminium Oxide Primarily Used for the Production of Aluminium—Determination of Fluorine Content—Alizarin Complexone and Lanthanum Chloride Spectrophotometric Method First Edition. 5 pp.

—Iron Content—Photometry
BSI BS 4140: Part 7-86. 1986 Methods of Test for Aluminium Oxide Part 7: Determination of Iron Content. 6 pp.
ISO 805-76. Aluminium Oxide Primarily Used for the Production of Aluminium—Determination of Iron Content—1,10-Phenanthroline Photometric Method First Edition. 6 pp.

—Loss of Mass
BSI BS 4140: Part 2-86. 1986 Methods of Test for Aluminium Oxide Part 2: Determination of Loss of Mass at 300 Degrees C. 4 pp.
BSI BS 4140: Part 3-86. 1986 Methods of Test for Aluminium Oxide Part 3: Determination of Loss of Mass at 1000 Degrees C and 1200 Degrees C. 5 pp.
ISO 803-76. Aluminium Oxide Primarily Used for the Production of Aluminium—Determination of Loss of Mass at 300 Degrees Celsius (Conventional Moisture) First Edition. 4 pp.
ISO 806-76. Aluminium Oxide Primarily Used for the Production of Aluminium—Determination of Loss of Mass at 1 000 Degrees Celsius and 1 200 Degrees Celsius First Edition. 5 pp.

—Manganese Content—Atomic Absorption Spectrophotometry
BSI BS 4140: Part 22-80. 1980 Methods of Test for Aluminium Oxide Part 22: Determination of Manganese Content. 8 pp.
ISO 3390-76. Aluminium Oxide Primarily Used for the Production of Aluminium—Determination of Manganese Content—Flame Atomic Absorption Method First Edition. 8 pp.

—Particle Size Analysis
BSI BS 4140: Part 21-80. 1980 Methods of Test for Aluminium Oxide Part 21: Particle Size Analysis. 4 pp.
BSI BS 4140: Part 24-87. 1987 Methods of Test for Aluminum Oxide Part 24: Determination of Fine Partical Size Distribution (Method Using Electroformed Sieves). 5 pp.
ISO 2926-74. Aluminium Oxide Primarily Used for the Production of Aluminium—Particle Size Analysis—Sieving Method First Edition. 4 pp.
ISO 8220-86. Aluminium Oxide Primarily Used for the Production of Aluminium—Determination of the Fine Particle Size Distribution (Less Than 60 Micrometer)—Method Using Electroformed Sieves First Edition. 5 pp.
SAA AS 2879.2-91. Alumina Determination of Particles Passing a 20 Milimeter Aperture Sieve. 4 pp.

—Pellets
CNS J3005-81. Nuclear-Grade Aluminium Oxide Pellets (Aug)(7801).

—Phosphorus Content—Spectrophotometry
BSI BS 4140: Part 18-86. 1986 Methods of Test for Aluminium Oxide Part 18: Determination of Phosphorus Content. 5 pp.
ISO 2829-73. Aluminium Oxide Primarily Used for the Production of Aluminium—Determination of Phosphorus Content—Reduced Phosphomolybdate Spectrophotometric Method First Edition. 5 pp.

—Physical Properties
JIS H 1902-90. Methods for Determination on Physical Properties of Aluminium Oxide. 15 pp.

Aluminum Oxide (Cont.)
—Sampling
BSI BS 4140: Part 20-80. 1980 Methods of Test for Aluminium Oxide Part 20: Sampling. 5 pp.
ISO 2927-73. Aluminium Oxide Primarily Used for the Production of Aluminium—Sampling First Edition. 5 pp.

—Silica Content—Spectrophotometry
BSI BS 4140: Part 5-86. 1986 Methods of Test for Aluminium Oxide Part 5: Determination of Silica Content. 7 pp.

—Sodium Content—Atomic Emission Spectrophotometry
BSI BS 4140: Part 11-86. 1986 Methods of Test for Aluminium Oxide Part 11: Determination of Sodium Content. 7 pp.
ISO 1617-76. Aluminium Oxide Primarily Used for the Production of Aluminium—Determination of Sodium Content—Flame Emission Spectrophotometric Method First Edition. 7 pp.

—Spectrophotometry
ISO 1232-76. Aluminium Oxide Primarily Used for the Production of Aluminium—Determination of Silica Content—Reduced Molybdosilicate Spectrophotometric Method First Edition. 7 pp.

—Test Specimens
BSI BS 4140: Part 1-86. 1986 Methods of Test for Aluminium Oxide Preparation and Storage of Test Samples. 4 pp.
BSI BS 4140: Part 4-86. 1986 Methods of Test for Aluminium Oxide Part 4: Preparation of Sample Solution by Alkaline Fusion. 5 pp.
BSI BS 4140: Part 13-86. 1986 Methods of Test for Aluminium Oxide Part 13: Preparation of Sample Solution by Treatment with Hydrochloric Acid Under Pressure. 5 pp.
ISO 804-76. Aluminium Oxide Primarily Used for the Production of Aluminium—Preparation of Solution for Analysis—Method by Alkaline Fusion First Edition. 5 pp.
ISO 2073-76. Aluminium Oxide Primarily Used for the Production of Aluminium—Preparation of Solution for Analysis—Method by Hydrochloric Acid Attack Under Pressure First Edition. 5 pp.

—Test Specimens—Storage
ISO 802-76. Aluminium Oxide Primarily Used for the Production of Aluminium—Preparation and Storage of Test Samples First Edition. 4 pp.

—Titanium Content—Photometry
BSI BS 4140: Part 6-80. 1980 Methods of Test for Aluminium Oxide Part 6: Determination of Titanium Content. 5 pp.
ISO 900-77. Aluminium Oxide Primarily Used for the Production of Aluminium—Determination of Titanium Content—Diantipyrylmethane Photometric Method First Edition. 6 pp.

—Vanadium Content—Photometry
BSI BS 4140: Part 12-86. 1986 Methods of Test for Aluminium Oxide Part 12: Determination of Vanadium Content. 6 pp.
ISO 1618-76. Aluminium Oxide Primarily Used for the Production of Aluminium—Determination of Vanadium Content—N-Benzoyl-N-Phenylhydroxylamine Photometric Method First Edition. 6 pp.

—Zinc Content—Atomic Absorption Spectrophotometry
BSI BS 4140: Part 16-86. 1986 Methods of Test for Aluminium Oxide Part 16: Determination of Zinc Content by Flame Atomic Absorption. 6 pp.
ISO 2071-76. Aluminium Oxide Primarily Used for the Production of Aluminium—Determination of Zinc Content—Flame Atomic Absorption Method First Edition. 6 pp.

—Zinc Content—Photometry
ISO 2072-81. Aluminium Oxide Primarily Used for the Production of Aluminium—Determination of Zinc Content—PAN Photometric Method First Edition. 7 pp.

Aluminum Oxide, Brockmann
Use: Activated Alumina

Aluminum Oxide Content Analysis
—Aluminum Sulfate
CPPA J.6-86. Analysis of Alum Reprinted—1988. 5 pp.

—Aluminum Sulfate—Gravimetric Analysis
CNS K6162-79. Method of Test for Aluminium Sulfate (Industrial Use) (Jun)(2073). 3 pp.

Aluminum Oxide Content Analysis (Cont.)
—Aluminum Sulfate—Volumetric Analysis
CNS K6162-79. Method of Test for Aluminium Sulfate (Industrial Use) (Jun)(2073). 3 pp.

—Calcium Carbonates
CNS K6721-82. Method of Test for Calcium Carbonate for Rubber (Jun)(9004). 4 pp.

—Cementitious Materials
SAA AS 3583.10-91. Methods of Test for Supplementary Cementitous Materials for Use with Portland Cement—Part 10: Determination of Alumina and Total Iron Content (In Professional Packages 30,58). 4 pp.

—Chromium Ores
CNS M3202-88. Method of Determination for Aluminium Oxide in Chrome Ores (May)(12319).
JIS M 8266-82. Method for Determination of Aluminium Oxide in Chrome Ores.

—Dolomite
ISO 10058-92. Magnesites and Dolomites—Chemical Analysis First Edition. 23 pp.

—Dolomite—Gravimetric Analysis
CNS M3104-82. Method for Determining Aluminium Oxide in Dolomite (Gravimetric Method) (Sep)(9422).

—Dolomite—Photometry
DIN ENGL 52241 Pt 4-86. Analysis of Raw Materials Used in Glass Production; Chemical Analysis of Dolomite Containing Not Less Than 95% of Calcium Magnesium Carbonate; Determination of Aluminium Oxide (Nov). 2 pp.

—Dolomite—Volumetric Analysis
CNS M3103-82. Method for Determining Aluminum Oxide in Dolomite (EDTA Titration Method) (Sep)(9421).

—Feldspars—Gravimetric Analysis
CNS M3084-82. Method for Determination of Aluminium Oxide in Feldspar (Oxine Gravimetric Method) (Jan)(8439).

—Feldspars—Volumetric Analysis
CNS M3085-82. Method for Determination of Aluminium Oxide in Feldspar (EDTA-Zinc Back Titration Method) (Jan)(8440).

—Glass—Molecular Absorption Spectrophotometry
BSI BS 7709: Part 4-93. 1993 Analysis of Extract Solutions of Glass Part 4: Method for Determination of Aluminium Oxide by Molecular Absorption Spectrometry (ISO 10136-4: 1993) (V). 11 pp.
ISO 10136 Pt 4-93. Glass and Glassware—Analysis of Extract Solutions—Part 4: Determination of Aluminium Oxide by Molecular Absorption Spectrometry First Edition. 9 pp.

—Glassware—Molecular Absorption Spectrophotometry
BSI BS 7709: Part 4-93. 1993 Analysis of Extract Solutions of Glass Part 4: Method for Determination of Aluminium Oxide by Molecular Absorption Spectrometry (ISO 10136-4: 1993) (V). 11 pp.
ISO 10136 Pt 4-93. Glass and Glassware—Analysis of Extract Solutions—Part 4: Determination of Aluminium Oxide by Molecular Absorption Spectrometry First Edition. 9 pp.

—Iron Ores—Atomic Absorption Spectrophotometry
JIS M 8220-83. Methods for Determination of Aluminium Oxide in Iron Ores. 20 pp.

—Iron Ores—Gravimetric Analysis
JIS M 8220-83. Methods for Determination of Aluminium Oxide in Iron Ores. 20 pp.

—Iron Ores—Volumetric Analysis
CNS M3128-83. Method for Determining Aluminium Oxide in Iron Ores (Zinc Titration Method) (Apr)(10183).
CNS M3129-83. Method for Determining Aluminium Oxide in Iron Ores (Oxine Method) (Apr) (10184).
JIS M 8220-83. Methods for Determination of Aluminium Oxide in Iron Ores. 20 pp.

—Limestone
CPPA J.4-92. Analysis of Limestone. 5 pp.

INDUSTRY STANDARDS

Aluminum Oxide Content Analysis (Cont.)

—Limestone—Photometry
DIN ENGL 52240 Pt 4-85. Analysis of Raw Materials Used in Glass Production; Chemical Analysis of Limestone Containing Not Less Than 95% of Calcium Carbonate; Determination of Aluminium Oxide (Sept). 2 pp.

—Magnesite
ISO 10058-92. Magnesites and Dolomites —Chemical Analysis First Edition. 23 pp.

—Manganese Ores—Atomic Absorption Spectrophotometry
JIS M 8239-82. Methods for Determination of Aluminium Oxide in Manganese Ores. 11 pp.

—Manganese Ores—Volumetric Analysis
CNS M3069-81. Method for Determination of Aluminum Oxide in Manganese Ores (Titration Method) (Aug) (7834).
JIS M 8239-82. Methods for Determination of Aluminium Oxide in Manganese Ores. 11 pp.

alpha-Aluminum Oxide Content Analysis

—Aluminum Oxide
SAA AS 2879.3-91. Alumina Determination of Alpha Alumina Content by X-Ray Diffraction. 3 pp.

Aluminum Oxide Grinding Wheels
Use: Grinding Wheels—Aluminum Oxide

Aluminum Paints (Made From Aluminum)
See Also: Aluminum; Aluminum Primers (Made From Aluminum); Paints; Varnishes
CNS K2129-86. Aluminum Paint (Jan)(8706). 2 pp.
CNS K6823-86. Method of Test for Aluminium Paint (Feb)(11481).
JIS K 5492-83. Aluminium Paint. 8 pp.

—Marine—Interior/Exterior
CGSB CAN/CGSB-1.93-92. Aluminum Marine Paint. 8 pp.

—Metal—Enamel—Interior/Exterior
CGSB CAN/CGSB-1.141-M89. Aluminum Alkyd Baking Enamel. 10 pp.

—Metal—Enamel—Interior/Exterior—Heat Resistant
CGSB CAN/CGSB-1.143-M90. Heat Resistant Aluminum Enamel, Silicone Alkyd. 10 pp.

—Metal—Interior/Exterior
CGSB CAN/CGSB-1.69-M89. Aluminum Paint. 10 pp.

—Metals
MOD UK DSTAN 80-31-91. Paint, Finishing, General Service, Aluminium Issue 2. 14 pp.
MOD UK DSTAN 80-31-01. Paint, Finishing, General Service, Aluminium Issue 2; Amendment 1. 14 pp.
MOD UK DSTAN 80-31-02. Paint, Finishing, General Service, Aluminium Issue 2; Amendment 2. 18 pp.

—Pipes—Heat Resistant
MOD UK DSTAN 80-131-89. Paint, Aluminium, Heat Resisting Type: Brushing Issue 1. 15 pp.
MOD UK TS 10194A. Paint, Aluminium Heat Resistant (Superseded by Def Stan 80-131).

—Tripolyphosphate Content
CNS K2200-88. Aluminum Tri-Polyphosphate for Paint (Apr)(12264). 1 p.

—Vehicles (Paints)—Varnishes
CGSB CAN/CGSB-1.19-M89. Varnish Vehicle for Aluminum Paint. 10 pp.

—Wood
MOD UK DSTAN 80-31-91. Paint, Finishing, General Service, Aluminium Issue 2. 14 pp.
MOD UK DSTAN 80-31-01. Paint, Finishing, General Service, Aluminium Issue 2; Amendment 1. 14 pp.
MOD UK DSTAN 80-31-02. Paint, Finishing, General Service, Aluminium Issue 2; Amendment 2. 18 pp.

—Wood—Interior/Exterior
CGSB CAN/CGSB-1.69-M89. Aluminum Paint. 10 pp.

Aluminum Panel Siding
Use: Aluminum Siding

Aluminum Pipes
Scope Note: Includes pipes made from aluminum alloys See Also: Pipes

Aluminum Pipes (Cont.)
BSI BS 5222: Part 1-75. (WITHDRAWN) 1975 Amd 1 Aluminium Piping Systems Part 1: Dimensions, Materials and Construction of Components. 36 pp.

—Aerospace—Seamless
DIN ENGL LN 9223-86. Aerospace; Tubes for Fluids in Wrought Aluminium Alloy, Seamless Drawn; Dimensions, Masses (Dec). 5 pp.

—Aircraft—Control Columns
SBAC AS 2255 ISSUE 8. Control Column Tube. 1 p.

—Aircraft—Control Columns—Plug Ends
SBAC AS 2257 ISSUE 4. Top Plug End for Control Column: Tube. 1 p.

—Aircraft—Control Columns—Sleeves
SBAC AS 2239 ISSUE 5. Sleeve for Control Column Tube. 1 p.

—Aircraft—Distance
SBAC AS 2307-2310 ISSUE 3. Distance Tubes— Aluminium Alloy. 1 p.

—Aircraft—Distance—Trim Controls
SBAC AS 879 ISSUE 3. Distance Tube—Trim Control. 1 p.

—Aircraft—Parachute Holders
SBAC AS 427 ISSUE 3(I). Tube—Parachute Stowage. 1 p.

—Aircraft—Seamless—Distance
SBAC AS 2809-2811 ISSUE 5. Seamless Distance Tubes. 1 p.

—Drainpipes
CSA CAN/CSA-B281-M90. Aluminum Drain, Waste, and Vent Pipe and Components; (Gen Instr 1). 27 pp.

—Gas—Line—Coils
CSA CAN/CSA-Z245.6-92. Coiled Aluminum Line Pipe and Accessories; (Gen Instr 1). 28 pp.

—Oil—Line—Coils
CSA CAN/CSA-Z245.6-92. Coiled Aluminum Line Pipe and Accessories; (Gen Instr 1). 28 pp.

—Sewer
CSA CAN/CSA-B281-M90. Aluminum Drain, Waste, and Vent Pipe and Components; (Gen Instr 1). 27 pp.

—Vent
CSA CAN/CSA-B281-M90. Aluminum Drain, Waste, and Vent Pipe and Components; (Gen Instr 1). 27 pp.

Aluminum Potassium
Use: Aluminum Potassium Sulfate

Aluminum Potassium Sulfate
Use For: Alum Potassium; Aluminum Potassium; Potash Alum; Potassium Alum
CNS K7038-82. Chemical Reagent (Potassium Alum) (May)(1538).
JIS K 1473-70. Potassium Alum (Aluminium Potassium Sulfate). 11 pp.
JIS K 8255-80. Aluminium Potassium Sulfate 12-Water.

—Photographic Chemicals
BSI BS 3312-78. 1978 Photographic Grade Aluminium Potassium Sulphate, Dodecahydrate (Potash Alum). 8 pp.
CNS Z9043-80. Aluminium Potassium Sulphate Dodecahydrate (Photographic Grade) (Mar)(5389).
CNS Z9044-80. Method of Test for Aluminium Potassium Sulphate Dodecahydrate of Photographic Grade (Mar)(5390).
ISO 3620-76. Photographic Grade Aluminium Potassium Sulphate Dodecahydrate—Specification First Edition. 5 pp.
JIS K 7730-87. Photographic Grade Aluminium Potassium Salphate Dodecahydrate.

Aluminum Powder
See Also: Metal Powders
CNS H3145-87. Aluminium Powder (Aug)(12049).
CNS K7598-88. Chemical Reagent (Aluminum Powder) (Oct)(8517).
MOD UK DSTAN 68-111-89. Aluminium Powder (Blown) Types A, B, C, D and E Issue 1. 12 pp.
MOD UK DSTAN 68-111-01. Aluminium Powder (Blown) Types A, B, C, D and E Issue 1; Amendment 1. 13 pp.
MOD UK TS 521A. Aluminium Powder, Heavy, Grade 1, Special (Superseded by Def Stan 68-111).

Aluminum Powder (Cont.)

—Explosives—Quality Assurance
NATO STANAG 4300 ED 1 AMD 0-00. (DRAFT) Test Procedures for Assessing the Quality of Aluminium Powder, for Use in Explosive Formulation, for Deliveries from One NATO Nation to Another. 13 pp.
NATO STANAG 4300 ED 1 AMD 0-93. Test Procedures for Assessing the Quality of Aluminium Powder, for Use in Explosive Formulation, for Deliveries from One NATO Nation to Another. 13 pp.

—Paints
JIS K 5906-91. Aluminium Pigments for Paints. 31 pp.

—Propellants—Quality Assurance
NATO STANAG 4300 ED 1 AMD 0-00. (DRAFT) Test Procedures for Assessing the Quality of Aluminium Powder, for Use in Explosive Formulation, for Deliveries from One NATO Nation to Another. 13 pp.
NATO STANAG 4300 ED 1 AMD 0-93. Test Procedures for Assessing the Quality of Aluminium Powder, for Use in Explosive Formulation, for Deliveries from One NATO Nation to Another. 13 pp.

—Pyrotechnics—Quality Assurance
NATO STANAG 4300 ED 1 AMD 0-00. (DRAFT) Test Procedures for Assessing the Quality of Aluminium Powder, for Use in Explosive Formulation, for Deliveries from One NATO Nation to Another. 13 pp.
NATO STANAG 4300 ED 1 AMD 0-93. Test Procedures for Assessing the Quality of Aluminium Powder, for Use in Explosive Formulation, for Deliveries from One NATO Nation to Another. 13 pp.

Aluminum Primers (Made From Aluminum)
See Also: Aluminum Coatings (Made From Aluminum); Aluminum Paints (Made From Aluminum)

—Wood
BSI BS 4756-71. 1971 Ready Mixed Aluminium Priming Paints for Woodwork. 8 pp.
MOD UK DEF-1418-63. Paint, Priming, Aluminum, for Wood. 4 pp.

Aluminum Primers (On Aluminum)
See Also: Aluminum Coatings (On Aluminum)
MOD UK DSTAN 80-77-79. Paint, Priming, Zinc Chrome, Fire-Retardant Type: Brushing Issue 1. 10 pp.
MOD UK CS 2625D. Solution Pretreatment for Aluminium.

—Alkyd Base
CNS K2201-88. Alkyd Resin Aluminum Tri-Polyphosphate Anti-Corrosive Primer (Apr)(12266). 2 pp.

—Alkyd Base—Tripolyphosphate Content
CNS K6952-88. Method of Test for Alkyd Resin Aluminum Tri-Polyphosphate Anticorrosive Primer (Apr)(12267). 3 pp.

—Chlorinated Rubber Base
CNS K2203-88. Chlorinated Rubber Aluminum Tri-Polyphosphate Anti-Corrosive Primer (Apr)(12270). 2 pp.

—Chlorinated Rubber—Tripolyphosphate Content
CNS K6954-88. Method of Test for Chlorinated Rubber Aluminum Tri-Polyphosphate Anti-Corrosive Primer (Apr)(12271). 3 pp.

—Epoxy Base
CNS K2202-88. Epoxy Resin Aluminum Tri-Polyphosphate Anti-Corrosive Primer (Apr)(12268). 2 pp.

—Epoxy—Tripolyphosphate Content
CNS K6953-88. Method of Test for Epoxy Resin Aluminum Tri-Polyphosphate Anti-Corrosive Primer (Apr)(12269). 3 pp.

—Iron Oxide—Alkyd Base
CGSB CAN/CGSB-1.81-M90. Air Drying and Baking Alkyd Primer for Vehicles and Equipment; (Corrigendum—Aug 1990). 11 pp.

Aluminum Products
See Also: Aluminum

—Inspection
CNS H2022-86. General Rules for Inspection of Household Aluminium Products (Jul) (760).

INTERNATIONAL AND NON-U.S. NATIONAL STANDARDS
SUBJECT INDEX
Aluminum

Aluminum Siding
Use For: Aluminum Panel Siding *See Also:* Siding
CGSB CAN/CGSB-93.2-M91. Prefinished Aluminum Siding, Soffits and Fascia, for Residential Use. 20 pp.

—Colors
BSI BS 4904-78. 1978 External Cladding Colours for Building Purposes. 10 pp.

—Installation
CGSB CAN/CGSB-93.5-92. Installation of Metal Residential Siding, Soffits and Fascia. 7 pp.

—Sheets—Corrugated
BSI CP 143: Part 1-58. 1958 Sheet Roof and Wall Coverings Part 1: Aluminium, Corrugated and Troughed. 22 pp.

Aluminum Silicate
—Pigments—Calcium Content
CGSB 1-GP-71 METH 50.15-81. Methods of Testing Paints and Pigments Pigment Analysis Soluble Calcium Compounds. 1 p.

Aluminum Silicon Alloys
See Also: Aluminum Alloys; Aluminum Silicon Bronze Alloys

—Heat Treated—Ingots
MOD UK DTD-722B-71. Ingots and Castings of Aluminium-Silicon-Magnesium Alloy (Precipitation Treated) (Si 5, Mg 0.5). 2 pp.
MOD UK DTD-727B-71. Ingots and Castings of Aluminium-Silicon-Magnesium Alloy (Heat Treated) (Si 5, Mg 0.5). 2 pp.
MOD UK DTD-735B-71. Ingots and Castings of Aluminium-Silicon-Magnesium Alloy (Solution Treated and Precipitation Treated) (Si 5, Mg 0.5). 2 pp.

Aluminum Silicon Bronze Alloys
See Also: Alloys; Aluminum Alloys; Aluminum Silicon Alloys

—Forgings—Marine
MOD UK NES 834: Part 2-91. Requirements for Aluminium Silicon Bronze Part 2: Forgings, Forging Stock, Rods and Sections Issue 2 (09.91). 31 pp.

—Ingots—Castings—Marine
MOD UK NES 834: Part 3-88. Requirements for Aluminium Silicon Bronze Part 3: Ingots and Castings Issue 1 (11.88). 21 pp.

—Plates—Marine
MOD UK NES 834: Part 1-88. Requirements for Aluminium Silicon Bronze Part 1: Sheet, Strip and Plate Issue 1 (11.88). 18 pp.

—Rods—Marine
MOD UK NES 834: Part 2-91. Requirements for Aluminium Silicon Bronze Part 2: Forgings, Forging Stock, Rods and Sections Issue 2 (09.91). 31 pp.

—Sections—Marine
MOD UK NES 834: Part 2-91. Requirements for Aluminium Silicon Bronze Part 2: Forgings, Forging Stock, Rods and Sections Issue 2 (09.91). 31 pp.

—Sheets—Marine
MOD UK NES 834: Part 1-88. Requirements for Aluminium Silicon Bronze Part 1: Sheet, Strip and Plate Issue 1 (11.88). 18 pp.

—Strips—Marine
MOD UK NES 834: Part 1-88. Requirements for Aluminium Silicon Bronze Part 1: Sheet, Strip and Plate Issue 1 (11.88). 18 pp.

Aluminum Sodium Sulfate
JIS K 8959-80. Aluminium Sodium Sulfate 12-Water.

Aluminum Stearate
—Pyrotechnics
CNS K1161-72. Aluminium Stearate (Industrial Grade) (Tentative) (Jun)(3380).
MOD UK TS 10073. Aluminium Stearate.

—Pyrotechnics—Ash Content
CNS K6320-72. Method of Test for Aluminum Stearate of Industrial Grade (Temporary Standard) (Jun)(3381). 1 p.

—Pyrotechnics—Density
CNS K6320-72. Method of Test for Aluminum Stearate of Industrial Grade (Temporary Standard) (Jun)(3381). 1 p.

—Pyrotechnics—Melting Points
CNS K6320-72. Method of Test for Aluminum Stearate of Industrial Grade (Temporary Standard) (Jun)(3381). 1 p.

Aluminum Stearate *(Cont.)*
—Pyrotechnics—Solubility
CNS K6320-72. Method of Test for Aluminum Stearate of Industrial Grade (Temporary Standard) (Jun)(3381). 1 p.

Aluminum Structures
See Also: Construction

—Coatings—Construction Contracts
DIN ENGL 18364-88. Tendering and Performance Stipulations in Contracts for Construction Works (VOB); Part C: General Technical Specifications in Contracts for Construction Works (ATV); Corrosion Protection of Steel and Aluminium Structures. 4 pp.

Aluminum Sulfate
Use For: Aluminum Sulphate; Alums (Aluminum Sulfate) *See Also:* Aluminum Iron Sulfate
CNS K7039-61. Chemical Reagent (Aluminum Sulfate) (Dec)(1539).
CNS K7632-82. Chemical Reagent (Aluminum Sulfate, Amhydrous) (Aug)(9289).
JIS K 1423-70. Aluminium Sulfate. 10 pp.
JIS K 8957-79. Aluminium Sulfate, 14-18 Water.

—Acidity
MOD UK M 9508/67. Examination of Aluminium Sulphate.

—Aluminum Oxide Content—Gravimetric Analysis
CNS K6162-79. Method of Test for Aluminium Sulfate (Industrial Use) (Jun)(2073). 3 pp.

—Aluminum Oxide Content—Volumetric Analysis
CNS K6162-79. Method of Test for Aluminium Sulfate (Industrial Use) (Jun)(2073). 3 pp.

—Cartridge Cases
MOD UK DSTAN 68-168-93. Aluminium Sulfate (Equivalent to 18% AL2O3) Issue 1. 14 pp.
MOD UK TS 10032. Aluminium Sulphate 18 Per Cent Al2O3 (Superseded by Def Stan 68-168).

—Content Analysis
MOD UK M 9508/67. Examination of Aluminium Sulphate.

—Ferric Oxide Content
CNS K6162-79. Method of Test for Aluminium Sulfate (Industrial Use) (Jun)(2073). 3 pp.

—Insoluble Matter Content
CNS K6162-79. Method of Test for Aluminium Sulfate (Industrial Use) (Jun)(2073). 3 pp.

—Papermaking—Acidity
CPPA J.6-86. Analysis of Alum Reprinted—1988. 5 pp.

—Papermaking—Aluminum Oxide Content
CPPA J.6-86. Analysis of Alum Reprinted—1988. 5 pp.

—Papermaking—Insoluble Matter Content
CPPA J.6-86. Analysis of Alum Reprinted—1988. 5 pp.

—Papermaking—Iron Content
CPPA J.6-86. Analysis of Alum Reprinted—1988. 5 pp.

—Papermaking—Soluble Matter Content
CPPA J.6-86. Analysis of Alum Reprinted—1988. 5 pp.

—Papermaking—Sulfate Content
CPPA J.6-86. Analysis of Alum Reprinted—1988. 5 pp.

—PH
CNS K6162-79. Method of Test for Aluminium Sulfate (Industrial Use) (Jun)(2073). 3 pp.

—Visual Inspection
MOD UK M 9508/67. Examination of Aluminium Sulphate.

—Water Treatment
CEN PREN 878-92. Aluminium Sulfate, Iron-Free Grade, Used for Water Intended for Human Consumption. 15 pp.
CEN PREN 879-92. Aluminium Sulfate, Low-Iron Grade, Used for Treatment of Water Intended for Human Consumption. 15 pp.
CEN PREN 880-92. Aluminium Sulfate, Special Grade, Used for Treatment of Water Intended for Human Consumption. 15 pp.
CNS K1068-74. Aluminum Sulfate for Water Purification (Jun)(2074).

Aluminum Sulfate *(Cont.)*
—Water Treatment *(Cont.)*
DIN ENGL 19600-87. Aluminium Sulfate for Water Treatment; Technical Delivery Conditions (May). 3 pp.
MOD UK TS 401C. Aluminium Sulphate, 14 per Cent Al2O3 (Withdrawn).
MOD UK TS 474A. Aluminium Sulphate 16% Al2O3 (Withdrawn).

—Waterworks
JIS K 1450-77. Aluminium Sulfate for Water Works.
JIS K 1450-68. Aluminium Sulfate for Water Works. 14 pp.

—Waterworks—Ammoniacal Nitrogen Content
CNS K6163-74. Method of Test for Aluminium Sulfate (for Water Works) (Jun)(2075). 4 pp.

—Waterworks—Arsenic Content
CNS K6163-74. Method of Test for Aluminium Sulfate (for Water Works) (Jun)(2075). 4 pp.

—Waterworks—Heavy Metals Content
CNS K6163-74. Method of Test for Aluminium Sulfate (for Water Works) (Jun)(2075). 4 pp.

Aluminum Sulphate
Use: Aluminum Sulfate

Aluminum Wire
Use For: Aluminum Alloy Wire
See Also: Aluminum Conductors; Electric Wire; Wire
BSI BS 1475-72. 1972 Amd 2 Wrought Aluminium and Aluminium Alloys for General Engineering Purposes; Wire. 26 pp.
BSI BS 4300/5-73. 1973 Amd 1 Specification (Supplementary Series) for Wrought Aluminium and Aluminium Alloys for General Engineering Purposes /5: 2011 Free-Cutting Bar and Wire-Alloy. 16 pp.
CNS H3048-84. Aluminium and Aluminium Alloy Rods, Bars and Wires (Sep)(3667).
DIN ENGL 1790 Pt 2-77. Wire of Aluminium and Wrought Aluminium Alloys; Technical Conditions of Delivery (May). 2 pp.
ISO 6365 Pt 1-88. Wrought Aluminium and Aluminium Alloy Cold-Drawn Wire—Part 1: Technical Conditions for Inspection and Delivery First Edition; (Supersedes 5191). 6 pp.
JIS H 4040-88. Aluminium and Aluminium Alloy Rods, Bars and Wires. 48 pp.

—Aerospace
AECMA PREN2615-86. Wire to Close Tolerance in Aluminium and Aluminium Alloys 1.6 Less Than or Equal to D Less Than or Equal to 9.6 mm Dimensions. 3 pp.

—Crimping Compounds
MOD UK DTD-5503-54. Cable Crimping Compound (for Aluminium Cables) (Reprinted December 1965). 1 p.

—Drawn
SAA AS 1865-86. Aluminium and Aluminium Alloys—Drawn Wire, Rod, Bar and Strip (This is a Joint Standard with SANZ NZS 1865:1986). 11 pp.
SNZ NZS/AS 1865-86. Aluminium and Aluminium Alloys—Drawn Wire, Rod, Bar and Strip (This is a Joint Standard with SAA AS 1865). 11 pp.

—Mechanical Properties
DIN ENGL 1790 Pt 1-83. Wrought Aluminium and Aluminium Alloy Wire; Properties (Feb). 5 pp.

—Rivet
CNS H3131-84. Aluminium and Aluminium Alloy Rivet Wires and Rods (Sep)(11016).
DIN ENGL 59675-57. High Grade Aluminium and Aluminium Wrought Alloy; Wire and Bars for Rivets; Drawn (May). 2 pp.
ISO R2101-71. Aluminium and Aluminium Alloys Shear Test for Rivet Wire and Rivets First Edition. 4 pp.

—Rivet—Aerospace
AECMA PREN2070-06-87. Aluminium and Aluminium Alloy Wrought Products—Technical Specification—Part 6—Rivet Wire. 6 pp.
AECMA PREN2114-85. Aluminium 1050A-H14 Wire for Solid Rivets D Less Than or Equal to 10 mm. 3 pp.
AECMA PREN2116-85. Aluminium Alloy 2017A-H13 Wire for Solid Rivets D Less Than or Equal to 10 mm. 3 pp.
AECMA PREN2117-85. Aluminium Alloy 5056A-H32 Wire for Solid Rivets D Less Than or Equal to 10 mm. 3 pp.
AECMA PREN2616-86. Wire for Rivets in Aluminium and Aluminium Alloys Large Tolerances D Less Than or Equal to 10 mm Dimensions. 3 pp.

INDUSTRY STANDARDS

Aluminum Wire (Cont.)
—Rivet—Aerospace (Cont.)
AECMA PREN2628-85. Aluminium Alloy 5056A-0 Wire for Solid Rivets D Less Than or Equal to 10 mm. 3 pp.

BSI BS EN 2070: Part 6-91. 1991 Aluminium and Aluminium Alloy Wrought Products Technical Specification Part 6: Rivet Wire.

BSI 5L 36-85. 1985 Wire for Solid, Cold Forged Rivets of 99.5% Aluminium (Not Exceeding 10 mm Diameter). 4 pp.

BSI 7L 37-89. 1989 Amd 1 Wire for Solid Cold-Forged Rivets of Aluminium-Copper-Magnesium-Silicon-Manganese Alloy (AMD 6625) December 21, 1990. 5 pp.

BSI 3L 58-71. 1971 Wire for Solid, Cold-Forged Rivets of Aluminium 5 Per Cent Magnesium Alloy (Not Exceeding 10 mm Diameter). 2 pp.

BSI 3L 86-71. 1971 Wire for Solid, Cold-Forged Rivets of Aluminium-Copper-Magnesium Alloy (Not Exceeding 10 mm Diameter) (Cu 2.5, Mg 0.3). 2 pp.

CEN EN 2070-6-89. Aerospace Series Aluminium and Aluminium Alloy Wrought Products Technical Specification Part 6: Rivet Wire. 8 pp.

Aluminum Zinc Alloys
See Also: Alloys; Aluminum Alloys; Aluminum Zinc Magnesium Alloys

—Heat Treated—Bars—Aerospace
MOD UK DTD-5114-73. Bars and Extruded Sections of Aluminium-Zinc-Magnesium-Copper-Chromium Alloy (Solution Treated and Precipitation Treated) (Zn 5.7, Mg 2.5, Cu 0.5, Cr 0.15). 2 pp.

MOD UK DTD-5124-73. Bars and Extruded Sections of Aluminium-Zinc-Magnesium-Copper-Chromium Alloy (Solution Treated and Precipitation Treated) (Zn 5.8, Mg 2.5, Cu 1.6, Cr 0.16). 2 pp.

—Heat Treated—Bars—Extruded—Aerospace
BSI L 170-89. 1989 Extruded Bars and Sections of Aluminium-Zinc-Copper-Chromium Alloy (Solution Treated, Controlled Stretched and Artif. Aged)(Not Exc. 150 mm Dia. or Minor Sect. Dim.)(Zn 5.6, Mg 2.5, Cu 1.6, Cr 0.23) (7075). 3 pp.

BSI L 170-01. 1989 Amd 1 Extruded Bars and Sections of Aluminium-Zinc-Copper-Chromium Alloy (Solution Treated, Controlled Stretched and Artif. Aged)(Not Exc. 150 mm Dia. or Minor Sect. Dim.)(Zn 5.6, Mg 2.5,Cu 1.6,Cr0.23)(7075) AMD 6838 November 29, 1991. 4 pp.

—Heat Treated—Plates—Aerospace
MOD UK DTD-5120B-81. Plate of Aluminium-Zinc-Magnesium-Copper-Zirconium Alloy (Solution Treated, Controlled Stretched and Artificially Aged) (Zn 6.2, Mg 2.4, Cu 1.7, Zr 0.13) (7010-T7651). 7 pp.

MOD UK DTD-5130A-81. Plate of Aluminium-Zinc-Magnesium-Copper-Zirconium Alloy (Solution Treated, Controlled Stretched and Artificially Aged) (Zn 6.2, Mg 2.4, Cu 1.7, Zr 0.13) (7010-T73651). 7 pp.

—Heat Treated—Sections—Extruded—Aerospace
BSI L 170-89. 1989 Extruded Bars and Sections of Aluminium-Zinc-Copper-Chromium Alloy (Solution Treated, Controlled Stretched and Artif. Aged)(Not Exc. 150 mm Dia. or Minor Sect. Dim.)(Zn 5.6, Mg 2.5, Cu 1.6, Cr 0.23) (7075). 3 pp.

BSI L 170-01. 1989 Amd 1 Extruded Bars and Sections of Aluminium-Zinc-Copper-Chromium Alloy (Solution Treated, Controlled Stretched and Artif. Aged)(Not Exc. 150 mm Dia. or Minor Sect. Dim.)(Zn 5.6, Mg 2.5,Cu 1.6,Cr0.23)(7075) AMD 6838 November 29, 1991. 4 pp.

MOD UK DTD-5114-73. Bars and Extruded Sections of Aluminium-Zinc-Magnesium-Copper-Chromium Alloy (Solution Treated and Precipitation Treated) (Zn 5.7, Mg 2.5, Cu 0.5, Cr 0.15). 2 pp.

MOD UK DTD-5124-73. Bars and Extruded Sections of Aluminium-Zinc-Magnesium-Copper-Chromium Alloy (Solution Treated and Precipitation Treated) (Zn 5.8, Mg 2.5, Cu 1.6, Cr 0.16). 2 pp.

—Ingots
BSI DD 139-86. 1986 Zinc-Aluminium Alloy Ingots and Casting. 13 pp.

—Plates—Aluminum Alloy Clad—Heat Treated—Aerospace
MOD UK DTD-5110-71. Aluminium-Alloy-Coated Plate of Aluminium-Zinc-Magnesium-Copper-Chromium Alloy (Solution Treated, Controlled Stretched and Precipitation Treated) (Zn 5.8, Mg 2.5, Cu 0.5, Cr 0.15). 2 pp.

Aluminum Zinc Alloys (Cont.)
—Precipitation Hardened—Ingots
MOD UK DTD-5008B-71. Ingots and Castings of Aluminium-Zinc-Magnesium-Chromium Alloy (Age Hardened) (Zn 5.2, Mg 0.6, Cr 0.5). 2 pp.

Aluminum Zinc Magnesium Alloys
Use For: Magnesium Aluminum Zinc Alloys
See Also: Aluminum Alloys

—Forgings—Aerospace
MOD UK DTD-5636-82. Die Forgings of Aluminium-Zinc-Magnesium-Copper-Zirconium Alloy (Solution Treated Water Quenched and Artificially Aged to an Overaged Condition) (Zn 6.2, Mg 2.4, Cu 1.7, Zr 0.13) (7010-T736). 3 pp.

MOD UK DTD-5636-01. Die Forgings of Aluminium-Zinc-Magnesium-Copper-Zirconium Alloy (Solution Treated Water Quenched and Artificially Aged to an Overaged Condition) (Zn 6.2, Mg 2.4, Cu 1.7, Zr 0.13) (7010-T736); Amendment 1. 4 pp.

—Forgings—Aircraft
MOD UK DTD-5024-58. Aluminium—Zinc—Magnesium—Copper—Manganese Alloy Forgings (Not Exceeding 10 Inches Diameter or Minor Sectional Dimension) (Solution Treated and Precipitation Treated) (Reprinted May 1966, Incorporating Amendment No. 1). 2 pp.

—Forgings—Extruded—Aerospace
BSI L 172-89. 1989 Amd 1 Extruded, Rolled on Cast Forging Stock of Aluminium-Zinc-Magnesium-Manganese-Copper Alloy (S) (AMD 6561) December 21, 1990. 5 pp.

BSI L 172-02. 1989 Amd 2 Extruded, Rolled or Cast Forging Stock of Aluminium-Zinc-Magnesium-Manganese-Copper Alloy (AMD 6840) November 29, 1991. 5 pp.

—Heat Treated—Bars—Aircraft
MOD UK DTD-5044-58. Aluminium—Zinc—Magnesium—Copper—Manganese Alloy Bars and Extruded Sections (Not Exceeding 10 Inches or Minor Sectional Dimension) (Solution Treated and Precipitation Treated) (Reprinted March 1961). 2 pp.

—Heat Treated—Forgings—Aerospace
MOD UK DTD-5094A-71. Forging Stock and Forgings of Aluminium-Zinc-Magnesium-Copper-Manganese Alloy (Heat Treated at a Ruling Thickness of Not More Than 75 mm) (Solution Treated, Step Quenched and Precipitation Treated for Low Internal Stress) (Zn 5.7, Mg 2.5, Cu 0.5, Mn 0.5). 2 pp.

MOD UK DTD-5104A-71. Forging Stock and Forgings of Aluminium-Zinc-Magnesium-Copper-Manganese Alloy (Solution Treated, Boiling Water Quenched and Duplex Precipitation Treated to Improve Stress Corrosion Resistance) (Zn 5.7, Mg 2.5, Cu 0.5, Mn 0.5). 2 pp.

—Heat Treated—Sections—Extruded—Aircraft
MOD UK DTD-5044-58. Aluminium—Zinc—Magnesium—Copper—Manganese Alloy Bars and Extruded Sections (Not Exceeding 10 Inches or Minor Sectional Dimension) (Solution Treated and Precipitation Treated) (Reprinted March 1961). 2 pp.

Alums (Aluminum Sulfate)
Use: Aluminum Sulfate

Alvis Leonides Engines
Use: Aircraft Engines—Alvis Leonides

AM Radio Receivers
Use: Radio Receivers—AM

Amalgamators
See Also: Dental Equipment

—Dental
ISO 7488-91. Dental Amalgamators First Edition. 8 pp.

Amalgams
Use: Mercury Amalgams

Amateur
Use: Amateur Radio Services

Amateur Aircraft
—Permit to Fly
CAA LEAFLET 11-2 07.90. Information and Advice on the Procedures for the Issue of a Permit to Fly for Amateur Constructed Aeroplanes. 3 pp.

Amateur Radio Equipment
Use For: Ham Radio

Amateur Radio Equipment (Cont.)
See Also: Amateur Radio Services; Radio Communications; Radio Receivers; Radio Transmitters

—Licenses
CEPT T/R 61-01 E-92. CEPT Radio Amateur License (Revised in Paris 1992 and by Correspondence August 1992). 7 pp.

CEPT T/R 61-01-85. Licence De Radioamateur CEPT. 4 pp.

CEPT T/R 61-01 E-85. CEPT Radio Amateur License. 5 pp.

Amateur Radio Services
Use For: Amateur; Amateur Services; Ham Radio Services
See Also: Amateur Radio Equipment; Radio Communications

CCIR Volume 8 Annex TOC-90. Table of Contents. 3 pp.

CCIR Volume 8 ANX NUMERIC-90. Numerical Index of Texts. 1 p.

CCIR Volume 8 ANX TEXTS-90. Index of Texts Deleted. 1 p.

CCIR Report 1154-90. Techniques and Frequency Usage in the Amateur and Amateur-Satellite Services—Section 8L—Amateur Service; Amateur-Satellite Service. 24 pp.

CCIR Volume 8 ANX 3 TOC-90. Table of Contents. 4 pp.

CCIR Volume 8 ANX 3 NUMERIC-90. Numerical Index of Texts. 1 p.

CCIR Volume 8 ANX 3 Index-90. Index of Texts Deleted. 1 p.

CCIR Volume XV-2 TOC-90. Table of Contents. 3 pp.

—Certificates
CEPT T/R 61-02 E-90. Harmonized Amateur Radio Examination Certificates. 16 pp.

—Communications Satellites
CCIR Report 1154-90. Techniques and Frequency Usage in the Amateur and Amateur-Satellite Services—Section 8L—Amateur Service; Amateur-Satellite Service. 24 pp.

—Frequency Bands
CCIR QUESTION 48-2/8-90. Techniques and Frequency Usage in the Amateur Service and Amateur-Satellite Service—Questions Concerning Study Group 8—Mobile, Radiodetermination, Amateur and Related Satellite Services. 1 p.

—Radio Link Protocols
CCIR Report 1154-90. Techniques and Frequency Usage in the Amateur and Amateur-Satellite Services—Section 8L—Amateur Service; Amateur-Satellite Service. 24 pp.

—Satellite
CCIR Volume 8 TOC-90. Table of Contents. 4 pp.

CCIR Volume 8 NUMERICAL-90. Numerical Index of Texts. 1 p.

CCIR Volume 8 INTRO-90. Terms of Reference of Study Group 8 and Introduction by the Chairman of Study Group 8. 9 pp.

CCIR Volume 8 Annex TOC-90. Table of Contents. 3 pp.

CCIR Volume 8 ANX NUMERIC-90. Numerical Index of Texts. 1 p.

CCIR Volume 8 ANX TEXTS-90. Index of Texts Deleted. 1 p.

CCIR Report 1154-90. Techniques and Frequency Usage in the Amateur and Amateur-Satellite Services—Section 8L—Amateur Service; Amateur-Satellite Service. 24 pp.

CCIR Volume 8 ANX 3 TOC-90. Table of Contents. 4 pp.

CCIR Volume 8 ANX 3 NUMERIC-90. Numerical Index of Texts. 1 p.

CCIR Volume 8 ANX 3 Index-90. Index of Texts Deleted. 1 p.

CCIR Volume XV-2 TOC-90. Table of Contents. 3 pp.

—Satellite—Frequency Bands
CCIR QUESTION 48-2/8-90. Techniques and Frequency Usage in the Amateur Service and Amateur-Satellite Service—Questions Concerning Study Group 8—Mobile, Radiodetermination, Amateur and Related Satellite Services. 1 p.

—Satellite—Radio Link Protocols
CCIR Report 1154-90. Techniques and Frequency Usage in the Amateur and Amateur-Satellite Services—Section 8L—Amateur Service; Amateur-Satellite Service. 24 pp.

—Satellite—UHF
CCIR Report 1154-90. Techniques and Frequency Usage in the Amateur and Amateur-Satellite Services—Section 8L—Amateur Service; Amateur-Satellite Service. 24 pp.

Amateur Radio Services (Cont.)
—Satellite—VHF
CCIR Report 1154-90. Techniques and Frequency Usage in the Amateur and Amateur-Satellite Services—Section 8L—Amateur Service; Amateur-Satellite Service. 24 pp.
—UHF
CCIR Report 1154-90. Techniques and Frequency Usage in the Amateur and Amateur-Satellite Services—Section 8L—Amateur Service; Amateur-Satellite Service. 24 pp.
—VHF
CCIR Report 1154-90. Techniques and Frequency Usage in the Amateur and Amateur-Satellite Services—Section 8L—Amateur Service; Amateur-Satellite Service. 24 pp.

Amateur Services
Use: Amateur Radio Services

Ambient Air
Use: Air

Ambient Air Monitoring
Use: Air Monitoring

Ambient Conditions Testing
Use: Environmental Testing

Ambient Temperature
See Also: Air Monitoring; Environmental Conditions; Environmental Testing; Temperature Testing
—Medical Electrical Equipment—Safety
CSA CAN/CSA-C22. 2NO601.1-M90. Medical Electrical Equipment, Part 1: General Requirements for Safety; (Gen Instr 1). 240 pp.

Amblyoscopes
JIS T 7307-88. Major Amblyoscopes. 6 pp.

Ambulances
See Also: Air Medical Transport Units; Automobiles; Motor Vehicles
—Design
MOD UK DSTAN 23-5-01. Medical Design Requirements for Military Motor Ambulances Issue 1; Amendment 1 (Withdrawn). 8 pp.
—Design—Military
NATO STANAG 2872 ED 2 AMD 1-82. Medical Design Requirements for Military Motor Ambulances. 7 pp.
NATO STANAG 2872 ED 3 AMD 0-89. Medical Design Requirements for Military Motor Ambulances. 8 pp.
—Medical Equipment
NATO STANAG 2342 ED 1 AMD 5-75. Minimum Essential Medical Equipment and Supplies for Motor Ambulances at All Levels. 6 pp.
NATO STANAG 2342 ED 1 AMD 6-75. Minimum Essential Medical Equipment and Supplies for Motor Ambulances at All Levels. 6 pp.

Americium-241 Content Analysis
Use: Americium Isotope Content Analysis

Americium Isotope Content Analysis
—Plutonium Nitrate—Extraction Analysis
CNS J2045-82. Method of Test for Americium-241 in Nuclear-Grade Plutonium Nitrate Solution by Extraction and Gamma Counting (Mar)(8592).
—Plutonium Nitrate—Gamma Counting
CNS J2045-82. Method of Test for Americium-241 in Nuclear-Grade Plutonium Nitrate Solution by Extraction and Gamma Counting (Mar)(8592).
CNS J2046-82. Method of Test for Americium-241 in Nuclear-Grade Plutonium Nitrate Solution by Gamma Counting (Mar)(8593).

Ametryn
See Also: Pesticides
SAA AS 1870.10D-78. Standard for Development —Pesticides for Agricultural Use—Part 10D: Atrazine, Simazine, Propazine, Prometryn, Methoprotryn, Ametryn and Tertbutryn. 35 pp.

Amides
Scope Note: Use a more specific term
See: Acetamides; Acetanilides; Asparagine Monohydrate; Diethyltoluamide; 2',4'-Dimethylacetoacetanilide; N,N-Dimethylformamide; Formamides; Sulfamic Acids; Sulfanilmide; Thiourea; o-Toluenesulfonamide; Urea

Amido Black Method
See Also: Dairy Products; Kjeldahl Method
—Milk
ISO 5542-84. Milk—Determination of Protein Content—Amido Black Dye-Binding Method (Routine Method) First Edition. 7 pp.

Amidoamines
—Amine Content—Potentiometric Analysis
CNS K6689-81. Total, Primary, Secondary and Tertiary Amine Values of Fatty Amines, Amidoamines and Diamines by Referee Potentiometric Method (Sep)(7913).
—Iodine Number
CNS K6682-81. Method of Test for Iodine Value of Fatty Amines, Amidoamines, and Diamines (Jul)(7726).

Amine Content Analysis
—Amidoamines—Potentiometric Analysis
CNS K6689-81. Total, Primary, Secondary and Tertiary Amine Values of Fatty Amines, Amidoamines and Diamines by Referee Potentiometric Method (Sep)(7913).
—Diamines—Potentiometric Analysis
CNS K6689-81. Total, Primary, Secondary and Tertiary Amine Values of Fatty Amines, Amidoamines and Diamines by Referee Potentiometric Method (Sep)(7913).
—Fatty Amines
CNS K6690-81. Total, Primary, Secondary and Tertiary Amine Value of Fatty Amines by Alternative Indicator Method (Sep)(7914).
CNS K6692-81. Calculation of Percent of Primary, Secondary and Tertiary Amines in Fatty Amines (Sep)(7916).
—Fatty Amines—Potentiometric Analysis
CNS K6689-81. Total, Primary, Secondary and Tertiary Amine Values of Fatty Amines, Amidoamines and Diamines by Referee Potentiometric Method (Sep)(7913).

Amines
Scope Note: Use a more specific term
See: Amidoamines; Aniline; Benzidine; N-Benzyl-N-Ethylaniline; Carbazoles; o-Chloroaniline; p-Chloroaniline; N,N'-di-beta-Naphthyl-p-Phenylenediamine; Diamines; Diethanolamine; Diethylamine; p-Dimethylaminoazobenzene; N,N-Dimethylaniline; 3,3'-Dimethylbenzidine Dihydrochloride; N,N-Dinitrosopentamethylenetetramine; Diphenylamine; Ethanolamine; N-Ethylaniline; Ethylenediamine; Fatty Amines; Hexamethylenetetramine; Histamine Content Analysis; N-Isopropyl-N'-Phenyl-p-Phenylenediamine; Isopropylamine; Melamines; Methyl Violet; Methylene Blue; Naphthylamine; 1-Naphthylamine; Nitramine; o-Nitrodiphenylamine; o-Phenylenediamine; p-Phenylenediamine; Phenylenediamines; Sulfanilamide; Triethanolamine; Triethylamine

Amino Acid Content Analysis
See Also: Acidity; Organic Acid Content Analysis
—Fruit Juices
CNS N6221-91. Method of Test for Fruit and Vegetable Juices and Drinks-Determination of Free Amino Acids (Jun)(12632). 3 pp.
—Vegetable Juices
CNS N6221-91. Method of Test for Fruit and Vegetable Juices and Drinks-Determination of Free Amino Acids (Jun)(12632). 3 pp.

5-Amino-2-Chlorotoluene-4-sulfonic Acid
JIS K 4147-88. 5-Amino-2-Chlorotoluene-4-Sulfonic Acid (C Acid). 6 pp.

4-Amino-3-hydroxy-1-napthalenesulfonic Acid
Use: 1-Amino-2-naphthol-4-sulfonic Acid

2-Amino-1-hydroxymethyl-1,3-propanediol
Use: Tromethamine

3-Amino-4-methoxytoluene
JIS K 4137-84. Anisidines (o-Anisidine, p-Anisidine, p-Phenetidine, 3-Amino--4-Methoxytoluene).

8-Amino-1-Naphthalenesulfonic Acid
Use: 1-Naphthylamine-8-sulfonic Acid

1-Amino-2-naphthol-4-sulfonic Acid
Use For: 4-Amino-3-hydroxy-1-napthalenesulfonic Acid
CNS K7043-61. Chemical Reagent (1,2,4-Aminonaphtol Sulfonic Acid) (Dec)(1543).
CNS K7594-81. Chemical Reagent (4-Amino-3-Hydroxy-1-Naphthalenesulfonic Acid) (Nov)(8139).
JIS K 8050-78. 4-Amino-3-Hydroxyl-1-Naphthalenesulfonic Acid.

2-Amino-5-naphthol-7-sulfonic Acid
Use For: J Acid
CNS K2229-88. J-Acid (Jun)(12346). 3 pp.
JIS K 4135-90. 7-Amino-4-Hydroxy-2-Naphthalene Sulfonic Acid (J-Acid).

Amino Nitrogen Content Analysis
Use For: Ammonia Nitrogen; Ammonia Nitrogen Content Analysis
—Waste Water
CNS K9015-76. Methods of Determination for Ammonia Nitrogen in Polluted and Waste Water (Jun)(3942).
—Water Pollution
CNS K9015-76. Methods of Determination for Ammonia Nitrogen in Polluted and Waste Water (Jun)(3942).

Amino Resins
Use For: Aminoplasts *See Also:* Melamine Formaldehyde Resins; Thermosetting Resins; Urea Formaldehyde Resins; Urea Resins
CNS K2098-80. Amino Resin (Feb)(5212). 1 p.
—Chemical Analysis
DIN ENGL 53749-70. Chemical Analysis of Urea-, Thiourea-and Melamine-Formaldehyde Resins and Aminoplastic Moulding Materials in the Moulded and Unmoulded State (July). 4 pp.
—Molding Materials
BSI BS 1322-92. 1992 Aminoplastic Moulding Materials (ISO 2112: 1990). 11 pp.
BSI BS 1322-81. 1981 Amd 1 Aminoplastic Moulding Materials. 11 pp.
DIN ENGL 7708 Suppl.-68. Types of Plastic Moulding Materials; Properties of Test Specimens from Phenolic, Aminoplastic and Aminoplastic/Phenolic Compression Moulding Materials (Oct). 6 pp.
DIN ENGL 7708 Pt 3-75. Types of Plastic Moulding Materials; Aminoplastic Moulding Materials; Aminoplastic/Phenolic Moulding Materials (Oct). 9 pp.
ISO 2112-90. Plastics—Aminoplastic Moulding Materials—Specification Second Edition. 8 pp.
—Molding Materials—Volatile Matter Content
BSI BS 2782:Pt4: METH 461A-78. 1978 Methods of Testing Plastics Part 4: Chemical Properties Method 461A: Determination of Volatile Matter in Aminoplastic Moulding Materials. 2 pp.
CNS K6618-80. Aminoplastic Moulding Materials Determination of Volatile Matter (Dec)(6684).
ISO 3671-76. Plastics—Aminoplastic Moulding Materials—Determination of Volatile Matter First Edition. 3 pp.

Aminoacetic Acid
Use: Glycine

p-Aminoacetophenone
CNS K7041-61. Chemical Reagent (P-Aminoacetophenone) (Dec)(1541).
CNS K7042-61. Chemical Reagent (P-Aminoacetophenone Acid) (Dec)(1542).
JIS K 8047-78. p-Aminoacetophenone.

Aminoalkyd Enamel Paints
See Also: Alkyd Enamel Paints; Enamel Paints
—Metal
CNS K2092-86. Aminoalkyd Resin Enamel (May)(4941). 4 pp.
CNS K6876-86. Method of Test for Aminoalkyd Resin Enamel (May)(11606).
JIS K 5651-92. Aminoalkyd Resin Paint. 29 pp.

4-Aminoantipyrine
Use: Ampyrone

Aminobenzene
Use: Aniline

4-Aminobenzenesulfonic Acid
Use: Sulfanilic Acid

o-Aminobenzoic Acid
Use For: Anthranilic Acid

o-Aminobenzoic Acid (Cont.)
CNS K2204-88. O-Aminobenzoic Acid (Anthranilic Acid) (Apr)(12272). 2 pp.
CNS K7555-81. Chemical Reagent (Anthranilic Acid) (Jul)(7713).
JIS K 4118-86. Substituted Benzoic Acids (Salicylic Acid (o-Hydroxybenzoic Acid), Anthranilic Acid (o-Aminobenzoic Acid)).

Aminoethanol
Use: Ethanolamine

2 Aminoethanol
Use: Ethanolamine

Aminoguanidinium Sulfate Monohydrate
Use For: Aminoguanidinium Sulphate Monohydrate
MOD UK DSTAN 68-32-91. Aminoguanidinium Sulphate Monohydrate Issue 2. 8 pp.

Aminoguanidinium Sulphate Monohydrate
Use: Aminoguanidinium Sulfate Monohydrate

2-Aminonaphthalene-1-sulfonic Acid
Use: 2-Naphthylamine-1-Sulfonic Acid

m-Aminophenol
CNS K2234-88. M-Aminophenol (Aug)(12402). 2 pp.
CNS K7551-81. Chemical Reagent (M-Aminophenol) (Jul)(7709).
JIS K 4144-86. Aminophenols (m-Aminophenol. p-Aminophenol).
JIS K 8052-64. m-Aminophenol.

p-Aminophenol
CNS K7633-82. Chemical Reagent (P-Aminophenol) (Aug)(9290).
JIS K 4144-86. Aminophenols (m-Aminophenol. p-Aminophenol).
JIS K 9059-87. p-Aminophenol.

Aminoplastic Adhesives
See Also: Adhesives

—Wood
BSI BS 1203-79. 1979 Amd 1 Synthetic Resin Adhesives (Phenolic and Aminoplastic) for Plywood (AMD 6284) November 30, 1990. 10 pp.
BSI BS 1204-93. 1993 Type MR Phenolic and Aminoplastic Synthetic Resin Adhesives for Wood (Supersedes BS 1204: Parts 1 & 2: 1979). 16 pp.
BSI BS 1204: Part 1-79. (WITHDRAWN) 1979 Amd 1 Synthetic Resin Adhesives (Phenolic and Aminoplastic) for Wood Part 1: Gap-Filling Adhesives (AMD 6502) December 21, 1990 (Superseded by BS EN 301: 1992 & BS EN 302: Parts 1-4: 1992). 14 pp.
BSI BS 1204: Part 2-79. (WITHDRAWN) 1979 Amd 1 Synthetic Resin Adhesives (Phenolic and Amino-Plastic) for Wood Part 2: Close-Contact Adhesives (AMD 6503) December 21, 1990 (Superseded by BS EN 301: 1992 & BS EN 302: Parts 1-4: 1992). 13 pp.

—Wooden Structures—Adhesive Strength
BSI BS EN 302-1-92. 1992 Adhesives for Load-Bearing Timber Structures: Test Methods Part 1: Determination of Bond Strength in Longitudinal Tensile Shear (Supersedes BS 1204: Part 1 & 2: 1979). 11 pp.
CEN PREN 302 (Part 1)-89. Adhesives for Load-Bearing Timber Structures: Polycondensation Adhesives of the Phenolic and Aminoplastic Types—Test Tethods—Part 1: Determination of Bond Strength in Longitudinal Shear. 12 pp.
CEN EN 302-1-92. Adhesives for Load-Bearing Timber Structures—Test Methods—Part 1: Determination of Bond Strength in Longitudinal Tensile Shear. 11 pp.

—Wooden Structures—Classification
BSI BS EN 301-92. 1992 Adhesives, Phenolic and Aminoplastic, for Load-Bearing Timber Structures: Classification and Performance Requirements (Supersedes BS 1204: Part 1&2: 1979). 11 pp.
CEN PREN 301-89. Adhesives for Load-Bearing Timber Structures: Polycondensation Adhesives of the Phenolic and Aminoplastic Types—Classification and Performance Requirements. 11 pp.
CEN EN 301-92. Adhesives, Phenolic and Aminoplastic, for Load Bearing Timber Structures: Classification and Performance Requirements. 12 pp.
DIN ENGL EN 301-92. Adhesives (Phenolic and Aminoplastic) for Loadbearing Timber Structures; Classification and Performance Requirements (Aug). 8 pp.

Aminoplastic Adhesives (Cont.)
—Wooden Structures—Delamination
BSI BS EN 302-2-92. 1992 Adhesives for Load-Bearing Timber Structures: Test Methods Part 2: Determination of Resistance to Delamination (Laboratory Method) (Supersedes BS 1204: Parts 1 & 2: 1979). 10 pp.
BSI BS EN 302-2-01. 1992 Amd 1 Adhesives for Load-Bearing Timber Structures: Test Methods Part 2: Determination of Resistance to Delamination (Laboratory Method) (AMD 7719) May 15, 1993 (W). 11 pp.
CEN PREN 302 (Part 2)-89. Adhesives for Load-Bearing Timber Structures: Polycondensation Adhesives of the Phenolic and Aminoplastic Types—Test Methods—Part 2: Determination of Resistance to Delamination (Laboratory Method). 10 pp.
CEN EN 302-2-92. Adhesives for Load-Bearing Timber Structures—Test Methods—Part 2: Determination of Resistance to Delamination (Laboratory Method). 9 pp.
DIN ENGL EN 302 Pt 2-92. Test Methods for Adhesives for Loadbearing Timber Structures; Determination of Resistance to Delamination (Laboratory Method) (Aug). 5 pp.

—Wooden Structures—Shear Strength
BSI BS EN 302-4-92. 1992 Adhesives for Load-Bearing Timber Structures: Test Methods Part 4: Determination of the Effects of Wood Shrinkage on the Shear Strength (Supersedes BS 1204: Part 1 & 2: 1979). 11 pp.
CEN PREN 302 (Part 4)-90. Adhesives for Load Bearing Timber Structures: Polycondensation Adhesives of the Phenolic and Aminoplastic Types—Test Methods—Part 4: Determination of the Influence of Shrinkage on the Sheer Strength. 9 pp.
CEN EN 302-4-92. Adhesives for Load-Bearing Timber Structures—Test Methods—Part 4: Determination of the Effects of Wood Shrinkage on the Shear Strength. 10 pp.
DIN ENGL EN 302 Pt 4-92. Test Methods for Adhesives for Loadbearing Timber Structures; Determination of the Effects of Wood Shrinkage on the Shear Strength (Aug). 6 pp.

—Wooden Structures—Tensile Testing
BSI BS EN 302-3-92. 1992 Adhesives for Load-Bearing Timber Structures: Test Methods Part 3: Determination of the Effect of Acid Damage to Wood Fibres by Temperature and Humidity Cycling on the Transverse Tensile Strength (Supersedes BS 1204: Part 1 & 2: 1979). 11 pp.
CEN PREN 302 (Part 3)-90. Adhesives for Load Bearing Timber Structures: Polycondensation Adhesives of the Phenolic and Aminoplastic Types—Test Methods—Part 3: Determination of the Influence of Cyclic Treatment on the Tensile Strength (Fibre Damage). 10 pp.
CEN EN 302-3-92. Adhesives for Load-Bearing Timber Structures—Test Methods—Part 3: Determination of the Effect of Acid Damage to Wood Fibres by Temperature and Humidity Cycling on the Transverse Tensile Strength. 10 pp.
DIN ENGL EN 302 Pt 3-92. Test Methods for Adhesives for Loadbearing Timber Structures; Determination of the Effect of Acid Damage to Wood Fibres by Temperature and Humidity Cycling on the Transverse Tensile Strength (Aug). 6 pp.

Aminoplasts
Use: Amino Resins

4-Aminotoluene-3-sulfonic Acid
Use For: p-Toluidine-m-sulfonic Acid
JIS K 4184-85. p-Toluidine-m-Sulfonic Acid (4-Aminotoluene-3-Sulfonic Acid).

4-Aminotoluene
Use: p-Toluidine

Ammeters
Use For: Ampere Meters See Also: Electric Measuring Instruments; Electrical Measurement; Ohmmeters; Voltmeters
BSI BS 89: Part 2-90. 1990 Direct Acting Indicating Analogue Electrical Measuring Instruments and Their Accessories Part 2: Special Requirements for Ammeters and Voltmeters. 16 pp.
CENELEC EN 60 051-2-89. Direct Acting Indicating Analogue Electrical Measuring Instruments and Their Accessories Part 2: Special Requirements for Ammeters and Voltmeters (Supersedes HD 233). 4 pp.
CNS C3163-82. Method of Test for Thermoelectric Type High-Frequency Ampere Meter for Panel (Oct)(9478).
CNS C4369-82. Thermoelectric Type High-Frequency Ampere Meter for Panel (Oct)(9477).
IEC 51 Pt 2-84. Direct Acting Indicating Analogue Electrical Measuring Instruments and Their Accessories Part 2: Special Requirements for Ammeters and Voltmeters Fourth Edition. 24 pp.

Ammeters (Cont.)
MOD UK DSTAN 66-7: Part 2-01. Instruments, Electrical Indicating (Sealed) Part 2: List of Instruments Issue 1; Amendment 6. 60 pp.

—AC
CSA CAN3-C17-M84. Alternating-Current Electricity Metering; (Gen Instr 1). 129 pp.

—AC—Symbols
BSI BS EN 60387-93. 1993 Symbols for Alternating-Current Electricity Meters (IEC 387: 1992) (G). 23 pp.
CENELEC EN 60387-92. Symbols for Alternating-Current Electricity Meters. 5 pp.
CENELEC EN 60387-92. Symbols for Alternating-Current Electricity Meters; (IEC 387: 1992). 18 pp.
IEC 387-92. Symbols for Alternating-Current Electricity Meters Second Edition; (CENELEC EN 60387:1992). 38 pp.

—Automotive
CNS D2040-87. Ammeters for Automobiles (Aug)(5796).
JIS D 5604-73. Ammeters for Automobiles.

—Construction Equipment
JIS A 8106-71. Ammeters for Construction Machinery.

Ammonia
Use For: Anhydrous Ammonia; Liquefied Ammonia
See Also: Ammonia Content Analysis; Ammonium Compounds; Ammonium Hydroxide; Hazardous Materials
SAA AS 2022-83. Anhydrous Ammonia—Storage and Handling (Known as the SAA Anhydrous Ammonia Code) Amdt 1 January 1985. 24 pp.

—Agricultural Spraying Equipment—Safety
ISO 4254 Pt 2-86. Tractors and Machinery for Agriculture and Forestry—Technical Means for Providing Safety—Part 2: Anhydrous Ammonia Applicators First Edition. 6 pp.

—Automotive—Safety Code
CNS Z1022-86. Safety Code for Liquefied Ammonia Tank on Automobiles (Feb)(7249).

—Carbon Dioxide Content
BSI BS 4431-89. 1989 Methods of Sampling and Test for Liquefied Anhydrous Ammonia. 16 pp.

—Cylinders—Construction
CNS Z3016-86. Construction Standard for Steel Cylinder of Liquid Ammonia (Oct)(5636).

—Cylinders—Safety
CNS Z1003-86. Safety Code for Steel Cylinder of Liquid Ammonia (Oct)(1325).

—Drying Agents
BSI BS 3482: Part 8-91. 1991 Methods of Test for Desiccants Part 8: Estimation of Ammonia and Ammonium Compounds Content. 6 pp.

—Hose Assemblies
ISO 5771-81. Rubber Hose and Hose Assemblies for Transferring Anhydrous Ammonia First Edition. 7 pp.

—Oil Content—Gravimetric Analysis
BSI BS 4431-89. 1989 Methods of Sampling and Test for Liquefied Anhydrous Ammonia. 16 pp.
ISO 7106-85. Liquefied Anhydrous Ammonia for Industrial Use—Determination of Oil Content—Gravimetric and Infra-Red Spectrometric Methods First Edition. 7 pp.

—Oil Content—Infrared Spectrophotometry
BSI BS 4431-89. 1989 Methods of Sampling and Test for Liquefied Anhydrous Ammonia. 16 pp.
ISO 7106-85. Liquefied Anhydrous Ammonia for Industrial Use—Determination of Oil Content—Gravimetric and Infra-Red Spectrometric Methods First Edition. 7 pp.

—Residue Content
BSI BS 4431-89. 1989 Methods of Sampling and Test for Liquefied Anhydrous Ammonia. 16 pp.

—Residue Content—Gravimetric Analysis
ISO 4276-78. Anhydrous Ammonia for Industrial Use—Evaluation of Residue on Evaporation—Gravimetric Method First Edition. 5 pp.

—Sampling
BSI BS 4431-89. 1989 Methods of Sampling and Test for Liquefied Anhydrous Ammonia. 16 pp.
ISO 7103-82. Liquefied Anhydrous Ammonia for Industrial Use—Sampling—Taking a Laboratory Sample First Edition; (Corrected and Reprinted -1983). 5 pp.

—Tanks (Containers)—Automotive—Safety
CNS Z1022-86. Safety Code for Liquefied Ammonia Tank on Automobiles (Feb)(7249).

INTERNATIONAL AND NON-U.S. NATIONAL STANDARDS
SUBJECT INDEX
Ammonium

Ammonia (Cont.)
—Water Content—Gas Chromatography
ISO 7104-85. Liquefied Anhydrous Ammonia for Industrial Use—Determination of Water Content—Gas Chromatographic Method First Edition. 6 pp.

—Water Content—Karl Fischer Method
BSI BS 4431-89. 1989 Methods of Sampling and Test for Liquefied Anhydrous Ammonia. 16 pp.
ISO 7105-85. Liquefied Anhydrous Ammonia for Industrial Use—Determination of Water Content—Karl Fischer Method First Edition. 6 pp.

Ammonia Content Analysis
Use For: Free Ammonia Content Analysis
See Also: Ammonia

—Ammonium Hydroxide—Volumetric Analysis
BSI BS 4651: Part 2-87. 1987 Ammonia Solution Part 2: Method for Determination of Ammonia Content. 4 pp.

—Cleaning Agents
CGSB 2-GP-11M METH 6.4-88. Methods of Testing and Analysis of Soaps and Detergents Free Ammonia; (Amendment 1 Nov 1988). 2 pp.
CGSB 31-GP-0A METH 68.2-62. Methods of Testing Corrosion-Prevention Materials and Processes Ammonia (Quantitative Determination); (Re-Edited April 1982). 2 pp.

—Coal Gas
CNS K6240-66. Method of Test for Ammonia in Coal Gas (Sep)(2634)(R 1971).

—Cosmetics
EC 83/514/EEC-83. Third Commission Directive on the Approximation of the Laws of the Member States Relating to Methods of Analysis Necessary for Checking the Composition of Cosmetic Products. 38 pp.

—Detergents—Volumetric Analysis
BSI BS 3762: Sec 3.12-85. 1985 Analysis of Formulated Detergents Part 3: Quantitative Test Methods Section 3.12: Method for Determination of Ammonia Content. 4 pp.

—Exhaust Gases
CNS K9099-82. Methods of Test for Ammonia in Exhaust Gas (Jul)(9186).
JIS K 0099-83. Methods for Determination of Ammonia in Exhaust Gas. 22 pp.

—Hexamethylenetetramine
MOD UK M 879/69. Examination of Hexamine.

—Olefins—Photometry
ISO 6192-81. Light Olefins for Industrial Use—Determination of Ammonia—Photometric Method First Edition. 5 pp.

—Phenol Formaldehyde Moldings
JIS K 7230-82. Determination of Free Ammonia in Phenol-Formaldehyde Moldings. 13 pp.

—Phenol Formaldehyde Resins
BSI BS 2782:Pt4: METH 451L-80. 1980 Methods of Testing Plastics Part 4: Chemical Properties Method 451L: Detection of Free Ammonia in Phenol-Formaldehyde Mouldings. 3 pp.
ISO 172-78. Plastics—Phenol—Formaldehyde Mouldings—Detection of Free Ammonia First Edition. 3 pp.

—Phenol Formaldehyde Resins—Colorimetric Analysis
BSI BS 2782:Pt4: METH 451D-78. 1978 Methods of Testing Plastics Part 4: Chemical Properties Method 451D: Determination of Free Ammonia and Ammonium Compounds in Phenol-Formaldehyde Mouldings (Colormetric Comparison Method). 4 pp.
ISO 120-77. Plastics—Phenol-Formaldehyde Mouldings—Determination of Free Ammonia and Ammonium Compounds—Colorimetric Comparison Method First Edition. 4 pp.

—Soldering Fluxes—Distillation Methods
ISO 9455 Pt 9-93. Soft Soldering Fluxes—Test Methods—Part 9: Determination of Ammonia Content First Edition. 8 pp.

—Urea
CNS K6158-75. Method of Test for Urea Technical Grade (Jan)(1426). 2 pp.

—Volumetric Analysis
ISO 7108-85. Ammonia Solution for Industrial Use—Determination of Ammonia Content—Titrimetric Method First Edition. 4 pp.

Ammonia Content Analysis (Cont.)
—Water
BSI BS 2690: Part 7-68. (WITHDRAWN) 1968 Water Used in Industry Part 7: Nitrite, Nitrate and Ammonia (Free, Saline and Albuminoid) (SUPERSEDED BY BS 6068: SECTIONS 2. 7, 2. 10, 2. 11, 2. 16, & 2. 36). 21 pp.

Ammonia Liquor
—Alkalinity
MOD UK M 2006/60. Examination of Ammonia Liquor.

—Evaporation Residue
MOD UK M 2006/60. Examination of Ammonia Liquor.

—Visual Inspection
MOD UK M 2006/60. Examination of Ammonia Liquor.

Ammonia Nitrogen
Use: Amino Nitrogen Content Analysis

Ammonia Nitrogen Content Analysis
Use: Amino Nitrogen Content Analysis

Ammonia Solution
Use: Ammonium Hydroxide

Ammonia Water
Use: Ammonium Hydroxide

Ammoniacal Copper Arsenate
Use: Copper Arsenate

Ammoniacal Nitrogen Content Analysis
—Aluminum Sulfate
CNS K6163-74. Method of Test for Aluminium Sulfate (for Water Works) (Jun)(2075). 4 pp.

—Ammonium Bicarbonate—Volumetric Analysis
ISO 2515-73. Ammonium Hydrogen Carbonate for Industrial Use (Including Foodstuffs)—Determination of Ammoniacal Nitrogen Content—Volumetric Method After Distillation First Edition. 5 pp.

—Ammonium Nitrate
CNS K6081-62. Method of Test for Ammonium Nitrate (for Industrial Use) (Dec)(998)(R 1971). 1 p.

—Ammonium Nitrate—Volumetric Analysis
BSI BS 4267: Part 1-88. 1988 Ammonium Nitrate Part 1: Method for Determination of Ammoniacal Nitrogen Content. 4 pp.
ISO 3330-75. Ammonium Nitrate for Industrial Use—Determination of Ammoniacal Nitrogen Content—Titrimetric Method After Distillation First Edition. 6 pp.

—Ammonium Sulfate
CNS K6075-74. Method of Test for Ammonium Sulfate (May)(919). 3 pp.

—Ammonium Sulfate—Volumetric Analysis
ISO 3332-75. Ammonium Sulphate for Industrial Use—Determination of Ammoniacal Nitrogen Content—Titrimetric Method After Distillation First Edition. 6 pp.

—Fertilizers
CNS N4027-70. Method of Test for Compound Fertilizer (May)(3077). 6 pp.

—Fertilizers—Volumetric Analysis
BSI BS 5551: SUBSEC 4.1.2-82. 1982 Fertilizers Part 4: Chemical Analysis Section 4.1: Determination of Nitrogen Subsection 4.1.2: Determination of Ammoniacal Nitrogen, Titrimetric Method After Distillations. 7 pp.
BSI BS 5551: SUBSEC 4.1.4-84. 1984 Fertilizers Part 4: Chemical Analysis Sec 4 Determination of Nitrogen Subsection 4.1.4: Method for Determination of Ammoniacal Nitrogen Content in the Presence of Other Substance Which Rel Ammonia When Treated with Sodium Hydroxide (Titrimetric Method). 6 pp.
ISO 5314-81. Fertilizers—Determination of Ammoniacal Nitrogen Content—Titrimetric Method After Distillation First Edition. 6 pp.
ISO 7408-83. Fertilizers—Determination of Ammoniacal Nitrogen Content in the Presence of Other Substances Which Release Ammonia When Treated with Sodium Hydroxide—Titrimetric Method First Edition. 7 pp.

Ammoniacal Nitrogen Content Analysis (Cont.)
—Nitric Acid—Spectrophotometry
ISO 2991-74. Nitric Acid for Industrial Use—Determination of Ammoniacal Nitrogen Content—Spectrophotometric Method First Edition. 7 pp.

—Photographic Processing Wastes—Kjeldahl Method
ISO 6851-87. Photography—Processing Waste—Determination of Total Amino Nitrogen—Microdiffusion Kjeldahl Method First Edition; (ANSI PH4.47-1991). 7 pp.
ISO 6853-87. Photography—Processing Waste—Determination of Ammoniacal Nitrogen Content—Microdiffusion Method First Edition; (ANSI PH4.48-1990). 6 pp.

—Sulfuric Acid—Spectrophotometry
ISO 2899-74. Sulphuric Acid and Oleums for Industrial Use—Determination of Ammoniacal Nitrogen Content—Spectrophotometric Method First Edition. 7 pp.

—Volumetric Analysis
BSI BS 6068: Sec 2.7-84. 1984 Water Quality Part 2: Physical, Chemical and Bio-Chemical Methods Section 2.7: Determination of Ammonium: Distillation and Titration Method (SUPERSEDES BS 2690; PART 7: 1968). 6 pp.

—Water—Distillation Methods
BSI BS 6068: Sec 2.7-84. 1984 Water Quality Part 2: Physical, Chemical and Bio-Chemical Methods Section 2.7: Determination of Ammonium: Distillation and Titration Method (SUPERSEDES BS 2690; PART 7: 1968). 6 pp.

—Water—Potentiometric Analysis
BSI BS 6068: Sec 2.10-84. 1984 Water Quality Part 2: Physical, Chemical and Bio-Chemical Methods Section 2.10: Determination of Ammonium: Potentiometric Methods (SUPERSEDES BS 2690: PART 7: 1968). 8 pp.
ISO 6778-84. Water Quality—Determination of Ammonium—Potentiometric Method First Edition. 7 pp.

—Water—Spectrometry
BSI BS 6068: Sec 2.11-84. 1984 Water Quality Part 2: Physical, Chemical and Bio-Chemical Methods Section 2.11: Determination of Ammonium: Manual Spectrometric Method (SUPERSEDES BS 2690: PART 7: 1968). 10 pp.
BSI BS 6068: Sec 2.33-87. 1987 Water Quality Part 2: Physical, Chemical and Bio-Chemical Methods Section 2.33: Method for the Determination of Ammonium: Automated Spectrometric Method. 10 pp.

Ammoniacal Silver Nitrate
Use: Silver Nitrate

Ammonium Acetate
See Also: Esters
CNS K7044-62. Chemical Reagent (Ammonium Acetate) (Jul)(1544).
JIS K 8359-79. Ammonium Acetate.
MOD UK CS 3125. Ammonium Acetate (Withdrawn).

—Drying Agents
BSI BS 3482: Part 8-91. 1991 Methods of Test for Desiccants Part 8: Estimation of Ammonia and Ammonium Compounds Content. 6 pp.

Ammonium Alum
Use: Aluminum Ammonium Sulfate

Ammonium Bicarbonate
Use For: Ammonium Hydrogen Carbonate
CNS K1013-81. Ammonium Bicarbonate for Industrial Use (Oct)(51).
CNS K6335-72. Method of Test for Ammonium Bicarbonate (for Industrial Use) (Oct)(3481).
JIS K 1451-83. Ammonium Hydrogen Carbonate. 10 pp.

—Alkalinity—Volumetric Analysis
ISO 2516-73. Ammonium Hydrogen Carbonate for Industrial Use (Including Foodstuffs)—Determination of Total Alkalinity—Volumetric Method First Edition. 3 pp.

—Ammoniacal Nitrogen Content—Volumetric Analysis
ISO 2515-73. Ammonium Hydrogen Carbonate for Industrial Use (Including Foodstuffs)—Determination of Ammoniacal Nitrogen Content—Volumetric Method After Distillation First Edition. 5 pp.

INDUSTRY STANDARDS

Ammonium Bicarbonate (Cont.)

—Arsenic Content—Photometry
ISO 4275-77. Ammonium Hydrogencarbonate for Industrial Use (Including Foodstuffs)—Determination of Arsenic Content—Silver Diethyldithiocarbamate Photometric Method First Edition; (Erratum—Sept 1978). 5 pp.

—Ash Content—Gravimetric Analysis
ISO 3420-75. Ammonium Hydrogen Carbonate for Industrial Use (Including Foodstuffs)—Determination of Ash—Gravimetric Method First Edition. 3 pp.

—Carbon Dioxide Content—Volumetric Analysis
ISO 3422-75. Ammonium Hydrogen Carbonate for Industrial Use (Including Foodstuffs—Determination of Total Carbon Dioxide Content—Titrimetric Method First Edition. 5 pp.

—Lead Content—Atomic Absorption Spectrometry
ISO 7110-85. Ammonium Bicarbonate (Ammonium Hydrogencarbonate) for Industrial Use (Including Foodstuffs)—Determination of Lead Content—Flame Atomic Absorption Method First Edition. 5 pp.

Ammonium Bichromate
Use: Ammonium Dichromate

Ammonium Bifluoride
Use For: Ammonium Hydrogen Fluoride
JIS K 1407-78. Ammonium Bifluoride. 7 pp.
JIS K 8817-79. Ammonium Hydrogen Fluoride.

Ammonium Bisulfate
Use For: Ammonium Hydrogen Sulfate
See Also: Ammonium Compounds; Sulfates
CNS K7045-62. Chemical Reagent (Ammonium Bisulfate) (Acid Ammonium Sulfate) (Jul)(1545).

Ammonium Bromide
CNS K7046-62. Chemical Reagent (Ammonium Bromide) (Jul)(1546).
JIS K 8502-75. Ammonium Bromide.

Ammonium Carbonate
CNS K7047-62. Chemical Reagent (Ammonium Carbonate) (Jul)(1547).
JIS K 8613-75. Ammonium Carbonate.

Ammonium Ceric Nitrate
Use For: Ceric Ammonium Nitrate
See Also: Ammonium Compounds
JIS K 8556-84. Cerium (IV) Diammonium Nitrate.

Ammonium Chloride
See Also: Quaternary Ammonium Chloride
CNS K1014-81. Ammonium Chloride for Industrial Use (Oct)(52).
CNS K6292-70. Method of Test for Ammonium Chloride of Industrial Grade (Dec)(3180).
CNS K7048-62. Chemical Reagent (Ammonium Chloride) (Jul)(1548).
JIS K 1441-86. Ammonium Chloride.
JIS K 1441-63. Ammonium Chloride. 8 pp.
JIS K 8116-75. Ammonium Chloride.

—Dry Cells
CNS K1152-70. Ammonium Chloride for Dry Cell (Dec)(3179).

—Fertilizers
CNS N3002-88. Ammonium Chloride, Fertilizer Grade (Aug)(53).

Ammonium Chromate
See Also: Ammonium Compounds
CNS K7049-62. Chemical Reagent (Ammonium Chromate) (Jul)(1549).

Ammonium Citrate, Dibasic
See Also: Ammonium Compounds
CNS K7050-62. Chemical Reagent (Ammonium Citrate, Dibasic) (Jul)(1550).

Ammonium Compound Content Analysis

—Dicyandiamide
MOD UK M 812/77. Examination of Dicyandiamide.

—Dyes
MOD UK M 818/63. Examination of Dyestuffs for Use in Pyrotechnic Compositions and HE Substitutes.

—Magnesium Carbonate
MOD UK M 870/67. Examination of Magnesium Carbonate, Heavy.

Ammonium Compound Content Analysis (Cont.)

—Milk
BSI BS 1741: Part 11-89. 1989 Methods for Chemical Analysis of Liquid Milk and Cream Part 11: Detection of Detergent/Disinfectant Residues. 4 pp.

—Phenol Formaldehyde Moldings
JIS K 7230-82. Determination of Free Ammonia in Phenol-Formaldehyde Moldings. 13 pp.

—Phenol Formaldehyde Resins—Colorimetric Analysis
BSI BS 2782:Pt4: METH 451D-78. 1978 Methods of Testing Plastics Part 4: Chemical Properties Method 451D: Determination of Free Ammonia and Ammonium Compounds in Phenol-Formaldehyde Mouldings (Colormetric Comparison Method). 4 pp.
ISO 120-77. Plastics—Phenol-Formaldehyde Mouldings—Determination of Free Ammonia and Ammonium Compounds—Colorimetric Comparison Method First Edition. 4 pp.

—Sodium Nitrate
MOD UK M 869/91. Examination of Sodium Nitrate, Grade 1.

—Strontium Nitrate
MOD UK M 867/92. Examination of Strontium Nitrate (Anhydrous).

Ammonium Compounds
Scope Note: For additional listings, use a more specific term *See Also:* Aluminum Ammonium Sulfate; Ammonia; Ammonium Acetate; Ammonium Bifluoride; Ammonium Bisulfate; Ammonium Bromide; Ammonium Carbonate; Ammonium Ceric Nitrate; Ammonium Chloride; Ammonium Chromate; Ammonium Citrate, Dibasic; Ammonium Compound Content Analysis; Ammonium Cupric Chloride; Ammonium Dichromate; Ammonium Dihydrogen Orthophosphate; Ammonium Ferric Citrate; Ammonium Ferric Sulfate; Ammonium Ferrous Sulfate; Ammonium Fluoride; Ammonium Formate; Ammonium Iodide; Ammonium Molybdate; Ammonium Nickel Sulfate; Ammonium Nickel Sulfate Hexahydrate; Ammonium Nitrate; Ammonium Oxalate; Ammonium Perchlorate; Ammonium Peroxydisulfate; Ammonium Phosphate; Ammonium Phosphate, Dibasic; Ammonium Phosphate, Monobasic; Ammonium Sodium Hydrogenphosphate Tetrahydrate; Ammonium Sodium Phosphate; Ammonium Sulfamate; Ammonium Sulfate; Ammonium Sulfide; Ammonium Sulfide Solution, Yellow; Ammonium Tartrate; Ammonium Thiocyanate; Ammonium Thiosulfate; Ammonium Vanadate; Ammonium Vanadate (V); Quaternary Ammonium Chloride

—Fertilizers
CNS N3063-88. Ammonium Humate Fertilizer (Aug)(11950).

Ammonium Content Analysis

—Ammonium Dihydrogen Orthophosphate
MOD UK M 9515/65. Examination of Ammonium Dihydrogen Orthophosphate (Withdrawn).

—Ammonium Peroxydisulfate
MOD UK M 9547/67. Examination of Ammonium Persulphate (Withdrawn).

—Soldering Fluxes—Distillation Methods
MOD UK M 9542/70. Examination of Flux, Soldering (Zinc Chloride Solution).

—Water—Distillation Methods
ISO 5664-84. Water Quality—Determination of Ammonium—Distillation and Titration Method First Edition. 5 pp.

—Water—Spectrometry
ISO 7150 Pt 1-84. Water Quality—Determination of Ammonium—Part 1: Manual Spectrometric Method First Edition. 9 pp.
ISO 7150 Pt 2-86. Water Quality—Determination of Ammonium—Part 2: Automated Spectrometric Method First Edition; (Corrected and Reprinted -1987). 9 pp.

Ammonium Cupric Chloride
Use For: Cupric Ammonium Chloride
See Also: Ammonium Compounds
CNS K7163-66. Chemical Reagent (Cupric Ammonium Chloride) (Mar)(1663).

Ammonium Dichromate
Use For: Ammonium Bichromate
CNS K7051-62. Chemical Reagent (Ammonium Dichromate) (Jul)(1551).
JIS K 8516-86. Ammonium Dichromate.

Ammonium Dihydrogen Orthophosphate

—Content Analysis
MOD UK M 9515/65. Examination of Ammonium Dihydrogen Orthophosphate (Withdrawn).

—Fire Retardants
MOD UK CS 3049A. Ammonium Dihydrogen Orthophosphate (Withdrawn).

—PH
MOD UK M 9515/65. Examination of Ammonium Dihydrogen Orthophosphate (Withdrawn).

Ammonium Dihydrogen Phosphate
Use: Ammonium Phosphate, Monobasic

Ammonium Dihydrogenphosphate
Use: Ammonium Phosphate, Monobasic

Ammonium Ferric Citrate
Use For: Ammonium Iron Citrate; Ferric Ammonium Citrate; Iron Ammonium Citrate
CNS K1174-75. Iron Ammonium Citrate (Industrial Grade) (Jan)(3726).
CNS K6373-75. Method of Test for Iron Ammonium Citrate of Industrial Grade (Jan)(3727).
JIS K 8287-81. Ammonium Iron (III) Citrate.

Ammonium Ferric Sulfate
Use For: Ammonium Iron Sulfate; Ferric Ammonium Sulfate
CNS K7222-66. Chemical Reagent (Ferric Ammonium Sulphate) (Jun)(1722).
JIS K 8982-80. Ammonium Iron (III) Sulfate 12-Water.

Ammonium Ferrous Sulfate
Use For: Ferrous Ammonium Sulfate; Mohr's Salt
CNS K7228-66. Chemical Reagent (Ferrous Ammonium Sulfate) (Jun)(1727).
JIS K 8979-76. Ferrous Ammonium Sulfate (Mohr's Salt).

Ammonium Fluoride
See Also: Ammonium Compounds; Hazardous Materials
CNS K7052-62. Chemical Reagent (Ammonium Fluoride) (Jul)(1552).

—Highway Transportation—Emergency Procedures
SAA AS 1678.6.1. 015-84. Emergency Procedure Guides—Transport—Part 6.1.015: Fluorides (Ammonium, Potassium or Sodium Fluoride).

Ammonium Formate
CNS K7611-82. Chemical Reagent (Ammonium Formate) (May)(8852).

Ammonium Hydrogen Carbonate
Use: Ammonium Bicarbonate

Ammonium Hydrogen Fluoride
Use: Ammonium Bifluoride

Ammonium Hydrogen Sulfate
Use: Ammonium Bisulfate

Ammonium Hydroxide
Use For: Ammonia Solution; Ammonia Water
See Also: Ammonia; Hazardous Materials
CGSB CAN/CGSB-15.30-92. Aqueous Ammonia Solution (Ammonium Hydroxide) Technical Grade. 7 pp.
CNS K1024-62. Anhydrous Ammonia Water (Dec)(263)(R 1968).
CNS K1081-75. Ammonium Hydroxide (Industrial Grade) (Aug)(2237).
CNS K6018-75. Method of Test for Ammonium Hydroxide (Aug)(280).
CNS K7053-62. Chemical Reagent (Ammonium Hydroxide, Ammonia Water, Aqua Ammonia) (Jul)(1553).
JIS K 8085-89. Ammonia Solution.

—Ammonia Content—Volumetric Analysis
BSI BS 4651: Part 2-87. 1987 Ammonia Solution Part 2: Method for Determination of Ammonia Content. 4 pp.

—Carbon Dioxide Content
BSI BS 4651: Part 7-88. 1988 Ammonia Solution Part 7: Methods for Determination of Carbon Content. 4 pp.

—Chloride Content
BSI BS 4651: Part 6-88. 1988 Ammonia Solution Part 6: Methods for Determination of Chloride Content. 4 pp.

Ammonium Hydroxide (Cont.)

—Density—Gravimetric Analysis
BSI BS 4651: Part 1-88. 1988 Ammonia Solution Part 1: Methods for Determination of Density at 20 Degrees Celsius. 4 pp.

—Iron Content—Spectrometry
BSI BS 4651: Part 5-88. 1988 Ammonia Solution Part 5: Methods for Determination of Iron Content. 4 pp.

—Residue Content—Gravimetric Analysis
BSI BS 4651: Part 3-87. 1987 Ammonia Solution Part 3: Method for Determination of Residue on Evaporation. 3 pp.
BSI BS 4651: Part 4-88. 1988 Ammonia Solution Part 4: Methods for Determination of Residue on Heating at 850 Degrees Celsius. 4 pp.
ISO 7109-85. Ammonia Solution for Industrial Use—Determination of Residue After Evaporation at 105 Degrees Celsius—Gravimetric Method First Edition. 3 pp.

—Sampling
BSI BS 4651: Part 0-87. 1987 Ammonia Solution Part 0: Methods for Sampling. 4 pp.

Ammonium Iodide
CNS K7054-62. Chemical Reagent (Ammonium Iodide) (Jul)(1554).
JIS K 8910-75. Ammonium Iodine.

Ammonium Ion Content Analysis

—Water
CNS K9067-81. Methods of Test for Ammonium Ion in Water (May)(7439).

Ammonium Iron Citrate
Use: Ammonium Ferric Citrate

Ammonium Iron Sulfate
Use: Ammonium Ferric Sulfate

Ammonium Metavanadate
Use: Ammonium Vanadate

Ammonium Molybdate
See Also: Molybdate
CNS K7055-62. Chemical Reagent (Ammonium Molybdate) (Jul)(1555).
JIS K 8905-76. Ammonium Molybdate.

Ammonium Nickel Sulfate
Use For: Nickel Ammonium Sulfate
See Also: Ammonium Compounds; Sulfates
CNS K7330-69. Chemical Reagent (Nickel Ammonium Sulfate) (Jul)(1829).

Ammonium Nickel Sulfate Hexahydrate
JIS K 8990-80. Ammonium Nickel (II) Sulfate Hexahydrate.

Ammonium Nitrate
See Also: Calcium Ammonium Nitrate; Hazardous Materials
BSI BS 4267: Part 0-87. 1987 Ammonium Nitrate Part 0: General Introduction. 4 pp.
CNS K1015-81. Ammonium Nitrate for Industrial Use (Oct)(54).
CNS K7056-62. Chemical Reagent (Ammonium Nitrate) (Jul)(1556).
JIS K 1424-85. Ammonium Nitrate.
JIS K 8545-75. Ammonium Nitrate.

—Acidity—Volumetric Analysis
ISO 2364-72. Ammonium Nitrate for Industrial Use—Determination of Free Acidity—Volumetric Method First Edition. 3 pp.

—Ammoniacal Nitrogen Content
CNS K6081-62. Method of Test for Ammonium Nitrate (for Industrial Use) (Dec)(998)(R 1971). 1 p.

—Ammoniacal Nitrogen Content—Volumetric Analysis
BSI BS 4267: Part 1-88. 1988 Ammonium Nitrate Part 1: Method for Determination of Ammoniacal Nitrogen Content. 4 pp.
ISO 3330-75. Ammonium Nitrate for Industrial Use—Determination of Ammoniacal Nitrogen Content—Titrimetric Method After Distillation First Edition. 6 pp.

—Chloride Content—Colorimetric Analysis
BSI BS 4267: Part 8-87. 1987 Ammonium Nitrate Part 8: Methods for Determination of Chloride Content. 4 pp.

Ammonium Nitrate (Cont.)

—Chloride Content—Potentiometric Analysis
BSI BS 4267: Part 8-87. 1987 Ammonium Nitrate Part 8: Methods for Determination of Chloride Content. 4 pp.

—Chloride Ion Content—Potentiometric Analysis
ISO 3695-77. Ammonium Nitrate for Industrial Use—Determination of Chloride Ions Content—Potentiometric Method First Edition. 6 pp.

—Explosives—Delivery Systems
NATO STANAG 4024 ED 2 AMD 2-62. Specification for Ammonium Nitrate Crystal and Prill-Type A and Type B (for Use in Explosives) for Deliveries from One NATO Nation to Another. 9 pp.

—Fertilizers
CNS N3003-88. Ammonium Nitrate, Fertilizer Grade (Jul)(55).
CNS N3076-88. Urea-Ammonium Nitrate Fertilizer, Fluid (Jul)(11963).
EC 80/876/EEC-80. Council Directive on the Approximation of the Laws of the Member States Relating to Straight Ammonium Nitrate Fertilizers of High Nitrogen Content. 5 pp.

—Fertilizers—Oil Retention
BSI BS 5551: Sec 3.4-87. 1987 Fertilizers Part 3: Physical Properties Section 3.4: Method for Determination of Oil Retention of High Nitrogen Content, Straight Ammonium Nitrate Fertilizers. 4 pp.
ISO 5313-86. High Nitrogen Content, Straight Ammonium Nitrate Fertilizers—Determination of Oil Retention First Edition. 4 pp.

—Highway Transportation—Emergency Procedures
SAA AS 1678.5.1. 002-91. Emergency Procedure Guides—Transport—Part 5.1.002: Ammonium Nitrate. 2 pp.

—Insoluble Matter Content
CNS K6081-62. Method of Test for Ammonium Nitrate (for Industrial Use) (Dec)(998)(R 1971). 1 p.

—Insoluble Matter Content—Gravimetric Analysis
BSI BS 4267: Part 5-87. 1987 Ammonium Nitrate Part 5: Method for Determination of Matter Insoluble in Water. 4 pp.
ISO 2995-74. Ammonium Nitrate for Industrial Use—Determination of Matter Insoluble in Water—Gravimetric Method First Edition. 4 pp.

—Nitric Acid Content
CNS K6081-62. Method of Test for Ammonium Nitrate (for Industrial Use) (Dec)(998)(R 1971). 1 p.

—Nitrite Content
BSI BS 4267: Part 7-87. 1987 Ammonium Nitrate Part 7: Method for Determination of Nitrite Content. 4 pp.

—Nitrogen Content
CNS K6081-62. Method of Test for Ammonium Nitrate (for Industrial Use) (Dec)(998)(R 1971). 1 p.

—Nitrogen Content—Volumetric Analysis
BSI BS 4267: Part 2-88. 1988 Ammonium Nitrate Part 2: Method for Determination of Total Nitrogen Content. 6 pp.
ISO 3331-75. Ammonium Nitrate for Industrial Use—Determination of Total Nitrogen Content—Titrimetric Method After Distillation First Edition. 6 pp.

—PH—Potentiometric Analysis
BSI BS 4267: Part 3-87. 1987 Ammonium Nitrate Part 3: Method of Measurement of pH Value. 3 pp.
ISO 2365-72. Ammonium Nitrate for Industrial Use—Measurement of pH Value—Potentiometric Method First Edition. 4 pp.

—Residue Content
CNS K6081-62. Method of Test for Ammonium Nitrate (for Industrial Use) (Dec)(998)(R 1971). 1 p.

—Sulfate Content—Turbidimetry
BSI BS 4267: Part 9-87. 1987 Ammonium Nitrate Part 9: Method for Determination of Sulphate Content. 4 pp.

—Sulfated Ash Content—Gravimetric Analysis
BSI BS 4267: Part 4-87. 1987 Ammonium Nitrate Part 4: Method for Determination of Sulphated Ash. 3 pp.

Ammonium Nitrate (Cont.)

—Sulfur Compound Content—Reduction
ISO 3329-75. Ammonium Nitrate for Industrial Use—Determination of Content of Sulphur Compounds—Method by Reduction and Titrimetry First Edition. 6 pp.

—Sulfur Compound Content—Volumetric Analysis
ISO 3329-75. Ammonium Nitrate for Industrial Use—Determination of Content of Sulphur Compounds—Method by Reduction and Titrimetry First Edition. 6 pp.

—Water Content
CNS K6081-62. Method of Test for Ammonium Nitrate (for Industrial Use) (Dec)(998)(R 1971). 1 p.

—Water Content—Karl Fischer Method
BSI BS 4267: Part 6-87. 1987 Ammonium Nitrate Part 6: Method for Determination of Water Content. 4 pp.
ISO 5791-78. Ammonium Nitrate for Industrial Use—Determination of Water Content—Karl Fischer Method First Edition. 4 pp.

Ammonium Nitrate Fuel Oil Explosives
Use For: ANFO See Also: Explosives
JIS K 4826-79. Ammonium Nitrate Fuel Oil Explosives (ANFO).

Ammonium Oxalate
See Also: Esters
CNS K7057-62. Chemical Reagent (Ammonium Oxalate) (Jul)(1557).
JIS K 8521-79. Ammonium Oxalate, Monohydrate.

Ammonium Perchlorate
See Also: Ammonium Compounds

—High Explosives
NATO STANAG 4299 ED 1 AMD 0-00. Specification Ammonium Perchlorate (NH4C104) for Deliveries from One NATO Nation to Another. 24 pp.

—Propellants
NATO STANAG 4299 ED 1 AMD 0-00. Specification Ammonium Perchlorate (NH4C104) for Deliveries from One NATO Nation to Another. 24 pp.

—Pyrotechnics
NATO STANAG 4299 ED 1 AMD 0-00. Specification Ammonium Perchlorate (NH4C104) for Deliveries from One NATO Nation to Another. 24 pp.

Ammonium Peroxydisulfate
Use For: Ammonium Persulfate; Ammonium Persulphate
CNS K7058-62. Chemical Reagent (Ammonium Persulfate) (Jul)(1558).
JIS K 8252-79. Ammonium Peroxydisulfate.
MOD UK CS 2522. Ammonium Persulphate (Withdrawn).

—Ammonium Content
MOD UK M 9547/67. Examination of Ammonium Persulphate (Withdrawn).

—Persulphate Content
MOD UK M 9547/67. Examination of Ammonium Persulphate (Withdrawn).

—Residue-On-Ignition Determination
MOD UK M 9547/67. Examination of Ammonium Persulphate (Withdrawn).

—Visual Inspection
MOD UK M 9547/67. Examination of Ammonium Persulphate (Withdrawn).

Ammonium Persulfate
Use: Ammonium Peroxydisulfate

Ammonium Persulphate
Use: Ammonium Peroxydisulfate

Ammonium Phosphate

—Fire Retardants
MOD UK DSTAN 80-2-69. Wood Preservative and Fire Retardant, Ammonium Phosphate Mixture. 14 pp.

—Wood Preservatives
MOD UK DSTAN 80-2-69. Wood Preservative and Fire Retardant, Ammonium Phosphate Mixture. 14 pp.

Ammonium Phosphate, Dibasic
Use For: Diammonium Hydrogenphosphate
CNS K7059-62. Chemical Reagent (Ammonium Phosphate, Dibasic) (Diammonium Hydrogen Phosphate) (Jul)(1559).
JIS K 9016-81. Diammonium Hydrogenphosphate.

Ammonium Phosphate, Monobasic
Use For: Ammonium Dihydrogen Phosphate; Ammonium Dihydrogenphosphate
See Also: Ammonium Compounds
CNS K7060-62. Chemical Reagent (Ammonium Phosphate, Monobasic)(Ammonium Dihydrogen Phosphate) (Jul)(1560).
JIS K 9006-81. Ammonium Dihydrogenphosphate.

Ammonium Salt Content Analysis
—**Reagents**
CNS K7619-82. Chemical Reagent (Ammonium Salt, Nitrates and Total Nitrogen Determination) (Jun)(8991).

Ammonium Sodium Hydrogenphosphate Tetrahydrate
JIS K 9013-81. Ammonium Sodium Hydrogenphosphate Tetrahydrate.

Ammonium Sodium Phosphate
Use For: Sodium Ammonium Phosphate
See Also: Ammonium Compounds
CNS K7443-76. Chemical Reagent (Sodium Ammonium Phosphate) (Sep)(1942).

Ammonium Sulfamate
CNS K7061-62. Chemical Reagent (Ammonium Sulfamate) (Jul)(1561).
JIS K 8588-75. Ammonium Sulfamate.

Ammonium Sulfate
CNS K1067-74. Ammonium Sulfate for Industrial Use (Jun)(2072).
CNS K7062-62. Chemical Reagent (Ammonium Sulfate) (Jul)(1562).
JIS K 8960-75. Ammonium Sulfate.
MOD UK TS 476. Ammonium Sulphate, Technical (Withdrawn).

—**Acidity—Volumetric Analysis**
ISO 2993-74. Ammonium Sulphate for Industrial Use—Determination of Free Acidity—Titrimetric Method First Edition. 4 pp.

—**Ammoniacal Nitrogen Content**
CNS K6075-74. Method of Test for Ammonium Sulfate (May)(919). 3 pp.

—**Ammoniacal Nitrogen Content—Volumetric Analysis**
ISO 3332-75. Ammonium Sulphate for Industrial Use—Determination of Ammoniacal Nitrogen Content—Titrimetric Method After Distillation First Edition. 6 pp.

—**Arsenic Content**
CNS K6075-74. Method of Test for Ammonium Sulfate (May)(919). 3 pp.

—**Arsenic Content—Photometry**
ISO 5786-78. Ammonium Sulphate for Industrial Use—Determination of Arsenic Content—Silver Diethyldithiocarbamate Photometric Method First Edition. 4 pp.

—**Chloride Ion Content—Potentiometric Analysis**
ISO 3694-77. Ammonium Sulphate for Industrial Use—Determination of Chloride Ions Content—Potentiometric Method First Edition. 6 pp.

—**Copper Content—Photometry**
ISO 3333-75. Ammonium Sulphate for Industrial Use—Determination of Copper Content—Zinc Dibenzyldithiocarbamate Photometric Method First Edition. 5 pp.

—**Fertilizers**
CNS N3004-88. Ammonium Sulfate, Fertilizer Grade (Aug)(262).

—**Insoluble Matter Content—Gravimetric Analysis**
ISO 2994-74. Ammonium Sulphate for Industrial Use—Determination of Matter Insoluble in Water—Gravimetric Method First Edition. 4 pp.

—**Iron Content—Spectrophotometry**
ISO 2992-74. Ammonium Sulphate for Industrial Use—Determination of Iron Content—2,2'-Bipyridyl Photometric Method First Edition. 5 pp.

Ammonium Sulfate (Cont.)
—**Sulfuric Acid Content**
CNS K6075-74. Method of Test for Ammonium Sulfate (May)(919). 3 pp.

—**Thiocyanate Content**
CNS K6075-74. Method of Test for Ammonium Sulfate (May)(919). 3 pp.

—**Water Content**
CNS K6075-74. Method of Test for Ammonium Sulfate (May)(919). 3 pp.

Ammonium Sulfide
CNS K7063-62. Chemical Reagent (Ammonium Sulfide Solution) (Jul)(1563).

Ammonium Sulfide Solution, Yellow
JIS K 8943-61. Ammonium Sulfide Solution, Colorless.

Ammonium Tartrate
See Also: Esters
CNS K7064-62. Chemical Reagent (Ammonium Tartrate) (Jul)(1564).
JIS K 8534-82. Ammonium Tartrate.

Ammonium Thiocyanate
CNS K7065-62. Chemical Reagent (Ammonium Thiocyanate) (Ammonium Sulfocyanide) (Jul)(1565).
JIS K 9000-75. Ammonium Thiocyanate.

—**Photographic Chemicals**
BSI BS 3750-78. 1978 Photographic Grade Ammonium Thiocyanate. 9 pp.
ISO 3622-76. Photographic Grade Ammonium Thiocyanate—Specification First Edition. 7 pp.
JIS K 7731-87. Photographic Grade Ammonium Thiocyanate.

Ammonium Thioglycolate
See Also: Hazardous Materials

—**Highway Transportation—Emergency Procedures**
SAA AS 1678.6.1. 014-84. Emergency Procedure Guides—Transport—Part 6.1.014: Ammonium Thioglycolate.

Ammonium Thiosulfate
Use For: Ammonium Thiosulphate

—**Photographic Chemicals**
BSI BS 3310-78. 1978 Photographic Grade Ammonium Thiosulphate Solution. 9 pp.
ISO 3619-76. Photographic Grade Ammonium Thiosulphate Solution—Specification First Edition. 7 pp.
JIS K 7721-86. Photographic Grade Ammonium Thiosulphate Solution.

Ammonium Thiosulphate
Use: Ammonium Thiosulfate

Ammonium Vanadate
See Also: Ammonium Compounds
CNS K7066-62. Chemical Reagent (Ammonium Vanadate) (Ammonium Metavanadate) (Jul)(1566).

Ammonium Vanadate (V)
JIS K 8747-78. Ammonium Vanadate (V).

Ammunition
See Also: Armor Piercing Ammunition; Bombs (Ordnance); Bullet Jackets; Cartridges (Explosives); Gilding Metals; Grenades; Knebworth Corvus Chaff System; Munitions; Ordnance; Projectiles; Shells (Ammunition); Small Arms Ammunition; Smoke Ammunition
MOD UK DSTAN 13-96: Part 2-01. Lotting and Batching of Ammunition Part 2: Land Service Ammunition Issue 2; Amendment 1. 64 pp.

—**Adhesives**
MOD UK DEF-71. Adhesive, Non-Corrosive and Waterproof (Superseded by Def Stan 80-119).

—**Baking Paints**
MOD UK DSTAN 80-174-93. Paint System Ammunition, Matt, Stoving Priming, Ammunition, Exterior, Stoving Finishing, Ammunition, Exterior Matt Stoving Type: Spraying Issue 1. 20 pp.

—**Baking Primers**
MOD UK DSTAN 80-174-93. Paint System Ammunition, Matt, Stoving Priming, Ammunition, Exterior, Stoving Finishing, Ammunition, Exterior Matt Stoving Type: Spraying Issue 1. 20 pp.

—**Bandoleers—Fabrics**
MOD UK DSTAN 83-62-01. Cloth, Drill, Cotton, Khaki, for Bandoliers Issue 1; Amendment 1. 15 pp.

Ammunition (Cont.)
—**Bituminous Coatings**
MOD UK CS 3004. Bitumen, Oxidised No. 2 (Superseded by Def Stan 68-119).

—**Calibers**
MOD UK DSTAN 05-15-01. Metric System Measurement of Weapon Calibres and Ammunition Issue 1; Amendment 1. 4 pp.

—**Capacitor Papers**
MOD UK DEF-49-A-67. Paper, Tissue, Condenser. 4 pp.

—**Cartridge Cases—Brasses**
MOD UK DSTAN 95-9-88. Cartridge Brass for Ammunition Components Issue 2. 14 pp.

—**Chargers—NATO 7.62mm**
NATO STANAG 2318 ED 2 AMD 0-70. Charger for NATO 7.62mm Ammunition. 7 pp.

—**Chipboard**
MOD UK DSTAN 81-40-80. Chipboard, Lined, and Chipboard, Unlined Issue 1. 10 pp.
MOD UK DEF-47-A-01. Paper, Chip, Patch and/or Bitumen Sized; Amendment 1. 7 pp.
MOD UK DEF-48-A-01. Board, Pitch and/or Bitumen Sized; Amendment 1. 6 pp.

—**Color Coding**
NATO STANAG 2321 ED 4 AMD 1-87. The NATO Code of Colours for the Indentification of Ammunition (Except Ammunition of a Calibre Below 20mm. 13 pp.
NATO STANAG 2953 ED 1 AMD 0-90. Method of Application of the NATO Code of Colours for the Identification of Ammunition (Except Ammunition of a Calibre Below 20 mm)—AOP-2(A). 4 pp.

—**Corrugated Board—Fire Resistant**
MOD UK TS 10280. Board, Corrugated, (Three Flute, Four Liner) Fire Resistant.

—**Decoy—100mm—Interchangeability**
NATO STANAG 1350 ED 2 AMD 0-92. 100mm Decoy System Ammunition Interchangeable Within NATO Naval Forces. 11 pp.
NATO STANAG 1350 ED 2 AMD 1-92. 100mm Decoy System Ammunition Interchangeable Within NATO Naval Forces. 12 pp.
NATO STANAG 1350 ED 2 AMD 2-92. 100mm Decoy System Ammunition Interchangeable Within NATO Naval Forces. 13 pp.

—**Design—Ballistics**
NATO STANAG 4110 ED 2 AMD 0-91. Definition of Pressure Terms and Their Inter-Relationship for Use in the Design and Proof of Cannons and Ammunition. 18 pp.

—**Design—Chemical Compatibility**
NATO STANAG 4147 ED 1 AMD 0-82. Chemical Compatibility of Ammunition Components with Explosives and Propellants (Non Nuclear Applications). 19 pp.
NATO STANAG 4147 ED 1 AMD 0-92. Chemical Compatibility of Ammunition Components with Explosives and Propellants (Non Nuclear Applications). 17 pp.

—**Fabrics—Cotton**
MOD UK DSTAN 83-36-79. Cloths, Plain Woven, Cotton, Type QX Issue 2. 10 pp.

—**Fabrics—Cotton/Polyester**
MOD UK DSTAN 83-43-01. Cloths, Plain Weave, Polyester and Cotton, Type QX Issue 2; Amendment 1. 15 pp.

—**Fabrics—Silk**
MOD UK DSTAN 83-45-01. Cloths, Silk, Type QX Issue 2; Amendment 1. 20 pp.

—**Fabrics—Wool**
MOD UK DSTAN 83-42-79. Cloths, Wool, Type QX Issue 2. 10 pp.
MOD UK DSTAN 83-44-80. Cloths, Melton, Wool, Type QX Issue 2. 12 pp.

—**Failure (Quality Control)**
NATO STANAG 2940 ED 2 AMD 2-85. Reporting of Major Ammunition Malfunctions. 14 pp.
NATO STANAG 2940 ED 2 AMD 3-85. Reporting of Major Ammunition Malfunctions. 16 pp.

—**Fiberboard**
MOD UK DEF-127-A-01. Board for Discs or Wads; Amendment 1. 5 pp.

—**Fiberboard—Glazed**
MOD UK DEF-43-A-67. Glazed Board and Glazed Board, Lead Free. 5 pp.

INTERNATIONAL AND NON-U.S. NATIONAL STANDARDS
SUBJECT INDEX
Ammunition

Ammunition *(Cont.)*

—**Identification Systems**
NATO STANAG 2316 ED 3 AMD 4-80. Marking of Ammunition (and its Packaging) of a Calibre Below 20 mm. 12 pp.
NATO STANAG 2316 ED 3 AMD 5-80. Marking of Ammunition (and Its Packaging) of a Calibre Below 20 mm. 12 pp.
NATO STANAG 2322 ED 2 AMD 4-80. Minimum Markings for the Indentification of Ammunition (and Its Packing). 13 pp.
NATO STANAG 2322 ED 2 AMD 5-80. Minimum Markings for the Indentification of Ammunition (and Its Packing). 14 pp.
NATO STANAG 2322 ED 2 AMD 6-80. Minimum Markings for the Indentification of Ammunition (and Its Packing). 13 pp.

—**Indirect Fire—Interchangeability**
NATO STANAG 4425 ED 1 AMD 0-00. A Procedure to Determine the Degree of Interchangeability of NATO Indirect Fire Ammunition. 29 pp.

—**Interchangeability**
NATO STANAG 2968 ED 1 AMD 0-90. Land Forces Catalogue of Bilateral Agreements on the Interchangeability and Safe Firing of Ammunition During Peacetime Training Exercises—AOP-14. 5 pp.

—**Interchangeability—Catalogs**
NATO STANAG 2928 ED 2 AMD 3-81. Land Forces Ammunition Interchangeability Catalogue in Wartime—AOP-6. 9 pp.

—**Interchangeability—Naval**
NATO STANAG 1346 ED 1 AMD 0-88. L-5 Mod 4 Torpedo Interchangeable Within NATO Naval Forces. 10 pp.
NATO STANAG 1346 ED 1 AMD 2-88. L-5 Mod 4 Torpedo Interchangeable Within NATO Naval Forces. 10 pp.
NATO STANAG 1359 ED 1 AMD 2-90. 20 mm F2 Gun Ammunition Interchangeable Within NATO Naval Forces. 28 pp.
NATO STANAG 1360 ED 2 AMD 0-91. 30mm Goalkeeper Ammunition Interchangeable Within NATO Naval Forces. 14 pp.
NATO STANAG 1386 ED 1 AMD 1-90. Breda Rockets Interchangeable Within NATO Naval Forces. 12 pp.

—**Interchangeability—Symbols**
NATO STANAG 2315 ED 3 AMD 4-80. Symbol to Denote the Operational Interchangeability of Ammunition and Demolition Accessories (Short Title: NATO Symbol of Inter-changeability). 9 pp.

—**Kraft Papers**
MOD UK DEF-99-A-64. Paper Kraft, Polythene Coated (Heat Sealable) (Reprinted January 1974, Incorporating Amendment No. 1). 6 pp.
MOD UK DEF-135-01. Paper, Kraft, Dry Paraffin-Waxed; Amendment 1. 5 pp.
MOD UK DEF-137-01. Paper, Laminated (Kraft-Polythene-Kraft) (Reprinted January 1963, Incorporating Correction); Amendment 1. 7 pp.
MOD UK TS 10084. Paper, Kraft Bleached (Withdrawn).

—**Lacquers**
MOD UK DSTAN 80-45-85. Lacquer, Shellac/Castor Oil Type QX Types: Brushing Dipping Issue 2. 11 pp.

—**Lacquers (Cellulose Nitrate)**
MOD UK TS 10226. Lacquer, Cellulose, Nitrate, Type QX (Withdrawn).

—**Lacquers (Ethyl Cellulose)**
MOD UK TS 492A. Lacquer, Ethyl Cellulose, LF Quality (Withdrawn).

—**Leather—Buffings**
MOD UK DSTAN 83-78-87. Leather Buffings, Vegetable-Tanned Issue 1. 8 pp.

—**Leatherboards**
MOD UK DEF-41-A-67. Imitation Leatherboard. 4 pp.

—**Lotting/Batching**
MOD UK DSTAN 13-96: Part 1-87. Lotting and Batching of Ammunition Part 1: General Requirements Issue 2. 11 pp.
MOD UK DSTAN 13-96: Part 2-90. Lotting and Batching of Ammunition Part 2: Land Service Ammunition Issue 2. 63 pp.

—**Lutes**
MOD UK DSTAN 68-173-93. Luting (Thick) MK 4 Luting (Thin) MK 5 Issue 1. 11 pp.
MOD UK CS 2677G. Luting Mark 8, LF Quality (Composition RD 1284) (Withdrawn).

Ammunition *(Cont.)*

—**Marking Inks**
MOD UK DSTAN 80-171-93. Ink, Liquid, Marking, Quick Drying Issue 1. 15 pp.
MOD UK TS 10157. Ink, Liquid, Marking, Quick Drying, LF Quality (Superseded by Def Stan 80-171).

—**Millboards**
MOD UK DEF-40-A-67. Millboard and Millboard, Lead Free. 5 pp.

—**NATO Design Mark**
NATO STANAG 2320 ED 3 AMD 4-80. NATO Design Mark. 6 pp.

—**Naval Operations—Interchangeability**
NATO STANAG 1245 ED 2 AMD 1-90. 40 MM/L60 Gun Ammunition Interchangeable Within NATO Naval Forces. 38 pp.

—**Naval Operations—Safety**
NATO STANAG 4224 ED 1 AMD 0-90. Safety and Suitability for Service Testing of Artillery and Naval Gun Ammunition—76 mm and Greater. 37 pp.

—**Netting—Polyester**
MOD UK DSTAN 83-71-01. Cloth, Netting, Polyester, Type QX Issue 1; Amendment 1. 13 pp.

—**Packaging Materials—Identification Systems**
NATO STANAG 2316 ED 3 AMD 4-80. Marking of Ammunition (and its Packaging) of a Calibre Below 20 mm. 12 pp.
NATO STANAG 2316 ED 3 AMD 5-80. Marking of Ammunition (and Its Packaging) of a Calibre Below 20 mm. 12 pp.
NATO STANAG 2322 ED 2 AMD 4-80. Minimum Markings for the Indentification of Ammunition (and Its Packing). 13 pp.
NATO STANAG 2322 ED 2 AMD 5-80. Minimum Markings for the Indentification of Ammunition (and Its Packing). 14 pp.
NATO STANAG 2322 ED 2 AMD 6-80. Minimum Markings for the Indentification of Ammunition (and Its Packing). 13 pp.

—**Packaging Papers**
MOD UK DEF-99-A-64. Paper Kraft, Polythene Coated (Heat Sealable) (Reprinted January 1974, Incorporating Amendment No. 1). 6 pp.
MOD UK DEF-135-01. Paper, Kraft, Dry Paraffin-Waxed; Amendment 1. 5 pp.
MOD UK DEF-137-01. Paper, Laminated (Kraft-Polythene-Kraft) (Reprinted January 1963, Incorporating Correction); Amendment 1. 7 pp.
MOD UK TS 10084. Paper, Kraft Bleached (Withdrawn).

—**Paints (Alkyd)**
MOD UK DSTAN 80-25-01. Paint, Finishing, Ammunition, Stoving Paint, Priming, Ammunition, Stoving Scheme A: System of Primer and Finish Scheme B: One-Coat Finish Types: Brushing Spraying Electrostatic Spraying Dipping Issue 2; Amendment 3. 15 pp.
MOD UK DSTAN 80-27-01. Paint, Finishing, Ammunition Paint, Priming, Ammunition Scheme A: System of Primer and Finish Scheme B: One-Coat Finish Types: Brushing Spraying Electrostatic Spraying Dipping Issue 2; Amendment 1. 13 pp.
MOD UK TS 10176A. Paint System, Ammunition, Matt, Stoving, LF Quality, Type: Spraying (Superseded by Def Stan 80-174).

—**Paints (Asphalt)—Acid/Alkali Resistant**
CGSB CAN/CGSB-1.108-M89. Bituminous Solvent Type Paint. 10 pp.

—**Paints (Nitrocellulose)**
MOD UK DSTAN 80-55-01. Paint, Finishing, General Service, Semi-Gloss, Cellulose Nitrate Types: Brushing Dipping Spraying Issue 1; Amendment 2. 9 pp.
MOD UK DSTAN 80-55-92. Paint, Finishing, General Service Semi-Gloss, Cellulose Nitrate Types: Brushing Spraying Dipping Issue 2. 17 pp.
MOD UK DSTAN 80-95-84. Paint, Bullet Tipping Issue 1. 11 pp.

—**Paints (Polyurethane)**
MOD UK TS 10243. Paint System for Aromatic Polyamide Fibre and Ammunition Components.

—**Papers**
MOD UK DSTAN 13-16-68. Paper, Kraft, Pure Issue 1. 8 pp.
MOD UK DSTAN 13-32-68. Paper, Pure, Friction Glazed, Lead Free Issue 1. 12 pp.
MOD UK DSTAN 81-97-91. Paper, Pure, Bleached Type QX Issue 1. 14 pp.

Ammunition *(Cont.)*

—**Papers** *(Cont.)*
MOD UK DEF-42-A. Paper, Pure, Bleached, Super-Calendered and Paper, Pure, Bleached, Super-Calendered, Lead Free (Superseded by Def Stan 81-97).
MOD UK DEF-86-A-01. Paper, Pure, Unbleached, Rolled; Amendment 1. 6 pp.

—**Polyester Laminates**
MOD UK TS 448. Laminate, Polyester Film/Glazed Board (Withdrawn).

—**Pressboard**
MOD UK DEF-45-A-67. Paper, Presspahn. 4 pp.

—**Primers (Zinc Chromate)**
MOD UK DSTAN 80-27-01. Paint, Finishing, Ammunition Paint, Priming, Ammunition Scheme A: System of Primer and Finish Scheme B: One-Coat Finish Types: Brushing Spraying Electrostatic Spraying Dipping Issue 2; Amendment 1. 13 pp.

—**Primers (Zinc Phosphate)**
MOD UK DSTAN 80-25-01. Paint, Finishing, Ammunition, Stoving Paint, Priming, Ammunition, Stoving Scheme A: System of Primer and Finish Scheme B: One-Coat Finish Types: Brushing Spraying Electrostatic Spraying Dipping Issue 2; Amendment 3. 15 pp.
MOD UK TS 10176A. Paint System, Ammunition, Matt, Stoving, LF Quality, Type: Spraying (Superseded by Def Stan 80-174).

—**Propellants**
BSI BS EN 268-91. 1991 Propellants for Commercial Ammunition—Requirements and Test Methods. 11 pp.
CEN PREN 268-87. Propellants for Commercial Ammunition; Requirements and Test Methods. 8 pp.
CEN EN 268-91. Propellants for Commercial Ammunition—Requirements and Test Methods. 9 pp.

—**Propellants—Identification Systems**
NATO STANAG 2374 ED 1 AMD 1-90. Method of Application of a Standard Marking System for the Identification of Propelling Charges. 19 pp.

—**Safe Firing**
NATO STANAG 2968 ED 1 AMD 0-90. Land Forces Catalogue of Bilateral Agreements on the Interchangeability and Safe Firing of Ammunition During Peacetime Training Exercises—AOP-14. 5 pp.

—**Shellac**
MOD UK DSTAN 80-44-76. Lacquer, Shellac, 40%, Type QX Lacquer, Shellac, 30%, Type QX Lacquer, Shellac, 20%, Type QX Lacquer, Shellac, 14.5%, Type QX Lacquer, Shellac, 11%, Type QX Lacquer, Shellac 6%, Type QX Issue 1. 8 pp.

—**Shellac Adhesives**
MOD UK DSTAN 80-46-76. Adhesive, Shellac, 50%, Type QX Adhesive, Shellac, 60%, Type QX Issue 1. 6 pp.
MOD UK DSTAN 80-47-76. Adhesive, Shellac, Special, Type QX Issue 1. 6 pp.

—**Soldering Fluxes**
MOD UK DSTAN 34-2-75. Flux, Soldering (Oleic Acid Based Paste) Issue 1. 8 pp.

—**Storage—Classification**
NATO STANAG 4123 ED 2 AMD 4-78. Methods to Determine and Classify the Hazards of Ammunition. 34 pp.
NATO STANAG 4123 ED 3 AMD 0-00. (DRAFT) Determination of the Classification of Military Ammunition and Explosives. 9 pp.

—**Storage—Quality Assurance**
MOD UK DSTAN 05-61: Part 8-01. Quality Assurance Procedural Requirements Part 8: Movement and Storage of Ammunition Under 'Red Card' Issue 1; Amendment 1. 10 pp.
MOD UK DSTAN 05-61: Part 8-92. Quality Assurance Procedural Requirements Part 8: Movement and Storage of Ammunition Under "Red Card" Restriction Issue 2. 15 pp.

—**Tanks (Combat Vehicles)—120 mm—Interoperability**
NATO STANAG 4385 ED 1 AMD 0-00. 120mm x 570 Ammunition for Smooth Bore Tank Guns. 16 pp.
NATO STANAG 4385 ED 1 AMD 0-93. 120mm x 570 Ammunition for Smooth Bore Tank Guns. 15 pp.

—**Textile Tapes**
MOD UK DSTAN 83-54-77. Tape, Textile, Cotton and Webbing, Textile, Cotton, Type QX Issue 1. 16 pp.

Ammunition (Cont.)

—Textile Tapes (Cont.)
MOD UK DSTAN 83-61-01. Tapes, Textile, Plain Weave, Polyester, Type QX Issue 3; Amendment 1. 15 pp.
MOD UK DSTAN 83-67-01. Tapes, Textile, Plain Weave, Polyester and Cotton, Tubular, Type QX Issue 1; Amendment 1. 14 pp.

—Thread (Textiles)
MOD UK DSTAN 83-46-01. Threads, Silk, Sewing, Type QX Issue 2; Amendment 1. 13 pp.
MOD UK DSTAN 83-51-75. Threads, Cotton, Type QX Issue 1. 11 pp.
MOD UK DSTAN 83-60-01. Threads, Polyester Cored, Cotton Sheathed, Type QX Issue 2; Amendment 2. 14 pp.

—Transportation—Classification
NATO STANAG 4123 ED 2 AMD 4-78. Methods to Determine and Classify the Hazards of Ammunition. 34 pp.
NATO STANAG 4123 ED 3 AMD 0-00. (DRAFT) Determination of the Classification of Military Ammunition and Explosives. 9 pp.

—Transportation—Safety
NATO STANAG 2890 ED 2 AMD 2-88. Regulations for Safety in the Transport of Ammunition and Explosives by Military Train. 10 pp.

—Varnishes (Baking)
MOD UK DSTAN 80-175-93. Varnish, Ammunition, Stoving, Type QX Class A Expoxy Ester B Silicone Alkyd C Tung Oil Phenolic Types: Brushing Spraying Dipping Issue 1. 16 pp.

—Vulcanized Rubber—Conductive
MOD UK TS 10057. Vulcanized Natural Rubber, Black Type QX, Electrically Conducting, Hardness 60 Degrees (Superseded by Def Stan 93-65).

—Wool Felts
CGSB 4-GP-34MA-89. Felt, Ammunition Filling. 18 pp.
MOD UK DSTAN 83-19: Part 4-79. Cloths, Compressed Felts, Wool, and Blended Wool Part 4: Type QX Issue 1. 15 pp.

—Writing Papers
MOD UK DEF-89-01. Paper, Writing; Amendment 1. 5 pp.

—20 mm—Ground/Air Defence
NATO STANAG 4136 ED 1 AMD 0-76. 20 mm x 139 Ammunition for Ground/Air Defence Weapons (Withdrawn). 17 pp.

—20 mm—Interchangeability—Aircraft—Weapons Systems
NATO STANAG 3585 ED 2 AMD 5-73. 20 mm Ammunition for Aircraft Guns. 11 pp.
NATO STANAG 3585 ED 6-73. AL/4 20mm Ammunition for Aircraft Guns. 12 pp.

—25 mm—Interchangeability
NATO STANAG 4173 ED 1 AMD 1-86. 25mm x 137 Ammunition. 15 pp.
NATO STANAG 4173 ED 1 AMD 2-86. 25mm x 137 Ammunition. 16 pp.

—27 mm x 145—Aircraft Guns
NATO STANAG 3820 ED 1 AMD 2-88. 27 mm X 145 Ammunition and Links for Aircraft Guns. 16 pp.

—7.62 mm—Disintegrating Belts—Links
NATO STANAG 2329 ED 2 AMD 2-82. Links for Disintegrating Belts for Use with NATO 7.62 mm Cartridges. 15 pp.

—7.62 mm—Small Arms
NATO STANAG 2310 ED 3 AMD 4-76. Small Arms Ammunition (7.62 mm.). 18 pp.

Amorphous Materials
Scope Note: Use a more specific term *See:* Glass; Glassy Alloys; Phosphorus; Plastics; Polymers

Amorphous Metals
Use: Glassy Alloys

Ampacity
Use: Current Ratings

Ampere Hour Meters
See Also: Electric Measuring Instruments; Measuring Instruments

—Call Duration
CCITT RECMN E.261-89. Devices for Measuring and Recording Call Durations—Telephone Network and ISDN—Operation, Numbering, Routing and Mobile Service (Study Group II) 3 pp. 3 pp.

Ampere Meters
Use: Ammeters

Amperometric Titration
See Also: Electrometric Analysis; Volumetric Analysis
CNS K0029-87. General Rules for Method of Potentiometric Amperometric and Coulometric Titrations (Oct)(12128).
JIS K 0113-90. General Rules for Methods of Potentiometric, Amperometric, Coulometric, and Karl-Fischer Titrations. 25 pp.

—Aircraft Fuels
BSI BS 2000: Part 342-93. 1993 Methods of Test for Petroleum and Its Products Part 342: Gasoline, Kerosine and Distillate Fuels—Determination of Mercaptan Sulfur—Potentiometric Method (ISO 3012: 1991) (W). 7 pp.
CNS K6545-80. Method of Test for Mercaptan Sulfur in Aviation Turbine Fuels (Amperometric Method) (Aug)(6190).
ISO 3012-91. Gasoline, Kerosene and Distillate Fuels—Determination of Mercaptan Sulfur—Potentiometric Method Second Edition. 8 pp.

—Nitric Acid
ISO 8298-87. Determination of Plutonium in Nitric Acid Solutions—Method by Oxidation by Cerium (IV), Reduction by Iron (II) Ammonium Sulfate and Amperometric Back-Titration with Potassium Dichromate First Edition. 10 pp.

Amperometry

—Sludge (Sewage)
DIN ENGL 38408 Pt 22 (S)-86. German Standard Methods for the Examination of Water, Waste Water and Sludge; Gaseous Constituents (Group G); Determination of Dissolved Oxygen Using the Membrane Oxygen Probe (G 22) (Nov) Superseded by EN 25814, November 1992 Edition. 8 pp.

—Waste Water
DIN ENGL 38408 Pt 22 (S)-86. German Standard Methods for the Examination of Water, Waste Water and Sludge; Gaseous Constituents (Group G); Determination of Dissolved Oxygen Using the Membrane Oxygen Probe (G 22) (Nov) Superseded by EN 25814, November 1992 Edition. 8 pp.
DIN ENGL 38408 Pt 23-87. German Standard Methods for the Examination of Water, Waste Water and Sludge; Gaseous Constituents (Group G); Determination of Oxygen Saturation Index (G 23) (Nov). 5 pp.
IEC 746 Pt 4-92. Expression of Performance of Electrochemical Analyzers Part 4: Dissolved Oxygen in Water Measured by Membrane Covered Amperometric Sensors First Edition. 59 pp.

—Water
BSI BS 2690: Part 101-84. (WITHDRAWN) 1984 Water Used in Industry Part 101: Dissolved Oxygen (Superseded by BS EN 25813). 10 pp.
DIN ENGL 38408 Pt 22 (S)-86. German Standard Methods for the Examination of Water, Waste Water and Sludge; Gaseous Constituents (Group G); Determination of Dissolved Oxygen Using the Membrane Oxygen Probe (G 22) (Nov) Superseded by EN 25814, November 1992 Edition. 8 pp.
DIN ENGL 38408 Pt 23-87. German Standard Methods for the Examination of Water, Waste Water and Sludge; Gaseous Constituents (Group G); Determination of Oxygen Saturation Index (G 23) (Nov). 5 pp.
IEC 746 Pt 4-92. Expression of Performance of Electrochemical Analyzers Part 4: Dissolved Oxygen in Water Measured by Membrane Covered Amperometric Sensors First Edition. 59 pp.

Amphibious Charts
See Also: Amphibious Operations; Charts; Combat Charts
NATO STANAG 1022 ED 5 AMD 8-78. Combat Charts, Amphibious Charts and Combat/Landing Charts. 24 pp.
NATO STANAG 1022 ED 5 AMD 9-78. Combat Charts, Amhibious Charts and Combat/Landing Charts. 29 pp.
NATO STANAG 1022 ED 5 AMD 10-78. Combat Charts, Amphibious Charts and Combat/Landing Charts. 31 pp.
NATO STANAG 1022 ED 5 AMD 11-78. Combat Charts, Amphibious Charts and Combat/Landing Charts. 32 pp.

Amphibious Operations
See Also: Amphibious Charts; Combat Charts; Naval Gunfire Support; Tactical Air Support; Warfare
NATO STANAG 1149 ED 14 AMD 0-91. Doctrine for Amphibious Operations—ATP-8(A). 5 pp.

—Command and Control
NATO STANAG 1356 ED 1 AMD 1-87. Guidance for the Command and Control/Coordination of Forces in the Amphibious Objective Area (AOA) Not Assigned to the Commander Amphibious Task Force (CATF) (Withdrawn). 10 pp.

—Command and Control—Termination
NATO STANAG 1303 ED 1 AMD 2-86. Guidance for Transitional Command and Control Procedures for Termination of Amphibious Operations. 12 pp.

—Embarkation
NATO STANAG 1195 ED 2 AMD 0-89. Amphibious Embarkation—ATP-39(A). 5 pp.
NATO STANAG 1195 ED 3 AMD 0-92. Amphibious Embarkation—ATP-39(A). 4 pp.

—Reconnaissance
NATO STANAG 1191 ED 2 AMD 1-87. Amphibious Reconnaissance—ATP-38. 5 pp.

—Ship to Shore
NATO STANAG 1180 ED 7 AMD 0-91. Amphibious Operations—Ship-to-Shore Movement—ATP-36. 5 pp.

—Supporting Arms
NATO STANAG 1181 ED 5 AMD 0-00. Supporting Arms in Amphibious Operations—ATP-37. 5 pp.
NATO STANAG 1181 ED 7 AMD 0-89. Supporting Arms in Amphibious Operations—ATP-37. 5 pp.
NATO STANAG 1181 ED 8 AMD 0-92. Supporting Arms in Amphibious Operations—ATP-37. 5 pp.

Amphibious Vehicles
See Also: Military Vehicles

—Paints
MOD UK DSTAN 80-163-93. Paint System Defence Equipment Heavy Duty (Two-Pack) Paint, Priming, Defence Equipment Heavy Duty, Epoxide Paint, Undercoat, Defence Equipment Epoxide/Pitch Paint, Finishing, Defence Equipment Heavy Duty, Epoxide Issue 1. 30 pp.

Amplification (Gain)
Use: Gain (Amplification)

Amplifier Klystrons
Use: Klystron Amplifiers

Amplifiers
Scope Note: For additional listings, use a more specific term *See Also:* Audio Amplifiers; Broadband Amplifiers; Cable Television Amplifiers; Circuits; DC Amplifiers; Differential Amplifiers; Differential Operational Amplifiers; Klystron Amplifiers; Line Amplifiers; Linear Amplifiers; Low Noise Amplifiers; Magnetic Amplifiers; Microwave Amplifiers; Operational Amplifiers; Parametric Amplifiers; Power Amplifiers; Preamplifiers; Pulse Amplifiers; Radio Frequency Amplifiers; Sense Amplifiers; Temperature Transmitters; Traveling Wave Amplifiers; Video Amplifiers; Wideband Amplifiers
BSI BS 6493: Sec 2.3-01. 1987 Amd 1 Semiconductor Devices Part 2: Integrated Circuits Section 2.3: Recommendations for Analogue Integrated Circuits (IEC 748-3: 1986) (AMD 7035) July 15, 1992. 190 pp.
CNS C6274-87. Method of Test for Amplifiers (Jan)(11812).
CSA 1423A Bull. Electrical Bulletin 1423A January 2, 1985 to C22.2 NO 1. 8 pp.
CSA 1434 Bull. Electrical Bulletin 1434 April 1, 1985 to C22.2 NO 1. 2 pp.
IEC 748 Pt 3-86. Semiconductor Devices Integrated Circuits Part 3: Analogue Integrated Circuits First Edition; (Amendment 1-1991). 334 pp.
MOD UK DSTAN 58-95: Part 2-76. Electronic Assemblies Part 2: Amplifier Assemblies Issue 1. 40 pp.

—Bipolar Transistors
BSI BS E9372-76. 1976 Amd 1 Blank Detail Specification: Ambient-Rated Bipolar-Transistors for Low and High Frequency Amplification. 16 pp.
BSI BS E9373-76. (WITHDRAWN) 1976 Amd 1 Blank Detail Specification: Case-Rated Bipolar Transistors for Low Frequency Amplification (Renumbered as BS EN 150003: 1993). 10 pp.
BSI BS E9377-76. (WITHDRAWN) 1976 Amd 1 Blank Detail Specification: Case-Rated Bipolar Transistors for High Frequency Amplification (Renumbered as BS EN 150007: 1993). 17 pp.
IEC 747 Pt 7-1-89. Semiconductor Devices Discrete Devices Part 7: Bipolar Transistors Section One—Blank Detail Specification for Ambient-Rated Bipolar Transistors for Low and High-Frequency Amplification First Edition (IECQ QC 750102). 35 pp.

INTERNATIONAL AND NON-U.S. NATIONAL STANDARDS
SUBJECT INDEX

Amplifiers (Cont.)
—Bipolar Transistors (Cont.)
IEC 747 Pt 7-2-89. Semiconductor Devices Discrete Devices Part 7: Bipolar Transistors Section Two—Blank Detail Specification for Case-Rated Bipolar Transistors for Low-Frequency Amplification First Edition (IECQ QC 750103). 31 pp.

IECQ QC 750102-89. Semiconductor Devices Discrete Devices Part 7: Bipolar Transistors Section One—Blank Detail Specification for Ambient-Rated Bipolar Transistors for Low and High-Frequency Amplification (IEC 747-7-1 ED 1). 35 pp.

IECQ QC 750102/CN 0001-88. Detail Specification for Electronic Components Bipolar Transistors for Ambient-Rated High-Frequency Amplification of Type 3DG130 Assessment Level II. 18 pp.

IECQ QC 750102/CN 0002-88. Detail Specification for Electronic Components Bipolar Transistors for Ambient-Rated, Forward A.G.C. Low-Noise, High-Frequency Amplification of Type 3DG79 Assessment Level II. 17 pp.

IECQ QC 750102/CN 0003-90. Detail Specification for Electronic Components Bipolar Transistors for Ambient-Rated High-Frequency Amplification of Type 3DG1815 Assessment Level II. 15 pp.

IECQ QC 750103-89. Semiconductor Devices Discrete Devices Part 7: Bipolar Transistors Section Two—Blank Detail Specification for Case-Rated Bipolar Transistors for Low-Frequency Amplification (IEC 747-7-2 ED 1). 31 pp.

IECQ QC 750103/CN 0001-92. Detail Specification for Electronic Components Case-Rated Bipolar Transistors for Low-Frequency Amplification Type 3DD870. 12 pp.

—Bipolar Transistors—Quality Assurance
BSI BS EN 150007-93. 1993 Amd 2 Blank Detail Specification: Case-Rated Bipolar Transistors for High Frequency Amplification (AMD 7598) February 1993 (T). 21 pp.

CENELEC EN 150 007-91. Blank Detail Specification: Case-Rated Bipolar Transistors for High Frequency Amplification. 11 pp.

—Integrated Circuits—Quality Assurance
BSI BS 9460-74. (OBSOLESCENT) 1974 Amd 2 Rules for the Preparation of Detail Specifications for Integrated Circuits of Assessed Quality: Differential Operational Amplifiers. General Application Category. 15 pp.

—Power Systems—Safety
IEC 65-85. Safety Requirements for Mains Operated Electronic and Related Apparatus for Household and Similar General Use Fifth Edition; (Amendment 2-1989, Incorporating Amendment 1) (Amendment 3-1992). 210 pp.

—Semiconductor Detectors
CENELEC HD 404-80. Test Procedures for Amplifiers and Preamplifiers for Semiconductor Detectors for Ionizing Radiation. 2 pp.

Amplitude Hits
See Also: Phase Hits

—Telephone Circuits
CCITT RECMN O.95-89. Phase and Amplitude Hit Counters for Telephone-Type Circuits—Specifications for Measuring Equipment (Study Group IV) 4 pp. 4 pp.

Amplitude Modulated Voice Frequency Telegraphy (AMVFT)
Use: Voice Frequency Telegraphy—Amplitude Modulated

Amplitude Response
See Also: Electrical Measurement

—Fiber Optic Cables
CCITT RECMN G.651-89. Characteristics of a 50/125 Micrometer Multimode Graded Index Optical Fibre Cable—Transmission Media Characteristics (Study Group XV) 30 pp. 30 pp.

Ampoules
Use: Ampules

Amprolium Content Analysis
—Animal Feed
CNS N4075-81. Method Test for Feed Additives: Determination of Amprolium (Aug) (7838). 2 pp.

Ampules
See Also: Injection Equipment (Medical)
BSI BS 795-83. 1983 Amd 1 Ampoules. 25 pp.
CNS R2057-80. Ampoules (Sep)(3062).
JIS R 3512-74. Ampoule.

Ampules (Cont.)
—Injection Equipment (Medical)
ISO 9187 Pt 1-91. Injection Equipment for Medical Use—Part 1: Ampoules for Injectables First Edition. 10 pp.

—One Point Cut—Injection Equipment (Medical)
ISO 9187 Pt 2-93. Injection Equipment for Medical Use—Part 2: One-Point-Cut (OPC) Ampoules First Edition. 8 pp.

—Tubes
CNS R2056-80. Ampoule Pipe (Sep)(3061).
JIS R 3511-74. Ampoule Pipe.

Ampyrone
Use For: 4-Aminoantipyrine
CNS K7558-88. Chemical Reagent (4-Aminoantipyrine) (Oct)(7716).
JIS K 8048-79. 4-Aminoantipyrine.

Amsonic Acid
Use For: 4,4'-Diamino-2,2-stilbenedisulfonic Acid; Diaminostilbenedisulfonic Acid; 4,4-Diaminostilbenedisulfonic Acid
CNS K1127-67. Diaminostilbene Disulfonic Acid (Aug)(2791)(R 1973).
CNS K2178-87. 4,4'-Diaminostilbene-2,2'-Disulfonic Acid (Dec)(12188). 5 pp.
CNS K6264-67. Method of Test for Diaminostilbene Disulfonic Acid (Aug)(2792)(R 1973).
JIS K 4158-88. 4,4'-Diaminostilbene-2,2'-Disulfonic Acid. 11 pp.

Amusement Rides
See Also: Ferris Wheels; Roller Coasters

—Inspection
CNS A3303-88. Inspection of Amusements (Airplane Tower) (Dec)(12472).
JIS A 1715-78. Inspection Standard of Amusements (Airplane Tower).
JIS A 1717-78. Inspection Standard of Amusements (Merry-Go-Round).

—Model Bylaws
SNZ NZS 9201: Chapter 10-72. Model General Bylaws Chapter 10: Amusement Devices and Shooting Galleries (Reconfirmed 1980). 11 pp.

—Safety
CSA Z267-M1983. Safety Code for Amusement Rides; (Gen Instr 1). 29 pp.

—Water Shoots
CEN PREN 1069-1-93. Water Slides over 2 m Height—Part 1: Specifications and Test Methods. 39 pp.
CEN PREN 1069-2-93. Water Slides over 2 m Height—Part 2: Instructions. 46 pp.
SNZ NZS 5838-86. Code of Practice for the Design, Contruction, Maintenance and Operation of Waterslides. 14 pp.

—Water Shoots—Inspection
CNS A3163-88. Inspection Standard of Amusements (Water Shoot) (Dec)(8916).
JIS A 1716-78. Inspection Standard of Amusements (Water Shoot).

Amyl Acetate
Use For: Amylacetic Ester; Banana Oil; Pear Oil; Pentyl Acetate See Also: Acetates; Hazardous Materials; Solvents
BSI BS 552-70. 1970 Amyl Acetate. 12 pp.
BSI BS 552-01. 1970 Amd 1 Amyl Acetate (AMD 7218) September 15, 1992. 14 pp.
CNS K1109-78. Amyl Acetate (98% Percent Grade) (May)(2556).
CNS K6228-66. Method of Test for Amyl Acetate of Industrial Grade (Sep)(2557)(R 1971).
JIS K 1515-57. Amyl Acetate.

—Storage and Handling—Information Cards
SAA AS 2508.3.00 2-82. Safe Storage and Handling Information Cards for Hazardous Materials—Part 3.002: Acetates (Ethyl Acetate, Propyl Acetates, Butyl Acetates, Amyl Acetate) Double Sided Card.

Amyl Alcohols
See Also: Alcohols; Methylamyl Alcohol
CNS K1225-80. Amyl Alcohol (Feb)(5231).

—Safety
SAA AS 2508.3.00 6-91. Safe Storage and Handling Information Cards for Hazardous Materials—Part 3.006: Butanols, Propanols, Amyl Alcohols (In Professional Package 38) Double Sided Card.

Amyl Ethyl Ketone
See Also: Ketones

Amyl Ethyl Ketone (Cont.)
—Alcohol Content—Volumetric Analysis
ISO 2501-74. Methyl Ethyl Ketone, isoButyl Methyl Ketone and isoAmyl Ethyl Ketone for Industrial Use—Determination of Alcoholic Impurities—Volumetric Method First Edition. 5 pp.

Amyl Nitrite
Use For: Isoamyl Nitrite; 3-Methylbutyl Nitrite
CNS K7572-88. Chemical Reagent (Isomyl Nitrite, 3-Methylbutyl Nitrite) (Oct)(7805).

Amylacetic Ester
Use: Amyl Acetate

Amylase
JIS K 7001-90. Amylase for Industrial Use. 35 pp.

Amylase Activity
See Also: Enzymatic Activity

—Cereal Products—Colorimetric Analysis
BSI BS 4317: Part 14-80. 1980 Methods of Test for Cereals and Pulses Part 14: Determination of Alpha-Amylase Activity in Cereals and Cereal Products (Colorimetric Method). 8 pp.
ISO 3983-77. Cereals and Cereal Products—Determination of Alpha-Amylase Activity—Colorimetric Method First Edition. 7 pp.

—Cereals—Colorimetric Analysis
BSI BS 4317: Part 14-80. 1980 Methods of Test for Cereals and Pulses Part 14: Determination of Alpha-Amylase Activity in Cereals and Cereal Products (Colorimetric Method). 8 pp.
ISO 3983-77. Cereals and Cereal Products—Determination of Alpha-Amylase Activity—Colorimetric Method First Edition. 7 pp.

—Cereals—Falling Number
BSI BS 4317: Part 9-82. 1982 Methods of Test for Cereals and Pulses Part 9: Determination of Falling Number of Cereals. 8 pp.
ISO 3093-82. Cereals—Determination of Falling Number Second Edition. 7 pp.

Amylographs
See Also: Viscosity

—Flours (Food)
BSI BS 4317: Part 28-93. 1993 Methods of Test for Cereals and Pulses Part 28: Determination of Viscosity of Flour Using an Amylograph (ISO 7973: 1992). 12 pp.
ISO 7973-92. Cereals and Milled Cereal Products—Determination of the Viscosity of Flour—Method Using an Amylograph First Edition. 11 pp.

Amylose Content Analysis
—Rice—Spectrometry
ISO 6647-87. Rice—Determination of Amylose Content First Edition. 7 pp.

Amyris Oil
See Also: Oils
CNS K5122-81. Oil of Amyris (Nov)(8134).

AN
Use: Access Networks

Anaerobic Adhesives
See Also: Adhesives; Anaerobic Polymers

—Aerospace—Adhesive Strength—Threaded Fasteners
CEN PREN 3794-90. Anaerobic Polymerisable Compounds Test Method Determination of Torque Strength on Threaded Fasteners. 6 pp.

—Aerospace—Cure Rate
CEN PREN 3795-90. Anaerobic Polymerisable Compounds Test Method Determination of Freedom from Excessive Cure Rate. 5 pp.

—Aerospace—Setting Rate
CEN PREN 3796-90. Anaerobic Polymerisable Compounds Test Method Determination of Ability of Anaerobic Adhesives to Set on Metal Surfaces. 6 pp.

—Aerospace—Shear Strength
CEN PREN 3793-90. Anaerobic Polymerisable Compounds Test Method Determination of Static Shear Strength. 15 pp.

—Quality Control
ISO 10123-90. Adhesives—Determination of Shear Strength of Anaerobic Adhesives Using Pin-and-Collar Specimens First Edition. 7 pp.

Anaerobic Adhesives *(Cont.)*
—**Shear Strength**
  ISO 10123-90. Adhesives—Determination of Shear Strength of Anaerobic Adhesives Using Pin-and-Collar Specimens First Edition. 7 pp.
—**Threaded Fasteners—Torque**
  ISO 10964-93. Adhesives—Determination of Torque Strength of Anaerobic Adhesives on Threaded Fasteners First Edition. 7 pp.

Anaerobic Compounds
Use: Anaerobic Polymers

Anaerobic Digestion
—**Sediments**
  DIN ENGL 38414 Pt 8-85. German Standard Methods for the Examination of Water, Waste Water and Sludge; Sludge and Sediments (Group S); Determination of the Amenability to Anaerobic Digestion (S 8) (June). 6 pp.
—**Sludge (Sewage)**
  DIN ENGL 38414 Pt 8-85. German Standard Methods for the Examination of Water, Waste Water and Sludge; Sludge and Sediments (Group S); Determination of the Amenability to Anaerobic Digestion (S 8) (June). 6 pp.

Anaerobic Polymers
Use For: Anaerobic Compounds
See Also: Adhesives; Anaerobic Adhesives
  CEN PREN 751-92. Hardening, Sealing Materials for Thread Joints in Contact with Combustible Gases. 12 pp.
  MOD UK DTD-5629-01. General Purpose Locking, Sealing and Retaining Materials: Anaerobic Polymerisable Compounds Giving Very Low Torque Strengths; Amendment 2. 7 pp.
  MOD UK DTD-5630-01. General Purpose Locking, Sealing and Retaining Materials: Anaerobic Polymerisable Compounds Giving Low Torque Strengths; Amendment 2. 7 pp.
  MOD UK DTD-5631-01. General Purpose Locking, Sealing and Retaining Materials: Anaerobic Polymerisable Compounds Giving Medium Torque Strengths; Amendment 2. 7 pp.
  MOD UK DTD-5632-01. General Purpose Locking, Sealing and Retaining Materials: Anaerobic Polymerisable Compounds Giving High Torque Strengths; Amendment 2. 7 pp.
  MOD UK DTD-5633-01. General Purpose Locking, Sealing and Retaining Materials: Anaerobic Polymerisable Compounds Giving Very High Torque Strengths; Amendment 2. 7 pp.
—**Aerospace**
  CEN PREN 3792-90. Anaerobic Polymerisable Compounds Technical Specification. 12 pp.
  CEN PREN 3797-90. Anaerobic Polymerisable Compounds Torque Strength 2Nm Viscosity 500 mm2s-1. 5 pp.
  CEN PREN 3798-90. Anaerobic Polymerisable Compounds Torque Strength 2Nm Viscosity 125 mm2s-1. 4 pp.
  CEN PREN 3799-90. Anaerobic Polymerisable Compounds Torque Strength 5Nm Viscosity 125 mm2s-1. 5 pp.
  CEN PREN 3800-90. Anaerobic Polymerisable Compounds Torque Strength 5Nm Viscosity 500 mm2s-1. 4 pp.
  CEN PREN 3801-90. Anaerobic Polymerisable Compounds Torque Strength 5Nm Thixotropic Index 8. 5 pp.
  CEN PREN 3802-90. Anaerobic Polymerisable Compounds Torque Strength 10Nm Viscosity 125 mm2s-1. 5 pp.
  CEN PREN 3803-90. Anaerobic Polymerisable Compounds Torque Strength 10Nm Viscosity 500 mm2s-1. 4 pp.
  CEN PREN 3804-90. Anaerobic Polymerisable Compounds Torque Strength 10Nm Thixotropic Index 3,5. 5 pp.
  CEN PREN 3805-90. Anaerobic Polymerisable Compounds Torque Strength 10Nm Thixotropic Index 8. 5 pp.
  CEN PREN 3806-90. Anaerobic Polymerisable Compounds Torque Strength 16Nm Viscosity < 50 mm2s-1. 5 pp.
  CEN PREN 3807-90. Anaerobic Polymerisable Compounds Torque Strength 16Nm Viscosity 500 mm2-1. 5 pp.
  CEN PREN 3808-90. Anaerobic Polymerisable Compounds Torque Strength 16Nm Thixotropic Index 8. 5 pp.
  CEN PREN 3809-90. Anaerobic Polymerisable Compounds Torque Strength 19Nm Viscosity 50 mm2s-1. 5 pp.
  CEN PREN 3810-90. Anaerobic Polymerisable Compounds Torque Strength 19Nm Viscosity 125 mm2s-1. 5 pp.
  CEN PREN 3811-90. Anaerobic Polymerisable Compounds Torque Strength 19Nm Viscosity 500 mm2s-1. 5 pp.
  CEN PREN 3812-90. Anaerobic Polymerisable Compounds Torque Strength 19Nm Thixotropic Index 8. 5 pp.
—**Mechanical Properties**
  BSI BS 6956: Part 7-92. 1992 Jointing Materials and Compounds Part 7: Specification for Anaerobic Jointing Compounds for Use with 1st, 2nd and 3rd Family Gases. 10 pp.
—**Procurement**
  MOD UK DTD-5628-01. Test Methods and Procurement Procedures for One Part Anaerobic Polymerisable Compounds for Locking, Sealing and Retaining; Amendment 1. 24 pp.

Anaesthesia Ventilators
Use: Anesthesia Ventilators

Anaesthesiology
Use: Anesthesiology

Anaesthetic Vaporizors
Use: Anesthetic Vaporizers

Anaesthetics
Use: Anesthetics

Analgesic Gas Machines
Use: Anesthesia Gas Machines

Analog Channels
Use: Communication Channels—Analog

Analog Circuits
Scope Note: For additional listings, use a more specific term *Use For:* Analogue Circuits
See Also: Analog Switches; Monostable Multivibrators; Operational Amplifiers; Voltage Regulators
  BSI BS 6493: Sec 2.3-87. 1987 Semiconductor Devices Part 2: Integrated Circuits Section 2.3: Recommendations for Analogue Integrated Circuits. 132 pp.
  BSI BS 6493: Sec 2.3-01. 1987 Amd 1 Semiconductor Devices Part 2: Integrated Circuits Section 2.3: Recommendations for Analogue Integrated Circuits (IEC 748-3: 1986) (AMD 7035) July 15, 1992. 190 pp.
  IEC 748 Pt 3-86. Semiconductor Devices Integrated Circuits Part 3: Analogue Integrated Circuits First Edition; (Amendment 1-1991). 334 pp.
  IECQ QC 790202-91. Semiconductor Devices Integrated Circuits Part 3: Analogue Integrated Circuits Section One—Blank Detail Specification for Monolithic Integrated Operational Amplifiers (IEC 748-3-1 ED 1). 37 pp.
—**Glossaries**
  JIS X 0019-87. Glossary of Terms Used in Information Processing (Analog Computing).
—**Quality Assurance**
  BSI BS 9491-75. 1975 Amd 3 Rules for the Preparation of Detail Specifications for Analogue Integrated Circuits of Assessed Quality: Full Assessment Level. 25 pp.
  BSI BS QC 790202-91. 1991 Harmonized System of Quality Assessment for Electronic Components. Semiconductor Devices. Integrated Circuits Blank Detail Specification Monolithic Integrated Operational Amplifiers (IEC 748-3-1: 1991). 17 pp.
  BSI BS CECC 90201-90. 1990 Integrated Voltage Regulators: Blank Detail Specification. 23 pp.
  BSI BS CECC 90203-85. 1985 Integrated Analogue Switching Circuits: Blank Detail Specification. 17 pp.
  IEC 748 Pt 3-1-91. Semiconductor Devices Integrated Circuits Part 3: Analogue Integrated Circuits Section One—Blank Detail Specification for Monolithic Integrated Operational Amplifiers First Edition (IECQ QC 790202). 37 pp.
—**Reliability Assured**
  JIS C 7410-81. General Rules for Reliability Assured Analogue Semiconductor Integrated Circuits. 16 pp.

Analog Computers
See Also: Computers
—**Glossaries**
  CNS C5157-90. Data Processing Vocabulary (Part 19: Analog Computing) (Nov)(10244).
  JIS X 0019-87. Glossary of Terms Used in Information Processing (Analog Computing).

Analog Data Selector/Multiplexers
Use: Analog Multiplexers

Analog Direct Current Signals
Use: DC Signals

Analog Direct Voltage Signals
Use: Direct Voltage Signals

Analog Electric Measuring Instruments
Use: Electric Measuring Instruments

Analog Elements
Use For: Analogue Elements
—**Symbols—Diagrams**
  BSI BS 3939: Part 13-85. (WITHDRAWN) 1985 Graphical Symbols for Electrical Power, Telecommunications and Electronics Diagrams Part 13: Analogue Elements (Superseded by BS EN 60617-13: 1993). 16 pp.
  BSI BS EN 60617-13-93. 1993 Graphical Symbols for Diagrams Part 13: Analogue Elements (IEC 617-13: 1993). 50 pp.
  CENELEC EN 60617-13-93. Graphical Symbols for Diagrams Part 13: Analogue Elements (IEC 617-13: 1993). 45 pp.
  IEC 617 Pt 13-93. Graphical Symbols for Diagrams Part 13: Analogue Elements Second Edition; (CENELEC EN 60617-13: 1993). 53 pp.
  SAA AS 1102.15-82. Graphical Symbols for Electrotechnology—Part 15: Analogue Elements. 12 pp.
  SNZ IEC 617: Part 13-78. Graphical Symbols for Diagrams Part 13: Analogue Elements. 16 pp.

Analog Links
Use: Data Links

Analog Multiplexers
Use For: Analog Data Selector/Multiplexers; Data Selector/Multiplexers, Analog *See Also:* Multiplexers
—**Preferred Products List**
  CECC CECC MUAHAG Vol 7 IS 8-92. Preferred Products List; Active Microcircuits (En, Fe, Ge). 89 pp.

Analog Ohmmeters
Use: Ohmmeters

Analog Phase Meters
Use: Phase Meters

Analog Pneumatic Signals
Use: Pneumatic Signals

Analog Subscriber Lines
Use: Subscriber Lines

Analog Switches
See Also: Analog Circuits; Electronic Switches; Interface Circuits; Semiconductor Devices; Silicon Bilateral Switches; Silicon Controlled Switches; Switches
  MOD UK DSTAN 59-62: Part 5-83. Microcircuits Electronic (Integrated Circuits) (Listed on EPIC Database) Part 5: Analogue Switch/Multiplexers Issue 1. 16 pp.
—**Preferred Products List**
  CECC CECC MUAHAG Vol 7 IS 8-92. Preferred Products List; Active Microcircuits (En, Fe, Ge). 89 pp.
—**Quad**
  CECC CECC 90 109-814 ISSUE 1-88. Digital Integrated Circuits in Accordance with FS 90 109; 54/74 HC 4016; Quad Bilateral Analog Switches (En, Fr). 13 pp.
—**Quality Assurance**
  BSI BS CECC 90200-88. 1988 Analogue Monolithic Integrated Circuits: Sectional Specification. 63 pp.
—**4 PST**
  CECC CECC 90 104-138 ISSUE 1-81. BS CECC 90 104-138; Silicon Complementary MOS with (B) Buffered Outputs and Cavity Packaging (En). 25 pp.
  CECC CECC 90 104-138 ISSUE 1-86. CEI CECC 90 104-138; Silicon Complementary MOS with (B) Buffered Outputs Cavity and Non Cavity Packaging (En). 2 pp.

Analog Switching Circuits
Use: Circuit Switching Systems—Analog

Analog to Digital Converters
Use For: A/D Converters *See Also:* Digital to Analog Converters; Integrated Circuits; Interface Circuits; Voltage Converters; Voltage to Frequency Converters

INTERNATIONAL AND NON-U.S. NATIONAL STANDARDS
SUBJECT INDEX
Anchor

Analog to Digital Converters (Cont.)
BSI BS 5704-79. 1979 Amd 1 Digital Electronic D.C. Voltmeters and D.C. Electronic Analogue-To-Digital Convertors (AMD 5231) June 30, 1988. 53 pp.
CNS C5056-80. Analog-to-Digital Conversion Equipment (Jul)(5766).
IEC 485-74. Digital Electronic D.C. Voltmeters and D.C. Electronic Analogue-to-Digital Convertors First Edition. 99 pp.

—Gateways—Military Communications
NATO STANAG 4209 ED 1 AMD 2-83. The NATO Multi-Channel Tactical Digital Gateway —Standards for Analogue to Digital Conversion of Speech Signals. 38 pp.

—Integrated Circuit—Binary Output—8-Bit—General Purpose—PPL
CECC CECC MUAHAG Vol 7 IS 8-92. Preferred Products List; Active Microcircuits (En, Fe, Ge). 89 pp.

—Integrated Circuit—Bipolar—6-Bit—Preferred Products List
CECC CECC MUAHAG Vol 7 IS 8-92. Preferred Products List; Active Microcircuits (En, Fe, Ge). 89 pp.

—Integrated Circuit—Video Equipment—8-Bit—Preferred Products List
CECC CECC MUAHAG Vol 7 IS 8-92. Preferred Products List; Active Microcircuits (En, Fe, Ge). 89 pp.

—Integrated Circuit—10-Bit—Preferred Products List
CECC CECC MUAHAG Vol 7 IS 8-92. Preferred Products List; Active Microcircuits (En, Fe, Ge). 89 pp.

—Integrated Circuit—8-Bit—Microprocessor Compatible—PPL
CECC CECC MUAHAG Vol 7 IS 8-92. Preferred Products List; Active Microcircuits (En, Fe, Ge). 89 pp.

—Integrated Circuit—8-Bit—Preferred Products List
CECC CECC MUAHAG Vol 7 IS 8-92. Preferred Products List; Active Microcircuits (En, Fe, Ge). 89 pp.

—Preferred Products List
CECC CECC MUAHAG Vol 7 IS 8-92. Preferred Products List; Active Microcircuits (En, Fe, Ge). 89 pp.

Analog Transmission
—Fixed Satellite Services—Hypothetical Reference Circuits
CCIR RECMN 352-4-82. Hypothetical Reference Circuit for Systems Using Analogue Transmission in the Fixed-Satellite Service—Section 4B2—Performance and Availability. 2 pp.

Analog Voltmeters
Use: Voltmeters—Analog

Analog Wattmeters
Use: Wattmeters

Analogue Circuits
Use: Analog Circuits

Analogue Elements
Use: Analog Elements

Analogue Switches
Use: Analog Switches

Analysers
Use: Analyzers

Analysis Methods
Scope Note: Use a more specific term
See: Absorptiometric Analysis; Acidimetry; Alpha Particle Spectrometry; Amido Black Method; Amperometric Titration; Amperometry; Atomic Absorption Analysis; Atomic Absorption Spectrometry; Atomic Absorption Spectrophotometry; Atomic Absorption Spectroscopy; Atomic Emission Spectrometry; Atomic Emission Spectrophotometry; Atomic Fluorescence Spectrometry; Azeotropic Distillation; Biological Analysis; Calcium Carbonate Saturation Analysis; California Bearing Ratio; Chemical Analysis; Chromatography; Conductimetric Method; Conductometric Analysis; Cost Analysis; Coulometric Analysis; Densitometry; Distillation Methods; Electrochemical Analysis; Electrogravimetric Analysis; Electrolytic Analysis; Electrometric

Analysis Methods (Cont.)
See: (Cont.)
Analysis; Enrichment; Enzyme Assay; Eschka Method; Evaporation Residue Analysis; Fault Tree Analysis; Float and Sink Analysis; Fluorometric Analysis; Gas Analysis; Harmonic Analysis; Hazard Analysis; Hydrometry; Infrared Analysis; Infrared Spectrometry; Infrared Spectrophotometry; Infrared Spectroscopy; Inoculation; Iodometry; Karl Fischer Method; Kjeldahl Method; Leachate Extraction; Manometry; Market Research; Mass Spectrometry; Mercurimetric Analysis; Metallography; Nuclear Magnetic Resonance Spectrometry; Oxidation; Ozonolysis; Paper Chromatography; Permeation Method; Petrography; Polarimetric Analysis; Polarographic Analysis; Potentiometric Analysis; Pycnometric Analysis; Quantitative Analysis; Radioimmunoassay; Reduction (Chemistry); Refractometry; Sedimentation; Sensory Analysis; Sieve Analysis; Soil Analysis; Spectrochemical Analysis; Spectrography; Spectrometry; Spectrophotometry; Spectrum Analysis; Statistical Analysis; Subsoil Analysis; Thermography (Temperature Measurement); Thermomechanical Analysis; Turbidimetry; Ultimate Analysis; Urinalysis; Viscometric Analysis; Volumetric Analysis

Analytical Chemistry
Use: Chemical Analysis

Analyzers
Scope Note: For additional listings, use a more specific term *Use For:* Analysers *See Also:* Carbon Monoxide Analyzers; Chlorine Analyzers; Comparators; Distortion Analyzers; Electrochemical Analyzers; Gas Analyzers; Harmonic Analyzers; Hydrocarbon Analyzers; Infrared Analyzers; Lactate Analyzers; Logic Analyzers; Monitors; Nitrogen Oxide Analyzers; Octave Band Analyzers; Oxygen Analyzers; Process Analyzers; Pulse Height Analyzers; Radiation Measuring Instruments; Radio Frequency Analyzers; Spectrum Analyzers; Sulfur Analyzers; Surface Area Analyzers; Vibration/Sound Analyzers; Water Analysis Equipment

—Automotive Engines
CSA CAN/CSA-C22. 2 NO 68-92. Motor-Operated Appliances (Household and Commercial); (Gen Instr 1 Thru 2). 115 pp.
CSA 1169 Bull. Electrical Bulletin 1169 June 27, 1978 to C22.2 NO 68. 2 pp.

—Olefins
ISO 8917-88. Light Olefins for Industrial Use—Determination of Water—Guidelines for Use of in-Line Analysers First Edition. 10 pp.

Anatomy
—Classification
BSI BS 1000: (6/611)-68. 1968 Amd 2 Universal Decimal Classification (UDC). English Full Edition (61/611): Applied Sciences in General: Anatomy. 26 pp.
SNZ NZS/BS 1000 (61/611)-68. Universal Decimal Classification Medical Sciences in General. Anatomy Amend: 1; 2. 28 pp.

Anchor Agitators
Use: Agitators

Anchor Bolts
Use For: Expansion Bolts; Foundation Bolts
See Also: Anchors (Fasteners); Bolts; Concrete Anchors
CNS B2246-81. Foundation Bolts (Apr)(4426).
DIN ENGL 529-86. Masonry and Foundation Bolts (Dec). 5 pp.
DIN ENGL 797-70. Special Foundation Bolts (Nov). 1 p.
JIS B 1178-76. Foundation Bolts. 17 pp.

—Aircraft
SBAC AS 4752 ISSUE 1. Single Anchor, Bolt B.A. and B.S.F.. 1 p.
SBAC AS 4753 ISSUE 1. Double Anchor, Bolt B.A. and B.S.F.. 1 p.
SBAC AS 6735 ISSUE 2. Single Anchor Bolt (Unified).
SBAC AS 6736 ISSUE 2. Double Anchor Bolt (Unified).

—Aircraft—Plates (Supports)
SBAC AS 4755 ISSUE 1. Plate for Single Anchor Bolt, B.A. and B.S.F.. 1 p.
SBAC AS 4756 ISSUE 1. Plate for Double Anchor Bolt, B.A. and B.S.F.. 1 p.
SBAC AS 6738 ISSUE 1. Plated for Single Anchor Bolt (Unified).
SBAC AS 6739 ISSUE 1. Plate for Double Anchor Bolt (Unified).

Anchor Buoys
See Also: Buoys
CNS F3065-80. Anchor Buoys (Oct)(6562).
JIS F 3308-77. Anchor Buoys (R 1984). 7 pp.

Anchor Capstans
Use: Capstans

Anchor Chains
Use: Chains—Anchor

Anchor Nuts
Use For: Stake Nuts *See Also:* Anchors (Fasteners); Fasteners; Locking Nuts; Nuts (Fasteners); Rivet Nuts; Screws
CNS B2221-83. Special Foundation Nuts (Feb)(4378).
DIN ENGL 798-72. Special Foundation Nuts for Special Foundation Bolts According to DIN 797 (Dec). 1 p.

—Aerospace
DIN ENGL LN 29638 Pt 1-69. Press Nuts; Self-Locking for Temperatures up to 260 Degrees C (Apr). 1 p.
DIN ENGL LN 29639 Pt 1-69. Press Nuts; Self-Locking, Floating for Temperatures up to 260 Degrees C (Apr). 1 p.
DIN ENGL LN 29679-77. Nuts Anchor; Self-Locking; Floating; Double Lug for Temperatures up to 235 Degrees C (July). 3 pp.
DIN ENGL LN 29791-80. Aerospace; Nuts, Anchor, Right Angle; Self-Locking, Floating for Temperatures up to 235 Degrees C (Aug). 3 pp.
DIN ENGL LN 29951-74. Nuts, Anchor; Self-Locking, Pressure Tight, Double Lug for Temperatures from-55 to +100 Degrees C (Dec). 2 pp.
DIN ENGL LN 29954-80. Aerospace; Nuts, Anchor, Self-Locking, Pressure Tight, Double Lug for Temperatures from-55 to +230 Degrees C (Mar). 3 pp.
DIN ENGL LN 29982-78. Aerospace; Nuts Anchor; Self-Locking; Deep Counterbore; Double Lug for Temperatures up to 235 Degrees C (Mar). 3 pp.

—Aerospace—Locking
CEN PREN 3653-92. Nuts, Anchor, Self-Locking, Floating, Self-Aligning, One Lug, in Steel, Cadmium Plated, MoS2 Lubricated Classificaton: 900 MPa (at Ambient Temperature) /235 Degrees C. 5 pp.
DIN ENGL LN 29671-77. Nuts Anchor; Self-Locking; Double Lug for Temperatures up to 235 Degrees C (May). 3 pp.

—Aerospace—Locking—Counterbore
CEN PREN 2862-90. Nuts Anchor, Self-Locking, Fixed, 90 Degrees Corner, with Counterbore, in Alloy Steel, Cadmium Plated, MO2 Lubricated Classification: 1100 MPa (at Ambient Temperature) /235 Degrees Celsius. 6 pp.
CEN PREN 2862-93. Aerospace Series Nuts, Anchor, Self-Locking, Fixed, 90 Degree Corner, with Counterbore, in Alloy Steel, Cadmium Plated, MoS2 Lubricated Classification: 1 100 MPa (at Ambient Temperature)/235 Degrees C. 5 pp.
CEN PREN 2862-93. CORRIGENDUM Aerospace Series Nuts, Anchor, Self-Locking, Fixed, 90 Degree Corner, with Counterbore, in Alloy Steel, Cadmium Plated, MoS2 Lubricated Classification: 1 100 MPa (at Ambient Temperature) /235 Degrees C. 1 p.
CEN PREN 2863-90. Nuts, Anchor Self-Locking, Fixed, 90 Degrees Corner, with Counterbore, in Heat Resisting Steel, Passivated, MOS2 Lubricated Classification: 1100 MPa (at Ambient Temperature) /315 Degrees Celsius. 6 pp.
CEN PREN 2863-93. Aerospace Series Nuts, Anchor, Self-Locking, Fixed, 90 Degree Corner, with Counterbore, in Heat Resisting Steel, MoS2 Lubricated Classification: 1 100 MPa (at Ambient Temperature)/315 Degrees C. 5 pp.
CEN PREN 2863-93. CORRIGENDUM Aerospace Series Nuts, Anchor, Self-Locking, Fixed, 90 Degree Corner, with Counterbore, in Heat Resisting Steel, MoS2 Lubricated Classification: 1 100 MPa (at Ambient Temperature) /315 Degrees C. 1 p.
CEN PREN 2865-93. Aerospace Series Nuts, Anchor, Self-Locking, Floating, Two Lug, with Counterbore, in Heat Resisting Steel, MoS2 Lubricated Classification: 1 100 MPa (at Ambient Temperature)/315 Degrees C. 5 pp.
CEN PREN 2865-93. CORRIGENDUM Aerospace Series Nuts, Anchor, Self-Locking, Floating, Two Lug, with Counterbore, in Heat Resisting Steel, MoS2 Lubricated Classification: 1 100 MPa (at Ambient Temperature) /315 Degrees C. 1 p.
CEN PREN 2866-92. Nuts, Anchor, Self-Locking, Floating, One Lug, with Counterbore, in Steel, Cadmium Plated, MoS2 Lubricated Classification: 1 110 MPa (at Ambient Temperature) /235 Degrees C. 5 pp.

INTERNATIONAL AND NON-U.S. NATIONAL STANDARDS
SUBJECT INDEX

Anchor

Anchor Nuts *(Cont.)*
—Aerospace—Locking—Counterbore *(Cont.)*

CEN PREN 2867-92. Nuts, Anchor, Self-Locking, Floating, One Lug, with Counterbore, in Heat Resisting Steel, MoS2 Lubricated Classification: 1 110 MPa (at Ambient Temperature) /315 Degrees C. 5 pp.

CEN PREN 3435-93. Aerospace Series Nuts, Anchor, Self-Locking, Floating, Two Lug, Reduced Series, with Counterbore, in Heat Resisting Steel, MoS2 Lubricated Classification: 1 100 MPa (at Ambient Temperature)/315 Degrees C. 5 pp.

CEN PREN 3435-93. CORRIGENDUM Aerospace Series Nuts, Anchor, Self-Locking, Floating, Two Lug, Reduced Series, with Counterbore, in Heat Resisting Steel, MoS2 Lubricated Classification: 1 100 MPa (at Ambient Temperature) /315 Degrees C. 1 p.

CEN PREN 3537-90. Nuts, Anchor, Self-Locking, Fixed, Two Lug, with Counterbore, in Heat Resisting Steel, Passivated, MoS2 Lubricated Classification:1100 MPa (at Ambient Temperature) /315 Degrees Celsius. 6 pp.

CEN PREN 3537-93. Aerospace Series Nuts, Anchor, Self-Locking, Fixed, Two Lug, with Counterbore, in Heat Resisting Steel, MoS2 Lubricated Classification: 1 100 MPa (at Ambient Temperature)/315 Degrees C. 5 pp.

CEN PREN 3537-93. CORRIGENDUM Aerospace Series Nuts, Anchor, Self-Locking, Fixed, Two Lug, with Counterbore, in Heat Resisting Steel, MoS2 Lubricated Classification: 1 100 MPa (at Ambient Temperature) /315 Degrees C. 1 p.

CEN PREN 3538-90. Nuts, Anchor, Self-Locking, Fixed, Two Lug, Reduced Series with Counterbore, in Heat Resisting Steel Passivated, MoS 2 Lubricated Classification: 100 MPa (at Ambient Temperature) /315 Degrees Celsius. 6 pp.

CEN PREN 3538-93. Aerospace Series Nuts, Anchor, Self-Locking, Fixed, Two Lug, Reduced Series, with Counterbore, in Heat Resisting Steel, MoS2 Lubricated Classification: 1 100 MPa (at Ambient Temperature)/315 Degrees C. 5 pp.

CEN PREN 3538-93. CORRIGENDUM Aerospace Series Nuts, Anchor, Self-Locking, Fixed, Two Lug, Reduced Series, with Counterbore, in Heat Resisting Steel, MoS2 Lubricated Classification: 1 100 MPa (at Ambient Temperature) /315 Degrees C. 1 p.

CEN PREN 3539-90. Nuts Anchor, Self-Locking, One Lug, Fixed, with Counterbore, in Heat Resisting Steel, MoS2 Lubricated Classification: 1100 MPa (at Ambient Temperature) /315 Degrees Celsius. 5 pp.

CEN PREN 3539-93. Aerospace Series Nuts, Anchor, Self-Locking, One Lug, Fixed, with Counterbore, in Heat Resisting Steel, MoS2 Lubricated Classification: 1 100 MPa (at Ambient Temperature) /315 Degrees C. 5 pp.

CEN PREN 3712-91. Nuts, Anchor, Self-Locking, One Lug Fixed, Reduced Series, with Counterbore, in Steel, Cadmium Plated, MoS2 Lubricated Classification: 1100 MPa (at Ambient Temperature)/ 235 Degrees C. 6 pp.

CEN PREN 3714-93. Aerospace Series Nuts, Anchor, Self-Locking, Floating, Two Lug, with Counterbore, in Heat Resisting Steel, Silver Plated Classification: 1 100 MPa (at Ambient Temperature)/425 Degrees C. 5 pp.

CEN PREN 3714-93. CORRIGENDUM Aerospace Series Nuts, Anchor, Self-Locking, Floating, Two Lug, with Counterbore, in Heat Resisting Steel, Silver Plated Classification: 1 100 MPa (at Ambient Temperature) /425 Degrees C. 1 p.

CEN PREN 3750-90. Nuts, Anchor, Self-Locking, Fixed 90 Degrees Corner, Reduced Series with Counterbore, in Heat Resisting Steel, Passivated, MoS2 Lubricated/Classification:100 MPa (at Ambient Temperature) /315 Degrees Celsius. 6 pp.

CEN PREN 3750-93. Aerospace Series Nuts, Anchor, Self-Locking, Fixed, 90 Degree Corner, Reduced Series, with Counterbore, in Heat Resisting Steel, MoS2 Lubricated Classification: 1 100 MPa (at Ambient Temperature)/315 Degrees C. 5 pp.

CEN PREN 3750-93. CORRIGENDUM Aerospace Series Nuts, Anchor, Self-Locking, Fixed, 90 Degree Corner, Reduced Series, with Counterbore, in Heat Resisting Steel, MoS2 Lubricated Classification: 1 100 MPa (at Ambient Temperature) /315 Degrees C. 1 p.

CEN PREN 3751-90. Nuts, Anchor, Self-Locking, Fixed, Closed Corner, Reduced Series, with Counterbore, in Heat Resisting Steel, Passivated, MoS2 Lubricated Classification: 1100 MPa (at Ambient Temperature)/315 Degrees Celsius. 6 pp.

CEN PREN 3751-93. Aerospace Series Nuts, Anchor, Self-Locking, Fixed, Closed Corner, Reduced Series, with Counterbore, in Heat Resisting Steel, MoS2 Lubricated Classification: 1 100 MPa (at Ambient Temperature)/315 Degrees C. 5 pp.

CEN PREN 3751-93. CORRIGENDUM Aerospace Series Nuts, Anchor, Self-Locking, Fixed, Closed Corner, Reduced Series, with Counterbore, in Heat Resisting Steel, MoS2 Lubricated Classification: 1 100 MPa (at Ambient Temperature) /315 Degrees C. 1 p.

Anchor Nuts *(Cont.)*
—Aerospace—Locking—Counterbore *(Cont.)*

CEN PREN 3753-90. Nuts, Anchor, Self-Locking, Fixed, Closed Corner, with Counterbore, in Alloy Steel, Cadmium Plated, MoS2 Lubricated Classification:1100 MPa (at Ambient Temperature) /235 Degrees Celsius. 6 pp.

CEN PREN 3753-93. Aerospace Series Nuts, Anchor, Self-Locking, Fixed, 60 Degree Corner, with Counterbore, in Alloy Steel, Cadmium Plated, MoS2 Lubricated Classification: 1 100 MPa (at Ambient Temperature)/235 Degrees C. 5 pp.

CEN PREN 3753-93. CORRIGENDUM Aerospace Series Nuts, Anchor, Self-Locking, Fixed, 60 Degree Corner, with Counterbore, in Alloy Steel, Cadmium Plated, MoS2 Lubricated Classification: 1 100 MPa (at Ambient Temperature) /235 Degrees C. 1 p.

CEN PREN 3754-90. Nuts, Anchor, Self-Locking, Fixed, Closed Corner, with Counterbore, in Heat Resisting Steel, Passivated, MoS2 Lubricated Classification: 100 MPa (at Ambient Temperature) /315 Degrees Celsius. 6 pp.

CEN PREN 3754-93. Aerospace Series Nuts, Anchor, Self-Locking, Fixed, 60 Degree Corner, with Counterbore, in Heat Resisting Steel, MoS2 Lubricated Classification: 1 100 MPa (at Ambient Temperature)/315 Degrees C. 5 pp.

CEN PREN 3754-93. CORRIGENDUM Aerospace Series Nuts, Anchor, Self-Locking, Fixed, 60 Degree Corner, with Counterbore, in Heat Resisting Steel, MoS2 Lubricated Classification: 1 100 MPa (at Ambient Temperature) /315 Degrees C. 1 p.

CEN PREN 3757-93. Aerospace Series Nuts, Anchor, Self-Locking, Floating, Self-Aligning, Two Lug, in Heat Resisting Steel, MoS2 Lubricated Classification: 900 MPa (at Ambient Temperature) /315 Degrees C. 5 pp.

CEN PREN 3757-93. CORRIGENDUM Aerospace Series Nuts, Anchor, Self-Locking, Floating, Self-Aligning, Two Lug, in Heat Resisting Steel, MoS2 Lubricated Classification: 900 MPa (at Ambient Temperature) /315 Degrees C. 1 p.

CEN PREN 3768-90. Nuts, Anchor, Self-Locking, One Lug, Fixed, Reduced Series, with Counterbore, in Heat Resisting Steel, MoS2 Lubricated Classification: 1100 MPa (at Ambient Temperature) /315 Degrees Celsius. 5 pp.

CEN PREN 3768-93. Aerospace Series Nuts, Anchor, Self-Locking, One Lug, Fixed, Reduced Series, with Counterbore, in Heat Resisting Steel, MoS2 Lubricated Classification: 1 100 MPa (at Ambient Temperature) /315 Degrees C. 5 pp.

CEN PREN 3834-91. Nuts, Anchor, Self-Locking, Floating, Two Lug, Incremental Counterbore, in Heat Resisting Steel, MoS2 Lubricated Classification: 900 MPa (at Ambient Temperature)/315 Degrees C. 7 pp.

CEN PREN 3834-93. Aerospace Series Nuts, Anchor, Self-Locking, Floating, Two Lug, Incremental Counterbore, in Heat Resisting Steel, MoS2 Lubricated Classification: 900 MPa (at Ambient Temperature) /315 Degrees C. 5 pp.

CEN PREN 4084-93. Aerospace Series Nuts, Anchor, Self-Locking, Fixed, Two Lug, with Counterbore, in Alloy Steel, Cadmium Plated, MoS2 Lubricated Classification: 1 100 MPa (at Ambient Temperature) /235 Degrees C. 5 pp.

DIN ENGL LN 29983-77. Nuts Anchor; Self-Locking; Deep Counterbore; Double Lug Reduced for Temperatures up to 235 Degrees C (July). 3 pp.

DIN ENGL LN 29984-77. Nuts Anchor; Self-Locking; Deep Counterbore; Single Lug for Temperatures up to 235 Degrees C (July). 3 pp.

DIN ENGL LN 29985-77. Nuts Anchor; Self-Locking; Floating; Deep Counterbore; Double Lug for Temperatures up to 235 Degrees C (Dec). 3 pp.

DIN ENGL LN 29987-77. Nuts Anchor; Self-Locking; Floating; Deep Counterbore; Single Lug for Temperatures up to 235 Degrees C (Dec). 3 pp.

DIN ENGL LN 29988-77. Nuts Anchor; Self-Locking; Deep Counterbore; Corner Lug for Temperatures up to 235 Degrees C (July). 3 pp.

DIN ENGL LN 29989-77. Nuts Anchor; Self-Locking; Deep Counterbore; Corner Lug Reduced for Temperatures up to 235 Degrees C (July). 3 pp.

DIN ENGL LN 29990-77. Anchor Nuts; Self-Locking; Deep Counterbore, Double Lug, for Temperatures up to 315 Degrees C and up to 425 Degrees C (July). 3 pp.

DIN ENGL LN 29992-77. Nuts, Anchor; Self-Locking, Deep Counterbore, Single Lug for Temperatures up to 315 Degrees C and up to 425 Degrees C (July). 3 pp.

DIN ENGL LN 29996-77. Nuts Anchor; Self-Locking; Deep Counterbore; Single Lug Reduced for Temperatures up to 315 Degrees C and up to 425 Degrees C (May). 3 pp.

ISO 3168-86. Aerospace—Self-Locking, Fixed, Single-Lug Anchor Nuts with Counterbore, Classification 1 100 MPa/235 Degrees Celsius First Edition. 5 pp.

Anchor Nuts *(Cont.)*
—Aerospace—Locking—Counterbore *(Cont.)*

ISO 3191-85. Aerospace—Self-Locking, Fixed, Single-Lug Anchor Nuts, Reduced Series, with Counterbore, Strength Classification 1 100 MPa and Maximum Operating Temperature 235 Degrees Celsius First Edition. 5 pp.

ISO 3209-89. Aerospace—Nuts, Anchor, Self-Locking, Floating, Two-Lug, with Counter-bore, with MJ Threads, Coated or Uncoated, Classification 1 100 MPa/235 Degrees Celsius, 1 100 MPa/315 Degrees Celsius or 1 100 MPa/425 Degrees Celsius—Dimensions First Edition. 6 pp.

ISO 3221-89. Aerospace—Nuts, Anchor, Self-Locking, Fixed, 90 Degree Corner, with Counterbore, with MJ Threads, Coated or Uncoated, Classification 1 100 MPa/235 Degrees Celsius, 1 100 MPa/315 Degrees Celsius or 1 100 MPa/425 Degrees Celsius—Dimensions First Edition. 5 pp.

ISO 3222-89. Aerospace—Nuts, Anchor, Self-Locking, Fixed, Closed Corner, Reduced Series, with Counterbore, with MJ Threads, Coated or Uncoated, Classification 1 100 MPa/235 Degrees C, 1 100 MPa/315 Degrees C or 1 100 MPa/425 Degrees C—Dimensions First Edition. 5 pp.

ISO 3223-89. Aerospace—Nuts, Anchor, Self-Locking, Fixed, Two-Lug, with Counter-bore, with MJ Threads, Coated or Uncoated, Classification 1 100 MPa/235 Degrees Celsius, 1 100 MPa/315 Degrees Celsius or 1 100 MPa/425 Degrees Celsius—Dimensions First Edition. 6 pp.

ISO 3224-85. Aerospace—Self-Locking, Floating, Single-Lug Anchor Nuts, with Counterbore, Classification 1 100 MPa/235 Degrees Celsius First Edition. 5 pp.

ISO 3225-85. Aerospace—Self-Locking, Fixed, Two Lug Anchor Nuts, Reduced Series, with Counterbore, Strength Classification 1 100 MPa and Maximum Operating Temperature 235 Degrees Celsius First Edition. 5 pp.

ISO 7332-83. Metric Fasteners for Aerospace Construction—Nuts, Anchor, Self-Locking, Floating, Two Lug, Reduced Series, with Counterbore—Strength Classification 1 100 MPa—Maximum Operating Temperature 235 Degrees Celsius First Edition. 5 pp.

ISO 8940-88. Aerospace—Nuts, Anchor, Self-Locking, Sealing, Floating, Two-Lug, with Counterbore, Classifications 900 MPa/ 120 Degrees Celsius, 900 MPa/175 Degrees Celsius and 900 MPa/235 Degrees Celsius Dimensions First Edition. 4 pp.

ISO 9156-89. Aerospace—Nuts, Anchor, Self-Locking, Fixed, 90 Degree Corner, Reduced Series, with Counter-bore, with MJ Threads, Coated or Uncoated, Classification 1 100 MPa/235 Degrees Celsius, 1 100 MPa/315 Degrees Celsius or 1 100 MPa/425 Degrees Celsius—Dimensions First Edition. 5 pp.

—Aerospace—Locking—Counterbore—Miniature

DIN ENGL LN 29986-77. Anchor Nuts; Self-Locking, Floating, Deep Counterbore, Double Lug, Miniature, for Temperatures up to 235 Degrees C (May). 3 pp.

DIN ENGL LN 29991-77. Anchor Nuts; Self-Locking, Deep Counterbore, Double Lug, Miniature for Temperatures up to 315 Degrees C and up to 425 Degrees C (July). 3 pp.

—Aerospace—Plate

AECMA PREN3019-89. Self-Locking Plate Nuts Floating, Two-Lug in Heat Resisting Steel FE-PA2HT (A286), Unplated Classification: 1100 MPa/ 650 Degrees Celsius. 5 pp.

DIN ENGL LN 29910-83. Aerospace; Nuts, Plate; Self-Locking, Floating, Double Lug, Reduced, for Temperatures up to 235 Degrees C (Oct). 3 pp.

—Aerospace—Plate—Locking

DIN ENGL LN 29680-77. Aerospace; Nuts, Plate; Self-Locking, Floating, Self-Aligning, Double Lug for Temperatures up to 235 Degrees C (July). 3 pp.

—Aircraft

BSI 2A122-2A124-81. 1981 Clinch Nuts (Unified Threads) for Aircraft. 15 pp.

SBAC TS 120 ISSUE 2. Technical Specifications for AGS 2000 Series Stiffnuts (Including Clinch Nuts) (BA and BSF Threads) for Aircraft.

—Aircraft—Blind

SBAC AGS 3037 ISSUE 1. Anchor, Blind, Avlok 4 BA and No.6. U.N.C. Blind Anchor Nuts (Steel).

SBAC AGS 3038 ISSUE 1. Avlok 2 BA, No.10 UNF/No.8 U.N.C. Blind Anchor (Steel).

SBAC AGS 3039 ISSUE 1. Avlok 1/4 Inch BSF and 1/4 Inch U.N.F. Blind Anchor (Steel).

SBAC AGS 3040 ISSUE 1. Avlok 5/16 Inch BSF and 5/16 Inch U.N.F. Blind Anchor (Steel).

—Aircraft—Bolt Assemblies

SBAC AS 4572 (V). Typical Assembly of Retained Bolts-Double Anchor Nuts. 2 pp.

INTERNATIONAL AND NON-U.S. NATIONAL STANDARDS
SUBJECT INDEX
Anchors

Anchor Nuts *(Cont.)*

—Aircraft—Bolts

SBAC AS 6752 (V). Typical Assembly of Retained Bolts—Double Anchor Nuts.

—Aircraft—Counterbore

AECMA PREN2264-82. Nuts, Anchor, Self Locking Two Lug, Floating Incremental Counterbore Classification 1100 MPa/235 Degrees Celsius. 6 pp.

AECMA PREN2752-86. Nuts, Anchor, Self Locking Fixed, Two Lug, Reduced Series, with Counterbore Classification 1100 MPa/235 Degrees Celsius. 4 pp.

AECMA PREN2753-87. Nuts, Anchor, Self Locking One Lug, with Counterbore Classification 1100 MPa/235 Degrees Celsius. 6 pp.

AECMA PREN2754-87. Nuts, Anchor, Self Locking Two Lug, Floating with Counterbore Classification 1100 MPa/235 Degrees Celsius. 6 pp.

AECMA PREN2855-89. Nuts, Anchor, Self-Locking Fixed, 90 Degrees Corner, Reduced Series, with Counterbore Classification: 1100 MPa/235 Degrees Celsius. 6 pp.

AECMA PREN2856-89. Nuts, Anchor, Self-Locking, Fixed, Closed Corner, Reduced Series, with Counterbore Classification: 1100 MPa/235 Degrees Celsius. 6 pp.

BSI BS EN 2752-90. 1990 Nuts, Anchor, Self Locking Fixed, Two Lug, Reduced Series with Counterbore. Classification: 1100 MPa/235 Degrees C. 8 pp.

CEN EN 2752-89. Nuts, Anchor, Self Locking Fixed, Two Lug, Reduced Series, with Counterbore Classification: 1100 MPa/235 Degrees C. 5 pp.

—Aircraft—Locking

SBAC AS 8602 ISSUE 4. Self-Locking Nuts, Double Lug Anchor, Steel, Cadmium Plated, 125,000 lbf/sq.in. 250 Degrees C.

SBAC AS 8603 ISSUE 4. Self-Locking Nuts, Miniature Double Lug Anchor Steel, Cadmium Plated, 125,000 lbf/sq. in. 250 Degrees C.

SBAC AS 8604 ISSUE 4. Self-Locking Nuts, Single Long Lug Anchor, Steel, Cadmium Plated, 125,000 lbf/sq.in. 250 Degrees C.

SBAC AS 8605 ISSUE 4. Self-Locking Nuts, Miniature, Single Long Lug Anchor, Steel, Cadmium Plated, 125,000 lbf/sq.in. 250 Degrees C.

SBAC AS 8606 ISSUE 4. Self-Locking Nuts, Miniature, Single Short Lug Anchor Steel, Cadmium Plated, 125,000 lbf/sq.in. 250 Degrees C.

SBAC AS 8607 ISSUE 4. Self-Locking Nuts, Corner Anchor Steel, Cadmium Plated, 125,000 lbf/sq. in. 250 Degrees C.

SBAC AS 8608 ISSUE 4. Self-Locking Nuts, Miniature Corner Anchor, Steel, Cadmium Plated, 125,000 lbf/sq.in. 250 Degrees C.

SBAC AS 8609 ISSUE 4. Self-Locking Nuts, Double Lug Floating Anchor, Steel, Cadmium Plated, 125,000 lbf/sq.in. 250 Degrees C.

SBAC AS 8611 ISSUE 4. Self-Locking Nuts, Single Long Lug Floating Anchor Steel, Cadmium Plated, 125,000 lbf/sq.in. 250 Degrees C.

SBAC AS 8625 ISSUE 2. Self-Locking Nuts, Double Lug Anchor, CR Steel, Silver Plated, lbf/in.2 450 Degrees C.

SBAC AS 8626 ISSUE 2. Self-Locking Nuts, Miniature, Double Lug Anchor, CR Steel, Silver Plated, 125,000 lbf/in.2 450 Degrees C.

SBAC AS 8627 ISSUE 2. Self-Locking Nuts,Single Long Lug Anchor, CR Steel, Silver Plated, 125,000 lbf/in.2 450 Degrees C.

SBAC AS 8628 ISSUE 2. Self-Locking Nuts, Miniature, Single, Long Lug Anchor, CR Steel, Silver Plated, 125,000 lbf/in.2 450 Degrees C.

SBAC AS 8629 ISSUE 2. Self-Locking Nuts, Miniature, Single, Short Lug Anchor, CR Steel, Silver Plated, 125,000 lbf/in.2 450 Degrees C.

SBAC AS 8630 ISSUE 2. Self-Locking Nuts, Corner Anchor, CR Steel, Silver Plated 125,000 lbf/in.2 450 Degrees C.

SBAC AS 8631 ISSUE 2. Self-Locking Nuts, Miniature, Corner Anchor, CR Steel, Silver Plated, 125,000 lbf/in.2 450 Degrees C.

SBAC AS 8632 ISSUE 2. Self-Locking Nuts, Double Lug Anchor, CR Steel, Silver Plated, 125,000 lbf/in.2.

SBAC AS 8633 ISSUE 3(I). Self-Locking Nuts, Miniature, Double Lug Floating Anchor C.R. Steel, Silver-Plated, 125,000 lbf/in2 450 Degrees C.

SBAC AS 8634 ISSUE 2. Self-Locking Nuts, Single Long Lug Floating Anchor, CR Steel, Silver Plated, 125,000 lbf/in.2 450 Degrees C.

SBAC AS 8652 ISSUE 4. Self-Locking Nuts, Double Lug Anchor, CR Steel, 125,000 lbf/in.2 250 Degrees C.

SBAC AS 8653 ISSUE 4. Self-Locking Nuts, Miniature, Double Lug Anchor, CR Steel, 125,000 lbf/in.2 250 Degrees C.

SBAC AS 8654 ISSUE 4. Self-Locking Nuts, Single, Long Lug Anchor, CR Steel, 125,000 lbf/in.2 250 Degrees C.

SBAC AS 8655 ISSUE 4. Self-Locking Nuts, Miniature, Single Long Lug Anchor, CR Steel, 125,000 lbf/in.2 Degrees C.

Anchor Nuts *(Cont.)*

—Aircraft—Locking *(Cont.)*

SBAC AS 8656 ISSUE 4. Self-Locking Nuts, Miniature, Single Short Lug Anchor, CR Steel, 125,000 lbf/in.2 250 Degrees C.

SBAC AS 8657 ISSUE 4. Self-Locking Nuts, Corner Anchor, CR Steel, 125,000 lbf/in2 250 Degrees C.

SBAC AS 8658 ISSUE 4. Self-Locking Nuts, Miniature, Corner Anchor, CR Steel, 125,000 lbf/in.2 250 Degrees C.

SBAC AS 8659 ISSUE 4. Self-Locking Nuts, Double Lug Floating Anchor, CR Steel, 125,000 lbf/in.2 250 Degrees C.

SBAC AS 8660 ISSUE 5(I). Self-Locking Nuts, Miniature, Double Lug Floating Anchor CR Steel, 125,000 lbf/in.2 250 Degrees C.

SBAC AS 8661 ISSUE 4. Self-Locking Nuts, Single, Long Lug Floating Anchor, CR Steel, 125,000 lbf/in.2 250 Degrees C.

SBAC AS 28003 ISSUE 2. Nut, Self-Locking, Non C.R. Steel, Long Lug Anchor, Cadmium Coated.

SBAC AS 28004 ISSUE 2. Nut—Self-Locking, Double Lug Anchor, Cadmium Coated.

SBAC AS 28005 ISSUE 2. Nut, Self-Locking, Non C.R. Steel, Corner Anchor, Cadmium Coated.

SBAC AS 28006 ISSUE 2. Nut, Self-Locking, Non C.R. Steel, Floating, Long Lug Anchor, Cadmium Coated.

SBAC AS 28007 ISSUE 4. Nut, Self-Locking, Non C.R. Steel, Floating, Double Lug Anchor, Cadmium Coated.

SBAC AS 28009 ISSUE 2. Nut, Self-Locking, Non C.R. Steel, Miniature, Single Long Lug Anchor, Cadmium Coated.

SBAC AS 28010 ISSUE 2. Nut, Self-Locking, Miniature, Double Lug Anchor, Cadmium Coated.

SBAC AS 28019 ISSUE 2. Nut, Self-Locking, Non C.R. Steel, Corner Anchor, Miniature, Cadmium Coated.

SBAC AS 46767 (V). Nut, Self Locking, Miniature, Double Lug Floating Anchor, Carbon Steel, Cadmium Coated (12500LBF/IN2) 235 Degrees Celsius.

SBAC AS 46768 (V). Nut, Self Locking, Miniature, Double Lug Floating Anchor, Cres. Steel, Silver Plated (125000 lbf/in2) 425 Degrees C.

SBAC AS 46769 (V). Nut, Self Locking, Miniature, Double Lug Floating Anchor, Cres. Steel, (125000 lbf/in2) 235 Degrees C.

SBAC AGS 2011 ISSUE 18(I). Stiffnuts—Clinch.

SBAC AGS 2023 ISSUE 11(I). Stiffnuts Double Anchor, with Translucent Ca

—Aircraft—Locking—Countersunk

SBAC AGS 2009 ISSUE 14(I). Stiffnuts—Double Anchor—Countersunk.

SBAC AGS 2014 ISSUE 14(I). Stiffnuts—Floating Anchor—Countersunk.

SBAC AGS 2020 ISSUE 13(I). Stiffnuts—Single Anchor—Countersunk.

—Aircraft—Locking—Thick

SBAC AGS 2007 ISSUE 14(I). Stiffnuts—Double Anchor—Thick.

SBAC AGS 2012 ISSUE 14(I). Stiffnuts—Floating Anchor—Thick.

SBAC AGS 2018 ISSUE 12(I). Stiffnuts—Single Anchor—Thick.

—Aircraft—Locking—Thin

SBAC AGS 2008 ISSUE 14(I). Stiffnuts—Double Anchor—Thin.

SBAC AGS 2013 ISSUE 15(I). Stiffnuts—Floating Anchor—Thin.

SBAC AGS 2019 ISSUE 12(I). Stiffnuts—Single Anchor—Thin.

—Aircraft—Packing Plates

SBAC AS 6753 ISSUE 1. Packing Plate for Steel Retaining Rings—Double Anchor Nuts.

SBAC AS 6754 ISSUE 1. Packing Plate for Nylon Retaining Rings—Double Anchor Nuts.

SBAC AS 6755 ISSUE 1. Packing Plate for Steel Retaining Rings—Single Anchor Nuts.

SBAC AS 6756 ISSUE 1. Packing Plate for Nylon Retaining Rings—Single Anchor Nuts.

SBAC AS 8431 ISSUE 1. Packing Plate—Plain for Use with B.S. A.125-201-Double Anchor Stiffnuts.

SBAC AS 8432 ISSUE 1. Packing Plate—Plain for Use with B.S. A.125-201-Single Anchor Stiffnuts.

—Aircraft—Self Aligning

SBAC AS 28008 ISSUE 2. Nut, Anchor, Double Lug, Floating, Self-Aligning, Cadmium Coated.

—Locking

AECMA PREN2680-87. Nuts, Anchor, Self Locking, Two Lug, Floating, Self Aligning Classification 900 MPa/235 Degrees Celsius. 4 pp.

Anchor Screws

See Also: Screws

Anchor Screws *(Cont.)*

—Tee Head

CNS B2382-79. T-Head Anchor Screws (Aug)(4947).

Anchor Stoppers

Use: Cable Stoppers

Anchor Studs

Use For: Foundation Studs *See Also:* Studs (Fasteners)

CNS B2247-81. Foundation Studs (Apr)(4427).

Anchor Windlasses

Use: Windlasses

Anchorages (Fasteners)

Use: Anchors (Fasteners)

Anchors (Fasteners)

Scope Note: For additional listings, use a more specific term *Use For:* Bolt Anchors
See Also: Anchor Bolts; Anchor Nuts; Bands; Bolts; Brackets; Clips; Concrete Anchors; Couplings; Fasteners; Ground Anchors; Holders; Ice Anchors; Nails (Fasteners); Nuts (Fasteners); Retaining Rings; Screws; Square Head Bolts; Straps (Fasteners); Structural Members; Studs (Fasteners)

CEN PREN 795-92. Protection Against Falls from a Height—Anchorage Devices—Requirements and Testing. 21 pp.

—Aircraft—Springs (Elastic)

SBAC AS 2802 ISSUE 1. Spring Anchorage. 1 p.
SBAC AS 2822 ISSUE 3. Spring Anchorage. 1 p.

—Automotive—Safety Belts

CSA CAN/CSA-M6683-92. Seat Belts and Anchorages—Machinery (EMM, FM) (ISO 6683-1981); (Gen Instr 1). 17 pp.

EC 76/115/EEC-75. Council Directive on the Approximation of the Laws of the Member States Relating to Anchorages for Motor-Vehicle Safety Belts. 15 pp.

EC 81/575/EEC-81. Council Directive Amending Council Directive 76/115/EEC on the Approximation of the Laws of the Member States Relating to Anchorages for Motor-Vehicle Safety Belts. 2 pp.

EC 82/318/EEC-82. Commission Directive Adapting to Technical Progress Council Directive 76/115/EEC on the Approximation of the Laws of the Member States Relating to Anchorages for Motor-Vehicle Safety Belts. 8 pp.

EC 90/629/EEC-90. Commission Directive Adapting to Technical Progress Council Directive 76/115/EEC on the Approximation of the Laws of the Member States Relating to Anchorages for Motor Vehicle Safety Belts. 6 pp.

ISO TR1417-74. Automobiles—Anchorages for Seat Belts First Edition. 8 pp.

ISO 6683-81. Earth-Moving Machinery—Seat Belts and Seat Belt Anchorages First Edition; (Amendment 1-1990) (CAN/CSA-M6683-92). 8 pp.

JIS D 4609-88. Seat Belt Installation and Anchorages for Passenger Cars. 16 pp.

—Automotive—Seats

EC 74/408/EEC-74. Council Directive on the Approximation of the Laws of the Member States Relating to the Interior Fittings of Motor Vehicles (Strength of Seats and of Their Anchorages). 9 pp.

EC 81/577/EEC-81. Council Directive Amending Council Directive 74/408/EEC on the Approximation of the Laws of the Member States Relating to the Interior Fittings of Motor Vehicles (Strength of Seats and of Their Anchorages). 1 p.

JIS D 4610-87. Strength of Seats and of Seat Anchorages for Passenger Cars. 10 pp.

—Automotive—Towing Cables

BSI BS AU 188-83. 1983 Anchorages for Towing Ropes, Cables or Bars for Road Vehicles. 4 pp.

ISO 5422-82. Road Vehicles—Anchorages for Towing Ropes, Cables or Bars First Edition. 4 pp.

—Forestry Equipment

CSA CAN/CSA-M6816-92. Machinery for Forestry—Winches—Classification and Nomenclature (ISO 6816-1984); (Gen Instr 1). 19 pp.

—Prestressing

BSI BS 4447-73. 1973 Amd 1 The Performance of Prestressing Anchorages for Post-Tensioned Construction (AMD 6536) December 21,1990. 13 pp.

SAA AS 1314-72. Prestressing Anchorages (Metric Units) (Incorporating Amdt 1). 12 pp.

—Rolling Stock—Vibration Isolators

CNS E1025-83. Rubber Vibration Isolators for Bilster Anchors of Railway Rolling Stock (Feb)(9995).

JIS E 4712-89. Rubber Vibration Isolators for Bolster Anchors of Railway Rolling Stock.

INDUSTRY STANDARDS

Anchors

Anchors (Fasteners) *(Cont.)*
—Trailers—Load Restraints
SNZ NZS 5444-89. Load Anchorage Points for Heavy Vehicles. 26 pp.
—Trucks—Load Restraints
SNZ NZS 5444-89. Load Anchorage Points for Heavy Vehicles. 26 pp.
—Wire Ropes
BSI BS 7166-89. 1989 Wedge and Socket Anchorages for Wire Ropes. 11 pp.

Anchors (Marine)
See Also: Cable Stoppers; Marine Equipment; Mooring Anchors; Mooring Hardware
CNS F1007-76. Method of Test for Anchors (Jun)(3938).
ISO 9437-86. Shipbuilding—Inland Vessels—Matrosov Anchors First Edition. 5 pp.
—Chains—Link
BSI BS 7160-90. 1990 Anchor Chains for Small Craft. 9 pp.
BSI BS MA 70: Part 1-75. 1975 Amd 1 Dimensions of Anchor Chain Cables Part 1: Stud Link Anchor Chain Cables. 12 pp.
CEN EN 24 565-89. Shipbuilding, Yachting, Ships, Ship Anchors, Chains, Specifications, Mechanical Properties, Breaking Load, Dimensions, Marking. 2 pp.
ISO 1704-91. Shipbuilding—Stud-Link Anchor Chains Second Edition. 16 pp.
ISO 4565-86. Small Craft—Anchor Chains First Edition. 5 pp.
JIS F 3303-82. Flash Butt Welded Anchor Chain Cables (R 1987). 16 pp.
MOD UK NES 171-01. Requirements for Copper—Based Alloy Stud Link Anchor Cables and Associated Items of Equipment Issue 1 (04.91); Amendment 1. 17 pp.
MOD UK NES 171-02. Requirements for Copper—Based Alloy Stud Link Anchor Cables and Associated Items of Equipment Issue 1 (04.91); Amendment 2. 18 pp.
MOD UK NES 172-01. Requirements for Forged Steel Stud Link Chain Cable Grades 1 and 2 Issue 1 (04.91); Amendment 1. 18 pp.
MOD UK NES 172-02. Requirements for Forged Steel Stud Link Chain Cable Grades 1, 2 and 3 Issue 2 (02.92); Amendment 1. 26 pp.
MOD UK NES 172-92. Requirements for Forged Steel Stud Link Chain Cable Grades 1, 2 and 3 Issue 2 (02.92). 19 pp.
MOD UK NES 175-91. Requirements for Equipment Associated with Forged Steel Chain Cable Issue 1 (06.91). 18 pp.
MOD UK NES 175-01. Requirements for Equipment Associated with Forged Steel Chain Cable Issue 1 (06.91); Amendment 1. 21 pp.
MOD UK NES 176-01. Requirements for Mooring Chain, Open—Link Chain Cable, Cast and Forged Steel and Associated Equipment Issue 1 (06.91); Amendment 1. 18 pp.
—Chains—Steel
JIS F 3302-90. Cast Steel Anchor Chain Cables.
JIS F 3302-75. Cast Steel Anchor Chain Cables. 13 pp.
—Dredges
JIS F 3991-75. Dredger's Anchors.
—Ground Effect Machines
CAA Chapter B5-9 04.79. Anchoring, Towing and Berthing. 4 pp.
CAA Chapter B5-9 App 04.79. Anchoring, Towing and Berthing. 2 pp.
—Identification Systems
MOD UK NES 174-01. Requirements for Ship and Mooring Anchors Issue 1 (06.91); Amendment 1. 24 pp.
MOD UK NES 174-02. Requirements for Ship and Mooring Anchors Issue 1 (06.91); Amendment 2. 29 pp.
—Ships
CNS F3014-76. Anchors (Jun)(3937).
JIS F 3301-90. Anchors.
JIS F 3301-75. Anchors. 18 pp.

Anchovies
See Also: Fish; Herrings
—Dried
CNS N5073-81. Dried Herring and Horse Mackerel (Aug)(2538). 3 pp.

AND Buffer Gates
See Also: AND Gates; Digital Circuits

—2-Input
CECC CECC 90 109-812 ISSUE 1-88. Digital Integrated Circuits; Silicon Monolithic C MOS, Cavity or Non-Cavity Packages; Type(s) 54/74 HC 808; Hex 2-Input AND Driver (En, Fr). 8 pp.

AND Gates
See Also: AND Buffer Gates; Digital Circuits; Gate Circuits
CECC CECC 90 106-008 ISSUE 1-85. UTE C 86-217; Digital Integrated Circuits in Accordance with FS 90 106; 54 ALS 08, 74 ALS 08; AND Gate (En, Fr) ADD 3 (En, Fr). 15 pp.
CECC CECC 90 106-013 ISSUE 1-85. UTE C 86-217; Digital Integrated Circuits in Accordance with FS 90 106; 54/74 ALS 11, 54/74 ALS 11; Positive AND Gate (En, Fr) ADD 3 (En, Fr). 15 pp.
CECC CECC 90 106-024 ISSUE 1-85. UTE C 86-217; Digital Integrated Circuits in Accordance with FS 90 106; 54/74 ALS 09, 54/72 ALS 09U; Positive AND Gates with Open Collector Outputs (En, Fr) ADD 3 (En, Fr). 15 pp.
CECC CECC 90 106-044 ISSUE 1-85. UTE C86-217; Digital Integrated Circuits in Accordance with FS 90 106; 54/74 ALS 15, 54/74 ALS 15U; Positive AND Gate with Open Collector Outputs (En, Fr) ADD 3 (En, Fr). 15 pp.
CECC CECC 90 106-045 ISSUE 1-85. UTE C 86-217; Digital Integrated Circuits in Accordance with FS 90 106; 54 ALS 21, 74 ALS 21; Positive AND Gate (En, Fr) ADD 3 (En, Fr). 15 pp.
CECC CECC 90 109-605 ISSUE 1-86. Digital Integrated Circuits in Accordance with FS 90 109; 54 HC 08, 74 HC 08; AND Gate (En, Fr). 5 pp.
CECC CECC 90 109-622 ISSUE 1-86. Digital Integrated Circuits in Accordance with FS 90 109; 54 HC 11, 74 HC 11; AND Gates (En, Fr). 5 pp.
CECC CECC 90 109-677 ISSUE 1-87. Digital Integrated Circuits in Accordance with FS 90 109; 54/74 HC 09; AND Gate with Open Drain Outputs (En, Fr). 6 pp.
CECC CECC 90 109-707 ISSUE 1-87. Digital Integrated Circuits in Accordance with FS 90 109; 54 HCT 11, 74 HCT 11; AND Gates (En, Fr). 6 pp.
CECC CECC 90 109-762 ISSUE 1-87. Digital Integrated Circuits in Accordance with FS 90 109; 54/74 HCT 08; AND Gates (En, Fr). 6 pp.

—2-Input
CECC CECC 90 101-007 ISSUE 1-87. UTE C 86-213/B 27; Digital Integrated Circuits in Accordance with FS 90 101; 54/64/74 F 08; Quadruple 2-Input AND Gates (En). 13 pp.
CECC CECC 90 101-008 ISSUE 1-87. UTE C 86-213/B 28; Digital Integrated Circuits in Accordance with FS 90 101; 54/64/74 F 09; Quad 2-Input AND Gates with Open Collector Outputs (En). 13 pp.
CECC CECC 90 102-014 ISSUE 1-81. BS CECC 90 102-014 Issue 1; Quadruple 2-Input Positive AND Gates (En). 16 pp.
CECC CECC 90 102-015 ISSUE 1-81. BS CECC 90102-015 Issue 1; Quad 2-Input Positive-AND Gates with Open Collector Outputs (En). 16 pp.
CECC CECC 90 103-007 ISSUE 1-81. BS CECC 90 103-007; Quad 2-Input Positive AND Gates (En) AMD 1 (En). 16 pp.
CECC CECC 90 103-008 ISSUE 1-81. BS CECC 90 103-008; Quad 2-Input Positive-AND Gates with Open Collector Outputs (En) AMD 1 (En). 17 pp.
CECC CECC 90 104-054 ISSUE 2. NL CECC 90 104-054 Issue 2; Digital Integrated Circuits in Accordance with FS 90 104; HEC/HEF 4081UB; Quadruple 2-Input AND Gate (En). 6 pp.
CECC CECC 90 104-148 ISSUE 1-81. BS CECC 90 104-148; Silicon Complementary MOS with (B) Buffered Outputs and Cavity Packaging (En). 22 pp.
CECC CECC 90 104-148 ISSUE 1-86. CEI CECC 90 104-148; Silicon Complementary MOS with (B) Buffered Outputs Cavity and Non Cavity Packaging (En). 2 pp.
CECC CECC 90 106-138 ISSUE 2-86. UTE C 86-217 ADD 2/GA 86; Digital Integrated Circuits in Accordance with FS 90 106; 54 ALS 808, 74 ALS 808; 2-Input AND Driver (En, Fr) ADD 3 (En, Fr). 15 pp.
CECC CECC 90 107-006 ISSUE 2-89. UTE C 86-218 ADD 2/FA 6; Digital Integrated Circuits in Accordance with FS 90 107; 54/74 F 08; AND Gate (En, Fr) ADD 3 (En, Fr). 9 pp.
CECC CECC 90 107-007 ISSUE 2-89. UTE C 86-218 ADD 2/FA 7; Digital Integrated Circuits in Accordance with FS 90 107; 54/74 F 11; AND Gate (En, Fr) ADD 3 (En, Fr). 9 pp.

—2-Input—Preferred Products List
CECC CECC MUAHAG Vol 7 IS 8-92. Preferred Products List; Active Microcircuits (En, Fe, Ge). 89 pp.

AND Gates *(Cont.)*
—3-Input
CECC CECC 90 102-016 ISSUE 1-81. BS CECC 90102-016 Issue 1; Triple 3-Input Positive AND Gate (En). 17 pp.
CECC CECC 90 102-017 ISSUE 1-81. BS CECC 90102-017 Issue 1; Triple Three-Input Positive-AND Gates with Open Collector Outputs (En). 16 pp.
CECC CECC 90 103-010 ISSUE 1-81. BS CECC 90 103-010; Triple 3-Input Positive AND Gates (En) AMD 1 (En). 17 pp.
CECC CECC 90 103-014 ISSUE 1-81. BS CECC 90 103-014; Triple 3-Input Positive AND Gates with Open Collector Outputs (En) AMD 1 (En). 17 pp.
CECC CECC 90 104-049 ISSUE 2. NL CECC 90 104-049 Issue 2; Digital Integrated Circuits in Accordance with FS 90 104; HEC/HEF 4073B; Triple 3-Input AND Gate (En). 6 pp.
CECC CECC 90 104-143 ISSUE 1-81. BS CECC 90 104-143; Silicon Complementary MOS with (B) Buffered Outputs and Cavity Packaging. 23 pp.
CECC CECC 90 104-143 ISSUE 1-86. CEI CECC 90 104-143; Silicon Complementary MOS with (B) Buffered Outputs Cavity and Non Cavity Packaging (En). 2 pp.

—3-Input—Preferred Products List
CECC CECC MUAHAG Vol 7 IS 8-92. Preferred Products List; Active Microcircuits (En, Fe, Ge). 89 pp.

—4-Input
CECC CECC 90 103-016 ISSUE 1-81. BS CECC 90 103-016; Dual 4-Input Positive AND Gates (En) AMD 1 (En). 17 pp.
CECC CECC 90 104-055 ISSUE 2. NL CECC 90 104-055 Issue 2; Digital Integrated Circuits in Accordance with FS 90 104; HEC/HEF 4082B; Dual 4-Input AND Gate (En). 6 pp.
CECC CECC 90 104-149 ISSUE 1-81. BS CECC 90 104-149; Dual 4 Input AND Gate (En). 23 pp.
CECC CECC 90 104-149 ISSUE 1-86. CEI CECC 90 104-149; Dual 4 Input AND Gate (En.). 2 pp.
CECC CECC 90 109-658 ISSUE 1-86. Digital Integrated Circuits in Accordance with FS 90 109; 54/74 HC 21; Dual 4-Input Positive AND Gates (En, Fr). 5 pp.
CECC CECC 90 109-766 ISSUE 1-87. Digital Integrated Circuits in Accordance with FS 90 109; 54/74 HCT 21; Dual 4-Input Positive AND Gate (En, Fr). 5 pp.

—4-Input—Preferred Products List
CECC CECC MUAHAG Vol 7 IS 8-92. Preferred Products List; Active Microcircuits (En, Fe, Ge). 89 pp.

AND NAND Gates
See Also: Gate Circuits
—8-Input
CECC CECC 90 104-151 ISSUE 1-81. BS CECC 50 104-151; Triple Gate Which May Be Combined to Give Dual 4-Input NAND Gate 2-Input NOR/OR Gate, 8-Input AND/NAND Gate (En). 29 pp.
CECC CECC 90 104-151 ISSUE 1-86. CEI CECC 90 104-149; Triple Gate: Dual 4-Input NAND Gate 2-Input NOR/OR Gate 8-Input AND/NAND Gate (En). 2 pp.

AND OR Circuits
Use: AND OR Gates

AND OR Gates
Use For: AND OR Circuits *See Also:* Digital Circuits; Gate Circuits

—2-2/3-3 Input
CECC CECC 90 109-805 ISSUE 1-88. Digital Integrated Circuits in Accordance with FS 90 109; 54/74 HC 58; AND-OR Gates (En, Fr). 6 pp.

AND OR INVERT Gates
Use For: AOI Gates *See Also:* Digital Circuits; Gate Circuits
CECC CECC 90 104-226 ISSUE 1-87. CEI-CECC 90 104-226; Silicon CMOS 4085 B with Buffered Outputs (En). 26 pp.
CECC CECC 90 107-009 ISSUE 2-89. UTE C 86-218 ADD 2/FA 9; Digital Integrated Circuits in Accordance with FS 90 107; 54/74 F 64; AND-OR Inverting Gate (En, Fr) ADD 3 (En, Fr). 9 pp.
CECC CECC 90 109-660 ISSUE 1-86. Digital Integrated Circuits in Accordance with FS 90 109; 54/74 HC 51; AND-OR-Inverting Gate (En, Fr). 5 pp.
MOD UK DSTAN 59-62: 90/072-68. Integrated Circuit Issue 1. 1 p.
MOD UK DSTAN 59-62: 90/074-68. Integrated Circuit Issue 1. 1 p.
MOD UK DSTAN 59-62: 90/083-68. Integrated Circuit Issue 1. 3 pp.

INTERNATIONAL AND NON-U.S. NATIONAL STANDARDS
SUBJECT INDEX
Anesthesia

AND OR INVERT Gates *(Cont.)*
MOD UK DSTAN 59-62: 90/084-68. Integrated Circuit Issue 1. 3 pp.
MOD UK DSTAN 59-62: 90/085-68. Integrated Circuit Issue 1. 3 pp.
MOD UK DSTAN 59-62: 90/086-68. Integrated Circuit Issue 1. 3 pp.

—**Expandable**
CECC CECC 90 104-154 ISSUE 1-81. BS CECC 90 104-154; Dual Expandable 'AND-OR-Invert' Gate with Three State Outputs (En). 28 pp.
CECC CECC 90 104-154 ISSUE 1-86. CEI CECC 90 104-154; Dual Expandable AND-OR-Invert Gate 3-State (En). 2 pp.

—**2-Input**
CECC CECC 90 102-063 ISSUE 1-83. BS CECC 90 102-063; Dual 2-Wide 2-Input AND-OR-Invert Gates (En). 17 pp.
CECC CECC 90 104-056 ISSUE 2. NL CECC 90 104-056 Issue 2; Digital Integrated Circuits in Accordance with FS 90 104; HEC/HEF 4085B; Dual 2-Wide 2-Input AND-OR-INVERT Gate (En). 6 pp.
CECC CECC 90 104-057 ISSUE 2. NL CECC 90 104-057 Issue 2; Digital Integrated Circuits in Accordance with FS 90 104; HEC/HEF 4086B; 4-Wide 2-Input AND-OR-INVERT Gate (En). 7 pp.

—**2-Input—Preferred Products List**
CECC CECC MUAHAG Vol 7 IS 8-92. Preferred Products List; Active Microcircuits (En, Fe, Ge). 89 pp.

—**2-3-2-2 Input**
CECC CECC 90 103-030 ISSUE 1-81. BS CECC 90 103-030 Issue 1; 2-Wide 3-Input, 2-Wide 2-Input AND-OR-Invert Gates (En) AMD 1 (En). 18 pp.

—**2-3/2-2 Input—Preferred Products List**
CECC CECC MUAHAG Vol 7 IS 8-92. Preferred Products List; Active Microcircuits (En, Fe, Ge). 89 pp.

—**4-Input**
CECC CECC 90 103-031 ISSUE 1-81. 4-Wide AND-OR Invert Gates (En) AMD 1 (En). 18 pp.
CECC CECC 90 103-032 ISSUE 1-81. BS CECC 90 103-032; 2-Wide 4-Input AND-OR-Invert Gates (En) AMD 1 (En). 17 pp.

—**4-Input—Preferred Products List**
CECC CECC MUAHAG Vol 7 IS 8-92. Preferred Products List; Active Microcircuits (En, Fe, Ge). 89 pp.

—**4-2-2-3 Input—Preferred Products List**
CECC CECC MUAHAG Vol 7 IS 8-92. Preferred Products List; Active Microcircuits (En, Fe, Ge). 89 pp.

—**4-2-3-2 Input**
CECC CECC 90 102-019 ISSUE 1-81. BS CECC 90102-019 Issue 1; 4-2-3-2 Input AND-OR-Invert Gates (En). 17 pp.
CECC CECC 90 102-020 ISSUE 1-81. BS 9405 F0607 to F0619 Issue 1; 4-2-3-2 Input AND-OR-Invert Gates with Open-Collector Outputs (En). 8 pp.

AND OR SELECT Gates
See Also: Digital Circuits; Gate Circuits

—**4-Bit**
CECC CECC 90 104-164 ISSUE 1-81. BS CECC 90 104-164; 4 Bit AND/OR Selector (En). 24 pp.
CECC CECC 90 104-164 ISSUE 1-86. CEI CECC 90 104-164; 4-Bit AND/OR Selector (En). 2 pp.

AND OR Selectors
Use: AND OR SELECT Gates

Anechoic Chambers
See Also: Acoustical Insulation; Acoustics; Test Chambers; Test Facilities
BSI BS 4196: Part 5-81. 1981 Sound Power Levels of Noise Sources Part 5: Precision Methods for Determination of Sound Power Levels for Sources in Anechoic and Semi-Anechoic Rooms. 26 pp.
ISO 3745-77. Acoustics—Determination of Sound Power Levels of Noise Sources—Precision Methods for Anechoic and Semi-Anechoic Rooms First Edition; (Amendment Slip-1982). 25 pp.
JIS Z 8732-86. Precision Methods for the Measurement of Sound Power Level in Anechoic and Hemi-Anechoic Rooms. 23 pp.

—**Noise**
BSI BS 4196: Part 8-91. 1991 Sound Power Levels of Noise Sources Part 8: Performance and Calibration of Reference Sound Sources. 12 pp.
ISO 6926-90. Acoustics—Determination of Sound Power Levels of Noise Sources—Requirements for the Performance and Calibration of Reference Sound Sources First Edition. 11 pp.

Anechoic Chambers *(Cont.)*
—**Noise** *(Cont.)*
SAA AS 1217.6-85. Acoustics—Determination of Sound Power Levels of Noise Sources—Part 6: Precision Methods for Anechoic and Hemi-Anechoic Rooms. 22 pp.

Anemometers
See Also: Measuring Instruments; Speedometers; Wind Velocity

—**Biram**
CNS M2016-80. Biram Type Anemometer (Feb)(5244).
JIS M 7604-53. Biram Type Anemometer.

—**Mining Equipment—Portable—Thermal**
CNS M2026-80. Thermal Anemometer for Mines (Portable Type) (Feb)(5254).
JIS M 7606-78. Thermal Anemometer for Mines (Portable Type).

—**Portable—Thermal**
CNS T2013-82. Portable Type Thermal Anemometer (Jan)(8455).
CNS T4002-82. Method of Test for Portable Type Thermal Anemometer (Jan)(8456).
JIS T 8202-76. Portable Type Thermal Anemometer.

Aneroid Barographs
Use: Barographs

Anesthesia
Use: Anesthesiology

Anesthesia Breathing Systems
See Also: Anesthesia Equipment; Anesthetics; Medical Electrical Equipment; Medical Equipment
CSA CAN/CSA-Z168.9-92. Breathing Systems for Use in Anaesthesia; (Gen Instr 1). 47 pp.

—**Breathing Tubes**
BSI BS 6151-92. 1992 Breathing Tubes for Use with Anaesthetic Apparatus and Ventilators (ISO 5367: 1991). 12 pp.
ISO 5367-91. Breathing Tubes Intended for Use with Anaesthetic Apparatus and Ventilators Third Edition. 9 pp.

—**Catheters—Mountings**
BSI BS 7143-89. 1989 Catheter Mounts (Flexible Adaptors) for Use with Medical Breathing Systems. 12 pp.
BSI BS 7143-01. 1989 Amd 1 Catheter Mounts (Flexible Adaptors) for Use with Medical Breathing Systems (AMD 7063) June 15, 1992. 13 pp.

—**Conical Connectors**
BSI BS 3849: Part 1-88. 1988 Amd 1 Conical Connectors for Anaesthetic and Respiratory Equipment Part 1: Cones and Sockets (AMD 6422) June 29, 1990. 21 pp.
ISO 5356 Pt 1-87. Anaesthetic and Respiratory Equipment—Conical Connectors—Part 1: Cones and Sockets First Edition; (Amendment 1-1993). 31 pp.

—**Firesafety**
SAA AS 1169-82. Minimizing of Combustion Hazards Arising from the Medical Use of Flammable Anaesthetic Agents. 23 pp.

—**Interruption Monitors**
CSA Z168.10-M1988. Testing of Breathing Gas Interruption Monitors for Use During Anaesthesia. 25 pp.

—**Reservoir Bags**
BSI BS 3353-87. 1987 Anaesthetic Reservoir Bags. 7 pp.
ISO 5362-86. Anaesthetic Reservoir Bags Second Edition. 5 pp.

—**Workstations**
CEN PREN 740-92. Medical Electrical Equipment—Anaesthetic Workstations and Their Modules—Particular Requirements. 135 pp.

—**Workstations—Modules**
CEN PREN 740-92. Medical Electrical Equipment—Anaesthetic Workstations and Their Modules—Particular Requirements. 135 pp.

Anesthesia Equipment
Scope Note: For additional listings, use a more specific term *See Also:* Anesthesia Breathing Systems; Anesthesia Gas Machines; Anesthesia Ventilators; Anesthesiology; Anesthetic Gas Scavenging Systems; Anesthetic Vaporizers; Anesthetics; Medical Electrical Equipment; Medical Equipment
JIS T 7201-90. Anesthetic Machines. 47 pp.
SAA AS 4059-92. Anaesthetic Machines—Non-Electrical—for Use with Humans (in Professional Package 17D). 18 pp.

Anesthesia Equipment *(Cont.)*
—**Alarm Signals**
BSI BS 7618: Part 1-92. 1992 Alarm Signals for Anaesthesia and Respiratory Care Part 1: Specification for Visual Alarm Signals (ISO 9703-1: 1992). 8 pp.
ISO 9703 Pt 1-92. Anaesthesia and Respiratory Care Alarm Signals—Part 1: Visual Alarm Signals First Edition. 7 pp.

—**Heat/Moisture Exchangers**
ISO 9360-92. Anaesthetic and Respiratory Equipment—Heat and Moisture Exchangers for Use in Humidifying Respired Gases in Humans First Edition. 14 pp.

—**Heat/Moisture Exchangers—Safety**
ISO 9360-92. Anaesthetic and Respiratory Equipment—Heat and Moisture Exchangers for Use in Humidifying Respired Gases in Humans First Edition. 14 pp.

—**Safety**
BSI BS 5724: Sec 2.13-90. 1990 Medical Electrical Equipment Part 2: Particular Requirements for Safety Section 2.13: Anaesthetic Machines. 24 pp.
BSI BS 5724: Sec 2.13-01. 1990 Amd 1 Medical Electrical Equipment Part 2: Particular Requirements for Safety Section 2.13: Specification for Anaesthetic Machines (IEC 601-2-13: 1989) (AMD 7447) January 15, 1993. 24 pp.
CENELEC HD 395.2.13 S1-89. Medical Electrical Equipment Part 2: Particular Requirements for the Safety of Anaesthetic Machines. 3 pp.
IEC 601 Pt 2-13-89. Medical Electrical Equipment Part 2: Particular Requirements for the Safety of Anaesthetic Machines First Edition. 43 pp.

Anesthesia Gas Machines
Use For: Analgesic Gas Machines; Continuous Flow Inhalation Anesthetic Apparatus
See Also: Anesthesia Equipment; Anesthetics; Medical Electrical Equipment; Medical Equipment
BSI BS 4272: Part 1-68. (OBSOLESCENT) 1968 Anaesthetic and Analgesic Machines Part 1: Anaesthetic Machines of the On-Demand Type Supplied with Nitrous Oxide and Oxygen from Separate Containers. 12 pp.
BSI BS 4272: Part 2-68. 1968 Anaesthetic and Analgesic Machines Part 2: Analgesic Machines of the On-Demand Type Supplied with Pre-Mixed Nitrous Oxide-Oxygen from a Single Container. 8 pp.
BSI BS 4272: Part 3-89. (WITHDRAWN) 1989 Anaesthetic and Analgesic Machines Part 3: Specification for Continuous Flow Anaesthetic Machines (N). 19 pp.
BSI BS 4272: Part 3-01. 1989 Amd 1 Anaesthetic and Analgesic Machines Part 3: Continuous Flow Anaesthetic Machines (AMD 6506) December 21, 1990. 21 pp.
CSA CAN3-Z168.3-M84. Anaesthetic Machines for Medical Use,; (Gen Instr 1) (Supplement 1 M1991). 48 pp.
ISO 5358-92. Anaesthetic Machines for Use with Humans Second Edition. 21 pp.

—**Conical Connectors**
BSI BS 3849: Part 2-88. 1988 Amd 1 Conical Connectors for Anaesthetic and Respiratory Equipment Part 2: Screw-Threaded Weight-Bearing Connectors (AMD 6460) June 29, 1990. 8 pp.
BSI BS 3849: Part 4-90. 1990 Conical Connectors for Anaesthetic and Respiratory Equipment Part 4: 8.5mm Cones and Sockets. 10 pp.
ISO 5356 Pt 2-87. Anaesthetic and Respiratory Equipment—Conical Connectors—Part 2: Screw-Threaded Weight-Bearing Connectors First Edition. 6 pp.

Anesthesia Ventilators
Use For: Anaesthesia Ventilators
See Also: Anesthesia Equipment; Anesthetics; Medical Electrical Equipment; Medical Equipment; Respirators; Ventilation Equipment
CSA CAN/CSA-Z168.5.1-M87. Anaesthesia Ventilators; (Gen Instr 1). 29 pp.

—**Breathing Tubes**
BSI BS 6151-92. 1992 Breathing Tubes for Use with Anaesthetic Apparatus and Ventilators (ISO 5367: 1991). 12 pp.
ISO 5367-91. Breathing Tubes Intended for Use with Anaesthetic Apparatus and Ventilators Third Edition. 9 pp.

—**Breathing Tubes—Veterinary**
BSI BS 6155-90. 1990 Tracheal Tubes for Large Animals in Veterinary Anaesthesia. 10 pp.

—**Catheters—Mountings**
BSI BS 7143-89. 1989 Catheter Mounts (Flexible Adaptors) for Use with Medical Breathing Systems. 12 pp.

INDUSTRY STANDARDS

Anesthesia Ventilators *(Cont.)*
—Catheters—Mountings *(Cont.)*
BSI BS 7143-01. 1989 Amd 1 Catheter Mounts (Flexible Adaptors) for Use with Medical Breathing Systems (AMD 7063) June 15, 1992. 13 pp.

—Conical Connectors
BSI BS 3849: Part 2-88. 1988 Amd 1 Conical Connectors for Anaesthetic and Respiratory Equipment Part 2: Screw-Threaded Weight-Bearing Connectors (AMD 6460) June 29, 1990. 8 pp.

BSI BS 3849: Part 4-90. 1990 Conical Connectors for Anaesthetic and Respiratory Equipment Part 4: 8.5mm Cones and Sockets. 10 pp.

ISO 5356 Pt 1-87. Anaesthetic and Respiratory Equipment—Conical Connectors—Part 1: Cones and Sockets First Edition; (Amendment 1-1993). 31 pp.

ISO 5356 Pt 2-87. Anaesthetic and Respiratory Equipment—Conical Connectors—Part 2: Screw-Threaded Weight-Bearing Connectors First Edition. 6 pp.

—Safety
BSI BS 5724: Sec 2.13-90. 1990 Medical Electrical Equipment Part 2: Particular Requirements for Safety Section 2.13: Anaesthetic Machines. 24 pp.

BSI BS 5724: Sec 2.13-01. 1990 Amd 1 Medical Electrical Equipment Part 2: Particular Requirements for Safety Section 2.13: Specification for Anaesthetic Machines (IEC 601-2-13: 1989) (AMD 7447) January 15, 1993. 24 pp.

IEC 601 Pt 2-13-89. Medical Electrical Equipment Part 2: Particular Requirements for the Safety of Anaesthetic Machines First Edition. 43 pp.

Anesthesiology
See Also: Anesthesia Equipment; Anesthetics

—Glossaries
BSI BS 6015-80. 1980 Amd 1 Glossary of Terms Used in Anaesthesiology. 13 pp.

ISO 4135-79. Anaesthesiology—Vocabulary First Edition; (Amendment 1-1986). 21 pp.

—Labels
CSA CAN/CSA-Z327-M91. Standard for User-Applied Drug Labels in Anaesthesia and Critical Care; (Gen Instr 1). 16 pp.

Anesthetic Gas Scavenging Systems
See Also: Anesthesia Equipment; Anesthetics; Medical Electrical Equipment; Medical Equipment
CSA CAN3-Z168.8-M82. Anaesthetic Gas Scavenging Systems; (Gen Instr 1). 28 pp.

—Conical Connectors
BSI BS 3849: Part 1-88. 1988 Amd 1 Conical Connectors for Anaesthetic and Respiratory Equipment Part 1: Cones and Sockets (AMD 6422) June 29, 1990. 21 pp.

ISO 5356 Pt 1-87. Anaesthetic and Respiratory Equipment—Conical Connectors—Part 1: Cones and Sockets First Edition; (Amendment 1-1993). 31 pp.

—Safety
BSI BS 6834-87. 1987 Active Anaesthetic Gas Scavenging Systems. 16 pp.

Anesthetic Mixtures
Use: Anesthetics

Anesthetic Vaporizers
Use For: Anaesthetic Vaporizors
See Also: Anesthesia Equipment; Anesthetics; Medical Electrical Equipment; Medical Equipment

—Conical Connectors
BSI BS 3849: Part 1-88. 1988 Amd 1 Conical Connectors for Anaesthetic and Respiratory Equipment Part 1: Cones and Sockets (AMD 6422) June 29, 1990. 21 pp.

—Filling Systems
ISO 5360-93. Anaesthetic Vaporizers—Agent-Specific Filling Systems First Edition. 21 pp.

—Keyed Fillers
CSA CAN3-Z168.4-M83. Keyed Filling Devices for Anaesthetic Vaporizers; (Gen Instr 1). 24 pp.

Anesthetics
Use For: Anaesthetics; Anesthetic Mixtures
See Also: Anesthesia Breathing Systems; Anesthesia Equipment; Anesthesia Gas Machines; Anesthesia Ventilators; Anesthesiology; Anesthetic Gas Scavenging Systems; Anesthetic Vaporizers; Drugs

—Flammable
CSA CAN/CSA-C22. 2NO601.1-M90. Medical Electrical Equipment, Part 1: General Requirements for Safety; (Gen Instr 1). 240 pp.

Anesthetics *(Cont.)*
—Flammable—Enclosure—Leakage
CSA CAN/CSA-C22. 2NO601.1-M90. Medical Electrical Equipment, Part 1: General Requirements for Safety; (Gen Instr 1). 240 pp.

—Flammable—Hazardous Environments
CSA CAN/CSA-C22. 2NO601.1-M90. Medical Electrical Equipment, Part 1: General Requirements for Safety; (Gen Instr 1). 240 pp.

—Flammable—Identification Systems
CSA CAN/CSA-C22. 2NO601.1-M90. Medical Electrical Equipment, Part 1: General Requirements for Safety; (Gen Instr 1). 240 pp.

—Flammable—Safety
CSA CAN/CSA-C22. 2NO601.1-M90. Medical Electrical Equipment, Part 1: General Requirements for Safety; (Gen Instr 1). 240 pp.

—Flammable—Veterinary
CSA CAN/CSA-C22. 2NO601.1-M90. Medical Electrical Equipment, Part 1: General Requirements for Safety; (Gen Instr 1). 240 pp.

ANFO
Use: Ammonium Nitrate Fuel Oil Explosives

Angle Brackets
Use For: Angle Connectors *See Also:* Brackets

—Masonry
CEN PREN 845-1-92. Specification for Ancillary Components for Masonry—Part 1: Ties, Straps, Hangers, Brackets and Support Angles. 26 pp.

Angle Check Valves
See Also: Angle Valves; Check Valves; Valves

—Bronze—Ships
CNS F3163-82. Marine Bronze 5 kgf/cm2 Screw-Down Check Angle Valves (Feb)(8493).

CNS F3205-82. Marine Bronze 5 kgf/cm2 Screw-Down Check Angle Valves (Union Bonnet Type) (Oct)(9488).

CNS F3207-82. Marine Bronze 16 kgf/cm2 Screw-Down Check Angle Valves (Union Bonnet Type) (Oct)(9490).

CNS F3209-82. Marine Bronze 5 kgf/cm2 Lift Check Angle Valves (Union Bonnet Type) (Oct)(9492).

JIS F 7352-87. Marine Bronze 5 K Screw-Down Check Angle Valves.

JIS F 7410-87. Marine Bronze 16 K Screw-Down Check Angle Valves.

JIS F 7412-88. Bronze 5 K Screw-Down Check Angle Valves for Marine Use (Union Bonnet Type).

JIS F 7414-88. Bronze 16 K Screw-Down Check Angle Valves for Marine Use (Union Bonnet Type).

JIS F 7416-88. Bronze 5 K Lift Check Angle Valves for Marine Use (Union Bonnet Type).

JIS F 7418-88. Bronze 16 K Lift Check Angle Valves for Marine Use (Union Bonnet Type).

—Cast Iron—Ships
CNS F3165-82. Marine Cast Iron 5 kgf/cm2 Screw-Down Check Angle Valves (Feb)(8495).

CNS F3174-82. Marine Cast Iron 5 kgf/cm2 Lift Check Angle Valves (Apr)(8690).

CNS F3186-82. Marine Cast Iron 16 kgf/cm2 Screw-Down Check Angle Valves (Jun)(8962).

JIS F 7354-89. Cast Iron 5 K Screw-Down Check Angle Valves for Marine Use.

JIS F 7359-89. Cast Iron 5 K Lift Check Angle Valves for Marine Use.

JIS F 7376-89. Cast Iron 10 K Screw-Down Check Angle Valves for Marine Use.

JIS F 7378-89. Cast Iron 16 K Screw-Down Check Angle Valves for Marine Use.

—Cast Steel—Ships
CNS F3008-86. Cast Iron Screw-Down Check Angle Valves for Marine Use (10kgf/cm2) (Mar)(3813).

JIS F 7472-88. Cast Steel 10 K Screw-Down Check Angle Valves for Marine Use.

JIS F 7474-88. Cast Steel 20 K Screw-Down Check Angle Valves for Marine Use.

—Steel
BSI BS 1873-75. 1975 Amd 1 Steel Globe and Globe Stop and Check Valves (Flanged and Butt-Welding Ends) for the Petroleum, Petrochemical and Allied Industries (AMD 6564) July 31, 1990. 34 pp.

Angle Cocks
Use: Angle Valves

Angle Connectors
Use: Angle Brackets

Angle Cutters
Use: Milling Cutters—Angle

Angle Gages
See Also: Gages
CNS B3209-71. Moveable Angle Gauge (Tentative) (Apr)(3224).

Angle Globe Valves
See Also: Angle Valves; Globe Valves

—Steel
BSI BS 1873-75. 1975 Amd 1 Steel Globe and Globe Stop and Check Valves (Flanged and Butt-Welding Ends) for the Petroleum, Petrochemical and Allied Industries (AMD 6564) July 31, 1990. 34 pp.

Angle Grinders
Use For: Right Angle Grinders; Small Angle Grinders *See Also:* Grinders

—Electric—Portable
MOD UK DSTAN 51-19: Part 2-92. Hand Tools, Powered Part 2: Electric Issue 1. 31 pp.

Angle Iron
See Also: Channels (Structural Shapes); Steels; Structural Members
CEN PREN 10056-91. Structural Steel Equal and Unequal Leg Angles—Tolerances on Shape and Dimensions. 9 pp.

CNS G2087-87. Method of Test for Hot Rolled Stainless Steel Equal Leg Angles (Mar)(7795).

—Cold Rolled
JIS G 4320-91. Cold Formed Stainless Steel Equal Leg Angles. 17 pp.

—Hot Rolled
ISO 657 Pt 1-89. Hot-Rolled Steel Sections—Part 1: Equal-Leg Angles—Dimensions First Edition. 7 pp.

ISO 657 Pt 2-89. Hot-Rolled Steel Sections—Part 2: Unequal-Leg Angles—Dimensions First Edition. 8 pp.

ISO 657 Pt V-76. Hot-Rolled Steel Sections—Part V: Equal-Leg Angles and Unequal-Leg Angles—Tolerances for Metric and Inch Series First Edition. 5 pp.

JIS G 4317-91. Hot Rolled Stainless Steel Equal Leg Angles. 17 pp.

Angle Joints
See Also: Joints

—Slotted
BSI BS 4345-68. 1968 Amd 2 Slotted Angles. 15 pp.

Angle Modulation
See Also: Modulation

—Carrier Systems—Fixed Satellite Services
CCIR RECMN 446-2-78. Carrier Energy Dispersal for Systems Employing Angle Modulation by Analogue Signals or Digital Modulation in the Fixed-Satellite Service—Section 4C—Earth Station and Baseband Characteristics —Earth Station Antennas —Maintenance of Earth Stations. 1 p.

CCIR RECMN 446-3-92. Carrier Energy Dispersal for Systems Employing Angle Modulation by Analogue Signals or Digital Modulation in the Fixed-Satellite Service—Section 4C—Earth Station and Baseband Characteristics —Earth Station Antennas —Maintenance of Earth Stations. 20 pp.

—Carrier Systems—Power Density—Fixed Satellite Services
CCIR Report 792-3-90. Calculation of the Maximum Power Density Averaged over 4 kHz of an Angle-Modulated Carrier—Section 4/9B—Co-Ordination and Interference Calculations. 6 pp.

Angle of Friction
Use: Angle of Repose

Angle of Repose
Use For: Angle of Friction *See Also:* Stability Testing

—Aluminum Oxide
BSI BS 4140: Part 9-86. 1986 Methods of Test for Aluminium Oxide Part 9: Measurement of the Angle of Repose. 6 pp.

ISO 902-76. Aluminium Oxide Primarily Used for the Production of Aluminium—Measurement of the Angle of Repose First Edition. 6 pp.

—Construction Materials
DIN ENGL 1055 Pt 1-78. Design Loads for Buildings; Stored Materials, Building Materials and Structural Members; Dead Load and Angle of Friction (July). 15 pp.

INTERNATIONAL AND NON-U.S. NATIONAL STANDARDS
SUBJECT INDEX

Angle of Repose (Cont.)
—Fertilizers
BSI BS 5551: Sec 3.6-89. 1989 Fertilizers Part 3: Physical Properties Section 3.6: Method for Determination of Static Angle of Repose of Solid Fertilizers. 6 pp.
ISO 8398-89. Solid Fertilizers—Measurement of Static Angle of Repose First Edition. 6 pp.

—Surfactants
ISO 4324-77. Surface Active Agents—Powders and Granules—Measurement of the Angle of Repose First Edition. 5 pp.

Angle of Slide
—Hexamethylenetetramine
MOD UK M 879/69. Examination of Hexamine.

Angle of Slip
Use: Angle of Slide

Angle Pins
See Also: Pins
—Dies
BSI BS 7564: Part 8-92. 1992 Tools for Moulding Part 8: Angle Pins—Basic Dimensions (ISO 8404: 1986). 7 pp.
CNS B2784-87. Angle Pin of Die Casting Molds (Sep)(12071).
ISO 8404-86. Angle Pins—Basic Dimensions First Edition. 4 pp.

—Molds (Casting)
CNS B2772-87. Angle Pins of Molds for Plastics (May)(11932).

Angle Plates
See Also: Plates (Supports)
BSI BS 5535-78. 1978 Right Angle and Box Angle Plates (Supersedes BS 3080: 1959). 8 pp.
SAA AS B263-68. Precision Angle Plates Reconfirmed 1987. 7 pp.

Angle Valves
Use For: Angle Cocks See Also: Angle Check Valves; Angle Globe Valves; Valves
CNS B2496-80. Face-to-Face and End-to-End Dimensions of Angle Valves (Jul)(5712).
CNS B2504-88. Bronze Screwed Angle Valves (Jul)(5965).
CNS B2509-85. Bronze Flanged Angle Valves (10kgf/cm2) (Mar)(5970).
CNS B2512-88. Cast Iron Flanged Angle Valves (10kgf/cm2) (Jul)(5973).
CNS B2536-82. Cast Steel Flanged Angle Valves (10 kgf/cm2) (Sep)(6883).
CNS B2550-82. Cast Steel Flanged Angle Valves (20 kgf/cm2) (Mar)(7113).
JIS B 2002-87. Face-to-Face and End-to-End Dimensions of Valves.
JIS B 2002-68. Face-to-Face and End-to-End Dimensions of Valves. 14 pp.
JIS B 2011-88. Bronze Gate, Globe, Angle and Check Valves. 41 pp.

—Compressed Air—Marine
CNS F3147-81. Marine Forged Steel Angle Valves for Compressed Air (Nov)(8111).
JIS F 7337-88. Forged Steel Angle Air Valves for Marine Use.
JIS F 7475-88. Cast Steel Angle Air Valves for Marine Use.

—Pneumatic—Ships
CNS F3147-81. Marine Forged Steel Angle Valves for Compressed Air (Nov)(8111).
JIS F 7337-88. Forged Steel Angle Air Valves for Marine Use.
JIS F 7475-88. Cast Steel Angle Air Valves for Marine Use.

—Rolling Stock
CNS E1051-85. Angle Cocks for Railway Rolling Stock (Nov)(11409).
JIS E 4102-91. Angle Cocks for Railway Rolling Stock.

—Ships
CNS F3004-86. Cast Iron Angle Valves for Marine Use (5kgf/cm2) (Mar)(3809).
CNS F3097-81. Marine Bronze 5 kgf/cm2 Angle Valves (Apr)(7240).
CNS F3099-81. Marine Bronze 16 kgf/cm2 Angle Valves (Apr)(7242).
CNS F3101-81. Marine Cast Iron 10 kgf/cm2 Angle Valves (Apr)(7244).
CNS F3103-81. Marine Cast Iron 16 kgf/cm2 Angle Valves (Apr)(7246).
CNS F3111-81. Marine Cast Steel 5 kgf/cm2 Angle Valves (Jul)(7683).
CNS F3113-81. Marine Cast Steel 20 kgf/cm2 Angle Valves (Jul)(7685).

Angle Valves (Cont.)
—Ships (Cont.)
CNS F3115-81. Marine Cast Steel 30 kgf/cm2 Angle Valves (Jul)(7687).
CNS F3117-81. Marine Cast Steel 40 kgf/cm2 Angle Valves (Jul)(7689).
CNS F3132-81. Marine Cast Steel 10 kgf/cm2 Angle Valves (Sep)(7899).
CNS F3134-81. Marine Malleable Iron 5 kgf/cm2 Angle Valves (Sep)(7901).
CNS F3136-81. Marine Malleable Iron 16 kgf/cm2 Angle Valves (Sep)(7903).
CNS F3138-81. Marine Forged Steel 40 kgf/cm2 Angle Valves (Sep)(7905).
CNS F3158-82. Marine Bronze 5 kgf/cm2 Angle Valves (Union Bonnet Type) (Jan)(8273).
CNS F3160-82. Marine Bronze 16 kgf/cm2 Angle Valves (Union Bonnet Type) (Jan)(8275).
CNS F3161-82. Marine Hull Cast Steel Angle Valves (Feb)(8491).
CNS F3191-82. Marine Bronze 20 kgf/cm2 Angle Valves (Aug)(9253).
CNS F3194-82. Marine Cast Iron 3 kgf/cm2 Angle Valves (Aug)(9256).
CNS F3196-82. Marine Bronze 3 kgf/cm2 Angle Valves (Aug)(9258).
JIS F 7302-87. Marine Bronze 5 K Angle Valves.
JIS F 7304-87. Marine Bronze 16 K Angle Valves.
JIS F 7306-89. Cast Iron 5 K Angle Valves for Marine Use.
JIS F 7308-89. Cast Iron 10 K Angle Valves for Marine Use.
JIS F 7310-88. Cast Iron 16 K Angle Valves for Marine Use.
JIS F 7312-88. Cast Steel 5 K Angle Valves for Marine Use.
JIS F 7314-89. Cast Steel 20 K Angle Valves for Marine Use.
JIS F 7316-88. Cast Steel 30 K Angle Valves for Marine Use.
JIS F 7318-88. Cast Steel 40 K Angle Valves for Marine Use.
JIS F 7320-88. Cast Steel 10 K Angle Valves for Marine Use.
JIS F 7322-88. Malleable Iron 5 K Angle Valves for Marine Use.
JIS F 7324-88. Malleable Iron 16 K Angle Valves for Marine Use.
JIS F 7330-88. Forged Steel 40 K Angle Valves for Marine Use.
JIS F 7347-88. Bronze 5 K Angle Valves (Union Bonnet Type) for Marine Use.
JIS F 7349-88. Bronze 16 K Angle Valves (Union Bonnet Type) for Marine Use.
JIS F 7350-86. Marine Hull Cast Steel Angle Valves.
JIS F 7389-88. Bronze 20 K Angle Valves for Marine Use.
JIS F 7404-87. Marine Hull Bronze Angle Valves.
JIS F 7422-89. Forged Steel 20 K Angle Valves for Marine Use.

—Waterworks
CNS B2617-84. Angle Valve for Water System (Jun)(8086). 3 pp.

Angles (Geometry)
—Cones
ISO 1119-75. Series of Conical Tapers and Taper Angles First Edition. 5 pp.

—Drills
CNS B1012-53. Drill Angles (Feb)(237) (R 1970).

—Extruded—Aircraft
SBAC RS 2-18 ISSUE 1. Angle Extrusions.
SBAC RS 31-42 ISSUE 1. Bulb Angle Extrusions.

—Prisms
ISO 2538-74. Limits and Fits—Series of Angles and Slopes on Wedges and Prisms First Edition. 7 pp.

—Tables (Data)
SAA AS 1377.1-73. Conversion Tables—Part 1: Length, Angle and Velocity Corrig. (2). 39 pp.

Anglo Polish Sailplanes SZD45A Ogar
See Also: Aircraft

—Foreign Airworthiness Directives
CAA. Anglo Polish Sailplanes SZD45A Ogar (Foreign Airworthiness Directives). 1 p.

Angular Contact Bearings
See Also: Antifriction Bearings; Bearings

—Needle
DIN ENGL 5429 Pt 2-87. Rolling Bearings; Needle Bearing Combinations; Needle Angular Contact Ball Bearings (Aug). 2 pp.

Angular Contact Bearings (Cont.)
—Needle—Designations
DIN ENGL 5429 Pt 2-87. Rolling Bearings; Needle Bearing Combinations; Needle Angular Contact Ball Bearings (Aug). 2 pp.

Angular Distance
See Also: Dimensions

—Geostationary Satellites—Satellite Networks
CCIR Report 772-2-86. Orbital Spacing Considerations for a Mobile-Satellite Service —Section 8H—Efficient Use of the Radio Spectrum Characteristics and Sharing of Frequency Resources. 13 pp.

Angular Position Detectors
Use: Position Detectors—Rotary—Angular

Angular Separation
Use: Angular Distance

Anhydrides
Scope Note: Use a more specific term See: Acetic Anhydride; Boric Anhydrides; Maleic Anhydride; Phthalic Anhydride; Phthalic Anhydride-13-8

Anhydrous Alcohol
Use: Ethanol

Anhydrous Ammonia
Use: Ammonia

Anhydrous Hydrogen Fluoride
See Also: Hydrogen Fluoride

—Safety—Information Cards
SAA AS 2508.8.01 2-91. Safe Storage and Handling Information Cards for Hazardous Materials—Part 8.012: Hydrogen Fluoride (Anhydrous) (in Professional Package 38).

Anhydrous Soaps
Use: Soaps

Aniline
See Also: N-Benzyl-N-Ethylaniline; o-Chloroaniline; p-Chloroaniline; Diethylaniline; N,N-Dimethylaniline; N-Ethylaniline; Hazardous Materials; Methyl Violet; N-Methylaniline; Sulfanilamide
CNS K2199-88. Aminobenzenes (Mar)(12234).
CNS K7067-62. Chemical Reagent (Aniline) (Jul)(1567).
JIS K 4109-84. Aminobenzenes (Aniline,o-Toluidin,p-Toluinene, m-Toluidine, N-Ethyl-m-Toluidine).
JIS K 8042-86. Aniline.

—Highway Transportation—Emergency Procedures
SAA AS 1678.6.0. 001-83. Emergency Procedure Guides—Transport—Part 6.0.001—Aniline (Amino benzene). 1 p.

Aniline Black
See Also: Aniline Hydrochloride; Pigments

—Pigments
CNS K2223-88. Aniline Black (May)(12311).
JIS K 5245-71. Aniline Black. 5 pp.

Aniline Hydrochloride
CNS K7068-62. Chemical Reagent (Aniline Chloride) (Aniline Hydrochloride) (Jul)(1568).

Aniline Point
Use For: Mixed Aniline Point

—Petroleum Products
BSI BS 2000: Part 2-82. (WITHDRAWN) 1982 Petroleum and Its Products Part 2: Aniline Point and Mixed Aniline Point of Petroleum Products and Hydrocarbon Solvents (Superseded by 7391: 1990). 12 pp.
BSI BS 7391-90. 1990 Method for Determination of Aniline Point and Mixed Aniline Point of Petroleum Products and Hydrocarbon Solvents. 16 pp.
CNS K6366-87. Method of Test for Aniline Point and Mixed Aniline Point of Petroleum Products and Hydrocarbon Solvents (Jul)(3575).
ISO 2977-89. Petroleum Products and Hydrocarbon Solvents—Determination of Aniline Point and Mixed Aniline Point Second Edition; (Corrigendum 1-1992). 13 pp.
JIS K 2256-85. Testing Methods for Aniline Point and Mixed Aniline Point of Petroleum Products.

Aniline Point (Cont.)
—Solvents
BSI BS 2000: Part 2-82. (WITHDRAWN) 1982 Petroleum and Its Products Part 2: Aniline Point and Mixed Aniline Point of Petroleum Products and Hydrocarbon Solvents (Superseded by 7391: 1990). 12 pp.
BSI BS 7391-90. 1990 Method for Determination of Aniline Point and Mixed Aniline Point of Petroleum Products and Hydrocarbon Solvents. 16 pp.
CNS K6366-87. Method of Test for Aniline Point and Mixed Aniline Point of Petroleum Products and Hydrocarbon Solvents (Jul)(3575).
ISO 2977-89. Petroleum Products and Hydrocarbon Solvents—Determination of Aniline Point and Mixed Aniline Point Second Edition; (Corrigendum 1-1992). 13 pp.

Aniline Sulfate
CNS K7069-62. Chemical Reagent (Aniline Sulfate) (Jul)(1569).

Animal Breeding
Use For: Breeding (Animals) *See Also:* Animal Diseases; Animal Husbandry
EC 89/608/EEC-89. Council Directive on Mutual Assistance Between the Administrative Authorities of the Member States and Cooperation Between the Latter and the Commision to Ensure the Correct Application of Legislation on Veterinary and. 4 pp.
EC 89/662/EEC-89. Council Directive Concerning Veterinary Checks in Intra-Community Trade with a View to the Completion of the International Market. 10 pp.

—Cattle
EC 87/328/EEC-87. Council Directive on the Acceptance for Breeding Purposes of Pure-Bred Breeding Animals of the Bovine Species. 2 pp.

—Cattle—Embryo Collection—Health Certification
EC COM(88) 785-89. Proposal for a Council Regulation (EEC) on Animal Health Conditions Governing Intra-Community Trade in and Importation from Third Countries of Embryos of Domestic Animals of the Bovine Species. 25 pp.
EC 89/556/EEC-89. Council Directive on Animal Health Conditions Governing Intra—Community Trade in and Importation from Third Countries of Embryos of Domestic Animals of the Bovine Species. 11 pp.

—Cattle—Semen—Bacteria Count Methods
ISO TR8607-91. Artificial Insemination of Animals—Frozen Semen of Breeding Bulls—Enumeration of Living Aerobic Micro-Organisms First Edition. 6 pp.

—Cattle—Semen Collection—Health Certification
EC COM(83) 512-83. Proposal for a Council Directive on Animal Health Problems Affecting Intra-Community Trade and Imports from Third Countries of Semen of Domestic Animals of the Bovine and Porcine Species. 37 pp.
EC COM(86) 657-86. Amendment to the Proposal for a Council Directive on Animal Health Problems Affecting Intra-Community Trade and Imports from Third Countries of Semen of Domestic Animals of the Bovine and Porcine Species. 4 pp.
EC 88/407/EEC-88. Council Directive Laying Down the Animal Health Requirements Applicable to Intra-Community Trade in the Imports of Deep Frozen Semen of Domestic Animals of the Bovine Species. 14 pp.
EC 92/571/EEC-92. Commission Decision Relating to New Transitional Measures Which are Necessary to Facilitate the Move to the System of Veterinary Checks Provided for in Council Directive 90/675/EEC. 3 pp.

—Goats
EC COM(87) 591-87. Proposal for a Council Directive Concerning Pure-Bred Breeding Sheep and Goats. 7 pp.
EC COM(88) 742-88. Proposal for a Council Regulation (EEC) on Animal Health Conditions Governing Intra-Community Trade in Ovine and Caprine Animals: Proposal for a Council Directive Amending Directive 72/462/EEC. 42 pp.
EC 89/361/EEC-89. Council Directive Concerning Pure-Bred Breeding Sheep and Goats. 2 pp.

—Pedigree Records
EC COM(88) 598-88. Proposal for a Council Regulation (EEC) Laying Down Zootechnical and Pedigree Requirements for the Marketing of Purebred Animals. 7 pp.
EC 91/174/EEC-91. Council Directive Laying Down Zootechnical and Pedigree Requirements for the Marketing of Pure-Bred Animals and Amending Directives 77/504/EEC and 90/425/EEC. 3 pp.

—Sheep
EC COM(87) 591-87. Proposal for a Council Directive Concerning Pure-Bred Breeding Sheep and Goats. 7 pp.
EC COM(88) 742-88. Proposal for a Council Regulation (EEC) on Animal Health Conditions Governing Intra-Community Trade in Ovine and Caprine Animals: Proposal for a Council Directive Amending Directive 72/462/EEC. 42 pp.
EC 89/361/EEC-89. Council Directive Concerning Pure-Bred Breeding Sheep and Goats. 2 pp.

—Swine
EC 88/661/EEC-88. Council Directive on the Zoo Technical Standards Applicable to Breeding Animals of the Porcine Species. 3 pp.

—Swine—Semen Collection—Health Certification
EC COM(83) 512-83. Proposal for a Council Directive on Animal Health Problems Affecting Intra-Community Trade and Imports from Third Countries of Semen of Domestic Animals of the Bovine and Porcine Species. 37 pp.
EC COM(86) 657-86. Amendment to the Proposal for a Council Directive on Animal Health Problems Affecting Intra-Community Trade and Imports from Third Countries of Semen of Domestic Animals of the Bovine and Porcine Species. 4 pp.
EC 90/429/EEC-90. Council Directive Laying Down the Animal Health Requirements Applicable to Intra-Community Trade in and Imparts of Semen of Domestic Animals of the Porcine Species. 12 pp.

Animal Diseases
Scope Note: For additional listings, use a more specific term *See Also:* Animal Breeding; Aujeszky's Disease; Brucellosis; Contagious Bovine Pleuropneumonia; Foot and Mouth Disease; Hog Cholera; Leukosis; Rabies; Swine Vesicular Disease; Tuberculosis

—Control Measures—Financial Aid
EC 90/424/EEC-90. Council Decision on Expenditure in the Veterinary Field. 9 pp.
EC 90/495/EEC-90. Council Decision Introducing a Community Financial Measure with a View to the Eradication of Infectious Haemopoietic Necrosis of Salmonids in the Community. 3 pp.

Animal Fats
See Also: Animal Feed; Fats; Lard; Oils; Tallow Fatty Acids
BSI BS 684: Part 0-89. 1989 Methods of Analysis of Fats and Fatty Oils Part 0: General Introduction. 8 pp.
BSI BS 684: Part 0-01. 1989 Amd 1 Methods of Analysis of Fats and Fatty Oils Part 0: General Introduction (AMD 7892) October 15, 1993 (W). 9 pp.
BSI BS 3919-87. 1987 Amd 1 Technical Tallow and Animal Grease. 9 pp.
CNS N5156-88. Edible Tallow (Dripping) (Apr)(4988). 2 pp.
CNS N5193-88. Refined Processed Oil and Fat (Edible) (Apr)(8156). 2 pp.

—Acidity
ISO 660-83. Animal and Vegetable Fats and Oils—Determination of Acid Value and of Acidity First Edition. 6 pp.

—Animal Feed
CNS N2032-84. Animal Fat (for Feeding) (Dec)(3400). 1 p.

—Anisidine Value
BSI BS 684: Sec 2.24-89. 1989 Methods of Analysis of Fats and Fatty Oils Part 2: Other Methods Section 2.24: Determination of Anisidine Value. 6 pp.
ISO 6885-88. Animal and Vegetable Fats and Oils—Determination of Anisidine Value First Edition. 6 pp.

—Antioxidant Content—Molecular Absorption Spectrophotometry
BSI BS 684: Sec 2.44-84. 1984 Methods of Analysis of Fats and Fatty Oils Part 2: Other Methods Section 2.44: Determination of Gallates. 4 pp.

—Antioxidant Content—Thin Layer Chromatography
BSI BS 684: Sec 2.33-83. 1983 Methods of Analysis of Fats and Fatty Oils Part 2: Other Methods Section 2.33: Detection and Identification of Antioxidants. 4 pp.
ISO 5558-82. Animal and Vegetable Fats and Oils—Detection and Identification of Antioxidants—Thin-Layer Chromatographic Method First Edition. 5 pp.

Animal Fats (Cont.)
—Ash Content
BSI BS 684: Sec 2.2-86. 1986 Methods of Analysis of Fats and Fatty Oils Part 2: Other Methods Section 2.2: Determination of Ash. 4 pp.
ISO 6884-85. Animal and Vegetable Fats and Oils—Determination of Ash First Edition. 4 pp.

—BHA Content—Gas Chromatography
BSI BS 684: Sec 2.36-83. 1983 Methods of Analysis of Fats and Fatty Oils Part 2: Other Methods Section 2.36: Determination of Butylhydroxyanisole (BHA) and Butylhydroxytoluene (BHT). 4 pp.
ISO 6463-82. Animal and Vegetable Fats and Oils—Determination of Butylhydroxyanisole (BHA) and Butylhydroxytoluene (BHT)—Gas-Liquid Chromatographic Method First Edition. 5 pp.

—BHT Content—Gas Chromatography
BSI BS 684: Sec 2.36-83. 1983 Methods of Analysis of Fats and Fatty Oils Part 2: Other Methods Section 2.36: Determination of Butylhydroxyanisole (BHA) and Butylhydroxytoluene (BHT). 4 pp.
ISO 6463-82. Animal and Vegetable Fats and Oils—Determination of Butylhydroxyanisole (BHA) and Butylhydroxytoluene (BHT)—Gas-Liquid Chromatographic Method First Edition. 5 pp.

—Bleaching
BSI BS 684: Sec 2.27-87. 1987 Methods of Analysis of Fats and Fatty Oils Part 2: Other Methods Section 2.27: Determination of Bleachability. 4 pp.

—Bomer Number
BSI BS 684: Sec 2.23-89. 1989 Methods of Analysis of Fats and Fatty Oils Part 2: Other Methods Section 2.23: Determination of Boumer Value. 5 pp.
ISO 3577-88. Animal and Vegetable Fats and Oils—Determination of Bomer Value Second Edition. 7 pp.

—Colorimetry
BSI BS 684: Sec 1.14-87. 1987 Methods of Analysis of Fats and Fatty Oils Part 1: Physical Methods Section 1.14: Determination of Colour. 4 pp.

—Copper Content—Colorimetry
BSI BS 684: Sec 2.16-76. 1976 Methods of Analysis of Fats and Fatty Oils Part 2: Other Methods Section 2.16: Determination of Copper. Colorimetric Method. 2 pp.

—Density
ISO 6883-87. Animal and Vegetable Fats and Oils—Determination of Mass per Unit Volume ("Litre Weight") in Air First Edition. 6 pp.

—Dilatation
ISO 8293-90. Animal and Vegetable Fats and Oils—Determination of Dilatation First Edition. 10 pp.

—Erucic Acid Content
BSI BS 684: Sec 2.41-87. 1987 Methods of Analysis of Fats and Fatty Oils Part 2: Other Methods Section 2.41: Determination of Erucic Acid. 6 pp.
ISO 8209-86. Animal and Vegetable Fats and Oils—Determination of Erucic Acid First Edition. 6 pp.

—Fatty Acid Content
BSI BS 684: Sec 2.12-84. 1984 Methods of Analysis of Fats and Fatty Oils Part 2: Other Methods Section 2.12: Determination of Oxidised Fatty Acids. 4 pp.
BSI BS 684: Sec 2.39-86. 1986 Methods of Analysis of Fats and Fatty Oils Part 2: Other Methods Section 2.39: Determination of the Composition of Fatty Acids in the 2-Position. 8 pp.
BSI BS 684: Sec 2.43-88. 1988 Methods of Analysis of Fats and Fatty Oils Part 2: Other Methods Section 2.43: Determination of Polyunsaturated Fatty Acids with a Cis, Cis 1,4-Diene Structure. 8 pp.
ISO 6800-85. Animal and Vegetable Fats and Oils—Determination of the Composition of Fatty Acids in the 2-Position First Edition. 7 pp.
ISO 7847-87. Animal and Vegetable Fats and Oils—Determination of Polyunsaturated Fatty Acids with a Cis,Cis 1,4-diene Structure First Edition. 8 pp.

—Gallates Content—Molecular Absorption Spectrophotometry
ISO 6464-83. Animal and Vegetable Fats and Oils—Determination of Gallates Content—Molecular Absorption Spectrometric Method First Edition. 5 pp.

—Glyceride Content
BSI BS 684: Sec 2.42-89. 1989 Methods of Analysis of Fats and Fatty Oils Section 2.42: Determination of 1-Monoglycerides and Free Glycerol Contents. 6 pp.
ISO 7366-87. Animal and Vegetable Fats and Oils—Determination of 1-Monoglycerides and Free Glycerol Contents First Edition. 6 pp.

INTERNATIONAL AND NON-U.S. NATIONAL STANDARDS
SUBJECT INDEX
Animal

Animal Fats (Cont.)

—**Glycerol Content**
BSI BS 684: Sec 2.42-89. 1989 Methods of Analysis of Fats and Fatty Oils Section 2.42: Determination of 1-Monoglycerides and Free Glycerol Contents. 6 pp.
ISO 7366-87. Animal and Vegetable Fats and Oils—Determination of 1-Monoglycerides and Free Glycerol Contents First Edition. 6 pp.

—**Impurities—Solvent Extraction**
BSI BS 684: Sec 2.3-93. 1993 Methods of Analysis of Fats and Fatty Oils Part 2: Other Methods Section 2.3: Determination of Insoluble Impurities Content (ISO 663: 1992). 6 pp.
BSI BS 684: Sec 2.3-83. 1983 Methods of Analysis of Fats and Fatty Oils Part 2: Other Methods Section 2.3: Determination of Insoluble Impurities. 4 pp.
ISO 663-92. Animal and Vegetable Fats and Oils—Determination of Insoluble Impurities Content Second Edition. 6 pp.

—**Iodine Number**
BSI BS 684: Sec 2.13-90. 1990 Methods of Analysis of Fats and Fatty Oils Part 2: Other Methods Section 2.13: Determination of Iodine Value. 4 pp.
ISO 3961-89. Animal and Vegetable Fats and Oils—Determination of Iodine Value Second Edition. 4 pp.

—**Iron Content—Colorimetric Analysis**
BSI BS 684: Sec 2.17-76. 1976 Methods of Analysis of Fats and Fatty Oils Part 2: Other Methods Section 2.17: Determination of Iron. Colorimetric Method. 2 pp.

—**Mass**
BSI BS 684: Sec 1.1-85. 1985 Methods of Analysis of Fats and Fatty Oils Part 1: Physical Methods Section 1.1: Determination of Litre Weight in Air. 4 pp.

—**Melting Points**
BSI BS 684: Sec 1.3-91. 1991 Methods of Analysis of Fats and Fatty Oils Part 1: Physical Methods Section 1.3: Determination of Melting Point (Slip Point) (ISO 6321: 1991) (W). 9 pp.
BSI BS 684: Sec 1.3-76. 1976 Methods of Analysis of Fats and Fatty Oils Part 1: Physical Methods Section 1.3: Determination of Melting Point (Slip Point). 4 pp.
ISO 6321-91. Animal and Vegetable Fats and Oils—Determination of Melting Point in Open Capillary Tubes (Slip Point) First Edition. 8 pp.

—**Moisture Content—Dehydration**
BSI BS 684: Sec 1.10-82. 1982 Methods of Analysis of Fats and Fatty Oils Part 1: Physical Methods Section 1.10: Determination of Moisture and Other Volatile Matter. 4 pp.
ISO 662-80. Animal and Vegetable Fats and Oils—Determination of Moisture and Volatile Matter Content First Edition; (Erratum—Nov 1981). 5 pp.

—**Peroxide Number**
BSI BS 684: Sec 2.14-87. 1987 Methods of Analysis of Fats and Fatty Oils Part 2: Other Methods Section 2.14: Determination of Peroxide Value. 4 pp.
ISO 3960-77. Animal and Vegetable Oils and Fats—Determination of Peroxide Value First Edition. 5 pp.

—**Polar Compounds Content Analysis**
BSI BS 684: Sec 2.45-91. 1991 Methods of Analysis of Fats and Fatty Oils Part 2: Other Methods Section 2.45: Determination of Polar Compounds Content (ISO 8420: 1990). 9 pp.
ISO 8420-90. Animal and Vegetable Fats and Oils—Determination of Polar Compounds Content First Edition. 8 pp.

—**Polyethylene Content**
ISO 6656-84. Animal and Vegetable Fats and Oils—Determination of Polyethylene-Type Polymers First Edition. 5 pp.

—**Refractive Index**
BSI BS 684: Sec 1.2-84. 1984 Amd 1 Methods of Analysis of Fats and Fatty Oils Part 1: Physical Methods Section 1.2: Determination of Refractive Index. 5 pp.
ISO 6320-85. Animal and Vegetable Fats and Oils—Determination of Refractive Index Second Edition. 4 pp.

—**Residue Content—Hexane**
BSI BS 684: Sec 2.37-93. 1993 Methods of Analysis of Fats and Fatty Oils Part 2: Other Methods Section 2.37: Determination of Residual Technical Hexane Content (ISO 9832: 1992). 9 pp.
ISO 9832-92. Animal and Vegetable Fats and Oils—Determination of Residual Technical Hexane Content First Edition. 8 pp.

Animal Fats (Cont.)

—**Sampling**
BSI BS 627-91. 1991 Methods for Sampling of Animal and Vegetable Fats and Oils (ISO 5555: 1991). 28 pp.
BSI BS 627-82. 1982 Sampling Animal and Vegetable Fats and Oils. 24 pp.
ISO 5555-91. Animal and Vegetable Fats and Oils—Sampling Second Edition. 28 pp.

—**Saponification Number**
BSI BS 684: Sec 2.15-86. 1986 Methods of Analysis of Fats and Fatty Oils Part 2: Other Methods Section 2.15: Calculation of Total Saponifiable Matter. 2 pp.
ISO 660-83. Animal and Vegetable Fats and Oils—Determination of Acid Value and of Acidity First Edition. 6 pp.
ISO 3657-88. Animal and Vegetable Fats and Oils—Determination of Saponification Value Second Edition. 4 pp.

—**Smoke Point**
BSI BS 684: Sec 1.8-76. 1976 Methods of Analysis of Fats and Fatty Oils Part 1: Physical Methods Section 1.8: Determination of Smoke Point. 4 pp.

—**Solid Fat Content—Nuclear Magnetic Resonance Spectrometry**
BSI BS 684: Sec 2.22-92. 1992 Methods of Analysis of Fats and Fatty Oils Part 2: Other Methods Section 2.22: Determination of Solid Fat Content (Pulsed NMR Method) (ISO 8292: 1991). 8 pp.
ISO 8292-91. Animal and Vegetable Fats and Oils—Determination of Solid Fat Content—Pulsed Nuclear Magnetic Resonance Method First Edition. 7 pp.

—**Solvent Content**
BSI BS 684: Sec 1.9-76. 1976 Methods of Analysis of Fats and Fatty Oils Part 1: Physical Methods Section 1.9: Determination of Water-Insoluble Solvents. 4 pp.

—**Sterol Content—Gas Chromatography**
BSI BS 684: Sec 2.38-92. 1992 Methods of Analysis of Fats and Fatty Oils Part 2: Other Methods Section 2.38: Determination of the Proportions of Individual Sterols in the Sterol Fraction (ISO 6799: 1991). 12 pp.
BSI BS 684: Sec 2.38-83. 1983 Methods of Analysis of Fats and Fatty Oils Part 2: Other Methods Section 2.38: Determination of the Proportions of Individual Sterols in the Sterol Fraction. 8 pp.
ISO 6799-83. Animal and Vegetable Fats and Oils—Determination of Composition of the Sterol Fraction—Method by Gas-Liquid Chromatography First Edition. 7 pp.

—**Test Specimens**
ISO 661-89. Animal and Vegetable Fats and Oils—Preparation of Test Sample Second Edition. 4 pp.

—**Ultraviolet Spectrophotometry**
BSI BS 684: Sec 1.15-90. 1990 Methods of Analysis of Fats and Fatty Oils Part 1: Physical Methods Section 1.15: Determi-nation of Ultraviolet Absorbance. 4 pp.
ISO 3656-89. Animal and Vegetable Fats and Oils—Determination of Ultraviolet Absorbance Second Edition. 6 pp.

—**Unsaponifiable Matter**
ISO 3596 Pt 1-88. Animal and Vegetable Fats and Oils—Determination of Unsaponifiable Matter—Part 1: Method Using Diethyl Ether Extraction (Reference Method) First Edition. 8 pp.
ISO 3596 Pt 2-88. Animal and Vegetable Fats and Oils—Determination of Unsaponifiable Matter—Part 2: Rapid Method Using Hexane Extraction First Edition. 6 pp.

—**Volatile Matter Content—Dehydration**
BSI BS 684: Sec 1.10-82. 1982 Methods of Analysis of Fats and Fatty Oils Part 1: Physical Methods Section 1.10: Determination of Moisture and Other Volatile Matter. 4 pp.
ISO 662-80. Animal and Vegetable Fats and Oils—Determination of Moisture and Volatile Matter Content First Edition; (Erratum—Nov 1981). 5 pp.

—**Water Content—Entrainment Method**
BSI BS 684: Sec 1.16-81. 1981 Amd 1 Methods of Analysis of Fats and Fatty Oils Part 1: Physical Methods Section 1.16: Determination of Water Content: Entrainment Method. 5 pp.
ISO 934-80. Animal and Vegetable Fats and Oils—Determination of Water Content—Entrainment Method First Edition. 4 pp.

Animal Feed

Use For: Feeds; Pet Foods *See Also:* Animal Fats; Aquaculture; Barley Hull Meal; Bone Meal; Buttermilk; Calcium Phosphates; Cornmeal; Feather Meal; Feathers; Fish Meal; Fish Paste; Food; Forage;

Animal Feed (Cont.)

See Also: (Cont.)
Leucaena Meal; Molasses; Napier Grass Meal; Oil Meal; Rice; Rice Bran; Shell Meal; Skim Milk; Wheat Bran; Whey; Yeasts
CNS N2026-91. Formula Feeds (for Livestock and Poultry) (Oct)(3027). 5 pp.
CNS N4024-86. Method of Test for Feeding: General Rule (Aug)(2770).
EC 90/44/EEC-90. Council Directive Amending Directive 79/373/EEC on the Marketing of Compound Feedingstuffs. 10 pp.

—**Additives**
EC 87/153/EEC-87. Council Directive Fixing Guidelines for the Assessment of Additives in Animal Nutrition. 10 pp.
EC 92/113/EEC-92. Commission Directive Amending Council Directive 70/524/EEC Concerning Additives in Feedingstuffs. 2 pp.

—**Aflatoxin Content**
BSI BS 5766: Part 7-88. 1988 Method for Analysis of Animal Feeding Stuffs Part 7: Determination of Aflatoxin B Less Than of 12 Greater Than 1. 12 pp.
CNS N4070-81. Methods for Determining Aflatoxins in Feedstuff (Feb)(7066). 6 pp.
ISO 6651-87. Animal Feeding Stuffs—Determination of Aflatoxin B1 Content Second Edition. 11 pp.

—**Amprolium Content**
CNS N4075-81. Method Test for Feed Additives: Determination of Amprolium (Aug) (7838). 2 pp.

—**Animal Waste—Health and Veterinary Inspection**
EC 90/667/EEC-90. Council Directive Laying down the Veterinary Rules for the Disposal and Processing of Animal Waste, for Its Placing on the Market and for the Prevention of Pathogens in Feedstuffs of Animal or Fish Origin and Amending Directive 90/425/EEC. 11 pp.

—**Anthracis**
CNS N4046-78. Methods of Test for Bacteria in Feeds (Mar) (3682). 3 pp.

—**Antibiotics Content**
CNS N4098-83. Method of Test for Feed Additives: General Rules for Microbiological Determination of Antibiotics (Aug)(9028).

—**Apramycin Content**
CNS N4124-84. Method of Test for Feed Additives: Determination of Apramycin (Nov)(11130).

—**Arsanilic Acid Content**
CNS N4100-82. Method of Test for Feed Additives: Determination of Arsanilic Acid (Aug)(9315).

—**Arsenic Content**
CNS N4024-17-86. Method of Test for Feeds: Determination of Arsenic (Aug)(2770-17).

—**Ash Content**
BSI BS 5766: Part 1-79. 1979 Amd 1 Methods for Analysis of Animal Feeding Stuffs Part 1: Determination of Crude Ash. 7 pp.
BSI BS 5766: Part 2-80. 1980 Amd 1 Methods for Analysis of Animal Feeding Stuffs Part 2: Determination of Ash Insoluble in Hydrochloric Acid. 7 pp.
CNS N4024-9-86. Method of Test for Feeds: Determination of Crude Ash (Aug)(2770-9).
CNS N4024-10-86. Method of Test for Feeds: Determination of Ash Insoluble in Hydrochloric Acid (Aug)(2770-10).
ISO 5984-78. Animal Feeding Stuffs—Determination of Crude Ash First Edition; (Erratum—Nov 1979). 5 pp.
ISO 5985-78. Animal Feeding Stuffs—Determination of Ash Insoluble in Hydrochloric Acid First Edition. 5 pp.

—**Avoparcin Content**
CNS N4121-84. Method of Test for Feed Additives: Determination of Avoparcin (Nov)(11127).

—**Bacitracin Content**
CNS N4102-82. Method of Test for Feed Additives: Determination of Bacitracin (Aug)(9317).

—**Bacteria Content**
BSI BS 5763: Part 2-91. 1991 Microbiological Examination of Food and Animal Feeding Stuffs Part 2: Enumeration of Coliforms—Colony Count Technique (ISO 4832: 1991). 11 pp.
BSI BS 5763: Part 3-91. 1991 Microbiological Examination of Food and Animal Feeding Stuffs Part 3: Enumeration of Coliforms—Most Probable Number Technique (ISO 4831: 1991). 19 pp.
BSI BS 5763: Part 4-90. 1990 Microbiological Examination of Food and Animal Feeding Stuffs Part 4: Detection of Salmonella. 22 pp.

INDUSTRY STANDARDS

Animal Feed (Cont.)

—Bacteria Content (Cont.)

BSI BS 5763: Part 7-83. 1983 Microbiological Examination of Food and Animal Feeding Stuffs Part 7: Enumeration of Staphylococcus Aureus by Colony Count Technique. 10 pp.

BSI BS 5763: Part 8-85. 1985 Microbiological Examination of Food and Animal Feeding Stuffs Part 8: Enumeration of Presumptive Escherichia Coli. 12 pp.

BSI BS 5763: Part 9-86. 1986 Microbiological Examination of Food and Animal Feeding Stuffs Part 9: Enumeration of Clostridium Perfringens. 10 pp.

BSI BS 5763: Part 10-86. 1986 Microbiological Examination of Food and Animal Feeding Stuffs Part 10: Enumeration of Enterobacteriaceae. 11 pp.

BSI BS 5763: Part 11-88. 1988 Microbiological Examination of Food and Animal Feeding Stuffs Part 11: Enumeration of Bacillus Cereus. 11 pp.

BSI BS 5763: Part 13-89. 1989 Microbiological Examination of Food and Animal Feeding Stuffs Part 13: Enumeration of Escherichia Coli: Colony Count Technique Using Membranes. 8 pp.

BSI BS 5763: Part 14-91. 1991 Methods for Microbiological Examination of Food and Animal Feeding Stuffs Part 14: Detection of Vibrio Parahaemolyticus. 17 pp.

ISO 4831-91. Microbiology—General Guidance for the Enumeration of Coliforms—Most Probable Number Technique Second Edition. 16 pp.

ISO 4832-91. Microbiology—General Guidance for the Enumeration of Coliforms—Colony Count Technique Second Edition. 9 pp.

ISO 6579-93. Microbiology—General Guidance on Methods for the Detection of Salmonella Third Edition. 21 pp.

ISO 6888-83. Microbiology—General Guidance for Enumeration of Staphylococcus Aureus—Colony Count Technique First Edition. 9 pp.

ISO 7251-84. Microbiology—General Guidance for Enumeration of Presumptive Escherichia Coli—Most Probable Number Technique First Edition. 11 pp.

ISO 7402-93. Microbiology—General Guidance for the Enumeration of Enterobacteriaceae Without Resuscitation—MPN Technique and Colony-Count Technique Second Edition. 14 pp.

ISO 7932-87. Microbiology—General Guidance for Enumeration of Bacillus Cereus—Colony Count Technique at 30 Degrees Celsius First Edition. 11 pp.

ISO 7937-85. Microbiology—General Guidance for Enumeration of Clostridium Perfringens—Colony Count Technique First Edition. 9 pp.

ISO 8914-90. Microbiology—General Guidance for the Detection of Vibrio Parahaemolyticus. 15 pp.

—Bambermycins Content

CNS N4131-84. Method of Test for Feed Additives: Determination of Flavomycin (Nov)(11137).

—Benzoic Acid Content

CNS N4116-84. Method of Test for Feed Additives: Determination of Benzoic Acid (May)(10885).

—BHA Content

CNS N4097-82. Method of Test for Feed Additives: Determination of Dibutyl Hydroixy Toluene (BHT) and Butyl Hydroxy Anisol (BHA) (Jun)(9027).

—BHT Content

CNS N4097-82. Method of Test for Feed Additives: Determination of Dibutyl Hydroixy Toluene (BHT) and Butyl Hydroxy Anisol (BHA) (Jun)(9027).

—Buquinolate Content

CNS N4080-82. Method of Test for Feed Additives: Determination of Buquinolate (Jan) (8316). 2 pp.

—Cadmium Content

CNS N4024-21-86. Method of Test for Feeds: Determination of Cadmium (Aug)(2770-21).

—Calcium Content

CNS N4024-15-86. Method of Test for Feeds: Determination of Calcium (Aug)(2770-15).

—Calcium Content—Atomic Absorption Spectrometry

BSI BS 5766: Part 6-84. 1984 Methods for Analysis of Animal Feeding Stuffs Part 6: Determination of Calcium by Atomic Absorption Spectrometry. 6 pp.

ISO 6490 Pt 2-83. Animal Feeding Stuffs—Determination of Calcium Content—Part 2: Atomic Absorption Spectrometric Method First Edition. 5 pp.

—Calcium Content—Volumetric Analysis

BSI BS 5766: Part 11-86. 1986 Methods for Analysis of Animal Feeding Stuffs Part 11: Determination of Calcium by Titration. 4 pp.

ISO 6490 Pt 1-85. Animal Feeding Stuffs—Determination of Calcium Content—Part 1: Titrimetric Method First Edition. 4 pp.

Animal Feed (Cont.)

—Calcium Phosphate Content

CNS N2045-82. Calcium Phosphates (for Feed) (Apr)(4526). 1 p.

CNS N4047-82. Methods of Test for Calcium Phosphates (for Feed)(Apr)(4527).

—Carbadox Content

CNS N4099-82. Method of Test for Feed Additives: Determination of Carbadox (Aug)(9314).

CNS N4107-83. Method of Test for Feed Additives (Identification of Carbadox) (May)(10289).

—Carotene Content

CNS N4062-80. Method of Test for Carotene in Feeds (Dec)(6710).

—Chloride Content

BSI BS 5766: Part 5-81. 1981 Amd 1 Methods for Analysis of Animal Feeding Stuffs Part 5: Determination of Water-Soluble Chlorides Content. 7 pp.

CNS N4024-13-86. Method of Test for Feeds: Determination of Water Soluble Chloride (Aug)(2770-13).

ISO 6495-80. Animal Feeding Stuffs—Determination of Water-Soluble Chlorides Content First Edition. 5 pp.

—Chlorobenzene Content

CNS N4068-81. Method of Test for BHC in Feeds (Jan)(6943). 2 pp.

—Chlortetracycline Content

CNS N4101-82. Method of Test for Feed Additives: Determination of Chloratetracycline (Aug)(9316).

—Chromium Content

CNS N4024-18-86. Method of Test for Feeds: Determination of Chromium (Aug)(2770-18).

—Classification

EC EEC/509/92-92. Commission Regulation Concerning the Classification of Certain Goods in the Combined Nomenclature. 2 pp.

—Coliform Bacteria

CNS N4046-78. Methods of Test for Bacteria in Feeds (Mar) (3682). 3 pp.

—Cyanide Content

CNS N4063-80. Method of Test for Cyanide in Feedstuff (Dec)(6711).

—Decoquinate Content

CNS N4082-82. Method of Test for Feed Additives: Determination of Decoquinate (Jan) (8318). 2 pp.

—Dibutyltin Dilaurate Content

CNS N4081-82. Method of Test for Butynorate in Feed (Jan)(8317). 2 pp.

—Dimetridazole Content

CNS N4122-84. Method of Test for Feed Additives: Determination of Dimetridazole (Nov)(11128).

—Dinitolmide Content

CNS N4094-82. Method of Test for Feed Additives: Determination of Zoalene (Apr)(8731).

—Enduracidin Content

CNS N4126-84. Method of Test for Feed Additives: Determination of Enramycin (Nov)(11132).

—Erythromycin Content

CNS N4103-82. Method of Test for Erythromycin in Feed (Oct)(9533).

—Ethopabate Content

CNS N4088-82. Method of Test for Feed Additives: Determination of Ethopabate (Feb)(8534).

—Ethoxyquin Content

CNS N4095-82. Method of Test for Feed Additives: Determination of Ethoxyguin (Jun)(9025).

—Fat Content

CNS N4024-4-86. Method of Test for Feeds: Determination of Crude Fat (Aug)(2770-4).

—Fiber Content

CNS N4024-8-86. Method of Test for Feeds: Determination of Crude Fiber (Aug)(2770-8).

—Fungus Content

BSI BS 5763: Part 12-88. 1988 Microbiological Examination of Food and Animal Feeding Stuffs Part 12: Enumeration of Yeasts and Moulds. 6 pp.

ISO 7954-87. Microbiology—General Guidance for Enumeration of Yeasts and Moulds—Colony Count Technique at 25 Degrees Celsius First Edition. 6 pp.

—Furazolidone Content

CNS N4138-86. Determination of Furazolidone in Feeds (Aug)(11663).

Animal Feed (Cont.)

—Gossypol Content

BSI BS 5766: Part 12-86. 1986 Methods for Analysis of Animal Feeding Stuffs Part 12: Determination of Gossypol. 6 pp.

CNS N4071-81. Determination of Free Gossypol in Cottonseed Meal of Cake (Feed Grade)(Feb)(7067).

ISO 6866-85. Animal Feeding Stuffs—Determination of Free and Total Gossypol First Edition. 5 pp.

—Halofuginone Content

CNS N4119-84. Method of Test for Feed Additives: Determination of Halofuginone (Nov)(11125).

—Hygromycin B Content

CNS N4115-84. Method of Test for Feed Additives: Determination of Hygromycin B (May)(10884).

—Industrial Plants—Dust Explosions—Safety

CNS Z1020-88. Safety Code of Preventing Dust Explosion for Grain Elevator (Oct)(3364).

—Kanamycin Content

CNS N4130-84. Method of Test for Feed Additives: Determination of Kanamycin (Nov)(11136).

—Lead Content

CNS N4024-20-86. Method of Test for Feeds: Determination of Lead (Aug)(2770-20).

—Leucomycins Content

CNS N4129-84. Method of Test for Feed Additives: Determination of Kitasamycin (Nov)(11135).

—Lincomycin Content

CNS N4104-82. Method of Test for Feed Additives (Determination of Lincomycin) (Oct)(9534).

—Lysine Content

ISO 5510-84. Animal Feeding Stuffs—Determination of Available Lysine First Edition. 7 pp.

—Medicated

EC COM(81) 795-82. Proposal for a Council Directive on the manufacture, putting into circulation and supply of medicated feeding-stuffs in the Community. 18 pp.

EC COM(83) 378-83. Amendment to the Proposal for a Council Directive on the Manufacture, Putting into Circulation and Supply of Medicated Feedingstuffs in the Community. 5 pp.

EC 90/167/EEC-90. Council Directive laying Down the Conditions Governing the Preparation, Placing on the Market and Use of Medicated Feeding Stuffs in the Community. 7 pp.

—Mercury Content

CNS N4024-19-86. Method of Test for Feeds: Determination of Mercury (Aug)(2770-19).

—Microbiological Analysis

BSI BS 5763: Part 0-86. (OBSOLESCENT) 1986 Microbiological Examination of Food and Animal Feeding Stuffs Part 0: General Laboratory Practices. 16 pp.

BSI BS 5763: Part 5-81. 1981 Microbiological Examination of Food and Animal Feeding Stuffs Part 5: Enumeration of Micro-Organisms—Colony Count at 30 Degrees Celsius (Surface Plate Technique). 7 pp.

BSI BS 5763: Part 6-83. 1983 Microbiological Examination of Food and Animal Feeding Stuffs Part 6: Preparation of Dilutions. 7 pp.

—Microorganism Content

BSI BS 5763: Part 1-91. 1991 Microbiological Examination of Food and Animal Feeding Stuffs Part 1: Enumeration of Micro-Organisms—Colony Count Technique at 30 Degrees C (ISO 4833: 1991). 11 pp.

BSI BS 5763: Part 5-81. 1981 Microbiological Examination of Food and Animal Feeding Stuffs Part 5: Enumeration of Micro-Organisms—Colony Count at 30 Degrees Celsius (Surface Plate Technique). 7 pp.

ISO 4833-91. Microbiology—General Guidance for the Enumeration of Micro-Organisms—Colony Count Technique at 30 Degrees Celsius Second Edition. 9 pp.

—Mineral Supplements

CNS N4064-80. Method of Test for the Purity of Copper Salts in Mineral Supplements (For Feeding) (Dec)(6712).

CNS N4069-81. Method of Test for Purity of Ferrous Salts in Mineral Premix (For Feeding) (Jan)(6944).

—Moisture Content

BSI BS 5766: Part 8-83. 1983 Method for Analysis of Animal Feeding Stuffs Part 8: Determination of Moisture. 6 pp.

CNS N4024-3-86. Method of Test for Feeds: Determination of Moisture (Aug)(2770-3).

INTERNATIONAL AND NON-U.S. NATIONAL STANDARDS
SUBJECT INDEX

Animal

Animal Feed (Cont.)
—Moisture Content (Cont.)
ISO 6496-83. Animal Feeding Stuffs—Determination of Moisture Content First Edition. 5 pp.

—Monensin Content
CNS N4089-82. Method of Test for Feed Additives: Determination of Monensin (Feb)(8535).

—Neomycin Content
CNS N4105-82. Method of Test for Feed Additives (Determination of Neomycin) (Oct)(9535).

—Nicarbazin Content
CNS N4090-82. Method of Test for Feed Additives: Determination of Nicarbazin (Feb)(8536).

—Nitrogen Content
CNS N4024-6-86. Method of Test for Feeds: Determination of Volatile Basic Nitrogen (Aug)(2770-6).

—Nitrogen Content—Kjeldahl Method
BSI BS 5766: Part 4-81. 1981 Amd 1 Methods for Analysis of Animal Feeding Stuffs Part 4: Determination of Nitrogen Content and Calculation of Crude Protein Content. 9 pp.
BSI BS 5766: Part 4-02. 1981 Amd 2 Methods for Analysis of Animal Feeding Stuffs Part 4: Determination of Nitrogen Content and Calculation of Crude Protein Content (ISO 5983: 1979) (AMD 7725) June 15, 1993 (W). 10 pp.
ISO 5983-79. Animal Feeding Stuffs—Determination of Nitrogen Content and Calculation of Crude Protein Content First Edition. 6 pp.

—Nitrovin Content
CNS N4120-84. Method of Test for Feed Additives: Determination of Nitrovin (Nov)(11126).

—Nutrition—Identification Systems
CGSB 146-GP-1M-87. Minimum Nutritional Testing, Labelling and Manufacturing Procedures for Pet Foods. 36 pp.

—Nystatin Content
CNS N4096-82. Method of Test for Feed Additives: Determination of Nystatin (Jun)(9026).

—Oils—Polychlorinated Biphenyl Content
CNS N4066-81. Method of Test for PCB in Oil (For Feeding) (Jan)(6941).

—Oilseed Residue Content—Microscopic Analysis
BSI BS 5766: Part 9-83. 1983 Method for Analysis of Animal Feeding Stuffs Part 9: Determination of Castor Oil Seed Husks. 9 pp.
ISO 5061-83. Animal Feeding Stuffs—Determination of Castor Oil Seed Husks—Microscopical Method First Edition. 8 pp.

—Olaquindox Content
CNS N4118-84. Method of Test for Feed Additives: Determination of Olaquindox (Nov)(11124).

—Oleandomycin Content
CNS N4106-82. Method of Test for Oleandomycin in Feed (Oct)(9536).

—Oxytetracycline Content
CNS N4108-83. Method of Test for Feed Additives (Determination of Oxytetracycline) (May)(10290).

—Penicillin Content
CNS N4109-83. Method of Test for Feed Additives (Determination of Penicillin) (Jul)(10463).

—Pesticide Residues
EC 87/519/EEC-87. Council Directive Amending Directive 74/63/EEC on Undesirable Substances and Products in Animal Nutrition. 2 pp.

—Phosphorus Content
CNS N4024-16-86. Method of Test for Feeds: Determination of Phosporus (Aug)(2770-16).

—Phosphorus Content—Spectrophotometry
BSI BS 5766: Part 3-81. 1981 Amd 1 Methods for Analysis of Animal Feeding Stuffs Part 3: Determination of Total Phosphorus Content (Spectrophotometric Method). 7 pp.
ISO 6491-80. Animal Feeding Stuffs—Determination of Total Phosphorus Content—Spectrophotometric Method First Edition. 5 pp.

—Polychlorinated Biphenyl Content
CNS N4065-81. Method of Test for PCB in Feeds (Jan)(6940).

—Propionic Acid Content
CNS N4114-84. Method of Test for Feed Additives: Determination of Propionic Acid (May)(10883).

Animal Feed (Cont.)
—Protein Content
BSI BS 5766: Part 4-81. 1981 Amd 1 Methods for Analysis of Animal Feeding Stuffs Part 4: Determination of Nitrogen Content and Calculation of Crude Protein Content. 9 pp.
BSI BS 5766: Part 4-02. 1981 Amd 2 Methods for Analysis of Animal Feeding Stuffs Part 4: Determination of Nitrogen Content and Calculation of Crude Protein Content (ISO 5983: 1979) (AMD 7725) June 15, 1993 (W). 10 pp.
CNS N4024-5-86. Method of Test for Feeds: Determination of Crude Protein (Aug)(2770-5).
ISO 5983-79. Animal Feeding Stuffs—Determination of Nitrogen Content and Calculation of Crude Protein Content First Edition. 6 pp.

—Pyrimethamine Content
CNS N4139-86. Determination of Pyrimethamine in Feeds (Aug)(11664).

—Roxarsone Content
CNS N4117-84. Method of Test for Feed Additives: Determination of Roxarsone (Nov)(11123).

—Salinomycin Content
CNS N4128-84. Method of Test for Feed Additives: Determination of Salinomycin (Nov)(11134).

—Salmonella
CNS N4046-78. Methods of Test for Bacteria in Feeds (Mar) (3682). 3 pp.

—Sampling
BSI BS 5766: Part 10-86. 1986 Methods for Analysis of Animal Feeding Stuffs Part 10: Preparation of Test Samples. 8 pp.
CNS N4024-1-86. Method of Test for Feeds: Sampling (Aug)(2770-1).

—Sorbic Acid Content
CNS N4113-84. Method of Test for Feed Additives: Determination of Sorbic Acid (May)(10882).

—Spectinomycin Content
CNS N4110-83. Method of Test for Feed Additives (Determination of Spectinomycin) (Jul)(10464).

—Starch Content
CNS N4024-12-86. Method of Test for Feeds: Determination of Starch (Aug)(2770-12).

—Streptomycin Content
CNS N4111-83. Method of Test for Feed Additives (Determination of Streptomycin) (Jul)(10465).

—Sugar Content
CNS N4024-11-86. Method of Test for Feeds: Determination of Total Sugars (Aug)(2770-11).

—Sulfadimethoxine Content
CNS N4091-82. Method of Test for Feed Additives: Determination of Sulfadimethoxine (Apr)(8728).

—Sulfamethazine Content
CNS N4123-84. Method of Test for Feed Additives: Determination of Sulfamethazine (Nov)(11129).

—Sulfanitran Content
CNS N4093-82. Method of Test for Sulfanitran in Feed (Apr)(8730).

—Sulfaquinoxaline Content
CNS N4092-82. Method of Test for Feed Additives: Determination of Sulfaquinoxaline (Apr)(8729).

—Test Specimens
CNS N4024-2-86. Method of Test for Feeds: Preparation of Test Samples (Aug)(2770-2).
ISO 6498-83. Animal Feeding Stuffs—Preparation of Test Samples First Edition. 8 pp.

—Thiopeptin Content
CNS N4125-84. Method of Test for Feed Additives: Determination of Thiopeptin (Nov)(11131).

—Tylosin Content
CNS N4112-83. Method of Test for Feed Additives (Determination of Tylosin) (Jul)(10466).

—Urea Content
CNS N4024-7-86. Method of Test for Feeds: Determination of Urea (Aug)(2770-7).

—Urea Content—Spectrometry
BSI BS 5766: Part 14-91. 1991 Analysis of Animal Feeding Stuffs Part 14: Determination of Urea Content. 7 pp.
ISO 6654-91. Animal Feeding Stuffs—Determination of Urea Content First Edition. 6 pp.

—Urease Activity
CNS N4024-14-86. Method of Test for Feeds: Urease Activity (Aug)(2770-14).

Animal Feed (Cont.)
—Virginiamycin Content
CNS N4127-84. Method of Test for Feed Additives: Determination of Virginiamycin (Nov)(11133).

—Vitamin Supplements
CNS N4083-82. Method of Test for Vitamin A in Premix of Feed (Jan)(8319).
CNS N4084-82. Method of Test for Vitamin A Material (For Feed (Jan)(8320).

—Zearalenone Content
BSI BS 5766: Part 13-86. (WITHDRAWN) 1986 Methods for Analysis of Animal Feeding Stuffs Part 13: Determination of Zearalenone. 8 pp.
ISO 6870-85. Animal Feeding Stuffs—Determination of Zearalenone Content First Edition. 7 pp.

Animal Feeding Equipment
Scope Note: Use a more specific term *See:* Animal Feed

Animal Glue
See Also: Adhesives
CNS K6174-63. Method of Test for Animal Glue (Jul)(2201).
CNS K8007-67. Animal Glue (Aug)(2200).
ISO 9665-93. Adhesives—Animal Glues —Methods for Sampling and Testing First Edition. 21 pp.
JIS K 6503-77. Animal Glues and Gelatins.
JIS K 6503-70. Animal Glues and Gelatins. 19 pp.

—Sampling
BSI BS 647-93. 1993 Adhesives—Methods for Sampling and Testing Animal Glues (ISO 9665: 1993) (W). 23 pp.
BSI BS 647-81. 1981 Sampling and Testing Glues (Bone, Skin and Fish Glues). 22 pp.
ISO 9665-93. Adhesives—Animal Glues —Methods for Sampling and Testing First Edition. 21 pp.

—Sizing (Surface Treatment)—Wall Coverings
BSI BS 3357-61. (WITHDRAWN) 1961 Glue Size for Decorator's Use. 9 pp.

—Wood
BSI BS 745-69. 1969 Amd 1 Animal Glue for Wood (Joiner's Glue) (Dry Glue; Jelly or Liquid Glue). 9 pp.

Animal Hair
Use: Hair (Animal)

Animal Husbandry
See Also: Animal Breeding; Farms; Veterinary Medicine

—Classification
BSI BS 1000: (636/639)-81. 1981 Universal Decimal Classification (UDC). English Full Edition (636/639): Animal Breeding. Animal Produce. Hunting. Fishing. 56 pp.
SNZ NZS/BS 1000 (636/639)-81. Universal Decimal Classification Animal Breeding. Animal Produce. Hunting. Fishing. 56 pp.

—Heaters—Safety
IEC 335 Pt 2-71-93. Safety of Household and Similar Electrical Appliances Part 2: Particular Requirements for Electrical Heating Appliances for Breeding and Rearing Animals First Edition. 40 pp.

Animal Oils
BSI BS 684: Part 0-89. 1989 Methods of Analysis of Fats and Fatty Oils Part 0: General Introduction. 8 pp.
BSI BS 684: Part 0-01. 1989 Amd 1 Methods of Analysis of Fats and Fatty Oils Part 0: General Introduction (AMD 7892) October 15, 1993 (W). 9 pp.

—Acidity
ISO 660-83. Animal and Vegetable Fats and Oils—Determination of Acid Value and of Acidity First Edition. 6 pp.

—Anisidine Value
BSI BS 684: Sec 2.24-89. 1989 Methods of Analysis of Fats and Fatty Oils Part 2: Other Methods Section 2.24: Determination of Anisidine Value. 6 pp.
ISO 6885-88. Animal and Vegetable Fats and Oils—Determination of Anisidine Value First Edition. 6 pp.

—Antioxidant Content—Thin Layer Chromatography
BSI BS 684: Sec 2.33-83. 1983 Methods of Analysis of Fats and Fatty Oils Part 2: Other Methods Section 2.33: Detection and Identification of Antioxidants. 4 pp.

INDUSTRY STANDARDS

Animal Oils (Cont.)

—Antioxidant Content—Thin Layer Chromatography (Cont.)
ISO 5558-82. Animal and Vegetable Fats and Oils—Detection and Identification of Antioxidants—Thin-Layer Chromatographic Method First Edition. 5 pp.

—Ash Content
BSI BS 684: Sec 2.2-86. 1986 Methods of Analysis of Fats and Fatty Oils Part 2: Other Methods Section 2.2: Determination of Ash. 4 pp.
ISO 6884-85. Animal and Vegetable Fats and Oils—Determination of Ash First Edition. 4 pp.

—BHA Content—Gas Chromatography
BSI BS 684: Sec 2.36-83. 1983 Methods of Analysis of Fats and Fatty Oils Part 2: Other Methods Section 2.36: Determination of Butylhydroxyanisole (BHA) and Butylhydroxytoluene (BHT). 4 pp.
ISO 6463-82. Animal and Vegetable Fats and Oils—Determination of Butylhydroxyanisole (BHA) and Butylhydroxytoluene (BHT)—Gas-Liquid Chromatographic Method First Edition. 5 pp.

—BHT Content—Gas Chromatography
BSI BS 684: Sec 2.36-83. 1983 Methods of Analysis of Fats and Fatty Oils Part 2: Other Methods Section 2.36: Determination of Butylhydroxyanisole (BHA) and Butylhydroxytoluene (BHT). 4 pp.
ISO 6463-82. Animal and Vegetable Fats and Oils—Determination of Butylhydroxyanisole (BHA) and Butylhydroxytoluene (BHT)—Gas-Liquid Chromatographic Method First Edition. 5 pp.

—Bleaching
BSI BS 684: Sec 2.27-87. 1987 Methods of Analysis of Fats and Fatty Oils Part 2: Other Methods Section 2.27: Determination of Bleachability. 4 pp.

—Colorimetry
BSI BS 684: Sec 1.14-87. 1987 Methods of Analysis of Fats and Fatty Oils Part 1: Physical Methods Section 1.14: Determination of Colour. 4 pp.

—Density
BSI BS 684: Sec 1.1-85. 1985 Methods of Analysis of Fats and Fatty Oils Part 1: Physical Methods Section 1.1: Determination of Litre Weight in Air. 4 pp.
ISO 6883-87. Animal and Vegetable Fats and Oils—Determination of Mass per Unit Volume ("Litre Weight") in Air First Edition. 6 pp.

—Dilatation
ISO 8293-90. Animal and Vegetable Fats and Oils—Determination of Dilatation First Edition. 10 pp.

—Erucic Acid Content
BSI BS 684: Sec 2.41-87. 1987 Methods of Analysis of Fats and Fatty Oils Part 2: Other Methods Section 2.41: Determination of Erucic Acid. 6 pp.
ISO 8209-86. Animal and Vegetable Fats and Oils—Determination of Erucic Acid First Edition. 6 pp.

—Fatty Acid Content
BSI BS 684: Sec 2.12-84. 1984 Methods of Analysis of Fats and Fatty Oils Part 2: Other Methods Section 2.12: Determination of Oxidised Fatty Acids. 4 pp.
BSI BS 684: Sec 2.39-86. 1986 Methods of Analysis of Fats and Fatty Oils Part 2: Other Methods Section 2.39: Determination of the Composition of Fatty Acids in the 2-Position. 8 pp.
BSI BS 684: Sec 2.43-88. 1988 Methods of Analysis of Fats and Fatty Oils Part 2: Other Methods Section 2.43: Determination of Polyunsaturated Fatty Acids with a Cis, Cis 1,4-Diene Structure. 8 pp.
ISO 6800-85. Animal and Vegetable Fats and Oils—Determination of the Composition of Fatty Acids in the 2-Position First Edition. 7 pp.
ISO 7847-87. Animal and Vegetable Fats and Oils—Determination of Polyunsaturated Fatty Acids with a Cis,Cis 1,4-diene Structure First Edition. 8 pp.

—Gallates Content—Molecular Absorption Spectrophotometry
BSI BS 684: Sec 2.44-84. 1984 Methods of Analysis of Fats and Fatty Oils Part 2: Other Methods Section 2.44: Determination of Gallates. 4 pp.

—Glycerides Content
BSI BS 684: Sec 2.42-89. 1989 Methods of Analysis of Fats and Fatty Oils Section 2.42: Determination of 1-Monoglycerides and Free Glycerol Contents. 6 pp.
ISO 7366-87. Animal and Vegetable Fats and Oils—Determination of 1-Monoglycerides and Free Glycerol Contents First Edition. 6 pp.

—Glycerol Content
BSI BS 684: Sec 2.42-89. 1989 Methods of Analysis of Fats and Fatty Oils Section 2.42: Determination of 1-Monoglycerides and Free Glycerol Contents. 6 pp.

Animal Oils (Cont.)

—Glycerol Content (Cont.)
ISO 7366-87. Animal and Vegetable Fats and Oils—Determination of 1-Monoglycerides and Free Glycerol Contents First Edition. 6 pp.

—Impurities Content
BSI BS 684: Sec 2.3-93. 1993 Methods of Analysis of Fats and Fatty Oils Part 2: Other Methods Section 2.3: Determination of Insoluble Impurities Content (ISO 663: 1992). 6 pp.
BSI BS 684: Sec 2.3-83. 1983 Methods of Analysis of Fats and Fatty Oils Part 2: Other Methods Section 2.3: Determination of Insoluble Impurities. 4 pp.
ISO 663-92. Animal and Vegetable Fats and Oils—Determination of Insoluble Impurities Content Second Edition. 6 pp.

—Iodine Number
BSI BS 684: Sec 2.13-90. 1990 Methods of Analysis of Fats and Fatty Oils Part 2: Other Methods Section 2.13: Determination of Iodine Value. 4 pp.
ISO 3961-89. Animal and Vegetable Fats and Oils—Determination of Iodine Value Second Edition. 4 pp.

—Melting Points
BSI BS 684: Sec 1.3-91. 1991 Methods of Analysis of Fats and Fatty Oils Part 1: Physical Methods Section 1.3: Determination of Melting Point (Slip Point) (ISO 6321: 1991) (W). 9 pp.
BSI BS 684: Sec 1.3-76. 1976 Methods of Analysis of Fats and Fatty Oils Part 1: Physical Methods Section 1.3: Determination of Melting Point (Slip Point). 4 pp.
ISO 6321-91. Animal and Vegetable Fats and Oils—Determination of Melting Point in Open Capillary Tubes (Slip Point) First Edition. 8 pp.

—Moisture Content—Dehydration
BSI BS 684: Sec 1.10-82. 1982 Methods of Analysis of Fats and Fatty Oils Part 1: Physical Methods Section 1.10: Determination of Moisture and Other Volatile Matter. 4 pp.
ISO 662-80. Animal and Vegetable Fats and Oils—Determination of Moisture and Volatile Matter Content First Edition; (Erratum—Nov 1981). 5 pp.

—Peroxide Number
BSI BS 684: Sec 2.14-87. 1987 Methods of Analysis of Fats and Fatty Oils Part 2: Other Methods Section 2.14: Determination of Peroxide Value. 4 pp.
ISO 3960-77. Animal and Vegetable Oils and Fats—Determination of Peroxide Value First Edition. 5 pp.

—Polar Compound Content
BSI BS 684: Sec 2.45-91. 1991 Methods of Analysis of Fats and Fatty Oils Part 2: Other Methods Section 2.45: Determination of Polar Compounds Content (ISO 8420: 1990). 9 pp.
ISO 8420-90. Animal and Vegetable Fats and Oils—Determination of Polar Compounds Content First Edition. 8 pp.

—Polyethylene Content
ISO 6656-84. Animal and Vegetable Fats and Oils—Determination of Polyethylene-Type Polymers First Edition. 5 pp.

—Refractive Index
BSI BS 684: Sec 1.2-84. 1984 Amd 1 Methods of Analysis of Fats and Fatty Oils Part 1: Physical Methods Section 1.2: Determination of Refractive Index. 5 pp.
ISO 6320-85. Animal and Vegetable Fats and Oils—Determination of Refractive Index Second Edition. 4 pp.

—Residue Content—Hexane
BSI BS 684: Sec 2.37-93. 1993 Methods of Analysis of Fats and Fatty Oils Part 2: Other Methods Section 2.37: Determination of Residual Technical Hexane Content (ISO 9832: 1992). 9 pp.
ISO 9832-92. Animal and Vegetable Fats and Oils—Determination of Residual Technical Hexane Content First Edition. 8 pp.

—Sampling
BSI BS 627-91. 1991 Methods for Sampling of Animal and Vegetable Fats and Oils (ISO 5555: 1991). 28 pp.
BSI BS 627-82. 1982 Sampling Animal and Vegetable Fats and Oils. 24 pp.
ISO 5555-91. Animal and Vegetable Fats and Oils—Sampling Second Edition. 28 pp.

—Saponification Number
BSI BS 684: Sec 2.15-86. 1986 Methods of Analysis of Fats and Fatty Oils Part 2: Other Methods Section 2.15: Calculation of Total Saponifiable Matter. 2 pp.
ISO 660-83. Animal and Vegetable Fats and Oils—Determination of Acid Value and of Acidity First Edition. 6 pp.

Animal Oils (Cont.)

—Saponification Number (Cont.)
ISO 3657-88. Animal and Vegetable Fats and Oils—Determination of Saponification Value Second Edition. 4 pp.

—Solid Fat Content—Nuclear Magnetic Resonance Spectrometry
BSI BS 684: Sec 2.22-92. 1992 Methods of Analysis of Fats and Fatty Oils Part 2: Other Methods Section 2.22: Determination of Solid Fat Content (Pulsed NMR Method) (ISO 8292: 1991). 8 pp.
ISO 8292-91. Animal and Vegetable Fats and Oils—Determination of Solid Fat Content—Pulsed Nuclear Magnetic Resonance Method First Edition. 7 pp.

—Sterol Content—Gas Chromatography
BSI BS 684: Sec 2.38-92. 1992 Methods of Analysis of Fats and Fatty Oils Part 2: Other Methods Section 2.38: Determination of the Proportions of Individual Sterols in the Sterol Fraction (ISO 6799: 1991). 12 pp.
BSI BS 684: Sec 2.38-83. 1983 Methods of Analysis of Fats and Fatty Oils Part 2: Other Methods Section 2.38: Determination of the Proportions of Individual Sterols in the Sterol Fraction. 8 pp.
ISO 6799-83. Animal and Vegetable Fats and Oils—Determination of Composition of the Sterol Fraction—Method by Gas-Liquid Chromatography First Edition. 7 pp.

—Test Specimens
ISO 661-89. Animal and Vegetable Fats and Oils—Preparation of Test Sample Second Edition. 4 pp.

—Ultraviolet Spectrophotometry
BSI BS 684: Sec 1.15-90. 1990 Methods of Analysis of Fats and Fatty Oils Part 1: Physical Methods Section 1.15: Determi-nation of Ultraviolet Absorbance. 4 pp.
ISO 3656-89. Animal and Vegetable Fats and Oils—Determination of Ultraviolet Absorbance Second Edition. 6 pp.

—Unsaponifiable Matter
ISO 3596 Pt 1-88. Animal and Vegetable Fats and Oils—Determination of Unsaponifiable Matter—Part 1: Method Using Diethyl Ether Extraction (Reference Method) First Edition. 8 pp.

—Volatile Matter Content—Dehydration
BSI BS 684: Sec 1.10-82. 1982 Methods of Analysis of Fats and Fatty Oils Part 1: Physical Methods Section 1.10: Determination of Moisture and Other Volatile Matter. 4 pp.
ISO 662-80. Animal and Vegetable Fats and Oils—Determination of Moisture and Volatile Matter Content First Edition; (Erratum—Nov 1981). 5 pp.

—Water Content—Entrainment Method
BSI BS 684: Sec 1.16-81. 1981 Amd 1 Methods of Analysis of Fats and Fatty Oils Part 1: Physical Methods Section 1.16: Determination of Water Content: Entrainment Method. 5 pp.
ISO 934-80. Animal and Vegetable Fats and Oils—Determination of Water Content—Entrainment Method First Edition. 4 pp.

Animal Products
See Also: Butter; Cheeses; Dairy Products; Goats Milk; Leather; Meat; Milk; Offals; Pelts

—Health Requirements
EC 92/120/EEC-92. Council Directive on the Conditions for Granting Temporary and Limited Derogations from Specific Community Health Rules on the Production and Marketing of Certain Products of Animal Origin. 2 pp.

Animal Traps
CGSB CAN/CGSB-144.1-M84. Animal Traps, Humane, Mechanically-Powered, Trigger-Activated. 14 pp.

Animal Welfare (Laboratory Animals)
Use: Laboratory Animals—Humane Treatment

Anion Exchangers
See Also: Ion Exchangers

—Exchange Capacity Measurement
DIN ENGL 54402-85. Testing of Ion Exchangers; Determination of the Total Capacity of Anion Exchangers (July). 9 pp.

Anionic Surfactant Content Analysis
See Also: Cationic Surfactant Content Analysis; Surfactant Content Analysis

Anionic Surfactant Content Analysis (Cont.)

—Detergents

BSI BS 3762: Sec 3.3-87. 1987 Analysis of Formulated Detergents Part 3: Quantitative Test Methods Section 3.3: Method for Determination of Hydrolizable and Non-Hydrolizable Anionic-Active Matter Content After Hydrolysis Under Acid Conditions. 5 pp.

ISO 2868-73. Surface Active Agents—Detergents—Anionic-Active Matter Stable to Acid Hydrolysis—Determination of Trace Amounts First Edition. 4 pp.

ISO 2869-73. Surface Active Agents—Detergents—Anionic-Active Matter Hydrolyzable Under Alkaline Conditions—Determination of Hydrolyzable and Non-Hydrolyzable Anionic-Active Matter First Edition. 4 pp.

ISO 2870-86. Surface Active Agents—Detergents—Determination of Anionic-Active Matter Hydrolyzable and Non-Hydrolyzable Under Acid Conditions Second Edition. 5 pp.

—Detergents—Volumetric Analysis

BSI BS 3762: Sec 3.1-90. 1990 Analysis of Formulated Detergents Part 3: Quantitative Test Methods Section 3.1: Method for Determination for Anionic-Active Matter Content. 8 pp.

BSI BS 3762: Sec 3.2-83. 1983 Analysis of Formulated Detergents Part 3: Quantitative Test Methods Section 3.2: Method for Determination of Hydrolizable and Non-Hydrolizable Anionic-Active Matter Content After Hydrolysis Under Alkaline. 5 pp.

CGSB 2-GP-11M METH 17.3-89. Methods of Testing and Analysis of Soaps and Detergents Anionic Active Ingredient (Cationic Titration Method); (Amendment 2 May 1989). 4 pp.

ISO 2271-89. Surface Active Agents—Detergents—Determination of Anionic-Active Matter by Manual or Mechanical Direct Two-Phase Titration Procedure Second Edition. 8 pp.

—Polyethoxylated Alcohol Sulfates

BSI BS 6829: Sec 5.1-90. 1990 Analysis of Surface Active Agents (Raw Materials) Part 5: Ethoxylated Alcohol and Alkylphenol Sulphates Section 5.1: Method for Determination of Total Active Matter Content. 4 pp.

ISO 6842-89. Surface Active Agents—Sulfated Ethoxylated Alcohols and Alkylphenols—Determination of Total Active Matter Content Second Edition. 4 pp.

—Polyethoxylated Alkylphenol Sulfates

BSI BS 6829: Sec 5.1-90. 1990 Analysis of Surface Active Agents (Raw Materials) Part 5: Ethoxylated Alcohol and Alkylphenol Sulphates Section 5.1: Method for Determination of Total Active Matter Content. 4 pp.

ISO 6842-89. Surface Active Agents—Sulfated Ethoxylated Alcohols and Alkylphenols—Determination of Total Active Matter Content Second Edition. 4 pp.

—Soaps

BSI BS 1715: Part 1-91. 1991 Analysis of Soaps Part 1: General Introduction, Sampling, and Test for Presence of Synthetic Anionic-Active Surface Active Agents. 8 pp.

BSI BS 1715: Part 1-89. 1989 Analysis of Soaps Part 1: General Introduction, Sampling, and Test for Presence of Synthetic Anionic-Active Surface Active Agents. 4 pp.

—Water—Spectrometry

BSI BS 6068: Sec 2.23-86. 1986 Water Quality Part 2: Physical, Chemical and Bio-Chemical Methods Section 2.23: Method for the Determination of Anionic Surfactants by the Methylene Blue Spectrometric Method. 9 pp.

ISO 7875 Pt 1-84. Water Quality—Determination of Surfactants—Part 1: Determination of Anionic Surfactants by the Methylene Blue Spectrometric Method First Edition. 8 pp.

Anionic Surfactants

See Also: Surfactants

—Solubility

BSI BS 6829: Sec 1.6-91. 1991 Analysis of Surface Active Agents (Raw Materials) Part 1: General Methods Section 1.6: Method for Determination of Solubility in Water (ISO 6839: 1982). 7 pp.

ISO 6839-82. Anionic Surface Active Agents—Determination of Solubility in Water First Edition. 5 pp.

Anionic Surfactants (Cont.)

—Surface Tension—Critical Micelle Concentration

ISO 4311-79. Anionic and Non-Ionic Surface Active Agents—Determination of the Critical Micellization Concentration—Method by Measuring Surface Tension with a Plate, Stirrup or Ring First Edition. 5 pp.

Anise Oil

Use For: Anise Seed Oil; Aniseed Oil
CNS K5086-80. Oil of Aniseed (Sep)(6457).
ISO 3475-75. Oil of Aniseed First Edition. 4 pp.

Anise Seed Oil

Use: Anise Oil

Aniseed

See Also: Herbs
ISO 7386-84. Aniseed (Pimpinella Anisum Linnaeus)—Specification First Edition. 5 pp.

Aniseed Oil

Use: Anise Oil

o-Anisidines

CNS K2232-88. Anisidines (Aug)(12400).
JIS K 4137-84. Anisidines (o-Anisidine, p-Anisidine, p-Phenetidine, 3-Amino-4-Methoxytoluene).

p-Anisidines

CNS K2232-88. Anisidines (Aug)(12400).
JIS K 4137-84. Anisidines (o-Anisidine, p-Anisidine, p-Phenetidine, 3-Amino-4-Methoxytoluene).

Anisochronous Communications

See Also: Telex Communications

—Data Networks—Data Terminal Equipment

CCITT RECMN X.52-89. Method of Encoding Anisochronous Signals Into a Synchronous User Bearer—Data Communication Networks—Transmission, Signalling and Switching, Network Aspects, Maintenance and Administrative Arrangements (Study Group VII) 3 pp. 3 pp.

—Data Networks—Quality Assurance

CCITT RECMN R.121-89. Standard Limits of Transmission Quality for Start-Stop User Classes of Service 1 and 2 on Anisochronous Data Networks—Telegraph Transmission (Study Group IX) 1 pp. 1 p.

Anisole

Use For: Methoxybenzene *See Also:* Ethers
CNS K7573-88. Chemical Reagent (Anisole) (Oct)(7806).
JIS K 8041-78. Anisole, Metoxybenzene.

Ankle Prostheses

See Also: Joint Prostheses; Leg Prostheses
BSI BS 2574: Part 1-91. 1991 Lower Limb Orthoses Part 1: Guide to the Design and Manufacture of Lower Limb Orthoses, Excluding Foot Orthoses (Supersedes BS 3330: 1961). 22 pp.
BSI BS 2931-57. 1957 Amd 2 Steel Ankle Joints for Steel Orthopaedic Appliances. 17 pp.
JIS T 9212-85. Artificial Feet and Ankle Joints.
JIS T 9214-91. Metallic Ankle Joints for Lower Extremity Orthoses. 13 pp.

Annealed Glass

Use: Heat Treated Glass

Annealing

Scope Note: See the subheading Annealing or Annealed under specific metals

Annealing Point

See Also: Glass; Thermodynamic Properties
BSI BS 7034: Part 7-88. 1988 Viscosity and Viscometric Fixed Points of Glass Part 7: Method for the Determination of Annealing Point and Strain Point by Beam Bending. 12 pp.
ISO 7884 Pt 7-87. Glass—Viscosity and Viscometric Fixed Points—Part 7: Determination of Annealing Point and Strain Point by Beam Bending First Edition. 12 pp.

Announcements (Telephone Services)

See Also: Telephone Services; Tones (Telephone Services)

—European Telephone Networks—Glossaries

CEPT T/CS 28-02-81. Denomination Et Signification Des Tonalites Et Designations Generales Pour Announces Parlees. 3 pp.
CEPT T/CS 28-02 E-86. Names and Meanings of Tones and General Designation for Verbal Announcements. 4 pp.

Announcements (Telephone Services) (Cont.)

—Telephone Networks

CEPT T/CS 20-15-81. Tonalites Et Announces Parlees. 10 pp.
CEPT T/CS 20-15 E-81. Tones and Announcements. 10 pp.

Annual Publications

Use: Yearbooks

Annual Rings

Use For: Growth Rings; Rings (Wood Structure)
See Also: Trees (Plants)

—Width

JIS Z 2102-57. Method of Measuring Average Width of Annual Rings, Moisture Content and Specific Gravity of Wood (R 1966). 4 pp.

Annunciators

See Also: Acoustic Signals; Audio Amplifiers; Display Devices; Indicator Lights

—Alarm Systems

CENELEC PREN 50136-4-93. Alarm Transmission System Part 4: Annunciation Equipment. 31 pp.

—Color Coding

BSI BS 4099: Part 2-77. (WITHDRAWN) 1977 Amd 1 Colours of Indicator Lights, Push Buttons, Annunciators and Digital Readouts: Flashing Lights, Annunciators and Digital Readouts (Superseded by BS EN 60073: 1993). 8 pp.

Anobiidae

Use: Coleoptera

Anodes

—Cadmium

BSI BS 2868-68. 1968 Cadmium Anodes and Cadmium Oxide for Electroplating. 8 pp.

—Nickel

BSI BS 558 & 564-70. 1970 Nickel Anodes, Anode Nickel and Nickel Salts for Electroplating. 28 pp.
DIN ENGL 1702-67. Nickel Anodes (Jan). 2 pp.

—Silver

BSI BS 1561-66. 1966 Silver Anodes and Silver Salts for Electroplating. 12 pp.

—Tin

BSI BS 1468-67. 1967 Amd 1 Tin Anodes and Tin Salts for Electroplating. 19 pp.

Anodic Coatings

Use For: Anodic Films *See Also:* Coatings; Conversion Coatings; Corrosion; Corrosion Prevention; Nonmetallic Coatings; Oxide Coatings; Paints; Passivation

—Aerospace—Codes

DIN ENGL LN 9368 Pt 4-83. Aerospace; Designation of Surface Treatments; Code Numbers for Anodic Treatments (Apr). 2 pp.

—Aluminum

BSI BS 1615-87. 1987 Methods for Specifying Anodic Oxidation Coating on Aluminium and Its Alloys. 14 pp.
BSI BS 4479: Part 5-90. 1990 Design of Articles That Are to Be Coated Part 5: Recommendations for Anodic Oxidation Coatings. 13 pp.
BSI BS 5599-78. 1978 Hard Anodic Oxide Coatings on Aluminium for Engineering Purposes. 8 pp.
CNS H3101-82. Coatings Combined with Anodic Oxidation and Organic Finishing on Aluminium and Aluminium Alloys (Jan)(8405).
CNS H3105-82. Anodic Oxidation Coatings on Aluminum and Aluminum Alloys (Feb)(8507).
CNS H3107-82. Anodic Oxidation Practice by Sulphuric Acid Process on Aluminium and Aluminium Alloys (Mar)(8590). 13 pp.
CNS H3108-82. Anodic Oxidation Practice by Oxalic Acid Process on Aluminum and Aluminum Alloys (Apr)(8703).
ISO 7599-83. Anodizing of Aluminium and Its Alloys—General Specifications for Anodic Oxide Coatings on Aluminium First Edition. 9 pp.
JIS H 8601-92. Anodic Oxide Coatings on Aluminium and Aluminium Alloys. 11 pp.
JIS H 8602-92. Combined Coatings of Anodic Oxide and Organic Coatings on Aluminium and Aluminium Alloys. 20 pp.
JIS H 9500-92. Recommended Practice for Anodizing on Aluminium and Aluminium Alloys. 11 pp.
JIS H 9502-92. Recommended Practice for Combined Coatings of Anodic Oxide and Organic Coatings on Aluminium and Aluminium Alloys. 11 pp.

Anodic Coatings (Cont.)

—Aluminum (Cont.)

MOD UK DSTAN 03-24-88. Chromic Acid Anodizing of Aluminium and Aluminium Alloys Issue 2. 25 pp.

MOD UK DSTAN 03-25-88. Sulphuric Acid Anodizing of Aluminium and Aluminium Alloys Issue 2. 18 pp.

MOD UK DSTAN 03-26-88. Hard Anodizing of Aluminium and Aluminium Alloys Issue 2. 15 pp.

—Aluminum—Abrasion Testing

BSI BS 6161: Part 9-87. 1987 Methods of Test for Anodic Oxidation Coatings on Aluminum and Its Alloys Part 9: Measurement of Wear Properties with an Abrasive Wheel Wear Test Apparatus. 8 pp.

BSI BS 6161: Part 10-87. 1987 Methods of Test for Anodic Oxidation Coatings on Aluminum and Its Alloys Part 10: Measurement of of Mean Specific Abrasion Resistance with an Abrasive Jet Test Apparatus. 12 pp.

BSI BS 6161: Part 18-91. 1991 Methods of Test for Anodic Oxidation Coatings on Aluminium and Its Alloys Part 18: Determination of Surface Abrasion Resistance. 8 pp.

CNS H2064-82. Methods of Test for Abrasion Resistance of Anodic Oxidation Coatings on Aluminium and Aluminium Alloys (Jan)(8411). 5 pp.

ISO 8251-87. Anodized Aluminium and Aluminium Alloys—Measurement of Wear Resistance and Wear Index of Anodic Oxide Coatings with an Abrasive Wheel Wear Test Apparatus First Edition; (Corrected and Reprinted -1989). 8 pp.

ISO 8252-87. Anodized Aluminium and Aluminium Alloys—Measurement of Mean Specific Abrasion Resistance of Anodic Oxidation Coatings with an Abrasive Jet Test Apparatus First Edition. 12 pp.

JIS H 8682-88. Test Methods for Abrasion Resistance of Anodic Oxidation Coatings on Aluminium and Aluminium Alloys. 18 pp.

SAA AS 2039.5.1-78. Methods for Testing Anodic Oxidation Coatings on Aluminium and Aluminium Alloys—Part 5.1: Abrasion Resistance Tests—Jet Test for Abrasion Resistance of Anodic Oxidation Coatings Reconfirmed 1990. 4 pp.

SAA AS 2039.5.2-78. Methods for Testing Anodic Oxidation Coatings on Aluminium and Aluminium Alloys—Part 5.2: Abrasion Resistance Tests—ASTM Test for Abrasion Resistance of Anodic Oxidation Coatings Reconfirmed 1990. 6 pp.

—Aluminum—Accelerated Testing

BSI BS 6161: Part 7-84. 1984 Methods of Test for Anodic Oxidation Coatings on Aluminium and Its Alloys Part 7: Accelerated Determination of Light Fastness of Coloured Anodic Oxidation Coatings Using Artificial Light. 4 pp.

ISO 2135-84. Anodizing of Aluminium and Its Alloys—Accelerated Test of Light Fastness of Coloured Anodic Oxide Coatings Using Artificial Light Second Edition. 4 pp.

—Aluminum—Acids

MOD UK DSTAN 68-109-89. Chromium Trioxide (Chromic Acid) Issue 1. 12 pp.

MOD UK CS 3119. Chromic Acid (Superseded by Def Stan 68-109).

—Aluminum—Aerospace

AECMA PREN2101-87. Chromic Acid Anodizing of Aluminium and Wrought Aluminium Alloys.

BSI BS EN 2101-91. 1991 Chromic Acid Anodizing of Aluminium and Wrought Aluminium Alloys. 19 pp.

CEN PREN 2101-89. Chromic Acid Anodizing of Aluminium and Wrought Aluminium Alloys. 14 pp.

CEN EN 2101-91. Chromic Acid Anodizing of Aluminium and Wrought Aluminium Alloys. 16 pp.

CEN PREN 2536-92. Aerospace Series Hard Anodizing of Aluminium Alloys. 11 pp.

ISO 8076-84. Aerospace Process—Anodic Treatment of Aluminium Alloys—Chromic Acid Process 40 V DC, Undyed Coating First Edition. 5 pp.

ISO 8077-84. Aerospace Process—Anodic Treatment of Aluminium Alloys—Chromic Acid Process 20 V DC, Undyed Coating First Edition. 5 pp.

ISO 8078-84. Aerospace Process—Anodic Treatment of Aluminium Alloys—Sulfuric Acid Process, Undyed Coating First Edition. 5 pp.

ISO 8079-84. Aerospace Process—Anodic Treatment of Aluminium Alloys—Sulfuric Acid Process, Dyed Coating First Edition. 5 pp.

—Aluminum—Aerospace—Adhesion Testing

AECMA PREN3002-90. Chromic Acid Anodizing Testing of Adhesives. 8 pp.

—Aluminum—Aerospace—Quality Assurance

AECMA PREN2101-87. Chromic Acid Anodizing of Aluminium and Wrought Aluminium Alloys.

AECMA PREN2284-87. Sulphuric Acid Anodizing of Aluminium and Wrought Aluminium Alloys.

Anodic Coatings (Cont.)

—Aluminum—Aerospace—Quality Assurance (Cont.)

BSI BS EN 2101-91. 1991 Chromic Acid Anodizing of Aluminium and Wrought Aluminium Alloys. 19 pp.

BSI BS EN 2284-91. 1991 Sulfuric Acid Anodizing of Aluminium and Wrought Aluminium Alloys. 14 pp.

CEN PREN 2101-89. Chromic Acid Anodizing of Aluminium and Wrought Aluminium Alloys. 14 pp.

CEN EN 2101-91. Chromic Acid Anodizing of Aluminium and Wrought Aluminium Alloys. 16 pp.

CEN PREN 2284-89. Sulphuric Acid Anodizing of Aluminium and Wrought Aluminium Alloys. 9 pp.

CEN EN 2284-91. Sulphuric Acid Anodizing of Aluminium and Wrought Aluminium Alloys. 11 pp.

—Aluminum—Aircraft

JIS W 1111-83. Anodic Coatings on Aluminum and Aluminum Alloys for Aircraft.

—Aluminum—Automotive

SAA AS 1956-76. Anodic Oxidation Coatings on Aluminium for Decorative and Automotive Applications. 16 pp.

—Aluminum—Buildings

BSI BS 3987-91. 1991 Anodic Oxidation Coatings on Wrought Aluminium for External Architectural Applications. 12 pp.

BSI BS 3987-74. 1974 Amd 1 Anodic Oxide Coatings on Wrought Aluminium for External Architectural Applications. 14 pp.

SAA AS 1231-85. Aluminium and Aluminium Alloys—Anodized Coatings for Architectural Applications (R 1993). 7 pp.

SNZ NZS/AS 1231-85. Aluminium and Aluminium Alloys—Anodized Coatings for Architectural Applications (This is a Joint Standard with SAA AS 1231). 7 pp.

—Aluminum—Clarity

BSI BS 6161: Part 13-87. 1987 Methods of Test for Anodic Oxidation Coatings on Aluminium and Its Alloys Part 13: Image Clarity Test Using the Gardam Grid. 4 pp.

BSI BS 6161: Part 19-93. 1993 Methods of Test for Anodic Oxidation Coatings on Aluminium and Its Alloys Part 19: Determination of Image Clarity—Instrumental Method (ISO 10216: 1992) (F). 11 pp.

ISO 10215-92. Anodized Aluminium and Aluminium Alloys—Visual Determination of Image Clarity of Anodic Oxidation Coatings—Chart Scale Method First Edition. 9 pp.

ISO 10216-92. Anodized Aluminium and Aluminium Alloys—Instrumental Determination of Image Clarity of Anodic Oxidation Coatings—Instrumental Method First Edition. 9 pp.

JIS H 8686-78. Test Methods for Image Clarity of Anodic Oxidation Coatings on Aluminium and Aluminium Alloys. 11 pp.

—Aluminum—Colorfastness Testing

BSI BS 6161: Part 8-81. 1981 Methods of Test for Anodic Oxidation Coatings on Aluminium and Its Alloys Part 8: Determination of the Fastness to Ultraviolet Light to Coloured Anodic Oxide Coatings. 4 pp.

CNS H2061-82. Methods of Accelerated Test for Light Fastness of Colored Anodic Oxidation Coatings on Aluminium and Aluminium Alloys (Jan)(8408). 3 pp.

ISO 6581-80. Anodizing of Aluminium and Its Alloys—Determination of the Fastness to Ultra-Violet Light of Coloured Anodic Oxide Coatings First Edition. 4 pp.

JIS H 8685-88. Accelerated Test Methods for Lightfastness of Coloured Anodic Oxidation Coatings on Aluminium and Aluminium Alloys. 11 pp.

—Aluminum—Colorimetry

ISO TR8125-84. Anodizing of Aluminium and Its Alloys—Determination of Colour and Colour Difference of Coloured Anodic Coatings First Edition. 33 pp.

—Aluminum—Comprehensive Testing

SAA AS 2039. Methods for Testing Anodic Oxidation Coatings on Aluminium and Aluminium Alloys Complete Set in Binder.

—Aluminum—Continuity Testing

ISO 2085-76. Anodizing of Aluminium and Its Alloys—Check of Continuity of Thin Anodic Oxide Coatings—Copper Sulphate Test First Edition. 3 pp.

—Aluminum—Corrosion Testing

CNS H2063-82. Methods of Test for Corrosion Resistance of Anodic Oxidation Coatings on Aluminium and Aluminium Alloys (Jan)(8410). 4 pp.

Anodic Coatings (Cont.)

—Aluminum—Corrosion Testing (Cont.)

SAA AS 2039.3.1-78. Methods for Testing Anodic Oxidation Coatings on Aluminium and Aluminium Alloys —Part 3.1: Corrosion Tests—Neutral Salt Spray Test (NSS Test) of Anodic Oxidation Coatings Reconfirmed 1990. 6 pp.

SAA AS 2039.3.2-78. Methods for Testing Anodic Oxidation Coatings on Aluminium and Aluminium Alloys —Part 3.2: Corrosion Tests—Copper Accelerated Acetic Acid Salt Spray Test (CASS Test) of Anodic Oxidation Coatings Reconfirmed 1990. 6 pp.

SAA AS 2039.3.3-78. Methods for Testing Anodic Oxidation Coatings on Aluminium and Aluminium Alloys—Part 3.3: Corrosion Tests—Mortar Test for Clear Organic Coatings on Anodic Oxidation Coatings Reconfirmed 1990. 2 pp.

—Aluminum—Cracking (Fracturing)

CNS H2062-89. Methods of Test for Resistance to Cracking by Deforming of Anodic Oxidation Coatings on Aluminium and Aluminium Alloys (Apr)(8409).

ISO 3211-77. Anodizing of Aluminium and Its Alloys—Assessment of Resistance of Anodic Oxide Coatings to Cracking by Deformation Second Edition. 4 pp.

JIS H 8684-77. Test Method for Resistance to Cracking by Deforming of Anodic Oxidation Coatings on Aluminium and Aluminium Alloys. 8 pp.

—Aluminum—Decorative

SAA AS 1956-76. Anodic Oxidation Coatings on Aluminium for Decorative and Automotive Applications. 16 pp.

—Aluminum—Dielectric Breakdown

BSI BS 6161: Part 15-87. 1987 Methods of Test for Anodic Oxidation Coatings on Aluminium and Its Alloys Part 15: Determination of Electrical Breakdown Potential. 4 pp.

ISO 2376-72. Anodization (Anodic Oxidation) of Aluminium and Its Alloys—Insulation Check by Measurement of Breakdown Potential First Edition. 3 pp.

—Aluminum—Dyeing Testing

BSI BS 6161: Part 5-82. 1982 Methods of Test for Anodic Oxidation Coatings on Aluminium and Its Alloys Part 5: Estimation of Loss of Absorptive Power of Sealed Coatings: Dye Spot Test with Prior Acid Treatment. 7 pp.

ISO 2143-81. Anodizing of Aluminium and Its Alloys—Estimation of Loss of Absorptive Power of Anodic Oxide Coatings After Sealing—Dye Spot Test with Prior Acid Treatment First Edition. 5 pp.

—Aluminum—Electrical Impedance

DIN ENGL 50949-84. Non-Destructive Testing of Anodic Oxidation Coatings on Pure Aluminium and Aluminium Alloys by Measurement of Admittance (Feb). 3 pp.

—Aluminum—Environmental Testing

ISO TR11728-93. Anodized Aluminium and Aluminium Alloys—Accelerated Test of Weather Fastness of Coloured Anodic Oxide Coatings Using Cyclic Artificial Light and Pollution Gas First Edition. 9 pp.

—Aluminum—Gloss

BSI BS 6161: Part 12-87. 1987 Methods of Test for Anodic Oxidation Coatings on Aluminium and Its Alloys Part 12: Measurement of Specular Reflectance and Specular Gloss at Angles of 20 Degrees, 45 Degrees, 60 Degrees or 85 Degrees. 14 pp.

ISO 7668-86. Anodized Aluminium and Aluminium Alloys—Measurement of Specular Reflectance and Specular Gloss at Angles of 20 Degrees, 45 Degrees, 60 Degrees or 85 Degrees First Edition. 12 pp.

—Aluminum—Glossaries

ISO 7583-86. Anodizing of Aluminium and Its Alloys—Vocabulary First Edition. 37 pp.

—Aluminum—Insulation Testing

SAA AS 2039.7.1-78. Methods for Testing Anodic Oxidation Coatings on Aluminium and Aluminium Alloys—Part 7.1: Insulation Tests—Measurement of Breakdown Potential of Anodic Oxidation Coatings Reconfirmed 1990. 2 pp.

—Aluminum—Light Testing

ISO TR11728-93. Anodized Aluminium and Aluminium Alloys—Accelerated Test of Weather Fastness of Coloured Anodic Oxide Coatings Using Cyclic Artificial Light and Pollution Gas First Edition. 9 pp.

INTERNATIONAL AND NON-U.S. NATIONAL STANDARDS
SUBJECT INDEX

Anodic Coatings (Cont.)
—Aluminum—Light Testing (Cont.)
SAA AS 2039.6.1-78. Methods for Testing Anodic Oxidation Coatings on Aluminium and Aluminium Alloys—Part 6.1: Light Fastness Tests—Grey Scale Method for Light Fastness of Anodic Oxidation Coatings Reconfirmed 1990. 6 pp.

—Aluminum—Pitting Testing
BSI BS 6161: Part 16-90. 1990 Methods of Test for Anodic Oxidation Coatings on Aluminium and Its Alloys Part 16: Rating System for the Evaluation of Pitting Corrosion—Chart Method. 12 pp.

ISO 8993-89. Anodized Aluminium and Aluminium Alloys—Rating System for the Evaluation of Pitting Corrosion—Chart Method First Edition. 12 pp.

—Aluminum—Pitting Testing—Rating System
BSI BS 6161: Part 17-90. 1990 Methods of Test for Anodic Oxidation Coatings on Aluminium and Its Alloys Part 17: Rating System for the Evaluation of Pitting Corrosion-Grid Method. 4 pp.

ISO 8994-89. Anodized Aluminium and Aluminium Alloys—Rating System for the Evaluation of Pitting Corrosion—Grid Method First Edition. 4 pp.

—Aluminum—Reflectance
BSI BS 6161: Part 11-85. 1985 Methods of Test for Anodic Oxidation Coatings on Aluminum and Its Alloys Part 11: Measurement of Total Reflectance Using a Photoelectric Reflectometer. 4 pp.

BSI BS 6161: Part 12-87. 1987 Methods of Test for Anodic Oxidation Coatings on Aluminium and Its Alloys Part 12: Measurement of Specular Reflectance and Specular Gloss at Angles of 20 Degrees, 45 Degrees, 60 Degrees or 85 Degrees. 14 pp.

BSI BS 6161: Part 14-87. 1987 Methods of Test for Anodic Oxidation Coatings on Aluminium and Its Alloys Part 14: Determination of Infra-Red Reflectance. 4 pp.

ISO 6719-86. Anodized Aluminium and Aluminium Alloys—Measurement of Reflectance Characteristics of Aluminium Surfaces Using Integrating-Sphere Instruments First Edition. 8 pp.

ISO 7668-86. Anodized Aluminium and Aluminium Alloys—Measurement of Specular Reflectance and Specular Gloss at Angles of 20 Degrees, 45 Degrees, 60 Degrees or 85 Degrees First Edition. 12 pp.

ISO 7759-83. Anodizing of Aluminium and Its Alloys—Measurement of Reflectivity Characteristics of Aluminium Surfaces Using Abridged Goniophotometer or Goniophotometer First Edition. 7 pp.

SAA AS 2039.4.1-78. Methods of Testing Anodic Oxidation Coatings on Aluminium and Aluminium Alloys—Part 4.1: Reflectivity Tests—Specular Reflectivity Test of Anodic Oxidation Coatings with the Gloss Head Reconfirmed 1990. 4 pp.

SAA AS 2039.4.2-78. Methods of Testing Anodic Oxidation Coatings on Aluminium and Aluminium Alloys—Part 4.2: Reflectivity Tests—Total Reflectivity Test of Anodic Oxidation Coatings with the PRS Head Reconfirmed 1990. 4 pp.

SAA AS 2039.4.3-78. Methods of Testing Anodic Oxidation Coatings on Aluminium and Aluminium Alloys—Part 4.3: Reflectivity Tests—Image Clarity Test of Anodic Oxidation Coatings with the Gardam Grid Reconfirmed 1990. 4 pp.

SAA AS 2039.4.4-78. Methods of Testing Anodic Oxidation Coatings on Aluminium and Aluminium Alloys—Part 4.4: Reflectivity Tests—Infared Reflectivity Test of Anodic Oxidation Coatings Reconfirmed 1990. 4 pp.

—Aluminum—Sealing
BSI BS 6161: Part 3-84. 1984 Methods of Test for Anodic Oxidation Coatings on Aluminium and Its Alloys Part 3: Assessment of Sealing Quality by Measurement of the Loss of Mass After Immersion in Phosphoric-Chromic Acid-Solution. 7 pp.

BSI BS 6161: Part 4-81. 1981 Methods of Test for Anodic Oxidaton Coatings on Aluminium and Its Alloys Part 4: Assessmentof Sealing Quality by Measurement of the Loss of Mass After Immersion in Acid Solution. 4 pp.

BSI BS 6161: Part 6-84. 1984 Methods of Test for Anodic Oxidation Coatings on Aluminium and Its Alloys Part 6: Assessment of Sealing Quality by Measurement of Admittance or Impedance. 7 pp.

CNS H2060-89. Methods of Test for Sealing Quality and Anodic Oxidation Coatings on Aluminum and Aluminum Alloys (Apr)(8407).

ISO 2931-83. Anodizing of Aluminium and Its Alloys—Assessment of Quality of Sealed Anodic Oxide Coatings by Measurement of Admittance or Impedance Second Edition. 5 pp.

ISO 3210-83. Anodizing of Aluminium and Its Alloys—Assessment of Quality of Sealed Anodic Oxide Coatings by Measurement of the Loss of Mass After Immersion in Phosphoric-Chromic Acid Solution Second Edition. 5 pp.

Anodic Coatings (Cont.)
—Aluminum—Sealing (Cont.)
JIS H 8683-79. Test Methods for Sealing Quality of Anodic Oxidation Coatings on Aluminium and Aluminium Alloys.

JIS H 8683-73. Test Methods for Sealing Quality of Anodic Oxidation Coatings on Aluminium and Aluminium Alloys. 6 pp.

SAA AS 2039.2.1-78. Methods for Testing Anodic Oxidation Coatings on Aluminium and Aluminium Alloys—Part 2.1: Sealing Property Tests—Measurement of Impedance of Clear Anodic Oxidation Coatings (Aztac Method) Reconfirmed 1990. 4 pp.

SAA AS 2039.2.2-78. Methods for Testing Anodic Oxidation Coatings on Aluminium and Aluminium Alloys—Part 2.2: Sealing Property Tests—Resistance of Anodic Oxidation Coatings to Staining Reconfirmed 1990. 2 pp.

SAA AS 2039.2.3-78. Methods for Testing Anodic Oxidation Coatings on Aluminium and Aluminium Alloys—Part 2.3: Sealing Property Tests—Acetic Acid Test of Anodic Oxidation Coatings Reconfirmed 1990. 2 pp.

SAA AS 2039.2.4-78. Methods for Testing Anodic Oxidation Coatings on Aluminium and Aluminium Alloys—Part 2.4: Sealing Property Tests—Acidified Sodium Sulphite (Kape) Test of Anodic Oxidaiton Coatings Reconfirmed 1990. 2 pp.

SAA AS 2039.2.5-78. Methods for Testing Anodic Oxidation Coatings on Aluminium and Aluminium Alloys—Part 2.5: Sealing Property Tests—Phosphoric-Chromic Acid Test of Anodic Oxidation Coatings Reconfirmed 1990. 2 pp.

—Aluminum—Semifinished Products
DIN ENGL 17611-85. Anodized Wrought Aluminium and Aluminium Alloy Semi-Finished Products with Coating Thicknesses of at Least 10 Glm; Technical Delivery Conditions (June). 5 pp.

—Aluminum—Surface Density—Gravimetric Analysis
BSI BS 6161: Part 1-84. 1984 Methods of Test for Anodic Oxidation Coatings on Aluminium and Its Alloys Part 1: Determination of Mass Per Unit Area (Surface Density) of Anodic Oxidation Coatings (Gravimetric Method). 4 pp.

ISO 2106-82. Anodizing of Aluminium and Its Alloys—Determination of Mass per Unit Area (Surface Density) of Anodic Oxide Coatings—Gravimetric Method Second Edition; (Amendment 1-1983). 5 pp.

—Aluminum—Thickness Measurement
BSI BS 6161: Part 2-81. 1981 Methods of Test for Anodic Oxidation Coatings on Aluminium and Its Alloys Part 2: Determination of Thickness of Anodic Oxide Coatings: Non-Destructive Measurement by Split-Beam Microscope. 4 pp.

CNS H2059-89. Methods of Test for Thickness of Anodic Oxidation Coatings on Aluminum and Aluminum Alloys (Apr)(8406).

ISO 2128-76. Anodizing of Aluminium and Its Alloys—Determination of Thickness of Anodic Oxide Coatings—Non-Destructive Measurement by Split-Beam Microscope First Edition. 4 pp.

JIS H 8680-79. Test Methods for Thickness of Anodic Oxidation Coatings on Aluminium and Aluminium Alloys (R 1984). 9 pp.

SAA AS 2039.1.1-78. Methods for Testing Anodic Oxidation Coatings on Aluminium and Aluminium Alloys—Part 1.1: Thickness and Related Property Tests—Local Thickness of Anodic Oxidation Coatings by Micrographic Examination of Cross-Sections Reconfirmed 1990. 4 pp.

SAA AS 2039.1.2-78. Methods for Testing Anodic Oxidation Coatings on Aluminium and Aluminium Alloys—Part 1.2: Thickness and Related Property Tests—Local Thickness of Anodic Oxidation Coatings by Optical Split-Beam Microscope Reconfirmed 1990. 4 pp.

SAA AS 2039.1.3-78. Methods for Testing Anodic Oxidation Coatings on Aluminium and Aluminium Alloys—Part 1.3: Thickness and Related Property Tests—Average Thickness of Anodic Oxidation Coatings by Strip and Weigh Method Reconfirmed 1990. 4 pp.

SAA AS 2039.1.4-78. Methods for Testing Anodic Oxidation Coatings on Aluminium and Aluminium Alloys—Part 1.4: Thickness and Related Property Tests—Average Thickness of Anodic Oxidation Coatings by Eddy Current Method Reconfirmed 1990. 4 pp.

SAA AS 2039.1.5-78. Methods for Testing Anodic Oxidation Coatings on Aluminium and Aluminium Alloys—Part 1.5: Thickness and Related Property Tests—Continuity of Thin Anodic Oxidation Coatings Reconfirmed 1990. 2 pp.

—Atmospheric Corrosion Testing
BSI BS 7561-92. 1992 Method for Corrosion Tests in Artificial Atmospheres at Very Low Concentrations of Polluting Gas(es) (ISO 10062: 1991) (V). 14 pp.

Anodic Coatings (Cont.)
—Atmospheric Corrosion Testing (Cont.)
ISO 10062-91. Corrosion Tests in Artificial Atmosphere at Very Low Concentrations of Polluting Gas(es) First Edition. 12 pp.

—Cadmium
BSI BS 2868-68. 1968 Cadmium Anodes and Cadmium Oxide for Electroplating. 8 pp.

—Corrosion Testing
BSI BS 5466: Part 10-92. 1992 Methods for Corrosion Testing of Metallic Coatings Part 10: Rating of Test Specimens with Coatings Anodic to the Substrate (ISO 8403: 1991). 14 pp.

ISO 8403-91. Metallic Coatings—Coatings Anodic to the Substrate—Rating of Test Specimens Subjected to Corrosion Tests First Edition. 11 pp.

—Magnesium
CGSB 31-GP-113M-89. Anodic Coatings for Magnesium Alloys (HAE). 15 pp.

—Oxalic Acid
CNS H2066-82. Analysis of Anodic Oxidation Solation-Oxalic Acid (Apr)(8704).

—Sulfuric Acid
CNS H2067-82. Analysis of Anodic Oxidation Solation-Sulphuric Acid (Apr) (8705).

—Sulfuric Acid—Aluminum—Aerospace
CEN PREN 2284-89. Sulphuric Acid Anodizing of Aluminium and Wrought Aluminium Alloys. 9 pp.

CEN EN 2284-91. Sulphuric Acid Anodizing of Aluminium and Wrought Aluminium Alloys. 11 pp.

—Titanium—Aerospace
BSI M 58-86. 1986 Anodic Coating of Titanium and Titanium Alloys by the Sulphuric Acid Process. 4 pp.

CEN PREN 2808-91. Anodizing of Titanium and Titanium Alloys. 9 pp.

ISO 8080-85. Aerospace—Anodic Treatment of Titanium and Titanium Alloys—Sulfuric Acid Process First Edition. 4 pp.

MOD UK DTD-942-73. Anodizing of Titanium and Titanium Alloys. 2 pp.

Anodic Films
Use: Anodic Coatings

Anodic Oxidation
Use: Anodic Coatings

Anodic Polarization
—Stainless Steels
CNS G2262-86. Method for Making Anode Polarization Measurement for Stainless Steel (Mar)(11521).

JIS G 0579-83. Method for Making Anodic Polarization Measurement for Stainless Steels. 14 pp.

Anodizing
Use: Anodic Coatings

Anodizing Processes
Use: Anodic Coatings

Anone
Use: Cyclohexanone

Anoraks
—Children's
BSI BS 5919-91. 1991 Children's Anoraks. 11 pp.

Anoscopes
See Also: Speculum
CNS T1018-81. Auoscopes (Jul)(6277).
JIS T 4403-53. Anoscopes.

Answer Signals
See Also: Signaling Systems
CCITT RECMN Q.27-89. Transmission of the Answer Signal—General Recommendations on Telephone Switching and Signalling—Functions and Information Flow for Services in the ISDN—Supplements (Study Group XI) 1 pp. 1 p.

CCITT RECMN Q.28-89. Determination of the Moment of the Called Subscriber's Answer in the Automatic Service—General Recommendations on Telephone Switching and Signalling—Functions and Information Flows for Services in the ISDN—Supplements (Study Group XI) 1 pp. 1 p.

—International Exchanges
CCITT RECMN Q.109-89. Transmission of the Answer Signal in International Exchanges—General Recommendations on Telephone Switching and Signalling—Functions and Information Flows for Services in the ISDN—Supplements (Study Group XI) 1 pp. 1 p.

Answerback

See Also: Codes; Information Interchange; Protocols

—Data Terminal Equipment—Telex Equipment

CCITT RECMN S.23-89. Automatic Request of the Answerback of the Terminal of the Calling Party, by the Telex Terminal of the Called Party or by the International Network—Telegraph Services Terminal Equipment (Study Group IX) 2 pp. 2 pp.

—Gentex Networks

CCITT RECMN F.1-89. Operational Provisions for the International Public Telegram Service—Telegraph and Mobile Services Operations and Quality of Service (Study Group I) 54 pp. 54 pp.

CCITT RECMN F.20-89. International Gentex Service—Telegraph and Mobile Services Operations and Quality of Service (Study Group I) 5 pp. 5 pp.

CCITT RECMN F.21-89. Composition of Answer-Back Codes for the International Gentex Service—Telegraph and Mobile Services Operations and Quality of Service (Study Group I) 3 pp. 3 pp.

CCITT RECMN S.6-89. Characteristics of Answerback Units (ITA2)—Telegraph Services Terminal Equipment (Study Group IX) 2 pp. 2 pp.

—Gentex Networks—Service Messages

CCITT RECMN F.21-89. Composition of Answer-Back Codes for the International Gentex Service—Telegraph and Mobile Services Operations and Quality of Service (Study Group I) 3 pp. 3 pp.

—Gentex Networks—Teleprinters

CCITT RECMN F.21-89. Composition of Answer-Back Codes for the International Gentex Service—Telegraph and Mobile Services Operations and Quality of Service (Study Group I) 3 pp. 3 pp.

—Maritime Mobile Services—Radioteletype Communications

CCITT RECMN F.130-89. Maritime Answer-Back Codes—Telegraph and Mobile Services Operations and Quality of Service (Study Group I) 2 pp. 2 pp.

—Maritime Mobile Services—Telex Communications

CCITT RECMN F.130-89. Maritime Answer-Back Codes—Telegraph and Mobile Services Operations and Quality of Service (Study Group I) 2 pp. 2 pp.

—Start-Stop Transmission

CCITT RECMN S.32-89. Answer-Back Units for 200- and 300-Baud Start-Stop Machines in Accordance with Recommendation S.30—Telegraph Services Terminal Equipment (Study Group IX) 2 pp. 2 pp.

—Telex Communications

CCITT RECMN F.60-92. Operational Provisions for the International Telex Service (Study Group I) 34 pp. 34 pp.

CCITT RECMN S.6-89. Characteristics of Answerback Units (ITA2)—Telegraph Services Terminal Equipment (Study Group IX) 2 pp. 2 pp.

—Telex Communications—Address Extraction

CCITT RECMN U.74-89. Extraction of Telex Selection Information from a Calling Telex Answerback—Telegraph Switching (Study Group IX) 3 pp. 3 pp.

—Telex Communications—Data Terminal Equipment

CCITT RECMN F.73-90. Operational Principles for Communication Between Terminals of the International Telex Service and Data Terminal Equipment on Packet Switched Public Data Networks (Study Group I) 9 pp. 9 pp.

CCITT RECMN F.73-89. Operational Principles for Communication Between Terminals on Telex Networks and Data Terminal Equipment on Packet Switched Public Data Networks—Telegraph and Mobile Services Operations and Quality of Service (Study Group I) 5 pp. 5 pp.

—Telex Communications—Electronic Mail Systems

CCITT RECMN F.74-92. Intermediate Storage Devices Accessed from the International Telex Service Using Single Stage Selection—Answerback Format (Study Group I) 5 pp. 5 pp.

CCITT RECMN F.74-89. Operational Provisions Relating to Mailbox Devices Connected to the Telex Network—Telegraph and Mobile Services Operations and Quality of Service (Study Group I) 3 pp. 3 pp.

—Telex Communications—Electronic Mail Systems—Overflow Traffic

CCITT RECMN F.74-89. Operational Provisions Relating to Mailbox Devices Connected to the Telex Network—Telegraph and Mobile Services Operations and Quality of Service (Study Group I) 3 pp. 3 pp.

Answerback (Cont.)

—Telex Communications—Message Switching

CCITT RECMN F.35-89. Provisions Applying to the Operation of an International Public Automatic Message Switching Service for Equipments Utilizing the International Telegraph Alphabet No. 2—Telegraph and Mobile Services Operations and Quality of Service (Study Group I) 7 pp. 7 pp.

—Telex Communications—Store and Forward Units

CCITT RECMN F.72-89. International Telex Store and Forward—General Principles and Operational Aspects—Telegraph and Mobile Services Operations and Quality of Service (Study Group I) 18 pp. 18 pp.

—Telex Equipment

CCITT RECMN F.60-89. Operational Provisions for the International Telex Service—Telegraph and Mobile Services Operations and Quality of Service (Study Group I) 17 pp. 17 pp.

—Telex Equipment—Data Terminal Equipment

CCITT RECMN U.75-89. Automatic Called Telex Answerback Check—Telegraph Switching (Study Group IX) 3 pp. 3 pp.

—Telex Packet Interworking Function

CCITT RECMN F.73-90. Operational Principles for Communication Between Terminals of the International Telex Service and Data Terminal Equipment on Packet Switched Public Data Networks (Study Group I) 9 pp. 9 pp.

CCITT RECMN F.73-89. Operational Principles for Communication Between Terminals on Telex Networks and Data Terminal Equipment on Packet Switched Public Data Networks—Telegraph and Mobile Services Operations and Quality of Service (Study Group I) 5 pp. 5 pp.

—Unit Simulators

CCITT RECMN S.17-89. Answer-Back Unit Simulators—Telegraph Services Terminal Equipment (Study Group IX) 1 pp. 1 pp.

Answering Machines

Use: Telephone Answering Equipment

Answering Time

Use For: Time to Answer
See Also: Telecommunication Equipment; Telephone Systems; Time

—Semiautomatic Demand Operating

CCITT RECMN E.142-89. Time-to-Answer by Operators—Telephone Network and ISDN—Operation, Numbering, Routing and Mobile Service (Study Group II) 1 pp. 1 pp.

—Service Observation

CCITT RECMN E.423-89. Observations on Traffic Set up by Operators—Telephone Network and ISDN—Quality of Service, Network Management and Traffic Engineering (Study Group II) 4 pp. 4 pp.

—Telex Communications—Electronic Mail Systems

CCITT RECMN F.74-89. Operational Provisions Relating to Mailbox Devices Connected to the Telex Network—Telegraph and Mobile Services Operations and Quality of Service (Study Group I) 3 pp. 3 pp.

—Telex Communications—International Exchanges

CCITT RECMN F.65-89. Time-to-Answer by Operators at International Telex Positions—Telegraph and Mobile Services Operations and Quality of Service (Study Group I) 1 pp. 1 pp.

Antenna Beams

Use: Antennas—Beams

Antenna Connectors

See Also: Antenna Couplers; Antennas; Connectors

—Automotive

ISO 10487 Pt 1-92. Passenger Car Radio Connections—Part 1: Dimensions and General Requirements First Edition. 6 pp.

Antenna Couplers

See Also: Antenna Connectors; Antennas; Couplers

—Capacitor Antennas

CNS C7150-82. Across-the-Line Capacitor Antenna-Coupling and Line-by-Pass Components for Radio and Television Type Appliances (Jan)(8382).

Antenna Masts

Use For: Masts (Antennas)
See Also: Communication Towers; Masts (Supports)

—Safety

MOD UK DSTAN 05-39-01. Safety Requirements for the Design, Erection and Maintenance of Land Based Masts, Towers, Turntables, Antenna Supporting, and Related Structures Issue 1; Amendment 2. 22 pp.

Antenna Positions

Use For: Aerial Positions *See Also:* Aircraft

CAA. Foreword (Approved Aerial Positions). 2 pp.

—Aircraft Types

CAA. Index (Approved Aerial Positions). 4 pp.
CAA. Aero/Super Aero 45/145 (Approved Aerial Positions). 1 p.
CAA. Aero Commander 200 (Approved Aerial Positions). 1 p.
CAA. Aero Commander 500/520/560 (Approved Aerial Positions). 1 p.
CAA. Aero Commander 680, 680E and 680F (Approved Aerial Positions). 1 p.
CAA. Aero Commander 680T (Approved Aerial Positions). 1 p.
CAA. Aero Commander 685, 690A, 690B, 690C, 690D, 695 and 695A (Approved Aerial Positions). 1 p.
CAA. Aeronca 15AC. 1 p.
CAA. Aerospatiale AS332L (Approved Aerial Positions). 1 p.
CAA. Aerospatiale AS350B, AS355, AS355F1 and AS355N (Approved Aerial Positions). 1 p.
CAA. Aerospatiale SA365C (Approved Aerial Positions). 1 p.
CAA. Aerospatiale SA365N (Approved Aerial Positions). 1 p.
CAA. Aerosport 2. 1 p.
CAA. Aerostar 601P (Approved Aerial Positions). 1 p.
CAA. Agusta A109A and 109C (Approved Aerial Positions). 1 p.
CAA. Airbus A300 (Approved Aerial Positions). 1 p.
CAA. Airbus A310 (Approved Aerial Positions). 1 p.
CAA. Airbus A320. 1 p.
CAA. Airship SKS500 and SKS600 (Approved Aerial Positions). 1 p.
CAA. Airtourer T6-24, 100 and 150 (Approved Aerial Positions). 1 p.
CAA. Alon Aircoupe and Alon A-2A. 1 p.
CAA. Aries (Approved Aerial Positions). 1 p.
CAA. ARV Aviation ARV-1/Super 2 (Approved Aerial Positions). 1 p.
CAA. ATL 98 (Approved Aerial Positions). 1 p.
CAA. Auster (All Series) and Beagle/Auster A61 (Approved Aerial Positions). 1 p.
CAA. Avro XIX (Approved Aerial Positions). 1 p.
CAA. Avro/HS 748 (Approved Aerial Positions). 1 p.
CAA. AWW 650 Argosy (Approved Aerial Positions). 1 p.
CAA. Ayres S2R (Approved Aerial Positions). 1 p.
CAA. BAC One-Eleven 200 and 400 Series (Approved Aerial Positions). 1 p.
CAA. BAC One-Eleven 500 Series. 1 p.
CAA. BAe ATP. 1 p.
CAA. BAe 146 (Approved Aerial Positions). 1 p.
CAA. Beagle B.121/125/(Bulldog) (Approved Aerial Positions). 1 p.
CAA. Beagle B.206/B.206Y (Approved Aerial Positions). 1 p.
CAA. Beagle/Auster A109 (Approved Aerial Positions). 1 p.
CAA. Bede BD4-150 (Approved Aerial Positions). 1 p.
CAA. Beechcraft D17S (Approved Aerial Positions). 1 p.
CAA. Beechcraft E18, H18 and 3TM (Approved Aerial Positions). 1 p.
CAA. Beechcraft 19, 23 and 24 (Approved Aerial Positions). 1 p.
CAA. Beech 33-33/F33A/35 Series/A36/58 Series (Approved Aerial Positions). 1 p.
CAA. Beechcraft 60 (Approved Aerial Positions). 1 p.
CAA. Beechcraft B60 (Approved Aerial Positions). 1 p.
CAA. Beechcraft 65, 70, 90 and 100 Series (Approved Aerial Positions). 1 p.
CAA. Beechcraft 76 (Approved Aerial Positions). 1 p.
CAA. Beechcraft D55, 56TC, 95, B95A and D95A (Approved Aerial Positions). 1 p.
CAA. Beechcraft 200 and B200 (Approved Aerial Positions). 1 p.
CAA. Beechcraft 300 (Approved Aerial Positions). 1 p.
CAA. Belfast SC5 (Approved Aerial Positions). 1 p.
CAA. Bell 47G and 47D (Approved Aerial Positions). 1 p.
CAA. Bell 47G5 (Approved Aerial Positions). 1 p.
CAA. Bell 47J (Approved Aerial Positions). 1 p.
CAA. Bell 206A, 206B and 206L (Approved Aerial Positions). 1 p.

INTERNATIONAL AND NON-U.S. NATIONAL STANDARDS
SUBJECT INDEX
Antenna

Antenna Positions *(Cont.)*
—*Aircraft Types* *(Cont.)*

CAA. Bell 212 and 214B (Approved Aerial Positions). 1 p.
CAA. Bell 214ST (Approved Aerial Positions). 1 p.
CAA. Bell 222 (Approved Aerial Positions). 1 p.
CAA. Bell 412. 1 p.
CAA. Bellanca (Champion) Series 7 and 8 (Approved Aerial Positions). 1 p.
CAA. Bellanca 17-30A (Approved Aerial Positions). 1 p.
CAA. BN2A B and T Series Islander (Approved Aerial Positions). 1 p.
CAA. BN2A MK.III Series Trislander (Approved Aerial Positions). 1 p.
CAA. Boeing A75 N1 and B75. 1 p.
CAA. Boeing Vertol 234 (Approved Aerial Positions). 1 p.
CAA. Boeing 707 and 720 (Approved Aerial Positions). 1 p.
CAA. Boeing 727-100 (Approved Aerial Positions). 1 p.
CAA. Boeing 727-200 Series (Approved Aerial Positions). 1 p.
CAA. Boeing 737 Series (Approved Aerial Positions). 1 p.
CAA. Boeing 747. 1 p.
CAA. Boeing 757-200 Series. 1 p.
CAA. Boeing 767-200 and-300 Series (Approved Aerial Positions). 1 p.
CAA. Bolkow BO.105 D/DB/DB4 (Approved Aerial Positions). 1 p.
CAA. Bolkow BO.105 DS/DBS/DBS4. 1 p.
CAA. Bolkow BO.208 (Approved Aerial Positions). 1 p.
CAA. Bolkow BO.209 (Approved Aerial Positions). 1 p.
CAA. Brantly B2 (Approved Aerial Positions). 1 p.
CAA. Brantly 305 (Approved Aerial Positions). 1 p.
CAA. Brasov IS 28M24 (Approved Aerial Positions). 1 p.
CAA. Bristol 170 MK.21 (Approved Aerial Positions). 1 p.
CAA. Bristol 170 MK.31 (Approved Aerial Positions). 1 p.
CAA. Bristol 170 MK.32 (Approved Aerial Positions). 1 p.
CAA. Bristol 171 (Approved Aerial Positions). 1 p.
CAA. Britannia 100 (Approved Aerial Positions). 1 p.
CAA. Britannia 300 (Approved Aerial Positions). 1 p.
CAA. Canadair CL44 (Approved Aerial Positions). 1 p.
CAA. CAP 10, CAP 20 and CAP 231 Series (Approved Aerial Positions). 1 p.
CAA. Catalina. 1 p.
CAA. Cessna 120 and 140 (Approved Aerial Positions). 1 p.
CAA. Cessna 150 and 152 (Approved Aerial Positions). 1 p.
CAA. Cessna 170/172/175/177/180/182/185/205/206/210 Series (Approved Aerial Positions). 1 p.
CAA. Cessna 207 (Approved Aerial Positions). 1 p.
CAA. Cessna T303 (Approved Aerial Positions). 1 p.
CAA. Cessna 310 and 320 (Approved Aerial Positions). 1 p.
CAA. Cessna 336/337 (Approved Aerial Positions). 1 p.
CAA. Cessna 335 and 340 (Approved Aerial Positions). 1 p.
CAA. Cessna 401/402/404/411/414/421/425/441 (Approved Aerial Positions). 1 p.
CAA. Cessna 406. 1 p.
CAA. Cessna 500 (Approved Aerial Positions). 1 p.
CAA. Cessna 550 (Approved Aerial Positions). 1 p.
CAA. Cessna 650 Serial No. 650-0179 and on. 1 p.
CAA. Chipmunk (Approved Aerial Positions). 1 p.
CAA. CIT/A1 (Approved Aerial Positions). 1 p.
CAA. Comet 4 (Approved Aerial Positions). 1 p.
CAA. Comet 4B and 4C (Approved Aerial Positions). 1 p.
CAA. Concorde (Approved Aerial Positions). 1 p.
GAL. Cygnet (Approved Aerial Positions). 1 p.
CAA. Dakota (Approved Aerial Positions). 1 p.
CAA. DC-4 (Approved Aerial Positions). 1 p.
CAA. DC-6B (Approved Aerial Positions). 1 p.
CAA. DC-7B (Approved Aerial Positions). 1 p.
CAA. DC-7C (Approved Aerial Positions). 1 p.
CAA. DC-8 (Approved Aerial Positions). 1 p.
CAA. DC-9 (Approved Aerial Positions). 1 p.
CAA. DC-10 (Approved Aerial Positions). 1 p.
CAA. Desford Trainer (Approved Aerial Positions). 1 p.
CAA. DHC-2 Beaver MK1. 1 p.
CAA. DHC-6 Twin Otter (Approved Aerial Positions). 1 p.
CAA. DHC-7 (Approved Aerial Positions). 1 p.
CAA. DH80A Puss Moth (Approved Aerial Positions). 1 p.
CAA. DH82A Tiger Moth (Approved Aerial Positions). 1 p.
CAA. DH85 Leopard Moth (Approved Aerial Positions). 1 p.
CAA. DH87B (Approved Aerial Positions). 1 p.

Antenna Positions *(Cont.)*
—*Aircraft Types* *(Cont.)*

CAA. DH89A Rapide (Approved Aerial Positions). 1 p.
CAA. DH90/DH90A Dragonfly (Approved Aerial Positions). 1 p.
CAA. DH104/Dove. 1 p.
CAA. DH 112/Venom MK1. 1 p.
CAA. Dornier DO 28. 1 p.
CAA. Dornier DO 228-100 and-200. 1 p.
CAA. Edgley EA7 Optica (Approved Aerial Positions). 1 p.
CAA. Embraer 110P1 and 110P2 (Approved Aerial Positions). 1 p.
CAA. Embraer 120 Brasilia. 1 p.
CAA. Embraer 121 (Approved Aerial Positions). 1 p.
CAA. Enstrom F28A Series and 280 Series (Approved Aerial Positions). 1 p.
CAA. EP9 (Approved Aerial Positions). 1 p.
CAA. Ercoupe 415 (Approved Aerial Positions). 1 p.
CAA. Fairchild 24R-46A Argus III (Approved Aerial Positions). 1 p.
CAA. Falco F8L (Approved Aerial Positions). 1 p.
CAA. Falcon 20 Series E, F and D (Approved Aerial Positions). 1 p.
CAA. FFA AS202/184A Bravo. 1 p.
CAA. Fokker F.27 Series 200/500 (Approved Aerial Positions). 1 p.
CAA. Fokker F.28 2000 and 4000 (Approved Aerial Positions). 1 p.
CAA. Forney F1A. 1 p.
CAA. Fournier RF.4 (Approved Aerial Positions). 1 p.
CAA. Fuji FA 200 (Approved Aerial Positions). 1 p.
CAA. Gardan GY.80 (Approved Aerial Positions). 1 p.
CAA. Gemini (Approved Aerial Positions). 1 p.
CAA. Gloster Meteor NFII (Approved Aerial Positions). 1 p.
CAA. Grob G109B (Approved Aerial Positions). 1 p.
CAA. Grob G115. 1 p.
CAA. Grumman American AA1 Series (Approved Aerial Positions). 1 p.
CAA. Grumman American AA5 Series/American General AG-5B (Approved Aerial Positions). 1 p.
CAA. Grumman American GA-7 (Approved Aerial Positions). 1 p.
CAA. Grumman G159 Gulfstream 1 (Approved Aerial Positions). 1 p.
CAA. Grumman G1159 Gulfstream 2 (Approved Aerial Positions). 1 p.
CAA. Grumman G1159A Gulfstream 3. 1 p.
CAA. Grumman G164A (Approved Aerial Positions). 1 p.
CAA. Gyroflug SCO1B 160. 1 p.
CAA. Harvard AT6D (Approved Aerial Positions). 1 p.
CAA. Helio H-295 and H-395 (Approved Aerial Positions). 1 p.
CAA. Herald (Dart) 100 Series (Approved Aerial Positions). 1 p.
CAA. Herald (Dart) 200 Series (Approved Aerial Positions). 1 p.
CAA. Heron (Approved Aerial Positions). 1 p.
CAA. Hiller UH12A/UH12C/UH12E and FH 1100 (Approved Aerial Positions). 1 p.
CAA. Hiller UH12E4 (Approved Aerial Positions). 1 p.
CAA. HP.137 Jetstream and BAe Jetstream 31 (Approved Aerial Positions). 1 p.
CAA. HS125-200 and 400 Series (Approved Aerial Positions). 1 p.
CAA. HS125-600 and 700 Series (Approved Aerial Positions). 1 p.
CAA. HS125-800 (Approved Aerial Positions). 1 p.
CAA. HS 125-1000. 1 p.
CAA. Hughes 269A (Approved Aerial Positions). 1 p.
CAA. Hughes 369 (Approved Aerial Positions). 1 p.
CAA. Jodel (CEA) 253, 315, 340, 360, 380 (Approved Aerial Positions). 1 p.
CAA. Jodel D.140 (Approved Aerial Positions). 1 p.
CAA. Jodel DR 100A, D 117, D 120, D 150, DR 200, DR 221, DR 250/160, DR 1050, DR 1051 (Approved Aerial Positions). 1 p.
CAA. Lake LA4 (Approved Aerial Positions). 1 p.
CAA. Learjet 25B (Approved Aerial Positions). 1 p.
CAA. Learjet 35A and 36A (Approved Aerial Positions). 1 p.
CAA. Learjet 55 (Approved Aerial Positions). 1 p.
CAA. Let Z.37 (Approved Aerial Positions). 1 p.
CAA. Lockheed L1011 (Approved Aerial Positions). 1 p.
CAA. Maule M5-235C and M6-235 (Approved Aerial Positions). 1 p.
CAA. Miles M14 (Hawk) (Approved Aerial Positions). 1 p.
CAA. Mitsubishi MU 300. 1 p.
CAA. Morane Saulnier MS880B/885/890/892/893/894 Rallye 100ST, 110ST and 150ST (Approved Aerial Positions). 1 p.
CAA. Morava L200A (Approved Aerial Positions). 1 p.
CAA. Nash Petrel (Approved Aerial Positions). 1 p.

Antenna Positions *(Cont.)*
—*Aircraft Types* *(Cont.)*

CAA. NDN1T Firecracker (Approved Aerial Positions). 1 p.
CAA. Nord 262 (Approved Aerial Positions). 1 p.
CAA. Nord 1002 and 1101 (Approved Aerial Positions). 1 p.
CAA. Norman NAC 6 Fieldmaster. 1 p.
CAA. Partenavia P64B (Approved Aerial Positions). 1 p.
CAA. Partenavia P68 Series (Approved Aerial Positions). 1 p.
CAA. Piaggio P.166 (Approved Aerial Positions). 1 p.
CAA. Pierre Robin CEA DR300 and DR400 Series (Approved Aerial Positions). 1 p.
CAA. Pierre Robin HR100/200, HR100/210, R1180 (T) and R2112 (Approved Aerial Positions). 1 p.
CAA. Pierre Robin HR200/100, R2100A and R2160 (Approved Aerial Positions). 1 p.
CAA. Pierre Robin R3000-120 (Approved Aerial Positions). 1 p.
CAA. Pilatus PC6 (Approved Aerial Positions). 1 p.
CAA. Twin Pioneer (Approved Aerial Positions). 1 p.
CAA. Piper J3C-65 (Approved Aerial Positions). 1 p.
CAA. Piper PA.12 (Approved Aerial Positions). 1 p.
CAA. Piper PA.18, PA.19, PA.20 and PA.22 (Approved Aerial Positions). 1 p.
CAA. Piper PA.23 (Approved Aerial Positions). 1 p.
CAA. Piper PA.24, PA.30 and PA.39 (Approved Aerial Positions). 1 p.
CAA. Piper PA.25/150 (Approved Aerial Positions). 1 p.
CAA. Piper PA.25/235 (Approved Aerial Positions). 1 p.
CAA. Piper PA.28—Short Fuselage (i.e. up to Con. No. 28—7305001) (Approved Aerial Positions). 1 p.
CAA. Piper PA.28R, 200-2, PA28-201T, PA.28-236 and PA.28—Long Fuselage (i.e. Con. No. 28-7305001 and Upwards) (Approved Aerial Positions). 1 p.
CAA. Piper PA.31, PA.31B and PA.31-325 (Approved Aerial Positions). 1 p.
CAA. Piper PA.31-350 and PA.31T3 (Approved Aerial Positions). 1 p.
CAA. Piper PA.31P and PA31T1 (Approved Aerial Positions). 1 p.
CAA. Piper PA32, PA32R, PA32RT and PA32-301T (Approved Aerial Positions). 1 p.
CAA. Piper PA.34 (Approved Aerial Positions). 1 p.
CAA. Piper PA-36-375 (Approved Aerial Positions). 1 p.
CAA. Piper PA.38 (Approved Aerial Positions). 1 p.
CAA. Piper PA.42 (Approved Aerial Positions). 1 p.
CAA. Piper PA.44 (Approved Aerial Positions). 1 p.
CAA. Piper PA46-310P (Approved Aerial Positions). 1 p.
CAA. Pitts S1 and S2 Series (Approved Aerial Positions). 1 p.
CAA. Prentice (Approved Aerial Positions). 1 p.
CAA. Prince (Approved Aerial Positions). 1 p.
CAA. Proctor (Approved Aerial Positions). 1 p.
CAA. Provost (Approved Aerial Positions). 1 p.
CAA. Robinson R22 (Approved Aerial Positions). 1 p.
CAA. Rockwell Commander 112 and 114 (Approved Aerial Positions). 1 p.
CAA. Rockwell 500S (Approved Aerial Positions). 1 p.
CAA. Rollason D62B Condor (Approved Aerial Positions). 1 p.
CAA. S.51 (Approved Aerial Positions). 1 p.
CAA. S.55 (Approved Aerial Positions). 1 p.
CAA. SA330J Puma (Approved Aerial Positions). 1 p.
CAA. Saab 91A/91/C Safir (Approved Aerial Positions). 1 p.
CAA. Saab SF 340A. 1 p.
CAA. Saro Skeeter (Approved Aerial Positions). 1 p.
CAA. Scheibe SF25E (Approved Aerial Positions). 1 p.
CAA. Schweizer 300C. 1 p.
CAA. Scintex, Linnet and Piel CP.301A (Approved Aerial Positions). 1 p.
CAA. Short SC.7 (Approved Aerial Positions). 1 p.
CAA. Short SD3-30 (Approved Aerial Positions). 1 p.
CAA. Short SD3-60 (Approved Aerial Positions). 1 p.
CAA. Siai Marchetti S.205 and S.205/22/R (Approved Aerial Positions). 1 p.
CAA. Siai Marchetti F.260 (Approved Aerial Positions). 1 p.
CAA. Sikorsky S58ET (Approved Aerial Positions). 1 p.
CAA. Sikorsky S61-N2 (Approved Aerial Positions). 1 p.
CAA. Sikorsky S76A and S76C (Approved Aerial Positions). 1 p.
CAA. Slingsby T61. 1 p.
CAA. Slingsby T67A.B, C and M (Approved Aerial Positions). 1 p.
CAA. Socata ST10 (Approved Aerial Positions). 1 p.
CAA. Socata TB9, TB10 and TB20 (Approved Aerial Positions). 1 p.
CAA. Spitfire (Approved Aerial Positions). 1 p.
CAA. Stampe SV4C. 1 p.
CAA. Stinson 108-3. 1 p.
CAA. Stinson V77 Reliant. 1 p.

INTERNATIONAL AND NON-U.S. NATIONAL STANDARDS
SUBJECT INDEX

Antenna

Antenna Positions (Cont.)
—Aircraft Types (Cont.)
CAA. Sud SA315/SA318 (Alouette) (Approved Aerial Positions). 1 p.
CAA. Swearingen SA226T (Approved Aerial Positions). 1 p.
CAA. Behev Mraz Sokol M.1C (Approved Aerial Positions). 1 p.
CAA. Meta/Sokol L.40 (Approved Aerial Positions). 1 p.
CAA. Thunder and Colt GA-42. 1 p.
CAA. Trident 1 (Approved Aerial Positions). 1 p.
CAA. Trident 2E (Approved Aerial Positions). 1 p.
CAA. Trident 3B (Approved Aerial Positions). 1 p.
CAA. Valentin Taifun 17E (Approved Aerial Positions). 1 p.
CAA. Vanguard 951 and 953 (Approved Aerial Positions). 1 p.
CAA. Varga 2150A (Approved Aerial Positions). 1 p.
CAA. VC.10 (Approved Aerial Positions). 1 p.
CAA. Super VC.10 (Approved Aerial Positions). 1 p.
CAA. Viscount 700 (Approved Aerial Positions). 1 p.
CAA. Viscount 800 (Approved Aerial Positions). 1 p.
CAA. Wassmer WA.40 and WA.41 (Approved Aerial Positions). 1 p.
CAA. Wassmer WA51/W52 (Approved Aerial Positions). 1 p.
CAA. Wessex 60 (Approved Aerial Positions). 1 p.
CAA. Westland Gazelle SA341-G (Approved Aerial Positions). 1 p.
CAA. Westland 30 (Approved Aerial Positions). 1 p.
CAA. Wilga 80. 1 p.
CAA. Wittman W8 Tailwind (Approved Aerial Positions). 1 p.
CAA. Zlin Z226T/Z326/Z526 (Approved Aerial Positions). 1 p.

Antenna Systems
Use: Antennas

Antenna Towers
Use: Communication Towers

Antenna Windows
Use: Radomes

Antennae
Use: Antennas

Antennas
Use For: Aerials; Radio Antennas
See Also: Adaptive Antenna Arrays; Antenna Connectors; Antenna Couplers; Capacitor Antennas; Communication Equipment; Communication Towers; Directional Antennas; Effective Isotropic Radiated Power; Electromagnetic Radiation; Exciters; Free Space; Radio Equipment; Radomes; Receiving Antennas; Reflecting Antennas; Splitters; Telecommunication Equipment; Telecommunication Systems; Television Antennas; Terrestrial Link Antennas; Transmitters; Transmitting Antennas; Wave Antennas; Wire Antennas
CCIR Volume 3 TOC-90. Table of Contents. 2 pp.
CCIR Volume 3 Index-90. Numerical Index of Texts. 1 p.
CCIR Volume 3 TERMS-90. Terms of Reference of Study Group 3. 7 pp.
CCIR Report 1129-90. Performance Over Real Ground of Antennas—Section 3Ab—Antennas Characteristics. 3 pp.
CENELEC HD 95.4-86. Aerials for the Reception of Sound and Television Broadcasting in the Frequency Range 30MHz to 1GHz Part 4: Guide for the Preparation of Aerial Performance Specifications—Detailed Specification Sheet Format. 2 pp.
CSA CAN/CSA-S37-M86. Antennas, Towers, and Antenna-Supporting Structures; (Amendments 1-8 March 1989) (Amendments 9-11 August 1990) (Amendments 12-15 September 1992). 89 pp.
DIN VDE 0855 Pt 2-75. VDE Regulation for Antenna Systems; Qualification Approval of Receiving Antenna Systems (Nov). 21 pp.
IEC 597 Pt 4-83. Aerials for the Reception of Sound and Television Broadcasting in the Frequency Range 30 MHz to 1 GHz Part 4: Guide for the Preparation of Aerial Performance Specifications, Detailed Specifications Sheet Format First Edition. 29 pp.

—Aircraft
CAA Chapter R4-4 04.74. Aerial Systems (Radio). 3 pp.

—Aircraft—Electrical Bonding
CAA Chapter R4-5 04.74. Bonding and Lightning Discharge Protection (Radio). 1 p.

—Aircraft—Lightning Protection
CAA Chapter R4-5 04.74. Bonding and Lightning Discharge Protection (Radio). 1 p.

Antennas (Cont.)
—Automotive
CSA CAN/CSA-C225-M88. Vehicle-Mounted Aerial Devices; (Gen Instr 1). 33 pp.

—Beams—Atmospheric Bending—Refraction
CCIR Report 393-4-90. Intersections of Radio-Relay Antenna Beams with Orbits Used by Space Stations in the Fixed-Satellite Service—Section 4/9A—Sharing Conditions. 34 pp.

—Beams—Orbits—Space Stations—Intersection
CCIR RECMN 765-92. Intersection of Radio-Relay Antenna Beams with Orbits Used by Space Stations in the Fixed-Satellite Service Section 4/9A—Sharing Conditions. 18 pp.
CCIR Report 393-4-90. Intersections of Radio-Relay Antenna Beams with Orbits Used by Space Stations in the Fixed-Satellite Service—Section 4/9A—Sharing Conditions. 34 pp.

—Directivity—Television Broadcasting
CCIR RECMN 419-2-90. Directivity of Antennas in the Reception of Television Broadcasting—Section 11E—Planning of Television Networks, Protection Ratios Television Receivers and Antennas. 1 p.
CCIR RECMN 419-3-92. Directivity and Polarization Discrimination of Antennas in the Reception of Television Broadcasting—Section 11E—Planning of Television Networks, Protection Ratios Television Receivers and Antennas. 3 pp.

—Discharge Units
CNS C7105-80. Antenna-Discharge Units (Feb)(5206).

—Earth Stations
CCIR Report 922-1-86. Reference Radiation Pattern for Ship Earth-Station Antennas—Section 8I—Technical and Operating Characteristics of Mobile Satellite Services. 10 pp.
DIN VDE 0855 Pt 12 (P)-88. Cabled Distribution Systems; Equipment for Receiving Aerial Systems with Satellite Receiving Equipment (Nov). 35 pp.

—Earth Stations—Coupling—Terrestrial Link Antennas
CCIR Report 709-1-82. Consideration of the Coupling Between an Earth-Station Antenna and a Terrestrial Link Antenna—Section 4/9B—Co-Ordination and Interference Calculations. 9 pp.

—Earth Stations—Cross Polarization
CCIR Report 1141-90. Polarization Discrimination in Interference Calculation—Section 4D1—Permissible Levels of Interference. 20 pp.

—Earth Stations—Fixed Satellite Services
CCIR Report 390-6-90. Earth-Station Antennas for the Fixed-Satellite Service—Section 4C—Earth Station and Baseband Characteristics —Earth Station Antennas —Maintenance of Earth Stations. 23 pp.
CCIR QUESTION 42/4-90. Characteristics of Antennas at Earth Stations in the Fixed-Satellite Service—Questions of Study Group 4 Fixed-Satellite Service. 1 p.

—Earth Stations—Fixed Satellites—Radiation Diagrams—Design
CCIR Report 391-6-90. Radiation Diagrams of Antennas for Earth Stations in the Fixed-Satellite Service for Use in Interference Studies and for the Determination of a Design Objective—Section 4C—Earth Station and Baseband Characteristics—Earth Station Antennas-. 16 pp.
CCIR QUESTION 40/4-90. Reference Radiation Diagram of Antennas at Earth Stations in the Fixed-Satellite Service—Questions of Study Group 4 Fixed-Satellite Service. 1 p.

—Earth Stations—Geostationary—Radiation Diagrams—Design
CCIR RECMN 580-2-90. Radiation Diagrams for Use as Design Objectives for Antennas of Earth Stations Operating with Geostationary Satellites —Section 4C—Earth Station and Baseband Characteristics—Earth Station Antennas—Maintenance of Earth Stations. 3 pp.
CCIR RECMN 580-3-92. Radiation Diagrams for Use as Design Objectives for Antennas of Earth Stations Operating with Geostationary Satellites —Section 4C—Earth Station and Baseband Characteristics—Earth Station Antennas—Maintenance of Earth Stations. 3 pp.

—Earth Stations—Mobile Satellite Communications
CCIR Report 1047-1-90. Compact Antennas for Mobile Satellite Communication—Section 8I—Technical and Operating Characteristics of Mobile Satellite Services. 18 pp.

Antennas (Cont.)
—Earth Stations—Multipath Fading
CCIR Report 1048-1-90. Fading Reduction Techniques Applicable to Ship Earth-Station Antennas—Section 8I—Technical and Operating Characteristics of Mobile Satellite Services. 13 pp.

—Earth Stations—Noise Temperature—Fixed Satellite Services
CCIR Report 390-6-90. Earth-Station Antennas for the Fixed-Satellite Service—Section 4C—Earth Station and Baseband Characteristics —Earth Station Antennas —Maintenance of Earth Stations. 23 pp.

—Earth Stations—Radiation
CCIR Report 390-6-90. Earth-Station Antennas for the Fixed-Satellite Service—Section 4C—Earth Station and Baseband Characteristics —Earth Station Antennas —Maintenance of Earth Stations. 23 pp.
CCIR RECMN 694-90. Reference Radiation Pattern for Ship Earth Station Antennas—Section 8I—Technical and Operating Characteristics of Mobile Satellite Services. 2 pp.
CCIR Report 922-1-86. Reference Radiation Pattern for Ship Earth-Station Antennas—Section 8I—Technical and Operating Characteristics of Mobile Satellite Services. 10 pp.

—Earth Stations—Radio Frequency Interference—Radiation Diagrams
CCIR Report 391-6-90. Radiation Diagrams of Antennas for Earth Stations in the Fixed-Satellite Service for Use in Interference Studies and for the Determination of a Design Objective—Section 4C—Earth Station and Baseband Characteristics—Earth Station Antennas-. 16 pp.
CCIR QUESTION 40/4-90. Reference Radiation Diagram of Antennas at Earth Stations in the Fixed-Satellite Service—Questions of Study Group 4 Fixed-Satellite Service. 1 p.

—Electric Wire
MOD UK DSTAN 61-12: Pt 16:Sec 1-01. Wires, Cords, and Cables, Electrical—Metric Units Part 16: Section 1: Wires Electrical (Aerial (Antenna) Wires) Issue 1; Amendment 1. 14 pp.
MOD UK DSTAN 61-12: Pt 16:Sec 2-80. Wires, Cords, and Cables, Electrical-Metric Units Part 16 Section 2: Wires, Electrical (for HF Aerial (Antenna) Feeder, or Earthing Systems) Issue 1. 11 pp.

—Electrical Measurement
BSI BS 5640: Part 1-78. 1978 Aerials for the Reception of Sound and Television Broadcasting in the Frequency Range to 30 MHz to 1 GHz Part 1: Specification for Electrical and Mechanical Characteristics. 10 pp.
BSI BS 5640: Part 2-78. 1978 Aerials for the Reception of Sound and Television Broadcasting in the Frequency Range 30 MHz to 1 GHz Part 2: Methods of Measurement of Electrical Performance Parameters. 15 pp.
CENELEC HD 95.1-78. Aerials for the Reception of Sound and Television Broadcasting in the Frequency Range 30 MHz to 2 GHz Part 1: Electrical and Mechanical Characteristics. 2 pp.
CENELEC HD 95.2-78. Aerials for the Reception of Sound and Television Broadcasting in the Frequency Range 30 MHz to 2 GHz Part 2: Methods of Measurement of Electrical Performance Parameters. 2 pp.
CENELEC HD 466.8-86. Methods of Measurement for Radio Equipment Used in the Mobil Services Part 8: Methods of Measurement for Antennas. 2 pp.
IEC 489 Pt 8-84. Methods of Measurement for Radio Equipment Used in the Mobile Services Part 8: Methods of Measurement for Antennas First Edition. 31 pp.
IEC 597 Pt 1-77. Aerials for the Reception of Sound and Television Broadcasting in the Frequency Range 30 MHz to 1 GHz Part 1: Electrical and Mechanical Characteristics First Edition. 18 pp.
IEC 597 Pt 2-77. Aerials for the Reception of Sound and Television Broadcasting in the Frequency Range 30 MHz to 1 GHz Part 2: Methods of Measurement of Electrical Performance Parameters First Edition. 27 pp.

—Environmental Testing
CENELEC HD 95.3-84. Aerials for the Reception of Sound and Television Broadcasting in the Frequency Range 30 MHz to 1GHz Part 3: Methods of Measurements of Mechanical Properties, Vibration and Environmental Tests. 2 pp.
CENELEC HD 95.3 S1-89. Aerials for the Reception of Sound and Television Broadcasting in the Frequency Range 30 MHz to 1 GHz-Part 3: Methods of Measurement of Mechanical Properties, Vibration and Environmental Tests. 3 pp.

INTERNATIONAL AND NON-U.S. NATIONAL STANDARDS
SUBJECT INDEX

Antennas

Antennas (Cont.)

—Environmental Testing (Cont.)

IEC 597 Pt 3-83. Aerials for the Reception of Sound and Television Broadcasting in the Frequency Range 30 MHz to 1 GHz Part 3: Methods of Measurement of Mechanical Properties, Vibration and Environmental Tests First Edition. 23 pp.

—Gain to Temperature Ratio—Earth Stations—Fixed Satellite Services

CCIR RECMN 733-92. Measurement of the G/T Ratio for Earth Stations Operating in the Fixed-Satellite Service—Section 4C—Earth Station and Baseband Characteristics—Earth Station Antennas—Maintenance of Earth Stations. 20 pp.

—Geostationary Satellites—Accurate Beam Pointing

CCIR Report 1136-90. Geostationary Satellite Antenna Pointing Accuracy—Section 4D3—Spacecraft Station Keeping—Satellite Antenna Radiation Pattern—Pointing Accuracy. 13 pp.

—Glossaries

BSI BS 4727:Pt3: Group 15-92. 1992 Electrotechnical, Power, Telecommunication, Electronics, Lighting and Colour Terms Part 3: Terms Particular to Telecommunications and Electronics Group 15: Antennas (IEC 50(712): 1992) (G). 158 pp.

CNS C5101-81. Standard Definitions of Terms on Antennas and Waveguides (Mar)(7021).

IEC 50 Chap 712-92. International Electrotechnical Vocabulary Chapter 712: Antennas First Edition. 164 pp.

—Installation

DIN VDE 0855 Pt 1-84. Antenna Systems; Installation and Operation (May). 34 pp.

—Ionospheric Propagation

CCIR QUESTION 40/6-90. Antenna Characteristics—Questions Concerning Study Group 6—Radio Wave Propagation in Ionized Media. 1 p.

—L Band Avionics—Aircraft

EUROCAE ED-25 07.76. MPS for Experimental AEROSAT L-Band Avionics. 75 pp.

—LF/MF—Electric Field Strength

CCIR Report 1117-90. Determination of the Electric Field Strength in the Near Field Zone of High-Power LF/MF Antennas—Section 1D—Spectrum Utilization and Applications. 12 pp.

—LF/MF/HF—Radiation Patterns—Sound Broadcasting

CCIR Decision 79-1-89. Calculation of the Radiation Patterns of LF, MF and HF Antennas—Volume X-1—Broadcasting Service (Sound). 3 pp.

—Magnetic Cores

CNS C1081-82. Dimension for Aerial Rod and Slab Made of Ferromagnetic Oxide (Jan)(8372).

CNS C5006-73. Dimensions of Aerial Rod Cores of Ferromagnetic Oxides (Jan)(3433).

—Mechanical Properties

BSI BS 5640: Part 1-78. 1978 Aerials for the Reception of Sound and Television Broadcasting in the Frequency Range to 30 MHz to 1 GHz Part 1: Specification for Electrical and Mechanical Characteristics. 10 pp.

BSI BS 5640: Part 2-78. 1978 Aerials for the Reception of Sound and Television Broadcasting in the Frequency Range 30 MHz to 1 GHz Part 2: Methods of Measurement of Electrical Performance Parameters. 15 pp.

CENELEC HD 95.1-78. Aerials for the Reception of Sound and Television Broadcasting in the Frequency Range 30 MHz to 2 GHz Part 1: Electrical and Mechanical Characteristics. 2 pp.

CENELEC HD 95.2-78. Aerials for the Reception of Sound and Television Broadcasting in the Frequency Range 30 MHz to 2 GHz Part 2: Methods of Measurement of Electrical Performance Parameters. 2 pp.

CENELEC HD 95.3-84. Aerials for the Reception of Sound and Television Broadcasting in the Frequency Range 30 MHz to 1GHz Part 3: Methods of Measurements of Mechanical Properties, Vibration and Environmental Tests. 2 pp.

CENELEC HD 95.3 S1-89. Aerials for the Reception of Sound and Television Broadcasting in the Frequency Range 30 MHz to 1GHz-Part 3: Methods of Measurement of Mechanical Properties, Vibration and Environmental Tests. 3 pp.

IEC 597 Pt 1-77. Aerials for the Reception of Sound and Television Broadcasting in the Frequency Range 30 MHz to 1 GHz Part 1: Electrical and Mechanical Characteristics First Edition. 18 pp.

Antennas (Cont.)

—Mechanical Properties (Cont.)

IEC 597 Pt 3-83. Aerials for the Reception of Sound and Television Broadcasting in the Frequency Range 30 MHz to 1 GHz Part 3: Methods of Measurement of Mechanical Properties, Vibration and Environmental Tests First Edition. 23 pp.

—Mobile Radio Equipment

CENELEC PRETS 300 296-93. Radio Equipment and Systems (RES); Land Mobile Service Technical Characteristics and Test Conditions for Radio Equipment Using Integral Antennas Intended Primarily for Analogue Speech. 70 pp.

ETSI PRETS 300 296-93. Radio Equipment and Systems (RES); Land Mobile Service Technical Characteristics and Test Conditions for Radio Equipment Using Integral Antennas Intended Primarily for Analogue Speech. 70 pp.

—Mobile Stations—Mobile Satellite Communications

CCIR Report 925-1-86. Factors Affecting the Choice of Antennas for Mobile Stations of the Land Mobile-Satellite Service—Section 8I—Technical and Operating Characteristics of Mobile Satellite Services. 10 pp.

CCIR Report 1047-1-90. Compact Antennas for Mobile Satellite Communication—Section 8I—Technical and Operating Characteristics of Mobile Satellite Services. 18 pp.

—Monitoring—Radio Spectra

CCIR QUESTION 31/1-70. Antennas for Monitoring Stations—Questions Concerning Study Group 1—Spectrum Management Techniques (Spectrum Engineering, Planning, Sharing, Monitoring and Utilization). 1 p.

—Polarization Discrimination—Television Broadcasting

CCIR RECMN 419-3-92. Directivity and Polarization Discrimination of Antennas in the Reception of Television Broadcasting—Section 11E—Planning of Television Networks, Protection Ratios Television Receivers and Antennas. 3 pp.

—Radiation Diagrams—Sound Broadcasting

CCIR RESOLUTION 76-1-90. Presentation of Antenna Diagrams—Volume X-1—Broadcasting Service (Sound). 1 p.

—Radiation Patterns—Earth Stations

CCIR Report 677-2-90. Radiation Patterns of Antennas for Space Research Earth Stations—Section 2A—Research in Space Technology. 5 pp.

—Radiation Patterns—Earth Stations—Fixed Satellite Services

CCIR RECMN 465-3-90. Reference Earth-Station Radiation Pattern for Use in Coordination and Interference Assessment in the Frequency Range from 2 to About 30 GHz—Section 4C—Earth Station and Baseband Characteristics—Earth Station Antennas—Maintenance of Earth Stations. 2 pp.

CCIR RECMN 465-4-92. Reference Earth-Station Radiation Pattern for Use in Coordination and Interference Assessment in the Frequency Range from 2 to About 30 GHz—Section 4C—Earth Station and Baseband Characteristics—Earth Station Antennas—Maintenance of Earth Stations. 2 pp.

CCIR RECMN 731-92. Reference Earth-Station Cross-Polarized Radiation Pattern for Use in Frequency Coordination and Interference Assessment in the Frequency Range from 2 to About 30 GHz—Section 4C—Earth Station and Baseband Characteristics—Earth Station Antennas-. 15 pp.

—Radiation Patterns—Earth Stations—Space Research Services

CCIR QUESTION 127/7-90. Radiation Patterns and Side Lobe Characteristics of Large Antennas Used for Space Research Earth Stations and Radioastronomy—Questions Concerning Study Group 7—Science Services. 1 p.

—Radiation Patterns—Fixed Services (Radio Communications)

CCIR QUESTION 155/9-90. Performance over Real Ground of Antennas Operating at Frequencies Below About 30 MHz in the Fixed Service—Questions of Study Group 9 Fixed Service. 1 p.

—Radiation Patterns—Frequency Management—Radio Relay Systems

CCIR RECMN 699-90. Reference Radiation Patterns for Line-of-Sight Radio-Relay System Antennas for Use in Coordination Studies and Interference Assessment in the Frequency Range from 1 to About 40 GHz—Section 9B2—System General Characteristics. 2 pp.

Antennas (Cont.)

—Radiation Patterns—Frequency Management—Radio Relay Systems (Cont.)

CCIR RECMN 699-1-92. Reference Radiation Patterns for Line-of-Sight Radio-Relay System Antennas for Use in Coordination Studies and Interference Assessment in the Frequency Range from 1 to About 40 GHz—Section 9B2—System General Characteristics. 2 pp.

—Radiation Patterns—Interference—Radio Relay Systems

CCIR RECMN 699-90. Reference Radiation Patterns for Line-of-Sight Radio-Relay System Antennas for Use in Coordination Studies and Interference Assessment in the Frequency Range from 1 to About 40 GHz—Section 9B2—System General Characteristics. 2 pp.

CCIR RECMN 699-1-92. Reference Radiation Patterns for Line-of-Sight Radio-Relay System Antennas for Use in Coordination Studies and Interference Assessment in the Frequency Range from 1 to About 40 GHz—Section 9B2—System General Characteristics. 2 pp.

—Radiation Patterns—Radio Relay Systems

CCIR Report 614-3-90. Reference Radiation Patterns for Radio-Relay System Antennas—Section 9B2—System General Characteristics. 7 pp.

CCIR QUESTION 110/9-90. Antenna Radiation Diagrams of Radio-Relay Stations for Use in Sharing Studies—Questions of Study Group 9 Fixed Service. 1 p.

—Radio Links—Interference Immunity

CCIR Report 831-1-90. Use of Special Screens to Improve Interference Immunity in Radio Links—Section 1D—Spectrum Utilization and Applications. 20 pp.

—Radio Links—Interference Screens

CCIR Report 831-1-90. Use of Special Screens to Improve Interference Immunity in Radio Links—Section 1D—Spectrum Utilization and Applications. 20 pp.

—Radio Receivers

SAA AS 1417.1-87. Receiving Antennas for Radio and Television in the Frequency Range 30 MHz to 1 GHz—Part 1: Construction and Installation. 10 pp.

SAA AS 1417.2-91. Receiving Antennas for Radio and Television in the Frequency Range 30 MHz to 1 GHz—Part 2: Performance. 28 pp.

SNZ NZS 6602-83. Compatibility of VHF-FM Sound Broadcast Receivers with Broadcast Signals. 14 pp.

—Safety

BSI BS 5373-77. 1977 Electrical Safety Requirements for Room Aerials. 7 pp.

—Satellites

CCIR Report 1172-90. Intersystem Frequency Sharing and Reuse in the Mobile-Satellite Services Operating at Mid to High Portions of Band 9—Section 8H—Efficient Use of the Radio Spectrum Characteristics and Sharing of Frequency Resources. 25 pp.

—Satellites—Fixed Satellite Services

CCIR Report 558-4-90. Satellite Antenna Patterns in the Fixed-Satellite Service—Section 4D3—Spacecraft Station Keeping—Satellite Antenna Radiation Pattern—Pointing Accuracy. 57 pp.

—Satellites—Radiation—Fixed Satellite Services

CCIR QUESTION 41/4-90. Radiation Characteristics of Satellite Antennas in the Fixed-Satellite Service—Questions of Study Group 4 Fixed-Satellite Service. 1 p.

—Satellites—Radiation Pattern—Design—Fixed Satellite Services

CCIR RECMN 672-90. Satellite Antenna Radiation Pattern for Use as a Design Objective in the Fixed-Satellite Service—Section 4D3—Spacecraft Station-Keeping—Satellite Antenna Radiation Pattern—Pointing Accuracy. 3 pp.

CCIR RECMN 672-1-92. Satellite Antenna Radiation Pattern for Use as a Design Objective in the Fixed-Satellite Service Employing Geostationary Satellites—Section 4D3—Spacecraft Station-Keeping—Satellite Antenna Radiation Pattern—Pointing Accuracy. 24 pp.

—Satellites—Shaped Beam

CCIR Report 558-4-90. Satellite Antenna Patterns in the Fixed-Satellite Service—Section 4D3—Spacecraft Station Keeping—Satellite Antenna Radiation Pattern—Pointing Accuracy. 57 pp.

INDUSTRY STANDARDS

Antennas (Cont.)

—Self Supporting—Ships
CCIR OPINION 43-2-78. Self-Supporting Antennas for Use on Board Ships Performance at 500 kHz—Volume VIII—Mobile, Radiodetermination, Amateur and Related Satellite Services. 1 p.

—Shaped Beam—Spacecraft
CCIR Report 676-1-90. Shaped Beam Antennas—Section 2A—Research in Space Technology. 3 pp.

—Ships—Electromagnetic Interference
CCIR RECMN 218-2-90. Prevention of Interference to Radio Reception on Board Ships —Section 8B—Maritime Mobile Service; Telegraphy and Related Subjects. 2 pp.

—Side Lobe Gain—Radio Astronomy
CCIR QUESTION 127/7-90. Radiation Patterns and Side Lobe Characteristics of Large Antennas Used for Space Research Earth Stations and Radioastronomy—Questions Concerning Study Group 7—Science Services. 1 p.

—Side Lobe Peak—Earth Stations—Fixed Satellite Services
CCIR RECMN 732-92. Method for Statistical Processing of Earth-Station Antenna Side-Lobe Peaks—Section 4C—Earth Station and Baseband Characteristics—Earth Station Antennas—Maintenance of Earth Stations. 2 pp.

—Signal Delay
CCIR Report 1017-86. Characterization of Signal Delays in Antennas—Section 7C—Systems for Dissemination and Comparison. 2 pp.

—Signal Delays—Time Transfer
CCIR QUESTION 111/7-90. Signal Delays in Antennas and Other Circuits for High-Precision Time Transfer—Questions Concerning Study Group 7—Science Services. 1 p.

—Sky Waves—Ionospheric Propagation
CCIR Report 891-2-90. Antenna Characteristics Important for the Analysis and Prediction of Sky-Wave Propagation Paths—Section 6E—Ionospheric Propagation Prediction at Frequencies Between About 1.6 and 30 MHz. 6 pp.

—Small—Earth Stations—Performance Measurement
CCIR Report 998-1-90. Performance of Small Earth-Station Antennas for the Fixed-Satellite Service—Section 4C—Earth Station and Baseband Characteristics —Earth Station Antennas—Maintenance of Earth Stations. 20 pp.

—Space Research Systems
CCIR QUESTION 123/7-90. Antennas for Space Research Systems—Questions Concerning Study Group 7—Science Services. 1 p.

—Spatial Radiation Diagrams
CCIR OPINION 64-82. Methods for Measuring the Spatial Radiation Diagram of Antennas—Volume I—Spectrum Utilization and Monitoring. 1 p.

—Symbols
CNS C5095-81. Graphical Symbols for Aerials (Mar)(7015).

—UHF—Mobile Satellite Communications
CCIR Report 925-1-86. Factors Affecting the Choice of Antennas for Mobile Stations of the Land Mobile-Satellite Service—Section 8I—Technical and Operating Characteristics of Mobile Satellite Services. 10 pp.
CCIR Report 1047-1-90. Compact Antennas for Mobile Satellite Communication—Section 8I—Technical and Operating Characteristics of Mobile Satellite Services. 18 pp.

Anthracene

See Also: Aromatic Hydrocarbons; Hydrocarbons
CNS K7556-81. Chemical Reagent (Anthracene) (Jul)(7714).
JIS K 2436-90. Naphthalene.Anthracene. Carbazole. 46 pp.
JIS K 8083-72. Anthracene.

Anthracis

—Animal Feed
CNS N4046-78. Methods of Test for Bacteria in Feeds (Mar) (3682). 3 pp.

Anthracite

Use For: Hard Coal *See Also:* Coal; Fossil Fuels; Rocks; Solid Mineral Fuels
CNS M1006-64. Low-Carbon Anthracite (for Export) (Jul) (1445)(R 1972). 1 p.
CNS M1013-66. Anthracite (for Taiwan Area) (Sep)(2644) (R 1972).

Anthracite (Cont.)

—Carbominerite Content—Petrography
BSI BS 6127: Part 4-90. 1990 Petrographic Analysis of Bituminous Coal and Anthracite Part 4: Method of Determining Microlithotype, Carbominerite and Minerite Composition. 10 pp.
ISO 7404 Pt 4-88. Methods for the Petrographic Analysis of Bituminous Coal and Anthracite—Part 4: Method of Determining Microlithotype, Carbominerite and Minerite Composition First Edition. 8 pp.

—Float and Sink Analysis
ISO 7936-92. Hard Coal—Determination and Presentation of Float and Sink Characteristics—General Directions for Apparatus and Procedures First Edition. 23 pp.

—Froth Flotation Testing
ISO 8858 Pt 1-90. Hard Coal—Froth Flotation Testing—Part 1: Laboratory Procedure First Edition. 14 pp.

—Glossaries—Petrography
BSI BS 6127: Part 1-81. 1981 Petrographic Analysis of Bituminous Coal and Anthracite Part 1: Glossary of Terms Relating to the Petrographic Analysis of Bituminous Coal and Anthracite. 7 pp.
ISO 7404 Pt 1-84. Methods for the Petrographic Analysis of Bituminous Coal and Anthracite—Part 1: Glossary of Terms First Edition. 5 pp.

—Maceral Content—Petrography
BSI BS 6127: Part 3-81. 1981 Petrographic Analysis of Bituminous Coal and Anthracite Part 3: Determining Maceral Group Composition of Bituminous Coal and Anthracite. 7 pp.
ISO 7404 Pt 3-84. Methods for the Petrographic Analysis of Bituminous Coal and Anthracite—Part 3: Method of Determining Maceral Group Composition First Edition. 6 pp.

—Microlithotype Content—Petrography
BSI BS 6127: Part 4-90. 1990 Petrographic Analysis of Bituminous Coal and Anthracite Part 4: Method of Determining Microlithotype, Carbominerite and Minerite Composition. 10 pp.
ISO 7404 Pt 4-88. Methods for the Petrographic Analysis of Bituminous Coal and Anthracite—Part 4: Method of Determining Microlithotype, Carbominerite and Minerite Composition First Edition. 8 pp.

—Minerite Content—Petrography
BSI BS 6127: Part 4-90. 1990 Petrographic Analysis of Bituminous Coal and Anthracite Part 4: Method of Determining Microlithotype, Carbominerite and Minerite Composition. 10 pp.
ISO 7404 Pt 4-88. Methods for the Petrographic Analysis of Bituminous Coal and Anthracite—Part 4: Method of Determining Microlithotype, Carbominerite and Minerite Composition First Edition. 8 pp.

—Moisture Content
ISO 589-81. Hard Coal—Determination of Total Moisture Second Edition. 8 pp.
ISO 1018-75. Hard Coal—Determination of Moisture-Holding Capacity First Edition. 8 pp.

—Sampling
ISO 1988-75. Hard Coal—Sampling First Edition; (Amendment-1975). 95 pp.

—Sampling—Petrography
BSI BS 6127: Part 2-82. 1982 Amd 2 Petrographic Analysis of Bituminous Coal and Anthracite Part 2: Method of Preparing Coal Samples for Petrographic Analysis. 12 pp.
ISO 7404 Pt 2-85. Methods for the Petrographic Analysis of Bituminous Coal and Anthracite—Part 2: Method of Preparing Coal Samples First Edition. 10 pp.

—Sieve Analysis
CNS M3157-84. Method of Test for Size of Anthracite (Jun) (10942).
ISO 1953-72. Hard Coals—Size Analysis First Edition. 19 pp.

—Swelling—Dilatometry
BSI BS 1016: Sec 107.3-90. 1990 Methods for the Analysis and Testing of Coal and Coke Part 107: Caking and Swelling Properties of Coal Section 107.3: Determination of Swelling Properties Using a Dilatometer. 14 pp.
ISO 8264-89. Hard Coal—Determination of the Swelling Properties Using a Dilatometer First Edition. 13 pp.

—Vitrinite—Reflectance—Petrography
ISO 7404 Pt 5-84. Methods for the Petrographic Analysis of Bituminous Coal and Anthracite—Part 5: Method of Determining Microscopically the Reflectance of Vitrinite First Edition. 13 pp.

Anthranilic Acid

Use: o-Aminobenzoic Acid

Anthraquinones

See Also: Ketones
JIS K 4145-87. Anthraquinone. 10 pp.

Anthrone

CNS K7070-62. Chemical Reagent (Athrone) (Jul)(1570).
JIS K 8082-64. Anthrone.

Anthropometric Characteristics

See Also: Ergonomics
BSI PP 7310-90. 1990 Anthropometrics An Introduction. 70 pp.
MOD UK DSTAN 00-12-81. Anthropometric Data Issue 3 (Withdrawn). 3 pp.

—Aging Persons
BSI BS 4467-91. 1991 Guide to Dimensions in Designing for Elderly People. 20 pp.

—Military—Combat Uniforms
NATO STANAG 2177 ED 1 AMD 0-00. Methodology for Anthropometric Data. 6 pp.

Anthropometric Data

Use: Anthropometric Characteristics

Anthropometric Studies

Use: Anthropometric Characteristics

Anti-Exposure Suits

Use: Exposure Suits

Anti-G Suits

Use: Antigravity Suits

Antiaircraft Ammunition

See Also: Projectiles; Small Arms Ammunition

—Boxes (Containers)—40/70
NATO STANAG 4038 ED 1 AMD 6-60. Specification for a 40/70 Ammunition Operational Box (Metal) for Naval Use. 23 pp.
NATO STANAG 4081 ED 1 AMD 1-62. Specification for a Lightweight 40/70 Ammunition Operational Box. 12 pp.

—Charger—40/70
NATO STANAG 4037 ED 2 AMD 0-63. Specification for a 40/70 Ammunition Charger. 17 pp.

—Design—40/70
NATO STANAG 4079 ED 1 AMD 1-62. Design of 40/70 Ammunition. 7 pp.

Antiarmor Ammunition

Use: Armor Piercing Ammunition

Antibiotic Content Analysis

See Also: Patulin Content Analysis

—Animal Feed
CNS N4098-83. Method of Test for Feed Additives: General Rules for Microbiological Determination of Antibiotics (Aug)(9028).

—Baby Foods
CNS N6169-82. Method of Test for the Residue of Anti-Biotic in Foodstuffs (Babyfood Use) (May)(8860). 2 pp.

—Food
CNS N6169-82. Method of Test for the Residue of Anti-Biotic in Foodstuffs (Babyfood Use) (May)(8860). 2 pp.

Anticollision Lights

See Also: Aircraft Lighting; Flasher Units; Warning Lights

—Aircraft
CAA Chapter G6-7 01.75. External Lights (Rotocraft). 6 pp.
CAA Chapter K6-7 10.92. External Lights (Light Aeroplanes). 6 pp.
CAA Chapter P6-7. External Lights (Provisional Airworthiness Requirements for Civil Powered Lift Aircraft). 7 pp.
CAA Chapter Q6-7 12.79. External Lights (Non-Rigid Airships). 5 pp.
NATO STANAG 3153 ED 6 AMD 5-73. Aircraft Navigation and Anti-Collision Lights. 11 pp.
NATO STANAG 3153 ED 7 AMD 0-93. Aircraft Navigation and Anti-Collision Lights. 14 pp.

Anticreepers

See Also: Railroad Equipment
CNS E1005-72. Antincreeper (for Railway) (Jun)(2665).

Anticreepers (Cont.)
JIS E 1111-78. Anti-Creepers.

Antidegradant Content Analysis
—Elastomers—Thin Layer Chromatography
DIN ENGL 53622 Pt 2-76. Testing of Rubber and Elastomers; Thin-Layer Chromatographic Analysis; Identification of Antidegradants in Rubber and Elastomers (June). 4 pp.

—Rubber—Thin Layer Chromatography
BSI BS 6630-85. 1985 Amd 1 Identification of Antidegradants in Rubber and Rubber Products by Thin Layer Chromatography (ISO 4645-1984) (AMD 6698) April 30, 1991. 12 pp.

DIN ENGL 53622 Pt 2-76. Testing of Rubber and Elastomers; Thin-Layer Chromatographic Analysis; Identification of Antidegradants in Rubber and Elastomers (June). 4 pp.

Antifogging Agents
—Automotive
JIS K 2399-87. Road Vehicles—Antifogging Agent. 23 pp.

—Eyeglasses
JIS S 4030-83. Methods of Antifogging Agent Test for Spectacle Lens.

Antifouling Coatings
Use For: Antifouling Paints
See Also: Biodegradability; Coatings; Marine Coatings; Metal Coatings (Made From Metal); Nonmetallic Coatings; Paints

MOD UK TS 10239. Paint, Antifouling, Black (317).
MOD UK TS 10240. Paint, Antifouling (161P).
SAA AS 1580.481. 5-93. Paints and Related Materials—Methods of Test—Part 481: Coatings—Part 481.5: Durability and Resistance to Fouling—Marine Underwater Paint Systems (in Professional Package 39). 8 pp.
SNZ NZS/AS 1580. 481.5-86. Methods of Test for Paints and Related Materials Part 481.5: Assessement of Durability and Resistance to Fouling of Marine Underwater Paint Systems and Degree of Protection of Substrate (This is a Joint Standard with SAA AS 1580.481.5). 2 pp.

—Chlorinated Rubber—Steel
CNS K2072-86. Chlorinated Rubber Antifouling Paint for Steel Ship (May)(4921). 2 pp.
CNS K6864-86. Method of Test for Chlorinated Rubber Antifouling Paint for Steel Ship (May)(11594).

—Copper Powder
CNS K2128-82. Copper Powder (Jan)(8424). 1 p.

—Oleoresinous—Steel
CNS K6859-86. Method of Test for Oleoresinous Antifouling Paint for Steel Ship Bottom (May)(11589).

—Pigments—Cuprous Oxide
MOD UK TS 10241. Cuprous Oxide for Antifouling Paints.

—PVC—Steel
CNS K2080-86. Vinyl Chloride Resin Antifouling Paint for Bottom of Steel Ship (May)(4929). 2 pp.
CNS K6869-86. Method of Test for Vinyl Chloride Resin Antifouling Paints for Bottom of Steel Ship (May)(11599).

—PVC—Wood
CNS K2083-86. Vinyl Chloride Resin Anti-Fouling Paint of Bottom for Wooden Ship (May)(4932). 2 pp.
CNS K6872-86. Method of Test for Vinyl Chloride Resin Antifouling Paint of Bottom for Wooden Ship (May)(11602).

—Steel
CNS K6894-86. Method of Test for Antifouling Properties of Steel Ships Bottom Paint (Dec)(11798).
JIS K 5630-83. Testing Method of Antifouling Properties for Steel Ships' Bottom Paint. 12 pp.

—Steel Panels—Salt Spray Testing
CGSB 1-GP-71 METH 113.1-79. Methods of Testing Paints and Pigments Corrosion Salt Water Resistance Vinyl Primer and Antifouling Paints; (Amended September 1980) (Re-Edited September 1982). 1 p.

—Vinyl
CGSB CAN/CGSB-1.123-92. Vinyl Antifouling Paint for Ships' Bottoms. 9 pp.

Antifouling Paints
Use: Antifouling Coatings

Antifreezes
See Also: Additives; Antiicing Additives
JIS K 2234-87. Engine Antifreeze Coolants. 24 pp.

—Airport Runways
MOD UK TS 10228A. Ice Control Agents for Aircraft Runways.

—Alkalinity
CNS K6481-80. Method of Test for Reserve Alkalinity of Engine Antifreezes Antirusts and Coolants (May)(5584).

—Ash Content
CNS K6484-80. Method of Test for Ash Content of Engine Antifreezes, Antirusts and Coolants (May)(5587).

—Boiling Points
BSI BS 5117: Sec 1.2-85. 1985 Testing Corrosion Inhibiting, Engine Coolant Concentrate ('Antifreeze') Part 1: Methods of Test for Determination of Physical and Chemical Properties Section 1.2: Determination of Boiling Point. 4 pp.
CNS K6482-80. Method of Test for Boiling Point of Engine Antifreezes (May)(5585).

—Corrosion Prevention
BSI BS 5117: Part 0-92. 1992 Testing Corrosion Inhibiting, Engine Coolant Concentrate ('Antifreeze') Part 0: General Introduction. 12 pp.
BSI BS 5117: Part 0-85. 1985 Testing Corrosion Inhibiting, Engine Coolant Concentrate ('Antifreeze') Part 0: General Introduction. 10 pp.
BSI BS 5117: Sec 1.1-85. 1985 Testing Corrosion Inhibiting, Engine Coolant Concentrate ('Antifreeze') Part 1: Methods of Test for Determination of Physical and Chemical Properties Section 1.1: Ancillary Procedures. 6 pp.
BSI BS 5117: Sec 2.1-85. 1985 Testing Corrosion Inhibiting, Engine Coolant Concentrate ('Antifreeze') Part 2: Methods of Test for Corrosion Inhibition Performance Section 2.1: General Procedures. 8 pp.
BSI BS 5117: Sec 2.2-85. 1985 Testing Corrosion Inhibiting, Engine Coolant Concentrate ('Antifreeze') Part 2: Methods of Test for Corrosion Inhibition Performance Section 2.2: Glassware Tests. 6 pp.
BSI BS 5117: Sec 2.3-85. 1985 Testing Corrosion Inhibiting, Engine Coolant Concentrate ('Antifreeze') Part 2: Methods of Test for Corrosion Inhibition Performance Section 2.3: Recirculating Rig Test. 23 pp.
BSI BS 5117: Sec 2.4-85. 1985 Testing Corrosion Inhibiting, Engine Coolant Concentrate ('Antifreeze') Part 2: Methods of Test for Corrosion Inhibition Performance Section 2.4: Static Engine Test. 8 pp.
BSI BS 5117: Sec 2.5-85. 1985 Testing Corrosion Inhibiting, Engine Coolant Concentrate ('Antifreeze') Part 2: Methods of Test for Corrosion Inhibition Performance Section 2.5: Field Test. 5 pp.
BSI BS 5117: Sec 2.6-92. 1992 Testing Corrosion Inhibiting, Engine Coolant Concentrate ('Antifreeze') Part 2: Methods of Test for Corrosion Inhibition Performance Section 2.6: Test for Corrosion of Cast Aluminium Alloys Under Heat-Transfer Conditions. 10 pp.
BSI BS 6580-92. 1992 Corrosion Inhibiting, Engine Coolant Concentrate ('Antifreeze'). 10 pp.
BSI BS 6580-85. 1985 Corrosion Inhibiting, Engine Coolant Concentrate. 8 pp.

—Density
CNS K6483-80. Method of Test for Specific Gravity of Engine Antifreezes (May)(5586).

—Ethylene Glycol
MOD UK DSTAN 68-127-89. Antifreeze, Inhibited Ethanediol NATO Code S-757 Joint Service Designation AL-39 Issue 1. 14 pp.

—Foaming Power
BSI BS 5117: Sec 1.4-85. 1985 Testing Corrosion Inhibiting, Engine Coolant Concentrate ('Antifreeze') Part 1: Methods of Test for Determination of Physical and Chemical Properties Section 1.4: Determination of Foaming Characteristics. 4 pp.

—Freezing Points
BSI BS 5117: Sec 1.3-85. 1985 Testing Corrosion Inhibiting, Engine Coolant Concentrate ('Antifreeze') Part 1: Methods of Test for Determination of Physical and Chemical Properties Section 1.3: Determination of Freezing Point. 4 pp.

—Stability Testing
BSI BS 5117: Sec 1.5-92. 1992 Testing Corrosion Inhibiting, Engine Coolant Concentrate ('Antifreeze') Part 1: Methods of Test for Determination of Physical and Chemical Properties Section 1.5: Coolant Hard Water Stability Test. 6 pp.

—Windshields
CGSB CAN2-3.532-M85. Antifreeze, Windshield Washer, Methanol Base. 10 pp.

Antifriction Alloys
Use: Bearing Alloys

Antifriction Bearings
Use For: Rolling Bearings; Rolling Contact Bearings
See Also: Airframe Bearings; Angular Contact Bearings; Ball Bearings; Bearing Caps; Bearings; Journal Bearings; Magnetic Bearings; Needle Roller Bearings; Plain Bearings; Precision Bearings; Radial Bearings; Roller Bearings; Sleeve Bearings; Taper Roller Bearings; Thrust Bearings

BSI BS 5988-80. 1980 Rolling Bearings, Metric Series; Minimum and Maximum Chamfer Dimensions of. 7 pp.
CNS B1169-80. Rolling Bearings; Component Parts and Spherical Plain Bearings; Summary of Standards (Jul)(5694).
CNS B1276-81. Rolling Bearings General Plans of Boundary Dimensions (Oct)(7947).
CNS B1287-82. Fitting Dimensions for Rolling Bearing (Aug)(9222).
CNS B1290-85. Tolerance of Rolling Bearings-Radial Bearings (May)(9572).
CNS B2477-82. Rolling Bearings; Bearing with Spherical Outside Surface and Extended Inner Ring Width, Insert Bearings (Dec)(5507).
CNS B7040-85. Method of Test for Tolerances of Rolling Bearings (May)(2863).
CNS G3060-88. High Carbon Chromium Bearing Steel (Mar)(3014).
DIN ENGL 616 (S)-73. Rolling Bearings; General Plans of Boundary; Dimensions (Feb) (Superseded by DIN EN 10203 and 10205). 12 pp.
DIN ENGL 620 Pt 2-88. Rolling Bearings; Tolerances for Rolling Bearings; Tolerances for Radial Bearings (Feb). 14 pp.
DIN ENGL 620 Pt 4-87. Rolling Bearings; Tolerances for Rolling Bearings; Radial Internal Clearance (Aug). 6 pp.
ISO 492-86. Rolling Bearings—Radial Bearings—Tolerances Second Edition; (Supersedes 2349). 15 pp.
ISO 582-79. Rolling Bearings—Metric Series—Chamfer Dimension Limits Second Edition. 6 pp.
ISO 1132-80. Rolling Bearings—Tolerances—Definitions First Edition. 14 pp.
ISO 3228-93. Rolling Bearings—Cast and Pressed Housings for Insert Bearings Third Edition. 11 pp.
JIS B 1511-86. General Code for Rolling Bearings.
JIS B 1511-75. General Rules for Rolling Bearings (R 1978). 23 pp.
JIS B 1512-86. Boundary Dimensions for Rolling Bearings. 65 pp.
JIS B 1514-86. Tolerances for Rolling Bearings.
JIS B 1514-75. Tolerances for Rolling Bearings (R 1978). 14 pp.
JIS B 1515-86. Measuring Methods for Rolling Bearings. 34 pp.
JIS B 1520-86. Radial Internal Clearance of Rolling Bearings. 9 pp.
JIS B 1557-65. Rolling Bearing Units (R 1968). 16 pp.
JIS G 4805-90. High Carbon Chromium Bearing Steels. 16 pp.
SAA AS 3890-91. Rolling Bearings—System Life and Reliability. 7 pp.

—Balls
CNS B2115-85. Parts of Rolling Bearings — Balls (Apr)(2861).
DIN ENGL 5401-78. Rolling Bearing Components; Balls (Jan). 3 pp.
ISO 3290-75. Rolling Bearings—Bearings Parts—Balls for Rolling Bearings First Edition. 10 pp.

—Control Cables—Fork Ends—Aircraft
AECMA PREN2360-85. Fork-Ends for Rolling Bearings in Corrosion Resisting Steel Swaged on Type, Control Cable—Dimensions and Loads. 6 pp.
BSI BS EN 2360-89. 1989 Fork-Ends for Rolling Bearings in Corrosion Resisting Steel Swaged on Type, Control Cable. Dimensions and Loads. 9 pp.
CEN EN 2360-88. Fork-Ends for Rolling Bearings in Corrosion Resisting Steel Swaged on Type, Control Cable Dimensions and Loads. 8 pp.

—Control Cables—Pulleys—Aircraft
AECMA PREN2062-79. Fully Non-Metallic Body Pulleys with Bearing for Control Cables—Technical Specification (C1/22-01). 19 pp.

—Control Cables—Turnbuckles—Fork Ends—Aircraft
AECMA PREN2356-86. Fork-Ends Threaded, Control Cable for Rolling Bearings in Corrosion Resisting Steel; Dimensions and Loads. 6 pp.
BSI BS EN 2356-89. 1989 Fork-Ends Threaded, Control Cable for Rolling Bearings in Corrosion Resisting Steel. Dimensions and Loads. 9 pp.
CEN EN 2356-88. Fork-Ends Threaded, Control Cable for Rolling Bearings in Corrosion Resisting Steel Dimensions and Loads. 8 pp.

Antifriction Bearings (Cont.)
—Cylindrical Rollers
CNS B2482-80. Rolling Bearing Components; Cylindrical Rollers (Jul)(5684).

—Drafting
JIS B 0005-73. Drawing Office Practice for Rolling Bearings. 16 pp.

—Dynamic Load Ratings
BSI BS 5512: Part 1-77. (WITHDRAWN) 1977 Rolling Bearings. Dynamic Load Ratings and Rating Life Part 1: Calculation Methods (Superseded by BS 5512: 1991). 13 pp.
BSI BS 5512-91. 1991 Dynamic Load Ratings and Rating Life of Rolling Bearings (ISO 281: 1990). 17 pp.
CNS B1168-80. Load Carrying Capacity of Rolling Bearings, Definitions, Load Rating, Methods of Load and Rating Life (Jul)(5683).
ISO 281-90. Rolling Bearings—Dynamic Load Ratings and Rating Life First Edition. 16 pp.
ISO TR8646-85. Explanatory Notes on ISO 281/1-1977 First Edition. 66 pp.
JIS B 1518-92. Dynamic Load Ratings and Rating Life for Rolling Bearings. 25 pp.

—Engineering Drawings—Symbols
ISO 8826 Pt 1-89. Technical Drawings—Rolling Bearings—Part 1: General Simplified Representation First Edition. 6 pp.

—Glossaries
BSI BS 6107: Part 1-81. 1981 Rolling Bearings: Tolerances Part 1: Glossary of Terms. 8 pp.
BSI BS 6560-84. 1984 Glossary of Terms for Rolling Bearings. 47 pp.
DIN ENGL 50282-79. Plain Bearings; the Tribiological Behaviour of Metallic Antifriction Materials; Significant Definitions (Feb). 2 pp.
ISO 5593-84. Rolling Bearings—Vocabulary First Edition. 83 pp.
JIS B 0104-91. Rolling Bearings—Vocabulary.
JIS B 0104-67. Glossary of Terms Relating to Rolling Bearings (R 1971). 87 pp.

—Greases
CGSB 3-GP-664MB-90. Heavy-Duty Water-Wash-Resistant Grease (NLGI Grades No. 1 and No. 2). 8 pp.
CGSB 3-GP-691MB-90. General Purpose Grease. 9 pp.
MOD UK DSTAN 91-55-01. Grease, Aircraft: Synthetic, High Temperature NATO Code No: G-372 Joint Service Designation: XG-300 Issue 1; Amendment 1. 12 pp.

—Greases—Corrosion Inhibitors
DIN ENGL 51802-90. Testing Lubricating Greases for Their Corrosion-Inhibiting Properties by the SKF Emcor Method (Apr). 5 pp.

—Housings—Pillow Blocks
BSI BS 5646: Part 4-79. 1979 Rolling Bearings—Accessories Part 4: Plummer Block Housings. 4 pp.
BSI BS 6010: Part 3-80. 1980 Rolling Bearings-Bearings with Spherical Outside Surface and Extended Inner Ring Width Part 3: Dimensions of Cast and Pressed Housings. 8 pp.
CNS B1169-80. Rolling Bearings; Component Parts and Spherical Plain Bearings; Summary of Standards (Jul)(5694).
CNS B1289-82. Felt Ring Groove; for Rolling Bearing Housing (Sep)(9336).
CNS B2478-80. Rolling Bearings; Bearings with Spherical Outside Surface and Extended Inner Ring Width, Steel Pressed Housings for Insert Bearings (May)(5508).
CNS B2479-80. Rolling Bearings; Bearings with Spherical Outside Surface and Extended Inner Ring Width, Cast Housings for Insert Bearings (May)(5509).
CNS B2699-82. Felt Ring for Rolling Bearing Housing (Sep)(9337).
CNS B2700-82. Felt Strip for Rolling Bearing Housing (Sep)(9338).
DIN ENGL 5419-59. Felt Rings; Felt Strips; Ring Grooves for Rolling Bearing Housings (Sept). 3 pp.
DIN ENGL 5425 Pt 1-84. Rolling Bearings; Mounting Tolerances; General Guidelines (Nov). 5 pp.
ISO 113 Pt II-79. Rolling Bearing Accessories—Part II: Plummer Block Housings First Edition. 4 pp.
JIS B 1559-76. Housings for Rolling Bearing Unit (R 1979). 22 pp.

—Identification Systems
CNS Z5064-80. Markings for Rolling Bearings and Their Packages (Apr)(5472).
JIS B 1513-87. Designation for Rolling Bearings. 13 pp.
JIS B 1516-87. Marking on Rolling Bearings and Packages. 10 pp.

Antifriction Bearings (Cont.)
—Loads (Forces)
CNS B1293-82. Load Carrying Capacity of Rolling Bearings (Foundation of Theory) (Nov)(9575).

—Lock Washers
BSI BS 5646: Part 1-78. 1978 Rolling Bearings—Accessories Part 1: Locknuts, Narrow Series, and Lockwashers with Straight Inner Tab. 4 pp.
BSI BS 5646: Part 2-78. 1978 Rolling Bearings—Accessories Part 2: Locknuts, Wide Series, and Lockwashers with Bent Inner Tab. 4 pp.
CNS B2720-82. Rolling Bearings Accessories (Lockwashers) (Nov)(9569).
DIN ENGL 5406-77. Rolling Bearing Accessories; Lockwashers (Apr). 2 pp.
ISO 2982-72. Rolling Bearings—Locknuts, Narrow Series, and Lockwashers with Straight Inner Tab First Edition. 4 pp.
ISO 2983-75. Rolling Bearings—Locknuts, Wide Series, and Lockwashers with Bent Inner Tab First Edition. 4 pp.
JIS B 1554-93. Rolling Bearings—Locknuts, Lockwashers and Lockplates. 20 pp.
JIS B 1555-79. Lock Washers and Lock Plates for Rolling Bearings. 10 pp.

—Locking Collars
BSI BS 6010: Part 2-80. (WITHDRAWN) 1980 Rolling Bearings-Bearings with Spherical Outside Surface and Extended Inner Ring Width Part 2: Dimensions and Tolerances of Eccentric Locking Collars (Superseded by BS 6010: Part 4: 1993). 6 pp.

—Locknuts
BSI BS 5646: Part 1-78. 1978 Rolling Bearings—Accessories Part 1: Locknuts, Narrow Series, and Lockwashers with Straight Inner Tab. 4 pp.
BSI BS 5646: Part 2-78. 1978 Rolling Bearings—Accessories Part 2: Locknuts, Wide Series, and Lockwashers with Bent Inner Tab. 4 pp.
CNS B2257-78. Rolling Bearing Accessories (Lock Nuts) (Aug)(4464).
DIN ENGL 981-83. Rolling Bearing Accessories; Locknuts (Jan). 2 pp.
ISO 2982-72. Rolling Bearings—Locknuts, Narrow Series, and Lockwashers with Straight Inner Tab First Edition. 4 pp.
ISO 2983-75. Rolling Bearings—Locknuts, Wide Series, and Lockwashers with Bent Inner Tab First Edition. 4 pp.
JIS B 1554-93. Rolling Bearings—Locknuts, Lockwashers and Lockplates. 20 pp.

—Lockplates
JIS B 1554-93. Rolling Bearings—Locknuts, Lockwashers and Lockplates. 20 pp.
JIS B 1555-79. Lock Washers and Lock Plates for Rolling Bearings. 10 pp.

—Long Rollers
CNS B2483-80. Rolling Bearing Components; Long Rollers (Jul)(5685).
DIN ENGL 5402 Pt 2-73. Rolling Bearing Components; Long Rollers (June). 2 pp.

—Mounting
CNS B1275-81. Tolerances for the Mounting of Rolling Bearings, General Guidelines (Oct)(7946).
DIN ENGL 5425 Pt 1-84. Rolling Bearings; Mounting Tolerances; General Guidelines (Nov). 5 pp.
JIS B 1566-89. Mounting Dimensions and Fits for Rolling Bearings. 33 pp.

—Packaging
JIS B 1517-84. Packaging of Rolling Bearings. 8 pp.
MOD UK DSTAN 81-32-01. Packaging of Ball and Roller Bearings Issue 3; Amendment 2. 26 pp.

—Packaging—Identification Systems
CNS Z5064-80. Markings for Rolling Bearings and Their Packages (Apr)(5472).
JIS B 1516-87. Marking on Rolling Bearings and Packages. 10 pp.

—Reliability
SAA AS 3890-91. Rolling Bearings—System Life and Reliability. 7 pp.

—Shafts (Machine Elements)
DIN ENGL 5425 Pt 1-84. Rolling Bearings; Mounting Tolerances; General Guidelines (Nov). 5 pp.

—Sleeves (Fittings)
BSI BS 5646: Part 3-78. 1978 Rolling Bearings—Accessories Part 3: Tapered Mounting Sleeves. 8 pp.
DIN ENGL 5415-77. Adapter Sleeves with Nut and Retainer for Rolling Bearings (Dec). 3 pp.
DIN ENGL 5416-90. Withdrawal Sleeves for Use with Rolling Bearings (Mar). 5 pp.
ISO 113 Pt I-79. Rolling Bearing Accessories—Part I: Tapered Sleeves First Edition. 6 pp.

Antifriction Bearings (Cont.)
—Sleeves (Fittings) (Cont.)
JIS B 1553-76. Tapered Adapter Sleeves for Rolling Bearings (R 1985). 20 pp.
JIS B 1556-93. Rolling Bearings—Withdrawal Sleeves and Nuts. 21 pp.

—Sleeves (Fittings)—Withdrawal
DIN ENGL 5416-90. Withdrawal Sleeves for Use with Rolling Bearings (Mar). 5 pp.
JIS B 1556-93. Rolling Bearings—Withdrawal Sleeves and Nuts. 21 pp.

—Snap Rings
BSI BS 5646: Part 5-80. 1980 Rolling Bearings—Accessories Part 5: Snap Ring Groove and Snap Ring Dimensions. 8 pp.
ISO 464-76. Rolling Bearings with Locating Snap Ring—Dimensions First Edition. 7 pp.
JIS B 1509-87. Locating Snap Rings for Rolling Bearings. 9 pp.

—Static Load Ratings
BSI BS 5645-87. 1987 Static Load Ratings for Roller Bearings. 8 pp.
CNS B1168-80. Load Carrying Capacity of Rolling Bearings, Definitions, Load Rating, Methods of Load and Rating Life (Jul)(5683).
ISO 76-87. Rolling Bearings—Static Load Ratings Second Edition. 8 pp.
ISO TR10657-91. Explanatory Notes on ISO 76 First Edition. 62 pp.
JIS B 1519-89. Static Load Ratings for Rolling Bearings. 12 pp.

—Symbols
CNS B1167-80. Symbols of Rolling Bearings (Jul)(5682).
JIS B 0124-65. Symbols for Quantities of Rolling Bearings (R 1971). 8 pp.

—Tribology
CNS B1284-82. Plain Bearings; the Tribological Behavior of Metallic Antifriction (Mar)(8556).

Antifungal Agents
Use: Fungicides

Antiglobulin Testing
—Antiserum
CGSB CAN/CGSB-106.2-M91. Polyspecific Anti-Human Globulin Reagent for the Antiglobulin Test (Monoclonal and/or Polyclonal). 9 pp.

Antigravity Suits
Use For: Anti-G Suits; G Suits *See Also:* Pressure Suits; Protective Clothing
NATO STANAG 3200 ED 4 AMD 1-80. Anti-G Suits. 11 pp.

Antiicing Additives
Use For: Icing Inhibitors *See Also:* Additives; Antifreezes; Antiicing Equipment; Deicer/Antiicing Fluids; Deicers; Ice Formation; Ice Prevention
MOD UK DSTAN 68-150-01. Mixture of Fuel System Icing Inhibitor and Corrosion Inhibitor/Lubricity Improving Additive Joint Service Designation: AL-48 Issue 1; Corrigendum. 12 pp.

—Ethylene Glycol Monomethyl Ether—Aircraft Fuels
CAA NOTICE #23 ISSUE 1. Fuel Additives—Health Hazards (Airworthiness Notices). 2 pp.

—Ethylene Glycol Monomethyl Ether—Jet Engine Fuels
CGSB CAN/CGSB-3.526-M87. Icing Inhibitor for Aviation Fuels. 9 pp.
MOD UK DERD 2451-80. Fuel System Icing Inhibitor Issue 3. 6 pp.

—Isopropyl Alcohol—Motor Vehicle Fuels
CGSB 3-GP-525MA-88. Isopropanol. 8 pp.

Antiicing/Deicing Fluids
Use: Deicer/Antiicing Fluids

Antiicing Equipment
See Also: Antiicing Additives; Deicer/Antiicing Fluids; Deicers

—Aircraft
BSI 3G 100:Pt 2: SUBSEC 3.9-72. 1972 General Requirements for Equipment for Use in Aircraft Part 2: All Equipment Section 3: Environmental Conditions Subsection 3.9: Ice Formation. 3 pp.
BSI 3G 100:Pt 2: SUBSEC 3.10-74. 1974 General Requirements for Equipment for Use in Aircraft Part 2: All Equipment Section 3: Environmental Conditions Subsection 3.10: Impact Icing. 12 pp.

Antiicing Equipment (Cont.)

—Aircraft—Self Propelled Machinery
ISO 11077-93. Aerospace—Self-Propelled De-Icing/Anti-Icing Vehicles—Functional Requirements First Edition. 9 pp.

Antiknock Ratings
Scope Note: Use a more specific term *See:* Cetane Number; Knock Rating; Octane Number

Antilock Brake Systems
Use: Antiskid Brake Systems

Antimicrobial Agents
See Also: Disinfectants

—Food Preparation Equipment
CGSB CAN/CGSB-2.161-M91. Assessment of Efficacy of Antimicrobial Agents for Use on Environmental Surfaces and Medical Devices. 8 pp.

—Food Processing Equipment
CGSB CAN/CGSB-2.161-M91. Assessment of Efficacy of Antimicrobial Agents for Use on Environmental Surfaces and Medical Devices. 8 pp.

—Food Storage
CGSB CAN/CGSB-2.161-M91. Assessment of Efficacy of Antimicrobial Agents for Use on Environmental Surfaces and Medical Devices. 8 pp.

—Health Care Facilities
CGSB CAN/CGSB-2.161-M91. Assessment of Efficacy of Antimicrobial Agents for Use on Environmental Surfaces and Medical Devices. 8 pp.

—Medical Equipment
CGSB CAN/CGSB-2.161-M91. Assessment of Efficacy of Antimicrobial Agents for Use on Environmental Surfaces and Medical Devices. 8 pp.

—Paperboard
CEN PREN 1104-93. Paper and Board Intended to Come into Contact with Foodstuff—Determination of Transfer of Antimicrobic Constituents. 9 pp.
DIN ENGL 54600 Pt 8-82. Testing of Paper and Board; Testing for Antimicrobial Additives; Determination of the Methylene-Bis-Thiocyanate Content (Mar). 3 pp.
DIN ENGL 54600 Pt 11-82. Testing of Paper and Board; Testing for Antimicrobial Additives; Determination of 2-Oxo-2 (4'-Hydroxyphenyl) Acetohydroximic Acid Chloride Content (Sept). 3 pp.

—Papers
CEN PREN 1104-93. Paper and Board Intended to Come into Contact with Foodstuff—Determination of Transfer of Antimicrobic Constituents. 9 pp.
DIN ENGL 54600 Pt 8-82. Testing of Paper and Board; Testing for Antimicrobial Additives; Determination of the Methylene-Bis-Thiocyanate Content (Mar). 3 pp.
DIN ENGL 54600 Pt 11-82. Testing of Paper and Board; Testing for Antimicrobial Additives; Determination of 2-Oxo-2 (4'-Hydroxyphenyl) Acetohydroximic Acid Chloride Content (Sept). 3 pp.

Antimony
See Also: Antimony Content Analysis; Metals; White Metals
JIS H 2112-58. Antimony Metal. 3 pp.
JIS K 8080-61. Antimony.

—Chemical Analysis
CNS H2013-53. Method of Chemical Test for Antimony (Feb)(276) (R 1973).
JIS H 1171-76. Methods for Chemical Analysis of Antimony Metal.

—Standard Solutions
CNS K0038-88. Antimony Standard Solution (Nov)(12468).
JIS K 0025-83. Antimony Standard Solution. 14 pp.

—Toys—Migration Testing
BSI BS 5665: Part 3-89. 1989 Safety of Toys Part 3: Migration of Certain Elements (Supersedes BS 3443: 1968). 11 pp.
CEN EN 71-3-88. Safety of Toys Part 3 Migration of Certain Elements. 14 pp.
CEN PREN 71-3-92. Safety of Toys—Part 3: Migration of Certain Elements. 24 pp.

Antimony Content Analysis
See Also: Antimony

—Coatings
CGSB CAN2-1.500-75 METH 4-73. Methods of Test for Toxic Trace Elements in Protective Coatings Determination of Leachable Antimony (Sb) in Low Concentration. 2 pp.

—Copper Alloys—Spectrophotometry
ISO 5956-84. Copper and Copper Alloys—Determination of Antimony Content—Rhodamine B Spectrometric Method First Edition. 4 pp.

—Copper—Spectrophotometry
ISO 5956-84. Copper and Copper Alloys—Determination of Antimony Content—Rhodamine B Spectrometric Method First Edition. 4 pp.

—Electrolytic Copper—Atomic Absorption Spectrophotometry
BSI BS 7317: Part 3-90. 1990 Methods for Analysis of High Purity Copper Cathode Cu-CATH-1 Part 3: Method for Determination of Antimony, Arsenic, Bismuth, Selenium, Tellurium and Tin by Hydride Generation and Atomic Absorption Spectrophotometry. 12 pp.
BSI BS 7317: Part 4-90. 1990 Methods for Analysis of High Purity Copper Cathode Cu-CATH-1 Part 4: Method for Determination of Antimony, Arsenic, Bismuth, Lead, Selenium, Tellurium and Tin by Electrothermal Atomization Atomic Absorption Spectrophotometry. 9 pp.
BSI DD 95: Part 3-84. (WITHDRAWN) 1984 Analysis of Higher Purity Copper Cathode Cu-CATH-1 Part 3: Methods for Determination of Antimony, Arsenic, Bismuth, Selenium, Tellurium and Tin by Hydride Generation and Atomic Absorption Spectrophotometry. 14 pp.
BSI DD 95: Part 4-86. (WITHDRAWN) 1986 Analysis of Higher Purity Copper Cathode CU-CATH-1 Part 4: Method for Determination of Antimony, Arsenic, Bismuth, Lead, Selenium, Tellurium and Tin by Electrothermal Atomization Atomic. 9 pp.

—Electrolytic Copper—Hydride Generation
BSI BS 7317: Part 3-90. 1990 Methods for Analysis of High Purity Copper Cathode Cu-CATH-1 Part 3: Method for Determination of Antimony, Arsenic, Bismuth, Selenium, Tellurium and Tin by Hydride Generation and Atomic Absorption Spectrophotometry. 12 pp.
BSI DD 95: Part 3-84. (WITHDRAWN) 1984 Analysis of Higher Purity Copper Cathode Cu-CATH-1 Part 3: Methods for Determination of Antimony, Arsenic, Bismuth, Selenium, Tellurium and Tin by Hydride Generation and Atomic Absorption Spectrophotometry. 14 pp.

—Ferrotungsten
CNS G2156-83. Methods of Chemical Analysis for Antimony in Ferrotungsten (Jan)(9867).

—Fluorspar—Atomic Absorption Spectrometry
ISO 9504-93. Metallurgical-Grade Fluorspar—Determination of Antimony Content—Solvent Extraction Atomic Absorption Spectrometric Method Second Edition. 10 pp.

—Iron
CNS G2257-85. Method of Determination for Antimony in Iron and Steel (Dec)(11449).
JIS G 1235-81. Methods for Determination of Antimony in Iron and Steel. 11 pp.

—Lead Alloys—Atomic Absorption Spectrometry
SAA AS 1671.1-87. Lead and Lead Alloys—Part 1: Determination of Antimony Content—Flame Atomic Absorption Spectrometric Method. 3 pp.

—Lead Alloys—Photometry
BSI BS 3908: Part 13-70. 1970 Methods for the Sampling and Analysis of Lead and Lead Alloys Part 13: Antimony in Lead and Lead Alloys (Low Contents) (Photometric Method). 8 pp.

—Lead Alloys—Volumetric Analysis
BSI BS 3908: Part 10-89. 1989 Sampling and Analysis of Lead and Lead Alloys Part 10: Antimony in Lead Alloys (Titrimetric Method). 6 pp.
SAA AS 1671.2-87. Lead Alloys—Part 2: Determination of Low Concentrations of Antimony in Lead and Lead Alloys Containing not More than 2.5 Percent Arsenic and 0.10 Percent Copper—Titrimetric Method. 4 pp.
SAA AS 1671.3-87. Lead and Lead Alloys—Part 3: Determination of High Concentrations of Antimony in Lead Alloys Containing not More than 2.5 Percent Arsenic and 1.0 Percent Copper-Titrimetric Method. 5 pp.

—Lead—Atomic Absorption Spectrometry
SAA AS 1671.1-87. Lead and Lead Alloys—Part 1: Determination of Antimony Content—Flame Atomic Absorption Spectrometric Method. 3 pp.

—Lead—Photometry
BSI BS 3908: Part 13-70. 1970 Methods for the Sampling and Analysis of Lead and Lead Alloys Part 13: Antimony in Lead and Lead Alloys (Low Contents) (Photometric Method). 8 pp.

—Lead—Volumetric Analysis
BSI BS 3908: Part 10-89. 1989 Sampling and Analysis of Lead and Lead Alloys Part 10: Antimony in Lead Alloys (Titrimetric Method). 6 pp.

—Nickel—Atomic Absorption Spectrometry
BSI BS 6783: Part 4-86. 1986 Sampling and Analysis of Nickel, Ferronickel and Nickel Alloys: Part 4: Method for Determination of Silver, Arsenic, Bismuth, Cadmium, Lead, Antimony, Selenium, Tin, Tellurium and Thallium in Nickel by Electrothermal Atomic Absorption Spectrometry. 11 pp.
ISO 7523-85. Nickel—Determination of Silver, Arsenic, Bismuth, Cadmium, Lead, Antimony, Selenium, Tin, Tellurium and Thallium Contents—Electrothermal Atomic Absorption Spectrometric Method First Edition. 10 pp.

—Ores
CNS M3058-81. Method for Determination of Antimony in Ores (Jun)(7520).
JIS M 8130-84. Methods for Determination of Antimony in Ores. 17 pp.

—Paints—Atomic Absorption Spectrometry
BSI BS 3900: Part B7-86. 1986 Methods of Test for Paints Group B: Tests Involving Chemical Examination of Liquid Paints and Dried Paint Films Part B7: Determination of 'Soluble' Antimony Content. 9 pp.
ISO 3856 Pt 2-84. Paints and Varnishes—Determination of "Soluble" Metal Content—Part 2: Determination of Antimony Content—Flame Atomic Absorption Spectrometric Method and Rhodamine B Spectrophotometric Method Second Edition. 8 pp.

—Paints—Spectrophotometry
ISO 3856 Pt 2-84. Paints and Varnishes—Determination of "Soluble" Metal Content—Part 2: Determination of Antimony Content—Flame Atomic Absorption Spectrometric Method and Rhodamine B Spectrophotometric Method Second Edition. 8 pp.

—Rubber—Iodometry
BSI BS 903: Part B16-67. (WITHDRAWN) 1967 Methods of Testing Vulcanized Rubber Part B16: Determination of Antimony. 7 pp.

—Solders—Volumetric Analysis
BSI BS 3338: Part 13-65. 1965 Methods for the Sampling and Analysis of Tin and Tin Alloys Part 13: Determination of Antimony in Solders and White Metal Bearing Alloys (Volumetric: Potassium Bromate Method). 8 pp.

—Steels
CNS G2257-85. Method of Determination for Antimony in Iron and Steel (Dec)(11449).
JIS G 1235-81. Methods for Determination of Antimony in Iron and Steel. 11 pp.

—Tin—Photometry
BSI BS 3338: Part 3-61. 1961 Amd 2 Methods for the Sampling and Analysis of Tin and Tin Alloys Part 3: Determination of Antimony in Ingot Tin (Photometric Method). 10 pp.

—Uranium Hexafluoride—Atomic Absorption Analysis
CNS J2062-82. Method for Determining Antimony in UF6 Atomic Absorption Method (Aug)(9280).

—White Metals—Volumetric Analysis
BSI BS 3338: Part 13-65. 1965 Methods for the Sampling and Analysis of Tin and Tin Alloys Part 13: Determination of Antimony in Solders and White Metal Bearing Alloys (Volumetric: Potassium Bromate Method). 8 pp.

Antimony Potassium Tartrate
Use For: Potassium Antimonyl Tartrate; Tartar Emetic *See Also:* Esters
JIS K 8533-61. Antimonyl Tartrate (Tartar Emetic).

Antimony Sulfides
MOD UK DSTAN 13-25-92. Antimony III Sulfide, Black Issue 2. 8 pp.
MOD UK M 811/91. Examination of Antimony Sulphide, Black.

Antimony Trichloride
CNS K7071-62. Chemical Reagent (Antimony Trichloride) (Antimonous Chloride) (Jul)(1571).
JIS K 8400-78. Antimony Trichloride.

Antimony Trioxide
CNS K7682-83. Chemical Reagent (Antimony Trioxide) (Dec)(10725).
JIS K 8407-61. Antimony Trioxide.

Antimonyl Potassium Tartrate
Use: Antimony Potassium Tartrate

Antioxidant Content Analysis
See Also: Antioxidants; BHA Content Analysis; BHT Content Analysis; Gallates Content Analysis

—Elastomers—Thin Layer Chromatography
DIN ENGL 53622 Pt 2-76. Testing of Rubber and Elastomers; Thin-Layer Chromatographic Analysis; Identification of Antidegradants in Rubber and Elastomers (June). 4 pp.

—Electrical Insulating Oils
BSI BS 5984-80. 1980 Detection and Determination of Specified Anti-Oxidant Additives in Insulating Oils. 15 pp.
CENELEC HD 415-80. Detection and Determination of Specified Anti-Oxydant Additives in Insulating Oils. 2 pp.
IEC 666-79. Detection and Determination of Specified Anti-Oxidant Additives in Insulating Oils First Edition. 27 pp.

—Fats—Thin Layer Chromatography
BSI BS 684: Sec 2.33-83. 1983 Methods of Analysis of Fats and Fatty Oils Part 2: Other Methods Section 2.33: Detection and Identification of Antioxidants. 4 pp.
ISO 5558-82. Animal and Vegetable Fats and Oils—Detection and Identification of Antioxidants—Thin-Layer Chromatographic Method First Edition. 5 pp.

—Oils—Thin Layer Chromatography
BSI BS 684: Sec 2.33-83. 1983 Methods of Analysis of Fats and Fatty Oils Part 2: Other Methods Section 2.33: Detection and Identification of Antioxidants. 4 pp.

—Olefin Resins—Spectrophotometry
BSI BS 2782:Pt4: METH 434D-75. 1975 Methods of Testing Plastics Part 4: Chemical Properties Method 434D: Determination of Antioxidants in Polyolefin Compounds by a Spectrophotometric Method. 3 pp.

—Olefin Resins—Thin Layer Chromatography
BSI BS 2782:Pt4: METH 434A-75. 1975 Methods of Testing Plastics Part 4: Chemical Properties Method 434A: The Identification of Antioxidants and Ultraviolet Absorbers in Polyolefin by Thin Layer Chromatography. 5 pp.

—Olefin Resins—Ultraviolet Absorption
BSI BS 2782:Pt4: METH 434B-77. 1977 Methods of Testing Plastics Part 4: Chemical Properties Method 434B: Determination of Antioxidants in Polyolefin Compounds by Ultraviolet Absorption of the Chloroform Extract. 3 pp.
BSI BS 2782:Pt4: METH 434C-77. 1977 Methods of Testing Plastics Part 4: Chemical Properties Method 434C: Determination of Anti-Oxidants in Polyolefin Compounds by Ultra-Violet Absorption of the Toluene/Ethanol Extract in Ethanol Solution. 4 pp.

—Rubber—Thin Layer Chromatography
DIN ENGL 53622 Pt 2-76. Testing of Rubber and Elastomers; Thin-Layer Chromatographic Analysis; Identification of Antidegradants in Rubber and Elastomers (June). 4 pp.

Antioxidants
See Also: Additives; Antioxidant Content Analysis; Corrosion Inhibitors; Retarders (Materials)

—Food—Purity
EC 78/664/EEC-78. Council Directive Laying Down Specific Criteria of Purity for Antioxidants Which May be Used in Foodstuffs Intended for Human Consumption. 18 pp.
EC 81/712/EEC-81. First Commission Directive Laying down Community Methods of Analysis for Verifying That Certain Additives Used in Foodstuffs Satisfy Criteria of Purity. 27 pp.
EC 82/712/EEC-82. Council Directive Amending Directive 78/664/EEC Laying down Specific Criteria of Purity for Antioxidants Which May be Used in Foodstuffs Intended for Human Consumption. 2 pp.

—Rubber
CNS K4043-87. Rubber Antioxidant DNPD (N, N-Di—B—Naphthyl P—Phenylenediamine) (Mar)(4861).
CNS K4082-87. Rubber Antioxidant TMDO (Poly-2,2, 4—Trimethyl—1,2—Dihydroquinoline) (Mar)(11879).
CNS K4083-87. Rubber Antioxidant IPPD (N—Isopropyl—N'-Phenyl—P Phenylenediamine) (Mar)(11881).

—Rubber (Cont.)
CNS K6900-87. Method of Test for Rubber Antioxidant TMDO (Poly-2,2,4-Trimethyl-1, 2-Dihydroquinoline) (Mar)(11880).
CNS K6901-87. Method of Test for Rubber Antioxidant IPPD (N-Isopropyl-N'-Phenyl P-Phenylenediamine) (Mar)(11882).
CNS K6902-87. Method of Test for Rubber Antioxidant DNPD Phenylenediamine) (Mar)(11883).
JIS K 6211-79. Antioxidants for Rubber.
MOD UK DSTAN 68-198-93. Antioxidant 2,2'-Methylene-Bis (4-Methyl-6 Tertiary Butyl Phenol) Issue 1. 11 pp.
MOD UK TS 10199A. Antioxidant 2-2-Methylene-Bis (4-Methyl-6-Tertiary Butyl Phenol).

Antiques
—Value-Added Tax
EC COM(88) 846-89. Proposal for a Council Directive Supplementing the Common System of Value-Added Tax and Amending Articles 32 and 28 of Directive 77/388/EEC—Special Arrangements for Second-Hand Goods, Works of Art, Antiques and Collectors' Items. 24 pp.

Antireflection Coatings
See Also: Coatings

—Glass—Aircraft Instruments
MOD UK DSTAN 66-26: Part 3-79. General Requirements for Aircraft Instruments and Displays Part 3: Reflection-Reducing Coatings Used on Glass Elements in Aircrew Station Displays (Specification) Issue 1. 7 pp.

—Glass—Display Devices—Aircraft Cabins
NATO STANAG 3643 ED 2 AMD 7-73. Coating, Reflection Reducing for Glass Elements Used in Aircrew Station Displays. 6 pp.
NATO STANAG 3643 ED 2 AMD 9-73. Coating, Reflection Reducing for Glass Elements Used in Aircrew Station Displays. 6 pp.

—Instrument Windows—Aircraft
BSI G 211-71. 1971 Amd 1 Reflection-Reducing Coating of Instrument Windows and Lighting Wedges. 6 pp.

—Lighting Wedges—Aircraft
BSI G 211-71. 1971 Amd 1 Reflection-Reducing Coating of Instrument Windows and Lighting Wedges. 6 pp.

Antiscatter Grids
Use: Radiographic Grids

Antiseize Compounds
Scope Note: Use a more specific term
See: Graphite-Petrolatum; Lubricants; Molybdenum Disulfide

Antiseptic
Use: Disinfectants

Antiseptics
Use: Disinfectants

Antiskid Brake Systems
Use For: Antilock Brake Systems *See Also:* Brakes

—Automotive—Connectors
BSI BS AU 196-84. 1984 Electrical Connections Between 24V Trailer Antilock Braking Systems and Towing Vehicles. 4 pp.
ISO 7638-85. Road Vehicles—Brake Anti-Lock Device Connector First Edition. 13 pp.

Antispreading Compounds
—Aircraft
MOD UK DTD-644-47. Anti-Spreading Composition (Reprinted June 1961). 1 p.

Antistatic Additives
Use: Antistatic Agents

Antistatic Agents
—Aircraft Fuels—Electrical Resistivity
CNS K6965-89. Method of Test for Electrical Conductivity of Aviation and Distillate Fuels Containing a Static Dissipator Additive (Oct)(12616).
ISO 6297-83. Petroleum Products—Aviation and Distillate Fuels Containing a Static Dissipator Additive—Determination of Electrical Conductivity First Edition. 6 pp.

Antistick Papers
Use: Release Papers

Antisubmarine Rockets
Use For: ASROC *See Also:* Antisubmarine Warfare; Rockets

—Interchangeability
NATO STANAG 1343 ED 1 AMD 2-90. ASROC System Interchangeable Within NATO Naval Forces. 12 pp.

Antisubmarine Warfare
See Also: Antisubmarine Rockets; Naval Mine Warfare; Naval Ships; Submarines; Warfare

—Steering
NATO STANAG 1041 ED 12 AMD 0-83. Anti-Submarine Evasive Steering. 4 pp.
NATO STANAG 1041 ED 12 AMD 1-83. Anti-Submarine Evasive Steering. 4 pp.

—Technical Manuals
NATO STANAG 1052 ED 22 AMD 0-88. Allied Submarine and Anti-Submarine Exercise Manual—AXP-1(B). 7 pp.
NATO STANAG 1052 ED 23 AMD 0-90. Allied Submarine and Anti-Submarine Exercise Manual—AXP-1(B). 5 pp.
NATO STANAG 1152 ED 12 AMD 0-89. Allied Anti-Submarine Warfare Manual—ATP-28 (A). 5 pp.
NATO STANAG 1152 ED 13 AMD 0-91. Allied Anti-Submarine Warfare Manual—ATP-28 (A). 5 pp.
NATO STANAG 1152 ED 14 AMD 0-93. Allied Anti-Submarine Warfare Manual—ATP-28 (A). 5 pp.

Antitheft Alarms
Use: Burglar Alarms

Antitransmit Receive Tubes
Use For: ATR Tubes *See Also:* Electron Tubes; Gas Filled Tubes; Transmit Receive Tubes

—Preferred Products List
CECC CECC MUAHAG Vol 12 IS 1-90. Preferred Products List; Microwave Components (En, Fr, Ge). 76 pp.

—Quality Assurance
BSI BS 9042-71. (WITHDRAWN) 1971 Rules for the Preparation of Detail Specifications for Gas-Filled Microwave Switching Tubes of Assessed Quality: ATR Tubes. 13 pp.

Antivacuum Valves
Use: Backflow Preventers

Anvils
See Also: Hammers
CNS B3345-80. Anvils, Blacksmith (Apr)(5393).
MOD UK DEF-1166-57. Anvils, Blacksmiths'. 5 pp.

AOI
Use: AND OR INVERT Gates

AOI Gates
Use: AND OR INVERT Gates

APC
Use: Aeronautical Passenger Communications

APC Coaxial Connectors
See Also: Coaxial Connectors

— 7mm—Rigid Cable
CENELEC HD 351.2-76. Rigid Precision Coaxial Lines and Their Associated Precision Connectors Part 2: 50ohm 7mm Rigid Precision Coaxial Line and Associated Hermaphroditic Precision Coaxial Connector. 2 pp.
IEC 457 Pt 2-74. Rigid Precision Coaxial Lines and Their Associated Precision Connectors Part 2: 50 Ohm 7 mm Rigid Precision Coaxial Line and Associated Hermaphroditic Precision Coaxial Connector First Edition. 15 pp.

—35mm—Rigid Cable
IEC 457 Pt 5-84. Rigid Precision Coaxial Lines and Their Associated Precision Connectors Part 5: 50 Ohms 3.5 mm Rigid Precision Coaxial Line with Provision for Mounting Connectors First Edition. 9 pp.

APC Connectors
Use: APC Coaxial Connectors

APD
Use: Avalanche Photodiodes

Aperture Cards
See Also: Microfilm; Punched Cards

Aperture Cards (Cont.)

BSI BS 4210: Part 3-77. 1977 35 mm Microcopying of Technical Drawings Part 3: Unitized Microfilm Carriers. 8 pp.
CGSB CAN/CGSB-72.14-M89. Unitized 35 mm Microfilm Carriers (Aperture Cards). 11 pp.
JIS Z 6006-89. Aperture Card for 35 mm Microfilm of Technical Drawings. 13 pp.
SAA AS 1717-75. Unitized Microfilm Carriers. 16 pp.

—Adhesion Testing—Protection Sheets

BSI BS 6247-82. 1982 Adhesion of Aperture-Card Protection Sheets. 8 pp.
ISO 6343-81. Micrographics—Unitized Microfilm Carrier (Aperture Card)—Determination of Adhesion of Protection Sheet to Aperture Adhesive First Edition. 6 pp.

—Aerospace

AECMA PREN2484-84. Microfilming of Drawings—Aperture Card for 35 mm Microfilm (CO/01A). 8 pp.
CEN EN 2484-89. Microfilming of Drawings Aperture Card for 35 mm Microfilm. 11 pp.

—Thickness Measurement

ISO 6342-93. Micrographics—Aperture Cards—Method of Measuring Thickness of Buildup Area First Edition. 7 pp.

API

Use: Application Program Interfaces

API Gravity

Use: Density

APL (A Programming Language)

Use For: A Programming Language (APL)
See Also: Programming Languages
BSI BS 7301-90. 1990 Programming Language: APL. 269 pp.
ISO 8485-89. Programming Languages—APL First Edition. 268 pp.
JTC1 8485-89. Programming Languages—APL First Edition. 268 pp.
OSI ISO 8485-89. Programming Languages—APL. 272 pp.

—Source Programs

ECMA ECMA 53-78. Representation of Source Program for Program Interchange-APL, COBOL, FORTRAN, Minimal BASIC and PL/1. 16 pp.

Apparel

Use: Clothing

Apparent Density

Use: Bulk Density

Apparent Porosity

Use: Porosity

Apparent Viscosity

Use: Viscosity

Apple Juice

See Also: Apples; Cider; Fruit Juices

—Concentrates—Patulin Content—Liquid Chromatography

ISO 8128 Pt 1-93. Apple Juice, Apple Juice Concentrates and Drinks Containing Apple Juice—Determination of Patulin Content—Part 1: Method Using High-Performance Liquid Chromatography First Edition. 7 pp.

—Concentrates—Patulin Content—Thin Layer Chromatography

ISO 8128 Pt 2-93. Apple Juice, Apple Juice Concentrates and Drinks Containing Apple Juice—Determination of Patulin Content—Part 2: Method Using Thin-Layer Chromatography First Edition. 8 pp.

—Patulin Content—Liquid Chromatography

ISO 8128 Pt 1-93. Apple Juice, Apple Juice Concentrates and Drinks Containing Apple Juice—Determination of Patulin Content—Part 1: Method Using High-Performance Liquid Chromatography First Edition. 7 pp.

—Patulin Content—Thin Layer Chromatography

ISO 8128 Pt 2-93. Apple Juice, Apple Juice Concentrates and Drinks Containing Apple Juice—Determination of Patulin Content—Part 2: Method Using Thin-Layer Chromatography First Edition. 8 pp.

Apples

Use For: Wax Apples *See Also:* Apple Juice; Food; Fruits

Apples (Cont.)

EC EEC/292/92. Commission Regulation Amending Regulation (EEC) No 920/89 Laying down Quality Standards for Carrots, Citrus Fruit and Dessert Apples and Pears as Regards the Tables Classifying Varieties of Apples. 2 pp.
EC EEC/2647/92-92. Commission Regulation Amending Regulation (EEC) No 1596/79 on Preventive Withdrawals of Apples and Pears. 1 p.
EC EEC/2648/92-92. Commission Regulation Enabling Member States to Authorize Preventive Withdrawals of Apples and Pears. 1 p.

—Boxes

BSI BS 5637-78. (WITHDRAWN) 1978 Non-Returnable Corrugated Cases for Apples. 9 pp.

—Canned

CNS N5161-85. Canned Apples (May)(5930). 6 pp.

—Dried

ISO 7701-86. Dried Apples—Specification First Edition. 8 pp.

—Grading

CNS N1119-83. Grades of Apple (Sep)(10575). 4 pp.
CNS N1120-83. Grades of Wax Apple (Sep)(10576). 3 pp.

—Storage

ISO 1212-76. Apples—Guide to Cold Storage First Edition. 8 pp.
ISO 8682-87. Apples—Storage in Controlled Atmospheres First Edition. 8 pp.

Appliance Wire

Use: Hookup Wire—Appliances

Appliances

Scope Note: For additional listings, use a more specific term *See Also:* Barbecues; Bed Warmers; Blenders; Clothes Dryers; Coffee Grinders; Coffee Makers; Convection Ovens; Cooking Appliances; Cooking Utensils; Corn Grinders; Curling Irons; Dishwashers; Food Cutters; Food Mixers; Food Preparation Equipment; Food Processing Equipment; Food Warmers; Foot Warmers (Electric); Freezers; Frying Utensils; Garbage Disposals; Garment Steamers; Griddles; Grills (Appliances); Hair Care Appliances; Hair Clippers; Hair Dryers; Hardware; Heaters; Hot Plates (Warmers); Ice Cream Makers; Ice Makers; Irons (Electric); Jugs; Juicers; Kettles; Massage Equipment; Meat Cutters; Meat Grinders; Meat Slicers; Microwave Ovens; Ovens; Pasta Machines; Peelers; Pots (Electric); Pressure Cookers; Range Hoods; Ranges; Refrigerator/Freezers; Refrigerators; Rice Cookers; Rice Dryers; Roasters; Sewing Machines; Shavers; Skin Care Appliances; Space Heaters; Steam Irons; Steam Tables; Steamers; Stoves; Teakettles; Toaster Ovens; Toasters; Toothbrushes; Towel Rails (Heated); Vacuum Cleaners; Waffle Irons; Washing Machines; Water Heaters

BSI PD 6434-69. 1969 Recommendations for the Design and Testing of Smoke Reducing Solid Fuel Burning Domestic Appliances. 12 pp.
CNS Z3024-83. Method of Test for Gas Appliances Used with Town Gas (Nov)(10671).
CSA C169 Bull. Electrical Bulletin 1169 June 27, 1978 to C22.2 NO 68. 2 pp.
DIN VDE 0720 Pt 1-72. Specification for Electrical Heating Appliances for Household and Similar Applications Part 1: General Provisions (Feb) (This Document is Not a Complete Translation and Must Only Be Used in Conjunction with the Following Standard CEE 11 (1964)). 10 pp.
DIN VDE 0720 Pt 1 AE-80. Electric Heating Appliances for Domestic and Similar Purposes; General Regulations: Amendment E (Mar). 10 pp.
DIN VDE 0730 Pt 1-72. Specifications for Electric Motor-Operated Appliances for Domestic and Similar Purposes; Part 1: General Specifications (Mar). 99 pp.
DIN VDE 0730 Pt 2 G/H-76. Regulations for Electrically Powered Appliances for Domestic Use and Similar Purposes; Part 2 G/H: Special Regulations for Kitchen Appliances (Sept). 34 pp.
DIN VDE 0730 Pt 2 ZR-80. Electric Motor-Operated Appliances for Domestic and Similar Purposes; Particular Requirements for Battery-Operated Appliances up to 20 VA Input and Their Charging and Battery Assemblies (VDE Specification) (May). 18 pp.
EC COM(88) 786-89. Proposal for a Council Directive on the Approximation of the Laws of the Member States Relating to Appliances Burning Gaseous Fuels. 51 pp.
JIS S 2092-91. General Constructions of Gas Burning Appliances for Domestic Use. 34 pp.
JIS S 2093-91. Test Methods of Gas Burning Appliances for Domestic Use. 98 pp.
JIS S 2121-79. Test Methods for Gas Appliances Used with Town Gas.

Appliances (Cont.)

JIS S 3031-89. General Rules for Test Methods of Oil Burning Appliances. 105 pp.
SAA AS 1375-85. Industrial Fuel-Fired Appliances (Known as the SAA Industrial Fuel-Fired Appliance Code). 44 pp.
SAA AS 2078-77. Domestic Oil-Fired Appliances (Quality and Performance) Corrig.. 40 pp.
SAA AS 3115-90. Approval and Test Specification—Motor Operated Appliances Amdt 1 December 1990 (This is a Joint Standard with SANZ NZS 3115). 9 pp.
SNZ NZS/AS 3115-90. Approval and Test Specification—Motor Operated Appliances (This is a Joint Standard with SAA AS 3115). 7 pp.

—Air Supply

BSI BS 5440: Part 1-78. (WITHDRAWN) 1978 Amd 1 Code of Practice for Flues and Air Supply for Gas Appliances of Rated Input Not Exceeding 60 kW (1st and 2nd Family Gases) Part 1: Flues. 42 pp.
BSI BS 5440: Part 2-89. 1989 Code of Practice for Flues and Air Supply for Gas Appliances of Rated Input Not Exceeding 60 kW (1st and 2nd Family Gases) Part 2: Installation of Ventilation for Gas Appliances. 14 pp.
BSI BS 5440: Part 2-76. (WITHDRAWN) 1976 Code of Practice for Flues and Air Supply for Gas Appliances of Rated Input Not Exceeding 60 kW (1st and 2nd Family Gases) Part 2: Air Supply. 11 pp.

—Aircraft

CAA 707 (G). Section G Rotorcraft Electrical Supply, Systems and Equipment (Blue Papers). 47 pp.
CAA NOTICE #99 ISSUE 1. Galley Equipment (Airworthiness Notices). 4 pp.
CAA Chap G6-14 App #4 11.85. Design and Installation of Galleys and Domestic Equipment (Rotorcraft). 2 pp.

—Aircraft—Documentation

CAA LEAFLET 11-1 07.90. Materials, Parts or Appliances Approved to USA Technical Standards Orders (TSO's) for Part 21 Sub-Part 0. 3 pp.

—Aquariums—Safety

BSI BS EN 60335-2-55-93. 1993 Safety of Household and Similar Electrical Appliances Part 2: Particular Requirements Section 2.55: Electrical Appliances for Use with Aquariums and Garden Ponds (L). 22 pp.
CENELEC EN 60335-2-55-92. Safety of Household and Similar Electrical Appliances Part 2: Particular Requirements for Electrical Appliances for Use with Aquariums and Garden Ponds. 16 pp.
CENELEC EN 60335-2-55-93. Safety of Household and Similar Electrical Appliances Part 2: Particular Requirements for Electric Appliances for Use with Aquariums and Garden Ponds (IEC 335-2-55: 1989, Modified). 14 pp.
IEC 335 Pt 2-55-89. Safety of Household and Similar Electrical Appliances Part 2: Particular Requirements for Electrical Appliances for Use with Aquariums and Garden Ponds First Edition (CENELEC EN 60335-2-55: 1993). 31 pp.

—Automatic Control Equipment

BSI BS 6795-87. 1987 Full Sequence Automatic Control Systems Used on Gas-Fired Appliances. 16 pp.
BSI BS EN 60730-1-92. 1992 Automatic Electrical Controls for Household And Similar Use Part 1: General Requirements. 229 pp.
BSI BS EN 60730-1-01. 1992 Amd 1 Automatic Electrical Controls for Household and Similar Use Part 1: General Requirements (AMD 7747) May 15, 1993 (L). 235 pp.
BSI BS EN 60730-2-1-92. 1992 Automatic Electrical Controls for Household and Similar Use Part 2: Particular Requirements Section 2.1: Electrical Controls for Electrical Household Appliances (L). 22 pp.
BSI BS EN 60730-2-1-01. 1992 Amd 1 Automatic Electrical Controls for Household and Similar Use Part 2: Particular Requirements Section 2.1: Electrical Controls for Electrical Household Appliances (AMD 7419) December 15, 1992. 28 pp.
CENELEC EN 60730-1-91. Automatic Electrical Controls for Household and Similar Use Part 1: General Requirements. 17 pp.
CENELEC EN 60730-1/A1-A11-91. AMD 1 Thru AMD 11 Automatic Electrical Controls for Household and Similar Use Part 1: General Requirements. 17 pp.
CENELEC EN 60730-1/A12-93. AMD 12 Automatic Electrical Controls for Household and Similar Use Part 1: General Requirements. 4 pp.
CENELEC EN 60730-2-1-91. Automatic Electrical Controls for Household and Similar Use Part 2: Particular Requirements for Electrical Controls for Electrical Household Appliances. 10 pp.

INTERNATIONAL AND NON-U.S. NATIONAL STANDARDS
SUBJECT INDEX

Appliances

Appliances (Cont.)

—Automatic Control Equipment (Cont.)
CENELEC EN 60730-2-1 /A11-92. AMD 11 Automatic Electrical Controls for Household and Similar Use Part 2: Particular Requirements for Electrical Controls for Electrical Household Appliances. 5 pp.
IEC 730 Pt 1-86. Automatic Electrical Controls for Household and Similar Use Part 1: General Requirements First Edition; (Amendment 1-1990) (Amendment 2-3:1991). 487 pp.
IEC 730 Pt 2-1-89. Automatic Electrical Controls for Household and Similar Use Part 2: Particular Requirements for Electrical Controls for Electrical Household Appliances First Edition. 28 pp.

—Backflow Testing
BSI BS 6280-82. 1982 Vacuum (Backsiphonage) Test for Water-Using Appliances. 4 pp.

—Battery Chargers
DIN VDE 0730 Pt 2 ZR-80. Electric Motor-Operated Appliances for Domestic and Similar Purposes; Particular Requirements for Battery-Operated Appliances up to 20 VA Input and Their Charging and Battery Assemblies (VDE Specification) (May). 18 pp.

—Burners—Automatic Control Equipment
CEN PREN 298-89. Automatic Burner Control Systems for Gas Burners and Gas Burning Appliances with or Without Fans. 39 pp.

—Cable Couplers
BSI BS 196-61. 1961 Amd 4 Protected-Type Non-Reversible Plugs, Socket-Outlets, Cable-Couplers and Appliance Couplers, with Earthing Contacts for Single Phase a.c. Circuits up to 250 Volts. 64 pp.

—Certification
EC 90/396/EEC-90. Council Directive on the Approximation of the Laws of the Member States Relating to Appliances Burning Gaseous Fuels (NEW APP). 16 pp.

—Combustion Control Equipment
BSI BS 6047: Part 1-81. 1981 Flame Supervision Devices for Domestic, Commercial and Catering Gas Appliances Part 1: Heat Sensitive Types. 11 pp.
BSI BS 6067-81. 1981 Amd 1 Multifunctional Gas Controls for Domestic, Commercial and Catering Appliances. 12 pp.

—Commercial
CSA CAN/CSA-C22. 2 NO 68-92. Motor-Operated Appliances (Household and Commercial); (Gen Instr 1 Thru 2). 115 pp.

—Connectors
BSI BS 3283: Part 1-80. 1980 Amd 2 Non-Reversible Connectors and Appliances Inlets for Portable Electrical Appliances(for Circuits up to 250 Volts) Part 1: 13A Connector and Appliance Inlet. 17 pp.
BSI BS EN 60320-2-2-92. 1992 Appliance Couplers for Household and Similar General Purposes Part 2: Specification for Individual Types of Coupler Section 2.2: Interconnection Couplers for Household and Similar Equipment. 37 pp.
CENELEC EN 60 320-87. Appliance Couplers for Household and Similar General Purposes. 115 pp.
CENELEC EN 60320-2-2-91. Appliance Couplers for Household and Similar General Purposes Part 2: Interconnection Couplers for Household and Similar Equipment. 7 pp.
CSA 655 Bull. Electrical Bulletin 655 October 12, 1966 to C22.2 NO 4. 4 pp.
CSA C22.2 NO 65-93. Wire Connectors; (Gen Instr 1). 67 pp.
CSA 655 Bull. Electrical Bulletin 655 October 12, 1966 to C22.2 NO 65. 4 pp.
CSA 812 Bull. Electrical Bulletin 812 February 4, 1971 to C22.2 NO 65. 5 pp.
CSA 1122A Bull. Electrical Bulletin 1122A January 10, 1979 to C22.2 NO 65. 3 pp.
CSA 1165B Bull. Electrical Bulletin 1165B June 25, 1984 to C22.2 NO 65. 5 pp.
CSA 1422 Bull. Electrical Bulletin 1422 July 6, 1984 to C22.2 NO 65. 3 pp.
CSA 1165C Bull. Electrical Bulletin 1165C December 4, 1984 to C22.2 NO 65. 2 pp.
CSA 1438 Bull. Electrical Bulletin 1438 October 23, 1985 to C22.2 NO 65. 20 pp.
CSA 1165D Bull. Electrical Bulletin 1165D April 23, 1986 to C22.2 NO 65. 2 pp.
DIN VDE 0627-86. Connectors and Plug-And-Socket Devices for Rated Voltages up to 1000 V a.c. and 1200 V d.c. and Rated Currents up to 500 A for Each Pole (June). 36 pp.
IEC 320 Pt 2-2-90. Appliance Couplers for Household and Similar General Purposes Part 2: Interconnection Couplers for Household and Similar Equipment First Edition; (Corrigendum—Aug 1990). 69 pp.

Appliances (Cont.)

—Connectors (Cont.)
IEC 884 Pt 2-2-89. Plugs and Socket-Outlets for Household and Similar Purposes Part 2: Particular Requirements for Socket-Outlets for Appliances First Edition. 31 pp.
IEC 884 Pt 2-3-89. Plugs and Socket-Outlets for Household and Similar Purposes Part 2: Particular Requirements for Switched Socket-Outlets Without Interlock for Fixed Installations First Edition. 35 pp.
SNZ NZS 1801-63. Specification for Domestic Appliance Connectors and Appliance Inlet Sockets. 17 pp.

—Construction
CNS C1074-80. General Rules for Insulation Construction of Class II Electrical Appliances (Dec)(6765).
JIS S 3030-89. General Rules for Construction of Oil Burning Appliances. 45 pp.

—Control Systems Equipment
BSI BS 3955-86. 1986 Amd 1 Electrical Controls for Household and Similar General Purposes. 84 pp.
CEN PREN 126-91. Multifunctional Controls for Gas Burning Appliances. 77 pp.
EC 84/530/EEC-84. Council Directive on the Approximation of the Laws of the Member States Relating to Common Provisions for Appliances Using Gaseous Fuels, Safety and Control Devices for These Appliances and Methods of Surveillance of Them. 11 pp.
EC 90/396/EEC-90. Council Directive on the Approximation of the Laws of the Member States Relating to Appliances Burning Gaseous Fuels (NEW APP). 16 pp.
SNZ NZS/AS 3197-93. Approval and Test Specifications—Portable Electric Control or Conditioning Devices (This is a Joint Standard with SAA AS 3197). 14 pp.

—Cord Switches
BSI BS EN 61058-2-1-93. 1993 Switches for Appliances Part 2: Particular Requirements Section 2.1: Cord Switches (L). 26 pp.
CENELEC EN 61058-2-1-93. Switches for Appliances Part 2-1: Particular Requirements for Cord Switches (IEC 1058-2-1:1992). 21 pp.
IEC 1058 Pt 2-1-92. Switches for Appliances Part 2-1: Particular Requirements for Cord Switches First Edition; (CENELEC EN 61058-2-1: 1993). 44 pp.

—Cords (Electric)
CSA CAN/CSA-C22. 2 NO 21-M90. Cord Sets and Power-Supply Cords; (Gen Instr 1 Thru 2). 74 pp.
CSA 895B Bull. Electrical Bulletin 895B May 12, 1976 to C22.2 NO 21. 2 pp.

—Couplers
BSI BS 196-61. 1961 Amd 4 Protected-Type Non-Reversible Plugs, Socket-Outlets, Cable-Couplers and Appliance Couplers, with Earthing Contacts for Single Phase a.c. Circuits up to 250 Volts. 64 pp.
BSI BS 4491: Part 1-89. 1989 Amd 1 Appliance Couplers for Household and Similar General Purposes Part 1: Specification of General Requirements (AMD 6462) April 30, 1991. 123 pp.
BSI BS 6991-90. 1990 6/10A, Two-Pole Weather-Resistant Couplers for Household, Commercial and Light Industrial Equipment. 32 pp.
BSI BS EN 60320-2-2-92. 1992 Appliance Couplers for Household and Similar General Purposes Part 2: Specification for Individual Types of Coupler Section 2.2: Interconnection Couplers for Household and Similar Equipment. 37 pp.
CENELEC EN 60320-2-2-91. Appliance Couplers for Household and Similar General Purposes Part 2: Interconnection Couplers for Household and Similar Equipment. 7 pp.
CNS C4293-90. Appliance Couplers for Domestic and Similar Use (Nov)(6797).
IEC 320-81. Appliance Couplers for Household and Similar General Purposes Second Edition; (Amendment 3-1987, Incorporating Amendment 1 and 2). 181 pp.
IEC 320 Pt 2-2-90. Appliance Couplers for Household and Similar General Purposes Part 2: Interconnection Couplers for Household and Similar Equipment First Edition; (Corrigendum—Aug 1990). 69 pp.
JIS C 8358-88. Appliance Couplers for Domestic and Similar Use. 19 pp.

—Diaphragms (Mechanics)
BSI BS EN 278-91. 1991 Rubber Materials for Diaphragms in Domestic Appliances Using Combustible Gases up to 200 mbar. 11 pp.
CEN PREN 278-88. Rubber Materials for Diaphragms in Domestic Appliances Using Combustible Gases. 10 pp.
CEN EN 278-91. Rubber Materials for Diaphragms in Domestic Appliances Using Combustible Gases up to 200 mbar. 12 pp.

Appliances (Cont.)

—Elastomer Sealants—Safety
DIN ENGL 3535 Pt 2-83. Sealants for Gas Supply; Homogeneous Elastomers for Sealants for Domestic Appliances; Safety Requirements, Testing (Apr). 5 pp.

—Electric Inlets
CSA CAN/CSA-C22. 2 NO 21-M90. Cord Sets and Power-Supply Cords; (Gen Instr 1 Thru 2). 74 pp.
CSA 895B Bull. Electrical Bulletin 895B May 12, 1976 to C22.2 NO 21. 2 pp.

—Electric Plugs—Heater Cords
CSA C22.2 NO 57-M1985. Appliance Plugs for Heater Cord Sets (R 1993); (Gen Instr 1 Thru 3). 26 pp.

—Electric Switches
BSI BS 3676-63. (WITHDRAWN) 1963 Amd 7 Switches for Domestic and Similar Purposes (For Fixed or Portable Mounting) (Superseded by BS 3676: Part 1: 1989). 49 pp.
BSI BS EN 61058-1-92. 1992 Switches for Appliances Part 1: General Requirements (E). 108 pp.
CENELEC EN 61058-1-92. Switches for Appliances Part 1: General Requirements. 7 pp.
CENELEC EN 61058-1-92. Switches for Appliances Part 1: General Requirements. 9 pp.
CENELEC EN 61058-1-92. CORRIGENDUM Switches for Appliances Part 1: General Requirements (IEC 1058-1:1990). 1 p.
DIN VDE 0630-86. Switches for Appliances for a Rated Voltage Not Exceeding 500 V and a Rated Current Not Exceeding 63 A (Apr). 23 pp.
DIN VDE 0630 A4 (D)-89. Switches for Appliances for a Rated Voltage Not Exceeding 500V and a Rated Current Not Exceeding 63A; Amendment 4: No-Load Switches (Apr). 11 pp.
DIN VDE 0630 Pt 10-88. Switches for Appliances for a Rated Voltage Not Exceeding 500 V and a Rated Current Not Exceeding 63 A (Sept). 17 pp.
IEC 1058 Pt 1-90. Switches for Appliances Part 1: General Requirements First Edition; (Amendment 1-1993) (CENELEC EN 61058-1: 1992). 211 pp.

—Electric Switches—Safety
DIN ENGL 3398 Pt 1-82. Pressure Cut-Off Switches for Gas Appliances; Safety Requirements, Testing (Nov). 4 pp.
DIN ENGL 3398 Pt 3-82. Pressure Cut-Off Switches for Gaseous Substances; Requirements and Testing (Nov). 8 pp.
DIN ENGL 3398 Pt 4-86. Pressure Cut-Off Switches for Liquid Fuels and Heat Transfer Oils (Oct). 8 pp.

—Electric Wire—Breakage
CNS C3170-83. Method of Test for Wires Breakage of Appliance (Jan)(9815).

—Electrical Codes
CSA 1020B Bull. Electrical Bulletin 1020B January 28, 1977 to C22.2 NO 68. 3 pp.
CSA 1164 Bull. Electrical Bulletin 1164 April 13, 1978 to C22.2 NO 68. 2 pp.

—Electrical Insulation
CNS C1035-67. Insulating Compound (for Common Electric Appliance) (Oct)(2801) (R 1973). 6 pp.
CNS C3172-83. Classification and Test Methods for Insulation of Household Appliances and Equipments (Jun)(10315).
JIS C 0702-74. General Rules for Insulation Construction of Class II Electrical Appliances (R 1980). 18 pp.
JIS C 0703-80. Classification and Test Methods for Insulation of Household Appliances and Equipments. 16 pp.

—Electromagnetic Interference
BSI BS 613-77. 1977 Amd 2 Components and Filter Units for Electromagnetic Interference Suppression. 24 pp.

—Electronic Switches
DIN VDE 0630 Pt 12-88. Switches for Appliances for a Rated Voltage Not Exceeding 500 V and a Rated Current Not Exceeding 63 A; Electronic Switches (Sept). 27 pp.

—Endurance Testing
JIS C 9101-87. General Rules of Endurance Test Methods of Household and Similar Electrical Appliances. 23 pp.

—Energy Consumption
CEN PREN 30-2-91. Gas Burning Domestic Cooking Appliances—Part 2: Rational Use of Energy. 9 pp.

—Flame Detectors—Thermoelectric
BSI BS EN 125-91. 1991 Flame Supervision Devices for Gas-Burning Appliances—Thermo-Electric Types. 27 pp.

INTERNATIONAL AND NON-U.S. NATIONAL STANDARDS
SUBJECT INDEX
Appliances

Appliances *(Cont.)*

—**Flame Detectors—Thermoelectric** *(Cont.)*
CEN PREN 125-90. Flame Supervision Devices for Gas Burning Appliances—Thermo-Electric Flame Supervision Devices. 38 pp.
CEN EN 125-91. Flame Supervision Devices for Gas Burning Appliances—Thermo-Electric Flame Supervision Devices. 38 pp.
DIN ENGL EN 125-91. Flame Supervision Devices for Gas Burning Appliances; Thermo-Electric Flame Supervision Devices (Sept). 38 pp.

—**Flues**
BSI BS 1181-89. 1989 Clay Flue Linings and Flue Terminals. 16 pp.
BSI BS 1289: Part 2-89. 1989 Flue Blocks and Masonry Terminals for Gas Appliances Part 2: Clay Flue Blocks and Terminals. 20 pp.
BSI BS 5440: Part 1-90. 1990 Code of Practice for Flues and Air Supply for Gas Appliances of Rated Input Not Exceeding 60 kW (1st and 2nd Family Gases) Part 1: Installation of Flues. 44 pp.
BSI BS 5440: Part 1-78. (WITHDRAWN) 1978 Amd 1 Code of Practice for Flues and Air Supply for Gas Appliances of Rated Input Not Exceeding 60 kW (1st and 2nd Family Gases) Part 1: Flues. 42 pp.
BSI BS 5440: Part 2-89. 1989 Code of Practice for Flues and Air Supply for Gas Appliances of Rated Input Not Exceeding 60 kW (1st and 2nd Family Gases) Part 2: Installation of Ventilation for Gas Appliances. 14 pp.
BSI BS 5440: Part 2-76. (WITHDRAWN) 1976 Code of Practice for Flues and Air Supply for Gas Appliances of Rated Input Not Exceeding 60 kW (1st and 2nd Family Gases) Part 2: Air Supply. 11 pp.

—**Fuel Pumps**
JIS S 2037-92. Filler Pumps for Oil Burning Appliances. 26 pp.

—**Garden Ponds—Safety**
BSI BS EN 60335-2-55-93. 1993 Safety of Household and Similar Electrical Appliances Part 2: Particular Requirements Section 2.55: Electrical Appliances for Use with Aquariums and Garden Ponds (L). 22 pp.
CENELEC EN 60335-2-55-92. Safety of Household and Similar Electrical Appliances Part 2: Particular Requirements for Electrical Appliances for Use with Aquariums and Garden Ponds. 16 pp.
CENELEC EN 60335-2-55-93. Safety of Household and Similar Electrical Appliances Part 2: Particular Requirements for Electric Appliances for Use with Aquariums and Garden Ponds (IEC 335-2-55: 1989, Modified). 14 pp.
IEC 335 Pt 2-55-89. Safety of Household and Similar Electrical Appliances Part 2: Particular Requirements for Electrical Appliances for Use with Aquariums and Garden Ponds First Edition (CENELEC EN 60335-2-55: 1993). 31 pp.

—**Glass Components**
SNZ NZS 7841-72. Specification for Glass Components for Domestic Appliances Amend: A, 1972; 1 & 1A, 1978. 12 pp.

—**Glossaries**
JIS S 2091-85. Glossary of Gas and Kerosine Appliances, and Accessories for Domestic Use.

—**Hard Water Testing**
BSI BS EN 60734-93. 1993 Hard Water to be Used for Testing the Performance of Some Household Electrical Appliances (IEC 734: 1993) (L). 15 pp.
CENELEC EN 60734-93. Hard Water to Be Used for Testing the Performance of Some Household Electrical Appliances (IEC 734: 1993). 10 pp.
IEC 734-93. Hard Water to be Used for Testing the Performance of Some Household Electrical Appliances Second Edition (CENELEC EN 60734: 1993). 23 pp.

—**Harmonics**
BSI BS 5406: Part 2-88. 1988 Disturbances in Supply Systems Caused by Household Appliances and Similar Electrical Equipment Part 2: Specification of Harmonics. 19 pp.
BSI BS 5406: Part 2-01. 1988 Amd 1 Disturbances in Supply Systems Caused by Household Appliances and Similar Electrical Equipment Part 2: Specification of Harmonics (AMD 7420) October 15, 1992 (F). 21 pp.
CENELEC EN 60 555 (Part 2)-87. Disturbances in Supply Systems Caused by Household Appliance and Similar Electrical Equipment—Part 2: Harmonics. 16 pp.
IEC 555 Pt 2-82. Disturbances in Supply Systems Caused by Household Appliances and Similar Electrical Equipment Part 2: Harmonics First Edition; (Second Impression, Incorporating Amendments 1 Thru 3). 44 pp.

Appliances *(Cont.)*

—**Harmonics** *(Cont.)*
SAA AS 2279.1-91. Disturbances in Mains Supply Networks—Part 1: Limitation of Harmonics Caused by Household and Similar Electrical Appliances. 8 pp.

—**Hookup Wire**
CSA CAN/CSA-C22. 2NO210.2-M90. Appliance Wiring Material Products (R 1992); (Gen Instr 1 Thru 4). 73 pp.

—**Hose Fittings**
BSI BS 669: Part 1-89. 1989 Flexible Hoses, End Fittings and Sockets for Gas Burning Appliances Part 1: Strip-Wound Metallic Flexible Hoses, Covers, End Fittings and Sockets for Domestic Appliances Burning 1st and 2nd Family Gases (L). 24 pp.

—**Hoses**
BSI BS 669: Part 1-89. 1989 Flexible Hoses, End Fittings and Sockets for Gas Burning Appliances Part 1: Strip-Wound Metallic Flexible Hoses, Covers, End Fittings and Sockets for Domestic Appliances Burning 1st and 2nd Family Gases (L). 24 pp.
JIS K 6348-80. Gas Tubing (Extruded Rubber Tubing). 19 pp.
JIS S 3022-91. Oil Discharge Rubber Hoses for Oil Burning Appliances. 14 pp.

—**Hoses—Rubber**
JIS K 6351-82. Wire Reinforced Rubber Hoses for Gaseous Fuels. 20 pp.

—**Household**
CSA CAN/CSA-C22. 2 NO 68-92. Motor-Operated Appliances (Household and Commercial); (Gen Instr 1 Thru 2). 115 pp.

—**Identification Systems**
EC 90/396/EEC-90. Council Directive on the Approximation of the Laws of the Member States Relating to Appliances Burning Gaseous Fuels (NEW APP). 16 pp.

—**Ignition Transformers**
JIS S 2142-85. Ignition Transformers for Burning Appliances.

—**Inspection**
EC 84/530/EEC-84. Council Directive on the Approximation of the Laws of the Member States Relating to Common Provisions for Appliances Using Gaseous Fuels, Safety and Control Devices for These Appliances and Methods of Surveillance of Them. 11 pp.

—**Installation**
BSI BS 5482: Part 1-79. 1979 Amd 1 Code of Practice for Domestic Butane-and Propane-Gas-Burning Installations Part 1: Installations in Permanent Dwellings (AMD 5474) March 31, 1989. 23 pp.
BSI BS 5871: Part 3-91. 1991 Installation of Gas Fires, Convector Heaters, Fire/Back Boilers and Decorative Fuel Effect Gas Appliances Part 3: Decorative Fuel Effect Gas Appliances of Heat Input Not Exceeding 15 kW (2nd and 3rd Family Gases) (L). 25 pp.
BSI BS 5871: Part 3-01. 1991 Amd 1 Installation of Gas Fires, Convector Heaters, Fire/Back Boilers and Decorative Fuel Effect Gas Appliances Part 3: Decorative Fuel Effect Gas Appliances of Heat Input Not Exceeding 15 kW (2nd and 3rd Family Gases). 27 pp.
SNZ NZS 5261-90. Code of Practice for the Installation of Gas Burning Appliances and Equipment Amend: 1, 1993. 164 pp.

—**Insulated Wire**
CNS C2010-80. 600 V Grade Rubber Insulated Wire (Tentative) (Jul)(675). 4 pp.
CNS C2048-80. Varnish Cloth Insulated Wire (Jul)(2698). 15 pp.

—**Kerosene Feeders**
JIS S 3026-82. Kerosene Feeders for Kerosene Combustion Appliances.

—**Liquefied Petroleum Gas Cylinders**
BSI BS EN 417-92. 1992 Non-Refillable Metallic Gas Cartridges for Liquefied Petroleum Gases, with or Without a Valve, for Use with Portable Appliances Construction, Inspection, Testing and Marking (E). 19 pp.
CEN PREN 417-90. Non-Refillable Metallic Gas Cartridges for Liquefied Petroleum Gases, with or Without a Valve, for Use with Portable Appliances—Construction, Inspection, Testing and Marking. 27 pp.

Appliances *(Cont.)*

—**Liquefied Petroleum Gas Cylinders** *(Cont.)*
CEN EN 417-92. Non-Refillable Metallic Gas Cartridges for Liquefied Petroleum Gases, with or Without a Valve, for Use with Portable Appliances—Construction, Inspection, Testing and Marking. 27 pp.
CEN PREN 521-91. Dedicated Liquefied Petroleum Gas Appliances—Portable Appliances Operating at Vapour Pressure from Liquefied Petroleum Gas Containers. 54 pp.

—**Lubricants**
BSI BS EN 377-93. 1993 Lubricants for Applications in Appliances and Associated Controls Using Combustible Gases Except Those Designed for Use in Industrial Processes (W). 15 pp.
CEN PREN 377-90. Lubricants for Applications in Domestic Appliances Using Combustible Gases. 20 pp.
CEN EN 377-93. Lubricants for Applications in Appliances and Associated Controls Using Combustible Gases Except Those Designed for Use in Industrial Processes. 10 pp.
DIN ENGL 3536-82. Lubricants for Gas Valves and Controls; Requirements, Testing (Nov). 5 pp.

—**Magnetic Valves**
DIN VDE 0730 Pt 2 ZB-81. Electric Motor-Operated Appliances for Domestic and Similar Purposes; Particular Requirements for Magnetic Devices; (VDE Specification) (June). 10 pp.

—**Membrane Switches**
DIN VDE 0630 Pt 13 (D). Switches for Appliances for a Rated Voltage Not Exceeding 500V and a Rated Current Not Exceeding 63A; Membrane Switches, Membrane-Switch Keyboard. 18 pp.

—**Mobile Homes—Installation**
BSI BS 5482: Part 2-77. 1977 Amd 4 Installations in Caravans and Non-Permanent Dwellings. 22 pp.

—**Name Plates**
JIS Z 8304-84. Design Standard for Nameplates. 11 pp.

—**Noise**
BSI BS 6686: Part 1-86. 1986 Methods for Determination of Airborne Acoustical Noise Emitted by Household and Similar Electrical Appliances Part 1: General Requirements for Testing. 32 pp.
CENELEC HD 423.100-83. Test Code for the Determination of Airborne Acoustical Noise Emitted by Household and Similar Electrical Appliances Part 1: General Requirements. 3 pp.
CSA Z107.71-M1981. Measurement and Rating of the Noise Output of Consumer Appliances; (Gen Instr 1). 31 pp.
EC 86/594/EEC-86. Council Directive on Airborne Noise Emitted by Household Appliances. 4 pp.
IEC 704 Pt 1-82. Test Code for the Determination of Airborne Acoustical Noise Emitted by Household and Similar Electrical Appliances Part 1: General Requirements First Edition. 58 pp.
IEC 704 Pt 3-92. Test Code for the Determination of Airborne Acoustical Noise Emitted by Household and Similar Electrical Appliances Part 3: Procedure for Determining and Verifying Declared Noise Emission Values First Edition. 37 pp.

—**Oil Tanks**
JIS S 3020-76. Oil Tanks for Oil Burning Appliances.

—**Panels—Glass**
BSI BS 3193-93. 1993 Thermally Toughened Glass Panels for Use in Domestic Appliances. 8 pp.
BSI BS 3193-89. 1989 Thermally Toughened Glass Panels for Use in Domestic Appliances. 7 pp.

—**Power Supplies—Safety**
DIN VDE 0700 Pt 207-82. Safety of Household and Similar Electrical Appliances; Particular Requirements for Power Supply Appliances up to 20 VA (Oct). 21 pp.
DIN VDE 0700 Pt 207 A1-85. Safety of Household and Similar Electrical Appliances; Power Supply Appliances up to 20 VA: Amendment 1 (May). 5 pp.

—**Power Supply Cords**
CNS C4412-15-85. Detachable Power-Supply Cords Having Appliance Plugs (Feb)(10917-15).
CNS C4412-17-85. Detachable Power-Supply Cords for Appliances Rated Not Greater Than 50 Watts (Feb)(10917-17).

—**Power System Disturbances**
CENELEC EN 50 006-75. The Limitation of Disturbances in Electricity Supply Networks Caused by Domestic and Similar Appliances Equipped with Electronic Devices. 29 pp.

INDUSTRY STANDARDS

Appliances *(Cont.)*

—Power System Disturbances *(Cont.)*

CENELEC EN 60 555 (Part 1)-87. Disturbances in Supply Systems Caused by Household Appliances and Similar Electrical Equipment—Part 1: Definitions. 15 pp.

IEC 555 Pt 1-82. Disturbances in Supply Systems Caused by Household Appliances and Similar Electrical Equipment Part 1: Definitions First Edition. 25 pp.

IEC 725-81. Considerations on Reference Impedances for Use in Determining the Disturbance Characteristics of Household Appliances and Similar Electrical Equipment First Edition. 13 pp.

SAA AS 2279.1-91. Disturbances in Mains Supply Networks—Part 1: Limitation of Harmonics Caused by Household and Similar Electrical Appliances. 8 pp.

SAA AS 2279.3-91. Disturbances in Mains Supply Networks—Part 3: Limitation of Voltage Fluctuations Caused by Household and Similar Electrical Appliances. 15 pp.

—Power System Disturbances—Glossaries

BSI BS 5406: Part 1-88. 1988 Disturbances in Supply Systems Caused by Household Appliances and Similar Electrical Equipment Part 1: Glossary of Terms. 17 pp.

—Pressure Regulators

BSI BS 6448: Part 1-87. (WITHDRAWN) 1987 Gas Governors Part 1: Pressure Governors with Nominal Connection Size up to 50mm for Gas Appliances with Inlet Pressures up to and Including 200mbar (Superseded by BS EN 88: 1991). 21 pp.

BSI BS EN 88-91. 1991 Pressure Governors for Gas Appliances for Inlet Pressures up to 200 mbar. 31 pp.

CEN PREN 88-90. Pressure Governors for Gas Appliances for Inlet Pressures up to 200 mbar. 44 pp.

CEN EN 88-91. Pressure Governors for Gas Appliances for Inlet Pressures up to 200 mbar. 43 pp.

—Pressure Transducers

IEC 730 Pt 2-6-91. Automatic Electrical Controls for Household and Similar Use Part 2: Particular Requirements for Automatic Electrical Pressure Sensing Controls Including Mechanical Requirements First Edition. 50 pp.

—Quick Disconnect Couplings

JIS S 2135-91. Quick Coupling Unit for Gas Appliances. 30 pp.

—Radio Frequency Interference

BSI BS 800-88. 1988 Amd 2 Specification for Limits and Methods of Measurement of Radio Interference Characteristics of Household Electrical Appliances, Portable Tools and Similar Electrical Apparatus (AMD 6578) June 28, 1991. 85 pp.

BSI BS 4809-72. (WITHDRAWN) 1972 Radio Interference Limits and Measurements for Radio Frequency Heating Equipment (Superseded by BS EN 55011: 1991). 17 pp.

BSI BS EN 55014-93. 1993 Limits and Methods of Measurement of Radio Disturbance Characteristics of Electrical Motor-Operated and Thermal Appliances for Household and Similar Purposes, Electric Tools and Similar Electric Apparatus (S). 73 pp.

CENELEC EN 55 014-87. Limits and Methods of Measurement of Radio Interference Characteristics of Household Electrical Appliances, Portable Tools and Similar Electrical Apparatus. 58 pp.

CENELEC EN 55 014/A1-88. AMD 1 Limits and Methods of Measurement of Radio Interference Characteristics of Household Electrical Appliances, Portable Tools and Similar Electrical Apparatus. 13 pp.

CENELEC EN 55014-93. Limits and Methods of Measurement of Radio Disturbance Characteristics of Electrical Motor-Operated and Thermal Appliances for Household and Similar Purposes, Electric Tools and Similar Electric Apparatus (CISPR 14: 1993). 67 pp.

DIN VDE 0875 Pt 208-89. Radio Interference Suppression of Electrical Appliances and Systems; Limits and Methods of Measurement of Radio Interference Characteristics of Household Electrical Appliances, Portable Tools and Similar Electrical Apparatus; Identical with. 13 pp.

EC 76/889/EEC-76. Council Directive on the Approximation of the Laws of the Member States Relating to Radio Interference Caused by Electrical Household Appliances, Portable Tools and Similar Equipment. 21 pp.

EC 82/499/EEC-82. Commission Directive Adapting to Technical Progress Council Directive 76/889/EEC on the Approximation of the Laws of the Member States Relating to Radio Interference Caused by Electrical Household Appliances, Portable Tools and Similar Equipment. 41 pp.

Appliances *(Cont.)*

—Radio Frequency Interference *(Cont.)*

EC 83/447/EEC-83. Commission Directive Adopting the Measures Provided for in Article 3 (3) of Directive 76/889/EEC on the Approximation of the Laws of the Member States Relating to Radio Interference Caused by Electrical Household Appliances, Portable Tools and Similar. 1 p.

EC 87/308/EEC-87. Commission Directive Adapting to Technical Progress Council Directive 76/889/EEC on the Approximation of the Laws of the Member States Relating to Radio Interference Caused by Electrical Household Appliances, Portable Tools and Similar Equipement. 2 pp.

IEC CISPR 14-93. Limits and Methods of Measurement of Radio Disturbance Characteristics of Electrical Motor-Operated and Thermal Appliances for Household and Similar Purposes, Electric Tools and Electric Apparatus Third Edition; (CEN EN 55014: 1993). 128 pp.

SAA AS/NZS 1044-92. Limits and Methods of Measurement of Radio Interference Characteristics of Household Electrical Appliances, Portable Tools and Similar Electrical Apparatus (IEC/CISPR 14:1985). 56 pp.

SNZ NZS/AS 1044-92. Limits and Methods of Measurement of Radio Interference Characteristics of Household Electrical Appliances, Portable Tools and Similar Electrical Apparatus (This is a Joint Standard with SAA AS 1044). 56 pp.

—Radio Frequency Interference—Suppression

DIN VDE 0875 Pt 1-88. Radio Interference Suppression of Elec. App-liances and Sys.; Limits and Meth. of Measurement of Radio Interference Char. of Household Elec. Appliances, Portable Tools and Similar Elec. Apparatus; (CISPR 14 (1985, 2nd Edn. Mod.)) German Version of EN 55 014: 1987 (Dec). 9 pp.

DIN VDE 0875 Pt 3-88. Radio Interference Suppression of Electrical Appliances and Systems; Radio Interference Suppression of Electrical Systems and Special Electrical Appliances (Dec) (Replaces in Parts DIN 57875/VDE 0875/06.77). 31 pp.

—Recreational Vehicles—Installation

BSI BS 5482: Part 2-77. 1977 Amd 4 Installations in Caravans and Non-Permanent Dwellings. 22 pp.

—Recreational Vehicles—Safety

ISO 7421-91. Leisure Accommodation Vehicles—Liquefied Petroleum Gas Systems First Edition. 7 pp.

—Repair/Maintenance

DIN VDE 0701 Pt 1-86. Repair, Modification and Testing of Electrical Appliances; General Requirements (Oct). 18 pp.

DIN VDE 0701 Pt 1 (D)-89. Repair, Modification and Testing of Electrical Appliances; General Requirements (Dec) (Intended to Replace DIN VDE 0701 Pt 1/10.86). 17 pp.

DIN VDE 0701 Pt 200-88. Repair, Modification and Testing of Electrical Appliances; Mains-Operated Electronic Appliances and Their Associated Equipment for Household and Similar General Use (June). 7 pp.

—Rubber Cork Seals—Safety

DIN ENGL 3535 Pt 5-83. Sealants for Gas Supply; Rubber-Cork and Rubber-Cork-Asbestos Sealants for Gas Valves and Gas Appliances, Safety Requirements, Testing (Apr). 6 pp.

—Safety

BSI BS 3456: Part 101-87. 1987 Amd 1 Safety of Household Electrical Appliances Part 101: General Requirements (AMD 6055) August 31, 1990. 85 pp.

BSI BS 3456: Part 201-90. 1990 Amd 3 Safety of Household and Similar Electrical Appliances Part 201: General Requirements (AMD 6494) November 30, 1990 (L). 196 pp.

BSI BS 3456: Part 201-06. 1990 Amd 6 Safety of Household and Similar Electrical Appliances Part 201: General Requirements (AMD 7347) November 15, 1992 (L). 216 pp.

BSI BS 3456: Part 201-07. 1990 Amd 7 Safety of Household and Similar Electrical Appliances Part 201: General Requirements (AMD 7564) March 15, 1993 (L). 225 pp.

BSI BS 3456: Part 201-08. 1990 Amd 8 Safety of Household and Similar Electrical Appliances Part 201: General Requirements (AMD 7887) August 15, 1993 (L). 230 pp.

BSI BS 3456: Part 1-69. 1969 Amd 5 Safety of Household Elecrical Appliances Part 1: General Requirements. 110 pp.

BSI BS 5258: Part 12-01. 1990 Amd 1 Safety of Domestic Gas Appliances Part 12: Specification for Decorative Fuel Effect Gas Appliances (2nd and 3rd Family Gases) (AMD 7507) January 15, 1993. 34 pp.

Appliances *(Cont.)*

—Safety *(Cont.)*

BSI BS 5415: Part 1-85. 1985 Amd 1 Safety of Electrical Motor-Operated Industrial Cleaning Appliances Part 1: General Requirements (AMD 6452) November 30, 1990. 73 pp.

BSI BS 7462-91. 1991 Electrical Safety of Domestic Gas Appliances. 52 pp.

BSI BS EN 60335-2-54-92. 1992 Safety of Household and Similar Electrical Appliances Part 2: Particular Requirements Section 2.54: General Purpose Cleaning Appliances. 18 pp.

BSI PP 888-82. (WITHDRAWN) 1982 Safety and Performance of Domestic Electrical Appliances. 33 pp.

CENELEC HD 250-74. General Specification for Electric Motor Operated Appliances for Household and Similar Purposes. 12 pp.

CENELEC HD 250.2-77. General Specification for Electric Motor Operated Appliances for Household and Similar Purposes. 2 pp.

CENELEC HD 251 S3-82. Safety of Household and Similar Electrical Appliances: General Requirements. 2 pp.

CENELEC HD 251 S3/A2-87. AMD 2 Safety of Household and Similar Electrical Appliances: General Requirements. 8 pp.

CENELEC HD 251 S3/A3-87. AMD 3 Safety of Household and Similar Electrical Appliances: General Requirements. 11 pp.

CENELEC HD 251 S3/A2 + Annexe-91. AMD 2 + Annexe Safety of Household and Similar Electrical Appliances: General Requirements. 15 pp.

CENELEC HD 289 S1-90. Safety of Household and Similar Appliances Particular Rules for Routine Tests Referring to Appliances Under the Scope of EN 60 335-1. 7 pp.

CENELEC HD 289 S1/A1-92. AMD 1 Safety of Household and Similar Appliances Particular Rules for Routine Tests Referring to Appliances Under the Scope of EN 60 335-1. 6 pp.

CENELEC EN 60 335-1-88. Safety of Household and Similar Electrical Appliances Part 1: General Requirements. 32 pp.

CENELEC EN 60 335-1/A2-88. AMD 2 Safety of Household and Similar Electrical Appliances Part 1: General Requirements. 8 pp.

CENELEC EN 60 335-1/A5-89. AMD 5 Safety of Household and Similar Electrical Appliances Part 1: General Requirements. 6 pp.

CENELEC EN 60 335-1/A6-89. AMD 6 Safety of Household and Similar Electrical Appliances Part 1: General Requirements. 7 pp.

CENELEC EN 60335-1/A51-91. AMD 51 Safety of Household and Similar Electrical Appliances Part 1: General Requirements. 7 pp.

CENELEC EN 60335-1/A52-92. AMD 52 Safety of Household and Similar Electrical Appliances Part 1: General Requirements. 7 pp.

CENELEC EN 60335-1/A53-92. AMD 53 Safety of Household and Similar Electrical Appliances Part 1: General Requirements. 6 pp.

CENELEC EN 60335-1/A54-92. AMD 54 Safety of Household and Similar Electrical Appliances Part 1: General Requirements. 5 pp.

CNS C4126-90. Safety Testing Standards for Household Electrical Appliances (Oct)(3766).

CNS C4126-81. Safety Testing Standards for Household Electrical Appliances (Jun)(3766). 3 pp.

CSA C22.2 NO 1335.1-93. Portable Electrical Motor-Operated and Heating Appliances: General Requirements; (Gen Instr 1). 208 pp.

DIN VDE 0700 Pt 1 A2-86. Safety of Household and Similar Electrical Appliances; General Requirements; Amendment 2 (Sept). 8 pp.

DIN VDE 0700 Pt 1 A3 (D)-81. Safety of Household and Similar Electrical Appliances; General Requirements (VDE Specification): Amendment 3 (July). 4 pp.

DIN VDE 0700 Pt 1 A7 (D)-82. Safety of Household and Similar Electrical Appliances; Part 1: General Requirements—Calibration of Spring-Operated Impact Test Apparatus: Amendment 7 (VDE Specification) (Feb). 10 pp.

DIN VDE 0700 Pt 1 A8 (D)-82. Safety of Household and Similar Electrical Appliances; General Requirements: Amendment 8 (VDE Specification) (Mar). 6 pp.

DIN VDE 0700 Pt 1 A11 (D)-83. (Amendment to IEC 335-1) Safety of Household and Similar Electrical Appliances; Part 1: General Requirements: Amendment 11 (VDE Specification) (DIN IEC 61(CO)330) (May). 9 pp.

DIN VDE 0700 Pt 1 A13 (D)-84. (Amendment to IEC 335-1) Safety of Household and Similar Electrical Appliances; Part 1: General Requirements: Amendment 13 (VDE Specification) (DIN IEC 61(CO)342) (Mar). 7 pp.

DIN VDE 0700 Pt 1 A14 (D)-84. (Amendment to IEC 335-1) Safety of Household and Similar Electrical Appliances; Part 1: General Requirements: Amendment 14 (VDE Specification) (DIN IEC 61(CO)346) (Mar). 5 pp.

INTERNATIONAL AND NON-U.S. NATIONAL STANDARDS
SUBJECT INDEX
Application

Appliances (Cont.)
—Safety (Cont.)
- DIN VDE 0700 Pt 1 A15 (D)-84. Safety of Household and Similar Electrical Appliances; General Requirements: Amendment 15 (VDE Specification) (Mar). 4 pp.
- DIN VDE 0700 Pt 1 A17 (D)-84. Safety of Household and Similar Electrical Appliances; General Requirements: Amendment 17 (VDE Specification) (Sept). 6 pp.
- DIN VDE 0700 Pt 1 A18 (D)-84. Safety of Household and Similar Electrical Appliances; General Requirements: Amendment 18 (VDE Specification) (Dec). 12 pp.
- DIN VDE 0700 Pt 1 A19 (D)-85. Safety of Household and Similar Electrical Appliances; General Requirements: Amendment 19 (June). 8 pp.
- DIN VDE 0700 Pt 1 A21 (D)-86. Safety of Household and Similar Electrical Appliances; General Requirements: Amendment 21; Identical to IEC 61(CO)458 (Sept). 7 pp.
- DIN VDE 0700 Pt 1 A24 (D)-86. Safety of Household and Similar Electrical Appliances; General Requirements: Amendment 24; Identical to IEC 61(CO)468 (Sept). 7 pp.
- DIN VDE 0700 Pt 1 A25 (D)-87. Safety of Household and Similar Electrical Appliances; General Requirements: Amendment 25 (Mar). 6 pp.
- DIN VDE 0700 Pt 1 A26 (D)-87. Safety of Household and Similar Electrical Appliances; General Requirements: Amendment 26; Identical to IEC 61(CO)484 (Sept) (Supersedes Draft 0700 Part 1 A20/5.86). 18 pp.
- DIN VDE 0700 Pt 1 A27 (D)-87. Safety of Household and Similar Electrical Appliances; General Requirements: Amendment 27 (Sept). 6 pp.
- DIN VDE 0700 Pt 1 A28 (D)-87. Safety of Household and Similar Electrical Appliances; General Requirements: Amendment 28 (Oct). 5 pp.
- DIN VDE 0700 Pt 1 A29 (D)-88. Safety of Household and Similar Electrical Appliances; General Requirements: Amendment 29 (Oct). 4 pp.
- DIN VDE 0700 Pt 1 A30 (D)-89. Safety of Household and Similar Electrical Appliances; General Requirements: Amendment 30 (Mar). 5 pp.
- EC 84/530/EEC-84. Council Directive on the Approximation of the Laws of the Member States Relating to Common Provisions for Appliances Using Gaseous Fuels, Safety and Control Devices for These Appliances and Methods of Surveillance of Them. 11 pp.
- EC 90/396/EEC-90. Council Directive on the Approximation of the Laws of the Member States Relating to Appliances Burning Gaseous Fuels (NEW APP). 16 pp.
- IEC 335 Pt 1-91. Safety of Household and Similar Electrical Appliances Part 1: General Requirements Third Edition; (I-SH 01—Feb 1993). 246 pp.
- IEC 335 Pt 2-18-84. Safety of Household and Similar Electrical Appliances Part 2: Guide for Preparing Safety Requirements for Battery-Powered Motor-Operated Appliances and Their Charging and Battery Assemblies Second Edition. 39 pp.
- SNZ NZS 6300-92. Approval and Test Specification—General Requirements for Household and Similar Electrical Appliances Amend: 1, 1992 (NZS 6300:1992/AS 3300-1992). 89 pp.

—Seals—Rubber Joints
- BSI BS EN 279-91. 1991 Homogenous Rubber Materials for Dynamic Seals in Domestic Appliances Using Combustible Gases up to 200 mbar. 11 pp.
- BSI BS EN 291-92. 1992 Rubber Seals—Static Seals in Domestic Appliances for Combustible Gas up to 200 mbar—Specification for Material. 11 pp.
- CEN PREN 279-88. Homogeneous Rubber for Dynamic Seals in Domestic Appliances Using Combustible Gases. 9 pp.
- CEN EN 279-91. Homogeneous Rubber Materials for Dynamic Seals in Domestic Appliances Using Combustible Gases up to 200 mbar. 11 pp.
- CEN PREN 291-89. Rubber Seals: Static Seals in Domestic Appliances for Combustible Gas up to 200 mbar: Specifications for Material. 7 pp.
- CEN EN 291-92. Rubber Seals—Static Seals in Domestic Appliances for Combustible Gas up to 200 mbar—Specifications for Material. 10 pp.
- CEN PREN 549-91. Rubber Materials for Seals and Diaphragms for Appliances and Associated Controls Using Combustible Gases Except Those Designed for Use in Industrial Processes. 17 pp.

—Selector Switches
- DIN VDE 0630 Pt 11-88. Switches for Appliances for a Rated Voltage Not Exceeding 500 V & a Rated Current Not Exceeding 63 A; Incorporated Programme Controls & Selector Switches. (Sept). 12 pp.

Appliances (Cont.)
—Ships—Installation
- BSI BS 5482: Part 3-79. 1979 Amd 3 Code of Practice for Domestic Butane-and Propane-Gas-Burning Installations Part 3: Installations in Boats, Yachts and Other Vessels (AMD 6816) June 28, 1991. 22 pp.

—Ships—Plastics
- CNS F5091-84. Criteria for Selection of Plastics Used for Marine Electrical Appliances (Apr)(10855).
- JIS F 0701-89. Criteria for Selection of Plastics Used for Marine Electrical Appliances.

—Ships—Stuffing Boxes
- JIS F 8801-86. Marine Cable Glands for Electric Watertight Box.

—Ships—Water Resistance Testing
- CNS F5014-82. Method of Watertight Testing for Marine Electric Appliances (Jan)(8266).
- JIS F 8001-76. Method of Watertight Testing for Marine Electric Appliances.

—Stop Valves
- BSI BS EN 161-91. 1991 Automatic Shut-Off Valves for Gas Burners and Gas Appliances. 27 pp.
- CEN EN 161-91. Automatic Shut-Off Valves for Gas Burners and Gas Appliances. 38 pp.

—Storage—Built In
- JIS A 0016-79. Modular Coordination-Coordinating Size of Opening for Built-in Appliances in Storage Furniture.

—Symbols
- SAA AS 1104-78. Informative Symbols for Use on Electrical and Electronic Equipment. 24 pp.

—Taps (Valves)
- CEN PREN 1106-93. Manually Operated Taps for Gas Burning Appliances. 32 pp.

—Temperature Controllers
- IEC 730 Pt 2-9-92. Automatic Electrical Controls for Household and Similar Use Part 2: Particular Requirements for Temperature Sensing Controls First Edition. 52 pp.

—Temperature Controllers—Safety
- IEC 730 Pt 2-9-92. Automatic Electrical Controls for Household and Similar Use Part 2: Particular Requirements for Temperature Sensing Controls First Edition. 52 pp.

—Test Gases
- BSI BS 4947-84. 1984 Test Gases for Gas Appliances. 11 pp.
- CEN PREN 437-90. Appliances Using Combustible Gases—Test Gases, Test Pressures and Catogories of Appliances. 35 pp.

—Testing Fingers
- CNS C4386-83. Testing Fingers for Electric Appliance (Jan)(9816).

—Thermal Cutouts
- DIN VDE 0631-83. Thermostats, Thermal Cut-Outs and Similar Devices (Dec). 37 pp.

—Thermostats
- BSI BS 4201-79. (WITHDRAWN) 1979 Amd 1 Thermostats for Gas-Burning Appliances (AMD 6268) November 30, 1989 (Superseded by BS EN 257: 1992). 24 pp.
- BSI BS EN 257-92. 1992 Mechanical Thermostats for Gas-Burning Appliances. 30 pp.
- CEN PREN 257-87. Mechanical Thermostats for Gas Burning Appliances. 37 pp.
- CEN EN 257-92. Mechanical Thermostats for Gas-Burning Appliances. 44 pp.
- DIN ENGL EN 257-92. Mechanical Thermostats for Gas Burning Appliances (Mar). 45 pp.
- DIN VDE 0631-83. Thermostats, Thermal Cut-Outs and Similar Devices (Dec). 37 pp.

—Valves
- BSI BS 1963-90. 1990 Pressure Operated Relay Valves for Domestic, Commercial and Catering Gas Appliances. 14 pp.
- BSI BS 1963-69. (WITHDRAWN) 1969 Amd 1 Pressure Operated Relay Valves for Domestic, Commercial and Catering Gas Appliances. 21 pp.
- BSI BS 7673-93. 1993 Domestic Gas Cooker Oven Control Valves with or Without Manual Override (L). 14 pp.
- JIS B 8414-90. Relief Valves for Hot Water Appliances. 14 pp.
- JIS S 2143-75. Solenoid Valves for Gas Appliances.

—Vitreous Enameled
- BSI BS 3831-64. 1964 Vitreous Enamel Finishes for Domestic and Catering Appliances. 8 pp.
- JIS R 4301-78. Quality Standard of Porcelain Enamelled Products. 18 pp.

Appliances (Cont.)
—Voltage Fluctuations
- BSI BS 5406: Part 3-88. 1988 Disturbances in Supply Systems Caused by Household Appliances and Similar Electrical Equipment Part 3: Voltage Fluctuations. 30 pp.
- BSI BS 5406: Part 3-01. 1988 Amd 1 Disturbances in Supply Systems Caused by Household Appliances and Similar Electrical Equipment Part 3: Specification of Voltage Fluctuations (IEC 555-3: 1987) (AMD 6960) September 15, 1992. 45 pp.
- CENELEC EN 60 555 (Part 3)-87. Disturbances in Supply Systems Caused by Household Appliances and Similar Electrical Equipment—Part 3: Voltage Fluctuations. 28 pp.
- CENELEC EN 60555-3/A1-91. AMD 1 Disturbances in Supply Systems Caused by Household Appliances and Similar Electrical Equipment—Part 3: Voltage Fluctuations. 6 pp.
- IEC 555 Pt 3-82. Disturbances in Supply Systems Caused by Household Appliances and Similar Electrical Equipment Part 3: Voltage Fluctuations First Edition; (Amendment 1-1990) (Corrigendum—June 1990). 58 pp.
- IEC 827-85. Guide to Voltage Fluctuation Limits for Household Appliances (Relating to IEC Publication 555-3) First Edition. 26 pp.
- SAA AS 2279.3-91. Disturbances in Mains Supply Networks—Part 3: Limitation of Voltage Fluctuations Caused by Household and Similar Electrical Appliances. 15 pp.

—Water Efficient—Quality Assurance
- SAA MP64-92. Manual of Assessment Procedure for Water Efficient Appliances. 7 pp.

—Welding—Electric
- DIN ENGL 44753-66. Electric Spot, Projection and Seam Welding Machines and Spot and Seam Welding Appliances; Definitions and Rating Details (Dec). 13 pp.

—Wicks
- JIS S 2038-84. Wicks for Oil Burning Appliances.

Application Layer (OSI)
See Also: Application Services (OSI); Message Transfer Layer; Open Systems Interconnection; Transaction Capabilities Application Part
- BSI BS ISO/IEC 9545-89. 1989 Information Technology—Open Systems: Interconnection—Application Layer Structure. 20 pp.
- IEC 9545-89. Information Technology—Open Systems Interconnection—Application Layer Structure First Edition. 18 pp.
- IEC 9545 Draft AMD 1. Information Technology—Open Systems Interconnection—Application Layer Structure Amendment 1: Extended Application Layer Structure; (1991) ***CD-ROM ONLY***. 47 pp.
- IEC 9545 Draft AMD 1.2. Information Technology—Open Systems Interconnection—Application Layer Structure Amendment 1: Extended Application Layer Structure; (1992) ***CD-ROM ONLY***. 34 pp.
- ISO 9545-89. Information Technology—Open Systems Interconnection—Application Layer Structure First Edition. 18 pp.
- ISO 9545 Draft AMD 1. Information Technology—Open Systems Interconnection—Application Layer Structure Amendment 1: Extended Application Layer Structure; (1991) ***CD-ROM ONLY***. 47 pp.
- ISO 9545 Draft AMD 1.2. Information Technology—Open Systems Interconnection—Application Layer Structure Amendment 1: Extended Application Layer Structure; (1992) ***CD-ROM ONLY***. 34 pp.
- JTC1 9545-89. Information Technology—Open Systems Interconnection—Application Layer Structure First Edition. 18 pp.
- JTC1 9545 Draft AMD 1. Information Technology—Open Systems Interconnection—Application Layer Structure Amendment 1: Extended Application Layer Structure; (1991) ***CD-ROM ONLY***. 47 pp.
- JTC1 9545 Draft AMD 1.2. Information Technology—Open Systems Interconnection—Application Layer Structure Amendment 1: Extended Application Layer Structure; (1992) ***CD-ROM ONLY***. 34 pp.
- JTC1 10040 Draft AMD 1. Information Technology—Open Systems Interconnection—Systems Management Overview Amendment 1: Management Knowledge Management Architecture; (1993) ***CD-ROM ONLY***. 3 pp.
- OSI ISO IEC 9545-89. Information Technology—Open Systems Interconnection—Application Layer Structure (ALS). 19 pp.
- OSI ISO DP 9545-87. Information Processing Systems-Open Systems Interconnection-Application Layer Structure (ALS). 19 pp.

Application
INTERNATIONAL AND NON-U.S. NATIONAL STANDARDS
SUBJECT INDEX

Application Layer (OSI) *(Cont.)*

SAA AS 4021-92. Information Technology—Open Systems Interconnection—Application Layer Structure (ISO/IEC 9545:1989) (in Professional Package 26A). 13 pp.

SNZ NZS/ISO-IEC 9545-89. Information Technology—Open Systems Interconnection—Application Layer Structure. 13 pp.

—Common Management Information Services and Protocol

BSI BS ISO/IEC 9595-91. 1991 Information Technology—Open Systems Interconnection—Common Management Information Service Definition. 34 pp.

BSI BS ISO/IEC 9595-01. 1991 Amd 1 Information Technology—Open Systems Interconnection—Common Management Information Service Definition (AMD 7306) September 15, 1992 (Technical Corr 1) (S). 37 pp.

BSI BS ISO/IEC 9595-04. 1991 Amd 4 Information Technology—Open Systems Interconnection—Common Management Information Service Definition (AMD 7476) November 15, 1992. 44 pp.

BSI BS ISO/IEC 9596-90. (WITHDRAWN) 1990 Information Technology—Open Systems Interconnection—Common Management Information Protocol Specification (S) (Superseded by BS ISO/IEC 9596-1: 1991). 34 pp.

BSI BS ISO/IEC 9596-1-91. 1991 Information Technology—Open Systems Interconnection—Common Management Information Protocol Part 1: Specification. 41 pp.

BSI BS ISO/IEC 9596-1-01. 1991 Amd 1 Information Technology—Open Systems Interconnection—Common Management Information Protocol Part 1: Specification (AMD 7305) September 15, 1992 (Technical Corr 1) (S). 44 pp.

BSI BS ISO/IEC 9596-1-03. 1991 Amd 3 Information Technology—Open Systems Interconnection—Common Management Information Protocol Part 1: Specification (AMD 7535) December 15, 1992 (Technical Corr 3). 49 pp.

BSI BS ISO/IEC 9596-1-04. 1991 Amd 4 Information Technology—Open Systems Interconnection—Common Management Information Protocol Part 1: Specification (AMD 7805) August 15, 1993 (Technical Corr 4) (S). 52 pp.

CCITT RECMN X.710-91. Common Management Information Service Definition for CCITT Applications (Study Group VII) 42 pp. 42 pp.

CCITT RECMN X.711-91. Common Management Information Protocol Specification for CCITT Applications (Study Group VII) 43 pp. 43 pp.

CSA CAN/CSA-Z243. 141-91. Information Technology—Open Systems Interconnection—Common Management Information Service Definition (ISO/IEC 9595:1991); (Gen Instr 1). 42 pp.

CSA CAN/CSA-Z243. 142-91. Information Technology—Open Systems Interconnection—Common Management Information Protocol—Part 1: Specification (ISO/IEC 9596-1:1991); (Gen Instr 1). 49 pp.

ECMA ECMA-TR 47-88. Configuration Management Service Definition. 48 pp.

IEC 9595-91. Information Technology—Open Systems Interconnection—Common Management Information Service Definition Amendment 4: Access Control Second Edition; (Corrigendum 1-2:1992) (Amendment 4-1992). 39 pp.

IEC 9596 Pt 1-91. Information Technology—Open Systems Interconnection—Common Management Information Protocol—Part 1: Specification Second Edition; (Corrigendum 1-3:1992) (Corrigendum 4-1993). 45 pp.

IEC DIS 10165 Pt 5-92. Information Technology—Open Systems Interconnection—Structure of Management Information: Generic Management Information ***CD-ROM ONLY***. 27 pp.

IEC DIS 10165 Pt 6-92. Information Technology—Open Systems Intercon-nection—Structure of Management Information—Requirements and Guidelines for Implementation Conformance Statement Proformas Associated with Management Information ***CD-ROM ONLY***. 41 pp.

ISO 9595-91. Information Technology—Open Systems Interconnection—Common Management Information Service Definition Amendment 4: Access Control Second Edition; (Corrigendum 1-2:1992) (Amendment 4-1992). 39 pp.

ISO 9596 Pt 1-91. Information Technology—Open Systems Interconnection—Common Management Information Protocol—Part 1: Specification Second Edition; (Corrigendum 1-3:1992) (Corrigendum 4-1993). 45 pp.

ISO DIS 10165 Pt 5-92. Information Technology—Open Systems Interconnection—Structure of Management Information: Generic Management Information ***CD-ROM ONLY***. 27 pp.

Application Layer (OSI) *(Cont.)*

—Common Management Information Services and Protocol *(Cont.)*

ISO DIS 10165 Pt 6-92. Information Technology—Open Systems Intercon-nection—Structure of Management Information—Requirements and Guidelines for Implementation Conformance Statement Proformas Associated with Management Information ***CD-ROM ONLY***. 41 pp.

JTC1 9595-91. Information Technology—Open Systems Interconnection—Common Management Information Service Definition Amendment 4: Access Control Second Edition; (Corrigendum 1-2:1992) (Amendment 4-1992). 4 pp.

JTC1 9596-91. Information Technology—Open Systems Interconnection—Common Management Information Protocol—Part 1: Specification Second Edition; (Corrigendum 1-3:1992) (Corrigendum 4-1993). 3 pp.

JTC1 DIS10165 Pt 5-92. Information Technology—Open Systems Interconnection—Structure of Management Information: Generic Management Information ***CD-ROM ONLY***. 27 pp.

JTC1 DIS10165 Pt 6-92. Information Technology—Open Systems Intercon-nection—Structure of Management Information—Requirements and Guidelines for Implementation Conformance Statement Proformas Associated with Management Information ***CD-ROM ONLY***. 41 pp.

OSI ISO IEC 9595 DAD 2-90. Information Technology-Open Systems Interconnection—Common Management Information Service Definition—Addendum 2. 4 pp.

OSI ISO IEC 9595 DAD 1-90. Information Technology—Open Systems Interconnection—Common Management Information Service Definition—Addendum 1. 5 pp.

OSI ISO/IEC 9595-90. Information Technology—Open Systems Interconnection—Common Management Information Service Definition. 21 pp.

OSI ISO DP 9595-1-87. Information Processing Systems—Open Systems Interconnection—Management Information Service Definition—Part 1: Overview. 16 pp.

OSI ISO IEC DIS 9595-2-89. Information Processing Systems—Open Systems Interconnection—Management Information Service Definition—Part 2: Common Management Information Service. 25 pp.

OSI ISO DP 9595-2-87. Information Processing Systems—Open Systems Interconnection—Management Information Service Definition—Part 2: Common Management Information Service Definition. 47 pp.

OSI ISO IEC 9596 DAD 2-90. Information Technology-Open Systems Interconnection—Common Management Information Protocol Specification-Addendum 2. 5 pp.

OSI ISO IEC 9596 DAD 1. Information Technology-Open Systems Interconnection—Common Management Informaiton Protocol Specification—Addendum 1. 5 pp.

OSI ISO/IEC 9596-90. Information Technology-Open Systems Interconnection—Common Management Information Protocol Specification. 31 pp.

OSI ISO DP 9596-1-86. Information Processing Systems—Open Systems Interconnection—Management Information Protocol Specification—Part 1: Overview. 5 pp.

OSI ISO DP 9596-2-86. Information Processing Systems—Open Systems Interconnection—Management Information Protocol Specification—Part 2: Common Management Information Protocol Specification. 33 pp.

SAA AS 3965-91. Information Technology—Open Systems Interconnection—Common Management Information Service Definition (ISO/IEC 9595:1990) (in Professional Packages 26A, 26C). 27 pp.

SAA AS 3964.1-92. Information Technology—Open Systems Interconnection—Common Management Information Protocol—Part 1: Specification (ISO/IEC 9596-1:1991) (in Professional Package 26A) (Supersedes AS 3964—1991). 33 pp.

SNZ NZS/ISO-IEC 9595-90. Information Technology—Open Systems Interconnection—Common Management Information Service Definition. 27 pp.

SNZ NZS/ISO-IEC 9596.1-91. Information Technology—Open Systems Interconnection—Common Management Information Protocol Part 1: Specification. 33 pp.

—Common Management Information Services and Protocol—Conformance

BSI BS ISO/IEC 9596-2-93. 1993 Information Technology—Open Systems Interconnection—Common Management Information Protocol: Protocol Implementation Conformance Statement (PICS) Proforma (S). 68 pp.

IEC 9596 Pt 2-93. Information Technology—Open Systems Interconnection—Common Management Information Protocol: Protocol Implementation Conformance Statement (PICS) Proforma First Edition; (CCITT RECMN X.712) (Corrigendum 1-1993). 67 pp.

Application Layer (OSI) *(Cont.)*

—Common Management Information Services and Protocol—Conformance *(Cont.)*

ISO 9596 Pt 2-93. Information Technology—Open Systems Interconnection—Common Management Information Protocol: Protocol Implementation Conformance Statement (PICS) Proforma First Edition; (CCITT RECMN X.712) (Corrigendum 1-1993). 67 pp.

JTC1 9596-93. Information Technology—Open Systems Interconnection—Common Management Information Protocol: Protocol Implementation Conformance Statement (PICS) Proforma First Edition; (CCITT RECMN X.712) (Corrigendum 1-1993). 67 pp.

—Common Management Information Services and Protocol—PICS

CCITT RECMN X.712-92. Information Technology—Open Systems Interconnection—Common Management Information Protocol: Protocol Implementation Conformance Statement Proforma 64 pp (ISO/CCIT Common Text). 66 pp.

—Distributed Office Application—Document Printing Application

IEC DIS 10175 Pt 2-91. Information Technology—Text and Office Systems—Document Printing Application—Part 2: Protocol Specification ***CD-ROM ONLY***. 22 pp.

ISO DIS 10175 Pt 2-91. Information Technology—Text and Office Systems—Document Printing Application—Part 2: Protocol Specification ***CD-ROM ONLY***. 22 pp.

—Distributed Office Application—Referenced Data Transfer

BSI BS ISO/IEC 10740-1-93. 1993 Information Technology—Text and Office Systems—Referenced Data Transfer—Part 1: Abstract Service Definition (S). 19 pp.

BSI BS ISO/IEC 10740-2-93. 1993 Information Technology—Text and Office Systems—Referenced Data Transfer—Part 2: Protocol Specification (S). 19 pp.

IEC 10740 Pt 1-93. Information Technology—Text and Office Systems—Referenced Data Transfer—Part 1: Abstract Service Definition First Edition. 16 pp.

IEC 10740 Pt 2-93. Information Technology—Text and Office Systems—Referenced Data Transfer—Part 2: Protocol Specification First Edition. 16 pp.

ISO 10740 Pt 1-93. Information Technology—Text and Office Systems—Referenced Data Transfer—Part 1: Abstract Service Definition First Edition. 16 pp.

ISO 10740 Pt 2-93. Information Technology—Text and Office Systems—Referenced Data Transfer—Part 2: Protocol Specification First Edition. 16 pp.

JTC1 10740 Pt 1-93. Information Technology—Text and Office Systems—Referenced Data Transfer—Part 1: Abstract Service Definition First Edition. 16 pp.

JTC1 10740 Pt 2-93. Information Technology—Text and Office Systems—Referenced Data Transfer—Part 2: Protocol Specification First Edition. 16 pp.

—Distributed Transaction Processing

BSI BS ISO/IEC 10026-1-92. 1992 Information Technology—Open Systems Interconnection—Distributed Transaction Processing—Part 1: OSI TP Model (S). 34 pp.

BSI BS ISO/IEC 10026-2-92. 1992 Information Technology—Open Systems Interconnection—Distributed Transaction Processing—Part 2: OSI TP Service. 62 pp.

IEC 10026 Pt 1-92. Information Technology—Open Systems Interconnection—Distributed Transaction Processing—Part 1: OSI TP Model First Edition. 34 pp.

IEC 10026 Pt 2-92. Information Technology—Open Systems Interconnection—Distributed Transaction Processing—Part 2: OSI TP Service First Edition. 61 pp.

IEC DIS 10026 Pt 5-93. Information Technology—Open Systems Interconnection—Distributed Transaction Processing—Part 5: Application Context Proforma and Guidelines When Using OSI TP ***CD-ROM ONLY***. 11 pp.

IEC DIS 10026 Pt 6-92. Information Technology—Open Systems Interconnection—Distributed Transaction Processing—Part 6: Unstructured Data Transfer ***CD-ROM ONLY***. 14 pp.

IEC DISP 12061 Pt 1-93. Information Technology—Open Systems Interconnection—International Standardized Profiles 12061: OSI Distributed Transaction Processing—Part 1: Introduction to the Transaction Processing Profiles ***CD-ROM ONLY***. 25 pp.

IEC DISP 12061 Pt 2-93. Information Technology—Open Systems Interconnection—International Standardized Profiles 12061: OSI Distributed Transaction Processing—Part 2: Support of OSI TP APDUs ***CD-ROM ONLY***. 46 pp.

INTERNATIONAL AND NON-U.S. NATIONAL STANDARDS
SUBJECT INDEX
Application

Application Layer (OSI) *(Cont.)*

—Distributed Transaction Processing *(Cont.)*

IEC DISP 12061 Pt 3-93. Information Technology—Open Systems Interconnection—International Standardized Profiles 12061: OSI Distributed Transaction Processing—Part 3: Support of CCR APDUs ***CD-ROM ONLY***. 30 pp.
IEC DISP 12061 Pt 4-93. Information Technology—Open Systems Interconnection—International Standardized Profiles 12061: OSI Distributed Transaction Processing—Part 4: Support of Session, Presentation and ACSE PDUs ***CD-ROM ONLY***. 51 pp.
ISO 10026 Pt 1-92. Information Technology—Open Systems Interconnection—Distributed Transaction Processing—Part 1: OSI TP Model First Edition. 34 pp.
ISO 10026 Pt 2-92. Information Technology—Open Systems Interconnection—Distributed Transaction Processing—Part 2: OSI TP Service First Edition. 61 pp.
ISO DIS 10026 Pt 5-93. Information Technology—Open Systems Interconnection—Distributed Transaction Processing—Part 5: Application Context Proforma and Guidelines When Using OSI TP ***CD-ROM ONLY***. 11 pp.
ISO DIS 10026 Pt 6-92. Information Technology—Open Systems Interconnection—Distributed Transaction Processing—Part 6: Unstructured Data Transfer ***CD-ROM ONLY***. 14 pp.
ISO DISP 12061 Pt 1-93. Information Technology—Open Systems Interconnection—International Standardized Profiles 12061: OSI Distributed Transaction Processing—Part 1: Introduction to the Transaction Processing Profiles ***CD-ROM ONLY***. 25 pp.
ISO DISP 12061 Pt 2-93. Information Technology—Open Systems Interconnection—International Standardized Profiles 12061: OSI Distributed Transaction Processing—Part 2: Support of OSI TP APDUs ***CD-ROM ONLY***. 46 pp.
ISO DISP 12061 Pt 3-93. Information Technology—Open Systems Interconnection—International Standardized Profiles 12061: OSI Distributed Transaction Processing—Part 3: Support of CCR APDUs ***CD-ROM ONLY***. 30 pp.
ISO DISP 12061 Pt 4-93. Information Technology—Open Systems Interconnection—International Standardized Profiles 12061: OSI Distributed Transaction Processing—Part 4: Supprot of Session, Presentation and ACSE PDUs ***CD-ROM ONLY***. 51 pp.
JTC1 10026 Pt 1-92. Information Technology—Open Systems Interconnection—Distributed Transaction Processing—Part 1: OSI TP Model First Edition. 34 pp.
JTC1 10026 Pt 2-92. Information Technology—Open Systems Interconnection—Distributed Transaction Processing—Part 2: OSI TP Service First Edition. 61 pp.
JTC1 DIS10026 Pt 5-93. Information Technology—Open Systems Interconnection—Distributed Transaction Processing—Part 5: Application Context Proforma and Guidelines When Using OSI TP ***CD-ROM ONLY***. 11 pp.
JTC1 DIS10026 Pt 6-92. Information Technology—Open Systems Interconnection—Distributed Transaction Processing—Part 6: Unstructured Data Transfer ***CD-ROM ONLY***. 14 pp.
JTC1 DISP 12061 Pt 1-93. Information Technology—Open Systems Interconnection—International Standardized Profiles 12061: OSI Distributed Transaction Processing—Part 1: Introduction to the Transaction Processing Profiles ***CD-ROM ONLY***. 25 pp.
JTC1 DISP 12061 Pt 2-93. Information Technology—Open Systems Interconnection—International Standardized Profiles 12061: OSI Distributed Transaction Processing—Part 2: Support of OSI TP APDUs ***CD-ROM ONLY***. 46 pp.
JTC1 DISP 12061 Pt 3-93. Information Technology—Open Systems Interconnection—International Standardized Profiles 12061: OSI Distributed Transaction Processing—Part 3: Support of CCR APDUs ***CD-ROM ONLY***. 30 pp.
JTC1 DISP 12061 Pt 4-93. Information Technology—Open Systems Interconnection—International Standardized Profiles 12061: OSI Distributed Transaction Processing—Part 4: Support of Session, Presentation and ACSE PDUs ***CD-ROM ONLY***. 51 pp.
OSI ISO IEC DIS 10026-1-90. Information Technology—Open Systems Interconnection—Distributed Transaction Processing—Part 1: Model. 35 pp.

—Distributed Transaction Processing—Conformance

IEC DIS 10026 Pt 4-92. Information Technology—Open Systems Interconnection—Distributed Transaction Processing—Part 4: Protocol Implementation Conformance Statement (PICS) Proforma ***CD-ROM ONLY***. 31 pp.

Application Layer (OSI) *(Cont.)*

—Distributed Transaction Processing—Conformance *(Cont.)*

ISO DIS 10026 Pt 4-92. Information Technology—Open Systems Interconnection—Distributed Transaction Processing—Part 4: Protocol Implementation Conformance Statement (PICS) Proforma ***CD-ROM ONLY***. 31 pp.
JTC1 DIS10026 Pt 4-92. Information Technology—Open Systems Interconnection—Distributed Transaction Processing—Part 4: Protocol Implementation Conformance Statement (PICS) Proforma ***CD-ROM ONLY***. 31 pp.

—Distributed Transaction Processing—Polarized Control

IEC DISP 12061 Pt 5-93. Information Technology—Open Systems Interconnection—International Standardized Profiles 12061: OSI Distributed Transaction Processing—Part 5: Application Supported Transactions—Polarized Control (ATP11) ***CD-ROM ONLY***. 21 pp.
IEC DISP 12061 Pt 7-93. Information Technology—Open Systems Interconnection—International Standardized Profiles 12061: OSI Distributed Transaction Processing—Part 7: Privoder Supported Unchained Transactions—Polarized Control (ATP21) ***CD-ROM ONLY***. 21 pp.
IEC DISP 12061 Pt 9-93. Information Technology—Open Systems Interconnection—International Standardized Profiles 12061: OSI Distributed Transaction Processing—Part 9: Provider Supported Chained Transactions—Polarized Control (ATP31) ***CD-ROM ONLY***. 21 pp.
ISO DISP 12061 Pt 5-93. Information Technology—Open Systems Interconnection—International Standardized Profiles 12061: OSI Distributed Transaction Processing—Part 5: Application Supported Transactions—Polarized Control (ATP11) ***CD-ROM ONLY***. 21 pp.
ISO DISP 12061 Pt 7-93. Information Technology—Open Systems Interconnection—International Standardized Profiles 12061: OSI Distributed Transaction Processing—Part 7: Provider Supported Unchained Transactions—Polarized Control (ATP21) ***CD-ROM ONLY***. 21 pp.
ISO DISP 12061 Pt 9-93. Information Technology—Open Systems Interconnection—International Standardized Profiles 12061: OSI Distributed Transaction Processing—Part 9: Provider Supported Chained Transactions—Polarized Control (ATP31) ***CD-ROM ONLY***. 21 pp.
JTC1 DISP 12061 Pt 5-93. Information Technology—Open Systems Interconnection—International Standardized Profiles 12061: OSI Distributed Transaction Processing—Part 5: Application Supported Transactions—Polarized Control (ATP11) ***CD-ROM ONLY***. 21 pp.
JTC1 DISP 12061 Pt 7-93. Information Technology—Open Systems Interconnection—International Standardized Profiles 12061: OSI Distributed Transaction Processing—Part 7: Provider Supported Unchained Transactions—Polarized Control (ATP21) ***CD-ROM ONLY***. 21 pp.
JTC1 DISP 12061 Pt 9-93. Information Technology—Open Systems Interconnection—International Standardized Profiles 12061: OSI Distributed Transaction Processing—Part 9: Provider Supported Chained Transactions—Polarized Control (ATP31) ***CD-ROM ONLY***. 21 pp.

—Distributed Transaction Processing—Protocol

BSI BS ISO/IEC 10026-3-92. 1992 Information Technology—Open Systems Interconnection—Distributed Transaction Processing—Part 3: Protocol Specification (S). 695 pp.
IEC 10026 Pt 3-92. Information Technology—Open Systems Interconnection—Distributed Transaction Processing—Part 3: Protocol Specification First Edition. 697 pp.
ISO 10026 Pt 3-92. Information Technology—Open Systems Interconnection—Distributed Transaction Processing—Part 3: Protocol Specification First Edition. 697 pp.
JTC1 10026 Pt 3-92. Information Technology—Open Systems Interconnection—Distributed Transaction Processing—Part 3: Protocol Specification First Edition. 697 pp.
OSI ISO/IEC DIS 10026-3-90. Information Technology—Open Systems Interconnection—Distributed Transaction Processing—Part 3: Protocol Specification. 455 pp.
OSI ISO DP 10026-3-88. Information Processing Systems—Open Systems Interconnection—Distributed Transaction Processing—Part 3: Transaction Processing Protocol Specification. 171 pp.

Application Layer (OSI) *(Cont.)*

—Distributed Transaction Processing—Service

OSI ISO/IEC DIS 10026-2-90. Information Technology—Open Systems Interconnection—Distributed Transaction Processing—Part 2: Service Definition. 364 pp.

—Distributed Transaction Processing—Shared Control

IEC DISP 12061 Pt 6-93. Information Technology—Open Systems Interconnection—International Standardized Profiles 12061: OSI Distributed Transaction Processing—Part 6: Application Supported Transactions—Shared Control (ATP12) ***CD-ROM ONLY***. 21 pp.
IEC DISP 12061 Pt 8-93. Information Technology—Open Systems Interconnection—International Standardized Profiles 12061: OSI Distributed Transaction Processing—Part 8: Provider Supported Unchained Transactions—Shared Control (ATP22) ***CD-ROM ONLY***. 21 pp.
IEC DISP 12061 Pt 10-93. Information Technology—Open Systems Interconnection—International Standardized Profiles 12061: OSI Distributed Transaction Processing—Part 10 Provider Supported Chained Transactions—Shared Control (ATP32) ***CD-ROM ONLY***. 21 pp.
ISO DISP 12061 Pt 6-93. Information Technology—Open Systems Interconnection—International Standardized Profiles 12061: OSI Distributed Transaction Processing—Part 6: Application Supported Transactions—Shared Control (ATP12) ***CD-ROM ONLY***. 21 pp.
ISO DISP 12061 Pt 8-93. Information Technology—Open Systems Interconnection—International Standardized Profiles 12061: OSI Distributed Transaction Processing—Part 8: Provider Supported Unchained Transactions—Shared Control (ATP22) ***CD-ROM ONLY***. 21 pp.
ISO DISP 12061 Pt 10-93. Information Technology—Open Systems Interconnection—International Standardized Profiles 12061: OSI Distributed Transaction Processing—Part 10: Provider Supported Chained Transactions—Shared Control (ATP32) ***CD-ROM ONLY***. 21 pp.
JTC1 DISP 12061 Pt 6-93. Information Technology—Open Systems Interconnection—International Standardized Profiles 12061: OSI Distributed Transaction Processing—Part 6: Application Supported Transactions—Shared Control (ATP12) ***CD-ROM ONLY***. 21 pp.
JTC1 DISP 12061 Pt 8-93. Information Technology—Open Systems Interconnection—International Standardized Profiles 12061: OSI Distributed Transaction Processing—Part 8: Provider Supported Unchained Transactions—Shared Control (ATP22) ***CD-ROM ONLY***. 21 pp.
JTC1 DISP 12061 Pt 10-93. Information Technology—Open Systems Interconnection—International Standardized Profiles 12061: OSI Distributed Transaction Processing—Part 10: Provider Supported Chained Transactions—Shared Control (ATP32) ***CD-ROM ONLY***. 21 pp.

—Document Filing and Retrieval—Abstract Service

BSI BS ISO/IEC 10166-1-91. 1991 Information Technology—Text and Office Systems—Document Filing and Retrieval (DFR)—Part 1: Abstract Service Definition and Procedures. 139 pp.
IEC 9593 Pt 4-91. Information Technology—Computer Graphics—Programmer's Hierarchical Interactive Graphics System (PHIGS) Language Bindings—Part 4: C First Edition. 316 pp.
IEC 10166 Pt 1-91. Information Technology—Text and Office Systems—Document Filing and Retrieval (DFR)—Part 1: Abstract Service Definition and Procedures First Edition. 135 pp.
ISO 10166 Pt 1-91. Information Technology—Text and Office Systems—Document Filing and Retrieval (DFR)—Part 1: Abstract Service Definition and Procedures First Edition. 135 pp.
JTC1 9593-91. Information Technology—Computer Graphics—Programmer's Hierarchical Interactive Graphics System (PHIGS) Language Bindings—Part 4: C First Edition. 316 pp.
JTC1 9593 Pt4 Draft AMD 1. Information Technology—Computer Graphics—Programmer's Hierarchical Interactive Graphics System (PHIGS) Language Bindings—Part 4: C; (1992) ***CD-ROM ONLY***. 167 pp.
JTC1 10166 Pt 1-91. Information Technology—Text and Office Systems—Document Filing and Retrieval (DFR)—Part 1: Abstract Service Definition and Procedures First Edition. 135 pp.
OSI ISO/IEC DIS 10166-1-89. Information Technology—Document Filing and Retrieval (DFR)—Part 1: Abstract Service Definition and Procedures. 138 pp.

INDUSTRY STANDARDS
197

Application
INTERNATIONAL AND NON-U.S. NATIONAL STANDARDS
SUBJECT INDEX

Application Layer (OSI) *(Cont.)*

—Document Filing and Retrieval—Protocol

BSI BS ISO/IEC 10166-2-91. 1991 Information Technology—Text and Office Systems—Document Filing and Retrieval (DFR)—Part 2: Protocol Specification. 19 pp.
ECMA ECMA 137-90. Document Filing and Retrieval (DFR) Part 1—Abstract-Service Definition and Procedures Part 2—Protocol Specification. 143 pp.
IEC 10166 Pt 2-91. Information Technology—Text and Office Systems—Document Filing and Retrieval (DFR)—Part 2: Protocol Specification First Edition. 16 pp.
ISO 10166 Pt 2-91. Information Technology—Text and Office Systems—Document Filing and Retrieval (DFR)—Part 2: Protocol Specification First Edition. 16 pp.
JTC1 10166 Pt 2-91. Information Technology—Text and Office Systems—Document Filing and Retrieval (DFR)—Part 2: Protocol Specification First Edition. 16 pp.
OSI ISO/IEC DIS 10166-2-89. Information Technology—Document Filing and Retrieval (DFR)—Part 2: Protocol Specification. 16 pp.

—Document Filing and Retrieval—Service

ECMA ECMA 137-90. Document Filing and Retrieval (DFR) Part 1—Abstract-Service Definition and Procedures Part 2—Protocol Specification. 143 pp.

—Document Printing Application—Abstract Service

ECMA ECMA-140-90. Document Printing Application (DPA). 221 pp.
IEC DIS 10175 Pt 1-91. Information Technology—Text and Office Systems—Document Printing Application—Part 1: Abstract Service Definition and Procedures ***CD-ROM ONLY***. 403 pp.
ISO DIS 10175 Pt 1-91. Information Technology—Text and Office Systems—Document Printing Application—Part 1: Abstract Service Definition and Procedures ***CD-ROM ONLY***. 403 pp.
OSI ISO IEC DP 10175-1-89. Information Processing—Text Communication—Document Printing Application—Part 1: Abstract Service Definition and Precedures. 199 pp.

—Document Printing Application—Protocol

ECMA ECMA-140-90. Document Printing Application (DPA). 221 pp.
OSI ISO IEC DP 10175-2-89. Information Processing—Text Communication—Document Printing Application—Part 2: Protocol Specification. 16 pp.

—File Transfer, Access, and Management

BSI BS 7090: Part 1-89. 1989 Amd 1 Open Systems Interconnection: File Transfer, Access and Management Part 1: General Introduction (ISO 8571-1: 1988) (AMD 6826) September 30, 1991. 33 pp.
BSI BS ISO/IEC ISP 10607-2-90. 1990 Amd 1 Information Technology—International Standardized Profiles AFTnn—File Transfer, Access and Management—Part 2: Definition of Document Types, Constraint Sets and Syntaxes (AMD 7153) July 15, 1992. 72 pp.
CEN ENV 41 204-88. Information Systems Interconnection: File Transfer, Access and Management; Simple File Transfer (Unstructured). 50 pp.
CEN PRENV 41 204-89. Review Information Systems Interconnection: File Transfer Access and Management (FTAM): Simple File Transfer (Unstructured). 55 pp.
CEN PRENV 41 206-89. Information Systems Interconnection; File Transfer Access and Management (FTAM); Positional File Transfer (flat). 98 pp.
CEN ENV 41206-91. Information Systems Interconnection—File Transfer Access and Management (FTAM)—Positional File Transfer. 102 pp.
CEN PRENV 41 207-89. Information Systems Interconnection; File Transfer Access and Management (FTAM): Positional Tile Access (Flat). 99 pp.
CEN ENV 41207-91. Information Systems Interconnection—File Transfer Access and Management (FTAM)—Positional File Access. 103 pp.
CSA CAN/CSA-Z243.180-89. Information Processing Systems—Open Systems Interconnection—File Transfer, Access and Management—Part 1: General Introduction (ISO 8571-1-1988). 39 pp.
IEC 8571 Pt 1 AMD 1. Amendment 1—Information Processing Systems—Open Systems Interconnection—File Transfer, Access and Management—Part 1: General Introduction Amendment 1: Filestore Management; (1992). 16 pp.
IEC 8571 Pt 1 COR 1. Corrigendum 1-Information Processing Systems—Open Systems Interconnection—File Transfer, Access and Management—Part 1: General Introduction; (1991). 2 pp.

Application Layer (OSI) *(Cont.)*

—File Transfer, Access, and Management *(Cont.)*

IEC 8571 Pt1 AMD 2. Information Processing Systems—Open Systems Interconnection—File Transfer, Access and Management—Part 1: General Introduction—Amendment 2: Overlapped Access. 14 pp.
IEC 8571 Pt 1 Draft AMD 3. Information Processing Systems—Open Systems Interconnection—File Transfer, Access and Management—Part 1: General Introduction Amendment 3: Service Enhancement; (1992) ***CD-ROM ONLY***. 13 pp.
IEC 8571 Pt 2 COR 1. Corrigendum 1-Information Processing Systems—Open Systems Interconnection—File Transfer, Access and Management—Part 2: Virtual Filestore Definition; (1991). 3 pp.
IEC ISP10607 Pt 2-90. Information Technology—International Standardized Profiles AFTnn—File Transfer, Access and Management—Part 2: Definition of Document Types, Constraint Sets and Syntaxes Amendment 1: Additional Definitions First Edition; (Amendment 1-1991). 45 pp.
IEC ISP10607 Pt 3-90. Information Technology—International Standardized Profiles AFTnn—File Transfer, Access and Management—Part 3: AFT 11—Simple File Transfer Service (Unstructured) First Edition; (Corrected and Reprinted -1991). 39 pp.
ISO 8571 Pt 1-88. Information Processing Systems—Open Systems Interconnection—File Transfer, Access and Management—Part 1: General Introduction Amendment 1: Filestore Management Amendment 2: Overlapped Access; First Edition; (Corrigendum 1-1991) (Amendment 1-1992 1992) (Amendment 2-1993). 63 pp.
ISO ISP10607 Pt 2-90. Information Technology—International Standardized Profiles AFTnn—File Transfer, Access and Management—Part 2: Definition of Document Types, Constraint Sets and Syntaxes Amendment 1: Additional Definitions First Edition; (Amendment 1-1991). 45 pp.
JTC1 8571-88. Information Processing Systems—Open Systems Interconnection—File Transfer, Access and Management—Part 1: General Introduction Amendment 1: Filestore Management First Edition; (Corrigendum 1-1991) (Amendment 1-1992) (Amendment 2-1993). 63 pp.
JTC1 ISP 10607 Pt 2-90. Information Technology—International Standardized Profiles AFTnn—File Transfer, Access and Management—Part 2: Definition of Document Types, Constraint Sets and Syntaxes Amendment 1: Additional Definitions First Edition; (Amendment 1-1991). 45 pp.
OSI ISO 8571-1 DAM 1-90. Information Processing Systems—Open Systems Interconnection—File Transfer, Access and Management—Part 1: General Introduction Amendment 1: Filestore Management. 14 pp.
OSI ISO DIS 8571-1-88. Final Text of DIS 8571-1: Information Processing Systems—Open Systems Interconnection—File Transfer, Access and Management Part 1: General Introduction.
OSI ISO DIS 8571-1 DAM 2-91. Information Processing Systems—Open Systems Interconnection—File Transfer, Access and Management—Part 1: General Introduction—Amendment 2: Overlapped Access.
OSI ISO 8571-1/DAM 1-90. Information Processig Systems—Open Systems Interconnection—File Transfer, Access and Management—Part 1: General Introduction Amendment 1: Filestore Management. 14 pp.
OSI ISO/IEC DISP 10607-2-90. Information Technology—International Standardized Profile AFTnn—File Transfer, Access and Management—Part 2: Definition of Document Types, Constraint Sets and Syntaxes. 34 pp.
OSI ISO/IEC ISP 10607-3-90. Information Technology—International Standard-ized Profile AFTnn—File Transfer, Access and Management—Part 3: AFT 11—Simple File Transfer Service (Unstructured). 39 pp.
OSI ISO/IEC DISP 10607-4-90. Information Technology—International Standard-ized Profiles AFTnn—File Transfer, Access and Management—Part 4: AFT12—Positional File Transfer Service (Flat). 56 pp.
OSI ISO/IEC DISP 10607-5-90. Information Technology—International Standard-ized Profiles AFTnn—File Transfer, Access and Management—Part 5: AFT 22—Positional File Access Service (FLAT). 57 pp.
OSI ISO/IEC DISP 10607-6-90. Information Technology—International Standard-ized Profiles AFTnn—File Transfer, Access and Management—Part 6: AFT3—File Management Service. 31 pp.
SAA MP67.2-92. Standardized Interoperability Tests in the OSI Environment—Part 2: FTAM Tests. 101 pp.

Application Layer (OSI) *(Cont.)*

—File Transfer, Access, and Management—COBOL

IEC DISP 10607 Pt1 DR AMD 1. Information Technology—International Standardized Profiles AFTnn—File Transfer, Access and Management—Part 1: Specification of ASCE, Presentation and Session Protocols for the Use by FTAM Amd. 1: Additional Specs. for COBOL Doc. Types; (1993) ***CD-ROM ONLY***. 10 pp.
IEC DISP 10607 Pt2 DR AMD 2. Information Technology—International Standardized Profiles AFTnn—File Transfer, Access and Management—Part 2: Definition of Document Types, Contraint Sets and Syntaxes Amendment 2: Additional Defs. for COBOL Doc. Types; (1993) ***CD-ROM ONLY***. 54 pp.
IEC DISP 10607 Pt4 DR AMD 1. Information Technology—International Standardized Profiles AFTnn—File Transfer, Access and Management—Part 4: AFT12—Positional File Transfer Service (Flat) Amendment 1: Additional Specifications for COBOL Document Types; (1993) ***CD-ROM ONLY***. 15 pp.
IEC DISP 10607 Pt5 DR AMD 1. Information Technology—International Standardized Profiles AFTnn—File Transfer, Access and Management—Part 5: AFT22—Positional File Access Service (Flat) Amendment 1: Additional Specifications for COBOL Document Types; (1993) ***CD-ROM ONLY***. 15 pp.
ISO DISP 10607 Pt1 DR AMD 1. Information Technology—International Standardized Profiles AFTnn—File Transfer, Access and Management—Part 1: Specification of ASCE, Presentation and Session Protocols for the Use by FTAM Amd. 1: Additional Specs. for COBOL Doc. Types; (1993) ***CD-ROM ONLY***. 10 pp.
ISO DISP 10607 Pt2 DR AMD 2. Information Technology—International Standardized Profiles AFTnn—File Transfer, Access and Management—Part 2: Definition of Document Types, Contraint Sets and Syntaxes Amendment 2: Additional Defs. for COBOL Doc. Types; (1993) ***CD-ROM ONLY***. 54 pp.
ISO DISP 10607 Pt4 DR AMD 1. Information Technology—International Standardized Profiles AFTnn—File Transfer, Access and Management—Part 4: AFT12—Positional File Transfer Service (Flat) Amendment 1: Additional Specifications for COBOL Document Types; (1993) ***CD-ROM ONLY***. 15 pp.
ISO DISP 10607 Pt5 DR AMD 1. Information Technology—International Standardized Profiles AFTnn—File Transfer, Access and Management—Part 5: AFT22—Positional File Access Service (Flat) Amendment 1: Additional Specifications for COBOL Document Types; (1993) ***CD-ROM ONLY***. 15 pp.
JTC1 DISP10607 Pt1 DR AMD 1. Information Technology—International Standardized Profiles AFTnn—File Transfer, Access and Management—Part 1: Specification of ASCE, Presentation and Session Protocols for the Use by FTAM Amd. 1: Additional Specs. for COBOL Doc. Types; (1993) ***CD-ROM ONLY***. 10 pp.
JTC1 DISP10607 Pt2 DR AMD 2. Information Technology—International Standardized Profiles AFTnn—File Transfer, Access and Management—Part 2: Definition of Document Types, Contraint Sets and Syntaxes Amendment 2: Additional Defs. for COBOL Doc. Types; (1993) ***CD-ROM ONLY***. 54 pp.
JTC1 DISP10607 Pt4 DR AMD 1. Information Technology—International Standardized Profiles AFTnn—File Transfer, Access and Management—Part 4: AFT12—Positional File Transfer Service (Flat) Amendment 1: Additional Specifications for COBOL Document Types; (1993) ***CD-ROM ONLY***. 15 pp.
JTC1 DISP10607 Pt5 DR AMD 1. Information Technology—International Standardized Profiles AFTnn—File Transfer, Access and Management—Part 5: AFT22—Positional File Access Service (Flat) Amendment 1: Additional Specifications for COBOL Document Types; (1993) ***CD-ROM ONLY***. 15 pp.

—File Transfer, Access, and Management—Protocol

BSI BS 7090: Part 4-89. 1989 Open Systems Interconnection: File Transfer, Access and Management Part 4: File Protocol Specification. 99 pp.
BSI BS 7090: Part 4-01. 1989 Amd 1 Open Systems Interconnection: File Transfer, Access and Management Part 4: File Protocol Specification (ISO 8571-4: 1988) (AMD 7604) February 15, 1993 (Technical Corr 1). 102 pp.
BSI BS 7090: Part 4-04. 1989 Amd 4 Open Systems Interconnection: File Transfer, Access and Management Part 4: File Protocol Specification (ISO 8571-4: 1988) (AMD 7669) Apirial 15, 1993 (S). 118 pp.

INTERNATIONAL AND NON-U.S. NATIONAL STANDARDS
SUBJECT INDEX
Application

Application Layer (OSI) *(Cont.)*

—File Transfer, Access, and Management—Protocol *(Cont.)*

BSI BS ISO/IEC 8571-5-90. 1990 Information Processing Systems—Open Systems Interconnection—File Transfer, Access and Management—Part 5: Protocol Implementation Conformance Statement Proforma. 49 pp.

BSI BS ISO/IEC ISP 10607-1-90. 1990 Amd 1 Information Technology—International Standardized Profiles AFTnn—File Transfer, Access and Management—Part 1: Specification of ACSE, Presentation and Session Protocols for the Use by FTAM (Technical Corr 1). 22 pp.

CSA CAN/CSA-Z243.183-89. Information Processing Systems—Open Systems Interconnection—File Transfer, Access and Management—Part 4: File Protocol Specification (ISO 8571-4-1988). 111 pp.

CSA CAN/CSA-Z243.184-92. Information Processing Systems—Open Systems Interconnection—File Transfer, Access and Management—Part 5: Protocol Implementation Conformance Statement Proforma (ISO/IEC 8571-5:1990); (Gen Instr 1). 56 pp.

IEC 8571 Pt 4 AMD 1. Amendment 1—Information Processing Systems—Open Systems Interconnection—File Transfer, Access and Management—Part 4: File Protocol Specification Amendment 1: Filestore Management (1992). 111 pp.

IEC 8571 Pt 4 AMD 4. Amendment 4-Information Processing Systems—Open Systems Interconnection—File Transfer, Access and Management—Part 4: File Protocol Specification; (1992). 55 pp.

IEC 8571 Pt 4 COR 1. Corrigendum 1-Information Processing Systems—Open Systems Interconnection—File Transfer, Access and Management—Part 4: File Protocol Specification; (1992). 2 pp.

IEC 8571 Pt4 AMD 2. Information Processing Systems—Open Systems Interconnection—File Transfer, Access and Management—Part 4: File Protocol Specification—Amendment 2: Overlapped Access. 99 pp.

IEC 8571 Pt 4 Draft AMD 3. Information Processing Systems—Open Systems Interconnection—File Transfer, Access and Management—Part 4: File Protocol Specification Amendment 3: Service Enhancement; (1992) ***CD-ROM ONLY***. 7 pp.

IEC 8571 Pt 5-90. Information Processing Systems—Open Systems Interconnection—File Transfer, Access and Management—Part 5: Protocol Implementation Conformance Statement Proforma First Edition. 46 pp.

IEC ISP10607 Pt 1-90. Information Technology—International Standardized Profiles AFTnn—File Transfer, Access and Management—Part 1: Specification of ACSE, Presentation and Session Protocols for the Use by FTAM First Edition; (Corrected and Reprinted -1991). 20 pp.

ISO 8571 Pt 4-88. Information Processing Systems—Open Systems Interconnection—File Transfer, Access and Management—Part 4: File Protocol Specification Amendment 1: Filestore Management Amendment 2: Overlapped Access First Edition; (Corrigendum 1-1992) (Amendment 4-1992). 370 pp.

ISO 8571 Pt 5-90. Information Processing Systems—Open Systems Interconnection—File Transfer, Access and Management—Part 5: Protocol Implementation Conformance Statement Proforma First Edition. 46 pp.

ISO ISP10607 Pt 1-90. Information Technology—International Standardized Profiles AFTnn—File Transfer, Access and Management—Part 1: Specification of ACSE, Presentation and Session Protocols for the Use by FTAM First Edition; (Corrected and Reprinted -1991). 20 pp.

JTC1 8571-88. Information Processing Systems—Open Systems Interconnection—File Transfer, Access and Management—Part 4: File Protocol Specification Amendment 1: Filestore Management Amendment 2: Overlapped Access First Edition; (Corrigendum 1-1992) (Amendment 4-1992). 370 pp.

JTC1 8571-90. Information Processing Systems—Open Systems Interconnection—File Transfer, Access and Management—Part 5: Protocol Implementation Conformance Statement Proforma First Edition. 46 pp.

JTC1 ISP 10607 Pt 1-90. Information Technology—International Standardized Profiles AFTnn—File Transfer, Access and Management—Part 1: Specification of ACSE, Presentation and Session Protocols for the Use by FTAM First Edition; (Corrected and Reprinted -1991). 20 pp.

OSI ISO DIS 8571-4-88. Final Text of DIS 8571-4: Information Processing Systems—Open Systems Interconnection—File Transfer, Access and Management Part 4: File Protocol Specification.

OSI ISO DIS 8571-4 DAM 2-91. Information Processing Systems—Open Systems Interconnection—File Transfer, Access and Management—Part 4: File Protocol Specification Amendment 2: Overlapped Access.

OSI ISO IEC DIS 8571-5-87. Information Processing Systems—Open Systems Interconnection—File Transfer, Access and Management Part 5: Protocol Implementation Conformance Statement Proforma. 47 pp.

OSI ISO/IEC 8571-5-90. Information Processing Systems—Open Systems Interconnection—File Transfer, Access and Management—Part 5: Protocol Implementation Conformance Statement Proforma. 46 pp.

OSI ISO/IEC DIS 8571-5-89. Information Processing Systems—Open Systems Interconnection—File Transfer, Access and Management—Part 5: Protocol Implementation Conformance Statement Proforma. 47 pp.

OSI ISO/IEC DIS 10035-89. Information Processing Systems—Open Systems Interconnection—Connectionless ACSE Protocol Specification. 13 pp.

OSI ISO/IEC DISP 10607-1-90. Information Technology—International Standardized Profile AFTnn—File Transfer, Access and Management—Part 1: Specification of ACSE, Presentation and Session Protocols for the Use by FTAM. 41 pp.

OSI ISO/IEC ISP 10607-1-90. Information Technology—International Standard-ized Profile AFTnn—File Transfer, Access and Management—Part 1: Specification of ACSE, Presentation and Session Protocols for the Use by FTAM. 20 pp.

—File Transfer, Access, and Management—Protocol—Conformance

BSI BS ISO/IEC 10170-1-93. 1993 Information Technology—Open Systems Interconnection—Conformance Test Suite for the FTAM Protocol—Part 1: Test Suite Structure and Test Purposes (S). 130 pp.

IEC 10170 Pt 1-93. Information Technology—Open Systems Interconnection—Conformance Test Suite for the FTAM Protocol—Part 1: Test Suite Structure and Test Purposes First Edition. 127 pp.

ISO 10170 Pt 1-93. Information Technology—Open Systems Interconnection—Conformance Test Suite for the FTAM Protocol—Part 1: Test Suite Structure and Test Purposes First Edition. 127 pp.

JTC1 10170 Pt 1-93. Information Technology—open Systems Interconnection—Conformance Test Suite for the FTAM Protocol—Part 1: Test Suite Structure and Test Purposes First Edition. 127 pp.

OSI ISO/IEC DIS 10170-1-90. Information Technology—Open Systems Interconnection—Conformance Test Suite for the FTAM Protocol—Part 1: Test Suite Structure and Test Purposes. 116 pp.

—File Transfer, Access, and Management—Service

BSI BS 7090: Part 3-89. 1989 Amd 1 Open Systems Interconnection: File Transfer, Access and Management Part 3: File Service Definition (ISO 8571-3: 1988) (AMD 6828) September 30, 1991. 77 pp.

BSI BS 7090: Part 3-02. 1989 Amd 2 Open Systems Interconnection: File Transfer, Access and Management Part 3: File Service Definition (ISO 8571-3: 1988) (AMD 7603) February 15, 1993 (Technical Corr 2). 79 pp.

BSI BS ISO/IEC ISP 10607-3-90. 1990 Amd 1 Information Technology—International Standardized Profiles AFTnn—File Transfer, Access and Management—Part 3: AFT 11—Simple File Transfer Service (Unstructured) (Technical Corr 1). 43 pp.

BSI BS ISO/IEC ISP 10607-4-91. 1991 Information Technology—International Standardized Profiles AFTnn—File Transfer, Access and Management—Part 4: AFT12—Positional File Transfer Service (Flat). 50 pp.

BSI BS ISO/IEC ISP 10607-5-91. 1991 Information Technology—International Standardized Profiles AFTnn—File Transfer, Access and Management—Part 5: AFT22—Positional File Access Service (Flat). 51 pp.

BSI BS ISO/IEC ISP 10607-6-91. 1991 Information Technology—International Standardized Profiles AFTnn—File Transfer, Access and Management—Part 6: AFT3—File Management Service. 26 pp.

CEN ENV 41205-91. Information Systems Interconnection—File Transfer Access and Management—File Management. 22 pp.

CSA CAN/CSA-Z243.182-89. Information Processing Systems—Open Systems Interconnection—File Transfer, Access and Management—Part 3: File Service Definition (ISO 8571-3-1988). 83 pp.

IEC 8571 Pt3 AMD 2. Information Processing Systems—Open Systems Interconnection—File Transfer, Access and Management—Part 3: File Service Definition Amendment 2: Overlapped Access. 83 pp.

Application Layer (OSI) *(Cont.)*

—File Transfer, Access, and Management—Service *(Cont.)*

IEC 8571 Pt 3 Draft AMD 3. Information Processing Systems—Open Systems Interconnection—File Transfer, Access and Management—Part 3: File Service Definition Amendment 3: Service Enhancement; (1992) ***CD-ROM ONLY***. 28 pp.

IEC 8571 Pt 3 COR 1. Corrigendum 1-Information Processing Systems—Open Systems Interconnection—File Transfer, Access and Management—Part 3: File Service Definition; (1991). 6 pp.

IEC 8571 Pt 3 COR 2. Corrigendum 2-Information Processing Systems—Open Systems Interconnection—File Transfer, Access and Management—Part 3: File Service Definition; (1992). 1 p.

IEC ISP10607 Pt 4-91. Information Technology—International Standardized Profiles AFTnn—File Transfer, Access and Management—Part 4: AFT 12—Positional File Transfer Service (Flat) First Edition. 47 pp.

IEC ISP10607 Pt 5-91. Information Technology—International Standardized Profiles AFTnn—File Transfer, Access and Management—Part 5: AFT22—Positional File Access Service (Flat) First Edition. 48 pp.

IEC ISP10607 Pt 6-91. Information Technology—International Standardized Profiles AFTnn—File Transfer, Access and Management—Part 6: AFT3—File Management Service First Edition. 23 pp.

ISO 8571 Pt 3-88. Information Processing Systems—Open Systems Interconnection—File Transfer, Access and Management—Part 3: File Service Definition Amd 1: Filestore Mgmt. Amd 2: Overlapped Access First Edition; (Corrigendum 1-1991) (Corrigendum 2-1992) (Amd 1-1992)(Amd 2-1992). 253 pp.

ISO ISP10607 Pt 3-90. Information Technology—International Standardized Profiles AFTnn—File Transfer, Access and Management—Part 3: AFT 11—Simple File Transfer Service (Unstructured) First Edition; (Corrected and Reprinted -1991). 39 pp.

ISO ISP10607 Pt 4-91. Information Technology—International Standardized Profiles AFTnn—File Transfer, Access and Management—Part 4: AFT 12—Positional File Transfer Service (Flat) First Edition. 47 pp.

ISO ISP10607 Pt 5-91. Information Technology—International Standardized Profiles AFTnn—File Transfer, Access and Management—Part 5: AFT22—Positional File Access Service (Flat) First Edition. 48 pp.

ISO ISP10607 Pt 6-91. Information Technology—International Standardized Profiles AFTnn—File Transfer, Access and Management—Part 6: AFT3—File Management Service First Edition. 23 pp.

JTC1 8571-88. Information Processing Systems—Open Systems Interconnection—File Transfer, Access and Management—Part 3: File Service Definition First Edition; (Corrigendum—1 June 1991) (Corrigendum—2 Oct 1992) (Amendment 1 1991) (Amendment 2 1993). 253 pp.

JTC1 ISP 10607 Pt 3-90. Information Technology—International Standardized Profiles AFTnn—File Transfer, Access and Management—Part 3: AFT 11—Simple File Transfer Service (Unstructured) First Edition; (Corrected and Reprinted -1991). 39 pp.

JTC1 ISP 10607 Pt 4-91. Information Technology—International Standardized Profiles AFTnn—File Transfer, Access and Management—Part 4: AFT 12—Positional File Transfer Service (Flat) First Edition. 47 pp.

JTC1 ISP 10607 Pt 5-91. Information Technology—International Standardized Profiles AFTnn—File Transfer, Access and Management—Part 5: AFT22—Positional File Access Service (Flat) First Edition. 48 pp.

JTC1 ISP 10607 Pt 6-91. Information Technology—International Standardized Profiles AFTnn—File Transfer, Access and Management—Part 6: AFT3—File Management Service First Edition. 23 pp.

OSI ISO 8571-3-88. Information Processing Systems—Open Systems Interconnection—File Transfer, Access and Management—Part 3: File Service Definition. 78 pp.

OSI ISO DIS 8571-3-83. Final Text of DIS 8571-3: Information Processing Systems—Open Systems Interconnection—File Transfer, Access and Management Part 3: File Service Definition.

OSI ISO DIS 8571-3 DAM 2-91. Information Processing Systems—Open Systems Interconnection—File Transfer, Access and Management—Part 3: File-Service Definition—Amendment 2: Overlapped Access.

OSI ISO 8571-4 DAM 1-90. Information Processig Systems—Open Systems Interconnection—File Transfer, Access and Management Part 4: File Protocol Specification Amendment 1: Filestore Management. 99 pp.

INDUSTRY STANDARDS 199

Application Layer (OSI) *(Cont.)*

—File Transfer, Access, and Management—Service *(Cont.)*

OSI ISO/IEC DISP 10607-3-90. Information Technology—International Standardized Profile AFTnn—File Transfer, Access and Management—Part 3: AFT 11—Simple File Transfer Service. 59 pp.

—File Transfer, Access, and Management—Virtual Files

ECMA ECMA 85-82. Virtual File Protocol. 120 pp.

—File Transfer, Access, and Management—Virtual Filestore

BSI BS 7090: Part 2-89. 1989 Amd 1 Open Systems Interconnection: File Transfer, Access and Management Part 2: Virtual Filestore Definition (ISO 8571-2: 1988) (AMD 6827) September 30, 1991. 55 pp.

CSA CAN/CSA-Z243.181-89. Information Processing Systems—Open Systems Interconnection—File Transfer, Access and Management—Part 2: Virtual Filestore Definition (ISO 8571-2-1988). 63 pp.

IEC 8571 Pt2 AMD 2. Information Processing Systems—Open Systems Interconnection—File Transfer, Access and Management—Part 2: Virtual Filestore Definition—Amendment 2: Overlapped Access. 8 pp.

ISO 8571 Pt 2-88. Information Processing Systems—Open Systems Interconnection—File Transfer, Access and ranagement—Part 2: Virtual Filestore Definition Amendment 1: Filestore Management Amendment 2: Overlapped Access First Edition; (Corrigendum 1-1991) (Amd 1-1992)(Amd 2-1993). 92 pp.

JTC1 8571-88. Information Processing Systems—Open Systems Interconnection—File Transfer, Access and Management—Part 2: Virtual Filestore Definition First Edition; (Corrigendum—1 June 1991) (Amendment 1 1992) (Amendment 2-1993). 92 pp.

OSI ISO 8571-2/DAM 1-90. Information Processig Systems—Open Systems Interconnection—File Transfer, Access and Management—Part 2: Virtual Filestore Definition Amendment 1: Filestore Management. 24 pp.

OSI ISO DIS 8571-2-88. Final Text of DIS 8571-2: Information Processing Systems—Open Systems Interconnection—File Transfer, Access and Management Part 2: Virtual Filestore Definition.

OSI ISO DIS 8571-2 DAM 2-91. Information Processing Systems—Open Systems Interconnection—File Transfer, Access and Management—Part 2: Virtual Filestore Definition—Amendment 2: Overlapped Access.

—Interlibrary Loans

BSI BS ISO 10160-93. 1993 Information and Documentation—Open Systems Interconnection—Interlibrary Loan Application Service Definition (S). 123 pp.

CSA Z243.45-88P. Interlibrary Loan (ILL) Service Definition. 67 pp.

CSA Z243.46-88P. Interlibrary Loan (ILL) Protocol Specification. 93 pp.

ISO 10160-93. Information and Documentation—Open Systems Interconnection—Interlibrary Loan Application Service Definition First Edition. 121 pp.

OSI ISO/DIS 10160-90. Information Documentation—Open Systems Interconnection—Interlibrary Loan Application Service Definition. 117 pp.

OSI ISO/DIS 10161-90. Information and Documentation—Open Systems Interconnection—Interlibrary Loan Application Protocol Specification. 152 pp.

—Job Transfer and Manipulation

BSI BS ISO 8831-89. (WITHDRAWN) 1989 Information Processing Systems—Open Systems Interconnection—Job Transfer and Manipulation Concepts and Services (Superseded by BS ISO/IEC 8831: 1992). 87 pp.

BSI BS ISO 8832-89. (WITHDRAWN) 1989 Information Processing Systems—Open Systems Interconnection—Specification of the Basic Class Protocol for Job Transfer and Manipulation (Superseded by BS ISO/IEC 8832: 1992). 71 pp.

BSI BS ISO/IEC 8831-92. 1992 Information Technology—Open Systems Interconnection—Job Transfer and Manipulation Concepts and Services. 123 pp.

BSI BS ISO/IEC 8832-92. 1992 Information Technology—Open Systems Interconnection—Specification of the Basic Class and Full Protocol for Job Transfer and Manipulation. 170 pp.

CSA CAN/CSA-Z243.173-90. Information Processing Systems—Open Systems Interconnection Transfer and Manipulation Concepts and Services (ISO 8831:1989). 93 pp.

CSA CAN/CSA-Z243.174-90. Information Processing Systems—Open Systems Interconnection Specification of the Basic Class Protocol for Job Transfer and Manipulation (ISO 8832:1989). 77 pp.

Application Layer (OSI) *(Cont.)*

—Job Transfer and Manipulation *(Cont.)*

IEC 8831-92. Information Technology—Open Systems Interconnection—Job Transfer and Manipulation Concepts and Services Second Edition. 120 pp.

IEC 8832-92. Information Technology—Open Systems Interconnection—Specification of the Basic Class and Full Protocol for Job Transfer and Manipulation Second Edition. 167 pp.

ISO 8831-92. Information Technology—Open Systems Interconnection—Job Transfer and Manipulation Concepts and Services Second Edition. 120 pp.

ISO 8832-92. Information Technology—Open Systems Interconnection—Specification of the Basic Class and Full Protocol for Job Transfer and Manipulation Second Edition. 167 pp.

JTC1 8831-92. Information Technology—Open Systems Interconnection—Job Transfer and Manipulation Concepts and Services Second Edition. 120 pp.

JTC1 8832-92. Information Technology—Open Systems Interconnection—Specification of the Basic Class and Full Protocol for Job Transfer and Manipulation Second Edition. 167 pp.

OSI ISO DIS 8831-87. Revised Text of 2nd ISO/ DP 8831, Information Processing Systems—Open Systems Interconnection—Job Transfer and Manipulation Concepts and Services.

OSI ISO 8831-89. Information Processing Systems—Open Systems Interconnection—Job Transfer and Manipulation Concepts and Services. 86 pp.

OSI ISO 8832/DAM 1-90. Information Processing Systems—Open Systems Interconnection—Specification of the Basic Class Protocol for Job Transfer and Manipulation Amendment 1: Extension to Specification of the Full Protocol. 86 pp.

OSI ISO 8832-89. Information Processing Systems—Open Systems Interconnection—Specification of the Basic Class Protocol for Job Transfer and Manipulation. 71 pp.

OSI ISO DIS 8832-87. Specification of the Basic Class Protocol for Job Transfer and Manipulation. 146 pp.

SAA AS 3697-91. Information Processing Systems—Open Systems Interconnection—Job Transfer and Manipulation Concepts and Services (ISO 8831: 1988) (In Professional Packages 26A, 26C). 78 pp.

SAA AS 3698-91. Information Processing Systems—Open Systems Interconnection—Specificstion of the Basic Class Protocol for Job Transfer and Manipulation (ISO 8832: 1989)(In Professional Packages 26A, 26C). 62 pp.

SNZ NZS/ISO 8831-88. Information Processing Systems—Open Systems Interconnection—Job Transfer and Manipulation Concepts and Services. 78 pp.

SNZ NZS/ISO 8832-89. Information Processing Systems—Open Systems Interconnection—Specification of the Basic Class Protocol for Job Transfer and Manipulation. 62 pp.

—Management Information Services

OSI ISO/IEC DIS 10165-1-90. Information Technology—Open Systems Interconnection—Management Information Services—Structure of Management Information—Part 1: Management Information Model. 34 pp.

OSI ISO IEC DP 10165-1-89. Information Processing Systems—Open Systems Interconnection—Management Information Services—Structure of Management Information—Part 1: Management Information Model. 31 pp.

OSI ISO/IEC DIS 10165-2-90. Information Technology—Open Systems Interconnection—Management Information Services—Structure of Management Information—Part 2: Definition of Management Information. 77 pp.

OSI ISO/IEC DIS 10165-4-90. Information Technology—Open Systems Interconnection—Management Information Services—Structure of Management Information—Part 4: Guidelines for the Definition of Managed Objects. 46 pp.

—Manufacturing Message Services and Protocol

BSI BS ISO/IEC 9506-1-90. 1990 Industrial Automation Systems—Manufacturing Message Specification—Part 1: Service Definition (S). 329 pp.

BSI BS ISO/IEC 9506-1-01. 1990 Amd 1 Industrial Automation Systems—Manufacturing Message Specification—Part 1: Service Definition (AMD 7991) November 15, 1993 (S). 339 pp.

BSI BS ISO/IEC 9506-2-90. 1990 Industrial Automation Systems—Manufacturing Message Specification—Part 2: Protocol Specification (S). 142 pp.

BSI BS ISO/IEC 9506-3-91. 1991 Industrial Automation Systems—Manufacturing Message Specification—Part 3: Companion Standard for Robotics. 129 pp.

BSI BS ISO/IEC 9506-4-92. 1992 Industrial Automation Systems—Manufacturing Message Specification—Part 4: Companion Standard for Numerical Control (S). 142 pp.

Application Layer (OSI) *(Cont.)*

—Manufacturing Message Services and Protocol *(Cont.)*

CEN ENV 40 003-90. Computer Integrated Manufacturing—Systems Architecture—Framework for Enterprise Modelling. 24 pp.

CEN ENV 40 003-91. AMD 1 Computer Integrated Manufacturing—Systems Architecture—Framework for Enterprise Modelling. 2 pp.

IEC 9506 Pt 1-90. Industrial Automation Systems—Manufacturing Message Specification—Part 1: Service Definition First Edition. 334 pp.

IEC 9506 Pt 1 Draft AMD 1. Industrial Automation Systems—Manufacturing Message Specification—Part 1: Service Definition Amendment 1: Data Exchange; (1992) ***CD-ROM ONLY***. 11 pp.

IEC 9506 Pt 2-90. Industrial Automation Systems—Manufacturing Message Specification—Part 2: Protocol Specification AMMENDMENT 1: Data Exchange First Edition; (Ammendment 1-1993). 147 pp.

IEC 9506 Pt 2 Draft AMD 1. Industrial Automation Systems—Manufacturing Message Specification—Part 2: Protocol Specification Amendment 1: Data Exchange; (1992) ***CD-ROM ONLY***. 7 pp.

IEC 9506 Pt 3-91. Industrial Automation Systems—Manufacturing Message Specification—Part 3: Companion Standard for Robotics First Edition; (Corrigendum 1-1993). 128 pp.

IEC 9506 Pt 4-92. Industrial Automation Systems—Manufacturing Message Specification—Part 4: Companion Standard for Numerical Control First Edition. 139 pp.

ISO 9506 Pt 1-90. Industrial Automation Systems—Manufacturing Message Specification—Part 1: Service Definition First Edition; (Amendment 1-1993). 334 pp.

ISO 9506 Pt 1 Draft AMD 1. Industrial Automation Systems—Manufacturing Message Specification—Part 1: Service Definition Amendment 1: Data Exchange; (1992) ***CD-ROM ONLY***. 11 pp.

ISO 9506 Pt 2-90. Industrial Automation Systems—Manufacturing Message Specification—Part 2: Protocol Specification AMENDMENT 1: Data Exchange First Edition; (Amendment 1-1993). 147 pp.

ISO 9506 Pt 2 Draft AMD 1. Industrial Automation Systems—Manufacturing Message Specification—Part 2: Protocol Specification Amendment 1: Data Exchange; (1992) ***CD-ROM ONLY***. 7 pp.

ISO 9506 Pt 3-91. Industrial Automation Systems—Manufacturing Message Specification—Part 3: Companion Standard for Robotics First Edition; (Corrigendum 1-1993). 128 pp.

ISO 9506 Pt 4-92. Industrial Automation Systems—Manufacturing Message Specification—Part 4: Companion Standard for Numerical Control First Edition. 139 pp.

JTC1 9506-90. Industrial Automation Systems—Manufacturing Message Specification—Part 1: Service Definition First Edition. 326 pp.

JTC1 9506 Pt1 Draft AMD 1. Industrial Automation Systems—Manufacturing Message Specification—Part 1: Service Definition Amendment 1: Data Exchange; (1992) ***CD-ROM ONLY***. 11 pp.

JTC1 9506-90. Industrial Automation Systems—Manufacturing Message Specification—Part 2: Protocol Specification First Edition. 139 pp.

JTC1 9506 Pt2 Draft AMD 1. Industrial Automation Systems—Manufacturing Message Specification—Part 2: Protocol Specification Amendment 1: Data Exchange; (1992) ***CD-ROM ONLY***. 7 pp.

JTC1 9506-91. Industrial Automation Systems—Manufacturing Message Specification—Part 3: Companion Standard for Robotics First Edition; (Corrigendum 1-1993). 128 pp.

JTC1 9506-92. Industrial Automation Systems—Manufacturing Message Specification—Part 4: Companion Standard for Numerical Control First Edition. 139 pp.

OSI ISO DIS 9506-1-88. Manufacturing Message Specification Part 1: Service Definition. 230 pp.

OSI ISO/IEC 9506-2-90. Industrial Automation Systems—Manufacturing Message Specification—Part 2: Protocol Specification. 139 pp.

OSI ISO IEC DIS 9506-2-88. Manufacturing Message Specification—Part 2: Protocol Specification. 90 pp.

OSI ISO DIS 9506-2-88. Manufacturing Message Specification Part 2: Protocol Specification. 101 pp.

OSI ISO/IEC DIS 9506-3-90. Manufacturing Message Specification—Part 3: Robot Specific Message System. 127 pp.

OSI ISO IEC DP 9506-5-89. Manufacturing Message Specification—Part 5: Programmable Controller Message Specification (PCMS Companion Standard). 90 pp.

OSI ISO IEC DP 9506-6-89. Manufacturing Message Specification—Part 6: Process Control Semantics for the Manufacturing Message Service and Protocol Standard. 119 pp.

SAA AS 4038.1-92. Industrial Automation Systems—Manufacturing Message Specification—Part 1: Service Definition (ISO/IEC 9506-1:1990). 316 pp.

INTERNATIONAL AND NON-U.S. NATIONAL STANDARDS
SUBJECT INDEX
Application

Application Layer (OSI) *(Cont.)*
—Manufacturing Message Services and Protocol *(Cont.)*
SAA AS 4038.2-92. Industrial Automation Systems—Manufacturing Message Specification—Part 2: Protocol Specification (ISO/IEC 9506-2:1990). 128 pp.
SAA AS 4038.3-92. Industrial Automation Systems—Manufacturing Message Specification—Part 3: Companion Standard for Robotics (ISO/IEC 9506-3:1991). 119 pp.

—Search and Retrieve—Protocol
BSI BS ISO 10163-1-93. 1993 Information and Documentation—Open Systems Interconnection—Search and Retrieve Application Protocol Specification—Part 1: Protocol Specification (S). 50 pp.
ISO 10163 Pt 1-93. Information and Documentation—Open Systems Interconnection—Search and Retrieve Application Protocol Specification—Part 1: Protocol Specification First Edition. 41 pp.
OSI ISO/DIS 10163-90. Documentation—Search and Retrieve Protocol Specification. 52 pp.

—Search and Retrieve—Service
BSI BS ISO 10162-93. 1993 Information and Documentation—Open Systems Interconnection—Search and Retrieve Application Service Definition (S). 28 pp.
ISO 10162-93. Information and Documentation—Open Systems Interconnection—Search and Retrieve Application Service Definition First Ediiton. 26 pp.
OSI ISO/DIS 10162-90. Documentation—Search and Retrieve Service Definition. 36 pp.

—SQL
IEC DIS 9579 Pt 2-91. Information Technology—Open Systems Interconnection—Remote Database Access—Part 2: SQL Specialization ***CD-ROM ONLY***. 55 pp.
ISO DIS 9579 Pt 2-91. Information Technology—Open Systems Interconnection—Remote Database Access—Part 2: SQL Specialization ***CD-ROM ONLY***. 55 pp.

—Systems Management
CCITT RECMN X.701-92. Information Technology—Open Systems Interconnection—Systems Management Overview (Study Group VII) 31 pp. 31 pp.
CCITT RECMN X.720-92. Information Technology—Open Systems Interconnection—Structure of Management Information: Management Information Model 32 pp (ISO/CCIT Common Text). 32 pp.
IEC 10040-92. Information Technology—Open Systems Interconnection—Systems Management Overview First Edition. 29 pp.
IEC 10040 Draft AMD 1. Information Technology—Open Systems Interconnection—Systems Management Overview Amendment 1: Management Knowledge Management Architecture; (1993) ***CD-ROM ONLY***. 3 pp.
IEC DIS 10164 Pt 11-92. Information Technology—Open Systems Interconnection—Systems Management—Part 11: Workload Monitoring Function ***CD-ROM ONLY***. 55 pp.
ISO 10040-92. Information Technology—Open Systems Interconnection—Systems Management Overview First Edition. 29 pp.
ISO 10040 Draft AMD 1. Information Technology—Open Systems Interconnection—Systems Management Overview Amendment 1: Management Knowledge Management Architecture; (1993) ***CD-ROM ONLY***. 3 pp.
ISO DIS 10164 Pt 11-92. Information Technology—Open Systems Interconnection—Systems Management—Part 11: Workload Monitoring Function ***CD-ROM ONLY***. 55 pp.
JTC1 10040-92. Information Technology—Open Systems Interconnection—Systems Management Overview First Edition. 29 pp.
JTC1 10040 Draft AMD 1. Information Technology—Open Systems Interconnection—Systems Management Overview Amendment 1: Management Knowledge Management Architecture; (1993) ***CD-ROM ONLY***. 3 pp.
OSI ISO/IEC DIS 10164-3-90. Information Technology—Open Systems Interconnection—Systems Management—Part 3: Attributes for Representing Relationships. 22 pp.

—Systems Management—Access Control
IEC DIS 10164 Pt 9-93. Information Technology—Open Systems Interconnection—Systems Management: Objects and Attributes for Access Control; (CCITT RECMN X.741) ***CD-ROM ONLY***. 97 pp.
JTC1 DIS10164 Pt 9-93. Information Technology—Open Systems Interconnection—Systems Management: Objects and Attributes for Access Control; (CCITT RECMN X.741) ***CD-ROM ONLY***. 97 pp.

Application Layer (OSI) *(Cont.)*
—Systems Management—Accounting
ISO DIS 10164 Pt 10-92. Information Technology—Open Systems Interconnection—Systems Management—Part 10: Accounting Metering Function ***CD-ROM ONLY***. 39 pp.
JTC1 DIS10164 Pt 10-92. Information Technology—Open Systems Interconnection—Systems Management—Part 10: Accounting Metering Function ***CD-ROM ONLY***. 39 pp.

—Systems Management—Alarm Reporting
BSI BS ISO/IEC 10164-4-92. 1992 Information Technology—Open Systems Interconnection—Systems Management: Alarm Reporting Function (S). 26 pp.
CCITT RECMN X.733-92. Information Technology—Open Systems Interconnection—Systems Management: Alarm Reporting Function (Study Group VII) 24 pp. 24 pp.
IEC 10164 Pt 4-92. Information Technology—Open Systems Interconnection—Systems Management: Alarm Reporting Function First Edition; (CCITT RECMN X.733). 24 pp.
IEC DISP 12059 Pt 4-92. Information Technology—International Standardized Profiles—Management Functions—Common Information for Management Functions—Part 4: Alarm Reporting ***CD-ROM ONLY***. 60 pp.
ISO 10164 Pt 4-92. Information Technology—Open Systems Interconnection—Systems Management: Alarm Reporting Function First Edition; (CCITT RECMN X.733). 24 pp.
ISO DISP 12059 Pt 4-92. Information Technology—International Standardized Profiles—Management Functions—Common Information for Management Functions—Part 4: Alarm Reporting ***CD-ROM ONLY***. 60 pp.
JTC1 10164 Pt 4-92. Information Technology—Open Systems Interconnection—Systems Management: Alarm Reporting Function First Edition; (CCITT RECMN X.733). 24 pp.
JTC1 DISP12059 Pt 4-92. Information Technology—International Standardized Profiles—Management Functions—Common Information for Management Functions—Part 4: Alarm Reporting ***CD-ROM ONLY***. 60 pp.
OSI ISO/IEC DIS 10164-4-90. Information Technology—Open Systems Interconnection—Systems Management—Part 4: Alarm Reporting Function. 18 pp.

—Systems Management—Log Control
CCITT RECMN X.735-92. Information Technology—Open Systems Interconnection—Systems Management: Log Control Function 22 pp (ISO/CCIT Common Text). 22 pp.
OSI ISO/IEC DIS 10164-6-90. Information Technology—Open Systems Interconnection—Systems Management—Part 6: Log Control Function. 23 pp.

—Systems Management—Object Management
BSI BS ISO/IEC 10164-1-93. 1993 Information Technology—Open Systems Interconnection—Systems Management: Object Management Function (S). 31 pp.
CCITT RECMN X.730-92. Information Technology—Open Systems Interconnection—Systems Management: Object Management Function 29 pp (ISO/CCIT Common Text). 29 pp.
IEC 10164 Pt 1-93. Information Technology—Open Systems Interconnection—Systems Management: Object Management Function First Edition; (CCITT RECMN X.730). 29 pp.
IEC DIS 10164 Pt 9-93. Information Technology—Open Systems Interconnection—Systems Management: Objects and Attributes for Access Control; (CCITT RECMN X.741) ***CD-ROM ONLY***. 97 pp.
ISO 10164 Pt 1-93. Information Technology—Open Systems Interconnection—Systems Management: Object Management Function First Edition; (CCITT RECMN X.730). 29 pp.
ISO DIS 10164 Pt 9-93. Information Technology—Open Systems Interconnection—Systems Management: Objects and Attributes for Access Control; (CCITT RECMN X.741) ***CD-ROM ONLY***. 97 pp.
JTC1 10164 Pt 1-93. Information Technology—Open Systems Interconnection—Systems Management: Object Management Function First Edition; (CCITT RECMN X.730). 29 pp.
JTC1 DIS10164 Pt 9-93. Information Technology—Open Systems Interconnection—Systems Management: Objects and Attributes for Access Control; (CCITT RECMN X.741) ***CD-ROM ONLY***. 97 pp.
OSI ISO/IEC DIS 10164-1-90. Information Technology—Open Systems Interconnection—Systems Management Part 1: Object Management Function. 30 pp.

Application Layer (OSI) *(Cont.)*
—Systems Management—Relationship Attributes
BSI BS ISO/IEC 10164-3-93. 1993 Information Technology—Open Systems Interconnection—Systems Management: Attributes for Representing Relationships (S). 28 pp.
CCITT RECMN X.732-92. Information Technology—Open Systems Interconnection—Systems Management: Attributes for Representing Relationships 22 pp (ISO/CCIT Common Text). 22 pp.
IEC 10164 Pt 3-93. Information Technology—Open Systems Interconnection—Systems Management: Attributes for Representing Relationships First Edition; (CCITT RECMN X.732). 22 pp.
ISO 10164 Pt 3-93. Information Technology—Open Systems Interconnection—Systems Management: Attributes for Representing Relationships First Edition; (CCITT RECMN X.732). 22 pp.
JTC1 10164 Pt 3-93. Information Technology—Open Systems Interconnection—Systems Management: Attributes for Representing Relationships First Edition; (CCITT RECMN X.732). 22 pp.

—Systems Management—Report Management
BSI BS ISO/IEC 10164-5-93. 1993 Information Technology—Open Systems Interconnection—Systems Management: Event Report Management Function (S). 26 pp.
CCITT RECMN X.734-93. Information Technology—Open Systems Interconnection—Systems Management: Event Report Management Function 22 pp (ISO/CCIT Common Text). 22 pp.
IEC 10164 Pt 5-93. Information Technology—Open Systems Interconnection—Systems Management: Event Report Management Function First Edition; (CCITT RECMN X.734). 24 pp.
ISO 10164 Pt 5-93. Information Technology—Open Systems Interconnection—Systems Management: Event Report Management Function First Edition; (CCITT RECMN X.734). 24 pp.
JTC1 10164 Pt 5-93. Information Technology—Open Systems Interconnection—Systems Management: Event Report Management Function First Edition; (CCITT RECMN X.734). 24 pp.
OSI ISO/IEC DIS 10164-5-90. Information Technology—Open Systems Interconnection—Systems Management—Part 5: Event Report Management Function. 24 pp.

—Systems Management—Scheduling
IEC DIS 10164 Pt 15-93. Information Technology—Open Systems Interconnection—Systems Management—Part 15: Scheduling Function ***CD-ROM ONLY***. 61 pp.
ISO DIS 10164 Pt 15-93. Information Technology—Open Systems Interconnection—Systems Management—Part 15: Scheduling Function ***CD-ROM ONLY***. 61 pp.
JTC1 DIS 10164-93. Information Technology—Open Systems Interconnection—Systems Management—Part 15: Scheduling Function ***CD-ROM ONLY***. 61 pp.

—Systems Management—Security Alarm Reporting
BSI BS ISO/IEC 10164-7-92. 1992 Information Technology—Open Systems Interconnection—Systems Management: Security Alarm Reporting Function (S). 20 pp.
CCITT RECMN X.736-92. Information Technology—Open Systems Interconnection—Systems Management: Security Alarm Reporting Function (Study Group VII) 18 pp. 18 pp.
IEC 10164 Pt 7-92. Information Technology—Open Systems Interconnection—Systems Management: Security Alarm Reporting Function First Edition; (CCITT RECMN X.736:1992). 18 pp.
IEC DISP 12060 Pt 2-92. Information Technology—International Standardized Profiles AOMnnn—OSI Management—Management Functions—Part 2: AOM 212—Alarm Reporting and State Management Capabilities ***CD-ROM ONLY***. 33 pp.
IEC DISP 12060 Pt 3-92. Information Technology—International Standardized Profiles AOMnnn—OSI Management—Management Functions—Part 3: AOM213—Alarm Reporting Capabilities ***CD-ROM ONLY***. 32 pp.
ISO 10164 Pt 7-92. Information Technology—Open Systems Interconnection—Systems Management: Security Alarm Reporting Function First Edition; (CCITT RECMN X.736). 18 pp.
ISO DISP 12060 Pt 2-92. Information Technology—International Standardized Profiles AOMnnn—OSI Management—Management Functions—Part 2: AOM 212—Alarm Reporting and State Management Capabilities ***CD-ROM ONLY***. 33 pp.

INDUSTRY STANDARDS

Application
INTERNATIONAL AND NON-U.S. NATIONAL STANDARDS
SUBJECT INDEX

Application Layer (OSI) *(Cont.)*

—Systems Management—Security Alarm Reporting *(Cont.)*

ISO DISP 12060 Pt 3-92. Information Technology—International Standardized Profiles AOMnnn—OSI Management—Management Functions—Part 3: AOM213—Alarm Reporting Capabilities ***CD-ROM ONLY***. 32 pp.
JTC1 10164 Pt 7-92. Information Technology—Open Systems Interconnection—Systems Management: Security Alarm Reporting Function First Edition; (CCITT RECMN X.736:1992). 18 pp.
JTC1 DISP12060 Pt 2-92. Information Technology—International Standardized Profiles AOMnnn—OSI Management—Management Functions—Part 2: AOM 212—Alarm Reporting and State Management Capabilities ***CD-ROM ONLY***. 33 pp.
JTC1 DISP12060 Pt 3-92. Information Technology—International Standardized Profiles AOMnnn—OSI Management—Management Functions—Part 3: AOM213—Alarm Reporting Capabilities ***CD-ROM ONLY***. 32 pp.
OSI ISO/IEC DIS 10164-7-90. Information Technology—Open Systems Interconnection—Systems Management—Part 7: Security Alarm Reporting Function. 15 pp.

—Systems Management—Security Audit Trail

CCITT RECMN X.740-92. Information Technology—Open Systems Interconnection—Systems Management: Security Audit Trail Function 30 pp (ISO/CCIT Common Text). 30 pp.

—Systems Management—State Management

BSI BS ISO/IEC 10164-2-93. 1993 Information Technology—Open Systems Interconnection—Systems Management: State Management Function (S). 31 pp.
CCITT RECMN X.731-92. Information Technology—Open Systems Interconnection—Systems Management: State Management Function 28 pp (ISO/CCIT Common Text). 28 pp.
IEC 10164 Pt 2-93. Information Technology—Open Systems Interconnection—Systems Management: State Management Function First Edition; (CCITT RECMN X.731). 28 pp.
ISO 10164 Pt 2-93. Information Technology—Open Systems Interconnection—Systems Management: State Management Function First Edition; (CCITT RECMN X.731). 28 pp.
JTC1 10164 Pt 2-93. Information Technology—Open Systems Interconnection—Systems Management: State Management Function First Edition; (CCITT RECMN X.731). 28 pp.
OSI ISO/IEC DIS 10164-2-90. Information Technology—Open Systems Interconnection—Systems Management Part 2: State Management Function. 24 pp.

—Systems Management—Summarization Function

IEC DIS 10164 Pt 13-92. Information Technology—Open Systems Interconnection—Systems Management—Part 13: Summarization Function ***CD-ROM ONLY***. 85 pp.
ISO DIS 10164 Pt 13-92. Information Technology—Open Systems Interconnection—Systems Management—Part 13: Summarization Function ***CD-ROM ONLY***. 85 pp.
JTC1 DIS10164 Pt 13-92. Information Technology—Open Systems Interconnection—Systems Management—Part 13: Summarization Function ***CD-ROM ONLY***. 85 pp.

—Systems Management—Test Management

IEC DIS 10164 Pt 12-91. Information Technology—Open Systems Interconnection—Systems Management—Part 12: Test Management Function ***CD-ROM ONLY***. 51 pp.
IEC DIS 10164 Pt 14-93. Information Technology—Open Systems Interconnection—Systems Management—Part 14: Confidence and Diagnostic Test Categories ***CD-ROM ONLY***. 104 pp.
ISO DIS 10164 Pt 12-91. Information Technology—Open Systems Interconnection—Systems Management—Part 12: Test Management Function ***CD-ROM ONLY***. 51 pp.
ISO DIS 10164 Pt 14-93. Information Technology—Open Systems Interconnection—Systems Management—Part 14: Confidence and Diagnostic Test Categories ***CD-ROM ONLY***. 104 pp.
JTC1 DIS10164 Pt 14-93. Information Technology—Open Systems Interconnection—Systems Management—Part 14: Confidence and Diagnostic Test Categories ***CD-ROM ONLY***. 104 pp.

—Virtual Terminal

BSI BS ISO 9040-90. 1990 Amd 1 Information Technology—Open Systems Interconnection—Virtual Terminal Basic Class Service (AMD 6773) August 30, 1991. 90 pp.

Application Layer (OSI) *(Cont.)*

—Virtual Terminal *(Cont.)*

BSI BS ISO 9040-02. 1990 Amd 2 Information Technology—Open Systems Interconnection—Virtual Terminal Basic Class Service (AMD 7620) February 15, 1993 (S). 117 pp.
BSI BS ISO 9041-1-90. 1990 Information Technology—Open Systems Interconnection—Virtual Terminal Basic Class Protocol—Part 1: Specification. 74 pp.
BSI BS ISO 9041-1-01. 1990 Amd 1 Information Technology—Open Systems Interconnection—Virtual Terminal Basic Class Protocol—Part 1: Specification (AMD 7680) April 15, 1993 (Technical Corr 1) (S). 81 pp.
BSI BS ISO 9041-1-02. 1990 Amd 2 Information Technology—Open Systems Interconnection—Virtual Terminal Basic Class Protocol—Part 1: Specification (AMD 7621) February 15, 1993 (S). 90 pp.
BSI DD ENV 41208-1-91. 1991 Information Systems Interconnection—Basic Class Virtual Terminal—S-Mode Forms Part 1: Virtual Terminal Service. 85 pp.
BSI DD ENV 41208-2-91. 1991 Information Systems Interconnection—Basic Class Virtual Terminal—S-Mode Forms Part 2: Check List. 46 pp.
BSI DD ENV 41208-3-91. 1991 Information Systems Interconnection—Basic Class Virtual Terminal—S-Mode Forms Part 3: Underlying Layers Check List. 38 pp.
BSI DD ENV 41209-91. 1991 Information Systems Interconnection—Basic Class Virtual Terminal—Common Control Objects. 13 pp.
BSI DD ENV 41211-1-93. 1993 Information Systems Interconnection—Virtual Terminal—Basic Class—A-Mode X.3 Part 1: Virtual Terminal Service. 19 pp.
BSI DD ENV 41211-2-93. 1993 Information Systems Interconnection—Virtual Terminal—Basic Class—A-Mode X.3 Part 2: Virtual Terminal Protocol Check List. 18 pp.
BSI DD ENV 41211-3-93. 1993 Information Systems Interconnection—Virtual Terminal—Basic Class—A-Mode X.3 Part 3: Underlying Layers Check List. 19 pp.
BSI DD ENV 41213-1-93. 1993 Information Systems Interconnection—Virtual Terminal—Basic Class—A-Mode "Telnet" Part 1: Virtual Terminal Service. 16 pp.
BSI DD ENV 41213-2-93. 1993 Information Systems Interconnection—Virtual Terminal—Basic Class—A-Mode "Telnet" Part 2: Virtual Terminal Protocol Check List. 20 pp.
BSI DD ENV 41213-3-93. 1993 Information Systems Interconnection—Virtual Terminal—Basic Class—A-Mode "Telnet" Part 3: Underlying Layers Check List. 19 pp.
BSI DD ENV 41513-93. 1993 Information Systems Interconnection—Virtual Terminal—Basic Class—VT Font Assignment Type 1. 21 pp.
CEN ENV 41208 (Part 1)-90. Information Systems Interconnections—Basic Class Virtual Terminal—S-Mode Forms—Part 1: Virtual Terminal Service. 83 pp.
CEN ENV 41208 (Part 2)-90. Information Systems Interconnections—Basic Class Virtual Terminal—S-Mode Forms—Part 2: Check List. 44 pp.
CEN ENV 41208 (Part 3)-90. Information Systems Interconnections—Basic Class Virtual Terminal—S-Mode Forms—Part 3: Underlying Layers Check List. 36 pp.
CEN ENV 41209-90. Information System Interconnection—Basic Class Virtual Terminal—Common Control Objects. 11 pp.
CEN ENV 41211-1-92. Information Systems Interconnection—Virtual Terminal—Basic Class—A-Mode X.3—Part 1: Virtual Terminal Service. 16 pp.
CEN ENV 41211-2-92. Information Systems Interconnection—Virtual Terminal Basic Class—A-Mode X.3—Part 2: Virtual Terminal Protocol Check List. 15 pp.
CEN ENV 41211-3-92. Information Systems Interconnection—Virtual Terminal Basic Class—A-Mode X.3—Part 3: Underlying Layers Checklist. 16 pp.
CEN ENV 41213-1-92. Information Systems Interconnection—Virtual Terminal—Basic Class—A-Mode "Telnet"—Part 1: Virtual Terminal Service. 13 pp.
CEN ENV 41213-2-92. Information Systems Interconnection—Virtual Terminal—Basic Class—A-Mode "Telnet"—Part 2: Virtual Terminal Protocol Check List. 17 pp.
CEN ENV 41213-3-92. Information Systems Interconnection—Virtual Terminal—Basic Class—A-Mode "Telnet"—Part 3: Underlying Layers Check List. 16 pp.
CSA CAN/CSA-Z243. 175-91. Information Technology—Open Syestems Interconnection—Virtual Terminal Basic Class Service (ISO 9040: 1990); (Gen Instr 1). 97 pp.
CSA CAN/CSA-Z243. 176-91. Information Technology—Open Systems Interconnection—Virtual Terminal Basic Class Protocol—Part 1: Specification (ISO 9041-1:1990); (Gen Instr 1). 83 pp.

Application Layer (OSI) *(Cont.)*

—Virtual Terminal *(Cont.)*

ECMA ECMA 87-83. Generic Virtual Terminal. Service and Protocol Description. 71 pp.
IEC 9040 COR 1. Corrigendum 1—Information Technology—Open Systems Interconnection—Virtual Terminal Basic Class Service; (1991). 1 p.
IEC 9040 COR 2. Corrigendum 2—Information Technology—Open Systems Interconnection—Virtual Terminal Basic Class Service; (1992). 3 pp.
IEC 9040 AMD 2. Amendment 2—Information Technology—Open Systems Interconnection—Virtual Terminal Basic Class Service Amendment 2: Additional Functional Units; (1992). 19 pp.
IEC 9041 Pt 1 AMD 2. Amendment 2—Information Technology—Open Systems Interconnection—Virtual Terminal Basic Class Protocol Part 1: Specification Amendment 2: Additional Functional Units; (1992). 12 pp.
IEC 9041 Pt 1 COR 1. Corrigendum 1-Information Technology—Open Systems Interconnection—Virtual Terminal Basic Class Protocol—Part 1: Specification; (1992). 3 pp.
IEC DIS 10739 Pt 1-91. Information Technology—Open Systems Interconnection—Virtual Terminal Basic Class Protocol—Part 1: Test Suite Structure and Test Purposes ***CD-ROM ONLY***. 151 pp.
ISO 9040-90. Information Technology—Open Systems Interconnection—Virtual Terminal Basic Class Service First Edition; (Corrigendum 1-1991) (Amendment 2-1992) (Corrigendum 2-1992). 109 pp.
ISO 9041 Pt 1-90. Information Technology—Open Systems Interconnection—Virtual Terminal Basic Class Protocol—Part 1: Specification First Edition; (Amendment 2-1992) (Corrigendum 1-1992). 87 pp.
ISO DIS 10739 Pt 1-91. Information Technology—Open Systems Interconnection—Virtual Terminal Basic Class Protocol—Part 1: Test Suite Structure and Test Purposes ***CD-ROM ONLY***. 151 pp.
JTC1 9040-90. Information Technology—Open Systems Interconnection—Virtual Terminal Basic Class Service First Edition; (Corrigendum 1-1991) (Amendment 2-1992) (Corrigendum 2-1992). 19 pp.
JTC1 9041-90. Information Technology—Open Systems Interconnection—Virtual Terminal Basic Class Protocol—Part 1: Specification First Edition; (Amendment 2-1992) (Corrigendum 1-1992). 12 pp.
OSI ISO 9040-90. Information Technology—Open Systems Interconnection—Virtual Terminal Basic Class Service. 86 pp.
OSI ISO 9040 DAD 1-88. Information Processing Systems—Open Systems Interconnection—Virtual Terminal Service —Basic Class—Addendum 1: Extended Facility Set. 89 pp.
OSI ISO 9040/DAM 2-90. Information Technology—Open Open Systems Interconnection—Virtual Terminal Basic Class Service-Amendment 2: Additional Functional Units. 14 pp.
OSI ISO DIS 9040-2-87. Information Processing Systems—Open Systems Interconnection—Virtual Terminal Service—Basic Class. 162 pp.
OSI ISO IEC 9041 DAD 1-88. Information Processing Systems—Open Systems Interconnection Virtual Terminal Protocol—Basic Class—Addendum 1: Extended Facility Set. 36 pp.
OSI ISO 9041-1/DAM 2-90. Information Technology—Open Systems Interconnection—Virtual Terminal Basic Class 1: Specification AMENDMENT 2: Additional Functional Units. 17 pp.
OSI ISO IEC 9041 DAD 1. Information Processing Systems—Open Systems Interconnection Virtual Terminal Protocol—Basic Class Addendum 1: Extended Facility Set.
OSI ISO DIS 9041-2-87. Information Processing Systems—Open Systems Interconnection—Virtual Terminal Protocol—Basic Class. 112 pp.
SAA AS 4017-92. Information Technology—Open Systems Interconnection—Virtual Terminal Basic Class Service (ISO 9040:1990) (in Professional Package 26A, 26C). 78 pp.
SAA AS 4018.1-92. Information Technology—Open Systems Interconnection—Virtual Terminal Basic Class Protocol—Part 1: Specification (ISO 9041-1:1990) (in Professional Packages 26A, 26C). 65 pp.

—Virtual Terminal—Conformance

IEC DIS 9041 Pt 2-91. Information Technology—Open Systems Interconnection—Basic Class Virtual Terminal Protocol—Part 2: Protocol Implementation Conformance Statement (PICS proforma) ***CD-ROM ONLY***. 66 pp.
ISO DIS 9041 Pt 2-91. Information Technology—Open Systems Interconnection—Basic Class Virtual Terminal Protocol—Part 2: Protocol Implementation Conformance Statement (PICS proforma) ***CD-ROM ONLY***. 66 pp.
JTC1 DIS9041 Pt 2-91. Information Technology—Open Systems Interconnection—Basic Class Virtual Terminal Protocol—Part 2: Protocol Implementation Conformance Statement (PICS proforma) ***CD-ROM ONLY***. 66 pp.

202
INDEX and DIRECTORY of

INTERNATIONAL AND NON-U.S. NATIONAL STANDARDS
SUBJECT INDEX Application

Application Layer (OSI) *(Cont.)*

—Virtual Terminal—Control Objects

IEC DISP 11185 Pt 1-93. Information Technology—International Standardized Profiles FVT2nn—Virtual Terminal Basic Class—Register of Control Object Type Definitions —Part 1: FVT211, FVT212—Sequenced and Unsequenced Application Control Objects ***CD-ROM ONLY***. 26 pp.

IEC DISP 11185 Pt 2-93. Information Technology—International Standardized Profiles FVT2nn—Virtual Terminal Basic Class—Register of Control Object Type Definitions —Part 2: FVT213, FVT214—Sequenced and Unsequenced Terminal Control Objects ***CD-ROM ONLY***. 26 pp.

IEC DISP 11185 Pt 3-93. Information Technology—International Standardized Profiles FVT2nn—Virtual Terminal Basic Class—Register of Control Object Type Definitions —Part 3: FVT215, FVT216—Application RIO Record Loading Control Object, Terminal RIO Record Notification. 31 pp.

IEC DISP 11185 Pt 4-93. Information Technology—International Standardized Profiles FVT2nn—Virtual Terminal Basic Class—Register of Control Object Type Definitions —Part 4: FVT217—Horizontal Tabulation Control Object ***CD-ROM ONLY***. 23 pp.

IEC DISP 11185 Pt 5-93. Information Technology—International Standardized Profiles FVT2nn—Virtual Terminal Basic Class—Register of Control Object Type Definitions —Part 5: FVT218—Logical Image Control Object ***CD-ROM ONLY***. 22 pp.

IEC DISP 11185 Pt 6-93. Information Technology—International Standardized Profiles FVT2nn—Virtual Terminal Basic Class—Register of Control Object Type Definitions —Part 6: FVT219—Status Message Control Object ***CD-ROM ONLY***. 24 pp.

IEC DISP 11185 Pt 7-93. Information Technology—International Standardized Profiles FVT2nn—Virtual Terminal Basic Class—Register of Control Object Type Definitions —Part 7: FVT2110—Entry-Control Control Object ***CD-ROM ONLY***. 23 pp.

IEC DISP 11185 Pt 8-93. Information Technology—International Standardized Profiles FVT2nn—Virtual Terminal Basic Class—Register of Control Object Type Definitions —Part 8: FVT221—Forms FEICO (Field Entry Instruction Control Object) No. 1 ***CD-ROM ONLY***. 36 pp.

IEC DISP 11185 Pt 9-93. Information Technology—International Standardized Profiles FVT2nn—Virtual Terminal Basic Class—Register of Control Object Type Definitions —Part 9: FVT222—Paged FEICO (Field Entry Instruction Control Object) No. 1 ***CD-ROM ONLY***. 33 pp.

IEC DISP 11185 Pt 10-93. Information Technology—International Standardized Profiles FVT2nn—Virtual Terminal Basic Class—Register of Control Object Type Definitions —Part 10: FVT231—Forms FEPCO (Field Entry Pilot Control Object) No. 1 ***CD-ROM ONLY***. 35 pp.

IEC DISP 11185 Pt 11-93. Information Technology—International Standardized Profiles FVT2nn—Virtual Terminal Basic Class—Register of Control Object Type Definitions —Part 11: FVT232—Paged FEPCO (Field Entry Pilot Control Object) No. 1 ***CD-ROM ONLY***. 33 pp.

ISO DISP 11185 Pt 1-93. Information Technology—International Standardized Profiles FVT2nn—Virtual Terminal Basic Class—Register of Control Object Type Definitions —Part 1: FVT211, FVT212—Sequenced and Unsequenced Application Control Objects ***CD-ROM ONLY***. 26 pp.

ISO DISP 11185 Pt 2-93. Information Technology—International Standardized Profiles FVT2nn—Virtual Terminal Basic Class—Register of Control Object Type Definitions —Part 2: FVT213, FVT214—Sequenced and Unsequenced Terminal Control Objects ***CD-ROM ONLY***. 26 pp.

ISO DISP 11185 Pt 3-93. Information Technology—International Standardized Profiles FVT2nn—Virtual Terminal Basic Class—Register of Control Object Type Definitions —Part 3: FVT215, FVT216—Application RIO Record Loading Control Object, Terminal RIO Record Notification. 31 pp.

ISO DISP 11185 Pt 4-93. Information Technology—International Standardized Profiles FVT2nn—Virtual Terminal Basic Class—Register of Control Object Type Definitions —Part 4: FVT217—Horizontal Tabulation Control Object ***CD-ROM ONLY***. 23 pp.

ISO DISP 11185 Pt 5-93. Information Technology—International Standardized Profiles FVT2nn—Virtual Terminal Basic Class—Register of Control Object Type Definitions —Part 5: FVT218—Logical Image Control Object ***CD-ROM ONLY***. 23 pp.

ISO DISP 11185 Pt 6-93. Information Technology—International Standardized Profiles FVT2nn—Virtual Terminal Basic Class—Register of Control Object Type Definitions —Part 6: FVT219—Status Message Control Object ***CD-ROM ONLY***. 24 pp.

ISO DISP 11185 Pt 7-93. Information Technology—International Standardized Profiles FVT2nn—Virtual Terminal Basic Class—Register of Control Object Type Definitions —Part 7: FVT2110—Entry-Control Control Object ***CD-ROM ONLY***. 23 pp.

Application Layer (OSI) *(Cont.)*

—Virtual Terminal—Control Objects *(Cont.)*

ISO DISP 11185 Pt 8-93. Information Technology—International Standardized Profiles FVT2nn—Virtual Terminal Basic Class—Register of Control Object Type Definitions —Part 8: FVT221—Forms FEICO (Field Entry Instruction Control Object) No. 1 ***CD-ROM ONLY***. 36 pp.

ISO DISP 11185 Pt 9-93. Information Technology—International Standardized Profiles FVT2nn—Virtual Terminal Basic Class—Register of Control Object Type Definitions —Part 9: FVT222—Paged FEICO (Field Entry Instruction Control Object) No. 1 ***CD-ROM ONLY***. 33 pp.

ISO DISP 11185 Pt 10-93. Information Technology—International Standardized Profiles FVT2nn—Virtual Terminal Basic Class—Register of Control Object Type Definitions —Part 10: FVT231—Forms FEPCO (Field Entry Pilot Control Object) No. 1 ***CD-ROM ONLY***. 35 pp.

ISO DISP 11185 Pt 11-93. Information Technology—International Standardized Profiles FVT2nn—Virtual Terminal Basic Class—Register of Control Object Type Definitions —Part 11: FVT232—Paged FEPCO (Field Entry Pilot Control Object) No. 1 ***CD-ROM ONLY***. 33 pp.

JTC1 DISP 11185 Pt 1-93. Information Technology—International Standardized Profiles FVT2nn—Virtual Terminal Basic Class—Register of Control Object Type Definitions —Part 1: FVT211, FVT212—Sequenced and Unsequenced Application Control Objects ***CD-ROM ONLY***. 26 pp.

JTC1 DISP 11185 Pt 2-93. Information Technology—International Standardized Profiles FVT2nn—Virtual Terminal Basic Class—Register of Control Object Type Definitions —Part 2: FVT213, FVT214—Sequenced and Unsequenced Terminal Control Objects ***CD-ROM ONLY***. 26 pp.

JTC1 DISP 11185 Pt 3-93. Information Technology—International Standardized Profiles FVT2nn—Virtual Terminal Basic Class—Register of Control Object Type Definitions —Part 3: FVT215, FVT216—Application RIO Record Loading Control Object, Terminal RIO Record Notification. 31 pp.

JTC1 DISP 11185 Pt 4-93. Information Technology—International Standardized Profiles FVT2nn—Virtual Terminal Basic Class—Register of Control Object Type Definitions —Part 4: FVT217—Horizontal Tabulation Control Object ***CD-ROM ONLY***. 23 pp.

JTC1 DISP 11185 Pt 5-93. Information Technology—International Standardized Profiles FVT2nn—Virtual Terminal Basic Class—Register of Control Object Type Definitions —Part 5: FVT218—Logical Image Control Object ***CD-ROM ONLY***. 22 pp.

JTC1 DISP 11185 Pt 6-93. Information Technology—International Standardized Profiles FVT2nn—Virtual Terminal Basic Class—Register of Control Object Type Definitions —Part 6: FVT219—Status Message Control Object ***CD-ROM ONLY***. 24 pp.

JTC1 DISP 11185 Pt 7-93. Information Technology—International Standardized Profiles FVT2nn—Virtual Terminal Basic Class—Register of Control Object Type Definitions —Part 7: FVT2110—Entry-Control Control Object ***CD-ROM ONLY***. 23 pp.

JTC1 DISP 11185 Pt 8-93. Information Technology—International Standardized Profiles FVT2nn—Virtual Terminal Basic Class—Register of Control Object Type Definitions —Part 8: FVT221—Forms FEICO (Field Entry Instruction Control Object) No. 1 ***CD-ROM ONLY***. 36 pp.

JTC1 DISP 11185 Pt 9-93. Information Technology—International Standardized Profiles FVT2nn—Virtual Terminal Basic Class—Register of Control Object Type Definitions —Part 9: FVT222—Paged FEICO (Field Entry Instruction Control Object) No. 1 ***CD-ROM ONLY***. 33 pp.

JTC1 DISP 11185 Pt 10-93. Information Technology—International Standardized Profiles FVT2nn—Virtual Terminal Basic Class—Register of Control Object Type Definitions —Part 10: FVT231—Forms FEPCO (Field Entry Pilot Control Object) No. 1 ***CD-ROM ONLY***. 35 pp.

JTC1 DISP 11185 Pt 11-93. Information Technology—International Standardized Profiles FVT2nn—Virtual Terminal Basic Class—Register of Control Object Type Definitions —Part 11: FVT232—Paged FEPCO (Field Entry Pilot Control Object) No. 1 ***CD-ROM ONLY***. 33 pp.

—Virtual Terminal—Fonts

CEN ENV 41513-92. Information Systems Interconnection—Virtual Terminal—Basic Class—VT Font Assignment Type 1. 18 pp.

Application Program Interfaces

See Also: Communication Interfaces; Peripheral Interfaces

IEC DIS 11685.2-91. Information Technology—IEEE Trial-Use Standard Specifications for Microprocessor Operating Systems Interfaces (MOSI) ***CD-ROM ONLY***. 250 pp.

Application Program Interfaces *(Cont.)*

ISO DIS 11685.2-91. Information Technology—IEEE Trial-Use Standard Specifications for Microprocessor Operating Systems Interfaces (MOSI) ***CD-ROM ONLY***. 250 pp.

—Microprocessors

IEC DIS 11685.2-91. Information Technology—IEEE Trial-Use Standard Specifications for Microprocessor Operating Systems Interfaces (MOSI) ***CD-ROM ONLY***. 250 pp.

ISO DIS 11685.2-91. Information Technology—IEEE Trial-Use Standard Specifications for Microprocessor Operating Systems Interfaces (MOSI) ***CD-ROM ONLY***. 250 pp.

Application Protocols

See Also: Protocols

—Association Control Service Element

BSI BS 7091-89. 1989 Amd 1 Open Systems Interconnection: Service Definition for the Association Control Service Element (ISO 8649: 1988) (AMD 6733) July 31, 1991 (Technical Corr 1) (AMD 6772) July 31, 1991. 34 pp.

BSI BS 7091-02. 1989 Amd 2 Open Systems Interconnection: Service Definition for the Association Control Service Element (ISO 8649: 1988) (AMD 6929) April 1, 1992. 46 pp.

BSI BS 7092-89. 1989 Amd 1 Open Systems Interconnection: Protocol Specification for the Association Control Service Element (ISO 8650: 1988) (AMD 6734) July 31, 1991 (Technical Corr 1) (AMD 6735) July 31, 1991. 54 pp.

BSI BS ISO/IEC 10035-91. 1991 Information Technology—Open Systems Interconnection—Connectionless ACSE Protocol Specification. 19 pp.

BSI BS ISO/IEC ISP 10607-1-90. 1990 Amd 1 Information Technology—International Standardized Profiles AFTnn—File Transfer, Access and Management—Part 1: Specification of ACSE, Presentation and Session Protocols for the Use by FTAM (Technical Corr 1). 22 pp.

CSA CAN/CSA-Z243.170-89. Information Processing Systems—Open Systems Interconnection—Service Definition for the Association Control Service Element (ISO 8649:1988). 23 pp.

CSA CAN/CSA-Z243.171-89. Information Processing Systems—Open Systems Interconnection—Protocol Specification for the Association Control Service Element (ISO 8650:1988). 39 pp.

IEC 8649 COR 1. Corrigendum 1—Information Processing Systems—Open Systems Interconnection—Service Definition for the Association Control Service Element; (1991). 2 pp.

IEC DIS 8650 Pt 2-92. Information Technology—Open Systems Interconnection—Protocol Specification for the Association Control Service Element—Part 2: Protocol Implementation Conformance Statement (PICS) Proforma ***CD-ROM ONLY***. 15 pp.

IEC 10035-91. Information Technology—Open Systems Interconnection—Connectionless ACSE Protocol Specification First Edition. 16 pp.

IEC ISP10607 Pt 1-90. Information Technology—International Standardized Profiles AFTnn—File Transfer, Access and Management—Part 1: Specification of ACSE, Presentation and Session Protocols for the Use by FTAM First Edition; (Corrected and Reprinted -1991). 20 pp.

ISO 8649-88. Information Processing Systems—Open Systems Interconnection—Ser-vice Definition for the Association Control Ser-vice Element Amd 1: Au-thentication During As-sociation Establishment Amd 2: Connectionless-Mode ACSE Service First Edition; (Amd 1-1990) (Cor 1-1991)(Amd 2-1991). 36 pp.

ISO 8650-88. Information Processing Systems—Open Systems Interconnection—Protocol Specification for the Association Control Service Element First Edition; (Amendment 1-1990) (Corrigendum 1-1990). 52 pp.

ISO DIS 8650 Pt 2-92. Information Technology—Open Systems Interconnection—Protocol Specification for the Association Control Service Element—Part 2: Protocol Implementation Conformance Statement (PICS) Proforma ***CD-ROM ONLY***. 15 pp.

ISO 10035-91. Information Technology—Open Systems Interconnection—Connectionless ACSE Protocol Specification First Edition. 16 pp.

ISO ISP10607 Pt 1-90. Information Technology—International Standardized Profiles AFTnn—File Transfer, Access and Management—Part 1: Specification of ACSE, Presentation and Session Protocols for the Use by FTAM First Edition; (Corrected and Reprinted -1991). 20 pp.

JIS X 5701-91. Information Processing Systems—Open Systems Interconnection—Service Definition for the Association Control Service Element (ISO 8649:1988).

INDUSTRY STANDARDS 203

Application
INTERNATIONAL AND NON-U.S. NATIONAL STANDARDS
SUBJECT INDEX

Application Protocols *(Cont.)*

—Association Control Service Element *(Cont.)*

JIS X 5702-91. Information Processing Systems—Open Systems Interconnection—Protocol Specification for the Association Control Service Element (ISO 8650:1988).
JTC1 DIS8650 Pt 2-92. Information Technology—Open Systems Interconnection—Protocol Specification for the Association Control Service Element—Part 2: Protocol Implementation Conformance Statement (PICS) Proforma ***CD-ROM ONLY***. 15 pp.
OSI ISO 8649 DAD 2-89. Information Processing Systems—Open Systems Interconnection—Service Definition for the Association Control Service Element ADDENDUM 2: Connectionless-Mode ACSE Service. 8 pp.
OSI ISO 8649-88. Information Processing Systems—Open Systems Interconnection—Service Definition for the Association Control Service Element (with Amendment 1). 30 pp.
OSI ISO IEC 8650 DAD 1-89. Revised Text of ISO 8650 PDAD 1: Information Processing Systems—Open Systems Interconnection—Protocol Specification for the Association Control Service Element—Addendum 1: Peer Entity Authentication During Association Establishment. 16 pp.
OSI ISO 8650 PDAD 2-89. Information Processing Systems—Open Systems Interconnection—Protocol Specification for the Association Control Service Element—Proposed Draft Addendum 2: Protocol Implementation Conformance Statement (PICS) Proforma. 12 pp.
OSI ISO 8650 DAD 1-89. Information Processing Systems—Open Systems Interconnection—Protocol Specification for the Association Control Service Element—Addendum 1: Peer-Entity Authentication During Association Establishment. 16 pp.
OSI ISO 8650-88. Information Processing Systems—Open Systems Interconnection—Protocol Specification for the Association Control Service Element (with Amendment 1). 58 pp.
OSI ISO/IEC DIS 10035-89. Information Processing Systems—Open Systems Interconnection—Connectionless ACSE Protocol Specification. 13 pp.
OSI ISO/IEC DISP 10607-1-90. Information Technology—International Standardized Profile AFTnn—File Transfer, Access and Management—Part 1: Specification of ACSE, Presentation and Session Protocols for the Use by FTAM. 41 pp.
OSI ISO/IEC ISP 10607-1-90. Information Technology—International Standard-ized Profile AFTnn—File Transfer, Access and Management—Part 1: Specification of ACSE, Presentation and Session Protocols for the Use by FTAM. 20 pp.
SAA AS 3683-91. Information Processing Systems—OpenSystems Interconnection—Service Definition for the AssociationControl Service Element (ISO 8649:1988) (In Professional Packages 26A,26C). 11 pp.
SAA AS 3684-91. Information Processing Systems—Open Systems Interconnection—Protocol Specification for the Association Control Service Element (ISO 8650:1988) (In Professional Packages 26A, 26C). 27 pp.
SNZ NZS/ISO 8649-88. Information Processing Systems—Open Systems Interconnection—Service Definition for the Association Control Service Element. 11 pp.
SNZ NZS/ISO 8650-88. Information Processing Systems—Open Systems Interconnection—Protocol Specification for the Association Control Service Element. 27 pp.

—Interlibrary Loans

BSI BS ISO 10161-1-93. 1993 Information and Documentation—Open Systems Interconnection—Interlibrary Loan Application Protocol Specification—Part 1: Protocol Specification (S). 183 pp.
CSA Z243.46-88P. Interlibrary Loan (ILL) Protocol Specification. 93 pp.
ISO 10161 Pt 1-93. Information and Documentation—Open Systems Interconnection—Interlibrary Loan Application Protocol Specification—Part 1: Protocol Specification First Edition. 181 pp.

Application Services (OSI)

See Also: Application Layer (OSI); Directory Services; Directory System Agents; Directory User Agents; Open Systems Interconnection

—Addressing Systems

BSI BS 7306-01. 1990 Amd 1 Procedures for the Operation of the UK Scheme for the Allocation of ISO DCC Format OSI NSAP Addresses (Including the Operation of the UK Registration Authority) (AMD 7652) May 15, 1993 (S). 17 pp.
BSI BS ISO 7498/3-89. 1989 Information Processing Systems—Open Systems Interconnection—Basic Reference Model 7498/3: Naming and Addressing. 28 pp.

Application Services (OSI) *(Cont.)*

—Addressing Systems *(Cont.)*

ECMA ECMA-TR 20-84. Layer 4 to 1 Addressing. 25 pp.
ISO 7498 Pt 3-89. Information Processing Systems—Open Systems Interconnection—Basic Reference Model—Part 3: Naming and Addressing First Edition. 26 pp.
JIS X 5005-90. Information Processing Systems—Open Systems Interconnection—Basic Reference Model—Part 3:Naming and Addressing (ISO 7498/3:1989).
JTC1 7498 Pt 3-89. Information Processing Systems—Open Systems Interconnection—Basic Reference Model—Part 3: Naming and Addressing First Edition. 26 pp.
OSI ISO 7498-3-89. Information Processing Systems—Open Systems Interconnection—Basic Reference Model—Part 3: Naming and Addressing. 26 pp.
OSI ISO/IEC DIS 7498-3-88. Information Processing Systems—Open Systems Interconnection—Basic Reference Model Part 3: Naming and Addressing.
SNZ NZS/ISO 7498. 3-88. Information Processing Systems—Open Systems Interconnection—Basic Reference Model Part 3: Naming and Addressing. 21 pp.

—Addressing Systems—NATO Reference Model

NATO STANAG 4250 Pt3 ED1 AMD0-00. (Draft) NATO Reference Model for Open System Interconnection Part 3: Addressing. 6 pp.

—Association Control Service Element—Protocols

BSI BS 7091-89. 1989 Amd 1 Open Systems Interconnection: Service Definition for the Association Control Service Element (ISO 8649: 1988) (AMD 6733) July 31, 1991 (Technical Corr 1) (AMD 6772) July 31, 1991. 34 pp.
BSI BS 7091-02. 1989 Amd 2 Open Systems Interconnection: Service Definition for the Association Control Service Element (ISO 8649: 1988) (AMD 6929) April 1, 1992. 46 pp.
BSI BS 7092-89. 1989 Amd 1 Open Systems Interconnection: Protocol Specification for the Association Control Service Element (ISO 8650: 1988) (AMD 6734) July 31, 1991 (Technical Corr 1) (AMD 6735) July 31, 1991. 54 pp.
BSI BS ISO/IEC 10035-91. 1991 Information Technology—Open Systems Interconnection—Connectionless ACSE Protocol Specification. 19 pp.
BSI BS ISO/IEC ISP 10607-1-90. 1990 Amd 1 Information Technology—International Standardized Profiles AFTnn—File Transfer, Access and Management—Part 1: Specification of ACSE, Presentation and Session Protocols for the Use by FTAM (Technical Corr 1). 22 pp.
CSA CAN/CSA-Z243.170-89. Information Processing Systems—Open Systems Interconnection—Service Definition for the Association Control Service Element (ISO 8649:1988). 23 pp.
CSA CAN/CSA-Z243.171-89. Information Processing Systems—Open Systems Interconnection—Protocol Specification for the Association Control Service Element (ISO 8650:1988). 39 pp.
IEC DIS 8650 Pt 2-92. Information Technology—Open Systems Interconnection—Protocol Specification for the Association Control Service Element—Part 2: Protocol Implementation Conformance Statement (PICS) Proforma ***CD-ROM ONLY***. 15 pp.
IEC 10035-91. Information Technology—Open Systems Interconnection—Connectionless ACSE Protocol Specification First Edition. 16 pp.
IEC ISP10607 Pt 1-90. Information Technology—International Standardized Profiles AFTnn—File Transfer, Access and Management—Part 1: Specification of ACSE, Presentation and Session Protocols for the Use by FTAM First Edition; (Corrected and Reprinted -1991). 20 pp.
ISO 8649-88. Information Processing Systems—Open Systems Interconnection—Ser-vice Definition for the Association Control Ser-vice Element Amd 1: Authentication During As-sociation Establishment Amd 2: Connectionless-Mode ACSE Service First Edition; (Amd 1-1990) (Cor 1-1991)(Amd 2-1991). 36 pp.
ISO 8650-88. Information Processing Systems—Open Systems Interconnection—Protocol Specification for the Association Control Service Element First Edition; (Amendment 1-1990) (Corrigendum 1-1990). 52 pp.
ISO DIS 8650 Pt 2-92. Information Technology—Open Systems Interconnection—Protocol Specification for the Association Control Service Element—Part 2: Protocol Implementation Conformance Statement (PICS) Proforma ***CD-ROM ONLY***. 15 pp.
ISO 10035-91. Information Technology—Open Systems Interconnection—Connectionless ACSE Protocol Specification First Edition. 16 pp.

Application Services (OSI) *(Cont.)*

—Association Control Service Element—Protocols *(Cont.)*

ISO ISP10607 Pt 1-90. Information Technology—International Standardized Profiles AFTnn—File Transfer, Access and Management—Part 1: Specification of ACSE, Presentation and Session Protocols for the Use by FTAM First Edition; (Corrected and Reprinted -1991). 20 pp.
JIS X 5701-91. Information Processing Systems—Open Systems Interconnection—Service Definition for the Association Control Service Element (ISO 8649:1988).
JIS X 5702-91. Information Processing Systems—Open Systems Interconnection—Protocol Specification for the Association Control Service Element (ISO 8650:1988).
JTC1 8649-88. Information Processing Systems—Open Systems Interconnection—Ser-vice Definition for the Association Control Ser-vice Element Amd 1: Authentication During As-sociation Establishment Amd 2: Connectionless-Mode ACSE Service First Edition; (Amd 1-1990) (Cor 1-1991)(Amd 2-1991). 36 pp.
JTC1 8650-88. Information Processing Systems—Open Systems Interconnection—Protocol Specification for the Association Control Service Element First Edition; (Amendment 1-1990) (Corrigendum 1-1990). 52 pp.
JTC1 DIS8650 Pt 2-92. Information Technology—Open Systems Interconnection—Protocol Specification for the Association Control Service Element—Part 2: Protocol Implementation Conformance Statement (PICS) Proforma ***CD-ROM ONLY***. 15 pp.
JTC1 10035-91. Information Technology—Open Systems Interconnection—Connectionless ACSE Protocol Specification First Edition. 16 pp.
JTC1 ISP 10607 Pt 1-90. Information Technology—International Standardized Profiles AFTnn—File Transfer, Access and Management—Part 1: Specification of ACSE, Presentation and Session Protocols for the Use by FTAM First Edition; (Corrected and Reprinted -1991). 20 pp.
OSI ISO 8649 DAD 2-89. Information Processing Systems—Open Systems Interconnection—Service Definition for the Association Control Service Element ADDENDUM 2: Connectionless-Mode ACSE Service. 8 pp.
OSI ISO 8649-88. Information Processing Systems—Open Systems Interconnection—Service Definition for the Association Control Service Element (with Amendment 1). 30 pp.
OSI ISO IEC 8650 DAD 1-89. Revised Text of ISO 8650 PDAD 1: Information Processing Systems—Open Systems Interconnection—Protocol Specification for the Association Control Service Element—Addendum 1: Peer Entity Authentication During Association Establishment. 16 pp.
OSI ISO 8650 PDAD 2-89. Information Processing Systems—Open Systems Interconnection—Protocol Specification for the Association Control Service Element—Proposed Draft Addendum 2: Protocol Implementation Conformance Statement (PICS) Proforma. 12 pp.
OSI ISO 8650 DAD 1-89. Information Processing Systems—Open Systems Interconnection—Protocol Specification for the Association Control Service Element—Addendum 1: Peer-Entity Authentication During Association Establishment. 16 pp.
OSI ISO 8650-88. Information Processing Systems—Open Systems Interconnection—Protocol Specification for the Association Control Service Element (with Amendment 1). 58 pp.
OSI ISO/IEC DIS 10035-89. Information Processing Systems—Open Systems Interconnection—Connectionless ACSE Protocol Specification. 13 pp.
OSI ISO/IEC DISP 10607-1-90. Information Technology—International Standardized Profile AFTnn—File Transfer, Access and Management—Part 1: Specification of ACSE, Presentation and Session Protocols for the Use by FTAM. 41 pp.
OSI ISO/IEC ISP 10607-1-90. Information Technology—International Standard-ized Profile AFTnn—File Transfer, Access and Management—Part 1: Specification of ACSE, Presentation and Session Protocols for the Use by FTAM. 20 pp.
SAA AS 3683-91. Information Processing Systems—OpenSystems Interconnection—Service Definition for the AssociationControl Service Element (ISO 8649:1988) (In Professional Packages 26A,26C). 11 pp.
SAA AS 3684-91. Information Processing Systems—Open Systems Interconnection—Protocol Specification for the Association Control Service Element (ISO 8650:1988) (In Professional Packages 26A, 26C). 27 pp.
SNZ NZS/ISO 8649-88. Information Processing Systems—Open Systems Interconnection—Service Definition for the Association Control Service Element. 11 pp.
SNZ NZS/ISO 8650-88. Information Processing Systems—Open Systems Interconnection—Protocol Specification for the Association Control Service Element. 27 pp.

INTERNATIONAL AND NON-U.S. NATIONAL STANDARDS
SUBJECT INDEX
Application

Application Services (OSI) *(Cont.)*

—Association Control Service Element—Service

BSI BS 7091-89. 1989 Amd 1 Open Systems Interconnection: Service Definition for the Association Control Service Element (ISO 8649: 1988) (AMD 6733) July 31, 1991 (Technical Corr 1) (AMD 6772) July 31, 1991. 34 pp.
BSI BS 7091-02. 1989 Amd 2 Open Systems Interconnection: Service Definition for the Association Control Service Element (ISO 8649: 1988) (AMD 6929) April 1, 1992. 46 pp.
BSI BS ISO/IEC 10035-91. 1991 Information Technology—Open Systems Interconnection—Connectionless ACSE Protocol Specification. 19 pp.
CSA CAN/CSA-Z243.170-89. Information Processing Systems—Open Systems Interconnection—Service Definition for the Association Control Service Element (ISO 8649:1988). 23 pp.
IEC 8649 AMD 1. Amendment 1—Authentication During Association Establishment Information Processing Systems—Open Systems Interconnection—Service Definition for the Association Control Service Element; (1990). 10 pp.
IEC 8649 AMD 2. Amendment 2—Connectionless-Mode ACSE Service and Corrigendum 1 Information Processing Systems—Open Systems Interconnection—Service Definition for the Association Control Service Element; (1991). 9 pp.
IEC 8649 COR 1. Corrigendum 1—Information Processing Systems—Open Systems Interconnection—Service Definition for the Association Control Service Element; (1991). 2 pp.
IEC 8650 AMD 1. Amendment 1—Authentication During Association Establishment Information Processing Systems—Open Systems Interconnection—Protocol Specification for the Association Control Service Element; (1990). 17 pp.
IEC 10035-91. Information Technology—Open Systems Interconnection—Connectionless ACSE Protocol Specification First Edition. 16 pp.
ISO 8649-88. Information Processing Systems—Open Systems Interconnection—Ser-vice Definition for the Association Control Ser-vice Element Amd 1: Authentication During As-sociation Establishment Amd 2: Connectionless-Mode ACSE Service First Edition; (Amd 1-1990) (Cor 1-1991)(Amd 2-1991). 36 pp.
ISO 10035-91. Information Technology—Open Systems Interconnection—Connectionless ACSE Protocol Specification First Edition. 16 pp.
JIS X 5701-91. Information Processing Systems—Open Systems Interconnection—Service Definition for the Association Control Service Element (ISO 8649:1988).
JTC1 8649-88. Information Processing Systems—Open Systems Interconnection—Ser-vice Definition for the Association Control Ser-vice Element Amd 1: Authentication During As-sociation Establishment Amd 2: Connectionless-Mode ACSE Service First Edition; (Amd 1-1990) (Cor 1-1991)(Amd 2-1991). 36 pp.
JTC1 10035-91. Information Technology—Open Systems Interconnection—Connectionless ACSE Protocol Specification First Edition. 16 pp.
OSI ISO IEC 8649 DAD 1-89. Revised Text of ISO 8649 PDAD 1: Information Processing Systems—Open Systems Interconnection—Service Definition for the Association Control Service Element—Addendum 1: Peer—Entity Authentication During Association Establishment. 27 pp.
OSI ISO 8649 DAD 2-89. Information Processing Systems—Open Systems Interconnection—Service Definition for the Association Control Service Element ADDENDUM 2: Connectionless-Mode ACSE Service. 8 pp.
OSI ISO 8649-88. Information Processing Systems—Open Systems Interconnection—Service Definition for the Association Control Service Element (with Amendment 1). 30 pp.
SAA AS 3683-91. Information Processing Systems—OpenSystems Interconnection—Service Definition for the AssociationControl Service Element (ISO 8649:1988) (In Professional Packages 26A,26C). 11 pp.
SNZ NZS/ISO 8649-88. Information Processing Systems—Open Systems Interconnection—Service Definition for the Association Control Service Element. 11 pp.

—Banking—Protocols

OSI ISO DP 9955-N205-87. Information Interchange—Methodology and Guidelines for the Development of Application Protocols for Banking Information Interchange. 145 pp.
OSI ISO DP 9955-N205-87. Information Interchange—Methodology and Guidelines for the Development of Application Protocols for Banking Information Interchange. 145 pp.

—Commitment, Concurrency and Recovery

BSI BS ISO/IEC 9804-90. 1990 Information Technology—Open Systems Interconnection—Service Definition for the Commitment, Concurrency and Recovery Service Element. 52 pp.

Application Services (OSI) *(Cont.)*

—Commitment, Concurrency and Recovery *(Cont.)*

BSI BS ISO/IEC 9804-01. 1990 Amd 1 Information Technology—Open Systems Interconnection—Service Definition for the Commitment, Concurrency and Recovery Service Element (AMD 7152) June 15, 1992. 54 pp.
BSI BS ISO/IEC 9804-02. 1990 Amd 2 Information Technology—Open Systems Interconnection—Service Definition for the Commitment, Concurrency and Recovery Service Element (AMD 7708) April 15, 1993 (S). 61 pp.
BSI BS ISO/IEC 9805-90. 1990 Information Technology—Open Systems Interconnection—Protocol Specification for the Commitment, Concurrency and Recovery Service Element. 44 pp.
BSI BS ISO/IEC 9805-01. 1990 Amd 1 Information Technology—Open Systems Interconnection—Protocol Specification for the Commitment, Concurrency and Recovery Service Element (AMD 7154) July 15, 1992 (Technical Corr 1) (S). 45 pp.
BSI BS ISO/IEC 9805-02. 1990 Amd 2 Information Technology—Open Systems Interconnection—Protocol Specification for the Commitment, Concurrency and Recovery Service Element (AMD 7681) April 15, 1993 (Technical Corr 2) (S). 83 pp.
CSA CAN/CSA-Z243. 178-91. Information Technology—Open Systems Interconnection—Service Definition for the Commitment, Concurrency and Recovery Service Element (ISO/IEC 9804:1990); (Gen Instr 1). 58 pp.
CSA CAN/CSA-Z243. 179-91. Information Technology—Open Systems Interconnection—Protocol Specification for the Commitment, Concurrency and Recovery Service Element (ISO/IEC 9805:1990); (Gen Instr 1). 49 pp.
IEC 9804-90. Information Technology—Open Systems Interconnection—Service Definition for the Commitment, Concurrency and Recovery Service Element Amendment 2: Session Mapping Changes First Edition; (Corrigendum 1-1991) (Amendment 2-1992). 58 pp.
IEC 9805-90. Information Technology—Open Systems Interconnection—Protocol Specification for the Commitment, Concurrency and Recovery Service Element First Edition; (Corrigendum 1-1991) (Amendment 2-1992) (Corrigendum 2-1992). 67 pp.
ISO 9804-90. Information Technology—Open Systems Interconnection—Service Definition for the Commitment, Concurrency and Recovery Service Element Amendment 2: Session Mapping Changes First Edition; (Corrigendum 1-1991) (Amendement 2-1992). 58 pp.
ISO 9805-90. Information Technology—Open Systems Interconnection—Protocol Specification for the Commitment, Concurrency and Recovery Service Element First Edition; (Corrigendum 1-1991) (Amendment 2-1992) (Corrigendum 2-1992). 67 pp.
JTC1 9804-90. Information Technology—Open Systems Interconnection—Service Definition for the Commitment, Concurrency and Recovery Service Element Amendment 2: Session Mapping Changes First Edition; (Corrigendum 1-1991) (Amendment 2-1992). 58 pp.
JTC1 9805-90. Information Technology—Open Systems Interconnection—Protocol Specification for the Commitment, Concurrency and Recovery Service Element First Edition; (Corrigendum 1-1991) (Amendment 2-1992) (Corrigendum 2-1992). 42 pp.
OSI ISO DIS 8649-3-85. Open Systems Interconnection: Definition of Common Application Service Elements: Commitment, Concurrency and Recovery. 58 pp.
OSI ISO DIS 8650-3-85. Open Systems Interconnection: Specification of Protocols Forcommon Application Service Elements: Commitment, Concurrency and Recovery. 54 pp.
OSI ISO/IEC 9804-90. Information Technology—Open Systems Interconnection—Service Definition for the Committment, Concurrency and Recovery Service Element. 50 pp.
OSI ISO IEC DIS 9804-2-88. Information Processing Systems—Open Systems Interconnection—Service Definition for the Commitment, Concurrency and Recovery Service Element. 37 pp.
OSI ISO/IEC 9805-90. Information Technology—Open Systems Interconnection—Protocol Specification for the Committment, Concurrency and Recovery Service Element. 40 pp.
OSI ISO IEC DIS 9805-2-88. Information Processing Systems—Open Systems Interconnection—Protocol Specification for the Commitment, Concurrency and Recovery Service Element. 39 pp.
SNZ NZS/AS 4105-93. Information Technology—Open Systems Interconnection—Protocol Specification for the Commitment, Concurrency and Recovery Service Element (This is a Joint Standard with SAA AS 4105). 33 pp.
SNZ NZS/AS 4106-93. Information Technology—Open Systems Interconnection—Service Definition for the Commitment, Concurrency and Recovery Service Element (This is a Joint Standard with SAA AS 4106). 42 pp.

Application Services (OSI) *(Cont.)*

—Commitment, Concurrency and Recovery—Conformance

IEC DIS 9805 Pt 2-92. Information Technology—Open Systems Interconnection—Commitment, Concurrency and Recovery Protocol—Part 2: Protocol Implementation Conformance Statement (PICS) Proforma ***CD-ROM ONLY***. 21 pp.
ISO DIS 9805 Pt 2-92. Information Technology—Open Systems Interconnection—Commitment, Concurrency and Recovery Protocol—Part 2: Protocol Implementation Conformance Statement (PICS) Proforma ***CD-ROM ONLY***. 21 pp.
JTC1 DIS9805 Pt 2-92. Information Technology—Open Systems Interconnection—Commitment, Concurrency and Recovery Protocol—Part 2: Protocol Implementation Conformance Statement (PICS) Proforma ***CD-ROM ONLY***. 21 pp.

—Libraries—Protocols

BSI BS ISO 10161-1-93. 1993 Information and Documentation—Open Systems Interconnection—Interlibrary Loan Application Protocol Specification—Part 1: Protocol Specification (S). 183 pp.
ISO 10161 Pt 1-93. Information and Documentation—Open Systems Interconnection—Interlibrary Loan Application Protocol Specification—Part 1: Protocol Specification First Edition. 181 pp.

—Reliable Transfer Service Element—Protocols

BSI BS ISO/IEC 9066-2-89. 1989 Protocol Specification. 71 pp.
IEC 9066 Pt 2-89. Information Processing Systems—Text Communication—Reliable Transfer—Part 2: Protocol Specification First Edition. 68 pp.
ISO 9066 Pt 2-89. Information Processing Systems—Text Communication—Reliable Transfer—Part 2: Protocol Specification First Edition. 68 pp.
JTC1 9066-89. Information Processing Systems—Text Communication—Reliable Transfer—Part 2: Protocol Specification First Edition. 68 pp.
OSI ISO IEC 9066-2-89. Information Processing Systems—Text Communication—Reliable Transfer—Part 2: Protocol Specification. 68 pp.
OSI ISO DP 9066-2-WD-86. Information Processing Systems—Text Communication—Motis—Reliable Transfer Part 2: Protocol Specification. 64 pp.
OSI ISO DP 9066 2-WD-86. Information Processing Systems—Text Communication—Motis—Reliable Transfer Part 2: Protocol Specification.
OSI ISO DP 9066 2 WD-86. Information Processing Systems-Text Communication-Motis-Reliable Transfer Part 2: Protocol Specification.
SAA AS 4016.2-92. Information Processing Systems—Text Communication—Reliable Transfer—Part 2: Protocol Specification (ISO/IEC 9066-2:1989) (in Professional Packages 26A, 26C). 64 pp.

—Reliable Transfer Service Element—Service

BSI BS ISO/IEC 9066-1-89. 1989 Model and Service Definition. 21 pp.
IEC 9066 Pt 1-89. Information Processing Systems—Text Communication—Reliable Transfer—Part 1: Model and Service Definition First Edition. 19 pp.
ISO 9066 Pt 1-89. Information Processing Systems—Text Communication—Reliable Transfer—Part 1: Model and Service Definition First Edition. 19 pp.
JTC1 9066-89. Information Processing Systems—Text Communication—Reliable Transfer—Part 1: Model and Service Definition First Edition. 19 pp.
OSI ISO IEC 9066-1-89. Information Processing Systems—Text Communication—Reliable Transfer—Part 1: Model and Service Definition. 19 pp.
OSI ISO DP 9066-1 WD-86. Information Processing Systems—Text Communication—Motis—Reliable Transfer—Part 1: Model and Service Definition. 29 pp.
OSI ISO DP 90661 WD-86. Information Processing Systems—Text Communication—Motis—Reliable Transfer (Part 1): Model and Service Definition.
SAA AS 4016.1-92. Information Processing Systems—Text Communication—Reliable Transfer—Part 1: Model and Service Definition (ISO/IEC 9066-1:1989) (in Professional Packages 26A, 26C). 14 pp.

—Remote Operation Service Element

ECMA ECMA 131-88. Referenced Data Transfer. 27 pp.
IEC DIS 9579 Pt 1-91. Information Technology—Open Systems Interconnection—Remote Database Access—Part 1: Generic Model, Service and Protocol ***CD-ROM ONLY***. 156 pp.
ISO DIS 9579 Pt 1-91. Information Technology—Open Systems Interconnection—Remote Database Access—Part 1: Generic Model, Service and Protocol ***CD-ROM ONLY***. 156 pp.
OSI ISO DP 9072-1-86. Information Processing—Message Oriented Text Interchange Systems—Remote Operation Service —Part 1: Concepts and Model.

INDUSTRY STANDARDS
205

INTERNATIONAL AND NON-U.S. NATIONAL STANDARDS
SUBJECT INDEX

Application

Application Services (OSI) *(Cont.)*
—Remote Operation Service Element *(Cont.)*
  OSI ISO DP 9072-2-86. Information Processing—Message Oriented Text Interchange Systems—Remote Operation Service —Part 2: Basic ROS. 56 pp.
  OSI ISO DP 9579-87. Information Processing Systems—Remote Database Access. 124 pp.
  OSI ISO/IEC DIS 10031-2-89. Information Technology—Text Communication—Distributed-Office—Applications Model—Part 2: Referenced Data Transfer. 21 pp.
  OSI ISO/IEC DIS 10148-88. Information Processing Systems—Basic Remote Procedure Call (RPC) Using OSI Remote Operations. 73 pp.

—Remote Operation Service Element—Protocols
  BSI BS ISO/IEC 9072-2-89. 1989 Protocol Specification. 35 pp.
  IEC 9072 Pt 2-89. Information Processing Systems—Text Communication—Remote Operations—Part 2: Protocol Specification First Edition. 32 pp.
  ISO 9072 Pt 2-89. Information Processing Systems—Text Communication—Remote Operations—Part 2: Protocol Specification First Edition. 32 pp.
  JTC1 9072-89. Information Processing Systems—Text Communication—Remote Operations—Part 2: Protocol Specification First Edition. 32 pp.
  OSI ISO IEC 9072-2-89. Information Processing Systems—Text Communication—Remote Operations—Part 2: Protocol Specification. 32 pp.
  OSI ISO/IEC DIS 10031-1-89. Information Technology—Text Communication—Distributed-Office—Applications Model—Part 1: General Model. 63 pp.
  SAA AS 3893.2-91. Information Processing Systems—Text Communication—Remote Operations—Part 2: Protocol Specification (ISO/IEC 9072-2:1989) (in Professional Packages 26A, 71A, 71B).
  SNZ NZS/ISO-IEC 9072.2-89. Information Processing Systems—Text Communication—Remote Operations Part 2: Protocol Specification. 28 pp.

—Remote Operation Service Element—Service
  BSI BS ISO/IEC 9072-1-89. 1989 Model, Notation and Service Definition. 42 pp.
  IEC 9072 Pt 1-89. Information Processing Systems—Text Communication—Remote Operations—Part 1: Model, Notation and Service Definition First Edition. 40 pp.
  ISO 9072 Pt 1-89. Information Processing Systems—Text Communication—Remote Operations—Part 1: Model, Notation and Service Definition First Edition. 40 pp.
  JTC1 9072-89. Information Processing Systems—Text Communication—Remote Operations—Part 1: Model, Notation and Service Definition First Edition. 40 pp.
  OSI ISO IEC 9072-1-89. Information Processing Systems—Text Communication—Remote Operations—Part 1: Model, Notation and Service Definition. 41 pp.
  SAA AS 3893.1-91. Information Processing Systems—Text Communication—Remote Operations—Part 1: Model, Notation and Service Definition (ISO/IEC 9072-1:1989) (in Professional Packages 26A, 71B). 34 pp.
  SNZ NZS/ISO-IEC 9072.1-89. Information Processing Systems—Text Communication—Remote Operations Part 1: Model, Notation and Service Definition. 34 pp.

Appointment Books
Use For: Datebooks
  CGSB CAN/CGSB-53.52-M89. Book, Appointment. 10 pp.

Approval Marks
Use: Standard Marks

Apramycin Content Analysis
—Animal Feed
  CNS N4124-84. Method of Test for Feed Additives: Determination of Apramycin (Nov)(11130).

Apricots
Use For: Chinese Apricots *See Also:* Food; Fruits
—Grading
  CNS N1054-64. Gradation of Chinese Plum and Chinese Apricot (for Export) (Tentative) (Jul)(2349)(R 1973). 2 pp.
—Kernels—Shelled
  ISO 6479-84. Shelled Sweet Kernels of Apricots—Specification First Edition. 9 pp.

Apricots *(Cont.)*
—Storage
  ISO 2826-74. Apricots—Guide to Cold Storage First Edition. 5 pp.

Apron Conveyors
Use: Slat Conveyors

Aprons
—Protective
  BSI BS 7185-89. (WITHDRAWN) 1989 Protective Aprons for Use with Hand Knives (Superseded by BS EN 412: 1993). 16 pp.
  BSI BS EN 412-93. 1993 Protective Aprons for Use with Hand Knives (Supersedes BS 7185: 1989) (N). 22 pp.
  CEN PREN 412-90. Protective Aprons for Use with Hand Knives. 16 pp.
  CEN EN 412-93. Protective Aprons for Use with Hand Knives. 17 pp.

—Protective—Wet Work
  BSI BS 3314-82. 1982 Protective Aprons for Wet Work. 4 pp.

—Protective—X-Ray Equipment
  CNS T2016-83. Medical X-Ray Protective Aprons (May)(10297).
  CNS T4005-83. Method of Test for X-Ray Protective Aprons (May)(10298).
  JIS Z 4803-91. Medical X-Ray Protective Aprons. 9 pp.

—Ring Spinning Frames
  BSI BS 2794: Part 3-78. 1978 Textile Machinery and Accessories. Drafting Equipment for Preparation and Spinning Machinery Part 3: Ring Spinning Frames and Speed Frames. Top and Bottom Aprons. 12 pp.
  ISO 5235-77. Textile Machinery and Accessories—Ring-Spinning Frames and Speedframes—Top and Bottom Aprons First Edition. 7 pp.

APU
Use: Auxiliary Power Units

Aqua Regia
Use For: Nitrohydrochloric Acid
See Also: Hydrochloric Acid
—Barium Peroxide
  MOD UK M 819/72. Examination of Barium Peroxide.
—Sludge Analysis
  DIN ENGL 38414 Pt 7-83. German Standard Methods for the Examination of Water, Waste Water and Sludge; Sludge and Sediments (Group S); Digestion Using Aqua Regia for Subsequent Determination of the Acid-Soluble Portion of Metals (S 7) (Jan). 4 pp.

Aquaculture
See Also: Animal Feed
  CNS N2043-86. Formula Feeds (for Aquaculture) (Mar) (4142). 2 pp.
—Health Inspection
  EC 91/67/EEC-91. Council Directive Concerning the Animal Health Conditions Governing the Placing on the Market of Aquaculture Animals and Products. 21 pp.

Aquaducts
Use: Aqueducts

Aquariums
  SAA AS 3192-92. Approval and Test Specification—Electrically Operated Aquarium Equipment Amdt 1 June 1993 Amdt 2 September 1993 (in Professional Package 28). 5 pp.
  SNZ NZS/AS 3192-86. Approval and Test Specification—Electrically Operated Aquarium Equipment (This is a Joint Standard with SAA AS 3192). 3 pp.

—Appliances—Safety
  BSI BS EN 60335-2-55-93. 1993 Safety of Household and Similar Electrical Appliances Part 2: Particular Requirements Section 2.55: Electrical Appliances for Use with Aquariums and Garden Ponds (L). 22 pp.
  CENELEC EN 60335-2-55-92. Safety of Household and Similar Electrical Appliances Part 2: Particular Requirements for Electrical Appliances for Use with Aquariums and Garden Ponds. 16 pp.
  CENELEC EN 60335-2-55-93. Safety of Household and Similar Electrical Appliances Part 2: Particular Requirements for Electric Appliances for Use with Aquariums and Garden Ponds (IEC 335-2-55: 1989, Modified). 14 pp.

Aquariums *(Cont.)*
—Appliances—Safety *(Cont.)*
  IEC 335 Pt 2-55-89. Safety of Household and Similar Electrical Appliances Part 2: Particular Requirements for Electrical Appliances for Use with Aquariums and Garden Ponds First Edition (CENELEC EN 60335-2-55: 1993). 31 pp.

Aquatic Organisms
See Also: Crustacea; Fish
—Sampling—Shallow Water
  BSI BS 6068: Sec 5.5-85. 1985 Water Quality Part 5: Methods for Biological Testing Section 5.5: Methods of Biological Sampling: Guidance on Handnet Sampling of Aquatic Benthic Macro-Invertebrates. 9 pp.
  BSI BS 6068: Sec 5.8-89. 1989 Water Quality Part 5: Biological Methods Sec 5.8: Methods of Biological Sampling: Guidance on the Design and Use of Quantitative Samplers for Benthic Macro-Invertebrates on Stony Substrata in Shallow Freshwaters (ISO 8265: 1988). 12 pp.
  ISO 7828-85. Water Quality—Methods of Biological Sampling—Guidance on Handnet Sampling of Aquatic Benthic Macro-Invertebrates First Edition. 8 pp.
  ISO 8265-88. Water Quality—Design and Use of Quantitative Samplers for Benthic Macro-Invertebrates on Stony Substrata in Shallow Freshwaters First Edition. 13 pp.

Aquatic Sports Equipment
See Also: Diving Equipment; Snorkels (Swimming); Sports Equipment
—Face Shields
  BSI BS 4532-69. 1969 Amd 1 Snorkels and Face Masks. 13 pp.

Aquatic Toys
See Also: Toys
—Flotation
  SAA AS 1900-91. Flotation Toys and Swimming Aids for Children Amdt 1 August 1993 (in Professional Package 45). 11 pp.

Aqueducts
Use For: Aquaducts *See Also:* Siphons; Water Pipelines
  DIN ENGL 19661 Pt 1-72. Guidelines for Hyraulic Structures; Crossing Works; Bridges, Aqueducts, Culverts, Inverted Syphons, Casings (Oct). 13 pp.

Aqueous Solutions
See Also: Solutions
—Dicyandiamide
  MOD UK M 812/77. Examination of Dicyandiamide.
—Kaolin
  MOD UK M 2002/59. Examination of Kaolin (Withdrawn).
—PH—Clear
  DIN ENGL 19268-85. Measurement of pH Value of Clear Aqueous Solutions (Feb). 4 pp.
—PH—Glass Electrodes
  CNS K0010-80. Method for Determination of PH of Aqueous Solutions (Sep)(6492).
  JIS Z 8802-84. Methods for Determination of pH of Aqueous Solutions. 13 pp.
—Polytetrafluoroethylene—Dispersion
  BSI BS 7485-91. 1991 Method of Specifying Aqueous Dispersions of Polytetrafluoroethylene (PTFE) Materials (V). 11 pp.
  JIS K 6893-76. Testing Methods for Polytetrafluoroethylene Aqueous Dispersion (R 1979). 10 pp.
—Reverse Osmosis
  JIS K 3805-90. Testing Methods for Solute Rejection and Water Flux of Reverse Osmosis Membrane Element and Module Using Aqueous Solution of Various Solutes. 27 pp.

Aquifers
See Also: Ground Water; Spring Water; Water Wells
—Water Analysis
  DIN ENGL 38402 Pt 13-85. German Standard Methods for the Examination of Water, Waste Water and Sludge; General Information (Group A); Sampling from Aquifers (A13) (Dec). 7 pp.

Arabic Characters
See Also: Graphic Characters

INTERNATIONAL AND NON-U.S. NATIONAL STANDARDS
SUBJECT INDEX

Arabic Characters (Cont.)
—Coded Character Sets
ECMA ECMA 114-86. 8-Bit Single Byte Coded Graphic Character Sets, Latin/Arabic Alphabet. 17 pp.
ISO 8859 Pt 6-87. Information Processing—8-Bit Single-Byte Coded Graphic Character Sets—Part 6: Latin/Arabic Alphabet First Edition. 9 pp.
ISO 9036-87. Information Processing—Arabic 7-Bit Coded Character Set for Information Interchange First Edition. 16 pp.
JTC1 8859 Pt 6-87. Information Processing—8-Bit Single-Byte Coded Graphic Character Sets—Part 6: Latin/Arabic Alphabet First Edition. 9 pp.
JTC1 9036-87. Information Processing—Arabic 7-Bit Coded Character Set for Information Interchange First Edition. 16 pp.
OSI ISO 8859-6-87. Information Processing—8-Bit Single-Byte Coded Graphic Character Sets—Part 6: Latin/Arabic Alphabet. 9 pp.
OSI ISO 9036-87. Information Processing—Arabic 7-Bit Coded Character Set for Information Interchange. 22 pp.
OSI ISO DIS 9036-86. Information Processing—Arabic 7-Bit Coded Character Set for Information Interchange. 22 pp.

—Engraving
JIS Z 8905-76. Standard Type of Letters Used in Mechanical Engraving (Arabic Figures and Roman Types).

—Romanization
BSI BS 4280-68. 1968 Amd 1 Transliteration of Arabic Characters. 10 pp.
ISO 233-84. Documentation—Transliteration of Arabic Characters into Latin Characters First Edition. 9 pp.
ISO 233 Pt 2-93. Information and Documentation—Transliteration of Arabic Characters—into Latin Characters—Part 2: Arabic Language—Simplified Transliteration First Edition. 10 pp.

Arabic Gum
See Also: Gums
CNS K4023-82. Arabic Gum (for General Uses) (Aug)(3281). 2 pp.

—Electric Primers
CNS K1272-82. Arabic Gum (for Electric Primer Uses) (Aug)(9301).
CNS K6725-82. Method of Test for Arabic Gum (for Electric Primer Use) (Aug)(9302).

Arabinose
Use For: L-Arabinose
CNS K7072-62. Chemical Reagent (L-Arabinose) (Jul)(1572).
JIS K 8054-91. L-Arabinose.

Arachis Oil
Use: Peanut Oil

Aramid Fiber Content Analysis
See Also: Aramid Fibers; Fiber Content Analysis

—Aerospace
CEN PREN 3783-90. Procedure for the Normalisation of Test Results of Fibre Dominated Composite Mechanical Properties. 6 pp.

Aramid Fibers
Use For: Aromatic Polyamide Resins; Para-Aramid Fibers *See Also:* Aramid Fiber Content Analysis; Fibers

—Aerospace—Filament Yarns—Sampling
CEN PREN 3675-90. Sampling Plan for Acceptance Testing of Aramid, Carbon Fibre and Textile Glass Filament Yarns. 8 pp.
CEN PREN 3675-90. Sampling Plan for Acceptance Testing of Aramid, Carbon Fibre and Textile Glass Filament Yarns. 7 pp.

—Aerospace—Joint Sealants
BSI F 130-87. 1987 Rubber Bonded Compressed Aramid Fibre Jointing Material for Aerospace Purposes. 12 pp.

—Aerospace—Prepreg—Fiber Content
CEN PREN 3783-90. Procedure for the Normalisation of Test Results of Fibre Dominated Composite Mechanical Properties. 6 pp.

—Electrical Insulating Papers
IEC 819 Pt 3-3-91. Specification for Non-Cellulosic Papers for Electrical Purposes Part 3: Specifications for Individual Materials Sheet 3: Unfilled Aramid (Aromatic Polyamide) Papers First Edition. 17 pp.

Arbors
See Also: Beams (Supports); Hobbing Arbors; Mandrels; Milling Cutter Arbors; Shafts (Machine Elements); Spindles

—Aligning
DIN ENGL 6381-62. Aligning Arbors with Taper Shank (Sept). 2 pp.

—Drill Chucks
DIN ENGL 238 Pt 1-62. Drill Chuck Mounting Arrangements; Taper Arbors (June). 1 p.

—Master Gears
DIN ENGL 3970 Pt 2-74. Master Gears for Checking Spur Gears; Receiving Arbors (Nov). 2 pp.

—Shell Drills
DIN ENGL 217-72. Arbors with Morse Taper for Shell Reamers and Shell Drills (Sept). 2 pp.

—Shell End Mills
CNS B3237-74. Arbors for Shell End Mills and Face Cutters (Jan)(3612). 5 pp.
JIS B 4216-80. Arbors for Shell End Mills (R 1986). 9 pp.

—Shell End Mills—Wrenches
DIN ENGL 6368-62. Wrenches for Arbors for Shell End Mills (Nov). 1 p.

—Shell Reamers
DIN ENGL 217-72. Arbors with Morse Taper for Shell Reamers and Shell Drills (Sept). 2 pp.
ISO 2402-72. Shell Reamers with Taper Bore (Taper Bore 1: 30 (Included)) with Slot Drive and Arbors for Shell Reamers First Edition. 7 pp.
JIS B 4407-64. Straight Shank Arbors for Shell Reamers (R 1984). 7 pp.
JIS B 4408-64. Taper Shank Arbors for Shell Reamers.

Arc Furnaces
Use: Electric Arc Furnaces

Arc Spraying Equipment
See Also: Arc Welding Equipment; Plasma Arc Cutting Equipment; Plasma Arc Welding Equipment; Thermal Spraying Equipment
DIN VDE 0544 Pt 2 (D)-78. Welding Equipment and Assemblies for Arc Welding and Allied Processes; Assemblies (VDE Specification) (Oct). 20 pp.

—Filler Metal
DIN ENGL 8566 Pt 2-84. Filler Metals for Thermal Spraying; Solid Wires for Arc Spraying; Technical Delivery Conditions (Dec). 8 pp.

Arc Suppression Coils
BSI BS 4944-73. 1973 Reactors, Arc-Suppression Coils and Earthing Transformers for Electric Power Systems. 19 pp.

Arc Testing
See Also: Testing

—Electrical Insulation
DIN VDE 0303 Pt 5-90. Testing of Electrical Insulating Materials; Low-Voltage High-Current Arc Test (July) (Replaces DIN 53484/10.55X). 11 pp.

Arc Welded Joints
Use: Arc Welds

Arc Welding
Scope Note: For additional listings, use a more specific term *See Also:* Arc Welding Equipment; Arc Welds; Braze Welding; Electric Welding; Electron Beam Welding; Electron Beam Welding Machines; Fusion Welding; Gas Filled Cables; Gas Metal Arc Welding; Gas Shielded Arc Welding; Gas Tungsten Arc Welding; Metal Arc Welding; Shielded Arc Welding; Shielded Metal Arc Welding; Stud Welding; Submerged Arc Welding; Welding
BSI BS 7384-91. 1991 Guide to Laboratory Methods for Sampling and Analysis of Particulate Matter Generated by Arc Welding Consumables. 17 pp.
JIS Z 3801-79. Standard Qualification Procedure for Welding Technique. 19 pp.
JIS Z 3841-79. Standard Qualification Procedure for Semi-Automatic Arc Welding Technique. 18 pp.

—Aluminum
BSI BS EN 288: Part 4-92. 1992 Specification and Approval of Welding Procedures for Metallic Materials Part 4: Welding Procedure Tests for the Arc Welding of Aluminium and its Alloys (Supersedes BS 4870: Part 2: 1982). 32 pp.
CEN PREN 288-4-91. Specification and Approval of Welding Procedures for Metallic Materials—Part 4: Welding Procedure Tests for the Arc Welding of Aluminium and its Alloys. 37 pp.

Arc Welding (Cont.)
—Aluminum (Cont.)
CEN EN 288-4-92. Specification and Approval of Welding Procedures for Metallic Materials—Part 4: Welding Procedure Tests for the Arc Welding of Aluminium and Its Alloys. 27 pp.
CEN EN 288-4-92. CORRIGENDUM Specification and Approval of Welding Procedures for Metallic Materials—Part 4: Welding Procedure Tests for the Arc Welding of Aluminium and its Alloys. 2 pp.
CEN EN 288-4-92. Specification and Approval of Welding Procedures for Metallic Materials—Part 4: Welding Procedure Tests for the Arc Welding of Aluminium and Its Alloys. 28 pp.
CEN PREN 630-92. Arc-Welded Joints in Aluminium and Its Weldable Alloys—Guidance on Quality Levels for Imperfections. 15 pp.
CSA W59.2-M1991. Welded Aluminum Construction; (Gen Instr 1) (Replaces CSA S244-1969). 76 pp.
DIN ENGL EN 288 Pt 4-92. Specification and Approval of Procedures for Welding Metallic Materials; Welding Procedure Tests for the Arc Welding of Aluminium and Its Alloys (Oct). 30 pp.

—Aluminum Alloys
BSI BS EN 288: Part 4-92. 1992 Specification and Approval of Welding Procedures for Metallic Materials Part 4: Welding Procedure Tests for the Arc Welding of Aluminium and its Alloys (Supersedes BS 4870: Part 2: 1982).
CEN PREN 288-4-91. Specification and Approval of Welding Procedures for Metallic Materials—Part 4: Welding Procedure Tests for the Arc Welding of Aluminium and its Alloys. 37 pp.
CEN EN 288-4-92. Specification and Approval of Welding Procedures for Metallic Materials—Part 4: Welding Procedure Tests for the Arc Welding of Aluminium and Its Alloys. 27 pp.
CEN EN 288-4-92. CORRIGENDUM Specification and Approval of Welding Procedures for Metallic Materials—Part 4: Welding Procedure Tests for the Arc Welding of Aluminium and Its Alloys. 2 pp.
CEN EN 288-4-92. Specification and Approval of Welding Procedures for Metallic Materials—Part 4: Welding Procedure Tests for the Arc Welding of Aluminium and Its Alloys. 28 pp.
CEN PREN 630-92. Arc-Welded Joints in Aluminium and Its Weldable Alloys—Guidance on Quality Levels for Imperfections. 15 pp.
DIN ENGL EN 288 Pt 4-92. Specification and Approval of Procedures for Welding Metallic Materials; Welding Procedure Tests for the Arc Welding of Aluminium and Its Alloys (Oct). 30 pp.

—Carbon Steels
BSI BS 5135-84. 1984 Amd 1 Process of Arc Welding of Carbon and Carbon Manganese Steels. 54 pp.

—Glossaries
DIN ENGL 1910 Pt 10-81. Welding; Mechanized Fusion Arc Welding Processes; Terms (July). 8 pp.

—Metals
BSI BS EN 288: Part 2-92. 1992 Specification and Approval of Welding Procedures for Metallic Materials Part 2: Welding Procedure Specification for Arc Welding. 11 pp.
CEN PREN 288-2-91. Specification and Approval of Welding Procedures for Metallic Materials—Part 2: Welding Procedure Specification for Arc Welding. 9 pp.
CEN EN 288-2-92. Specification and Approval of Welding Procedures for Metallic Materials—Part 2: Welding Procedure Specification for Arc Welding. 8 pp.
DIN ENGL EN 288 Pt 2-92. Specification and Approval of Procedures for Welding Metallic Materials; Welding Procedure Specification for Arc Welding (Apr). 8 pp.

—Particulates—Chemical Analysis
BSI BS 7384-91. 1991 Guide to Laboratory Methods for Sampling and Analysis of Particulate Matter Generated by Arc Welding Consumables. 17 pp.

—Power Supplies
BSI BS 7418-91. 1991 Isolation of the Welding Circuit in Arc Welding Plant. 8 pp.
IEC 974 Pt 1-89. Safety Requirements for Arc Welding Equipment Part 1: Welding Power Sources First Edition. 110 pp.
SAA AS 1966.1-85. Electric Arc Welding Power Sources—Part 1: Transformer Type. 21 pp.
SAA AS 1966.2-85. Electric Arc Welding Power Sources—Part 2: Rotary Type. 16 pp.
SAA AS 1966.3-90. Electric Arc Welding Power Sources—Part 3: Plasma Arc Cutting and Welding Types. 11 pp.

—Primers (Coatings)—Porosity
BSI BS 6084-81. 1981 Method of Test for Comparison of Prefabrication Primers by Porosity Rating in Arc Welding. 7 pp.

Arc Welding (Cont.)

—Quality Assurance

CEN PREN 288-7-93. Specification and Approval of Welding Procedures for Metallic Materials—Part 7: Approval by a Standard Welding Procedure for Arc Welding. 11 pp.

—Stainless Steels

BSI BS 7475-91. 1991 Fusion Welding of Austenitic Stainless Steels (Supersedes BS 3019: Part 2: 1960). 16 pp.

BSI BS 7475-01. 1991 Amd 1 Fusion Welding of Austenitic Stainless Steels (AMD 7642) April 15, 1993 (Supersedes BS 3019: Part 2: 1960) (F). 17 pp.

JIS Z 3821-89. Standard Qualification Procedure for Welding Technique of Stainless Steel. 14 pp.

—Steel Pipes

BSI BS 2633-87. 1987 Amd 1 Class 1 Arc Welding of Ferritic Steel Pipework for Carrying Fluids (Amd 5798) February 29, 1988. 56 pp.

BSI BS 2633-02. 1987 Amd 2 Class I Arc Welding of Ferritic Steel Pipework for Carrying Fluids (Amd 6969) July 15, 1992. 60 pp.

BSI BS 2971-91. 1991 Class II Arc Welding of Carbon Steel Pipework for Carrying Fluids. 37 pp.

BSI BS 2971-77. 1977 Amd 3 Class II Arc Welding of Carbon Steel Pipework for Carrying Fluids (AMD 5167) July 31, 1986. 25 pp.

BSI BS 4515-84. 1984 Amd 1 Process of Welding of Steel Pipelines on Land and Offshore (AMD 5810) December, 1987. 45 pp.

BSI BS 6990-89. 1989 Code of Practice for Welding on Steel Pipes Containing Process Fluids or Their Residuals. 22 pp.

SNZ NZS/BS 2971-91. Specification for Class II Arc Welding of Carbon Steel Pipework for Carrying Fluids. 36 pp.

—Steel Pipes—Stainless

BSI BS 4677-84. 1984 Amd 1 Arc Welding of Austenitic Stainless Steel Pipework for Carrying Fluids. 46 pp.

—Steels

BSI BS EN 288: Part 3-92. 1992 Specification and Approval of Welding Procedures for Metallic Materials Part 3: Welding Procedure Tests for the Arc Welding of Steels. 41 pp.

CEN PREN 288-3-91. Specification and Approval of Welding Procedures for Metallic Materials—Part 3: Welding Procedure Tests for the Arc Welding of Steels. 43 pp.

CEN EN 288-3-92. Specification and Approval of Welding Procedures for Metallic Materials—Part 3: Welding Procedure Tests for the Arc Welding of Steels. 40 pp.

CEN PREN 1011-93. Recommendations for Arc Welding of Ferritic Steels. 67 pp.

DIN ENGL EN 288 Pt 3-92. Specification and Approval of Procedures for Welding Metallic Materials; Welding Procedure Tests for the Arc Welding of Steel (Apr). 39 pp.

—Training

BSI BS 1295-87. 1987 Code of Practice for Training in Arc Welding Skills. 10 pp.

—Tubes

BSI BS 4870: Part 3-85. 1985 Amd 1 Approval Testing of Welding Procedures Part 3: Arc Welding of Tube to Tube-Plate Joints in Metallic Materials (AMD 6292) February 28, 1990. 15 pp.

—Welders (Personnel)—Acceptance Testing

BSI BS 4871: Part 3-85. 1985 Approval Testing of Welders Working to Approved Welding Procedures Part 3: Arc Welding of Tube to Tube-Plate Joints in metallic Materials. 14 pp.

Arc Welding Cables

See Also: Arc Welding Equipment; Cables (Electric)

BSI BS 638: Part 4-79. 1979 Amd 1 Arc Welding Power Sources, Equipment and Accessories Part 4: Welding Cables. 8 pp.

CENELEC HD 22.6 S1-90. Rubber Insulated Cables of Rated Voltages up to and Including 450/750 V Part 6: Arc Welding Cables. 9 pp.

CENELEC HD 22.6 S1/A1-92. AMD 1 Rubber Insulated Cables of Rated Voltages up to and Including 450/750 V Part 6: Arc Welding Cables. 7 pp.

DIN VDE 0250 Pt 803-86. Cables, Wires and Flexible Cords for Power Installation: Arc Welding Cable (July) (Replaces Parts of VDE 0250 and VDE 0250C—but Note Transitional Period.). 4 pp.

DIN VDE 0282 Pt 803-91. Rubber Insulated Cables, Wires and Flexible Cords for Power Installation: Arc Welding Cable (Apr) (Supersedes DIN VDE 0250 Part 803, but Note Transitional Period.). 7 pp.

IEC 245 Pt 6-80. Rubber Insulated Cables of Rated Voltages up to and Including 450/750 V Part 6: Arc Welding Electrode Cables First Edition; (Amendment 1-1985). 17 pp.

Arc Welding Cables (Cont.)

—Couplings

IEC 974 Pt 12-92. Arc-Welding Equipment Part 12: Coupling Devices for Welding Cables First Edition. 26 pp.

—Electrical Codes

CSA 632 Bull. Electrical Bulletin 632 March 29, 1966 to C22.2 NO 38. 2 pp.

—Safety

IEC 501-75. Safety Requirements for Arc Welding Equipment—Plugs, Socket-Outlets and Couplers for Welding Cables First Edition. 13 pp.

Arc Welding Equipment

See Also: Arc Spraying Equipment; Arc Welding; Arc Welding Cables; Arc Welding Machines; Arc Welds; Braze Welding Equipment; Flux Cored Arc Welding Equipment; Gas Metal Arc Welding Equipment; Gas Shielded Arc Welding Equipment; Inert Gas Welding Equipment; Submerged Arc Welding Equipment; Welding Equipment

BSI BS 7570-92. 1992 Validation of Arc Welding Equipment. 34 pp.

CENELEC HD 24-73. No—Load Voltage of Arc —Welding Equipment (Superseded by EN 60974-1-1990). 7 pp.

CNS C4072-81. Alternating Current Electric Arc Welding Machine (Jun)(2786).

CNS C4160-84. Rectifier Type DC Arc Welding Machines with Drooping Characteristic (Mar)(5028).

CNS C4161-84. Light Duty AC Arc Welding Machine (Mar)(5029).

DIN VDE 0544 Pt 2 (D)-78. Welding Equipment and Assemblies for Arc Welding and Allied Processes; Assemblies (VDE Specification) (Oct). 20 pp.

JIS C 9301-81. AC Arc Welding Machines. 12 pp.

JIS C 9306-81. Rectifier Type DC Arc Welding Machines with Drooping Characteristic. 11 pp.

JIS C 9314-81. Light Duty AC Arc Welding Machines. 10 pp.

JIS C 9322-88. Engine-Driven Arc Welding Machines with Drooping Characteristic. 16 pp.

—Accessories

BSI BS 638: Part 5-88. 1988 Amd 1 Arc Welding Power Sources, Equipment and Accessories Part 5: Accessories (AMD 6178) June 30, 1989. 27 pp.

—Automated

CSA C22.2 NO 60-M1990. Arc Welding Equipment; (Gen Instr 1). 59 pp.

CSA 1169 Bull. Electrical Bulletin 1169 June 27, 1978 to C22.2 NO 60. 2 pp.

—Electric Converters

DIN VDE 0540-65. Regulations for Direct Current Arc-Welding; Generators and Convertors (Feb). 19 pp.

—Electrode Holders

IEC 974 Pt 11-92. Arc-Welding Equipment Part 11: Electrode Holders First Edition. 28 pp.

JIS C 9302-91. Welding Electrode Holders for Arc Welding. 15 pp.

—Electrodes

BSI BS 7084-89. 1989 Amd 1 Carbon and Carbon-Manganese Steel Tubular Cored Welding Electrodes (F) (AMD 6281) November 30, 1990. 23 pp.

BSI BS 7084-02. 1989 Amd 2 Carbon and Carbon-Manganese Steel Tubular Cored Welding Electrodes (AMD 6814) November 29, 1991. 24 pp.

CNS C4031-85. Covered Electrodes for Mild Steel (Dec)(1215).

ISO 10446-90. Welding—All-Weld Metal Test Assembly for the Classification of Corrosion-Resisting Chromium and Chromium-Nickel Steel Covered Arc Welding Electrodes First Edition; (Corrected and Reprinted -1991). 6 pp.

JIS Z 3211-91. Covered Electrodes for Mild Steel. 11 pp.

JIS Z 3212-90. Covered Electrodes for High Tensile Strength Steel. 14 pp.

JIS Z 3214-91. Covered Electrodes for Atmospheric Corrosion Resisting Steel. 18 pp.

JIS Z 3221-89. Stainless Steel Covered Electrodes. 20 pp.

JIS Z 3223-87. Molybdenum Steel and Chromium Molybdenum Steel Covered Electrodes. 15 pp.

JIS Z 3224-91. Nickel and Nickel-Alloy Covered Electrodes. 15 pp.

JIS Z 3225-90. Covered Electrodes for 9% Nickel Steel. 9 pp.

JIS Z 3231-89. Copper and Copper Alloy Covered Electrodes. 15 pp.

JIS Z 3241-88. Covered Electrodes for Low Temperature Service Steel. 13 pp.

JIS Z 3252-92. Covered Electrodes for Cast Iron. 9 pp.

Arc Welding Equipment (Cont.)

—Electrodes—Core Wire

CNS G3033-88. Wire Rods for Core Wire of Covered Electrode (May)(2067). 3 pp.

JIS G 3503-80. Wire Rods for Core Wire of Covered Electrode. 6 pp.

JIS G 3523-80. Core Wires for Covered Electrode. 7 pp.

—Electrodes—Symbols

ISO 1071-83. Covered Electrodes for Manual Arc Welding of Cast Iron—Symbolization First Edition. 7 pp.

ISO 3580-75. Covered Electrodes for Manual Arc Welding of Creep-Resisting Steels—Code of Symbols for Identification First Edition. 5 pp.

ISO 3581-76. Covered Electrodes for Manual Arc Welding of Stainless and Other Similar High Alloy Steels—Code of Symbols for Identification First Edition. 7 pp.

—Generators

DIN VDE 0540-65. Regulations for Direct Current Arc-Welding; Generators and Convertors (Feb). 19 pp.

—Power Supplies

IEC 974 Pt 1-89. Safety Requirements for Arc Welding Equipment Part 1: Welding Power Sources First Edition. 110 pp.

JIS C 9300-92. General Rules for Arc Welding Power Sources. 16 pp.

—Power Supplies—Safety

CENELEC EN 60 974-1-90. Safety Requirements for Arc Welding Equipment Part 1: Welding Power Sources; (Supersedes HD 24:1973). 56 pp.

—Rectifiers

DIN VDE 0542-65. Regulations for Rectifier-Type Arc Welders (July). 38 pp.

DIN VDE 0542A-76. Standards for Arc Welding Rectifiers: Amendment A (Feb). 8 pp.

—Safety

BSI BS 638: Part 6-84. 1984 Arc Welding Power Sources, Equipment and Accessories Part 6: Safety Requirements for Construction. 8 pp.

BSI BS 638: Part 7-84. 1984 Amd 3 Arc Welding Power Sources, Equipment and Accessories Part 7: Safety Requirements for Installation and Use (AMD 6661) March 28, 1991. 21 pp.

BSI BS 638: Part 10-90. 1990 Arc Welding Power Sources, Equipment and Accessories Part 10: Specification for Safety Requirements for Arc Welding Equipment: Welding Power Sources. 32 pp.

CENELEC HD 362-76. Safety Rules for the Construction of Equipment for Electric Arc Welding and Allied Processes. 8 pp.

CENELEC HD 362/A1-78. AMD 1 Safety Rules for the Construction of Equipment for Electric Arc Welding and Allied Processes. 3 pp.

CENELEC HD 407-79. Safety Rules for the Use of Equipment for Electric arc Welding and Allied Processes. 7 pp.

CENELEC HD 427-80. Specific Safety Rules for the Installation of Equipment for Electric Arc Welding and Allied Processes. 10 pp.

CENELEC HD 433-83. Safety Requirements for Arc Welding Equipment Coupling Devices for Welding Cables. 10 pp.

CENELEC EN 60 974-1-90. CORRIGENDUM Safety Requirements for Arc Welding Equipment Part 1: Welding Power Sources; (Supersedes HD 24:1973). 4 pp.

IEC 501-75. Safety Requirements for Arc Welding Equipment—Plugs, Socket-Outlets and Couplers for Welding Cables First Edition. 13 pp.

IEC 974 Pt 1-89. Safety Requirements for Arc Welding Equipment Part 1: Welding Power Sources First Edition. 110 pp.

—Screens (Protective)

DIN ENGL 32504 Pt 1-83. Translucent Screens at Welders' Working Positions; Arc Welding Processes; Safety Requirements, Testing and Marking (Oct). 3 pp.

—Symbols

DIN ENGL 32520 Pt 3-89. Graphical Symbols for Welding; Graphical Symbols for Use on Arc Welding Equipment (Feb). 6 pp.

—Voltage Reducing Devices

CNS C4147-84. Voltage Reducing Device for AC Arc Welding (Jul)(4782).

Arc Welding Machines

See Also: Arc Welding Equipment; Welding Machines

INTERNATIONAL AND NON-U.S. NATIONAL STANDARDS
SUBJECT INDEX
Architectural

Arc Welding Machines *(Cont.)*
—Portable
SAA AS 3195-90. Approval and Test Specifications—Portable Machines for Electric Arc Welding and Allied Processes (This is a Joint Standard with SANZ NZS 3195) Amdt September 1993. 9 pp.

SNZ NZS/AS 3195-90. Approval and Test Specification—Portable Machines for Electric Arc Welding and Allied Processes (This is a Joint Standard with SAA AS 3195). 9 pp.

—Voltage Reducing Devices
JIS C 9311-89. Voltage Reducing Devices for AC Arc Welding Machines. 16 pp.

Arc Welds
See Also: Arc Welding; Arc Welding Equipment; Welded Joints

—Aluminum Alloys—Defects—Classification
ISO 10042-92. Arc-Welded Joints in Aluminium and Its Weldable Alloys—Guidance on Quality Levels for Imperfections First Edition. 14 pp.

—Aluminum—Defects—Classification
ISO 10042-92. Arc-Welded Joints in Aluminium and Its Weldable Alloys—Guidance on Quality Levels for Imperfections First Edition. 14 pp.

—Rolling Stock—Design
CNS E3001-82. Design Method of Arc Welded Joints for Railway Rolling Stock (Feb)(8487).

JIS E 4047-88. Design Methods of Arc Welded Joints for Railway Rolling Stock.

—Steels—Cracking (Fracturing)
JIS Z 3159-93. Method of H-Type Restrained Weld Cracking Test. 7 pp.

—Steels—Defects—Classification
BSI BS EN 25817-92. 1992 Acr-Welded Joints in Steel—Guidance on Quality Levels for Imperfections (ISO 5817: 1992). 18 pp.

CEN EN 25817-92. Arc-Welded Joints in Steel—Guidance on Quality Levels for Imperfections. 13 pp.

DIN ENGL EN 25817-92. Arc-Welded Joints in Steel; Guidance on Quality Levels for Imperfections (ISO 5817: 1992) (Sep) (Supersedes DIN 8563 Part 3, October 1985 Edition). 15 pp.

ISO 5817-92. Arc-Welded Joints in Steel—Guidance on Quality Levels for Imperfections First Edition; (Corrected and Reprinted -1992). 14 pp.

Arch Clipboard Files
Use For: Archfiles *See Also:* Clipboards
CGSB CAN/CGSB-53.28-M88. Arch Clip Board Files. 8 pp.

CGSB CAN/CGSB-53.92-M89. Arch Board File and Box. 8 pp.

Arches
Use For: Pipe Arches *See Also:* Structural Forms; Structural Members
SAA AS 2041-84. Corrugated Steel Pipes, Pipe-Arches, and Arches. 11 pp.

—Design
SAA AS 2042-84. Corrugated Steel Pipes, Pipe-Arches and Arches—Design and Installation Amdt 1 June 1988. 24 pp.

—Installation
SAA AS 2042-84. Corrugated Steel Pipes, Pipe-Arches and Arches—Design and Installation Amdt 1 June 1988. 24 pp.

Archfiles
Use: Arch Clipboard Files

Architectural Barriers
Use: Barrier Free Design

Architectural Concrete
See Also: Concretes; Plaster; Precast Concrete

—Manufacturers—Certification
CSA A251-M1982. Qualification Code for Manufacturers of Architectural and Structural Precast Concrete; (Gen Instr 1). 64 pp.

Architectural Drawings
Use For: Building Drawings; Civil Engineering Drawings *See Also:* Architecture; Drawings; Engineering Drawings
BSI BS 1192: Part 2-87. 1987 Constructon Drawing Practice Part 2: Recommendations for Architectural and Engineering Drawings. 58 pp.

Architectural Drawings *(Cont.)*
BSI BS 6100: SUBSEC 1.5.7-01. 1988 Amd 1 Glossary of Building and Civil Engineering Terms Part 1: General and Miscellaneous Section 1.5: Operations; Associated Plant and Equipment Subsection 1.5.7: Drawings (AMD 7245) August 15, 1992. 8 pp.

BSI PP 7319-89. 1989 Construction Drawing Practice for Universities, Polytechnics and Colleges. 80 pp.

BSI PP 7320-88. 1988 Construction Drawing Practice for Schools. 45 pp.

CNS A1042-87. Architectural Drawing (Dec)(11567).

CNS B1001-88. Engineering Drawing General Principle of Presentation (Dec)(3).

CSA CAN3-B78.3-M77. Building Drawings (R 1991); (Amd 1-3 September 1979). 51 pp.

ISO 128-82. Technical Drawings—General Principles of Presentation First Edition. 18 pp.

ISO 406-87. Technical Drawings—Tolerancing of Linear and Angular Dimensions Second Edition. 7 pp.

ISO 2594-72. Building Drawings—Projection Methods First Edition; (Erratum—Oct 1972). 5 pp.

ISO 6284-85. Tolerances for Building—Indication of Tolerances on Building and Construction Drawings First Edition. 7 pp.

ISO TR7084-81. Technical Drawings—Coding and Referencing Systems for Building and Civil Engineering Drawings and Associated Documents First Edition. 11 pp.

ISO 7519-91. Technical Drawings—Construction Drawings—General Principles of Presentation for General Arrangement and Assembly Drawings First Edition. 11 pp.

JIS A 0150-78. Drawing Office Practice for Architecture and Building (General Rules).

SAA AS 1100.301-85. Technical Drawing—Part 301: Architectural Drawing. 19 pp.

SAA AS 1100.301 Supp 1-86. Technical Drawing—Part 301 Supplement 1: Architectural Drawings (Supplement to AS 1100. 301—1985). 6 pp.

SNZ NZS/AS 1100: Part 301-85. Technical Drawing Part 301: Architectural Drawing. 19 pp.

SNZ NZS/AS 1100: Pt301: Supp1-86. Technical Drawing Part 301: Architectural Drawing Supplement 1: Architectural Drawings. 6 pp.

SNZ NZS 5902: Part 0-86. Building and Civil Engineering Drawing Practice Part 0: General (Superseded by NZS/AS 1100 Parts 101-501).

SNZ NZS 5902: Part 1-86. Building and Civil Engineering Drawing Practice Part 1: Architectural (Superseded by NZS/AS 1100 Parts 101-501).

SNZ NZMP 5905-81. Standard Technical Drawing Practice for School Certificate Courses in New Zealand. 31 pp.

SNZ NZMP 5906-84. Standard Technical Drawing Practice for Senior Courses in Schools and Technical Institutes. 74 pp.

—CAD
CSA B78.5-93. Computer-Aided Design Drafting (Buildings). 123 pp.

ISO TR10127-90. Computer-Aided Design (CAD) Technique—Use of Computers for the Preparation of Construction Drawings First Edition. 20 pp.

JTC1 TR10127-90. Computer-Aided Design (CAD) Technique—Use of Computers for the Preparation of Construction Drawings First Editon. 20 pp.

OSI ISO TR 10127-90. Computer-Aided Design (CAD) Technique—Use of Computers for the Preparation of Construction Drawings. 20 pp.

—Computer Graphics—Information Interchange
SAA AS 3643.2-92. Computer Graphics—Initial Graphics Exchange Specification (IGES) for Digital Exchange of Product Definition Data—Part 2: Subset of AS 3643.1—Two-Dimensional Drawings for Architectural, Engineering and Construction (AEC) Industries. 37 pp.

—Demolition/Rebuilding
ISO 7518-83. Technical Drawings—Construction Drawings—Simplified Representation of Demolition and Rebuilding First Edition. 4 pp.

—Electric Wiring
SNZ NZS 5902: Part 4-76. Building and Civil Engineering Drawing Practice Part 4: Services—Electrical Amend: 1, 1984; 2, 1985 (Superseded bu NZS/AS 1100 Parts 101-501).

—Electric Wiring—Symbols
CNS C1090-87. Graphical Symbols of Interior Wiring Diagram for Architectural Plans (Dec)(9101).

CNS C1091-87. Symbols of Switch of Interior Wiring Diagram for Architectural Plans (Dec)(9102).

CNS C1092-87. Symbols of Relays of Interior Wiring Diagram for Architectural Plans (Dec)(9103).

CNS C1093-87. Symbols of Meters of Interior Wiring Diagram for Architectural Plans (Dec)(9104).

CNS C1094-87. Symbols of Electric Machines of Interior Wiring Diagram for Architectural Plans (Dec)(9105).

Architectural Drawings *(Cont.)*
—Electric Wiring—Symbols *(Cont.)*
CNS C1095-87. Symbols of Potential Transformers and Current Transformers of Interior Wiring Diagrams for Architectural Plans (Dec)(9106).

CNS C1096-87. Symbols of Connecting of Interior Wiring Diagrams for Architectural Plans (Dec)(9107).

CNS C1097-87. Symbols of Distribution of Interior Wiring Diagram for Architectural Plans (Dec)(9108).

CNS C1098-87. Symbols of Wiring of Interior Wiring Diagram for Architectural Plans (Dec)(9109).

CNS C1099-87. Symbols of Bus-Bar of Interior Wiring Diagram for Architectural Plans (Dec)(9110).

CNS C1100-87. Symbols of Lamps and Socket of Interior Wiring Diagram for Architectural Plans (Dec)(9111).

CNS C1101-87. Symbols of Telephone and Bell of Interior Wiring Diagram for Architectural Plans (Dec)(9112).

CNS C1102-87. Symbols of Fire Alarm System of Interior Wiring Diagrams for Architectural Plans (Dec)(9113).

CNS C1103-87. Symbols of Television and Broadcast Equipment of Interior Wiring Diagram for Architectural Plans (Dec)(9114).

CSA Z99.3-1979. Graphical Electrical Symbols for Architectural Plans (R 1989). 32 pp.

JIS C 0303-84. Graphical Symbols of Interior Wiring Diagram for Architectural Plans. 52 pp.

—Glossaries
BSI BS 6100: SUBSEC 1.5.7-88. 1988 Glossary of Building and Civil Engineering Terms Part 1: General and Miscellaneous Section 1.5: Operations: Associated Plant and Equipment Subsection 1.5.7: Drawings. 7 pp.

BSI BS 6100: SUBSEC 1.5.7-01. 1988 Amd 1 Glossary of Building and Civil Engineering Terms Part 1: General and Miscellaneous Section 1.5: Operations; Associated Plant and Equipment Subsection 1.5.7: Drawings (AMD 7245) August 15, 1992. 8 pp.

—Installation Plans—Symbols
BSI BS 3939: Part 11-85. 1985 Graphical Symbols for Electrical Power, Telecommunications and Electronics Diagrams Part 11: Architectural and Topographical Installation Plans and Diagrams (G). 16 pp.

IEC 617 Pt 11-83. Graphical Symbols for Diagrams Part 11: Architectural and Topographical Installation Plans and Diagrams First Edition. 33 pp.

SNZ IEC 617: Part 11-83. Graphical Symbols for Diagrams Part 11: Architectural and Topographical Installation Plans and Diagrams. 30 pp.

—Modular Construction
ISO 8560-86. Technical Drawings—Construction Drawings—Representation of Modular Sizes, Lines and Grids First Edition. 8 pp.

—Prefabricated Buildings
ISO 4172-91. Technical Drawings—Construction Drawings—Drawings for the Assembly of Prefabricated Structures Second Edition. 14 pp.

ISO 7437-90. Technical Drawings—Construction Drawings—General Rules for Execution of Production Drawings for Prefabricated Structural Components First Edition. 6 pp.

—Reference Lines
ISO 4068-78. Building and Civil Engineering Drawings—Reference Lines First Edition. 4 pp.

—Reinforcing Steels
SNZ NZS 5902: Part 2-76. Building and Civil Engineering Drawing Practice Part 2: Structural—Concrete, Steel, and Timber Amend: 1, 1979; 2, 1984; 3, 1985 (Superseded by NZS/AS 1100 Parts 101-501).

—Reinforcing Steels—Symbols
ISO 3766-77. Building and Civil Engineering Drawings—Symbols for Concrete Reinforcement First Edition. 7 pp.

—Sanitary Facilities
BSI BS 1192: Part 3-87. 1987 Amd 1 Construction Drawing Practice Part 3: Recommendations for Symbols and Other Graphic Conventions (AMD 5882) February 28, 1989. 52 pp.

ISO 4067 Pt 2-80. Building and Civil Engineering Drawings—Installations—Part 2: Simplified Representation of Sanitary Appliances First Edition. 4 pp.

SNZ NZS 5902: Part 3-76. Building and Civil Engineering Drawing Practice Part 3: Services—Mechanical and Sanitary Amend: 1, 1979; 2, 1981; 3, 1985 (Superseded by NZS/AS 1100 Parts 101-501).

—Steel Bars
ISO 4066-77. Building and Civil Engineering Drawings—Bar Scheduling First Edition. 9 pp.

INDUSTRY STANDARDS

Architectural Drawings (Cont.)
—Steel Structures
SNZ NZS 5902: Part 2-76. Building and Civil Engineering Drawing Practice Part 2: Structural—Concrete, Steel, and Timber Amend: 1, 1979; 2, 1984; 3, 1985 (Superseded by NZS/AS 1100 Parts 101-501).

—Surface Representation
ISO 4069-77. Building and Civil Engineering Drawings—Representation of Areas on Sections and Views—General Principles First Edition. 4 pp.
ISO 8048-84. Technical Drawings—Construction Drawings—Representation of Views, Sections and Cuts First Edition. 7 pp.

—Symbols
ISO 4157 Pt 1-80. Building Drawings—Part 1: Designation of Buildings and Parts of Buildings First Edition. 5 pp.
ISO 4157 Pt 2-82. Technical Drawings—Construction Drawings—Designation of Buildings and Parts of Buildings—Part 2: Designation of Rooms and Other Areas First Edition. 4 pp.
SNZ NZS 5902: Part 1-86. Building and Civil Engineering Drawing Practice Part 1: Architectural (Superseded by NZS/AS 1100 Parts 101-501).

—Technical Manuals
SNZ NZS 5902: Part 5-81. Building and Civil Engineering Drawing Practice Part 5: Recommendations for Drawings Associated with Engineering Services Operating Manuals and Maintenance Manuals Amend: 1, 1985. 24 pp.

—Wooden Structures
SNZ NZS 5902: Part 2-76. Building and Civil Engineering Drawing Practice Part 2: Structural—Concrete, Steel, and Timber Amend: 1, 1979; 2, 1984; 3, 1985 (Superseded by NZS/AS 1100 Parts 101-501).

Architectural Lighting
Use: Decorative Lighting

Architectural Plans
Use: Architectural Drawings

Architecture
Scope Note: For additional listings, use a more specifc term *See Also:* Acoustics; Architectural Drawings; Arts; Buildings; Civil Engineering; Construction; Structures

—Classification
BSI BS 1000: (72)-75. 1975 Universal Decimal Classification (UDC). English Full Edition (72): Architecture. 41 pp.
SNZ NZS/BS 1000 (72)-75. Universal Decimal Classification Architecture. 44 pp.

Archival Storage
BSI BS 5454-89. 1989 Recommendations for the Storage and Exhibition of Archival Documents. 23 pp.

—Binding
BSI BS 4971: Part 2-80. 1980 Repair and Allied Processes for the Conservation of Documents Part 2: Archival Binding. 8 pp.

—Bookbinding—Leather
BSI BS 7451-91. 1991 Archival Quality Bookbinding Leather. 11 pp.

—Microfilming—Newspapers
BSI BS 5847-80. (WITHDRAWN) 1980 35 mm Microcopying of Newspapers for Archival Purposes (Superseded by BS ISO 4087: 1991). 11 pp.
BSI BS ISO 4087-91. 1991 Micrographics—Microfilming of Newspapers for Archival Purposes on 35mm Microfilm (H). 14 pp.
ISO 4087-91. Micrographics—Microfilming of Newspapers for Archival Purposes on 35 mm Microfilm Second Edition. 11 pp.

—Photographic Films
BSI BS 5699: Part 1-79. 1979 Processed Photographic Film for Archival Records Part 1: Specifications for Silver-Gelatine Type on Cellulose Ester Base. 12 pp.
BSI BS 5699: Part 2-79. 1979 Processed Photographic Film for Archival Records Part 2: Specifications for Silver-Gelatin Type on Poly (Ethylene Terephthalate) Base. 14 pp.
ISO 4331-86. Photography—Processed Photographic Black-and-White Film for Archival Records—Silver-Gelatin Type on Cellulose Ester Base—Specifications Second Ediiton. 12 pp.
ISO 4332-86. Photography—Processed Photographic Black-and-White Film for Archival Records—Silver-Gelatin Type on Poly(Ethylene Terephthalate) Base—Specifications Second Edition. 13 pp.

Archival Storage (Cont.)
—Photographic Films (Cont.)
ISO 5466-92. Photography—Processed Safety Photographic Films—Storage Practices Third Edition. 16 pp.

—Photographic Films—Containers
ISO 10214-91. Photography—Processed Photographic Materials—Filing Enclosures for Storage First Edition; (Corrigendum 1-1992). 14 pp.

Archives
—Directories
ISO 2146-88. Documentation—Directories of Libraries, Archives, Information and Documentation Centres, and Their Data Bases Second Edition. 28 pp.

Area
Use For: Surface Area *See Also:* Coordination Areas

—Aluminum Oxide—Adsorption
BSI BS 4140: Part 23-87. 1987 Methods of Test for Aluminium Oxide Part 23: Determination of Specific Surface Area by Nitrogen Absorption (Single-Point Method). 11 pp.
ISO 8008-86. Aluminium Oxide Primarily Used for the Production of Aluminium—Determination of Specific Surface Area by Nitrogen Adsorption—Single-Point Method First Edition. 11 pp.
SAA AS 2879.4-91. Alumina Determination of Specific Surface Area by Nitrogen Adsorption. 2 pp.

—Carbon Black—Adsorption
BSI BS 5293: Part 12-88. 1988 Sampling and Testing Carbon Black for Use in the Rubber Industry Part 12: Methods for Determination of Surface Area by Surfactant Absorption. 15 pp.
ISO 4652-81. Rubber Compounding Ingredients—Carbon Black—Determination of Specific Surface Area—Nitrogen Adsorption Methods First Edition. 17 pp.
ISO 6810-85. Rubber Compounding Ingredients—Carbon Black—Determination of Surface Area—Surfactant Adsorption Methods First Edition. 13 pp.

—Ceramic Powders—Nitrogen Absorption
CEN PREN 725-6-92. Advanced Technical Ceramics—Methods of Test for Ceramic Powders—Part 6: Determination of the Specific Surface Area. 17 pp.

—Gas Permeability
MOD UK M 825/62. Specific Surface Area and Surface Mean Diameter Determination by Air Permeability.

—Metal Powders
ISO 10070-91. Metallic Powders—Determination of Envelope-Specific Surface Area from Measurements of the Permeability to Air of a Powder Bed Under Steady-State Flow Conditions First Edition. 16 pp.

—Powders—Gas Adsorption
JIS Z 8830-90. Determination of the Specific Surface Area of Powders by Gas Adsorption Methods. 13 pp.

—Tables (Data)
CNS Z7137-82. Square Foot-Square Metre Conversion Tables (Dec)(9779).
CNS Z7138-82. Square Yard-Square Metre Conversion Tables (Dec)(9780).
CNS Z7139-82. Square Inch-Square Centimetre Conversion Tables (Dec)(9781).
CNS Z7201-86. Square Millimeter-Circular Mile Conversion Tables (Jun)(11623).
JIS Z 8423-71. Square Inch-Square Centimetre Conversion Tables.
JIS Z 8424-71. Square Foot-Square Metre Conversion Tables.
JIS Z 8425-71. Square Yard-Square Metre Conversion Tables.
SAA AS 1377.2-73. Conversion Tables—Part 2: Area. 21 pp.

Area Navigation Computing Instruments
See Also: Navigation

—Accuracy Testing
EUROCAE ED-28 07.82. MPS for Airborne Area Navigation Computing Equipment Based on VOR and DME as Sensors. 31 pp.
EUROCAE ED-40 06.84. MPS for Airborne Computing Equipment for Area Navigation System Using Two DME as Sensors. 33 pp.

—Design
EUROCAE ED-28 07.82. MPS for Airborne Area Navigation Computing Equipment Based on VOR and DME as Sensors. 31 pp.

Area Navigation Computing Instruments (Cont.)
—Design (Cont.)
EUROCAE ED-40 06.84. MPS for Airborne Computing Equipment for Area Navigation System Using Two DME as Sensors. 33 pp.

—Electrical Properties
EUROCAE ED-28 07.82. MPS for Airborne Area Navigation Computing Equipment Based on VOR and DME as Sensors. 31 pp.
EUROCAE ED-40 06.84. MPS for Airborne Computing Equipment for Area Navigation System Using Two DME as Sensors. 33 pp.

—Environmental Testing
EUROCAE ED-28 07.82. MPS for Airborne Area Navigation Computing Equipment Based on VOR and DME as Sensors. 31 pp.
EUROCAE ED-40 06.84. MPS for Airborne Computing Equipment for Area Navigation System Using Two DME as Sensors. 33 pp.

Area Navigation Systems
See Also: Navigation

—Accuracy Testing
EUROCAE ED-27 10.79. MOPR for Airborne Area Navigation Systems, Based on VOR and DME as Sensors. 63 pp.
EUROCAE ED-39 06.84. MOPR for Airborne Area Navigation Systems, Based on Two DME as Sensors. 67 pp.

—Design
EUROCAE ED-27 10.79. MOPR for Airborne Area Navigation Systems, Based on VOR and DME as Sensors. 63 pp.
EUROCAE ED-39 06.84. MOPR for Airborne Area Navigation Systems, Based on Two DME as Sensors. 67 pp.

—Environmental Testing
EUROCAE ED-27 10.79. MOPR for Airborne Area Navigation Systems, Based on VOR and DME as Sensors. 63 pp.

—Failure (Quality Control)
EUROCAE ED-27 10.79. MOPR for Airborne Area Navigation Systems, Based on VOR and DME as Sensors. 63 pp.
EUROCAE ED-39 06.84. MOPR for Airborne Area Navigation Systems, Based on Two DME as Sensors. 67 pp.

—Quality Assurance
EUROCAE ED-27 10.79. MOPR for Airborne Area Navigation Systems, Based on VOR and DME as Sensors. 63 pp.

Area Rugs
Use: Rugs—Area

Arginine Monohydrochloride
JIS K 9046-72. L-Arginine Monohydrochloride.

Argon
See Also: Hazardous Materials; Noble Gases; Nonmetals
CNS K1054-61. Argon, High Purity (Jun)(1374)(R 1968).
CNS K1131-69. Argon for Welding (Jun)(2983).

—Gas Cylinders
JIS K 1105-84. Argon. 9 pp.

—Storage and Handling—Information Cards
SAA AS 2508.2.01 2-85. Safe Storage and Handling Information Cards for Hazardous Materials—Part 2.012: Non-Flammable, Inert, Compressed Gases (Includes: Argon, Helium, Krypton, Neon, Nitrogen and Xenon) Double Sided Card.

Argon Content Analysis
—Butadienes—Gas Chromatography
ISO 6792-82. Butadiene for Industrial Use—Determination of Oxygen and Argon in the Gaseous Phase Above Liquid Butadiene—Gas Chromatographic Method First Edition. 4 pp.

Aries Aircraft
See Also: Aircraft

—Antenna Positions
CAA. Aries (Approved Aerial Positions). 1 p.

ARIS
Use: Automatic Reporting and Identification System

Arithmetic
Scope Note: Use a more specific term *See:* Floating Point Arithmetic; Mathematics

INTERNATIONAL AND NON-U.S. NATIONAL STANDARDS
SUBJECT INDEX
Armored

Arithmetic Circuits
Scope Note: Use a more specific term
See: Arithmetic Logic Unit/Function Generators; Decade Counter/Dividers; Digital Circuits; Octal Counter/Dividers; Ripple Counter/Dividers; Ripple Counters

Arithmetic Logic Unit/Function Generators
See Also: Arithmetic Logic Units; Digital Circuits; Function Generators; Generators; Look Ahead Carry Generators
  CECC CECC 90 102-032 ISSUE 1-81. BS CECC 90 102-032; Arithmetic Logic Units/Function Generators (En) AMD 1 (En). 26 pp.
  CECC CECC 90 109-832 ISSUE 1-89. Digital Integrated Circuits; Silicon Monolithic C MOS, Cavity or Non-Cavity Packages; Type(s) 54/74 HC 181 Arithmetic Logic Unit /Function Generator Assessment Levels P, Y, L (En, Fr, Ge). 14 pp.

—**4-Bit—Preferred Products List**
  CECC CECC MUAHAG Vol 7 IS 8-92. Preferred Products List; Active Microcircuits (En, Fe, Ge). 89 pp.

Arithmetic Logic Units
See Also: Arithmetic Logic Unit/Function Generators

—**NMOS—Microprocessors—Preferred Products List**
  CECC CECC MUAHAG Vol 7 IS 8-92. Preferred Products List; Active Microcircuits (En, Fe, Ge). 89 pp.

—**4-Bit**
  CECC CECC 90 103-078 ISSUE 1-81. BS CECC 90 103-078; Arithmetic Logic Units/Function Generators (En) AMD 1 (En). 25 pp.

Arm Prostheses
See Also: Hand Prostheses; Joint Prostheses; Prosthetic Devices; Wrist Prostheses

—**Cosmetic Hand Attachment**
  JIS T 9224-90. Cosmetic Hands for Arm Prostheses. 10 pp.

—**Stumps**
  ISO 8548 Pt 3-93. Prosthetics and Orthotics—Limb Deficiencies—Part 3: Method of Describing Upper Limb Amputation Stumps First Edition. 19 pp.

Armaments
Use: Weapons

Armature Relays
See Also: Induction Loops; Relays
  MOD UK DSTAN 59-7: Part 2-77. Relays, Armature, and Relays, Thermal Part 2: Relays, Armature, Sealed Issue 2. 40 pp.
  MOD UK DSTAN 59-7: Part 4-69. Relays, Armature, and Relays, Thermal Part 4: Relays, Armature, Non-Sealed, Other Than 3000, 600, and 500 Types. 3 pp.
  MOD UK DSTAN 59-7: Part 5-69. Relays, Armature, and Relays, Thermal Part 5: Relays, Armature K600, 600, and 500 High and Low Voltage Types. 3 pp.

Armatures
See Also: Rotating Machines; Rotors (Machine Elements)

—**Piano Wire**
  CNS C2070-80. Tin Coated Piano Wire for Armature Binding (Aug)(6061). 4 pp.

—**Spindles**
  CNS B2588-81. Remote Control Gear with Transmission Tubing for Manual Operation; Lengthening for Spindles of Armatures (Jun)(7585).

—**Steel Wire**
  CNS C2071-90. Tin Coated Non-Magnetic Steel Wire for Armature Binding (Aug)(6062).
  CNS C2071-80. Tin Coated Non-Magnetic Steel Wire for Armature Binding (Aug)(6062). 4 pp.
  JIS C 2507-90. Tin Coated Non-Magnetic Steel Wire for Armature Binding. 9 pp.

Armbands
See Also: Bands

—**Inflatable**
  BSI BS 7661-93. 1993 Inflatable Armbands Worn as Flotation Aids (N). 11 pp.

Armchairs
Use For: Easy Chairs   *See Also:* Chairs (Seats)
  CNS S1047-68. Arm Chairs (Apr)(2838).

Armed Forces
Use: Military Personnel

Arming Devices
See Also: Bombs (Ordnance); Intervalometers

—**Aircraft Stores**
  NATO STANAG 3605 ED 2 AMD 1-86. Compatibility of Mechanical Fuzing Systems and Arming Devices for Expendable Aircraft Stores for Fixed Wing Aircraft. 12 pp.
  NATO STANAG 3605 ED 3 AMD 0-92. Compatibility of Arming Units and Expendable Aircraft Stores. 9 pp.

Armor, Body
Use: Body Armor

Armor Piercing Ammunition
Use For: Antiarmor Ammunition
See Also: Ammunition; Grenades; Guns; Projectiles; Propellants; Rockets; Small Arms Ammunition

—**Penetration Resistance**
  NATO STANAG 4190 ED 1 AMD 0-87. Test Procedures for Measuring Behind-Armour Effects of Anit-Armour Ammunition. 20 pp.

—**Perforation Tests**
  NATO STANAG 4164 ED 1 AMD 0-85. Test Procedures for Armour Perforation Tests of Direct Fire Armour Defeating Ammunition. 30 pp.
  NATO STANAG 4164 ED 1 AMD 1-85. Test Procedures for Armour Perforation Tests of Direct Fire Armour Defeating Ammunition. 28 pp.

Armor Plates
  MOD UK DSTAN 95-13-81. Armour Plates for Proof of Projectile Purposes Issue 1. 17 pp.

Armored Cables
See Also: Cables (Electric); Telephone Cables; Transmission Lines
  CNS C2081-90. Steel Tape Armour and Anti-Corrosive Layer for Electric Cable (Aug)(6074).
  CNS C2081-80. Steel Tape Armour and Anti-Corrosive Layer for Electric Cable (Aug)(6074). 5 pp.
  CNS C2106-82. 2P x 1.2mm PE Insulated Star Quad Armoured Cable (Mar)(8560).
  JIS C 3610-87. Steel Tape Armor and Anti-Corrosive Layer for Cables. 9 pp.

—**Armor Thickness**
  DIN VDE 0472 Pt 402-84. Testing of Insulated Cables, Wires and Flexible Cords; Thickness of Insulation and Armouring (May). 12 pp.

—**Bushings**
  CSA CAN/CSA-C22. 2 NO 18-92. Outlet Boxes, Conduit Boxes, and Fittings; (Gen Instr 1 Thru 2). 118 pp.

—**Cable Clamps**
  CSA CAN/CSA-C22. 2 NO 18-92. Outlet Boxes, Conduit Boxes, and Fittings; (Gen Instr 1 Thru 2). 118 pp.

—**Connectors**
  CSA CAN/CSA-C22. 2 NO 18-92. Outlet Boxes, Conduit Boxes, and Fittings; (Gen Instr 1 Thru 2). 118 pp.

—**Electrical Codes**
  CSA 632 Bull. Electrical Bulletin 632 March 29, 1966 to C22.2 NO 38. 2 pp.
  CSA 671 Bull. Electrical Bulletin 671 March 16, 1967 to C22.2 NO 38. 3 pp.
  CSA 851 Bull. Electrical Bulletin 851 November 17, 1971 to C22.2 NO 38. 2 pp.
  CSA 983C Bull. Electrical Bulletin 983C November 14, 1978 to C22.2 NO 38. 3 pp.
  CSA 632 Bull. Electrical Bulletin 632 March 29, 1966 to C22.2 NO 51. 2 pp.
  CSA 671 Bull. Electrical Bulletin 671 March 16, 1967 to C22.2 NO 51. 3 pp.
  CSA 851 Bull. Electrical Bulletin 851 November 17, 1971 to C22.2 NO 51. 2 pp.
  CSA 983C Bull. Electrical Bulletin 983C November 14, 1978 to C22.2 NO 51. 3 pp.

—**Identification Systems**
  CSA 1368 Bull. Electrical Bulletin 1368 April 7, 1982 to C22.2 NO 38. 2 pp.
  CSA 1368 Bull. Electrical Bulletin 1368 April 7, 1982 to C22.2 NO 51. 2 pp.
  CSA 1368-51 Bull. Electrical Bulletin 1368-51 September 14, 1983 to C22.2 NO 51. 2 pp.

—**Mechanical Protection**
  CCITT RECMN L.3-89. Armouring of Cables—Construction, Installation and Protection of Cable and Other Elements of Outside Plant (Study Group VI) 3 pp. 3 pp.

Armored Cables *(Cont.)*
—**Multiconductor—Thermoset Insulated**
  CSA 983B Bull. Electrical Bulletin 983B April 25, 1977 to C22.2 NO 38. 39 pp.
  CSA 1213A Bull. Electrical Bulletin 1213A October 25, 1979 to C22.2 NO 38. 3 pp.
  CSA CAN/CSA-C22. 2 NO 51-M89. Armoured Cable; (Gen Instr 1 Thru 2). 39 pp.
  CSA 983B Bull. Electrical Bulletin 983B April 25, 1977 to c22.2 NO 51. 39 pp.
  CSA 1240 Bull. Electrical Bulletin 1240 September 24, 1979 to C22.2 NO 51. 2 pp.
  CSA 1213A Bull. Electrical Bulletin 1213A October 25, 1979 to C22.2 NO 51. 3 pp.

—**Paper Insulated—Lead Sheathed—Steel Wire—Submarines**
  CNS C2057-80. Eight Pair Submarine Cable (Star Quad, Paper-Insulated, Lead-Sheathed Steel Wire Armoured) (Tentative) (Jul)(3262). 3 pp.

—**Staples**
  CSA CAN/CSA-C22. 2 NO 18-92. Outlet Boxes, Conduit Boxes, and Fittings; (Gen Instr 1 Thru 2). 118 pp.

—**Steel Wire—Galvanized**
  BSI BS 1442-69. 1969 Amd 3 Galvanized Mild Steel Wire for Armouring Cables. 14 pp.
  CCITT RECMN L.3-89. Armouring of Cables—Construction, Installation and Protection of Cable and Other Elements of Outside Plant (Study Group VI) 3 pp. 3 pp.
  CNS C2102-81. Galvanized Low Carbon Steel Wires for Armouring Cable (Oct)(7968).
  CNS G2195-83. Method of Test for Galvanized Low Carbon Steel Wires for Armoring Cables (Dec)(10715).
  CNS G3212-83. Galvanized Low Carbon Steel Wires for Armouring Cables (Dec) (10714).
  SAA AS 3863-91. Galvanized Mild Steel Wire for Armouring Cables. 5 pp.

—**Steel Wire—Galvanized—Submarines**
  BSI BS 1441-69. 1969 Galvanized Steel Wire for Armouring Submarine Cables. 13 pp.
  SAA AS 3863-91. Galvanized Mild Steel Wire for Armouring Cables. 5 pp.

—**Straps (Fasteners)**
  CSA CAN/CSA-C22. 2 NO 18-92. Outlet Boxes, Conduit Boxes, and Fittings; (Gen Instr 1 Thru 2). 118 pp.

—**Thermoset Insulated**
  BSI BS 6724-90. 1990 Armoured Cables for Electricity Supply Having Thermosetting Insulation with Low Emission of Smoke and Corrosive Gases When Affected by Fire (Q). 44 pp.
  BSI BS 6724-01. 1990 Amd 1 Armoured Cables for Electricity Supply Having Thermosetting Insulation with Low Emission of Smoke and Corrosive Gases When Affected by Fire (AMD 6832) December 24, 1991. 46 pp.
  BSI BS 6724-02. 1990 Amd 2 Armoured Cables for Electricity Supply Having Thermosetting Insulation with Low Emission of Smoke and Corrosive Gases When Affected by Fire (AMD 7569) July 15, 1993 (Q). 53 pp.
  CSA 983B Bull. Electrical Bulletin 983B April 25, 1977 to C22.2 NO 38. 39 pp.
  CSA 1213A Bull. Electrical Bulletin 1213A October 25, 1979 to C22.2 NO 38. 3 pp.
  CSA CAN/CSA-C22. 2 NO 51-M89. Armoured Cable; (Gen Instr 1 Thru 2). 39 pp.
  CSA 983B Bull. Electrical Bulletin 983B April 25, 1977 to c22.2 NO 51. 39 pp.
  CSA 1240 Bull. Electrical Bulletin 1240 September 24, 1979 to C22.2 NO 51. 2 pp.
  CSA 1213A Bull. Electrical Bulletin 1213A October 25, 1979 to C22.2 NO 51. 3 pp.

Armored Vehicles
See Also: Motor Vehicles; Tanks (Combat Vehicles); Trucks

—**Electronic Equipment**
  MOD UK DSTAN 00-32: Part 1-85. Systematic Approach to Vehicle Electronics (SAVE) Part 1: General Requirements Issue 1. 7 pp.

—**Lighting**
  MOD UK DSTAN 62-7-81. Lights, Light Units and Reflectors for Service Vehicles Issue 2. 20 pp.

—**Spring Steels**
  MOD UK DEF-103-60. Steel for Hardened and Tempered High Quality Coil Springs for Guns and Armoured Fighting Vehicles (Reprinted December 1971). 4 pp.
  MOD UK DEF-104-60. Hardened and Tempered Coil Springs of High Quality for Guns and Armoured Fighting Vehicles (Reprinted May 1973, Incorporating Amendment No. 1). 4 pp.

INDUSTRY STANDARDS

Armored Vehicles (Cont.)
—Track Pins
MOD UK DSTAN 95-17-91. Track Pins, Induction Hardened Issue 1. 10 pp.

Armoured Cables
Use: Armored Cables

Aromatic Compound Content Analysis
—Aromatic Hydrocarbons—Infrared Spectrophotometry
MOD UK DSTAN 05-50: Part 60-91. Methods for Testing Fuels, Lubricants and Associated Products Part 60: Aromatics Content in Hydrocarbon Solvents Issue 1. 7 pp.

Aromatic Hydrocarbon Content Analysis
See Also: Aromatic Hydrocarbons
—Air—Gas Chromatography
BSI BS 6069: Sec 3.4-91. 1991 Characterization of Air Quality Part 3: Workplace Atmospheres Section 3.4: Method for the Determination of Vaporous Aromatic Hydrocarbons by Charcoal Tube/Solvent Desorption/ Gas Chromatography (ISO 9487. 1991). 18 pp.
ISO 9487-91. Workplace Air—Determination of Vaporous Aromatic Hydrocarbons—Charcoal Tube/Solvent Desorption/ Gas Chromatographic Method First Edition. 15 pp.
—Electrical Insulating Oils
CENELEC HD 382-78. Determination of the Aromatic Hydorcarbon Content of New Mineral Insulating Oils. 2 pp.
IEC 590-77. Determination of the Aromatic Hydrocarbon Content of New Mineral Insulating Oils First Edition; (Corrigendum—July 1978). 21 pp.
—Gasoline—Gas Chromatography
CGSB CAN/CGSB-3.0 NO.14.3-M91. Methods of Testing Petroleum and Associated Products Standard Test Method for the Identification of Hydrocarbon Components in Automotive Gasoline Using Gas Chromatography. 32 pp.
—Paint Thinners
CGSB 1-GP-71 METH 66.3-82. Methods of Testing Paints and Pigments Hydrocarbons and Esters Aromatic Hydrocarbons. 1 p.
—Petroleum Products—Gas Chromatography
DIN ENGL 51413 Pt 3-88. Testing of Liquid Petroleum Products; Determination of Aromatics Content by Gas Chromatography (Oct). 3 pp.

Aromatic Hydrocarbons
See Also: Anthracene; Aromatic Hydrocarbon Content Analysis; Aromatic Polycyclic Hydrocarbon Content Analysis; Benzene; Benzopyrene Content Analysis; Hydrocarbons; Naphthalene; Solvents; Styrene; Toluene; Xylene
—Acidity
CNS K6257-74. Method of Test for Acidity and Alkalinity of Aromatic Hydrocarbons (Oct)(2758).
—Alkalinity
CNS K6257-74. Method of Test for Acidity and Alkalinity of Aromatic Hydrocarbons (Oct)(2758).
—Alkanes Content
CNS K6260-67. Method of Test for Paraffins in Aromatic Hydrocarbons (Mar)(2761)(R 1973).
—Aromatic Compound Content—Infrared Spectrophotometry
MOD UK DSTAN 05-50: Part 60-91. Methods for Testing Fuels, Lubricants and Associated Products Part 60: Aromatics Content in Hydrocarbon Solvents Issue 1. 7 pp.
—Bromine Number—Coulometric Titration
CNS K6521-80. Method of Test for Bromine Index of Aromatic Hydrocarbons by Coulometric Titration (Jul)(5838).
—Bromine Number—Potentiometric Analysis
CNS K6520-80. Method of Test for Bromine Index of Aromatic Hydrocarbons by Potentiometric Titration (Jul)(5837).
—Chemical Neutrality—Color Testing
ISO 5276-79. Aromatic Hydrocarbons—Test for Neutrality First Edition. 3 pp.

Aromatic Hydrocarbons (Cont.)
—Color Testing
CNS K6254-74. Method of Test for Colors of Aromatic Hydrocarbons (Jun)(2755).
CNS K6256-74. Method of Test for Acid-Wash Colors of Aromatic Hydrocarbons (Oct)(2757).
ISO 5274-79. Aromatic Hydrocarbons—Acid-Wash Test First Edition. 4 pp.
—Corrosion Testing—Copper
CNS K6259-74. Method of Test for Copper Corrosion of Aromatic Hydrocarbons (Jun)(2760).
—Distillation Methods
CNS K6255-74. Method of Test for Distillation of Aromatic Hydrocarbons (Jun)(2756).
—Glossaries
JIS K 2410-88. Glossary of Terms Used in Aromatic Hydrocarbons and Tar Products. 36 pp.
—Hydrocarbon Content—Gas Chromatography
CNS K6497-80. Nonaromatic Hydrocarbon in Monocyclic Aromatic Hydrocarbon by Gas Chromatography (May)(5603).
—Hydrogen Sulfide Content—Qualitative Analysis
CNS K6258-67. Method of Test for Hydrogen Sulfide and Sulfur Dioxide Contents (Qualitative) of Aromatic Hydrocarbons (Mar)(2759)(R 1973).
—Insulating Liquids
BSI BS 6802-87. 1987 Unused Insulating Liquids Based on Synthetic Aromatic Hydrocarbons. 12 pp.
CENELEC HD 497 S1-87. Specifications for Unused Insulating Liquids Based on Synthetic Aromatic Hydrocarbons. 3 pp.
CENELEC HD 497 S1-88. Specifications for Unused Insulating Liquids Based on Synthetic Aromatic Hydrocarbons. 3 pp.
CENELEC PREN 60867-92. Insulating Liquids—Specifications for Unused Liquids Based on Synthetic Aromatic Hydrocarbons. 15 pp.
IEC 867-93. Insulating Liquids—Specifications for Unused Liquids Based on Synthetic Aromatic Hydrocarbons Second Edition. 31 pp.
—Residue Content—Evaporation
ISO 5277-81. Aromatic Hydrocarbons—Determination of Residue on Evaporation of Products Having Boiling Points up to 150 Degrees Celsius First Edition. 4 pp.
—Sampling
ISO 1995-81. Aromatic Hydrocarbons—Sampling First Edition. 15 pp.
JIS K 2420-85. Method of Sampling for Aromatic Hydrocarbons and Tar Products. 15 pp.
—Sulfur Content
DIN ENGL 51764-55. Testing of Liquid Fuels; Determination of the Active Sulphur Reacting with Mercury (Feb). 1 p.
—Sulfur Content—Spectrophotometry
ISO 5282-82. Aromatic Hydrocarbons—Determination of Sulphur Content—Pitt-Ruprecht Reduction and Spectrophotometric Method First Edition. 8 pp.
—Sulfur Dioxide Content—Qualitative Analysis
CNS K6258-67. Method of Test for Hydrogen Sulfide and Sulfur Dioxide Contents (Qualitative) of Aromatic Hydrocarbons (Mar)(2759)(R 1973).
—Thiol Content—Doctor Testing
ISO 5275-79. Aromatic Hydrocarbons—Test for Presence of Mercaptans (Thiols)—Doctor Test First Edition. 4 pp.

Aromatic Polyamide Resins
Use: Aramid Fibers

Aromatic Polycyclic Hydrocarbon Content Analysis
See Also: Aromatic Hydrocarbons
—Waste Water—Thin Layer Chromatography
DIN ENGL 38409 Pt 13-81. German Standard Methods for the Analysis of Water, Waste Water and Sludge; Summary Action and Material; Characteristic Parameters (Group H); Determination of Polycyclic Aromatic Hydrocarbons (PAH) in Drinking Water (H 13—1 to 3) (June). 10 pp.

Aromatic Polycyclic Hydrocarbon Content Analysis (Cont.)
—Water—Thin Layer Chromatography
DIN ENGL 38409 Pt 13-81. German Standard Methods for the Analysis of Water, Waste Water and Sludge; Summary Action and Material; Characteristic Parameters (Group H); Determination of Polycyclic Aromatic Hydrocarbons (PAH) in Drinking Water (H 13—1 to 3) (June). 10 pp.

ARPA
Use: Automatic Radar Plotting Aids

ARQ
Use: Automatic Request-Repeat Mode

ARQ Equipment
Use: Automatic Request-Repeat Equipment

Arrester Hooks
See Also: Aircraft Equipment
—Manual Controls
SBAC AS 3195 ISSUE 4. Knob (Arrestor Hook Control). 1 p.

Arresters
Use: Surge Arresters

Arresting Systems, Aircraft
Use: Aircraft Arresting Systems

Arrows
See Also: Hunting
—Symbols
ISO 4196-84. Graphical Symbols—Use of Arrows First Edition. 7 pp.
ISO TR10488-91. Graphical Symbols Incorporating Arrows—Synopsis First Edition. 26 pp.

Arsanilic Acid Content Analysis
—Animal Feed
CNS N4100-82. Method of Test for Feed Additives: Determination of Arsanilic Acid (Aug)(9315).

Arsenates
Scope Note: Use a more specific term *See:* Copper Arsenate; Lead Arsenate; Sodium Arsenate, Dibasic

Arsenazo III
JIS K 9524-76. Arsenazo III.

Arsenic
See Also: Arsenic Content Analysis; Hazardous Materials; Insecticides; Pesticides
SAA AS N25-55. Arsenical Preparations for Agricultural Insecticides and Pest Destroyers. 24 pp.
—Standard Solutions
CNS K0039-88. Arsenic Standard Solution (Nov)(12469).
JIS K 0026-83. Arsenic Standard Solution. 11 pp.
—Toys—Migration Testing
BSI BS 5665: Part 3-89. 1989 Safety of Toys Part 3: Migration of Certain Elements (Supersedes BS 3443: 1968). 11 pp.
CEN EN 71-3-88. Safety of Toys Part 3 Migration of Certain Elements. 14 pp.
CEN PREN 71-3-92. Safety of Toys—Part 3: Migration of Certain Elements. 24 pp.
—Wood Preservatives
BSI BS 4072: Part 1-87. 1987 Amd 1 Wood Preservation by Means of Copper/Chromium /Arsenic Compositions Part 1: Preservatives (AMD 6200) June 30, 1989. 5 pp.

Arsenic Acid
Use For: Ortho-Arsenic Acid; Othoarsenic Acid
See Also: Hazardous Materials
—Highway Transportation—Emergency Procedures
SAA AS 1678.6.1. 010-83. Emergency Procedure Guides—Transport—Part 6.1.010: Arsenic Acid—Ortho-Arsenic Acid (Liquid) Meta-Arsenic Acid (Solid).

Arsenic Content Analysis
See Also: Arsenic
—Aluminum Sulfate
CNS K6163-74. Method of Test for Aluminium Sulfate (for Water Works) (Jun)(2075). 4 pp.
MOD UK M 9508/67. Examination of Aluminium Sulphate.

INTERNATIONAL AND NON-U.S. NATIONAL STANDARDS
SUBJECT INDEX
Arsenic

Arsenic Content Analysis (Cont.)

—Ammonium Bicarbonate—Photometry
ISO 4275-77. Ammonium Hydrogencarbonate for Industrial Use (Including Foodstuffs)—Determination of Arsenic Content—Silver Diethyldithiocarbamate Photometric Method First Edition; (Erratum—Sept 1978). 5 pp.

—Ammonium Sulfate
CNS K6075-74. Method of Test for Ammonium Sulfate (May)(919). 3 pp.

—Ammonium Sulfate—Photometry
ISO 5786-78. Ammonium Sulphate for Industrial Use—Determination of Arsenic Content—Silver Diethyldithiocarbamate Photometric Method First Edition. 4 pp.

—Animal Feed
CNS N4024-17-86. Method of Test for Feeds: Determination of Arsenic (Aug)(2770-17).

—Bearing Alloys—White Metals—Photometry
BSI BS 3338: Part 9-65. 1965 Methods for the Sampling and Analysis of Tin and Tin Alloys Part 9: Determination of Arsenic in Ingot Tin Solders and White Metal Bearing Alloys (Photometric Method). 8 pp.

—Carbonated Beverage Bottles
CNS S2111-83. Method of Test for Lead, Arsenic and Alkali Content Extracted from Carbonated Beverage Bottles (Oct)(10633).

—Ceramic Ware
CNS R3138-86. Method of Test for Arsenic Extracted from Glazed Ceramic or Porcelain Enamel Surfaces of Food Utensils and Containers (Apr)(11555).

—Coal—Oxidation Method
BSI BS 1016: Part 10-77. 1977 Methods for the Analysis and Testing of Coal and Coke Part 10: Arsenic in Coal and Coke. 8 pp.
SAA AS 1038.10-80. Methods for the Analysis and Testing of Coal and Coke—Part 10: Arsenic in Coal and Cokes. 16 pp.

—Coatings
CGSB CAN2-1.500-75 METH 7-73. Methods of Test for Toxic Trace Elements in Protective Coatings Determination of Leachable Arsenic (As) in Low Concentration. 2 pp.

—Coke—Oxidation Method
BSI BS 1016: Part 10-77. 1977 Methods for the Analysis and Testing of Coal and Coke Part 10: Arsenic in Coal and Coke. 8 pp.
SAA AS 1038.10-80. Methods for the Analysis and Testing of Coal and Coke—Part 10: Arsenic in Coal and Cokes. 16 pp.

—Copper
JIS H 1059-87. Methods for Determination of Arsenic in Copper and Copper Alloys. 10 pp.

—Copper Alloys
JIS H 1059-87. Methods for Determination of Arsenic in Copper and Copper Alloys. 10 pp.

—Copper Alloys—Photometry
ISO 3220-75. Copper and Copper Alloys—Determination of Arsenic—Photometric Method First Edition. 4 pp.

—Copper—Photometry
ISO 3220-75. Copper and Copper Alloys—Determination of Arsenic—Photometric Method First Edition. 4 pp.

—Cosmetics
CNS S2087-82. Methods of Hygienic Test for Cosmetics — Arsenic (Oct)(9541).

—Electrolytic Copper—Atomic Absorption Spectrophotometry
BSI BS 7317: Part 3-90. 1990 Methods for Analysis of High Purity Copper Cathode Cu-CATH-1 Part 3: Method for Determination of Antimony, Arsenic, Bismuth, Selenium, Tellurium and Tin by Hydride Generation and Atomic Absorption Spectrophotometry. 12 pp.
BSI BS 7317: Part 4-90. 1990 Methods for Analysis of High Purity Copper Cathode Cu-CATH-1 Part 4: Method for Determination of Antimony, Arsenic, Bismuth, Lead, Selenium, Tellurium and Tin by Electrothermal Atomiza-tion Atomic Absorption Spectrophotometry. 9 pp.
BSI DD 95: Part 3-84. (WITHDRAWN) 1984 Analysis of Higher Purity Copper Cathode Cu-CATH-1 Part 3: Methods for Determination of Antimony, Arsenic, Bismuth, Selenium, Tellurium and Tin by Hydride Generation and Atomic Absorption Spectrophotometry. 14 pp.

Arsenic Content Analysis (Cont.)

—Electrolytic Copper—Atomic Absorption Spectrophotometry (Cont.)
BSI DD 95: Part 4-86. (WITHDRAWN) 1986 Analysis of Higher Purity Copper Cathode CU-CATH-1 Part 4: Method for Determination of Antimony, Arsenic, Bismuth, Lead, Selenium, Tellurium and Tin by Electrothermal Atomization Atomic. 9 pp.

—Electrolytic Copper—Hydride Generation
BSI BS 7317: Part 3-90. 1990 Methods for Analysis of High Purity Copper Cathode Cu-CATH-1 Part 3: Method for Determination of Antimony, Arsenic, Bismuth, Selenium, Tellurium and Tin by Hydride Generation and Atomic Absorption Spectrophotometry. 12 pp.
BSI DD 95: Part 3-84. (WITHDRAWN) 1984 Analysis of Higher Purity Copper Cathode Cu-CATH-1 Part 3: Methods for Determination of Antimony, Arsenic, Bismuth, Selenium, Tellurium and Tin by Hydride Generation and Atomic Absorption Spectrophotometry. 14 pp.

—Enamel Ware
CNS R3138-86. Method of Test for Arsenic Extracted from Glazed Ceramic or Porcelain Enamel Surfaces of Food Utensils and Containers (Apr)(11555).

—Ferrotungsten
CNS G2158-83. Methods of Chemical Analysis for Arsenic in Ferrotungsten (Jan)(9869).

—Fertilizers
CNS N4028-86. Method of Test for Hazardous Components in Compound Fertilizer (Jan)(3080). 4 pp.

—Flue Gases
JIS K 0221-88. Methods of Determination of Arsenic in Flue Gas. 16 pp.

—Fluorite—Spectrometry
ISO 9505-92. All Grades of Fluorspar—Determination of Arsenic Content—Silver Diethyldithiocarbamate Spectrometric Method First Edition. 9 pp.

—Fruits—Spectrophotometry
ISO 6634-82. Fruits, Vegetables and Derived Products Determination of Arsenic Content—Silver Diethyldithiocarbamate Spectrophotometric Method First Edition. 7 pp.

—Glucose
CNS N6053-73. Method of Test for Dried Glucose and Glucose Syrup (Jan)(3218). 9 pp.
CNS N6055-72. Method of Test for Powdered Glucose (Jun)(3351). 3 pp.

—Glucose Syrups
CNS N6053-73. Method of Test for Dried Glucose and Glucose Syrup (Jan)(3218). 9 pp.

—Glycerol—Photometry
BSI BS 5711: Part 10-79. 1979 Methods of Sampling and Test for Glycerol Part 10: Determination of Arsenic Content: Silver Diethyldithiocarbamate Method. 4 pp.
ISO 2465-74. Glycerols for Industrial Use—Determination of Arsenic Content—Silver Diethyldithiocarbamate Photometric Method First Edition. 5 pp.

—Ground Water—Atomic Absorption Spectrometry
DIN ENGL 38405 Pt 18-85. German Standard Methods for the Examination of Water, Waste Water and Sludge; Anions (Group D); Determination of Arsenic by Atomic Absorption Spectrometry (D 18) (Sept). 5 pp.

—Ground Water—Photometry
DIN ENGL 38405 Pt 12 (S)-81. German Standard Methods for the Analysis of Water, Waste Water and Sludge; Anions (Group D); Determination of Arsenic (D 12) (June) (Superseded by DIN EN 26595). 5 pp.

—Hydrochloric Acid—Photometry
ISO 5785-78. Hydrochloric Acid for Industrial Use—Determination of Arsenic Content—Silver Diethyldithiocarbamate Photometric Method First Edition. 4 pp.

—Iron
CNS G2243-85. Method of Determination for Arsenic in Iron and Steel (Apr)(11245).
JIS G 1225-81. Methods for Determination of Arsenic in Iron and Steel.

—Iron Ores—Spectrophotometry
BSI BS 7020: Part 13-88. 1988 Analysis of Iron Ores Part 13: Method for the Determination of Arsenic Content: Spectrophotometric Method. 12 pp.

Arsenic Content Analysis (Cont.)

—Iron Ores—Spectrophotometry (Cont.)
ISO 7834-87. Iron Ores—Determination of Arsenic Content—Molybdenum Blue Spectrophotometric Method First Edition. 12 pp.
JIS M 8226-83. Methods for Determination of Arsenic in Iron Ores.
JIS M 8226-71. Methods for Determination of Arsenic in Iron Ores. 10 pp.

—Iron Ores—Volumetric Analysis
JIS M 8226-83. Methods for Determination of Arsenic in Iron Ores.
JIS M 8226-71. Methods for Determination of Arsenic in Iron Ores. 10 pp.

—Iron—Spectrophotometry
CEN PREN 10212-91. Chemical Analysis of Ferrous Materials—Determination of Arsenic in Steel and Iron—Spectrophotometric Method. 12 pp.

—Lead Alloys—Photometry
BSI BS 3908: Part 2-67. 1967 Methods for the Sampling and Analysis of Lead and Lead Alloys Part 2: Arsenic inLead and Lead Alloys (Photometric Method). 8 pp.

—Lead—Photometry
BSI BS 3908: Part 2-67. 1967 Methods for the Sampling and Analysis of Lead and Lead Alloys Part 2: Arsenic inLead and Lead Alloys (Photometric Method). 8 pp.

—Manganese Ores—Spectrometry
ISO 317-84. Manganese Ores and Concentrates—Determination of Arsenic Content—Spectrometric Method First Edition. 6 pp.

—Metal Conditioners
CGSB 31-GP-0A METH 35.1-57. Methods of Testing Corrosion-Prevention Materials and Processes Determination of Arsenic. 1 p.

—Monosodium Glutamate
CNS N6007-82. Method of Test for Monosodium L-Glutamate (May)(765). 4 pp.

—Nickel—Atomic Absorption Spectrometry
BSI BS 6783: Part 4-86. 1986 Sampling and Analysis of Nickel, Ferronickel and Nickel Alloys: Part 4: Method for Determination of Silver, Arsenic, Bismuth, Cadmium, Lead, Antimony, Selenium, Tin, Tellurium and Thallium in Nickel by Electrothermal Atomic Absorption Spectrometry. 11 pp.
ISO 7523-85. Nickel—Determination of Silver, Arsenic, Bismuth, Cadmium, Lead, Antimony, Selenium, Tin, Tellurium and Thallium Contents—Electrothermal Atomic Absorption Spectrometric Method First Edition. 10 pp.

—Ores
JIS M 8132-86. Methods for Determination of Arsenic in Ores. 17 pp.

—Ores—Volumetric Analysis
CNS M3094-82. Method for Determining Arsenic in Ores (Titration Method After Separating Arsenic by Distillation) (May) (8854).
CNS M3095-82. Method for Determining Arsenic in Ores (Titration Method after Separating Arsenic with Hydrogen Sulfide) (May) (8855).
CNS M3096-82. Method for Determining Arsenic in Ores (Titration Method After Separating Arsenic as Silver Arsenate) (May) (8856).

—Paints
CNS K6491-80. Method of Test for Arsenic in Paints (May)(5595).

—Phosphates—Photometry
BSI BS 4427: Part 12-79. 1979 Methods of Test for Sodium Tripolyphosphate (Pentasodium Triphosphate) and Sodium Pyro-phosphate (Tetrasodium Pyrophosphate) for Industrial Use Part 12: Condensed Phosphates for Industrial Use (Including Foodstuffs),. 6 pp.
ISO 5372-78. Condensed Phosphates for Industrial Use (Including Foodstuffs)—Determination of Arsenic Content—Silver Diethyldithiocarbamate Photometric Method First Edition. 4 pp.

—Phosphoric Acids
CNS K6235-72. Method of Test for Phosphoric Acid (Industrial Use) (Jun)(2620). 2 pp.

—Phosphoric Acids—Photometry
BSI BS 4258: Part 6-76. 1976 Methods of Test for Phosphoric Acid (Orthophosphoric Acid) for Industrial Use Part 6: Determinaton of Arsenic Content. 5 pp.
ISO 3359-75. Phosphoric Acid for Industrial Use—Determination of Arsenic Content—Silver Diethyldithiocarbamate Photometric Method First Edition. 4 pp.

INDUSTRY STANDARDS

Arsenic Content Analysis (Cont.)

—Photometry
ISO 2590-73. General Method for the Determination of Arsenic—Silver Diethyldithiocarbamate Photometric Method First Edition. 7 pp.

—Rust Removers
CGSB 31-GP-0A METH 35.1-57. Methods of Testing Corrosion-Prevention Materials and Processes Determination of Arsenic. 1 p.

—Sea Water—Atomic Absorption Spectrometry
DIN ENGL 38405 Pt 18-85. German Standard Methods for the Examination of Water, Waste Water and Sludge; Anions (Group D); Determination of Arsenic by Atomic Absorption Spectrometry (D 18) (Sept). 5 pp.

—Sea Water—Photometry
DIN ENGL 38405 Pt 12 (S)-81. German Standard Methods for the Analysis of Water, Waste Water and Sludge; Anions (Group D); Determination of Arsenic (D 12) (June) (Superseded by DIN EN 26595). 5 pp.

—Sodium Hexametaphosphate
MOD UK M 9520/66. Examination of Sodium Hexametaphosphate (Withdrawn).

—Solders—Tin—Photometry
BSI BS 3338: Part 9-65. 1965 Methods for the Sampling and Analysis of Tin and Tin Alloys Part 9: Determination of Arsenic in Ingot Tin Solders and White Metal Bearing Alloys (Photometric Method). 8 pp.

—Solid Fuels—Photometry
ISO 601-81. Solid Mineral Fuels—Determination of Arsenic Content Using the Standard Silver Diethyldithiocarbamate Photometric Method of ISO 2590 First Edition. 6 pp.

—Spectrophotometry
BSI BS 4404-68. 1968 Determination of Arsenic (Silver Diethyldithiocarbamate Procedure). 12 pp.

—Steels
CNS G2243-85. Method of Determination for Arsenic in Iron and Steel (Apr)(11245).
JIS G 1225-81. Methods for Determination of Arsenic in Iron and Steel.

—Steels—Spectrophotometry
CEN PREN 10212-91. Chemical Analysis of Ferrous Materials—Determination of Arsenic in Steel and Iron—Spectrophotometric Method. 12 pp.

—Sulfur—Photometry
ISO 3705-76. Sulphur for Industrial Use—Determination of Arsenic Content—Silver Diethyldithiocarbamate Photometric Method First Edition. 4 pp.

—Sulfuric Acid—Photometry
ISO 5792-78. Sulphuric Acid for Industrial Use—Determination of Arsenic Content—Silver Diethyldithiocarbamate Photometric Method First Edition. 5 pp.

—Surface Water—Atomic Absorption Spectrometry
DIN ENGL 38405 Pt 18-85. German Standard Methods for the Examination of Water, Waste Water and Sludge; Anions (Group D); Determination of Arsenic by Atomic Absorption Spectrometry (D 18) (Sept). 5 pp.

—Surface Water—Photometry
DIN ENGL 38405 Pt 12 (S)-81. German Standard Methods for the Analysis of Water, Waste Water and Sludge; Anions (Group D); Determination of Arsenic (D 12) (June) (Superseded by DIN EN 26595). 5 pp.

—Vegetables—Spectrophotometry
ISO 6634-82. Fruits, Vegetables and Derived Products Determination of Arsenic Content—Silver Diethyldithiocarbamate Spectrophotometric Method First Edition. 7 pp.

—Waste Water—Atomic Absorption Spectrometry
DIN ENGL 38405 Pt 18-85. German Standard Methods for the Examination of Water, Waste Water and Sludge; Anions (Group D); Determination of Arsenic by Atomic Absorption Spectrometry (D 18) (Sept). 5 pp.

—Waste Water—Photometry
DIN ENGL 38405 Pt 12 (S)-81. German Standard Methods for the Analysis of Water, Waste Water and Sludge; Anions (Group D); Determination of Arsenic (D 12) (June) (Superseded by DIN EN 26595). 5 pp.

—Water
CNS K9038-80. Method of Test for Arsenic in Water (Silver Diethyldithiocarbamate Method) (May)(5579).
CNS K9059-80. Method of Test for Arsenic in Water (Dec)(6676).

—Water—Spectrophotometry
BSI BS 6068: Sec 2.1-93. 1993 Amd 1 Water Quality—Determination of Total Aresenic—Silver Diethyldithiocarbamate Spectrophotometric Method (ISO 6695: 1992) (AMD 7427) May 15, 1993 (N). 15 pp.
BSI BS 6068: Sec 2.1-83. 1983 Water Quality Part 2: Physical, Chemical and Bio-Chemical Methods Section 2.1: Determination of Total Arsenic Silver Diethyldithiocarbamate Spectrophotometric Method (Renumbered as BS EN 26595: 1993). 7 pp.
BSI BS EN 26595-93. 1993 Amd 1 Water Quality—Determination of Total Arsenic—Silver Diethyldithiocarbamate Spectrophotometric Method (ISO 6695: 1982) (AMD 7427) May 15, 1993 (N). 14 pp.
CEN EN 26595-92. Water Quality—Determination of Total Arsenic—Silver Diethyldithiocarbamate Spectrophotometric Method (ISO 6595:1982). 8 pp.
DIN ENGL EN 26595-93. Water Quality; Determination of Total Arsenic; by the Silver Diethyldithiocarbamate Spectrometric Method (ISO 6595: 1982) (Jan) (Supersedes DIN 38405 Part 12, June 1981 Edition). 9 pp.
ISO 6595-82. Water Quality—Determination of Total Arsenic—Silver Diethyldithiocarbamate Spectrophotometric Method First Edition; (CEN EN 26595:1992). 7 pp.

—Wood Preservatives—Atomic Absorption Spectrophotometry
BSI BS 5666: Part 3-91. 1991 Methods of Analysis of Wood Preservatives and Treated Timber Part 3: Quantitative Analysis of Preservatives and Treated Timber Containing Copper/Chromium/Arsenic Formulations. 14 pp.

—Wood Preservatives—Colorimetric Analysis
BSI BS 5666: Part 3-91. 1991 Methods of Analysis of Wood Preservatives and Treated Timber Part 3: Quantitative Analysis of Preservatives and Treated Timber Containing Copper/Chromium/Arsenic Formulations. 14 pp.

—Zinc Alloys—Photometry
BSI BS 3630: Part 12-70. 1970 Methods for the Sampling and Analysis of Zinc and Zinc Alloys Part 12: Arsenic in Ingot Zinc and Zinc Alloys (Photometric Method). 11 pp.

—Zinc Ingots—Photometry
BSI BS 3630: Part 12-70. 1970 Methods for the Sampling and Analysis of Zinc and Zinc Alloys Part 12: Arsenic in Ingot Zinc and Zinc Alloys (Photometric Method). 11 pp.

Arsenic Inorganic Compounds
Scope Note: Use a more specific term *See:* Arsenic Pentoxide; Arsenic Trioxide; Gallium Arsenide; Sodium Arsenite

Arsenic Pentoxide
CNS K7073-62. Chemical Reagent (Arsenic Pentoxide) (Arsenic Acid Anhydride) (Jul)(1573).

Arsenic Trioxide
CNS K7074-62. Chemical Reagent (Arsenic Trioxide, Primary Standard) (Jul)(1574).
CNS K7075-62. Chemical Reagent (Arsenic Trioxide) (Arsenous Acid) (Jul)(1575).
JIS K 8044-80. Arsenic (III) Trioxide.

Arsine
—Standard Gases
JIS K 0040-90. Arsine (Standard Reference Gas). 9 pp.

Art Papers
See Also: Drawing Papers; Fine Papers; Mian Papers; Papers; Shiuan Papers
CNS P2019-84. Art Paper (Oct)(1464). 2 pp.
JIS P 3105-89. Art Papers. 6 pp.

—Printability
CNS P3060-83. Method of Test for IGT-Printibility of Art Paper (Sep)(10579).

Art Works
Use: Works of Art

Artemisia Oil
—Thujone Content—Gas Chromatography
ISO 7356-85. Oil of Thujone-Containing Artemisia and Oil of Sage (Salvia Officinalis Linnaeus)—Determination of Alpha Particle and Beta Ray-Thujone Content—Gas Chromatographic Method on Packed Columns First Edition. 6 pp.

Articulated Vehicles
See Also: Motor Vehicles

—Length Measurement
EC 89/461/EEC-89. Council Directive Amending, with a View to Fixing Certain Maximum Authorized Dimensions for Articulated Vehicles, Directive 85/3/EEC on the Weights, Dimensions and Certain Other Technical Characteristics of Certain Road Vehicles. 1 p.

Articulation (Speech)
Use: Speech

Artificial Abrasives
Use: Abrasives

Artificial Ears
See Also: Acoustic Measuring Instruments; Earphones; Occluded Ear Simulators

—Earphone Calibration
BSI BS 4669-71. 1971 An Artificial Ear of the Wide Band Type for the Calibration of Earphones Audiometry. 12 pp.
IEC 318-70. An IEC Artificial Ear, of the Wide Band Type, for the Calibration of Earphones Used in Audiometry First Edition. 17 pp.
SAA AS 1591.5-87. Acoustics—Instrumentation for Audiomerty—Part 5: Wide Band Artifical Ear. 7 pp.

—Telephone Transmission
CCITT RECMN P.51-89. Artificial Ear and Artificial Mouth—Telephone Transmission Quality (Study Group XII) 8 pp. 8 pp.

Artificial Fibers
Use: Synthetic Fibers

Artificial Horizons
See Also: Aircraft Instruments
BSI G 152-56. 1956 A.C. Electrically Operated Artificial Horizons for Aircraft. 5 pp.

Artificial Leather
Use: Leathercloth

Artificial Light Testing
Use: Light Testing—Artificial

Artificial Mouths
See Also: Acoustic Measuring Instruments

—Telephone Transmission
CCITT RECMN P.51-89. Artificial Ear and Artificial Mouth—Telephone Transmission Quality (Study Group XII) 8 pp. 8 pp.

Artificial Plasmas
Use: Plasmas (Physics)

Artificial Reverberation Amplifiers
Use: Audio Amplifiers—Artificial Reverberation

Artificial Satellites
Use: Satellites

Artificial Speech
Use For: Artificial Voice *See Also:* Speech

—Telephone Transmission
CCITT RECMN P.50-89. Artificial Voices—Telephone Transmission Quality (Study Group XII) 13 pp. 13 pp.

Artificial Sweeteners Content Analysis
—Beverages
CNS N6091-91. Method of Test for Fruit and Vegetable Juices and Drinks (General Rules) (Jun)(3736). 1 p.

—Food—Gas Chromatography
CNS N6191-86. Method of Test for Artificial Sweeteners in Food (Sep)(10950). 4 pp.

Artificial Sweeteners Content Analysis (Cont.)

—Food—Liquid Chromatography
CNS N6191-86. Method of Test for Artificial Sweeteners in Food (Sep)(10950). 4 pp.

—Food—Paper Chromatography
CNS N6191-86. Method of Test for Artificial Sweeteners in Food (Sep)(10950). 4 pp.

—Fruit Juices
CNS N6091-91. Method of Test for Fruit and Vegetable Juices and Drinks (General Rules) (Jun)(3736). 1 p.

—Vegetable Juices
CNS N6091-91. Method of Test for Fruit and Vegetable Juices and Drinks (General Rules) (Jun)(3736). 1 p.

Artificial Teeth
Use: Dentures

Artificial Ventilation
Use: Ventilation

Artificial Voice
Use: Artificial Speech

Artificial Weathering Tests
Use: Environmental Testing

Artillery
See Also: Artillery Computers; Field Artillery; Guns; Mortars (Weapons); Munitions; Weapons

—Ballistics—Meteorological Data
NATO STANAG 4061 ED 3 AMD 2-69. Adoption of a Standard Ballistic Meteorological Message. 27 pp.

—Connectors
NATO STANAG 4007 ED 2 AMD 0-92. (Draft) Electrical Connectors Between Prime Movers, Trailers and Towed Artillery. 17 pp.
NATO STANAG 4007 ED 2 AMD 0-00. Electrical Connectors Between Prime Movers, Trailers and Towed Artillery. 21 pp.

—Fuses (Ordnance)—Interchangeable
NATO STANAG 2916 ED 1 AMD 0-89. Nose Fuze Contours and Matching Projectile Cavities for Artillary and Mortar Projectiles. 31 pp.

—Graphite Oils
MOD UK DSTAN 91-30-77. Lubricating Oil, Colloidal Graphite NATO Code No: O-218 Joint Service Designation: OX-320 Issue 1 (Withdrawn). 14 pp.

—Greases
MOD UK DSTAN 91-17-90. Grease, Calcium Base Joint Service Designation LG-280 Issue 2. 12 pp.
MOD UK DSTAN 91-27-76. Grease, Automotive and Artillery NATO Code No: G-403 Joint Service Designation: XG-279 Issue 1. 13 pp.

—Procedures
NATO STANAG 2934 ED 1 AMD 0-89. Artillary Procedures. 5 pp.
NATO STANAG 2934 ED 1 AMD 1-89. Artillary Procedures. 5 pp.
NATO STANAG 2934 ED 1 AMD 2-89. Artillary Procedures. 5 pp.

—Safety
NATO STANAG 4224 ED 1 AMD 0-90. Safety and Suitability for Service Testing of Artillery and Naval Gun Ammunition—76 mm and Greater. 37 pp.

—Survey Accuracy
NATO STANAG 2373 ED 1 AMD 0-89. Survey Accuracy Requirements for Surface to Surface Artillary. 9 pp.
NATO STANAG 2373 ED 1 AMD 1-89. Survey Accuracy Requirements for Surface to Surface Artillary. 12 pp.

—Survey Control Points
NATO STANAG 2865 ED 1 AMD 5-76. Recording of Data for Artillary Survey Control Points. 8 pp.
NATO STANAG 2865 ED 1 AMD 6-76. Recording of Data for Artillary Survey Control Points. 7 pp.

Artillery Computers
See Also: Artillery; Computers; Weapons Systems

—Meteorological Data
NATO STANAG 4082 ED 1 AMD 3-69. Adoption of a Standard Artillary Computer Meteorological Message. 14 pp.

Artillery Shells
Use: Projectiles

Artists' Brushes
Use: Paint Brushes

Artists' Colors
Use: Paints

Arts
See Also: Architecture; Handicrafts; Literature (Fine Arts); Music; Printing

—Classification
BSI BS 1000: (7/7.0)-76. 1976 Universal Decimal Classification (UDC). English Full Edition (7/7.0): The Arts in General: Theory, History, Techniques. 18 pp.
BSI BS 1000: (73/76)-77. 1977 Universal Decimal Classification (UDC). English Full Edition (73/76): Various Arts and Crafts. 23 pp.
SNZ NZS/BS 1000 (7/7.0)-76. Universal Decimal Classification Arts in General: Theory, History, Techniques. 24 pp.
SNZ NZS/BS 1000 (73/76)-77. Universal Decimal Classification Various Arts and Crafts. 24 pp.

ARV Aviation ARV 1 Aircraft
See Also: Aircraft

—Antenna Positions
CAA. ARV Aviation ARV-1/Super 2 (Approved Aerial Positions). 1 p.

ARV Aviation Super 2 Aircraft
See Also: Aircraft

—Data Sheets
CAA BA22 ISSUE 2. ARV Aviation Ltd, Super 2. 2 pp.

—Mandatory Aircraft Modifications and Inspections Summaries
CAA. ARV Super 2 Series. 1 p.

AS Resins
Use: Acrylic Styrene Resins

AS 332L2 Super Puma Helicopters
Use For: Eurocopter France AS 332L2 Super Puma Helicopters *See Also:* Helicopters

—Data Sheets
CAA FR20 ISSUE 1. Eurocopter France AS332L2 Super Puma Mk.2. 2 pp.

ASA Resins
Use: Acrylic Styrene Acrylonitrile Resins

Asarone Content Analysis
—Calamus Oil—Gas Chromatography
ISO 7357-85. Oil of Calamus—Determination of cis-Beta Ray-Asarone Content—Gas Chromatographic Method on Packed Columns First Edition. 5 pp.

Asbestos
Scope Note: For additional listings, see specific products made from asbestos *See Also:* Asbestos Board; Asbestos Cements; Asbestos Fabrics; Asbestos Millboards; Asbestos Papers; Asbestos Sealants (Made From Asbestos); Asbestos Sealants (On Asbestos); Asbestos Yarns; Fibers; Minerals; Ores; Serpentine
CNS K6004-72. Method of Analysis for Asbestos (Jun)(58).
CNS K7076-62. Chemical Reagent (Asbestos)(Acid Washed) (Jul)(1576).
MOD UK DSTAN 53-99-85. Ropes, Asbestos and Packing (Braids), Asbestos: Dust Suppressed and Fibres, Asbestos Issue 1. 21 pp.

—Electrical Insulation
BSI BS 7018-88. 1988 Guide to Use of Electrical Insulating Materials Containing Asbestos. 11 pp.

—Fiber Emissions—Industrial Plants
ISO 10397-93. Stationary Source Emissions—Determination of Asbestos Plant Emissions—Method by Fibre Count Measurement First Edition. 26 pp.

—Identification Systems
EC 83/478/EEC-83. Council Directive Amending for the Fifth Time (Asbestos) Directive 76/769/EEC on the Approximation of the Laws, Regs. and Ad. Provs. of the Member States Relating to Restrictions on the Mktg. and Use of Certain Dangerous Substances and Preparations. 4 pp.

Asbestos (Cont.)
—Restricted Use
EC 83/478/EEC-83. Council Directive Amending for the Fifth Time (Asbestos) Directive 76/769/EEC on the Approximation of the Laws, Regs. and Ad. Provs. of the Member States Relating to Restrictions on the Mktg. and Use of Certain Dangerous Substances and Preparations. 4 pp.
EC 85/610/EEC-85. Council Directive Amending for the Seventh Time (Asbestos) Directive 76/769/EEC on the Approximation of the Laws, Regs. and Ad. Provisions of the Member States Relating to Restrictions on the Mktg. and Use of Certain Dangerous Substances and Preparations. 2 pp.

—Sheets
CGSB 34-GP-18M-77. Sheets, Asbestos, Low Density, Standard for; (Amendment 1 Feb 1983). 10 pp.
CNS R3105-86. Method of Test for Compressed Asbestos Sheet Packing (May)(9327). 7 pp.

Asbestos Board
Use For: Asbestos Cement Board
See Also: Asbestos; Building Board

—Calcium Silicate—Autoclaved
CNS A2174-84. Autoclave Asbestos Cement Silicate Boards (Apr)(10843).
CNS A3205-84. Method of Test for Autoclave Asbestos Cement Silicate Boards (Apr)(10844).
JIS A 5418-89. Silica-Asbestos-Cement Boards. 19 pp.

—Calcium Silicate—Autoclaved—Decorative
CNS A2175-84. Decorated Autoclave Asbestos Cement Calcium Silicate Boards (Apr)(10845).
CNS A3206-84. Method of Test for Decorated Autoclave Asbestos Cement Calcium Silicate Boards (Apr)(10846).

—Electrical Equipment
BSI BS 3497-79. 1979 Unimpregnated Asbestos Cement Boards for Electrical Purposes. 14 pp.

—Impact Testing
CNS A3178-83. Method of Impact Test for Boards of Buildings (Feb)(9961).
JIS A 1421-81. Method of Impact Test for Boards of Buildings. 9 pp.

—Perlite
CNS A2216-87. Asbestos Cement Perlite Boards (Sep)(12064).
CNS A3268-87. Method of Test for Asbestos Cement Perlite Boards (Sep)(12065).
JIS A 5413-89. Asbestos Cement Perlite Boards. 15 pp.

—Plywood Laminated
CNS A2197-86. Asbestos Cement Boards with Plywood Lamina (Nov)(11756).
CNS A3246-86. Method of Test for Asbestos Cement Boards with Plywood Lamina (Nov)(11757).

—Sandwich Panels
CNS A2198-86. Flat Sandwich Boards in Asbestos Cement Sheets—Cemented Excelsior Boards (Nov)(11758).
CNS A3247-86. Method of Test for Flat Sandwich Boards in Asbestos Cement Sheets—Cemented Excelsior Boards (Nov)(11759).
JIS A 5426-89. Flat Sandwich Boards in Asbestos Cement Sheets-Wood-Wool Cement Boards. 14 pp.

—Siding
CNS A2193-86. Asbestos Cement Sidings (Sep)(11699).
CNS A3242-86. Method of Test for Asbestos Cement Sidings (Sep)(11700).
JIS A 5422-87. Asbestos Cement Sidings.
JIS A 5422-83. Asbestos Cement Sidings. 17 pp.

Asbestos Cement Board
Use: Asbestos Board

Asbestos Cement Chimneys
Use: Chimneys—Asbestos Cement

Asbestos Cement Claddings
Use: Asbestos Cement Siding

Asbestos Cement Panel Siding
Use: Asbestos Cement Siding

Asbestos Cement Pipes
See Also: Pipes
DIN ENGL 2410 Pt 4-78. Pipes; Survey of Standards for Asbestos Cement Pipes (Feb). 1 p.

—Chimney
JIS A 5405-82. Asbestos Cement Pipes. 10 pp.

—Drainpipes
BSI BS 569-73. 1973 Amd 1 Asbestos-Cement Rainwater Goods (AMD 5755) October 31, 1988. 35 pp.

INTERNATIONAL AND NON-U.S. NATIONAL STANDARDS
SUBJECT INDEX
Asbestos

Asbestos Cement Pipes (Cont.)
—Drainpipes (Cont.)
- BSI BS 3656-81. 1981 Amd 1 Asbestos-Cement Pipes, Joints and Fittings for Sewerage and Drainage. 13 pp.
- BSI BS 4624-81. 1981 Amd 1 Asbestos-Cement Building Products. 13 pp.
- BSI BS 4624-02. 1981 Amd 2 Methods of Test for Asbestos-Cement Building Products (AMD 7307) October 15, 1992. 14 pp.
- CGSB CAN/CGSB-34.22-M87. Pipe Asbestos-Cement, Drain; (Corrigendum). 12 pp.
- CGSB CAN/CGSB-34.23-M87. Pipe, Asbestos-Cement, Sewer, House Connection. 11 pp.
- CSA B127.1-M1977. Components for Use in Asbestos Cement Drain, Waste and Vent Systems; (Amd 1-3 March 1988) (Amd 4-11 December 1990). 23 pp.
- CSA B127.11-M1982. Recommended Practice for the Installation of Asbestos Cement Drain, Waste, and Vent Pipe and Pipe Fittings; (Gen Instr 1). 15 pp.
- ISO 881-80. Asbestos-Cement Pipes, Joints and Fittings for Sewerage and Drainage First Edition. 13 pp.
- ISO 4488-79. Asbestos-Cement Pipes and Joints for Thrust-Boring and Pipe Jacking First Edition. 13 pp.

—Flues
- BSI BS 4624-81. 1981 Amd 1 Asbestos-Cement Building Products. 13 pp.
- BSI BS 4624-02. 1981 Amd 2 Methods of Test for Asbestos-Cement Building Products (AMD 7307) October 15, 1992. 14 pp.

—Pressure
- BSI BS 486-81. 1981 Amd 1 Asbestos-Cement Pressure Pipes and Joints (AMD 5532) March 31, 1989. 13 pp.
- CGSB CAN/CGSB-34.1-M87. Pipe, Asbestos-Cement, Pressure. 11 pp.
- CNS A3091-82. Method of Test for Asbestos Cement Pressure Pipes (Feb)(5180).
- DIN ENGL 19800 Pt 1-73. Asbestos Cement Pipes and Fittings for Pressure Pipelines; Pipes; Dimensions (Jan). 3 pp.
- DIN ENGL 19800 Pt 2-73. Asbestos Cement Pipes and Fittings for Pressure Pipelines; Pipes, Joints and Fittings; Technical Conditions of Delivery (Jan). 6 pp.
- DIN ENGL 19800 Pt 3-79. Asbestos-Cement Pipes and Fittings for Pressure Pipelines; Joint Assemblies; Dimensions (Mar). 3 pp.
- ISO 160-80. Asbestos-Cement Pressure Pipes and Joints First Edition. 12 pp.

—Pressure—Gas
- CNS A2077-81. Asbestos Cement Pressure Pipes (Apr)(5179). 6 pp.

—Pressure—Water
- CNS A2077-81. Asbestos Cement Pressure Pipes (Apr)(5179). 6 pp.
- DIN ENGL 4279 Pt 6-75. Testing of Pressure Pipelines for Water by Internal Pressure; Asbestos-Cement Pressure Pipes (Nov). 3 pp.

—Sewer
- BSI BS 3656-81. 1981 Amd 1 Asbestos-Cement Pipes, Joints and Fittings for Sewerage and Drainage. 13 pp.
- CGSB CAN/CGSB-34.9-M87. Pipe Asbestos-Cement, Sewer. 11 pp.
- CGSB CAN/CGSB-34.22-M87. Pipe Asbestos-Cement, Drain; (Corrigendum). 12 pp.
- CGSB CAN/CGSB-34.23-M87. Pipe, Asbestos-Cement, Sewer, House Connection. 11 pp.
- CSA B127.1-M1977. Components for Use in Asbestos Cement Drain, Waste and Vent Systems; (Amd 1-3 March 1988) (Amd 4-11 December 1990). 23 pp.
- CSA B127.2-M1977. Components for Use in Asbestos Cement Building Sewer Systems. 13 pp.
- CSA B127.11-M1982. Recommended Practice for the Installation of Asbestos Cement Drain, Waste, and Vent Pipe and Pipe Fittings; (Gen Instr 1). 15 pp.
- ISO 391-82. Building and Sanitary Pipes in Asbestos-Cement First Edition. 11 pp.
- ISO 881-80. Asbestos-Cement Pipes, Joints and Fittings for Sewerage and Drainage First Edition. 13 pp.
- ISO 4488-79. Asbestos-Cement Pipes and Joints for Thrust-Boring and Pipe Jacking First Edition. 13 pp.

—Thrust Boring
- ISO 4488-79. Asbestos-Cement Pipes and Joints for Thrust-Boring and Pipe Jacking First Edition. 13 pp.

—Underground—Selection Criteria
- ISO 2785-86. Directives for Selection of Asbestos-Cement Pipes Subject to External Loads with or Without Internal Pressure Second Edition. 41 pp.

Asbestos Cement Pipes (Cont.)
—Vent
- CGSB CAN/CGSB-34.10-M87. Pipe, Asbestos-Cement, Industrial Vent. 9 pp.
- CGSB CAN/CGSB-34.22-M87. Pipe Asbestos-Cement, Drain; (Corrigendum). 12 pp.
- CSA B127.1-M1977. Components for Use in Asbestos Cement Drain, Waste and Vent Systems; (Amd 1-3 March 1988) (Amd 4-11 December 1990). 23 pp.
- CSA B127.11-M1982. Recommended Practice for the Installation of Asbestos Cement Drain, Waste, and Vent Pipe and Pipe Fittings; (Gen Instr 1). 15 pp.

Asbestos Cement Roofing
See Also: Roofing

—Fittings
- BSI BS 690: Part 6-76. 1976 Amd 1 Asbestos-Cement Slates and Sheets Part 6: Fittings for Use with Corrugated Sheets (AMD 5487) July 31, 1989. 17 pp.
- ISO 393 Pt 5-87. Asbestos-Cement Products—Part 5: Short Corrugated and Asymmetrical Section Sheets and Fittings for Roofing First Edition. 19 pp.

—Panels
- CGSB CAN/CGSB-34.21-M89. Panels, Sandwich, Asbestos-Cement, with Insulating Cores. 12 pp.

—Sheets
- BSI BS 5247: Part 14-75. 1975 Amd 2 Sheet Roof and Wall Coverings Part 14: Corrugated Asbestos Cement. 15 pp.
- CEN PREN 494-91. Fibre-Cement Profiled Sheets and Fittings for Roofing—Product Specification. 46 pp.
- ISO 393 Pt 1-83. Asbestos-Cement Products—Part 1: Corrugated Sheets and Fittings for Roofing and Cladding First Edition. 14 pp.
- ISO 393 Pt 2-86. Asbestos-Cement Products—Part 2: Asbestos-Cement-Cellulose Corrugated Sheets and Fittings for Roofing and Cladding First Edition. 13 pp.
- ISO 393 Pt 3-84. Asbestos-Cement Products—Part 3: Asymmetrical Section Corrugated Sheets and Fittings for Roofing and Cladding First Edition. 16 pp.
- ISO 393 Pt 4-86. Asbestos-Cement Products—Part 4: Trapezoidal Section Sheets for Roofing and Cladding First Edition. 14 pp.
- ISO 393 Pt 5-87. Asbestos-Cement Products—Part 5: Short Corrugated and Asymmetrical Section Sheets and Fittings for Roofing First Edition. 19 pp.
- ISO R394-64. Asymmetrical Section Corrugated Sheets in Asbestos-Cement for Roofing and Cladding First Edition. 15 pp.
- ISO 8108-86. Directives for Fixing Asbestos-Cement Corrugated and Asymmetrical Section Sheets and Fittings for Roofing First Edition. 12 pp.

—Sheets—Corrugated
- BSI BS 690: Part 3-73. 1973 Amd 3 Asbestos-Cement Slates and Sheets Part 3: Corrugated Sheets (AMD 5674) July 31, 1989. 16 pp.

—Sheets—Corrugated—Linings
- BSI BS 690: Part 5-75. 1975 Amd 1 Asbestos-Cement Slates and Sheets Part 5: Lining Sheets and Panels (AMD 5486) September 29, 1989. 13 pp.

—Shingles
- CNS A2192-86. Asbestos Cement Shingles for Roofing (Sep)(11697).
- CNS A3241-86. Method of Test for Asbestos Cement Shingles for Roofing (Sep)(11698).
- JIS A 5423-89. Decorated Asbestos Cement Shingles for Roofing. 16 pp.

—Slates
- BSI BS 690: Part 4-74. 1974 Amd 2 Asbestos-Cement Slates and Sheets Part 4: Slates (AMD 5675) September 29, 1989. 12 pp.
- CEN PREN 492-91. Fiber-Cement Slates and Their Fittings for Roofing. 27 pp.
- ISO 395-83. Asbestos-Cement Slates First Edition. 10 pp.
- ISO 9125-90. Fibre-Cement Slates and Fittings First Edition; (Corrigendum 1-1993). 14 pp.

Asbestos Cement Siding
Use For: Asbestos Cement Panel Siding
See Also: Siding
- BSI BS 690: Part 3-73. 1973 Amd 3 Asbestos-Cement Slates and Sheets Part 3: Corrugated Sheets (AMD 5674) July 31, 1989. 16 pp.
- BSI BS 690: Part 5-75. 1975 Amd 1 Asbestos-Cement Slates and Sheets Part 5: Lining Sheets and Panels (AMD 5486) September 29, 1989. 13 pp.
- CNS A2193-86. Asbestos Cement Sidings (Sep)(11699).
- CNS A3242-86. Method of Test for Asbestos Cement Sidings (Sep)(11700).
- JIS A 5422-87. Asbestos Cement Sidings.
- JIS A 5422-83. Asbestos Cement Sidings. 17 pp.

—Board
- CNS A2193-86. Asbestos Cement Sidings (Sep)(11699).

Asbestos Cement Siding (Cont.)
—Board (Cont.)
- CNS A3242-86. Method of Test for Asbestos Cement Sidings (Sep)(11700).
- JIS A 5422-87. Asbestos Cement Sidings.
- JIS A 5422-83. Asbestos Cement Sidings. 17 pp.

—Clapboards
- CGSB CAN/CGSB-34.4-M89. Siding, Asbestos-Cement, Shingles and Clapboards. 12 pp.

—Colors
- BSI BS 4904-78. 1978 External Cladding Colours for Building Purposes. 10 pp.

—Fittings
- BSI BS 690: Part 6-76. 1976 Amd 1 Asbestos-Cement Slates and Sheets Part 6: Fittings for Use with Corrugated Sheets (AMD 5487) July 31, 1989. 17 pp.

—Paints (Latex)—Exterior
- CGSB CAN/CGSB-1.138-93. Exterior Latex Type, Flat Paint. 9 pp.

—Sheets
- ISO 393 Pt 1-83. Asbestos-Cement Products—Part 1: Corrugated Sheets and Fittings for Roofing and Cladding First Edition. 14 pp.
- ISO 393 Pt 5-87. Asbestos-Cement Products—Part 5: Short Corrugated and Asymmetrical Section Sheets and Fittings for Roofing First Edition. 19 pp.

—Sheets—Corrugated
- BSI BS 690: Part 3-73. 1973 Amd 3 Asbestos-Cement Slates and Sheets Part 3: Corrugated Sheets (AMD 5674) July 31, 1989. 16 pp.
- BSI BS 5247: Part 14-75. 1975 Amd 2 Sheet Roof and Wall Coverings Part 14: Corrugated Asbestos Cement. 15 pp.
- ISO 393 Pt 5-87. Asbestos-Cement Products—Part 5: Short Corrugated and Asymmetrical Section Sheets and Fittings for Roofing First Edition. 19 pp.

—Sheets—Corrugated—Linings
- BSI BS 690: Part 5-75. 1975 Amd 1 Asbestos-Cement Slates and Sheets Part 5: Lining Sheets and Panels (AMD 5486) September 29, 1989. 13 pp.

—Shingles
- CGSB CAN/CGSB-34.4-M89. Siding, Asbestos-Cement, Shingles and Clapboards. 12 pp.
- ISO 880-81. Asbestos-Cement Siding Shingles First Edition. 9 pp.

—Slates
- BSI BS 690: Part 4-74. 1974 Amd 2 Asbestos-Cement Slates and Sheets Part 4: Slates (AMD 5675) September 29, 1989. 12 pp.
- CEN PREN 492-91. Fiber-Cement Slates and Their Fittings for Roofing. 27 pp.
- DIN ENGL 18517 Pt 1-85. Wall Cladding with Small-Size Facade Slates; Asbestos-Cement Slates (Nov). 16 pp.
- ISO 9125-90. Fibre-Cement Slates and Fittings First Edition; (Corrigendum 1-1993). 14 pp.

Asbestos Cements
See Also: Asbestos; Cements; Construction Materials
- ISO 7337-84. Asbestos Reinforced Cement Products—Guidelines for on-Site Work Practices First Edition. 12 pp.

—Fittings—Roofing
- BSI BS 690: Part 6-76. 1976 Amd 1 Asbestos-Cement Slates and Sheets Part 6: Fittings for Use with Corrugated Sheets (AMD 5487) July 31, 1989. 17 pp.
- ISO 393 Pt 5-87. Asbestos-Cement Products—Part 5: Short Corrugated and Asymmetrical Section Sheets and Fittings for Roofing First Edition. 19 pp.

—Gutters
- BSI BS 569-73. 1973 Amd 1 Asbestos-Cement Rainwater Goods (AMD 5755) October 31, 1988. 35 pp.

—Inspection
- ISO 390-93. Products in Fibre-Reinforced Cement—Sampling and Inspection Second Edition. 22 pp.

—Panels—Acoustical Insulation
- CGSB CAN/CGSB-34.19-M89. Tiles or Panels, Asbestos-Cement, Acoustical. 10 pp.

—Plates—Electrical Insulation
- CNS C4352-82. Asbestos Cement Plate for Electric Insulation (Jul)(9086).
- JIS C 2210-75. Asbestos Cement Plate for Electric Insulation (R 1983). 9 pp.

—Sampling
- ISO 390-93. Products in Fibre-Reinforced Cement—Sampling and Inspection Second Edition. 22 pp.

INTERNATIONAL AND NON-U.S. NATIONAL STANDARDS
SUBJECT INDEX

Ash

Asbestos Cements (Cont.)
—Sheets
BSI BS 690: Part 2-81. 1981 Amd 1 Asbestos-Cement Slates and Sheets Part 2: Asbestos-Cement and Cellulose-Asbestos-Cement Flat Sheets (AMD 5673) July 31, 1989. 9 pp.
BSI BS 4624-81. 1981 Amd 1 Asbestos-Cement Building Products. 13 pp.
BSI BS 4624-02. 1981 Amd 2 Methods of Test for Asbestos-Cement Building Products (AMD 7307) October 15, 1992. 14 pp.
CGSB CAN/CGSB-34.16-M89. Sheets, Asbestos-Cement, Flat, Fully Compressed. 12 pp.
CGSB CAN/CGSB-34.17-M89. Sheets, Asbestos-Cement, Flat, Semicompressed. 10 pp.
CNS R2011-88. Flat Asbestos Cement Sheet (Nov)(473). 2 pp.
CNS R3003-88. Method of Test for Flat Asbestos Cement Sheet (Nov)(474). 2 pp.
ISO 396 Pt 1-80. Products in Fibre Reinforced Cement—Part 1: Asbestos-Cement Flat Sheets First Edition. 11 pp.
ISO 396 Pt 2-80. Products in Fibre Reinforced Cement—Part 2: Silica-Asbestos-Cement Flat Sheets First Edition. 11 pp.
ISO 396 Pt 3-80. Products in Fibre Reinforced Cement—Part 3: Cellulose-Asbestos-Cement Flat Sheets First Edition. 11 pp.
JIS A 5403-89. Asbestos Cement Sheets. 29 pp.

—Sheets—Compressed
CNS R2161-86. Compressed Asbestos Sheet Packing (May)(9326). 2 pp.
DIN ENGL 3754 Pt 1-84. Gasket Materials; Compressed Asbestos Fibre Sheets; Dimensions, Requirements, Testing (May). 5 pp.
JIS R 3453-85. Compressed Asbestos Sheets.
JIS R 3453-79. Compressed Asbestos Sheets. 11 pp.

—Sheets—Corrugated
BSI BS 690: Part 3-73. 1973 Amd 3 Asbestos-Cement Slates and Sheets Part 3: Corrugated Sheets (AMD 5674) July 31, 1989. 16 pp.
BSI BS 4624-81. 1981 Amd 1 Asbestos-Cement Building Products. 13 pp.
BSI BS 4624-02. 1981 Amd 2 Methods of Test for Asbestos-Cement Building Products (AMD 7307) October 15, 1992. 14 pp.
BSI BS 5247: Part 14-75. 1975 Amd 2 Sheet Roof and Wall Coverings Part 14: Corrugated Asbestos Cement. 15 pp.
CGSB CAN/CGSB-34.5-M89. Sheets, Asbestos-Cement, Corrugated. 12 pp.
CNS R2012-88. Corrugated Asbestos Cement Sheets (Nov)(475). 4 pp.
CNS R3004-88. Method of Test for Corrugated Asbestos Cement Sheets (Nov)(476). 3 pp.
ISO 393 Pt 1-83. Asbestos-Cement Products—Part 1: Corrugated Sheets and Fittings for Roofing and Cladding First Edition. 14 pp.
ISO 393 Pt 2-86. Asbestos-Cement Products—Part 2: Asbestos-Cement-Cellulose Corrugated Sheets and Fittings for Roofing and Cladding First Edition. 13 pp.
ISO 393 Pt 3-84. Asbestos-Cement Products—Part 3: Asymmetrical Section Corrugated Sheets and Fittings for Roofing and Cladding First Edition. 16 pp.
ISO 393 Pt 5-87. Asbestos-Cement Products—Part 5: Short Corrugated and Asymmetrical Section Sheets and Fittings for Roofing First Edition. 19 pp.
ISO R394-64. Asymmetrical Section Corrugated Sheets in Asbestos-Cement for Roofing and Cladding First Edition. 15 pp.

—Sheets—Corrugated—Linings
BSI BS 690: Part 5-75. 1975 Amd 1 Asbestos-Cement Slates and Sheets Part 5: Lining Sheets and Panels (AMD 5486) September 29, 1989. 13 pp.

—Sheets—Decorative
CGSB CAN/CGSB-34.14-M89. Sheets, Asbestos-Cement, Decorative. 11 pp.
CNS A2191-86. Decorated Asbestos Cement Sheets (Sep)(11695).
CNS A3240-86. Method of Test for Decorated Asbestos Cement Sheets (Sep)(11696).

—Sheets—Perforated—Acoustic Insulated
CNS A2155-83. Perforated Asbestos Cement Flat Sheets for Acoustic Use (May) (10211).
JIS A 6302-75. Perforated Asbestos Cement Flat Sheets for Acoustic Use. 18 pp.

—Shingles
ISO 880-81. Asbestos-Cement Siding Shingles First Edition. 9 pp.

—Slates
BSI BS 690: Part 2-81. 1981 Amd 1 Asbestos-Cement Slates and Sheets Part 2: Asbestos-Cement and Cellulose-Asbestos-Cement Flat Sheets (AMD 5673) July 31, 1989. 9 pp.
BSI BS 690: Part 4-74. 1974 Amd 2 Asbestos-Cement Slates and Sheets Part 4: Slates (AMD 5675) September 29, 1989. 12 pp.

Asbestos Cements (Cont.)
—Slates (Cont.)
BSI BS 4624-81. 1981 Amd 1 Asbestos-Cement Building Products. 13 pp.
BSI BS 4624-02. 1981 Amd 2 Methods of Test for Asbestos-Cement Building Products (AMD 7307) October 15, 1992. 14 pp.
CEN PREN 492-91. Fiber-Cement Slates and Their Fittings for Roofing. 27 pp.
DIN ENGL 18517 Pt 1-85. Wall Cladding with Small-Size Facade Slates; Asbestos-Cement Slates (Nov). 16 pp.
ISO 395-83. Asbestos-Cement Slates First Edition. 10 pp.
ISO 9125-90. Fibre-Cement Slates and Fittings First Edition; (Corrigendum 1-1993). 14 pp.

—Tiles—Acoustical Insulation
CGSB CAN/CGSB-34.19-M89. Tiles or Panels, Asbestos-Cement, Acoustical. 10 pp.

Asbestos Cloths
Use: Asbestos Fabrics

Asbestos Fabrics
See Also: Asbestos; Fabrics
CNS R2177-85. Asbestos Cloths (May) (11266).
JIS R 3451-79. Asbestos Cloths. 10 pp.
MOD UK DSTAN 83-74-83. Cloths, Asbestos and Webbings and Tapes, Asbestos: Dust Suppressed Issue 1. 26 pp.
MOD UK DSTAN 83-74-01. Cloths, Asbestos and Webbings and Tapes, Asbestos: Dust Suppressed Issue 1; Amendment 1. 28 pp.

—Diaphragms (Mechanics)—Electrolytic Analysis—Water
CNS R2180-85. Asbestos Cloth Diaphragms for Electrolysis of Water (May)(11269).
JIS R 3456-79. Asbestos Cloth Diaphragms for Electrolysis of Water. 8 pp.

Asbestos Filled Asphalt Coatings
Use: Asphalt Coatings (Made From Asphalt)

Asbestos Millboards
See Also: Asbestos; Building Board
MOD UK TS 426. Asbestos Millboard, General Purpose.

—Electrical Insulation
CNS R2178-85. Asbestos Mill Boards (May)(11267).
JIS R 3454-79. Asbestos Mill Boards. 6 pp.

—Gaskets
CNS R2178-85. Asbestos Mill Boards (May)(11267).
JIS R 3454-79. Asbestos Mill Boards. 6 pp.

—Thermal Insulation
CNS R2178-85. Asbestos Mill Boards (May)(11267).
JIS R 3454-79. Asbestos Mill Boards. 6 pp.

Asbestos Papers
See Also: Asbestos; Papers

—Electrical Insulation
BSI BS 3057-58. 1958 Amd 2 Untreated Asbestos Paper for Electrical Purposes. 29 pp.

Asbestos Ropes
Use: Ropes—Asbestos

Asbestos Sealants (Made From Asbestos)
See Also: Asbestos; Sealants
BSI BS 1832-91. 1991 Compressed Asbestos Fibre Jointing (Supersedes BS 2815: 1973). 20 pp.
BSI BS 1832-01. 1991 Amd 1 Compressed Asbestos Fibre Jointing (AMD 7558) March 15, 1993 (Supersedes BS 2815: 1973). 21 pp.
BSI BS 2815-73. (WITHDRAWN) 1973 Compressed Asbestos Fiber Jointing (Superseded by BS 1832: 1991). 19 pp.

—Oil Resistant
BSI BS 1832-72. 1972 Oil Resistant Compressed Asbestos Fibre Jointing. 20 pp.

Asbestos Sealants (On Asbestos)
See Also: Asbestos; Sealants
CGSB 1-GP-205M-84. Standard for: Sealer for Application to Asbestos—Fibre—Releasing Materials; (QPL (Type 2 Class A) Apr 1986). 13 pp.

Asbestos Yarns
See Also: Asbestos; Yarns

—Duct Insulation
MOD UK DSTAN 53-99-85. Ropes, Asbestos and Packing (Braids), Asbestos: Dust Suppressed and Fibres, Asbestos Issue 1. 21 pp.

Asbestos Yarns (Cont.)
—Electric Wire
CNS R2176-85. Asbestos Yarns and Twisted Ropes (May) (11265).
JIS R 3450-79. Asbestos Yarns and Twisted Ropes. 10 pp.

—Pipe Insulation
MOD UK DSTAN 53-99-85. Ropes, Asbestos and Packing (Braids), Asbestos: Dust Suppressed and Fibres, Asbestos Issue 1. 21 pp.

—Pipes
CNS R2176-85. Asbestos Yarns and Twisted Ropes (May) (11265).
JIS R 3450-79. Asbestos Yarns and Twisted Ropes. 10 pp.

ASCC
Use For: Air Standardization Coordinating Committee

—Standards Preparation
MOD UK NES 5-01. Guide to Implementation & Data Sheets for ABCA NAVSTAGS, Army QSTAGS and ASCC (Air Standards) Issue 2 (04.84); Amendment 1. 47 pp.

Ascorbic Acid
Use For: Vitamin C *See Also:* Ascorbic Acid Content Analysis
CNS K7077-62. Chemical Reagent (Ascorbic Acid) (L-Ascorbic Acid;Vitamin C L-Xyloascorbic Acid) (Jul)(1577).
JIS K 9502-88. L-Ascorbic Acid.

Ascorbic Acid Content Analysis
See Also: Ascorbic Acid

—Fruits
ISO 6557 Pt 1-86. Fruits, Vegetables and Derived Products—Determination of Ascorbic Acid—Part 1: Reference Method First Edition. 5 pp.
ISO 6557 Pt 2-84. Fruits, Vegetables and Derived Products—Determination of Ascorbic Acid Content—Part 2: Routine Methods First Edition. 6 pp.

—Vegetables
ISO 6557 Pt 1-86. Fruits, Vegetables and Derived Products—Determination of Ascorbic Acid—Part 1: Reference Method First Edition. 5 pp.
ISO 6557 Pt 2-84. Fruits, Vegetables and Derived Products—Determination of Ascorbic Acid Content—Part 2: Routine Methods First Edition. 6 pp.

Aseptic Valves
Use: Sanitary Valves

Aseptors
Use: Sterilizers

Ash Cans
Use: Waste Containers

Ash Content Analysis
Use For: Ash Determination *See Also:* Residue Content Analysis; Residue-on-Ignition Determination; Sulfated Ash Content Analysis

—Activated Carbon
CNS K6478-80. Method of Test for Total Ash Content of Activated Carbon (May)(5581).

—Adipates—Gravimetric Analysis
ISO 2526-74. Adipate Esters for Industrial Use—Determination of Ash—Gravimetric Method First Edition. 4 pp.

—Agar
CNS N6113-79. Method of Test for Agar-Agar (Oct)(5006). 2 pp.

—Alcohols
ISO 1843 Pt VI-77. Higher Alcohols for Industrial Use—Methods of Test—Part VI: Determination of Ash First Edition. 3 pp.

—Alcohols—Gravimetric Analysis
CNS K6413-78. Determination of Ash of Higher Alcohols for Industrial Use (Gravimetric Method) (Mar)(4282).

—Algae
CNS N6106-78. Method of Test for Edible Chlorella (Oct)(4597). 9 pp.

—Aluminum Stearate
CNS K6320-72. Method of Test for Aluminum Stearate of Industrial Grade (Temporary Standard) (Jun)(3381). 1 p.

INDUSTRY STANDARDS

INTERNATIONAL AND NON-U.S. NATIONAL STANDARDS
SUBJECT INDEX — Ash

Ash Content Analysis (Cont.)

—**Ammonium Bicarbonate—Gravimetric Analysis**
ISO 3420-75. Ammonium Hydrogen Carbonate for Industrial Use (Including Foodstuffs)—Determination of Ash—Gravimetric Method First Edition. 3 pp.

—**Animal Feed**
BSI BS 5766: Part 1-79. 1979 Amd 1 Methods for Analysis of Animal Feeding Stuffs Part 1: Determination of Crude Ash. 7 pp.
BSI BS 5766: Part 2-80. 1980 Amd 1 Methods for Analysis of Animal Feeding Stuffs Part 2: Determination of Ash Insoluble in Hydrochloric Acid. 7 pp.
CNS N4024-9-86. Method of Test for Feeds: Determination of Crude Ash (Aug)(2770-9).
CNS N4024-10-86. Method of Test for Feeds: Determination of Ash Insoluble in Hydrochloric Acid (Aug)(2770-10).
ISO 5984-78. Animal Feeding Stuffs—Determination of Crude Ash First Edition; (Erratum—Nov 1979). 5 pp.
ISO 5985-78. Animal Feeding Stuffs—Determination of Ash Insoluble in Hydrochloric Acid First Edition. 5 pp.

—**Antifreezes**
CNS K6484-80. Method of Test for Ash Content of Engine Antifreezes, Antirusts and Coolants (May)(5587).

—**Asphalts**
CNS K6205-65. Method of Test for Ash of Asphalt and Coal Tar (Sep)(2487)(R 1973).

—**Beeswax**
MOD UK M 2012/71. Examination of Beeswax GS and Beeswax, Lead Free.
MOD UK M 9517/65. Examination of Beeswax and Beeswax, Lead Free (LAG 931A, No Information).

—**Beverages**
CNS N6091-91. Method of Test for Fruit and Vegetable Juices and Drinks (General Rules) (Jun)(3736). 1 p.

—**Boiler Water Treatment**
MOD UK M 9501/59. Determination of Solution Boiler Treatment (No Information) (Withdrawn).

—**Bunker Fuel Oils**
CPPA J.24P(C)-74. Ash Content. 1 p.

—**Cane Sugar**
CNS N6026-79. Method of Test for Cane Sugar (Jan)(1338). 7 pp.

—**Carbon Black**
BSI BS 5293: Part 7-91. 1991 Sampling and Testing Carbon Black for Use in the Rubber Industry Part 7: Method for Determination of Ash Content (ISO 1125: 1990). 6 pp.
ISO 1125-90. Rubber Compounding Ingredients—Carbon Black—Determination of Ash Content Second Edition. 4 pp.

—**Caseins**
BSI BS 6248: Part 2-82. 1982 Caseins and Caseinates Part 2: Method for Determination of 'Fixed Ash' of Caseins (Reference Method). 4 pp.
BSI BS 6248: Part 3-82. 1982 Caseins and Caseinates Part 3: Method for Determination of Ash of Rennet Caseins and Caseinates (Reference Method). 4 pp.

—**Cellulose Acetate**
BSI BS 2782:Pt4: METH 470C-91. 1991 Methods of Testing Plastics Part 4: Chemical Properties Method 470C: Determination of Ash of Unplasticized Cellulose Acetate (ISO 3451/3: 1984). 6 pp.
ISO 3451 Pt 3-84. Plastics—Determination of Ash—Part 3: Unplasticized Cellulose Acetate First Edition. 4 pp.

—**Cellulose Nitrate**
CNS K6185-87. Method of Test for Nitrocellulose for Industrial Use (Jun)(2358). 7 pp.

—**Cereal Products**
BSI BS 4317: Part 10-93. 1993 Methods of Test for Cereals and Pulses Part 10: Determination of Ash of Cereals and Milled Cereal Products (ISO 2171: 1993) (W). 11 pp.
BSI BS 4317: Part 10-81. 1981 Methods of Test for Cereals and Pulses Part 10: Determination of Ash in Cereals, Pulses and Their Derived Products. 7 pp.
ISO 2171-93. Cereals and Milled Cereal Products—Determination of Total Ash Third Edition. 8 pp.

—**Cereals**
BSI BS 4317: Part 10-93. 1993 Methods of Test for Cereals and Pulses Part 10: Determination of Ash of Cereals and Milled Cereal Products (ISO 2171: 1993) (W). 11 pp.

Ash Content Analysis (Cont.)

—**Cereals** (Cont.)
BSI BS 4317: Part 10-81. 1981 Methods of Test for Cereals and Pulses Part 10: Determination of Ash in Cereals, Pulses and Their Derived Products. 7 pp.
ISO 2171-93. Cereals and Milled Cereal Products—Determination of Total Ash Third Edition. 8 pp.

—**Cheeses**
CNS N6089-74. Methods of Test for Milk and Milk Products (Determination of Ash of Edible Cheese) (Apr)(3722). 1 p.

—**Chemical Products**
CNS K0019-82. Method for Determining Loss and Residue of Chemical Products (May)(8838).
JIS K 0067-92. Test Methods for Loss and Residue of Chemical Products. 18 pp.

—**Coal**
BSI BS 1016: Part 3-73. (WITHDRAWN) 1973 Amd 2 Methods for the Analysis and Testing of Coal and Coke Part 3: Proximate Analysis of Coal (Superseded by BS 1016: Sections 104.1, 104.2, 104.3, and 104.4: 1991). 23 pp.
CNS M3141-84. Method for Determination of Ash of Coal and Coke (Mar)(10822).
CNS M3155-84. Method for Analysis of Coal Ash and Coke Ash (Mar)(10836).
JIS M 8812-84. Methods for Proximate Analysis of Coal and Coke. 28 pp.
SAA AS 1038.3-89. Methods for the Analysis and Testing of Coal and Coke—Part 3: Proximate Analysis of Higher Rank Coal. 12 pp.
SAA AS 1038.14.1-81. Methods for the Analysis and Testing of Coal and Coke—Part 14.1: Analysis of Coal Ash, Coke Ash and Mineral Matter (Borate Fusion—Flame Atomic Absorption Spectrometric Method). 18 pp.
SAA AS 1038.14.2-85. Methods for the Analysis and Testing of Coal and Coke—Part 14.2: Analysis of Higher Rank Coal Ash, Coke Ash (Acid Digestion—Flame Atomic Absorption Spectrometric Method). 11 pp.
SAA AS 1038.15-87. Methods for the Analysis and Testing of Coal and Coke—Part 15: Fusibility of Higher Rank Coal Ash and Coke Ash. 6 pp.

—**Coal—Reference Method**
BSI BS 1016: Sec 104.4-91. 1991 Analysis and Testing of Coal and Coke Part 104: Proximate Analysis Section 104.4: Determination of Ash. 9 pp.

—**Coal Tar**
CNS K6205-65. Method of Test for Ash of Asphalt and Coal Tar (Sep)(2487)(R 1973).

—**Coal Tar Pitch**
BSI BS 6043: Sec 1.8-86. 1986 Methods of Sampling and Test for Carbonaceous Materials Used in Aluminium Manufacture Part 1: Electrode Pitch Section 1.8: Determination of Ash. 7 pp.
ISO 8006-85. Carbonaceous Materials Used in the Production of Aluminium—Pitch for Electrodes—Determination of Ash First Edition. 5 pp.

—**Coatings**
CGSB 1-GP-71 METH 69.3-82. Methods of Testing Paints and Pigments Ash Content of Varnish, Tar-Base and Other Coatings. 1 p.

—**Coke**
BSI BS 1016: Part 4-73. (WITHDRAWN) 1973 Amd 2 Methods for the Analysis and Testing of Coal and Coke Part 4: Moisture, Volatile Matter and Ash in the Analysis Sample of Coke (Superseded by BS 1016: Sections 104.1, 104.2, 104.3 and 104.4: 1991). 18 pp.
BSI BS 6043: Sec 2.2-85. 1985 Methods of Sampling and Test for Carbonaceous Materials Used in Aluminium Manufacture Part 2: Electrode Coke Section 2.2: Determination of Ash Content of Green and Calcined Cokes. 6 pp.
CNS M3141-84. Method for Determination of Ash of Coal and Coke (Mar)(10822).
CNS M3155-84. Method for Analysis of Coal Ash and Coke Ash (Mar)(10836).
ISO 8005-84. Carbonaceous Materials Used in the Production of Aluminium—Green and Calcined Coke—Determination of Ash Content First Edition. 4 pp.
JIS M 8812-84. Methods for Proximate Analysis of Coal and Coke. 28 pp.
SAA AS 1038.4-79. Methods for the Analysis and Testing of Coal and Coke—Part 4: Proximate Analysis of Coke. 16 pp.
SAA AS 1038.14.1-81. Methods for the Analysis and Testing of Coal and Coke—Part 14.1: Analysis of Coal Ash, Coke Ash and Mineral Matter (Borate Fusion—Flame Atomic Absorption Spectrometric Method). 18 pp.

Ash Content Analysis (Cont.)

—**Coke** (Cont.)
SAA AS 1038.14.2-85. Methods for the Analysis and Testing of Coal and Coke—Part 14.2: Analysis of Higher Rank Coal Ash, Coke Ash (Acid Digestion—Flame Atomic Absorption Spectrometric Method). 11 pp.
SAA AS 1038.15-87. Methods for the Analysis and Testing of Coal and Coke—Part 15: Fusibility of Higher Rank Coal Ash and Coke Ash. 6 pp.

—**Coke—Reference Method**
BSI BS 1016: Sec 104.4-91. 1991 Analysis and Testing of Coal and Coke Part 104: Proximate Analysis Section 104.4: Determination of Ash. 9 pp.

—**Comprehensive Testing**
SAA AS 1038.16-86. Methods for the Analysis and Testing of Coal and Coke—Part 16: Acceptance and Reporting of Results. 15 pp.

—**Condensed Milk**
BSI BS 1742: Part 4-91. 1991 Chemical Analysis of Condensed Milks Part 4: Determination of Ash from Condensed Milks. 8 pp.

—**Condiments**
BSI BS 4585: Part 3-81. 1981 Methods of Test for Spices and Condiments Part 3: Determination of Total Total Ash. 4 pp.
BSI BS 4585: Part 9-81. 1981 Methods of Test for Spices and Condiments Part 9: Determination of Acid-Insoluble Ash. 4 pp.
BSI BS 4585: Part 10-81. (WITHDRAWN) 1981 Methods of Test for Spices and Condiments Part 10: Determination of Water-Soluble Ash. 4 pp.
CNS N6153-81. Spices and Condiments—Determination of Total Ash (Sep)(7919). 3 pp.
CNS N6154-81. Spices and Condiments—Determination of Water-Insoluble Ash (Sep)(7920). 2 pp.
CNS N6155-81. Spices and Condiments—Determination of Acid-Insoluble Ash (Sep)(7921). 2 pp.
ISO 928-80. Spices and Condiments—Determination of Total Ash First Edition. 4 pp.
ISO 929-80. Spices and Condiments—Determination of Water-Soluble Ash First Edition; (Erratum—Dec 1981). 5 pp.
ISO 930-80. Spices and Condiments—Determination of Acid-Insoluble Ash First Edition. 4 pp.

—**Corrosion Inhibitors**
CNS K6484-80. Method of Test for Ash Content of Engine Antifreezes, Antirusts and Coolants (May)(5587).

—**Crude Oils**
CEN EN 7-74. Determination of Ash from Petroleum Products. 5 pp.
JIS K 2272-85. Testing Methods for Ash and Sulfated Ash of Crude Oil and Petroleum Products.

—**Dairy Products**
BSI BS 1741: Part 9-88. 1988 Methods for Chemical Analysis of Liquid Milk and Cream Part 9: Determination of Ash from Liquid Milk. 4 pp.

—**Dairy Products—Gravimetric Analysis**
ISO 5544-78. Caseins—Determination of "Fixed Ash" (Reference Method) First Edition. 4 pp.
ISO 5545-78. Rennet Caseins and Caseinates—Determination of Ash (Reference Method) First Edition. 4 pp.

—**Detergents**
MOD UK M 9521/65. Examination of Detergent Solution for Plastic Food Utensils.

—**Dicyandiamide**
MOD UK M 812/77. Examination of Dicyandiamide.

—**Drying Oils**
CNS K6631-81. Method of Test for Ash in Drying Oils and Fatty Acid (Mar)(7040).

—**Eggs—Frozen**
CNS N6050-80. Method of Test for Frozen Eggs (Jul)(2923). 3 pp.

—**Engine Coolants**
CNS K6484-80. Method of Test for Ash Content of Engine Antifreezes, Antirusts and Coolants (May)(5587).

—**Fabrics**
BSI BS 6646-85. 1985 Amd 1 Method for Determination of Ash of Textiles (AMD 5720) July 29, 1988. 6 pp.
CNS L3166-82. Determination of the Contents of Ash and Sulfur in Textile Materials (Apr)(8709).

INTERNATIONAL AND NON-U.S. NATIONAL STANDARDS
SUBJECT INDEX

Ash Content Analysis (Cont.)

—Fats
BSI BS 684: Sec 2.2-86. 1986 Methods of Analysis of Fats and Fatty Oils Part 2: Other Methods Section 2.2: Determination of Ash. 4 pp.
ISO 6884-85. Animal and Vegetable Fats and Oils—Determination of Ash First Edition. 4 pp.

—Fatty Acids
CNS K6631-81. Method of Test for Ash in Drying Oils and Fatty Acid (Mar)(7040).

—Flammable Materials—Solids
CPPA G.6U-77. Ashing by Oxygen Flask Combustion. 2 pp.

—Floor Coverings
CEN PREN 670-92. Identification and Composition of Linoleum—Determination of Cement Content and Ash Residue. 6 pp.

—Floor Waxes
MOD UK M 9503/59. Examination of Wax/Water Emulsion Floor Polish.

—Flours (Food)
CNS N6028-73. Method of Test for Soybean Powder (Nov)(1418). 3 pp.

—Food
CNS N6115-84. Method of Test for Ash in Food (Jan)(5034). 2 pp.

—Formaldehyde
BSI BS 2942-57. (WITHDRAWN) 1957 Amd 2 Formaldehyde Solution. 12 pp.
ISO 2224-72. Formaldehyde Solutions for Industrial Use—Determination of Ash First Edition. 3 pp.

—Fructose Syrups
CNS N6200-85. Method of Test for High Fructose Syrup (Sep)(11370). 3 pp.

—Fruit Juices
CNS N6091-91. Method of Test for Fruit and Vegetable Juices and Drinks (General Rules) (Jun)(3736). 1 p.
CNS N6217-91. Method of Test for Fruit and Vegetable Juices and Drinks-Determination of Ash (Jun)(12571). 1 p.

—Fruits
CNS N6166-91. Method of Test for Fruit and Vegetable Products-Determination of Ash Insoluble in Hydrochloric Acid (Jun)(8625). 2 pp.
ISO 763-82. Fruit and Vegetable Products—Determination of Ash Insoluble in Hydrochloric Acid First Edition. 4 pp.

—Fruits—Dried
CNS N6049-85. Method of Test for Dehydrated Fruits and Vegetables (Mar)(2896). 2 pp.

—Fuel Oils
CEN EN 7-74. Determination of Ash from Petroleum Products. 5 pp.

—Glucose
CNS N6053-73. Method of Test for Dried Glucose and Glucose Syrup (Jan)(3218). 9 pp.
CNS N6054-72. Method of Test for Solid Glucose (Jun)(3349). 5 pp.
CNS N6055-72. Method of Test for Powdered Glucose (Jun)(3351). 3 pp.

—Glucose Syrups
CNS N6053-73. Method of Test for Dried Glucose and Glucose Syrup (Jan)(3218). 9 pp.

—Glycerol—Gravimetric Analysis
BSI BS 5711: Part 6-79. 1979 Methods of Sampling and Test for Glycerol Part 6: Determination of Ash: Gravimetric Method. 2 pp.
ISO 2098-72. Glycerols for Industrial Use—Determinaton of Ash—Gravimetric Method First Edition. 3 pp.

—Graphite
CNS K6806-84. Method of Test for Ash in Graphite (Nov)(11114).
JIS R 7223-79. Chemical Analysis of High Purity Graphite Material (R 1985). 32 pp.

—Hexachloroethane
BSI BS 577-66. (WITHDRAWN) 1966 Amd 1 Hexachlorethane. 17 pp.

—Hexamethylenetetramine
MOD UK M 879/69. Examination of Hexamine.

—Honey
CNS N6027-74. Method of Test for Honey (Oct)(1344). 8 pp.

—Lactose Monohydrate
MOD UK M 821/71. Examination of Lactose.

Ash Content Analysis (Cont.)

—Leather
CNS K6130-81. Method of Test for Soluble Ash of Leather (Aug)(1286).
CNS K6134-81. Method of Test for Total Ash of Leather (Aug)(1290).

—Legumes
BSI BS 4317: Part 10-93. 1993 Methods of Test for Cereals and Pulses Part 10: Determination of Ash of Cereals and Milled Cereal Products (ISO 2171: 1993) (W). 11 pp.
BSI BS 4317: Part 10-81. 1981 Methods of Test for Cereals and Pulses Part 10: Determination of Ash in Cereals, Pulses and Their Derived Products. 7 pp.
ISO 2171-93. Cereals and Milled Cereal Products—Determination of Total Ash Third Edition. 8 pp.

—Lubricants
CNS K6002-47. Method of Test for Lubricants (Mar)(41). 4 pp.

—Lubricating Oils
CEN EN 7-74. Determination of Ash from Petroleum Products. 5 pp.

—Maleic Anhydride
CNS K6569-80. Method of Test for Ash of Maleic Anhydride for Industrial Use (Aug)(6214).
ISO 1390 Pt V-77. Maleic Anhydride for Industrial Use—Methods of Test—Part V: Determination of Ash First Edition. 3 pp.

—Malononitrile
MOD UK M 866/66. Examination of Malononitrile.

—Meat
BSI BS 4401: Part 1-80. 1980 Methods of Test for Meat and Meat Products Part 1: Determination of Ash (Reference Method). 4 pp.
CNS N6134-80. Method of Test for Meat and Meat Products—Determination of Ash (Aug)(6259). 2 pp.
ISO 936-78. Meat and Meat Products—Determination of Ash (Reference Method) First Edition. 4 pp.

—Meat—Dried
CNS N6036-87. Method of Test for Fried and Dried Shredded Meat (Jul)(2168). 2 pp.

—Musk
CNS N4029-70. Method of Test for Musk (Jan)(3083). 2 pp.

—Nitrodiphenylamine
MOD UK M 824/92. Examination of 2-Nitrodiphenylamine.

—Oilseeds
BSI BS 4325: Part 2-78. 1978 Analysis of Oilseed Residues Part 2: Determination of Total Ash. 4 pp.
BSI BS 4325: Part 3-78. 1978 Analysis of Oilseed Residues Part 3: Determination of Ash Insoluble in Hydrochloric Acid. 4 pp.
ISO 735-77. Oilseed Residues—Determination of Ash Insoluble in Hydrochloric Acid First Edition. 4 pp.
ISO 749-77. Oilseed Residues—Determination of Total Ash First Edition. 4 pp.

—Oleic Acid
MOD UK M 9545/68. Examination of Oleic Acid, Special, Lead Free.

—Organic Coatings
CNS K6804-5-84. Method of Test for Organic Coating Ash Content (Jun)(10880-5).

—Packaging Papers
MOD UK DSTAN 13-10-68. Paper, Wrapping, Unglazed, and Paper, Wrapping, Unglazed, Lead Free Issue 1. 9 pp.
MOD UK M 6512/61. Examination of Paper Wrapping, Unglazed and Paper, Wrapping, Unglazed, LF Quality (Superseded by Def Stan 13-10).

—Paints—Spectrography
MOD UK M 9511/65. Spectrographic Analysis of Inorganic Materials by the Barium Chloride Flux Method (No Information).
MOD UK M 9512/65. Spectrographic Determination of Lead in Paint Ashes and Similar Materials (No Information).

—Paperboard
BSI BS 3631-84. 1984 Determination of Ash of Paper and Board. 7 pp.
CNS P3044-81. Method of Test for Ash Content in Paper and Paperboard (Jan)(6950). 2 pp.
CPPA G.11-86. Ash of Paper and Paperboard. 1 p.
ISO 2144-87. Paper and Board—Determination of Ash Third Edition. 6 pp.
JIS P 8128-76. Testing Method for Ash in Paper and Paperboard.

Ash Content Analysis (Cont.)

—Paperboard (Cont.)
SAA AS 1301.P418 S-78. Methods of Test for Pulp and Paper (Metric Units)—Part P418s: Ash Content of Paper and Paperboard (This is a Joint Standard with SANZ NZS 1301).
SNZ NZS/AS 1301. P418S-78. Methods of Test for Pulp and Paper Ash Content of Paper and Paperboard (This is a Joint Standard with SAA AS 1301.P418S). 2 pp.

—Papers
BSI BS 3631-84. 1984 Determination of Ash of Paper and Board. 7 pp.
CNS P3044-81. Method of Test for Ash Content in Paper and Paperboard (Jan)(6950). 2 pp.
CPPA G.11-86. Ash of Paper and Paperboard. 1 p.
ISO 2144-87. Paper and Board—Determination of Ash Third Edition. 6 pp.
JIS P 8128-76. Testing Method for Ash in Paper and Paperboard.
SAA AS 1301.P418 S-78. Methods of Test for Pulp and Paper (Metric Units)—Part P418s: Ash Content of Paper and Paperboard (This is a Joint Standard with SANZ NZS 1301).
SNZ NZS/AS 1301. P418S-78. Methods of Test for Pulp and Paper Ash Content of Paper and Paperboard (This is a Joint Standard with SAA AS 1301.P418S). 2 pp.

—Paraffin Wax
MOD UK M 2009/67. Examination of Paraffin Wax.

—Paraformaldehyde
BSI BS 2941-57. (WITHDRAWN) 1957 Amd 1 Paraformaldehyde. 14 pp.
CNS K6549-80. Method of Test for Ash of Paraformaldehyde for Industrial Use (Aug)(6194).
ISO 1391 Pt II-76. Paraformaldehyde for Industrial Use—Methods of Test—Part II: Determination of Ash First Edition; (Corrected and Reprinted -1977). 3 pp.

—Pasta
CNS N6170-82. Method of Test for Macaroni, Spaghetti and Vermicelli (Aug)(9320). 3 pp.

—Pastes
JIS R 7231-78. Testing Method for Solderberg Paste.
JIS R 7231-71. Testing Method for Solderberg Paste. 11 pp.

—Peat
BSI BS 4156-90. 1990 Recommendations for Peat for Horticultural and Landscape Use. 15 pp.
BSI BS 4156-67. 1967 Peat. 11 pp.

—Petroleum Products
BSI BS 2000: Part 223-93. 1993 Methods of Test for Petroleum and Its Products Part 223: Determination of Ash of Petroleum Products Containing Mineral Matter (W). 3 pp.
BSI BS 2000: Part 223-84. 1984 Petroleum and Its Products Part 223: Ash from Petroleum Products Containing Mineral Matter. 4 pp.
BSI BS 4450-76. 1976 Determination of Ash from Petroleum Products. 7 pp.
CEN EN 7-74. Determination of Ash from Petroleum Products. 5 pp.
CNS K6367-73. Method of Test for Ash of Petroleum Products (Nov)(3576).
ISO 6245-93. Petroleum Products—Determination of Ash Second Edition. 6 pp.
JIS K 2272-85. Testing Methods for Ash and Sulfated Ash of Crude Oil and Petroleum Products.

—Phthalate Esters
ISO 1385 Pt III-77. Phthalate Esters for Industrial Use—Methods of Test—Part III: Determination of Ash First Edition. 3 pp.

—Phthalic Anhydride
ISO 1389 Pt VIII-77. Phthalic Anhydride for Industrial Use—Methods of Test—Part VIII: Determination of Ash First Edition. 4 pp.

—Plastic Pipes
CNS K6198-65. Method of Test for Polyethylene Pipe for Chemical Industrial and General Uses (Apr)(2459) (R 1971). 3 pp.

—Plastics
BSI BS 2782:Pt4: METH 470A-91. 1991 Methods of Testing Plastics Part 4: Chemical Properties Method 470A: Determination of Ash (General Methods) (ISO 3451/1: 1981) (Supersedes BS 2782: Method: 454A & 454B: 1978). 6 pp.
ISO 3451 Pt 1-81. Plastics—Determination of Ash—Part 1: General Methods Second Edition. 5 pp.

—Polyalkylene Terephthalates
BSI BS 2782:Pt4: METH 470B-91. 1991 Methods of Testing Plastics Part 4: Chemical Properties Method 470B: Determination of Ash of Polyalkylene Terephthalates (ISO 3451/2: 1984). 6 pp.

Ash Content Analysis (Cont.)

—Polyalkylene Terephthalates (Cont.)
ISO 3451 Pt 2-84. Plastics—Determination of Ash—Part 2: Polyalkylene Terephthalates First Edition. 4 pp.

—Polyamide Resins
BSI BS 2782:Pt4: METH 470D-91. 1991 Methods of Testing Plastics Part 4: Chemical Properties Method 470D: Determination of Ash of Polyamides (ISO 3451/4: 1986). 5 pp.
ISO 3451 Pt 4-86. Plastics—Determination of Ash—Part 4: Polyamides First Edition. 4 pp.

—Polybasic Organic Acids
CNS K6701-81. Method of Test for Ash in Polybasic Acids (Oct)(8033).

—Polyurethane Foams
SAA AS 2282.16-91. Methods for Testing Flexible Cellular Polyurethane—Part 16: Determination of Ash Content. 2 pp.

—Pulps
BSI BS 4496-84. 1984 Determination of Acid-Insoluble Ash in Pulp. 4 pp.
BSI BS 4896-73. 1973 Determination of Ash Pulp. 7 pp.
CNS P3006-91. Method of Test for Ash Content in Pulp (Jan)(1356). 2 pp.
CNS P3055-82. Method of Test for Acid Insoluble Ash Pulp (Aug)(9323).
CPPA G.10-73. Ash in Pulp and Wood. 1 p.
CPPA G.33P-84. Acid-Insoluble Ash in Pulp. 1 p.
ISO 776-82. Pulps—Determination of Acid-Insoluble Ash Second Edition. 4 pp.
ISO 1762-74. Pulps—Determination of Ash First Edition. 4 pp.
JIS P 8204-76. Method of Testing Ash Content for Paper Pul
JIS P 8204-61. Method of Determining Ash Content in Pulp for Paper. 4 pp.
SAA AS 1301.P3S-78. Methods of Test for Pulp and Paper (Metric Units)—Part P3s: Ash Content of Wood and Pulp (This is a Joint Standard with SANZ NZS 1301).
SNZ NZS/AS 1301. P3S-78. Methods of Test for Pulp and Paper Ash Content of Wood and Pulp (This is a Joint Standard with SAA AS 1301.P3s). 1 p.

—Pulpwood
CNS O2016-84. Method of Test for Ash in Wood (May)(3084). 2 pp.
JIS P 8003-76. Testing Method for Ash of Pulpwood (R 1985). 5 pp.

—PVC
BSI BS 2782:Pt4: METH 454A&B-78. (WITHDRAWN) 1978 Methods of Testing Plastics Part 4: Chemical Properties Method 454A: Determination of Ash Method 454B: Determina-tion of Sulphated Ash (Superseded by BS 2782: Method 470A:1991). 4 pp.
BSI BS 2782:Pt4: METH 470E-91. 1991 Methods of Testing Plastics Part 4: Chemical Properties Method 470E: Determination of Ash of Poly(Vinyl Chloride) (ISO 3451/5: 1989). 7 pp.
ISO 1270-75. Plastics—PVC Resins—Determination of Ash and Sulphated Ash First Edition. 4 pp.
ISO 3451 Pt 5-89. Plastics—Determination of Ash—Part 5: Poly(Vinyl Chloride) First Edition. 5 pp.

—Quaternary Ammonium Chloride
CNS K6680-81. Method of Test for Ash in Fatty Quaternary Ammonium Chlorides (Jul)(7724).

—Refractory Materials
BSI BS 6043: Sec 4.13-91. 1991 Methods of Sampling and Test for Carbonaceous Materials Used in Aluminium Manufacture Part 4: Cold Ramming Pastes Section 4.13: Determination of Ash of Baked Rammed Paste. 8 pp.

—Rosins
CNS K6466-80. Method of Test for Ash in Rosin (Mar)(5351).

—Rubber
BSI BS 5923: Part 1-80. (WITHDRAWN) 1980 Chemical Analysis of Rubber Part 1: Determination of Ash (Superseded by BS 7164: Part 5: 1991). 7 pp.
BSI BS 7164: Part 5-91. 1991 Chemical Tests for Raw and Vulcanized Rubber Part 5: Methods for Determination of Ash Content (ISO 247: 1990). 7 pp.
CNS K6358-73. Chemical Analysis of Ash Content in Natural Rubber (May)(3567).
ISO 247-90. Rubber—Determination of Ash Third Edition. 6 pp.

—Soldering Fluxes
MOD UK M 9524/66. Examination of Flux Soldering Solution (Withdrawn).

Ash Content Analysis (Cont.)

—Solid Fuels
ISO 1171-81. Solid Mineral Fuels—Determination of Ash Second Edition. 4 pp.

—Soybeans
CNS N6009-83. Method of Test for Soybean Paste (Dec)(956). 1 p.

—Spent Liquors
CPPA G.6U-77. Ashing by Oxygen Flask Combustion. 2 pp.

—Spices
BSI BS 4585: Part 3-81. 1981 Methods of Test for Spices and Condiments Part 3: Determination of Total Total Ash. 4 pp.
BSI BS 4585: Part 9-81. 1981 Methods of Test for Spices and Condiments Part 9: Determination of Acid-Insoluble Ash. 4 pp.
BSI BS 4585: Part 10-81. (WITHDRAWN) 1981 Methods of Test for Spices and Condiments Part 10: Determination of Water-Soluble Ash. 4 pp.
CNS N6153-81. Spices and Condiments—Determination of Total Ash (Sep)(7919). 3 pp.
CNS N6154-81. Spices and Condiments—Determination of Water-Insoluble Ash (Sep)(7920). 2 pp.
CNS N6155-81. Spices and Condiments—Determination of Acid-Insoluble Ash (Sep)(7921). 2 pp.
ISO 928-80. Spices and Condiments—Determination of Total Ash First Edition. 4 pp.
ISO 929-80. Spices and Condiments—Determination of Water-Insoluble Ash First Edition; (Erratum—Dec 1981). 5 pp.
ISO 930-80. Spices and Condiments—Determination of Acid-Insoluble Ash First Edition. 4 pp.

—Starches
ISO 3593-81. Starch—Determination of Ash Second Edition. 4 pp.

—Sulfur—High Temperature Testing
ISO 3425-75. Sulphur for Industrial Use—Determination of Ash at 850-900 Degrees Celsius and of Residue at 200 Degrees Celsius First Edition. 4 pp.

—Sulfuric Acid—Gravimetric Analysis
ISO 913-77. Sulphuric Acid and Oleum for Industrial Use—Determination of Ash—Gravimetric Method First Edition. 4 pp.

—Tar Coatings
CGSB 1-GP-71 METH 69.3-82. Methods of Testing Paints and Pigments Ash Content of Varnish, Tar-Base and Other Coatings. 1 p.

—Tea
BSI BS 6049: Part 4-88. 1988 Methods of Test for Tea Part 4: Determination of Total Ash. 4 pp.
BSI BS 6049: Part 5-89. 1989 Methods of Test for Tea Part 5: Determination of Water-Soluble Ash and Water-Insoluble Ash. 4 pp.
BSI BS 6049: Part 6-88. 1988 Methods of Test for Tea Part 6: Determination of Acid-Insoluble Ash. 4 pp.
BSI BS 6049: Part 7-81. 1981 Methods of Test for Tea Part 7: Determination of Alkalinity of Water-Soluble Ash. 4 pp.
BSI BS 6986: Part 4-90. 1990 Analysis of Instant Tea Part 4: Method for Determination of Total Ash. 7 pp.
ISO 1575-87. Tea—Determination of Total Ash Third Edition. 4 pp.
ISO 1576-88. Tea—Determination of Water-Soluble Ash and Water-Insoluble Ash Second Edition. 4 pp.
ISO 1577-87. Tea—Determination of Acid-Insoluble Ash Second Edition. 4 pp.
ISO 1578-75. Tea—Determination of Alkalinity of Water-Soluble Ash First Edition. 4 pp.
ISO 7514-90. Instant Tea in Solid Form—Determination of Total Ash First Edition. 4 pp.

—Toluenesulfonamide
CNS K6202-65. Method of Test for Para-Toluene-Sulfonyl Amide (Mar)(2484) (R 1971). 1 p.

—Urea—Industrial
ISO 1594-77. Urea for Industrial Use—Determination of Ash First Edition. 4 pp.

—Varnishes
CGSB 1-GP-71 METH 69.3-82. Methods of Testing Paints and Pigments Ash Content of Varnish, Tar-Base and Other Coatings. 1 p.

—Vegetable Juices
CNS N6091-91. Method of Test for Fruit and Vegetable Juices and Drinks (General Rules) (Jun)(3736). 1 p.
CNS N6217-91. Method of Test for Fruit and Vegetable Juices and Drinks-Determination of Ash (Jun)(12571). 1 p.

Ash Content Analysis (Cont.)

—Vegetables
CNS N6166-91. Method of Test for Fruit and Vegetable Products-Determination of Ash Insoluble in Hydrochloric Acid (Jun)(8625). 2 pp.
ISO 763-82. Fruit and Vegetable Products—Determination of Ash Insoluble in Hydrochloric Acid First Edition. 4 pp.

—Vegetables—Dried
CNS N6049-85. Method of Test for Dehydrated Fruits and Vegetables (Mar)(2896). 2 pp.

—Waxes
CEN EN 7-74. Determination of Ash from Petroleum Products. 5 pp.
CGSB 25-GP-1M M 17.1, 17.2-84. Methods of Sampling and Testing Waxes and Polishes Ash Content. 1 p.

—Wood
CPPA G.10-73. Ash in Pulp and Wood. 1 p.
SAA AS 1301.P3S-78. Methods of Test for Pulp and Paper (Metric Units)—Part P3s: Ash Content of Wood and Pulp (This is a Joint Standard with SANZ NZS 1301).
SNZ NZS/AS 1301. P3S-78. Methods of Test for Pulp and Paper Ash Content of Wood and Pulp (This is a Joint Standard with SAA AS 1301.P3s). 1 p.

Ash Determination
Use: Ash Content Analysis

Ash Receivers
Use: Ashtrays

Ash Wood
Scope Note: For additional listings, see the subheading Ash Wood under specific products made from Ash Wood *See Also:* Hardwoods; Wood

—Aircraft
BSI 3V 4-29. 1929 Amd 3 Ash. 4 pp.

Ashes
See Also: Clinkers; Fertilizers; Fly Ash

—Aggregates—Concretes
BSI BS 1165-85. (WITHDRAWN) 1985 Clinker and Furnace Bottom Ash Aggregates for Concrete (R) (Superseded by BS 3797: 1990). 4 pp.

—Metallic Element Content—Spectrography
MOD UK M 9510/65. Spectrographic Analysis of Inorganic Non-Metallic Materials by the Iron Flux Method (No Information).
MOD UK M 9548/70. Emission Spectrographic Analysis of Inorganic Materials Using a Lithium Fluoride-Graphite Flux (No Information).

Ashtrays

—Metal—Floor Type
CGSB CAN/CGSB-44.118-M89. Floor Type Tobacco Ash Receiver, Metal or Plastic. 14 pp.

—Plastic
CNS Z7120-82. Plastics Ash-Tray (Jun)(9050).
CNS Z8034-82. Method of Test for Plastics Ash-Tray (Jun)(9051).

—Plastic—Floor Type
CGSB CAN/CGSB-44.118-M89. Floor Type Tobacco Ash Receiver, Metal or Plastic. 14 pp.

Askarels
See Also: Electrical Insulating Oils

—Capacitors
IEC 588 Pt 1-77. Askarels for Transformers and Capacitors Part 1: General First Edition. 12 pp.
IEC 588 Pt 2-78. Askarels for Transformers and Capacitors Part 2: Test Methods First Edition. 48 pp.
IEC 588 Pt 3-77. Askarels for Transformers and Capacitors Part 3: Specifications for New Askarels First Edition. 22 pp.

—Capacitors—Screening Testing
IEC 588 Pt 6-79. Askarels for Transformers and Capacitors Part 6: Screening Test for Effects of Materials on Capacitor Askarels First Edition. 17 pp.

—Transformers
IEC 588 Pt 1-77. Askarels for Transformers and Capacitors Part 1: General First Edition. 12 pp.
IEC 588 Pt 2-78. Askarels for Transformers and Capacitors Part 2: Test Methods First Edition. 48 pp.
IEC 588 Pt 3-77. Askarels for Transformers and Capacitors Part 3: Specifications for New Askarels First Edition. 22 pp.

INTERNATIONAL AND NON-U.S. NATIONAL STANDARDS
SUBJECT INDEX

Askarels *(Cont.)*
—Transformers—Maintenance
IEC 588 Pt 4-79. Askarels for Transformers and Capacitors Part 4: Guide for Maintenance of Transformer Askarel in Equipment First Edition. 34 pp.

—Transformers—Screening Testing
IEC 588 Pt 5-79. Askarels for Transformers and Capacitors Part 5: Screening Test for Compatibility of Materials and Transformer Askarels First Edition. 14 pp.

ASN.1
Use: Open Systems Interconnection—Abstract Syntax Notation One

Asparagine
CNS K7078-62. Chemical Reagent (Asparagin) (L-Aminosuccinamic Acid) (Jul)(1578).

Asparagine Monohydrate
JIS K 8021-72. L-Asparagine, Monohydrate.

Asparagus
See Also: Food; Vegetables
EC EEC/454/92-92. Commission Regulation Laying down Quality Standards for Asparagus. 5 pp.

—Canned
CNS N5063-89. Canned Asparagus (Jan)(2348). 9 pp.

—Frozen
CNS N5099-85. Frozen Green Asparagus (Mar)(3286). 4 pp.
CNS N5139-85. Frozen White Asparagus (Mar)(4528). 4 pp.

—Refrigerated Transportation
ISO 6882-81. Asparagus—Guide to Refrigerated Transport First Edition. 4 pp.

—Storage
ISO 4186-80. Asparagus—Guide to Storage First Edition. 3 pp.

Aspartic Acid
JIS K 9045-72. L-Aspartic Acid.

Aspect Ratio
See Also: Peripheral Equipment; Television Equipment

—Normal—Enhanced Quality Television Systems
CCIR Report 1077-1-90. Enhanced 4:3 Aspect Ratio Television Systems—Section 11A—Characteristics of Systems for Monochrome and Colour Television. 9 pp.

—Wider—Television Systems
CCIR Report 1220-91. Wider Aspect Ratio Television Systems—Section 11A—Characteristics of Systems for Monochrome and Colour Television. 4 pp.

Asphalt Cements
Use For: Cutback Asphalt Cements
See Also: Cements; Lap Cements
CGSB CAN/CGSB-37.5-M89. Cutback Asphalt Plastic, Cement. 10 pp.
CGSB 37-GP-11M-76. Application of Cutback Asphalt Plastic Cement, Standard for (R 1984). 6 pp.

—Paving Materials
CGSB CAN/CGSB-16.3-M90. Asphalt Cements for Road Purposes. 12 pp.

Asphalt Coatings (Made From Asphalt)
Use For: Asphalt Paints (Made From Asphalt)
See Also: Asphalt Primers; Bituminous Coatings; Coal Tar Coatings; Coatings; Nonmetallic Coatings; Tar Coatings; Waterproof Coatings

—Adhesion—Immersion Testing
DIN ENGL 52006 Pt 2-80. Testing of Bituminous Binders; Effect of Water on Binder Coatings; Rapid Curing Cutback or Cold Tar Binder Coating (Dec). 2 pp.

—Concrete
CGSB CAN/CGSB-37.1-M89. Chemical Emulsifer Type, Emulsified Asphalt for Damp-Proofing. 10 pp.
CGSB CAN/CGSB-37.2-M88. Emulsified Asphalt, Mineral—Colloid Type, Unfilled, for Dampproofing and Waterproofing and for Roof Coatings. 13 pp.
CGSB CAN/CGSB-37.3-M89. Application of Emulsified Asphalts for Dampproofing or Waterproofing. 9 pp.
CGSB 37-GP-6MA-83. Asphalt, Cutback, Unfilled, for Dampproofing, Standard for. 9 pp.

Asphalt Coatings (Made From Asphalt) *(Cont.)*
—Concrete *(Cont.)*
CGSB CAN/CGSB-37.16-M89. Filled, Cutback Asphalt for Dampproofing and Waterproofing. 10 pp.
CGSB CAN/CGSB-37.28-M89. Reinforced, Mineral Colloid Type, Emulsified Asphalt for Roof Coatings and Waterproofing. 14 pp.
CGSB 37-GP-36M-76. Application of Filled Cutback Asphalts for Dampproofing and Waterproofing, Standard for. 7 pp.
CGSB 37-GP-37M-77. Application of Hot Asphalt for Dampproofing or Waterproofing, Standard for. 8 pp.
CGSB CAN/CGSB-37.50-M89. Hot-Applied, Rubberized Asphalt for Roofing and Waterproofing. 23 pp.
CGSB CAN/CGSB-37.51-M90. Application for Hot-Applied Rubberized Asphalt for Roofing and Waterproofing. 21 pp.
CGSB CAN/CGSB-37.65-M88. Mastic Asphalt (Hot Process) for Flooring. 19 pp.

—Masonry
CGSB CAN/CGSB-37.1-M89. Chemical Emulsifer Type, Emulsified Asphalt for Damp-Proofing. 10 pp.
CGSB CAN/CGSB-37.2-M88. Emulsified Asphalt, Mineral—Colloid Type, Unfilled, for Dampproofing and Waterproofing and for Roof Coatings. 13 pp.
CGSB CAN/CGSB-37.3-M89. Application of Emulsified Asphalts for Dampproofing or Waterproofing. 9 pp.
CGSB 37-GP-6MA-83. Asphalt, Cutback, Unfilled, for Dampproofing, Standard for. 9 pp.
CGSB CAN/CGSB-37.16-M89. Filled, Cutback Asphalt for Dampproofing and Waterproofing. 10 pp.
CGSB CAN/CGSB-37.28-M89. Reinforced, Mineral Colloid Type, Emulsified Asphalt for Roof Coatings and Waterproofing. 14 pp.
CGSB 37-GP-36M-76. Application of Filled Cutback Asphalts for Dampproofing and Waterproofing, Standard for. 7 pp.
CGSB 37-GP-37M-77. Application of Hot Asphalt for Dampproofing or Waterproofing, Standard for. 8 pp.

—Pipes—Chemical Resistant
CGSB CAN/CGSB-1.108-M89. Bituminous Solvent Type Paint. 10 pp.

—Roofing
CGSB CAN/CGSB-37.2-M88. Emulsified Asphalt, Mineral—Colloid Type, Unfilled, for Dampproofing and Waterproofing and for Roof Coatings. 13 pp.
CGSB CAN/CGSB-37.8-M88. Asphalt, Cutback, Filled, for Roof Coating. 9 pp.
CGSB 37-GP-14M-76. Application of Filled Cutback Asphalt Roof Coating, Standard for (R 1984). 7 pp.
CGSB CAN/CGSB-37.28-M89. Reinforced, Mineral Colloid Type, Emulsified Asphalt for Roof Coatings and Waterproofing. 14 pp.
CGSB CAN/CGSB-37.50-M89. Hot-Applied, Rubberized Asphalt for Roofing and Waterproofing. 23 pp.
CGSB CAN/CGSB-37.51-M90. Application for Hot-Applied Rubberized Asphalt for Roofing and Waterproofing. 21 pp.

—Steel Pipes—Water
JIS G 3491-77. Asphalt Protective Coatings for Steel Water Pipe (R 1983). 28 pp.

Asphalt Content Analysis
—Lubricants
CNS K6002-47. Method of Test for Lubricants (Mar)(41). 4 pp.

Asphalt Emulsion Coatings (Made From Asphalt)
Use: Asphalt Coatings (Made From Asphalt)

Asphalt Emulsions
Use: Emulsified Asphalts

Asphalt Finishers
Use: Finishing Machinery—Asphalt

Asphalt Mastics
Use For: Mastic Asphalts

—Binder Content
BSI BS 5284-76. 1976 Sampling and Testing Mastic Asphalt and Pitchmastic Used in Building. 22 pp.

—Buildings
DIN ENGL 18195 Pt 2-83. Waterproofing of Buildings and Structures; Materials (Aug). 7 pp.
DIN ENGL 18195 Pt 3-83. Waterproofing of Buildings and Structures; Processing of Materials (Aug). 5 pp.

Asphalt Mastics *(Cont.)*
—Buildings *(Cont.)*
DIN ENGL 18195 Pt 5-84. Waterproofing of Buildings and Structures; Waterproofing Against Water That Exerts No Hydrostatic Pressure; Design and Workmanship (Feb). 6 pp.

—Compression Testing
DIN ENGL 1996 Pt 12-85. Testing of Asphalt; Compression Testing of Mastic Asphalt (Feb). 3 pp.

—Floors
BSI BS 988,1076, 1097,1451-73. 1973 Mastic Asphalt for Building (Limestone Aggregate). 15 pp.
BSI BS 6577-85. 1985 Mastic Asphalt for Building (Natural Rock Asphalt Aggregate). 13 pp.
BSI BS 6925-88. 1988 Mastic Asphalt for Building and Civil Engineering (Limestone Aggregate). 12 pp.
BSI BS 6925-01. 1988 Amd 1 Mastic Asphalt for Building and Civil Engineering (Limestone Aggregate) (AMD 7150) July 15, 1992 (R). 12 pp.
CGSB CAN/CGSB-37.26-M89. Mineral Colloid Type, Emulsified Asphalt for Mastic Flooring. 12 pp.
CGSB CAN/CGSB-37.27-M89. Chemical Emulsifier Type, Emulsified Asphalt for Mastic Flooring. 9 pp.
CGSB CAN/CGSB-37.38-M89. Application of Mineral-Colloid Type, Emulsified Asphalt Mixes for Cold Mastic Flooring. 10 pp.
CGSB CAN/CGSB-37.41-M89. Application of Chemical Type, Emulsified Asphalt Mixes for Cold Mastic Flooring. 11 pp.
SNZ NZS/BS 6925-88. Specification for Mastic Asphalt for Building and Civil Engineering (Limestone Aggregate). 12 pp.

—Roofing
BSI BS 988,1076, 1097,1451-73. 1973 Mastic Asphalt for Building (Limestone Aggregate). 15 pp.
BSI BS 6577-85. 1985 Mastic Asphalt for Building (Natural Rock Asphalt Aggregate). 13 pp.
BSI BS 6925-88. 1988 Mastic Asphalt for Building and Civil Engineering (Limestone Aggregate). 12 pp.
BSI BS 6925-01. 1988 Amd 1 Mastic Asphalt for Building and Civil Engineering (Limestone Aggregate) (AMD 7150) July 15, 1992 (R). 12 pp.
BSI CP 144: Part 4-70. 1970 Amd 1 Roof Coverings Part 4: Mastic Asphalt (AMD 6161) December 21, 1990. 48 pp.
CGSB CAN/CGSB-37.67-M87. Mastic Asphalt (Hot Process) for Roofing. 13 pp.
SNZ NZS/BS 6925-88. Specification for Mastic Asphalt for Building and Civil Engineering (Limestone Aggregate). 12 pp.

—Sampling
BSI BS 5284-93. 1993 Methods of Sampling and Testing Mastic Asphalt Used in Building and Civil Engineering (R). 24 pp.
BSI BS 5284-76. 1976 Sampling and Testing Mastic Asphalt and Pitchmastic Used in Building. 22 pp.

—Segregation
DIN ENGL 1996 Pt 16-75. Testing of Bituminous Materials for Road Building and Related Purposes; Determination of Segregation Tendency (Dec). 4 pp.

—Ships
MOD UK NES 805: Part 2-88. Requirements for Thermal Insulation Material Part 2: Mastic and Sealing Products Flexible Mastic Sealers Issue 2 (11.88). 9 pp.

—Softening Points
DIN ENGL 1996 Pt 15-75. Testing of Bituminous Materials for Road Building and Related Purposes; Determination of Softening Point in Accordance with Wilhelmi (Dec). 4 pp.

—Submarines
MOD UK NES 805: Part 2-88. Requirements for Thermal Insulation Material Part 2: Mastic and Sealing Products Flexible Mastic Sealers Issue 2 (11.88). 9 pp.

—Water Content
BSI BS 5284-76. 1976 Sampling and Testing Mastic Asphalt and Pitchmastic Used in Building. 22 pp.

—Waterproofing
BSI BS 988,1076, 1097,1451-73. 1973 Mastic Asphalt for Building (Limestone Aggregate). 15 pp.
BSI BS 6925-88. 1988 Mastic Asphalt for Building and Civil Engineering (Limestone Aggregate). 12 pp.
BSI BS 6925-01. 1988 Amd 1 Mastic Asphalt for Building and Civil Engineering (Limestone Aggregate) (AMD 7150) July 15, 1992 (R). 12 pp.
CGSB CAN/CGSB-37.68-M87. Mastic Asphalt (Hot Process) for Waterproofing. 13 pp.
DIN ENGL 18195 Pt 2-83. Waterproofing of Buildings and Structures; Materials (Aug). 7 pp.
DIN ENGL 18195 Pt 3-83. Waterproofing of Buildings and Structures; Processing of Materials (Aug). 5 pp.

INTERNATIONAL AND NON-U.S. NATIONAL STANDARDS
SUBJECT INDEX

Asphalt

Asphalt Mastics (Cont.)
—Waterproofing (Cont.)
DIN ENGL 18195 Pt 5-84. Waterproofing of Buildings and Structures; Waterproofing Against Water That Exerts No Hydrostatic Pressure; Design and Workmanship (Feb). 6 pp.
SNZ NZS/BS 6925-88. Specification for Mastic Asphalt for Building and Civil Engineering (Limestone Aggregate). 12 pp.

Asphalt Paints (Made From Asphalt)
Use: Asphalt Coatings (Made From Asphalt)

Asphalt Papers
See Also: Asphalts; Impregnated Papers; Papers
CNS P2025-84. Asphalt Laminating Paper (Jan)(2305). 2 pp.
CNS P2036-72. Asphalt Paper, Double Coated (Tentative Standard) (Jun)(2597). 2 pp.

Asphalt Pavements
Use For: Asphalt Paving Materials; Bituminous Pavements; Flexible Pavements *See Also:* Pavements; Paving Materials
BSI BS 594: Part 1-92. 1992 Hot Rolled Asphalt for Roads and Other Paved Areas Part 1: Specification for Constituent Materials and Asphalt Mixtures (R). 20 pp.
BSI BS 594: Part 1-85. 1985 Amd 1 Hot Rolled Asphalt for Roads and Other Paved Areas Part 1: Constituent Materials and Asphalt Mixtures (AMD 5196) July 31, 1986. 19 pp.
BSI BS 594: Part 2-92. 1992 Hot Rolled Asphalt for Roads and Other Paved Areas Part 2: Specification for the Transport, Laying and Compaction of Rolled Asphalt (R). 11 pp.
BSI BS 594: Part 2-85. 1985 Hot Rolled Asphalt for Roads and Other Paved Areas Part 2: Transport, Laying and Compactiion of Rolled Asphalt. 10 pp.
BSI BS 1446-73. 1973 Mastic Asphalt (Natural Rock Asphalt Fine Aggregate) for Roads and Footways. 12 pp.
BSI BS 1447-88. 1988 Mastic Asphalt (Limestone Fine Aggregate) for Roads, Footways and Pavings in Building. 10 pp.
CGSB CAN/CGSB-16.1-M89. Cutback Asphalts for Road Purposes. 13 pp.
CGSB CAN/CGSB-16.2-M89. Emulsified Asphalts, Anionic Type, for Road Purposes. 12 pp.
CGSB CAN/CGSB-16.4-M89. Emulsified Asphalts, Cationic Type, for Road Purposes. 10 pp.
CGSB CAN/CGSB-16.5-M84. Asphalt, Emulsified, High-Float Type, for Road Purposes. 15 pp.

—Air Voids
CNS A3148-88. Method of Test for Percent Air Voids in Compacted Dense and Open Bituminous Paving Mixtures (Aug)(8756).

—Airports—Static Loads
CNS A3290-88. Method of Test for Nonrepetitive Static Plate Load of Soils and Flexible Pavement Components, for Use in Evaluation and Design of Airport and Highway Pavements (Aug)(12392).
CNS A3291-88. Method of Test for Repetitive Static Plate Load of Soils and Flexible Pavement Components, for Use in Evaluation and Design of Airport and Highway Pavements (Aug)(12393).

—Calcium Silicate
BSI BS 6677: Part 1-86. 1986 Clay and Calcium Silicate Pavers for Flexible Pavements Part 1: Specification for Pavers. 10 pp.
BSI BS 6677: Part 2-86. 1986 Clay and Calcium Silicate Pavers for Flexible Pavements Part 2: Code of Practice for Design of Lightly Trafficked Pavements. 10 pp.
BSI BS 6677: Part 3-86. 1986 Clay and Calcium Silicate Pavers for Flexible Pavements Part 3: Method for Construction of Pavements. 8 pp.

—Clay
BSI BS 6677: Part 1-86. 1986 Clay and Calcium Silicate Pavers for Flexible Pavements Part 1: Specification for Pavers. 10 pp.
BSI BS 6677: Part 2-86. 1986 Clay and Calcium Silicate Pavers for Flexible Pavements Part 2: Code of Practice for Design of Lightly Trafficked Pavements. 10 pp.
BSI BS 6677: Part 3-86. 1986 Clay and Calcium Silicate Pavers for Flexible Pavements Part 3: Method for Construction of Pavements. 8 pp.

—Coal Tar Coatings
CGSB 37-GP-39M-77. Pitch, Emulsified Coal Tar, Mineral Colloid Type Filled, for Coating Bituminous Pavements, Standard for. 12 pp.
CGSB 37-GP-40M-77. Application of Coal Tar Pitch Emulsion as a Bituminous Pavement Coating, Standard for. 7 pp.

Asphalt Pavements (Cont.)
—Compacting
CNS A3288-88. Method of Test for Determining Degree of Pavement Compaction of Bituminous Aggregate Mixtures (Aug)(12390).

—Construction Contracts
DIN ENGL 18317-88. Tendering and Performance Stipulations in Contracts for Construction Works (VOB); Part C: General Technical Specifications in Contracts for Construction Works (ATV); Pavements—Bituminous Surfacings (Sept) (This Standard, Together with DIN 18299,. 6 pp.

—Density
CNS A3150-87. Method of Test for Theoretical Maximum Specific Gravity of Bituminous Paving Mixtures (Aug)(8758).

—Highways—Static Loads
CNS A3290-88. Method of Test for Nonrepetitive Static Plate Load of Soils and Flexible Pavement Components, for Use in Evaluation and Design of Airport and Highway Pavements (Aug)(12392).
CNS A3291-88. Method of Test for Repetitive Static Plate Load of Soils and Flexible Pavement Components, for Use in Evaluation and Design of Airport and Highway Pavements (Aug)(12393).

—Limestone Fillers
CNS A2094-85. Limestone Filler for Bituminous Paving Mixtures (Oct)(6990). 3 pp.
JIS A 5008-76. Limestone Filler for Bituminous Paving Mixtures (R 1984). 7 pp.

—Sampling
CNS A3286-88. Method of Practice for Sampling Bituminous Paving Mixtures (Aug)(12388).

—Thickness Measurement
CNS A3147-87. Method of Test for Thickness or Height of Compacted Bituminous Paving Mixture Specimens (Aug)(8755).

Asphalt Paving Materials
Use: Asphalt Pavements

Asphalt Plants
See Also: Industrial Plants
JIS A 8703-85. Test Code of Asphalt Mixing Plants. 18 pp.

—Aggregate Dryers
BSI BS 2096-54. 1954 Method of Testing Oil-Fired Rotary Dryers for Use in Asphalt and Coated Macadam Plant. 12 pp.

—Batch Type
JIS A 8704-91. Standard Form of Specifications of Batch Type Asphalt Mixing Plant. 19 pp.

Asphalt Primers
See Also: Asphalt Coatings (Made From Asphalt)

—Concrete
CGSB 37-GP-9MA-83. Primer, Asphalt, Unfilled, for Asphalt Roofing, Dampproofing and Waterproofing, Standard for. 8 pp.
CGSB 37-GP-15M-76. Application of Asphalt Primer for Asphalt Roofing, Dampproofing and Waterproofing, Standard for (R 1984). 7 pp.

—Masonry
CGSB 37-GP-9MA-83. Primer, Asphalt, Unfilled, for Asphalt Roofing, Dampproofing and Waterproofing, Standard for. 8 pp.
CGSB 37-GP-15M-76. Application of Asphalt Primer for Asphalt Roofing, Dampproofing and Waterproofing, Standard for (R 1984). 7 pp.

Asphalt Roofing
See Also: Roofing; Shingles
CGSB CAN/CGSB-37.31-M89. Application of Emulsified Asphalt for Asphalt Roofing. 9 pp.
SNZ NZS 4408-88. Specification for Asphalt Roofing Shingles Made from Glass Felt and Surfaced with Mineral Granules Amend: A, 1988. 3 pp.

—Asphalt/Aluminum Coatings
CGSB CAN/CGSB-37.42-M89. Aluminum Cutback Asphalt Roof Coatings. 12 pp.

—Asphalt Coatings
CGSB CAN/CGSB-37.8-M88. Asphalt, Cutback, Filled, for Roof Coatings. 9 pp.

—Built Up—Felts
CSA A123.17-1963. Asphalt-Saturated Felted Glass-Fibre Mat for Use in Construction of Built-Up Roofs (R 1992); (Rev 1 November 1964). 8 pp.

Asphalt Roofing (Cont.)
—Felts
CSA A123.3-M1979. Asphalt or Tar Saturated Roofing Felt (R 1992); (Amd 1-3 November 1985). 13 pp.
CSA CAN/CSA-A123.5-M90. Asphalt Shingles Made from Glass Felt and Surfaced with Mineral Granules; (Gen Instr 1). 18 pp.
CSA CAN/CSA-A123.16-M88. Asphalt-Coated Glass-Base Sheet; (Gen Instr 1). 17 pp.
JIS A 6005-91. Asphalt Roofing Felts. 12 pp.
JIS A 6022-91. Stretchy Asphalt Roofing Felts (Synthetic Fiber Base). 12 pp.
JIS A 6023-91. Perforated Asphalt Roofing Felts. 11 pp.

—Lap Cements
CGSB CAN/CGSB-37.4-M89. Fibrated, Cutback Asphalt, Lap Cement for Asphalt Roofing. 9 pp.
CGSB 37-GP-10MA-84. Application of Asphalt Lap Cement, Standard for. 6 pp.

—Primers (Asphalt)
CGSB 37-GP-9MA-83. Primer, Asphalt, Unfilled, for Asphalt Roofing, Dampproofing and Waterproofing, Standard for. 8 pp.
CGSB 37-GP-15M-76. Application of Asphalt Primer for Asphalt Roofing, Dampproofing and Waterproofing, Standard for (R 1984). 7 pp.

—Sheets
CSA A123.2-M1979. Asphalt Coated Roofing Sheets (R 1992). 12 pp.
CSA CAN/CSA-A123.16-M88. Asphalt-Coated Glass-Base Sheet; (Gen Instr 1). 17 pp.

—Shingles
CSA A123.1-M1979. Asphalt Shingles Surfaced with Mineral Granules (R 1992); (Amd 1 May 1985). 14 pp.
CSA CAN/CSA-A123.5-M90. Asphalt Shingles Made from Glass Felt and Surfaced with Mineral Granules; (Gen Instr 1). 18 pp.
CSA CAN3-A123.51-M85. Asphalt Shingle Application on Roof Slopes 1:3 and Steeper (R 1992); (Gen Instr 1 Thru 2). 26 pp.
CSA CAN3-A123.52-M85. Asphalt Shingle Application on Roof Slopes 1:6 to Less Than 1:3 (R 1992); (Gen Instr 1 Thru 2). 27 pp.

—Woven Fabric
JIS A 6012-77. Woven Fabrics Asphalt Roofings.

Asphalt Sealants
See Also: Asphalts; Sealants

—Adhesion Testing
DIN ENGL 1996 Pt 19-84. Testing of Asphalt; Determination of Extensibility and Adhesion Using a Rabe Joint Model (May). 6 pp.

—Extensibility
DIN ENGL 1996 Pt 19-84. Testing of Asphalt; Determination of Extensibility and Adhesion Using a Rabe Joint Model (May). 6 pp.

Asphalt Shingles
Use: Shingles—Asphalt

Asphaltene Content Analysis
—Bitumens
BSI BS 2000: Part 143-93. 1993 Methods of Test for Petroleum and Its Products Part 143: Determination of Asphaltenes (Heptane Insolubles) (W). 6 pp.
BSI BS 2000: Part 143-85. 1985 Petroleum and Its Products Part 143: Asphaltenes in Petroleum Products (Precipitation with Normal Heptane). 8 pp.

—Crude Oil
BSI BS 2000: Part 143-93. 1993 Methods of Test for Petroleum and Its Products Part 143: Determination of Asphaltenes (Heptane Insolubles) (W). 6 pp.
BSI BS 2000: Part 143-85. 1985 Petroleum and Its Products Part 143: Asphaltenes in Petroleum Products (Precipitation with Normal Heptane). 8 pp.

—Diesel Fuels
BSI BS 2000: Part 143-93. 1993 Methods of Test for Petroleum and Its Products Part 143: Determination of Asphaltenes (Heptane Insolubles) (W). 6 pp.
BSI BS 2000: Part 143-85. 1985 Petroleum and Its Products Part 143: Asphaltenes in Petroleum Products (Precipitation with Normal Heptane). 8 pp.

—Fuel Oils
BSI BS 2000: Part 143-93. 1993 Methods of Test for Petroleum and Its Products Part 143: Determination of Asphaltenes (Heptane Insolubles) (W). 6 pp.
BSI BS 2000: Part 143-85. 1985 Petroleum and Its Products Part 143: Asphaltenes in Petroleum Products (Precipitation with Normal Heptane). 8 pp.

Asphaltene Content Analysis (Cont.)
—Gas Oils
BSI BS 2000: Part 143-93. 1993 Methods of Test for Petroleum and Its Products Part 143: Determination of Asphaltenes (Heptane Insolubles) (W). 6 pp.
BSI BS 2000: Part 143-85. 1985 Petroleum and Its Products Part 143: Asphaltenes in Petroleum Products (Precipitation with Normal Heptane). 8 pp.

—Lubricating Oils
BSI BS 2000: Part 143-93. 1993 Methods of Test for Petroleum and Its Products Part 143: Determination of Asphaltenes (Heptane Insolubles) (W). 6 pp.
BSI BS 2000: Part 143-85. 1985 Petroleum and Its Products Part 143: Asphaltenes in Petroleum Products (Precipitation with Normal Heptane). 8 pp.

Asphalts
Scope Note: For additional listings, use a more specific term *See Also:* Asphalt Coatings (Made From Asphalt); Asphalt Mastics; Asphalt Papers; Asphalt Primers; Asphalt Roofing; Asphalt Sealants; Emulsified Asphalts; Petroleum Asphalts; Petroleum Products; Pitch (Material); Tars
CNS K5030-83. Asphalt (Jul)(2260). 2 pp.
MOD UK CS 1540B. Asphaltic Bitumen, LF Quality (Withdrawn).

—Ash Content
CNS K6205-65. Method of Test for Ash of Asphalt and Coal Tar (Sep)(2487)(R 1973).

—Benzene Content
CNS K6206-65. Determination of Benzene-Insoluble Substance in Asphalt and Coal Tar (Sep)(2488)(R 1973).

—Binder Content
DIN ENGL 1996 Pt 6-88. Testing of Asphalt; Determination of Binder Content and Recovery of Binder (Oct). 11 pp.

—Bulk Density
DIN ENGL 1996 Pt 7-83. Testing of Bituminous Materials for Road Building and Related Purposes; Determination of Density and Voids (Jan). 8 pp.

—Calorific Value
CNS K6209-65. Determination of Calorific Value of Asphalt (Sep)(2491)(R 1971).

—Carbon Content
CNS K6207-65. Determination of Fixed Carbon in Asphalt and Coal Tar (Sep)(2489)(R 1971).

—Compactibility
DIN ENGL 1996 Pt 7-83. Testing of Bituminous Materials for Road Building and Related Purposes; Determination of Density and Voids (Jan). 8 pp.

—Indentation
DIN ENGL 1996 Pt 13-84. Testing of Asphalt; Indentation Testing Using a Flat-Ended Indentor Pin (July). 7 pp.

—Insoluble Matter Content
CNS K6206-65. Determination of Benzene-Insoluble Substance in Asphalt and Coal Tar (Sep)(2488)(R 1973).

—Loss on Heating
CNS K6758-83. Method of Test for Loss on Heating of Oil and Asphaltic Compounds (Mar)(10093).

—Moisture Content
CNS K6208-65. Determination of Moisture Contents in Asphalt, Coal Tar and Phenols (Sep)(2490)(R 1971).

—Rolled
DIN ENGL 1996 Pt 20-84. Testing of Asphalt; Laboratory Preparation of Bituminous Mixtures; Rolled Asphalt Mixture (Nov). 3 pp.

—Rolled—Binder Content
BSI BS 598: Part 107-90. 1990 Sampling and Examination of Bituminous Mixtures for Roads and Other Paved Areas Part 107: Method of Test for the Determination of the Composition of Design Wearing Course Rolled Asphalt (Supersedes BS 598: Part 3: 1985). 20 pp.

—Softening Points
CNS K6203-65. Determination of Softening Point for Asphalt (Mercury Method) (Sep)(2485)(R 1973).

—Softening Points—Ring and Ball Testing
CNS K6204-65. Determination of Softening Point for Asphalt (Ring and Ball Method) (Sep)(2486)(R 1973).

—Solubility
CNS K6757-83. Method of Test for Solubility of Asphalt Materials in Trichloroethylene (Mar)(10092).

Asphalts (Cont.)
—Staining
CGSB 1-GP-71 METH 114.3-79. Methods of Testing Paints and Pigments Cleansability Asphalt Stain Test. 1 p.

—Sulfur Content
CNS K6210-65. Determination of Total Sulfur in Asphalt (Sep)(2492)(R 1973).

—Test Specimens
DIN ENGL 1996 Pt 4-84. Testing of Asphalt; Moulding of Specimens from Bituminous Mixtures (Nov). 8 pp.
DIN ENGL 1996 Pt 18-89. Testing of Asphalt; Herrmann Falling Ball Test (Jan). 3 pp.

—Voids
DIN ENGL 1996 Pt 7-83. Testing of Bituminous Materials for Road Building and Related Purposes; Determination of Density and Voids (Jan). 8 pp.

ASROC
Use: Antisubmarine Rockets

Assaying
See Also: Chemical Analysis; Cupellation; Mineral Deposits; Volumetric Analysis

—Maleic Anhydride—Volumetric Analysis
CNS K6568-80. Determination of the Assay of Maleic Anhydride for Industrial Use (Titrimetric Method) (Aug)(6213).

—Potassium Hydroxide
ISO 990-73. Potassium Hydroxide for Industrial Use—Method of Assay First Edition. 4 pp.

—Sodium Hydroxide
BSI BS 6075: Part 1-81. 1981 Sampling and Test for Sodium Hydroxide for Industrial Use Part 1: Determination of Sodium Hydroxide Content. 4 pp.
ISO 979-74. Sodium Hydroxide for Industrial Use—Method of Assay First Edition. 4 pp.

Assembly Languages
See Also: Programming Languages

—Microprocessors
OSI ISO/IEC DIS 11198-91. Information Technology—Assembly Language Mnemonics for Microprocessors. 44 pp.

Assembly Machinery
Scope Note: Use a more specific term
See: Industrial Equipment; Machinery; Stitching Machinery

Assembly Tools
See Also: Setting Punches; Tools

—Fasteners—Glossaries
CNS B1096-78. Assembly Tools for Screws and Nut Nomenclature (Oct)(4585).
CNS B3304-79. Assembly Tools for Bolts and Screws Drive Ends for Power Tools (Oct)(5003).
ISO 1703-83. Assembly Tools for Screws and Nuts—Nomenclature Second Edition. 37 pp.

Assignment of Frequencies
Use: Frequency Assignment

Association Europeene des Constructeurs de Materiel Aerospatial
Use: AECMA

Association for the Promotion of Electrotechnical Standardization
Use: FEN

Association pour Promotion de la Normalisation Electrotechnique
Use: FEN

Astronautics
Use For: Space Technology

—Glossaries
BSI BS 185: Sec 1-69. 1969 Amd 1 Glossary of Aeronautical and Astronautical Terms Section 1: General (Including Main Types of Aircraft). 7 pp.
BSI BS 185: Sec 13-72. 1972 Glossary of Aeronautical and Astronautical Terms Section 13: Air-Traffic and Ground Services (General, Air-Traffic Control, Ground Services, Naval Terms). 12 pp.
BSI BS 185: Sec 18-70. 1970 Glossary of Aeronautical and Astronautical Terms Section 18: Astronautics (Vehicles, Components, Launching, Re-Entry and Orbits). 5 pp.

Astronautics (Cont.)
—Indexes
BSI BS 185 Index-73. Obsolescent 1973 Glossary of Aeronautical and Astronautical Terms: Index. 38 pp.

Astronomical Telescopes
See Also: Optical Equipment
JIS B 7155-88. Astronomical Telescopes. 20 pp.

Astronomy
See Also: Extraterrestrial Intelligence; Radar Astronomy; Radio Astronomy

—Classification
BSI BS 1000: (52)-77. 1977 Universal Decimal Classification (UDC). English Full Edition (52): Astronomy. Astrophysics. Space Research. Geodesy. 47 pp.
SNZ NZS/BS 1000 (52)-77. Universal Decimal Classification Astronomy. Astrophysics. Space Research. Geodesy. 48 pp.

Asymmetric Bars
Use: Uneven Bars (Gymnastic Equipment)

Asynchronous Communications Interfaces
See Also: Communication Interfaces
BSI BS 6514-84. (WITHDRAWN) 1984 Implementation of V24 or RS232 as an Asynchronous Local Interface. 4 pp.

—Numerical Control—Industrial Equipment
ISO 8867 Pt 1-88. Industrial Asynchronous Data Link and Physical Layer—Part 1: Physical Interconnection and Two-Way Alternate Communication First Edition. 17 pp.
JTC1 8867 Pt 1-88. Industrial Asynchronous Data Link and Physical Layer—Part 1: Physical Interconnection and Two-Way Alternate Communication First Edition. 17 pp.
OSI ISO 8867-1-88. Industrial Asynchronous Data Link and Physical Layer—Part 1: Physical Interconnection and Two-Way Alternate Communication. 17 pp.
OSI ISO DIS 8867-2-88. Industrial Asynchronous Data Link and Physical Layer—Part 2: Two-Way Simultaneous Communication. 16 pp.

Asynchronous Time Division Multiplexers
Use: Time Division Multiplexers—Asynchronous

Asynchronous to Synchronous Converters
See Also: Converters

—Data Circuit Terminating Equipment—Error Correction
CCITT RECMN V.42-89. Error-Correcting Procedures for DCEs Using Asynchronous-to-Synchronous Conversion—Data Communication over the Telephone Network (Study Group XVII) 75 pp. 75 pp.

Asynchronous Transfer Mode
See Also: ATM Adaptation Layer (OSI); ATM Layer (OSI); Telecommunication Services

—B-ISDN
CCITT RECMN I.150-91. B-ISDN Asynchronous Transfer Mode Functional Characteristics (Study Group XVIII) 12 pp. 12 pp.
CENELEC PRETS 300 298-1-93. Network Aspects (NA); Basic Characteristics and Functional Specification of Asynchronous Transfer Mode (ATM) Part 1: B-ISDN ATM Functional Specification. 15 pp.
CENELEC PRETS 300 298-2-93. Network Aspects (NA); Basic Characteristics and Functional Specification of Asynchronous Transfer Mode (ATM) Part 2: B-ISDN ATM Layer Specification. 16 pp.
ETSI PRETS 300 298-1-93. Network Aspects (NA); Basic Characteristics and Functional Specification of Asynchronous Transfer Mode (ATM) Part 1: B-ISDN ATM Functional Specification. 15 pp.
ETSI PRETS 300 298-2-93. Network Aspects (NA); Basic Characteristics and Functional Specification of Asynchronous Transfer Mode (ATM) Part 2: B-ISDN ATM Layer Specification. 16 pp.

—B-ISDN—Television Transmission
CCIR Report 1240-90. Television Transmission in an ATM-Based Network—Section CMTT A—Television Transmission Standards and Performance Objectives. 5 pp.

Asynchronous Transfer Mode (Cont.)
—B-ISDN—Television Transmission (Cont.)
CCIR QUESTION 38/CMTT-90. Service Requirements for Long-Distance Digital Television Connections on the Integrated Services Digital Network (ISDN)—Questions Concerning the CMTT CCIR/CCITT Joint Study Group for Television and Sound Transmission. 1 p.

Athletic Fields
Use For: Baseball Fields; Hockey Fields; Soccer Fields; Sports Grounds; Synthetic Sporting Surfaces
See Also: Recreational Facilities; Sports Facilities
BSI BS 7044: Sec 2.5-91. 1991 Artificial Sports Surfaces Part 2: Methods of Test Section 2.5: Miscellaneous. 14 pp.
BSI BS 7044: Part 4-91. 1991 Artificial Sports Surfaces Part 4: Surfaces for Multi-Sports Use (L). 13 pp.
BSI BS 7044: Part 4-01. 1991 Amd 1 Artificial Sports Surfaces Part 4: Specification for Surfaces for Mulit-Sports Use (AMD 7426) February 15, 1993. 13 pp.
DIN ENGL 18035 Pt 6-78. Sports Grounds; Synthetic Surfacings; Requirements, Test, Maintenance (Apr). 28 pp.

—Abrasion Resistance
BSI BS 7044: Sec 2.3-90. 1990 Artificial Sports Surfaces Part 2: Methods of Test Section 2.3: Methods of Determination of Durability. 12 pp.

—Classification
BSI BS 7044: Part 1-90. 1990 Artificial Sports Surfaces Part 1: Classification and General Introduction. 16 pp.

—Drainage
DIN ENGL 18035 Pt 3-78. Sports Grounds; Drainage (Dec). 8 pp.

—Environmental Testing
BSI BS 7044: Sec 2.4-89. 1989 Artificial Sports Surfaces Part 2: Methods of Test Section 2.4: Methods for Determination of Environmental Resistance. 8 pp.

—Lighting
JIS Z 9120-88. Lighting for Outdoor Sports (Tennis Courts, Baseball Fields). 15 pp.
JIS Z 9121-89. Lighting for Outdoor Sports (Track and Field, Soccer Field, Rugby Field). 14 pp.

—Polyurethane Resins
CNS K3046-85. Polyurethane (PU) Athletic Installation Material (Feb)(6482). 2 pp.

—Polyurethane Resins—Physical Properties
CNS K6591-86. Method of Test for Polyurethane (PU) Athletic Installation Material (May)(6483). 6 pp.

—Resilience
BSI BS 7044: Sec 2.1-89. 1989 Artificial Sports Surfaces Part 2: Methods of Test Section 2.1: Methods of Determination of Ball/Surface Interaction. 11 pp.

—Sprinklers
DIN ENGL 18035 Pt 2-79. Sports Grounds; Watering of Turfed Areas and Tamped Areas (Jan). 4 pp.

—Traction
BSI BS 7044: Sec 2.2-90. 1990 Artificial Sports Surfaces Part 2: Methods of Test Section 2.2: Methods for Determination of Person/Surface Interaction. 10 pp.

Athletic Shoes
Use For: Sport Shoes See Also: Shoes (Footwear)
CNS S2009-82. Method of Test for Sport Shoes (Oct)(742).

—Fabric
CNS S1136-82. Textile Fabrics Sport Shoes (Oct)(8632). 2 pp.

—Leather
CNS S1138-82. Leather Sport Shoes (Oct)(8634). 2 pp.

—Plastic
CNS S1137-82. Plastic Sport Shoes (Oct)(8633). 2 pp.

—Rubber
CNS S1139-82. Rubber Sport Shoes (Oct)(8635). 3 pp.

ATIC
Use: Time Assignment Systems

ATIS
Use: Automatic Transmitter Identification Systems

ATL 98 Aircraft
See Also: Aircraft

ATL 98 Aircraft (Cont.)
—Antenna Positions
CAA. ATL 98 (Approved Aerial Positions). 1 p.

ATLAS
Use For: Abbreviated Test Language for Avionic Systems
MOD UK DSTAN 00-14-87. Use of ATLAS in the Defence Industry Issue 2. 26 pp.
NATO STANAG 3957 ED 2 AMD 1-91. C/Atlas Hol for Application in Avionics Test Systems. 7 pp.

Atlases
—Ground Conductivity
CCIR Report 717-3-90. World Atlas of Ground Conductivities (Published Separately)—Section 5B—Effects of the Ground (Including Ground-Wave Propagation). 9 pp.
CCIR RECMN 832-92. World Atlas of Ground Conductivities. 48 pp.

ATM (Advanced Testing Methods)
Use: Advanced Testing Methods (OSI)

ATM Adaptation Layer (OSI)
Use For: AAL (OSI) See Also: Asynchronous Transfer Mode; ATM Layer (OSI); Open Systems Interconnection

—B-ISDN
CCITT RECMN I.362-91. B-ISDN ATM Adaptation Layer (AAL) Functional Description (Study Group XVIII) 7 pp. 7 pp.
CCITT RECMN I.363-91. B-ISDN ATM Adaptation Layer (AAL) Specification (Study Group XVIII) 21 pp. 21 pp.

—Information Interchange
CCITT RECMN I.363-91. B-ISDN ATM Adaptation Layer (AAL) Specification (Study Group XVIII) 21 pp. 21 pp.

ATM Layer (OSI)
See Also: Asynchronous Transfer Mode; ATM Adaptation Layer (OSI); Open Systems Interconnection

—B-ISDN
CCITT RECMN I.361-91. B-ISDN ATM Layer Specification (Study Group XVIII) 10 pp. 10 pp.

—B-ISDN—Maintenance
CCITT RECMN I.610-91. OAM Principles of the B-ISDN Access (Study Group XVIII) 18 pp. 18 pp.

ATME
Use: Automatic Transmission Measuring and Signaling Testing Equipment

Atmospheres (Standard)
Use: Standard Atmospheres

Atmospheric Attenuation
See Also: Acoustic Absorption; Attenuation; Rain Attenuation

—Frequency Bands—Radio Spectra
CCIR Report 1100-90. Impact of Atmospheric Attenuation on Spectrum Management Above 20 GHz—Section 1D—Spectrum Utilization and Applications. 6 pp.

—Infrared Radiation
CCIR Report 883-2-90. Attenuation of Visible and Infra-Red Radiation—Section 5C—Effects of the Atmosphere (Radiometeorology). 8 pp.

—Light (Visible Radiation)
CCIR Report 883-2-90. Attenuation of Visible and Infra-Red Radiation—Section 5C—Effects of the Atmosphere (Radiometeorology). 8 pp.

—Radio Wave Propagation
CCIR RECMN 676-90. Attenuation by Atmospheric Gases—Section 5C—Effects of the Atmosphere (Radiometeorology). 1 p.
CCIR Report 719-3-90. Attenuation by Atmospheric Gases—Section 5C—Effects of the Atmosphere (Radiometeorology). 16 pp.
CCIR Report 721-3-90. Attenuation by Hydrometeors, in Particular Precipitation, and Other Atmospheric Particles—Section 5C—Effects of the Atmosphere (Radiometeorology). 20 pp.
CCIR RECMN 676-1-92. Attenuation by Atmospheric Gases in the Frequency Range 1-350 GHz. 11 pp.

Atmospheric Attenuation (Cont.)
—Radio Wave Propagation—Radio Spectra
CCIR QUESTION 79/1-90. Propagation Models for Spectrum Management and Planning Above 20 GHz—Questions Concerning Study Group 1—Spectrum Management Techniques (Spectrum Engineering, Planning, Sharing, Monitoring and Utilization). 1 p.

—Standard Atmospheres
CCIR RECMN 835-92. Reference Standard Atmosphere for Gaseous Attenuation. 3 pp.

Atmospheric Corrosion Testing
See Also: Corrodkote Testing; Corrosion Testing; Humidity; Immersion Testing; Salt Spray Testing
DIN ENGL 50018-88. Testing in a Saturated Atmosphere in the Presence of Sulfur Dioxide (June). 4 pp.
ISO 7384-86. Corrosion Tests in Artificial Atmosphere—General Requirements First Edition. 6 pp.

—Accelerated
IEC 355-71. Appraisal of the Problems of Accelerated Testing for Atmospheric Corrosion First Edition. 21 pp.

—Alloys
BSI BS 6917-87. 1987 Method for Corrosion Testing in Artificial Atmospheres: General Principles. 7 pp.
ISO 8565-92. Metals and Alloys—Atmospheric Corrosion Testing—General Requirements for Field Tests First Edition. 12 pp.
ISO 9224-92. Corrosion of Metals and Alloys—Corrosivity of Atmospheres—Guiding Values for the Corrosivity Categories First Edition. 7 pp.

—Alloys—Atmosphere Classification
ISO 9223-92. Corrosion of Metals and Alloys—Corrosivity of Atmospheres—Classification First Edition. 18 pp.
ISO 9226-92. Corrosion of Metals and Alloys—Corrosivity of Atmospheres—Determination of Corrosion Rate of Standard Specimens for the Evaluation of Corrosivity First Edition. 8 pp.

—Aluminum
JIS H 0521-68. Testing Method for Atmospheric Corrosion of Aluminium and Aluminium Alloys. 12 pp.

—Aluminum Alloys
JIS H 0521-68. Testing Method for Atmospheric Corrosion of Aluminium and Aluminium Alloys. 12 pp.

—Anodic Coatings
BSI BS 7561-92. 1992 Method for Corrosion Tests in Artificial Atmospheres at Very Low Concentrations of Polluting Gas(es) (ISO 10062: 1991) (V). 14 pp.
ISO 10062-91. Corrosion Tests in Artificial Atmosphere at Very Low Concentrations of Polluting Gas(es) First Edition. 12 pp.

—Coatings
BSI BS 5466: Part 7-82. 1982 Methods for Corrosion Testing of Metallic Coatings Part 7: Guidance on Stationary Outdoor Exposure Corrosion Tests. 12 pp.
BSI BS 6682-86. 1986 Method for Determination of Bimetallic Corrosion in Outdoor Exposure Corrosion Tests. 15 pp.
ISO 7441-84. Corrosion of Metals and Alloys—Determination of Bimetallic Corrosion in Outdoor Exposure Corrosion Tests First Edition. 14 pp.

—Connectors
BSI BS 2011: Part 2.1KD-77. 1977 Basic Environmental Testing Procedures Part 2.1: Tests Part 2.1KD: Test Kd: Hydrogen Sulfide Test for Contacts and Connections. 8 pp.
BSI BS 2011: Part 2.2KD-84. 1984 Basic Environmental Testing Procedures Part 2.2: Guidance Part 2.2KD: Test Kd. Guidance on Test Kd: Hydrogen Sulfide Test for Contacts and Connections. 12 pp.
CENELEC HD 323.2.46 S1-88. Basic Environmental Testing Procedures—Part 2: Tests. Guidance to Test Kd: Hydrogen Sulphide Test for Contacts and Connections. 3 pp.
IEC 68 Pt 2-42-82. Basic Environmental Testing Procedures Part 2: Tests Test Kc: Sulphur Dioxide Test for Contacts and Connections Second Edition. 18 pp.
IEC 68 Pt 2-43-76. Basic Environmental Testing Procedures Part 2: Tests Test Kd: Hydrogen Sulphide Test for Contacts and Connections First Edition. 16 pp.
IEC 68 Pt 2-46-82. Basic Environmental Testing Procedures Part 2: Tests Guidance to Test Kd: Hydrogen Sulphide Test for Contacts and Connections First Edition. 27 pp.

Atmospheric Corrosion Testing (Cont.)

—Connectors (Cont.)
IEC 68 Pt 2-49-83. Basic Environmental Testing Procedures Part 2: Tests Guidance to Test Kc: Sulphur Dioxide Test for Contacts and Connections First Edition. 23 pp.
IEC 68 Pt 2-60 TTD-89. Environmental Testing Part 2: Tests—Test Ke: Corrosion Tests in Artificial Atmosphere at Very Low Concentration of Polluting Gas(es). 25 pp.
SAA AS 1099.2KC-81. Basic Environmental Testing Procedures for Electrotechnology—Part 2: Tests—Part 2Kc: Sulphur Dioxide Test for Contacts and Connections. 9 pp.
SAA AS 1099.2KD-81. Basic Environmental Testing Procedures for Electrotechnology—Part 2: Tests—Part 2Kd: Hydrogen Sulphide Test for Contacts and Connections. 6 pp.
SNZ IEC 68: Part 2-42-82. Basic Environmental Testing Procedures Part 2-42: Test Kc: Sulphur Dioxide Test for Controls and Connections. 14 pp.
SNZ IEC 68: Part 2-43-76. Basic Environmental Testing Procedures Part 2-43: Test Kd: Hydrogen Sulphide Test for Contacts and Connections. 14 pp.
SNZ IEC 68: Part 2-46-82. Basic Environmental Testing Procedures Part 2-46: Guidance to Test Kd: Hydrogen Sulphide Test for Contacts and Connections. 23 pp.
SNZ IEC 68: Part 2-49-83. Basic Environmental Testing Procedures Part 2-49: Guidance to Test Kc: Sulphur Dioxide for Contacts and Connections. 19 pp.

—Conversion Coatings
BSI BS 7561-92. 1992 Method for Corrosion Tests in Artificial Atmospheres at Very Low Concentrations of Polluting Gas(es) (ISO 10062: 1991) (V). 14 pp.
ISO 10062-91. Corrosion Tests in Artificial Atmosphere at Very Low Concentrations of Polluting Gas(es) First Edition. 12 pp.

—Electric Contacts
BSI BS 2011: Part 2.1KD-77. 1977 Basic Environmental Testing Procedures Part 2.1: Tests Part 2.1KD: Test Kd: Hydrogen Sulfide Test for Contacts and Connections. 8 pp.
BSI BS 2011: Part 2.2KD-84. 1984 Basic Environmental Testing Procedures Part 2.2: Guidance Part 2.2KD: Test Kd. Guidance on Test Kd: Hydrogen Sulfide Test for Contacts and Connections. 12 pp.
CENELEC HD 323.2.46 S1-88. Basic Environmental Testing Procedures—Part 2: Tests. Guidance to Test Kd: Hydrogen Sulphide Test for Contacts and Connections. 3 pp.
IEC 68 Pt 2-42-82. Basic Environmental Testing Procedures Part 2: Tests Test Kc: Sulphur Dioxide Test for Contacts and Connections Second Edition. 18 pp.
IEC 68 Pt 2-43-76. Basic Environmental Testing Procedures Part 2: Tests Test Kd: Hydrogen Sulphide Test for Contacts and Connections First Edition. 16 pp.
IEC 68 Pt 2-46-82. Basic Environmental Testing Procedures Part 2: Tests Guidance to Test Kd: Hydrogen Sulphide Test for Contacts and Connections First Edition. 27 pp.
IEC 68 Pt 2-49-83. Basic Environmental Testing Procedures Part 2: Tests Guidance to Test Kc: Sulphur Dioxide Test for Contacts and Connections First Edition. 23 pp.
IEC 68 Pt 2-60 TTD-89. Environmental Testing Part 2: Tests—Test Ke: Corrosion Tests in Artificial Atmosphere at Very Low Concentration of Polluting Gas(es). 25 pp.
SAA AS 1099.2KC-81. Basic Environmental Testing Procedures for Electrotechnology—Part 2: Tests—Part 2Kc: Sulphur Dioxide Test for Contacts and Connections. 9 pp.
SAA AS 1099.2KD-81. Basic Environmental Testing Procedures for Electrotechnology—Part 2: Tests—Part 2Kd: Hydrogen Sulphide Test for Contacts and Connections. 6 pp.
SNZ IEC 68: Part 2-42-82. Basic Environmental Testing Procedures Part 2-42: Test Kc: Sulphur Dioxide Test for Controls and Connections. 14 pp.
SNZ IEC 68: Part 2-43-76. Basic Environmental Testing Procedures Part 2-43: Test Kd: Hydrogen Sulphide Test for Contacts and Connections. 14 pp.
SNZ IEC 68: Part 2-46-82. Basic Environmental Testing Procedures Part 2-46: Guidance to Test Kd: Hydrogen Sulphide Test for Contacts and Connections. 23 pp.
SNZ IEC 68: Part 2-49-83. Basic Environmental Testing Procedures Part 2-49: Guidance to Test Kc: Sulphur Dioxide for Contacts and Connections. 19 pp.

—Electrical Components
IEC 68 Pt 2-60 TTD-89. Environmental Testing Part 2: Tests—Test Ke: Corrosion Tests in Artificial Atmosphere at Very Low Concentration of Polluting Gas(es). 25 pp.
SAA AS 1099.3.7-81. Basic Environmental Testing Procedures for Electrotechnology—Part 3: Background Information—Part 3.7: Section 7—Appraisal of the Problems of Accelerated Testing for Atmospheric Corrosion. 4 pp.

—Lubricating Oils
DIN ENGL 51386 Pt 1-86. Testing of Corrosion Preventive Oils in a Condensation Water Alternating Atmosphere (Mar). 5 pp.

—Metal Coatings (Made From Metal)
ISO 8565-92. Metals and Alloys—Atmospheric Corrosion Testing—General Requirements for Field Tests First Edition. 12 pp.
ISO 9224-92. Corrosion of Metals and Alloys—Corrosivity of Atmospheres—Guiding Values for the Corrosivity Categories First Edition. 7 pp.

—Metals
BSI BS 6917-87. 1987 Method for Corrosion Testing in Artificial Atmospheres: General Principles. 7 pp.
BSI BS 7479-91. 1991 Method for Salt Spray Corrosion Tests in Artificial Atmospheres (ISO 9227: 1990). 15 pp.
BSI BS 7561-92. 1992 Method for Corrosion Tests in Artificial Atmospheres at Very Low Concentrations of Polluting Gas(es) (ISO 10062: 1991) (V). 14 pp.
ISO 8565-92. Metals and Alloys—Atmospheric Corrosion Testing—General Requirements for Field Tests First Edition. 12 pp.
ISO 9224-92. Corrosion of Metals and Alloys—Corrosivity of Atmospheres—Guiding Values for the Corrosivity Categories First Edition. 7 pp.
ISO 9227-90. Corrosion Tests in Artificial Atmospheres—Salt Spray Tests First Edition. 13 pp.
ISO 10062-91. Corrosion Tests in Artificial Atmosphere at Very Low Concentrations of Polluting Gas(es) First Edition. 12 pp.

—Metals—Atmosphere Classification
ISO 9223-92. Corrosion of Metals and Alloys—Corrosivity of Atmospheres—Classification First Edition. 18 pp.
ISO 9226-92. Corrosion of Metals and Alloys—Corrosivity of Atmospheres—Determination of Corrosion Rate of Standard Specimens for the Evaluation of Corrosivity First Edition. 8 pp.

—Organic Coatings
BSI BS 7561-92. 1992 Method for Corrosion Tests in Artificial Atmospheres at Very Low Concentrations of Polluting Gas(es) (ISO 10062: 1991) (V). 14 pp.
ISO 10062-91. Corrosion Tests in Artificial Atmosphere at Very Low Concentrations of Polluting Gas(es) First Edition. 12 pp.

—Plastics
ISO 291-77. Plastics—Standard Atmospheres for Conditioning and Testing First Edition. 4 pp.

—Precious Metals
IEC 68 Pt 2-42-82. Basic Environmental Testing Procedures Part 2: Tests Test Kc: Sulphur Dioxide Test for Contacts and Connections Second Edition. 18 pp.
IEC 68 Pt 2-49-83. Basic Environmental Testing Procedures Part 2: Tests Guidance to Test Kc: Sulphur Dioxide Test for Contacts and Connections First Edition. 23 pp.
SAA AS 1099.2KC-81. Basic Environmental Testing Procedures for Electrotechnology—Part 2: Tests—Part 2Kc: Sulphur Dioxide Test for Contacts and Connections. 9 pp.
SNZ IEC 68: Part 2-42-82. Basic Environmental Testing Procedures Part 2-42: Test Kc: Sulphur Dioxide Test for Controls and Connections. 14 pp.
SNZ IEC 68: Part 2-49-83. Basic Environmental Testing Procedures Part 2-49: Guidance to Test Kc: Sulphur Dioxide for Contacts and Connections. 19 pp.

—Salinity—Deposition Rates
ISO 9225-92. Corrosion of Metals and Alloys—Corrosivity of Atmospheres—Measurement of Pollution First Edition. 14 pp.

—Sulfur Dioxide—Deposition Rates
ISO 9225-92. Corrosion of Metals and Alloys—Corrosivity of Atmospheres—Measurement of Pollution First Edition. 14 pp.

Atmospheric Noise
See Also: Noise (Spurious Signals); Radio Frequency Interference

—Maritime Mobile Services
CCIR Report 1032-1-90. Radio Noise Environment on Board Vessels—Section 8B—Maritime Mobile Service: Telegraphy and Related Subjects. 9 pp.

—Radiotelegraphy—Receivers
CCIR Report 183-3-82. Usable Sensitivity of Radiotelegraphy Receivers in the Presence of Quasi-Impulsive Interference—Section 3Cd—Performance of Digital Transmission Systems. 6 pp.

—Sound Broadcasting—Receivers—Tropical Regions
CCIR Report 303-3-86. Determination of the Effects of Atmospheric Noise on the Grade of Reception in the Tropical Zone—Section 10A-2—Sound Broadcasting in the Tropical Zone. 2 pp.
CCIR QUESTION 96/10-90. Determination of the Effects of Atmospheric Noise on the Grade of Reception in the Tropical Zone—Questions Concerning Study Group 10—Broadcasting Service (Sound). 1 p.

Atmospheric Pressure
Use For: Barometric Pressure
See Also: Environmental Conditions; Flight Levels; Pressure; Pressure Gages; Pressure Measurement

—Medical Electrical Equipment—Safety
CSA CAN/CSA-C22. 2NO601.1-M90. Medical Electrical Equipment, Part 1: General Requirements for Safety; (Gen Instr 1). 240 pp.

Atmospheric Scattering

—Precipitation
CCIR Report 882-2-90. Scattering by Precipitation—Section 5C—Effects of the Atmosphere (Radiometeorology). 5 pp.

—Precipitation—Earth Stations—Radio Wave Propagation
CCIR Report 1010-86. Propagation Data for Bi-Directional Coordination of Earth Stations—Section 5G—Propagation Data Required for the Evaluation of Interference: Space and Terrestrial Systems. 5 pp.

Atmospheric Sounding

—Ionospheric Propagation
CCIR Report 249-7-90. Use of Oblique Sounding for Propagation Analysis and Optimization—Section 6C—Ionospheric Propagation and Operational Forecasting. 6 pp.

—Optimum Working Frequencies
CCIR Report 249-7-90. Use of Oblique Sounding for Propagation Analysis and Optimization—Section 6C—Ionospheric Propagation and Operational Forecasting. 6 pp.

ATN
Use: Aeronautical Telecommunication Network

Atomic Absorption Analysis
See Also: Chemical Analysis; Photometry; Spectrochemical Analysis; Spectrophotometry

—Coal
SAA AS 1038.14.2-85. Methods for the Analysis and Testing of Coal and Coke—Part 14.2: Analysis of Higher Rank Coal Ash, Coke Ash (Acid Digestion—Flame Atomic Absorption Spectrometric Method). 11 pp.

—Coke
SAA AS 1038.14.2-85. Methods for the Analysis and Testing of Coal and Coke—Part 14.2: Analysis of Higher Rank Coal Ash, Coke Ash (Acid Digestion—Flame Atomic Absorption Spectrometric Method). 11 pp.

—Cryolite
CNS M3118-83. Method for Determining Aluminium Content in Natural and Artificial Cryolite (Atomic Absorption Method) (Mar)(10117).

—Ores
CNS M3077-81. Method for Determination of Copper in Ores (Atomic Absorption Method) (Oct)(8045).

—Uranium Hexafluoride
CNS J2062-82. Method for Determining Antimony in UF6 Atomic Absorption Method (Aug)(9280).
CNS J2068-82. Method for Determining Ruthenium in UF6 Atomic Absorption Method (Sept)(9401).
CNS J2078-82. Method for Determining Metallic Impurities in UF6 Method (Dec)(9711).
CNS J2087-83. Method for Determining Chromium Soluble in UF6 Method (Jan)(9886).
CNS J2088-83. Method for Determining Chromium Insoluble in UF6 Method (Jan)(9887).

Atomic Absorption Spectrochemical Analysis
Use: Spectrochemical Analysis

Atomic Absorption Spectrometry

Use For: Electrothermal Atomization Atomic Absorption Spectrometry; Flame Atomic Absorption Spectrometry

ISO 78 Pt 4-83. Layouts for Standards—Part 4: Standard for Atomic Absorption Spectrometric Analysis First Edition. 7 pp.

—**Air**

BSI BS 6069: Sec 3.2-91. 1991 Characterization of Air Quality Part 3: Workplace Atmospheres Section 3.2: Method for the Determination of Particulate Lead and Lead Compounds by Flame Atomic Absorption Spectrometry (ISO 8518: 1990). 15 pp.

ISO 8518-90. Workplace Air—Determination of Particulate Lead and Lead Compounds—Flame Atomic Absorption Spectrometric Method First Edition. 13 pp.

—**Alloy Steels**

CEN EN 10 177-89. Chemical Analysis of Ferrous Materials Determination of Calcium in Steels Flame Atomic Absorption Spectrometric Method. 6 pp.

CEN PREN 10211-91. Chemical Analysis of Ferrous Materials—Determination of Titanium in Steel and Iron—Flame Atomic Absorption Spectrometric Method. 16 pp.

DIN ENGL EN 10177-90. Chemical Analysis of Ferrous Materials; Determination of Calcium in Steels; Flame Atomic Absorption Spectrometric Method (Apr). 7 pp.

ISO 10138-91. Steel and Iron—Determination of Chromium Content—Flame Atomic Absorption Spectrometric Method First Edition. 12 pp.

—**Aluminum**

ISO 4192-81. Aluminium and Aluminium Alloys—Determination of Lead Content—Flame Atomic Absorption Spectrometric Method First Edition. 5 pp.

ISO 4193-81. Aluminium and Aluminium Alloys—Determination of Chromium Content—Flame Atomic Absorption Spectrometric Method First Edition. 5 pp.

ISO 5194-81. Aluminium and Aluminium Alloys—Determination of Zinc Content—Flame Atomic Absorption Spectrometric Method First Edition. 6 pp.

—**Aluminum Alloys**

ISO 4192-81. Aluminium and Aluminium Alloys—Determination of Lead Content—Flame Atomic Absorption Spectrometric Method First Edition. 5 pp.

ISO 4193-81. Aluminium and Aluminium Alloys—Determination of Chromium Content—Flame Atomic Absorption Spectrometric Method First Edition. 5 pp.

ISO 5194-81. Aluminium and Aluminium Alloys—Determination of Zinc Content—Flame Atomic Absorption Spectrometric Method First Edition. 6 pp.

—**Aluminum Fluoride**

BSI BS 4993: Part 10-83. 1983 Methods of Test for Aluminium Fluoride for Industrial Use Parts 10 & 13: Determination for Phosphorus Content. 7 pp.

BSI BS 5050: Part 13-83. 1983 Methods of Test for Cryolite Part 13: Determination of Phosphorus Content. 7 pp.

ISO 6374-81. Cryolite, Natural and Artificial, and Aluminium Fluoride for Industrial Use—Determination of Phosphorus Content—Atomic Absorption Spectrometric Method After Extraction First Edition. 6 pp.

—**Aluminum Oxide**

BSI BS 4140: Part 14-90. 1990 Methods of Test for Aluminium Oxide Part 14: Determination of Calcium Content (W). 6 pp.

BSI BS 4140: Part 14-01. 1990 Amd 1 Methods of Test for Aluminium Oxide Part 14: Determination of Calcium Content (AMD 6893) February 28, 1992 (W). 7 pp.

—**Ammonium Bicarbonate**

ISO 7110-85. Ammonium Bicarbonate (Ammonium Hydrogencarbonate) for Industrial Use (Including Foodstuffs)—Determination of Lead Content—Flame Atomic Absorption Spectrometric Method First Edition. 5 pp.

—**Animal Feed**

BSI BS 5766: Part 6-84. 1984 Methods for Analysis of Animal Feeding Stuffs Part 6: Determination of Calcium by Atomic Absorption Spectrometry. 6 pp.

ISO 6490 Pt 2-83. Animal Feeding Stuffs—Determination of Calcium Content—Part 2: Atomic Absorption Spectrometric Method First Edition. 5 pp.

Atomic Absorption Spectrometry (Cont.)

—**Carbon Steels**

BSI BS 6200: SUBSEC 3.1.4-90. 1990 Sampling and Analysis of Iron, Steel and Other Ferrous Metals Part 3: Methods of Analysis Section 3.1: Determination of Aluminium Subsection 3.1.4: Non-Alloyed Steel: Flame Atomic Absorption Spectrometric Method. 13 pp.

BSI BS 6200: SUB SEC3.1.4-01. 1990 Amd 1 Sampling and Analysis of Iron, Steel and Other Ferrous Metals Part 3: Methods of Analysis Section 3.1: Determination of Aluminium Subsection 3.1.4: Non-Alloyed Steel: Flame Atomic Absorption Spectrometric Method. 18 pp.

CEN PREN 10211-91. Chemical Analysis of Ferrous Materials—Determination of Titanium in Steel and Iron—Flame Atomic Absorption Spectrometric Method. 16 pp.

DIN ENGL EN 29658-92. Determination of Aluminium Content of Steel by Flame Atomic Absorption Spectrometry; (ISO 9658: 1990) (July). 14 pp.

ISO 9658-90. Steel—Determination of Aluminium Content—Flame Atomic Absorption Spectrometric Method First Edition. 14 pp.

—**Cast Iron**

BSI BS 6200: SUB SEC 3.12.3-86. 1986 Sampling and Analysis of Iron, Steel and Other Ferrous Metals Part 3: Methods of Analysis Section 3.12: Determination of Copper Subsection 3.12.3: Steel and Cast Iron: Flame Atomic Absorption Spectrometric Method. 8 pp.

BSI BS 6200: SUB SEC 3.12.3-01. 1986 Amd 1 Sampling and Analysis of Iron, Steel and Other Ferrous Metals Part 3: Methods of Analysis Sec 3.12: Determination of Copper Subsec 3.12.3: Steel and Cast Iron: Flame Atomic Absorption Spectrometric Method (ISO 4943: 1985) (AMD 7071). 13 pp.

CEN EN 24 943-90. Chemical Analysis of Ferrous Metal—Determination of Copper Content—Flame Atomic Absorption Spectrometric Method. 2 pp.

DIN ENGL EN 24943-92. Determination of Copper Content of Steel and Cast Iron by Flame Atomic Absorption Spectrometry; (ISO 4943: 1985) (Oct). 9 pp.

ISO 4940-85. Steel and Cast Iron—Determination of Nickel Content—Flame Atomic Absorption Spectrometric Method First Edition. 9 pp.

ISO 4943-85. Steel and Cast Iron—Determination of Copper Content—Flame Atomic Absorption Spectrometric Method First Edition. 8 pp.

SAA AS 1050.33-88. Methods for the Analysis of Iron and Steel—Part 33: Determination of Nickel Content—Flame Atomic Absorption Spectrometric Method. 4 pp.

—**Ceramics**

CNS R3073-86. Method of Test for Lead and Cadmium Extracted from Glazed Ceramic Surfaces (Apr)(3503). 4 pp.

—**Coal**

SAA AS 1038.10.1-86. Methods for the Analysis and Testing of Coal and Coke—Part 10.1: Determination of Trace Elements—Determination of Eleven Trace Elements in Coal, Coke and Fly-Ash—Flame Atomic Absorption Spectrometric Method (R 1992). 10 pp.

SAA AS 1038.14.1-81. Methods for the Analysis and Testing of Coal and Coke—Part 14.1: Analysis of Coal Ash, Coke Ash and Mineral Matter (Borate Fusion—Flame Atomic Absorption Spectrometric Method). 18 pp.

—**Coke**

BSI BS 6043: Sec 2.3-89. 1989 Methods of Sampling and Test for Carbonaceous Materials Used in Aluminium Manufacture Part 2: Electrode Coke Section 2.3: Determination of Trace Elements by Flame Atomic Absorption Spectrometry. 10 pp.

SAA AS 1038.10.1-86. Methods for the Analysis and Testing of Coal and Coke—Part 10.1: Determination of Trace Elements—Determination of Eleven Trace Elements in Coal, Coke and Fly-Ash—Flame Atomic Absorption Spectrometric Method (R 1992). 10 pp.

SAA AS 1038.14.1-81. Methods for the Analysis and Testing of Coal and Coke—Part 14.1: Analysis of Coal Ash, Coke Ash and Mineral Matter (Borate Fusion—Flame Atomic Absorption Spectrometric Method). 18 pp.

—**Copper**

DIN ENGL 38406 Pt 7-91. German Standard Methods for the Examination of Water, Waste Water and Sludge; Cations (Group E); Determination of Copper by Atomic Absorption Spectrometry (AAS) (E 7) (Sept). 5 pp.

Atomic Absorption Spectrometry (Cont.)

—**Copper** (Cont.)

ISO 4740-85. Copper and Copper Alloys—Determination of Zinc Content—Flame Atomic Absorption Spectrometric Method First Edition. 6 pp.

ISO 4744-84. Copper and Copper Alloys—Determination of Chromium Content—Flame Atomic Absorption Spectrometric Method First Edition. 5 pp.

JIS H 1060-89. Methods for Determination of Cobalt in Copper and Copper Alloys. 12 pp.

—**Copper Alloys**

ISO 4740-85. Copper and Copper Alloys—Determination of Zinc Content—Flame Atomic Absorption Spectrometric Method First Edition. 6 pp.

ISO 4744-84. Copper and Copper Alloys—Determination of Chromium Content—Flame Atomic Absorption Spectrometric Method First Edition. 5 pp.

ISO 4749-84. Copper Alloys—Determination of Lead Content—Flame Atomic Absorption Spectrometric Method First Edition. 6 pp.

ISO 5960-84. Copper Alloys—Determination of Cadmium Content—Flame Atomic Absorption Spectrometric Method First Edition. 6 pp.

JIS H 1060-89. Methods for Determination of Cobalt in Copper and Copper Alloys. 12 pp.

JIS H 1062-89. Methods for Determination of Zinc in Copper Alloys. 16 pp.

JIS H 1063-89. Methods for Determination of Beryllium in Copper Alloys. 12 pp.

SAA AS K209.1-70. Methods for the Analysis of Copper Alloys—Part 1: Lead in Copper Alloys (Atomic Absorption Spectrometric Method). 5 pp.

SAA AS K209 Part 2-71. Methods for the Analysis of Copper Alloys—Part 2: Manganese in Copper Alloys (Atomic Absorption Spectrometric Method). 5 pp.

SAA AS 1515.3-89. Copper Alloys—Part 3: Determination of Silver Content—Flame Atomic Absorption Spectrometric Method. 3 pp.

SAA AS 1515.5-87. Copper Alloys—Part 5: Determination of Cadmium—Flame Atomic Absorption Spectrometric Method. 3 pp.

—**Cryolite**

BSI BS 5050: Part 13-83. 1983 Methods of Test for Cryolite Part 13: Determination of Phosphorus Content. 7 pp.

ISO 6374-81. Cryolite, Natural and Artificial, and Aluminium Fluoride for Industrial Use—Determination of Phosphorus Content—Atomic Absorption Spectrometric Method After Extraction First Edition. 6 pp.

—**Dolomite**

DIN ENGL 52241 Pt 7-86. Analysis of Raw Materials Used in Glass Production; Chemical Analysis of Dolomite Containing Not Less Than 95% of Calcium Magnesium Carbonate; Determination of Manganese Content Calculated as MnO (Nov). 2 pp.

—**Drinking Water**

DIN ENGL 38406 Pt 3-82. German Standard Methods for the Examination of Water, Waste Water and Sludge; Cations (Group E); Determination of Calcium and Magnesium (E3) (Sept). 7 pp.

DIN ENGL 38406 Pt 18-90. German Standard Methods for the Examination of Water, Waste Water and Sludge; Cations (Group E); Determination of Dissolved Silver by Atomic Absorption Spectrometry Using Electrothermal Atomization (E 18) (May). 5 pp.

ISO 9965-93. Water Quality—Determination of Selenium—Atomic Absorption Spectrometric Method (Hydride Technique) First Edition. 8 pp.

—**Ferronickel**

BSI BS 6783: Part 3-86. 1986 Sampling and Analysis of Nickel, Ferronickel and Nickel Alloys Part 3: Method for Determination of Cobalt in Ferronickel by Flame Atomic Absorption Spectrometry. 6 pp.

BSI BS 6783: Part 3-01. 1986 Amd 1 Sampling and Analysis of Nickel, Ferronickel and Nickel Alloys Part 3: Method for Determination of Cobalt in Ferronickel by Flame Atomic Absorption Spectrometry (ISO 7520: 1985) (AMD 6992) February 28, 1992 (V). 10 pp.

CEN EN 27520-91. Ferronickel—Determination of Cobalt Content—Flame Atomic Absorption Spectrometric Method. 3 pp.

DIN ENGL EN 27520-92. Determination of Cobalt Content of Ferronickel; Flame Atomic Absorption Spectrometric Method (ISO 7520: 1985) (Feb). 6 pp.

ISO 7520-85. Ferronickel—Determination of Cobalt Content—Flame Atomic Absorption Spectrometric Method First Edition. 5 pp.

INTERNATIONAL AND NON-U.S. NATIONAL STANDARDS
SUBJECT INDEX

Atomic Absorption Spectrometry (Cont.)

—Ferrosilicon

BSI BS 6200: SUBSEC 3.1.5-85. 1985 Sampling and Analysis of Iron, Steel and Other Ferrous Metals Part 3: Methods of Analysis Section 3.1: Determination of Aluminium Subsection 3.1.5: Ferro-Silicon: Atomic Absorption Spectrometric Method. 6 pp.

ISO 4139-79. Ferrosilicon—Determination of Aluminium Content—Flame Atomic Absorption Spectrometric Method First Edition. 5 pp.

—Fluorite

ISO 9779-93. Metallurgical-Grade Fluorspar—Determination of Lead Content—Solvent Extraction Atomic Absorption Spectrometric Method Second Edition. 9 pp.

—Fluorspar

ISO 9504-93. Metallurgical-Grade Fluorspar—Determination of Antimony Content—Solvent Extraction Atomic Absorption Spectrometric Method Second Edition. 10 pp.

—Fly Ash

SAA AS 1038.10.1-86. Methods for the Analysis and Testing of Coal and Coke—Part 10.1: Determination of Trace Elements—Determination of Eleven Trace Elements in Coal, Coke and Fly-Ash—Flame Atomic Absorption Spectrometric Method (R 1992). 10 pp.

—Fruits

ISO 6561-83. Fruits, Vegetables and Derived Products—Determination of Cadmium Content—Flameless Atomic Absorption Spectrometric Method First Edition. 5 pp.

ISO 6633-84. Fruits, Vegetables and Derived Products—Determination of Lead Content—Flameless Atomic Absorption Spectrometric Method First Edition. 5 pp.

ISO 6636 Pt 2-81. Fruits, Vegetables and Derived Products—Determination of Zinc Content—Part 2: Atomic Absorption Spectrometric Method First Edition. 6 pp.

ISO 6637-84. Fruits, Vegetables and Derived Products—Determination of Mercury Content—Flameless Atomic Absorption Method First Edition. 7 pp.

ISO 9526-90. Fruits, Vegetables and Derived Products—Determination of Iron Content by Flame Atomic Absorption Spectrometry First Edition. 6 pp.

—Fuel Oils

DIN ENGL 51416 Pt 1-86. Testing of Liquid Fuels; Determination of the Aluminium Content of Fuel Oils; Determination by Atomic Absorption Spectrometry (AAS) After Ashing (Apr). 4 pp.

—Gasoline

CNS K6912-87. Method of Test for Lead in Gasoline by Atomic Absorption Spectrometry (Jul)(12013).

—Glass

BSI BS 3473: Sec 4.2-89. 1989 Chemical Resistance of Glass Used in the Pro-duction of Laboratory Glassware: Part 4: Methods for determination of Hydrolytic Resistance of the Interior Surfaces of Glass Containers: Section 4.2:. 13 pp.

BSI BS 3473: Part 5-87. 1987 Chemical Resistance of Glass Used in the Production of Laboratory Glassware Part 5: Method for Determination of Resistance of Glass to Attack by 6 mol/L Hydrochloric Acid at 100 Degrees Celcius. 10 pp.

BSI BS 7709: Part 5-93. 1993 Analysis of Extract Solutions of Glass Part 5: Method for Determination of Iron(III) Oxide by Molecular Absorption Spectrometry and Flame Atomic Absorption Spectrometry (ISO 10136-5: 1993) (V). 13 pp.

ISO 1776-85. Glass—Resistance to Attack by Hydrochloric Acid at 100 Degrees Celsius—Flame Emission or Flame Atomic Absorption Spectrometric Method First Edition. 7 pp.

ISO 4802 Pt 2-88. Glassware—Hydrolytic Resistance of the Interior Surfaces of Glass Containers—Part 2: Determination by Flame Spectrometry and Classification First Edition. 11 pp.

ISO 10136 Pt 3-93. Glass and Glassware—Analysis of Extract Solutions—Part 3: Determination of Calcium Oxide and Magnesium Oxide by Flame Atomic Absorption Spectrometry First Edition. 10 pp.

ISO 10136 Pt 5-93. Glass and Glassware—Analysis of Extract Solutions—Part 5: Determination of Iron(III) Oxide by Molecular Absorption Spectrometry and Flame Atomic Absorption Spectrometry First Edition. 10 pp.

Atomic Absorption Spectrometry (Cont.)

—Glassware

BSI BS 7709: Part 5-93. 1993 Analysis of Extract Solutions of Glass Part 5: Method for Determination of Iron(III) Oxide by Molecular Absorption Spectrometry and Flame Atomic Absorption Spectrometry (ISO 10136-5: 1993) (V). 13 pp.

ISO 10136 Pt 3-93. Glass and Glassware—Analysis of Extract Solutions—Part 3: Determination of Calcium Oxide and Magnesium Oxide by Flame Atomic Absorption Spectrometry First Edition. 10 pp.

ISO 10136 Pt 5-93. Glass and Glassware—Analysis of Extract Solutions—Part 5: Determination of Iron(III) Oxide by Molecular Absorption Spectrometry and Flame Atomic Absorption Spectrometry First Edition. 10 pp.

—Glossaries

ISO 6955-82. Analytical Spectroscopic Methods—Flame Emmission, Atomic Absorption, and Atomic Fluorescence—Vocabulary First Edition. 23 pp.

—Greases

DIN ENGL 51827-89. Determination of Lead Content of Lubricating Greases by Atomic Absorption Spectrometry After Incineration (Jan). 3 pp.

—Ground Water

DIN ENGL 38405 Pt 18-85. German Standard Methods for the Examination of Water, Waste Water and Sludge; Anions (Group D); Determination of Arsenic by Atomic Absorption Spectrometry (D 18) (Sept). 5 pp.

DIN ENGL 38406 Pt 18-90. German Standard Methods for the Examination of Water, Waste Water and Sludge; Cations (Group E); Determination of Dissolved Silver by Atomic Absorption Spectrometry Using Electrothermal Atomization (E 18) (May). 5 pp.

ISO 9965-93. Water Quality—Determination of Selenium—Atomic Absorption Spectrometric Method (Hydride Technique) First Edition. 8 pp.

—Hard Metals

BSI BS 5600: SUB SEC 4.17.1-86. (WITHDRAWN) 1986 Powder Metallurgical Materials and Products Part 4: Methods of Testing and Chemical Analysis of Hardmetals Section 4.17: Chemical Analysis by Flame Atomic Absorption Spectrometry Subsection 4.17.1: General Requirements. 4 pp.

BSI BS 5600: SUB SEC 4.17.2-86. (WITHDRAWN) 1986 Part 4: Methods of Testing and Chemical Analysis of Hardmetals Section 4.17: Chemical Analysis by Flame Atomic Absorption Spectrometry Subsection 4.17.2: De-termination of Calcium, Potassium, Magnesium and Sodium in Contents from 0.001% to 0.02% (m/m). 4 pp.

BSI BS 5600: SUB SEC 4.17.3-86. (WITHDRAWN) 1986 Part 4: Methods of Testing and Chemical Analysis of Hardmetals Section 4.17: Chemical Analysis by Flame Atomic Absorption Spectrometry Subsection 4.17.3: De-termination of Cobalt, Iron, Manganese and Nickel in Contents from 0.01% to 0.05% (m/m). 4 pp.

BSI BS 5600: SUB SEC 4.17.4-86. (WITHDRAWN) 1986 Part 4: Methods of Testing and Chemical Analysis of Hardmetals Section 4.17: Chemical Analysis by Flame Atomic Absorption Spectrometry Subsection 4.17.4: De-termination of Malybdenum, Titanium and Vanadium in Contents From 0.01% to 0.05% m/m. 4 pp.

BSI BS 5600: SUB SEC 4.17.5-86. (WITHDRAWN) 1986 Part 4: Methods of Testing and Chemical Analysis of Hardmetals Section 4.17: Chemical Analysis by Flame Atomic Absorption Spectrometry Sub 4.17.5:Determination of Cobalt, Iron, Manganese, Molybdenum, Nickel Titanium and Vanadium in Cont from 0.5% to 2% m/m. 4 pp.

BSI BS 5600: SUB SEC 4.17.6-86. (WITHDRAWN) 1986 Part 4: Methods of Testing and Chemical Analysis of Hardmetals Section 4.17: Chemical Analysis by Flame Atomic Absorption Spectrometry Subsection 4.17.6: Determination of Chromium in Contents from 0.01% to 2% (m/m). 4 pp.

BSI BS EN 27627-1-93. 1993 Hardmetals—Chemical Analysis by Flame Atomic Absorption Spectrometry Part 1: General Requirements (ISO 7627-1: 1983) (V). 9 pp.

BSI BS EN 27627-2-93. 1993 Hardmetals—Chemical Analysis by Flame Atomic Absorption Spectrometry Part 2: Determination of Calcium, Potassium, Magnesium and Sodium, in Contents from 0.001 to 0.02 % (m/m) (ISO 7627-2: 1983) (V). 10 pp.

BSI BS EN 27627-3-93. 1993 Hardmetals—Chemical Analysis by Flame Atomic Absorption Spectrometry Part 3: Determination of Cobalt, Iron, Manganese and Nickel in Contents from 0.01 to 0.5 % (m/m) (ISO 7627-3: 1983) (V). 10 pp.

Atomic Absorption Spectrometry (Cont.)

—Hard Metals (Cont.)

BSI BS EN 27627-4-93. 1993 Hardmetals—Chemical Analysis by Flame Atomic Absorption Spectrometry Part 4: Determination of Molybdenum, Titanium and Vanadium in Contents from 0.01 to 0.5 % (m/m) (ISO 7627-4: 1983) (V). 10 pp.

BSI BS EN 27627-5-93. 1993 Hardmetals—Chemical Analysis by Flame Atomic Absorption Spectrometry Part 5: Determination of Cobalt, Iron, Manganese, Molybdenum, Nickel, Titanium and Vanadium in Contents from 0.5 to 2 % (m/m) (ISO 7627-5: 1985) (V). 10 pp.

BSI BS EN 27627-6-93. 1993 Hardmetals—Chemical Analysis by Flame Atomic Absorption Spectrometry Part 6: Determination of Chromium in Contents from 0.01 to 2 % (m/m) (ISO 7627-6: 1985) (V). 10 pp.

CEN EN 27627-1-93. Hardmetals—Chemical Analysis by Flame Atomic Absorption Spectrometry—Part 1: General Requirements (ISO 7627-1: 1983). 4 pp.

CEN EN 27627-2-93. Hardmetals—Chemical Analysis by Flame Atomic Absorption Spectrometry—Part 2: Determination of Calcium, Potassium, Magnesium and Sodium in Contents from 0,001 to 0,02 % (m/m) (ISO 7627-2: 1983). 5 pp.

CEN EN 27627-3-93. Hardmetals—Chemical Analysis by Flame Atomic Absorption Spectrometry—Part 3: Determination of Cobalt, Iron, Manganese and Nickel in Contents from 0,01 to 0,5 % (m/m) (ISO 7627-3: 1983). 5 pp.

CEN EN 27627-4-93. Hardmetals—Chemical Analysis by Flame Atomic Absorption Spectrometry—Part 4: Determination of Molybdenum, Titanium and Vanadium in Contents from 0,01 to 0,5 % (m/m) (ISO 7627-4: 1983). 5 pp.

CEN EN 27627-5-93. Hardmetals—Chemical Analysis by Flame Atomic Absorption Spectrometry—Part 5: Determination of Cobalt, Iron, Manganese, Molybdenum, Nickel, Titanium and Vanadium in Contents from 0,5 to 2 % (m/m) (ISO 7627-5: 1983). 5 pp.

CEN EN 27627-6-93. Hardmetals—Chemical Analysis by Flame Atomic Absorption Spectrometry—Part 6: Determination of Chromium in Contents from 0,01 to 2 % (m/m) (ISO 7627-6: 1985). 5 pp.

ISO 7627 Pt 1-83. Hardmetals—Chemical Analysis by Flame Atomic Absorption Spectrometry—Part 1: General Requirements First Edition (CEN EN 27627-1: 1993). 4 pp.

ISO 7627 Pt 2-83. Hardmetals—Chemical Analysis by Flame Atomic Absorption Spectrometry—Part 2: Determination of Calcium, Potassium, Magnesium and Sodium in Contents from 0,001 to 0,02 % (m/m) First Edition (CEN EN 27627-2: 1993). 4 pp.

ISO 7627 Pt 3-83. Hardmetals—Chemical Analysis by Flame Atomic Absorption Spectrometry—Part 3: Determination of Cobalt, Iron, Manganese and Nickel in Contents from 0,01 to 0,5 % (m/m) First Edition (CEN EN 27627-3: 1993). 4 pp.

ISO 7627 Pt 4-83. Hardmetals—Chemical Analysis by Flame Atomic Absorption Spectrometry—Part 4: Determination of Molybdenum, Titanium and Vanadium in Contents from 0,01 to 0,5 % (m/m) First Edition (CEN EN 27627-4: 1993). 4 pp.

ISO 7627 Pt 5-83. Hardmetals—Chemical Analysis by Flame Atomic Absorption Spectrometry—Part 5: Determination of Cobalt, Iron, Manganese, Molybdenum, Nickel, Titanium and Vanadium in Contents from 0,5 to 2 % (m/m) First Edition (CEN EN 27627-5: 1993). 4 pp.

ISO 7627 Pt 6-85. Hardmetals—Chemical Analysis by Flame Atomic Absorption Spectrometry—Part 6: Determination of Chromium in Contents from 0,01 to 2 % (m/m) First Edition (CEN EN 27627-6: 1993). 4 pp.

—Iron

BSI BS 6200: SUB SEC 3.10.2-89. 1989 Sampling and Analysis of Iron, Steel and Other Ferrous Metals Part 3: Methods of Analysis Section 3.10 Determination of Chromium Subsection 3.10.2: Steel and Cast Iron: Flame Atomic Absorption Spectrometric Method. 12 pp.

BSI BS 6200: SUB SEC 3.34.3-90. 1990 Sampling and Analysis of Iron, Steel and Other Ferrous Metals Part 3: Methods of Analysis Section 3.34: Determination of Vanadium Subsection 3. 34.3: Steel and Cast. Iron: Flame Atomic Absorption Spectrometric Method. 13 pp.

BSI BS 6200: Sec 6.1-90. 1990 Sampling and Analysis of Iron, Steel and Other Ferrous Metals: Part 6: Guidelines on Atomic Absorption Spectrometric Techniques: Section 6.1: Recommendations for the Drafting of Standard Methods for the Chemical Anal. of Iron and Steel by Flame Atomic Absorp. 21 pp.

Atomic Absorption Spectrometry (Cont.)

—Iron (Cont.)

BSI BS 6200: Sec 6.2-90. 1990 Sampling and Analysis of Iron, Steel and Other Ferrous Metals: Part 6: Guidelines on Atomic Absorption Spectrometric Techniques: Section 6.2: Recom. for the Appl. of Flame Atomic Absorp. Spectrometry in Standard Methods for the Chemical Anal. of Iron and Steel. 17 pp.

BSI BS EN 10 136-91. 1991 Chemical Analysis of Ferrous Materials. Determination of Nickel in Steel and Irons. Flame Atomic Absorption Spectrometric Method (V). 11 pp.

CEN EN 10 136-89. Chemical Analysis of Ferrous Materials Determination of Nickel in Steels and Irons Flame Atomic Absorption Spectrometric Method. 7 pp.

CEN EN 10 188-89. Chemical Analysis of Ferrous Materials Determination of Chromium in Steels and Irons Flame Atomic Absorption Spectrometic Method. 9 pp.

CEN PREN 10 201-88. Chemical Analysis of Ferrous Materials Determination of Silicon in Steel and Iron Flame Atomic Absorbption Spectrometric Method. 13 pp.

CEN PREN 10211-91. Chemical Analysis of Ferrous Materials—Determination of Titanium in Steel and Iron—Flame Atomic Absorption Spectrometric Method. 16 pp.

DIN ENGL EN 10188-90. Chemical Analysis of Ferrous Materials; Determination of Chromium in Steels and Iron; Flame Atomic Absorption Spectrometric Method (Apr). 10 pp.

ISO 9647-89. Steel and Iron—Determination of Vanadium Content—Flame Atomic Absorption Spectrometric Method First Edition. 11 pp.

ISO 10138-91. Steel and Iron—Determination of Chromium Content—Flame Atomic Absorption Spectrometric Method First Edition. 12 pp.

ISO TR10281-90. Steel and Iron—Determination of Manganese Content—Flame Atomic Absorption Spectrometric Method First Edition. 14 pp.

SAA AS 1050.20-83. Methods for the Analysis of Iron and Steel—Part 20: Determination of Magnesium in Iron and Steel (Flame Atomic Absorption Spectrometric Method) Amdt 1 March 1983 Reconfirmed 1989. 6 pp.

SAA AS 1050.25-86. Methods for the Analysis of Iron and Steel—Part 25: Determination of Lead Content (Flame Atomic Absorption Spectrometric Method). 3 pp.

SAA AS 1050.29-89. Methods for the Analysis of Iron and Steel—Part 29: Determination of Cobalt Content—Flame Atomic Absorption Spectrometric Method. 3 pp.

—Iron Ores

BSI BS 7020: Sec 8.2-93. 1993 Analysis of Iron Ores Part 8: Methods for the Determination of Aluminium Content Section 8.2: Flame Atomic Absorption Spectrometric Method (ISO 4688-1: 1990) (Q). 15 pp.

BSI BS 7020: Sec 9.2-93. 1993 Analysis of Iron Ores Part 9: Methods for the Determination of Manganese Content Section 9.2: Flame Atomic Absorption Spectrometric Method (ISO 9682-1: 1991) (Q). 16 pp.

BSI BS 7020: Sec 10.2-88. 1988 Analysis of Iron Ores Part 10: Methods for the Determination of Copper Content Section 10.2: Flame Atomic Absorption Spectrometric Method. 12 pp.

BSI BS 7020: Part 14-88. 1988 Analysis of Iron Ores Part 14: Method for the Determination of Sodium and/or Potassium Contents: Flame Atomic Absorption Spectrometric Method. 12 pp.

BSI BS 7020: Part 16-93. 1993 Analysis of Iron Ores Part 16: Method for the Determination of Nickel and/or Chromium Contents: Flame Atomic Absorption Spectrometric Method (ISO 9685: 1991) (Q). 16 pp.

BSI BS 7020: Part 17-88. 1988 Analysis of Iron Ores Part 17: Method for the Determination of Lead and/or Zinc Contents: Flame Atomic Absorption Spectrometric Method. 14 pp.

ISO 4688 Pt 1-92. Iron Ores—Determination of Aluminium Content—Part 1: Flame Atomic Absorption Spectrometric Method First Edition. 12 pp.

ISO 4693-86. Iron Ores—Determination of Copper Content—Flame Atomic Absorption Spectrometric Method First Edition. 10 pp.

ISO 6831-86. Iron Ores—Determination of Sodium and/or Potassium Contents—Flame Atomic Absorption Spectrometric Method First Edition. 11 pp.

ISO 8753-87. Iron Ores—Determination of Lead and/or Zinc Content—Flame Atomic Absorption Spectrometric Method First Edition. 12 pp.

ISO 9682 Pt 1-91. Iron Ores—Determination of Manganese Content—Part 1: Flame Atomic Absorption Spectrometric Method First Edition. 14 pp.

ISO 9684-91. Iron Ores—Determination of Vanadium Content—Flame Atomic Absorption Spectrometric Methods First Edition. 14 pp.

ISO 9685-91. Iron Ores—Determination of Nickel and/or Chromium Contents—Flame Atomic Absorption Spectrometric Method First Edition. 14 pp.

ISO 10203-92. Iron Ores—Determination of Calcium Content—Flame Atomic Absorption Spectrometric Method First Edition. 12 pp.

ISO 10204-92. Iron Ores—Determination of Magnesium Content—Flame Atomic Absorption Spectrometric Method First Edition. 12 pp.

—Lead

SAA AS 1671.1-87. Lead and Lead Alloys—Part 1: Determination of Antimony Content—Flame Atomic Absorption Spectrometric Method. 3 pp.

SAA AS 1671.4-88. Lead and Lead Alloys—Part 4: Determination of Tin in Antimonial Lead—Flame Atomic Absorption Spectrometric Method. 3 pp.

—Lead Alloys

SAA AS 1671.1-87. Lead and Lead Alloys—Part 1: Determination of Antimony Content—Flame Atomic Absorption Spectrometric Method. 3 pp.

—Limestone

DIN ENGL 52240 Pt 7-85. Analysis of Raw Materials Used in Glass Production; Chemical Analysis of Limestone Containing Not Less Than 95% of Calcium Carbonate; Determination of Manganese Content Expressed as MnO (Sept). 2 pp.

DIN ENGL 52240 Pt 8-85. Analysis of Raw Materials Used in Glass Production; Chemical Analysis of Limestone Containing Not Less Than 95% of Calcium Carbonate; Determination of Calcium Oxide and Magnesium Oxide (Sept). 4 pp.

—Liquid Fuels

DIN ENGL 51404 Pt 2-85. Testing of Lubricants and Fuels; Determination of Copper Content of Lubricating Oils and Liquid Fuels; Direct Determination by Atomic Absorption Spectrometry (May). 4 pp.

DIN ENGL 51416 Pt 1-86. Testing of Liquid Fuels; Determination of the Aluminium Content of Fuel Oils; Determination by Atomic Absorption Spectrometry (AAS) After Ashing (Apr). 4 pp.

—Lubricating Oils

DIN ENGL 51379 Pt 3-85. Testing of Lubricants; Determination of the Molybdenum Content of Lubricating Oils by Atomic Absorption Spectrometry (AAS) (July). 3 pp.

DIN ENGL 51391 Pt 1-85. Testing of Lubricants; Determination of the Barium, Calcium and Zinc Content of Lubricating Oils; Direct Determination by Atomic Absorption Spectrometry (Dec). 4 pp.

DIN ENGL 51404 Pt 2-85. Testing of Lubricants and Fuels; Determination of Copper Content of Lubricating Oils and Liquid Fuels; Direct Determination by Atomic Absorption Spectrometry (May). 4 pp.

DIN ENGL 51431-86. Testing of Lubricants; Determination of the Magnesium Content of Lubricating Oils; Direct Determination by Atomic Absorption Spectrometry (AAS) (Feb). 4 pp.

—Magnesium Alloys

ISO 4194-81. Magnesium Alloys—Determination of Zinc Content—Flame Atomic Absorption Spectrometric Method First Edition. 6 pp.

JIS H 1341-90. Method for Determination of Calcium in Magnesium Alloys. 6 pp.

—Manganese Ores

ISO 315-84. Manganese Ores and Concentrates—Determination of Nickel Content—Dimethylglyoxime Spectrometric Method and Flame Atomic Absorption Spectrometric Method First Edition. 8 pp.

ISO 5889-83. Manganese Ores and Concentrates—Determination of Aluminium, Copper, Lead and Zinc Contents—Flame Atomic Absorption Spectrometric Method First Edition. 6 pp.

ISO 7953-85. Manganese Ores and Concentrates—Determination of Calcium and Magnesium Contents—Flame Atomic Absorption Spectrometric Method First Edition. 6 pp.

ISO 7969-85. Manganese Ores and Concentrates—Determination of Sodium and Potassium Contents—Flame Atomic Absorption Spectrometric Method First Edition. 7 pp.

ISO 9681-90. Manganese Ores and Concentrates—Determination of Iron Content—Flame Atomic Absorption Spectrometric Method First Edition. 7 pp.

JIS M 8242-89. Methods for Determination of Copper in Manganese Ores. 11 pp.

—Mineral Water

DIN ENGL 38406 Pt 18-90. German Standard Methods for the Examination of Water, Waste Water and Sludge; Determination of Dissolved Silver by Atomic Absorption Spectrometry Using Electrothermal Atomization (E 18) (May). 5 pp.

—Nickel

BSI BS 3727: Part 21-66. 1966 Methods for the Analysis of Nickel for Use in Electronic Tubes and Valves Part 21: Determination of Magnesium (Atomic Absorption Method). 8 pp.

BSI BS 3727: Part 22-66. 1966 Methods for the Analysis of Nickel for Use in Electronic Tubes and Valves Part 22: Determination of Zinc (Atomic Absorption Method). 8 pp.

BSI BS 6783: Part 1-86. 1986 Sampling and Analysis of Nickel, Ferronickel and Nickel Alloys Part 1: Method for Determination of Silver, Bismuth, Cadmium, Cobalt, Copper, Iron, Manganese, Lead and Zinc in Nickel by Flame Atomic Absorption Spectrometry. 15 pp.

BSI BS 6783: Part 4-86. 1986 Sampling and Analysis of Nickel, Ferronickel and Nickel Alloys: Part 4: Method for Determination of Silver, Arsenic, Bismuth, Cadmium, Lead, Antimony, Selenium, Tin, Tellurium and Thallium in Nickel by Electrothermal Atomic Absorption Spectrometry. 11 pp.

DIN ENGL 38406 Pt 11-91. German Standard Methods for the Examination of Water, Waste Water and Sludge; Cations (Group E); Determination of Nickel by Atomic Absorption Spectrometry (AAS) (E 11) (Sept). 5 pp.

ISO 6351-85. Nickel—Determination of Silver, Bismuth, Cadmium Cobalt, Copper, Iron, Manganese, Lead and Zinc Contents—Flame Atomic Absorption Spectrometric Method First Edition. 13 pp.

ISO 7523-85. Nickel—Determination of Silver, Arsenic, Bismuth, Cadmium, Lead, Antimony, Selenium, Tin, Tellurium and Thallium Contents—Electrothermal Atomic Absorption Spectrometric Method First Edition. 10 pp.

—Nickel Alloys

BSI BS 7455: Part 1-91. 1991 Analysis of Nickel Alloys by Flame Atomic Absorption Spectrometry Part 1: General Requirements and Sample Dissolution (ISO 7530-1: 1990). 12 pp.

BSI BS 7455: Part 1-01. 1991 Amd 1 Analysis of Nickel Alloys by Flame Atomic Absorption Spectrometry Part 1: General Requirements and Sample Dissolution (ISO 7530-1: 1990) (AMD 7216) August 15, 1992. 13 pp.

BSI BS 7455: Part 2-91. 1991 Analysis of Nickel Alloys by Flame Atomic Absorption Spectrometry Part 2: Method for the Determination of Cobalt (ISO 7530-2: 1990). 8 pp.

BSI BS 7455: Part 3-91. 1991 Analysis of Nickel Alloys by Flame Atomic Absorption Spectrometry Part 3: Method for the Determination of Chromium (ISO 7530-3: 1990). 8 pp.

BSI BS 7455: Part 5-91. 1991 Analysis of Nickel Alloys by Flame Atomic Absorption Spectrometry Part 5: Method for the Determination of Iron (ISO 7530-5: 1990). 8 pp.

BSI BS 7455: Part 6-91. 1991 Analysis of Nickel Alloys by Flame Atomic Absorption Spectrometry Part 6: Method for the Determination of Manganese (ISO 7530-6: 1990). 8 pp.

BSI BS 7455: Part 7-92. 1992 Analysis of Nickel Alloys by Flame Atomic Absorption Spectrometry Part 7: Method for Determination of Aluminium (ISO 7530-7: 1992). 10 pp.

BSI BS 7455: Part 8-92. 1992 Analysis of Nickel Alloys by Flame Atomic Absorption Spectrometry Part 8: Method for Determination of Silicon (ISO 7530-8: 1992). 8 pp.

ISO 7530 Pt 1-90. Nickel Alloys—Flame Atomic Absorption Spectrometric Analysis—Part 1: General Requirements and Sample Dissolution First Edition; (Corrected and Reprinted -1992). 11 pp.

ISO 7530 Pt 7-92. Nickel Alloys—Flame Atomic Absorption Spectrometric Analysis—Part 7: Determination of Aluminium Content First Edition. 7 pp.

ISO 7530 Pt 8-92. Nickel Alloys—Flame Atomic Absorption Spectrometric Analysis—Part 8: Determination of Silicon Content First Edition. 6 pp.

ISO 7530 Pt 9-93. Nickel Alloys—Flame Atomic Absorption Spectrometric Analysis—Part 9: Determination of Vanadium Content First Edition. 6 pp.

INTERNATIONAL AND NON-U.S. NATIONAL STANDARDS
SUBJECT INDEX

Atomic Absorption Spectrometry (Cont.)

—Paints
BSI BS 3900: Part B3-83. 1983 Group B: Tests Involving Chemical Examination of Liquid Paints and Dried Paint Films: Part B3: Determination of 'Soluble' Lead in Solid Matter in Liquid Paints: Meth for Use in Conj with the Control of Lead at Work Regul, 1980 (S.I. 1980 No. 1248). 8 pp.

BSI BS 3900: Part B4-86. 1986 Methods of Test for Paints Group B: Tests Involving Chemical Examination of Liquid Paints and Dried Paint Films Part B4: Determination of Total Lead in Paints and Similar Materials. 8 pp.

BSI BS 3900: Part B6-86. 1986 Methods of Test for Paints Group B: Tests Involving Chemical Examination of Liquid Paints and Dried Paint Films Part B6: Determination of Content 'Soluble' Lead Content. 9 pp.

BSI BS 3900: Part B7-86. 1986 Methods of Test for Paints Group B: Tests Involving Chemical Examination of Liquid Paints and Dried Paint Films Part B7: Determination of 'Soluble'Antimony Content. 9 pp.

BSI BS 3900: Part B9-86. 1986 Methods of Test for Paints Group B: Tests Involving Chemical Examination of Liquid Paints and Dried Paint Films Part B9: Determination of 'Soluble' Cadmium Content. 9 pp.

BSI BS 3900: Part B11-86. 1986 Methods of Test for Paints Group B: Tests Involving Chemical Examination of Liquid Paints and Dried Paint Films Part B11: Determination of Total Chromium Content of Liquid Matter. 6 pp.

BSI BS 3900: Part B12-86. 1986 Methods of Test for Paints Group B: Tests Involving Chemical Examination of Liquid Paints and Dried Paint Films Part B12: Determination of 'Soluble'Mercury Content. 11 pp.

BSI BS 3900: Part B16-90. 1990 Methods of Test for Paints Group B: Tests Involving Chemical Examination of Liquid Paints and Dried Paint Films Part B16: Determination of Total Mercury. 10 pp.

ISO 3856 Pt 1-84. Paints and Varnishes—Determination of "Soluble" Metal Content—Determination of Lead Content—Flame Atomic Absorption Spectrometric Method and Dithizone Spectrophotometric Method Third Edition. 8 pp.

ISO 3856 Pt 2-84. Paints and Varnishes—Determination of "Soluble" Metal Content—Part 2: Determination of Antimony Content—Flame Atomic Absorption Spectrometric Method and Rhodamine B Spectrophotometric Method Second Edition. 8 pp.

ISO 3856 Pt 4-84. Paints and Varnishes—Determination of "Soluble" Metal Content—Part 4: Determination of Cadmium Content—Flame Atomic Absorption Spectrometric Method and Polarographic Method Second Edition. 8 pp.

ISO 3856 Pt 6-84. Paints and Varnishes—Determination of "Soluble" Metal Content—Part 6: Determination of Total Chromium Content of the Liquid Portion of the Paint—Flame Atomic Absorption Spectrometric Method Second Edition. 5 pp.

ISO 3856 Pt 7-84. Paints and Varnishes—Determination of "Soluble" Metal Content—Part 7: Determination of Mercury Content of the Pigment Portion of the Paint and of the Liquid Portion of Water-Dilutable Paints—Flameless Atomic Absorption Spectrometric Method First Edition. 10 pp.

ISO 6503-84. Paints and Varnishes—Determination of Total Lead—Flame Atomic Absorption Spectrometric Method First Edition. 9 pp.

ISO 7252-84. Paints and Varnishes—Determination of Total Mercury—Flameless Atomic Absorption Spectrometric Method First Edition. 8 pp.

—Petroleum Products
CEN PREN 237-85. Liquid Petroleum Products, Determination of Low Lead Concentrations; Atomic Absorption Spectronmetric Method. 8 pp.

—Phosphates
BSI BS 4427: Part 13-82. 1982 Methods of Test for Sodium Tripolyphosphate (Pentasodium Triphosphate) and Sodium Pyro-phosphate (Tetrasodium Pyrophosphate) for Industrial Use Part 13: Condensed Phosphates for Industrial Use (Including Foodstuffs). 7 pp.

ISO 5373-81. Condensed Phosphates for Industrial Use (Including Foodstuffs)—Determination of Calcium Content—Flame Atomic Absorption Spectrometric Method First Edition. 6 pp.

—Phosphoric Acids
BSI BS 4258: Part 11-82. 1982 Methods of Test for Phosphoric Acid (Orthophosphoric Acid) for Industrial Use Part 11: Determination of Lead Content (Atomic Absorption Spectrometric Method). 4 pp.

ISO 6678-81. Phosphoric Acid for Industrial Use—Determination of Lead Content—Atomic Absorption Spectrometric Method First Edition. 4 pp.

Atomic Absorption Spectrometry (Cont.)

—Pulps
BSI BS 4897: Part 1-83. 1983 Trace Metal Contents of Pulps Part 1: Method for Determination of Calcium Content by Edta Titrimetric and Flame Atomic Absorption Spectrometric Methods. 8 pp.

BSI BS 4897: Part 2-83. 1983 Trace Metal Contents of Pulps Part 2: Method for Determination of Copper Content by Extraction—Photometric and Flame Atomic Absorption Spectrometric Methods. 8 pp.

BSI BS 4897: Part 3-83. 1983 Trace Metal Contents of Pulps Part 3: Method for Determination of Iron Content by 1,10-Phenanthroline Photometric and Flame Atomic Absorption Spectrometric Methods. 8 pp.

BSI BS 4897: Part 4-83. 1983 Trace Metal Contents of Pulps Part 4: Method for Determination of Maganese Content by Sodium Peridate Photometric and Flame Atomic Absorption Spectrometric Methods. 8 pp.

ISO 777-82. Pulps—Determination of Calcium Content—EDTA Titrimetric and Flame Atomic Absorption Spectrometric Methods First Edition. 7 pp.

ISO 778-82. Pulps—Determination of Copper Content—Extraction-Photometric and Flame Atomic Absorption Spectrometric Methods First Edition. 6 pp.

ISO 779-82. Pulps—Determination of Iron Content—1, 10-Phenanthroline Photometric and Flame Atomic Absorption Spectrometric Methods First Edition. 6 pp.

ISO 1830-82. Pulps—Determination of Manganese Content—Sodium Periodate Photometric and Flame Atomic Absorption Spectrometric Methods First Edition. 6 pp.

ISO 9668-90. Pulps—Determination of Magnesium Content—Flame Atomic Absorption Spectrometric Method First Edition. 6 pp.

—Rubber
BSI BS 7164: Sec 27.1-91. 1991 Chemical Tests for Raw and Vulcanized Rubber Part 27: Methods for Determination of Iron Content Section 27.1: Atomic Absorption Spectrometry (ISO 6101-5: 1990). 9 pp.

BSI BS 7164: Sec 29.1-92. 1992 Chemical Tests for Raw and Vulcanized Rubber Part 29: Methods for Determination of Zinc Content Section 29.1: Atomic Absorption Spectrometry (ISO 6101-1: 1991) (V). 12 pp.

BSI BS 7165-91. 1991 Recommendations for Achievement of Quality in Software. 30 pp.

ISO 6101 Pt 1-91. Rubber—Determination of Metal Content by Atomic Absorption Spectrometry—Part 1: Determination of Zinc Content Second Edition. 9 pp.

ISO 6101 Pt 2-86. Rubber—Determination of Metal Content—Flame Atomic Absorption Spectrometric Method—Part 2: Determination of Lead Content First Edition. 8 pp.

ISO 6101 Pt 3-88. Rubber—Determination of Metal Content by Atomic Absorption Spectrometry—Part 3: Determination of Copper Content First Edition. 8 pp.

ISO 6101 Pt 4-88. Rubber—Determination of Metal Content by Atomic Absorption Spectrometry—Part 4: Determination of Manganese Content First Edition. 8 pp.

ISO 6101 Pt 5-90. Rubber—Determination of Metal Content by Atomic Absorption Spectrometry—Part 5: Determination of Iron Content First Edition. 7 pp.

—Sea Water
DIN ENGL 38405 Pt 18-85. German Standard Methods for the Examination of Water, Waste Water and Sludge; Anions (Group D); Determination of Arsenic by Atomic Absorption Spectrometry (D 18) (Sept). 5 pp.

—Silicon
CNS G2137-82. Method of Chemical Analysis for Ferrite in Metallic Silicon (Atomic Absorptimetric Method) (Oct)(9496).

CNS G2140-82. Method of Chemical Analysis for Aluminum in Metallic Silicon (Atomic Absorptimetric Method) (Oct)(9499).

CNS G2143-82. Method of Chemical Analysis for Calcium in Metallic Silicon (Atomic Absorptimetric Method) (Oct)(9502).

—Sodium Borates
BSI BS 5688: Part 28-86. 1986 Methods of Test for Orthoboric Acid (Boric Acid), Diboron Trioxide (Boric Oxide), Disodium Tetraborates, Sodium Perborates and Crude Sodium Borates for Industrial Use Part 28: Determination of Total and AlkaliSoluble Calcium and Magnesium. 7 pp.

Atomic Absorption Spectrometry (Cont.)

—Sodium Borates (Cont.)
ISO 6918-84. Crude Sodium Borates for Industrial Use—Determination of Total and Alkali-Soluble Calcium and Magnesium Contents—Flame Atomic Absorption Spectrometric Method First Edition. 6 pp.

—Sodium Chloride
BSI BS 7319: Part 6-90. 1990 Analysis of Sodium Chloride for Industrial Use Part 6: Method for Determination of Cadmium Content. 7 pp.

BSI BS 7319: Part 8-90. 1990 Analysis of Sodium Chloride for Industrial Use Part 8: Method for Determination of Lead Content. 7 pp.

BSI BS 7319: Part 9-90. 1990 Analysis of Sodium Chloride for Industrial Use Part 9: Method for Determination of Mercury Content. 8 pp.

—Sodium Hydroxide
BSI BS 6075: Part 11-81. 1981 Sampling and Test for Sodium Hydroxide for Industrial Use Part 11: Determination of Mercury Content (Flameless Atomic Absorption Method). 10 pp.

ISO 5993-79. Sodium Hydroxide for Industrial Use—Determination of Mercury Content—Flameless Atomic Absorption Spectrometric Method First Edition. 8 pp.

—Sodium Sulfate
ISO 5994-79. Sodium Sulphate for Industrial Use—Determination of Calcium Content—Flame Atomic Absorption Spectrometric Method First Edition. 6 pp.

—Sorghum
ISO 9648-88. Sorghum—Determination of Tannin Content First Edition. 6 pp.

—Steels
BSI BS 6200: SUBSEC 3.7.1-87. 1987 Sampling and Analysis of Iron, Steel and Other Ferrous Metals Part 3: Methods of Analysis Section 3.7: Determination of Calcium Subsection 3.7.1: Steel: Flame Atomic Absorption Spectrometric Method. 8 pp.

BSI BS 6200: SUBSEC 3.7.1-01. 1987 Amd 1 Sampling and Analysis of Iron, Steel and Other Ferrous Metals Part 3: Methods of Analysis Sec 3.7: Determination of Calcium Subsec 3.7.1: Steel: Flame Atomic Absorption Spectrometric Method (AMD 7069) February 28, 1992 (V). 13 pp.

BSI BS 6200: SUB SEC 3.10.2-89. 1989 Sampling and Analysis of Iron, Steel and Other Ferrous Metals Part 3: Methods of Analysis Section 3.10 Determination of Chromium Subsection 3.10.2: Steel and Cast Iron: Flame Atomic Absorption Spectrometric Method. 12 pp.

BSI BS 6200: SUB SEC 3.12.3-86. 1986 Sampling and Analysis of Iron, Steel and Other Ferrous Metals Part 3: Methods of Analysis Section 3.12: Determination of Copper Subsection 3.12.3: Steel and Cast Iron: Flame Atomic Absorption Spectrometric Method. 8 pp.

BSI BS 6200: SUB SEC 3.12.3-01. 1986 Amd 1 Sampling and Analysis of Iron, Steel and Other Ferrous Metals Part 3: Methods of Analysis Sec 3.12: Determination of Copper Subsec 3.12.3: Steel and Cast Iron: Flame Atomic Absorption Spectrometric Method (ISO 4943: 1985) (AMD 7071). 13 pp.

BSI BS 6200: SUB SEC 3.16.4-87. 1987 Sampling and Analysis of Iron, Steel and Other Ferrous Metals Part 3: Methods of Analysis Section 3.16: Determination of Lead Subsection 3.16.4: Steel: Flame Atomic Absorption Spectrometric Method. 8 pp.

BSI BS 6200: SUB SEC 3.16.4-01. 1987 Amd 1 Sampling and Analysis of Iron, Steel and Other Ferrous Metals Part 3: Methods of Analysis Sec 3.16: Determination of Lead Subsec 3.16.4: Steel: Flame Atomic Absorption Spectrometric Method (AMD 7072) February 28, 1992 (V). 13 pp.

BSI BS 6200: SUB SEC 3.34.3-90. 1990 Sampling and Analysis of Iron, Steel and Other Ferrous Metals Part 3: Methods of Analysis Section 3.34: Determination of Vanadium Subsection 3. 34.3: Steel and Cast. Iron: Flame Atomic Absorption Spectrometric Method. 13 pp.

BSI BS 6200: Sec 6.1-90. 1990 Sampling and Analysis of Iron, Steel and Other Ferrous Metals: Part 6: Guidelines on Atomic Absorption Spectrometric Techniques: Section 6.1: Recommendations for the Drafting of Standard Methods for the Chemical Anal. of Iron and Steel by Flame Atomic Absorp. 21 pp.

BSI BS 6200: Sec 6.2-90. 1990 Sampling and Analysis of Iron, Steel and Other Ferrous Metals: Part 6: Guidelines on Atomic Absorption Spectrometric Techniques: Section 6.2: Recom. for the Appl. of Flame Atomic Absorp. Spectrometry in Standard Methods for the Chemical Anal. of Iron and Steel. 17 pp.

Atomic
INTERNATIONAL AND NON-U.S. NATIONAL STANDARDS
SUBJECT INDEX

Atomic Absorption Spectrometry *(Cont.)*

—Steels *(Cont.)*

BSI BS EN 10 136-91. 1991 Chemical Analysis of Ferrous Materials. Determination of Nickel in Steel and Irons. Flame Atomic Absorption Spectrometric Method (V). 11 pp.
CEN EN 10 136-89. Chemical Analysis of Ferrous Materials Determination of Nickel in Steels and Irons Flame Atomic Absorption Spectrometric Method. 7 pp.
CEN EN 10 177-89. Chemical Analysis of Ferrous Materials Determination of Calcium in Steels Flame Atomic Absorption Spectrometric Method. 6 pp.
CEN EN 10 181-89. Chemical Analysis of Ferrous Materials Determination of Lead in Steels Flame Atomic Absorption Spectrometric Method. 6 pp.
CEN EN 10 188-89. Chemical Analysis of Ferrous Materials Determination of Chromium in Steels and Irons Flame Atomic Absorption Spectrometic Method. 9 pp.
CEN PREN 10 201-88. Chemical Analysis of Ferrous Materials Determination of Silicon in Steel and Iron Flame Atomic Absorbption Spectrometric Method. 13 pp.
CEN EN 24 943-90. Chemical Analysis of Ferrous Metal—Determination of Copper Content—Flame Atomic Absorption Spectrometric Method. 2 pp.
CENELEC PREN 10 206-89. Chemical Analysis of Ferrous Materials: Determination of Manganese in Steel: Flame Atomic Absorption Spectrometric Method. 16 pp.
DIN ENGL EN 10136-90. Chemical Analysis of Ferrous Materials; Determination of Nickel in Steel and Iron; Flame Atomic Absorption Spectrometric Method (Apr). 8 pp.
DIN ENGL EN 10177-90. Chemical Analysis of Ferrous Materials; Determination of Calcium in Steels; Flame Atomic Absorption Spectrometric Method (Apr). 7 pp.
DIN ENGL EN 10181-90. Chemical Analysis of Ferrous Materials; Determination of Lead in Steels; Flame Atomic Absorption Spectrometric Method (Apr). 7 pp.
DIN ENGL EN 10188-90. Chemical Analysis of Ferrous Materials; Determination of Chromium in Steels and Iron; Flame Atomic Absorption Spectrometric Method (Apr). 10 pp.
DIN ENGL EN 24943-92. Determination of Copper Content of Steel and Cast Iron by Flame Atomic Absorption Spectrometry; (ISO 4943: 1985) (Oct). 9 pp.
ISO 4940-85. Steel and Cast Iron—Determination of Nickel Content—Flame Atomic Absorption Spectrometric Method First Edition. 9 pp.
ISO 4943-85. Steel and Cast Iron—Determination of Copper Content—Flame Atomic Absorption Spectrometric Method First Edition. 8 pp.
ISO 9647-89. Steel and Iron—Determination of Vanadium Content—Flame Atomic Absorption Spectrometric Method First Edition. 11 pp.
ISO 10138-91. Steel and Iron—Determination of Chromium Content—Flame Atomic Absorption Spectrometric Method First Edition. 12 pp.
ISO TR10281-90. Steel and Iron—Determination of Manganese Content—Flame Atomic Absorption Spectrometric Method First Edition. 14 pp.
ISO 10697 Pt 1-92. Steel—Determination of Calcium Content by Flame Atomic Absorption Spectrometry—Part 1: Determination of Acid-Soluble Calcium Content First Edition. 13 pp.
SAA AS 1050.20-83. Methods for the Analysis of Iron and Steel—Part 20: Determination of Magnesium in Iron and Steel (Flame Atomic Absorption Spectrometric Method) Amdt 1 March 1983 Reconfirmed 1989. 6 pp.
SAA AS 1050.23-90. Methods for the Analysis of Iron and Steel—Part 23: Determination of Molybdenum Content—Flame Atomic Absorption Spectrometric Method. 4 pp.
SAA AS 1050.25-86. Methods for the Analysis of Iron and Steel—Part 25: Determination of Lead Content (Flame Atomic Absorption Spectrometric Method). 3 pp.
SAA AS 1050.29-89. Methods for the Analysis of Iron and Steel—Part 29: Determination of Cobalt Content—Flame Atomic Absorption Spectrometric Method. 3 pp.
SAA AS 1050.33-88. Methods for the Analysis of Iron and Steel—Part 33: Determination of Nickel Content—Flame Atomic Absorption Spectrometric Method. 4 pp.

—Surface Water

DIN ENGL 38405 Pt 18-85. German Standard Methods for the Examination of Water, Waste Water and Sludge; Anions (Group D); Determination of Arsenic by Atomic Absorption Spectrometry (D 18) (Sept). 5 pp.
ISO 9965-93. Water Quality—Determination of Selenium—Atomic Absorption Spectrometric Method (Hydride Technique) First Edition. 8 pp.

Atomic Absorption Spectrometry *(Cont.)*

—Vegetables

ISO 6561-83. Fruits, Vegetables and Derived Products—Determination of Cadmium Content—Flameless Atomic Absorption Spectrometric Method First Edition. 5 pp.
ISO 6633-84. Fruits, Vegetables and Derived Products—Determination of Lead Content—Flameless Atomic Absorption Spectrometric Method First Edition. 5 pp.
ISO 6636 Pt 2-81. Fruits, Vegetables and Derived Products—Determination of Zinc Content—Part 2: Atomic Absorption Spectrometric Method First Edition. 6 pp.
ISO 6637-84. Fruits, Vegetables and Derived Products—Determination of Mercury Content—Flameless Atomic Absorption Method First Edition. 7 pp.
ISO 9526-90. Fruits, Vegetables and Derived Products—Determination of Iron Content by Flame Atomic Absorption Spectrometry First Edition. 6 pp.

—Vulcanized Rubber

BSI BS 7164: Sec 26.1-90. 1990 Chemical Tests for Raw and Vulcanized Rubber Part 26: Methods for Determination of Manganese Content Section 26.1: Atomic Absorption Spectrometry (ISO 6101-4: 1988). 11 pp.
BSI BS 7164: Sec 28.1-90. 1990 Chemical Tests for Raw and Vulcanized Rubber Part 28: Methods for Determination of Copper Content Section 28.1: Atomic Absorption Spectrometry (ISO 6101-3: 1988). 11 pp.
BSI BS 7164: Sec 30.1-90. 1990 Chemical Tests for Raw and Vulcanized Rubber Part 30: Methods for Determination of Lead Content Section 30.1: Atomic Absorption Spectrometry. 12 pp.

—Waste Water

DIN ENGL 38405 Pt 18-85. German Standard Methods for the Examination of Water, Waste Water and Sludge; Anions (Group D); Determination of Arsenic by Atomic Absorption Spectrometry (D 18) (Sept). 5 pp.
DIN ENGL 38406 Pt 3-82. German Standard Methods for the Examination of Water, Waste Water and Sludge; Cations (Group E); Determination of Calcium and Magnesium (E3) (Sept). 7 pp.
DIN ENGL 38406 Pt 6-81. German Standard Methods for the Analysis of Water, Waste Water and Sludge; Cations (Group E); Determination of Lead (E 6) (May). 7 pp.
DIN ENGL 38406 Pt 8-80. German Standard Methods for the Examination of Water, Waste Water and Sludge; Cations (Group E); Determination of Zinc (E8) (Oct). 6 pp.
DIN ENGL 38406 Pt 10-85. German Standard Methods for the Examination of Water, Waste Water and Sludge; Cations (Group E); Determination of Chromium (E 10) (June). 6 pp.
DIN ENGL 38406 Pt 12-80. German Standard Methods for Examination of Water, Waste Water and Sludge; Cations (Group E); Determination of Mercury (E 12) (July). 5 pp.
DIN ENGL 38406 Pt 19-80. German Standard Methods for the Analysis of Water, Waste Water and Sludge; Cations (Group E); Determination of Cadmium (E 19) (July). 8 pp.

—Water

BSI BS 6068: Sec 2.4-84. 1984 Water Quality: Part 2: Physical, Chemical and Bio-chemical Methods: Section 2.4: Determination of Total Mercury by Flameless Atomic Absorption Spectrometry: Method After Digestion with Permanganate-Peroxodisulphate. 9 pp.
BSI BS 6068: Sec 2.5-84. 1984 Water Quality Part 2: Physical, Chemical and Bio-Chemical Methods Section 2.5: Determination of Total Mercury of Flameless Atomic Absorption Spectrometry: Method After Pretreatment with Ultraviolet Radiation. 11 pp.
BSI BS 6068: Sec 2.6-84. 1984 Water Quality Part 2: Physical, Chemical and Bio-Chemical Methods Section 2.6: Determination of Total Mercury by Flameless Atomic Absorption Spectometry: Method After Digestion with Bromine. 9 pp.
BSI BS 6068: Sec 2.21-85. 1985 Water Quality Part 2: Physical, Chemical and Bio-Chemical Methods Section 2.21: Determination of Cadmium: Flame Atomic Absorption Spectrometric Methods. 12 pp.
BSI BS 6068: Sec 2.29-87. 1987 Water Quality Part 2: Physical, Chemical and Bio-Chemical Methods Section 2.29: Determination of Cobalt, Nickel, Copper, Zinc, Cadmium and Lead: Flame Atomic Absorption Spectrometric Methods. 13 pp.
BSI BS 6068: Sec 2.30-87. 1987 Water Quality Part 2: Physical, Chemical and Bio-Chemical Methods Section 2.30: Methods for Determination of Calcium and Magnesium by Atomic Absorption Spectrometric Method. 6 pp.

Atomic Absorption Spectrometry *(Cont.)*

—Water *(Cont.)*

BSI BS 6068: Sec 2.38-90. 1990 Water Quality Part 2: Physical, Chemical and Bio-Chemical Methods Section 2.38: Methods for the Determination of Total Chromium by Atomic Absorption Spectrometry. 11 pp.
BSI BS 6068: Sec 2.42-93. 1993 Water Quality Part 2: Physical, Chemical and Biochemical Methods Section 2.42: Determination of Sodium and Potassium: Determination of Sodium by Atomic Absorption Spectrometry (ISO 9964-1: 1993) (N). 21 pp.
BSI BS 6068: Sec 2.43-93. 1993 Water Quality Part 2: Physical, Chemical and Biochemical Methods Section 2.43: Determination of Sodium and Potassium: Determination of Patassium by Atomic Absorption Spectrometry (ISO 9964-2: 1993) (N). 9 pp.
CNS K9024-80. Method of Test for Calcium and Magnesium in Water (Atomic Absorption, Direct) (Feb)(5222).
CNS K9025-80. Method of Test for Zinc in Water (Atomic Absorption, Direct) (Feb)(5223).
CNS K9026-80. Method of Test for Nickel in Water (Atomic Absorption, Direct) (Feb)(5224).
CNS K9027-80. Method of Test for Manganese in Water (Atomic Absorption, Direct) (Feb)(5225).
CNS K9028-80. Method of Test for Lead in Water (Atomic Absorption, Direct) (Apr)(5462).
CNS K9029-80. Method of Test for Cadmium in Water (Atomic Absorption, Direct) (Apr)(5463).
CNS K9030-80. Method of Test for Copper in Water (Atomic Absorption, Direct) (Apr)(5464).
CNS K9031-80. Method of Test for Cobalt in Water (Atomic Absorption, Direct) (Apr)(5465).
CNS K9032-80. Method of Test for Iron in Water (Atomic Absorption, Direct) (Apr)(5466).
CNS K9034-80. Method of Test for Lead in Water (Atomic Absorption, Chelation-Extraction) (Apr)(5468).
DIN ENGL 38406 Pt 3-82. German Standard Methods for the Examination of Water, Waste Water and Sludge; Cations (Group E); Determination of Calcium and Magnesium (E3) (Sept). 7 pp.
DIN ENGL 38406 Pt 6-81. German Standard Methods for the Analysis of Water, Waste Water and Sludge; Cations (Group E); Determination of Lead (E 6) (May). 7 pp.
DIN ENGL 38406 Pt 8-80. German Standard Methods for the Examination of Water, Waste Water and Sludge; Cations (Group E); Determination of Zinc (E8) (Oct). 6 pp.
DIN ENGL 38406 Pt 10-85. German Standard Methods for the Examination of Water, Waste Water and Sludge; Cations (Group E); Determination of Chromium (E 10) (June). 6 pp.
DIN ENGL 38406 Pt 12-80. German Standard Methods for Examination of Water, Waste Water and Sludge; Cations (Group E); Determination of Mercury (E 12) (July). 5 pp.
DIN ENGL 38406 Pt 19-80. German Standard Methods for the Analysis of Water, Waste Water and Sludge; Cations (Group E); Determination of Cadmium (E 19) (July). 8 pp.
ISO 5666 Pt 1-83. Water Quality—Determination of Total Mercury by Flameless Atomic Absorption Spectrometry—Part 1: Method After Digestion with Permanganate-Peroxodisulfate First Edition. 8 pp.
ISO 5666 Pt 2-83. Water Quality—Determination of Total Mercury by Flameless Atomic Absorption Spectrometry—Part 2: Method After Pretreatment with Ultraviolet Radiation First Edition. 10 pp.
ISO 5666 Pt 3-84. Water Quality—Determination of Total Mercury by Flameless Atomic Absorption Spectrometry—Part 3: Method After Digestion with Bromine First Edition. 8 pp.
ISO 5961-85. Water Quality—Determination of Cadmium—Flame Atomic Absorption Spectrometric Methods First Edition. 11 pp.
ISO 7980-86. Water Quality—Determination of Calcium and Magnesium—Atomic Absorption Spectrometric Method First Edition. 5 pp.
ISO 8288-86. Water Quality—Determination of Cobalt, Nickel, Copper, Zinc, Cadmium and Lead—Flame Atomic Absorption Spectrometric Methods First Edition. 13 pp.
ISO 9174-90. Water Quality—Determination of Total Chromium—Atomic Absorption Spectrometric Methods First Edition. 10 pp.
ISO 9964 Pt 1-93. Water Quality—Determination of Sodium and Potassium—Part 1: Determination of Sodium by Atomic Absorption Spectrometry First Edition. 7 pp.
ISO 9964 Pt 2-93. Water Quality—Determination of Sodium and Potassium—Part 2: Determination of Potassium by Atomic Absorption Spectrometry First Edition. 7 pp.
JIS K 0553-90. Testing Methods for Determination of Metallic Elements in Highly Purified Water. 27 pp.

INTERNATIONAL AND NON-U.S. NATIONAL STANDARDS
SUBJECT INDEX
Atomic

Atomic Absorption Spectrometry (Cont.)

—Zinc
DIN ENGL 50551-90. Determination of the Lead, Cadmium and Copper Content of Zinc and Zinc Alloys by Atomic Absorption Spectrometry (Oct). 5 pp.

—Zinc Alloys
DIN ENGL 50551-90. Determination of the Lead, Cadmium and Copper Content of Zinc and Zinc Alloys by Atomic Absorption Spectrometry (Oct). 5 pp.
SAA AS 1329.2-73. Methods for the Analysis of Zinc and Zinc Alloys—Part 2: Magnesium in Zinc Alloys (Atomic Absorption Spectrometric Method). 6 pp.
SAA AS 1329.3-73. Methods for the Analysis of Zinc and Zinc Alloys—Part 3: Aluminium in Zinc Alloys (Atomic Absorption Spectrometric Method). 6 pp.
SAA AS 1329.5-80. Methods for the Analysis of Zinc and Zinc Alloys—Part 5: Determination of Copper Content (0.0001 Percent to 0.0025 Percent)—Flame Atomic Absorption Spectrometric Method. 10 pp.
SAA AS 1329.6-81. Methods for the Analysis of Zinc and Zinc Alloys—Part 6: Determination of Copper Content (0.25 Percent to 1.25 Percent)—Flame Atomic Absorption Spectrometric Method. 5 pp.
SAA AS 1329.7-80. Methods for the Analysis of Zinc and Zinc Alloys—Part 7: Determination of Lead Content—Flame Atomic Absorption Spectrometric Method. 10 pp.
SAA AS 1329.8-80. Methods for the Analysis of Zinc and Zinc Alloys—Part 8: Determination of Cadmium Content—Flame Atomic Absorption Spectrometric Method. 10 pp.

Atomic Absorption Spectrophotometry
Use For: AAS; Electrothermal Atomic Absorption Spectrophotometry; Flame Atomic Absorption Spectrophotometry *See Also:* Molecular Absorption Spectrophotometry
CNS K0027-85. General Rules for Chemical Analysis by Atomic Absorption Spectrophotometry (Feb)(11209).
JIS K 0121-82. General Rules for Atomic Absorption Spectrochemical Analysis. 20 pp.

—Air
CSA CAN/CSA-Z223.26-M87. Measurement of Total Mercury in Air—Cold Vapour Atomic Absorption Spectrophotometric Method; (Gen Instr 1). 35 pp.

—Aluminum
BSI BS 1728: Part 19-71. 1971 Methods for the Analysis of Aluminium and Aluminium Alloys: Part 19: Magnesium (Atomic Absorption Method). 9 pp.
BSI BS 1728: Part 20-71. 1971 Methods for the Analysis of Aluminium and Aluminium Alloys: Part 20: Lead (Atomic Absorption Method). 9 pp.
BSI BS 1728: Part 21-73. 1973 Methods for the Analysis of Aluminium and Aluminium Alloys: Part 21: Zinc (Atomic Absorption Method). 8 pp.
BSI BS 1728: Part 23-76. 1976 Methods for the Analysis of Aluminium and Aluminium Alloys: Part 23: Copper (Atomic Absorption Method). 4 pp.
BSI BS 1728: Part 24-76. 1976 Methods for the Analysis of Aluminium and Aluminium Alloys: Part 24: Nickel (Atomic Absorption Method). 4 pp.
ISO 3256-77. Aluminium and Aluminium Alloys—Determination of Magnesium—Atomic Absorption Spectrophotometric Method First Edition. 6 pp.
ISO 3980-77. Aluminium and Aluminium Alloys—Determination of Copper—Atomic Absorption Spectrophotometric Method First Edition. 5 pp.
ISO 3981-77. Aluminium and Aluminium Alloys—Determination of Nickel—Atomic Absorption Spectrophotometric Method First Edition. 5 pp.

—Aluminum Alloys
BSI BS 1728: Part 19-71. 1971 Methods for the Analysis of Aluminium and Aluminium Alloys: Part 19: Magnesium (Atomic Absorption Method). 9 pp.
BSI BS 1728: Part 20-71. 1971 Methods for the Analysis of Aluminium and Aluminium Alloys: Part 20: Lead (Atomic Absorption Method). 9 pp.
BSI BS 1728: Part 21-73. 1973 Methods for the Analysis of Aluminium and Aluminium Alloys: Part 21: Zinc (Atomic Absorption Method). 8 pp.
BSI BS 1728: Part 23-76. 1976 Methods for the Analysis of Aluminium and Aluminium Alloys: Part 23: Copper (Atomic Absorption Method). 4 pp.
BSI BS 1728: Part 24-76. 1976 Methods for the Analysis of Aluminium and Aluminium Alloys: Part 24: Nickel (Atomic Absorption Method). 4 pp.
ISO 3256-77. Aluminium and Aluminium Alloys—Determination of Magnesium—Atomic Absorption Spectrophotometric Method First Edition. 6 pp.
ISO 3980-77. Aluminium and Aluminium Alloys—Determination of Copper—Atomic Absorption Spectrophotometric Method First Edition. 5 pp.
ISO 3981-77. Aluminium and Aluminium Alloys—Determination of Nickel—Atomic Absorption Spectrophotometric Method First Edition. 5 pp.

Atomic Absorption Spectrophotometry (Cont.)

—Aluminum Oxide
BSI BS 4140: Part 16-86. 1986 Methods of Test for Aluminium Oxide Part 16: Determination of Zinc Content by Flame Atomic Absorption. 6 pp.
ISO 2069-76. Aluminium Oxide Primarily Used for the Production of Aluminium—Determination of Calcium Content—Flame Atomic Absorption Method First Edition. 6 pp.
ISO 2070-81. Aluminium Oxide Primarily Used for the Production of Aluminium—Determination of Calcium Content—Naphthalhydroxamic Acid Spectrophotometric Method First Edition. 6 pp.
ISO 2071-76. Aluminium Oxide Primarily Used for the Production of Aluminium—Determination of Zinc Content—Flame Atomic Absorption Method First Edition. 6 pp.

—Cast Iron
BSI BS 6200: SUB SEC 3.20.4-86. (WITHDRAWN) 1986 Sampling and Analysis of Iron, Steel and Other Ferrous Metals: Part 3: Methods of Anal. Sec. 3.20: Determ. of Nickel: Subsec 3.20.4:Steel & Cast Iron: FlameAtomic Absp.Spectrphotometric (Superseded by BS EN 10 136: 1991). 10 pp.

—Chemical Analysis
CNS H2082-87. Method of Chemical Analysis by Atomic Absorption Spectrophotometry (Oct)(12126).

—Copper
BSI BS 6721: Part 4-86. 1986 Sampling and Analysis of Copper and Copper Alloys Part 4: Method for Determination of Chromium in Copper and Copper Alloys by Flame Atomic Absorption Spectrophotometry. 7 pp.
BSI BS 6721: Part 5-86. 1986 Sampling and Analysis of Copper and Copper Alloys Part 5: Method for Determination of Zinc in Copper and Copper Alloys by Flame Atomic Absorption Spectrophotometry. 7 pp.

—Copper Alloys
BSI BS 6721: Part 4-86. 1986 Sampling and Analysis of Copper and Copper Alloys Part 4: Method for Determination of Chromium in Copper and Copper Alloys by Flame Atomic Absorption Spectrophotometry. 7 pp.
BSI BS 6721: Part 5-86. 1986 Sampling and Analysis of Copper and Copper Alloys Part 5: Method for Determination of Zinc in Copper and Copper Alloys by Flame Atomic Absorption Spectrophotometry. 7 pp.

—Cryolite
BSI BS 5050-74. 1974 Methods of Test for Cryolite. 20 pp.
BSI BS 5050: Part 8-80. 1980 Methods of Test for Cryolite Part 8: Determination of Calcium Content. 7 pp.
CNS M3116-83. Method for Determining Sodium Content in Natural and Artificial Cryolite (Flame Emission and Atomic Absorption Spectrophotometric Method) (Mar)(10115).
CNS M3119-83. Method for Determining Calcium Content in Natural and Artificial Cryolite (Flame Atomic Absorption Method) (Mar)(10118).
ISO 2366-74. Cryolite, Natural and Artificial—Determination of Sodium Content—Flame Emission and Atomic Absorption Spectrophotometric Methods First Edition. 6 pp.
ISO 2830-73. Cryolite, Natural and Artificial—Determination of Aluminium Content—Atomic Absorption Method First Edition. 5 pp.
ISO 3391-76. Cryolite, Natural and Artificial—Determination of Calcium Content—Flame Atomic Absorption Method First Edition. 6 pp.

—Electrolytic Copper
BSI BS 7317: Part 1-90. 1990 Methods for Analysis of High Purity Copper Cathode Cu-CATH-1 Part 1: Method for Determination of Cadmium Manganese and Silver (Screening Procedure for Chromium, Cobalt, Iron, Nickel and Zinc) by Atomic Absorption. 9 pp.
BSI BS 7317: Part 2-90. 1990 Methods for Analysis of High Purity Copper Cathode Cu-CATH-1 Part 2: Method for Determination of Chromium, Cobalt, Iron, Nickel and Zinc by Discrete Volume Nebulization Atomic Absorption Spectrophotometry. 9 pp.
BSI BS 7317: Part 3-90. 1990 Methods for Analysis of High Purity Copper Cathode Cu-CATH-1 Part 3: Method for Determination of Antimony, Arsenic, Bismuth, Selenium, Tellurium and Tin by Hydride Generation and Atomic Absorption Spectrophotometry. 12 pp.
BSI BS 7317: Part 4-90. 1990 Methods for Analysis of High Purity Copper Cathode Cu-CATH-1 Part 4: Method for Determination of Antimony, Arsenic, Bismuth, Lead, Selenium, Tellurium and Tin by Electrothermal Atomiza-tion Atomic Absorption Spectrophotometry. 9 pp.

Atomic Absorption Spectrophotometry (Cont.)

—Electrolytic Copper *(Cont.)*
BSI BS 7317: Part 7-90. 1990 Methods for Analysis of High Purity Copper Cathode Cu-CATH-1 Part 7: Method for Determination of Lead by Lanthanum Hydroxide Separation and Atomic Absorption Spectrophotometry. 9 pp.
BSI DD 95: Part 1-84. (WITHDRAWN) 1984 Amd 1 Analysis of Higher Purity Copper Cathode Cu-CATH-1 Part 1: Method for Determination of Cadmium, Manganese and Silver (Screening Procedures for Chromium, Cobalt, Iron, Nickel and Zinc) by Atomic Absorption. 10 pp.
BSI DD 95: Part 2-84. (WITHDRAWN) 1984 Amd 1 Analysis of Higher Purity Copper Cathode Cu-CATH-1 Part 2: Method for Determination of Chromium, Cobalt, Iron, Nickel, and Zinc by Discrete Volume Nebulization Atomic Absorption Spectrophotometry. 10 pp.
BSI DD 95: Part 3-84. (WITHDRAWN) 1984 Analysis of Higher Purity Copper Cathode Cu-CATH-1 Part 3: Methods for Determination of Antimony, Arsenic, Bismuth, Selenium, Tellurium and Tin by Hydride Generation and Atomic Absorption Spectrophotometry. 14 pp.
BSI DD 95: Part 4-86. (WITHDRAWN) 1986 Analysis of Higher Purity Copper Cathode CU-CATH-1 Part 4: Method for Determination of Antimony, Arsenic, Bismuth, Lead, Selenium, Tellurium and Tin by Electrothermal Atomization Atomic. 9 pp.
BSI DD 95: Part 7-86. (WITHDRAWN) 1986 Analysis of Higher Purity Copper Cathode Cu-CATH-1 Part 7: Method for Determination of Lead by Lanthanum Hydroxide Separation and Atomic Absorption Spectrophotometry (Superseded by BS 7317: Part 7: 1990). 8 pp.

—Fabrics
CGSB CAN/CGSB-4.2 NO.42-M91. Textile Test Methods Copper Content of Textiles. 6 pp.

—Flue Gases
JIS K 0083-76. Method for Determination of Vanadium in Stack Gas. 12 pp.

—Iron Ores
CNS M3016-80. Method for Determination of Cobalt in Iron Ores (Mar)(5378).
JIS M 8210-83. Methods for Determination of Cobalt in Iron Ores. 11 pp.
JIS M 8218-83. Methods for Determination of Copper in Iron Ores. 14 pp.
JIS M 8220-83. Methods for Determination of Aluminium Oxide in Iron Ores. 20 pp.
JIS M 8221-83. Methods for Determination of Calcium Oxide in Iron Ores. 20 pp.
JIS M 8222-83. Methods for Determination of Magnesium Oxide in Iron Ores. 14 pp.
JIS M 8223-83. Methods for Determination of Nickel in Iron Ores. 16 pp.
JIS M 8227-83. Methods for Determination of Tin in Iron Ores. 15 pp.
JIS M 8229-83. Methods for Determination of Lead in Iron Ores. 15 pp.
JIS M 8230-83. Methods for Determination of Bismuth in Iron Ores. 14 pp.

—Magnesium
BSI BS 3907: Part 15-76. 1976 Methods for the Analysis of Magnesium and Magnesium Alloys Part 15: Lead in Magnesium and Magnesium Alloys (Atomic Absorption Method). 7 pp.

—Magnesium Alloys
BSI BS 3907: Part 15-76. 1976 Methods for the Analysis of Magnesium and Magnesium Alloys Part 15: Lead in Magnesium and Magnesium Alloys (Atomic Absorption Method). 7 pp.

—Manganese Ores
CNS M3066-81. Method for Determination of Iron in Manganese Ores (o-Phenanthroline Absorptiometric Method) (Aug)(7831).
JIS M 8234-92. Manganese Ores—Determination of Iron Content. 16 pp.
JIS M 8239-82. Methods for Determination of Aluminium Oxide in Manganese Ores. 11 pp.
JIS M 8240-82. Methods for Determination of Calcium Oxide in Manganese Ores. 10 pp.

—Nickel Castings
JIS H 1279-88. Methods for Determination of Chromium in Nickel Alloy Castings. 12 pp.

—Paints
ISO 3856 Pt 1-84. Paints and Varnishes—Determination of "Soluble" Metal Content—Determination of Lead Content—Flame Atomic Absorption Spectrometric Method and Dithizone Spectrophotometric Method Third Edition. 8 pp.

INDUSTRY STANDARDS

Atomic Absorption Spectrophotometry (Cont.)

—Paperboard
BSI BS 7427-91. 1991 Method for Determination of Titanium Dioxide Content of Paper and Board. 10 pp.
ISO 5647-90. Paper and Board—Determination of Titanium Dioxide Content First Edition. 7 pp.

—Papers
BSI BS 7427-91. 1991 Method for Determination of Titanium Dioxide Content of Paper and Board. 10 pp.
CPPA G.34P-92. Determination of Sodium, Magnesium, Calcium, Manganese, Iron, Copper and Cadmium in Wood, Pulp or Paper by Atomic Absorption Spectrophotometry. 3 pp.
ISO 5647-90. Paper and Board—Determination of Titanium Dioxide Content First Edition. 7 pp.

—Phosphoric Acids
BSI BS 4258: Part 8-78. 1978 Methods of Test for Phosphoric Acid (Orthophosphoric Acid) for Industrial Use Part 8: Determination of Calcium Content (Flame Atomic Absorption Method). 7 pp.
ISO 3707-76. Phosphoric Acid for Industrial Use (Including Foodstuffs)—Determination of Calcium Content—Flame Atomic Absorption Method First Edition. 6 pp.

—Potassium Hydroxides
ISO 3698-76. Potassium Hydroxide for Industrial Use—Determination of Calcium and Magnesium Contents—Flame Atomic Absorption Method First Edition. 6 pp.

—Pulps
CPPA G.34P-92. Determination of Sodium, Magnesium, Calcium, Manganese, Iron, Copper and Cadmium in Wood, Pulp or Paper by Atomic Absorption Spectrophotometry. 3 pp.

—Sodium Hydroxide
BSI BS 6075: Part 9-81. 1981 Amd 1 Sampling and Test for Sodium Hydroxide for Industrial Use Part 9: Determination of Calcium and Magnesium Contents. 7 pp.
ISO 3697-76. Sodium Hydroxide for Industrial Use—Determination of Calcium and Magnesium Contents—Flame Atomic Absorption Method First Edition. 6 pp.

—Steels
BSI BS 6200: SUB SEC 3.20.4-86. (WITHDRAWN) 1986 Sampling and Analysis of Iron, Steel and Other Ferrous Metals: Part 3: Methods of Anal. Sec. 3.20: Determ. ofNickel: Subsec 3.20.4:Steel & Cast Iron: FlameAtomic Absp.Spectrphotometric (Superseded by BS EN 10 136: 1991). 10 pp.

—Urea
ISO 4274-77. Urea for Industrial Use—Determination of Biuret Content—Flame Atomic Absorption and Photometric Absorption Methods First Edition. 8 pp.

—Vegetable Oils
CNS N6099-77. Method of Test for Edible Vegetable Oils (Determination of Mercury) (Jun)(4143). 4 pp.

—Waste Water
CNS K9089-82. Method of Test for Copper in Industrial Waste Water Absorption Spectrophotometric Method (Jun)(8998).

—Water
CNS K9033-80. Method of Test for Molybdenum in Water (Atomic Absorption Spectrophotometry) (Apr)(5467).

—Wood
CPPA G.34P-92. Determination of Sodium, Magnesium, Calcium, Manganese, Iron, Copper and Cadmium in Wood, Pulp or Paper by Atomic Absorption Spectrophotometry. 3 pp.

—Wood Preservatives
BSI BS 5666: Part 3-91. 1991 Methods of Analysis of Wood Preservatives and Treated Timber Part 3: Quantitative Analysis of Preservatives and Treated Timber Containing Copper/Chromium/Arsenic Formulations. 14 pp.
BSI BS 5666: Part 4-79. 1979 Amd 1 Wood Preservatives and Treated Timber Part 4: Quantitative Analysis of Preservatives and Treated Timber Containing Copper Naphthenate. 5 pp.
BSI BS 5666: Part 5-86. 1986 Wood Preservatives and Treated Timber Part 5: Determination of Zinc Naphthenate in Preservative Solutions and Treated Timber. 8 pp.
BSI BS 5666: Part 7-91. 1991 Methods of Analysis of Wood Preservatives and Treated Timber Part 7: Quantitative Analysis of Preservatives Containing Bis(tri-n-butyltin)oxide: Determination of Total Tin. 12 pp.

—Zinc Alloys
BSI BS 3630: Part 15-76. 1976 Methods for the Sampling and Analysis of Zinc and Zinc Alloys Part 15: Magnesium in Zinc Alloys (Atomic Absorption Method). 7 pp.
ISO 3750-76. Zinc Alloys—Determination of Magnesium Content—Atomic Absorption Method First Edition. 4 pp.

Atomic Absorption Spectroscopy
Use For: Flame Atomic Absorption Spectroscopy

—Crude Oil
BSI BS 2000: Part 288-93. 1993 Methods of Test for Petroleum and Its Products Part 288: Determination of Nickel, Sodium and Vanadium—Atomic Absorption Spectroscopy Method (W). 5 pp.
BSI BS 2000: Part 288-83. 1983 Petroleum and Its Products Part 288: Sodium, Nickel and Vanadium in Fuel Oils and Crude Oils by Atomic Absorption Spectroscopy. 6 pp.

—Ebonite
DIN ENGL 53599 Pt 1-78. Testing of Rubber and Elastomers; Determination of the Lead Content; Determination by Atomic Absorption Spectroscopy for Lead Contents up to 1000 mg/kg (0.1%) (May). 4 pp.

—Effluents—Photographic Processing
ISO 10348-93. Photography—Processing Wastes—Determination of Silver Content First Edition. 20 pp.

—Elastomers
DIN ENGL 53569 Pt 2-72. Testing of Rubber and Elastomers; Determination of Copper Content; Determination by Atomic Absorption Spectroscopy (Nov). 7 pp.
DIN ENGL 53599 Pt 1-78. Testing of Rubber and Elastomers; Determination of the Lead Content; Determination by Atomic Absorption Spectroscopy for Lead Contents up to 1000 mg/kg (0.1%) (May). 4 pp.

—Fuel Oils
BSI BS 2000: Part 288-93. 1993 Methods of Test for Petroleum and Its Products Part 288: Determination of Nickel, Sodium and Vanadium—Atomic Absorption Spectroscopy Method (W). 5 pp.
BSI BS 2000: Part 288-83. 1983 Petroleum and Its Products Part 288: Sodium, Nickel and Vanadium in Fuel Oils and Crude Oils by Atomic Absorption Spectroscopy. 6 pp.

—Gas Turbine Fuels
DIN ENGL 51790 Pt 3-78. Testing of Liquid Fuels; Determination of Vanadium Content; Vanadium Content in the Range of 0.4 to 4.0 mg/kg; Determination by Flameless Atomic Absorption Spectroscopy After Incineration (Apr). 6 pp.

—Gold
SAA AS 3895.1-91. Methods for the Analysis of Copper, Lead, Zinc, Gold and Silver Ores—Part 1: Determination of Gold (Fire Assay—Flame AAS Method). 9 pp.

—Liquid Fuels
DIN ENGL 51790 Pt 3-78. Testing of Liquid Fuels; Determination of Vanadium Content; Vanadium Content in the Range of 0.4 to 4.0 mg/kg; Determination by Flameless Atomic Absorption Spectroscopy After Incineration (Apr). 6 pp.

—Rubber
DIN ENGL 53569 Pt 2-72. Testing of Rubber and Elastomers; Determination of Copper Content; Determination by Atomic Absorption Spectroscopy (Nov). 7 pp.
DIN ENGL 53599 Pt 1-78. Testing of Rubber and Elastomers; Determination of the Lead Content; Determination by Atomic Absorption Spectroscopy for Lead Contents up to 1000 mg/kg (0.1%) (May). 4 pp.

Atomic Emission Spectrometry
Use For: Flame Atomic Emission Spectrometry
See Also: Spectrometry

—Glass
BSI BS 3473: Sec 4.2-89. 1989 Chemical Resistance of Glass Used in the Pro-duction of Laboratory Glassware: Part 4: Methods for determination of Hydrolytic Resistance of the Interior Surfaces of Glass Containers: Section 4.2:. 13 pp.
ISO 4802 Pt 2-88. Glassware—Hydrolytic Resistance of the Interior Surfaces of Glass Containers—Part 2: Determination by Flame Spectrometry and Classification First Edition. 11 pp.
ISO 10136 Pt 2-93. Glass and Glassware—Analysis of Extract Solutions—Part 2: Determination of Sodium Oxide and Potassium Oxide by Flame Spectrometric Methods First Edition. 10 pp.

—Glassware
ISO 10136 Pt 2-93. Glass and Glassware—Analysis of Extract Solutions—Part 2: Determination of Sodium Oxide and Potassium Oxide by Flame Spectrometric Methods First Edition. 10 pp.

—Glossaries
ISO 6955-82. Analytical Spectroscopic Methods—Flame Emmission, Atomic Absorption, and Atomic Fluorescence—Vocabulary First Edition. 23 pp.

—Manganese Ores
ISO 4571-81. Manganese Ores and Concentrates—Determination of Potassium and Sodium Content—Flame Atomic Emission Spectrometric Method First Edition. 5 pp.

—Paints
BSI BS 3900: Part B8-86. 1986 Methods of Test for Paints Group B: Tests Involving Chemical Examination of Liquid Paints and Dried Paint Films Part B8: Determination of 'Soluble' Barium Content. 7 pp.
ISO 3856 Pt 3-84. Paints and Varnishes—Determination of "Soluble" Metal Content—Part 3: Determination of Barium Content—Flame Atomic Emission Spectrometric Method Second Edition. 6 pp.

—Powdered Milk
BSI BS 1743: Part 20-88. 1988 Analysis of Dried Milk and Dried Milk Products Part 20: Determination of Sodium and Potassium Contents of Dried Milk. 6 pp.
ISO 8070-87. Dried Milk—Determination of Sodium and Potassium Contents—Flame Emission Spectrometric Method First Edition. 6 pp.

—Surface Waters
DIN ENGL 38406 Pt 22-88. German Standard Methods for the Examination of Water, Waste Water and Sludge; Cations (Group E); Determination of Ag, Al, As, B, Ba, Be, Bi, Ca, Cd, Co, Cr, Cu, Fe, K, Li, Mg, Mn, Mo, Na, Ni, P, Pb, S, Sb, Se, Si, Sn, Sr, Ti, V, W, Zn, and Zr by Inductively Coupled. 12 pp.

—Waste Water
DIN ENGL 38406 Pt 22-88. German Standard Methods for the Examination of Water, Waste Water and Sludge; Cations (Group E); Determination of Ag, Al, As, B, Ba, Be, Bi, Ca, Cd, Co, Cr, Cu, Fe, K, Li, Mg, Mn, Mo, Na, Ni, P, Pb, S, Sb, Se, Si, Sn, Sr, Ti, V, W, Zn, and Zr by Inductively Coupled. 12 pp.

—Water
BSI BS 6068: Sec 2.44-93. 1993 Water Quality Part 2: Physical, Chemical and Biochemical Methods Section 2.44: Determination of Sodium and Potassium: Determination of Sodium and Potassium by Flame Emission Spectrometry (ISO 9964-3: 1993) (N). 9 pp.
ISO 9964 Pt 3-93. Water Quality—Determination of Sodium and Potassium—Part 3: Determination of Sodium and Potassium by Flame Emission Spectrometry First Edition. 8 pp.

Atomic Emission Spectrophotometry
Use For: Flame Atomic Emission Spectrophotometry; Flame Emission Spectrophotometry
See Also: Spectrophotometry

—Aluminum Fluoride
BSI BS 4993: Part 7-80. 1980 Methods of Test for Aluminium Fluoride for Industrial Use Part 7: Determination of Sodium Content. 8 pp.
ISO 4279-77. Aluminium Fluoride for Industrial Use—Determination of Sodium Content—Flame Emission Spectrophotometric Method First Edition; (Erratum—Sept 1978). 7 pp.

—Aluminum Oxide
BSI BS 4140: Part 11-86. 1986 Methods of Test for Aluminium Oxide Part 11: Determination of Sodium Content. 7 pp.

INTERNATIONAL AND NON-U.S. NATIONAL STANDARDS
SUBJECT INDEX
Attenuation

Atomic Emission Spectrophotometry (Cont.)
—Aluminum Oxide (Cont.)
ISO 1617-76. Aluminium Oxide Primarily Used for the Production of Aluminium—Determination of Sodium Content—Flame Emission Spectrophotometric Method First Edition. 7 pp.

—Cryolite
CNS M3116-83. Method for Determining Sodium Content in Natural and Artificial Cryolite (Flame Emission and Atomic Absorption Spectrophotometric Method) (Mar)(10115).

ISO 2366-74. Cryolite, Natural and Artificial—Determination of Sodium Content—Flame Emission and Atomic Absorption Spectrophotometric Methods First Edition. 6 pp.

—Magnesium Carbonate
MOD UK M 870/67. Examination of Magnesium Carbonate, Heavy.

—Potassium Chloride
ISO 2050-76. Potassium Chloride for Industrial Use—Determination of Potassium Content—Flame Emission Spectrophotometric Method First Edition. 6 pp.

—Potassium Hydroxides
ISO 1550-73. Potassium Hydroxide for Industrial Use—Determination of Sodium Content—Flame Emission Spectrophotometric Method First Edition. 5 pp.

—Potassium Sulfate
ISO 2484-73. Potassium Sulphate for Industrial Use—Determination of Potassium Content—Flame Emission Spectrophotometric Method First Edition. 5 pp.

—Waste Water
CNS K9086-82. Method of Test for Potassium in Industrial Waste Water Emission Spectrophotometric Method (May)(8842).

CNS K9087-82. Method of Test for Sodium in Industrial Waste Water Emission Spectrophotometric Method (May)(8843).

Atomic Energy
Use: Nuclear Power

Atomic Fluorescence Spectrometry
Use For: AFS *See Also:* Specific Rotation; Spectrometry

—Glossaries
ISO 6955-82. Analytical Spectroscopic Methods—Flame Emmission, Atomic Absorption, and Atomic Fluorescence—Vocabulary First Edition. 23 pp.

Atomic Physics
—Units of Measurement
BSI BS 5775: Part 9-93. 1993 Quantities, Units and Symbols Part 9: Atomic and Nuclear Physics (ISO 31-9: 1992) (G). 29 pp.

BSI BS 5775: Part 9-82. 1982 Amd 1 Specification for Quantities, Units and Symbols Part 9: Atomic and Nuclear Physics (AMD 5854) July 31, 1989. 17 pp.

CNS Z7196-9-85. Quantities and Units of Atomic and Nuclear Physics (Dec)(11296-9).

ISO 31 Pt 9-92. Quantities and Units Part 9: Atomic and Nuclear Physics Third Edition. 27 pp.

Atomic Power Stations
Use: Nuclear Power Plants

Atomic Reactors
Use: Nuclear Reactors

Atomic Time
Use: Atomic Time Scales

Atomic Time Scales
Use For: Atomic Time; TA *See Also:* International Atomic Time; Time Scales

CCIR RECMN 458-2-90. International Comparisons of Atomic Time Scales—Section 7B—Specifications for the Standard-Frequency and Time-Signal Services. 1 p.

—Abbreviations
CCIR RECMN 536-78. Time-Scale Notations—Section 7B—Specifications for the Standard-Frequency and Time-Signal Services. 2 pp.

Atomic Weapons Research Establishments
Use For: AWRE *See Also:* Weapons

—Defense Contracts
MOD UK DEFCON 112CE-83. Basic Set of Conditions of Contract for AWRE Aldermaston 9/83. 1 p.

—Service Contracts
MOD UK DEFCON 112CZ-83. Conditions of Contract for Miscellaneous On-Site Services for AWRE Aldermaston 9/83. 2 pp.

ATR Tubes
Use: Antitransmit Receive Tubes

ATR 42 Aircraft
See Also: Aircraft

—Accidents
CAA. ATR 42 and 72. 1 p.

—Certification
CAA. ATR 42. 12 pp.

—Data Sheets
CAA FA47 ISSUE 1. ATR 42, Model 300. 4 pp.

ATR 72 Aircraft
See Also: Aircraft

—Accidents
CAA. ATR 42 and 72. 1 p.

Atrazine
See Also: Pesticides

SAA AS 1870.10D-78. Standard for Development—Pesticides for Agricultural Use—Part 10D: Atrazine, Simazine, Propazine, Prometryn, Methoprotryn, Ametryn and Tertbutryn. 35 pp.

Atriums
See Also: Buildings; Windows

—Firesafety
SNZ NZS 4238-91. Code of Practice for Fire Safety in Atrium Buildings. 16 pp.

Attache' Cases
Use: Briefcases

Attachment Plugs
Use: Electric Plugs

Attack Blocking Glazings
Use: Security Glazing

Attenuation
See Also: Atmospheric Attenuation; Attenuation Distortion; Attenuators; Diffraction; Fading; Light (Visible Radiation); Rain Attenuation; Signal Processing; Signal to Noise Ratio; Sound Transmission

—Connectors
CEN PREN 2591-F2-93. Aerospace Series Elements of Electrical and Optical Connection Test Methods Part F2—Optical Elements Variation of Attenuation and Optical Discontinuity. 3 pp.

—Fiber Optic Cables
CCITT RECMN G.651-89. Characteristics of a 50/125 Micrometer Multimode Graded Index Optical Fibre Cable—Transmission Media Characteristics (Study Group XV) 30 pp. 30 pp.

CCITT RECMN G.652-89. Characteristics of a Single-Mode Optical Fibre Cable—Transmission Media Characteristics (Study Group XV) 34 pp. 34 pp.

CCITT RECMN G.653-89. Characteristics of a Dispersion-Shifted Single-Mode Optical Fibre Cable—Transmission Media Characteristics (Study Group XV) 5 pp. 5 pp.

CCITT RECMN G.654-89. Characteristics of a 1550 mm Wavelength Loss-Minimized Single-Mode Optical Fibre Cable—Transmission Media Characteristics (Study Group XV) 3 pp. 3 pp.

CCITT RECMN L.14-92. Measurement Method to Determine the Tensile Performance of Optical Fibre Cables Under Load (Study Group VI) 8 pp. 8 pp.

—Free Space Propagation
CCIR RECMN 525-1-82. Calculation of Free-Space Attenuation—Section 5A—Texts of General Interest. 5 pp.

—Frequency Bands—Space Communications
CCIR Report 1119-90. Method of Calculating Attenuation, Noise Temperature, and Telecommunication Link Performance for the Selection of Preferred Frequency Bands—Section 2B—Topics of General Interest. 20 pp.

Attenuation (Cont.)
—Groups—Carrier Systems
CCITT RECMN G.313-89. Open-Wire Lines for Use with 12-Channel Carrier Systems—International Analogue Carrier Systems (Study Group XV) 3 pp. 3 pp.

—Optical Fibers
CNS C6277-86. Method of Test for Fiber Optic Devices (FOTP-46 Spectral Attennation Measurement for Long-Length, Graded-Index Optical Fibers) (Dec)(11789).

CNS C6287-87. Method of Test for Fiber Optic Device (FOTP—50 Light Launch Conditions for Long Length Graded Index Optical Fiber Spectral Attenuation Measurements) (Jun)(11993).

CNS C6324-88. Method of Test for Fiber Optic Devices (FOTP-53 Attenuation by Substitution Measurement for Multimode Graded-Index Optical Fibers or Fiber Assemblies Used in Long Length Communications System) (Jul)(12372).

JIS C 6826-89. Test Methods for Attenuation of Single-Mode Optical Fibers. 9 pp.

JIS C 6863-90. Test Methods for Attenuation of All Plastic Multimode Optical Fibers. 7 pp.

—Optical Fibers—Temperature
CNS C6280-87. Method of Test for Fiber Optic Devices (FOTP-52 Temperature Dependence of Attenuation for Optical Fibers) (Apr)(11902).

—Telephone Cables
CCITT RECMN G.612-89. Characteristics of Symmetric Cable Pairs Designed for the Transmission of Systems with Bit Rates of the Order of 6 to 34 Mbit/s—Transmission Media Characteristics (Study Group XV) 4 pp. 4 pp.

CCITT RECMN G.613-89. Characteristics of Symmetric Cable Pairs Usable Wholly for the Transmission of Digital Systems with a Bit Rate of up to 2 Mbits—Transmission Media Characteristics (Study Group XV) 4 pp. 4 pp.

CCITT RECMN G.614-89. Characteristics of Symmetric Pair Star-Quad Cables Designed Earlier for Analogue Transmission Systems and Being Used Now for Digital System Transmission at Bit Rates of 6 to 34 Mbit/s—Transmission Media Characteristics (Study Group XV) 5 pp. 5 pp.

CCITT RECMN G.621-89. Characteristics of 0.7/2.9 mm Coaxial Cable Pairs—Transmission Media Characteristics (Study Group XV) 4 pp. 4 pp.

CCITT RECMN G.622-89. Characteristics of 1.2/4.4 mm Coaxial Cable Pairs—Transmission Media Characteristics (Study Group XV) 8 pp. 8 pp.

CCITT RECMN G.623-89. Characteristics of 2.6/9.5 mm Coaxial Cable Pairs—Transmission Media Characteristics (Study Group XV) 8 pp. 8 pp.

DIN VDE 0472 Pt 515-85. Testing of Cables, Wires and Flexible Cords; Attenuation (May). 7 pp.

—Telephone Repeaters
CCITT RECMN G.323-89. Typical Transistorized System on Symmetric Cable Pairs—International Analogue Carrier Systems (Study Group XV) 3 pp. 3 pp.

CCITT RECMN G.611-89. Characteristics of Symmetric Cable Pairs for Analogue Transmission—Transmission Media Characteristics (Study Group XV) 5 pp. 5 pp.

—VHF/UHF—Woodlands
CCIR RECMN 833-92. Attenuation in Vegetation. 1 p.

Attenuation Distortion
See Also: Attenuation; Communication Circuits; Distortion (Electrical)

—Communication Terminal Equipment—Groups
CCITT RECMN G.232-89. 12-Channel Terminal Equipments—International Analogue Carrier Systems (Study Group XV) 13 pp. 13 pp.

CCITT RECMN G.235-89. 16-Channel Terminal Equipments—International Analogue Carrier Systems (Study Group XV) 4 pp. 4 pp.

—Digital Transmission Systems
CCITT RECMN G.622-89. Characteristics of 1.2/4.4 mm Coaxial Cable Pairs—Transmission Media Characteristics (Study Group XV) 8 pp. 8 pp.

CCITT RECMN G.623-89. Characteristics of 2.6/9.5 mm Coaxial Cable Pairs—Transmission Media Characteristics (Study Group XV) 8 pp. 8 pp.

—Echo Cancellers
CCITT RECMN G.165-89. Echo Cancellers—General Characteristics of International Telephone Connections and Circuits (Study Groups XII and XV) 23 pp. 23 pp.

INDUSTRY STANDARDS

Attenuation

Attenuation Distortion (Cont.)

—Echo Suppressors
CCITT RECMN G.164-89. Echo Suppressors—General Characteristics of International Telephone Connections and Circuits (Study Groups XII and XV) 36 pp. 36 pp.

—Maritime Mobile Satellite Communications—Automatic Services
CCITT RECMN G.473-89. Interconnection of a Maritime Mobile Satellite System with the International Automatic Switched Telephone Service; Transmission Aspects—International Analogue Carrier Systems (Study Group XV) 10 pp. 10 pp.

—Speech Circuits—Loudness Ratings
CCITT RECMN G.111-89. Loudness Ratings (LRs) in an International Connection—General Characteristics of International Telephone Connections and Circuits (Study Groups XII and XV) 22 pp. 22 pp.

—Telephone Circuits
CCITT RECMN G.113-89. Transmission Impairments—General Characteristics of International Telephone Connections and Circuits (Study Groups XII and XV) 22 pp. 22 pp.
CCITT RECMN G.131-89. Stability and Echo—General Characteristics of International Telephone Connections and Circuits (Study Groups XII and XV) 13 pp. 13 pp.
CCITT RECMN G.132-89. Attenuation Distortion—General Characteristics of International Telephone Connections and Circuits (Study Groups XII and XV) 2 pp. 2 pp.
CCITT RECMN G.141-89. Attenuation Distortion—General Characteristics of International Telephone Connections and Circuits (Study Groups XII and XV) 2 pp. 2 pp.
CCITT RECMN G.151-89. General Performance Objectives Applicable to All Modern International Circuits and National Extension Circuits—General Characteristics of International Telephone Connections and Circuits (Study Groups XII and XV) 6 pp. 6 pp.
CCITT RECMN G.153-89. Characteristics Appropriate to International Circuits More Than 2500 km in Length—General Characteristics of International Telephone Connections and Circuits (Study Groups XII and XV) 4 pp. 4 pp.
CCITT RECMN Q.44-89. Attenuation Distortion—General Recommendations on Telephone Switching and Signalling—Functions and Information Flows for Services in the ISDN—Supplements (Study Group XI) 2 pp. 2 pp.

—Telephone Circuits—Carrier Systems
CCITT RECMN G.125-89. Characteristics of Circuits on Carrier Systems—General Characteristics of International Telephone Connections and Circuits (Study Groups XII and XV) 1 pp. 1 pp.
CCITT RECMN G.141-89. Attenuation Distortion—General Characteristics of International Telephone Connections and Circuits (Study Groups XII and XV) 2 pp. 2 pp.

—Telephone Circuits—Public Switched Telephone Networks
CCITT RECMN G.113-89. Transmission Impairments—General Characteristics of International Telephone Connections and Circuits (Study Groups XII and XV) 22 pp. 22 pp.

—Telephone Connections
CCITT RECMN G.113-89. Transmission Impairments—General Characteristics of International Telephone Connections and Circuits (Study Groups XII and XV) 22 pp. 22 pp.

—Telephone Connections—Pulse Code Modulation
CCITT RECMN G.141-89. Attenuation Distortion—General Characteristics of International Telephone Connections and Circuits (Study Groups XII and XV) 2 pp. 2 pp.

—Telephone Exchanges—Public Switched Telephone Networks
CCITT RECMN G.113-89. Transmission Impairments—General Characteristics of International Telephone Connections and Circuits (Study Groups XII and XV) 22 pp. 22 pp.

—Telephones
CCITT RECMN P.62-89. Measurements on Subscribers' Telephone Equipment—Telephone Transmission Quality (Study Group XII) 2 pp. 2 pp.

—Transmission Systems
CCITT RECMN G.622-89. Characteristics of 1.2/4.4 mm Coaxial Cable Pairs—Transmission Media Characteristics (Study Group XV) 8 pp. 8 pp.

Attenuation Distortion (Cont.)

—Transmission Systems (Cont.)
CCITT RECMN G.623-89. Characteristics of 2.6/9.5 mm Coaxial Cable Pairs—Transmission Media Characteristics (Study Group XV) 8 pp. 8 pp.

—Underwater Telephone Cables
CCITT RECMN G.153-89. Characteristics Appropriate to International Circuits More Than 2500 km in Length—General Characteristics of International Telephone Connections and Circuits (Study Groups XII and XV) 4 pp. 4 pp.

—Voice Band Data Transmission
CCITT RECMN G.113-89. Transmission Impairments—General Characteristics of International Telephone Connections and Circuits (Study Groups XII and XV) 22 pp. 22 pp.

Attenuator Sets
Use: Attenuators

Attenuators
Use For: Attenuator Sets *See Also:* Absorbents; Attenuation; Electronic Components; Equalizers (Electrical); Microwave Attenuators; Mufflers; Silencers

—Fiber Optic
IEC 869 Pt 1-88. Fibre Optic Attenuators Part 1: Generic Specification First Edition (IECQ QC 8000000). 80 pp.
IECQ QC 800000-88. Fibre Optic Attenuators Part 1: Generic Specification (IEC 869-1 ED 1). 81 pp.
JIS C 5920-88. General Rules of Optical Attenuator. 10 pp.

Attic Fans
Use: Fans—Attic

Attic Ventilation Equipment
Use: Ventilation Equipment

Attitude Control
See Also: Attitude Indicators

—Geostationary Satellites
CCIR Report 546-4-90. Space Systems Technology: Attitude Control Technology—Section 2A—Research in Space Technology. 22 pp.

—Satellites
CCIR Report 546-4-90. Space Systems Technology: Attitude Control Technology—Section 2A—Research in Space Technology. 22 pp.

—Space Communications
CCIR Report 546-4-90. Space Systems Technology: Attitude Control Technology—Section 2A—Research in Space Technology. 22 pp.

—Spacecraft
CCIR Report 546-4-90. Space Systems Technology: Attitude Control Technology—Section 2A—Research in Space Technology. 22 pp.

Attitude Director Indicators
Use: Attitude Indicators

Attitude Indicators
Use For: ADI; Attitude Director Indicators; Attitude Reference Indicators; EADI; Electronic Attitude Direction Indicators *See Also:* Aircraft Instruments; Attitude Control; Gyrohorizons; Horizontal Indicators; Indicating Instruments; Turn and Bank Indicators

CAA Chapter P6-10. Attitude Displays (Provisional Airworthiness Requirements for Civil Powered Lift Aircraft). 3 pp.
MOD UK DSTAN 66-43-87. General Requirements for Aircraft Instruments and Displays Attitude Indicators (Self-Contained and Remote Driven) Issue 1. 10 pp.
MOD UK DSTAN 66-44-87. General Requirements for Aircraft Instruments and Displays Attitude Director Indicators Issue 1. 13 pp.
NATO STANAG 3637 ED 2 AMD 9-72. Attitude Indicators (Self-Contained and Remote Driven). 8 pp.
NATO STANAG 3637 ED 2 AMD 10-92. Attitude Indicators (Self-Contained and Remote Driven). 9 pp.
NATO STANAG 3640 ED 3 AMD 6-76. Attitude Director Indicators. 11 pp.
NATO STANAG 3640 ED 3 AMD 7-76. Attitude Director Indicators. 12 pp.
NATO STANAG 3640 ED 3 AMD 8-92. Attitude Director Indicators. 13 pp.

—Gliders
CAA JAR-22 SUBPART F. Equipment (Joint Airworthiness Requirements). 4 pp.

Attitude Indicators (Cont.)

—Rotary Wing Aircraft
CAA Chapter G6-10 08.82. Attitude Display Systems (Rotorcraft). 2 pp.
CAA Chapter G6-10 App 08.82. Attitude Display Systems (Rotorcraft). 2 pp.

Attitude Reference Indicators
Use: Attitude Indicators

Attrition Rate

—Sodium Perborates
ISO 5937-80. Sodium Perborates for Industrial Use—Determination of Degree of Attrition First Edition. 7 pp.

Audio Amplifiers
See Also: Amplifiers; Annunciators; Audio Equipment; Communication Equipment; Electroacoustics; Hearing Aids; Power Amplifiers; Preamplifiers; Sound Recording and Reproduction Equipment; Speakers

BSI BS 5428: Part 2-79. (WITHDRAWN) 1979 Methods for Specifying and Measuring the Characteristics of Sound System Equipment Part 2: Amplifiers (Superseded by BS 6840: Part 3: 1989). 31 pp.
BSI BS 5942: Part 6-80. 1980 High Fidelity Audio Equipment and Systems; Minimum Performance Requirements Part 6: Amplifiers. 10 pp.
BSI BS 6493: Sec 2.3-87. 1987 Semiconductor Devices Part 2: Integrated Circuits Section 2.3: Recommendations for Analogue Integrated Circuits. 132 pp.
BSI BS 6493: Sec 2.3-01. 1987 Amd 1 Semiconductor Devices Part 2: Integrated Circuits Section 2.3: Recommendations for Analogue Integrated Circuits (IEC 748-3: 1986) (AMD 7035) July 15, 1992. 190 pp.
BSI BS 6840: Part 3-92. 1992 Sound System Equipment Part 3: Methods for Specifying and Measuring the Characteristics of Sound System Amplifiers (IEC 268-3: 1988). 60 pp.
BSI BS 6840: Part 3-89. 1989 Amd 1 Sound System Equipment Part 3: Methods for Specifying and Measuring the Characteristics of Sound System Amplifiers (AMD 6719) September 30, 1991. 53 pp.
CENELEC HD 483.3 S1-90. Sound System Equipment Part 3: Amplifiers. 3 pp.
IEC 268 Pt 3-88. Sound System Equipment Part 3: Amplifiers Second Edition; (Amendment 1-1990) (Amendment 2-1991). 117 pp.
IEC 581 Pt 6-79. High Fidelity Audio Equipment and Systems; Minimum Performance Requirements Part 6: Amplifiers First Edition. 22 pp.
IEC 748 Pt 3-86. Semiconductor Devices Integrated Circuits Part 3: Analogue Integrated Circuits First Edition; (Amendment 1-1991). 334 pp.
SAA AS 1127.6-90. Sound System Equipment—Part 6: Amplifiers. 45 pp.

—Aircraft
EUROCAE ED-18 07.85. Audio Systems Characteristics and MPS Aircraft Microphones (Except Carbon), Aircraft Headsets, Handsets and Loudspeakers, Aircraft Audio Selector Panels and Amplifiers. 108 pp.

—Aircraft—Environmental Testing
EUROCAE ED-18 07.85. Audio Systems Characteristics and MPS Aircraft Microphones (Except Carbon), Aircraft Headsets, Handsets and Loudspeakers, Aircraft Audio Selector Panels and Amplifiers. 108 pp.

—Artificial Reverberation
BSI BS 6840: Part 9-87. 1987 Sound System Equipment Part 9: Methods for Specifying and Measuring the Characteristics of Artificial Reverberation Time Delay and Frequency Shift Equipment. 14 pp.
IEC 268 Pt 9-77. Sound System Equipment Part 9: Artificial Reverberation, Time Delay and Frequency Shift Equipment First Edition. 29 pp.

—Frequency Shift
BSI BS 6840: Part 9-87. 1987 Sound System Equipment Part 9: Methods for Specifying and Measuring the Characteristics of Artificial Reverberation Time Delay and Frequency Shift Equipment. 14 pp.
IEC 268 Pt 9-77. Sound System Equipment Part 9: Artificial Reverberation, Time Delay and Frequency Shift Equipment First Edition. 29 pp.

—Time Delay
BSI BS 6840: Part 9-87. 1987 Sound System Equipment Part 9: Methods for Specifying and Measuring the Characteristics of Artificial Reverberation Time Delay and Frequency Shift Equipment. 14 pp.

INTERNATIONAL AND NON-U.S. NATIONAL STANDARDS
SUBJECT INDEX

Audio

Audio Amplifiers (Cont.)
—Time Delay (Cont.)
IEC 268 Pt 9-77. Sound System Equipment Part 9: Artificial Reverberation, Time Delay and Frequency Shift Equipment First Edition. 29 pp.

Audio Cables
Scope Note: Use a more specific term
See: Microphone Cables; Telephone Cables

Audio Cassettes
Use For: Cassette Tapes *See Also:* Audio Equipment; Audio Tapes

—Digital Recording—Professional
IEC 1119 Pt 5-93. Digital Audio Tape Cassette System (DAT) Part 5: DAT for Professional Use First Edition. 40 pp.

—Digital Recording—Sound-Program Exchange
CCIR RECMN 777-92. International Exchange of Digital Audio Recordings. 2 pp.

Audio Connectors
See Also: Circular Connectors; Connectors; Phone Jacks; Phone Plugs

BSI BS 5817: Part 3-89. 1989 Audio-Visual, Video and Television Equipment and Systems Part 3: Connectors for the Interconnection of Equipment in Audio-Visual Systems. 15 pp.

BSI BS 6840: Part 11-88. 1988 Sound System Equipment Part 11: Application of Connectors for the Interconnection of Sound System Components. 21 pp.

BSI BS 6840: Part 11-01. 1988 Amd 1 Sound System Equipment Part 11: Specification for Application of Connectors for the Interconnection of Sound System Components (AMD 6410) October 31, 1991. 24 pp.

BSI BS 6840: Part 12-93. 1993 Sound System Equipment Part 12: Specification for Application of Connectors for Broadcast and Similar Use (IEC 268-12: 1987) (S). 15 pp.

BSI BS 6840: Part 12-87. 1987 Amd 1 Sound System Equipment Part 12: Specification for Application of Connectors for Broadcast and Similar Use (IEC 268-12: 1987) (AMD 6409) August 30, 1991. 12 pp.

BSI BS 6840: Part 15-92. 1992 Sound System Equipment Part 15: Specification for Matching Values for the Interconnection of Sound System Components (IEC 268-15: 1987). 27 pp.

BSI BS 6840: Part 15-88. 1988 Sound System Equipment Part 15: Matching Values for the Interconnection of Sound System Components. 21 pp.

CENELEC HD 369.3-86. Audio-Visual, Video and Television Equipment and Systems Part 3: Connectors for the Interconnection of Equipment in Audio-Visual Systems. 2 pp.

CENELEC HD 369.18 S1-89. Audiovisual, Video and Television Equipment and Systems Part 18: Connectors for Automatic Slide Projectors with Built-in Triacs for Audiovisual Application. 3 pp.

CENELEC HD 483.11 S1-90. Sound System Equipment Part 11: Application of Connectors for the Interconnection of Sound System Components. 3 pp.

CENELEC HD 483.11 S2-91. Sound System Equipment Part 11: Application of Connectors for the Interconnection of Sound System Components. 5 pp.

CENELEC HD 483.12 S1-90. Sound System Equipment Part 12: Application of Connectors for Broadcast and Similar Use. 3 pp.

CENELEC HD 483.12 S2-93. Sound System Equipment Part 12: Application of Connectors for Broadcast and Similar Use (IEC 268-12:1987 + A1: 1991). 5 pp.

CENELEC HD 483.15 S1-89. Sound System Equipment-Part 15: Prefered Matching Valves for the Interconnection of Sound System Components. 3 pp.

CENELEC HD 483.15 S2-90. Sound System Equipment Part 15: Preferred Matching Values for the Interconnection of Sound System Components. 3 pp.

CENELEC HD 483.15 S3-91. Sound System Equipment Part 15: Preferred Matching Values for the Interconnection of Sound System Components. 7 pp.

CENELEC HD 483.15 S4-92. Sound System Equipment Part 15: Preferred Matching Values for the Interconnection of Sound System Components (IEC 268-15:1987 + A1: 1989 + A2:1990 + A3: 1991). 6 pp.

IEC 130 Pt 9-89. Connectors for Frequencies Below 3 MHz Part 9: Circular Connectors for Radio and Associated Sound Equipment Second Edition; (Amendment 1-1993). 78 pp.

IEC 268 Pt 11-87. Sound System Equipment Part 11: Application of Connectors for the Interconnection of Sound System Components Second Edition; (Amendment 1-1989) (Amendment 2-1991). 58 pp.

Audio Connectors (Cont.)
IEC 268 Pt 12-87. Sound System Equipment Part 12: Application of Connectors for Broadcast and Similar Use Second Edition; (Amendment 1-1991) (CENELEC HD 483.12 S2: 1993). 21 pp.

IEC 268 Pt 15-87. Sound System Equipment Part 15: Preferred Matching Values for the Interconnection of Sound System Components Second Edition; (Amendment 1-1989) (Amendment 2-1990) (Amendment 3-1991) (CENELEC HD 483.15 S4: 1992). 58 pp.

IEC 574 Pt 3-83. Audio-Visual, Video and Television Equipment and Systems Part 3: Connectors for the Interconnection of Equipment in Audio-Visual Systems First Edition. 27 pp.

—Identification Systems
IEC 1062-91. Audiovisual Equipment and Systems—Rating Plates—Marking of Electricity Supply First Edition (ANSI IT7.105-1992). 12 pp.

—Printed Circuit Boards
IEC 603 Pt 11-92. Connectors for Frequencies Below 3 MHz for Use with Printed Boards Part 11: Detail Specification for Concentric Connectors (Dimensions for Free Connectors and Fixed Connectors) First Edition. 50 pp.

—Quick Disconnect—Headsets—Preferred Products List
CECC CECC MUAHAG Vol 3A IS 4-91. Preferred Products List; Connectors; L. F. (En, Fr, Ge). 46 pp.

—Quick Disconnect—Microphones—Preferred Products List
CECC CECC MUAHAG Vol 3A IS 4-91. Preferred Products List; Connectors; L. F. (En, Fr, Ge). 46 pp.

Audio Detectors
Use For: Acoustic Intrusion Detectors; Sound Sensing and Detection Systems *See Also:* Burglar Alarms

—Burglar
BSI BS 4737: Sec 3.6-78. 1978 Amd 1 Intruder Alarm Systems in Buildings Part 3: Components Section 3.6: Acoustic Detectors. 4 pp.

Audio Equipment
Use For: Audio Systems and Equipment; Sound System Equipment *See Also:* Audio Amplifiers; Audio Cassettes; Audio Selector Panels; Audio Systems; Audio Tapes; Audiovisual Equipment; Compact Disk Players; Compact Disks; Conference Communication Systems; Earphones; Entertainment Equipment; Handsets; Headphones; Intercom Equipment; Microphones; Phonographs; Program Level Meters; Sound Recording and Reproduction Equipment; Sound Studios; Speakers; Stylus; Tape Players (Audio); Tape Recorders; Tuners

BSI BS 5942: Part 1-80. 1980 High Fidelity Audio Equipment and Systems; Minimum Performance Requirements Part 1: General Requirements. 5 pp.

BSI BS 5942: Part 8-87. 1987 High Fidelity Audio Equipment and Systems; Minimum Preformance Requirements Part 8: Combination Equipment. 22 pp.

BSI BS 6840: Part 1-87. 1987 Sound System Equipment Part 1: Methods for Specifying and Measuring General Characteristics Used for Equipment Performance (IEC 268-1: 1985). 20 pp.

BSI BS 6840: Part 1-01. 1987 Amd 1 Sound System Equipment Part 1: Methods for Specifying and Measuring General Characteristics Used for Equipment Performance (IEC 268-1: 1985) (AMD 6921) May 1, 1992. 23 pp.

CENELEC HD 483.1 S2-89. Sound System Equipment—Part 1: General. 3 pp.

CENELEC HD 483.2 S1-89. Sound System Equipment-Part 2: Explanation of General Terms and Calculation Methods. 3 pp.

CENELEC HD 483.2 S2-93. Sound System Equipment Part 2: Explanation of General Terms and Calculation Methods (IEC 268-2:1987 + A1: 1991). 5 pp.

CNS C6122-81. Sound Systems (Jul)(7644).

IEC 268 Pt 1-85. Sound System Equipment Part 1: General Second Edition; (Amendment 1-2:1988). 47 pp.

IEC 268 Pt 2-87. Sound System Equipment Part 2: Explanation of General Terms and Calculation Methods Second Edition (Amendment 1-1991) (CENELEC HD 483.2 S2: 1993). 40 pp.

IEC 581 Pt 1-77. High Fidelity Audio Equipment and Systems; Minimum Performance Requirements Part 1: General First Edition. 10 pp.

IEC 581 Pt 8-86. High Fidelity Audio Equipment and Systems; Minimum Performance Requirements Part 8: Combination Equipment First Edition. 47 pp.

SAA AS 1127.1-89. Sound System Equipment—Part 1: General (IEC 268-1:1985). 16 pp.

Audio Equipment (Cont.)
SAA AS 1127.2-89. Sound System Equipment—Part 2: Explanation of General Terms (IEC 268-2:1987). 13 pp.

—Auditoriums
BSI BS 6840: Part 16-89. 1989 Sound System Equipment Part 16: Guide to the 'RASTI' Method for the Objective Rating of Speech Intelligibility in Auditoria. 16 pp.

IEC 268 Pt 16-88. Sound System Equipment Part 16: The Objective Rating of Speech Intelligibility in Auditoria by the "RASTI" Method First Edition. 31 pp.

—Automatic Gain Control
BSI BS 6840: Part 8-88. 1988 Sound System Equipment Part 8: Methods for Specifying and Measuring the Characteristics of Automatic Gain Control Devices. 16 pp.

IEC 268 Pt 8-73. Sound System Equipment Part 8: Automatic Gain Control Devices First Edition. 27 pp.

—Automotive
BSI BS 5942: Part 11-89. 1989 High Fidelity Audio Equipment and Systems; Minimum Performance Requirements Part 11: High Fidelity Systems for Use in Vehicles (For Example, Motor Cars); General Requirements. 7 pp.

BSI BS 5942: Part 13-89. 1989 High Fidelity Audio Equipment and Systems; Minimum Performance Requirements Part 13: High Fidelity Systems for Use in Vehicles (For Example, Motor Cars); FM Radio Tuner Units. 10 pp.

IEC 581 Pt 11-81. High Fidelity Audio Equipment and Systems; Minimum Performance Requirements Part 11: High Fidelity Systems for Use in Vehicles (for Example, Motor Cars) First Edition. 13 pp.

IEC 581 Pt 13-88. High Fidelity Audio Equipment and Systems; Minimum Performance Requirements Part 13: High Fidelity Systems for Use in Vehicles (for Example, Motor Cars); FM Radio Tuner Units First Edition. 19 pp.

—Broadcast Receivers—Interfaces
CCIR OPINION 74-1-90. Systems for Signal Interface Connection Between Sound-Broadcasting Receivers and Associated Equipment—Volume X-1—Broadcasting Service (Sound). 1 p.

—Cassette—Automobiles
CNS C6268-86. Method of Test for Cassette Players for Automobile Use (Oct)(11717).

—Compact Disks—Address Codes
BSI BS 6288: Part 11-89. 1989 Magnetic Tape Sound Recording and Reproducing Systems Part 11: Guide to an Address Code for Compact Cassettes. 11 pp.

—Digital—Interfaces—Data Channels—Formats
CCIR RECMN 776-92. Format for User Data Channel of the Digital Audio Interface—Section 10C—Audio-Frequency Characteristics of Sound-Broadcasting Signals. 18 pp.

—Digital—Quality Assurance
CCIR Report 1070-86. Quality Measurement Methods for Digital Sound—Section 10C—Audio-Frequency Characteristics of Sound-Broadcasting Signals. 2 pp.

—Digital—Serial Interfaces—Sound Broadcasting
CCIR RECMN 647-1-90. Digital Audio Interface for Broadcasting Studios —Section 10C—Audio-Frequency Characteristics of Sound-Broadcasting Signals. 14 pp.

CCIR RECMN 647-2-92. Digital Audio Interface for Broadcasting Studios —Section 10C—Audio-Frequency Characteristics of Sound-Broadcasting Signals. 21 pp.

—Glossaries
BSI BS 6840: Part 2-88. 1988 Amd 1 Sound System Equipment Part 2: Glossary of General Terms and Calculation Methods (AMD 6412) September 30, 1991. 19 pp.

—Interconnection—Matching Values
BSI BS 6840: Part 15-92. 1992 Sound System Equipment Part 15: Specification for Matching Values for the Interconnection of Sound System Components (IEC 268-15: 1987). 27 pp.

BSI BS 6840: Part 15-01. 1992 Amd 1 Sound System Equipment Part 15: Specification for Matching Values for the Interconnection of Sound System Components (IEC 268-15: 1987) (AMD 7700) July 15, 1993 (S). 32 pp.

BSI BS 6840: Part 15-88. 1988 Sound System Equipment Part 15: Matching Values for the Interconnection of Sound System Components. 21 pp.

INDUSTRY STANDARDS

INTERNATIONAL AND NON-U.S. NATIONAL STANDARDS
SUBJECT INDEX
Audio

Audio Equipment (Cont.)
—Interconnection—Matching Values (Cont.)
CENELEC HD 483.15 S1-89. Sound System Equipment-Part 15: Prefered Matching Valves for the Interconnection of Sound System Components. 3 pp.
CENELEC HD 483.15 S2-90. Sound System Equipment Part 15: Preferred Matching Values for the Interconnection of Sound System Components. 3 pp.
CENELEC HD 483.15 S3-91. Sound System Equipment Part 15: Preferred Matching Values for the Interconnection of Sound System Components. 7 pp.
CENELEC HD 483.15 S4-92. Sound System Equipment Part 15: Preferred Matching Values for the Interconnection of Sound System Components (IEC 268-15:1987 + A1: 1989 + A2:1990 + A3: 1991). 6 pp.
IEC 268 Pt 15-87. Sound System Equipment Part 15: Preferred Matching Values for the Interconnection of Sound System Components Second Edition; (Amendment 1-1989) (Amendment 2-1990) (Amendment 3-1991) (CENELEC HD 483.15 S4-1992). 58 pp.

—Interfaces
BSI BS 7239-89. 1989 Digital Audio Interface (S) (IEC 958: 1989). 26 pp.
BSI BS 7239-01. 1989 Amd 1 Digital Audio Interface (IEC 958: 1989) (AMD 6413) October 31, 1991. 31 pp.
CCIR OPINION 90-90. Equipment Interconnection in Professional Programme Production Installations—Volumes X and XI—Part 3—Sound and Television Recording. 1 p.
CENELEC EN 60 958-90. Digital Audio Interface. 4 pp.
IEC 958-89. Digital Audio Interface First Edition; (Amendment 1-1993). 91 pp.

—Packaging
CNS C5069-80. Packing Tests for Radio, Radio-Phonograph, High Fidelity Equipments and Recorders (Aug)(6318).
CNS C6302-87. Method of Test for Packaging and Appearance Audio Product (Oct)(12119).

—Passive Elements
BSI BS 6840: Part 6-87. 1987 Sound System Equipment Part 6: Methods for Specifying and Measuring the Characteristics of Auxilary Passive Elements. 39 pp.
IEC 268 Pt 6-71. Sound System Equipment Part 6: Auxiliary Passive Elements First Edition. 69 pp.

—Receivers—Intermediate Frequencies
CNS C5040-80. Intermediate Frequencies for Entertainment Receivers (May)(5552).

—Reliability
CNS C6303-87. Method of Test for Reliability of Household Audio Product (Oct)(12120).

—Symbols
CNS C1143-8-89. Graphical Symbols for Use on Electrical and Electronic Equipment (Sound, Speak, Listen) (Feb)(12491-8).

—Visual Inspection
CNS C6302-87. Method of Test for Packaging and Appearance Audio Product (Oct)(12119).

—Volume Indicators
BSI BS 6840: Part 17-91. 1991 Sound System Equipment Part 17: Methods for Specifying and Measuring the Characteristics of Standard Volume Indicators (IEC 268-17: 1990). 14 pp.
BSI BS 6840: Part 17-01. 1991 Amd 1 Sound System Equipment Part 17: Methods for Specifying and Measuring the Characteristics of Standard Volume Indicators (IEC 268-17: 1990) (AMD 7701) July 15, 1993 (S). 23 pp.
CENELEC HD 483.17 S1-92. Sound System Equipment Part 17: Standard Volume Indicators. 5 pp.
IEC 268 Pt 17-90. Sound System Equipment Part 17: Standard Volume Indicators First Edition. 25 pp.

Audio Frequencies
See Also: Radio Frequencies; Sound Recording; Sound Recording and Reproduction Equipment
ISO 266-75. Acoustics—Preferred Frequencies for Measurements First Edition. 4 pp.
SNZ NZS/ISO 266-75. Acoustics. Preferred Frequencies for Measurements. 2 pp.

—Noise Voltage—Sound Broadcasting
CCIR RECMN 468-4-86. Measurement of Audio-Frequency Noise Voltage Level in Sound Broadcasting—Section 10C—Audio-Frequency Characteristics of Sound-Broadcasting Signals. 7 pp.

Audio Frequencies (Cont.)
—Radio Relay Systems—Telephony—Interfaces
CCIR RECMN 268-1-70. Interconnection at Audio Frequencies of Radio-Relay Systems for Telephony—Section 9C—Interconnection Characteristics (Baseband and Intermediate Frequency). 1 p.

—Symbols
CNS C5132-82. Graphic Symbols for Frequencies, Bands, Modulation and Frequency Diagrams (Apr)(8664).

—Telecommunication Circuits
CCITT FASCICLE III.2-88. International Analogue Carrier Systems Recommendations G.211-G.544. 226 pp.

Audio Frequency Transformers
Use For: Audio Transformers See Also: Output Transformers; Transformers
CNS C7129-80. Audio Transformers for Electrical Equipment (Sep)(6433).

—Telephone Cables
CCITT RECMN K.1-89. Connection to Earth of an Audio-Frequency Telephone Line in Cable—Protection Against Interference (Study Group V) 1 pp. 1 p.

Audio Pages
Scope Note: A sheet having a printing/writing surface on one side and a magnetic coating suitable for the recording and reproduction of audio information on the other side See Also: Audiovisual Equipment
BSI BS 5817: Part 15-85. 1985 Audio-Visual, Video and Television Equipment and Systems Part 15: Audio Pages. 8 pp.
IEC 574 Pt 15-84. Audio-Visual, Video and Television Equipment and Systems Part 15: Audio Pages First Edition. 13 pp.

Audio Selector Panels
See Also: Audio Equipment

—Aircraft
EUROCAE ED-18 07.85. Audio Systems Characteristics and MPS Aircraft Microphones (Except Carbon), Aircraft Headsets, Handsets and Loudspeakers, Aircraft Audio Selector Panels and Amplifiers. 108 pp.

—Aircraft—Environmental Testing
EUROCAE ED-18 07.85. Audio Systems Characteristics and MPS Aircraft Microphones (Except Carbon), Aircraft Headsets, Handsets and Loudspeakers, Aircraft Audio Selector Panels and Amplifiers. 108 pp.

Audio Signals
Use: Acoustic Signals

Audio Systems
Use For: Sound Systems See Also: Audio Equipment; Phonograph Pickups; Radio Paging Systems; Sound-Program Signals; Sound Recording
BSI BS 5942: Part 1-80. 1980 High Fidelity Audio Equipment and Systems; Minimum Performance Requirements Part 1: General Requirements. 5 pp.
BSI BS 5942: Part 8-87. 1987 High Fidelity Audio Equipment and Systems; Minimum Performance Requirements Part 8: Combination Equipment. 22 pp.

—Aircraft
EUROCAE ED-18 07.85. Audio Systems Characteristics and MPS Aircraft Microphones (Except Carbon), Aircraft Headsets, Handsets and Loudspeakers, Aircraft Audio Selector Panels and Amplifiers. 108 pp.

—Automotive
BSI BS 5942: Part 11-89. 1989 High Fidelity Audio Equipment and Systems; Minimum Performance Requirements Part 11: High Fidelity Systems for Use in Vehicles (For Example, Motor Cars); General Requirements. 7 pp.
BSI BS 5942: Part 13-89. 1989 High Fidelity Audio Equipment and Systems; Minimum Performance Requirements Part 13: High Fidelity Systems for Use in Vehicles (For Example, Motor Cars); FM Radio Tuner Units. 10 pp.
IEC 581 Pt 11-81. High Fidelity Audio Equipment and Systems; Minimum Performance Requirements Part 11: High Fidelity Systems for Use in Vehicles (for Example, Motor Cars) First Edition. 13 pp.
IEC 581 Pt 13-88. High Fidelity Audio Equipment and Systems; Minimum Performance Requirements Part 13: High Fidelity Systems for Use in Vehicles (for Example, Motor Cars); FM Radio Tuner Units First Edition. 19 pp.

Audio Systems (Cont.)
—Buildings
BSI BS 6259-82. 1982 Planning and Installation of Sound Systems. 29 pp.

—Emergency
BSI BS 7443-91. 1991 Sound Systems for Emergency Purposes. 13 pp.
BSI BS 7443-01. 1991 Amd 1 Sound Systems for Emergency Purposes (AMD 7675) June 15, 1993 (S). 15 pp.

—Enhanced Quality Television Systems
CCIR QUESTION 79/10-90. Suitable Sound Systems to Accompany High-Definition Television and Enhanced Television Systems—Questions Concerning Study Group 10—Broadcasting Service (Sound). 1 p.

—High Definition Television Systems
CCIR Report 1072-1-90. Suitable Sound Systems to Accompany High-Definition and Enhanced Television Systems—Section 10C—Audio-Frequency Characteristics of Sound-Broadcasting Signals. 9 pp.
CCIR QUESTION 79/10-90. Suitable Sound Systems to Accompany High-Definition Television and Enhanced Television Systems—Questions Concerning Study Group 10—Broadcasting Service (Sound). 1 p.

—Mining Equipment
BSI BS 6353-83. 1983 Underground Loudspeaker Communication Systems in Coal Mines. 8 pp.

—Multichannel—High Definition Television Systems
CCIR Decision 94-89. Multi-Channel Sound Systems (Especially Suited to Accompany High-Definition and Enhanced Television Systems)—Volume X-1—Broadcasting Service (Sound). 3 pp.

—Multichannel—Stereophonic
CCIR RECMN 775-92. Multi-Channel Stereophonic Sound System with and Without Accompanying Picture—Section 10C—Audio-Frequency Characteristics of Sound-Broadcasting Signals. 9 pp.

—Stadiums
BSI BS 6259-82. 1982 Planning and Installation of Sound Systems. 29 pp.

—Warning Systems
IEC 849-89. Sound Systems for Emergency Purposes First Edition. 25 pp.

Audio Systems and Equipment
Use: Audio Equipment

Audio Tape Recorders
Use: Tape Recorders

Audio Tapes
See Also: Audio Cassettes; Audio Equipment; Magnetic Tapes
BSI BS 5817: Part 10-92. 1992 Audiovisual, Video and Television Equipment and Systems Part 10: Specification for Audio Cassette Systems (IEC 574-10: 1983). 18 pp.
BSI BS 5817: Part 10-89. 1989 Audio-Visual, Video and Television Equipment and Systems Part 10: Audio Cassette Systems. 14 pp.
BSI BS 5942: Part 4-81. 1981 High Fidelity Audio Equipment and Systems; Minimum Performance Requirements Part 4: Magnetic Recording and Reproducing Equipment. 8 pp.
BSI BS 6288: Part 3-90. 1990 Magnetic Tape Sound Recording and Reproducing Systems Part 3: Methods of Measuring the Characteristics of Recording and Reproducing Equipment for Sound on Magnetic Tape. 27 pp.
BSI BS 6288: Part 5-89. 1989 Amd 1 Magnetic Tape Sound Recording and Reproducing Systems Part 5: Specification for Electrical Magnetic Tape Properties (AMD 6407) September 30, 1991. 33 pp.
BSI BS 6288: Part 6-87. 1987 Amd 1 Magnetic Tape Sound Recording and Reproducing Systems Part 6: Specification for Reel-to-Reel Systems (AMD 6406) September 30, 1991. 30 pp.
BSI BS 6288: Part 7-87. 1987 Magnetic Tape Sound Recording and Reproducing Systems Part 7: Cassettes for Commercial Tape Records and Domestic Use. 19 pp.
BSI BS 6288: Part 9-89. 1989 Magnetic Tape Sound Recording and Reproducing Systems Part 9: Magnetic Tape Cartridge for Professional Use. 18 pp.
CCIR RECMN 649-86. Measuring Methods for Analogue Audio Disk and Tape Recordings—Section 10/11F—Exchange of Recorded Sound Programmes. 1 p.
CCIR RECMN 649-1-92. Measuring Methods for Analogue Audio Tape Recordings—Section 10/11F—Exchange of Recorded Sound Programmes. 1 p.

INTERNATIONAL AND NON-U.S. NATIONAL STANDARDS
SUBJECT INDEX
Audiometry

Audio Tapes *(Cont.)*

CENELEC HD 311.3 S2-89. Magnetic Tape Sound Recording and Reproducing Systems Part 3: Methods of Measuring the Characteristics of Recording and Reproducing Equipment for Sound on Magnetic Tape. 3 pp.

CENELEC HD 311.6 S1-88. Magnetic Tape Sound Recording and Reproducing Systems-Part 6: Reel-to-Reel Systems. 3 pp.

CENELEC HD 311.7 S1-88. Magnetic Tape Sound Recording and Reproducing Systems-Part 7: Cassette for Commercial Tape Records and Domestic Use. 3 pp.

CENELEC HD 369.10 S2-88. Audiovisual, Video and Television Equipment and Systems—Part 10: Audio Cassette Systems. 3 pp.

CENELEC HD 369.10 S3-89. Audiovisual, Video and Television Equipment and Systems Part 10: Audio Cassette Systems. 3 pp.

CNS C1040-74. Standard Dimensions of Endless Loop Magnetic Tape Cartridges; Type I, II, III (Mar)(3620).

CNS C5026-79. Magnetic Tape Recording and Reproducing System: Dimensions and Characteristic (Apr)(4783).

CNS C5034-85. Cassette for Commercial Tape Records and Domestic Use Dimensions and Characteristics (Sep)(4953).

CNS C5089-81. Standard Dimensions for Unrecorded Magnetic Sound Recording Tape (Jan)(6898).

CNS C5129-82. Standard for Magnetic Tape Records: Four-Track Open-Reel Stereophonic Records at 9.5 and 19cm/s (3.75 and 7.5 in/s) (Jan)(8385).

CNS C6269-86. Method of Test for Magnetic Sound Recording Tapes (Nov)(11765).

CNS C7183-86. Reference Tape for Magnetic Sound Recording (Nov)(11764).

CNS C7184-86. Test Tape Records for Magnetic Sound Recording and Reproducing Equipment (Dec)(11784).

CNS S1178-84. Cartridge Tape Records (Jul)(10973).

CNS S1179-84. Cassette Tape Records (Jul)(10975).

CNS S2117-84. Method of Test for Cartridge Tape Records (Jul)(10974).

CNS S2118-84. Method of Test for Cassette Tape Records (Jul)(10976).

IEC 94 Pt 3-79. Magnetic Tape Sound Recording and Repro-ducing Systems Part 3: Methods of Measuring the Characteristics of Recording and Reproducing Equipment for Sound on Magnetic Tape First Edition; (Amendment 2-1988). 70 pp.

IEC 94 Pt 5-88. Magnetic Tape Sound Recording and Reproducing Systems Part 5: Electrical Magnetic Tape Properties First Edition. 66 pp.

IEC 94 Pt 6-85. Magnetic Tape Sound Recording and Reproducing Systems Part 6: Reel-to-Reel Systems First Edition. 38 pp.

IEC 94 Pt 7-86. Magnetic Tape Sound Recording and Reproducing Systems Part 7: Cassette for Commercial Tape Records and Domestic Use First Edition. 30 pp.

IEC 94 Pt 8-87. Magnetic Tape Sound Recording and Reproducing Systems Part 8: Eight-Track Magnetic Tape Cartridge for Commercial Tape Records and Domestic Use First Edition. 27 pp.

IEC 94 Pt 9-88. Magnetic Tape Sound Recording and Reproducing Systems Part 9: Magnetic Tape Cartridge for Professional Use First Edition. 34 pp.

IEC 574 Pt 10-83. Audiovisual, Video and Television Equipment and Systems Part 10: Audio Cassette Systems Second Edition; (Amendment 2-1989 Incoporating Amendment 1). 32 pp.

IEC 581 Pt 4-79. High Fidelity Audio Equipment and Systems; Minimum Performance Requirements Part 4: Magnetic Recording and Reproducing Equipment First Edition. 17 pp.

JIS C 5564-91. Magnetic Tape Sound Recording and Reproducing Systems Part 3: Methods of Measuring the Characteristics of Recording and Reproducing Equipment for Sound on Magnetic Tape; (IEC 94-3).

JIS C 5566-89. Magnetic Tape Sound Recording and Reproducing Systems Part 5: Electrical Magnetic Tape Properties; (IEC 94-5-1988).

JIS C 5568-89. Magnetic Tape Sound Recording and Reproducing Systems Part 7: Cassette for Commercial Tape Record and Domestic Use; (IEC 94-7-1986).

JIS S 8603-85. Cartridge Tape Records. 26 pp.

JIS S 8604-86. Cassette Tape Records.

—Digital

CCIR Report 950-2-90. Digital Recording of Audio Signals—Section 10/11F—Exchange of Recorded Sound Programmes. 2 pp.

—Handling

BSI BS 4783: Part 4-88. 1988 Storage, Transportation and Maintenance of Magnetic Media for Use in Data Processing and Information Storage Part 4: Recommendations for Magnetic Tape Cartridges and Cassettes. 10 pp.

Audio Tapes *(Cont.)*
—Handling *(Cont.)*

BSI BS 4783: Part 4-01. 1988 Amd 1 Storage, Transportation and Maintenance of Media for Use in Data Processing and Information Storage Part 4: Recommendations for Magnetic Tape Cartridges and Cassettes (AMD 6890) February 28, 1992. 11 pp.

—Identification Systems

BSI BS 5817: Part 16-87. 1987 Audio-Visual, Video and Television Equipment and Systems Part 16: Labelling for Educational Audio Cassettes. 7 pp.

IEC 574 Pt 16-87. Audio-Visual, Video and Television Equipment and Systems Part 16: Labelling for Educational Audio Cassettes First Edition; (ANSI IT7.401-89). 13 pp.

—Reel to Reel

JIS C 5567-89. Magnetic Tape Sound Recording and Reproducing Systems Part 6: Reel-to-Reel Systems; (IEC 94-6-1985).

—Storage

BSI BS 4783: Part 4-88. 1988 Storage, Transportation and Maintenance of Magnetic Media for Use in Data Processing and Information Storage Part 4: Recommendations for Magnetic Tape Cartridges and Cassettes. 10 pp.

BSI BS 4783: Part 4-01. 1988 Amd 1 Storage, Transportation and Maintenance of Media for Use in Data Processing and Information Storage Part 4: Recommendations for Magnetic Tape Cartridges and Cassettes (AMD 6890) February 28, 1992. 11 pp.

Audio Track 2
Use: Sound Tracks

Audio Transformers
Use: Audio Frequency Transformers

Audio Visual Equipment
Use: Audiovisual Equipment

Audiographic Conferencing Services
Use For: AGC Services *See Also:* Audiovisual Services; Teleconferencing Services; Telephone Services

CCITT RECMN F.701-89. Teleconference Service—Telematic, Data Transmission and Teleconference Services —Operations and Quality of Service (Study Group I) 13 pp (Renumbered from Recmn F.710). 13 pp.

CCITT RECMN F.710-91. General Principles for Audiographic Conference Service—Telematic, Data Transmission and Teleconference Services Operations and Quality of Service (Study Group I) 15 pp. 15 pp.

—Integrated Services Digital Networks

CENELEC PRETS 300 101-90. Integrated Services Digital Network (ISDN) International Digital Audiographic Teleconference (T/N 33-01). 185 pp.

CENELEC PRI-ETS 300 101-92. Integrated Services Digital Network (ISDN); International Digital Audiographic Teleconference. 179 pp.

CENELEC I-ETS 300 101-93. Integrated Services Digital Network (ISDN); International Digital Audiographic Teleconference. 179 pp.

ETSI PRETS 300 101-90. Integrated Services Digital Network (ISDN) International Digital Audiographic Teleconference (T/N 33-01). 181 pp.

ETSI I-ETS 300 101-93. Integrated Services Digital Network (ISDN); International Digital Audiographic Teleconference. 179 pp.

ETSI PRI-ETS 300 101-92. Integrated Services Digital Network (ISDN); International Digital Audiographic Teleconference. 179 pp.

ETSI PRETS 300 101-90. Integrated Services Digital Network (ISDN) International Digital Audiographic Teleconference (T/N 33-01). 185 pp.

Audiometers
See Also: Acoustic Measuring Instruments; Audiometry; Medical Electrical Equipment; Medical Equipment; Occluded Ear Simulators

BSI BS 5966-80. 1980 Audiometers. 22 pp.

CSA CAN3-Z107.4-M86. Pure Tone Air Conduction Audiometers for Hearing Conservation and for Screening; (Gen Instr 1). 27 pp.

IEC 645 Pt 1-92. Audiometers Part 1: Pure-Tone Audiometers First Edition; (Corrigendum—Feb 1993). 59 pp.

—Calibration

BSI BS 2497-92. 1992 Standard Reference Zero for the Calibration of Pure Tone Air Conduction Audiometers (ISO 389: 1991) (Supersedes BS 2497: Parts 5,6 &7: 1980) (N). 14 pp.

BSI BS 2497: Part 5-88. (WITHDRAWN) 1988 Reference Zero for the Calibration of Pure-Tone Audiometers Part 5: Standard Reference Zero Using an Acoustic Coupler Complying with BS 4668 (Superseded by BS 2497: 1992). 8 pp.

Audiometers *(Cont.)*
—Calibration *(Cont.)*

BSI BS 2497: Part 7-88. (WITHDRAWN) 1988 Reference Zero for the Calibration of Pure-Tone Audiometers Part 7: Standard Reference Zero at Frequencies Intermediate Between Those Given in Parts 5 and 6 of BS 2497 (Superseded by BS 2497: 1992). 7 pp.

ISO 389-91. Acoustics—Standard Reference Zero for the Calibration of Pure-Tone Air Conduction Audiometers Third Edition; (Incorporating Addenda 1 and 2). 11 pp.

SAA AS 1591.2-87. Acoustics—Instrumentation for Audiometry—Part 2: Reference Zero for the Calibration of Pure-Tone Audiometers. 2 pp.

—Calibration—Artificial Ears

BSI BS 2497: Part 6-88. (WITHDRAWN) 1988 Reference Zero for the Calibration of Pure-Tone Audiometers Part 6: Standard Reference Zero Using an Artificial Ear Complying with BS 4669 (Superseded by BS 2497: 1992). 7 pp.

IEC 318-70. An IEC Artificial Ear, of the Wide Band Type, for the Calibration of Earphones Used in Audiometry First Edition. 17 pp.

—Calibration—Bone Vibrators

BSI BS 4009-91. 1991 Artificial Mastoids for the Calibration of Bone Vibrators Used in Hearing Aids and Audiometers (IEC 373: 1990). 16 pp.

BSI BS 6950-88. 1988 Standard Reference Zero for the Calibration of Pure Tone Bone Conduction Audiometers. 13 pp.

BSI BS 6950-01. 1988 Amd 1 Standard Reference Zero for the Calibration of Pure Tone Bone Conduction Audiometers (ISO 7566: 1987) (AMD 7340) October 15, 1992. 17 pp.

CEN EN 27566-91. Acoustics—Standard Reference Zero for the Calibration of Pure-Tone Bone Conduction Audiometers. 3 pp.

CENELEC HD 590 S1-91. Mechanical Coupler for Measurements on Bone Vibrators. 5 pp.

IEC 373-90. Mechanical Coupler for Measurements on Bone Vibrators Second Edition. 29 pp.

ISO 7566-87. Acoustics—Standard Reference Zero for the Calibration of Pure-Tone Bone Conduction Audiometers First Edition. 12 pp.

SAA AS 1591.4-74. Acoustics—Instrumentation for Audiometry—Part 4: Mechanical Coupler for Calibration of Bone Vibrators Used in Hearing Aids and Audiometers. 11 pp.

—Calibration—Earphones

BSI BS 4668-71. 1971 An Acoustic Coupler (IEC Reference Type) for the Calibration of Earphones Used in Audiometry. 11 pp.

IEC 303-70. IEC Provisional Reference Coupler for the Calibration of Earphones Used in Audiometry First Edition. 12 pp.

—Calibration—Masking Noise

BSI BS 7113-89. 1989 Reference Levels for Narrow-Band Masking Noise. 8 pp.

BSI BS 7113-01. 1989 Amd 1 Reference Levels for Narrow-Band Masking Noise (ISO 8798: 1987) (AMD 7342) October 15, 1992. 13 pp.

CEN EN 28798-91. Acoustics—Reference Levels for Narrow-Band Masking Noise. 3 pp.

ISO 8798-87. Acoustics—Reference Levels for Narrow-Band Masking Noise First Edition. 8 pp.

—Diagnostic

JIS T 1201-82. Diagnostic Audiometers (R 1987). 17 pp.

Audiometric Testing
Use: Audiometry

Audiometry
Use For: Auditory Threshold *See Also:* Audiometers

BSI BS 6951-88. 1988 Threshold of Hearing by Air Conduction as a Function of Age and Sex for Otologically Normal Persons. 11 pp.

BSI BS 6951-01. 1988 Amd 1 Threshold of Hearing by Air Conduction as a Function of Age and Sex for Otologically Normal Persons (ISO 7029: 1984) (AMD 7341) October 15, 1992. 16 pp.

CEN EN 27029-91. Acoustics—Threshold of Hearing by Air Conduction as a Function of Age and Sex Otologically Persons. 3 pp.

ISO 7029-84. Acoustics—Threshold of Hearing by Air Conduction as a Function of Age and Sex for Otologically Normal Persons First Edition. 10 pp.

SAA AS 1591. Acoustics—Instrumentation for Audiometry See Also Z43.

—Pure Tone Air Conduction Threshold

BSI BS 6655-01. 1986 Amd 1 Pure Tone Air Conduction Threshold Audiometry for Hearing Conservation Purposes (AMD 7339) October 15, 1992. 17 pp.

CEN EN 26189-91. Acoustics—Pure Tone Air Conduction Threshold Audiometry for Hearing Conservation Purposes. 11 pp.

INDUSTRY STANDARDS

Audiometry (Cont.)
—Pure Tone Air Conduction Threshold (Cont.)
CSA CAN/CSA-Z107.6-M90. Pure Tone Air Conduction Threshold Audiometry for Hearing Conservation; (Gen Instr 1). 30 pp.
ISO 6189-83. Acoustics—Pure Tone Air Conduction Threshold Audiometry for Hearing Conservation Purposes First Edition. 11 pp.
ISO 8253 Pt 1-89. Acoustics—Audiometric Test Methods—Part 1: Basic Pure Tone Air and Bone Conduction Threshold Audiometry First Edition. 18 pp.

—Pure Tone Bone Conduction Threshold
ISO 8253 Pt 1-89. Acoustics—Audiometric Test Methods—Part 1: Basic Pure Tone Air and Bone Conduction Threshold Audiometry First Edition. 18 pp.

—Test Signals—Narrowband
BSI BS 7636-93. 1993 Method for Determination of Thresholds of Hearing Using Sound Field Audiometry with Pure Tone and Narrow-Band Test Signals (ISO 8253-2 1992) (S). 20 pp.
ISO 8253 Pt 2-92. Acoustics—Audiometric Test Methods—Part 2: Sound Field Audiometry with Pure Tone and Narrow-Band Test Signals First Edition. 16 pp.

—Test Signals—Pure Tone
BSI BS 7636-93. 1993 Method for Determination of Thresholds of Hearing Using Sound Field Audiometry with Pure Tone and Narrow-Band Test Signals (ISO 8253-2 1992) (S). 20 pp.
ISO 8253 Pt 2-92. Acoustics—Audiometric Test Methods—Part 2: Sound Field Audiometry with Pure Tone and Narrow-Band Test Signals First Edition. 16 pp.

Audiovisual Documents
—Glossaries
ISO 5127 Pt 11-87. Documentation and Information—Vocabulary—Part 11: Audio-Visual Documents Second Edition. 15 pp.

Audiovisual Equipment
Use For: Audio Visual Equipment; Audiovisual Systems and Equipment *See Also:* Audio Equipment; Audio Pages; Charts; Education; Filmstrips; Magnetic Cards; Maps; Phonographs; Projectors; Screens (Projection); Slide Projectors; Slides (Photographic); Tape Recorders; Television Equipment; Training Films; Transparencies; Video Disks

BSI BS 5817: Part 1-80. 1980 Audio-Visual, Video and Television Equipment and Systems Part 1: General. 10 pp.
CENELEC HD 369.1-78. Audio-Visual, Video and Television Equipment and Systems Part 1: General. 2 pp.
IEC 574 Pt 1-77. Audio-Visual, Video and Television Equipment and Systems Part 1: General First Edition. 18 pp.

—Broadcasting—Interfaces
CCIR QUESTION 115/11-90. Interconnection Specifications for Audiovisual Equipment Related to Broadcasting—Questions Concerning Study Group 11—Broadcasting Service (Television). 1 p.

—Connectors
BSI BS 5817: Part 3-89. 1989 Audio-Visual, Video and Television Equipment and Systems Part 3: Connectors for the Interconnection of Equipment in Audio-Visual Systems. 15 pp.
CENELEC HD 369.3-86. Audio-Visual, Video and Television Equipment and Systems Part 3: Connectors for the Interconnection of Equipment in Audio-Visual Systems. 2 pp.
IEC 574 Pt 3-83. Audio-Visual, Video and Television Equipment and Systems Part 3: Connectors for the Interconnection of Equipment in Audio-Visual Systems First Edition. 27 pp.

—Connectors—Identification Systems
IEC 1062-91. Audiovisual Equipment and Systems—Rating Plates—Marking of Electricity Supply First Edition (ANSI IT7.105-1992). 12 pp.

—Connectors—Preferred Matching Values
BSI BS 5817: Part 4-89. 1989 Audio-Visual, Video and Television Equipment and Systems Part 4: Matching Values for the Interconnection of Equipment in a System. 14 pp.
BSI BS 5817: Part 4-01. 1989 Amd 1 Audiovisual, Video and Television Equipment and Systems Part 4: Specification for Matching Values for the Interconnection of Equipment in a System (IEC 574-4: 1982) (AMD 7857) September 15, 1993 (S). 25 pp.

—Connectors—Preferred Matching Values (Cont.)
CENELEC HD 369.4-84. Audio-Visual, Video and Television Equipment and Systems Part 4: Preferred Matching Valves for the Interconnection of Equipment in a System. 2 pp.
CENELEC HD 369.4 S2-93. Audiovisual, Video and Television Equipment and Systems Part 4: Preferred Matching Values for the Interconnection of Equipment in a System (IEC 574-4:1982 + A1: 1991). 6 pp.
IEC 574 Pt 4-82. Audiovisual, Video and Television Equipment and Systems Part 4: Preferred Matching Values for the Interconnection of Equipment in a System First Edition; (Amendment 1-1991) (CENELEC HD 369.4 S2: 1993). 33 pp.
IEC 933 Pt 2-91. Audio, Video and Audiovisual Systems—Interconnections and Matching Values Part 2: 21-Pin Connector for Video Systems—Application No.2 First Edition. 18 pp.

—Control Equipment
BSI BS 5817: Sec 5.1-81. 1981 Amd 1 Audio-Visual, Video and Television Equipment and Systems Part 5: Control, Synchronization and Address Codes Section 5.1: Synchronize Tape/Visual Operating Practice (AMD 6286) December 21, 1990. 10 pp.
CENELEC HD 369.5-87. Audio-Visual, Video and Television Equipment and Systems Part 5: Control, Synchronization and Address Codes Chapter 1: Synchronized Tape/Visual Operating Practice. 2 pp.
IEC 574 Pt 5-80. Audio-Visual, Video and Television Equipment and Systems Part 5: Control, Synchronization and Address Codes Chapter I: Synchronized Tape/Visual Operating Practice First Edition. 18 pp.

—Data Terminal Equipment—Digital Channels
CCITT RECMN H.242-90. System for Establishing Communication Between Audiovisual Terminals Using Digital Channels up to 2 Mbit/s (Study Group XV) 35 pp. 35 pp.

—Data Transmission—Message Channel
CEPT T/N 34-02 E-88. Message Channel Specification, Layers 1-6 for Use in Audiovisual Communication. 7 pp.

—Digital Counters
BSI BS 5817: Part 13-90. 1990 Audio-Visual, Video and Television Equipment and Systems Part 13: Digital Counters for Audio Cassette Systems. 8 pp.
CENELEC HD 369.13-84. Audio-Visual Video and Television Equipment and Systems Part 13: Digital Counter for Audio Cassette Systems. 2 pp.
IEC 574 Pt 13-82. Audio-Visual, Video and Television Equipment and Systems Part 13: Digital Counter for Audio Cassette Systems First Edition. 10 pp.

—Educational—Design
BSI BS 8205-85. 1985 Determining the Design of Learning Spaces Where Audio-Visual Equipment Will Be Used. 43 pp.

—Glossaries
CENELEC HD 369.2-78. Audio-Visual, Video and Television Equipment and Systems Part 2: Explanation of General Terms. 2 pp.
IEC 574 Pt 2-92. Audiovisual, Video and Television Equipment and Systems Part 2: Definition of General Terms Second Edition. 33 pp.

—Handling
BSI BS 5817: Part 7-89. 1989 Audio-Visual, Video and Television Equipment and Systems Part 7: Guide for Safe Handling and Operation of Audiovisual Equipment. 11 pp.
IEC 574 Pt 7-87. Audiovisual, Video and Television Equipment and Systems Part 7: Safe Handling and Operation of Audiovisual Equipment First Edition. 23 pp.

—Identification Systems
BSI BS 5817: Part 8-90. 1990 Audio-Visual, Video and Television Equipment and Systems Part 8: Symbols and Identification. 12 pp.
CENELEC HD 369.8 S1-88. Audio-Visual, Video and Television Equipment and Systems-Part 8: Symbols and Identification. 3 pp.
IEC 574 Pt 8-79. Audio-Visual, Video and Television Equipment and Systems Part 8: Symbols and Identification First Edition; (Amendment 1-1988). 23 pp.

—Instructional Systems—Language
BSI BS 5817: Part 17-90. 1990 Audio-Visual, Video and Television Equipment and Systems Part 17: Methods for Specifying and Measuring the Performance Characteristics of Audio Learning Systems. 22 pp.

—Instructional Systems—Language (Cont.)
IEC 574 Pt 17-89. Audiovisual, Video and Television Equipment and Systems Part 17: Audio-Learning Systems First Edition. 41 pp.

—Integrated Services Digital Networks
CENELEC PRETS 300 142-91. Integrated Services Digital Network (ISDN) and Other Digital Telecommunications Networks Audiovisual Teleservices Video Codec for Audiovisual Services at p* 64 Kbit/s (T/N 31-04). 6 pp.
CENELEC PRETS 300 143-91. Integrated Services Digital Network (ISDN) and Other Digital Telecommunications Networks Audiovisual Teleservices System for Establishing Communication Between Audiovisual Terminals Using Digital Channels up to 2048 Kbit/s. 10 pp.
CENELEC PRETS 300 143-93. Integrated Services Digital Network (ISDN); Audiovisual Services Inband Signalling Procedures for Audiovisual Terminals Using Digital Channels up to 2 048 kbit/s. 27 pp.
CENELEC PRETS 300 144-91. Integrated Services Digital Network (ISDN) and Other Digital Telecommunications Networks Audiovisual Teleservices Frame Structure for a 64 to 1920 Kbit/s Channel (T/N 32-04). 8 pp.
CENELEC PRETS 300 144-93. Integrated Services Digital Network (ISDN); Audiovisual Services Frame Structure for a 64 kbit/s to 1 920 kbit/s Channel and Associated Syntax for Inband Signalling. 51 pp.
CENELEC PRETS 300 145-91. Integrated Services Digital Network (ISDN) and Other Digital Telecommunications Networks Audiovisual Teleservices Narrowband Visual Telephone Systems (T/N 32-05). 28 pp.
CENELEC PRETS 300 145-93. Integrated Services Digital Network (ISDN); Audiovisual Services Videotelephone Systems and Terminal Equipment Operating on One or Two 64 kbit/s Channels. 24 pp.
CENELEC PRETS 300 146-91. Integrated Services Digital Network (ISDN) and Other Digital Telecommunications Networks Audiovisual Teleservices Frame Synchronous Control and Indication Signals for Audiovisual Systems (T/N 32-06). 9 pp.
ETSI PRETS 300 142-91. Integrated Services Digital Network (ISDN) and Other Digital Telecommunications Networks—Audiovisual Teleservices—Video Codec for Audiovisual Services at p* 64 kbit/s (T/N 31-04).
ETSI PRETS 300 143-91. Integrated Services Digital Network (ISDN) and Other Digital Telecommunications Networks—Audiovisual Teleservices—System for Establishing Communication Between Audiovisual Terminals Using Digital Channels up to 2048 kbit/s (T/N 32-03).
ETSI PRETS 300 144-91. Integrated Services Digital Network (ISDN) and Other Digital Telecommunications Networks—Audiovisual Teleservices—Frame Structure for a 64 to 1920 kbit/s Channel (T/N 32-04).
ETSI PRETS 300 145-91. Integrated Services Digital Network (ISDN) and Other Digital Telecommunications Networks—Audiovisual Teleservices—Narrowband Visual Telephone Systems (T/N 32-05).
ETSI PRETS 300 146-91. Integrated Services Digital Network (ISDN) and Other Digital Telecommunications Networks—Audiovisual Teleservices—Frame Synchronous Control and Indication Signals for Audiovisual Systems (T/N 32-06).
ETSI PRETS 300 142-91. Integrated Services Digital Network (ISDN) and Other Digital Telecommunications Networks Audiovisual Teleservices Video Codec for Audiovisual Services at p* 64 Kbit/s (T/N 31-04). 6 pp.
ETSI PRETS 300 143-91. Integrated Services Digital Network (ISDN) and Other Digital Telecommunications Networks Audiovisual Teleservices System for Establishing Communication Between Audiovisual Terminals Using Digital Channels up to 2048 Kbit/s. 10 pp.
ETSI PRETS 300 143-93.
ETSI PRETS 300 144-91. Integrated Services Digital Network (ISDN) and Other Digital Telecommunications Networks Audiovisual Teleservices Frame Structure for a 64 to 1920 Kbit/s Channel (T/N 32-04). 8 pp.
ETSI PRETS 300 144-93. Integrated Services. 51 pp.
ETSI PRETS 300 145-91. Integrated Services Digital Network (ISDN) and Other Digital Telecommunications Networks Audiovisual Teleservices Narrowband Visual Telephone Systems (T/N 32-05). 28 pp.
ETSI PRETS 300 145-93.

INTERNATIONAL AND NON-U.S. NATIONAL STANDARDS
SUBJECT INDEX

Audiovisual Equipment (Cont.)
—Integrated Services Digital Networks (Cont.)
ETSI PRETS 300 146-91. Integrated Services Digital Network (ISDN) and Other Digital Telecommunications Networks Audiovisual Teleservices Frame Synchronous Control and Indication Signals for Audiovisual Systems (T/N 32-06). 9 pp.

—Interfaces
CCIR OPINION 90-90. Equipment Interconnection in Professional Programme Production Installations—Volumes X and XI—Part 3—Sound and Television Recording. 1 p.

—Magnetic Cards
BSI BS 5817: Part 14-85. 1985 Amd 1 Audio-Visual, Video and Television Equipment and Systems Part 14: Audio Striped Card System (AMD 6570) December 21, 1990. 24 pp.
CENELEC HD 369.14 S2-89. Audiovisual, Video and Television Equipment on Systems Part 14: Audio Striped Card System. 3 pp.
IEC 574 Pt 14-83. Audio-Visual, Video and Television Equipment and Systems Part 14: Audio Striped Card System First Edition; (Amendment 1-1988). 21 pp.

—Medical Equipment—Symbols
BSI BS 7139-89. 1989 Guide to Graphical Symbols for Use on Medical Electrical Equipment. 51 pp.

—Medical—Symbols
IEC 878-88. Graphical Symbols for Electrical Equipment in Medical Practice First Edition. 63 pp.

—Operation
BSI BS 5817: Part 7-89. 1989 Audio-Visual, Video and Television Equipment and Systems Part 7: Guide for Safe Handling and Operation of Audiovisual Equipment. 11 pp.
IEC 574 Pt 7-87. Audiovisual, Video and Television Equipment and Systems Part 7: Safe Handling and Operation of Audiovisual Equipment First Edition. 23 pp.

—Symbols
BSI BS 5817: Part 8-90. 1990 Audio-Visual, Video and Television Equipment and Systems Part 8: Symbols and Identification. 12 pp.
BSI BS 7139-89. 1989 Guide to Graphical Symbols for Use on Medical Electrical Equipment. 51 pp.
CENELEC HD 369.8 S1-88. Audiovisual, Video and Television Equipment and Systems-Part 8: Symbols and Identification. 3 pp.
IEC 574 Pt 8-79. Audiovisual, Video and Television Equipment and Systems Part 8: Symbols and Identification First Edition; (Amendment 1-1988). 23 pp.
IEC 878-88. Graphical Symbols for Electrical Equipment in Medical Practice First Edition. 63 pp.

Audiovisual Services
Use For: Audiovisual Teleservices
See Also: Audiographic Conferencing Services; Telecommunication Services; Teleconferencing Services; Video Telephone Services; Videoconferencing Services
CCIR Report 1232-90. Release of Programmes in a Multimedia Environment —Section 10/11G—Exchange of Recorded Television Programmes. 2 pp.
CCITT RECMN F.701-89. Teleconference Service—Telematic, Data Transmission and Teleconference Services —Operations and Quality of Service (Study Group I) 13 pp (Renumbered from Recmn F.710). 13 pp.
CCITT RECMN F.710-91. General Principles for Audiographic Conference Service—Telematic, Data Transmission and Teleconference Services Operations and Quality of Service (Study Group I) 15 pp. 15 pp.

—Encoder/Decoders—384 kbit/s
CCITT RECMN H.261-89. Codec for Audiovisual Services at n x 384 kBit/s Transmission of Non-Telephone Signals—Transmission of Sound-Programme and Television Signals (Study Group XV) 9 pp. 9 pp.

—Encoder/Decoders—64 kbit/s
CCITT RECMN G.725-89. System Aspects for the Use of the 7 kHz Audio Codec Within 64 kbit/s—General Aspects of Digital Transmission Systems; Terminal Equipments (Study Groups XV and XVIII) 11 pp. 11 pp.

—Encoders/Decoders—1920 kbit/s
CCITT RECMN H.261-90. Video Codec for Audiovisual Services at p x 64 kBit/s (Study Group XV) 30 pp. 30 pp.

Audiovisual Services (Cont.)
—Frame Structure—384-1920 kbit/s
CCITT RECMN H.222-89. Frame Structure for 384-1920 kBit/s Channels in Audiovisual Teleservices—Transmission of Non-Telephone Signals—Transmission of Sound-Programme and Television Signals (Study Group XV) 2 pp. 2 pp.

—Frame Structure—64 kbit/s
CCITT RECMN H.221-90. Frame Structure for a 64 to 1920 kBit/s Channnel in Audiovisual Teleservices (Study Group XV) 32 pp. 32 pp.
CCITT RECMN H.221-89. Frame Structure for a 64 kBit/s Channel in Audiovisual Teleservices—Line Transmission of Non-Telephone Signals—Transmission of Sound-Programme and Television Signals (Study Group XV) 16 pp. 16 pp.

—Framework
CCITT RECMN H.200-89. Framework for Recommendations for Audiovisual Services—Line Transmission of Non-Telephone Signals—Transmission of Sound-Programme and Television Signals (Study Group XV) 4 pp. 4 pp.

—Transmission Systems—Control and Indication Signals
CCITT RECMN H.230-90. Frame-Synchronous Control and Indication Signals for Audiovisual Systems (Study Group XV) 8 pp. 8 pp.

Audiovisual Systems and Equipment
Use: Audiovisual Equipment

Audiovisual Teleservices
Use: Audiovisual Services

Auditoriums
See Also: Theaters
—Acoustics
BSI BS 5363-76. 1976 Measurement of Reverberation Time in Auditoria. 6 pp.
ISO 3382-75. Acoustics—Measurement of Reverberation Time in Auditoria First Edition. 5 pp.

—Sound Systems and Equipment
BSI BS 6840: Part 16-89. 1989 Sound System Equipment Part 16: Guide to the 'RASTI' Method for the Objective Rating of Speech Intelligibility in Auditoria. 16 pp.
IEC 268 Pt 16-88. Sound System Equipment Part 16: The Objective Rating of Speech Intelligibility in Auditoria by the "RASTI" Method First Edition. 31 pp.

Auditors (Personnel)
See Also: Personnel
—Quality Audit—Qualification
BSI BS 7229: Part 2-91. 1991 Quality Systems Auditing Part 2: Qualification Criteria for Auditors (G) (ISO 10011-2: 1991). 11 pp.
CSA CAN/CSA-Q10011-91. Guidelines for Auditing Quality Systems (ISO 10011-1:1990, 10011-2:1991, 10011-3: 1991); (Gen Instr 1). 31 pp.
ISO 10011 Pt 2-91. Guidelines for Auditing Quality Systems—Part 2: Qualification Criteria for Quality Systems Auditors First Edition; (Corrected and Reprinted -1993). 10 pp.

Auditory Masking
—Calibration
BSI BS 7113-89. 1989 Reference Levels for Narrow-Band Masking Noise. 8 pp.
BSI BS 7113-01. 1989 Amd 1 Reference Levels for Narrow-Band Masking Noise (ISO 8798: 1987) (AMD 7342) October 15, 1992. 13 pp.
CEN EN 28798-91. Acoustics—Reference Levels for Narrow-Band Masking Noise. 3 pp.
ISO 8798-87. Acoustics—Reference Levels for Narrow-Band Masking Noise First Edition. 8 pp.

Auditory Threshold
Use: Audiometry

Auger Bits
See Also: Augers; Bits (Tools); Drill Bits
BSI BS 2054-53. 1953 Amd 1 Augers and Auger Bits. 26 pp.

Auger Drills
Use: Augers

Augers
Scope Note: For additional listings, use a more specific term See Also: Auger Bits; Drills; Earth Augers; Hand Tools; Screw Conveyors
BSI BS 2054-53. 1953 Amd 1 Augers and Auger Bits. 26 pp.

Augers (Cont.)
—Air
CNS M3028-80. Methods of Performance Test for Air Auger Drills (Dec)(6708).
JIS M 2003-77. Method of Performance Test for Air Auger Drills.

Aujeszky's Disease
See Also: Animal Diseases
—Control Measures
EC COM(82) 529-82. Proposal for a Council Directive amending Directives 64/432/EEC 72/461/EEC as regards certain measures relating to foot-and-mouth disease, Aujeszky's disease and swine vesicular disease. 5 pp.

Auoscopes
CNS T1050-81. Auoscopes Optical Light Guide System (Jul)(7754).

Auramine
CNS K7587-81. Auramine (Aug)(7823).
JIS K 8215-72. Auramine.

Aurin
Use For: Pararosolic Acid; Rosolic Acid
CNS K7420-75. Chemical Reagent (Rosolic Acid) (Aug)(1919).
JIS K 9037-61. Rosolic Acid (Pararosolic Acid).

Aurintricarboxylic Acid Triammonium Salt
Use: Aluminon

Austenitic Cast Iron
Use For: Austenitic Iron See Also: Cast Iron
ISO 2892-73. Austenitic Cast Iron First Edition. 11 pp.
SAA AS 1833-86. Iron Castings—Austenitic Cast Iron. 13 pp.
SNZ NZS/AS 1833-86. Iron Castings—Austenitic Cast Iron (This is a Joint Standard with SAA AS 1833). 13 pp.

—Chemical Analysis
BSI BS 3468-86. 1986 Austenitic Cast Iron. 18 pp.
DIN ENGL 1694-81. Austenitic Cast Iron (Sept). 9 pp.

—Mechanical Properties
BSI BS 3468-86. 1986 Austenitic Cast Iron. 18 pp.
DIN ENGL 1694-81. Austenitic Cast Iron (Sept). 9 pp.
DIN ENGL 1694 Suppl. 1-81. Austenitic Cast Iron; Reference Data on Mechanical and Physical Properties (Sept). 7 pp.

—Physical Properties
DIN ENGL 1694-81. Austenitic Cast Iron (Sept). 9 pp.
DIN ENGL 1694 Suppl. 1-81. Austenitic Cast Iron; Reference Data on Mechanical and Physical Properties (Sept). 7 pp.

Austenitic Iron
Use: Austenitic Cast Iron

Austenitic Iron Castings
See Also: Castings; Iron Castings
SAA AS 1833-86. Iron Castings—Austenitic Cast Iron. 13 pp.
SNZ NZS/AS 1833-86. Iron Castings—Austenitic Cast Iron (This is a Joint Standard with SAA AS 1833). 13 pp.

Austenitic Stainless Steels
Scope Note: For additional listings, see also specific products made from austenitic stainless steel
See Also: Alloy Steels; Austenitic Steels; Corrosion Resistant Steels; Ferritic Stainless Steels; Ferroalloys; Free Machining Steels; Heat Resistant Steels; Martensitic Stainless Steels; Nickel Chromium Molybdenum Steels; Stainless Steels; Steels

—Acid Resistance Testing
CNS G2053-86. Method of Ferric Sulfate-Sulfuric Acid Test for Stainless Steels (Jul)(4763).
CNS G2055-86. Method of Nitro-Hydrofluoric Acid Test for Stainless Steels (Jul)(4765).
CNS G2056-86. Method of Copper Sulfate-Sulfuric Acid Test for Stainless Steels (Jul)(4766).
JIS G 0572-84. Method of Ferric Sulfate-Sulfuric Acid Test for Stainless Steels. 6 pp.
JIS G 0574-80. Method of Nitric-Hydrofluoric Acid Test for Stainless Steels. 6 pp.
JIS G 0575-80. Method of Copper Sulfate-Sulfuric Acid Test for Stainless Steels. 7 pp.

INDUSTRY STANDARDS

Austenitic Stainless Steels (Cont.)

—Aerospace—Intergranular Corrosion Testing
AECMA PREN2003-11-86. Test Methods for Austenitic Stainless Steels—Part 11—Determination of Resistance to Intergranular Corrosion by the Huey Method. 1 p.
AECMA PREN2003-12-86. Test Methods for Austenitic Stainless Steels—Part 12—Determination of Resistance to Intergranular Corrosion by the Monypenny-Strauss Method. 1 p.

—Aerospace—Surface Finishing
ISO 8074-85. Aerospace—Surface Treatment of Austenitic Stainless Steel Parts First Edition. 7 pp.

—Bars
BSI BS 6258-88. 1988 Hollow Steel Bars for Machining. 14 pp.
BSI BS 6258-01. 1988 Amd 1 Hollow Steel Bars for Machining (AMD 7278) September 15, 1992. 15 pp.

—Chemical Analysis
CEN PREN 10088-1-93. Stainless Steels—Part 1: List of Stainless Steels. 19 pp.

—Electrolytic Etching
JIS G 0571-80. Method of 10 Per Cent Oxalic Acid Etch Test for Stainless Steels. 10 pp.

—Forgings
BSI BS 1503-89. 1989 Amd 1 Steel Forgings for Pressure Purposes (AMD 6739) September 30, 1991. 35 pp.
BSI BS 1503-02. 1989 Amd 2 Steel Forgings for Pressure Purposes (AMD 7744) July 15, 1993 (Q). 36 pp.
SNZ NZS/BS 1503-89. Steel Forgings for Pressure Purposes Amend: 1, 1991. 32 pp.

—Intergranular Corrosion Testing
BSI BS 5903-80. 1980 Determination of Resistance to Intergranular Corrosion of Austenitic Stainless Steels: Copper Sulphate-Sulphuric Acid Method (Moneypenny Strauss Test). 4 pp.
ISO 3651 Pt II-76. Austenitic Stainless Steels—Determination of Resistance to Intergranular Corrosion—Part II: Corrosion Test in a Sulphuric Acid/Copper Sulphate Medium in the Presence of Copper Turnings (Monypenny Strauss Test) First Edition. 5 pp.
SAA AS 2038-77. Methods for Detecting the Susceptibility of Austenitic Stainless Steel to Intergranular Corrosion Reconfirmed 1986. 8 pp.

—Loss of Mass—Corrosion Testing
DIN ENGL 50921-84. Corrosion of Metals; Testing of Austenitic Stainless Steels for Resistance to Local Corrosion in Highly Oxidizing Acids; Corrosion Test in Nitric Acid Medium by Measurement of Loss in Mass (Huey Test) (Oct). 3 pp.
ISO 3651 Pt I-76. Austenitic Stainless Steels—Determination of Resistance to Intergranular Corrosion—Part I: Corrosion Test in Nitric Acid Medium by Measurement of Loss in Mass (Huey Test) First Edition. 4 pp.

—Physical Properties
CEN PREN 10088-1-93. Stainless Steels—Part 1: List of Stainless Steels. 19 pp.

—Weld Metal—Ferrite Content
JIS Z 3119-88. Methods of Measurement for Ferrite Content in Austenitic Stainless Steel Deposited Metal. 11 pp.

Austenitic Steels
Scope Note: For additional listings, see also products made from austenitic steel See Also: Austenitic Stainless Steels; Heat Resistant Steels; Manganese Steels; Steels

—Bars
DIN ENGL 17460-92. High-Temperature Austenitic Steel Plate and Sheet, Cold and Hot Rolled Strip, Bars and Forgings; Technical Delivery Conditions (Sept). 20 pp.

—Flats—Ultrasonic Testing
BSI BS 5996-93. 1993 Acceptance Levels for Internal Imperfections in Steel Plate, Strip and Wide Flats, Based on Ultrasonic Testing (SUPERSEDES BS 4336: 1980) (V). 13 pp.

—Forgings
DIN ENGL 17460-92. High-Temperature Austenitic Steel Plate and Sheet, Cold and Hot Rolled Strip, Bars and Forgings; Technical Delivery Conditions (Sept). 20 pp.

—Grain Size Analysis
BSI BS 4490-89. 1989 Amd 1 Methods for Micrographic Determination of the Grain Size of Steel (AMD 6597) January 31, 1991. 36 pp.
CNS G2177-83. Method of Austenite Grain Size Test for Steel (Jul)(10436).
DIN ENGL 50601-85. Metallographic Examination; Determination of the Ferritic or Austenitic Grain Size of Steel and Ferrous Materials (Aug). 13 pp.
ISO 643-83. Steels—Micrographic Determination of the Ferritic or Austenitic Grain Size First Edition. 31 pp.
JIS G 0551-77. Method of Austenite Grain Size Test for Steel. 16 pp.

—Heat Treated—Gage Lengths
BSI BS 3894: Part 2-85. 1985 Conversion for Elongation Values for Steel Part 2: Method of Conversion for Application to Austenitic Steels. 32 pp.
ISO 2566 Pt 2-84. Steel—Conversion of Elongation Values—Part 2: Austenitic Steels First Edition. 30 pp.

—Plates
ISO 9328 Pt 1-91. Steel Plates and Strips for Pressure Purposes—Technical Delivery Conditions—Part 1: General Requirements First Edition. 24 pp.
ISO 9328 Pt 5-91. Steel Plates and Strips for Pressure Purposes—Technical Delivery Conditions—Part 5: Austenitic Steels First Edition. 15 pp.

—Plates—Hot Rolled
DIN ENGL 17460-92. High-Temperature Austenitic Steel Plate and Sheet, Cold and Hot Rolled Strip, Bars and Forgings; Technical Delivery Conditions (Sept). 20 pp.

—Plates—Ultrasonic Testing
BSI BS 5996-93. 1993 Acceptance Levels for Internal Imperfections in Steel Plate, Strip and Wide Flats, Based on Ultrasonic Testing (SUPERSEDES BS 4336: 1980) (V). 13 pp.

—Strips
ISO 9328 Pt 1-91. Steel Plates and Strips for Pressure Purposes—Technical Delivery Conditions—Part 1: General Requirements First Edition. 24 pp.
ISO 9328 Pt 5-91. Steel Plates and Strips for Pressure Purposes—Technical Delivery Conditions—Part 5: Austenitic Steels First Edition. 15 pp.

—Strips—Cold Rolled
DIN ENGL 17460-92. High-Temperature Austenitic Steel Plate and Sheet, Cold and Hot Rolled Strip, Bars and Forgings; Technical Delivery Conditions (Sept). 20 pp.

—Strips—Hot Rolled
DIN ENGL 17460-92. High-Temperature Austenitic Steel Plate and Sheet, Cold and Hot Rolled Strip, Bars and Forgings; Technical Delivery Conditions (Sept). 20 pp.

—Strips—Ultrasonic Testing
BSI BS 5996-93. 1993 Acceptance Levels for Internal Imperfections in Steel Plate, Strip and Wide Flats, Based on Ultrasonic Testing (SUPERSEDES BS 4336: 1980) (V). 13 pp.

—Tubes—Color Coding
ISO 9095-90. Steel Tubes—Continuous Character Marking and Colour Coding for Material Identification First Edition. 9 pp.

—Tubes—Identification Systems
ISO 9095-90. Steel Tubes—Continuous Character Marking and Colour Coding for Material Identification First Edition. 9 pp.

—Tubes—Seamless
DIN ENGL 17459-92. Seamless Circular High-Temperature Austenitic Steel Tubes; Technical Delivery Conditions (Sept). 18 pp.

Auster Aircraft
See Also: Aircraft

—Antenna Positions
CAA. Auster (All Series) and Beagle/Auster A61 (Approved Aerial Positions). 1 p.

Auster Beagle A 61 Aircraft
See Also: Aircraft

—Mandatory Aircraft Modifications and Inspections Summaries
CAA. Auster 6A and Beagle A61 Series. 2 pp.

Auster 3 Aircraft
See Also: Aircraft

—Mandatory Aircraft Modifications and Inspections Summaries
CAA. Auster 3, 4 and 5 Variants J and D Series and Taylor Craft Plus C and D. 8 pp.

Auster 4 Aircraft
See Also: Aircraft

—Mandatory Aircraft Modifications and Inspections Summaries
CAA. Auster 3, 4 and 5 Variants J and D Series and Taylor Craft Plus C and D. 8 pp.

Auster 5 Aircraft
See Also: Aircraft

—Mandatory Aircraft Modifications and Inspections Summaries
CAA. Auster 3, 4 and 5 Variants J and D Series and Taylor Craft Plus C and D. 8 pp.

Auster 6A Aircraft
See Also: Aircraft

—Mandatory Aircraft Modifications and Inspections Summaries
CAA. Auster 6A and Beagle A61 Series. 2 pp.

Australian Garment Marks
See Also: Clothing
SAA MP63-92. Australian GarmentMark—Specifications and Procedures Amdt 1 June 1992. 19 pp.

Auto Call
Use: Automatic Calling

Auto Starter Transformers
Use: Autotransformers

Autocatalytic Coatings
Use: Electroless Coatings

Autoclaves
See Also: Containers; Kettles; Laboratory Equipment; Pressure Vessels; Sterilizers
BSI BS 2646: Part 1-93. 1993 Autoclaves for Sterilization in Laboratories Part 1: Specification for Design, Construction, Safety and Performance (SUPERSEDES BS 2646: 1955). 40 pp.
BSI BS 2646: Part 1-88. 1988 Autoclaves for Sterilization in Laboratories Part 1: Specification for Design and Construction. 27 pp.
BSI BS 2646: Part 2-90. 1990 Autoclaves for Sterilization in Laboratories Part 2: Guide to Planning and Installation. 11 pp.
BSI BS 2646: Part 5-93. 1993 Autoclaves for Sterilization in Laboratories Part 5: Methods of Test for Function and Performance. 15 pp.

—Maintenance
BSI BS 2646: Part 4-91. 1991 Autoclaves for Sterilization in Laboratories Part 4: Guide to Maintenance. 9 pp.

—Tapes—Tray Wrap Material—Adhesion Testing
BSI BS 6988-89. 1989 Method for Determination of the Adhesion of Autoclave Tape to Tray Wrap Material. 4 pp.

Autoclaving

—Bricks—Water Resistance Testing
CNS R3078-77. Method of Test for Hydration Resistance of Basic Brick (Apr)(4093).
JIS R 2211-91. Testing Method for Hydration Resistance of Basic Bricks. 6 pp.

Autocollimators
See Also: Collimators; Optical Equipment
CNS B6064-82. Autocollimators (Jan)(8345).
JIS B 7538-92. Autocollimators. 7 pp.

Automated Guided Vehicle Systems
Use: Automatic Guided Vehicles

Automated Warehouses
Use: Warehouses

Automatic Answering
See Also: Automatic Calling; Automatic Clearing; Telecommunication Services; Telephone Services

—Public Switched Telephone Networks
BSI BS 6789: Sec 3.2-87. 1987 Apparatus with One or More Particular Functions for Connection to the British Telephone Network Part 3: Apparatus with Auto-Calling, Auto-Answering and Auto-Clearing Facilities Section 3.2: Auto-Answering and Auto-Clearing Facilities. 16 pp.
BSI PD 6563-92. 1992 Draft Amendment to BS 6789: Section 3.2: 1987 'Apparatus with One or More Particular Functions for Connection to Certain Public Switched Telephone Networks, Part 3: Apparatus with Auto-Calling, Auto-Answering and. 6 pp.

INTERNATIONAL AND NON-U.S. NATIONAL STANDARDS
SUBJECT INDEX

Automatic

Automatic Answering (Cont.)
—**Public Switched Telephone Networks** (Cont.)

BSI PD 6569: Sec 3.2-92. 1992 Apparatus with One or More Particular Functions for Connection to Public Switched Telephone Networks Run by Certain Public Telecommunications Operators Part 3: Apparatus with Auto-Calling, Auto-Answering and. 10 pp.

Automatic Calling
Use For: Auto Call *See Also:* Automatic Answering; Automatic Clearing; Telecommunication Services; Telephone Services

—**Public Switched Telephone Networks**

BSI BS 6789: Sec 3.1-85. 1985 Apparatus with One or More Particular Functions for Connection to the British Tele-communications Public Switched Telephone Net-work: Part 3: Apparatus with Auto-Calling, Auto-Answering and Auto-Clearing Fac. Sec 3.1: Auto-Calling Facilities. 12 pp.

BSI PD 6562-92. 1992 Draft Amendment to BS 6789: Section 3.1: 1985 'Apparatus with One or More Particular Functions for Connection to the British Telecommunications Public Switched Network, Part 3: Apparatus with Auto-Calling, Auto-Answering and. 6 pp.

BSI PD 6569: Sec 3.1-92. 1992 Apparatus with One or More Particular Functions for Connection to Public Switched Telephone Networks Run by Certain Public Telecommunications Operators Part 3: Apparatus with Auto-Calling, Auto-Answering and. 10 pp.

Automatic Clearing
See Also: Automatic Answering; Automatic Calling; Telecommunication Services; Telephone Services

—**Public Switched Telephone Networks**

BSI BS 6789: Sec 3.2-87. 1987 Apparatus with One or More Particular Functions for Connection to the British Telephone Network Part 3: Apparatus with Auto-Calling, Auto-Answering and Auto-Clearing Facilities Section 3.2: Auto-Answering and Auto-Clearing Facilities. 16 pp.

BSI PD 6563-92. 1992 Draft Amendment to BS 6789: Section 3.2: 1987 'Apparatus with One or More Particular Functions for Connection to Certain Public Switched Telephone Networks, Part 3: Apparatus with Auto-Calling, Auto-Answering and. 6 pp.

BSI PD 6569: Sec 3.2-92. 1992 Apparatus with One or More Particular Functions for Connection to Public Switched Telephone Networks Run by Certain Public Telecommunications Operators Part 3: Apparatus with Auto-Calling, Auto-Answering and. 10 pp.

Automatic Control
Scope Note: For additional listings, use a more specific term *See Also:* Automatic Control Equipment; Automatic Gain Control; Flight Dynamics; Numerical Control; Remote Control Equipment; Speed Control Testing

—**Glossaries**

CNS B8002-89. Glossary of Terms Used in Automatic Control (General) (Jul)(11057).

JIS Z 8116-72. Glossary of Terms Used in Automatic Control (General).

Automatic Control Equipment
Scope Note: For additional listings, use a more specific term *Use For:* Automatic Control Systems *See Also:* Actuators; Automatic Control; Automatic Gain Control; Automation; Building Automation Systems; Control Systems Equipment; Pneumatic Servomechanisms; Radar Equipment; Remote Control Equipment; Robots; Safety Valves; Servomotors; Servovalves; Speed Control Testing; Speed Controls; Thermostatic Valves; Thermostats; Voltage Regulators

—**Aircraft**

CAA Chapter K6-9 10.92. Flight Directors (Light Aeroplanes). 2 pp.

CAA Chapter P6-9. Directors for Take-Off, Discontinued Approach and Balked Landing (Provisional Airworthiness Requirements for Civil Powered Lift Aircraft). 7 pp.

—**Aircraft—Tubing (Flexible)—Rubber**

MOD UK DTD-329D-53. Rubber Tubing for Automatic Controls (Reprinted February 1966). 2 pp.

—**Appliances**

BSI BS 6795-87. 1987 Full Sequence Automatic Control Systems Used on Gas-Fired Appliances. 16 pp.

Automatic Control Equipment (Cont.)
—**Appliances** (Cont.)

BSI BS EN 60730-1-92. 1992 Automatic Electrical Controls for Household And Similar Use Part 1: General Requirements. 229 pp.

BSI BS EN 60730-1-01. 1992 Amd 1 Automatic Electrical Controls for Household and Similar Use Part 1: General Requirements (AMD 7747) May 15, 1993 (L). 235 pp.

BSI BS EN 60730-2-1-92. 1992 Automatic Electrical Controls for Household and Similar Use Part 2: Particular Requirements Section 2.1: Particular Controls for Electrical Household Appliances (L). 22 pp.

BSI BS EN 60730-2-1-01. 1992 Amd 1 Automatic Electrical Controls for Household and Similar Use Part 2: Particular Requirements Section 2.1: Electrical Controls for Electrical Household Appliances (AMD 7419) December 15, 1992. 28 pp.

CENELEC EN 60730-1-91. Automatic Electrical Controls for Household and Similar Use Part 1: General Requirements. 17 pp.

CENELEC EN 60730-1/A1-A11-91. AMD 1 Thru AMD 11 Automatic Electrical Controls for Household and Similar Use Part 1: General Requirements. 17 pp.

CENELEC EN 60730-1/A12-93. AMD 12 Automatic Electrical Controls for Household and Similar Use Part 1: General Requirements. 4 pp.

CENELEC EN 60730-2-1-91. Automatic Electrical Controls for Household and Similar Use Part 2: Particular Requirements for Electrical Controls for Electrical Household Appliances. 10 pp.

CENELEC EN 60730-2-1 /A11-92. AMD 11 Automatic Electrical Controls for Household and Similar Use Part 2: Particular Requirements for Electrical Controls for Electrical Household Appliances. 5 pp.

IEC 730 Pt 1-86. Automatic Electrical Controls for Household and Similar Use Part 1: General Requirements First Edition; (Amendment 1-1990) (Amendment 2-3:1991). 487 pp.

IEC 730 Pt 2-1-89. Automatic Electrical Controls for Household and Similar Use Part 2: Particular Requirements for Electrical Controls for Electrical Household Appliances First Edition. 28 pp.

IEC 730 Pt 2-6-91. Automatic Electrical Controls for Household and Similar Use Part 2: Particular Requirements for Automatic Electrical Pressure Sensing Controls Including Mechanical Requirements First Edition. 50 pp.

—**Appliances—Gas**

CEN PREN 298-89. Automatic Burner Control Systems for Gas Burners and Gas Burning Appliances with or Without Fans. 39 pp.

—**Automotive—Carburetors**

CNS D3100-89. Automatic Devices Test Code of Carburetors for Automobiles (Jan)(8257).

—**Classification**

BSI BS 1000: (681.5)-76. 1976 Amd 2 Universal Decimal Classification (UDC). English Full Edition (681.5): Automatic Control Engineering. 18 pp.

SNZ NZS/BS 1000 (681.5)-76. Universal Decimal Classification Automatic Control Engineering Amend: 1. 16 pp.

—**Engineering Drawings—Symbols**

ISO TR8545-84. Technical Drawings—Installations—Graphical Symbols for Automatic Control First Edition. 6 pp.

—**Fire Alarm and Extinguishing Equipment**

SAA AS 1603.4-87. Automatic Fire Detection and Alarm Systems—Part 4: Control and Indicating Equipment Amdt 1 June 1988 Amdt 2 October 1989. 32 pp.

—**Gas Burners**

BSI BS EN 60730-2-5-92. 1992 Automatic Electrical Controls for Household and Similar Use Part 2: Particular Requirements Section 2.5: Automatic Electrical Burner Control Systems. 30 pp.

CEN PREN 298-89. Automatic Burner Control Systems for Gas Burners and Gas Burning Appliances with or Without Fans. 39 pp.

CENELEC EN 60730-2-5-91. Automatic Electrical Controls for Household and Similar Use Part 2: Particular Requirements for Automatic Electrical Burner Control Systems. 7 pp.

IEC 730 Pt 2-5-90. Automatic Electrical Controls for Household and Similar Use Part 2: Particular Requirements for Automatic Electrical Burner Control Systems First Edition. 48 pp.

—**Glossaries**

IEC 50 Chap 351-75. International Electrotechnical Vocabulary Chapter 351: Automatic Control First Edition; (Amendment 1-1978) (Erratum—April 1979). 74 pp.

SAA AS 1852.351-78. International Electrotechnical Vocabulary—Part 351: Automatic Control Being IEC 50(351). 66 pp.

Automatic Control Equipment (Cont.)
—**Glossaries** (Cont.)

SNZ IEC 50: 50(351)-75. International Electrotechnical Vocabulary 50(351): Automatic Control Amend: 1. 58 pp.

—**Motor Compressors—Thermal Protection**

BSI BS EN 60730-2-4-93. 1993 Automatic Electrical Controls for Household and Similar Use Part 2: Particular Requirements Section 2.4: Thermal Motor Protectors for Motor-Compressors of Hermetic and Semi-Hermetic Type (L). 28 pp.

IEC 730 Pt 2-4-90. Automatic Electrical Controls for Household and Similar Use Part 2: Particular Requirements for Thermal Motor Protectors for Motor-Compressors of Hermetic and Semi-Hermetic Type First Edition. 40 pp.

—**Oil Burners**

BSI BS 799: Part 3-81. 1981 Oil Burning Equipment Part 3: Automatic and Semi-Automatic Atomizing Burners up to 36 Litres per Hour. 8 pp.

BSI BS 799: Part 6-79. (WITHDRAWN) 1979 Oil Burning Equipment Part 6: Safety Times and Safety Control and Monitoring Devices for Atomizing Oil Burners of the Monobloc Type (Superseded by BS EN 230: 1991). 13 pp.

BSI BS EN 230-91. 1991 Monobloc Oil Burners—Safety, Control and Regulation Devices and Safety Times. 22 pp.

BSI BS EN 60730-2-5-92. 1992 Automatic Electrical Controls for Household and Similar Use Part 2: Particular Requirements Section 2.5: Automatic Electrical Burner Control Systems. 30 pp.

CEN EN 230-90. Monobloc Oil Burners—Safety, Control and Regulation Devices and Safety Times. 31 pp.

CENELEC EN 60730-2-5-91. Automatic Electrical Controls for Household and Similar Use Part 2: Particular Requirements for Automatic Electrical Burner Control Systems. 7 pp.

IEC 730 Pt 2-5-90. Automatic Electrical Controls for Household and Similar Use Part 2: Particular Requirements for Automatic Electrical Burner Control Systems First Edition. 48 pp.

ISO 3544-78. Atomizing Oil Burners of the Monobloc Type—Safety Times and Safety, Control and Monitoring Devices First Edition. 12 pp.

—**Oil Burners—Safety**

BSI BS EN 230-91. 1991 Monobloc Oil Burners—Safety, Control and Regulation Devices and Safety Times. 22 pp.

CEN EN 230-90. Monobloc Oil Burners—Safety, Control and Regulation Devices and Safety Times. 31 pp.

—**Pressure Transducers**

IEC 730 Pt 2-6-91. Automatic Electrical Controls for Household and Similar Use Part 2: Particular Requirements for Automatic Electrical Pressure Sensing Controls Including Mechanical Requirements First Edition. 50 pp.

—**Radio Equipment—Emergency Relief Operations**

CCIR Decision 63-2-89. Use of Automatically Controlled Radio Systems Below 30 MHz to Provide Service to Sparsely Populated Areas and for Emergency Relief Operations—Annex to Volume III—Fixed Service at Frequencies Below About 30 MHz. 2 pp.

—**Radio Equipment—Sparsely Populated Areas**

CCIR Decision 63-2-89. Use of Automatically Controlled Radio Systems Below 30 MHz to Provide Service to Sparsely Populated Areas and for Emergency Relief Operations—Annex to Volume III—Fixed Service at Frequencies Below About 30 MHz. 2 pp.

—**Rotary Wing Aircraft**

CAA Chapter G6-4 08.82. Automatic Flight Control and Stability Augmentation Systems (Rotorcraft). 2 pp.

—**Rotary Wing Aircraft—Safety**

CAA Chapter G6-4 App 08.82. Automatic Flight Control and Stability Augmentation Systems—Safety Assessment (Rotorcraft). 2 pp.

—**Safety**

DIN VDE 0411 Pt 2-69. VDE Specifications for Electronic Measuring Instruments and Automatic Controls (Dec). 5 pp.

—**Semiconductor Devices**

OSI ISO DTR 10450-89. Industrial Automation Systems—Operating Coditions of Discrete Part Manufacturing Equipments in Industrial Environment. 9 pp.

INDUSTRY STANDARDS

Automatic Control Equipment (Cont.)

—Ships—Fuel Systems
JIS F 0803-79. Methods of Onboard Test on Automatic Control of Fuel Oil System for Smaller Ships.

—Ships—Lubricating Systems
JIS F 0804-79. Methods of Onboard Test on Automatic Control of Lubricating Oil System for Smaller Ships.

—Ships—Pneumatic Systems
JIS F 0806-79. Methods of Onboard Test on Automatic Control of Compressed Air, Bilge, Fresh Water and Sanitary Systems for Smaller Ships.

—Ships—Sanitary Systems
JIS F 0806-79. Methods of Onboard Test on Automatic Control of Compressed Air, Bilge, Fresh Water and Sanitary Systems for Smaller Ships.

—Ships—Water Systems
JIS F 0805-79. Methods of Onboard Test on Automatic Control of Cooling Water System for Smaller Ships.
JIS F 0806-79. Methods of Onboard Test on Automatic Control of Compressed Air, Bilge, Fresh Water and Sanitary Systems for Smaller Ships.

—Thermal Protectors
BSI BS EN 60730-2-2-92. 1992 Automatic Electrical Controls for Household and Similar Use Part 2: Particular Requirements Section 2.2: Thermal Motor Protectors. 34 pp.
CENELEC EN 60730-2-2-91. Automatic Electrical Controls for Household and Similar Use Part 2: Particular Requirements for Thermal Motor Protectors. 7 pp.
IEC 730 Pt 2-2-90. Automatic Electrical Controls for Household and Similar Use Part 2: Particular Requirements for Thermal Motor Protectors First Edition. 54 pp.

—Time Switches
BSI BS EN 60730-2-7-92. 1992 Automatic Electrical Controls for Household and Similar Use Part 2: Particular Requirements Section 2.7: Timers and Time Switches. 39 pp.
CENELEC EN 60730-2-7-91. Automatic Electrical Controls for Household and Similar Use Part 2: Particular Requirements for Timers and Time Switches. 8 pp.
IEC 730 Pt 2-7-90. Automatic Electrical Controls for Household and Similar Use Part 2: Particular Requirements for Timers and Time Switches First Edition. 52 pp.

—Timers
BSI BS EN 60730-2-7-92. 1992 Automatic Electrical Controls for Household and Similar Use Part 2: Particular Requirements Section 2.7: Timers and Time Switches. 39 pp.
CENELEC EN 60730-2-7-91. Automatic Electrical Controls for Household and Similar Use Part 2: Particular Requirements for Timers and Time Switches. 8 pp.
IEC 730 Pt 2-7-90. Automatic Electrical Controls for Household and Similar Use Part 2: Particular Requirements for Timers and Time Switches First Edition. 52 pp.

Automatic Control Systems
Use: Automatic Control Equipment

Automatic Controllers
Use: Automatic Control Equipment

Automatic Couplings
See Also: Couplings
CNS E1018-82. Automatic Couplers (Dec)(9699).
CNS E1019-82. Alloy Steel Casting for Automatic Couplers (Dec)(9700).
JIS E 4201-89. Automatic Couplers.

Automatic Dependent Surveillance

—Air Traffic Control
ICAO Circular 226. Automatic Dependent Surveillance—1990. 30 pp.

Automatic Direction Finders
Use For: ADF *See Also:* Aircraft Instruments; Direction Finders

—Aircraft
CAA NOTICE #85 ISSUE 2. (This Notice Gives Details of a Mandatory Modification) Automatic Direction Finding Equipment on Turbine-Engined Aeroplanes and Helicopters (Airworthiness Notices). 2 pp.
CAA Chapter R4-6 App #3 04.74. Tests for ADF Systems (Radio). 3 pp.

Automatic Direction Finders (Cont.)

—Aircraft (Cont.)
CCIR RECMN 487-74. Use of Radio-Beacon Stations for Communications—Section 8D—Radiodetermination, Global Maritime Distress and Safety System and Related Subjects. 1 p.

—Aircraft—Medium Frequency
CAA Part 5 CAP 208. MF Radio Receiving and Direction Finding Apparatus (Civil Air Publications: Airborne Radio Apparatus). 11 pp.

—Radio Frequency Interference—NAVTEX Services
CCIR Report 910-1-86. Sharing Between the Maritime Mobile Service and the Aeronautical Radionavigation Service in the Band 415-526.5 kHz—Section 8C—Maritime Mobile Service; Telephony and Related Subjects. 12 pp.

Automatic Door Openers
Use: Door Openers—Automatic

Automatic Error Correction
See Also: Error Correction

—Automatic Repeat Request—Telegraphy Systems
CCIR RECMN 518-78. Single-Channel Simplex ARQ Telegraph System—Section 3Ca—Radiotelegraph Circuits. 1 p.

—Automatic Request-Repeat Mode—Telegraphy Systems
CCIR RECMN 519-78. Single-Channel Duplex ARQ Telegraph System—Section 3Ca—Radiotelegraph Circuits. 1 p.

Automatic Exchanges
See Also: Private Automatic Branch Exchanges; Telephone Exchanges

—Circuit Noise
CCITT RECMN G.123-89. Circuit Noise in National Networks—General Characteristics of International Telephone Connections and Circuits (Study Groups XII and XV) 6 pp. 6 pp.

—Noise
CCITT RECMN Q.31-89. Noise in a National 4-Wire Automatic Exchange—General Recommendations on Telephone Switching and Signalling—Functions and Information Flows for Services in the ISDN—Supplements (Study Group XI) 1 pp. 1 p.

—Remote Control Equipment—Radiotelephone Circuits
CCIR RECMN 480-74. Semi-Automatic Operation on HF Radiotelephone Circuits Devices for Remote Connection to an Automatic Exchange by Radiotelephone Circuits—Section 3B—Radiotelephony. 2 pp.

—Transit—Routing
CCITT RECMN E.148-89. Routing of Traffic by Automatic Transit Exchanges—Telephone Network and ISDN—Operation, Numbering, Routing and Mobile Service (Study Group II) 1 pp. 1 p.

Automatic Flight Control Systems
Use: Automatic Control Equipment—Aircraft

Automatic Frequency Control
Use For: AFC *See Also:* Circuits; Frequencies

—Fixed Services (Radio Communications)—Frequency Stability
CCIR RECMN 349-4-86. Frequency Stability Required for Systems Operating in the HF Fixed Service to Make the Use of Automatic Frequency Control Superfluous—Section 3Aa—Technical Characteristics. 3 pp.

—Reference Receivers—Sound Broadcasting
CCIR RECMN 704-90. Characteristics of FM Sound Broadcasting Reference Receivers for Planning Purposes—Section 10B—Frequency-Modulation Sound Broadcasting in Bands 8 (VHF) and 9 (UHF). 4 pp.

Automatic Gain Control
See Also: Automatic Control; Automatic Control Equipment

—Audio Equipment
BSI BS 6840: Part 8-88. 1988 Sound System Equipment Part 8: Methods for Specifying and Measuring the Characteristics of Automatic Gain Control Devices. 9 pp.
IEC 268 Pt 8-73. Sound System Equipment Part 8: Automatic Gain Control Devices First Edition. 27 pp.

Automatic Gain Control (Cont.)

—Bridges (Electrical)—Conference Calling
CCITT RECMN G.172-89. Transmission Plan Aspects of International Conference Calls—General Characteristics of International Telephone Connections and Circuits (Study Groups XII and XV) 3 pp. 3 pp.

—Hearing Aids
BSI BS 6083: Part 2-84. (WITHDRAWN) 1984 Hearing Aids Part 2: Methods for Measurement of Electroacoustical Characteristics of Hearing Aids with Automatic Gain Control Circuits. 11 pp.
BSI BS 6083: Part 2-01. 1984 Amd 1 Hearing Aids Part 2: Methods for Measurement of Electroacoustical Characteristics of Hearing Aids with Automatic Gain Control Circuits (AMD 6181) October 31, 1990. 12 pp.
CENELEC HD 450.2-84. Hearing Aids Part 2: Hearing Aids with Automatic Gain Control-Circuits. 2 pp.
IEC 118 Pt 2-83. Hearing Aids Part 2: Hearing Aids with Automatic Gain Control Circuits Second Edition; (Amendment 1-1993). 36 pp.
SAA AS 1088.2-87. Hearing Aids—Part 2: Hearing Aids with Automatic Gain Control Circuits. 7 pp.

Automatic Gain Control Circuits
Use: Automatic Gain Control

Automatic Guided Vehicles
Use For: Automated Guided Vehicle Systems
See Also: Electric Vehicles; Industrial Trucks; Robots; Self Propelled Machinery; Tracked Vehicles; Vehicles (Transportation)

—Fork Trucks—Safety
JIS D 6802-90. Safety Standards of Automatic Guided Vehicles. 11 pp.

—Glossaries
JIS D 6801-90. Glossary of Terms Relating to Automatic Guided Vehicles.

—Safety
JIS D 6802-90. Safety Standards of Automatic Guided Vehicles. 11 pp.
SNZ NZS/ASME/ANSI B56.5-88. Safety Standard for Guided Industrial Vehicles. 16 pp.

—Tractors—Safety
JIS D 6802-90. Safety Standards of Automatic Guided Vehicles. 11 pp.

Automatic Number Identification
See Also: Telephone Services
CEPT T/CS 20-06-81. Identification Automatique Du Numero D'Abonne. 2 pp.
CEPT T/CS 20-06 E-81. Automatic Number Identification. 2 pp.

Automatic Pilots
See Also: Aircraft Instruments; Avionics; Pilots (Aircraft)

—Aerostats
CAA Chapter Q6-4 12.79. Automatic Pilots (Non-Rigid Airships). 1 p.

—Aircraft
CAA Chapter K7-2 App 10.72. Operating Information (Light Aeroplanes). 1 p.

—Aircraft—Design
CAA 741 (K). Section K Light Aeroplanes Autopilots and Flight Directors (Blue Papers). 15 pp.
CAA Chapter K6-4 10.92. Automatic Pilots (Light Aeroplanes). 4 pp.
CAA Chapter K6-4 App #2 10.92. Flight Demonstrations (Light Aeroplanes). 4 pp.

—Aircraft—Failure (Quality Control)
CAA 741 (K). Section K Light Aeroplanes Autopilots and Flight Directors (Blue Papers). 15 pp.
CAA Chapter K6-4 App #1 10.92. Failure Characteristics (Light Aeroplanes). 4 pp.

Automatic Radar Plotting Aids
Use For: ARPA *See Also:* Navigational Plotters; Radar Equipment
BSI BS 6919-87. (WITHDRAWN) 1987 Marine Automatic Radar Plotting Aids (ARPA) (Superseded by BS EN 60872: 1993). 22 pp.
BSI BS EN 60872-93. 1993 Marine Automatic Radar Plotting Aids (ARPA) Operational Requirements—Methods of Testing and Test Results (IEC 872: 1987) (Q). 40 pp.
CENELEC HD 526 S1-89. Marine Automatic Radar Plotting Aids (ARPA). Operational Requirements-Methods of Testing and Test Results. 3 pp.

Automatic Radar Plotting Aids (Cont.)

CENELEC EN 60872-93. Marine Automatic Radar Plotting Aids (ARPA) Operational Requirements—Methods of Testing and Test Results (IEC 872:1987) (Supersedes HD 526 S1: 1989). 4 pp.

CENELEC EN 60872/A1-93. AMD 1 Marine Automatic Radar Plotting Aids (ARPA) Operational Requirements—Methods of Testing and Test Results (IEC 872:1987/A1:1991). 4 pp.

IEC 872-87. Marine Automatic Radar Plotting Aids (ARPA) Operational Requirements—Methods of Testing and Test Results First Edition; (Amendment 1-1991) (CENELEC EN 60872/A1: 1993). 75 pp.

Automatic Repeat Attempt

See Also: Call Set Up

CCITT RECMN Q.12-89. Overflow—Alternative Routing—Rerouting—Automatic Repeat Attempt—General Recommendations on Telephone Switching and Signalling—Functions and Information Flows for Services in the ISDN —Supplements (Study Group XI) 1 pp. 1 p.

—CCITT No. 6 Signaling Systems

CCITT RECMN Q.264-89. 4.4 Potential for Automatic Repeat Attempt and Re-Routing—Specifications of Signalling System No. 6 (Study Group XI) 1 pp. 1 p.

Automatic Repeat Request Equipment

Use: Automatic Request-Repeat Equipment

Automatic Repeat Request Mode

Use: Automatic Request-Repeat Mode

Automatic Reporting and Identification System

Use For: ARIS *See Also:* Identification Systems; Radio Navigation; Radio Navigation Equipment

—Ships

CCIR Report 1164-90. Frequency Requirements for Shipborne Transponders Operating in the Frequency Bands Allocated to the Maritime Mobile Service—Section 8D—Radiodetermination, Global Maritime Distress and Safety System and Related Subjects. 7 pp.

Automatic Request-Repeat Equipment

Use For: ARQ Equipment; Automatic Repeat Request Equipment *See Also:* Automatic Request-Repeat Mode; Data Transmission; Error Control

—Radiotelegraph Circuits

CCITT RECMN F.68-89. Establishment of the Automatic Intercontinental Telex Network—Telegraph and Mobile Services Operations and Quality of Service (Study Group I) 6 pp. 6 pp.

—Radiotelegraph Circuits—Distortion

CCIR QUESTION 144/9-90. Efficiency Factor and Telegraph Distortion on ARQ Circuits—Questions of Study Group 9 Fixed Service. 1 p.

—Radiotelegraph Circuits—Efficiency Factor

CCIR QUESTION 144/9-90. Efficiency Factor and Telegraph Distortion on ARQ Circuits—Questions of Study Group 9 Fixed Service. 1 p.

—Radiotelegraphy—Maritime Mobile Services

CCIR RECMN 476-4-86. Direct-Printing Telegraph Equipment in the Maritime Mobile Service—Section 8B—Maritime Mobile Service; Telegraphy and Related Subjects. 10 pp.

—Telegraph Printers—Maritime Mobile Services

CCIR RECMN 492-4-90. Operational Procedures for the Use of Direct-Printing Telegraph Equipment in the Maritime Mobile Service—Section 8B—Maritime Mobile Service; Telegraphy and Related Subjects. 7 pp.

CCIR RECMN 492-5-92. Operational Procedures for the Use of Direct-Printing Telegraph Equipment in the Maritime Mobile Service—Section 8B—Maritime Mobile Service; Telegraphy and Related Subjects. 7 pp.

CCIR RECMN 625-1-90. Direct-Printing Telegraph Equipment Employing Automatic Identification in the Maritime Mobile Service—Section 8B—Maritime Mobile Service; Telegraphy and Related Subjects. 55 pp.

CCIR RECMN 625-2-92. Direct-Printing Telegraph Equipment Employing Automatic Identification in the Maritime Mobile Service; Telegraphy and Related Subjects. 59 pp.

Automatic Request-Repeat Mode

Use For: ARQ; Automatic Repeat Request Mode
See Also: Automatic Request-Repeat Equipment; Data Transmission; Error Control

—Automatic Error Correction—Telegraphy Systems

CCIR RECMN 518-78. Single-Channel Simplex ARQ Telegraph System—Section 3Ca—Radiotelegraph Circuits. 1 p.

CCIR RECMN 519-78. Single-Channel Duplex ARQ Telegraph System—Section 3Ca—Radiotelegraph Circuits. 1 p.

—Radiotelegraph Channels—Telex Communications

CCIR Report 436-70. Efficient Use of HF Radiotelegraph Channels in the Telex Network by Means of Automatic Selection and Allocation Procedures—Section 3Ca —Radiotelegraph Circuits. 3 pp.

—Telegraph Printers—Adaptive Coding

CCIR Report 1027-1-90. Adaptive Coding/Decoding Methods for Narrow-Band Direct-Printing Equipment—Section 8B—Maritime Mobile Service: Telegraphy and Related Subjects. 6 pp.

—Telegraphy Systems

CCIR RECMN 518-78. Single-Channel Simplex ARQ Telegraph System—Section 3Ca—Radiotelegraph Circuits. 1 p.

CCIR RECMN 519-78. Single-Channel Duplex ARQ Telegraph System—Section 3Ca—Radiotelegraph Circuits. 1 p.

—Telex Communications—Radiotelegraph Circuits

CCITT RECMN U.23-89. Use of Radiotelegraph Circuits with ARQ Equipment for Fully Automatic Telex Calls Charged on the Basis of Elapsed Time—Telegraph Switching (Study Group IX) 4 pp. 4 pp.

Automatic Services (Telephone)

Use For: International Automatic Telephone Service
See Also: Recorded Information Services (Telephone); Telecommunication Services; Telephone Services

CCITT RECMN E.105-92. International Telephone Service (Study Group I) 8 pp. 8 pp.

CCITT RECMN E.145-89. Advantages of International Automatic Service—Telephone Network and ISDN—Operation, Numbering, Routing and Mobile Service (Study Group II) 1 pp. 1 p.

—Accounting

CCITT RECMN D.140-92. Accounting Rate Principles for International Telephone Services (Study Group III) 6 pp. 6 pp.

—Accounting—Europe

CCITT RECMN D.390R-89. Accounting System in the International Automatic Telephone Service—General Tariff Principles—Charging and Accounting in International Telecommunications Services (Study Group III) 2 pp. 2 pp.

—Accounting—Mediterranean Basin

CCITT RECMN D.390R-89. Accounting System in the International Automatic Telephone Service—General Tariff Principles—Charging and Accounting in International Telecommunications Services (Study Group III) 2 pp. 2 pp.

—Call Duration—Accounting

CCITT RECMN E.260-89. Basic Technical Problems Concerning the Measurement and Recording of Call Durations—Telephone Network and ISDN—Operation, Numbering, Routing and Mobile Service (Study Group II) 4 pp. 4 pp.

—Call Duration—Data Logging

CCITT RECMN E.260-89. Basic Technical Problems Concerning the Measurement and Recording of Call Durations—Telephone Network and ISDN—Operation, Numbering, Routing and Mobile Service (Study Group II) 4 pp. 4 pp.

—Call Duration—Measurement

CCITT RECMN E.260-89. Basic Technical Problems Concerning the Measurement and Recording of Call Durations—Telephone Network and ISDN—Operation, Numbering, Routing and Mobile Service (Study Group II) 4 pp. 4 pp.

—Circuit Groups

CCITT RECMN E.520-89. Number of Circuits to be Provided in Automatic and/or Semiautomatic Operation, Without Overflow Facilities—Telephone Network and ISDN—Quality of Service, Network Management and Traffic Engineering (Study Group II) 3 pp. 3 pp.

Automatic Services (Telephone) (Cont.)

—Grade of Service

CCITT RECMN E.541-89. Overall Grade of Service for International Connections (Subscriber-to-Subscriber)—Telephone Network and ISDN—Quality of Service, Network Management and Traffic Engineering (Group Study II) 3 pp. 3 pp.

—Grade of Service—Traffic Units

CCITT RECMN E.541-89. Overall Grade of Service for International Connections (Subscriber-to-Subscriber)—Telephone Network and ISDN—Quality of Service, Network Management and Traffic Engineering (Group Study II) 3 pp. 3 pp.

—International Prefixes

CCITT RECMN E.122-89. Measures to Reduce Customer Difficulties in the International Telephone Service—Telephone Network and ISDN—Operation, Numbering, Routing and Mobile Service (Study Group II) 2 pp. 2 pp.

—Maritime Mobile Satellite Communications

CCITT RECMN G.473-89. Interconnection of a Maritime Mobile Satellite System with the International Automatic Switched Telephone Service; Transmission Aspects—International Analogue Carrier Systems (Study Group XV) 10 pp. 10 pp.

—Maritime Mobile Satellite Communications—Attenuation Distortion

CCITT RECMN G.473-89. Interconnection of a Maritime Mobile Satellite System with the International Automatic Switched Telephone Service; Transmission Aspects—International Analogue Carrier Systems (Study Group XV) 10 pp. 10 pp.

—Maritime Mobile Satellite Communications—Bandwidth

CCITT RECMN G.473-89. Interconnection of a Maritime Mobile Satellite System with the International Automatic Switched Telephone Service; Transmission Aspects—International Analogue Carrier Systems (Study Group XV) 10 pp. 10 pp.

—Maritime Mobile Satellite Communications—Crosstalk

CCITT RECMN G.473-89. Interconnection of a Maritime Mobile Satellite System with the International Automatic Switched Telephone Service; Transmission Aspects—International Analogue Carrier Systems (Study Group XV) 10 pp. 10 pp.

—Maritime Mobile Satellite Communications—Echo Control Equipment

CCITT RECMN G.473-89. Interconnection of a Maritime Mobile Satellite System with the International Automatic Switched Telephone Service; Transmission Aspects—International Analogue Carrier Systems (Study Group XV) 10 pp. 10 pp.

—Maritime Mobile Satellite Communications—EDD

CCITT RECMN G.473-89. Interconnection of a Maritime Mobile Satellite System with the International Automatic Switched Telephone Service; Transmission Aspects—International Analogue Carrier Systems (Study Group XV) 10 pp. 10 pp.

—Maritime Mobile Satellite Communications—Noise

CCITT RECMN G.473-89. Interconnection of a Maritime Mobile Satellite System with the International Automatic Switched Telephone Service; Transmission Aspects—International Analogue Carrier Systems (Study Group XV) 10 pp. 10 pp.

—Maritime Mobile Satellite Communications—Transmission Loss

CCITT RECMN G.473-89. Interconnection of a Maritime Mobile Satellite System with the International Automatic Switched Telephone Service; Transmission Aspects—International Analogue Carrier Systems (Study Group XV) 10 pp. 10 pp.

—Maritime Mobile Satellite Communications—VHF/UHF—Sidetone

CCITT RECMN 549-1-82. Side Tone Reference Equivalent of Handset Used on Board a Ship in the Maritime Mobile-Satellite Service and in Automated VHF/UHF Mari-time Mobile Radiotele-phone Systems—Section 8G—Availability, Performance Objectives and Interworking with Terrestrial Networks. 1 p.

Automatic Services (Telephone) (Cont.)

—Maritime Mobile Services—MF/HF
CCIR QUESTION 93/8-88. Automation of MF/HF Maritime Mobile Communications—Questions Concerning Study Group 8—Mobile, Radiodetermination, Amateur and Related Satellite Services. 1 p.

—Maritime Mobile Services—VHF
CCIR RECMN 689-90. Operational Procedures for an International Maritime VHF Radiotelephone System with Automatic Facilities Based on DSC Signalling Format—Section 8C—Maritime Mobile Service; Telephony and Related Subjects. 8 pp.

CCIR RECMN 689-1-92. Operational Procedures for an International Maritime VHF Radiotelephone System with Automatic Facilities Based on DSC Signalling Format—Section 8C—Maritime Mobile Service; Telephony and Related Subjects. 9 pp.

CCIR Report 1033-1-90. VHF Radiotelephone Systems with Automatic Facilities for the Maritime Mobile Service—Section 8C—Maritime Mobile Service; Telephony and Related Subjects. 6 pp.

—Maritime Mobile Services—VHF/UHF
CCIR RECMN 586-1-86. Automated VHF/UHF Maritime Mobile Telephone System—Section 8C—Maritime Mobile Service; Telephony and Related Subjects. 44 pp.

CCIR RECMN 587-1-86. Coast Station Identities and Initiation of Location Registration in an Automated VHF/UHF Maritime Mobile Telephone System—Section 8C—Maritime Mobile Service; Telephony and Related Subjects. 1 p.

—Quality of Service—Statistical Analysis
CCITT RECMN E.427-89. Collection and Statistical Analysis of Special Quality of Service Observation Data for Measurements of Customer Difficulties in the International Automatic Service—Telephone Network and ISDN—Quality of Service, Network Management and Traffic. 3 pp.

—Routing—Documentation
CCITT RECMN E.149-89. Presentation of Routing Data—Telephone Network and ISDN—Operation, Numbering, Routing and Mobile Service (Study Group II) 4 pp. 4 pp.

—Tariffs
CCITT RECMN D.101-89. Charging in Automatic International Telephone Service—General Tariff Principles—Charging and Accounting in International Telecommunications Services (Study Group III) 1 pp. 1 p.

—Tariffs—Cancelled Subscribers
CCITT RECMN D.103-89. Charging in Automatic Service for Calls Terminating on Special Services for Suspended, Cancelled or Transferred Subscribers—General Tariff Principles—Charging and Accounting in International Telecommunications Services (Study Group III) 1 pp. 1 p.

CCITT RECMN E.231-89. Charging in Automatic Service for Calls Terminating on Special Services for Suspended, Cancelled or Transferred Subscribers—Telephone Network and ISDN—Operation, Numbering, Routing and Mobile Service (Study Group II) 1 pp (Same as Recmn D.103). 1 p.

—Tariffs—Recorded Information Services
CCITT RECMN D.103 (REV 1)-92. Charging in Automatic Service for Calls Terminating on a Recorded Announcement Stating the Reason for the Call Not Being Completed (Study Group III) 4 pp (Same as Recmn E.231). 4 pp.

—Tariffs—Suspended Subscribers
CCITT RECMN D.103-89. Charging in Automatic Service for Calls Terminating on Special Services for Suspended, Cancelled or Transferred Subscribers—General Tariff Principles—Charging and Accounting in International Telecommunications Services (Study Group III) 1 pp. 1 p.

CCITT RECMN E.231-89. Charging in Automatic Service for Calls Terminating on Special Services for Suspended, Cancelled or Transferred Subscribers—Telephone Network and ISDN—Operation, Numbering, Routing and Mobile Service (Study Group II) 1 pp (Same as Recmn D.103). 1 p.

—Tariffs—Transferred Subscribers
CCITT RECMN D.103-89. Charging in Automatic Service for Calls Terminating on Special Services for Suspended, Cancelled or Transferred Subscribers—General Tariff Principles—Charging and Accounting in International Telecommunications Services (Study Group III) 1 pp. 1 p.

Automatic Services (Telephone) (Cont.)

—Tariffs—Transferred Subscribers (Cont.)
CCITT RECMN E.231-89. Charging in Automatic Service for Calls Terminating on Special Services for Suspended, Cancelled or Transferred Subscribers—Telephone Network and ISDN—Operation, Numbering, Routing and Mobile Service (Study Group II) 1 pp (Same as Recmn D.103). 1 p.

—Telephone Circuits
CCITT RECMN E.520-89. Number of Circuits to be Provided in Automatic and/or Semiautomatic Operation, Without Overflow Facilities—Telephone Network and ISDN—Quality of Service, Network Management and Traffic Engineering (Study Group II) 3 pp. 3 pp.

—Telephone Circuits—Telecommunication Traffic—Tables (Data)
CCITT FASCICLE II.3-88. Telephone Network and ISDN Quality of Service, Network Management and Traffic Engineering. Recommendations E.401-E.880. 368 pp.

—Telephone Networks
CCITT RECMN E.110-89. Organization of the International Telephone Network—Telephone Network and ISDN—Operation, Numbering, Routing and Mobile Service (Study Group II) 1 pp.
1 p.

—Time Consistent Busy Hour
CCITT RECMN E.520-89. Number of Circuits to be Provided in Automatic and/or Semiautomatic Operation, Without Overflow Facilities—Telephone Network and ISDN—Quality of Service, Network Management and Traffic Engineering (Study Group II) 3 pp. 3 pp.

Automatic Sprinklers (Fire Protection)
Use: Fire Alarm and Extinguishing Equipment

Automatic Standby Power Plants
Use: Emergency and Standby Power Supplies

Automatic Terminal Equipment
Use: Data Terminal Equipment

Automatic Test Equipment
Use For: Automatic Test Systems *See Also:* Testing Equipment

—Aircraft—Software
CAA Chapter B7-4 App 1 07.89. Automatic Test Equipment Software. 3 pp.

—Exhaust Gases—Sulfur Dioxide Content
BSI BS 6069: Sec 4.4-93. 1993 Characterization of Air Quality Part 4: Stationary Source Emissions Section 4.4: Determination of the Mass Concentration of Sulfur Dioxide—Performance Characteristics of Auto-mated Measuring Methods (ISO 7935: 1992). 16 pp.

ISO 7935-92. Stationary Source Emissions—Determination of the Mass Concentration of Sulphur Dioxide—Performance Characteristics of Automated Measuring Methods First Edition. 16 pp.

—Printed Circuit Boards
NATO STANAG 4288 ED 1 AMD 0-00. Design Criteria to Facilitate Test Capability for NATO Commuications and Associated Electronic Sub-Assemblies in NATO-Depots. 18 pp.

—Sound-Program Signals
CCIR Report 822-2-86. Transmission of Sound-Programme Signals over Long Distances Routine Automatic Testers for Sound-Programme Circuits—Section CMTT D—Methods of Operation and Assessment of Performance of Sound-Programme Transmission Channels. 7 pp.

—Telegraph Circuits
CCITT RECMN R.79-89. Automatic Tests of Transmission Quality on Telegraph Circuits Between Switching Centres—Telegraph Transmission (Study Group IX) 8 pp. 8 pp.

Automatic Test Systems
Use: Automatic Test Equipment

Automatic Transfer Switches
See Also: Electric Switches; Emergency and Standby Power Supplies

—Bypass—Isolating—120-240 V AC
CSA C22.2 NO 178-1978. Automatic Transfer Switches (R 1992); (Amd 1-2 July 1982) (Amd 3 June 1987). 51 pp.

Automatic Transfer Switches (Cont.)

—Bypass—Isolating—120-240 V DC
CSA C22.2 NO 178-1978. Automatic Transfer Switches (R 1992); (Amd 1-2 July 1982) (Amd 3 June 1987). 51 pp.

—Bypass—Isolating—250-500 V AC
CSA C22.2 NO 178-1978. Automatic Transfer Switches (R 1992); (Amd 1-2 July 1982) (Amd 3 June 1987). 51 pp.

—Bypass—Isolating—250-500 V DC
CSA C22.2 NO 178-1978. Automatic Transfer Switches (R 1992); (Amd 1-2 July 1982) (Amd 3 June 1987). 51 pp.

—Bypass—Isolating—550-1000 V AC
CSA C22.2 NO 178-1978. Automatic Transfer Switches (R 1992); (Amd 1-2 July 1982) (Amd 3 June 1987). 51 pp.

—Bypass—Isolating—550-1000 V DC
CSA C22.2 NO 178-1978. Automatic Transfer Switches (R 1992); (Amd 1-2 July 1982) (Amd 3 June 1987). 51 pp.

—Emergency and Standby Power Supplies—1000-1500 A AC
BSI BS EN 60947-6-1-92. 1992 Low-Voltage Switchgear and Controlgear Part 6: Multiple Function Equipment Section 1: Automatic Transfer Switching Equipment. 46 pp.

BSI BS EN 60947-6-1-01. 1992 Amd 1 Low-Voltage Switchgear and Controlgear Part 6: Multiple Function Equipment Section 1: Automatic Transfer Switching Equipment (AMD 7851) July 15, 1993 (E). 47 pp.

CENELEC EN 60947-6-1-91. Low-Voltage Switchgear and Controlgear Part 6: Multiple Function Equipment Section One—Automatic Transfer Switching Equipment. 6 pp.

IEC 947 Pt 6-1-89. Low-Voltage Switchgear and Controlgear Part 6: Multiple Function Equipment Section One—Automatic Transfer Switching Equipment First Edition. 81 pp.

—Service Entrance Equipment
CSA C22.2 NO 178-1978. Automatic Transfer Switches (R 1992); (Amd 1-2 July 1982) (Amd 3 June 1987). 51 pp.

—Static
CSA CAN/CSA-C22. 2NO107.1-M91. Commercial and Industrial Power Supplies; (Gen Instr 1 Thru 2). 112 pp.

Automatic Transmission Measuring and Signaling Testing Equipment
Use For: ATME *See Also:* Measuring Instruments; Telephone Equipment; Testing Equipment; Transmission Measuring Equipment

CCITT RECMN Q.49-89. Specification for the CCITT Automatic Transmission Measuring and Signalling Testing Equipment ATME No. 2—General Recommendations on Telephone Switching and Signalling—Functions and Information Flows for Services in the ISDN—Supplements. 1 p.

—CCITT R1 Signaling Systems
CCITT RECMN Q.330-89. Automatic Transmission and Signalling Testing—Specifications of Signalling Systems R1 and R2 (Study Group XI) 1 pp. 1 p.

—Telephone Circuits
CCITT RECMN O.22-89. CCITT Automatic Transmission Measuring and Signalling Testing Equipment ATME No. 2—Specifications for Measuring Equipment (Study Group IV) 28 pp. 28 pp.

Automatic Transmitter Identification Systems
Use For: ATIS *See Also:* Identification Systems; Radio Transmitters; Signaling Systems

—Maritime Mobile Services—UHF—Radio Telephones
CCIR Report 1159-90. Characteristics of an Automatic Identification System for VHF and UHF Transmitting Stations in the Maritime Mobile Service—Section 8C—Maritime Mobile Service; Telephony and Related Subjects. 9 pp.

—Maritime Mobile Services—VHF—Radio Telephones
CCIR Report 1159-90. Characteristics of an Automatic Identification System for VHF and UHF Transmitting Stations in the Maritime Mobile Service—Section 8C—Maritime Mobile Service; Telephony and Related Subjects. 9 pp.

INTERNATIONAL AND NON-U.S. NATIONAL STANDARDS
SUBJECT INDEX

Automobiles

Automatic Vehicle Location Systems
Use For: AVL *See Also:* Early Mobile Data Services; Radio Equipment; Vehicles (Transportation)

—Cellular Mobile Radio Services
CCIR Report 904-2-90. Automatic Determination of Location and Guidance in the Land Mobile Service—Section 8A—Land Mobile Service and Related Subjects. 20 pp.

—Land Mobile Services
CCIR Report 904-2-90. Automatic Determination of Location and Guidance in the Land Mobile Service—Section 8A—Land Mobile Service and Related Subjects. 20 pp.
CCIR QUESTION 51-2/8-90. Automatic Determination of Location and Guidance in the Land Mobile Service—Questions Concerning Study Group 8 —Mobile, Radiodetermination, Amateur and Related Satellite Services. 2 pp.

—Spread Spectrum Modulation
CCIR Report 904-2-90. Automatic Determination of Location and Guidance in the Land Mobile Service—Section 8A—Land Mobile Service and Related Subjects. 20 pp.

—UHF—Land Mobile Satellite Communications
CCIR Report 770-3-90. Technical and Operating Considerations for a Land Mobile-Satellite Service Operating in Band 9—Section 8F—Frequencies, Orbits and Systems. 27 pp.

Automation
See Also: Automatic Control Equipment; Building Automation Systems; CAM; CIM; Control Systems Equipment; Controllers; Data Processing; Process Control Systems
CNS C5092-81. Standard Dimensional System for Automation Requirements (Mar)(7010).

—Flight Decks—Ergonomics
ICAO Circular 234. Human Factors Digest No. 5 Operational Implications of Automation in Advanced Technology Flight Decks—1992. 46 pp.

—Glossaries
ISO TR11065-92. Industrial Automation Glossary First Edition. 162 pp.
JIS B 3000-90. Glossary of Terms Used in Factory Automation.

—Office—Glossaries
IEC DIS 2382 Pt 27-92. Information Technology—Vocabulary—Part 27: Office Automation ***CD-ROM ONLY***. 22 pp.
ISO DIS 2382 Pt 27-92. Information Technology—Vocabulary—Part 27: Office Automation ***CD-ROM ONLY***. 22 pp.
JTC1 DIS2382 Pt 27-92. Information Technology—Vocabulary—Part 27: Office Automation ***CD-ROM ONLY***. 22 pp.

Automobile Accidents
Use: Collision Research

Automobile Engines
Use: Automotive Engines

Automobile Roofs
See Also: Automotive Equipment

—Canvas
CNS L3133-81. Method of Test for Linen or Jute Canvas for Waterproof Roofing of Car (Jul)(7741).
CNS L4136-81. Linen or Tute Canvas for Waterproof Roofing of Car (Jul)(7740).

Automobiles
Scope Note: See also the subheading Automotive under specific types of equipment. For additional listing, consult the following list *Use For:* Passenger Cars *See Also:* Ambulances; Buses (Vehicles); Electric Vehicles; Ground Vehicles; Ignition Systems; Motor Vehicles; Starters; Steering Gear; Trailers; Trucks; Vehicles (Transportation)
BSI BS AU 179-81. 1981 Dimensional Codes for Passenger Cars. 20 pp.
CNS D1020-82. Standard Form of Specification for Passenger Cars and Motor Trucks (Mar)(8568).
ISO 4131-79. Road Vehicles—Dimensional Codes for Passenger Cars First Edition; (Erratum—July 1980). 27 pp.
JIS D 0001-67. Standard Form of Specification for Passenger Cars and Motor Trucks (R 1970). 22 pp.

—Acceleration Testing
CNS D3016-89. Method of Acceleration Test for Automobiles (Apr)(2735).
JIS D 1014-82. Acceleration Test Method of Automobiles. 6 pp.

Automobiles *(Cont.)*

—Air Cargo—Unit Loads
ISO 8268-87. Air Cargo Equipment—Automobile Transport Devices—Basic Requirements First Edition. 15 pp.

—Air Pollution
EC COM(87) 706-88. Proposal for a Council Directive Amending Directive 70/220/EEC on the Approximation of the Laws of Member States Relating to Measures to be Taken Against air Pollution by Gases from the Engines of Motor Vehicles—European Emission Standard for Cars Below 1.4 Litres. 13 pp.
EC COM(88) 675-88. Amendment to the Proposal for a Council Directive amending Directive 70/220/EEC on the Approximation of the Laws of the Member States Relating to Measures to be Taken Against Air Pollution by Gases from the Engines of Motor Vehicles (European Emission). 4 pp.

—Driver's Visual Field
BSI BS AU 176-80. 1980 Method for Establishment of Eyellipses for Driver's Eye Location. 14 pp.
EC 81/643/EEC-81. Commission Directive Adapting to Technical Progress Directive 77/649/EEC on the Approximation of the Laws of the Member States Relating to the Field of Vision of Motor Vehicle Drivers. 2 pp.
EC 90/630/EEC-90. Commission Directive Adapting to Technical Progress Council Directive 77/649/EEC on the Approximation of the Laws of the Member States Relating to the Field of Vision of Motor Vehicle Drivers. 10 pp.
ISO 4513-78. Road Vehicles—Visibility—Method for Establishment of Eyellipses for Driver's Eye Location First Edition. 13 pp.
ISO 7397 Pt 2-93. Passenger Cars—Verification of Driver's Direct Field of View—Part 2: Test Method First Edition. 18 pp.

—Driver's Visual Field—Vehicle Positioning
ISO 7397 Pt 1-93. Passenger Cars—Verification of Driver's Direct Field of View—Part 1: Vehicle Positioning for Static Measurement First Edition. 4 pp.

—Electroplated Coatings
JIS D 0201-76. General Rules of Electroplating for Automobile Parts (R 1979). 30 pp.

—Exteriors
CNS D3073-81. Measuring Method of Automobile Exterior Dimensions (Jul)(7673).
JIS D 0302-67. Measuring Method of Automobile Exterior Dimensions.

—Fiducial Points
JIS D 0030-82. Three-Dimensional Reference System for Automobiles.

—Fuel Consumption
SAA AS 2077-82. Methods of Test for Fuel Consumption of Passenger Cars, Their Derivatives and Multi-Purpose Passenger Cars. 32 pp.
SNZ NZS 5420-80. Methods of Test for Petrol Consumption of Passenger Cars. 19 pp.

—Fuel Consumption—Identification Systems
SNZ NZS 5421-81. Specification for Fuel Consumption Labelling of Passenger Cars. 7 pp.

—Glossaries
CNS D1037-82. Glossary of Terms Relating to Automobiles (Kinds of Automobiles) (Nov)(9590).
CNS D1038-82. Glossary of Terms Relating to Automobiles (General Terms of Automobiles) (Nov)(9591). 1 p.
CNS D1039-82. Glossary of Terms Relating to Automobiles (Dimensions of Automobiles) (Nov)(9592). 16 pp.
CNS D1040-82. Glossary of Terms Relating to Automobiles (Weights of Automobiles) (Nov)(9593). 3 pp.
CNS D1041-82. Glossary of Terms Relating to Automobiles (Performances of Automobiles) (Nov)(9594). 4 pp.
CNS D1049-83. Glossary of Terms Relating to Equipment and Decoration of Automobiles (Mar)(10065).
CNS D1057-83. Glossary of Terms Relating to the Body of Passenger Cars (May)(10248).
JIS D 0101-82. Glossary of Terms Relating to Kinds of Automobiles.
JIS D 0102-79. Glossary of Terms Relating to Automobiles-Dimensions, Masses, Weights and Performances of Automobiles (R 1984). 62 pp.
JIS D 0104-86. Glossary of Terms Relating to Main System of Automobiles.
JIS D 0110-88. Glossary of Terms Relating to Equipments of Automobiles. 20 pp.

—Inspection
CNS D3025-89. Method of Disassembly Inspection for Automobiles (Apr)(2744).

Automobiles *(Cont.)*

—Inspection *(Cont.)*
JIS D 1023-82. Overhauled Inspection Method of Automobiles (R 1987). 6 pp.

—Interior Protusions—Control Devices
ISO 3208-74. Road Vehicles—Evaluation of Protrusions Inside Passenger Cars First Edition; (Amendment-1974). 7 pp.

—Interiors
CNS D3072-81. Measuring Method of Automobile Interior Dimensions (Jul)(7672).
JIS D 0301-82. Measuring Method of Automobile Body Interior Dimensions. 21 pp.

—Loads (Forces)
ISO 2416-92. Passenger Cars—Mass Distribution Third Edition. 5 pp.
JIS D 0050-83. Load Distribution for Passenger Cars. 6 pp.

—Noise
CNS D3058-87. Method of Test for Noise Emission for Automobiles (Sep)(5799). 8 pp.
CNS D3146-82. Method of Test for Tire Noise for Automobiles (Oct)(9480).
ISO 7188-85. Acoustics—Measurement of Noise Emitted by Passenger Cars Under Conditions Representative of Urban Driving First Edition. 7 pp.
JIS D 1024-89. Measurement of Noise Emitted by Accelerating Road Vehicles. 13 pp.

—Road Holding—Stability
BSI BS AU 189-83. 1983 Steady State Cornering Behaviour for Road Vehicles. 16 pp.
ISO 4138-82. Road Vehicles—Steady State Circular Test Procedure First Edition. 16 pp.
ISO TR8726-88. Road Vehicles—Transient Open-Loop Response Test Method with Pseudo-Random Steering Input First Edition. 22 pp.

—Road Holding—Stability—Glossaries
CNS D1059-83. Glossary of Terms Relating to Automobile Stability and Controllability (Aug)(10493).

—Road Testing
CNS D3013-86. General Rules of Running Test Method of Automobiles (Jun)(2731).
CNS D3017-89. Method of Coasting Test for Automobiles (Apr)(2736).
CNS D3019-89. Method of Steep Hill Climbing Test for Automobiles (Apr)(2738).
CNS D3020-89. Method of Long Hill Climbing Test for Automobiles (Apr)(2739).
CNS D3024-89. Method of Driving Test for Automobiles (Apr)(2743).
CNS D3088-88. Method of Incline Test of Carburetors for Automobiles (Dec)(8245).
CNS D3095-89. Cornering Test Code of Carburetors for Automobiles (Jan)(8252).
CNS D3119-82. Test Procedure for Cornering Performance for Passenger Car (May)(8809).
CNS D3131-82. Pylon Course Slalom Test Procedure for Automobiles (Aug)(9244).
CNS D3134-82. Pylon Slalom Test Procedure for Passenger Car and Light Trailer Combinations (Aug)(9247).
JIS D 1010-82. General Rules of Running Test Method of Automobiles. 9 pp.
JIS D 1015-82. Coasting Test Method of Automobiles (R 1987). 7 pp.
JIS D 1017-82. Steep Hill Climbing Test Method of Automobiles (R 1987). 6 pp.
JIS D 1018-82. Long Hill Climbing Test Method of Automobiles (R 1987). 5 pp.
JIS D 1022-82. Driving Test Method of Automobiles (R 1987). 6 pp.

—Road Testing—Traction
CNS D3021-89. Method of Traction Test for Automobiles (Apr)(2740).
JIS D 1019-82. Traction Test Method of Automobiles (R 1987). 5 pp.

—Sand Ground Testing
CNS D3022-89. Method of Sand Ground Test for Automobiles (Apr)(2741).
JIS D 1020-82. Sand Ground Test Method of Automobiles (R 1987). 7 pp.

—Sealants
JIS K 6830-77. Testing Methods for Sealant for Automobiles.

—Seating—Mannequins
CNS D2182-84. Three Dimensional Mannequins for Use in Defining Automobile Seating Accommodations (Oct)(11059).
JIS D 4607-77. Three Dimensional Manikins for Use in Defining Automobile Seating Accommodations. 16 pp.

INDUSTRY STANDARDS

INTERNATIONAL AND NON-U.S. NATIONAL STANDARDS
SUBJECT INDEX

Automobiles

Automobiles (Cont.)

—**Speed**
CNS D3018-89. Method of Maximum Speed Test for Automobiles (Apr)(2737).
JIS D 1016-82. Maximum Speed Test Method of Automobiles (R 1987). 5 pp.

—**Standard Voltages**
CNS D1007-80. Standard Voltages for Automobiles (Jul)(5783).

—**Starting Testing**
CNS D3023-89. Method of Starting Test for Automobiles (Apr)(2742).
JIS D 1021-82. Starting Test Method of Automobiles.

—**Steady State Circulation Testing**
JIS D 1070-86. Steady State Circular Test Procedure for Passenger Cars. 12 pp.

—**Testing Fluids—Polymer Materials**
DIN ENGL 51604 Pt 1-82. FAM Testing Fluid for Polymer Materials; Composition and Requirements (Nov). 2 pp.
DIN ENGL 51604 Pt 2-84. Methanolic FAM Testing Fluid for Polymer Materials; Composition and Requirements (Mar). 2 pp.
DIN ENGL 51604 Pt 3-84. Methanolic Lower Layer FAM Testing Fluid for Polymer Materials; Composition and Requirements (Mar). 2 pp.

—**Three Dimensional Display Systems**
CNS D1064-84. Three-Dimensional Reference System for Automobiles (Mar)(10799).

—**Transient Response**
ISO 7401-88. Road Vehicles—Lateral Transient Response Test Methods First Edition. 19 pp.
ISO TR8725-88. Road Vehicles—Transient Open-Loop Response Test Method with One Period of Sinusoidal Input First Edition. 19 pp.

—**Turning Radius**
CNS D3118-82. Test Procedure for Minimum Turning Radius for Automobiles (May)(8808).
JIS D 1025-85. Minimum Turning Radius Test Procedure of Automobiles. 8 pp.

Automotive Air Conditioning Equipment
Use: Air Conditioning Equipment—Automotive

Automotive Diagnostic Testing
Use: Diagnostic Testing—Automotive

Automotive Engines
Use For: Automobile Engines *See Also:* Internal Combustion Engines
CNS D3070-81. Test Code of Air Cooled Gasoline Engines for Automobiles (Jul)(7665).
CNS D3071-81. Test Code of Water Cooled Gasoline Engines for Automobiles (Jul)(7666).

—**Analyzers**
CSA CAN/CSA-C22. 2 NO 68-92. Motor-Operated Appliances (Household and Commercial); (Gen Instr 1 Thru 2). 115 pp.
CSA 1169 Bull. Electrical Bulletin 1169 June 27, 1978 to C22.2 NO 68. 2 pp.

—**Cooling Systems—Pressure Caps**
ISO 9133-88. Passenger Cars—Engine Cooling Systems—Threaded Pressure Caps and Their Seats on Filler Necks First Edition. 4 pp.
ISO 9817-91. Passenger Cars—Engine Cooling Systems—Dimensions of Pressure Caps and Their Ramp Seats on Filler Necks First Edition. 6 pp.
ISO 9818-91. Passenger Cars—Engine Cooling Systems—Test Methods and Marking of Pressure Caps First Edition. 7 pp.

—**Cooling Systems—Pressure Caps—Identification Systems**
ISO 9818-91. Passenger Cars—Engine Cooling Systems—Test Methods and Marking of Pressure Caps First Edition. 7 pp.

—**Diesel Fuels**
MOD UK DSTAN 91-7-01. Reference Gasoline, High Lead Reference Gasoline, Low Lead Reference Diesel Fuel, High Sulphur Issue 3 Amendment 2. 12 pp.
MOD UK DSTAN 91-7-93. Reference Diesel, High Sulphur Issue 4. 12 pp.

—**Fuel Consumption**
CNS D3014-87. Method of Test for Fuel Consumption for Passenger Cars (Sep)(2733). 15 pp.
EC 80/1268/EEC-80. Council Directive on the Approximation of the Laws of the Member States Relating to the Fuel Consumption of Motor Vehicles. 10 pp.

Automotive Engines (Cont.)

—**Fuel Consumption** (Cont.)
ISO 1585-92. Road Vehicles—Engine Test Code—Net Power Third Edition. 29 pp.
ISO 2534-74. Road Vehicles—Engine Test Code—Gross Power First Edition; (Erratum—Aug 1974). 17 pp.
JIS D 1012-83. Fuel Consumption Test Methods of Automobiles. 26 pp.
SAA AS 2077-82. Methods of Test for Fuel Consumption of Passenger Cars, Their Derivatives and Multi-Purpose Passenger Cars. 32 pp.
SNZ NZS 5420-80. Methods of Test for Petrol Consumption of Passenger Cars. 19 pp.
SNZ NZS 5421-81. Specification for Fuel Consumption Labelling of Passenger Cars. 7 pp.

—**Pistons**
CNS D2003-82. Piston Pins for Automobile Engines (Jan)(2935).
CNS D2008-86. Piston for Automobile Engines (Jun)(3211). 10 pp.
JIS D 3104-76. Pistons for Automobile Engines.

—**Power**
DIN ENGL 70020 Pt 6-76. Automotive Engineering; Power (Nov). 2 pp.
EC 80/1269/EEC-80. Council Directive on the Approximation of the Laws of the Member States Relating to the Engine Power of Motor Vehicles. 22 pp.
EC 88/195/EEC-88. Commission Directive Adapting to Technical Progress Council Directive 80/1269/EEC on the Approximation of the Laws of the Member States Relating to the Engine Power of Motor Vehicles. 10 pp.
ISO 1585-92. Road Vehicles—Engine Test Code—Net Power Third Edition. 29 pp.
ISO 2534-74. Road Vehicles—Engine Test Code—Gross Power First Edition; (Erratum—Aug 1974). 17 pp.

—**Sleeve Bearings**
CNS D2007-86. Sleeve Type Half Bearing for the Crank-Shaft of Automobile Engines (Aug)(3102). 7 pp.
JIS D 3102-87. Sleeve Type Half Bearings for Automobile Engines. 12 pp.
JIS D 3106-88. Sleeve Type Flanged Half Bearings for Automobile Engines. 15 pp.

—**Tachometers**
CNS D2158-82. Electric Engine Tachometers for Automobiles (Nov)(9585).

Automotive Equipment
Scope Note: See also the subheading Automotive under specific types of equipment. For additional listings, consult the following list
See Also: Accelerators (Mechanical); Air Brakes; Automobile Roofs; Automotive Lifts; Automotive Locks; Automotive Trim Materials; Backup Lights; Brake Cylinders; Brakes; Bumpers; Car Washes; Chocks; Defrosters; Demisters; Disk Brakes; Drum Brakes; Flasher Units; Hydraulic Brakes; Jumper Cables; Motor Vehicle Lighting; Motor Vehicle Operators; Motor Vehicles; Mufflers; Power Steering; Radiators (Automotive); Rear View Mirrors; Safety Belts; Steering Wheels; Tire Chains; Tire Cords; Tires; Towbars; Turn Signals; Warning Triangles; Wheel Ramps; Wheels; Windshield Washers (Mechanical); Windshield Wipers; Windshields

—**Door Hinges**
CNS D2098-89. Door Hinges for Automobiles (Jul)(7681).
JIS D 1621-88. Test Method of Side Door Hinge Systems for Automobiles. 7 pp.

—**Door Latches**
CNS D3162-83. Method of Test for Side Door Lock System of Automobiles (Dec)(10708).
JIS D 1620-88. Test Method of Side Door Latch Systems for Automobiles. 9 pp.

—**Gearshifts**
JIS D 0012-87. Transmission Shift Positions for Passenger Cars. 6 pp.

—**Luggage Compartments—Capacity Measurement**
ISO 3832-91. Passenger Cars—Luggage Compartments—Method of Measuring Reference Volume Second Edition; (Corrected and Reprinted -1991). 5 pp.
JIS D 0303-82. Method of Measuring the Reference Volume for the Luggage Compartments of Passenger Cars.

—**Manual Controls—Disabled Persons**
SNZ NZS 5832: Part 1-88. Driving Controls for People with Disabilities Part 1: Hand Controls. 24 pp.

Automotive Equipment (Cont.)

—**Manual Transmissions**
JIS D 0012-87. Transmission Shift Positions for Passenger Cars. 6 pp.

—**Safety Glass**
BSI BS AU 178-80. 1980 Road Vehicle Safety Glass. 32 pp.
BSI BS AU 178A-92. 1992 Road Vehicle Safety Glass. 41 pp.

—**Tachographs**
CNS D2063-85. Tachographs for Automobiles (Jul)(6816).
JIS D 5607-82. Tachographs for Automobiles (R 1987). 11 pp.

—**Tachographs—Flexible Shafts**
CNS D2062-85. Flexible Shafts for Speedometers and Tachographs of Automobiles or Motorcycles (Jul)(6815).
JIS D 5602-91. Flexible Shafts for Speedometers and Tachographs of Automobiles. 16 pp.

Automotive Fuels
Use For: Motor Fuels; Motor Vehicle Fuels
See Also: Diesel Fuels; Gasoline; Kerosene
BSI BS EN 228-93. 1993 Unleaded Petrol (Gasoline) for Motor Vehicles (W). 16 pp.
CEN EN 228-87. Liquid Petroleum Products: Unleaded Petrol; Specification. 5 pp.
CEN PREN 228-91. Automotive Fuels—Unleaded Petrol—Requirements and Methods of Test. 20 pp.
CNS K6910-87. Method of Test for Knock Characteristics of Motor Fuels by the Research Method (Jul)(12011).
MOD UK DSTAN 91-13-01. Gasoline, Automotive: Military (91 RON) NATO Code No: F-46 Joint Service Designation: COMBATGAS Gasoline, Automotive: (91 RON) NATO Code No: F-50 Joint Service Designation: CIVGAS Gasoline, Automotive: Combat, Sub-Zero (95 RON) Joint Service. 12 pp.

—**Antiicing Additives**
CGSB 3-GP-525MA-88. Isopropanol. 8 pp.

—**Import Tariffs—Exemptions**
EC 85/347/EEC-85. Council Directive Amending Directive 68/297/EEC on the Standardization of Provisions Regarding the Duty-Free Admission of Fuel Contained in the Fuel Tanks of Commercial Motor Vehicles. 2 pp.

—**Knock Ratings**
CNS K6910-87. Method of Test for Knock Characteristics of Motor Fuels by the Research Method (Jul)(12011).

—**Liquefied Petroleum Gas**
BSI BS 4250: Part 2-87. (WITHDRAWN) 1987 Amd 1 Liquified Petroleum Gas Part 2: Automotive LPG (AMD 6230) February 28, 1990 (Superseded by BS EN 589: 1993). 9 pp.

—**Octane Number**
BSI BS 2637-91. 1991 Amd 1 Determination of Knock Characteristics of Motor and Aviation-Type Fuels (Motor Method). 7 pp.
BSI BS 2638-91. 1991 Determination of Knock Characteristics of Motor Fuels (Research Method) (ISO 5164: 1990). 7 pp.
ISO 5163-90. Motor and Aviation-Type Fuels—Determination of Knock Characteristics—Motor Method Second Edition. 6 pp.
ISO 5164-90. Motor Fuels—Determination of Knock Characteristics—Research Method Second Edition. 6 pp.

Automotive Lifts
See Also: Automotive Equipment
JIS D 8108-87. Lifts Above Ground for Automobiles. 20 pp.

Automotive Lighting
Use: Motor Vehicle Lighting

Automotive Locks
See Also: Automotive Equipment; Locks (Security)
BSI BS AU 209: Part 5A-88. 1988 Amd 1 Vehicle Security Part 5A: Central Power Locking Systems for Passenger Cars and Car Derived Vehicles. 7 pp.
BSI BS AU 209: Part 6-88. 1988 Vehicle Security Part 6: Dead Locking Systems for Passenger Cars and Car Derived Vehicles. 4 pp.

—**Doors**
BSI BS AU 209: Part 1A-92. 1992 Vehicle Security Part 1a: Specification for Locking Systems for Passenger Cars and Car Derived Vehicles (E). 9 pp.
BSI BS AU 209: Part 7-93. 1993 Vehicle Security Part 7: Specification for Locking Systems for Goods Vehicle Driver's Compartments (E). 9 pp.

INTERNATIONAL AND NON-U.S. NATIONAL STANDARDS
SUBJECT INDEX

Automotive Locks *(Cont.)*
—**Luggage Compartments**
BSI BS AU 209: Part 1A-92. 1992 Vehicle Security Part 1a: Specification for Locking Systems for Passenger Cars and Car Derived Vehicles (E). 9 pp.

—**Windows**
BSI BS AU 209: Part 1-86. (WITHDRAWN) 1986 Vehicle Security Part 1: Mechanical Locking System for Passenger Cars and Car Derived Vehicles (Superseded by BS AU 209: Part 1a: 1992). 7 pp.

Automotive Maintenance Equipment
Use: Maintenance Equipment—Automotive

Automotive Mirrors
See Also: Mirrors; Rear View Mirrors
CNS D2050-86. Rear View Mirrors for Automobiles (Sep)(6172). 10 pp.
JIS D 5705-87. Mirrors for Automobiles. 23 pp.

Automotive Servicing Equipment
Use: Maintenance Equipment—Automotive

Automotive Trim Materials
See Also: Automotive Equipment

—**Adhesives—Staining**
SAA AS 1937.5-77. Methods of Test for Sealers and Adhesives for Automotive Purposes—Part 5: Determination of Staining of Trim Materials by Adhesives Reconfirmed 1989. 4 pp.

—**Flammability Testing**
JIS D 1201-77. Test Method for Flammability of Organic Interior Materials for Automobiles (R 1983). 38 pp.

Autopatrols
Use: Motor Graders

Autopilots
Use: Automatic Pilots

Autotransformer Starters
See Also: Starters
BSI BS 587-57. 1957 Amd 6 Motor Starters and Controllers. 62 pp.
BSI BS 4941: Part 4-77. (WITHDRAWN) 1977 Motor Starters for Voltages up to and Including 1000 V a.c. and 1200 V d.c. Part 4: Reduced Voltage a.c. Starters, Two-Step Auto-Transformer Starters (Superseded by BS EN 60947-4-1: 1992). 25 pp.
SAA AS 1202.3-76. A.C. Motor Starters (up to and Including 1000 V) —Part 3: Autotransformer Starters. 28 pp.

Autotransformers
Use For: Auto Starter Transformers
See Also: Power Dividers; Power Transformers; Transformers
BSI BS 171: Part 1-78. 1978 Amd 1 Power Transformers Part 1: General. 23 pp.
BSI BS 171-70. 1970 Amd 3 Power Transformers. 115 pp.
BSI BS 7452-91. 1991 Separating Transformers, Autotransformers, Variable Transformers and Reactors (IEC 989: 1991). 130 pp.
DIN VDE 0550 Pt 1 (D)-87. Separating Transformers, Autotransformers, Variable Transformers and Reactors. Identical with IEC 14D(CO)29 (Nov) (Partially Supersedes 0550 Part 1/12.69, Part 2/ 12.69 and Part 6/4.66) (Supersedes 0550 Part 4/ 4.66, Part 5/9.72 and 0552/5.69). 117 pp.
DIN VDE 0550 Pt 4 (S)-66. Regulations for Small Transformers; Part 4: Special Regulations for Autotransformers (Apr). 5 pp.
IEC 989-91. Separating Transformers, Autotransformers, Variable Transformers and Reactors First Edition. 241 pp.

—
CSA C22.2 NO 66-1988. Speciality Transformers; (Gen Instr 1). 53 pp.
CSA 1169 Bull. Electrical Bulletin 1169 June 27, 1978 to C22.2 NO 66. 2 pp.
CSA 1440 Bull. Electrical Bulletin 1440 June 23, 1986 to C22.2 NO 66. 2 pp.

— **601-999 V**
CSA C22.2 NO 66-1988. Speciality Transformers; (Gen Instr 1). 53 pp.
CSA 1169 Bull. Electrical Bulletin 1169 June 27, 1978 to C22.2 NO 66. 2 pp.
CSA 1440 Bull. Electrical Bulletin 1440 June 23, 1986 to C22.2 NO 66. 2 pp.

—**Connectors**
BSI BS 171: Part 4-78. 1978 Power Transformers Part 4: Tappings and Connections. 18 pp.

Autotransformers *(Cont.)*
—**Electrical Insulation**
BSI BS 171: Part 3-87. 1987 Power Transformers Part 3: Insulation Levels and Dielectric Tests. 35 pp.

—**Short Circuits**
BSI BS 171: Part 5-78. 1978 Power Transformers Part 5: Ability to Withstand Short Circuit. 12 pp.

—**Temperature Rise**
BSI BS 171: Part 2-78. 1978 Amd 1 Power Transformers Part 2: Temperature Rise. 15 pp.

—**Variable—0-600 V**
CSA C22.2 NO 66-1988. Speciality Transformers; (Gen Instr 1). 53 pp.
CSA 1169 Bull. Electrical Bulletin 1169 June 27, 1978 to C22.2 NO 66. 2 pp.
CSA 1440 Bull. Electrical Bulletin 1440 June 23, 1986 to C22.2 NO 66. 2 pp.

—**Variable—601-999 V**
CSA C22.2 NO 66-1988. Speciality Transformers; (Gen Instr 1). 53 pp.
CSA 1169 Bull. Electrical Bulletin 1169 June 27, 1978 to C22.2 NO 66. 2 pp.
CSA 1440 Bull. Electrical Bulletin 1440 June 23, 1986 to C22.2 NO 66. 2 pp.

Auxiliary Gutters
See Also: Electric Raceways; Electrical Enclosures
CSA C22.2 NO 26-1952. Construction and Test of Wireways, Auxiliary Gutters, and Associated Fittings (R 1993); (Rev 1 March 1955) (Rev 2-3 May 1960) (Amd 4-6 July 1985) (Amd 7 June 1992). 16 pp.

—**Construction**
CSA C22.2 NO 26-1952. Construction and Test of Wireways, Auxiliary Gutters, and Associated Fittings (R 1993); (Rev 1 March 1955) (Rev 2-3 May 1960) (Amd 4-6 July 1985) (Amd 7 June 1992). 16 pp.

—**Meter Centers**
CSA C22.2 NO 26-1952. Construction and Test of Wireways, Auxiliary Gutters, and Associated Fittings (R 1993); (Rev 1 March 1955) (Rev 2-3 May 1960) (Amd 4-6 July 1985) (Amd 7 June 1992). 16 pp.

—**Switchboards (Electrical)**
CSA C22.2 NO 26-1952. Construction and Test of Wireways, Auxiliary Gutters, and Associated Fittings (R 1993); (Rev 1 March 1955) (Rev 2-3 May 1960) (Amd 4-6 July 1985) (Amd 7 June 1992). 16 pp.

Auxiliary Power Units
Use For: APU *See Also:* Emergency and Standby Power Supplies; Generators; Power Supplies

—**Aircraft—Testing Equipment**
MOD UK DERD 2083-90. Design, Development and Production of Test Equipment (GSE) for Aircraft Propulsion and Auxiliary Power Unit Systems Issue 2. 9 pp.

—**Connectors—Combat Vehicles**
NATO STANAG 4074 ED 1 AMD 3-69. Auxiliary Power Unit Connections for Starting Combat and Tactical Vehicles. 11 pp.
NATO STANAG 4074 ED 2 AMD 0-00. (Draft) Auxilliary Power Unit Connections for Starting Tactical Land Vehicles. 15 pp.

—**Gas Turbine—Aircraft—Certificate of Airworthiness**
CAA. Contents (Joint Airworthiness Requirements). 1 p.
CAA. Foreword. 1 p.
CAA. Check List of Pages (Joint Airworthiness Requirements). 1 p.
CAA JAR-APU Section 1. Regulations (Joint Airworthiness Requirements). 4 pp.
CAA JAR-APU Section 2. Acceptable Means of Compliance and Interpretations (Joint Airworthiness Requirements). 4 pp.
CAA JAR-APU Section 3. Advisory Material Joint—AMJ (Joint Airworthiness Requirements). 5 pp.
CAA JAR-APU Section 4. Approval of Gas Turbine Auxiliary Power Units and Associated Equipment. 5 pp.

—**Ground Support Equipment**
MOD UK DERD 2080-90. General Procedures for Design and Development of Ground Support Equipment for Aircraft Propulsion and Auxiliary Power Unit Systems Issue 6. 14 pp.

—**Ground Support Equipment—Design**
MOD UK DERD 2081-90. Design, Development and Production of Tools, Ground Support Equipment (GSE) for Aircraft Propulsion and Auxiliary Power Unit Systems Issue 5. 8 pp.

Auxiliary Power Units *(Cont.)*
—**Rotary Wing Aircraft—Fire Testing**
CAA Chapter G5-8 06.76. Fire Precautions (Rotocraft). 8 pp.

—**Stands (Supports)—Design**
MOD UK DERD 2084-90. Design, Development of Stands for Aircraft Propulsion and Auxiliary Power Unit Systems Issue 4. 12 pp.

Avalanche Diodes
Use For: Breakdown Diodes *See Also:* Avalanche Photodiodes; Avalanche Rectifier Diodes; IMPATT Diodes; Semiconductor Devices; Voltage Reference Diodes; Voltage Regulators; Zener Diodes
BSI BS 6493: Sec 1.4-92. 1992 Semiconductor Devices Part 1: Discrete Devices Section 1.4: Recommendations for Microwave Diodes and Transistors (IEC 747-4: 1991). 83 pp.
IEC 747 Pt 4-91. Semiconductor Devices Discrete Devices Part 4: Microwave Diodes and Transistors First Edition. 163 pp.
IECQ QC 750101-86. Semiconductor Devices Discrete Devices Part 3: Signal (Including Switching) and Regulator Diodes, Section One—Blank Detail Specification for Signal Diodes, Switching Diodes and Controlled-Avalanche Diodes (IEC 747-3-1 ED 1). 27 pp.
MOD UK DSTAN 59-61: 90/079-01. Semiconductor Devices Diode Issue 1; Amendment 1. 9 pp.
MOD UK DSTAN 59-61: 90/085-72. Semiconductor Devices Diode Issue 1. 19 pp.
MOD UK DSTAN 59-61: 90/197-80. Semiconductor Device, Diode Issue 1. 10 pp.

—**Preferred Products List**
CECC CECC MUAHAG Vol 9 IS 3-90. Preferred Products List; Semiconductors (En, Fr, Ge) AMD 1 (En, Fr, Ge). 51 pp.

—**Quality Assurance**
IEC 747 Pt 3-1-86. Semiconductor Devices Discrete Devices Part 3: Signal (Including Switching) and Regulator Diodes Section One—Blank Detail Specification for Signal Diodes, Switching Diodes and Controlled-Avalanche Diodes First Edition (IECQ QC 750101). 27 pp.

Avalanche Oscillators
See Also: Microwave Equipment; Microwave Oscillators; Semiconductor Devices

—**Quality Assurance**
BSI BS 9327-77. 1977 Rules for the Preparation of Detail Specifications for Semiconductor Devices of Assessed Quality Microwave Avalanche Oscillators (C.W. Operation). 8 pp.
BSI BS 9328-77. 1977 Rules for the Preparation of Detail Specifications for Semiconductor Devices of Assessed Quality: Microwave Avalanche Oscillators (Pulse Operation). 8 pp.

Avalanche Photodiodes
Use For: APD *See Also:* Avalanche Diodes; Photodiodes
MOD UK DSTAN 59-61: 90/139-01. Semiconductor Device—Photodiode Issue 1; Amendment 2. 19 pp.
MOD UK DSTAN 59-61: 90/182-01. Semiconductor Device—Photodiode Issue 1; Amendment 1. 21 pp.

Avalanche Rectifier Diodes
See Also: Avalanche Diodes; Diodes; Rectifier Diodes
IECQ PQC 75-91. Semiconductor Devices—Discrete Devices—Rectifier Diodes Blank Detail Specification: Rectifier Diodes (Including Avalanche Rectifier Diodes), Ambient and Case-Rated, Greater Than 100 A. 19 pp.

—**Quality Assurance**
BSI BS 9300 C476-73. (OBSOLESCENT) 1973 Detail Requirements for Silicon Avalanche Rectifier Diode. 12 pp.
BSI BS 9300 C667-668-71. (OBSOLESCENT) 1971 Detail Requirements for Silicon Avalanche Rectifier Diodes. 15 pp.
IEC 747 Pt 2-1-89. Semiconductor Devices Discrete Devices Part 2: Rectifier Diodes Section One—Blank Detail Specification for Rectifier Diodes (Including Avalanche Rectifier Diodes), Ambient and Case-Rated, up to 100A First Edition (IECQ. 27 pp.
IECQ QC 750108-89. Semiconductor Devices Discrete Devices Part 2: Rectifier Diodes Section One—Blank Detail Specification for Rectifier Diodes (Including Avalanche Rectifier Diodes), Ambient and Case-Rated, up to 100A (IEC 747-2-1 ED 1). 27 pp.

Avalanches
—**Search and Rescue—Transceivers—Safety**
CEN PREN 282-88. Avalanche Beacons Transmitter/ Receiver Systems Safety Requirements and Testing. 9 pp.

INTERNATIONAL AND NON-U.S. NATIONAL STANDARDS
SUBJECT INDEX

Avalanches (Cont.)
—Search and Rescue—Transceivers—Safety (Cont.)
  CEN EN 282-91. Avalanche Beacons—Transmitter/Receiver Systems—Safety Requirements and Testing. 9 pp.

Avco Lycoming Engines
Use: Aircraft Engines—Avco Lycoming

Average Daily Peak Hour
See Also: Busy Hour; Fixed Daily Measurement Hour; Fixed Daily Measurement Period; Telecommunication Circuits; Telecommunication Traffic; Time; Time Consistent Busy Hour; Units of Measurement
  CCITT RECMN E.500 (REV 1)-92. Traffic Intensity Measurement Principles (Study Group II) 17 pp. 17 pp.
  CCITT RECMN E.500-89. Traffic Intensity Measurement Principles —Telephone Network and ISDN—Quality of Service, Network Management and Traffic Engineering (Study Group II) 12 pp. 12 pp.

Avia Propellers
Use: Aircraft Propellers—Avia

Aviation Accidents
Use: Aircraft Accidents

Aviation Fuels
Use: Aircraft Fuels

Aviation Gasoline
Use: Aircraft Gasoline

Aviation Lighting
Use: Airport Lighting

Aviation Medicine
Use: Aerospace Medicine

Aviation Meteorology
Use: Aeronautical Meteorology

Aviation Personnel
See Also: Air Traffic Controllers; Air Transportation; Flight Crews; Pilots (Aircraft)
—Emergency Planning—Training
  CAA Part 8 CAP 168. Rescue and Fire Fighting Services. 29 pp.
—Licensing
  ICAO Annex 1. Personnel Licensing Eighth Edition—July 1988 (Incorporating Amendments 1-159); (Amendment 160 7/26/93). 86 pp.
  ICAO 7192 Part B-2. Training Manual Part B-2 Pilots—Helicopter Licenses, CVFR-Rating and Instrument Flight Rating First Edition—1977. 90 pp.
  ICAO 9379. Manual of Procedures for Establishment and Management of a State's Personnel Licensing System First Edition—1983. 188 pp.
—Radiation—Earth Stations
  CCIR Report 682-1-86. Probability of Hazards to Personnel Within Aircraft Due to Radiation from Deep-Space Earth Stations—Section 2A—Research in Space Technology. 2 pp.
—Training
  ICAO Annex 18. Safe Transport of Dangerous Goods by Air Second Edition—July 1989; (Supplement 8/1/90). 40 pp.
  ICAO 9284 Supplement. Technical Instructions for the Safe Transport of Dangerous Goods by Air—1986 Edition. 117 pp.
  ICAO 9375 Book 1. Dangerous Goods Training Programme Book 1 Shippers and Packers Second Edition—October 1985. 102 pp.
  ICAO 9375 Book 2. Dangerous Goods Training Programme Book 2 Cargo Agents Second Edition—October 1985. 91 pp.
  ICAO 9375 Book 3. Dangerous Goods Training Programme Book 3 Operator's Cargo Acceptance Staff Second Edition—October 1985. 89 pp.
  ICAO 9375 Book 4. Dangerous Goods Training Programme Book 4 Load Planners and Flight Crew Second Edition—January 1986. 36 pp.
  ICAO 9375 Book 5. Dangerous Goods Training Programme Book 5 Passenger Handling Staff and Flight Attendants Second Edition—January 1986. 19 pp.
  ICAO 9375 Book 6. Dangerous Goods Training Programme Book 6 Loading and Warehouse Personnel Second Edition—January 1986. 35 pp.

Aviation Personnel (Cont.)
—Training Centers
  ICAO 9401. Manual on Establishment and Operation of Aviation Training Centres First Edition—1983. 94 pp.
—Training Manuals
  ICAO 7192 Part A-1. Training Manual Part A-1 General Considerations First Edition—1975; (Amendment 1 02/15/88). 61 pp.
  ICAO 7192 Part A-3. Training Manual Part A-3 Composite Ground Subject Curriculum First Edition—1975. 100 pp.
  ICAO 7192 Part B-2. Training Manual Part B-2 Pilots—Helicopter Licenses, CVFR-Rating and Instrument Flight Rating First Edition—1977. 90 pp.
  ICAO 7192 Part B-5 Vol 1. Training Manual Part B-5 Integrated Commercial Pilot Course Volume 1—Course Details First Edition—1985. 294 pp.
  ICAO 7192 Part B-5 Vol 2. Training Manual Part B-5 Integrated Commercial Pilot Course Volume 2—Instructor Briefing Sheets First Edition—1985. 299 pp.
  ICAO 7192 Part C-3. Training Manual Part C-3 Flight Engineer First Edition—1977 (OUT OF PRINT). 57 pp.
  ICAO 7192 Part D-1. Training Manual Part D-1 Aircraft Maintenance Technician Type II and Type I First Edition—1976. 132 pp.
  ICAO 7192 Part D-2. Training Manual Part D-2 Air Traffic Controller First Edition—1977 (OUT OF PRINT). 89 pp.
  ICAO 7192 Part D-3. Training Manual Part D-3 Flight Operations Officer First Edition—1975 (OUT OF PRINT). 39 pp.
  ICAO 7192 Part E-1. Training Manual Part E-1 Cabin Personnel First Edition—1976. 42 pp.
  ICAO 7192 Part E-3. Training Manual Part E-3 Aeronautical Information Services Personnel First Edition—1980. 35 pp.

Aviation Traders ATL 98 Carvair Aircraft
See Also: Aircraft
—Accidents
  CAA. Aviation Traders ATL 98 Carvair (World Airline Accident Summary). 1 p.
—Mandatory Aircraft Modifications and Inspections Summaries
  CAA. Aviation Traders ATL98 Carvair. 1 p.

Aviation Training Centers
Use: Aviation Personnel—Training Centers

Aviation Turbine Fuels
Use: Jet Engine Fuels

Avionic Computers
Use For: Airborne Computers See Also: Avionics; Computers; Navigation Computers
—Dead Reckoning—Electrical Properties
  EUROCAE WG7C/2-74 08.74. MPS for Airborne Automatic Dead Reckoning Computer Equipment Utilizing Aircraft Heading and Doppler Obtained Velocity Vector Data. 19 pp.
—Dead Reckoning—Environmental Testing
  EUROCAE WG7C/2-74 08.74. MPS for Airborne Automatic Dead Reckoning Computer Equipment Utilizing Aircraft Heading and Doppler Obtained Velocity Vector Data. 19 pp.
—Software
  NATO STANAG 3094 ED 1 AMD 1-86. 16 Bit Computer Instruction Set Architecture. 9 pp.
  NATO STANAG 3094 ED 1 AMD 2-86. 16 Bit Computer Instruction Set Architecture. 10 pp.
—16-Bit—Instruction Set
  NATO STANAG 3094 ED 1 AMD 1-86. 16 Bit Computer Instruction Set Architecture. 9 pp.
  NATO STANAG 3094 ED 1 AMD 2-86. 16 Bit Computer Instruction Set Architecture. 10 pp.

Avionics
Use For: Airborne Electronics See Also: Aircraft; Aircraft Safety; Automatic Pilots; Avionic Computers; Electronic Equipment; Inertial Navigation Systems; Search and Rescue Avionic Systems
  BSI G 244-89. 1989 Code of Practice for Installation/Repair of Aircraft Electrical Interconnection Systems. 21 pp.
  CAA NOTICE #39 ISSUE 4. Selection and Procurement of Electronic Components (Airworthiness Notices). 4 pp.
  JIS W 7202-83. Installation and Test of Electronic Equipment in Aircraft, General Specification for.

Avionics (Cont.)
—Aerospace
  ISO TR10201-91. Aerospace—Standards for Electronic Instruments and Systems Second Edition. 17 pp.
  JIS W 7006-85. Electronic Equipment, Aerospace, General Specification for.
—ATLAS
  NATO STANAG 3957 ED 2 AMD 1-91. C/Atlas Hol for Application in Avionics Test Systems. 7 pp.
—Data Transmission—Interfaces
  MOD UK DSTAN 00-18: Pt 1:Sec 1-89. Avionic Data Transmission Interface Systems Part 1: Guide to Interface Systems Section 1: Introduction Issue 2. 15 pp.
  MOD UK DSTAN 00-18: Pt 1:Sec 3. Avionic Data Transmission Interface Systems Part 1: Guide to Interface Systems Section 3: Guide to the Simplex and Half Duplex Serial Digital Transmission Interface Systems Standard Issue 2. 69 pp.
  MOD UK DSTAN 00-18: Pt 1:Sec 4-91. Avionic Data Transmission Interface Systems Part 1: Guide to Interface Systems Section 4: Guide to the Discrete Signal Interfaces Standard Issue 2. 40 pp.
  MOD UK DSTAN 00-18: Pt 1:Sec 5-91. Avionic Data Transmission Interface Systems Part 1: Guide to Interface Systems Section 5: Guide to the Fibre Optic Interface Standardization Issue 2. 156 pp.
  MOD UK DSTAN 00-18: Pt 1:Sec 6-91. Avionic Data Transmission Interface Systems Part 1: Guide to Interface Systems Section 6: Guide to the Analogue Video Standard for Aircraft System Applications Issue 1. 57 pp.
  MOD UK DSTAN 00-18: Pt 1:Sec 6-01. Avionic Data Transmission Interface Systems Part 1: Guide to Interface Systems Section 6: Guide to the Analogue Video Standard for Aircraft System Applications Issue 1; Amendment 1. 58 pp.
  MOD UK DSTAN 00-18: Part 2-01. Avionic Data Transmission Interface Systems Part 2: Serial, Time Division, Command/Response Multiplex Data Bus Standard Issue 2; Amendment 1. 54 pp.
  MOD UK DSTAN 00-18: Pt 2:Supp A-90. Avionic Data Transmission Interface Systems Part 2: Serial, Time Division, Command/Response Multiplex Data Bus Supplement A: Fibre Optic Supplement for a Point to Point Link Issue 1. 15 pp.
  MOD UK DSTAN 00-18: Pt 2:Supp B-90. Avionic Data Transmission Interface Systems Part 2: Serial, Time Division, Command/Response Multiplex Data Bus Supplement B: Fibre Optic Supplement for a Single Transmissive Star Issue 1. 15 pp.
  MOD UK DSTAN 00-18: Pt 2:Supp C-90. Avionic Data Transmission Interface Systems Part 2: Serial, Time Division, Command/Response Multiplex Data Bus Supplement C: Fibre Optic Supplement for a Single Reflective Star Issue 1. 15 pp.
  MOD UK DSTAN 00-18: Pt 2:Supp D-90. Avionic Data Transmission Interface Systems Part 2: Serial, Time Division, Command/Response Multiplex Data Bus Supplement D: Fibre Optic Supplement for a Multi-Local Transmissive Star Issue 1. 20 pp.
  MOD UK DSTAN 00-18: Part 3-90. Avionic Data Transmission Interface Systems Part 3: Simplex and Half Duplex Serial Digital Transmission Interface Systems Issue 3. 43 pp.
  MOD UK DSTAN 00-18: Pt 3:Supp B-91. Avionic Data Transmission Interface Systems Part 3: Simplex and Half Duplex Serial Digital Interface System Supplement B: Fibre Optic Supplement for a Single Transmissive Star Issue 1. 21 pp.
  MOD UK DSTAN 00-18: Pt 3:Supp D-91. Avionic Data Transmission Interface Systems Part 3: Simplex and Half Duplex Serial Digital Interface System Supplement D: Fibre Optic Supplement for a Multi-Local Transmissive Star Issue 1. 23 pp.
  MOD UK DSTAN 00-18: Part 4-90. Avionic Data Transmission Interface Systems Part 4: Discrete Signal Interfaces Issue 3. 29 pp.
  MOD UK DSTAN 00-18: Part 6-90. Avionic Data Transmission Interface Systems Part 6: Analogue Video Standard for Aircraft System Applications Issue 2. 30 pp.
  MOD UK DSTAN 00-18: Part 6-01. Avionic Data Transmission Interface Systems Part 6: Analogue Video Standard for Aircraft System Applications Issue 2; Amendment 1. 31 pp.
  MOD UK DSTAN 00-18: Pt 6:Supp A-90. Avionic Data Transmission Interface Systems Part 6: Analogue Video Standard for Aircraft System Applications Supplement A: Fibre Optic Supplement for a Point to Point Single Way Link Issue 1. 13 pp.
—Discrete Signal Interfaces
  NATO STANAG 3909 ED 2 AMD 0-90. Discrete Signal Interfaces. 23 pp.
  NATO STANAG 3909 ED 2 AMD 1-90. Discrete Signal Interfaces. 24 pp.

Avionics (Cont.)

—Display Devices
EUROCAE ED-64 12.88. Transport Category Airplane Electronic Display Systems Changes to Be Applied to FAA Advisory Circular No. 25-11 for Adoption as Jar AC. 17 pp.

—Display Devices—Color
NATO STANAG 3940 ED 1 AMD 0-91. Aircraft Electronic Colour Display Systems. 7 pp.

—Electromagnetic Radiation—Glossaries
NATO STANAG 3968 ED 1 AMD 1-85. NATO Glossary of Electromagnetic Terms and Definitions. 11 pp.
NATO STANAG 3968 ED 2 AMD 0-92. NATO Glossary of Electromagnetic Terminology. 7 pp.
NATO STANAG 3968 ED 2 AMD 1-92. NATO Glossary of Electromagnetic Terminology. 8 pp.

—Environmental Testing
EUROCAE ED-25 07.76. MPS for Experimental AEROSAT L-Band Avionics. 75 pp.
JIS W 7004-81. Testing, Environmental, Airborne Electronic and Associated Equipment.

—Glossaries
NATO STANAG 3908 ED 1 AMD 2-84. Standardized Avionics Terminology and Abbreviations. 29 pp.

—L Band Aerosat Experimental
EUROCAE ED-25 07.76. MPS for Experimental AEROSAT L-Band Avionics. 75 pp.

—Maintainability
EUROCAE ED-25 07.76. MPS for Experimental AEROSAT L-Band Avionics. 75 pp.

—Microprocessors
MOD UK DSTAN 00-38-91. Guidelines for the Evaluation of Microprocessors for Avionics Applications Issue 1. 36 pp.

—Real Time Software—ADA
NATO STANAG 3912 ED 1 AMD 1-85. Standardization of a Real-Time High Order Computer Programming Language for Avionics Systems Applications. 8 pp.
NATO STANAG 3912 ED 1 AMD 2-85. Standardization of a Real-Time High Order Computer Programming Language for Avionics Systems Applications. 8 pp.

—Reliability
EUROCAE ED-25 07.76. MPS for Experimental AEROSAT L-Band Avionics. 75 pp.

—Safety
MOD UK DSTAN 00-31-87. Development of Safety Critical Software for Airborne Systems Issue 1. 12 pp.

—Soldering
JIS W 7204-84. Method of Soldering for Aircraft Electrical and Electronic Equipments.

Avions Mudry/CAARP CAP 10B Aircraft
See Also: Aircraft

—Foreign Airworthiness Directives
CAA. Avions Mudry/CAARP CAP 10B and CAP 20 L/S 200 (Foreign Airworthiness Directives). 1 p.

Avions Mudry/CAARP CAP 20 L/S 200 Aircraft
See Also: Aircraft

—Foreign Airworthiness Directives
CAA. Avions Mudry/CAARP CAP 10B and CAP 20 L/S 200 (Foreign Airworthiness Directives). 1 p.

AVL
Use: Automatic Vehicle Location Systems

Avocados
See Also: Food; Fruits

—Refrigerated Transportation
ISO 2295-74. Avocados—Guide for Storage and Transport First Edition. 6 pp.

—Storage
ISO 2295-74. Avocados—Guide for Storage and Transport First Edition. 6 pp.

Avoparcin Content Analysis

—Animal Feed
CNS N4121-84. Method of Test for Feed Additives: Determination of Avoparcin (Nov)(11127).

Avro Aircraft
See Also: Aircraft

—Antenna Positions
CAA. Avro/HS 748 (Approved Aerial Positions). 1 p.

Avro Anson Aircraft
See Also: Aircraft

—Mandatory Aircraft Modifications and Inspections Summaries
CAA. Avro Anson and Avro 19. 6 pp.

Avro Lancaster Aircraft
See Also: Aircraft

—Accidents
CAA. Avro Lancaster/Lancastrian (World Airline Accident Summary). 1 p.

Avro Lancastrian Aircraft
See Also: Aircraft

—Accidents
CAA. Avro Lancaster/Lancastrian (World Airline Accident Summary). 1 p.

Avro Tudor Aircraft
See Also: Aircraft

—Accidents
CAA. Avro Tudor. 1 p.

Avro XIX Aircraft
See Also: Aircraft

—Antenna Positions
CAA. Avro XIX (Approved Aerial Positions). 1 p.

Avro York Aircraft
See Also: Aircraft

—Accidents
CAA. Avro York (World Airline Accidence Summary). 1 p.

Avro 19 Aircraft
See Also: Aircraft

—Mandatory Aircraft Modifications and Inspections Summaries
CAA. Avro Anson and Avro 19. 6 pp.

AW 650 Argosy Aircraft
See Also: Aircraft

—Antenna Positions
CAA. AWW 650 Argosy (Approved Aerial Positions). 1 p.

Awnings
See Also: Fabrics; Roofing

—Construction Contracts
DIN ENGL 18358-88. Tendering and Performance Stipulations in Contracts for Construction Works (VOB); Part C: General Technical Specifications in Contracts for Construction Works (ATV); Installation of Roller Shutters and Similar Equipment (Sept). 3 pp.

—Installation—Construction Contracts
DIN ENGL 18358-88. Tendering and Performance Stipulations in Contracts for Construction Works (VOB); Part C: General Technical Specifications in Contracts for Construction Works (ATV); Installation of Roller Shutters and Similar Equipment (Sept). 3 pp.

—Recreational Vehicles
ISO 8937-91. Caravan Awnings—Functional Requirements and Test Methods First Edition. 11 pp.

—Recreational Vehicles—Safety
BSI BS 5576-85. 1985 Safety Features of Camping Tents, Awnings, Trailer Tents and Caravan Awnings. 4 pp.

—Safety
BSI BS 5576-85. 1985 Safety Features of Camping Tents, Awnings, Trailer Tents and Caravan Awnings. 4 pp.

—Ships
BSI BS MA 42-75. 1975 Ships' Awnings and Supports. 5 pp.
MOD UK NES 727-88. Guide to the Design and Manufacture of Awnings and Associated Screens and Covers Issue 1 (12.88). 28 pp.

AWRE
Use: Atomic Weapons Research Establishments

Axes
See Also: Cutting Tools; Hand Tools; Hatchets; Mattocks; Tools
BSI BS 2945-58. 1958 Amd 1 Axes and Hatchets. 20 pp.
CGSB 39-GP-3M-79. Axes, Standard for. 15 pp.
CNS B3221-72. Axes (Single Blade, Double Blades) (Jun)(3513).

—Fire Fighting Equipment
BSI BS 2957-58. 1958 Fireman's Axe with Ash Handle. 11 pp.
BSI BS 3054-59. 1959 Fireman's Axe with Rubber Insulated Handle. 12 pp.

—Fire Fighting Equipment—Ships
CNS F3085-83. Fire Axes for Ships (Mar)(6915).
JIS F 3610-89. Ships' Fire Axes.

—Handles
CGSB 39-GP-10M-79. Handle, Hickory, Striking Tool (Metric), Standard for; (Amendment 2 Jan 1984). 15 pp.

Axial Fans
Use For: Axial Flow Fans See Also: Fans; Mining Equipment
CNS C4148-79. Axial Flow Fan for Industry Use (Jun)(4872).

—Mines
CNS M2068-80. Axial Flow Type Local Fans (Oct)(6592).
JIS M 7612-77. Axial Flow Type Local Fans for Mine (R 1987). 21 pp.

Axial Flow Fans
Use: Axial Fans

Axial Flow Pumps
Use For: Propeller Pumps See Also: Centrifugal Pumps; Pumps; Rotary Pumps; Screw Pumps
BSI BS 5316: Part 3-88. 1988 Acceptance Tests for Centifugal Mixed Flow and Axial Pumps Part 3: Precision Class Tests. 87 pp.
ISO 5198-87. Centrifugal, Mixed Flow and Axial Pumps—Code for Hydraulic Performance Tests—Precision Class First Edition. 86 pp.
JIS B 8301-90. Testing Methods for Centrifugal Pumps, Mixed Flow Pumps and Axial Flow Pumps. 33 pp.
JIS B 8327-89. Testing Methods for Performance of Pump, Using Model Pump. 38 pp.

—Acceptance Testing
BSI BS 5316: Part 1-76. 1976 Acceptance Tests for Centifugal Mixed Flow and Axial Pumps Part 1: Class C Tests. 36 pp.
BSI BS 5316: Part 2-77. 1977 Acceptance Tests for Centifugal Mixed Flow and Axial Pumps Part 2: Class B Tests. 40 pp.
ISO 2548-73. Centrifugal, Mixed Flow and Axial Pumps—Code for Acceptance Tests—Class C First Edition; (Amendment Slip-1980). 39 pp.
ISO 3555-77. Centrifugal, Mixed Flow and Axial Pumps—Code for Acceptance Tests—Class B First Edition. 40 pp.

Axial Leads
See Also: Electrical Leads

—Tape Packaging
BSI BS 6062: Part 1-81. 1981 Packaging of Electronic Components for Automatic Handling Part 1: Specification for Tape Packaging of Components with Axial Leads on Continous Tapes. 8 pp.
CENELEC HD 143.1-83. Packaging of Components for Automatic Handling Part 1: Tape Packaging of Components with Axial Leads on Continuous Tapes. 2 pp.
CNS Z5077-81. Reel Packaging of Components with Axial Leads (Jan)(6981).
IEC 286 Pt 1-80. Packaging of Components for Automatic Handling Part 1: Tape Packaging of Components with Axial Leads on Continuous Tapes First Edition. 14 pp.
JIS C 0805-89. Packaging of Electronic Components on Continuous Tapes (Components with Axial Type and Radial Wire Lead Terminals). 20 pp.

Axial Load Fatigue Testing
Use: Fatigue Testing

Axial Loads
See Also: Loads (Forces)

—Connectors
CEN PREN 2591 (Part D5)-92. Elements of Electrical and Optical Connection Test Methods Part D5—Axial Load. 4 pp.

Axilla Crutches
See Also: Crutches

Axilla Crutches *(Cont.)*
—Axilla
SNZ NZS 5831: Part 3-89. Walking Aids for the Disabled Part 3: Specification for Axilla Crutches. 12 pp.

Axle Boxes
See Also: Railroad Equipment; Rolling Stock
—Rolling Stock
CNS E1037-85. Axle Boxes for Railway Rolling Stock (Feb)(11202).
CNS E1038-85. Dust Guards for Plain Bearing Axle Boxes of Railway Rolling Stock (Feb)(11203).
JIS E 4701-77. Axle Boxes for Railway Rolling Stock.
JIS E 4702-89. Dust Guards for Plain Bearing Axle Boxes of Railway Rolling Stock.

Axle Power
See Also: Axles
—Tractors
BSI BS 6347: Part 7-93. 1993 Performance Assessment of Agricultural Tractors Part 7: Method for Determination of Axle Power (ISO 789-7: 1991) (E). 16 pp.
ISO 789 Pt 7-91. Agricultural Tractors—Test Procedures—Part 7: Axle Power Determination First Edition. 15 pp.

Axle Shafts
Use: Axles

Axles
See Also: Axle Power; Bearings; Drive Axles; Gear Reducers; Keyways; Machine Keys; Mandrels; Running Gear; Shafts (Machine Elements); Spindles; Wheels
—Automotive—Rigidity
CNS D3149-82. Method of Rigidity Test for Front Axles for Automobiles (Nov)(9598).
—Bicycles
CNS B2036-53. Bracket Axle for Bicycles (Jul)(346). 2 pp.
CNS B2037-75. Bracket Axle For Bicycle (Dec)(347). 3 pp.
CNS B7053-75. Method of Test for the Bracket Axle of Bicycles (Dec)(3866). 2 pp.
—Bicycles—Pedals
BSI BS 6102: Part 14-91. 1991 Cycles Part 14: Specification for Dimensions of Bottom Bracket Axle and Crank Assembly with Square End Fittings (ISO 6695: 1991). 7 pp.
ISO 6695-91. Cycles—Pedal Axle and Crank Assembly with Square End Fitting—Assembly Dimensions First Edition. 4 pp.
—Cranes—Seal Covers
DIN ENGL 15092-82. Cranes; Driving Wheel Units and Idler Wheel Units; Sealing Covers (July). 6 pp.
—Dollies—Springs
CNS B2606-81. Industrial Trucks up to 20 km/h; Rubberboned-Springs for Axels of Handcarts and Trailers (Sep)(7867).
—Gears—Lubricants
CNS K6940-88. Method of Test for Performance of Gear Lubricants in Axles at High Speed, Low Torque, Followed by Low Speed, High Torque (Jan)(12213).
CNS K6941-88. Method of Test for Performance of Gear Lubricants in Axles Under High Speed and Shock Loading (Jan)(12214).
—Motorcycles—Pedals
BSI BS 6102: Part 14-91. 1991 Cycles Part 14: Specification for Dimensions of Bottom Bracket Axle and Crank Assembly with Square End Fittings (ISO 6695: 1991). 7 pp.
ISO 6695-91. Cycles—Pedal Axle and Crank Assembly with Square End Fitting—Assembly Dimensions First Edition. 4 pp.
—Rolling Stock
BSI BS 5892: Part 1-92. 1992 Railway Rolling Stock Materials Part 1: Specification for Axles for Traction and Trailing Stock. 24 pp.
BSI BS 5892: Part 1-83. 1983 Railway Rolling Stock Material (Metric) Part 1: Axles for Traction and Trailing Stock. 11 pp.
CNS E1043-85. Axles for Railway Rolling Stock (Jun)(11280).
JIS E 4502-89. Axles for Railway Rolling Stock.
JIS E 4502-85. Axles for Railway Rolling Stock. 16 pp.

Axles *(Cont.)*
—Rolling Stock—Acceptance Testing—Magnetic Particle Testing
ISO 1005 Pt 3-82. Railway Rolling Stock Material—Part 3: Axles for Tractive and Trailing Stock—Quality Requirements First Edition. 12 pp.
ISO 1005 Pt 9-86. Railway Rolling Stock Material—Part 9: Axles for Tractive and Trailing Stock—Dimensional Requirements First Edition. 9 pp.
ISO 6933-86. Railway Rolling Stock Material—Magnetic Particle Acceptance Testing First Edition. 8 pp.
SNZ NZS/ISO 1005: Part 3-82. Railway Rolling Stock Material Part 3: Axles for Tractive and Trailing Stock. Quality Requirements. 10 pp.
—Rolling Stock—Acceptance Testing—Ultrasonic Testing
ISO 5948-81. Railway Rolling Stock Material—Ultrasonic Acceptance Testing First Edition. 5 pp.
—Rolling Stock—Center Holes
DIN ENGL 332 Pt 4-90. Centre Holes for Rail Vehicle Axles (June). 2 pp.
—Rolling Stock—Strength
CNS E3004-82. Strength Design Methods of Axles for Railway Rolling Stock (Feb)(8490).
JIS E 4501-89. Design Methods for Strength of Axles of Railway Rolling Stock.
—Rolling Stock—Vibration Isolators
CNS E1024-83. Rubber Vibration Isolators for Axle and Bolster Springs of Railway Rolling Stock (Feb)(9994).
JIS E 4711-89. Rubber Vibration Isolators for Axle and Bolster Springs of Railway Rolling Stock.
—Ships—Pins—Fasteners
DIN ENGL 82242-63. Pin Fastener for Axle Pins from 20 to 80 mm Diameter (May). 1 p.
—Trailers
CNS B4037-80. Axles with Brakes of Trailer for Power Tiller (Aug)(6012).
JIS B 9208-76. Axles with Brakes of Trailer for Power Tiller (R 1979). 9 pp.
—Trailers—Springs
CNS B2606-81. Industrial Trucks up to 20 km/h; Rubberboned-Springs for Axels of Handcarts and Trailers (Sep)(7867).

Axonometric Projections
Use For: Dimetric Projections; Isometric Projections
See Also: Engineering Drawings
DIN ENGL 5 Pt 1-70. Drawing Practice; Axonometric Projections; Isometric Projection (Dec). 4 pp.
DIN ENGL 5 Pt 2-70. Drawing Practice; Axonometric Projections; Dimetric Projection (Dec). 1 p.
—Pipelines
DIN ENGL 2428-68. Drawings for Pipelines; Forms for Isometric Representation (Dec). 5 pp.
ISO 6412 Pt 2-89. Technical Drawings—Simplified Representation of Pipelines—Part 2: Isometric Projection First Edition. 13 pp.

Ayres S2R Aircraft
See Also: Aircraft
—Antenna Positions
CAA. Ayres S2R (Approved Aerial Positions). 1 p.
—Foreign Airworthiness Directives
CAA. Ayres S-2R (Foreign Airworthiness Directives). 1 p.

Azeotropic Distillation
See Also: Distillation Methods
—Fruits
ISO 1026-82. Fruit and Vegetable Products—Determination of Dry Matter Content by Drying Under Reduced Pressure and of Water Content by Azeotropic Distillation First Edition. 6 pp.
—Refractory Materials
BSI BS 6043: Sec 4.5-91. 1991 Methods of Sampling and Test for Carbonaceous Materials Used in Aluminium Manufacture Part 4: Cold Ramming Pastes Section 4.5: Determination of Water Content of Unbaked Pastes. 9 pp.
—Soaps
ISO 4318-89. Surface Active Agents and Soaps—Determination of Water Content—Azeotropic Distillation Method Second Edition. 5 pp.

Azeotropic Distillation *(Cont.)*
—Surfactants
ISO 4318-89. Surface Active Agents and Soaps—Determination of Water Content—Azeotropic Distillation Method Second Edition. 5 pp.
—Vegetables
ISO 1026-82. Fruit and Vegetable Products—Determination of Dry Matter Content by Drying Under Reduced Pressure and of Water Content by Azeotropic Distillation First Edition. 6 pp.

Azimuth Indicators
See Also: Indicating Instruments
—Ships
BSI BS MA 2: Part 1-85. 1985 Magnetic Compasses and Binnacles Part 1: Class A Magnetic Compasses, Their Binnacles and Azimuth Reading Devices (Including Test Procedures). 19 pp.
ISO 449-79. Shipbuilding—Magnetic Compasses and Binnacles, Class A First Edition. 8 pp.
ISO 613-82. Shipbuilding—Magnetic Compasses, Binnacles and Azimuth Reading Devices—Class B First Edition; (Corrigendum 1-1991). 7 pp.

Azinphos-Ethyl
See Also: Pesticides
SAA AS 1870.4D-76. Standard for Development—Pesticides for Agricultural Use—Part 4D: Azinphos-Methyl and Azinphos-Ethyl. 21 pp.

Azinphos-Methyl
See Also: Pesticides
SAA AS 1870.4D-76. Standard for Development—Pesticides for Agricultural Use—Part 4D: Azinphos-Methyl and Azinphos-Ethyl. 21 pp.

Azobenzene
CNS K7574-81. Chemical Reagent (Azobenzene) (Aug)(7807).

Azolitmin
CNS K7079-62. Chemical Reagent (Azolitmin) (Jul)(1579).

Babbitt
Use: Bearing Alloys

Babbitt Metals
Use: Bearing Alloys

Babies' Dummies
Use: Pacifiers

Babies' Napkins
Use: Diapers

Baby Beds
Use: Cribs (Furniture)

Baby Bottles
Use: Feeding Bottles (Babies)

Baby Carriages
Use For: Perambulators; Prams
—Mattresses
BSI BS 1877: Part 10-82. 1982 Amd 2 Domestic Bedding Part 10: Specification for Mattresses and Bumpers for Children's Cots, Perambulators and Similar Domestic Articles (AMD 6593) April 30, 1991. 12 pp.
BSI BS 1877: Part 10-03. 1982 Amd 3 Domestic Bedding Part 10: Specification for Mattresses and Bumpers for Children's Cots, Perambulators and Similar Domestic Articles (AMD 6949) February 28, 1992. 13 pp.
—Safety
BSI BS 4139-86. (WITHDRAWN) 1986 Amd 1 Safety Requirements for Perambulators (Baby Carriages) (AMD 6474) March 30, 1990 (Superseded by BS 7409: 1991). 22 pp.
BSI BS 7409-91. 1991 Safety Requirements for Wheeled Child Conveyances. 32 pp.
BSI BS 7409-01. 1991 Amd 1 Safety Requirements for Wheeled Child Conveyances (AMD 7177) July 15, 1992. 34 pp.
SAA AS 2088-89. Prams and Strollers—Safety Requirements Amdt 1 May 1990 Amdt 2 January 1991 Amdt 3 June 1991 (Superseded by AS/NZS 2088:1993).
SNZ NZS/AS 2088-93. Prams and Strollers—Safety Requirements (This is a Joint Standard with SAA AS 2088). 43 pp.
SNZ NZS 5804-86. Safety Requirements for Baby Carriages Amend: 1, 1987. 26 pp.

Baby Carriages (Cont.)

—Safety Belts

BSI BS 3785-64. 1964 Amd 3 Webbing Safety Harness for Baby Carriages and Chairs and Walking Reins. 13 pp.

BSI BS 6684-89. 1989 Amd 1 Safety Harnesses (Including Detachable Walking Reins) for Restraining Children When in Perambulators (Baby Carriages), Pushchairs and High Chairs and When Walking (AMD 6531) June 29, 1990. 16 pp.

SAA AS 3747-89. Harnesses for Use in Prams, Strollers, and High Chairs (Including a Detachable Walking Rein) (This is a Joint Standard with SANZ NZS 3747). 8 pp.

SNZ NZS/AS 3747-89. Harnesses for Use in Prams, Strollers and High Chairs, (Including a Detachable Walking Rein) (This is a Joint Standard with SAA AS 3747). 8 pp.

Baby Corn

Use: Corn

Baby Cucumbers

Use: Cucumbers

Baby Foods

Use For: Infant Foods *See Also:* Food

EC COM(86) 91-86. Proposal for a Council Directive on the Approximation of the Laws of the Member States Relating to Foodstuffs Intended for Particular Nutritional Uses. 13 pp.

—Antibiotics Content

CNS N6169-82. Method of Test for the Residue of Anti-Biotic in Foodstuffs (Babyfood Use) (May)(8860). 2 pp.

—Canned

CNS N5206-84. Canned Baby Food (Mar)(10838). 4 pp.

—Cereal Based

CNS N5201-83. Cereal Based Foods for Infant and Children (Jan)(9906). 4 pp.

—Formulas

CNS N5174-91. Infant Formula (Oct)(6849). 6 pp.

EC COM(84) 703-84. Proposal for a Council Directive on the Approximation of the Laws of the Member States Relating to Infant Formulae and Follow-up Milks. 60 pp.

EC COM(86) 91-86. Proposal for a Council Directive on the Approximation of the Laws of the Member States Relating to Foodstuffs Intended for Particular Nutritional Uses. 13 pp.

EC COM(86) 564-86. Modified Proposal for a Council Directive on the Approximation of the Laws of the Member States Relating to Infant Formulae and Follow-up Milks. 38 pp.

EC 91/321/EEC-91. Commission Directive on Infant Formulae and Follow-on Formulae. 15 pp.

—Formulas—Compositional Requirements

EC 91/321/EEC-91. Commission Directive on Infant Formulae and Follow-on Formulae. 15 pp.

—Formulas—Identification Systems

EC 91/321/EEC-91. Commission Directive on Infant Formulae and Follow-on Formulae. 15 pp.

—Formulas—Mineral Content

CNS N6231-91. Method of Test for Minerals in Infant Formula (May)(12869). 6 pp.

—Formulas—Test Specimens

CNS N6056-80. Method of Test for Milk and Milk Products (General Rules) (Aug)(3440). 2 pp.

—Formulas—Vitamin A Content

CNS N6230-90. Method of Test for Determining of Vitamin A Content for Infant Formula (May)(12725). 2 pp.

—Formulas—Vitamin E Content

CNS N6228-90. Method of Test for Determining of Vitamin E Content for Infant Formula (May)(12723). 2 pp.

—Milk Based—Fat Content—Gravimetric Analysis

BSI BS 7142: Part 1-89. 1989 Analysis of Milk-Based Products Part 1: Determination of Fat Content of Milk-Based Infant Foods by the Rose-Gottlieb Gravimetric Method. 12 pp.

BSI BS 7142: Part 2-89. 1989 Analysis of Milk-Based Products Part 2: Determination of Fat Content of Milk-Based Infant Foods by the Weibull-Berntrop Gravimetric Method. 8 pp.

ISO 8262 Pt 1-87. Milk Products and Milk-Based Foods—Determination of Fat Content by the Weibull-Berntrop Gravimetric Method (Reference Method)—Part 1: Infant Foods First Edition. 8 pp.

Baby Foods (Cont.)

—Milk Based—Fat Content—Gravimetric Analysis (Cont.)

ISO 8381-87. Milk-Based Infant Foods—Determination of Fat Content—Rose-Gottlieb Gravimetric Method (Reference Method) First Edition. 12 pp.

Baby Nests

Use: Buntings (Clothing)

Baby Safety Seats

Use: Child Safety Seats

BAC/SNIAS Concorde Aircraft

See Also: Aircraft

—Data Sheets

CAA BA10 ISSUE 10. BAC/SNIA Concorde Type 1 Variant 102. 4 pp.

—Mandatory Aircraft Modifications and Inspections Summaries

CAA. British Aircraft Corporation/SNIAS Concorde Type 1, Variant 102. 26 pp.

BAC Super VC 10 Aircraft

See Also: Aircraft

—Mandatory Aircraft Modifications and Inspections Summaries

CAA. British Aircraft Corporation VC10 and Super VC10. 15 pp.

BAC 1-11 Aircraft

See Also: Aircraft

—Accidents

CAA. British Aircraft Corporation BAC 1-11 (World Airline Accident Summary). 1 p.

CAA. British Aircraft Corporation BAC 1-11. 1 p.

—Antenna Positions

CAA. BAC One-Eleven 200 and 400 Series (Approved Aerial Positions). 1 p.

CAA. BAC One-Eleven 500 Series. 1 p.

—Data Sheets

CAA BA3 ISSUE 10. BAC One Eleven 475 and 500 Series, 475EZ, 476FM, 479FU, 481FW, 485GD, 487GK, 488GH, 492GM, 500EN, 501EX, 509EW, 501EW, 510ED, 515FB, 516FP, 517FE, 518FG, 520FN, 521FH, 523FJ, 524FF, 525FT, 525/1FT, 527FK, 528FL, 529FR, 530FX, 531FS, 537GF, 539GL, 560RB. 5 pp.

—Mandatory Aircraft Modifications and Inspections Summaries

CAA. BAC One-Eleven. 30 pp.

Bacillaceae

Scope Note: Use a more specific term
See: Anthracis; Bacillus Cereus; Bacteria; Clostridium

Bacillus Cereus

See Also: Bacteria

CNS N6212-89. Method of Test for Microbiology—Test of Bacillus Cereus (Jun)(12540). 9 pp.

—Count Methods

BSI BS 5763: Part 11-88. 1988 Microbiological Examination of Food and Animal Feeding Stuffs Part 11: Enumeration of Bacillus Cereus. 11 pp.

ISO 7932-87. Microbiology—General Guidance for Enumeration of Bacillus Cereus—Colony Count Technique at 30 Degrees Celsius First Edition. 11 pp.

—Dairy Products

BSI BS 4285: Sec 3.12-89. 1989 Microbiological Examination for Dairy Purposes Part 3: Methods for Detection and/or Enumeration of Specific Groups of Microorganisms Section 3.12: Enumeration of Bacillus Cereus. 6 pp.

—Food

SAA AS 1766.2.6-91. Methods for the Microbiological Examination of Food—Part 2: Examination for Specific Organisms—Part 2.6: Bacillus Cereus (Supersedes AS 1766.2.1.6 Addendum 1—1976).

Bacitracin Content Analysis

—Animal Feed

CNS N4102-82. Method of Test for Feed Additives: Determination of Bacitracin (Aug)(9317).

Back Shells (Electric)

Use: Shells (Electric)

Back Up Alarms

Use: Backup Alarms

Backflow Gates

Use: Backwater Valves

Backflow Preventers

Use For: Antivacuum Valves; Vacuum Breakers
See Also: Backwater Valves; Check Valves; Foot Valves

BSI BS 6282: Part 2-82. 1982 Devices with Moving Parts for the Prevention of Contamination of Water by Backflow Part 2: Terminal Anti-Vacuum Valves of Nominal Size up to and Including DN 54. 8 pp.

BSI BS 6282: Part 3-82. 1982 Devices with Moving Parts for the Prevention of Contamination of Water by Backflow Part 3: In-Line Anti-Vacuum Valves of Nominal Size up to and Including DN 42. 8 pp.

CSA CAN/CSA-B64 Series-M88. Backflow Preventers and Vacuum Breakers (Including CAN/CSA-B64.10-M88 on Selection, Installation, Maintenance, and Field Testing); (Gen Instr 1 Thru 3). 110 pp.

—Air Gaps

BSI BS 6281: Part 1-92. 1992 Devices Without Moving Parts for the Prevention of Contamination of Water by Backflow Part 1: Specification for Type A Air Gaps. 12 pp.

BSI BS 6281: Part 1-82. 1982 Devices Without Moving Parts for the Prevention of Contamination of Water by Backflow Part 1: Type A Air Gaps. 4 pp.

BSI BS 6281: Part 2-82. 1982 Devices Without Moving Parts for the Prevention of Contamination of Water by Backflow Part 2: Type B Air Gaps. 4 pp.

—Pipe Interrupters

BSI BS 6281: Part 3-82. 1982 Devices Without Moving Parts for the Prevention of Contamination of Water by Backflow Part 3: Pipe Interrupters of Nominal Size up to and Including DN 42. 8 pp.

—Water Pipes

DIN ENGL 3266 Pt 1-86. Valves for Drinking Water Installations on Private Premises; PN 10 Pipe Interrupters, Pipe Disconnectors, Anti-Vacuum Valves (July). 8 pp.

DIN ENGL 3266 Pt 2-87. Valves for Drinking Water Installations on Private Premises; PN 10 Pipe Interrupters, Pipe Disconnectors, Anti-Vacuum Valves; Testing (Dec). 6 pp.

SAA AS 2845.1-91. Water Supply—Mechanical Backflow Prevention Devices—Part 1: Materials, Design and Performance Requirements. 64 pp.

Backhoes

See Also: Earthmoving Equipment; Excavating Equipment; Loaders

—Booms—Hydraulic Valves

BSI BS 6912: Part 1-88. 1988 Safety of Earth-Moving Machinery Part 1: Hydraulic Excavator and Backhoe Loader Boom Lowering Control Device. 6 pp.

ISO 8643-88. Earth-Moving Machinery—Hydraulic Excavator and Backhoe Loader Boom Lowering Control Device—Requirements and Tests First Edition. 6 pp.

—Manual Controls

JIS A 8405-90. Excavators—Operator's Controls. 16 pp.

—Roll Over Protective Structures

BSI BS 5527: Part 1-87. 1987 Roll-Over Protective Structures on Earth-Moving Machinery: Laboratory Tests and Preformance Requirements Part 1: Crawler, Wheel Loaders and Tractors, Backhoe Loaders, Graders, Tractor Scrapers and Articulated Steel Dumpers. 19 pp.

ISO 3471 Pt 1-86. Earth-Moving Machinery—Roll-over Protective Structures—Laboratory Tests and Performance Requirements—Part 1: Crawler, Wheel Loaders and Tractors, Backhoe Loaders, Graders, Tractor Scrapers, Articulated Steer Dumpers First Edition. 18 pp.

—Safety

CEN PREN 474-4-93. Earth-Moving Machinery—Safety—Part 4: Requirements for Backhoe Loaders. 16 pp.

Backing Boards

Use: Plasterboard

Backing Plates

See Also: Plates (Supports)

—Molds (Casting)

DIN ENGL 16760 Pt 1-90. Compression, Injection and Die-Casting Moulds; Machined Plates, Undrilled (Sept). 9 pp.

Backlash

See Also: Gears; Resolution Guides

Backlash (Cont.)

—Gears
DIN ENGL 3967-78. System of Gear Fits; Backlash; Tooth Thickness Allowances; Tooth Thickness Tolerances; Principles (Aug). 24 pp.

—Gears—Bevel
JIS B 1705-73. Backlashes for Bevel Gears (R 1979). 18 pp.

—Gears—Cylindrical—Glossaries
CNS B1189-84. Definitions and Denominations for Glossary of Gear Terms — Errors for the Axial Positions and Backlash for a Cylindrical Gear Teeth (May)(5731).

—Gears—Glossaries
CNS B1184-84. Definitions and Denomination of Glossary of Gear Terms — Crest, Root, Surface, Tooth Flank, Tooth Profiles, Backlash, and Clearance (May)(5726).

—Gears—Helical
JIS B 1703-76. Backlash for Spur and Helical Gears (R 1979). 22 pp.

—Gears—Spur
JIS B 1703-76. Backlash for Spur and Helical Gears (R 1979). 22 pp.

Backplanes

See Also: Data Buses; Printed Circuit Boards
MOD UK DSTAN 59-86-90. Generic Specification for Press-In Backplane Assemblies and Capability Approval Procedures Issue 1. 53 pp.

— 8-Bit—Interfaces
OSI ISO/IEC DIS 10859-90. Information Technology—8-Bit Backplane Interface STEbus. 60 pp.

—Data Buses—VICbus Compatible
IEC DIS 11458-92. Information Technology—VICbus: an Inter-Crate Bus for the IEC 821 Bus (VMEbus) ***CD-ROM ONLY***. 103 pp.
ISO DIS 11458-92. Information Technology—VICbus: an Inter-Crate Bus for the IEC 821 Bus (VMEbus) ***CD-ROM ONLY***. 103 pp.
JTC1 DIS11458-92. Information Technology—VICbus: an Inter-Crate Bus for the IEC 821 Bus (VMEbus) ***CD-ROM ONLY***. 103 pp.

—Data Buses—VMEbus Compatible
BSI BS IEC 821-91. 1991 IEC 821 VMEbus—Microprocessor System Bus for 1 Byte to 4 Byte Data. 564 pp.
IEC 821-91. IEC 821 VMEbus—Microprocessor System Bus for 1 Byte to 4 Byte Data Second Edition. 563 pp.
ISO 821-91. IEC 821 VMEbus—Microprocessor System Bus for 1 Byte to 4 Byte Data Second Edition. 563 pp.
JTC1 821-91. IEC 821 VMEbus—Microprocessor System Bus for 1 Byte to 4 Byte Data Second Edition. 563 pp.
OSI ISO/IEC DIS 821-89. IEC 821 VME Bus—Microprocessor System Bus for 1 to 4 Byte Data. 282 pp.

—32-Bit—Interfaces
OSI ISO/IEC DIS 10860-90. Information Technology—Simple 32-Bit Backplane Bus: NuBus. 70 pp.

Backsaws

See Also: Saws

—Woodworking
CNS B3137-59. Back Saw for Carpenter (Apr)(1101)(R 1970). 1 p.

Backscattering

See Also: Scatter Propagation; Sidescattering

—Ground Wave Propagation
CCIR Report 726-2-90. Ground and Ionospheric Side- and Back-Scatter—Section 6C—Ionospheric Propagation and Operational Forecasting. 7 pp.

—Ionospheric Propagation
CCIR Report 726-2-90. Ground and Ionospheric Side- and Back-Scatter—Section 6C—Ionospheric Propagation and Operational Forecasting. 7 pp.
CCIR Report 890-2-90. Operational Use of Side-Scatter and Back-Scatter—Section 6C—Ionospheric Propagation and Operational Forecasting. 3 pp.
CCIR QUESTION 32/6-90. Propagation Via Side-and Back-Scatter—Questions Concerning Study Group 6—Radio Wave Propagation in Ionized Media. 1 p.

—Scattering Ratio
JIS Z 4508-80. Testing Method of Scattering-Ratio of Low-Backscattering Materials.

Backshells (Electric)

Use: Shells (Electric)

Backup Alarms

Use For: Back Up Alarms; Forward and Reverse Audible Warning Alarms *See Also:* Alarm Systems; Construction Equipment; Warning Systems

—Automotive
CNS D2170-86. Backup Alarm for Automobiles (Nov)(10155).
CNS D3153-86. Method of Test for Backup Alarm for Automobiles (Nov)(10156).

—Earthmoving Equipment—Acoustic Testing
BSI BS 6912: Part 3-90. 1990 Safety of Earth Moving Machinery Part 3: Sound Test Method for Machine-Mounted Forward and Reverse Warning Alarm. 12 pp.
ISO 9533-89. Earth-Moving Machinery—Machine-Mounted Forward and Reverse Audible Warning Alarm—Sound Test Method First Edition. 8 pp.

Backup Lights

Use For: Reversing Lights *See Also:* Automotive Equipment; Motor Vehicle Lighting; Signal Lights

—Automobiles—Electric Switches
CNS D2056-87. Switches of Back-Up Lamps for Automobiles (Jan)(6324).
JIS D 5810-74. Back-up-Lamp Switches for Automobiles.

—Automotive
EC 77/539/EEC-77. Council Directive on the Approximation of the Laws of the Member States Relating to Reversing Lamps for Motor Vehicles and Their Trailers. 11 pp.

—Trailers
EC 77/539/EEC-77. Council Directive on the Approximation of the Laws of the Member States Relating to Reversing Lamps for Motor Vehicles and Their Trailers. 11 pp.

Backup Power Supplies

Use: Emergency and Standby Power Supplies

Backward Interworking Telephone Events

Use: BITES

Backward Signals

See Also: Signaling Systems

—Glossaries—CCITT R2 Signaling Systems
CCITT RECMN Q.400-89. Definitions and Functions of Signals—Forward Line Signals. Backward Line Signals. Forward Register Signals. Backward Register Signals—Specifications of Signalling Systems R1 and R2 (Study Group XI) 4 pp. 4 pp.

—Pulse Transmission—Interregister Signaling—CCITT R2 Signaling
CCITT RECMN Q.442-89. Interregister Signalling —Pulse Transmission of Backward Signals A-3, A-4, A-6 or A-15. Multifrequency Signalling Equipment—Specifications of Signalling Systems R1 and R2 (Study Group XI) 2 pp. 2 pp.

Backward Wave Oscillators

Use For: Carcinatorn Oscillators; Carcinatrons
See Also: Oscillators

—Preferred Products List
CECC CECC MUAHAG Vol 12 IS 1-90. Preferred Products List; Microwave Components (En, Fr, Ge). 76 pp.

Backward Wave Tubes

See Also: Electron Tubes; Microwave Equipment; Microwave Tubes; Thermionic Tubes

—Electrical Measurement
IEC 235 Pt 8-72. Measurement of the Electrical Properties of Microwave Tubes Part 8: Backward-Wave Oscillator Tubes—"O" Type First Edition; (Supplement A-1974). 17 pp.

Backwater Valves

Use For: Backflow Gates *See Also:* Backflow Preventers; Check Valves; Valves
BSI BS 6282: Part 4-82. 1982 Devices with Moving Parts for the Prevention of Contamination of Water by Backflow Part 4: Combined Check and Anti-Vacuum Valves of Nominal Size up to and Including DN 42. 4 pp.
CSA CAN/CSA-B70-M91. Cast Iron Soil Pipe, Fittings, and Means of Joining; (Gen Instr 1). 27 pp.
CSA CAN/CSA-B181.1-M90. ABS Drain, Waste, and Vent Pipe and Pipe Fittings; (Gen Instr 1). 50 pp.
CSA CAN/CSA-B181.2-M90. PVC Drain, Waste, and Vent Pipe and Pipe Fittings; (Gen Instr 1). 56 pp.
DIN ENGL 1997 Pt 1-84. Shut-Off Valves for Site Drainage Systems; Backflow Gates for Non-Faecal Sewage; Requirements, Design Principles and Materials (May). 3 pp.
DIN ENGL 1997 Pt 2-84. Shut-Off Valves for Site Drainage Systems; Backflow Gates for Non-Faecal Sewage; Testing (May). 4 pp.

—Inspection
DIN ENGL 1986 Pt 32-86. Site Drainage Systems; Backflow Gates for Non-Faecal Sewage; Inspection and Maintenance (June). 2 pp.
DIN ENGL 1986 Pt 33-87. Site Drainage Systems; Backflow Gates for Faecal Sewage; Inspection and Maintenance (Oct). 2 pp.

—Maintenance
DIN ENGL 1986 Pt 32-86. Site Drainage Systems; Backflow Gates for Non-Faecal Sewage; Inspection and Maintenance (June). 2 pp.
DIN ENGL 1986 Pt 33-87. Site Drainage Systems; Backflow Gates for Faecal Sewage; Inspection and Maintenance (Oct). 2 pp.

—Sewers
DIN ENGL 19578 Pt 1-88. Stop Valves for Site Drainage Systems; Anti-Flooding Valves for Faecal Sewage Systems; Requirements (Feb). 3 pp.

—Sewers—Inspection
DIN ENGL 19578 Pt 2-88. Stop Valves for Site Drainage Systems; Anti-Flooding Valves for Faecal Sewage Systems; Testing and Inspection (Feb). 3 pp.

Bacon

See Also: Food; Pork
CGSB 32.61M-90. Back Bacon and Side Bacon. 8 pp.

Bacteria

Scope Note: For additional listings, use a more specific term *See Also:* Aerobic Bacteria; Bacillus Cereus; Campylobacter; Clostridium; Clostridium Botulinum; Clostridium Perfringens; Coliform Bacteria; Enterobacteriaceae; Escherichia Coli; Legionella Pneumophilla; Listeria Monocytogenes; Nitrobactericeae; Photobacteria; Pseudomonas; Salmonella; Shigella; Staphylococcus; Streptococcus; Vibrio Parahaemolyticus

—Cellular Plastics
ISO 846-78. Plastics—Determination of Behavior Under the Action of Fungi and Bacteria—Evaluation by Visual Examination or Measurement of Change in Mass or Physical Properties First Edition. 12 pp.

—Dairy Products
SAA AS 1095.3.7-79. Microbiological Methods for the Dairy Industry—Part 3: Methods of Examination for Specific Groups of Microorganisms —Part 3.7: Bacterial Spores. 2 pp.

—Fabrics
JIS L 1902-90. Testing Method for Antibacterial of Textiles. 13 pp.

—Membrane Filters
JIS K 3824-90. Testing Methods for Endotoxin Rejection of Ultrafiltration Modules. 15 pp.
JIS K 3835-90. Testing Methods for Determining Bacterial Retention of Membrane Filters. 17 pp.

—Plastics
DIN ENGL 53739-84. Testing of Plastics; Influence of Fungi and Bacteria; Visual Evaluation; Change in Mass or Physical Properties (Nov). 11 pp.

Bacteria Count Methods

Use For: Colony Count Technique; Most Probable Number Technique *See Also:* Microbiological Analysis

—Aerobic Bacteria—Animal Feed
BSI BS 5763: Part 1-91. 1991 Microbiological Examination of Food and Animal Feeding Stuffs Part 1: Enumeration of Micro-Organisms—Colony Count Technique at 30 Degrees C (ISO 4833: 1991). 11 pp.

—Aerobic Bacteria—Cattle Semen
ISO TR8607-91. Artificial Insemination of Animals—Frozen Semen of Breeding Bulls—Enumeration of Living Aerobic Micro-Organisms First Edition. 6 pp.

—Aerobic Bacteria—Dairy Products
BSI BS 4285: Sec 3.3-86. 1986 Microbiological Examination for Dairy Purposes Part 3: Methods for Detection and/or Enumeration of Specific Groups of Microorganisms Section 3.3: Enumeration of Aerobic Bacterial Spores. 4 pp.

INTERNATIONAL AND NON-U.S. NATIONAL STANDARDS
SUBJECT INDEX
Bacteria

Bacteria Count Methods (*Cont.*)

—**Aerobic Bacteria—Food**
BSI BS 5763: Part 1-91. 1991 Microbiological Examination of Food and Animal Feeding Stuffs Part 1: Enumeration of Micro-Organisms—Colony Count Technique at 30 Degrees C (ISO 4833: 1991). 11 pp.
BSI BS 5763: Part 5-81. 1981 Microbiological Examination of Food and Animal Feeding Stuffs Part 5: Enumeration of Micro-Organisms—Colony Count at 30 Degrees Celsius (Surface Plate Technique). 7 pp.
ISO 4833-91. Microbiology—General Guidance for the Enumeration of Micro-Organisms—Colony Count Technique at 30 Degrees Celsius Second Edition. 9 pp.

—**Aerobic Bacteria—Water**
ISO 6222-88. Water Quality—Enumeration of Viable Micro-Organisms—Colony Count by Inoculation in or on a Nutrient Agar Culture Medium First Edition. 4 pp.

—**Aerobic Microorganisms—Meat**
ISO 2293-88. Meat and Meat Products—Enumeration of Micro-Organisms—Colony Count Technique at 30 Degrees Celsius (Reference Method) Second Edition. 7 pp.

—**Bacillus Cereus—Dairy Products**
BSI BS 4285: Sec 3.12-89. 1989 Microbiological Examination for Dairy Purposes Part 3: Methods for Detection and/or Enumeration of Specific Groups of Microorganisms Section 3.12: Enumeration of Bacillus Cereus. 6 pp.

—**Bacillus Cereus—Food**
BSI BS 5763: Part 11-88. 1988 Microbiological Examination of Food and Animal Feeding Stuffs Part 11: Enumeration of Bacillus Cereus. 11 pp.
ISO 7932-87. Microbiology—General Guidance for Enumeration of Bacillus Cereus—Colony Count Technique at 30 Degrees Celsius First Edition. 11 pp.

—**Cereals**
ISO 7698-90. Cereals, Pulses and Derived Products—Enumeration of Bacteria, Yeasts and Moulds First Edition. 10 pp.

—**Clostridium—Animal Feed**
BSI BS 5763: Part 9-86. 1986 Microbiological Examination of Food and Animal Feeding Stuffs Part 9: Enumeration of Clostridium Perfringens. 10 pp.

—**Clostridium—Dairy Products**
BSI BS 4285: Sec 3.13-91. 1991 Microbiological Examination for Dairy Purposes Part 3: Methods for Detection and/or Enumeration of Specific Groups of Microorganisms Section 3.13: Enumeration of Clostridium Perfringens. 6 pp.

—**Clostridium—Food**
BSI BS 5763: Part 9-86. 1986 Microbiological Examination of Food and Animal Feeding Stuffs Part 9: Enumeration of Clostridium Perfringens. 10 pp.
ISO 7937-85. Microbiology—General Guidance for Enumeration of Clostridium Perfringens—Colony Count Technique First Edition. 9 pp.

—**Coliforms—Animal Feed**
BSI BS 5763: Part 2-91. 1991 Microbiological Examination of Food and Animal Feeding Stuffs Part 2: Enumeration of Coliforms—Colony Count Technique (ISO 4832: 1991). 11 pp.

—**Coliforms—Dairy Products**
BSI BS 4285: Sec 3.7-87. 1987 Microbiological Examination for Dairy Purposes Part 3: Methods for Detection and/or Enumeration of Specific Groups of Microorganisms Section 3.7: Enumeration of Coliform Bacteria. 4 pp.
ISO 5541 Pt 1-86. Milk and Milk Products—Enumeration of Coliforms—Part 1: Colony Count Technique at 30 Degrees Celsius First Edition. 9 pp.
ISO 5541 Pt 2-86. Milk and Milk Products—Enumeration of Coliforms—Part 2: Most Probable Number Technique at 30 Degrees Celsius First Edition. 10 pp.

—**Coliforms—Food**
BSI BS 5763: Part 2-91. 1991 Microbiological Examination of Food and Animal Feeding Stuffs Part 2: Enumeration of Coliforms—Colony Count Technique (ISO 4832: 1991). 11 pp.
BSI BS 5763: Part 3-91. 1991 Microbiological Examination of Food and Animal Feeding Stuffs Part 3: Enumeration of Coliforms—Most Probable Number Technique (ISO 4831: 1991). 9 pp.
ISO 4831-91. Microbiology—General Guidance for the Enumeration of Coliforms —Most Probable Number Technique Second Edition. 16 pp.

Bacteria Count Methods (*Cont.*)

—**Coliforms—Food** (*Cont.*)
ISO 4832-91. Microbiology—General Guidance for the Enumeration of Coliforms —Colony Count Technique Second Edition. 9 pp.

—**Coliforms—Water**
DIN ENGL 38411 Pt 6-91. German Standard Methods for the Examination of Water, Waste Water and Sludge; Microbiological Methods (Group K); Determination of Escherichia Coli and Coliform Organisms (K 6) (June). 6 pp.
ISO 9308 Pt 1-90. Water Quality—Detection and Enumeration of Coliform Organisms, Thermotolerant Coliform Organisms and Presumptive Escherichia Coli—Part 1: Membrane Filtration Method First Edition. 14 pp.
ISO 9308 Pt 2-90. Water Quality—Detection and Enumeration of Coliform Organisms, Thermotolerant Coliform Organisms and Presumptive Escherichia Coli—Part 2: Multiple Tube (Most Probable Number) Method First Edition. 13 pp.
SAA AS 1095.4.1. 3-81. Microbiological Methods for the Dairy Industry—Part 4: Methods for the Examination of Water and Air—Part 4.1.3: Microbiological Examination of Water—Coliforms by Multiple Tube Dilution. 3 pp.
SAA AS 1095.4.1. 5-81. Microbiological Methods for the Dairy Industry—Part 4: Methods for the Examination of Water and Air—Part 4.1.5: Microbiological Examination of Water—Coliforms by Membrane Filtration. 4 pp.

—**Cork Stoppers**
ISO 10718-93. Cork Stoppers—Enumeration of Colony-Forming Units of Yeasts, Moulds and Bacteria Capable of Growth in an Alcoholic Medium First Edition. 5 pp.

—**Dairy Products**
BSI BS 4285: Sec 2.5-89. 1989 Microbiological Examination for Dairy Purposes Part 2: Methods of General Application for Enumeration of Microoganisms Section 2.5: Enumeration of Bacteria by Direct Microscopic Counts. 6 pp.
DIN ENGL 10192 Pt 1-84. Microbiological Analysis of Milk; Determination of Bacterial Count; Reference Method (Apr). 4 pp.
DIN ENGL 10192 Pt 2-83. Microbiological Analysis of Milk; Determination of Bacterial Count; Simplified Plate Count Method According to Koch (Feb). 2 pp.
DIN ENGL 10192 Pt 3-83. Microbiological Analysis of Milk; Determination of Bacterial Count; Plate-Loop Method (Feb). 3 pp.
DIN ENGL 10195 Pt 1-85. Microbiological Analysis of Milk; Microcolony Count; Enumeration of Microcolonies by Stereomicroscopic Counts (Dec). 3 pp.
ISO 6610-92. Milk and Milk Products—Enumeration of Colony-Forming Units of Micro-Organisms—Colony-Count Technique at 30 Degrees Celsius First Edition. 8 pp.
ISO 6730-92. Milk—Enumeration of Colony-Forming Units of Psychrotrophic Micro-Organisms—Colony-Count Technique at 6,5 Degrees C First Edition. 9 pp.

—**Enterobacteriaceae—Animal Feed**
BSI BS 5763: Part 10-86. 1986 Microbiological Examination of Food and Animal Feeding Stuffs Part 10: Enumeration of Enterobacteriaceae. 11 pp.

—**Enterobacteriaceae—Food**
BSI BS 5763: Part 10-86. 1986 Microbiological Examination of Food and Animal Feeding Stuffs Part 10: Enumeration of Enterobacteriaceae. 11 pp.
ISO 7402-93. Microbiology—General Guidance for the Enumeration of Enterobacteriaceae Without Resuscitation—MPN Technique and Colony-Count Technique Second Edition. 14 pp.

—**Escherichia Coli—Dairy Products**
BSI BS 4285: Sec 3.8-88. 1988 Microbiological Examination for Dairy Purposes Part 3: Methods for Detection and/or Enumeration of Specific Groups of Microorganisms Section 3.8: Enumeration of Presumptive Esherichia Coli. 6 pp.

—**Escherichia Coli—Food**
BSI BS 5763: Part 8-85. 1985 Microbiological Examination of Food and Animal Feeding Stuffs Part 8: Enumeration of Presumptive Escherichia Coli. 12 pp.
BSI BS 5763: Part 13-89. 1989 Microbiological Examination of Food and Animal Feeding Stuffs Part 13: Enumeration of Escherichia Coli: Colony Count Technique Using Membranes. 8 pp.
ISO 7251-84. Microbiology—General Guidance for Enumeration of Presumptive Escherichia Coli—Most Probable Number Technique First Edition. 11 pp.
SAA AS 1766.2.12-84. Method for the Microbiological Examination of Food—Part 2: Examination for Specific Organisms—Part 2.12: Escherichia Coli—Direct Plate Method. 3 pp.

Bacteria Count Methods (*Cont.*)

—**Escherichia Coli—Water**
DIN ENGL 38411 Pt 6-91. German Standard Methods for the Examination of Water, Waste Water and Sludge; Microbiological Methods (Group K); Determination of Escherichia Coli and Coliform Organisms (K 6) (June). 6 pp.
ISO 9308 Pt 1-90. Water Quality—Detection and Enumeration of Coliform Organisms, Thermotolerant Coliform Organisms and Presumptive Escherichia Coli—Part 1: Membrane Filtration Method First Edition. 14 pp.
ISO 9308 Pt 2-90. Water Quality—Detection and Enumeration of Coliform Organisms, Thermotolerant Coliform Organisms and Presumptive Escherichia Coli—Part 2: Multiple Tube (Most Probable Number) Method First Edition. 13 pp.
SAA AS 1095.4.1. 4-81. Microbiological Methods for the Dairy Industry—Part 4: Methods for the Examination of Water and Air—Part 4.1.4: Microbiological Examination of Water—Escherichia Coli by Multiple Tube Dilution. 3 pp.
SAA AS 1095.4.1. 6-81. Microbiological Methods for the Dairy Industry—Part 4: Methods for the Examination of Water and Air—Part 4.1.6: Microbiological Examination of Water—Escherichia Coli by Membrane Filtration. 4 pp.

—**Food**
CNS N6186-91. Method of Test for Food Microbiology-Test of Standard Plate Count (Feb)(10890). 5 pp.
SAA AS 1766.1.3-91. Methods for the Microbiological Examination of Food—Part 1: General Procedures and Techniques—Part 1.3: Colony Count—Pour Plate Method. 3 pp.
SAA AS 1766.1.4-91. Methods for the Microbiological Examination of Food—Part 1: General Procedures and Techniques—Part 1.4: Colony Count—Surface Spread Method. 2 pp.
SAA AS 1766.1.5-91. Methods for the Microbiological Examination of Food—Part 1: General Procedures and Techniques—Part 1.5: Colony Count—Membrane Filtration Method. 4 pp.
SAA AS 1766.1.6-91. Methods for the Microbiological Examination of Food—Part 1: General Procedures and Techniques—Part 1.6: Estimation of Most Probable Number of Micro-Organisms (MPN). 4 pp.
SAA AS 1766.2.1-91. Methods for the Microbiological Examination of Food—Part 2: Examination for Specific Organisms—Part 2.1: Standard Plate Count. 2 pp.

—**Fungi—Animal Feed**
BSI BS 5763: Part 12-88. 1988 Microbiological Examination of Food and Animal Feeding Stuffs Part 12: Enumeration of Yeasts and Moulds. 6 pp.

—**Fungi—Dairy Products**
BSI BS 4285: Sec 3.6-86. 1986 Microbiological Examination for Dairy Purposes Part 3: Methods Part 3: Methods for Detection and/or Enumeration of Specific Groups of Microorganisms Section 3.6: Enumeration of Yeasts and Moulds. 4 pp.

—**Fungi—Food**
BSI BS 5763: Part 12-88. 1988 Microbiological Examination of Food and Animal Feeding Stuffs Part 12: Enumeration of Yeasts and Moulds. 6 pp.
ISO 7954-87. Microbiology—General Guidance for Enumeration of Yeasts and Moulds—Colony Count Technique at 25 Degrees Celsius First Edition. 6 pp.

—**Fungi—Water**
ISO 6222-88. Water Quality—Enumeration of Viable Micro-Organisms—Colony Count by Inoculation in or on a Nutrient Agar Culture Medium First Edition. 4 pp.

—**Legumes**
ISO 7698-90. Cereals, Pulses and Derived Products—Enumeration of Bacteria, Yeasts and Moulds First Edition. 10 pp.

—**Milk**
SAA AS 1766.3.10-91. Methods for the Microbiological Examination of Food—Part 3: Examination of Specific Products—Part 3.10: Liquid Milks—Direct Microscopic Count. 4 pp.

—**Paperboard**
ISO 8784 Pt 1-87. Paper and Board—Determination of Microbiological Properties—Part 1: Total Bacterial Count First Edition. 8 pp.

—**Papers**
ISO 8784 Pt 1-87. Paper and Board—Determination of Microbiological Properties—Part 1: Total Bacterial Count First Edition. 8 pp.

Bacteria Count Methods (Cont.)

—Pseudomonas—Water
SAA AS 1095.4.1. 10-81. Microbiological Methods for the Dairy Industry—Part 4: Methods for the Examination of Water and Air—Part 4.1.10: Microbiological Examination of Water—Pseudomonas by Multiple Tube Dilution. 3 pp.
SAA AS 1095.4.1. 11-81. Microbiological Methods for the Dairy Industry—Part 4: Methods for the Examination of Water and Air—Part 4.1.11: Microbiological Examination of Water—Pseudomonas Aeruginosa by Multiple Tube Dilution. 3 pp.
SAA AS 1095.4.1. 12-81. Microbiological Methods for the Dairy Industry—Part 4: Methods for the Examination of Water and Air—Part 4.1.12: Microbiological Examination of Water—Pseudomonads by Membrane Filtration. 3 pp.
SAA AS 1095.4.1. 13-81. Microbiological Methods for the Dairy Industry—Part 4: Methods for the Examination of Water and Air—Part 4.1.13: Microbiological Examination of Water—Pseudomonas Aeruginosa by Membrane Filtration. 4 pp.

—Salmonella—Dairy Products
BSI BS 4285: Sec 3.9-87. 1987 Microbiological Examination for Dairy Purposes Part 3: Methods for Detection and/or Enumeration of Specific Groups of Microorganisms Section 3.9: Detection of Salmonella. 15 pp.
BSI BS 4285: SUB SEC 3.9.1-01. 1987 Amd 1 Microbiological Examination for Dairy Purposes Part 3: Methods for Detection and/or Enumeration of Specific Groups of Microorganisms Section 3.9: Detection of Salmonella Subsection 3.9.1: Reference Method. 17 pp.
BSI BS 4285: SUB SEC 3.9.2-92. 1992 Microbiological Examination for Dairy Purposes Part 3: Methods for Detection and/or Enumeration of Specific Groups of Microorganisms Section 3.9: Dectection of Salmonella Subsec 3.9.2: Screening Method Using Electrical Conductance. 14 pp.

—Salmonella—Water
CEN PREN 26340-93. Water Quality—Detection of Salmonella. 16 pp.
SAA AS 1095.4.1. 9-81. Microbiological Methods for the Dairy Industry—Part 4: Methods for the Examination of Water and Air—Part 4.1.9: Microbiological Examination of Water—Salmonellae. 5 pp.

—Staphylococcus—Animal Feed
BSI BS 5763: Part 7-83. 1983 Microbiological Examination of Food and Animal Feeding Stuffs Part 7: Enumeration of Staphylococcus Aureus by Colony Count Technique. 10 pp.

—Staphylococcus—Dairy Products
BSI BS 4285: SUB SEC 3.10.1-89. 1989 Microbiological Examination for Dairy Purposes Part 3: Methods for Detection and/or Enumeration of Specific Groups of Microorganisms Sec 3.10: Staphylococcus Aureus Subsection 3.10.1: Enumeration Using the Colony Count Technique. 6 pp.
BSI BS 4285: SUB SEC 3.10.2-89. 1989 Microbiological Examination for Dairy Purposes Part 3: Methods for Detection and/or Enumeration of Specific Groups of Microorganisms Sec 3.10: Staphylococcus Aureus Subsection 3.10.2: Detection. 6 pp.
BSI BS 4285: SUB SEC 3.10.3-89. 1989 Microbiological Examina-tion for Dairy Purposes Part 3: Methods for Detection and/or Enumeration of Specific Groups of Microorganisms Section 3.10: Staphylo-Coccus Aureus Subsection 3.10.3: Enumeration Using the Most Probable Number Technique. 6 pp.
DIN ENGL 10178 Pt 3-88. Microbiological Analysis of Milk; Determination of Coagulase-Positive Staphylococci; Colony Count Technique (Apr). 3 pp.

—Staphylococcus—Food
BSI BS 5763: Part 7-83. 1983 Microbiological Examination of Food and Animal Feeding Stuffs Part 7: Enumeration of Staphylococcus Aureus by Colony Count Technique. 10 pp.
ISO 6888-83. Microbiology—General Guidance for Enumeration of Staphylococcus Aureus—Colony Count Technique First Edition. 9 pp.

—Streptococcus—Dairy Products
BSI BS 4285: Sec 3.11-85. 1985 Microbiological Examination for Dairy Purposes Part 3: Methods for Detection and/or Enumeration of Specific Groups of Microorganisms Section 3.11: Detection and Enumeration of Faecal Streptococci. 4 pp.

—Streptococcus—Water
SAA AS 1095.4.1. 7-81. Microbiological Methods for the Dairy Industry—Part 4: Methods for the Examination of Water and Air—Part 4.1.7: Microbiological Examination of Water—Enterococci by Multiple Tube Dilution. 3 pp.

Bacteria Count Methods (Cont.)

—Streptococcus—Water (Cont.)
SAA AS 1095.4.1. 8-81. Microbiological Methods for the Dairy Industry—Part 4: Methods for the Examination of Water and Air—Part 4.1.8: Microbiological Examination of Water—Enterococci by Membrane Filtration. 3 pp.

—Thermoduric Bacteria—Dairy Products
BSI BS 4285: Sec 3.2-91. 1991 Microbiological Examination for Dairy Purposes Part 3: Methods for Detection and/or Enumeration of Specific Groups of Microorganisms Sec 3.2: Enumeration of Thermoduric Bacteria (W). 5 pp.
BSI BS 4285: Sec 3.2-01. 1991 Amd 1 Microbiological Examination for Dairy Purposes Part 3: Methods for Detection and/or Enumeration of Specific Groups of Microorganisms Sec 3.2: Enumeration of Thermoduric Bacteria (AMD 6985) February 28, 1992 (W). 6 pp.

—Urinalysis—Dip Slides
DIN ENGL 58942 Pt 3-85. Medical Microbiology; Culture Media Used in Bacteriology; Dip Slides for Bacteriological Urine Testing (Sept). 3 pp.

—Water
SAA AS 1095.4.1. 2-81. Microbiological Methods for the Dairy Industry—Part 4: Methods for the Examination of Water and Air—Part 4.1.2: Microbiological Examination of Water—Colony Count by the Pour Plate Method. 2 pp.

—Water—Solid Media
ISO 9998-91. Water Quality—Practices for Evaluating and Controlling Microbiological Colony Count Media Used in Water Quality Tests First Edition. 26 pp.

—Yeasts—Animal Feed
BSI BS 5763: Part 12-88. 1988 Microbiological Examination of Food and Animal Feeding Stuffs Part 12: Enumeration of Yeasts and Moulds. 6 pp.

—Yeasts—Dairy Products
BSI BS 4285: Sec 3.6-86. 1986 Microbiological Examination for Dairy Purposes Part 3: Methods Part 3: Methods for Detection and/or Enumeration of Specific Groups of Microorganisms Section 3.6: Enumeration of Yeasts and Moulds. 4 pp.

—Yeasts—Food
BSI BS 5763: Part 12-88. 1988 Microbiological Examination of Food and Animal Feeding Stuffs Part 12: Enumeration of Yeasts and Moulds. 6 pp.
ISO 7954-87. Microbiology—General Guidance for Enumeration of Yeasts and Moulds—Colony Count Technique at 25 Degrees Celsius First Edition. 6 pp.

—Yeasts—Water
ISO 6222-88. Water Quality—Enumeration of Viable Micro-Organisms—Colony Count by Inoculation in or on a Nutrient Agar Culture Medium First Edition. 4 pp.

Bactericide Content Analysis
Use For: Germicide Content Analysis

—Halophenol—Cosmetics
CNS S2081-82. Methods of Hygienic Test for Cosmetics Halophenol Germicide (Sep)(9441).

Badges
See Also: Identification Systems

—Machine Embroidered
CGSB CAN/CGSB-38.41-M91. Machine Embroidered Badges and Insignia (Schiffli Type). 12 pp.
CGSB 38-GP-43M-78. Construction Details for Uniform Coats, Jackets, and Tunics (Fusible Interlining), Standard for. 29 pp.

Badminton Equipment
—Nets
CNS S2027-80. Method of Test for Badminton Nets (May)(5632).

—Rackets
CNS S1089-77. Badminton Racket (Apr)(4094).
CNS S2024-77. Method of Test for Badminton Racket (Apr)(4095).
JIS S 7006-77. Badminton Racket.

BAe ATP Aircraft
See Also: Aircraft

—Antenna Positions
CAA. BAe ATP. 1 p.

BAe Jetstream Series Aircraft
See Also: Aircraft

BAe Jetstream Series Aircraft (Cont.)

—Accidents
CAA. British Aerospace Jetstream Series Aircraft. 21 pp.

BAe Jetstream 31 Aircraft
See Also: Aircraft

—Antenna Positions
CAA. HP.137 Jetstream and BAe Jetstream 31 (Approved Aerial Positions). 1 p.

BAe 125 Aircraft
See Also: Aircraft

—Accidents
CAA. Hawker Siddely/BAe 125 Series. 1 p.

—Mandatory Aircraft Modifications and Inspections Summaries
CAA. British Aerospace 125 Series Aircraft. 28 pp.

BAe 146 Aircraft
See Also: Aircraft

—Accidents
CAA. British Aerospace 146. 1 p.

—Antenna Positions
CAA. BAe 146 (Approved Aerial Positions). 1 p.

—Data Sheets
CAA BA16 ISSUE 11. British Aerospace BAe 146 Series 100, Series 200, Series 300. 5 pp.

—Mandatory Aircraft Modifications and Inspections Summaries
CAA. British Aerospace 146 Series. 18 pp.

BAe 748 Aircraft
See Also: Aircraft

—Accidents
CAA. Hawker Siddely/BAe 748 Series. 1 p.

Baffle Plates
—Aircraft—Filter Packs
SBAC AS 3318 ISSUE 3. Baffle Plate Filter Pack. 1 p.

Bag Filters
—Glass Fabric
JIS R 3421-91. Textile Finished Glass Fabrics for Bag Filter. 7 pp.

Bag Machinery
Use: Bag Machines

Bag Machines
Use For: Bag Machinery; Bag Making Machinery
See Also: Bags; Packaging Machines

—Woven
CNS B7193-82. Method of Test for PE/PP Woven Bag Making Machine (Jul)(9069).

Bag Making Machinery
Use: Bag Machines

Bag Sealers
—Industrial
CNS B4051-85. Single Thread Chain Stitch, Filled Bag Closing Machine for Industrial Use (Oct)(8344). 5 pp.

Bagasse Pulps
See Also: Pulps
CNS P2005-85. Bleached Bagasse Pulp (Dec)(1317). 1 p.

Baggage Compartments
Use: Aircraft Compartments—Baggage

Bags
Use For: Sacks *See Also:* Bag Machines; Containers; Garbage Bags; Gunny Sacks; Hot Water Bottles; Ice Water Bags; Luggage; Mailbags; Ostomy Collection Bags; Packaging; Packaging Materials; Paper Bags; Paper Tubes; Pouches; Shipping Containers; Urinary Collection Bags

—Bulk Foods—Export
CEN PREN 1086-93. Sacks for the Transport of Food Aid—Recommendations on the Selection of Type of Sack and the Liner in Relation to the Product to Be Packed. 9 pp.
CGSB CAN/CGSB-43.44-M91. Bags for Transport of Food Aid. 9 pp.

INTERNATIONAL AND NON-U.S. NATIONAL STANDARDS
SUBJECT INDEX

Baking

Bags (Cont.)

—Bulk Foods—Export (Cont.)
EC SM 2. Sacks for the Transport of Food Aid. 1 p.

—Cement
CNS Z6003-70. Method of Test for Cement Bags for Domestic Market (Mar)(2109).

—Fabric—Hazardous Materials
CGSB CAN/CGSB-43.67-93. Lined Cloth and Paper Bags (TC-45B). 7 pp.

—Insulated—Food
BSI BS 6672: Part 2-86. 1986 Insulated Domestic Food Containers Part 2: Insulated Bags and Boxes. 10 pp.
BSI BS 6672: Part 2-01. 1986 Amd 1 Insulated Domestic Food Containers Part 2: Specification for Insulated Bags and Boxes (AMD 7200) June 15, 1992. 11 pp.

—Jute
CEN PREN 766-92. Sacks for the Transport of Food Aid—Sacks Made of Jute Fabric. 8 pp.
CNS L3018-88. Method of Test for Bast Fiber Bag (Sep)(1389).
CNS L4005-88. Bast Fiber Bag (Sep)(1388).

—Packaging Machinery
CNS B4051-85. Single Thread Chain Stitch, Filled Bag Closing Machine for Industrial Use (Oct)(8344). 5 pp.

—Plastic
BSI BS EN 26591-2-93. 1993 Packaging—Sacks—Description and Method of Measurement Part 2: Empty Sacks Made from Thermoplastic Flexible Film (H). 13 pp.
CEN EN 26591-2-92. Packaging—Sacks—Description and Method of Measurement—Part 2: Empty Sacks Made from Thermoplastic Flexible Film (ISO 6591-2: 1985). 8 pp.
ISO 6591 Pt 2-85. Packaging—Sacks—Description and Method of Measurement—Part 2: Empty Sacks Made from Thermoplastic Flexible Film First Edition; (CEN EN 26591-2:1992). 6 pp.

—Plastic—Glossaries
BSI BS 3130: Part 8-87. (WITHDRAWN) 1987 Glossary of Packaging Terms Part 8: Plastic Sacks (Superseded by BS EN 26590-2: 1993). 15 pp.
BSI BS EN 26590-2-93. 1993 Packaging—Sacks—Vocabulary and Types Part 2: Sacks Made from Thermoplastic Flexible Film (Supersedes BS 3130: Part 8: 1987) (H). 20 pp.
CEN EN 26590-2-92. Packaging—Sacks—Vocabulary and Types—Part 2: Sacks Made from Thermoplastic Flexible Film (ISO 6590-2: 1986). 15 pp.
ISO 6590 Pt 2-86. Packaging—Sacks—Vocabulary and Types—Part 2: Sacks Made from Thermoplastic Flexible Film First Edition; (CEN EN 26590-2:1992). 18 pp.

—Plastic—Sterilization
BSI BS 6871-89. 1989 Heat-Sealable Pouches and Tube Material Converted from Transparent Plastics Film and Paper from Steam Sterilization for Medical Use. 11 pp.

—Polyethylene
BSI BS 4932-73. (WITHDRAWN) 1973 Amd 1 Heavy Duty Low Density Polyethylene Sacks. 17 pp.
CNS Z5125-83. Size for Heavy Duty Polyethylene Bags (Mar)(10133).
JIS Z 1534-78. Size for Heavy Duty Polyethylene Bags. 6 pp.
JIS Z 1711-86. Polyethylene Film Bags. 14 pp.

—Polyethylene—High Density
CNS K3096-90. U-Shaped High Density Polyethylene Bag (May)(12721). 3 pp.
CNS K6971-90. Method of Test for U-Shaped High Density Polyethylene Bag (May)(12722). 3 pp.

—Polyolefin
BSI BS 6162-81. 1981 Open-Mouth Sacks Manufactured from Woven Polyolefin Tape Yarn. 11 pp.
CNS Z5106-82. Polyolefine Stretched Tape Yarn for Woven Bags (Apr)(8751).
JIS Z 1533-76. Polyolefine Stretched Tape Yarn for Woven Bags (R 1980). 6 pp.

—Polypropylene—Bulk Foods
BSI BS 7303: Part 1-90. 1990 Sacks for the Transport of Polypropylene Fabric Part 1: Sacks Made of Polypropylene Fabric. 10 pp.
CEN EN 277-89. Sacks for the Transport of Food Aid—Sacks Made of Polypropylene Fabric. 9 pp.
CEN PREN 277-93. Sacks for the Transport of Food Aid—Sacks Made of Woven Polypropylene Fabric. 8 pp.

Bags (Cont.)

—Polypropylene—Sugar
CNS Z5032-70. PP Seamless Bags for Packing of 50kg Crystal Sugar (Dec)(3223).

—Powdered Milk
CGSB CAN/CGSB-43.41-M86. Bags for Milk Powder. 13 pp.

—Sampling
BSI BS 6428-83. (WITHDRAWN) 1983 Sampling Empty Sacks for Testing (Superseded by BS EN 27023: 1993). 4 pp.
BSI BS EN 27023-93. 1993 Packaging—Sacks—Method of Sampling Empty Sacks for Testing (ISO 7023: 1983). 9 pp.
CEN EN 27023-92. Packaging—Sacks—Method of Sampling Empty Sacks for Testing (ISO 7023: 1983). 4 pp.
ISO 7023-83. Packaging—Sacks—Method of Sampling Empty Sacks for Testing First Edition; (CEN EN 27023:1992). 4 pp.

Bains Marie
Use: Steam Tables

Bake Ovens
Use: Ovens

Baked Finishes
Use: Baking Finishes

Bakery Products
See Also: Bread; Cereal Products; Food; Pies

—Fat Content
ISO 7302-82. Cereals and Cereal Products—Determination of Total Fat Content First Edition. 5 pp.

Baking Enamels
See Also: Baking Finishes; Enamels

—Aluminum Panels
CGSB 1-GP-71 METH 112.4-82. Methods of Testing Paints and Pigments Baking Resistance Enamels. 1 p.

—Metal—Aircraft
MOD UK DTD-56D-54. Stoving Enamel (Reprinted April 1954). 3 pp.

Baking Finishes
Use For: Stove Coatings *See Also:* Baking Enamels; Baking Lacquers; Baking Paints; Baking Varnishes; Coatings; Finishes

—Metal—Aeronautical
BSI X 31-67. 1967 Low Temperature Stoving Scheme for Aeronautical Purposes. 5 pp.

—Powder
MOD UK DSTAN 80-122-01. Powder Coating High Durability, Stoving Types: Near Matt Semi-Gloss and Glossy Issue 1; Amendment 1. 19 pp.

—Reflectance
CGSB 1-GP-71 METH 39.3-80. Methods of Testing Paints and Pigments Reflectance on Baked Finishes. 1 p.

Baking Lacquers
See Also: Baking Finishes; Lacquers

—Aluminum Panels
CGSB 1-GP-71 METH 112.3-82. Methods of Testing Paints and Pigments Baking Resistance Lacquers. 1 p.

Baking Paints
Use For: Stoving Paints *See Also:* Baking Finishes; Baking Primers; Paints
CGSB 1-GP-71 METH 44.1-81. Methods of Testing Paints and Pigments Baking Properties. 1 p.
CGSB 1-GP-71 METH 103.2-79. Methods of Testing Paints and Pigments Drying Conditions Oven Drying. 1 p.
SAA AS 1580.101. 3-92. Paints and Related Materials—Methods of Test—Part 101.3: Standard Procedure for Stoving (in Professional Packages 30, 39). 2 pp.
SNZ NZS/AS 1580. 101.3-92. Methods of Test for Paints and Related Materials Part 101.3: Standard Procedure for Storing (This is a Joint Standard with SAA AS 1580.101.3). 2 pp.

—Aluminum Coatings (Made From Aluminum)
MOD UK DSTAN 80-11-01. Paint, Aluminium, Flame Resisting, Stoving Issue 1; Amendment 1. 12 pp.
MOD UK DSTAN 80-11-92. Paint, Aluminium, Flame Resisting, Stoving, Spraying Issue 2. 16 pp.

Baking Paints (Cont.)

—Ammunition
MOD UK TS 10176A. Paint System, Ammunition, Matt, Stoving, LF Quality, Type: Spraying (Superseded by Def Stan 80-174).

—Ammunition—Exterior
MOD UK DSTAN 80-174-93. Paint System Ammunition, Matt, Stoving Priming, Ammunition, Exterior, Stoving Finishing, Ammunition, Exterior Matt Stoving Type: Spraying Issue 1. 20 pp.

—Containers—Gasoline
MOD UK DSTAN 80-65-76. Paint, Finishing, Stoving, Jerricans Paint, Finishing, Stoving, Interior, Jerricans Types: Spraying Flowcoating Paint, Finishing, Stoving, Exterior, Jerricans Types: Spraying Flowcoating Dipping Issue 1. 11 pp.

—Epoxy Coatings—Powder
MOD UK DSTAN 80-169-93. Powder Coating, Stoving Types: Matt, Semi-Gloss and Glossy Issue 1. 17 pp.

—Grenades
MOD UK DSTAN 80-116-91. Paint, System, Stoving, Textured Issue 1. 11 pp.
MOD UK TS 10183. Paint System, Stoving Textured for Grenades, LF Quality.

—Instruments
MOD UK DEF-1059-A. Stoving Paint System for Instruments (Superseded by Def Stan 80-152).

—Metal Coatings (On Metal)
MOD UK DSTAN 80-60-77. Paint System, Hammer Finish, Stoving Paint, Hammer Finish, Stoving Paint, Filler, for Paint, Hammer Finish, Stoving Type: Spraying Issue 1. 15 pp.
MOD UK DSTAN 80-60-92. Paint, System, Spraying, Hammer Finish, Epoxide, Stoving Paint, Filler, Spraying, Expoxide, Stoving, Two-Pack Paint, Finishing, Spraying, Hammer Finish, Expoxide, Stoving Issue 2. 19 pp.
MOD UK DSTAN 80-152-91. Paint, System, Spraying, Epoxide, Stoving Paint, Priming, Spraying, Expoxide, Stoving Paint, Filler, Spraying, Epoxide, Stoving Paint, Finishing, Straying, Epoxide, Stoving Types: Semi-Gloss. 18 pp.

—Military Vehicles
MOD UK DSTAN 80-42-82. Paint System, Defence Equipment, IRR, Matt, Stoving Types: Spraying Dipping Issue 1. 19 pp.

—Military Vehicles—Exterior
MOD UK DSTAN 80-51-01. Paint, System, Defence Equipment, Gloss, Stoving Paint, Undercoat, Defence Equipment, Stoving Paint, Finishing, Defence Equipment, Gloss, Stoving Types: Dipping and Spraying Issue 3; Amendment 1. 22 pp.

—Optical Equipment
MOD UK DSTAN 80-8-70. 1. Paint, Finishing, for the Interior of Optical Instruments, Stoving, Matt Black 2. Paint, Priming, for Use Under Paint, Finishing, for the Interior of Optical Instruments, Stoving, Light Bronze Green Types: Brushing Spraying Issue 1. 10 pp.
MOD UK DSTAN 80-8-01. Paint, System, Optical Instruments, Matt Black, Stoving Paint, Priming, Optical Instruments, Stoving Paint, Finishing, Optical Instruments, Matt Black, Stoving Types: Brushing Spraying Issue 2; Amendment 1. 16 pp.

—Small Arms
MOD UK DSTAN 80-56-01. Paint, Finishing, Epoxide, Stoving, Black, for Small Arms Type: Spraying Issue 1; Amendment 1. 11 pp.

Baking Primers
See Also: Baking Paints; Coatings; Primers (Coatings)
CNS K6835-86. Method of Test for Baking Primer (Apr)(11548).
MOD UK DSTAN 80-124-89. Paint, Priming, Stoving, Defence Equipment Types: Spraying Dipping Issue 1. 15 pp.

—Ammunition—Exterior
MOD UK DSTAN 80-174-93. Paint System Ammunition, Matt, Stoving Priming, Ammunition, Exterior, Stoving Finishing, Ammunition, Exterior Matt Stoving Type: Spraying Issue 1. 20 pp.

—Military Vehicles—Exterior
MOD UK DSTAN 80-51-01. Paint, System, Defence Equipment, Gloss, Stoving Paint, Undercoat, Defence Equipment, Stoving Paint, Finishing, Defence Equipment, Gloss, Stoving Types: Dipping and Spraying Issue 3; Amendment 1. 22 pp.

—Steel
CGSB 1-GP-71 METH 130.4-79. Methods of Testing Paints and Pigments Behavior Towards Topcoats Baking Primers. 1 p.

INDUSTRY STANDARDS

Baking Primers (Cont.)

—Zinc Chromate
MOD UK DSTAN 80-8-70. 1. Paint, Finishing, for the Interior of Optical Instruments, Stoving, Matt Black 2. Paint, Priming, for Use Under Paint, Finishing, for the Interior of Optical Instruments, Stoving, Light Bronze Green Types: Brushing Spraying Issue 1. 10 pp.

—Zinc Chromate—Steel
MOD UK DSTAN 80-3-70. Paint, Priming, Zinc Chrome Stoving Types: Brushing Spraying Issue 1. 7 pp.

—Zinc Phosphate
MOD UK TS 10115B. Paint, Priming, Zinc Phosphate, Stoving (Superseded by Def Stan 80-123).

Baking Resins
Use: Baking Varnishes

Baking Varnishes
Use For: Baking Resins See Also: Baking Finishes; Varnishes
CNS K2018-70. Baking Varnish (Jan)(770). 2 pp.
CNS K2019-70. Clear Varnish for Baking Varnish (Jan)(772). 1 p.
CNS K6056-70. Method of Test for Baking Varnish (Jan)(771). 6 pp.
CNS K6057-70. Method of Test for Clear Varnish for Baking Varnish (Jan)(773). 3 pp.

—Chemical Resistance
CNS K6056-70. Method of Test for Baking Varnish (Jan)(771). 6 pp.
CNS K6057-70. Method of Test for Clear Varnish for Baking Varnish (Jan)(773). 3 pp.
CNS K6103-59. Method of Test for Low Temperature Baking Varnish (Apr)(1113). 3 pp.

—Colorimetric Analysis
CNS K6057-70. Method of Test for Clear Varnish for Baking Varnish (Jan)(773). 3 pp.

—Electrical Insulating—Windings
MOD UK DEF-31-A-64. Varnish, Impregnating, for Electrical Purposes, Baking (Reprinted June, 1968). 5 pp.

—Epoxy—Steel—Ammunition
MOD UK DSTAN 80-175-93. Varnish, Ammunition, Stoving, Type QX Class A Expoxy Ester B Silicone Alkyd C Tung Oil Phenolic Types: Brushing Spraying Dipping Issue 1. 16 pp.

—Heat Resistance
CNS K6103-59. Method of Test for Low Temperature Baking Varnish (Apr)(1113). 3 pp.

—Light Testing
CNS K6056-70. Method of Test for Baking Varnish (Jan)(771). 6 pp.
CNS K6103-59. Method of Test for Low Temperature Baking Varnish (Apr)(1113). 3 pp.

—Magnesium—Aircraft
MOD UK DTD-5562-63. Clear Baking Resin for Surface Sealing Magnesium (Reprinted March 1965). 3 pp.

—Melamine Content
CNS K6103-59. Method of Test for Low Temperature Baking Varnish (Apr)(1113). 3 pp.

—Phenolic Base—Steel—Ammunition
MOD UK DSTAN 80-175-93. Varnish, Ammunition, Stoving, Type QX Class A Expoxy Ester B Silicone Alkyd C Tung Oil Phenolic Types: Brushing Spraying Dipping Issue 1. 16 pp.

—Phthalic Acid Content
CNS K6103-59. Method of Test for Low Temperature Baking Varnish (Apr)(1113). 3 pp.

—Physical Properties
CNS K6056-70. Method of Test for Baking Varnish (Jan)(771). 6 pp.
CNS K6057-70. Method of Test for Clear Varnish for Baking Varnish (Jan)(773). 3 pp.
CNS K6103-59. Method of Test for Low Temperature Baking Varnish (Apr)(1113). 3 pp.

—Saponification Number
CNS K6103-59. Method of Test for Low Temperature Baking Varnish (Apr)(1113). 3 pp.

—Silicone—Steel—Ammunition
MOD UK DSTAN 80-175-93. Varnish, Ammunition, Stoving, Type QX Class A Expoxy Ester B Silicone Alkyd C Tung Oil Phenolic Types: Brushing Spraying Dipping Issue 1. 16 pp.

—Urea Content
CNS K6103-59. Method of Test for Low Temperature Baking Varnish (Apr)(1113). 3 pp.

Baking Varnishes (Cont.)

—Water Resistance Testing
CNS K6056-70. Method of Test for Baking Varnish (Jan)(771). 6 pp.
CNS K6057-70. Method of Test for Clear Varnish for Baking Varnish (Jan)(773). 3 pp.

Balance of Payments
See Also: Commerce; Credit Institutions

—Credit Institutions
EC EEC/1969/88-88. Council Regulation Establishing a Single Facility Providing Medium-Term Financial Assistance for Member States' Balances of Payments. 4 pp.

Balance of Trade
See Also: Commerce

—Aircraft
EURO ECO/4690/F & E And 4675/-75. Demand Prospects for Civil Transport Aircraft Currency Balance of European Airlines Purchases and European Industry Sales. 41 pp.

Balance Weights
Use: Weights

Balancing
Use For: Equilibration See Also: Balancing Machines; Weight and Balance Systems

—Air Cushion Vehicles
CAA Chapter A5-1 App 08.81. Weight and Balance Data. 2 pp.

—Aircraft
CAA Chapter A6-4 09.91. Weight and Balance of Aircraft. 2 pp.
CAA Chapter A6-4 App 1 07.89. Weight and Balance of Aircraft—Fleet Mean Weight and Fleet Mean Centre-of-Gravity. 2 pp.
CAA Chapter B5-4 07.89. Weight and Balance of Aircraft. 2 pp.
CAA Chapter B6-4 07.89. Weight and Balance of Aircraft. 2 pp.
CAA Chapter B7-10 07.89. Weight and Balance Report. 2 pp.
CAA LEAFLET 1-4 07.90. Weight and Balance of Aircraft. 27 pp.

—Aircraft—Logs (Reports)
CAA Chapter A7-10 07.89. Weight and Balance Report. 5 pp.

—Chain Saws
BSI BS 6916: Part 10-88. 1988 Chain Saws Part 10: Specification for Longitudinal Balance (ISO 8334: 1985). 3 pp.
ISO 8334-85. Forestry Machinery—Portable Chain-Saws—Determination of Balance First Edition. 3 pp.

—Glossaries
BSI BS 3851-90. 1990 Terms Used in the Mechanical Balancing of Rotating Machinery. 35 pp.
ISO 1925-90. Mechanical Vibration—Balancing—Vocabulary Third Edition. 37 pp.

—Rotating Machines—Ships
NATO STANAG 4139 ED 1 AMD 5-76. Criteria for Balancing of Shipboard Rotating Machinery. 16 pp.

—Rotors (Machine Elements)
BSI BS 5265: Part 2-81. 1981 Mechanical Balancing of Rotating Bodies Part 2: Methods for the Mechanical Balancing of Flexible Rotors. 34 pp.
BSI BS 5265: Part 3-84. 1984 Mechanical Balancing of Rotating Bodies Part 3: Recommendations for Criteria for Evaluating Flexible Rotor Balance. 12 pp.
BSI BS 6861: Part 1-87. 1987 Amd 1 Balance Quality Requirements of Rigid Rotors Part 1: Method for Determination of Permissible Residual Unbalance (AMD 6365) June 28, 1991. 20 pp.
BSI BS 7130-89. 1989 Mechanical Balancing of Rotating Machinery: Shaft and Fitment Key Convention. 16 pp.
ISO 1940 Pt 1-86. Mechanical Vibration—Balance Quality Requirements of Rigid Rotors—Part 1: Determination of Permissible Residual Unbalance First Edition. 17 pp.
ISO 5343-83. Criteria for Evaluating Flexible Rotor Balance First Edition. 11 pp.
ISO 5406-80. Mechanical Balancing of Flexible Rotors First Edition. 32 pp.
ISO 8821-89. Mechanical Vibration—Balancing—Shaft and Fitment Key Convention First Edition. 15 pp.
JIS B 0905-92. Rotating Machines—Balance Quality Requirements of Rigid Motors. 16 pp.
SNZ NZS/ISO 1940. Part 1-86. Mechanical Vibration—Balance Quality Requirements of Rigid Rotors Part 1: Determination of Permissible Residual Unbalance. 15 pp.

Balancing (Cont.)

—Rotors (Machine Elements) (Cont.)
SNZ NZS/ISO 5343-83. Criteria for Evaluating Flexible Rotor Balance. 9 pp.
SNZ NZS/ISO 5406-80. Mechanical Balancing of Flexible Rotors. 30 pp.
SNZ NZS/ISO 8821-89. Mechanical Vibration—Balancing—Shaft and Fitment Key Convention. 12 pp.

—Shafts (Machine Elements)
BSI BS 7130-89. 1989 Mechanical Balancing of Rotating Machinery: Shaft and Fitment Key Convention. 16 pp.
ISO 8821-89. Mechanical Vibration—Balancing—Shaft and Fitment Key Convention First Edition. 15 pp.
SNZ NZS/ISO 8821-89. Mechanical Vibration—Balancing—Shaft and Fitment Key Convention. 12 pp.

—Transport Aircraft
ISO 6702 Pt 1-91. Aircraft—Requirements for on-Board Weight and Balance Systems—Part 1: General First Edition. 17 pp.
ISO 6702 Pt 2-91. Aircraft—Requirements for on-Board Weight and Balance Systems—Part 2: Design, Performance and Interface Characteristics First Edition. 4 pp.

Balancing Beams (Gymnastic Equipment)
CNS S1193-85. Balancing Beams (Dec)(11469).
CNS S2125-85. Method of Test for Balancing Beams (Dec)(11470).
JIS S 7015-83. Balancing Beams.

Balancing Equipment
Use: Balancing Machines

Balancing Machinery
Use: Balancing Machines

Balancing Machines
Use For: Balancing Equipment See Also: Balancing
BSI BS 3852: Part 1-79. 1979 Balancing Machines Part 1: Description and Evaluation. 24 pp.
ISO 2953-85. Balancing Machines—Description and Evaluation Second Edition. 38 pp.
JIS B 7737-84. Balancing Machines; (Erratum). 22 pp.

—Field
ISO 2371-74. Field Balancing Equipment—Description and Evaluation First Edition. 8 pp.

—Safety
ISO 7475-84. Balancing Machines—Enclosures and Other Safety Measures First Edition. 10 pp.

—Symbols
ISO 3719-82. Balancing Machines—Symbols for Front Panels First Edition. 21 pp.

Balconies
JIS A 6601-92. Metal Components for Balcony and Handrails of Dwellings. 32 pp.

Balers
Use For: Pickup Balers See Also: Agricultural Equipment

—Safety
CEN PREN 704-92. Safety Requirements for Agricultural and Forestry Machinery—Pick-Up Balers. 12 pp.
DIN ENGL 11001 Pt 4-80. Agricultural Machines and Tractors; Pick-Up Balers; Special Technical Safety Requirements and Testing (Aug). 3 pp.

Bales
See Also: Containers

—Cotton—Density
BSI BS 6874-87. 1987 Dimensions and Density of Cotton Bales. 4 pp.
ISO 8115-86. Cotton Bales—Dimensions and Density First Edition. 4 pp.

—Rubber—Coatings
ISO 1434-75. Natural Rubber in Bales—Amount of Bale Coating—Specification and Determination First Edition. 4 pp.

Ball Bearing Slides
Use: Mechanical Guides

Ball Bearings
See Also: Antifriction Bearings; Ball Races; Ball Thrust Bearings; Bearings; Precision Bearings; Radial Bearings; Roller Bearings; Self Aligning Bearings; Sleeve Bearings; Taper Roller Bearings; Thrust Bearings

INTERNATIONAL AND NON-U.S. NATIONAL STANDARDS
SUBJECT INDEX

Ball

Ball Bearings *(Cont.)*

BSI BS 292: Part 1-82. 1982 Amd 1 Rolling Bearings: Ball Bearings, Cylindrical and Spherical Roller Bearings Part 1: Dimensions of Ball Bearings, Cylindrical and Spherical Roller Bearings (Metric Series). 36 pp.

BSI BS 292: Part 2-82. 1982 Amd 1 Rolling Bearings: Ball Bearings, Cylindrical and Spherical Roller Bearings Part 2: Dimensions of Ball Bearings and Cylindrical Roller Bearings (Inch Series). 17 pp.

BSI BS 6010: Part 1-80. (WITHDRAWN) 1980 Rolling Bearings-Bearings with Spherical Outside Surface and Extended Inner Ring Width Part 1: Characteristics, Dimensions and Bore Tolerances (Superseded byBS 6010: Part 4: 1993). 9 pp.

CNS B1292-85. Tolerance of Ball and Roller Bearings (Radial Clearance) (May)(9574).

CNS B2116-85. Deep Groove Ball Bearings: Single-Row Without Filling Notch (May)(2862). 7 pp.

CNS B2469-85. Angular Contact Ball Bearings; Non Self Retaining Single-Row (Mar)(5498).

CNS B2470-85. Deep Groove Ball Bearings (Double-Row, with Filling Notch) (Mar)(5499). 2 pp.

CNS B2634-82. Deep Groove Ball Bearings; Single-Row, Without Filling Notch, with Tapered Bore (Jan)(8204).

CNS B2635-82. Deep Groove Ball Bearings; Single-Row, Without Filling Notch, with Ring Groove (Jan)(8205).

CNS B2636-82. Deep Groove Ball Bearings; Single-Row, Without Filling Notch, with Filling Discs (Jan)(8206).

CNS B2637-82. Deep Groove Ball Bearings; Single-Row, Without Filling Notch, with Shields (Jan)(8207).

DIN ENGL 628 Pt 1-73. Angular Contact Ball Bearings; Single-Row and Double-Row (Mar). 5 pp.

JIS B 1521-86. Deep Groove Ball Bearings.

JIS B 1521-76. Deep Groove Ball Bearings. 14 pp.

JIS B 1522-86. Angular Contact Ball Bearings.

JIS B 1522-65. Angular Contact Ball Bearings (R 1968). 7 pp.

JIS B 1558-76. Ball Bearings for Rolling Bearing Unit (R 1979). 13 pp.

SNZ NZS/BS 292: Part 1-82. Rolling Bearings: Ball Bearings, Cylindrical and Spherical Roller Bearings Part 1: Specification for Dimensions of Ball Bearings, Cylindrical and Spherical Roller Bearings (Metric Series) Amend: 1. 33 pp.

SNZ NZS/BS 292: Part 2-82. Rolling Bearings: Ball Bearings, Cylindrical and Spherical Roller Bearings Part 2: Specification for Dimensions of Ball Bearings and cylindrical Roller Bearings (Inch Series Amend: 1. 12 pp.

—Aerospace

CEN PREN 2130-91. Corrosion Resisting Steel Precision Ball Bearings for Instruments and Equipments Technical Specification. 16 pp.

CEN PREN 3281-93. Aerospace Series Bearings, Airframe Rolling—Rigid Single Row Ball Bearings in Steel—Diameter Series 8 and 9—Dimensions and Loads. 7 pp.

CEN PREN 3446-93. Aerospace Series Bearings, Precision Ball Without Flange in Corrosion Resisting Steel, for Instruments and Equipment Dimensions and Loads. 9 pp.

MOD UK DSTAN 31-4: Part 1-01. Bearings, Ball, Roller and Needle Roller, for Aerospace Use Part 1: Bearings—General Purpose Issue 1; Amendment 1. 28 pp.

MOD UK DSTAN 31-4: Part 2A-01. Bearings, Ball, Roller and Needle Roller, for Aerospace Use Part 2A: Rolling Bearings for Inch Based Airframe Applications (Excluding Tapered Roller Bearings) Issue 1; Amendment 3. 15 pp.

—Aerospace—Precision

AECMA PREN2079-86. Bearings, Precision Ball, in Corrosion Resisting Steel, for Instruments and Equipments, with Flange Dimensions and Loads. 10 pp.

CEN PREN2008-76. Precision Ball Bearings for Instruments and Equipment—Without Flange—Dimensions (C1/ 02-01). 9 pp.

—Aerospace—Pulleys

DIN ENGL LN 29637-87. Aerospace; Rolling Bearings with Seals, Rigid Ball Bearings; Dimensions, Masses (June). 4 pp.

—Aerospace—Self Aligning—Control Rods

AECMA PREN2067-89. Rod-Ends with Self-Aligning Ball Bearings Technical Specification. 18 pp.

AECMA PREN2492-80. Rod-Ends Self Aligning Ball Bearing with Threaded Shank—Dimensions and Loads (C1/15-01). 7 pp.

AECMA PREN3541-90. Rod Ends, Adjustable Self-Aligning Ball Bearing with Threaded Shank Dimensions, Torques, Clearances and Loads. 6 pp.

BSI BS EN 3541-92. 1992 Rod Ends, Adjustable, Self-Aligning Ball Bearing with Threaded Shank—Dimensions, Torques and Loads. 11 pp.

Ball Bearings *(Cont.)*

—Aerospace—Self Aligning—Control Rods *(Cont.)*

CEN EN 3541-92. Rod Ends, Adjustable, Self-Aligning Ball Bearing with Threaded Shank—Dimensions, Torques, Clearances and Loads. 7 pp.

—Aircraft—Shafts (Machine Elements)

SBAC RS 182 ISSUE 1. Fits for Ball Bearings.

—Airframe—Alloy Housings

SBAC RS 182 ISSUE 1. Fits for Ball Bearings.

—Airframe—Control Cables—Pulleys

AECMA PREN2081-77. Non-Metallic Bodied Pulleys with Ball Bearing—for Control Cables—Dimensions and Loads (C1/22-02). 5 pp.

AECMA PREN2496-80. Bearing, Ball for Control Cable pulleys—Dimensions and Load (C1/22). 5 pp.

AECMA PREN3182-89. Ball Bearings, Rigid in Corrosion Resisting Steel Cadmium Plated, for Control Cable Pulleys Dimensions and Loads. 5 pp.

AECMA PREN3629-89. Ball Bearings, Rigid in Corrosion Resisting Steel for Control Cable Pulleys Dimensions and Loads. 5 pp.

ISO 7938-86. Aircraft—Ball Bearings for Control Cable Pulleys—Dimensions and Loads First Edition. 6 pp.

ISO 7939-88. Aircraft—Non-Metallic Pulleys with Ball Bearings for Control Cables—Dimensions and Loads First Edition. 6 pp.

—Airframe—Loads (Forces)

AECMA PREN3045-87. Bearings-Airframe Rolling Rigid Single Row Ball Bearings in Steel Diameter Series 0 and 2 Reduced Clearance Category Dimensions and Loads. 5 pp.

AECMA PREN3046-87. Bearings-Airframe Rolling Rigid Single Row Ball Bearings in Steel Cadmium Plated Diameter Series 0 and 2 Reduced Clearance Category Dimensions and Loads. 5 pp.

AECMA PREN3047-87. Bearings-Airframe Rolling Rigid Single Row Ball Bearings in Corrosion Resisting Steel Diameter Series 0 and 2 Reduced Clearance Category Dimensions and Loads. 5 pp.

AECMA PREN3056-87. Bearings-Airframe Rolling Rigid Two Row Ball Bearings in Steel Dimensions and Loads. 5 pp.

AECMA PREN3057-87. Bearings-Airframe Rolling Rigid Two Row Ball Bearings in Steel Cadmium Plated Dimensions and Loads. 5 pp.

AECMA PREN3058-87. Bearings-Airframe Rolling Rigid Two Row Ball Bearings in Corrosion Resisting Steel Dimensions and Loads. 5 pp.

AECMA PREN3281-87. Bearings-Airframe Rolling Rigid, Single Row Ball Bearings in Steel Diameter Series 8 and 9 Dimensions, Torques, Clearances and Loads. 7 pp.

AECMA PREN3282-87. Bearings-Airframe Rolling Rigid, Single Row Ball Bearings in Steel, Cadmium Plated Diameter Series 8 and 9 Dimensions, Torques, Clearances and Loads. 6 pp.

AECMA PREN3283-87. Bearings-Airframe Rolling Rigid, Single Row Ball Bearings in Corrosion Resisting Steel Diameter Series 8 and 9 Dimensions, Torques, Clearances and Loads. 6 pp.

AECMA PREN3284-87. Bearings-Airframe Rolling Rigid, Single Row Ball Bearings in Steel Diameter Series 0 and 2 Normal Clearance Category Dimensions, Torques, and Loads. 6 pp.

AECMA PREN3285-87. Bearings-Airframe Rolling Rigid, Single Row Ball Bearings in Steel, Cadmium Plated Diameter Series 0 and 2 Normal Clearance Category Dimensions, Torques, and Loads. 6 pp.

AECMA PREN3286-87. Bearings-Airframe Rolling Rigid, Single Row Ball Bearings in Corrosion Resisting Steel Diameter Series 0 and 2 Normal Clearance Category Dimensions, Torques, and Loads. 6 pp.

BSI BS EN 2009-92. 1992 Bearings—Airframe Rolling Rigid, Single Row Ball Bearings in Steel Diameter Series 8 and 9 Dimensions and Loads. 10 pp.

BSI BS EN 2012-89. 1989 Bearings—Airframe Rolling Rigid, Single Row Ball Bearings in Steel Diameter Series 0 and 2. Dimensions and Loads. 8 pp

BSI BS EN 2014-89. 1989 Bearings—Airframe Rolling Rigid Single Row Ball Bearings in Corrosion Resisting Steel Diameter Series 0 and 2. Dimensions and Loads. 8 pp.

CEN PREN2009-82. Bearings Airframe Rolling Rigid Single Row Ball Bearings in Steel—Diameter Series 8 and 9—Dimensions and Loads (C1/04).

CEN EN2009-84. Bearings-Airframe Rolling Rigid, Single Roll Ball Bearings in Steel Diameter Series 8 and 9 Dimensions and Loads. 4 pp.

CEN PREN2010-82. Bearings Airframe Rolling Rigid Single Row Ball Bearings in Steel Cadmium Plated—Diameters Series 8 and 9 —Dimensions and Loads (C1/04). 5 pp.

CEN PREN2011-82. Bearings Airframe Rolling Rigid Single Row Ball Bearings in Corrosion Resisting Steel—Diameter Series 8 and 9—Dimensions and Loads (C1/04).

Ball Bearings *(Cont.)*

—Airframe—Loads (Forces) *(Cont.)*

CEN EN 2011-84. Bearings-Airframe Rolling Rigid Single Row Ball Bearings in Corrosion Resisting Steel Diameter Series 8 and 9 Dimensions and Loads. 5 pp.

CEN PREN2012-82. Bearings Airframe Rolling Rigid Single Row Ball Bearings in Steel—Diameter Series 0 and 2—Dimensions and Loads (C1/05).

CEN EN 2012-84. Bearings-Airframe Rolling Rigid, Single Row Ball Bearings in Steel Diameter Series 0 and 2 Dimensions and Loads. 5 pp.

CEN PREN2013-82. Bearings Airframe Rolling Rigid Single Row Ball Bearings in Steel Cadmium Plated—Diameter Series 0 and 2—Dimensions and Loads (C1/05). 5 pp.

CEN PREN2014-82. Bearings Airframe Rolling Rigid Single Row Ball Bearings in Corrosion Resisting Steel—Diameter Series 0 and 2—Dimensions and Loads (C1/05).

CEN EN 2014-84. Bearings—Airframe Rolling Rigid Single Row Ball Bearings in Corrosion Resisting Steel Diametes Series 0 and 2 Dimensions and Loads Aerospace Series. 7 pp.

CEN PREN 3059-91. Bearings-Airframe Rolling Rigid Single Row Ball Bearings in Steel with Flanged Alignment Bush Dimensions and Loads. 7 pp.

CEN PREN 3060-91. Bearings-Airframe Rolling Rigid Single Row Ball Bearings in Steel Cadmium Plated with Flanged Alignment Bush Dimensions and Loads. 7 pp.

CEN PREN 3061-91. Bearings-Airframe Rolling Rigid Single Row Ball Bearings in Corrosion Resisting Steel with Flanged Alignment Bush Dimensions and Loads. 7 pp.

—Airframe—Pulleys

SBAC AS 103 ISSUE 6. Pulley Ballbearing 2.0 Diameter. 1 p.

SBAC AS 104 ISSUE 7. Pulley Ballbearing 3.0 Diameter. 1 p.

SBAC AS 105 ISSUE 7. Pulley Ballbearing 4.0 Diameter. 1 p.

SBAC AS 106 ISSUE 7. Pulley Ballbearing 5.0 Diameter. 1 p.

SBAC AS 111 ISSUE 11. Pulley Ballbearing 1.15 Diameter. 1 p.

SBAC AS 1151 ISSUE 5. Pulley, (Ball Bearing) Shrouded, Single. 1 p.

SBAC AS 1152 ISSUE 5. Pulleys, (Ball Bearing) Shrouded, Twin. 1 p.

SBAC AS 1153 ISSUE 5. Pulleys, (Ball Bearing) Shrouded, Triple. 1 p.

SBAC AS 1154 ISSUE 5. Pulleys, (Ball Bearing) Shrouded, Quadruple. 1 p.

—Airframe—Steel Housings

SBAC RS 182 ISSUE 1. Fits for Ball Bearings.

—Automotive—Clutch Release

CNS D2085-81. Clutch Release Ball Bearings for Automobiles (Mar)(7133).

—Automotive—Water Pumps

CNS D2017-86. Water Pump Ball Bearings for Automobiles (Aug)(5323).

—Balls

JIS B 1501-88. Steel Balls for Ball Bearings. 12 pp.

—Bearing Alloys

DIN ENGL 17230-80. Ball and Roller Bearing Steels; Technical Conditions of Delivery (Sept). 18 pp.

—Bearing Boxes

CNS B3249-76. Bearing Box (Ball Bearing, Assembly 2, Light Grade, Cone Form) (Dec)(4030).

CNS B3250-76. Bearing Box (Ball Bearing, Assembly 3, Medium Grade, Cone Form) (Dec)(4031).

CNS B3251-76. Bearing Box (Ball Bearing, Assembly 2, Light Grade, Round Form) (Dec)(4032).

CNS B3252-76. Bearing Box (Ball Bearing, Assembly 3, Medium, Grade, Round Form) (Dec)(4033).

—Chamfers

CNS B1349-85. Tolerances of Ball and Roller Bearings-Chamfer Dimension Limits (May)(11259).

—Cranes—Covers

CNS B1288-82. Cranes; Covers for Balland Roller Bearings; Connecting Dimensions (Aug)(9223).

—Door Hinges

JIS A 5516-72. Door Hinges (with Ball Bearings).

—Dynamic Load Ratings

ISO TR8646-85. Explanatory Notes on ISO 281/1-1977 First Edition. 66 pp.

—Electrical Equipment

CNS B1161-80. Ball Bearings; Conrad Type, for Electrical Machines; Tolerances and Radial Clearance (May)(5506).

INDUSTRY STANDARDS

Ball Bearings (Cont.)

—Flanged
BSI BS 5646: Part 7-86. 1986 Amd 1 Rolling Bearings—Accessories Part 7: Specification for Flange Dimensions of Radial Ball Bearings with Flanged Outer Ring (AMD 6349) July 31, 1991. 9 pp.
ISO 8443-85. Radial Ball Bearings with Flanged Outer Ring—Flange Dimensions First Edition; (Corrigendum 1-1990). 7 pp.

—Greases
MOD UK DSTAN 91-28-76. Grease, Naval: General Purpose, Ball and Roller Bearing NATO Code No: G-450 Joint Service Designation: XG-274 Issue 1. 11 pp.

—Internal Clearance
BSI BS 6107: Part 3-92. 1992 Rolling Bearings: Tolerances Part 3: Specification for Radial Internal Clearance (ISO 5753: 1991). 12 pp.
BSI BS 6107: Part 3-81. 1981 Rolling Bearings: Tolerances Part 3: Radial Internal Clearance. 9 pp.
ISO 5753-91. Rolling Bearings—Radial Internal Clearance Second Edition. 9 pp.

—Journal
CNS B2472-85. Self-Aligning Ball Bearing; Wide Inner Race; Inner Race with Clamping Sleeve (Mar)(5501).
DIN ENGL 630 Pt 1-60. Self-Aligning Ball (Journal) Bearings; Parallel and Tapered Bore (May). 2 pp.

—Life Testing—Hydraulic Fluids
BSI PD 6487-79. 1979 Calculating the Life of Ball Bearings When Used in Contact with Fire Resistant Hydraulic Fluids. 7 pp.

—Linear—Recirculating
BSI BS 7637-93. 1993 Linear Motion, Recirculating Ball, Sleeve Type Rolling Bearings (Metric Series) (ISO 10285: 1992) (E). 15 pp.
ISO 10285-92. Rolling Bearings, Linear Motion, Recirculating Ball, Sleeve Type—Metric Series First Edition. 13 pp.

—Magnetic
CNS B2476-85. Magneto Ball Bearings (Mar)(5505).
JIS B 1538-76. Magneto Ball Bearings. 9 pp.

—Packaging
MOD UK DSTAN 81-32-01. Packaging of Ball and Roller Bearings Issue 3; Amendment 2. 26 pp.

—Pulleys—Cranes
DIN ENGL 15062 Pt 2-82. Cranes; Rope Pulleys; Dimensions of Hubs and Bearings (July). 4 pp.

—Radial
CNS B2475-85. Radial-Angular Contact Ball Bearings; Single and Double-Row (Mar)(5504).
DIN ENGL 625 Pt 1-89. Rolling Bearings; Single Row Radial Contact Ball Bearings (Apr). 9 pp.
DIN ENGL 625 Pt 3-90. Rolling Bearings; Double Row Radial Contact Ball Bearings (Mar). 2 pp.

—Radial—Flanged
DIN ENGL 625 Pt 4-87. Rolling Bearings; Radial Contact Groove Ball Bearings with Flanged Outer Ring (Aug). 4 pp.

—Radial—Self Aligning
CNS B2471-85. Radial Self-Aligning Ball Bearings; Parallel and Tapered Bore (Mar)(5500).

—Radial—Separable
DIN ENGL 615-59. Separable Ball Bearings (Radial) (Apr). 1 p.

—Roving Frames
JIS L 5166-75. Ball and Roller Bearings Top Rollers for Roving Frames.

—Self Aligning
DIN ENGL 630 Pt 1-60. Self-Aligning Ball (Journal) Bearings; Parallel and Tapered Bore (May). 2 pp.
JIS B 1523-86. Self-Aligning Ball Bearings.
JIS B 1523-65. Self-Aligning Ball Bearings (R 1977). 6 pp.

—Self Aligning—Airframe—Loads (Forces)
AECMA PREN3287-87. Bearings-Airframe Rolling, Double Row Self-Aligning Ball Bearings in Steel Diameter Series 2 Dimensions, Torques, Clearances and Loads. 6 pp.
AECMA PREN3288-87. Bearings-Airframe Rolling Double Row, Self-Aligning Ball Bearings in Steel, Cadmium Plated Diameter Series 2 Dimensions, Torques, Clearances and Loads. 6 pp.
AECMA PREN3289-87. Bearings-Airframe Rolling Double Row, Self-Aligning Ball Bearings in Corrosion Resisting Steel Diameter Series 2 Dimensions, Torques, Clearances and Loads. 6 pp.

Ball Bearings (Cont.)

—Self Aligning—Airframe—Loads (Forces) (Cont.)
BSI BS EN 2015-89. 1989 Bearings—Airframe Rolling Double Row, Self Aligning Ball Bearings in Steel Diameter Series 2. Dimensions and Loads. 8 pp.
BSI BS EN 2017-89. 1989 Bearings—Airframe Rolling Double Row Self Aligning Ball Bearings in Corrosion Resisting Steel Diameter Series 2. Dimensions and Loads. 8 pp.
CEN PREN2015-82. Bearing—Airframe Rolling Double Row Self Aligning Ball Bearings in Steel—Diameter Series—Dimensions and Loads (C1/06).
CEN EN 2015-84. Bearings-Airframe Rolling Double Row, Self Aligning Ball Bearings in Steel Diameter Series 2 Dimensions and Loads. 5 pp.
CEN PREN2016-82. Bearings—Airframe Rolling Double Row Self Aligning Ball Bearings in Steel Cadmium Plated—Diameter Series 2—Dimensions and Loads (C1/06). 5 pp.
CEN PREN2017-82. Bearings—Airframe Rolling Double Row Self Aligning Ball Bearings in Corrosion Resisting Steel—Diameter Series 2—Dimensions and Loads (C1/06).
CEN EN 2017-84. Bearings-Airframe Rolling Double Row Self Aligning Ball Bearings in Corrosion Resisting Steel Diameter Series 2 Dimensions and Loads. 5 pp.
CEN PREN2018-82. Bearings—Airframe Rolling Single Row Self Aligning Roller Bearings in Steel—Diameter Series 3 and 4—Dimensions and Loads (C1/07).
CEN EN 2018-84. Bearings-Airframe Rolling Single Row, Self Aligning Roller Bearings in Steel Diameter Series 3 and 4 Dimensions and Loads. 5 pp.
CEN PREN2019-82. Bearings—Airframe Rolling Single Row Self Aligning Roller Bearings in Steel Cadmium Plated—Diameter Series 3 and 4—Dimensions and Loads (C1/07). 5 pp.

—Sleeves
CNS B3192-76. Adapter Sleeves with Nut and Retainer for Ball and Roller Bearings (Dec)(3007).
CNS B3193-76. Withdrawal Sleeves for Ball and Roller Bearings (Dec)(3008).

—Sound Level Measurement
JIS B 1548-76. Measuring Method of Sound Pressure Levels of Ball and Roller Bearings (R 1985). 14 pp.

Ball Cage Bushes
Use: Guide Bushings—Ball Cage

Ball Couplings
See Also: Couplings; Towing Attachments

—Automotive
SAA AS D18-68. Ball Couplings (for Automotive Purposes) Amdt 1 August 1972. 11 pp.

—Recreational Vehicles
BSI BS AU 113C-78. 1978 Amd 2 Dimensional Characteristics of Coupling Balls for Caravans and Light Trailers (AMD 5212) October 31, 1986. 7 pp.
BSI BS AU 114B-79. 1979 Amd 1 Strength Requirements of Towing Brackets and Coupling Balls for Caravans and Light Trailers (AMD 5211) October 31, 1986. 7 pp.
ISO 1103-76. Road Vehicles—Caravans and Light Trailers—Coupling Ball—Dimensional Characteristics Second Edition. 4 pp.
ISO 8704-91. Caravans and Light Trailers—Couplings—Strength Tests First Edition. 6 pp.
ISO 10061-91. Road Vehicles—Trailers, Including Caravans—Height of Coupling Head First Edition. 5 pp.

—Recreational Vehicles—Dynamic Loads
ISO 3853-77. Road Vehicles—Caravans and Light Trailers—Towing Brackets and Coupling Balls—Strength Test First Edition. 6 pp.

—Recreational Vehicles—Static Loads
ISO TR4114-79. Road Vehicles—Caravans and Light Trailers—Static Load on Ball Couplings First Edition. 3 pp.

—String Insulators
IEC 120-84. Dimensions of Ball and Socket Couplings of String Insulator Units Third Edition; (Corrigendum—May 1989). 38 pp.

—String Insulators—Locks
BSI BS 3288: Part 4-89. 1989 Insulator and Conductor Fittings for Overhead Power Lines Part 4: Locking Devices for Ball and Socket Couplings of String Insulator Units: Dimensions and Tests. 18 pp.
IEC 372-84. Locking Devices for Ball and Socket Couplings of String Insulator Units: Dimensions and Tests Third Edition; (Amendment 1-1991). 34 pp.

Ball Couplings (Cont.)

—Trailers
BSI BS AU 113C-78. 1978 Amd 2 Dimensional Characteristics of Coupling Balls for Caravans and Light Trailers (AMD 5212) October 31, 1986. 7 pp.
BSI BS AU 114B-79. 1979 Amd 1 Strength Requirements of Towing Brackets and Coupling Balls for Caravans and Light Trailers (AMD 5211) October 31, 1986. 7 pp.
ISO 8704-91. Caravans and Light Trailers—Couplings—Strength Tests First Edition. 6 pp.
ISO 10061-91. Road Vehicles—Trailers, Including Caravans—Height of Coupling Head First Edition. 5 pp.

Ball Cup Testing
Use For: Erichsen Cup Testing See Also: Ductility
JIS Z 2247-77. Method of Erichsen Cupping Test. 6 pp.

—Metal Sheets
BSI BS 3855-65. 1965 Modified Erichsen Cupping Test for Sheet and Strip Metal. 8 pp.
DIN ENGL 50101 Pt 1-79. Testing of Metals; Erichsen Cupping Test on Sheet and Strip Metal Having a Width of Greater Than or Equal to 90 mm; Thickness Range: 0,2 mm to 2 mm (Sept). 4 pp.
DIN ENGL 50101 Pt 2-79. Testing of Metals; Erichsen Cupping Test on Sheet and Strip Metal Having a Width of Greater Than or Equal to 90 mm; Thickness Range: over 2 mm up to 3 mm (Sept). 3 pp.
DIN ENGL 50102-79. Testing of Metals; Erichsen Cupping Test on Narrow Strip Metal; Width Range: 30 mm to Less Than 90 mm (Sept). 4 pp.
ISO 8490-86. Metallic Materials—Sheet and Strip—Modified Erichsen Cupping Test First Edition. 6 pp.

—Metal Strips
BSI BS 3855-65. 1965 Modified Erichsen Cupping Test for Sheet and Strip Metal. 8 pp.
DIN ENGL 50101 Pt 1-79. Testing of Metals; Erichsen Cupping Test on Sheet and Strip Metal Having a Width of Greater Than or Equal to 90 mm; Thickness Range: 0,2 mm to 2 mm (Sept). 4 pp.
DIN ENGL 50101 Pt 2-79. Testing of Metals; Erichsen Cupping Test on Sheet and Strip Metal Having a Width of Greater Than or Equal to 90 mm; Thickness Range: over 2 mm up to 3 mm (Sept). 3 pp.
DIN ENGL 50102-79. Testing of Metals; Erichsen Cupping Test on Narrow Strip Metal; Width Range: 30 mm to Less Than 90 mm (Sept). 4 pp.
ISO 8490-86. Metallic Materials—Sheet and Strip—Modified Erichsen Cupping Test First Edition. 6 pp.

Ball Handles
Use: Handles—Ball

Ball Joints
See Also: Joints; Universal Joints
JIS B 1454-88. Universal Ball Joints. 11 pp.

—Aircraft
SBAC AGS 385 (V). Ball Joint.
SBAC AGS 3022 ISSUE 1. Ball Joint.

—Aircraft—Balls
SBAC AGS 3024 ISSUE 1. Ball for Ball Joint.

—Aircraft—Plugs
SBAC AGS 3025 ISSUE 1. Plug for Ball Joint.

—Aircraft—Sockets
SBAC AGS 3023 ISSUE 1. Socket for Ball Joint.

—Automotive—Commercial—Steering
BSI BS AU 225-88. 1988 Dimensions of Steering Ball Joints for Commercial Road Vehicles. 4 pp.
ISO 7803-87. Commercial Road Vehicles—Steering Ball Joints First Edition. 4 pp.

Ball Knobs
Use: Handles—Ball

Ball Mill Grindability Testing
JIS M 4002-76. Testing Method of Grinding Work Index (R 1984). 7 pp.

—Abrasives
CNS R3107-82. Method of Test for Toughness of Artificial Abrasives (Ball Mill Method) (Sep)(9437).
JIS R 6128-87. Ball Mill Test for Toughness of Artificial Abrasives. 7 pp.

Ball Nuts
See Also: Nuts (Fasteners)

—Machine Tools—Ball Screws
BSI BS 6101: Part 6-90. 1990 Machine Tool Ball Screws Part 6: Ball Nuts and Principal Dimensions of Ball Screws. 13 pp.

INTERNATIONAL AND NON-U.S. NATIONAL STANDARDS
SUBJECT INDEX

Ball Powders
MOD UK M 106/93. Examination of Ball Powders Containing Nitroglycerine.

Ball Races
See Also: Ball Bearings

—Aircraft—Control Columns
SBAC AS 867 ISSUE 3. Ball Race and Housing—Control Column. 1 p.

—Aircraft—Housings—Control Columns
SBAC AS 868 ISSUE 4. Ballrace Housing—Control Column. 1 p.

—Aircraft—Spigots—Control Columns
SBAC AS 855 ISSUE 7. Rear Ball Race Spigot—Control Column. 1 p.
SBAC AS 856 ISSUE 7. Front Ball Race Spigot Control Column. 1 p.

Ball Screws
Use For: Ballscrews See Also: Screws
BSI BS 6101: Part 6-90. 1990 Machine Tool Ball Screws Part 6: Ball Nuts and Principal Dimensions of Ball Screws. 13 pp.
CNS B2737-83. Ball Screws for General Use (Oct)(9786).
DIN ENGL 69051 Pt 5-82. Machine Tools; Ball Screws; Connecting Dimensions for Ball Nuts (Sept). 4 pp.
JIS B 1191-87. Ball Screws for General Use.
JIS B 1191-80. Ball Screws for General Use. 21 pp.

—Acceptance Testing
ISO 3408 Pt 3-92. Ball Screws—Part 3: Acceptance Conditions and Acceptance Tests First Edition. 25 pp.

—Accuracy Testing
BSI BS 6101: Part 2-91. 1991 Machine Tool Ball Screws Part 2: Specification for Accuracy, Including Geometrical Tests. 19 pp.

—Ball Nuts
BSI BS 6101: Part 6-90. 1990 Machine Tool Ball Screws Part 6: Ball Nuts and Principal Dimensions of Ball Screws. 13 pp.

—Dynamic Load Ratings
BSI BS 6101: Part 1-81. 1981 Machine Tool Ball Screws Part 1: Methods of Calculating Dynamic Load and Life Ratings. 6 pp.

—Glossaries
BSI BS 6101: Part 3-92. 1992 Machine Tool Ball Screws Part 3: Glossary of Terms (ISO 3408-1: 1991). 16 pp.
BSI BS 6101: Part 3-84. 1984 Machine Tool Ball Screws Part 3: Glossary of Terms. 11 pp.
ISO 3408 Pt 1-91. Ball Screws—Part 1: Vocabulary and Designation First Edition. 15 pp.

—Leads
BSI BS 6101: Part 5-92. 1992 Machine Tool Ball Screws Part 5: Specification for Nominal Diameters and Nominal Leads (ISO 3408-2: 1991). 7 pp.
BSI BS 6101: Part 5-84. 1984 Machine Tool Ball Screws Part 5: Nominal Diameters and Nominal Leads. 6 pp.
ISO 3408-75. Ball Screws—Nominal Diameters and Basic Leads—Metric and Inch Series First Edition. 5 pp.
ISO 3408 Pt 2-91. Ball Screws—Part 2: Nominal Diameters and Nominal Leads—Metric Series First Edition. 4 pp.

—Life (Durability)
BSI BS 6101: Part 1-81. 1981 Machine Tool Ball Screws Part 1: Methods of Calculating Dynamic Load and Life Ratings. 6 pp.

—Nominal Diameters
BSI BS 6101: Part 5-92. 1992 Machine Tool Ball Screws Part 5: Specification for Nominal Diameters and Nominal Leads (ISO 3408-2: 1991). 7 pp.
BSI BS 6101: Part 5-84. 1984 Machine Tool Ball Screws Part 5: Nominal Diameters and Nominal Leads. 6 pp.
ISO 3408-75. Ball Screws—Nominal Diameters and Basic Leads—Metric and Inch Series First Edition. 5 pp.
ISO 3408 Pt 2-91. Ball Screws—Part 2: Nominal Diameters and Nominal Leads—Metric Series First Edition. 4 pp.

—Precision
CNS B2738-83. Precision Ball Screws (Oct)(9787).
JIS B 1192-87. Precision Ball Screws.
JIS B 1192-80. Precision Ball Screws. 28 pp.

Ball Screws (Cont.)
—Rigidity
BSI BS 6101: Part 4-87. 1987 Amd 1 Machine Tool Ball Screws Part 4: Method for Calculation of Static Axial Rigidity (AMD 6264) November 30, 1989. 23 pp.

—Static Load Ratings
BSI BS 6101: Part 1-81. 1981 Machine Tool Ball Screws Part 1: Methods of Calculating Dynamic Load and Life Ratings. 6 pp.

Ball Thrust Bearings
See Also: Ball Bearings; Thrust Bearings
BSI BS 5989: Part 2-80. 1980 Rolling Bearings: Metric Thrust Ball Bearings Part 2: Tolerances of Thrust Ball Bearings. 7 pp.
CNS B1291-85. Tolerance of Rolling Bearings-Thrust Ball Bearings (May)(9573).
CNS B2117-85. Deep Groove Thrust Ball Bearings, Single-Thrust, with Flat Seating (Mar)(2864).
CNS B2623-81. Deep Groove Thrust Ball Bearings; with Aligning Seat and Housing Washer with Aligning Seat (Nov)(8095).
CNS B2624-81. Deep Groove Thrust Ball Bearings; Double-Thrust with Flat Seatings (Nov)(8096).
CNS B2625-81. Deep Groove Thrust Ball Bearings; Double-Thrust with Aligning Seat and Housing Washer with Aligning Seat (Nov)(8097).
CNS B2628-81. Deep Groove Full Type Thrust Ball Bearings; Single-Thrust, with Retaining Shield (No Cage) (Nov)(8100).
DIN ENGL 711-88. Rolling Bearings; Single Direction Thrust Ball Bearings (Feb). 8 pp.
DIN ENGL 715-87. Rolling Bearings; Double Direction Thrust Ball Bearings (Aug). 6 pp.
DIN ENGL 5429 Pt 1-87. Rolling Bearings; Needle Bearing Combinations; Needle Roller Thrust Bearings and Needle Ball Thrust Bearings (Aug). 3 pp.
ISO 199-79. Rolling Bearings—Thrust Ball Bearings—Tolerances First Edition. 5 pp.
JIS B 1532-86. Thrust Ball Bearings with Flat Back Faces. 14 pp.

—Automotive—Kingpins
CNS D2018-80. Kingpin Thrust Ball Bearing for Automobiles (Mar)(5324).

—Designations
DIN ENGL 5429 Pt 1-87. Rolling Bearings; Needle Bearing Combinations; Needle Roller Thrust Bearings and Needle Ball Thrust Bearings (Aug). 3 pp.

Ball Valves
See Also: Check Valves; Cocks; Float Valves; Gas Valves; Globe Valves; Hydraulic Valves; Pneumatic Valves; Valves
BSI BS 5159-74. 1974 Amd 2 Cast Iron and Carbon Steel Ball Valves for General Purposes. 22 pp.
BSI BS 5351-86. 1986 Amd 3 Steel Ball Valves for the Petroleum, Petrochemical and Allied Industries (AMD 6271) September 28, 1990. 36 pp.
CNS B2499-80. Face-to-Face and End-to-End Dimensions of Ball Valves (Jul)(5715).
CNS B2769-89. Bronze Screwed Ball Valves (10 kgf/cm2) (Apr)(11355). 5 pp.
DIN ENGL 3357 Pt 1-89. Metal Ball Valves; General Requirements and Methods of Test (Oct). 7 pp.
DIN ENGL 3357 Pt 2-81. Full Bore Steel Ball Valves (Dec). 7 pp.
DIN ENGL 3357 Pt 3-81. Reduced Bore Steel Ball Valves (Dec). 7 pp.
DIN ENGL 3357 Pt 4-81. Full Bore Nonferrous Metal Ball Valves (Dec). 5 pp.
DIN ENGL 3357 Pt 5-81. Reduced Bore Nonferrous Metal Ball Valves (Dec). 4 pp.
DIN ENGL 3357 Pt 6-81. Full Bore Cast Iron Ball Valves (Dec). 3 pp.
DIN ENGL 3357 Pt 7-81. Reduced Bore Cast Iron Ball Valves (Dec). 3 pp.
DIN ENGL 3441 Pt 2-84. Unplasticized Polyvinyl Chloride (PVC-U) Valves; Ball Valves; Dimensions (Aug). 3 pp.
DIN ENGL 3442 Pt 2-80. Fittings of PP (Polypropylene); Ball Valves; Dimensions (Oct). 2 pp.
ISO 5752-82. Metal Valves for Use in Flanged Pipe Systems—Face-to-Face and Centre-to-Face Dimensions Second Edition. 13 pp.
ISO 7121-86. Flanged Steel Ball Valves First Edition. 12 pp.
JIS B 2002-87. Face-to-Face and End-to-End Dimensions of Valves.
JIS B 2002-68. Face-to-Face and End-to-End Dimensions of Valves. 14 pp.
SNZ NZS 1822-64. Half-Inch High-Pressure Ball Valves (New Zealand Type), Excluding Floats. 14 pp.
SNZ NZS/BS 5351-86. Specification for Steel Ball Valves for the Petroleum, Petrochemical and Allied Industries Amend: 1; 2. 28 pp.

Ball Valves (Cont.)
—Aircraft—Float Operated
SBAC AS 6605 ISSUE 1. Float Operated Cut-Off Valve.

—Floats
BSI BS 1968-53. 1953 Amd 2 Floats for Ballvalves (Copper). 16 pp.
BSI BS 2456-90. 1990 Floats (Plastics) for Float Operated Valves for Cold Water Services. 13 pp.

—Gas
CEN PREN 331-93. Manually Operated Ball Valves and Closed Bottom Taper Plug Valves for Gas Installations for Buildings. 31 pp.

—Gas Industry
CSA CAN/CSA-Z245.15-M91. Steel Valves; (Gen Instr 1). 81 pp.

—Petroleum Industry
CSA CAN/CSA-Z245.15-M91. Steel Valves; (Gen Instr 1). 81 pp.

—Pipelines—Flanged
CEN PREN 558-1-91. Metal Valves for Use in Flanged Pipe Systems—Face-to-Face and Centre-to-Face Dimensions—Part 1: General. 5 pp.
CEN PREN 558-2-91. Metal Valves for Use in Flanged Pipe Systems—Face-to-Face and Centre-to-Face Dimensions—Part 2: PN Designated Valves. 13 pp.
CEN PREN 558-3-91. Metal Valves for Use in Flanged Pipe Systems—Face-to-Face and Centre-to-Face Dimensions—Part 3: Class-Designation Valves. 17 pp.

—Piston Type
BSI BS 1212: Part 1-90. 1990 Float Operated Valves (Excluding Floats) Part 1: Piston Type Float Operated Valves (Copper Alloy Body) (Excluding Floats). 36 pp.
BSI BS 1212: Part 1-53. (WITHDRAWN) 1953 Amd 13 Ballvalves (Excluding Floats) Part 1: Piston Type. 46 pp.

—Stop—Gas
CEN PREN 333-90. Manually Operated Metallic Ball Shut-Off Valves for Domestic Gas Installations up to PN 4 and with DN 15 to DN 50. 13 pp.

Ballast Regulating Tubes
Use: Rheostats

Ballast Tanks
See Also: Tanks (Containers)

—Paints (Aluminum)—Marine—Interior/Exterior
CGSB CAN/CGSB-1.93-92. Aluminum Marine Paint. 8 pp.

Ballasts (Electric)
See Also: Capacitors; Lamps; Lighting; Power Reducers; Resistors; Transformers
IEC 259-68. Miscellaneous Lamps and Ballasts First Edition; (Supplement A-1972). 37 pp.

—Discharge Lamps
BSI BS 4782-71. (WITHDRAWN) 1971 Amd 3 Ballasts for Discahrge Lamps (Excluding Ballasts for Tubular Flourescent Lamps) (Superseded by BS EN 60922: 1991 and BS EN 60923: 1991). 67 pp.
BSI BS EN 60922-91. 1991 General and Safety Requirements for Ballasts for Discharge Lamps (Excluding Tubular Fluorescent Lamps) (E). 38 pp.
BSI BS EN 60922-01. 1991 Amd 1 General and Safety Requirements for Ballasts for Discharge Lamps (Excluding Tubular Fluorescent Lamps) (AMD 7173) September 15, 1992. 50 pp.
CENELEC EN 60922-91. Ballasts for Discharge Lamps (Excluding Tubular Fluorescent Lamps) General and Safety Requirements. 6 pp.
CENELEC EN 60922/A1-92. AMD 1 Ballasts for Discharge Lamps (Excluding Tubular Fluorescent Lamps) General and Safety Requirements. 6 pp.
CENELEC EN 60923/A1-92. AMD 1 Ballasts for Discharge Lamps (Excluding Tubular Fluorescent Lamps)—Performance Requirements. 5 pp.
CSA CAN/CSA-C22. 2 NO 74-92. Equipment for Use with Electric Discharge Lamps (Incorporating Electrical Bulletin Nos. 523F, 753, 846, 1124A, 1125A, 1325, and 1326); (Gen Instr 1). 82 pp.
DIN VDE 0712 Pt 2-71. Accessories for Discharge Lamps with Nominal Voltage up to 1000 V; Part 2: Special Specifications for Ballasts (June) (Valid Only in Conjunction with 0712 Part 1/6.71). 50 pp.
DIN VDE 0712 Pt 2 A2 (D)-83. Equipment for Discharge Lamps for a Supply Voltage up to 1000V; Ballasts: Amendment 2 (VDE Specifications) (Oct). 9 pp.

INDUSTRY STANDARDS

INTERNATIONAL AND NON-U.S. NATIONAL STANDARDS
SUBJECT INDEX
Ballasts

Ballasts (Electric) *(Cont.)*

—Discharge Lamps *(Cont.)*
DIN VDE 0712 Pt 2A (D)-78. Equipment for Discharge Lamps for a Supply Voltage up to 1000V: Particular Specification for Ballasts (VDE Specification) (Oct). 16 pp.

IEC 922-89. Ballasts for Discharge Lamps (Excluding Tubular Fluorescent Lamps) General and Safety Requirements First Edition; (Amendment 2-1992, Includes Amendment 1). 89 pp.

—Fluorescent Lighting
BSI BS 2818: Part 1-85. (WITHDRAWN) 1985 Amd 1 Ballasts for Tubular Fluorescent Lamps Part 1: Ballasts for Use Internationally (Superseded by BS EN 60920: 1991 and BS EN 60921: 1991). 73 pp.

BSI BS 2818: Part 2-85. (WITHDRAWN) 1985 Amd 1 Ballasts for Tubular Fluorescent Lamps Part 2: Ballasts for Use with Lamps Used in the UK Not Included in BS 1853 'Tubular Fluorescent Lamps for General Lighting Service 'Part 1' Specif for Internationally Specified Lamps'. 11 pp.

BSI BS 5717-84. (WITHDRAWN) 1984 Transistorized Ballasts for Tubular Fluorescent Lamps (Superseded by BS EN 60924: 1991). 30 pp.

BSI BS EN 60730-2-3-92. 1992 Automatic Electrical Controls for Household and Similar Use Part 2: Particular Requirements Section 2.3: Thermal Protectors for Ballasts for Tubular Fluorescent Lamps. 19 pp.

BSI BS EN 60920-91. 1991 Ballasts for Tubular Fluorescent Lamps. General and Safety Requirements (Supersedes BS 2818: Part 1 & 2: 1985). 57 pp.

BSI BS EN 60921-91. 1991 Ballasts for Tubular Fluorescent Lamps. Performance Requirements. 34 pp.

BSI BS EN 60921-01. 1991 Amd 1 Ballasts for Tubular Fluorescent Lamps. Performance Requirements (AMD 7433) December 15, 1992 (Supersedes BS 2818: Parts 1 & 2: 1985). 45 pp.

BSI BS EN 60924-91. 1991 General and Safety Requirements for D.C. Supplied Electronic Ballasts for Tubular Fluorescent Lamps. 58 pp.

BSI BS EN 60925-91. 1991 Performance Requirements for D.C. Supplied Electronic Ballasts for Tubular Fluorescent Lamps. 26 pp.

BSI BS EN 60928-91. 1991 A.C. Supplied Electronic Ballasts for Tubular Fluorescent Lamps. General and Safety Requirements. 31 pp.

BSI BS EN 60929-92. 1992 A.C.-Supplied Electronic Ballasts for Tubular Fluorescent Lamps Performance Requirements. 46 pp.

CENELEC HD 302-75. Transistorized Ballast for Fluorescent Lamps. 4 pp.

CENELEC EN 60730-2-3-92. Automatic Electrical Controls for Household and Similar Use Part 2: Particular Requirements for Thermal Protectors for Ballasts for Tubular Fluorescent Lamps. 7 pp.

CENELEC EN 60920-91. Ballasts for Tubular Fluorescent Lamps General and Safety Requirements. 7 pp.

CENELEC EN 60921-91. Ballasts for Tubular Fluorescent Lamps Performance Requirements. 6 pp.

CENELEC EN 60921/A1-92. AMD 1 Ballasts for Tubular Fluorescent Lamps Performance Requirements. 5 pp.

CENELEC EN 60924-91. D.C. Supplied Electronic Ballasts for Tubular Fluorescent Lamps General and Safety Requirements. 6 pp.

CENELEC EN 60925-91. D.C. Supplied Electronic Ballasts for Tubular Fluorescent Lamps—Performance Requirements. 5 pp.

CENELEC EN 60928-91. A.C. Supplied Electronic Ballasts for Tubular Fluorescent Lamps—General and Safety Requirements. 6 pp.

CENELEC EN 60929-92. A.C.-Supplied Electronic Ballasts for Tubular Fluorescent Lamps—Performance Requirements. 5 pp.

CNS C3041-85. Method of Test for Ballasts for Fluorescent Lamp (Oct)(3888). 5 pp.

CNS C4020-86. Ballasts for Fluorescent Lamp (Oct)(927).

CSA CAN/CSA-C.2 NO 74-92. Equipment for Use with Electric Discharge Lamps (Incorporating Electrical Bulletin Nos. 523F, 753, 846, 1124A, 1125A, 1325, and 1326); (Gen Instr 1). 82 pp.

CSA CAN/CSA-C654-M91. Fluorescent Lamp Ballast Efficacy Measurements; (Gen Instr 1 Thru 2). 54 pp.

IEC 82-84. Ballasts for Tubular Fluorescent Lamps Fifth Edition; (Amendment 1-1986). 131 pp.

IEC 730 Pt 2-3-90. Automatic Electrical Controls for Household and Similar Use Part 2: Particular Requirements for Thermal Protectors for Ballasts for Tubular Fluorescent Lamps First Edition. 30 pp.

IEC 920-90. Ballasts for Tubular Fluorescent Lamps General and Safety Requirements First Edition; (Amendment 1-1993). 121 pp.

IEC 921-88. Ballasts for Tubular Fluorescent Lamps Performance Requirements First Edition; (Corrigendum—April 1989) (Amendment 1-1990). 64 pp.

Ballasts (Electric) *(Cont.)*

—Fluorescent Lighting *(Cont.)*
IEC 924-90. D.C. Supplied Electronic Ballasts for Tubular Fluorescent Lamps General and Safety Requirements First Edition; (Amendment 1-1993). 107 pp.

IEC 925-89. D.C. Supplied Electronic Ballasts for Tubular Fluorescent Lamps Performance Requirements First Edition. 40 pp.

IEC 928-90. A.C. Supplied Electronic Ballasts for Tubular Fluorescent Lamps General and Safety Requirements First Edition; (Amendment 1-1992) (Amendment 2-1993). 71 pp.

IEC 929-90. A.C.-Supplied Electronic Ballast for Tubular Fluorescent Lamps-Performance Requirements First Edition; (Corrigendum—June 1991). 77 pp.

JIS C 8108-91. Ballasts for Fluorescent Lamps. 43 pp.

JIS C 8117-92. AC Supplied Electronic Ballasts for Fluorescent Lamps. 45 pp.

SAA AS 3134-92. Approval and Test Specification—a.c. Supplied Electronic Ballasts for Tubular Fluorescent Lamps (in Professional Package 28) (Supersedes AS 3134(Int)—1992) (This is a Joint Standard with SNZ NZS 3134). 15 pp.

SAA AS 3963-91. a.c. Supplied Electronic Ballasts for Tubular Fluorescent Lamps—Performance Requirements. 18 pp.

SNZ NZS 6210-84. Ballasts for Tubular Fluorescent Lamps Amend: A, 1984. 104 pp.

—Fluorescent Lighting—Ships
CNS F5034-87. Marine Ballasts for Fluorescent Lamp (Nov)(9390).

JIS F 8431-85. Marine Ballasts for Fluorescent Lam

—Fluorescent Lighting—Thermal Cutoffs
CSA C22.2 NO 209-M1985. Thermal Cut-Offs (R 1992); (Gen Instr 1 Thru 2). 28 pp.

—High Intensity Discharge Lamps—Branch Circuits
CSA CAN/CSA-C22. 2 NO 74-92. Equipment for Use with Electric Discharge Lamps (Incorporating Electrical Bulletin Nos. 523F, 753, 846, 1124A, 1125A, 1325, and 1326); (Gen Instr 1). 82 pp.

—Lead Wire
CSA C22.2 NO 116-1980. Coil-Lead Wires (R 1992); (Gen Instr 1 Thru 8). 52 pp.

CSA 1227 Bull. Electrical Bulletin 1227 May 16, 1979 to C22.2 NO 116. 9 pp.

CSA 1236 Bull. Electrical Bulletin 1236 September 12, 1979 to C22.2 NO 116. 9 pp.

CSA 1333 Bull. Electrical Bulletin 1333 August 17, 1981 to C22.2 NO 116. 9 pp.

CSA 1383 Bull. Electrical Bulletin 1383 November 29, 1982 to C22.2 NO 116. 9 pp.

CSA 1409 Bull. Electrical Bulletin 1409 July 28, 1983 to C22.2 NO 116. 9 pp.

—Mercury Vapor Lamps
BSI BS 3677-89. 1989 Amd 1 Specification for High-Pressure Mercury Vapour Lamps (AMD 6714) June 28, 1991. 44 pp.

BSI BS 3677-82. 1982 High Pressure Mercury Vapour Lamps. 35 pp.

BSI BS EN 60923-91. 1991 Performance Requirements for Ballasts for Discharge Lamps (Excluding Tubular Fluorescent Lamps). 26 pp.

BSI BS EN 60923-01. 1991 Amd 1 Performance Requirements for Ballasts for Discharge Lamps (Excluding Tubular Fluorescent Lamps) (AMD 7434) December 15, 1992 (E). 34 pp.

BSI BS EN 60923-02. 1991 Amd 2 Performance Requirements for Ballasts for Discharge Lamps (Excluding Tubular Fluorescent Lamps) (AMD 7749) May 15, 1993. 35 pp.

CENELEC EN 60923-91. Ballasts for Discharge Lamps (Excluding Tubular Fluorescent Lamps)—Performance Requirements. 6 pp.

CNS C3043-75. Method of Test for Ballasts for High Pressure Mercury Vapor Lamps (Dec)(3890). 4 pp.

CNS C4069-79. Ballasts for High Pressure Mercury Vapour Lamps (Jan)(2729). 5 pp.

IEC 262-69. Ballasts for High Pressure Mercury Vapour Lamps Second Edition; (Amendment 1-1974) (Amendment 2-1976) (Amendment 3-1978). 83 pp.

IEC 923-88. Ballasts for Discharge Lamps (Excluding Tubular Fluorescent Lamps) Performance Requirements First Edition; (Corrigendum—April 1989) (Corrigendum—Oct 1989) (Amendment 1-1990). 44 pp.

JIS C 8110-87. Ballasts for High Pressure Mercury Vapor Lamps and Ballasts for Low Pressure Sodium Vapour Lamps. 21 pp.

SAA AS 1468-74. Ballasts for High Pressure Mercury Vapour and Low Pressure Sodium Vapour Discharge Lamps. 40 pp.

—Mercury Vapor Lamps—Ships
CNS F5121-87. Marine Ballasts for High Pressure Mercury Vapour Lamp (Nov)(12155).

Ballasts (Electric) *(Cont.)*

—Mercury Vapor Lamps—Ships *(Cont.)*
JIS F 8447-85. Marine Ballasts for High Pressure Mercury Vapour Lam

—Metal Halide Lamps
BSI BS EN 60923-91. 1991 Performance Requirements for Ballasts for Discharge Lamps (Excluding Tubular Fluorescent Lamps). 26 pp.

BSI BS EN 60923-01. 1991 Amd 1 Performance Requirements for Ballasts for Discharge Lamps (Excluding Tubular Fluorescent Lamps) (AMD 7434) December 15, 1992 (E). 34 pp.

BSI BS EN 60923-02. 1991 Amd 2 Performance Requirements for Ballasts for Discharge Lamps (Excluding Tubular Fluorescent Lamps) (AMD 7749) May 15, 1993. 35 pp.

CENELEC EN 60923-91. Ballasts for Discharge Lamps (Excluding Tubular Fluorescent Lamps)—Performance Requirements. 6 pp.

IEC 923-88. Ballasts for Discharge Lamps (Excluding Tubular Fluorescent Lamps) Performance Requirements First Edition; (Corrigendum—April 1989) (Corrigendum—Oct 1989) (Amendment 1-1990). 44 pp.

—Sodium Vapor Lamps
BSI BS EN 60923-91. 1991 Performance Requirements for Ballasts for Discharge Lamps (Excluding Tubular Fluorescent Lamps). 26 pp.

BSI BS EN 60923-01. 1991 Amd 1 Performance Requirements for Ballasts for Discharge Lamps (Excluding Tubular Fluorescent Lamps) (AMD 7434) December 15, 1992 (E). 34 pp.

BSI BS EN 60923-02. 1991 Amd 2 Performance Requirements for Ballasts for Discharge Lamps (Excluding Tubular Fluorescent Lamps) (AMD 7749) May 15, 1993. 35 pp.

CENELEC EN 60923-91. Ballasts for Discharge Lamps (Excluding Tubular Fluorescent Lamps)—Performance Requirements. 6 pp.

IEC 459-74. Ballasts for Low Pressure Sodium Vapour Lamps First Edition; (Amendment 1-1976) (Amendment 2-1978) (Amendment 3-1980). 64 pp.

IEC 923-88. Ballasts for Discharge Lamps (Excluding Tubular Fluorescent Lamps) Performance Requirements First Edition; (Corrigendum—April 1989) (Corrigendum—Oct 1989) (Amendment 1-1990). 44 pp.

JIS C 8110-87. Ballasts for High Pressure Mercury Vapor Lamps and Ballasts for Low Pressure Sodium Vapour Lamps. 21 pp.

SAA AS 1468-74. Ballasts for High Pressure Mercury Vapour and Low Pressure Sodium Vapour Discharge Lamps. 40 pp.

—Sodium Vapor Lamps—Branch Circuits
CSA CAN/CSA-C22. 2 NO 74-92. Equipment for Use with Electric Discharge Lamps (Incorporating Electrical Bulletin Nos. 523F, 753, 846, 1124A, 1125A, 1325, and 1326); (Gen Instr 1). 82 pp.

—Starters—Discharge Lamps
CSA CAN/CSA-C22. 2 NO 74-92. Equipment for Use with Electric Discharge Lamps (Incorporating Electrical Bulletin Nos. 523F, 753, 846, 1124A, 1125A, 1325, and 1326); (Gen Instr 1). 82 pp.

—Thermal Protectors
CSA CAN/CSA-C22. 2 NO 74-92. Equipment for Use with Electric Discharge Lamps (Incorporating Electrical Bulletin Nos. 523F, 753, 846, 1124A, 1125A, 1325, and 1326); (Gen Instr 1). 82 pp.

Ballasts (Weights)

—Lead—Aircraft
SBAC AS 4589 ISSUE 1. Ballast Weight 10lb. 1 p.
SBAC AS 4590 ISSUE 2. Ballast Weight 17.5lb. 1 p.
SBAC AS 8666 ISSUE 1. Ballast, Weight 5lb.

—Lead—Ships
MOD UK NES 168-90. Requirements for Polyvinyl Chloride (PVC) Coated Lead Ballast Blocks Issue 2 (07.90). 10 pp.

—Tractors
EC 88/410/EEC-88. Commission Directive Adapting to Technical Progress Council Directive 74/151/EEC on the Approximation of the Laws of the Member States Relating to Certain Components or Characteristics of Wheeled Agricultural or Forestry Tractors. 3 pp.

Ballistic Attack Testing
Use For: Bullet Attack Testing *See Also:* Testing

—Munitions
NATO STANAG 4241 ED 1 AMD 0-91. Bullet Attack Test for Munitions. 11 pp.

—Weapons Systems
NATO STANAG 4241 ED 1 AMD 0-91. Bullet Attack Test for Munitions. 11 pp.

Ballistic Missiles
—Glossaries
BSI BS 185: Sec 6-70. 1970 Amd 1 Glossary of Aeronautical and Astronautical Terms Section 6: Ballistic and Guided Missiles (Propulsion, Launching, Control and Guidance). 8 pp.

Ballistics
See Also: Weapons Systems
—Artillery—Meteorological Data
NATO STANAG 4061 ED 3 AMD 2-69. Adoption of a Standard Ballistic Meteorological Message. 27 pp.
—Design—Ammunition
NATO STANAG 4110 ED 2 AMD 0-91. Definition of Pressure Terms and Their Inter-Relationship for Use in the Design and Proof of Cannons and Ammunition. 18 pp.
—Design—Guns
NATO STANAG 4110 ED 2 AMD 0-91. Definition of Pressure Terms and Their Inter-Relationship for Use in the Design and Proof of Cannons and Ammunition. 18 pp.
—Design—Small Arms Ammunition
NATO STANAG 4110 ED 2 AMD 0-91. Definition of Pressure Terms and Their Inter-Relationship for Use in the Design and Proof of Cannons and Ammunition. 18 pp.
—Firing Table Format—Guns
NATO STANAG 4144 ED 1 AMD 1-84. Dynamic Firing Techniques to Determine Ballistic Data for Cannon Artillary Firing Tables and Associated Fire Control Equipment. 13 pp.
—Gun Propellants
NATO STANAG 4115 ED 1 AMD 0-70. Definition and Determination of Ballistic Properties of Gun Propellants. 16 pp.
—Meteorological Data
NATO STANAG 4103 ED 2 AMD 1-76. Format of Requests for Meteorological Messages for Ballistic and Special Purposes. 13 pp.
NATO STANAG 4103 ED 3 AMD 0-90. Format of Requests for Meteorological Messages for Ballistic and Special Purposes. 12 pp.
—Procedures—Projectiles
NATO STANAG 4106 ED 2 AMD 1-86. Procedure for the Statistical Comparison of External Ballistic Performance of Shell. 29 pp.
—Procedures—Small Arms Ammunition
NATO STANAG 4097 ED 1 AMD 3-66. Procedures with Respect to Charge Adjustment for Velocity for Standardized Gun Ammunition. 12 pp.
NATO STANAG 4098 ED 3 AMD 0-84. Procedures for the Determination of Propellant Charge Temperature Corrections. 11 pp.
—Thermochemical Values
NATO STANAG 4400 ED 1 AMD 0-00. (Draft) Derivation of Thermochemical Values for Interior Ballistics Calculations. 44 pp.
NATO STANAG 4400 ED 1 AMD 0-93. Derivation of Thermochemical Values for Interior Ballistic Calculation. 44 pp.
—Trajectory Models
NATO STANAG 4355 ED 1 AMD 0-92. Modified Point Mass Trajectory Model. 28 pp.
NATO STANAG 4355 ED 2 AMD 0-92. (Draft) Modified Point Mass Trajectory Model. 40 pp.
—Trajectory Models—Thermodynamic
NATO STANAG 4367 ED 1 AMD 0-92. Thermodynamic Interior Ballistic Model with Global Parameters. 30 pp.

Balloons
Use For: Hot Air Balloons See Also: Aircraft; Cameron DG 14 Balloons; Cameron Skystar Balloons
BSI 3F 57-75. (WITHDRAWN) 1975 Scoured Cotton Fabric for Inflatable Equipments (Superseded by BS 4F 57: 1990). 3 pp.
BSI 4F 57-90. 1990 Scoured Cotten Fabric for Inflatable Equipment for Aerospace Purposes. 4 pp.
—Hydrogen Generators
MOD UK TS 10188. Methanol/Water Hydrogen Generators AL-40 (Superseded by Def Stan 68-129).
NATO STANAG 4168 ED 1 AMD 1-85. Characteristics of Hydrogen Generating Equipment. 14 pp.
—Mandatory Aircraft Modifications and Inspections Summaries
CAA. Hot Air Balloons and Hot-Air Airships. 2 pp.

Balls (Sports Equipment)
See Also: Golf Balls
CNS S1152-82. Athletic Balls (Dec)(9748). 4 pp.
CNS S2092-82. Method of Test for Athletic Balls (Dec)(9749). 1 p.
—Rubber
CNS S1143-82. Sporting Rubber Balls (Jun)(9038). 3 pp.
CNS S2075-82. Method of Test for Sporting Rubber Ball (Jun)(9039). 1 p.

Ballscrews
Use: Ball Screws

Balsam
—Canada
CNS K7635-88. Chemical Reagent (Canada Balsam) (Oct)(9292).

Bambermycins Content Analysis
Use For: Flavomycin Content Analysis
—Animal Feed
CNS N4131-84. Method of Test for Feed Additives: Determination of Flavomycin (Nov)(11137).

Bamboo
Use For: Ma Chu Sprouts
—Poles (Supports)—Preservation
CNS O1019-85. Preservative Treatment of Bamboo (by Pressure) (May)(3219). 4 pp.
—Shoots—Canned
CNS N5019-89. Canned Bamboo Shoots (Feb)(1253). 6 pp.
—Shoots—Grading
CNS N1043-63. Grades of Ma-Chu Sprout (for Export) (Mar) (2128)(R 1973). 2 pp.
CNS N1048-63. Grades of Winter Bamboo Shoots (for Export) (Mar)(2133)(R 1973). 2 pp.
—Shoots—Pickled
CNS N5176-84. Brine-Cured Bamboo Shoots (Jul)(7069). 2 pp.

Banana Oil
Use: Amyl Acetate

Bananas
See Also: Food; Fruits
CNS N1086-73. Bananas (Packed in Carton) (for Export) (Sep)(2951). 3 pp.
—Grading
CNS N1091-72. Grades and Packing of Bananas (Packed in Carton) (Oct)(3496). 4 pp.
—Marine Transportation
ISO 931-80. Green Bananas—Guide to Storage and Transport First Edition. 5 pp.
—Ripening
ISO 3959-77. Green Bananas—Ripening Conditions First Edition. 8 pp.
—Storage
ISO 931-80. Green Bananas—Guide to Storage and Transport First Edition. 5 pp.

Band Clamps
See Also: Clamps
—Aircraft—Worms (Mechanical Drives)
BSI 2SP 91-92-73. 1973 Amd 3 Carbon Steel Worm Drive Clamps for Aircraft. 17 pp.

Band Dryers
Use: Dryers

Band Elimination Filters
Use: Band Reject Filters

Band Pass Filters
Use: Bandpass Filters

Band Reject Filters
Use For: Band Elimination Filters; Band Stop Filters
See Also: Electric Filters
—Circuit Noise—Measurement
CCITT RECMN G.228-89. Measurement of Circuit Noise in Cable Systems Using a Uniform-Spectrum Random Noise Loading—International Analogue Carrier Systems (Study Group XV) 11 pp. 11 pp.
—Noise—Measurement
CCITT RECMN G.230-89. Measuring Methods for Noise Produced by Modulating Equipment and Through-Connection Filters—International Analogue Carrier Systems (Study Group XV) 4 pp. 4 pp.

Band Resaws
Use: Band Saws

Band Saw Blades
See Also: Band Saws; Saw Blades
CGSB 46-GP-3-61. Blades; Band Saw, Spring Tempered. 6 pp.
—High Carbon Steel
CGSB 46-GP-2-61. Blades; Band Saw, Flexible Back. 6 pp.
—Metal Cutting
BSI BS 3877-91. 1991 Metal Cutting Bandsaw Blades. 19 pp.
BSI BS 3877-79. 1979 Specification for Metal Cutting Bandsaw Blades. 14 pp.
ISO 4875 Pt 1-78. Metal Cutting Band Saw Blades—Part 1: Definitions and Terminology First Edition. 10 pp.
ISO 4875 Pt 2-78. Metal Cutting Band Saw Blades—Part 2: Basic Dimensions and Tolerances First Edition. 4 pp.
ISO 4875 Pt 3-78. Metal Cutting Band Saw Blades—Part 3: Characteristics Relating to Each Type of Blade First Edition. 8 pp.
—Woodworking
BSI BS 4411-69. 1969 Wood Cutting Bandsaw Blades. 21 pp.
CNS B2695-82. Saw Blades for Woodworking; Technical Conditions of Delivery for Bandsaw Blades (Aug)(9224).
CNS B3297-85. Band Saw Blades for Woodworking Machines (Oct)(4969).
DIN ENGL 5134 Pt 1-75. Saw Blades for Woodworking; Technical Conditions of Delivery for Bandsaw Blades (Aug). 2 pp.
ISO 3295-75. Narrow Bandsaw Blades for Woodworking—Dimensions First Edition. 3 pp.
JIS B 4803-73. Wood Band Saw Blades.
—Woodworking—Sharpeners—Glossaries
BSI BS 4361: Part 26-90. 1990 Woodworking Machines Part 26: Nomenclature for Bandsaw Blade Sharpening Machines. 8 pp.
ISO 9267-88. Woodworking Machines—Bandsaw Blade Sharpening Machines—Nomenclature First Edition. 7 pp.

Band Saws
See Also: Band Saw Blades; Cutting Tools; Machine Tools; Saws; Woodworking Equipment
EC 89/392/EEC-89. Council Directive on the Approximation of the Laws of the Member States Relating to Machinery (NEW APP) (for Rel Stds See SM 1) (for Rel Stds See TEST/89/392, TEST/89/392 A1, TEST/89/392 A2). 24 pp.
EC TEST/89/392-89. Commission Communication in the Framework of the Implementation of Council Directive 89/392/EEC of 14 June 1989 in Relation to Machinery, as Amended by Directive 91/368/EEC of 20 June 1991. 3 pp.
EC TEST/89/392 A1-89. AMD 1 Commission Communication in the Framework of the Implementation of Council Directive 89/392/EEC of 14 June 1989 in Relation to Machinery, as Amended by Council Directive 91/368/EEC of 20 June 1991. 1 p.
EC TEST/89/392 A2-89. AMD 2 Commission Communication in the Framework of the Implementation of Council Directive 89/392/EEC of 14 June 1989 in Relation to Machinery, as Amended by Council Directive 91/368/EEC of 20 June 1991. 1 p.
EC TEST/89/398-89. List of Competent Authorities of the Member States Within the Meaning of Article 9 of Council Directive 89/398/EEC on the Approximation of the Laws of the Member States Relating to Foodstuffs Intended for Foods for Particular Nutritional Uses. 2 pp.
EC 91/368/EEC-91. Council Directive Amending Directive 89/392/EEC on the Approximation of the Laws of the Member States Relating to Machinery (NEW APP) (for Rel Stds See SM 1). 18 pp.
EC 93/68/EEC-93. Council Directive Amending Directives 87/404/EEC (Simple Pressure Vessels), 88/378/EEC (Safety of Toys), 89/106/EEC (Construction Products), 89/336/EEC (Electromagnetic Compatibility), 89/392/EEC (Machinery), 89/686/ EEC (Personal Protective. 22 pp.
—Accuracy Testing
CNS B7154-86. Test Code for Accuracy of Feed Carriages of Band Saw Mills (Feb)(6305).
—Glossaries
ISO 4875 Pt 1-78. Metal Cutting Band Saw Blades—Part 1: Definitions and Terminology First Edition. 10 pp.
—Inspection
CNS B7153-80. Running Inspection Method of Band Saw Mills and Their Feed Carriages (Aug)(6304).

Band Saws (Cont.)

—Meat Cutting
CGSB CAN/CGSB-52.29-M87. Saw, Band, Meat Cutting, Electric. 12 pp.

—Metal Cutting
CNS B7116-82. Accuracy Inspection of Metal Cutting Vertical Band Sawing Machines (Dec)(4896). 3 pp.
CNS B7117-82. Accuracy Inspection of Metal Cutting Horizontal Band Sawing Machines (Dec)(4897). 3 pp.

—Safety
IEC 1029 Pt 2-5-93. Safety of Transportable Motor-Operated Electric Tools Part 2: Particular Requirements for Band Saws First Edition. 24 pp.
JIS B 6605-83. Safety Standards for Construction of Table Band Resaws. 10 pp.
JIS B 6606-83. Safety Standards for Construction of Band Resaws with Feed Rollers. 11 pp.
JIS B 6607-83. Safety Standards for Construction of Band Saw Machines with Feed Carriages. 13 pp.

—Scroll—Woodworking
JIS B 6509-90. Test Methods for Performance and Accuracy of Band Saw Machines and Feed Equipment. 18 pp.

—Scroll—Woodworking—Accuracy Testing
JIS B 6519-90. Test Methods for Performance and Accuracy of Band Scroll Saws. 11 pp.

—Scroll—Woodworking—Rigidity
JIS B 6519-90. Test Methods for Performance and Accuracy of Band Scroll Saws. 11 pp.

—Sharpeners
CNS B7160-80. Band Saw Sharpeners (Dec)(6739).
JIS B 6556-90. Test Methods for Performance and Accuracy of Bandsaw Sharpening Machines. 11 pp.

—Stretchers
CNS B7158-80. Method of Test for Band Saw Roll Stretcher (Dec)(6737).
JIS B 6555-90. Test Methods for Performance and Accuracy of Band Saw Stretcher. 10 pp.

—Woodworking
BSI BS 4361: Part 4-88. 1988 Woodworking Machines Part 4: Table Bandsawing Machines. 12 pp.
BSI BS 6854: Part 5-89. 1989 Code of Practice for Safeguarding Woodworking Machines Part 5: Narrow Band Sawing Machines. 28 pp.
ISO 7007-83. Woodworking Machines—Table Bandsawing Machines—Nomenclature and Acceptance Conditions First Edition. 11 pp.

Band Sharing
Use: Frequency Band Sharing

Band Stop Filters
Use: Band Reject Filters

Bandpass Filters
Scope Note: For additional listings, use a more specific term *See Also:* Crystal Filters; Dielectric Filters; Microwave Filters; Radio Frequency Filters; Through Connection Filters; Through Group Filters; Through Mastergroup Filters; Through Supergroup Filters; Through Supermastergroup Filters; Through 15 Supergroup Assembly Filters

BSI BS 2475-64. 1964 Octave and One Third Octave Band-Pass Filters. 13 pp.
IEC 225-66. Octave, Half-Octave and Third-Octave Band Filters Intended for the Analysis of Sounds and Vibrations First Edition. 17 pp.
SNZ NZS 1499-65. Specification for Octave and One-Third Octave Band-Pass Filters Amend: 1, 1974. 12 pp.

—Circuit Noise—Measurement
CCITT RECMN G.228-89. Measurement of Circuit Noise in Cable Systems Using a Uniform-Spectrum Random Noise Loading—International Analogue Carrier Systems (Study Group XV) 11 pp. 11 pp.

—Noise—Measurement
CCITT RECMN G.230-89. Measuring Methods for Noise Produced by Modulating Equipment and Through-Connection Filters—International Analogue Carrier Systems (Study Group XV) 4 pp. 4 pp.
SAA AS Z41-69. Octave, Half Octave and One-Third Octave Band Pass Filters Intended for the Analysis of Sound and Vibrations. 15 pp.

Bands
See Also: Anchors (Fasteners); Armbands; Clamps; Clips; Fasteners; Holders; Packaging Materials; Retaining Rings; Rubber Bands; Straps (Fasteners); Wrist Bands

—Copper
DIN ENGL 46415-58. Copper and Copper Alloy Band; Cold Rolled with Rounded (Lightly Rolled) Edges; Dimensions (Oct). 2 pp.

—Copper Alloys
DIN ENGL 46415-58. Copper and Copper Alloy Band; Cold Rolled with Rounded (Lightly Rolled) Edges; Dimensions (Oct). 2 pp.

—Glass Fabric—Aircraft—Direction Indicator—Pipes
SBAC AGS 2094 ISSUE 4. Flow Direction Indicator, Bands for Use in Hot Zones.

—Glass Fabric—Aircraft—Identification Systems—Pipes
SBAC AGS 2088 ISSUE 4. Pipe Identification Band for Use in Hot Zones (1 1/4 Inch Wide).
SBAC AGS 2090 ISSUE 4. Pipe Identification Band for Use in Hot Zones (1.0 Inches Wide).
SBAC AGS 2092 ISSUE 4. Pipe Identification Band for Use in Hot Zones (3/4 Inches Wide).
SBAC AGS 3441 ISSUE 2. Pipe Identification Band for Use in Hot Zones (1/2 Inch Wide).

—Glass Fabric—Aircraft—Symbols—Pipes
SBAC AGS 3442 ISSUE 2. Pipe Warning Symbol, Bands for Use in Hot Zones (1/2 Inch Wide).

—Ignition Coils
CNS D2022-1-86. Ignition Coils Fixing Band for Automobiles (Nov)(5561-1).

—Paper String
JIS Z 1517-76. Paper String Bands (R 1985). 6 pp.

—Polypropylene
CNS Z5037-81. Polypropylene Band (Aug)(3465).
JIS Z 1527-76. Polypropylene Band (R 1980). 6 pp.

—Steel—Aircraft—Identification Systems—Clips
SBAC AGS 2089 ISSUE 4. Clip Identification Band.
SBAC AGS 2091 ISSUE 4. Clip Identification Band.
SBAC AGS 2093 ISSUE 4. Clip—Identification Band.
SBAC AGS 2095 ISSUE 4. Clip—Identification Band.
SBAC AGS 2114-17 ISSUE 4. Clip Identification Band.

Bandwidth
See Also: Frequencies; Out of Band Emissions; Radio Frequency Bandwidth; Resonance Frequency; Video Bandwidth

—Earth Exploration Satellites
CCIR RECMN 514-1-90. Telecommunication Links for Earth Exploration Satellites Frequencies, Bandwidths and Criteria for Protection from Interference—Section 2F—Earth Exploration Satellites. 2 pp.

—Expansion—Spread Spectrum Modulation
CCIR Report 651-2-86. Spread Spectrum Techniques—Section 1D—Spectrum Utilization and Applications. 13 pp.
CCIR QUESTION 71/1-90. Bandwidth Expansion Techniques and Spectrum Sharing—Questions Concerning Study Group 1 —Spectrum Management Techniques (Spectrum Engineering, Planning, Sharing, Monitoring and Utilization). 1 p.

—Fixed Services (Radio Communications)
CCIR QUESTION 143/9-90. Signal-to-Noise Ratios and Protection Ratios; Bandwidth, Adjacent Channel Spacing and Frequency Stability—Questions of Study Group 9 Fixed Service. 3 pp.

—Glossaries
CCIR RECMN 573-3-90. Radiocommunication Vocabulary—Section A—Terminology. 60 pp.
CCIR RECMN 662-1-90. Terms and Definitions—Section A—Terminology. 19 pp.

—Land Mobile Services
CCIR Decision 69-2-89. Future Public Land Mobile Telecommunication Systems (FPLMTS)—Annex 1 to Volume VIII—Land Mobile Service—Amateur Service—Amateur-Satellite Service. 4 pp.

—Maritime Mobile Satellite Communications—Automatic Services
CCITT RECMN G.473-89. Interconnection of a Maritime Mobile Satellite System with the International Automatic Switched Telephone Service; Transmission Aspects—International Analogue Carrier Systems (Study Group XV) 10 pp. 10 pp.

—Multimode Distortion—Fiber Optic Cables
CCITT RECMN G.651-89. Characteristics of a 50/125 Micrometer Multimode Graded Index Optical Fibre Cable—Transmission Media Characteristics (Study Group XV) 30 pp. 30 pp.

—Radar Spectra
CCIR Report 837-82. Methods for Calculating Pulsed Radar Emission Spectrum Bandwidth—Section 1A—Spectrum Engineering and Computer-Aided Principles and Techniques. 11 pp.

—Radio Altimeters
CCIR QUESTION 94/8-90. Necessary Bandwidth Required for Radio Altimeters Operating in the Band 4200-4400 MHz—Questions Concerning Study Group 8—Mobile, Radiodetermination, Amateur and Related Satellite Services. 1 p.

—Radio Determination Satellite Services
CCIR Report 1050-1-90. Technical and Operational Considerations for a Radiodetermination Satellite Service in Bands 9 and 10—Section 8J—Technical and Oper-ating Characteristics of Radiocommunications Using Satellite Distress and Safety Operation and of Radio Determination Satellite Services. 31 pp.

—Radio Equipment
CCIR RECMN 339-6-86. Bandwidths, Signal-to-Noise Ratios and Fading Allowances in Complete Systems—Section 3Aa—Technical Characteristics. 3 pp.

—Radio Frequency Emissions
CCIR RECMN 328-7-90. Spectra and Bandwidth of Emissions—Section 1A—Spectrum Engineering and Computer-Aided Principles and Techniques. 16 pp.
CCIR Report 324-3-90. Methods of Measuring the Bandwidths of Emissions—Section 1A—Spectrum Engineering and Computer-Aided Principles and Techniques. 5 pp.
CCIR Report 836-2-90. Necessary Bandwidth Calculations—Section 1A—Spectrum Engineering and Computer-Aided Principles and Techniques. 7 pp.
CCIR Report 977-1-90. Spectra and Bandwidth of Emissions—Section 1A—Spectrum Engineering and Computer-Aided Principles and Techniques. 25 pp.
CCIR QUESTION 60/1-82. Spectra and Bandwidths of Emissions—Questions Concerning Study Group 1 —Spectrum Management Techniques (Spectrum Engineering, Planning, Sharing, Monitoring and Utilization). 1 p.
CCIR QUESTION 76/1-90. Spectra and Bandwidths of Emissions—Questions Concerning Study Group 1 —Spectrum Management Techniques (Spectrum Engineering, Planning, Sharing, Monitoring and Utilization). 1 p.
CCIR RECMN 853-92. Necessary Bandwidth. 8 pp.

—Radio Frequency Emissions—Formulas
CCIR QUESTION 77/1-90. Formulae and Examples for the Calculation of Necessary Bandwidths—Questions Concerning Study Group 1—Spectrum Management Techniques (Spectrum Engineering, Planning, Sharing, Monitoring and Utilization). 1 p.

—Radio Spectra—Monitoring
CCIR RECMN 443-1-78. Bandwidth Measurements at Monitoring Stations—Section 1C—Spectrum Monitoring Techniques. 1 p.
CCIR Report 275-5-90. Bandwidth Measurement at Monitoring Stations—Section 1C—Spectrum Monitoring Techniques. 9 pp.
CCIR Report 668-3-90. Automatic Monitoring and Measurement in the Radio-Frequency Spectrum—Section 1C—Spectrum Monitoring Techniques. 17 pp.
CCIR QUESTION 26-2/1-90. Bandwidth Measurements at Monitoring Stations—Questions Concerning Study Group 1—Spectrum Management Techniques (Spectrum Engineering, Planning, Sharing, Monitoring and Utilization). 1 p.

—Radio Transmitters
CENELEC HD 236.2-75. Methods of Measurement for Radio Transmitters Part 2: Bandwidth, Out-of-Band Power and Power of Non-Essential Oscillations. 2 pp.
IEC 244 Pt 2-69. Methods of Measurement for Radio Transmitters Part 2: Bandwidth, out-of-Band Power and Power of Non-Essential Oscillations First Edition; (Supplement A-1969) (Amendment 1-1973 to Supplement A) (Amendment 1-1974). 149 pp.
SAA AS 1174.2-71. Methods of Measurement for Radio Transmitters—Part 2: Bandwidth, Out-of-Band Power and Power of Non-Essential Oscillations Being IEC Publications 244-2 (1969), 244-2A (1969) and 244-2B (1969) Endor-sed as the Australian Standard Without Australian Amendment. 189 pp.

INTERNATIONAL AND NON-U.S. NATIONAL STANDARDS SUBJECT INDEX

Bandwidth (Cont.)

—Radio Transmitters—Single Sideband
BSI BS 6160: Part 4-81. 1981 Measurement for Radio the Equipment Used in Mobile Services Part 4: Transmitters Emloying Single Sideband Emissions (A3A, A3H, or A3J). 34 pp.

IEC 489 Pt 4-91. Methods of Measurement for Radio Equipment Used in the Mobile Services Part 4: Transmitters Employing Single-Sideband Emissions (R3E, H3E or J3E) Second Edition. 113 pp.

—Radiotelegraph Systems—Maritime Mobile Services
CCIR RECMN 627-86. Technical Characteristics for HF Maritime Radio Equipment Using Narrow-Band Phase-Shift Keying (NBPSK) Telegraphy—Section 8B—Maritime Mobile Service; Telegraphy and Related Subjects. 2 pp.

—Satellite Communication Equipment
CCIR RECMN 363-4-90. Space Operation Systems Frequencies, Bandwidths and Protection Criteria—Section 2C—Space Operations. 2 pp.

—Satellite Communications—Formulas
CCIR QUESTION 77/1-90. Formulae and Examples for the Calculation of Necessary Bandwidths—Questions Concerning Study Group 1—Spectrum Management Techniques (Spectrum Engineering, Planning, Sharing, Monitoring and Utilization). 1 p.

—Sound Broadcasting
CCIR RECMN 639-86. Necessary Bandwidth of Emission in LF, MF and HF Broadcasting—Section 10A-1—Amplitude-Modulation Sound Broadcasting in Bands 5 (LF), 6 (MF) and 7 (HF). 7 pp.

—Space Operation Services
CCIR Report 845-1-86. Space Operation Systems Frequencies, Bandwidths and Protection Criteria—Section 2C—Space Operations. 12 pp.

—Space Operation Systems
CCIR RECMN 363-4-90. Space Operation Systems Frequencies, Bandwidths and Protection Criteria—Section 2C—Space Operations. 2 pp.

—Telegraph Receivers
CCIR RECMN 338-2-70. Bandwidth Required at the Output of a Telegraph or Telephone Receiver—Section 3Aa—Technical Characteristics. 1 p.

—Telegraphy Systems
CCIR Report 195-63. Prediction of the Performance of Telegraph Systems in Terms of Bandwidth and Signal-to-Noise Ratio in Complete Systems—Section 3Cd—Performance of Digital Transmission Systems. 8 pp.

—Telephone Receivers
CCIR RECMN 338-2-70. Bandwidth Required at the Output of a Telegraph or Telephone Receiver—Section 3Aa—Technical Characteristics. 1 p.

—Television Signals—Encoding
CCIR Report 962-2-90. Filtering, Sampling and Multiplexing for Digital Encoding of Colour Television Signals—Section 11F—Digital Methods of Transmitting Television Information. 10 pp.

Banjo Bolts
See Also: Bolts

—Adapters—Aircraft
SBAC AS 6758 (V). Pipe Couplings (Unified Adaptor) Steel Banjo Bolt with Hole for Bleeder Screw.

SBAC AS 6759 (V). Pipe Couplings (Unified Adaptor) Aluminium Alloy Banjo Bolt with Hole for Bleeder Screw.

—Cone Banjos—Aircraft
SBAC AS 6727 (V). Single End Cone Banjo Body (3/16 Inch to 1/2 Inch) for Use with Banjo Bolts AS 6721-6.

SBAC AS 6728 (V). Double End Cone Banjo Body (3/16 Inch to 1/2 Inch) for Use with Banjo Bolts AS 6721-6.

SBAC AS 6729 (V). Double R.A. Cone Banjo Body (3/16 Inch to 1/2 Inch) for Use with Banjo Bolts AS 6721-6.

—Pipe Couplings—Aircraft
SBAC AS 6721 (V). Pipe Couplings (Unified Adaptor) Aluminium Alloy Banjo Bolt.

SBAC AS 6722 (V). Pipe Couplings (Unified Adaptor) Banjo Bolt Steel.

SBAC AS 6723 (V). Pipe Couplings (Unified Adaptor) Aluminium Alloy Banjo Bolt with Cone Head Connection.

SBAC AS 6724 (V). Pipe Couplings (Unified Adaptor) Steel Banjo Bolt with Cone Head Connection.

SBAC AS 6725 (V). Pipe Couplings (Unified Adaptor) Aluminium Alloy Banjo Bolt with Union Head Connection.

Banjo Bolts (Cont.)

—Pipe Couplings—Aircraft (Cont.)
SBAC AS 6726 (V). Pipe Couplings (Unified Adaptor) Steel, Banjo Bolt with Union Head Connection.

—Pipe Fittings—Aircraft
SBAC AGS 1135 (V). Pipe Couplings (for Aluminium Alloy Range) Banjo Bolt (3/16 to 1).

SBAC AGS 1136 (V). Pipe Couplings (for Aluminium Alloy Range) Banjo Bolt with Cone Head Connection.

SBAC AGS 1137 (V). Pipe Couplings (Aluminium Alloy Range) Banjo Bolt with Union Head Connection.

SBAC AGS 1173 (V). Pipe Couplings for Aluminium Alloy Range, Banjo Bolt 3/16" to 1" B.S.P.F. with Hole for Bleeder Screw.

SBAC AGS 1213 (V). Pipe Couplings (Aluminium Alloy Range) Aluminium Alloy Banjo Bolt.

SBAC AGS 1214 (V). Pipe Couplings (Aluminium Alloy Range) Aluminium Alloy Banjo Bolt with Cone Head Connection.

SBAC AGS 1220 (V). Pipe Couplings for Aluminium Alloy Range, Aluminium Alloy Banjo Bolt 3/16" to 1" B.S.P.F. with Hole for Bleeder Screw.

—Pitot Static Tubes—Aircraft
SBAC AS 8680 ISSUE 1. Banjo Bolt, Pitot, Pitot Static System.

SBAC AS 8682 ISSUE 1. Banjo Bolt, Pitot, Pitot Static System.

SBAC AS 8683 ISSUE 1. Banjo Bolt, Static, Pitot Static System.

—Ring Nipples
DIN ENGL 7644-85. Compression Couplings with Spherical Liner; Male Studed Banjo Bolts for Ring Type Nipples (Jan). 4 pp.

—Union Banjos—Aircraft
SBAC AS 6730 (V). Single End Union Banjo Body (3/16 Inch to 1 Inch) for Use with Banjo Bolts AS 6721-6.

SBAC AS 6731 (V). Double End Union Banjo Body (3/16 Inch to 1 Inch) for Use with Banjo Bolts AS 6721-6.

SBAC AS 6732 (V). Double R.A. Union Banjo Body (3/16 Inch to 1/2 Inch) for Use with Banjo Bolts AS 6721-6.

Banjo Fittings
See Also: Fittings

—Pitot Static Tubes—Aircraft
SBAC AS 8679 ISSUE 2. Arrangement of Banjo Fittings, Pitot Static Systems.

Bank Cards
Use For: Financial Transaction Cards; Transaction Cards *See Also:* Banking; Banking Documents; Cards

BSI BS 7108-90. (WITHDRAWN) 1990 Guide to Design and Use of Identification Cards as Financial Transaction Cards (Superseded by BS EN 27813: 1992). 11 pp.

BSI BS EN 27813-92. 1992 Identification Cards—Financial Transaction Cards (ISO/IEC 7813: 1990) (G). 12 pp.

CEN EN 27 813-89. Identification Cards Financial Transaction Cards. 9 pp.

CEN EN 27813-92. Identification Cards—Financial Transaction Cards. 9 pp.

CNS C5246-90. Identification Cards—Financial Transaction Cards (Dec)(12820).

IEC 7813-90. Identification Cards—Financial Transaction Cards Third Edition. 7 pp.

IEC DIS 7813-92. Identification Cards—Financial Transaction Cards (Revision of Third Edition (ISO 7813:1990)) ***CD-ROM ONLY***. 11 pp.

ISO 7813-90. Identification Cards—Financial Transaction Cards Third Edition. 7 pp.

ISO DIS 7813-92. Identification Cards—Financial Transaction Cards (Revision of Third Edition (ISO 7813:1990)) ***CD-ROM ONLY***. 11 pp.

JTC1 7813-90. Identification Cards—Financial Transaction Cards Third Edition. 7 pp.

JTC1 DIS7813-92. Identification Cards—Financial Transaction Cards (Revision of Third Edition (ISO 7813:1990)) ***CD-ROM ONLY***. 11 pp.

OSI ISO 7813-90. Identification Cards—Financial Transaction Cards. 7 pp.

—Integrated Circuit
BSI BS EN 29992-1-93. 1993 Amd 1 Financial Transaction Cards—Messages Between the Integrated Circuit Card and the Card Accepting Device Part 1: Concepts and Structures (ISO 9992-1: 1990) (AMD 7943) September 15, 1993 (S). 19 pp.

CEN EN 29992-1-93. Financial Transaction Cards—Messages Between the Integrated Circuit Card and the Card Accepting Device—Part 1: Concepts and Structures (ISO 9992-1: 1990). 11 pp.

Bank Cards (Cont.)

—Integrated Circuit (Cont.)
ISO 9992 Pt 1-90. Financial Transaction Cards—Messages Between the Integrated Circuit Card and the Card Accepting Device—Part 1: Concepts and Structures First Edition (CEN EN 29992-1: 1993). 9 pp.

—Integrated Circuit—Security
BSI BS EN 30202-1-93. 1993 Amd 1 Financial Transaction Cards—Security Architecture of Financial Transaction Systems Using Integrated Circuit Cards—Part 1: Card Life Cycle (ISO 10202-1: 1991) (AMD 7944) September 15, 1993 (S). 23 pp.

BSI BS ISO 10202-1-91. 1991 Financial Transaction Cards—Security Architecture of Financial Transaction Systems Using Integrated Circuit Cards—Part 1: Card Life Cycle (Renumbered as BS EN 30202-1: 1993). 15 pp.

CEN EN 30202-1-93. Financial Transaction Cards—Security Architecture of Financial Transaction Systems Using Integrated Circuit Cards Part 1: Card Life Cycle (ISO 10202-1: 1991). 15 pp.

ISO 10202 Pt 1-91. Financial Transaction Cards—Security Architecture of Financial Transaction Systems Using Integrated Circuit Cards—Part 1: Card Life Cycle First Edition (CEN EN 30202-1: 1993). 12 pp.

JTC1 10202 Pt 1-91. Financial Transaction Cards—Security Architecture of Financial Transaction Systems Using Integrated Circuit Cards—Part 1: Card Life Cycle First Edition. 12 pp.

OSI ISO DIS 10202-1-89. Financial Transaction Cards—Security Architectures of Financial Transaction Systems Using Integrated Circuit Cards—Part 1: Card Life Cycle. 19 pp.

—Magnetic
BSI BS 7096-90. 1990 Guide to Design and Use of Bank Cards with a Magnetic Stripe That Employs Track 3. 17 pp.

CEN EN 24 909-90. Bank Cards Magnetic Stripe Data Content for Track 3. 3 pp.

ISO 4909-87. Bank Cards—Magnetic Stripe Data Content for Track 3 Second Edition. 13 pp.

JTC1 4909-87. Bank Cards—Magnetic Stripe Data Content for Track 3 Second Edition. 13 pp.

OSI ISO 4909-87. Bankcards—Magnetic Stripe Data Content for Track 3. 15 pp.

—Magnetic—Prepaid
JIS X 6311-90. Prepaid Cards—Physical Characteristics and Dimensions. 7 pp.

JIS X 6313-91. Prepaid Cards—Magnetic Recording Characteristics and Information Interchange. 10 pp.

JIS X 6314-92. Prepaid Cards—Magnetic Recording Format. 12 pp.

—Magnetic—Prepaid—Identification Systems
JIS X 6312-91. Prepaid Cards—Required Marking. 5 pp.

—Message Interchange Format
BSI BS 7100-90. (WITHDRAWN) 1990 Method of Formatting Electronic Messages Relating to Financial Transactions for Exchange Between Financial Institutions (Superseded by BS EN 28583: 1991). 38 pp.

BSI BS EN 28583-91. 1991 Bank Card Originated Messages—Interchange Message Specifications—Content for Financial Transactions (ISO 8583: 1987). 42 pp.

BSI BS ISO 9992-1-90. 1990 Concepts and Structures. 11 pp.

CEN EN 28 583-91. Bank Card Originated Messages Message Specifications—Content for Financial Transactions. 3 pp.

ISO 7580-87. Identification Cards—Card Originated Messages—Content for Financial Transactions First Edition. 16 pp.

ISO 8583-87. Bank Card Originated Messages—Interchange Message Specifications—Content for Financial Transactions First Edition. 37 pp.

JTC1 7580-87. Identification Cards—Card Originated Messages—Content for Financial Transactions First Edition. 16 pp.

JTC1 8583-87. Bank Card Originated Messages—Interchange Message Specifications—Content for Financial Transactions First Edition. 37 pp.

OSI ISO 8583-87. Bank Card Originated Messages—Interchange Message Specifications—Content for Financial Transactions. 37 pp.

Banking
Use For: Banks *See Also:* Accounting; Bank Cards; Banking Documents; Capital Movements; Credit Institutions; Currencies; Electronic Funds Transfer; Financial Institutions; Securities; Telecommunication Equipment

OSI ISO 8732-88. Banking—Key Management (Wholesale). 84 pp.

Banking (Cont.)

OSI ISO DIS 8732-87. Banking—Key Management (Wholesale). 146 pp.
OSI ISO DIS 8732-87. Banking—Key Management (Wholesale). 146 pp.

—Annual Accounts—Information Disclosure
EC 86/635/EEC-86. Council Directive on the Annual Accounts and Consolidated Accounts of Banks and Other Financial Institutions. 17 pp.

—Consolidated Accounts—Information Disclosure
EC 86/635/EEC-86. Council Directive on the Annual Accounts and Consolidated Accounts of Banks and Other Financial Institutions. 17 pp.

—Data Transmission—Bank Identifier Codes
BSI BS 7104-90. 1990 Procedure for Allocating a Bank Identifier Code (BIC). 11 pp.
ISO 9362-87. Banking—Banking Telecommunication Messages—Bank Indentifier Codes First Edition. 7 pp.
ISO 10383-92. Codes for Exchanges and Regulated Markets—Market Identifier Codes (MIC) First Edition. 6 pp.
JTC1 9362-87. Banking—Banking Telecommunication Messages—Bank Indentifier Codes First Edition. 7 pp.
OSI ISO 9362-87. Banking—Banking Telecommunication Messages—Banker Indentifier Codes. 5 pp.

—Data Transmission—Data Encryption
BSI BS ISO 10126-1-91. 1991 Banking—Procedures for Message Encipherment (Wholesale) Part 1: General Principles. 26 pp.
BSI BS ISO 10126-2-91. 1991 Banking—Procedures for Message Encipherment (Wholesale)—Part 2: DEA Algorithm (G). 10 pp.
ISO 10126 Pt 1-91. Banking—Procedures for Message Encipherment (Wholesale)—Part 1: General Principles First Edition. 23 pp.
ISO 10126 Pt 2-91. Banking—Procedures for Message Encipherment (Wholesale)—Part 2: DEA Algorithm First Edition. 7 pp.
JTC1 10126 Pt 1-91. Banking—Procedures for Message Encipherment (Wholesale)—Part 1: General Principles First Edition. 23 pp.
JTC1 10126 Pt 2-91. Banking—Procedures for Message Encipherment (Wholesale)—Part 2: DEA Algorithm First Edition. 7 pp.

—Data Transmission—Funds Transfer
BSI BS 7099: Part 1-89. (WITHDRAWN) 1989 Procedures for Telecommunication of Funds Transfer Messages Between Banks Part 1: Vocabulary and Data Elements (Superseded by BS EN 27982-1: 1991). 23 pp.
BSI BS EN 27982: Part 1-91. 1991 Bank Telecommunication—Funds Transfer Messages Part 1: Vocabulary and Data Elements (ISO 7982-1: 1987). 28 pp.
CEN EN 27 982 (Part 1)-91. Bank Telecommunication Funds Transfer Messages — Part 1 Vocabulary and Data Elements. 2 pp.
CEPT T/SF 57 E-87. Electronic Funds Transfer Telecommunications Services. 17 pp.
ISO 7982 Pt 1-87. Bank Telecommunication—Funds Transfer Messages—Part 1: Vocabulary and Data Elements First Edition. 23 pp.
JTC1 7982 Pt 1-87. Bank Telecommunication—Funds Transfer Messages—Part 1: Vocabulary and Data Elements First Edition. 23 pp.

—Data Transmission—Glossaries
OSI ISO/DIS 8908. Banking and Related Financial Services—Vocabulary and Data Elements. 135 pp.

—Data Transmission—Key Management
BSI BS 7205-90. 1990 Procedures for Key Management to Achieve Security for Financial Institutions Engaged in Financial Transactions (Wholesale). 86 pp.
ISO 8732-88. Banking—Key Management (Wholesale) First Edition. 84 pp.

—Data Transmission—Market Identifier Codes
ISO 10383-92. Codes for Exchanges and Regulated Markets—Market Identifier Codes (MIC) First Edition. 6 pp.

—Data Transmission—Message Authentication Codes
BSI BS 7101-89. (WITHDRAWN) 1989 Procedures for Protecting Authentic Wholesale Messages Between Financial Institutions (Superseded by BS ISO 8730: 1990). 11 pp.
BSI BS 7102: Part 1-89. 1989 Recommendations for Algorithms for Use in Banking Message Authentication Part 1: Data Encryption Algorithm (DEA). 6 pp.

Banking (Cont.)

—Data Transmission—Message Authentication Codes (Cont.)
BSI BS 7102: Part 2-89. (WITHDRAWN) 1989 Recommendations for Algorithms for Use in Banking Message Authentication Part 2: Message Authenticator Algorithm (MAA) (Superseded by BS ISO 8731-2: 1992). 12 pp.
BSI BS ISO 8730-90. 1990 Banking—Requirements for Message Authentication (Wholesale). 37 pp.
BSI BS ISO 9807-91. 1991 Banking and Related Financial Services—Requirements for Message Authentication (Retail) (G). 19 pp.
ISO 8730-90. Banking—Requirements for Message Authentication (Wholesale) Second Edition. 31 pp.
ISO 8731 Pt 1-87. Banking—Approved Algorithms for Message Authentication—Part 1: DEA First Edition. 4 pp.
ISO 9807-91. Banking and Related Financial Services—Requirements for Message Authentication (Retail) First Edition. 16 pp.
JTC1 8730-90. Banking—Requirements for Message Authentication (Wholesale) Second Edition. 31 pp.
JTC1 8731 Pt 1-87. Banking—Approved Algorithms for Message Authentication—Part 1: DEA First Edition. 4 pp.
JTC1 9807-91. Banking and Related Financial Services—Requirements for Message Authentication (Retail) First Edition. 16 pp.
OSI ISO 8730-90. Banking—Requirements for Message Authentication (Wholesale). 31 pp.
OSI ISO 8730-86. Banking—Requirements for Message Authentication (Wholesale). 9 pp.
OSI ISO 8730-86. Banking—Requirements for Message Authentication (Wholesale).

—Data Transmission—Message Authentication Codes—Algorithms
BSI BS ISO 9807-91. 1991 Banking and Related Financial Services—Requirements for Message Authentication (Retail) (G). 19 pp.
ISO 9807-91. Banking and Related Financial Services—Requirements for Message Authentication (Retail) First Edition. 16 pp.
JTC1 9807-91. Banking and Related Financial Services—Requirements for Message Authentication (Retail) First Edition. 16 pp.

—Data Transmission—Personal Identification Numbers
BSI BS ISO 9564-1-91. 1991 Banking—Personal Identification Number Management and Security—Part 1: PIN Protection Principles and Techniques (G). 38 pp.
ISO 9564 Pt 1-91. Banking—Personal Identification Number Management and Security—Part 1: PIN Protection Principles and Techniques First Edition. 35 pp.
JTC1 9564 Pt 1-91. Banking—Personal Identification Number Management and Security—Part 1: PIN Protection Principles and Techniques First Edition. 35 pp.

—Data Transmission—Personal Identification Numbers—Algorithms
BSI BS ISO 9564-2-91. 1991 Banking—Personal Identification Number Management and Security—Part 2: Approved Algorithm(s) for PIN Encipherment (G). 7 pp.
ISO 9564 Pt 2-91. Banking—Personal Identification Number Management and Security—Part 2: Approved Algorithm(s) for PIN Encipherment First Edition. 5 pp.
JTC1 9564 Pt 2-91. Banking—Personal Identification Number Management and Security—Part 2: Approved Algorithm(s) for PIN Encipherment First Edition. 5 pp.

—Data Transmission—Sign-On Authentication
ISO 11131-92. Banking and Related Financial Services—Sign-on Authentication First Edition. 15 pp.

—Data Transmission—TELEX
BSI BS 7098-89. 1989 Method of Formatting Inter-Bank Telexes for Messages Relating to the Transfer of Funds or Other Financial Messages. 55 pp.
ISO 7746-88. Banking—Telex Formats for Inter-Bank Messages First Edition. 52 pp.
JTC1 7746-88. Banking—Telex Formats for Inter-Bank Messages First Edition. 52 pp.
OSI ISO 7746-88. Banking—Telex Formats for Interbank Messages. 52 pp.

—Open Systems Interconnection
OSI ISO DP 9955-N205-87. Information Interchange—Methodology and Guidelines for the Development of Application Protocols for Banking Information Interchange. 145 pp.

Banking (Cont.)

—Open Systems Interconnection (Cont.)
OSI ISO DP 9955-N205-87. Information Interchange—Methodology and Guidelines for the Development of Application Protocols for Banking Information Interchange. 145 pp.

Banking Documents
See Also: Bank Cards; Banking; Checks; Securities

—Account Statements
BSI BS 6695-86. 1986 Nostro Accounts Reconciliation in Banking. 14 pp.
ISO 7341-85. Banking—Nostro Accounts Reconciliation First Edition. 13 pp.
OSI ISO 7341-85. Banking—Nostro Accounts Reconciliation. 13 pp.

—Drawing Lists
ISO 6536-81. Bank Operations—Standard Scheme for Drawing Lists First Edition; (Erratum—Jan 1982). 11 pp.
OSI ISO 6536-81. Bank Operations—Standard Scheme for Drawing Lists. 10 pp.

—Foreign Exchange
OSI ISO DP 9777-87. Forms for Confirming Foreign Exchange Deals. 14 pp.

—Loan/Deposit Contracts
OSI ISO DP 9778-87. Forms for Confirming Loan/Deposit Contracts. 21 pp.

—Microfiche—Signature Lists
ISO 6234-81. Bank Operations—Authorized Signature Lists and Their Representation on Microfiche First Edition. 9 pp.
OSI ISO 6234-81. Bank Operations—Authorized Lists and Their Representation on Microfiche. 9 pp.

—Payment Orders
BSI BS 6601-85. 1985 Mail Payment Orders. 8 pp.
ISO 6260-84. Mail Payment Orders First Edition. 7 pp.
OSI ISO 6260-84. Mail Payment Orders. 7 pp.

—Savings Books—Magnetic
BSI BS 7110-90. 1990 Design and Use of Savings Books with Magnetic Stripes for Information Interchange. 13 pp.
ISO 8484-87. Magnetic Stripes on Savingsbooks First Edition. 9 pp.
JTC1 8484-87. Magnetic Stripes on Savingsbooks First Edition. 9 pp.
OSI ISO 8484-87. Magnetic Stripes on Savings Books. 9 pp.

—Signature Lists
BSI BS 6321-82. 1982 Authorized Signature Lists and Their Representation on Microfiche in Bank Operations. 11 pp.
ISO 6234-81. Bank Operations—Authorized Signature Lists and Their Representation on Microfiche First Edition. 9 pp.
OSI ISO 6234-81. Bank Operations—Authorized Lists and Their Representation on Microfiche. 9 pp.

Banks
Use: Banking

Banners
See Also: Flags; Signs

—Aircraft
CAA Chapter K4-11 10.92. Banner Towing Installations. 2 pp.
CAA Chapter K4-11 App 10.92. Banner Towing Installations. 1 p.

Bar Codes
See Also: Codes

—Symbols
CEN PREN 796-92. Bar Coding—Symbology Identifiers. 8 pp.
CEN PREN 797-92. Bar Coding—Symbology Specifications—EAN/UPC. 20 pp.
CEN PREN 798-92. Bar Coding—Symbology Specifications—Codabar. 10 pp.
CEN PREN 799-92. Bar Coding—Symbology Specifications—Code 128. 15 pp.
CEN PREN 800-92. Bar Coding—Symbology Specifications—Code 39. 10 pp.
CEN PREN 801-92. Bar Coding—Symbology Specifications—Interleaved 2 of 5. 10 pp.
MOD UK DSTAN 00-34-01. Standard Symbology for Automatic Identification Marking—Bar Code 39 Issue 1; Amendment 1. 20 pp.
NATO STANAG 4329 ED 1 AMD 0-92. NATO Standard Bar Code Symbology. 26 pp.

—Symbols—Format
CEN PREN 841-92. Bar Coding—Symbology Specifications—Format Description. 4 pp.

Bar Codes *(Cont.)*

—Uniform Commodity Codes
CNS Z5143-88. Commodity Bar Code (Aug)(12408).
JIS X 0501-85. Bar Code Symbols for Uniform Commodity Code.

—Unit Loads
CNS Z5144-88. Despatch Unit Bar Code (Aug)(12409).
JIS X 0502-87. Bar Code Symbol for Dispatch Unit Code. 16 pp.

Barbecues
See Also: Appliances; Cooking Appliances; Grills (Appliances)

—Electric—Household—Safety
CSA C22.2 NO 1335.2.9-93. Portable Electrical Motor-Operated and Heating Appliances: Particular Requirements for Portable Electric Cooking Appliances; (Gen Instr 1). 40 pp.

—Liquefied Petroleum Gas
CEN PREN 498-91. Barbecues Burning Liquefied Petroleum Gases for Outdoor Use. 44 pp.

—Rotisseries
CSA CAN/CSA-C22. 2 NO 68-92. Motor-Operated Appliances (Household and Commercial); (Gen Instr 1 Thru 2). 115 pp.
CSA 1169 Bull. Electrical Bulletin 1169 June 27, 1978 to C22.2 NO 68. 2 pp.

—Safety
BSI BS 5258: Part 14-84. 1984 Safety of Domestic Gas Appliances Part 14: Specification for Barbecues (3rd Family Gas). 19 pp.

Barbed Wire
See Also: Wire
CNS G3037-86. Barbed Wires (Jun)(2323).
JIS G 3533-88. Barbed Wires. 8 pp.
SNZ NZS 3471-74. Specification for Galvanized Steel Fencing Wire. Plain and Barbed. 12 pp.

Barbeques
Use: Barbecues

Bare Conductors
See Also: Electric Conductors; Electric Wire

—Color Coding
CENELEC HD 324-76. Identification of Insulated and Bare Conductors by Colours. 2 pp.
IEC 446-89. Identification of Conductors by Colours or Numerals Second Edition. 18 pp.

—Short Circuits
IEC 865-86. Calculation of the Effects of Short-Circuit Currents First Edition. 52 pp.

Bare Electrodes
See Also: Electrodes; Welding Electrodes
CNS Z8010-73. Testing Standard for Bared Electrodes of Nickel and Nickel Alloys (May)(3593).

Barges
See Also: Boats (Marine); Cargo Ships; Tugboats
BSI BS 7472-91. 1991 Shipborne Barges of All Series (ISO 9382: 1990). 8 pp.
BSI BS MA 87: Part 1-80. 1980 Amd 1 Main Dimensions of Shipborne Barges Part 1: Specification for Main Dimensions (AMD 4781) February 28, 1985. 6 pp.
BSI BS MA 87: Part 2-85. 1985 Main Dimensions of Shipborne Barges Part 2: Principal Technical Requirements for Shipborne Barges. 3 pp.
BSI BS MA 99: Part 1-85. 1985 Series 4 Shipborne Barges Part 1: Main Dimensions. 4 pp.
BSI BS MA 104: Part 1-87. 1987 Shipborne Barges Part 1: Main Dimensions. 6 pp.
BSI BS MA 105: Part 1-87. 1987 Shipborne Barges Part 1: Main Dimensions. 6 pp.
BSI BS MA 105: Part 2-87. 1987 Shipborne Barges Part 2: Main Operational and Technical Requirements. 3 pp.
ISO 4175-79. Shipbuilding—Shipborne Barges, Series 1—Main Dimensions First Edition. 5 pp.
ISO 6765-85. Shipbuilding—Shipborne Barges, Series 3—Main Dimensions First Edition. 6 pp.
ISO 6766-84. Shipbuilding—Shipborne Barges, Series 4—Main Dimensions First Edition. 4 pp.
ISO 7221-84. Shipbuilding and Marine Structures—Shipborne Barges, Series 1, on Barge Carriers—Principal Technical Requirements First Edition. 3 pp.
ISO 7222-85. Shipbuilding—Shipborne Barges, Series 2—Main Dimensions First Edition. 6 pp.
ISO 8303-85. Shipbuilding—Shipborne Barges, Series 3—Main Operational and Technical Requirements First Edition. 3 pp.

Barges *(Cont.)*
ISO 9382-90. Shipborne Barges, All Series—Classification and Main Requirements First Edition. 6 pp.

—Classification
BSI BS 7472-91. 1991 Shipborne Barges of All Series (ISO 9382: 1990). 8 pp.
ISO 9382-90. Shipborne Barges, All Series—Classification and Main Requirements First Edition. 6 pp.

—Lifting Equipment
BSI BS MA 87: Part 3-87. 1987 Main Dimensions of Shipborne Barges Part 3: The Arrangement, Dimensions and Method of Test for Lifting Post Castings. 7 pp.
ISO 6764-85. Shipbuilding—Shipborne Barges, Series 1—Lifting Post Casting—Arrangement, Dimensions and Method of Testing First Edition. 7 pp.

—Ventilation Ducts
BSI BS MA 105: Part 3-87. 1987 Shipborne Barges Part 3: Principal Mating Dimensions for the Ventilating Systems. 4 pp.
ISO 8304-84. Shipbuilding—Shipborne Barges, Series 3—Ventilation System—Principal Mating Dimensions First Edition. 4 pp.

Barite
Use For: Barytes *See Also:* Pigments
MOD UK DSTAN 68-55-78. Barytes, LF Quality Issue 1. 5 pp.

—Pigments
CNS K2154-87. Precipitated Barium Sulfate and Ground Barite (Pigment) (May)(11944). 2 pp.
CNS K6904-87. Method of Test for Precipitated Barium Sulfate and Ground Barite (Pigment) (May)(11945). 3 pp.
JIS K 5115-65. Precipitated Barium Sulfate and Ground Barite (Pigment). 6 pp.

Barium
See Also: Metals; Radioactive Barium Content Analysis

—Toys—Migration Testing
BSI BS 5665: Part 3-89. 1989 Safety of Toys Part 3: Migration of Certain Elements (Supersedes BS 3443: 1968). 11 pp.
CEN EN 71-3-88. Safety of Toys Part 3 Migration of Certain Elements. 14 pp.
CEN PREN 71-3-92. Safety of Toys—Part 3: Migration of Certain Elements. 24 pp.

Barium Acetate
See Also: Acetates
CNS K7080-62. Chemical Reagent (Barium Acetate) (Jul)(1580).
JIS K 8376-61. Barium Acetate.

—Explosives
MOD UK DSTAN 68-41-91. Barium Acetate Anhydrous and Monohydrate Issue 2. 8 pp.

Barium Carbonates
CNS K7081-62. Chemical Reagent (Barium Carbonate) (Jul)(1581).
JIS K 1415-61. Barium Carbonate. 7 pp.
JIS K 8626-75. Barium Carbonate.

Barium Chlorides
CNS K7082-62. Chemical Reagent (Barium Chloride) (Jul)(1582).
JIS K 1414-61. Barium Chloride. 6 pp.
JIS K 8155-91. Barium Chloride Dihydrate.

—Initiators (Ordnance)
MOD UK DSTAN 68-34-91. Barium Chloride Dihydrate Issue 2. 8 pp.

Barium Chromates
See Also: Pigments
CNS K7083-62. Chemical Reagent (Barium Chromate) (Jul)(1583).
CNS K7585-81. Barium Chloride (Aug)(7821).

—Pigments
ISO 2068-72. Barium Chromate Pigments for Paints First Edition. 7 pp.

Barium Content Analysis

—Barium Peroxide
MOD UK M 819/72. Examination of Barium Peroxide.

—Coatings
CGSB CAN2-1.500-75 METH 3-73. Methods of Test for Toxic Trace Elements in Protective Coatings Determination of Leachable Barium (Ba) in Low Concentration. 2 pp.

Barium Content Analysis *(Cont.)*

—Lubricating Oils—Atomic Absorption Spectrometry
DIN ENGL 51391 Pt 1-85. Testing of Lubricants; Determination of the Barium, Calcium and Zinc Content of Lubricating Oils; Direct Determination by Atomic Absorption Spectrometry (Dec). 4 pp.

—Paints—Atomic Emission Spectrometry
BSI BS 3900: Part B8-86. 1986 Methods of Test for Paints Group B: Tests Involving Chemical Examination of Liquid Paints and Dried Paint Films Part B8: Determination of 'Soluble' Barium Content. 7 pp.
ISO 3856 Pt 3-84. Paints and Varnishes—Determination of "Soluble" Metal Content—Part 3: Determination of Barium Content—Flame Atomic Emission Spectrometric Method Second Edition. 6 pp.

—Strontium Nitrate
MOD UK M 867/92. Examination of Strontium Nitrate (Anhydrous).

—Strontium Peroxide
MOD UK M 884/70. Examination of Strontium Peroxide (Anhydrous).

Barium Hydroxide
CNS K7084-62. Chemical Reagent (Barium Hydroxide) (Jul)(1584).
CNS K7085-62. Chemical Reagent (Barium Hydroxide, Anhydrous) (Jul)(1585).
JIS K 1417-61. Barium Hydroxide. 7 pp.
JIS K 8577-91. Barium Hydroxide Octahydrate.

Barium Inorganic Compounds
Scope Note: Use a more specific term *See:* Barium Acetate; Barium Carbonates; Barium Chlorides; Barium Hydroxide; Barium Nitrate; Barium Oxides; Barium Perchlorate; Barium Peroxide; Barium Sulfate

Barium Nitrate
CNS K7086-62. Chemical Reagent (Barium Nitrate) (Jul)(1586).
JIS K 1416-61. Barium Nitrate. 6 pp.
JIS K 8565-75. Barium Nitrate.

—Explosives
MOD UK DSTAN 68-74-01. Barium Nitrate Grades 1 and 1A Issue 1; Amendment 1. 8 pp.

—Pyrotechnics
MOD UK DSTAN 68-74-01. Barium Nitrate Grades 1 and 1A Issue 1; Amendment 1. 8 pp.

Barium Oxide Content Analysis

—Manganese Ores—Gravimetric Analysis
ISO 548-81. Manganese Ores—Determination of Barium Oxide Content—Barium Sulphate Gravimetric Method Second Edition. 4 pp.

Barium Oxides
JIS K 8428-61. Barium Oxide.

Barium Perchlorate
CNS K1168-74. Barium Perchlorate (Tentative) (Mar)(3702).
JIS K 9551-76. Barium Perchlorate.

Barium Peroxide
CNS K7088-62. Chemical Reagent (Barium Peroxide) (Barium Dioxide) (Jul)(1588).

—Content Analysis
MOD UK M 819/72. Examination of Barium Peroxide.

—Sieve Analysis
MOD UK M 819/72. Examination of Barium Peroxide.

—Visual Inspection
MOD UK M 819/72. Examination of Barium Peroxide.

Barium Sulfate
Use For: Barium Sulphates *See Also:* Pigments
CNS K7087-62. Chemical Reagent (Barium Sulfate) (Jul)(1587).
MOD UK DSTAN 68-38-75. Barium Sulphate, Precipitated Issue 1. 6 pp.

—Pigments
CNS K2118-86. Barium Sulfate Pigment (Feb)(5883). 1 p.
CNS K2154-87. Precipitated Barium Sulfate and Ground Barite (Pigment) (May)(11944). 2 pp.
CNS K6584-86. Method of Test for Barium Sulfate Pigments (Feb)(6365).

Barium Sulfate (Cont.)
—Pigments (Cont.)
CNS K6904-87. Method of Test for Precipitated Barium Sulfate and Ground Barite (Pigment) (May)(11945). 3 pp.
JIS K 5115-65. Precipitated Barium Sulfate and Ground Barite (Pigment). 6 pp.

Barium Sulfate Content Analysis
—Fluorite—Gravimetric Analysis
BSI BS 5659: Part 8-89. (WITHDRAWN) 1989 Acid-Grade Fluorspar Part 8: Determination of Barium Sulphate Content. 6 pp.
ISO 5437-92. Acid-Grade and Ceramic-Grade Fluorspar—Determination of Barium Sulfate Content—Gravimetric Method Second Edition. 6 pp.

Barium Sulphates
Use: Barium Sulfate

Barley
See Also: Cereals
CNS N1062-89. Barley (Nov)(2428). 5 pp.

Barley Hull Meal
See Also: Animal Feed; Bone Meal; Cornmeal; Oil Meal
—Animal Feed
CNS N2033-81. Barley Hull Meal (for Feeding) (Jun)(3401). 1 p.

Barographs
Use For: Aneroid Barographs; Recording Barometers
See Also: Barometers; Meteorology
CNS B6071-82. Barographs (Jul)(9061).
JIS B 7307-89. Barographs. 7 pp.

Barometers
See Also: Absolute Pressure Gages; Altimeters; Barographs; Manometers; Pressure Measurement
—Tables (Data)
BSI BS 2520-83. 1983 Barometer Conventions and Tables, Their Application and Use. 32 pp.

Barometric Altimeters
Use: Pressure Altimeters

Barometric Pressure
Use: Atmospheric Pressure

Barrels (Containers)
See Also: Containers; Drums (Containers); Shipping Containers
—Shipping Containers—Hazardous Materials—Steel
CGSB CAN/CGSB-43.96-M89. Steel Drums, Removable, or Non-Removable Head (TC-5B). 12 pp.
CGSB CAN/CGSB-43.97-M90. Stainless Steel Drums, Non-Removable Head (TC-5C). 12 pp.
CGSB CAN/CGSB-43.99-M88. Barrels or Drums, Steel (TC-6B). 11 pp.

Barricades
Use: Barriers

Barrier Free Design
CSA CAN/CSA-B651-M90. Barrier-Free Design; (Gen Instr 1). 122 pp.

Barrier Materials
See Also: Packaging Papers; Vapor Barriers
CNS Z5127-83. Corrosion Preventive Greaseproofed Barried Materials (Jun)(10396).
CNS Z6065-83. Method of Test for Corrosion Preventive Greaseproofed Barried Materials (Jun)(10397).
JIS Z 1705-76. Corrosion Preventive Greaseproofed Barrier Materials. 13 pp.

Barriers
Use For: Obstacles; Protective Barriers; Safety Barriers *See Also:* Child Safety Barriers; Dams; Fire Barriers; Guardrails; Parapets; Seals; Stiles; Stoppers; Termite Protection Barriers; Traffic Barriers; Walls
—Automotive—Collision Research
JIS D 1060-82. Frontal and Rear Vehicle Collision Test Procedure. 7 pp.
—Buildings
BSI BS 6180-82. 1982 Amd 1 Protective Barriers in and About Buildings (AMD 4858) August 30, 1985. 25 pp.
—Military Operations—Transfer
NATO STANAG 2989 ED 1 AMD 3-85. Transfer of Barriers. 39 pp.

Barriers (Cont.)
—Military Operations—Transfer (Cont.)
NATO STANAG 2989 ED 1 AMD 4-85. Transfer of Barriers. 40 pp.
—Minefields
NATO STANAG 2123 ED 2 AMD 3-84. Obstacle Folder. 45 pp.
—Railroad Equipment
JIS E 3016-76. Performance Test Methods for Motor-Operated Barriers.

Bars (Metallurgy)
Scope Note: See the subheading Bars under the specific metal or alloy

Bars (Textile Machinery)
See Also: Textile Machinery
BSI BS 4977-73. 1973 Amd 1 Terminology and Dimensions of Serrated Bars for Warp Stop Motions. 12 pp.
ISO 1865-77. Textile Machinery and Accessories—Serrated Bars for Mechanical Warp Stop Motions—Designations of Dimensions, and Dimensions of Cross-Section First Edition. 14 pp.
—Dyeing Equipment—Winding
BSI BS ISO 10458-93. 1993 Textile Machinery—Square Bars for Winding Devices Relating to Dyeing and Finishing Machines—Dimensions (L). 7 pp.
ISO 10458-93. Textile Machinery—Square Bars for Winding Devices Relating to Dyeing and Finishing Machines—Dimensions First Edition. 5 pp.
—Textile Finishing Machinery—Winding
BSI BS ISO 10458-93. 1993 Textile Machinery—Square Bars for Winding Devices Relating to Dyeing and Finishing Machines—Dimensions (L). 7 pp.
ISO 10458-93. Textile Machinery—Square Bars for Winding Devices Relating to Dyeing and Finishing Machines—Dimensions First Edition. 5 pp.

Barytes
Use: Barite

Base Capacity
—Water—Volumetric Analysis
DIN ENGL 38409 Pt 7-79. German Standard Methods for the Examination of Water, Waste Water and Sludge; Summary Indices of Actions and Substances (Group H); Determination of Acid and Base Capacity (H7) (May). 5 pp.

Base Metals
See Also: Metals
—Inspection
CNS H2046-82. General Rules for Tests and Inspection of Base Metals (Oct)(4196).
JIS H 0301-52. General Rules for Tests and Inspection of Base Metals (R 1964). 5 pp.

Base Number
—Petroleum Products—Potentiometric Analysis
ISO 3771-77. Petroleum Products—Total Base Number—Perchloric Acid Potentiometric Titration Method First Edition. 8 pp.

Base Plates
Use: Plates (Supports)

Base Rings
Use: Rings—Base

Base Station Systems
See Also: Repeater Stations
CENELEC GSM 11.20-93. GSM Base Station System Equipment Specification. 425 pp.
CENELEC GSM 11.20-93. GSM Base Station System Equipment Specification. 53 pp.
ETSI GSM 11.20-93. GSM Base Station System Equipment Specification. 425 pp.
ETSI GSM 11.20-93. GSM Base Station System Equipment Specification. 53 pp.
—Controllers—Interfaces—Transceivers
CENELEC GSM 08.51-92. BSC-BTS Interface General Aspects. 7 pp.
CENELEC GSM 08.52-92. BSC-BTS Interface Principles. 7 pp.
ETSI GSM 08.51-92. BSC-BTS Interface General Aspects. 7 pp.
—Data Link Layer (OSI)—Interfaces
CENELEC GSM 04.05-92. MS—BSS Data Link Layer General Aspects. 27 pp.

Base Station Systems (Cont.)
—Data Link Layer (OSI)—Interfaces (Cont.)
CENELEC GSM 08.56-92. BSC-BTS Specification. 14 pp.
ETSI GSM 04.05-92. MS—BSS Data Link Layer General Aspects. 27 pp.
ETSI GSM 08.56-92. BSC-BTS Specification. 14 pp.
—Land Mobile Services—Radio Frequency Interference
CCIR Report 1019-86. Sources of Unwanted Signals in Multiple Base Station Sites in the Land Mobile Service—Section 8A—Land Mobile Service and Related Subjects. 5 pp.
—Mobile Stations (Communications)—Data Link Layer (OSI)
CENELEC PRI-ETS 300 021-90. European Digital Cellular Telecommunications System (Phase 1); Mobile Station-Base Station System (MS-BSS) Interface Data Link Layer Specification. 83 pp.
CENELEC PRI-ETS 300 021-91. European Digital Cellular Telecommunications System (Phase 1); Mobile Station—Base Station System (MS-BSS) Interface Data Link Layer Specification. 86 pp.
CENELEC PRI-ETS 300 021-92. European Digital Cellular Telecommunications System (Phase 1); Mobile Station—Base Station System (MS-BSS) Interface Data Link Layer Specification. 86 pp.
CENELEC I-ETS 300 021-92. European Digital Cellular Telecommunications System (Phase 1); Mobile Station—Base Station System (MS-BSS) Interface Data Link Layer Specification (GSM 04.06). 86 pp.
CENELEC GSM 04.06-92. See PRI-ETS 300 021. 87 pp.
ETSI PRI-ETS 300 021-90. European Digital Cellular Telecommunications System (Phase 1); Mobile Station—Base Station System (MS-BSS) Interface Data Link Specification (GSM 04. 06). 83 pp.
ETSI I-ETS 300 021-92. European Digital Cellular Telecommunications System (Phase 1); Mobile Station—Base Station System (MS-BSS) Interface Data Link Layer Specification (GSM 04.06). 86 pp.
ETSI PRI-ETS 300 021-92. European Digital Cellular Telecommunications System (Phase 1); Mobile Station—Base Station System (MS-BSS) Interface Data Link Layer Specification. 86 pp.
ETSI GSM 04.06-92. See PRI-ETS 300 021. 87 pp.
—Network Layer (OSI)—Interfaces
CENELEC GSM 08.58-92. BSC-BTS Specification. 83 pp.
CENELEC GSM 08.58-DCS-92. European Digital Cellular Telecommunication System (Phase 1); Base Station Controller (BSC) to Base Station Tranceiver (BTS) Interface Layer 3 Specification. 5 pp.
ETSI GSM 08.58-92. BSC-BTS Specification. 83 pp.
ETSI GSM 08.58-DCS-92. European Digital Cellular Telecommunication System (Phase 1); Base Station Controller (BSC) to Base Station Tranceiver (BTS) Interface Layer 3 Specification. 5 pp.
—Operations and Maintenance Signaling Transport
CENELEC GSM 08.59-92. BSC—BTS O&M Signalling Transport. 9 pp.
ETSI GSM 08.59-92. BSC—BTS O&M Signalling Transport. 9 pp.
—Physical Layer (OSI)—Interfaces
CENELEC PRI-ETS 300 078-90. European Digital Cellular Telecommunications System (Phase 1): Layer 1—General Requirements. 21 pp.
CENELEC PRI-ETS 300 078-92. European Digital Cellular Telecommunications System (Phase 1); MS-BSS Layer 1—General Requirements. 24 pp.
CENELEC I-ETS 300 078-92. European Digital Cellular Telecommunications System (Phase 1); MS-BSS Layer 1—General Requirements (GSM 04.04). 24 pp.
CENELEC GSM 04.04-92. See PRI-ETS 300 078. 25 pp.
CENELEC GSM 08.54-92. BSC-BTS: Layer 1 Structure of Physical Circuits. 6 pp.
ETSI PRI-ETS 300 078-90. European Digital Cellular Telecommunications System (Phase 1); Layer 1—General Requirements (GSM 04-04). 21 pp.
ETSI I-ETS 300 078-92. European Digital Cellular Telecommunications System (Phase 1); MS-BSS Layer 1—General Requirements (GSM 04.04). 24 pp.
ETSI PRI-ETS 300 078-92. European Digital Cellular Telecommunications System (Phase 1); MS-BSS Layer 1—General Requirements. 24 pp.
ETSI GSM 04.04-92. See PRI-ETS 300 078. 25 pp.
ETSI GSM 08.54-92. BSC-BTS: Layer 1 Structure of Physical Circuits. 6 pp.

INTERNATIONAL AND NON-U.S. NATIONAL STANDARDS
SUBJECT INDEX

Base Station Systems (Cont.)
—Radio Paging Systems
CENELEC PRETS 300 133-6-91. Paging Systems; European Radio Message System (ERMES) Part 6: Base Station Conformance Specification. 11 pp.
CENELEC ETS 300 133-6-92. Paging Systems (PS); European Radio Message System (ERMES) Part 6: Base Station Conformance Specification. 10 pp.
ETSI PRETS 300 133-6-91. Paging Systems; European Radio Message System (ERMES) Part 6: Base Station Conformance Specification. 10 pp.
ETSI ETS 300 133-6-92. Paging Systems (PS); European Radio Message System (ERMES) Part 6: Base Station Conformance Specification. 10 pp.
ETSI PRETS 300 133-6-91. Paging Systems; European Radio Message System (ERMES) Part 6: Base Station Conformance Specification. 11 pp.

Baseball Fields
Use: Athletic Fields

Baseballs
CNS S1080-77. Hard Type Baseballs (Apr)(3462).

Basebands
See Also: Frequencies; Frequency Bands; Sidebands
—Frequency Division Multiplexers—Radio Relay Systems
CCITT RECMN G.423-89. Interconnection at the Baseband Frequencies of Frequency-Division Multiplex Radio-Relay Systems—International Analogue Carrier Systems (Study Group XV) 8 pp. 8 pp.
—Frequency Response—Fiber Optic Cables
CCITT RECMN G.651-89. Characteristics of a 50/125 Micrometer Multimode Graded Index Optical Fibre Cable—Transmission Media Characteristics (Study Group XV) 30 pp. 30 pp.
—Meteor Burst Radio Equipment
CCIR QUESTION 157/9-90. Radio Systems Employing Meteor-Burst Propagation—Questions of Study Group 9 Fixed Service. 1 p.
—Radio Relay Systems—Interfaces
CCIR Report 938-82. Baseband Interconnection of Digital Radio-Relay Systems—Section 9C—Interconnection Characteristics (Baseband and Intermediate Frequency). 4 pp.
CCIR QUESTION 101/9-90. Analogue Radio-Relay Systems Using Amplitude or Frequency Modulation—Questions of Study Group 9 Fixed Service. 1 p.
CCIR QUESTION 137/9-90. Interconnection at Baseband and Intermediate Frequencies for Digital Radio-Relay Systems—Questions of Study Group 9 Fixed Service. 2 pp.
—Radio Relay Systems—Telephony—Interfaces
CCIR RECMN 380-4-86. Interconnection at Baseband Frequencies of Radio-Relay Systems for Telephony Using Frequency-Division Multiplex—Section 9C—Interconnection Characteristics (Baseband and Intermediate Frequency). 4 pp.

Baseboard Heaters
See Also: Baseboards; Heaters
CSA C273.2-1971. Performance Standard for Residential Electric Baseboard Heaters; (Rev 1 March 1973) (Amd 2-9 October 1981). 26 pp.
JIS A 4004-87. Convectors, Baseboard-Heaters and Panel-Radiators. 24 pp.
—Electric—Portable
CSA 1063A Bull. Electrical Bulletin 1063A June 7, 1977 to C22.2 NO 46. 4 pp.

Baseboards
Use For: Skirtboards; Skirting (Walls); Skirting Boards (Wall); Wall Bases *See Also:* Baseboard Heaters; Construction Materials
—Resilient
CSA CAN/CSA-A126.5-87. Resilient Wall Base; (Gen Instr 1). 14 pp.
CSA A126.10-M1984. Test Methods for Resilient Flooring; (Gen Instr 1 Thru 2). 26 pp.

Bases (Supports)
Use: Platforms

BASIC
See Also: Programming Languages
BSI BS 7149-90. 1990 Programming Language Minimal BASIC. 39 pp.
BSI BS ISO/IEC 10279-91. 1991 Information Technology—Programming Languages—Full BASIC. 19 pp.

BASIC (Cont.)
CEN EN 26 373-89. Data Processing Programming Languages—Minimal Basic. 38 pp.
ECMA ECMA 55-78. Minimal BASIC. 38 pp.
ECMA ECMA 116-86. BASIC: ECMA Basic 1, ECMA Basic 2, ECMA Graphics Module. 202 pp.
ISO 6373-84. Data Processing—Programming Languages—Minimal BASIC First Edition. 36 pp.
ISO 10279-91. Information Technology—Programming Languages—Full BASIC First Edition; (Amendment 1-1993). 21 pp.
JIS X 3003-82. Programming Language Minimal BASIC.
JTC1 6373-84. Data Processing—Programming Languages—Minimal BASIC First Edition. 36 pp.
OSI ISO 6373-84. Data Processing—Programming Languages—Minimal BASIC. 36 pp.
OSI ISO/IEC DIS 10279-90. Information Technology—Programming Languages—Full BASIC. 17 pp.
—CAMAC
IEC 775-83. Real-Time BASIC for CAMAC First Edition. 41 pp.
—Man-Machine Language—Format Layout
CCITT RECMN Z.312-89. Basic Format Layout—Man-Machine Language (MML) (Study Group X) 2 pp. 2 pp.
—Source Programs
ECMA ECMA 53-78. Representation of Source Program for Program Interchange-APL, COBOL, FORTRAN, Minimal BASIC and PL/1. 16 pp.

Basic Bismuth Nitrate
Use: Bismuth Subnitrate

Basic Cupric Salicylate
Use: Cupric Salicylate

Basic Data Link Control Procedures
Use: Basic Mode Control Procedures

Basic Fuchsin
Use: Fuchsin

Basic Lead Carbonates
Use For: Carbonate White Lead; Lead Carbonates; White Lead *See Also:* Pigments
CNS K7271-68. Chemical Reagent (Lead Carbonate) (Nov)(1770).
MOD UK DSTAN 68-123-89. White Lead, Special Issue 1. 10 pp.
—Pastes—Oil Base
CGSB 1-GP-9MA-90. White Basic Lead Carbonate in Oil. 7 pp.
—Pigments
CNS K2001-84. Lead Carbonate for Paint (Feb)(13). 2 pp.
CNS K6090-84. Method of Test for Lead Carbonate for Paint (Mar)(1038).
ISO 511-74. White Lead Pigments for Paints First Edition. 6 pp.
JIS K 5103-65. White Lead (Pigment). 9 pp.

Basic Mode Control Procedures
See Also: Control Procedures; Link Control Procedures
BSI BS 4505: Part 1-81. 1981 Digital Data Transimission Part 1: Basic Mode Control Procedures. 22 pp.
BSI BS 4505: Part 4-71. (OBSOLESCENT) 1971 Amd 1 Digital Data Transmission Part 4: Basic Mode Control Procedures; Code Independent Information Transfer. 10 pp.
BSI BS 4505: Part 6-76. (OBSOLESCENT) 1976 Digital Data Transmission Part 6: Complements to the Basic Mode Control Procedures Recovery, Abort and Interrupt, Multiple Station Selection. 10 pp.
BSI BS 4505: Part 7-76. (OBSOLESCENT) 1976 Digital Data Transmission Part 7: Basic Mode Control Procedures; Conversational Information Message Transfer. 4 pp.
CNS C5242-90. Basic Mode Control Procedures—Complements (Nov)(12798).
ISO 1745-75. Information Processing—Basic Mode Control Procedures for Data Communication Systems First Edition. 22 pp.
ISO 2111-85. Data Communication—Basic Mode Control Procedures—Code Independent Information Transfer Second Edition. 4 pp.
ISO 2628-73. Basic Mode Control Procedures—Complements First Edition. 10 pp.
ISO 2629-73. Basic Mode Control Procedures—Conversational Information Message Transfer First Edition. 4 pp.
JIS X 5002-75. Basic Mode Data Transmission Control Procedures.
JTC1 1745-75. Information Processing—Basic Mode Control Procedures for Data Communication Systems First Edition. 22 pp.

Basic Mode Control Procedures (Cont.)
JTC1 2111-85. Data Communication—Basic Mode Control Procedures—Code Independent Information Transfer Second Edition. 4 pp.
JTC1 2628-73. Basic Mode Control Procedures—Complements First Edition. 10 pp.
JTC1 2629-73. Basic Mode Control Procedures—Conversational Information Message Transfer First Edition. 4 pp.
OSI ISO 1745-75. Information Processing—Basic Mode Control Procedures for Data Communication Systems. 23 pp.
OSI ISO 2111-85. Data Communication—Basic Mode Control Procedures—Code Independent Information Transfer. 4 pp.
OSI ISO 2628-73. Basic Mode Control Procedures—Complements. 10 pp.
OSI ISO 2629-73. Basic Mode Control Procedures—Conversational Information Message Transfer. 4 pp.
SAA AS 1484.5-76. Digital Data Transmission—Part 5: Basic Mode Control Procedures. 20 pp.
SNZ NZS/ISO 211, 2628, 2629-73. Digital Data Transmission. Extensions to Basic Mode Control Procedures. 20 pp.
SNZ NZS/ISO 1745-75. Information Processing—Basic Mode Control Procedures for Data Communication Systems. 19 pp.
—Data Communication
CNS C5240-90. Information Processing—Basic Mode Control Procedures for Data Communication Systems (Oct)(12796).
—Data Communication—Code Independent Information
CNS C5241-90. Data Communication—Basic Mode Control Procedures—Code Independent Information (Nov)(12797).
—Message Handling Systems
CNS C5243-90. Basic Mode Control Procedures—Conversational Information Message Transfer (Nov)(12799).

Basicity
Use: Alkalinity

Basidiomycetes
See Also: Fungi; Plants (Botany)
—Construction Materials—Fungus Resistance Testing
BSI BS 1982: Part 1-90. 1990 Fungal Resistance of Panel Products Made of or Containing Materials of Organic Origin Part 1: Method for Determination of Resistance to Wood-Rotting Basidiomycetes. 19 pp.
BSI BS 1982: Part 1-01. 1991 Amd 1 Fungal Resistance of Panel Products Made of or Containing Materials of Organic Origin Part 1: Method for Determination of Resistance to Wood-Rotting Basidiomycetes (AMD 7780) June 15, 1993 (R). 20 pp.
BSI BS 1982: Part 2-90. 1990 Fungal Resistance of Panel Products Made of or Containing Materials of Organic Origin Part 2: Method for Determination of Resistance to Cellulose-Decomposing Microfungi. 16 pp.
—Toxicity—Wood Preservatives
BSI BS 6009-82. 1982 Amd 2 Wood Preservatives. Determination of the Toxic Values Against Wood Destroying Basidiomycetes Cultured on an Agar Medium (AMD 6257) May 31, 1990. 20 pp.
CEN EN 113-80. Wood Preservatives: Determination of Toxic Values Against Wood Destroying Basidiomycetes Cultured on an Agar Medium. 17 pp.
CEN PREN 113-89. Wood Preservatives: Determination of the Toxic Valves Against Wood Destroying Basidiomycetes Cultured on an Agar Medium (Supersedes EN 113 Per Index). 18 pp.

Basil
See Also: Spices
BSI BS 7087: Part 12-92. 1992 Herbs and Spices Ready for Food Use Part 12: Specification for Dried Basil (Bouquets, Rubbed and Ground). 9 pp.

Basins (Containers)
See Also: Pus Basins; Retarding Basins; Storage Pools; Tanks (Containers); Wash Basins
—Aluminum
CNS H3013-86. Aluminium Basins, Shallow Shape (Jul)(755).
CNS H3014-86. Aluminium Basins, Deep Shape (Jul)(756).
—Enameled
CNS R2016-58. Enamel Basins, Shallow (Nov)(893)(R 1973).

INDUSTRY STANDARDS

Basins (Containers) (Cont.)

—Enameled (Cont.)
CNS R2017-58. Enamel Basins, Deep (Nov)(894)(R 1973).
CNS R2022-58. Enamel Bowls (Nov)(899) (R 1973).

—Platinum
CNS K0024-82. Platinum Basins (Nov)(9622).

—Porcelain—Chemical Analysis
CNS R2103-85. Porcelain Basins for Chemical Analysis (May)(6517).
JIS R 1302-80. Porcelain Basins for Chemical Analysis.

—Stands—Medical Equipment
BSI BS 3475-76. 1976 Amd 1 Bowl Stands for Use In Hospitals. 9 pp.

—Steel
CNS S1141-88. Steel Basins (Jan)(8741).
CNS S2071-88. Method of Test for Steel Basins (Jan)(8742).

Basins (Laboratory)
Use: Laboratory Dishes

Basins (Sinks)
Use: Sinks

Basis Weight Measurement
See Also: Weight Measurement

—Paperboard
CNS P3002-91. Method of Test for Basis Weight of Paper and Paperboard (Jun)(1352). 2 pp.

—Papers
CNS P3002-91. Method of Test for Basis Weight of Paper and Paperboard (Jun)(1352). 2 pp.
JIS P 8124-76. Determination of Basis Weight of Paper. 4 pp.

Basis Weight Scales
See Also: Papers; Weight Indicators

—Toledo
CPPA M-8-71. Toledo Basis Weight Scale—Models 9210 & 9212. 1 p.

Basketball Equipment
See Also: Sports Equipment
BSI BS 1892: Sec 2.7-86. 1986 Gymnasium Equipment Part 2: Particular Requirements Section 2.7: Basketball and Mini-Basketball Equipment. 12 pp.

—Nets
CNS S2028-80. Method of Test for Basketball Nets (May)(5633).

Baskets
See Also: Containers

—Bicycles
CNS S2031-80. Method of Test for Basket for Bicycles & Motor-Cycles (Aug)(6261).

—Motorcycles
CNS S2031-80. Method of Test for Basket for Bicycles & Motor-Cycles (Aug)(6261).

Bassinets
Use For: Carry Cots *See Also:* Cribs (Furniture)

—Mattresses
BSI BS 1877: Part 10-82. 1982 Amd 2 Domestic Bedding Part 10: Specification for Mattresses and Bumpers for Children's Cots, Perambulators and Similar Domestic Articles (AMD 6593) April 30, 1991. 12 pp.
BSI BS 1877: Part 10-03. 1982 Amd 3 Domestic Bedding Part 10: Specification for Mattresses and Bumpers for Children's Cots, Perambulators and Similar Domestic Articles (AMD 6949) February 28, 1992. 13 pp.

—Portable—Safety
BSI BS 3881-86. (WITHDRAWN) 1986 Amd 1 Safety Requirements for Carry Cots. Carry Cot Stands and Carry Cot Transporters (AMD 6473) April 30, 1990 (Superseded by BS 7409: 1991 & BS 7551: 1992). 15 pp.
BSI BS 7551-92. 1992 Safety Requirements for Carry Cots and Similar Handled Products and Stands. 13 pp.

Bath Boilers
Use: Boilers—Bath

Bath Salts
MOD UK DSTAN 68-187-93. Salts for Salt Baths Issue 1. 18 pp.

Bath Salts (Cont.)
MOD UK TS 307A. Salts for Salt Baths (Superseded by Def Stan 68-187).

—Moisture Content
MOD UK M 9531/77. Examination of Salts for Salt Baths (Superseded by Def Stan 68-187).

—Nitrate Content
MOD UK M 9531/77. Examination of Salts for Salt Baths (Superseded by Def Stan 68-187).

—Nitrite Content
MOD UK M 9531/77. Examination of Salts for Salt Baths (Superseded by Def Stan 68-187).

Bath Towels
BSI BS 5815: Part 2-88. 1988 Sheets, Sheeting, Pillowslips, Towels, Napkins, Counterpanes and Continental Quilt Secondary Covers Suitable for Use in the Public Sector Part 2: Towels and Napkins. 8 pp.
BSI BS 7311: Part 1-90. 1990 Cabinet Towels Part 1: Specification. 8 pp.
CGSB 4-GP-31MA-87. Towel, Terry. 10 pp.
CNS L4087-89. Woven Towel and Woven Towel Blankets (Jun) (5899). 3 pp.
JIS L 4105-87. YOKUYO Towel. 7 pp.

—Cleaning
BSI BS 7311: Part 2-93. 1993 Cabinet Towels Part 2: Specification for Processing. 7 pp.

—Cotton
BSI BS 5815-79. (WITHDRAWN) 1979 Cotton and Man-Made Fibre Blend Sheeting, Sheets, Pillowslips, Towels and Napkins for Use in the Public Sector (Superseded by BS 5815: Part 1: 1989, Part 2: 1988, and Part 3: 1991). 11 pp.

—Polyester/Cotton
BSI BS 5815-79. (WITHDRAWN) 1979 Cotton and Man-Made Fibre Blend Sheeting, Sheets, Pillowslips, Towels and Napkins for Use in the Public Sector (Superseded by BS 5815: Part 1: 1989, Part 2: 1988, and Part 3: 1991). 11 pp.

Bathing Pools
Use: Swimming Pools

Bathing Suits
Use: Swim Suits

Bathrobes
See Also: Clothing

—Children's
CGSB CAN/CGSB-49.44-M80. Dusters and Robes, Little Girls' and Girls', Regular Range—Dimensions; (Amendment 2 Nov 1988). 14 pp.

Bathroom Cabinets
Use For: Bathroom Vanities *See Also:* Medicine Cabinets
CSA CAN3-A278-M82. Kitchen Cabinets and Bathroom Vanities; (Gen Instr 1 Thru 3). 23 pp.
JIS A 4401-84. Vanities and Medicine Cabinets.

Bathroom Vanities
Use: Bathroom Cabinets

Bathrooms
See Also: Bathtubs; Showers; Sinks; Toilet Facilities

—Fans
CNS C4394-86. Ventilating Fan for Bath Room Use (Nov)(10597). 6 pp.

—Panels—Plastic
JIS K 6786-85. Plastic Panels for Bath Room. 11 pp.

—Pans
JIS A 4419-91. Prefabricative Pans for Bathrooms. 19 pp.

—Ships
MOD UK NES 120-01. Requirements for WCs, Urinals, Bathrooms and Washing Facilities Issue 3 (05.86); Amendment 1. 58 pp.

Baths
See Also: Cleaning Agents; Cleaning Equipment and Supplies; Water Baths

—Ablutionary
SAA AS 2023-89. Baths for Ablutionary Purposes Amdt 1 September 1991.

Bathtubs
See Also: Bathrooms; Hydrobaths; Plumbing Fixtures; Saunas; Shower Bases; Toilet Facilities; Whirlpool Baths
BSI BS EN 232-92. 1992 Baths—Connecting Dimensions. 13 pp.

Bathtubs (Cont.)
CEN EN 232-90. Baths—Connecting Dimensions. 11 pp.
DIN ENGL EN 232-91. Baths; Connecting Dimensions (Feb). 10 pp.
JIS A 4416-80. Bath Unit for Dwellings.

—Acrylic
BSI BS 4305: Part 2-89. 1989 Baths for Domestic Purposes Made of Acrylic Material Part 2: Connecting Dimensions (Supersedes BS 4305: 1972). 8 pp.
CEN EN 198-87. Specification for Finished Baths for Domestic Purposes: Made of Acrylic Material. 18 pp.

—Boilers
JIS A 5713-80. Bathtubs Connected with Boiler for Dwellings.

—Cast Iron—Porcelain Enamel
BSI BS 1189-86. 1986 Baths Made from Porcelain Enamelled Cast Iron. 12 pp.
BSI BS 1189-72. 1972 Amd 6 Cast Iron Baths (AMD 4157) April 29, 1983. 24 pp.

—Covers—Plastic
JIS A 5708-76. Plastic Covers for Bathtubs (R 1979). 12 pp.

—Drainage Systems
CEN PREN 128-89. Performance Criteria and Requirements for Sanitary Appliances. 4 pp.

—Drainpipe Fittings
BSI BS 3380-82. 1982 Wastes (Excluding Skeleton Sink Wastes) and Bath Overflows. 12 pp.
BSI BS EN 274-93. 1993 Sanitary Tapware—Waste Fittings for Basins, Bidets and Baths —General Technical Specifications. 22 pp.
CEN PREN 274-88. Sanitary Tapware: Waste Fittings for Basins Bidets and Baths; General Technical Specifications. 29 pp.
CEN EN 274-92. Sanitary Tapware—Waste Fittings for Basins, Bidets and Baths—General Technical Specifications. 28 pp.
CEN EN 274-92. Sanitary Tapware—Waste Fittings for Basins, Bidets and Baths—General Technical Specifications. 17 pp.
JIS A 5711-78. Waste Fittings for Bathtubs (R 1983). 10 pp.

—Electrical Installations
DIN VDE 0100 Pt 701-84. Installation of Power Plant with Rated Voltages up to 1000 V; Locations Containing a Bath or Shower (May). 19 pp.
IEC 364 Pt 7-701-84. Electrical Installations of Buildings Part 7: Requirements for Special Installations or Locations Section 701: Locations Containing a Bath Tub or Shower Basin First Edition. 20 pp.

—Fiberglass Reinforced Plastic
CNS A2107-85. Glassfiber Reinforced Plastic Bathtubs (Jan)(7612). 6 pp.
CNS A2139-82. Glassfiber Reinforced Plastic Bathtubs Connected with Floor (Jun)(8913).
CNS A3124-85. Method of Test for Glassfiber Reinforced Plastic Bathtubs (Jan)(7613).
CNS A3161-82. Method of Test for Glassfiber Reinforced Plastic Bathtubs Connected with Floor (Jun)(8914).
JIS A 5704-79. Glassfiber Reinforced Plastic Bathtubs. 19 pp.
JIS A 5712-79. Glassfiber Reinforced Plastic Bathtubs Connected with Floor. 18 pp.

—Joint Sealants
CGSB CAN/CGSB-19.22-M89. Mildew-Resistant Sealing Compound for Tubs and Tiles. 9 pp.

—Plastic Sheets
BSI BS 7015-89. 1989 Cast Acrylic Sheet for Baths and Shower Trays for Domestic Purposes. 11 pp.
CEN EN 263-87. Specification for Cast Acrylic Sheet for Baths and Shower Trays for Domestic Purposes. 7 pp.

—Sizes
CNS A2106-85. Standard Sizes for Bathtubs (Jan)(7611).
JIS A 0061-79. Standard Sizes for Bathtubs.

—Stainless Steel
JIS A 5710-83. Stainless Steel Bathtubs.

—Steel
BSI BS 1390-90. 1990 Baths Made from Vitreous Enamelled Sheet Steel. 22 pp.
BSI BS 1390-72. 1972 Amd 6 Sheet Steel Baths (AMD 4156) April 29, 1983. 28 pp.

INTERNATIONAL AND NON-U.S. NATIONAL STANDARDS
SUBJECT INDEX

Bathtubs (Cont.)

—Symbols—Architectural Drawings
BSI BS 1192: Part 3-87. 1987 Amd 1 Construction Drawing Practice Part 3: Recommendations for Symbols and Other Graphic Conventions (AMD 5882) February 28, 1989. 52 pp.
ISO 4067 Pt 2-80. Building and Civil Engineering Drawings—Installations—Part 2: Simplified Representation of Sanitary Appliances First Edition. 4 pp.

—Thermoplastic
JIS A 5709-79. Thermoplastic Bathtubs. 17 pp.

—Vitreous Enamel
JIS A 5532-79. Porcelain Enameled Bathtubs (R 1989). 13 pp.

—Water Pipe Fittings
JIS A 5527-78. Fittings of Circulating Pipe for Japanese Bathtub.

Bats (Sports Equipment)

—Wooden
CNS S1079-77. Wooden Bat for Baseball (for Youth)(Apr)(3461).

Batteries

Scope Note: For additional listings, use a more specific term *Use For:* Electric Batteries
See Also: Air Batteries; Aircraft Batteries; Alkaline Batteries; Battery Chargers; Battery Connectors; Battery Electrolytes; Battery Eliminators; Battery Separators; Battery Terminals; Battery Warmers; Jumper Cables; Lead Acid Batteries; Leclanche Batteries; Lithium Batteries; Lithium Manganese Dioxide Batteries; Lithium Thionyl Chloride Batteries; Manganese Batteries; Nickel Cadmium Batteries; Power Supplies; Primary Batteries; Starter Batteries; Storage Batteries; Torpedo Batteries; Traction Batteries; Water Activated Batteries
DIN ENGL 43539 Pt 1-85. Storage Cells and Batteries; Testing; General Information and General Test Methods (May). 6 pp.
MOD UK DSTAN 61-17-86. Users' Guide to the Selection of Batteries for Portable Equipment Issue 2. 22 pp.
MOD UK DSTAN 61-17-92. Requirements for the Selection of Batteries for Service Equipment Issue 3. 34 pp.

—Automotive—Guides
SAA AS MP50-82. Public Information Guide to the Use of Automotive Batteries. 10 pp.

—Connectors—Plug In Pin
IEC 130 Pt 3-65. Connectors for Frequencies Below 3 MHz (Mc/s) Part 3: Battery Connectors First Edition; (Corrigendum—Jan 1993). 27 pp.

—Connectors—Snap Fastener
IEC 130 Pt 3-65. Connectors for Frequencies Below 3 MHz (Mc/s) Part 3: Battery Connectors First Edition; (Corrigendum—Jan 1993). 27 pp.

—Defense Contracts
MOD UK DEFCON 112E-90. Basic Set of Conditions of Contract—Batteries 10/90. 2 pp.

—Fire Fighting Equipment
CNS Z2050-83. Batteries Device for Fire Fighting Emergency Use (Apr)(10205).

—Hazardous Materials—Disposal
EC 91/157/EEC-91. Council Directive on Batteries and Accumulators Containing Certain Dangerous Substances. 4 pp.

—Housings—Design—Marine
MOD UK NES 2055-88. Specification for Battery Box NSN 5850-99-523-3078 for Use with 2 1/2 Inch Intermediate and 5 Inch Hand Signalling Lanterns Issue 1 (11.88) (Withdrawn). 6 pp.

—Installation
DIN VDE 0510-77. VDE Regulation for Storage; Batteries and Battery Installations (Jan). 76 pp.

—Rotary Wing Aircraft
CAA 707 (G). Section G Rotorcraft Electrical Supply, Systems and Equipment (Blue Papers). 47 pp.
CAA Chapter G6-13 11.85. Electrical Generation, Supply and Distribution (Rotocraft). 11 pp.

—Watches
ISO TR10219-89. Batteries for Watches—Dimensions, Requirements and Marking First Edition. 10 pp.

—Watches—Identification Systems
ISO TR10219-89. Batteries for Watches—Dimensions, Requirements and Marking First Edition. 10 pp.

Batteries (Cont.)

—Watches—Leakage
ISO TR10220-89. Batteries for Watches—Leakage Tests First Edition. 6 pp.

Battery Chargers

See Also: Batteries; Storage Batteries
CNS C3119-81. Method of Test for Chargers for Cells (May)(7358).
CSA CAN/CSA-C22. 2NO107.2-M89. Battery Chargers; (Gen Instr 1 Thru 3). 88 pp.
CSA CAN/CSA-C22. 2 NO 223-M91. Power Supplies with Extra-Low-Voltage Class 2 Outputs; (Gen Instr 1). 53 pp.

—Aircraft
CAA LEAFLET 9-2 07.90. Charging Rooms for Aircraft Batteries. 5 pp.

—Appliances
DIN VDE 0730 Pt 2 ZR-80. Electric Motor-Operated Appliances for Domestic and Similar Purposes; Particular Requirements for Battery-Operated Appliances up to 20 VA Input and Their Charging and Battery Assemblies (VDE Specification) (May). 18 pp.

—Automotive
CSA CAN/CSA-C22. 2NO107.2-M89. Battery Chargers; (Gen Instr 1 Thru 3). 88 pp.
JIS D 5304-84. Household Battery Chargers for Automobiles.

—Commercial
CSA CAN/CSA-C22. 2NO107.2-M89. Battery Chargers; (Gen Instr 1 Thru 3). 88 pp.

—Electric Vehicles
IEC 718-92. Electrical Equipment for the Supply of Energy to Battery-Powered Road Vehicles Second Edition. 59 pp.

—Fixed—Indoor
CSA CAN/CSA-C22. 2NO107.2-M89. Battery Chargers; (Gen Instr 1 Thru 3). 88 pp.

—Fixed—Marine
CSA CAN/CSA-C22. 2NO107.2-M89. Battery Chargers; (Gen Instr 1 Thru 3). 88 pp.

—Household
CSA CAN/CSA-C22. 2NO107.2-M89. Battery Chargers; (Gen Instr 1 Thru 3). 88 pp.

—Industrial
CSA CAN/CSA-C22. 2NO107.2-M89. Battery Chargers; (Gen Instr 1 Thru 3). 88 pp.

—Lead Acid Batteries
NATO STANAG 4247 ED 1 AMD 0-00. (DRAFT) Battery Chargers, Non-Rotating, for Lead/Acid and Nickel/Cadmium Batteries. 7 pp.

—Medical Equipment
CSA CAN/CSA-C22. 2NO107.2-M89. Battery Chargers; (Gen Instr 1 Thru 3). 88 pp.

—Mobile—Indoor/Outdoor
CSA CAN/CSA-C22. 2NO107.2-M89. Battery Chargers; (Gen Instr 1 Thru 3). 88 pp.

—Nickel Cadmium Batteries
NATO STANAG 4247 ED 1 AMD 0-00. (DRAFT) Battery Chargers, Non-Rotating, for Lead/Acid and Nickel/Cadmium Batteries. 7 pp.

—Portable—Indoor/Outdoor
CSA CAN/CSA-C22. 2NO107.2-M89. Battery Chargers; (Gen Instr 1 Thru 3). 88 pp.

—Safety
BSI BS 3456: Sec 3.25-81. (WITHDRAWN) 1981 Amd 1 1972-1981 Edition. Specification for Safety of Household Electrical Appliances Part 3: Complete Particular Specifications Section 3.25: Battery Chargers (Superseded by BS EN 60335-2-29: 1991). 56 pp.
BSI BS EN 60335-2-29-91. 1991 Safety of Household and Similar Electrical Appliances Part 2: Particular Requirements Section 2.29: Battery Chargers. 23 pp.
CENELEC EN 60 335-2-29-91. Safety of Household and Similar Electrical Appliances Part 2: Particular Requirements for Battery Chargers. 9 pp.
DIN VDE 0700 Pt 207-82. Safety of Household and Similar Electrical Appliances; Particular Requirements for Power Supply Appliances up to 20 VA (Oct). 21 pp.
IEC 335 Pt 2-29-87. Safety of Household and Similar Electrical Appliances Part 2: Particular Requirements for Battery Chargers Second Edition; (Amendment 1-1989) (Amendment 2-1991). 62 pp.
SNZ NZS 6329-90. Safety of Household and Similar Electrical Appliances. Particular Requirements for Battery Chargers. 20 pp.

Battery Chargers (Cont.)

—Selenium Rectifiers
CNS C7007-64. Selenium Rectifiers for Battery Charging (Aug)(2317)(R 1973).

—Stationary
SAA AS 4044-92. Battery Chargers for Stationary Batteries (in Professional Package 44). 16 pp.

—Stationary—Indoor/Outdoor
CSA CAN/CSA-C22. 2NO107.2-M89. Battery Chargers; (Gen Instr 1 Thru 3). 88 pp.

—Transformerless
CSA 1442 Bull. Electrical Bulletin 1442 May 25, 1988 to C22.2 NO 223. 13 pp.

—Wheelchairs
CSA CAN/CSA-C22. 2NO107.2-M89. Battery Chargers; (Gen Instr 1 Thru 3). 88 pp.

Battery Connectors

Use For: Battery Disconnects *See Also:* Batteries; Connectors

—External—Aircraft
IEC 952 Pt 3-93. Aircraft Batteries Part 3: External Electrical Connectors First Edition. 15 pp.

—Two Pore—Preferred Products List
CECC CECC MUAHAG Vol 3A IS 4-91. Preferred Products List; Connectors; L. F. (En, Fr, Ge). 46 pp.

Battery Disconnects

Use: Battery Connectors

Battery Electrolytes

See Also: Batteries; Electrically Conductive Liquids

—Hydrometers
MOD UK DSTAN 66-3: Part 1-01. Testers, Specific Gravity Part 1: Tester, Battery Electrolyte Solution Issue 2; Amendment 1. 9 pp.

—Lead Acid Batteries—Submarines
NATO STANAG 4287 ED 1 AMD 2-88. Electrolyte for Submarine Main Lead Acid Batteries. 21 pp.

—Nickel Cadmium Batteries—Marine
MOD UK NES 2062-90. Specification for Electrolyte Potassium Hydroxide (Caustic Potash) Solid NSN 6810-99-224-7619 for Use with Nickel-Cadmium and Silver-Zinc Type Batteries Issue 1 (02.90). 7 pp.

—Nickel Cadmium Cells
CENELEC HD 585 S1-91. Electrolyte for Vented Nickel-Cadmium Cells. 5 pp.
IEC 993-89. Electrolyte for Vented Nickel-Cadmium Cells First Edition. 26 pp.

—Perchloric Acid
MOD UK TS 10128A. Electrolyte, Perchloric Acid Type (Withdrawn).

—Silver Zinc Batteries—Marine
MOD UK NES 2062-90. Specification for Electrolyte Potassium Hydroxide (Caustic Potash) Solid NSN 6810-99-224-7619 for Use with Nickel-Cadmium and Silver-Zinc Type Batteries Issue 1 (02.90). 7 pp.

—Sulfuric Acid
CGSB 15-GP-8M-83. Sulfuric Acid, Battery Electrolyte Grade, Standard for. 14 pp.
CNS K1175-77. Sulfuric Acid for Storage Batteries (May)(3753). 2 pp.

Battery Eliminators

See Also: Batteries

—Power Systems—Safety
IEC 65-85. Safety Requirements for Mains Operated Electronic and Related Apparatus for Household and Similar General Use Fifth Edition; (Amendment 2-1989, Incorporating Amendment 1) (Amendment 3-1992). 210 pp.

Battery Locomotives

See Also: Locomotives; Railroad Equipment

—Explosion Proof
CNS M2086-80. Explosion-Proof Type Battery Locomotives for Mine (Dec)(6847).

—Explosion Proof—Electric Plugs
CNS M2079-80. Plug Connection Devices for Explosion-Proof Type Battery Locomotives in Mines (Dec)(6703).

—Explosion Proof—Traction Batteries
CNS M2078-80. Traction Batteries for Explosion-Proof Type Locomotives for Mine (Dec)(6702).
JIS M 7632-75. Traction Batteries for Explosion-Proof Type Locomotives.

Battery Locomotives (Cont.)

—Explosion Proof—Traction Battery Boxes
CNS M2080-80. Traction Battery Boxes for Explosion-Proof Type Battery Locomotives in Mines (Dec)(6704).

Battery Powered Appliances
Use: Appliances

Battery Powered Electric Vehicles
Use: Electric Vehicles—Battery Powered

Battery Powered Emergency Lighting
Use: Emergency Lighting

Battery Separators
See Also: Batteries
JIS C 2313-90. Separators for Lead-Acid Batteries. 11 pp.

Battery Terminals
See Also: Batteries; Electric Terminals

—Automotive—Connectors
BSI BS AU 91-65. 1965 Dimensions of Electrical Cable Connectors for Starter Battery Taper Terminal Posts (Cars and Light Commercial Vehicles Using Positive Earth System). 2 pp.

—Lead Acid Batteries
CENELEC HD 465.2 S1-89. Lead-Acid Traction Batteries-Part 2: Dimensions of Cells and Terminals and Marking of Polarity on Cells. 3 pp.
IEC 254 Pt 2-85. Lead-Acid Traction Batteries Part 2: Dimensions of Cells and Terminals and Marking of Polarity on Cells Second Edition. 12 pp.

Battery Warmers
See Also: Batteries

—Automotive
CSA CAN/CSA-C22. 2 NO 191-M89. Engine Heaters and Battery Warmers; (Gen Instr 1 Thru 3). 54 pp.

—Trucks
CSA CAN/CSA-C22. 2 NO 191-M89. Engine Heaters and Battery Warmers; (Gen Instr 1 Thru 3). 54 pp.

Batting (Fibers)
See Also: Fibers

—Steaming
CGSB CAN/CGSB-4.2 NO.33.3-M86. Textile Test Methods Method of Pressing Steaming. 6 pp.

Battle Tanks
Use: Tanks (Combat Vehicles)

Battlefield Illumination
See Also: Flares (Pyrotechnic); Searchlights; Warfare
NATO STANAG 2088 ED 6 AMD 0-88. Battlefield Illumination. 11 pp.

Baumann Testing
Use For: Sulfur Print Testing

—Steels
AECMA PREN2003-13-86. Test Methods for Steel Products—Part 13—Macrographic Examination by Sulphur Print (Baumann Method). 1 p.
BSI BS 6285-82. 1982 Macrographic Examination of Steel by Sulphur Print (Baumann Method). 4 pp.
CNS G2032-86. Method of Sulphur Print Test for Steel (Mar)(3939).
ISO 4968-79. Steel—Macrographic Examination by Sulphur Print (Baumann Method) First Edition. 5 pp.
JIS G 0560-83. Method of Sulphur Print Test for Steel. 8 pp.

Bauxite
See Also: Aluminum Ores

—Chemical Analysis
CNS K6025-69. Method of Chemical Test for Bauxite (Jun)(288)(R 1973).
JIS M 8361-68. Methods for Chemical Analysis of Bauxite Ores (R 1984). 14 pp.

—Sampling
JIS M 8110-75. Method for Sampling of Bauxite Ores. 39 pp.

Bay Oil
Use For: Myrica Oil See Also: Essential Oils
CNS K5074-80. Oil of Bay (Aug)(6353).
ISO 3045-74. Oil of Bay First Edition. 3 pp.

Bayonet Base Lamps
Use: Bayonet Lamps

Bayonet Connectors
Scope Note: See the subheading Quick Disconnect under specific Connectors

Bayonet Lamps
Use For: Bayonet Base Lamps See Also: Lamps; Lighting

—Lampholders
BSI BS 5042-87. 1987 Bayonet Lampholder. 28 pp.
SAA AS 3117-88. Approval and Test Specification—Bayonet Lampholders Amdt 1 1990 (This is a Joint Standard with SANZ NZS 3117). 12 pp.
SNZ NZS/AS 3117-88. Approval and Test Specification—Bayonet Lampholders Amend: 1, 1990 (This is a Joint Standard with SAA AS 3117). 12 pp.

BDT
Use For: Telecommunications Development Bureau

—Cooperation—CCIR—Radio Wave Propagation—Developing Countries
CCIR RESOLUTION 101-90. Cooperation Between the CCIR and BDT in Radio Propagation Measurement Campaigns in the Developing Countries—Volume XIV—Administrative Texts of the CCIR. 2 pp.

Beaches
Use For: Landing Beaches See Also: Parks; Recreational Facilities

—Military Geographic Documentation
NATO STANAG 2263 ED 4 AMD 4-76. Military Geograghic Documentation—Coastal Areas and Landing Beaches. 18 pp.
NATO STANAG 2263 ED 4 AMD 5-76. MGD—Coastal Areas and Landing Beaches. 28 pp.

—Model Bylaws
SNZ NZS 9201: Chapter 19-72. Model General Bylaws Chapter 19: Beaches: Bathing and Control. 11 pp.

Beacon Lights
See Also: Beacons; Emergency Lighting; Lighthouses; Lighting; Navigational Aids; Projectors; Searchlights; Signal Lights

—Naval Ships
MOD UK NES 148-92. Requirements for Lifesaving Equipment Issue 1 (06.92). 68 pp.

Beacons
Scope Note: For additional terms, use a more specific term See Also: Beacon Lights; Crash Locator Beacons; Lighthouses; Marker Beacons; Radar Transponders; Sonar Beacons; Warning Lights

—Spacecraft—Frequency Bands
CCIR Report 456-3-86. Preferred Frequency Bands for Spacecraft Transmitters Used as Beacons—Section 2E—Space Research. 5 pp.

Beading Machines
DIN ENGL 55211-80. Machine Tools; Beading Machines with Swivelling Top Shaft; Sizes (Mar). 2 pp.
DIN ENGL 55212-80. Machine Tools; Beading Machines with Parallel Adjustable Top Shaft; Sizes (Mar). 2 pp.

—Acceptance Testing
DIN ENGL 55801-79. Machine Tools; Beading Machines; Acceptance Conditions (Dec). 3 pp.

Beads, Glass
Use: Glass Beads

Beads (Plastering)
See Also: Plastering

—Galvanized Steel
BSI BS 6452: Part 1-84. 1984 Beads for Internal Plastering and Dry Lining Part 1: Galvanized Steel Beads. 8 pp.

Beagle A 109 Aircraft
See Also: Aircraft

—Mandatory Aircraft Modifications and Inspections Summaries
CAA. Beagle A109. 1 p.

Beagle-Auster A 61 Aircraft
See Also: Aircraft

—Antenna Positions
CAA. Auster (All Series) and Beagle/Auster A61 (Approved Aerial Positions). 1 p.

Beagle-Auster A109 Aircraft
See Also: Aircraft

—Antenna Positions
CAA. Beagle/Auster A109 (Approved Aerial Positions). 1 p.

Beagle B. (Bulldog) Aircraft
See Also: Aircraft

—Antenna Positions
CAA. Beagle B.121/125/(Bulldog) (Approved Aerial Positions). 1 p.

Beagle B 121 Aircraft
See Also: Aircraft

—Antenna Positions
CAA. Beagle B.121/125/(Bulldog) (Approved Aerial Positions). 1 p.

—Mandatory Aircraft Modifications and Inspections Summaries
CAA. British Aerospace Beagle B121 Pup Series 1, 2 and 3 Aircraft. 5 pp.

Beagle B 125 Aircraft
See Also: Aircraft

—Antenna Positions
CAA. Beagle B.121/125/(Bulldog) (Approved Aerial Positions). 1 p.

Beagle B 206 Aircraft
See Also: Aircraft

—Antenna Positions
CAA. Beagle B.206/B.206Y (Approved Aerial Positions). 1 p.

Beagle B 206Y Aircraft
See Also: Aircraft

—Antenna Positions
CAA. Beagle B.206/B.206Y (Approved Aerial Positions). 1 p.

Beagle 121 Aircraft
See Also: Aircraft

—Data Sheets
CAA BA1 ISSUE 7. Beagle 121 Series 1, Series 2 and Series 3. 4 pp.

Beagle 206 Aircraft
See Also: Aircraft

—Mandatory Aircraft Modifications and Inspections Summaries
CAA. Beagle 206. 2 pp.

Beakers
Scope Note: For additional listings, use a more specific term See Also: Containers; Flasks; Laboratory Dishes; Laboratory Ware

—Enameled
CNS R2158-85. Enamel Beaker for Laboratory (Jun)(9033).

—Glass
BSI BS 6523-84. 1984 Glass Beakers for Laboratory Use. 4 pp.
DIN ENGL 12331-88. Laboratory Glassware; Beakers (Oct). 3 pp.
ISO 3819-85. Laboratory Glassware—Beakers First Edition. 4 pp.
JIS R 3505-83. Volumetric Glassware. 16 pp.

—Glass—Chemical Analysis
CNS R2124-86. Glass Beakers for Chemical Analysis (Mar) (7300).

—Glass—Medicine Measures
BSI BS 1922-87. 1987 Glass Dispensing Measures for Pharmaceutical Purposes. 12 pp.

—Plastic
BSI BS 5404: Part 1-76. 1976 Plastics Laboratory Ware Part 1: Beakers. 8 pp.
ISO 7056-81. Plastics Laboratory Ware—Beakers First Edition. 8 pp.

—Porcelain—Chemical Analysis
CNS R2114-85. Porcelain Beaker for Chemical Analysis (May)(6613).
JIS R 1303-80. Porcelain Beakers for Chemical Analysis (R 1985). 5 pp.

—Volumetric
JIS R 3505-83. Volumetric Glassware. 16 pp.

Beam Bridges
—Concrete Beams
BSI JIS A 5316-91. Prestressed Concrete Beams for Beam Bridges. 44 pp.

Beam Interruption Detectors
Use: Photoelectric Alarm Systems

Beam Saddles
Use: Vaulting Horses

Beam Sensors
Use: Photoelectric Sensors

Beams (Supports)
Use For: Composite Beams *See Also:* Arbors; Columns (Supports); Concrete Beams; Footings; Girders; H Beams; Headers (Supports); I Beams; Joists; Lintels; Purlins; Rafters; Structural Members; T Beams

—Identification Systems
BSI BS 4263-67. 1967 Marking of Hatchway Beams. 8 pp.
CNS F4010-82. Marking of Hatchway Beam (Apr)(8688).

—Steel
BSI BS 5950: Sec 3.1-90. 1990 Structural Use of Steelwork in Building Part 3: Design in Composite Construction Section 3.1: Code of Practice for Design of Simple and Continous Composite Beams. 36 pp.
BSI CP 117: Part 1-65. (WITHDRAWN) 1965 Composite Construction in Structural Steel and Concrete Part 1: Simply-Supported Beams in Building (Superseded by BS 5950: Section 3.1: 1990). 24 pp.
SNZ NZS 1545-60. Specification for the Design and Testing of Steel Overhead Runway Beams Amend: 1A, 1960. 32 pp.

—Steel—Bridges (Structures)
BSI CP 117: Part 2-67. 1967 Composite Construction in Structural Steel and Concrete Part 2: Beams for Bridges. 33 pp.

—Steel—Hot Rolled
BSI BS 4: Part 1-80. 1980 Amd 5 Structural Steel Sections Part 1: Hot-Rolled Sections. 30 pp.
ISO 657 Pt 13-81. Hot-Rolled Steel Sections—Part 13: Tolerances on Sloping Flange Beam, Column and Channel Sections First Edition. 6 pp.
ISO 657 Pt 15-80. Hot-Rolled Steel Sections—Part 15: Sloping Flange Beam Sections (Metric Series)—Dimensions and Sectional Properties First Edition. 4 pp.
SNZ NZS/BS 4: Part 1-80. Structural Steel Sections Part 1: Specification for Hot-Rolled Sections. 27 pp.

Beams (Textile Machinery)
Use For: Sectional Beams (Textile Machinery)
See Also: Knitting Beams; Textile Machinery; Warp Beams; Weavers' Beams
BSI BS 6547-84. (WITHDRAWN) 1984 Methods for Measurement of Variations of CF Form and Position of Beams in Textile Machinery. 6 pp.
ISO 2013-83. Textile Machinery and Accessories—Beams—Method of Measuring Variations of Form and Position Second Edition. 5 pp.

—Dyeing Equipment
BSI BS ISO 8116-9-91. 1991 Textile Machinery and Accessories—Beams for Winding—Part 9: Dyeing Beams for Textile Fabrics (L). 7 pp.
ISO 8116 Pt 9-91. Textile Machinery and Accessories—Beams for Winding—Part 9: Dyeing Beams for Textile Fabrics First Edition. 5 pp.

—Dyeing Equipment—Glossaries
ISO 1037-82. Textile Machinery and Accessories—Beams for Dyeing Slivers and Yarn—Terminology and Main Dimensions First Edition. 5 pp.

—Flanges
BSI BS 3225: Part 5-86. 1986 Beams for Wraping, Sizing and Weaving of Textile Yarns art 5: Method of Classification of Flanges for Weaver's Beams, Warper's Beams and Sectional Beams. 8 pp.
ISO 8116 Pt 4-85. Textile Machinery and Accessories—Beams for Winding—Part 4: Quality Classification of Flanges for Weaver's Beams, Warper's Beams and Sectional Beams First Edition. 6 pp.

—Glossaries
BSI BS 3225: Part 4-86. 1986 Beams for Warping, Sizing and Weaving of Textile Yarns Part 4: Glossary of Terms for Winding Beams. 6 pp.
ISO 8116 Pt 1-85. Textile Machinery and Accessories—Beams for Winding—Part 1: Vocabulary First Edition. 5 pp.

Beams (Textile Machinery) *(Cont.)*
—Glossaries *(Cont.)*
ISO 8116 Pt 6-90. Textile Machinery and Accessories—Beams for Winding—Part 6: Beams for Ribbon Weaving and Ribbon Knitting—Terminology and Main Dimensions First Edition. 8 pp.

—Looms
CNS L2043-82. Yarn Beams for Woollen and Worsted Loom (Apr)(8718).

—Perforated
BSI BS 4753: Part 2-72. (WITHDRAWN) 1972 Amd 1 Dyeing and Finishing Machines Part 2: Beams for Silver and Yarn Dyeing. 8 pp.

—Warp Knitting Machines
BSI BS 3945-82. (WITHDRAWN) 1982 Main Dimensions of Sectional Beams for Warp Knitting Machines. 4 pp.
ISO 8116 Pt 5-88. Textile Machinery and Accessories—Beams for Winding—Part 5: Sectional Beams for Warp Knitting Machines—Terminology and Main Dimensions First Edition; (Supersedes 1025). 6 pp.

Bean Curd
Use: Tofu

Bean Sprouts
See Also: Beans; Vegetables

—Canned
CNS N5100-85. Canned Bean Sprout (Nov)(3314). 2 pp.

Beans
See Also: Bean Sprouts; Food; Green Beans; Kidney Beans; Vegetables
CNS N1064-90. Beans and Peas (Except Soybean) (Sep)(2430). 2 pp.
EC EEC/452/92-92. Commission Regulation Adding a Temporary Provision to the Detailed Rules for the Application of the Special Measures for Peas, Field Beans and Sweet Lupins. 1 p.

—Canned
CNS N5023-86. Canned Beans (Sep)(1257). 3 pp.

—Dried
CGSB 32.262M-89. Dried Beans and Peas (Supersedes 32-GP-261A Jan 1968). 7 pp.

Bearded Needles
Use: Knitting Needles

Bearer Circuits
Use: Telephone Circuits

Bearer Services
See Also: Connectionless Broadband Data Service; Integrated Services Digital Networks; Network Services (OSI); Telecommunication Services; Telephone Services; Teleservices

—Accounting
CEPT T/PGT 41 E-91. Charging and Accounting Principles for the International 2 Mbit/s Switched Bearer Service. 1 p.

—Application Layer—Protocols
CENELEC PRETS 300 223-92. Integrated Services Digital Network (ISDN); Syntax-Based Videotex Common End-to-End Protocols. 103 pp.
CENELEC PRETS 300 223-92. Terminal Equipment (TE); Syntax-Based Videotex Common End-to-End Protocols. 102 pp.
CENELEC ETS 300 223-93. Terminal Equipment (TE); Syntax-Based Videotex Common End-to-End Protocols. 102 pp.
ETSI ETS 300 223-93. Terminal Equipment (TE); Syntax-Based Videotex Common End-to-End Protocols. 102 pp.
ETSI PRETS 300 223-92. Terminal Equipment (TE); Syntax-Based Videotex Common End-to-End Protocols. 102 pp.
ETSI PRETS 300 223-92. Integrated Services Digital Network (ISDN); Syntax-Based Videotex Common End-to-End Protocols. 103 pp.

—Cellular Mobile Radio Equipment
CCIR Report 1156-90. Digital Cellular Public Land Mobile Telecommunication Systems (DCPLMTS)—Section 8A—Land Mobile Service and Related Subjects. 28 pp.

—Circuit Mode—ISDN
CCITT RECMN F.353-89. Provision of Telematic and Data Transmission Services on Integrated Services Digital Network (ISDN)—Telematic, Data Transmission and Teleconference Services—Operations and Quality of Service (Study Group I) 5 pp. 5 pp.

Bearer Services *(Cont.)*
—Circuit Mode—ISDN *(Cont.)*
CCITT RECMN I.231-89. Circuit-Mode Bearer Service Categories—Integrated Services Digital Network (ISDN)—General Structure and Service Capabilities (Study Group XVIII) 25 pp. 25 pp.
CCITT RECMN I.231. 1-88. Circuit-Mode 64 kbit/s Unrestricted, 8 kHz Structured Bearer Service Category (Note 1)—Integrated Services Digital Network (ISDN)—General Structure and Service Capabilities (Study Group XVIII) 5 pp. 5 pp.
CCITT RECMN I.231. 5-88. Circuit-Mode 2 X 64 kbit/s Unrestricted, 8 kHz Structured Bearer Service Category—Integrated Services Digital Network (ISDN)—General Structure and Service Capabilities (Study Group XVIII) 2 pp. 2 pp.
CCITT RECMN I.231. 6-88. Circuit-Mode 384 kbit/s Unrestricted, 8 kHz Structured Bearer Service Category—Integrated Services Digital Network (ISDN)—General Structure and Service Capabilities (Study Group XVIII) 3 pp. 3 pp.
CCITT RECMN I.231. 7-88. Circuit-Mode 1536 kbit/s Unrestricted, 8 kHz Structured Bearer Service Category—Integrated Services Digital Network (ISDN)—General Structure and Service Capabilities (Study Group XVIII) 3 pp. 3 pp.
CCITT RECMN I.231. 8-88. Circuit-Mode 1920 kbit/s Unrestricted, 8 kHz Structured Bearer Service Category—Integrated Services Digital Network (ISDN)—General Structure and Service Capabilities (Study Group XVIII) 3 pp. 3 pp.
CENELEC PRETS 300 108-90. Integrated Services Digital Network (ISDN) Circuit-Mode 64Kbit/s Unrestricted SKHz Structured Bearer Service Category (T/NA1 (89)35) Service Description. 6 pp.
CENELEC PRETS 300 108-92. Integrated Services Digital Network (ISDN); Circuit-Mode 64 kbit/s Unrestricted 8 kHz Structured Bearer Service Category Service Description. 15 pp.
CENELEC ETS 300 108-92. Integrated Services Digital Network (ISDN); Circuit-Mode 64 kbit/s Unrestricted 8 kHz Structured Bearer Service Category Service Description. 14 pp.
CENELEC PRETS 300 109-90. Integrated Services Digital Network (ISDN) Circuit-Mode 64Kbit/s 8KHz Structured Bearer Service Category (T/NA1 (89)36) Service Description. 6 pp.
CENELEC PRETS 300 109-92. Integrated Services Digital Network (ISDN); Circuit-Mode 64 kbit/s 8 kHz Structured Bearer Service Category Usable for Speech Information Transfer Service Description. 15 pp.
CENELEC ETS 300 109-92. Integrated Services Digital Network (ISDN); Circuit-Mode 64 kbit/s 8 kHz Structured Bearer Service Category Usable for Speech Information Transfer Service Description. 13 pp.
CENELEC PRETS 300 110-90. Integrated Services Digital Network (ISDN) Circuit-Mode 64 Kbit/s 8 KHz Structured Bearer Service Category (T/NA1 (89)37) Usable for 3.1 KHz Audio Information Transfer Service Description. 7 pp.
CENELEC PRETS 300 110-92. Integrated Services Digital Network (ISDN); Circuit-Mode 64 kbit/s 8 kHz Structured Bearer Service Category Usable for 3,1 kHz Audio Information Transfer Service Description. 15 pp.
CENELEC ETS 300 110-92. Integrated Services Digital Network (ISDN); Circuit-Mode 64 kbit/s 8 kHz Structured Bearer Service Category Usable for 3,1 kHz Audio Information Transfer Service Description. 14 pp.
ETSI PRETS 300 108-92. Integrated Services Digital Network (ISDN); Circuit-Mode 64 kbit/s Unrestricted 8 kHz Structured Bearer Service Category Service Description. 15 pp.
ETSI ETS 300 109-92. Integrated Services Digital Network (ISDN); Circuit-Mode 64 kbit/s 8 kHz Structured Bearer Service Category Usable for Speech Information Transfer Service Description. 13 pp.
ETSI PRETS 300 109-92. Integrated Services Digital Network (ISDN); Circuit-Mode 64 kbit/s 8 kHz Structured Bearer Service Category Usable for Speech Information Transfer Service Description. 15 pp.
ETSI ETS 300 110-92. Integrated Services Digital Network (ISDN); Circuit-Mode 64 kbit/s 8 kHz Structured Bearer Service Category Usable for 3,1 kHz Audio Information Transfer Service Description. 14 pp.
ETSI PRETS 300 110-92. Integrated Services Digital Network (ISDN); Circuit-Mode 64 kbit/s 8 kHz Structured Bearer Service Category Usable for 3,1 kHz Audio Information Transfer Service Description. 15 pp.

—Circuit Mode—ISDN—Accounting
CCITT RECMN D.220-91. Charging and Accounting Principles to Be Applied to International Circuit-Mode Demand Bearer Services Provided over the Integrated Services Digital Network (ISDN) (Study Group III) 5 pp. 5 pp.

Bearer Services (Cont.)

—Circuit Mode—ISDN—Accounting (Cont.)
CCITT RECMN D.220-89. Charging and Accounting Principles to Be Applied to International Circuit Mode Demand Bearer Services Provided over the Integrated Services Digital Network (ISDN)—General Tariff Principles—Charging and Accounting of Intl Telecom. Serv. (Study Group III) 2 pp. 2 pp.

—Circuit Mode—ISDN—Audio
CCITT RECMN I.231. 3-88. Circuit-Mode 64 kbit/s, 8 kHz Structured Bearer Service Category Usable for 3.1 kHz Audio Information Transfer—Integrated Services Digital Network (ISDN)—General Structure and Service Capabilities (Study Group XVIII) 5 pp. 5 pp.

—Circuit Mode—ISDN—Speech Transmission
CCITT RECMN I.231. 2-88. Circuit-Mode 64 kbit/s, 8 kHz Structured Bearer Service Category Usable for Speech Information Transfer—Integrated Services Digital Network (ISDN)—General Structure and Service Capabilities (Study Group XVIII) 5 pp. 5 pp.

CCITT RECMN I.231. 4-88. Circuit-Mode, Alternate Speech /64 kbit/s Unrestricted, 8 kHz Structured Bearer Service Category—Integrated Services Digital Network (ISDN)—General Structure and Service Capabilities (Study Group XVIII) 5 pp. 5 pp.

—Circuit Mode—ISDN—Tariffs
CCITT RECMN D.220-91. Charging and Accounting Principles to Be Applied to International Circuit-Mode Demand Bearer Services Provided over the Integrated Services Digital Network (ISDN) (Study Group III) 5 pp. 5 pp.

CCITT RECMN D.220-89. Charging and Accounting Principles to Be Applied to International Circuit Mode Demand Bearer Services Provided over the Integrated Services Digital Network (ISDN)—General Tariff Principles—Charging and Accounting of Intl Telecom. Serv. (Study Group III) 2 pp. 2 pp.

—Circuit Mode—Land Mobile Services
CCIR RECMN 687-90. Future Public Land Mobile Telecommunication Systems (FPLMTS)—Section 8A—Land Mobile Service and Related Subjects. 11 pp.

CCIR RECMN 687-1-92. Future Public Land Mobile Telecommunication Systems (FPLMTS)—Section 8A—Land Mobile Service and Related Subjects. 21 pp.

CCIR Report 1153-90. Future Public Land Mobile Telecommunication Systems—Section 8A—Land Mobile Service and Related Subjects. 62 pp.

—Circuit Mode—Multiple Rate—ISDN
CCITT RECMN I.231. 10-92. Circuit-Mode Multiple-Rate Unrestricted 8 kHz Structured Bearer Service Category (Study Group I) 8 pp. 8 pp.

—Circuit Mode—PISN
IEC DIS 11574-92. Information Technology—Telecommunications and Information Exchange Between Systems—Private Integrated Services Network—Circuit-Mode 64 kbit/s Bearer Services—Service Description, Functional Capabilities and Information Flows ***CD-ROM ONLY***. 73 pp.

ISO DIS 11574-92. Information Technology—Telecommunications and Information Exchange Between Systems—Private Integrated Services Network—Circuit-Mode 64 kbit/s Bearer Services—Service Description, Functional Capabilities and Information Flows ***CD-ROM ONLY***. 73 pp.

JTC1 DIS11574-92. Information Technology—Telecommunications and Information Exchange Between Systems—Private Integrated Services Network—Circuit-Mode 64 kbit/s Bearer Services—Service Description, Functional Capabilities and Information Flows ***CD-ROM ONLY***. 73 pp.

—Circuit Mode—Private Switched Networks
CENELEC PRETS 300 171-91. Specification, Functional Model and Information Flows for Control Aspects of Circuit Mode Basic Services in Private Telecommunications Networks. 103 pp.

CENELEC PRETS 300 171-92. Private Telecommunications Network (PTN); Specification, Functional Model and Information Flows Control Aspects of Circuit Mode Basic Services. 96 pp.

CENELEC ETS 300 171-92. Private Telecommunication Network (PTN); Specification, Functional Models and Information Flows Control Aspects of Circuit Mode Basic Services. 95 pp.

ECMA ECMA 143-90. Layer 3 Protocol for Signalling Between Exchanges of Private Telecommunication Networks for the Control of Circuit-Switched Calls. 164 pp.

Bearer Services (Cont.)

—Circuit Mode—Private Switched Networks (Cont.)
ECMA ECMA 143-92. Private Telecommunication Network (PTN)—Inter-Exchange Signalling Protocol—Circuit Mode Basic Services (QSIG-BC). 157 pp.

ETSI PRETS 300 171-91. Specification, Functional Model and Information Flows for Control Aspects of Circuit Mode Basic Services in Private Telecommunications Networks (Standard ECMA-142). 102 pp.

ETSI ETS 300 171-92. Private Telecommunication Network (PTN); Specification, Functional Models and Information Flows Control Aspects of Circuit Mode Basic Services. 95 pp.

ETSI PRETS 300 171-92. Private Telecommunications Network (PTN); Specification, Functional Model and Information Flows Control Aspects of Circuit Mode Basic Services. 96 pp.

ETSI PRETS 300 171-91. Specification, Functional Model and Information Flows for Control Aspects of Circuit Mode Basic Services in Private Telecommunications Networks. 103 pp.

—Circuit Mode—PTN—Signaling Protocols—S Reference Point
CENELEC PRETS 300 192-91. Layer 3 Protocol for Signalling over the D-Channel of Interfaces at the S Reference Point Between Terminal Equipment and Private Telecommunication Networks for the Control of Circuit-Switched Calls. 41 pp.

CENELEC PRETS 300 192-92. Private Telecommunications Network (PTN); Signalling Protocol at the S-Reference Point Circuit Mode Basic Services. 37 pp.

CENELEC ETS 300 192-92. Private Telecommunication Network (PTN); Signalling Protocol at the S-Reference Point Circuit Mode Basic Services. 37 pp.

ETSI ETS 300 192-92. Private Telecommunication Network (PTN); Signalling Protocol at the S-Reference Point Circuit Mode Basic Services. 37 pp.

ETSI PRETS 300 192-92. Private Telecommunications Network (PTN); Signalling Protocol at the S-Reference Point Circuit Mode Basic Services. 37 pp.

ETSI PRETS 300 192-91. Layer 3 Protocol for Signalling over the D-Channel of Interfaces at the S Reference Point Between Terminal Equipment and Private Telecommunication Networks for the Control of Circuit-Switched Calls. 41 pp.

—Circuit Mode—PTNX—Signaling Protocols—Q Reference Point
CENELEC PRETS 300 172-91. Layer 3 Protocol for Signalling Between Exchanges of Private Telecommunication Networks for the Control of Circuit-Switched Calls. 167 pp.

CENELEC PRETS 300 172-92. Private Telecommunications Network (PTN); Inter-Exchange Signalling Protocol Circuit Mode Basic Services. 149 pp.

CENELEC ETS 300 172-92. Private Telecommunication Network (PTN); Inter-Exchange Signalling Protocol Circuit Mode Basic Services. 148 pp.

CENELEC PRETS 300 172-93. Private Telecommunication Network (PTN); Inter-Exchange Signalling Protocol Circuit Mode Basic Services. 159 pp.

ETSI PRETS 300 172-91. Layer 3 Protocol for Signalling Between Exchanges of Private Telecommunication Networks for the Control of Circuit-Switched Calls (Standard ECMA-143). 166 pp.

ETSI ETS 300 172-92. Private Telecommunication Network (PTN); Inter-Exchange Signalling Protocol Circuit Mode Basic Services. 148 pp.

ETSI PRETS 300 172-92. Private Telecommunications Network (PTN); Inter-Exchange Signalling Protocol Circuit Mode Basic Services. 149 pp.

ETSI PRETS 300 172-91. Layer 3 Protocol for Signalling Between Exchanges of Private Telecommunication Networks for the Control of Circuit-Switched Calls. 167 pp.

ETSI PRETS 300 172-93. Private Telecommunication Network (PTN): Interexchange Signalling Protocol Circuit Mode Basic Services. 159 pp.

—Circuit Mode—Telecommunication Services
CCITT RECMN I.220-89. Common Dynamic Description of Basic Telecommunication Services—Integrated Services Digital Network (ISDN)—General Structure and Service Capabilities (Study Group XVIII) 9 pp. 9 pp.

Bearer Services (Cont.)

—Communication Terminal Equipment—ISDN
CCITT RECMN I.333-89. Terminal Selection in ISDN—Integrated Services Digital Network (ISDN)—Overall Network Aspects and Functions, ISDN User-Network Interfaces (Study Group XVIII) 18 pp. 18 pp.

CENELEC ETR 026-92. Network Aspects (NA); Terminal Selection Principles for Priority 1 and 2 Services of MoU—ISDN Applicable in Multi-Terminal Environments at Customer Premises. 21 pp.

ETSI ETR 026-92. Network Aspects (NA); Terminal Selection Principles for Priority 1 and 2 Services of MoU—ISDN Applicable in Multi-Terminal Environments at Customer Premises. 21 pp.

—Connection Oriented Broadband Service
CCITT RECMN F.811-92. Broadband Connection-Oriented Bearer Service (Study Group I) 12 pp. 12 pp.

—Connectionless Broadband Data Service
CCITT RECMN F.812-92. Broadband Connectionless Data Bearer Service (Study Group I) 8 pp. 8 pp.

CENELEC PRETS 300 217-2-91. Network Aspects (NA); Connectionless Broadband Data Service (CBDS) Part 2: Basic Bearer Service Definition (DE/NA-53201-2). 22 pp.

CENELEC PRETS 300 217-2-92. Network Aspects (NA); Connectionless Broadband Data Service (CBDS) Part 2: Basic Bearer Service Definition. 20 pp.

CENELEC ETS 300 217-2-92. Network Aspects (NA); Connectionless Broadband Data Service (CBDS) Part 2: Basic Bearer Service Definition. 20 pp.

ETSI ETS 300 217-2-92. Network Aspects (NA); Connectionless Broadband Data Service (CBDS) Part 2: Basic Bearer Service Definition. 20 pp.

ETSI PRETS 300 217-2-92. Network Aspects (NA); Connectionless Broadband Data Service (CBDS) Part 2: Basic Bearer Service Definition. 20 pp.

ETSI PRETS 300 217-2-91. Network Aspects (NA); Connectionless Broadband Data Service (CBDS) Part 2: Basic Bearer Service Definition (DE/NA-53201-2). 22 pp.

—Digital Line Systems—Frequency Division Multiplexers
CCITT RECMN G.941-89. Digital Line Systems Provided by FDM Transmission Bearers—Digital Networks, Digital Sections and Digital Line Systems (Study Groups XV and XVIII) 4 pp. 4 pp.

—Frame Mode—Data Link Layer—ISDN—DSS1
CCITT RECMN Q.922-92. ISDN Data Link Layer Specification for Frame Mode Bearer Services (Study Group XI) 112 pp. 112 pp.

—Frame Relaying—ISDN
CCITT RECMN I.233-92. Frame Mode Bearer Services (Study Group XVIII) 49 pp. 49 pp.

—Frame Relaying—ISDN—Congestion Management
CCITT RECMN I.370-91. Congestion Management for the ISDN Frame Relaying Bearer Service (Study Group XVIII) 13 pp. 13 pp.

—ISDN
CCITT RECMN D.210-89. General Charging and Accounting Principles for International Telecommunication Services Provided over the Integrated Services Digital Network (ISDN)—General Tariff Principles—Charging and Accounting in Intl Telecom. Serv. (Study Group III) 4 pp. 4 pp.

CCITT RECMN E.711-92. User Demand Modelling Study Group II) 11 pp. 11 pp.

CCITT RECMN I.230-89. Definition of Bearer Service Categories—Integrated Services Digital Network (ISDN)—General Structure and Service Capabilities (Study Group XVIII) 5 pp. 5 pp.

CCITT RECMN I.241-89. Teleservices Supported by an ISDN—Integrated Services Digital Network (ISDN)—General Structure and Service Capabilities (Study Group XVIII) 28 pp. 28 pp.

CCITT RECMN I.320-89. ISDN Protocol Reference Model—Integrated Services Digital Network (ISDN)—Overall Network Aspects and Functions, ISDN User-Network Interfaces (Study Group XVIII) 10 pp. 10 pp.

CENELEC PRETS 300 083-90. Integrated Services Digital Network (ISDN): Circuit Mode Structural Bearer Service Category Usable for Speech Information Transfer End-to-End Compatibility (T/TE 12-07). 7 pp.

Bearer Services (Cont.)

—ISDN (Cont.)

CENELEC PRETS 300 083-92. Integrated Services Digital Network (ISDN); Circuit Mode Structured Bearer Service Category Usable for Speech Information Transfer Terminal Requirements Necessary for End-to-End Compatibility. 9 pp.

CENELEC ETS 300 083-93. Integrated Services Digital Network (ISDN); Circuit Mode Structured Bearer Service Category Usable for Speech Information Transfer Terminal Requirements Necessary for End-to-End Compatibility. 9 pp.

CENELEC PRETS 300 084-90. Integrated Services Digital Network (ISDN); Circuit Mode Structural Bearer Service Category Usable for 3.1 KHz Audio Information Transfer End-to-End Compatibility (T/TE 12-08). 7 pp.

CENELEC PRETS 300 084-92. Integrated Services Digital Network (ISDN); Circuit Mode Structured Bearer Service Category Usable for 3,1 kHz Audio Information Transfer Terminal Requirements Necessary for End-to-End Compatibility. 9 pp.

CENELEC ETS 300 084-93. Integrated Services Digital Network (ISDN); Circuit Mode Structured Bearer Service Category Usable for 3,1 kHz Audio Information Transfer Terminal Requirements Necessary for End-to-End Compatibility. 9 pp.

CEPT T/CAC S 10.2 E-92. Bearer Services to Be Provided by an Integrated Services Digital Network (ISDN) (Revised in Odense 1986, Vienna 1989 (CAC) and Athens 1992). 70 pp.

CEPT T/CAC S 10.4 E-92. Intercommunication Aspects of an Integrated Services Digital Network (ISDN) (Revised in Odense 1986, Vienna 1989 (CAC) and Athens 1992). 11 pp.

CEPT T/GSI 02-02-87. Services Supports. 16 pp.

CEPT T/GSI 02-02 E-87. Bearer Services. 16 pp.

CEPT T/GSI 02-07-87. Services Supplementaires Associes Aux Services Supports. 5 pp.

CEPT T/GSI 02-07 E-87. Supplementary Services Associated with Bearer Services. 5 pp.

CEPT T/N 13-04 E-89. Supplementary Services Associated with Bearer Services. 5 pp.

CEPT T/SF 31-02 E-89. Bearer Services to Be Provided by Integrated Services Digital Network (ISDN) (Revised in Odense 1986 and Vienna 1989 (CAC)). 38 pp.

CEPT T/SF 69 E-91. International Switched 2 Mbit/s Bearer Service. 4 pp.

CSA CAN/CSA-T543-91. Integrated Services Digital Network (ISDN)—Minimal Set of Bearer Services for the Primary Rate Interface (ANSI T1.603-1990); (Gen Instr 1). 30 pp.

CSA CAN/CSA-T544-91. Integrated Services Digital Network (ISDN)—Minimal Set of Bearer Services for the Basic Rate Interface (ANSI T1.604-1990); (Gen Instr 1). 34 pp.

ETSI PRETS 300 083-90. Integrated Services Digital Network (ISDN); Circuit Mode Structured Bearer Service Category Usable for Speech Information Transfer—End-To-End Compatibility (T/TE 12-07). 7 pp.

ETSI PRETS 300 084-90. Integrated Services Digital Network (ISDN); Circuit Mode Structured Bearer Service Category Usable for 3.1 KHz Audio Information Transfer—End-To-End Compatibility (T/TE 12-08). 7 pp.

ETSI PRETS 300 108-90. Integrated Services Digital Network (ISDN); Circuit-Mode 64 Kbit/s Unrestricted 8kHz Structed Bearer Service Category (T/NA1 (89)35) Service Description. 10 pp.

ETSI PRETS 300 109-90. Integrated Services Digital Network (ISDN) Circuit-Mode 64 Kbit/s 8 KHz Structured Bearer Service Category (T/NA1 (89)36) Service Description. 10 pp.

ETSI PRETS 300 110-90. Integrated Services Digital Network (ISDN) Circuit-Mode 64 Kbit/s 8 KHz Structured Bearer Service Category (T/NAI (89)37) Usable for 3.1 KHz Audio Information Transfer—Service Description. 10 pp.

ETSI ETS 300 083-93. Integrated Services Digital Network (ISDN); Circuit Mode Structured Bearer Service Category Usable for Speech Information Transfer Terminal Requirements Necessary for End-to-End Compatibility. 9 pp.

ETSI PRETS 300 083-92. Integrated Services Digital Network (ISDN); Circuit Mode Structured Bearer Service Category Usable for Speech Information Transfer Terminal Requirements Necessary for End-to-End Compatibility. 9 pp.

ETSI PRETS 300 083-90. Integrated Services Digital Network (ISDN): Circuit Mode Structural Bearer Service Category Usable for Speech Information Transfer End-to-End Compatibility (T/TE 12-07). 7 pp.

ETSI ETS 300 084-93. Integrated Services Digital Network (ISDN); Circuit Mode Structured Bearer Service Category Usable for 3,1 kHz Audio Information Transfer Terminal Requirements Necessary for End-to-End Compatibility. 9 pp.

ETSI PRETS 300 084-92. Integrated Services Digital Network (ISDN); Circuit Mode Structured Bearer Service Category Usable for 3,1 kHz Audio Information Transfer Terminal Requirements Necessary for End-to-End Compatibility. 9 pp.

ETSI PRETS 300 084-90. Integrated Services Digital Network (ISDN): Circuit Mode Structural Bearer Service Category Usable for 3.1 KHz Audio Information Transfer End-to-End Compatibility (T/TE 12-08). 7 pp.

ETSI PRETS 300 108-90. Integrated Services Digital Network (ISDN) Circuit-Mode 64Kbit/s Unrestricted 8KHz Structured Bearer Service Category (T/NA1 (89)35) Service Description. 6 pp.

ETSI PRETS 300 109-90. Integrated Services Digital Network (ISDN) Circuit-Mode 64Kbit/s 8KHz Structured Bearer Service Category (T/NA1 (89)36) Service Description. 6 pp.

ETSI PRETS 300 110-90. Integrated Services Digital Network (ISDN) Circuit-Mode 64 Kbit/s 8 KHz Structured Bearer Service Category (T/NA1 (89)37) Usable for 3.1 KHz Audio Information Transfer Service Description. 7 pp.

—ISDN—Classifications

CCITT RECMN I.210-89. Principles of Telecommunication Services Supported by an ISDN and the Means to Describe Them—Integrated Services Digital Network (ISDN)—General Structure and Service Capabilities (Study Group XVIII) 19 pp. 19 pp.

—ISDN—Functions/Information Flows

CCITT RECMN Q.71-89. ISDN 64 kBit/s Circuit Mode Switched Bearer Services—General Recommendations on Telephone Switching and Signalling—Functions and Information Flows for Services in the ISDN—Supplements (Study Group XI) 44 pp. 44 pp.

—ISDNUP—Preference Indicators

CCITT RECMN E.172-89. Call Routing in the ISDN Era—Telephone Network and ISDN—Operation, Numbering, Routing and Mobile Service (Study Group II) 14 pp. 14 pp.

—Packet Mode—Connectionless—ISDN

CCITT RECMN I.232. 2-88. Connectionless Bearer Service Category—Integrated Services Digital Network (ISDN)—General Structure and Service Capabilities (Study Group XVIII) 1 pp. 1 p.

—Packet Mode—ISDN

CCITT RECMN F.353-89. Provision of Telematic and Data Transmission Services on Integrated Services Digital Network (ISDN)—Telematic, Data Transmission and Teleconference Services—Operations and Quality of Service (Study Group I) 5 pp. 5 pp.

CCITT RECMN I.122-89. Framework for Providing Additional Packet Mode Bearer Services—Integrated Services Digital Network (ISDN)—General Structure and Service Capabilities (Study Group XVIII) 17 pp. 17 pp.

CCITT RECMN I.232-89. Packet-Mode Bearer Services Categories—Integrated Services Digital Network (ISDN)—General Structure and Service Capabilities (Study Group XVIII) 5 pp. 5 pp.

—Packet Mode—Land Mobile Services

CCIR RECMN 687-90. Future Public Land Mobile Telecommunication Systems (FPLMTS)—Section 8A—Land Mobile Service and Related Subjects. 11 pp.

CCIR RECMN 687-1-92. Future Public Land Mobile Telecommunication Systems (FPLMTS)—Section 8A—Land Mobile Service and Related Subjects. 21 pp.

CCIR Report 1153-90. Future Public Land Mobile Telecommunication Systems—Section 8A—Land Mobile Service and Related Subjects. 62 pp.

—Packet Mode—User Signaling—ISDN

CCITT RECMN I.232. 3-88. User Signalling Bearer Service Category—Integrated Services Digital Network (ISDN)—General Structure and Service Capabilities (Study Group XVIII) 1 pp. 1 p.

—Packet Mode—Virtual Call—ISDN

CCITT RECMN I.232. 1-88. Virtual Call and Permanent Virtual Circuit Bearer Service Category—Integrated Services Digital Network (ISDN)—General Structure and Service Capabilities (Study Group XVIII) 5 pp. 5 pp.

—Public Land Mobile Networks

CENELEC GSM 02.02-92. European Digital Cellular Telecommunication System (Phase 1); Bearer Services Supported by a GSM PLMN. 40 pp.

ETSI GSM 02.02-92. European Digital Cellular Telecommunication System (Phase 1); Bearer Services Supported by a GSM PLMN. 40 pp.

—Symbols—Start-Stop

CCITT RECMN V.14-89. Transmission of Start-Stop Characters over Synchronous Bearer Channels—Data Communication over the Telephone Network (Study Group XVII) 4 pp. 4 pp.

—Tariffs (Telecommunications)

CEPT T/PGT 41 E-91. Charging and Accounting Principles for the International 2 Mbit/s Switched Bearer Service. 1 p.

—Telegraph Circuits—Dual Diversity

CCITT RECMN R.150-89. Automatic Protection Switching of Dual Diversity Bearers—Telegraph Transmission (Study Group IX) 4 pp. 4 pp.

Bearers

Use: Bearing Plates

Bearing Alloys

Use For: Antifriction Alloys; Babbitt; Babbitt Metals
See Also: Alloys; Bearings; Ferroalloys; Tin Alloys; White Metal Bearing Alloys; Zinc Alloys

—Aluminum

BSI BS 5812: Part 1-79. 1979 Materials for Plain Bearings Part 1: Specification for Aluminium Alloy for Solid Bearings. 4 pp.

ISO 6279-79. Plain Bearings—Aluminium Alloy for Solid Bearings First Edition; (Corrected and Reprinted -1979). 4 pp.

—Carbide Content—Photomicrography

ISO 5949-83. Tool Steels and Bearing Steels—Micrographic Method for Assessing the Distribution of Carbides Using Reference Photomicrographs First Edition. 12 pp.

—Chemical Analysis

CNS G2004-53. Method of Chemical Test for Antifriction Alloys (Jul)(293)(R 1973).

—Chromium Steel

CNS G3060-88. High Carbon Chromium Bearing Steel (Mar)(3014).

JIS G 4805-90. High Carbon Chromium Bearing Steels. 16 pp.

—Compression Testing

ISO 4385-81. Plain Bearings—Compression Testing of Metallic Bearing Materials First Edition. 6 pp.

—Copper

CNS H3130-84. Copper-Lead Alloy Castings for Bearing (Aug)(10983).

ISO 4382 Pt 1-91. Plain Bearings—Copper Alloys—Part 1: Cast Copper Alloys for Solid and Multilayer Thick-Walled Plain Bearings Second Edition. 11 pp.

ISO 4382 Pt 2-91. Plain Bearings—Copper Alloys—Part 2: Wrought Copper Alloys for Solid Plain Bearings Second Edition. 7 pp.

—Crown Metal—Marine

MOD UK NES 836-88. Requirements for Crown Metal Cast Bearings Issue 1 (08.88). 16 pp.

—Hardness Testing

ISO 4384 Pt 1-82. Plain Bearings—Hardness Testing of Bearing Metals—Part 1: Compound Materials First Edition. 4 pp.

ISO 4384 Pt 2-82. Plain Bearings—Hardness Testing of Bearing Metals—Part 2: Solid Materials First Edition. 3 pp.

—Lead

BSI BS 3332-87. 1987 White Metal Bearing Alloy Ingots. 7 pp.

ISO 4381-91. Plain Bearings—Lead and Tin Casting Alloys for Multilayer Plain Bearings Second Edition. 8 pp.

—Lubricants—Corrosion Testing

ISO TR10129-93. Plain Bearings—Testing of Bearing Metals—Resistance to Corrosion by Lubricants Under Static Conditions First Edition. 7 pp.

—Mild Steel—Aerospace

BSI 3S 91-77. 1977 Mild Steel Billets, Bars, Forgings and Parts (Suitable for Bearing Shells). 2 pp.

—Multilayer

ISO 4381-91. Plain Bearings—Lead and Tin Casting Alloys for Multilayer Plain Bearings Second Edition. 8 pp.

ISO 4382 Pt 1-91. Plain Bearings—Copper Alloys—Part 1: Cast Copper Alloys for Solid and Multilayer Thick-Walled Plain Bearings Second Edition. 11 pp.

ISO 4383-91. Plain Bearings—Multilayer Materials for Thin-Walled Plain Bearings Second Edition. 12 pp.

Bearing Alloys (Cont.)

—Sintered

BSI BS 5600: Sec 5.1-88. 1988 Powder Metallurgical Materials and Products Part 5: Material Specifications for Sintered Metal Products, Excluding Hardmetals Section 5.1: Materials, for Bearings, Impregnated with Liquid Lubricant. 4 pp.

BSI BS 5600: Sec 5.2-88. 1988 Powder Metallurgical Materials and Products Part 5: Material Specs for Sintered Metal Products, Excluding Hardmetals Section 5.2: Sintered Iron and Sint. Steel Containing One or Both of the Elements Carbon and Copper, Used for Structural Parts. 6 pp.

BSI BS 5600: Sec 5.3-88. 1988 Powder Metallurgical Materials and Products Part 5: Material Specifications for Sintered Metal Products, Excluding Hardmetals Section 5.3: Sintered Alloyed and Sintered Stainless Steels Used for Structural Parts. 6 pp.

ISO 5755 Pt 1-87. Sintered Metal Materials—Specifications—Part 1: Materials, for Bearings, Impregnated with Liquid Lubricant Third Edition. 4 pp.

ISO 5755 Pt 3-87. Sintered Metal Materials—Specifications—Part 3: Sintered Alloyed and Sintered Stainless Steels Used for Structural Parts First Edition. 7 pp.

—Steel

DIN ENGL 17230-80. Ball and Roller Bearing Steels; Technical Conditions of Delivery (Sept). 18 pp.

ISO 683 Pt 17-76. Heat-Treated Steels, Alloy Steels and Free-Cutting Steels—Part 17: Ball and Roller Bearing Steels First Edition. 11 pp.

—Tin

BSI BS 3332-87. 1987 White Metal Bearing Alloy Ingots. 7 pp.

ISO 4381-91. Plain Bearings—Lead and Tin Casting Alloys for Multilayer Plain Bearings Second Edition. 8 pp.

Bearing Boxes

Use For: Bearing Liners   *See Also:* Bearing Housings; Bearings

—Ball Bearings

CNS B3249-76. Bearing Box (Ball Bearing, Assembly 2, Light Grade, Cone Form) (Dec)(4030).

CNS B3250-76. Bearing Box (Ball Bearing, Assembly 3, Medium Grade, Cone Form) (Dec)(4031).

CNS B3251-76. Bearing Box (Ball Bearing, Assembly 2, Light Grade, Round Form) (Dec)(4032).

CNS B3252-76. Bearing Box (Ball Bearing, Assembly 3, Medium, Grade, Round Form) (Dec)(4033).

Bearing Capacity

Use For: Load Capacity   *See Also:* Cohesion; Foundations; Loads (Forces); Stability Testing

—Foot Bridges

DIN ENGL 1072-85. Road and Foot Bridges; Design Loads (Dec). 14 pp.

—High Strength Steels

DIN ENGL 50969-90. Testing of High-Strength Steel Building Elements for Resistance to Hydrogen-Induced Brittle Fracture and Advice on the Prevention of Such Fracture (Dec). 5 pp.

—Highway Bridges

DIN ENGL 1072-85. Road and Foot Bridges; Design Loads (Dec). 14 pp.

—Sheeting

DIN ENGL 18807 Pt 1-87. Trapezoidal Sheeting in Building; Trapezoidal Steel Sheeting; General Requirements and Determination of Loadbearing Capacity by Calculation (June). 16 pp.

DIN ENGL 18807 Pt 2-87. Trapezoidal Sheeting in Building; Trapezoidal Steel Sheeting; Determination of Loadbearing Capacity by Testing (June). 10 pp.

—Transport Aircraft—Information Interchange

NATO STANAG 3767 ED 1 AMD 10-75. Exchange of Data on Load Capabilities of Transport Aircraft. 20 pp.

NATO STANAG 3767 ED 1 AMD 11-75. Exchange of Data on Load Capabilities of Transport Aircraft. 20 pp.

Bearing Caps

See Also: Antifriction Bearings

—Textile Machinery—Roller Bearings

BSI BS 5509-77. 1977 Textile Machinery and Accessories. Bearings for Bottom Rollers Allied Dimensions. Caps with Central Nose and Caps with Side Lugs. 7 pp.

ISO 3464-77. Textile Machinery and Accessories—Bearings for Bottom Rollers and Allied Dimensions—Caps with Central Nose and Caps with Side Lugs First Edition. 8 pp.

Bearing Greases

See Also: Bearings; Greases

CGSB 3-GP-664MB-90. Heavy-Duty Water-Wash-Resistant Grease (NLGI Grades No. 1 and No. 2). 8 pp.

DIN ENGL 51825-90. Type K Lubricating Greases (Aug) (This Standard Supersedes the June 1981 Edition of DIN 51825 Part 1, the December 1979 Edition of DIN 51825 Part 2, and the June 1981 Edition of DIN 51825 Part 3 Withdrawn in January 1986). 5 pp.

MOD UK DSTAN 91-17-90. Grease, Calcium Base Joint Service Designation LG-280 Issue 2. 12 pp.

MOD UK DSTAN 91-28-76. Grease, Naval: General Purpose, Ball and Roller Bearing NATO Code No: G-450 Joint Service Designation: XG-274 Issue 1. 11 pp.

—Corrosion Prevention

DIN ENGL 51802-90. Testing Lubricating Greases for Their Corrosion-Inhibiting Properties by the SKF Emcor Method (Apr). 5 pp.

—Helicopters

MOD UK DSTAN 91-51-01. Grease, Aircraft: Helicopter Oscillating Bearing NATO Code No: G-366 Joint Service Designation: XG-284 Issue 1; Amendment 1. 13 pp.

—Naval

CGSB 3-GP-691MB-90. General Purpose Grease. 9 pp.

—Roller Bearings

BSI BS 2000: Part 168-93. 1993 Methods of Test for Petroleum and Its Products Part 168: Determination of Rolling Bearing Performance of Lubricating Grease (W). 8 pp.

BSI BS 2000: Part 168-87. 1987 Petroleum and Its Products Part 168: Rolling Bearing Performance Test for Lubricating Greases. 10 pp.

Bearing Housings

See Also: Bearing Boxes; Bearings; Housings; Pillow Blocks

BSI BS 5646: Part 4-79. 1979 Rolling Bearings—Accessories Part 4: Plummer Block Housings. 4 pp.

CNS B1169-80. Rolling Bearings; Component Parts and Spherical Plain Bearings; Summary of Standards (Jul)(5694).

CNS B1210-80. Roller Bearings for Electric Traction, Selection of Fit for Shafts and Housings (Aug)(6310).

CNS B2478-80. Rolling Bearings; Bearings with Spherical Outside Surface and Extended Inner Ring Width, Steel Pressed Housings for Insert Bearings (May)(5508).

CNS B2479-80. Rolling Bearings; Bearings with Spherical Outside Surface and Extended Inner Ring Width, Cast Housings for Insert Bearings (May)(5509).

CNS B2677-82. Pedestal Plain Bearings; Assembly, Housing (Jul)(9062).

CNS B2678-82. Pedestal Plain Bearings; Bearing Shells (Jul)(9063).

DIN ENGL 5425 Pt 1-84. Rolling Bearings; Mounting Tolerances; General Guidelines (Nov). 5 pp.

ISO 113 Pt II-79. Rolling Bearing Accessories—Part II: Plummer Block Housings First Edition. 4 pp.

JIS B 1551-88. Plummer Block Housings for Rolling Bearings. 18 pp.

JIS B 1559-76. Housings for Rolling Bearing Unit (R 1979). 22 pp.

SAA AS 1911-84. Solid Bearing Housings—Dimensions and Tolerances. 4 pp.

—Aircraft—Alloy

SBAC RS 182 ISSUE 1. Fits for Ball Bearings.

—Aircraft—Steel

SBAC RS 182 ISSUE 1. Fits for Ball Bearings.

—Felt Rings

CNS B1289-82. Felt Ring Groove; for Rolling Bearing Housing (Sep)(9336).

CNS B2699-82. Felt Ring for Rolling Bearing Housing (Sep)(9337).

DIN ENGL 5419-59. Felt Rings; Felt Strips; Ring Grooves for Rolling Bearing Housings (Sept). 3 pp.

—Felt Strips

CNS B1289-82. Felt Ring Groove; for Rolling Bearing Housing (Sep)(9336).

CNS B2700-82. Felt Strip for Rolling Bearing Housing (Sep)(9338).

DIN ENGL 5419-59. Felt Rings; Felt Strips; Ring Grooves for Rolling Bearing Housings (Sept). 3 pp.

—Plain Bearings

DIN ENGL 31693-90. Pedestal Plain Bearings; Side Flange Bearings (Sept). 7 pp.

Bearing Housings (Cont.)

—Ring Grooves

CNS B1289-82. Felt Ring Groove; for Rolling Bearing Housing (Sep)(9336).

DIN ENGL 5419-59. Felt Rings; Felt Strips; Ring Grooves for Rolling Bearing Housings (Sept). 3 pp.

—Washers

CNS B2623-81. Deep Groove Thrust Ball Bearings; with Aligning Seat and Housing Washer with Aligning Seat (Nov)(8095).

CNS B2625-81. Deep Groove Thrust Ball Bearings; Double-Thrust with Aligning Seat and Housing Washer with Aligning Seat (Nov)(8097).

Bearing Liners

Use: Bearing Boxes

Bearing Lubricants

See Also: Bearing Oils; Bearings; Gear Lubricants; Lubricants

BSI BS 6413: Part 2-83. 1983 Lubricants, Industrial Oils and Related Products (Class L) Part 2: Classification for Family F (Spindle Bearings, Bearings and Associated Clutches). 3 pp.

ISO 6743 Pt 2-81. Lubricants, Industrial Oils and Related Products (Class L)—Classification—Part 2: Family F (Spindle Bearings, Bearings and Associated Clutches) First Edition. 3 pp.

ISO 6743 Pt 13-89. Lubricants, Industrial Oils and Related Products (Class L)—Classification—Part 13: Family G (Slideways) First Edition. 4 pp.

JIS K 2239-83. Bearing Oils. 7 pp.

MOD UK DSTAN 91-51-01. Grease, Aircraft: Helicopter Oscillating Bearing NATO Code No: G-366 Joint Service Designation: XG-284 Issue 1; Amendment 1. 13 pp.

—Corrosion Testing

ISO TR10129-93. Plain Bearings—Testing of Bearing Metals—Resistance to Corrosion by Lubricants Under Static Conditions First Edition. 7 pp.

Bearing Metals

Use: Bearing Alloys

Bearing Oils

See Also: Bearing Lubricants; Lubricating Oils; Oils

JIS K 2239-83. Bearing Oils. 7 pp.

Bearing Plates

Use For: Bearers   *See Also:* Plates (Supports)

—Rail Tracks

BSI BS 751-92. 1992 Steel Baseplates for Flat Bottom Railway Rails (Q). 11 pp.

BSI BS 751-59. 1959 Amd 1 Steel Bearing Plates for Flat Bottom Railway Rails. 13 pp.

—Spring Link Washer—Suspension Systems (Vehicles)

CNS B2603-81. Laminated Spring; Spring Link Washer Bearing Plates for Spring Suspension (Sep)(7864). 3 pp.

Bearing Pressure

See Also: Pressure

—Foundations—Subsoil

DIN ENGL 4018-74. Subsoil; Calculation of the Bearing Pressure; Distribution Under Spread Foundations (Sept). 3 pp.

Bearing Rings

See Also: Bearings; Rings; Roller Bearings

—Driving Wheels—Cranes

DIN ENGL 15094-82. Cranes; Driving Wheel Units and Idler Wheel Units; Bearing Cage Rings (July). 3 pp.

—Idlers—Cranes

DIN ENGL 15094-82. Cranes; Driving Wheel Units and Idler Wheel Units; Bearing Cage Rings (July). 3 pp.

Bearing Sleeves

Use: Sleeve Bearings

Bearing Steels

Use: Bearing Alloys

Bearing Strength

See Also: Loads (Forces)

—Reinforced Plastics

JIS K 7080-91. Testing Methods for Bearing Strength of Carbon Fiber Reinforced Plastics. 13 pp.

Bearing Strength (Cont.)

—Soil Analysis

SNZ NZS 4402: Pt 6:TEST 6.1.1-86. Methods of Testing Soils for Civil Engineering Purposes Part 6: Soil Strength Tests. Section 6.1: Determination of the California Bearing Ratio (CBR). Test 6.1.1: Standard Laboratory Method for Remoulded Specimens. 9 pp.

SNZ NZS 4402: Pt 6:TEST 6.1.2-86. Methods of Testing Soils for Civil Engineering Purposes Part 6: Soil Strength Tests. Section 6.1: Determination of the California Bearing Ratio (CBR). Test 6.1.2: Standard Laboratory Method for Undisturbed Specimens. 7 pp.

SNZ NZS 4402: Pt 6:TEST 6.1.3-86. Methods of Testing Soils for Civil Engineering Purposes Part 6: Soil Strength Tests. Section 6.1: Determination of the California Bearing Ratio (CBR). Test 6.1.3: Standard Method for In Situ Tests. 5 pp.

Bearings

Scope Note: For additional listings, use a more specific term *See Also:* Airframe Bearings; Angular Contact Bearings; Antifriction Bearings; Axles; Ball Bearings; Bearing Alloys; Bearing Boxes; Bearing Greases; Bearing Housings; Bearing Lubricants; Bearing Rings; Bridge Bearings; Bushings; Crown Metal Bearings; Gear Boxes; Idlers; Jewel Bearings; Journal Bearings; Lubrication; Magnetic Bearings; Needle Roller Bearings; Pillow Blocks; Piston Pins; Pivots; Plain Bearings; Precision Bearings; Radial Bearings; Roller Bearings; Roller Thrust Bearings; Self Aligning Bearings; Self Lubricating Bearings; Shafts (Machine Elements); Sleeve Bearings; Slide Bearings; Split Bearings; Structural Bearings; Thrust Bearings; Wheels; Wrapped Bushes

CNS B2587-81. Remote Control Gear with Transmission Tubing for Manual Operation; Bearings (Jun)(7584).

JIS B 1581-76. Oil Impregnated Sintered Bearings (R 1979). 13 pp.

—Aircraft—Design

CAA LEAFLET 11-14. Design Requirements and Quality Assurance Associated with Bearing Assemblies. 2 pp.

—Aircraft—Quality Assurance

CAA LEAFLET 11-14. Design Requirements and Quality Assurance Associated with Bearing Assemblies. 2 pp.

—Compression Testing

CNS Z8043-82. Method for Determining Radial Crushing Strength Constant of Metal Powder Sintered Bearing, Oil Impregnated (Jul)(9207).

JIS Z 2507-89. Method for Determination of Radial Crushing Strength Constant of Metal Powder Sintered Bearing, Oil Impregnated. 7 pp.

—Eye Mounted

CNS B2640-82. Eye-Mounted Bearings (Jan)(8210).

DIN ENGL 504-73. Driving Components; Eye-Mounted Bearings (July). 2 pp.

—Flanged

CNS B2638-82. Flange-Mounted Bearings; Fastened with 2 Screws (Jan)(8208).

CNS B2639-82. Flange-Mounted Bearings; Fastened with 4 Screws (Jan)(8209).

DIN ENGL 502-73. Driving Components; Flanged Bearings; Fastened with 2 Screws (July). 2 pp.

DIN ENGL 503-73. Driving Components; Flanged Bearings; Fastened with 4 Screws (July). 2 pp.

—Flanged—Multispindle Heads

DIN ENGL 69001 Pt 37-81. Machine Tools; Multi-Spindle Heads; Bearing Flanges; Type A (Oct). 1 p.

—Friction—Glossaries

CNS B1280-82. Friction in Bearing: Types; Physical Quantities (Feb)(8467).

DIN ENGL 50281-77. Friction in Bearings; Definitions, Types, Conditions, Physical Quantities (Oct). 2 pp.

—Motorcycles—Steering

CNS B2644-83. Steering Bearings for Motorcycles (Jan)(8215).

—Oil Content

CNS Z8037-82. Method for Determining Oil Content in Metal Powder Sintered Bearing Oil Impregnated (Jul)(9201).

JIS Z 2501-89. Method for Determination of Oil Content in Metal Powder Sintered Bearing, Oil Impregnated. 7 pp.

—Preservation—Ships

MOD UK NES 733-83. Requirements for Surface Ships in Long Term Preservation—Protection of Propellers, Shafts & Bearings Issue 1 (07.83). 43 pp.

Bearings (Cont.)

—Ships—Derricks—Goose Necks

ISO 6045-87. Shipbuilding and Marine Structures—Bearings for Derrick Goosenecks—Assemblies and Components First Edition. 19 pp.

—Ships—Derricks—Trunnions

ISO 8314-87. Shipbuilding and Marine Structures—Trunnion Pieces for Span Bearings and Lead Block Bearings First Edition. 7 pp.

—Snap Rings

CNS B2683-82. Snap Rings for Bearings with Ring Groove (Jul)(9072).

—Soleplates

DIN ENGL 189-77. Driving Elements; Sole Plates; Main Dimensions (July). 1 p.

Beaters (Textile Machinery)

See Also: Textile Machinery

—Lags

JIS L 5163-60. Spiked Wood Lag for Kirschner Beater.

Beating

—Pulps—Jokro Mill

DIN ENGL 54360-77. Testing of Pulp; Laboratory Beating of Pulp with the Jokro Mill (Nov). 4 pp.

ISO 5264 Pt III-79. Pulps—Laboratory Beating—Part III: Jokro Mill Method First Edition. 7 pp.

—Pulps—Pebble Mill

CPPA C.3H-80. Laboratory Processing of Pulp (Pebble Mill Method). 2 pp.

—Pulps—PFI Mill

BSI BS 6094: Part 2-81. 1981 Methods for Laboratory Beating of Pulp Part 2: PFI Mill Method. 8 pp.

CPPA C.7-72. Laboratory Processing of Pulp (PFI Mill). 4 pp.

ISO 5264 Pt 2-79. Pulps—Laboratory Beating—Part 2: PFI Mill Method First Edition. 6 pp.

—Pulps—Valley Beater

BSI BS 6094: Part 1-81. 1981 Methods for Laboratory Beating of Pulp Part 1: Valley Beater Method. 10 pp.

CPPA C.2-90. Laboratory Beating of Pulp (Valley Beater Method). 4 pp.

ISO 5264 Pt 1-79. Pulps—Laboratory Beating—Part 1: Valley Beater Method First Edition. 8 pp.

Beaubey Papers

Use For: Chinese Backing Papers *See Also:* Mian Papers; Papers; Shiuan Papers

CNS P2068-85. Beaubey Paper (Chinese Backing Paper) (Apr)(11248). 2 pp.

Beauty Cases

Use: Overnight Cases

Bed Joints

See Also: Joints; Masonry

—Reinforcement—Bonding Strength

CEN PREN 846-2-92. Methods of Test for Ancillary Components for Masonry—Part 2: Determination of Bond Strength of Bed Joint Reinforcement in Mortar Joints. 12 pp.

—Reinforcement—Welded Joints—Shear Strength

CEN PREN 846-3-92. Methods of Test for Ancillary Components for Masonry—Part 3: Determination of Shear Strength of Welds in Bed Joint Reinforcement. 9 pp.

—Shear Strength

CEN PREN 1052-3-93. Methods of Test for Masonry—Part 3: Determination of Initial Shear Strength. 10 pp.

CEN PREN 1052-4-93. Methods of Test for Masonry—Part 4: Determination of Shear Strength Including Damp Proof Course. 11 pp.

Bed Pans

See Also: Toilet Facilities

BSI BS 2588-55. (WITHDRAWN) 1955 Amd 1 Bed Pans of Perfection Type (Adult Size) (AMD 2708) July 31, 1986 (Superseded by BS 2588: Part 1: 1992). 8 pp.

—Portable

BSI BS 2588: Part 1-92. 1992 Portable Sanitary Pans for Use in Healthcare Establishments Part 1: Specification for Reusable Bed Pans. 14 pp.

BSI BS 2588: Part 2-92. 1992 Portable Sanitary Pans for Use in Healthcare Establishments Part 2: Specification for Reusable Pans for Commode Chairs. 13 pp.

Bed Rugs

Use: Blankets

Bed Sheets

Use For: Bedsheets *See Also:* Bedspreads; Fabrics; Sheets

BSI BS 5815: Part 1-01. 1989 Amd 1 Sheets, Sheeting, Pillowslips, Towels, Napkins, Counterpanes and Continental Quilt Secondary Covers Suitable for Use in the Public Sector Part 1: Specif for Sheeting, Sheets and Pillowslips (AMD 6806) December 24, 1991. 11 pp.

—Cotton

BSI BS 5815-79. (WITHDRAWN) 1979 Cotton and Man-Made Fibre Blend Sheeting, Sheets, Pillowslips, Towels and Napkins for Use in the Public Sector (Superseded by BS 5815: Part 1: 1989, Part 2: 1988, and Part 3: 1991). 11 pp.

—Cotton/Polyester—Flannelette

CGSB 4-GP-6MA-84. Sheeting, Sheets and Blankets, Cotton Polyester Flannelette, Standard for. 7 pp.

—Electric—Safety

CSA C22.2 No. 101-M1984. Electrically Heated Bedding Appliances for Household Use (R 1992); (Gen Instr 1 Thru 4). 61 pp.

CSA 1169 Bull. Electrical Bulletin 1169 June 27, 1978 to C22.2 NO 101. 2 pp.

—Medical—Rubber

CNS T2001-80. Rubber Sheet for Medical Purposes (Jul)(5942).

JIS T 9101-75. Rubber Sheet for Medical Purposes.

—Polyester/Cotton

BSI BS 5815-79. (WITHDRAWN) 1979 Cotton and Man-Made Fibre Blend Sheeting, Sheets, Pillowslips, Towels and Napkins for Use in the Public Sector (Superseded by BS 5815: Part 1: 1989, Part 2: 1988, and Part 3: 1991). 11 pp.

BSI BS 5815: Part 1-89. 1989 Sheets, Sheeting, Pillowslips, Towels, Napkins, Counterpanes and Continental Quilt Secondary Covers Suitable for Use in the Public Sector Part 1: Sheeting, Sheets and Pillowslips. 10 pp.

BSI BS 5815: Part 1-01. 1989 Amd 1 Sheets, Sheeting, Pillowslips, Towels, Napkins, Counterpanes and Continental Quilt Secondary Covers Suitable for Use in the Public Sector Part 1: Specif for Sheeting, Sheets and Pillowslips (AMD 6806) December 24, 1991. 11 pp.

CGSB 4-GP-133M-77. Sheeting and Sheets, Polyester/Cotton, Bleached or Dyed; (Amendment 2 March 1984 Incorporates Amendment 1). 10 pp.

—Sampling

CNS L4048-73. Bedspreads (Jan)(2811). 3 pp.

—Shrinkage

CNS L4048-73. Bedspreads (Jan)(2811). 3 pp.

—Visual Inspection

CNS L4048-73. Bedspreads (Jan)(2811). 3 pp.

Bed Side Cabinets

Use: Night Tables

Bed Warmers

See Also: Appliances

—Electric

SAA AS 3148-91. Approval and Test Specification—Electric Bed-Warmers Amdt 1 October 1992 (in Professional Package 28). 7 pp.

Bedding

Scope Note: For additional listings, use a more specific term *See Also:* Air Mattresses; Bed Sheets; Beds; Bedspreads; Blankets; Futons; Mattress Covers; Mattresses; Pillowcases; Pillows; Quilts

SAA AS 3789.1-91. Textiles for Health Care Facilities and Institutions—Part 1: General Ward Linen Amdt 1 January/February 1992 (in Professional Package 17G).

—Electric—Safety

CSA C22.2 NO 101-M1984. Electrically Heated Bedding Appliances for Household Use (R 1992); (Gen Instr 1 Thru 4). 61 pp.

CSA 1169 Bull. Electrical Bulletin 1169 June 27, 1978 to C22.2 NO 101. 2 pp.

—Filling Materials—Dust

BSI BS 3400-67. (WITHDRAWN) 1967 Methods of Test for Dust in Filling Materials. 20 pp.

—Flammability Testing

BSI BS 7175-89. 1989 Methods of Test for the Ignitability of Bedcovers and Pillows by Smouldering and Flaming Ignition Sources. 20 pp.

Bedding (Cont.)

—Health Care Facilities
SAA AS 3789.2-91. Textiles for Health Care Facilities and Institutions—Part 2: Theatre Linen and Pre-Packs Amdt 1 January/February 1992 (in Professional Package 17G). 71 pp.

—Identification Systems
SNZ NZS 8721-88. Care Labelling of Clothing, Household Textiles, Furnishings, Upholstered Furniture, Bedding, Piece Goods and Yarns Amend: A, 1988. 7 pp.
SNZ NZS 8722-88. Care Labelling—Guide to the Selection of Correct Care Labelling Instructions from NZS 8721 Amend: A, 1988. 15 pp.

—Labels
SAA AS 1957-87. Care Labelling of Clothing, Household Textiles, Furnishings, Upholstered Furniture, Bedding, Piece Goods and Yarns. 7 pp.

—Stuffings—Cleanliness
BSI BS 1425-60. 1960 Amd 14 Cleanliness of Fillings and Stuffings for Bedding, Upholstery, Toys and Other Domestic Articles. 108 pp.
BSI BS 1425: Supp 1-66. 1966 Amd 1 Cleanliness of Fillings and Stuffings for Bedding, Upholstery, Toys and Other Domestic Articles: Supplement 1: Appendices. 44 pp.
BSI BS 1425: Part 1-91. 1991 Cleanliness of Fillings and Stuffings for Bedding, Upholstery and Other Domestic Articles Part 1: Specification for Fillings and Stuffings Other Than Feather and/or Down. 15 pp.
BSI BS 1425: Part 1-01. 1991 Amd 1 Cleanliness of Fillings and Stuffings for Bedding, Upholstery and Other Domestic Articles Part 1: Specification for Fillings and Stuffings Other Than Feather and/or Down (AMD 7788) July 15, 1993 (L). 15 pp.
BSI BS 1425: Part 2-91. 1991 Cleanliness of Fillings and Stuffings for Bedding, Upholstery and Other Domestic Articles Part 2: Specification for Feather and/or Down Fillings and Stuffings. 5 pp.

—Stuffings—Feathers
BSI PD 6522-88. 1988 Terms Relevant to the Feather Industry for Fillings for Bedding, Upholstery and Other Domestic Articles. 7 pp.

—Stuffings—Flock
BSI BS 3474: Parts 1-3-62. (WITHDRAWN) 1962 Amd 4 Quality of Fillings for Bedding, Upholstery and Domestic Articles Part 1: Layered Washed Fillings Containing Wool Part 2: Wased Flock (Looe) for Use in Bedding Part 3: Curled Hair. 21 pp.

—Stuffings—Glossaries
BSI BS 2005-66. 1966 Amd 1 Glossary of Terms for Fillings for Bedding, Upholstery and Other Domestic Articles. 35 pp.
BSI PD 6522-88. 1988 Terms Relevant to the Feather Industry for Fillings for Bedding, Upholstery and Other Domestic Articles. 7 pp.

—Stuffings—Hair
BSI BS 3474: Parts 1-3-62. (WITHDRAWN) 1962 Amd 4 Quality of Fillings for Bedding, Upholstery and Domestic Articles Part 1: Layered Washed Fillings Containing Wool Part 2: Wased Flock (Looe) for Use in Bedding Part 3: Curled Hair. 21 pp.
BSI BS 3474: Part 4-65. (WITHDRAWN) 1965 Amd 1 Quality of Fillings for Bedding, Upholstery and Domestic Articles Part 4: Recleansed Hair. 9 pp.
BSI BS 3474: Part 5-71. (WITHDRAWN) 1971 Quality of Fillings for Bedding, Upholstery and Domestic Articles Part 5: Rubberized Curled Hair for Use in Bedding. 12 pp.

—Stuffings—Wool
BSI BS 3474: Parts 1-3-62. (WITHDRAWN) 1962 Amd 4 Quality of Fillings for Bedding, Upholstery and Domestic Articles Part 1: Layered Washed Fillings Containing Wool Part 2: Wased Flock (Looe) for Use in Bedding Part 3: Curled Hair. 21 pp.

—Ticking
BSI BS 4820-72. (WITHDRAWN) 1972 Performance Requirements for Coverings for Domestic Bedding (Superseded by BS 7337: 1990). 8 pp.

Bedding Compounds
See Also: Sealants

—Bleeding
CGSB CAN2-19.0-M77METH 9.4-78. Methods of Testing Putty, Caulking and Sealing Compounds Oil Bleeding and Spotting of Bedding Compounds. 1 p.

—Wrinkling
CGSB CAN2-19.0-M77METH 9.4-78. Methods of Testing Putty, Caulking and Sealing Compounds Oil Bleeding and Spotting of Bedding Compounds. 1 p.

Bede BD 4-150 Aircraft
See Also: Aircraft

—Antenna Positions
CAA. Bede BD4-150 (Approved Aerial Positions). 1 p.

Beds
See Also: Bedding; Bedsteads; Bunk Beds; Cots; Cribs (Furniture); Divans; Furniture; Hospital Beds

—Bases—Flammability Testing
BSI BS 6807-90. 1990 Methods of Test for Assessment of the Ignitability of Mattresses, Divans and Bed Bases with Primary and Secondary Sources of Ignition. 22 pp.
BSI BS 6807-01. 1990 Amd 1 Methods of Test for Assessment of the Ignitability of Mattresses, Divans and Bed Bases with Primary and Secondary Sources of Ignition (AMD 7551) January 15, 1993. 23 pp.
CEN PREN 597-1-91. Furniture—Assessment of the Ignitability of Mattresses and Bed Bases—Part 1: Ignition Source: Smouldering Cigarette. 15 pp.
CEN PREN 597-2-91. Furniture—Assessment of the Ignitability of Mattresses and Bed Bases—Part 2: Ignition Source: Match Flame Equivalent. 16 pp.

—Flammability Testing
SNZ NZS 8720-92. Methods of Test for Assessment of the Ignitability of Mattresses, Divans and Bed Bases with Primary and Secondary Sources of Ignition Amend: A, 1992. 24 pp.

—Folding—Steel
CNS S2040-80. Method of Test for Steel Folding Bed (Aug)(6270).

—Household
CNS S1062-68. Bed, Wooden, Single (Apr)(2853).
CNS S1124-81. Beds for Domestic Use (May)(7449). 3 pp.
CNS S1125-82. Nominal Dimensions of Domestic Beds (Jun)(7451). 5 pp.
CNS S2053-81. Method of Test for Beds for Domestic Use (May)(7450). 2 pp.
JIS S 1102-88. Beds for Domestic Use. 19 pp.

—Medical
SNZ NZS 7121-79. Specification for Hospital Beds. 20 pp.

—Medical—Accessories
SNZ NZS 7122-80. Specification for Accessories for Use with Hospital Beds. 12 pp.

—Motor Operated
CSA CAN/CSA-C22. 2 NO 68-92. Motor-Operated Appliances (Household and Commercial); (Gen Instr 1 Thru 2). 11 pp.
CSA 1169 Bull. Electrical Bulletin 1169 June 27, 1978 to C22.2 NO 68. 2 pp.

—Springs
CNS S1123-82. Bedspring, Box Type (Aug)(7447).
CNS S2052-81. Method of Test for Bedspring, Box Type (May)(7448). 3 pp.

—Test Methods
BSI BS 7397-91. 1991 Methods of Test for Mattresses, Bed Bases and Bed Sets. 15 pp.

—Ticking
BSI BS 7337-90. 1990 Tickings (Supersedes BS 2036: 1962 & BS 2732: 1961 & BS 4820: 1972). 11 pp.

—Webbing
BSI BS 5127-74. 1974 Polyolefin Webbing for Divans and Bed Bases. 11 pp.

Bedsheets
Use: Bed Sheets

Bedspreads
Use For: Counterpanes See Also: Bed Sheets; Bedding; Blankets; Fabrics

—Flammability Testing
BSI BS 5815: Part 3-91. 1991 Sheets, Sheeting, Pillowslips, Towels, Napkins, Counterpanes and Continental Quilt Secondary Covers Suitable for Use in the Public Sector Part 3: Counterpanes and Continental Quilt Sec. Covers Including Flammability Performance. 12 pp.

—Sampling
CNS L4048-73. Bedspreads (Jan)(2811). 3 pp.

—Shrinkage
CNS L4048-73. Bedspreads (Jan)(2811). 3 pp.

—Visual Inspection
CNS L4048-73. Bedspreads (Jan)(2811). 3 pp.

Bedsteads
See Also: Beds

Bedsteads (Cont.)

—Medical
BSI BS 1979-85. 1985 Hospital Ward Bedsteads for Psychiatric Hospitals. 12 pp.
BSI BS 4886-88. 1988 Hospital Bedsteads. 20 pp.

—Oak Panelled
MOD UK DEF-1209-59. Bedsteads, Panelled, Oak. 12 pp.

Beech 3TM Aircraft
See Also: Aircraft; Beech Aircraft

—Antenna Positions
CAA. Beechcraft E18, H18 and 3TM (Approved Aerial Positions). 1 p.

Beech 19 Aircraft
See Also: Aircraft; Beech Aircraft

—Antenna Positions
CAA. Beechcraft 19, 23 and 24 (Approved Aerial Positions). 1 p.

Beech 23 Aircraft
See Also: Aircraft; Beech Aircraft

—Antenna Positions
CAA. Beechcraft 19, 23 and 24 (Approved Aerial Positions). 1 p.

Beech 24 Aircraft
See Also: Aircraft; Beech Aircraft

—Antenna Positions
CAA. Beechcraft 19, 23 and 24 (Approved Aerial Positions). 1 p.

Beech 33-33 Aircraft
See Also: Aircraft; Beech Aircraft

—Antenna Positions
CAA. Beech 33-33/F33A/35 Series/A36/58 Series (Approved Aerial Positions). 1 p.

Beech 35 Aircraft
See Also: Aircraft; Beech Aircraft

—Antenna Positions
CAA. Beech 33-33/F33A/35 Series/A36/58 Series (Approved Aerial Positions). 1 p.

Beech 56TC Aircraft
See Also: Aircraft; Beech Aircraft

—Antenna Positions
CAA. Beechcraft D55, 56TC, 95, B95A and D95A (Approved Aerial Positions). 1 p.

Beech 58 Aircraft
See Also: Aircraft; Beech Aircraft

—Antenna Positions
CAA. Beech 33-33/F33A/35 Series/A36/58 Series (Approved Aerial Positions). 1 p.

Beech 60 Aircraft
See Also: Aircraft; Beech Aircraft

—Antenna Positions
CAA. Beechcraft 60 (Approved Aerial Positions). 1 p.

—Data Sheets
CAA FA4 ISSUE 3. Beech 60. 2 pp.

Beech 65 Aircraft
See Also: Aircraft; Beech Aircraft

—Antenna Positions
CAA. Beechcraft 65, 70, 90 and 100 Series (Approved Aerial Positions). 1 p.

Beech 65-A90 King Air Aircraft
See Also: Aircraft; Beech Aircraft

—Certification
CAA. Beech 65-A90 King Air. 11 pp.

Beech 65-B90 King Air Aircraft
See Also: Aircraft; Beech Aircraft

—Certification
CAA. Beech 65-B90 and King Air. 11 pp.

Beech 65-C90 King Air Aircraft
See Also: Aircraft; Beech Aircraft

—Certification
CAA. Beech 65-C90 and C90A. 14 pp.

Beech 65-C90A King Air Aircraft
See Also: Aircraft; Beech Aircraft

Beech 65-C90A King Air Aircraft (Cont.)
—Certification
CAA. Beech 65-C90 and C90A. 14 pp.

Beech 65-E90 King Air Aircraft
See Also: Aircraft; Beech Aircraft
—Certification
CAA. Beech 65—E90 King Air. 15 pp.

Beech 65-90 King Air Aircraft
See Also: Aircraft; Beech Aircraft
—Certification
CAA. Beech 65-90 King Air. 11 pp.

Beech 70 Aircraft
See Also: Aircraft; Beech Aircraft
—Antenna Positions
CAA. Beechcraft 65, 70, 90 and 100 Series (Approved Aerial Positions). 1 p.

Beech 76 Aircraft
See Also: Aircraft; Beech Aircraft
—Antenna Positions
CAA. Beechcraft 76 (Approved Aerial Positions). 1 p.

Beech 90 Aircraft
See Also: Aircraft; Beech Aircraft
—Antenna Positions
CAA. Beechcraft 65, 70, 90 and 100 Series (Approved Aerial Positions). 1 p.

Beech 95 Aircraft
See Also: Aircraft; Beech Aircraft
—Antenna Positions
CAA. Beechcraft D55, 56TC, 95, B95A and D95A (Approved Aerial Positions). 1 p.

Beech 99 Airliner Aircraft
See Also: Aircraft; Beech Aircraft
—Certification
CAA. Beech 99. 8 pp.
—Data Sheets
CAA FA32 ISSUE 1. Beech Model 99 Airliner. 2 pp.

Beech 100 Aircraft
See Also: Aircraft; Beech Aircraft
—Antenna Positions
CAA. Beechcraft 65, 70, 90 and 100 Series (Approved Aerial Positions). 1 p.
—Certification
CAA. Beech 100 and A100 King Air. 14 pp.
—Data Sheets
CAA FA6 ISSUE 2. Beech 100. 3 pp.

Beech 200 Aircraft
See Also: Aircraft; Beech Aircraft
—Antenna Positions
CAA. Beechcraft 200 and B200 (Approved Aerial Positions). 1 p.
—Certification
CAA. Beech 200 and B200 Super King Air. 12 pp.

Beech 200 Super King Air Aircraft
See Also: Aircraft; Beech Aircraft
—Data Sheets
CAA FA15 ISSUE 2. Beech Super King Air Model 200. 5 pp.

Beech 300 Aircraft
See Also: Aircraft; Beech Aircraft
—Antenna Positions
CAA. Beechcraft 300 (Approved Aerial Positions). 1 p.
—Certification
CAA. Beech Model 300 and 300 LW. 12 pp.

Beech 300LW Aircraft
See Also: Aircraft; Beech Aircraft
—Certification
CAA. Beech Model 300 and 300 LW. 12 pp.

Beech A 36 Aircraft
See Also: Aircraft; Beech Aircraft
—Antenna Positions
CAA. Beech 33-33/F33A/35 Series/A36/58 Series (Approved Aerial Positions). 1 p.

Beech A 100 King Air Aircraft
See Also: Aircraft; Beech Aircraft
—Certification
CAA. Beech 100 and A100 King Air. 14 pp.

Beech Aircraft
Use For: Beechcraft Aircraft See Also: Aircraft; Beech; Beech; Beech; Beech; Beech; Beech; Beech; Beech; Beech; Beech; Beech; Beech; Beech; Beech; Beech; Beech; Beech 100 Aircraft; Beech 200 Aircraft; Beech 300 Aircraft; Beech 300LW Aircraft; Beech A 36 Aircraft; Beech A 100 King Air Aircraft; Beech B 60 Aircraft; Beech B 95A Aircraft; Beech B 200 Aircraft; Beech B 200 Super King Air Aircraft; Beech D 17S Aircraft; Beech D 55 Aircraft; Beech D 95A Aircraft; Beech E 18 Aircraft; Beech F 33A Aircraft; Beech F 90 Super King Air Aircraft; Beech H 18 Aircraft; Beech Super King Air 200 Aircraft; Beech Super King Air 300 Aircraft
—Foreign Airworthiness Directives
CAA. Beech Series Aircraft (Foreign Airworthiness Directives). 9 pp.

Beech B 60 Aircraft
See Also: Aircraft; Beech Aircraft
—Antenna Positions
CAA. Beechcraft B60 (Approved Aerial Positions). 1 p.

Beech B 95A Aircraft
See Also: Aircraft; Beech Aircraft
—Antenna Positions
CAA. Beechcraft D55, 56TC, 95, B95A and D95A (Approved Aerial Positions). 1 p.

Beech B 200 Aircraft
See Also: Aircraft; Beech Aircraft
—Antenna Positions
CAA. Beechcraft 200 and B200 (Approved Aerial Positions). 1 p.

Beech B 200 Super King Air Aircraft
See Also: Aircraft; Beech Aircraft
—Certification
CAA. Beech 200 and B200 Super King Air. 12 pp.

Beech D 17S Aircraft
See Also: Aircraft; Beech Aircraft
—Antenna Positions
CAA. Beechcraft D17S (Approved Aerial Positions). 1 p.

Beech D 55 Aircraft
See Also: Aircraft; Beech Aircraft
—Antenna Positions
CAA. Beechcraft D55, 56TC, 95, B95A and D95A (Approved Aerial Positions). 1 p.

Beech D 95A Aircraft
See Also: Aircraft; Beech Aircraft
—Antenna Positions
CAA. Beechcraft D55, 56TC, 95, B95A and D95A (Approved Aerial Positions). 1 p.

Beech E 18 Aircraft
See Also: Aircraft; Beech Aircraft
—Antenna Positions
CAA. Beechcraft E18, H18 and 3TM (Approved Aerial Positions). 1 p.

Beech F 33A Aircraft
See Also: Aircraft; Beech Aircraft
—Antenna Positions
CAA. Beech 33-33/F33A/35 Series/A36/58 Series (Approved Aerial Positions). 1 p.

Beech F 90 Super King Air Aircraft
See Also: Aircraft; Beech Aircraft
—Certification
CAA. Beech F90 Super King Air. 12 pp.

Beech H 18 Aircraft
See Also: Aircraft; Beech Aircraft
—Antenna Positions
CAA. Beechcraft E18, H18 and 3TM (Approved Aerial Positions). 1 p.

Beech Super King Air 200 Aircraft
See Also: Aircraft; Beech Aircraft
—Accidents
CAA. Beachcraft 200 and 300 Super Kingair. 1 p.

Beech Super King Air 300 Aircraft
See Also: Aircraft; Beech Aircraft
—Accidents
CAA. Beachcraft 200 and 300 Super Kingair. 1 p.
—Certification
CAA. Beech Model B300. 8 pp.

Beech Wood
Scope Note: For additional listings, see the subheading Beech Wood under specific products made from Beech Wood See Also: Hardwoods; Wood
—Visual Inspection
CEN PREN 975-1-92. Wood—Hardwood Sawn Timber—Visual Grading—Part 1: Oak and Beech. 22 pp.

Beechcraft Aircraft
Use: Beech Aircraft

Beechjet 400 Aircraft
See Also: Aircraft
—Certification
CAA. Beechjet 400. 13 pp.

Beef
Use For: Minced Beef See Also: Food
CGSB 32.44-92. Beef Cuts (Supersedes 32-GP-44M and 32-GP-45M). 25 pp.
—Dried
CNS N5068-87. Dried, Sliced Meat (Jul)(2419). 2 pp.
—Organ Meat
CGSB 32.56M-90. Fresh or Frozen Organs. 9 pp.
—Organ Meat—Frozen
CGSB 32.56M-90. Fresh or Frozen Organs. 9 pp.

Beef Cattle
See Also: Dairy Cattle
—Glossaries
ISO 3973-77. Definitions of Living Animals for Slaughter—Bovines First Edition. 8 pp.

Beer
See Also: Beverages
—Bottles
BSI BS 6118-81. 1981 Amd 3 Multi-Trip Glass Bottles for Beer and Cider. 19 pp.
—Excise Taxes
EC COM(87) 328-87. Proposal for a Council Directive Concerning Approximation of the Rates of Excise Duty on Alcoholic Beverages and on the Alcohol Contained in Other Products. 18 pp.

Bees
See Also: Insect Contamination
—Model Bylaws
SNZ NZS 9201: Chapter 13-72. Model General Bylaws Chapter 13: The Keeping of Animals, Poultry and Bees. 12 pp.

Beeswax
Use For: White Wax See Also: Waxes
CNS K1133-69. Beeswax (Jun)(3015) (R 1973). 1 p.
CNS K5003-47. White Wax ((Chinese) Insect Wax) (Mar)(81). 1 p.
CNS K6276-69. Method of Test for Bee Wax (Jul)(3016)(R 1973).
MOD UK DSTAN 91-24-93. Beeswax GS and Beeswax, LF Quality Issue 1. 21 pp.
MOD UK CS 2177A. Beeswax GS and Beeswax, LF Quality.
—Acidity
MOD UK M 2012/71. Examination of Beeswax GS and Beeswax, Lead Free.
MOD UK M 9517/65. Examination of Beeswax and Beeswax, Lead Free (LAG 931A, No Information).

Beeswax (Cont.)

—Cloud Point
MOD UK M 2012/71. Examination of Beeswax GS and Beeswax, Lead Free.

—Content Analysis
MOD UK M 2012/71. Examination of Beeswax GS and Beeswax, Lead Free.
MOD UK M 9517/65. Examination of Beeswax and Beeswax, Lead Free (LAG 931A, No Information).

—Density
MOD UK M 2012/71. Examination of Beeswax GS and Beeswax, Lead Free.
MOD UK M 9517/65. Examination of Beeswax and Beeswax, Lead Free (LAG 931A, No Information).

—Ester Value
MOD UK M 2012/71. Examination of Beeswax GS and Beeswax, Lead Free.

—Saponification Number
MOD UK M 9517/65. Examination of Beeswax and Beeswax, Lead Free (LAG 931A, No Information).

Beets
See Also: Food; Vegetables

—Harvesting Equipment—Safety
DIN ENGL 11001 Pt 7-80. Agricultural Machines and Tractors; Beet-Harvesters; Special Safety Requirements and Testing (Dec). 3 pp.

—Seeds—Marketing
EC 88/380/EEC-88. Council Directive Amending Directives 66/400/EEC, 66/401/EEC, 66/ 402/EEC, 66/403/EEC, 69/ 208/EEC, 70/457/EEC AND 70/458/EEC on the Marketing of Beet Seed, Fodder Plant Seed, Cereal Seed, Seed Potatoes, Seed of Oil and Fibre Plants and Vegetable Seed. 18 pp.

Behez Aircraft
See Also: Aircraft

—Antenna Positions
CAA. Behez Mraz Sokol M.1C (Approved Aerial Positions). 1 p.

Bekk Testers
Use For: Smoothness Testers (Bekk)
See Also: Manometers; Testing Equipment
CPPA M-2-70. Bekk Smoothness Tester. 1 p.

Belfast SC 5 Aircraft
See Also: Aircraft

—Antenna Positions
CAA. Belfast SC5 (Approved Aerial Positions). 1 p.

Bell and Spigot Joints
See Also: Pipe Joints

—Cast Iron
CNS B5018-57. Spigot and Bell End of Cast-Iron Pipes (May)(828).
CNS B5062-67. Cast-Iron Pipes, Bell and Spigot Lead Calking Joint, Class A (Aug)(2798).
CNS B5063-67. Cast-Iron Pipes, Bell and Spigot Lead Calking Joint, Class B (Aug)(2799).

Bell Helicopters
Scope Note: Use a more specific term *See:* Bell 205 Helicopters; Bell 206A Helicopters; Bell 206B Helicopters; Bell 206L Helicopters; Bell 206L1 Helicopters; Bell 206L3 Helicopters; Bell 212 Helicopters; Bell 214 Helicopters; Bell 214B Helicopters; Bell 214ST Helicopters; Bell 222 Helicopters; Bell 412 Helicopters; Bell 47 Helicopters; Bell 47D Helicopters; Bell 47G Helicopters; Bell 47G5 Helicopters; Bell 47J Helicopters

Bell Housings
Use For: Clutch Housings *See Also:* Housings

—Automotive—Flanges
CNS D1016-81. Dimension of Flanges to Be Fitted up Clutch Housing for Automobiles (Oct)(7970).
JIS D 4304-56. Dimension of Flanges to Be Fitted up Clutch Housing for Automobiles (R 1983). 4 pp.

—Internal Combustion Piston Engines
BSI BS AU 249-93. 1993 Clutch Housings for Reciprocating Internal Combustion Engines—Nominal Dimensions and Tolerances (ISO 7649:1991) (E). 9 pp.
ISO 7649-91. Clutch Housings for Reciprocating Internal Combustion Engines—Nominal Dimensions and Tolerances First Edition. 7 pp.

Bell Mouths

—Oil Suction—Ships
JIS F 3020-85. Ships' Oil Suction Bellmouths.

Bell Mouths (Cont.)

—Oil Suction—Ships (Cont.)
JIS F 3020-69. Ships' Oil Suction Bellmouths. 7 pp.

Bell 205 Helicopters
See Also: Helicopters

—Accidents
CAA. Bell 205 Series (World Helicopter Accident Summary). 1 p.

Bell 206A Helicopters
See Also: Helicopters

—Antenna Positions
CAA. Bell 206A, 206B and 206L (Approved Aerial Positions). 1 p.

—Foreign Airworthiness Directives
CAA. Bell 206A and B Series Helicopters (Foreign Airworthiness Directives). 6 pp.

Bell 206B Helicopters
See Also: Helicopters

—Antenna Positions
CAA. Bell 206A, 206B and 206L (Approved Aerial Positions). 1 p.

—Foreign Airworthiness Directives
CAA. Bell 206A and B Series Helicopters (Foreign Airworthiness Directives). 6 pp.

Bell 206L Helicopters
See Also: Helicopters

—Antenna Positions
CAA. Bell 206A, 206B and 206L (Approved Aerial Positions). 1 p.

—Certification
CAA. Bell 206L, L1 and L3. 3 pp.

—Data Sheets
CAA FR7 ISSUE 3. Bell 206L. 3 pp.

—Foreign Airworthiness Directives
CAA. Bell 206 L Series Helicopters (Foreign Airworthiness Directives). 2 pp.

Bell 206L1 Helicopters
See Also: Helicopters

—Certification
CAA. Bell 206L, L1 and L3. 3 pp.

Bell 206L3 Helicopters
See Also: Helicopters

—Certification
CAA. Bell 206L, L1 and L3. 3 pp.

Bell 212 Helicopters
See Also: Helicopters

—Accidents
CAA. Bell/Augusta Bell 212 Series (World Helicopter Accident Summary). 3 pp.

—Antenna Positions
CAA. Bell 212 and 214B (Approved Aerial Positions). 1 p.

—Certification
CAA. Bell 212. 6 pp.

—Data Sheets
CAA FR1 ISSUE 2. Bell Model 212. 2 pp.

—Foreign Airworthiness Directives
CAA. Bell 212 Series Helicopters. 8 pp.

Bell 214 Helicopters
See Also: Helicopters

—Accidents
CAA. Bell 214 Series (World Helicopter Accident Summary). 1 p.

—Foreign Airworthiness Directives
CAA. Bell 214 Series Helicopters (Foreign Airworthiness Directives). 3 pp.

Bell 214B Helicopters
See Also: Helicopters

—Antenna Positions
CAA. Bell 212 and 214B (Approved Aerial Positions). 1 p.

Bell 214ST Helicopters
See Also: Helicopters

Bell 214ST Helicopters (Cont.)

—Antenna Positions
CAA. Bell 214ST (Approved Aerial Positions). 1 p.

—Certification
CAA. Bell 214ST. 11 pp.

—Data Sheets
CAA FR17 ISSUE 1. Bell Model 214ST. 2 pp.

Bell 222 Helicopters
See Also: Helicopters

—Antenna Positions
CAA. Bell 222 (Approved Aerial Positions). 1 p.

—Data Sheets
CAA FR12 ISSUE 1. Bell Model 222. 2 pp.

—Foreign Airworthiness Directives
CAA. Bell 222 Series Helicopters (Foreign Airworthiness Directives). 2 pp.

Bell 412 Helicopters
See Also: Helicopters

—Accidents
CAA. Bell 412 Series (World Helicopter Accident Summary). 1 p.

—Antenna Positions
CAA. Bell 412. 1 p.

—Certification
CAA. Bell 412. 7 pp.

—Data Sheets
CAA FR14 ISSUE 3. Model 412. 2 pp.

Bell 47 Helicopters
See Also: Helicopters

—Foreign Airworthiness Directives
CAA. Bell 47 Series Helicopters. 3 pp.

Bell 47D Helicopters
See Also: Helicopters

—Antenna Positions
CAA. Bell 47G and 47D (Approved Aerial Positions). 1 p.

Bell 47G Helicopters
See Also: Helicopters

—Antenna Positions
CAA. Bell 47G and 47D (Approved Aerial Positions). 1 p.

Bell 47G5 Helicopters
See Also: Helicopters

—Antenna Positions
CAA. Bell 47G5 (Approved Aerial Positions). 1 p.

Bell 47J Helicopters
See Also: Helicopters

—Antenna Positions
CAA. Bell 47J (Approved Aerial Positions). 1 p.

Bellanca (Champion) 7 Aircraft
See Also: Aircraft

—Antenna Positions
CAA. Bellanca (Champion) Series 7 and 8 (Approved Aerial Positions). 1 p.

Bellanca (Champion) 8 Aircraft
See Also: Aircraft

—Antenna Positions
CAA. Bellanca (Champion) Series 7 and 8 (Approved Aerial Positions). 1 p.

Bellanca 17-30A Aircraft
See Also: Aircraft

—Antenna Positions
CAA. Bellanca 17-30A (Approved Aerial Positions). 1 p.

Bellows
See Also: Bellows Expansion Joints; Expansion Joints

—Rubber—Air Springs
CNS D3080-81. Testing Method of Rubber Bellows for Air Springs (Oct)(7977).
JIS D 4101-77. Testing Method of Rubber Bellows for Air Springs.

INTERNATIONAL AND NON-U.S. NATIONAL STANDARDS
SUBJECT INDEX
Belt

Bellows *(Cont.)*
—Rubber—Identification Systems
MOD UK DEFCON 67-90. Specifications Covering Identification, Marking and Age on Delivery of Rubber Materiel, Assemblies and Rubber Containing Composites 3/90. 2 pp.

MOD UK DEFCON 67-92. Specifications Covering Identification, Marking and Age on Delivery of Rubber Materiel, Assemblies and Rubber Containing Composites (Navy Contracts Only) 2/92. 2 pp.

—Rubber—Marine
MOD UK NES 345-92. Requirements for Flexible Rubber Pipe Assemblies and Bellows for Use in Systems from Vacuum to 10 Bar Issue 1 (10.92). 73 pp.

—Rubber—Submarines
MOD UK NES 374-82. Rubber Flexible Pipes and Bellows for Use in Submarines Issue 1 (02.82). 28 pp.

Bellows Expansion Compensations
Use: Bellows Expansion Joints

Bellows Expansion Joints
Use For: Bellows Expansion Compensations; Expansion Compensators, Bellows *See Also:* Bellows; Expansion Joints; Pipe Joints
JIS B 2352-77. Bellows Type Expansion Pipe Joints (R 1980). 12 pp.

—Metal
BSI BS 6129: Part 1-81. 1981 Selection and Application of Bellows Expansion Joints for Use in Pressure Systems Part 1: Metallic Bellows Expansion Joints. 26 pp.

Bells
Use For: Electric Bells *See Also:* Acoustic Signals; Electroacoustic Transducers; Warning Systems
CNS C4345-90. Electric Bell (Dec)(8799).
CSA C22.2 NO 205-M1983. Signal Equipment (R 1992); (Gen Instr 1). 32 pp.

—Bicycles
CNS B2139-73. Bells for Bicycles (Jan)(3515). 2 pp.
ISO 7636-84. Bells for Bicycles and Mopeds—Technical Specifications First Edition. 4 pp.
JIS D 9451-83. Bells for Bicycles.

—Fire Detection Systems
SAA AS 1603.6-87. Automatic Fire Detection and Alarm Systems—Part 6: Fire Alarm Bells. 9 pp.

—Marine
CNS F5077-87. Marine Electric Bells (Watertight Type) (Mar)(10624).
CNS F5122-87. Marine Electric Bells for Accommodation Spaces (Nov)(12156).
JIS F 8501-83. Marine Watertight Electric Bells.
JIS F 8506-85. Marine Electric Bells for Accommodation Spaces.

—Mopeds
ISO 7636-84. Bells for Bicycles and Mopeds—Technical Specifications First Edition. 4 pp.

—Ships
CNS F3088-81. Ships' Bell (Jan)(6918).
JIS F 3614-76. Ships' Bells (R 1984). 5 pp.
MOD UK NES 739-82. Sound Signal Appliances (Whistles, Gongs, Bells Etc) Issue 1 (10.82). 14 pp.

—Submarines
MOD UK NES 739-82. Sound Signal Appliances (Whistles, Gongs, Bells Etc) Issue 1 (10.82). 14 pp.

—Symbols—Architectural Drawings
CNS C1101-87. Symbols of Telephone and Bell of Interior Wiring Diagram for Architectural Plans (Dec)(9112).

Belt Conveyors
Use For: Belt Feeders; Drum Conveyors; Slide Conveyors *See Also:* Belt Scales; Belts (Machinery); Conveyor Belts; Conveyors; Idlers; Materials Handling Equipment
ISO 2109-75. Continuous Mechanical Handling Equipment—Light Duty Belt Conveyors for Loose Bulk Materials First Edition. 4 pp.
JIS B 8805-92. Rubber Belt Conveyors with Carrying Idlers—Calculation of Operating Power and Tensile Forces. 18 pp.
SAA AS 1333-88. Conveyor Belting of Elastomeric and Steel Cord Construction. 19 pp.

—Adhesion Testing
SAA AS 1334.7-82. Methods of Testing Conveyor and Elevator Belting—Part 7: Determination of Ply Adhesion of Conveyor Belting. 2 pp.

Belt Conveyors *(Cont.)*
—Comprehensive Testing
SAA AS 1334. Methods of Testing Conveyor and Elevator Belting Complete Set in Binder.

—Design
DIN ENGL 22101-82. Continuous Mechanical Handling Equipment; Belt Conveyors for Bulk Materials; Bases for Calculations and Design (Feb). 23 pp.

—Ducts
BSI BS 4286: Part 2-70. 1970 Steel Ducting for Grain and Fodder Conveying Part 2: Metric Units. 11 pp.

—Electrical Resistance
BSI BS 490: Part 3-91. 1991 Conveyor and Elevator Belting Part 3: Specification for Flammability and Anti-Static Properties of Rubber and of Plastics Conveyor Belting of Textile Construction for General Use. 7 pp.
SAA AS 1334.9-82. Methods of Testing Conveyor and Elevator Belting—Part 9: Determination of Electrical Resistance of Conveyor Belting. 3 pp.
SNZ NZS/BS 490: Part 3-91. Conveyor and Elevator Belting Part 3: Specification for Flammability and Anti-Static Properties of Rubber and of Plastics Conveyor Belting of Textile Construction for General Use. 8 pp.

—Elongation
SAA AS 1334.3-82. Methods of Testing Conveyor and Elevator Belting—Part 3: Determination of Full Thickness Tensile Strength and Elongation of Conveyor Belting. 4 pp.

—Fastenings—Dynamic Testing
BSI BS 4890-73. 1973 Test Methods for Mechanical Joints in Conveyor Belting. 14 pp.

—Fastenings—Static Testing
BSI BS 4890-73. 1973 Test Methods for Mechanical Joints in Conveyor Belting. 14 pp.
ISO 1120-84. Conveyor Belts—Determination of Strength of Mechanical Fastenings—Static Test Method Second Edition. 5 pp.

—Fire Testing
BSI BS 490: Sec 11.2-91. 1991 Conveyor and Elevator Belting Part 11: Methods of Test for Safety Section 11.2: Low Energy and High Energy Propane Gallery Tests. 12 pp.
SNZ NZS/BS 490: Pt11:Sec11.2-91. Conveyor and Elevator Belting Part 11: Methods of Test for Safety Section 11.2: Low Energy and High Energy Propane Gallery Tests. 12 pp.

—Flammability Testing
BSI BS 490: Part 3-91. 1991 Conveyor and Elevator Belting Part 3: Specification for Flammability and Anti-Static Properties of Rubber and of Plastics Conveyor Belting of Textile Construction for General Use. 7 pp.
SAA AS 1334.10-82. Methods of Testing Conveyor and Elevator Belting—Part 10: Determination of Ignitability and Flame Propagation Characteristics of Conveyor Belting. 3 pp.
SAA AS 1334.12-86. Methods of Testing Conveyor and Elevator Belting—Part 12: Determination of Combustion Propagation Characteristics of Conveyor Belting Amdt 1 June 1988 Amdt 2 May 1993. 6 pp.
SNZ NZS/BS 490: Part 3-91. Conveyor and Elevator Belting Part 3: Specification for Flammability and Anti-Static Properties of Rubber and of Plastics Conveyor Belting of Textile Construction for General Use. 8 pp.

—Identification Systems
ISO 433-91. Conveyor Belts—Marking Second Edition. 7 pp.

—Idlers
BSI BS 5934-80. 1980 Calculation of Operating Power and Tensile Forces in Belt Conveyors with Carrying Idlers on Continuous Mechanical Handling Equipment. 18 pp.
DIN ENGL 15207 Pt 1-88. Continuous Mechanical Handling Equipment; Idlers for Belt Conveyors Which Handle Loose Bulk Materials; Principal Dimensions (July). 4 pp.
DIN ENGL 22107-84. Continuous Mechanical Handling Equipment; Idler Sets for Belt Conveyors for Loose Bulk Materials; Principal Dimensions (Aug). 4 pp.
ISO 1537-75. Continuous Mechanical Handling Equipment for Loose Bulk Materials—Troughed Belt Conveyors (Other Than Portable Conveyors)—Idlers First Edition. 7 pp.
ISO 5048-89. Continuous Mechanical Handling Equipment—Belt Conveyors with Carrying Idlers—Calculation of Operating Power and Tensile Forces Second Edition. 17 pp.

Belt Conveyors *(Cont.)*
—Length Measurement
SAA AS 1334.1-82. Methods of Testing Conveyor and Elevator Belting—Part 1: Determination of Length of Endless Belting. 1 p.

—Mining Equipment
BSI BS 3289-90. 1990 Textile Carcase Conveyor Belting for Use in Underground Mines (Including Fire Performance). 34 pp.

—Portable
BSI BS 4531-86. 1986 Portable and Mobile Troughed Belt Conveyors. 14 pp.
CNS K4077-83. Rubber Belts for Portable Conveyors (Feb)(10025). 4 pp.
JIS B 8808-76. Portable Belt Conveyors. 10 pp.

—Pulleys
BSI BS 5934-80. 1980 Calculation of Operating Power and Tensile Forces in Belt Conveyors with Carrying Idlers on Continuous Mechanical Handling Equipment. 18 pp.
ISO 1536-75. Continuous Mechanical Handling Equipment for Loose Bulk Materials—Troughed Belt Conveyors (Other Than Portable Conveyors)—Belt Pulleys First Edition. 3 pp.
ISO 1816-75. Continuous Mechanical Handling Equipment for Loose Bulk Materials and Unit Loads—Belt Conveyors—Basic Characteristics of Motorized Driving Pulleys First Edition. 3 pp.
ISO 3684-90. Conveyors Belts—Determination of Minimum Pulley Diameters Second Edition. 8 pp.
ISO 5048-89. Continuous Mechanical Handling Equipment—Belt Conveyors with Carrying Idlers—Calculation of Operating Power and Tensile Forces Second Edition. 17 pp.
JIS B 8814-92. Pulleys for Belt Conveyors. 12 pp.

—Rollers
CNS B4049-81. Rollers for Belt Conveyors (Sep)(7879).
JIS B 8803-90. Rollers for Belt Conveyor. 19 pp.

—Safety
BSI BS 5667: Part 19-80. 1980 Continuous Mechanical Handling Equipment. Safety Requirements Part 19: Belt Conveyors: Examples for Guarding of Nip Points. 18 pp.
BSI BS 7300-90. 1990 Code of Practice for Safeguarding of the Hazard Points on Troughed Belt Conveyors. 30 pp.
ISO TR5045-79. Continuous Mechanical Handling Equipment—Safety Code for Belt Conveyors—Examples for Guarding of Nip Points First Edition. 15 pp.
ISO TR8435-84. Continuous Mechanical Handling Equipment—Safety Code for Belt Conveyors—Examples for Protection of Pinch Points on Idlers First Edition. 13 pp.

—Tensile Strength
ISO 283-90. Conveyor Belts—Full Thickness Tensile Strength and Elongation—Specifications and Method of Test Second Edition. 11 pp.

—Tensile Testing
BSI BS 490: Sec 11.3-91. 1991 Conveyor and Elevator Belting Part 11: Methods of Test for Safety Section 11.3: Drum Friction Tests. 11 pp.
BSI BS 5934-80. 1980 Calculation of Operating Power and Tensile Forces in Belt Conveyors with Carrying Idlers on Continuous Mechanical Handling Equipment. 18 pp.
ISO 5048-89. Continuous Mechanical Handling Equipment—Belt Conveyors with Carrying Idlers—Calculation of Operating Power and Tensile Forces Second Edition. 17 pp.
JIS B 8805-92. Rubber Belt Conveyors with Carrying Idlers—Calculation of Operating Power and Tensile Forces. 18 pp.
SAA AS 1334.3-82. Methods of Testing Conveyor and Elevator Belting—Part 3: Determination of Full Thickness Tensile Strength and Elongation of Conveyor Belting. 4 pp.
SNZ NZS/BS 490: Pt11:Sec11.3-91. Conveyor and Elevator Belting Part 11: Methods of Test for Safety Section 11.3: Drum Friction Tests. 12 pp.

—Thickness Measurement
SAA AS 1334.2-82. Methods of Testing Conveyor and Elevator Belting—Part 2: Determination of Thickness of Belting and Rubber Covers Across the Width. 2 pp.
SAA AS 1334.2A-84. Methods of Testing Conveyor and Elevator Belting—Part 2A: Determination of Thickness of Cover Using an Optical Magnifier. 2 pp.

—Troughed
BSI BS 2890-89. 1989 Troughed Belt Conveyors. 26 pp.
BSI BS 2890-01. 1989 Amd 1 Troughed Belt Conveyors (AMD 6779) November 29, 1991. 27 pp.

INDUSTRY STANDARDS

Belt Conveyors (Cont.)

—Troughed (Cont.)
BSI BS 4531-86. 1986 Portable and Mobile Troughed Belt Conveyors. 14 pp.
ISO 1535-75. Continuous Mechanical Handling Equipment for Loose Bulk Materials—Troughed Belt Conveyors (Other Than Portable Conveyors)—Belts First Edition. 4 pp.

—Troughed—Flexibility
SAA AS 1334.4-82. Methods of Testing Conveyor and Elevator Belting—Part 4: Determination of Troughability of Conveyor Belting Reconfirmed 1989. 2 pp.

Belt Drives
Use For: Wedge Belt Drives *See Also:* Belts (Machinery); Pulleys; Rope Drives; Sprockets; Synchronous Belt Drives; V Belt Drives; V Belts
BSI BS 3790-81. 1981 Endless Wedge Belt Drives and Endless V-Belt Drives. 34 pp.
ISO 9608-88. V-Belts—Uniformity of Belts—Centre Distance Variation—Specifications and Test Method First Edition. 6 pp.

—Electrical Resistance
ISO 9563-90. Belt Drives—Electrical Conductivity of Antistatic Endless Synchronous Belts—Characteristics and Test Method First Edition. 7 pp.

—Glossaries
ISO 5288-82. Synchronous Belt Drives—Vocabulary First Edition. 29 pp.

—Pulleys
ISO 155-89. Belt Drives—Pulleys—Limiting Values for Adjustment of Centres Second Edition. 7 pp.
ISO 254-90. Belt Drives—Pulleys—Quality, Finish and Balance Second Edition. 6 pp.
ISO 5294-89. Synchronous Belt Drives—Pulleys Second Edition. 12 pp.

—Pulleys—Racks
BSI BS 6889-87. 1987 Generating Racks for Synchronous Belt Drive Pulleys. 7 pp.

Belt Feeders
Use: Belt Conveyors

Belt Sanders
See Also: Sanders (Tools); Tools; Woodworking Equipment
CNS B1304-83. Nominal Dimension of Wide Belt Sanders (Mar)(10056).
CNS B7246-83. Test Code for Accuracy of Wild Belt Sanders (May)(10226).
CNS B7264-84. Test Code for Performance of Wide Belt Sanders (Mar)(10787).
JIS B 6546-91. Wide Belt Sanders—Test and Inspection Methods. 15 pp.

—Electric
MOD UK DSTAN 51-19: Part 2-92. Hand Tools, Powered Part 2: Electric Issue 1. 31 pp.

—Glossaries
BSI BS 4361: Part 23-90. 1990 Woodworking Machines Part 23: Nomenclature for Narrow Belt Sanding Machines with Sliding Table or Frame. 9 pp.
BSI BS 4361: Part 29-90. 1990 Woodworking Machines Part 29: Nomenclature for Wide Belt Sanding Machines. 11 pp.
ISO 9264-88. Woodworking Machines—Narrow Belt Sanding Machines with Sliding Table or Frame—Nomenclature First Edition. 8 pp.
ISO 9415-89. Woodworking Machines—Wide Belt Sanding Machines—Nomenclature First Edition. 9 pp.

Belt Scales
See Also: Belt Conveyors; Weight Indicators

—Conveyors
JIS B 7606-91. Belt Weighers. 14 pp.

Belts (Clothing)
See Also: Belts (Machinery); Clothing

—Cotton
CNS L4049-70. Cotton Belt for Pants (Jan)(3087).

—Cotton Canvas
CNS S1181-85. Cotton Canvas Waist Belt (Feb)(11213).

—Leather
CNS K8003-60. Leather (for Waist Band) (May)(1270).
MOD UK DSTAN 83-14-88. Leather, Cattlehide, Hydraulic Packing Leather, Cattlehide, Sole Leather, Cattlehide, for Hoses, Bellows, Straps, Harness and Saddlery, Leather Belting, Leather Butts, Special Issue 3. 19 pp.

Belts (Clothing) (Cont.)

—Leather (Cont.)
SAA AS B4-39. Vegetable Tanned Leather Belting. 7 pp.

Belts (Machinery)
See Also: Abrasive Belts; Belt Conveyors; Belt Drives; Belts (Clothing); Buckles; Cables (Mechanical); Conveyor Belts; Elevator Belts; Flat Belts; Hexagonal Belts; Pulleys; Safety Belts; Synchronous Belts; Transmission Belts; V Belts

—Clamping Plates
CNS Z7083-81. Clamping Plate for Straps and Belts (May)(7472).

—Dental Engines
CNS T3017-80. Belts for Electric Dental Engine (Dec)(6721).
JIS T 5106-76. Belts for Electric Dental Engines (R 1986). 5 pp.

—Eyes
CNS Z7082-81. Eyes for Straps and Belts (May)(7471).

—Fasteners
CNS B2113-61. Belt Fastener (Jun)(1362)(R 1973).

—Rubber—Cotton Fabrics
CNS L3135-81. Method of Test for Cotton Fabrics for Rubber Belt (Jul)(7745).
CNS L4138-81. Cotton Fabrics for Rubber Belt (Jul)(7744).
JIS L 3104-59. Cotton Fabrics for Rubber Belt.

Bench Grinders
See Also: Grinders; Machine Tools; Tools
CNS C3130-83. Method of Test for Electric Bench Grinders (Oct)(7965).
CNS C4327-83. Electric Bench Grinders (Oct)(7964).
JIS C 9621-76. Electric Bench Grinders. 16 pp.

—Electric
CGSB 45-GP-17B-72. Grinder, Bench, Electric, Heavy Duty. 9 pp.

—Safety
IEC 1029 Pt 2-4-93. Safety of Transportable Motor-Operated Electric Tools Part 2: Particular Requirements for Bench Grinders First Edition. 38 pp.

Bench Lathes
See Also: Lathes
CNS B1099-79. Specified Items of Machine Tools Bench Lathe (Jan)(4670-2). 1 p.

Bench Tools
Use: Tools

Bench Tops
Use: Countertops

Bench Vises
See Also: Vises
CGSB 39-GP-54B-84. Vises. 56 pp.
MOD UK DSTAN 51-11: Part 21-76. Hand Tools, General Purpose Part 21: Vices, Engineer's and Vice, Bench, Clamp Base Issue 1. 16 pp.

—Parallel
CNS B3254-83. Parallel Bench Vices (with Square Chamber and Solid Base) (Mar)(4037). 2 pp.
CNS B3255-83. Parallel Bench Vices (with Round Chamber and Solid Base) (Mar)(4038). 2 pp.
JIS B 4620-80. Parallel Bench Vices. 8 pp.
JIS B 4621-75. Parallel Bench Vices. 7 pp.

Bend Properties
Use: Bend Testing

Bend Radius
See Also: Bending

—Steel Pipes—Aerospace
CEN PREN 3658-92. Tube Bend Radii, for Engine Application Design Standard. 6 pp.

—Titanium Pipes—Aerospace
CEN PREN 3658-92. Tube Bend Radii, for Engine Application Design Standard. 6 pp.

Bend Resistance
Use: Bend Testing

Bend Testing
See Also: Fatigue Testing; Flexural Strength; Flexure; High Temperature Testing; Impact Testing; Low Temperature Testing; Reverse Bend Testing; Static Testing; Stiffness Testing

Bend Testing (Cont.)

—Aluminum
ISO 2142-81. Wrought Aluminium, Magnesium and Their Alloys—Selection of Specimens and Test Pieces for Mechanical Testing First Edition. 5 pp.

—Aluminum Alloys
ISO 2142-81. Wrought Aluminium, Magnesium and Their Alloys—Selection of Specimens and Test Pieces for Mechanical Testing First Edition. 5 pp.

—Bituminous Coatings
MOD UK M 3004/67. Protective PX-2 and Composition Rust Preventative Type B.

—Bone Plates
BSI BS 3531: Sec 23.1-91. 1991 Implants for Osteosynthesis Part 23: Bone Plates Section 23.1: Method for Determination of Bending Strength and Stiffness (ISO 9585: 1990). 11 pp.
ISO 9585-90. Implants for Surgery—Determination of Bending Strength and Stiffness of Bone Plates First Edition. 7 pp.

—Building Board
CNS A3064-86. Method of Bending Test for Boards for Buildings (Feb)(3904). 2 pp.
JIS A 1408-77. Method of Bending Test for Boards of Buildings.
JIS A 1408-64. Method of Bending Test for Boards of Buildings (R 1974). 6 pp.

—Butt Welded Joints
CEN PREN 910-92. Welding—Welded Butt Joints in Metallic Materials—Bend Tests. 16 pp.
ISO 5173-81. Fusion Welded Butt Joints in Steel—Transverse Root and Face Bend Test First Edition. 6 pp.
ISO 5177-81. Fusion Welded Butt Joints in Steel—Transverse Side Bend Test First Edition. 6 pp.
JIS Z 3122-90. Methods of Bend Test for Butt Welded Joint. 11 pp.
JIS Z 3125-62. Method of Notched Bend Test for Butt Welded Joint (R 1980). 5 pp.

—Cables (Electric)
DIN VDE 0472 Pt 610-85. Testing of Cables, Wires and Flexible Cords; Bending Test at Low Temperature (Jan). 6 pp.

—Cables (Electric)—Aircraft
CEN PREN 3475 (Part 405)-92. Cables, Electrical, Aircraft Use Test Methods Part 405—Bending at Ambient Temperature. 2 pp.
CEN PREN 3475 (Part 406)-92. Cables, Electrical, Aircraft Use Test Methods Part 406—Cold Bend Test. 2 pp.

—Cables (Electric)—Telecommunication Equipment
DIN VDE 0472 Pt 603-89. Testing of Cables, Wires and Flexible Cords; Bending Behaviour (VDE Specification) (July). 17 pp.

—Cellular Plastics
ISO 1209 Pt 1-90. Cellular Plastics, Rigid —Flexural Tests—Part 1: Bending Test First Edition. 6 pp.
JIS K 7221-84. Testing Method for Flexural Properties of Rigid Cellular Plastics. 12 pp.

—Coated Fabrics
ISO 4675-90. Rubber-or Plastics-Coated Fabrics—Low-Temperature Bend Test Second Edition. 8 pp.
SAA AS 1441.6-73. Methods of Test for Coated Fabrics—Part 6: Method for Determination of Resistance to Flex Cracking.

—Coatings
BSI BS 3900: Part E11-85. 1985 Methods of Test for Paints Group E: Mechanical Tests on Paint Films Part E11: Bend Test (Conical Mandrel). 5 pp.
ISO 6860-84. Paints and Varnishes—Bend Test (Conical Mandrel) First Edition. 5 pp.

—Coatings—Bicycles
CNS B7072-75. Method of Test for Surface Treatment of Bicycle Parts (Dec)(3885). 2 pp.

—Coatings—Cans
CNS Z6070-83. Method of Test for Coating Materials of Metal Cans for Foods—Impact Bent Test of Lacquered Film (Sep)(10586). 2 pp.

—Connectors
CEN PREN 2591 (Part D4)-92. Elements of Electrical and Optical Connection Test Methods Part D4—Transverse Load (External Bending Moment). 4 pp.
CEN PREN 2591 (Part D20)-92. Elements of Electrical and Optical Connection Test Methods Part D20—Mechanical Strength of Rear Accessories. 5 pp.
CNS C6145-87. Method of Test for Low Frequency (Below 3 MHz) Electrical Connectors (TP-41 Circular Jacket Cable Flexing Test) (Oct)(8223).

INTERNATIONAL AND NON-U.S. NATIONAL STANDARDS
SUBJECT INDEX

Bend Testing (Cont.)

—Connectors (Cont.)
CNS C6183-88. Method of Test for Low Frequency (Below 3 MHz) Electrical Connectors (TP-43 Bending Moment Test Connector) (May)(9243).
SBAC TS 356 ISSUE 1. Test for Electrical Connectors Contact Bending Strength.

—Cords (Electric)
DIN VDE 0472 Pt 610-85. Testing of Cables, Wires and Flexible Cords; Bending Test at Low Temperature (Jan). 6 pp.

—Cords (Electric)—Telecommunication Equipment
DIN VDE 0472 Pt 603-89. Testing of Cables, Wires and Flexible Cords; Bending Behaviour (VDE Specification) (July). 17 pp.

—Corkboard
ISO 2077-79. Pure Expanded Corkboard—Determination of the Modulus of Rupture by Bending First Edition; (Amendment Slip-1980). 4 pp.

—Curbs—Concrete
CNS A3066-85. Method of Test for Concrete Curbs (Feb)(3931). 2 pp.

—Electric Contacts
CEN PREN 2591 (Part D16)-92. Elements of Electrical and Optical Connection Test Methods Part D16—Contact Bending Strength. 4 pp.

—Electric Wire
DIN VDE 0472 Pt 610-85. Testing of Cables, Wires and Flexible Cords; Bending Test at Low Temperature (Jan). 6 pp.

—Enamels
CNS K6030-72. Method of Test for Enamel (Jun)(627). 7 pp.

—Fiber Optic Cables
CNS C6276-86. Method of Test for Fiber Optic Devices (FOTP-37 Cable Bend Test Low and High Temperature) (Dec)(11788).
CNS C6297-87. Method of Test for Fiber Optic Devices (FOTP-33 Fiber Optic Cable Tensile Loading and Bending Test) (Sep)(12083).
CNS C6315-88. Method of Test for Fiber Optic Devices (FOTP-91 Fiber Optic Cable Twist-Bend Test) (Jul)(12362).
DIN VDE 0472 Pt 232-85. Testing of Cables, Wires and Flexible Cords; Attenuation Changes in Optical Fibre Cables During Bending (Jan). 4 pp.

—Fiberboard
BSI BS EN 310-93. 1993 Wood-Base Panels—Determination of Modulus of Elasticity in Bending and of Bending Strength (R). 12 pp.
CEN PREN 310-90. Wood—Based Panels—Determination of Moduls of Elasticity—in Beding and Bending Strength. 7 pp.
CEN EN 310-93. Wood-Based Panels—Determination of Modulus of Elasticity in Bending and of Bending Strength. 8 pp.
CNS A3064-86. Method of Bending Test for Boards for Buildings (Feb)(3904). 2 pp.
ISO 768-72. Fibre Building Boards—Determination of Bending Strength First Edition. 5 pp.

—Fillet Welds
JIS Z 3134-65. Method of Bend Test for T Type Fillet Welded Joint (R 1980). 6 pp.
JIS Z 3135-71. Method of Soundness Test for Fillet Welds. 5 pp.

—Glass
BSI BS 7604: Part 2-92. 1992 Method for Determination of Stress-Optical Coefficient of Glass Part 2: Bending Test (ISO 10345-2: 1992). 11 pp.
DIN ENGL 52292 Pt 1-84. Testing of Glass and Glass Ceramics; Determination of Bending Strength; Coaxial Double Ring Bending Test on Flat Specimens with Small Test Area (Apr). 6 pp.
DIN ENGL 52303 Pt 1-84. Methods of Testing Flat Glass for Use in Buildings; Determination of Bending Strength; Test with Specimen Supported at Two Points (Aug). 3 pp.
DIN ENGL 52303 Pt 2-83. Testing of Glass; Determination of Bending Strength; Testing of Profile Glass (Mar). 3 pp.
ISO 10345 Pt 2-92. Glass—Determination of Stress-Optical Coefficient—Part 2: Bending Test First Edition. 8 pp.

—Gutters—Concrete
CNS A3068-86. Method of Test for Reinforced Concrete Ditches (U Type) (Mar)(4062). 2 pp.

—Gutters—Covers—Concrete
CNS A3069-86. Method of Test for Reinforced Concrete Covers for Ditches (U Type) (Mar)(4064). 2 pp.

Bend Testing (Cont.)

—Hoses
BSI BS 5173: Sec 103.5-85. (WITHDRAWN) 1985 Rubber and Plastics Hoses and Hose Assemblies Part 103: Physical Tests Section 103.5: Bending Tests (Renumbered as BS EN 21746: 1993). 7 pp.
BSI BS 5173: Sec 106.1-89. (WITHDRAWN) 1989 Rubber and Plastics Hoses and Hose Assemblies Part 106: Environmental Tests Section 106.1: Determination of Low Temperature Flexibility (Renumbered as BS EN 24672: 1993). 6 pp.
BSI BS EN 21746-93. 1993 Amd 1 Rubber and Plastics Hoses and Tubing—Bending Tests (ISO 1746: 1983) (AMD 7523) April 15, 1993. 14 pp.
BSI BS EN 24672-93. 1993 Amd 1 Rubber and Plastics Hoses—Sub-Ambient Temperature Flexibility Tests (ISO 4672: 1988) (AMD 7524) April 15, 1993. 13 pp.
CEN EN 21746-93. Rubber and Plastics Hoses and Tubing—Bending Tests (ISO 1746: 1983). 8 pp.
CEN EN 24672-93. Rubber and Plastics Hoses—Sub-Ambient Temperature Flexibility Tests (ISO 4672: 1988). 6 pp.
ISO 1746-83. Rubber or Plastics Hoses and Tubing—Bending Tests Second Edition (CEN EN 21746:1993). 6 pp.
ISO 4672-88. Rubber and Plastics Hoses—Sub-Ambient Temperature Flexibility Tests Second Edition (CEN EN 24672: 1993). 6 pp.
SAA AS 1180.8A-72. Methods of Test for Hose Made from Elastomeric Materials—Part 8A: Resistance to Cold Flexing of Hose Assembly Corrig. Reconfirmed 1988.
SAA AS 1180.8B-72. Methods of Test for Hose Made from Elastomeric Materials—Part 8B: Resistance to Cold Flexing of Hose Lining and Cover (Bound Together) Reconfirmed 1988. 42 pp.
SAA AS 1180.14-91. Methods of Test for Hose Made from Elastomeric Materials—Part 14: Determination of Reeling Properties of Non-Collapsible Hose. 3 pp.

—Lacquers
SAA AS 1580.402. 1-92. Paints and Related Materials—Methods of Test—Part 402.1: Bend Test (in Professional Packages 30, 39). 3 pp.
SNZ NZS/AS 1580. 402.1-92. Methods of Test for Paints and Related Materials Part 402.1: Bend Test (This is a Joint Standard with SAA AS 1580.402.1). 3 pp.

—Leather
BSI BS 3144: Add 1-81. 1981 Methods of Sampling and Physical Testing of Leather Addendum 1: Determination of Resistance to Bending and Abrasion of Heavy Leather. 12 pp.
CNS K6120-60. Method of Test for Leather (Creasing Test) (May)(1276)(R 1973).

—Magnesium
ISO 2142-81. Wrought Aluminium, Magnesium and Their Alloys—Selection of Specimens and Test Pieces for Mechanical Testing First Edition. 5 pp.

—Magnesium Alloys
ISO 2142-81. Wrought Aluminium, Magnesium and Their Alloys—Selection of Specimens and Test Pieces for Mechanical Testing First Edition. 5 pp.

—Metal Tubes
ISO 8491-86. Metallic Materials—Tube (in Full Section)—Bend Test First Edition. 4 pp.

—Metals
AECMA PREN2002-06-87. Test Methods for Metallic Materials—Part 6—Bend Testing. 10 pp.
BSI BS 1639-64. 1964 Methods for Bend Testing of Metals. 20 pp.
BSI 4A 4-66. 1966 Amd 1 Test Pieces and Test Methods for Metallic Materials for Aircraft. 13 pp.
CNS G2033-84. Pieces for Impact Test for Metallic Materials (Aug)(3033). 3 pp.
CNS G2034-84. Method of Bend Test for Metallic Materials (Sep)(3941).
DIN ENGL 50111-87. Testing of Metallic Materials; Bend Test (Sept). 4 pp.
DIN ENGL 50151-87. Testing of Metallic Materials; Bend Testing of 0,05 to 1,0 mm Thick Sheet and Strip Used for Manufacturing Springs (July). 5 pp.
ISO 7438-85. Metallic Materials—Bend Test First Edition. 7 pp.
JIS Z 2204-69. Bend Test Pieces for Metallic Materials (R 1971). 5 pp.
JIS Z 2248-75. Method of Bend Test for Metallic Materials. 6 pp.

—Optical Fibers
CNS C6331-89. Method of Test for Fiber Optic Devices (FOTP-103 Buffered Fiber Bend Test) (Jul)(12562).

—Paints
BSI BS 3900: Part E1-70. 1970 Methods of Test for Paints Group E: Mechanical Tests on Paint Films Part E1: Bend Test (Cylindrical Mandrel). 4 pp.

Bend Testing (Cont.)

—Paints (Cont.)
ISO 1519-73. Paints and Varnishes—Bend Test (Cylindrical Mandrel) First Edition. 7 pp.
SAA AS 1580.402. 1-92. Paints and Related Materials—Methods of Test—Part 402.1: Bend Test (in Professional Packages 30, 39). 3 pp.
SNZ NZS/AS 1580. 402.1-92. Methods of Test for Paints and Related Materials Part 402.1: Bend Test (This is a Joint Standard with SAA AS 1580.402.1). 3 pp.

—Pallets
CNS Z6041-81. Method of Test for Flat Pallet (Nov)(8170).
JIS Z 0602-88. Test Methods for Flat Pallets. 10 pp.

—Paperboard
BSI BS 3748-92. 1992 Method for Determination of Resistance to Bending of Paper and Board (ISO 2493: 1992). 7 pp.
BSI BS 7424-91. 1991 Determination of the Bending Stiffness of Paper and Board by Static Methods (ISO 5628: 1990). 11 pp.
CNS P3076-85. Method of Test for Stiffness of Paperboard by Bending Load (Oct)(11395).
DIN ENGL 53123 Pt 1-78. Testing of Paper and Board; Determination of Bending Stiffness; Resonant Length Method (Jan). 4 pp.
ISO 2493-92. Paper and Board—Determination of Resistance to Bending Second Edition. 6 pp.
ISO 5628-90. Paper and Board—Determination of Bending Stiffness by Static Methods—General Principles First Edition. 9 pp.
ISO 5629-83. Paper and Board—Determination of Bending Stiffness—Resonance Method First Edition. 8 pp.
JIS P 8125-76. Testing Method for Stiffness of Paperboard by Bending Load (R 1984). 7 pp.
SAA AS 1301.P406 M-86. Methods of Test for Pulp and Paper (Metric Units)—Part 406m: Bending Quality of Paperboard (This is a Joint Standard with SANZ NZS 1301).
SAA AS 1301.431R P-89. Methods of Test for Pulp and Paper (Metric Units)—Part 431rp: Taber Bending Resistance of Paper and Paperboard (This is a Joint Standard with SANZ NZS 1301). 5 pp.
SAA AS 1301.453S-91. Methods of Test for Pulp and Paper—Part 453S: Bending Resistance of Paper and Paperboard—Constant Rate of Deflection (This is a Joint Standard with SANZ NZS 1301.453S). 3 pp.
SNZ NZS/AS 1301. P406M-86. Methods of Test for Pulp and Paper Bending Quality of Paperboard (This is a Joint Standard with SAA AS 1301.P406M). 3 pp.
SNZ NZS/AS 1301. 431RP-89. Methods of Test for Pulp and Paper Taber Bending Resistance of Paper and Paperboard (This is a Joint Standard with SAA AS 1301.431RP). 5 pp.
SNZ NZS/AS 1301. 453S-91. Methods of Test for Pulp and Paper Bending Resistance of Paper and Paperboard—Constant Rate of Deflection (This is a Joint Standard with SAA AS 1301.453S). 3 pp.

—Papers
BSI BS 3748-92. 1992 Method for Determination of Resistance to Bending of Paper and Board (ISO 2493: 1992). 7 pp.
BSI BS 7424-91. 1991 Determination of the Bending Stiffness of Paper and Board by Static Methods (ISO 5628: 1990). 11 pp.
DIN ENGL 53123 Pt 1-78. Testing of Paper and Board; Determination of Bending Stiffness; Resonant Length Method (Jan). 4 pp.
ISO 2493-92. Paper and Board—Determination of Resistance to Bending Second Edition. 6 pp.
ISO 5628-90. Paper and Board—Determination of Bending Stiffness by Static Methods—General Principles First Edition. 9 pp.
ISO 5629-83. Paper and Board—Determination of Bending Stiffness—Resonance Method First Edition. 8 pp.
JIS P 8143-67. Testing Method for Stiffness of Paper by Selfbending (Clark Method).
SAA AS 1301.431R P-89. Methods of Test for Pulp and Paper (Metric Units)—Part 431rp: Taber Bending Resistance of Paper and Paperboard (This is a Joint Standard with SANZ NZS 1301). 5 pp.
SAA AS 1301.453S-91. Methods of Test for Pulp and Paper—Part 453S: Bending Resistance of Paper and Paperboard—Constant Rate of Deflection (This is a Joint Standard with SANZ NZS 1301.453S). 3 pp.
SNZ NZS/AS 1301. 431RP-89. Methods of Test for Pulp and Paper Taber Bending Resistance of Paper and Paperboard (This is a Joint Standard with SAA AS 1301.431RP). 5 pp.
SNZ NZS/AS 1301. 453S-91. Methods of Test for Pulp and Paper Bending Resistance of Paper and Paperboard—Constant Rate of Deflection (This is a Joint Standard with SAA AS 1301.453S). 3 pp.

Bend Testing (Cont.)

—Particle Board
BSI BS EN 310-93. 1993 Wood-Base Panels—Determination of Modulus of Elasticity in Bending and of Bending Strength (R). 12 pp.
CEN PREN 310-90. Wood—Based Panels—Determination of Moduls of Elasticity—in Beding and Bending Strength. 7 pp.
CEN EN 310-93. Wood-Based Panels—Determination of Modulus of Elasticity in Bending and of Bending Strength. 8 pp.

—Phenolic Resins
CNS K6273-74. Method of Test for Phenol Formaldehyde Resin Molding Compounds (Oct)(2988). 11 pp.

—Plastic Pipes
CNS K6641-81. Method of Test for Rigid Polyvinyl Chloride Corrugated Pipes (Mar)(7057). 6 pp.

—Plastics
DIN ENGL 53435-83. Testing of Plastics; Bending Test and Impact Test on Dynstat Test Pieces (July). 4 pp.
DIN ENGL 53457-87. Testing of Plastics; Determination of the Elastic Modulus by Tensile, Compression and Bend Testing (Oct). 7 pp.
ISO 6721-83. Plastics—Determination of Damping Properties and Complex Modulus by Bending Vibration First Edition. 7 pp.
JIS K 7055-87. Testing Method for Flexural Properties of Glass Fiber Reinforced Plastics. 12 pp.

—Plywood
BSI BS EN 310-93. 1993 Wood-Base Panels—Determination of Modulus of Elasticity in Bending and of Bending Strength (R). 12 pp.
CEN PREN 310-90. Wood—Based Panels—Determination of Moduls of Elasticity—in Beding and Bending Strength. 7 pp.
CEN EN 310-93. Wood-Based Panels—Determination of Modulus of Elasticity in Bending and of Bending Strength. 8 pp.
CNS O2040-87. Method of Test for Bending Strength of Plywood (May)(8538).
CNS O2041-82. Method of Test for Bending Stiffness of Plywood (Feb)(8539). 3 pp.

—Reinforcing Steels
ISO 10065-90. Steel Bars for Reinforcement of Concrete—Bend and Rebend Tests First Edition. 6 pp.

—Sandwich Structures
DIN ENGL 53293-82. Testing of Sandwiches; Bending Test (Feb). 5 pp.

—Shutters
CNS A3071-85. Method of Test for Light-Weight Shutters for Buildings (Apr)(4167). 2 pp.

—Silicon Wafers
JIS H 0611-71. Methods of Measurement of Thickness, Taper and Bow for Silicon Wafers (R 1983). 7 pp.

—Skis
ISO 6267-80. Alpine Skis—Measurement of Bending Vibrations First Edition. 4 pp.

—Spokes—Bicycles
CNS B7067-75. Method of Test for the Spokes of Bicycles (Dec)(3880). 2 pp.

—Steels—Aerospace
BSI BS EN 2831-93. 1993 Hydrogen Embrittlement of Steels—Test by Slow Bending (S). 10 pp.
CEN PREN 2831-90. Hydrogen Embrittlement of Steels Test by Slow Bending. 4 pp.
CEN EN 2831-93. Aerospace Series—Hydrogen Embrittlement of Steels—Test by Slow Bending. 5 pp.

—Stud Welds
JIS Z 3145-81. Method of Bend Test for Stud Welds. 5 pp.

—Tent Frames
CNS S2145-90. Method of Test for Tent: Testing of Bending Resistance of the Framework (Mar)(12690). 2 pp.

—Thermoplastic Resins—Molding Materials
SAA AS 1370-73. Method for the Determination of the Softening Point of Thermoplastic Moulding Materials (Bending Test) Reconfirmed 1986. 5 pp.

—Urea Resins
CNS K6272-74. Method of Test for Urea Resin Molding Compounds (Oct)(2986). 9 pp.

Bend Testing (Cont.)

—Varnishes
BSI BS 3900: Part E1-70. 1970 Methods of Test for Paints Group E: Mechanical Tests on Paint Films Part E1: Bend Test (Cylindrical Mandrel). 4 pp.
CNS K6056-70. Method of Test for Baking Varnish (Jan)(771). 6 pp.
CNS K6057-70. Method of Test for Clear Varnish for Baking Varnish (Jan)(773). 3 pp.
ISO 1519-73. Paints and Varnishes—Bend Test (Cylindrical Mandrel) First Edition. 7 pp.
SAA AS 1580.402. 1-92. Paints and Related Materials—Methods of Test—Part 402.1: Bend Test (in Professional Packages 30, 39). 3 pp.
SNZ NZS/AS 1580. 402.1-92. Methods of Test for Paints and Related Materials Part 402.1: Bend Test (This is a Joint Standard with SAA AS 1580.402.1). 3 pp.

—Vulcanized Rubber
BSI BS 903: Part A49-84. 1984 Methods of Testing Vulcanized Rubber Part A49: Determination of Temperature Rise and Resistance to Fatigue in Flexometer Testing (Basic Principles). 10 pp.
BSI BS 903: Part A50-84. 1984 Methods of Testing Vulcanized Rubber Part A50: Determination of Temperature Rise and Resistance to Fatigue in Flexometer Testing (Compression Flexometer). 11 pp.
CNS K6745-83. Method of Test for Bending Fatigue of Vulcanized Rubber (Feb)(10017).
ISO 4666 Pt 1-82. Rubber, Vulcanized—Determination of Temperature Rise and Resistance to Fatigue in Flexometer Testing—Part 1: Basic Principles First Edition. 8 pp.
ISO 4666 Pt 2-82. Rubber, Vulcanized—Determination of Temperature Rise and Resistance to Fatigue in Flexometer Testing—Part 2: Rotary Flexometer First Edition. 9 pp.
ISO 4666 Pt 3-82. Rubber, Vulcanized—Determination of Temperature Rise and Resistance to Fatigue in Flexometer Testing—Part 3: Compression Flexometer First Edition. 9 pp.

—Weld Platings
DIN ENGL 50121 Pt 3-78. Testing of Metallic Materials; Technological Bending Test on Welded Joints and Weld Platings; Fusion Welded Platings (Jan). 5 pp.

—Welded Joints
DIN ENGL 50121 Pt 1-78. Testing of Metallic Materials; Technological Bending Test on Welded Joints and Weld Platings; Fusion Welded Joints (Jan). 9 pp.

—Welded Joints—Test Specimens
DIN ENGL 50127-76. Testing of Steel; Bendover Test Specimen, Wedge Test Specimen, Notched Pipe Tension Test Specimen for Assessment of Fusion-Welded Fillet or Butt Joints (Aug). 4 pp.

—Wire
CNS H2021-55. Testing Standard for Bending Strength of Metallic Wires (Oct)(658). 1 p.

—Wood
CNS O2005-81. Method of Test for Static Bending of Wood (Mar)(454). 2 pp.
CNS O2008-81. Method of Test for Impact Bending of Wood (Mar)(457). 1 p.
DIN ENGL 52189 Pt 1-81. Testing of Wood; Impact Bending Test; Determination of Impact Bending Strength (Dec). 3 pp.
ISO 3133-75. Wood—Determination of Ultimate Strength in Static Bending First Edition. 4 pp.
ISO 3348-75. Wood—Determination of Impact Bending Strength First Edition. 4 pp.
ISO 3349-75. Wood—Determination of Modulus of Elasticity in Static Bending First Edition. 5 pp.
JIS Z 2113-77. Method of Bending Test for Wood. 3 pp.
JIS Z 2113-63. Method of Bending Test for Wood. 4 pp.
JIS Z 2116-77. Method of Impact Bending Test for Wood.
JIS Z 2116-63. Method of Impact Bending Test for Wood (R 1970). 3 pp.

Bend Testing Equipment
See Also: Stiffness Testers; Testing Equipment
DIN ENGL 51227-77. Material Testing Machines; Bending Test Machines (Dec). 4 pp.
DIN ENGL 51230-77. Material Testing Machines; Dynstat Apparatus for the Determination of Bending Strength and Impact Strength of Small Specimens (Dec). 2 pp.
DIN ENGL 51302 Pt 1-85. Materials Testing Machines; Verification of Tensile, Compression and Bend Testing Machines; Supplementary Verification of Compression Testing Machines for Building Materials (Aug). 6 pp.
DIN ENGL 51302 Pt 2-86. Materials Testing Machines; Verification of Tensile, Compression and Bend Testing Machines; Supplementary Verification of Compression Testing Machines for Building Materials (Aug). 6 pp.

Benders
Use: Bending Equipment

Bending
Use For: Bowing *See Also:* Bend Radius; Deflection; Deformation; Modulus of Elasticity

—Glossaries
DIN ENGL 9870 Pt 3-72. Terms for Stamping Practice; Production Processes and Tools for Forming by Bending (Oct). 6 pp.

—Pipes—Filling Compounds
MOD UK TS 10142. Composition, Filling, for Pipe Bending (Withdrawn).

Bending Equipment
Use For: Benders; Bending Machines

—Bars
CNS B3200-85. Bender for Steel Bars (Dec)(3099).

—Roll
DIN ENGL 55805-79. Machine Tools; Sheet Metal Bending Rolls; Acceptance Conditions (Dec). 8 pp.

—Sheet
DIN ENGL 55802-79. Machine Tools; Folding Machines; Acceptance Conditions (Dec). 5 pp.

—Swivel
DIN ENGL 55220-80. Machine Tools; Swivel Bending Machine; Sizes (Mar). 2 pp.

Bending Machines
Use: Bending Equipment

Bending Strength
Use: Flexural Strength

Bending Stress
See Also: Flexibility; Stresses

—Ebonite—Deflection Temperature
BSI BS 2782:Pt1: METH 121A-B-91. 1991 Methods of Testing Plastics Part 1: Thermal Properties: Methods 121A and 121B: Determination of Temperature of Deflection of Plastics Under a Bending Stress (ISO 75: 1987) (V). 8 pp.
BSI BS 2782:Pt1: METH 121A-C-76. (WITHDRAWN) 1976 Methods of Testing Plastics Part 1: Thermal Properties: Method 121A-C: (A) Determ of Temp of Deflection Under a Bending Stress of 1.8 MPa of Plastics and Ebonite.(B) Determ of Temp of Deflec Under a Bending Stress of 0.45 MPa of Plastic & Ebonite. 4 pp.
CNS K6617-80. Plastics and Ebonite Determination of Temperature of Deflection Under Load (Dec)(6683).
ISO 75 Pt 2-93. Plastics—Determination of Temperature of Deflection Under Load—Part 2: Plastics and Ebonite First Edition; (Cancels and Replaces ISO 75:1987). 6 pp.

—Plastics—Deflection Temperature
BSI BS 2782:Pt1: METH 121A-B-91. 1991 Methods of Testing Plastics Part 1: Thermal Properties: Methods 121A and 121B: Determination of Temperature of Deflection of Plastics Under a Bending Stress (ISO 75: 1987) (V). 8 pp.
BSI BS 2782:Pt1: METH 121C-92. 1992 Methods of Testing Plastics Part 1: Thermal Properties: Method 121C: Determination of Temperature of Deflection Under a Bending Stress for Rigid Thermosetting Resin Bonded Laminated Plastics Sheet (V). 7 pp.
BSI BS 2782:Pt1: METH 121A-C-76. (WITHDRAWN) 1976 Methods of Testing Plastics Part 1: Thermal Properties: Method 121A-C: (A) Determ of Temp of Deflection Under a Bending Stress of 1.8 MPa of Plastics and Ebonite.(B) Determ of Temp of Deflec Under a Bending Stress of 0.45 MPa of Plastic & Ebonite. 4 pp.
CNS K6617-80. Plastics and Ebonite Determination of Temperature of Deflection Under Load (Dec)(6683).
DIN ENGL 53462-87. Testing of Plastics; Martens Method of Determining the Temperature of Deflection Under a Bending Stress (Jan) (Supersedes March 1965 Edition and July 1968 Edition of DIN 53458). 5 pp.
ISO 75 Pt 1-93. Plastics—Determination of Temperature of Deflection Under Load—Part 1: General Test Method First Edition; (Cancels and Replaces ISO 75:1987). 8 pp.
ISO 75 Pt 2-93. Plastics—Determination of Temperature of Deflection Under Load—Part 2: Plastics and Ebonite First Edition; (Cancels and Replaces ISO 75:1987). 6 pp.

—Reinforced Plastics—Deflection Temperature
ISO 75 Pt 3-93. Plastics—Determination of Temperature of Deflection Under Load—Part 3: High Strength Thermosetting Laminates and Long-Fibre-Reinforced Plastics First Edition; (Cancels and Replaces ISO 75:1987). 6 pp.

INTERNATIONAL AND NON-U.S. NATIONAL STANDARDS
SUBJECT INDEX

Bending Stress (Cont.)
—Thermosetting Resins—Deflection Temperature
ISO 75 Pt 3-93. Plastics—Determination of Temperature of Deflection Under Load—Part 3: High Strength Thermosetting Laminates and Long-Fibre-Reinforced Plastics First Edition; (Cancels and Replaces ISO 75:1987). 6 pp.

Bending Test Machines
Use: Bend Testing Equipment

Bending Testing
Use: Bend Testing

Bends (Fittings)
Scope Note: See specific types of fittings

Bendtsen Testers
Use For: Smoothness Testers (Bendtsen)
See Also: Testing Equipment
CPPA M-5-71. Bendtsen Smoothness Tester—Model 5. 1 p.

—Paperboard
BSI BS 6538: Part 2-92. 1992 Air Permeance of Paper and Board Part 2: Method for Determination of Air Permeance Using the Bendtsen Apparatus (ISO 5636-3: 1992) (Supersedes BS 2925: 1958) (H). 15 pp.

BSI BS 6538: Part 2-85. 1985 Air Permeance of Paper and Board Part 2: Method for Determination of Air Permeance Using the Bendtsen Apparatus. 11 pp.

CPPA D.15U-86. Roughness of Paper & Paperboard—Bendtsen. 1 p.

ISO 5636 Pt 3-92. Paper and Board—Determination of Air Permeance (Medium Range)—Part 3: Bendtsen Method Second Edition. 12 pp.

ISO 8791 Pt 2-90. Paper and Board—Determination of Roughness/Smoothness (Air Leak Methods)—Part 2: Bendtsen Method First Edition. 11 pp.

—Papers
BSI BS 6538: Part 2-92. 1992 Air Permeance of Paper and Board Part 2: Method for Determination of Air Permeance Using the Bendtsen Apparatus (ISO 5636-3: 1992) (Supersedes BS 2925: 1958) (H). 15 pp.

BSI BS 6538: Part 2-85. 1985 Air Permeance of Paper and Board Part 2: Method for Determination of Air Permeance Using the Bendtsen Apparatus. 11 pp.

CPPA D.15U-86. Roughness of Paper & Paperboard—Bendtsen. 1 p.

ISO 5636 Pt 3-92. Paper and Board—Determination of Air Permeance (Medium Range)—Part 3: Bendtsen Method Second Edition. 12 pp.

ISO 8791 Pt 2-90. Paper and Board—Determination of Roughness/Smoothness (Air Leak Methods)—Part 2: Bendtsen Method First Edition. 11 pp.

Benzaldehyde
CNS K2194-88. Benzaldehyde (Mar)(12226). 2 pp.
CNS K6944-88. Method of Test for Benzaldehyde (Mar)(12227). 3 pp.
CNS K7089-62. Chemical Reagent (Benzaldehyde) (Jul)(1589).
JIS K 4126-87. Benzaldehyde.
JIS K 8857-81. Benzaldehyde.

Benzene
Use For: Benzol See Also: Aromatic Hydrocarbons; Benzene Content Analysis; Chlorobenzene; Chloronitrobenzenes; o,p-Dihydroxyazo-p-Nitrobenzene; Dinitrobenzene; m-Dinitrobenzene; Dodecylbenzene; Hazardous Materials; Hexanitrostilbene; Hydrocarbons; 1,2,4-Trichlorobenzene

CNS K1065-80. Benzene (Industrial Grade and Nitration Grade) (Apr)(1473). 2 pp.
CNS K1241-80. Refined Benzene (Apr)(5461).
CNS K7090-62. Chemical Reagent (Benzene) (Benzol) (Jul)(1590).
DIN ENGL 51633-86. Benzene and Benzene Homologues; Requirements (Nov). 4 pp.
ISO 5271-79. Benzene for Industrial Use—Specification First Edition. 3 pp.
JIS K 2435-85. Benzenes (Benzene-Toluene-Xylene-Solventnaphtha). 6 pp.
JIS K 8858-80. Benzene.

—Bromine Assimilation—Iodometry
DIN ENGL 51774 Pt 3-75. Testing of Liquid Fuels; Determination of Bromine Assimilation by the Iodometric Method (BC) (Aug). 2 pp.

—Carbon Disulfide Content
CNS K6620-80. Method of Test for Carbon Disulfide in Benzene (Dec)(6835).

Benzene (Cont.)
—Crystallization
ISO 5278-80. Benzene—Determination of Crystallizing Point First Edition; (Erratum—Dec 1981). 6 pp.

—Density—Pycnometric Analysis
ISO 5281-80. Aromatic Hydrocarbons—Benzene, Xylene and Toluene—Determination of Density at 20 Degrees Celsius First Edition. 11 pp.

—Freezing Points
CNS K6261-74. Method of Test for Freezing Point of Benzene (Jun)(2762).

—Glossaries
ISO 1543-81. Benzole Industry—Vocabulary First Edition. 9 pp.

—Restricted Use
EC 82/806/EEC-82. Council Directive Amending, for the Second Time (Benzene), Directive 76/769/EEC on the Approximation of the Laws, Regs. and Ad. Provs. of the Member Sts. Relating to Restrictions on the Mktg. and Use of Certain Dangerous Substances and Preparations. 2 pp.

—Sulfuric Acid Reaction
DIN ENGL 51762-74. Testing of Liquid Fuels; Determination of Reaction to Sulphuric Acid (Sept). 2 pp.

—Thiophene Content
CNS K6621-80. Method of Test for Thiophene in Benzene (Dec)(6836).

—Thiophene Content—Spectrophotometry
CNS K6622-80. Method of Test for Traces of Thiophene in Benzene Using Isatin and Spectrophotometry (Dec)(6837).

Benzene Carbonyl Chloride
Use: Benzoyl Chloride

Benzene Carboxylic Acid
Use: Benzoic Acid

Benzene Chloride
Use: Chlorobenzene

Benzene Content Analysis
See Also: Benzene

—Asphalts
CNS K6206-65. Determination of Benzene-Insoluble Substance in Asphalt and Coal Tar (Sep)(2488)(R 1973).

—Coatings
CGSB 1-GP-71 METH 65.1-82. Methods of Testing Paints and Pigments Detection of Benzene (Benzol). 2 pp.

—Creosole Oil
CNS K6071-57. Method of Test for Benzene Insoluble Substances of Creosote Oil (Jul)(914)(R 1971).

—Cyclohexane—Gas Chromatography
CNS K6651-81. Method of Test for Purity and Benzene Content of Cyclohexane by Gas Chromatography (Mar)(7170).

—Exhaust Gases
CNS K9108-90. Method for Determination of Benzene in Exhaust Gas (Mar)(12688).
JIS K 0088-83. Methods for Determination of Benzene in Exhaust Gas. 13 pp.

—Gasoline—Gas Chromatography
DIN ENGL 51413 Pt 2-84. Analysis of Liquid Petroleum Products by Gas Chromatography; Determination of Benzene Content (Nov). 2 pp.

—Gasoline—Infrared Spectroscopy
DIN ENGL 51414-85. Testing of Petroleum Products; Determination of the Benzene Content of Gasolines by Infrared Spectroscopy (June). 3 pp.

—Petroleum Products—Infrared Spectroscopy
CEN PREN 238-85. Liquid Petroleum Products Determination of the Benzene Content; Infrared Spectrometric Method. 7 pp.

—Polishes
CGSB 25-GP-1M METH 18.1-84. Methods of Sampling and Testing Waxes and Polishes Benzol. 1 p.

Benzene Content Analysis (Cont.)
—Sludge (Sewage)—Gas Chromatography
DIN ENGL 38407 Pt 9-91. German Standard Methods for the Examination of Water, Waste Water and Sludge; Substance Group Analysis (Group F); Determination of Benzene and Some of Its Derivatives by Gas Chromatography (F9) (May). 16 pp.

—Waste Water—Gas Chromatography
DIN ENGL 38407 Pt 9-91. German Standard Methods for the Examination of Water, Waste Water and Sludge; Substance Group Analysis (Group F); Determination of Benzene and Some of Its Derivatives by Gas Chromatography (F9) (May). 16 pp.

—Water—Gas Chromatography
DIN ENGL 38407 Pt 9-91. German Standard Methods for the Examination of Water, Waste Water and Sludge; Substance Group Analysis (Group F); Determination of Benzene and Some of Its Derivatives by Gas Chromatography (F9) (May). 16 pp.

—Waxes
CGSB 25-GP-1M METH 18.1-84. Methods of Sampling and Testing Waxes and Polishes Benzol. 1 p.

Benzene Hexachloride
Use For: gamma-Benzene Hexachloride; BHC; Hexachlorocyclohexane; 1,2,3,4,5,6-Hexachlorocyclohexane; Lindane
See Also: Cyclohexane; Pesticides

CNS K1040-57. BHC (Industrial Grade) (Oct)(636)(R 1973).
CNS K6078-57. Method of Test for BHC of Industrial Grade and Fine Grade (Dec)(960)(R 1973).
SAA AS 1870.2D-76. Standard for Development—Pesticides for Agricultural Use—Part 2D: BHC and BHC-DDT. 30 pp.

Benzene Hexachloride Content Analysis
Use For: gamma-Benzene Hexachloride Content Analysis; Lindane Content Analysis

—Animal Feed
CNS N4068-81. Method of Test for BHC in Feeds (Jan)(6943). 2 pp.

—Wood Preservatives—Colorimetric Analysis
BSI BS 5666: Part 6-83. 1983 Amd 1 Wood Preservatives and Treated Timber: Part 6: Quantitative Analysis or Preservative Solution and Treated Timber Containing Pentachloro-phenol, Phenachloro-phenyl Laurate, GS Hexachlorocyclohexane and Dieldrin (AMD 6224). 11 pp.

—Wood Preservatives—Gas Chromatography
BSI BS 5666: Part 6-83. 1983 Amd 1 Wood Preservatives and Treated Timber: Part 6: Quantitative Analysis or Preservative Solution and Treated Timber Containing Pentachloro-phenol, Phenachloro-phenyl Laurate, GS Hexachlorocyclohexane and Dieldrin (AMD 6224). 11 pp.

gamma-Benzene Hexachloride Content Analysis
Use: Benzene Hexachloride Content Analysis

gamma-Benzene Hexachloride
Use: Benzene Hexachloride

Benzenearsonic Acid
Use For: Phenylarsonic Acid
CNS K7355-69. Chemical Reagent (Phenylarsonic Acid) (Jul)(1854).

1,3-Benzenediol
Use: Resorcinol

Benzenemethanol
Use: Benzyl Alcohol

1,2-Benzenodiol,4-(1,1-Dimethylethyl) Content Analysis
Use: tert-Butylcatechol Content Analysis

Benzidine
CNS K7091-62. Chemical Reagent (Benzidine) (P,P-Diaminodiphenyl) (Jul)(1591).

Benzidine Dihydrochloride
CNS K7092-62. Chemical Reagent (Benzidine Dihydrochloride) (Jul)(1592).

Benzidine Orange
See Also: Pigments
—**Pigments**
CNS K2217-88. Benzidine Orange (May)(12305).
JIS K 5221-71. Benzidine Orange. 4 pp.

Benzidine Yellow
See Also: Pigments
—**Pigments**
CNS K2216-88. Benzidine Yellow (May)(12304).
JIS K 5220-71. Benzidine Yellow. 4 pp.

Benzoates
Scope Note: Use a more specific term *See:* Lead-2, 4-Dihydroxybenzoate

Benzoic Acid
Use For: Benzene Carboxylic Acid *See Also:* Gallic Acid; Salicylic Acid; Tannic Acid
CNS K2195-88. Benzoic Acid (Mar)(12228). 1 p.
CNS K6945-88. Method of Test for Benzoic Acid (Mar)(12229). 2 pp.
CNS K7093-62. Chemical Reagent (Benzoic Acid) (Jul)(1593).
CNS K7094-62. Chemical Reagent (Benzoic Acid, Primary Standard) (Jul)(1594).
JIS K 4127-86. Benzoic Acid.
JIS K 8073-79. Benzoic Acid.

Benzoic Acid Content Analysis
—**Animal Feed**
CNS N4116-84. Method of Test for Feed Additives: Determination of Benzoic Acid (May)(10885).
—**Food—Gas Chromatography**
CNS N6190-84. Method of Test for Preservatives in Food (Jun)(10949). 8 pp.
—**Food—Molecular Absorption Spectrophotometry**
ISO 6560-83. Fruit and Vegetable Products—Determination of Benzoic Acid Content (Benzoic Acid Contents Greater Than 200 mg per Litre or per Kilogram)—Molecular Absorption Spectrometric Method First Edition. 5 pp.
—**Food—Thin Layer Chromatography**
CNS N6190-84. Method of Test for Preservatives in Food (Jun)(10949). 8 pp.
—**Fruits—Molecular Absorption Spectrophotometry**
ISO 6560-83. Fruit and Vegetable Products—Determination of Benzoic Acid Content (Benzoic Acid Contents Greater Than 200 mg per Litre or per Kilogram)—Molecular Absorption Spectrometric Method First Edition. 5 pp.
—**Fruits—Spectrophotometry**
ISO 5518-78. Fruits, Vegetables and Derived Products—Determination of Benzoic Acid Content—Spectrophotometric Method First Edition; (Erratum—Aug 1978). 6 pp.
—**Vegetables—Molecular Absorption Spectrophotometry**
ISO 6560-83. Fruit and Vegetable Products—Determination of Benzoic Acid Content (Benzoic Acid Contents Greater Than 200 mg per Litre or per Kilogram)—Molecular Absorption Spectrometric Method First Edition. 5 pp.
—**Vegetables—Spectrophotometry**
ISO 5518-78. Fruits, Vegetables and Derived Products—Determination of Benzoic Acid Content—Spectrophotometric Method First Edition; (Erratum—Aug 1978). 6 pp.

alpha-Benzoin Oxime
CNS K7095-62. Chemical Reagent (Alpha-Benzoin Oxime) (Jul)(1595).
JIS K 8860-78. Benzoin-Alpha Particle-Oxime.

Benzol
Use: Benzene

Benzophenones
Use For: Diphenyl Ketone; Phenyl Ketone
See Also: Ketones
CNS K7654-82. Chemical Reagent (Benzophenone) (Oct)(9511).
JIS K 8861-61. Benzophenone.

Benzopurpurine 4B
JIS K 8862-64. Benzopurpurine 4B.

Benzopyrene Content Analysis
See Also: Aromatic Hydrocarbons
—**Creosote—Wood Preservatives**
CEN PREN 1014-3-93. Wood Preservation—Creosote and Creosoted Timber—Methods of Sampling and Analysis—Part 3: Method for the Determination of the Benzo a Pyrene Content of Creosote. 11 pp.

Benzothiazyl Disulfide
Use: 2,2'-Dithiobis(benzothiazole)

Benzotriazole
—**Photographic Chemicals**
BSI BS 3309-78. 1978 Photographic Grade Benzotriazole. 8 pp.
CNS Z9041-80. Benzotriazole (Photographic Grade) (Mar)(5387).
CNS Z9042-80. Method of Test for Benzotriazole (Photographic Grade) (Mar)(5388).
ISO 3618-76. Photographic Grade Benzotriazole—Specification First Edition. 6 pp.
JIS K 7734-89. Photographic Grade Benzotriazole.

Benzoyl Chloride
Use For: Benzene Carbonyl Chloride
CNS K7096-62. Chemical Reagent (Benzyl Chloride) (Sep)(1596).
JIS K 8158-80. Benzoyl Chloride.

Benzoyl Peroxide
Use For: Dibenzbyl Peroxides *See Also:* Organic Peroxides
—**Safety—Information Cards**
SAA AS 2508.5.00 4-91. Safe Storage and Handling Information Cards for Hazardous Materials—Part 5.004: Benzoyl Peroxide (in Professional Package 38). 2 pp.

N-Benzoyl-N-phenylhydroxylamine
JIS K 9569-76. N-Benzoyl-N-phenylhydroxylamine.

Benzyl Alcohol
Use For: Benzenemethanol; alpha-Hydroxytoluene; Phenyl Carbinol; Phenyl Methanol; Phenylmethanol
See Also: Alcohols
CNS K7097-62. Chemical Reagent (Benzyl Alcohol) (Sep)(1597).
JIS K 8854-78. Benzyl Alcohol.

Benzyl Benzoate
CNS K7559-88. Chemical Reagent (Benzyl Benzoate) (Oct)(7717).
JIS K 8079-80. Benzyl Benzoate.

Benzyl Chloride
Use For: (Chloromethyl) Benzene; a-Chlorotoluene; alpha-Chlorotoluene *See Also:* Halogenated Hydrocarbons
CNS K2193-88. Benzyl Chloride (Mar)(12224). 2 pp.
CNS K6943-88. Method of Test for Benzyl Chloride (Mar)(12225).
CNS K7098-62. Chemical Reagent (Benzyl Chloride) (Sep)(1598).
ISO 3362-76. Benzyl Chloride for Industrial Use—Methods of Test First Edition. 5 pp.
JIS K 4125-83. Benzyl Chloride.
JIS K 8157-81. Benzyl Chloride.

N-Benzyl-N-Ethylaniline
See Also: Aniline
CNS K2184-87. N-Benzyl-N-Ethylaniline (Dec)(12200). 5 pp.
JIS K 4112-81. N-Substituted Anilines (N, N-Dimethylaniline.N, N-Diethylaniline. N-Ethylaniline. N-Methylaniline. N-Benzyl-N-Ethylaniline).

BER
Use: Bit Error Rate

Bergamot Oil
See Also: Essential Oils
CNS K5119-81. Oil of Bergamot Italy (Nov)(8131).
ISO 3520-80. Oil of Bergamot, Italy First Edition; (Erratum—Aug 1981). 5 pp.
ISO 8900-87. Oil of Bergamot Petitgrain (Citrus Aurantium (Linnaeus) Ssp. Bergamia (Wight Et Arnott) Engler) First Edition. 6 pp.

Bergapten Content Analysis
—**Cosmetics**
CNS S2095-82. Methods of Hygienic Test for Cosmetics Bergapten (Dec)(9753).

Berths
See Also: Aircraft; Mooring Hardware

Berths (Cont.)
—**Aircraft—Design**
CAA Chapter P4-4. Seats and Berths: Safety Belts and Harnesses (Provisional Airworthiness Requirements for Civil Powered Lift Aircraft). 6 pp.
—**Design**
BSI BS 6349: Part 3-88. 1988 Code of Practice for Maritime Structures; Part 3: Design of Dry Docks, Locks, Slipways and Shipbuilding Berths, Shiplifts and Dock and Lock Gates. 75 pp.
—**Ground Effect Machines**
CAA Chapter B5-9 04.79. Anchoring, Towing and Berthing. 4 pp.
—**Ships—Fluorescent Lighting**
CNS F5039-82. Fluorescent Berth Lights with Spare Light for Marine Use (Sep)(9395).

Beryllium
See Also: Hazardous Materials; Metals
—**Highway Transportation—Emergency Procedures**
SAA AS 1678.6.0. 015-84. Emergency Procedure Guides—Transport—Part 6.0.015: Beryllium and Compounds.

Beryllium Content Analysis
—**Copper Alloys—Absorptiometric Analysis**
JIS H 1063-89. Methods for Determination of Beryllium in Copper Alloys. 12 pp.
—**Copper Alloys—Atomic Absorption Spectrometry**
JIS H 1063-89. Methods for Determination of Beryllium in Copper Alloys. 12 pp.
—**Copper Alloys—Emission Spectrometry**
JIS H 1063-89. Methods for Determination of Beryllium in Copper Alloys. 12 pp.
—**Flue Gases**
JIS K 0224-83. Methods for Determination of Beryllium in Stack Gas (R 1988). 11 pp.
—**Magnesium Alloys**
JIS H 1339-87. Methods for Determination of Beryllium in Magnesium Alloys. 9 pp.

Beryllium Copper
Use: Beryllium Copper Alloys

Beryllium Copper Alloys
Scope Note: For additional listings, see also specific products made from copper beryllium alloys
Use For: Beryllium Copper *See Also:* Alloys; Copper Alloys
—**Bars**
CNS H3127-85. Copper-Beryllium Alloy Plate, Sheet, Strip and Rolled Bar (Jan)(10879).
JIS H 3270-92. Copper Beryllium Alloy, Phosphor Bronze and Nickel Silver Rods, Bars and Wires. 22 pp.
—**Chemical Analysis**
CNS H2072-85. Method of Chemical Analysis for Copper-Beryllium Alloy (Jan)(11189).
JIS H 1261-76. Methods for Chemical Analysis of Beryllium Copper.
JIS H 1553-76. Methods for Chemical Analysis of Beryllium Copper Master Alloys.
—**Master Alloy**
JIS H 2504-63. Copper Beryllium Master Alloy. 4 pp.
—**Plates**
CNS H3127-85. Copper-Beryllium Alloy Plate, Sheet, Strip and Rolled Bar (Jan)(10879).
JIS H 3130-92. Copper Beryllium Alloy, Phosphor Bronze and Nickel Silver Sheets, Plates and Strips for Springs. 19 pp.
—**Rods**
JIS H 3270-92. Copper Beryllium Alloy, Phosphor Bronze and Nickel Silver Rods, Bars and Wires. 22 pp.
—**Sheets**
CNS H3127-85. Copper-Beryllium Alloy Plate, Sheet, Strip and Rolled Bar (Jan)(10879).
JIS H 3130-92. Copper Beryllium Alloy, Phosphor Bronze and Nickel Silver Sheets, Plates and Strips for Springs. 19 pp.

Beryllium Oxides
—**Nuclear Grade**
CNS J3008-81. Specification for Nuclear-Grade Beryllium Oxide Powder (Nov)(8126).

Beta Contamination Meters
Use: Radiation Meters

Beta Contamination Monitors
Use: Radiation Monitors

Beta Dosimeters
Use: Dosimeters—Beta Radiation

Beta Particles
See Also: Radiation Detection and Measurement

—**Radiation Detection and Measurement—Surface Contamination**
ISO 7503 Pt 1-88. Evaluation of Surface Contamination—Part 1: Beta-Emitters (Maximum Beta Energy Greater Than 0.15 MeV) and Alpha-Emitters First Edition. 15 pp.

—**Radioactive Isotopes—Measurement**
CNS J2011-81. Method for Analysis of Radioisotopes Gross Beta Particle Count (May)(7388).

—**Water—Measurement**
BSI BS 6068: Sec 3.2-93. 1993 Water Quality Part 3: Radiological Methods Section 3.2: Measurement of Gross Beta Activity in Non-Saline Water (ISO 9697: 1992). 14 pp.
CNS J2033-82. Method of Test for Beta Particle Radioactivity of Water (Jan)(8301).
ISO 9697-92. Water Quality—Measurement of Gross Beta Activity in Non-Saline Water First Edition. 13 pp.

Beta Radiation Monitors
See Also: Radiation Measuring Instruments; Radiation Monitors

—**Calibration**
ISO 8769-88. Reference Sources for the Calibration of Surface Contamination Monitors—Beta-Emitters (Maximum Beta Energy Greater Than 0,15 MeV) and Alpha-Emitters First Edition. 10 pp.
JIS Z 4334-92. Reference Sources for the Calibration of Surface Contamination Monitors. 10 pp.

—**Effluents**
IEC 861-87. Equipment for Continuously Monitoring for Beta and Gamma Emitting Radionuclides in Liquid Effluents First Edition. 71 pp.

—**Personnel Exposure**
IEC 1098-92. Installed Personnel Surface Contamination Monitoring Assemblies for Alpha and Beta Emitters First Edition. 77 pp.

Bevel Gear Pairs
Use: Bevel Gears

Bevel Gears
See Also: Gear Reducers; Gears; Pinions; Ring Gears
BSI BS 545-82. 1982 Amd 2 Bevel Gears (Machine Cut). 30 pp.
BSI BS 978: Part 3-52. 1952 Amd 1 Fine Pitch Gears Part 3: Bevel Gears. 9 pp.
CNS B2558-81. Remote Control Gear with Transmission Tubing for Manual Operation; Bevel Gear Units (Apr)(7189).
DIN ENGL 3965 Pt 1-86. Tolerancing of Bevel Gears; Basic Concepts (Aug). 4 pp.
DIN ENGL 3965 Pt 2-86. Tolerancing of Bevel Gears; Tolerances for Individual Parameters (Aug). 16 pp.
DIN ENGL 3965 Pt 3-86. Tolerancing of Bevel Gears; Tolerances for Tangential Composite Errors (Aug). 16 pp.
DIN ENGL 3965 Pt 4-86. Tolerancing of Bevel Gears; Tolerances for Shaft Angle Errors and Axes Intersection Point Deviations (Aug). 3 pp.
JIS B 1704-78. Accuracy for Bevel Gears. 21 pp.

—**Backlash**
JIS B 1705-73. Backlashes for Bevel Gears (R 1979). 18 pp.

—**Engineering Drawings**
ISO 1341-76. Straight Bevel Gears—Information to Be Given to the Manufacturer by the Purchaser in Order to Obtain the Gear Required First Edition. 4 pp.

—**Gear Cutters**
JIS B 4351-85. Straight Bevel Gear Generating Cutters (Type G). 15 pp.

—**Gear Teeth**
DIN ENGL 3966 Pt 2-78. Information on Gear Teeth in Drawings; Information on Straight Bevel Gear Teeth (Aug). 6 pp.
JIS B 1741-77. Tooth Contact Marking of Gears. 10 pp.

Bevel Gears *(Cont.)*
—**Gear Teeth—Loads (Forces)**
CNS B1204-84. Calculation of Load Capacity on Tooth Root for Bevel Gears (Aug)(5984).
CNS B1205-84. Calculation of Load Capacity on Tooth Flank for Bevel Gears (Aug)(5985).
CNS B1206-84. Calculation of Load Capacity for Cylindrical and Bevel Gears-Tooth Form Factor YF (Aug)(5986).

—**Glossaries**
CNS B1177-84. Glossary of Gears Geometrical Definitions (Bevel and Hypoid Gears and Gear Pairs) (Feb)(5719).
CNS B1190-84. Definition and Denominations for Glossary of Gear Terms — Reference Cone of Bevel Gears (May)(5732).
CNS B1192-84. Definition and Denominations for Glossary of Gear Terms — Special Additional Definition for Bevel Gears (May)(5734).
DIN ENGL 3971-80. Definitions and Parameters for Bevel Gears and Bevel Gear Pairs (July). 24 pp.
DIN ENGL 3998 Pt 3-76. Denominations on Gears and Gear Pairs; Bevel and Hypoid Gears and Gear Pairs (Sept). 14 pp.

—**Loads (Forces)**
CNS B2519-84. Calculation of Load Capacity for Cylindrical and Bevel Gears-Load Sharing Factor Y Sub Epsilon and Overlap Ration Sub Epsilon Sub Beta (Aug)(6407).
CNS B2520-84. Calculation of Load Capacity of Spur, Helical and Bevel Gears; Zone Factor ZH (Aug)(6408).
CNS B2521-84. Calculation of Load Capacity of Spur, Helical and Bevel Gears; Material Factor ZM (Aug)(6409).
CNS B2522-84. Calculation of Load Capacity for Cylindrical and Bevel Gears-Pinion Single Tooth Contact Factor ZB Wheel Single Tooth Contact Factor ZD and Traverse Contact Ratio Sub Epsilon Sub Beta (Aug)(6410).
CNS B2523-84. Calculation of Load Capacity for Cylindrical and Bevel Gears-Contact Ration Factor Z Sub Epsilon (Aug)(6411).
CNS B2524-84. Calculation of Load Capacity for Cylindrical and Bevel Gears-Helix Angle Factor Y Beta Ray (Aug)(6412).

—**Marine**
JIS F 7451-78. Marine Bevel Gears.

—**Modules**
CNS B2527-84. Series of Modules for Straight Bevel Gears (Aug)(6415).

—**Pitch**
ISO 678-76. Straight Bevel Gears for General Engineering and Heavy Engineering—Modules and Diametral Pitches First Edition. 3 pp.

—**Rack**
CNS B2526-84. Basic Rack of Straight Bevel Gears for General and Heavy Engineering (Aug)(6414).
ISO 677-76. Straight Bevel Gears for General Engineering and Heavy Engineering—Basic Rack First Edition. 4 pp.

—**Symbols**
DIN ENGL 3971-80. Definitions and Parameters for Bevel Gears and Bevel Gear Pairs (July). 24 pp.

—**Trim Controls—Aircraft**
SBAC AS 888 ISSUE 6. Bevel and Shaft (Sprocket), for Trim Control. 1 p.
SBAC AS 889 ISSUE 7. Bevel and Shaft (Handwheel), for Trim Control. 1 p.

Bevel Protractors
Use: Protractors

Bevel Washers
Use For: Tapered Square Washers
See Also: Washers (Fasteners)

—**Aerospace**
DIN ENGL LN 9016-80. Aerospace; Washers; Bevelled (Apr). 3 pp.

Bevelled Washers
Use: Bevel Washers

Bevels
See Also: Hand Tools; Squares (Instruments); Woodworking Equipment
BSI BS 3322-81. 1981 Carpenters' Squares and Bevels. 8 pp.

—**Carpenters'**
MOD UK DEF-1207-58. Bevels, Carpenters', and Squares, Carpenters'. 6 pp.

Beverage Coolers
See Also: Beverages; Coolers; Milk Coolers

Beverage Coolers *(Cont.)*
BSI BS 6671-86. 1986 Thermal Packs and Drink Coolers. 12 pp.
BSI BS 6671-01. 1986 Amd 1 Thermal Packs and Drink Coolers (AMD 7198) June 15, 1992. 13 pp.

Beverage Urns
BSI BS 4167: Part 6-69. 1969 Amd 1 Electrically-Heated Catering Equipment Part 6: Bulk Liquid Heaters. 46 pp.
CGSB CAN/CGSB-52.8-M86. Urns, Beverage, Food Service. 12 pp.

Beverages
Scope Note: For additional listings, use a more specific term *See Also:* Alcoholic Beverages; Beer; Beverage Coolers; Buttermilk; Carbonated Beverage Bottles; Carbonated Beverages; Cider; Coffee; Condensed Milk; Cream (Dairy Products); Evaporated Milk; Food; Fruit Juices; Milk; Powdered Milk; Skim Milk; Tea; Vegetable Juices; Wines

—**Acidity**
CNS N6091-91. Method of Test for Fruit and Vegetable Juices and Drinks (General Rules) (Jun)(3736). 1 p.
CNS N6176-82. Method of Test for Beverage: Determination of Acidity (Sep)(9434). 2 pp.

—**Artificial Sweeteners Content**
CNS N6091-91. Method of Test for Fruit and Vegetable Juices and Drinks (General Rules) (Jun)(3736). 1 p.

—**Ash Content**
CNS N6091-91. Method of Test for Fruit and Vegetable Juices and Drinks (General Rules) (Jun)(3736). 1 p.

—**Brix Testing**
CNS N6091-91. Method of Test for Fruit and Vegetable Juices and Drinks (General Rules) (Jun)(3736). 1 p.

—**Caffeine Content**
CNS N6174-82. Method of Test for Beverage: Determination of Caffeine (Sep)(9432). 2 pp.

—**Cans—Capacity Measurement**
BSI BS 6992: Part 3-88. (OBSOLESCENT) 1988 Capacities and Related Cross Sections of Open-Top Cans for Food and Drinks Part 3: Recommendations for Cans for Drinks. 7 pp.

—**Classification**
BSI BS 1000: (663)-85. 1985 Universal Decimal Classification (UDC). English Full Edition (663): Industrial Microbiology. Industrial Mycology. Fermentation Industry. Beverage Industry. Stimulant Industry. 30 pp.
SNZ NZS/BS 1000 (663)-85. Universal Decimal Classification Industrial Microbiology. Industrial Mycology. Fermentation Industry. Beverage Industry. Stimulant Industry.. 32 pp.

—**Essential Oil Content**
CNS N6175-82. Method of Test for Beverage: Determination of Essential Oil (Sep)(9433). 1 p.

—**Ethanol Content—Gas Chromatography**
CNS N6181-82. Method of Test for Beverage—Determination of Ethanol Content (May)(10292). 1 p.

—**Impurities Content**
CNS N6091-91. Method of Test for Fruit and Vegetable Juices and Drinks (General Rules) (Jun)(3736). 1 p.

—**Insoluble Matter Content**
CNS N6091-91. Method of Test for Fruit and Vegetable Juices and Drinks (General Rules) (Jun)(3736). 1 p.
CNS N6173-82. Method of Test for Beverage: Determination of Water-Insoluble Solids (Sep)(9431). 1 p.

—**Methanol Content—Gas Chromatography**
CNS N6180-83. Method of Test for Beverage: Determination of Methanol Content (May)(10291). 1 p.

—**Moisture Content**
CNS N6172-82. Method of Test for Beverage: Determination of Total Solids and Moisture (Sep)(9430). 1 p.

—**Nitrogen Content**
CNS N6091-91. Method of Test for Fruit and Vegetable Juices and Drinks (General Rules) (Jun)(3736). 1 p.

—**Powdered—Fruit Flavored**
CGSB 32.283M-87. Beverage Powders, Fruit-Flavoured. 7 pp.

Beverages (Cont.)
—Powdered—Fruit Flavored (Cont.)
CNS N5104-72. Citrus Essence Powder (Tentative) (Jun)(3438). 1 p.
CNS N5105-72. Citrus Juice Powder (Tentative) (Jun)(3439). 1 p.

—Saponin Content
CNS N6171-82. Method of Test for Beverage: Determination of Saponin (Sep)(9429). 1 p.

—Solids Content
CNS N6172-82. Method of Test for Beverage: Determination of Total Solids and Moisture (Sep)(9430). 1 p.

—Soluble Matter Content
CNS N6177-82. Method of Test for Beverage: Determination of Water-Soluble Solids (Sep)(9435). 2 pp.

—Soybean
CNS N5210-84. Soybean Drink (Packaged) (Nov)(11138). 2 pp.
CNS N5211-84. Formulated Soymilk (Packaged) (Nov)(11139). 2 pp.
CNS N5212-84. Soymilk (Packaged) (Nov)(11140). 2 pp.

—Sport Drinks
CNS N5223-87. Sport Drinks (Packaged) (Oct)(12149). 2 pp.

—Sport Drinks—Electrolytes Content
CNS N6208-87. Method of Test for Sport Drinks (Oct)(12150). 3 pp.

—Sport Drinks—Impurities Content
CNS N6208-87. Method of Test for Sport Drinks (Oct)(12150). 3 pp.

—Sport Drinks—PH
CNS N6208-87. Method of Test for Sport Drinks (Oct)(12150). 3 pp.

BHA Content Analysis
Use For: Butylated Hydroxyanisole Content Analysis
See Also: Antioxidant Content Analysis

—Animal Feed
CNS N4097-82. Method of Test for Feed Additives: Determination of Dibutyl Hydroixy Toluene (BHT) and Butyl Hydroxy Anisol (BHA) (Jun)(9027).

—Fats—Gas Chromatography
BSI BS 684: Sec 2.36-83. 1983 Methods of Analysis of Fats and Fatty Oils Part 2: Other Methods Section 2.36: Determination of Butylhydroxyanisole (BHA) and Butylhydroxytoluene (BHT). 4 pp.
ISO 6463-82. Animal and Vegetable Fats and Oils—Determination of Butylhydroxyanisole (BHA) and Butylhydroxytoluene (BHT)—Gas-Liquid Chromatographic Method First Edition. 5 pp.

—Oils—Gas Chromatography
BSI BS 684: Sec 2.36-83. 1983 Methods of Analysis of Fats and Fatty Oils Part 2: Other Methods Section 2.36: Determination of Butylhydroxyanisole (BHA) and Butylhydroxytoluene (BHT). 4 pp.
ISO 6463-82. Animal and Vegetable Fats and Oils—Determination of Butylhydroxyanisole (BHA) and Butylhydroxytoluene (BHT)—Gas-Liquid Chromatographic Method First Edition. 5 pp.

BHC
Use: Benzene Hexachloride

BHT Content Analysis
Use For: Butylated Hydroxytoluene Content Analysis
See Also: Antioxidant Content Analysis

—Animal Feed
CNS N4097-82. Method of Test for Feed Additives: Determination of Dibutyl Hydroxy Toluene (BHT) and Butyl Hydroxy Anisol (BHA) (Jun)(9027).

—Fats—Gas Chromatography
BSI BS 684: Sec 2.36-83. 1983 Methods of Analysis of Fats and Fatty Oils Part 2: Other Methods Section 2.36: Determination of Butylhydroxyanisole (BHA) and Butylhydroxytoluene (BHT). 4 pp.
ISO 6463-82. Animal and Vegetable Fats and Oils—Determination of Butylhydroxyanisole (BHA) and Butylhydroxytoluene (BHT)—Gas-Liquid Chromatographic Method First Edition. 5 pp.

—Oils—Gas Chromatography
BSI BS 684: Sec 2.36-83. 1983 Methods of Analysis of Fats and Fatty Oils Part 2: Other Methods Section 2.36: Determination of Butylhydroxyanisole (BHA) and Butylhydroxytoluene (BHT). 4 pp.
ISO 6463-82. Animal and Vegetable Fats and Oils—Determination of Butylhydroxyanisole (BHA) and Butylhydroxytoluene (BHT)—Gas-Liquid Chromatographic Method First Edition. 5 pp.

Bias (Statistical Analysis)
Use: Statistical Analysis—Bias

Biaxial Compression Testing
Use: Compression Testing

Biblid
Use: Bibliographic Identification

Bibliographic Identification
Use For: Biblid *See Also:* Bibliographic References

—Books
BSI BS 7187-89. 1989 Code for Bibliographic Identification (Biblid) of Contributions in Serials and Books. 8 pp.
ISO 9115-87. Documentation—Bibliographic Identification (Biblid) of Contributions in Serials and Books First Edition. 7 pp.

—Music Scores
BSI BS 4754-82. 1982 Presentation of Music Scores and Parts. 15 pp.

—Periodicals
BSI BS 7187-89. 1989 Code for Bibliographic Identification (Biblid) of Contributions in Serials and Books. 8 pp.
ISO 9115-87. Documentation—Bibliographic Identification (Biblid) of Contributions in Serials and Books First Edition. 7 pp.

Bibliographic References
Use For: Citations *See Also:* Bibliographic Identification
BSI BS 1629-89. 1989 Recommendations for References to Published Materials. 22 pp.
BSI BS 5605-90. 1990 Recommendations for Citing and Referencing Published Material (G). 11 pp.
ISO 690-87. Documentation—Bibliographic References—Content, Form and Structure Second Edition; (Supersedes 3388). 15 pp.

—Abbreviations
ISO 832-75. Documentation—Bibliographical References—Abbreviations of Typical Words First Edition. 43 pp.

—Coded Character Sets
OSI ISO DIS 6862.2-90. Documentation—Mathematical Coded Character Set for Bibliographic Information Interchange. 18 pp.

—Coded Character Sets—African
ISO 6438-83. Documentation—African Coded Character Set for Bibliographic Information Interchange First Edition. 8 pp.
OSI ISO 6438-83. Documentation—African Coded Character Set for Bibliographic Information Interchange. 8 pp.

—Coded Character Sets—Cyrillic
BSI BS 6474: Part 4-86. 1986 Coded Character Sets for Bibliographic Information Interchange Part 4: Extension of the Cyrillic Alphabet Coded Character Set. 7 pp.
ISO 5427-84. Extension of the Cyrillic Alphabet Coded Character Set for Bibliographic Information Interchange First Edition. 6 pp.

—Coded Character Sets—Greek
BSI BS 6474: Part 2-85. 1985 Coded Character Sets for Bibliographic Information Interchange Part 2: Greek Alphabet Coded Character Set. 8 pp.
ISO 5428-84. Greek Alphabet Coded Character Set for Bibliographic Information Interchange Second Edition. 7 pp.
OSI ISO 5428-84. Greek Alphabet Coded CharacterSet for Bibliographic Information Interchange. 7 pp.

—Coded Character Sets—Roman
BSI BS 6474: Part 1-84. 1984 Coded Character Sets for Bibliographic Information Interchange Part 1: Specification for Extension of the Latin Alphabet Coded Character Set. 9 pp.
ISO 5426-83. Extension of the Latin Alphabet Coded Character Set for Bibliographic Information Interchange Second Edition. 8 pp.
OSI ISO 5426-83. Extension of the Latin Alphabet Coded Character Set for Bibliographic Information Interchange. 8 pp.

—Control Characters
ISO 6630-86. Documentation—Bibliographic Control Characters First Edition. 8 pp.
JTC1 6630-86. Documentation—Bibliographic Control Characters First Edition. 8 pp.

Bibliographic References (Cont.)
—Control Characters (Cont.)
OSI ISO 6630-86. Documentation—Bibliographic Control Characters. 8 pp.

—Filing Systems
BSI BS 6478-84. 1984 Filing Bibliographic Information in Libraries and Documentation. 11 pp.
ISO 7154-83. Documentation—Bibliographic Filing Principles First Edition. 10 pp.
ISO TR8393-85. Documentation—ISO Bibliographic Filing Rules (International Standard Bibliographic Filing Rules)—Exemplification of Bibliographic Filing Principles in a Model Set of Rules First Edition. 26 pp.

—Magnetic Tapes—Format
BSI BS 4748-82. 1982 Bibliographic Information Interchange Format for Magnetic Tape. 8 pp.
ISO 2709-81. Documentation—Format for Bibliographic Information Interchange on Magnetic Tape Second Edition. 7 pp.
JTC1 2709-81. Documentation—Format for Bibliographic Information Interchange on Magnetic Tape Second Edition. 7 pp.

—Unpublished Documents
BSI BS 6371-83. 1983 Citation of Unpublished Documents. 10 pp.

Bibliographies
Scope Note: See also the subheading Bibliographies under specific topics *See Also:* Bibliographic References

—Electromagnetic Compatibility
NATO STANAG 3731 ED 1 AMD 1-88. Bibliography on Electromagnetic Compatibility (EMC). 22 pp.
NATO STANAG 3731 ED 2 AMD 0-90. Bibliography on Electromagnetic Compatibility (EMC). 22 pp.
NATO STANAG 3731 ED 2 AMD 1-90. Bibliography on Electromagnetic Compatibility (EMC). 23 pp.
NATO STANAG 3731 ED 2 AMD 2-90. Bibliography on Electromagnetic Compatibility (EMC). 25 pp.

—Filing Systems
BSI BS 6478-84. 1984 Filing Bibliographic Information in Libraries and Documentation. 11 pp.
ISO 7154-83. Documentation—Bibliographic Filing Principles First Edition. 10 pp.
ISO TR8393-85. Documentation—ISO Bibliographic Filing Rules (International Standard Bibliographic Filing Rules)—Exemplification of Bibliographic Filing Principles in a Model Set of Rules First Edition. 26 pp.

Bibulous Papers
Use: Absorbent Papers

Bicycle Helmets
Use: Helmets—Bicycling

Bicycle Lanes
Use: Bicycle Paths

Bicycle Paths
Use For: Bicycle Lanes *See Also:* Roads

—Traffic Control
SAA AS 1742.9-86. Manual of Uniform Traffic Control Devices—Part 9: Bicycle Facilities. 12 pp.

Bicycle Pumps
See Also: Bicycles; Pumps
CNS B7137-80. Method of Test for Air Pump for Bicycles (Jul)(5654). 2 pp.
JIS D 9455-85. Air Pumps for Bicycles.

Bicycle Racks
Use For: Bicycle Stands (Storage); Bike Racks; Bike Stands (Storage); Cycle Holders; Cycle Racks
See Also: Bicycles; Luggage Racks; Racks (Storage)
JIS D 9453-84. Luggage Carriers and Stands for Bicycles.

Bicycle Seats
See Also: Bicycles; Seats
CNS B2055-75. Saddles for Bicycles (Dec)(365). 3 pp.

Bicycle Stands (Storage)
Use: Bicycle Racks

Bicycles
See Also: Bicycle Pumps; Bicycle Racks; Bicycle Seats; Cycles; Exercise Bicycles; Ground Vehicles; Mopeds; Motorcycles; Vehicles (Transportation)
CNS B1340-84. Classification and Essential Characteristics of Bicycles (Feb)(10764).

INTERNATIONAL AND NON-U.S. NATIONAL STANDARDS
SUBJECT INDEX

Bicycles

Bicycles (Cont.)
CNS B7011-90. Inspection Standard of Test Code of Bicycles (Sep)(366).
CNS B7139-80. Method of Test for Infant Bicycle (Jul)(5656). 2 pp.
CNS B7140-80. Method of Test for Two Wheel Boy Bicycles (Jul)(5657). 2 pp.
JIS D 9301-92. Bicycles for General Use. 25 pp.
JIS D 9302-92. Bicycles for Young Children.

—**Assembly**
JIS D 9311-88. Assembling of Bicycles. 52 pp.
SNZ NZS 5455-89. Requirements for the Safe Assembly of Pedal Bicycles for Road Use. 12 pp.

—**Axles**
CNS B2036-53. Bracket Axle for Bicycles (Jul)(346). 2 pp.
CNS B2037-75. Bracket Axle For Bicycle (Dec)(347). 3 pp.
CNS B7053-75. Method of Test for the Bracket Axle of Bicycles (Dec)(3866). 2 pp.

—**Balls—Carbon Steel**
CNS B2141-74. Carbon Steel Ball for Bicycles (Jan)(3616). 2 pp.
JIS D 9404-88. Carbon Steel Balls for Bicycles.

—**Baskets**
CNS S2031-80. Method of Test for Basket for Bicycles & Motor-Cycles (Aug)(6261).

—**Bells**
CNS B2139-73. Bells for Bicycles (Jan)(3515). 2 pp.
ISO 7636-84. Bells for Bicycles and Mopeds—Technical Specifications First Edition. 4 pp.

—**Brakes**
CNS B2044-90. Hand Brakes for Bicycles (Nov)(354). 6 pp.
CNS B7058-75. Method of Test for the Rear Brakes of Bicycles (Dec)(3871). 2 pp.
CNS K4060-81. Brake Rubber for Bicycles (Jan)(6926). 2 pp.
JIS D 9201-90. Bicycles—Method of Braking Test. 12 pp.
JIS D 9414-88. Brakes for Bicycles.
JIS K 6310-77. Brake Rubber for Bicycle.

—**Cables—Brakes**
CNS B2676-82. Brake Cable for Bicycle (Jun)(8934). 9 pp.

—**Carburizing**
CNS B7073-75. Standard for Heat Treatment of Bicycle Parts (Dec)(3886). 2 pp.

—**Chains**
CNS B2047-75. Chain for Bicycles (Dec)(357). 2 pp.
CNS B7062-91. Method of Test for Chain of Bicycles (Jun)(3875). 1 p.
JIS D 9417-76. Chains for Bicycles.
JIS D 9432-80. Chain Adjusters and Crank Cotter Pins for Bicycles.
JIS D 9454-83. Chain Cases for Bicycles.

—**Classification**
CNS B1340-84. Classification and Essential Characteristics of Bicycles (Feb)(10764).

—**Coatings**
CNS B7069-75. Standard for Coating of Bicycle Parts (Dec)(3882). 1 p.
CNS B7071-75. Standard for Surface Treatment of Bicycle Parts (Dec)(3884). 3 pp.

—**Coatings—Bend Testing**
CNS B7072-75. Method of Test for Surface Treatment of Bicycle Parts (Dec)(3885). 2 pp.

—**Coatings—Impact Testing**
CNS B7070-75. Method of Test for Coating of Bicycle Parts (Dec)(3883). 1 p.

—**Coatings—Light Testing**
CNS B7072-75. Method of Test for Surface Treatment of Bicycle Parts (Dec)(3885). 2 pp.

—**Coatings—Salt Spray Testing**
CNS B7070-75. Method of Test for Coating of Bicycle Parts (Dec)(3883). 1 p.
CNS B7072-75. Method of Test for Surface Treatment of Bicycle Parts (Dec)(3885). 2 pp.

—**Cotter Pins**
BSI BS 6102: Part 7-82. 1982 Cycles Part 7: Cotter Pins and Assembly of the Axle/Cotter Pin/Crank. 6 pp.
ISO 6693-81. Cycles—Cotter Pin and Assembly of the Axle/Cotter Pin/Crank First Edition. 5 pp.
JIS D 9432-80. Chain Adjusters and Crank Cotter Pins for Bicycles.

—**Cranks**
CNS B2045-75. Chain Wheels and Cranks for Bicycles (Dec)(355). 5 pp.

Bicycles (Cont.)
—**Cranks (Cont.)**
CNS B7059-75. Method of Test for Chain Wheels and Cranks for Bicycles (Dec)(3872). 2 pp.
JIS D 9415-83. Chainwheels and Cranks for Bicycles.

—**Derailleurs**
JIS D 9428-87. Derailleur for Bicycles.

—**Frames**
CNS B2033-75. Frames for Bicycles (Dec)(343). 6 pp.
CNS B7051-91. Method of Test for Frames of Bicycles (Jun)(3864). 1 p.
JIS D 9401-90. Frame-Fork Assembly for Bicycles. 18 pp.
JIS D 9402-90. Front Forks for Bicycles. 9 pp.
JIS D 9403-90. Frame Parts for Bicycles. 13 pp.

—**Glossaries**
BSI BS 6102: Part 4-91. 1991 Cycles Part 4: Glossary of Terms (ISO 8090: 1990). 35 pp.
ISO 8090-90. Cycles—Terminology First Edition. 35 pp.
JIS D 9101-91. Cycles—Terminology.

—**Handles**
CNS B2040-75. Handles for Bicycles (Dec)(350). 5 pp.
CNS B7056-90. Method of Test for the Handles of Bicycles (Nov)(3869).
CNS B7056-75. Method of Test for the Handles of Bicycles (Dec)(3869). 3 pp.
JIS D 9412-90. Handlebars for Bicycles. 16 pp.

—**Handles—Grips**
CNS B2041-75. Handle Grips for Bicycles (Dec)(351). 2 pp.
CNS B7057-90. Method of Test for the Handle Grips of Bicycles (Nov)(3870).
CNS B7057-75. Method of Test for the Handle Grips of Bicycles (Dec)(3870). 2 pp.
CNS K4055-81. Grip Rubber for Bicycles (Jan)(6921). 2 pp.
JIS D 9413-90. Handle Grips for Bicycles. 10 pp.

—**Handles—Stems**
BSI BS 6102: Part 11-91. 1991 Cycles Part 11: Specification for Assembly Dimensions for Stem and Handlebar Bend (ISO 6699: 1990). 7 pp.
BSI BS 6102: Part 12-91. 1991 Cycles Part 12: Specification for Stem Wedge Angle (ISO 8562: 1990). 7 pp.
CNS B2035-75. Handle Stems and Parts for Bicycles (Dec)(345). 3 pp.
CNS B7052-90. Method of Test for the Handle Stems and Parts of Bicycles (Nov)(3865).
CNS B7052-75. Method of Test for the Handle Stems and Parts of Bicycles (Dec)(3865). 2 pp.
ISO 6699-90. Cycles—Stem and Handlebar Bend—Assembly Dimensions First Edition. 4 pp.
ISO 8562-90. Cycles—Stem Wedge Angle First Edition. 4 pp.

—**Identification Systems**
ISO 6692-81. Cycles—Marking of Cycle Components First Edition. 4 pp.

—**Lamps**
BSI BS 3648-63. 1963 Amd 2 Cycle Rear Lamps. 21 pp.
BSI BS 6102: Part 3-86. 1986 Amd 1 Cycles Part 3: Photometric and Physical Requirements of Lighting Equipment. 23 pp.
CNS C3013-84. Method of Test for Miniature Lamps (Jun)(720). 5 pp.
CNS C3014-71. Testing Standard for Illuminating Equipments of Bicycles (Apr)(804). 3 pp.
CNS C4089-84. Electrical Bulbs for Dynamo Lamps of Bicycles (Jun)(2944). 4 pp.
ISO 6742 Pt 1-87. Cycles—Lighting and Retro Reflective Devices—Photometric and Physical Requirements—Part 1: Lighting Equipment Second Edition. 19 pp.
JIS C 7510-92. Incandescent Lamps for Bicycle Dynamo Lamps. 11 pp.
JIS C 9502-81. Dynamo Lamps for Bicycles.
SNZ NZS 5441: Part 1-87. Lighting and Retroreflectors for Pedal Cycles Part 1: Specification for Lamp Units and Retroreflectors Suitable for Fitting to Pedal Cycles. 13 pp.

—**Lamps—Installation**
SNZ NZS 5441: Part 2-87. Lighting and Retroreflectors for Pedal Cycles Part 2: Code of Practice for the Fitting of Lamp Units and Retroreflectors to Pedal Cycles. 7 pp.

—**Locks**
JIS D 9456-88. Locks for Bicycles.
JIS D 9456-61. Locks for Bicycles (R 1973). 5 pp.

—**Mudguards**
CNS B2039-75. Mudguards for Bicycles (Dec)(349). 3 pp.
CNS B7055-90. Method of Test for the Mudguards of Bicycles (Nov)(3868).

Bicycles (Cont.)
—**Mudguards (Cont.)**
CNS B7055-75. Method of Test for the Mud-guards of Bicycles (Dec)(3868). 2 pp.
CNS K4061-81. Rubber Mudguard Flapper for Bicycle (Jan)(6927). 1 p.
JIS D 9411-92. Mudguards for Bicycles. 11 pp.

—**Packaging—Export**
CNS Z5092-81. Packaging and Packing of Bicycle for Export (Jun)(7544).

—**Pedals**
CNS B2046-91. Pedals for Bicycles (Jun)(356). 6 pp.
CNS B7063-75. Method of Test for Pedal of Bicycles (Dec)(3876). 2 pp.
CNS K4059-81. Pedal Rubbers for Bicycles (Jan)(6925). 4 pp.
JIS D 9416-88. Pedals for Bicycles.

—**Pedals—Cranks**
BSI BS 6102: Part 14-91. 1991 Cycles Part 14: Specification for Dimensions of Bottom Bracket Axle and Crank Assembly with Square End Fittings (ISO 6695: 1991). 7 pp.
ISO 6695-91. Cycles—Pedal Axle and Crank Assembly with Square End Fitting—Assembly Dimensions First Edition. 4 pp.

—**Reflectors**
BSI BS 6102: Part 2-82. 1982 Amd 1 Cycles Part 2: Photometric and Physical Requirements of Reflective Devices. 11 pp.
CNS B2385-79. Reflex Reflectors for Pedal Bicycles (Oct)(5004). 7 pp.
ISO 6742 Pt 2-85. Cycles—Lighting and Retro-Reflective Devices—Photometric and Physical Requirements—Part 2: Retro-Reflective Devices Second Edition. 11 pp.
JIS D 9452-90. Reflex Reflectors for Bicycles. 11 pp.
SNZ NZS 5441: Part 1-87. Lighting and Retroreflectors for Pedal Cycles Part 1: Specification for Lamp Units and Retroreflectors Suitable for Fitting to Pedal Cycles. 13 pp.

—**Reflectors—Installation**
SNZ NZS 5441: Part 2-87. Lighting and Retroreflectors for Pedal Cycles Part 2: Code of Practice for the Fitting of Lamp Units and Retroreflectors to Pedal Cycles. 7 pp.

—**Safety**
BSI BS 6101: Part 1-81. 1981 Machine Tool Ball Screws Part 1: Methods of Calculating Dynamic Load and Life Ratings. 6 pp.
BSI BS 6102: Part 1-92. 1992 Cycles Part 1: Specification for Safety Requirements for Bicycles. 37 pp.
BSI BS 6102: Part 1-81. 1981 Amd 4 Cycles Part 1: Safety Requirements for Bicycles (AMD 5540) May 29, 1987. 45 pp.
CSA CAN3-D113.1-M80. Bicycles. 34 pp.
ISO 4210-89. Cycles—Safety Requirements of Bicycles Third Edition. 34 pp.
ISO 8098-89. Cycles—Safety Requirements for Bicycles for Young Children First Edition; (Amendment 1-1992). 24 pp.
JIS D 9102-90. Bicycles for General Use —Safety Requirements. 19 pp.
SAA AS 1927-89. Pedal Bicycles for Normal Road Use—Safety Requirements (This is a Joint Standard with SANZ NZS 1927:1989). 31 pp.
SNZ NZMP 205-91. Guide to New Zealand Product Safety Standard: Pedal Bicycles for Normal Use—Safety Requirements. 36 pp.
SNZ NZS/AS 1927-89. Pedal Bicycles for Normal Road Use—Safety Requirements (This is a Joint Standard with SAA AS 1927). 31 pp.

—**Screw Threads**
BSI BS 811-50. 1950 Amd 1 Cycle Threads. 27 pp.
BSI BS 6102: Part 8-90. 1990 Cycles Part 8: Screw Threads Used to Assemble Free Wheels on Bicycle Hubs. 13 pp.
BSI BS 6102: Part 9-90. 1990 Cycles Part 9: Screw Threads Used in Bottom Bracket Assemblies. 13 pp.
BSI BS 6102: Part 10-87. 1987 Cycles Part 10: Screw Threads Used to Assemble Head Fittings on Bicycle Forks. 9 pp.
CNS B2032-75. Screw Threads for Bicycles (Dec)(341). 5 pp.
DIN ENGL 79012-60. Screw Threads for Bicycles and Mopeds; Theoretical Values; Limiting Screw Thread Dimensions (Oct). 1 p.
ISO 6696-89. Cycles—Screw Threads Used in Bottom Bracket Assemblies First Edition. 10 pp.
ISO 6698-89. Cycles—Screw Threads Used to Assemble Freewheels on Bicycle Hubs Second Edition. 10 pp.
ISO 8488-86. Cycles—Screw Threads Used to Assemble Head Fittings on Bicycle Forks First Edition. 8 pp.
JIS B 0225-60. Cycle Threads (R 1973). 5 pp.

INDUSTRY STANDARDS

INTERNATIONAL AND NON-U.S. NATIONAL STANDARDS
SUBJECT INDEX

Bicycles (Cont.)
—Seats
 CNS B2055-75. Saddles for Bicycles (Dec)(365). 3 pp.
 JIS D 9431-88. Saddles for Bicycles.
—Seats—Posts
 CNS B2038-91. Saddle Post (Pillar) for Bicycles (Jun)(348). 2 pp.
 CNS B7066-91. Method of Test for Saddles Post (Pillar) of Bicycles (Jun)(3879). 2 pp.
—Spokes
 CNS B2052-75. Spokes for Bicycles (Dec)(362). 4 pp.
 JIS D 9420-88. Spokes for Bicycles.
—Spokes—Bend Testing
 CNS B7067-75. Method of Test for the Spokes of Bicycles (Dec)(3880). 2 pp.
—Spokes—Nipples
 BSI BS 6102: Part 15-91. 1991 Cycles Part 15: Specification for Dimensions of Spoke Nipples (ISO 6701: 1991). 7 pp.
 ISO 6701-91. Cycles—External Dimensions of Spoke Nipples First Edition. 4 pp.
—Spokes—Tensile Testing
 CNS B7067-75. Method of Test for the Spokes of Bicycles (Dec)(3880). 2 pp.
—Sprockets
 BSI BS 6102: Part 13-91. 1991 Cycles Part 13: Specification for Mating Dimensions for Splined Hub and Sprocket (ISO 10230: 1990). 9 pp.
 CNS B2045-75. Chain Wheels and Cranks for Bicycles (Dec)(355). 5 pp.
 CNS B7059-75. Method of Test for Chain Wheels and Cranks for Bicycles (Dec)(3872). 2 pp.
 ISO 10230-90. Cycles—Splined Hub and Sprocket—Mating Dimensions First Edition. 4 pp.
 JIS D 9415-83. Chainwheels and Cranks for Bicycles.
—Stability Testing
 JIS D 9203-76. Method of Stability Test for Bicycles.
—Stands
 JIS D 9453-84. Luggage Carriers and Stands for Bicycles.
—Tire Tubes
 CNS K4002-83. Rubber Inner Tubes for Bicycles (Dec)(738). 2 pp.
 CNS K4056-81. Rubber Valve Tubing (Jan)(6922). 3 pp.
 CNS K6050-83. Method of Test for Rubber Inner Tubes of Bicycles (Dec)(739).
 JIS K 6304-82. Rubber Inner Tubes for Bicycles.
 JIS K 6304-76. Rubber Inner Tubes for Bicycles. 7 pp.
 JIS K 6307-77. Valve Tubing (R 1980). 6 pp.
—Tire Valves
 CNS B2054-84. Tire Valves for Bicycles (Sep)(364). 9 pp.
 CNS B7061-75. Method of Test for the Tire Valves of Bicycles (Dec)(3874). 1 p.
 JIS D 9422-92. Tire Valves for Bicycles. 18 pp.
—Tires
 BSI BS 6102: Part 5-90. 1990 Cycles Part 5: Bicycle Tyre Designations and Dimensions. 14 pp.
 CNS K4001-83. Rubber Pneumatic Tires for Bicycles (Dec)(736). 5 pp.
 CNS K4058-81. Solid Tires to Insert Bicycles Rim (Jan)(6924). 2 pp.
 CNS K4079-83. Dimensions of Tires for Bicycles (Dec)(10724). 4 pp.
 CNS K6049-83. Method of Test for Rubber Pneumatic Tires of Bicycles (Dec)(737).
 ISO 5775 Pt 1-88. Bicycle Tyres and Rims—Part 1: Tyre Designations and Dimensions Third Edition. 11 pp.
 JIS D 9112-91. Cycle—Tires—Dimensions. 9 pp.
 JIS K 6302-82. Rubber Pneumatic Tires for Bicycles.
 JIS K 6306-77. Solid Tire to Insert Bicycles Rim.
—Tubes—Steel
 BSI BS 1717-83. 1983 Steel Tubes for Cycle and Motor-Cycle Purposes. 16 pp.
—Wheels
 BSI BS 6102: Part 6-90. 1990 Cycles Part 6: Bicycle Rims. 19 pp.
 CNS B2048-75. Free Wheels for Bicycles (Dec)(358). 5 pp.
 CNS B2049-75. Hub Cogs for Bicycles (Dec)(359). 2 pp.
 CNS B2050-75. Front Hubs for Bicycles (Dec)(360). 3 pp.
 CNS B2051-75. Rear Hubs for Bicycles (Dec)(361). 13 pp.
 CNS B2053-75. Rims for Bicycles (Dec)(363). 5 pp.
 CNS B2757-84. Plastic Rims of Bicycles (Jan)(10735).
 CNS B7054-90. Method of Test for the Front Hubs of Bicycles (Nov)(3867).

Bicycles (Cont.)
—Wheels (Cont.)
 CNS B7054-75. Method of Test for the Front Hubs of Bicycles (Dec)(3867). 2 pp.
 CNS B7060-75. Method of Test for the Rear Hubs of Bicycles (Dec)(3873). 2 pp.
 CNS B7064-75. Method of Test for the Hub Cogs of Bicycles (Dec)(3877). 1 pp.
 CNS B7065-75. Method of Test for the Free Wheels of Bicycles (Dec)(3878). 2 pp.
 CNS B7068-75. Method of Test for the Rims of Bicycles (Dec)(3881). 2 pp.
 CNS B7262-84. Method of Test for the Plastic Rims of Bicycles (Jan)(10734).
 ISO 5775 Pt 2-89. Bicycle Tyres and Rims—Part 2: Rims First Edition. 15 pp.
 JIS D 9418-87. Free Wheels and Hub Cogs for Bicycles.
 JIS D 9419-88. Hubs for Bicycles.

Bidets
See Also: Toilet Facilities
 BSI BS 5505: Part 3-77. 1977 Bidets Part 3: Vitreous China Bidets, over Rim Supply Only. Quality, Workmanship and Functional Dimensions Other Than Connecting Dimensions. 4 pp.
 CNS R2061-6-82. Sanitary Ceramic Ware Bidet (Sep)(3220-6).
 JIS A 4422-86. Toilet Seat with Douche to Wash Anus with Warm Water. 22 pp.
—Drainage Systems
 CEN PREN 128-89. Performance Criteria and Requirements for Sanitary Appliances. 4 pp.
—Drainpipe Fittings
 BSI BS 3380-82. 1982 Wastes (Excluding Skeleton Sink Wastes) and Bath Overflows. 12 pp.
 BSI BS EN 274-93. 1993 Sanitary Tapware—Waste Fittings for Basins, Bidets and Baths—General Technical Specifications. 22 pp.
 CEN PREN 274-88. Sanitary Tapware: Waste Fittings for Basins Bidets and Baths; General Technical Specifications. 29 pp.
 CEN EN 274-92. Sanitary Tapware—Waste Fittings for Basins, Bidets and Baths—General Technical Specifications. 28 pp.
 CEN EN 274-92. Sanitary Tapware—Waste Fittings for Basins, Bidets and Baths—General Technical Specifications. 17 pp.
—Pedestal
 BSI BS 5505: Part 1-77. 1977 Bidets Part 1: Pedestal Bidets, over Rim Supply Only. Connecting Dimensions. 8 pp.
 BSI BS 5505: Part 1-01. 1977 Amd 1 Bidets Part 1: Pedestal Bidets over Rim Supply Only. Connecting Dimensions (AMD 7308) September 15, 1992. 14 pp.
 CEN EN 35-77. Pedestal Bidets Over Rim Supply Only: Connecting Dimensions. 6 pp.
—Wall Hung
 BSI BS 5505: Part 2-77. 1977 Bidets Part 2: Wall Hung Bidets, over Rim Supply Only. Connecting Dimensions. 8 pp.
 BSI BS 5505: Part 2-01. 1977 Amd 1 Bidets Part 2: Wall Hung Bidets over Rim Supply Only. Connecting Dimensions (AMD 7309) September 15, 1992. 14 pp.
 CEN EN 36-77. Wall Hung Bidets over Rim Supply Only: Connecting Dimensions. 6 pp.

Bidirectional Bus Drivers
Use: Bus Transceivers

Bidirectional Triode Thyristors
Use For: Triac (TM) See Also: Thyristors
 BSI BS QC 750111-91. 1991 Bidirectional Triode Thyristors (Triacs), Ambient or Case-Rated up to 100A (IEC 747-6-2: 1991). 14 pp.
 IECQ QC 750111-91. Semiconductor Devices Discrete Devices Part 6: Thyristors Section Two—Blank Detail Specification for Bidirectional Triode Thyristors (Triacs), Ambient or Case-Rated, up to 100A (IEC 747-6-2 ED 1). 33 pp.
—Quality Assurance
 BSI BS 9343-77. (WITHDRAWN) 1977 Amd 1 Rules for the Preparation of Detail Specifications for Semiconductor Devices of Assessed Quality: Case-Rated Bi-Directional Thyristors (Triacs) (AMD 6171) December 31, 1989. 12 pp.
—Reliability Assured
 CNS C6194-88. Method of Test for Reliability Assured Bi-Directional Triode Thyristors Rule (Dec)(9834).
 CNS C7154-88. Reliability Assured Bi-Directional Triode Thyristors (Low Current) (Dec)(9832).
 CNS C7155-88. Reliability Assured Bi-Directional Triode Thyristors (High and Medium Current) (Dec)(9833).

Bidirectional Triode Thyristors (Cont.)
—Reliability Assured (Cont.)
 JIS C 7233-78. Reliability Assured Bi-Directional Triode Thyristors (Low Current).
 JIS C 7234-78. Reliability Assured Bi-Directional Triode Thyristors (High and Medium Current).
—Reliability Assured—Current Measurement
 CNS C6196-88. Method of Test for Reliability Assured Bi-Directional Triode Thyristors Trigger Voltage and Current (Dec)(9836).
 CNS C6198-88. Method of Test for Reliability Assured Bi-Directional Triode Thyristors Current (Dec)(9838).
 CNS C6200-88. Method of Test for Reliability Assured Bi-Directional Triode Thyristors Test for Reverse Current (Dec)(9840).
 CNS C6203-88. Method of Test for Reliability Assured Bi-Directional Triode Thyristors Current and Surge on Current (Dec)(9843).
—Reliability Assured—Life (Durability)
 CNS C6204-88. Method of Test for Reliability Assured Bi-Directional Triode Thyristors Durability (Dec)(9844).
—Reliability Assured—Thermal Resistance
 CNS C6202-88. Method of Test for Reliability Assured Bi-Directional Triode Thyristors Resistance (Dec)(9842).
—Reliability Assured—Turn On Time
 CNS C6201-88. Method of Test for Reliability Assured Bi-Directional Triode Thyristors Time (Dec)(9841).
—Reliability Assured—Voltage Measurement
 CNS C6195-88. Method of Test for Reliability Assured Bi-Directional Triode Thyristors Voltage (Dec)(9835).
 CNS C6196-88. Method of Test for Reliability Assured Bi-Directional Triode Thyristors Trigger Voltage and Current (Dec)(9836).
 CNS C6197-88. Method of Test for Reliability Assured Bi-Directional Triode Thyristors Nontrigger Voltage (Dec)(9837).
 CNS C6199-88. Method of Test for Reliability Assured Bi-Directional Triode Thyristors of Critical off Voltage (Dec)(9839).

Bids
See Also: Competitive Bidding; Invitation for Bids; Procurement
—Certificates
 MOD UK DEFCON 48-81. Bona-Fide Tendering Certificate (See Leaflet N3 of DCH) 8/81. 1 p.
—European Communities Official Journal
 MOD UK DEFCON 49-89. Tender Notice for the Official Journal—EEC 12/89. 2 pp.
 MOD UK DEFCON 49A-89. Request for Certificate of Posting DEFCON 49 12/89. 1 p.

Bihexagonal Head Bolts
Use: Double Hexagonal Bolts

Bihexagonal Nuts
Use: Double Hexagonal Nuts

Bike Racks
Use: Bicycle Racks

Bike Stands (Storage)
Use: Bicycle Racks

Bilberries
See Also: Food; Fruits
—Storage
 ISO 6664-83. Bilberries and Blueberries—Guide to Cold Storage First Edition. 5 pp.

Bilge Pumps
See Also: Marine Equipment; Pumps; Sump Pumps
 MOD UK NES 717-84. Requirements for Bilge, Sullage and Drain Tank Systems for Surface Ships Issue 2 (05.84). 40 pp.
—Boats
 BSI BS EN 28849-93. 1993 Small Craft—Electrically Operated Bilge-Pumps (ISO 8849: 1990) (Q). 10 pp.
 ISO 8849-90. Small Craft—Electrically Operated Bilge-Pumps First Edition. 4 pp.
—Mud Boxes
 ISO 5621-84. Shipbuilding—Bilge Mud Boxes for Machinery Spaces and Tunnels—General Design Characteristics First Edition. 5 pp.

INTERNATIONAL AND NON-U.S. NATIONAL STANDARDS
SUBJECT INDEX

Binary

Bilge Water
See Also: Water
—Automatic Control Equipment
JIS F 0806-79. Methods of Onboard Test on Automatic Control of Compressed Air, Bilge, Fresh Water and Sanitary Systems for Smaller Ships.

Bilge Wells
Use: Bilges

Bilges
Use For: Bilge Wells *See Also:* Ships
MOD UK NES 717-84. Requirements for Bilge, Sullage and Drain Tank Systems for Surface Ships Issue 2 (05.84). 40 pp.
—Cleaning Agents
MOD UK DSTAN 79-13-89. Quick-Break Type Cleaners for Ships' Bilges Issue 1. 19 pp.
—Covers
BSI BS MA 61-74. 1974 Bilge Drain and Suction Hats. 5 pp.

Bill Hooks
See Also: Cutting Tools
MOD UK DEF-1188. Hooks, Bill and Slashers (Withdrawn).

Billboards
See Also: Posters; Signs
—Buildings—Model Bylaws
SNZ NZS 9201: Chapter 8-75. Model General Bylaws Chapter 8: Control of Advertising Signs (Reconfirmed 1980). 8 pp.

Billets
Scope Note: For additional listings, see the subheading Billets under the specific metal or alloy
See Also: Metal Products

Billiard Rooms
Use For: Pool Halls *See Also:* Buildings; Recreational Facilities
—Model Bylaws
SNZ NZS 9201: Chapter 18-72. Model General Bylaws Chapter 18: Billiard Rooms (Reconfirmed 1980). 11 pp.

Billing, Telecommunications
Use: Tariffs (Telecommunications)

Bills (Invoices)
Use: Invoices

Bimetal Thermometers
Use: Bimetallic Thermometers

Bimetallic Thermometers
Use For: Bimetal Thermometers; Differential Thermometers *See Also:* Thermometers
CGSB CAN/CGSB-14.5-M88. Thermometers, Bimetallic, Self-Indicating, Commercial/Industrial Type. 11 pp.
CGSB 14.6M-89. Thermometers, Bimetallic, Self-Indicating, Shipboard Use. 17 pp.
DIN ENGL 16203-88. Filled-System and Bimetallic Dial Thermometers; Requirements and Testing (Feb). 4 pp.
DIN ENGL 16204-88. Bimetallic Dial Thermometers, Accuracy Classes 1 and 2, with Case Diameters of 63, 80, 100 and 160 mm; Dimensions and Nominal Ranges (Feb). 4 pp.
JIS B 7542-79. Bimetallic Thermometers. 18 pp.
—Temperature Measurement
JIS Z 8707-92. Method of Temperature Measurement by Filled-System Thermometers and Bimetallic Thermometers. 15 pp.

Binapacryl
—Plants (Botany)—Prohibited Use
EC 90/533/EEC-90. Council Directive Amending the Annex to Directive 79/117/EEC Prohibiting the Placing on the Market and Use of Plant Protection on the Market and Use of Plant Protection Product Containing Certain Active Substances. 1 p.

Binary Counter/Dividers
See Also: Binary Counters; Digital Counters
—4-Bit
CECC CECC 90 104-206 ISSUE 1-84. BS CECC 90 104-206; Silicon Complementary MOS. with (B) Buffered Outputs and Cavity Packaging (En). 25 pp.
CECC CECC 90 104-206 ISSUE 1-86. CEI CECC 90 104-206; Silicon Complementary MOS. with (B) Buffered Outputs Cavity and Non Cavity Packaging (En). 2 pp.

Binary Counter/Latches
See Also: Binary Counters; Latches (Circuits)
CECC CECC 90 102-058 ISSUE 1-82. BS CECC 90 102-058; Presettable Binary Counters/Latches (En). 20 pp.
CECC CECC 90 103-142 ISSUE 1-82. BS CECC 90 103-142; Presettable Binary Counters/Latches (En). 22 pp.

Binary Counter/Registers
See Also: Binary Counters
—Asynchronous—4-Bit
CECC CECC 90 109-837 ISSUE 1-89. Digital Integrated Circuits; Silicon Monolithic C MOS, Cavity or Non-Cavity Packages; Type(s) 54/74 HC 691 Asynchronous Clear 4-Bit Binary Counter /Register with 3-State Outputs Assessment Levels P, Y, L (En, Fr, Ge). 11 pp.
—Synchronous/Asynchronous—Count Up/Down
CECC CECC 90 109-849 ISSUE 1-90. Digital Integrated Circuits; Silicon Monolithic C MOS, Cavity or Non-Cavity Packaging; Type(s) 54/74 HC 697 Synchronous Up/Down Binary Counter/Register with Asynchronous Clear and 3-State Outputs Assessment Levels P, Y, L (En, Fr, Ge). 11 pp.
—Synchronous—Count Up/Down
CECC CECC 90 109-851 ISSUE 1-90. Digital Integrated Circuits; Silicon Monolithic C MOS, Cavity or Non-Cavity Packages; Type(s) 54/74 HC 699 Synchronous Up/Down Binary Counter/Register with Synchronous Clear and 3-State Outputs Assessment Levels P, Y, L (En, Fr, Ge). 11 pp.
—Synchronous—4-Bit
CECC CECC 90 109-809 ISSUE 1-88. Digital Integrated Circuits in Accordance with FS 90 109; 54/74 HC 693; Synchronous Clear 4-Bit Binary Counter/Register with 3-State Outputs (En, Fr). 8 pp.
—8-Bit
CECC CECC 90 109-856 ISSUE 1-90. Digital Integrated Circuits; Silicon Monolithic C MOS, Cavity or Non-Cavity Packages; Type(s) 54/74 HCT 166 Parallel Load 8-Bit Shift Register Assessment Levels P, Y, L (En, Fr, Ge). 10 pp.

Binary Counters
Use For: Binary Dividers; Binary Scalers; Scale of Two Circuits *See Also:* Binary Counter/Dividers; Binary Counter/Latches; Binary Counter/Registers; Binary/Decade Counters; Digital Circuits; Ripple Counters
CECC CECC 90 104-222 ISSUE 1-85. BS CECC 90 104-222; Three-Digit BCD Counter (En). 35 pp.
CECC CECC 90 104-222 ISSUE 1-86. CEI-CECC 90 104-222; Silicon with (B) Buffered Outputs Three-Digit BCD Counter (En). 2 pp.
CECC CECC 90 109-629 ISSUE 1-86. Digital Integrated Circuits in Accordance with FS 90 109; 54/74 HC 4020; 14 Stage Binary Counter with Clear (En, Fr). 6 pp.
—Asynchronous
CECC CECC 90 109-816 ISSUE 1-88. Digital Integrated Circuits in Accordance with FS 90 109; 54/74 HC 4520; Dual Binary Counter with Asynchronous Clear (En, Fr). 6 pp.
—Asynchronous—Count Up/Down—Preferred Products List
CECC CECC MUAHAG Vol 7 IS 8-92. Preferred Products List; Active Microcircuits (En, Fe, Ge). 89 pp.
—Count Up/Down
CECC CECC 90 104-124 ISSUE 1-81. BS CECC 90 104-124; Silicon Complementary MOS with (B) Buffered Outputs and Cavity Packaging (En). 14 pp.
CECC CECC 90 104-124 ISSUE 1-86. CEI CECC 90 104-124; Silicon Complementary MOS with (B) Buffered Outputs Cavity and Non Cavity Packaging (En). 2 pp.
CECC CECC 90 104-161 ISSUE 1-81. BS CECC 90 104-161; Binary Up/Down Counter (En). 28 pp.
CECC CECC 90 104-161 ISSUE 1-86. CEI CECC 90 104-161; Binary Up /Down Counter (En). 2 pp.
—Oscillators—14 Stage—Preferred Products List
CECC CECC MUAHAG Vol 7 IS 8-92. Preferred Products List; Active Microcircuits (En, Fe, Ge). 89 pp.
—Synchronous/Asynchronous
CECC CECC 90 109-631 ISSUE 1-86. Digital Integrated Circuits in Accordance with FS 90 109; 54/74 HC 161; Synchronous Binary Counter with Asynchronous Clear (En, Fr). 6 pp.

Binary Counters *(Cont.)*
—Synchronous/Asynchronous *(Cont.)*
CECC CECC 90 109-785 ISSUE 1-87. Detail Specification: Digital Integrated Circuits; Silicon Monolithic C MOS, Cavity or Non-Cavity Packages; Type(s) 54/74 HCT 161 Synchronous Binary Counter with Asynchronous Clear; Assessment Levels P, Y, L (En, Fr). 9 pp.
—Synchronous/Asynchronous—4-Bit
CECC CECC 90 104-094 ISSUE 2. NL CECC 90 104-094 Issue 2; Digital Integrated Circuits in Accordance with FS 90 104; HEC/HEF 40161B; 4-Bit Synchronous Binary Counter with Asynchronous Reset (En). 14 pp.
CECC CECC 90 104-190 ISSUE 1-81. BS CECC 90 104-190 Issue 1; 4 Bit Synchronous Binary Counter with Asynchronous Clear (En). 26 pp.
—Synchronous—Count Up/Down
CECC CECC 90 104-064 ISSUE 2. NL CECC 90 104-064 Issue 2; Digital Integrated Circuits in Accordance with FS 90 104; HEC/HEF 4510B; BCD Up/Down Counter (En). 13 pp.
CECC CECC 90 109-729 ISSUE 1-87. Digital Integrated Circuits in Accordance with FS 90 109; 54 HC 193, 74 HC 193; Synchronous Up/Down Binary Counter (En, Fr). 7 pp.
CECC CECC 90 109-795 ISSUE 1-87. Digital Integrated Circuits; Silicon Monolithic C MOS, Cavity or Non-Cavity Packages; Type(s) 54/74 HCT 193; Synchronous Up/Down Binary Counter (En, Fr). 9 pp.
—Synchronous—Count Up/Down—Preferred Products List
CECC CECC MUAHAG Vol 7 IS 8-92. Preferred Products List; Active Microcircuits (En, Fe, Ge). 89 pp.
—Synchronous—4-Bit
CECC CECC 90 102-028 ISSUE 1-81. BS CECC 90 102-028; Fully Synchronous 4-Bit Binary Counters (En) AMD 1 (En). 20 pp.
CECC CECC 90 103-070 ISSUE 1-82. BS CECC 90 103-070; Binary Synchronous Clear 4-Bit Direct Counters (En). 21 pp.
CECC CECC 90 104-071 ISSUE 2. NL CECC 90 104-071 Issue 2; Digital Integrated Circuits in Accordance with FS 90 104; HEC/HEF 4518B; Dual BCD Counter (En). 9 pp.
CECC CECC 90 104-073 ISSUE 2. NL CECC 90 104-073 Issue 2; Digital Integrated Circuits in Accordance with FS 90 104; HEC/HEF 4520B; Dual Binary Counter (En). 9 pp.
CECC CECC 90 104-096 ISSUE 2. NL CECC 90 104-096 Issue 2; Digital Integrated Circuits in Accordance with FS 90 104; HEC/HEF 40163B; 4-Bit Synchronous Binary Counter with Synchronous Reset (En). 14 pp.
CECC CECC 90 104-192 ISSUE 1-81. BS CECC 90 104-192 Issue 1; Synchronous 4-Bit Binary Counter with Synchronous Clear (En). 27 pp.
CECC CECC 90 104-192 ISSUE 1-86. CEI CECC 90 104-192 Issue 1; Synchronous 4-Bit Binary Counter with Synchronous Clear (En). 2 pp.
CECC CECC 90 106-004 ISSUE 2-86. UTE C 86-217; Digital Integrated Circuits in Accordance with FS 90 106; 54/74 ALS 161, 54/74 ALS 161B; Synchronous 4-Bit Binary Counters with Asynchronous Clear (En, Fr) ADD 3 (En, Fr). 17 pp.
CECC CECC 90 106-037 ISSUE 2-86. UTE C 86-217 ADD 2/GA 37; Digital Integrated Circuits in Accordance with FS 90 106; 54/74 ALS 561, 54/74 ALS 561A; Synchronous 4-Bit Binary Counter with 3 State Outputs (En, Fr) ADD 3 (En, Fr). 18 pp.
CECC CECC 90 106-070 ISSUE 1-85. UTE C 86-217; Digital Integrated Circuits in Accordance with FS 90 106; 54/74 ALS 163, 54/74 ALS 163A; Synchronous 4-Bit Binary Counters with Synchronous Clear (En, Fr) ADD 3 (En, Fr). 17 pp.
CECC CECC 90 106-075 ISSUE 1-85. UTE C 86-217; Digital Integrated Circuits in Accordance with FS 90 106; 54/74 ALS 169/169B, 54/74 ALS 169U; Synchronous 4-Bit Up/Down Binary Counter (En, Fr) ADD 3 (En, Fr). 17 pp.
CECC CECC 90 107-020 ISSUE 2-89. UTE C 86-218 ADD 2/FA 20; Digital Integrated Circuits in Accordance with FS 90 107; 54/74 F 161A; Synchronous 4-Bit Binary Counters with Asynchronous Clear (En, Fr) ADD 3 (En, Fr). 11 pp.
CECC CECC 90 107-021 ISSUE 2-89. UTE C 86-218 ADD 2/FA 21; Digital Integrated Circuits in Accordance with FS 90 107; 54/74 F 162A; Synchronous 4-Bit Decade Counters with Synchronous Clear (En, Fr) ADD 3 (En, Fr). 11 pp.

INDUSTRY STANDARDS

Binary Counters (Cont.)

—Synchronous—4-Bit (Cont.)
CECC CECC 90 107-022 ISSUE 2-89. UTE C 86-218 ADD 3/FA 22; Digital Integrated Circuits in Accordance with FS 90 107; 54/74 F 163A; Synchronous 4-Bit Binary Counters with Synchronous Clear (En, Fr) ADD 3 (En, Fr). 12 pp.
CECC CECC 90 109-646 ISSUE 1-86. Digital Integrated Circuits in Accordance with FS 90 109; 54/74 HC 163; Synchronous 4-Bit Binary Counter with Synchronous Clear (En, Fr). 7 pp.
CECC CECC 90 109-787 ISSUE 1-87. Detail Spacification: Digital Integrated Circuits Silicon Monolithic C MOS, Cavity or Non-Cavity Packages; Type(s) 54/74 HCT 163 Synchronous Binary Counter with Synchronous Clear Assessment Levels P, Y, L (En, Fr, Ge). 9 pp.

—Synchronous—4-Bit—Count Down
CECC CECC 90 104-075 ISSUE 2. NL CECC 90 104-075 Issue 2; Digital Integrated Circuits in Accordance with FS 90 104; HEC/HEF 4522B; Programmable 4-Bit BCD Down Counter (En). 14 pp.
CECC CECC 90 104-076 ISSUE 2. NL CECC 90 104-076 Issue 2; Digital Integrated Circuits in Accordance with FS 90 104; HEC/HEF 4526B; Programmable 4-Bit Binary Down Counter (En). 14 pp.

—Synchronous—4-Bit—Count Up/Down
CECC CECC 90 102-030 ISSUE-81. BS CECC 90 102-030; 4-Bit Up/Down Synchronous Binary Counter (En) AMD 1 (En). 21 pp.
CECC CECC 90 103-074 ISSUE 1-81. BS CECC 90 103-074; Synchronous 4-Bit Up/Down Binary Counter (En) AMD 1 (En). 24 pp.
CECC CECC 90 103-139 ISSUE 1-81. BS CECC 90 103-139; Synchronous Up/Down 4-Bit Binary Counters with Mode Control AMD 1 (En). 21 pp.
CECC CECC 90 103-173 ISSUE 1-81. BS 90 103-173; Synchronous 4-Bit Up/Down Binary Counter (En) AMD 1 (En). 21 pp.
CECC CECC 90 104-069 ISSUE 2. NL CECC 90 104-069 Issue 2; Digital Integrated Circuits in Accordance with FS 90 104; HEC/HEF 4516B; Binary Up/Down Counter (En). 13 pp.
CECC CECC 90 106-030 ISSUE 1-85. UTE C 86-217; Digital Integrated Circuits in Accordance with FS 90 106; 54 ALS 191, 74 ALS 191; Synchronous 4-Bit Up/Down Binary Counter (En, Fr) ADD 3 (En, Fr). 17 pp.
CECC CECC 90 106-032 ISSUE 1-85. UTE C 86-217; Digital Integrated Circuits in Accordance with FS 90 106; 54/74 ALS 193, 54/74 ALS 193U; Synchronous 4-Bit Up/Down Binary Counter (En, Fr) ADD 3 (En, Fr). 18 pp.
CECC CECC 90 106-039 ISSUE 2-86. UTE C 86-217 ADD 2/GA 39; Digital Integrated Circuits in Accordance with FS 90 106; 54/74 ALS 569, 54/74 ALS 569A; Synchronous 4-Bit Up/Down Decade Counter with 3 State Outputs (En, Fr) ADD 3 (En, Fr). 18 pp.
CECC CECC 90 107-028 ISSUE 2-89. UTE C 86-218 ADD 2/FA 28; Digital Integrated Circuits in Accordance with FS 90 107; 54/74 F 193; Synchronous 4-Bit Up/Down Binary Counter (En, Fr) ADD 3 (En, Fr). 12 pp.
CECC CECC 90 109-727 ISSUE 1-87. Digital Integrated Circuits in Accordance with FS 90 109; 54 HC 191, 74 HC 191; Synchronous 4-Bit Up/Down Binary Counter (En, Fr). 7 pp.
CECC CECC 90 109-793 ISSUE 1-87. Digital Integrated Circuits; Silicon Monolithic C MOS, Cavity or Non-Cavity Packages; Type(s) 54/74 HCT 191; Synchronous 4-Bit Up/Down Binary Counter (En, Fr). 10 pp.

—Synchronous—8-Bit—Count Down
CECC CECC 90 109-886 ISSUE 1-90. Digital Integrated Circuits; Silicon Monolithic C MOS, Cavity or Non-Cavity Packages; Type(s); 54/74 HC 40103 8-Bit Synchronous Binary Down Counter; Assessment Levels P, Y, L (En, Fr). 10 pp.
CECC CECC 90 109-888 ISSUE 1-90. Digital Integrated Circuits Silicon Monolithic C MOS, Cavity or Non-Cavity Packages; Type(s); 54/74 HCT 40103 8-Bit Synchronous Binary down Counter; Assessment Levels P, Y, L (En, Fr). 10 pp.

—Synchrounous—4-Bit—Preferred Products List
CECC CECC MUAHAG Vol 7 IS 8-92. Preferred Products List; Active Microcircuits (En, Fe, Ge). 89 pp.

—12 Stage—Preferred Products List
CECC CECC MUAHAG Vol 7 IS 8-92. Preferred Products List; Active Microcircuits (En, Fe, Ge). 89 pp.

Binary Counters (Cont.)

—14-Bit
CECC CECC 90 104-116 ISSUE 1-81. BS CECC 90 104-116; 14-Bit Binary Counter (En). 12 pp.
CECC CECC 90 104-116 ISSUE 1-86. CEI CECC 90 104-116; 14-Bit Binary Counter (En). 2 pp.
CECC CECC 90 104-213 ISSUE 1-86. CEI-CECC 90 104-213; 14-Bit Binary Counter and Oscillator (En). 2 pp.
CECC CECC 90 104-213 ISSUE 1-85. BS CECC 90 104-213; 14-Bit Binary Counter and Oscillator (En). 33 pp.

—14 Stage—Preferred Products List
CECC CECC MUAHAG Vol 7 IS 8-92. Preferred Products List; Active Microcircuits (En, Fe, Ge). 89 pp.

—4-Bit
CECC CECC 90 101-031 ISSUE 1-87. UTE C 86-213/B 51; Digital Integrated Circuits in Accordance with FS 90 101; 54/64/74 93A; 4-Bit Binary Counters (En). 16 pp.
CECC CECC 90 101-047 ISSUE 1-87. UTE C 86-213/B 67; Digital Integrated Circuits in Accordance with FS 90 101; 54/64/74 161; 4-Bit Binary Counter (En). 16 pp.
CECC CECC 90 101-048 ISSUE 1-87. UTE C 86-213/B 68; Digital Integrated Circuits in Accordance with FS 90 101 54/64/74 163; 4-Bit Binary Counter (En). 16 pp.
CECC CECC 90 103-044 ISSUE 1-81. BS CECC 90 103-044; 4-Bit Binary Counter (En) AMD 1 (En). 19 pp.
CECC CECC 90 103-108 ISSUE 1-82. BS CECC 90 103-108; 4-Bit Binary Counter with Supply to Pins 7 and 14 (En). 21 pp.
CECC CECC 90 104-216 ISSUE 1-85. BS CECC 90 104-216; Programmable BCD Divide-By-N 4-Bit Counter (En). 37 pp.
CECC CECC 90 104-216 ISSUE 1-86. CEI-CECC 90 104-216; Silicon with (B) Buffered Outputs Programmable BCD Divide-By-N 4-Bit Counter (En). 2 pp.

—4-Bit—Count Up/Down
CECC CECC 90 104-100 ISSUE 2. NL CECC 90 104-100 Issue 2; Digital Integrated Circuits in Accordance with FS 90 104; HEC/HEF 40193B; 4-Bit Up/Down Binary Counter (En). 13 pp.

—4-Bit—Dual
CECC CECC 90 103-129 ISSUE 1-81. BS CECC 90 103-129; Dual 4-Bit Binary Counter with Clear (En) AMD 1 (En). 19 pp.

—4-Bit—Dual—Preferred Products List
CECC CECC MUAHAG Vol 7 IS 8-92. Preferred Products List; Active Microcircuits (En, Fe, Ge). 89 pp.

—4-Bit—Preferred Products List
CECC CECC MUAHAG Vol 7 IS 8-92. Preferred Products List; Active Microcircuits (En, Fe, Ge). 89 pp.

—7 Stage
CECC CECC 90 109-655 ISSUE 1-86. Digital Integrated Circuits in Accordance with FS 90 109; 54/74 HC 4024; 7-Stage Binary Counter with Clear (En, Fr). 6 pp.
CECC CECC 90 109-871 ISSUE 1-90. Digital Integrated Circuits; Silicon Monolithic C MOS, Cavity or Non-Cavity Packages; Type(s); 54/74 HCT 4024 7-Stage Binary Counter with Clear; Assessment Levels P, Y, L (En, Fr). 8 pp.

—7 Stage—Preferred Products List
CECC CECC MUAHAG Vol 7 IS 8-92. Preferred Products List; Active Microcircuits (En, Fe, Ge). 89 pp.

Binary/Decade Counters
See Also: Binary Counters; Decade Counters

—Synchronous—Count Up/Down
CECC CECC 90 104-026 ISSUE 3. NL CECC 90 104-026 Issue 3; Digital Integrated Circuits in Accordance with FS 90 104; HEC/HEF 4029B; Synchronous Up/Down Counter, Binary/Decade Counter (En). 16 pp.

Binary Direct Voltage Signals
Use: Direct Voltage Signals

Binary Dividers
Use: Binary Counters

Binary Encoding
Use: Encoding

Binary Floating Point Arithmetic
Use: Floating Point Arithmetic

Binary Logic Elements
Use: Logic Elements

Binary Notation
Use: Binary Number Systems

Binary Number Systems
Use For: Binary Notation; Binary System
See Also: Numbering Systems

—Two Condition—Codes—Data Transmission
CCITT RECMN V.1-89. Equivalence Between Binary Notation Symbols and the Significant Conditions of a Two-Condition Code—Data Communication Over the Telephone Network (Study Group XVII) 2 pp. 2 pp.

Binary Scalers
Use: Binary Counters

Binary System
Use: Binary Number Systems

Binder Content Analysis

—Asphalt Mastics
BSI BS 5284-76. 1976 Sampling and Testing Mastic Asphalt and Pitchmastic Used in Building. 22 pp.

—Asphalts
BSI BS 598: Part 107-90. 1990 Sampling and Examination of Bituminous Mixtures for Roads and Other Paved Areas Part 107: Method of Test for the Determination of the Composition of Design Wearing Course Rolled Asphalt (Supersedes BS 598: Part 3: 1985). 20 pp.
DIN ENGL 1996 Pt 6-88. Testing of Asphalt; Determination of Binder Content and Recovery of Binder (Oct). 11 pp.

—Refractory Materials
BSI BS 6043: Sec 4.2-91. 1991 Methods of Sampling and Test for Carbonaceous Materials Used in Aluminium Manufacture Part 4: Cold Ramming Pastes Section 4.2: Determination of Effective Binder Content and Aggregate Content by Extraction with Quinoline; Determination. 9 pp.

Binder Discs
Use: Gaskets

Binders (Adhesives)
Use: Binders (Materials)

Binders (Agricultural)
See Also: Agricultural Equipment

—Finger Liners
CNS B2632-82. Finger Liners for Agricultural Binders and Rice Combines (Jan)(8202).

—Knife Sections
CNS B2563-81. Knife Section for Agricultural Binders and Rice Combines (Jun)(7551).
JIS B 9213-89. Knife Section and Ledger Plate of Agricultural Binders and Head-Feeding Combines. 7 pp.

—Twines
JIS B 9211-78. Binder's Twines.

Binders (Files)
Use For: Loose Leaf Binders *See Also:* Bookbinding; Notebooks; Paper Fasteners
BSI BS 5641-82. 1982 Loose-Leaf Publications. 7 pp.
JIS S 5507-83. Multi-Prong Binders.

—Holes
SAA AS P5-69. Punching Patterns for Round Holes Used in Files and Loose Leaf Binders Reconfirmed 1986. 4 pp.

—Note Pad
CGSB CAN/CGSB-53.128-93. Notepad and Binder. 11 pp.

—Printouts
CGSB CAN/CGSB-53.115-92. Binder for Computer Printout Forms. 8 pp.

—Prong
CGSB CAN/CGSB-53.41-M89. Loose-Leaf Binder, Flexible Prong Fastener. 10 pp.

—Ring
BSI BS 5097: Part 1-74. 1974 Amd 1 Loose Leaf Binders Part 1: Ring Binders with Metal Mechanisms. 19 pp.
CGSB CAN/CGSB-53.23-93. Ring Type Loose-Leaf Binder. 10 pp.

Binders (Materials)

Use For: Binders (Adhesives) *See Also:* Adhesives; Clays; Cohesion; Masonry Sealants; Sealants; Sizing (Surface Treatment); Sizing Compounds

—Aggregates
BSI BS 598: Part 108-90. 1990 Sampling and Examination of Bituminous Mixtures for Roads and Other Paved Areas Part 108: Methods for Determination of the Condition of the Binder on Coated Chippings and for Measurement of the Rate of Spread of Coated Chippings. 8 pp.

—Bituminous
DIN ENGL 1995 Pt 1-89. Bituminous Binders; Road Bitumen; Requirements (Oct). 3 pp.
DIN ENGL 1995 Pt 2-89. Bituminous Binders; Cutback Bitumen; Requirements (Oct). 2 pp.
DIN ENGL 1995 Pt 3-89. Bituminous Binders; Bitumen Emulsions; Requirements (Oct). 3 pp.
DIN ENGL 1995 Pt 4-89. Bituminous Binders; Dissolved Bitumen; Requirements (Oct). 2 pp.
DIN ENGL 1995 Pt 5-89. Bituminous Binders; Road Tars, Pitch-Bitumen Mixtures and Cold Pitch Solutions; Requirements (Oct). 5 pp.

—Bituminous—Flow Rate
DIN ENGL 52023 Pt 1-80. Testing of Bituminous Binders; Determination of Flow Time by the Tar Efflux Viscometer; Method of Measurement (Dec). 4 pp.

—Bituminous—Paraffin Wax Content
DIN ENGL 52015-80. Testing of Bituminous Binders; Determination of Paraffin Wax Content (Dec). 4 pp.

—Bituminous—Recovery
BSI BS 2000: Part 105-91. 1991 Petroleum and Its Products Part 105: Recovery of Bituminous Binders by Dichloromethane Extraction (Identical with IP 105/89) (W) (Supersedes BS 598: Part 2: 1974). 8 pp.

—Bituminous—Sampling
BSI BS 3195: Part 3-87. 1987 Sampling Petroleum Products Part 3: Method for Sampling Bituminous Binders. 15 pp.
BSI DD 193-91. 1991 Recovery of Bitumen Binders by Dichloromethane Extraction (Rotary Film Evaporator Method). 10 pp.
CEN EN 58-84. Sampling Bituminous Binders. 13 pp.

—Bituminous—Solubility
BSI BS 2000: Part 47-93. 1993 Methods of Test for Petroleum and Its Products Part 47: Solubility of Bituminous Binders. 4 pp.
BSI BS 2000: Part 47-83. 1983 Petroleum and Its Products Part 47: Solubility of Bituminous Binders. 4 pp.

—Bituminous—Stability Testing
BSI BS 598: Part 106-90. 1990 Sampling and Examination of Bituminous Mixtures for Roads and Other Paved Areas Part 106: Method for the Determination of the Stability Index of Pitch-Bitumen Binders (Supersedes BS 598: Part 3: 1985). 7 pp.

—Bituminous—Viscosity
DIN ENGL 52007 Pt 1-80. Testing of Bituminous Binders; Determination of Viscosity; General Principles and Evaluation (Dec). 5 pp.
DIN ENGL 52007 Pt 2-80. Testing of Bituminous Binders; Determination of Viscosity; Measurement by Drawn-Sphere Viscometer (Dec). 5 pp.

—Coatings
DIN ENGL 55928 Pt 9-91. Corrosion Protection of Steel Structures by the Application of Organic or Metallic Coatings; Composition of Binders and Pigments for Coating Materials (May). 6 pp.

—Colorimetry
BSI BS 6782: Part 5-87. 1987 Binders for Paints Part 5: Method for Estimation of Colour of Clear Liquids by the Gardner Colour Scale. 8 pp.
ISO 4630-81. Binders for Paints and Varnishes—Estimation of Colour of Clear Liquids by the Gardner Colour Scale First Edition. 7 pp.

—Concretes
SAA AS 1012. Methods of Testing Concrete Complete Set in Binder.

—Distributors
BSI BS 1707-89. 1989 Hot Binder Distributors for Road Surface Dressing. 14 pp.
BSI BS 1707-70. (WITHDRAWN) 1970 Hot Binder Distributors for Road Surface Dressing. 20 pp.

—Fatty Acid Content—Gas Chromatography
DIN ENGL 55957-83. Binders for Paints and Varnishes; Determination of Fatty Acids by Gas Chromatography (Oct). 4 pp.

Binders (Materials) (Cont.)

—Fiberglass Reinforced Plastics—Mats
ISO 2558-74. Textile Glass Chopped-Strand Mats for Reinforcement of Plastics—Determination of Time of Dissolution of the Binder in Styrene First Edition. 5 pp.

—Glossaries
BSI BS 6100: Sec 6.1-84. 1984 Glossary of Building and Civil Engineering Terms Part 6: Concrete and Plaster Section 6.1: Binders. 10 pp.
BSI BS 6100: Sec 6.1-01. 1984 Amd 1 Glossary of Building and Civil Engineering Terms Part 6: Concrete and Plaster Section 6.1: Binders (AMD 7263) August 15, 1992. 12 pp.

—Gypsum Deposits
ISO 1587-75. Gypsum Rock for the Manufacture of Binders—Specifications First Edition. 5 pp.

—Hydroxyl Value—Volumetric Analysis
BSI BS 6782: Part 4-87. 1987 Binders for Paints Part 4: Method for Determination of Hydroxyl Value (Titrimetric Method). 6 pp.
ISO 4629-78. Paint Media—Determination of Hydroxyl Value—Titrimetric Method First Edition; (Erratum—Nov 1979). 6 pp.

—Latex
MOD UK TS 10255. Acrylonitrile/Butadiene/Styrene Latex, Type QX.

—Melting Points
DIN ENGL 53181-81. Binders for Paints and Varnishes; Determination of the Melting Range of Resins by the Capillary Method (Mar). 2 pp.

—Paints
BSI BS 6782: Part 1-87. 1987 Binders for Paints Part 1: Method for Determination of Volatile and Non-Volatile Content. 4 pp.
BSI BS 6782: Part 2-87. 1987 Binders for Paints Part 2: Method for Determination of Saponification Valve (Titrimetric Method). 6 pp.
BSI BS 6782: Part 3-87. 1987 Binders for Paint Part 3: Method for Determination of Acid Valve (Titrimetric Method). 4 pp.
BSI BS 6782: Part 5-87. 1987 Binders for Paints Part 5: Method for Estimation of Colour of Clear Liquids by the Gardner Colour Scale. 8 pp.
BSI BS 6782: Part 6-87. 1987 Binders for Paints Part 6: Method for Determination of Softening Point (Ring and Ball Method). 10 pp.
DIN ENGL 53181-81. Binders for Paints and Varnishes; Determination of the Melting Range of Resins by the Capillary Method (Mar). 2 pp.
DIN ENGL 55952-81. Binders for Paints and Varnishes; Cellulose Ethers; Testing (June). 5 pp.
DIN ENGL 55953-81. Binders for Paints and Varnishes; Cellulose Esters of Organic Acids; Testing (May). 6 pp.
ISO 3681-83. Binders for Paints and Varnishes—Determination of Saponification Value—Titrimetric Method Second Edition. 5 pp.
ISO 3682-83. Binders for Paints and Varnishes—Determination of Acid Value—Titrimetric Method Second Edition. 4 pp.
ISO 4625-80. Binders for Paints and Varnishes—Determination of Softening Point—Ring-and-Ball Method First Edition. 9 pp.
ISO 4630-81. Binders for Paints and Varnishes—Estimation of Colour of Clear Liquids by the Gardner Colour Scale First Edition. 7 pp.
ISO 6744-84. Binders for Paints and Varnishes—Alkyd Resins—General Methods of Test First Edition. 9 pp.
ISO 7142-84. Binders for Paints and Varnishes—Epoxy Resins—General Methods of Test First Edition. 6 pp.
ISO 7143-82. Binders for Paints and Varnishes—Aqueous Dispersions of Polymers and Copolymers—General Methods of Test First Edition. 3 pp.

—Saponification Number—Volumetric Analysis
BSI BS 6782: Part 2-87. 1987 Binders for Paints Part 2: Method for Determination of Saponification Valve (Titrimetric Method). 6 pp.
BSI BS 6782: Part 3-87. 1987 Binders for Paint Part 3: Method for Determination of Acid Valve (Titrimetric Method). 4 pp.
ISO 3681-83. Binders for Paints and Varnishes—Determination of Saponification Value—Titrimetric Method Second Edition. 5 pp.
ISO 3682-83. Binders for Paints and Varnishes—Determination of Acid Value—Titrimetric Method Second Edition. 4 pp.

—Softening Points
BSI BS 6782: Part 6-87. 1987 Binders for Paints Part 6: Method for Determination of Softening Point (Ring and Ball Method). 10 pp.

Binders (Materials) (Cont.)

—Softening Points (Cont.)
ISO 4625-80. Binders for Paints and Varnishes—Determination of Softening Point—Ring-and-Ball Method First Edition. 9 pp.

—Varnishes
BSI BS 6782: Part 2-87. 1987 Binders for Paints Part 2: Method for Determination of Saponification Valve (Titrimetric Method). 6 pp.
BSI BS 6782: Part 3-87. 1987 Binders for Paint Part 3: Method for Determination of Acid Valve (Titrimetric Method). 4 pp.
BSI BS 6782: Part 5-87. 1987 Binders for Paints Part 5: Method for Estimation of Colour of Clear Liquids by the Gardner Colour Scale. 8 pp.
BSI BS 6782: Part 6-87. 1987 Binders for Paints Part 6: Method for Determination of Softening Point (Ring and Ball Method). 10 pp.
DIN ENGL 53181-81. Binders for Paints and Varnishes; Determination of the Melting Range of Resins by the Capillary Method (Mar). 2 pp.
DIN ENGL 55952-81. Binders for Paints and Varnishes; Cellulose Ethers; Testing (June). 5 pp.
DIN ENGL 55953-81. Binders for Paints and Varnishes; Cellulose Esters of Organic Acids; Testing (May). 6 pp.
ISO 3681-83. Binders for Paints and Varnishes—Determination of Saponification Value—Titrimetric Method Second Edition. 5 pp.
ISO 3682-83. Binders for Paints and Varnishes—Determination of Acid Value—Titrimetric Method Second Edition. 4 pp.
ISO 4625-80. Binders for Paints and Varnishes—Determination of Softening Point—Ring-and-Ball Method First Edition. 9 pp.
ISO 4630-81. Binders for Paints and Varnishes—Estimation of Colour of Clear Liquids by the Gardner Colour Scale First Edition. 7 pp.
ISO 6744-84. Binders for Paints and Varnishes—Alkyd Resins—General Methods of Test First Edition. 9 pp.
ISO 7142-84. Binders for Paints and Varnishes—Epoxy Resins—General Methods of Test First Edition. 6 pp.
ISO 7143-82. Binders for Paints and Varnishes—Aqueous Dispersions of Polymers and Copolymers—General Methods of Test First Edition. 3 pp.

Binding Agents

Use: Binders (Materials)

Binding Posts

See Also: Connectors; Electric Terminals

—Screw Type
CSA 655 Bull. Electrical Bulletin 655 October 12, 1966 to C22.2 NO 4. 4 pp.
CSA 655 Bull. Electrical Bulletin 655 October 12, 1966 to C22.2 NO 65. 4 pp.
CSA 655A Bull. Electrical Bulletin 655A April 26, 1977 to C22.2 NO 65. 1 p.

Binding Tapes

See Also: Adhesive Tapes

—Amalgamative
MOD UK TS 10215. Self Amalgamating Tape.

—Nylon Film—Mailbags
CGSB CAN/CGSB-4.146-M90. Nylon Binding Textile Tape. 8 pp.

—PTFE Film—Pipe Threads
BSI BS 6974-91. 1991 Unsintered PTFE Tape for Thread Sealing Applications (Coarse Threads). 10 pp.
BSI BS 6974-01. 1991 Amd 1 Unsintered PTFE Tape for Thread Sealing Applications (Coarse Threads) (AMD 6776) February 28, 1992. 11 pp.

Bindings, Ski

Use: Ski Bindings

Binnacles

BSI BS MA 2: Part 1-85. 1985 Magnetic Compasses and Binnacles Part 1: Class A Magnetic Compasses, Their Binnacles and Azimuth Reading Devices (Including Test Procedures). 19 pp.
BSI BS MA 2: Part 2-69. 1969 Magnetic Compasses and Binnacles Part 2: Class B Instruments. 6 pp.
ISO 449-79. Shipbuilding—Magnetic Compasses and Binnacles, Class A First Edition. 8 pp.
ISO 613-82. Shipbuilding—Magnetic Compasses, Binnacles and Azimuth Reading Devices—Class B First Edition; (Corrigendum 1-1991). 7 pp.

—Glossaries
BSI BS MA 2: Part 4-85. 1985 Magnetic Compasses and Binnacles Part 4: Glossary of Terms. 10 pp.
ISO 1069-73. Magnetic Compasses and Binnacles for Sea Navigation—Vocabulary First Edition. 40 pp.

Binnacles (Cont.)

—Positioning
BSI BS MA 2: Part 3-69. 1969 Amd 1 Magnetic Compasses and Binnacles Part 3: Recommendations for the Positioning of Compasses (AMD 5029) March 31, 1986. 8 pp.
ISO R694-68. Positioning of Magnetic Compasses in Ships First Edition; (Erratum—Aug 1969). 9 pp.

Binoculars
See Also: Optical Equipment

—Galilean
JIS B 7122-88. Galilean Binoculars. 12 pp.

—Packaging—Export
CNS Z5075-81. Packaging and Packing of Binoculars for Export (Jan)(6979).

—Prism
JIS B 7121-85. Prism Binoculars. 26 pp.

Binomial Distributions

—Confidence Limits—Tables (Data)
MOD UK DSTAN 05-49-78. Tables of the Lower Single Sided 95% Confidence Limit for the Binomial Distribution Sample Sizes 5 to 2020 Issue 1. 174 pp.

Bioacoustics
Scope Note: Methods of measurement and testing in the fields of psychological and physiological acoustics; includes general acoustics, shock, and vibration pertaining to safety, tolerance, and comfort
See Also: Acoustics; Noise; Vibration
BSI BS 3045-81. 1981 Method of Expression of Physical and Subjective Magnitudes of Sound or Noise in Air. 7 pp.
BSI BS 3383-88. 1988 Normal Equal-Loudness Contours for Pure Tones and Normal Threshold of Hearing Under Free-Field Listening Conditions. 12 pp.
BSI BS 4198-67. 1967 Method for Calculating Loudness. 25 pp.
BSI BS 5330-76. 1976 Method of Test for Estimating the Risk of Hearing Handicap Due to Noise Exposure. 8 pp.
BSI BS 6655-86. 1986 Puretone Air Conduction Threshold Audiometry for Hearing Conservation Purposes. 12 pp.
BSI BS 6655-01. 1986 Amd 1 Pure Tone Air Conduction Threshold Audiometry for Hearing Conservation Purposes (AMD 7339) October 15, 1992. 17 pp.
CEN EN 26189-91. Acoustics—Pure Tone Air Conduction Threshold Audiometry for Hearing Conservation Purposes. 11 pp.
CSA CAN/CSA-Z107.6-M90. Pure Tone Air Conduction Threshold Audiometry for Hearing Conservation; (Gen Instr 1). 30 pp.
ISO 131-79. Acoustics—Expression of Physical and Subjective Magnitudes of Sound or Noise in Air First Edition. 5 pp.
ISO 226-87. Acoustics—Normal Equal-Loudness Level Contours First Edition; (Supersedes 454). 11 pp.
ISO 1999-90. Acoustics—Determination of Occupational Noise Exposure and Estimation of Noise-Induced Hearing Impairment Second Edition. 22 pp.
ISO 2204-79. Acoustics—Guide to International Standards on the Measurement of Airborne Acoustical Noise and Evaluation of Its Effects on Human Beings Second Edition. 11 pp.
ISO TR3352-74. Acoustics—Assessment of Noise with Respect to Its Effect on the Intelligibility of Speech First Edition. 3 pp.
ISO 6189-83. Acoustics—Pure Tone Air Conduction Threshold Audiometry for Hearing Conservation Purposes First Edition. 11 pp.
JIS Z 8731-83. Methods for Measurement and Description of A-Weighted Sound Pressure Level. 13 pp.
SNZ NZS/ISO 131-79. Method of Expression of Physical and Subjective Magnitudes of Sound or Noise in Air. 3 pp.
SNZ NZS/ISO 226-87. Acoustics. Normal Equal-Loudness Level Contours. 8 pp.

Bioassay
Use For: Biological Assay

—Surface Waters—Spectrometry
ISO 10260-92. Water Quality—Measurement of Biochemical Parameters—Spectrometric Determination of the Chlorophyll-a Concentration First Edition. 10 pp.

Biochemical Oxygen Demand
Use For: Biological Oxygen Demand
See Also: Chlorination; Filtration; Oxygen Demand; Sedimentation

Biochemical Oxygen Demand (Cont.)

—Effluents
CPPA H.2-91. Determination of Biochemical Oxygen Demand. 5 pp.

—Waste Water—Dilution
DIN ENGL 38409 Pt 51-87. German Standard Methods for the Examination of Water, Waste Water and Sludge; Parameters Characterizing Effects and Substances (Group H) Determination of Biochemical Oxygen Demand After n Days, BODn, Using the Dilution Method (H 51) (May). 9 pp.

—Waste Water—Microbial Sensors
JIS K 3602-90. Apparatus for the Estimation of Biochemical Oxygen Demand (BODs) with Microbial Sensor. 10 pp.

—Water
CPPA H.2-91. Determination of Biochemical Oxygen Demand. 5 pp.

—Water—Dilution
BSI BS 6068: Sec 2.14-90. 1990 Water Quality Part 2: Physical, Chemical and Bio-Chemical Methods Section 2.14: Determination of Biochemical Oxygen Demand After 5 Days (BOD): Dilution and Seeding Method. 12 pp.
DIN ENGL 38409 Pt 51-87. German Standard Methods for the Examination of Water, Waste Water and Sludge; Parameters Characterizing Effects and Substances (Group H) Determination of Biochemical Oxygen Demand After n Days, BODn, Using the Dilution Method (H 51) (May). 9 pp.
ISO 5815-89. Water Quality—Determination of Biochemical Oxygen Demand After 5 Days (BOD5)—Dilution and Seeding Method Second Edition. 8 pp.

—Water—Inoculation
BSI BS 6068: Sec 2.14-90. 1990 Water Quality Part 2: Physical, Chemical and Bio-Chemical Methods Section 2.14: Determination of Biochemical Oxygen Demand After 5 Days (BOD): Dilution and Seeding Method. 12 pp.

Biocompatibility
Use For: Cytocompatibility

—Surgical Implants
CSA CAN3-Z310.6-M84. Testing for Biocompatibility; (Gen Instr 1). 51 pp.
ISO TR9966-89. Implants for Surgery—Biocompatibility—Selection of Biological Test Methods for Materials and Devices First Edition. 9 pp.

Biodegradability
Use For: Biodeterioration See Also: Antifouling Coatings

—Aerobic—Organic Compounds—Water
BSI BS 6068: Sec 5.11-93. 1993 Water Quality Evaluation in an Aqueous Medium of the 'Ultimate' Aerobic Biodegradability of Organic Compounds: Method by Determining the Oxygen Demand in a Closed Respirometer (ISO 9408: 1991) (N). 17 pp.
BSI BS 6068: Sec 5.12-93. 1993 Water Quality Evaluation in an Aqueous Medium of the 'Ultimate' Aerobic Biodegradability of Organic Compounds: Method by Analysis of Released Carbon Dioxide (ISO 9439: 1990) (N). 16 pp.
BSI BS EN 29408-93. 1993 Water Quality Evaluation in an Aqueous Medium of the 'Ultimate' Aerobic Biodegradability of Organic Compounds: Method by Determining the Oxygen Demand in a Closed Respirometer (ISO 9408: 1991) (N). 17 pp.
BSI BS EN 29439-93. 1993 Water Quality Evaluation in an Aqueous Medium of the 'Ultimate' Aerobic Biodegradability of Organic Compounds: Method by Analysis of Released Carbon Dioxide (ISO 9439: 1990) (N). 16 pp.
CEN EN 29408-93. Water Quality—Evaluation in an Aqueous Medium of the 'Ultimate' Aerobic Biodegradability of Organic Compounds—Method by Determining the Oxygen Demand in a Closed Respirometer (ISO 9408: 1991). 12 pp.
CEN EN 29439-93. Water Quality—Evaluation in an Aqueous Medium of the 'Ultimate' Aerobic Biodegradability of Organic Compounds—Method by Analysis of Released Carbon Dioxide (ISO 9439: 1990). 11 pp.
CENELEC PREN 210634-92. Water Quality—Guidance for the Evaluation in an Aqueous Medium of the "Ultimate" Biodegradability of Poorly Soluble Organic Compounds. 14 pp.
ISO 9408-91. Water Quality—Evaluation in an Aqueous Medium of the "Ultimate" Aerobic Biodegradability of Organic Compounds—Method by Determining the Oxygen Demand in a Closed Respirometer First Edition; (Corrigendum 1-1992) (CEN EN 29408: 1993). 14 pp.

Biodegradability (Cont.)

—Aerobic—Organic Compounds—Water (Cont.)
ISO 9439-90. Water Quality—Evaluation in an Aqueous Medium of the "Ultimate" Aerobic Biodegradability of Organic Compounds—Method by Analysis of Released Carbon Dioxide First Edition; (Corrigendum 1-1992) (CEN EN 29439: 1993). 13 pp.

—Aerobic—Organic Compounds—Water—Activated Sludge Method
ISO 9887-92. Water Quality—Evaluation of the Aerobic Biodegradability of Organic Compounds in an Aqueous Medium—Semi-Continuous Activated Sludge Method (SCAS) First Edition. 11 pp.

—Anaerobic—Plastics—Sludge (Sewage)
CSA Z218.0-93. Test Method for Determining the Anaerobic Biodegradability of Plastic Materials; (Gen Instr 1). 25 pp.

—Detergents
CNS K6436-79. Method of Test for Biodegradability of Synthetic Detergents (Sep)(4984).
JIS K 3363-90. Testing Method for Biodegradability of Synthetic Detergent. 14 pp.

—Nonionic Surfactants
CNS K6437-79. Method of Test for Biodegradability of Nonionic Surface Active Agent (Sep)(4985).

—Organic Compounds—Waste Water
BSI BS 6068: Sec 5.13-93. 1993 Water Quality—Evaluation of the Aerobic Biodegradability of Organic Compounds in an Aqueous Medium—Static Test (Zahn-Wellens Method) (ISO 9888: 1991) (N). 14 pp.
BSI BS EN 29888-93. 1993 Water Quality—Evaluation of the Aerobic Biodegradability of Organic Compounds in an Aqueous Medium Static Test (Zahn-Wellens Method) (ISO 9888: 1991) (N). 14 pp.
CEN EN 29888-93. Water Quality—Evaluation of the Aerobic Biodegradability of Organic Compounds in an Aqueous Medium—Static Test (Zahn-Wellens Method) (ISO 9888: 1991). 9 pp.
ISO 9888-91. Water Quality—Evaluation of the Aerobic Biodegradability of Organic Compounds in an Aqueous Medium—Static Test (Zahn-Wellens Method) First Edition; (CEN EN 29888:1993). 10 pp.

—Organic Compounds—Water
BSI BS 6068: Sec 5.6-85. 1985 Water Quality Part 5: Methods for Biological Testing Section 5.6: Evaluation in an Aqueous Medium of the 'Ultimate' Aerobic Biodegradability of Organic Compounds: Method by Analysis of Dissolved Organic Carbon (DOC). 8 pp.
BSI BS 6068: Sec 5.13-93. 1993 Water Quality—Evaluation of the Aerobic Biodegradability of Organic Compounds in an Aqueous Medium—Static Test (Zahn-Wellens Method) (ISO 9888: 1991) (N). 14 pp.
BSI BS EN 29888-93. 1993 Water Quality—Evaluation of the Aerobic Biodegradability of Organic Compounds in an Aqueous Medium Static Test (Zahn-Wellens Method) (ISO 9888: 1991) (N). 14 pp.
CEN EN 29888-93. Water Quality—Evaluation of the Aerobic Biodegradability of Organic Compounds in an Aqueous Medium—Static Test (Zahn-Wellens Method) (ISO 9888: 1991). 9 pp.
ISO 7827-84. Water Quality—Evaluation in an Aqueous Medium of the "Ultimate" Aerobic Biodegradability of Organic Compounds—Method by Analysis of Dissolved Organic Carbon (DOC) First Edition. 7 pp.
ISO 9888-91. Water Quality—Evaluation of the Aerobic Biodegradability of Organic Compounds in an Aqueous Medium—Static Test (Zahn-Wellens Method) First Edition; (CEN EN 29888:1993). 10 pp.

—Sewage
DIN ENGL 38412 Pt 25-84. German Standard Methods for the Examination of Water, Waste Water and Sludge; Test Methods Using Water Organisms (Group L); Determination of Biodegradability Static Test (L 25) (Jan). 8 pp.

—Surfactants
SAA AS 1792-76. Method for Determining the Biodegradability of Surfactants. 40 pp.

—Waste Water
DIN ENGL 38412 Pt 24-81. German Standard Methods for the Analysis of Water, Waste Water and Sludge; Bio-Assays (Group L); Determination of Biodegradability by Use of Special Methods of Analysis (24) (Apr). 6 pp.

Biodegradability (Cont.)

—Waste Water (Cont.)
DIN ENGL 38412 Pt 25-84. German Standard Methods for the Examination of Water, Waste Water and Sludge; Test Methods Using Water Organisms (Group L); Determination of Biodegradability Static Test (L 25) (Jan). 8 pp.

—Water
DIN ENGL 38412 Pt 24-81. German Standard Methods for the Analysis of Water, Waste Water and Sludge; Bio-Assays (Group L); Determination of Biodegradability by Use of Special Methods of Analysis (24) (Apr). 6 pp.
DIN ENGL 38412 Pt 25-84. German Standard Methods for the Examination of Water, Waste Water and Sludge; Test Methods Using Water Organisms (Group L); Determination of Biodegradability Static Test (L 25) (Jan). 8 pp.

Biodeterioration
Use: Biodegradability

Biogas
See Also: Agricultural Products; Methane
SNZ NZS 5228: Part 1-87. Code of Practice for the Production and Use of Biogas, Farm Scale Operation Part 1: Production of Biogas. 120 pp.
SNZ NZS 5228: Part 2-87. Code of Practice for the Production and Use of Biogas, Farm Scale Operation Part 2: Uses of Biogas. 24 pp.
SNZ NZS 5442-90. Specification for Reticulated Natural Gas Amend: 1, 1993. 13 pp.

Biological Analysis
See Also: Microbiological Analysis

—Waste Water
DIN ENGL 38412 Pt 1-82. German Standard Methods for the Examination of Water, Waste Water and Sludge; Test Methods Using Water Organisms (Group L); General Guideline for the Planning, Performance and Evaluation of Biological Test (L 1) (June). 5 pp.

—Water
DIN ENGL 38412 Pt 1-82. German Standard Methods for the Examination of Water, Waste Water and Sludge; Test Methods Using Water Organisms (Group L); General Guideline for the Planning, Performance and Evaluation of Biological Test (L 1) (June). 5 pp.

Biological Assay
Use: Bioassay

Biological Containment Cabinets
Use: Containment Cabinets

Biological Hazards
See Also: Hazardous Materials; Pathogens

—Air Filters
JIS K 3801-89. Penetration Test Method of HEPA Filters for Microbiological Filtration. 10 pp.

—Medical Equipment
BSI BS 5736: Part 1-89. 1989 Evaluation of Medical Devices for Biological Hazards Part 1: Guide for the Selection of Biological Methods of Test. 8 pp.
BSI BS 5736: Part 2-90. 1990 Evaluation of Medical Devices for Biological Hazards Part 2: Method of Testing by Tissue Implantation. 8 pp.
BSI BS 5736: Part 2-81. 1981 Amd 1 Evaluation of Medical Devices for Biological Hazards Part 2: Method of Testing by Tissue Implantation (N) (AMD 6207) July 31, 1989. 9 pp.
BSI BS 5736: Part 3-81. 1981 Amd 1 Evaluation of Medical Devices for Biological Hazards Part 3: Method of Test for Systemic Toxicity; Assessment of Acute Toxicity of Extracts from Medical Devices (AMD 6208) July 31, 1989. 8 pp.
BSI BS 5736: Part 4-81. 1981 Amd 1 Evaluation of Medical Devices for Biological Hazards Part 4: Method of Test for Intracutaneous Reactivity of Extracts from Medical Devices (AMD 6209) July 31, 1989. 9 pp.
BSI BS 5736: Part 5-82. 1982 Amd 1 Evaluation of Medical Devices for Biological Hazards Part 5: Method of Test for Systemic Toxicity; Assessment of Pyrogenicity in Rabbits of Extracts from Medical Devices (AMD 6210) July 31, 1989. 9 pp.
BSI BS 5736: Part 6-83. 1983 Amd 1 Evaluation of Medical Devices for Biological Hazards Part 6: Methods of Test for Sensitization; Assmt. of the Potential of Medical Devices to Produce Delayed Contact Dermatitis (AMD 6211) July 31, 1989. 9 pp.
BSI BS 5736: Part 7-83. 1983 Amd 1 Evaluation of Medical Devices for Biological Hazards Part 7: Method of Test for Skin Irritation of Extracts from Medical Devices (AMD 6212) July 31, 1989. 9 pp.

Biological Hazards (Cont.)

—Medical Equipment (Cont.)
BSI BS 5736: Part 8-84. 1984 Amd 1 Evaluation of Medical Devices for Biological Hazards Part 8: Method of Test for Skin Irritation of Solid Medical Devices (AMD 6213) July 31, 1989. 9 pp.
BSI BS 5736: Part 9-86. 1986 Amd 1 Evaluation of Medical Devices for Biological Hazards Part 9: Method of Test for Eye Irritation (AMD 6214) July 31, 1989. 9 pp.
BSI BS 5736: Part 10-88. 1988 Amd 1 Evaluation of Medical Devices for Biological Hazards Part 10: Method of Test for Toxicity to Cells in Culture of Extracts from Medical Devices (AMD 6215) July 31, 1989. 10 pp.
BSI BS 5736: Part 11-90. 1990 Evaluation of Medical Devices for Biological Hazards Part 11: Method of Test for Haemolysis. 10 pp.
CEN PREN 30993-6-92. Evaluation of Medical Devices for Biological Hazards New Part: Tests for Local Effects After Implantation. 27 pp.
CEN PREN 30993-7-93. Biological Testing of Medical and Dental Materials and Devices—Part 7: Ethylene Oxide Sterilization Residuals (ISO/DIS 10993-7). 63 pp.
ISO 10993 Pt 1-92. Biological Evaluation of Medical Devices—Part 1: Guidance on Selection of Tests First Edition; (Corrigendum 1-1992). 18 pp.
ISO 10993 Pt 3-92. Biological Evaluation of Medical Devices—Part 3: Tests for Genotoxicity, Carcinogenicity, and Reproductive Toxicity First Edition. 12 pp.
ISO 10993 Pt 4-92. Biological Evaluation of Medical Devices—Part 4: Selection of Tests for Interactions with Blood First Edition. 24 pp.
ISO 10993 Pt 5-92. Biological Evaluation of Medical Devices—Part 5: Tests for Cytotoxicity: in Vitro Methods First Edition. 12 pp.

—Packaging
CGSB CAN/CGSB-43.125-M90. Packaging of Infectious Substances and Diagnostic Specimens. 20 pp.

—Wood
BSI BS EN 335-1-92. 1992 Hazard Classes of Wood and Wood-Based Products Against Biological Attack Part 1: Classification of Hazard Classes. 11 pp.
BSI BS EN 335-2-92. 1992 Hazard Classes of Wood and Wood-Based Products Against Biological Attack Part 2: Guide to the Application of Hazard Classes to Solid Wood. 13 pp.
CEN PREN 335 (Part 1)-90. Wood and Wood Based Products—Definition of Hazard Classes of Biological Attack—Part 1: Solid Wood. 17 pp.
CEN EN 335-1-92. Durability of Wood and Derived Materials—Definition of Hazard Classes of Biological Attack—Part 1: General. 10 pp.
CEN EN 335-2-92. Durability of Wood and Wood-Based Products-Definition of Hazard Classes of Biological Attack—Part 2: Application to Solid Wood. 13 pp.
DIN ENGL EN 335 Pt 1-92. Durability of Wood and Wood-Based Panel Products; Definition of Hazard Classes of Biological Attack; General (Sept). 5 pp.
DIN ENGL EN 335 Pt 2-92. Durability of Wood and Wood-Based Panel Products; Definition of Hazard Classes of Biological Attack; Application to Solid Wood (Oct). 8 pp.

—Wood Preservatives
CEN PREN 599-1-91. Durability of Wood and Derived Materials—Performances of Wood Preservative as Determined by Biological Tests—Part 1: Specifications According to Hazard Class. 25 pp.
CEN PREN 599-2-91. Durability of Wood and Derived Materials—Performances of Wood Preservative as Determined by Biological Tests—Part 2: Classification and Labelling. 16 pp.

Biological Oxygen Demand
Use: Biochemical Oxygen Demand

Biology
Scope Note: Use a more specific term
See: Anatomy; Bioassay; Biological Analysis; Biological Hazards; Biotechnology; Botany; Fungi; Microbiology; Physiology; Plants (Botany); Zoology

Biomechanical Effects (Human Body)
Use For: Impact Effects (Human Body); Mechanical Shock (Human Body) *See Also:* Ergonomics; Occupational Safety and Health; Vibration

—Electric Measuring Instruments
BSI DD ENV 28041-93. 1993 Human Response to Vibration Measuring Instrumentation (ISO 8041: 1990) (E). 32 pp.
CEN ENV 28041-93. Human Response to Vibration—Measuring Instrumentation (ISO 8041: 1990). 26 pp.

Biomechanical Effects (Human Body) (Cont.)

—Electric Measuring Instruments (Cont.)
ISO 8041-90. Human Response to Vibration—Measuring Instrumentation First Edition; (Corrigendum 1-1993) (CEN EN 28041: 1993). 30 pp.

—Electric Shock
BSI PD 6519: Part 1-88. 1988 Guide to Effects of Current Passing Through the Human Body Part 1: General Aspects. 21 pp.
BSI PD 6519: Part 2-88. 1988 Guide to Effects of Current Passing Through the Human Body Part 2: Special Aspects. 24 pp.
IEC 479 Pt 1-84. Effects of Current Passing Through the Human Body Part 1: General Aspects Chapter 1: Electrical Impedance of the Human Body Chapter 2: Effects of Alternating Current in the Range of 15 Hz to 100 Hz Chapter 3: Ef-fects of Direct Current Second Edition. 40 pp.
IEC 479 Pt 2-87. Effects of Current Passing Through the Human Body Part 2: Special Aspects Chapter 4: Effects of Alternating Current with Frequencies Above 100 Hz Chapter 5: Effects of Special Waveforms of Current Chapter 6: Effects of Unidirectional Single. 48 pp.
IEC 990-90. Methods of Measurement of Touch-Current and Protection Conductor Current First Edition; (Corrigendum—Aug 1991). 73 pp.
SAA AS 3859-91. Guide to the Effects of Current Passing Through the Human Body (IEC 479-1:1984, IEC 479-2:1987) (In Professional Package 47). 39 pp.

—Environmental Testing Equipment
BSI DD ENV 28041-93. 1993 Human Response to Vibration Measuring Instrumentation (ISO 8041: 1990) (E). 32 pp.
CEN ENV 28041-93. Human Response to Vibration—Measuring Instrumentation (ISO 8041: 1990). 26 pp.
ISO 8041-90. Human Response to Vibration—Measuring Instrumentation First Edition; (Corrigendum 1-1993) (CEN EN 28041: 1993). 30 pp.

—Glossaries
ISO 5805-81. Mechanical Vibration and Shock Affecting Man—Vocabulary First Edition. 11 pp.

—Safety
BSI BS 7085-89. 1989 Safety Aspects of Experiments in Which People are Exposed to Mechanical Vibration and Shock. 19 pp.

—Vibration
BSI BS 6472-92. 1992 Evaluation of Human Exposure to Vibration in Buildings (1Hz to 80Hz). 21 pp.
BSI BS 6472-84. 1984 Evaluation of Human Exposure to Vibration and Shock in Buildings (1Hz to 80Hz). 16 pp.
BSI BS 6611-85. 1985 Evaluation of the Response of Occupants of Fixed Structures, Especially Buildings and Offshore Structures, to Low-Frequency Horizontal Motion (0.063 Hz to 1 Hz). 12 pp.
BSI BS 6841-87. 1987 Measurement and Evaluation of Human Exposure to Whole-Body Mechanical Vibration and Repeated Shock. 24 pp.
BSI BS 6842-87. 1987 Measurement and Evaluation of Human Exposure to Vibration Transmitted to the Hand. 18 pp.
BSI BS 7439-91. 1991 Exchange of Information for the Analytical Assessment of Shock Resistance of Mechanical Systems (ISO 9688: 1990). 12 pp.
BSI BS 7482: Part 1-91. 1991 Instrumentation for the Measurement of Vibration Exposure of Human Beings Part 1: Specification for General Requirements for Instrumentation for Measuring the Vibration Applied to Human Beings. 11 pp.
BSI BS 7482: Part 2-91. 1991 Instrumentation for the Measurement of Vibration Exposure of Human Beings Part 2: Specification for Instrumentation for Measuring Vibration Transmitted to the Hand. 11 pp.
BSI BS 7482: Part 3-91. 1991 Instrumentation for the Measurement of Vibration Exposure of Human Beings Part 3: Specification for Instrumentation for Measuring Vibration Exposure to the Whole Body. 13 pp.
BSI DD ENV 25349-93. 1993 Mechanical Vibration—Guidelines for the Measurement and the Assessment of Human Esposure to Hand-Transmitted Vibration (ISO 5349: 1986) (F). 22 pp.
CEN PREN 1031-93. Measurement and Evaluation of Whole-Body Vibration—General Requirements. 11 pp.
CEN ENV 25349-92. Mechanical Vibration—Guidelines for the Measurement and the Assessment of Human Exposure to Hand-Transmitted Vibration; (ISO 5349: 1986). 15 pp.
DIN ENGL 45675 Pt 1-87. Exposure to Mechanical Vibration Transmitted to the Hand-Arm System; Principles of Measurement (Sept) (Superseded in Part by DIN EN 28662 Part 1). 8 pp.

Biomechanical Effects (Human Body) (Cont.)

—Vibration (Cont.)

ISO 2631 Pt 1-85. Evaluation of Human Exposure to Whole-Body Vibration—Part 1: General Requirements First Edition. 19 pp.

ISO 2631 Pt 2-89. Evaluation of Human Exposure to Whole-Body Vibration—Part 2: Continuous and Shock-Induced Vibration in Buildings (1 to 80 Hz) First Edition. 20 pp.

ISO 2631 Pt 3-85. Evaluation of Human Exposure to Whole-Body Vibration—Part 3: Evaluation of Exposure to Whole-Body Z-Axis Vertical Vibration in the Frequency Range 0,1 to 0,63 Hz First Edition. 6 pp.

ISO 5349-86. Mechanical Vibration—Guidelines for the Measurement and the Assessment of Human Exposure to Hand-Transmitted Vibration First Edition; (CEN ENV 25349:1992). 14 pp.

ISO 5805-81. Mechanical Vibration and Shock Affecting Man—Vocabulary First Edition. 11 pp.

ISO 5982-81. Vibration and Shock—Mechanical Driving Point Impedance of the Human Body First Edition. 10 pp.

ISO 6897-84. Guidelines for the Evaluation of the Response of Occupants of Fixed Structures, Especially Buildings and off-Shore Structures, to Low-Frequency Horizontal Motion (0,063 to 1 Hz) First Edition. 10 pp.

ISO 7962-87. Mechanical Vibration and Shock—Mechanical Transmissibility of the Human Body in the Z Direction First Edition. 8 pp.

ISO 9688-90. Mechanical Vibration and Shock—Analytical Methods of Assessing Shock Resistance of Mechanical Systems—Information Exchange Between Suppliers and Users of Analyses First Edition. 10 pp.

JIS B 4900-86. Method of Measurement and Description of Hand-Transmitted Vibration Level. 9 pp.

SNZ NZS/ISO 2631: Part 1-85. Evaluation of Human Exposure to Whole-Body Vibration Part 1: General Requirements. 17 pp.

SNZ NZS/ISO 2631: Part 2-89. Evaluation of Human Exposure to Whole-Body Vibration Part 2: Continuous and Shock-Induced Vibration in Buildings (1 to 80 Hz). 18 pp.

SNZ NZS/ISO 2631: Part 3-85. Evaluation of Human Exposure to Whole-Body Vibration Part 3: Evaluation of Exposure to Whole-Body Z-Axis Vertical Vibration in the Frequency Range 0,1 to 0, 63 Hz. 4 pp.

—Vibration—Machinery

CEN PREN 1032-93. Testing of Machinery in Order to Measure the Whole-Body Vibration Emission Value—General Requirements. 8 pp.

—Vibration—Machinery Design

CEN PREN 1030-1-93. Hand-Arm Vibration—Guidelines for Vibration Hazards Reduction—Part 1: Engineering Methods by Design of Machinery. 10 pp.

—Vibration—Safety

BSI BS 7085-89. 1989 Safety Aspects of Experiments in Which People are Exposed to Mechanical Vibration and Shock. 19 pp.

—Vibration Testers

BSI DD ENV 28041-93. 1993 Human Response to Vibration Measuring Instrumentation (ISO 8041: 1990) (E). 32 pp.

CEN ENV 28041-93. Human Response to Vibration—Measuring Instrumentation (ISO 8041: 1990). 26 pp.

ISO 8041-90. Human Response to Vibration—Measuring Instrumentation First Edition; (Corrigendum 1-1993) (CEN EN 28041: 1993). 30 pp.

Biomedical Water
Use: High Purity Water

Biotechnical Medicinal Products
Use: Medicinal Products—Biotechnical

Biotechnology

—Glossaries

JIS K 3600-89. Technical Terms for Biotechnology.

—Inventions—Patents

EC COM(88) 496-88. Proposal for a Council Directive on the Legal Protection of Biotechnological Inventions. 68 pp.

—Vials—Frozen Storage

JIS K 3603-90. Plastic Vials for Frozen Storage and Ultra Low-Temperature Preservation. 13 pp.

Bipolar Digital Circuits
Use: Digital Circuits—Bipolar

Bipolar Transistors

See Also: Germanium Transistors; NPN/PNP Transistors; NPN Transistors; PNP Transistors; Silicon Transistors; Transistors

BSI BS 6493: Sec 1.4-92. 1992 Semiconductor Devices Part 1: Discrete Devices Section 1.4: Recommendations for Microwave Diodes and Transistors (IEC 747-4: 1991). 83 pp.

CECC CECC 50 002 ISSUE 3-80. Blank Detail Specification: Ambient-Rated Bipolar Transistors for Low and High Frequency Amplification (En, Fr, Ge) AMD 1 (En, Ge). 39 pp.

CNS C6029-76. Method of Test for Transistors (Sep)(3997). 28 pp.

IEC 747 Pt 4-91. Semiconductor Devices Discrete Devices Part 4: Microwave Diodes and Transistors First Edition. 163 pp.

IECQ QC 750102/SU 0001-90. Detail Specification for Electronic Components High-Frequency Epitaxy-Planar Ambient-Rated Bipolar Transistors Type KT3I17A. 17 pp.

IECQ QC 750107-91. Semiconductor Devices Discrete Devices Part 7: Bipolar Transistors Section Four—Blank Detail Specification for Case-Rated Transistors for High-Frequency Amplification (IEC 747-7-4 ED 1). 37 pp.

—Amplifiers

BSI BS E9372-76. 1976 Amd 1 Blank Detail Specification: Ambient-Rated Bipolar-Transistors for Low and High Frequency Amplification. 16 pp.

BSI BS E9373-76. (WITHDRAWN) 1976 Amd 1 Blank Detail Specification: Case-Rated Bipolar Transistors for Low Frequency Amplification (Renumbered as BS EN 150003: 1993). 10 pp.

BSI BS E9377-76. (WITHDRAWN) 1976 Amd 1 Blank Detail Specification: Case-Rated Bipolar Transistors for High Frequency Amplification (Renumbered as BS EN 150007: 1993). 17 pp.

IEC 747 Pt 7-1-89. Semiconductor Devices Discrete Devices Part 7: Bipolar Transistors Section One—Blank Detail Specification for Ambient-Rated Bipolar Transistors for Low and High-Frequency Amplification First Edition (IECQ QC 750102). 35 pp.

IEC 747 Pt 7-2-89. Semiconductor Devices Discrete Devices Part 7: Bipolar Transistors Section Two—Blank Detail Specification for Case-Rated Bipolar Transistors for Low-Frequency Amplification First Edition (IECQ QC 750103). 31 pp.

IECQ QC 750102-89. Semiconductor Devices Discrete Devices Part 7: Bipolar Transistors Section One—Blank Detail Specification for Ambient-Rated Bipolar Transistors for Low and High-Frequency Amplification (IEC 747-7-1 ED 1). 35 pp.

IECQ QC 750102/CN 0001-88. Detail Specification for Electronic Components Bipolar Transistors for Ambient-Rated High-Frequency Amplification of Type 3DG130 Assessment Level II. 18 pp.

IECQ QC 750102/CN 0002-88. Detail Specification for Electronic Components Bipolar Transistors for Ambient-Rated, Forward A.G.C. Low-Noise, High-Frequency Amplification of Type 3DG79 Assessment Level II. 17 pp.

IECQ QC 750102/CN 0003-90. Detail Specification for Electronic Components Bipolar Transistors for Ambient-Rated High-Frequency Amplification of Type 3DG1815 Assessment Level II. 15 pp.

IECQ QC 750103-89. Semiconductor Devices Discrete Devices Part 7: Bipolar Transistors Section Two—Blank Detail Specification for Case-Rated Bipolar Transistors for Low-Frequency Amplification (IEC 747-7-2 ED 1). 31 pp.

IECQ QC 750103/CN 0001-92. Detail Specification for Electronic Components Case-Rated Bipolar Transistors for Low-Frequency Amplification Type 3DD870. 12 pp.

—Amplifiers—Quality Assurance

BSI BS EN 150003-93. 1993 Amd 2 Blank Detail Specification:u Case-Rated Bipolar Transistors for Low Frequency Amplification (AMD 7595) February 1993 (T). 14 pp.

BSI BS EN 150007-93. 1993 Amd 2 Blank Detail Specification: Case-Rated Bipolar Transistors for High Frequency Amplification (AMD 7598) February 1993 (T). 21 pp.

CENELEC EN 150 007-91. Blank Detail Specification: Case-Rated Bipolar Transistors for High Frequency Amplification. 11 pp.

—Current Gain

CNS C5068-80. Ranges and Conditions for Specifying Beta for Low Power, Audio Frequency Transistor (Aug)(6317).

—Electrical Impedance

CNS C6105-81. Method of Test for the Collector-Base Time Constant and for the Resistive Part of the Common-Emitter Input Impedance (Mar)(7012).

—Microwave—Preferred Products List

CECC CECC MUAHAG Vol 9 IS 3-90. Preferred Products List; Semiconductors (En, Fr, Ge) AMD 1 (En, Fr, Ge). 51 pp.

Bipolar Transistors (Cont.)

—Power

BSI BS 6493: Sec 1.7-89. 1989 Semiconductor Devices Part 1: Discrete Devices Section 1.7: Recommendations for Bipolar Transistors. 116 pp.

BSI BS 6493: Sec 1.7-01. 1989 Amd 1 Semiconductor Devices Part 1: Discrete Devices Section 1.7: Recommendations for Bipolar Transistors (IEC 747-7: 1988) (AMD 7206) September 15, 1992. 122 pp.

IEC 747 Pt 7-88. Semiconductor Discrete Devices and Integrated Circuits Part 7: Bipolar Transistors First Edition; (Amendment 1-1991). 242 pp.

MOD UK DSTAN 59-61: 80/013-78. Semiconductor Device Transistor Issue 1. 12 pp.

MOD UK DSTAN 59-61: 90/032-80. Semiconductor Device Transistor Issue 2. 22 pp.

MOD UK DSTAN 59-61: 90/075-01. Transistor Issue 1; Amendment 1. 13 pp.

MOD UK DSTAN 59-61: 90/076-01. Transistor Issue 1; Amendment 1. 13 pp.

MOD UK DSTAN 59-61: 90/078-01. Semiconductor Device Transistor Issue 1; Amendment 1. 14 pp.

MOD UK DSTAN 59-61: 90/080-01. Semiconductor Devices Transistors Issue 1; Amendment 1. 18 pp.

MOD UK DSTAN 59-61: 90/084-72. Transistor Issue 1. 9 pp.

MOD UK DSTAN 59-61: 90/097-72. Semiconductor Devices, Transistor Issue 1. 13 pp.

MOD UK DSTAN 59-61: 90/111-72. Semiconductor Device, Transistor Issue 1. 14 pp.

MOD UK DSTAN 59-61: 90/150-73. Semiconductor Device, Transistor Issue 1. 8 pp.

MOD UK DSTAN 59-61: 90/152-74. Transistors Issue 1. 18 pp.

MOD UK DSTAN 59-61: 90/156-74. Transistor Issue 1. 17 pp.

MOD UK DSTAN 59-61: 90/162-80. Semiconductor Device, Transistor Issue 1. 15 pp.

—Preferred Products List

CECC CECC MUAHAG Vol 9 IS 3-90. Preferred Products List; Semiconductors (En, Fr, Ge) AMD 1 (En, Fr, Ge). 51 pp.

—Quality Assurance

BSI BS E9372-76. 1976 Amd 1 Blank Detail Specification: Ambient-Rated Bipolar-Transistors for Low and High Frequency Amplification. 16 pp.

BSI BS E9373-76. (WITHDRAWN) 1976 Amd 1 Blank Detail Specification: Case-Rated Bipolar Transistors for Low Frequency Amplification (Renumbered as BS EN 150003: 1993). 10 pp.

BSI BS E9377-76. (WITHDRAWN) 1976 Amd 1 Blank Detail Specification: Case-Rated Bipolar Transistors for High Frequency Amplification (Renumbered as BS EN 150007: 1993). 17 pp.

BSI BS QC 750107-91. 1991 Case-Rated Bipolar Transistors for High-Frequency Amplifications (IEC 747-7-4: 1991) (T). 17 pp.

CENELEC EN 150 003-91. Blank Detail Specification: Case-Rated Bipolar Transistors for Low Frequency Amplification. 5 pp.

CENELEC EN 150 004-91. Blank Detail Specification: Bipolar Transistors for Switching Applications. 10 pp.

IEC 747 Pt 7-1-89. Semiconductor Devices Discrete Devices Part 7: Bipolar Transistors Section One—Blank Detail Specification for Ambient-Rated Bipolar Transistors for Low and High-Frequency Amplification First Edition (IECQ QC 750102). 35 pp.

IEC 747 Pt 7-2-89. Semiconductor Devices Discrete Devices Part 7: Bipolar Transistors Section Two—Blank Detail Specification for Case-Rated Bipolar Transistors for Low-Frequency Amplification First Edition (IECQ QC 750103). 31 pp.

IEC 747 Pt 7-4-91. Semiconductor Devices Discrete Devices Part 7: Bipolar Transistors Section Four—Blank Detail Specification for Case-Rated Bipolar Transistors for High-Frequency Amplification First Edition (IECQ QC 750107). 37 pp.

—Saturation Voltage

CNS C6136-81. Test Method for Transistor Collector Emitter Saturation Voltage (Nov)(8105).

—Switching

BSI BS 6493: Sec 1.7-89. 1989 Semiconductor Devices Part 1: Discrete Devices Section 1.7: Recommendations for Bipolar Transistors. 116 pp.

BSI BS 6493: Sec 1.7-01. 1989 Amd 1 Semiconductor Devices Part 1: Discrete Devices Section 1.7: Recommendations for Bipolar Transistors (IEC 747-7: 1988) (AMD 7206) September 15, 1992. 122 pp.

IEC 747 Pt 7-88. Semiconductor Discrete Devices and Integrated Circuits Part 7: Bipolar Transistors First Edition; (Amendment 1-1991). 242 pp.

IECQ QC 750104-91. Semiconductor Devices Discrete Devices Part 7: Bipolar Transistors Section Three—Blank Detail Specification for Bipolar Transistors for Switching Applications (IEC 747-7-3 ED 1). 39 pp.

INTERNATIONAL AND NON-U.S. NATIONAL STANDARDS
SUBJECT INDEX
Bistable

Bipolar Transistors (Cont.)
—Switching (Cont.)
IECQ QC 750104/CN 0001-89. Detail Specification for Electronic Components Bipolar Transistors for Switching Application of Type 3DK 106 Assessment Level II. 16 pp.
MOD UK DSTAN 59-61: 90/111-72. Semiconductor Device, Transistor Issue 1. 14 pp.

—Switching—Quality Assurance
BSI BS E9374-76. (WITHDRAWN) 1976 Amd 1 Blank Detail Specification: Bipolar Transistors for Switching Applications (Renumbered as BS EN 150004: 1993). 15 pp.
BSI BS QC 750104-91. 1991 Bipolar Transistors for Switching Applications (IEC 747-7-3: 1991) (T). 17 pp.
BSI BS EN 150004-93. 1993 Amd 2 Blank Detail Specification: Bipolar Transistors for Switching Applications (AMD 7596) February 1993 (T). 19 pp.
IEC 747 Pt 7-3-91. Semiconductor Devices Discrete Devices Part 7: Bipolar Transistors Section Three—Blank Detail Specification for Bipolar Transistors for Switching Applications First Edition (IECQ QC 750104). 39 pp.

—Time Constants
CNS C6105-81. Method of Test for the Collector-Base Time Constant and for the Resistive Part of the Common-Emitter Input Impedance (Mar)(7012).

2,2'-Bipyridine
Use: alpha,alpha'-Dipyridyl

Bird Strikes, Aircraft
Use: Aircraft Safety—Bird Strikes

2,2-Bis(Ethyl Ferrocenyl) Propane
—Propellants
MOD UK TS 10293. 2,2-Bis (Ethyl Ferrocenyl) Propane.

Bis(2-Ethylhexyl) Phthalate
Use: Di(2-Ethylhexyl) Phthalate

Bis (2-Ethylhexyl) Sebacate
Use For: Di (2-Ethylhexyl) Sebacate
CNS K1232-80. Di-(2-Ethyl-Hexyl) Sebacate (Apr)(5445).
CNS K6470-80. Method of Test for Di-(2 Ethyl-Hexyl) Sebacate (Apr)(5446).

Bis(1-Phenyl-3-methyl-5-pyrazolone)
JIS K 9545-78. Bis (1-phenyl-3-methyl-5-pyrazolone).

Biscuits
See Also: Bread
CNS N5218-86. Biscuits (Packaged) (Jul)(11635). 2 pp.

Bismuth
See Also: Bismuth Content Analysis; Metals

—Standard Solutions
JIS K 0017-83. Bismuth Standard Solution. 13 pp.

Bismuth Chloride
Use For: Bismuth Trichloride
CNS K7099-62. Chemical Reagent (Bismuth Chloride) (Bismuth Trichloride) (Sep)(1599).
JIS K 8156-78. Bismuth Chloride.

Bismuth Content Analysis
See Also: Bismuth

—Aluminum Alloys
JIS H 1364-71. Methods for Determination of Bismuth and Lead in Aluminium Alloy.

—Bearing Alloys—White Metals—Photometry
BSI BS 3338: Part 8-61. 1961 Methods for the Sampling and Analysis of Tin and Tin Alloys Part 8: Determination of Bismuth in Ingot Tin, Tin-Lead Solder and White Metal Bearing Alloys (Photometric Method). 8 pp.

—Copper Alloys—Spectrometry
ISO 5959-84. Copper and Copper Alloys—Determination of Bismuth Content—Diethyldithiocarbamate Spectrometric Method First Edition. 5 pp.

—Copper—Spectrometry
ISO 5959-84. Copper and Copper Alloys—Determination of Bismuth Content—Diethyldithiocarbamate Spectrometric Method First Edition. 5 pp.

Bismuth Content Analysis (Cont.)
—Electrolytic Copper—Atomic Absorption Spectrophotometry
BSI BS 7317: Part 3-90. 1990 Methods for Analysis of High Purity Copper Cathode Cu-CATH-1 Part 3: Method for Determination of Antimony, Arsenic, Bismuth, Selenium, Tellurium and Tin by Hydride Generation and Atomic Absorption Spectrophotometry. 12 pp.
BSI BS 7317: Part 4-90. 1990 Methods for Analysis of High Purity Copper Cathode Cu-CATH-1 Part 4: Method for Determination of Antimony, Arsenic, Bismuth, Lead, Selenium, Tellurium and Tin by Electrothermal Atomiza-tion Atomic Absorption Spectrophotometry. 9 pp.
BSI DD 95: Part 3-84. (WITHDRAWN) 1984 Analysis of Higher Purity Copper Cathode Cu-CATH-1 Part 3: Methods for Determination of Antimony, Arsenic, Bismuth, Selenium, Tellurium and Tin by Hydride Generation and Atomic Absorption Spectrophotometry. 14 pp.
BSI DD 95: Part 4-86. (WITHDRAWN) 1986 Analysis of Higher Purity Copper Cathode CU-CATH-1 Part 4: Method for Determination of Antimony, Arsenic, Bismuth, Lead, Selenium, Tellurium and Tin by Electrothermal Atomization Atomic. 9 pp.

—Electrolytic Copper—Hydride Generation
BSI BS 7317: Part 3-90. 1990 Methods for Analysis of High Purity Copper Cathode Cu-CATH-1 Part 3: Method for Determination of Antimony, Arsenic, Bismuth, Selenium, Tellurium and Tin by Hydride Generation and Atomic Absorption Spectrophotometry. 12 pp.
BSI DD 95: Part 3-84. (WITHDRAWN) 1984 Analysis of Higher Purity Copper Cathode Cu-CATH-1 Part 3: Methods for Determination of Antimony, Arsenic, Bismuth, Selenium, Tellurium and Tin by Hydride Generation and Atomic Absorption Spectrophotometry. 14 pp.

—Ferrotungsten
CNS G2155-83. Methods of Chemical Analysis for Bismuth in Ferrotungsten (Jan)(9866).

—Iron Ores—Absorptiometric Analysis
JIS M 8230-83. Methods for Determination of Bismuth in Iron Ores. 14 pp.

—Iron Ores—Atomic Absorption Spectrophotometry
JIS M 8230-83. Methods for Determination of Bismuth in Iron Ores. 14 pp.

—Lead Alloys—Photometry
BSI BS 3908: Part 3-67. 1967 Methods for the Sampling and Analysis of Lead and Lead Alloys Part 3: Bismuth in Lead and Lead Alloys (Photometric Method). 8 pp.

—Lead—Photometry
BSI BS 3908: Part 3-67. 1967 Methods for the Sampling and Analysis of Lead and Lead Alloys Part 3: Bismuth in Lead and Lead Alloys (Photometric Method). 8 pp.

—Lead—Polarographic Analysis
MOD UK M 9331/63. Determination of Bismuth, Copper, Cadmium, and Zinc in Pure Lead (Polarographic Method) (No Information) (Withdrawn).

—Nickel—Atomic Absorption Spectrometry
BSI BS 6783: Part 1-86. 1986 Sampling and Analysis of Nickel, Ferronickel and Nickel Alloys Part 1: Method for Determination of Silver, Bismuth, Cadmium, Cobalt, Copper, Iron, Manganese, Lead and Zinc in Nickel by Flame Atomic Absorption Spectrometry. 16 pp.
BSI BS 6783: Part 4-86. 1986 Sampling and Analysis of Nickel, Ferronickel and Nickel Alloys: Part 4: Method for Determination of Silver, Arsenic, Bismuth, Cadmium, Lead, Antimony, Selenium, Tin, Tellurium and Thallium in Nickel by Electro-thermal Atomic Absorption Spectrometry. 11 pp.
ISO 6351-85. Nickel—Determination of Silver, Bismuth, Cadmium Cobalt, Copper, Iron, Manganese, Lead and Zinc Contents—Flame Atomic Absorption Spectrometric Method First Edition. 13 pp.
ISO 7523-85. Nickel—Determination of Silver, Arsenic, Bismuth, Cadmium, Lead, Antimony, Selenium, Tin, Tellurium and Thallium Contents—Electrothermal Atomic Absorption Spectrometric Method First Edition. 10 pp.

—Ores
JIS M 8133-83. Methods for Determination of Bismuth in Ores.

—Ores—Gravimetric Analysis
CNS M3097-82. Method for Determining Bismuth in Ores (Gravimetric Method) (May)(8857).

Bismuth Content Analysis (Cont.)
—Ores—Volumetric Analysis
CNS M3098-82. Method for Determining Bismuth in Ores (Titration Method) (May)(8858).

—Solders—Photometry—Tin-Lead
BSI BS 3338: Part 8-61. 1961 Methods for the Sampling and Analysis of Tin and Tin Alloys Part 8: Determination of Bismuth in Ingot Tin, Tin-Lead Solder and White Metal Bearing Alloys (Photometric Method). 8 pp.

—Tin—Photometry
BSI BS 3338: Part 8-61. 1961 Methods for the Sampling and Analysis of Tin and Tin Alloys Part 8: Determination of Bismuth in Ingot Tin, Tin-Lead Solder and White Metal Bearing Alloys (Photometric Method). 8 pp.

Bismuth Ethyl Camphorate Content Analysis
—Cosmetics
CNS S2103-83. Method of Hygienic Test for Cosmetics Bismuth Salts (Jun)(10381).

Bismuth Inorganic Compounds
Scope Note: Use a more specific term See: Bismuth Chloride; Bismuth Nitrate; Bismuth Subcarbonate; Bismuth Subnitrate; Bismuth Trioxide

Bismuth Nitrate
See Also: Bismuth Subnitrate
CNS K7100-62. Chemical Reagent (Bismuth Nitrate) (Bismuth Trinitrate) (Sep)(1600).
JIS K 8566-61. Bismuth Nitrate.

Bismuth Salt Content Analysis
Use: Bismuth Ethyl Camphorate Content Analysis

Bismuth Sodium Trioxide
JIS K 8770-80. Bismuth Sodium Trioxide.

Bismuth Subcarbonate
CNS K7101-62. Chemical Reagent (Bismuth Subcarbonate) (Basic Bismuth Carbonate Bismuth Oxycarbonate) (Sep)(1601).

Bismuth Subnitrate
Use For: Basic Bismuth Nitrate See Also: Bismuth Nitrate
CNS K7102-62. Chemical Reagent (Bismuth Subnitrate) (Basic Bismuth Nitrate Bismuth Oxynitrate) (Sep)(1602).
JIS K 8169-61. Bismuth Nitrate, BASIC.

Bismuth Trichloride
Use: Bismuth Chloride

Bismuth Trioxide
MOD UK DSTAN 13-52-92. Bismuth (III) Oxide Issue 2. 8 pp.

Bistable Latches
See Also: Flip Flops; Latches (Circuits)

—2-Bit
CECC CECC 90 103-035 ISSUE 1-81. BS CECC 90 103-035; Dual 2-Bit Bistable Latches (En) AMD 1 (En). 20 pp.
CECC CECC MUAHAG Vol 7 IS 8-92. Preferred Products List; Active Microcircuits (En, Fe, Ge). 89 pp.

—4-Bit
CECC CECC 90 103-181 ISSUE 1-82. BS 90 103-181; 4-Bit Bistable Latches (En). 19 pp.
CECC CECC 90 109-678 ISSUE 1-87. Digital Integrated Circuits in Accordance with FS 90 109; 54/74 HC 75; 4 Bit Bistable Latch (En, Fr). 6 pp.
CECC CECC 90 109-743 ISSUE 1-87. Digital Integrated Circuits in Accordance with FS 90 109; 54 HCT 75, 74 HCT 75; 4-Bit Bistable Latch (En, Fr). 6 pp.
CECC CECC 90 109-806 ISSUE 1-88. Digital Integrated Circuits in Accordance with FS 90 109; 54/74 HC 77; 4-Bit Bistable Latch (En, Fr). 6 pp.
CECC CECC 90 109-880 ISSUE 1-90. Digital Integrated Circuits; Silicon Monolithic C MOS, Cavity or Non-Cavity Packages; Type(s); 54/74 HC 375 4-Bit Bistable Latch; Assessment Levels P, Y, L (En, Fr). 9 pp.

Bistable Multivibrators
Use: Flip Flops

Bistable Relays
Use For: Electromechanical Flip Flops; Flip Flop Relays; Relay Flip Flops See Also: Latching Relays; Relays

INDUSTRY STANDARDS

Bistable Relays (Cont.)

—All-Or-Nothing—Sensitive—Crystal Can—Severe Environments—PPL

CECC CECC MUAHAG Vol 5 IS 3-88. Preferred Products List; Relays (En, Fr, Ge). 45 pp.

—All-Or-Nothing—0-49 V DC—Crystal Can—Severe Environments—PPL

CECC CECC MUAHAG Vol 5 IS 3-88. Preferred Products List; Relays (En, Fr, Ge). 45 pp.

—All-Or-Nothing—0-49 V DC—Severe Environments—PPL

CECC CECC MUAHAG Vol 5 IS 3-88. Preferred Products List; Relays (En, Fr, Ge). 45 pp.

—All-Or-Nothing—50-599 V DC—Crystal Can—Severe Environments—PPL

CECC CECC MUAHAG Vol 5 IS 3-88. Preferred Products List; Relays (En, Fr, Ge). 45 pp.

—All-Or-Nothing—50-599 V DC—Severe Environments—PPL

CECC CECC MUAHAG Vol 5 IS 3-88. Preferred Products List; Relays (En, Fr, Ge). 45 pp.

Bistable Sequential Circuits

Use: Flip Flops

Bistables

Use: Flip Flops

Bisulfite Liquors

See Also: Papermaking Equipment

—Sulfite Pulps—Volumetric Analysis

CPPA J-1-74. Analysis of Bisulphite Liquor. 2 pp.

Bit Blanks

See Also: Bits (Tools)

—Core Drills

CNS M2030-80. Tungsten Carbide Bits and Bit Blanks for Core Drills (Mar)(5357).

JIS M 1403-82. Tungsten Carbide Bits and Bit Blanks for Core Drills.

Bit Error

See Also: Bit Error Rate; Bit Rates; Data Communication

—Data Circuits

CCITT RECMN O.153-89. Basic Parameters for the Measurement of Error Performance at Bit Rates Below the Primary Rate—Specifications for Measuring Equipment (Study Group IV) 5 pp. 5 pp.

—Digital Paths

CCITT RECMN O.152-89. Error Performance Measuring Equipment for 64 kBit/s Paths—Specifications for Measuring Equipment (Study Group IV) 3 pp. 3 pp.

—Digital Transmission Systems

CCITT RECMN O.151-89. Error Performance Measuring Equipment for Digital Systems at the Primary Bit Rate and Above—Specifications for Measuring Equipment (Study Group IV) 5 pp. 5 pp.

—Hypothetical Reference Circuits—Mobile Satellite Communications

CCIR QUESTION 86-1/8-90. Performance Objectives for Mobile-Satellite Services—Questions Concerning Study Group 8 —Mobile, Radiodetermination, Amateur and Related Satellite Services. 1 p.

—Sound-Program Signals

CCIR Report 968-82. Methods of Measuring the Bit-Error Distortion of Digital Coded Sound-Programme Signals—Section CMTT D—Methods of Operation and Assessment of Performance of Sound-Programme Transmission Channels. 2 pp.

Bit Error Rate

Use For: BER *See Also:* Bit Error; Bit Rates

CCIR Report 865-82. Improvement in Bit Error-Rate by the Use of Spread Spectrum Techniques—Section 3Cb —Data Transmission. 3 pp.

CCIR RECMN 700-90. Error Performance and Availability Measurement Algorithm for Digital Radio-Relay Links at the System Bit Rate Interface—Section 9D1—Digital Systems. 3 pp.

CCIR RECMN 700-1-91. Error Performance and Availability Measurement Algorithm for Digital Radio-Relay Links at the Bit Rate Interface—Section 9D1—Digital Systems. 4 pp.

Bit Error Rate (Cont.)

—Hypothetical Reference Digital Path—Fixed Satellite Services

CCIR RECMN 522-3-90. Allowable Bit Error Ratios at the Output of the Hypothetical Reference Digital Path for Systems in the Fixed-Satellite Service Using Pulse-Code Modulation for Telephony —Section 4B2—Performance and Availability. 2 pp.

CCIR RECMN 522-4-92. Allowable Bit Error Ratios at the Output of the Hypothetical Reference Digital Path for Systems in the Fixed-Satellite Service Using Pulse-Code Modulation for Telephony —Section 4B2—Performance and Availability. 8 pp.

—Hypothetical Reference Digital Paths—Radio Relay Systems—ISDN

CCIR RECMN 594-2-90. Allowable Bit Error Ratios at the Output of the Hypothetical Reference Digital Path for Radio-Relay Systems Which May Form Part of an Integrated Services Digital Network—Section 9A—Performance Objectives, Propagation and Interference Effects. 2 pp.

CCIR RECMN 594-3-91. Allowable Bit Error Ratios at the Output of the Hypothetical Reference Digital Path for Radio-Relay Systems Which May Form Part of an Integrated Services Digital Network—Section 9A—Performance Objectives, Propagation and Interference Effects. 4 pp.

CCIR RECMN 696-90. Error Performance and Availability Objectives for Hypothetical Reference Digital Sect. Utilizing Digital Radio-Relay Systems Forming Part or All of the Medium Grade Portion of an ISDN Connection—Section 9A—Performance Objectives, Propagation and Interference Effects. 3 pp.

CCIR RECMN 696-1-91. Error Performance and Availability Objectives for Hypothetical Reference Digital Sect. Utilizing Digital Radio-Relay Systems Forming Part or All of the Medium-Grade Portion of an ISDN Connection—Section 9A—Performance Objectives, Propagation and Interference Effects. 4 pp.

—Radio Links—Radio Relay Systems—ISDN

CCIR RECMN 634-1-90. Error Performance Objectives for Real Digital Radio-Relay Links Forming Part of a High-Grade Circuit Within an Integrated Services Digital Network—Section 9A—Performance Objectives, Propagation and Interference Effects. 2 pp.

CCIR RECMN 634-2-91. Error Performance Objectives for Real Digital Radio-Relay Links Forming Part of a High-Grade Circuit Within an Integrated Services Digital Network—Section 9A—Performance Objectives, Propagation and Interference Effects. 12 pp.

—Radio Relay Systems

CCIR Report 930-2-90. Performance Objectives for Digital Radio-Relay Systems—Section 9A—Performance Objectives, Propagation and Interference Effects. 15 pp.

CCIR Report 613-4-90. Performance Measurements for Digital Radio-Relay Systems—Section 9D—Maintenance. 9 pp.

—Radio Relay Systems—Integrated Services Digital Networks

CCIR Report 1052-1-90. Error Performance and Availability Objectives for Digital Radio-Relay Systems Used in the "Medium Grade" Portion of an ISDN Connection—Section 9A—Performance Objectives, Propagation and Interference Effects. 6 pp.

CCIR Report 1053-1-90. Error Performance and Availability Objectives for Digital Radio-Relay Systems Used in the Local-Grade Portion of an ISDN Connection—Section 9A—Performance Objectives, Propagation and Interference Effects. 5 pp.

—Radio Relay Systems—ISDN

CCIR RECMN 697-90. Error Performance Objectives for the Local-Grade Portion at Each End of an ISDN Connection Utilizing Digital Radio-Relay Systems—Section 9A—Performance Objectives, Propagation and Interference Effects. 2 pp.

CCIR RECMN 697-1-91. Error Performance and Availability Objectives for the Local-Grade Portion at Each End of an ISDN Connection Utilizing Digital Radio-Relay Systems—Section 9A—Performance Objectives, Propagation and Interference Effects. 5 pp.

Bit Rate Converters

Use: Demultiplexers

Bit Rates

See Also: Bit Error; Bit Error Rate; Data Communication; Data Links; Data Transfer Rates; Data Transmission

Bit Rates (Cont.)

—Digital Hierarchy

CCITT RECMN G.702-89. Digital Hierarchy Bit Rates—General Aspects of Digital Transmission Systems; Terminal Equipments (Study Groups XV and XVIII) 4 pp. 4 pp.

—Digital Hierarchy—Interfaces—Electrical Characteristics

CCITT RECMN G.703-91. Physical/Electrical Characteristics of Hierarchical Digital Interfaces (Study Group XVIII) 42 pp. 42 pp.

CCITT RECMN G.703-89. Physical/Electrical Characteristics of Hierarchical Digital Interfaces—General Aspects of Digital Transmission Systems; Terminal Equipments (Study Groups XV and XVIII) 31 pp. 31 pp.

—Digital Hierarchy—Interfaces—Physical Characteristics

CCITT RECMN G.703-91. Physical/Electrical Characteristics of Hierarchical Digital Interfaces (Study Group XVIII) 42 pp. 42 pp.

CCITT RECMN G.703-89. Physical/Electrical Characteristics of Hierarchical Digital Interfaces—General Aspects of Digital Transmission Systems; Terminal Equipments (Study Groups XV and XVIII) 31 pp. 31 pp.

—Digital Line Sections—3152 kbit/s

CCITT RECMN G.931-89. Digital Line Sections at 3152 kBit/s—Digital Networks, Digital Sections and Digital Line Systems (Study Groups XV and XVIII) 2 pp. 2 pp.

—Digital Multiplexers—139,264 kbit/s

CCITT RECMN G.751-89. Digt. Multiplex Eqpts. Operating at the Third Order Bit Rate of 34 368 kbit/s and the Fourth Order Bit Rate of 139 264 kbit/s and Using Positive Justification—General Aspects of Digital Transmission Systems; Terminal Equipments (Study Groups XV and XVIII) 13 pp. 13 pp.

CCITT RECMN G.754-89. Fourth Order Digital Multiplex Equipment Operating at 139 264 kbit/s and Using Positive/Zero/Negative Justification—General Aspects of Digital Transmission Systems; Terminal Equipments (Study Groups XV and XVIII) 3 pp. 3 pp.

CCITT RECMN G.755-89. Digital Multiplex Eqpt. Operating at 139 264 kBit/s and Multiplexing Three Tributaries at 44 736 kBit/s—General Aspects of Digital Transmission Systems; Terminal Equipments (Study Groups XV and XVIII) 7 pp. 7 pp.

—Digital Multiplexers—2048 kbit/s

CCITT RECMN G.747-89. Second Order Digital Multiplex Equipment Operating at 6312 kbit/s and Multiplexing Three Tributaries at 2048 kbit/s—General Aspects of Digital Transmission Systems; Terminal Equipments (Study Groups XV and XVIII) 7 pp. 7 pp.

—Digital Multiplexers—34,368 kbit/s

CCITT RECMN G.751-89. Digt. Multiplex Eqpts. Operating at the Third Order Bit Rate of 34 368 kbit/s and the Fourth Order Bit Rate of 139 264 kbit/s and Using Positive Justification—General Aspects of Digital Transmission Systems; Terminal Equipments (Study Groups XV and XVIII) 13 pp. 13 pp.

CCITT RECMN G.753-89. Third Order Digital Multiplex Equipment Operating at 34 368 kbit/s and Using Positive/Zero/Negative Justification—General Aspects of Digital Transmission Systems; Terminal Equipments (Study Groups XV and XVIII) 4 pp. 4 pp.

—Digital Multiplexers—44,736 kbit/s

CCITT RECMN G.755-89. Digital Multiplex Eqpt. Operating at 139 264 kBit/s and Multiplexing Three Tributaries at 44 736 kBit/s—General Aspects of Digital Transmission Systems; Terminal Equipments (Study Groups XV and XVIII) 7 pp. 7 pp.

—Digital Multiplexers—6312 kbit/s

CCITT RECMN G.747-89. Second Order Digital Multiplex Equipment Operating at 6312 kbit/s and Multiplexing Three Tributaries at 2048 kbit/s—General Aspects of Digital Transmission Systems; Terminal Equipments (Study Groups XV and XVIII) 7 pp. 7 pp.

CCITT RECMN G.752-89. Characteristics of Digital Multiplex Equipments Based on a Second Order Bit Rate of 6312 kbit/s and Using Positive Justification—General Aspects of Digital Transmission Systems; Terminal Equipments (Study Groups XV and XVIII) 10 pp. 10 pp.

—Digital Sections—2048 kbit/s

CCITT RECMN G.921-89. Digital Sections Based on the 2048 kBit/s Hierarchy—Digital Networks, Digital Sections and Digital Line Systems (Study Groups XV and XVIII) 5 pp. 5 pp.

Bit Rates (Cont.)

—Land Mobile Services
CCIR RECMN 623-86. Data Transmission Bit Rates and Modulation Techniques in the Land Mobile Service—Section 8A—Land Mobile Service and Related Subjects. 1 p.

—Pulse Code Modulation Multiplexers
CCITT RECMN G.741-89. General Considerations on Second Order Multiplex Equipments—General Aspects of Digital Transmission Systems; Terminal Equipments (Study Groups XV and XVIII) 1 pp. 1 p.

—Pulse Code Modulation Multiplexers—6312 kbit/s
CCITT RECMN G.743-89. Second Order Digital Multiplex Equipment Operating at 6312 kbit/s and Using Positive Justification—General Aspects of Digital Transmission Systems; Terminal Equipments (Study Groups XV and XVIII) 5 pp. 5 pp.
CCITT RECMN G.746-89. Characteristics of Second Order PCM Multiplex Equipment Operating at 6312 kbit/s—General Aspects of Digital Transmission Systems; Terminal Equipments (Study Groups XV and XVIII) 3 pp. 3 pp.

—Pulse Code Modulation Multiplexers—8448 kbit/s
CCITT RECMN G.742-89. Second Order Digital Multiplex Equipment Operating at 8448 kbit/s and Using Positive Justification—General Aspects of Digital Transmission Systems; Terminal Equipments (Study Groups XV and XVIII) 5 pp. 5 pp.
CCITT RECMN G.744-89. Second Order PCM Multiplex Equipment Operating at 8448 kbit/s—General Aspects of Digital Transmission Systems; Terminal Equipments (Study Groups XV and XVIII) 5 pp. 5 pp.
CCITT RECMN G.745-89. Second Order Digital Multiplex Equipment Operating at 8448 kbit/s and Using Positive/Zero/Negative Justification—General Aspects of Digital Transmission Systems; Terminal Equipments (Study Groups XV and XVIII) 4 pp. 4 pp.

—Reduction—Television Signals
CCIR Report 1089-1-90. Bit-Rate Reduction for Digital Television Signals—Section 11F—Digital Methods of Transmitting Television Information. 34 pp.
CCIR QUESTION 44/11-90. Bit-Rate Reduction and Associated Quality Parameters for Digital Television Signals—Questions Concerning Study Group 11—Broadcasting Service (Television). 1 p.

—Reduction—Television Signals—Encoding
CCIR QUESTION 81/11-90. Reduction in the Bit Rate in the Digital Coding of Colour Television Signals—Questions Concerning Study Group 11—Broadcasting Service (Television). 1 p.

—Reduction—Television Signals—HDTV—Encoding
CCIR RECMN 788-92. Coding Rate for the Wide RF-Band HDTV Broadcasting-Satellite Service. 1 p.

—Reduction—Television Signals—Quality Assurance
CCIR QUESTION 82/11-90. Subjective and Objective Quality Parameters for the Bit Rate Reduction of Digital Television Signals—Questions Concerning Study Group 11—Broadcasting Service (Television). 1 p.

—Synchronous Digital Hierarchy
CCITT RECMN G.707-91. Synchronous Digital Hierarchy Bit Rates (Study Group XVIII) 5 pp. 5 pp.
CCITT RECMN G.707-89. Synchronous Digital Hierarchy Bit Rates—General Aspects of Digital Transmission Systems; Terminal Equipments (Study Groups XV and XVIII) 2 pp. 2 pp.

—Television Transmission—B-ISDN
CCIR QUESTION 39/CMTT-90. Requirements for Channel Capacities Above the H4 Rate for High-Bit-Rate Television Services—Questions Concerning the CMTT CCIR/CCITT Joint Study Group for Television and Sound Transmission. 1 p.

—Television Transmission—140 Mbit/s
CCIR RECMN 721-90. Transmission of Component-Coded Digital Television Signals for Contribution-Quality Applications at Bit Rates Near 140 Mbit/s—Section CMTT A—Television Transmission Standards and Performance Objectives. 12 pp.
CCIR RECMN 721-1-92. Transmission of Component-Coded Digital Television Signals for Contribution-Quality Applications at Bit Rates Near 140 Mbit/s—Section CMTT A—Television Transmission Standards and Performance Objectives. 14 pp.

Bit Rates (Cont.)

—Television Transmission—140 Mbit/s (Cont.)
CCIR Report 1234-90. Digital Transmission of Component-Coded Television Signals at Bit Rates near 68 Mbit/s and 140 Mbit/s—Section CMTT A—Television Transmission Standards and Performance Objectives. 12 pp.

—Television Transmission—30-34 Mbit/s
CCIR Report 1235-90. Digital Transmission of Component-Coded Television Signals at 30-34 Mbit/s and 45 Mbit/s—Section CMTT A—Television Transmission Standards and Performance Objectives. 43 pp.

—Television Transmission—30-45 Mbit/s
CCIR RECMN 723-90. Transmission of Component-Coded Digital Television Signals for Contribution-Quality Applications at the Third Hierarchical Level of CCITT Recommendation G.702—Section CMTT A—Television Transmission Standards and Performance Objectives. 14 pp.
CENELEC PRETS 300 174-91. Network Aspects Digital Coding of Component Television Signals for Contribution Quality Applications in the Range 34-45 Mbit/s. 84 pp.
CENELEC ETS 300 174-92. Network Aspects (NA); Digital Coding of Component Television Signals for Contribution Quality Applications in the Range 34-45 Mbit/s. 83 pp.
ETSI PRETS 300 174-91. Network Aspects Digital Coding of Component Television Signals for Contribution Quality Applications in the Range 34-45 Mbit/s.
ETSI ETS 300 174-92. Network Aspects (NA); Digital Coding of Component Television Signals for Contribution Quality Applications in the Range 34-45 Mbit/s. 83 pp.
ETSI PRETS 300 174-91. Network Aspects Digital Coding of Component Television Signals for Contribution Quality Applications in the Range 34-45 Mbit/s. 84 pp.

—Television Transmission—45 Mbit/s
CCIR Report 1235-90. Digital Transmission of Component-Coded Television Signals at 30-34 Mbit/s and 45 Mbit/s—Section CMTT A—Television Transmission Standards and Performance Objectives. 43 pp.

—Television Transmission—68 Mbit/s
CCIR Report 1234-90. Digital Transmission of Component-Coded Television Signals at Bit Rates near 68 Mbit/s and 140 Mbit/s—Section CMTT A—Television Transmission Standards and Performance Objectives. 12 pp.

BITES

Use For: Backward Interworking Telephone Events
See Also: Signaling Systems
CCITT FASCICLE VI.6-88. Interworking of Signalling Systems Recommendations Q.601—Q.699. 259 pp.
CCITT RECMN Q.603-89. Events—Interworking of Signalling Systems (Study Group XI) 1 pp. 1 p.

Bits (Tools)

Scope Note: For additional listings, use a more specific term *See Also:* Auger Bits; Bit Blanks; Core Bits; Cutting Tools; Drill Bits; Drill Pipes; Drill Steels; Drilling Equipment; Drills; Hand Tools; Insert Bits; Percussion Drills; Reamers; Screwdriver Bits; Tools
CNS B1009-53. Diameters of Drill Heads (for Drilling Bolt Holes) (Feb)(211) (R 1970).

—High Speed—Steel—Tool Holders
DIN ENGL 4964 Pt 1-62. High Speed Steel Tool Holder Bits (Sept). 3 pp.

—High Speed Steels
BSI BS 1296: Part 4-78. 1978 Single Point Cutting Tools Part 4: High Speed Tool Bits. 6 pp.
CNS B3309-79. High Speed Steel Tool Bits (Nov)(5016).
ISO 5421-77. Ground High Speed Steel Tool Bits First Edition. 4 pp.
JIS B 4151-87. Ground High Speed Steel Tool Bits. 13 pp.

—Socket Spanners
MOD UK DSTAN 51-11: Part 26-80. Hand Tools, General Purpose Part 26: Bits and Attachments for Socket Spanners and Screwdriver Type Tools, and Bits for Spiral Ratchet Screwdrivers Issue 1. 18 pp.

Bitter Melons

See Also: Food; Fruits; Watermelons

—Grading
CNS N1102-82. Grades of Bitter Melon (Nov)(9630). 2 pp.

Bitter Orange Oil

See Also: Essential Oils

—Quality Assurance
ISO 9844-91. Oil of Bitter Orange (Citrus Aurantium Linnaeus ssp. Aurantium) First Edition. 7 pp.

Bitter Orange Petitgrain Oil

Use: Petitgrain Oil

Bitts

Use For: Cross Bitts *See Also:* Mooring Hardware; Ships
JIS F 1012-84. Yacht's Cross Bitts.
JIS F 2052-87. Recess Type Tug Bitts.
JIS F 2804-76. Ships' Cross Bitts.

—Tugboats
CNS F3126-81. Double Type Cross Bitts for Tugboat (Aug)(7788).
JIS F 2051-76. Double Type Cross Bitts for Tugboat. 5 pp.

Bitumen Content Analysis

—Asphalt Paving Materials
SAA AS 2891.3.1-91. Methods of Sampling and Testing Asphalt—Part 3: Bitumen Content and Aggregate Grading—Part 3.1: Reflux Method. 4 pp.
SAA AS 2891.3.2-91. Methods of Sampling and Testing Asphalt—Part 3: Bitumen Content and Aggregate Grading—Part 3.2: Centrifugal Extraction. 4 pp.
SAA AS 2891.3.3-91. Methods of Sampling and Testing Asphalt—Part 3: Bitumen Content and Aggregate Grading—Part 3.3: Pressure Filter Method. 4 pp.

Bitumens

Use For: Bituminous Materials; Cutback Bitumens; Fluxed Bitumens *See Also:* Bituminous Cements; Bituminous Coatings; Coal Tar; Construction Materials; Emulsified Asphalts; Ethylene Copolymer Bitumen Coatings; Paving Materials; Tars
BSI BS 434: Part 1-84. 1984 Amd 3 Bitumen Road Emulsions (Anionic and Cationic) Part 1: Specification for Bitumen Road Emulsions (AMD 6376) June 28, 1991. 15 pp.
BSI BS 434: Part 2-84. 1984 Amd 1 Bitumen Road Emulsions (Anionic and Cationic) Part 2: Code of Practice for Use of Bitumen Road Emulsions. 16 pp.
BSI BS 3690: Part 1-89. 1989 Bitumens for Building and Civil Engineering Part 1: Bitumens for Roads and Other Paved Areas. 8 pp.
BSI BS 3690: Part 1-01. 1989 Amd 1 Bitumens for Building and Civil Engineering Part 1: Specification for Bitumens for Roads and Other Paved Areas (AMD 7316) November 15, 1992. 9 pp.
BSI BS 3690: Part 1-82. (WITHDRAWN) 1982 Bitumens for Building and Civil Engineering Part 1: Specification for Bitumens for Road Purposes. 7 pp.
BSI BS 3690: Part 3-90. 1990 Bitumens for Building and Civil Engineering Part 3: Mixtures of Bitumen with Pitch, Tar and Trinidad Lake Asphalt. 10 pp.
BSI BS 3690: Part 3-83. (WITHDRAWN) 1983 Bitumens for Building and Civil Engineering Part 3: Specification for Bitumen Mixtures. 10 pp.
MOD UK DSTAN 68-138-92. Bitumen Grades 1, 2 & 3 Issue 1. 13 pp.
MOD UK DSTAN 68-138-01. Bitumen Grades 1, 2 & 3 Issue 1; Amendment 1. 14 pp.
MOD UK CS 1471D. Bitumen, Special LF Quality (Withdrawn).
SAA AS 1160-88. Bitumen Emulsions for Construction and Maintenance of Pavements (superseded by AS 4131(INT)—1993, but will Remain Current). 20 pp.

—Asphaltene Content
BSI BS 2000: Part 143-93. 1993 Methods of Test for Petroleum and Its Products Part 143: Determination of Asphaltenes (Heptane Insolubles) (W). 6 pp.
BSI BS 2000: Part 143-85. 1985 Petroleum and Its Products Part 143: Asphaltenes in Petroleum Products (Precipitation with Normal Heptane). 8 pp.

—Binders
DIN ENGL 1995 Pt 1-89. Bituminous Binders; Road Bitumen; Requirements (Oct). 3 pp.
DIN ENGL 1995 Pt 2-89. Bituminous Binders; Cutback Bitumen; Requirements (Oct). 2 pp.
DIN ENGL 1995 Pt 4-89. Bituminous Binders; Dissolved Bitumen; Requirements (Oct). 2 pp.
DIN ENGL 1995 Pt 5-89. Bituminous Binders; Road Tars, Pitch-Bitumen Mixtures and Cold Pitch Solutions; Requirements (Oct). 5 pp.

—Binders (Materials)—Sampling
BSI BS 3195: Part 3-87. 1987 Sampling Petroleum Products Part 3: Method for Sampling Bituminous Binders. 15 pp.
BSI DD 193-91. 1991 Recovery of Bitumen Binders by Dichloromethane Extraction (Rotary Film Evaporator Method). 10 pp.

Bitumens (Cont.)

—Binders (Materials)—Solubility
BSI BS 2000: Part 47-93. 1993 Methods of Test for Petroleum and Its Products Part 47: Solubility of Bituminous Binders. 4 pp.

BSI BS 2000: Part 47-83. 1983 Petroleum and Its Products Part 47: Solubility of Bituminous Binders. 4 pp.

—Binders (Materials)—Stability Testing
BSI BS 598: Part 106-90. 1990 Sampling and Examination of Bituminous Mixtures for Roads and Other Paved Areas Part 106: Method for the Determination of the Stability Index of Pitch-Bitumen Binders (Supersedes BS 598: Part 3: 1985). 7 pp.

—Binders (Materials—Viscosity
DIN ENGL 52007 Pt 1-80. Testing of Bituminous Binders; Determination of Viscosity; General Principles and Evaluation (Dec). 5 pp.

DIN ENGL 52007 Pt 2-80. Testing of Bituminous Binders; Determination of Viscosity; Measurement by Drawn-Sphere Viscometer (Dec). 5 pp.

—Binders—Emulsions
DIN ENGL 1995 Pt 3-89. Bituminous Binders; Bitumen Emulsions; Requirements (Oct). 3 pp.

—Binders—Flow Rate
DIN ENGL 52023 Pt 1-80. Testing of Bituminous Binders; Determination of Flow Time by the Tar Efflux Viscometer; Method of Measurement (Dec). 4 pp.

—Binders—Paraffin Wax Content
DIN ENGL 52015-80. Testing of Bituminous Binders; Determination of Paraffin Wax Content (Dec). 4 pp.

—Bulk Density
BSI BS 598: Part 104-89. 1989 Amd 1 Sampling and Examination of Bituminous Mixtures for Roads and Other Paved Areas Part 104: Method of Test for the Determination of Density and Compaction (AMD 6738) September 30, 1991 (Supersedes BS 598: Part 3: 1985). 17 pp.

CNS A3149-87. Method of Test for Bulk Specific Gravity and Density of Compacted Bituminous Mixtures Using Paraffin-Coated Specimens (Aug)(8757).

—Comprehensive Testing
BSI BS 598: Part 2-88. 1988 Amd 3 Sampling and Examination of Bituminous Mixtures for Roads and Other Paved Areas: Part 2: Methods for Analytical Testing (AMD 6053) March 30, 1990. 28 pp.

BSI BS 598: Part 2-74. (WITHDRAWN) 1974 Amd 3 Sampling and Examination of Bituminous Mixtures for Roads and Other Paved Areas Part 2: Methods for Analytical Testing (AMD 6053) March 30, 1990 (Superseded by BS 598: Part 101:87, 102:89, BS 2000: Part 105: 1991). 34 pp.

BSI BS 598: Part 102-89. 1989 Amd 1 Sampling and Examination of Bituminous Mixtures for Roads and Other Paved Areas Part 102: Analytical Test Methods (AMD 6585) November 30, 1990. 28 pp.

BSI BS 598: Part 102-02. 1989 Amd 2 Sampling and Examination of Bituminous Mixtures for Roads and Other Paved Areas Part 102: Analytical Test Methods (AMD 7534) December 15, 1992 (Supersedes BS 598: part 2: 1974). 29 pp.

DIN ENGL 1996 Pt 1-74. Testing of Bituminous Materials for Road Building and Related Purposes; General, Synopsis and Indications Relating to the Evaluation of the Tests (Dec). 8 pp.

—Compression Testing
BSI BS 598: Part 104-89. 1989 Amd 1 Sampling and Examination of Bituminous Mixtures for Roads and Other Paved Areas Part 104: Method of Test for the Determination of Density and Compaction (AMD 6738) September 30, 1991 (Supersedes BS 598: Part 3: 1985). 17 pp.

—Compression Testing—Marshall Stability
DIN ENGL 1996 Pt 11-81. Testing of Bituminous Materials for Road Building and Related Purposes; Determination of Marshall Stability and of Marshall Flow Value (July). 4 pp.

—Deformation
BSI DD 185-90. 1990 Method for Determination of Creep Stiffness of Bitumen Aggregate Mixtures Subject to Confined Uniaxial Loading. 12 pp.

—Deformation—Marshall Flow Value
DIN ENGL 1996 Pt 11-81. Testing of Bituminous Materials for Road Building and Related Purposes; Determination of Marshall Stability and of Marshall Flow Value (July). 4 pp.

Bitumens (Cont.)

—Density
CNS A3149-87. Method of Test for Bulk Specific Gravity and Density of Compacted Bituminous Mixtures Using Paraffin-Coated Specimens (Aug)(8757).

CNS A3151-87. Method of Test for Bulk Specific Gravity and Density of Compacted Bituminous Mixtures Using Saturated Surface-Dry Specimens (Aug)(8759).

—Dimensional Stability
CEN PREN 1107-93. Bitumen Sheets for Waterproofing—Determination of Dimensional Stability at Elevated Temperature. 11 pp.

DIN ENGL 1996 Pt 17-90. Testing of Asphalt; Determination of Dimensional Stability When Heated (Nussel Deformation Index) (Nov). 2 pp.

—Distillation Methods
BSI BS 2000: Part 27-93. 1993 Methods of Test for Petroleum and Its Products Part 27: Determination of Distillation Characteristics of Cutback Bitumen (Supersedes BS 4453: 1976). 6 pp.

BSI BS 2000: Part 27-82. 1982 Petroleum and Its Products Part 27: Distillation of Cut-Back Asphaltic (Bituminous) Products. 8 pp.

—Ductility
CNS K6756-83. Method of Test for Ductility of Bituminous Materials (Mar)(10091).

DIN ENGL 52013-85. Testing of Bitumen; Determination of Ductility (July). 5 pp.

—Fillers—Electrical Equipment
BSI BS 1858-73. 1973 Amd 1 Bitumen Based Filing Compounds for Electrical Purposes. 24 pp.

—Fillers—Softening Points
DIN ENGL 52096-87. Testing of Fillers for Road Construction; Testing the Stiffening Action of Fillers on Bitumen (Apr). 3 pp.

—Fillers—Stiffness Testing
DIN ENGL 52096-87. Testing of Fillers for Road Construction; Testing the Stiffening Action of Fillers on Bitumen (Apr). 3 pp.

—Flash Point—Closed Cup Method
BSI BS 2000: Part 113-82. (WITHDRAWN) 1982 Petroleum and Its Products Part 113: Flash Point (Closed) of Cutback Bitumen. 4 pp.

—Flexibility
CEN PREN 1109-93. Bitumen Sheets for Waterproofing—Determination of Flexibility at Low Temperature. 11 pp.

—Float Testing
CNS K6798-83. Method of Test for Float Test of Bituminous Materials (Jul)(10459).

—Form Stability
CEN PREN 1108-93. Bitumen Sheets for Waterproofing—Determination of Formstability Under Cyclical Temperature Changes. 11 pp.

—Heaters
BSI BS 1676-70. 1970 Heaters for Tar and Bitumen (Mobile and Transportable). 12 pp.

—Highway Transportation—Emergency Procedures
SAA AS 1678.3T1-87. Emergency Procedure Guide —Transport Group Text EPGs for Class 3 Substances—Flammable Liquids—Part 3T1: Bitumen Products.

—Hoses
BSI BS 6130-81. 1981 Amd 1 Hose and Hose Assemblies for Asphalt and Bitumen. 5 pp.

—Immersion Testing
DIN ENGL 1996 Pt 10-77. Testing of Bituminous Materials for Road Building and Related Purposes; Testing of Behaviour of Mix in Respect of Immersion in Water (Dec). 3 pp.

—Industrial
BSI BS 3690: Part 2-89. 1989 Bitumens for Building and Civil Engineering Part 2: Bitumens for Industrial Purposes. 9 pp.

BSI BS 3690: Part 2-82. (WITHDRAWN) 1982 Bitumens for Building and Civil Engineering Part 2: Specification for Bitumens for Industrial Purposes. 8 pp.

—Loss of Mass—High Temperature Testing
BSI BS 2000: Part 45-93. 1993 Methods of Test for Petroleum and Its Products Part 45: Determination of Loss on Heating of Bitumen and Flux Oil. 4 pp.

BSI BS 2000: Part 45-82. 1982 Petroleum and Its Products Part 45: Loss on Heating of Bitumen and Flux Oil. 4 pp.

Bitumens (Cont.)

—Moisture Content—Distillation Methods
CNS K6339-73. Method of Test for Moisture Content in Petroleum Products and Bituminous Materials by Distillation (Feb)(3517). 4 pp.

—Neutralization Number
BSI BS 2000: Part 213-93. 1993 Methods of Test for Petroleum and Its Products Part 213: Determination of Neutralization Value of Bitumen—Colour Indicator Titration Method (W). 3 pp.

BSI BS 2000: Part 213-83. 1983 Petroleum and Its Products Part 213: Acidity of Bitumen (Neutralization Value). 4 pp.

—Pavements
SAA AS 2008-80. Residual Bitumen for Pavements. 8 pp.

—Penetration Resistance
BSI BS 2000: Part 49-93. 1993 Methods of Test for Petroleum and Its Products Part 49: Determination of Needle Penetration of Bituminous Material. 6 pp.

BSI BS 2000: Part 49-83. 1983 Petroleum and Its Products Part 49: Penetration of Bituminous Materials. 7 pp.

CNS K6755-83. Method of Test for Penetration of Bituminous Materials (Mar)(10090).

DIN ENGL 52010-83. Testing of Bitumen; Determination of Needle Penetration (Dec). 4 pp.

—Permittivity
BSI BS 2000: Part 357-83. (OBSOLESCENT) 1983 Petroleum and Its Products Part 357: Permittivity of Bitumen. 14 pp.

—Physical Testing
BSI BS 598: Part 3-85. (WITHDRAWN) 1985 Amd 2 Sampling and Examination of Bituminous Mixtures for Roads and Other Paved Areas Part 3: Methods for Design and Physical Testing (AMD 6121) September 29, 1988 (Sup'd by BS 598: Part 104: 1989, 105, 106, 107, 108, & 109: 1990). 33 pp.

—Plastic Properties—Fraass Breaking Point
DIN ENGL 52012-85. Testing of Bitumen; Determination of the Fraass Breaking Point (Aug). 6 pp.

—Sampling
BSI BS 598: Part 100-87. 1987 Amd 1 Sampling and Examination of Bituminous Mixtures for Roads and Other Paved Areas Part 100: Methods for Sampling for Analysis (AMD 6122) May 31, 1989 (Superseding BS 598: Part 1). 11 pp.

BSI BS 598: Part 101-87. 1987 Sampling and Examination of Bituminous Mixtures for Roads and other Paved Areas Part 101: Methods for Preparatory Treatment of Samples for Analysis (Supersedes BS 598: Part 2: 1974). 8 pp.

DIN ENGL 1996 Pt 2-71. Testing of Bituminous Materials for Road Building and Related Purposes; Sampling (Oct). 8 pp.

—Segregation
DIN ENGL 1996 Pt 16-75. Testing of Bituminous Materials for Road Building and Related Purposes; Determination of Segregation Tendency (Dec). 4 pp.

—Softening Points
DIN ENGL 1996 Pt 15-75. Testing of Bituminous Materials for Road Building and Related Purposes; Determination of Softening Point in Accordance with Wilhelmi (Dec). 4 pp.

DIN ENGL 52011-86. Determination of the Ring-and-Ball Softening Point of Bitumen (Oct). 5 pp.

—Softening Points—Ring and Ball Testing
BSI BS 2000: Part 58-93. 1993 Methods of Test for Petroleum and Its Products Part 58: Determination of Softening Point of Bitumen—Ring and Ball Method. 5 pp.

BSI BS 2000: Part 58-88. 1988 Petroleum and its Products Part 58: Softening Point of Bitumen (Ring and Ball). 8 pp.

—Swelling
DIN ENGL 1996 Pt 9-81. Testing of Bituminous Materials for Road Building and Related Purposes; Swelling Test (May). 4 pp.

—Temperature Measurement
BSI BS 598: Part 109-90. 1990 Sampling and Examination of Bituminous Mixtures for Roads and Other Paved Areas Part 109: Methods for the Assessment of the Compaction Performance of a Roller and Recm. Proc. for the Mst. of the Temp of Bituminous Mixtures. 7 pp.

—Test Specimens
DIN ENGL 1996 Pt 3-90. Testing of Asphalt; Sample Preparation (May). 4 pp.

INTERNATIONAL AND NON-U.S. NATIONAL STANDARDS
SUBJECT INDEX
Bituminous

Bitumens (Cont.)

—Texture Depth
BSI BS 598: Part 105-90. 1990 Sampling and Examination of Bituminous Mixtures for Roads and Other Paved Areas Part 105: Methods of Test for the Determination of Texture Depth. 18 pp.

BSI BS 598: Part 105-01. 1990 Amd 1 Sampling and Examination of Bituminous Mixtures for Roads and Other Paved Areas Part 105: Methods of Test for the Determination of Texture Depth (AMD 7294) November 15, 1992 (Supersedes BS 598: Part 3: 1985). 19 pp.

—Thermal Stability
DIN ENGL 52016-88. Testing the Thermal Stability of Bitumen in a Rotating Flask (Dec) (Supersedes the December 1980 Edition of DIN 52017). 3 pp.

—Tracking
BSI DD 184-90. 1990 Method for the Determination of the Wheel Tracking Rate of Cores of Bituminous Wearing Courses. 10 pp.

—Viscosity
BSI BS 2000: Part 72-93. 1993 Methods of Test for Petroleum and Its Products Part 72: Determination of Viscosity of Cutback Bitumen. 5 pp.

BSI BS 2000: Part 72-88. 1988 Petroleum and its Products Part 72: Viscosity of Cutback Bitumen. 7 pp.

CEN PREN 1110-93. Bitumen Sheets for Waterproofing—Determination of Flow Properties at Elevated Temperature. 11 pp.

—Water Absorption
DIN ENGL 1996 Pt 8-77. Testing of Bituminous Materials for Road Building and Related Purposes; Determination of Water Absorption (Sept). 3 pp.

—Water Content
DIN ENGL 1996 Pt 5-83. Testing of Bituminous Materials for Road Building and Related Purposes; Determination of Water Content (Apr). 4 pp.

—Water Content—Distillation Methods
BSI BS 4385-80. (WITHDRAWN) 1980 Determination of Water in Petroleum Products and Bituminous Materials (Distillation Method) (Superseded by BS 4385: Part 1: 1991). 8 pp.

Bituminous Board

—Test Facilities
MOD UK DEF-149-65. Board, Bituminous (for Use at Proof & Experimental Establishments). 5 pp.

Bituminous Cements
See Also: Cements

CGSB 37-GP-19M-76. Cement, Plastic, Cutback Tar, Standard for (R 1985). 9 pp.

Bituminous Coal
See Also: Coal; Fossil Fuels; Rocks; Sub-Bituminous Coal

CGSB 18-GP-7-69. Coal, Bituminous, Subbituminous and Lignite; (Amendment 1 Feb 1980). 11 pp.

CNS M1002-63. Bituminous Coal (for Export from Taiwan Area) (Sep)(205) (R 1972). 1 p.

—Carbominerite Content—Petrography
BSI BS 6127: Part 4-90. 1990 Petrographic Analysis of Bituminous Coal and Anthracite Part 4: Method of Determining Microlithotype, Carbominerite and Minerite Composition. 10 pp.

ISO 7404 Pt 4-88. Methods for the Petrographic Analysis of Bituminous Coal and Anthracite—Part 4: Method of Determining Microlithotype, Carbominerite and Minerite Composition First Edition. 8 pp.

—Coking
CNS M1001-65. Bituminous Coal for Coking (for Export from Taiwan Area) (Sep)(204) (R 1972). 1 p.

—Maceral Content—Petrography
BSI BS 6127: Part 3-81. 1981 Petrographic Analysis of Bituminous Coal and Anthracite Part 3: Determining Maceral Group Composition of Bituminous Coal and Anthracite. 7 pp.

ISO 7404 Pt 3-84. Methods for the Petrographic Analysis of Bituminous Coal and Anthracite—Part 3: Method of Determining Maceral Group Composition First Edition. 6 pp.

—Microlithotype Content—Petrography
BSI BS 6127: Part 4-90. 1990 Petrographic Analysis of Bituminous Coal and Anthracite Part 4: Method of Determining Microlithotype, Carbominerite and Minerite Composition. 10 pp.

ISO 7404 Pt 4-88. Methods for the Petrographic Analysis of Bituminous Coal and Anthracite—Part 4: Method of Determining Microlithotype, Carbominerite and Minerite Composition First Edition. 8 pp.

Bituminous Coal (Cont.)

—Minerite Content—Petrography
BSI BS 6127: Part 4-90. 1990 Petrographic Analysis of Bituminous Coal and Anthracite Part 4: Method of Determining Microlithotype, Carbominerite and Minerite Composition. 10 pp.

ISO 7404 Pt 4-88. Methods for the Petrographic Analysis of Bituminous Coal and Anthracite—Part 4: Method of Determining Microlithotype, Carbominerite and Minerite Composition First Edition. 8 pp.

—Petrography—Glossaries
BSI BS 6127: Part 1-81. 1981 Petrographic Analysis of Bituminous Coal and Anthracite Part 1: Glossary of Terms Relating to the Petrographic Analysis of Bituminous Coal and Anthracite. 7 pp.

ISO 7404 Pt 1-84. Methods for the Petrographic Analysis of Bituminous Coal and Anthracite—Part 1: Glossary of Terms First Edition. 5 pp.

—Petrography—Sampling
BSI BS 6127: Part 2-82. 1982 Amd 2 Petrographic Analysis of Bituminous Coal and Anthracite Part 2: Method of Preparing Coal Samples for Petrographic Analysis. 12 pp.

ISO 7404 Pt 2-85. Methods for the Petrographic Analysis of Bituminous Coal and Anthracite—Part 2: Method of Preparing Coal Samples First Edition. 10 pp.

—Sieve Analysis
CNS M3158-84. Method for Sieve Analysis of Crushed Bituminous Coal (Jun)(10943).

—Vitrinite—Reflectance—Petrography
ISO 7404 Pt 5-84. Methods for the Petrographic Analysis of Bituminous Coal and Anthracite—Part 5: Method of Determining Microscopically the Reflectance of Vitrinite First Edition. 13 pp.

—Weight Measurement
CNS M3160-84. Method of Test for Cubic Meter Weight of Crushed Bituminous Coal (Jun) (10945).

Bituminous Coatings
Use For: Bituminous Paints *See Also:* Asphalt Coatings (Made From Asphalt); Bituminous Primers; Coal Tar Coatings; Nonmetallic Coatings; Organic Coatings; Paving Materials; Waterproof Coatings

BSI BS 3416-91. 1991 Bitumen-Based Coatings for Cold Application, Suitable for Use in Contact with Potable Water. 11 pp.

BSI BS 3416-88. 1988 Bitumen-Based Coatings for Cold Application, Suitable for use in Contact with Potable Water (W). 10 pp.

BSI BS 3416-01. 1991 Amd 1 Bitumen-Based Coatings for Cold Application, Suitable for Use in Contact with Potable Water (AMD 7288) September 15, 1992. 12 pp.

BSI BS 6949-91. 1991 Bitumen-Based Coatings for Cold Application, Excluding Use in Contact with Potable Water. 11 pp.

BSI BS 6949-01. 1991 Bitumen-Based Coatings for Cold Application, Excluding Use in Contact with Potable Water (AMD 7287) September 15, 1992. 12 pp.

BSI BS 6949-88. 1988 Bitumen-Based Coatings for Cold Application, Excluding Use in Contact with Potable Water. 10 pp.

MOD UK DSTAN 80-23-72. Paint, Finishing, Bituminous, Aluminium Types: Brushing Spraying Issue 1. 6 pp.

MOD UK CS 2297D. Bitumen, Oxidized No.1 (Withdrawn).

—Adhesion—Immersion Testing
DIN ENGL 52006 Pt 1-80. Testing of Bituminous Binders; Effect of Water on Binder Coatings; Bitumen Emulsion Binder Coating (Dec). 2 pp.

DIN ENGL 52006 Pt 3-85. Bitumen and Coal Tar Pitch; Effect of Water on Binder Coatings; Testing of Fluxed Bitumen Binder Coatings (Sept). 2 pp.

—Adhesion Testing
MOD UK M 3004/67. Protective PX-2 and Composition Rust Preventative Type B.

—Aggregates
CNS A3287-88. Method of Test for Determining Degree of Particle Coating of Bituminous—Aggregate Mixtures (Aug)(12389).

CNS A3289-88. Method of Test for Effect of Water on Bituminous—Coated Aggregate Quick Field (Aug)(12391).

CNS A3292-88. Method of Test for Coating and Stripping of Bituminous—Aggregate Mixtures (Aug)(12394).

—Aggregates—Stripping
CNS A3292-88. Method of Test for Coating and Stripping of Bituminous—Aggregate Mixtures (Aug)(12394).

Bituminous Coatings (Cont.)

—Ammunition
MOD UK CS 3004. Bitumen, Oxidised No. 2 (Superseded by Def Stan 68-119).

—Bombs (Ordnance)
MOD UK DSTAN 80-82-81. Composition, Bituminous, RD 1083 Issue 1. 8 pp.

—Buildings
DIN ENGL 18195 Pt 2-83. Waterproofing of Buildings and Structures; Materials (Aug). 7 pp.

DIN ENGL 18195 Pt 3-83. Waterproofing of Buildings and Structures; Processing of Materials (Aug). 5 pp.

DIN ENGL 18195 Pt 5-84. Waterproofing of Buildings and Structures; Waterproofing Against Water That Exerts No Hydrostatic Pressure; Design and Workmanship (Feb). 6 pp.

DIN ENGL 18195 Pt 6-83. Waterproofing of Buildings and Structures; Waterproofing Against Water That Exerts Hydrostatic Pressure from the Outside; Design and Workmanship (Aug). 6 pp.

—Concrete
DIN ENGL 18195 Pt 4-83. Waterproofing of Buildings and Structures; Damp-Proofing Against Moisture from the Ground; Design and Workmanship (Aug). 8 pp.

—Dean Stark Method
MOD UK M 3004/67. Protective PX-2 and Composition Rust Preventative Type B.

—Drying
MOD UK M 3004/67. Protective PX-2 and Composition Rust Preventative Type B.

—Ductile Iron Pipes
DIN ENGL 30674 Pt 4-83. Coating of Ductile Cast Iron Pipes; Bitumen Coating (May). 3 pp.

—Humidity
MOD UK M 3004/67. Protective PX-2 and Composition Rust Preventative Type B.

—Iron
BSI BS 4147-80. 1980 Amd 1 Bitumen-Based Hot-Applied Coating Material for Protecting Iron and Steel, Including Suitable Primers Where Required. 11 pp.

—Masonry
DIN ENGL 18195 Pt 4-83. Waterproofing of Buildings and Structures; Damp-Proofing Against Moisture from the Ground; Design and Workmanship (Aug). 8 pp.

—Pipe Fittings
DIN ENGL 30673-86. Bitumen Coatings and Linings for Steel Pipes, Fittings and Vessels (Dec). 6 pp.

—Ships
JIS K 5850-79. Testing Method for Ship's Bitumen Paints.

JIS K 5851-57. Ship's Bitumen Solution.

MOD UK TS 10195A. Paint Systems, Bitumen, for Defence Equipment Types I, II and III.

—Steel
BSI BS 4147-80. 1980 Amd 1 Bitumen-Based Hot-Applied Coating Material for Protecting Iron and Steel, Including Suitable Primers Where Required. 11 pp.

—Steel—Pipe Fittings
DIN ENGL 30673-86. Bitumen Coatings and Linings for Steel Pipes, Fittings and Vessels (Dec). 6 pp.

—Steel Pipes
ISO 5256-85. Steel Pipes and Fittings for Buried or Submerged Pipelines—External and Internal Coating by Bitumen or Coal Tar Derived Materials First Edition. 57 pp.

—Storage Tanks
MOD UK TS 10195A. Paint Systems, Bitumen, for Defence Equipment Types I, II and III.

—Sulfur Content Analysis
MOD UK M 3004/67. Protective PX-2 and Composition Rust Preventative Type B.

—Vehicles
MOD UK TS 310B. Compound, Protective for Vehicle Under-Bodies Solvent-Based.

MOD UK TS 10065A. Compound, Protective for Vehicle Underbodies, Water-Based (Superseded by Def Stan 80-168).

—Water Tanks
BSI BS 3416-91. 1991 Bitumen-Based Coatings for Cold Application, Suitable for Use in Contact with Potable Water. 11 pp.

INDUSTRY STANDARDS

Bituminous Coatings (Cont.)
—Water Tanks (Cont.)
BSI BS 3416-88. 1988 Bitumen-Based Coatings for Cold Application, Suitable for use in Contact with Potable Water (W). 10 pp.
BSI BS 3416-01. 1991 Amd 1 Bitumen-Based Coatings for Cold Application, Suitable for Use in Contact with Potable Water (AMD 7288) September 15, 1992. 12 pp.

Bituminous Concretes
See Also: Plastic Properties
—Plastic Properties
CNS A3293-88. Method of Test for Resistance to Plastic Flow of Bituminous Mixtures Using Marshall Apparatus (Aug)(12395).

Bituminous Materials
Use: Bitumens

Bituminous Paints
Use: Bituminous Coatings

Bituminous Pavements
Use: Asphalt Pavements

Bituminous Pipes
Use For: Bitumized Fiber Pipes See Also: Pipes
CNS A3179-89. Method of Test for Homogeneous Bituminized Fiber Pipe (Jul)(10041).
CNS A3180-89. Method of Test for Laminated-Wall Bituminized Fiber Pipe (Jul)(10044).
—Drainpipes
CNS A2146-89. Homogeneous Bituminized Fiber Drain and Sewer Pipe (Jul)(10039).
CNS A2148-89. Laminated-Wall Bituminized Fiber Drain and Sewer Pipe (Jul)(10042).
—Drainpipes—Perforated
CNS A2147-89. Perforated Homogeneous Bituminized Fiber Pipe for General Drainage (Jul)(10040).
CNS A2149-89. Perforated, Laminated-Wall Bituminized Fiber Pipe for General Drainage (Jul)(10043).
—Sewer
CNS A2146-89. Homogeneous Bituminized Fiber Drain and Sewer Pipe (Jul)(10039).
CNS A2148-89. Laminated-Wall Bituminized Fiber Drain and Sewer Pipe (Jul)(10042).

Bituminous Primers
See Also: Bituminous Coatings
—Iron
BSI BS 4147-80. 1980 Amd 1 Bitumen-Based Hot-Applied Coating Material for Protecting Iron and Steel, Including Suitable Primers Where Required. 11 pp.
—Steel
BSI BS 4147-80. 1980 Amd 1 Bitumen-Based Hot-Applied Coating Material for Protecting Iron and Steel, Including Suitable Primers Where Required. 11 pp.

Bituminous Varnishes
See Also: Varnishes
MOD UK DSTAN 80-39-01. Varnish, Oil Bitumen, Anchoring, Type QX Types: Brushing Spraying Issue 1; Corrigendum. 9 pp.
MOD UK DSTAN 80-104-85. Varnish Oil Bitumen Unpigmented Type QX (RD1219) Soft Type QX (1322) Pigmented Type QX (RD 1229) Issue 1. 10 pp.

Bitumized Fiber Pipes
Use: Bituminous Pipes

Biuret Content Analysis
—Fertilizers
CNS N4028-86. Method of Test for Hazardous Components in Compound Fertilizer (Jan)(3080). 4 pp.
—Urea—Atomic Absorption Spectrophotometry
ISO 4274-77. Urea for Industrial Use—Determination of Biuret Content—Flame Atomic Absorption and Photometric Absorption Methods First Edition. 8 pp.
—Urea—Photometry
ISO 2754-73. Urea for Industrial Use—Determination of Biuret Content—Photometric Method First Edition. 5 pp.

Biuret Content Analysis (Cont.)
—Urea—Spectrophotometry
ISO 4274-77. Urea for Industrial Use—Determination of Biuret Content—Flame Atomic Absorption and Photometric Absorption Methods First Edition. 8 pp.

Black Boxes
Use: Flight Recorders

Black Light
See Also: Ultraviolet Radiation
—Illumination—Output
ISO 3059-74. Non-Destructive Testing—Method for Indirect Assessment of Black Light Sources First Edition. 8 pp.
—Optical Density
BSI BS 4489-84. 1984 Method for Measurement of UV-A Radiation (Black Light) Used in Non-Destructive Testing. 4 pp.

Black Liquors
See Also: Green Liquors; Industrial Wastes; Kraft Liquors; Tall Oil; White Liquors
—Sulfate Process
CPPA J.15P-81. Analysis of Sulphate Process Black Liquors. 8 pp.

Black Oxide Coatings
See Also: Coatings; Nonmetallic Coatings; Oxide Coatings
—Copper
CGSB CAN/CGSB-31.112-M90. Black Finish for Copper Alloys. 6 pp.
—Ferroalloys—Qualitative Analysis
CGSB 31-GP-0A METH 61.2-62. Methods of Testing Corrosion-Prevention Materials and Processes Oxalic Acid Spot Test. 1 p.
—Iron
CGSB CAN/CGSB-31.114-M91. Black Oxide Conversion Coatings for Ferrous Metals. 8 pp.
DIN ENGL 50938-87. Alkaline Blackening Treatment of Ferrous Components; Principles and Methods of Test (Nov). 3 pp.

Black Pepper
Use: Pepper

Black Plate
See Also: Steels
CEN PREN 10 205-90. Cold Reduced Blackplate in Coil Form for the Production of Tinplate or Electrolytic Chromium/Chromium Oxide Coated Steel. 17 pp.
CNS G2040-87. Method of Test for Tinplate and Blackplate (Dec)(4156). 15 pp.
CNS G3097-87. Tinplate and Blackplate (Dec)(4155).
DIN ENGL 1616 (S)-84. Tinplate and Blackplate Sheet; Grades, Dimensions and Permissible Deviations (Oct) (Superseded DIN EN 10203 August 1991 and DIN EN 10205 January 1992). 12 pp.
ISO 1111 Pt 1-83. Single Cold-Reduced Tinplate and Single Cold-Reduced Blackplate—Part 1: Electrolytic and Hot-Dipped Tinplate Sheet and Blackplate Sheet First Edition. 16 pp.
ISO 1111 Pt 2-83. Single Cold-Reduced Tinplate and Single Cold-Reduced Blackplate—Part 2: Electrolytic Tinplate Coil and Blackplate Coil for Subsequent Cutting into Sheet Form Second Edition. 17 pp.
JIS G 3303-87. Tinplate and Blackplate. 40 pp.
SAA AS 1517-91. Tinplate and Blackplate. 19 pp.
—Coils—Cold Reduced
BSI BS EN 10205-92. 1992 Cold Reduced Blackplate in Coil Form for the Production of Tinplate or Electrolytic Chromium/Chromium Oxide Coated Steel (V). 15 pp.
CEN EN 10205-91. Cold Reduced Blackplate in Coil Form for the Production of Tinplate or Electrolytic Chromium/Chromium Oxide Coated Steel. 18 pp.
DIN ENGL EN 10205-92. Cold Reduced Blackplate in Coil Form for the Production of Tinplate or Electrolytic Chromium/Chromium Oxide Coated Steel (Jan) (This Standard, Together with DIN EN 10203, Supersedes DIN 1616, October 1984 Edition). 10 pp.
—Cold Reduced
BSI BS 2920-73. (WITHDRAWN) 1973 Cold-Reduced Tinplate and Cold-Reduced Blackplate (Superseded by BS EN 10203: 1991). 29 pp.

Black Powder
See Also: Explosives
JIS K 4805-68. Black Powder.

Black Smoke Index
See Also: Air Pollution; Particulates; Smoke
BSI BS 1747: Part 11-93. 1993 Methods for Measurement of Air Pollution Part 11: Determination of a Black Smoke Index in Ambient Air (ISO 9835: 1993) (N). 14 pp.
ISO 9835-93. Ambient Air—Determination of a Black Smoke Index First Edition. 12 pp.

Black Wax
See Also: Waxes
CNS K1266-81. Black Wax (Oct)(8014).
CNS K6695-81. Method of Test for Black Wax (Oct)(8015).

Blackboards
Use: Chalkboards

Blackheart Malleable Iron
Use: Malleable Cast Iron

Blackout Lamps
Use: Blackout Lights

Blackout Lights
Use For: Blackout Lamps See Also: Motor Vehicle Lighting
—Combat Vehicles
NATO STANAG 4381 ED 1 AMD 0-00. (Draft) Blackout Lighting Systems for Tactical Land Vehicles. 18 pp.

Blackplate
Use: Black Plate

Blacksmiths' Tongs
Use: Tongs—Forging

Blades (Tools)
Scope Note: See the subheading Blades under the specific tool

Blanik Gliders
See Also: Aircraft; Gliders
—Foreign Airworthiness Directives
CAA. Blanik Gliders (Foreign Airworthiness Directives). 3 pp.

Blanket Insulation
Use For: Quilt Insulation; Thermal Insulation Pads
See Also: Acoustical Insulation; Electrical Insulation; Pipe Insulation; Thermal Insulation
—Ceramic Fiber—Marine
MOD UK NES 806-89. Requirements for Ceramic Thermal Insulation Material Ceramic Products Ceramic Rope and Ceramic Blanket Issue 2 (04.89). 11 pp.
—Electrical
IEC 1112-92. Blankets of Insulating Material for Electrical Purposes First Edition. 80 pp.
—Fiber
CNS R2188-86. High-Temperature Fiber Blanket Thermal Insulation (Mar)(11522).
CNS R3136-86. Method of Test for High-Temperature Fiber Blanket Thermal Insulation (Apr)(11523).
—Glass Wool
CNS R2059-85. Glass Fiber Heat Insulation Materials (May)(3065). 10 pp.
JIS A 9505-89. Thermal Insulation Material Made of Glass Wool. 13 pp.
—Mineral Wool—Boilers
CGSB CAN/CGSB-51.11-92. Mineral Fibre Thermal Insulation Blanket. 8 pp.
—Mineral Wool—Ducts
CGSB CAN/CGSB-51.11-92. Mineral Fibre Thermal Insulation Blanket. 8 pp.
—Mineral Wool—Machinery
CGSB CAN/CGSB-51.11-92. Mineral Fibre Thermal Insulation Blanket. 8 pp.
—Rock Wool
CNS R2080-73. Rock Wool for Heat Insulation (Nov)(3657).
JIS A 9504-89. Thermal Insulation Material Made of Rock Wool. 14 pp.
—Silica Fabrics—Naval Ships
MOD UK NES 807-82. Inorganic Insulation Material Silica Fabric for Welding and Burning Blankets Issue 1 (01.82). 11 pp.

Blanket Insulation (Cont.)
—Silica Fabrics—Naval Ships (Cont.)
MOD UK NES 807: Part 1-89. Requirements for the Production of Welding and Burning Blankets and Hanger Fire Curtains Part 1: Silica Cloth for Welding and Burning Blankets Issue 2 (04.89). 13 pp.
MOD UK NES 807: Part 1-01. Requirements for the Production of Welding and Burning Blankets and Hanger Fire Curtains Part 1: Silica Cloth for Welding and Burning Blankets Issue 2 (04.89); Amendment 1. 15 pp.

—Silica Fabrics—Submarines
MOD UK NES 807-82. Inorganic Insulation Material Silica Fabric for Welding and Burning Blankets Issue 1 (01.82). 11 pp.
MOD UK NES 807: Part 1-89. Requirements for the Production of Welding and Burning Blankets and Hanger Fire Curtains Part 1: Silica Cloth for Welding and Burning Blankets Issue 2 (04.89). 13 pp.
MOD UK NES 807: Part 1-01. Requirements for the Production of Welding and Burning Blankets and Hanger Fire Curtains Part 1: Silica Cloth for Welding and Burning Blankets Issue 2 (04.89); Amendment 1. 15 pp.

Blankets
Use For: Bed Rugs *See Also:* Bedspreads; Electric Blankets; Fire Blankets; Quilts
CEN EN 14-82. Dimensions of Bed Blankets. 9 pp.
CEN PREN 14-93. Dimensions of Bed Blankets. 7 pp.
JIS L 1098-92. Testing Methods for Blankets. 9 pp.

—Cellular Fiber
BSI BS 5866: Part 3-91. 1991 Blankets Suitable for Use in the Public Sector Part 3: Specification for Synthetic Fibre Cellular Blankets. 9 pp.

—Ceramic Fiber
BSI BS 7225: Sec 3.1-90. 1990 Classification of Refractories Part 3: Section 3.1: Blankets, Mats, Felts and Paper. 8 pp.
JIS R 3311-91. Ceramic Fiber Blanket. 10 pp.

—Cotton/Cellular
BSI BS 5866: Part 2-91. 1991 Blankets Suitable for Use in the Public Sector Part 2: Specification for Cotton Leno Cellular Blankets. 9 pp.

—Cotton—Grading
CNS L4017-63. Cotton Blankets and Viscose Rayon Blankets (for Export) (Mar) (1480)(R 1973). 4 pp.

—Cotton/Polyester—Flannelette
CGSB 4-GP-6MA-84. Sheeting, Sheets and Blankets, Cotton Polyester Flannelette, Standard for. 7 pp.

—Flammability Testing
BSI BS 5866: Part 4-01. 1991 Amd 1 Blankets Suitable for Use in the Public Sector Part 4: Specification for Flammability Performance (AMD 7065) May 1, 1992. 9 pp.

—Rayon—Grading
CNS L4017-63. Cotton Blankets and Viscose Rayon Blankets (for Export) (Mar) (1480)(R 1973). 4 pp.

—Repair
SNZ NZCP 28-62. Code of Recommended Practice for the Repair of Electrically Heated Blankets for Domestic Use. 15 pp.

—Wool
BSI BS 5866: Part 1-90. 1990 Amd 1 Blankets Suitable for Use in the Public Sector Part 1: Specification for Wool and Wool/Polyamide Blankets (AMD 6732) August 31, 1991. 9 pp.
CGSB CAN/CGSB-4.67-M91. Wool Blankets. 10 pp.
CGSB 4-GP-111MA-90. Wool Blankets (RCMP); (Corrigendum—August 1990). 10 pp.

—Wool/Nylon
BSI BS 5866: Part 1-90. 1990 Amd 1 Blankets Suitable for Use in the Public Sector Part 1: Specification for Wool and Wool/Polyamide Blankets (AMD 6732) August 31, 1991. 9 pp.

Blankets (Fire)
Use: Fire Blankets

Blanking Caps
See Also: Caps (Lids)

—Aircraft—Bolts
SBAC AS 2942 ISSUE 3. Bolt. 1 p.

—Aircraft—Brackets
SBAC AS 2977 ISSUE 3. Bracket. 1 p.

—Aircraft—Chains
SBAC AS 2944 ISSUE 2. Chain Assembly. 1 p.
SBAC AS 2949 ISSUE 2. Chain Assembly. 1 p.

—Aircraft—Flanges
SBAC AS 2973 ISSUE 2. Locking Flange. 1 p.

—Aircraft—Fuel Couplings
SBAC AS 2941 ISSUE 2. Blanking Cap for 1 1/2 Inch Fuel Coupling. 1 p.
SBAC AS 2946 ISSUE 2. Blanking Cap for 2 1/2 Inch Fuel Coupling. 1 p.
SBAC AS 6544 ISSUE 1. Blanking Cap for 1 1/2 Inch Fuel Coupling. 1 p.
SBAC AS 6552 ISSUE 1. Blanking Cap for 2 1/2 Inch Fuel Coupling. 1 p.

—Aircraft—Rivets
SBAC AS 2976 ISSUE 4. Chain Rivet. 1 p.

—Aircraft—Sealing Caps
SBAC AS 2948 ISSUE 3. Sealing Cap. 1 p.

—Aircraft—Springs (Elastic)
SBAC AS 2943 ISSUE 3. Locking Spring. 1 p.
SBAC AS 2947 ISSUE 3. Locking Spring. 1 p.

Blanks (Fasteners)
See Also: Fasteners

—Nipples—Aerospace
SBAC AS 62352 ISSUE 1. Nipple-Blanking Short (Cres). 2 pp.

—Pipe Couplings—Bulkhead—Adapter—Aerospace
SBAC AS 62373 ISSUE 1. Bulkhead Adaptor-Forged Blank (Cres) Swaged Pipe Coupling. 2 pp.

—Pipe Couplings—Bulkhead—Elbow—Aerospace
SBAC AS 62374 ISSUE 1. Bulkhead Elbow-Forged Blank (Cres) Swaged Pipe Coupling. 2 pp.

—Pipe Couplings—Cross—Aerospace
SBAC AS 62371 ISSUE 1. Cross-Forged Blank (Cres) Swaged Pipe Coupling. 2 pp.
SBAC AS 62372 ISSUE 1. Cross-Forged Blank (Cres) Swaged Pipe Coupling. 2 pp.

—Pipe Couplings—Elbow—Aerospace
SBAC AS 62369 ISSUE 1. Elbow-Forged Blank (Cres). 2 pp.

—Pipe Couplings—Tee—Aerospace
SBAC AS 62370 ISSUE 1. Tee-Forged Blank (Cres) Swaged Pipe Coupling. 2 pp.

Blanks (Forms)
Use: Forms (Paper)

Blast Cleaning
Use For: Abrasive Blast Cleaning *See Also:* Metal Cleaning; Metal Finishing; Sandblasting; Surface Finishing

—Steels
BSI BS 7079: Part D2-93. 1993 Preparation of Steel Substrates Before Application of Paints and Related Products Group D: Methods for Surface Preparation Part D2: Abrasive Blast-Cleaning (ISO 8504-2: 1992) (V). 15 pp.
CEN PREN 10238-93. Automatically Blast Cleaned and Primed Steel Products. 12 pp.
ISO 8504 Pt 2-92. Preparation of Steel Substrates Before Application of Paints and Related Products—Surface Preparation Methods—Part 2: Abrasive Blast-Cleaning First Edition. 14 pp.

—Steels—Corrosion Testing
BSI DD 207-92. 1992 Method of Test for Soluble Iron Corrosion Products to Assess Surface Cleanliness of Steel Substrates Before Application of Paints and Related Products (ISO/TR 8502-1: 1991) (V). 12 pp.
ISO TR8502 Pt 1-91. Preparation of Steel Substrates Before Application of Paints and Related Products—Tests for the Assessment of Surface Cleanliness—Part 1: Field Test for Soluble Iron Corrosion Products First Edition. 10 pp.

Blast Furnace Slag Cements
Use: Slag Cements

Blast Furnace Slag Content Analysis
Use: Slag Content Analysis

Blast Furnaces
See Also: Furnaces

—Firebricks
BSI BS 3056: Part 5-85. (WITHDRAWN) 1985 Sizes of Refractory Bricks Part 5: Bricks for Blast Furnace Walls. 3 pp.

Blasters
See Also: Blasting; Blasting Caps; Explosives
CNS M3010-73. Method of Test for Dynamo Type Blasting Machine (Dec)(3681).
JIS M 2505-62. Condenser Discharge Type Blasting Machines.

Blasting
See Also: Blasters; Blasting Caps; Mining Equipment; Shot Blasting

—Abrasives
DIN ENGL 50315-88. Testing the Wear and Effectiveness of Metallic Abrasives by Centrifugal Blasting (Oct). 10 pp.

—Noise
CSA CAN3-Z107.54-M85. Procedure for Measurement of Sound and Vibration Due to Blasting Operations; (Gen Instr 1). 15 pp.

—Vibration
CSA CAN3-Z107.54-M85. Procedure for Measurement of Sound and Vibration Due to Blasting Operations; (Gen Instr 1). 15 pp.

Blasting Caps
Use For: Caps (Explosive) *See Also:* Blasters; Blasting; Explosives; Initiators (Explosives)
JIS K 4806-78. Blasting Caps.
JIS K 4807-81. Electric Blasting Caps.

—Glass Powder
MOD UK DSTAN 13-76-91. Glass Powder Issue 2. 7 pp.

—Shotguns
JIS K 4818-74. Percussion Caps for Shot Gun.

—Single Shot
JIS K 4852-66. Single Shot Caps and Tape Caps.

—Sporting Paper
JIS K 4853-66. Sporting Paper Caps.

—Tape
JIS K 4852-66. Single Shot Caps and Tape Caps.

Blasting Machines
Use: Blasters

Blazers
See Also: Clothing; Jackets
BSI BS 3838-80. 1980 Amd 1 Blazer Cloths. 9 pp.

—Boys'
CGSB CAN/CGSB-49.65-M91. Canada Standard Children's Sizes 2 to 6X, Girls' Sizes 7 to 16 and Boys' Sizes 7 to 20 Blazers, Dress Jackets and Sport Jackets, Regular Range—Dimensions. 19 pp.

—Children's
CGSB CAN/CGSB-49.65-M91. Canada Standard Children's Sizes 2 to 6X, Girls' Sizes 7 to 16 and Boys' Sizes 7 to 20 Blazers, Dress Jackets and Sport Jackets, Regular Range—Dimensions. 19 pp.

—Girls'
CGSB CAN/CGSB-49.65-M91. Canada Standard Children's Sizes 2 to 6X, Girls' Sizes 7 to 16 and Boys' Sizes 7 to 20 Blazers, Dress Jackets and Sport Jackets, Regular Range—Dimensions. 19 pp.

—Women's
CGSB CAN/CGSB-49.214-M89. Junior, Misses and Women's Canada Standard Sizes Blazers—Dimensions. 19 pp.

Bleach Liquors
See Also: Bleaching Agents; Calcium Hypochlorite; Hypochlorites; Spent Liquors

—Alkali Content
CPPA J.22P-86. Analysis of Chlorine Solutions, Hypochlorite Bleach Liquors, and Spent Bleach Liquors. 3 pp.

—Alkalinity
CPPA J.22P-86. Analysis of Chlorine Solutions, Hypochlorite Bleach Liquors, and Spent Bleach Liquors. 3 pp.

—Chlorine Content
CPPA J.22P-86. Analysis of Chlorine Solutions, Hypochlorite Bleach Liquors, and Spent Bleach Liquors. 3 pp.

Bleach Liquors *(Cont.)*
—Residue Content
CPPA J.22P-86. Analysis of Chlorine Solutions, Hypochlorite Bleach Liquors, and Spent Bleach Liquors. 3 pp.

Bleached Shellac
See Also: Finishes; Nonmetallic Coatings; Shellacs
JIS K 5911-60. Bleached Shellac.

Bleached Shellac Varnishes
Use: Shellac Varnishes

Bleaches
Use: Bleaching Agents

Bleaching
See Also: Chlorination; Colorfastness Testing; Papermaking Equipment; Pulp Bleaching
—Fats
BSI BS 684: Sec 2.27-87. 1987 Methods of Analysis of Fats and Fatty Oils Part 2: Other Methods Section 2.27: Determination of Bleachability. 4 pp.
CNS K6762-83. Method of Bleaching Test for Fats and Oils (Mar)(10097).
JIS K 3501-62. Method of Bleaching Test for Fats and Oils.
—Fats—Alkali Refining
JIS K 3502-62. Method of Alkali Refining of Fats and Oils for Bleaching Test.
—Hair (Animal)
CNS N4025-85. Method of Test for Animal Hairs (May)(2826). 7 pp.
—Oils
BSI BS 684: Sec 2.27-87. 1987 Methods of Analysis of Fats and Fatty Oils Part 2: Other Methods Section 2.27: Determination of Bleachability. 4 pp.
CNS K6762-83. Method of Bleaching Test for Fats and Oils (Mar)(10097).
JIS K 3501-62. Method of Bleaching Test for Fats and Oils.
—Oils—Alkali Refining
JIS K 3502-62. Method of Alkali Refining of Fats and Oils for Bleaching Test.

Bleaching Agents
See Also: Bleach Liquors; Bleaching Powder; Cleaning Agents; Fabric Sizings
—Calcium Hypochlorites
CGSB CAN/CGSB-15.32-93. Calcium Hypochlorite. 8 pp.
CNS K1162-74. Calcium Hypochlorite Bleaching Solution (Sep)(3395).
CNS K6331-74. Method of Test for Calcium Hypochlorite Bleaching Solution (Oct)(3396).
—Calcium Hypochlorites—Chemical Analysis
CPPA J.2H-64. Analysis of Calcium Hypochlorite Bleaching Powder, Bleach Liquor, and Bleach Sludge. 4 pp.
—Chlorine Dioxide
CPPA J.14P-84. Chlorine Dioxide Plant Analyses. 5 pp.
—Hydrogen Peroxide—Pulps
CPPA J.16P-92. Analysis of Peroxides. 3 pp.
—Hydrosulfites
JIS K 1476-80. Hydrosulfites for Discharging and Bleaching Agents. 9 pp.
—Sodium Hydrosulfite—Pulps
CPPA J.17P-92. Analysis of Hydrosulphites. 3 pp.
—Sodium Hypochlorite
CGSB CAN/CGSB-15.31-93. Sodium Hypochlorite. 7 pp.
CNS K1163-74. Sodium Hypochlorite Bleaching Solution (Sep)(3397).
CNS K6332-74. Method of Test for Sodium Hypochlorite Bleaching Solution (Sep)(3398).
MOD UK CS 3137. Sodium Hypochlorite Solution (Withdrawn).
—Sodium Peroxide—Pulps
CPPA J.16P-92. Analysis of Peroxides. 3 pp.
—Zinc Hydrosulfite—Pulps
CPPA J.17P-92. Analysis of Hydrosulphites. 3 pp.

Bleaching Powder
Use For: Chlorinated Lime *See Also:* Bleaching Agents
CNS K1028-70. Bleaching Powder (Jan)(377).

Bleaching Powder *(Cont.)*
CNS K6038-74. Method of Test for Bleaching Powder (Sep)(725).
JIS K 1425-59. Bleaching Powder and High Test Hypochlorite.
JIS K 8388-61. Chlorinated Lime.
—Calcium Hypochlorites—Chemical Analysis
CPPA J.2H-64. Analysis of Calcium Hypochlorite Bleaching Powder, Bleach Liquor, and Bleach Sludge. 4 pp.

Bleed Screws
See Also: Screws
—Hydraulic Cylinders
CNS D2179-84. Bleed Screw of Hydraulic Brake Wheel Cylinders for Automobiles (Jun)(10934).

Bleeder Screws
See Also: Screws
—Aircraft—Pipe Fittings
SBAC AGS 1174 ISSUE 10. Pipe Couplings, Bleeder Screw, 1/4" B.S.F..
SBAC AGS 3049 ISSUE 5. Pipe Couplings Bleeder Screw 1/4 Inch U.N.F..

Bleeding
—Bedding Compounds
CGSB CAN2-19.0-M77METH 9.4-78. Methods of Testing Putty, Caulking and Sealing Compounds Oil Bleeding and Spotting of Bedding Compounds. 1 p.
—Concretes
CEN PREN 480 (Part 4)-91. Admixtures for Concrete, Mortar and Grout—Test Methods—Part 4: Determination of Bleeding. 5 pp.
CNS A3048-86. Method of Test for Bleeding Water from Concrete (Dec)(1235).
JIS A 1123-75. Method of Test for Bleeding of Concrete. 5 pp.
SAA AS 1012.6-83. Methods of Testing Concrete—Part 6: Method for the Determination of Bleeding of Concrete. 9 pp.
SNZ NZS 3112: Part 1-86. Methods of Test for Concrete Part 1: Tests Relating to Fresh Concrete. 20 pp.
—Enamels
CGSB 1-GP-71 METH 37.3-80. Methods of Testing Paints and Pigments Bleeding Resistance. 1 p.
—Glazing Compounds
CGSB CAN2-19.0-M77METH 9.3-78. Methods of Testing Putty, Caulking and Sealing Compounds Oil Bleeding and Spotting of Face Glazing Compounds. 1 p.
—Grout
SNZ NZS 3112: Part 4-86. Methods of Test for Concrete Part 4: Tests Relating to Grout. 10 pp.
—Pigments
BSI BS 3483: Part C7-80. 1980 Methods for Testing Pigments in Paints Part C7: Comparison of Resistance to Bleeding. 4 pp.
CNS K6589-80. Method of Test for Bleeding of Pigments (Sep)(6474).
—Sealing Materials
CGSB CAN2-19.0-M77METH 9.2-78. Methods of Testing Putty, Caulking and Sealing Compounds Bleeding of Sealing Compounds. 1 p.

Bleeding Resistance
Use: Bleeding

Blended Cements
See Also: Cements; Hydraulic Cements
SAA AS 3972-91. Portland and Blended Cements (Supersedes AS 1315—1982 and AS 1317—1982) (in Professional Package 58). 2 pp.
—Calcium Oxide Content
SAA AS 2350.10-91. Methods of Testing Portland and Blended Cements—Part 10: Calcium Oxide Content of Blended Cement (in Professional Package 58). 4 pp.
—Sampling
SAA AS 2349-91. Method of Sampling Portland and Blended Cements (in Professional Package 58). 2 pp.

Blenders
Use For: Continuous Blenders *See Also:* Appliances; Blending; Cooking Appliances; Food Processing Equipment
CNS C4066-86. Electric Juice Blenders (Mar)(2661). 3 pp.

Blenders *(Cont.)*
—Commercial
CSA C22.2 NO 195-M1987. Motor Operated Food Processing Appliances (Household and Commercial); (Gen Instr 1 Thru 2). 55 pp.
—Household
CSA C22.2 NO 195-M1987. Motor Operated Food Processing Appliances (Household and Commercial); (Gen Instr 1 Thru 2). 55 pp.
JIS C 9609-90. Electric Blenders and Electric Juicers for Household Use. 29 pp.

Blending
See Also: Blenders
—Fuel Oils
JIS F 7010-85. Application of Fuel Oil Blending System.

Blending Valves
Use: Mixing Valves

Blind Fasteners
Scope Note: Use a more specific term *See:* Blind Nuts; Blind Rivets; Grommet Nuts

Blind Nuts
See Also: Grommet Nuts; Nuts (Fasteners)
—Aircraft—Cages
SBAC AGS 3041 ISSUE 1. Cage for Blind Nuts.
SBAC AGS 3042 ISSUE 1. Cage for Blind Nuts.
SBAC AGS 3043 (V). Cage for Blind Nuts.

Blind Persons
Use: Visually Impaired Persons

Blind Rivets
Use For: A Rivets; C Rivets *See Also:* Rivets
DIN ENGL 7337-85. Blind Rivets with Breaking Mandrel (July). 9 pp.
—Aircraft
SBAC AS 46784-785 (V). Rivet, Blind, Self-Plugging, Flush Break, Mechanically Locked Spindle, Aluminium Alloy, Universal Head.
SBAC AGS 2039 ISSUE 3. Rivets, Blind, A.G.S. (Chobert and Tucker), Application.
SBAC RS 699 ISSUE 1. Installation Information for Rivet Blind, (AGS 3920 To 3923).
SBAC TS 38 ISSUE 2. Rivet, Blind, Self-Plugging, Flush Break, Mechanically Locked Stem, Corrosion-Resisting Steel, Plain and Cadmium Plated AS 46788 to AS 46791.
SBAC TS 39 ISSUE 2. Rivet, Blind, Self-Plugging, Flush Break, Mechanically Locked Stem, Aluminum Alloys AS 46784 to AS 46787.
SBAC TS 53 ISSUE 1. Rivet, Blind, Self-Plugging, Interference Lock (Audel) Corrosion Resisting Steel.
SBAC TS 117 ISSUE 2. SBAC Technical Specification for Rivet, Blind, Self-Plugging, Flush Break, Mechanically Locked Spindle (Audel MBC) Aluminium Alloy.
SBAC TS 154 ISSUE 1. Rivet, Blind, Self Plugging, Flush Break, Visible Mechanically Locked Stem, Aluminium Alloys.
SBAC TS 155 ISSUE 1. Rivet, Blind, Self Plugging, Flush Break, Visible Mechanically Locked Stem, Corrosion Resisting Steel.
—Aircraft—Countersunk Head
SBAC AS 46786-787 (V). Rivet, Blind, Self-Plugging, Flush Break, Mechanically Locked Spindle, Aluminium Alloy, 100 Degree CSK Head.
SBAC AS 46790 ISSUE 2. Rivet, Blind, Self-Plugging, Flush Break, Mechanically Locked Stem, Corrosion-Resisting Steel, 100 Degree CSK Head, Cadmium Plated.
SBAC AS 46791 ISSUE 1. Rivet, Blind, Self-Plugging, Flush Break, Mechanically Locked Stem, Corrosion-Resisting Steel, 100 Degree CSK Head, Unplated.
SBAC AS 63250-255 ISSUE 1. Rivet Blind, Self-Plugging, Flush Break, Visible Mechanically Locked Stem, Aluminium Alloys 100 Degree Countersunk Head.
SBAC AS 63270-271 ISSUE 1. Rivet Blind, Self-Plugging, Flush Break, Visible Mechanically Locked Stem, Corrosion Resisting Steel 100 Degree Countersunk Head.
SBAC AGS 3921 ISSUE 2. Rivet Blind, 100 Degree Countersunk Head, Self Plugging, Interference Lock (Avdel), C.R. Steel.
SBAC AGS 3923 (V). Rivet Blind, 100 Degree Countersunk Head, Self Plugging, Interference Lock(Avdel), C.R. Steel, Cadmium Coated.
SBAC AGS 4720 (V). Rivet, Blind, Self-Plugging, Flush Break, Mechanically Locked Spindle, (Avdel MBC), Aluminium Alloy, 100 Degrees GSE.
—Aircraft—Snap Head
SBAC AGS 2040 ISSUE 7. Rivets, Chobert, Steel D.T.D. 720 Snap Head.

Blind Rivets (Cont.)

—Aircraft—Snap Head (Cont.)
SBAC AGS 3920 ISSUE 2. Rivets, Blind, Snap Head, Self Plugging, Interference Lock (Avdel), C.R. Steel.
SBAC AGS 3922 (V). Rivet Blind, Snap Head, Self Plugging, Interference Lock (Avdel), C.R. Steel, Cadium Coated.

—Aircraft—Universal Head
SBAC AS 46788 ISSUE 2. Rivet, Blind, Self-Plugging, Flush Break, Mechanically Locked Stem, Corrosion-Resisting Steel, Universal Head Cadmium Plated.
SBAC AS 46789 ISSUE 2. Rivet, Blind, Self-Plugging, Flush Break, Mechanically Locked Stem, Corrosion-Resisting Steel, Universal Head, Unplated.
SBAC AS 63240-245 ISSUE 1. Rivet Blind, Self-Plugging, Flush Break, Visible Mechanically Locked Stem, Aluminium Alloys Universal Head.
SBAC AS 63260-261 ISSUE 1. Rivet Blind, Self-Plugging, Flush Break, Visible Mechanically Locked Stem, Corrosion Resisting Steel Universal Head.
SBAC AGS 4719 (V). Rivet, Blind, Self Plugging, Flush Break, Mech. Locked Spindle (Avdel MBC), Alloy, Head.

—Automobiles
CNS D2167-83. Blind Rivets for Automobiles (Feb)(9986).
CNS D3151-83. Method of Test for Blind Rivets for Automobiles (Feb)(9987).

—Break Mandrel
BSI BS 7349: Part 1-93. 1993 Blind Rivets with Break Mandrel Part 1: Specification for Closed End Blind Rivets (E). 17 pp.
BSI BS 7349: Part 2-93. 1993 Blind Rivets with Break Mandrel Part 2: Specification for Open End Blind Rivets (E). 24 pp.

Blinds (Windows)
Scope Note: Use a more specific term See: Barriers; Draperies; Shutters (Blinds); Venetian Blinds

Blinker Units
Use: Flasher Units

Blister Copper
See Also: Copper

—Copper Content
CNS M3043-81. Methods for Determination of Copper in Blister Coppers (Apr)(7291).
JIS M 8125-76. Method for Determination of Copper in Blister Copper (R 1984). 6 pp.

—Gold Content
CNS M3041-81. Methods for Determination of Gold and Silver in Blister Coppers (Apr) (7289).
JIS M 8114-50. Determination of Gold and Silver in Blister Copper.

—Moisture Content
CNS M3044-81. Methods for Sampling and Determination of Moisture Content of Blister Copper (May) (7441).
JIS M 8102-76. Methods of Sampling and Determination of Moisture Content of Blister Copper.

—Sampling
CNS M3044-81. Methods for Sampling and Determination of Moisture Content of Blister Copper (May) (7441).
JIS M 8102-76. Methods of Sampling and Determination of Moisture Content of Blister Copper.

—Silver Content
CNS M3041-81. Methods for Determination of Gold and Silver in Blister Coppers (Apr) (7289).
JIS M 8114-50. Determination of Gold and Silver in Blister Copper.

Blistering (Coatings)
Scope Note: See the subheading Blistering under specific type of coating

Block Formats
See Also: Record Layouts; Variable Block Formats

—Numerical Control—Cutting Equipment—Ships
ISO 6582-83. Shipbuilding—Numerical Control of Machines—ESSI Format First Edition. 11 pp.
JTC1 6582-83. Shipbuilding—Numerical Control of Machines—ESSI Format First Edition. 11 pp.

Block Valves
Use: Isolation Valves

Blocking Resistance Testing

—Cables (Electric)—Aircraft
CEN PREN 3475 (Part 403)-92. Cables, Electrical, Aircraft Use Test Methods Part 403—Delamination and Blocking. 3 pp.

—Coated Fabrics
BSI BS 3424: Part 11-82. 1982 Testing Coated Fabrics Part 11: Method 13. Method for Determination of Resistance to Blocking. 4 pp.
ISO 5978-90. Rubber-or Plastics-Coated Fabrics—Determination of Blocking Resistance Second Edition. 6 pp.

—Latex Paints
SAA AS 1580.409. 3-92. Paints and Related Materials—Methods of Test—Part 409.3: Blocking Resistance of Latex Paint Finishes (in Professional Packages 30, 39). 2 pp.
SNZ NZS/AS 1580. 409.3-92. Methods of Test for Paints and Related Materials Part 409.3: Blocking Resistance to Latex Paint Finished (This is a Joint Standard with SAA AS 1580.409.3). 2 pp.

Blocking Signals
Use For: Unblocking Signals See Also: Signaling Systems

—CCITT No. 4 Signaling Systems
CCITT RECMN Q.129-89. Maximum Duration of a Blocking Signal—Specifications of Signalling Systems Nos. 4 and 5 (Study Group XI) 1 pp. 1 p.
CCITT RECMN Q.130-89. Special Arrangements in Case of Failures in the Sequence of Signals—Specifications of Signalling Systems Nos. 4 and 5 (Study Group XI) 1 pp. 1 p.

—CCITT No. 6 Signaling Systems
CCITT RECMN Q.266-89. 4.6 Blocking and Unblocking Sequences and Control of Quasi-Associated Signalling—Specifications of Signalling System No. 6 (Study Group XI) 3 pp. 3 pp.

—Telecommunication Traffic—Telex Communications
CCITT RECMN U.6-89. Prevention of Fraudulent Transit Traffic in the Fully Automatic International Telex Service—Telegraph Switching (Study Group IX) 1 pp. 1 p.

Blocks
Scope Note: For additional listings, use a more specific term See Also: Brake Blocks; Building Blocks; Building Board; Calcium Silicate Blocks; Concrete Blocks; Pillow Blocks; Pulley Blocks; Sheave Blocks; V Blocks; Wood Blocks

—Chains
CNS B4048-81. Chain Blocks (Sep)(7878).

—Chains—Aircraft
SBAC AS 195-198 ISSUE 3. Bi-Planar Block, Chain. 1 p.

—Chains—Aircraft—Assembly
SBAC AS 173 ISSUE 5. Assembly of Chain to Bi-Planar Block. 1 p.

—Setup
CNS B3316-83. Setup Blocks Adjustable Type (Mar)(5058).

—Ships
CNS F3084-81. Ships' Small Size Steel Blocks (Jan)(6914).

—Shore Hardness Testing
CNS B6094-89. Standard Blocks of Shore Hardness (Jan)(12481).

—Tenon
CNS B3314-83. Loose Tenon Blocks (Mar)(5056).

Blood
Use For: Blood Platelets See Also: Blood Culture Media; Blood Refrigerators; Blood Serum; Blood Storage; Serum Albumin

—Cyanmethemoglobin—Photometry
BSI BS 3985-78. 1978 Specification for Cyanmethaemoglobin (Haemiglobincyanide) Solution for Photometric Haemoglobinometry. 7 pp.

—Medical Equipment—Chemical Reactions
ISO 10093 Pt 4-92. Medical Devices—Part 4: Selection of Tests for Interactions with Blood First Edition. 24 pp.

—Refrigerated Containers
SAA AS 3864-91. Medical Refrigeration Equipment—for the Storage of Blood and Blod Products, and Containers for the Transport of Blood and Blood Products (In Professional Package 17F). 30 pp.

Blood (Cont.)

—Test Specimens—Containers
BSI BS 4851-82. 1982 Single Use Labelled Medical Specimen Containers for Haematology and Biochemistry. 7 pp.
BSI BS 6242-82. 1982 Single Use Unlabelled Medical Specimen Containers for Haematology and Biochemistry. 4 pp.

Blood Cells
Scope Note: Use a more specific term See: Blood

Blood Culture Media
See Also: Blood

—Bottles
DIN ENGL 58942 Pt 2-85. Medical Microbiology; Culture Media Used in Bacteriology; Ready-to-Use Blood Culture Bottles; Physical, Chemical and Biological Requirements (Sept). 4 pp.

Blood Donors
See Also: Blood Transfusion Equipment; Personnel

—Medical Requirements
NATO STANAG 2939 ED 2 AMD 1-87. Medical Requirements for Blood, Blood Donors and Associated Equipment. 8 pp.

Blood Groupings

—Serum
CGSB CAN/CGSB-106.1-M86. Blood Grouping Sera. 11 pp.
CGSB CAN/CGSB-106.3-M86. Anti-Rh Typing Sera, Anti-D. 11 pp.
CGSB CAN/CGSB-106.4-M86. Anti-Rh Typing Sera, (Other Than Anti-D). 12 pp.

Blood Meal
Use: Bone Meal

Blood Platelets
Use: Blood

Blood Pressure Apparatus
Use: Sphygmomanometers

Blood Pumps
See Also: Medical Electrical Equipment; Medical Equipment; Pumps

—Cardiotomy
JIS T 1603-89. Electric Motor Driven Blood Pump for Cardiopulmonary Bypass. 12 pp.

Blood Refrigerators
See Also: Blood; Refrigerators
BSI BS 4376-82. 1982 Amd 1 Electrically Operated Blood Storage Refrigerators (AMD 6326) April 30, 1991. 8 pp.
BSI BS 4376: Part 1-91. 1991 Electrically Operated Blood Storage Refrigerators Part 1: Specification for Closed Reach-in Types. 12 pp.

Blood Serum
See Also: Blood; Serum Albumin

—Anti-A Grouping
CGSB CAN/CGSB-106.1-M86. Blood Grouping Sera. 11 pp.

—Anti-AB Grouping
CGSB CAN/CGSB-106.1-M86. Blood Grouping Sera. 11 pp.

—Anti-B Grouping
CGSB CAN/CGSB-106.1-M86. Blood Grouping Sera. 11 pp.

—Anti-Rh
CGSB CAN/CGSB-106.4-M86. Anti-Rh Typing Sera, (Other Than Anti-D). 12 pp.

—Anti-Rh Grouping—Anti-D
CGSB CAN/CGSB-106.3-M86. Anti-Rh Typing Sera, Anti-D. 11 pp.

Blood Storage
See Also: Blood; Blood Transfusion Equipment

—Packs
SAA AS 3787.2-92. General Requirements for Single-Use, Sterile, Plasticized Polyvinyl Chloride (PVC) Blood Packs for Whole Blood and Blood Components—Part 2: Multiple Blood Pack Systems (in Professional Package 17B). 54 pp.

Blood Transfusion Equipment
Use For: Transfusion Equipment See Also: Blood Donors; Blood Storage; Blood Transfusion Sets; Medical Electrical Equipment; Medical Equipment

Blood Transfusion Equipment (Cont.)
BSI BS 2463-62. (WITHDRAWN) 1962 Amd 1 Transfusion Equipment for Medical Use (N) (Superseded by BS 2463: Part 1: 1990 & Part 2: 1989). 43 pp.
MOD UK DSTAN 65-11-83. Blood Transfusion Equipment Issue 2. 10 pp.

—Bottles—Glass
ISO 1135 Pt 1-87. Transfusion Equipment for Medical Use—Part 1: Glass Transfusion Bottles, Closures and Caps Second Edition. 13 pp.

—Containers
BSI BS 2463: Part 1-90. 1990 Transfusion Equipment for Medical Use Part 1: Collapsible Containers for Blood and Blood Components. 15 pp.

—Containers—Single Use
ISO 4822-81. Single Use Blood Specimen Containers up to 25 ml Capacity First Edition. 6 pp.

—Interdepartmental Procurement
NATO STANAG 2939 ED 2 AMD 1-87. Medical Requirements for Blood, Blood Donors and Associated Equipment. 8 pp.

—Lids
ISO 1135 Pt 1-87. Transfusion Equipment for Medical Use—Part 1: Glass Transfusion Bottles, Closures and Caps Second Edition. 13 pp.

—Plastic Containers—Collapsible
ISO 3826-93. Plastics Collapsible Containers for Human Blood and Blood Components First Edition. 21 pp.

—Stoppers
ISO 1135 Pt 1-87. Transfusion Equipment for Medical Use—Part 1: Glass Transfusion Bottles, Closures and Caps Second Edition. 13 pp.

Blood Transfusion Sets
See Also: Blood Transfusion Equipment

—Blood Taking—Single Use
ISO 1135 Pt 3-86. Transfusion Equipment for Medical Use—Part 3: Blood-Taking Set First Edition. 11 pp.

—Single Use
BSI BS 2463: Part 2-89. 1989 Amd 1 Transfusion Equipment for Medical Use Part 2: Specification for Administration Sets (AMD 6332) September 30, 1991. 17 pp.
ISO 1135 Pt 4-87. Transfusion Equipment for Medical Use—Part 4: Transfusion Sets for Single Use First Edition. 13 pp.

Blood Typing
Use: Blood Groupings

Blotters, Desk
Use: Desk Blotters

Blotting Papers
See Also: Absorbent Papers; Papers
CGSB 53.34M-88. Paper, Blotting. 7 pp.

—Dry Batteries
CNS C3034-85. Method of Test for Blotting Paper for Dry Batteries (Jun)(3257).
CNS C4100-85. Blotting Paper for Dry Batteries (Jun)(3256). 1 p.

—Ink Absorption
CNS P3062-83. Method of Test for Ink Absorbency of Blotting Paper (Dec)(10727). 2 pp.
CPPA F.4-92. Absorption of Water and Ink by Bibulous and Blotting Paper. 2 pp.

Blouses
See Also: Clothing
JIS L 4114-80. Blouses.

—Dust
BSI BS 4771-71. 1971 Positive Pressure, Powered Dust Hoods and Blouses. 10 pp.

—Girls'
CGSB CAN/CGSB-49.45-M80. Blouses, Little Girls' and Girls', Regular and Chubby Ranges—Dimensions; (Amendment 2 Oct 1988). 14 pp.

—Women's
CGSB CAN/CGSB-49.210-M84. Blouses, Junior, Misses and Women's Sizes—Dimensions. 14 pp.

Blower Cleaners
Use: Vacuum Blowers

Blower/Vacs
Use: Vacuum Blowers

Blowers (Cleaning Equipment)
Scope Note: Use a more specific term See: Cleaning Equipment; Sprayers; Vacuum Blowers; Vacuum Cleaners

Blowers (Ventilation)
See Also: Air Handling Equipment; Centrifugal Fans; Cooling Systems; Exhaust Blowers; Exhaust Systems; Fans; Materials Handling Equipment; Nozzles; Pneumatic Conveyors; Refrigeration Equipment; Rotary Blowers; Rotary Compressors; Sprayers; Ventilation Equipment
CNS B7038-73. Method of Test for Blowers (Aug)(2726).

—Glossaries
JIS B 0132-84. Glossary of Terms for Fans, Blowers and Compressors.

—Mist
CNS B7046-81. Inspection Standard of Mist Blowers and Desters (Jan)(3374).

—Noise
CNS Z8024-82. Method of Noise Level Measurement for Fans Blowers and Compressors (Apr)(8753).
CNS Z8025-82. Method of Noise Spectrum Measurement for Fans Blowers and Compressors (Apr)(8754).

—Sound Pressure
JIS B 8346-91. Fans, Blowers and Compressors—Determination of A-Weighted Sound Pressure Level. 60 pp.

—Turbo
ISO 5389-92. Turbocompressors—Performance Test Code First Edition. 177 pp.

Blowing Agents
Use For: Foam Concentrates See Also: Cellular Plastics; Foam Rubber

—N,N-Dinitrosopentamethylenetetramine
CNS K6403-77. Method of Test for N,N'-Dinitrosopentameth-ylenete-Tramine Blowing Agent (Jun)(4141).

Blowpipes
See Also: Pneumatic Conveyors; Welding Equipment
BSI BS 6503-84. 1984 Handheld Blowpipes and Nozzles, Using Fuel Gas and Oxygen, for Gas Welding, Cutting and Related Processes. 20 pp.
CEN PREN 731-92. Air-Aspirated Hand Blowpipes—Specifications and Tests. 14 pp.
CEN PREN 874-92. Welding—Oxygen/Fuel Gas Blowpipes (Cutting Machine Type) of Cylindrical Barrel—Type of Construction, General Specifications, Test Methods. 25 pp.
DIN ENGL 8543 Pt 1-81. Blowpipes for Oxy-Fuel Gas Processes; Manual Blowpipes for Fuel Gas/Oxygen and Fuel Gas/Compressed Air; Types, Concepts, Requirements, Marking (June). 7 pp.
DIN ENGL 8543 Pt 2-86. Blowpipes for Oxy-Fuel Gas Processes; Manual Blowpipes for Fuel Gas/Oxygen and Fuel Gas/Compressed Air; Testing (Sept). 12 pp.
DIN ENGL 8543 Pt 4-81. Blowpipes for Oxy-Fuel Gas Processes; Manual Blowpipes for Fuel Gas/Aspirated Air; Types, Concepts, Requirements, Testing, Marking (Nov). 7 pp.
DIN ENGL 8543 Pt 5-86. Blowpipes for Oxy-Fuel Gas Processes; Machine Cutting Blowpipes for Fuel Gas/Oxygen; Types, Concepts, Requirements, Marking, Testing (Sept). 11 pp.
ISO 5172-77. Manual Blowpipes for Welding and Cutting First Edition. 13 pp.
ISO 9012-88. Air-Aspirated Hand Blowpipes—Specifications First Edition. 10 pp.

—Cutting Equipment
BSI BS 6503-84. 1984 Handheld Blowpipes and Nozzles, Using Fuel Gas and Oxygen, for Gas Welding, Cutting and Related Processes. 20 pp.
CEN PREN 874-92. Welding—Oxygen/Fuel Gas Blowpipes (Cutting Machine Type) of Cylindrical Barrel—Type of Construction, General Specifications, Test Methods. 25 pp.
ISO 5172-77. Manual Blowpipes for Welding and Cutting First Edition. 13 pp.

—Oxyacetylene Welding Equipment
JIS B 6801-91. Manual Blowpipes for Oxyacetylene Welding. 15 pp.

Blue Lead, Basic Sulfate
Use: Lead Sulfate, Blue Basic

Blueberries
See Also: Food; Fruits

—Storage
ISO 6664-83. Bilberries and Blueberries—Guide to Cold Storage First Edition. 5 pp.

Bluefish
Use For: Skipjack See Also: Fish; Mackerel

—Dried—Inspection
CNS N6045-81. Method of Test for Dried Shipjacks and Mackerels (Aug)(2417). 1 p.

—Dried—Sampling
CNS N6045-81. Method of Test for Dried Shipjacks and Mackerels (Aug)(2417). 1 p.

Blushing

—Cellulose Nitrate Lacquers
CGSB 1-GP-71 METH 38.4-81. Methods of Testing Paints and Pigments Blushing Resistance Cellulose Nitrate Lacquer at High Humidity. 1 p.

—Lacquer Thinners
CGSB 1-GP-71 METH 38.1-80. Methods of Testing Paints and Pigments Blushing Resistance Lacquer Thinner. 1 p.

—Lacquers
CGSB 1-GP-71 METH 38.2-80. Methods of Testing Paints and Pigments Blushing Resistance Lacquers at Normal Humidity. 1 p.
CGSB 1-GP-71 METH 38.3-80. Methods of Testing Paints and Pigments Blushing Resistance Lacquers at High Temperature and Humidity. 1 p.

Blushing Resistance
Use: Blushing

BMC
Use: Bulk Molding Compounds

BN 2A B and T Islander Aircraft
See Also: Aircraft

—Antenna Positions
CAA. BN2A B and T Series Islander (Approved Aerial Positions). 1 p.

BN 2A MK III Trislander Aircraft
See Also: Aircraft

—Antenna Positions
CAA. BN2A MK.III Series Trislander (Approved Aerial Positions). 1 p.

BNC Coaxial Connectors
Scope Note: See also the subheading BNC under specific coaxial components See Also: Coaxial Connectors

— 50 Ohms
BSI BS 9210 N004-81. 1981 Detail Specification for Radio Frequency Coaxial Connector (Series BNC), Unsealed, Soldered, Captive Contact 50 ohms, Basic Assessement Level. 17 pp.
BSI BS CECC 22120-93. 1993 Sectional Specification: Radio Frequency Coaxial Connectors Series BNC (T). 40 pp.
BSI BS CECC 22120-81. 1981 Radio Frequency Coaxial Connectors, Series BNC: Sectional Specification. 31 pp.
BSI BS CECC 22121-81. 1981 Radio Frequency Coaxial Connectors, Series BNC: Blank Detail Specification. 11 pp.
CECC CECC 22 120 ISSUE 2-92. Sectional Specification: Radio Frequency Coaxial Connectors; Series BNC (En, Fr, Ge). 112 pp.
CECC CECC 22 121 ISSUE 1-79. Blank Detail Specification: Radio Frequency Coaxial Connectors; Series BNC (En, Fr, Ge). 31 pp.
IEC 169 Pt 8-78. Radio-Frequency Connectors Part 8: R.F. Coaxial Connectors with Inner Diameter of Outer Conductor 6.5 mm (0.256 in) with Bayonet Lock—Characteristic Impedance 50 Ohms (Type BNC) First Edition. 51 pp.
IEC 313-83. Coaxial Cable Connectors Used in Nuclear Instrumentation Second Edition. 9 pp.

— 50 Ohms—Adapters—Preferred Products List
CECC CECC MUAHAG Vol 3B IS 4-91. Preferred Products List; Connectors; R.F. and Fibre Optics (En, Fr, Ge). 65 pp.

— 50 Ohms—Adapters—Tee—Preferred Products List
CECC CECC MUAHAG Vol 3B IS 4-91. Preferred Products List; Connectors; R.F. and Fibre Optics (En, Fr, Ge). 65 pp.

— 50 Ohms—Crimp—Plugs
CECC CECC 22 121-802 ISSUE 1-85. Detail Specification: Radio Frequency Coaxial Connectors; Series BNC (En, Fr, Ge). 21 pp.

INTERNATIONAL AND NON-U.S. NATIONAL STANDARDS
SUBJECT INDEX

Boats

BNC Coaxial Connectors *(Cont.)*

— 50 Ohms—Crimp—Plugs—Preferred Products List
CECC CECC MUAHAG Vol 3B IS 4-91. Preferred Products List; Connectors; R.F. and Fibre Optics (En, Fr, Ge). 65 pp.

— 50 Ohms—Crimp—Receptacles—Preferred Products List
CECC CECC MUAHAG Vol 3B IS 4-91. Preferred Products List; Connectors; R.F. and Fibre Optics (En, Fr, Ge). 65 pp.

— 50 Ohms—Hermetically Sealed—Receptacles
CECC CECC 22 121-810 ISSUE 1-85. Detail Specification: Radio Frequency Coaxial Connectors; Series BNC (En, Fr, Ge). 21 pp.

— 50 Ohms—Plugs
CECC CECC 22 121-007 ISSUE 1-82. BS CECC 22 121-007; Series BNC Connector, Unsealed, Captive Contact 50 Ohms; Free Plug, Cable Outlet (En). 6 pp.
CECC CECC 22 121-801 ISSUE 1-85. Detail Specification: Radio Frequency Coaxial Connectors; Series BNC (En, Fr, Ge). 21 pp.
MOD UK DSTAN 59-35: Pt 4:Sec 1-82. Connectors, Electrical Part 4: Radio Frequency Coaxial Connectors (See Also EPIC Database) Section 1: Pattern 9210 N0004 (Series BNC Unsealed) Issue 1. 12 pp.
MOD UK DSTAN 59-35: Pt 4:Sec 2-01. Connectors, Electrical Part 4: Radio Frequency Coaxial Connectors (See Also EPIC Database) Section 2: Pattern 9210 F0013 (Series BNC Sealed) List of Items Conforming to BS 9210 F0013 Issue 1; Amendment 1. 12 pp.

— 50 Ohms—Preferred Products List
CECC CECC MUAHAG Vol 3B IS 4-91. Preferred Products List; Connectors; R.F. and Fibre Optics (En, Fr, Ge). 65 pp.

— 50 Ohms—Quality Assurance
BSI BS 9210-84. 1984 Amd 3 Radio Frequency Connectors of Assessed Quality: Generic Data and Methods of Test (AMD 6629) September 30, 1991. 93 pp.
BSI BS 9210 N001: Pt 1-75. 1975 Radio Frequency Connector (Type BNC), Sealed, Soldered, Captive Content, 50 ohms Part 1: Detail Specification. Full Assessment Level. 23 pp.
BSI BS 9210 N003-76. 1976 Detail Specification for Radio Frequency Connectors (Type BNC), Unsealed, Soldered, Captive Contact, 50 ohms Full Assessment Level. 21 pp.

— 50 Ohms—Receptacles
CECC CECC 22 121-001 ISSUE 1-82. BS CECC 22 121-001; Series BNC Connector, Unsealed, Captive Contact 50 Ohms; Fixed Straight Socket, Single Hole Rear Mounting, Cable Outlet (En). 6 pp.
CECC CECC 22 121-002 ISSUE 1-82. BS CECC 22 121-002; Series BNC Connector, Usealed, Captive Contact 50 Ohms; Fixed Straight Socket, 4-Hole Fixing, Cable Outlet (En). 6 pp.
CECC CECC 22 121-003 ISSUE 1-82. BS CECC 22 121-003; Series BNC Connector, Unsealed, Captive Contact 50 Ohms, Fixed Straight Socket, 4 Hole Fixing, Solder Bucket (En). 6 pp.
CECC CECC 22 121-005 ISSUE 1-82. BS CECC 22 121-005; Series BNC Connector, Unealed, Captive Contact 50 Ohms; Free Socket, Cable Outlet (En). 6 pp.
CECC CECC 22 121-006 ISSUE 1-82. BS CECC 22 121-006; Series BNC Connector, Unsealed, Captive Contact 50 Ohms; Fixed Straight Socket, Single Hole Mounting, Solder Bucket (En). 6 pp.
CECC CECC 22 121-007 ISSUE 1-82. BS CECC 22 121-007; Series BNC Connector, Unsealed, Captive Contact 50 Ohms; Free Plug, Cable Outlet (En). 6 pp.
CECC CECC 22 121-804 ISSUE 1-85. Detail Specification: Radio Frequency Coaxial Connectors; Series BNC (En, Fr, Ge). 21 pp.
CECC CECC 22 121-805 ISSUE 1-85. Detail Specification: Radio Frequency Coaxial Connectors; Series BNC (En, Fr, Ge). 21 pp.
CECC CECC 22 121-806 ISSUE 1-85. Detailed Specification: Radio Frequency Coaxial Connectors; Series BNC (En, Fr, Ge). 21 pp.
CECC CECC 22 121-806 ISSUE 1-85. Detail Specification: Radio Ferquency Coaxial Connectors; Series BNC (En, Fr, Ge). 21 pp.
CECC CECC 22 121-807 ISSUE 1-85. Detail Specification: Radio Frequency Coaxial Connectors; Series BNC (En, Fr, Ge). 21 pp.
CECC CECC 22 121-808 ISSUE 1-85. Detail Specification: Radio Frequency Coaxial Connectors; Series BNC (En, Fr, Ge). 21 pp.
CECC CECC 22 121-809 ISSUE 1-85. Detail Specification: Radio Frequency Coaxial Connectors; Series BNC (En, Fr, Ge). 21 pp.

BNC Coaxial Connectors *(Cont.)*

— 50 Ohms—Receptacles *(Cont.)*
MOD UK DSTAN 59-35: Pt 4:Sec 1-82. Connectors, Electrical Part 4: Radio Frequency Coaxial Connectors (See Also EPIC Database) Section 1: Pattern 9210 N0004 (Series BNC Unsealed) Issue 1. 12 pp.
MOD UK DSTAN 59-35: Pt 4:Sec 2-01. Connectors, Electrical Part 4: Radio Frequency Coaxial Connectors (See Also EPIC Database) Section 2: Pattern 9210 F0013 (Series BNC Sealed) List of Items Conforming to BS 9210 F0013 Issue 1; Amendment 1. 12 pp.

— 50 Ohms—Receptacles—Preferred Products List
CECC CECC MUAHAG Vol 3B IS 4-91. Preferred Products List; Connectors; R.F. and Fibre Optics (En, Fr, Ge). 65 pp.

— 50 Ohms—Right Angle—Plugs
CECC CECC 22 121-004 ISSUE 1-82. BS CECC 22 121-004; Series BNC Connector, Unsealed, Captive Contact 50 Ohms; Right Angle Free Plug, Cable Outlet (En). 6 pp.
CECC CECC 22 121-803 ISSUE 1-85. Detail Specification: Radio Frequency Coaxial Connectors; Series BNC (En, Fr, Ge) ERRATUM (Fr). 22 pp.

— 50 Ohms—Right Angle—Plugs—Preferred Products List
CECC CECC MUAHAG Vol 3B IS 4-91. Preferred Products List; Connectors; R.F. and Fibre Optics (En, Fr, Ge). 65 pp.

— 50 Ohms—Right Angle—Receptacles
CECC CECC 22 121-004 ISSUE 1-82. BS CECC 22 121-004; Series BNC Connector, Unsealed, Captive Contact 50 Ohms; Right Angle Free Plug, Cable Outlet (En). 6 pp.
CECC CECC 22 121-811 ISSUE 1-85. Detail Specification: Radio Frequency Coaxial Connectors; Series BNC (En, Fr, Ge). 21 pp.

—Engineering Drawings—Quality Assurance
BSI BS 9210 N001: Pt 2-82. 1982 Radio Frequency Connector (Type BNC), Sealed, Soldered, Captive Content, 50 ohms Part 2: Control Drawing, Mating Face Details and Gauge Information. 6 pp.

—Quality Assurance
BSI BS CECC 22120-93. 1993 Sectional Specification: Radio Frequency Coaxial Connectors Series BNC (T). 40 pp.
BSI BS CECC 22120-81. 1981 Radio Frequency Coaxial Connectors, Series BNC: Sectional Specification. 31 pp.
BSI BS CECC 22121-81. 1981 Radio Frequency Coaxial Connectors, Series BNC: Blank Detail Specification. 11 pp.

Board (Building)
Use: Building Board

Board (Paper)
Use: Paperboard

Boardroom Tables
Use: Conference Room Tables

Boards (Gymnastic Equipment)
Use: Mats (Gymnastic Equipment)

Boats (Chemistry)
Scope Note: Use a more specific term
See: Combustion Boats

Boats (Marine)
Scope Note: For additional listings, use a more specific term *See Also:* Barges; Canoes; Dinghies; Hulls; Lifeboats; Marine Engines; Motor Boats; Pilot Vessels; Pleasure Boats; Rafts; Sailboats; Ships; Survival Craft; Tugboats; Vehicles (Transportation); Warships; Yachts

—Alarm Systems—Panels
JIS F 0415-89. Arrangement of Main Engine Control Stand in Wheelhouse and Local Alarm Panel for Small Shi

—Aluminum—Preservation
MOD UK NES 742: Part 3-89. Preservation of Small Craft and Boats Part 3: Aluminium Construction Issue 1 (07.89). 21 pp.

—Bilge Pumps
BSI BS EN 28849-93. 1993 Small Craft—Electrically Operated Bilge-Pumps (ISO 8849: 1990) (Q). 10 pp.
ISO 8849-90. Small Craft—Electrically Operated Bilge-Pumps First Edition. 4 pp.

Boats (Marine) *(Cont.)*

—Control Consoles—Positioning
JIS F 0415-89. Arrangement of Main Engine Control Stand in Wheelhouse and Local Alarm Panel for Small Shi

—Electrical Installations
CSA C22.2 NO 183.1-M1982. Alternating-Current (AC) Electrical Installations on Boats (R 1992); (Gen Instr 1). 27 pp.
CSA C22.2 NO 183.2-M1983. DC Electrical Installations on Boats (R 1992); (Gen Instr 1). 31 pp.
DIN VDE 0100 Pt 721-84. Erection of Power Installations with Rated Voltages up to 1000 V; Caravans, Boats and Yachts, as Well as Power Supply Thereof at Camping Sites and Berths (Apr). 10 pp.

—Electrical Protection Equipment
BSI BS 7489-91. 1991 Protection of Electrical Devices Used on Small Craft to Prevent Ignition of Surrounding Flammable Gases. 12 pp.
BSI BS EN 28846-93. 1993 Small Craft—Electrical Devices—Protection Against Ignition of Surrounding Flammable Gases (ISO 8846: 1990). 14 pp.
ISO 8846-90. Small Craft—Electrical Devices—Protection Against Ignition of Surrounding Flammable Gases First Edition. 10 pp.

—Fans—Air Flow
ISO 9097-91. Small Craft—Electric Fans First Edition. 6 pp.

—Fiberglass Reinforced Plastics
MOD UK NES 166-01. Requirements for Glass Woven Roving Fabrics for Ship Structures Issue 1 (11.83); Amendment 1. 32 pp.
MOD UK NES 166-93. Requirements for Glass Reinforcing Fabrics for Ships, Boats and Craft Structures Issue 2 (02.93). 53 pp.
MOD UK NES 742: Part 2-88. Preservation of Small Craft and Boats Part 2: GRP Construction Issue 3 (11.88). 13 pp.
MOD UK NES 742: Part 2-91. Requirements for the Preservation of Small Craft and Boats Part 2: G R P Construction Issue 4 (12.91). 14 pp.
MOD UK NES 752-81. GRP Ships and Boats Maintenance, Survey and Repair Issue 2 (04.81). 48 pp.

—Fuel Systems
ISO 10088-92. Small Craft—Permanently Installed Fuel Systems and Fixed Fuel Tanks First Edition. 12 pp.

—Fuel Tanks
ISO 10088-92. Small Craft—Permanently Installed Fuel Systems and Fixed Fuel Tanks First Edition. 12 pp.

—Hoisting Slings—Repair
MOD UK NES 744-90. Requirements for the Manufacture and Repair of Rings and Links Issue 2 (10.90). 11 pp.

—Hydrometric—Positioning—Open Channel Flow Measurement
ISO 6420-84. Liquid Flow Measurement in Open Channels—Position Fixing Equipment for Hydrometric Boats First Edition. 13 pp.
SAA AS 3778.6.8-92. Measurement of Water Flow in Open Channels—Part 6: Measuring Devices, Instruments and Equipment—Part 6.8: Position Fixing Equipment for Hydrometric Boats (ISO 6420:1984). 9 pp.

—Inflatable—Safety
BSI BS MA 16-71. 1971 Inflatable Boats (Manual or Motor Propelled). 12 pp.
ISO 6185-82. Shipbuilding and Marine Structures—Inflatable Boats—Boats Made of Reinforced Elastomers or Plastomers First Edition; (Corrected and Reprinted -1983). 9 pp.
SNZ NZS 5829: Part 2-87. Code of Practice for the Safe Design and Use of Motor Powered Pleasure Boats Part 2: Inflatable Boats. 8 pp.

—Lightning Protection Equipment
ISO 10134-93. Small Craft—Electrical Devices—Lightning Protection First Edition. 9 pp.

—Liquefied Petroleum Gas—Cylinders
SNZ NZS 5428-83. Code of Practice for the Use of LPG for Domestic Purposes in Caravans and Boats (Reconfirmed 1988). 27 pp.

—Sanitary Facilities
SNZ NZS 5465-90. Self-Containment of Caravans, Motor Caravans and Boats. 12 pp.

—Space Heaters—Safety
CEN PREN 624-91. LPG Space Heating Equipment in Vehicles and Boats. 60 pp.

INDUSTRY STANDARDS

INTERNATIONAL AND NON-U.S. NATIONAL STANDARDS
SUBJECT INDEX

Boats (Marine) (Cont.)
—Toilet Facilities—Sewage Treatment
 BSI BS MA 101-86. 1986 Toilet Retention and Recirculation Systems for the Treatment of Toilet Waste on Small Craft. 5 pp.
 ISO 8099-85. Small Craft—Toilet Retention and Recirculating Systems for the Treatment of Toilet Waste First Edition. 5 pp.
—Wood—Preservation
 MOD UK NES 742: Part 1-91. Requirements for the Preservation of Small Craft and Boats Part 1: Wood Construction Issue 4 (10.91). 22 pp.

Bobbin Cores
See Also: Bobbins (Electric); Magnetic Cores
 MOD UK DEF-5195-62. Bobbins, Moulded (for Use with Cores, Magnetic, Ferrite, or Laminations, Magnetic). 19 pp.
—Magnetic Cores
 MOD UK DEF-5195-62. Bobbins, Moulded (for Use with Cores, Magnetic, Ferrite, or Laminations, Magnetic). 19 pp.

Bobbins (Electric)
See Also: Bobbin Cores; Bobbins (Spools)
—Magnetic Wire
 CNS C2096-90. Plastic Bobbins for Magnet Wires (Jun)(7780).

Bobbins (Spools)
See Also: Bobbins (Electric); Bobbins (Thread); Spinning Frames; Spinning Machinery; Textile Machinery; Winders
—Aircraft—Parachutes
 SBAC AS 478 ISSUE 2(I). Bobbin Anti-Spin Parachute Jettison Slip. 1 p.
—Cheese
 CNS L2019-81. Cheese Bobbin (Mar)(595).
 JIS L 5115-75. Cheese Bobbins.
—Cigarette Papers
 CNS P2024-84. Cigarette Paper Bobbins (Feb)(2304). 2 pp.
—Classification
 JIS L 0311-83. Classification of Tubes and Bobbins.
—Magnet Wire
 JIS C 3201-82. Plastic Bobbins for Magnet Wires. 11 pp.

Bobbins (Thread)
See Also: Bobbins (Spools); Cones (Bobbins)
—Looms
 JIS L 6416-92. Leno Selvage Bobbins for Looms. 6 pp.
—Ring Spinning Frames
 CNS L2026-81. Plastic Bobbins for Ring Spinning Frame (Apr)(7263).
 CNS L3111-81. Method of Test for Plastic Bobbins for Ring Spinning Frame (Apr) (7264).
 JIS L 5302-92. Plastic Bobbins for Ring Spinning Frames. 13 pp.
—Roving
 CNS L2011-81. Roving Bobbins (Mar)(587).
—Sewing Machines—Household
 CNS B2404-90. Shuttle Bobbin Cases for Household Sewing Machines (Mar)(5185).
 CNS B2405-90. Bobbins for Household Sewing Machines (Mar)(5186).
 JIS B 9024-79. Bobbins for Household Sewing Machines (R 1984). 6 pp.
—Sewing Machines—Industrial
 CNS B2453-89. Bobbins of Sewing Machines for Industrial Use (Jan)(5410).
 CNS B2454-89. Bobbin Cases of Sewing Machines for Industrial Use (Jan)(5411).
 JIS B 9073-64. Bobbins of Sewing Machines for Industrial Use. 4 pp.
 JIS B 9074-64. Bobbin Cases of Sewing Machines for Industrial Use. 4 pp.
—Spinning Machinery
 BSI BS 1836: Part 1-82. 1982 Flyer Bobbins Part 1: Dimensions for Plastics Flyer Bobbins for Speed Frames. 4 pp.
 BSI BS 5508-77. 1977 Textile Machinery and Accessories. Condenser Bobbins for Woolen Spinning. Dimensions. 3 pp.
 ISO 344-81. Textile Machinery and Accessories—Spinning Machines—Flyer Bobbins First Edition. 4 pp.

Bobbins (Thread) (Cont.)
—Spinning Machinery (Cont.)
 ISO 1946-76. Textile Machinery and Accessories—Condenser Bobbins for Woollen Spinning—Dimensions First Edition. 3 pp.
—Textile Machinery
 BSI BS 1809-83. 1983 Flanged Bobbins for Doubling and Twisting. 14 pp.
 ISO 6169-82. Textile Machinery and Accessories—Flanged Bobbins for Doubling and Twisting First Edition. 12 pp.
 JIS L 5116-83. Flyer Bobbins.
—Twisting Machinery
 CNS L2012-81. Twisting Bobbins (Mar)(588).
 CNS L2024-81. Twisting Bobbins for Italian Type Twisters (Mar)(7172).
 JIS L 5111-75. Twisting Bobbins.
 JIS L 6102-62. Twisting Bobbins for Italian Type Twisters. 5 pp.
—Warping—Cotton
 CNS L2013-81. Warping Bobbin for Cotton (Mar)(589).
 JIS L 6407-58. Warping Bobbin for Cotton. 5 pp.
—Weft
 CNS L2010-81. Weft Bobbins for Shuttle (Mar)(586).
 CNS L2014-81. Weft Bobbin for Cop Change (Mar)(590).
 JIS L 5113-76. Weft Bobbins for Cop Change.
 JIS L 6502-79. Weft Bobbins for Shuttles.
—Winding
 CNS L2023-81. Winding Bobbins for Italian Type Twisters (Mar)(7171).
 JIS L 6103-62. Winding Bobbins for Italian Type Twisters. 4 pp.
—Winding—Spindles
 CNS L2039-82. Spindles for Winding Bobbin (Italian Type) (Mar)(8615).
 JIS L 6106-79. Spindles for Winding Bobbin (Italian Type).

BOD
Use: Biochemical Oxygen Demand

Body Armor
—Aircraft
 SBAC RS 119 (V). Armour Protection.
—Ballistic Limit Protection
 NATO STANAG 2920 ED 1 AMD 0-00. Ballistic Test Method for Personal Armours. 9 pp.

Body Belts, Safety
Use: Safety Belts

Boeing A75 N1 Aircraft
See Also: Aircraft
—Antenna Positions
 CAA. Boeing A75 N1 and B75. 1 p.

Boeing Kawasaki 107 Aircraft
See Also: Aircraft
—Accidents
 CAA. Boeing Vertol/Kawasaki 107 Series (World Helicopter Accident Summary). 1 p.

Boeing Vertol 44 Aircraft
See Also: Aircraft
—Accidents
 CAA. Boeing Vertol 44 Series (World Helicopter Accident Summary). 1 p.

Boeing Vertol Helicopters
See Also: Helicopters
—Foreign Airworthiness Directives
 CAA. Boeing Vertol Series Helicopters (Foreign Airworthiness Directives). 2 pp.

Boeing Vertol 107 Aircraft
See Also: Aircraft
—Accidents
 CAA. Boeing Vertol/Kawasaki 107 Series (World Helicopter Accident Summary). 1 p.

Boeing Vertol 234 Aircraft
See Also: Aircraft
—Antenna Positions
 CAA. Boeing Vertol 234 (Approved Aerial Positions). 1 p.

Boeing Vertol 234 Chinook Aircraft
See Also: Aircraft

Boeing Vertol 234 Chinook Aircraft (Cont.)
—Accidents
 CAA. Boeing Vertol 234 Chinook Series (World Helicopter Accident Summary). 1 p.

Boeing Vertol 234LR Aircraft
See Also: Aircraft
—Data Sheets
 CAA FR13 ISSUE 2. Boeing Vertol 234LR. 2 pp.

Boeing 307 Stratoline Aircraft
See Also: Aircraft
—Accidents
 CAA. Boeing 307 Stratoline (World Airline Accident Summary). 1 p.

Boeing 377 Stratocruiser Aircraft
See Also: Aircraft
—Accidents
 CAA. British 377 Stratocruiser (World Airline Accident Summary). 1 p.

Boeing 707 Aircraft
See Also: Aircraft
—Accidents
 CAA. Boeing 707/720 Series (World Airline Accident Summary). 3 pp.
 CAA. Boeing 707/720 Series. 2 pp.
—Antenna Positions
 CAA. Boeing 707 and 720 (Approved Aerial Positions). 1 p.
—Certification
 CAA. Boeing 707-138B and 720-051B: Explanatory Notes. 3 pp.
 CAA. Boeing 707-100B Series: Amplification of Special Conditions No's 17 and 18. 2 pp.
 CAA. Boeing 707-300 Series. 3 pp.
 CAA. Boeing 707-321 and 331: Amplification of Special Conditions No's 17 and 18. 4 pp.
 CAA. Boeing 707-300B and 300C: Amplification of Special Conditions No's 18 and 19. 3 pp.
 CAA. Boeing 707-400 Series. 7 pp.
 CAA. Boeing 707-436 Appendix 1: UK Validation Design Appraisal. 26 pp.
—Foreign Airworthiness Directives
 CAA. Boeing 707/720 Series Aircraft (Foreign Airworthiness Directives). 10 pp.

Boeing 720 Aircraft
See Also: Aircraft
—Accidents
 CAA. Boeing 707/720 Series (World Airline Accident Summary). 3 pp.
 CAA. Boeing 707/720 Series. 2 pp.
—Antenna Positions
 CAA. Boeing 707 and 720 (Approved Aerial Positions). 1 p.
—Foreign Airworthiness Directives
 CAA. Boeing 707/720 Series Aircraft (Foreign Airworthiness Directives). 10 pp.

Boeing 720B Aircraft
See Also: Aircraft
—Certification
 CAA. Boeing 720B: Amplification of Special Conditions No's 17 and 18. 2 pp.

Boeing 727 Aircraft
See Also: Aircraft
—Accidents
 CAA. Boeing 727 Series (World Airline Accident Summary). 1 p.
 CAA. Boeing 727. 4 pp.
—Antenna Positions
 CAA. Boeing 727-100 (Approved Aerial Positions). 1 p.
 CAA. Boeing 727-200 Series (Approved Aerial Positions). 1 p.
—Certification
 CAA. Boeing 727 Series 100, 100C and 200. 10 pp.
—Data Sheets
 CAA FA11 ISSUE 2. Boeing 727-100 Series, Model 727-46. 2 pp.
—Foreign Airworthiness Directives
 CAA. Boeing 727 Series Aircraft. 9 pp.

Boeing 737 Aircraft
See Also: Aircraft

—Accidents
CAA. Boeing 737 (World Airline Accident Summary). 1 p.
CAA. Boeing 737. 2 pp.

—Antenna Positions
CAA. Boeing 737 Series (Approved Aerial Positions). 1 p.

—Certification
CAA. Boeing 737-200 and 200 ADV. 7 pp.
CAA. Boeing 737-300 and 737-400. 17 pp.

—Data Sheets
CAA FA2 ISSUE 1. Boeing 737-200 Series, Model 204, 204C, 222, 204 (ADV), 219, 236,2S3, 2S3, 2K9, 2L9, 2M8, 2Q8, 2T5, 2T7, 2U4 (ADV), 219DA, 2T4, Boeing 737-300 Series Model 3T5. 8 pp.

—Foreign Airworthiness Directives
CAA. Boeing 737 Series Aircraft (Foreign Airworthiness Directives). 28 pp.

Boeing 747 Aircraft
See Also: Aircraft

—Accidents
CAA. Boeing 747 (World Airline Accident Summary). 1 p.
CAA. Boeing 747. 2 pp.

—Antenna Positions
CAA. Boeing 747. 1 p.

—Certification
CAA. Boeing 747—100,-100SF,-200B and-200F Series. 12 pp.

—Data Sheets
CAA FA5 ISSUE 1. Boeing 747-100 & 200 Series, Models 136 and 236B. 2 pp.

—Foreign Airworthiness Directives
CAA. Boeing 747 Series Aircraft (Foreign Airworthiness Directives). 13 pp.

Boeing 747-400 Aircraft
See Also: Aircraft

—Certification
CAA. Boeing 747-400 Series. 12 pp.

Boeing 757 Aircraft
See Also: Aircraft

—Accidents
CAA. Boeing 757 (World Airline Accident Summary). 1 p.

—Certification
CAA. Boeing 757-200. 10 pp.

—Data Sheets
CAA FA28 ISSUE 1. Boeing 757 Model 236, 2T7. 3 pp.

—Foreign Airworthiness Directives
CAA. Boeing 757 Series Aircraft (Foreign Airworthiness Directives). 7 pp.

Boeing 757-200 Aircraft
See Also: Aircraft

—Antenna Positions
CAA. Boeing 757-200 Series. 1 p.

Boeing 767 Aircraft
See Also: Aircraft

—Accidents
CAA. Boeing 767 (World Airline Accident Summary). 1 p.

—Antenna Positions
CAA. Boeing 767-200 and-300 Series (Approved Aerial Positions). 1 p.

—Certification
CAA. Boeing 767-200 with GE GE CF6-80A Engines. 20 pp.

—Data Sheets
CAA FA33 ISSUE 2. Boeing 767-200 Model 204. 3 pp.

—Foreign Airworthiness Directives
CAA. Boeing 767 Series Aircraft (Foreign Airworthiness Directives). 8 pp.

Boeing 767-300ER Aircraft
See Also: Aircraft

—Certification
CAA. Boeing 767-300ER. 22 pp.

Boiled Oil
See Also: Drying Oils; Linseed Oil; Oils

—Iodine Number
CNS K6804-7-84. Method of Test for Organic Coating Iodine Value of Boiled Oil (Jun)(10880-7).

Boiler Codes
See Also: Boilers; Codes

CNS B1023-89. Boiler Code (Code for Boiler Manufacture) (Apr)(2139).
CNS B1024-89. Boiler Code (Performance Test Code for General Instructions) (Apr)(2140).
CNS B1025-89. Boiler Code (Calculation Method of Boiler Efficiency for Land Use) (Apr)(2141).
CNS B1027-89. Boiler Code (Boiler Material and Material Test Code) (Apr)(2143).
CNS B1028-89. Boiler Code (Part 6, Boiler Operation and Maintenance Code) (Apr)(2144).
CNS B1341-84. Code for Miniature Boiler Manufacture (Jun)(10897).
CSA B51-M1991. Boiler, Pressure Vessel, and Pressure Piping Code; (Gen Instr 1). 96 pp.

—Gas Burners
CNS B1356-89. Boiler Code (Gas Burning Equipment) (Jul)(12554).

—Safety
CNS B1026-89. Boiler Code (Part 4, Boiler Safety Equipment Test Code) (Apr)(2142).

Boiler Feed Pumps
Use: Feed Pumps

Boiler Feedwater
Use: Feedwater

Boiler Furnaces
Use: Furnaces

Boiler Tubes
See Also: Boilers

—Alloy Steel
BSI BS 3059: Part 2-90. 1990 Steel Boiler and Superheater Tubes Part 2: Carbon, Alloy and Austenitic Stainless Steel Tubes with Specified Elevated Temperature Properties. 32 pp.
CNS G2082-87. Method of Test for Alloy Steel Tube for Boiler and Heat Exchanger (Aug)(7382).
CNS G3141-87. Alloy Steel Tube for Boiler and Heat Exchanger (Aug)(7381).
JIS G 3462-88. Alloy Steel Boiler and Heat Exchanger Tubes. 21 pp.

—Austenitic Stainless Steel
BSI BS 3059: Part 2-90. 1990 Steel Boiler and Superheater Tubes Part 2: Carbon, Alloy and Austenitic Stainless Steel Tubes with Specified Elevated Temperature Properties. 32 pp.

—Carbon Steel
BSI BS 3059: Part 1-87. 1987 Steel Boiler and Superheater Tubes Part 1: Low Tensile Carbon Steel Tubes Without Specified Elevated Temperature Properties. 15 pp.
BSI BS 3059: Part 2-90. 1990 Steel Boiler and Superheater Tubes Part 2: Carbon, Alloy and Austenitic Stainless Steel Tubes with Specified Elevated Temperature Properties. 32 pp.
CNS G2081-87. Method of Test for Carbon Steel Tubes for Boiler and Heat Exchanger (Aug)(7380).
CNS G3140-87. Carbon Steel Tubes for Boiler and Heat Exchanger (Aug)(7379). 12 pp.
JIS G 3461-88. Carbon Steel Boiler and Heat Exchanger Tubes. 24 pp.

—Deposits—Sampling
BSI BS 2455: Part 2-83. 1983 Methods of Sampling and Examining Deposits from Boilers and Associated Industrial Plant Part 2: Methods for Sampling and Examining Free-Side Deposits. 41 pp.
BSI BS 2455: Part 2-01. 1983 Amd 1 Sampling and Examining Deposits from Boilers and Associated Industrial Plant Part 2: Methods for Sampling and Examining Fire-Side Deposits (AMD 7751) October 15, 1993 (Q). 42 pp.

—Stainless Steel
CNS G2083-88. Method of Test for Stainless Steel Boiler and Heat Exchanger Tubes (Mar)(7384).
CNS G3142-88. Stainless Steel Boiler and Heat Exchanger Tubes (Mar)(7383). 23 pp.
JIS G 3463-88. Stainless Steel Boiler and Heat Exchanger Tubes. 33 pp.

—Steel
ISO 1129-80. Steel Tubes for Boilers, Superheaters and Heat Exchangers—Dimensions, Tolerances and Conventional Masses per Unit Length Second Edition. 6 pp.

—Steel—Marine
MOD UK DSTAN 44-1. Naval Boiler Tubes—Specifications and Sizes (Guide) Issue 1. 11 pp.

Boiler Water
See Also: Boilers; Feedwater; Industrial Water; Water
CNS B1312-89. Boiler Code of Boiler Feed Water and Boiler Water (Apr)(10231).
JIS B 8224-86. Testing Methods for Boiler Feed Water and Boiler Water. 93 pp.

Boiler Water Treatment
See Also: Feedwater Treatment; Water Treatment
BSI BS 2486-78. 1978 Recommendations for Treatment of Water for Land Boilers. 60 pp.
JIS B 8223-89. Water Conditioning for Boiler Feed Water and Boiler Water. 15 pp.

—Alkalinity
MOD UK M 9501/59. Determination of Solution Boiler Treatment (No Information) (Withdrawn).

—Ash Content
MOD UK M 9501/59. Determination of Solution Boiler Treatment (No Information) (Withdrawn).

—Marine
BSI BS 1170-83. 1983 Amd 1 Treatment of Water for Marine Boilers. 62 pp.

—Oxygen Content—Volumetric Analysis
MOD UK M 9501/59. Determination of Solution Boiler Treatment (No Information) (Withdrawn).

—Sodium Nitrate
CNS K1146-71. Sodium Nitrate for Boiler Water Treatment (Jan)(3166).

—Sodium Sulfite Anhydrous
CNS K1145-71. Anhydrated Sodium Sulfite for Boiler Water Treatment (Jan)(3165).

—Solids Content
MOD UK M 9501/59. Determination of Solution Boiler Treatment (No Information) (Withdrawn).

—Tannic Acid Content—Colorimetric Analysis
MOD UK M 9501/59. Determination of Solution Boiler Treatment (No Information) (Withdrawn).

Boilers
Use For: Gas Boilers; Heating Boilers; Hot Water Boilers; Hot Water Heating Boilers; Hot Water Space Heating Boilers; Hot Water Supply Boilers; Water Boilers See Also: Boiler Codes; Boiler Tubes; Boiler Water; Cast Iron Boilers; Chimneys; Electric Boilers; Fire Tube Boilers; Furnaces; Generators; Heat Balance; Heating, Ventilating and Air Conditioning Equipment; Heating Equipment; Oil Burners; Safety Plugs; Steam; Steam Boilers; Steam Condensers; Steel Boilers; Stokers; Waste Heat Boilers; Water Heaters; Watertube Boilers
BSI BS 1113: ENQ CASE 6-90. 1990 Design and Manufacture of Water-Tube Steam Generating Plant (Including Superheaters, Reheaters and Steel Tube Economizers) Enquiry Case 6: Use of BS 3604 591 Material. 1 p.
BSI BS 2790: CASE 15-90. 1990 Design and Manufacture of Shell Boilers of Welded Construction Enquiry Cases: Enquiry Case 15: Allowable Stresses in Shells Under Hydraulic Test Conditions at Boiler Supports. 1 p.
BSI BS 4167: Part 6-69. 1969 Amd 1 Electrically-Heated Catering Equipment Part 6: Bulk Liquid Heaters. 46 pp.
BSI BS 4167: Part 7-69. 1969 Amd 1 Electrically-Heated Catering Equipment Part 7: Water Boilers. 46 pp.
BSI BS 5978: Part 1-89. 1989 Safety and Performance of Gas-Fired Hot Water Boilers (60kW to 2MW Input) Part 1: General Requirements. 14 pp.
CEN PREN 656-92. Gas-Fired Hot Water Boilers with Atmospheric Burners—Type B Boilers of Nominal Heat Input Exceeding 70 kW but Not Exceeding 300 kW. 112 pp.
CSA CAN/CSA-B415.1-92. Performance Testing of Solid-Fuel-Burning Stoves, Inserts, and Low-Burn-Rate Factory-Built Fireplaces; (Gen Instr 1 Thru 2). 89 pp.
CSA CAN/CSA-C22. 2 NO 165-92. Electric Boilers; (Gen Instr 1). 61 pp.
JIS S 3021-82. Oil Burning Water Boilers.
SAA AS 1200-88. Boilers and Pressure Vessels (Known as the SAA Boiler Code). 6 pp.

Boilers (Cont.)

SAA AS 1200 Supp 1-88. Boilers and Pressure Vessels (Known as the SAA Boiler Code)—Supplement 1: Rulings to SAA Boiler Code (Supplement to AS 1200—1988) Amdt 1 January 1990 Amdt 2 April 1991 (in Professional Packages 19-21, 61-69).

SAA AS 1200 Supp 1.2-88. Boilers and Pressure Vessls (Known as the SAA Boiler Code)—Supplement 1.2: Supplement 1.2—1988). 1 p.

SAA AS 1200 Supp 1.3-90. Boilers and Pressure Vessels (Known as the SAA Boiler Code)—Supplement 1.3: Rulings to the SAA Boiler Code (Supplement to AS 1200—1988). 1 p.

SAA AS 1200 Supp 1.4-91. Boilers and Pressure Vessels (Known as the SAA Boiler Code)—Supplement 1.4: Rulings to the SAA Boiler Code (Supplement to AS 1200—1988). 1 p.

SAA AS 1797-86. Boilers—Fire-Tube, Shell, and Miscellaneous Amdt 1 November 1987. 152 pp.

SNZ NZS 5351-72. Code of Practice for the Installation, Operation and Maintenance of Building Service and Small Industrial Boilers (Reconfirmed 1989) Amend: 1, 1976. 46 pp.

—**Acceptance Testing**

BSI BS 2885-74. 1974 Amd 1 Code for Acceptance Tests on Stationary Steam Generators of the Power Station Type. 51 pp.

—**Atmospheric Burners**

BSI BS 5978: Part 2-89. 1989 Amd 1 Safety and Performance of Gas-Fired Hot Water Boilers (60kW to 2MW Input) Part 2: Additional Requirements for Boilers with Atmospheric Burners. 16 pp.

CEN PREN 297-89. Gas-Fired Central Heating Boilers Fitted with Atmospheric Burners; Type B11 Boilers of Nominal Heat Input Not Exceeding 70kw. 105 pp.

CEN PREN 297-91. AMD 1 Gas-Fired Central Heating Boilers Fitted with Atmospheric Burners; Type B11 Boilers of Nominal Heat Input Not Exceeding 70kw. 8 pp.

CEN PREN 297/A2-92. AMD 2 Gas-Fired Central Heating Boilers Fitted with Atmospheric Burners; Type B11 Boilers of Nominal Heat Input Not Exceeding 70kw. 10 pp.

CEN PREN 297/PRA5-92. AMD 5 Gas-Fired Central Heating Boilers Fitted with Atmospheric Burners; Type B11 Boilers of Nominal Heat Input Not Exceeding 70kw. 5 pp.

CEN PREN 483-91. Gas-Fired Central Heating Boilers Fitted with Atmospheric Burners—Type C Boilers of Nominal Heat Input Not Exceeding 70kw. 158 pp.

CEN PREN 625-92. Gas-Fired Central Heating Boilers Fitted with Atmospheric Burners—Specific Requirements for Combination Boilers of a Nominal Heat Input Not Exceeding 70 kW. 25 pp.

CEN PREN 677-92. Gas-Fired Central Heating Boilers—Specific Requirements for Condensing Boilers with a Nominal Heat Input Not Exceeding 70 kW. 16 pp.

—**Atmospheric Burners—Safety**

BSI BS 5978: Part 2-89. 1989 Amd 1 Safety and Performance of Gas-Fired Hot Water Boilers (60kW to 2MW Input) Part 2: Additional Requirements for Boilers with Atmospheric Burners. 16 pp.

CEN PREN 297-89. Gas-Fired Central Heating Boilers Fitted with Atmospheric Burners; Type B11 Boilers of Nominal Heat Input Not Exceeding 70kw. 105 pp.

CEN PREN 483-91. Gas-Fired Central Heating Boilers Fitted with Atmospheric Burners—Type C Boilers of Nominal Heat Input Not Exceeding 70kw. 158 pp.

—**Bath**

JIS A 5713-80. Bathtubs Connected with Boiler for Dwellings.

JIS S 2111-82. Bath Boilers Used with Liquefied Petroleum Gas.

JIS S 3017-82. Wick Feed Type Oil Burning Bath Boilers.

JIS S 3018-82. Vaporizing Pot Type Oil Firing Bath Boilers.

JIS S 3023-82. Pressure Type Oil Burning Bath Boilers.

JIS S 3027-87. Oil Burning Bath Boilers with Water Heaters for Domestic Use. 34 pp.

—**Brazing**

SAA AS 3992-92. Boilers and Pressure Vessels—Welding and Brazing Qualification Amdt 1 June 1993 (in Professional Package 19). 58 pp.

—**Breathing Spaces**

BSI BS 2790: CASE 24-89. 1989 Design and Manufacture of Shell Boilers of Welded Construction Enquiry Cases: Case 24: Breathing Space for Reverse Flame Boilers. 1 p.

—**Cisterns**

JIS A 4006-76. Cisterns for Hot-Water Boiler.

Boilers (Cont.)

—**Classification**

BSI BS 1000: (621.1/22)-84. 1984 Universal Decimal Classificaton (UDC). English Full Edition (621.1/22): UDC 621.1 Heat Engines in General. Steam Power. Steam Engines. Boilers UDC 621.22 Water Power. Hydraulic Energy. 22 pp.

SNZ NZS/BS 1000 (621.1/.22)-84. Universal Decimal Classification Heat Engines in General. Steam Power. Steam Engines. Boilers. Water Power. Hydraulic Energy. 24 pp.

—**Comprehensive Testing**

SAA AS 4037-92. Boilers and Pressure Vessels—Examination and Testing (in Professional Package 19). 51 pp.

—**Eddy Current Testing**

BSI BS 1113: ENQ CASE 1-86. 1986 Design and Manufacture of Water-Tube Steam Generating Plant (Including Superheaters, Reheaters and Steel Tube Economizers) Enquiry Case 1: Eddy Current Testing in Lieu of Hydraulic Testing. 1 p.

—**Feed Pumps**

JIS B 8303-90. Testing Methods for Boiler Feed Pumps. 10 pp.

—**Firebricks**

CGSB CAN/CGSB-10.1-92. Fireclay Refractory Brick for Stationary Boiler Service. 7 pp.

—**Fittings**

BSI BS 759: Part 1-84. 1984 Amd 1 Valves, Gauges and Other Safety Fittings for Application to Boilers and to Piping Installations for and in Connection with Boilers Part 1: Specification for Valves, Mounting and Fittings. 15 pp.

—**Flues**

BSI BS 5440: Part 1-90. 1990 Code of Practice for Flues and Air Supply for Gas Appliances of Rated Input Not Exceeding 60 kW (1st and 2nd Family Gases) Part 1: Installation of Flues. 44 pp.

—**Forced Draft Burners**

BSI BS 5978: Part 3-89. 1989 Safety and Performance of Gas-Fired Hot Water Boilers (60 kW to 2MW Input) Part 3: Additional Requirements for Boilers with Forced or Induced Draught Burners. 4 pp.

—**Forced Draft Burners—Safety**

BSI BS 5978: Part 3-89. 1989 Safety and Performance of Gas-Fired Hot Water Boilers (60 kW to 2MW Input) Part 3: Additional Requirements for Boilers with Forced or Induced Draught Burners. 4 pp.

—**Gage Glasses**

CNS B2743-83. Gage Glasses for Boilers (Feb)(9970).

JIS B 8211-75. Gauge Glasses for Boilers.

—**Gages**

CNS B2744-83. Water Gages for Boilers — Reflex Type (Feb)(9971).

CNS B2745-83. Water Gages for Boilers — Transparent Type (Feb)(9972).

CNS B2746-83. Water Gages for Boilers — with 10 kgf/cm2 Cocks Tubular Type (Feb)(9973).

CNS B2747-83. Water Gages for Boilers — with 10 kgf/cm2 Valves Tubular Type (Feb)(9974).

JIS B 8213-74. Reflex Type Water Gauges for Boilers (R 1983). 6 pp.

JIS B 8215-74. Transparent Type Water Gauges for Boilers (R 1983). 11 pp.

JIS B 8216-77. 10 kgf/cm2 Tubular Type Water Gauges with Cocks for Boilers (R 1982). 8 pp.

JIS B 8217-77. 10 kgf/cm2 Tubular Type Water Gauges with Valves for Boilers (R 1982). 8 pp.

SAA AS 1271-90. Safety Valves, Other Valves, Liquid Level Gauges, and Other Fittings for Boilers and Unfired Pressure Vessels Amdt 1 April 1991 (in Professional Package 19).

—**Glossaries**

BSI BS 1846: Part 2-68. 1968 Glossary of Terms Relating to Solid Fuel Burning Equipment Part 2: Industrial Water Heating and Steam Raising Installations. 19 pp.

JIS B 0126-86. Glossary of Terms for Thermal Power Plant Use (Boilers and Auxiliary Equipments).

—**Heat Balance**

JIS B 8222-86. Heat Balancing of Boilers for Land Use.

—**Induced Draft Burners**

BSI BS 5978: Part 3-89. 1989 Safety and Performance of Gas-Fired Hot Water Boilers (60 kW to 2MW Input) Part 3: Additional Requirements for Boilers with Forced or Induced Draught Burners. 4 pp.

Boilers (Cont.)

—**Induced Draft Burners—Safety**

BSI BS 5978: Part 3-89. 1989 Safety and Performance of Gas-Fired Hot Water Boilers (60 kW to 2MW Input) Part 3: Additional Requirements for Boilers with Forced or Induced Draught Burners. 4 pp.

—**Installation**

BSI BS 5871-80. (WITHDRAWN) 1980 Amd 2 Installation of Gas Fires, Convectors and Fire/Back Boilers (2nd Family Gas) (AMD 4638) March 29, 1985 (Superseded by BS 5871: Parts 1&2: 1991). 22 pp.

BSI BS 5871: Part 1-91. 1991 Installation of Gas Fires, Convector Heaters, Fire/Back Boilers and Decorative Fuel Effect Gas Appliances Part 1: Gas Fires, Convector Heaters and Fire/Back Boilers (1st, 2nd and 3rd Family Gases). 32 pp.

BSI BS 6644-91. 1991 Installation of Gas-Fired Hot Water Boilers of Rated Inputs Between 60 kW and 2 MW (2nd and 3rd Family Gases). 25 pp.

BSI BS 6644-01. 1991 Amd 1 Installation of Gas-Fired Hot Water Boilers of Rated Inputs Between 60 kW and 2 MW (2nd and 3rd Family Gases) (AMD 7640) June 15, 1993 (F). 26 pp.

BSI BS 6798-87. 1987 Installation of Gas-Fired Hot Water Boilers of Rated Input Not Exceeding 60 kW. 14 pp.

BSI CP 403-74. 1974 Installation of Domestic Heating and Cooking Appliances Burning Solid Fuel. 37 pp.

SAA AS 3892-91. Boilers and Pressure Vessels—Installation (in Professional Package 19). 18 pp.

—**Installation Codes**

CSA CAN/CSA-B365-M91. Installation Code for Solid-Fuel-Burning Appliances and Equipment; (Gen Instr 1 Thru 5). 60 pp.

—**Loads (Forces)**

BSI BS 1113: ENQ CASE 3-87. 1987 Design and Manufacture of Water-Tube Steam Generating Plant (Including Superheaters, Reheaters and Steel Tube Economizers) Enquiry Case 3: Local Loads. 1 p.

—**Log Sheets**

BSI BS 1374-72. (WITHDRAWN) 1972 Recommendations on the Use of British Standard Log Sheets for Steam and Hot Water Boiler Plants. 21 pp.

—**Maintenance**

SAA AS 3873-91. Boiles and Pressure Vessels—Operation and Maintenance (in Professional Package 19). 32 pp.

—**Marine—Fuel Oils**

CGSB 3-GP-12MA-88. Fuel Oil, Marine Boiler. 9 pp.

—**Marine—Gages**

JIS F 5609-88. Forged Steel 20 K Reflex Type Water Gauges with Cocks for Marine Boilers.

JIS F 5610-88. Forged Steel 20 K Reflex Type Water Gauges with Valves for Marine Boilers.

JIS F 5611-88. Forged Steel 63 K Transparent Type Water Gauges with Valves for Marine Boilers.

—**Metals—Inspection**

BSI BS 7339-90. 1990 Code of Practice for Supplementary Checking of Metallic Materials of Construction in Pressurized Systems. 11 pp.

—**Mountings**

BSI BS 759: Part 1-84. 1984 Amd 1 Valves, Gauges and Other Safety Fittings for Application to Boilers and to Piping Installations for and in Connection with Boilers Part 1: Specification for Valves, Mounting and Fittings. 15 pp.

—**Oil Burners—Atomizing**

BSI BS EN 303: Part 1-92. 1992 Heating Boilers. Heating Boilers with Forced Draught Burners Part 1: Terminology, General Requirements, Testing and Marking. 26 pp.

BSI BS EN 303: Part 2-92. 1992 Heating Boilers—Heating Boilers with Forced Draught Burners Part 2: Special Requirements for Boilers with Atomizing Oil Burners. 10 pp.

BSI BS EN 304-92. 1992 Heating Boilers. Test Code for Heating Boilers for Atomizing Oil Burners (F). 27 pp.

CEN PREN 303-90. Central Heating Boilers—Special Heating Boilers for Liquid Fuels—Terminology, Requirements, Testing and Marketing. 32 pp.

CEN EN 303-1-92. Heating Boilers—Heating Boilers with Forced Draught Burners—Part 1: Terminology, General Requirements, Testing and Marking. 20 pp.

CEN EN 303-2-92. Heating Boilers—Heating Boilers with Forced Draught Burners—Part 2: Special Requirements for Boilers with Atomizing Oil Burners. 5 pp.

CEN PREN 304-90. Central Heating Boilers—Test Code for Special Heating Boilers for Liquid Fuels. 29 pp.

CEN EN 304-92. Heating Boilers—Test Code for Heating Boilers for Atomizing Oil Burners. 22 pp.

INTERNATIONAL AND NON-U.S. NATIONAL STANDARDS
SUBJECT INDEX

Boilers

Boilers (Cont.)

—Oil—Energy Consumption
CSA B212-93. Seasonal Energy Utilization Efficiencies of Oil-Fired Furnaces and Boilers; (Gen Instr 1). 72 pp.

—Oil—Industrial
CSA B140.7.2-1967. Oil-Fired Steam and Hot-Water Boilers for Commercial and Industrial Use (R 1991); (Rev 1 August 1967). 23 pp.

—Oil—Marine—Firebricks
CGSB CAN/CGSB-10.4-92. Fireclay Refractory Brick for Marine Oil-Fired Boilers. 7 pp.

—Oil—Residential
CSA B140.7.1-1976. Oil-Fired Steam and Hot-Water Boilers for Residential Use (R 1991); (Rev 1-3 September 1977). 30 pp.

—Operation
SAA AS 3873-91. Boiles and Pressure Vessels—Operation and Maintenance (in Professional Package 19). 32 pp.

—Pipelines—Construction
BSI BS 806-90. 1990 Amd 2 Design and Construction of Ferrous Piping Installations for and in Connection with Land Boilers (AMD 6742) July 31, 1991. 109 pp.

BSI BS 806-03. 1990 Amd 3 Design and Construction of Ferrous Piping Installations for and in Connection with Land Boilers (AMD 6952) December 24, 1991 (Q). 107 pp.

BSI BS 806-04. 1990 Amd 4 Design and Construction of Ferrous Piping Installations for and in Connection with Land Boilers (AMD 7416) December 15, 1992 (Q). 123 pp.

BSI BS 806: ENQ CASE 1-92. 1992 Design and Construction of Ferrous Piping Installations for and in Connection with Land Boilers Enquiry Cases: Enquiry Case 1: Use of Materials Complying with API 5L, ASTM A 106, A 182, A 312 and A 335. 6 pp.

BSI BS 806: ENQ CASE 1-91. 1991 Design and Construction of Ferrous Piping Installations for and in Connection with Land Boilers Enquiry Cases: Enquiry Case 1: Use of Materials Complying with API 5L, ASTM A 106, A 182, A 312 and A 335 (FCP). 6 pp.

SNZ NZS/BS 806-90. Specification for Design and Construction of Ferrous Piping Installations for and in Connection with Land Boilers Amend: 1, 2, 3. 128 pp.

—Pipelines—Design
BSI BS 806-90. 1990 Amd 2 Design and Construction of Ferrous Piping Installations for and in Connection with Land Boilers (AMD 6742) July 31, 1991. 109 pp.

BSI BS 806-03. 1990 Amd 3 Design and Construction of Ferrous Piping Installations for and in Connection with Land Boilers (AMD 6952) December 24, 1991 (Q). 107 pp.

BSI BS 806-04. 1990 Amd 4 Design and Construction of Ferrous Piping Installations for and in Connection with Land Boilers (AMD 7416) December 15, 1992 (Q). 123 pp.

BSI BS 806: ENQ CASE 1-92. 1992 Design and Construction of Ferrous Piping Installations for and in Connection with Land Boilers Enquiry Cases: Enquiry Case 1: Use of Materials Complying with API 5L, ASTM A 106, A 182, A 312 and A 335. 6 pp.

BSI BS 806: ENQ CASE 1-91. 1991 Design and Construction of Ferrous Piping Installations for and in Connection with Land Boilers Enquiry Cases: Enquiry Case 1: Use of Materials Complying with API 5L, ASTM A 106, A 182, A 312 and A 335 (FCP). 6 pp.

SNZ NZS/BS 806-90. Specification for Design and Construction of Ferrous Piping Installations for and in Connection with Land Boilers Amend: 1, 2, 3. 128 pp.

—Plates—Carbon Manganese Steel
SAA AS 1548-88. Steel Plates for Boilers and Pressure Vessels. 25 pp.

—Plates—Carbon Steel
CNS G3169-82. Carbon Steel and Molybdenum Alloy Steel Plates for Boilers and Other Pressure Vessels (Apr)(8696).

JIS G 3103-87. Carbon Steel and Molybdenum Alloy Steel Plates for Boilers and Other Pressure Vessels.

JIS G 3103-77. Carbon Steel and Molybdenum Alloy Steel Plates for Boilers and Other Pressure Vessels. 12 pp.

—Plates—Copper and Copper Alloys
ISO 1634 Pt 2-87. Wrought Copper and Copper Alloy Plate, Sheet and Strip Part 2: Technical Conditions of Delivery for Plate and Sheet for Boilers, Pressure Vessels and Heat-Exchangers First Edition. 6 pp.

Boilers (Cont.)

—Plates—High Strength Steel
CNS G2232-84. Method of Test for High Strength Steel Plates for Pressure Vessel for Intermediate and Moderate Temperature Service (Nov)(11108).

CNS G3223-84. High Strength Steel Plates for Pressure Vessel for Intermediate and Moderate Temperature Service (Nov)(11107).

JIS G 3124-87. High Strength Steel Plates for Pressure Vessel for Intermediate and Moderate Temperature Service.

JIS G 3124-83. High Strength Steel Plates for Pressure Vessel for Intermediate and Moderate Temperature Service. 17 pp.

—Plates—Manganese Steel
CNS G2125-82. Method of Test for Manganese-Molybdenum and Manganese-Molybdenum-Nickel Alloy Steel Plates for Boilers and Other Pressure Vessels (Jun)(8972).

CNS G3183-82. Manganese-Molybdenum and Manganese-Molybdenum-Nickel Alloy Steel Plates for Boilers and Other Pressure Vessels (Jun)(8971).

JIS G 3119-87. Manganese-Molybdenum Alloy and Manganese-Molybdenum-Nickel Alloy Steel Plates forBoilers and Other Pressure Vessels.

JIS G 3119-77. Manganese-Molybdenum and Manganese-Molybdenum-Nickel Alloy Steel Plates for Boilers and Other Pressure Vessels. 14 pp.

—Plates—Molybdenum Steel
CNS G3169-82. Carbon Steel and Molybdenum Alloy Steel Plates for Boilers and Other Pressure Vessels (Apr)(8696).

JIS G 3103-87. Carbon Steel and Molybdenum Alloy Steel Plates for Boilers and Other Pressure Vessels.

JIS G 3103-77. Carbon Steel and Molybdenum Alloy Steel Plates for Boilers and Other Pressure Vessels. 12 pp.

—Plates—Steel
CNS G3169-82. Carbon Steel and Molybdenum Alloy Steel Plates for Boilers and Other Pressure Vessels (Apr)(8696).

JIS G 3103-87. Carbon Steel and Molybdenum Alloy Steel Plates for Boilers and Other Pressure Vessels.

JIS G 3103-77. Carbon Steel and Molybdenum Alloy Steel Plates for Boilers and Other Pressure Vessels. 12 pp.

JIS G 4109-87. Chromium-Molybdenum Alloy Steel Plates for Boilers and Pressure Vessels. 18 pp.

—Safety
BSI BS 5258: Part 1-86. 1986 Safety of Domestic Gas Appliances Part 1: Central Heating Boilers and Circulators. 43 pp.

BSI BS 5258: Pt 1: Supp 1-83. 1983 Safety of Domestic Gas Appliances Part 1: Central Heating Boilers and Circulators Supplement 1: Fan-Powered Appliances. 13 pp.

BSI BS 5258: Part 8-80. 1980 Safety of Domestic Gas Appliances Part 8: Combined Appliances: Gas Fire/Back Boiler. 24 pp.

BSI BS 5258: Part 15-90. 1990 Safety of Domestic Gas Appliances Part 15: Combination Boilers. 52 pp.

BSI BS 5978: Part 1-89. 1989 Safety and Performance of Gas-Fired Hot Water Boilers (60kW to 2MW Input) Part 1: General Requirements. 14 pp.

BSI DD 189-90. 1990 Safety of Condensing Boilers (2nd and 3rd Family Gases). 9 pp.

—Safety Plugs
SAA AS 1732-75. Fusible Plugs for Boilers Reconfirmed 1982. 4 pp.

—Safety Valves
BSI BS 6759: Part 1-84. 1984 Amd 1 Safety Valves Part 1: Specification for Safety Valves for Steam and Hot Water (AMD 5493) March 31, 1987. 25 pp.

CNS B2742-83. Spring Loaded Safety Valves for Steam Boilers and Pressure Vessels (Feb)(9969).

—Sheets—Copper and Copper Alloys
ISO 1634 Pt 2-87. Wrought Copper and Copper Alloy Plate, Sheet and Strip Part 2: Technical Conditions of Delivery for Plate and Sheet for Boilers, Pressure Vessels and Heat-Exchangers First Edition. 6 pp.

—Ships—Fuel Oils
MOD UK DSTAN 91-5-01. Fuel, Residual: Light Viscosity, Boiler NATO Code No: F-77 Joint Service Designation: 50/50 FFO Fuel Residual: Medium Viscosity, Boiler NATO Code No: F-82 Joint Service Designation: 75/50 FFO Issue 3 Amendment 2. 28 pp.

—Ships—Glossaries
JIS F 0022-85. Glossary of Terms for Shipbuilding (Machinery Part—Propulsion Machinery, Boilers, Generator Engines and Prime Mover for Auxiliary Machinery).

Boilers (Cont.)

—Solid Fuels
BSI BS 3376-91. 1991 Open Fires Burning Solid Mineral Fuels with Convection, with or Without Boilers. 37 pp.

BSI BS 3376-82. (WITHDTRAWN) 1982 Solid Mineral Fuel Open Fires with Convection, with or Without Boilers. 20 pp.

BSI BS 3377-85. 1985 Amd 1 Boilers for Use with Domestic Solid Mineral Fuel Appliances (AMD 6138) May 31, 1990. 9 pp.

—Steam Pipes—Ships
MOD UK NES 344-87. Feed Steam and Drain Systems in Ships Fitted with Auxiliary Boilers Issue 2 (11.87). 59 pp.

—Steels—Creep Rupture Strength
ISO TR7468-81. Summary of Average Stress Rupture Properties of Wrought Steels for Boilers and Pressure Vessels First Edition. 50 pp.

—Steels—Rolled
CNS G3072-72. Rolled Steel for Boilers and Other Pressure Vessels for High Temperature Services (Jun)(3331).

—Storage Water Heaters
CEN PREN 303-3-93. Gas-Fired Central Heating Boilers—Assembly Comprising a Boiler Body and a Forced Draught Burner. 51 pp.

—Stresses
BSI BS 2790: CASE 15-90. 1990 Design and Manufacture of Shell Boilers of Welded Construction Enquiry Cases: Enquiry Case 15: Allowable Stresses in Shells Under Hydraulic Test Conditions at Boiler Supports. 1 p.

—Temperature Controllers
JIS B 8413-84. Temperature Limiters and Thermostats for Hot Water Boilers.

—Thermal Efficiency
BSI BS 845: Part 1-87. 1987 Assessing Thermal Performance of Boilers for Steam, Hot Water and High Temperature Heat Transfer Fluids Part 1: Concise Procedure. 18 pp.

BSI BS 845: Part 2-87. 1987 Assessing Thermal Performance of Boilers for Steam, Hot Water and High Temperature Heat Transfer Fluids Part 2: Comprehensive Procedure. 41 pp.

BSI BS 6332: Part 1-88. 1988 Thermal Performance of Domestic Gas Appliances Part 1: Thermal Performance of Central Heating Boilers and Circulators. 12 pp.

BSI BS 6332: Part 3-84. 1984 Thermal Performance of Domestic Gas Appliances Part 3: Specification for Thermal Performance of Combined Appliances: Gas Fire/Back Boilers. 15 pp.

BSI BS 7190-89. 1989 Method for Assessing Thermal Performance of Low Temperature Hot Water Boilers Using a Test Rig. 24 pp.

—Thermal Insulation
CGSB CAN/CGSB-51.2-M88. Thermal Insulation, Calcium Silicate, for Piping, Machinery and Boilers; (Amendment 1 December 1991). 11 pp.

CGSB CAN/CGSB-51.5-92. Thermal Insulation, Block or Blanket, Elevated Temperature. 8 pp.

CGSB CAN/CGSB-51.10-92. Mineral Fibre Board Thermal Insulation. 8 pp.

CGSB CAN/CGSB-51.11-92. Mineral Fibre Thermal Insulation Blanket. 8 pp.

—Tube Plate Temperature
BSI BS 2790: CASE 21-88. (WITHDRAWN) 1988 Design and Manufacture of Shell Boilers of Welded Construction Enquiry Cases: Case 21: Tube Plate Metal Temperatures. 1 p.

—Valves
BSI BS 759: Part 1-84. 1984 Amd 1 Valves, Gauges and Other Safety Fittings for Application to Boilers and to Piping Installations for and in Connection with Boilers Part 1: Specification for Valves, Mounting and Fittings. 15 pp.

SAA AS 1271-90. Safety Valves, Other Valves, Liquid Level Gauges, and Other Fittings for Boilers and Unfired Pressure Vessels Amdt 1 April 1991 (in Professional Package 19).

—Water Formed Deposits—Sampling
BSI BS 2455: Part 1-73. 1973 Methods of Sampling and Examining Deposits from Boilers and Associated Industrial Plant Part 1: Water-Side Deposits. 31 pp.

BSI BS 2455: Part 1-01. 1973 Amd 1 Methods of Sampling and Examining Deposits from Boilers and Associated Industrial Plant Part 1: Water-Side Deposits (AMD 7753) October 15, 1993 (Q). 32 pp.

INDUSTRY STANDARDS

Boilers

Boilers (Cont.)
—Welded Joints
BSI BS 1113: ENQ CASE 4-89. 1989 Design and Manufacture of Water-Tube Steam Generating Plant (Including Superheaters, Reheaters and Steel Tube Economizers) Enquiry Case 4: Methods for Manual Examination of Fusion Welds in Ferritic Steels. 1 p.
BSI BS 1113: ENQ CASE 7-92. 1992 Design and Manufacture of Water-Tube Steam Generating Plant (Including Superheaters, Reheaters and Steel Tube Economizers) Enquiry Case 7: Use of Butt Welds in Tubes for Manufacture of Coils for Coil Type Boilers/Superheaters (Q). 1 p.

—Welding
JIS Z 3040-88. Method of Qualification Test for Welding Procedure. 50 pp.
SAA AS 3992-92. Boilers and Pressure Vessels—Welding and Brazing Qualification Amdt 1 June 1993 (in Professional Package 19). 58 pp.

Boiling Flasks
Use: Flasks—Boiling

Boiling Pans
See Also: Cooking Pans; Cooking Utensils
BSI BS 4167: Part 9-69. (WITHDRAWN) 1969 Amd 1 Electrically-Heated Catering Equipment Part 9: Boiling Pans (Superseded by BS EN 60335-2-47: 1992). 45 pp.
BSI BS 5314: Part 9-79. (WITHDRAWN) 1979 Amd 1 Gas Heated Catering Equipment Part 9: Boiling Pans (AMD 5405) July 31, 1987 (Superseded by BS EN 203-1: 1993). 22 pp.

Boiling Points
Use For: Boiling Range; Bubble Points
See Also: Ebulliometry; Flash Point; Freezing Points; High Temperature Testing; Melting Points; Softening Points; Temperature; Thermodynamic Properties; Vapor Pressure

—Antifreezes
CNS K6482-80. Method of Test for Boiling Point of Engine Antifreezes (May)(5585).

—Chlorofluorocarbons
BSI BS 5598: Part 9-81. 1981 Methods of Sampling and Test for Halogenated Hydrocarbons Part 9: Determination of Boiling Range of Chlorofluorinated Hydrocarbons. 6 pp.
ISO 5917-80. Chlorofluorinated Hydrocarbons—Determination of Boiling Range—Test for Product Characterization First Edition; (Amendment Slip-1981). 6 pp.

—Engine Coolants
BSI BS 5117: Sec 1.2-85. 1985 Testing Corrosion Inhibiting, Engine Coolant Concentrate ('Antifreeze') Part 1: Methods of Test for Determination of Physical and Chemical Properties Section 1.2: Determination of Boiling Point. 4 pp.

—Membrane Filters
JIS K 3832-90. Testing Methods for Bubble Point of Membrane Filters. 11 pp.

—Organic Solvents
ISO 4626-80. Volatile Organic Liquids—Determination of Boiling Range of Organic Solvents Used as Raw Materials First Edition. 14 pp.

—Petroleum Products—Gas Chromatography
ISO 3924-77. Petroleum Products—Determination of Boiling Range Distribution—Gas Chromatography Method First Edition. 11 pp.

—Reagents
CNS K7006-82. Chemical Reagent Boiling Point and Distillate Determination (Feb)(1506).

—Solvents
JIS K 1545-78. Testing Methods for High Boiling Point Solvents.

Boiling Range
Use: Boiling Points

Boiling Tubes
Use: Test Tubes—Boiling

Boiling Water Resistance Testing
Use: Water Resistance Testing

Bolkow BO 105 Aircraft
See Also: Aircraft

—Antenna Positions
CAA. Bolkow BO.105 D/DB/DB4 (Approved Aerial Positions). 1 p.
CAA. Bolkow BO.105 DS/DBS/DBS4. 1 p.

Bolkow BO 208 Aircraft
See Also: Aircraft

—Antenna Positions
CAA. Bolkow BO.208 (Approved Aerial Positions). 1 p.

Bolkow BO 209 Aircraft
See Also: Aircraft

—Antenna Positions
CAA. Bolkow BO.209 (Approved Aerial Positions). 1 p.

Bolkow 107 Aircraft
See Also: Aircraft

—Foreign Airworthiness Directives
CAA. Bolkow 107 and 207 Series (Foreign Airworthiness Directives). 2 pp.

Bolkow 207 Aircraft
See Also: Aircraft

—Foreign Airworthiness Directives
CAA. Bolkow 107 and 207 Series (Foreign Airworthiness Directives). 2 pp.

Bolkow 208 Junior Aircraft
See Also: Aircraft

—Foreign Airworthiness Directives
CAA. Bolkow 208 Junior (Foreign Airworthiness Directives). 1 p.

Bolkow 209 Aircraft
See Also: Aircraft

—Foreign Airworthiness Directives
CAA. Bolkow 209 Series (Foreign Airworthiness Directives). 3 pp.

Bollards
See Also: Marine Equipment; Mooring Hardware
CNS F3026-80. Bollards (Mar)(5334).
CNS F3027-80. Bollards (Mar)(5335).
JIS F 2001-90. Bollards.
JIS F 2001-75. Bollards. 4 pp.
JIS F 2018-76. Bollards (Small Type).
JIS F 2018-65. Simple Type Bollards (R 1968). 4 pp.

—Marine—Steel
BSI BS MA 12: Part 1-83. 1983 Welded Steel Bollards Part 1: Vertical Type. 7 pp.
BSI BS MA 12: Part 2-83. 1983 Welded Steel Bollards Part 2: Cruciform Type. 7 pp.
ISO 3913-77. Shipbuilding—Welded Steel Bollards First Edition; (Addendum 1-1980). 14 pp.

—Traffic
BSI BS 873: Part 3-80. 1980 Road Traffic Signs and Internally Illuminated Bollards Part 3: Internally Illuminated Bollards. 10 pp.

Bolsters (Pillows)
Use: Pillows

Bolt Anchors
Use: Anchors (Fasteners)

Bolt Rods
See Also: Rods

—Aircraft
SBAC AS 2535 ISSUE 5. Long Bolt Rod. 1 p.
SBAC AS 2536 ISSUE 5. Short Bolt Rod. 1 p.

—Aircraft Door—Locks
SBAC AS 6529 ISSUE 1. Short Bolt Rod.
SBAC AS 6530 ISSUE 1. Long Bolt Rod.

Bolted Connections
Use: Bolted Joints

Bolted Joints
Use For: Bolted Connections *See Also:* Butt Joints; Joints; Pipe Joints
CNS B1094-83. Bolted Connections with Reduced Body (Survey and Installation) (Mar)(4576).
CNS B1095-78. Bolted Connections with Reduced Body (Studies Relating to the Calculation of Bolted Connection) (Oct)(4577).

—Flanged—Pressure Vessels
CNS B5092-83. Construction of Pressure Vessels (9) (Bolted Flange Connection) (Jan)(9796).

Bolts
Scope Note: For additional listings, use a more specific term *See Also:* Aerospace Bolts; Aircraft Bolts; Anchor Bolts; Anchors (Fasteners); Banjo Bolts; Buffer Bolts; Clamping Bolts; Countersunk Head Bolts; Couplings; Crown Head Bolts; Cup

Bolts (Cont.)
See Also: (Cont.)
Head Bolts; Dee Head Bolts; Door Bolts; Double Hexagonal Bolts; Eyebolts; Fasteners; Flat Head Bolts; Friction Grip Bolts; Hexagonal Bolts; High Strength Bolts; Holders; Hub Bolts; Machine Bolts; Mounting Bolts; Nuts (Fasteners); Oval Head Bolts; Pan Head Bolts; Phillips Bolts; Plow Bolts; Recessed Head Bolts; Ringbolts; Rock Bolts; Roof Bolts; Round Head Bolts; Screws; Shear Bolts; Sliding Bolts; Slotted Head Bolts; Socket Head Bolts; Spring Bolts; Square Head Bolts; Square Neck Bolts; Stop Bolts; Structural Bolts; Studs (Fasteners); T Slot Bolts; Taper Bolts; Tee Head Bolts; Threaded Fasteners; Triangular Head Bolts; U Bolts; Washer Head Bolts; Weld Bolts; Wing Bolts

BSI BS 4882-90. 1990 Bolting for Flanges and Pressure Containing Purposes. 42 pp.
BSI BS 6565-85. (WITHDRAWN) 1985 Method for Dimensioning and Designating Bolts, Screws, Studs and Nuts (Superseded by BS EN 20225: 1992). 22 pp.
BSI BS 7345-90. 1990 Nominal Lengths of Bolts, Screws and Studs, and Thread Lengths for General Purpose Bolts. 8 pp.
BSI BS EN 20225-92. 1992 Fasteners—Bolts, Screws, Studs and Nuts—Symbols and Designations of Dimensions (ISO 225: 1983). 26 pp.
CEN EN 20225-91. Fasteners—Bolts, Screws, Studs and Nuts—Symbols and Designations of Dimensions. 3 pp.
CNS B1030-78. Nominal Lengths and Thread Lengths of Bolts, Screws and Studs (Mar)(3126).
CNS B1054-81. Finishes and Tolerances of Bolts, Screws, Studs and Nuts (Jan)(4238).
CNS B1057-78. Widths Across Flats and Widths Across Corners for Bolts, Screws and Nuts (Mar)(4242).
CNS B1058-78. Radius Under Head for Bolts, Screws (Mar)(4243).
CNS B2143-78. Mechanical Properties of Bolts, Screws and Studs (Mar)(3934).
CNS B2169-88. Bolts and Nuts, Made of Stainless Steels (Apr)(4234). 8 pp.
DIN ENGL 267 Pt 3-83. Fasteners; Technical Delivery Conditions; Property Classes for Carbon Steel and Alloy Steel Bolts and Screws; Conversion of Property Classes (Aug). 2 pp.
DIN ENGL 267 Pt 19 (S)-84. Fasteners; Technical Delivery Conditions, Surface Discontinuities on Bolts and Screws (Oct) (Superseded by DIN EN 26157 Parts 1 and 3). 15 pp.
DIN ENGL 267 Pt 25 (P)-84. Fasteners; Technical Delivery Conditions; Torsion Testing of M 1 to M 10 Bolts and Screws (Nov). 4 pp.
DIN ENGL 267 Pt 27-90. Fasteners; Adhesive-Coated Steel Screws, Bolts and Studs; Technical Delivery Conditions (Mar). 4 pp.
DIN ENGL 267 Pt 28-90. Fasteners; Steel-Screws, Bolts and Studs with Locking Coating; Technical Delivery Conditions (Mar). 3 pp.
DIN ENGL 475 Pt 1-84. Widths Across Flats for Bolts, Screws, Valves and Fittings (Jan). 4 pp.
DIN ENGL 17111-80. Low Carbon Unalloyed Steels for Bolts, Nuts and Rivets; Technical Conditions of Delivery (Sept). 10 pp.
DIN ENGL EN 20225-92. Fasteners—Bolts, Screws, Studs and Nuts; Symbols and Designations for Dimensioning (ISO 225: 1983) (Feb) (Supersedes DIN ISO 225, January 1984 Edition). 22 pp.
ISO 225-83. Fasteners—Bolts, Screws, Studs and Nuts—Symbols and Designations of Dimensions Second Edition. 22 pp.
ISO 885-76. General Purpose Bolts and Screws—Metric Series—Radii Under the Head First Edition. 3 pp.
ISO 888-76. Bolts, Screws and Studs—Nominal Lengths, and Thread Lengths for General Purpose Bolts First Edition. 4 pp.
ISO 8992-86. Fasteners—General Requirements for Bolts, Screws, Studs and Nuts First Edition. 4 pp.
MOD UK DSTAN 53-11-78. Guide to Selection of Bolts, Screws, and Nuts for General Purpose Use Issue 2. 10 pp.

—Clip
DIN ENGL 6378-74. Clip Bolts (July). 2 pp.

—Corrosion Resistant
BSI BS 6105-81. 1981 Corrosion-Resistant Stainless Steel Fasteners. 18 pp.
ISO 3506-79. Corrosion-Resistant Stainless Steel Fasteners—Specifications First Edition. 17 pp.

—Counterbores
JIS B 1001-85. Diameter of Clearance Holes and Counterbores for Bolts and Screws. 6 pp.

—Cover
CNS B2307-81. Cover Bolts (Apr)(4574).

—Drafting
DIN ENGL 30-70. Drawing Practice; Simplified Presentations (Dec). 2 pp.

INTERNATIONAL AND NON-U.S. NATIONAL STANDARDS
SUBJECT INDEX
Bolts

Bolts (Cont.)

—**Drawbars—Overrun Devices—Trailers**
BSI BS AU 190-83. 1983 Interface Dimensions and Fixing Bolts for the Mounting of Overrun Devices on Delta Shaped Drawbars of Trailers up to 2000 Kg. 4 pp.

—**Electroplated Coatings**
CNS B1056-80. Electroplated Coating of Bolts, Screws and Nuts (May)(4240).
CNS B7085-78. Method of Test for Electroplated Coatings of Bolts, Screws and Nuts (Mar)(4241).

—**Fish Plates**
BSI BS 64-92. 1992 Normal and High Strength Steel Bolts and Nuts for Railway Rail Fishplates (Q). 16 pp.
BSI BS 64-46. 1946 Amd 2 Steel Fishbolts and Nuts for Railway Rails. 16 pp.
CNS B2350-83. Fish Bolt with Round Head and Oval Neck (Mar)(4693).
CNS B2351-79. Fish Bolt with Square Head (Jan)(4694).
ISO 6305 Pt 4-85. Railway Components—Technical Delivery Requirements—Part 4: Untreated Steel Nuts and Bolts and High-Strength Nuts and Bolts for Fish-Plates and Fastenings First Edition. 10 pp.
JIS E 1107-65. Fishbolts and Nuts (R 1971). 6 pp.
JIS E 1113-88. Heat-Treated Fish Bolts and Nuts. 7 pp.
JIS E 1114-78. Fish Bolts and Nuts for N Type Rails and 60kg Rails.
JIS E 1114-65. Fishbolts and Nuts for N Rails. 5 pp.

—**Glossaries**
BSI BS 6040-81. 1981 Nomenclature for Bolts, Screws, Nuts and Accessories. 58 pp.
CNS B1078-78. Terms of Fasteners for Portions of Screw, Bolt and Stud (Oct)(4534).
ISO 1891-79. Bolts, Screws, Nuts and Accessories—Terminology and Nomenclature First Edition; (Amendment Slip-1980). 63 pp.

—**Heat Resistant**
CNS B1093-78. Finishes and Tolerance of Bolts, Screws and Nuts Primarily of Steels Exhibiting Toughness at Subzero Temperature and High Temperature Steels (Oct)(4575). 12 pp.
DIN ENGL 17240-76. Heat Resisting and Highly Heat Resisting Materials for Bolts and Nuts; Quality Specifications (July). 23 pp.

—**High Torque**
CNS B2791-88. Sets of High Strength Torque Control Bolt, Hexagon Nut and Plain Washers (Jan)(12209). 12 pp.

—**Holes**
BSI BS 4186-84. (WITHDRAWN) 1984 Clearance Holes for Metric Bolts and Screws (Superseded by BS EN 20273: 1992). 3 pp.
BSI BS 6487-84. 1984 Split Pin Holes and Wire Holes in Bolts, Screws and Studs. 4 pp.
BSI BS EN 20273-92. 1992 Fasteners—Clearance Holes for Bolts and Screws (ISO 273: 1979). 9 pp.
CEN EN 20273-91. Fasteners—Clearance Holes for Bolts and Screws. 3 pp.
CNS B1060-78. Clearance Holes for Metric Bolts (l.6mm up to and Including 150mm Thread Diameter) (Mar)(4245).
CNS B5012-57. Distribution of Bolt Holes (Mar)(796).
DIN ENGL EN 20273-92. Fasteners; Clearance Holes for Bolts and Screws (ISO 273:1979) (Supersedes DIN ISO 273, September 1979 Edition). 5 pp.
ISO 273-79. Fasteners—Clearance Holes for Bolts and Screws First Edition; (Amendment Slip-1979). 5 pp.
ISO 7378-83. Fasteners—Bolts, Screws and Studs—Split Pin Holes and Wire Holes First Edition. 5 pp.
JIS B 1001-85. Diameter of Clearance Holes and Counterbores for Bolts and Screws. 6 pp.
JIS B 1008-88. Split Pin Holes and Wire Holes for Bolts, Screws and Studs. 6 pp.
SNZ NZS/ISO 273-79. Fasteners. Clearance Holes for Bolts and Screws. 2 pp.

—**Hot Dip Coatings**
CNS B2171-78. Hot Dip Galvanized Bolts and Nuts (Mar)(4237). 4 pp.

—**Identification Systems**
CNS B1055-78. Marking of Bolts, Screws, Studs and Nuts (Mar)(4239). 4 pp.

—**Leaf Springs**
CNS B2599-81. Center Bolts for Laminated (Leaf) Springs (Sep)(7860). 3 pp.

—**Low Temperature**
CNS B1093-78. Finishes and Tolerance of Bolts, Screws and Nuts Primarily of Steels Exhibiting Toughness at Subzero Temperature and High Temperature Steels (Oct)(4575). 12 pp.

Bolts (Cont.)

—**Marine—Pipe Flanges**
BSI BS MA 9-70. 1970 Amd 3 Flanges, Bolting and Gaskets for Exhaust Gas Piping for Diesel Engines and Boiler Uptakes. 10 pp.

—**Mechanical Properties**
BSI BS 6104: Part 1-81. (WITHDRAWN) 1981 Amd 1 Mechanical Properties of Fasteners Part 1: Bolts, Screws and Studs (AMD 4724) December 31, 1984 (Superseded by BS EN 20898-1: 1992). 20 pp.
BSI BS EN 20898-1-92. 1992 Mechanical Properties of Fasteners Part 1: Bolts, Screws and Studs (ISO 898-1: 1988) (E). 26 pp.
BSI BS EN 20898-1-01. 1992 Amd 1 Mechanical Properties of Fasteners Part 1: Bolts, Screws and Studs (ISO 898-1: 1988) (AMD 7300) July 15, 1992 (W). 28 pp.
BSI BS EN 28839-92. 1992 Mechanical Properties of Fasteners—Bolts, Screws, Studs and Nuts Made of Non-Ferrous Metals (ISO 8839: 1986) (E). 10 pp.
CEN EN 20898-1-91. Mechanical Properties of Fasteners—Part 1: Bolts, Screws and Studs. 21 pp.
CEN EN 28839-91. Mechanical Properties of Fasteners—Part 1: Bolts, Screws and Studs Made of Non-Ferrous Metals. 3 pp.
CNS B7076-78. Method of Test for Mechanical Properties of Bolts, Screws and Studs (Mar)(3935). 5 pp.
DIN ENGL EN 20898 Pt 1-92. Mechanical Properties of Fasteners; Bolts, Screws and Studs (ISO 898-1: 1988) (Apr) (Supersedes DIN ISO 898 Part 1, January 1989 Edition). 21 pp.
DIN ENGL EN 28839-91. Mechanical Properties of Fasteners; Nonferrous Metal Bolts, Screws, Studs and Nuts; (ISO 8839: 1986) (Dec) (Supersedes February 1981 Edition of DIN 267 Part 18). 8 pp.
ISO 898 Pt 1-88. Mechanical Properties of Fasteners—Part 1: Bolts, Screws and Studs Second Edition. 22 pp.
ISO 8839-86. Mechanical Properties of Fasteners—Bolts, Screws, Studs and Nuts Made of Non-Ferrous Metals First Edition. 6 pp.
JIS B 1051-91. Mechanical Properties for Steel Bolts and Screws. 40 pp.
JIS B 1057-89. Mechanical Properties of Non-Ferrous Metal Fasteners. 16 pp.

—**Metal Coatings (Made From Metal)**
BSI BS 7371: Part 1-91. 1991 Coatings on Metal Fasteners Part 1: Specification for General Requirements and Selection Guidelines. 17 pp.

—**Nut Assemblies—Friction**
DIN ENGL 946-91. Determination of Coefficient of Friction of Bolt/Nut Assemblies Under Specified Conditions (Oct). 4 pp.

—**Pipes**
CNS B5059-72. Bolts and Nuts for Bolt-Gland Flexible Joint Cast-Iron Pipes and Fittings (Jun)(2795).
DIN ENGL 2507-86. Bolts, Screws and Nuts for Pipelines (Feb). 3 pp.

—**Quality Assurance**
CNS B1053-78. Acceptable Quality Level for Screws, Bolts, Nuts (Mar)(4235).

—**Screw Threads**
BSI BS 1580: Part 3-65. 1965 Amd 3 Unified Screw Threads Part 3: Diameters Below Quarter of an Inch. 19 pp.
BSI BS 4827-72. 1972 Amd 1 ISO Miniature Screw Threads (Supersedes BS 3369: 1961). 24 pp.
BSI BS 6322: Part 1-82. 1982 Tolerances for Fasteners Part 1: Tolerances of Bolts, Screws and Nuts with Thread Diameters MW1.6 mm and MV150 mm and Product Grades A, B and C. 21 pp.
BSI BS 6322: Part 2-82. 1982 Tolerances for Fasteners Part 2: Tolerances of Bolts, Screws and Nuts with Thread Diameters from 1 mm up to 3 mm and Product Grade F, for Fine Mechanics. 7 pp.
CNS B1020-78. ISO Metric Screw Threads Limits of Size for Commercial Bolt and Nut Threads Quality (Jul)(530). 4 pp.
CNS B1062-78. ISO Miniature Screw Threads—Tolerances (Jun)(4318). 6 pp.
CNS B1063-78. ISO Miniature Screw Threads—Limits of Size (Jun)(4319). 3 pp.
CNS B2165-78. Metric Screw Threads Selected Sizes for Screws, Bolts and Nuts (Jun)(4229).
DIN ENGL 13 Pt 51-88. ISO Metric Screw Threads; External Threads for Transition Fits; Tolerances, Limit Deviations, Limits of Size (Dec). 7 pp.
ISO 262-73. ISO General Purpose Metric Screw Threads—Selected Sizes for Screws, Bolts and Nuts First Edition. 3 pp.
ISO 263-73. ISO Inch Screw Threads—General Plan and Selection for Screws, Bolts and Nuts—Diameter Range 0,06 to 6 in First Edition. 7 pp.
ISO 965 Pt 2-80. ISO General Purpose Metric Screw Threads—Tolerances, Part 2: Limits of Sizes for General Purpose Bolt and Nut Threads—Medium Quality Second Edition. 8 pp.

Bolts (Cont.)

—**Screw Threads (Cont.)**
ISO R1501-70. ISO Miniature Screw Threads First Edition. 15 pp.
ISO 4759 Pt 1-78. Tolerances for Fasteners—Part I: Bolts, Screws and Nuts with Thread Diameters Greater Than or Equal to 1,6 and Less Than or Equal to 150 mm and Product Grades A, B and C First Edition. 20 pp.
ISO 4759 Pt 2-79. Tolerances for Fasteners—Part 2: Bolts, Screws and Nuts with Thread Diameters from 1 up to 3 mm and Product Grade F, for Fine Mechanics First Edition; (Amendment Slip-1980). 7 pp.
JIS B 0201-73. Miniature Screw Threads. 14 pp.
SNZ NZS/BS 4827-72. Specification for ISO Miniature Screw Threads. Metric Series Amend: 1. 24 pp.

—**Screw Threads—Nominal Sizes**
CNS B1030-78. Nominal Lengths and Thread Lengths of Bolts, Screws and Studs (Mar)(3126).
DIN ENGL 103 Pt 7-72. ISO Metric Trapezoidal Screw Threads; Limiting Sizes for Bolt Threads from 8 to 100 mm Nominal Diameter (Oct). 9 pp.
DIN ENGL 103 Pt 8-72. ISO Metric Trapezoidal Screw Threads; Limiting Sizes for Bolt Threads from 105 to 300 mm Nominal Diameter (Oct). 8 pp.
JIS B 1009-91. Bolts, Screws and Studs—Nominal Lengths, and Thread Lengths for General Purpose Bolts. 7 pp.

—**Screw Threads—Projections**
DIN ENGL 78-83. Thread Ends and Lengths of Projection of Bolt Ends for ISO Metric Threads in Accordance with DIN 13 (Dec). 5 pp.

—**Ships**
MOD UK NES 730: Part 1. Fasteners Part 1: Non-Ferrous Bolts, Screws, Studs, Studbolts, Nuts, Wing Nuts and Washers (Unified) Issue 2 (04.84).
MOD UK NES 730: Part 2-88. Fasteners Part 2: Steel Bolts, Stud Bolts, Studs and Screws for Use in Temperatures 0 Degrees—520 Degrees C (Metric) Issue 2 (01.88). 15 pp.
MOD UK NES 730: Part 5-01. Fasteners Part 5: Non-Ferrous Bolts, Screws, Studs, Studbolts, Nuts, Wing-Nuts and Washers (Metric) Issue 2 (11.84); Amendment 1. 22 pp.
MOD UK NES 730: Part 8-88. Fasteners Part 8: Steel Bolts, Studbolts, Studs and Screws, for Use in Temperatures 60 Degrees—520 Degrees C (Imperial) Issue 2 (01.88). 15 pp.
MOD UK NES 730: Part 12-84. Fasteners Part 12: Non-Ferrous Bolts, Screws, Studs, Studbolts, Nuts, Wing Nuts and Washers (Unified) Issue 2 (04.84). 20 pp.

—**Shot**
BSI BS 6753-86. 1986 Shotbolts (Solenoid Operated) for Guarding Machinery. 4 pp.

—**Submarines**
MOD UK NES 730: Part 2-88. Fasteners Part 2: Steel Bolts, Stud Bolts, Studs and Screws for Use in Temperatures 0 Degrees—520 Degrees C (Metric) Issue 2 (01.88). 15 pp.
MOD UK NES 730: Part 5-01. Fasteners Part 5: Non-Ferrous Bolts, Screws, Studs, Studbolts, Nuts, Wing-Nuts and Washers (Metric) Issue 2 (11.84); Amendment 1. 22 pp.
MOD UK NES 730: Part 7-88. Fasteners Part 7: Steel Fasteners Submarine Grade (Metric) Issue 1 (05.88). 26 pp.
MOD UK NES 730: Part 7-01. Fasteners Part 7: Steel Fasteners Submarine Grade (Metric) Issue 1 (05.88); Amendment 1. 28 pp.
MOD UK NES 730: Part 8-88. Fasteners Part 8: Steel Bolts, Studbolts, Studs and Screws, for Use in Temperatures 60 Degrees—520 Degrees C (Imperial) Issue 2 (01.88). 15 pp.
MOD UK NES 730: Part 11-88. Fasteners Part 11: Steel Fasteners Submarine Grade (Imperial) Issue 1 (05.88). 26 pp.
MOD UK NES 730: Part 11-01. Fasteners Part 11: Steel Fasteners Submarine Grade (Imperial) Issue 1 (05.88); Amendment 1. 28 pp.

—**Surface Defects**
BSI BS EN 26157-1-92. 1992 Fasteners—Surface Discontinuities Part 1: Bolts, Screws and Studs for General Requirements (ISO 6157-1: 1988). 16 pp.
BSI BS EN 26157-3-92. 1992 Fasteners—Surface Discontinuities Part 3: Bolts, Screws and Studs for Special Requirements (ISO 6157-3: 1988). 20 pp.
CEN EN 26157-1-91. Fasteners—Surface Discontinuities—Part 1: Bolts, Screws and Studs for General Requirements. 12 pp.
CEN EN 26157-3-91. Fasteners—Surface Discontinuities—Part 3: Bolts, Screws and Studs for Special Requirements. 4 pp.

Bolts (Cont.)
—Surface Defects (Cont.)
DIN ENGL EN 26157 Pt 1-91. Fasteners; Surface Discontinuities; Bolts, Screws and Studs Subject to General Requirements (ISO 6157-1: 1988) (Dec) (This Standard, Together with the December 1991 Edition of DIN EN 26157 Part 3, Supersedes October 1984 Edition of DIN 267 Part 19). 13 pp.
DIN ENGL EN 26157 Pt 3-91. Fasteners; Surface Discontinuities; Bolts, Screws and Studs Subject to Special Requirements (ISO 6157-3: 1988) (Dec) (This Standard, Together with the December 1991 Edition of DIN EN 26157 Part 1, Supersedes October 1984 Edition of DIN 267 Part 19). 17 pp.
ISO 6157 Pt 1-88. Fasteners—Surface Discontinuities—Part 1: Bolts, Screws and Studs for General Requirements First Edition. 12 pp.
ISO 6157 Pt 3-88. Fasteners—Surface Discontinuities—Part 3: Bolts, Screws and Studs for Special Requirements First Edition. 16 pp.

—Symbols
BSI BS 6565-85. (WITHDRAWN) 1985 Method for Dimensioning and Designating Bolts, Screws, Studs and Nuts (Superseded by BS EN 20225: 1992). 22 pp.
BSI BS EN 20225-92. 1992 Fasteners—Bolts, Screws, Studs and Nuts—Symbols and Designations of Dimensions (ISO 225: 1983). 26 pp.
CEN EN 20225-91. Fasteners—Bolts, Screws, Studs and Nuts—Symbols and Designations of Dimensions. 3 pp.
DIN ENGL EN 20225-92. Fasteners Bolts, Screws, Studs and Nuts; Symbols and Designations for Dimensioning (ISO 225: 1983) (Feb) (Supersedes DIN ISO 225, January 1984 Edition). 22 pp.
ISO 225-83. Fasteners—Bolts, Screws, Studs and Nuts—Symbols and Designations of Dimensions Second Edition. 22 pp.

—Torque
DIN ENGL 267 Pt 25 (P)-84. Fasteners; Technical Delivery Conditions; Torsion Testing of M 1 to M 10 Bolts and Screws (Nov). 4 pp.

—Torsional Strength
ISO 898 Pt 7-92. Mechanical Properties of Fasteners—Part 7: Torsional Test and Minimum Torques for Bolts and Screws with Nominal Diameters 1 mm to 10 mm First Editon. 7 pp.

—Tower
SAA AS 1559-86. Fasteners—Bolts, Nuts and Washers for Tower Construction. 24 pp.
SNZ NZS/AS 1559-86. Fasteners—Bolts, Nuts and Washers for Tower Construction (This is a Joint Standard with SAA AS 1559:1986). 24 pp.

—Turnbuckles
CNS A2075-89. Bolts of Turnbuckle for Building (Jan)(5045).
JIS A 5542-82. Bolts of Turnbuckle for Building. 12 pp.

—Washers (Fasteners)
BSI BS 6322: Part 3-82. 1982 Tolerances for Fasteners Part 3: Tolerances of Washers for Metric Bolts, Screws and Nuts with Thread Diameters from 1 mm up to and Including 150 mm. Product Grades A and C. 6 pp.
CNS B2010-78. Washers Medium Type Primarily for Bolts and Nuts (Mar)(150). 3 pp.
CNS B2012-47. Semi-Finished Washer for Bolts and Nuts of Cylindrical and Pan Head (Mar)(152). 2 pp.
CNS B2014-47. Large Unfinished Washers for Metric Bolts and Nuts (M6—M52) (Mar)(154). 2 pp.
CNS B2015-78. Unfinished Square Washer for Metric Bolts and Nuts (Mar)(155).
CNS B2018-47. Rectangular Lock Washers for Metric Bolts and Nuts (M3-M52) (Mar)(158)(R 1970).
CNS B2241-81. Seating Bolts and Conical Spring Washers for Clamping (Apr)(4421).
CNS B2326-78. Washers for Bolt (with Heavy Type Spring Pins) (Nov)(4648).
CNS B2399-80. Washers, Type Medium, for Bolts (Jan)(5114).
CNS B2400-80. Washers, Type Coarse, for Bolts (Jan)(5115).
DIN ENGL 128-87. Curved and Wave Spring Lock Washers (Oct). 2 pp.
DIN ENGL 267 Pt 26-87. Fasteners; Technical Delivery Conditions; Steel Spring Washers for Bolt/Nut Assemblies (Oct). 10 pp.
DIN ENGL 1440 (S)-74. Washers; Type Medium for Bolts (July) (Superseded by DIN EN 28738). 1 p.
DIN ENGL 1441-74. Washers; Type Coarse for Bolts (July). 1 p.
DIN ENGL 6796-87. Conical Spring Washers for Bolt/Nut Assemblies (Oct). 2 pp.
DIN ENGL 7349-74. Plain Washers for Bolts with Heavy Clamping Sleeves (July). 1 p.

Bolts (Cont.)
—Welding
BSI BS 2996-58. 1958 Amd 1 Projection Welding of Low Carbon Wrought Steel Studs, Bosses, Bolts, Nuts and Similar Rings. 12 pp.

—Wire Rods
DIN ENGL 59115-72. Steel Wire Rod for Bolts, Nuts and Rivets; Dimensions, Permissible Variations, Weights (Nov). 2 pp.

—Wooden Structures
BSI BS EN 383-93. 1993 Timber Structures—Test Methods—Determination of Embedding Strength and Foundation Values for Dowel Type Fasteners (V). 13 pp.
CEN PREN 383-90. Timber Structures—Determination of Embedding Strength. 12 pp.

Bomb Calorimeters
See Also: Calorimeters
BSI BS 4791-85. 1985 Calorimeter Bombs. 12 pp.

—Liquid Fuels
DIN ENGL 51900 Pt 1-89. Determination of Gross Calorific Value of Solid and Liquid Fuels by the Bomb Calorimeter and Calculation of Net Calorific Value; General Information (Nov). 11 pp.
DIN ENGL 51900 Pt 2-77. Testing of Solid and Liquid Fuels; Determination of the Gross Calorific Value by the Bomb Calorimeter and Calculation of the Net Calorific Value; Method Using Isothermal Water Jacket (Aug). 8 pp.
DIN ENGL 51900 Pt 3-77. Testing of Solid and Liquid Fuels; Determination of the Gross Calorific Value by the Bomb Calorimeter and Calculation of the Net Calorific Value; Method Using Adiabatic Jacket (Aug). 5 pp.

—Solid Fuels
CNS M3154-84. Method of Test for Gross Calorific Value of Solid Mineral Fuels by the Calorimeter Bomb Method, and Calculation of Net Calorific Value (Mar) (10835).
DIN ENGL 51900 Pt 1-89. Determination of Gross Calorific Value of Solid and Liquid Fuels by the Bomb Calorimeter and Calculation of Net Calorific Value; General Information (Nov). 11 pp.
DIN ENGL 51900 Pt 2-77. Testing of Solid and Liquid Fuels; Determination of the Gross Calorific Value by the Bomb Calorimeter and Calculation of the Net Calorific Value; Method Using Isothermal Water Jacket (Aug). 8 pp.
DIN ENGL 51900 Pt 3-77. Testing of Solid and Liquid Fuels; Determination of the Gross Calorific Value by the Bomb Calorimeter and Calculation of the Net Calorific Value; Method Using Adiabatic Jacket (Aug). 5 pp.
ISO 1928-76. Solid Mineral Fuels—Determination of Gross Calorific Value by the Calorimeter Bomb Method, and Calculation of Net Calorific Value First Edition; (Erratum—June 1979). 18 pp.

Bomb Sights
—Tubing (Flexible)—Rubber—Aircraft
MOD UK DTD-373A-50. Braided Rubber Tubing for Bomb Sight Installations (Reprinted May 1955, Incorporating Amendments Nos. 1 and 2). 4 pp.

Bombardier-Rotax Engines
Use: Aircraft Engines—Bombardier-Rotax

Bombs (Ordnance)
See Also: Ammunition; Arming Devices; Dispenser Weapons; Explosives; Initiators (Explosives); Mortars (Weapons); Ordnance; Projectiles; Pyrotechnics; Weapons

—Bituminous Coatings
MOD UK DSTAN 80-82-81. Composition, Bituminous, RD 1083 Issue 1. 8 pp.

Bomer Number
—Animal Fats
BSI BS 684: Sec 2.23-89. 1989 Methods of Analysis of Fats and Fatty Oils Part 2: Other Methods Section 2.23: Determination of Boumer Value. 5 pp.
ISO 3577-88. Animal and Vegetable Fats and Oils—Determination of Bomer Value Second Edition. 7 pp.

Bomer Value
Use: Bomer Number

Bon Maroon L
See Also: Pigments

—Pigments
JIS K 5222-71. Bon Maroon L. 5 pp.

Bon Maroon M
See Also: Pigments

Bon Maroon M (Cont.)
—Pigments
CNS K2218-88. Bon Maroon M (May)(12306).
JIS K 5223-71. Bon Maroon M. 4 pp.

Bond Papers
See Also: Manifold Papers; Papers; Printing Papers; Writing Papers
CGSB 9-GP-1MA-82. Paper, Bond, Standard for. 12 pp.
CGSB 9-GP-50M-79. Paper, Register Bond, Standard for. 8 pp.

—Check Printing—Magnetic Ink Character Recognition
CGSB 9-GP-47M-79. Paper, Bond, for Magnetic Ink Character Recognition Cheque Printing, Standard for; (Amendment 1 Mar 1980). 9 pp.

—Lithography
CGSB 9-GP-47M-79. Paper, Bond, for Magnetic Ink Character Recognition Cheque Printing, Standard for; (Amendment 1 Mar 1980). 9 pp.

Bonded Coatings
See Also: Coatings; Metal Coatings (Made From Metal); Nonmetallic Coatings

—Epoxy
CSA CAN/CSA-Z245.20-M92. External Fusion Bond Epoxy Coating for Steel Pipe; (Gen Instr 1 Thru 2). 87 pp.

Bonded Fabrics
See Also: Fabrics; Fused Fabrics; Laminated Fabrics

—Bonding Strength
CGSB CAN/CGSB-4.2 NO.65-M91. Textile Test Methods Determination of Strength of Bonds of Bonded, Laminated and Fused Fabrics. 9 pp.

—Dry Cleaning—Dimensional Stability
CGSB CAN/CGSB-4.2 NO.66-M91. Textile Test Methods Dimensional Change and Appearance After Dry Cleaning of Coated, Bonded, Laminated and Fused Fabrics. 9 pp.

Bonded Joints
Use For: Adhesive Joints *See Also:* Construction Joints; Joints

—Aerospace—Aging Testing
BSI BS EN 2243-5-92. 1992 Test Methods for Structural Adhesives Part 5: Ageing Tests. 13 pp.
CEN EN 2243-5-92. Structural Adhesives—Test Methods—Part 5: Ageing Tests. 8 pp.

—Aging Testing
BSI BS EN 29142-93. 1993 Adhesives—Guide to the Selection of Standard Laboratory Ageing Conditions for Testing Bonded Joints (ISO 9142: 1990) (W). 28 pp.
CEN EN 29142-93. Adhesives—Guide to the Selection of Standard Laboratory Ageing Conditions for Testing Bonded Joints (ISO 9142: 1990). 22 pp.
ISO 9142-90. Adhesives—Guide to the Selection of Standard Laboratory Ageing Conditions for Testing Bonded Joints First Edition (CEN EN 29142: 1993). 24 pp.

Bonding
Scope Note: Use a more specific term *See:* Adhesive Bonding; Bonded Joints; Bonding Equipment; Bonding Strength; Cohesion; Electrical Bonding; Surface Strength

Bonding Agents
Use For: Bonding Films *See Also:* Adhesive Bonding; Adhesives; Admixtures

—Plaster
CSA A261-1970. Liquid Bonding Agents for Interior Plasters (R 1992). 12 pp.

—Propellants
MOD UK TS 10276. Tris (2-Methylaziridin-1-YL) Phosphine Oxide (MAPO).

—Rubber
MOD UK TS 10246. Rubber to Metal Bonding Agent, Type QX.

Bonding Clamps
See Also: Cable Clamps; Clamps
SAA AS 1882-76. Earth and Bonding Clamps Reconfirmed 1985. 12 pp.

Bonding Compounds
Use: Bonding Agents

INTERNATIONAL AND NON-U.S. NATIONAL STANDARDS
SUBJECT INDEX

Bonding Conductors
See Also: Electrical Bonding; Grounding Conductors; Grounding Connectors

—Installation
DIN VDE 0100 Pt 540-91. Installation of Power Plant with Rated Voltages Not Exceeding 1000 V; Selection and Insallation of Electrical Equipment; Earthing Arrangements, Protective Conductors and Equipotential Bonding Conductors (Nov). 37 pp.

DIN VDE 0100 Pt 540 A2 (D)-92. Erection of Power Installations with Nominal Voltages up to 1000 V; Selection and Erection of Equipment; Earthing Arrangements, Protective Conductors, Equipotential Bonding Conductors; Amendment 2 Identical with IEC 64 (Sec)570 (Jan). 8 pp.

Bonding Equipment
See Also: Woodworking Equipment

—Glossaries
BSI BS 4361: Part 32-90. 1990 Woodworking Machines Part 32: Nomenclature for Single-End Edge Bonding Machines. 12 pp.

ISO 9537-89. Woodworking Machines—Single-End Edge Bonding Machines—Nomenclature First Edition. 11 pp.

Bonding Films
Use: Bonding Agents

Bonding Flexibles
Use: Cordage

Bonding Hardware (Electric)
Scope Note: Use a more specific term *See:* Bonding Clamps

Bonding Strength
See Also: Adhesive Strength; Adhesives; Cements; Shear Testing

—Adhesive Tapes
SAA AS 1635.16.1-74. Methods of Testing Pressure Sensitive Adhesive Tape—Part 16.1: Bond Strength at Elevated Temperature. 1 p.

—Adhesives
SAA AS 1937.2-77. Methods of Test for Sealers and Adhesives for Automotive Purposes—Part 2: Determination of Initial Bond Strength of Adhesives Reconfirmed 1989. 4 pp.

—Adhesives—Ceilings
CNS A3311-89. Method of Test for Bonding Strength of Ceiling Boards Adhesives (Sep)(12605).

—Adhesives—Footwear
CEN PREN 522-91. Adhesives for Leather and Footwear Materials—Bond Strength—Minimum Requirements and Adhesive Classification. 17 pp.

—Adhesives—Leather
CEN PREN 522-91. Adhesives for Leather and Footwear Materials—Bond Strength—Minimum Requirements and Adhesive Classification. 17 pp.

—Adhesives—Wallboard
CNS A3309-89. Method of Test for Bonding Strength of Wall Boards Adhesives (Sep)(12602).

JIS A 1613-84. Testing Methods for Bonding Strength of Wall Boards Adhesives. 11 pp.

—Bed Joints—Reinforcement
CEN PREN 846-2-92. Methods of Test for Ancillary Components for Masonry—Part 2: Determination of Bond Strength of Bed Joint Reinforcement in Mortar Joints. 12 pp.

—Concretes—Reinforcing Steels
CEN PREN 989-92. Determination of the Bond Behaviour Between Reinforcing Steel and Autoclaved Aerated Concrete by the "Push-Out" Test. 5 pp.

CNS A3219-84. Method of Test for Comparing Concrete on the Basis of the Bond Developed with Reinforcing Steel (Dec)(11152).

—Conveyor Belts
ISO 7623-84. Steel Cord Conveyor Belts—Cord-to-Coating Bond Test First Edition. 4 pp.

—Corrugated Fiberboard—Water Resistance Testing
ISO 3038-75. Corrugated Fibreboard—Determination of the Water Resistance of the Glue Bond by Immersion First Edition. 6 pp.

—Electrical Insulating Varnishes
IEC 699-81. Test Method for the Evaluation of Bond Strength of Impregnating Varnishes by the Wire Bundle Test First Edition. 17 pp.

Bonding Strength (Cont.)

—Fabrics
CGSB CAN/CGSB-4.2 NO.65-M91. Textile Test Methods Determination of Strength of Bonds of Bonded, Laminated and Fused Fabrics. 9 pp.

—Grinding Wheels
CNS R3077-87. Method of Test for Grinding Wheels (Nov) (3786).

JIS R 6240-86. Testing Methods of Grinding Wheels. 18 pp.

—Impregnating Agents—Enameled Electric Wire
IEC 1033-91. Test Methods for the Determination of Bond Strength of Impregnating Agents to an Enamelled Wire Substrate First Edition. 30 pp.

—Metal Adhesives
DIN ENGL 54454-84. Testing of Adhesives for Metals and of Bonded Metal Joints; Initial Break-Away Test on Bonded Screw Threads (June). 3 pp.

DIN ENGL 54455-84. Testing of Adhesives for Metals and of Bonded Metal Joints; Torsional Shear Test (May). 3 pp.

—Mortars
DIN ENGL 18555 Pt 6-87. Testing of Mortars Containing Mineral Binders; Determination of Bond Strength of Hardened Mortar (Nov). 3 pp.

—Mortars—Shear Stress
DIN ENGL 18555 Pt 5-86. Testing of Mortars Containing Mineral Binders; Hardened Mortars; Determination of Bond Shear Strength of Masonry Mortars (Mar). 4 pp.

—Paperboard
CPPA D.37P-80. Z—Directional Strength of Paper and Paperboard. 1 p.

—Papers
CPPA D.37P-80. Z—Directional Strength of Paper and Paperboard. 1 p.

—Plain Bearings
BSI BS 7585: Part 1-92. 1992 Metallic Multilayer Plain Bearings Part 1: Method for Non-Destructive Ultrasonic Testing of Bond (ISO 4386-1: 1992). 16 pp.

BSI BS 7585: Part 2-92. 1992 Metallic Multilayer Plain Bearings Part 2: Method for Destructive Testing of Bond for Bearing Metal Layer Thicknesses Greater Than or Equal to 2 mm (ISO 4386-2: 1992). 10 pp.

BSI BS 7585: Part 3-92. 1992 Metallic Multilayer Plain Bearings Part 3: Method for Non-Destructive Die Penetrant Testing (ISO 4386-3: 1992). 16 pp.

ISO 4386 Pt 1-92. Plain Bearings—Metallic Multilayer Plain Bearings—Part 1: Non-Destructive Ultrasonic Testing of Bond Second Edition. 13 pp.

ISO 4386 Pt 2-82. Plain Bearings—Metallic Multilayer Plain Bearings—Part 2: Destructive Testing of Bond for Bearing Metal Layer Thicknesses Greater Than or Equal to 2 mm First Edition. 6 pp.

ISO 4386 Pt 3-92. Plain Bearings—Metallic Multilayer Plain Bearings—Part 3: Non-Destructive Penetrant Testing First Edition. 13 pp.

—Plywood
BSI BS 6566: Part 8-85. 1985 Amd 1 Plywood Part 8: Bond Performance of Veneer Plywood. 16 pp.

BSI BS EN 314-1-93. 1993 Plywood—Bonding Quality Part 1: Test Methods (R). 14 pp.

BSI BS EN 314-2-93. 1993 Plywood—Bonding Quality Part 2: Requirements (R). 10 pp.

CEN PREN 314-89. Plywood—Bonding—Tests. 4 pp.

CEN EN 314-1-93. Plywood—Bonding Quality —Part 1: Test Methods. 11 pp.

CEN PREN 314 (Part 2)-90. Plywood—Bonding Quality—Part 2: Requirements. 6 pp.

CEN EN 314-2-93. Plywood—Bonding Quality —Part 2: Requirements. 6 pp.

CNS O2032-81. Method of Test for Bonding Strength of Plywood (Oct)(8060). 3 pp.

—Refractory Mortars
SAA AS 1774.22-85. Refractories and Refractory Materials—Physical Test Methods—Part 22: Bonding Strength of Refractory Mortars. 4 pp.

—Resins—Slant Shear Method
BSI BS 6319: Part 4-84. 1984 Testing of Resin Compositions for Use in Construction Part 4: Method for Measurement of Bond Strength (Slant Shear Method). 12 pp.

—Sprayed Coatings
BSI DD 156-87. 1987 Method of Test for Bond Strength of Sprayed Mineral Coatings. 9 pp.

Bonding Strength (Cont.)

—Textile Floor Coverings
SAA AS 2111.16-91. Methods of Test for Textile Floor Coverings—Part 16: Determination of Bond Strength Between Backing Components of a Textile Floor Covering. 2 pp.

—Welding Rods—Braze Welding Equipment
ISO 698-75. Filler Rods for Braze Welding—Determination of Conventional Bond Strength on Steel, Cast Iron and Other Metals First Edition. 6 pp.

—Wood Adhesives
CNS A3204-89. Method of Test for Bonding Strength of Anchoring Wooden Block Adhesives (Sep)(10842).

JIS A 1611-84. Testing Methods for Bonding Strength of Anchoring Wooden Block Adhesives.

SAA AS 1321.3-76. Methods for the Sampling and Testing of Adhesives—Part 3: Bond Strength of Cured Wood-to-Wood Adhesives in Shear Corrig. Reconfirmed 1989. 28 pp.

Bonding Wire
See Also: Electric Conductors; Wire
BSI BS 7197-90. 1990 Performance of Bonds for Electric Power Cable Terminations and Joints for System Voltages up to 36 kV. 18 pp.

Bonds (Securities)
Use: Securities

Bone Black

—Pigments
CNS K2104-80. Bone Black (Jul)(5869). 1 p.

Bone Cements
See Also: Prosthetic Devices; Surgical Implants

—Acrylic Resins
BSI BS 7253: Part 1-93. 1993 Non-Metallic Materials for Surgical Implants Part 1: Specification for Acrylic Resin Cement (ISO 5833: 1992) (Supersedes BS 3531: Part 7: 1981) (SUPERSEDES BS 3531: PART 7: 1981). 22 pp.

BSI BS 7253: Part 1-90. 1990 Non-Metallic Materials for Surgical Implants Part 1: Acrylic Bone Cement. 12 pp.

ISO 5833-92. Implants for Surgery—Acrylic Resin Cements First Edition. 21 pp.

Bone China
Use: Tableware—Bone China

Bone Chisels
Use: Osteotomes

Bone Files
See Also: Files (Tools); Surgical Instruments
CNS T1044-81. Bone Files (Jan)(6975).

Bone Meal
Use For: Blood Meal; Flesh Meal *See Also:* Animal Feed; Barley Hull Meal; Feather Meal; Fertilizers; Fish Meal; Leucaena Meal; Oil Meal; Shell Meal
CNS N1070-66. Raw Bone Powder (Nov)(2714). 1 p.

—Animal Feed
CNS N2017-80. Bone Meal (Feed Grade) (Apr)(2297). 2 pp.

CNS N2019-80. Blood Meal (Feed Grade) (Apr)(2594). 1 p.

CNS N2021-84. Meat and Bone Meal (for Feeding) (Oct)(2710). 1 p.

—Fertilizers
CNS N3016-88. Steamed Bone Meal for Fertilizer (Aug)(2596).

CNS N3046-88. Flesh Meal Fertilizer (Jul)(11912).

CNS N3047-88. Flesh-Bone Meal Fertilizer (Jun)(11913).

CNS N3048-88. Raw Bone Meal Fertilizer (Jun)(11914).

Bone Nails
See Also: Bone Pins; Nails (Fasteners); Prosthetic Devices; Surgical Implants
BSI BS 3531: Part 15-90. 1990 Surgical Implants Part 15: Devices for the Fixation of the Ends of the Femur in Adults. 17 pp.

—Intramedullary
BSI BS 3531: Sec 14.1-90. 1990 Surgical Implants Part 14: Intramedullary Nailing Systems Section 14.1: Dimensions of Nails of Clover Leaf and V-Shaped Cross Section and Dimensions of Extraction Hooks. 12 pp.

ISO 5837 Pt 1-85. Implants for Surgery—Intramedullary Nailing Systems—Part 1: Intramedullary Nails with Cloverleaf or V-Shaped Cross-Section First Edition. 7 pp.

INDUSTRY STANDARDS

Bone Nails (Cont.)

—Intramedullary—Introducers/Extractors
BSI BS 3531: Sec 14.1-90. 1990 Surgical Implants Part 14: Intramedullary Nailing Systems Section 14.1: Dimensions of Nails of Clover Leaf and V-Shaped Cross Section and Dimensions of Extraction Hooks. 12 pp.

—Medullary
BSI BS 3531: Sec 14.2-90. 1990 Surgical Implants Part 14: Intramedullary Nailing Systems Section 14.2: Medullary Pins (Rush Type) and Introducer/Extractors. 7 pp.

—Medullary—Introducers/Extractors
BSI BS 3531: Sec 14.2-90. 1990 Surgical Implants Part 14: Intramedullary Nailing Systems Section 14.2: Medullary Pins (Rush Type) and Introducer/Extractors. 7 pp.

Bone Pins
See Also: Bone Nails; Bone Screws; Bone Wire; Pins; Prosthetic Devices; Surgical Implants
BSI BS 3531: Sec 6.1-90. 1990 Surgical Implants Part 6: Skeletal Pins and Wires Section 6.1: General Requirements. 8 pp.
BSI BS 3531: Sec 6.2-91. 1991 Implants for Osteosynthesis Part 6: Skeletal Pins and Wires Section 6.2: Specification for Dimensions of Steinmann Skeletal Pins (ISO 5838-2: 1991). 8 pp.
ISO 5838 Pt 1-83. Implants for Surgery—Skeletal Pins and Wires—Part 1: Material and Mechanical Requirements First Edition. 4 pp.
ISO 5838 Pt 2-91. Implants for Surgery—Skeletal Pins and Wires—Part 2: Steinmann Skeletal Pins—Dimensions First Edition. 4 pp.

—Intramedullary
ISO 5837 Pt 2-80. Implants for Surgery—Intramedullary Nailing Systems—Part 2: Medullary Pins First Edition. 4 pp.

Bone Plates
See Also: Plates (Supports); Prosthetic Devices; Surgical Implants
BSI BS 3531: Part 15-90. 1990 Surgical Implants Part 15: Devices for the Fixation of the Ends of the Femur in Adults. 17 pp.

—Bend Testing
BSI BS 3531: Sec 23.1-91. 1991 Implants for Osteosynthesis Part 23: Bone Plates Section 23.1: Method for Determination of Bending Strength and Stiffness (ISO 9585: 1990). 11 pp.
ISO 9585-90. Implants for Surgery—Determination of Bending Strength and Stiffness of Bone Plates First Edition. 7 pp.

—Holes
BSI BS 3531: Sec 23.2-93. 1993 Implants for Osteosynthesis Part 23: Bone Plates Section 23.2: Specification for Holes and Slots for Use with Screws of 4.5 mm, 4.2 mm, 4.0 mm, 3.9 mm, 3.5 mm and 2.9 mm Nominal Sizes (ISO 9269: 1988) (SUP'S BS 3531: PART 5: 1982). 12 pp.
BSI BS 3531: Sec 23.2-90. 1990 Surgical Implants Part 23: Bone Plates Section 23.2: Holes and Slots for Use with Screws of 4mm, 3.5mm and 3mm Nominal Sizes. 8 pp.
BSI BS 3531: Sec 23.3-91. 1991 Implants for Osteosynthesis Part 23: Bone Plates Section 23.3: Specification for Holes Corresponding to Screws with Asymmetrical Thread and Spherical Undersurfaces (ISO 5836: 1988). 12 pp.
ISO 5836-88. Implants for Surgery—Metal Bone Plates—Holes Corresponding to Screws with Asymmetrical Thread and Spherical Under-Surface First Edition. 8 pp.
ISO 9269-88. Implants for Surgery—Metal Bone Plates—Holes and Slots Corresponding to Screws with Conical Under-Surface First Edition; (Supersedes 5836 Pt 4). 8 pp.

—Stiffness Testing
ISO 9585-90. Implants for Surgery—Determination of Bending Strength and Stiffness of Bone Plates First Edition. 7 pp.

Bone Powder
Use: Bone Meal

Bone Screws
See Also: Bone Pins; Prosthetic Devices; Surgical Implants
BSI BS 3531: Part 15-90. 1990 Surgical Implants Part 15: Devices for the Fixation of the Ends of the Femur in Adults. 17 pp.

—Bits (Tools)
BSI BS 3531: Sec 5.7-91. 1991 Implants for Osteosynthesis Part 5: Bone Screws and Auxiliary Equipment Section 5.7: Specification for Drill Bits, Taps and Countersink Cutters (ISO 9714-1: 1991). 10 pp.
ISO 9714 Pt 1-91. Orthopaedic Drilling Instruments—Part 1: Drill Bits, Taps and Countersink Cutters First Edition. 8 pp.

—Conical Undersurface Head
ISO 9268-88. Implants for Surgery—Metal Bone Screws with Conical Under-Surface of Head—Dimensions First Edition; (Supersedes 5835 Pt 4). 12 pp.

—Countersinks
BSI BS 3531: Sec 5.7-91. 1991 Implants for Osteosynthesis Part 5: Bone Screws and Auxiliary Equipment Section 5.7: Specification for Drill Bits, Taps and Countersink Cutters (ISO 9714-1: 1991). 10 pp.
ISO 9714 Pt 1-91. Orthopaedic Drilling Instruments—Part 1: Drill Bits, Taps and Countersink Cutters First Edition. 8 pp.

—Hexagonal Head—Socket—Keys
BSI BS 3531: Sec 5.1-90. 1990 Surgical Implants Part 5: Bone Screws and Auxiliary Equipment Section 5.1: Keys for Use with Screws Having Heads with Hexagon Socket. 8 pp.
ISO 8319 Pt 1-86. Orthopaedic Instruments—Drive Connections—Part 1: Keys for Use with Screws with Hexagon Socket Heads First Edition. 6 pp.

—Phillips—Screwdrivers
BSI BS 3531: Sec 5.2-90. 1990 Surgical Implants Part 5: Bone Screws and Auxiliary Equipment Section 5.2: Screwdrivers for Screws Having Heads with Single Slot, Cruciate Slot or Cross-Recess. 12 pp.
ISO 8319 Pt 2-86. Orthopaedic Instruments—Drive Connections—Part 2: Screwdrivers for Single Slot Head Screws, Screws with Cruciate Slot and Cross-Recessed Head Screws First Edition. 10 pp.

—Screw Threads
BSI BS 3531: Sec 5.3-91. 1991 Implants for Osteosynthesis Part 5: Bone Screws and Auxiliary Equipment Section 5.3: Spec. for the Dimensions of Screws Having Hexagonal Drive Connection, Spherical Under-Surfaces and Asymmetrical Thread (ISO 5835: 1991). 16 pp.
BSI BS 3531: Sec 5.4-93. 1993 Implants for Osteosynthesis Part 5: Bone Screws and Auxiliary Equipment Section 5.4: Specification for Screws Having Symmetrical and Asymmetrical Thread and Conical Under-Surfaces (ISO 9268: 1988) (SUP'S BS 3456: PART 5: 1982). 15 pp.
BSI BS 3531: Sec 5.4-90. 1990 Surgical Implants Part 5: Bone Screws and Auxiliary Equipment Section 5.4: Screws Having Symmetrical Thread and Conical Under-Surfaces. 8 pp.
BSI BS 3531: Sec 5.6-91. 1991 Implants for Osteosynthesis Part 5: Bone Screws and Auxiliary Equipment Section 5.6: Spec. for the Mechanical Requir. and Test Meth for Screws Having Asymmetrical Thread and Spherical Undersurface (ISO 6475: 1989). 11 pp.
ISO 5835-91. Implants for Surgery—Metal Bone Screws with Hexagonal Drive Connection, Spherical Under-Surface of Head, Asymmetrical Thread—Dimensions First Edition. 14 pp.
ISO 6475-89. Implants for Surgery—Metal Bone Screws with Asymmetrical Thread and Spherical Under-Surface—Mechanical Requirements and Test Methods First Edition. 9 pp.

—Slotted Head—Screwdrivers
BSI BS 3531: Sec 5.2-90. 1990 Surgical Implants Part 5: Bone Screws and Auxiliary Equipment Section 5.2: Screwdrivers for Screws Having Heads with Single Slot, Cruciate Slot or Cross-Recess. 12 pp.
ISO 8319 Pt 2-86. Orthopaedic Instruments—Drive Connections—Part 2: Screwdrivers for Single Slot Head Screws, Screws with Cruciate Slot and Cross-Recessed Head Screws First Edition. 10 pp.

—Taps
BSI BS 3531: Sec 5.5-90. 1990 Surgical Implants Part 5: Bone Screws and Auxiliary Equipment Section 5.5: Taps and Drills. 4 pp.
BSI BS 3531: Sec 5.7-91. 1991 Implants for Osteosynthesis Part 5: Bone Screws and Auxiliary Equipment Section 5.7: Specification for Drill Bits, Taps and Countersink Cutters (ISO 9714-1: 1991). 10 pp.
ISO 9714 Pt 1-91. Orthopaedic Drilling Instruments—Part 1: Drill Bits, Taps and Countersink Cutters First Edition. 8 pp.

—Twist Drills
BSI BS 3531: Sec 5.5-90. 1990 Surgical Implants Part 5: Bone Screws and Auxiliary Equipment Section 5.5: Taps and Drills. 4 pp.

Bone Staples
See Also: Prosthetic Devices; Staples; Surgical Implants
BSI BS 3531: Part 11-91. 1991 Implants for Osteosynthesis Part 11: Specification for Staples with Parallel Legs (ISO 8827: 1988). 11 pp.
BSI BS 3531: Part 11-90. 1990 Amd 1 Surgical Implants Part 11: Staples for Use in Orthopaedic Surgery. 8 pp.
ISO 8827-88. Implants for Surgery—Staples with Parallel Legs for Orthopaedic Use—General Requirements First Edition. 8 pp.

Bone Vibrators
See Also: Vibration Equipment

—Audiometers
BSI BS 4009-91. 1991 Artificial Mastoids for the Calibration of Bone Vibrators Used in Hearing Aids and Audiometers (IEC 373: 1990). 16 pp.
BSI BS 6950-88. 1988 Standard Reference Zero for the Calibration of Pure Tone Bone Conduction Audiometers. 13 pp.
BSI BS 6950-01. 1988 Amd 1 Standard Reference Zero for the Calibration of Pure Tone Bone Conduction Audiometers (ISO 7566: 1987) (AMD 7340) October 15, 1992. 17 pp.
CEN EN 27566-91. Acoustics—Standard Reference Zero for the Calibration of Pure-Tone Bone Conduction Audiometers. 3 pp.
CENELEC HD 590 S1-91. Mechanical Coupler for Measurements on Bone Vibrators. 5 pp.
IEC 373-90. Mechanical Coupler for Measurements on Bone Vibrators Second Edition. 29 pp.
ISO 7566-87. Acoustics—Standard Reference Zero for the Calibration of Pure-Tone Bone Conduction Audiometers First Edition. 12 pp.
SAA AS 1591.4-74. Acoustics—Instrumentation for Audiometry—Part 4: Mechanical Coupler for Calibration of Bone Vibrators Used in Hearing Aids and Audiometers. 11 pp.

—Hearing Aids
BSI BS 6083: Part 9-86. 1986 Amd 1 Hearing Aids Part 9: Methods for Measurement of Characteristics of Hearing Aids with Bone Vibrator Output (AMD 6186) October 31, 1990. 16 pp.
CENELEC HD 450.9-87. Hearing Aids Part 9: Methods of Measurement of Characteristics of Hearing Aids with Bone Vibrator Output. 3 pp.
IEC 118 Pt 9-85. Hearing Aids Part 9: Methods of Measurement of Characteristics of Hearing Aids with Bone Vibrator Output First Edition. 27 pp.

Bone Wire
See Also: Bone Pins; Prosthetic Devices; Surgical Implants; Wire
BSI BS 3531: Sec 6.1-90. 1990 Surgical Implants Part 6: Skeletal Pins and Wires Section 6.1: General Requirements. 8 pp.
BSI BS 4106-67. (OBSOLESCENT) 1967 Surgical Stainless Steel Monofilament Wire. Fully Softened. 12 pp.
ISO 5838 Pt 1-83. Implants for Surgery—Skeletal Pins and Wires—Part 1: Material and Mechanical Requirements First Edition. 4 pp.
ISO 5838 Pt 3-93. Implants for Surgery—Skeletal Pins and Wires—Part 3: Kirschner Skeletal Wires First Edition. 7 pp.

Bonito
See Also: Fish; Mackerel

—Canned
CNS N5138-91. Canned Tuna and Bonito (Feb)(4456). 5 pp.

—Dried
CNS N5066-81. Dried Bonito and Mackerel (Aug)(2379). 3 pp.

Book Cases
Use: Bookcases

Book Ends
Use: Bookends

Book Papers
See Also: Papers; Printing Papers
BSI BS 1413-89. 1989 Page Sizes for Books. 4 pp.
CGSB 9-GP-29MA-82. Paper, Book, No. 1 Book and No. 1 Opaque Book, Standard for. 12 pp.

Bookbinding
See Also: Binders (Files); Bookbinding Machinery; Sewing

—Classification
BSI BS 1000: (686)-71. 1971 Universal Decimal Classification (UDC). English Full Edition (686): Bookbinding. Metallizing. Mirror Making. Stationery. 21 pp.
SNZ NZS/BS 1000 (686)-71. Universal Decimal Classification Bookbinding. Metallizing. Mirror Making. Stationery. 24 pp.

Bookbinding (Cont.)

—Leather
BSI BS 7451-91. 1991 Archival Quality Bookbinding Leather. 11 pp.

—Photoelectric Sensors—Sensing Marks
JIS B 9607-91. Book Making Machine—Photoelectrical Sensing Marks for Signature. 7 pp.

Bookbinding Machinery
See Also: Bookbinding; Machinery

—Adhesives—Hot Melt
CGSB CAN/CGSB-21.12-M91. Hot Melt Adhesives. 7 pp.

—Wire—Stitching
CGSB CAN/CGSB-21.14-M91. Stitching Wire. 7 pp.

Bookcases
See Also: Bookends; Books; Furniture; Office Furniture; Shelves

—Modular—With Credenza
CGSB 44.153-85. Credenza, Modular, Single Case with Bookcase. 36 pp.

—Office—Modular
CGSB 44.170-85. Bookcase Modular, Wood, Metal Legs. 13 pp.

Bookends
Use For: Book Ends *See Also:* Bookcases; Books

—Steel
CGSB 44.11A-93. Metal Book Ends. 7 pp.

Books
Use For: Textbooks *See Also:* Bookcases; Bookends; Documentation; International Standard Book Numbering; Monographs; Publications; Title Pages
ISO 6716-83. Graphic Technology—Text-Books and Periodicals—Sizes of Untrimmed Sheets and Trimmed Pages First Edition. 4 pp.

—Bibliographic Identification
BSI BS 7187-89. 1989 Code for Bibliographic Identification (Biblid) of Contributions in Serials and Books. 8 pp.
ISO 9115-87. Documentation—Bibliographic Identification (Biblid) of Contributions in Serials and Books First Edition. 7 pp.

—Dating
SNZ NZSR 11-65. Recommendation for Book Sizes and Dating of Books. 8 pp.

—Libraries—Price Indexes
BSI BS ISO 9230-91. 1991 Information and Documentation—Determination of Price Indexes for Books and Serials Purchased by Libraries. 15 pp.
ISO 9230-91. Information and Documentation—Determination of Price Indexes for Books and Serials Purchased by Libraries First Edition. 13 pp.

—Military—Medical—Multilingual
NATO STANAG 2131 ED 3 AMD 3-82. Multilingual Phrase Book for Use by the NATO Medical Services (AMedP-5(A)). 7 pp.

—Sizes
SNZ NZSR 11-65. Recommendation for Book Sizes and Dating of Books. 8 pp.

—Statistics
BSI BS ISO 9707-91. 1991 Information and Documentation—Statistics on the Production and Distribution of Books, Newspapers, Periodicals and Electronic Publications. 19 pp.
ISO 9707-91. Information and Documentation—Statistics on the Production and Distribution of Books, Newspapers, Periodicals and Electronic Publications First Edition. 17 pp.

—Telecommunication Services—Abbreviations
CCITT RECMN F.92-89. Service Codes—Telegraph and Mobile Services Operations and Quality of Service (Study Group I) 3 pp. 3 pp.

—Telecommunication Services—Codes
CCITT RECMN F.92-89. Service Codes—Telegraph and Mobile Services Operations and Quality of Service (Study Group I) 3 pp. 3 pp.

Bookshelves
Use: Bookcases

Booms (Equipment)
See Also: Cranes; Derricks; Hoists

Booms (Equipment) (Cont.)

—Excavating Equipment—Hydraulic Valves
BSI BS 6912: Part 1-88. 1988 Safety of Earth-Moving Machinery Part 1: Hydraulic Excavator and Backhoe Loader Boom Lowering Control Device. 6 pp.
ISO 8643-88. Earth-Moving Machinery—Hydraulic Excavator and Backhoe Loader Boom Lowering Control Device—Requirements and Tests First Edition. 6 pp.

Boots (Footwear)
See Also: Clothing; Footwear; Overshoes; Shoes (Footwear)
BSI BS 1870: Part 1-88. 1988 Amd 1 Safety Footwear Part 1: Safety Footwear Other Than All-Rubber and All-Plastics Moulded Types (AMD 6273) February 28, 1990. 24 pp.

—Canvas
JIS S 5002-92. Canvas Boots and Shoes. 18 pp.

—Chain Saw Users
CEN PREN 381 (Part 3)-90. Protective Clothing for Users of Hand Held Chain Saws—Part 3: Test Method for Boots. 14 pp.
CEN PREN 381 (Part 6)-90. Protective Clothing for Users of Hand Held Chain Saws—Part 6: Requirements for Boots. 8 pp.

—Conductive
BSI BS 7193-89. 1989 Amd 1 Lined Lightweight Rubber Overshoes and Overboots (AMD 6606) December 21, 1990. 11 pp.

—Flying
MOD UK DTD-1250-01. Boots Flying 1965 Pattern; Amendment 2. 16 pp.

—Leather
CNS S2014-72. Testing Standard for Leather Boots (Tentative) (Jun)(888). 2 pp.

—Leather—Firefighters'
BSI BS 2723-56. 1956 Amd 1 Firemen's Leather Boots. 18 pp.

—Mountaineering
CNS S1098-79. Mountaineering Boots with Light Outfit (May) (4871). 4 pp.
JIS S 5035-92. Mountaineering Boots with Light Outfit. 9 pp.

—Polyurethane
ISO 5423-92. Moulded Plastics Footwear—Lined or Unlined Polyurethane Boots for General Industrial Use—Specification First Edition. 18 pp.
ISO 6910-92. Moulded Plastics Footwear—Lined or Unlined Polyurethane Industrial Boots with General-Purpose Resistance to Animal Fats and Vegetable Oils—Specification First Edition. 5 pp.

—Protective
BSI BS 1870: Part 1-88. 1988 Amd 1 Safety Footwear Part 1: Safety Footwear Other Than All-Rubber and All-Plastics Moulded Types (AMD 6273) February 28, 1990. 24 pp.
BSI BS 1870: Part 2-76. 1976 Amd 4 Safety Footwear Part 2: Lined Rubber Safety Boots. 8 pp.
ISO 2023-73. Lined Industrial Rubber Footwear First Edition. 12 pp.
ISO 4643-92. Moulded Plastics Footwear—Lined or Unlined Poly (Vinyl Chloride) Boots for General Industrial Use—Specification Second Edition. 17 pp.
JIS T 8117-79. Protective Boots for Occupational Health.

—Protective—Antistatic
BSI BS 7193-89. 1989 Amd 1 Lined Lightweight Rubber Overshoes and Overboots (AMD 6606) December 21, 1990. 11 pp.

—PVC
BSI BS 6159: Part 1-87. 1987 Amd 1 Polyvinyl Chloride Boots Part 1: General Industrial Lined or Unlined Boots. 13 pp.
BSI BS 6159: Part 1-02. 1987 Amd 2 Polyvinyl Chloride Boots Part 1: Specification for General and Industrial Lined or Unlined Boots (AMD 7053) May 1, 1992. 14 pp.
ISO 4643-92. Moulded Plastics Footwear—Lined or Unlined Poly (Vinyl Chloride) Boots for General Industrial Use—Specification Second Edition. 17 pp.

—PVC—Chemical Resistant
ISO 6110-92. Moulded Plastics Footwear—Lined or Unlined Poly (Vinyl Chloride) Industrial Boots with Chemical Resistance—Specification Second Edition. 5 pp.

—PVC—Oil Resistant
ISO 6112-92. Moulded Plastics Footwear—Lined or Unlined Poly (Vinyl Chloride) Industrial Boots with General-Purpose Resistance to Animal Fats and Vegetable Oils—Specification Second Edition. 5 pp.

Boots (Footwear) (Cont.)

—Rubber
BSI BS 1870: Part 2-76. 1976 Amd 4 Safety Footwear Part 2: Lined Rubber Safety Boots. 8 pp.
BSI BS 5145-89. 1989 Amd 1 Lined Industrial Vulcanized Rubber Boots (AMD 6596) April 30, 1991. 13 pp.
CNS S1097-79. Rubber Boots and Shoes (May)(4870). 4 pp.
ISO 2023-73. Lined Industrial Rubber Footwear First Edition. 12 pp.
ISO 3910-83. Rubber Boots, Unlined Moulded Second Edition. 6 pp.
JIS S 5005-92. High Boots. 11 pp.

—Rubber—Antistatic
ISO 2251-91. Lined Antistatic Rubber Footwear—Specification Second Edition. 9 pp.

—Rubber—Chemical Resistant
ISO 6111-82. Rubber Footwear—Lined or Unlined Rubber Industrial Boots with Chemical Resistance First Edition. 3 pp.

—Rubber—Cold Resistant
ISO 2252-83. Rubber Footwear, Lined Industrial, for Use at Low Temperatures Third Edition. 6 pp.

—Rubber—Oil Resistant
ISO 2025-72. Lined Industrial Rubber Boots with General Purpose Oil Resistance First Edition. 4 pp.

—Rugby—Studs
BSI BS 6366-83. 1983 Studs for Rugby Football Boots. 12 pp.
SNZ NZS 5836-85. Specification for Studs for Rugby Football Boots Amend: A, 1985. 12 pp.

—Safety
JIS T 8117-79. Protective Boots for Occupational Health.

—Ski—Alpine
ISO 5355-91. Alpine Ski-Boots—Safety Requirements and Test Methods Second Edition. 23 pp.

—Ski—Cross Country
ISO 6959-83. Cross-Country Ski Boots with Three Pin Holes—Dimensions, Interface and Design First Edition. 5 pp.

BOPP
Use: Polypropylene Films

Boracic Acid
Use: Boric Acid

Borate Content Analysis

—Cosmetics
CNS S2080-82. Method of Hygienic Test for Cosmetics Boric Acid and Borate (Sep)(9440).

—Water—Spectrometry
BSI BS 6068: Sec 2.40-91. 1991 Water Quality Part 2: Physical, Chemical and Biochemical Methods Section 2.40: Method for the Determination of Borate by Spectrometry Using Azomethine-H (ISO 9390: 1990). 10 pp.
ISO 9390-90. Water Quality—Determination of Borate—Spectrometric Method Using Azomethine-H First Edition. 7 pp.

Borate Ion Content Analysis

—Waste Water—Photometry
DIN ENGL 38405 Pt 17-81. German Standard Methods for the Analysis of Water, Waste Water and Sludge; Anions (Group D); Determination of Borate Ions (D 17) (Mar). 3 pp.

—Water—Photometry
DIN ENGL 38405 Pt 17-81. German Standard Methods for the Analysis of Water, Waste Water and Sludge; Anions (Group D); Determination of Borate Ions (D 17) (Mar). 3 pp.

Borax
Use: Sodium Borates

Bordeaux 5B
See Also: Pigments

—Pigments
CNS K2139-83. Lithol Red B (Jun)(10360). 3 pp.
JIS K 5214-71. Bordeaux 5 B. 5 pp.

Bordeaux Connections

—Wire Rope
BSI BS 461-70. (WITHDRAWN) 1970 Bordeaux Connections (Superseded by BS 7167: 1990). 25 pp.
BSI BS 7167-90. 1990 Bordeaux Connections. 20 pp.

INTERNATIONAL AND NON-U.S. NATIONAL STANDARDS
SUBJECT INDEX

Bordeaux

Bordeaux 10B
See Also: Pigments
—Pigments
CNS K2140-83. Bordeaux 10B (Jun)(10361). 3 pp.
JIS K 5215-71. Bordeaux 10 B. 5 pp.

Border Control
Use For: Frontier Control
EC COM(88) 800-89. Proposal for a Council Regulation (EEC) on the Elimination of Controls Performed at the Frontiers of Member States in the Field of Road and Inland Waterway Transport. 16 pp.
EC EEC/4060/89-89. Council Regulation on the Elimination of Controls Performed at the Frontiers of Member States in the Field of Road and Inland Waterway Transport. 5 pp.
EC EEC/474/90-90. Council Regulation Amending, with a View to Abolishing Lodgement of the Transit Advice Note on Crossing an Internal Frontier of the Community, Regulation (EEC) No 222/77 on Community Transit. 2 pp.
—Goods
EC EEC/3690/86-86. Council Regulation Concerning the Abolition Within the Framework of the TIR Convention of Customs Formalities on Exit from a Member State at a Frontier between Two Member States. 2 pp.
EC EEC/4283/88-88. Council Regulation on the Abolition of Certain Exit Formalities at Internal Community Frontiers-Introduction of Common Border Posts. 2 pp.
—Intra-Community Nationals
EC COM(84) 749-85. Proposal for a Council Directive on the Easing of Controls and Formalities Applicable to Nationals of the Member States When Crossing Intra-Community Borders. 9 pp.
EC COM(85) 224-85. Amendment to the Proposal for a Council Directive on the Easing of Controls and Formalities Applicable to Nationals of the Member States When Crossing Intra-Community Borders. 13 pp.
—Military—Roads
NATO STANAG 2176 ED 2 AMD 1-81. Procedures for Military Road Movement Across National Frontiers. 10 pp.
NATO STANAG 2176 ED 2 AMD 2-81. Procedures for Military Road Movement Across National Frontiers. 12 pp.
—Military—Rolling Stock
NATO STANAG 2171 ED 5 AMD 3-84. Procedures for Military Trains Crossing Frontiers. 19 pp.
NATO STANAG 2173 ED 3 AMD 2-83. Regulations for the Securing of Military Tracked and Wheeled Vehicles on Railway Wagons. 26 pp.
NATO STANAG 2832 ED 2 AMD 1-80. Restrictions for the Transport of Military Equipment by Rail on European Railways. 34 pp.
—Weapons
EC COM(87) 383-87. Proposal for a Council Directive on the Control of the Acquisition and Possession of Weapons. 26 pp.
EC 91/477/EEC-92. Council Directive on Control of the Acquisition and Possession of Weapons. 7 pp.

Border Pens
Use: Pens (Writing Implements)—Border

Borehole Loggers
Use For: Borehole Logging Equipment
See Also: Borehole Logs; Boreholes; Testing Equipment
—Radiation Measuring Instruments—Portable
BSI BS 5547-78. 1978 Portable Bore-Hole Logging Equipment (Down to 300 m): General Characteristics. 30 pp.
CENELEC HD 371-78. Portable Bore-Hole Logging Equipment (Down to 300 m) General Characteristics. 2 pp.
IEC 576-77. Portable Bore-Hole Logging Equipment (down to 300 m): General Characteristics First Edition. 40 pp.

Borehole Logging Equipment
Use: Borehole Loggers

Borehole Logs
See Also: Borehole Loggers
BSI BS 7022-88. 1988 Guide for Geophysical Logging of Boreholes for Hydrogeological Purposes. 25 pp.
DIN ENGL 4022 Pt 1-87. Subsoil and Groundwater; Classification and Description of Soil and Rock; Borehole Logging of Soil and Rock Not Involving Continuous Core Sample Recovery (Sept). 21 pp.

Borehole Logs *(Cont.)*
DIN ENGL 4022 Pt 3-82. Subsoil and Groundwater; Designation and Description of Soil Types and Rock; Borehole Log for Boring in Soil (Loose Rock) by Continuous Extraction of Cores (May). 16 pp.
—Symbols
DIN ENGL 4023-84. Borehole Logging; Graphical Representation of the Results (Mar). 11 pp.

Boreholes
Use For: Wellbores *See Also:* Borehole Loggers
—Construction—Safety
BSI BS 5573-78. 1978 Code of Practice for Safety Precautions in the Construction of Large Diameter Boreholes for Piling and Other Purposes. 11 pp.

Borer Rigs
Use: Drilling Rigs

Borers (Drills)
Use: Drills

Bores
—Aircraft—Stoppers
SBAC AGS 2108 ISSUE 2. Protective Plugs for Plain Bores 5/8 Inch to 2 1/2 Inch Diameter.

Boric Acid
Use For: Boracic Acid; Orthoboric Acid
See Also: Fluoboric Acid
BSI BS 3476-80. (WITHDRAWN) 1980 Boric Acid and Boric Oxide, Technical Grades. 4 pp.
CNS K1101-65. Boric Acid (Industrial Grade) (Aug)(2481)(R 1971).
CNS K6201-65. Method of Test for Boric Acid of Industrial Grade (Aug)(2482)(R 1971).
CNS K7103-64. Chemical Reagent (Boric Acid) (May)(1603).
JIS K 8863-91. Boric Acid.
—Boric Acid Content
BSI BS 5688: Part 1-79. 1979 Methods of Test for Orthoboric Acid (Boric Acid), Diboron Trioxide (Boric Oxide), Disodium Tetraborates, Sodium Perborates and Crude Sodium Borates for In-dustrial Use Part 1: Determination of Boric Acid Content of Boric Acid. 3 pp.
—Chloride Content—Mercurimetric Analysis
BSI BS 5688: Part 16-79. 1979 Orthoboric Acid (Boric Acid), Diboron Trioxide (Boric Oxide), Disodium Tetraborates, Sodium Perborates and Crude Sodium Borates for Industrial Use Part 16: Determin. of Chloride Content of Boric Acid, Boric Oxide and Disodium Tetraborates. 4 pp.
ISO 3121-76. Boric Acid, Boric Oxide and DiSodium Tetraborates for Industrial Use—Determination of Chloride Content—Mercurimetric Method First Edition. 5 pp.
—Chromium Content—Photometry
BSI BS 5688: Part 14-79. 1979 Orthoboric Acid (Boric Acid), Diboron Trioxide (Boric Oxide), Disodium Tetraborates, Sodium Perborates and Crude Sodium Borates for Industrial Use Part 14: Determ. of Chromium Content of Boric Acid, Boric Oxide and Disodium Tetraborates. 4 pp.
ISO 3119-76. Boric Acid, Boric Oxide and diSodium Tetraborates for Industrial Use—Determination of Chromium Content—Diphenylcarbazide Photometric Method First Edition. 5 pp.
—Cobalt Content—Photometry
ISO 5932-80. Boric Acid, Boric Oxide and DiSodium Tetraborates for Industrial Use—Determination of Cobalt Content—2-Nitroso-1-Naphthol Photometric Method First Edition. 6 pp.
—Comprehensive Testing
BSI BS 5688: Part 0-86. 1986 Methods of Test for Orthoboric Acid (Boric Acid), Diboron Trioxide (Boric Oxide), Disodium Tetraborates, Sodium Perborates and Crude Sodium Borates for Industrial Use Part 0: General Introduction. 3 pp.
—Copper Content—Photometry
BSI BS 5688: Part 7-79. 1979 Orthoboric Acid (Boric Acid), Diboron Trioxide (Boric Oxide), Disodium Tetraborates, Sodium Perborates and Crude Sodium Borates for Industrial Use Part 7: Determination of Copper Content of Boric Acid, Boric Oxide and Disodium Tetraborates. 4 pp.
ISO 2215-72. Boric Acid, Boric Oxide and Disodium Tetraborates for Industrial Use—Determination of Copper Content—Zinc Dibenzyldithiocarbamate Photometric Method First Edition. 4 pp.
—Fertilizers
CNS N3040-88. Boric Acid, Fertilizer Grade (Jul)(11863).

Boric Acid *(Cont.)*
—Iron Content—Photometry
BSI BS 5688: Part 17-79. 1979 Orthoboric Acid, Diboron Trioxide, Disodium Tetraborates, Sodium Perborates and Crude Sodium Borates for Industrial Use Part 17: Deter of Iron Content of Boric Acid, Boric Oxide, Disodium Tetraborates, Sodium Perborates and Crude Sodium Borates. 5 pp.
ISO 3122-76. Boric Acid, Boric Oxide, DiSodium Tetraborates, Sodium Perborates and Crude Sodium Borates for Industrial Use—Determination of Iron Content—2,2'-Bipyridyl Photometric Method First Edition. 6 pp.
—Manganese Content—Photometry
BSI BS 5688: Part 6-79. 1979 Orthoboric Acid (Boric Acid), Diboron Trioxide (Boric Oxide), Disodium Tetraborates, Sodium Perborates and Crude Sodium Borates for Ind. Use Part 6: Detn. of Manganese Content of Boric Acid, Boric Oxide and Disodium Tetraborates. 4 pp.
ISO 2214-72. Boric Acid, Boric Oxide and Disodium Tetraborates for Industrial Use—Determination of Manganese Content—Formaldehyde Oxime Photometric Method First Edition. 5 pp.
—Nickel Content—Photometry
ISO 5933-80. Boric Acid, Boric Oxide, DiSodium Tetraborates and Crude Sodium Borates for Industrial Use—Determination of Total Nickel Content of Boric Acid, Boric Oxide and DiSodium Tetraborates and the Alkali-Soluble Nickel Content of Crude Sodium Borates—Furil a-Dioxime Photometric. 7 pp.
—Photographic Chemicals
BSI BS 5565-78. 1978 Photographic Grade Boric Acid. 7 pp.
ISO 3628-76. Photographic Grade Boric Acid—Specification First Edition. 5 pp.
JIS K 7724-86. Photographic Grade Boric Acid.
—Sulfur Compound Content—Volumetric Analysis
BSI BS 5688: Part 5-79. 1979 Orthoboric Acid (Boric Acid), Diboron Trioxide (Boric Oxide), Disodium Tetraborates, Sodium Perborates and Crude Sodium Borates for Ind. Use Part 5: Detn. of Sulphur Compounds in Boric Acid, Boric Oxide, Disodium Tetraborates and Crude Sodium Borates. 4 pp.
ISO 1918-72. Boric Acid, Boric Oxide, Disodium Tetraborates and Crude Sodium Borates for Industrial Use—Determination of Sulphur Compounds—Volumetric Method First Edition. 4 pp.
—Volumetric Analysis
ISO 1914-72. Boric Acid for Industrial Use—Determination of Boric Acid Content—Volumetric Method First Edition. 3 pp.

Boric Acid Content Analysis
—Boric Acids
BSI BS 5688: Part 1-79. 1979 Methods of Test for Orthoboric Acid (Boric Acid), Diboron Trioxide (Boric Oxide), Disodium Tetraborates, Sodium Perborates and Crude Sodium Borates for In-dustrial Use Part 1: Determination of Boric Acid Content of Boric Acid. 3 pp.
—Cosmetics
CNS S2080-82. Method of Hygienic Test for Cosmetics Boric Acid and Borate (Sep)(9440).
—Dry Noodles
CNS N6112-79. Method of Test for Dry Noodles (Sep)(4992). 2 pp.
—Latex
BSI BS 6057: Sec 3.12-92. 1992 Rubber Latices Part 3: Methods of Test Section 3.12: Determination of Boric Acid Content of Natural Rubber Latex Concentrate (ISO 1802: 1992). 6 pp.
BSI BS 6057: Sec 3.12-88. 1988 Rubber Latices Part 3: Methods of Test Section 3.12: Determination of Boric Acid Content of Natural Rubber Latex Concentrate. 4 pp.
DIN ENGL 53605-77. Testing of Latex; Determination of Boric Acid Content in Natural Rubber Latex (Feb). 3 pp.
ISO 1802-92. Natural Rubber Latex Concentrate—Determination of Boric Acid Content Third Edition. 5 pp.

Boric Anhydrides
Use For: Boron Trioxide; di-Boron Trioxide
See Also: Boron Oxides
CNS K7104-64. Chemical Reagent (Boric Anhydride) (Boron Trioxide) (May)(1604).

Boric Oxides
Use: Boron Oxides

Boring Bars

—Indexable Inserts

BSI BS 4193: Part 14-84. 1984 Hardmetal Insert Tooling Part 14: Designation of Boring Bars (Tool Holders with Cylindrical Shank) for Indexable Inserts. 10 pp.

BSI BS 4193: Part 18-90. 1990 Hardmetal Insert Tooling Part 18: Dimensions of Boring Bars for Indexable Inserts. 10 pp.

ISO 5609-89. Boring Bars for Indexable Inserts—Dimensions AMENDMENT 1: Boring Bars Style Q and U, with Rhombic V-Shape Indexable Inserts Second Edition; (Amendment 1-1993). 11 pp.

ISO 6261-84. Boring Bars (Tool Holders with Cylindrical Shank) for Indexable Inserts—Designation First Edition. 8 pp.

—Probes—Lathes

CNS B3460-87. Boring Bar Probes Used in CNC Lathe (Aug)(12042).

Boring Machines

See Also: Boring Spindles; Earth Augers; Excavating Equipment; Machine Tools; Tools; Turret Lathes; Woodworking Equipment

CNS B7258-83. Test Code for Performance of Wood Borer and Hollow Chisel Mortiser (Sep)(10541).

—Accuracy Testing

CNS B7007-85. Test Code for Accuracy of Table Type Horizontal Boring Machines (Jan)(100).

CNS B7008-85. Test Code for Accuracy of Floor Type Horizontal Boring Machines (Jan)(101).

CNS B7098-82. Accuracy Inspection of Vertical Boring and Turning Mills (Aug)(4350).

—Glossaries

BSI BS 4361: Part 11-88. 1988 Woodworking Machines Part 11: Single Spindle Boring Machines. 10 pp.

BSI BS 4361: Part 24-90. 1990 Woodworking Machines Part 24: Nomenclature for Multi-Spindle Boring Machines. 8 pp.

CNS B1222-80. Glossary of Terms Relating to Parts of Boring Machine (Dec)(6746).

ISO 7945-85. Woodworking Machines—Single Spindle Boring Machines—Nomenclature and Acceptance Conditions First Edition. 8 pp.

ISO 9265-88. Woodworking Machines—Multi-Spindle Boring Machines—Nomenclature First Edition. 7 pp.

—Horizontal

BSI BS 4656: Part 21-88. 1988 Amd 1 Accuracy or Machine Tools and Methods of Test Part 21: Boring and Milling Machines, Horizontal Spindle Table Type and Rotary Table Type. 37 pp.

BSI BS 4656: Part 32-83. 1983 Accuracy of Machine Tools and Methods of Test Part 32: Boring and Milling Machines, Planer Type. 30 pp.

CNS B1107-79. Specified Items for Machine Tools (Horizontal Boring Machine) (Table Type) (Jan)(4670-10).

CNS B1108-79. Specified Items for Machine Tools (Horizontal Boring Machine) (Floor Type) (Jan)(4670-11).

CNS B7007-85. Test Code for Accuracy of Table Type Horizontal Boring Machines (Jan)(100).

CNS B7008-85. Test Code for Accuracy of Floor Type Horizontal Boring Machines (Jan)(101).

CNS B7156-85. Test Code for Performance of Table Type Horizontal Boring Machines (Jan)(6552).

CNS B7173-86. Test Code for Performance of Numerically Controlled Table Type Horizontal Boring Machines (Feb)(7962).

CNS B7174-86. Test Code for Accuracy of Numerically Controlled Table Type Horizontal Boring Machines (Feb)(7963).

ISO 3070 Pt 0-82. Acceptance Conditions for Boring and Milling Machines with Horizontal Spindle—Testing of the Accuracy—Part 0: General Introduction Second Edition. 7 pp.

ISO 3070 Pt 1-87. Acceptance Conditions for Boring and Milling Machines with Horizontal Spindle—Testing of the Accuracy—Part 1: Table-Type Machines Second Edition; (Addendum 1-1987). 35 pp.

ISO 3070 Pt II-78. Test Conditions for Boring and Milling Machines with Horizontal Spindle—Testing of the Accuracy—Part II: Floor Type Machines First Edition. 16 pp.

JIS B 6210-86. Test Code for Performance and Accuracy of Horizontal Boring Machines (Table Type). 47 pp.

JIS B 6222-85. Test Code for Performance and Accuracy of Horizontal Boring Machines (Floor Type). 34 pp.

—Horizontal—Acceptance Testing

DIN ENGL 8620 Pt 1-78. Machine Tools; Boring (Drilling) and Milling Machines with Horizontal Spindle; Acceptance Conditions; General Introduction (May). 6 pp.

DIN ENGL 8620 Pt 2-78. Machine Tools; Boring and Milling Machines with Horizontal Spindle with Table and Fixed Column Acceptance Conditions (May). 13 pp.

DIN ENGL 8620 Pt 3-80. Machine Tools; Boring and Milling Machines with Horizontal Spindle; Table-Type Machines; Rotary Tables; Acceptance Conditions (July). 6 pp.

DIN ENGL 8620 Pt 4-80. Machine Tools; Boring and Milling Machines with Horizontal Spindle; Floor-Type Machines; Acceptance Conditions (July). 18 pp.

—Horizontal—Safety

CEN PREN 815-92. Safety of Unshielded Tunnel Boring Machines and Rodless Shaft Boring Machines for Rock. 40 pp.

—Inspection

CNS B7241-83. Test Code for Accuracy of Wood Borer and Hollow Chisel Mortiser (May)(10221).

JIS B 6514-89. Test Methods for Performance and Accuracy of Hollow Chisel Mortising Machines. 12 pp.

—Jig

CNS B1110-79. Specified Items for Machine Tools (Jig Borer) (Double Housing Type) (Jan)(4670-13).

CNS B1111-79. Specified Items for Machine Tools (Jig Borer) (Single Column Type) (Jan)(4670-14).

CNS B7236-83. Test Code for Accuracy of Jig Boring Machines (Apr)(10152).

JIS B 6224-78. Test Code for Performance and Accuracy of Jig Boring Machines. 27 pp.

—Jig—Reamers

DIN ENGL 1862-62. End Mill Reamers for Jig Boring; Machines (Jan). 4 pp.

—Jig—Twist Drills

DIN ENGL 1861-62. Twist Drills for Jig Boring; Machines (Jan). 2 pp.

—Machine Tool Tables

DIN ENGL 8620 Pt 5-85. Machine Tools; Boring and Milling Machines with Horizontal Spindle; Work-Holding Fixed Tables; Acceptance Conditions (Feb). 6 pp.

—Multiple

BSI BS 4361: Part 24-90. 1990 Woodworking Machines Part 24: Nomenclature for Multi-Spindle Boring Machines. 8 pp.

ISO 9265-88. Woodworking Machines—Multi-Spindle Boring Machines—Nomenclature First Edition. 7 pp.

—Single

BSI BS 4361: Part 11-88. 1988 Woodworking Machines Part 11: Single Spindle Boring Machines. 10 pp.

ISO 3686-76. Test Conditions for Turret and Single Spindle Co-Ordinate Drilling and Boring Machines with Table of Fixed Height with Vertical Spindle—High Accuracy Machines—Testing of the Accuracy First Edition. 16 pp.

ISO 7945-85. Woodworking Machines—Single Spindle Boring Machines—Nomenclature and Acceptance Conditions First Edition. 8 pp.

JIS B 6517-89. Test Methods for Performance and Accuracy of Single Spindle Wood Boring Machines. 12 pp.

—Vertical

CNS B1109-79. Specified Items for Machine Tools (Vertical Boring Machine) (Jan)(4670-12).

CNS B7098-82. Accuracy Inspection of Vertical Boring and Turning Mills (Aug)(4350).

ISO 3686-76. Test Conditions for Turret and Single Spindle Co-Ordinate Drilling and Boring Machines with Table of Fixed Height with Vertical Spindle—High Accuracy Machines—Testing of the Accuracy First Edition. 16 pp.

—Vertical—Safety

CEN PREN 815-92. Safety of Unshielded Tunnel Boring Machines and Rodless Shaft Boring Machines for Rock. 40 pp.

Boring Spindles

See Also: Boring Machines; Multispindle Heads; Spindles

—Type A

DIN ENGL 69001 Pt 33-81. Machine Tools; Multi-Spindle Heads; Boring Spindles; Types A and B (Oct). 5 pp.

—Type B

DIN ENGL 69001 Pt 33-81. Machine Tools; Multi-Spindle Heads; Boring Spindles; Types A and B (Oct). 5 pp.

Boring Tools

See Also: Cutting Tools; Drills; Single Point Cutting Tools; Tools

—Indexable Inserts

BSI BS 4193: Part 2-86. 1986 Hardmetal Insert Tooling Part 2: Dimensions for Indexable Hardmetal (Carbide) Inserts with Rounded Corners, Without Fixing Hole. 10 pp.

BSI BS 4193: Part 3-86. 1986 Hardmetal Insert Tooling Part 3: Dimensions of Indexable Hardmetal (Carbide) Inserts with Rounded Corners and with Cylindrical Fixing Hole. 12 pp.

BSI BS 4193: Part 13-84. 1984 Hardmetal Insert Tooling Part 13: Dimensions of Indexable Hardmetal Inserts Having No Normal Clearance Rounded Corners and Partly Cylindrical Fixing Hole. 14 pp.

BSI BS 4193: Part 19-93. 1993 Hardmetal Insert Tooling Part 19: Specification for Dimensions of Indexable Ceramic Inserts with Rounded Corners Without Fixing Hole (ISO 9361-1: 1991). 18 pp.

BSI BS 4193: Part 20-93. 1993 Hardmetal Insert Tooling Part 20: Specification for Dimensions of Ceramic Inserts with Rounded Corners with Fixing Hole (ISO 9361-2: 1991). 16 pp.

ISO 883-85. Indexable Hardmetal (Carbide) Inserts with Rounded Corners, Without Fixing Hole—Dimensions Third Edition. 9 pp.

ISO 3364-85. Indexable Hardmetal (Carbide) Inserts with Rounded Corners, with Cylindrical Fixing Hole—Dimensions Second Edition. 11 pp.

ISO 6987 Pt 1-83. Indexable Hardmetal (Carbide) Inserts with Rounded Corners, with Partly Cylindrical Fixing Hole—Part 1: Dimensions of Inserts with 7 Degrees Normal Clearance First Edition. 9 pp.

ISO 9361 Pt 1-91. Indexable Inserts for Cutting Tools—Ceramic Inserts with Rounded Corners—Part 1: Dimensions of Inserts Without Fixing Hole First Edition. 15 pp.

ISO 9361 Pt 2-91. Indexable Inserts for Cutting Tools—Ceramic Inserts with Rounded Corners—Part 2: Dimensions of Inserts with Cylindrical Fixing Hole First Edition. 15 pp.

Boron

See Also: Boron Content Analysis; Boron Steels

MOD UK DSTAN 13-128-92. Boron, Amorphous, Grades 1 and 2 Issue 1. 8 pp.

—Pellets—Nuclear Grade

CNS J3003-81. Nuclear-Grade Boron Pellets (Aug)(7799).

Boron Carbide

—Powder—Nuclear Grade

CNS J3002-81. Nuclear-Grade Boron Carbide Powder (Aug)(7798).

Boron Content Analysis

See Also: Boron

—Aluminum

JIS H 1365-73. Method for Determination of Boron in Aluminium Alloy.

—Aluminum Oxide—Spectrophotometry

BSI BS 4140: Part 19-86. 1986 Methods of Test for Aluminium Oxide Part 19: Determinaton of Boron Content. 7 pp.

BSI BS 4140: Part 19-01. 1986 Amd 1 Methods of Test for Aluminium Oxide Part 19: Determination of Boron Content (ISO 2865: 1973) (AMD 6896) April 1, 1992. 8 pp.

ISO 2865-73. Aluminium Oxide Primarily Used for the Production of Aluminium—Determination of Boron Content—Curcumin Spectrophotometric Method First Edition; (Corrigendum 1-1991). 8 pp.

—Carbon Steels—Spectrophotometry

CEN PREN 10 200-88. Chemical Analysis of Ferrous Materials Determination of Boron in Steel Spectrophotometric Method. 14 pp.

—Coal

SAA AS 1038.10.3-88. Methods for the Analysis and Testing of Coal and Coke—Part 10.3: Determination of Trace Elements—Coal, Coke and Fly-Ash—Determination of Boron Content—Spectrophotometric Method. 5 pp.

—Coke

SAA AS 1038.10.3-88. Methods for the Analysis and Testing of Coal and Coke—Part 10.3: Determination of Trace Elements—Coal, Coke and Fly-Ash—Determination of Boron Content—Spectrophotometric Method. 5 pp.

Boron

Boron Content Analysis (Cont.)

—Ferroboron—Volumetric Analysis
BSI BS 6200: SUBSEC 3.5.2-91. 1991 Sampling and Analysis of Iron, Steel and Other Ferrous Metals Part 3: Methods of Analysis Section 3.5: Determination of Boron Subsection 3.5.2: Ferroboron: Volumetric Method. 7 pp.

—Fertilizers
EC COM(88) 562-88. Proposal for a Council Directive on the Approximation of the Laws of the Member States in Respect of the Trace Elements Boron, Cobalt, Copper, Iron, Manganese, Molybdenum and Zinc Contained in Fertilizers. 20 pp.
EC 89/530/EEC-89. Council Directive Supplementing and Amending Directive 76/116/EEC in Respect of the Trace Elements Boron, Cobalt, Copper Iron, Manganese, Molybdenum and Zinc Contained in Fertilizers. 9 pp.

—Fly Ash
SAA AS 1038.10.3-88. Methods for the Analysis and Testing of Coal and Coke—Part 10.3: Determination of Trace Elements—Coal, Coke and Fly-Ash—Determination of Boron Content—Spectrophotometric Method. 5 pp.

—Gallium Arsenide
JIS H 1191-91. Methods for Chemical Analysis of Gallium Arsenide. 57 pp.

—Graphite
JIS R 7223-79. Chemical Analysis of High Purity Graphite Material (R 1985). 32 pp.

—Iron
CNS G2246-85. Method of Determination for Boron in Iron and Steel (Jul)(11303).
JIS G 1227-92. Methods for Determination of Boron in Iron and Steel. 19 pp.

—Magnesite
ISO 10058-92. Magnesites and Dolomites —Chemical Analysis First Edition. 23 pp.

—Magnesite Refractories—Spectrophotometry
BSI BS 1902: Sec 2.3:Add1-76. 1976 Amd 1 Methods for Testing Refractory Materials Part 2: Chemical Analysis (Wet Methods) Section 2.3: Chemical Analysis of Magnesites and Dolomites Addendum 1: Determination of Boron in Magnesites. 5 pp.

—Neutron Cross Sections
CNS J2080-82. Method for Determining Boron Neutron Cross Section (Dec)(9713).

—Nickel Alloys—Molecular Absorption Spectrophotometry
ISO 11436-93. Nickel and Nickel Alloys —Determination of Total Boron Content—Curcumin Molecular Absorption Spectrometric Method First Edition. 11 pp.

—Nickel—Molecular Absorption Spectrophotometry
ISO 11436-93. Nickel and Nickel Alloys —Determination of Total Boron Content—Curcumin Molecular Absorption Spectrometric Method First Edition. 11 pp.

—Nickel—Photometry
BSI BS 3727: Part 2-68. 1968 Methods for the Analysis of Nickel for Use in Electronic Tubes and Valves Part 2: Determination of Boron (Photometric Method). 8 pp.

—Silicon
JIS H 0608-78. Estimation of Boron Content in Silicon (R 1983). 7 pp.

—Silicon—Infrared Spectroscopy
DIN ENGL 50438 Pt 3-84. Testing of Materials for Use in Semiconductor Technology; Determination of Interstitial Atomic Boron and Phosphorus Content of Silicon by Infrared Absorption Spectroscopy (Feb). 4 pp.

—Steels
CNS G2246-85. Method of Determination for Boron in Iron and Steel (Jul)(11303).
JIS G 1227-92. Methods for Determination of Boron in Iron and Steel. 19 pp.

—Steels—Spectrophotometry
BSI BS EN 10200-92. 1992 Chemical Analysis of Ferrous Materials—Determination of Boron in Steel—Spectrophotometric Method. 13 pp.
CEN EN 10200-91. Chemical Analysis of Ferrous Materials—Determination of Boron in Steel—Spectrophotometric Method. 16 pp.
DIN ENGL EN 10200-92. Chemical Analysis of Ferrous Materials; Determination of Boron Content of Steel by Spectrophotometry; (May). 9 pp.

Boron Content Analysis (Cont.)

—Steels—Spectrophotometry (Cont.)
ISO 10153-91. Steel—Determination of Boron Content—Curcumin Spectrophotometric Method First Edition. 12 pp.
SAA AS 1050.30-78. Methods for the Analysis of Iron and Steel—Part 30: Determination of Boron in Steel (Spectrophotometric Method Using Curcumin) Reconfirmed 1989. 10 pp.

—Uranium Hexafluoride—Spectrography
CNS J2067-82. Method for Determining Boron and Silicon in UF6 Method (Sept)(9400).

—Washing Powders—Volumetric Analysis
BSI BS 3762: Sec 3.13-83. 1983 Analysis of Formulated Detergents Part 3: Quantitative Test Methods Section 3.13: Method of Determination of Total Boron Content. 6 pp.
ISO 6835-81. Surface Active Agents—Washing Powders—Determination of Total Boron Content—Titrimetric Method First Edition. 5 pp.

—Water
CNS K9035-80. Method of Test for Boron in Water (Curcumin Method) (May)(5576).

—Water—Colorimetric Analysis
CNS K9036-80. Method of Test for Boron in Water (Carmine Colorimetric Method) (May)(5577).

—Zirconium
JIS H 1670-82. Method for Determination of Boron in Zirconium and Zirconium Alloys (R 1988). 7 pp.

—Zirconium Alloys
JIS H 1670-82. Method for Determination of Boron in Zirconium and Zirconium Alloys (R 1988). 7 pp.

Boron Inorganic Compounds
See Also: Boric Acid; Boron Nitride; Boron Oxides; Fluoboric Acid; Fluoroborates; Sodium Perborates

—Fertilizers
CNS N3041-88. Fused Boron Fertilizer (Jul)(11864).
CNS N3042-88. Processed Boron Fertilizer (Jul)(11865).

Boron Nitride
Scope Note: Includes boron nitride to be used in the manufacturing process. For finished products manufactured from boron nitride, see the specific product

—Grain Size Analysis
BSI BS 5851-80. 1980 Grain Sizes of Diamond or Cubic Boron Nitride. 18 pp.
ISO 6106-79. Abrasive Products—Grain Sizes of Diamond or Cubic Boron Nitride First Edition. 8 pp.
JIS B 4130-82. Grain Sizes of Diamond or Cubic Boron Nitride.

Boron Nitride Grinding Wheels
Use: Grinding Wheels—Boron Nitride

Boron Oxide Content Analysis
See Also: Boron Oxides

—Glass—Molecular Absorption Spectrophotometry
BSI BS 7709: Part 6-93. 1993 Analysis of Extract Solutions of Glass Part 6: Method for Determination of Boron(III) Oxide by Molecular Absorption Spectrometry (ISO 10136-6: 1993) (V). 11 pp.
ISO 10136 Pt 6-93. Glass and Glassware—Analysis of Extract Solutions—Part 6: Determination of Boron(III) Oxide by Molecular Absorption Spectrometry First Edition. 9 pp.

—Glassware—Molecular Absorption Spectrophotometry
BSI BS 7709: Part 6-93. 1993 Analysis of Extract Solutions of Glass Part 6: Method for Determination of Boron(III) Oxide by Molecular Absorption Spectrometry (ISO 10136-6: 1993) (V). 11 pp.
ISO 10136 Pt 6-93. Glass and Glassware—Analysis of Extract Solutions—Part 6: Determination of Boron(III) Oxide by Molecular Absorption Spectrometry First Edition. 9 pp.

—Sodium Borates—Volumetric Analysis
BSI BS 5688: Part 8-79. 1979 Orthoboric Acid (Boric Acid), Diboron Trioxide (Boric Oxide), Disodium Tetraborates, Sodium Perborates and Crude Sodium Borates for Industrial Use Part 8: Determination of Sodium Oxide and Boric Oxide Contents of Crude Sodium Borates. 4 pp.

Boron Oxide Content Analysis (Cont.)

—Sodium Perborates—Ignition Loss
BSI BS 5688: Part 3-79. 1979 Orthoboric Acid (Boric Acid), Diboron Trioxide (Boric Oxide), Disodium Tetraborates, Sodium Perborates and Crude Sodium Borates for Ind. Use Part 3: Detn. of Sodium Oxide and Boric Oxide Contents and Loss on Ingintion of Disodium Tetraborates. 4 pp.
ISO 1916-72. Disodium Tetraborates for Industrial Use—Determination of Sodium Oxide and Boric Oxide Contents and Loss on Ignition First Edition. 4 pp.

—Sodium Perborates—Volumetric Analysis
BSI BS 5688: Part 4-79. 1979 Orthoboric Acid (Boric Acid), Diboron Trioxide (Boric Oxide), Disodium Tetraborates, Sodium Perborates and Crude Sodium Borates for Ind. Use Part 4: Detn. of Sodium Oxide, Boric Oxide and Available Oxygen Cont. Hydrated Sodium Perborates. 4 pp.
ISO 1917-72. Hydrated Sodium Perborates for Industrial Use—Determination of Sodium Oxide, Boric Oxide and Available Oxygen Contents—Volumetric Methods First Edition. 4 pp.

Boron Oxides
Use For: Boric Oxides; Boron Trioxide
See Also: Boric Anhydrides; Boron Oxide Content Analysis

BSI BS 3476-80. (WITHDRAWN) 1980 Boric Acid and Boric Oxide, Technical Grades. 4 pp.
BSI BS 5688: Part 0-86. 1986 Methods of Test for Orthoboric Acid (Boric Acid), Diboron Trioxide (Boric Oxide), Disodium Tetraborates, Sodium Perborates and Crude Sodium Borates for Industrial Use Part 0: General Introduction. 3 pp.
BSI BS 5688: Part 2-79. 1979 Methods of Test for Orthoboric Acid (Boric Acid), Diboron Trioxide (Boric Oxide), Disodium Tetraborates, Sodium Perborates and Crude Sodium Borates for Industrial Use Part 2: Determination of Boric Oxide Content of Boric Oxide. 4 pp.

—Chloride Content—Mercurimetric Analysis
BSI BS 5688: Part 16-79. 1979 Orthoboric Acid (Boric Acid), Diboron Trioxide (Boric Oxide), Disodium Tetraborates, Sodium Perborates and Crude Sodium Borates for Industrial Use Part 16: Determin. of Chloride Content of Boric Acid, Boric Oxide and Disodium Tetraborates. 4 pp.
ISO 3121-76. Boric Acid, Boric Oxide and DiSodium Tetraborates for Industrial Use—Determination of Chloride Content—Mercurimetric Method First Edition. 5 pp.

—Chromium Content—Photometry
BSI BS 5688: Part 14-79. 1979 Orthoboric Acid (Boric Acid), Diboron Trioxide (Boric Oxide), Disodium Tetraborates, Sodium Perborates and Crude Sodium Borates for Industrial Use Part 14: Determ. of Chromium Content of Boric Acid, Boric Oxide and Disodium Tetraborates. 4 pp.
ISO 3119-76. Boric Acid, Boric Oxide and diSodium Tetraborates for Industrial Use—Determination of Chromium Content—Diphenylcarbazide Photometric Method First Edition. 5 pp.

—Cobalt Content—Photometry
ISO 5932-80. Boric Acid, Boric Oxide and DiSodium Tetraborates for Industrial Use—Determination of Cobalt Content—2-Nitroso-1-Naphthol Photometric Method First Edition. 6 pp.

—Copper Content—Photometry
BSI BS 5688: Part 7-79. 1979 Orthoboric Acid (Boric Acid), Diboron Trioxide (Boric Oxide), Disodium Tetraborates, Sodium Perborates and Crude Sodium Borates for Industrial Use Part 7: Determination of Copper Content of Boric Acid, Boric Oxide and Disodium Tetraborates. 4 pp.
ISO 2215-72. Boric Acid, Boric Oxide and Disodium Tetraborates for Industrial Use—Determination of Copper Content—Zinc Dibenzyldithiocarbamate Photometric Method First Edition. 4 pp.

—Iron Content—Photometry
BSI BS 5688: Part 17-79. 1979 Orthoboric Acid, Diboron Trioxide, Disodium Tetraborates, Sodium Perborates and Crude Sodium Borates for Industrial Use Part 17: Deter of Iron Content of Boric Acid, Boric Oxide, Disodium Tetraborates, Sodium Perborates and Crude Sodium Borates. 5 pp.
ISO 3122-76. Boric Acid, Boric Oxide, DiSodium Tetraborates, Sodium Perborates and Crude Sodium Borates for Industrial Use—Determination of Iron Content—2,2'-Bipyridyl Photometric Method First Edition. 6 pp.

Boron Oxides (Cont.)

—Manganese Content—Photometry
BSI BS 5688: Part 6-79. 1979 Orthoboric Acid (Boric Acid), Diboron Trioxide (Boric Oxide), Disodium Tetraborates, Sodium Perborates and Crude Sodium Borates for Ind. Use Part 6: Detn. of Manganese Content of Boric Acid, Boric Oxide and Disodium Tetraborates. 4 pp.
ISO 2214-72. Boric Acid, Boric Oxide and Disodium Tetraborates for Industrial Use—Determination of Manganese Content—Formaldehyde Oxime Photometric Method First Edition. 5 pp.

—Nickel Content—Photometry
ISO 5933-80. Boric Acid, Boric Oxide, DiSodium Tetraborates and Crude Sodium Borates for Industrial Use—Determination of Total Nickel Content of Boric Acid, Boric Oxide and DiSodium Tetraborates and the Alkali-Soluble Nickel Content of Crude Sodium Borates—Furil a-Dioxime Photometric. 7 pp.

—Sulfur Compound Content—Volumetric Analysis
BSI BS 5688: Part 5-79. 1979 Orthoboric Acid (Boric Acid), Diboron Trioxide (Boric Oxide), Disodium Tetraborates, Sodium Perborates and Crude Sodium Borates for Ind. Use Part 5: Detn. of Sulphur Compounds in Boric Acid, Boric Oxide, Disodium Tetraborates and Crude Sodium Borates. 4 pp.
ISO 1918-72. Boric Acid, Boric Oxide, Disodium Tetraborates and Crude Sodium Borates for Industrial Use—Determination of Sulphur Compounds—Volumetric Method First Edition. 4 pp.

—Volumetric Analysis
ISO 1915-72. Boric Oxide for Industrial Use—Determination of Boric Oxide Content—Volumetric Method First Edition. 4 pp.

Boron Steels
See Also: Alloy Steels; Boron

—Bars—Quenched
CEN PREN 10083-3-91. Quenched and Tempered Steels—Part 3: Technical Delivery Conditions for Boron Steels. 31 pp.

—Bars—Tempered
CEN PREN 10083-3-91. Quenched and Tempered Steels—Part 3: Technical Delivery Conditions for Boron Steels. 31 pp.

—Flats—Quenched
CEN PREN 10083-3-91. Quenched and Tempered Steels—Part 3: Technical Delivery Conditions for Boron Steels. 31 pp.

—Flats—Tempered
CEN PREN 10083-3-91. Quenched and Tempered Steels—Part 3: Technical Delivery Conditions for Boron Steels. 31 pp.

—Forgings—Quenched
CEN PREN 10083-3-91. Quenched and Tempered Steels—Part 3: Technical Delivery Conditions for Boron Steels. 31 pp.

—Forgings—Tempered
CEN PREN 10083-3-91. Quenched and Tempered Steels—Part 3: Technical Delivery Conditions for Boron Steels. 31 pp.

—Plates—Quenched
CEN PREN 10083-3-91. Quenched and Tempered Steels—Part 3: Technical Delivery Conditions for Boron Steels. 31 pp.

—Plates—Tempered
CEN PREN 10083-3-91. Quenched and Tempered Steels—Part 3: Technical Delivery Conditions for Boron Steels. 31 pp.

—Rods
JIS G 3508-91. Boron Steel Wire Rods for Cold Heading and Cold Forging. 23 pp.

—Rods—Quenched
CEN PREN 10083-3-91. Quenched and Tempered Steels—Part 3: Technical Delivery Conditions for Boron Steels. 31 pp.

—Rods—Tempered
CEN PREN 10083-3-91. Quenched and Tempered Steels—Part 3: Technical Delivery Conditions for Boron Steels. 31 pp.

—Sheets—Quenched
CEN PREN 10083-3-91. Quenched and Tempered Steels—Part 3: Technical Delivery Conditions for Boron Steels. 31 pp.

Boron Steels (Cont.)

—Sheets—Tempered
CEN PREN 10083-3-91. Quenched and Tempered Steels—Part 3: Technical Delivery Conditions for Boron Steels. 31 pp.

—Strips—Quenched
CEN PREN 10083-3-91. Quenched and Tempered Steels—Part 3: Technical Delivery Conditions for Boron Steels. 31 pp.

—Strips—Tempered
CEN PREN 10083-3-91. Quenched and Tempered Steels—Part 3: Technical Delivery Conditions for Boron Steels. 31 pp.

Boron Trioxide
Use: Boric Anhydrides; Boron Oxides

Borosilicate Glass
See Also: Glass; Glassware
BSI BS 2598: Part 1-91. 1991 Glass Plant, Pipeline and Fittings Part 1: Specification for Properties of Borosilicate Glass 3.3 (ISO 3585: 1991). 8 pp.
BSI BS 2598: Part 1-80. 1980 Glass Plant, Pipeline and Fittings Part 1: Properties of Borosilicate Glass 3.3. 4 pp.
BSI BS 2598: Part 3-80. 1980 Glass Plant, Pipeline and Fittings Part 3: Pipeline and Fittings of Nominal Bore 15 to 150 mm: Compatibility and Interchangeability. 21 pp.
ISO 3585-91. Borosilicate Glass 3.3—Properties Second Edition. 7 pp.
ISO 3587-76. Glass Plant, Pipeline and Fittings—Pipeline and Fittings of Nominal Bore 15 to 150 mm—Compatibility and Interchangeability First Edition; (Erratum—Aug 1979). 20 pp.

—Chemical Analysis
CNS R3134-86. Method for Chemical Analysis of Borosilicate Glasses (Feb)(11510).
JIS R 3105-81. Methods for Chemical Analysis of Borosilicate Glasses. 26 pp.

—Optical—Crown
CNS R2147-86. Borosilicate Crown Optical Glass (Feb) (8541).

—Tubes—Laboratory
BSI BS 5895-80. 1980 Amd 1 Borosilicate Glass Tubing for Laboratory Apparatus. 5 pp.
BSI BS EN 28362-1-93. 1993 Injection Containers for Inectables and Accessories Part 1: Injection Vials Made of Glass Tubing (ISO 8362-1: 1989) (N). 12 pp.
CEN EN 28362-1-93. Injection Containers for Injectables and Accessories—Part 1. Injectable Vials Made of Glass Tubing (ISO 8362-1: 1989). 7 pp.
ISO 4803-78. Laboratory Glassware—Borosilicate Glass Tubing First Edition. 4 pp.
ISO 8362 Pt 1-89. Injection Containers for Injectables and Accessories—Part 1: Injection Vials Made of Glass Tubing First Edition; (CEN EN 28362-1: 1993). 8 pp.

Boss Pipes
Use: Pipe Bosses

Bosses (Fasteners)

—Aerospace—Fluid Power Equipment
CEN PREN 3633-92. Installation Hole for Fluid Fittings, Flanged. 5 pp.

—Fittings—Aerospace
DIN ENGL LN 29858-74. Internal Thread Bosses for Fitting Assemblies, Ring Locked; Design Dimensions (Nov). 2 pp.

Botany

—Classification
BSI BS 1000: (58)-79. 1979 Universal Decimal Classification (UDC). English Full Edition (58): Botany. 34 pp.
SNZ NZS/BS 1000 (58)-79. Universal Decimal Classification Botany. 36 pp.

Bottle Caps
See Also: Bottles; Caps (Lids)
CNS Z5036-72. Caps for Bottles, Crown Type (Jun)(3365).
CNS Z6015-73. Method of Test for Caps of Bottles of Crown Type (May)(3366).

—Aluminum
BSI BS EN 28872-93. 1993 Aluminium Caps for Transfusion, Infusion and Injection Bottles—General Requirements and Test Methods (ISO 8872: 1988) (N). 13 pp.
CEN EN 28872-93. Aluminium Caps for Transfusion, Infusion and Injection Bottles—General Requirements and Test Methods (ISO 8872: 1988). 8 pp.

Bottle Caps (Cont.)

—Aluminum (Cont.)
ISO 8872-88. Aluminium Caps for Transfusion, Infusion and Injection Bottles—General Requirements and Test Methods First Edition; (CEN EN 28872: 1993). 10 pp.

—Syrup—Glass
CNS Z5005-53. Caps of Cylindrical Glass Bottle for Syrup Storage (Jul)(340)(R 1973).

Bottle Openers
Use For: Crown Bottle Openers *See Also:* Bottles
BSI BS 3414-61. 1961 Crown Bottle Openers. 8 pp.
JIS S 9023-81. Bottle Openers.

Bottle Warmers
See Also: Food Warmers; Medical Equipment
CSA CAN/CSA-C22. 2 NO 64-M91. Household Cooking and Liquid-Heating Appliances; (Gen Instr 1 Thru 2). 91 pp.

Bottles
Scope Note: For additional listings, use a more specific term *See Also:* Bottle Caps; Bottle Openers; Cans; Carbonated Beverage Bottles; Containers; Feeding Bottles (Babies); Flasks; Glass Containers; Laboratory Ware; Plastic Containers; Reagent Bottles; Shipping Containers; Suction Bottles; Thermos Bottles; Weighing Bottles

—Beer
BSI BS 6118-81. 1981 Amd 3 Multi-Trip Glass Bottles for Beer and Cider. 19 pp.

—Blood Culture
DIN ENGL 58942 Pt 2-85. Medical Microbiology; Culture Media Used in Bacteriology; Ready-to-Use Blood Culture Bottles; Physical, Chemical and Biological Requirements (Sept). 4 pp.

—Chemicals—Glass
CNS R2074-72. Glass Bottles for Chemicals and Agricultural Drugs (Oct) (3502).

—Cider—Glass
BSI BS 6118-81. 1981 Amd 3 Multi-Trip Glass Bottles for Beer and Cider. 19 pp.

—Classification
BSI BS 1000: (682/683)-73. 1973 Universal Decimal Classification (UDC). English Full Edition (682/683): Smithery. Farriery. Hand-Forged Ironwork. Ironmongery. Hardware. Locksmithing. Gunsmithing. Bottling. Lamps. Stoves. 27 pp.
SNZ NZS/BS 1000 (682/683)-73. Universal Decimal Classification Smithery. Farriery. Hand-Forged Ironwork. Ironmongery. Hardware. Locksmithing. Gunsmithing. Bottling. Lamps. Stoves. 28 pp.

—Crowns
JIS S 9017-79. Crowns of Bottles.

—Drugs—Agricultural
CNS R2074-72. Glass Bottles for Chemicals and Agricultural Drugs (Oct) (3502).

—Glass
ISO 8162-85. Glass Containers—Tall Crown Finishes—Dimensions First Edition. 4 pp.
ISO 8163-85. Glass Containers—Shallow Crown Finishes—Dimensions First Edition; (Corrected and Reprinted -1985). 4 pp.
ISO 8164-90. Glass Containers—520 ml Euro-Form Bottles—Dimensions First Edition. 4 pp.

—Glass—Acid Resistant
CNS R2171-83. Acid Proof Glass Bottles (Mar)(10127).
JIS R 3801-78. Acid Proof Glass Bottles. 10 pp.

—Glass—Aerosol—Impact Testing
CNS Z6057-82. Method of Drop Testing of Glass Aerosol Bottles (Nov)(9650).

—Glass—Chemical Analysis
CNS R2129-86. Glass Bottles for Chemical Analysis (Mar) (7305).

—Glass—Continuous Threads
DIN ENGL 6094 Pt 12-83. Packaging; Finishes; 7,5 R Continuous Thread Finishes for Bottles Subjected to Internal Pressure (Jan). 2 pp.

—Glass—Handling
BSI BS 7488-91. 1991 Handling Glass Bottles, Containing Carbonated Soft Drinks or Carbonated Water, in Filling Plants, Warehouses, and Retailers' Premises (H). 8 pp.

—Glass Joints—Conical
JIS R 3650-81. Interchangeable Conical Ground Glass Joints for Narrow Mouth Bottle.

INTERNATIONAL AND NON-U.S. NATIONAL STANDARDS
SUBJECT INDEX

Bottles

Bottles (Cont.)
—Glass—Verticality
ISO 9008-91. Glass Bottles—Verticality—Test Method First Edition. 5 pp.

—Hazardous Materials—Polyethylene
CGSB 43-GP-57M-81. Polyethylene, Inside Containers, (TC-2E, TC-2S, TC-2SL, TC-2T, TC-2TL, TC-2U), Standard for. 12 pp.

—Infusion—Stoppers
ISO 8536 Pt 2-92. Infusion Equipment for Medical Use—Part 2: Closures for Infusion Bottles First Edition. 13 pp.
ISO 8536 Pt 3-92. Infusion Equipment for Medical Use—Part 3: Aluminium Caps for Infusion Bottles First Edition. 10 pp.
ISO 8536 Pt 7-92. Infusion Equipment for Medical Use—Part 7: Caps Made of Aluminium-Plastics Combinations for Infusion Bottles First Edition. 8 pp.

—Intravenous Medical Equipment—Caps (Lids)
ISO 10985-92. Caps Made of Aluminium-Plastics Combinations for Infusion Bottles and Injection Vials—Requirements and Test Methods First Edition. 8 pp.

—Intravenous Medical Equipment—Glass
ISO 8536 Pt 1-91. Infusion Equipment for Medical Use—Part 1: Infusion Glass Bottles First Edition. 9 pp.

—Kerosene—Polyethylene
JIS Z 1710-77. Blow Moulded Polyethylene Containers for Kerosine (R 1980). 11 pp.

—Laboratory—Glass
BSI BS 2461-83. 1983 Gas Washing Bottles. 7 pp.

—Laboratory—Plastic
BSI BS 5404: Part 4-79. 1979 Plastics Laboratory Ware Part 4: Wash Bottles. 4 pp.

—Medical—Eye Dropper
BSI BS 1679: Part 5-73. 1973 Amd 1 Containers for Pharmaceutical Dispensing Part 5: Eye Dropper Bottles. 10 pp.

—Medical—Glass
BSI BS 1679: Part 6-84. 1984 Containers for Pharmaceutical Dispensing Part 6: Specification for Glass Medicine Bottles. 8 pp.
BSI BS 1679: Part 7-68. 1968 Containers for Pharmaceutical Dispensing Part 7: Ribbed Oval Glass Bottles. 13 pp.
BSI BS 1679: Part 8-92. 1992 Containers for Pharmaceutical Dispensing Part 8: Specification for Glass and Plastics Containers for Solid Dosage Forms, Semi-Solids and Powders (Supersedes BS 1679: 1965). 14 pp.
BSI BS 1679: Part 7-01. 1968 Amd 1 Containers for Pharmaceutical Dispensing Part 7: Ribbed Oval Glass Bottles (AMD 7203) August 15, 1992. 13 pp.
CNS R2170-83. Tube Bottle for Injections (Mar)(10126).
CNS T2030-85. Feeding Bottles (Aug)(11348).
CNS T4016-85. Method of Test for Feeding Bottles (Aug)(11349).
ISO 1135 Pt 1-87. Transfusion Equipment for Medical Use—Part 1: Glass Transfusion Bottles, Closures and Caps Second Edition. 13 pp.
JIS R 3522-77. Glass Bottles for Drug.
JIS R 3523-78. Tube Bottle for Injections.
JIS T 9112-75. Feeding Bottles.

—Medical—Plastic
BSI BS 1679: Part 8-92. 1992 Containers for Pharmaceutical Dispensing Part 8: Specification for Glass and Plastics Containers for Solid Dosage Forms, Semi-Solids and Powders (Supersedes BS 1679: 1965). 14 pp.

—Milk
SAA AS 1382-74. Glass Milk and Cream Bottles of the Metal Foil Cap Type (200-600 mL Capacity) Reconfirmed 1990. 12 pp.
SNZ NZS 7842-73. Specification for Glass Milk-Bottles of the Metal Foil Cap Type. 12 pp.

—Milk—Glass
BSI BS 6106-81. 1981 1 Pint (568 ml) Multi-Trip Glass Bottles for Pasteurized Milk. 8 pp.

—Milk—Labels—Papers
CNS Z5027-66. Paper Slips for Milk Bottles (Mar)(2651) (R 1972).

—Plastic—Buttress Threads
DIN ENGL 6063 Pt 1-89. Screw Threads Used Mainly for Plastic Containers; Dimensions of Buttress Threads (May). 2 pp.

Bottles (Cont.)
—Polyethylene
CNS Z5024-82. Polyethylene Bottle (Tentative) (Sep)(2444).
CNS Z5131-83. Plastic Films for Food Packaging (Jul)(10481). 6 pp.
CNS Z6005-82. Method of Test for Polyethlene Bottle (Sep)(2445).
JIS Z 1703-76. Polyethylene Bottles (R 1980). 8 pp.
JIS Z 1706-76. Containers Blow-Molded from Polyethylene (R 1980). 10 pp.

—Syrup—Glass
CNS Z5004-53. Cylindrical Glass Bottle for Syrup Storage (Jul)(339).

—Wine—Glass
BSI BS 6117-81. 1981 Amd 1 Glass Bottles for Light Wine (AMD 4886) July 31, 1985. 25 pp.

Bottom Dump Haulers
Use: Dump Trucks

Bougies (Medical)
CNS T1009-80. Standard Scale for Medical Bougie and Catheter (Jul)(5939).
JIS T 4202-53. Standard Scale for Medical Bougie and Catheter.

—Urethral
CNS T1001-81. Urethral Bougies (Jul)(5931).
JIS T 3202-53. Urethral Bougies.

Bounce Buffers
Use For: Bounce Eliminators *See Also:* Buffers (Data)

—Hex
CECC CECC 90 104-174 ISSUE 1-81. BS CECC 90 104-174; Hex Contact Bounce Eliminator (En). 24 pp.
CECC CECC 90 104-174 ISSUE 1-86. CEI CECC 90 104-174; Hex Contact Bounce Eliminator (En). 2 pp.

Bounce Eliminators
Use: Bounce Buffers

Bound Styrene Content Analysis
Use: Styrene Content Analysis

Boundaries
See Also: Airspace

—Mapping
NATO STANAG 2029 ED 7 AMD 1-89. Method of Describing Ground Locations, Areas and Boundaries. 8 pp.
NATO STANAG 2029 ED 7 AMD 2-89. Method of Describing Ground Locations, Areas and Boundaries. 8 pp.

Bourdon Pressure Gages
Use: Bourdon Tubes

Bourdon Tubes
Use For: Bourdon Pressure Gages *See Also:* Gages; Pressure Gages
BSI BS 1780-85. 1985 Amd 1 Bourdon Tube Pressure and Vacuum Gauges (AMD 6124) July 31, 1989. 29 pp.
BSI BS 3127-79. 1979 Amd 1 Ferrous and Non-Ferrous Bourdon Tubing. 9 pp.
BSI BS 6752-86. 1986 Bourdon Tube Pressure Gauges Used in Welding, Cutting and Related Processes. 12 pp.
CEN PREN 562-91. Pressure Gauges Used in Welding, Cutting and Allied Processess. 14 pp.
CEN PREN 837-1-92. Pressure Gauges—Part 1: Bourdon Tube Pressure Gauges—Dimensions, Metrology, Requirements and Testing. 27 pp.
DIN ENGL 8549-86. Bourdon-Tube Pressure Gauges Used in Welding, Cutting and Related Processes with 50 mm or 63 mm Diameter Casing (Dec). 8 pp.
DIN ENGL 16005-87. General Purpose Pressure Gauges with Elastic Pressure-Responsive Elements; Requirements and Testing (Feb). 7 pp.
JIS B 7505-80. Bourdon Tube Pressure Gauges. 46 pp.
MOD UK DSTAN 66-2-01. Gauges, Pressure, Dial Indicating (Bourdon Tube Type) Issue 3; Amendment 3. 56 pp.
SNZ NZS/BS 1780-85. Specification for Bourdon Tube Pressure and Vacuum Gauges Amend: 1. 28 pp.

—Cutting Equipment
BSI BS 6752-86. 1986 Bourdon Tube Pressure Gauges Used in Welding, Cutting and Related Processes. 12 pp.
ISO 5171-80. Pressure Gauges Used in Welding, Cutting and Related Processes First Edition; (Erratum—Dec 1981). 10 pp.

Bourdon Tubes (Cont.)
—Gas Cutting Equipment
CEN PREN 562-91. Pressure Gauges Used in Welding, Cutting and Allied Processess. 14 pp.
DIN ENGL 8549-86. Bourdon-Tube Pressure Gauges Used in Welding, Cutting and Related Processes with 50 mm or 63 mm Diameter Casing (Dec). 8 pp.

—Rolling Stock
CNS E3007-85. Bourdon Tube Pressure Gauges for Railway Rolling Stock (Feb)(11201).
JIS E 4118-78. Bourdon Tube Pressure Gauges for Railway Rolling Stock.

—Welding Equipment
BSI BS 6752-86. 1986 Bourdon Tube Pressure Gauges Used in Welding, Cutting and Related Processes. 12 pp.
CEN PREN 562-91. Pressure Gauges Used in Welding, Cutting and Allied Processess. 14 pp.
DIN ENGL 8549-86. Bourdon-Tube Pressure Gauges Used in Welding, Cutting and Related Processes with 50 mm or 63 mm Diameter Casing (Dec). 8 pp.
ISO 5171-80. Pressure Gauges Used in Welding, Cutting and Related Processes First Edition; (Erratum—Dec 1981). 10 pp.

Bourdon Vacuum Gages
Use: Pressure Gages

Bovines
Use: Cattle

Bow Compasses
See Also: Compasses (Drawing)
CNS Z7109-82. Spring Bow Compass with Interchangeable Points for Precision Drawing Instruments (P) (May)(8890).
CNS Z7128-82. Drawing Instruments (Big Bow Combination with Interchangeable Parts) (Dec)(9770).

—Pivot Needles
CNS Z7117-82. Pivots Needles for Fixed Center Spring Bow (Jun)(9047).

—School
CNS Z7148-83. Fixed Centre Spring Bow with Interchangeable Point for School Drawing Instruments (Jan)(9924).

Bow Nuts
See Also: Nuts (Fasteners)
CNS B2219-87. Bow Nuts (Jul)(4376).

—Screw Threads
DIN ENGL 80704-68. Bow Nuts; Metric Screw Thread (Aug). 2 pp.

Bowing
Use: Bending

Bowl Stands
Use: Basins (Containers)—Stands

Box Spanners
See Also: Hand Tools; Railroad Equipment; Spanners
BSI BS 2558-54. 1954 Amd 3 Tubular Box Spanners. 26 pp.
BSI BS 2558: Supp 1-68. 1968 Tubular Box Spanners Supplement 1: Metric Sizes. 3 pp.
MOD UK DSTAN 51-11: Part 3-88. Hand Tools, General Purpose Part 3: Spanners, Box (Tubular) and Tommy Bars Issue 2. 13 pp.

—Double Ended—Gap Combinations
CNS B3281-78. Combinations of Double Ended Wrench Gaps (Mar)(4295). 2 pp.

Box Wrenches
See Also: Flare Nut Wrenches; Wrenches
BSI BS 2090-54. 1954 Amd 3. Spanners, Peg Spanners, Coupling Wrenches and the Related Slots, Holes and Horns. 27 pp.
BSI BS 3555-88. 1988 Ring Wrenches. 11 pp.
CGSB 39-GP-11C-86. Wrenches, Box, Open End, and Combination, (Inch Series). 32 pp.
CGSB 39-GP-11M-79. Wrenches, Box, Open End, Combination and Flare Nut, Standard for; (Amendment 1 Apr 1984). 29 pp.
CNS B3282-78. Combination Wrenches (Long Series) (Mar)(4296). 2 pp.
CNS B3283-78. Combination Wrenches (Short Series) (Mar)(4297). 3 pp.
CNS B3347-80. Assembly Tools for Screws and Nuts (Ring), Wrench, Test Torques Series B (Jul)(5658).
ISO 3318-90. Assembly Tools for Screws and Nuts—Double-Headed Open-Ended Wrenches, Double-Headed Ring Wrenches and Combination Wrenches—Maximum Widths of Heads Third Edition. 4 pp.

INTERNATIONAL AND NON-U.S. NATIONAL STANDARDS
SUBJECT INDEX

Box Wrenches *(Cont.)*

—**Double Ended**
CNS B3217-78. Double Ended Box Wrenches (Mar)(3413). 3 pp.

—**Double Ended—Flat**
ISO 10103-90. Assembly Tools for Srews and Nuts—Double-Headed, Flat and Offset, Box Wrenches First Edition. 6 pp.

—**Double Ended—Offset**
CNS B3218-78. Double Ended Offset Box Wrenches (Mar)(3414). 4 pp.
CNS B3349-80. Assembly Tools for Screws and Nuts Wrench Double End, Modified Offset, Test Torques Series A (Jul)(5660).
ISO 10103-90. Assembly Tools for Srews and Nuts—Double-Headed, Flat and Offset, Box Wrenches First Edition. 6 pp.
ISO 10104-90. Assembly Tools for Srews and Nuts—Double-Headed, Deep Offset and Modified Offset, Box Wrenches First Edition. 6 pp.

—**Double Ended—Offset—Aerospace**
CEN PREN 3711-90. Wrench-Double Ended, Bi-Hexagonal Straight, Cranked, Offset. 5 pp.
CEN PREN 3819-90. Clearance for Wrenchs and Sockets. 6 pp.
CEN PREN 3819-92. Clearance for Wrenchs and Sockets. 5 pp.

—**Single Ended**
CNS B3021-80. Assembly Tools for Screws and Nuts—Box (Ring) Wrench (Single End) Test Torques Series A (Jul)(142). 3 pp.
CNS B3022-80. Box (Ring) Wrench, Single End, Square, Test Torques Series A. Assembly Tools for Screws and Nuts (Jul)(143). 3 pp.

Boxboard
See Also: Fiberboard; Packaging Materials; Paperboard; Papers
CGSB 9-GP-24M-78. Paperboard, Boxboard Grades, Standard for. 11 pp.

—**Puncture Resistance**
CPPA D.22P-84. Puncture Test of Containerboard—Beach Method. 3 pp.

—**Weight Measurement**
CGSB 9-GP-24M-78. Paperboard, Boxboard Grades, Standard for. 11 pp.

Boxes (Containers)
Scope Note: For additonal listings, use a more specific term *Use For:* Tote Boxes
See Also: Cartons; Containers; Crates (Shipping Containers); Fire Hydrant Boxes; Gear Boxes; Lunch Boxes; Mud Boxes; Shipping Containers

—**Bulk Butter**
CGSB CAN/CGSB-43.42-M86. Boxes, Bulk Butter and Similar Products. 20 pp.

—**Bulk Cheeses—Liners**
CGSB CAN/CGSB-43.43-M86. Boxes, Bulk Cheese, and Liners. 13 pp.

—**Butter**
DIN ENGL 10079-81. Means of Packaging; One-Way Consignment Boxes in One and Two Pieces for Butter Pats and Butter Bars (Apr). 2 pp.
DIN ENGL 10089-81. Means of Packaging; Folding Boxes for Butter Samples (Mar). 2 pp.

—**Corrugated Board**
CNS Z6013-88. Method of Test for Adhesion of Corrugated Board by Selective Separation (Sep)(3327).
MOD UK DSTAN 81-28-89. Cartons, Fibreboard, Fixed Joint, Multi-Wall, Corrugated Issue 3. 11 pp.
MOD UK TS 415C. Board, Corrugated (Two Flutes, Three Linears) Type 1, Type 2 and Type 3.
MOD UK TS 417D. Board, Corrugated (Three -Flute, Four-Liner) Heavy Duty Grade.
MOD UK TS 10015. Board Corrugated, Double Faced, 'E' Flute.

—**Corrugated Fiberboard**
CGSB CAN/CGSB-43.22-92. Corrugated Fibreboard Products. 9 pp.
JIS Z 1507-89. Types of Corrugated Fibreboard Boxes. 16 pp.

—**Corrugated—Glossaries**
CNS Z5085-81. Terminology of Materials and Accessories for Corrugated Boxes (May)(7458).
CNS Z5096-81. Terminology of Various Types of Corrugated Boxes (Oct)(8071).

—**Corrugated—Waterproof**
CGSB CAN/CGSB-43.33-92. Water Resistant Corrugated Fibreboard Boxes. 13 pp.

Boxes (Containers) *(Cont.)*

—**Fiberboard**
BSI BS 1133: SUBSEC 7.5-90. 1990 Packaging Code Section 7: Paper and Board Wrappers, Bags and Containers Subsection 7.5: Fibreboard Cases and Fitments. 22 pp.
CGSB CAN/CGSB-43.21-M91. Fibreboard Boxes. 23 pp.
CNS Z5117-82. Fiberboard Boxes for Outer Packing (Dec)(9758).
JIS Z 1408-89. Crates, Slatted, Wirebound Boxes. 21 pp.
JIS Z 1507-89. Types of Corrugated Fibreboard Boxes. 16 pp.
MOD UK DSTAN 81-15-01. Cartons and Boxes, Fibreboard Issue 3; Amendment 3. 21 pp.
SAA AS 1048-82. International Fibreboard Box Code. 32 pp.

—**Fiberboard—Compression Testing**
SAA AS 1301.800S-87. Methods of Test for Pulp and Paper (Metric Units)—Part 800s: Compression Resistance of Fibreboard Boxes (Cases) (This is a Joint Standard with SANZ NZS 1301). 2 pp.
SNZ NZS/AS 1301. 800S-87. Methods of Test for Pulp and Paper Compression Resistance of Fibreboard Boxes (Cases) (This is a Joint Standard with SAA AS 1301.800S). 2 pp.

—**Fiberboard—Food—Export**
CGSB CAN/CGSB-43.162-M91. Fibreboard Boxes and Partitions for Transport of Food Aid. 18 pp.

—**Fiberboard—Hazardous Materials**
CGSB CAN/CGSB-43.84-M89. Fibreboard Boxes for Inside Plastic Containers (TC-12P). 14 pp.
CGSB CAN/CGSB-43.90-M89. Nonreusable Fibreboard Boxes (TC-12A). 14 pp.
CGSB CAN/CGSB-43.91-M89. Fibreboard Boxes (TC-12B). 21 pp.
CGSB CAN/CGSB-43.93-M89. Fibreboard Boxes (TC-12D). 16 pp.
CGSB CAN/CGSB-43.95-M89. Fibreboard Boxes (TC-12H). 19 pp.

—**Fiberboard—Wirebound**
JIS Z 1407-89. Wirebound Boxes. 16 pp.

—**Food—Export—Pallets**
CGSB CAN/CGSB-43.164-M90. Palletizing of Food Containers for Export as Aid. 16 pp.

—**Insulated—Food**
BSI BS 6672: Part 2-86. 1986 Insulated Domestic Food Containers Part 2: Insulated Bags and Boxes. 10 pp.
BSI BS 6672: Part 2-01. 1986 Amd 1 Insulated Domestic Food Containers Part 2: Specification for Insulated Bags and Boxes (AMD 7200) June 15, 1992. 11 pp.

—**Marine**
JIS F 0902-78. Spare Part Boxes for Marine Use.

—**Marine—Electrical Equipment**
CNS F5109-86. Spare-Part Boxes of Electrical Equipment for Marine Use (Dec)(11792).

—**Metal—Antiaircraft Ammunition**
NATO STANAG 4038 ED 1 AMD 6-60. Specification for a 40/70 Ammunition Operational Box (Metal) for Naval Use. 23 pp.
NATO STANAG 4081 ED 1 AMD 1-62. Specification for a Lightweight 40/70 Ammunition Operational Box. 12 pp.

—**Paperboard**
BSI BS 1133: SUBSEC 7.3-86. 1986 Packaging Code Section 7: Paper and Board Wrappers, Bags and Containers Subsection 7.3: Cartons and Boxes. 18 pp.
CGSB CAN/CGSB-43.17-M91. Folding Paperboard Boxes. 20 pp.

—**Plants (Botany)**
BSI BS 3936: Part 7-89. 1989 Nursery Stock Part 7: Bedding Plants. 7 pp.

—**Plywood—Hazardous Materials**
CGSB CAN/CGSB-43.80-M90. Plywood Boxes (TC-19A and TC-19B). 15 pp.

—**Wirebound**
CNS Z5114-82. Wirebound Boxes (Dec)(9755).
CNS Z5115-82. Wirebound Crated Boxes (Dec)(9756).

—**Wooden**
BSI BS 1133: Sec 8-91. 1991 Packaging Code Section 8: Wooden Boxes, Cases and Crates (H). 89 pp.
BSI BS 1133: Sec 8-81. 1981 Packaging Code Section 8: Wooden Containers. 40 pp.
CNS Z5123-83. Wooden Box for Export Packing (Feb)(10035).
CNS Z5136-86. Wooden Framed Boxes for Export Packing (Sep)(11714).

Boxes (Containers) *(Cont.)*

—**Wooden** *(Cont.)*
JIS Z 1402-84. Wooden Boxes for Export Packing. 53 pp.
JIS Z 1403-84. Wooden Framed Boxes for Export Packing. 80 pp.
JIS Z 1406-78. Cleated Plywood Boxes.

—**Wooden—Glossaries**
JIS Z 0107-74. Glossary of Terms Used in Wooden Box for Packing.

—**Wooden—Wirebound**
CGSB CAN/CGSB-43.23-92. Wirebound Wooden Boxes. 47 pp.

Boxes (Electrical)
Use: Electrical Enclosures

Boxing Equipment
See Also: Sports Equipment
BSI BS 1892: Sec 2.6-86. 1986 Gymnasium Equipment Part 2: Particular Requirements Section 2.6: Boxing Rings. 10 pp.

Boxing Rings
Use: Boxing Equipment

Brackets
See Also: Anchors (Fasteners); Angle Brackets; Fasteners; Hangers (Fasteners); Holders; Pipe Brackets

—**Aircraft—Blanking Caps**
SBAC AS 2977 ISSUE 3. Bracket. 1 p.

—**Aircraft—Connections**
SBAC AS 8665 ISSUE 1. Bracket.

—**Aircraft—Hinges**
SBAC AS 3290 ISSUE 2. Hinge Bracket. 1 p.

—**Aircraft—Parachute Holders**
SBAC AS 424 ISSUE 4(I). Top Bracket—Parachute Stowage. 1 p.
SBAC AS 425 ISSUE 4(I). Bottom Bracket—Parachute Stowage. 1 p.

—**Aircraft—Pedal—Rudder Bars**
SBAC AS 2217-2218 ISSUE 7. Bracket for Two-Position Pedal (Rudder Bar). 1 p.

—**Aircraft—Pedal—Rudder Bars—Castings**
SBAC AS 2216 ISSUE 5. Casting for Pedal Bracket (Rudder Bar). 1 p.

—**Aircraft—Stop Valves**
SBAC AS 4896 ISSUE 3. Bracket. 1 p.

—**Aircraft—Stop Valves—Floats**
SBAC AS 4887 ISSUE 1. Bracket for Float. 1 p.

—**Aircraft—Stop Valves—Floats—Tab Washers**
SBAC AS 4904 ISSUE 3. Tab Washer—Float Bracket. 1 p.

—**Automotive—Ignition Coils**
ISO 3285-86. Road Vehicles—Ignition Coil Mounting Brackets Second Edition. 3 pp.

—**Brake Blocks**
DIN ENGL 37081-82. Adjusting Equipment for Brake Blocks; Brackets; Spring Plates (July). 2 pp.

—**Containers**
DIN ENGL 28083 Pt 1-87. Bracket Supports; Dimensions and Maximum Loads (Jan). 8 pp.
DIN ENGL 28145 Pt 7-87. Welded Attachments for Steel, Glass-Lined Agitator Vessels; Bracket Supports Welded to Jackets; Dimensions, Maximum Permissible Loads (Nov). 9 pp.

—**Containers—Bending Moments**
DIN ENGL 28083 Pt 2-87. Bracket Supports; Maximum Bending Moments Imposed on the Vessel Wall Resulting from Type A Bracket Support Loads (Jan). 23 pp.

—**Containers—Loads (Forces)**
DIN ENGL 28083 Pt 1-87. Bracket Supports; Dimensions and Maximum Loads (Jan). 8 pp.
DIN ENGL 28145 Pt 7-87. Welded Attachments for Steel, Glass-Lined Agitator Vessels; Bracket Supports Welded to Jackets; Dimensions, Maximum Permissible Loads (Nov). 9 pp.

—**Flywheels—Housings**
BSI BS 3529-62. 1962 Dimensions of Flywheel Housing Mounting Pads and Brackets for I.C. Engines. 12 pp.

INDUSTRY STANDARDS

Brackets (Cont.)

—Furniture
ISO 8555 Pt 6-87. Hardware for Furniture—Terms for Furniture Fittings—Part 6: Shelf Supports, Hanging Rails, Cabinet Suspension Brackets First Edition. 10 pp.

—Lamps
SAA AS/NZS 3128-93. Approval and Test Specification—Portable Lamp Standards and Brackets (in Professional Package 28). 7 pp.
SNZ NZS/AS 3128-90. Approval and Test Specification—Portable Lamp Standards and Brackets Amend: 4, 1992 (This is a Joint Standard with SAA AS 3128). 6 pp.

—Lighting Poles
SAA AS 1798-92. Lighting Poles and Bracket Arms—Preferred Dimensions. 18 pp.

—Masonry
CEN PREN 845-1-92. Specification for Ancillary Components for Masonry—Part 1: Ties, Straps, Hangers, Brackets and Support Angles. 26 pp.

—Masonry—Flexural Strength
CEN PREN 846-10-92. Methods of Test for Ancillary Components for Masonry—Part 10: Determination of Flexural Resistance and Stiffness of Brackets and Support Angles. 9 pp.

—Masonry—Stiffness Testing
CEN PREN 846-10-92. Methods of Test for Ancillary Components for Masonry—Part 10: Determination of Flexural Resistance and Stiffness of Brackets and Support Angles. 9 pp.

—Modular Units
DIN ENGL 69524-79. Machine Tools; Modular Units; Support Brackets, Constructural Sizes (Jan). 3 pp.
ISO 3610-76. Modular Units for Machine Tool Construction—Support Brackets First Edition; (Amendment Slip-1981). 5 pp.

—Pipes
DIN ENGL 3567-63. Pipe Brackets for NW 20 to 500 (Aug). 2 pp.

—Ships—Derricks
CNS F3149-81. Ships' Derrick Topping Brackets (Nov)(8113).
CNS F3150-81. Ships' Derrick Gooseneck Brackets (Nov)(8114).
CNS F3152-81. Ships' Light Load Derrick Topping Brackets (Nov)(8116).
CNS F3153-81. Ships' Light Load Derrick Gooseneck Brackets (Nov)(8117).
JIS F 2202-90. Ships' Derrick Topping Brackets.
JIS F 2202-76. Ships' Derrick Topping Brackets. 8 pp.
JIS F 2203-90. Ships' Derrick Gooseneck Brackets.
JIS F 2203-73. Ships' Derrick Gooseneck Brackets. 6 pp.
JIS F 2206-76. Ships' Light Load Derrick Topping Brackets.
JIS F 2206-70. Ships' Light Load Derrick Topping Brackets. 4 pp.
JIS F 2207-76. Ships' Light Load Derrick Gooseneck Brackets (R 1984). 7 pp.

—Ships—Moorings
CNS F3124-81. Ships' Towing and Mooring Brackets (Jul)(7696).
JIS F 2029-78. Ships' Towing and Mooring Brackets. 6 pp.

—Ships—Towing
CNS F3124-81. Ships' Towing and Mooring Brackets (Jul)(7696).
JIS F 2029-78. Ships' Towing and Mooring Brackets. 6 pp.

Braided Wire
Use For: Wire Braids

—Copper
BSI BS 4801: Part 2-72. 1972 Varnish-Bonded Glass Braided Copper Conductors Part 2: Rectangular Conductors. 18 pp.
MOD UK DSTAN 61-12: Part 20-80. Wires, Cords, and Cables, Electrical-Metric Units Part 20: Braids, Wire Issue 1. 7 pp.

Braiding Machinery
See Also: Textile Machinery

—Glossaries
JIS L 0307-85. Glossary of Terms Used in Knitting, Braiding and Related Machinery.

Braids (Uniform)
Use: Fourrageres

Brake Blocks
See Also: Brakes

—Adjusting Equipment
DIN ENGL 37080-75. Adjusting Equipment for Brake Blocks; Survey and Arrangement (Sept). 2 pp.
DIN ENGL 37081-82. Adjusting Equipment for Brake Blocks; Brackets; Spring Plates (July). 2 pp.
DIN ENGL 37082-75. Adjusting Equipment for Brake Blocks; Compression Springs; Spring Bolt (Sept). 1 p.

—Emergency—Cable Railways
JIS E 9201-68. Emergency Brake Blocks for Cable Cars.

Brake Cables
See Also: Brakes; Control Cables (Electric)

—Bicycles
CNS B2676-82. Brake Cable for Bicycle (Jun)(8934). 9 pp.

Brake Cylinders
See Also: Automotive Equipment; Brakes

—Hydraulic—Automotive
CNS D2011-85. Hydraulic Brake Master Cylinders for Automobiles (Jun)(3474). 14 pp.
JIS D 2603-88. Hydraulic Brake Master Cylinders for Automotive Hydraulic Brake Systems Using a Non-Petroleum Base Hydraulic Brake Fluid. 21 pp.

—Hydraulic—Bleed Screws—Automotive
CNS D2179-84. Bleed Screw of Hydraulic Brake Wheel Cylinders for Automobiles (Jun)(10934).

—Hydraulic—Boots—Automotive
CNS D2009-85. Rubber Cups of Hydraulic Brake Cylinders for Automobiles (Jun)(3472).
CNS D2175-83. Rubber Boots of Hydraulic Brake Wheel Cylinders for Automobiles (Dec)(10710).

—Hydraulic—Cups
ISO 4928-80. Road Vehicles—Elastomeric Cups and Seals for Cylinders for Hydraulic Braking Systems Using a Non-Petroleum Base Hydraulic Brake Fluid (Service Temperature 120 Degrees Celsius Maximum) Second Edition. 14 pp.
ISO 7631-85. Road Vehicles—Elastomeric Cups and Seals for Cylinders for Hydraulic Braking Systems Using a Petroleum Base Hydraulic Brake Fluid (Service Temperature 120 Degrees Celsius Max.) First Edition. 13 pp.

—Hydraulic—Cups—Automotive
JIS D 2605-88. Rubber Cups for Hydraulic Cylinders for Automotive Hydraulic Brake Systems Using a Non-Petroleum Base Hydraulic Brake Fluid. 35 pp.

—Hydraulic—Disc—Seals
ISO 4930-78. Road Vehicles—Elastomeric Seals for Hydraulic Disc Brake Cylinders Using a Non-Petroleum Base Hydraulic Brake Fluid (Service Temperature 150 Degrees Celsius Maximum) First Edition; (Erratum—June 1979). 9 pp.
ISO 6119-80. Road Vehicles—Elastomeric Seals for Hydraulic Disc Brake Cylinders Using a Petroleum Base Hydraulic Brake Fluid (Service Temperature 120 Degrees Celcius Maximum) First Edition. 8 pp.

—Hydraulic—Disc—Seals—Automotive
CNS D2176-83. Elastomeric Seals of Hydraulic Disc Brake Cylinders for Automobiles (Dec)(10711).
JIS D 2609-82. Elastomeric Seals of Hydraulic Disc Brake Cylinders for Automobiles.

—Hydraulic—Drum—Boots
ISO 4927-78. Road Vehicles—Elastomeric Boots for Drum Type Hydraulic Brake Wheel Cylinders Using a Non-Petroleum Base Hydraulic Brake Fluid (Service Temperature 120 Degrees Celsius Maximum) First Edition. 6 pp.
ISO 6117-80. Road Vehicles—Elastomeric Boots for Drum Type Hydraulic Brake Wheel Cylinders Using A Non-Petroleum Base Hydraulic Brake Fluid (Service Temperature 100 Degrees Celsius Maximum) First Edition. 6 pp.
ISO 7633-85. Road Vehicles—Elastomeric Boots for Drum Type Hydraulic Brake Wheel Cylinders Using a Petroleum Base Hydraulic Brake Fluid (Service Temperature 120 Degrees Celsius Max.) First Edition. 6 pp.

—Hydraulic—Drum—O Rings
ISO 7630-85. Road Vehicles—Elastomeric O-Rings for Hydraulic Drum Brake Wheel Cylinders Using a Petroleum Base Hydraulic Brake Fluid (Service Temperature 120 Degrees Celsius Max.) First Edition. 7 pp.

—Hydraulic—Gaskets—Automotive
CNS D2177-83. Diaphragm Gaskets of Hydraulic Brake Master Cylinder Reservoirs for Automobiles (Dec)(10712).
JIS D 2610-82. Diaphragm Gaskets of Hydraulic Brake Master Cylinder Reservoirs for Automobiles.

—Hydraulic—Seals
ISO 4928-80. Road Vehicles—Elastomeric Cups and Seals for Cylinders for Hydraulic Braking Systems Using a Non-Petroleum Base Hydraulic Brake Fluid (Service Temperature 120 Degrees Celsius Maximum) Second Edition. 14 pp.
ISO 7631-85. Road Vehicles—Elastomeric Cups and Seals for Cylinders for Hydraulic Braking Systems Using a Petroleum Base Hydraulic Brake Fluid (Service Temperature 120 Degrees Celsius Max.) First Edition. 13 pp.
ISO 7632-85. Road Vehicles—Elastomeric Seals for Hydraulic Disc Brake Cylinders Using a Petroleum Base Hydraulic Brake Fluid (Service Temperature 120 Degrees Celsius Max.) First Edition. 8 pp.

—Hydraulic—Wheel—Automotive
CNS D2010-85. Hydraulic Brake Wheel Cylinders for Automobiles (Jun)(3473). 6 pp.
JIS D 2604-89. Hydraulic Brake Wheel Cylinders for Automotive Hydraulic Brake Systems Using a Non-Petroleum Base Brake Fluid.

Brake Cylinders, Hydraulic
Use: Hydraulic Cylinders

Brake Fluids
See Also: Hydraulic Fluids

BSI BS AU 174: Part 1-79. 1979 Hydraulic Brake Systems, Using Non-Petroleum Base Brake Fluids, for Road Vehicles Part 1: Non-petroleum Base Reference Fluids. 4 pp.
BSI BS AU 174: Part 2-85. 1985 Hydraulic Brake Systems, Using Non-Petroleum Base Brake Fluids, for Road Vehicles Part 2: Petroleum Base Brake Fluid. 25 pp.
BSI BS AU 216: Part 1-87. 1987 Hydraulic Brake Systems Using Petroleum Base Brake Fluids for Road Vehicles Part 1: Petroleum Base Reference Fluid. 3 pp.
CNS K5009-83. Hydraulic Brake Fluid (Dec)(261). 2 pp.
ISO 4925-78. Road Vehicles—Non-Petroleum Base Brake Fluid First Edition; (Erratum—June 1979). 27 pp.
ISO 4926-78. Road Vehicles—Hydraulic Brake Systems—Non-Petroleum Base Reference Fluids First Edition. 4 pp.
ISO 7308-87. Road Vehicles—Petroleum-Based Brake-Fluid for Stored-Energy Hydraulic Brakes First Edition. 25 pp.
ISO 7309-85. Road Vehicles—Hydraulic Braking Systems—ISO Reference Petroleum Base Fluid First Edition. 3 pp.
JIS K 2228-88. Rubber Lubricant of Non-Petroleum Base for Motor Vehicle Brake Systems. 28 pp.
JIS K 2233-89. Non-Petroleum Base Motor Vehicle Brake Fluids. 41 pp.
MOD UK TS 10145. Brake Fluid, Automotive OX-8 NATO H-542.
SAA AS 1960-83. Motor Vehicle Brake Fluids (Non-Petroleum Type). 4 pp.
SNZ NZS/AS 1960-83. Motor Vehicle Brake Fluids (Non-Petroleum Type) (This is a Joint Standard with SAA AS 1960). 4 pp.

—Containers—Identification Systems
ISO 3871-80. Road Vehicles—Labelling of Containers for Petroleum or Non-Petroleum Base Brake Fluid Third Edition. 3 pp.

—Resistance Testing—Brake Linings
BSI BS AU 180: Part 5-83. 1983 Brake Linings Part 5: Method for Determining Resistance to Water, Saline Solution, Oil and Brake Fluid of Brake Lining Material. 4 pp.
ISO 6314-80. Road Vehicles—Brake Linings—Resistance to Water, Saline Solution, Oil and Brake Fluid—Test Procedure First Edition. 4 pp.
JIS D 4420-86. Test Procedure of Deterioration by Water, Salt Water, Oil and Brake Fluid for Brake Linings and Pads of Automobiles. 7 pp.

—Resistance Testing—Brake Pads
JIS D 4420-86. Test Procedure of Deterioration by Water, Salt Water, Oil and Brake Fluid for Brake Linings and Pads of Automobiles. 7 pp.

—Symbols
BSI BS AU 229-89. 1989 Graphical Symbols to Designate Brake Fluid Types in Road Vehicles. 4 pp.
ISO 9128-87. Road Vehicles—Graphical Symbols to Designate Brake Fluid Types First Edition. 6 pp.

INTERNATIONAL AND NON-U.S. NATIONAL STANDARDS
SUBJECT INDEX
Brake

Brake Horsepower
—Internal Combustion Piston Engines
BSI BS 5514: Part 7-88. 1988 Reciprocating Internal Combustion Engines: Performance Part 7: Codes for Engine Power. 8 pp.
ISO 3046 Pt 7-87. Reciprocating Internal Combustion Engines—Performance—Part 7: Codes for Engine Power First Edition. 7 pp.

Brake Lights
Use For: Stop Lamps *See Also:* Indicator Lights; Motor Vehicle Lighting
EC 76/758/EEC-76. Council Directive on the Approximation of the Laws of the Member States Relating to the End-Outline Marker Lamps, Front Position (Side) Lamps, Rear Position (Side) Lamps and Stop Lamps for Motor Vehicles and Their Trailers. 17 pp.
EC 89/516/EEC-89. Commission Directive Adapting to Technical Progress Council Directive 76/758/EEC on the Approximation of the Laws of the Member Sts. Rel. to the End-Outline Marker Lamps, Front Pos. (Side) Lamps, Rear Pos. (Side) Lamps and Stop Lamps for Motor Vehicles and Their Trailers. 14 pp.
SNZ NZS 5461-89. High Level Stop Light. 12 pp.

—Switches
CNS D2026-87. Mechanical Stop Lamp Switches for Automobiles (Aug)(5780).
CNS D2052-87. Switches of Hydraulic Stop Lamp for Automobiles (Jan)(6174).
JIS D 5808-76. Mechanical Stop Lamp Switches for Automobiles.
JIS D 5808-72. Mechanical Stop Lamp Switches for Automobiles (R 1976). 9 pp.

—Trailers
EC 76/758/EEC-76. Council Directive on the Approximation of the Laws of the Member States Relating to the End-Outline Marker Lamps, Front Position (Side) Lamps, Rear Position (Side) Lamps and Stop Lamps for Motor Vehicles and Their Trailers. 17 pp.
EC 89/516/EEC-89. Commission Directive Adapting to Technical Progress Council Directive 76/758/EEC on the Approximation of the Laws of the Member Sts. Rel. to the End-Outline Marker Lamps, Front Pos. (Side) Lamps, Rear Pos. (Side) Lamps and Stop Lamps for Motor Vehicles and Their Trailers. 14 pp.

Brake Linings
See Also: Brake Pads; Brake Shoes; Brakes; Linings
BSI BS AU 142-68. 1968 Methods of Test for Brake Linings Materials. 16 pp.

—Automobiles
CNS D2002-85. Brake Linings for Automobiles (Jun)(2586). 9 pp.
JIS D 4411-75. Brake Linings for Automobiles (R 1983). 15 pp.

—Automobiles—Corrosion Testing
CNS D3178-87. Method of Test of Seizure to Ferrous Mating Surface Due to Corrosion for Brake Linings and Pads of Automobiles (Dec)(12174).
JIS D 4414-86. Test Procedure of Seizure to Ferrous Mating Surface Due to Corrosion for Brake Linings and Pads of Automobiles. 8 pp.

—Automobiles—Density
CNS D3182-87. Method of Test of Specific Gravity for Brake Linings and Pads of Automobiles (Dec)(12178).
JIS D 4417-86. Test Procedure of Specific Gravity for Brake Linings and Pads of Automobiles. 7 pp.

—Automobiles—Hardness Testing
JIS D 4421-87. Method of Hardness Test for Brake Linings, Pads and Clutch Facings of Automobiles. 11 pp.

—Automobiles—Porosity
CNS D3183-87. Method of Porosity for for Brake Linings and Pads of Automobiles (Dec)(12179).
CNS K5139-87. Oils for Testing of Porosity for Brake Linings and Pads of Automobiles (Dec)(12180).
JIS D 4418-86. Test Procedure of Porosity for Brake Linings and Pads of Automobiles. 13 pp.

—Automobiles—Rivets
CNS D2006-85. Rivets for Brake Linings and Clutch Facings for Automobiles (Jun)(3012).
JIS D 4312-90. Rivets of Brake Linings and Clutch Facings for Automobiles. 8 pp.

—Automobiles—Shear Strength
CNS D3179-87. Method of Test of Shear Strength for Brake Linings and Pads of Automobiles (Dec)(12175).
JIS D 4415-86. Test Procedure of Shear Strength for Brake Linings and Pads of Automobiles. 8 pp.

Brake Linings (Cont.)
—Automobiles—Thermal Expansion
CNS D3180-87. Method of Test of Thermal Expansion for Brake Linings and Pads of Automobiles (Dec)(12176).
CNS D3181-87. Method of Test of Thermal Expansion with Heating Plate for Disc Brake Pads of Automobiles (Dec)(12177).
JIS D 4416-86. Test Procedure of Thermal Expansion for Brake Linings and Pads of Automobiles. 13 pp.

—Automotive—Thermal Conductivity
ISO TR7882-86. Road Vehicles—Brake Linings—Determination of Thermal Conductivity by Guarded Hot-Plate Apparatus First Edition. 8 pp.

—Compression Testing
BSI BS AU 180: Part 1-83. 1983 Brake Linings Part 1: Method for Measurement of Compressibility of Brake Lining Material. 5 pp.
ISO 6310-81. Road Vehicles—Brake Linings—Compressibility—Test Procedure First Edition. 6 pp.

—Corrosion Testing
BSI BS AU 180: Part 6-82. 1982 Brake Linings Part 6: Method for Assessment of Seizure to Ferrous Mating Surface Due to Corrosion. 4 pp.
ISO 6315-80. Road Vehicles—Brake Linings—Seizure to Ferrous Mating Surface Due to Corrosion—Test Procedure First Edition. 4 pp.

—Friction
ISO 7881-87. Road Vehicles—Brake Linings—Evaluation of Friction Material Characteristics—Small Sample Bench Test Procedure First Edition. 12 pp.

—Industrial Equipment
CNS R2179-85. Asbestos Brake Linings for Industrial Machines (May)(11268).
JIS R 3455-79. Asbestos Brake Linings for Industrial Machines. 15 pp.

—Resistance Testing
BSI BS AU 180: Part 5-83. 1983 Brake Linings Part 5: Method for Determining Resistance to Water, Saline Solution, Oil and Brake Fluid of Brake Lining Material. 4 pp.
ISO 6314-80. Road Vehicles—Brake Linings—Resistance to Water, Saline Solution, Oil and Brake Fluid—Test Procedure First Edition. 4 pp.
JIS D 4420-86. Test Procedure of Deterioration by Water, Salt Water, Oil and Brake Fluid for Brake Linings and Pads of Automobiles. 7 pp.

—Rivets
BSI BS 3575-63. 1963 Rivets for the Attachment of Friction Linings. 8 pp.
CNS B2582-81. Rivets for Brake and Clutch Lining (Jun)(7570).
DIN ENGL 7338-83. Rivets for Brake Linings and Clutch Linings (Dec). 5 pp.

—Shear Strength
BSI BS AU 180: Part 2-82. 1982 Brake Linings Part 2: Method for Measurement of Internal Shear Strength of Lining Material. 4 pp.
ISO 6311-80. Road Vehicles—Brake Linings—Internal Shear Strength of Lining Material—Test Procedure First Edition. 4 pp.

Brake Pads
See Also: Brake Linings; Brakes; Disk Brakes

—Automobiles—Compression Testing
CNS D3177-87. Method of Test of Compressibility for Disc Brake Pads of Automobiles (Dec)(12173).
JIS D 4413-86. Test Procedure of Compressibility for Disc Brake Pads of Automobiles. 9 pp.

—Automobiles—Corrosion Testing
CNS D3178-87. Method of Test of Seizure to Ferrous Mating Surface Due to Corrosion for Brake Linings and Pads of Automobiles (Dec)(12174).
JIS D 4414-86. Test Procedure of Seizure to Ferrous Mating Surface Due to Corrosion for Brake Linings and Pads of Automobiles. 8 pp.

—Automobiles—Density
CNS D3182-87. Method of Test of Specific Gravity for Brake Linings and Pads of Automobiles (Dec)(12178).
JIS D 4417-86. Test Procedure of Specific Gravity for Brake Linings and Pads of Automobiles. 7 pp.

—Automobiles—Dimensional Stability—High Temperature Testing
BSI BS AU 180: Part 4-82. 1982 Brake Linings Part 4: Method for Determining Effects of Heat on Dimensions and Form of Disc Brake Pads. 6 pp.
ISO 6313-80. Road Vehicles—Brake Linings—Effects of Heat on Dimensions and Form of Disc Brake Pads—Test Procedure First Edition. 6 pp.

Brake Pads (Cont.)
—Automobiles—Hardness Testing
JIS D 4421-87. Method of Hardness Test for Brake Linings, Pads and Clutch Facings of Automobiles. 11 pp.

—Automobiles—Porosity
CNS D3183-87. Method of Porosity for for Brake Linings and Pads of Automobiles (Dec)(12179).
CNS K5139-87. Oils for Testing of Porosity for Brake Linings and Pads of Automobiles (Dec)(12180).
JIS D 4418-86. Test Procedure of Porosity for Brake Linings and Pads of Automobiles. 13 pp.

—Automobiles—Rusting
CNS D3176-87. Method of Test of Rusting at Material Interfaces for Disc Brake Pads of Automobiles (Dec)(12172).
JIS D 4419-86. Test Procedure of Rusting at Material Interfaces for Disc Brake Pads of Automobiles. 7 pp.

—Automobiles—Shear Strength
CNS D3179-87. Method of Test of Shear Strength for Brake Linings and Pads of Automobiles (Dec)(12175).
JIS D 4415-86. Test Procedure of Shear Strength for Brake Linings and Pads of Automobiles. 8 pp.

—Automobiles—Thermal Expansion
CNS D3180-87. Method of Test of Thermal Expansion for Brake Linings and Pads of Automobiles (Dec)(12176).
CNS D3181-87. Method of Test of Thermal Expansion with Heating Plate for Disc Brake Pads of Automobiles (Dec)(12177).
JIS D 4416-86. Test Procedure of Thermal Expansion for Brake Linings and Pads of Automobiles. 13 pp.

—Compression Testing
BSI BS AU 180: Part 1-83. 1983 Brake Linings Part 1: Method for Measurement of Compressibility of Brake Lining Material. 5 pp.

—Resistance Testing
JIS D 4420-86. Test Procedure of Deterioration by Water, Salt Water, Oil and Brake Fluid for Brake Linings and Pads of Automobiles. 7 pp.

—Shear Strength
BSI BS AU 180: Part 3-82. 1982 Brake Linings Part 3: Method for Measurement of Shear Strength of Disc Brake Pad and Drum Brake Shoe Assemblies. 5 pp.
ISO 6312-81. Road Vehicles—Brake Linings—Shear Strength of Disc Brake Pad and Drum Brake Shoe Assemblies—Test Procedure First Edition. 6 pp.
JIS D 4422-90. Shear Strength Test Produce for Drum Brake Shoe Assemblies and Parts of Automobiles. 10 pp.

—Surface Defects
BSI BS AU 222-88. 1988 Method of Evaluating Surface and Material Flaws of Disc Brake Pads. 6 pp.
ISO 7629-87. Road Vehicles—Brake Linings—Disc Brake Pads—Evaluation of Surface and Material Flaws After Testing First Edition. 7 pp.

Brake Pipes
See Also: Air Brakes; Brakes

—Air Strainers—Rolling Stock
CNS E1016-82. Brake Pipe Air Strainers for Railway Rolling Stock (Dec)(9697).
JIS E 4308-87. Air Filters for Railway Rolling Stock.

—Automobiles—Engineering Drawings
CNS D1032-82. Simplified Drawing Methods for Brake Pipes of Automobiles (Sep)(9385).

Brake Shoes
See Also: Brake Linings; Brakes; Drum Brakes
SAA AS D20-72. Bonded Brake Shoe Assemblies. 22 pp.

—Compression Testing
BSI BS AU 180: Part 1-83. 1983 Brake Linings Part 1: Method for Measurement of Compressibility of Brake Lining Material. 5 pp.

—Motorcycles
CNS D2129-82. Brake Shoe Assembly for Motorcycles (May)(8814). 2 pp.
CNS D3122-82. Method of Test for Brake Shoe Assembly of Motorcycles (May)(8813). 2 pp.

—Rolling Stock
CNS E3003-82. Composition Brake Shoes for Railway Rolling Stock (Feb)(8489).
JIS E 4309-91. Composition Brake Shoes for Railway Rolling Stock.
JIS E 7501-75. Cast Iron Brake Shoes for Railway Rolling Stock.

INDUSTRY STANDARDS

Brake

Brake Shoes (Cont.)

—Shear Strength

BSI BS AU 180: Part 3-82. 1982 Brake Linings Part 3: Method for Measurement of Shear Strength of Disc Brake Pad and Drum Brake Shoe Assemblies. 5 pp.

ISO 6312-81. Road Vehicles—Brake Linings—Shear Strength of Disc Brake Pad and Drum Brake Shoe Assemblies—Test Procedure First Edition. 6 pp.

JIS D 4422-90. Shear Strength Test Produce for Drum Brake Shoe Assemblies and Parts of Automobiles. 10 pp.

Brakes

Scope Note: For additional listings, use a more specific term Use For: Braking Systems
See Also: Air Brakes; Aircraft Brakes; Antiskid Brake Systems; Automotive Equipment; Brake Blocks; Brake Cables; Brake Cylinders; Brake Linings; Brake Pads; Brake Pipes; Brake Shoes; Braking Testing; Disk Brakes; Drum Brakes; Hydraulic Brakes; Hydraulic Cylinders; Landing Gear; Nose Wheels; Retarders (Devices); Vacuum Brakes; Wheels

IEC 631-78. Characteristics and Tests of Electrodynamic and Electromagnetic Braking Systems First Edition. 22 pp.

—Agricultural Equipment

BSI BS 6384-83. 1983 Determination of Braking Performance of Agricultural and Forestry Vehicles. 18 pp.

ISO 5697-82. Agricultural and Forestry Vehicles—Determination of Braking Performance First Edition. 18 pp.

—Agricultural Equipment—Hydraulic Couplings

BSI BS 4742: Part 4-85. 1985 Hydraulic Equipment for Agricultural Machinery Part 4: Hydraulic Couplings on Braking Systems for Trailers and Trailed Equipment. 4 pp.

ISO 5676-83. Tractors and Machinery for Agriculture and Forestry—Hydraulic Coupling—Braking Circuit First Edition. 4 pp.

—Automotive

CNS D3015-89. Method of Brake Test for Automobiles (Apr)(2734).

CNS D3106-82. Method of Road Test for Simulated Mountain Brake Fade for Automobiles (Mar)(8563).

CNS D3163-83. Method of Service Brake Test for Automobiles — on Brake Tester (Dec)(10713).

CNS K5139-87. Oils for Testing of Porosity for Brake Linings and Pads of Automobiles (Dec)(12180).

EC 91/422/EEC-91. Commission Directive Adapting to Technical Progress Council Directive 71/320/EEC on the Approximation of the Laws of the Member States Relating to the Braking Devices of Certain Categories of Motor Vehicles and Their Trailers. 11 pp.

EC 92/54/EEC-92. Council Directive Amending Directive 77/143/EEC on the Approximation of the Laws of the Member States Relating to Roadworthiness Tests for Motor Vehicles and Their Trailers (Brakes). 5 pp.

ISO 6597-91. Road Vehicles—Hydraulic Braking Systems—Measurement of Braking Performance Second Edition. 23 pp.

JIS D 0210-85. General Rules of Brake Test Method of Automobiles and Motor Cycles. 13 pp.

JIS D 1013-82. Brake Test Method of Automobiles. 9 pp.

JIS D 2607-80. Vacuum Brake Hoses for Automobiles (R 1985). 11 pp.

—Automotive—Diagrams—Symbols

CNS D1033-82. Graphical Symbols for Braking Diagram of Automobiles (Sep)(9386).

—Automotive—Dynamometer Testing

CNS D3107-82. Method of Dynamometer Test for Simulated Mountain Fade for Automobiles (Mar)(8564).

CNS D3108-82. Method of Braking Device Dynamometer Test for Passenger Car (Mar)(8565).

—Automotive—Glossaries

CNS D1047-83. Glossary of Terms Relating to Braking Equipment of Automobiles (Feb)(9992).

CNS D1054-83. Glossary of Terms Relating to Brake Types, Braking and Brake Operation of Automobiles (Apr)(10163).

ISO 611-80. Road Vehicles—Braking of Automotive Vehicles and Their Trailers—Vocabulary Second Edition. 10 pp.

JIS D 0106-84. Glossary of Terms Relating to Brake Types, Braking Mechanics and Brake Operation of Automobiles. 41 pp.

JIS D 0107-84. Glossary of Terms Relating to Braking Equipments of Automobiles. 95 pp.

Brakes (Cont.)

—Bicycles

CNS B2044-90. Hand Brakes for Bicycles (Nov)(354). 6 pp.

CNS B7058-75. Method of Test for the Rear Brakes of Bicycles (Dec)(3871). 2 pp.

CNS K4060-81. Brake Rubber for Bicycles (Jan)(6926). 2 pp.

JIS D 9201-90. Bicycles—Method of Braking Test. 12 pp.

JIS K 6310-77. Brake Rubber for Bicycle.

—Chain Saws

BSI BS 6916: Part 5-88. 1988 Chain Saws Part 5: Methods of Evaluating Chain Brake Performance (ISO 6535: 1983). 4 pp.

ISO 6535-91. Portable Chain-Saws—Chain Brake Performance Second Edition. 5 pp.

—Combat Vehicles—Antilock Devices—Connectors

NATO STANAG 4395 ED 1 AMD 0-00. Connector for Tactical Land Vehicles with Anti-Lock Braking Systems. 4 pp.

—Cultivators

CNS B4037-80. Axles with Brakes of Trailer for Power Tiller (Aug)(6012).

JIS B 9208-76. Axles with Brakes of Trailer for Power Tiller (R 1979). 9 pp.

—Earthmoving Equipment

BSI BS 6824-87. 1987 Performance and Test of Braking Systems for Wheeled Earth-Moving Machinery. 10 pp.

CNS A4017-88. Off-Highway Earth-Moving Machinery Performance Criteria for Brake Systems (Dec)(9948).

CSA CAN/CSA-M3450-M92. Braking Systems—Performance Requirements and Test Procedures—Wheeled Machinery; (Gen Instr 1). 21 pp.

ISO 3450-85. Earth-Moving Machinery—Wheeled Machines—Performance Requirements and Test Procedures for Braking Systems Second Edition. 8 pp.

—Forestry Equipment

BSI BS 6384-83. 1983 Determination of Braking Performance of Agricultural and Forestry Vehicles. 18 pp.

BSI BS 7705-93. 1993 Brake Systems for Special Wheeled Forestry Machines (ISO 11169: 1993) (E). 12 pp.

ISO 5697-82. Agricultural and Forestry Vehicles—Determination of Braking Performance First Edition. 18 pp.

ISO 11169-93. Machinery for Forestry—Wheeled Special Machines —Vocabulary, Performance Test Methods and Criteria for Brake Systems First Edition. 10 pp.

—Forestry Equipment—Glossaries

BSI BS 7705-93. 1993 Brake Systems for Special Wheeled Forestry Machines (ISO 11169: 1993) (E). 12 pp.

ISO 11169-93. Machinery for Forestry—Wheeled Special Machines —Vocabulary, Performance Test Methods and Criteria for Brake Systems First Edition. 10 pp.

—Forestry Equipment—Hydraulic Couplings

BSI BS 4742: Part 4-85. 1985 Hydraulic Equipment for Agricultural Machinery Part 4: Hydraulic Couplings on Braking Systems for Trailers and Trailed Equipment. 4 pp.

ISO 5676-83. Tractors and Machinery for Agriculture and Forestry—Hydraulic Coupling—Braking Circuit First Edition. 4 pp.

—Fork Trucks

JIS D 6023-85. Brake Performance and Brake Tests for Fork Lift Trucks.

—Glossaries

JIS B 0152-73. Glossary of Terms Relating to Clutches and Brakes.

—Hoses—Rolling Stock

BSI BS 3682-69. 1969 Amd 2 Railway Brake Hose. 14 pp.

—Industrial Trucks

ISO 6292 Pt 1-81. Powered Industrial Trucks—Brake Performance—Part 1: High-Lift, Low-Lift and Non-Lifting First Edition. 5 pp.

ISO 6500-80. Powered Industrial Trucks—Service Brakes—Component Strength-Performance Requirements First Edition; (Amendment Slip-1980). 4 pp.

—Mining Equipment

CSA CAN/CSA-M424.3-M90. Braking Performance—Rubber-Tired, Self-Propelled Underground Mining Machines; (Gen Instr 1). 18 pp.

Brakes (Cont.)

—Mopeds

EC 93/14/EEC-93. Council Directive on the Braking of Two or Three-Wheel Motor Vehicles. 19 pp.

—Motorcycles

CNS D3030-88. Method of Brake Test for Motorcycles (May)(3106).

JIS D 0210-85. General Rules of Brake Test Method of Automobiles and Motor Cycles. 13 pp.

JIS D 1034-90. Method of Brake Test for Mopeds and Motorcycles. 9 pp.

—Railroad Equipment

JIS E 9310-76. Braking Test Methods of Lifts.

—Recreational Vehicles

BSI BS AU 213-87. 1987 Calculation of Forces on Control Devices for Inertia Braking of Caravans and Light Trailers. 4 pp.

ISO 7642-91. Caravans and Light Trailers—Trailers of Categories O1 and O2 with Overrun Brakes—Inertia Bench Test Methods for Brakes Second Edition. 7 pp.

ISO 7643-91. Caravans and Light Trailers—Trailers of Categories O1 and O2 with Overrun Brakes—Linear Bench Test Methods for Brake Controls Second Edition. 8 pp.

ISO 8703-86. Road Vehicles—Caravans and Light Trailers—Control Device for Inertia Braking—Calculation of Forces First Edition. 4 pp.

—Rollers (Compactors)—Safety

SAA AS 2958.3-92. Earth-Moving Machinery—Safety—Part 3: Roller Compactors—Brake Systems. 3 pp.

—Rolling Stock—Hoses

BSI BS 3682-69. 1969 Amd 2 Railway Brake Hose. 14 pp.

—Semitrailers—Couplings

BSI BS AU 5-63. 1963 Dimensions of a Contact Type Coupling for Vacuum and Pressure Braking Systems on Trailers and Semi-Trailers. 3 pp.

—Semitrailers—Military

NATO STANAG 2604 ED 3 AMD 0-86. Breaking Systems Between Tractor, Draw-Bar Trailer and Semi-Trailer Equipment Combinations for Military Use. 24 pp.

NATO STANAG 2604 ED 3 AMD 2-86. Breaking Systems Between Tractor, Draw-Bar Trailer and Semi-Trailer Equipment Combinations for Military Use. 28 pp.

—Towed Vehicles

BSI BS 4639-87. 1987 Brakes and Braking Systems for Towed Agricultural Vehicles. 8 pp.

BSI BS AU 196-84. 1984 Electrical Connections Between 24V Trailer Antilock Braking Systems and Towing Vehicles. 4 pp.

EC 89/173/EEC-88. Council Directive on the Approximation of the Laws of the Member States Relating to Certain Components and Characteristics of Wheeled Agricultural or Forestry Tractors. 119 pp.

ISO 5696-84. Trailed Agricultural Vehicles—Brakes and Braking Devices—Laboratory Test Method First Ediiton. 15 pp.

—Tractors

EC 76/432/EEC-76. Council Directive on the Approximation of the Laws of the Member States Relating to the Braking Devices of Wheeled Agricultural or Forestry Tractors. 14 pp.

EC 89/173/EEC-88. Council Directive on the Approximation of the Laws of the Member States Relating to Certain Components and Characteristics of Wheeled Agricultural or Forestry Tractors. 119 pp.

—Trailers

EC 91/422/EEC-91. Commission Directive Adapting to Technical Progress Council Directive 71/320/EEC on the Approximation of the Laws of the Member States Relating to the Braking Devices of Certain Categories of Motor Vehicles and Their Trailers. 11 pp.

EC 92/54/EEC-92. Council Directive Amending Directive 77/143/EEC on the Approximation of the Laws of the Member States Relating to Roadworthiness Tests for Motor Vehicles and Their Trailers (Brakes). 5 pp.

—Trailers—Couplings

BSI BS AU 5-63. 1963 Dimensions of a Contact Type Coupling for Vacuum and Pressure Braking Systems on Trailers and Semi-Trailers. 3 pp.

—Trailers—Military

NATO STANAG 2604 ED 3 AMD 0-86. Breaking Systems Between Tractor, Draw-Bar Trailer and Semi-Trailer Equipment Combinations for Military Use. 24 pp.

Brakes (Cont.)
—Trailers—Military (Cont.)
NATO STANAG 2604 ED 3 AMD 2-86. Breaking Systems Between Tractor, Draw-Bar Trailer and Semi-Trailer Equipment Combinations for Military Use. 28 pp.

—Wheelchairs
CSA CAN/CSA-Z323.4.4-M89. Wheelchairs—Determination of Brake Effectiveness (ISO 7176-3:1988); (Gen Instr 1). 17 pp.
ISO 7176 Pt 3-88. Wheelchairs—Part 3: Determination of Efficiency of Brakes First Edition. 6 pp.

Braking Systems
Use: Brakes

Braking Testing
See Also: Brakes; Deceleration Testing; Road Testing

—Automobiles
CNS D3106-82. Method of Road Test for Simulated Mountain Brake Fade for Automobiles (Mar)(8563).

—Automobiles—Dynamometers
CNS D3107-82. Method of Dynamometer Test for Simulated Mountain Fade for Automobiles (Mar)(8564).
CNS D3108-82. Method of Braking Device Dynamometer Test for Passenger Car (Mar)(8565).

—Automotive
BSI BS AU 205-86. 1986 Effect of Braking on Steady State Cornering Behaviour of Road Vehicles. 14 pp.
CNS D3143-82. Service Brake Curved Road Test Procedure for Combinations (Sep)(9387).
CNS D3144-82. Service Brake Road Test Procedure for Combinations (Sep)(9388).
CNS D3145-82. Service Brake Road Test Performance Requirements for Combinations (Sep)(9389).
CNS D3157-83. Air Servo and Vacuum Servo Brake Test Procedure (Nov)(10653).
ISO 6597-91. Road Vehicles—Hydraulic Braking Systems—Measurement of Braking Performance Second Edition. 23 pp.
ISO 7635-91. Road Vehicles—Air or Air over Hydraulic Braking Systems—Measurement of Braking Performance First Edition. 22 pp.
ISO 7975-85. Road Vehicles—Braking in a Turn—Open Loop Test Procedure First Edition. 14 pp.

—Buses (Vehicles)
CNS D3114-82. Method of Service Brake Road Test for Trucks and Buses (Apr)(8678).
CNS D3115-82. Method of Service Brake Road Test for Performance Requirements for Trucks and Buses (Apr)(8679).

—Buses (Vehicles)—Dynamometers
CNS D3116-82. Method of Braking Device Dynamometer Test for Trucks and Buses (Apr)(8680).

—Earthmoving Equipment
CSA CAN/CSA-M3450-M92. Braking Systems—Performance Requirements and Test Procedures—Wheeled Machines; (Gen Instr 1). 21 pp.
ISO 3450-85. Earth-Moving Machinery—Wheeled Machines—Performance Requirements and Test Procedures for Braking Systems Second Edition. 8 pp.

—Motorcycles
CNS D3060-88. Method of Brake Test for Motorcycles (May)(3106). 3 pp.

—Recreational Vehicles
ISO 7642-91. Caravans and Light Trailers—Trailers of Categories O1 and O2 with Overrun Brakes—Inertia Bench Test Methods for Brakes Second Edition. 7 pp.
ISO 7643-91. Caravans and Light Trailers—Trailers of Categories O1 and O2 with Overrun Brakes—Linear Bench Test Methods for Brake Controls Second Edition. 8 pp.

—Trucks
CNS D3114-82. Method of Service Brake Road Test for Trucks and Buses (Apr)(8678).
CNS D3115-82. Method of Service Brake Road Test for Performance Requirements for Trucks and Buses (Apr)(8679).

—Trucks—Dynamometers
CNS D3116-82. Method of Braking Device Dynamometer Test for Trucks and Buses (Apr)(8680).

Branch Circuits
See Also: Circuits; Lighting Circuits

Branch Circuits (Cont.)
—Ballasts—High Intensity Discharge Lights
CSA CAN/CSA-C22. 2 NO 74-92. Equipment for Use with Electric Discharge Lamps (Incorporating Electrical Bulletin Nos. 523F, 753, 846, 1124A, 1125A, 1325, and 1326); (Gen Instr 1). 82 pp.

—Ballasts—Sodium Vapor Lamps
CSA CAN/CSA-C22. 2 NO 74-92. Equipment for Use with Electric Discharge Lamps (Incorporating Electrical Bulletin Nos. 523F, 753, 846, 1124A, 1125A, 1325, and 1326); (Gen Instr 1). 82 pp.

—Panel Boards (Electrical)
CSA 655 Bull. Electrical Bulletin 655 October 12, 1966 to C22.2 NO 4. 4 pp.
CSA C22.2 NO 29-M1989. Panelboards and Enclosed Panelboards; (Gen Instr 1 Thru 6). 100 pp.
CSA 397 Bull. Electrical Bulletin 397 May 27, 1957 to C22.2 NO 29. 2 pp.
CSA 655 Bull. Electrical Bulletin 655 October 12, 1966 to C22.2 NO 29. 4 pp.
CSA 1120 Bull. Electrical Bulletin 1120 May 26, 1977 to C22.2 NO 29. 8 pp.
CSA 1169 Bull. Electrical Bulletin 1169 June 27, 1978 to C22.2 NO 29. 2 pp.

Branching Networks
See Also: Communication Networks; Signaling Networks

—Digital Microwave Radios—Relay Systems
BSI BS 7573: Sec 2.3-93. 1993 Methods of Measurement for Equipment Used in Digital Microwave Radio Transmission Systems Part 2: Measurements on Terrestrial Radio-Relay Systems Section 2.3: RF Branching Networks (IEC 835-2-3: 1992) (S). 11 pp.
IEC 835 Pt 2-3-92. Methods of Measurement for Equipment Used in Digital Microwave Radio Transmission Systems Part 2: Measurements on Terrestrial Radio-Relay Systems Section 3: RF Branching Networks First Edition. 20 pp.

—Radio Relay Systems
IEC 487 Pt 2-3-84. Methods of Measurement for Equipment Used in Terrestrial Radio-Relay Systems Part 2: Measurements for Sub-Systems Section Three—R.F. Branching Networks First Edition. 19 pp.

Branding
Use: Identification Systems

Brantly B2 Aircraft
See Also: Aircraft

—Antenna Positions
CAA. Brantly B2 (Approved Aerial Positions). 1 p.

Brantly 305 Aircraft
See Also: Aircraft

—Antenna Positions
CAA. Brantly 305 (Approved Aerial Positions). 1 p.

Brasov IS 28M24 Aircraft
See Also: Aircraft

—Antenna Positions
CAA. Brasov IS 28M24 (Approved Aerial Positions). 1 p.

Brass Castings
See Also: Brasses; Castings; Metal Products
CNS H3064-87. Brass Castings (May)(4336).
CNS H3068-88. High-Strength Brass Castings (Dec)(4386).
JIS H 5101-88. Brass Castings. 8 pp.
JIS H 5102-88. High Strength Brass Castings. 10 pp.

—Ingots
BSI DD 187-90. 1990 Dezincification Resistant Brass Die Castings and Ingots for Die Castings. 11 pp.
CNS H3055-88. High Strength Brass Ingots for Castings (Dec)(4082).
JIS H 2202-85. Brass Ingots for Castings. 6 pp.
JIS H 2205-85. High Strength Brass Ingots for Castings. 5 pp.

Brass Coatings (On Brasses)
See Also: Coatings

—Chromate—Passivation
MOD UK DSTAN 03-12-76. Chromate Passivation of Brass Articles Issue 1. 12 pp.

Brass Coatings (On Brasses) (Cont.)
—Chromium Coatings—Electroplated
BSI BS 3382: Parts 3 & 4-65. 1965 Amd 1 Electroplated Coatings on Threaded Components Part 3: Nickel or Nickel Plus Chromium on Steel Components Part 4: Nickel or Nickel Plus Chromium on Copper and Copper Alloy (Including Brass) Components. 33 pp.
JIS H 9121-53. Standard Operation of Nickel Plating and Chromium Plating. 9 pp.

—Nickel Coatings—Electroplated
BSI BS 3382: Parts 3 & 4-65. 1965 Amd 1 Electroplated Coatings on Threaded Components Part 3: Nickel or Nickel Plus Chromium on Steel Components Part 4: Nickel or Nickel Plus Chromium on Copper and Copper Alloy (Including Brass) Components. 33 pp.
JIS H 9121-53. Standard Operation of Nickel Plating and Chromium Plating. 9 pp.

—Silver Coatings—Electroplated
BSI BS 3382: Parts 5 & 6-67. 1967 Electroplated Coatings on Threaded Components Part 5: Tin on Copper and Copper Alloy (Including Brass) Components Part 6: Silveron Copper and Copper Alloy (Including Brass) Components. 33 pp.

—Tin Coatings—Electroplated
BSI BS 3382: Parts 5 & 6-67. 1967 Electroplated Coatings on Threaded Components Part 5: Tin on Copper and Copper Alloy (Including Brass) Components Part 6: Silveron Copper and Copper Alloy (Including Brass) Components. 33 pp.

Brass Wire
CNS H3113-85. Brass Wires (Oct)(9505).

—Aircraft
MOD UK DTD-627-44. Brass Rod or Wire for Machined Components Subject to a Riveting Operation (Not Suitable for the Manufacture of Rivets) (Reprinted September 1963). 1 p.

—Springs
BSI BS 2786-63. 1963 Brass Wire for Springs, 2/1 Brass. 12 pp.

Brasses
Scope Note: For additional listings, see also specific products made from brass See Also: Aluminum Brasses; Brass Castings; Copper Alloys; Copper Zinc Alloys; Leaded Brasses; Silicon Brasses

—Bars
CNS H3050-85. Brass Rods and Bars (Oct)(4008).

—Bars—Aircraft
BSI 3B 11-33. (WITHDRAWN) 1933 Brass Bars Suitable to Be Brazed or Silver Soldered (Superseded by BS 2874: 1986). 4 pp.

—Chemical Analysis
CNS H2008-87. Method of Chemical Analysis for Brass (May) (271).
CNS H2077-87. Method of Chemical Analysis for Special Brass (May)(11943).
DIN ENGL 17660-83. Wrought Copper Alloys; Copper-Zinc Alloys; (Brass); (Special Brass); Composition (Dec). 9 pp.
JIS H 1211-77. Methods for Chemical Chemical Analysis of Brass.
JIS H 1222-77. Methods for Chemical Analysis of Special Brass.

—Dezincification Resistance
BSI DD 187-90. 1990 Dezincification Resistant Brass Die Castings and Ingots for Die Castings. 11 pp.
ISO 6509-81. Corrosion of Metals and Alloys—Determination of Dezincification Resistance of Brass First Edition. 5 pp.

—Plates
CNS H3065-85. Brass Sheets, Plates and Strips (Oct)(4383).

—Rods
CNS H3050-85. Brass Rods and Bars (Oct)(4008).

—Rods—Aircraft
MOD UK DTD-627-44. Brass Rod or Wire for Machined Components Subject to a Riveting Operation (Not Suitable for the Manufacture of Rivets) (Reprinted September 1963). 1 p.

—Sheets
CNS H3065-85. Brass Sheets, Plates and Strips (Oct)(4383).

—Sintered—Structural
CNS H3150-89. Sintered Brass Structural Parts (Jan)(12486).

INDUSTRY STANDARDS

Brasses (Cont.)

—Stampings
BSI BS 3885-65. 1965 Tolerances for Hot Brass Stampings. 8 pp.

—Strips
CNS H3065-85. Brass Sheets, Plates and Strips (Oct)(4383).
MOD UK NES 2044-92. Specification for Formed Naval Brass Strip for Making Retaining Straps Issue 3 (12.92). 8 pp.

Brat Pans
Use: Deep Fat Fryers

Bratt Pans
Use: Deep Fat Fryers

Braze Welding
See Also: Arc Welding; Brazing; Brazing Alloys; Fusion Welding; Gas Welding; Welding

—Glossaries
BSI BS EN 24063-92. 1992 Welding, Brazing, Soldering and Braze Welding of Metals—Nomenclature of Processes and Reference Numbers for Symbolic Representation on Drawings (ISO 4063: 1990). 15 pp.
CEN EN 24063-92. Welding, Brazing, Soldering and Braze Welding of Metals—Nomenclature of Processes and Reference Numbers for Symbolic Representation on Drawings. 10 pp.
DIN ENGL EN 24063-92. Welding, Brazing, Soldering and Braze Welding of Metals; List of Process Names and Reference Numbers for Use in Technical Documentation (ISO 4063: 1990) (Sept) (Supersedes DIN ISO 4063, July 1981 Edition). 11 pp.
ISO 4063-90. Welding, Brazing, Soldering and Braze Welding of Metals—Nomenclature of Processes and Reference Numbers for Symbolic Representation on Drawings Second Edition. 13 pp.

Braze Welding Equipment
See Also: Arc Welding Equipment; Brazing Equipment; Welding Equipment

—Welding Rods—Bonding Strength
ISO 698-75. Filler Rods for Braze Welding—Determination of Conventional Bond Strength on Steel, Cast Iron and Other Metals First Edition. 6 pp.

Brazed Joints
See Also: Joints; Lap Joints; Pipe Joints
DIN ENGL 32515-91. Rating Levels for Brazed Joints (June). 6 pp.

—Copper—Pipe
JIS F 7436-89. Copper Tube 10 K Brazed Unions for Marine Use.
JIS F 7440-89. Copper Pipe 20 K Brazed Type Umions Marine Use.

—Corrosion Testing
JIS Z 3195-71. Method of Wet Corrosion Test for Brazed Joint. 6 pp.
JIS Z 3196-72. Method of Gaseous Corrosion Test for Brazed Joint (R 1979). 9 pp.

—Defects—Classification
DIN ENGL 8515 Pt 1-79. Defects in Metallic Solder Joints; Brazing Joints and High-Temperature Solder Joints; Classification; Designations; Explanatory Notes (June). 8 pp.

—Glossaries
DIN ENGL 1912 Pt 1-76. Graphical Representation of Welded, Soldered and Brazed Joints; Definitions and Terms for Welding Joints, Edges and Welds (June). 11 pp.
DIN ENGL 1912 Pt 4-81. Graphical Representation of Welded, Soldered and Brazed Joints; Concepts and Terms for Soldered and Brazed Joints and Seams (May). 3 pp.

—Shear Testing
DIN ENGL 8525 Pt 2-77. Testing of Brazing Joints; Gap Brazed Joints; Shear Test (Nov). 4 pp.
JIS Z 3192-88. Methods of Tension and Shear Tests for Brazed Joint. 13 pp.

—Symbols
DIN ENGL 1912 Pt 2-77. Graphical Representation of Welded, Soldered and Brazed Joints; Working Positions; Slope, Rotation (Sept). 4 pp.
DIN ENGL 1912 Pt 5-87. Symbolic Representation of Welded, Soldered and Brazed Joints; Symbols, Dimensioning (Modified Version of ISO 2553, 1984 Edition) (Dec). 19 pp.
ISO 2553-92. Welded, Brazed and Soldered Joints—Symbolic Representation on Drawings Third Edition; (Includes Draft Addendum 1). 55 pp.

Brazed Joints (Cont.)

—Tensile Strength
DIN ENGL 8525 Pt 1-77. Testing of Brazing Joints; Gap Brazed Joints; Tensile Test (Nov). 4 pp.
DIN ENGL 8525 Pt 3-86. Testing of Brazed Joints; Tensile Testing of High Temperature Brazed Close Joints (July). 4 pp.

Brazing
See Also: Braze Welding; Brazing Alloys; Brazing Fluxes; Filler Metal; Gas Welding; Oxyacetylene Welding; Sealing; Soldering; Welding
BSI BS 1723: Part 1-86. 1986 Brazing Part 1: Specification for Brazing. 8 pp.
BSI BS 1723: Part 2-86. 1986 Amd 1 Brazing Part 2: Guide to Brazing. 29 pp.
BSI BS 1723: Part 3-88. 1988 Methods for Non-Destructive and Destructive Testing. 53 pp.
BSI BS 1723: Part 4-88. 1988 Methods for Specifying Brazing Procedure and Operator Approval Testing. 19 pp.
ISO 5179-83. Investigation of Brazeability Using a Varying Gap Test Piece First Edition. 7 pp.
JIS Z 3621-92. Recommended Practice for Brazing. 9 pp.
JIS Z 3891-90. Standard Qualification Procedure for Silver Brazing Technique. 9 pp.
MOD UK DSTAN 03-22-83. Guide to Soldering and Brazing Issue 1. 20 pp.

—Classification
DIN ENGL 8505 Pt 2-79. Soldering and Brazing; Classification of Processes, Terms (May). 2 pp.
DIN ENGL 8505 Pt 3-83. Soldering and Brazing; Classification of Processes According to Energy Transfer Media; Description of Processes (Jan). 12 pp.

—Glossaries
BSI BS 499: Part 1-91. 1991 Welding Terms and Symbols Part 1: Glossary for Welding, Brazing and Thermal Cutting. 116 pp.
BSI BS EN 24063-92. 1992 Welding, Brazing, Soldering and Braze Welding of Metals—Nomenclature of Processes and Reference Numbers for Symbolic Representation on Drawings (ISO 4063: 1990). 15 pp.
CEN EN 24063-92. Welding, Brazing, Soldering and Braze Welding of Metals—Nomenclature of Processes and Reference Numbers for Symbolic Representation on Drawings. 10 pp.
DIN ENGL 8505 Pt 1-79. Soldering and Brazing; General; Terms (May). 3 pp.
DIN ENGL 8505 Pt 2-79. Soldering and Brazing; Classification of Processes, Terms (May). 2 pp.
DIN ENGL EN 24063-92. Welding, Brazing, Soldering and Braze Welding of Metals; List of Process Names and Reference Numbers for Use in Technical Documentation (ISO 4063: 1990) (Sept) (Supersedes DIN ISO 4063, July 1981 Edition). 11 pp.
ISO 857-90. Welding, Brazing and Soldering Processes—Vocabulary Second Edition. 33 pp.
ISO 4063-90. Welding, Brazing, Soldering and Braze Welding of Metals—Nomenclature of Processes and Reference Numbers for Symbolic Representation on Drawings Second Edition. 13 pp.
SNZ NZS/AS 2812-85. Welding, Brazing and Cutting of Metals—Glossary of Terms (This is a Joint Standard with SAA AS 2812). 98 pp.

—Pressure Pipes—Plain Flanges
DIN ENGL 2573-75. Plain Face Flanges for Brazing or Welding; Nominal Pressure 6 (Mar). 2 pp.

—Pressure Pipes—Slip-On Flanges
DIN ENGL 2576-75. Flanges, Slip-On Type for Brazing or Welding; Nominal Pressure 10 (Mar). 2 pp.

—Safety
CSA CAN/CSA-W117.2-M87. Safety in Welding, Cutting, and Allied Processes; (Gen Instr 2 Thru 3). 63 pp.

Brazing Alloys
Use For: Brazing Filler Metals *See Also:* Alloys; Aluminum Alloys; Braze Welding; Brazing; Copper Alloys; Filler Metal; Solders
BSI BS 1845-84. 1984 Filler Metals for Brazing. 14 pp.
CEN PREN 1044-93. Welding—Brazing Filler Metals. 16 pp.
SAA AS 1167.1-93. Welding and Brazing—Filler Metals—Part 1: Filler Metal for Brazing and Braze Welding (in Professional Package 36). 7 pp.

—Aluminum
DIN ENGL 8513 Pt 4-81. Brazing and Braze Weld Filler Metals; Aluminium-Base Brazing Alloys; Composition; Use; Technical Conditions of Delivery (Feb). 3 pp.

Brazing Alloys (Cont.)

—Aluminum Alloys
DIN ENGL 8513 Pt 4-81. Brazing and Braze Weld Filler Metals; Aluminium-Base Brazing Alloys; Composition; Use; Technical Conditions of Delivery (Feb). 3 pp.
JIS Z 3263-92. Aluminium Alloy Brazing Filler Metals and Brazing Sheets. 23 pp.

—Brass—Chemical Analysis
JIS Z 3902-84. Methods for Chemical Analysis of Brass Brazing Filler Metals. 46 pp.

—Copper
JIS Z 3262-86. Copper and Copper—Zinc Brazing Filler Metals. 6 pp.

—Copper Alloys
DIN ENGL 8513 Pt 1-79. Brazing and Braze Weld Filler Metals; Copper Base Brazing Alloys; Composition, Use, Technical Conditions of Delivery (Oct). 6 pp.

—Copper Phosphorus
JIS Z 3264-85. Copper Phosphorus Brazing Filler Metals. 7 pp.

—Copper Phosphorus—Chemical Analysis
JIS Z 3903-88. Methods for Chemical Analysis of Copper Phosphorus Brazing Filler Metals. 18 pp.

—Copper Zinc
JIS Z 3262-86. Copper and Copper—Zinc Brazing Filler Metals. 6 pp.

—Dental Materials
BSI BS EN 29333-92. 1992 Dental Brazing Materials (ISO 9333: 1990). 11 pp.
CEN PREN 29 333-90. Dental Brazing Materials. 3 pp.
CEN EN 29333-91. Dental Brazing Materials. 6 pp.
CSA CAN/CSA-Z349.45-93. Dental Brazing Materials (ISO 9333-1990); (Gen Instr 1). 31 pp.
ISO 9333-90. Dental Brazing Materials First Edition. 7 pp.

—Gold
JIS Z 3266-85. Gold Brazing Filler Metals. 7 pp.

—Gold—Chemical Analysis
JIS Z 3904-79. Methods for Chemical Analysis of Gold Brazing Filler Metals.

—Nickel
DIN ENGL 8513 Pt 5-83. Brazing Filler Metals; Nickel Base Filler Metals for High Temperature Brazing; Application, Composition, Technical Delivery Conditions (Feb). 4 pp.
JIS Z 3265-86. Nickel Brazing Filler Metals. 7 pp.

—Nickel—Chemical Analysis
JIS Z 3905-76. Methods for Chemical Analysis of Nickel Brazing Filler Metals.
JIS Z 3905-70. Methods for Chemical Analysis of Nickel Brazing Filler Metals. 22 pp.

—Palladium
JIS Z 3267-86. Palladium Brazing Filler Metals. 7 pp.

—Palladium—Chemical Analysis
JIS Z 3906-88. Methods for Chemical Analysis of Palladium Brazing Filler Metals. 21 pp.

—Powder—Chemical Analysis
CNS K6028-74. Method of Chemical Test for Brazing Powder (Oct)(295).

—Precious Metals
JIS Z 3268-88. Vacuum Grade Precious Brazing Filler Metals. 8 pp.

—Precious Metals—Sampling
JIS Z 3900-74. Methods for Sampling of Precious Brazing Filler Metals.

—Silver
DIN ENGL 8513 Pt 2-79. Brazing and Braze Weld Filler Metals; Silver-Bearing Brazing Alloys with Less Than 20% by Wt. Silver; Composition, Use, Technical Conditions of Delivery (Oct). 6 pp.
DIN ENGL 8513 Pt 3-86. Brazing Filler Metals; Silver Brazing Filler Metals Containing Not Less Than 20% of Silver; Composition, Application, Technical Delivery Conditions (July). 7 pp.
JIS Z 3261-85. Silver Brazing Filler Metals. 7 pp.

—Silver Alloys
DIN ENGL 8513 Pt 2-79. Brazing and Braze Weld Filler Metals; Silver-Bearing Brazing Alloys with Less Than 20% by Wt. Silver; Composition, Use, Technical Conditions of Delivery (Oct). 6 pp.

—Silver—Chemical Analysis
JIS Z 3901-88. Methods for Chemical Analysis of Silver Brazing Filler Metals. 35 pp.

Brazing Alloys (Cont.)
—Spreading Rate
JIS Z 3191-63. Method of Spreading Test for Brazing. 4 pp.

Brazing Equipment
Use For: Hard Soldering Equipment
See Also: Braze Welding Equipment; Brazing Fluxes; Gas Welding Equipment; Welding Equipment
—Hoses
CEN PREN 559-91. Rubber Hoses for Welding, Cutting and Allied Processes. 12 pp.
ISO 3821-92. Welding—Rubber Hoses for Welding, Cutting and Allied Processes Second Edition. 12 pp.

Brazing Filler Metals
Use: Brazing Alloys

Brazing Fluxes
See Also: Brazing; Brazing Equipment
DIN ENGL 8511 Pt 1-85. Fluxes for Brazing and Soldering Metallic Materials; Fluxes for Brazing (July). 2 pp.
—Classification
CEN PREN 1045-93. Welding—Fluxes for Brazing—Classification. 5 pp.

Bread
See Also: Bakery Products; Biscuits; Flours (Food); Food
CNS N5120-87. Breads (May)(3899). 3 pp.
—Rye Flour
CGSB 32.3M-88. Rye Bread. 7 pp.
—Rye Flour—Bread Making Testing
ISO 6820-85. Wheat Flour and Rye Flour—General Guidance on the Drafting of Bread-Making Tests First Edition. 6 pp.
—Wheat Flour
CGSB 32.1M-87. Bread, White Enriched White and Whole Wheat. 8 pp.
—Wheat Flour—Bread Making Testing
ISO 6820-85. Wheat Flour and Rye Flour—General Guidance on the Drafting of Bread-Making Tests First Edition. 6 pp.
—Wheat Flour—Fruit
CGSB 32.2M-88. Bread, Fruit and Raisin. 7 pp.
—Wheat Flour—Raisin
CGSB 32.2M-88. Bread, Fruit and Raisin. 7 pp.
—Wheat Flour—Rolls
CGSB 32.4M-88. Bread Rolls. 8 pp.
—Whole Wheat Flour
CGSB 32.1M-87. Bread, White Enriched White and Whole Wheat. 8 pp.

Bread Toasters
Use: Toasters

Breakdown (Electrical)
Use: Electrical Faults

Breakdown (Metal Working)
Use: Metal Working

Breakdown Diodes
Use: Avalanche Diodes

Breakdown Voltage
See Also: Dielectric Breakdown
—Electrical Insulating Liquids
IEC 897-87. Methods for the Determination of the Lightning Impulse Breakdown Voltage of Insulating Liquids First Edition. 30 pp.
—Electrical Insulation
DIN VDE 0303 Pt 2-74. Electrical Tests of Insulating Materials; Breakdown Voltage; Electric Strength (Nov). 23 pp.

Breaker Points
Use For: Contact Breakers See Also: Distributors (Electrical); Electric Contacts; Ignition Systems
—Automotive
CNS D1004-87. Configuration and Mounting Dimension of Distributor Contact Breakers for Automobiles (Oct)(5559). 1 p.
JIS D 5206-71. Shape and Mounting Dimensions of Distributor Contact Breakers for Automobiles. 3 pp.

Breakers
—Safety
CEN PREN 792-4-92. Handheld Non-Electric Power Tools—Safety Requirements—Part 4: Percussive Non-Rotary Power Tools. 8 pp.

Breakfast Cereals
See Also: Food
—Ready-to-Eat
CGSB 32.13M-88. Cereals, Breakfast, Prepared, Ready-to-Eat. 14 pp.
—Uncooked—Rolled Oats
CGSB 32.14M-87. Cereal, Rolled Oats. 9 pp.
—Uncooked—Wheat
CGSB 32.15M-87. Wheat Cereal, Uncooked. 9 pp.

Breaking Force
Use: Breaking Load

Breaking Load
Use For: Breaking Force; Breaking Strength
See Also: Tensile Testing
—Adhesive Tapes
SAA AS 1635.5.1-74. Methods of Testing Pressure Sensitive Adhesive Tape—Part 5.1: Breaking Strength. 1 p.
—Aggregates
CNS A3031-85. Method of Test for Soundness of Aggregates by Use of Sodium Sulfate or Magnesium Sulfate (Jun)(1167).
JIS A 1122-89. Method of Test for Soundness of Aggregates by Use of Sodium Sulfate. 11 pp.
—Cellular Plastics
BSI BS 4370: Part 1-88. 1988 Methods of Test for Rigid Cellular Materials Part 1: Methods 1-5. 12 pp.
—Coated Fabrics
BSI BS 3424: Part 4-82. 1982 Testing Coated Fabrics Part 4: Method 6. Method for Determination of Breaking Strength and Elongation at Break. 4 pp.
ISO 1421-77. Fabrics Coated with Rubber or Plastics—Determination of Breaking Strength and Elongation at Break First Edition. 4 pp.
—Cordage
CGSB 40-GP-1M METH 5-78. Methods of Sampling and Testing Cordage Breaking Strength. 1 p.
—Cotton Fibers
BSI BS 5116-74. 1974 Method of Test for the Determination of Breaking Tenacity of Flat Bundles of Cotton Fibres. 11 pp.
ISO 3060-74. Cotton Fibres—Determination of Breaking Tenacity of Flat Bundles First Edition. 8 pp.
—Crepe Papers
CPPA D.32P-90. Dry and Wet Tensile Breaking Strength and Stretch of Creped Papers. 2 pp.
—Ebonite
CNS K6450-80. Method of Test for Cross-Breaking Strength of Ebonite (Jan)(5140).
—Fabrics
BSI BS 2576-86. 1986 Amd 1 Method for Determination of Breaking Strength and Elongation (Strip Method) of Woven Fabrics (AMD 6199) September 29, 1989. 11 pp.
CGSB CAN/CGSB-4.2 NO.9.1-M90. Textile Test Methods Breaking Strength of Fabrics—Strip Method—Constant-Time-to-Break Principle. 11 pp.
CGSB CAN/CGSB-4.2 NO.9.2-M90. Textile Test Methods Breaking Strength of Fabrics—Grab Method—Constant-Time-to-Break Principle. 10 pp.
CGSB CAN/CGSB-4.2 NO.9.3-M90. Textile Test Methods Breaking Strength of High-Strength Fabrics—Constant-Time-to-Break Principle. 13 pp.
ISO 5081-77. Textiles—Woven Fabrics—Determination of Breaking Strength and Elongation (Strip Method) First Edition. 10 pp.
ISO 5082-82. Textiles—Woven Fabrics—Determination of Breaking Strength—Grab Method First Edition. 9 pp.
SAA AS 2001.2.3-88. Methods of Test for Textiles—Part 2: Physical Tests—Part 2.3: Determination of Breaking Force and Extension of Textile Fabrics. 4 pp.
SAA AS 2001.2.7-87. Methods of Test for Textiles—Part 2: Physical Tests—Part 2.7: Determination of Breaking Force and Extension of Yarns. 7 pp.
SAA AS 2001.2.20-86. Methods of Test for Textiles—Part 2: Physical Tests—Part 2.20: Seam Breaking Force. 3 pp.

Breaking Load (Cont.)
—Fabrics (Cont.)
SAA AS 2001.2.21-89. Methods of Test for Textiles—Part 2: Physical Tests—Part 2.21: Determination of Seam Opening Due to the Application of Force in the Transverse Direction. 5 pp.
SAA AS 2001.2.22-86. Methods of Test for Textiles—Part 2: Physical Tests—Part 2.22: Determination of Yarn Slippage in Woven Fabrics at a Standard Stitched Seam. 4 pp.
—Fibers
ISO 5079-77. Textiles—Man-Made Fibres—Determination of Breaking Strength and Elongation of Individual Fibres First Edition. 6 pp.
—Fishing Nets
BSI BS 4650-70. 1970 Amd 1 Determination of the Mesh Breaking Load of Netting for Fishing. 12 pp.
BSI BS 4674-71. 1971 Amd 1 Determination of the Breaking Load and Knot Breaking Load of Netting Yarns for Fishing Nets. 12 pp.
ISO 1805-73. Fishing Nets—Determination of Breaking Load and Knot Breaking Load of Netting Yarns First Edition. 6 pp.
ISO 1806-73. Fishing Nets—Determination of Mesh Breaking Load of Netting First Edition. 6 pp.
—Geotextiles
CGSB CAN/CGSB-148. 1 NO.7.3-92. Methods of Testing Geotextiles and Geomembranes Grab Tensile Test for Geotextiles. 9 pp.
—Glass Fabrics
BSI BS 3396: Part 2-87. 1987 Woven Glass Fibre Fabrics for Plastics Reinforcement Part 2: Desized Fabrics. 4 pp.
BSI BS 3396: Part 3-87. 1987 Woven Glass Fibre Fabrics for Plastics Reinforcement Part 3: Finished Fabrics for Use with Polyester Resin Systems. 8 pp.
ISO 4606-79. Textile Glass—Woven Fabric—Determination of Tensile Breaking Force and Breaking Elongation by the Strip Method First Edition. 7 pp.
—Intrauterine Devices
ISO 7857 Pt 1-83. Intra-Uterine Devices—Part 1: Determination of Breaking Force First Edition. 3 pp.
—Laces (Footwear)
BSI BS 5131: Sec 3.7-91. 1991 Methods of Test for Footwear and Footwear Materials Part 3: Uppers, Textiles and Threads Section 3.7: Breaking Strength of Shoe Laces. 4 pp.
—Nonwoven Fabrics
CGSB CAN/CGSB-4.2 NO.9.6-93. Textile Test Methods Breaking Strength of Nonwoven Textiles. 8 pp.
—Paperboard
CNS P3004-90. Method of Test for Tensile Breaking Strength and Elongation of Paper and Paperboard (Jul)(1354). 5 pp.
CPPA D.6H-84. Tensile Breaking Strength of Paper and Paperboard (Using Constant Rate of Loading Apparatus). 2 pp.
—Papers
CNS P3004-90. Method of Test for Tensile Breaking Strength and Elongation of Paper and Paperboard (Jul)(1354). 5 pp.
CPPA D.6H-84. Tensile Breaking Strength of Paper and Paperboard (Using Constant Rate of Loading Apparatus). 2 pp.
—Plastic Sheets
BSI BS 2782:Pt1: METH 150C-83. 1983 Methods of Testing Plastics Part 1: Thermal Properties Method 150C: Determination of Low Temperature Extensibility of Flexible Polyvinyl Chloride Sheet. 4 pp.
DIN ENGL 53372-70. Testing of Plastic Films; Determination of Low-Temperature Breakage of Non-Rigid Polyvinyl Chloride (PVC) Films (Dec). 2 pp.
—Seams
CGSB CAN/CGSB-4.2 NO.32.2-M89. Textile Test Methods Breaking Strength of Seams in Woven Fabrics. 9 pp.
—Skis—Alpine
ISO 6265-92. Alpine Skis—Determination of Deformation Load and Breaking Load Second Edition. 7 pp.
—Skis—Cross Country
ISO 7797-85. Cross-Country Skis—Determination of Breaking Load and Deflection at Break with Quasistatic Load First Edition. 5 pp.

Breaking Load (Cont.)

—Wire Rope
BSI BS 302: Part 8-89. 1989 Stranded Steel Wire Ropes Part 8: Higher Breaking Load Ropes. 7 pp.
ISO 3108-74. Steel Wire Ropes for General Purposes—Determination of Actual Breaking Load First Edition. 3 pp.
ISO 10092-90. High Breaking Load Steel Wire Ropes—Specifications First Edition. 6 pp.
SNZ NZS/BS 302: Part 8-89. Stranded Steel Wire Ropes Part 8: Specification for Higher Breaking Load Ropes. 4 pp.

—Yarns
BSI BS 1932: Part 1-89. 1989 Amd 1 Testing the Strength of Yarns and Threads from Packages Part 1: Methods for Determination of Breaking Strength and Breaking Extension (AMD 6339) June 28, 1991. 9 pp.
BSI BS 1932: Part 2-89. 1989 Methods of Testing the Strength of Yarns from Packages Part 2: Determination of Knot Strength and Loop Strength. 7 pp.
BSI BS 6372-83. 1983 Determination of Breaking Strength of Yarn From Packages: Skein Method. 8 pp.
CGSB CAN/CGSB-4.2 NO.9.4-M91. Textile Test Methods Breaking Strength of Yarns—Single Strand Method. 8 pp.
CGSB CAN/CGSB-4.2 NO.9.5-M89. Textile Test Methods Breaking Strength of Yarns—Skein Method. 10 pp.
ISO 2062-72. Textiles—Yarn from Packages—Method for Determination of Breaking Load and Elongation at the Breaking Load of Single Strands—(CRL, CRE and CRT Testers) First Edition. 11 pp.
ISO 6939-88. Textiles—Yarns from Packages—Method of Test for Breaking Strength of Yarn by the Skein Method Second Edition. 7 pp.

—Yarns—Glass Fabrics
ISO 3341-84. Textile Glass—Yarns—Determination of Breaking Force and Breaking Elongation Second Edition. 7 pp.

Breaking Strength
Use: Breaking Load

Breakwaters
See Also: Shore Protection

—Construction
BSI BS 6349: Part 7-91. 1991 Maritime Structures Part 7: Guide to the Design and Construction of Breakwaters. 92 pp.

—Design
BSI BS 6349: Part 7-91. 1991 Maritime Structures Part 7: Guide to the Design and Construction of Breakwaters. 92 pp.

Breast Drills
See Also: Drills; Hand Tools
BSI BS 2556-73. 1973 Amd 2 Hand and Breast Drills. 19 pp.
BSI BS 2556-03. 1973 Amd 3 Hand and Breast Drills (AMD 7729) July 15, 1993 (F). 20 pp.
CNS B3264-77. Breast Drills (May)(4099). 2 pp.

Breathing Apparatus
Use For: Escape Breathing Apparatus; Respiratory Protective Equipment *See Also:* Diving Equipment; Face Shields; Fire Protection; Heat/Moisture Exchangers; Helmets; Hoses; Mouthpieces; Oxygen Masks; Oxygen Supply Equipment; Protective Clothing; Protective Masks; Respirators; Self Rescuers; Survival Kits

BSI BS 3806-64. (WITHDRAWN) 1964 Amd 2 Breathing Machines for Medical Use (PD 6056) March 3, 1967. 36 pp.
BSI BS 4275-74. 1974 Recommendations for the Selection, Use and Maintenance of Respiratory Protective Equipment. 25 pp.
BSI BS 4667: Part 1-74. 1974 Breathing Apparatus Part 1: Closed-Circuit Breathing Apparatus. 16 pp.
BSI BS 4667: Part 2-74. 1974 Amd 1 Breathing Apparatus Part 2: Open-Circuit Breathing Apparatus. 14 pp.
BSI BS 4667: Part 3-74. 1974 Amd 1 Breathing Apparatus Part 3: Fresh Air Hose and Compressed Air Line Breathing Apparatus. 6 pp.
BSI BS 4667: Part 4-89. 1989 Breathing Apparatus Part 4: Specification for Open-Circuit Escape Breathing Apparatus. 19 pp.
BSI BS 4667: Part 5-90. 1990 Breathing Apparatus Part 5: Closed Circuit Escape Breathing Apparatus. 22 pp.
BSI BS 7004-88. (WITHDRAWN) 1988 Respiratory Protective Devices: Self-Contained Open-Circuit Compressed Air Breathing Apparatus (Superseded by BS EN 137: 1993). 12 pp.
BSI BS EN 137-93. 1993 Respiratory Protective Devices: Self-Contained Open-Circuit Compressed Air Breathing Apparatus (Supersedes BS 7004: 1988) (N). 21 pp.
CEN EN 137-86. Respiratory Protective Devices: Self-Contained Open-Circuit Compressed Air Breathing Apparatus; Requirements, Testing, Marking. 11 pp.
CEN PREN 137-90. Respiratory Protective Devices—Self Contained Open-Circuit Compressed Air Breathing Apparatus—Requirements, Testing, Marking. 35 pp.
CEN EN 137-93. Respiratory Protective Devices—Self-Contained Open-Circuit Compressed Air Breathing Apparatus—Requirements, Testing, Marking. 16 pp.
CEN PREN 138-89. Respiratory Protective Devices; Fresh Air Hose Breathing Apparatus: Requirements, Testing, Marking. 19 pp.
CEN PREN 139-89. Respiratory Protective Devices; Compressed Air Line Breathing Apparatus: Requirements, Testing, Marking. 19 pp.
CEN PREN 269-89. Respiratory Protective Devices; Power Assisted Air Hose Breathing Apparatus Incorporating a Hood: Requirements, Testing Marking. 27 pp.
CEN PREN 270-89. Respiratory Protective Devices: Compressed Air Line Breathing Apparatus Incorporating a Hood; Requirements, Testing, Marking. 26 pp.
CEN PREN 400-90. Respiratory Protective Devices-Self Contained Closed-Circuit Breathing Apparatus for Self-Rescue-Compressed Oxygen Self-Rescuer-Requirements, Testing, Marking. 40 pp.
CEN EN 400-92. Respiratory Protective Devices for Self-Rescue Self-Contained Closed-Circuit Breathing Apparatus Compressed Oxygen Escape Apparatus Requirements, Testing, Marking. 21 pp.
CEN PREN 401-90. Respiratory Protective Devices-Self Contained Closed-Circuit for Breathing Apparatus for Self-Rescue-Chemical Oxygen Self-Rescuer-Requirements, Testing Marking. 36 pp.
CEN EN 401-92. Respiratory Protective Devices for Self-Rescue Self-Contained Closed-Circuit Breathing Apparatus Chemical Oxygen (KO2) Escape Apparatus Requirements, Testing, Marking. 20 pp.
CEN PREN 1061-93. Respiratory Protective Devices for Escape—Self-Contained Closed-Circuit Breathing Apparatus—Chemical Oxygen (NaC1O3) Escape Apparatus—Requirements, Testing, Marking. 39 pp.
CEN PREN 1146-93. Respiratory Protective Devices for Escape—Self-Contained Open-Circuit Compressed Air Breathing Apparatus Incorporating a Hood (Compressed Air Escape Apparatus with Hood)—Requirements, Testing, Marking. 29 pp.
CSA CAN3-Z180.1-M85. Compressed Breathing Air and Systems; (Gen Instr 1). 34 pp.
JIS M 7600-87. Compressed Oxygen Open Circuit Self-Contained Breathing Apparatus. 13 pp.
JIS M 7601-87. Compressed Oxygen Closed Circuit Self-Contained Breathing Apparatus. 18 pp.
JIS T 8155-83. Self-Contained Breathing Apparatus (Compressed Air Type). 14 pp.
JIS T 8156-88. Oxygen-Generating Closed-Circuit Breathing Apparatus. 12 pp.
MOD UK DSTAN 42-11-01. Respirators and Breathing Apparatus Issue 2; Amendment 1. 14 pp.
SAA AS 1715-91. Selection, Use and Maintenance of Respiratory Protective Devices (in Professional Packages 32, 36, 47) (This is a Joint Standard with SANZ NZS 1715:1991). 62 pp.
SAA AS 1716-91. Respiratory Protective Devices (in Professional Packages 32, 36, 47) (This is a Joint Standard with SANZ NZS 1716:1991). 65 pp.
SNZ NZS/AS 1715-91. Selection Use and Maintenance of Respiratory Protective Devices. 62 pp.
SNZ NZS/AS 1716-91. Respiratory Protective Devices. 65 pp.

—Abrasive Blasting
CEN PREN 271-89. Respiratory Protective Devices: Compressed Air Line Breathing Apparatus for Use in Abrasive Blasting Operations; Requirements, Testing, Marking. 8 pp.

—Alarm Signals
BSI BS 7618: Part 1-92. 1992 Alarm Signals for Anaesthesia and Respiratory Care Part 1: Specification for Visual Alarm Signals (ISO 9703-1: 1992). 8 pp.
ISO 9703 Pt 1-92. Anaesthesia and Respiratory Care Alarm Signals—Part 1: Visual Alarm Signals First Edition. 7 pp.

—Classification
BSI BS EN 133-91. 1991 Respiratory Protective Devices—Classification. 11 pp.
CEN EN 133-86. Respiratory Protective Devices: Classification. 8 pp.
CEN EN 133-90. Respiratory Protective Devices-Classification. 6 pp.

—Compressed Oxygen
BSI BS EN 145-2-93. 1993 Respiratory Protective Devices. Self-Contained Closed-Circuit Compressed Oxygen Breathing Apparatus Part 2: Specification for Apparatus for Special Use (N). 9 pp.
BSI BS EN 250-93. 1993 Respiratory Equipment—Open-Circuit Self-Contained Compressed Air Diving Apparatus—Requirements, Testing, Marking (N). 20 pp.
BSI BS EN 400-93. 1993 Respiratory Protective Devices for Self-Rescue—Self-Contained Closed-Circuit Breathing Apparatus—Compressed Oxygen Escape Apparatus—Requirements, Testing, Marking (N). 27 pp.
BSI BS EN 401-93. 1993 Respiratory Protective Devices for Self-Rescue—Self-Contained Closed-Circuit Breathing Apparatus—Chemical Oxygen (KO2) Escape Apparatus—Requirements, Testing, Marking (N). 26 pp.
CEN PREN 145 (Part 2)-91. Respiratory Protective Devices—Self-Contained Closed-Circuit Compresses Oxygen Breathing Apparatus for Special Use—Safety Requirements, Testing Marking. 8 pp.
CEN PREN 250-86. Breathing Apparatus, Open Circuit, Self-Contained, Compressed—Air Diving Apparatus; Safety Requirements, Testing, Marking. 16 pp.
CEN EN 250-93. Respiratory Equipment—Open-Circuit, Self-Contained, Compressed Air Diving Apparatus—Requirements, Testing, Marking. 15 pp.
CEN PREN 400-90. Respiratory Protective Devices-Self Contained Closed-Circuit Breathing Apparatus for Self-Rescue-Compressed Oxygen Self-Rescuer-Requirements, Testing, Marking. 40 pp.
CEN EN 400-92. Respiratory Protective Devices for Self-Rescue Self-Contained Closed-Circuit Breathing Apparatus Compressed Oxygen Escape Apparatus Requirements, Testing, Marking. 21 pp.
CEN EN 400-93. Respiratory Protective Devices for Self-Rescue Self-Contained Closed-Circuit Breathing Apparatus Compressed Oxygen Escape Apparatus Requirements, Testing, Marking. 20 pp.
CEN PREN 401-90. Respiratory Protective Devices-Self Contained Closed-Circuit for Breathing Apparatus for Self-Rescue-Chemical Oxygen Self-Rescuer-Requirements, Testing Marking. 36 pp.
CEN EN 401-92. Respiratory Protective Devices for Self-Rescue Self-Contained Closed-Circuit Breathing Apparatus Chemical Oxygen (KO2) Escape Apparatus Requirements, Testing, Marking. 20 pp.
CEN EN 401-93. Respiratory Protective Devices for Self-Rescue Self-Contained Closed-Circuit Breathing Apparatus Chemical Oxygen (KO2) Escape Apparatus Requirements, Testing, Marking. 20 pp.
CNS Z2026-80. Self-Contained Compressed Air Breathing Apparatus (Dec)(6860).
CNS Z2027-80. Self-Contained Compressed Oxygen Open Circuit Breathing Apparatus (Dec)(6861).
CNS Z2028-80. Self-Contained Compressed Oxygen Closed Circuit Breathing Apparatus (Dec)(6862).

—Diving Equipment
BSI BS 4001: Part 1-81. 1981 Recommendations for the Care and Maintenance of Underwater Part 1: Breathing Apparatus Compressed Air Open Circuit Type. 8 pp.
BSI BS 4001: Part 2-67. 1967 Amd 1 Recommendations for the Care and Maintenance of Underwater Breathing Apparatus Part 2: Standard Diving Equipment. 38 pp.

—Facepieces—Screw Threads
BSI BS 7156: Part 1-90. 1990 Respiratory Protective Devices: Threads for Facepieces Part 1: Standard Thread Connection. 10 pp.
BSI BS 7156: Part 1-01. 1990 Amd 1 Respiratory Protective Devices: Threads for Facepieces Part 1: Specification for Standard Thread Connection (AMD 7406) January 15, 1993. 11 pp.
BSI BS 7156: Part 2-90. 1990 Respiratory Protective Devices: Threads for Facepieces Part 2: Centre Thread Connection. 8 pp.
BSI BS 7156: Part 2-01. 1990 Amd 1 Respiratory Protective Devices: Threads for Facepieces Part 2: Specification for Centre Thread Connection (AMD 7407) January 15, 1993. 9 pp.
BSI BS EN 148-3-92. 1992 Respiratory Protective Devices: Threads for Facepieces Part 3: Specification for Thread Connection M 45 x 3. 10 pp.
CEN EN 148 (Part 1)-87. Respiratory Protective Devices; Threads for Facepieces; Standard Thread Connection. 6 pp.
CEN EN 148-1-92. CORRIGENDUM Respiratory Protective Devices; Threads for Facepieces; Standard Thread Connection. 1 p.
CEN EN 148 (Part 2)-87. Respiratory Protective Devices; Threads for Facepieces; Centre Thread Connection. 4 pp.
CEN EN 148 (Part 2)/A1-87. AMD 1 Respiratory Protective Devices; Threads for Facepieces; Centre Thread Connection. 4 pp.

INTERNATIONAL AND NON-U.S. NATIONAL STANDARDS
SUBJECT INDEX
Breathing

Breathing Apparatus *(Cont.)*

—Facepieces—Screw Threads *(Cont.)*

CEN EN 148-2-92. CORRIGENDUM Respiratory Protective Devices; Threads for Facepieces; Centre Thread Connection. 1 p.

CEN EN 148-3-92. Respiratory Protective Devices—Threads for Facepieces—Thread Connection M 45 x 3. 4 pp.

CEN PREN 148 (Part 10)-90. Respiratory Protective Devices—Threads for Facepieces—Thread Connection M 45X3. 4 pp.

—Filters

BSI BS EN 141-91. 1991 Gas Filters and Combined Filters Used in Respiratory Protective Equipment. 13 pp.

BSI BS EN 143-91. 1991 Particle Filters Used in Respiratory Protective Equipment (N). 30 pp.

BSI BS EN 371-92. 1992 AX Gas Filters and Combined Filters Against Low Boiling Organic Compounds Used in Respiratory Protective Equipment. 11 pp.

BSI BS EN 372-92. 1992 SX Gas Filters and Combined Filters Against Specific Named Compounds Used in Respiratory Protective Equipment. 11 pp.

BSI BS EN 404-93. 1993 Respiratory Protective Devices for Self-Rescue: Filter Self-Rescuer (N). 28 pp.

CEN EN 141-90. Respiratory Protective Devices—Gas Filters and Combined Filters—Requirements, Testing, Marking. 13 pp.

CEN EN 143-90. Respiratory Protective Devices—Particle Filters—Requirements, Testing, Marking. 34 pp.

CEN PREN 371-90. Respiratory Protective Devices: AX Gas Filters and Combined Filters Against Low Boiling Organic Compounds: Requirements, Testing, Marking. 10 pp.

CEN EN 371-92. Respiratory Protective Devices—AX Gas Filters and Combined Filters Against Low Boiling Organic Compounds—Requirements, Testing, Marking. 14 pp.

CEN PREN 372-90. Respiratory Protective Devices: SX Gas Filters and Combined Filters Against Specific Named Compounds; Requirements, Testing, Marking. 11 pp.

CEN EN 372-92. Respiratory Protective Devices—SX Gas Filters and Combined Filters Against Specific Named Compounds—Requirements, Testing, Marking. 14 pp.

CEN PREN 404-90. Respiratory Protective Devices for Self-Rescue-Filter Self-Rescuer-Requirements, Testing, Marking. 30 pp.

—Filters—Identification Systems

BSI BS EN 141-91. 1991 Gas Filters and Combined Filters Used in Respiratory Protective Equipment. 13 pp.

BSI BS EN 143-91. 1991 Particle Filters Used in Respiratory Protective Equipment (N). 30 pp.

BSI BS EN 371-92. 1992 AX Gas Filters and Combined Filters Against Low Boiling Organic Compounds Used in Respiratory Protective Equipment. 11 pp.

BSI BS EN 372-92. 1992 SX Gas Filters and Combined Filters Against Specific Named Compounds Used in Respiratory Protective Equipment. 11 pp.

BSI BS EN 403-93. 1993 Filtering Respiratory Protective Devices with Hood for Self-Rescue From Fire (N). 26 pp.

BSI BS EN 404-93. 1993 Respiratory Protective Devices for Self-Rescue: Filter Self-Rescuer (N). 28 pp.

CEN EN 141-90. Respiratory Protective Devices—Gas Filters and Combined Filters—Requirements, Testing, Marking. 13 pp.

CEN EN 143-90. Respiratory Protective Devices—Particle Filters—Requirements, Testing, Marking. 34 pp.

CEN PREN 149-89. Respiratory Protective Devices; Filtering Jalf Masks to Protect Against Particles; Requirements, Testing, Marking. 49 pp.

CEN PREN 371-90. Respiratory Protective Devices: AX Gas Filters and Combined Filters Against Low Boiling Organic Compounds: Requirements, Testing, Marking. 10 pp.

CEN EN 371-92. Respiratory Protective Devices—AX Gas Filters and Combined Filters Against Low Boiling Organic Compounds—Requirements, Testing, Marking. 14 pp.

CEN PREN 372-90. Respiratory Protective Devices: SX Gas Filters and Combined Filters Against Specific Named Compounds; Requirements, Testing, Marking. 11 pp.

CEN EN 372-92. Respiratory Protective Devices—SX Gas Filters and Combined Filters Against Specific Named Compounds—Requirements, Testing, Marking. 14 pp.

CEN PREN 403-90. Respiratory Protective Devices: Filtering Devices with Hood for Self-Rescue from Fire-Requirements, Testing, Marking. 11 pp.

Breathing Apparatus *(Cont.)*

—Filters—Identification Systems *(Cont.)*

CEN PREN 404-90. Respiratory Protective Devices for Self-Rescue-Filter Self-Rescuer-Requirements, Testing, Marking. 30 pp.

—Fire Fighting Equipment—Nuclear, Biological, and Chemical Warfare

NATO STANAG 7049 ED 1 AMD 0-93. Personal Protective Equipment Requirements Together with Standard Operating Procedures for CFR Operations in an NBC Environment. 6 pp.

—Flight Crews

BSI N5-92. 1992 Crew Portable Protective Breathing Equipment for Use During Aircraft Emergencies. 11 pp.

—Full Mask

BSI BS EN 402-93. 1993 Respiratory Protective Devices for Escape: Self-Contained Open-Circuit Compressed Air Breathing Apparatus with Full Face Mask or Mouthpiece Assembly (N). 21 pp.

CEN PREN 147-84. Respiratory Protective Equipment Power Assisted Particle Filtering Devices Incorporating Full Face Masks-Half Masks and Quarter Masks. 16 pp.

CEN EN 147-91. Respiratory Protective Devices—Assisted Particle Filtering Devices Full Face Incorporating Full Face Masks, Half Masks—Requirements, Testing, Marking. 14 pp.

CEN PREN 402-90. Respiratory Protective Devices-for Self-Rescue-Self-Contained Open-Circuit Compressed Air Breathing Apparatus with Full Face Mask or Mouthpiece Assembly. 10 pp.

CEN EN 402-93. Respiratory Protective Devices for Escape—Self-Contained Open-Circuit Compressed Air Breathing Apparatus with Full Face Mask or Mouthpiece Assembly—Requirements, Testing, Marking. 32 pp.

—Full Mask—Identification Systems

BSI BS 7355-90. 1990 Full Face Masks for Respiratory Protective Devices. 26 pp.

BSI BS EN 136: Part 10-92. 1992 Parts for Full Face Masks for Respiratory Protective Devices Part 10: Specification for Full Face Masks for Special Use. 11 pp.

BSI BS EN 147-92. 1992 Respiratory Protective Devices Specification for Power Assisted Particle Filtering Devices Incorporating Full Face Masks, Half Masks or Quarter Masks. 19 pp.

BSI BS EN 402-93. 1993 Respiratory Protective Devices for Escape: Self-Contained Open-Circuit Compressed Air Breathing Apparatus with Full Face Mask or Mouthpiece Assembly (N). 21 pp.

CEN EN 136-89. Respiratory Protective Devices: Full Face Tasks; Requirements, Testing, Making. 51 pp.

CEN PREN 136-90. Respiratory Equipment—Parts for Respiratory Protective Devices—Full Face Masks for Special Use—Requirements, Testing, Masking. 9 pp.

CEN EN 136-10-92. Respiratory Protective Devices—Full Face Masks for Special Use—Requirements, Testing, Marking. 6 pp.

CEN EN 136-10-92. Respiratory Protective Devices—Full Face Masks for Special Use—Requirements, Testing, Marking. 12 pp.

CEN EN 147-91. Respiratory Protective Devices—Assisted Particle Filtering Devices Full Face Incorporating Full Face Masks, Half Masks—Requirements, Testing, Marking. 29 pp.

CEN PREN 402-90. Respiratory Protective Devices-for Self-Rescue-Self-Contained Open-Circuit Compressed Air Breathing Apparatus with Full Face Mask or Mouthpiece Assembly. 10 pp.

CEN EN 402-93. Respiratory Protective Devices for Escape—Self-Contained Open-Circuit Compressed Air Breathing Apparatus with Full Face Mask or Mouthpiece Assembly—Requirements, Testing, Marking. 32 pp.

—Gas Cylinders—Valve Fittings

BSI BS 341: Part 2-63. 1963 Amd 1 Valve Fittings for Compressed Gas Cylinders Part 2: Valves with Taper Stems for Use with Breathing Apparatus (Excluding Medical Gas Cylinders to BS 1319). 29 pp.

BSI BS EN 144: Part 1-91. 1991 Respiratory Protective Devices—Gas Cylinder Valves Part 1: Specification for Thread Connection for Insert Connector (N). 12 pp.

CEN EN 144-01-91. Respiratory Protective Devices—Gas Cylinder Valves—Thread Connection for Insert Connector. 9 pp.

—Gas Cylinders—Valves

CEN PREN 144 (Part 1)-88. Respiratory Protective Devices; Gas Cylinder Valves; Thread Connection for Insert Connector. 5 pp.

CEN PREN 144-2-90. Respiratory Protective Devices; Gas Cylinder Valves; Thread Connection for Side Connector. 4 pp.

Breathing Apparatus *(Cont.)*

—Glossaries

BSI BS EN 132-91. 1991 Respiratory Protective Devices—Definitions. 11 pp.

BSI BS EN 134-91. 1991 Respiratory Protective Devices—Nomenclature of Components. 25 pp.

BSI BS EN 135-91. 1991 Respiratory Protective Devices—List of Equivalent Terms. 20 pp.

CEN EN 132-86. Respiratory Protective Devices: Definitions. 7 pp.

CEN EN 132-90. Respiratory Protective Devices-Definitions. 11 pp.

CEN EN 134-86. Respiratory Protective Devices: Nomenclature of Components. 23 pp.

CEN EN 134-90. Respiratory Protective Devices—Nomenclature of Components. 20 pp.

CEN EN134-90. Respiratory Protective Devices-Nomenclature of Components. 20 pp.

CEN EN 135-86. Respiratory Protective Devices List of Equivalent Terms. 14 pp.

CEN EN 135-90. Respiratory Protective Devices—List of Equivalent Terms. 18 pp.

JIS T 8001-85. Glossary of Terms Relating to Respiratory Protective Equipments.

—Half Mask

BSI BS EN 149-92. 1992 Filtering Half Masks to Protect Against Particles. 27 pp.

CEN EN 140-89. Respiratory Protective Devices; Half-Masks and Quarter-Masks; Requirements, Testing, Making. 43 pp.

CEN EN 140/A1-90. AMD 1 Respiratory Protective Devices; Half-Masks and Quarter-Masks; Requirements, Testing, Marking. 4 pp.

CEN EN 140/A1-92. AMD A1 Respiratory Protective Devices; Half-Masks and Quarter-Masks; Requirements, Testing, Marking. 3 pp.

CEN PREN 147-84. Respiratory Protective Equipment Power Assisted Particle Filtering Devices Incorporating Full Face Masks-Half Masks and Quarter Masks. 16 pp.

CEN EN 147-91. Respiratory Protective Devices—Assisted Particle Filtering Devices Full Face Incorporating Full Face Masks, Half Masks—Requirements, Testing, Marking. 14 pp.

CEN EN 149-91. Respiratory Protective Devices—Filtering Half Masks to Protect Against Particles—Requirements Testing Marking. 49 pp.

—Half Mask—Identification Systems

BSI BS 7356-90. 1990 Half Masks and Quarter Masks for Respiratory Protective Devices. 23 pp.

BSI BS 7356-01. 1990 Amd 1 Half Masks and Quarter Masks for Respiratory Protective Devices (AMD 7113) November 15, 1992. 29 pp.

BSI BS EN 147-92. 1992 Respiratory Protective Devices Specification for Power Assisted Particle Filtering Devices Incorporating Full Face Masks, Half Masks or Quarter Masks. 19 pp.

BSI BS EN 149-92. 1992 Filtering Half Masks to Protect Against Particles. 27 pp.

BSI BS EN 405-93. 1993 Respiratory Protective Devices Valved Filtering Half Masks to Protect Against Gases or Gases and Particles (N). 26 pp.

CEN EN 147-91. Respiratory Protective Devices—Assisted Particle Filtering Devices Full Face Incorporating Full Face Masks, Half Masks—Requirements, Testing, Marking. 29 pp.

CEN EN 149-91. Respiratory Protective Devices—Filtering Half Masks to Protect Against Particles—Requirements Testing Marking. 49 pp.

CEN PREN 405-90. Respiratory Protective Devices-Filtering Half Masks to Protect Against Gases or Gases and Particles-Requirements, Testing, Marking. 48 pp.

CEN EN 405-92. Respiratory Protective Devices—Valved Filtering Half Masks to Protect Against Gases or Gases and Particles—Requirements, Testing, Marking. 20 pp.

—Heat/Moisture Exchangers

ISO 9360-92. Anaesthetic and Respiratory Equipment—Heat and Moisture Exchangers for Use in Humidifying Respired Gases in Humans First Edition. 14 pp.

—Heat/Moisture Exchangers—Safety

ISO 9360-92. Anaesthetic and Respiratory Equipment—Heat and Moisture Exchangers for Use in Humidifying Respired Gases in Humans First Edition. 14 pp.

—Hoses

MOD UK DEF-1282-65. Air Hose 1 Inch Internal Diameter for Remote Breathing Apparatus. 3 pp.

—Identification Systems

BSI BS 7170-90. 1990 Respiratory Protective Devices: Self-Contained Closed-Circuit Compressed Oxygen Breathing Apparatus. 19 pp.

BSI BS EN 137-93. 1993 Respiratory Protective Devices: Self-Contained Open-Circuit Compressed Air Breathing Apparatus (Supersedes BS 7004: 1988) (N). 21 pp.

INDUSTRY STANDARDS

Breathing Apparatus (Cont.)

—Identification Systems (Cont.)
CEN EN 137-86. Respiratory Protective Devices: Self-Contained Open-Circuit Compressed Air Breathing Apparatus; Requirements, Testing, Marking. 11 pp.
CEN PREN 137-90. Respiratory Protective Devices—Self Contained Open-Circuit Compressed Air Breathing Apparatus—Requirements, Testing, Marking. 35 pp.
CEN EN 137-93. Respiratory Protective Devices—Self-Contained Open-Circuit Compressed Air Breathing Apparatus —Requirements, Testing, Marking. 16 pp.
CEN PREN 138-89. Respiratory Protective Devices; Fresh Air Hose Breathing Apparatus: Requirements, Testing, Marking. 19 pp.
CEN PREN 139-89. Respiratory Protective Devices: Compressed Air Line Breathing Apparatus: Requirements, Testing, Marking. 19 pp.
CEN EN 145/A1-88. Respiratory Protective Devices; Self-Contained Closed—Circuit Breathing Apparatus, Compressed Oxygen Type,—Requirements, Testing, Marking. 37 pp.
CEN PREN 269-89. Respiratory Protective Devices; Power Assisted Air Hose Breathing Apparatus Incorporating a Hood: Requirements, Testing Marking. 27 pp.
CEN PREN 270-89. Respiratory Protective Devices: Compressed Air Line Breathing Apparatus Incorporating a Hood; Requirements, Testing, Marking. 26 pp.
CEN PREN 1061-93. Respiratory Protective Devices for Escape-Self-Contained Closed-Circuit Breathing Apparatus—Chemical Oxygen (NaC103) Escape Apparatus—Requirements, Testing, Marking. 39 pp.
CEN PREN 1146-93. Respiratory Protective Devices for Escape—Self-Contained Open-Circuit Compressed Air Breathing Apparatus Incorporating a Hood (Compressed Air Escape Apparatus with Hood)—Requirements, Testing, Marking. 29 pp.
DIN ENGL 23320 Pt 5-88. Flameproof Clothing for the Mining Industry; Protective Hoods (May). 3 pp.

—Miners'
JIS M 7623-87. Escape Breathing Apparatus for Miners. 10 pp.

—Mouthpieces
BSI BS 7309-90. 1990 Mouthpiece Assemblies for Respiratory Protective Devices. 12 pp.
BSI BS EN 402-93. 1993 Respiratory Protective Devices for Escape: Self-Contained Open-Circuit Compressed Air Breathing Apparatus with Full Face Mask or Mouthpiece Assembly (N). 21 pp.
CEN EN 142-89. Respiratory Protective Devices; Mouthpiece Assemblies; Requirements, Testing, Marking. 15 pp.
CEN EN 142/A1-89. AMD 1 Respiratory Protective Devices; Mouthpiece Assemblies; Requirements, Testing, Marking. 2 pp.
CEN PREN 402-90. Respiratory Protective Devices-for Self-Rescue-Self-Contained Open-Circuit Compressed Air Breathing Apparatus with Full Face Mask or Mouthpiece Assembly. 10 pp.
CEN EN 402-93. Respiratory Protective Devices for Escape—Self-Contained Open-Circuit Compressed Air Breathing Apparatus with Full Face Mask or Mouthpiece Assembly—Requirements, Testing, Marking. 32 pp.

—Mouthpieces—Identification Systems
BSI BS EN 402-93. 1993 Respiratory Protective Devices for Escape: Self-Contained Open-Circuit Compressed Air Breathing Apparatus with Full Face Mask or Mouthpiece Assembly (N). 21 pp.
CEN PREN 402-90. Respiratory Protective Devices-for Self-Rescue-Self-Contained Open-Circuit Compressed Air Breathing Apparatus with Full Face Mask or Mouthpiece Assembly. 10 pp.
CEN EN 402-93. Respiratory Protective Devices for Escape—Self-Contained Open-Circuit Compressed Air Breathing Apparatus with Full Face Mask or Mouthpiece Assembly—Requirements, Testing, Marking. 32 pp.

—Particle Filters—Identification Systems
BSI BS EN 146-92. 1992 Respiratory Protective Devices Specification for Powered Particle Filtering Devices Incorporating Helmets or Hoods. 22 pp.
CEN PREN 146-89. Respiratory Protective Devices Powered Particle Filtering Devices Incorporating Helmets and Hoods Requirements, Testing, Marking. 28 pp.
CEN EN 146-91. Respiratory Protective Devices—Powered Particle Filtering Devices Incorporating Helmets or Hoods—Requirements, Testing, Marking. 16 pp.
CEN EN 146-91. Respiratory Protective Devices—Powered Particle Filtering Devices Incorporating Helmets or Hoods—Requirements, Testing, Marking. 30 pp.

Breathing Apparatus (Cont.)

—Quarter Mask
CEN EN 140-89. Respiratory Protective Devices; Half-Masks and Quarter-Masks; Requirements, Testing, Making. 43 pp.
CEN EN 140/A1-90. AMD 1 Respiratory Protective Devices; Half-Masks and Quarter-Masks; Requirements, Testing, Marking. 4 pp.
CEN EN 140/A1-92. AMD A1 Respiratory Protective Devices; Half-Masks and Quarter-Masks; Requirements, Testing, Marking. 3 pp.
CEN PREN 147-84. Respiratory Protective Equipment Power Assisted Particle Filtering Devices Incorporating Full Face Masks-Half Masks and Quarter Masks. 16 pp.
CEN EN 147-91. Respiratory Protective Devices—Assisted Particle Filtering Devices Full Face Incorporating Full Face Masks, Half Masks—Requirements, Testing, Marking. 14 pp.

—Quarter Mask—Identification Systems
BSI BS 7356-90. 1990 Half Masks and Quarter Masks for Respiratory Protective Devices. 23 pp.
BSI BS 7356-01. 1990 Amd 1 Half Masks and Quarter Masks for Respiratory Protective Devices (AMD 7113) November 15, 1992. 29 pp.
BSI BS EN 147-92. 1992 Respiratory Protective Devices Specification for Power Assisted Particle Filtering Devices Incorporating Full Face Masks, Half Masks or Quarter Masks. 19 pp.
CEN EN 147-91. Respiratory Protective Devices—Assisted Particle Filtering Devices Full Face Incorporating Full Face Masks, Half Masks—Requirements, Testing, Marking. 29 pp.

—Safety
EUROCAE ED-65 04.91. MOPS for Passenger Protective Breathing Equipment. 62 pp.
JIS T 8150-85. Guidance for Selection, Use and Maintenance of Respiratory Protective Equipments. 4 pp.

Breathing Systems, Anesthesia
Use: Anesthesia Breathing Systems

Breech Mechanisms

—Lubricants
MOD UK DSTAN 91-42-78. Lubricating Oil, Petroleum: Compressor, Light Joint Service Designation: OM-58 Lubricating Oil, Petroleum: Compressor, Medium Joint Service Designation: OM-160 Issue 1. 6 pp.

Breeding (Animals)
Use: Animal Breeding

Brewing Equipment

—Flasks
BSI BS 701-53. 1953 Amd 1 Brewers' Mash Flask. 10 pp.

—Glass Lined Steel
JIS R 4201-83. Glass-Lined Apparatus and Equipment for Industry. 24 pp.

—Hoses
CNS K4068-82. Brewing Hoses (Nov)(9617).
JIS K 6344-82. Rubber Hoses for Brewing.

Bricks
Use For: Building Bricks; Clay Bricks See Also: Air Bricks; Building Stones; Calcium Silicate Bricks; Ceramics; Clays; Compass Bricks; Concrete Blocks; Construction Materials; Facing Bricks; Firebricks; Hollow Bricks; Magnesia Bricks; Masonry; Silica Bricks

BSI BS 3921-85. 1985 Clay Bricks. 22 pp.
BSI BS 3921-74. 1974 Clay Bricks and Blocks. 35 pp.
BSI BS 4729-90. 1990 Dimensions of Bricks of Special Shapes and Sizes. 68 pp.
BSI BS 4729-71. (WITHDRAWN) 1971 Amd 1 Shapes and Dimensions of Special Bricks. 68 pp.
BSI BS 6649-85. 1985 Clay and Calcium Silicate Modular Bricks. 4 pp.
CEN PREN 771-1-92. Specification for Masonry Units—Part 1: Clay Masonry Units. 21 pp.
CNS R2002-78. Bricks for Building (Oct)(382). 1 p.
CNS R3042-59. Method of Test for General Type Bricks for Building (Apr)(1127) (R 1973). 2 pp.
CSA CAN/CSA-A82.1-M87. Burned Clay Brick (Solid Masonry Units Made from Clay or Shale) (R 1992). 20 pp.
CSA CAN3-A82.2-M78. Methods of Sampling and Testing Brick (R 1992). 26 pp.
DIN ENGL 105 Pt 3-84. Clay Bricks; High Strength Bricks and High Strength Engineering Bricks (May). 6 pp.
DIN ENGL 105 Pt 4-84. Clay Bricks; Ceramic Engineering Bricks (May). 4 pp.
DIN ENGL 105 Pt 5-84. Clay Bricks; Lightweight Horizontally Perforated Bricks and Lightweight Horizontally Perforated Brick Panels (May). 6 pp.

Bricks (Cont.)
DIN ENGL 18152-87. Lightweight Concrete Solid Bricks and Blocks (Apr). 7 pp.
JIS A 5213-87. Clay Bricks for Building. 11 pp.
JIS R 1250-91. Common Bricks. 7 pp.
SAA AS 1225-84. Clay Building Bricks. 4 pp.
SAA AS 1653-85. Calcium Silicate Building Bricks. 23 pp.
SNZ NZS 366-63. Clay Building Bricks. 11 pp.

—Acid Resistant
CNS R2122-81. Acid-Proof Brick for Chemical Industry (Jan) (6959).

—Cleaning
BSI BS 6270: Part 1-82. 1982 Amd 2 Cleaning and Surface Repair of Buildings Part 1: Natural Stone, Cast Stone and Clay and Calcium Silicate Brick Masonry (AMD 5605) April 28, 1989. 46 pp.

—Comprehensive Testing
SAA AS 1226. Methods of Sampling and Testing Clay Building Bricks (Complete Set in Binder).

—Compression Testing
CEN PREN 772-1-92. Methods of Test for Masonry Units—Part 1: Determination of Compressive Strength. 13 pp.
SAA AS 1226.4-84. Methods of Sampling and Testing Clay Building Bricks—Part 4: Method for Determining Compressive Strength. 3 pp.
SNZ NZS 366-63. Clay Building Bricks. 11 pp.

—Decks
DIN ENGL 4159-78. Structurally Cooperating Bricks for Floors and Wall Panels (Apr). 9 pp.
DIN ENGL 4160-78. Structurally Non-Cooperating Bricks for Floors (Aug). 6 pp.

—Density
DIN ENGL 105 Pt 1-89. Clay Bricks; Solid Bricks and Vertically Perforated Bricks (Aug). 9 pp.
DIN ENGL 105 Pt 2-89. Clay Bricks; Lightweight Vertically Perforated Bricks (Aug). 6 pp.

—Design
DIN ENGL 105 Pt 1-89. Clay Bricks; Solid Bricks and Vertically Perforated Bricks (Aug). 9 pp.
DIN ENGL 105 Pt 2-89. Clay Bricks; Lightweight Vertically Perforated Bricks (Aug). 6 pp.

—Effluorescence
SAA AS 1226.6-84. Methods of Sampling and Testing Clay Building Bricks—Part 6: Method of Test for Efflorescence. 2 pp.

—Expansion
SAA AS 1226.5-84. Methods of Sampling and Testing Clay Building Bricks—Part 5: Method for Determining Characteristic Expansion. 4 pp.

—Floors
DIN ENGL 4159-78. Structurally Cooperating Bricks for Floors and Wall Panels (Apr). 9 pp.
DIN ENGL 4160-78. Structurally Non-Cooperating Bricks for Floors (Aug). 6 pp.

—Glossaries
BSI BS 6100: Sec 5.3-84. 1984 Glossary of Building and Civil Engineering Terms Part 5: Masonry Section 5.3: Bricks and Blocks. 7 pp.
BSI BS 6100: Sec 5.3-01. 1984 Amd 1 Glossary of Building and Civil Engineering Terms Part 5: Masonry Section 5.3: Bricks and Blocks (AMD 7262) August 15, 1992. 9 pp.

—High Strength
DIN ENGL 105 Pt 3-84. Clay Bricks; High Strength Bricks and High Strength Engineering Bricks (May). 6 pp.

—Inspection
DIN ENGL 105 Pt 1-89. Clay Bricks; Solid Bricks and Vertically Perforated Bricks (Aug). 9 pp.
DIN ENGL 105 Pt 2-89. Clay Bricks; Lightweight Vertically Perforated Bricks (Aug). 6 pp.

—Modular Construction
CNS R2065-72. General Type Bricks for Building (Modulus Coordination) (Jan) (3319). 1 p.

—Modulus of Rupture
SAA AS 1226.11-84. Methods of Sampling and Testing Clay Building Bricks—Part 11: Method for Determining Lateral Modulus of Rupture. 3 pp.

—Painting—Interior/Exterior
CGSB 85-GP-31M-79. Painting Stucco, Masonry and Brick Surfaces. 16 pp.

—Pitting
SAA AS 1226.7-84. Methods of Sampling and Testing Clay Building Bricks—Part 7: Method of Examination for Pitting Due to Lime Particles. 1 p.

INTERNATIONAL AND NON-U.S. NATIONAL STANDARDS
SUBJECT INDEX
Bridges

Bricks (Cont.)
—**Prefabricated**
DIN ENGL 1053 Pt 4-78. Masonry; Buildings of Prefabricated Brickwork Components (Sept). 31 pp.

—**Repair**
BSI BS 6270: Part 1-82. 1982 Amd 2 Cleaning and Surface Repair of Buildings Part 1: Natural Stone, Cast Stone and Clay and Calcium Silicate Brick Masonry (AMD 5605) April 28, 1989. 46 pp.

—**Salt Spray Testing**
SAA AS 1226.10-84. Methods of Sampling and Testing Clay Building Bricks—Part 10: Method for Determining Resistance to Salt Attack. 2 pp.

—**Sampling**
CSA CAN3-A82.2-M78. Methods of Sampling and Testing Brick (R 1992). 26 pp.
SAA AS 1226. Methods of Sampling and Testing Clay Building Bricks (Complete Set in Binder).
SAA AS 1226.1-84. Methods of Sampling and Testing Clay Building Bricks—Part 1: Methods of Sampling. 2 pp.
SNZ NZS 366-63. Clay Building Bricks. 11 pp.

—**Sizes**
CEN PREN 772-16-92. Methods of Test for Masonry Units—Part 16: Determination of Size and Dimensions (Excluding Natural Stone Masonry Units). 6 pp.
DIN ENGL 105 Pt 1-89. Clay Bricks; Solid Bricks and Vertically Perforated Bricks (Aug). 9 pp.
DIN ENGL 105 Pt 2-89. Clay Bricks; Lightweight Vertically Perforated Bricks (Aug). 6 pp.
SNZ NZS 366-63. Clay Building Bricks. 11 pp.

—**Sizes—Measurement**
SAA AS 1226.2-84. Methods of Sampling and Testing Clay Building Bricks—Part 2: Method of Measurement of Dimensions. 2 pp.

—**Soluble Matter Content—Extraction Analysis**
CEN PREN 772-5-92. Methods of Test for Masonry Units—Part 5: Determination of Soluble Salt Content of Clay Masonry Units. 11 pp.

—**Stoneware—Acid Resistant**
JIS R 1535-91. Acid Proof Porcelain Bricks for Chemical Industry. 4 pp.

—**Thickness Measurement**
CEN PREN 772-16-92. Methods of Test for Masonry Units—Part 16: Determination of Size and Dimensions (Excluding Natural Stone Masonry Units). 6 pp.

—**Transverse Strength**
SAA AS 1226.3-84. Methods of Sampling and Testing Clay Building Bricks—Part 3: Method for Determining Transverse Strength. 3 pp.

—**Voids—Sand Filling**
CEN PREN 772-9-92. Methods of Test for Masonry Units—Part 9: Determination of Volume of Voids and Net Volume of Clay and Calcium Silicate Masonry Units by Sand Filling. 6 pp.

—**Volume**
CEN PREN 772-3-92. Methods of Test for Masonry Units—Part 3: Determination of Net Volume of Clay and Calcium Silicate Masonry Units by Weighing. 5 pp.

—**Walls**
DIN ENGL 4159-78. Structurally Cooperating Bricks for Floors and Wall Panels (Apr) 9 pp.

—**Water Absorption**
CEN PREN 772-7-92. Methods of Test for Masonry Units—Part 7: Determination of Water Absorption of Clay Masonry Units by Boiling in Water. 7 pp.
CEN PREN 772-11-92. Methods of Test for Masonry Units—Part 11: Determination of Water Absorption of Clay and Aggregate Concrete Masonry Units Due to Capillary Action. 7 pp.
SAA AS 1226.8-84. Methods of Sampling and Testing Clay Building Bricks—Part 8: Method for Determining Initial Rate of Absorption (Suction). 2 pp.
SAA AS 1226.9-84. Methods of Sampling and Testing Clay Building Bricks—Part 9: Method for Determining Water Absorption Properties. 2 pp.

—**Water Resistance Testing—Autoclaving**
CNS R3078-77. Method of Test for Hydration Resistance of Basic Brick (Apr)(4093).
JIS R 2211-91. Testing Method for Hydration Resistance of Basic Bricks. 6 pp.

Bridge Bearings
See Also: Bearings; Bridges (Structures); Highway Bridges; Structural Bearings

Bridge Bearings (Cont.)
BSI BS 5400: Sec 9.1-83. 1983 Steel, Concrete and Composite Bridges Part 9: Bridge Bearings Section 9.1: Code of Practice for Design of Bridge Bearings. 20 pp.
BSI BS 5400: Sec 9.2-83. 1983 Steel, Concrete and Composite Bridges Part 9: Bridge Bearings Section 9.2: Materials, Manufacture and Installation of Bridge Bearings. 16 pp.
DIN ENGL 4141 Pt 2-84. Structural Bearings; Bearing Systems for Civil Engineering Structures Forming Part of Traffic Routes (Bridges) (Sept). 5 pp.
DIN ENGL 4141 Pt 4-87. Structural Bearings; Transport, Storage on Site and Installation (Oct). 7 pp.

Bridge Cranes
Use For: Goligth Cranes; Overhead Cranes; Overhead Traveling Cranes; Staddle Hoists; Transporter Bridges; Transporter Cranes; Transporters; Traveling Bridge Cranes; Traveling Bridges *See Also:* Cranes
BSI BS 466-84. 1984 Power Driven Overhead Travelling Cranes Semi-Goliath and Goliath Cranes for General Use. 36 pp.
CNS B4039-84. Overhead Travelling Cranes (Sep)(6543).
CSA B167-1964. General Purpose Electric Overhead Travelling Cranes; (Erratum November 1966) (Erratum May 1967) (Erratum November 1969). 43 pp.
ISO 4310-81. Cranes—Test Code and Procedures First Edition. 6 pp.
ISO 8306-85. Cranes—Overhead Travelling Cranes and Portal Bridge Cranes—Tolerances for Cranes and Tracks First Edition. 9 pp.
ISO 9374 Pt 5-91. Cranes—Information to Be Provided—Part 5: Overhead Travelling Cranes and Portal Bridge Cranes First Edition. 11 pp.
JIS B 8801-74. Electric Overhead Travelling Cranes. 14 pp.

—**Cabins**
ISO 8566 Pt 5-92. Cranes—Cabins—Part 5: Overhead Travelling and Portal Bridge Cranes First Edition. 5 pp.

—**Classification**
ISO 4301 Pt 5-91. Cranes—Classification—Part 5: Overhead Travelling and Portal Bridge Cranes First Edition. 5 pp.

—**Control Systems Equipment**
ISO 7752 Pt 5-85. Lifting Appliances—Controls—Layout and Characteristics—Part 5: Overhead Travelling Cranes and Portal Bridge Cranes First Edition. 5 pp.

—**Design Loads**
ISO 8686 Pt 5-92. Cranes—Design Principles for Loads and Load Combinations—Part 5: Overhead Travelling and Portal Bridge Cranes First Edition. 7 pp.

—**Marine**
JIS F 0802-89. Shop Test Code for Marine AC Electric Overhead Travelling Cranes in Engine Room.

—**Sheaves**
CNS B4042-80. Rope Sheaves for Overhead Traveling Cranes (Dec)(6641).
JIS B 8807-76. Rope Sheaves for Electric Overhead Travelling Cranes. 9 pp.

—**Wheels**
CNS B4040-83. Cast Steel Traveling Wheel for Overhead Traveling Cranes (Jan)(6544).
CNS B4041-80. Wrought Steel Traveling Wheel for Overhead Traveling Cranes (Dec)(6615).
JIS B 8806-92. Cast Steel Wheels and Wrought Steel Wheels for Cranes. 10 pp.

Bridge Pieces
Use: Holders

Bridge Plates
See Also: Materials Handling Equipment
MOD UK DSTAN 39-4-01. Materials Handling Equipment Bridge Plates Lifting Slings Vehicle Loading Ramps Issue 1; Amendment 2. 10 pp.

Bridge Rectifiers
See Also: Rectifiers
MOD UK DSTAN 59-61: 90/207-01. Semiconductor Device Rectifier Bridges Issue 1; Amendment 2. 16 pp.
MOD UK DSTAN 59-61: 90/208-01. Semiconductor Device Rectifier Bridges Issue 1; Amendment 2. 18 pp.
MOD UK DSTAN 59-61: 90/213-01. Semiconductor Devices Rectifier Bridges Issue 1; Amendment 2. 22 pp.

Bridge Rectifiers (Cont.)
MOD UK DSTAN 59-61: 90/214-01. Semiconductor Devices Rectifier Bridges Issue 1; Amendment 2. 22 pp.

—**Single Phase—Preferred Products List**
CECC CECC MUAHAG Vol 9 IS 3-90. Preferred Products List; Semiconductors (En, Fr, Ge) AMD 1 (En, Fr, Ge). 51 pp.

—**Three Phase—Preferred Products List**
CECC CECC MUAHAG Vol 9 IS 3-90. Preferred Products List; Semiconductors (En, Fr, Ge) AMD 1 (En, Fr, Ge). 51 pp.

Bridges (Electrical)
See Also: Measuring Instruments

—**Conference Calling**
CCITT RECMN G.172-89. Transmission Plan Aspects of International Conference Calls—General Characteristics of International Telephone Connections and Circuits (Study Groups XII and XV) 3 pp. 3 pp.

—**Conference Calling—Automatic Gain Control**
CCITT RECMN G.172-89. Transmission Plan Aspects of International Conference Calls—General Characteristics of International Telephone Connections and Circuits (Study Groups XII and XV) 3 pp. 3 pp.

—**Conference Calling—Echo**
CCITT RECMN G.172-89. Transmission Plan Aspects of International Conference Calls—General Characteristics of International Telephone Connections and Circuits (Study Groups XII and XV) 3 pp. 3 pp.

—**Conference Calling—Noise**
CCITT RECMN G.172-89. Transmission Plan Aspects of International Conference Calls—General Characteristics of International Telephone Connections and Circuits (Study Groups XII and XV) 3 pp. 3 pp.

—**Conference Calling—Transmission Loss**
CCITT RECMN G.172-89. Transmission Plan Aspects of International Conference Calls—General Characteristics of International Telephone Connections and Circuits (Study Groups XII and XV) 3 pp. 3 pp.

—**Media Access Control—Local Area Networks**
IEC DIS 10038-92. Information Technology—Telecommunications and Information Exchange Between Systems—Local Area Networks—Media Access Control (MAC) Bridges ***CD-ROM ONLY***. 163 pp.
IEC DISP 10612 Pt 1-92. Information Technology—International Standardized Profile RD5p.5q—Relaying the MAC Service Using Transparent Bridging—Part 1: Subnetwork-Independent Requirements ***CD-ROM ONLY***. 31 pp.
ISO DIS 10038-92. Information Technology—Telecommunications and Information Exchange Between Systems—Local Area Networks—Media Access Control (MAC) Bridges ***CD-ROM ONLY***. 163 pp.
ISO DISP 10612 Pt 1-92. Information Technology—International Standardized Profile RD5p.5q—Relaying the MAC Service Using Transparent Bridging—Part 1: Subnetwork-Independent Requirements ***CD-ROM ONLY***. 31 pp.
JTC1 DISP10612 Pt 1-92. Information Technology—International Standardized Profile RD5p.5q—Relaying the MAC Service Using Transparent Bridging—Part 1: Subnetwork-Independent Requirements ***CD-ROM ONLY***. 31 pp.

—**Media Access Control—Local Area Networks—CSMA/CD**
IEC DISP 10612 Pt 2-92. Information Technology—International Standardized Profile RD5p.5q—Relaying the MAC Service Using Transparent Bridging—Part 2: CSMA/CD LAN Subnetwork-Dependent, Media-Dependent Requirements ***CD-ROM ONLY***. 32 pp.
IEC DISP 10612 Pt 4-92. Information Technology—International Standardized Profile RD5p.5q—Relaying the MAC Service Using Transparent Bridging—Part 4: Profile RD51.51 (CSMA/CD-CSMA/CD) ***CD-ROM ONLY***. 28 pp.
ISO DISP 10612 Pt 2-92. Information Technology—International Standardized Profile RD5p.5q—Relaying the MAC Service Using Transparent Bridging—Part 2: CSMA/CD LAN Subnetwork-Dependent, Media-Dependent Requirements ***CD-ROM ONLY***. 32 pp.

INTERNATIONAL AND NON-U.S. NATIONAL STANDARDS
SUBJECT INDEX

Bridges

Bridges (Electrical) (Cont.)
—**Media Access Control—Local Area Networks—CSMA/CD** (Cont.)
ISO DISP 10612 Pt 4-92. Information Technology—International Standardized Profile RD5p.5q—Relaying the MAC Service Using Transparent Bridging—Part 4: Profile RD51.51 (CSMA/CD-CSMA/CD) ***CD-ROM ONLY***. 28 pp.
JTC1 DISP10612 Pt 2-92. Information Technology—International Standardized Profile RD5p.5q—Relaying the MAC Service Using Transparent Bridging—Part 2: CSMA/CD LAN Subnetwork-Dependent, Media-Dependent Requirements ***CD-ROM ONLY***. 32 pp.
JTC1 DISP10612 Pt 4-92. Information Technology—International Standardized Profile RD5p.5q—Relaying the MAC Service Using Transparent Bridging—Part 4: Profile RD51.51 (CSMA/CD-CSMA/CD) ***CD-ROM ONLY***. 28 pp.

—**Media Access Control—Local Area Networks—Source Routing**
IEC DIS 10038 Draft AMD 2. Information Technology—Telecommunications and Information Exchange Between Systems—Local Area Networks—Media Access Control (MAC) Bridges Amendment 2: Source Routeing Supplement; (1992) ***CD-ROM ONLY***. 44 pp.
ISO DIS 10038 Draft AMD 2. Information Technology—Telecommunications and Information Exchange Between Systems—Local Area Networks—Media Access Control (MAC) Bridges Amendment 2: Source Routeing Supplement; (1992) ***CD-ROM ONLY***. 44 pp.

Bridges (Structures)
Scope Note: For additional listings, use a more specific term *Use For:* Composite Bridges
See Also: Bridge Bearings; Concrete Bridges; Foot Bridges; Highway Bridges; Ramps (Loading); Sheeting; Slab Bridges; Steel Bridges; Structures; Superstructures; Trestles (Supports)
BSI BS 5400: Part 1-88. 1988 Steel, Concrete and Composite Bridges Part 1: General Statement. 10 pp.
DIN ENGL 19661 Pt 1-72. Guidelines for Hyraulic Structures; Crossing Works; Bridges, Aqueducts, Culverts, Inverted Syphons, Casings (Oct). 13 pp.

—**Beams (Supports)**
BSI CP 117: Part 2-67. 1967 Composite Construction in Structural Steel and Concrete Part 2: Beams for Bridges. 33 pp.

—**Design**
BSI BS 5400: Part 5-79. 1979 Amd 1 Steel, Concrete and Composite Bridges Part 5: Code of Practice for Design of Composite Bridges. 40 pp.

—**Fatigue (Materials)**
BSI BS 5400: Part 10-80. 1980 Steel, Concrete and Composite Bridges Part 10: Code of Practice for Fatigue. 55 pp.
BSI BS 5400: Part 10C-80. 1980 Steel, Concrete and Composite Bridges Part 10C: Charts for Classification of Details for Fatigue. 2 pp.

—**Glossaries**
BSI BS 6100: Sec 2.3-92. 1992 Glossary of Building and Civil Engineering Terms Part 2: Civil Engineering Section 2.3: Superstructures. Bridges. Large Span Structures. 16 pp.
BSI BS 6100: Sec 2.3-89. 1989 Glossary of Building and Civil Engineering Terms Part 2: Civil Engineering Section 2.3: Superstructures. Bridges. Large Span Structures. 14 pp.

—**Lighting**
BSI BS 5489: Part 6-92. 1992 Road Lighting Part 6: Code of Practice for Lighting for Bridges and Elevated Roads. 11 pp.
BSI BS 5489: Part 6-90. 1990 Road Lighting Part 6: Code of Practice for Bridges and Elevated Roads (Group D). 11 pp.
BSI BS 5489: Part 6-67. (WITHDRAWN) 1967 Amd 1 Road Lighting Part 6: Lighting for Bridges and Elevated Roads. 17 pp.

—**Loads (Forces)**
BSI BS 5400: Part 2-78. 1978 Amd 1 Steel, Concrete and Composite Bridges Part 2: Specification for Loads. 45 pp.

—**Loads (Forces)—Classification**
NATO STANAG 2021 ED 4 AMD 0-87. Computation of Bridge, Ferry, Raft and Vehicle Classifications. 31 pp.
NATO STANAG 2021 ED 5 AMD 0-90. Military Computation of Bridge, Ferry, Raft and Vehicle Classifications. 39 pp.

Bridges (Structures) (Cont.)
—**Loads (Forces)—Identification Systems**
MOD UK DSTAN 23-2-01. Computation and Marking of Bridge and Raft Load Classification Numbers for Military Vehicles, Using SI Units Issue 2; Amendment 1 (Withdrawn). 36 pp.
NATO STANAG 2010 ED 5 AMD 5-80. Military Load Classification Markings (New Title). 13 pp.
NATO STANAG 2010 ED 5 AMD 6-80. Military Load Classification Markings (New Title). 16 pp.

—**Structural Bearings**
DIN ENGL 4141 Pt 4-87. Structural Bearings; Transport, Storage on Site and Installation (Oct). 7 pp.

—**Tendons**
BSI BS 5400: Part 7-78. 1978 Steel, Concrete and Composite Bridges Part 7: Specification for Materials and Workmanship: Concrete, Reinforcement and Prestressing Tendons. 15 pp.
BSI BS 5400: Part 8-78. 1978 Steel, Concrete and Composite Bridges Part 8: Recommendations for Materials and Workmanship: Concrete, Reinforcement and Prestressing Tendons. 23 pp.

Bridging (Coatings)
Scope Note: See the subheading Bridging under specific coatings

Bridle Gates
Use: Gates (Barriers)—Equestrian

Briefcases
See Also: Luggage
—**ABS Resin**
CNS S1108-81. ABS Attache Case (Nov)(6858). 4 pp.
CNS S2060-81. Method of Test for ABS Attache Case (Nov) (8159). 4 pp.

—**Leather**
CGSB 113-GP-3M-79. Leather, Cowhide, Vegetable or Chrome Tanned, Case, Standard for (R 1983). 9 pp.
CGSB 113-GP-4M-79. Portfolios, Leather, Slide Fastener, Standard for (R 1983); (Amendment 1 Feb 1982). 10 pp.
CGSB 113-GP-5MA-83. Briefcase, Leather, Frame Style, Standard for. 11 pp.
CGSB 113-GP-6MA-83. Briefcase, Leather, Inspectors, Solid Compartment Style, Standard for. 11 pp.

—**Plastic**
CGSB 113-GP-1MA-81. Case, Dispatch, Molded Plastic (Metal Frame), Standard for. 15 pp.

—**Vinyl**
CGSB CAN/CGSB-53.125-M89. Portfolio, Vinyl. 8 pp.
CGSB 113-GP-8M-84. Case, Dispatch Vinyl-Covered (Wood Frame). 11 pp.

Bright Steels
Scope Note: See the subheading Bright under the specific type of steel

Brightness Testing
—**Paperboard**
CNS P3010-89. Method of Test for Brightness of Pulp, Paper and Paperboard (Aug)(1466). 2 pp.
CPPA E.1-90. Brightness of Pulp, Paper and Paperboard. 3 pp.
SAA AS 1301.446S-92. Methods of Test for Pulp and Paper—Part 446s: Measurement of Diffuse Blue Reflectance Factor (Brightness) of Pulp, Paper and Paperboard (Supersedes AS 1301.P446s—1982).
SNZ NZS/AS 1301. 446S-92. Methods of Test for Pulp and Paper Measurement of Diffuse Blue Reflectance Factor (Brightness) of Pulp, Paper and Paperboard (This is a Joint Standard with SAA AS 1301.446S). 4 pp.

—**Papers**
CNS P3010-89. Method of Test for Brightness of Pulp, Paper and Paperboard (Aug)(1466). 2 pp.
CPPA E.1-90. Brightness of Pulp, Paper and Paperboard. 3 pp.
JIS P 8123-61. Testing Methods for Brightness by Hunter of Paper and Pulp. 8 pp.
SAA AS 1301.446S-92. Methods of Test for Pulp and Paper—Part 446s: Measurement of Diffuse Blue Reflectance Factor (Brightness) of Pulp, Paper and Paperboard (Supersedes AS 1301.P446s—1982).
SNZ NZS/AS 1301. 446S-92. Methods of Test for Pulp and Paper Measurement of Diffuse Blue Reflectance Factor (Brightness) of Pulp, Paper and Paperboard (This is a Joint Standard with SAA AS 1301.446S). 4 pp.

—**Pulps**
CNS P3010-89. Method of Test for Brightness of Pulp, Paper and Paperboard (Aug)(1466). 2 pp.

Brightness Testing (Cont.)
—**Pulps** (Cont.)
CPPA E.1-90. Brightness of Pulp, Paper and Paperboard. 3 pp.
CPPA E.4P-88. Accelerated Aging Test for Brightness Reversion of Pulp. 2 pp.
JIS P 8123-61. Testing Methods for Brightness by Hunter of Paper and Pulp. 8 pp.
SAA AS 1301.446S-92. Methods of Test for Pulp and Paper—Part 446s: Measurement of Diffuse Blue Reflectance Factor (Brightness) of Pulp, Paper and Paperboard (Supersedes AS 1301.P446s—1982).
SNZ NZS/AS 1301. 446S-92. Methods of Test for Pulp and Paper Measurement of Diffuse Blue Reflectance Factor (Brightness) of Pulp, Paper and Paperboard (This is a Joint Standard with SAA AS 1301.446S). 4 pp.

Brilliant Carmine 6B
See Also: Pigments
—**Pigments**
CNS K2135-83. Brilliant Carmine 6B (Jun)(10356). 3 pp.
JIS K 5206-71. Brilliant Carmine 6 B. 5 pp.

Brilliant Fast Scarlet
See Also: Pigments
—**Pigments**
CNS K2219-88. Brilliant Fast Scarlet (May)(12307).
JIS K 5224-71. Brilliant Fast Scarlet. 4 pp.

Brilliant Scarlet G
See Also: Pigments
—**Pigments**
CNS K2133-83. Brilliant Scarlet G (Jun)(10354). 3 pp.
JIS K 5203-71. Brilliant Scarlet G. 5 pp.

Brinell Hardness
Use: Brinell Hardness Testing

Brinell Hardness Testers
See Also: Hardness Testers
BSI BS 240-86. 1986 Method for Brinell Hardness Test and for Verification of Brinell Hardness Testing Machines. 42 pp.
CEN PREN 10003-3-91. Metallic Materials—Hardness Test—Brinell—Part 3: Calibration of Standardized Blocks to be Used for Brinell Hardness Testing Machines. 7 pp.
CNS B6077-82. Brinell Hardness Testing Machines (Oct)(9472).
CNS B7212-82. Method of Test for Brinell Hardness Testing Machines (Oct)(9473).
DIN ENGL 51225 Pt 1-85. Materials Testing Machines; Hardness Testing Machines with an Optical Indentation-Measuring Device; Test Forces from 49,03 to 29 420 N (Jan). 6 pp.
ISO 156-82. Metallic Materials—Hardness Test—Verification of Brinell Hardness Testing Machines First Edition. 5 pp.
JIS B 7724-86. Brinell Hardness Testing Machines. 10 pp.

—**Blocks—Calibration**
ISO 726-82. Metallic Materials—Hardness Test—Calibration of Standardized Blocks to Be Used for Brinell Hardness Testing Machines First Edition. 5 pp.

—**Metals**
CEN PREN 10003-1-91. Metallic Materials—Hardness Test—Brinell—Part 1: Test Method. 7 pp.

—**Test Blocks**
CEN PREN 10003-2-91. Metallic Materials—Hardness Test—Brinell—Part 2: Verification of Brinell Hardness Testing Machines. 8 pp.

Brinell Hardness Testing
See Also: Hardness Testing; Indentation Hardness Testing
BSI BS 240-86. 1986 Method for Brinell Hardness Test and for Verification of Brinell Hardness Testing Machines. 42 pp.

—**Blocks**
JIS B 7736-83. Standardized Blocks of Brinell Hardness. 9 pp.

—**Cast Iron**
DIN ENGL 1691-85. Flake Graphite Cast Iron (Grey Cast Iron); Properties (May). 7 pp.

—**Metals**
CNS Z8002-87. Method of Brimell Hardness Test (Jun)(2113).
DIN ENGL 50351-85. Testing of Metallic Materials; Brinell Hardness Test (Feb). 7 pp.
ISO 6506-81. Metallic Materials—Hardness Test—Brinell Test First Edition. 6 pp.

INTERNATIONAL AND NON-U.S. NATIONAL STANDARDS
SUBJECT INDEX

Brinell Hardness Testing (Cont.)
—Metals (Cont.)
- JIS Z 2243-92. Method of Brinell Hardness Test. 9 pp.
- SAA AS 1816-90. Metallic Materials—Brinell Hardness Test. 29 pp.

—Steel—Structural Members
- ISO TR10108-89. Steel—Conversion of Hardness Values to Tensile Strength Values First Edition. 9 pp.

—Tables (Data)
- BSI BS 860-67. 1967 Tables for Comparison of Hardness Scales. 13 pp.
- DIN ENGL 50150-76. Testing of Steel and Cast Steel; Conversion Table for Vickers Hardness, Brinell Hardness, Rockwell Hardness and Tensile Strength (Dec). 5 pp.
- ISO 410-82. Metallic Materials—Hardness Test—Tables of Brinell Hardness Values for Use in Tests Made on Flat Surfaces First Edition. 11 pp.
- ISO TR10108-89. Steel—Conversion of Hardness Values to Tensile Strength Values First Edition. 9 pp.

—Titanium
- JIS H 0511-90. Testing Methods for Brinell Hardness of Titanium Sponge. 11 pp.

Brines
See Also: Salt Water

—Sampling
- DIN ENGL 19570 Pt 1-83. Analysis of Waste Brine and Salt-Laden Bodies of Water; Sampling (Apr). 3 pp.

Briquets
See Also: Pellets

—Ash Content
- JIS M 8812-84. Methods for Proximate Analysis of Coal and Coke. 28 pp.

—Calorific Value
- JIS M 8814-85. Determination of Calorific Value of Coal and Coke; (Erratum). 20 pp.

—Fixed Carbon Content
- JIS M 8812-84. Methods for Proximate Analysis of Coal and Coke. 28 pp.

—Moisture Content
- JIS M 8812-84. Methods for Proximate Analysis of Coal and Coke. 28 pp.

—Volatile Matter Content
- JIS M 8812-84. Methods for Proximate Analysis of Coal and Coke. 28 pp.

Briquettes
Use: Briquets

Bristles
See Also: Brushes (Cleaning/Polishing); Fibers

—Hog
- CNS N1003-85. Hog Bristles (May)(84). 2 pp.
- CNS N4010-85. Method of Test for Hog Bristles (May)(1482).

—Horse
- CNS N1076-85. Horse Bristle and Tail Hair (May)(2821).

Bristol Board
See Also: Paperboard
- CGSB 9-GP-36M-79. Paper, Printing, Bristol, Standard for. 9 pp.
- CGSB 9-GP-37M-79. Paper, Index Bristol, Standard for. 10 pp.
- CGSB 9-GP-48M-79. Paper, Announcement Bristol, Kid and Plate Finish, Standard for; (Amendment 1 Feb 1982). 10 pp.

—Weight Measurement
- CGSB CAN/CGSB-9.63-M78. Grammage of Fine Papers. 6 pp.

Bristol Britannia Aircraft
See Also: Aircraft

—Accidents
- CAA. Bristol Britannia (World Airline Accident Summary). 1 p.
- CAA. Bristol Britannia. 1 p.

—Mandatory Aircraft Modifications and Inspections Summaries
- CAA. Bristol Britannia. 21 pp.

Bristol 170 Mark 31 Aircraft
See Also: Aircraft

—Antenna Positions
- CAA. Bristol 170 MK.31 (Approved Aerial Positions). 1 p.

—Mandatory Aircraft Modifications and Inspections Summaries
- CAA. Bristol 170 Mark 31. 6 pp.

Bristol 170 MK 21 Aircraft
See Also: Aircraft

—Antenna Positions
- CAA. Bristol 170 MK.21 (Approved Aerial Positions). 1 p.

Bristol 170 MK 32 Aircraft
See Also: Aircraft

—Antenna Positions
- CAA. Bristol 170 MK.32 (Approved Aerial Positions). 1 p.

Bristol 170 Wayfarer Aircraft
See Also: Aircraft

—Accidents
- CAA. Bristol 170 Wayfarer/Freighter (World Airline Accident Summary). 1 p.
- CAA. Bristol 170 Wayfarer/Freighter. 1 p.

Bristol 171 Aircraft
See Also: Aircraft

—Antenna Positions
- CAA. Bristol 171 (Approved Aerial Positions). 1 p.

Britannia 100 Aircraft
See Also: Aircraft

—Antenna Positions
- CAA. Britannia 100 (Approved Aerial Positions). 1 p.

Britannia 300 Aircraft
See Also: Aircraft

—Antenna Positions
- CAA. Britannia 300 (Approved Aerial Positions). 1 p.

British Aerospace ATP Aircraft
See Also: Aircraft

—Data Sheets
- CAA BA23 ISSUE 2. British Aerospace ATP. 4 pp.

—Mandatory Aircraft Modifications and Inspections Summaries
- CAA. British Aerospace ATP Series Aircraft. 12 pp.

British Aerospace Bulldog Series 100/120 Aircraft
See Also: Aircraft

—Mandatory Aircraft Modifications and Inspections Summaries
- CAA. British Aerospace Bulldog Series 100/120 Aircraft. 2 pp.

British Aerospace C1 Chipmunk Aircraft
See Also: Aircraft

—Mandatory Aircraft Modifications and Inspections Summaries
- CAA. British Aerospace C1 Chipmunk. 4 pp.

British Aerospace DH 82 Tiger Moth Aircraft
See Also: Aircraft

—Mandatory Aircraft Modifications and Inspections Summaries
- CAA. British Aerospace DH 82 Tiger Moth. 1 p.

British Aerospace HS 125 Aircraft
See Also: Aircraft

—Data Sheets
- CAA BA9 ISSUE 11. Corporate Jets Ltd, HS 125 Series 600B Model F600B; HS Series 400B Models 401B, 403B, F400B, HS 125 Series 700B, BAe 125 Series 800B, 1000B. 8 pp.

British Aerospace PLC Jetstream Aircraft
See Also: Aircraft

British Aerospace PLC Jetstream Aircraft (Cont.)
—Data Sheets
- CAA BA15 ISSUE 10. British Aerospace PLC Jetstream Series 3100 Jetstream Series 3200. 11 pp.

British Aerospace Scottish Division Bulldog Series Aircraft
See Also: Aircraft

—Data Sheets
- CAA BA7 ISSUE 8. British Aerospace Scottish Division Bulldog Series 100, Models 101, 102, 103, 104 Series 120, Models 122, 123, 125, 126, 127, 128, 129, 1210. 5 pp.

British Aircraft Corporation/Aerospatiale Concorde Aircraft
See Also: Aircraft

—Accidents
- CAA. British Aircraft Corporation/Aerospatiale Concorde (World Airline Accident Summary). 1 p.

British Aircraft Corporation VC 10 Aircraft
See Also: Aircraft

—Accidents
- CAA. British Aircraft Corporation VC10 Series (World Airline Accident Summary). 1 p.

—Mandatory Aircraft Modifications and Inspections Summaries
- CAA. British Aircraft Corporation VC10 and Super VC10. 15 pp.

British Association Screw Threads
Use: Screw Threads

British Standards Institution
Use: BSI

British Thermal Units
Use: Thermal Units

Britten-Norman Trislander BN 2A Mark III Aircraft
See Also: Aircraft

—Data Sheets
- CAA BA6 ISSUE 15. Britten-Norman Trislander BN2A Mark III, BN2A Mark III-1 and BN2A Mark III-2. 4 pp.

Britten-Norman Trislander BN 2A Mark III-1 Aircraft
See Also: Aircraft

—Data Sheets
- CAA BA6 ISSUE 15. Britten-Norman Trislander BN2A Mark III, BN2A Mark III-1 and BN2A Mark III-2. 4 pp.

Britten-Norman Trislander BN 2A Mark III-2 Aircraft
See Also: Aircraft

—Data Sheets
- CAA BA6 ISSUE 15. Britten-Norman Trislander BN2A Mark III, BN2A Mark III-1 and BN2A Mark III-2. 4 pp.

Brittleness
Use: Brittleness Testing

Brittleness Testing
Use For: Wedge Brittleness Testing
See Also: Ductility; Hardness Testing; High Temperature Testing; Impact Testing; Knife Testing; Low Temperature Testing; Tensile Testing; Weldability

—Photographic Film
- ISO 6077-93. Photography—Photographic Films and Papers—Wedge Test for Brittleness Second Edition. 11 pp.

—Photographic Papers
- ISO 6077-93. Photography—Photographic Films and Papers—Wedge Test for Brittleness Second Edition. 11 pp.

—Plastics
- ISO 974-80. Plastics—Determination of the Brittleness Temperature by Impact First Edition. 9 pp.
- JIS K 7216-80. Testing Method for Brittleness Temperature of Plastics. 10 pp.

Brittleness Testing (Cont.)

—Vulcanized Rubber
BSI BS 903: Part A25-92. 1992 Physical Testing of Rubber Part A25: Determination of Low-Temperature Brittleness (ISO 812: 1991) (V). 11 pp.
BSI BS 903: Part A25-68. 1968 Amd 1 Methods of Testing Vulcanized Rubber Part A25: Determination of Impact Brittleness Temperature. 11 pp.
CNS K6355-85. Method of Test for Brittleness Temperature of Vulcanized Rubber by Impact (Dec)(3564).
ISO 812-91. Rubber, Vulcanized—Determination of Low-Temperature Brittleness First Edition. 10 pp.

Broaches
See Also: Broaching Machines; Cutting Tools; High Speed Steel Tools; Planers (Tools); Reamers; Scrapers (Earthmoving Equipment); Tools
BSI BS 5326-76. 1976 Internal Broaching Tools and Pull Chuck Locations. 4 pp.
JIS B 4237-87. Dimensions of Pull and Follower Ends of Broach. 10 pp.
MOD UK DEF-1198. Broaches and Burnishers for Hand Use (Withdrawn).

—Dental
CNS T3024-81. Dental Barbed Broaches (Jan)(6964).
CNS T3025-81. Dental Smooth Broaches (Jan)(6965).
JIS T 5206-88. Dental Broaches. 7 pp.

—Dental—Holders
CNS T3038-81. Dental Broach Holders (Jul)(7758).
JIS T 5409-78. Dental Broach Holders (R 1987). 6 pp.

—Glossaries
JIS B 0175-79. Glossary of Terms Relating for Broaches. 37 pp.

—Involute Splines
CNS B1336-83. Involute Splines Hobs, Pinion-Type Cutters, Broaches (Sep)(10538).
DIN ENGL 5480 Pt 16-86. Involute Splines with 30 Degree Pressure Angle; Hobs; Pinion Type Cutters; Broaches (Mar). 11 pp.
JIS B 4239-88. Involute Spline Broaches. 14 pp.

—Keyway
JIS B 4238-88. Keyway Broaches. 11 pp.

Broaching Machines
Use For: Continuous Broaching Machines
See Also: Broaches; Cutting Tools; Grinders; Machine Tools; Milling Machines

—Chucks
BSI BS 5326-76. 1976 Internal Broaching Tools and Pull Chuck Locations. 4 pp.

—Glossaries
CNS B1227-80. Glossary of Terms Relating to Parts of Broaching Machine (Dec)(6751).

—Horizontal
BSI BS 4656: Part 18-81. 1981 Amd 1 Accuracy of Machine Tools and Methods of Test Part 18: Broaching Machines, Horizontal Internal Type. 11 pp.
DIN ENGL 55142-82. Machine Tools; Horizontal Surface Broaching Machines; Sizes (Aug). 3 pp.
DIN ENGL 55144-82. Machine Tools; Horizontal Internal Broaching Machines; Sizes (Aug). 3 pp.
DIN ENGL 55145-82. Machine Tools; Continuously Working Horizontal Surface Broaching Machines (Chain Broaching Machines); Sizes (Aug). 2 pp.
ISO 6480-83. Conditions of Acceptance for Horizontal Internal Broaching Machines—Testing of the Accuracy Second Edition. 11 pp.

—Vertical
BSI BS 4656: Part 13-83. 1983 Accuracy of Machine Tools and Methods of Test Part 13: Broaching Machines, Vertical Surface Type. 12 pp.
BSI BS 4656: Part 14-83. 1983 Accuracy of Machine Tools and Methods of Test Part 14: Broaching Machines, Vertical Internal Type. 11 pp.
BSI BS 4656: Part 17-73. 1973 Accuracy of Machine Tools and Methods of Test Part 17: Broaching Machines, Vertical Universal Type. 14 pp.
CNS B7251-89. Method of Test for Performance of Vertical Internal Type Broaching Machines (Jun)(10232).
CNS B7252-89. Test Code for Accuracy of Vertical Internal Type Broaching Machines (Jun)(10233).
DIN ENGL 55141-82. Machine Tools; Vertical Surface Broaching Machines; Sizes (Aug). 3 pp.
DIN ENGL 55143-82. Machine Tools; Vertical Internal Broaching Machines; Sizes (Aug). 4 pp.
ISO 6481-81. Acceptance Conditions for Vertical Surface Type Broaching Machines—Testing of Accuracy First Edition. 11 pp.
ISO 6779-81. Acceptance Conditions for Broaching Machines of Vertical Internal Type—Testing of Accuracy First Edition. 9 pp.

Broaching Machines (Cont.)

—Vertical (Cont.)
JIS B 6227-82. Test Code for Performance and Accuracy of Vertical Internal Type Broaching Machines (R 1987). 12 pp.

Broadband Amplifiers
See Also: Amplifiers; Data Communication Equipment; Integrated Circuits; Video Amplifiers; Wideband Amplifiers
MOD UK DSTAN 59-62: 90/042-69. Integrated Circuit Issue 1. 12 pp.
MOD UK DSTAN 59-62: 90/165-71. Integrated Circuit Issue 1. 12 pp.

—Cabled Distribution Systems
CENELEC PREN 50083-3-91. Cabled Distribution Systems for Television and Sound Signals: Part 3: Active Coaxial Wideband Distribution Components. 36 pp.

Broadcast Receivers

—Audio Equipment—Interfaces
CCIR OPINION 74-1-90. Systems for Signal Interface Connection Between Sound-Broadcasting Receivers and Associated Equipment—Volume X-1—Broadcasting Service (Sound). 1 p.

—Automotive—Sound Broadcasting—Digital Transmission Systems
CCIR RECMN 774-92. Digital Sound Broadcasting to Vehicular, Portable and Fixed Receivers Using Terrestrial Transmitters in the VHF/UHF Bands—Section 10B—Frequency-Modulation Sound Broadcasting in Bands 8 (VHF) and 9 (UHF). 7 pp.
CCIR RECMN 789-92. Digital Sound Broadcasting to Vehicular, Portable and Fixed Receivers for BSS (Sound) in the Frequency Range 500-3 000 MHz. 7 pp.

—Broadcasting Satellite Services
CCIR Report 473-5-90. Characteristics of Receiving Equipment for the Broadcasting-Satellite Service—Section 10/11C—Technology. 28 pp.
CCIR QUESTION 97/11-90. Characteristics of Receiving Systems of the Broadcasting-Satellite Service (Sound and Television)—Questions Concerning Study Group 11—Broadcasting Service (Television). 1 p.

—Figure of Merit—Broadcasting Satellite Services
CCIR RECMN 790-92. Characteristics of Receiving Equipment and Calculation of Receiver Figure-of-Merit (G/T) for the Broadcasting-Satellite Service. 3 pp.

—Fixed—Sound Broadcasting—Digital Transmission Systems
CCIR RECMN 774-92. Digital Sound Broadcasting to Vehicular, Portable and Fixed Receivers Using Terrestrial Transmitters in the VHF/UHF Bands—Section 10B—Frequency-Modulation Sound Broadcasting in Bands 8 (VHF) and 9 (UHF). 7 pp.
CCIR Report 1203-90. Digital Sound Broadcasting to Mobile, Portable and Fixed Receivers Using Terrestrial Transmitters —Section 10B—Frequency-Modulation Sound Broadcasting in Bands 8 (VHF) and 9 (UHF). 14 pp.
CCIR RECMN 789-92. Digital Sound Broadcasting to Vehicular, Portable and Fixed Receivers for BSS (Sound) in the Frequency Range 500-3 000 MHz. 7 pp.

—FM—Interference Immunity
CCIR QUESTION 75/10-90. Immunity of FM Broadcast Receivers Against Interference—Questions Concerning Study Group 10—Broadcasting Service (Sound). 2 pp.

—Low Cost—Sound Broadcasting
CCIR RECMN 415-2-86. Minimum Performance Specifications for Low-Cost Sound-Broadcasting Receivers—Section 10A-2—Sound Broadcasting in the Tropical Zone. 3 pp.

—Mobile—Broadcasting Satellite Services
CCIR QUESTION 93/10-90. Characteristics of Broadcasting-Satellite Systems (Sound) for Individual Reception by Means of Transportable, Portable and/or Mobile Receivers—Questions Concerning Study Group 10—Broadcasting Service (Sound). 2 pp.

—Mobile—Sound Broadcasting—Digital Transmission Systems
CCIR Report 1203-90. Digital Sound Broadcasting to Mobile, Portable and Fixed Receivers Using Terrestrial Transmitters —Section 10B—Frequency-Modulation Sound Broadcasting in Bands 8 (VHF) and 9 (UHF). 14 pp.

Broadcast Receivers (Cont.)

—Portable—Broadcasting Satellite Services
CCIR QUESTION 93/10-90. Characteristics of Broadcasting-Satellite Systems (Sound) for Individual Reception by Means of Transportable, Portable and/or Mobile Receivers—Questions Concerning Study Group 10—Broadcasting Service (Sound). 2 pp.

—Portable—Sound Broadcasting—Digital Transmission Systems
CCIR RECMN 774-92. Digital Sound Broadcasting to Vehicular, Portable and Fixed Receivers Using Terrestrial Transmitters in the VHF/UHF Bands—Section 10B—Frequency-Modulation Sound Broadcasting in Bands 8 (VHF) and 9 (UHF). 7 pp.
CCIR Report 1203-90. Digital Sound Broadcasting to Mobile, Portable and Fixed Receivers Using Terrestrial Transmitters —Section 10B—Frequency-Modulation Sound Broadcasting in Bands 8 (VHF) and 9 (UHF). 14 pp.
CCIR RECMN 789-92. Digital Sound Broadcasting to Vehicular, Portable and Fixed Receivers for BSS (Sound) in the Frequency Range 500-3 000 MHz. 7 pp.

—Radio—Electromagnetic Interference—Measurement
CSA CAN/CSA-C108.9-M91. Sound and Television Broadcasting Receivers and Associated Equipment —Limits and Methods of Measurement of Immunity Characteristics; (Gen Instr 1). 64 pp.

—Radio Frequency Interference
CCIR QUESTION 10-1/1-86. Radiated and Conducted Interference from Receivers—Questions Concerning Study Group 1—Spectrum Management Techniques (Spectrum Engineering, Planning, Sharing, Monitoring and Utilization). 1 p.

—Radio—Spurious Radiation
CCIR RECMN 239-2-78. Spurious Emissions from Sound and Television Broadcast Receivers—Section 1A—Spectrum Engineering and Computer-Aided Principles and Techniques. 1 p.

—Sound Broadcasting
CCIR QUESTION 49-1/10-90. Receivers for Sound Broadcasting—Questions Concerning Study Group 10—Broadcasting Service (Sound). 1 p.

—Sound Broadcasting—Atmospheric Noise—Tropical Regions
CCIR Report 303-3-86. Determination of the Effects of Atmospheric Noise on the Grade of Reception in the Tropical Zone—Section 10A-2—Sound Broadcasting in the Tropical Zone. 2 pp.
CCIR QUESTION 96/10-90. Determination of the Effects of Atmospheric Noise on the Grade of Reception in the Tropical Zone—Questions Concerning Study Group 10—Broadcasting Service (Sound). 1 p.

—Sound Broadcasting—Electrical Measurement
IEC 1079 Pt 3-93. Methods of Measurement on Receivers for Satellite Broadcast Transmissions in the 12 GHz Band Part 3: Electrical Measurements of Overall Performance of Receiver Systems Comprising an Outdoor Unit and a DBS Tuner Unit First Edition. 52 pp.

—Sound Broadcasting—Radio Frequency Interference
IEC CISPR 13-90. Limits and Methods of Measurement of Radio Interference Characteristics of Sound and Television Broadcast Receivers and Associated Equipment Second Edition; (Amendment 1-1992) (Amendment 2-1993). 54 pp.
SAA AS/NZS 1053-92. Limits and Methods of Measurement of Radio Interference Characteristics of Sound and Television Broadcast Receivers and Associated Equipment (IEC/CISPR 13:1990). 24 pp.
SNZ NZS/AS 1053-92. Limits and Methods of Measurement of Radio Interference Characteristics of Sound and Television Broadcast Receivers and Associated Equipment (This is a Joint Standard with SAA AS 1053). 24 pp.

—Sound Broadcasting—Radio Frequency Measurement
IEC 1079 Pt 1-92. Methods of Measurement on Receivers for Satellite Broadcast Transmissions in the 12 GHz Band Part 1: Radio-Frequency Measurements on Outdoor Units First Edition. 91 pp.

—Sound Broadcasting—Signal Interface Connections
CCIR OPINION 74-1-90. Systems for Signal Interface Connection Between Sound-Broadcasting Receivers and Associated Equipment—Volumes X and XI—Part 3—Sound and Television Recording. 1 p.

INTERNATIONAL AND NON-U.S. NATIONAL STANDARDS
SUBJECT INDEX
Broadcasting

Broadcast Receivers *(Cont.)*

—Sound/Data Decoders—Satellite Communications—Electric Measurement
IEC 1079 Pt 4-93. Methods of Measurement on Receivers for Satellite Broadcast Transmissions in the 12 GHz Band Part 4: Electrical Measurements on Sound/Data Decoder Units for the Digital Subcarrier NTSC System. 43 pp.

—Television—Electromagnetic Interference—Measurement
CSA CAN/CSA-C108.9-M91. Sound and Television Broadcasting Receivers and Associated Equipment —Limits and Methods of Measurement of Immunity Characteristics; (Gen Instr 1). 64 pp.

—Television—Radio Frequency Interference
IEC CISPR 13-90. Limits and Methods of Measurement of Radio Interference Characteristics of Sound and Television Broadcast Receivers and Associated Equipment Second Edition; (Amendment 1-1992) (Amendment 2-1993). 54 pp.
SAA AS/NZS 1053-92. Limits and Methods of Measurement of Radio Interference Characteristics of Sound and Television Broadcast Receivers and Associated Equipment (IEC/CISPR 13:1990). 24 pp.
SNZ NZS/AS 1053-92. Limits and Methods of Measurement of Radio Interference Characteristics of Sound and Television Broadcast Receivers and Associated Equipment (This is a Joint Standard with SAA AS 1053). 24 pp.

—Television—Signal Interface Connections
CCIR OPINION 75-1-90. Systems for Signal Interface Connection Between Television Receivers and Associated Equipment—Volumes X and XI—Part 3—Sound and Television Recording. 1 p.

—Television—Spurious Radiation
CCIR RECMN 239-2-78. Spurious Emissions from Sound and Television Broadcast Receivers—Section 1A—Spectrum Engineering and Computer-Aided Principles and Techniques. 1 p.

Broadcast Stations
Use For: Broadcasting Stations
See Also: Broadcasting; Radio Stations; Television Stations

—Aeronautical Mobile Services—Radio Navigation—Compatibility
CCIR Decision 71-1-89. Continuation of Studies on Compatibility Between the Aeronautical Radionavigation Service in the Band 108-117.975 MHz, the Aeronautical Mobile (R) Service in the Band 117.975-137 MHz and the FM Sound Broadcasting Stations in the Band About 87-108 MHz—Annex 3 to Volume. 1 p.

—Sound Broadcasting—Lightning Protection
CCIR Report 943-1-86. Protection of Sound-Broadcasting Stations Against Atmospheric Electricity—Section 10A-1—Amplitude-Modulation Sound Braodcasting in Bands 5 (LF), 6 (MF) and 7 (HF). 4 pp.
CCIR QUESTION 48-1/10-86. Protection of Sound-Broadcasting Stations Against Atmospheric Electricity—Questions Concerning Study Group 10—Broadcasting Service (Sound). 1 p.

—Standard Frequency Dissemination
CCIR Report 576-3-86. Standard-Frequency Dissemination Via Stabilized Broadcast Station Carriers—Section 7C—Systems for Dissemination and Comparison. 1 p.

—Television Broadcasting—Lightning Protection
CCIR QUESTION 38/11-82. Protection of Television Broadcasting Stations Against Lightning—Questions Concerning Study Group 11—Broadcasting Service (Television). 1 p.

Broadcast Transmitters
See Also: Transmitters

—Aircraft
CEPT T/R 51-01-85. Mesures A Prendre Pour Empecher Le Fonctionne-ment De Stations De Radiodiffusion A Bord De Navires Ou D'Aeronefs Hors Des Limites Des Territoires Nationaux. 1 p.

—AM—Sound Broadcasting—Energy Consumption
CCIR Report 1060-1-90. Energy Saving Methods in Amplitude Modulation Broadcasting and Their Influence on Reception Quality—Section 10A-1— Amplitude-Modulation Sound Broadcasting in Bands 5 (LF), 6 (MF) and 7 (HF). 4 pp.

Broadcast Transmitters *(Cont.)*

—FM—Radio Data Systems—Radio Paging Services
CCIR Report 900-2-90. Radio-Paging Systems Standardization of Code and Format—Section 8A— Land Mobile Service and Related Subjects. 22 pp.

—FM—Sound Broadcasting—Radio Spectra
CCIR Report 1065-86. RF Spectrum of Frequency-Modulation Sound-Broadcasting Transmitters— Section 10B—Frequency-Modulation Sound Broadcasting in Bands 8 (VHF) and 9 (UHF). 6 pp.

—FM—Terrestrial—Sound Broadcasting
CCIR RECMN 774-92. Digital Sound Broadcasting to Vehicular, Portable and Fixed Receivers Using Terrestrial Transmitters in the VHF/UHF Bands— Section 10B—Frequency-Modulation Sound Broadcasting in Bands 8 (VHF) and 9 (UHF). 7 pp.

—HF—Synchronized—Sound Broadcasting
CCIR QUESTION 62/10-90. Synchronized Transmitters in HF Sound Broadcasting —Questions Concerning Study Group 10—Broadcasting Service (Sound). 1 p.

—Interfaces—Supervisory Control Systems
CENELEC HD 577-90. Standardization of Interconnections Between Broadcasting Transmitters or Transmitter Systems and Supervisory Equipment Part 1: Interface Standards for Systems Using Dedicated Interconnections. 3 pp.
IEC 864 Pt 1-86. Standardization of Interconnections Between Broadcasting Transmitters or Transmitter Systems and Supervisory Equipment Part 1: Interface Standards for Systems Using Dedicated Interconnections First Edition; (Amendment 1-1987). 60 pp.

—Radio—Phase Modulation—Time Signals
CCIR Report 577-3-90. Dissemination of Time Signals and Time Codes by Addition of Phase Modulation on Amplitude-Modulated Sound Broadcasting Transmitters—Section 7C—Systems for Dissemination and Comparison. 3 pp.
CCIR QUESTION 102/7-90. Dissemination of Standard Frequencies by Carrier-Frequency Stabilization of Broadcasting Emissions and Dissemination of Time Signals by Addition of Phase Modulation on Amplitude-Modulated Sound Broadcasting Transmitters—Questions Concerning Study Group. 1 p.

—Radio—Spurious Radiation
CCIR RECMN 329-6-90. Spurious Emissions—Section 1A—Spectrum Engineering and Computer-Aided Principles and Techniques. 8 pp.

—Ships
CEPT T/R 51-01-85. Mesures A Prendre Pour Empecher Le Fonctionne-ment De Stations De Radiodiffusion A Bord De Navires Ou D'Aeronefs Hors Des Limites Des Territoires Nationaux. 1 p.

—Sound Broadcasting—Carrier Power— Tropical Regions
CCIR RECMN 215-2-82. Maximum Transmitter Powers for Broadcasting in the Tropical Zone— Section 10A-2—Sound Broadcasting in the Tropical Zone. 1 p.

—Sound Broadcasting—Communication Channels
CCIR Report 1201-90. Number of HF Sound Broadcasting Transmitters Using a Single Channel— Section 10A-1—Amplitude-Modulation Sound Broadcasting in Bands 5 (LF), 6 (MF) and 7 (HF). 2 pp.

—Supervisory Equipment—Interfaces
CENELEC HD 577-90. Standardization of Interconnections Between Broadcasting Transmitters or Transmitter Systems and Supervisory Equipment Part 1: Interface Standards for Systems Using Dedicated Interconnections. 3 pp.
IEC 864 Pt 1-86. Standardization of Interconnections Between Broadcasting Transmitters or Transmitter Systems and Supervisory Equipment Part 1: Interface Standards for Systems Using Dedicated Interconnections First Edition; (Amendment 1-1987). 60 pp.

—Television Broadcasting—Television Signals
CCIR QUESTION 49/11-90. Characteristics of Television Signals Radiated in Band 10 (SHF) from Terrestrial Broadcasting Transmitters—Questions Concerning Study Group 11—Broadcasting Service (Television). 1 p.

Broadcasting
Scope Note: For additional listings, use a more specific term *See Also:* Broadcast Stations; Data Broadcasting; European Broadcasting Union; Integrated Service Digital Broadcasting; Luminance Signal; Rebroadcasting; Sound Broadcasting; Television Broadcasting
CCIR Report 929-2-90. Compatibility Between the Broadcasting Service in the Band of About 87-108 MHz and the Aeronautical Services in the Band 108-136 MHz—Section 8K—Aeronautical Mobile Service (Terrestrial) (This Report is Published Separately). 1 p.

— 30 MHz+—Design—Radio Wave Propagation
CCIR QUESTION 11-1/5-90. Propagation Data and Prediction Methods for the Terrestrial Broadcasting and Terrestrial Mobile Services in the Frequency Range Above 30 MHz—Questions Concerning Study Group 5 —Radio Wave Propagation in Non-Ionized Media. 1 p.

—Adjacent Channel Interference
CCIR Report 485-1-82. Contribution to the Planning of Broadcasting Services—Section 11E—Planning of Television Networks, Protection Ratios, Television Receivers and Antennas. 5 pp.

—Aircraft—Territorial Limits
CEPT T/R 51-01 E-88. Measures to Be Taken to Prevent the Operation of Broadcasting Stations on Board Ships or Aircraft Outside National Territorial Limits. 2 pp.

—Broadcasting Satellite Services— Frequency Band Sharing
CCIR Report 631-4-90. Frequency Sharing Between the Broadcasting-Satellite Service (Sound and Television) and Terrestrial Services—Section 10/ 11E—Sharing. 49 pp.

—CCIR—Liaison
CCIR RESOLUTION 80-1-90. CCIR Liasion with Organizations Dealing with Radiocommunications Including Broadcasting—Volume XIV— Administrative Texts of the CCIR. 1 p.

—Co-Channel Interference
CCIR Report 485-1-82. Contribution to the Planning of Broadcasting Services—Section 11E—Planning of Television Networks, Protection Ratios, Television Receivers and Antennas. 5 pp.

—Conditional Access
CCIR QUESTION 77/11-90. Conditional-Access Broadcasting Systems—Questions Concerning Study Group 11—Broadcasting Service (Television). 1 p.
CCIR RECMN 810-92. Conditional-Access Broadcasting Systems. 13 pp.

—Directional Antennas
CCIR Report 356-3-90. Use of Directional Antennas in the Bands 4 to 28 MHz—Section 3Ab—Antennas Characteristics. 11 pp.

—Fixed Services (Radio Communications)— Frequency Band Sharing
CCIR QUESTION 2/12-90. Sharing Between the Broadcasting Service and the Fixed and/or Mobile Services in VHF and UHF Bands—Questions Concerning Study Group 12—Inter-Service Sharing and Compatibility. 2 pp.
CCIR QUESTION 131/9-90. Criteria for Frequency Sharing Between the Fixed and Broadcasting Services—Questions of Study Group 9 Fixed Service. 1 p.
CCIR RECMN 851-92. Sharing Between the Broadcasting Service and the Fixed and/or Mobile Services in the VHF and UHF Bands. 26 pp.

—Glossaries
BSI BS 4727:Pt3: Group 04-76. 1976 Electrotechnical, Power, Telecommunication, Electronics, Lighting and Colour Terms Part 3: Terms Particular to Telecommunications and Electronics Group 04: Broadcasting and Television Terminology. 36 pp.

—HF
CCIR Decision 85-89. Studies of the Propagation Prediction Method for HF Broadcasting—Annex to Volume VI—Propagation in Ionized Media. 1 p.

—HF—Field Strength
CCIR Report 1149-90. HF Field Strength Measurements Specifications for a Field Strength Measurement Campaign Intended for Future Improvements in Prediction Methods, Particularly Those Used for HF Broadcasting—Section 6G— Ionospheric Propagation Measurements. 17 pp.

—HF—Frequency Band Sharing
CCIR RECMN 831-92. Frequency Sharing Between Services in the Band 4-30 MHz. 3 pp.

INDUSTRY STANDARDS

Broadcasting (Cont.)

—HF—Radio Wave Propagation
CCIR Report 562-4-90. Propagation Data Required for Terrestrial Broadcasting and Point-to-Multipoint Communication Systems in the Frequency Bands Above 10 GHz—Section 5D—Aspects Relative to the Terrestrial Broadcasting and Mobile Services. 3 pp.

—HF—World Administrative Radio Conference
CCIR RESOLUTION 98-90. CCIR Studies to Be Carried out for Submission to the World Administrative Radio Conference (WARC-93) Dealing with Matters Connected with the HF Broadcasting Service—Volume XIV—Administrative Texts of the CCIR. 3 pp.

—LF/MF—Sky Wave Field Strength—Forecasting
CCIR RECMN 435-7-92. Sky-Wave Field-Strength Prediction Method for the Broadcasting Service in the Frequency Range 150 to 1 600 kHz Section 6D—Ionospheric Propagation Prediction at Frequencies Below About 1.6 MHz. 25 pp.

—MF/HF—Fixed Services—Frequency Band Sharing
CCIR Decision 97-89. Sharing Criteria Between Broadcasting and Fixed and Mobile Services in the Band 2-30 MHz—Annex to Volume III—Fixed Service at Frequencies Below About 30 MHz (See Annex to Volume X-1). 1 p.
CCIR Decision 97-89. Sharing Criteria Between Broadcasting and Fixed and Mobile Services in the Band 2-30 MHz—(See Annex to Vol. X-1)—Annex 1 to Volume VIII —Land Mobile Service—Amateur Service—Amateur-Satellite Service. 1 p.

—MF/HF—Mobile Radio Services—Frequency Band Sharing
CCIR Decision 97-89. Sharing Criteria Between Broadcasting and Fixed and Mobile Services in the Band 2-30 MHz—Annex to Volume III—Fixed Service at Frequencies Below About 30 MHz (See Annex to Volume X-1). 1 p.

—Mobile Radio Services—Frequency Band Sharing
CCIR Report 1098-90. Sharing Between the Land Mobile Service and the Broadcasting Service in the VHF and UHF Bands Where the Service Areas Are Geographically Seperated—Section 1B—Spectrum Sharing and Planning Principles and Techniques. 7 pp.
CCIR QUESTION 2/12-90. Sharing Between the Broadcasting Service and the Fixed and/or Mobile Services in VHF and UHF Bands—Questions Concerning Study Group 12—Inter-Service Sharing and Compatibility. 2 pp.
CCIR RECMN 851-92. Sharing Between the Broadcasting Service and the Fixed and/or Mobile Services in the VHF and UHF Bands. 26 pp.

—Protection Ratio
CCIR Report 485-1-82. Contribution to the Planning of Broadcasting Services—Section 11E—Planning of Television Networks, Protection Ratios, Television Receivers and Antennas. 5 pp.

—Radio Frequency Interference
CEPT T/R 70-01-72. Possibilite D'Attenuer Certains Brouillages Mutuels Entre Le Service De Radiodiffusion Et Le Service Mobile. 1 p.
CEPT T/R 70-01 E-72. Possible Reduction in Certain Types of Mutual Interference Between the Broadcasting Service and the Mobile Service. 2 pp.

—SHF/EHF—Design—Radio Wave Propagation
CCIR QUESTION 13/5-90. Propagation Data Required for Terrestrial Broadcasting Above 10 GHz—Questions Concerning Study Group 5—Radio Wave Propagation in Non-Ionized Media. 1 p.

—Ships—Territorial Limits
CEPT T/R 51-01 E-88. Measures to Be Taken to Prevent the Operation of Broadcasting Stations on Board Ships or Aircraft Outside National Territorial Limits. 2 pp.

—Standard Frequency Dissemination
CCIR QUESTION 102/7-90. Dissemination of Standard Frequencies by Carrier-Frequency Stabilization of Broadcasting Emissions and Dissemination of Time Signals by Addition of Phase Modulation on Amplitude-Modulated Sound Broadcasting Transmitters—Questions Concerning Study Group. 1 p.

Broadcasting (Cont.)

—Time Codes
CCIR RECMN 808-92. Broadcasting of Time and Date Information in Coded Form. 5 pp.

—VHF—Radio Frequency Interference—Aeronautical Mobile Services
CCIR RECMN 591-1-86. Compatibility Between the Broadcasting Service in the Band of About 87-108 MHz and the Aeronautical Services in the Band 108-136 MHz—Section 8K —Aeronautical Mobile Service (Terrestrial). 1 p.
CCIR QUESTION 1/12-90. Compatibility Between the Broadcasting Service in the Band of About 87-108 MHz and the Aeronautical Services in the Band 108-137 MHz—Questions Concerning Study Group 12—Inter-Service Sharing and Compatibility. 2 pp.

—VHF/UHF—Field Strength
CCIR Report 228-3-86. Measurement of Field Strength for VHF (Metric) and UHF (Decimetric) Broadcast Services, Including Television—Section 5D—Aspects Relative to the Terrestrial Broadcasting and Mobile Services. 8 pp.

—VHF/UHF—Radio Wave Propagation—Frequency Curves
CCIR RECMN 370-5-86. VHF and UHF Propagation Curves for the Frequency Range from 30 MHz to 1000 MHz. Broadcasting Services—Section 5D—Aspects Relative to the Terrestrial Broadcasting and Mobile Services. 28 pp.
CCIR Report 239-7-90. Propagation Statistics Required for Broadcasting Services Using the Frequency Range 30 to 1000 MHz—Section 5D—Aspects Relative to the Terrestrial Broadcasting and Mobile Services. 19 pp.

Broadcasting Satellite Services

Use For: BSS *See Also:* Fixed Satellite Services; Satellite Communications; Satellites; Sound Broadcasting; Television Broadcasting

—Acoustic Signals—Multiplexing
CCIR Report 954-2-90. Multiplexing Methods for the Emission of Several Digital Audio Signals and Also Data Signals in Broadcasting—Section 10/11B—Systems. 8 pp.

—Antennas—Electrical Measurement
IEC 1114 Pt 1-92. Methods of Measurement on Receiving Antennas for Satellite Broadcast Transmissions in the 12 GHz Band First Edition; (CENELEC EN 61114-1: 1993). 72 pp.

—Antennas—Reference Patterns
CCIR RECMN 652-86. 12 GHz Receiving Earth-Station Antenna and Satellite Transmitting Antenna Reference Patterns for the Broadcasting-Satellite Service—Section 10/11C —Technology. 6 pp.
CCIR RECMN 652-1-92. Reference Patterns for Earth-Station and Satellite Antennas for the Broadcasting-Satellite Service in the 12 GHz Band and for the Associated Feeder Links in the 14 GHz and 17 GHz Bands—Section 10/11C—Technology. 17 pp.
CCIR Report 810-3-90. Broadcasting-Satellite Service (Sound and Television) Reference Patterns and Technology for Transmitting and Receiving Antennas—Section 10/11C—Technology. 22 pp.

—Broadcast Receivers
CCIR Report 473-5-90. Characteristics of Receiving Equipment for the Broadcasting-Satellite Service—Section 10/11C—Technology. 28 pp.

—Broadcast Receivers—Figure of Merit
CCIR RECMN 790-92. Characteristics of Receiving Equipment and Calculation of Receiver Figure-of-Merit (G/T) for the Broadcasting-Satellite Service. 3 pp.

—Data Transmission—Multiplexing
CCIR Report 954-2-90. Multiplexing Methods for the Emission of Several Digital Audio Signals and Also Data Signals in Broadcasting—Section 10/11B—Systems. 8 pp.

—Design—Radio Wave Propagation
CCIR QUESTION 16-1/5-90. Propagation Data and Prediction Methods for Fixed-Satellite and Broadcasting-Satellite Services—Questions Concerning Study Group 5—Radio Wave Propagation in Non-Ionized Media. 2 pp.

—Digital Audio Coding
CCIR RECMN 651-86. Digital PCM Coding for the Emission of High-Quality Sound Signals in Satellite Broadcasting (15 kHz Nominal Bandwidth)—Section 10/11B—Systems. 2 pp.
CCIR Report 953-2-90. Digital Coding for the Emission of High-Quality Sound Signals in Satellite Broadcasting (15 kHz Nominal Bandwidth)—Section 10/11B—Systems. 18 pp.

Broadcasting Satellite Services (Cont.)

—Earth Stations—Fixed Satellites—Radio Frequency Emissions
CCIR Report 712-1-82. Factors Concerning the Protection of Fixed-Satellite Earth Stations Operating in Adjacent Frequency Band Allocations Against Unwanted Emissions from Broadcasting Satellites Operating in Frequency Bands Around 12 GHz—Section 4E—Frequency Sharing Between Networks. 12 pp.

—Feeder Links—Downlinks—Noise Partioning
CCIR RECMN 793-92. Partitioning of Noise Between Feeder Links for the Broadcasting-Satellite Service (BSS) and BSS Down Links. 5 pp.

—Feeder Links—Earth Stations—Fixed Satellite Services
CCIR QUESTION 44/4-90. Use of Transportable Transmitting Earth Stations in the Fixed-Satellite Service Including Use for Feeder Links to Broadcasting Satellites—Questions of Study Group 4 Fixed-Satellite Service. 1 p.

—Feeder Links—Fixed Services—EIRP
CCIR Report 1006-86. Fixed Service e.i.r.p. Limits for the Protection of the Broadcasting-Satellite Feeder Links Around 18 GHz—Section 4/9A—Sharing Conditions. 2 pp.

—Feeder Links—Radio Frequency Interference
CCIR RECMN 795-92. Techniques for Alleviating Mutual Interference Between Feeder Links to the BSS. 3 pp.

—Feeder Links—Rain Fading
CCIR RECMN 794-92. Techniques for Minimizing the Impact on the Overall BSS System Performance Due to Rain Along the Feeder-Link Path. 11 pp.

—Fixed Satellite Services—Frequency Band Sharing
CCIR Report 809-3-90. Inter-Regional Sharing of the 11.7 to 12.75 GHz Frequency Band Between the Broadcasting-Satellite Service and the Fixed-Satellite Service—Section 10/11E —Sharing. 6 pp.
CCIR QUESTION 66/4-90. Frequency Sharing Between the Broadcasting-Satellite Service and the Broadcasting-Satellite Service—Questions of Study Group 4 Fixed-Satellite Service. 1 p.

—Fixed Satellite Services—Radio Frequency Interference
CCIR Report 873-2-90. Analysis of the Interfer-ence from the Broadcast-ing-Satellite Service of One Region into the Fixed-Satellite Service of Another Region Around 12 GHz—Section 4E—Frequency Sharing Between Networks of the Fixed-Satellite Service and Those of Other Space Radiocommunications Sys.. 7 pp.

—Fixed Services (Radio Communications)—Frequency Band Sharing
CCIR QUESTION 111/9-90. Protection Criteria Between the Broadcasting-Satellite Service and the Fixed Service—Questions of Study Group 9 Fixed Service. 1 p.

—Glossaries
CCIR RECMN 566-3-90. Terminology Relating to the Use of Space Communication Techniques for Broadcasting—Section 10/11A—Terminology. 5 pp.

—High Definition Television Systems
CCIR QUESTION 100/1-90. Satellite Broadcasting of High Definition Television (HDTV)—Questions Concerning Study Group 11—Broadcasting Service (Television). 1 p.

—High Definition Television Systems—Bit Rate Reduction
CCIR RECMN 788-92. Coding Rate for the Wide RF-Band HDTV Broadcasting-Satellite Service. 1 p.

—High Definition Television Systems—Frequency Assignment
CCIR Decision 93-89. Preparatory Work for the WARC 1992—Annex to Volumes X and XI—Part 2—Broadcasting-Satellite Service (Sound and Television). 3 pp.

—High Definition Television Systems—Frequency Band Sharing
CCIR QUESTION 89/11-90. Sharing Studies Between High-Definition Television (HDTV) in the Broadcasting-Satellite Service and Other Services—Questions Concerning Study Group 11—Broadcasting Service (Television). 1 p.

INTERNATIONAL AND NON-U.S. NATIONAL STANDARDS
SUBJECT INDEX
Broadcasting

Broadcasting Satellite Services (Cont.)

—High Definition Television Systems—MAC/Packet
CCIR RECMN 787-92. MAC/Packet Based System for HDTV Broadcasting-Satellite Services. 18 pp.

—High Definition Television Systems—MUSE
CCIR RECMN 786-92. MUSE System for HDTV Broadcasting-Satellite Services. 19 pp.

—Integrated Service Digital Broadcasting
CCIR Report 1227-90. Satellite Broadcasting Systems for ISDB (Integrated Service Digital Broadcasting)—Section 10/11B—Systems. 4 pp.

—Interference—Radio Relay Systems
CCIR RECMN 760-92. Protection of Terrestrial Line-of-Sight Radio-Relay Systems Against Interference from the Broadcasting-Satellite Service in the Band 22.5-23 GHz—Section 9F—Frequency Sharing with Other Services. 3 pp.
CCIR Report 789-1-82. Protection of Terrestrial Line-of-Sight Radio-Relay Systems Against Interference from the Broadcasting-Satellite Service in the Band 11.7 to 12.75 GHz—Section 9F—Frequency Sharing with Other Services. 2 pp.
CCIR Report 1189-90. Protection of Terrestrial Line-of-Sight Radio-Relay Systems Against Interference from the Broadcasting-Satellite Service in the Band 22.5 GHz—Section 9F—Frequency Sharing with Other Services. 4 pp.

—Interference—Tropospheric Propagation
CCIR Decision 4-8-89. Tropospheric Propagation Data for Planning Space and Point-to-Point Terrestrial Telecommunication Systems and Determining Likely Interference Between Systems—Volume V—Propagation in Non-Ionized Media. 2 pp.

—Intersatellite Services—Frequency Band Sharing
CCIR Report 951-82. Sharing Between the Inter-Satellite Service and the Broadcasting-Satellite Service in the Vicinity of 23 GHz—Section 10/11E—Sharing. 7 pp.

—Planning—Computer Programs
CCIR Report 812-3-90. Computer Programs for Planning Broadcasting-Satellite Services in the 12 GHz Band—Section 10/11D—Planning. 7 pp.

—Polarization
CCIR Report 814-2-86. Factors to be Considered in the Choice of Polarization for Planning the Broadcasting-Satellite Service—Section 10/11D—Planning. 6 pp.
CCIR RECMN 791-92. Choice of Polarization for the Broadcasting-Satellite Service. 7 pp.

—SHF—Feeder Links—Frequency Assignment
CCIR Report 952-2-90. Technical Characteristics of Feeder Links to Broadcasting Satellites Elements Required for the Establishment of Plans of Frequency Assignments and Orbital Positions for the Broadcasting-Satellite Service and the Associated Feeder Links—Sharing in the Feeder-. 56 pp.

—SHF—Feeder Links—Orbits
CCIR Report 952-2-90. Technical Characteristics of Feeder Links to Broadcasting Satellites Elements Required for the Establishment of Plans of Frequency Assignments and Orbital Positions for the Broadcasting-Satellite Service and the Associated Feeder Links—Sharing in the Feeder-. 56 pp.

—SHF—Frequency Assignment
CCIR Report 811-2-86. Broadcasting-Satellite Service Planning Elements Including Those Used in the Establishment of Plans of Frequency Assignments and Orbital Positions for the Broadcasting-Satellite Service in the 12 GHz Band—Section 10/11D—Planning. 7 pp.

—SHF—Frequency Planning
CCIR Report 633-3-86. Orbit and Frequency Planning in the Broadcasting-Satellite Service—Section 10/11D—Planning. 24 pp.

—SHF—Orbits
CCIR Report 633-3-86. Orbit and Frequency Planning in the Broadcasting-Satellite Service—Section 10/11D—Planning. 24 pp.
CCIR Report 811-2-86. Broadcasting-Satellite Service Planning Elements Including Those Used in the Establishment of Plans of Frequency Assignments and Orbital Positions for the Broadcasting-Satellite Service in the 12 GHz Band—Section 10/11D—Planning. 7 pp.

Broadcasting Satellite Services (Cont.)

—Sound
CCIR Volume X&XI TOC-90. Table of Contents. 1 p.
CCIR Volume X&XI NUM IND TXTS-90. Numerical Index of Texts. 1 p.
CCIR Volume X&XI INTRO Chrmn-90. Introduction by the Chairman. 3 pp.
CCIR Volume X&XI Pt2/ANX TOC-90. Table of Contents. 3 pp.
CCIR Volume X&XI Pt2/ANX IND-90. Numerical Index of Texts. 1 p.
CCIR QUESTION 84/11-90. System Characteristics for the Broadcasting-Satellite Service (Sound and Television)—Questions Concerning Study Group 11—Broadcasting Service (Television). 2 pp.
DIN VDE 0855 Pt 11 (P)-88. Cabled Distribution Systems; Receiving Aerial Systems with Satellite Receiving Equipment (Nov). 10 pp.

—Sound—Acoustic Signals—Pulse Code Modulation
CCIR RECMN 651-86. Digital PCM Coding for the Emission of High-Quality Sound Signals in Satellite Broadcasting (15 kHz Nominal Bandwidth)—Section 10/11B—Systems. 2 pp.

—Sound—Broadcast Receivers
CCIR Report 955-2-90. Satellite Sound Broadcasting with Portable Receivers and Receivers in Automobiles—Section 10/11B—Systems. 79 pp.
CCIR Decision 43-5-89. Satellite Sound Broadcasting for Portable and Vehicle Receivers and Sharing and Spectrum Aspects of Wide RF-Band HDTV Satellite Broadcasting—Annex to Volumes X and XI—Part 2—Broadcasting-Satellite Service (Sound and Television). 5 pp.
CCIR QUESTION 93/10-90. Characteristics of Broadcasting-Satellite Systems (Sound) for Individual Reception by Means of Transportable, Portable and/or Mobile Receivers—Questions Concerning Study Group 10—Broadcasting Service (Sound). 2 pp.
CCIR QUESTION 97/11-90. Characteristics of Receiving Systems of the Broadcasting-Satellite Service (Sound and Television)—Questions Concerning Study Group 11—Broadcasting Service (Television). 1 p.

—Sound—Broadcast Receivers—Electrical Measurement
IEC 1079 Pt 3-93. Methods of Measurement on Receivers for Satellite Broadcast Transmissions in the 12 GHz Band Part 3: Electrical Measurements of Overall Performance of Receiver Systems Comprising an Outdoor Unit and a DBS Tuner Unit First Edition. 52 pp.

—Sound—Broadcast Receivers—Radio Frequency Measurement
IEC 1079 Pt 1-92. Methods of Measurement on Receivers for Satellite Broadcast Transmissions in the 12 GHz Band Part 1: Radio-Frequency Measurements on Outdoor Units First Edition. 91 pp.

—Sound/Data
CCIR Decision 51-4-89. Satellite Broadcasting of High Definition Television (HDTV) Signals and Accomodation of Several Audio and/or Data Signals Either Associated with Television Signals or for Sound/Data Broadcasting in Terrestrial and Satellite Broadcasting. 6 pp.

—Sound/Data Decoders—Electrical Measurement
IEC 1079 Pt 4-93. Methods of Measurement on Receivers for Satellite Broadcast Transmissions in the 12 GHz Band Part 4: Electrical Measurements on Sound/Data Decoder Units for the Digital Subcarrier NTSC System. 43 pp.

—Sound/Data—SHF
CCIR RECMN 712-90. High-Quality Sound/Data Standards for the Broadcasting-Satellite Service in the 12 GHz Band—Section 10/11B—Systems. 1 p.
CCIR RECMN 712-1-92. High-Quality Sound/Data Standards for the Broadcasting-Satellite Service in the 12 GHz Band—Section 10/11B—Systems. 48 pp.
CCIR Report 1228-90. High Quality Sound/Data Standards for the Broadcasting Satellite Service in the 12 GHz Band—Section 10/11B—Systems. 34 pp.

—Sound—Design
CCIR Report 215-7-90. Systems for the Broadcasting Satellite Service (Sound and Television)—Section 10/11B—Systems. 27 pp.
CCIR Report 632-4-90. Broadcasting-Satellite Service (Sound and Television) Technically Suitable Methods of Modulation—Section 10/11B—Systems. 32 pp.

Broadcasting Satellite Services (Cont.)

—Sound—Design (Cont.)
CCIR QUESTION 90/11-90. Possible Broadcasting-Satellite Systems (Sound and Television) and Their Relative Acceptability—Questions Concerning Study Group 11—Broadcasting Service (Television). 1 p.

—Sound—Digital Modulation
CCIR QUESTION 92/11-90. Digital Techniques in the Broadcasting-Satellite Service (Sound and Television)—Questions Concerning Study Group 11—Broadcasting Service (Television). 1 p.

—Sound—Feeder Links—Frequency Assignment
CCIR QUESTION 86/11-90. Frequencies for the Feeder Links to a Broadcasting Satellite (Sound and Television)—Questions Concerning Study Group 11—Broadcasting Service (Television). 2 pp.

—Sound—Feeder Links—Sound-Program Signals
CCIR QUESTION 59/CMTT-90. Transmission of Digital Sound-Programme Signals over Broadcasting Satellite Feeder Links—Questions Concerning the CMTT CCIR/CCITT Joint Study Group for Television and Sound Transmission. 1 p.

—Sound—Fixed Satellite Services—Frequency Band Sharing
CCIR Report 665-2-86. New Technologies for Off-Set Reflector Multiple-Beam Antennas—Section 1D—Spectrum Utilization and Applications. 10 pp.

—Sound—Fixed Services—Frequency Band Sharing
CCIR Report 631-4-90. Frequency Sharing Between the Broadcasting-Satellite Service (Sound and Television) and Terrestrial Services—Section 10/11E—Sharing. 49 pp.
CCIR QUESTION 112/9-90. Frequency Sharing Between the Broadcasting-Satellite Service (Sound) and the Fixed Service in the Band 0.5 to 3 GHz—Questions of Study Group 9 Fixed Service. 1 p.

—Sound—Frequency Assignment
CCIR Decision 93-89. Preparatory Work for the WARC 1992—Annex to Volumes X and XI—Part 2—Broadcasting-Satellite Service (Sound and Television). 3 pp.

—Sound—Frequency Band Sharing
CCIR Report 631-4-90. Frequency Sharing Between the Broadcasting-Satellite Service (Sound and Television) and Terrestrial Services—Section 10/11E—Sharing. 49 pp.

—Sound—Integrated Service Digital Broadcasting
CCIR QUESTION 101/11-90. Integrated Service Digital Broadcasting in the Broadcasting-Satellite Service (Sound and Television)—Questions Concerning Study Group 11—Broadcasting Service (Television). 1 p.

—Sound—Mobile Radio Services—Frequency Band Sharing
CCIR Report 631-4-90. Frequency Sharing Between the Broadcasting-Satellite Service (Sound and Television) and Terrestrial Services—Section 10/11E—Sharing. 49 pp.

—Sound—Modulation
CCIR Report 632-4-90. Broadcasting-Satellite Service (Sound and Television) Technically Suitable Methods of Modulation—Section 10/11B—Systems. 32 pp.

—Sound—Radio Frequency Interference
CCIR QUESTION 83/11-90. Broadcasting-Satellite Service (Sound and Television) Protection from Interference—Questions Concerning Study Group 11—Broadcasting Service (Television). 2 pp.

—Sound—Radio Relay Systems—Radio Frequency Interference
CCIR Report 941-82. Protection of Terrestrial Line-of-Sight Radio-Relay Systems Against Interference from the Broadcasting-Satellite Service (Sound) in the Band 1427 to 1530 MHz—Section 9F—Frequency Sharing with Other Services. 5 pp.

—Sound—Receiving Antennas
CCIR QUESTION 97/11-90. Characteristics of Receiving Systems of the Broadcasting-Satellite Service (Sound and Television)—Questions Concerning Study Group 11—Broadcasting Service (Television). 1 p.

INDUSTRY STANDARDS

Broadcasting

Broadcasting Satellite Services (Cont.)

—Sound—SHF
CCIR QUESTION 85/11-90. Broadcasting-Satellite Service (Sound and Television) Use of the 12 GHz Band—Questions Concerning Study Group 11—Broadcasting Service (Television). 2 pp.

—Sound—Space Stations—Spurious Radiation
CCIR QUESTION 94/11-90. Radiation of Unwanted Emissions from Space Stations by the Broadcasting-Satellite Service (Sound and Television)—Questions Concerning Study Group 11—Broadcasting Service (Television). 1 p.

—Sound Transmission—Television Channels
CCIR QUESTION 78/10-90. Standards for the Transmission of Several Sound Signals in One Television Channel in Terrestrial or Satellite Broadcasting—Questions Concerning Study Group 10—Broadcasting Service (Sound). 1 p.

—Sound—Transmitting Antennas
CCIR QUESTION 93/11-90. Transmitting Antennas for the Broadcasting-Satellite Service (Sound and Television)—Questions Concerning Study Group 11—Broadcasting Service (Television). 1 p.

—Sound—Tuners
IEC 1079 Pt 2-92. Methods of Measurement on Receivers for Satellite Broadcast Transmissions in the 12 GHz Band Part 2: Electrical Measurements on DBS Tuner Units First Edition. 111 pp.

—Space Segment Technology
CCIR Report 808-3-90. Broadcasting-Satellite Service Space-Segment Technology—Section 10/11C—Technology. 13 pp.

—Space Stations—Feeder Links
CCIR Report 561-4-90. Feeder Links to Space Stations in the Broadcasting-Satellite Service—Section 4E—Frequency Sharing Between Networks of the Fixed-Satellite Service and Those of Other Space Radiocommunications Systems. 13 pp.

CCIR QUESTION 54/4-90. Feeder Links for the Space Stations in the Broadcasting-Satellite Service—Questions of Study Group 4 Fixed-Satellite Service. 1 p.

—Space Stations—Frequency Band Sharing
CCIR Report 1076-86. Considerations Affecting the Accommodation of Spacecraft Service Functions (TTC) Within the Broadcasting-Satellite and Feeder-Link Service Bands—Section 10/11E—Sharing. 14 pp.

—Space Stations—Noise
CCIR Report 807-3-90. Unwanted Emissions from Broadcasting-Satellite Space Stations—Section 10/11E—Sharing. 11 pp.

—Television
CCIR Volume X&XI TOC-90. Table of Contents. 1 p.

CCIR Volume X&XI NUM IND TXTS-90. Numerical Index of Texts. 1 p.

CCIR Volume X&XI INTRO Chrmn-90. Introduction by the Chairman. 3 pp.

CCIR RECMN 650-2-92. Standards for Conventional Television Systems for Satellite Broadcasting in the Channels Defined by Appendix 30 of the Radio Regulations—Section 10/11B—Systems. 19 pp.

CCIR Volume X&XI Pt2/ANX TOC-90. Table of Contents. 3 pp.

CCIR Report 1073-1-90. Television Standards for the Broadcasting-Satellite Service—Section 10/11B—Systems. 16 pp.

CCIR QUESTION 84/11-90. System Characteristics for the Broadcasting-Satellite Service (Sound and Television)—Questions Concerning Study Group 11—Broadcasting Service (Television). 2 pp.

EC 92/38/EEC-92. Council Directive on the Adoption of Standards for Satellite Broadcasting of Television Signals. 4 pp.

—Television—Broadcast Receivers
CCIR QUESTION 97/11-90. Characteristics of Receiving Systems of the Broadcasting-Satellite Service (Sound and Television)—Questions Concerning Study Group 11—Broadcasting Service (Television). 1 p.

—Television Broadcasting—Frequency Band Sharing
CCIR Report 665-2-86. New Technologies for Off-Set Reflector Multiple-Beam Antennas—Section 1D—Spectrum Utilization and Applications. 10 pp.

Broadcasting Satellite Services (Cont.)

—Television Broadcasting—MAC/Packet
EC 86/529/EEC-86. Council Directive on the Adoption of Common Technical Specifications of the MAC/Packet Family of Standards for Direct Satellite Television Broadcasting. 2 pp.

—Television Channels
CCIR RECMN 650-1-90. Television Standards for Satellite Broadcasting in the Channels Defined by the WARC BS-77 and the RARC SAT-83—Section 10/11B—Systems. 2 pp.

CCIR RECMN 650-2-92. Standards for Conventional Television Systems for Satellite Broadcasting in the Channels Defined by Appendix 30 of the Radio Regulations—Section 10/11B—Systems. 19 pp.

—Television—Conditional Access
CCIR Report 1079-1-90. General Characteristics of a Conditional-Access Broadcasting System—Section 11B—Ancillary Television Services. 24 pp.

—Television—Design
CCIR Report 215-7-90. Systems for the Broadcasting Satellite Service (Sound and Television)—Section 10/11B—Systems. 27 pp.

CCIR Report 632-4-90. Broadcasting-Satellite Service (Sound and Television) Technically Suitable Methods of Modulation—Section 10/11B—Systems. 32 pp.

CCIR QUESTION 90/11-90. Possible Broadcasting-Satellite Systems (Sound and Television) and Their Relative Acceptability—Questions Concerning Study Group 11—Broadcasting Service (Television). 1 p.

—Television—Digital Modulation
CCIR QUESTION 92/11-90. Digital Techniques in the Broadcasting-Satellite Service (Sound and Television)—Questions Concerning Study Group 11—Broadcasting Service (Television). 1 p.

—Television—Earth Stations—Outside Broadcasts
CCIR Report 1090-1-90. Use of Transportable Transmitting Earth Stations for the Transmission of Television Outside Broadcasts over Satellites—Section CMTT A—Television Transmission Standards and Performance Objectives. 6 pp.

CCIR QUESTION 32/CMTT-90. Use of Portable and Transportable Transmitting Earth Stations for Outside Television Broadcasts (Analogue) over Satellites—Questions Concerning the CMTT CCIR/CCITT Joint Study Group for Television and Sound Transmission. 1 p.

—Television—Feeder Links—Frequency Assignment
CCIR QUESTION 86/11-90. Frequencies for the Feeder Links to a Broadcasting Satellite (Sound and Television)—Questions Concerning Study Group 11—Broadcasting Service (Television). 2 pp.

—Television—Fixed Satellite Services—Frequency Band Sharing
CCIR Report 665-2-86. New Technologies for Off-Set Reflector Multiple-Beam Antennas—Section 1D—Spectrum Utilization and Applications. 10 pp.

CCIR QUESTION 88/11-90. Broadcasting-Satellite Service (Television) Criteria to Be Applied for Frequency Sharing Between the Broadcasting-Satellite Service and the Terrestrial and Space Services in the Frequency Range 2500 MHz to 2690 MHz—Questions Concerning. 1 p.

—Television—Fixed Services—Frequency Band Sharing
CCIR Report 665-2-86. New Technologies for Off-Set Reflector Multiple-Beam Antennas—Section 1D—Spectrum Utilization and Applications. 10 pp.

—Television—High Definition—Earth Stations—Outside Broadcasts
CCIR QUESTION 74/CMTT-90. Use of Portable and Transportable Satellite Earth Stations for the Transmission of High Definition Television Outside Broadcasts over Satellites—Questions Concerning the CMTT CCIR/CCITT Joint Study Group for Television and Sound Transmission. 1 p.

—Television—High Definition Television Systems
CCIR Report 1075-1-90. High Definition Television Broadcasting by Satellite—Section 10/11B—Systems. 98 pp.

CCIR Decision 51-4-89. Satellite Broadcasting of High Definition Television (HDTV) Signals and Accomodation of Several Audio and/or Data Signals Either Associated with Television Signals or for Sound/Data Broadcasting in Terrestrial and Satellite Broadcasting. 6 pp.

Broadcasting Satellite Services (Cont.)

—Television—High Definition Television Systems—Frequency Sharing
CCIR Decision 43-5-89. Satellite Sound Broadcasting for Portable and Vehicle Receivers and Sharing and Spectrum Aspects of Wide RF-Band HDTV Satellite Broadcasting—Annex to Volumes X and XI—Part 2—Broadcasting-Satellite Service (Sound and Television). 5 pp.

—Television—Integrated Service Digital Broadcasting
CCIR QUESTION 101/11-90. Integrated Service Digital Broadcasting in the Broadcasting-Satellite Service (Sound and Television)—Questions Concerning Study Group 11—Broadcasting Service (Television). 1 p.

—Television—Mobile Radio Services—Frequency Band Sharing
CCIR Report 665-2-86. New Technologies for Off-Set Reflector Multiple-Beam Antennas—Section 1D—Spectrum Utilization and Applications. 10 pp.

—Television—Modulation
CCIR Report 632-4-90. Broadcasting-Satellite Service (Sound and Television) Technically Suitable Methods of Modulation—Section 10/11B—Systems. 32 pp.

—Television—Monitoring
CCIR QUESTION 64/11-90. Quality Parameters and Measurement and Monitoring Methods to Be Used in the Studio Complex and in Direct Broadcasting from Terrestrial Transmitters and from Satellites Using Digital or Analogue-and-Digital Modulation—Questions Concerning Study Group. 1 p.

—Television—Multiplexed Analog Component
CCIR Report 1074-1-90. Satellite Transmission of Multiplexed Analogue Component (MAC) Vision Signals—Section 10/11B—Systems. 26 pp.

—Television—Protection Ratio
CCIR RECMN 600-1-86. Standardized Set of Test Conditions and Measurement Procedures for the Subjective and Objective Determination of Protection Ratios for Television in the Terrestrial Broadcasting and the Broadcasting-Satellite Services—Section 10/11E—Sharing. 4 pp.

CCIR Report 634-4-90. Broadcasting-Satellite Service (Sound and Television) Measured Interference Protection Ratios for Planning Television Broadcasting Systems—Section 10/11E—Sharing. 71 pp.

—Television—Quality Assurance
CCIR QUESTION 64/11-90. Quality Parameters and Measurement and Monitoring Methods to Be Used in the Studio Complex and in Direct Broadcasting from Terrestrial Transmitters and from Satellites Using Digital or Analogue-and-Digital Modulation—Questions Concerning Study Group. 1 p.

—Television—Radio Frequency Interference
CCIR QUESTION 83/11-90. Broadcasting-Satellite Service (Sound and Television) Protection from Interference—Questions Concerning Study Group 11—Broadcasting Service (Television). 2 pp.

—Television—Receiving Antennas
CCIR QUESTION 97/11-90. Characteristics of Receiving Systems of the Broadcasting-Satellite Service (Sound and Television)—Questions Concerning Study Group 11—Broadcasting Service (Television). 1 p.

—Television—SHF
CCIR QUESTION 85/11-90. Broadcasting-Satellite Service (Sound and Television) Use of the 12 GHz Band—Questions Concerning Study Group 11—Broadcasting Service (Television). 2 pp.

—Television—SHF—Feeder Links—Frequency Assignment
CCIR QUESTION 98/11-90. Technical Characteristics of and Planning Methods for Feeder Links to Broadcasting Satellites (Television) in the 12 GHz Band in Regions 1 and 3—Questions Concerning Study Group 11—Broadcasting Service (Television). 1 p.

—Television—SHF—Protection Ratio
CCIR RECMN 792-92. Interference Protection Ratios for the Broadcasting-Satellite Service (Television) in the 12 GHz Band. 5 pp.

—Television Signals—Composite—625 Lines
CCIR QUESTION 96/11-90. Composite 625-Line Signal for Television Broadcasting from Satellites—Questions Concerning Study Group 11—Broadcasting Service (Television). 1 p.

INTERNATIONAL AND NON-U.S. NATIONAL STANDARDS
SUBJECT INDEX
Bromine

Broadcasting Satellite Services (Cont.)

—**Television—Space Stations—Spurious Radiation**
CCIR QUESTION 94/11-90. Radiation of Unwanted Emissions from Space Stations by the Broadcasting-Satellite Service (Sound and Television)—Questions Concerning Study Group 11—Broadcasting Service (Television). 1 p.

—**Television Standards**
CCIR QUESTION 95/11-90. Broadcasting-Satellite Service—Television Standards—Questions Concerning Study Group 11—Broadcasting Service (Television). 1 p.

—**Television—Television Broadcasting—Frequency Band Sharing**
CCIR QUESTION 87/11-90. Broadcasting-Satellite Service (Television) Criteria to Be Applied for Frequency Sharing Between the Broadcasting-Satellite Service and the Terrestrial Broadcasting Service in the Frequency Range 620 to 790 MHz—Questions Concerning Study Group 11-. 2 pp.

—**Television—Transmitting Antennas**
CCIR QUESTION 93/11-90. Transmitting Antennas for the Broadcasting-Satellite Service (Sound and Television)—Questions Concerning Study Group 11—Broadcasting Service (Television). 1 p.

—**Television—Uplinks**
CCIR Report 815-1-82. Transmission Performance of Satellite up Links for Television Broadcasting—Section CMTT A—Television Transmission Standards and Performance Objectives. 2 pp.

—**UHF—Land Mobile Services—Frequency Band Sharing**
CCIR Report 770-3-90. Technical and Operating Considerations for a Land Mobile-Satellite Service Operating in Band 9—Section 8F—Frequencies, Orbits and Systems. 27 pp.

Broadcasting Satellite Systems
See Also: Communications Satellites; Satellite Communications; Satellites

—**Design—Radio Wave Propagation**
CCIR RECMN 679-90. Propagation Data Required for the Design of Broadcasting-Satellite Systems—Section 5F—Aspects Relative to Space Telecommunication Systems. 1 p.
CCIR RECMN 679-1-92. Propagation Data Required for the Design of Broadcasting-Satellite Systems. 5 pp.

—**Radio Frequency Characteristics—Command Signals**
CCIR QUESTION 99/11-90. Telemetry, Tracking and Command Signals and Test Signals for Maintenance Testing of Broadcasting-Satellite Radio-Frequency Characteristics—Questions Concerning Study Group 11—Broadcasting Service (Television). 1 p.

—**Radio Frequency Characteristics—Telemetry Signals**
CCIR QUESTION 99/11-90. Telemetry, Tracking and Command Signals and Test Signals for Maintenance Testing of Broadcasting-Satellite Radio-Frequency Characteristics—Questions Concerning Study Group 11—Broadcasting Service (Television). 1 p.

—**Radio Frequency Characteristics—Test Signals**
CCIR QUESTION 99/11-90. Telemetry, Tracking and Command Signals and Test Signals for Maintenance Testing of Broadcasting-Satellite Radio-Frequency Characteristics—Questions Concerning Study Group 11—Broadcasting Service (Television). 1 p.

—**Radio Frequency Characteristics—Tracking Signals**
CCIR QUESTION 99/11-90. Telemetry, Tracking and Command Signals and Test Signals for Maintenance Testing of Broadcasting-Satellite Radio-Frequency Characteristics—Questions Concerning Study Group 11—Broadcasting Service (Television). 1 p.

—**Radio Wave Propagation**
CCIR Report 565-4-90. Propagation Data for Broadcasting from Satellites—Section 5F—Aspects Relative to Space Telecommunication Systems. 9 pp.

—**Sound—Television Reception**
CCIR QUESTION 91/11-90. Technical Characteristics of Broadcasting-Satellite Systems (Sound and Television) for Community and Individual Reception—Questions Concerning Study Group 11—Broadcasting Service (Television). 1 p.

Broadcasting Satellite Systems (Cont.)

—**Television—Television Reception**
CCIR QUESTION 91/11-90. Technical Characteristics of Broadcasting-Satellite Systems (Sound and Television) for Community and Individual Reception—Questions Concerning Study Group 11—Broadcasting Service (Television). 1 p.

Broadcasting Services
Use: Broadcasting

Broadcasting Stations
Use: Broadcast Stations

Broadcasting Transmitter Networks
Use: Transmitter Networks

Broadleaved Woods
Use: Hardwoods

Broilers
See Also: Food Warmers
CSA CAN/CSA-C22. 2 NO 64-M91. Household Cooking and Liquid-Heating Appliances; (Gen Instr 1 Thru 2). 91 pp.

—**Commercial—Electric**
CSA C22.2 NO 109-M1981. Commercial Cooking Appliances; (Gen Instr 1 Thru 3). 36 pp.
CSA 1169 Bull. Electrical Bulletin 1169 June 27, 1978 to C22.2 NO 109. 2 pp.

Bromates
Scope Note: Use a more specific term *See:* Sodium Bromate

Bromcrepsol Purple
Use For: Bromocresol Purple
CNS K7106-64. Chemical Reagent (Bromcresol Purple) (Dibromo-O-Cresol Sulfonphthalein) (May)(1606).
JIS K 8841-72. Bromcresol Purple.

Bromcresol Green
Use For: Bromocresol Green
CNS K7105-64. Chemical Reagent (Bromcresol Green) (Tetrabromo-m-Cresolsulfonphthalein; Bromcresol Blue) (May)(1605).
JIS K 8840-72. Bromcresol Green.

Bromide Content Analysis

—**Water—Colorimetric Analysis**
CNS K9046-80. Method of Test for Bromide in Water (Colorimetric Method) (Aug)(6228).

Bromide Ion Content Analysis

—**Waste Water—Ion Exchange Chromatography**
DIN ENGL 38405 Pt 20-91. German Standard Methods for the Examination of Water, Waste Water and Sludge; Anions (Group D); Determination of Dissolved Bromide, Chloride, Nitrate, Nitrite, (Ortho) Phosphate and Sulfate Anions in Waste Water by Ion Chromatography (D 20) (Sept). 11 pp.

—**Water—Ion Exchange Chromatography**
JIS K 0556-90. Testing Methods for Determination of Anions in Highly Purified Water. 18 pp.

—**Water—Liquid Chromatography**
DIN ENGL 38405 Pt 19-88. German Standard Methods for the Examination of Water, Waste Water and Sludge; Anions (Group D) Determination of Fluo-ride, Chloride, Nitrite, (Ortho)Phosphate, Brom-ide, Nitrate and Sulfate Anions in Water with a Low Pollution Level by Ion Exchange Chromatography (D 19) (Feb). 11 pp.
ISO 10304 Pt 1-92. Water Quality—Determination of Dissolved Fluoride, Chloride, Nitrite, Orthophosphate, Bromide, Nitrate and Sulfate Ions, Using Liquid Chromatography of Ions—Part 1: Method for Water with Low Contamination First Edition. 16 pp.

Bromides
Scope Note: Use a more specific term
See: Ammonium Bromide; Cadmium Bromide; Cupric Bromide; Cyanogen Bromide; Ethyl Bromide; Halides; Hydrobromic Acid; Mercuric Bromide; Potassium Bromide; Silver Bromide; Sodium Bromide

Bromination

—**Phenols**
ISO 1904-72. Liquefied Phenol for Industrial Use—Determination of Phenols Content—Bromination Method First Edition. 5 pp.

Bromine
See Also: Bromine Content Analysis; Bromine Number
CNS K7107-64. Chemical Reagent (Bromine) (May)(1607).
JIS K 8529-80. Bromine.

Bromine Content Analysis
See Also: Bromine

—**Exhaust Gases**
JIS K 0085-83. Methods for Determination of Bromine in Exhaust Gas. 13 pp.

—**Rubber—Combustion**
BSI BS 7164: Sec 22.2-92. 1992 Chemical Tests for Raw and Vulcanized Rubber Part 22: Methods for Determination of Bromine and Chlorine Content Section 22.2: Oxygen Flask Combustion Technique for Determination of Bromine and Chlorine (ISO 7725: 1991) (V). 14 pp.
ISO 7725-91. Rubber and Rubber Products—Determination of Bromine and Chlorine Content—Oxygen Flask Combustion Technique First Edition. 11 pp.

—**Uranium Oxide Fluoride—Spectrophotometry**
CNS J2063-82. Method for Determining Bromine in UO2F2 Spectrophoto-Metric Method (Aug)(9281).

Bromine Number
See Also: Bromine
BSI BS 4523-70. (WITHDRAWN) 1970 Determination of Bromine Index. 8 pp.

—**Acetic Anhydride**
ISO 761-77. Acetic Anhydride and Butan-1-ol for Industrial Use—Determination of Bromine Number First Edition. 4 pp.

—**Alcohols—Volumetric Analysis**
ISO 1843 Pt IV-77. Higher Alcohols for Industrial Use—Methods of Test—Part IV: Determination of Bromine Number—Titrimetric Method in the Presence of Mercury (II) Chloride First Edition. 4 pp.

—**Aromatic Hydrocarbons—Coulometric Titration**
CNS K6521-80. Method of Test for Bromine Index of Aromatic Hydrocarbons by Coulometric Titration (Jul)(5838).

—**Aromatic Hydrocarbons—Potentiometric Analysis**
CNS K6520-80. Method of Test for Bromine Index of Aromatic Hydrocarbons by Potentiometric Titration (Jul)(5837).

—**Butanols**
ISO 761-77. Acetic Anhydride and Butan-1-ol for Industrial Use—Determination of Bromine Number First Edition. 4 pp.

—**Olefins—Electrometric Analysis**
BSI BS 5089-80. 1980 Amd 1 Determination of Bromine Number of Petroleum Distillates and Commercial Aliphatic Olefins by Electrometric Method. 16 pp.
CNS K6523-80. Method of Test for Bromine Number of Petroleum Distillates and Commerical Aliphatic Olefins by Electrometric Titration (Jul)(5840).
ISO 3839-78. Petroleum Distillates and Commercial Aliphatic Olefins—Determination of Bromine Number—Electrometric Method First Edition; (Erratum—Nov 1979) (Erratum—June 1980). 16 pp.

—**Petroleum Products**
CNS K6268-68. Method of Test for Unsaturated Hydrocarbons in Petroleum Distillates by Bromine Number (Oct)(2915)(R 1973).

—**Petroleum Products—Electrometric Analysis**
BSI BS 5089-80. 1980 Amd 1 Determination of Bromine Number of Petroleum Distillates and Commercial Aliphatic Olefins by Electrometric Method. 16 pp.
CNS K6522-80. Method of Test for Bromine Index of Petroleum Hydrocarbons by Electrometric Titration (Jul)(5839).
CNS K6523-80. Method of Test for Bromine Number of Petroleum Distillates and Commerical Aliphatic Olefins by Electrometric Titration (Jul)(5840).
ISO 3839-78. Petroleum Distillates and Commercial Aliphatic Olefins—Determination of Bromine Number—Electrometric Method First Edition; (Erratum—Nov 1979) (Erratum—June 1980). 16 pp.
JIS K 2605-80. Testing Method for Bromine Number of Petroleum Products (Electrometric Titration Method).

INDUSTRY STANDARDS

Bromine Number (Cont.)
—Petroleum Products—Volumetric Analysis
BSI BS 2000: Part 129-93. 1993 Methods of Test for Petroleum and Its Products Part 129: Determination of Bromine Number—Colour Indicator Titration Method. 5 pp.
BSI BS 2000: Part 129-82. 1982 Petroleum and Its Products Part 129: Bromine Number (Colour Indicator Method). 7 pp.

—Vinyl Resins
ISO 3499-76. Plastics—Aqueous Dispersions of Homopolymers and Copolymers of Vinyl Acetate—Determination of Bromine Number First Edition. 4 pp.

Bromocresol Green
Use: Bromcresol Green

Bromocresol Purple
Use: Bromcrepsol Purple

Bromophenol Blue
Use: Bromphenol Blue

N-Bromosuccinimide
Use For: NBS
CNS K7661-82. Chemical Reagent (N-Bromosuccinimide) (Oct)(9518).
JIS K 9553-78. N-Bromosuccinimide.

Bromothymol Blue
Use: Bromthymol Blue

Bromphenol Blue
Use For: Bromophenol Blue
CNS K7108-64. Chemical Reagent (Bromphenol Blue) (Tetrabromphenolsul-fonphthalein) (May)(1608).
JIS K 8844-72. Bromophenol Blue.

Bromthymol Blue
Use For: Bromothymol Blue
CNS K7109-64. Chemical Reagent (Bromthymol Blue) (May)(1609).
JIS K 8842-72. Bromothymol Blue.

Bronze Castings
See Also: Aluminum Bronze Castings; Bronzes; Castings; Gunmetal Castings; Leaded Bronze Castings; Leaded Tin Bronze Castings; Metal Products; Phosphor Bronze Castings; Silicon Bronze Castings; Tin Bronze Castings
CNS H3057-89. Bronze Castings (Jan)(4125).
JIS H 5111-88. Bronze Castings. 10 pp.

—Ingots
CNS H3053-89. Bronze Ingots for Castings (Jan)(4080).
JIS H 2203-85. Bronze Ingots for Castings. 6 pp.

Bronze Powder
See Also: Metal Powders
—Insoluble Matter Content
BSI BS 5600: Sec 2.9-80. (WITHDRAWN) 1980 Powder Metallurgical Materials and Products Part 2: Methods of Sampling and Testing Metallic Powders Section 2.9: Determination of Acid-Insoluble Content in Iron, Copper, Tin and Bronze Powders. 4 pp.
BSI BS EN 24496-93. 1993 Metallic Powders—Determination of Acid Insoluble Content in Iron, Copper, Tin and Bronze Powders (ISO 4496: 1978) (V). 10 pp.
CEN EN 24496-93. Metallic Powders—Determination of Acid Insoluble Content in Iron, Copper, Tin and Bronze Powders (ISO 4496: 1978). 5 pp.
ISO 4496-78. Metallic Powders—Determination of Acid-Insoluble Content in Iron, Copper, Tin and Bronze Powders First Edition (CEN EN 24496: 1993). 5 pp.

Bronzes
Scope Note: For additional listings, see also specific products made from bronzes See Also: Aluminum Bronzes; Bronze Castings; Copper Alloys; Phosphor Bronzes; Tin Bronzes
—Chemical Analysis
CNS H2001-88. Method for Chemical Analysis of Bronze (Apr)(198).
JIS H 1251-77. Methods for Chemical Analysis of Bronze.

—Sintered—Structural
CNS H3151-89. Sintered Bronze Structural Parts (Jan)(12487).

Brooders
See Also: Agricultural Equipment

Brooders (Cont.)
—Electric
CSA 1169 Bull. Electrical Bulletin 1169 June 27, 1978 to C22.2 NO 102. 2 pp.

—Electric—Construction
CSA C22.2 NO 102-1958. Construction and Test of Brooders and Incubators (R 1992). 41 pp.

—Electric—Dielectric Strength
CSA C22.2 NO 102-1958. Construction and Test of Brooders and Incubators (R 1992). 41 pp.

—Electric—Grounding
CSA 1374 Bull. Electrical Bulletin 1374 June 28, 1982 to C22.2 NO 102. 1 p.

—Electric—Portable
CSA 1374 Bull. Electrical Bulletin 1374 June 28, 1982 to C22.2 NO 102. 1 p.

—Electric—Portable—Construction
CSA C22.2 NO 102-1958. Construction and Test of Brooders and Incubators (R 1992). 41 pp.

—Electric—Portable—Dielectric Strength
CSA C22.2 NO 102-1958. Construction and Test of Brooders and Incubators (R 1992). 41 pp.

—Electric—Portable—Temperature Testing
CSA C22.2 NO 102-1958. Construction and Test of Brooders and Incubators (R 1992). 41 pp.

—Electric—Temperature Testing
CSA C22.2 NO 102-1958. Construction and Test of Brooders and Incubators (R 1992). 41 pp.

Brookfield Viscosity
Use: Viscosity

Brooklands A07 Optica Aircraft
See Also: Aircraft
—Mandatory Aircraft Modifications and Inspections Summaries
CAA. Brooklands Aerospace 0A7 Optica Series Aircraft. 2 pp.

Brooms
See Also: Brushes (Cleaning/Polishing); Floor Cleaning Equipment
—Monofilaments
CGSB CAN/CGSB-41.21-M85. Monofilaments, Synthetic, for Brushes. 11 pp.

—Push
CGSB 22-GP-13M-83. Brush, Floor Sweeping (With Brace) 90 and 120 cm, Standard for (R 1989). 11 pp.
CGSB 22-GP-14M-83. Brush, Floor Sweeping 35 cm, 45 cm and 60 cm, Standard for (R 1989). 11 pp.
CGSB 22-GP-15M-82. Broom, Push, 35 cm, Standard for (R 1989). 10 pp.

—Upright
CGSB 22-GP-2M-81. Brooms, Upright, (Corn) Household (R 1989). 13 pp.
CGSB 22-GP-10M-81. Brooms, Upright, (Corn) Warehouse, Standard for (R 1989). 14 pp.

Brown Coal
Use: Lignite

Brown Rice
Use: Rice

Brown Stock
Use: Unbleached Pulps

Brucellosis
—Cattle—Control Measures
EC 87/58/EEC-86. Council Decision Introducing a Supplementary Community Measure for the Eradication of Brucellosis, Tuberculosis and Leucosis in Cattle. 3 pp.

—Goats—Control Measures
EC 90/242/EEC-90. Council Decision Introducing a Community Financial Measure for the Eradication of Brucellosis in Sheep and Goats. 9 pp.

—Sheep—Control Measures
EC 90/242/EEC-90. Council Decision Introducing a Community Financial Measure for the Eradication of Brucellosis in Sheep and Goats. 9 pp.

Brucine
CNS K7110-64. Chemical Reagent (Brucine) (May)(1610).

Brucine Dihydrate
JIS K 8832-79. Birucine Dihydrate.

Brucine Sulfate
JIS K 9527-76. Brucine Sulfate.

Brush Coatings
—Spreading Rate
BSI BS 3900: Part A16-86. 1986 Methods of Test for Paints Group A: Tests on Liquid Paints (Excluding Chemical Tests) Part A16: Determination of Natural Spreading Rate by Brush Application. 4 pp.
BSI BS 3900: Part A17-86. 1986 Methods of Test for Paints Group A: Tests on Liquid Paints (Excluding Chemical Tests) Part A17: Coating of Test Panels at a Spreading Rate. 4 pp.
ISO 7254-84. Paints and Varnishes—Assessment of Natural Spreading Rate—Brush Application First Edition. 4 pp.

—Wood—Test Panels
CGSB 1-GP-71 METH 97.2-82. Methods of Testing Paints and Pigments Brush Application Preparation of Wooden Panels. 1 p.

Brush Cutters
Use: Brush Saws

Brush Holders (Machine Components)
Use For: Carbon Brush Holders See Also: Brushes (Machine Components)
BSI BS 4999: Part 147-88. 1988 General Requirements for Rotating Electrical Machines Part 147: Dimensions of Brushes and Brush-Holders for Electrical Machinery. 36 pp.
IEC 136-86. Dimensions of Brushes and Brush-Holders for Electrical Machinery Second Edition. 66 pp.
IEC 778-84. Brush-Holders for Slip-Rings Group R—Type RA First Edition. 25 pp.
SAA AS 1359.47-88. Rotating Electrical Machines—General Requirements—Part 47: Brushes and Brush-Holders—Dimensions. 31 pp.

—Glossaries
BSI BS 4999: Part 146-88. 1988 General Requirements for Rotating Electrical Machines Part 146: Definitions and Nomenclature for Carbon Brushes, Brush Holders Commutators and Slip-Rings. 36 pp.
BSI BS 4999: Part 146-01. 1988 Amd 1 General Requirements for Rotating Electrical Machines Part 146: Definitions and Nomenclature for Carbon Brushes, Brush-Holders, Commutators and Slip-Rings (IEC 276: 1968) (AMD 7014) September 15, 1992. 40 pp.
CENELEC HD 56 S2-91. Definitions and Nomenclature for Carbon Brushes, Brush-Holders, Commutators and Slip-Rings. 4 pp.
IEC 276-68. Definitions and Nomenclature for Carbon Brushes, Brush-Holders, Commutators and Slip-Rings First Edition; (Amendment 1-1987). 41 pp.
IEC 560-77. Definitions and Terminology of Brush-Holders for Electrical Machines First Edition. 25 pp.
SAA AS 1359.46-88. Rotating Electrical Machines—General Requirements—Part 46: Brushes, Brush-Holders, and Slip-Rings—Glossary of Terms. 19 pp.

—Pressure Measurement
IEC 1015-90. Brush-Holders for Electrical Machines Guide to the Measurement of the Static Thrust Applied to Brushes First Edition. 17 pp.

Brush Saws
See Also: Forestry Equipment; Saws
CNS B2564-81. Bush Cutter Saws (Jun)(7552).
JIS B 9212-88. Bush Cutter Saws. 7 pp.

—Blade Guards
ISO 7918-85. Forestry Machinery—Portable Brush-Saws—Circular Saw-Blade Guard—Dimensions First Edition. 5 pp.
ISO 8380-85. Forestry Machinery—Portable Brush-Saws—Circular Saw-Blade Guard—Strength First Edition. 4 pp.

—Blades
ISO 7113-91. Forestry—Machinery Portable Brush-Saws—Saw Blades Second Edition. 7 pp.

—Fuel Consumption
ISO 8893-89. Forestry Machinery—Portable Brush-Saws—Engine Performance and Fuel Consumption First Edition. 6 pp.

—Glossaries
ISO 7112-82. Machinery for Forestry—Portable Brush Saws—Vocabulary First Edition. 10 pp.

—Noise
BSI BS 7136: Part 1-89. 1989 Brush Saws Part 1: Method of Measurement at the Operator's Position of Emitted Airborne Noise. 8 pp.

INTERNATIONAL AND NON-U.S. NATIONAL STANDARDS
SUBJECT INDEX

Brush Saws (Cont.)
—Noise (Cont.)
BSI BS 7136: Part 1-01. 1989 Amd 1 Brush Saws Part 1: Method of Measurement at the Operator's Position of Emitted Airborne Noise (ISO 7917: 1987) (AMD 6979) April 1, 1992. 12 pp.

CEN EN 27917-91. Acoustics—Measurement at the Operator's Position of Airborne Noise Emitted by Brush Saws. 3 pp.

DIN ENGL EN 27917-91. Acoustics; Measurement, at Operator's Position, of Airborne Noise Emitted by Hand-Held Brush Saws (ISO 7917: 1987) (Nov). 8 pp.

ISO 7917-87. Acoustics—Measurement at the Operator's Position of Airborne Noise Emitted by Brush Saws First Edition. 7 pp.

—Safety
CEN PREN 31806-92. Safety Requirements for Agricultural and Forestry Machinery—Brush Cutters and Grass Trimmers. 28 pp.

—Sound Pressure Levels
CEN EN 27917-91. Acoustics—Measurement at the Operator's Position of Airborne Noise Emitted by Brush Saws. 3 pp.

DIN ENGL EN 27917-91. Acoustics; Measurement, at Operator's Position, of Airborne Noise Emitted by Hand-Held Brush Saws (ISO 7917: 1987) (Nov). 8 pp.

ISO 7917-87. Acoustics—Measurement at the Operator's Position of Airborne Noise Emitted by Brush Saws First Edition. 7 pp.

—Vibration
DIN ENGL 45675 Pt 3-87. Exposure to Mechanical Vibration Transmitted to the Hand-Arm System; Measurement of Vibration Induced by Portable, Internal Combustion Engine Brush Saws (Sept). 3 pp.

ISO 7916-89. Forestry Machinery—Portable Brush-Saws—Measurement of Hand-Transmitted Vibration First Edition. 10 pp.

Brushes (Cleaning/Polishing)
Scope Note: For additional listings, use a more specific term *See Also:* Bristles; Brooms; Dusting Brushes; Paint Brushes; Scrub Brushes; Toilet Bowl Brushes; Toothbrushes; Window Washing Brushes; Wire Brushes

CNS S1023-83. Brushes (Tentative) (Feb)(2308).

CNS Z7010-83. Row of Brushes (Tentative) (Feb)(2307).

MOD UK DL-7. Brushware.

—Monofilaments
CGSB CAN/CGSB-22.50-92. Synthetic Monofilaments for Brushes (Supersedes CAN/CGSB-41.21-M85). 8 pp.

CGSB CAN/CGSB-41.21-M85. Monofilaments, Synthetic, for Brushes. 11 pp.

—Power—Glossaries
CEN PREN 1083-1-93. Power-Driven Brushes—Part 1: Terminology. 22 pp.

—Power—Safety
BSI BS 2939-74. (WITHDRAWN) 1974 Wire Brushes (Superseded by BS 6694: 1986). 21 pp.

BSI BS 6694-86. 1986 Safety of Power-Driven Industrial Brushes (Supersedes BS 2939: 1974). 24 pp.

CEN PREN 1083-2-93. Power-Driven Brushes—Part 2: Safety Requirements. 5 pp.

Brushes (Machine Components)
Use For: Carbon Brushes *See Also:* Brush Holders (Machine Components); Commutators; Electric Contacts; Slip Rings

BSI BS 4999: Part 147-88. 1988 General Requirements for Rotating Electrical Machines Part 147: Dimensions of Brushes and Brush-Holders for Electrical Machinery. 36 pp.

CNS C1131-84. Dimensions of Brushes for Electric Machines (Jun)(10912).

CNS C4051-64. Dimensions of Brushes for Electric Machines (Jul)(2322)(R 1971).

CNS C4391-83. Brushes for Electric Machines (Jun)(10327).

IEC 136-86. Dimensions of Brushes and Brush-Holders for Electrical Machinery Second Edition. 66 pp.

IEC 773-83. Test Methods and Apparatus for the Measurement of the Operational Characteristics of Brushes First Edition. 63 pp.

JIS C 2802-84. Dimensions of Brushes for Electric Machines. 13 pp.

SAA AS 1359.47-88. Rotating Electrical Machines—General Requirements—Part 47: Brushes and Brush-Holders—Dimensions. 31 pp.

Brushes (Machine Components) (Cont.)
—Glossaries
BSI BS 4999: Part 146-88. 1988 General Requirements for Rotating Electrical Machines Part 146: Definitions and Nomenclature for Carbon Brushes, Brush Holders Commutators and Slip-Rings. 36 pp.

BSI BS 4999: Part 146-01. 1988 Amd 1 General Requirements for Rotating Electrical Machines Part 146: Definitions and Nomenclature for Carbon Brushes, Brush-Holders, Commutators and Slip-Rings (IEC 276: 1968) (AMD 7014) September 15, 1992. 40 pp.

CENELEC HD 56 S2-91. Definitions and Nomenclature for Carbon Brushes, Brush-Holders, Commutators and Slip-Rings. 4 pp.

IEC 276-68. Definitions and Nomenclature for Carbon Brushes, Brush-Holders, Commutators and Slip-Rings First Edition; (Amendment 1-1987). 41 pp.

SAA AS 1359.46-88. Rotating Electrical Machines—General Requirements—Part 46: Brushes, Brush-Holders, Commutators, and Slip-Rings—Glossary of Terms. 19 pp.

—Physical Properties
IEC 413-72. Test Procedures for Determining Physical Properties of Brush Materials for Electrical Machines First Edition. 43 pp.

—Pressure Measurement
IEC 1015-90. Brush-Holders for Electrical Machines Guide to the Measurement of the Static Thrust Applied to Brushes First Edition. 17 pp.

BSI
Use For: British Standards Institution
See Also: Standards

—Compendiums
BSI PP 7302-87. 1987 Compendium of British Standards for Design and Technology in Schools. 139 pp.

—Compendiums—Construction
BSI HANDBOOK NO. 3-85. 1992 Ad 41 Summaries of British Standards for Building (R). 2371 pp.

BSI HANDBOOK NO. 20-85. (WITHDRAWN) 1985 Building Regulations 1985 British Standards Summarized. 239 pp.

—Standards Preparation
BSI BS 0: Part 1-91. 1991 A Standard for Standards Part 1: Guide to General Principles of Standardization (G). 15 pp.

BSI BS 0: Part 1-81. 1981 A Standard for Standards Part 1: General Principles of Standardization. 12 pp.

BSI BS 0: Part 2-91. 1991 A Standard for Standards Part 2: Guide to BSI Committee Procedures (G). 22 pp.

BSI BS 0: Part 2-81. 1981 Amd 2 A Standard for Standards Part 2: BSI and Its Committee Procedures. 38 pp.

BSI BS 0: Part 3-91. 1991 A Standard for Standards Part 3: Guide to Drafting and Presentation of British Standards (G). 64 pp.

BSI BS 0: Part 3-81. 1981 Amd 1 A Standard for Standards Part 3: Drafting and Presentation of British Standards. 42 pp.

—Standards Preparation—Construction/Civil Engineering
BSI PD 6501: Part 1-82. 1982 Amd 1 Preparation of British Standards for Building and Civil Engineering Part 1: Guide to the Types of British Standard, Their Aims, Relationship, Content and Application. 20 pp.

BSI PD 6501: Part 2-84. 1984 Preparation of British Standards for Building and Civil Engineering Part 2: Guide to Presentation. 34 pp.

BSI PD 6501: Part 3-85. 1985 Preparation of British Standards for Building and Civil Engineering Part 3: Guidance to BSI Committees on the Preparation of British Standards for Use with the Building Act 1984. 47 pp.

BSI PD 6501: Part 5-90. 1990 Preparation of British Standards for Building and Civil Engineering Part 5: Guide to Technical Committees on the Preparation of Performance Specifications in British Standards for Construction Products. 24 pp.

BSS
Use: Broadcasting Satellite Services

Bubble Leak Testing
Use: Leakage—Bubble Emission

Bubble Points
Use: Boiling Points

Buchholz Indentation Testing
Use: Indentation Hardness Testing

Bucket Conveyors
See Also: Bucket Elevators

—Belts
BSI BS 490: Part 2-75. 1975 Conveyor and Elevator Belting Part 2: Rubber and Plastics Belting of Textile Construction for Use on Bucket Elevators. 18 pp.

SNZ NZS/BS 490: Part 2-75. Conveyor and Elevator Belting Part 2: Rubber and Plastics Belting of Textile Construction for Use on Bucket Elevators. 15 pp.

—Link Chains—Chain Ends
CNS B2608-81. Round Steel Link Chains for Conveyer, Chain-Ends for Bucket Conveyers, Half Long Link (Oct)(7941).

—Mining Equipment
BSI BS 2969-80. 1980 High-Tensile Steel Chains (Round Link) for Chain Conveyors and Coal Ploughs. 22 pp.

ISO 610-90. High-Tensile Steel Chains (Round Link) for Chain Conveyors and Coal Ploughs Second Edition. 21 pp.

Bucket Elevators
See Also: Bucket Conveyors; Conveyors; Elevators (Conveyors)

BSI BS 6318: Part 3-82. 1982 Bucket Elevators Part 3: Dimensions of Deep Elevator Bucket with Flat Rear Wall. 3 pp.

ISO 5050-81. Continuous Mechanical Handling Equipment—Vertical Bucket Elevators with Calibrated Round Steel Link Chains—General Characteristics First Edition. 5 pp.

ISO 5051-81. Continuous Mechanical Handling Equipment—Deep Elevator Bucket with Flat Rear Wall—Main Dimensions First Edition. 4 pp.

ISO 7190-81. Continuous Mechanical Handling Equipment—Bucket Elevators—Classification First Edition. 9 pp.

—Belts
BSI BS 490: Part 2-75. 1975 Conveyor and Elevator Belting Part 2: Rubber and Plastics Belting of Textile Construction for Use on Bucket Elevators. 18 pp.

SNZ NZS/BS 490: Part 2-75. Conveyor and Elevator Belting Part 2: Rubber and Plastics Belting of Textile Construction for Use on Bucket Elevators. 15 pp.

—Classification
BSI BS 6318: Part 1-82. 1982 Bucket Elevators Part 1: Classification of Bucket Elevators. 6 pp.

—Link Chains
BSI BS 6318: Part 2-82. 1982 Bucket Elevators Part 2: Dimensions of Vertical Bucket Eelevators with Calibrated Round Steel Link Chains. 4 pp.

—Mining Equipment
CNS M2082-80. Bucket Elevator for Coal Washing (Dec)(6843).

JIS M 4601-52. Bucket Elevator for Coal Washing. 17 pp.

Buckets (Pails)
Use For: Pails *See Also:* Containers; Laboratory Buckets

JIS Z 1620-85. Pails. 14 pp.

—Aerial Cableways
CNS M2083-80. Bucket for Rope Ways (Dec)(6844).

—Aluminum
CNS S1090-88. Aluminium Pail (20 L) (May)(4154).

—Canvas
BSI BS 6170-81. 1981 Amd 1 Flax Canvas Water Buckets (AMD 5594) July 29, 1988. 6 pp.

—Gullies
DIN ENGL 1236 Pt 1-81. Class A and Class B Concrete Elements and Buckets for Gullies; Design, Installation and Assemblies (Nov). 3 pp.

DIN ENGL 1236 Pt 3-81. Classes A and B Concrete Elements and Buckets for Gullies; Buckets (Nov). 4 pp.

—Metal
CNS Z5006-84. Metal Bucket (Aug)(802).

CNS Z6001-84. Method of Test for Metal Bucket (Aug)(803).

—Mop
CGSB 44.32M-89. Bucket, Mop. 6 pp.

—Plastic
CNS S1105-86. Plastic Bucket (Feb)(6855).

JIS S 2032-82. Plastics Buckets. 9 pp.

JIS S 2040-82. Plastics Lockable Containers Household Wares. 11 pp.

INDUSTRY STANDARDS

Buckets (Pails) (Cont.)

—Shipping Containers—Hazardous Materials—Metal
CGSB CAN/CGSB-43.124-M89. Inside Metal Containers and Liners (TC-2F and TC-2N). 9 pp.

—Shipping Containers—Hazardous Materials—Steel
CGSB CAN/CGSB-43.117-M89. Single-Trip Steel Drums, Removable or Non-Removable Head (TC-17C). 12 pp.
CGSB CAN/CGSB-43.118-M89. Single-Trip Steel Drums, Non-Removable Head (TC-17E). 12 pp.
CGSB CAN/CGSB-43.119-M89. Single-Trip Steel Drums, Non-Removable Head (TC-17F). 12 pp.
CGSB CAN/CGSB-43.121-M90. Single-Trip Steel Drums, Removable Head (TC-37A). 11 pp.
CGSB CAN/CGSB-43.122-M89. Single-Trip Steel Drums, Non-Removable Head (TC-37B). 11 pp.

—Steel
CNS S1140-88. Steel Bucket (Jan)(8739).
CNS S2070-88. Method of Test for Steel Bucket (Jan)(8740).
JIS S 2009-76. Steel Made Buckets.

Buckles
Scope Note: For additional listings, use a more specific term *See Also:* Belts (Machinery); Fasteners; Spring Buckles; Straps (Fasteners)
CNS Z7063-79. 25mm Width Buckle (Apr)(4794).

—Straps
CNS Z7081-81. Strap Buckles (May)(7470).

Bucks (Vaulting)
Use: Vaulting Horses

Buechner Funnels
Use: Funnels—Filter

Buffer Bolts
See Also: Bolts; Door Bolts

—Doors
CNS A2102-81. Flag-Form Hinges, Buffer and Flush Bolts for Steel Doors (Apr)(7185). 8 pp.

Buffer Coefficient
Use: Buffer Value

Buffer Gates
Scope Note: Use a more specific term *See:* AND Buffer Gates; Buffer/Line Drivers; Buffers (Data); Gate Circuits; Interface Circuits; Inverting Buffer Gates; NAND Buffer Gates; Non-Inverting Buffer Gates; NOR Buffer Gates; OR Buffer Gates

Buffer/Line Drivers
See Also: Buffers (Data); Integrated Circuits; Interface Circuits; Line Drivers
CECC CECC 90 101-005 ISSUE 1-87. UTE C 86-213/B 25; Digital Integrated Circuits in Accordance with FS 90 101; 54/64/74 06; Hex Inverter Buffer/Drivers with Open Collector High Voltage Outputs (En). 16 pp.
CECC CECC 90 101-006 ISSUE 1-87. UTE C 86-213/B 26; Digital Integrated Circuits in Accordance with FS 90 101; 54/64/74 07; Hex Buffers/Drivers with Open Collector High Voltage Outputs (En). 16 pp.
CECC CECC 90 103-143 ISSUE 1-81. BS CECC 90 103-143; Hex Buffer Drivers with Three Level Outputs (En) AMD 1 (En). 21 pp.
CECC CECC 90 103-146 ISSUE 1-82. BS CECC 90 103-146; Octal Buffers and Line Drivers with Three-State Inverted Outputs (En). 19 pp.

—Hex
CECC CECC 90 101-059 ISSUE 1-87. UTE C 86-213/B 79; Digital Integrated Circuits in Accordance with FS 90 101 54/64/74 365A; Hex Buffer Drivers with 3-State Outputs (En). 19 pp.

—Hex—Inverting
CECC CECC 90 101-011 ISSUE 1-87. UTE C 86-213/B 31; Digital Integrated Circuits in Accordance with FS 90 101; 54/64/74 16; Hex Inverter Buffer/Drivers with Open Collector Outputs (En). 16 pp.
CECC CECC 90 101-012 ISSUE 1-87. UTE C 86-213/B 32; Digital Integrated Circuits in Accordance with FS 90 101; 54/64/74 17; Hex Buffers/Drivers with Open Collector High Voltage Outputs (En). 16 pp.
CECC CECC 90 103-120 ISSUE 1-81. BS CECC 90 103-120; Hex Inverting Buffer Drivers with Three Level Outputs (En) AMD 1 (En). 20 pp.

Buffer/Line Drivers (Cont.)

—Hex—Noninverting
CECC CECC 90 103-121 ISSUE 1-81. BS CECC 90 103-121; Non-Inverting Hex Buffers 2-Line and 4-Line Enable Inputs 3-Level Outputs (En) AMD 1 (En). 21 pp.

—Hex—Preferred Products List
CECC CECC MUAHAG Vol 7 IS 8-92. Preferred Products List; Active Microcircuits (En, Fe, Ge). 89 pp.

—Octal
CECC CECC 90 102-042 ISSUE 1-82. BS CECC 90 102-042; Octal Buffer and Line-Driver with Three Level Outputs (En). 20 pp.
CECC CECC 90 102-065 ISSUE 1-89. UTE C 86-216/F01; BS CECC 90 102-034 Issue A; Octal Buffers/Line Drivers with Inverted 3-State Outputs (En, Fr). 9 pp.
CECC CECC 90 102-066 ISSUE 1-89. UTE C 86-216/F02 BS CECC 90 102-035 Issue 1; Octal Buffer and Line Driver with Complementary G and Bar G Inputs and 3-State Outputs (En, Fr). 10 pp.
CECC CECC 90 103-087 ISSUE 1-82. BS CECC 90 103-087; Octal Buffers/Line Drivers with Inverted 3-Level Outputs (En) AMD 1 (En). 20 pp.
CECC CECC 90 103-088 ISSUE 1-82. BS CECC 90 103-088; Octal Buffers/Line Drivers with 3-Level Outputs (En). 20 pp.
CECC CECC 90 103-091 ISSUE 1-81. BS CECC 90 103-091; Octal Buffers/Level Drivers Non-Inverted 3-Level Outputs (En) AMD 1 (En). 21 pp.
CECC CECC 90 106-005 ISSUE 2-86. UTE C 86-217 ADD 2/GA 5; Digital Integrated Circuits in Accordance with FS 90 106; 54/74 ALS 244, 54/74 ALS 244A; Octal Buffer/Line Drivers with 3 State Non Inverting Outputs (En, Fr) ADD 3 (En, Fr). 16 pp.
CECC CECC 90 106-010 ISSUE 2-86. UTE C 86-217 ADD 2/GA 10; Digital Integrated Circuits in Accordance with FS 90 106; 54/74 ALS 240, 54/74 ALS 240A; Octal Buffer/Line Drivers with 3 State Inverting Outputs (En, Fr) ADD 3 (En, Fr). 16 pp.
CECC CECC 90 106-014 ISSUE 2-86. UTE C 86-217 ADD 2/GA 14; Digital Integrated Circuits in Accordance with FS 90 106; 54/74 ALS 241, 54/74 ALS 241A; Octal Buffer/Line Drivers with 3 State Inverting Non Outputs (En, Fr) ADD 3 (En, Fr). 16 pp.
CECC CECC 90 106-170 ISSUE 1-86. UTE C 86-217 ADD 3/GA 96 Digital Integrated Circuits in Accordance with FS 90 106; 54 ALS 1244A, 74 ALS 1244A; Octal Buffer/Line Driver with 3 State Non Inverting Outputs (En, Fr) ADD 3 (En, Fr). 15 pp.
CECC CECC 90 107-030 ISSUE 2-89. UTE C 86-218 ADD 3/FA 30; Digital Integrated Circuits in Accordance with FS 90 107; 54/74 F 240; Octal Buffer/Line Driver with 3-State Inverting Outputs (En, Fr) ADD 3 (En, Fr). 9 pp.
CECC CECC 90 107-031 ISSUE 2-89. UTE C 86-218 ADD 3/FA 31; Digital Integrated Circuits in Accordance with FS 90 107; 54/74 F 241; Octal Buffer/Line Driver with 3-State Non Inverting Outputs (En, Fr) ADD 3 (En, Fr). 9 pp.
CECC CECC 90 107-032 ISSUE 2-89. UTE C 86-218 ADD 3/FA 32; Digital Integrated Circuits in ACcordance with FS 90 107; 54/74 F 244; Octal Buffer/Line Driver with 3-State Non Inverting Outputs (En, Fr) ADD 3 (En, Fr). 9 pp.

—Octal—Inverting
CECC CECC 90 109-757 ISSUE 1-87. Digital Integrated Circuits in Accordance with FS 90 109; 54/74 HC 540; Inverting Octal 3-State Buffer Line Driver (En, Fr). 6 pp.
CECC CECC 90 109-759 ISSUE 1-87. Digital Integrated Circuits in Accordance with FS 90 109; 54/74 HCT 540; Inverting Octal 3-State Buffer Line Driver (En, Fr). 6 pp.

—Octal—Noninverting
CECC CECC 90 109-758 ISSUE 1-87. Digital Integrated Circuits in Accordance with FS 90 109; 54/74 HC 541; Non Inverting Octal 3-State Buffer Line Driver (En, Fr). 6 pp.
CECC CECC 90 109-760 ISSUE 1-87. Digital Integrated Circuits in Accordance with FS 90 109; 54/74 HCT 541; Non Inverting Octal 3-State Buffer Line Driver (En, Fr). 6 pp.

—Octal—Preferred Products List
CECC CECC MUAHAG Vol 7 IS 8-92. Preferred Products List; Active Microcircuits (En, Fe, Ge). 89 pp.

Buffer Storage Devices
Use: Buffers (Data)

Buffer Value

—Urea—Potentiometric Analysis
ISO 2751-73. Urea for Industrial Use—Determination of the Buffer Coefficient—Potentiometric Method First Edition. 4 pp.

Buffering
Use: Bumpers

Buffers (Chemical)
See Also: Alkalinity; pH
CNS K7011-82. Chemical Reagent (Buffer Solution) (Jun)(1511).

Buffers (Data)
Use For: Buffer Storage Devices; Data Buffers
See Also: Bounce Buffers; Buffer/Line Drivers; Line Buffers

—Data Buses—Integrated Circuit—Octal—Preferred Products List
CECC CECC MUAHAG Vol 7 IS 8-92. Preferred Products List; Active Microcircuits (En, Fe, Ge). 89 pp.

—Data Buses—Integrated Circuit—Quad—Preferred Products List
CECC CECC MUAHAG Vol 7 IS 8-92. Preferred Products List; Active Microcircuits (En, Fe, Ge). 89 pp.

—Integrated Circuit
CECC CECC 90 109-610 ISSUE 1-86. Digital Integrated Circuits in Accordance with FS 90 109; 54 HC 244, 74 HC 244; Octal 3-State Buffer (En, Fr). 6 pp.
CECC CECC 90 109-647 ISSUE 1-86. Digital Integrated Circuits in Accordance with FS 90 109; 54/74 HC 241; Octal 3-State Buffer (En, Fr). 6 pp.
CECC CECC 90 109-665 ISSUE 1-87. Digital Integrated Circuits in Accordance with FS 90 109; 54/74 HC 125; Quad 3-State Buffer (En, Fr). 6 pp.
CECC CECC 90 109-666 ISSUE 1-87. Digital Integrated Circuits in Accordance with FS 90 109; 54/74 HC 126; Quad 3-State Buffer (En, Fr). 6 pp.
CECC CECC 90 109-668 ISSUE 1-87. Digital Integrated Circuits in Accordance with FS 90 109; 54/74 HC 365; Hex 3-State Buffer (En, Fr). 6 pp.
CECC CECC 90 109-670 ISSUE 1-87. Digital Integrated Circuits in Accordance with FS 90 109; 54/74 HC 367; Hex 3-State Buffer (En, Fr). 6 pp.
CECC CECC 90 109-688 ISSUE 1-87. Digital Integrated Circuits in Accordance with FS 90 109; 54 HCT 240, 74 HCT 240; Inverting Octal 3-State Buffer (En, Fr). 6 pp.

—Integrated Circuit—Hex
CECC CECC 90 106-163 ISSUE 1-85. UTE C 86-217; Digital Integrated Circuits in Accordance with FS 90 106; 54 ALS 1034, ALS 1034; Buffer with Non Inverting Outputs (En, Fr) ADD 3 (En, Fr). 15 pp.
CECC CECC 90 109-844 ISSUE 1-90. Digital Integrated Circuits; Silicon Monolithic C MOS, Cavity or Non-Cavity Packages; Type(s) 54/74 HCT 365 Hex 3-State Buffer Assessment Levels P, Y, L (En, Fr, Ge). 9 pp.
CECC CECC 90 109-845 ISSUE 1-90. Digital Integrated Circuits; Silicon Monolithic C MOS, Cavity or Non-Cavity Packages; Type(s) 54/74 HCT 367 Hex 3-State Buffer Assessment Levels P, Y, L (En, Fr, Ge). 9 pp.
CECC CECC MUAHAG Vol 7 IS 8-92. Preferred Products List; Active Microcircuits (En, Fe, Ge). 89 pp.

—Integrated Circuit—Hex—Preferred Products List
CECC CECC MUAHAG Vol 7 IS 8-92. Preferred Products List; Active Microcircuits (En, Fe, Ge). 89 pp.

—Integrated Circuit—Inverting
CECC CECC 90 109-669 ISSUE 1-87. Digital Integrated Circuits in Accordance with FS 90 109; 54/74 HC 366; Hex Inverting 3-State Buffer (En, Fr). 6 pp.
CECC CECC 90 109-671 ISSUE 1-87. Digital Integrated Circuits in Accordance with FS 90 109; 54/74 HC 368; Hex Inverting 3-State Buffer (En, Fr). 6 pp.
CECC CECC 90 109-688 ISSUE 1-87. Digital Integrated Circuits in Accordance with FS 90 109; 54 HCT 240, 74 HCT 240; Inverting Octal 3-State Buffer (En, Fr). 6 pp.

—Integrated Circuit—Noninverting
CECC CECC 90 104-228 ISSUE 1. NL CECC 90 104-228 Issue 1; Digital Integrated Circuits in Accordance with FS 90 104; HEC/HEF 40244B; Octal Buffers with 3-State Outputs (En). 9 pp.

Buffers (Data) (Cont.)

—Integrated Circuit—Noninverting (Cont.)
CECC CECC 90 109-733 ISSUE 1-87. Digital Integrated Circuits in Accordance with FS 90 109; 54 HCT 7007 74 HCT 7007; Non Inverting Buffer (En, Fr). 5 pp.

—Integrated Circuit—Octal
CECC CECC 90 109-689 ISSUE 1-87. Digital Integrated Circuits in Accordance with FS 90 109; 54/74 HCT 241; Octal 3-State Buffer (En, Fr). 6 pp.
CECC CECC 90 109-690 ISSUE 1-87. Digital Integrated Circuits in Accordance with FS 90 190; 54 HCT 244, 74 HCT 244; Octal 3-State Buffer (En, Fr). 6 pp.

—Integrated Circuit—Quad
CECC CECC 90 104-031 ISSUE 3. NL CECC 90 104-031 Issue 3; Digital Integrated Circuits in Accordance with FS 90 104; HEC/HEF 4041B; Quadruple True/Complement Buffer (En). 6 pp.
CECC CECC 90 109-774 ISSUE 1-87. Digital Integrated Circuits in Accordance with FS 90 109; 54/74 HCT 125; Quad 3-State Buffer (En, Fr). 7 pp.

Buffers (Mechanical)
Use: Bumpers

Buffers (Polishers)
Use: Buffing Equipment

Buffing Equipment
Use For: Buffers (Polishers) *See Also:* Polishers

—Electric—Portable
CSA CAN/CSA-C22. 2 NO71.1-M89. Portable Electric Tools; (Gen Instr 1 Thru 4). 89 pp.

Bug Contamination (Insects)
Use: Insect Contamination

Bug Zappers
Use: Insect Electrocution Devices

Builders' Steps
Use: Stepladders

Building Automation Systems
Use For: Building Management Systems
See Also: Automatic Control Equipment; Automation; Buildings; Control Systems Equipment; Controllers; Environmental Engineering

—Residential
CENELEC PREN 50090-1-90. Home Electronic System, HES Part 1: Standardization Structure. 16 pp.
IEC DIS 10192 Pt 1.2-91. Information Technology—Home Electronic System (HES)—Part 1: Introduction ***CD-ROM ONLY***. 19 pp.
ISO DIS 10192 Pt 1.2-91. Information Technology—Home Electronic System (HES)—Part 1: Introduction ***CD-ROM ONLY***. 19 pp.

—Residential—Keyboards
BSI BS 7245-89. 1989 Numeric Keyboard for Home Electronic Systems (HES). 4 pp.
CENELEC EN 60 948-90. Numeric Keyboard for Home Electronic Systems (HES). 5 pp.
CNS C7209-90. Numeric Keyboard for Home Electronic Systems (May)(12710).
IEC 948-88. Numeric Keyboard for Home Electronic Systems (HES) First Edition. 12 pp.
JTC1 948-88. Numeric Keyboard for Home Electronic Systems (HES) First Edition. 12 pp.

Building Blocks
See Also: Building Stones; Concrete Blocks; Construction Materials; Glass Blocks

—Clay
JIS A 5210-75. Hollow Clay Building Blocks.

—Glossaries
BSI BS 6100: Sec 5.3-84. 1984 Glossary of Building and Civil Engineering Terms Part 5: Masonry Section 5.3: Bricks and Blocks. 7 pp.
BSI BS 6100: Sec 5.3-01. 1984 Amd 1 Glossary of Building and Civil Engineering Terms Part 5: Masonry Section 5.3: Bricks and Blocks (AMD 7262) August 15, 1992. 9 pp.

—Stone—Sampling
CNS A3004-59. Method of Sampling for Block Stone, Crushed Stone, Gravel Sand and Furnace Slag (Oct)(485) (R 1970).

Building Board
Scope Note: For additional listings, use a more specific term

See Also: Asbestos Board; Asbestos Millboards; Blocks; Construction Materials; Fiberboard; Insulating Board; Paperboard; Particle Board; Pressboard; Pulpboards; Wallboard; Weatherboards

—Cellular Plastics—Adhesives
CNS A2238-89. Adhesives for Plastic Foam Boards (Oct)(12609).
CNS A3312-89. Method of Test for Adhesives for Plastic Foam Boards (Oct)(12610).
JIS A 5547-81. Adhesives for Plastic Foam Boards. 17 pp.

—Modular Construction
ISO 2777-74. Modular Co-Ordination—Co-Ordinating Sizes for Rigid Flat Sheet Boards Used in Building First Edition. 3 pp.
SNZ NZS 4207-75. Modular Co-Ordination. Preferred Co-Ordinating Sizes for Rigid Flat Sheet Boards Used in Building. 1 p.

—Pine Wood
SAA AS 1489-73. Sawn Boards from Radiata Pine (Metric Units) Amdt 1 April 1977. 13 pp.
SAA AS 1493-73. Shelving and Dressed Boards Milled from Radiata Pine Amdt 1 April 1977.

—Polyester Coated
CNS O1016-68. Unsaturated Polyester Decorative Boards (Jan)(2870). 2 pp.

—Puncture Resistance
BSI BS 4816-72. 1972 Amd 1 Determination of the Puncture Resistance of Board. 13 pp.

—PVC
CNS K3021-70. Polyvinyl Chloride Board (Mar)(3142). 3 pp.

—Sizes
CNS A1020-78. Standard Size of Boards for Buildings (May)(4348).
JIS A 0006-66. Standard Size of Boards for Buildings.

—Slag-Gypsum Cement
CNS A2199-86. Slag—Gypsum Cement Boards (Nov)(11760).
CNS A3248-86. Method of Test for Slag—Gypsum Cement Boards (Nov)(11761).
JIS A 5429-83. Slag-Gypsum Cement Boards. 22 pp.

—Water Resistance Testing
JIS A 1437-92. Test Methods for Moisture Resistance of Boards for Interior System of Building. 10 pp.

—Wood Chip Cement
CNS A2163-86. Wood Chip Cement Boards (Apr)(10483).
CNS A3195-86. Method of Test for Wood Chip Cement Boards (Apr)(10484).

—Wood-Wool Cement
CNS A2142-86. Wood-Wool Cement Board (Apr)(9456). 3 pp.
CNS A3057-86. Method of Test for Wood-Wool Cement Board (Apr)(2465). 2 pp.
JIS A 5404-87. Wood-Wool Cement Boards. 14 pp.

Building Bricks
Use: Bricks

Building Codes
Use For: Building Regulations *See Also:* Mechanical Codes; Standards

—Agricultural Buildings—Models
SNZ NZS 1900: Chapter 11.2-85. Model Building Bylaw Chapter 11.2: Special Structures. Division 11.2 Farm Buildings Amend: 1, 1992. 20 pp.

—Australian—Limit Design Method
SAA AS MP38-79. Proposed Timetable for the Coordinated Revision of Australian Structural Codes into Limit State Terms. 4 pp.

—Cited Standards—BSI
BSI HANDBOOK NO. 20-85. (WITHDRAWN) 1985 Building Regulations 1985 British Standards Summarized. 239 pp.

—Protection—Ground Water
BSI CP 102-73. 1973 Amd 3 Protection of Buildings Against Water from the Ground. 30 pp.

—Structural Timber—Framing
SAA AS 1684-92. National Timber Framing Code (in Professional Packages 20, 21, 30, 62-69) Bound Together with Supplements 2, 3, 4, 5, 9, 15 Amdt 1—1993. 249 pp.
SAA AS 1684 Supp 0-92. National Timber Framing Code—Supplement 0: General Introduction and Index (Supplement to AS 1684—1992) (Supersedes AS 1684 Supplements 1 to 20—1975 (in Part) and Supplements 21 to 22—1978 (in Part)) (in Professional Packages 20, 21, 30,). 27 pp.
SAA AS 1684 Supp 1-92. National Timber Framing Code—Supplement 1: Timber Framing Span Tables—Unseasoned Timber—Stress Grade F4 (Supplement to AS 1684—1992) (Supersedes AS 1684 Supplement 1—1975 (in Part)) (in Professional Packages 20, 21, 30, 62-69). 27 pp.
SAA AS 1684 Supp 7-92. National Timber Framing Code—Supplement 7: Timber Framing Span Tables—Unseasoned Timber—Stress Grade F17 (Supplement to AS 1684—1992) (Supersedes AS 1684 Supplement 7—1975 (in Part)) (in Professional Packages 20, 21, 30, 62-69). 27 pp.
SAA AS 1684 Supp 8-92. National Timber Framing Code—Supplement 8: Timber Framing Span Tables—Unseasoned Timber—Stress Grade F22 (Supplement to AS 1684—1992) (Supersedes AS 1684 Supplement 8—1975 (in Part)) (in Professional Packages 20, 21, 30, 62-69). 27 pp.
SAA AS 1684 Supp 9-92. National Timber Framing Code—Supplement 9: Timber Framing Span Tables—Seasoned Softwood—Stress Grade F5 (Supplement to AS 1684—1992) (Supersedes AS 1684 Supplement 9—1975 (in Part)) (in Professional Packages 20, 21, 30, 62-69) Bound Together. 27 pp.
SAA AS 1684 Supp 10-92. National Timber Framing Code—Supplement 10: Timber Framing Span Tables—Seasoned Softwood—Stress Grade F7 (Supplement to AS 1684—1992) (Supersedes AS 1684 Supplement 10—1975 (in Part)) (in Professional Packages 20, 21, 30, 62-69). 27 pp.
SAA AS 1684 Supp 11-92. National Timber Framing Code—Supplement 11: Timber Framing Span Tables—Seasoned Softwood—Stress Grade F8 (Supplement to AS 1684—1992) (Supersedes AS 1684 Supplement 11—1975 (in Part)) (in Professional Packages 20, 21, 30, 62-69). 27 pp.
SAA AS 1684 Supp 12-92. National Timber Framing Code—Supplement 12: Timber Framing Span Tables—Seasoned Softwood—Stress Grade F11 (Supplement to AS 1684—1992) (Supersedes AS 1684 Supplement 12—1975 (in Part)) (in Professional Packages 20, 21, 30, 62-69). 27 pp.
SAA AS 1684 Supp 13-75. Supplement to AS 1684 SAA Timber Framing Code—Supplement 13: Light Timber Framing Span Tables—Seasoned Hardwood—Stress Grade F11 Amdt 1 March 1991 (in Professional Package 30) (Superseded (in Part) by AS 1684 Supplement 0 and AS 1684 Supplement 13—1992). 15 pp.
SAA AS 1684 Supp 13-92. National Timber Framing Code—Supplement 13: Timber Framing Span Tables—Seasoned Hardwood—Stress Grade F11 (Supplement to AS 1684—1992) (Supersedes AS 1684 Supplement 13—1975 (in Part)) (in Professional Packages 20, 21, 30, 62-69). 27 pp.
SAA AS 1684 Supp 14-75. Supplement to AS 1684 SAA Timber Framing Code—Supplement 14: Light Timber Framing Span Tables—Seasoned Hardwood—Stress Grade F14 Amdt 1 March 1991 (in Professional Package 30) (Superseded (in Part) by AS 1684 Supplement 0 and AS 1684 Supplement 14—1992). 15 pp.
SAA AS 1684 Supp 14-92. National Timber Framing Code—Supplement 14: Timber Framing Span Tables—Seasoned Hardwood—Stress Grade F14 (Supplement to AS 1684—1992) (Supersedes AS 1684 Supplement 14—1975 (in Part)) (in Professional Packages 20, 21, 30, 62-69). 27 pp.
SAA AS 1684 Supp 15-75. Supplement to AS 1684 SAA Timber Framing Code—Supplement 15: Light Timber Framing Span Tables—Seasoned Hardwood—Stress Code F17 Amdt 1 March 1991 (in Professional Package 30) (Superseded (in Part) by AS 1684 Supplement 0 and AS 1684 Supplement 15—1992). 15 pp.
SAA AS 1684 Supp 15-92. National Timber Framing Code—Supplement 15: Timber Framing Span Tables—Seasoned Hardwood—Stress Grade F17 (Supplement to AS 1684—1992) (Supersedes AS 1684 Supplement 15—1975 (in Part)) (in Professional Packages 20, 21, 30, 62-69) Bound Together. 27 pp.
SAA AS 1684 Supp 16-75. Supplement to AS 1684 SAA Timber Framing Code—Supplement 16: Light Timber Framing Span Tables—Seasoned Hardwood—Stress Grade F27 Amdt 1 March 1991 (in Professional Package 30) (Superseded (in Part) by AS 1684 Supplement 0 and AS 1684 Supplement 16—1992). 15 pp.
SAA AS 1684 Supp 16-92. National Timber Framing Code—Supplement 16: Timber Framing Span Tables—Seasoned Hardwood—Stress Grade F27 (Supplement to AS 1684—1992) (Supersedes AS 1684 Supplement 16—1975 (in Part)) (in Professional Packages 20, 21, 30, 62-69). 27 pp.

Building Codes (Cont.)
—Structural Timber—Framing (Cont.)
SAA AS 1684 Supp 17-75. Supplement to AS 1684 SAA Timber Framing Code—Supplement 17: Light Timber Framing Span Tables—Unseasoned Timber (Alternative Sizes)—Stress Grade F4 (Superseded (in Part) by AS 1684 Supplement 0 and AS 1684 Supplement 17—1992). 15 pp.

SAA AS 1684 Supp 17-92. National Timber Framing Code—Supplement 17: Timber Framing Span Tables—Unseasoned Timber (Alternative Sizes)—Stress Grade F4 (Supplement to AS 1684—1992) (Supersedes AS 1684 Supplement 17—1975 (in Part)) (in Professional Packages 20, 21, 30, 62-69). 27 pp.

SAA AS 1684 Supp 18-75. Supplement to AS 1684 SAA Timber Framing Code—Supplement 18: Light Timber Framing Span Tables—Unseasoned Timber (Alternative Sizes)—Stress Grade F5 (Superseded (in Part) by AS 1684 Supplement 0 and AS 1684 Supplement 18—1992). 15 pp.

SAA AS 1684 Supp 18-92. National Timber Framing Code—Supplement 18: Timber Framing Span Tables—Unseasoned Timber (Alternative Sizes)—Stress Grade F5 (Supplement to AS 1684—1992) (Supersedes AS 1684 Supplement 18—1975 (in Part)) (in Professional Packages 20, 21, 30, 62-69). 27 pp.

SAA AS 1684 Supp 19-75. Supplement to AS 1684 SAA Timber Framing Code—Supplement 19: Light Timber Framing Span Tables—Unseasoned Timber (Alternative Sizes)—Stress Grade F8 (Superseded (in Part) by AS 1684 Supplement 0 and AS 1684 Supplement 19—1992). 15 pp.

SAA AS 1684 Supp 19-92. National Timber Framing Code—Supplement 19: Timber Framing Span Tables—Unseasoned Timber (Alternative Sizes)—Stress Grade F8 (Supplement to AS 1684—1992) (Supersedes AS 1684 Supplement 19—1975 (in Part)) (in Professional Packages 20, 21, 30, 62-69). 27 pp.

SAA AS 1684 Supp 20-92. National Timber Framing Code—Supplement 20: Timber Framing Span Tables—Unseasoned Timber (Alternative Sizes)—Stress Grade F11 (Supplement to AS 1684—1992) (Supersedes AS 1684 Supplement 20—1975 (in Part)) (in Professional Packages 20, 21, 30, 62-69). 27 pp.

—Temporary Structures
DIN ENGL 4112-83. Temporary Structures; Code of Practice for Design and Construction (Feb). 32 pp.

Building Construction
Use: Construction

Building Contracts
Use: Construction Contracts

Building Costs
Use: Construction Costs

Building Drawings
Use: Architectural Drawings

Building Insulation
Use: Thermal Insulation

Building Insulation (Thermal)
Use: Thermal Insulation

Building Joint Sealants
Use: Joint Sealants

Building Management Systems
Use: Building Automation Systems

Building Materials
Use: Construction Materials

Building Papers
See Also: Asphalt Papers; Construction Materials; Papers

BSI BS 1521-72. 1972 Amd 1 Waterproof Building Papers. 13 pp.

BSI BS 4016-72. 1972 Building Papers (Breather Type). 9 pp.

SNZ NZS/BS 1521-72. Specification for Waterproof Building Papers Amend: 1. 12 pp.

SNZ NZS 2295-88. Building Papers (Breather Type) Amend: A, 1988. 12 pp.

Building Plaster
Use: Gypsum Plaster

Building Regulations
Use: Building Codes

Building Services
See Also: Air Handling Equipment; Buildings; Construction; Power Systems; Water Supply Installations

BSI BS 5997-80. 1980 Guide to British Standard Codes of Practice for Building Services. 49 pp.

Building Sites
Use: Construction Sites

Building Stones
See Also: Bricks; Building Blocks; Construction Materials; Masonry; Rocks; Slates

—Abrasion Testing
CNS A3227-85. Method of Test for Abrasion Resistance of Stone Subjected to Foot Traffic (Aug)(11320).

—Bulk Density
CNS A3228-85. Method of Test for Absorption and Bulk Specific Gravity of Natural Building Stone (Aug)(11321).

—Compression Testing
CNS A3226-85. Method of Test for Compressive Strength of Natural Building Stone (Aug)(11319).

—Glossaries
CNS A1041-85. Terms Relating to Natural Building Stones (Aug)(11318).

—Modulus of Rupture
CNS A3229-85. Method of Test for Modulus of Rupture of Natural Building Stone (Aug)(11322).

—Water Absorption
CNS A3228-85. Method of Test for Absorption and Bulk Specific Gravity of Natural Building Stone (Aug)(11321).

Buildings
Scope Note: For specific materials or equipment used in buildings, see the equipment or material. For additional references, consult the following list
Use For: Industrial Buildings; Public Buildings
See Also: Agricultural Buildings; Architecture; Atriums; Auditoriums; Balconies; Billiard Rooms; Building Automation Systems; Building Services; Carports; Ceilings; Construction; Crematoria; Facades; Floors; Handrails; Houses; Libraries; Mobile Homes; Office Buildings; Prefabricated Buildings; Residential Buildings; Restaurants; School Buildings; Shelters; Shopping Malls; Shops; Sports Facilities; Stairways; Structures; Terraces; Theaters; Transportable Shelters; Walls; Warehouses

CNS A1013-75. Preferred Horizontal Dimensions for Industrial Buildings (Feb)(3538).

CNS A1014-75. General Rules for Buildings (Application of Tolerances) (Feb)(3539).

—Access Doors
SAA AS 1428.1-93. Design for Access and Mobility—Part 1: General Requirements for Access—Buildings (in Professional Packages 17, 20, 21, 30, 55, 61, 62, 63, 64, 65, 66, 67, 68, 69) Amdt 1 October 1993. 59 pp.

SAA AS 1428.1 Supp 1-90. Design for Access and Mobility—Part 1: General Requirements for Access—Buildings—Supplement 1: Commentary (Supplement to AS 1428.1 —1988). 18 pp.

—Acoustical Insulation
BSI BS 2750: Part 1-80. 1980 Measurement of Sound Insulation in Buildings and of Building Elements Part 1: Recommendations for Laboratories. 6 pp.

BSI BS 2750: Part 2-93. 1993 Acoustics—Measurements of Sound Insulation in Buildings and of Building Elements Part 2: Determination, Verification and Application of Precision Data (F). 21 pp.

BSI BS 2750: Part 2-80. 1980 Measurement of Sound Insulation in Buildings and of Building Elements Part 2: Statement of Precision Requirements. 10 pp.

BSI BS 2750: Part 3-80. 1980 Measurement of Sound Insulation in Buildings and of Building Elements Part 3: Laboratory Measurements of Airborne Sound Insulation of Building Elements. 8 pp.

BSI BS 2750: Part 4-80. 1980 Measurement of Sound Insulation in Buildings and of Building Elements Part 4: Field Measurements of Airborne Sound Insulation Between Rooms. 8 pp.

BSI BS 8233-87. 1987 Amd 1 Sound Insulation and Noise Reduction for Buildings. 59 pp.

BSI BS EN 20140-10-92. 1992 Acoustics—Measurement of Sound Insulation in Buildings and of Building Elements Part 10: Laboratory Measurement of Airborne Sound Insulation of Small Building Elements (ISO 140-10: 1991) (F). 14 pp.

Buildings (Cont.)
—Acoustical Insulation (Cont.)
BSI BS EN 20140-2-93. 1993 Acoustics—Measurement 4f Sound Insulation in Buildings and of Building Elements Part 2: Determination, Verification and and Application of Precision Data (ISO 140-2: 1991) (Supersedes BS 2750: Part 2: 1980). 21 pp.

CEN EN 20140-2-93. Acoustics—Measurement of Sound Insulation in Buildings and of Building Elements—Part 2: Determination, Verification and Application of Precision Data (ISO 140-2: 1991). 15 pp.

CEN PREN 20140-3-92. Methods of Measurement of Sound Insulation in Buildings and of Building Elements: Part 3: Laboratory Measurements of Airborne Sound Insulation of Building Elements. 23 pp.

CEN PREN 20140-10-91. Acoustics—Measurement of Sound Insulation in Buildings and of Building Elements—Part 10: Laboratory Measurement of Airborne Sound Insulation of Small Building Elements. 16 pp.

CEN EN 20140-10-92. Acoustics—Measurement of Sound Insulation in Buildings and of Building Elements—Part 10: Laboratory Measurement of Airborne Sound Insulation of Small Building Elements. 9 pp.

CNS A1031-87. Classification of Air-Borne and Impact Sound Insulation for Buildings (Mar)(8465).

DIN ENGL 4109-89. Sound Insulation in Buildings; Requirements and Testing (Nov). 25 pp.

DIN ENGL 52210 Pt 3-87. Testing of Acoustics in Buildings; Airborne and Impact Sound Insulation; Laboratory Measurements of Sound Insulation of Building Elements and Field Measurements Between Rooms (Feb). 12 pp.

DIN ENGL 52210 Pt 5-85. Testing in Building Acoustics; Airborne and Impact Sound Insulation; Field Measurements of Airborne Sound Insulation of Exterior Building Elements (July). 10 pp.

DIN ENGL 52210 Pt 6-89. Testing of Acoustics in Buildings; Airborne Impact and Sound Insulation; Measurement of Level Difference (May). 4 pp.

DIN ENGL 52210 Pt 7-89. Testing of Acoustics in Buildings; Airborne Impact and Sound Insulation; Measurement of Lateral Sound Reduction Index (May). 4 pp.

DIN ENGL EN 20140 Pt 10-92. Acoustics; Measurement of Sound Insulation in Buildings and of Building Elements; Laboratory Measurement of Airborne Sound Insulation of Small Building Elements (ISO 140-10:1991) (Sept). 10 pp.

ISO 140 Pt I-78. Acoustics—Measurement of Sound Insulation in Buildings and of Building Elements—Part I: Requirements for Laboratories First Edition. 4 pp.

ISO 140 Pt 2-91. Acoustics—Measurement of Sound Insulation in Buildings and of Building Elements—Part 2: Determination, Verification and Application of Precision Data Second Edition; (CEN EN 20140-2: 1993). 19 pp.

ISO 140 Pt 3-78. Acoustics—Measurement of Sound Insulation in Buildings and of Building Elements—Part 3: Laboratory Measurements of Airborne Sound Insulation of Building Elements First Edition; (Amendment 1-1990). 15 pp.

ISO 140 Pt IV-78. Acoustics—Measurement of Sound Insulation in Buildings and of Building Elements—Part IV: Field Measurements of Airborne Sound Insulation Between Rooms First Edition. 7 pp.

ISO 140 Pt 10-91. Acoustics—Measurement of Sound Insulation in Buildings and of Building Elements—Part 10: Laboratory Measurement of Airborne Sound Insulation of Small Building Elements First Edition. 11 pp.

JIS A 1419-92. Classification of Air-Borne and Impact Sound Insulation for Buildings. 7 pp.

SNZ NZS/ISO 140: Part 1-90. Acoustics Part 1: Requirements for Laboratories. 4 pp.

SNZ NZS/ISO 140: Part 2-78. Acoustics Part 2: Statement of Precision Requirements. 5 pp.

SNZ NZS/ISO 140: Part 3-78. Acoustics Part 3: Laboratory Measurements of Airborne Sound Insulation. 5 pp.

SNZ NZS/ISO 140: Part 4-78. Acoustics Part 4: Field Measurements of Airborne Sound Insulation Between Rooms. 5 pp.

—Acoustical Insulation—Airborne
BSI BS EN 20140-10-92. 1992 Acoustics—Measurement of Sound Insulation in Buildings and of Building Elements Part 10: Laboratory Measurement of Airborne Sound Insulation of Small Building Elements (ISO 140-10: 1991) (F). 14 pp.

CEN PREN 20140-3-92. Methods of Measurement of Sound Insulation in Buildings and of Building Elements: Part 3: Laboratory Measurements of Airborne Sound Insulation of Building Elements. 23 pp.

CEN PREN 20140-10-91. Acoustics—Measurement of Sound Insulation in Buildings and of Building Elements—Part 10: Laboratory Measurement of Airborne Sound Insulation of Small Building Elements. 16 pp.

INTERNATIONAL AND NON-U.S. NATIONAL STANDARDS
SUBJECT INDEX
Buildings

Buildings *(Cont.)*

—Acoustical Insulation—Airborne *(Cont.)*

CEN EN 20140-10-92. Acoustics—Measurement of Sound Insulation in Buildings and of Building Elements—Part 10: Laboratory Measurement of Airborne Sound Insulation of Small Building Elements. 9 pp.
DIN ENGL EN 20140 Pt 10-92. Acoustics; Measurement of Sound Insulation in Buildings and of Building Elements; Laboratory Measurement of Airborne Sound Insulation of Small Building Elements (ISO 140-10:1991) (Sept). 10 pp.
ISO 140 Pt 10-91. Acoustics—Measurement of Sound Insulation in Buildings and of Building Elements—Part 10: Laboratory Measurement of Airborne Sound Insulation of Small Building Elements First Edition. 11 pp.

—Acoustics

SAA AS 1191-85. Acoustics—Method for Laboratory Measurement of Airborne Sound Transmission Loss of Building Partitions Amdt 1 January 1987. 16 pp.

—Acoustics—Ergonomics

BSI BS 7643: Part 3-93. 1993 Building Construction—Expression of Users' Requirements Part 3: Acoustical Requirements (ISO 6242-3: 1992). 12 pp.
ISO 6242 Pt 3-92. Building Construction—Expression of Users' Requirements—Part 3: Acoustical Requirements First Edition. 9 pp.

—Air Leakage

CGSB CAN/CGSB-149.10-M86. Determination of the Airtightness of Building Envelopes by the Fan Depressurization Method. 39 pp.

—Air Pollution—Ergonomics

BSI BS 7643: Part 2-93. 1993 Building Construction—Expression of Users' Requirements Part 2: Air Purity Requirements (ISO 6242-2: 1992). 9 pp.
ISO 6242 Pt 2-92. Building Construction—Expression of Users' Requirements—Part 2: Air Purity Requirements First Edition. 8 pp.

—Billboards—Model Bylaws

SNZ NZS 9201: Chapter 8-75. Model General Bylaws Chapter 8: Control of Advertising Signs (Reconfirmed 1980). 8 pp.

—Cable Television—Distribution Systems

SNZ NZS 6603-84. Guide for the Design and Installation of Multiple Outlet Distribution Systems for Sound and Television Signals. 19 pp.

—Cabling Systems (Electric Power)

CENELEC PRHD 384.5. 52 S1-92. Electrical Installations of Buildings Part 5: Selection and Erection of Equipment Chapter 52: Selection and Erection of Wiring Systems. 26 pp.

—Commercial—Communication Transmission Lines

CSA CAN/CSA-T529-M91. Design Guidelines for Telecommunications Wiring Systems in Commercial Buildings; (Gen Instr 1). 76 pp.

—Commercial—Power Supplies

CSA CAN/CSA-C22. 2NO107.1-M91. Commercial and Industrial Power Supplies; (Gen Instr 1 Thru 2). 112 pp.

—Commercial—Telecommunication Systems—Management

CSA CAN/CSA-T528-93. Design Guidelines for Administration of Telecommunications Infrastructure in Commercial Buildings; (Gen Instr 1). 108 pp.

—Condensation Control

BSI BS 5250-89. 1989 Code of Practice for Control of Condensation in Buildings. 83 pp.
BSI BS 5250-75. (WITHDRAWN) 1975 Amd 2 Code of Basic Data for the Design of Buildings: The Control of Condensation in Dwellings. 37 pp.

—Design

BSI DD ENV 1993-1-1-92. 1992 Eurocode 3: Design of Steel Structures Part 1.1: General Rules and Rules for Buildings (Together with United Kingdom National Application Document). 364 pp.
CEN ENV 1993-1-1-92. Eurocode 3: Design of Steel Structures—Part 1-1: General Rules and Rules for Buildings. 345 pp.
CEN ENV 1993-1-1 /AC-92. Eurocode 3 Design of Steel Structures Part 1.1: General Rules and Rules for Buildings. 28 pp.
CEN ENV 1994-1-1-92. Design of Composite Steel and Concrete Structures—Part 1-1: General Rules and Rules for Buildings. 180 pp.
MOD UK DSTAN 00-57-91. Design of Buildings Used for Electrical and Mechanical Metrology Issue 1. 22 pp.
SNZ NZS 4203-84. Code of Practice for General Structural Design and Design Loadings for Buildings. 100 pp.

Buildings *(Cont.)*

—Design *(Cont.)*

SNZ NZS 4203-92. General Structural Design and Design Loadings for Buildings. 218 pp.

—Design—Access

SAA AS 1428.1-93. Design for Access and Mobility—Part 1: General Requirements for Access—Buildings (in Professional Packages 17, 20, 21, 30, 55, 61, 62, 63, 64, 65, 66, 67, 68, 69) Amdt 1 October 1993. 59 pp.
SAA AS 1428.1 Supp 1-90. Design for Access and Mobility—Part 1: General Requirements for Access—Buildings—Supplement 1: Commentary (Supplement to AS 1428.1 —1988). 18 pp.

—Design—Handicapped Persons

BSI BS 5619-78. 1978 Amd 1 Design of Housing for the Convenience of Disabled People. 9 pp.
BSI BS 5810-79. 1979 Code of Practice for Access for the Disabled to Buildings. 15 pp.
BSI PD 6523-89. 1989 Information on Access to and Movement Within and Around Buildings and on Certain Facilities for Disabled People. 49 pp.
SAA AS 1428.2-92. Design for Access and Mobility—Part 2: Enhanced and Additional Requirements—Buildings and Facilities (Supersedes AS 1428 Supplement 1—1988) (in Professional Packages 17F, 20, 21, 55, 62-69). 49 pp.
SNZ NZS 4121-85. Code of Practice for Design for Access and Use of Buildings and Facilities by Disabled Persons. 37 pp.
SNZ NZMP 4122-89. Guide to Approachability, Accessibility and Usability of Buildings. 60 pp.

—Dimensions

SAA AS 1234-72. Recommendations for Coordinated Preferred Dimensions in Building (Bound with AS 1233—1972). 16 pp.

—Durability

BSI BS 7543-92. 1992 Durability of Buildings and Building Elements, Products and Components. 45 pp.

—Electric Wiring

SAA AS 3000-91. Electrical Installations —Buildings, Structures and Premises (Known as the SAA Wiring Rules) Amdt 1 April 1992 Amdt 2 January/February 1993 Amdt 3 May 1993 (in Professional Packages 17F, 21, 55) 68). 23 pp.
SAA AS 3000 DOC R/1-91. Electrical Installations —Buildings, Structures and Premises—Part R/1: Rulings to SAA Wiring Rules (AS 3000—1991) (Supersedes AS Doc 3000 R/1—1986, AS Doc 3000 R/2—1987, AS Doc 3000 R/3—1988, AS Doc 3000 R/4—1989, AS Doc 3000 R/5—1989, AS Doc 3000 R/6—1990 and. 61 pp.
SAA AS DOC 3000 R/7-91. Electrical Installations —Buildings, Structures and Premises—Part R/7: Rulings to SAA Wiring Rules (AS 3000—1986) Seventh Group (Superseded by AS 3000 Doc R/1—1991).

—Electrical Bonding—Telecommunications

CCITT RECMN K.27-91. Bonding Configurations and Earthing Inside a Telecommunication Building (Study Group V) 25 pp. 25 pp.
CENELEC PRETS 300 253-92. Equipment Engineering (EE); Earthing and Bonding of Telecommunication Equipment in Telecommunication Centres. 21 pp.
ETSI PRETS 300 253-92. Equipment Engineering (EE); Earthing and Bonding of Telecommunication Equipment in Telecommunication Centres. 21 pp.

—Electrical Grounding—Telecommunications

CCITT RECMN K.27-91. Bonding Configurations and Earthing Inside a Telecommunication Building (Study Group V) 25 pp. 25 pp.
CENELEC PRETS 300 253-92. Equipment Engineering (EE); Earthing and Bonding of Telecommunication Equipment in Telecommunication Centres. 21 pp.
ETSI PRETS 300 253-92. Equipment Engineering (EE); Earthing and Bonding of Telecommunication Equipment in Telecommunication Centres. 21 pp.

—Electrical Installations

BSI BS 4363-68. 1968 Amd 2 Distribution Units for Electricity Supplies for Construction and Building Sites. 26 pp.
CENELEC HD 384.1-78. Electrical Installations of Buildings Part 1: Scope. 2 pp.
CENELEC HD 384.3-85. Electrical Installations of Buildings Part 3: Assessment of General Characteristics. 21 pp.
CENELEC PRHD 384.3 S2-92. Electrical Installations of Buildings Part 3: Assessment of General Characteristics of Installations. 27 pp.
CENELEC HD 384.5.51-85. Electrical Installations of Buildings Part 5: Selection and Erection of Electrical Equipment Chapter 51: Common Rules. 9 pp.

Buildings *(Cont.)*

—Electrical Installations *(Cont.)*

CENELEC HD 384.5.54 S1-88. Electrical Installatioans of Buildings—Part 5: Selection and Erection of Electrical Equipment. Chapter 54: Earthing Arrangements and Protective Conductors. 7 pp.
CENELEC HD 384.5.523-91. Electrical Installations of Buildings Part 5: Selection and Erection of Electrical Equipment Chapter 52: Wiring Systems Section 523—Current-Carrying Capacities. 10 pp.
CENELEC HD 384.6.61 S1-92. Electrical Installations of Buildings Part 6: Verification Chapter 61: Initial Verification. 7 pp.
CENELEC PRHD 384.7. 704 S1-91. Electrical Installations of Buildings Part 7: Requirements for Special Installations of Locations Section 704—Construction and Demolition Site Installations. 7 pp.
CENELEC HD 384.7.705 S1-91. Electrical Installations of Buildings Part 7: Requirements for Special Installations or Locations Section 705—Electrical Installations of Agricultural and Horticultural Premises. 5 pp.
CENELEC HD 384.7.706 S1-91. Electrical Installations of Buildings Part 7: Requirements for Special Installations or Locations Section 706—Restrictive Conducting Locations. 7 pp.
DIN ENGL 18015 Pt 3-90. Electrical Installations in Dwellings; Circuit Arrangement and Location of Electrical Equipment (July). 3 pp.
DIN VDE 0100 Pt 701-84. Installation of Power Plant with Rated Voltages up to 1000 V; Locations Containing a Bath or Shower (May). 19 pp.
DIN VDE 0108 Pt 1-89. Power Installations and Safety Power Supply in Communal Buildings; General (Oct) (VDE 0107/ 11.89 with DIN VDE 0108 Part 1 to Part 8 (Issues of 10.89) Supersedes DIN 57108/VDE 0108/12.79). 76 pp.
DIN VDE 0108 Pt 1 Suppl 1-89. Power Installations and Safety Power Supply in Communal Buildings; Building Regulations (Oct). 36 pp.
DIN VDE 0108 Pt 2-89. Power Installations and Safety Power Supply in Communal Buildings; Communal Facilities (Oct) (VDE 0107/11.89 with DIN VDE 0108 Part 1 to Part 8 (Issues of 10.89) Supersedes DIN 57 108/VDE 0108/12.79). 20 pp.
DIN VDE 0108 Pt 4-89. Power Installations and Safety Power Supply in Communal Buildings; High-Rise Buildings (Oct) (VDE 0107/11.89 with DIN VDE 0108 Part 1 to Part 8 (Issues of 10.89) Supersedes DIN 57108/VDE 0108/12.79). 5 pp.
DIN VDE 0108 Pt 7-89. Power Installations and Safety Power Supply in Communal Buildings; Working and Business Premises (Oct) (VDE 0107/ 11.89 with DIN VDE 0108 Part 1 to Part 8 (Issues of 10.89) Supersedes DIN 57108/VDE 0108/12.79). 9 pp.
IEC 364 Pt 1-92. Electrical Installations of Buildings Part 1: Scope, Object and Fundamental Principles Third Edition. 35 pp.
IEC 364 Pt 3-93. Electrical Installations of Buildings Part 3: Assessment of General Characteristics Second Edition. 63 pp.
IEC 364 Pt 5-51-79. Electrical Installations of Buildings Part 5: Selection and Erection of Electrical Equipment Chapter 51: Common Rules First Edition; (Amendment 1-1982) (Amendment 2-1993). 33 pp.
IEC 364 Pt 5-54-80. Electrical Installations of Buildings Part 5: Selection and Erection of Electrical Equipment Chapter 54: Earthing Arrangements and Protective Conductors First Edition; (Amendment 1-1982). 35 pp.
IEC 364 Pt 5-523-83. Electrical Installations of Buildings Part 5: Selection and Erection of Electrical Equipment Chapter 52: Wiring Systems Section 523—Current-Carrying Capacities First Edition. 72 pp.
IEC 364 Pt 6-61-86. Electrical Installations of Buildings Part 6: Verification Chapter 61: Initial Verification First Edition; (Amendment 1-1993). 47 pp.
IEC 364 Pt 7-701-84. Electrical Installations of Buildings Part 7: Requirements for Special Installations or Locations Section 701: Locations Containing a Bath Tub or Shower Basin First Edition. 20 pp.
IEC 364 Pt 7-704-89. Electrical Installations of Buildings Part 7: Requirements for Special Installations or Locations Section 704—Construction and Demolition Site Installations First Edition. 15 pp.
IEC 364 Pt 7-705-84. Electrical Installations of Buildings Part 7: Requirements for Special Installations or Locations Section 705—Electrical Installations of Agricultural and Horticultural Premises First Edition. 13 pp.
IEC 364 Pt 7-706-83. Electrical Installations of Buildings Part 7: Requirements for Special Installations or Locations Section 706—Restrictive Conducting Locations First Edition. 12 pp.
SAA AS 1852.826-83. International Electrotechnical Vocabulary—Part 826: Electrical Installations of Buildings. 13 pp.

INDUSTRY STANDARDS
345

Buildings (Cont.)

—Electrical Installations (Cont.)
SAA AS 3000-91. Electrical Installations —Buildings, Structures and Premises (Known as the SAA Wiring Rules) Amdt 1 April 1992 Amdt 2 January/February 1993 Amdt 3 May 1993 (in Professional Packages 17F, 21, 55) 68). 23 pp.
SAA AS 3000 DOC R/1-91. Electrical Installations —Buildings, Structures and Premises—Part R/1: Rulings to SAA Wiring Rules (AS 3000—1991) (Supersedes AS Doc 3000 R/1—1986, AS Doc 3000 R/2—1987, AS Doc 3000 R/3—1988, AS Doc 3000 R/4—1989, AS Doc 3000 R/5—1989, AS Doc 3000 R/6—1990 and. 61 pp.
SAA AS DOC 3000 R/7-91. Electrical Installations —Buildings, Structures and Premises—Part R/7: Rulings to SAA Wiring Rules (AS 3000—1986) Seventh Group (Superseded by AS 3000 Doc R/1—1991).
SNZ IEC 50: 50(826)-82. International Electrotechnical Vocabulary 50(826): Electrical Installations of Buildings. 24 pp.

—Electrical Installations—Glossaries
CENELEC HD 384.2-86. International Electrotechnical Vocabulary Chapter 826: Electrical Installations of Buildings. 2 pp.
CENELEC HD 384.2 S1/A1-92. AMD 1 International Electrotechnical Vocabulary (IEV) Chapter 826: Electrical Installations of Buildings (IEC 50(826):1982/A1: 1990). 4 pp.
IEC 50 Chap 826-82. International Electrotechnical Vocabulary Chapter 826: Electrical Installations of Buildings First Edition; (Amendment 1-1990) (CENELEC HD 384.2 S1/A1:1992). 48 pp.
IEC 364 Pt 2-21-93. Electrical Installations of Buildings Part 2: Definitions Chapter 21: Guide to General Terms First Edition. 19 pp.

—Emergency and Standby Power Supplies
DIN VDE 0108 Pt 1-89. Power Installations and Safety Power Supply in Communal Buildings; General (Oct) (VDE 0107/ 11.89 with DIN VDE 0108 Part 1 to Part 8 (Issues of 10.89) Supersedes DIN 57108/VDE 0108/12.79). 76 pp.
DIN VDE 0108 Pt 1 Suppl 1-89. Power Installations and Safety Power Supply in Communal Buildings; Building Regulations (Oct). 36 pp.
DIN VDE 0108 Pt 2-89. Power Installations and Safety Power Supply in Communal Buildings; Communal Facilities (Oct) (VDE 0107/11.89 with DIN VDE 0108 Part 1 to Part 8 (Issues of 10.89) Supersedes DIN 57 108/VDE 0108/12.79). 20 pp.
DIN VDE 0108 Pt 4-89. Power Installations and Safety Power Supply in Communal Buildings; High-Rise Buildings (Oct) (VDE 0107/11.89 with DIN VDE 0108 Part 1 to Part 8 (Issues of 10.89) Supersedes DIN 57108/VDE 0108/12.79). 5 pp.
DIN VDE 0108 Pt 7-89. Power Installations and Safety Power Supply in Communal Buildings; Working and Business Premises (Oct) (VDE 0107/ 11.89 with DIN VDE 0108 Part 1 to Part 8 (Issues of 10.89) Supersedes DIN 57108/VDE 0108/12.79). 9 pp.

—Enclosures—Design
BSI BS 8200-85. 1985 Code of Practice for Design of Non-Load Bearing External Vertical Enclosures of Buildings. 73 pp.

—Energy Conservation
SNZ NZS 4220-82. Code of Practice for Energy Conservation in Non-Residential Buildings. 30 pp.

—Energy Consumption
BSI BS 8207-85. 1985 Code of Practice for Energy Efficiency in Buildings. 39 pp.
BSI BS 8207: Supplement-85. 1985 Code of Practice for Engery Efficiency in Buildings: Supplement: Energy Design Guide. 40 pp.

—Fire Protection
BSI BS 5588: Part 2-85. (WITHDRAWN) 1985 Amd 1 Fire Precautions in the Design and Construction of Buildings Part 2: Code of Practice for Shops. 62 pp.
BSI BS 5588: Part 2-03. 1985 Amd 3 Fire Precautions in the Design and Construction of Buildings Part 2: Code of Practice for Shops (AMD 6478) August 31, 1990. 62 pp.
BSI BS 5588: Part 6-91. 1991 Fire Precautions in the Design, Construction and Use of Buildings Part 6: Code of Practice for Places of Assembly. 67 pp.

—Fire Protection—Handicapped Persons
BSI BS 5588: Part 8-88. 1988 Fire Precautions in the Design and Construction of Buildings Part 8: Code of Practice for Means of Escape for Disabled People. 24 pp.

—Fire Testing
BSI BS 476: Part 1-53. 1953 Amd 3 Fire Tests on Building Materials and Structures Part 1: (AMD 686) February 18, 1971. 27 pp.

Buildings (Cont.)

—Fire Testing (Cont.)
BSI BS 476: Part 32-89. 1989 Fire Tests on Building Materials and Structures Part 32: Guide to Full Scale Fire Tests Within Buildings. 16 pp.

—Firesafety—Model Bylaws
SNZ NZS 9232-91. Model Bylaw for Precautions Against Fire and Panic in Cinemas, Theatres and Places of Assembly. 20 pp.

—Foundations—Construction
BSI CP 101-72. 1972 Foundations and Substructures for Non-Industrial Buildings of Not More Than Four Storeys. 29 pp.

—Glossaries
BSI BS 6100: Sec 1.1-87. 1987 Glossary of Building and Civil Engineering Terms Part 1: General and Miscellaneous Section 1.1: Types of Building. 9 pp.
BSI BS 6100: Sec 1.1-01. 1987 Amd 1 Glossary of Building and Civil Engineering Terms Part 1: General and Miscellaneous Section 1.1: Types of Building (AMD 7230) August 15, 1992. 10 pp.
SNZ NZMP 4212-79. Glossary of Building Terminology. 160 pp.

—Lighting—Design
BSI BS 8206: Part 1-85. 1985 Lighting for Buildings Part 1: Code of Practice for Artifical Lighting. 40 pp.
BSI BS 8206: Part 2-92. 1992 Lighting for Buildings Part 2: Code of Practice for Daylighting (R). 43 pp.
BSI BS 8206: Part 2-01. 1992 Amd 1 Lighting for Buildings Part 2: Code of Practice for Daylighting (AMD 7391) October 15, 1992. 45 pp.
BSI DD 73-82. (WITHDRAWN) 1982 Basic Data for the Design of Buildings: Daylight (Superseded by BS 8206: Part 2: 1992). 103 pp.

—Loads (Forces)
SNZ NZS 4203-92. General Structural Design and Design Loadings for Buildings. 218 pp.

—Maintenance
BSI BS 8210-86. 1986 Guide to Building Maintenance Management. 46 pp.

—Measurement
SNZ NZS 4202-86. Standard Method of Measurement of Building Works. 98 pp.

—Model Bylaws
SNZ NZS 9201: Chapter 1-72. Model General Bylaws Chapter 1: Introductory (Reconfirmed 1980). 12 pp.
SNZ NZS 9201: Chapter 2-72. Model General Bylaws Chapter 2: Public Places. 20 pp.

—Noise—Aircraft
SAA AS 2021-85. Acoustics—Aircraft Noise Intrusion—Building Siting and Construction. 42 pp.

—Painting
BSI BS 6150-91. 1991 Code of Practice for Painting of Buildings. 133 pp.
BSI BS 6150-82. 1982 Code of Practice for Painting of Buildings. 83 pp.
BSI BS 8000: Part 12-89. 1989 Workmanship on Building Sites Part 12: Code of Practice for Decorative Wallcoverings and Painting. 21 pp.
SNZ NZS 7703-85. Painting of Buildings. 51 pp.

—Protection
DIN ENGL 4123-72. Protection of Buildings in the Area of Excavations, Foundations and Underpinnings (May). 5 pp.

—Protection—Ground Water
BSI BS 8102-90. 1990 Code of Practice for Protection of Structures Against Water from the Ground. 40 pp.
BSI CP 102-73. 1973 Amd 3 Protection of Buildings Against Water from the Ground. 30 pp.

—Radio Wave Propagation
CCIR Report 880-2-90. Short Distance Radio-Wave Propagation in Special Environments Buildings, Tunnels, Mines, Etc.—Section 5D—Aspects Relative to the Terrestrial Broadcasting and Mobile Services. 2 pp.

—Scaffolds—Model Bylaws
SNZ NZS 9201: Chapter 9-72. Model General Bylaws Chapter 9: Scaffolding and Deposit of Building Materials. 8 pp.

—Seismic Design—Engineering Systems
SNZ NZS 4219-83. Specification for Seismic Resistance of Engineering Systems in Buildings Amend: 1, 1990; 2, 1992. 37 pp.

Buildings (Cont.)

—Smoke Control
BSI BS 5588: Part 4-78. 1978 Amd 1 Fire Precautions in the Design and Construction of Buildings Part 4: Code of Practice for Smoke Control in Protected Escape Routes Using Pressurization. 35 pp.
SAA AS 1668.1-91. Use of Mechanical Ventilation and Air-Conditioning in Buildings—Part 1: Fire and Smoke Control (Supersedes SAA MP47.C1—1980) (in Professional Packages 20, 21, 44, 50, 62-69). 53 pp.

—Soil Treatment—Termites
SAA AS 2057-86. Protection of Buildings from Subterranean Termites—Chemical Treatment of Soil for Buildings Under (Withdrawn).
SAA AS 2057A-86. Protection of Buildings from Subterranean Termites—Chemical Treatment of Soil for Buildings Under Con-struction—Treatment Cetrtification Pads of 20 Certificates in Triplicate (Withdrawn).

—Solar Radiation—Design
BSI DD 67-80. (WITHDRAWN) 1980 Basic Data for the Design of Buildings: Sunlight (Superseded by BS 8206: Part 2: 1992). 78 pp.

—Structural Timber
SNZ NZMP 3600: Part 1-92. Builders' Guide to NZS 3604. 148 pp.
SNZ NZS 3604-90. Code of Practice for Light Timber Frame Buildings Not Requiring Specific Design. 224 pp.

—Surface Area Analyzers
BSI BS 7641-93. 1993 Performance Standards in Building—Definition and Calculation of Area and Space Indicators (ISO 9836: 1992). 15 pp.
ISO 9836-92. Performance Standards in Building— Definition and Calculation of Area and Space Indicators First Edition. 13 pp.

—Termite Protection Barriers
SAA AS 1694-74. Code of Practice for Physical Barriers Used in the Protection of Buildings Against Subterranean Termites (Metric Units) (Withdrawn).

—Thermal Environments—Ergonomics
BSI BS 7643: Part 1-93. 1993 Building Construction— Expression of Users' Requirements Part 1: Thermal Requirements (ISO 6242-1: 1992). 11 pp.
ISO 6242 Pt 1-92. Building Construction—Expression of Users' Requirements—Part 1: Thermal Requirements First Edition. 9 pp.

—Thermal Insulation
SNZ NZS 4222-92. Specification for Materials for the Thermal Insulation of Buildings. 28 pp.

—Thermal Insulation—Glossaries
SNZ NZS/AS 2352-80. Glossary of Terms for Thermal Insulation of Buildings (This is a Joint Standard with SAA AS 2352:1980). 8 pp.

—Thermal Resistance
SNZ NZS 4214-77. Methods of Determining the Total Thermal Resistance of Parts of Buildings. 20 pp.

—Thermography (Temperature Measurement)
ISO 6781-83. Thermal Insulation—Qualitative Detection of Thermal Irregularities in Building Envelopes—Infrared Method First Edition. 15 pp.

—Thermography (Temperature Measurement)—Training Manuals
CGSB 149-GP-2MP-86. Manual for Thermographic Analysis of Building Enclosures. 125 pp.

—Units of Measurement
SNZ NZS 4202-86. Standard Method of Measurement of Building Works. 98 pp.

—Ventilation
SNZ NZS 4303-90. Ventilation for Acceptable Indoor Air Quality. 34 pp.

—Vibration
BSI BS 6472-92. 1992 Evaluation of Human Exposure to Vibration in Buildings (1Hz to 80Hz). 21 pp.
BSI BS 6472-84. 1984 Evaluation of Human Exposure to Vibration and Shock in Buildings (1Hz to 80Hz). 16 pp.
BSI BS 7385: Part 1-90. 1990 Evaluation and Measurement of Vibration in Buildings Part 1: Guide for Measurement of Vibrations and Evaluation of Their Effects on Buildings. 26 pp.
DIN ENGL 4150 Pt 3-86. Structural Vibration in Buildings; Effects on Structures (May). 6 pp.
ISO 4866-90. Mechanical Vibration and Shock— Vibration of Buildings—Guidelines for the Measurement of Vibrations and Evaluation of Their Effects on Buildings First Edition. 23 pp.

INTERNATIONAL AND NON-U.S. NATIONAL STANDARDS
SUBJECT INDEX

Bulk

Buildings *(Cont.)*

—Vibration *(Cont.)*
ISO 10137-92. Bases for Design of Structures—Serviceability of Buildings Against Vibration First Edition. 38 pp.

—Volume Indicators
BSI BS 7641-93. 1993 Performance Standards in Building—Definition and Calculation of Area and Space Indicators (ISO 9836: 1992). 15 pp.
ISO 9836-92. Performance Standards in Building—Definition and Calculation of Area and Space Indicators First Edition. 13 pp.

—Water Supply—Model Bylaws
SNZ NZS 9201: Chapter 6-72. Model General Bylaws Chapter 6: Removal of Refuse (Reconfirmed 1980). 8 pp.
SNZ NZS 9201: Chapter 7-73. Model General Bylaws Chapter 7: Water Supply Amend: 1, 1981. 31 pp.

—Waterproof Coatings
CNS A3145-88. Method of Test for Coating for Waterproofing of Buildings (Jan)(8645).

Bulk Density
Use For: Apparent Density
See Also: Compactibility; Density; Metal Powders; Tap Density

—Abrasives
BSI BS 7425: Part 1-91. 1991 Abrasive Grains Part 1: Method for the Determination of the Bulk Density of Macrograins. 8 pp.
CNS R3101-81. Method of Test for Bulk Density of Artificial Abrasives (Aug)(7841).
ISO 9136-89. Abrasive Macrograins—Determination of Bulk Density First Edition. 6 pp.
JIS R 6126-70. Testing Method for Bulk Density of Artificial Abrasives. 5 pp.

—Activated Carbon
CNS K6477-80. Method of Test for Apparent Density of Activation Carbon (May)(5580).

—Aggregates
BSI BS 812: Part 2-75. 1975 Amd 1 Testing Aggregates Part 2: Physical Properties. 24 pp.
JIS A 1134-89. Methods of Test for Bulk Specific Gravity and Absorption of Light Weight Fine Aggregates for Structural Concrete. 9 pp.
JIS A 1135-89. Methods of Test for Bulk Specific Gravity and Absorption of Light Weight Coarse Aggregates for Structural Concrete. 6 pp.

—Aggregates—Concretes
BSI BS 3681: Part 2-73. (WITHDRAWN) 1973 Methods for the Sampling and Testing of Lightweight Aggregates for Concrete Part 2: Metric Units (Superseded by BS 3797: 1990). 19 pp.
ISO 6782-82. Aggregates for Concrete—Determination of Bulk Density First Edition. 4 pp.

—Asphalts
DIN ENGL 1996 Pt 7-83. Testing of Bituminous Materials for Road Building and Related Purposes; Determination of Density and Voids (Jan). 8 pp.

—Bitumens
BSI BS 598: Part 104-89. 1989 Amd 1 Sampling and Examination of Bituminous Mixtures for Roads and Other Paved Areas Part 104: Methods of Test for the Determination of Density and Compaction (AMD 6738) September 30, 1991 (Supersedes BS 598: Part 3: 1985). 17 pp.
CNS A3149-87. Method of Test for Bulk Specific Gravity and Density of Compacted Bituminous Mixtures Using Paraffin-Coated Specimens (Aug)(8757).
CNS A3151-87. Method of Test for Bulk Specific Gravity and Density of Compacted Bituminous Mixtures Using Saturated Surface-Dry Specimens (Aug)(8759).

—Building Stones
CNS A3228-85. Method of Test for Absorption and Bulk Specific Gravity of Natural Building Stone (Aug)(11321).

—Calcium Carbonates
BSI BS 1460-67. 1967 Amd 1 Determination of Apparent Density After Compaction of Precipitated Calcium Carbonate. 13 pp.
CNS K6721-82. Method of Test for Calcium Carbonate for Rubber (Jun)(9004). 4 pp.

—Calcium Silicide
MOD UK M 805/84. Examination of Calcium Silicide.

—Carbon Black
DIN ENGL 53600-76. Testing of Carbon Blacks; Determination of the Bulk Density of Pelletized Carbon Blacks for Use in the Rubber Industry (May). 2 pp.

Bulk Density *(Cont.)*

—Cellular Materials
DIN ENGL 53420-78. Testing of Cellular Materials; Determination of Apparent Density (Dec). 2 pp.

—Cellular Plastics
CNS K6667-81. Determination of Apparent Density of Cellular Rubbers and Plastics (May)(7407).
ISO 845-88. Cellular Plastics and Rubbers—Determination of Apparent (Bulk) Density Second Edition. 6 pp.
ISO 9054-90. Cellular Plastics, Rigid —Test Methods for Self-Skinned, High-Density Materials First Edition. 8 pp.
JIS K 7222-85. Methods for Determining the Apparent Density of Rigid Cellular Plastics. 7 pp.

—Ceramic Powders
CEN PREN 725-8-93. Advanced Technical Ceramics—Methods of Test for Ceramic Powders—Part 8: Determination of Tapped Bulk Density. 7 pp.
CEN PREN 725-9-93. Advanced Technical Ceramics—Methods of Test for Ceramic Powders—Part 9: Determination of Untamped Bulk Density. 7 pp.

—Ceramics
CNS C6068-80. Method of Test for Apparent Density of Ceramics for Electron Device and Semiconductor Application (Jul)(5751). 3 pp.
DIN ENGL 51065 Pt 1-85. Testing of Ceramic Materials; Determination of Bulk Density of Shaped Products and Broken Pieces (Aug). 2 pp.

—Cereals
BSI BS 4317: Part 23-90. 1990 Methods of Test for Cereals and Pulses Part 23: Determination of Bulk Density of Cereals, Called 'Mass per Hectolitre'. 8 pp.
ISO 7971-86. Cereals—Determination of Bulk Density, Called "Mass per Hectolitre" (Reference Method) First Edition. 7 pp.
ISO 8840-87. Refractory Materials—Determination of Bulk Density of Granular Materials (Grain Density) First Edition. 8 pp.

—Coal
SAA AS 3899-91. Higher Rank Coal and Coke—Bulk Density (in Professional Package 32). 7 pp.

—Coal Tar Pitch
BSI BS 6043: Sec 1.7-88. 1988 Methods of Sampling and Test for Carbonaceous Materials Used in Aluminium Manufacture Part 1: Electrode Pitch Section 1.7: Determination of Apparent Density (Bouyancy Method). 7 pp.

—Coffee
BSI BS 5752: Part 11-87. 1987 Coffee and Coffee Products Part 11: Instant Coffee: Determination of Free-Flow and Compacted Bulk Densities. 9 pp.
ISO 8460-87. Instant Coffee—Determination of Free-Flow and Compacted Bulk Densities First Edition. 9 pp.

—Coke
BSI BS 1016: Sec 108.3-91. 1991 Analysis and Testing of Coal and Coke Part 108: Tests Special to Coke Section 108.3: Determination of Bulk Density (Small Container). 7 pp.
BSI BS 1016: Sec 108.4-91. 1991 Analysis and Testing of Coal and Coke Part 108: Tests Special to Coke Section 108.4: Determination of Bulk Density (Large Container). 6 pp.
CNS M3173-84. Method of Test for Bulk Density of Coke in a Larger Container (Nov) (11115).
CNS M3174-84. Method of Test for Bulk Density of Coke in a Small Container (Nov) (11116).
ISO 567-74. Coke—Determination of the Bulk Density in a Small Container First Edition. 4 pp.
ISO 1013-75. Coke—Determination of Bulk Density in a Large Container First Edition. 4 pp.
SAA AS 3899-91. Higher Rank Coal and Coke—Bulk Density (in Professional Package 32). 7 pp.

—Concretes
JIS A 1161-73. Testing Methods for Bulk Specific Gravity, Water Content, Absorption and Compressive Strength of Cellular Concrete. 7 pp.

—Cork
ISO 2031-91. Granulated Cork—Determination of Bulk Density Second Edition. 5 pp.
ISO 2189-86. Expanded Pure Agglomerated Cork—Determination of Bulk Density Second Edition. 3 pp.
ISO 9727-91. Cylindrical Stoppers of Natural Cork—Physical Tests—Reference Methods First Edition. 12 pp.

—Drying Agents
BSI BS 3482: Part 10-91. 1991 Methods of Test for Desiccants Part 10: Determination of Bulk Density (Dry Basis). 6 pp.

Bulk Density *(Cont.)*

—Extenders
BSI BS 3483: Part B10-82. 1982 Methods for Testing Pigments in Paints Part B10: Determination of Tamped Volume and Apparent Density After Tamping. 6 pp.
ISO 787 Pt 11-81. General Methods of Test for Pigments and Extenders—Part 11: Determination of Tamped Volume and Apparent Density After Tamping First Edition. 6 pp.

—Fertilizers
BSI BS 5551: Sec 3.1-93. 1993 Fertilizers Part 3: Physical Properties Section 3.1: Method for Determination of Bulk Density (Loose) (ISO 3944: 1992). 7 pp.
BSI BS 5551: Sec 3.1-81. 1981 Fertilizers Part 3: Physical Properties Section 3.1: Method for the Determination of Bulk Density (Loose). 4 pp.
BSI BS 5551: Sec 3.2-93. 1993 Fertilizers Part 3: Physical Properties Section 3.2: Method for Determination of Bulk Density (Tapped) (ISO 5311: 1992). 10 pp.
BSI BS 5551: Sec 3.2-83. 1983 Fertilizers Part 3: Physical Properties Section 3.2: Method for the Determination of Bulk Density (Tapped). 5 pp.
BSI BS 5551: Sec 3.3-92. 1992 Fertilizers Part 3: Physical Properties Section 3.3: Method for Determination of Bulk Density (Loose) of Fine-Grained Fertilizers (ISO 7837: 1992). 7 pp.
BSI BS 5551: Sec 3.3-84. 1984 Fertilizers Part 3: Physical Properties Section 3.3: Method for Determination of Bulk Density (Loose) of Fine-Grained Fertilizers. 5 pp.
ISO 3944-92. Fertilizers—Determination of Bulk Density (Loose) Third Edition. 7 pp.
ISO 5311-92. Fertilizers—Determination of Bulk Density (Tapped) Third Edition. 8 pp.
ISO 7837-92. Fertilizers—Determination of Bulk Density (Loose) of Fine-Grained Fertilizers Second Edition. 7 pp.

—Firebricks
CNS R3008-78. Method of Test for Size and Bulk Density of Refractory Brick and Insulating Firebrick (Nov)(614).
CNS R3013-86. Method of Test for Apparant Porosity Water Absorption and Specific Gravity of Refractory Bricks (Oct)(619).
CNS R3141-86. Method of Test for Specific Gravity and True Porosity of Insulating Fire Bricks (Oct)(11738).
JIS R 2205-92. Testing Method for Apparent Porosity, Water Absorption and Specific Gravity of Refractory Bricks. 8 pp.
JIS R 2614-85. Testing Method for Specific Gravity and True Porosity of Insulating Fire Bricks. 6 pp.

—Foam Rubber
CNS K6667-81. Determination of Apparent Density of Cellular Rubbers and Plastics (May)(7407).
ISO 845-88. Cellular Plastics and Rubbers—Determination of Apparent (Bulk) Density Second Edition. 6 pp.

—Germanium Dioxide
CNS C6069-80. Method of Test for Bulk Density of Germanium Dioxide (Jul)(5753).

—Iron Ores
BSI BS 6147-89. 1989 Methods for Determination of Bulk Density of Iron Ores. 7 pp.
ISO 3852-88. Iron Ores—Determination of Bulk Density Second Edition; (Supersedes 5464). 6 pp.
JIS M 8716-90. Iron Ore Pellets—Determination of Apparent Density and Porosity. 5 pp.

—Leather
ISO 2420-72. Leather—Determination of Apparent Density First Edition. 4 pp.

—Magnesium Carbonate
MOD UK M 870/67. Examination of Magnesium Carbonate, Heavy.

—Magnesium Oxides
MOD UK M 9541/69. Examination of Magnesium Oxide and Magnesium Oxide, Special, Lead Free (Withdrawn).

—Masonry—Stone
CEN PREN 772-4-92. Methods of Test for Masonry Units—Part 4: Determination of Real Density and Bulk Density for Natural Stone Masonry Units. 7 pp.

—Metal Powders
BSI BS 5600: SUBSEC 2.2.1-89. 1989 Powder Metallurgical Materials and Products Part 2: Methods of Sampling and Testing Metallic Powders Section 2.2: Determination of Apparent Density Subsection 2.2.1: Funnel Method. 7 pp.

INDUSTRY STANDARDS

Bulk Density (Cont.)

—Metal Powders (Cont.)

BSI BS 5600: SUBSEC 2.2.1-01. (WITHDRAWN) 1989 Amd 1 Powder Metallurgical Materials and Products Part 2: Methods of Sampling and Testing Metallic Powders Sec 2.2: Determination of Apparent Density Subsection 2.2.1: Funnel Method (AMD 7032) May 1, 1992. 7 pp.

BSI BS 5600: SUBSEC 2.2.2-84. (WTIHDRAWN) 1984 Powder Metallurgical Materials and Products Part 2: Methods of Sampling and Testing Metallic Powders Section 2.2: Determination of Apparent Density by the Cup and Funnel Method Subsection 2.2.2: Scott Volumeter Method. 6 pp.

BSI BS 5600: SUBSEC 2.2.3-84. 1984 Powder Metallurgical Materials and Products Part 2: Methods of Sampling and Testing Metallic Powders Section 2.2: Determination of Apparent Density by the Cup and Funnel Method Subsec 2.2.3: Oscillating Funnel Method. 4 pp.

BSI BS EN 23923-1-93. 1993 Metallic Powders—Determination of Apparent Density Part 1: Funnel Method (ISO 3923-1: 1979) (V). 11 pp.

BSI BS EN 23923-2-93. 1993 Metallic Powders—Determination of Apparent Density Part 2: Scott Volumeter Method (ISO 3923-2: 1981) (V). 12 pp.

CEN EN 23923-1-93. Metallic Powders—Determination of Apparent Density—Part 1: Funnel Method (ISO 3923-1: 1979). 6 pp.

CEN EN 23923-2-93. Metallic Powders—Determination of Apparent Density—Part 2: Scott Volumeter Method (ISO 3923-2: 1981). 7 pp.

CNS Z8040-82. Method for Determining Apparent Density of Metal Powders (Jul)(9204).

ISO 3923 Pt 1-79. Metallic Powders—Determination of Apparent Density—Part 1: Funnel Method Second Edition (CEN EN 23923-1: 1993). 5 pp.

ISO 3923 Pt 2-81. Metallic Powders—Determination of Apparent Density—Part 2: Scott Volumeter Method First Edition (CEN EN 23923-2: 1993). 6 pp.

JIS Z 2504-79. Method for Determination of Apparent Density of Metal Powders. 5 pp.

—Mineral Aggregates

DIN ENGL 52102-88. Determination of Absolute Density, Dry Density, Compactness and Porosity of Natural Stone and Mineral Aggregates (Aug). 10 pp.

—Molding Materials

CNS K6558-80. Determination of Apparent Density of Material That Can Be Poured from a Specified Funnel (Aug)(6203).

CNS K6559-80. Determination of Apparent Density of Moulding Material That Cannot Be Poured from a Specified Funnel (Aug)(6204).

—Mortars

CEN PREN 1015-6-93. Methods of Test for Mortar for Masonry—Part 6: Determination of Bulk Density of Fresh Mortar. 7 pp.

CEN PREN 1015-10-93. Methods of Test for Mortar for Masonry—Part 10: Determination of Bulk Density of Hardened Mortar. 7 pp.

DIN ENGL 18555 Pt 2-82. Testing of Mortars Containing Mineral Binders; Freshly Mixed Mortars Containing Aggregates of Dense Structure (Heavy Aggregates); Determination of Consistence, Bulk Density and Air Content (Sept). 4 pp.

DIN ENGL 18555 Pt 3-82. Testing of Mortars Containing Mineral Binders; Hardened Mortars; Determination of Flexural Strength, Compressive Strength and Bulk Density (Sept). 4 pp.

—Paperboard

CNS P3028-89. Method of Test for Thickness and Apparent Density of Paper and Paperboard (Aug)(3685). 3 pp.

JIS P 8118-76. Testing Method for Thickness and Bulk Density of Paper and Paperboard. 6 pp.

SAA AS 1301.427S-88. Methods of Test for Pulp and Paper (Metric Units)—Part 427s: Bulking Thickness of Paper and Paperboard (This is a Joint Standard with SANZ NZS 1301). 3 pp.

SNZ NZS/AS 1301. 427S-88. Methods of Test for Pulp and Paper Bulking Thickness of Paper and Paperboard (This is a Joint Standard with SAA AS 1301.427S). 3 pp.

—Papers

BSI BS 7387-91. 1991 Method for Determination of the Bulking Thickness, Apparent Bulk Density, Compressibility and Compressibility Index of Soft Creped Tissue Paper. 12 pp.

CNS P3028-89. Method of Test for Thickness and Apparent Density of Paper and Paperboard (Aug)(3685). 3 pp.

JIS P 8118-76. Testing Method for Thickness and Bulk Density of Paper and Paperboard. 6 pp.

SAA AS 1301.427S-88. Methods of Test for Pulp and Paper (Metric Units)—Part 427s: Bulking Thickness of Paper and Paperboard (This is a Joint Standard with SANZ NZS 1301). 3 pp.

Bulk Density (Cont.)

—Papers (Cont.)

SNZ NZS/AS 1301. 427S-88. Methods of Test for Pulp and Paper Bulking Thickness of Paper and Paperboard (This is a Joint Standard with SAA AS 1301.427S). 3 pp.

—Phenolic Resins

CNS K6273-74. Method of Test for Phenol Formaldehyde Resin Molding Compounds (Oct)(2988). 11 pp.

—Pigments

BSI BS 3483: Part B10-82. 1982 Methods for Testing Pigments in Paints Part B10: Determination of Tamped Volume and Apparent Density After Tamping. 6 pp.

ISO 787 Pt 11-81. General Methods of Test for Pigments and Extenders—Part 11: Determination of Tamped Volume and Apparent Density After Tamping First Edition. 6 pp.

—Plastics

BSI BS 2782:Pt6: METH 621A-78. 1978 Methods of Testing Plastics Part 6: Dimensional Properties Method 621A: Determination of Apparent Density of Moulding Material That Can Be Poured from a Funnel. 4 pp.

BSI BS 2782:Pt6: METH 621B-78. 1978 Methods of Testing Plastics Part 6: Dimensional Properties Method 621B: Determination of Apparent Density of Moulding Material Which Cannot Be Poured from a Funnel. 4 pp.

BSI BS 2782:Pt6: METH 621C-83. 1983 Methods of Testing Plastics Part 6: Dimensional Properties Method 621C: Determination of the Bulk Factor of Moulding Materials. 4 pp.

ISO 60-77. Plastics—Determination of Apparent Density of Material That Can Be Poured from a Specified Funnel Second Edition. 4 pp.

ISO 61-76. Plastics—Determination of Apparent Density of Moulding Material That Cannot Be Poured from a Specified Funnel First Edition. 3 pp.

ISO 171-80. Plastics—Determination of Bulk Factor of Moulding Materials First Edition. 3 pp.

—Powdered Milk

BSI BS 1743: Part 5-92. 1992 Methods for Analysis of Dried Milk and Dried Milk Products Part 5: Determination of Bulk Density (ISO 8967: 1992). 10 pp.

ISO 8967-92. Dried Milk and Dried Milk Products—Determination of Bulk Density First Edition. 8 pp.

—PVC

JIS K 6721-77. Testing Methods for Polyvinyl Chloride (R 1980). 10 pp.

JIS K 6722-77. Testing Methods for Polyvinyliden Chloride (R 1980). 5 pp.

—PVC Resins

BSI BS 2782:Pt6: METH 621D-78. 1978 Methods of Testing Plastics Part 6: Dimensional Properties Method 621D: Determination of Compacted Apparent Bulk Density of PVC Resins. 4 pp.

ISO 1068-75. Plastics—PVC Resins—Determination of Compacted Apparent Bulk Density First Edition; (Erratum—Nov 1980). 5 pp.

—Refractory Materials

BSI BS 1902: Sec 3.6-84. 1984 Methods for Testing Refractory Materials Part 3: General and Textural Properties Section 3.6: Determination of Grain Density (Methods 1902-306). 14 pp.

BSI BS 1902: Sec 3.7-89. 1989 Methods for Testing Refractory Materials Part 3: General and Textural Properties Section 3.7: Determination of Bulk Density and True Porosity of Shaped Insulating Products (Method 1902-307). 4 pp.

BSI BS 1902: Sec 3.8-89. 1989 Methods for Testing Refractory Materials Part 3: General and Textural Properties Section 3.8: Determination of Bulk Density, True Porosity and Apparent Porosity of Dense Shaped Products (Methods 1902-308). 7 pp.

BSI BS 1902: Sec 3.17-90. 1990 Methods for Testing Refractory Materials Part 3: General and Textural Properties Section 3.17: Determination of Volume and Bulk Density of Dense Shaped Products (Methods 1902-317). 11 pp.

BSI BS 6043: Sec 4.6-91. 1991 Methods of Sampling and Test for Carbonaceous Materials Used in Aluminium Manufacture Part 4: Cold Ramming Pastes Section 4.6: Determination of Maximum Rammed Apparent Density of Unbaked Paste. 9 pp.

BSI BS 6043: Sec 4.7-91. 1991 Methods of Sampling and Test for Carbonaceous Materials Used in Aluminium Manufacture Part 4: Cold Ramming Pastes Section 4.7: Preparation of Baked Rammed Test Pieces and Deter. of Baked Apparent Density, Change in Vol. and Loss in Mass. 10 pp.

CEN PREN 993-1-93. Methods of Test for Dense Shaped Refractory Products—Part 1: Determination of Bulk Density, Apparent Porosity and True Porosity. 9 pp.

Bulk Density (Cont.)

—Refractory Materials (Cont.)

CEN PREN 1094-4-93. Insulating Refractory Products—Part 4: Determination of Bulk Density and True Porosity. 7 pp.

CNS R3082-78. Method of Test for Bulk Density and Porosity of Granular Refractory Materials by Mercury Displacement (Nov)(4682).

ISO 5016-86. Shaped Insulating Refractory Products—Determination of Bulk Density and True Porosity First Edition. 4 pp.

ISO 5017-88. Dense Shaped Refractory Products—Determination of Bulk Density, Apparent Porosity and True Porosity First Edition. 6 pp.

JIS R 2655-85. Testing Method for Bulk Specific Gravity of Light Weight Castable Refractories (Method for Molded Goods). 5 pp.

SAA AS 1774.2-91. Refractories and Refractory Materials—Physical Test Methods—Part 2: Determination of Bulk Density of Granular Materials. 3 pp.

—Rocks

DIN ENGL 52102-88. Determination of Absolute Density, Dry Density, Compactness and Porosity of Natural Stone and Mineral Aggregates (Aug). 10 pp.

DIN ENGL 52110-85. Testing of Natural Stone; Determination of Bulk Density of Stone Gradings (Aug). 3 pp.

—Sands

DIN ENGL 52110-85. Testing of Natural Stone; Determination of Bulk Density of Stone Gradings (Aug). 3 pp.

—Sodium Perborates

BSI BS 5688: Part 21-79. 1979 Orthoboric Acid (Boric Acid), Diboron Trioxide (Boric Oxide), Disodium Tetraborates, Sodium Perborates and Crude Sodium Borates for Industrial Use Part 21: Determination of Bulk Density of Sodium Perborates. 4 pp.

ISO 3424-75. Sodium Perborates for Industrial Use—Determination of Bulk Density First Edition. 4 pp.

—Soil Analysis

DIN ENGL 19683 Pt 12-73. Methods of Soil Analysis for Water Management for Agricultural Purposes; Physical Laboratory Tests; Determination of Bulk Density (Apr). 2 pp.

—Surfactants—Pastes

ISO 1064-74. Surface Active Agents—Determination of Apparent Density of Pastes on Filling First Edition. 5 pp.

—Tea

BSI BS 6986: Part 2-88. 1988 Analysis of Instant Tea Part 2: Methods for Determination of Free-Flow and Compacted Bulk Densities. 9 pp.

ISO 6770-82. Instant Tea—Determination of Free-Flow and Compacted Bulk Densities First Edition. 8 pp.

—Uranium Dioxide Pellets

ISO 9278-92. Uranium Dioxide Pellets—Determination of Density and Amount of Open and Closed Porosity—Boiling Water Method and Penetration Immersion Method First Edition. 11 pp.

—Urea Resins

CNS K6272-74. Method of Test for Urea Resin Molding Compounds (Oct)(2986). 9 pp.

—Washing Powders

BSI BS 3762: Sec 4.2-86. 1986 Analysis of Formulated Detergents Part 4: Physical Test Methods Section 4.2: Method for Determination of Apparent Bulk Density. 5 pp.

ISO 697-81. Surface Active Agents—Washing Powders—Determination of Apparent Density—Method by Measuring the Mass of a Given Volume Second Edition. 5 pp.

Bulk Flotation

See Also: Flotation

—Sampling

CNS M3195-88. Method of Sampling, for Non-Ferrous Flotation Concentrates in Bulk (May)(12312).

JIS M 8083-84. Methods for Sampling of Non-Ferrous Flotation Concentrates in Bulk. 21 pp.

Bulk Handling

See Also: Bulk Materials; Packaging

—Classification

ISO 3435-77. Continuous Mechanical Handling Equipment—Classification and Symbolization of Bulk Materials First Edition. 4 pp.

INTERNATIONAL AND NON-U.S. NATIONAL STANDARDS
SUBJECT INDEX

Bulk Handling (Cont.)
—Document Transfer and Manipulation
CCITT RECMN T.521-89. Communication Application Profile BT0 for Document Bulk Transfer Based on the Session Service—Terminal Equipment and Protocols for Telematic Services (Study Group VIII) 13 pp. 13 pp.
CCITT RECMN T.522-92. Communication Application Profile BT1 for Document Bulk Transfer (Study Group VIII) 7 pp. 7 pp.

—Symbols
ISO 3435-77. Continuous Mechanical Handling Equipment—Classification and Symbolization of Bulk Materials First Edition. 4 pp.

Bulk Materials
Scope Note: For additional listings, use a more specific term *See Also:* Bulk Handling; Bulk Storage

—Quality Assurance
ISO 9004 Pt 3-93. Quality Management and Quality System Elements—Part 3: Guidelines for Processed Materials First Edition. 27 pp.

—Sampling
CNS M4008-85. General Rules for Method of Sampling of Bulk Materials (Dec) (11456).
JIS M 8100-92. Particulate Materials—General Rules for Methods of Sampling. 121 pp.

Bulk Molding Compounds
Use For: BMC; DMC; Dough Molding Compounds
See Also: Molding Materials; Prepreg; Sheet Molding Compounds
ISO 8606-90. Plastics—Prepregs—Bulk Moulding Compound (BMC) and Dough Moulding (DMC)—Basis for a Specification First Edition. 10 pp.

—Density
CNS K6558-80. Determination of Apparent Density of Material That Can Be Poured from a Specified Funnel (Aug)(6203).
CNS K6559-80. Determination of Apparent Density of Moulding Material That Cannot Be Poured from a Specified Funnel (Aug)(6204).

Bulk Storage
See Also: Bulk Materials; Containers; Fuel Storage Tanks; Hoppers; Oil Storage Tanks

—Safety
BSI BS 6989-89. 1989 Guide to Safety of Storage Equipment for Loose Bulk Materials. 8 pp.
ISO 8456-85. Storage Equipment for Loose Bulk Materials—Safety Code First Edition. 6 pp.

Bulk Throwing Loaders
Use: Bulk Throwing Machines

Bulk Throwing Machines
Use For: Bulk Throwing Loaders

—Safety
BSI BS 5667: Part 9-79. 1979 Continuous Mechanical Handling Equipment. Safety Requirements Part 9: Loose Bulk Materials: Bulk Throwing Machines. 3 pp.

Bulk Transporters
Scope Note: Excludes long distance freight transportation *See Also:* Bulk Handling; Bulk Throwing Machines; Containers; Conveyors; Dollies; Forklifts; Front End Loaders; Ground Vehicles; Lift Trucks; Loaders; Materials Handling Equipment; Mine Cars; Shipping; Trailers; Trucks

—Fuels
MOD UK DSTAN 47-18: Part 1-01. Equipment for Receipt and Delivery of Liquid Fuels Part 1: Bulk Transportation and Transfer Issue 1; Amendment 1. 19 pp.

Bulkhead Connectors
Use: Feedthrough Connectors

Bulkhead Fittings
Scope Note: See also the subheading Bulkhead under specific types of fittings *See Also:* Pipe Fittings
CNS B2609-81. Elbow, Flared and Bulkhead Universal, 90, Precision Type (Oct)(7944).
JIS F 3009-87. Ships' 5 K and 10 K Deck and Bulkhead Pieces for Pipe Connection. 6 pp.

—Aircraft
SBAC AGS 1110 ISSUE 4. Pipe, Couplings Arrangement of Flanged Bulkhead Union Bodies.
SBAC AGS 1144 ISSUE 3. Arrangement of Hexagon Bulkhead Unions.

—Aircraft—Aluminum Alloy
SBAC AS 43097 ISSUE 1. Union—Hexagon, Bulkhead (Aluminium Alloy).
SBAC AS 43102 ISSUE 1. Union—Bulkhead, Triangular Flange (Aluminium Alloy).
SBAC AS 43104 ISSUE 2. Union—Bulkhead, Sealed Flange (Aluminium Alloy).
SBAC AS 43142 ISSUE 1. Elbow—135 Degree—Bulkhead (Aluminium Alloy).
SBAC AS 43184 ISSUE 1. Elbow 90 Degree—Bulkhead (Aluminium Alloy).
SBAC AS 43185 ISSUE 1. Tee-Branch—Bulkhead (Aluminium Alloy).
SBAC AS 43186 ISSUE 1. Tee-End—Bulkhead (Aluminium Alloy).
SBAC AGS 1111 (V). Pipe, Couplings (Aluminium Alloy) Flanged Bulkhead Cone Union Body 3/16 Inch to 1/2 Inch.
SBAC AGS 1112 (V). Pipe, Couplings (Aluminium Alloy) Flanged Bulkhead Union Body 3/16 Inch to 1 1/2 Inches.
SBAC AGS 1113 (V). Pipe, Couplings (Aluminium Alloy) Flanged Bulkhead Union Body, (Cone To Nipple Type).
SBAC AGS 1145 (V). Pipe Couplings Aluminium Alloy) Hexagon Bulkhead Cone Union Body (3/16" to 1/2").
SBAC AGS 1146 (V). Pipe Couplings (Aluminium Alloy) Hex, Bulkhead Union Body (3/16" to 1/2").
SBAC AGS 1147 (V). Pipe Couplings Aluminium Alloy Hexagon Bulkhead Union Body (3/16" to 1/2") Cone to Nipple Type.

—Aircraft—Fluid Power Equipment—Titanium Alloy
AECMA PREN3246-89. Pipe coupling 8 Degrees 30' in Titanium Alloy Union, Bulkhead. 5 pp.
AECMA PREN3253-89. Pipe Coupling 8 Degrees 30' in Titanium Alloy Elbow 90 Degrees Swivel Nut, Bulkhead. 5 pp.
AECMA PREN3257-90. Pipe coupling 8 Degrees 30' in Titanium Alloy Elbow 45 Degrees, Bulk Head. 5 pp.
AECMA PREN3262-90. Pipe Coupling 8 Degrees 30' in Titanium Alloy Tee Bulkhead Branch. 5 pp.
AECMA PREN3263-90. Pipe Coupling 8 Degress 30' in Titanium Alloy Bulkhead End. 5 pp.
AECMA PREN3266-90. Nut, Bulkhead in Titanium. 6 pp.
AECMA PREN3267-90. Washer, Bulkhead in Titanium Alloy. 6 pp.
CEN PREN3564-90. Pipe Coupling 8 Degrees 30 in Titanium Alloy Tee, Bulkhead Fitting on Limb. 5 pp.
CEN PREN 3690-91. Pipe Coupling 8 Degree 30' in Titanium Alloy Union Bulkhead Long. 6 pp.
CEN PREN 3691-91. Pipe Coupling 8 Degree 30' in Titanium Alloy Union—Bulkhead Long Welded. 7 pp.
CEN PREN 3692-91. Pipe Coupling 8 Degree 30' in Titanium Alloy Elbow 90 Degree Bulkhead Long. 6 pp.
CEN PREN 3693-91. Pipe Coupling 8 Degree 30' in Titanium Alloy Elbow 90 Degree—Bulkhead Long Welded. 7 pp.
CEN PREN 3694-91. Pipe Coupling 8 Degree 30' in Titanium Alloy Tee—Bulkhead Long Branch. 6 pp.

—Aircraft—Fluid Power Equipment—Welded—Titanium Alloy
AECMA PREN3247-89. Pipe Coupling 8 Degree 30' in Titanium Alloy Union, Bulkhead, Welded. 6 pp.
AECMA PREN3254-89. Pipe Coupling 8 Degree 30' in Titanium Alloy Elbow 90 Degrees Swivel Nut, Bulkhead, Welded. 6 pp.

—Aircraft—Sleeves—Aluminum Alloy
SBAC AGS 1114 (V). Pipe, Couplings (Aluminium Alloy) Flanged Bulkhead Screwed Sleeve.

—Aircraft—Steel
SBAC AS 43027 ISSUE 1. Union—Hexagon, Bulkhead (Steel).
SBAC AS 43032 ISSUE 1. Union—Bulkhead, Triangular Flange (Steel).
SBAC AS 43034 ISSUE 2. Union—Bulkhead, Sealed Flanged (Steel).
SBAC AS 43069 ISSUE 1. Elbow—90 Degree—Bulkhead (Steel).
SBAC AS 43070 ISSUE 1. Tee-Branch—Bulkhead (Steel).
SBAC AS 43071 ISSUE 1. Tee-End—Bulkhead (Steel).
SBAC AS 43072 ISSUE 1. Elbow—135 Degree—Bulkhead (Steel).

—Aircraft—Swaged—Aluminum Alloy
SBAC AS 43090 ISSUE 1. Swaged Couplings—Typical Joints (Aluminium Alloy).
SBAC AS 43152 ISSUE 2. Elbow—Forged Blank—Swaged Pipe Coupling Aluminium Alloy.
SBAC AS 43155 ISSUE 2. Bulkhead Adaptor—Forged Blank Swaged Pipe Coupling (Aluminium Alloy).
SBAC AS 43159 ISSUE 2. Bulkhead Elbow 135 Degree—Forged Blank—Swaged Pipe Coupling (Aluminium Alloy).
SBAC AS 43176 ISSUE 1. Bulkhead Elbow—Forged Blank—Aluminium Alloy Swaged Pipe Coupling.
SBAC AS 43177 ISSUE 1. Bulkhead Tee—Forged Blank—Aluminium Alloy Swaged Pipe Coupling.
SBAC AS 43178 ISSUE 1. Bulkhead Tee—Forged Blank—Aluminium Alloy Swaged Pipe Coupling.

—Aircraft—Swaged—Steel
SBAC AS 43020 ISSUE 1. Swaged Couplings—Typical Joints (Steel). 37 pp.
SBAC AS 43085 ISSUE 2. Bulkhead Adaptor—Forged Blank, Steel Swaged Pipe Coupling.
SBAC AS 43086 ISSUE 2. Bulkhead Elbow—Forged Blank, Steel Swaged Pipe Coupling.
SBAC AS 43087 ISSUE 2. Bulkhead Tee—Forged Blank, Steel Swaged Pipe Coupling.
SBAC AS 43088 ISSUE 2. Bulkhead Tee—Forged Blank, Steel Swaged Pipe Coupling.
SBAC AS 43089 ISSUE 2. Bulkhead Elbow 135 Degree—Forged—Blank Steel Swaged Pipe Coupling.

—Aircraft—Washer Plates—Aluminum
SBAC AGS 1149 ISSUE 3. Pipe Couplings, Washer Plates for Hexagon Bulkhead Fittings.

—Aircraft—Washer Plates—Steel
SBAC AGS 1115 ISSUE 3. Pipe Couplings, Washer Plates for Flanged Bulkhead Fittings.

—Compression
DIN ENGL 3910-84. Solderless Compression Couplings with Olive; Male Stud Bulkhead Compression Couplings (Apr). 3 pp.
DIN ENGL 3911-84. Solderless Compression Couplings with Olive; Elbow Bulkhead Compression Couplings (Apr). 3 pp.
DIN ENGL 3912-84. Solderless Compression Couplings with Olive; Straight Bulkhead Compression Couplings (Apr). 2 pp.

—Fluid Power Equipment—Titanium Alloy
CEN PREN 3784-91. Pipe Coupling 8 Degree 30' in Titanium Alloy Tee Reduced Bulkhead Branch Long. 8 pp.
CEN PREN 3785-91. Pipe Coupling 8 Degree 30' in Titanium Alloy Tee Reduced Bulkhead Branch. 8 pp.
CEN PREN 3786-91. Pipe Coupling 8 Degree 30' in Titanium Alloy Tee Reduced Bulkhead Long. 8 pp.
CEN PREN 3787-91. Pipe Coupling 8 Degree 30' in Titanium Alloy Tee Reduced Bulkhead. 8 pp.

Bulkhead Penetrators
Use: Feedthrough Connectors

Bulkheads
See Also: Hulls; Revetments; Ships; Walls; Wharves

—Flanges—Pipes
BSI BS MA 75: Part 1-85. 1985 Bulkhead Pieces and Tank Pads Part 1: Bulkhead Pieces of Fabricated Steel or Copper Alloy and Steel Pad Pieces. 13 pp.
BSI BS MA 75: Part 2-78. 1978 Bulkhead Pieces and Tank Pads Part 2: Bulkhead Pieces of Ductile Iron (Spheroidal Graphite Iron). 7 pp.
ISO 5625-78. Shipbuilding—Welded Bulkhead Pieces with Flanges for Steel Pipework—PN 6, PN 10 and PN 16 First Edition. 4 pp.

—Incandescent Lighting—Fire Resistant
CNS F5083-84. Explosion-Proof Bulkhead Lights for Marine Use (Feb)(10766).

—Pipes—Mountings
CNS F3062-80. Ships' Deck and Bulkhead Pieces for Small Size Copper Tubes (Sep)(6442).

—Shafts (Machine Elements)—Mountings
JIS F 3008-77. Deck and Bulkhead Pieces for Transmission Shaft (R 1984). 8 pp.

—Ships
MOD UK NES 130: Part 2-91. Requirements for Decorative Linings and Composite Bulkheads Part 2: Composite Bulkheads Issue 3 (07.91). 10 pp.

—Stuffing Boxes
CNS F5045-83. Marine Cable Glands for Bulkhead and Deck (Jan)(9854).
JIS F 8802-87. Marine Cable Glands for Bulkhead and Deck.

Bulking
See Also: Crimping

—Aggregates
BSI BS 812: Part 2-75. 1975 Amd 1 Testing Aggregates Part 2: Physical Properties. 24 pp.

INDUSTRY STANDARDS

Bull Dozers
Use: Bulldozers

Bulldozers
Use For: Bull Dozers; Dozers; Dozers (Earthmoving Equipment) *See Also:* Construction Equipment; Crawler Tractors; Excavating Equipment; Tractors
JIS D 6507-76. Test Method for Bulldozer on Crawler Tractors (R 1979). 17 pp.

—Blades—Capacity Measurement
BSI BS 6911: Part 1-88. 1988 Testing Earth-Moving Machinery Part 1: Method for Determination of Volumetric Rating of Crawler and Wheel Tractor Dozer Blades. 7 pp.
ISO 9246-88. Earth-Moving Machinery—Crawler and Wheel Tractor Dozer Blades—Volumetric Ratings First Edition. 7 pp.

—Cutting Edges
BSI BS 6913: Part 3-89. 1989 Operation and Maintenance of Earth-Moving Machinery Part 3: Principal Shapes and Basic Dimensions for Cutting Edges for Tractors with Dozer, Graders and Tractor Scrapers. 15 pp.
ISO 7129-89. Earth-Moving Machinery—Tractors with Dozer, Graders, Tractor Scrapers—Cutting Edges—Principal Shapes and Basic Dimensions Second Edition. 14 pp.
JIS D 6101-86. Shapes and Dimensions of Cutting Edges for Bulldozers. 6 pp.

—Cutting Edges—End Bits—Holes
BSI BS 6544-84. 1984 Holes in Dozer End Bits for Earth-Moving Machinery. 11 pp.
ISO 7891-84. Earth-Moving Machinery—Dozer End Bits—Hole Specification First Edition. 9 pp.
JIS D 6110-90. Earth-Moving Machinery—Dozer End Bits—Hole Specification; (ISO 7891-1984).

—Noise—Limitation
EC COM(93) 154-93. Commission Proposal for a Council Directive to Amend Council Directive 86/662/EEC on the Limitation of Noise Emitted by Earth-Moving Machinery (93/C 157/11). 4 pp.
EC 86/662/EEC-86. Council Directive on the Limitation of Noise Emited by Hydraulic Excavators, Rope-Operated Excavators, Dozers, Loaders and Excavator-Loaders. 11 pp.

—Roll Over Protective Structures
CNS A3126-81. Earth-Moving Machinery-Roll-Over Protective Structures for Driver Laboratory Tests and Performance Requirements (Sep)(7849).
JIS A 8910-89. Earth-Moving Machinery—Roll-Over Protective Structures—Laboratory Tests and Performance Requirements. 25 pp.

Bullet Attack Testing
Use: Ballistic Attack Testing

Bullet Envelopes
Use: Bullet Jackets

Bullet Jackets
See Also: Ammunition

—Gilding Metals
MOD UK DSTAN 95-11-80. Bullet Envelope Materials Issue 1. 14 pp.

Bullet Resistant Glazing
See Also: Glazing Materials; Security Glazing
CEN PREN 1063-93. Specification for Security Glazing—Bullet-Resistant Glazing—Classification and Test Method. 10 pp.
DIN ENGL 52290 Pt 2-88. Testing of Security Glazing for Bullet Resistance (Nov). 3 pp.

—Classification
CEN PREN 1063-93. Specification for Security Glazing—Bullet-Resistant Glazing—Classification and Test Method. 10 pp.

Bulletins
See Also: Documents; Newsletters

—European Standards
EURO 1992-05 Bulletin-92. Bulletin of the European Standards Organizations. 19 pp.
EURO 1992-06 Bulletin-92. Bulletin of the European Standards Organizations. 24 pp.
EURO 1992-07 Bulletin-92. Bulletin of the European Standards Organizations. 26 pp.

—Quality Assurance
MOD UK Special BTN 11.81. Special Bulletin. 1 p.
MOD UK Special BTN 6.82. Special Bulletin (2). 2 pp.
MOD UK Special BTN 10.82. Special Bulletin (3). 1 p.
MOD UK Special BTN 1.83. Special Bulletin (4). 1 p.
MOD UK Special BTN 2.83. Special Bulletin (5). 1 p.
MOD UK Special BTN 5.83. Special Bulletin (6). 1 p.
MOD UK Special BTN 11.83. Special Bulletin (7). 1 p.
MOD UK Special BTN 11.84. Special Bulletin (1/84). 6 pp.
MOD UK Special BTN 2.84. Special Bulletin (8). 4 pp.
MOD UK Special BTN 3.84. Special Bulletin (9). 3 pp.
MOD UK Special BTN 4.84. Special Bulletin (10); Amendment 1. 3 pp.
MOD UK Special BTN 8.85. Special Bulletin (1/85). 4 pp.
MOD UK Special BTN 8.85. Special Bulletin (2/85). 2 pp.
MOD UK Special BTN 8.85. Special Bulletin (3/85); Amendment 1. 4 pp.
MOD UK Special BTN 9.86. Special Bulletin (1/86). 3 pp.
MOD UK Special BTN 12.87. Special Bulletin (1/87). 1 p.
MOD UK Special BTN 4.88. Special Bulletin (1/88). 3 pp.
MOD UK Special BTN 6.89. Special Bulletin (1/89). 1 p.
MOD UK Special BTN 7.90. Special Bulletin (1/90). 4 pp.
MOD UK Special BTN 9.91. Special Bulletin (2/91). 3 pp.

Bullets
Use: Small Arms Ammunition

Bullion

—Gold Content
CNS M3042-81. Methods for Determination of Gold and Silver in Crude Bullions (Apr) (7290).
JIS M 8115-50. Determination of Gold and Silver in Crude Bullion.

—Sampling
CNS M3046-81. Methods for Sampling of Crude Bullion (May) (7443).
JIS M 8104-77. Method for Sampling of Crude Bullion.

—Silver Content
CNS M3042-81. Methods for Determination of Gold and Silver in Crude Bullions (Apr)(7290).
JIS M 8115-50. Determination of Gold and Silver in Crude Bullion.

Bulwarks
See Also: Ships

—Ladders
CNS F3225-83. Bulwark Ladders (Jun)(10340).
JIS F 2614-67. Bulwark Ladders. 7 pp.

—Rail Sections—Steel
DIN ENGL 80284-78. Steel Bars; Hot Rolled Bulwark Rail Section Steel; Dimensions, Weights, Permissible Deviations (Aug). 2 pp.

Bump Caps
Use: Headgear

Bump Testing
Use: Mechanical Shock

Bumpers
Use For: Buffers (Mechanical) *See Also:* Automotive Equipment; Shock Absorbers

—Automobiles
CNS D1061-83. Bumper Heights for Automobiles (Oct)(10622).
CNS D2173-83. Plastic Bumpers for Passenger Cars (Dec)(10705). 2 pp.
CNS D3160-83. Method of Test for the Plastic Bumpers of Passenger Cars (Dec)(10706).

—Automobiles—Collision Research
ISO 2958-73. Road Vehicles—Exterior Protection for Passenger Cars First Edition. 6 pp.

—Automobiles—Impact Testing
CNS D3158-83. Method of Impact Test for Bumper of Passenger Cars (Dec)(10703).

—Buses (Vehicles)—Doors
CNS D2093-87. Rubber Bumpers of Entrance Door for Buses (Jul)(7676).
JIS D 4706-88. Rubber Bumpers of Entrance Doors for Buses. 9 pp.

—Hospital Trolleys
BSI BS 4322-68. 1968 Recommendations for Buffering on Hospital Vehicles Such as Trolleys. 8 pp.

Bunching Onions
Use: Scallions

Bungee Jumping
Use For: Bungy Jumping *See Also:* Recreation
SNZ NZS/AS 5848-92. Code of Practice for Bungy Jumping. 68 pp.

Bungy Jumping
Use: Bungee Jumping

Bunk Beds
See Also: Beds; Furniture
BSI BS 6998-88. (WITHDRAWN) 1988 AMD 1 Bunk Beds (SUPERSEDED By BS EN 747: PARTS 1 & 2; 1993). 17 pp.
CNS S1063-68. Bed, Wooden, Double Bunks Single (Apr)(2854).

—Household
JIS S 1104-84. Bunk Beds for Domestic Use.

—Household—Safety
BSI BS EN 747-1-93. 1993 Furniture—Bunk Beds for Domestic Use Part 1: Safety Requirements (L). 10 pp.
CEN EN 747-1-93. Furniture—Bunk Beds for Domestic Use—Part 1: Safety Requirements. 7 pp.

—Safety
BSI BS EN 747-2-93. 1993 Furniture—Bunk Beds for Domestic Use Part 2: Test Methods (L). 15 pp.
CEN EN 747-2-93. Furniture—Bunk Beds for Domestic Use—Part 2: Test Methods. 12 pp.

Bunker C Oil
Use: Bunker Fuel Oil—Type C

Bunker Fuel Oil
See Also: Crude Oil; Fuel Oils; Fuels; Liquid Fuels; Oils; Petroleum Products

—Type C—Ash Content
CPPA J.24P(C)-74. Ash Content. 1 p.

—Type C—Density
CPPA J.24(D)P-77. API Gravity of Bunker "C" Oil. 1 p.

—Type C—Heat of Combustion
CPPA J-24(B)P-75. Thermal Value of Bunker "C" Oil. 3 pp.

—Type C—Saybolt Viscosity
CPPA J.24(E)P-77. Saybolt Viscosity of Bunker "C" Oil. 2 pp.

—Type C—Sediment Content—Centrifuging
CPPA J.24(F)-77. Bunker "C" Oil—Water and Sediment by Centrifuge. 2 pp.

—Type C—Sulfur Content
CPPA J.24(G)P-80. Sulphur Content. 2 pp.

—Type C—Water Content—Centrifuging
CPPA J.24(F)-77. Bunker "C" Oil—Water and Sediment by Centrifuge. 2 pp.

Bunsen Funnels
Use: Funnels—Glass—Long Stem

Buntings (Clothing)
Use For: Baby Nests

—Infants'—Safety
BSI BS 6595-85. 1985 Safety Requirements for Baby Nests. 4 pp.

Buoyancy

—Ground Effect Machines
CAA Chapter B5-8 08.83. Buoyancy and Displacement Mode. 3 pp.

Buoyancy Compensators
Use For: Buoyancy Control Devices
See Also: Diving Equipment
JIS S 7305-88. SCUBA Diving Goods—Buoyancy Compensators. 11 pp.

Buoyancy Control Devices
Use: Buoyancy Compensators

Buoys
See Also: Anchor Buoys; Marine Equipment; Navigational Aids
CNS F4021-90. Buoyant Marine Use (Aug)(12769).

INTERNATIONAL AND NON-U.S. NATIONAL STANDARDS
SUBJECT INDEX

Buoys (Cont.)

—Chains
CNS B2753-83. Buoy Chains (Oct)(10592).
CNS B2754-83. Buoy Chains (Shackles, Swivels, Swivel Hooks) (Oct)(10593).

—Identification Systems—Submarines
NATO STANAG 1335 ED 1 AMD 0-88. Numerical Block Allocation for Identification Figures Transmitted by Submarine Indicator Buoys. 6 pp.
NATO STANAG 1335 ED 1 AMD 1-88. Numerical Block Allocation for Identification Figures Transmitted by Submarine Indicator Buoys. 5 pp.

—Monitors
CCIR Report 1168-90. Beacon and Buoy Remote-Monitoring System—Section 8D—Radiodetermination, Global Maritime Distress and Safety System and Related Subjects. 4 pp.

—Shackles
CNS B2754-83. Buoy Chains (Shackles, Swivels, Swivel Hooks) (Oct)(10593).
JIS F 3306-90. Buoy Shackles.
JIS F 3306-58. Buoy Shackles (R 1976). 9 pp.

—Submarines—Identification Systems
NATO STANAG 1083 ED 4 AMD 2-82. Tethered Submarine Marker Buoys. 7 pp.

—Swivel Hooks
CNS B2754-83. Buoy Chains (Shackles, Swivels, Swivel Hooks) (Oct)(10593).

—Swivels
CNS B2754-83. Buoy Chains (Shackles, Swivels, Swivel Hooks) (Oct)(10593).

—Telephone Answering Equipment
CCIR Report 1168-90. Beacon and Buoy Remote-Monitoring System—Section 8D—Radiodetermination, Global Maritime Distress and Safety System and Related Subjects. 4 pp.

Buquinolate Content Analysis

—Animal Feed
CNS N4080-82. Method of Test for Feed Additives: Determination of Buquinolate (Jan) (8316). 2 pp.

Bureaufax Communications

Use For: Bureaufax Services See Also: FAX Communications; Information Interchange; Telephone Services
CCITT RECMN A.21-89. Collaboration with Other International Organizations on CCITT-Defined Telematic Services—Terminal Equipment and Protocols for Telematic Services (Study Group VIII) 2 pp. 2 pp.
CCITT RECMN F.160-89. General Operational Provisions for the International Public Facsimile Services—Telematic, Data Transmission and Teleconference Services —Operations and Quality of Service (Study Group I) 4 pp. 4 pp.
CCITT RECMN F.170-92. Operational Provisions for the International Public Facsimile Service Between Public Bureaux (Bureaufax) (Study Group I) 9 pp. 9 pp.
CCITT RECMN F.170-89. Operational Provisions for the International Public Facsimile Service Between Public Bureaux (Bureaufax)—Telematic, Data Transmission and Teleconference Services —Operations and Quality of Service—(Study Group I) 8 pp. 8 pp.
CCITT RECMN F.190-92. Operational Provisions for the International Facsimile Service Between Public Bureaus and Subscriber Stations and vice versa (Bureaufax—telefax and vice versa) (Study Group I) 5 pp. 5 pp.
CCITT RECMN F.190-89. Operational Provisions for the International Facsimile Service Between Public Bureaus and Subscriber Stations & Vice Versa (Bureaufax—Telefax and Vice-Versa)—Telematic, Data Transmission and Teleconference Serv.—Operations & Quality of Serv. (Study Gr. I) 3 pp. 3 pp.

—Accounting
CCITT RECMN D.70 (REV 1)-92. General Tariff Principles for the International Public Facsimile Service Between Public Bureaux (Bureaufax Service) (Study Group III) 6 pp. 6 pp.
CCITT RECMN D.70-89. General Tariff Principles for the International Public Facsimile Service Between Public Bureaux (Bureaufax Service)—General Tariff Principles—Charging and Accounting in International Telecommunications Services (Study Group III) 3 pp. 3 pp.

—Document Formats
CCITT RECMN F.170-92. Operational Provisions for the International Public Facsimile Service Between Public Bureaux (Bureaufax) (Study Group I) 9 pp. 9 pp.

Bureaufax Communications (Cont.)

—Document Formats (Cont.)
CCITT RECMN F.170-89. Operational Provisions for the International Public Facsimile Service Between Public Bureaux (Bureaufax)—Telematic, Data Transmission and Teleconference Services —Operations and Quality of Service—(Study Group I) 8 pp. 8 pp.

—Documentation—Tables (Data)
CCITT RECMN F.170-89. Operational Provisions for the International Public Facsimile Service Between Public Bureaux (Bureaufax)—Telematic, Data Transmission and Teleconference Services —Operations and Quality of Service—(Study Group I) 8 pp. 8 pp.

—FAX Communications—Interworking
CCITT RECMN D.73 (REV 1)-92. General Tariff and Intl Accounting Principles for Interworking Between the International Bureaufax and Telefax Services (Study Group III) 5 pp. 5 pp.
CCITT RECMN D.73-89. General Tariff and Intl Accounting Principles for Interworking Between the International Bureaufax and Telefax Services—General Tariff Principles—Charging and Accounting in International Telecommunications Services (Study Group III) 2 pp. 2 pp.
CCITT RECMN F.190-92. Operational Provisions for the International Facsimile Service Between Public Bureaus and Subscriber Stations and vice versa (Bureaufax—telefax and vice versa) (Study Group I) 5 pp. 5 pp.
CCITT RECMN F.190-89. Operational Provisions for the International Facsimile Service Between Public Bureaus and Subscriber Stations & Vice Versa (Bureaufax—Telefax and Vice-Versa)—Telematic, Data Transmission and Teleconference Serv.—Operations & Quality of Serv. (Study Gr. I) 3 pp. 3 pp.

—Integrated Services Digital Networks
CCITT RECMN F.170-92. Operational Provisions for the International Public Facsimile Service Between Public Bureaux (Bureaufax) (Study Group I) 9 pp. 9 pp.
CCITT RECMN F.170-89. Operational Provisions for the International Public Facsimile Service Between Public Bureaux (Bureaufax)—Telematic, Data Transmission and Teleconference Services —Operations and Quality of Service—(Study Group I) 8 pp. 8 pp.

—Public Data Networks
CCITT RECMN F.170-92. Operational Provisions for the International Public Facsimile Service Between Public Bureaux (Bureaufax) (Study Group I) 9 pp. 9 pp.
CCITT RECMN F.170-89. Operational Provisions for the International Public Facsimile Service Between Public Bureaux (Bureaufax)—Telematic, Data Transmission and Teleconference Services —Operations and Quality of Service—(Study Group I) 8 pp. 8 pp.

—Public Switched Telephone Networks
CCITT RECMN F.170-92. Operational Provisions for the International Public Facsimile Service Between Public Bureaux (Bureaufax) (Study Group I) 9 pp. 9 pp.
CCITT RECMN F.170-89. Operational Provisions for the International Public Facsimile Service Between Public Bureaux (Bureaufax)—Telematic, Data Transmission and Teleconference Services —Operations and Quality of Service—(Study Group I) 8 pp. 8 pp.

—Quality of Service
CCITT RECMN F.160-89. General Operational Provisions for the International Public Facsimile Services—Telematic, Data Transmission and Teleconference Services —Operations and Quality of Service (Study Group I) 4 pp. 4 pp.

—Service Messages
CCITT RECMN D.70 (REV 1)-92. General Tariff Principles for the International Public Facsimile Service Between Public Bureaux (Bureaufax Service) (Study Group III) 6 pp. 6 pp.
CCITT RECMN D.70-89. General Tariff Principles for the International Public Facsimile Service Between Public Bureaux (Bureaufax Service)—General Tariff Principles—Charging and Accounting in International Telecommunications Services (Study Group III) 3 pp. 3 pp.
CCITT RECMN F.160-89. General Operational Provisions for the International Public Facsimile Services—Telematic, Data Transmission and Teleconference Services —Operations and Quality of Service (Study Group I) 4 pp. 4 pp.
CCITT RECMN F.170-92. Operational Provisions for the International Public Facsimile Service Between Public Bureaux (Bureaufax) (Study Group I) 9 pp. 9 pp.

Bureaufax Communications (Cont.)

—Service Messages (Cont.)
CCITT RECMN F.170-89. Operational Provisions for the International Public Facsimile Service Between Public Bureaux (Bureaufax)—Telematic, Data Transmission and Teleconference Services —Operations and Quality of Service—(Study Group I) 8 pp. 8 pp.

—Store and Forward Mode—Message Switching
CCITT RECMN F.171-89. Operational Provisions Relating to the Use of Store-and-Forward Switching Nodes Within the Bureaufax Service—Telematic, Data Transmission and Teleconference Services —Operations and Quality of Service—(Study Group I) 3 pp. 3 pp.

—Tariffs
CCITT RECMN D.70-89. General Tariff Principles for the International Public Facsimile Service Between Public Bureaux (Bureaufax Service)—General Tariff Principles—Charging and Accounting in International Telecommunications Services (Study Group III) 3 pp. 3 pp.

—Tariffs (Telecommunications)
CCITT RECMN D.70 (REV 1)-92. General Tariff Principles for the International Public Facsimile Service Between Public Bureaux (Bureaufax Service) (Study Group III) 6 pp. 6 pp.

—Tariffs—Refunds
CCITT RECMN D.70 (REV 1)-92. General Tariff Principles for the International Public Facsimile Service Between Public Bureaux (Bureaufax Service) (Study Group III) 6 pp. 6 pp.
CCITT RECMN D.70-89. General Tariff Principles for the International Public Facsimile Service Between Public Bureaux (Bureaufax Service)—General Tariff Principles—Charging and Accounting in International Telecommunications Services (Study Group III) 3 pp. 3 pp.

—Trademark Rights—Exceptions
CCITT FASCICLE I.2-88. Opinions and Resolutions Recommendations on the Organisation and Working Procedures of CCITT (Series A). 64 pp.
CCITT FASCICLE II.4-88. Telegraph and Mobile Service—Operations and Quality of Service Recommendations F.1—F.140. 293 pp.
CCITT FASCICLE II.5-88. Telematic, Data Transmission and Teleconference Service—Operations and Quality of Service Recommendations F.160—F.353, F.600, F.601, F.710—F.730. 154 pp.
CCITT FASCICLE II.6-88. Message Handling and Directory Services—Operations and Definition of Service Recommendations F.400—F.422, F.500. 184 pp.

Bureaufax Equipment

—Terminal Identification
CCITT RECMN F.160-89. General Operational Provisions for the International Public Facsimile Services—Telematic, Data Transmission and Teleconference Services —Operations and Quality of Service (Study Group I) 4 pp. 4 pp.

Bureaufax Services
Use: Bureaufax Communications

Burets
See Also: Containers; Laboratory Ware
BSI BS 846-85. 1985 Burettes. 14 pp.
BSI BS 846-01. 1985 Amd 1 Burettes (AMD 7387) January 15, 1993 (H). 15 pp.
ISO 385 Pt 1-84. Laboratory Glassware—Burettes—Part 1: General Requirements First Edition. 10 pp.
ISO 385 Pt 2-84. Laboratory Glassware—Burettes—Part 2: Burettes for Which No Waiting Time is Specified First Edition. 4 pp.
ISO 385 Pt 3-84. Laboratory Glassware—Burettes—Part 3: Burettes for Which a Waiting Time of 30 s is Specified First Edition. 4 pp.
JIS R 3505-83. Volumetric Glassware. 16 pp.

—Volumetric
BSI BS 1428: Part D1-65. 1965 Microchemical Apparatus: Group D: Volumetric Apparatus Part D1: Burettes with Pressure Filling Device and Automatic Zero. 16 pp.
CNS R2149-82. Glass Burette for Chemical Analysis (May)(8863).
JIS R 3505-83. Volumetric Glassware. 16 pp.

Burettes
Use: Burets

Burglar Alarms
Use For: Intruder Alarms; Security Alarms; Theft Alarms

Burglar Alarms (Cont.)

See Also: Alarm Systems; Audio Detectors; Burglar Detectors; Capacitance Operated Intrusion Detectors; Floor Mat Detectors; Foil on Glass Intruder Systems; Photoelectric Alarm Systems; Security Systems; Vibration Detection Systems; Warning Systems

BSI BS 4737: Sec 3.0-88. 1988 Intruder Alarm Systems in Buildings Part 3: Components Section 3.0: General Requirements. 8 pp.

BSI BS 4737: Sec 4.1-87. 1987 Intruder Alarm Systems in Buildings Part 4: Codes of Practice Section 4.1: Code of Practice for Planning and Installation. 24 pp.

BSI BS 4737: Sec 4.3-88. 1988 Intruder Alarm Systems in Buildings Part 4: Codes of Practice Section 4.3: Code of Practice for Exterior Alarm Systems. 11 pp.

BSI BS 7042-88. 1988 High Security Intruder Alarm Systems in Buildings. 20 pp.

BSI BS 7150-89. 1989 Code of Practice for Intruder Alarm Systems with Mains Wiring Communication. 8 pp.

BSI BS 7230-89. 1989 Code of Practice for Article Theft Detection Systems. 12 pp.

BSI BS 8220: Part 2-87. 1987 Security of Buildings Against Crime Part 2: Offices and Shops. 58 pp.

BSI BS 8220: Part 3-90. 1990 Security of Buildings Against Crime Part 3: Warehouses and Distribution Units. 137 pp.

DIN VDE 0833 Pt 1-89. Alarm Systems for Fire, Intrusion and Hold-Up; General Requirements (Jan). 25 pp.

DIN VDE 0833 Pt 3-82. Alarm Systems for Fire, Burglary and Hold-Ups; Requirements for Burglar and Hold-Up Alarm Systems (VDE Specification) (Aug). 14 pp.

SNZ NZS 4301-83. Intruder Alarm Systems. 32 pp.

—Acoustic Signals
BSI BS 4737: Part 1-86. 1986 Amd 1 Intruder Alarm Systems in Building Part 1: Installed Systems with Local Audible and/or Remote Signalling (AMD 5804) December 23, 1987. 16 pp.

—Central Stations—Signaling Links
SNZ NZS/AS 2201. 2-86. Intruder Alarm Systems Part 2: Central Stations and Signalling Links Amend: 1, 1988 (This is a Joint Standard with SAA AS 2201.2). 7 pp.

—Connectors
BSI BS 4166-67. (WITHDRAWN) 1967 Amd 2 Automatic Intruder Alarm Terminating Equipment in Police Stations (AMD 5802) May 31, 1989. 28 pp.

—Detectors
SAA AS 2201.3-91. Intruder Alarm Systems—Part 3: Detection Devices for Internal Use. 30 pp.

SNZ NZS 4301: Part 3-93. Intruder Alarm Systems Part 3: Detection Devices for Internal Use Amend: A, 1992. 30 pp.

—Electric Wiring
BSI BS 4737: Sec 3.1-77. 1977 Intruder Alarm Systems in Buildings Part 3: Components Section 3.1: Continous Wiring. 4 pp.

—Installation
BSI BS 4737: Sec 4.1-01. 1987 Amd 1 Intruder Alarm Systems in Buildings Part 4: Codes of Practice Section 4.1: Code of Practice for Planning and Installation (AMD 7192) May 1, 1992. 25 pp.

BSI BS 6707-86. 1986 Intruder Alarm Systems for Consumer Installation. 11 pp.

SNZ NZS/AS 2201. 4-90. Intruder Alarm Systems Part 4: Wire-Free Systems Installed in Client's Premises Amend: 1, 1990 (This is a Joint Standard with SAA AS 2201.4). 6 pp.

SNZ NZS 4301: Part 1-93. Intruder Alarm Systems Part 1: Systems Installed in Client's Premises. 48 pp.

—Insulated Cables
BSI BS 4737: Sec 3.30-86. 1986 Intruder Alarm Systems in Buildings Part 3: Components Section 3.30: PVC Insulated Cables for Interconnecting Wiring. 7 pp.

—Maintenance
BSI BS 4737: Sec 4.2-86. 1986 Intruder Alarm Systems in Buildings Part 4: Codes of Practice Section 4.2: Code of Practice for Maintenance and Records. 8 pp.

—Manually Operated
BSI BS 4737: Part 2-86. 1986 Amd 1 Intruder Alarm Systems in Buildings Part 2: Installed Systems for Deliberate Operation. 14 pp.

BSI BS 4737: Sec 3.14-86. 1986 Intruder Alarm Systems in Buildings Part 3: Components Section 3.14: Deliberately-Operated Devices. 4 pp.

Burglar Alarms (Cont.)

—Protective Switches
BSI BS 4737: Sec 3.3-77. 1977 Intruder Alarm Systems in Buildings Part 3: Components Section 3.3: Protective Switches. 2 pp.

—Remote Stations
BSI BS 5979-87. 1987 Remote Centres for Intruder Alarm Systems. 10 pp.

—Remote Stations—Signaling Links
SNZ NZS/AS 2201. 2-86. Intruder Alarm Systems Part 2: Central Stations and Signalling Links Amend: 1, 1988 (This is a Joint Standard with SAA AS 2201.2). 7 pp.

—Residential Buildings
BSI BS 6800-86. 1986 Home and Personal Security Devices. 8 pp.

BSI BS 8220: Part 1-86. 1986 Security of Buildings Against Crime Part 1: Dwellings. 48 pp.

—Symbols—Diagrams
BSI BS 4737: Sec 5.2-88. 1988 Intruder Alarm Systems in Buildings Part 5: Terms and Symbols Section 5.2: Recommendations for Symbols for Diagrams. 20 pp.

—Wire Free
BSI BS 6799-86. 1986 Code of Practice for Wire-Free Intruder Alarm Systems. 8 pp.

Burglar Detectors

Use For: Intruder Alarm Systems; Intruder Detectors
See Also: Burglar Alarms; Capacitance Operated Intrusion Detectors; Floor Mat Detectors; Foil on Glass Intruder Systems; Photoelectric Alarm Systems; Ultrasonic Motion Detectors; Vibration Detection Systems

CENELEC PREN 50131-2-1-93. Alarm System—Intrusion System Part 2-1: Intrusion Detectors—Common Requirements. 77 pp.

CENELEC PREN 50131-2-2-93. Alarm System—Intrusion System Part 2-2: Intrusion Detectors—Volume Detectors. 39 pp.

CEPT T/R 60-01-76. Appareils De Radiolocalisation De Faible Puissance Pour La Detection De Mouvements Et Pour L'Alerte. 10 pp.

DIN VDE 0830 Pt 221 (D)-87. Alarm Systems; Intruder Alarm Systems; General Requirements for Detectors for Use in Intruder Alarm Systems; Identical with IEC 79(CO)6 and IEC 79(CO)12 (June). 14 pp.

IEC 839 Pt 2-2-87. Alarm Systems Part 2: Requirements for Intruder Alarm Systems Section Two—Requirements for Detectors—General. 20 pp.

SNZ NZS 4301: Part 3-93. Intruder Alarm Systems Part 3: Detection Devices for Internal Use Amend: A, 1992. 30 pp.

—Audio
BSI BS 4737: Sec 3.6-78. 1978 Amd 1 Intruder Alarm Systems in Buildings Part 3: Components Section 3.6: Acoustic Detectors. 4 pp.

—Capacitance Operated
BSI BS 4737: Sec 3.8-78. 1978 Amd 1 Intruder Alarm Systems in Buildings Part 3: Requirements for Detection Devices Section 3.8: Volumetric Capacitive Detectors. 4 pp.

BSI BS 4737: Sec 3.13-78. 1978 Amd 1 Intruder Alarm Systems in Buildings Part 3: Components Section 3.13: Capacitive Proximity Detectors. 4 pp.

—Floor Mat
BSI BS 4737: Sec 3.9-78. 1978 Amd 1 Intruder Alarm Systems in Buildings Part 3: Components Section 3.9: Pressure Mats. 5 pp.

—Foil on Glass
BSI BS 4737: Sec 3.2-77. 1977 Intruder Alarm Systems in Buildings Part 3: Components Section 3.2: Foil on Glass. 4 pp.

—Microwave
IEC 839 Pt 2-5-90. Alarm Systems Part 2: Requirements for Intruder Alarm Systems Section 5—Microwave Doppler Detectors for Use in Buildings First Edition. 22 pp.

—Passive Infrared
BSI BS 4737: Sec 3.7-78. 1978 Amd 1 Intruder Alarm Systems in Buildings Part 3: Components Section 3.7: Passive Infra Red Detectors. 4 pp.

IEC 839 Pt 2-6-90. Alarm Systems Part 2: Requirements for Intruder Alarm Systems Section Six—Passive Infra-Red Detectors for Use in Buildings First Edition. 26 pp.

—Photoelectric
BSI BS 4737: Sec 3.12-78. 1978 Amd 1 Intruder Alarm Systems in Buildings Part 3: Components Section 3.12: Beam Interruption Detectors. 5 pp.

Burglar Detectors (Cont.)

—Photoelectric (Cont.)
IEC 839 Pt 2-3-87. Alarm Systems Part 2: Requirements for Intruder Alarm Systems Section Three—Requirements for Infrared-Beam Interruption Detectors in Buildings First Edition. 16 pp.

—Printed Circuit Wiring
BSI BS 4737: Sec 3.11-78. 1978 Amd 1 Intruder Alarm Systems in Buildings Part 3: Components Section 3.11: Rigid Printed Circuit Wiring. 4 pp.

—Radio Frequency
BSI BS 4737: Sec 3.4-78. 1978 Amd 2 Intruder Alarm Systems in Buildings Part 3: Components Section 3.4: Radiowave Doppler Detectors. 5 pp.

—Ultrasonic
BSI BS 4737: Sec 3.5-78. 1978 Amd 2 Intruder Alarm Systems in Buildings Part 3: Components Section 3.5: Ultrasonic Movement Detectors. 4 pp.

IEC 839 Pt 2-4-90. Alarm Systems Part 2: Requirements for Intruder Alarm Systems Section 4—Ultrasonic Doppler Detectors for Use in Buildings First Edition. 22 pp.

—Vibration
BSI BS 4737: Sec 3.10-78. 1978 Amd 1 Intruder Alarm Systems in Buildings Part 3: Components Section 3.10: Vibration Detectors. 4 pp.

Burlaps

Use For: Gunny Cloth; Hessian *See Also:* Fabrics; Gunny Sacks

CNS L3200-83. Method of Test for Gunny Cloths (Aug) (10377).

CNS L4149-83. Gunny Cloths (Jun) (10376).

JIS L 3405-87. Hessian Cloths. 5 pp.

JIS L 3414-78. Gunny Cloths.

—Tubing (Flexible)
BSI BS 6894-87. 1987 Jute Hessian Tubing. 8 pp.

MOD UK DSTAN 81-34-80. Tubing, Hessian, Jute Issue 2. 8 pp.

Burnerd Chucks

Use: Scroll Chucks

Burners

Use For: Fuel Burners; Spirit Burners
See Also: Combustion Chambers; Furnaces; Gas Burners; Oil Burners

—Laboratory
BSI BS 5865-80. (OBSOLESCENT) 1980 Construction and Operation of Spirit Burner for Small-Scale Laboratory Flame Tests. 8 pp.

Burning Quality

See Also: Burning Rate; Combustion; Fire Testing

—Kerosene
BSI BS 2000: Part 10-93. 1993 Methods of Test for Petroleum and Its Products Part 10: Determination of Kerosine Burning Characteristics—24 Hour Method. 5 pp.

BSI BS 2000: Part 10-82. 1982 Petroleum and Its Products Part 10: Burning Test (24-Hour) for Kerosine. 7 pp.

CNS K6957-88. Method of Test for Burning Quality of Kerosine (Jul)(12377).

Burning Rate

See Also: Burning Quality; Combustion; Explosives; Fire Testing; Flame Propagation

—Propellants—Iron Oxide Yellow
MOD UK TS 10254. Iron Oxide, Synthetic, Yellow.

Burnishers

See Also: Tools

MOD UK DEF-1198. Broaches and Burnishers for Hand Use (Withdrawn).

Burns (Injuries)

Scope Note: Use a more specific term *See:* Fires; Thermal Burns

Burnt Sienna

Use: Sienna

Burnt Umber

Use: Umber

Burrs (Tools)

Use For: Burs *See Also:* Cutting Tools; High Speed Steel Tools; Milling Cutters; Riveting Burrs; Tools

ISO 7755 Pt 1-84. Hardmetal Burrs—Part 1: General Specifications First Edition. 6 pp.

INTERNATIONAL AND NON-U.S. NATIONAL STANDARDS
SUBJECT INDEX

Burrs (Tools) (Cont.)
—Arch
ISO 7755 Pt 6-84. Hardmetal Burrs—Part 6: Arch Round-(Ball-) Nose Burrs (Style F) First Edition. 3 pp.
ISO 7755 Pt 7-84. Hardmetal Burrs—Part 7: Arch Pointed-Nose Burrs (Style G) First Edition. 3 pp.

—Cone
ISO 7755 Pt 9-84. Hardmetal Burrs—Part 9: 60 Degrees and 90 Degrees Cone Burrs (Styles J and K) First Edition. 3 pp.
ISO 7755 Pt 10-84. Hardmetal Burrs—Part 10: Conical Round-(Ball-) Nose Burrs (Style L) First Edition. 3 pp.
ISO 7755 Pt 11-84. Hardmetal Burrs—Part 11: Conical Pointed-Nose Burrs (Style M) First Edition. 3 pp.
ISO 7755 Pt 12-84. Hardmetal Burrs—Part 12: Inverted Cone Burrs (Style N) First Edition. 3 pp.

—Cylindrical
ISO 7755 Pt 2-84. Hardmetal Burrs—Part 2: Cylindrical Burrs (Style A) First Edition. 3 pp.
ISO 7755 Pt 3-84. Hardmetal Burrs—Part 3: Cylindrical Round-(Ball-) Nose Burrs (Style C) First Edition. 3 pp.

—Dental Instruments
BSI BS 6828: Part 3-87. 1987 Dental Rotary Instruments Part 3: Steel and Carbide Burs (ISO 3823/1: 1986). 16 pp.
BSI BS 6828: Part 3-01. 1987 Amd 1 Dental Rotary Instruments Part 3: Specification for Steel and Carbide Burs (ISO 3823/1: 1986) (AMD 7022) May 1, 1992. 21 pp.
BSI BS 6828: Part 4-87. 1987 Dental Rotary Instruments Part 4: Steel and Carbide Finishing Burs (ISO 3823/2: 1986). 17 pp.
BSI BS 6828: Part 4-01. 1987 Amd 1 Dental Rotary Instruments Part 4: Specification for Steel and Carbide Finishing Burs (ISO 3823/2: 1986) (AMD 7020) May 1, 1992. 23 pp.
CEN PREN 23 823-1-90. Dental Rotary Instruments—Part 1: Steel and Carbide Burs. 3 pp.
CEN EN 23823-1-91. Dental Rotary Instruments—Part 1: Steel and Carbide Burs. 2 pp.
CEN PREN 23 823-2-90. Dental Rotary Instruments—Part 2: Steel and Carbide Finishing Burs. 3 pp.
CEN EN 23823-2-91. Dental Rotary Instruments—Part 2: Steel and Carbide Finishing Burs. 3 pp.
ISO 3823 Pt 1-86. Dental Rotary Instruments—Part 1: Steel and Carbide Burs First Edition. 15 pp.
ISO 3823 Pt 2-86. Dental Rotary Instruments—Part 2: Steel and Carbide Finishing Burs First Edition. 16 pp.

—Flame
ISO 7755 Pt 8-84. Hardmetal Burrs—Part 8: Flame Burrs (Style H) First Edition. 3 pp.

—Oval
ISO 7755 Pt 5-84. Hardmetal Burrs—Part 5: Oval Burrs (Style E) First Edition. 3 pp.

—Spherical
ISO 7755 Pt 4-84. Hardmetal Burrs—Part 4: Spherical Burrs (Style D) First Edition. 3 pp.

Burs
Use: Burrs (Tools)

Burst Disks
Use For: Bursting Disks; Rupture Disks
See Also: Aircraft Safety; Valves
BSI BS 2915-90. 1990 Amd 1 Bursting Discs and Bursting Disc Devices (AMD 6781) July 31, 1991. 30 pp.
ISO 6718-91. Bursting Discs and Bursting Disc Devices Second Edition. 36 pp.
JIS B 8226-86. Bursting Discs and Bursting Disc Assemblies. 2 pp.
JIS B 8226-81. Bursting Discs and Bursting Disc Assemblies. 25 pp.
SAA AS 1358-89. Bursting Discs and Bursting Disc Devices—Guide to Application, Selection, and Installation Amdt 1 April 1991 (in Professional Package 19).
SNZ NZS/BS 2915-90. Specification for Bursting Discs and Bursting Disc Devices Amend: 1, 1991. 30 pp.

Burst Strength Testers
Use For: Mullen Testers See Also: Bursting Strength; Testing Equipment
—Corrugated Board
CPPA M-18-84. Perkins Model 'AH' Mullen Tester. 1 p.
—Fiberboard
CPPA M-18-84. Perkins Model 'AH' Mullen Tester. 1 p.

Burst Strength Testers (Cont.)
—Liner Boards
CPPA M-18-84. Perkins Model 'AH' Mullen Tester. 1 p.
—Paperboard
CPPA M-18-84. Perkins Model 'AH' Mullen Tester. 1 p.
CPPA D.8U-77. Plybond Strength of Paper Board (Using Jumbo Mullen Tester). 2 pp.
—Papers
CPPA M-17-77. Perkins Model "C" Mullen Tester. 2 pp.
CPPA D.5U-77. Determination of Bursting Strength of Paper Using Thibeault Instrument. 2 pp.

Burst Testing
Use: Bursting Strength

Bursting Disks
Use: Burst Disks

Bursting Strength
See Also: Burst Strength Testers; Tensile Testing
—Coated Fabrics
ISO 3303-90. Rubber-or Plastics-Coated Fabrics—Determination of Bursting Strength Second Edition. 7 pp.
—Condoms
CEN PREN 600-6-91. Latex Rubber Condoms—Part 6: Determination Bursting Volume and Pressure. 13 pp.
ISO 4074 Pt 6-84. Rubber Condoms—Part 6: Determination of Bursting Volume and Pressure Second Edition. 4 pp.
—Corrugated Board
CPPA D.19P-90. Bursting Strength of Board. 2 pp.
SAA AS 1301.438S-89. Methods of Test for Pulp and Paper (Metric Units)—Part 438s: Bursting Strength of Paperboard and Corrugated Fibreboard (This is a Joint Standard with SANZ NZS 1301). 7 pp.
SNZ NZS/AS 1301. 438S-89. Methods of Test for Pulp and Paper Bursting Strength of Paperboard and Corrugated Fibreboard (This is a Joint Standard with SAA AS 1301.438S). 7 pp.
—Fabrics
BSI BS 3424: Part 6-82. 1982 Testing Coated Fabrics Part 6: Methods 8A and 8B. Methods for Determination of Bursting Strength. 6 pp.
BSI BS 4768-72. 1972 Amd 1 Determination of the Bursting Strength and Bursting Distension of Fabrics (AMD 6340) June 28, 1991. 9 pp.
CGSB CAN/CGSB-4.2 NO.11.1-M88. Textile Test Methods Bursting Strength—Diaphragm Pressure Test. 9 pp.
CGSB CAN/CGSB-4.2 NO.11.2-M89. Textile Test Methods Bursting Strength—Ball Burst Test. 9 pp.
CNS L3082-80. Method of Test for Bursting Strength of Fabrics (May)(5613). 3 pp.
ISO 2960-74. Textiles—Determination of Bursting Strength and Bursting Distension—Diaphragm Method First Edition. 4 pp.
SAA AS 2001.2.4-90. Methods of Test for Textiles—Part 2: Physical Tests—Part 2.4: Determination of Bursting Pressure of Textile Fabrics—Hydraulic Diaphragm Method. 7 pp.
SAA AS 2001.2.19-88. Methods of Test for Textiles—Part 2: Physical Tests—Part 2.19: Determination of Bursting Force of Textile Fabrics—Ball Burst Method. 2 pp.
—Fiberboard
CPPA D.19P-90. Bursting Strength of Board. 2 pp.
—Glass Mats
CNS K6706-82. Method of Test of Glass Fiber Mat (Jan)(8428). 2 pp.
—Hydraulic Fluid Filters
BSI BS 6275: Part 2-84. 1984 Hydraulic Fluid Power Filter Elements Part 2: Method of Test to Verify Structural Integrity. 8 pp.
ISO 2941-74. Hydraulic Fluid Power—Filter Elements—Verification of Collapse/Burst Resistance First Edition. 4 pp.
—Hydraulic Systems—Tubes
ISO 8574-90. Aerospace—Hydraulic System Tubing—Qualification Tests First Edition. 8 pp.
—Leather
ISO 3379-76. Leather—Determination of Distension and Strength of Grain—Ball Burst Test First Edition. 5 pp.
—Leathercloth
CNS K6771-83. Method of Test for Polyurethane Leather (May)(10270). 3 pp.

Bursting Strength (Cont.)
—Liner Boards
CPPA D.19P-90. Bursting Strength of Board. 2 pp.
—Oil Filters
BSI BS 7403: Part 6-91. 1991 Full-Flow Lubricating Oil Filters for Internal Combustion Engines Part 6: Method of Test for Static Burst Pressure (ISO 4548-6: 1985). 8 pp.
ISO 4548 Pt 6-85. Methods of Test for Full-Flow Lubricating Oil Filters for Internal Combustion Engines—Part 6: Static Burst Pressure Test First Edition; (Corrigendum 1-1990). 5 pp.
—Packaging Papers
MOD UK DSTAN 13-10-68. Paper, Wrapping, Unglazed, and Paper, Wrapping, Unglazed, Lead Free Issue 1. 9 pp.
MOD UK M 6512/61. Examination of Paper Wrapping, Unglazed and Paper, Wrapping, Unglazed, LF Quality (Superseded by Def Stan 13-10).
—Paper Products
CPPA D.8-64. Bursting Strength of Paper. 4 pp.
—Paperboard
BSI BS 2922: Part 1-85. 1985 Strength of Wet Paper and Board Part 1: Method for Determinationof the Bursting Strengthof Paper and Board After Immersion in Water. 5 pp.
BSI BS 3137-72. 1972 Methods for Determining the Bursting Strength of Paper and Board. 27 pp.
CNS P3003-86. Method of Test for Low Bursting Strength of Paper and Paperboard (Jan)(1353). 2 pp.
CNS P3011-89. Method of Test for High Bursting Strength of Paper and Paperboard (Aug)(2054). 5 pp.
CPPA D.19P-90. Bursting Strength of Board. 2 pp.
CPPA D.8U-77. Plybond Strength of Paper Board (Using Jumbo Mullen Tester). 2 pp.
ISO 2759-83. Board—Determination of Bursting Strength Second Edition. 11 pp.
ISO 3689-83. Paper and Board—Determination of Bursting Strength After Immersion in Water Second Edition. 4 pp.
JIS P 8112-76. Testing Method for Bursting Strength of Paper and Paperboard by Mullen Low-Pressure Tester.
JIS P 8131-77. Testing Method for Bursting Strength of Paper and Paperboard by Mullen High-Pressure Tester. 7 pp.
SAA AS 1301.438S-89. Methods of Test for Pulp and Paper (Metric Units)—Part 438s: Bursting Strength of Paperboard and Corrugated Fibreboard (This is a Joint Standard with SANZ NZS 1301). 7 pp.
SNZ NZS/AS 1301. 438S-89. Methods of Test for Pulp and Paper Bursting Strength of Paperboard and Corrugated Fibreboard (This is a Joint Standard with SAA AS 1301.438S). 7 pp.
—Papers
BSI BS 2922: Part 1-85. 1985 Strength of Wet Paper and Board Part 1: Method for Determinationof the Bursting Strengthof Paper and Board After Immersion in Water. 5 pp.
BSI BS 3137-72. 1972 Methods for Determining the Bursting Strength of Paper and Board. 27 pp.
CNS P3003-86. Method of Test for Low Bursting Strength of Paper and Paperboard (Jan)(1353). 2 pp.
CNS P3011-89. Method of Test for High Bursting Strength of Paper and Paperboard (Aug)(2054). 5 pp.
CPPA D.8-64. Bursting Strength of Paper. 4 pp.
CPPA D.2U-77. Quick Method for Bursting Strength. 1 p.
CPPA D.5U-77. Determination of Bursting Strength of Paper Using Thibeault Instrument. 2 pp.
DIN ENGL 53113-90. Determining the Bursting Strength of Paper (June). 2 pp.
ISO 2758-83. Paper—Determination of Bursting Strength Second Edition. 11 pp.
ISO 3689-83. Paper and Board—Determination of Bursting Strength After Immersion in Water Second Edition. 4 pp.
JIS P 8112-76. Testing Method for Bursting Strength of Paper and Paperboard by Mullen Low-Pressure Tester.
JIS P 8131-77. Testing Method for Bursting Strength of Paper and Paperboard by Mullen High-Pressure Tester. 7 pp.
SAA AS 1301.403S-89. Methods of Test for Pulp and Paper (Metric Units)—Part 403s: Bursting Strength of Paper (This is a Joint Standard with SANZ NZS 1301). 7 pp.
SNZ NZS/AS 1301. 403S-89. Methods of Test for Pulp and Paper Bursting Strength of Paper (This is a Joint Standard with SAA AS 1301.403S). 7 pp.
—Pipes
BSI BS 4728-71. 1971 Determination of the Resistance to Constant Internal Pressure of Thermoplastics Pipe. 12 pp.

INDUSTRY STANDARDS

INTERNATIONAL AND NON-U.S. NATIONAL STANDARDS
SUBJECT INDEX

Bursting Strength (Cont.)
—Pipes (Cont.)
ISO 1167-73. Plastics Pipes for the Transport of Fluids—Determination of the Resistance to Internal Pressure First Edition. 6 pp.

—Plastic Pipes
CNS K3040-80. Requirements for Burst Test of Polypropylene Pipes (Jul)(5851). 1 p.
ISO 3212-75. Polypropylene Pipes—Burst Test Requirements First Edition. 3 pp.

Bus Arbiters
See Also: Microprocessors

—Microprocessors—Preferred Products List
CECC CECC MUAHAG Vol 7 IS 8-92. Preferred Products List; Active Microcircuits (En, Fe, Ge). 89 pp.

Bus Conductors
See Also: Busbars; Electric Conductors

—Aluminum
JIS H 4180-90. Aluminium and Aluminium Alloy Bus Conductors. 25 pp.

—Aluminum Alloy
JIS H 4180-90. Aluminium and Aluminium Alloy Bus Conductors. 25 pp.

Bus Connectors
Use: Busway Fittings

Bus Controllers
See Also: Computers; Controllers; Interface Circuits; Microprocessors; Multiplexers

—Microprocessors—Preferred Products List
CECC CECC MUAHAG Vol 7 IS 8-92. Preferred Products List; Active Microcircuits (En, Fe, Ge). 89 pp.

Bus Drivers
See Also: Bus Transceivers; Data Buses; Integrated Circuits; Interface Circuits; Microprocessors
CECC CECC 90 101-061 ISSUE 1-87. UTE C 86-213/B 81; Digital Integrated Circuits in Accordance with FS 90 101 54/64/74 368A; Bus Driver with 3 State Outputs (En). 19 pp.

—Latching—Microprocessors—Preferred Products List
CECC CECC MUAHAG Vol 7 IS 8-92. Preferred Products List; Active Microcircuits (En, Fe, Ge). 89 pp.

Bus Fittings
Use: Busway Fittings

Bus Registers
See Also: Digital Circuits; Integrated Circuits

—8 Stage
CECC CECC 90 104-127 ISSUE 1-81. BS CECC 90 104-127; Silicon Complementary MOS with (B) Buffered Outputs and Cavity Packaging (En). 26 pp.
CECC CECC 90 104-127 ISSUE 1-86. CEI CECC 90 104-127; Silicon Complementary MOS with (B) Buffered Outputs Cavity and Non Packaging (En). 2 pp.
CECC CECC 90 104-188 ISSUE 1-81. BS CECC 90 104-188 Issue 1; 8-Stage Shift-and-Store Bus Register with Three State Outputs (En). 28 pp.
CECC CECC 90 104-188 ISSUE 1-86. CEI CECC 90 104-188 Issue 1; 8-Stage Shift-and-Store Bus Register with Three State Outputs (En). 2 pp.

Bus Transceiver/Registers
See Also: Bus Transceivers; Digital Circuits; Microprocessors; Transceivers

—Octal
CECC CECC 90 109-881 ISSUE 1-90. Digital Integrated Circuits; Silicon Monolithic C MOS, Cavity or Non-Cavity Packages; Type(s); 54/74 HC 651 Inverting Octal Bus Transceiver/Register with 3-State Outputs; Assessment Levels P, Y, L (En, Fr). 11 pp.
CECC CECC 90 109-882 ISSUE 1-90. Digital Integrated Circuits; Silicon Monolithic C MOS, Cavity or Non-Cavity Packages; Type(s); 54/74 HC 652 Non-Inverting Octal Bus Transceiver/Register with 3-State Outputs; Assessment Levels P, Y, L (En, Fr). 11 pp.
CECC CECC 90 109-883 ISSUE 1-90. Digital Integrated Circuits; Silicon Monolithic C MOS, Cavity or Non-Cavity Packages; Type(s); 54/74 HCT 651 Inverting Octal Bus Transceiver/Register with 3-State Outputs; Assessment Levels P, Y, L (En, Fr). 11 pp.
CECC CECC 90 109-884 ISSUE 1-90. Digital Integrated Circuits; Silicon Monolithic C MOS, Cavity or Non-Cavity Packages; Type(s); 54/74 HCT 652 Non-Inverting Octal Bus Transceiver/Register with 3-State Outputs; Assessment Levels P, Y, L (En, Fr). 17 pp.
CECC CECC 90 109-890 ISSUE 1-90. Digital Integrated Circuits; Silicon Monolithic C MOS, Cavity or Non-Cavity Packages; Type(s); 54/74 HCT 648 Inverting Octal Bus Transceiver/Register with 3-State Outputs; Assessment Levels P, Y, L (En, Fr). 10 pp.

—Octal—Inverting
CECC CECC 90 109-703 ISSUE 1-87. Digital Integrated Circuits in Accordance with FS 90 109; 54/74 HC 648; Inverting Octal Bus Transceiver/Register with 3 State Outputs (En, Fr). 7 pp.

—Octal—Noninverting
CECC CECC 90 109-702 ISSUE 1-87. Digital Integrated Circuits in Accordance with FS 90 109; 54/74 HC 646; Non-Inverting Octal Bus Transceiver/Register with 3 State Outputs (En, Fr). 8 pp.

Bus Transceivers
Use For: Bidirectional Bus Drivers See Also: Bus Drivers; Bus Transceiver/Registers; Digital Circuits; Interface Circuits; Microprocessors; Transceivers

—Inverting
CECC CECC 90 109-648 ISSUE 1-86. Digital Integrated Circuits in Accordance with FS 90 109; 54/74 HC 242; 3-State Inverting Bus Transceiver (En, Fr). 6 pp.
CECC CECC 90 109-819 ISSUE 1-88. Digital Integrated Circuits in Accordance with FS 90 109; 54/74 HCT 242; 3-State Inverting Bus Transceiver (En, Fr). 6 pp.

—Noninverting
CECC CECC 90 109-649 ISSUE 1-86. Digital Integrated Circuits in Accordance with FS 90 109; 54/74 HC 243; 3-State Non Inverting Bus Transceiver (En, Fr). 6 pp.
CECC CECC 90 109-820 ISSUE 1-88. Digital Integrated Circuits in Accordance with FS 90 109; 54/74 HCT 243; 3-State Non Inverting Bus Transceiver (En, Fr). 6 pp.

—Octal
CECC CECC 90 103-092 ISSUE 1-82. BS CECC 90 103-092; Octal Bus Transceivers with 3-Level Outputs (En) AMD 1 (En). 18 pp.
CECC CECC 90 103-168 ISSUE 1-82. BS CECC 90 103-168; Octal Bus Transceivers with 3-State Non-Inverting Outputs (En). 18 pp.
CECC CECC 90 104-229 ISSUE 1. NL CECC 90 104-228 Issue 1; Digital Integrated Circuits in Accordance with FS 90 104; HEC/HEF 40245B; Octal Bus Transceiver with 3-State Outputs (En). 9 pp.
CECC CECC 90 106-006 ISSUE 2-86. UTE C 86-217 ADD 2/GA 6; Digital Integrated Circuits in Accordance with FS 90 106; 54/74 ALS 245, 54/74 ALS 245A; Octal Bus Transceivers with 3 State Non Inverting Outputs (En, Fr) ADD 3 (En, Fr). 17 pp.
CECC CECC 90 106-106 ISSUE 2-86. UTE C 86-217 ADD 2/GA 73; Digital Integrated Circuits in Accordance with FS 90 106; 54/74 ALS 620, 54/74 ALS 620A; Octal Bus Transceivers with 3 State Inverting Outputs (En, Fr) ADD 3 (En, Fr). 17 pp.
CECC CECC 90 106-107 ISSUE 2-86. UTE C 86-217 ADD 2/GA 74; Digital Integrated Circuits in Accordance with FS 90 106; 54/74 ALS 621, 54/74 ALS 621A; Octal Bus Transceivers with Non Inverting Open Collector Outputs (En, Fr) ADD 3 (En, Fr). 17 pp.
CECC CECC 90 106-108 ISSUE 2-86. UTE C 86-217 ADD 2/GA 75; Digital Integrated Circuits in Accordance with FS 90 106; 54/74 ALS 622, 54/74 ALS 622A; Octal Bus Transceivers with Inverting Open Collector Outputs (En, Fr) ADD 3 (En, Fr). 17 pp.
CECC CECC 90 106-109 ISSUE 2-86. UTE C 86-217 ADD 2/GA 76; Digital Integrated Circuits in Accordance with FS 90 106; 54/74 ALS 623, 54/74 ALS 623A; Octal Bus Transceivers with 3 State Non Inverting Outputs (En, Fr) ADD 3 (En, Fr). 17 pp.
CECC CECC 90 106-116 ISSUE 2-86. UTE C 86-217 ADD 2/GA 77; Digital Integrated Circuits in Accordance with FS 90 106; 54/74 ALS 640, 54/74 ALS 640A; Octal Bus Transceivers with Inverting 3 State Outputs (En, Fr) ADD 3 (En, Fr). 17 pp.
CECC CECC 90 106-117 ISSUE 2-86. UTE C 86-217 ADD 2/GA 78; Digital Integrated Circuits in Accordance with FS 90 106; 54/74 ALS 641, 54/74 ALS 641A; Octal Bus Transceivers with Non Inverting Open Collector Outputs (En, Fr) ADD 3 (En, Fr). 17 pp.
CECC CECC 90 106-118 ISSUE 2-86. UTE C 86-217 ADD 2/GA 79; Digital Integrated Circuits in Accordance with FS 90 106; 54/74 ALS 642, 54/74 ALS 642A; Octal Bus Transceivers with Inverting Open Collector Outputs (En, Fr) ADD 3 (En, Fr). 17 pp.
CECC CECC 90 106-119 ISSUE 2-86. UTE C 86-217 ADD 2/GA 80; Digital Integrated Circuits in Accordance with FS 90 106; 54/74 ALS 643, 54/74 ALS 643A; Octal Bus Transceivers with Non Inverting and Inverting 3 State Outputs (En, Fr) ADD 3 (En, Fr). 17 pp.
CECC CECC 90 106-120 ISSUE 2-86. UTE C 86-217 ADD 2/GA 81; Digital Integrated Circuits in Accordance with FS 90 106; 54/74 ALS 644, 54/74 ALS 644A; Octal Bus Transceivers with Non Inverting and Inverting Open Collector Outputs (En, Fr) ADD 3 (En, Fr). 17 pp.
CECC CECC 90 106-121 ISSUE 2-86. UTE C 86-217 ADD 2/GA 82; Digital Integrated Circuits in Accordance with FS 90 106; 54/74 ALS 645, 54/74 ALS 645A; Octal Bus Transceivers with Non Inverting 3 State Outputs (En, Fr) ADD 3 (En, Fr). 17 pp.
CECC CECC 90 107-034 ISSUE 2-89. UTE C 86-218 ADD 3/FA 34; Digital Integrated Circuits in Accordance with FS 90 107; 54/74 F 245; Octal Bidirectional Transceiver with 3-State Inputs/Outputs (En, Fr) ADD 3 (En, Fr). 11 pp.
CECC CECC 90 109-746 ISSUE 1-87. Digital Integrated Circuits in Accordance with FS 90 109; 54/74 HC 620; Octal Bus Transceiver with Inverting 3-State Outputs (En, Fr). 6 pp.
CECC CECC 90 109-747 ISSUE 1-87. Digital Integrated Circuits in Accordance with FS 90 109; 54/74 HC 623; Octal Bus Transceiver with Non Inverting 3-State Outputs (En, Fr). 6 pp.

—Octal—Preferred Products List
CECC CECC MUAHAG Vol 7 IS 8-92. Preferred Products List; Active Microcircuits (En, Fe, Ge). 89 pp.

—Quad
CECC CECC 90 103-089 ISSUE 1-82. BS CECC 90 103-089; Quadruple Bus Transceivers with Inverted 3-Level Outputs (En). 19 pp.
CECC CECC 90 103-090 ISSUE 1-81. BS CECC 90 103-090; Quadruple Bus Transceiver Non-Inverted with 3-Level Outputs (En) AMD 1 (En). 21 pp.
CECC CECC 90 106-011 ISSUE 1-85. UTE C 86-217; Digital Integrated Circuits in Accordance with FS 90 106; 54/74 ALS 243, 54/74 ALS 243; Quadruple Bus Transceivers with 3 State Non Inverting Outputs (En, Fr) ADD 3 (En, Fr). 16 pp.
CECC CECC 90 106-015 ISSUE 1-85. UTE C 86-217; Digital Integrated Circuits in Accordance with FS 90 106; 54/74 ALS 24254/74 ALS 242A; Quadruple Bus Transceivers with 3 State Inverting Outputs (En, Fr) ADD 3 (En, Fr). 16 pp.
CECC CECC 90 107-033 ISSUE 2-89. UTE C 86-218 ADD 3/FA 33; Digital Integrated Circuits in Accordance with FS 90 107; 54/74 F 243; Quad Bus Transceiver with 3-State Non Inverting Outputs (En, Fr) ADD 3 (En, Fr). 10 pp.

—Quad—Preferred Products List
CECC CECC MUAHAG Vol 7 IS 8-92. Preferred Products List; Active Microcircuits (En, Fe, Ge). 89 pp.

Busbar Enclosures
Use: Busways

Busbar Trunking
Use: Busways

Busbar Trunking Systems
Use: Busways

Busbars
See Also: Air Switches; Bus Conductors; Busways; Cables (Electric); Data Buses; Electric Wire; Power Lines; Power Systems; Stranded Conductors; Switchgear; Transmission Lines
SNZ NZS 1998-65. Specification for Neutral and Earth Busbars Amend: 1, 1967. 8 pp.

—Aluminum
BSI BS 159-92. 1992 High-Voltage Busbars and Busbar Connections. 16 pp.
BSI BS 159-57. 1957 Amd 3 Busbars and Busbar Connections. 37 pp.

INTERNATIONAL AND NON-U.S. NATIONAL STANDARDS
SUBJECT INDEX

Busbars *(Cont.)*
—Aluminum *(Cont.)*
BSI BS 2898-70. 1970 Amd 1 Wrought Aluminium and Aluminium Alloys for Electrical Purposes. Bars, Extruded Round Tube and Sections. 32 pp.
DIN ENGL 46424-73. Switchgear; Extruded Channel Sections of Aluminium for Busbars (Aug). 2 pp.
IEC 105-58. Recommendation for Commercial-Purity Aluminium Busbar Material First Edition. 5 pp.
SNZ NZS 211-61. Busbars and Busbar Connections Amend: 1A, 1961; 3A, 1979 (Reconfirmed 1991). 36 pp.

—Aluminum Alloy
BSI BS 2898-70. 1970 Amd 1 Wrought Aluminium and Aluminium Alloys for Electrical Purposes. Bars, Extruded Round Tube and Sections. 32 pp.

—Aluminum Alloy—Heat Treated
IEC 114-59. Recommendation for Heat-Treated Aluminium Alloy Busbar Material of the Aluminium-Magnesium-Silicon Type First Edition; (Corrigendum). 6 pp.

—Connectors
BSI BS 159-92. 1992 High-Voltage Busbars and Busbar Connections. 16 pp.
BSI BS 159-57. 1957 Amd 3 Busbars and Busbar Connections. 37 pp.

—Copper
BSI BS 159-92. 1992 High-Voltage Busbars and Busbar Connections. 16 pp.
BSI BS 159-57. 1957 Amd 3 Busbars and Busbar Connections. 37 pp.
BSI BS 1432-87. 1987 Copper for Electrical Purposes: High Conductivity Copper Rectangular Conductors with Drawn or Rolled Edges. 18 pp.
BSI BS 1433-70. 1970 Amd 3 Copper for Electrical Purposes. Rod and Bar. 34 pp.
JIS H 3140-92. Copper Bus Bars. 10 pp.
SNZ NZS 211-61. Busbars and Busbar Connections Amend: 1A, 1961; 3A, 1979 (Reconfirmed 1991). 36 pp.

—Drilled Holes
DIN ENGL 43673 Pt 1-82. Drilled Holes and Screw Connections for Busbars; Rectangular Cross Section Busbars (Feb). 4 pp.
DIN ENGL 43673 Pt 2-82. Drilled Holes and Screw Connections for Busbars; Channel Section Busbars (Feb). 4 pp.

—Identification Systems
SNZ NZS 73-62. Specification for the Marking and Arrangement of Switchgear, Busbars, Main Connections and Small Wiring (Reconfirmed 1979) Amend: 1, 1971. 18 pp.

—Installation
DIN VDE 0100 Pt 520-85. Erection of Power Installations with Rated Voltages of up to 1000 V; Selection and Erection of Electrical Apparatus; Cables, Conductors and Busbars (Nov). 50 pp.

—Interfaces
BSI BS 6475-84. 1984 Processor System Bus Interface (Eurobus A). 59 pp.
ISO 6951-86. Information Processing—Processor System Bus Interface (Eurobus A) First Edition. 57 pp.
JTC1 6951-86. Information Processing—Processor System Bus Interface (Eurobus A) First Edition. 57 pp.
OSI ISO 6951-86. Information Processing—Processor System Bus Interface (Eurobus A). 57 pp.

—Local Area Networks
BSI DD 100: Part 1-85. (WITHDRAWN) 1985 Token Bus Local Area Networks Part 1: Technical Specification (Superseded by BS ISO/IEC 8802-4: 1990). 287 pp.

—Screw Connections
DIN ENGL 43673 Pt 1-82. Drilled Holes and Screw Connections for Busbars; Rectangular Cross Section Busbars (Feb). 4 pp.
DIN ENGL 43673 Pt 2-82. Drilled Holes and Screw Connections for Busbars; Channel Section Busbars (Feb). 4 pp.

—Subsystems
BSI BS 7241-89. 1989 IEC 822 SB. Parallel Sub-System Bus of the IEC 821 VME Bus. 159 pp.
CENELEC HD 576 S1-90. Parallel Sub-System Bus of the IEC 821 VME Bus. 3 pp.
IEC 822-88. IEC 822 VSB Parallel Sub-System Bus of the IEC 821 VME Bus First Edition. 315 pp.
JTC1 822-88. IEC 822 VSB Parallel Sub-System Bus of the IEC 821 VME Bus First Edition. 315 pp.

—Symbols—Architectural Drawings
CNS C1099-87. Symbols of Bus-Bar of Interior Wiring Diagram for Architectural Plans (Dec)(9110).

Buses (Data)
Use: Data Buses

Buses (Vehicles)
See Also: Automobiles; Ground Vehicles; Motor Vehicles; Vehicles (Transportation)
CSA D250-M1985. School Buses; (Gen Instr 1). 29 pp.

—Air Compressors—Flanges
BSI BS AU 211-87. 1987 Dimensions of Four-Hole Flanges for Gear-Driven Air Compressors for Commercial Vehicles and Buses. 5 pp.
ISO 8719-86. Commercial Vehicles and Buses—Four-Hole Flanges for Gear-Driven Air Compressors First Edition. 5 pp.

—Alternators
BSI BS AU 250-93. 1993 Mounting Dimensions of Alternators, Types 1, 2 and 3 for Commercial Vehicles and Buses (ISO 7651: 1991) (E). 9 pp.
ISO 7651-91. Commercial Vehicles and Buses—Mounting Dimensions for Alternators of Types 1, 2 and 3 First Edition. 8 pp.

—Alternators—Shaft Ends (Hubs)
BSI BS AU 227-88. 1988 Dimensions of Cylindrical Shaft Ends and Hubs for Alterations for Commercial Vehicles and Buses. 4 pp.
ISO 7467-87. Commercial Vehicles and Buses—Cylindrical Shaft Ends and Hubs for Alternators First Edition; (Corrigendum 1-1990). 7 pp.

—Brakes
CNS D3114-82. Method of Service Brake Road Test for Trucks and Buses (Apr)(8678).
CNS D3115-82. Method of Service Brake Road Test for Performance Requirements for Trucks and Buses (Apr)(8679).
CNS D3116-82. Method of Braking Device Dynamometer Test for Trucks and Buses (Apr)(8680).

—Chassis
BSI BS AU 116-65. (OBSOLESCENT) 1965 Dimensions of Chassis for 27 ft Public Service Vehicle (Double Deck). 2 pp.
BSI BS AU 117-65. (OBSOLESCENT) 1965 Dimensions of Chassis for 30 ft Public Service Vehicle (Double Deck). 2 pp.

—Destination Signs
CNS D2094-87. Destination Indicators for Buses (Jul)(7677).
JIS D 4705-88. Destination Indicators for Buses. 9 pp.

—Door Bumpers
CNS D2093-87. Rubber Bumpers of Entrance Door for Buses (Jul)(7676).
JIS D 4706-88. Rubber Bumpers of Entrance Doors for Buses. 9 pp.

—Door Locks
CNS D2107-87. Door Locks for Buses (Jul)(7969).

—Door Stops
CNS D2092-87. Stoppers of Entrance Door for Buses (Jul)(7675).
JIS D 4707-76. Stoppers of Entrance Door for Buses (R 1984). 4 pp.

—Gear Boxes—Flanges
BSI BS AU 214A: Part 1-87. 1987 Amd 1 Gearbox Flanges Part 1: Dimensions of Cross-Tooth Gearbox Flanges, Type T, for Commercial Vehicles and Buses. 5 pp.
BSI BS AU 214: Part 2-88. 1988 Gearbox Flanges Part 2: Dimensions for Type A Gearbox Flanges for Commercial Vehicles and Buses. 5 pp.
BSI BS AU 214: Part 3-88. 1988 Gearbox Flanges Part 3: Dimensions for Type S Gearbox Flanges for Commercial Vehicles and Buses. 5 pp.
ISO 7646-86. Commercial Vehicles and Buses—Gearbox Flanges—Type A First Edition. 5 pp.
ISO 7647-86. Commercial Vehicles and Buses—Gearbox Flanges—Type S First Edition. 5 pp.
ISO 8667-92. Commercial Vehicles and Buses—Cross-Tooth Gearbox Flanges, Type T Second Edition. 7 pp.

—Gearshifts
CNS D1014-81. Transmission Shift Positions for Trucks and Buses (Sep)(7888).
JIS D 0011-79. Transmission Shift Positions for Trucks and Buses (R 1983). 4 pp.

—Incandescent Lighting
CNS D2041-87. Incandescent Lamps for Buses (Aug)(5797).

—Motor Starters—Mounting
ISO 7650-87. Commercial Vehicles and Buses—Mounting Dimensions for Starter Motors of Types 1, 2, 3 and 4 First Edition. 8 pp.

Buses (Vehicles) *(Cont.)*
—Seats
CNS D2097-89. Shape and Dimension of Seats for Buses (Oct)(7680).

—Tickets—Documentation
SAA AS 4039-92. Tender Documentation for Urban Transit Buses. 75 pp.

—Tires
BSI BS AU 50:Pt 1:SUBSEC 2.2-91. 1991 Tyres and Wheels Part 1: Tyres Section 2: Commercial Vehicle Tyres Subsection 2.2: Method of Measuring Tyre Rolling Loaded New Tyres (ISO 9112: 1991). 6 pp.
CGSB 20-GP-7-80. Tires, for Trucks and Buses, Standard for. 8 pp.
CNS K4029-89. Dimensions of Tires for Truck and Bus (Oct)(3672).
CNS K4029-85. Dimensions of Tires for Truck and Bus (May)(3672). 12 pp.
ISO 4209 Pt 1-88. Truck and Bus Tyres and Rims (Metric Series)—Part 1: Tyres Fourth Edition. 14 pp.
ISO 9112-91. Truck and Bus Tyres—Method of Measuring Tyre Rolling Circumference—Loaded New Tyres First Edition. 6 pp.
ISO 10454-93. Truck and Bus Tyres—Verifying Tyre Capabilites—Laboratory Test Methods First Edition. 11 pp.
SNZ NZS 5464-90. Pneumatic Tyres—Light Truck and Truck/Bus—New (NZS 5464:1990/AS 2230:1990). 12 pp.

—Tires—Rolling Resistance
BSI BS AU 50:Pt 1:SUBSEC 2.3-93. 1993 Tyres and Wheels Part 1: Tyres Section 2: Commercial Vehicle Tyres Subsection 2.3: Method of Measuring Rolling Resistance (ISO 9948: 1992). 15 pp.
ISO 9948-92. Truck and Bus Tyres—Methods of Measuring Rolling Resistance First Edition. 14 pp.

—Towing Attachments—Front
BSI BS AU 245-92. 1992 Front Towing Attachments for Commercial Road Vehicles and Buses over 3.5 t (ISO 8035: 1991) (E). 6 pp.
ISO 8035-91. Commercial Road Vehicles and Buses over 3,5 t—Front Towing Attachments First Edition. 5 pp.

—Ventilation Equipment
CNS D2096-89. Inlet Type Ventilators for Buses (Jul)(7679).
JIS D 4703-76. Inlet Type Ventilators for Buses (R 1984). 6 pp.

—Wheels
ISO 4209 Pt 2-87. Truck and Bus Tyres and Rims (Metric Series)—Part 2: Rims First Edition. 10 pp.

—Windows
CNS D2095-87. Sash Locks for Buses (Jul)(7678).
JIS D 4704-76. Sash Locks for Buses (R 1984). 7 pp.

—Windshield Defrosters
CNS D3125-82. Method of Performance Test for Windshield Defrosting and Defogging Systems for Passenger Trucks (Jun)(8946).
JIS D 4503-85. Windscreen Defrosting Systems for Trucks and Buses. 9 pp.

—Windshield Demisters
JIS D 4504-85. Windscreen Demisting Systems for Trucks and Buses. 10 pp.

Bush Chains
See Also: Chains
—Breaking Load
DIN ENGL 8164-90. Bush Chains (July). 3 pp.

—Motorcycles—Dynamic Testing
BSI BS 7615-92. 1992 Motor Cycle Chains (ISO 10190: 1992) (E). 14 pp.
ISO 10190-92. Motor Cycle Chains—Characteristics and Test Methods First Edition. 11 pp.

—Motorcycles—Length Measurement
BSI BS 7615-92. 1992 Motor Cycle Chains (ISO 10190: 1992) (E). 14 pp.
ISO 10190-92. Motor Cycle Chains—Characteristics and Test Methods First Edition. 11 pp.

—Motorcycles—Mechanical Properties
BSI BS 7615-92. 1992 Motor Cycle Chains (ISO 10190: 1992) (E). 14 pp.
ISO 10190-92. Motor Cycle Chains—Characteristics and Test Methods First Edition. 11 pp.

—Motorcycles—Proof Testing
BSI BS 7615-92. 1992 Motor Cycle Chains (ISO 10190: 1992) (E). 14 pp.
ISO 10190-92. Motor Cycle Chains—Characteristics and Test Methods First Edition. 11 pp.

Bush Chains (Cont.)

—Motorcycles—Tensile Testing
BSI BS 7615-92. 1992 Motor Cycle Chains (ISO 10190: 1992) (E). 14 pp.
ISO 10190-92. Motor Cycle Chains—Characteristics and Test Methods First Edition. 11 pp.

—Short Pitch
CNS B2155-77. Short Pitch Transmission Precision Bush Chains and Chain Wheels (Oct)(4172).
ISO 1395-77. Short Pitch Transmission Precision Bush Chains and Chain Wheels First Edition; (Amendment 1-1982). 15 pp.

Bush Cutters
Use: Mowers

Bush Saws
Use: Brush Saws

Bushes (Mechanical Components)
Use: Bushings

Bushings
See Also: Bearings; Chuck Bushings; Die Bushings; Electric Raceways; Electrical Insulating Bushings; Grommets; Guide Bushings; Indexable Inserts; Inserts (Fasteners); Jig Bushings; Linings; Shafts (Machine Elements); Spacers (Electric); Spacers (Mechanical); Spring Collets; Sprue Bushings; Wrapped Bushes
CSA CAN/CSA-C22. 2 NO 18-92. Outlet Boxes, Conduit Boxes, and Fittings; (Gen Instr 1 Thru 2). 118 pp.

—Aerospace
CEN EN 2285-89. Bushes, Plain Aluminium Alloy with Self-Lubricating Liner Dimensions and Loads. 6 pp.

—Aircraft
SBAC AS 144 ISSUE 4. Bush. 1 p.

—Alternating Voltage
BSI BS 223-85. 1985 Bushings for Alternating Voltages Above 1000V. 32 pp.
IEC 137-84. Bushings for Alternating Voltages Above 1000 V Third Edition. 63 pp.
SAA AS 1265-90. Bushings for Alternating Voltages Above 1000 V Amdt 1 July 1992. 32 pp.

—Armored Cables
CSA CAN/CSA-C22. 2 NO 18-92. Outlet Boxes, Conduit Boxes, and Fittings; (Gen Instr 1 Thru 2). 118 pp.

—Bearings—Aerospace
AECMA PREN2285-78. Bushes Plain Aluminium Alloy with Self Lubricating Liner Dimensions (C1/24-01). 7 pp.
AECMA PREN2286-78. Bushes Flanged Aluminium Alloy with Self-Lubricating Liner-Dimensions (C1/24-01). 7 pp.
AECMA PREN2287-78. Bushes Plain Corrosion Resisting Steel with Self-Lubricating Liner—Dimensions (C1/24-01). 7 pp.
AECMA PREN2288-78. Bushes Flanged Corrosion Resisting Steel with Self-Lubricating Liner—Dimensions (C1/24-01). 7 pp.
AECMA PREN2311-87. Bushes with Self-Lubricating Liner Technical Specification.
BSI BS EN 2285-90. 1990 Bushes Plain, Aluminium Alloy with Self-Lubricating Liner. Dimensions and Loads. 9 pp.
BSI BS EN 2286-90. 1990 Bushes Flanged, Aluminium Alloy with Self-Lubricating Liner Dimensions and Loads. 9 pp.
BSI BS EN 2287-90. 1990 Bushes Plain, Corrosion Resisting Steel with Self-Lubricating Liner. Dimensions and Loads. 9 pp.
BSI BS EN 2288-90. 1990 Bushes Flanged, Corrosion Resisting Steel with Self-Lubricating Liner. Dimensions and Loads. 9 pp.
BSI BS EN 2311-90. 1990 Bushes with Self-Lubricating Liner. Technical Specification. 21 pp.
CEN EN 2286-89. Bushes, Flanged Aluminium Alloy with Self-Lubricating Liner Dimensions and Loads. 6 pp.
CEN EN 2287-89. Bushes, Plain Corrosion Resisting Steel with Self-Lubricating Liner Dimensions and Loads. 7 pp.
CEN EN 2288-89. Bushes, Flanged Corrosion Resisting Steel with Self-Lubricating Liner Dimensions and Loads. 6 pp.
CEN EN 2311-87. Bushes with Self-Lubricating Liner Technical Specification. 18 pp.

—Bronze—Oil Retaining—Aircraft
BSI SP 94-64. 1964 Amd 1 Bronze Oil-Retaining Bushes and Thrust Washers for Aircraft. 7 pp.

—Cable Support—Aircraft
SBAC ESC 53 ISSUE 3. Connector Accessory: Bush (Cable Support). 3 pp.

Bushings (Cont.)

—Compressive Properties
BSI BS 5600: Sec 3.9-79. 1979 Powder Metallurgical Materials and Products Part 3: Methods of Testing Sintered Metal Materials and Products, Excluding Hardmetals Section 3.9: Determination of Radial Crushing Strength for Sintered Metal Bushes. 4 pp.
ISO 2739-73. Sintered Metal Bushes—Determination of Radial Crushing Strength First Edition. 4 pp.

—Control Columns—Aircraft
SBAC AS 860 ISSUE 6. Combined Dust Washer and Bush Control Column. 1 p.

—Copper Alloys
CNS B2708-82. Bushes for Plain Bearings Made from Copper Alloys, Solid (Sep)(9349).
DIN ENGL 1850 Pt 1-76. Bushes for Plain Bearings; Made from Copper Alloys, Solid (Oct). 3 pp.

—Cords (Electric)
CNS C4214-80. Drop Cord Bushing (Aug)(6047).

—Door Hinges
CNS A2087-86. Door Hinges (with Bushings or Washers) (Dec)(6538). 5 pp.
JIS A 5511-72. Door Hinges (with Bushing or Washers).

—Driving Elements
CNS B2665-82. Drivings Elements, Bushes for Plain Bearings (Jun)(8922).
DIN ENGL 8221-73. Driving Elements; Bushes for Plain Bearings According to DIN 502, DIN 503 and DIN 504 (July). 2 pp.

—Electric Conduits
CNS C4226-86. Bushings for Rigid Steel Conduit (Oct)(6082). 2 pp.

—Electrical Bonding
CSA C22.2 NO 41-M1987. Grounding and Bonding Equipment (R 1993); (Gen Instr 1). 30 pp.

—Electrical Grounding
CSA C22.2 NO 41-M1987. Grounding and Bonding Equipment (R 1993); (Gen Instr 1). 30 pp.

—Ferrule Welding—Aircraft
SBAC AS 15760-761 ISSUE 3. Bushing (for Use With Ferrule Welding Pipe Fitting).

—Flanged—Aircraft
SBAC AS 4670-4675 ISSUE 3(I). Flanged Bush. 1 p.

—Journal Bearings
JIS B 1582-76. Bushes for Journal Bearings. 11 pp.

—Linkages—Rudder Bars—Aircraft
SBAC AS 2207 ISSUE 5. Bush for Link-Bar (Rudder Bar). 1 p.

—Lubrication Feed
DIN ENGL 1850 Pt 2-82. Plain Bearings; Bushes with Lubrication Holes, Lubrication Grooves and Lubrication Bore Reliefs (Nov). 4 pp.

—Multispindle Heads
DIN ENGL 69001 Pt 51-81. Machine Tools; Multi-Spindle Heads; Bearing Bushes; Type A (Oct). 1 p.

—Pedal Pivots—Rudder Bars—Aircraft
SBAC AS 2211 ISSUE 5. Bush for Pedal Pivot (Rudder Bar). 1 p.

—Pipe Joints
DIN ENGL 16966 Pt 7-88. Glass Fibre Reinforced Polyester Resin (UP-GF) Pipe Fittings and Joint Assemblies; Bushings, Flanges, Flanged and Laminated Joints; General Quality Requirements and Testing (Nov). 6 pp.

—Pivots—Pedal—Rudder Bars—Aircraft
SBAC AS 2211 ISSUE 5. Bush for Pedal Pivot (Rudder Bar). 1 p.

—Plain Bearings
BSI BS 4480: Part 2-74. 1974 Plain Bearings. Metric Series Part 2: Solid Plastic Bushes. 15 pp.
BSI BS 4480: Part 3-93. 1993 Plain Bearings: Metric Series Part 3: Copper Alloy Bushes (ISO 4379: 1993) (E). 12 pp.
BSI BS 4480: Part 3-79. 1979 Plain Bearings. Metric Series Part 3: Dimensions and Tolerances of Solid Copper Alloy Bushes. 8 pp.
CNS B2709-82. Bushes for Plain Bearings with Lubrication Holes and Grooves (Sep)(9350).
CNS B2710-82. Bushes for Plain Bearings Made of Artificial Carbon (Sep)(9351).
CNS B2711-82. Bushes for Plain Bearings Made from Dura Plastics (Sep)(9352).
CNS B7202-82. Method of Test for Bushes of Plain Bearings Made from Dura Plastics (Sep)(9353).
ISO 4379-93. Plain Bearings—Copper Alloy Bushes Second Edition. 10 pp.

Bushings (Cont.)

—Plain Bearings (Cont.)
SNZ BS 4480: Part 2-74. Specification for Plain Bearings: Metric Series Part 2: Solid Plastics Bushes Amend: 1. 16 pp.

—Plain Bearings—Quality Assurance
DIN ENGL 31670 Pt 4-77. Quality Assurance of Plain Bearings; Quality Characteristics and Testing of the Dimensions of Metallic Bearing Bushes (Dec). 12 pp.
DIN ENGL 31670 Pt 7-81. Quality Assurance of Plain Bearings; Quality Characteristics and Measurement of the Dimensions of Bearing Bushes Made from Thermoplastics (Jan). 10 pp.

—Plastics—Aircraft
SBAC AS 12870-889 ISSUE 4. Bushing Plastics.
SBAC AS 12890-939 ISSUE 4. Bushing Plastics.

—Pressure Pipe Fittings
DIN ENGL 8063 Pt 4-83. Pipe Joint Assemblies and Fittings for Unplasticized Polyvinyl Chloride (PVC-U) Pressure Pipes; Bushings, Flanges, Seals; Dimensions (Sept). 12 pp.
DIN ENGL 16962 Pt 12-80. Pipe Joints and Elements for Polypropylene (PP) Pressure Pipelines; Types 1 and 2; Bushes, Flanges and Seals for Socket-Welding; Dimensions (Aug). 8 pp.
DIN ENGL 16963 Pt 11-80. Pipe Joints and Elements for High Density Polyethylene (HDPE) Pressure Pipelines; Types 1 and 2; Bushes, Flanges and Seals for Socket-Welding; Dimensions (Aug). 8 pp.

—Rolling Stock
CNS E1011-82. Bushes for Rail Vehicles; Press-Fit Bushes (Pressed on) Steel (Jun)(8959).
CNS E1012-82. Bushes for Rail Vehicles; Press-Fit Bushes of Steel (Jun)(8960).
CNS E1045-85. Bushes for Railway Rolling Stock (Jun)(11282).
JIS E 4112-89. Steel Bushes and Sintered Metallic Bushes for Railway Rolling Stock.
JIS E 4119-89. Polyurethane Bushes for Railway Rolling Stock.

—Rubber—Pipe Fittings—Aircraft
SBAC AS 44442 ISSUE 2. Bush—for 3/8 Inch Diameter Pipe Union Nut.

—Rudder Bars—Aircraft
SBAC AS 2180 ISSUE 6. Bush—(Rudder Bar). 1 p.
SBAC AS 2226 ISSUE 8. Bush—(Rudder Bar). 1 p.

—Shear—Aircraft
SBAC AS 484 ISSUE 5(I). Shear Bushes—Taper. 1 p.

—Sintered
CNS B2707-82. Bushes for Plain Bearings Made from Sintered Material (Sep)(9348).
DIN ENGL 1850 Pt 3-90. Sintered Metal Bushes for Use with Plain Bearings (June). 3 pp.

—Sockets—Aircraft
SBAC AS 6325 ISSUE 3. Bush for Bonding Socket.

—Spools—Rudder Bars—Aircraft
SBAC AS 2222 ISSUE 3. Bush for Pedal Spool (Rudder Bar). 1 p.

—Steel—Aircraft
SBAC AS 3119 ISSUE 2. Shear Bushes Parallel. 1 p.

—Steel—Bearings—Control Wheels—Aircraft
SBAC AS 3121 ISSUE 2. Bearing Bush for Pilot's Control Wheel. 1 p.

—Steel—Forgings
DIN ENGL 7527 Pt 4-72. Steel Forgings; Machining Allowances and Permissible Variations for Seamless Open-Die Forged Bushes (Jan). 3 pp.

—Tension
CNS B2663-82. Tension Bush for External Application (Jun)(8920).
CNS B2664-82. Tension Bush for Internal Application (Jun)(8921).

—Thermoplastic
DIN ENGL 1850 Pt 6-79. Bushes for Plain Bearings; Press Bushes Made from Thermoplastics (Feb). 3 pp.

—Threaded—Drawing Equipment
CNS Z7155-83. Threaded Bushing for Drawing Instruments (Jan)(9931).

—Trim Controls—Aircraft
SBAC AS 898 ISSUE 5. Bush—Trim Control. 1 p.

—Tubes—Aircraft
SBAC RS 684 ISSUE 1. Spring Clips and Bushes for Use with Rigid Tubes.

INTERNATIONAL AND NON-U.S. NATIONAL STANDARDS
SUBJECT INDEX

Bushings (Cont.)
—Union Nuts—Assembly—Aircraft
SBAC AGS 3412 ISSUE 2. Assy. of Union Nut and Bush—Unified.
—Union Nuts—Pipe Fittings—Aircraft
SBAC AGS 3411 ISSUE 3. Bush for Use with 5/16 Inch Pipe Union Nut.
—Valve Guides—Aircraft
SBAC AS 2563 ISSUE 2. Bush for Valve Guide. 1 p.
—Windshield Wipers—Automobiles
CNS D2137-82. Windscreen Wiper for Automobiles, Bushing (Jun)(8954).

Business Cards
Use For: Visiting Cards
CNS S1091-79. Dimension for Visiting Card (Jan)(4722).

Business Facilities
Use: Office Buildings

Business Forms
See Also: Continuous Forms; Forms (Paper)
BSI BS 1808-85. 1985 Amd 1 Cut Business Forms and Letterheads (AMD 5612) July 31, 1987. 10 pp.
SAA AS P4-68. Commercial Forms Reconfirmed 1986. 19 pp.
—Banking
BSI BS 6601-85. 1985 Mail Payment Orders. 8 pp.
BSI BS 6695-86. 1986 Nostro Accounts Reconciliation in Banking. 14 pp.
ISO 6234-81. Bank Operations—Authorized Signature Lists and Their Representation on Microfiche First Edition. 9 pp.
ISO 6260-84. Mail Payment Orders First Edition. 7 pp.
ISO 6536-81. Bank Operations—Standard Scheme for Drawing Lists First Edition; (Erratum—Jan 1982). 11 pp.
ISO 7341-85. Banking—Nostro Accounts Reconciliation First Edition. 13 pp.
OSI ISO 6234-81. Bank Operations—Authorized Lists and Their Representation on Microfiche. 9 pp.
OSI ISO 6260-84. Mail Payment Orders. 7 pp.
OSI ISO 6536-81. Bank Operations—Standard Scheme for Drawing Lists. 10 pp.
OSI ISO 7341-85. Banking—Nostro Accounts Reconciliation. 13 pp.
OSI ISO DP 9777-87. Forms for Confirming Foreign Exchange Deals. 14 pp.
OSI ISO DP 9778-87. Forms for Confirming Loan/Deposit Contracts. 21 pp.
OSI ISO IEC DIS 9796-90. Information Technology—Security Techniques—Digital Signature Scheme Giving Message Recovery. 14 pp.
—Continuous—Printing Papers
CNS P2073-90. Base Paper for Continuous Business Forms (Jul)(12747). 2 pp.
—Design
BSI BS ISO 8439-90. 1990 Forms Design—Basic Layout. 9 pp.
ISO 8439-90. Forms Design—Basic Layout First Edition. 7 pp.
JIS Z 8303-90. Design for Business Forms.
—Design—EDIFACT
SAA AS 3805-91. Electronic Data Interchange for Administration, Commerce and Transport (EDIFACT)—Forms Design—Basic Layout (ISO 8439:1990). 3 pp.
—Securities
BSI BS 6636: Part 1-85. (WITHDRAWN) 1985 Message Types for Securities Part 1: Messages for Receipt/Delivery (Superseded by BS ISO 7775: 1991). 23 pp.
BSI BS 6636: Part 2-87. (WITHDRAWN) 1987 Message Types for Securities Part 2: Messages for Orders to Buy/Sell (Superseded by BS ISO 7775: 1991). 19 pp.
BSI BS 6822-87. 1987 Format for Transmission of Certificate Numbers of Securities. 7 pp.
BSI BS ISO 7775-91. 1991 Securities—Scheme for Message Types (Supersedes BS 6636: Part 1: 1985 & part 2: 1987) (G). 58 pp.
ISO 7775-91. Securities—Scheme for Message Types First Edition. 55 pp.
ISO 8532-86. Securities—Format for Transmission of Certificate Numbers First Edition. 6 pp.
JTC1 8532-86. Securities—Format for Transmission of Certificate Numbers First Edition. 6 pp.
OSI ISO 7775-1-84. Securities—Standard Scheme for Message Types—Part: Receipt: Delivery. 23 pp.
OSI ISO 7775-2-86. Securities—Scheme for Message Types—Part 2: Order to Buy-Sell. 17 pp.
OSI ISO 8532-86. Securities—Format for Transmission of Certificate Numbers. 6 pp.

Business Machines
Use: Office Machines

Business Practices
Use For: Competitive Business Practices
—Defense Contracts
MOD UK DEFCON 176B-90. MOD Requirements for Competition in Sub-Contracting (Competitive Main Contract) 9/90. 1 p.
MOD UK DEFCON Guide no. 8-86. Code of Practice for Competitive Sub-Contracting 5/86. 8 pp.

Business Services by Satellite (BSS)
Use: Satellite Communications—Business Services

Busway Connectors
Use: Busway Fittings

Busway Fittings
Use For: Bus Connectors; Bus Fittings; Busway Connectors; Power Connectors *See Also:* Busways; Electric Conduit Fittings; Electric Raceway Fittings; Fittings; Wireway Fittings
CSA C22.2 NG 27-1973. Busways; (Amd 1 June 1987). 19 pp.
CSA 1169 Bull. Electrical Bulletin 1169 June 27, 1978 to C22.2 NO 27. 2 pp.
—Aerospace—Electric Terminals—Impedance Measurement
CEN PREN 2591 (Part G7)-92. Elements of Electrical and Optical Connection Test Methods Part G7—Measurement of Characteristic Impedance of a Bus or a Stub Terminator. 3 pp.

Busways
Use For: Busbar Enclosures; Conductor Busways; Feeder Busways; Plug In Bushways; Power Distribution Busways *See Also:* Busbars; Busway Fittings; Electric Conduits; Electric Raceways; Lighting Busways; Panel Boards (Electrical); Trolley Busways
CSA C22.2 NO 201-M1984. Metal-Enclosed High Voltage Busways (R 1992); (Gen Instr 1 Thru 2). 24 pp.
JIS C 8364-90. Busways. 23 pp.
—Controlgear—Low Voltage
BSI BS EN 60439-2-93. 1993 Low-Voltage Switchgear and Controlgear Assemblies Part 2: Particular Requirements for Busbar Trunking Systems (Busways) (AMD 7775) July 15, 1993 (E). 27 pp.
CENELEC EN 60439-2-93. Low-Voltage Switchgear and Controlgear Assemblies Part 2: Particular Requirements for Busbar Trunking Systems (Busways) (IEC 439-2: 1987 + A1: 1991, Modified). 8 pp.
CENELEC EN 60439-2-93. Low-Voltage Switchgear and Controlgear Assemblies Part 2: Particular Requirements for Busbar Trunking Systems (Busways) (IEC 439-2: 1987 + A1: 1991, Modified). 15 pp.
IEC 439 Pt 2-87. Low-Voltage Switchgear and Controlgear Assemblies Part 2: Particular Requirements for Busbar Trunking Systems (Busways) Second Edition; (Amendment 1-1991) (CENELEC EN 60439-2: 1993). 42 pp.
—Fasteners—Corrosion Prevention
CSA C22.2 NO 27-1973. Busways; (Amd 1 June 1987). 19 pp.
CSA 1091 Bull. Electrical Bulletin 1091 February 8, 1977 to C22.2 NO 27. 1 p.
CSA 1169 Bull. Electrical Bulletin 1169 June 27, 1978 to C22.2 NO 27. 2 pp.
—Fasteners—Plating Protection
CSA C22.2 NO 27-1973. Busways; (Amd 1 June 1987). 19 pp.
CSA 1091 Bull. Electrical Bulletin 1091 February 8, 1977 to C22.2 NO 27. 1 p.
CSA 1169 Bull. Electrical Bulletin 1169 June 27, 1978 to C22.2 NO 27. 2 pp.
—Isolated Phase
CSA C22.2 NO 201-M1984. Metal-Enclosed High Voltage Busways (R 1992); (Gen Instr 1 Thru 2). 24 pp.
—Low Voltage
BSI BS 5486: Part 2-88. (WITHDRAWN) 1988 Low-Voltage Switchgear and Controlgear Assemblies Part 2: Particular Requirements for Busbar Trunking Systems (Busways) (Renumbered as BS EN 60439-2: 1993). 15 pp.
—Segregated Phase
CSA C22.2 NO 201-M1984. Metal-Enclosed High Voltage Busways (R 1992); (Gen Instr 1 Thru 2). 24 pp.

Busways (Cont.)
—Splitters
CSA 1169 Bull. Electrical Bulletin 1169 June 27, 1978 to C22.2 NO 76. 2 pp.
—Switchgear—Low Voltage
BSI BS EN 60439-2-93. 1993 Low-Voltage Switchgear and Controlgear Assemblies Part 2: Particular Requirements for Busbar Trunking Systems (Busways) (AMD 7775) July 15, 1993 (E). 27 pp.
CENELEC EN 60439-2-93. Low-Voltage Switchgear and Controlgear Assemblies Part 2: Particular Requirements for Busbar Trunking Systems (Busways) (IEC 439-2: 1987 + A1: 1991, Modified). 8 pp.
CENELEC EN 60439-2-93. Low-Voltage Switchgear and Controlgear Assemblies Part 2: Particular Requirements for Busbar Trunking Systems (Busways) (IEC 439-2: 1987 + A1: 1991, Modified). 15 pp.
IEC 439 Pt 2-87. Low-Voltage Switchgear and Controlgear Assemblies Part 2: Particular Requirements for Busbar Trunking Systems (Busways) Second Edition; (Amendment 1-1991) (CENELEC EN 60439-2: 1993). 42 pp.
—Unventilated
CSA C22.2 NO 27-1973. Busways; (Amd 1 June 1987). 19 pp.
CSA 1169 Bull. Electrical Bulletin 1169 June 27, 1978 to C22.2 NO 27. 2 pp.
—Ventilated
CSA C22.2 NO 27-1973. Busways; (Amd 1 June 1987). 19 pp.
CSA 1169 Bull. Electrical Bulletin 1169 June 27, 1978 to C22.2 NO 27. 2 pp.

Busy Hour
See Also: Average Daily Peak Hour; Fixed Daily Measurement Hour; Fixed Daily Measurement Period; Telecommunication Circuits; Telecommunication Traffic; Time; Time Consistent Busy Hour; Units of Measurement
CCITT RECMN E.500 (REV 1)-92. Traffic Intensity Measurement Principles (Study Group II) 17 pp. 17 pp.
CCITT RECMN E.500-89. Traffic Intensity Measurement Principles —Telephone Network and ISDN—Quality of Service, Network Management and Traffic Engineering (Study Group II) 12 pp. 12 pp.
—Circuit Groups—Grade of Service
CCITT RECMN E.521-89. Calculation of the Number of Circuits in a Group Carrying Overflow Traffic—Telephone Network and ISDN—Quality of Service, Network Management and Traffic Engineering (Study Group II) 11 pp. 11 pp.
—Connection Accessibility—Telecommunication Networks
CCITT RECMN E.845-89. Connection Accessibility Objective for the International Telephone Service—Telephone Network and ISDN—Quality of Service, Network Management and Traffic Engineering (Study Group II) 7 pp. 7 pp.
—Manual Demand Operating
CCITT RECMN E.510-89. Determination of the Number of Circuits in Manual Operation —Telephone Network and ISDN—Quality of Service, Network Management and Traffic Engineering (Study Group II) 5 pp. 5 pp.
—Network Congestion—Telecommunication Networks
CCITT RECMN E.845-89. Connection Accessibility Objective for the International Telephone Service—Telephone Network and ISDN—Quality of Service, Network Management and Traffic Engineering (Study Group II) 7 pp. 7 pp.

Busy Signals
Use: Busy Tones

Busy Tones
See Also: Tones (Telephone Services)
CCITT RECMN E.182-89. Application of Tones and Recorded Announcements in Telephone Services—Telephone Network and ISDN—Operation, Numbering, Routing and Mobile Service (Study Group II) 8 pp. 8 pp.
CCITT RECMN Q.35-89. Technical Characteristics of Tones for the Telephone Service—General Recommendations on Telephone Switching and Signalling—Functions and Information Flows for Services in the ISDN —Supplements (Study Group XI) 9 pp. 9 pp.
CENELEC PRETS 300 295-92. Human Factors (HF); Specification of Characteristics of Telephone Services Tones When Locally Generated in Terminals. 9 pp.

INDUSTRY STANDARDS

Busy Tones (Cont.)
ETSI PRETS 300 295-92. Human Factors (HF); Specification of Characteristics of Telephone Services Tones When Locally Generated in Terminals. 9 pp.

—**DSS1—Integrated Services Digital Networks User Part—Interworking**
CCITT RECMN Q.699-89. Interworking Between the Digital Subscriber Signalling System Layer 3 Protocol and the Signalling System No. 7 ISDN User Part—Interworking of Signalling Systems (Study Group XI) 66 pp. 66 pp.

—**Foreign Countries**
CCITT RECMN Q.36-89. Customer Recognition of Foreign Tones—General Recommendations on Telephone Switching and Signalling—Functions and Information Flows for Services in the ISDN—Supplements (Study Group XI) 2 pp. 2 pp.

—**Frequency Bands**
CCITT RECMN E.180-89. Technical Characteristics of Tones for the Telephone Service—Telephone Network and ISDN—Operation, Numbering, Routing and Mobile Service (Study Group II) 9 pp. 9 pp.

—**Integrated Services Digital Networks—Supplementary Services**
CCITT RECMN I.221-89. Common Specific Characteristics of Services—Integrated Services Digital Network (ISDN)—General Structure and Service Capabilities (Study Group XVIII) 3 pp. 3 pp.

—**Integrated Services Digital Networks—Telecommunication Services**
CCITT RECMN I.221-89. Common Specific Characteristics of Services—Integrated Services Digital Network (ISDN)—General Structure and Service Capabilities (Study Group XVIII) 3 pp. 3 pp.

—**Periodicity**
CCITT RECMN E.180-89. Technical Characteristics of Tones for the Telephone Service—Telephone Network and ISDN—Operation, Numbering, Routing and Mobile Service (Study Group II) 9 pp. 9 pp.

But-2-ene-1,4-diol
Use: 2-Butene-1,4-Diol

Buta-1,3-Diene
Use: 1,3-Butadiene

Butadiene
Use: 1,3-Butadiene

Butadiene Acrylonitrile Rubber
Use: Nitrile Rubber

Butadiene Rubber
Use: Polybutadiene

1,3-Butadiene
Use For: Butadiene See Also: Butadienes; Hazardous Materials; Nitrile Rubber; Polybutadiene
MOD UK TS 10288. Buta-1, 3-Diene (1:3 Butadiene).

Butadienes
Use For: Vinyl Ethylene See Also: 1,3-Butadiene; Hydrocarbons
CNS K1169-74. Butadiene (Mar)(3703).
CNS K6370-74. Method of Test for Butadiene (Mar)(3704).
JIS K 1533-80. Butadiene.

—**Argon Content—Gas Chromatography**
ISO 6792-82. Butadiene for Industrial Use—Determination of Oxygen and Argon in the Gaseous Phase Above Liquid Butadiene—Gas Chromatographic Method First Edition. 4 pp.

—**Butylcatechol Content—Liquid Chromatography**
ISO 8176-86. Butadiene for Industrial Use—Determination of Active Tert-Butyl-Catechol (TBC) (4-(1,1-Dimethylethyl)-1,2-Benzenediol)—High Performance Liquid Chromatographic Method First Edition. 6 pp.

—**Butylcatechol Content—Spectrometry**
ISO 6684-82. Butadiene for Industrial Use—Determination of Tert-Butyl-Catechol (TBC) (4-(1,1-Dimethylethyl)-1,2 Benzenediol)—Spectrometric Method First Edition. 5 pp.

Butadienes (Cont.)
—**Highway Transportation—Emergency Procedures**
SAA AS 1678.2.1. 010-86. Emergency Procedure Guide—Transport—Part 2.1.010: Butadiene (Inhibited).

—**Hydrocarbon Content—Gas Chromatography**
ISO 6378-81. Butadiene for Industrial Use—Determination of Hydrocarbon Impurities—Gas Chromatographic Method First Edition. 9 pp.

—**Oligomer Content—Gas Chromatography**
ISO 7381-86. Butadiene for Industrial Use—Determination of Oligomers—Gas Chromatographic Method First Edition. 10 pp.

—**Oxygen Content—Gas Chromatography**
ISO 6792-82. Butadiene for Industrial Use—Determination of Oxygen and Argon in the Gaseous Phase Above Liquid Butadiene—Gas Chromatographic Method First Edition. 4 pp.

—**Sampling**
ISO 8563-87. Propylene and Butadiene for Industrial Use—Sampling in the Liquid Phase First Edition. 8 pp.

Butan-1-ol
Use: n-Butyl Alcohol

Butane-1,4-diol
Use: 1,4-Butylene Glycol

n-Butane
Use: Butanes

2,3-Butanedione Monoxime
Use For: Diacetyl Monoxime
CNS K7181-65. Chemical Reagent (Diacetyl Monoxime) (Jan)(1681).

Butanes
Use For: n-Butane See Also: 1,4-Butylene Glycol; Gases; Hydrocarbons; Liquefied Petroleum Gas
BSI BS 4250: Part 1-87. 1987 Liquified Petroleum Gas Part 1: Commercial Butane and Propane. 12 pp.
CGSB CAN/CGSB-3.13-M88. Liquefied Petroleum Gas (Butanes). 10 pp.
CNS K1171-74. N-Butane (Industrial Grade) (Jun)(3718).
DIN ENGL 51622-85. Liquefied Petroleum Gases; Propane, Propene, Butane, Butene and Their Mixtures; Requirements (Dec). 4 pp.
SNZ NZS 5435-84. Specification for Liquefied Petroleum Gas (LPG). 11 pp.

—**Automatic Changeover Devices**
BSI BS 3016-89. 1989 Pressure Regulators and Automatic Changeover Devices for Liquefied Petroleum Gases. 24 pp.

—**Gas Chromatography**
DIN ENGL 51619-85. Testing of Petroleum Products; Determination of the Composition of Liquefied Petroleum Gas; Analysis by Gas Chromatography (Aug). 2 pp.

—**Hydrocarbon Content—Gas Chromatography**
BSI BS 7276-90. 1990 Method for Gas Chromatographic Analysis of Liquified Petroleum Gas. 16 pp.
CEN PREN 242-85. Commercial Propane and Butane (LPG); Determination of Hydrocarbons Content by Gas Chromatographic Method. 24 pp.
ISO 7941-88. Commercial Propane and Butane—Analysis by Gas Chromatography First Edition. 15 pp.

—**Pressure Regulators**
BSI BS 3016-89. 1989 Pressure Regulators and Automatic Changeover Devices for Liquefied Petroleum Gases. 24 pp.

—**Residue Content—Evaporation**
DIN ENGL 51613-66. Testing of Liquefied Petroleum Gases; Determination of Elementary Sulphur and Residue on Evaporation (Apr). 3 pp.

—**Sulfur Content**
DIN ENGL 51613-66. Testing of Liquefied Petroleum Gases; Determination of Elementary Sulphur and Residue on Evaporation (Apr). 3 pp.

1-Butanol
Use: n-Butyl Alcohol

2-Butanol
Use: sec-Butyl Alcohol

2-Butanone
Use: Methyl Ethyl Ketone

2-Butene-1,4-Diol
Use For: But-2-ene-1,4-diol

—**Iodine Number**
ISO 6793-81. But-2-Ene-1,4-Diol for Industrial Use—Determination of Iodine Value First Edition. 5 pp.

2,3-Butenediondioxime
Use: Dimethylglyoxime

Butenes
Use: Butylene

2-Butoxyethanol
Use: Ethylene Glycol Monobutyl Ether

Butt Hinges
See Also: Hinges

—**Doors**
CGSB CAN/CGSB-69.18-M90. Butts and Hinges (ANSI/BHMA A156.1-1981) (Supersedes 69-GP-1M); (Amendment 1 July 1993). 45 pp.
CNS A2085-86. Loose Pin Butt Hinges (Dec)(6536). 4 pp.

—**Doors—Stainless Steel**
CNS A2007-89. Steel and Stainless Steel Butt Hinges (Jun)(857).
CNS A3084-86. Method of Test for Steel and Stainless Steel Butt Hinges (Dec)(5084).
JIS A 5501-75. Wrought Steel and Wrought Stainless Steel Butt Hinges (R 1983). 11 pp.
JIS A 5510-71. Wrought Steel and Wrought Stainless Steel Loose Pin Butt Hinges (R 1987). 9 pp.

—**Doors—Steel**
CNS A2007-89. Steel and Stainless Steel Butt Hinges (Jun)(857).
CNS A3084-86. Method of Test for Steel and Stainless Steel Butt Hinges (Dec)(5084).
JIS A 5501-75. Wrought Steel and Wrought Stainless Steel Butt Hinges (R 1983). 11 pp.
JIS A 5510-71. Wrought Steel and Wrought Stainless Steel Loose Pin Butt Hinges (R 1987). 9 pp.

Butt Joints
See Also: Bolted Joints; Butt Strap Joints; Butt Welded Joints; Joints; Lap Joints; Pipe Joints; Soldered Joints
BSI BS 6223-90. 1990 Friction Welding of Butt Joints in Metals for High Duty Applications. 26 pp.

—**Adhesive Strength—Tensile Testing**
BSI BS 5350: Part C3-89. (WITHDRAWN) 1989 Methods of Test for Adhesives Group C: Adhesively Bonded Joints Part C3: Determination of Bond Strength in Direct Tension (Renumbered as BS EN 26922: 1993). 6 pp.
BSI BS EN 26922-93. 1993 Amd 1 Adhesives—Determination of Tensile Strength of Butt Joints (ISO 6922: 1987) (AMD 7613) August 15, 1993 (W). 9 pp.
CEN EN 26922-93. Adhesives—Determination of Tensile Strength of Butt Joints (ISO 6922: 1987). 6 pp.
ISO 6922-87. Adhesives—Determination of Tensile Strength of Butt Joints First Edition (CEN EN 26922: 1993). 6 pp.

—**Hexagonal Nuts**
DIN ENGL 3872-85. Solderless Pipe Couplings with Cutting Ring; Coupling Nuts for Butt Joints (June). 2 pp.

—**Radiography**
BSI BS 2600: Part 1-83. 1983 Radiographic Examination of Fusion Welded Butt Joints in Steel Part 1: Methods for Steel 2 mm up to and Including 50 mm Thick. 15 pp.
BSI BS 2600: Part 2-73. 1973 Amd 2 Radiographic Examination of Fusion Welded Butt Joints in Steel Part 2: Over 50 mm up to and Including 200 mm Thick. 15 pp.

—**Thrust Collars**
DIN ENGL 3867-82. Solderless Pipe Unions with Cutting Ring; Thrust Collars for Butt Joints (Nov). 2 pp.

Butt Strap Joints
See Also: Butt Joints; Joints

—**Pipe**
DIN ENGL 4922 Pt 1-78. Steel Filter Pipes for Drilled Wells with Slot Perforation and Fishing; (Butt Strap Joint) (Feb). 4 pp.

Butt Welded Joints

See Also: Butt Joints; Joints; Lap Joints; Pipe Joints; Welded Joints

BSI BS 6944-88. 1988 Flash Welding of Butt Joints in Ferrous Metals (Excluding Pressure Piping Applications). 15 pp.

ISO 9692-92. Metal-Arc Welding with Covered Electrode, Gas-Shielded Metal-Arc Welding and Gas Welding—Joint Preparations for Steel First Edition. 17 pp.

—Bend Testing

CEN PREN 910-92. Welding—Welded Butt Joints in Metallic Materials—Bend Tests. 16 pp.

DIN ENGL 50121 Pt 1-78. Testing of Metallic Materials; Technological Bending Test on Welded Joints and Weld Platings; Fusion Welded Joints (Jan). 9 pp.

ISO 5173-81. Fusion Welded Butt Joints in Steel—Transverse Root and Face Bend Test First Edition. 6 pp.

ISO 5177-81. Fusion Welded Butt Joints in Steel—Transverse Side Bend Test First Edition. 6 pp.

JIS Z 3122-90. Methods of Bend Test for Butt Welded Joint. 11 pp.

JIS Z 3125-62. Method of Notched Bend Test for Butt Welded Joint (R 1980). 5 pp.

JIS Z 3126-64. Reverse Bend Testing of Butt Welded Joint (R 1980). 6 pp.

—Boilers

BSI BS 1113: ENQ CASE 7-92. 1992 Design and Manufacture of Water-Tube Steam Generating Plant (Including Superheaters, Reheaters and Steel Tube Economizers) Enquiry Case 7: Use of Butt Welds in Tubes for Manufacture of Coils for Coil Type Boilers/Superheaters (Q). 1 p.

—Fatigue Testing

JIS Z 3103-87. Method of Repeated Tension Fatigue Testing for Fusion Welded Joints. 14 pp.

—Impact Testing

CEN PREN 875-92. Welding—Welded Joints in Metallic Materials—Specimen Location and Notch Orientation for Impact Tests. 8 pp.

JIS Z 3128-86. Method of Impact Test for Welded Joint. 7 pp.

—Inspection

JIS Z 3120-80. Method of Inspection for Gas Pressure Welded Joint of Steel Bars for Concrete Reinforcement. 6 pp.

JIS Z 3143-61. Inspection for Flash Weld (Steel) (R 1978). 8 pp.

—Pipe

DIN ENGL 2559 Pt 1-73. Edge Preparation for Welding; Directions Regarding Edge Forms; Fusion Welding of Butt Joints in Steel Tubes (May). 3 pp.

DIN ENGL 2559 Pt 2-84. Edge Preparation for Welding; Matching of Inside Diameter for Circumferential Welds on Seamless Pipes (Feb). 4 pp.

JIS B 2312-91. Steel Butt-Welding Pipe Fittings. 36 pp.

JIS B 2313-91. Steel Plate Butt-Welding Pipe Fittings. 39 pp.

—Pipe—Nondestructive Testing

JIS Z 3050-78. Method of Nondestructive Inspection for Weld of Pipeline. 30 pp.

—Pipe—Radiography

BSI BS 2910-86. 1986 Radiographic Examination of Fusion Welded Circumferential Butt-Joints in Steel Pipes. 28 pp.

JIS Z 3105-84. Methods of Radiographic Test and Classification of Radiographs for Aluminium Welds. 16 pp.

JIS Z 3108-86. Methods of Radiographic Test for Circumferential Butt Welds of Aluminium Pipes and Tubes. 22 pp.

—Quality Assurance

DIN ENGL 8563 Pt 11-87. Quality Assurance of Welding Operations; Electron Beam Welded Joints in Steel; Specification of Quality Classes (Oct). 4 pp.

—Radiography

CGSB 48-GP-2M-79. Spot Radiography of Welded Butt Joints in Ferrous Materials, Standard for. 9 pp.

ISO 1106 Pt 1-84. Recommended Practice for Radiographic Examination of Fusion Welded Joints—Part 1: Fusion Welded Butt Joints in Steel Plates up to 50 mm Thick First Edition. 10 pp.

ISO 1106 Pt 2-85. Recommended Practice for Radiographic Examination of Fusion Welded Joints—Part 2: Fusion Welded Butt Joints in Steel Plates Thicker Than 50 mm and up to and Including 200 mm in Thickness First Edition. 7 pp.

ISO 2437-72. Recommended Practice for the X-Ray Inspection of Fusion Welded Butt Joints in Aluminium and Its Alloys and Magnesium and Its Alloys 5 to 50 mm Thick First Edition. 8 pp.

Butt Welded Joints (Cont.)

—Tensile Testing

CNS Z8072-88. Method of Tension Test for Butt Welded Joint (Oct)(12455).

DIN ENGL 50120 Pt 1-75. Testing of Steel; Tensile Test on Welded Joints; Fusion Welded Butt Joints (Sept). 6 pp.

DIN ENGL 50120 Pt 2-78. Testing of Steel; Tensile Test on Welded Joints; Pressure Welded Butt Joints (Aug). 3 pp.

DIN ENGL 50123-77. Testing of Non-Ferrous Metals; Tensile Test on Welded Joints; Fusion Welded Butt Joints (Apr). 3 pp.

ISO 4136-89. Fusion-Welded Butt Joints in Steel—Transverse Tensile Test First Edition. 6 pp.

JIS Z 3121-84. Method of Tension Test for Butt Welded Joint. 9 pp.

JIS Z 3127-77. Method of Wide Plate Tension Test for Butt Welded Joint (R 1980). 6 pp.

—Test Specimens

DIN ENGL 50127-76. Testing of Steel; Bendover Test Specimen, Wedge Test Specimen, Notched Pipe Tension Test Specimen for Assessment of Fusion-Welded Fillet or Butt Joints (Aug). 4 pp.

—Test Specimens—Tensile Testing

CEN PREN 895-92. Welding—Welded Butt Joints in Metallic Materials—Transverse Tensile Tests. 12 pp.

—Ultrasonic Testing

BSI BS 1113: ENQ CASE 8-92. 1992 Design and Manufacture of Water-Tube Steam Generating Plant (Including Superheaters, Reheaters and Steel Tube Economizers) Enquiry Case 8: Ultrasonic Acceptance Criteria (Q). 2 pp.

BSI BS 3923: Part 1-86. 1986 Methods for Ultrasonic Examination of Welds Part 1: Methods for Manual Examination of Fusion Welds in Ferritic Steels. 94 pp.

BSI BS 3923: Part 2-72. 1972 Methods for Ultrasonic Examination of Welds Part 2: Automatic Examination of Fusion Welded Butt Joints in Ferritic Steels. 19 pp.

CGSB 48-GP-6A-75. Recommended Practices for Ultrasonic Inspection of Structural Welds. 58 pp.

CNS Z8063-85. Method of Ultrasonic Test for Steel Butt Weld (Oct)(11401).

JIS Z 3080-88. Method of Ultrasonic Angle Beam Testing and Classification of Test Results for Butt Welds in Aluminium Plates. 17 pp.

Butt Welding

See Also: Seam Welding; Welding

—Steels—Nickel Coated

JIS Z 3044-91. Method of Welding Procedure Qualification Test for Nickel and Nickel Alloy Clad Steels. 19 pp.

Butt Welding Equipment

See Also: Welding Equipment

—Glossaries

DIN ENGL 44752-67. Electric Butt Welding Machines; Definitions and Rating Details (June). 11 pp.

Butter

See Also: Animal Products; Dairy Products; Food; Margarine; Milk

CGSB 32.161M-89. Butter. 7 pp.

CNS N5085-86. Edible Butter (Apr)(2877). 2 pp.

CNS N6074-83. Methods of Test for Edible Oils and Fats (General Rules) (Mar)(3639). 2 pp.

EC 88/570/EEC-93. CORRIGENDUM to Commision Regulation (EEC) No 1813 /93 of 7 July 1993 Amending Regulation (EEC) No 570/88 on the Sale of Butter and Concentrated Butter for Use in the Manufacture of Pastry Products, Ice-Cream and Other Foodstuffs. 1 p.

EC EEC/1813/93-93. Commission Regulation Amending Regulation (EEC) No 570/88 on the Sale of Butter at Reduced Prices and the Grant of Aid for Cream, Butter and Concentrated Butter for Use in the Manufacture of Pastry Products, Ice-Cream and Other Foodstuffs. 2 pp.

—Boxes

CGSB CAN/CGSB-43.42-M86. Boxes, Bulk Butter and Similar Products. 20 pp.

DIN ENGL 10079-81. Means of Packaging; One-Way Consignment Boxes in One and Two Pieces for Butter Pats and Butter Bars (Apr). 2 pp.

DIN ENGL 10089-81. Means of Packaging; Folding Boxes for Butter Samples (Mar). 2 pp.

—Curd Content

SAA AS 2300.7.2-91. Methods of Chemical and Physical Testing for the Dairying Industry—Part 7: Butter—Part 7.2: Determination of Fat and Curd (Supersedes AS 1739 —1975 (in Part). 2 pp.

Butter (Cont.)

—Dispersion Index—Water

DIN ENGL 10311-85. Determination of the Water Dispersion in Butter; Indicator Paper Method (Aug). 2 pp.

ISO 7586-85. Butter—Determination of Water Dispersion Value First Edition; (Corrected and Reprinted -1985). 4 pp.

—Dry Matter Content

BSI BS 5086: Part 2-84. 1984 Analysis of Butter Part 2: Methods for Determination of Water, Solids-Non-Fat and Fat Contents (Reference Method). 4 pp.

ISO 3727-77. Butter—Determination of Water, Solids-Not-Fat and Fat Contents on the Same Test Portion (Reference Method) First Edition. 5 pp.

ISO 3728-77. Ice-Cream and Milk Ice—Determination of Total Solids Content (Reference Method) First Edition. 5 pp.

—Fat Content

BSI BS 5086: Part 2-84. 1984 Analysis of Butter Part 2: Methods for Determination of Water, Solids-Non-Fat and Fat Contents (Reference Method). 4 pp.

CNS N6072-86. Method of Test for Butter (Apr)(3528). 3 pp.

ISO 3727-77. Butter—Determination of Water, Solids-Not-Fat and Fat Contents on the Same Test Portion (Reference Method) First Edition. 5 pp.

SAA AS 2300.7.2-91. Methods of Chemical and Physical Testing for the Dairying Industry—Part 7: Butter—Part 7.2: Determination of Fat and Curd (Supersedes AS 1739 —1975 (in Part)). 2 pp.

—Fat Content—Peroxide Number

ISO 3976-77. Anhydrous Milk Fat—Determination of Peroxide Value (Reference Method) First Edition. 5 pp.

—Fat Content—Refractive Index

ISO 1739-75. Butter—Determination of the Refractive Index of the Fat (Reference Method) First Edition. 3 pp.

—Fat Content—Saponification Number

BSI BS 5086: Part 6-92. 1992 Analysis of Butter Part 6: Method for Determination of Fat Acidity and Free Fatty Acids in the Fat of Butter and Milk Fat Products (Reference Method) (ISO 1740: 1991). 8 pp.

ISO 1740-91. Milk Fat Products and Butter—Determination of Fat Acidity (Reference Method) Second Edition. 8 pp.

—Fatty Acid Content

BSI BS 5086: Part 6-85. 1985 Analysis of Butter Part 6: Method for Determination of Free Free Fatty Acids in the Fat (Reference Method). 4 pp.

—Hardness Testing

DIN ENGL 10331-85. Determination of the Hardness of Butter (Aug). 3 pp.

—Margarine Content

CNS N6072-86. Method of Test for Butter (Apr)(3528). 3 pp.

—Microbiological Analysis

SAA AS 1095.2.3-83. Microbiological Methods for the Dairy Industry—Part 2: Methods for the Examination of Specific Dairy Products—Part 2. 3: Butter and Related Products. 3 pp.

—Moisture Content

CNS N6072-86. Method of Test for Butter (Apr)(3528). 3 pp.

—Salt Content

BSI BS 5086: Part 4-85. 1985 Analysis of Butter Part 4: Method for Determination of Salt Content (Reference Method). 4 pp.

BSI BS 5086: Part 5-84. 1984 Analysis of Butter Part 5: Methods for Determination of Salt Content (Routine Method). 4 pp.

CNS N6072-86. Method of Test for Butter (Apr)(3528). 3 pp.

ISO 1738-80. Butter—Determination of Salt Content (Reference Method) First Edition. 4 pp.

SAA AS 2300.7.3-91. Methods of Chemical and Physical Testing for the Dairying Industry—Part 7: Butter—Part 7.3: Determination of Salt (Supersedes AS 1739—1975 (in Part)). 2 pp.

—Sampling

BSI BS 5086: Part 1-84. 1984 Analysis of Butter Part 1: General Introduction Including Preparations of Samples. 4 pp.

CNS N6072-86. Method of Test for Butter (Apr)(3528). 3 pp.

—Sensory Analysis

CNS N6072-86. Method of Test for Butter (Apr)(3528). 3 pp.

Butter (Cont.)

—Serum-Fat Content—Gravimetric Analysis
BSI BS 1743: Part 11-86. 1986 Analysis of Dried Milk and Dried Milk Products Part 11: Determination of Fat Content (Gravimetric Reference Method). 10 pp.
ISO 1736-85. Dried Milk, Dried Whey, Dried Buttermilk and Dried Butter Serum—Determination of Fat Content—Gravimetric Method (Reference Method) Second Edition. 10 pp.

—Serum-PH
BSI BS 5086: Part 7-91. 1991 Analysis of Butter Part 7: Method for Determination of pH of Butter Serum (W). 6 pp.
BSI BS 5086: Part 7-85. 1985 Analysis of Butter Part 7: Method for Determination of pH of Butter Serum. 4 pp.

—Serum-PH—Potentiometric Analysis
ISO 7238-83. Butter—Determination of pH of the Serum—Potentiometric Method First Edition. 4 pp.

—Sodium Chloride
BSI BS 998-69. (WITHDRAWN) 1969 Amd 1 Vacuum Salt for Butter and Cheese Making and Other Food Uses (Superseded by BS 998: 1990 & BS 7319: Parts 1 to 12: 1990). 22 pp.
BSI BS 998-90. 1990 Vacuum Salt for Food Use. 9 pp.

—Solids Content
CNS N6072-86. Method of Test for Butter (Apr)(3528). 3 pp.

—Test Specimens
CNS N6056-80. Method of Test for Milk and Milk Products (General Rules) (Aug)(3440). 2 pp.

—Test Specimens—Microbiological Analysis
DIN ENGL 10191 Pt 3-86. Microbiological Examination of Milk; Preparation of Samples of Milk Fat Products (Oct). 2 pp.

—Water Content
BSI BS 5086: Part 2-84. 1984 Analysis of Butter Part 2: Methods for Determination of Water, Solids-Non-Fat and Fat Contents (Reference Method). 4 pp.
BSI BS 5086: Part 3-84. 1984 Analysis of Butter Part 3: Method for Determination of Water Content (Routine Method). 4 pp.
ISO 3727-77. Butter—Determination of Water, Solids-Not-Fat and Fat Contents on the Same Test Portion (Reference Method) First Edition. 5 pp.
ISO 3728-77. Ice-Cream and Milk Ice—Determination of Total Solids Content (Reference Method) First Edition. 5 pp.

Butterfat
Use: Butter—Fat Content

Butterfly Angle Disk Valves
Use: Butterfly Valves

Butterfly Dampers
Use: Butterfly Valves

Butterfly Nuts
Use: Wing Nuts

Butterfly Valves
Use For: Butterfly Dampers; Disk Wafer Valves; Wafer Valves (Butterfly) See Also: Gas Valves; Hydraulic Valves; Pneumatic Valves; Swing Check Valves; Thermostatic Valves; Valves
BSI BS 5155-02. 1984 Amd 2 Butterfly Valves (AMD 6990) May 1, 1992. 19 pp.
CNS B2801-90. Wafer-Type Rubber-Seated Butterfly Valves (Jul)(12744).
DIN ENGL 3354 Pt 1-82. Butterfly Valves; General Data (June). 6 pp.
DIN ENGL 3354 Pt 2-82. Butterfly Valves; Soft Material Seat Seal Cast Iron Tight Butterfly Shutoff Valves with Flanged Ends (June). 4 pp.
DIN ENGL 3354 Pt 3-82. Butterfly Valves; Soft Material Seat Seal Steel or Cast Steel Tight Butterfly Shutoff Valves with Flanged Ends or Weld-on Ends (June). 4 pp.
DIN ENGL 3354 Pt 4-82. Butterfly Valves; Metallic Seat Seal Steel or Cast Steel Tight Butterfly Shutoff Valves with Flanged Ends or Weld-on Ends (June). 4 pp.
DIN ENGL 3441 Pt 5-84. Unplasticized Polyvinyl Chloride (PVC-U) Valves; PN 6 and PN 10 Wafer Type Butterfly Valves; Dimensions (Jan). 2 pp.
ISO 5752-82. Metal Valves for Use in Flanged Pipe Systems—Face-to-Face and Centre-to-Face Dimensions Second Edition. 13 pp.
JIS B 2032-87. Wafer-Type Rubber-Seated Butterfly Valves. 20 pp.

—Flanged
BSI BS 5155-84. 1984 Amd 1 Butterfly Valves. 18 pp.

Butterfly Valves (Cont.)

—Flanged (Cont.)
BSI BS 5155-02. 1984 Amd 2 Butterfly Valves (AMD 6990) May 1, 1992. 19 pp.
SNZ NZS/BS 5155-84. Specification for Butterfly Valves Amend: 1; 2, 1992. 16 pp.

—Marine
JIS F 7480-91. Rubber-Seat Butterfly Valves for Marine Use.

—Pipelines—Flanged
CEN PREN 558-1-91. Metal Valves for Use in Flanged Pipe Systems—Face-to-Face and Centre-to-Face Dimensions—Part 1: General. 5 pp.
CEN PREN 558-2-91. Metal Valves for Use in Flanged Pipe Systems—Face-to-Face and Centre-to-Face Dimensions—Part 2: PN Designated Valves. 13 pp.
CEN PREN 558-3-91. Metal Valves for Use in Flanged Pipe Systems—Face-to-Face and Centre-to-Face Dimensions—Part 3: Class-Designation Valves. 17 pp.

—Pressure Pipes
CEN PREN 593-91. Industrial Valves—Metallic Butterfly Valves for General Purposes. 17 pp.

—Waterworks
CNS B2798-90. Butterfly Valves for Water Works (Short Body Type) (Jul)(12741).
CNS B2799-90. Butterfly Valves for Water Works (Long Body Type) (Jul)(12742).
CNS B2800-90. Butterfly Valves for Water Works (Wafer Type) (Jul)(12743).
JIS B 2064-84. Butterfly Valves for Water Works. 16 pp.

Buttermilk
See Also: Animal Feed; Beverages; Dairy Products; Milk; Whey

—Animal Feed
CNS N2048-84. Buttermilk, Dried (for Animal Feed Purpose) (May)(10881). 1 p.

—Fat Content—Gravimetric Analysis
BSI BS 1743: Part 11-86. 1986 Analysis of Dried Milk and Dried Milk Products Part 11: Determination of Fat Content (Gravimetric Reference Method). 10 pp.
ISO 1736-85. Dried Milk, Dried Whey, Dried Buttermilk and Dried Butter Serum—Determination of Fat Content—Gravimetric Method (Reference Method) Second Edition. 10 pp.
ISO 7208-84. Skimmed Milk, Whey and Buttermilk—Determination of Fat Content—Gravimetric Method (Reference Method) First Edition. 9 pp.

—Phosphatase Activity
ISO 3356-75. Milk and Dried Milk, Buttermilk and Buttermilk Powder, Whey and Whey Powder—Determination of Phosphatase Activity (Reference Method) First Edition. 5 pp.

—Solubility
BSI BS 1743: Part 2-80. 1980 Analysis of Dried Milk and Dried Milk Products Part 2: Determination of the Solubility of Dried Milk, Dried Whey and Dried Buttermilk (Reference Method). 4 pp.

Buttler Paper
Use: Greaseproof Papers

Button Head Screws
See Also: Screws

—Hexagonal—Socket
BSI BS 4168: Part 6-82. 1982 Amd 1 Hexagon Socket Screws and Wrench Keys; Metric Series Part 6: Specification for Hexagon Socket Button Head Screws. 8 pp.
ISO 7380-83. Hexagon Socket Button Head Screws—Metric Series First Edition. 6 pp.
JIS B 1174-89. Hexagon Socket Button Head Screws. 11 pp.
SNZ NZS/BS 4168: Part 6-82. Hexagon Socket Screws and Wrench Keys: Metric Series Part 6: Specification for Hexagon Socket Button Head Screws Amend: 1. 8 pp.

Buttons (Fasteners)
See Also: Clothing; Fasteners; Zippers
BSI BS 3866-79. 1979 Amd 1 Holes and Shanks in Buttons. 9 pp.
BSI BS 4162-83. 1983 Methods of Test for Buttons. 12 pp.
CNS Z7077-81. Blech Knob with Fastened Plate (May)(7466).
JIS S 4024-77. Snap Buttons.

—Aircraft
SBAC AS 2533 ISSUE 3. Button for Catch. 1 p.

Buttons (Fasteners) (Cont.)

—Double Rivet
CNS Z7092-81. Double Rivet Button (Jul)(7769).
JIS S 9022-71. Double Tubular Rivets or Leather Buttons.

—Leather
JIS S 9022-71. Double Tubular Rivets or Leather Buttons.

—Metal
CNS Z7093-81. Sheet-Metal Four Hole Button (Jul)(7770).

—Shirts
JIS S 4025-82. Buttons for Dress-Shirts.

Buttress Threads
See Also: Screw Threads
BSI BS 1657-50. 1950 Buttress Threads. 31 pp.
CNS B2095-78. Metric Buttress Screw Thread (General Plan) (Aug)(516).
CNS B2096-78. Metric Buttress Screw Thread (Allowance and Tolerance) (Aug)(517).
DIN ENGL 76 Pt 3-77. Runouts; Undercuts for Trapezoidal Threads, Buttress Threads and Knuckle Threads and Other Threads of Coarse Pitch (Jan). 3 pp.
DIN ENGL 513 Pt 2-85. Metric Buttress Threads; General Plan (Apr). 3 pp.
DIN ENGL 513 Pt 3-85. Metric Buttress Threads; Deviations and Tolerances (Apr). 8 pp.
MOD UK DSTAN 53-55-72. Buttress Threads-Metric Issue 1. 19 pp.
SAA AS B182-61. Buttress Threads Being BS 1657:1950, Endorsed Without Amendment Reconfirmed 1984 1987. 30 pp.

—Hydraulic Presses
CNS B2261-78. Buttress Thread 45 degree (for Hydraulic Presses) (Aug)(4468).

—Plastic Bottles
DIN ENGL 6063 Pt 1-89. Screw Threads Used Mainly for Plastic Containers; Dimensions of Buttress Threads (May). 2 pp.

—Profiles
CNS B2094-78. Metric Buttress Screw Thread (Profiles) (Aug)(515).
DIN ENGL 513 Pt 1-85. Metric Buttress Threads; Thread Profiles (Apr). 4 pp.

—Runouts
CNS B1066-78. Runout and Undercut for Metric Trapezoidal Screw Threads, Kunkle Screw Threads, Bultress Screw Threads and Other Screw Threads of Coarse Pitches (Mar)(4325).

—Undercuts
CNS B1066-78. Runout and Undercut for Metric Trapezoidal Screw Threads, Kunkle Screw Threads, Bultress Screw Threads and Other Screw Threads of Coarse Pitches (Mar)(4325).

Butts, Rifle
Use: Rifles—Butts

n-Butyl Acetate
Use For: Acetic Acid Butyl Ester
See Also: Acetates; sec-Butyl Acetate; tert-Butyl Acetate; Esters; Hazardous Materials; Isobutyl Acetate
BSI BS 551-90. 1990 Butyl Acetate for Industrial Use. 8 pp.
CNS K1118-78. Normal Butyl Acetate (May)(2617).
JIS K 1514-57. n-Butyl Acetate.
JIS K 8377-88. Butyl Acetate (n-Butyl Acetate).

—Storage and Handling—Information Cards
SAA AS 2508.3.00 2-82. Safe Storage and Handling Information Cards for Hazardous Materials—Part 3.002: Acetates (Ethyl Acetate, Propyl Acetates, Butyl Acetates, Amyl Acetate) Double Sided Card.

sec-Butyl Acetate
See Also: Acetates; n-Butyl Acetate; tert-Butyl Acetate; Hazardous Materials; Isobutyl Acetate
CNS K1211-79. Secondary Butyl Acetate (85-88% Grade) (May)(4864).

—Storage and Handling—Information Cards
SAA AS 2508.3.00 2-82. Safe Storage and Handling Information Cards for Hazardous Materials—Part 3.002: Acetates (Ethyl Acetate, Propyl Acetates, Butyl Acetates, Amyl Acetate) Double Sided Card.

tert-Butyl Acetate
See Also: Acetates; n-Butyl Acetate; sec-Butyl Acetate; Hazardous Materials; Isobutyl Acetate

INTERNATIONAL AND NON-U.S. NATIONAL STANDARDS
SUBJECT INDEX

tert-Butyl Acetate *(Cont.)*
—Storage and Handling—Information Cards
SAA AS 2508.3.00 2-82. Safe Storage and Handling Information Cards for Hazardous Materials—Part 3.002: Acetates (Ethyl Acetate, Propyl Acetates, Butyl Acetates, Amyl Acetate) Double Sided Card.

n-Butyl Acrylate
See Also: Hazardous Materials
CNS K1252-80. N-Butyl Acrylate (Jul)(5831).

n-Butyl Alcohol
Use For: Butan-1-ol; 1-Butanol *See Also:* Alcohols
BSI BS 508: Part 1-86. 1986 Butan-1-ol for Industrial Use Part 1: Butan-1-ol. 4 pp.
BSI BS 508: Part 2-84. 1984 Butan-1-ol for Industrial Use Part 2: Methods of Test. 4 pp.
CNS K1020-84. Normal Butyl Alcohol, Industrial Grade (Sep)(194).
CNS K6155-62. Method of Test for Butanol (May)(1420)(R 1973).
CNS K7111-64. Chemical Reagent (N-Butyl Alcohol) (N-Butanol) (May)(1611).
ISO 755 Pt 1-81. Butan-1-ol for Industrial Use—Methods of Test—Part 1: General First Edition. 5 pp.
JIS K 1504-59. Butanol.
JIS K 8810-79. 1-Butanol.
—Acidity—Volumetric Analysis
ISO 755 Pt 2-81. Butan-1-ol for Industrial Use—Methods of Test—Part 2: Determination of Acidity—Titrimetric Method First Edition. 4 pp.
—Bromine Number
ISO 761-77. Acetic Anhydride and Butan-1-ol for Industrial Use—Determination of Bromine Number First Edition. 4 pp.
—Safety
SAA AS 2508.3.00 6-91. Safe Storage and Handling Information Cards for Hazardous Materials—Part 3.006: Butanols, Propanols, Amyl Alcohols (In Professional Package 38) Double Sided Card.
—Sulfuric Acid Content—Color Testing
ISO 755 Pt 3-81. Butan-1-ol for Industrial Use—Methods of Test—Part 3: Sulphuric Acid Colour Test First Edition; (Amendment Slip-1982). 6 pp.

sec-Butyl Alcohol
Use For: 2-Butanol *See Also:* Alcohols
CNS K1210-79. Secondary Butyl Alcohol (Industrial Grade) (May)(4863).
CNS K7651-82. Chemical Reagent (2-Butanol) (Oct)(9508).
ISO 2496-73. secButyl Alcohol for Industrial Use—List of Methods of Test First Edition. 4 pp.
JIS K 1523-78. 2-Butanol (Secondary Butyl Alcohol).
JIS K 1523-63. 2-Butanol (Secondary Butyl Alcohol). 8 pp.
JIS K 8812-91. 2-Butanol.
—Acidity—Volumetric Analysis
ISO 2887-73. secButyl Alcohol, Methyl Ethyl Ketone, isoButyl Methyl Ketone, isoAmyl Ethyl Ketone, Diacetone Alcohol, and Hexylene Glycol for Industrial Use—Determination of Acidity to Phenolphthalein—Volumetric Method First Edition. 4 pp.

tert-Butyl Alcohol
Use For: 2-Methyl-2-propanol
CNS K7652-82. Chemical Reagent (2-Methyl-2-Propanol) (Oct)(9509).
JIS K 8813-91. 2-Methyl-2-Propanol.

Butyl Methyl Ketone
See Also: Ketones
—Alcohol Content—Volumetric Analysis
ISO 2501-74. Methyl Ethyl Ketone, isoButyl Methyl Ketone and isoAmyl Ethyl Ketone for Industrial Use—Determination of Alcoholic Impurities—Volumetric Method First Edition. 5 pp.

Butyl Rubber
Use For: Isobutylene Isoprene Rubber
See Also: Chlorobutyl Rubber; Latex; Rubber
BSI BS 3227-90. 1990 Butyl Rubber Compounds (Including Halobutyl Compounds). 10 pp.
MOD UK DSTAN 93-4-01. Butyl Rubber Type QX Issue 2; Amendment 3. 11 pp.
—Molding Materials
MOD UK DSTAN 93-22-84. Butyl Rubber, Black, Type MD Issue 1. 8 pp.
—Sheets
MOD UK DSTAN 93-53-88. Butyl Rubber, Calendered Sheet for Flexible Containers Issue 1. 13 pp.

Butyl Rubber *(Cont.)*
—Sheets *(Cont.)*
MOD UK TS 10161A. Butyl Rubber, Calendered Sheet for Flexible Containers (Superseded by Def Stan 93-53).
MOD UK TS 10260. Butyl Rubber, Scrim Reinforced Sheet.
—Shoes
MOD UK TS 10233. Halogenated Butyl Rubber, Type NBC.
—Thermogravimetric Analysis
ISO 9924 Pt 1-93. Rubber and Rubber Products—Determination of the Composition of Vulcanizates and Uncured Compounds by Thermogravimetry—Part 1: Butadiene, Ethylene-Propylene Copolymer and Terpolymer, Isobutene-Isoprene, Isoprene and Styrene-Butadiene. 8 pp.
—Vulcanization
BSI BS 4470-87. 1987 Evaluation of Vulcani-zation of Isobutene-Isoprene Rubber (IIR). 7 pp.
BSI BS 6845-87. 1987 Evaluation of Vulcanization Characteristics of Halogenated Isobutene-Isoprene Rubber (BIIR and CIIR). 4 pp.
ISO 7663-85. Rubber, Halogenated Isobutene-Isoprene (BIIR and CIIR)—Test Recipe and Evaluation of Vulcanization Characteristics First Edition. 4 pp.
—Vulcanized—Hoses
MOD UK DSTAN 93-71-93. Vulcanized Butyl Rubber for Tubing of Baby Viper Hose Type QX Issue 1. 13 pp.
MOD UK CS 3109. Vulcanized Butyl Rubber for Tubing of Baby Viper Hose (Superseded by Def Stan 93-71).

Butyl Rubber Content Analysis
—Elastomers
DIN ENGL 53621 Pt 8-77. Testing of Rubber and Elastomers; Quantitative Determination of Polymers; Determination of the Content of Butyl Rubber and of Polyisobutylene (July). 4 pp.
—Polyethylene
BSI BS 2782:Pt4: METH 452C-79. (OBSOLESCENT) 1979 Methods of Testing Plastics Part 4: Chemical Properties Method 452C: Determination of Butyl Content of Low Density Polythene Compounds. 2 pp.
—Rubber
DIN ENGL 53621 Pt 8-77. Testing of Rubber and Elastomers; Quantitative Determination of Polymers; Determination of the Content of Butyl Rubber and of Polyisobutylene (July). 4 pp.

Butylated Hydroxyanisole Content Analysis
Use: BHA Content Analysis

Butylated Hydroxytoluene Content Analysis
Use: BHT Content Analysis

tert-Butylcatechol Content Analysis
Use For: 1,2-Benzenodiol,4-(1,1-Dimethylethyl) Content Analysis
—Butadienes—Liquid Chromatography
ISO 8176-86. Butadiene for Industrial Use—Determination of Active Tert-Butyl-Catechol (TBC) (4-(1,1-Dimethylethyl)-1,2-Benzenediol)—High Performance Liquid Chromatographic Method First Edition. 6 pp.
—Butadienes—Spectrometry
ISO 6684-82. Butadiene for Industrial Use—Determination of Tert-Butyl-Catechol (TBC) (4-(1,1-Dimethylethyl)-1,2 Benzenediol)—Spectrometric Method First Edition. 5 pp.

Butylene
Use For: Butenes *See Also:* Hydrocarbons; Liquefied Petroleum Gas
CGSB CAN/CGSB-3.13-M88. Liquefied Petroleum Gas (Butanes). 10 pp.
DIN ENGL 51622-85. Liquefied Petroleum Gases; Propane, Propene, Butane, Butene and Their Mixtures; Requirements (Dec). 4 pp.
—Gas Chromatography
DIN ENGL 51619-85. Testing of Petroleum Products; Determination of the Composition of Liquefied Petroleum Gas; Analysis by Gas Chromatography (Aug). 2 pp.
—Residue Content—Evaporation
DIN ENGL 51613-66. Testing of Liquefied Petroleum Gases; Determination of Elementary Sulphur and Residue on Evaporation (Apr). 3 pp.

Butylene *(Cont.)*
—Sulfur Content
DIN ENGL 51613-66. Testing of Liquefied Petroleum Gases; Determination of Elementary Sulphur and Residue on Evaporation (Apr). 3 pp.

1,4-Butylene Glycol
Use For: Butane-1,4-diol *See Also:* Butanes
—Unsaturation—Volumetric Analysis
ISO 6794-81. Butane-1,4-Diol for Industrial Use—Determination of Degree of Unsaturation First Edition. 4 pp.

p-tert-Butylphenol-98
See Also: Phenol
CNS K1248-80. P-Tert-Butylphenol-98 (Jul)(5827).

Butynorate Content Analysis
Use: Dibutyltin Dilaurate Content Analysis

Butyrometers
—Cheeses
ISO 3432-75. Cheese—Determination of Fat Content—Butyrometer for Van Gulik Method First Edition. 6 pp.
SNZ NZS 501: Part 1-57. Gerber Method for the Determination of Fat in Milk and Milk Products Part 1: Apparatus Amend: 2, 1971; 3, 1982; 4, 1982; 4A, 1982. 32 pp.
—Cream
SNZ NZS 501: Part 1-57. Gerber Method for the Determination of Fat in Milk and Milk Products Part 1: Apparatus Amend: 2, 1971; 3, 1982; 4, 1982; 4A, 1982. 32 pp.
—Milk
BSI BS 696: Part 1-89. 1989 Gerber Method for the Determination of Fat in Milk and Milk Products Part 1: Apparatus. 23 pp.
ISO 488-83. Milk—Determination of Fat Content—Gerber Butyrometers First Edition. 12 pp.
SNZ NZS 501: Part 1-57. Gerber Method for the Determination of Fat in Milk and Milk Products Part 1: Apparatus Amend: 2, 1971; 3, 1982; 4, 1982; 4A, 1982. 32 pp.
—Milk—Pipettes
SNZ NZS 501: Part 1-57. Gerber Method for the Determination of Fat in Milk and Milk Products Part 1: Apparatus Amend: 2, 1971; 3, 1982; 4, 1982; 4A, 1982. 32 pp.

Buyers Guides
See Also: Documents
SNZ NZMP 2100-92. New Zealand Buyers Guide 1992/93. 68 pp.

Buzzers
See Also: Acoustic Signals; Signal Devices
CNS C4179-90. Dry Battery Type Buzzers (Dec)(5427).
CSA C22.2 NO 205-M1983. Signal Equipment (R 1992); (Gen Instr 1). 32 pp.
JIS C 9701-76. Dry Battery Type Buzzers (R 1980). 8 pp.
—Automotive
CNS D2057-86. Warning Buzzers for Automobiles (Jun)(6436). 5 pp.
JIS D 5712-73. Warning Buzzers for Automobiles. 8 pp.
—Ships
CNS F5078-87. Marine Electric Buzzers (Mar)(10625).
JIS F 8502-84. Marine Electric Buzzers.
JIS F 8502-58. Marine Electric Buzzers (R 1970). 10 pp.

By Pass Switches
Use: Bypass Switches

By-Pass Valves
Use: Bypass Valves

Bypass Disconnecting Switches
Use: Bypass Switches

Bypass Switches
Use For: By Pass Switches; Bypass Disconnecting Switches *See Also:* Disconnecting Switches; Electric Switches; Isolating Switches; Switches
—Automatic Transfer—Isolating—120-240 V AC
CSA C22.2 NO 178-1978. Automatic Transfer Switches (R 1992); (Amd 1-2 July 1982) (Amd 3 June 1987). 51 pp.

INDUSTRY STANDARDS

Bypass Switches *(Cont.)*

—Automatic Transfer—Isolating—120-240 V DC
- CSA C22.2 NO 178-1978. Automatic Transfer Switches (R 1992); (Amd 1-2 July 1982) (Amd 3 June 1987). 51 pp.

—Automatic Transfer—Isolating—250-500 V AC
- CSA C22.2 NO 178-1978. Automatic Transfer Switches (R 1992); (Amd 1-2 July 1982) (Amd 3 June 1987). 51 pp.

—Automatic Transfer—Isolating—250-500 V DC
- CSA C22.2 NO 178-1978. Automatic Transfer Switches (R 1992); (Amd 1-2 July 1982) (Amd 3 June 1987). 51 pp.

—Automatic Transfer—Isolating—550-1000 V AC
- CSA C22.2 NO 178-1978. Automatic Transfer Switches (R 1992); (Amd 1-2 July 1982) (Amd 3 June 1987). 51 pp.

—Automatic Transfer—Isolating—550-1000 V DC
- CSA C22.2 NO 178-1978. Automatic Transfer Switches (R 1992); (Amd 1-2 July 1982) (Amd 3 June 1987). 51 pp.

Bypass Valves
Use For: By-Pass Valves *See Also:* Directional Control Valves; Hydraulic Valves; Valves

—Electrohydraulic—Proportional
- JIS B 8656-89. Test Methods for Electro-Hydraulic Proportional Bypass Flow Control Valves. 30 pp.
- JIS B 8657-89. Test Methods for Electro-Hydraulic Proportional Directional Bypass Flow Control Valves. 36 pp.

—Meters—Indicator Plates
- BSI BS 3251-76. 1976 Amd 1 Indicator Plates for Fire Hydrants and Emergency Water Supplies (AMD 6736) September 30, 1991. 14 pp.

C Acid
- CNS K2236-88. C-Acid (Sep)(12418). 2 pp.

C Coaxial Connectors
See Also: Coaxial Connectors
- CENELEC HD 134.7-76. Radio-Frequency Connector Part 7: R.F. Coxial Connectors with Inner Diameter of Outer Conductor 9.5mm (0.374 in) with Bayonet Lock-Characteristic Impedance 50 Ohms (Type C). 2 pp.
- CNS C7075-86. CO1 Type Connectors for Radio Frequency Coaxial Cables (Jun)(4739).
- CNS C7076-86. CO2 Type Connectors for Radio Frequency Coaxial Cables (Jun)(4740).
- CNS C7077-86. CO3 Type Connectors for Radio Frequency Coaxial Cables (Jun)(4741).
- CNS C7078-86. CO4 Type Connectors for Radio Frequency Coaxial Cables (Jun)(4742).
- CNS C7079-86. CO5 Type Connectors for Radio Frequency Coaxial Cables (Jun)(4743).
- CNS C7080-86. C11 Type Connectors for Radio Frequency Coaxial Cables (Jun)(4744).
- JIS C 5411-76. C0 1 Type Connectors for Radio Frequency Coaxial Cables (R 1985). 22 pp.
- JIS C 5412-76. C0 2 Type Connectors for Radio Frequency Coaxial Cables (R 1985). 22 pp.
- JIS C 5413-76. C 03 Type Connectors for Radio Frequency Coaxial Cables (R 1985). 22 pp.
- JIS C 5414-76. C 04 Type Connectors for Radio Frequency Coaxial Cables (R 1985). 21 pp.
- JIS C 5415-76. C 05 Type Connectors for Radio Frequency Coaxial Cables (R 1985). 16 pp.
- JIS C 5419-76. C 11 Type Connectors for Radio Frequency Coaxial Cables (R 1985). 15 pp.

— 50 Ohms
- IEC 169 Pt 7-75. Radio-Frequency Connectors Part 7: R.F. Coaxial Connectors with Inner Diameter of Outer Conductor 9,5 mm (0.374 in) with Bayonet Lock—Characteristic Impedance 50 ohms (Type C) First Edition; (Amendment 1-1993). 51 pp.

—Plugs
- MOD UK DSTAN 59-35: Pt 4:Sec 6-76. Plugs and Sockets, Electrical Part 4: Radio Frequency Coaxial Connectors (See Also EPIC Database) Section 6: Detail Specification Series C Issue 1. 46 pp.

—Receptacles
- MOD UK DSTAN 59-35: Pt 4:Sec 6-76. Plugs and Sockets, Electrical Part 4: Radio Frequency Coaxial Connectors (See Also EPIC Database) Section 6: Detail Specification Series C Issue 1. 46 pp.

C Connectors
Use: C Coaxial Connectors

C Language
See Also: Programming Languages
- BSI BS EN 29899-93. 1993 Amd 0 Programming Languages—C (ISO 9899: 1990) (AMD 7933) September 15, 1993 (S). 233 pp.
- BSI BS ISO/IEC 9899-90. 1990 Programming Languages—C (Renumbered as BS EN 29899: 1993). 230 pp.
- CEN EN 29899-93. Programming Languages—C (ISO/IEC 9899: 1990). 228 pp.
- IEC 9899-90. Programming Languages—C First Edition (CEN EN 29899: 1993). 227 pp.
- ISO 9899-90. Programming Languages—C First Edition (CEN EN 29899: 1993). 227 pp.
- JTC1 9899-90. Programming Languages—C First Edition. 227 pp.
- OSI ISO/IEC 9899-90. Programming Languages—C. 226 pp.
- OSI ISO IEC DIS 9899-89. Programming Languages—C. 225 pp.
- SAA AS 3955-91. Programming Languages—C (ISO/IEC 9899:1990) (in Professional Package 26A). 219 pp.
- SNZ NZS/ISO-IEC 9899-90. Programming Languages—C. 219 pp.

—Language Binding—Graphical Kernel System
- BSI BS ISO/IEC 8651-4-91. 1991 Information Technology—Computer Graphics—Graphical Kernel System (GKS) Language Bindings—Part 4: C. 205 pp.
- IEC 8651 Pt 4-91. Information Technology—Computer Graphics—Graphical Kernel System (GKS) Language Bindings—Part 4: C First Edition. 202 pp.
- ISO 8651 Pt 4-91. Information Technology—Computer Graphics—Graphical Kernel System (GKS) Language Bindings—Part 4: C First Edition. 202 pp.
- JTC1 8651-91. Information Technology—Computer Graphics—Graphical Kernel System (GKS) Language Bindings—Part 4: C First Edition. 202 pp.
- OSI ISO/IEC DIS 8651-4-90. Information Technology—Computer Graphics—Graphic Kernel System (GKS) Language Bindings—Part 4: C. 189 pp.
- OSI ISO/IEC DIS 8806-4-90. Information Technology—Computer Graphics—Graphical Kernel System for Three Dimensions (GKS-3D) Language Bindings—Part 4: C. 253 pp.

—Language Binding—Portable Common Tool Environment
- ECMA ECMA 158-93. Portable Common Tool Environment (PCTE)—C Programming Language Binding. 164 pp.
- IEC DIS 13719 Pt 2-93. Information Technology—Portable Common Tool Environment (PCTE)—Part 2: C Programming Language Binding ***CD-ROM ONLY***. 157 pp.
- ISO DIS 13719 Pt 2-93. Information Technology—Portable Common Tool Environment (PCTE)—Part 2: C Programming Language Binding ***CD-ROM ONLY***. 157 pp.
- JTC1 DIS 13719 Pt 2-93. Information Technology—Portable Common Tool Environment (PCTE)—Part 2: C Programming Language Binding ***CD-ROM ONLY***. 157 pp.

—Language Binding—Programmers' Hierarchical Interactive Graphics
- BSI BS EN 28806-4-93. 1993 Amd 0 Information Technology—Computer Graphics—Graphical Kernel System for Three Dimensions (GKS-3D) Language Bindings Part 4: C (ISO/IEC 8806-4: 1991) (AMD 7932) September 15, 1993 (S). 276 pp.
- BSI BS ISO/IEC 8806-4-91. 1991 Information Technology—Computer Graphics—Graphical Kernel System for Three Dimensions (GKS-3D) Language Bindings Part 4: C (Renumbered as BS EN 28806-4: 1993). 273 pp.
- CEN EN 28806-4-93. Information Technology—Computer Graphics—Graphical Kernel System for Three Dimensions (GKS-3D) Language Bindings—Part 4: C (ISO/IEC 8806-4: 1991). 271 pp.
- IEC 8806 Pt 4-91. Information Technology—Computer Graphics—Graphical Kernel System for Three Dimensions (GKS-3D) Language Bindings—Part 4: C First Edition (CEN EN 28806-4: 1993). 270 pp.
- IEC 9593 Pt 4 Draft AMD 1. Information Technology—Computer Graphics—Programmer's Hierarchical Interactive Graphics System (PHIGS) Language Bindings—Part 4: C; (1992) ***CD-ROM ONLY***. 167 pp.
- ISO 8806 Pt 4-91. Information Technology—Computer Graphics—Graphical Kernel System for Three Dimensions (GKS-3D) Language Bindings—Part 4: C First Edition (CEN EN 28806-4: 1993). 270 pp.

C Language *(Cont.)*

—Language Binding—Programmers' Hierarchical Interactive Graphics *(Cont.)*
- ISO 9593 Pt 4 Draft AMD 1. Information Technology—Computer Graphics—Programmer's Hierarchical Interactive Graphics System (PHIGS) Language Bindings—Part 4: C; (1992) ***CD-ROM ONLY***. 167 pp.
- JTC1 8806-91. Information Technology—Computer Graphics—Graphical Kernel System for Three Dimensions (GKS-3D) Language Bindings—Part 4: C First Edition. 270 pp.
- OSI ISO/IEC DIS 9593-4-90. Information Technology—Computer Graphics—Programmer's Hierarchical Interactive Graphics System (PHIGS) Language Bindings—Part 4: C. 298 pp.

—POSIX
- BSI BS EN 29945-1-93. 1993 Amd 0 Information Technology—Portable Operating System Interface (POSIX)—Part 1: System Application Program Interface (API) (C Language) (ISO/IEC 9945-1: 1990) (AMD 7934) September 15,1993 (S). 373 pp.
- BSI BS ISO/IEC 9945-1-90. 1990 Information Technology—Portable Operating System Interface (POSIX)—Part 1: System Application Program Interface (API) (C Language) (Renumbered as BS EN 29945-1: 1990). 369 pp.
- CEN EN 29945-1-93. Information Technology—Portable Operating System Interface (POSIX)—Part 1: System Application Program Interface (API) (C Language) (ISO/IEC 9945-1: 1990). 369 pp.
- OSI ISO/IEC 9945-1-90. Information Technology—Portable Operating System Interface (POSIX)—Part 1: System Application Program Interface (API) (C Language). 367 pp.
- SAA AS 2976.1-91. Information Processing—Portable Operating System Interface (POSIX)—Part 1: System Application Program Interface (API)—C Language (ISO/IEC 9945-1:1990) (in Professional Package 26A).
- SAA AS 3976.1-91. Informaiton Processing—Portable Operating System Interface (POSIX) —Part 1: System Application Program Interface (API)—C Language (ISO/IEC 9945-1:1990) (in Professional Package 26A). 340 pp.
- SNZ NZS/ISO-IEC 9945.1-90. Information Processing—Portable Operating System Interface (POSIX) Part 1: System Application Program Interface (API)—Language. 340 pp.

C Rivets
Use: Blind Rivets

CAA
Use For: Civil Aviation Authority

—Air Navigation—Orders and Regulations
- CAA. Contents 11.91. 1 p.
- CAA. Foreword 11.89. 1 p.
- CAA Section 3 CAP 393. References in the Air Navigation (General) Regulations 1981 to the Air Navigation Order. 34 pp.
- CAA Section 5 CAP 393. CAA Regulations 1991. 31 pp.

—Approved Authorities
- CAA Chapter N5-1 08.90. Approved Organisations (Noise). 1 p.
- CAA. Name Changes. 1 p.
- CAA. Approval Cancellations. 2 pp.
- CAA Part 1 03.93 CAP 475. Organisations Approved in Accordance with BCAR, Section A, Subsection A8. 159 pp.
- CAA Part 2 03.93 CAP .475. Organisations Approved in Accordance with BHSR, Section A, Subsection A2. 5 pp.
- CAA Part 3 03.93 CAP 475. Organisations Approved in Accordance with JAR, Part 145. 5 pp.
- CAA Part 4 03.93 CAP 475. Organisations Conducting Approved AB-Initio Maintenance Engineer Training. 1 p.
- CAA App 1 03.93 CAP 475. List of CAA Safety Regulation Group Supervilsing Area Offices (UK). 1 p.
- CAA App 2 03.93 CAP 475. List of CAA Safety Regulation Group Supervising Area Offices (Overseas). 1 p.
- CAA App 3 03.93 CAP 475. List of CAA Safety Regulation Group Supervising HQ (Gatwick) Offices. 1 p.
- CAA INFORMATION Sheet. JAR 145 Approved Maintenance Organisations. 2 pp.

—Approved Authorities—Ground Effect Machines
- CAA Chapter A2-1 08.81. General. 4 pp.
- CAA. Introductory Note 4.79.

—British Civil Airworthiness Requirements (BCAR)
- CAA. General Contents Issue 97. 2 pp.
- CAA. General Forward Issue 56. 3 pp.

INTERNATIONAL AND NON-U.S. NATIONAL STANDARDS
SUBJECT INDEX

CAA *(Cont.)*
—**British Civil Airworthiness Requirements (BCAR)** *(Cont.)*
CAA. Cross-Reference Index 06.90. 4 pp.

—**British Civil Airworthiness Requirements (BCAR)—Amendment List**
CAA. Current List 05.93 (Blue Papers). 4 pp.

—**Bureau Veritas Agreement**
CAA NOTICE #34 ISSUE 6. Civil Aviation Authority—Bureau Veritas Agreement (Airworthiness Notices). 4 pp.

—**Regulations**
CAA Section 5 CAP 393. CAA Regulations 1991. 31 pp.

Cabbages
See Also: Food; Vegetables

—**Grading**
CNS N1026-82. Grades and Packaging of Cabbages (Jan)(2093). 3 pp.

—**Packaging**
CNS N1026-82. Grades and Packaging of Cabbages (Jan)(2093). 3 pp.

—**Refrigerated Transportation**
ISO 2167-91. Round-Headed Cabbage—Guide to Cold Storage and Refrigerated Transport Second Edition. 7 pp.

—**Spiced**
CNS N5040-82. Spice Cabbage (Jun)(2135). 1 p.

—**Storage**
ISO 2167-91. Round-Headed Cabbage—Guide to Cold Storage and Refrigerated Transport Second Edition. 7 pp.
ISO 6000-81. Round-Headed Cabbage—Storage in the Open First Edition. 7 pp.
ISO 6822-84. Potatoes, Root Vegetables and Round-Headed Cabbages—Guide to Storage in Silos Using Forced Ventilation First Edition. 6 pp.

Cabinet Locks
Use: Locks (Security)—Cabinet

Cabinets (Electrical)
See Also: Cabinets (Furniture); Electrical Enclosures; Racks (Electrical)
BSI BS 5954: Part 2-85. 1985 Dimensions of Mechanical Structures of the 482.6mm (19in) Series Part 2: Cabinets and Pitches of Rack Structures. 6 pp.
CENELEC HD 493.2 S1-88. Dimensions of Mechanical Structures of the 482.6 mm (19 in) Series—Part 2: Cabinets and Pitches of Rack Structures. 3 pp.
IEC 297 Pt 2-82. Dimensions of Mechanical Structures of the 482.6 mm (19 Inch) Series Part 2: Cabinets and Pitches of Rack Structures First Edition. 13 pp.

—**Cable Distribution**
DIN VDE 0660 Pt 503-86. Switchgear and Control Gear; Low Voltage Switchgear and Control Gear Assemblies; Additional Specification for Cable Distribution Equipment (July). 11 pp.

—**Slides—Glossaries**
IEC 916-88. Mechanical Structures for Electronic Equipment Terminology First Edition. 16 pp.

—**Swing Frame—Glossaries**
IEC 916-88. Mechanical Structures for Electronic Equipment Terminology First Edition. 16 pp.

—**Telecommunication Equipment**
CENELEC PRETS 300 119-2-90. European Telecommunication Standard for Equipment Practice Part 2: Engineering Requirements for Racks. 14 pp.
CENELEC PRETS 300 119-2-92. Equipment Engineering (EE); European Telecommunication Standard for Equipment Practice Part 2: Engineering Requirements for Racks and Cabinets. 14 pp.
CENELEC PRETS 300 119-2-92. Equipment Engineering (EE); European Telecommunication Standard for Equipment Practice Part 2: Engineering Requirements for Racks and Cabinets. 13 pp.
CENELEC PRETS 300 119-3-92. Equipment Engineering (EE); European Telecommunication Standard for Equipment Practice Part 3: Engineering Requirements for Miscellaneous Racks and Cabinets. 15 pp.
ETSI PRETS 300 119-2-90. European Telecommunication Standard for Equipment Practice Part 2: Engineering Requirements for Racks (T/TM 02-13 Part B) (T/TM 02-13 Part A). 13 pp.

Cabinets (Electrical) *(Cont.)*
—**Telecommunication Equipment** *(Cont.)*
ETSI PRETS 300 119-2-92. Equipment Engineering (EE); European Telecommunication Standard for Equipment Practice Part 2: Engineering Requirements for Racks and Cabinets. 13 pp.
ETSI PRETS 300 119-3-92. Equipment Engineering (EE); European Telecommunication Standard for Equipment Practice Part 3: Engineering Requirements for Miscellaneous Racks and Cabinets. 15 pp.

—**Wooden—Finishes**
CNS C6107-81. Finish Test of Wooden Cabinets for Electronic Equipments (Apr)(7228).

Cabinets (Furniture)
Scope Note: For additional listings, use a more specific term *Use For:* Storage Cabinets
See Also: Cabinets (Electrical); Card Cabinets; Containment Cabinets; Countertops; Cupboards; Filing Cabinets; Furniture; Kitchen Cabinets; Medicine Cabinets; Office Furniture; Warming Cabinets (Medical)
BSI BS 6250: Part 3-91. 1991 Domestic and Contract Furniture Part 3: Specification for Performance Requirements for Cabinet Furniture. 16 pp.
BSI BS 6250: Part 3-01. 1991 Amd 1 Domestic and Contract Furniture Part 3: Specification for Performance Requirements for Cabinet Furniture (AMD 6951) February 28, 1992 (L). 17 pp.
BSI BS 6250: Part 3-02. 1991 Amd 2 Domestic and Contract Furniture Part 3: Specification for Performance Requirements for Cabinet Furniture (AMD 7056) May 1, 1992. 18 pp.

—**Guns**
BSI BS 7558-92. 1992 Gun Cabinets. 11 pp.

—**Hardware**
CGSB CAN/CGSB-69.25-M90. Cabinet Hardware (ANSI/BHMA A156.9-1982); (Amendment 1 March 1993). 58 pp.

—**Office—Steel**
CNS S1067-84. Office Steel Storage Cabinets (Feb)(2996).
JIS S 1034-91. Office Furniture—Steel Storage Cabinets. 25 pp.

—**Office—Wooden**
JIS S 1024-89. Office Wooden Furniture (Cabinet).

—**Schools—Cleaning Equipment**
JIS S 1085-81. School Furnitures (Cabinet for Cleaning—Tools).

—**Steel**
CGSB CAN/CGSB-44.17-M89. Stationery Storage Cabinet, Steel. 12 pp.

Cabins (Ships)
—**Doors**
JIS F 2334-87. Ships' Cabin Hollow Doors.
JIS F 2334-73. Ships' Cabin Hollow Door Units. 10 pp.

Cable Assemblies (Electric)
See Also: Cables (Electric); Cabling Systems (Electric Power); Coaxial Cable Assemblies; Fiber Optic Cable Assemblies
MOD UK DSTAN 59-82-86. Wiring Harness and Connector Cable Assembles Generic Data and Methods of Test for Capability Approval Issue 1. 50 pp.
MOD UK DSTAN 59-95: Part 1-01. Connector/Cable Assemblies General Requirements and Method of Test Issue 1; Amendment 1 (Obsolescent). 18 pp.

—**Aircraft—Naval Ships**
MOD UK NES 642-89. Requirements to Aircraft Starting and Servicing Cable Assemblies Issue 4 (04.89). 21 pp.
MOD UK NES 642-92. Requirements for Aircraft Starting and Servicing Cable Assemblies Issue 5 (10.92). 21 pp.

—**Construction—Marine**
MOD UK NES 2067-90. Specification for High Strength AHG Buoyant Wire Aerial Cable Assembly Issue 1 (03.90). 25 pp.
MOD UK NES 2068-90. Specification for 0020 Towed Buoy Cable Assembly Issue 1 (06.90). 25 pp.
MOD UK NES 2069-90. Specification for 0030 Towed Buoy Cable Assembly Issue 1 (04.90). 24 pp.

—**Impulse Voltage Testing**
DIN VDE 0472 Pt 511-85. Testing of Cables, Wires and Flexible Cords; Impulse Test (Aug). 7 pp.

Cable Assemblies (Electric) *(Cont.)*
—**Packaging**
MOD UK DSTAN 81-66-90. Packaging of Electrical Cable Assemblies and Wiring Harnesses Issue 1. 12 pp.

—**Quality Assurance**
BSI BS 9550-89. (OBSOLESCENT) 1989 Capability Approval Procedures for d.c. and Low Frequency Connector Cable Assemblies and Wiring Harnesses: Generic Data. 24 pp.

Cable Assemblies (Mechanical)
See Also: Cables (Mechanical)

—**Aircraft**
SBAC AS 2818 ISSUE 4. Cable Assembly. 1 p.
SBAC AS 41092 ISSUE 8. Cable Assembly, Retaining.

—**Splicing—Aircraft**
CAA LEAFLET 2-12 07.90. Cable—Splicing and Swaging. 10 pp.

—**Swaging—Aircraft**
CAA LEAFLET 2-12 07.90. Cable—Splicing and Swaging. 10 pp.

Cable Breakers
Use: Cable Cutters

Cable Cars
See Also: Rolling Stock

—**Brake Blocks—Emergency**
JIS E 9201-68. Emergency Brake Blocks for Cable Cars.

—**Mechanical Guides**
JIS E 9202-72. Guide Rollers for Cable Cars.

—**Wire Rope—Sockets—Attachment**
JIS E 9210-73. Fixating Method of Rope Sockets on Wire Rope for Cable Cars.

Cable Clamps
Use For: Cord Clamps; Wire Clamps
See Also: Bonding Clamps; Cable Supports; Cable Ties; Cables (Electric); Clamps
CSA CAN/CSA-C22. 2 NO 126-M91. Cable Tray Systems; (Gen Instr 1). 28 pp.
DIN VDE 0220 Pt 3-77. VDE Regulations for Single and Multiple Cable Clamps with Insulated Parts in Electrical Power Cable Installations up to 1000 V (Oct). 19 pp.
IEC 998 Pt 2-1-90. Connecting Devices for Low Voltage Circuits for Household and Similar Purposes Part 2-1: Particular Requirements for Connecting Devices as Seperate Entities with Screw-Type Clamping Units First Edition. 45 pp.
IEC 999-90. Connecting Devices—Safety Requirements for Screw-Type and Screwless -Type Clamping Units for Electrical Copper Conductors First Edition. 59 pp.
MOD UK DSTAN 53-15-01. Clamps, Loop (Rigid) Issue 2; Amendment 1. 11 pp.

—**Aircraft**
SBAC TS 376 ISSUE 1. Cable Clamp Effectiveness.
SBAC ESC 50 ISSUE 2. Connector Assembly: End Fittings (Cable Clamps Without Strain Relief). 2 pp.
SBAC ESC 51 ISSUE 2. Connector Accessory: End Fittings (Cable Clamps, Straight Strain Relief). 2 pp.
SBAC ESC 52 ISSUE 2. Connector Accessory: End Fittings (Cable Clamps, 90 Degrees Strain Relief). 3 pp.

—**Aircraft—Accessories**
SBAC TS 374 ISSUE 1. Test for Accessory Thread Strength.

—**Armored Cables**
CSA CAN/CSA-C22. 2 NO 18-92. Outlet Boxes, Conduit Boxes, and Fittings; (Gen Instr 1 Thru 2). 118 pp.

—**Electrical Flexible Tubing**
CSA CAN/CSA-C22. 2 NO 18-92. Outlet Boxes, Conduit Boxes, and Fittings; (Gen Instr 1 Thru 2). 118 pp.

—**Flexible Metal Conduits**
CSA CAN/CSA-C22. 2 NO 18-92. Outlet Boxes, Conduit Boxes, and Fittings; (Gen Instr 1 Thru 2). 118 pp.

—**Insulation Piercing**
IEC 998 Pt 2-3-91. Connecting Devices for Low Voltage Circuits for Household and Similar Puposes Part 2-3: Particular Requirements for Connecting Devices as Separate Entities with Insulation Piercing Clamping Units First Edition. 39 pp.

INDUSTRY STANDARDS

Cable Clamps (Cont.)

—Loop
- DIN VDE 0220 Pt 3-77. VDE Regulations for Single and Multiple Cable Clamps with Insulated Parts in Electrical Power Cable Installations up to 1000 V (Oct). 19 pp.
- MOD UK DSTAN 53-15-01. Clamps, Loop (Rigid) Issue 2; Amendment 1. 11 pp.

—Loop—Aerospace
- AECMA PREN2901-89. Clamps, Loop (P Type) with Rubber Cushioning in Corrosion Resisting Steel Dimensions-Masses. 6 pp.
- AECMA PREN2902-89. Clamps, Loop (P type) with Rubber Cushioning in Aluminium Alloy Dimensions-Masses. 6 pp.

—Screwless
- IEC 998 Pt 2-2-91. Connecting Devices for Low Voltage Circuits for Household and Similar Puposes Part 2-2: Particular Requirements for Connecting Devices as Separate Entities with Screwless-Type Clamping Units First Edition. 34 pp.

—Ships
- CNS F5116-87. Cable Clamping Devices for Motor Starters and Controllers in Ships (Jun)(11994).

—Wire Rope
- CNS B5070-70. Cable Clamp (14mm) (Tentative) (Dec)(3039).

Cable Clenches
- CNS F3121-81. Cable Clenches (Jul)(7693).
- JIS F 2025-76. Cable Clenches. 11 pp.

Cable Clips
See Also: Cable Supports; Clips

—Filler Caps—Aircraft
- SBAC AS 2813 ISSUE 2. Cable Clip (Filler Cap). 1 p.

—Ships
- CNS F5046-86. Electric Cable Clips for Marine Use (Jan)(9855).

—Turnbuckles—Aircraft
- AECMA PREN2363-81. Locking Clips for Turnbuckles of Control Cables—Dimensions. 4 pp.
- BSI BS EN 2363-89. 1989 Locking Clips for Turnbuckles of Control Cables. Dimensions. 7 pp.
- CEN EN 2363-89. Locking Clips for Turnbuckles of Control Cables Dimensions. 7 pp.

Cable Connectors (Electric)
Use: Connectors

Cable Couplers
Use For: Cable Couplings *See Also:* Cables (Electric); Connectors; Couplers
- BSI BS 196-61. 1961 Amd 4 Protected-Type Non-Reversible Plugs, Socket-Outlets, Cable-Couplers and Appliance Couplers, with Earthing Contacts for Single Phase a.c. Circuits up to 250 Volts. 64 pp.
- IEC 501-75. Safety Requirements for Arc Welding Equipment—Plugs, Socket-Outlets and Couplers for Welding Cables First Edition. 13 pp.

—Aerospace—Common Mode Rejection
- CEN PREN 2591 (Part G3)-92. Elements of Electrical and Optical Connection Test Methods Part G3—Common Mode Rejection. 4 pp.

—Aerospace—Distortion (Electrical)
- CEN PREN 2591 (Part G2)-92. Elements of Electrical and Optical Connection Test Methods Part G2—Measurement of Signal Distortion. 3 pp.

—Aerospace—Endurance Testing
- CEN PREN 2591 (Part CG1)-92. Elements of Electrical and Optical Connection Test Methods Part CG1—Endurance Test. 4 pp.

—Aerospace—Impedance Measurement
- CEN PREN 2591 (Part G1)-92. Elements of Electrical and Optical Connection Test Methods Part G1—Measurement of Open Circuit Impedance. 3 pp.

—Aerospace—Input Impedance
- CEN PREN 2591 (Part G5)-92. Elements of Electrical and Optical Connection Test Methods Part G5—Measurement of Stub Input Impedance. 3 pp.

—Aerospace—Tensile Testing
- CEN PREN 2591 (Part G9)-92. Elements of Electrical and Optical Connection Test Methods Part G9—Tensile Strength. 3 pp.

—Aerospace—Thermal Stresses
- CEN PREN 2591 (Part G9)-92. Elements of Electrical and Optical Connection Test Methods Part G9—Tensile Strength. 3 pp.

Cable Couplers (Cont.)

—Aerospace—Transfer Impedance
- CEN PREN 2591 (Part G8)-92. Elements of Electrical and Optical Connection Test Methods Part G8—Measurement of Surface Transfer Impedance. 4 pp.

—Aerospace—Turns Ratio
- CEN PREN 2591 (Part G4)-92. Elements of Electrical and Optical Connection Test Methods Part G4—Measurement of Turns Ratio. 3 pp.

—Appliances
- BSI BS 196-61. 1961 Amd 4 Protected-Type Non-Reversible Plugs, Socket-Outlets, Cable-Couplers and Appliance Couplers, with Earthing Contacts for Single Phase a.c. Circuits up to 250 Volts. 64 pp.

—Arc Welding Cables
- IEC 501-75. Safety Requirements for Arc Welding Equipment—Plugs, Socket-Outlets and Couplers for Welding Cables First Edition. 13 pp.

—Industrial
- DIN VDE 0623 & 0623A-77. Regulations for Plugs, Socket-Outlets, Couplers and Connectors for Industrial Purposes up to 200 A and up to 750 V (Mar). 68 pp.

—Mining Equipment
- BSI BS 3454-84. 1984 1.9/3.3 kV, 300 A Bolted Flameproof Cable Couplers and Adaptors (Including 380/660 V and 640/1100 V, 300 A Adaptors), Primarily for Use in Mines. 25 pp.
- BSI BS 3905-84. 1984 3.8/6.6 kV, 300 A Bolted Flameproof Cable Couplers and Adapters, Primarily for Use in Mines. 24 pp.
- BSI BS 7019-89. 1989 Fully Screened 400 A Bolted Flameproof Cable Couplers and Adaptors for Use on Systems up to and Including 6.6 kV, Primarily for Use in Mines. 26 pp.
- SAA AS 1300-89. Electrical Equipment for Coal Mines—Bolted Flameproof Cable Coupling. 12 pp.

—Motor Vehicles—Safety
- BSI BS EN 50066-93. 1993 Mini-Couplers for the Interconnection of Electrical Mains Supplied Equipment in Road Vehicles (E). 23 pp.
- CENELEC EN 50066-92. Mini-Couplers for the Interconnection of Electrical Mains Supplied Equipment in Road Vehicles. 26 pp.

Cable Couplings
Use: Cable Couplers

Cable Cranes
See Also: Cranes
- ISO 4310-81. Cranes—Test Code and Procedures First Edition. 6 pp.

Cable Cutters
Use For: Cable Breakers; Cable Shears; Cable Slicers; Swagers *See Also:* Cutting Tools; Tools

—Safety
- CEN PREN 792-10-92. Handheld Non-Electric Power Tools—Safety Requirements—Part 10: Compression Power Tools, Squeeze Riveters, Presses, Punches. 6 pp.

Cable Distribution Cabinets
Use: Cabinets (Electrical)—Cable Distribution

Cable Distribution Networks
Use: Cable Networks

Cable Ducts
Use: Electric Conduits

Cable Fittings
See Also: Cable Glands; Stuffing Tubes
- CSA CAN/CSA-C22. 2 NO 126-M91. Cable Tray Systems; (Gen Instr 1). 28 pp.
- DIN VDE 0278 Pt 1-91. Power Cable Accessories with Nominal Voltages U up to 30 kV (Um up to 36 kV); Requirements and Test Methods (Feb) (Replacement for DIN 57278 Part 1/VDE 0278 Part 1/06.80 and DIN 57278 Part 100/VDE 0278 Part 100/10.82). 30 pp.

—Aircraft
- BSI SP 49-50-53. 1953 Ball Ends for Swaging on Cable for Aircraft. 2 pp.
- SBAC AS 2826 ISSUE 3. End for Cable. 1 p.
- SBAC AS 2902-2903 ISSUE 2. Fits for Ball Bearing. 1 p.
- SBAC AS 2904-2905 ISSUE 2. 1/4 Inch Ball End. 1 p.
- SBAC AGS 1713 ISSUE 4. L.T. Cable End.
- SBAC AGS 1767 ISSUE 2. End Fitting Cable, for Magneto Low Tension Connection.

Cable Fittings (Cont.)

—Aircraft—Identification Systems
- BSI SP 51-52-53. 1953 Amd 2 Identification Tags for Swaged Cable-End Assemblies for Aircraft. 1 p.

—Control Cables—Aircraft
- BSI SP 33-39-51. 1951 Amd 3 Turnbarrels, Tension Rods and Swaged Cable-End Connections for Aircraft. 23 pp.
- BSI SP 101-104-56. 1956 Amd 1 Turnbarrels, Tension Rods and Swaged Cable-End Connections (Unified Threads) for Aircraft. 12 pp.

—Crosses
- CSA CAN/CSA-C22. 2 NO 126-M91. Cable Tray Systems; (Gen Instr 1). 28 pp.

—Elbows
- CSA CAN/CSA-C22. 2 NO 126-M91. Cable Tray Systems; (Gen Instr 1). 28 pp.

—Mining Equipment
- BSI BS 3454-84. 1984 1.9/3.3 kV, 300 A Bolted Flameproof Cable Couplers and Adaptors (Including 380/660 V and 640/1100 V, 300 A Adaptors), Primarily for Use in Mines. 25 pp.
- DIN VDE 0279-82. Accessories for Underground Mining Cables; Joints Uo/U = 0.6/1 kV (VDE Specification) (Oct). 11 pp.

—Reducers
- CSA CAN/CSA-C22. 2 NO 126-M91. Cable Tray Systems; (Gen Instr 1). 28 pp.

—Steel Wire Rope—Aircraft
- BSI SP 53-53. 1953 Amd 4 Swaged Cable-End Assemblies (B.A. and B.S.F. Threads) for Preformed Steel Wire Rope for Aircraft. 10 pp.
- BSI SP 54-53. 1953 Amd 4 Swaged Cable-End Assemblies (B.A. and B.S.F. Threads) for Preformed Non-Corrodible Steel Wire Rope. 10 pp.
- BSI 2SP 105-55. 1955 Amd 3 Swaged Cable-End Assemblies (UNF Threads) for Preformed Steel Wire Rope (W.9) for Aircraft. 13 pp.
- BSI 2SP 106-55. 1955 Amd 3 Swaged Cable-End Assemblies (UNF Threads) for Preformed Non-Corrodible Steel Wire Rope (W.11) for Aircraft. 10 pp.

—Tees
- CSA CAN/CSA-C22. 2 NO 126-M91. Cable Tray Systems; (Gen Instr 1). 28 pp.

Cable Glands
Use For: Glands (Cables) *See Also:* Cable Fittings; Glands (Seals); Stuffing Boxes
- BSI BS 6121: Part 1-89. 1989 Mechanical Cable Glands Part 1: Metallic Glands. 19 pp.
- BSI BS 6121: Part 1-87. (WITHDRAWN) 1987 Mechanical Cable Glands Part 1: Specification for Metallic Glands. 19 pp.
- BSI BS 6121: Part 2-89. 1989 Mechanical Cable Glands Part 2: Polymeric Glands. 14 pp.

—Appliance—Ships
- JIS F 8801-86. Marine Cable Glands for Electric Watertight Box.

—Bulkheads
- CNS F5045-83. Marine Cable Glands for Bulkhead and Deck (Jan)(9854).
- JIS F 8802-87. Marine Cable Glands for Bulkhead and Deck.

—Corrosion Resistant
- BSI BS 6121: Part 3-90. 1990 Mechanical Cable Glands Part 3: Special Corrosion Resistant Glands. 24 pp.

—Deck Equipment
- CNS F5045-83. Marine Cable Glands for Bulkhead and Deck (Jan)(9854).
- JIS F 8802-87. Marine Cable Glands for Bulkhead and Deck.

—Explosive Atmospheres
- SAA AS 1828-84. Electrical Equipment for Explosive Atmospheres—Cable Glands Amdt 1 November 1985. 11 pp.

—Hazardous Locations
- CSA C22.2 NO 174-M1984. Cables and Cable Glands for Use in Hazardous Locations (R 1992); (Gen Instr 1 Thru 5). 31 pp.

—Inspection
- BSI BS 6121: Part 5-93. 1993 Mechanical Cable Glands Part 5: Code of Practice for Selection, Installation and Inspection of Cable Glands Used in Electrical Installations (E). 10 pp.

INTERNATIONAL AND NON-U.S. NATIONAL STANDARDS
SUBJECT INDEX

Cable

Cable Glands (Cont.)
—Installation
BSI BS 6121: Part 5-93. 1993 Mechanical Cable Glands Part 5: Code of Practice for Selection, Installation and Inspection of Cable Glands Used in Electrical Installations (E). 10 pp.

—Mining Equipment
BSI BS 542-73. (OBSOLESCENT) 1973 Cable Glands and Sealing Boxes for Association with Apparatus for Use at Mines and Quarries. 15 pp.
CSA C22.2 NO 174-M1984. Cables and Cable Glands for Use in Hazardous Locations (R 1992); (Gen Instr 1 Thru 5). 31 pp.

—Ships
CNS F5004-87. Marine Cable Glands for Electric Watertight Box (Nov)(7906).
CNS F5115-87. Conformation Between the Cable Glands and the Cable Specified in CNS 9124 (Jan)(11817).
MOD UK NES 512: Part 11-01. Guide to Cables, Electrical and Associated Items Part 11: Glands, Grommets and Deck Tubes Issue 2 (03.91); Amendment 1. 81 pp.

—Submarines
MOD UK NES 512: Part 11-01. Guide to Cables, Electrical and Associated Items Part 11: Glands, Grommets and Deck Tubes Issue 2 (03.91); Amendment 1. 81 pp.
MOD UK NES 524: Part 1-88. Guide to the Selection of Cable Glands in Submarines Part 1: Pressure Hull Glands Issue 2 (08.88). 69 pp.
MOD UK NES 524: Part 1-91. Guide to the Selection of Cable Glands in Submarines Part 1: Pressure Hull Glands Issue 3 (12.91). 70 pp.
MOD UK NES 524: Part 2-88. Guide to the Selection of Cable Glands in Submarines Part 2: Pressure Equipment and Disc Seal Glands Issue 1 (08.88). 52 pp.
MOD UK NES 524: Part 2-91. Guide to the Selection of Cable Glands in Submarines Part 2: Pressure Equipment and Disc Seal Glands Issue 2 (12.91). 53 pp.
MOD UK NES 524: Part 3-90. Guide to the Selection of Cable Glands in Submarines Part 3: Bulkhead Penetrators and Gland Tubes Issue 1 (11.90). 19 pp.

—Washers
CNS B2401-80. Washers for Cable Glands (Jan)(5116).

Cable Hangers
See Also: Cable Supports; Cables (Electric); Power Line Hardware

—Plenum Cables
CSA CAN/CSA-C22. 2 NO 18-92. Outlet Boxes, Conduit Boxes, and Fittings; (Gen Instr 1 Thru 2). 118 pp.

—Ships
CNS F5047-83. Electric Cable Hangers and Saddles for Marine Use (Jan)(9856).

Cable Harnesses
Use: Wiring Harnesses

Cable Insulation
Use For: Cable Oversheaths; Cable Sheaths; Sheaths (Cables) *See Also:* Cable Strippers; Electrical Insulation; Sleeving

BSI BS 6469: Sec 99.1-92. 1992 Insulating and Sheathing Materials of Electric Cables Part 99: Test Methods Used in the United Kingdom but Not Specified in Parts 1 to 5 Section 99: Non-Electrical Tests. 14 pp.
BSI BS 6469: Sec 99.2-92. 1992 Insulating and Sheathing Materials of Electric Cables Part 99: Test Methods Used in the United Kingdom but Not Specified in Parts 1 to 5 Section 99: Electrical Tests. 10 pp.
BSI BS 7655: Part 0:-93. 1993 Insulating and Sheathing Materials for Cables Part 0: General Introduction. 6 pp.
DIN VDE 0207 Pt 1-82. Insulating and Sheathing Compounds for Cables and Flexible Cords; List of the Standards in the DIN 57207/VDE 0201 Series (July). 4 pp.
DIN VDE 0299 Pt 1-86. Calculation Method on the Basis of Fictitious Diameters for the Determination of Dimensions of Protective Coverings for Cables and Flexible Cords for Power Installations; Part 1: Power Cables (Mar). 9 pp.
IEC 229-82. Tests on Cable Oversheaths Which Have a Special Protective Function and Are Applied by Extrusion Second Edition. 17 pp.

—Adhesion Testing
DIN VDE 0472 Pt 606-83. Testing of Cables, Wires and Flexible Cords. Wipe Resistance and Absence of Sticking (Aug). 8 pp.

Cable Insulation (Cont.)
—Aerial—Fiber Optic Cables
CCITT RECMN L.10-89. Optical Fibre Cables for Duct, Tunnel, Aerial and Buried Application—Construction, Installation and Protection of Cable and Other Elements of Outside Plant (Study Group VI) 8 pp. 8 pp.

—Aircraft—Adhesion Testing
CEN PREN 3475 (Part 701)-92. Cables, Electrical, Aircraft Use Test Methods Part 701—Strippability and Adherence of Insulation to the Conductor. 3 pp.

—Aircraft—Stripping
CEN PREN 3475 (Part 701)-92. Cables, Electrical, Aircraft Use Test Methods Part 701—Strippability and Adherence of Insulation to the Conductor. 3 pp.

—Aluminum
CCITT RECMN L.4-89. Aluminium Cable Sheaths—Construction, Installation and Protection of Cable and Other Elements of Outside Plant (Study Group VI) 4 pp. 4 pp.

—Clay
BSI BS 2484-85. (OBSOLESCENT) 1985 Straight Concrete Clayware Cable Covers. 12 pp.

—Color Coding
BSI BS 6746C-69. 1969 Amd 1 Colour Chart for Insulation and Sheath of Electric Cables. 4 pp.
CENELEC HD 402 S2-85. Standard Colours for Thermoplastic Materials Used for the Insulation for Low-Frequency Cables and Wires. 2 pp.
DIN VDE 0206-64. Recommendations on Colours for the Overall Sheaths and Coverings of Plastics or Rubber for Cables and Flexible Cords (June). 5 pp.
IEC 304-82. Standard Colours for Insulation for Low-Frequency Cables and Wires Third Edition. 11 pp.

—Concrete
BSI BS 2484-85. (OBSOLESCENT) 1985 Straight Concrete Clayware Cable Covers. 12 pp.

—Density
DIN VDE 0472 Pt 601-83. Testing of Cables, Wires and Flexible Cords; Density (Aug). 8 pp.

—Density—Pycnometric Analysis
BSI BS 6469: Sec 1.3-92. 1992 Insulation and Sheathing Materials of Electric Cables Part 1: Methods of Test for General Application Section 1.3: Methods for Determining the Density—Water Absorption Test—Shrinkage Test. 20 pp.
CENELEC HD 505.1.3 S2-91. Common Test Methods for Insulating and Sheathing Materials of Electric Cables Part 1: Methods for General Application Section Three—Methods for Determining the Density Water Absorption Tests-. 5 pp.

—Direct Burial—Fiber Optic Cables
CCITT RECMN L.10-89. Optical Fibre Cables for Duct, Tunnel, Aerial and Buried Application—Construction, Installation and Protection of Cable and Other Elements of Outside Plant (Study Group VI) 8 pp. 8 pp.

—Elastomeric
CENELEC HD 385 S2-86. Test Methods for Insulations and Sheaths of Electric Cables and Cords (Elastomeric and Thermoplastic Compounds). 2 pp.

—Elastomeric—Electrical Properties
BSI BS 7655: Sec 1.1-93. 1993 Insulating and Sheathing Materials for Cables Part 1: Elastomeric Insulating Compounds Section 1.1: Harmonized Types. 7 pp.
BSI BS 7655: Sec 1.2-93. 1993 Insulating and Sheathing Materials for Cables Part 1: Elastomeric Insulating Compounds Section 1.2: General 90 Degrees Celsius Application. 7 pp.
BSI BS 7655: Sec 1.3-93. 1993 Insulating and Sheathin Cables Part 1: Elastomeric Insulating Compounds Section 1.3: XLPE. 6 pp.

—Elastomeric—Heat Resistant—Electrical Properties
BSI BS 7655: Sec 1.5-93. 1993 Insulating and Sheathing Materials for Cables Part 1: Elastomeric Insulating Compounds Section 1.5: Flame Retardant Composites. 8 pp.

—Elastomeric—Heat Resistant—Physical Properties
BSI BS 7655: Sec 1.5-93. 1993 Insulating and Sheathing Materials for Cables Part 1: Elastomeric Insulating Compounds Section 1.5: Flame Retardant Composites. 8 pp.

Cable Insulation (Cont.)
—Elastomeric—Heat Resistant—Physical Properties (Cont.)
BSI BS 7655: Sec 2.2-93. 1993 Insulating and Sheathing Materials for Cables Part 2: Elastomeric Sheathing Compounds Section 2.2: Heat Resisting Types. 6 pp.
BSI BS 7655: Sec 2.5-93. 1993 Insulating and Sheathing Materials for Cables Part 2: Elastometric Insulating Compounds Section 2.5: Sheathing Compounds for Low Smoke and Acid Gas Emission for General Application. 6 pp.

—Elastomeric—Hot Set Testing
BSI BS 6469: Sec 2.1-92. 1992 Insulation and Sheathing Materials of Electric Cables Part 2: Methods of Test Specific to Elastomeric Compounds Section 2.1: Ozone Resistance Test—Hot Set Test—Mineral Oil Immersion Test. 19 pp.
CENELEC HD 505.2.1 S1-88. Common Test Methods for Insulating and Sheathing Materials of Electric Cables Part 2: Methods Specific to Elastomeric Compounds Section One—Ozone Resistance Test—Hot Set Test—Mineral Oil Immersion Test. 3 pp.
IEC 811 Pt 2-1-86. Common Test Methods for Insulating and Sheathing Materials of Electric Cables Part 2: Methods Specific to Elastomeric Compounds Section 1: Ozone Resistance Test—Hot Set Test—Mineral Oil Immersion Test First Edition; (Amendment 1-1992) (Amendment 2-1993). 39 pp.

—Elastomeric—Immersion Testing
IEC 811 Pt 2-1-86. Common Test Methods for Insulating and Sheathing Materials of Electric Cables Part 2: Methods Specific to Elastomeric Compounds Section 1: Ozone Resistance Test—Hot Set Test—Mineral Oil Immersion Test First Edition; (Amendment 1-1992) (Amendment 2-1993). 39 pp.

—Elastomeric—Ozone Resistance
DIN VDE 0472 Pt 805-91. Testing of Cables, Wires and Flexible Cords; Ozone Resistance (Jan). 13 pp.
DIN VDE 0472 Pt 805 A1 (D)-91. Testing of Cables, Wires and Flexible Cords; Ozone Resistance; Amendment 1 (Sept). 5 pp.
IEC 811 Pt 2-1-86. Common Test Methods for Insulating and Sheathing Materials of Electric Cables Part 2: Methods Specific to Elastomeric Compounds Section 1: Ozone Resistance Test—Hot Set Test—Mineral Oil Immersion Test First Edition; (Amendment 1-1992) (Amendment 2-1993). 39 pp.

—Elastomeric—Physical Properties
BSI BS 7655: Sec 1.1-93. 1993 Insulating and Sheathing Materials for Cables Part 1: Elastomeric Insulating Compounds Section 1.1: Harmonized Types. 7 pp.
BSI BS 7655: Sec 1.2-93. 1993 Insulating and Sheathing Materials for Cables Part 1: Elastomeric Insulating Compounds Section 1.2: General 90 Degrees Celsius Application. 7 pp.
BSI BS 7655: Sec 1.3-93. 1993 Insulating and Sheathin Cables Part 1: Elastomeric Insulating Compounds Section 1.3: XLPE. 6 pp.
BSI BS 7655: Sec 1.6-93. 1993 Insulating and Sheathing Materials for Cables Part 1: Elastomeric Insulating Compounds Section 1.6: Coil End Lead Types. 6 pp.
BSI BS 7655: Sec 2.1-93. 1993 Insulating and Sheathing Materials for Cables Part 2: Elastomeric Insulating Compounds Section 2.1: Harmonized Types. 7 pp.
BSI BS 7655: Sec 2.3-93. 1993 Insulating and Sheathing Materials for Cables Part 2: Elastomeric Sheathing Compounds Section 2.3: General Application. 6 pp.
BSI BS 7655: Sec 2.4-93. 1993 Insulating and Sheathing Materials for Cables Part 2: Elastomeric Insulating Compounds Section 2.4: Welding Cable Covering. 6 pp.
BSI BS 7655: Sec 2.6-93. 1993 Insulating and Sheathing Materials for Cables Part 2: Elastomeric Insulating Compounds Section 2.6: Sheathing Compounds for Ships Wiring and Offshore Applications. 7 pp.

—Electrical Faults
DIN VDE 0472 Pt 514-85. Testing of Cables, Wires and Flexible Cords; Faults on Insulation (May). 7 pp.

—Electrical Resistivity
DIN VDE 0472 Pt 502-91. Testing of Cables & Insulated Flexible Cords; Insulation Resistance and Volume Resistivity (Nov) (Supersedes DIN 57 472 Part 502). 15 pp.

—Elongation
DIN VDE 0472 Pt 602-83. Testing of Cables and Insulated Cords; Tensile Strength and Elongation at Break (Aug). 13 pp.
DIN VDE 0472 Pt 616-84. Testing of Cables, Wires and Flexible Cords. Elongation Test at Low Temperature (Oct). 7 pp.

INDUSTRY STANDARDS

Cable Insulation (Cont.)

—Elongation—Fiber Optic Cables
CNS C6233-84. Method of Test for Fiber Optic Devices (Cable Jacket Elongation) (Jun)(10933).

—Environmental—Fiber Optic Cables
CCITT RECMN L.10-89. Optical Fibre Cables for Duct, Tunnel, Aerial and Buried Application—Construction, Installation and Protection of Cable and Other Elements of Outside Plant (Study Group VI) 8 pp. 8 pp.

—Evaporation Loss
DIN VDE 0472 Pt 612-83. Testing of Cables and Insulated Cords; Weight Loss by Evaporation (Aug). 8 pp.

—Fiber Optic Cables—Adhesion Testing
CNS C6284-87. Method of Test for Fiber Optic Devices (FOTP-84 Jacket Self-Adhesion (Blocking) Test for Fiber Optic Cable) (Apr)(11906).

—Fiber Optic Cables—Shrinkage
CNS C6232-84. Method of Test for Fiber Optic Devices (Cable Jacket Shrinkage) (Jun)(10932).

—Flammability Testing—Communication Cables
JIS C 3521-86. Flame Test Method for Flame Retardant Sheath of Telecommunication Cables. 8 pp.

—Fluorocarbon Polymers
DIN VDE 0207 Pt 6-89. Insulating and Sheathing Compounds for Cables and Flexible Cords; Fluorinated Polymers (Feb). 4 pp.

—Halogen Free
DIN VDE 0207 Pt 23-86. Insulating and Sheathing Compounds for Cables and Flexible Cords; Halogen-Free Insulating Compounds (Feb). 5 pp.
DIN VDE 0207 Pt 24-86. Insulating and Sheathing Compounds for Cables and Flexible Cords; Halogen-Free Sheathing Compounds (Feb). 4 pp.

—Halogen Free—Combustion—Corrosion Testing
DIN VDE 0472 Pt 813-83. Testing of Cables, Wires and Flexible Cords; Corrosivity of Combustion Gases (Aug). 6 pp.

—Heat Shrink
MOD UK DSTAN 59-42: Part 1-90. Heat Shrink Solder Sheaths Part 1: Requirements and Test Methods for Capability Approval Issue 1. 30 pp.

—High Temperature Testing
DIN VDE 0472 Pt 632-87. Testing of Cables and Insulated Cords; Behaviour at High Temperatures (Oct) (Partially Supersedes VDE 0271/03.69 Which Was Withdrawn in December 1988). 4 pp.

—Hot Set Testing
DIN VDE 0472 Pt 615-84. Testing of Cables, Wires and Flexible Cords; Hot Set Test) (Oct). 6 pp.

—Identification Systems—Wipe Resistance
DIN VDE 0472 Pt 606-83. Testing of Cables, Wires and Flexible Cords. Wipe Resistance and Absence of Sticking (Aug). 8 pp.

—Indentation Hardness Testing
DIN VDE 0472 Pt 619-83. Testing of Cables, Wires and Flexible Cords; Indentation Strength (Jan). 4 pp.

—Insulation Resistance
DIN VDE 0282 Pt 1000-83. Rubber Cables, Wires and Flexible Cords for Power Installation; Surface Resistance of Sheath (Feb). 4 pp.
DIN VDE 0472 Pt 503-85. Testing of Cables and Insulated Flexible Cords; Surface Insulation Resistance of Sheath (May). 6 pp.

—Joints—Fiber Optic Cables
CCITT RECMN L.13-92. Sheath Joints and Organizers of Optical Fibre Cables in the Outside Plant (Study Group VI) 6 pp. 6 pp.

—Lead
CNS H3142-86. Lead and Lead Alloys for Electric Cable Sheathing (Dec)(11797).

—Lead Alloys
CNS H3142-86. Lead and Lead Alloys for Electric Cable Sheathing (Dec)(11797).

—Lead Alloys—Chemical Analysis
BSI BS 3908: Part 1-65. 1965 Amd 1 Methods for the Sampling and Analysis of Lead and Lead Alloys Part 1: Sampling of Ingot Lead, Lead Alloy Ingots, Sheet Pipe, and Cable Sheathing Alloys. 8 pp.
DIN ENGL 17640 Pt 2-86. Lead Alloys for Cable Sheathing (Jan). 3 pp.

Cable Insulation (Cont.)

—Lead Alloys—Sampling
BSI BS 3908: Part 1-65. 1965 Amd 1 Methods for the Sampling and Analysis of Lead and Lead Alloys Part 1: Sampling of Ingot Lead, Lead Alloy Ingots, Sheet Pipe, and Cable Sheathing Alloys. 8 pp.

—Lead—Chemical Analysis
BSI BS 801-84. 1984 Composition of Lead and Lead Alloy Sheaths of Electric Cables. 4 pp.
BSI BS 3908: Part 1-65. 1965 Amd 1 Methods for the Sampling and Analysis of Lead and Lead Alloys Part 1: Sampling of Ingot Lead, Lead Alloy Ingots, Sheet Pipe, and Cable Sheathing Alloys. 8 pp.

—Lead—Sampling
BSI BS 3908: Part 1-65. 1965 Amd 1 Methods for the Sampling and Analysis of Lead and Lead Alloys Part 1: Sampling of Ingot Lead, Lead Alloy Ingots, Sheet Pipe, and Cable Sheathing Alloys. 8 pp.

—Low Temperature Testing
BSI BS 6469: Sec 1.4-92. 1992 Insulation and Sheathing Materials of Electric Cables Part 1: Methods of Test for General Application Section 1.4: Test at Low Temperature. 22 pp.
CENELEC HD 505.1.4 S1-88. Common Test Methods for Insulating and Sheathing Materials of Electric Cables Part 1 Methods for General Application Section Four—Tests at Low Temperature. 3 pp.
IEC 811 Pt 1-4-85. Common Test Methods for Insulating and Sheathing Materials of Electric Cables Part 1: Methods for General Application Section Four—Tests at Low Temperature First Edition; (Corrigendum—May 1986) (Amendment 1-1993). 40 pp.

—Mechanical Testing
BSI BS 6469: Sec 1.1-92. 1992 Insulation and Sheathing Materials of Electric Cables Part 1: Methods of Test for General Application Section 1.1: Measurement of Thickness and Overall Dimensions—Tests for Determining the Mechanical Properties. 33 pp.
CENELEC HD 505.1.1 S1-88. Common Test Methods for Insulating and Sheathing Materials of Electric Cables Part 1: Methods for General Application Section One—Measurement of Thickness and Overall Dimension Tests for Determining the Mechanical Properties. 3 pp.
CENELEC HD 505.1.1 S2-90. Common Test Methods for Insulating and Sheathing Materials of Electric Cables Part 1: Methods for General Application Section One—Measurement of Thickness and Overall Dimensions—Tests for Determining the Mechanical Properties. 3 pp.
CENELEC HD 505.1.1 S3-91. Common Test Methods for Insulating and Sheathing Materials of Electric Cables Part 1: Methods for General Application Section One—Measurement of Thickness and Overall Dimensions—Tests for Determining the Mechanical Properties. 5 pp.
IEC 811 Pt 1-1-85. Common Test Meth. for Ins. and Sheathing Mat. of Electric Cables Part 1: Meth. for Gen. Application Sec. One—Meas. of Thickness and Overall Dimensions—Tests for Determining the Mech. Properties First Edition; (Corrigendum—May 1986) (Amd 1-1988)(Amd 2-1989). 51 pp.

—Melting Points
DIN VDE 0472 Pt 621-83. Testing of Cables and Insulated Cords; Crystallite Melting Point (Jan). 4 pp.

—Paper
CNS C4209-85. Insulating Papers for Electric Power Cables (Nov)(6042).
JIS C 2307-84. Insulating Papers for Electric Power Cables. 7 pp.
JIS C 2308-84. Insulating Papers for Communication Cables. 7 pp.

—Polyethylene
BSI BS 6469-90. (WITHDRAWN) 1990 Insulation and Sheaths of Electric Cables (Superseded by Various Sections of Parts 1-5 of BS 6469). 43 pp.
DIN VDE 0207 Pt 2-90. Insulating and Sheathing Compounds for Cables and Flexible Cords; Polyethylene Insulating Compounds (Feb). 4 pp.
DIN VDE 0207 Pt 2 (S)-82. Insulating and Sheathing Compounds for Cables and Flexible Cords; Polyethylene Insulating Compounds (July). 6 pp.
DIN VDE 0207 Pt 3-90. Insulating and Sheathing Compounds for Cables and Flexible Cords; Polyethylene Sheathing Compounds (Feb). 4 pp.
DIN VDE 0207 Pt 22-82. Insulating and Sheathing Compounds for Cables and Flexible Cords; XLPE Insulating Compounds (July). 6 pp.
DIN VDE 0472 Pt 404-87. Testing of Cables, Insulated Wires and Flexible Cords; Irregularities of Inner Conductive Layers and Insulating Sleeve (Oct). 6 pp.

Cable Insulation (Cont.)

—Polyethylene—Adhesive Strength
DIN VDE 0472 Pt 618-83. Testing of Cables, Wires and Flexible Cords; Adhesive Strength (Jan). 6 pp.

—Polyethylene—Carbon Black Content Analysis
DIN VDE 0472 Pt 702-89. Testing of Cables, Wires and Flexible Cords; Carbon Black and Filler Content of PE Sheaths (Feb). 8 pp.

—Polyethylene—Environmental Stress Cracking
DIN VDE 0472 Pt 810-85. Testing of Cables, Wires and Flexible Cords; Resistance to Environmental Stress Cracking (Sept). 11 pp.

—Polyethylene—Environmental Testing
BSI BS 6469: Sec 4.1-92. 1992 Insulation and Sheathing Materials of Electric Cables Part 4: Methods of Test Specific to Polyethylene and Polypropylene Compounds Section 4.1: Tesistance to Environmental Stress Cracking—Wrapping Test After Thermal Ageing in. 27 pp.
CENELEC HD 505.4.1 S1-88. Test Methods—Insulating/Sheathing Materials of Electric Cables Part 4: Polyethylene/Polyphlylene Compounds Section One—Resistance —Environmental Stress Cracking—Wrapping Test Thermal Ageing in Air—Measurement of Melt Flow Index—Carbon Black. 3 pp.
CENELEC HD 505.4.1 S2-90. Common Test Methods-Insulating/Sheathing Materials Electric Cables Part 4: Methods Polyethylene/Polypropylene Compounds Sec 1-Resistance Environmental Stress-Cracking/Wrapping Test After Thermal Ageing in Air-Measurement of Melt Flow Index-Carbon Black. 3 pp.
IEC 811 Pt 4-1-85. Common Test Methods for Insulating and Sheathing Materials of Electric Cables Part 4: Methods Specific to Polyethylene and Polypropylene Compounds Section One—Resistance to Environmental Stress Cracking—Wrapping Test After Thermal Ageing in Air—Measurement of the. 52 pp.

—Polyethylene—Hardness Testing
DIN VDE 0472 Pt 631-87. Testing of Cables, Wires and Flexible Cords; Shore D Hardness Test (Oct). 7 pp.

—Polyethylene—Mass Increase
CENELEC HD 505.4.2 S1-92. Common Test Methods for Insulating and Sheathing Materials of Electric Cables Part 4: Methods Specific to Polyethylene and Polypropylene Compounds Section Two—Elongation at Break After Pre-Conditioning—Wrapping Test After Thermal Ageing in Air—Measurement of Mass. 6 pp.
IEC 811 Pt 4-2-90. Common Test Methods for Insulating and Sheathing Materials of Electric Cables Part 4: Methods Specific to Polyethylene and Polypropylene Compounds Section Two—Elongation at Break After Pre-Conditioning—Wrapping Test After Pre-Conditioning—Wrapping Test After. 38 pp.

—Polyethylene—Melt Flow Index
DIN VDE 0472 Pt 607-84. Testing of Cables, Wires and Flexible Cords; Melt Flow Index (Oct). 7 pp.

—Polyethylene—Mineral Fiber Content
DIN VDE 0472 Pt 702-89. Testing of Cables, Wires and Flexible Cords; Carbon Black and Filler Content of PE Sheaths (Feb). 8 pp.

—Polyethylene—Shrinkage
DIN VDE 0472 Pt 630-87. Testing of Cables, Wires and Flexible Cords; Longitudinal Shrinkage of PE Sheaths (Oct). 4 pp.

—Polyethylene—Stability
CENELEC HD 505.4.2 S1-92. Common Test Methods for Insulating and Sheathing Materials of Electric Cables Part 4: Methods Specific to Polyethylene and Polypropylene Compounds Section Two—Elongation at Break After Pre-Conditioning—Wrapping Test After Thermal Ageing in Air—Measurement of Mass. 6 pp.
IEC 811 Pt 4-2-90. Common Test Methods for Insulating and Sheathing Materials of Electric Cables Part 4: Methods Specific to Polyethylene and Polypropylene Compounds Section Two—Elongation at Break After Pre-Conditioning—Wrapping Test After Pre-Conditioning—Wrapping Test After. 38 pp.

INTERNATIONAL AND NON-U.S. NATIONAL STANDARDS
SUBJECT INDEX

Cable Insulation (Cont.)

—Polyethylene—Thermal Aging Testing
CENELEC HD 505.4.2 S1-92. Common Test Methods for Insulating and Sheathing Materials of Electric Cables Part 4: Methods Specific to Polyethylene and Polypropylene Compounds Section Two—Elongation at Break After Pre-Conditioning—Wrapping Test After Thermal Ageing in Air—Measurement of Mass. 6 pp.

DIN VDE 0472 Pt 617-85. Testing of Cables, Wires and Flexible Cords; Mechanical Behaviour of Polyethylene After Thermal Ageing (Nov). 9 pp.

IEC 811 Pt 4-2-90. Common Test Methods for Insulating and Sheathing Materials of Electric Cables Part 4: Methods Specific to Polyethylene and Polypropylene Compounds Section Two—Elongation at Break After Pre-Conditioning—Wrapping Test After Pre-Conditioning—Wrapping Test After. 38 pp.

—Polyethylene—Winding Testing
CENELEC HD 505.4.2 S1-92. Common Test Methods for Insulating and Sheathing Materials of Electric Cables Part 4: Methods Specific to Polyethylene and Polypropylene Compounds Section Two—Elongation at Break After Pre-Conditioning—Wrapping Test After Thermal Ageing in Air—Measurement of Mass. 6 pp.

IEC 811 Pt 4-2-90. Common Test Methods for Insulating and Sheathing Materials of Electric Cables Part 4: Methods Specific to Polyethylene and Polypropylene Compounds Section Two—Elongation at Break After Pre-Conditioning—Wrapping Test After Pre-Conditioning—Wrapping Test After. 38 pp.

—Polyolefin—Elongation
CENELEC HD 505.4.2 S1-92. Common Test Methods for Insulating and Sheathing Materials of Electric Cables Part 4: Methods Specific to Polyethylene and Polypropylene Compounds Section Two—Elongation at Break After Pre-Conditioning—Wrapping Test After Thermal Ageing in Air—Measurement of Mass. 6 pp.

IEC 811 Pt 4-2-90. Common Test Methods for Insulating and Sheathing Materials of Electric Cables Part 4: Methods Specific to Polyethylene and Polypropylene Compounds Section Two—Elongation at Break After Pre-Conditioning—Wrapping Test After Pre-Conditioning—Wrapping Test After. 38 pp.

—Polyolefin—Oxidation Resistance
BSI BS 6469: Sec 4.2-92. 1992 Insulating and Sheathing Materials of Electric Cables Part 4: Methods of Test Specific to Polyethylene and Polypropylene Compounds Section 4.2: Elongation at Break After Pre-Conditioning—Wrapping Test After Pre-Conditioning-Wrapping. 24 pp.

CENELEC HD 505.4.2 S1-92. Common Test Methods for Insulating and Sheathing Materials of Electric Cables Part 4: Methods Specific to Polyethylene and Polypropylene Compounds Section Two—Elongation at Break After Pre-Conditioning—Wrapping Test After Thermal Ageing in Air—Measurement of Mass. 6 pp.

IEC 811 Pt 4-2-90. Common Test Methods for Insulating and Sheathing Materials of Electric Cables Part 4: Methods Specific to Polyethylene and Polypropylene Compounds Section Two—Elongation at Break After Pre-Conditioning—Wrapping Test After Pre-Conditioning—Wrapping Test After. 38 pp.

—Polypropylene
DIN VDE 0207 Pt 7-90. Insulating and Sheathing Compounds for Cables and Flexible Cords; Polypropylene Compounds (Feb). 4 pp.

—Polypropylene—Environmental Testing
BSI BS 6469: Sec 4.1-92. 1992 Insulation and Sheathing Materials of Electric Cables Part 4: Methods of Test Specific to Polyethylene and Polypropylene Compounds Section 4.1: Resistance to Environmental Stess Cracking—Wrapping Test After Thermal Ageing. 27 pp.

CENELEC HD 505.4.1 S1-88. Test Methods—Insulating/Sheathing Materials of Electric Cables Part 4: Polyethylene/Polyphlylene Compounds Section One—Resistance—Environmental Stress Cracking—Wrapping Test Thermal Ageing in Air—Measurement of Melt Flow Index—Carbon Black. 3 pp.

CENELEC HD 505.4.1 S2-90. Common Test Methods-Insulating/Sheathing Materials Electric Cables Part 4: Methods Polyethylene/Polypropylene Compounds Sec 1-Resistance Environmental Stress-Cracking/Wrapping Test After Thermal Ageing in Air-Measurement of Melt Flow Index-Carbon Black. 3 pp.

Cable Insulation (Cont.)

—Polypropylene—Environmental Testing (Cont.)
IEC 811 Pt 4-1-85. Common Test Methods for Insulating and Sheathing Materials of Electric Cables Part 4: Methods Specific to Polyethylene and Polypropylene Compounds Section One—Resistance to Environmental Stress Cracking—Wrapping Test After Thermal Ageing in Air—Measurement of the. 52 pp.

—Polypropylene—Mass Increase
CENELEC HD 505.4.2 S1-92. Common Test Methods for Insulating and Sheathing Materials of Electric Cables Part 4: Methods Specific to Polyethylene and Polypropylene Compounds Section Two—Elongation at Break After Pre-Conditioning—Wrapping Test After Thermal Ageing in Air—Measurement of Mass. 6 pp.

IEC 811 Pt 4-2-90. Common Test Methods for Insulating and Sheathing Materials of Electric Cables Part 4: Methods Specific to Polyethylene and Polypropylene Compounds Section Two—Elongation at Break After Pre-Conditioning—Wrapping Test After Pre-Conditioning—Wrapping Test After. 38 pp.

—Polypropylene—Winding Testing
CENELEC HD 505.4.2 S1-92. Common Test Methods for Insulating and Sheathing Materials of Electric Cables Part 4: Methods Specific to Polyethylene and Polypropylene Compounds Section Two—Elongation at Break After Pre-Conditioning—Wrapping Test After Thermal Ageing in Air—Measurement of Mass. 6 pp.

IEC 811 Pt 4-2-90. Common Test Methods for Insulating and Sheathing Materials of Electric Cables Part 4: Methods Specific to Polyethylene and Polypropylene Compounds Section Two—Elongation at Break After Pre-Conditioning—Wrapping Test After Pre-Conditioning—Wrapping Test After. 38 pp.

—Porosity
CSA C22.2 NO 16-M1986. Insulated Conductors for Power-Operated Electronic Devices (R 1992); (Gen Instr 1 Thru 4). 64 pp.

CSA 1334 Bull. Electrical Bulletin 1334 August 28, 1981 to C22.2 NO 16. 1 p.

CSA C22.2 NO 35-M1987. Extra-Low-Voltage Control Circuit Cables, Low-Energy Control Cable, and Extra-Low-Voltage Control Cable (R 1993); (Gen Instr 1). 28 pp.

CSA 1334 Bull. Electrical Bulletin 1334 August 28, 1981 to C22.2 NO 35. 1 p.

CSA C22.2 NO 38-M1986. Thermoset Insulated Wires and Cables; (Gen Instr 1 Thru 3). 64 pp.

CSA 1334 Bull. Electrical Bulletin 1334 August 28, 1981 to C22.2 NO 38. 1 p.

CSA 1334 Bull. Electrical Bulletin 1334 August 28, 1981 to C22.2 NO 48. 1 p.

CSA CAN/CSA-C22. 2 NO 49-92. Flexible Cords and Cables; (Gen Instr 1). 121 pp.

CSA 1334 Bull. Electrical Bulletin 1334 August 28, 1981 to C22.2 NO 49. 1 p.

CSA C22.2 NO 75-M1983. Thermoplastic Insulated Wires and Cables (R 1992); (Gen Instr 1 Thru 6). 48 pp.

CSA 1334 Bull. Electrical Bulletin 1334 August 28, 1981 to C22.2 NO 75. 1 p.

CSA 1334 Bull. Electrical Bulletin 1334 August 28, 1981 to C22.2 NO 96. 1 p.

CSA C22.2 NO 116-1980. Coil-Lead Wires (R 1992); (Gen Instr 1 Thru 8). 52 pp.

CSA 1334 Bull. Electrical Bulletin 1334 August 28, 1981 to C22.2 NO 116. 1 p.

CSA C22.2 NO 127-1988. Equipment Wires; (Gen Instr 1 Thru 3). 47 pp.

CSA 1334 Bull. Electrical Bulletin 1334 August 28, 1981 to C22.2 NO 127. 1 p.

—Pressure Measurement
DIN VDE 0472 Pt 609-85. Testing of Cables and Insulated Flexible Cords; Pressure Test at High Temperature (June). 11 pp.

—PVC
BSI BS 6469-90. (WITHDRAWN) 1990 Insulation and Sheaths of Electric Cables (Superseded by Various Sections of Parts 1-5 of BS 6469). 43 pp.

CNS K3001-89. Vinyl Chloride Resin Compounds (for Electrical Wires and Cables) (Nov)(1087). 2 pp.

DIN VDE 0207 Pt 4-86. Insulating and Sheathing and Compounds for Cables and Flexible Cords; PVC Insulating and Compounds (June). 8 pp.

DIN VDE 0207 Pt 5-86. Insulating and Sheathing Compounds for Cables and Flexible Cords; PVC Sheathing Compounds (July). 8 pp.

JIS K 6723-83. Plasticized Polyvinyl Chloride Compounds. 14 pp.

Cable Insulation (Cont.)

—PVC—Deformation
BSI BS 6469: Sec 99.1-92. 1992 Insulating and Sheathing Materials of Electric Cables Part 99: Test Methods Used in the United Kingdom but Not Specified in Parts 1 to 5 Section 99: Non-Electrical Tests. 14 pp.

—PVC—Electrical Properties
BSI BS 7655: Sec 3.1-93. 1993 Insulating and Sheathing Materials for Cables Part 3: PVC Insulating Compounds Section 3.1: Harmonized Types. 7 pp.

BSI BS 7655: Sec 3.2-93. 1993 Insulating and Sheathing Materials for Cables Part 3: PVC Insulating Compounds Section 3.2: Hard Grade Types. 6 pp.

BSI BS 7655: Sec 4.2-93. 1993 Insulating and Sheathing Materials for Cables Part 4: PVC Sheathing Compounds Section 4.2: General Application. 7 pp.

BSI BS 7655: Sec 4.3-93. 1993 Insulation and Sheathing Materials for Cables Part 4: PVC Sheathing Compounds Section 4.3: Special Applications-RF Cables. 6 pp.

—PVC—Loss of Mass
BSI BS 6469: Sec 3.2-92. 1992 Insulation and Sheathing Materials of Electric Cables Part 3: Methods of Test Specific to PVC Compounds Section 3.2: Loss of Mass Test—Thermal Stability Test. 19 pp.

CENELEC HD 505.3.2 S1-88. Common Test Methods for Insulating and Sheathing Materials of Electric Cables Part 3: Methods Specific to PVC Compounds Section Two—Loss of Mass Test—Thermal Stability Test. 3 pp.

IEC 811 Pt 3-2-85. Common Test Methods for Insulating and Sheathing Materials of Electric Cables Part 3: Methods Specific to PVC Compounds Section Two—Loss of Mass Test—Thermal Stability Test First Edition; (Corrigendum—May 1986) (Amendment 1-1993). 36 pp.

—PVC—Physical Properties
BSI BS 7655: Sec 3.1-93. 1993 Insulating and Sheathing Materials for Cables Part 3: PVC Insulating Compounds Section 3.1: Harmonized Types. 7 pp.

BSI BS 7655: Sec 3.2-93. 1993 Insulating and Sheathing Materials for Cables Part 3: PVC Insulating Compounds Section 3.2: Hard Grade Types. 6 pp.

BSI BS 7655: Sec 4.1-93. 1993 Insulating and Sheathing Materials for Cables Part 4: PVC Sheathing Compounds Section 4.1: Harmonized Types. 6 pp.

BSI BS 7655: Sec 4.2-93. 1993 Insulating and Sheathing Materials for Cables Part 4: PVC Sheathing Compounds Section 4.2: General Application. 7 pp.

BSI BS 7655: Sec 4.3-93. 1993 Insulating and Sheathing Materials for Cables Part 4: PVC Sheathing Compounds Section 4.3: Special Applications-RF Cables. 6 pp.

—PVC—Pressure Measurement
BSI BS 6469: Sec 3.1-92. 1992 Insulating and Sheathing Materials of Electric Cables Part 3: Methods of Test Specific to PVC Compounds Section 3.1: Pressure Test at High Temperature—Tests for Resistance to Cracking. 21 pp.

CENELEC HD 505.3.1 S1-88. Common Test Methods for Insulating and Sheathing Materials of Electric Cables Part 2: Methods Specific to PVC Compounds Section one—Pressure Test at High Temperature—Tests for Resistance to Cracking. 3 pp.

IEC 811 Pt 3-1-85. Common Test Methods for Insulating and Sheathing Materials of Electric Cables Part 3: Methods Specific to PVC Compounds Section One—Pressure Test At High Temperature—Tests for Resistance to Cracking First Edition; (Corrigendum—May 1986). 31 pp.

—PVC—Splitting Tests
IEC 811 Pt 3-1-85. Common Test Methods for Insulating and Sheathing Materials of Electric Cables Part 3: Methods Specific to PVC Compounds Section One—Pressure Test At High Temperature—Tests for Resistance to Cracking First Edition; (Corrigendum—May 1986). 31 pp.

—PVC—Thermal Testing
BSI BS 6469: Sec 3.2-92. 1992 Insulation and Sheathing Materials of Electric Cables Part 3: Methods of Test Specific to PVC Compounds Section 3.2: Loss of Mass Test—Thermal Stability Test. 19 pp.

CENELEC HD 505.3.2 S1-88. Common Test Methods for Insulating and Sheathing Materials of Electric Cables Part 3: Methods Specific to PVC Compounds Section Two—Loss of Mass Test—Thermal Stability Test. 3 pp.

INTERNATIONAL AND NON-U.S. NATIONAL STANDARDS
SUBJECT INDEX

Cable

Cable Insulation (Cont.)
—PVC—Thermal Testing (Cont.)
IEC 811 Pt 3-2-85. Common Test Methods for Insulating and Sheathing Materials of Electric Cables Part 3: Methods Specific to PVC Compounds Section Two—Loss of Mass Test—Thermal Stability Test First Edition; (Corrigendum—May 1986) (Amendment 1-1993). 36 pp.

—Rubber
DIN VDE 0207-82. Insulating and Sheathing Compounds for Cables and Flexible Cords; Rubber Sheathing Compounds (July). 8 pp.
DIN VDE 0207 Pt 20-82. Insulating and Sheathing Compounds for Cables and Flexible Cords; Rubber Insulating Compounds (July). 9 pp.
DIN VDE 0207 Pt 20 (D)-87. Insulating and Sheathing Compounds for Cables and Flexible Cords; Rubber Insulating Compounds (Oct). 9 pp.
DIN VDE 0207 Pt 21-82. Insulating and Sheathing Compounds for Cables and Flexible Cords; Rubber Sheathing Compounds (VDE Specification) (July). 9 pp.

—Ships
IEC 92 Pt 351-83. Electrical Installations in Ships Part 351: Insulating Materials for Shipboard Power Cables First Edition; (Amendment 1-1992). 23 pp.
IEC 92 Pt 359-87. Electrical Installations in Ships Part 359: Sheathing Materials for Shipboard Power and Telecommunication Cables First Edition. 15 pp.
JIS F 8806-85. Protective Rubber-Like Sheaths of Portable Cord for Marine Use. 7 pp.

—Ships—Acceptance Testing
MOD UK NES 518. Limited Fire Hazard Sheathing for Electric Cables Issue 3 (02.89) (Superseded by Def Stan 61-12: Part 31).

—Shrinkage
BSI BS 6469: Sec 1.3-92. 1992 Insulation and Sheathing Materials of Electric Cables Part 1: Methods of Test for General Application Section 1.3: Methods for Determining the Density—Water Absorption Test—Shrinkage Test. 20 pp.
CENELEC HD 505.1.3 S2-91. Common Test Methods for Insulating and Sheathing Materials of Electric Cables Part 1: Methods for General Application Section Three—Methods for Determining the Density Water Absorption Tests-. 5 pp.
DIN VDE 0472 Pt 628-86. Testing of Cables, Wires and Flexible Cords; Longitudinal Shrinkage of Insulation (Apr). 5 pp.

—Spark Testing
BSI BS 5099-92. 1992 Spark Testing of Electric Cables. 12 pp.
BSI BS 5099-74. 1974 Amd 2 Spark Testing of Electric Cables. 9 pp.

—Steel—Corrugated
CCITT RECMN L.5-89. Cable Sheaths Made of Metals Other Than Lead or Aluminium—Construction, Installation and Protection of Cable and Other Elements of Outside Plant (Study Group VI) 2 pp. 2 pp.

—Tear Strength
BSI BS 6469: Sec 99.1-92. 1992 Insulating and Sheathing Materials of Electric Cables Part 99: Test Methods Used in the United Kingdom but Not Specified in Parts 1 to 5 Section 99: Non-Electrical Tests. 14 pp.

—Telephone Cables—Pressurization
CCITT RECMN L.6-89. Methods of Keeping Cables Under Gas Pressure—Construction, Installation and Protection of Cable and Other Elements of Outside Plant (Study Group VI) 1 pp. 1 p.

—Tensile Testing
DIN VDE 0472 Pt 602-83. Testing of Cables and Insulated Cords; Tensile Strength and Elongation at Break (Aug). 13 pp.

—Thermal Aging Testing
BSI BS 6469: Sec 1.2-92. 1992 Insulation and Sheathing Materials of Electric Cables Part 1: Methods of Test for General Application Section 1.2: Thermal Ageing Methods. 23 pp.
IEC 811 Pt 1-2-85. Common Test Methods for Insulating and Sheathing Materials of Electric Cables Part 1: Methods for General Application Section Two—Thermal Ageing Methods First Edition; (Corrigendum—May 1986) (Amendment 1-1989). 41 pp.

—Thermal Shock
DIN VDE 0472 Pt 608-85. Testing of Cables, Wires and Flexible Cords; Heat Shock Behaviour (Nov). 7 pp.

—Thermal Stability
DIN VDE 0472 Pt 614-85. Testing of Cables, Wires and Flexible Cords; Thermal Stability (Nov). 5 pp.

Cable Insulation (Cont.)
—Thermoplastic
CENELEC HD 385 S2-86. Test Methods for Insulations and Sheaths of Electric Cables and Cords (Elastomeric and Thermoplastic Compounds). 2 pp.

—Thickness Measurement
CSA C22.2 NO 16-M1986. Insulated Conductors for Power-Operated Electronic Devices (R 1992); (Gen Instr 1 Thru 4). 64 pp.
CSA 916 Bull. Electrical Bulletin 916 July 3, 1973 to C22.2 NO 16. 2 pp.

—Tightness Testing
DIN VDE 0472 Pt 604-85. Testing Cables and Insulated Wires; Tightness of Cable Sheaths (May). 4 pp.

—Water Absorption
BSI BS 6469: Sec 99.2-92. 1992 Insulating and Sheathing Materials of Electric Cables Part 99: Test Methods Used in the United Kingdom but Not Specified in Parts 1 to 5 Section 99: Electrical Tests. 10 pp.
CENELEC HD 505.1.3 S2-91. Common Test Methods for Insulating and Sheathing Materials of Electric Cables Part 1: Methods for General Application Section Three—Methods for Determining the Density Water Absorption Tests-. 5 pp.
DIN VDE 0472 Pt 802-86. Testing of Cables, Wires and Flexible Cords; Water Absorption (Apr). 5 pp.

—Weight Increase
DIN VDE 0472 Pt 809-84. Testing of Cables, Wires and Flexible Cords; Weight Increase (Oct). 5 pp.

Cable Joints
See Also: Cables (Electric); Connectors; Joints; Splices (Electrical); Splicing

—Cable Insulation—Fiber Optic Cables
CCITT RECMN L.13-92. Sheath Joints and Organizers of Optical Fibre Cables in the Outside Plant (Study Group VI) 6 pp. 6 pp.

—Cathodic Protection Equipment—Underground
CCITT RECMN L.7-89. Application of Joint Cathodic Protection—Construction, Installation and Protection of Cable and Other Elements of Outside Plant (Study Group VI) 2 pp. 2 pp.

—Electrical Insulating Tapes
CSA C22.2 NO 197-M1983. PVC Insulating Tape (R 1992); (Gen Instr 1). 15 pp.

—Fiber Optic Cables—Telecommunication Equipment
CCITT RECMN L.12-92. Optical Fibre Joints (Study Group VI) 10 pp. 10 pp.
CCITT RECMN L.13-92. Sheath Joints and Organizers of Optical Fibre Cables in the Outside Plant (Study Group VI) 6 pp. 6 pp.

Cable Junctions
Use For: Junctions (Cables Electric)
See Also: Cables (Electric); Power Line Hardware; Transmission Lines
DIN VDE 0619-87. Cable Entries for Cables and Cords; Requirements, Tests (Sept). 20 pp.
IEC 685 Pt 2-3-83. Connecting Devices (Junction and/or Tapping) for Household and Similar Fixed Electrical Installations Part 2: Particular Requirements—Insulation Piercing Connecting Devices for Insulated Copper Conductors First Edition. 34 pp.

Cable Lifters
Use: Windlasses

Cable Networks
See Also: Cable Television; Cable Television Equipment; Communication Networks; Television Transmission

—Radio Frequency Interference
CCIR QUESTION 76/11-90. Radiation from Cable Distribution Networks—Questions Concerning Study Group 11—Broadcasting Service (Television). 1 p.

Cable Outlets
Use: Electric Outlets

Cable Oversheaths
Use: Cable Insulation

Cable Penetration
—Ships
CNS F5103-86. Basic Design for Cable Penetration of "A" Class Division (Jun)(11615).

Cable Penetration (Cont.)
—Ships (Cont.)
JIS F 8051-84. Basic Design for Cable Penetration of "A" Class Division.

Cable Plugs
Use: Electric Plugs

Cable Protectors
Use: Current Limiting Fuses

Cable Racks
See Also: Cable Supports; Racks (Electrical)

—Carrier Systems—Telephony
CCITT RECMN G.231-89. Arrangement of Carrier Equipment—International Analogue Carrier Systems (Study Group XV) 2 pp. 2 pp.

Cable Reels
Use For: Coil Loader Reels; Cord Reels
See Also: Cables (Electric); Hose Reels; Reels
MOD UK DSTAN 81-20-01. Drums, Cable—Metric Units Issue 2; Amendment 1. 14 pp.
MOD UK DSTAN 81-42-82. Reeling Machines Cable Hand; Apparatus Cable Laying; Reels Cable and Drums Cable for Field Use Issue 1. 17 pp.

—Cables (Electric)—Ships
MOD UK NES 113: Part 6-89. Requirements for Mechanical Handling Part 6: Winches, Reels and Windlasses Issue 2 (11.89). 16 pp.

—Cables (Electric)—Submarines
MOD UK NES 113: Part 6-89. Requirements for Mechanical Handling Part 6: Winches, Reels and Windlasses Issue 2 (11.89). 16 pp.

—Cords (Electric)
CSA CAN/CSA-C22. 2 NO 21-M90. Cord Sets and Power-Supply Cords; (Gen Instr 1 Thru 2). 74 pp.

Cable Saddles
See Also: Cables (Electric)

—Ships
CNS F5047-83. Electric Cable Hangers and Saddles for Marine Use (Jan)(9856).

Cable Shears
Use: Cable Cutters

Cable Sheaths
Use: Cable Insulation

Cable Ships
See Also: Ships

—Tables (Data)
CCITT FASCICLE III.3. Transmission Media Characteristics Recommendations G.601—G.654. 134 pp.

Cable Slicers
Use: Cable Cutters

Cable Splices
Use: Splices (Electrical)

Cable Splicing Systems
Use: Splices (Electrical)

Cable Stoppers
Use For: Anchor Stoppers; Chain Stoppers
See Also: Anchors (Marine); Mooring Anchors; Stoppers; Windlasses
BSI BS MA 56-74. 1974 Ships' Deck Machinery-Anchor Cable Stoppers. 6 pp.
CNS F3066-80. Anchor Stopper (Small Size) (Oct)(6563).
CNS F3069-80. Chain Stoppers (Oct)(6566).
ISO 6325-87. Shipbuilding—Cable Stoppers Second Edition. 7 pp.
JIS F 3307-78. Anchor Stoppers.
JIS F 3307-70. Anchor Stoppers. 9 pp.
JIS F 3310-90. Anchor Stoppers (Small Size).
JIS F 3310-76. Anchor Stoppers (Small Size) (R 1984). 8 pp.
JIS F 3406-90. Chain Stoppers.

—Bar
JIS F 2028-87. Roller Bar Type Chain Cable Stoppers for Grade 2 Chain Cable.
JIS F 2033-87. Roller Bar Type Chain Cable Stoppers for Grade 3 Chain Cable.

—Dog
CNS F3090-81. Cast Iron Dog Type Chain Cable Compressors (Apr)(7233).
CNS F3105-81. Cast Steel Dog Type Chain Stoppers (May)(7370).

Cable Stoppers (Cont.)
—Dog (Cont.)
CNS F3119-81. Ships' Cast Steel Dog Type Chain Cable Compressor (Small Size) (Jul)(7691).
JIS F 2002-76. Cast Iron Dog Type Chain Cable Compressors (R 1984). 8 pp.
JIS F 2015-87. Cast Steel Bar Type Chain Cable Stoppers.
JIS F 2015-75. Cast Steel Dog Type Anchor Chain Cable Stoppers. 7 pp.
JIS F 2023-76. Ships' Small Size Cast Steel Dog Type Anchor Chain Cable Stoppers. 7 pp.
JIS F 2028-76. Rollered Dog-Type Anchor Chain Cable Stoppers for Grade 2 Anchor Chain Cable. 9 pp.

—Pawl
JIS F 2016-87. Cast Steel Pawl Type Chain Cable Stoppers for Grade 2 Chain Cable.
JIS F 2016-76. Cast Steel Pawl-Type Anchor Chain Cable Stoppers for Grade 2 Anchor Chain Cable. 8 pp.
JIS F 2027-87. Roller Pawl Type Chain Cable Stoppers for Grade 2 Chain Cable. 16 pp.
JIS F 2031-87. Cast Steel Pawl Type Chain Cable Stoppers for Grade 3 Chain Cable.
JIS F 2032-87. Roller Pawl Type Chain Cable Stoppers for Grade 3 Chain Cable.

—Tongue
CNS F3106-81. Cast Steel Tongue Type Chain Cable Stoppers for Grade 2 Chain Cable (May)(7371).
CNS F3123-81. Roller Tongue Type Chain Cable Stoppers for Grade 2 Anchor Chain Cable (Jul)(7695).

Cable Straps
See Also: Cable Ties; Cables (Electric); Straps (Fasteners)

—Aircraft
SBAC AS 2508 ISSUE 3. Strap for Cable Fastening. 1 p.

Cable Strippers
See Also: Cable Insulation; Cables (Electric); Tools

—Quality Assurance
CEN PREN 2812-91. Stripping of Electric Cables as Defined by EN 2084, EN 2234, EN 2235 and EN 2346. 10 pp.

Cable Supports
Use For: Power Line Supports *See Also:* Cable Clamps; Cable Clips; Cable Hangers; Cable Racks; Communication Transmission Line Hardware; Power Line Hardware

—Plenum Cables
CSA CAN/CSA-C22. 2 NO 18-92. Outlet Boxes, Conduit Boxes, and Fittings; (Gen Instr 1 Thru 2). 118 pp.

Cable Systems (Electric Power)
Use: Cabling Systems (Electric Power)

Cable Systems (Telephony)
Use: Telephone Lines

Cable Television
Use For: CATV; Community Antenna Television
See Also: Cable Networks; Television Systems

—MAC/Packet System—Channel Rasters
CCIR RECMN 806-92. Common Channel Raster for the Distribution of D-MAC, D2-MAC and HD-MAC Signals in Collective Antenna and Cable Distribution Systems. 1 p.

Cable Television Amplifiers
Use For: CATV Amplifiers *See Also:* Amplifiers; Cable Television Equipment

—HF—Hybrid Circuits
CECC CECC 63 101-007 ISSUE 1-88. NL-CECC 63 101-007 Issue 1; CATV Hybrid Modules Manufacturer's Type Numbers BGY 84 H, BGY 85 H (En). 12 pp.
CECC CECC 63 101-009 ISSUE 2-91. NL CECC 63 101-009 Issue 1; CATV Hybrid Module Manufacterer's Type Number BGD 502 (En). 13 pp.
CECC CECC 63 101-026 ISSUE 2-91. 9NL CECC 63 101-026 Issue 2; CATV Hybrid Module Manufacturer's Type Number BGY585A (En). 11 pp.
CECC CECC 63 101-027 ISSUE 2-91. NL CECC 63 101-027 Issue 2; CATV Hybrid Module Manufacturer's Type Number BGY586 (En). 11 pp.
CECC CECC 63 101-028 ISSUE 2-91. NL CECC 63 101-028 Issue 2; CATV Hybrid Module Manufacturer's Type Number BGY587 (En). 11 pp.

Cable Television Amplifiers (Cont.)
—HF—Hybrid Circuits (Cont.)
CECC CECC 63 101-029 ISSUE 2-91. NL CECC 63 101-029 Issue 2; CATV Hybrid Module Manufacturer's Type Number BGY588 (En). 11 pp.

Cable Television Cables
See Also: Cable Television Equipment; Coaxial Cables; Communication Cables
DIN VDE 0887 Pt 3-84. Coaxial Radio Frequency Cable, Z = 75 Ohm for Receiving and Transmission Systems Including CATV Systems; Outdoor Cable with Sealed Outer Conductor (Dec). 10 pp.
JIS C 3503-90. CATV Aluminium Pipe Coaxial Cables. 10 pp.

— 75 Ohms
DIN VDE 0887 Pt 1-80. Radio Frequency Cables; Coaxial, Z=75 Ohms for Television Antenna Systems; General Requirements; (VDE Specifications) (Dec). 17 pp.
DIN VDE 0887 Pt 3-84. Coaxial Radio Frequency Cable, Z = 75 Ohm for Receiving and Transmission Systems Including CATV Systems; Outdoor Cable with Sealed Outer Conductor (Dec). 10 pp.

—Ships
IEC 92 Pt 373-77. Shipboard Telecommunication Cables and Radio-Frequency Cables Shipboard Flexible Coaxial Cables First Edition. 4 pp.

Cable Television Equipment
See Also: Cable Networks; Cable Television Amplifiers; Cable Television Cables; Television Equipment

—Data Processing Equipment
CSA CAN/CSA-C22. 2 NO 220-M91. Information Processing and Business Equipment; (Gen Instr 1). 119 pp.

—Distribution Systems—Buildings
SNZ NZS 6603-84. Guide for the Design and Installation of Multiple Outlet Distribution Systems for Sound and Television Signals. 19 pp.

—Office Machines
CSA CAN/CSA-C22. 2 NO 220-M91. Information Processing and Business Equipment; (Gen Instr 1). 119 pp.

Cable Television Networks
Use: Cable Networks

Cable Television Systems
Use: Cable Television

Cable Terminations
Use: Electric Terminals

Cable Ties
Use For: Lashing Ties *See Also:* Cable Clamps; Cable Straps; Cables (Electric); Power Line Hardware; Ties (Fasteners)

—Electrical Installations
CENELEC PREN 50146-93. Specification for Cable Ties for Electrical Installations. 17 pp.

Cable Trays
Use For: Cabletroughs *See Also:* Cables (Electric); Electric Raceways

—Metal
CSA CAN/CSA-C22. 2 NO 126-M91. Cable Tray Systems; (Gen Instr 1). 28 pp.

—Nonmetallic
CSA CAN/CSA-C22. 2 NO 126-M91. Cable Tray Systems; (Gen Instr 1). 28 pp.

—Reinforced Concrete
CNS A2186-87. Reinforced Concrete Cable Troughs (Dec)(11325).
CNS A3231-87. Method of Test for Reinforced Concrete Cable Troughs (Dec)(11326).
JIS A 5321-90. Reinforced Concrete Cable Troughs. 18 pp.

Cable Trunking
Use: Electric Raceways

Cabled Distribution Systems
See Also: Sound Broadcasting; Television Equipment
BSI BS 6513: Part 5-87. 1987 Wideband Cabled Distribution Systems Part 5: Recommendations for One-Way and Interactive Data Services. 45 pp.
DIN VDE 0855 Pt 11 (P)-88. Cabled Distribution Systems; Receiving Aerial Systems with Satellite Receiving Equipment (Nov). 10 pp.

Cabled Distribution Systems (Cont.)
IEC 728 Pt 1-86. Cabled Distribution Systems Part 1: Systems Primarily Intended for Sound and Television Signals Operating Between 30 MHz and 1 GHz Second Edition; (Corrigendum—Nov 1986) (Amendment 1-1992). 242 pp.
SAA AS 1367-85. Multiple Outlet Distribution Systems—Sound and Vision. 15 pp.

—Broadband Amplifiers—Measurement
CENELEC PREN 50083-3-91. Cabled Distribution Systems for Television and Sound Signals: Part 3: Active Coaxial Wideband Distribution Components. 36 pp.

—Components
CENELEC PREN 50083-4-91. Cabled Distribution Systems for Television and Sound Signals: Part 4: Passive Coaxial Wideband Distribution Components. 12 pp.

—Electromagnetic Interference
CENELEC PREN 50083-2-91. Cable Distribution Systems ENC Requirements Radiation and Immunity Performance Requirements. 72 pp.

—Equipment
CENELEC PREN 50083-5-91. Cabled Distribution Systems for Television and Sound Signals: Part 5: Headend. 43 pp.
DIN VDE 0855 Pt 10 (P)-88. Cabled Distribution Systems; Equipment for Reception and/or Distribution of Sound and Television Signals (Nov). 15 pp.

—Glossaries
BSI BS 6513: Part 1-84. 1984 Wideband Cabled Distribution Systems Part 1: Glossary of Terms. 11 pp.

—Measurement
BSI BS 6513: Part 2-84. 1984 Amd 1 Wideband Cabled Distribution Systems Part 2: Methods of Measurement (AMD 5065) December 31, 1985. 41 pp.
CENELEC PREN 50083-7-91. Cabled Distribution Systems for Television and Sound Signals: Part 7: System Performance. 64 pp.

—Radiation Detection and Measurement
CENELEC PREN 50083-2-91. Cable Distribution Systems ENC Requirements Radiation and Immunity Performance Requirements. 72 pp.

—Safety
BSI BS 6513: Part 3-84. 1984 Amd 1 Wideband Cabled Distribution Systems Part 3: Specification for Performance Requirements, Including System Safety, for Downstream Television and F.M. Sound Radio Signals (AMD 5793) December 23, 1987. 13 pp.
BSI BS 6513: Part 4-84. 1984 Amd 1 Wideband Cabled Distribution Systems Part 4: Specification for Performance Requirements, Including System Safety, for Upstream Television (AMD 5794) December 23, 1987. 9 pp.
BSI BS 6513: Part 6-87. 1987 Wideband Cabled Distribution Systems Part 6: Safety Requirements. 7 pp.
CENELEC PREN 50083-1-91. Cabled Distribution Systems for Television and Sound Signals Part 1: Safety Requirements. 22 pp.

—Television Signals
CCIR QUESTION 31-1/11-90. Performance and Testing of Cabled Distribution Systems for Television Signals—Questions Concerning Study Group 11—Broadcasting Service (Television). 1 p.

Cables (Electric)
Scope Note: For additional listings, use a more specific term *Use For:* Flexible Cables; Insulated Cables; Radio Frequency Cables; Sheathed Cables *See Also:* Alarm Cables; Arc Welding Cables; Armored Cables; Busbars; Cable Assemblies (Electric); Cable Clamps; Cable Couplers; Cable Hangers; Cable Joints; Cable Junctions; Cable Reels; Cable Saddles; Cable Straps; Cable Strippers; Cable Ties; Cable Trays; Cabling Systems (Electric Power); Coaxial Cables; Communication Cables; Control Cables (Electric); Cords (Electric); Electric Wire; Electric Wiring; Elevator Cables; Flat Cables; Gas Filled Cables; Harnesses (Electric); Heat Shrink Tubing; Heating Cables; Ignition Cables; Instrumentation Cables; Jumper Cables; Local Area Network Cables; Local Cables; Marine Cables (Electric); Microphone Cables; Neutral Cables; Oil Filled Cables; Paired Cables; Plenum Cables; Power Cables; Resistance Cables; Ribbon Cables; Service Entrance Cables; Shielded Conductor Cables; Stuffing Tubes; Telephone Cables; Television Cables; Thermocouple Cables; Trailing Cables; Transmission Lines; Tray Cables; Twisted Pair Cables; Wiring Harnesses

Cables (Electric) (Cont.)

BSI BS 3988-70. 1970 Amd 3 Wrought Aluminium for Electrical Purposes: Solid Conductors for Insulated Cables. 23 pp.
BSI BS 6141-91. 1991 Insulated Cables and Flexible Cords for Use in High Temperature Zones. 26 pp.
BSI BS 6141-01. 1991 Amd 1 Insulated Cables and Flexible Cords for Use in High Temperature Zones (AMD 7567) May 15, 1993 (F). 32 pp.
BSI BS 6141-81. 1981 Amd 2 Insulated Cables and Flexible Cords for Use in High Temperature Zones (AMD 6228) April 30, 1990. 27 pp.
BSI BS 6195-69. 1969 Amd 4 Insulated Flexible Cables and Cords for Coil Leads (AMD 4929) September 30, 1985. 40 pp.
BSI BS 7450-91. 1991 Determination of Economic Optimization of Power Cable Size (Implementation of CENELEC HD 558 S1) (IEC 1059: 1991). 34 pp.
CENELEC HD 383 S2-86. Conductors of Insulated Cables Guide to the Dimensional Limits of Circular Conductors. 2 pp.
CENELEC HD 383 S2/A1-89. AMD 1 Conductors of Insulated Cables Guide to the Dimensional Limits of Circular Conductors. 4 pp.
CENELEC HD 383 S2/A2-93. Conductors of Insulated Cables First Supplement: Guide to the Dimensional Limits of Circular Conductors (IEC 228:1978 + IEC 228A: 1982, Modified). 6 pp.
CNS C2050-87. Aluminum Stranded Wires for Insulated Cables (Dec)(2728). 4 pp.
CNS C4077-69. Fixtures for Aluminium Conducting Wires and Cables (Jul)(2831) (R 1973).
CNS C7022-89. Flexible R.F. Twin Cables (VHF) (May)(3306).
CNS C7133-83. Flexible R.F. Twin Cables (UHF) (Feb)(6655).
CSA CAN/CSA-C22. 2 NO 0.3-92. Test Methods for Electrical Wires and Cables; (Gen Instr 1). 121 pp.
CSA 1334 Bull. Electrical Bulletin 1334 August 28, 1981 to C22.2 NO 16. 1 p.
CSA 1334 Bull. Electrical Bulletin 1334 August 28, 1981 to C22.2 NO 96. 1 p.
CSA CAN/CSA-C22. 2 NO 131-M89. Type TECK 90 Cable; (Gen Instr 1 Thru 2). 44 pp.
CSA 1368-131 Bull. Electrical Bulletin 1368-131 September 20, 1985 to C22.2 NO 131. 2 pp.
DIN VDE 0100 Pt 520-85. Erection of Power Installations with Rated Voltages of up to 1000 V; Selection and Erection of Electrical Apparatus; Cables, Conductors and Busbars (Nov). 50 pp.
DIN VDE 0298 Pt 1-82. Application of Cables and Flexible Cords in Power Installations; General Requirement for Cables with Rated Voltages Uo/U up to 18/30 kV (Nov). 29 pp.
DIN VDE 0298 Pt 3-83. Application of Cables and Flexible Cords in Power Installations; General Information on Cables (Aug). 37 pp.
DIN VDE 0472 Suppl. 1-87. Testing of Cables, Wires and Flexible Cords; List of Standards in the DIN VDE 0472 Series (June). 18 pp.
DIN VDE 0472 Pt 1-87. Testing of Cables, Wires and Flexible Cords; General Requirements (June). 13 pp.
DIN VDE 0472 Pt 401-84. Testing of Cables and Insulated Flexible Cords; Outer Dimensions (June). 8 pp.
IEC 96 Pt 1-86. Radio-Frequency Cables Part 1: General Requirements and Measuring Methods Fourth Edition; (Amendment 2-1993, Incorporating Amendment 1). 143 pp.
IEC 96 Pt 4-1-90. Radio-Frequency Cables Part 4: Specification for Superscreened Cables Section 1—General Requirements and Test Methods First Edition. 38 pp.
IEC 183-84. Guide to the Selection of High-Voltage Cables Second Edition. 23 pp.
IEC 228-78. Conductors of Insulated Cables Second Edition; (Errata—Sept 1979) (Supplement A-1982) (Amendment 1-1993) (CENELEC HD 383 S2/A2: 1993). 52 pp.
IEC 649-79. Calculation of Maximum External Diameter of Cables for Indoor Installations First Edition. 19 pp.
IEC 719-92. Calculation of the Lower and Upper Limits for the Average Outer Dimensions of Cables with Circular Copper Conductors and of Rated Voltages up to and Including 450/750 V Second Edition. 19 pp.
IEC 1059-91. Economic Optimization of Power Cable Size First Edition. 65 pp.
JIS C 3330-87. Flexible R.F. Twin Cables. 9 pp.
MOD UK DSTAN 59-82-86. Wiring Harness and Connector Cable Assembles Generic Data and Methods of Test for Capability Approval Issue 1. 50 pp.
MOD UK DSTAN 61-12: Part 1-01. Wires, Cords, and Cables, Electrical-Metric Units Part 1: Cables, Electrical (Insulated Flexible Cords and Flexible Cables) Issue 4; Amendment 5. 47 pp.
MOD UK DSTAN 61-12: Part 9-90. Wires, Cords and Cables Electrical—Metric Units Part 9: Cables, Radio Frequency Including Limited Fire Hazard (LFH) Variants Issue 3. 36 pp.

Cables (Electric) (Cont.)

MOD UK DSTAN 61-12: Part 9-93. Wires, Cords and Cables Electrical—Metric Units Part 9: Cables, Radio Frequency Including Limited Fire Hazard (LFH) Variants Issue 4. 49 pp.
MOD UK DSTAN 61-12: Part 11-01. Wires, Cords, and Cables, Electrical-Metric Units Part 11: Imperial/Metric Cross Reference Index Issue 2; Amendment 1. 89 pp.
SAA AS 1125-93. Conductors in Insulated Electric Cables and Flexible Cords. 14 pp.
SAA AS 1660. Methods of Test for Electric Cables, Cords and Conductors Complete Set in Binder.

—300 V

CSA 983B Bull. Electrical Bulletin 983B April 25, 1977 to C22.2 NO 38. 39 pp.
CSA CAN/CSA-C22. 2 NO 49-92. Flexible Cords and Cables; (Gen Instr 1). 121 pp.
CSA 983B Bull. Electrical Bulletin 983B April 25, 1977 to C22.2 NO 49. 39 pp.
CSA 1378 Bull. Electrical Bulletin 1378 August 23, 1982 to C22.2 NO 49. 4 pp.

—600 V

CSA 983B Bull. Electrical Bulletin 983B April 25, 1977 to C22.2 NO 38. 39 pp.
CSA CAN/CSA-C22. 2 NO 49-92. Flexible Cords and Cables; (Gen Instr 1). 121 pp.
CSA 983B Bull. Electrical Bulletin 983B April 25, 1977 to C22.2 NO 49. 39 pp.
CSA 1378 Bull. Electrical Bulletin 1378 August 23, 1982 to C22.2 NO 49. 4 pp.

—Abrasion Testing

DIN VDE 0472 Pt 605-85. Testing of Cables, Insulated Wires and Flexible Cords; Abrasion (Jan) (Partially Supersedes 0472/09.71 and VDE 0472 d/12.77). 8 pp.

—Adhesive Strength

DIN VDE 0472 Pt 618-83. Testing of Cables, Wires and Flexible Cords; Adhesive Strength (Jan). 6 pp.

—Aerospace

BSI G 231-83. (WITHDRAWN) 1983 Conductors for General Purpose Aircraft Electrical Cables and Aerospace Applications (Superseded by 2G 231: 1990). 6 pp.
ISO 2635-79. Aircraft—Conductors for General Purpose Aircraft Electrical Cables and Aerospace Applications—Dimensions and Characteristics First Edition. 4 pp.

—Aging Testing

DIN VDE 0472 Pt 303-90. Testing of Cables, Wires and Flexible Cords; Ageing Procedures (May). 15 pp.

—Aircraft

AECMA PREN2084-80. Electric Cables for General Purpose with Conductors in Copper or Copper Alloy—Technical Specification (C2/02). 25 pp.
AECMA PREN2265-86. Electrical Cables for General Purpose Operating Temperatures Between-55 Degrees Celsius and +150 Degrees Celsius. 5 pp.
AECMA PREN2266-86. Electrical Cables for General Purpose Operating Temperatures Between-55 Degrees Celsius and +200 Degrees Celsius. 5 pp.
AECMA PREN2267-86. Electrical Cables for General Purpose Operating Temperatures Between-55 Degrees Celsius and +260 Degrees Celsius. 5 pp.
BSI G 212-71. 1971 Amd 8 General Requirements for Aircraft Electrical Cables. 43 pp.
BSI G 222-76. 1976 Amd 3 Efglas Type Electric Cables (Metric Units) (AMD 6188) December 21, 1990. 19 pp.
BSI G 230-84. (WITHDRAWN) 1984 Amd 2 General Requirements for Aircraft Electrical Cables (Second Series) (AMD 6253) December 22, 1989 (Superseded by 2G 230: 1991). 48 pp.
BSI 2G 230-91. 1991 General Requirements for Aircraft Electrical Cables (Second Series) (V). 39 pp.
BSI 2G 230-01. 1991 Amd 1 General Requirements for Aircraft Electrical Cables (Second Series) (AMD 7066) April 1, 1992. 49 pp.
BSI G 231-83. (WITHDRAWN) 1983 Conductors for General Purpose Aircraft Electrical Cables and Aerospace Applications (Superseded by 2G 231: 1990). 6 pp.
BSI 2G 231-90. 1990 Conductors for General-Purpose Aircraft Electrical Cables and Aerospace Applications. 8 pp.
BSI 2G 232-87. 1987 Electrical Cables for General Airframe or Equipment Interconnect Use (135 Degrees C), Wrapped Insulation. 18 pp.
BSI 2G 233-87. (WITHDRAWN) 1987 Electric Cables for General Airframe or Equipment Interconnect Use (135 Degrees C), Extruded Insulation (Superseded by 3G 233: 1990). 20 pp.
BSI 3G 233-90. 1990 Electric Cables for General Airframe or Equipment Interconnect Use (135 Degrees Celsius), Extruded Insulation. 28 pp.

Cables (Electric) (Cont.)

—Aircraft (Cont.)

BSI G 235-87. 1987 Electric Cables for General Airframe or Equipment Interconnect Use (150 Degrees Celsius) Wrapped Insulation with Silver Plated Conductors. 20 pp.
BSI G 236-87. 1987 Electric Cables for General Airframe or Equipment Interconnect Use (200 Degrees Celsius), Wrapped Insulation with Nickel Plated Conductors. 18 pp.
BSI G 237-87. 1987 Electric Cables for General Airframe or Equipment Interconnect Use (200 Degrees Celsius), Extruded Insulation with Nikel Plated Conductors. 19 pp.
BSI G 238-87. 1987 Electric Cables for General Airframe or Equipment Interconnect Use (260 Degrees Celsius), Wrapped Insulation with Nickel Plated Conductors. 18 pp.
CAA 706 (K). Section K Light Aeroplanes Electrical Supply, Systems and Equipment (Blue Papers). 54 pp.
CAA 707 (G). Section G Rotorcraft Electrical Supply, Systems and Equipment (Blue Papers). 47 pp.
CAA Chapter J3-2 09.66. Equipment Including Cables. 2 pp.
CAA Chapter J3-2 App 09.66. Cables. 4 pp.
CAA Chap G6-14 App #3 11.85. Cables and Cable Installations (Rotocraft). 1 p.
CAA CAP 482 SUB-Part F 03.83. Equipment (Small Light Aeroplanes). 3 pp.
CAA LEAFLET 9-3 07.90. Cables—Installation and Maintenance. 13 pp.
CAA LEAFLET 11-5 12.90. Aircraft Electrical Cables. 19 pp.
CEN PREN 2235-91. Single and Multicore Electrical Cables Screened and Sheathed Technical Specification. 14 pp.
CEN PREN 2265-002-91. Electrical Cables for General Use Operating Temperatures Between-55 and +150 Degrees C Product Standard: General. 5 pp.
CEN PREN 2265-003-91. Electrical Cables for General Use Operating Temperatures Between-55 and +150 Degrees C Product Standard. 8 pp.
CEN PREN 2265-004-91. Electrical Cables for General Use Operating Temperatures Between-55 and +150 Degrees C CO2-Laser Printable Product Standard. 8 pp.
CEN PREN 2265-005-91. Electrical Cables for General Use Operating Temperatures Between-55 and +150 Degrees C UV-Laser Printable Product Standard. 8 pp.
CEN PREN 2265-006-91. Electrical Cables for General Use Operating Temperatures Between-55 and +150 Degrees C YAG-X3-Laser Printable Product Standard. 8 pp.
CEN PREN 2266-002-91. Electrical Cables for General Use Operating Temperatures Between-55 and +200 Degrees C Product Standard: General. 5 pp.
CEN PREN 2266-003-91. Electrical Cables for General Use Operating Temperatures Between-55 and +200 Degrees C Product Standard. 8 pp.
CEN PREN 2266-004-91. Electrical Cables for General Use Operating Temperatures Between-55 and +200 Degrees C CO2-Laser Printable Product Standard. 8 pp.
CEN PREN 2266-005-91. Electrical Cables for General Use Operating Temperatures Between-55 and +200 Degrees C UV-Laser Printable Product Standard. 8 pp.
CEN PREN 2266-006-91. Electrical Cables for General Use Operating Temperatures Between-55 and +200 Degrees C YAG-X3 Laser Printable Product Standard. 8 pp.
CEN PREN 2267-002-91. Electrical Cables for General Use Operating Temperatures Between-55 and +260 Degrees C Product Standard: General. 5 pp.
CEN PREN 2267-003-91. Electrical Cables for General Use Operating Temperatures Between-55 and +260 Degrees C Product Standard. 8 pp.
CEN PREN 2267-004-91. Electrical Cables for General Use Operating Temperatures Between-55 and +260 Degrees C CO2-Laser Printable Product Standard. 9 pp.
CEN PREN 2267-005-91. Electrical Cables for General Use Operating Temperatures Between-55 and +260 Degrees C UV-Laser Printable Product Standard. 9 pp.
CEN PREN 2267-006-91. Electrical Cables for General Use Operating Temperatures Between-55 and +260 Degrees C YAG-X3 Laser Printable Product Standard. 10 pp.
CEN PREN 2712-002-91. Single and Multicore Electrical Cables Screened and Jacketed for General Use Operating Temperatures Between-55 and +150 Degrees C Product Standard: General. 7 pp.
CEN PREN 2712-003-91. Single and Multicore Electrical Cables Screened (Spiral) and Jacketed for General Use Operating Temperatures Between-55 and +150 Degrees C Product Standard. 8 pp.

INTERNATIONAL AND NON-U.S. NATIONAL STANDARDS
SUBJECT INDEX
Cables

Cables (Electric) (Cont.)
—Aircraft (Cont.)

CEN PREN 2712-004-91. Single and Multicore Electrical Cables Screened (Braided) and Jacketed for General Use Operating Temperatures Between -55 and +150 Degrees C Product Standard. 8 pp.

CEN PREN 2712-005-91. Single and Multicore Electrical Cables Screened (Spiral) and Jacketed for General Use Operating Temperatures Between -55 and +150 Degrees C CO2 Laser Printable Product Standard. 8 pp.

CEN PREN 2712-007-91. Single and Multicore Electrical Cables Screened (Spiral) and Jacketed for General Use Operating Temperatures Between -55 and +150 Degrees C UV Laser Printable Product Standard. 8 pp.

CEN PREN 2712-008-91. Single and Multicore Electrical Cables Screened (Braided) and Jacketed for General Use Operating Temperatures Between -55 and +150 Degrees C UV Laser Printable Product Standard. 8 pp.

CEN PREN 2712-009-91. Single and Multicore Electrical Cables Screened (Spiral) and Jacketed for General Use Operating Temperatures Between -55 and +150 Degrees C YAG Laser Printable Product Standard. 8 pp.

CEN PREN 2712-010-91. Single and Multicore Electrical Cables Screened (Braided) and Jacketed for General Use Operating Temperatures Between -55 and +150 Degrees C YAG-X3 Laser Printable Product Standard. 8 pp.

CEN PREN 2713-002-91. Single and Multicore Electrical Cables Screened and Jacketed for General Use Operating Temperatures Between -55 and +200 Degrees C Product Standard: General. 7 pp.

CEN PREN 2713-003-91. Single and Multicore Electrical Cables Screened (Spiral) and Jacketed for General Use Operating Temperatures Between -55 and +200 Degrees C Product Standard. 8 pp.

CEN PREN 2713-004-91. Single and Multicore Electrical Cables Screened (Braided) and Jacketed for General Use Operating Temperatures Between -55 and +200 Degrees C Product Standard. 8 pp.

CEN PREN 2713-005-91. Single and Multicore Electrical Cables Screened (Spiral) and Jacketed for General Use Operating Temperatures Between -55 and +200 Degrees C CO2 Laser Printable Product Standard. 8 pp.

CEN PREN 2713-006-91. Single and Multicore Electrical Cables Screened (Braided) and Jacketed for General Use Operating Temperatures Between -55 and +200 Degrees C CO2 Laser Printable Product Standard. 8 pp.

CEN PREN 2713-008-91. Single and Multicore Electrical Cables Screened (Braided) and Jacketed for General Use Operating Temperatures Between -55 and +200 Degrees C UV Laser Printable Product Standard. 8 pp.

CEN PREN 2713-010-91. Single and Multicore Electrical Cables Screened (Braided) and Jacketed for General Use Operating Temperatures Between -55 and +200 Degrees C YAG-X3 Laser Printable Product Standard. 8 pp.

CEN PREN 2714-002-91. Single and Multicore Electrical Cables Screened and Jacketed for General Use Operating Temperatures Between -55 and +260 Degrees C Product Standard: General. 7 pp.

CEN PREN 2714-003-91. Single and Multicore Electrical Cables Screened (Spiral) and Jacketed for General Use Operating Temperatures Between -55 and +260 Degrees C Product Standard. 8 pp.

CEN PREN 2714-004-91. Single and Multicore Electrical Cables Screened (Braided) and Jacketed for General Use Operating Temperatures Between -55 and +260 Degrees C Product Standard. 8 pp.

CEN PREN 2714-005-91. Single and Multicore Electrical Cables Screened (Spiral) and Jacketed for General Use Operating Temperatures Between -55 and +260 Degrees C CO2 Laser Printable Product Standard. 8 pp.

CEN PREN 2714-006-91. Single and Multicore Electrical Cables Screened (Braided) and Jacketed for General Use Operating Temperatures Between -55 and +260 Degrees C CO2 Laser Printable Product Standard. 8 pp.

CEN PREN 2714-007-91. Single and Multicore Electrical Cables Screened (Spiral) and Jacketed for General Use Operating Temperatures Between -55 and +260 Degrees C UV Laser Printable Product Standard. 8 pp.

CEN PREN 2714-008-91. Single and Multicore Electrical Cables Screened (Braided) and Jacketed for General Use Operating Temperatures Between -55 and +260 Degrees C UV Laser Printable Product Standard. 8 pp.

CEN PREN 2714-009-91. Single and Multicore Electrical Cables Screened (Spiral) and Jacketed for General Use Operating Temperatures Between -55 and +260 Degrees C YAG-X3 Laser Printable Product Standard. 8 pp.

Cables (Electric) (Cont.)
—Aircraft (Cont.)

CEN PREN 2714-010-91. Single and Multicore Electrical Cables Screened (Braided) and Jacketed for General Use Operating Temperatures Between -55 and +260 Degrees C YAG-X3 Laser Printable Product Standard. 8 pp.

CEN PREN 3475 (Part 100)-92. Cables, Electrical, Aircraft Use Test Methods Part 100—General. 7 pp.

CEN PREN 3475 (Part 203)-92. Cables, Electrical, Aircraft Use Test Methods Part 203—Dimensions. 3 pp.

ISO 1967-74. Aircraft—Fire-Resisting Electrical Cables—Dimensions, Conductor Resistance and Mass First Edition. 3 pp.

ISO 2635-79. Aircraft—Conductors for General Purpose Aircraft Electrical Cables and Aerospace Applications—Dimensions and Characteristics First Edition. 4 pp.

MOD UK DSTAN 61-12: Part 21-02. Wires, Cords, and Cables, Electrical-Metric Units Part 21: Wires and Cables, Lightweight, Extruded Insulation, 135 Degrees C Tin Plated, and 150 Degrees C Silver Plated Conductors Issue 2; Amendment 3 (Superseded in Part by Def Stan 61-12: Part 33). 64 pp.

MOD UK DSTAN 61-12: Part 24-01. Wires, Cords, and Cables, Electrical—Metric Units Part 24: Wires and Cables, Lightweight, Taped Insulation, 150 Degrees C Silver Plated and 260 Degrees C Nickel Plated Conductors Issue 1; Amendment 2 (Superseded by Def Stan 61-12: Part 33). 90 pp.

MOD UK DSTAN 61-12: Part 33-92. Wires Cords and Cables Electrical—Metric Units Part 33: Airframe Wires and Cables in Temperature Categories 135 Degrees C, 150 Degrees C, 200 Degrees C and 260 Degrees C Sectional Specification Issue 1 (Supersedes 61-12: Part 21 and Part 24). 40 pp.

SBAC AS 3360 ISSUE 6(I). Bonding Flexibles—General. 1 p.

—Aircraft—Abrasion Testing

CEN PREN 3475 (Part 503)-92. Cables, Electrical, Aircraft Use Test Methods Part 503—Scrape Abrasion. 3 pp.

—Aircraft—Aging Testing

CEN PREN 3475 (Part 401)-92. Cables, Electrical, Aircraft Use Test Methods Part 401—Accelerated Ageing. 3 pp.

CEN PREN 3475 (Part 409)-92. Cables, Electrical, Aircraft Use Test Methods Part 409—Air-Excluded Ageing. 2 pp.

—Aircraft—Bend Testing

CEN PREN 3475 (Part 405)-92. Cables, Electrical, Aircraft Use Test Methods Part 405—Bending at Ambient Temperature. 2 pp.

CEN PREN 3475 (Part 406)-92. Cables, Electrical, Aircraft Use Test Methods Part 406—Cold Bend Test. 2 pp.

—Aircraft—Blocking Resistance Testing

CEN PREN 3475 (Part 403)-92. Cables, Electrical, Aircraft Use Test Methods Part 403—Delamination and Blocking. 3 pp.

—Aircraft—Braid Screen—Pushback Testing

CEN PREN 3475 (Part 702)-92. Cables, Electrical, Aircraft Use Test Methods Part 702—Braid Screen Pushback Capability. 2 pp.

—Aircraft—Bridge Pieces

SBAC AS 2507 ISSUE 1. Bridge Piece. 1 p.

—Aircraft—Crimp Contacts

BSI 5G 178: Part 1-93. 1993 Crimped Joints for Aircraft Electrical Cables and Wires Part 1: Specification for Design Requirements (Including Tests) for Components and Tools (S). 15 pp.

BSI 4G 178: Part 1-84. (WITHDRAWN) 1984 Amd 1 Crimped Joints for Aircraft Electrical Cables and Wires Part 1: Design Requirements (Including Tests) for Components and Tools (Superseded by 5G 178: Part 1: 1993). 11 pp.

BSI 4G 178: Part 2-86. 1986 Crimped Joints for Aircraft Electrical Cables and Wires Part 2: Control of Crimping (Including User Control Tests). 10 pp.

BSI G 184-G188-64. (OBSOLESCENT) 1964 Amd 1 Aluminium Terminal Ends and Inline Connectors for Hexagonal Crimping to Aircraft Aluminium Electric Cables. 17 pp.

BSI G 204-67. 1967 Amd 1 Copper Terminal Ends for Crimping to Electric Cables with Copper Conductors. 13 pp.

BSI G 220-76. (WITHDRAWN) 1976 Amd 1 Crimped Terminations to Stranded Flexible Thermocouple Cables of the Nickel-Chromium and Nickel-Aluminium Types (Superseded by BS 2G 215: Part 2: 1989). 5 pp.

ISO 1965-73. Aluminium Terminal Ends for Crimping to Aircraft Aluminium Electrical Cables First Edition. 4 pp.

Cables (Electric) (Cont.)
—Aircraft—Crimp Contacts (Cont.)

ISO 1966-73. Crimped Joints for Aircraft Electrical Cables First Edition. 9 pp.

—Aircraft—Delamination

CEN PREN 3475 (Part 402)-92. Cables, Electrical, Aircraft Use Test Methods Part 402—Shrinkage and Delamination. 2 pp.

CEN PREN 3475 (Part 403)-92. Cables, Electrical, Aircraft Use Test Methods Part 403—Delamination and Blocking. 3 pp.

—Aircraft—Electrical Resistance

CEN PREN 3475 (Part 301)-92. Cables, Electrical, Aircraft Use Test Methods Part 301—Electrical Resistance per Unit Length. 2 pp.

—Aircraft—Endurance Testing

CEN PREN 3475 (Part 410)-92. Cables, Electrical, Aircraft Use Test Methods Part 410—Thermal Endurance. 2 pp.

—Aircraft—Fastening Strips

SBAC AS 2506 ISSUE 4. Flexible Strip for Cable Fastening. 1 p.

SBAC AS 4528 ISSUE 3. Flexible, Strip for Cable Fastening. 1 p.

—Aircraft—Fire Resistance

AECMA PREN2346-86. Fire Resistant Electrical Cables Dimensions, Conductor Resistance and Mass. 3 pp.

AECMA PREN2623-86. Electrical Cables for General Purpose Operating Temperatures Between -55 Degrees Celsius and +135 Degrees Celsius. 5 pp.

—Aircraft—Fire Resistant

BSI G 241-88. 1988 Fire Proof Electric Cables for Engine Fire Zone and Airframe Use. 18 pp.

CEN PREN 2234-91. Fire-Resistant Electrical Cables Technical Specification. 12 pp.

ISO 2155-74. Aircraft—Fire-Resisting Electrical Cables—Performance Requirements First Edition. 5 pp.

ISO 2156-74. Aircraft—Fire-Resisting Electrical Cables—Methods of Test First Edition. 9 pp.

—Aircraft—Flammability Testing

CEN PREN 3475 (Part 407)-92. Cables, Electrical, Aircraft Use Test Methods Part 407—Flammability. 3 pp.

—Aircraft—Harnesses—Ties—Design

BSI G 246-91. 1991 Cable Ties for Installation Under Controlled Tension on Aircraft Electric Cable Harnesses. 17 pp.

—Aircraft—Harnesses—Ties—Quality Assurance

BSI G 246-91. 1991 Cable Ties for Installation Under Controlled Tension on Aircraft Electric Cable Harnesses. 17 pp.

—Aircraft—Identification Systems

ISO 2574-74. Aircraft—Electrical Cables—Identification First Edition; (Amendment-1974). 4 pp.

NATO STANAG 3794 ED 1 AMD 3-76. Identification of Aircraft Electrical Cables. 5 pp.

SAA AS 3891.1-91. Air Navigation—Cables and Their Supporting Structures—Mapping and Marking—Part 1: Permanent Marking of Overhead Cables and their Supporting Structures. 5 pp.

—Aircraft—Identification Systems—Quality Assurance

CEN PREN 3838-92. Requirements and Tests on User-Applied Markings on Aircraft Electrical Cables. 11 pp.

—Aircraft—Insulation Resistance

CEN PREN 3475 (Part 303)-92. Cables, Electrical, Aircraft Use Test Methods Part 303—Insulation Resistance. 3 pp.

CEN PREN 3475 (Part 501)-92. Cables, Electrical, Aircraft Use Test Methods Part 501—Dynamic Cut-Through. 3 pp.

CEN PREN 3475 (Part 502)-92. Cables, Electrical, Aircraft Use Test Methods Part 502—Notch Propagation. 3 pp.

—Aircraft—Mass

CEN PREN 3475 (Part 202)-92. Cables, Electrical, Aircraft Use Test Methods Part 202—Mass. 2 pp.

ISO 1967-74. Aircraft—Fire-Resisting Electrical Cables—Dimensions, Conductor Resistance and Mass First Edition. 3 pp.

—Aircraft—Overload Testing

CEN PREN 3475 (Part 305)-92. Cables, Electrical, Aircraft Use Test Methods Part 305—Overload Resistance. 3 pp.

INDUSTRY STANDARDS

INTERNATIONAL AND NON-U.S. NATIONAL STANDARDS
SUBJECT INDEX

Cables

Cables (Electric) (Cont.)

—Aircraft—Screen Terminators
BSI 2G 223-89. 1989 Screen Terminating Devices for Aircraft Electric Cables. 9 pp.

—Aircraft—Shrinkage
CEN PREN 3475 (Part 402)-92. Cables, Electrical, Aircraft Use Test Methods Part 402—Shrinkage and Delamination. 2 pp.

—Aircraft—Sleeves
BSI 2G 198: Part 3-85. 1985 Sleeves for Aircraft Electric Cables and Equipment Wires Part 3: Heat Shrinkable Sleeving for Binding and Insulation. 27 pp.
BSI 2G 198: Part 4-90. 1990 Sleeves for Aircraft Electric Cables and Equipment Wires Part 4: Semi-Rigid High Performance Heat Shrinkable Sleeving for Binding and Insulation. 16 pp.

—Aircraft—Sleeves—Identification Systems
BSI 2G 198: Part 2-82. 1982 Amd 1 Sleeves for Aircraft Electric Cables and Equipment Wires Part 2: Slip-On Sleeves for Identification Purposes. 13 pp.
BSI 3G 198: Part 1-89. 1989 Amd 1 Sleeves for Aircraft Electric Cables and Equipment Wires Part 1: Specification for Elastomeric Sleeves for Binding and Identification (AMD 6389) September 30, 1991. 17 pp.

—Aircraft—Splices (Electrical)
BSI 2G 180-74. 1974 Permanent Splicing of Aircraft Electrical Cables. 6 pp.

—Aircraft—Studs (Fasteners)
SBAC AS 2505 ISSUE 3. Stud for Cable Fastening. 1 p.
SBAC AS 2509 ISSUE 4. Stud for Cable Fastening. 1 p.
SBAC AS 4527 ISSUE 3. Stud for Cable Fastening. 1 p.
SBAC AS 4714 ISSUE 3. Cable Stud. 1 p.
SBAC AS 4715 ISSUE 2. Cable Stud. 1 p.

—Aircraft—Surface Resistance
CEN PREN 3475 (Part 304)-92. Cables, Electrical, Aircraft Use Test Methods Part 304—Surface Resistance. 4 pp.

—Aircraft—Thermal Shock
CEN PREN 3475 (Part 404)-92. Cables, Electrical, Aircraft Use Test Methods Part 404—Thermal Shock. 3 pp.

—Aircraft—Thimbles
BSI SP 66-67-55. 1955 Cable Thimbles for Aircraft. 2 pp.
SBAC AGS 1531 ISSUE 6. Thimble for Cambric Covered H.T. Cables.

—Aircraft—Torsion
CEN PREN 3475 (Part 504)-92. Cables, Electrical, Aircraft Use Test Methods Part 504—Torsion. 3 pp.

—Aircraft—Union Nuts
SBAC AGS 1634 ISSUE 6. Nut, Union, for H.T. Cable.
SBAC AGS 1635 ISSUE 5. Nut, Union, for L.T. Cable.

—Aircraft—Visual Inspection
CEN PREN 3475 (Part 201)-92. Cables, Electrical, Aircraft Use Test Methods Part 201—Visual Examination. 2 pp.

—Aircraft—Voltage Measurement
CEN PREN 3475 (Part 302)-92. Cables, Electrical, Aircraft Use Test Methods Part 302—Voltage Proof Test. 4 pp.

—Airport Lighting—5000 V
CSA C22.2 NO 179-M1987. Airport Series Lighting Cables (R 1993); (Gen Instr 1). 21 pp.

—Aluminum Conductors—Aerospace
CEN PREN 3719-91. Aluminium or Aluminium Alloy Conductors for Electrical Cables Product Standard. 7 pp.

—Aluminum Sheathed
CSA 983B Bull. Electrical Bulletin 983B April 25, 1977 to C22.2 NO 38. 39 pp.
CSA 1368 Bull. Electrical Bulletin 1368 April 7, 1982 to C22.2 NO 38. 2 pp.
CSA 983B Bull. Electrical Bulletin 983B April 25, 1977 to C22.2 NO 123. 39 pp.
CSA 1368 Bull. Electrical Bulletin 1368 April 7, 1982 to C22.2 NO 123. 2 pp.

—Aluminum Sheathed—Electrical Codes
CSA 632 Bull. Electrical Bulletin 632 March 29, 1966 to C22.2 NO 38. 2 pp.
CSA 671 Bull. Electrical Bulletin 671 March 16, 1967 to C22.2 NO 38. 3 pp.

Cables (Electric) (Cont.)

—Aluminum Sheathed—Electrical Codes (Cont.)
CSA 983C Bull. Electrical Bulletin 983C November 14, 1978 to C22.2 NO 38. 3 pp.
CSA 1213A Bull. Electrical Bulletin 1213A October 25, 1979 to C22.2 NO 38. 3 pp.
CSA 1240 Bull. Electrical Bulletin 1240 September 24, 1979 to C22.2 NO 51. 2 pp.
CSA 632 Bull. Electrical Bulletin 632 March 29, 1966 to C22.2 NO 123. 2 pp.
CSA 671 Bull. Electrical Bulletin 671 March 16, 1967 to C22.2 NO 123. 3 pp.
CSA 983C Bull. Electrical Bulletin 983C November 14, 1978 to C22.2 NO 123. 3 pp.
CSA 1240 Bull. Electrical Bulletin 1240 September 24, 1979 to C22.2 NO 123. 2 pp.
CSA 1213A Bull. Electrical Bulletin 1213A October 25, 1979 to C22.2 NO 123. 3 pp.

—Automotive
BSI BS 6862: Part 1-71. 1971 Amd 2 Cables for Vehicles Part 1: Cables with Copper Conductors. 25 pp.
CNS C2088-86. High-Voltage Cables for Automobiles (Nov)(6646). 3 pp.
CNS C2089-86. Low-Voltage Cables for Automobile (Dec)(6648).
CNS C3103-86. Method of Test for High-Voltage Cables for Automobiles (Nov)(6647).
CNS C3104-86. Method of Test for Low-Voltage Cables for Automobiles (Dec)(6649).
ISO 6722 Pt 1-84. Road Vehicles—Unscreened Low-Tension Cables—General Requirements and Test Methods Second Edition. 12 pp.
ISO 6722 Pt 2-85. Road Vehicles—Unscreened Low-Tension Cables—Part 2: Cable Classes, Applicable Tests and Special Requirements Second Edition. 5 pp.
ISO 6722 Pt 3-93. Road Vehicles—Unscreened Low-Tension Cables—Part 3: Conductor Sizes and Dimensions for Thick-Wall Insulated Cables Second Edition. 6 pp.

—Automotive—Color Coding
BSI BS AU 7A-83. 1983 Colour Code for Road Vehicle Electrical Cables. 11 pp.

—Automotive—Current Rating
BSI BS AU 88A-85. 1985 Recommendations for Ratings for Light Duty Cables for Automobile Use. 4 pp.
CNS D1051-83. Current Capacity of Low Tension Cables for Automobiles (Apr)(10158).

—Automotive—Sizes
ISO 6722 Pt 4-93. Road Vehicles—Unscreened Low-Tension Cables—Part 4: Conductor Sizes and Dimensions for Thin-Wall Insulated Cables First Edition. 6 pp.

—Automotive—Starting
MOD UK DSTAN 25-8-71. Cable Assemblies, Intervehicle, Electrical, for Starting Purposes Issue 2. 6 pp.

—Automotive—Towing Attachments
BSI BS AU 231-89. 1989 Seven-Core Connecting Cable for Road Vehicles. 4 pp.
ISO 4141-88. Road Vehicles—Seven-Core Connecting Cables Second Edition. 4 pp.

—Bend Testing
DIN VDE 0472 Pt 610-85. Testing of Cables, Wires and Flexible Cords; Bending Test at Low Temperature (Jan). 6 pp.

—Breaking Length
DIN VDE 0472 Pt 626-83. Testing of Cables, Wires and Flexible Cords; Breaking Length (Jan). 5 pp.

—Buildings—Construction Contracts
DIN ENGL 18382-88. Tendering and Performance Stipulations in Contracts for Construction Works (VOB); Part C: General Technical Specifications in Contracts for Construction Works (ATV); Electrical Installations in Buildings (Sept). 3 pp.

—Cable Reels—Ships
MOD UK NES 113: Part 6-89. Requirements for Mechanical Handling Part 6: Winches, Reels and Windlasses Issue 2 (11.89). 16 pp.

—Cable Reels—Submarines
MOD UK NES 113: Part 6-89. Requirements for Mechanical Handling Part 6: Winches, Reels and Windlasses Issue 2 (11.89). 16 pp.

—Channels (Structural Shapes)—Naval Ships
MOD UK NES 2046-90. Specification for Channel Perforated Steel for Electric Cable Issue 2 (06.90). 10 pp.

Cables (Electric) (Cont.)

—Channels (Structural Shapes)—Naval Ships (Cont.)
MOD UK NES 2047-90. Specification for Channel Perforated Aluminium Alloy for Electric Cable Issue 2 (06.90). 10 pp.

—Channels (Structural Shapes)—Submarines
MOD UK NES 2046-90. Specification for Channel Perforated Steel for Electric Cable Issue 2 (06.90). 10 pp.
MOD UK NES 2047-90. Specification for Channel Perforated Aluminium Alloy for Electric Cable Issue 2 (06.90). 10 pp.

—Classification
BSI BS 1000: (679)-83. 1983 Universal Decimal Classification (UDC). English Full Edition (679): Industries Based on Various Processable Materials. Cable and Cordage Industries. Stone Industries. 15 pp.
SNZ NZS/BS 1000 (679)-83. Universal Decimal Classification Industries Based on Various Processable Materials. Cable and Cordage Industries. Stone Industry. 16 pp.

—Color Coding
BSI BS 1843-52. 1952 Amd 1 Colour Code for Twin Compensating Cables for Thermocouples. 8 pp.
BSI PD 2379-90. 1990 Register of Colours of Manufacturers' Identification Threads for Electric Cables and Cords. 25 pp.
CENELEC HD 27-78. Colours of the Cores of Flexible Cables and Cords. 7 pp.
CENELEC HD 308-76. Identification and Use of Cores of Flexible Cables. 4 pp.
DIN VDE 0293-90. Identification of Cores and Cables and Flexible Cords Used in Power Installations with Voltage Ratings up to 1000 V (Jan). 14 pp.
IEC 173-64. Colours of the Cores of Flexible Cables and Cords First Edition. 5 pp.
SNZ NZS/BSPD 2379-90. Register of Colours of Manufacturers' Identification Threads for Electric Cables and Cords. 26 pp.

—Combustion
BSI BS 6425: Part 1-90. 1990 Gases Evolved During Combustion of Electric Cables Part 1: Method for Determination of Amount of Halogen Acid Gas Evolved During Combustion of Polymeric Materials Taken from Cables. 4 pp.
IEC 754 Pt 1-82. Test on Gases Evolved During Combustion of Electric Cables Part 1: Determination of the Amount of Halogen Acid Gas Evolved During the Combustion of Polymeric Materials Taken from Cables First Edition. 11 pp.

—Combustion Gases—Acidity
BSI BS 6425: Part 2-93. 1993 Test on Gases Evolved During the Combustion of Materials from Cables Part 2: Determination of Degree of Acidity (Corrosivity) of Gases by Measuring pH and Conductivity (E). 19 pp.
CENELEC HD 602 S1-92. Test on Gases Evolved During Combustion of Electric Cables Part 2: Determination of Degree of Acidity of Gases Evolved During the Combustion of Materials Taken from Electric Cables by Measuring pH and Conductivity. 4 pp.
CENELEC HD 602 S1-92. Test on Gases Evolved During Combustion of Materials from Cables Determination of Degree of Acidity (Corrosivity) of Gases by Measuring pH and Conductivity. 4 pp.
IEC 754 Pt 2-91. Test on Gases Evolved During Combustion of Electric Cables Part 2: Determination of Degree of Acidity of Gases Evolved During the Combustion of Materials Taken from Electric Cables by Measuring pH and Conductivity First Edition. 33 pp.

—Combustion Testing
SAA AS 1660.5.3-88. Methods of Test for Electrical Cables, Cords and Conductors—Part 5: Fire Tests—Part 5.3: Determination of the Amount of Halogen Acid Gas Evolved During Combustion of Polymeric Materials Taken from Cables. 2 pp.

—Concrete Embedded
CNS C1076-80. Installation Methods of Cables Embedded Direct in Concrete (Dec)(6769).
CNS C2091-80. Cables Embedded Direct in Concrete (Dec)(6770).
JIS C 3650-87. Installation Methods of Cables Embedded Direct in Concrete. 19 pp.

—Conformance
DIN VDE 0287-85. Technical Procedures for Determining the Conformity of Harmonized Cables and Cords (Apr). 14 pp.

INTERNATIONAL AND NON-U.S. NATIONAL STANDARDS
SUBJECT INDEX

Cables

Cables (Electric) (Cont.)

—**Cracking (Fracturing)**
BSI BS 6469: Sec 99.1-92. 1992 Insulating and Sheathing Materials of Electric Cables Part 99: Test Methods Used in the United Kingdom but Not Specified in Parts 1 to 5 Section 99: Non-Electrical Tests. 14 pp.

—**Crimping Tools—Quality Assurance**
CEN PREN 2242-91. Control of Tools Used for Crimping of Electric Cables with Conductors Defined by EN 2083 and EN 2346. 10 pp.

—**Cross Reference Index**
MOD UK DSTAN 61-12: Part 90-89. Wires, Cords and Cables of Assessed Quality Part 90: Detail Specifications Issue 1. 5 pp.
MOD UK DSTAN 61-12: Part 90-93. Wires, Cords and Cables Electrical Metric Units Part 90: Detail Specification Cross Reference Index Issue 2. 5 pp.
MOD UK DSTAN 61-12: Part 90-01. Wires, Cords and Cables Electrical Metric Units Part 90: Detail Specification Cross Reference Index Issue 2; Amendment 1. 6 pp.

—**Current Ratings**
DIN VDE 0100 Pt 523-81. Installation of Power Plant with Rated Voltages up to 1000 V; Dimensioning of Cables and Cords; Mechanical Strength, Voltage Drop and Current Carrying Capacity (June). 10 pp.
DIN VDE 0298 Pt 2-79. Application of Cables and Flexible Cords in Power Installations; Recommended Values for the Current Carrying of Cables with Rated Voltages Uo/U up to 18/30 kV (Nov). 73 pp.
DIN VDE 0298 Pt 2 (D)-90. Application of Cables and Leads in Power Installations; Recommended Current-Carrying Capacities for Cables with Rated Voltages V/V up to 18/30 kV (July). 54 pp.
DIN VDE 0298 Pt 4-88. Application of Cables and Insulated Conductors in Power Plant (Feb) (Partially Supersedes: DIN 57100 Part 523/VDE 0100 Part 523/06.81). 43 pp.
IEC 287-82. Calculation of the Continuous Current Rating of Cables (100% Load Factor) Second Edition; (Amendment 3-1993, Including Amendments 1 and 2). 156 pp.
IEC 853 Pt 1-85. Calculation of the Cyclic and Emergency Current Rating of Cables Part 1: Cyclic Rating Factor for Cables up to and Including 18/30 (36) kV First Edition. 42 pp.
IEC 853 Pt 2-89. Calculation of the Cyclic and Emergency Current Rating of Cables Part 2: Cyclic Rating of Cables Greater Than 18/30 (36) kV and Emergency Ratings for Cables of All Voltages First Edition. 108 pp.
IEC 1042-91. Method for Calculating Reduction Factors for Groups of Cables in Free Air, Protected from Solar Radiation First Edition. 26 pp.

—**Data Processing Equipment**
DIN VDE 0813-88. Distribution Wires for Telecommunication and Data Processing Systems (Nov). 25 pp.
DIN VDE 0815-85. Wiring Cables for Telecommunication and Information Processing Systems (Sept). 51 pp.
DIN VDE 0815 A1-88. Wiring Cables for Telecommunication and Data Processing Systems; Amendment 1 (May). 6 pp.

—**Data Processing Equipment—Wiring**
DIN VDE 0891 Pt 5-85. Use of Cables and Insulated Cords for Telecommunications and Information Processing Installations. Special Guidance on Wiring Cables and Cords According to DIN/VDE 0815 (Nov). 11 pp.

—**Designations**
CENELEC HD 361-78. System for Cable Designation. 11 pp.
CENELEC HD 361 S2-86. System for Cable Designation. 11 pp.
CENELEC HD 361 S2/A1-88. AMD 1 System for Cable Designation. 3 pp.
CENELEC HD 361 S2/A2-90. AMD 2 System for Cable Designation. 5 pp.
CENELEC HD 361 S2/A3-90. AMD 3 System for Cable Designation. 5 pp.
CENELEC HD 361 S2/A4-91. AMD 4 System for Cable Designation. 5 pp.
CENELEC HD 361 S2/A5-92. AMD 5 System for Cable Designation. 4 pp.
CENELEC HD 361 S2/A6-93. System for Cable Designation. 4 pp.

—**Dielectric Loss Factor**
DIN VDE 0472 Pt 505-83. Testing of Cables and Insulated Flexible Cords; Loss Factor, Dielectric Loss Coefficient and Leakage (Apr). 9 pp.

Cables (Electric) (Cont.)

—**Dielectric Strength**
DIN VDE 0472 Pt 508-86. Testing of Insulated Cables and Flexible Cords; Dielectric Strength of Insulated Cables and Flexible Cords for Power Installations (May). 8 pp.

—**Digital Communications**
BSI BS 7630-93. 1993 Symmetric Pair/Quad and Multicore Cables for Digital Communication (S). 21 pp.
CENELEC HD 608 S1-92. Generic Specification for Symmetric Pair/Quad and Multicore Cables for Digital Communication. 35 pp.

—**Earthmoving Equipment—Identification Systems**
BSI BS 6913: Part 6-91. 1991 Operation and Maintenance of Earth-Moving Machinery Part 6: Specification for Identification and Marking of Electrical Wires and Cables (ISO 9247: 1990). 7 pp.
ISO 9247-90. Earth-Moving Machinery—Electrical Wires and Cables—Principles of Identification and Marking First Edition. 6 pp.

—**Elastomer Insulated**
BSI BS 5055-91. 1991 Amd 1 Elastomer-Insulated Cables for Electric Signs and High-Voltage Luminous-Discharge-Tube Installations (AMD 6768) September 30, 1991. 18 pp.
BSI BS 5055-02. 1991 Amd 2 Elastomer-Insulated Cables for Electric Signs and High-Voltage Luminous-Discharge-Tube Installations (AMD 7565) May 15, 1993. 20 pp.

—**Elastomer Insulated—Mining Equipment**
BSI BS 6116-70. (WITHDRAWN) 1970 Amd 1 Elastomer-Insulated Flexible Trailing Cables for Quarries and Miscellaneous Mines (Superseded by BS 6708: 1991). 22 pp.

—**Elastomer Insulated—Ships**
SAA AS 1168-73. Elastomer-Insulated Electric Cables for Fixed Wiring in Ships Amdt 1 July 1986. 63 pp.

—**Elastomer Sheathed—Fire Resistant**
MOD UK DSTAN 61-12: Part 27-87. Wires, Cords and Cables Electrical—Metric Units Part 27: Cables and Wires, Electrical (for Power Connections to Mobile Equipment) Issue 1. 34 pp.

—**Electric Discharges**
DIN VDE 0472 Pt 513-82. Testing of Cables, Wires and Flexible Cords; Partial Discharge Measurements (July). 28 pp.

—**Electric Terminals**
DIN VDE 0220 Pts 1 & 1A-71. Specifications for Detachable Cable Terminals in Heavy-Current Cable Installations up to 1000 V (Nov). 22 pp.

—**Electric Wire**
DIN VDE 0472 Pt 605-85. Testing of Cables, Insulated Wires and Flexible Cords; Abrasion (Jan) (Partially Supersedes 0472/09.71 and VDE 0472 d/12.77). 8 pp.

—**Electrical Codes**
CSA 916 Bull. Electrical Bulletin 916 July 3, 1973 to C22.2 NO 16. 2 pp.
CSA 632 Bull. Electrical Bulletin 632 March 29, 1966 to C22.2 NO 38. 2 pp.
CSA 671 Bull. Electrical Bulletin 671 March 16, 1967 to C22.2 NO 38. 3 pp.
CSA 851 Bull. Electrical Bulletin 851 November 17, 1971 to C22.2 NO 38. 2 pp.
CSA 983B Bull. Electrical Bulletin 983B April 25, 1977 to C22.2 NO 38. 39 pp.
CSA 983C Bull. Electrical Bulletin 983C November 14, 1978 to C22.2 NO 38. 3 pp.
CSA 1213A Bull. Electrical Bulletin 1213A October 25, 1979 to C22.2 NO 38. 3 pp.
CSA 675A Bull. Electrical Bulletin 675A January 2, 1968 to C22.2 NO 49. 1 p.
CSA 675B Bull. Electrical Bulletin 675B June 20, 1968 to C22.2 NO 49. 1 p.
CSA 851 Bull. Electrical Bulletin 851 November 17, 1971 to C22.2 NO 49. 2 pp.
CSA 959X Bull. Electrical Bulletin 959X May 14, 1976 to C22.2 NO 49. 1 p.
CSA 1293 Bull. Electrical Bulletin 1293 November 7, 1980 to C22.2 NO 49. 3 pp.
CSA 1240 Bull. Electrical Bulletin 1240 September 24, 1979 to C22.2 NO 51. 2 pp.
CSA 851 Bull. Electrical Bulletin 851 November 17, 1971 to C22.2 NO 96. 2 pp.
CSA 916 Bull. Electrical Bulletin 916 July 3, 1973 to C22.2 NO 96. 2 pp.
CSA 983B Bull. Electrical Bulletin 983B April 25, 1977 to C22.2 NO 96. 39 pp.
CSA 983C Bull. Electrical Bulletin 983C November 14, 1978 to C22.2 NO 96. 3 pp.
CSA 1293 Bull. Electrical Bulletin 1293 November 7, 1980 to C22.2 NO 96. 3 pp.

Cables (Electric) (Cont.)

—**Electrical Codes** (Cont.)
CSA 632 Bull. Electrical Bulletin 632 March 29, 1966 to C22.2 NO 131. 2 pp.
CSA 671 Bull. Electrical Bulletin 671 March 16, 1967 to C22.2 NO 131. 3 pp.
CSA 851 Bull. Electrical Bulletin 851 November 17, 1971 to C22.2 NO 131. 2 pp.
CSA 983C Bull. Electrical Bulletin 983C November 14, 1978 to C22.2 NO 131. 3 pp.
CSA 1240 Bull. Electrical Bulletin 1240 September 24, 1979 to C22.2 NO 131. 2 pp.
CSA 1213A Bull. Electrical Bulletin 1213A October 25, 1979 to C22.2 NO 131. 3 pp.

—**Electrical Installations**
DIN VDE 0100 Pt 730-86. Erection of Power Installations with Nominal Voltages up to 1000 V; Laying of Cables and Cords in Hollow Walls and in Buildings and Structures Made of Chiefly Combustible Building Materials According to DIN 4102 (Feb). 7 pp.
DIN VDE 0100 Pt 732-90. Erection of Power Installations with Nominal Voltages up to 1000 V; Cable Entries into Buildings in Public Cable Networks (Nov) (With DIN VDE 0211/12.85 Supersedes DIN 57 100 Part 732/VDE 0100 Part 732/03.83). 10 pp.
DIN VDE 0245 Pt 1 (D)-90. Cables & Cords for Electrical & Electronic Equipment in Power Installations; General Requirements (Oct). 15 pp.
DIN VDE 0298 Pt 1-82. Application of Cables and Flexible Cords in Power Installations; General Requirement for Cables with Rated Voltages Uo/U up to 18/30 kV (Nov). 29 pp.
SAA AS 3000 Supp 1-91. Electrical Installations—Buildings, Structures and Premises—Supplement 1: Cable and Conductor Tables (Imperial Units) (Supplement to AS 3000—1991). 13 pp.

—**Electrical Properties**
DIN VDE 0282 Pt 1000-83. Rubber Cables, Wires and Flexible Cords for Power Installation; Surface Resistance of Sheath (Feb). 4 pp.
IEC 885 Pt 1-87. Electrical Test Methods for Electric Cables Part 1: Electrical Tests for Cables, Cords and Wires for Voltages up to and Including 450/750 V First Edition. 13 pp.
ISO 1967-74. Aircraft—Fire-Resisting Electrical Cables—Dimensions, Conductor Resistance and Mass First Edition. 3 pp.
SAA AS 1660.3-93. Electrical Tests. 22 pp.

—**Electrical Protection Equipment**
DIN VDE 0100 Pt 430-81. Installation of Power Plant with Rated Voltages up to 1000 V; Protection of Cables and Cords Against Undue Temperature Rise (June). 22 pp.
DIN VDE 0100 Pt 430 Suppl. 1-91. Erection of Power Installations with Nominal Voltages up to 1000V; Protection of Cables and Cords Against Overcurrent; Recommended Values for Current-Carrying Capacity Iz and the Allocation of Overcurrent. 17 pp.

—**Electrical Resistance**
DIN VDE 0472 Pt 501-83. Testing of Cables and Insulated Cords; Conductor Resistance (Apr). 7 pp.
DIN VDE 0472 Pt 510-84. Testing of Cables and Insulated Flexible Cords; Resistance to Direct Current (Sept). 6 pp.
DIN VDE 0472 Pt 512-85. Testing of Cables & Insulated Conductors; Resistance Between Protective Conductor and Conductive Layer (May). 5 pp.

—**Elevators (Lifts)**
CNS C2109-90. Travelling Cables for Elevators (Mar)(8936).
CNS C3150-90. Method for Test for Traveling Cables for Elevators (Mar)(8937).
JIS C 3408-87. Travelling Cables for Elevators. 22 pp.
SAA AS 1979-93. Electric Cables—Lifts—Flexible Travelling (in Professional Package 55). 9 pp.

—**Elongation**
DIN VDE 0472 Pt 625-83. Testing of Cables and Insulated Conductors; Tensile and Elongation Characteristics of Strain Bearing Elements (Jan). 5 pp.

—**Ethylene Propylene Rubber Insulated**
BSI BS 6622-91. 1991 Cables with Extruded Cross-Linked Polyethylene or Ethylene Propylene Rubber Insulation for Rated Voltages from 3800/6600 V up to 19000/33000 V. 43 pp.
BSI BS 6622-01. 1991 Amd 1 Cables with Extruded Cross-Linked Polyethylene or Ethylene Propylene Rubber Insulation for Rated Voltages from 3800/6600 V up to 19000/33000 V (AMD 7568) May 15, 1993. 47 pp.

INDUSTRY STANDARDS

INTERNATIONAL AND NON-U.S. NATIONAL STANDARDS
SUBJECT INDEX

Cables (Electric) (Cont.)

—Ethylene Propylene Rubber Insulated—Short Circuiting/Grounding
IEC 1138-92. Cables for Portable Earthing and Short-Circuiting Equipment First Edition. 29 pp.

—Ethylene Propylene Rubber Insulated—600 V
CNS C2159-83. 600V Grade Ethylene-Propylene Rubber Insulated Cables (Oct)(10599).
JIS C 3621-87. 600 V Grade Ethylene Propylene Rubber Insulated Cables. 15 pp.

—Ethylenetetrafluoroethylene Insulated
DIN VDE 0250 Pt 106-82. Insulated Power Cables, ETFE Single-Core Non-Sheathed Cables for Internal Wiring; (VDE Specification) (Oct). 8 pp.
MOD UK DSTAN 61-12: Part 29-01. Wires, Cords, and Cables, Electrical-Metric Units Part 29: Wires and Cables, Extruded Insulation, Types ETFE and ETFE Metsheath 135 Degrees C Tin Plated, and 150 Degrees C Silver Plated Conductors Issue 1; Amendment 2. 40 pp.

—Fillers—Mining Equipment
BSI BS 7383-90. 1990 Cold-Pour Resin-Based Compound for Use as a Filling Medium in Terminating Cables in Enclosures for Voltages Not Exceeding 11 kV for Use in Coal Mines. 14 pp.

—Filling Compounds
BSI BS 6469: Sec 5.1-92. 1992 Insulating and Sheathing Materials of Electric Cables Part 5: Methods of Test Specific to Filling Compounds Section 5.1: Drop-Point—Separation of Oil—Lower Temperature Brittleness—Total Acid Number—Absence of Corrosive Components. 24 pp.
CENELEC HD 505.5.1 S1-92. Common Test Methods for Insulating and Sheathing Materials of Electric Cables Part 5: Methods Specific to Filling Compounds Section One: Drop-Point—Separation of Oil Lower Temperature Brittleness—Total Acid Number Absence of Corrosive Components—Permittivity at 23. 6 pp.
IEC 811 Pt 5-1-90. Common Test Methods for Insulating and Sheathing Materials of Electric Cables Part 5: Methods Specific to Filling Compounds Section One—Drop-Point—Separation of Oil—Lower Temperature Brittleness—Total Acid Number—Absence of Corrosive Components-. 36 pp.

—Fire Hazards
MOD UK DSTAN 61-12: Part 9-90. Wires, Cords and Cables Electrical—Metric Units Part 9: Cables, Radio Frequency Including Limited Fire Hazard (LFH) Variants Issue 3. 36 pp.
MOD UK DSTAN 61-12: Part 9-93. Wires, Cords and Cables Electrical—Metric Units Part 9: Cables, Radio Frequency Including Limited Fire Hazard (LFH) Variants Issue 4. 49 pp.
MOD UK DSTAN 61-12: Part 18-90. Wires Cords and Cables, Metric Units Part 18: Equipment Wires, Limited Fire Hazard Issue 3. 41 pp.
MOD UK DSTAN 61-12: Part 18-01. Wires Cords and Cables, Metric Units Part 18: Equipment Wires, Limited Fire Hazard Issue 3; Amendment 2. 43 pp.
MOD UK DSTAN 61-12: Part 18-02. Wires Cords and Cables, Metric Units Part 18: Equipment Wires, Limited Fire Hazard Issue 3; Amendment 3. 44 pp.
MOD UK DSTAN 61-12: Part 25-01. Wires, Cords, and Cables, Electrical Metric Units Part 25: Cables, Electrical Limited Fire Hazard, up to Size 2.5mm2 Cross-Sectional Area Issue 1; Amendment 1. 45 pp.
MOD UK DSTAN 61-12: Part 25-92. Wires, Cords and Cables Electrical Metric Units Part 25: Cables Electrical Limited Fire Hazard, up to Conductor Size 2.5sq mm Cross Sectional Area Issue 2. 43 pp.
MOD UK DSTAN 61-12: Part 25-02. Wires, Cords and Cables Electrical Metric Units Part 25: Cables Electrical Limited Fire Hazard, up to Conductor Size 2.5sq mm Cross Sectional Area Issue 2; Amendment 1. 44 pp.
MOD UK DSTAN 61-12: Part 25-03. Wires, Cords and Cables Electrical Metric Units Part 25: Cables Electrical Limited Fire Hazard, up to Conductor Size 2.5sq mm Cross Sectional Area Issue 2; Amendment 2. 45 pp.
MOD UK DSTAN 61-12: Part 25-04. Wires, Cords and Cables Electrical Metric Units Part 25: Cables Electrical Limited Fire Hazard, up to Conductor Size 2.5sq mm Cross Sectional Area Issue 2; Amendment 3. 46 pp.

—Fire Resistant
IEC 331-70. Fire-Resisting Characteristics of Electric Cables First Edition. 10 pp.
ISO 1967-74. Aircraft—Fire-Resisting Electrical Cables—Dimensions, Conductor Resistance and Mass First Edition. 3 pp.

—Fire Testing
BSI BS 4066: Part 1-80. 1980 Tests on Electric Cables Under Fire Conditions Part 1: Method of Test on a Single Vertical Insulated Wire or Cable. 8 pp.

Cables (Electric) (Cont.)

—Fire Testing (Cont.)
BSI BS 4066: Part 2-89. 1989 Tests on Electric Cables Under Fire Conditions Part 2: Method of Test on a Single Small Vertical Insulated Wire or Cable. 8 pp.
BSI BS 4066: Part 3-86. 1986 Tests on Electric Cables Under Fire Conditions Part 3: Method for Classification of Flame Propagation Characteristics of Bunched Cables. 14 pp.
BSI BS 6387-83. 1983 Amd 2 Performance Requirements for Cables Required to Maintain Circuit Integrity Under Fire Conditions (AMD 5615) July 29, 1988. 26 pp.
CENELEC HD 405.1-85. Test on Electric Cables Under Fire Conditions Part 1: Test on a Single Vertical Insulated Wire or Cable. 3 pp.
CENELEC HD 405.2 S1-91. Tests on Electric Cables Under Fire Conditions Part 2: Test on a Single Small Vertical Insulated Copper Wire or Cable. 7 pp.
DIN ENGL 4102 Pt 9-90. Fire Behaviour of Building Materials and Elements; Seals for Cable Penetrations; Concepts, Requirements and Testing (May). 11 pp.
DIN VDE 0472 Pt 804-89. Testing of Cables, Wires and Flexible Cords; Burning. 17 pp.
DIN VDE 0472 Pt 814-91. Testing of Cables and Flexible Cords; Fire Resisting Characteristics of Electric Cables (Jan). 10 pp.
IEC 332 Pt 1-93. Tests on Electric Cables Under Fire Conditions Part 1: Test on a Single Vertical Insulated Wire or Cable Third Edition; (Corrigendum—May 1993). 23 pp.
IEC 332 Pt 2-89. Tests on Electric Cables Under Fire Conditions Part 2: Test on a Single Small Vertical Insulated Copper Wire or Cable First Edition. 13 pp.
IEC 332 Pt 3-92. Tests on Electric Cables Under Fire Conditions Part 3: Tests on Bunched Wires or Cables Second Edition. 62 pp.

—Flammability Testing
BSI BS 5866: Part 4-91. 1991 Blankets Suitable for Use in the Public Sector Part 4: Specification for Flammability Performance (L). 8 pp.

—Flammability Testing—Identification Systems
CSA CAN/CSA-C22. 2 NO 49-92. Flexible Cords and Cables; (Gen Instr 1). 121 pp.

—Fuses (Electric)
CNS C1072-84. General Requirements for Fuses for the Protection of Low-Voltage Cables and Lines (Dec)(6056).

—Gateways—Military Communications
NATO STANAG 4210 ED 1 AMD 2-83. The NATO Multi-Channel Tactical Digital Gateway —Cable Link Standards. 18 pp.
NATO STANAG 4210 ED 2 AMD 0-89. (DRAFT) The NATO Multi-Channel Tactical Digital Gateway—Cable Link Standards. 17 pp.

—Glass Insulated
BSI BS 6500-90. 1990 Amd 1 Insulated Flexible Cords and Cables (AMD 6759) September 30, 1991 (F). 51 pp.
BSI BS 6500-02. 1990 Amd 2 Insulated Flexible Cords and Cables (AMD 6866) April 1, 1992. 56 pp.
MOD UK DSTAN 61-12: Part 30-87. Wires, Cords, and Cables, Electrical—Metric Units Part 30: Wires and Cables, Taped and Extruded PTFE/Glass Insulation, Types Uniflexefglas and Uniflexefglasmet 150 Degrees C Silver Plated Conductors Issue 1. 23 pp.

—Glossaries
BSI BS 4727:Pt2: Group 08-86. 1986 Electrotechnical, Power, Telecommunication, Electronics, Lighting and Colour Terms Part 2: Terms Particular to Power Engineering Group 08: Electric Cable Terminology. 15 pp.
DIN VDE 0289 Pt 1-88. Definition for Insulated Cables and Cords; General Definitions (Mar) (Together with DIN VDE 0289 P.2/03.88 and DIN VDE 0289 P.4/03.88 Supersedes DIN 57289 Part 100/VDE 0289 Part 100/11.79). 11 pp.
DIN VDE 0289 Pt 4-88. Definitions for Cables, Wires and Flexible Cords for Power Installations Testing (Mar) (Together with DIN VDE 0289 Part 1 and DIN VDE 0289 Part 2 Supersedes DIN 57 289 Part 100/VDE 0289 Part 100). 6 pp.
DIN VDE 0289 Pt 5-88. Definitions for Cables, Wires and Flexible Cords for Power Installations; Lengths (Mar). 3 pp.
IEC 50 Chap 461-84. International Electrotechnical Vocabulary Chapter 461: Electric Cables. 53 pp.
SAA AS 1852.461-88. International Electrotechnical Vocabulary—Part 461: Electric Cables. 26 pp.
SNZ IEC 50: 50(461)-84. International Electrotechnical Vocabulary 50(461): Electric Cables. 43 pp.

Cables (Electric) (Cont.)

—Halogen Free
DIN VDE 0472 Pt 815-89. Testing of Cables, Wires and Flexible Cords; Non-Halogen Verification (Mar). 14 pp.

—Halogen Free—Fire Resistant
DIN VDE 0250 Pt 503-89. Cables, Wires and Flexible Cords for Power Installations; Halogenfree Single-Core Non-Sheathed Cable with Improved Characteristics in Case of Fire; Nominal Voltages Uo/U 450/750 V (Mar). 14 pp.
DIN VDE 0266 Pt 1 (D)-90. Halogen-Free Cables with Improved Characteristics in the Case of Fire: Nominal Voltages: Uo/UYO.6/1 kV; General Requirements (July) (Intended as Partial Replacement for DIN VDE 0266/02.85). 13 pp.
DIN VDE 0266 Pt 2 (D)-90. Halogen-Free Cables with Improved Characteristics in Case of Fire: Nominal Voltages: Uo/UYO.6/1 kV; Halogen-Free, Low Fuming Cables with Reduced Fire Propagation (July) (Intended as Partial Replacement for DIN VDE 0266/02.85). 12 pp.
DIN VDE 0266 Pt 3 (D)-90. Halogen-Free Cables with Improved Characteristics in Case of Fire: Nominal Voltages: Uo/UYO.6/1kV; Halogen-Free, Low Fuming Cables with Reduced Fire Propagation and Continuance of Isolation (July) (Intended as Partial Replacement for DIN VDE 0266/02.85). 14 pp.

—Halogen Free—Polymer Insulated
DIN VDE 0266-85. Halogen-Free Cables with Improved Characteristics in Case of Fire; Nominal Voltages: Uo/U 0.6/1 kV (Feb). 18 pp.

—Hazardous Locations
CSA C22.2 NO 174-M1984. Cables and Cable Glands for Use in Hazardous Locations (R 1992); (Gen Instr 1 Thru 5). 31 pp.

—High Voltage—X-Ray Equipment
CNS C2090-85. High Tension Cables for X-Ray Apparatus (May)(6550).
JIS C 3407-87. High Tension Cables for X-Ray Apparatus. 9 pp.
JIS Z 4732-88. High Voltage Cables with Plugs for Medical X-Ray Equipment. 11 pp.

—Identification Systems
CEN PREN 3475 (Part 703)-92. Cables, Electrical, Aircraft Use Test Methods Part 703—Permanence of Manufacturer's Marking. 3 pp.
CENELEC HD 186 S2-89. Marking by Inscription for the Identification of Cores of Electric Cables Having More Than 5 Cores. 8 pp.
CENELEC HD 186 S2/A1-92. AMD 1 Marking by Inscription Identification of Cores of Electric Cables Having More Than 5 Cores. 5 pp.
CSA 1368 Bull. Electrical Bulletin 1368 April 7, 1982 to C22.2 NO 38. 2 pp.
CSA 1368 Bull. Electrical Bulletin 1368 April 7, 1982 to C22.2 NO 48. 2 pp.
CSA 1368 Bull. Electrical Bulletin 1368 April 7, 1982 to C22.2 NO 49. 2 pp.
CSA C22.2 NO 75-M1983. Thermoplastic Insulated Wires and Cables (R 1992); (Gen Instr 1 Thru 6). 48 pp.
CSA 1368 Bull. Electrical Bulletin 1368 April 7, 1982 to C22.2 NO 75. 2 pp.
CSA 1110 Bull. Electrical Bulletin 1110 March 17, 1977 to C22.2 NO 96. 1 p.
CSA 1368 Bull. Electrical Bulletin 1368 April 7, 1982 to C22.2 NO 96. 2 pp.
CSA C22.2 NO 123-M1985. Aluminum Sheathed Cables (R 1992); (Gen Instr 1). 26 pp.
CSA 1368 Bull. Electrical Bulletin 1368 April 7, 1982 to C22.2 NO 123. 2 pp.
CSA 1368 Bull. Electrical Bulletin 1368 April 7, 1982 to C22.2 NO 131. 2 pp.
DIN VDE 0293-90. Identification of Cores and Cables and Flexible Cords Used in Power Installations with Voltage Ratings up to 1000 V (Jan). 14 pp.

—Immersion Testing
DIN VDE 0472 Pt 803-86. Testing of Cables, Wires and Flexible Cords; Oil Immersion (Apr). 7 pp.

—Impact Testing
DIN VDE 0472 Pt 611-85. Testing of Cables and Insulated Flexible Cords; Cold Impact Test (Jan). 7 pp.

—Impulse Voltage Testing
CENELEC HD 48 S1-88. Impulse Tests on Cables and Their Accessories. 3 pp.
DIN VDE 0472 Pt 511-85. Testing of Cables, Wires and Flexible Cords; Impulse Test (Aug). 7 pp.
IEC 230-66. Impulse Tests on Cables and Their Accessories First Edition. 11 pp.

INDEX and DIRECTORY of

INTERNATIONAL AND NON-U.S. NATIONAL STANDARDS
SUBJECT INDEX

Cables

Cables (Electric) (Cont.)

—Impulse Voltage Testing (Cont.)
IEC 840-88. Tests for Power Cables with Extruded Insulation for Rated Voltages Above 30 kV (Um = 36 kV) up to 150 kV (Um = 170 kV) First Edition; (Corrigendum—Nov 1988) (Amendment 2-1993, Including Amendment 1) (Corrigendum—July 1993). 71 pp.

—Insulation Resistance
CSA 1433 Bull. Electrical Bulletin 1433 April 8, 1985 to C22.2 NO 38. 3 pp.

—Insulation Thickness
DIN VDE 0472 Pt 402-84. Testing of Insulated Cables, Wires and Flexible Cords; Thickness of Insulation and Armouring (May). 12 pp.

—Invitation for Bids
MOD UK DEFCON 47B-91. Invitation to Tender Special Notices and Instructions—Wire, Cables Etc. 9/91. 1 p.

—Leakage
DIN VDE 0472 Pt 505-83. Testing of Cables and Insulated Flexible Cords; Loss Factor, Dielectric Loss Coefficient and Leakage (Apr). 9 pp.

—Length Measurement
DIN VDE 0100 Suppl. 5 (D)-88. Erection of Power Installations with Rated Voltages up to 1000 V; Maximum Permissible Lengths of Cables and Cords Taking into Consideration Protection Against Electric Shock in the Event of Fault, Short Circuit of Voltage Drop (Mar). 54 pp.

—Lighting
CSA C22.2 NO 17-1973. Cable for Luminous-Tube Signs and for Oil-and Gas-Burner Ignition Equipment (R 1992); (Gen Instr 1) (Amd 1 February 1981) (Amd 2-8 March 1988). 28 pp.
MOD UK DSTAN 61-12: Part 2-01. Wires, Cords, and Cables, Electrical—Metric Units Part 2: Cables, Electrical (for Power and Lighting) Issue 4; Amendment 1. 19 pp.

—Lighting Chains—300 V
BSI BS 6726-91. 1991 Festoon and Temporary Lighting Cables and Cords. 27 pp.
BSI BS 6726-01. 1991 Amd 1 Festoon and Temporary Lighting Cables and Cords (AMD 6939) May 1, 1992. 31 pp.
BSI BS 6726-02. 1991 Amd 2 Festoon and Temporary Lighting Cables and Cords (AMD 7888) September 15, 1993 (F). 37 pp.

—Loss Factor
DIN VDE 0472 Pt 505-83. Testing of Cables and Insulated Flexible Cords; Loss Factor, Dielectric Loss Coefficient and Leakage (Apr). 9 pp.

—Marine—Flammability Testing
MOD UK NES 641-91. Determination of the Vertical Flammability of Electric Cables Issue 2 (07.91). 13 pp.

—Marine—Packings (Seals)
MOD UK NES 2063-90. Specification for Plastic Compound NSN 5975-99-735-5042 for Cable Glands Issue 1 (10.90). 8 pp.

—Mechanical Testing
BSI BS 6387-83. 1983 Amd 2 Performance Requirements for Cables Required to Maintain Circuit Integrity Under Fire Conditions (AMD 5615) July 29, 1988. 26 pp.

—Mica Glass Insulated
CSA 1315A Bull. Electrical Bulletin 1315A February 24, 1989 to C22.2 NO 210.2. 6 pp.

—Microphone Cables
CNS C5052-80. Basic Requirements for Broadcast Microphone Cables (Jul)(5762).

—Mineral Insulated
IEC 702 Pt 1-88. Mineral Insulated Cables and Their Terminations with a Rated Voltage Not Exceeding 750 V Part 1: Cables Second Edition; (Amendment 1-1992). 42 pp.

—Mineral Insulated—Copper Sheathed
BSI BS 6207-91. 1991 Mineral-Insulated Copper-Sheathed Cables with Copper Conductors. 18 pp.
CSA C22.2 NO 124-M1986. Mineral-Insulated Cable (R 1992); (Gen Instr 1 Thru 4). 28 pp.

—Mineral Insulated—Metal Sheathed
SAA AS 3187-86. Approval and Test Specifications—Mineral-Insulated Metal-Sheathed Cables (This is a Joint Standard with SANZ NZS 3187). 18 pp.
SNZ NZS/AS 3187-86. Approval and Test Specification—Mineral-Insulated Metal-Sheathed Cables (This is a Joint Standard with SAA AS 3187). 18 pp.

Cables (Electric) (Cont.)

—Mineral Insulated—Stainless Steel Sheathed
CSA C22.2 NO 124-M1986. Mineral-Insulated Cable (R 1992); (Gen Instr 1 Thru 4). 28 pp.

—Mineral Insulated—Terminations
BSI BS 6081-89. 1989 Amd 2 Terminations for Mineral Insulated Cables. 8 pp.
IEC 702 Pt 1-88. Mineral Insulated Cables and Their Terminations with a Rated Voltage Not Exceeding 750 V Part 1: Cables Second Edition; (Amendment 1-1992). 42 pp.
IEC 702 Pt 2-86. Mineral Insulated Cables and Their Terminations with a Rated Voltage Not Exceeding 750 V Part 2: Terminations First Edition. 23 pp.

—Miniature—Naval Ships
MOD UK NES 512: Part 6-91. Guide to Cables, Electrical and Associated Items Part 6: Cables, Electrical, Miniature and Sub-Miniature and Equipment Wires Issue 2 (05.91). 100 pp.
MOD UK NES 512: Part 6-01. Guide to Cables, Electrical and Associated Items Part 6: Cables, Electrical, Miniature and Sub-Miniature and Equipment Wires Issue 2 (05.91); Amendment 2. 116 pp.

—Miniature—Submarines
MOD UK NES 512: Part 6-91. Guide to Cables, Electrical and Associated Items Part 6: Cables, Electrical, Miniature and Sub-Miniature and Equipment Wires Issue 2 (05.91). 100 pp.
MOD UK NES 512: Part 6-01. Guide to Cables, Electrical and Associated Items Part 6: Cables, Electrical, Miniature and Sub-Miniature and Equipment Wires Issue 2 (05.91); Amendment 2. 116 pp.

—Mining Equipment
BSI BS 6708-91. 1991 Flexible Cables for Use at Mines and Quarries. 65 pp.
BSI BS 6708-77. 1977 Trailing Cables for Mining Purposes. 23 pp.
CSA C22.2 NO 174-M1984. Cables and Cable Glands for Use in Hazardous Locations (R 1992); (Gen Instr 1 Thru 5). 31 pp.
SAA AS 1747-93. Reeling, Trailing and Feeder Cables Used for Mining—Repair and Testing (in Professional Package 32A). 40 pp.
SAA AS 1766.0-91. Methods for the Microbiological Examination of food—Part 0: General Introduction and List of Methods. 1 p.
SAA AS 1802-92. Electric Cables—Reeling and Trailing—for Underground Coal Mining Purposes (in Professional Package 32). 41 pp.
SAA AS 1972-91. Electric Cables—Underground Coal Mines—Other Than Reeling Trailing Cables (in Professional Package 32). 14 pp.

—Multiconductor
CSA CAN/CSA-C22. 2 NO 131-M89. Type TECK 90 Cable; (Gen Instr 1 Thru 2). 44 pp.
CSA 1368-131 Bull. Electrical Bulletin 1368-131 September 20, 1985 to C22.2 NO 131. 2 pp.

—Naval Ships
MOD UK NES 503-88. Requirements for Electrical Installation Cabling Diagrams and Associated Data Issue 2 (08.88). 89 pp.
MOD UK NES 503-92. Requirements for Electrical Installation Cabling Diagrams and Associated Data Issue 3 (01.92). 94 pp.
MOD UK NES 512: Part 7-91. Guide to Cables, Electrical and Associated Items Part 7: Cables and Wires, Electrical, Miscellaneous for Special Services Issue 3 (04.91). 90 pp.
MOD UK NES 512: Part 7-01. Guide to Cables, Electrical and Associated Items Part 7: Cables and Wires, Electrical, Miscellaneous for Special Services Issue 3 (04.91); Amendment 1. 91 pp.
MOD UK NES 512: Part 7-93. Guide to Cables, Electrical and Associated Items Part 7: Cables and Wires, Electrical, Miscellaneous for Special Services Issue 4 (02.93). 84 pp.
MOD UK NES 512: Part 13-91. Guide to Cables, Electrical and Associated Items Part 13: Cables (Imperial) Nearest Metric Equivalents Issue 2 (03.91). 66 pp.
MOD UK NES 513-01. Guide to the Design of System Cabling Issue 1 (06.81); Amendment 1 (Superseded by NES 502). 42 pp.
MOD UK NES 517: Part 5-92. Requirements for Cables, Electric Part 5: Miscellaneous Issue 1 (07.92). 53 pp.
MOD UK NES 523-88. Guide to the Installation of Thin-Wall Insulated, Limited Fire Hazard Equipment Wire and Electric Cable Issue 2 (05.88). 51 pp.
MOD UK NES 523-92. Guide to the Installation of Thin-Wall Insulated, Limited Fire Hazard Equipment Wire and Electric Cable Issue 3 (01.92). 52 pp.

Cables (Electric) (Cont.)

—Naval Ships (Cont.)
MOD UK NES 525-91. Requirements for Electric Cables Thin-Wall Insulated, Limited Fire Hazard Issue 3 (09.91). 36 pp.
MOD UK NES 525-01. Requirements for Electric Cables Thin—Wall Insulated, Limited Fire Hazard Issue 3 (09.91); Amendment 1. 47 pp.

—Neoprene Coated
CENELEC HD 22.8 S1-89. Polychloroprene or Equivalent Synthetics Elastomer Sheathed Cables for Use as a Decorative Chain. 10 pp.
CENELEC HD 22.8 S1-92. Rubber Insulated Cables of Rated Voltages up to and Including 450/750 V—Part 8: Polychloroprene or Equivalent Synthetic Elastomer Sheathed Cables for Use as Decorative Chains (Reprint Incorporating A1). 10 pp.
CENELEC HD 22.8 S1/A2-92. AMD 2 Rubber Insulated Cables of Rated Voltages up to and Including 450/750 V—Part 8: Polychloroprene or Equivalent Synthetic Elastomer Sheathed Cables for Use as Decorative Chains. 4 pp.
CNS C2081-90. Steel Tape Armour and Anti-Corrosive Layer for Electric Cable (Aug)(6074).
CNS C2081-80. Steel Tape Armour and Anti-Corrosive Layer for Electric Cable (Aug)(6074). 5 pp.
JIS C 3610-87. Steel Tape Armor and Anti-Corrosive Layer for Cables. 9 pp.

—Neoprene Sheathed—Flat
DIN VDE 0250 Pt 809-85. Cables, Wires and Flexible Cords for Power Installations; Flat Ordinary Tough-Polychloroprene-Sheathed Flexible Cable (May). 11 pp.

—Nonmetallic Sheathed
CSA 983B Bull. Electrical Bulletin 983B April 25, 1977 to C22.2 NO 38. 39 pp.
CSA CAN/CSA-C22. 2 NO 48-M90. Nonmetallic Sheathed Cable; (Gen Instr 1 Thru 2). 55 pp.
CSA 983B Bull. Electrical Bulletin 983B April 25, 1977 to C22.2 NO 48. 39 pp.

—Paper Insulated
SAA AS 1026-92. Electric Cables—Impregnated Paper Insulated—Working Voltages up to and Including 33 kV Amdt 1 August 1993. 29 pp.

—Paper Insulated—Aluminum Sheathed
BSI BS 5593-78. 1978 Amd 2 Impregnated Paper-Insulated Cables With Aluminium Sheath/Neutral Conductor and Three Shaped Solid Aluminium Phase Conductors (CONSAC) 500/1000 V, for Electricity Supply (AMD 5519) February 29, 1988. 14 pp.

—Paper Insulated—Lead Sheathed
BSI BS 6480-88. 1988 Impregnated Paper-Insulated Lead or Lead Alloy Sheathed Electric Cables of Rated Voltages up to and Including 33 000 V. 56 pp.
SNZ BS 6480-88. Specification for Impregnated Paper-Insulated Lead or Lead Alloy Sheathed Electric Cables of Rated Voltages up to and Including 33 000 V. 52 pp.

—Paper Insulated—Metal Sheathed
DIN VDE 0255-72. Regulations for Mass-Impregnated Paper-Insulated Metal-Sheathed Cables for Electricity Supply (Except External Gas-Pressure and Oil-Filled Cables) (Nov). 65 pp.
DIN VDE 0255 A4-81. Impregnated Paper-Insulated Metal-Sheathed Cables for Power Plant (Excluding Gas-Pressure and Oil Filled Cables); Amendment 4 (Oct). 3 pp.
IEC 55 Pt 1-78. Paper-Insulated Metal-Sheathed Cables for Rated Voltages up to 18/30 kV (with Copper or Aluminium Conductors and Excluding Gas-Pressure and Oil-Filled Cables) Part 1: Tests Fourth Edition; (Amendment 1-1989). 32 pp.
IEC 55 Pt 2-81. Paper-Insulated Metal-Sheathed Cables for Rated Voltages up to 18/30 kV (with Copper or Aluminium Conductors and Excluding Gas-Pressure and Oil-Filled Cables) Part 2: General and Construction Requirements First Edition; (Amendment 1-1989). 85 pp.

—Partial Discharges
IEC 885 Pt 2-87. Electrical Test Methods for Electric Cables Part 2: Partial Discharge Tests First Edition. 13 pp.
IEC 885 Pt 3-88. Electrical Test Methods for Electric Cables Part 3: Test Methods for Partial Discharge Measurements on Lengths of Extruded Power Cable First Edition. 56 pp.

—Plastic Insulated
CNS C3011-88. Methods of Test for Plastic Insulated Wires and Cables (Mar)(689). 21 pp.
JIS C 3005-91. Testing Methods for Rubber or Plastic Insulated Wires and Cables. 39 pp.

INDUSTRY STANDARDS

Cables (Electric) (Cont.)

—Polyethylene Coated
CNS C2081-90. Steel Tape Armour and Anti-Corrosive Layer for Electric Cable (Aug)(6074).

CNS C2081-80. Steel Tape Armour and Anti-Corrosive Layer for Electric Cable (Aug)(6074). 5 pp.

JIS C 3610-87. Steel Tape Armor and Anti-Corrosive Layer for Cables. 9 pp.

—Polyethylene Insulated
BSI BS 6195-69. 1969 Amd 4 Insulated Flexible Cables and Cords for Coil Leads (AMD 4929) September 30, 1985. 40 pp.

BSI BS 6622-91. 1991 Cables with Extruded Cross-Linked Polyethylene or Ethylene Propylene Rubber Insulation for Rated Voltages from 3800/6600 V up to 19000/33000 V. 43 pp.

BSI BS 6622-01. 1991 Amd 1 Cables with Extruded Cross-Linked Polyethylene or Ethylene Propylene Rubber Insulation for Rated Voltages from 3800/6600 V up to 19000/33000 V (AMD 7568) May 15, 1993. 47 pp.

CNS C2107-82. 2P x 1.2mm PE Insulated Star Quad Pipe Guide Cable (Mar)(8561).

CSA 1433 Bull. Electrical Bulletin 1433 April 8, 1985 to C22.2 NO 38. 3 pp.

DIN VDE 0263-91. Cables with Cross-Linked Polyethylene Insulation and Their Accessories; Nominal Voltages Uo/U > 18/30 kV up to 87/150 kV (Feb). 17 pp.

DIN VDE 0273-87. Crosslinked Polyethylene Insulated Cables; Nominal Voltages: Uo/U 6/10, 12/20 and 18/30 kV (Dec). 27 pp.

DIN VDE 0273 A1 (D)-91. Crosslinked Polyethylene Insulated Cables, Nominal Voltages: Uo/U 6/10, 12/20 and 18/30 kV; Amendment 1 (Feb). 6 pp.

DIN VDE 0273 A2 (D)-91. Cross-Linked Polyethylene Insulated Cables; Nominal Voltages: Uo/U 6/10, 12/20 and 18/30 kV; Amendment 2 (Oct). 6 pp.

DIN VDE 0472 Pt 404-87. Testing of Cables, Insulated Wires and Flexible Cords; Irregularities of Inner Conductive Layers and Insulating Sleeve (Oct). 6 pp.

MOD UK DSTAN 61-12: Part 5-01. Wires, Cords, and Cables, Electrical-Metric Units Part 5: Cables, Special Purpose, Electrical and Cables, Power, Electrical (Small Multi-Core Cables) Issue 3; Amendment 2. 29 pp.

SAA AS 1049-86. Telecommunication Cables—Insulation and Sheath—Polyethylene. 64 pp.

—Polyethylene Insulated—Accessories
DIN VDE 0263-91. Cables with Cross-Linked Polyethylene Insulation and Their Accessories; Nominal Voltages Uo/U > 18/30 kV up to 87/150 kV (Feb). 17 pp.

—Polyethylene Insulated—Aluminum Sheathed
CSA C22.2 NO 123-M1985. Aluminum Sheathed Cables (R 1992); (Gen Instr 1). 26 pp.

—Polyethylene Insulated—High Voltage
JIS C 3606-87. High-Voltage Cross-Linked Polyethylene Insulated Cables. 16 pp.

—Polyethylene Insulated—Identification Systems
CSA 1213A Bull. Electrical Bulletin 1213A October 25, 1979 to C22.2 NO 38. 3 pp.

—Polyethylene Insulated—Polyethylene Sheathed
BSI BS 6234-87. 1987 Amd 1 Polyethylene Insulation and Sheath of Electric Cables (AMD 6683) April 30, 1991. 18 pp.

DIN VDE 0272 (D)-91. Cross Linked Polyethylene Insulated Cables with Nominal Voltage Uo/U 0.6/1 kV (Jan). 25 pp.

—Polyethylene Insulated—Polyethylene Sheathed—Data Processing
DIN VDE 0891 Pt 6-90. Use of Cables and Insulated Cords for Telecommunications and Information Processing Installations; Special Requirements for Outside Cables According to DIN VDE 0816 Part 1 to Part 3 (May). 31 pp.

—Polyethylene Insulated—PVC Sheathed
CNS C2047-87. Cross-Linked Polyethylene Insulated Polyvinyl Chloride Jacketed Cable (Nov)(2655).

DIN VDE 0272-89. Cables with Crosslinked Polyethylene Insulation and Thermoplastic PVC Sheaths, Rated Voltage Uo/U 0.6/1 kV (Sept). 25 pp.

DIN VDE 0272 (D)-91. Cross Linked Polyethylene Insulated Cables with Nominal Voltage Uo/U 0.6/1 kV (Jan). 25 pp.

Cables (Electric) (Cont.)

—Polyethylene Insulated—PVC Sheathed (Cont.)
MOD UK DSTAN 61-12: Part 5-01. Wires, Cords, and Cables, Electrical-Metric Units Part 5: Cables, Special Purpose, Electrical and Cables, Power, Electrical (Small Multi-Core Cables) Issue 3; Amendment 2. 29 pp.

—Polyethylene Insulated—PVC Sheathed—Fire Resistant
CNS C2167-85. Polyethylene (Cross-Linked Polyethylene) Insulated and Polyvinyl Chloride (Polyethylene) Sheathed Fire-Resistant Cables (Sep)(11359).

—Polyethylene Insulated—PVC Sheathed—Telecommunication Systems
DIN VDE 0891 Pt 7-84. Use of Cables and Insulated Cords for Telecommunications Systems and Data Processing Systems; Special Guidelines for Cables with Stranded Conductors for Increased Mechanical Stress for Telecommunications Systems and Data Processing Systems. 8 pp.

—Polyethylene Insulated—Water Resistance Testing
DIN VDE 0472 Pt 811-87. Testing of Cables, Wires and Flexible Cords; Longitudinal Watertightness (Oct). 8 pp.

—Polyethylene Insulated—0.6/1 kV
SAA AS 3198-90. Approval and Test Specification—Electric Cables—XLPE Insulated—for Working Voltages up to and Including 0.6/1 kV (This is a Joint Standard with SANZ NZS 3198). 25 pp.

SAA AS 3560-91. Electric Cables—XLPE Insulated—Aerial Bundled—for Working Voltages up to and Including 0.6/1 kV Amdt 1 July 1991 Amdt 2 August 1992. 22 pp.

SNZ NZS/AS 3198-90. Approval and Test Specification—XLPE Insulated—for Working Voltages up to and Including 0.6/1 kV (This is a Joint Standard with SAA AS 3198). 25 pp.

SNZ NZS/AS 3560-91. Electric Cables—XLPE—Aerial Bundled—for Working Voltages up to and Including 0.6/1 kV Amend: 1, 1991 (This is a Joint Standard with SAA AS 3560). 21 pp.

—Polyethylene Insulated—600 V
JIS C 3605-87. 600 V Polyethylene Insulated Cables and 600 V Cross Linked Polyethylene Insulated Cables. 15 pp.

—Polymer Insulated—12.7/22(24) kV
SAA AS 3599.2-91. Electric Cables—Aerial Bundled—Polymeric Insulated—Voltages 6.35/11(12) kV and 12.7/22(24) kV—Part 2: Non-Metallic Screened. 16 pp.

—Polymer Insulated—19.33 kV
SAA AS 1429-85. Polymeric Insulated Cables for Electricity Supply at Working Voltages 1.9/3.3 kV up to and Including 19.33 kV. 31 pp.

—Polymer Insulated—6.35/11(12) kV
SAA AS 3599.2-91. Electric Cables—Aerial Bundled—Polymeric Insulated—Voltages 6.35/11(12) kV and 12.7/22(24) kV—Part 2: Non-Metallic Screened. 16 pp.

—Polyolefin Insulated—Polyolefin Sheathed
IEC 708 Pt 1-81. Low-Frequency Cables with Polyolefin Insulation and Moisture Barrier Polyolefin Sheath Part 1: General Design Details and Requirements First Edition; (Amendment 3-1988, Contains Amendment 1 and 2). 87 pp.

IEC 708 Pt 2-81. Low-Frequency Cables with Polyolefin Insulation and Moisture Barrier Polyolefin Sheath Part 2: Unit Type, Filled, Moisture Barrier Polyethylene Sheathed Cables with Copper Conductors and Solid or Cellular Insulation First Edition; (Amendment 1-1983). 19 pp.

IEC 708 Pt 3-81. Low-Frequency Cables with Polyolefin Insulation and Moisture Barrier Polyolefin Sheath Part 3: Unit Type, Unfilled, Moisture Barrier Polyethylene Sheathed Cables with Copper Conductors and Solid or Cellular Insulation First Edition; (Amendment 1-1983). 19 pp.

IEC 708 Pt 4-81. Low-Frequency Cables with Polyolefin Insulation and Moisture Barrier Polyolefin Sheath Part 4: Unit Type, Unfilled, Moisture Barrier Poly-ethylene Sheathed Cables with Copper Conductors, Solid Insulation and In-tegral Suspension Strand First Edition; (Amendment 1-1983). 17 pp.

Cables (Electric) (Cont.)

—Polyolefin Insulated—Polyolefin Sheathed—Tensile Testing
IEC 771-83. Calculation of Maximum Overall Diameter of Cables and Specification of Minimum Tensile Strength of Suspension Strand for Low-Frequency Cables with Polyolefin Insulation and Moisture Barrier Polyolefin Sheath First Edition. 13 pp.

—Polyolefin Insulated—Polyvinylidene Fluoride Sheathed
MOD UK DSTAN 61-12: Part 26-01. Wires, Cords, and Cables, Electrical—Metric Units Part 26: Wires and Cables, Dual Extruded Insulation. Types OVF and OVFMETSHEATH. 140 Degrees C Conductors Issue 1; Amendment 3. 47 pp.

—Polytetrafluoroethylene Insulated
MOD UK DSTAN 61-12: Part 30-87. Wires, Cords, and Cables, Electrical—Metric Units Part 30: Wires and Cables, Taped and Extruded PTFE/Glass Insulation, Types Uniflexefglas and Uniflexefglasmet 150 Degrees C Silver Plated Conductors Issue 1. 23 pp.

—Polyurethane Sheathed
DIN VDE 0250 Pt 818-85. Cables, Wires and Flexible Cords for Power Installations; Polyurethane-Sheathed Flexible Cable (May). 12 pp.

—Portable Instruments
CSA 1110 Bull. Electrical Bulletin 1110 March 17, 1977 to C22.2 NO 96. 1 p.

—Power Distribution Systems—0.6/1.2 kV
CENELEC PRHD 626 S1-93. Bundle Assembled Cores for Overhead Distribution and Service. 76 pp.

—PVC Coated
CNS C2081-90. Steel Tape Armour and Anti-Corrosive Layer for Electric Cable (Aug)(6074).

CNS C2081-80. Steel Tape Armour and Anti-Corrosive Layer for Electric Cable (Aug)(6074). 5 pp.

JIS C 3610-87. Steel Tape Armor and Anti-Corrosive Layer for Cables. 9 pp.

—PVC Insulated
BSI BS 6004-91. 1991 PVC-Insulated Cables (Non-Armoured) for Electric Power and Lighting. 27 pp.

BSI BS 6004-01. 1991 Amd 1 PVC-Insulated Cables (Non-Armoured) for Electric Power and Lighting (AMD 7898) (E). 30 pp.

BSI BS 6004-84. (WITHDRAWN) 1984 Amd 2 PVC-Insulated Cables (Non-Armoured) for Electric Power and Lighting (AMD 6227) January 31, 1990 (E). 22 pp.

BSI BS 6195-69. 1969 Amd 4 Insulated Flexible Cables and Cords for Coil Leads (AMD 4929) September 30, 1985. 40 pp.

BSI BS 6500-90. 1990 Amd 1 Insulated Flexible Cords and Cables (AMD 6759) September 30, 1991 (F). 51 pp.

BSI BS 6500-02. 1990 Amd 2 Insulated Flexible Cords and Cables (AMD 6866) April 1, 1992. 56 pp.

BSI BS 6977-91. 1991 Insulated Flexible Cables for Lifts and for Other Flexible Connections. 27 pp.

BSI BS 6977-01. 1991 Amd 1 Insulated Flexible Cables for Lifts and for Other Flexible Connections (AMD 6934) May 1, 1992. 32 pp.

CENELEC HD 21.3-76. Polyvinyl Chloride (PVC) Insulated Cables and Flexible Cords of Rated Voltage Uo/U up to and Including 450/750v. 20 pp.

CENELEC HD 21.3 S2-83. Polyvinyl Chloride Insulated Cables of Rated Voltages up to and Including 450/750 V—Part 3: Non Sheathed Cables for Fixed Wiring. 16 pp.

CENELEC HD 21.3 S2-90. Polyvinyl Chloride Insulated Cables of Rated Voltages up to and Including 450/750 V Part 3: Single Core Non-Sheathed Cables for Fixed Wiring. 5 pp.

CENELEC HD 21.4 S2-83. Polyvinyl Chloride Insulated Cables of Rated Voltages up to and Including 450/750 V—Part 4: Sheathed Cables for Fixed Wiring. 10 pp.

CENELEC HD 21.5 S2-81. Polyvinyl Chloride Insulated Cables of Rated Voltages up to and Including 450/750 V—Part 5: Flexible Cables. 19 pp.

CENELEC HD 21.5 S2/A2-89. AMD 2 Polyvinyl Chloride Insulated Cables of Rated Voltages up to and Including 450/750 V—Part 5: Flexible Cables. 4 pp.

CENELEC HD 21.5 S2/A3-90. AMD 3 Polyvinyl Chloride Insulated Cables of Rated Voltages up to and Including 450/750 V—Part 5: Flexible Cables. 4 pp.

CENELEC HD 21.5 S2-90. Polyvinyl Chloride Insulated Cables of Rated Voltages up to and Including 450/750 v Part 5: Flexible Cables (Cords). 17 pp.

INTERNATIONAL AND NON-U.S. NATIONAL STANDARDS
SUBJECT INDEX

Cables

Cables (Electric) *(Cont.)*
—PVC Insulated *(Cont.)*
CENELEC HD 21.5 S2/A4-91. AMD 4 Polyvinyl Chloride Insulated Cables of Rated Voltages up to and Including 450/750 V Part 5: Flexible Cables (Cords). 6 pp.
CENELEC HD 21.8 S1-89. Polyvinyl Chloride Insulated Cables of Rated Voltages up to and Including 450/750V Part 8: Non Sheathed Cable for Decorative Chain. 8 pp.
CENELEC HD 21.8 S1-90. Polyvinyl Chloride Insulated Cables of Rated Voltages up to and Including 450/750 V Single Core Non-Sheathed Cable for Decorative Chains. 10 pp.
CENELEC HD 359-76. Flat Polyvinylchloride Sheated Cables for Lifts and Similar Applications. 8 pp.
CENELEC HD 359 S2-90. Flat Polyvinylchloride Sheathed Lift Cables. 17 pp.
CENELEC HD 359.2-79. Flat Polyvinyl Chloride Sheathed Flexible Cables for Lifts and Similar Applications. 1 p.
DIN VDE 0250 Pt 1-81. Insulated Power Cables; General Specifications (Oct). 25 pp.
DIN VDE 0250 Pt 102-91. Insulated Power Cables PVC Insulated Twin Cables for Internal Wiring (July). 7 pp.
DIN VDE 0281 Pt 102-90. PVC Cables, Wires and Flexible Cords for Power Installation; PVC Single-Core Non-Sheathed Cable for Internal Wiring (Nov) (Partially Supersedes DIN 57 250 Part 102/VDE 0250 Part 102/11.83). 8 pp.
IEC 227 Pt 2-79. Polyvinyl Chloride Insulated Cables of Rated Voltages up to and Including 450/750 V Part 2: Test Methods First Edition; (Amendment 1-1985). 25 pp.
IEC 227 Pt 3-93. Polyvinyl Chloride Insulated Cables of Rated Voltages up to and Including 450/750 V Part 3: Non-Sheathed Cables for Fixed Wiring Second Edition. 41 pp.
IEC 227 Pt 4-92. Polyvinyl Chloride Insulated Cables of Rated Voltages up to and Including 450/750 V Part 4: Sheathed Cables for Fixed Wiring Second Edition. 21 pp.
IEC 227 Pt 5-79. Polyvinyl Chloride Insulated Cables of Rated Voltages up to and Including 450/750 V Part 5: Flexible Cables (Cords) First Edition; (Amendment 1-1987). 38 pp.
IEC 227 Pt 6-85. Polyvinyl Chloride Insulated Cables of Rated Voltages up to and Including 450/750 V Part 6: Lift Cables and Cables for Flexible Connections Second Edition. 25 pp.
MOD UK DSTAN 61-12: Part 4-01. Wires, Cords, and Cables, Electrical-Metric Units Part 4: Cables, Special Purpose, Electrical (Sub-Miniature Electric Cables) Issue 3; Amendment 2. 30 pp.
MOD UK DSTAN 61-12: Part 13-01. Wires, Cords, and Cables, Electrical-Metric Units Part 13: Wires, Electrical (PVC Insulated Non-Sheathed Cables for Switchgear and Controlgear Wiring) Issue 2; Amendment 1. 22 pp.
MOD UK DSTAN 61-12: 90/001-88. Cables, Special Purpose, Electrical (Digital Data Transmission) Issue 2. 10 pp.
SAA AS 1178-86. Concentric Wire Neutral Cables—XLPE Insulated—for Electricity Supply at Working Voltages of 0.6/1 KV. 11 pp.
SNZ NZS 6401-73. Specification for PVC-Insulated Cables for Electric Power and Lighting (Reconfirmed 1991). 24 pp.

—PVC Insulated—Automotive
MOD UK DSTAN 61-12: Part 10-01. Wires, Cords and Cables, Electrical—Metric Units Part 10: PVC Insulated Cables for Road Vehicles Issue 1; Amendment 2. 18 pp.

—PVC Insulated—Cold Resistant
DIN VDE 0281 Pt 107-90. PVC Cables, Wires and Flexible Cords for Power Installation; PVC Single-Core Non-Sheathed Cable Cold-Resisting (May). 10 pp.

—PVC Insulated—Controlgear
BSI BS 6231-90. 1990 PVC—Insulated Cables for Switchgear and Controlgear Wiring. 13 pp.
BSI BS 6231-01. 1990 Amd 1 PVC-Insulated Cables for Switchgear and Controlgear Wiring (AMD 6936) May 1, 1992. 18 pp.
BSI BS 6231-81. (WITHDRAWN) 1981 PVC—Insulated Cables for Switchgear and Controlgear Wiring. 12 pp.

—PVC Insulated—PVC Sheathed
BSI BS 6004-01. 1991 Amd 1 PVC-Insulated Cables (Non-Armoured) for Electric Power and Lighting (AMD 7898) (E). 30 pp.
BSI BS 6500-90. 1990 Amd 1 Insulated Flexible Cords and Cables (AMD 6759) September 30, 1991 (F). 51 pp.
BSI BS 6500-02. 1990 Amd 2 Insulated Flexible Cords and Cables (AMD 6866) April 1, 1992. 56 pp.
BSI BS 6746-90. 1990 PVC Insulation and Sheath of Electric Cables. 12 pp.

Cables (Electric) *(Cont.)*
—PVC Insulated—PVC Sheathed *(Cont.)*
BSI BS 6746-01. 1990 Amd 1 PVC Insulation and Sheath of Electric Cables (AMD 6938) February 28, 1992. 14 pp.
BSI BS 6977-91. 1991 Insulated Flexible Cables for Lifts and for Other Flexible Connections. 27 pp.
BSI BS 6977-01. 1991 Amd 1 Insulated Flexible Cables for Lifts and for Other Flexible Connections (AMD 6934) May 1, 1992. 32 pp.
DIN VDE 0250 Pt 406-82. Insulated Power Cables; PVC Sheathed Cables NYMHYV (Nov). 10 pp.
DIN VDE 0250 Pt 819 (D). Cables, Wires and Flexible Cords for Power Installation; Cross-Linked PVC Flexible Cable. 10 pp.
DIN VDE 0281 Pt 403-89. PVC Cables, Wires and Flexible Cords for Power Installation; Flat PVC Sheathed Flexible Cable 05VVH6 (June) (Partially Supersedes DIN 57 281 Part 403/VDE 0281 Part 402/11.81). 11 pp.
DIN VDE 0281 Pt 404-89. PVC Cables, Wires and Flexible Cords for Power Installation; Flat PVC Sheathed Flexible Cable 07VVH6 (June) (Partially Supersedes DIN 57281 Part 404/VDE 0281 Part 404/11.81). 12 pp.
IEC 189 Pt 1-86. Low-Frequency Cables and Wires with PVC Insulation and PVC Sheath Part 1: General Test and Measuring Methods Second Edition; (Amendment 3-1992, Incorporating Amendment 1 and 2). 47 pp.
IEC 189 Pt 2-81. Low-Frequency Cables and Wires with PVC Insulation and PVC Sheath Part 2: Cables in Pairs, Triples, Quads and Quintuples for Inside Installations Third Edition; (Amendment 1-1989). 49 pp.
IEC 189 Pt 5-80. Low-Frequency Cables and Wires with PVC Insulation and PVC Sheath Part 5: Equipment Wires and Cables with Solid or Stranded Conductors, PVC Insulated, Screened, Single or One Pair Second Edition; (Amendment 2-1992, Incorporating Amd 1). 43 pp.
IEC 189 Pt 6-82. Low-Frequency Cables and Wires with PVC Insulation and PVC Sheath Part 6: Signalling Cables in Singles for Telecommunication Equipment and Installation Second Edition; (Amendment 1-1989). 33 pp.

—PVC Insulated—PVC Sheathed—Aerial
DIN VDE 0250 Pt 206-83. Cables, Wires and Flexible Cords for Power Installations; PVC-Sheathed Cable with Suspension Strand (Oct) (Partially Supersedes VDE 0250/03.69 and VDE 0250c/08.75). 12 pp.

—PVC Insulated—PVC Sheathed—Flameproof—Telecommunication Systems
DIN VDE 0891 Pt 7-84. Use of Cables and Insulated Cords for Telecommunications Systems and Data Processing Systems; Special Guidelines for Cables with Stranded Conductors for Increased Mechanical Stress for Telecommunications Systems and Data Processing Systems. 8 pp.

—PVC Insulated—PVC Sheathed—Insulation Resistance
BSI BS 6469: Sec 99.2-92. 1992 Insulating and Sheathing Materials of Electric Cables Part 99: Test Methods Used in the United Kingdom but Not Specified in Parts 1 to 5 Section 99: Electrical Tests. 10 pp.

—PVC Insulated—PVC Sheathed—Lighting
DIN VDE 0250 Pt 211-86. Cables, Wires and Flexible Cords for Power Installation; PVC—Lighting Cable with Sheath (June) (Partially Supersedes VDE 0250/03.69). 10 pp.

—PVC Insulated—PVC Sheathed—600 V
CNS C2058-89. 600V Grade Polyvinyl Chloride Insulated and Sheathed Cable (Nov)(3301).

—PVC Insulated—Sheathed—Junction Boxes
JIS C 8365-88. Junction Box for Indoor Wiring (for 600 V Grade Polyvinyl Chloride Insulated and Sheathed Cables: VVF). 12 pp.

—PVC Insulated—Short Circuiting/Grounding
IEC 1138-92. Cables for Portable Earthing and Short-Circuiting Equipment First Edition. 29 pp.

—PVC Insulated—Switchgear
BSI BS 6231-90. 1990 PVC—Insulated Cables for Switchgear and Controlgear Wiring. 13 pp.
BSI BS 6231-01. 1990 Amd 1 PVC-Insulated Cables for Switchgear and Controlgear Wiring (AMD 6936) May 1, 1992. 18 pp.
BSI BS 6231-81. (WITHDRAWN) 1981 PVC—Insulated Cables for Switchgear and Controlgear Wiring. 12 pp.

—PVC Insulated—1900/3300 V
BSI BS 6346-89. 1989 PVC-Insulated Cables for Electricity Supply. 40 pp.

Cables (Electric) *(Cont.)*
—PVC Insulated—1900/3300 V *(Cont.)*
BSI BS 6346-01. 1989 Amd 1 PVC-Insulated Cables for Electricity Supply (AMD 6831) February 28, 1992. 48 pp.
BSI BS 6346-02. 1989 Amd 2 PVC-Insulated Cables for Electricity Supply (AMD 7636) June 15, 1993. 49 pp.
SNZ BS 6346-87. Specification for PVC-Insulated Cables for Electricity Supply. 40 pp.

—PVC Insulated—450/750 V
BSI BS 7540-92. 1992 Use of Cables with a Rated Voltage Not Exceeding 450/750 V (E). 24 pp.
BSI BS 7540-01. 1992 Amd 1 Use of Cables with a Rated Voltage Not Exceeding 450/750 V (AMD 7293) August 15, 1992. 27 pp.
CENELEC HD 21.1 S2-90. Polyvinyl Chloride Insulated Cables of Rated Voltages up to and Including 450/750 V Part 1: General Requirements. 27 pp.
CENELEC HD 21.1 S2-90. Polyvinyl Chloride Insulated Cables of Rated Voltages up to and Including 450/750 V Part 1: General Requirements (Reprint). 28 pp.
CENELEC HD 21.1 S2/A1-88. AMD 1 Polyvinyl Chloride Insulated Cables of Rated Voltages up to and Including 450/750 V Part 1: General Requirements. 3 pp.
CENELEC HD 21.1 S2/A3-89. AMD 3 Polyvinyl Chloride Insulated Cables of Rated Voltages up to and Including 450/750 V Part 1: General Requirements. 4 pp.
CENELEC HD 21.1 S2/A4-88. AMD 4 Polyvinyl Chloride Insulated Cables of Rated Voltages up to and Including 450/750 V Part 1: General Requirements. 5 pp.
CENELEC HD 21.1 S2/A5-90. AMD 5 Polyvinyl Chloride Insulated Cables of Rated Voltages up to and Including 450/750 V Part 1: General Requirements. 4 pp.
CENELEC HD 21.1 S2/A6-91. AMD 6 Polyvinyl Chloride Insulated Cables of Rated Voltages up to and Including 450/750 V Part 1: General Requirements. 7 pp.
CENELEC HD 21.1 S2/A7-92. AMD 7 Polyvinyl Chloride Insulated Cables of Rated Voltages up to and Including 450/750 V Part 1: General Requirements. 5 pp.
CENELEC HD 21.1 S2/A9-93. AMD 9 Polyvinyl Chloride Insulated Cables of Rated Voltages up to and Including 450/750 V Part 1: General Requirements. 6 pp.
CENELEC HD 21.2 S2-90. Polyvinyl Chloride Insulated Cables of Rated Voltages up to and Including 450/750 V—Part 2: Test Methods. 18 pp.
CENELEC HD 21.2 S2/A2-90. AMD 2 Polyvinyl Chloride Insulated Cables of Rated Voltages up to and Including 450/750 V—Part 2: Test Methods. 6 pp.
CENELEC HD 21.2 S2/A3-93. AMD 3 Polyvinyl Chloride Insulated Cables of Rated Voltages up to and Including 450/750 V Part 2: Test Methods. 8 pp.
CENELEC HD 21.7 S1-90. Polyvinyl Chloride Insulated Cables of Rated Voltages up to and Including 450/750V Part 7: Single Core Non-Sheated Cables for Internal Wiring for a Conductor Temperature of 90 Degrees Celsius. 8 pp.
CENELEC HD 21.7 S1/A1-92. AMD 1 Polyvinyl Chloride Insulated Cables of Rated Voltages up to and Including 450/750 V Part 7: Single Core Non-Sheated Cables for Internal Wiring for a Conductor Temperature of 90 Degrees Celsius. 4 pp.
CENELEC HD 21.9 S1-88. Single Core Non Sheathed Cable for Installation at Low Temperatures. 11 pp.
CENELEC HD 21.9 S1-90. Polyvinyl Chloride Insulated Cables of Rated Voltages Single Core Non-Sheathed Cable for Installation at Low Temperature. 11 pp.
CENELEC HD 21.9 S1/A1-90. AMD 1 Single Core Non Sheathed Cable for Installation at Low Temperatures. 4 pp.
CENELEC HD 516 S1-90. Guide to Use of Low Voltage Harmonized Cables. 26 pp.
CENELEC HD 516 S1/A1-91. AMD 1 Guide to Use of Low Voltage Harmonized Cables. 5 pp.
CENELEC HD 516 S1/A2-92. AMD 2 Guide to Use of Low Voltage Harmonized Cables. 6 pp.
CENELEC HD 516 S1/A3-93. AMD 3 Guide to Use of Low Voltage Harmonized Cables. 10 pp.
CENELEC HD 516 S1/A4-92. AMD 4 Guide to Use of Low Voltage Harmonized Cables. 7 pp.
CENELEC HD 516 S1/A5-93. AMD 5 Guide to Use of Low Voltage Harmonized Cables. 6 pp.
CENELEC HD 516 S1/A6-93. AMD 6 Guide to Use of Low Voltage Harmonized Cables. 6 pp.
IEC 227 Pt 1-93. Polyvinyl Chloride Insulated Cables of Rated Voltages up to and Including 450/750 V Part 1: General Requirements Second Edition. 47 pp.

INDUSTRY STANDARDS

INTERNATIONAL AND NON-U.S. NATIONAL STANDARDS
SUBJECT INDEX

Cables

Cables (Electric) (Cont.)

—PVC Insulated—450/750 V—Electrical Leads
  CENELEC HD 21.10 S1-93. Polyvinyl Chloride Insulated Cables of Rated Voltages up to and Including 450/750 V Part 10: Extensible Leads. 14 pp.

—PVC Insulated—600/1000 V
  BSI BS 4553-92. 1992 600/1000 V PVC-Insulated Single-Phase Split Concentric Cables With Copper Conductors for Electricity Supply. 12 pp.
  BSI BS 4553-88. 1988 Amd 3 600/1000 PVC-Insulated Single-Phase Split Concentric Cables With Copper Conductors for Electricity Supply. 10 pp.
  BSI BS 6346-89. 1989 PVC-Insulated Cables for Electricity Supply. 40 pp.
  BSI BS 6346-01. 1989 Amd 1 PVC-Insulated Cables for Electricity Supply (AMD 6831) February 28, 1992. 48 pp.
  BSI BS 6346-02. 1989 Amd 2 PVC-Insulated Cables for Electricity Supply (AMD 7636) June 15, 1993. 49 pp.
  SNZ BS 6346-87. Specification for PVC-Insulated Cables for Electricity Supply. 40 pp.

—Quality Assurance
  MOD UK DSTAN 61-12: Part 0-92. Wires Cords and Cables Electrical—Metric Units Part 0: General Requirements and Test Methods for Qualification Approval Generic Specification Issue 1. 62 pp.
  MOD UK DSTAN 61-12: Part 0-01. Wires Cords and Cables Electrical—Metric Units Part 0: General Requirements and Test Methods for Qualification Approval Generic Specification Issue 1; Amendment 1. 65 pp.

—Reels
  SAA AS 3983-91. Metal Drums for Insulated Electric Cables and Bare Conductors (Supersedes AS C365.2—1970). 10 pp.

—Resistance Welding Equipment
  ISO 5828-83. Resistance Welding Equipment—Secondary Connecting Cables with Terminals Connected to Water-Cooled Lugs—Dimensions and Characteristics First Edition. 6 pp.

—Rolling Stock—Jumpers
  JIS E 6801-90. Jumper Cables for Railway Rolling Stock.

—Rubber Insulated
  BSI BS 6195-93. 1993 Insulated Flexible Cables and Cords for Coil Leads (E). 19 pp.
  BSI BS 6195-69. 1969 Amd 4 Insulated Flexible Cables and Cords for Coil Leads (AMD 4929) September 30, 1985. 40 pp.
  BSI BS 6500-90. 1990 Amd 1 Insulated Flexible Cords and Cables (AMD 6759) September 30, 1991 (F). 51 pp.
  BSI BS 6500-02. 1990 Amd 2 Insulated Flexible Cords and Cables (AMD 6866) April 1, 1992. 56 pp.
  BSI BS 6977-91. 1991 Insulated Flexible Cables for Lifts and for Other Flexible Connections. 27 pp.
  BSI BS 6977-01. 1991 Amd 1 Insulated Flexible Cables for Lifts and for Other Flexible Connections (AMD 6934) May 1, 1992. 32 pp.
  CENELEC HD 22.4 S2-83. Rubber Insulated Cables of Rated Voltages up to and Including 450/750 V—Part 4: Cords and Flexible Cables. 22 pp.
  CENELEC HD 22.4 S2/A1-89. AMD 1 Rubber Insulated Cables of Rated Voltages up to and Including 450/750 V—Part 4: Cords and Flexible Cables. 3 pp.
  CENELEC HD 22.4 S2/A2-89. AMD 2 Rubber Insulated Cables of Rated Voltages up to and Including 450/750 V—Part 4: Cords and Flexible Cables. 3 pp.
  CENELEC HD 22.4 S2/A3-89. AMD 3 Rubber Insulated Cables of Rated Voltages up to and Including 450/750 V—Part 4: Cords and Flexible Cables. 4 pp.
  CENELEC HD 22.4 S2/A4-89. AMD 4 Rubber Insulated Cables of Rated Voltages up to and Including 450/750 V—Part 4: Cords and Flexible Cables. 4 pp.
  CENELEC HD 360-76. Rubber Insulated Lift Cables for Normal Use. 9 pp.
  CENELEC HD 360 S2-90. Circular Rubber Insulated Lift Cables for Normal Use. 17 pp.
  CENELEC HD 360 S2/A1-91. AMD 1 Circular Rubber Insulated Lift Cables for Normal Use. 6 pp.
  CENELEC HD 360.2-79. Rubber Insulated Lift Cables for Normal Use. 3 pp.
  CNS C3009-87. Method of Test for Rubber-Insulated Wires & Cables (Sep)(687).
  CSA C22.2 NO 38-M1986. Thermoset Insulated Wires and Cables; (Gen Instr 1 Thru 3). 64 pp.
  CSA 983B Bull. Electrical Bulletin 983B April 25, 1977 to C22.2 NO 38. 39 pp.

Cables (Electric) (Cont.)

—Rubber Insulated (Cont.)
  DIN VDE 0250 Pt 602-85. Cables, Wires and Flexible Cords for Power Installation: Special Rubber Insulated Single Core Cables (Mar) (Replaces Parts of VDE 0250 and DIN VDE 0250C—but Note Transitional Period.). 9 pp.
  DIN VDE 0250 Pt 812-85. Cables, Wires and Flexible Cords for Power Installation: Rubber Insulated Flexible Cable NSSHOU (May) (Replaces Part of VDE 0250—but Note Transitional Period.). 13 pp.
  DIN VDE 0250 Pt 814-85. Cables, Wires and Flexible Cords for Power Installation: Rubber Insulated Flexible Cable NSHTOU (Feb) (Replaces Part of VDE 0250C—but Note Transitional Period.). 10 pp.
  DIN VDE 0282 Pt 1-85. Rubber Cables, Wires and Flexible Cords for Power Installation; General Requirements (Apr). 16 pp.
  IEC 245 Pt 4-80. Rubber Insulated Cables of Rated Voltage up to and Including 450/750 V Part 4: Cords and Flexible Cables First Edition; (Amendment 2-1988, Incorporating Amendment 1). 56 pp.
  JIS C 3005-91. Testing Methods for Rubber or Plastic Insulated Wires and Cables. 39 pp.
  JIS C 3327-87. 600 V Rubber Insulated Flexible Cables. 22 pp.
  MOD UK DSTAN 61-12: 90/001-88. Cables, Special Purpose, Electrical (Digital Data Transmission) Issue 2. 10 pp.

—Rubber Insulated—Aluminum Sheathed
  CSA C22.2 NO 123-M1985. Aluminum Sheathed Cables (R 1992); (Gen Instr 1). 26 pp.

—Rubber Insulated—Arc Welding Equipment
  BSI BS 638: Part 4-79. 1979 Amd 1 Arc Welding Power Sources, Equipment and Accessories Part 4: Welding Cables. 8 pp.
  IEC 245 Pt 6-80. Rubber Insulated Cables of Rated Voltages up to and Including 450/750 V Part 6: Arc Welding Electrode Cables First Edition; (Amendment 1-1985). 17 pp.

—Rubber Insulated—Heat Resistant
  DIN VDE 0282 Pt 501-90. Rubber Cables, Wires and Flexible Cords for Power Installation; Heat-Resistant Rubber-Insulated Cable (Jan) (Supersedes DIN VDE 0250 Part 501/03.85). 11 pp.

—Rubber Insulated—High Voltage—Automotive
  JIS C 3405-87. High-Voltage Cables for Automobile. 8 pp.
  JIS C 3409-87. High-Voltage Resistance Cables for Automobile. 13 pp.

—Rubber Insulated—Identification Systems
  CSA C22.2 NO 38-M1986. Thermoset Insulated Wires and Cables; (Gen Instr 1 Thru 3). 64 pp.

—Rubber Insulated—Lead Sheathed—600 V
  CNS C2037-87. 600 V Grade Rubber Insulated Lead Sheathed Cable (Oct)(2146). 9 pp.
  JIS C 3310-87. 600V Grade Rubber Insulated Lead Sheathed Cables. 11 pp.

—Rubber Insulated—Neoprene Sheathed
  CNS C2075-87. Rubber Insulated Poly-Chloroprene Sheathed Cables (Dec)(6068). 11 pp.

—Rubber Insulated—Rubber Sheathed
  BSI BS 6899-91. 1991 Specification for Rubber Insulation and Sheath of Electric Cables. 19 pp.
  BSI BS 6899-01. 1991 Amd 1 Rubber Insulation and Sheath of Electric Cables (AMD 6937) April 1, 1992. 22 pp.
  BSI BS 6977-91. 1991 Insulated Flexible Cables for Lifts and for Other Flexible Connections. 27 pp.
  BSI BS 6977-01. 1991 Amd 1 Insulated Flexible Cables for Lifts and for Other Flexible Connections (AMD 6934) May 1, 1992. 32 pp.
  DIN VDE 0282 Pt 804-85. Rubber Cables, Wires and Flexible Cords for Power Installation; Ordinary Tough Rubber-Sheathed Flexible Cable 05RR (Apr). 8 pp.
  DIN VDE 0282 Pt 810-85. Rubber Cables, Wires and Flexible Cords for Power Installation; Rubber-Sheathed Flexible Cable 07RN (Apr). 11 pp.
  DIN VDE 0282 Pt 817-85. Rubber Cables, Wires and Flexible Cords for Power Installation; Rubber-Sheathed Flexible Cable 05RN (Apr). 7 pp.

—Rubber Insulated—Rubber Sheathed—Lighting
  DIN VDE 0282 Pt 604-90. Rubber Cables, Wires and Flexible Cords for Power Installation; Flexible Cable for Illumination Chains (Jan) (Supersedes DIN 57250 Part 604/VDE 0250 Part 604/10.83). 8 pp.

—Rubber Insulated—Welding Equipment
  JIS C 3404-87. Welding Cables. 9 pp.

Cables (Electric) (Cont.)

—Rubber Insulated—450/750 V
  BSI BS 7540-92. 1992 Use of Cables with a Rated Voltage Not Exceeding 450/750 V (E). 24 pp.
  BSI BS 7540-01. 1992 Amd 1 Use of Cables with a Rated Voltage Not Exceeding 450/750 V (AMD 7293) August 15, 1992. 27 pp.
  CENELEC HD 22.1 S2-83. Rubber Insulated Cables of Rated Voltages up to and Including 450/750 V—Part 1: General Requirements. 27 pp.
  CENELEC HD 22.1 S2/A1-89. AMD 1 Rubber Insulated Cables of Rated Voltages up to and Including 450/750 V—Part 1: General Requirements. 4 pp.
  CENELEC HD 22.1 S2/A2-89. AMD 2 Rubber Insulated Cables of Rated Voltages up to and Including 450/750 V—Part 1: General Requirements. 8 pp.
  CENELEC HD 22.1 S2/A3-89. AMD 3 Rubber Insulated Cables of Rated Voltages up to and Including 450/750 V—Part 1: General Requirements. 4 pp.
  CENELEC HD 22.1 S2/A4-89. AMD 4 Rubber Insulated Cables of Rated Voltages up to and Including 450/750 V—Part 1: General Requirements. 4 pp.
  CENELEC HD 22.1 S2/A5-89. AMD 5 Rubber Insulated Cables of Rated Voltages up to and Including 450/750 V—Part 1: General Requirements. 4 pp.
  CENELEC HD 22.1 S2/A6-90. AMD 6 Rubber Insulated Cables of Rated Voltages up to and Including 450/750 V—Part 1: General Requirements. 5 pp.
  CENELEC HD 22.1 S2/A7-90. AMD 7 Rubber Insulated Cables of Rated Voltages up to and Including 450/750 V—Part 1: General Requirements. 4 pp.
  CENELEC HD 22.1 S2/A8-91. AMD 8 Rubber Insulated Cables of Rated Voltages up to and Including 450/750 V—Part 1: General Requirements. 5 pp.
  CENELEC HD 22.1 S2/A9-91. AMD 9 Rubber Insulated Cables of Rated Voltages up to and Including 450/750 V—Part 1: General Requirements. 4 pp.
  CENELEC HD 22.1 S2-92. Rubber Insulated Cables of Rated Voltages up to and Including 450/750 V Part 1: General Requirements (Reprint Incorporating A1 to A10). 31 pp.
  CENELEC HD 22.1 S2/A11-92. AMD 11 Rubber Insulated Cables of Rated Voltages up to and Including 450/750 V Part 1: General Requirements. 4 pp.
  CENELEC HD 22.1 S2/A12-92. AMD 12 Rubber Insulated Cables of Rated Voltages up to and Including 450/750 V Part 1: General Requirements. 4 pp.
  CENELEC HD 22.1 S2/A13-92. AMD 13 Rubber Insulated Cables of Rated Voltages up to and Including 450/750 V Part 1: General Requirements. 9 pp.
  CENELEC HD 22.2 S2-82. Rubber Insulated Cables or Rated Voltages up to and Including 450/750 V—Part 2: Test Methods. 21 pp.
  CENELEC HD 22.2 S2-92. Rubber Insulated Cables of Rated Voltages up to and Including 450/750 V Part 2: Test Methods (Reprint Incorporating A1 to A4). 26 pp.
  CENELEC HD 22.2 S2/A1-89. AMD 1 Rubber Insulated Cables of Rated Voltages up to and Including 450/750 V—Part 2: Test Methods. 21 pp.
  CENELEC HD 22.2 S2/A2-90. AMD 2 Rubber Insulated Cables of Rated Voltages up to and Including 450/750 V—Part 2: Test Methods. 7 pp.
  CENELEC HD 22.2 S2/A3-90. AMD 3 Rubber Insulated Cables of Rated Voltages up to and Including 450/750 V—Part 2: Test Methods. 7 pp.
  CENELEC HD 22.2 S2/A5-92. AMD 5 Rubber Insulated Cables of Rated Voltages up to and Including 450/750 V Part 2: Test Methods. 6 pp.
  CENELEC HD 22.2 S2/A6-92. AMD 6 Rubber Insulated Cables of Rated Voltages up to and Including 450/750 V Part 2: Test Methods. 6 pp.
  CENELEC HD 22.2 S2/A7-92. AMD 7 Rubber Insulated Cables of Rated Voltages up to and Including 450/750 V Part 2: Test Methods. 5 pp.
  CENELEC HD 22.4 S2-92. Rubber Insulated Cables of Rated Voltages up to and Including 450/750 V Part 4: Cords and Flexible Cables (Reprint Incorporating A1 to A5). 26 pp.
  CENELEC HD 22.4 S2/A6-92. AMD 6 Rubber Insulated Cables of Rated Voltages up to and Including 450/750 V Part 4: Cords and Flexible Cables. 4 pp.
  CENELEC HD 22.7 S1-89. Rubber Insulated Cables of Rated Voltage up to and Including 450/750V Part 7: Cables with Increased Heat Resistance for Internal Wiring for a Conductor Temperature of 110 Ddegrees Celsius. 9 pp.

INDEX and DIRECTORY of

INTERNATIONAL AND NON-U.S. NATIONAL STANDARDS
SUBJECT INDEX

Cables

Cables (Electric) *(Cont.)*

—Rubber Insulated—450/750 V *(Cont.)*
CENELEC HD 22.7 S1-92. Rubber Insulated Cables of Rated Voltages up to and Including 450/750 V Part 7: Cables with Increased Heat Resistance for Internal Wiring for a Conductor Temperature of 110 Degrees C (Reprint Incorporating A1). 12 pp.
CENELEC HD 22.9 S1-92. Rubber Insulated Cables of Rated Voltages up to and Including 450/750V—Part 9: Single Core Non-Sheathed Cables for Fixed Wiring Low Emission of Smoke and Corrosive Gases. 15 pp.
CENELEC HD 516 S1-90. Guide to Use of Low Voltage Harmonized Cables. 26 pp.
CENELEC HD 516 S1/A1-91. AMD 1 Guide to Use of Low Voltage Harmonized Cables. 5 pp.
CENELEC HD 516 S1/A2-92. AMD 2 Guide to Use of Low Voltage Harmonized Cables. 6 pp.
CENELEC HD 516 S1/A3-93. AMD 3 Guide to Use of Low Voltage Harmonized Cables. 10 pp.
CENELEC HD 516 S1/A4-92. AMD 4 Guide to Use of Low Voltage Harmonized Cables. 7 pp.
CENELEC HD 516 S1/A5-93. AMD 5 Guide to Use of Low Voltage Harmonized Cables. 6 pp.
CENELEC HD 516 S1/A6-93. AMD 6 Guide to Use of Low Voltage Harmonized Cables. 6 pp.
IEC 245 Pt 1-85. Rubber Insulated Cables of Rated Voltages up to and Including 450/750 V Part 1: General Requirements Second Edition; (Corrigendum—Jan 1988). 41 pp.
IEC 245 Pt 2-80. Rubber Insulated Cables of Rated Voltages up to and Including 450/750 V Part 2: Test Methods First Edition; (Amendment 1-1985). 36 pp.

—Saponification Number
DIN VDE 0472 Pt 704-89. Testing of Insulated Cables and Cords; Saponification Value (Mar) (Supersedes DIN 57 472 Part 704). 6 pp.

—Sheathed—Fire Hazards
CSA 1334 Bull. Electrical Bulletin 1334 August 28, 1981 to C22.2 NO 16. 1 p.
CSA 632 Bull. Electrical Bulletin 632 March 29, 1966 to C22.2 NO 38. 2 pp.
CSA 671 Bull. Electrical Bulletin 671 March 16, 1967 to C22.2 NO 38. 3 pp.
CSA 1368 Bull. Electrical Bulletin 1368 April 7, 1982 to C22.2 NO 38. 2 pp.
CSA 632 Bull. Electrical Bulletin 632 March 29, 1966 to C22.2 NO 48. 2 pp.
CSA 671 Bull. Electrical Bulletin 671 March 16, 1967 to C22.2 NO 48. 3 pp.
CSA 1334 Bull. Electrical Bulletin 1334 August 28, 1981 to C22.2 NO 48. 1 p.
CSA 1368 Bull. Electrical Bulletin 1368 April 7, 1982 to C22.2 NO 48. 2 pp.
MOD UK DSTAN 61-12: Part 31-90. Wires, Cords, and Cables, Electrical, Metric Units Part 31: Sheaths Limited Fire Hazard Issue 1. 20 pp.

—Sheathed—Halogen Free
DIN VDE 0250 Pt 214-87. Cables, Wires and Flexible Cords for Power Installations; Halogen Free Light Sheathed Cable with Improved Fire Behaviour (Feb). 14 pp.
DIN VDE 0266 Pt 4 (D)-90. Halogen-Free Cables with Improved Characteristics in Case of Fire: Nominal Voltage: Uo/UO.6/1 kV; Halogen-Free, Low Fuming Cables with Reduced Fire Propagation for Use in the Containment of Nuclear Power Plants (July) (Intended as Partial Replacement for DIN VDE 0266/02.85). 12 pp.

—Ships
IEC 92 Pt 3-65. Electrical Installations in Ships Part 3: Cables (Construction, Testing and Installations) Second Edition; (Amendment 1-1969) (Amendment 2-1971) (Amendment 3-1973) (Amendment 4-1974) (Amendment 5-1979) (Amendment 6-1984). 266 pp.
IEC 92 Pt 350-88. Electrical Installations in Ships Part 350: Low-Voltage Shipboard Power Cables General Construction and Test Requirements First Edition. 79 pp.
IEC 92 Pt 352-79. Electrical Installations in Ships Choice and Installation of Cables for Low-Voltage Power Systems First Edition; (Amendment 1-1987). 59 pp.
JIS F 8071-86. Electrical Installations in Ships Part 352 Choice and Installation of Cables for Low-Voltage Power Systems (IEC 92-352-1979).
SAA AS 1168-73. Elastomer-Insulated Electric Cables for Fixed Wiring in Ships Amdt 1 July 1986. 63 pp.

—Ships—Multiconductor—AC Power Supplies
NATO STANAG 4143 ED 1 AMD 4-76. Ship/Shore Connection Terminals for 3-Phase AC Power. 6 pp.

—Ships—PVC Insulated
SAA AS 1695-75. PVC Insulated Fixed Electric Cables for Ships Amdt 1 July 1986. 64 pp.

Cables (Electric) *(Cont.)*

—Signal Circuits
CSA C22.2 NO 208-M1986. Fire Alarm and Signal Cable (R 1992); (Gen Instr 1 Thru 5). 28 pp.

—Signal Circuits—Multiconductor
CSA C22.2 NO 208-M1986. Fire Alarm and Signal Cable (R 1992); (Gen Instr 1 Thru 5). 28 pp.

—Signs
CSA C22.2 NO 17-1973. Cable for Luminous-Tube Signs and for Oil-and Gas-Burner Ignition Equipment (R 1992); (Gen Instr 1) (Amd 1 February 1981) (Amd 2-8 March 1988). 28 pp.

—Silicone Rubber Insulated—Aluminum Sheathed
CSA C22.2 NO 123-M1985. Aluminum Sheathed Cables (R 1992); (Gen Instr 1). 26 pp.

—Silicone Rubber Insulated—Heat Resistant
CENELEC HD 22.3 S2-83. Rubber Insulated Cables of Rated Voltages up to and Including 450/750 V—Part 3: Heat Resistant Silicone Rubber Insulated Cables. 8 pp.
CENELEC HD 22.3 S2/A1-90. AMD 1 Rubber Insulated Cables of Rated Voltages up to and Including 450/750 V—Part 3: Heat Resistant Silicone Rubber Insulated Cables. 6 pp.
CENELEC HD 22.3 S2-92. Rubber Insulated Cables of Rated Voltages up to and Including 450/750 V Part 3: Heat Resistant Silicone Rubber Insulated Cables (Reprint Incorporating A1). 12 pp.
DIN VDE 0250 Pt 816-82. Insulated Power Cables; Heat Resistant Silicone Rubber Insulated Flexible Cable (June). 8 pp.
DIN VDE 0282 Pt 601-85. Rubber Cables, Wires and Flexible Cords for Power Installation; Heat-Resistant Silicon Rubber Insulated Cable (Apr). 8 pp.
IEC 245 Pt 3-80. Rubber Insulated Cables of Rated Voltages up to and Including 450/750 V Part 3: Heat Resistant Silicone Insulated Cables First Edition; (Amendment 1-1985). 14 pp.

—Sleeves—High Temperature Testing
DIN VDE 0472 Pt 632-87. Testing of Cables and Insulated Cords; Behaviour at High Temperatures (Oct) (Partially Supersedes VDE 0271/03.69 Which Was Withdrawn in December 1988). 4 pp.

—Smoke Density
BSI BS 7622: Part 2-93. 1993 Measurement of Smoke Density of Electric Cables Buning Under Defined Conditions Part 2: Test Procedure and Requirements (E). 14 pp.
CENELEC HD 606.2 S1-92. Measurement of Smoke Density of Electric Cables Burning Under Defined Conditions Part 2: Test Procedure and Requirements. 10 pp.
IEC 1034 Pt 2-91. Measurement of Smoke Density of Electric Cables Burning Under Defined Conditions Part 2: Test Procedure and Requirements First Edition. 16 pp.

—Smoke Density—Testing Equipment
BSI BS 7622: Part 1-93. 1993 Measurement of Smoke Density of Electric Cables Buning Under Defined Conditions Part 1: Test Apparatus (E). 17 pp.
CENELEC HD 606.1 S1-92. Measurement of Smoke Density of Electric Cables Burning Under Defined Conditions Part 1: Test Apparatus. 4 pp.
IEC 1034 Pt 1-90. Measurement of Smoke Density of Electric Cables Burning Under Defined Conditions Part 1: Test Apparatus First Edition. 24 pp.

—Solderability
DIN VDE 0472 Pt 808-84. Testing of Cables, Wires and Flexible Cords; Tinning, Solderability and Soldering Shrinkage (VDE Specification) (Feb). 7 pp.

—Spot Welding Equipment
CNS C2062-84. Secondary Cables for Portable Spot Welding Machine (Jul)(4618).
CNS C3058-84. Method of Test for Secondary Cable for Portable Spot Welding Machine (Jul)(4619).
JIS C 9318-90. Water-Cooled Secondary Cables for Portable Spot Welding Machines. 20 pp.

—Stripping
CEN PREN 2812-91. Stripping of Electric Cables as Defined by EN 2084, EN 2234, EN 2235 and EN 2346. 10 pp.

—Submarines
MOD UK NES 503-88. Requirements for Electrical Installation Cabling Diagrams and Associated Data Issue 2 (08.88). 89 pp.
MOD UK NES 503-92. Requirements for Electrical Installation Cabling Diagrams and Associated Data Issue 3 (01.92). 94 pp.

Cables (Electric) *(Cont.)*

—Submarines *(Cont.)*
MOD UK NES 512: Part 7-91. Guide to Cables, Electrical and Associated Items Part 7: Cables and Wires, Electrical, Miscellaneous for Special Services Issue 3 (04.91). 90 pp.
MOD UK NES 512: Part 7-01. Guide to Cables, Electrical and Associated Items Part 7: Cables and Wires, Electrical, Miscellaneous for Special Services Issue 3 (04.91); Amendment 1. 91 pp.
MOD UK NES 512: Part 7-93. Guide to Cables, Electrical and Associated Items Part 7: Cables and Wires, Electrical, Miscellaneous for Special Services Issue 4 (02.93). 84 pp.
MOD UK NES 512: Part 13-91. Guide to Cables, Electrical and Associated Items Part 13: Cables (Imperial) Nearest Metric Equivalents Issue 2 (03.91). 66 pp.
MOD UK NES 513-01. Guide to the Design of System Cabling Issue 1 (06.81); Amendment 1 (Superseded by NES 502). 42 pp.
MOD UK NES 517: Part 5-92. Requirements for Cables, Electric Part 5: Miscellaneous Issue 1 (07.92). 53 pp.
MOD UK NES 523-88. Guide to the Installation of Thin-Wall Insulated, Limited Fire Hazard Equipment Wire and Electric Cable Issue 2 (05.88). 51 pp.
MOD UK NES 523-92. Guide to the Installation of Thin—Wall Insulated, Limited Fire Hazard Equipment Wire and Electric Cable Issue 3 (01.92). 52 pp.
MOD UK NES 525-91. Requirements for Electric Cables Thin-Wall Insulated, Limited Fire Hazard Issue 3 (09.91). 36 pp.
MOD UK NES 525-01. Requirements for Electric Cables Thin—Wall Insulated, Limited Fire Hazard Issue 3 (09.91); Amendment 1. 47 pp.

—Subminiature—Naval Ships
MOD UK NES 512: Part 6-91. Guide to Cables, Electrical and Associated Items Part 6: Cables, Electrical, Miniature and Sub—Miniature and Equipment Wires Issue 2 (05.91). 100 pp.
MOD UK NES 512: Part 6-01. Guide to Cables, Electrical and Associated Items Part 6: Cables, Electrical, Miniature and Sub-Miniature and Equipment Wires Issue 2 (05.91); Amendment 2. 116 pp.
MOD UK NES 643-92. Requirements for Cables Electrical, Sub-Miniature and Small Multicore, Limited Fire Hazard for Special Purposes Issue 1 (12.92). 42 pp.

—Subminiature—Submarines
MOD UK NES 512: Part 6-91. Guide to Cables, Electrical and Associated Items Part 6: Cables, Electrical, Miniature and Sub—Miniature and Equipment Wires Issue 2 (05.91). 100 pp.
MOD UK NES 512: Part 6-01. Guide to Cables, Electrical and Associated Items Part 6: Cables, Electrical, Miniature and Sub-Miniature and Equipment Wires Issue 2 (05.91); Amendment 2. 116 pp.
MOD UK NES 643-92. Requirements for Cables Electrical, Sub-Miniature and Small Multicore, Limited Fire Hazard for Special Purposes Issue 1 (12.92). 42 pp.

—Symbols
CNS C5113-81. Graphic Symbols for Conductor, Cable and Wiring (May)(7362).

—Tear Strength
DIN VDE 0472 Pt 613-85. Testing of Cables, Wires and Flexible Cords; Tear Resistance (Nov) (Partial Replacement for VDE 0472/09.71 and VDE 0472d/12.77). 5 pp.

—Telecommunication Equipment
DIN VDE 0813-88. Distribution Wires for Telecommunication and Data Processing Systems (Nov). 25 pp.

—Telecommunication Equipment—Bend Testing
DIN VDE 0472 Pt 603-89. Testing of Cables, Wires and Flexible Cords; Bending Behaviour (VDE Specification) (July). 17 pp.

—Telecommunication Systems—Wiring
DIN VDE 0891 Pt 5-85. Use of Cables and Insulated Cords for Telecommunications and Information Processing Installations. Special Guidance on Wiring Cables and Cords According to DIN/VDE 0815 (Nov). 11 pp.

—Temperature Limits
IEC 724-84. Guide to the Short-Circuit Temperature Limits of Electric Cables with a Rated Voltage Not Exceeding 0,6/1,0 kV Second Edition; (Amendment 1-1993). 35 pp.

INDUSTRY STANDARDS

Cables (Electric) (Cont.)

—Temperature Limits (Cont.)
IEC 986-89. Guide to the Short-Circuit Temperature Limits of Electric Cables with a Rated Voltage from 1,8/3 (3,6) kV to 18/30 (36) kV First Edition; (Amendment 1-1993). 34 pp.

—Tensile Testing
DIN VDE 0472 Pt 625-83. Testing of Cables and Insulated Conductors; Tensile and Elongation Characteristics of Strain Bearing Elements (Jan). 5 pp.

—Testing
SAA AS 1660.4-86. Methods of Test for Electrical Cables, Cords and Conductors—Part 4: Complete Cable and Flexible Cord. 6 pp.

—Testing Equipment
CSA CAN/CSA-C22. 2 NO 0.3-92. Test Methods for Electrical Wires and Cables; (Gen Instr 1). 121 pp.

—Thermoplastic Insulated
CSA 1334 Bull. Electrical Bulletin 1334 August 28, 1981 to C22.2 NO 16. 1 p.
CSA 632 Bull. Electrical Bulletin 632 March 29, 1966 to C22.2 NO 38. 2 pp.
CSA 671 Bull. Electrical Bulletin 671 March 16, 1967 to C22.2 NO 38. 3 pp.
CSA 730A Bull. Electrical Bulletin 730A May 31, 1974 to C22.2 NO 38. 2 pp.
CSA 983B Bull. Electrical Bulletin 983B April 25, 1977 to C22.2 NO 38. 39 pp.
CSA 983C Bull. Electrical Bulletin 983C November 14, 1978 to C22.2 NO 38. 3 pp.
CSA 1368 Bull. Electrical Bulletin 1368 April 7, 1982 to C22.2 NO 38. 2 pp.
CSA 1403 Bull. Electrical Bulletin 1403 July 5, 1983 to C22.2 NO 38. 2 pp.
CSA 1316A Bull. Electrical Bulletin 1316A February 28, 1985 to C22.2 NO 38. 2 pp.
CSA 1240 Bull. Electrical Bulletin 1240 September 24, 1979 to C22.2 NO 51. 2 pp.
CSA 632 Bull. Electrical Bulletin 632 March 29, 1966 to C22.2 NO 75. 2 pp.
CSA 671 Bull. Electrical Bulletin 671 March 16, 1967 to C22.2 NO 75. 3 pp.
CSA 730A Bull. Electrical Bulletin 730A May 31, 1974 to C22.2 NO 75. 2 pp.
CSA 983B Bull. Electrical Bulletin 983B April 25, 1977 to C22.2 NO 75. 39 pp.
CSA 983C Bull. Electrical Bulletin 983C November 14, 1978 to C22.2 NO 75. 3 pp.
CSA 1240 Bull. Electrical Bulletin 1240 September 24, 1979 to C22.2 NO 75. 2 pp.
CSA 1334 Bull. Electrical Bulletin 1334 August 28, 1981 to C22.2 NO 75. 1 p.
CSA 1368 Bull. Electrical Bulletin 1368 April 7, 1982 to C22.2 NO 75. 2 pp.
CSA 1403 Bull. Electrical Bulletin 1403 July 5, 1983 to C22.2 NO 75. 4 pp.
CSA 1316A Bull. Electrical Bulletin 1316A February 28, 1985 to C22.2 NO 75. 2 pp.

—Thermoplastic Insulated—Well Pumps
CSA C22.2 NO 75-M1983. Thermoplastic Insulated Wires and Cables (R 1992); (Gen Instr 1 Thru 6). 48 pp.

—Thermoplastic Insulated—600 V
CSA C22.2 NO 75-M1983. Thermoplastic Insulated Wires and Cables (R 1992); (Gen Instr 1 Thru 6). 48 pp.

—Thermoset Insulated
CSA 916 Bull. Electrical Bulletin 916 July 3, 1973 to C22.2 NO 16. 2 pp.
CSA 1334 Bull. Electrical Bulletin 1334 August 28, 1981 to C22.2 NO 16. 1 p.
CSA C22.2 NO 38-M1986. Thermoset Insulated Wires and Cables; (Gen Instr 1 Thru 3). 64 pp.
CSA 632 Bull. Electrical Bulletin 632 March 29, 1966 to C22.2 NO 38. 2 pp.
CSA 671 Bull. Electrical Bulletin 671 March 16, 1967 to C22.2 NO 38. 3 pp.
CSA 851 Bull. Electrical Bulletin 851 November 17, 1971 to C22.2 NO 38. 2 pp.
CSA 916 Bull. Electrical Bulletin 916 July 3, 1973 to C22.2 NO 38. 2 pp.
CSA 730A Bull. Electrical Bulletin 730A May 31, 1974 to C22.2 NO 38. 2 pp.
CSA 983B Bull. Electrical Bulletin 983B April 25, 1977 to C22.2 NO 38. 39 pp.
CSA 983C Bull. Electrical Bulletin 983C November 14, 1978 to C22.2 NO 38. 3 pp.
CSA 1213A Bull. Electrical Bulletin 1213A October 25, 1979 to C22.2 NO 38. 3 pp.
CSA 1334 Bull. Electrical Bulletin 1334 August 28, 1981 to C22.2 NO 38. 1 p.
CSA 1368 Bull. Electrical Bulletin 1368 April 7, 1982 to C22.2 NO 38. 2 pp.
CSA 1403 Bull. Electrical Bulletin 1403 July 5, 1983 to C22.2 NO 38. 4 pp.
CSA 1316A Bull. Electrical Bulletin 1316A February 28, 1985 to C22.2 NO 38. 2 pp.

Cables (Electric) (Cont.)

—Thermoset Insulated (Cont.)
CSA 1316A Bull. Electrical Bulletin 1316A February 28, 1985 to C22.2 NO 51. 2 pp.
CSA C22.2 NO 52-M1989. Service-Entrance Cables; (Gen Instr 1 Thru 3). 34 pp.
CSA 1213A Bull. Electrical Bulletin 1213A October 25, 1979 to C22.2 NO 52. 3 pp.
CSA 1316A Bull. Electrical Bulletin 1316A February 28, 1985 to C22.2 NO 52. 2 pp.
CSA 1316A Bull. Electrical Bulletin 1316A February 28, 1985 to C22.2 NO 131. 2 pp.

—Thermoset Insulated— 600/1000 V
BSI BS 5467-89. 1989 Amd 1 Cables with Thermosetting Insulation for Electricity Supply for Rated Voltages of up to and Including 600/1000 V and up to and Including 1900/3300 V (AMD 6425) March 30, 1990. 36 pp.
BSI BS 5467-02. 1989 Amd 2 Cables with Thermosetting Insulation for Electricity Supply for Rated Voltages of up to and Including 600/1000 V and up to and Including 1900/3300 V (AMD 6829) February 28, 1992. 44 pp.
BSI BS 5467-03. 1989 Amd 3 Cables with Thermosetting Insulation for Electricity Supply for Rated Voltages of up to and Including 600/1000 V and up to and Including 1900/3300 V (AMD 7611) May 15, 1993. 45 pp.

—Thermoset Insulated—Deep Well Pumps
CSA 1403 Bull. Electrical Bulletin 1403 July 5, 1983 to C22.2 NO 38. 4 pp.

—Thermoset Insulated—Identification Systems
CSA C22.2 NO 38-M1986. Thermoset Insulated Wires and Cables; (Gen Instr 1 Thru 3). 64 pp.

—Thermoset Insulated—Multiconductor
CSA C22.2 NO 38-M1986. Thermoset Insulated Wires and Cables; (Gen Instr 1 Thru 3). 64 pp.
CSA 983B Bull. Electrical Bulletin 983B April 25, 1977 to C22.2 NO 38. 39 pp.

—Thermoset Insulated—Multiconductor—Identification Systems
CSA C22.2 NO 38-M1986. Thermoset Insulated Wires and Cables; (Gen Instr 1 Thru 3). 64 pp.

—Thermoset Insulated—Submersible Pumps
CSA 1403 Bull. Electrical Bulletin 1403 July 5, 1983 to C22.2 NO 38. 4 pp.

—Thermoset Insulated—1900/3300 V
BSI BS 5467-89. 1989 Amd 1 Cables with Thermosetting Insulation for Electricity Supply for Rated Voltages of up to and Including 600/1000 V and up to and Including 1900/3300 V (AMD 6425) March 30, 1990. 36 pp.
BSI BS 5467-02. 1989 Amd 2 Cables with Thermosetting Insulation for Electricity Supply for Rated Voltages of up to and Including 600/1000 V and up to and Including 1900/3300 V (AMD 6829) February 28, 1992. 44 pp.
BSI BS 5467-03. 1989 Amd 3 Cables with Thermosetting Insulation for Electricity Supply for Rated Voltages of up to and Including 600/1000 V and up to and Including 1900/3300 V (AMD 7611) May 15, 1993. 45 pp.

—Thermoset Insulated—300/500 V—Fire Testing
BSI BS 7629-93. 1993 Thermosetting Insulated Cables with Limited Circuit Integrity When Affected by Fire (V). 17 pp.

—Tin Coatings
DIN VDE 0472 Pt 808-84. Testing of Cables, Wires and Flexible Cords; Tinning, Solderability and Soldering Shrinkage (VDE Specification) (Feb). 7 pp.

—Underground—Power Distribution Systems
SAA AS 4026-92. Electric Cables—for Underground Residential Distribution Systems. 16 pp.

—Underground—Splicing
CSA C22.2 NO 198.2-M1986. Underground Cable Splicing Kits (R 1992); (Gen Instr 1). 18 pp.

—Vinyl Coated—Automotive
JIS C 3406-87. Low-Voltage Cables for Automobile. 12 pp.

—Wax Coated Nylon
CNS Z7203-87. Wax Coated Nylon Band for Electric Wire and Cable (Jan)(11822).

—Welding
MOD UK DSTAN 61-12: Part 19-01. Wires, Cords, and Cables, Electrical—Metric Units Part 19: Wires, Electrical (Single Core Flexible Cables for Welding Circuits) Issue 1; Amendment 2. 10 pp.

Cables (Electric) (Cont.)

—Welding (Cont.)
SAA AS 1995-77. Welding Cables. 12 pp.

—Welding Equipment
CNS C2108-89. Welding Cables (May)(8935).

Cables (Fiber Optics)
Use: Fiber Optic Cables

Cables (Heating)
Use: Heating Cables

Cables (Mechanical)
See Also: Belts (Machinery); Cable Assemblies (Mechanical); Control Cables (Mechanical); Cordage; Elevator Cables; Lanyards; Strands (Wire); Transmission Lines; Wire; Wire Ropes

—Aircraft—Drum Assemblies
SBAC AS 2824 ISSUE 6. Drum Assembly. 1 p.

—Aircraft—Drums (Containers)
SBAC AS 2825 ISSUE 6. Drum. 1 p.

—Aircraft—Sleeves
SBAC AGS 1511 ISSUE 9. Sleeves, Bonding, Metal Braided Cable, H.T. Inner, Straight, Long.
SBAC AGS 1512 ISSUE 12. Sleeves, Bonding, Metal Braided Cable, H.T. Outer, Straight.
SBAC AGS 1513 ISSUE 10. Sleeves, Bonding, Metal Braided Cable, H.T. Inner, Straight, Medium.
SBAC AGS 1514 ISSUE 9. Sleeves, Identification Cable, H.T. Black and Red.
SBAC AGS 1520 ISSUE 9. Sleeves, Bonding, Metal Braided Cable, L.T. Inner, Type A.
SBAC AGS 1521 ISSUE 11. Sleeves, Bonding, Metal Braided Cable, L.T. Outer, Type A.
SBAC AGS 1522 ISSUE 5. Sleeves, Insulating, Cable, L.T. Black and Red.
SBAC AGS 1523 ISSUE 8. Sleeves, Identification, Cable, L.T. Black and Red.
SBAC AGS 1651 ISSUE 12. Sleeves, Plain Cable, L.T. Type 1. Circular No.1.
SBAC AGS 1652 ISSUE 13. Sleeves, Plain Cable, L.T. Type 1. Oval No.1..
SBAC AGS 1653 ISSUE 13. Sleeves, Bonding, Metal Braided Cable L.T. Type 2, Circ. Inner.
SBAC AGS 1654 ISSUE 14. Sleeves, Bonding, Metal Braided Cable L.T. Type 2, Circ. Outer.
SBAC AGS 1655 ISSUE 9. Sleeves, Bonding, Metal Braided Cable L.T. Type 2, Oval Inner.
SBAC AGS 1656 ISSUE 10. Sleeves, Bonding, Metal Braided Cable L.T. Type 2, Oval Outer.
SBAC AGS 1716 ISSUE 2. Insulating, Sleeve (For H.T. Ignition Cable Sparking Plug Terminal).
SBAC AGS 1723 ISSUE 5(I). Sleeves, Bonding, Metal Braided Cable, R.T. Type 5, Circular, Inner.
SBAC AGS 1724 ISSUE 5(I). Sleeves, Bonding, Metal Braided Cable, R.T. Type 5, Circular, Outer.
SBAC AGS 1726 ISSUE 6(I). Sleeves, Bonding, Metal Braided Cable, L.T. Type 6, Circular, Inner and Outer.
SBAC AGS 1727 ISSUE 6(I). Sleeves, Bonding, Metal Braided Cable, L.T. Type 7, Circular, Inner.
SBAC AGS 1731 ISSUE 4(I). Sleeve, Outer, Circular (For Cable Electric, L.T. Metal Braided).
SBAC AGS 1732 ISSUE 6(I). Sleeves, Bonding, Metal Braided Cable, L.T. Type 8, Circular, Inner and Outer.
SBAC AGS 1733 ISSUE 8(I). Plain Cable, Sleeves, L.T. Type 1, Circular, No. 3.
SBAC AGS 1734 ISSUE 6(I). Sleeves, Bonding, Metal Braided Cable, L.T. Type 9, Circular, Inner.
SBAC AGS 1735 ISSUE 5(I). Sleeves, Bonding, Metal Braided Cable, L.T. Type 9, Circular, Outer.

—Aircraft—Union Nuts
SBAC AGS 1517 ISSUE 16. H.T. Cable Union Nut.
SBAC AGS 1519 ISSUE 14. L.T. Cable Union Nut.

—Lubricating Oils
CNS K5057-88. Steel Cable Oil (Jan)(2982)(R 1973). 2 pp.

Cabletroughs
Use: Cable Trays

Cabling Systems (Electric Power)
Use For: Wiring Systems *See Also:* Cable Assemblies (Electric); Cables (Electric); Electric Raceways; Wireways; Wiring Devices

IEC 1200 Pt 52-93. Electrical Installation Guide—Part 52: Selection and Erection of Electrical Equipment—Wiring Systems First Edition. 19 pp.

—Buildings
CENELEC PRHD 384.5. 52 S1-92. Electrical Installations of Buildings Part 5: Selection and Erection of Equipment Chapter 52: Selection and Erection of Wiring Systems. 26 pp.

INTERNATIONAL AND NON-U.S. NATIONAL STANDARDS
SUBJECT INDEX
Cadmium

Cabling Systems (Electric Power) (Cont.)

—**Electrical Equipment**
DIN VDE 0100 Pt 520 A1 (D)-86. Erection of Power Station Installations with Nominal Voltages up to 1000 Volts; Selection and Erection of Equipment Wiring Systems: Amendment 1. Identical with IEC 64 (Sec) 430 and 431 (Feb). 44 pp.

—**Fire Testing—Construction Materials**
DIN ENGL 4102 Pt 12-91. Fire Behaviour of Building Materials and Elements; Fire Resistance of Electrical Cable Systems; Requirements and Testing (Jan). 6 pp.

—**Flat Cables—Under Carpet—Safety**
CSA C22.2 NO 222-M1986. Type FCC Under-Carpet Wiring System (R 1992); (Gen Instr 1). 25 pp.

—**Military—Identification Systems**
NATO STANAG 2028 ED 4 AMD 1-83. System for Field Cable or Field Wire Labelling. 7 pp.

—**Office Furniture—Safety**
CSA CAN/CSA-C22. 2 NO 203-M91. Modular Wiring Systems for Office Furniture; (Gen Instr 1 Thru 2). 39 pp.

Cacotheline
CNS K7112-62. Chemical Reagent (Cacotheline) (Sep)(1612).

CAD
Use For: Computer Aided Design *See Also:* CAD/CAM; CAM

—**Architectural Drawings**
CSA B78.5-93. Computer-Aided Design Drafting (Buildings). 123 pp.
ISO TR10127-90. Computer-Aided Design (CAD) Technique—Use of Computers for the Preparation of Construction Drawings First Edition. 20 pp.
JTC1 TR10127-90. Computer-Aided Design (CAD) Technique—Use of Computers for the Preparation of Construction Drawings First Editon. 20 pp.
OSI ISO TR 10127-90. Computer-Aided Design (CAD) Technique—Use of Computers for the Preparation of Construction Drawings. 20 pp.

—**Documentation**
ISO 11442 Pt 3-93. Technical Product Documentation—Handling of Computer-Based Technical Information—Part 3: Phases in the Product Design Process First Edition. 5 pp.

—**Engineering Drawings**
BSI BS 1192: Part 5-90. 1990 Construction Drawing Practice Part 5: Guide to Structuring of Computer Graphic Information. 22 pp.
ISO TR10127-90. Computer-Aided Design (CAD) Technique—Use of Computers for the Preparation of Construction Drawings First Edition. 20 pp.
JIS B 3402-89. Technical Drawing—Drawing by CAD. 12 pp.
JTC1 TR10127-90. Computer-Aided Design (CAD) Technique—Use of Computers for the Preparation of Construction Drawings First Editon. 20 pp.
OSI ISO TR 10127-90. Computer-Aided Design (CAD) Technique—Use of Computers for the Preparation of Construction Drawings. 20 pp.
SAA AS 3883-91. Computer Graphics—Computer Aided Design (CAD)—Guide for Structuring of Computer Graphic Information (BS 1192:5:1990). 16 pp.

—**Engineering Drawings—Circuit Diagrams—Aircraft**
EURO CTI/79/9343-79. Recommendation for the Automation of Wiring Diagrams. 50 pp.

—**Engineering Drawings—Plotters**
DIN ENGL 6774 Pt 10-84. Technical Drawings; Principles for the Computer Aided Preparation of Drawings (Dec). 2 pp.

—**Glossaries**
CNS B8004-89. Glossary of Terms for Computer-Aided Design (CAD) (Jul)(11275).
ISO TR10623-91. Technical Product Documentation—Requirements for Computer-Aided Design and Draughting—Vocabulary First Edition. 30 pp.
JIS B 3401-89. Glossary of Terms Used in CAD. 14 pp.
JTC1 TR10623-91. Technical Product Documentation—Requirements for Computer-Aided Design and Draughting—Vocabulary First Edition. 30 pp.

—**Information Interchange—Geometry—Aerospace**
EURO CTI/80/10832-87. Recommendation for Geometry Data Exchange. 48 pp.

CAD (Cont.)

—**Information Interchange—Geometry—Aerospace (Cont.)**
EURO CTI/85/15617-85. Report of Geometry Data Exchange Study Group. 88 pp.

—**Information Security**
ISO 11442 Pt 1-93. Technical Product Documentation—Handling of Computer-Based Technical Information—Part 1: Security Requirements First Edition. 7 pp.

—**Procurement**
MOD UK DSTAN 00-37-88. Guide to the Procurement of a Computer Aided Design (CAD) System Issue 1. 33 pp.

CAD/CAM
Use For: Computer Aided Design/Computer Aided Manufacturing *See Also:* CAD; CAM; CIM

—**Aircraft**
SBAC TS 119 ISSUE 2. Guidelines for the Introduction and Operation of a CAD/CAM System Within a CAE Environment.

—**Information Interchange—Aircraft**
SBAC TS 133 ISSUE 1. Transfer of Information Between Different CAD/CAM Systems.

Cadmium
See Also: Cadmium Coatings (Made From Cadmium); Cadmium Coatings (On Cadmium); Cadmium Content Analysis; Hazardous Materials; Metals; Pigments
JIS H 2113-61. Cadmium Metal (R 1968). 3 pp.

—**Chemical Analysis**
JIS H 1161-91. Methods for Chemical Analysis of Cadmium Metal. 42 pp.

—**Pigments**
BSI BS 6857-87. 1987 Cadmium Pigments for Paints. 13 pp.
ISO 4620-86. Cadmium Pigments—Specifications and Methods of Test First Edition. 11 pp.

—**Spectrochemical Analysis**
JIS H 1163-91. Method for Photoelectric Emission Spectrochemical Analysis of Cadmium Metal. 8 pp.

—**Standard Solutions**
JIS K 0012-83. Cadmium Standard Solution (R 1988). 14 pp.

—**Toys—Migration Testing**
BSI BS 5665: Part 3-89. 1989 Safety of Toys Part 3: Migration of Certain Elements (Supersedes BS 3443: 1968). 11 pp.
CEN EN 71-3-88. Safety of Toys Part 3 Migration of Certain Elements. 14 pp.
CEN PREN 71-3-92. Safety of Toys—Part 3: Migration of Certain Elements. 24 pp.

Cadmium Acetate
See Also: Acetates; Cadmium Compounds
CNS K7113-62. Chemical Reagent (Cadmium Acetate) (Sep)(1613).
JIS K 8362-78. Cadmium Acetate, Dihydrate.

Cadmium Bromide
CNS K7114-62. Chemical Reagent (Cadmium Bromide) (Sep)(1614).

Cadmium Cells
BSI BS 5142-74. 1974 Standard Cells. 15 pp.
CENELEC HD 612 S1-92. Standard Cells; (IEC 428:1973). 4 pp.
CNS C3085-80. Method of Test for Standard Cells (Aug)(6027). 4 pp.
IEC 428-73. Standard Cells First Edition; (CENELEC HD 612 S1:1992). 25 pp.
JIS C 1021-69. Standard Cells. 10 pp.

Cadmium Chloride
CNS K7115-62. Chemical Reagent (Cadmium Chloride) (Sep)(1615).
JIS K 8120-72. Cadmium Chloride.

Cadmium Coatings (Made From Cadmium)
Scope Note: Includes coatings made from cadmium alloys *See Also:* Cadmium; Cadmium Coatings (On Cadmium); Coatings; Galvanized Materials; Metal Coatings (Made From Metal); Plating

—**Aircraft—Electrodeposited**
JIS W 1114-84. Plating Cadmium (Electrodeposited for Aircraft).

Cadmium Coatings (Made From Cadmium) (Cont.)

—**Copper Alloys—Electroplated—Aerospace**
CEN PREN 2133-92. Cadmium Plating of Steels with Maximum Specified Tensile Strength Equal to or Less Than 1450 MPa Copper, Copper Alloys and Nickel Alloys. 16 pp.

—**Copper—Electrodeposited**
MOD UK DTD-904C-63. Cadmium Plating (Reprinted December 1966, Incorporating Amendment No. 1) (Superseded by Def Stan 03-19). 5 pp.

—**Copper—Electrodeposited—Aircraft**
AECMA PREN2133-81. Cadmium Plating of Steels with Maximum Specified Tensile Strength Less Than or Equal to 1450 MPa and Copper and Copper Alloys (C7/SC4/D). 14 pp.

—**Copper—Electroplated—Aerospace**
CEN PREN 2133-92. Cadmium Plating of Steels with Maximum Specified Tensile Strength Equal to or Less Than 1450 MPa Copper, Copper Alloys and Nickel Alloys. 16 pp.

—**Electroplated**
BSI BS 2868-68. 1968 Cadmium Anodes and Cadmium Oxide for Electroplating. 8 pp.
BSI BS 6338-82. 1982 Chromate Conversion Coatings on Electroplated Zinc and Cadmium Coatings. 4 pp.
BSI BS 7371: Part 3-93. 1993 Coatings on Metal Fasteners Part 3: Specification for Electroplated Zinc and Cadmium Coatings (F). 14 pp.
ISO 4520-81. Chromate Conversion Coatings on Electroplated Zinc and Cadmium Coatings First Edition. 4 pp.

—**Electroplated—Impact Testing**
CNS H2048-79. Dropping Test for Electroplated Coatings of Cadmium (Apr)(4830).

—**Iron—Electroplated**
BSI BS 1706-90. 1990 Amd 1 Electroplated Coatings of Zinc and Cadmium on Iron and Steel (AMD 6731) May 31, 1991. 17 pp.
CNS H3080-87. Electroplated Coatings of Cadmium on Iron and Steel (Sep)(4828).
ISO 2082-86. Metallic Coatings—Electroplated Coatings of Cadmium on Iron or Steel Second Edition. 7 pp.
JIS H 8611-93. Electroplated Coatings of Cadmium on Steel. 8 pp.
SAA AS 1790-84. Electroplated Coatings—Cadmium on Iron or Steel (R 1993). 5 pp.

—**Nickel Alloys—Electroplated—Aerospace**
CEN PREN 2133-92. Cadmium Plating of Steels with Maximum Specified Tensile Strength Equal to or Less Than 1450 MPa Copper, Copper Alloys and Nickel Alloys. 16 pp.

—**Steel—Electrodeposited**
MOD UK DSTAN 03-19-81. Electro-Deposition of Cadmium Issue 1. 15 pp.
MOD UK DSTAN 03-19-01. Electro-Deposition of Cadmium Issue 1; Amendment 1. 16 pp.
MOD UK DTD-904C-63. Cadmium Plating (Reprinted December 1966, Incorporating Amendment No. 1) (Superseded by Def Stan 03-19). 5 pp.

—**Steel—Electrodeposited—Aircraft**
AECMA PREN2133-81. Cadmium Plating of Steels with Maximum Specified Tensile Strength Less Than or Equal to 1450 MPa and Copper and Copper Alloys (C7/SC4/D). 14 pp.

—**Steel—Electroplated**
BSI BS 1706-90. 1990 Amd 1 Electroplated Coatings of Zinc and Cadmium on Iron and Steel (AMD 6731) May 31, 1991. 17 pp.
BSI BS 3382: Parts 1 & 2-61. 1961 Amd 2 Electroplated Coatings on Threaded Components Part 1: Cadmium on Steel Components Part 2: Zinc on Steel Components. 34 pp.
CNS H3080-87. Electroplated Coatings of Cadmium on Iron and Steel (Sep)(4828).
ISO 2082-86. Metallic Coatings—Electroplated Coatings of Cadmium on Iron or Steel Second Edition. 7 pp.
JIS H 8611-93. Electroplated Coatings of Cadmium on Steel. 8 pp.
SAA AS K132.1-73. Electroplated Coatings on Threaded Components—Part 1: Cadmium on Steel. 23 pp.
SAA AS 1790-84. Electroplated Coatings—Cadmium on Iron or Steel (R 1993). 5 pp.

—**Steel—Electroplated—Aerospace**
CEN PREN 2133-92. Cadmium Plating of Steels with Maximum Specified Tensile Strength Equal to or Less Than 1450 MPa Copper, Copper Alloys and Nickel Alloys. 16 pp.

Cadmium Coatings (Made From Cadmium) (Cont.)

—Steel—Vacuum Deposited
MOD UK DSTAN 03-28: Part 2-01. Physical Vapour Deposition of Metals Part 2: Ion Vapour Deposition of Cadmium for Protection Against Corrosion Issue 1; Amendment 1. 19 pp.

—Steel—Vacuum Evaporated
MOD UK DTD-940-68. Cadmium Coating of Very Strong Steel Parts by Vacuum Evaporation. 4 pp.

Cadmium Coatings (On Cadmium)
Scope Note: Includes coatings on cadmium alloys
See Also: Cadmium; Cadmium Coatings (Made From Cadmium); Coatings; Metal Coatings (On Metal)

—Chromate
BSI BS 5411: Part 13-82. 1982 Amd 1 Method of Test for Metallic and Related Coatings Part 13: Chromate Conversion Coatings on Zinc and Cadmium. 9 pp.
CGSB 31-GP-115MA-87. Coating, Conversion, Chromate and Phosphate, for Zinc and Cadmium Surfaces. 8 pp.
MOD UK DEF-130-61. Chromate Passivation of Cadmium and Zinc Surfaces (Reprinted March 1977, Incorporating Amendment No. 1). 4 pp.

—Chromate—Electroplated
ISO 3613-80. Chromate Conversion Coatings on Zinc and Cadmium—Test Methods First Edition; (Erratum—Aug 1982). 7 pp.
JIS H 8625-93. Chromate Conversion Coatings on Electroplated Zinc and Cadmium Coatings. 14 pp.
SAA AS 1791-86. Chromate Conversion Coatings—Zinc and Cadmium (R 1993). 5 pp.

—Chromate—Surface Density
BSI BS 5411: Part 14-82. 1982 Method of Test for Metallic and Related Coatings Part 14: Gravimetric Method for Determination of Coating Mass per Unit Area of Conversion Coatings on Metallic Materials. 7 pp.
ISO 3892-80. Conversion Coatings on Metallic Materials—Determination of Coating Mass per Unit Area—Gravimetric Methods First Edition. 5 pp.

—Phosphate
CGSB 31-GP-115MA-87. Coating, Conversion, Chromate and Phosphate, for Zinc and Cadmium Surfaces. 8 pp.
ISO 9717-90. Phosphate Conversion Coatings for Metals—Method of Specifying Requirements First Edition. 19 pp.

—Phosphate—Surface Density
BSI BS 5411: Part 14-82. 1982 Method of Test for Metallic and Related Coatings Part 14: Gravimetric Method for Determination of Coating Mass per Unit Area of Conversion Coatings on Metallic Materials. 7 pp.
ISO 3892-80. Conversion Coatings on Metallic Materials—Determination of Coating Mass per Unit Area—Gravimetric Methods First Edition. 5 pp.

Cadmium Compounds
See Also: Cadmium Acetate; Hazardous Materials

—Highway Transportation—Emergency Procedures
SAA AS 1678.6.1. 017-84. Emergency Procedure Guides—Transport—Part 6.1.017: Cadmium Compounds (Other Than Selenide and Sulphide).

Cadmium Content Analysis
See Also: Cadmium

—Algae
CNS N6106-78. Method of Test for Edible Chlorella (Oct)(4597). 9 pp.

—Animal Feed
CNS N4024-21-86. Method of Test for Feeds: Determination of Cadmium (Aug)(2770-21).

—Bearing Alloys—Photometry
BSI BS 3338: Part 17-65. 1965 Methods for the Sampling and Analysis of Tin and Tin Alloys Part 17: Determination of Cadmium in Solders and White Metal Bearing Alloys (Photometric Method). 9 pp.

—Ceramic Ware
BSI BS 6748-86. 1986 Limits of Metal Release from Ceramic Ware, Glassware, Glass Ceramic Ware and Vitreous Enamel Ware. 8 pp.
CNS R2081-74. Lead and Cadmium Content from Glazed Ceramic Surface for Table Wares (Jun)(3725). 1 p.
ISO 6486 Pt 1-81. Ceramic Ware in Contact with Food—Release of Lead and Cadmium—Part 1: Method of Test First Edition. 6 pp.

Cadmium Content Analysis (Cont.)

—Ceramic Ware (Cont.)
ISO 6486 Pt 2-81. Ceramic Ware in Contact with Food—Release of Lead and Cadmium—Part 2: Permissible Limits First Edition. 4 pp.
ISO 7086 Pt 1-82. Glassware and Glass Ceramic Ware in Contact with Food—Release of Lead and Cadmium—Part 1: Method of Test First Edition. 7 pp.
ISO 7086 Pt 2-82. Glassware and Glass Ceramic Ware in Contact with Food—Release of Lead and Cadmium—Part 2: Permissible Limits First Edition. 3 pp.
ISO 8391 Pt 1-86. Ceramic Cookware in Contact with Food—Release of Lead and Cadmium—Part 1: Method of Test First Edition. 6 pp.
ISO 8391 Pt 2-86. Ceramic Cookware in Contact with Food—Release of Lead and Cadmium—Part 2: Permissible Limits First Edition. 4 pp.

—Ceramics—Atomic Absorption Spectrometry
CNS R3073-86. Method of Test for Lead and Cadmium Extracted from Glazed Ceramic Surfaces (Apr)(3503). 4 pp.

—Coatings
CGSB CAN2-1.500-75 METH 2-73. Methods of Test for Toxic Trace Elements in Protective Coatings Determination of Leachable Cadmium (Cd) in Low Concentration. 2 pp.

—Control Rods
CNS J2050-82. Method for Determining of Silver, Indium, and Cadmium in Silver-Indium-Cadmium Alloys Control Rod (May)(8832).

—Copper Alloys—Atomic Absorption Spectrometry
ISO 5960-84. Copper Alloys—Determination of Cadmium Content—Flame Atomic Absorption Spectrometric Method First Edition. 6 pp.
SAA AS 1515.5-87. Copper Alloys—Part 5: Determination of Cadmium—Flame Atomic Absorption Spectrometric Method. 3 pp.

—Copper Alloys—Electrolytic Analysis
BSI BS 1825-52. 1952 Determination of Cadmium in Copper-Cadmium Alloys (Electrolytic Method). 8 pp.

—Cosmetics
CNS S2089-83. Methods of Hygienic Test for Cosmetics — Cadmium (Jun)(9543).

—Drinking Water—Voltametry
DIN ENGL 38406 Pt 16-90. German Standard Methods for the Examination of Water, Waste Water and Sludge; Cations (Group E); Determination of Zinc, Cadmium, Lead, Copper, Thallium, Nickel, Cobalt by Voltammetry (E16) (Mar). 8 pp.

—Electrolytic Copper—Atomic Absorption Spectrophotometry
BSI BS 7317: Part 1-90. 1990 Methods for Analysis of High Purity Copper Cathode Cu-CATH-1 Part 1: Method for Determination of Cadmium Manganese and Silver (Screening Procedure for Chromium, Cobalt, Iron, Nickel and Zinc) by Atomic Absorption. 9 pp.
BSI DD 95: Part 1-84. (WITHDRAWN) 1984 Amd 1 Analysis of Higher Purity Copper Cathode Cu-CATH-1 Part 1: Method for Determination of Cadmium, Manganese and Silver (Screening Procedures for Chromium, Cobalt, Iron, Nickel and Zinc) by Atomic Absorption. 10 pp.

—Enameled Surfaces
BSI BS 6748-86. 1986 Limits of Metal Release from Ceramic Ware, Glassware, Glass Ceramic Ware and Vitreous Enamel Ware. 8 pp.
CNS R3137-86. Method of Test for Lead and Cadmium Release from Porcelain Enamel Surface (Apr)(11554).

—Enameled Surfaces—Food Contamination
DIN ENGL 51032-86. Ceramics, Glass, Glass Ceramics, Vitreous Enamels; Permissible Limits for the Release of Lead and Cadmium from Articles Intended for Use in Contact with Foodstuffs (Feb). 4 pp.

—Flue Gases
JIS K 0097-79. Method for Determination of Cadmium and Lead in Stack Gas (R 1984). 22 pp.

—Fruits—Atomic Absorption Spectrometry
ISO 6561-83. Fruits, Vegetables and Derived Products—Determination of Cadmium Content—Flameless Atomic Absorption Spectrometric Method First Edition. 5 pp.

Cadmium Content Analysis (Cont.)

—Glassware
BSI BS 6748-86. 1986 Limits of Metal Release from Ceramic Ware, Glassware, Glass Ceramic Ware and Vitreous Enamel Ware. 8 pp.
ISO 7086 Pt 1-82. Glassware and Glass Ceramic Ware in Contact with Food—Release of Lead and Cadmium—Part 1: Method of Test First Edition. 7 pp.
ISO 7086 Pt 2-82. Glassware and Glass Ceramic Ware in Contact with Food—Release of Lead and Cadmium—Part 2: Permissible Limits First Edition. 3 pp.

—Ground Water—Voltametry
DIN ENGL 38406 Pt 16-90. German Standard Methods for the Examination of Water, Waste Water and Sludge; Cations (Group E); Determination of Zinc, Cadmium, Lead, Copper, Thallium, Nickel, Cobalt by Voltammetry (E16) (Mar). 8 pp.

—Lead—Polarographic Analysis
MOD UK M 9331/63. Determination of Bismuth, Copper, Cadmium, and Zinc in Pure Lead (Polarographic Method) (No Information) (Withdrawn).

—Nickel—Atomic Absorption Spectrometry
BSI BS 6783: Part 1-86. 1986 Sampling and Analysis of Nickel, Ferronickel and Nickel Alloys Part 1: Method for Determination of Silver, Bismuth, Cadmium, Cobalt, Copper, Iron, Manganese, Lead and Zinc in Nickel by Flame Atomic Absorption Spectrometry. 15 pp.
BSI BS 6783: Part 4-86. 1986 Sampling and Analysis of Nickel, Ferronickel and Nickel Alloys: Part 4: Method for Determination of Silver, Arsenic, Bismuth, Cadmium, Lead, Antimony, Selenium, Tin, Tellurium and Thallium in Nickel by Electrothermal Atomic Absorption Spectrometry. 11 pp.
ISO 6351-85. Nickel—Determination of Silver, Bismuth, Cadmium Cobalt, Copper, Iron, Manganese, Lead and Zinc Contents—Flame Atomic Absorption Spectrometric Method First Edition. 13 pp.
ISO 7523-85. Nickel—Determination of Silver, Arsenic, Bismuth, Cadmium, Lead, Antimony, Selenium, Tin, Tellurium and Thallium Contents—Electrothermal Atomic Absorption Spectrometric Method First Edition. 10 pp.

—Ores
JIS M 8135-81. Methods for Determination of Cadmium in Ores.

—Painted Surfaces—Food Contamination—Permissible Limits
BSI BS 7557-92. 1992 Limits of Metal Release from Painted Surfaces of Articles, Liable to Come into Contact with Foodstuffs (W). 8 pp.

—Paints—Atomic Absorption Spectrometry
BSI BS 3900: Part B9-86. 1986 Methods of Test for Paints Group B: Tests Involving Chemical Examination of Liquid Paints and Dried Paint Films Part B9: Determination of 'Soluble' Cadmium Content. 9 pp.
ISO 3856 Pt 4-84. Paints and Varnishes—Determination of "Soluble" Metal Content—Part 4: Determination of Cadmium Content—Flame Atomic Absorption Spectrometric Method and Polarographic Method Second Edition. 8 pp.

—Paints—Polarographic Analysis
BSI BS 3900: Part B9-86. 1986 Methods of Test for Paints Group B: Tests Involving Chemical Examination of Liquid Paints and Dried Paint Films Part B9: Determination of 'Soluble' Cadmium Content. 9 pp.
ISO 3856 Pt 4-84. Paints and Varnishes—Determination of "Soluble" Metal Content—Part 4: Determination of Cadmium Content—Flame Atomic Absorption Spectrometric Method and Polarographic Method Second Edition. 8 pp.

—Papers—Atomic Absorption Spectrophotometry
CPPA G.34P-92. Determination of Sodium, Magnesium, Calcium, Manganese, Iron, Copper and Cadmium in Wood, Pulp or Paper by Atomic Absorption Spectrophotometry. 3 pp.

—Precipitation—Voltametry
DIN ENGL 38406 Pt 16-90. German Standard Methods for the Examination of Water, Waste Water and Sludge; Cations (Group E); Determination of Zinc, Cadmium, Lead, Copper, Thallium, Nickel, Cobalt by Voltammetry (E16) (Mar). 8 pp.

Cadmium Content Analysis (Cont.)

—Pulps—Atomic Absorption Spectrophotometry
CPPA G.34P-92. Determination of Sodium, Magnesium, Calcium, Manganese, Iron, Copper and Cadmium in Wood, Pulp or Paper by Atomic Absorption Spectrophotometry. 3 pp.

—Silicate Surfaces—Food Contamination
DIN ENGL 51031-86. Testing of Articles Intended for Use in Contact with Foodstuffs; Determination of Release of Lead and Cadmium from Silicate Surfaced Articles Intended for Use in Contact with Foodstuffs (Feb). 5 pp.

—Sodium Chloride—Atomic Absorption Spectrometry
BSI BS 7319: Part 6-90. 1990 Analysis of Sodium Chloride for Industrial Use Part 6: Method for Determination of Cadmium Content. 7 pp.

—Solders
BSI BS 3338: Part 21-83. 1983 Methods for the Sampling and Analysis of Tin and Tin Alloys Part 21: Determination of High Cadium Content of Soft Solders. 4 pp.

—Solders—Photometry
BSI BS 3338: Part 17-65. 1965 Methods for the Sampling and Analysis of Tin and Tin Alloys Part 17: Determination of Cadmium in Solders and White Metal Bearing Alloys (Photometric Method). 9 pp.

—Surface Waters—Voltametry
DIN ENGL 38406 Pt 16-90. German Standard Methods for the Examination of Water, Waste Water and Sludge; Cations (Group E); Determination of Zinc, Cadmium, Lead, Copper, Thallium, Nickel, Cobalt by Voltammetry (E16) (Mar). 8 pp.

—Varnishes—Polarographic Analysis
ISO 3856 Pt 4-84. Paints and Varnishes—Determination of "Soluble" Metal Content—Part 4: Determination of Cadmium Content—Flame Atomic Absorption Spectrometric Method and Polarographic Method Second Edition. 8 pp.

—Vegetables—Atomic Absorption Spectrometry
ISO 6561-83. Fruits, Vegetables and Derived Products—Determination of Cadmium Content—Flameless Atomic Absorption Spectrometric Method First Edition. 5 pp.

—Waste Water—Atomic Absorption Spectrometry
DIN ENGL 38406 Pt 19-80. German Standard Methods for the Analysis of Water, Waste Water and Sludge; Cations (Group E); Determination of Cadmium (E 19) (July). 8 pp.

—Water
CNS K9037-80. Method of Test for Cadmium in Water (Dithizone Method) (May)(5578).

—Water—Atomic Absorption Spectrometry
BSI BS 6068: Sec 2.21-85. 1985 Water Quality Part 2: Physical, Chemical and Bio-Chemical Methods Section 2.21: Determination of Cadmium: Flame Atomic Absorption Spectrometric Methods. 12 pp.
BSI BS 6068: Sec 2.29-87. 1987 Water Quality Part 2: Physical, Chemical and Bio-Chemical Methods Section 2.29: Determination of Cobalt, Nickel, Copper, Zinc, Cadmium and Lead: Flame Atomic Absorption Spectrometric Methods. 13 pp.
CNS K9029-80. Method of Test for Cadmium in Water (Atomic Absorption, Direct) (Apr)(5463).
DIN ENGL 38406 Pt 19-80. German Standard Methods for the Analysis of Water, Waste Water and Sludge; Cations (Group E); Determination of Cadmium (E 19) (July). 8 pp.
ISO 5961-85. Water Quality—Determination of Cadmium—Flame Atomic Absorption Spectrometric Methods First Edition. 11 pp.
ISO 8288-86. Water Quality—Determination of Cobalt, Nickel, Copper, Zinc, Cadmium and Lead—Flame Atomic Absorption Spectrometric Methods First Edition. 13 pp.

—White Metals—Photometry
BSI BS 3338: Part 17-65. 1965 Methods for the Sampling and Analysis of Tin and Tin Alloys Part 17: Determination of Cadmium in Solders and White Metal Bearing Alloys (Photometric Method). 9 pp.

—Wood—Atomic Absorption Spectrophotometry
CPPA G.34P-92. Determination of Sodium, Magnesium, Calcium, Manganese, Iron, Copper and Cadmium in Wood, Pulp or Paper by Atomic Absorption Spectrophotometry. 3 pp.

—Zinc Alloys—Atomic Absorption Spectrometry
DIN ENGL 50551-90. Determination of the Lead, Cadmium and Copper Content of Zinc and Zinc Alloys by Atomic Absorption Spectrometry (Oct). 5 pp.
SAA AS 1329.8-80. Methods for the Analysis of Zinc and Zinc Alloys—Part 8: Determination of Cadmium Content—Flame Atomic Absorption Spectrometric Method. 10 pp.

—Zinc Alloys—Polarographic Analysis
BSI BS 3630: Part 8-71. 1971 Methods for the Sampling and Analysis of Zinc and Zinc Alloys: Lead and Cadmium in Zinc (Grades Zn1 and Zn2) and Zinc Alloys (Polarographic Method). 11 pp.
ISO 2576-72. Chemical Analysis of Zinc Alloys—Polarographic Determination of Lead and Cadmium in Zinc Alloys Containing Copper First Edition. 5 pp.

—Zinc—Atomic Absorption Spectrometry
DIN ENGL 50551-90. Determination of the Lead, Cadmium and Copper Content of Zinc and Zinc Alloys by Atomic Absorption Spectrometry (Oct). 5 pp.

—Zinc—Polarographic Analysis
BSI BS 3630: Part 8-71. 1971 Methods for the Sampling and Analysis of Zinc and Zinc Alloys: Lead and Cadmium in Zinc (Grades Zn1 and Zn2) and Zinc Alloys (Polarographic Method). 11 pp.
BSI BS 3630: Part 14-72. 1972 Methods for the Sampling and Analysis of Zinc and Zinc Alloys Part 14: Cadmium in Zinc (Grades Zn3 and Zn4) (Polarographic Method). 8 pp.
ISO 713-75. Zinc—Determination of Lead and Cadmium Contents—Polarographic Method First Edition. 4 pp.
ISO 1054-75. Zinc—Determination of Cadmium Content—Polarographic Method First Edition. 4 pp.

—Zinc—Spectrochemical Analysis
JIS H 1110-89. Methods for Determination of Cadmium in Zinc Metal. 10 pp.

—Zirconium
JIS H 1671-82. Methods for Determination of Cadmium in Zirconium and Zirconium Alloys (R 1988). 7 pp.

—Zirconium Alloys
JIS H 1671-82. Methods for Determination of Cadmium in Zirconium and Zirconium Alloys (R 1988). 7 pp.

Cadmium Cyanide
CNS K1095-73. Cadmium Cyanide (Industrial Grade) (Nov)(2368).
CNS K6189-73. Method of Test for Cadmium Cyanide of Industrial Grade (Nov)(2369).

Cadmium Inorganic Compounds
Scope Note: Use a more specific term
See: Cadmium Chloride; Cadmium Compounds; Cadmium Iodide; Cadmium Sulfate; Inorganic Compounds

Cadmium Iodide
CNS K7116-62. Chemical Reagent (Cadmium Iodide) (Sep)(1616).
JIS K 8921-88. Cadmium Iodide.

Cadmium Nitrate
CNS K7117-62. Chemical Reagent (Cadmium Nitrate) (Sep)(1617).

Cadmium Oxide
CNS K1110-82. Cadmium Oxide (Industrial Grade) (Jul)(2561).

Cadmium Sulfate
CNS K7118-62. Chemical Reagent (Cadmium Sulfate) (Sep)(1618).
JIS K 8961-80. Cadmium Sulfate N Hydrate.

Cady Micrometers
Use: Micrometers—Papers

CAEE (ICAO)
Use: ICAO Committee on Aircraft Engine Emissions

CAEP (ICAO)
Use: ICAO Committee on Aviation Environmental Protection

Cafes
Use: Restaurants

Caffeine Content Analysis

—Beverages
CNS N6174-82. Method of Test for Beverage: Determination of Caffeine (Sep)(9432). 2 pp.

—Coffee
BSI BS 5752: Part 3-83. 1983 Coffee and Coffee Products Part 3: Coffee: Determination of Caffein Content (Reference Method). 9 pp.
CNS N6180-83. Method of Test for Beverage: Determination of Methanol Content (May)(10291). 1 p.
ISO 4052-83. Coffee—Determination of Caffeine Content (Reference Method) First Edition. 8 pp.

—Coffee Extracts
EC 79/1066/EEC-79. First Commission Directive Laying Down Community Methods of Analysis for Testing Coffee Extracts and Chicory Extracts. 12 pp.

—Coffee—Liquid Chromatography
BSI BS 5752: Part 12-92. 1992 Methods of Test for Coffee and Coffee Products Part 12: Coffee: Determination of Caffeine Content (Routine Method by HPLC) (ISO 10095: 1992). 12 pp.
ISO 10095-92. Coffee—Determination of Caffeine Content—Method Using High-Performance Liquid Chromatography First Edition. 9 pp.

Cages
See Also: Materials Handling Equipment

—Aircraft—Nuts (Fasteners)
SBAC AGS 3041 ISSUE 1. Cage for Blind Nuts.
SBAC AGS 3042 ISSUE 1. Cage for Blind Nuts.
SBAC AGS 3043 (V). Cage for Blind Nuts.
SBAC AGS 3044 ISSUE 1. Nut for AGS 3041 Cage.
SBAC AGS 3045 ISSUE 1. Nut for AGS 3042 Cage.
SBAC AGS 3046 ISSUE 1. Nut for AGS 3043 Cage.
SBAC AGS 3047 ISSUE 1. Nut for AGS 3043 Cage.

Cake Mixes
See Also: Food
CGSB 32.9M-88. Cake Mixes, Prepared, Complete. 7 pp.

Cakes
Scope Note: See the subheading Cakes under the specific metal or alloy

Caking
See Also: Compacting; Drying

—Cleaning Agents—Granular
CGSB 31-GP-0A METH 47.1-62. Methods of Testing Corrosion-Prevention Materials and Processes Caking Characteristics. 1 p.

—Coal
BSI BS 1016: Part 12-80. (WITHDRAWN) 1980 Methods for the Analysis and Testing of Coal and Coke Part 12: Caking and Swelling Properties of Coal (Superseded by BS 1016: Sections 107.1, 107.2: 1991 and 107.3: 1990). 23 pp.
BSI BS 1016: Sec 107.2-91. 1991 Analysis and Testing of Coal and Coke Part 107: Caking and Swelling Properties of Coal Section 107.2: Assessment of Caking Power by Gray-King Coke Test. 17 pp.
CNS M3168-84. Method for Determination of Caking Power of Hard Coal-Roga Test (Sep) (11021).
CNS M3169-84. Method for Determination of Caking Power of Coal-Gray-King Coke Test (Sep)(11022).
ISO 335-74. Hard Coal—Determination of Caking Power—Roga Test First Edition. 6 pp.
ISO 502-82. Coal—Determination of Caking Power—Gray-King Coke Test Second Edition. 11 pp.

Calabash
Use: Gourds

Calamus Oil

—Asarone Content—Gas Chromatography
ISO 7357-85. Oil of Calamus—Determination of cis-Beta Ray-Asarone Content—Gas Chromatographic Method on Packed Columns First Edition. 5 pp.

Calcination

—Pipe Fittings
ISO 6964-86. Polyolefin Pipes and Fittings—Determination of Carbon Black Content by Calcination and Pyrolysis—Test Method and Basic Specification First Edition. 4 pp.

Calcination (Cont.)
—Plastic Pipes
ISO 6964-86. Polyolefin Pipes and Fittings—Determination of Carbon Black Content by Calcination and Pyrolysis—Test Method and Basic Specification First Edition. 4 pp.

Calcined Coke
See Also: Coke
—Ash Content
BSI BS 6043: Sec 2.2-85. 1985 Methods of Sampling and Test for Carbonaceous Materials Used in Aluminium Manufacture Part 2: Electrode Coke Section 2.2: Determination of Ash Content of Green and Calcined Cokes. 6 pp.
ISO 8005-84. Carbonaceous Materials Used in the Production of Aluminium—Green and Calcined Coke—Determination of Ash Content First Edition. 4 pp.
—Coking
BSI BS 6043: Sec 1.6-84. 1984 Methods of Sampling and Test for Carbonaceous Materials Used in Aluminium Manufacture Part 1: Electrode Pitch Section 1.6: Determination of Coking Value. 7 pp.
BSI BS 6043: Sec 1.6-01. 1984 Amd 1 Methods of Sampling and Test for Carbonaceous Materials Used in Aluminium Manufacture Part 1: Electrode Pitch Section 1.6: Determination of Coking Value (AMD 7662) May 15, 1993 (ISO 6998: 1984) (V). 8 pp.
ISO 6998-84. Carbonaceous Materials for the Production of Aluminium—Pitch for Electrodes—Determination of Coking Value First Edition. 5 pp.
—Oil Content—Gravimetric Analysis
BSI BS 6043: Sec 2.7-86. 1986 Methods of Sampling and Test for Carbonaceous Materials Used in Aluminium Manufacture Part 2: Electrode Coke Section 2.7: Determination of the Apparent Oil Content of Calcined Cokes (Gravimetric Method). 7 pp.
ISO 6997-85. Carbonaceous Materials for the Production of Aluminium—Calcined Coke—Determination of Apparent Oil Content—Heating Method First Edition. 5 pp.
—Oil Content—Solvent Extraction
BSI BS 6043: Sec 2.8-87. 1987 Methods of Sampling and Test for Carbonaceous Materials Used in Aluminium Manufacture Part 2: Electrode Coke Section 2.8: Determination of Oil Content by Solvent Extraction of Calcined Coke. 7 pp.
ISO 8723-86. Carbonaceous Materials for the Production of Aluminium—Calcined Coke—Determination of Oil Content—Method by Solvent Extraction First Edition. 5 pp.
—Xylene—Density—Pycnometric Analysis
BSI BS 6043: Sec 2.6-86. 1986 Methods of Sampling and Test for Carbonaceous Materials Used in Aluminium Manufacture Part 2: Electrode Coke Section 2.6: Determination of the Density in Xylene of Calcined Coke. 8 pp.
ISO 8004-85. Carbonaceous Materials for the Production of Aluminium—Calcined Coke and Calcined Carbon Products—Determination of the Density in Xylene—Pyknometric Method First Edition. 6 pp.

Calcined Magnesium Oxide
Use: Magnesium Oxide

Calcium Acetate
See Also: Acetates
CNS K7119-62. Chemical Reagent (Cadmium Acetate) (Sep)(1619).
JIS K 8364-76. Calcium Acetate.

Calcium Ammonium Nitrate
See Also: Ammonium Nitrate
—Fertilizers
CNS N3066-88. Calcium Ammonium Nitrate Fertilizer (Jun)(11953).

Calcium Carbide
See Also: Carbides; Hazardous Materials
CNS K1025-75. Calcium Carbide (Jan)(264).
CNS K6012-75. Method of Test for Calcium Carbide (Apr)(187). 7 pp.
DIN ENGL 53922-79. Calcium Carbide (July). 10 pp.
JIS K 1901-83. Calcium Carbide. 37 pp.

Calcium Carbonate
See Also: Chalk; Marble; Precipitated Calcium Carbonate
CNS K1057-72. Calcium Carbonate Precipitated, Industrial Grade (Oct)(1383). 2 pp.
CNS K6150-72. Method of Test for Calcium Carbonate of Industrial Use (Oct)(1384).
JIS K 8617-75. Calcium Carbonate.

Calcium Carbonate (Cont.)
—Bulk Density
BSI BS 1460-67. 1967 Amd 1 Determination of Apparent Density After Compaction of Precipitated Calcium Carbonate. 13 pp.
—Chemical Analysis
BSI BS 6463: Part 2-84. 1984 Quicklime, Hydrated Lime and Natural Calcium Carbonate Part 2: Methods of Chemical Analysis. 14 pp.
BSI BS 6463: Part 2-01. 1984 Amd 1 Quicklime, Hydrated Lime and Natural Calcium Carbonate Part 2: Methods of Chemical Analysis (AMD 7586) May 15, 1993 (W). 15 pp.
—Fertilizers
CNS N3024-88. Calcium Carbonate, Fertilizer Grade (Aug)(11847).
—Rubber
CNS K4062-82. Calcium Carbonate for Rubber (Jun)(9003).
ISO 5796-90. Rubber Compounding Ingredients—Natural Calcium Carbonate—Test Methods First Edition. 11 pp.
JIS K 6223-76. Calcium Carbonate for Rubber.
—Rubber—Aluminum Oxide Content
CNS K6721-82. Method of Test for Calcium Carbonate for Rubber (Jun)(9004). 4 pp.
—Rubber—Bulk Density
CNS K6721-82. Method of Test for Calcium Carbonate for Rubber (Jun)(9004). 4 pp.
—Rubber—Ferric Oxide Content
CNS K6721-82. Method of Test for Calcium Carbonate for Rubber (Jun)(9004). 4 pp.
—Rubber—Ignition Loss
CNS K6721-82. Method of Test for Calcium Carbonate for Rubber (Jun)(9004). 4 pp.
—Rubber—Insoluble Matter Content
CNS K6721-82. Method of Test for Calcium Carbonate for Rubber (Jun)(9004). 4 pp.
—Rubber—Moisture Content
CNS K6721-82. Method of Test for Calcium Carbonate for Rubber (Jun)(9004). 4 pp.
—Rubber—Oil Absorption
CNS K6721-82. Method of Test for Calcium Carbonate for Rubber (Jun)(9004). 4 pp.
—Rubber—Residue Content
CNS K6721-82. Method of Test for Calcium Carbonate for Rubber (Jun)(9004). 4 pp.
—Rubber—Soluble Matter Content
CNS K6721-82. Method of Test for Calcium Carbonate for Rubber (Jun)(9004). 4 pp.
—Sampling
BSI BS 6463: Part 1-84. 1984 Quicklime, Hydrated Lime and Natural Calcium Carbonate Part 1: Methods of Sampling. 8 pp.
—Water Treatment
CEN PREN 1018-93. Calcium Carbonate Used for Treatment of Water Intended for Human Consumption. 12 pp.

Calcium Carbonate Saturation Analysis
—Waste Water—PH
DIN ENGL 38404 Pt 10-79. German Standard Method for the Analysis of Water, Waste and Water and Sludge; Physical and Physical-Chemical Characteristics; (Group C); Calcium Carbonate Saturation of a Water (C 10) (May). 6 pp.
—Water—PH
DIN ENGL 38404 Pt 10-79. German Standard Method for the Analysis of Water, Waste and Water and Sludge; Physical and Physical-Chemical Characteristics; (Group C); Calcium Carbonate Saturation of a Water (C 10) (May). 6 pp.

Calcium Chloride
BSI BS 3587-63. (WITHDRAWN) 1963 Amd 2 Calcium Chloride (Technical). 13 pp.
CGSB CAN/CGSB-15.1-92. Calcium Chloride. 8 pp.
CNS K1030-84. Calcium Chloride, for Industrial Use (Dec) (380). 1 p.
CNS K6076-84. Method of Test for Calcium Chloride (Dec)(920).
CNS K7122-64. Chemical Reagent (Calcium Chloride, Anhydrous) (Jul)(1622).
CNS K7124-64. Chemical Reagent (Calcium Chloride, Dehydrate) (Jul)(1624).
JIS K 8122-75. Calcium Chloride, Dihydrate.
JIS K 8123-75. Calcium Chloride, Anhydrous.

Calcium Chloride (Cont.)
—Drying Agents
CNS K7123-64. Chemical Reagent (Calcium Chloride, Anhydrous)(for Drying) (Jul)(1623).
JIS K 8124-75. Calcium Chloride (for Drying).
—Manometers
CNS K7561-88. Chemical Reagent (Calcium Chloride for Moisture Testing) (Oct)(7719).
JIS K 8125-75. Calcium Chloride (for U-Tube).

Calcium Chloride Content Analysis
—Herbicides—Volumetric Analysis
MOD UK M 9529/66. Examination of Solution Weedkiller Chlorate.

Calcium Content Analysis
—Alloy Steels—Atomic Absorption Spectrometry
CEN EN 10 177-89. Chemical Analysis of Ferrous Materials Determination of Calcium in Steels Flame Atomic Absorption Spectrometric Method. 6 pp.
DIN ENGL EN 10177-90. Chemical Analysis of Ferrous Materials; Determination of Calcium in Steels; Flame Atomic Absorption Spectrometric Method (Apr). 7 pp.
—Aluminum Oxide—Atomic Absorption Spectrometry
BSI BS 4140: Part 14-90. 1990 Methods of Test for Aluminium Oxide Part 14: Determination of Calcium Content (W). 6 pp.
BSI BS 4140: Part 14-01. 1990 Amd 1 Methods of Test for Aluminium Oxide Part 14: Determination of Calcium Content (AMD 6893) February 28, 1992 (W). 7 pp.
—Aluminum Oxide—Atomic Absorption Spectrophotometry
ISO 2069-76. Aluminium Oxide Primarily Used for the Production of Aluminium—Determination of Calcium Content—Flame Atomic Absorption Method First Edition. 6 pp.
—Aluminum Oxide—Spectrophotometric Analysis
ISO 2070-81. Aluminium Oxide Primarily Used for the Production of Aluminium—Determination of Calcium Content—Naphthalhydroxamic Acid Spectrophotometric Method First Edition. 6 pp.
—Animal Feed
CNS N4024-15-86. Method of Test for Feeds: Determination of Calcium (Aug)(2770-15).
—Animal Feed—Atomic Absorption Spectrometry
BSI BS 5766: Part 6-84. 1984 Methods for Analysis of Animal Feeding Stuffs Part 6: Determination of Calcium by Atomic Absorption Spectrometry. 6 pp.
ISO 6490 Pt 2-83. Animal Feeding Stuffs—Determination of Calcium Content—Part 2: Atomic Absorption Spectrometric Method First Edition. 5 pp.
—Animal Feed—Volumetric Analysis
BSI BS 5766: Part 11-86. 1986 Methods of Analysis of Animal Feeding Stuffs Part 11: Determination of Calcium by Titration. 4 pp.
ISO 6490 Pt 1-85. Animal Feeding Stuffs—Determination of Calcium Content—Part 1: Titrimetric Method First Edition. 4 pp.
—Calcium Silicide
MOD UK M 805/84. Examination of Calcium Silicide.
—Calcium Silicon
CNS G2104-81. Method of Chemical Analysis for Calcium and Silica in Calcium-Silicon (Nov)(8123).
—Chromium Ores—Volumetric Analysis
ISO 5975-83. Chromium Ores—Determination of Calcium and Magnesium Contents—EDTA Titrimetric Method First Edition. 6 pp.
—Cryolite—Atomic Absorption Spectrophotometry
BSI BS 5050: Part 8-80. 1980 Methods of Test for Cryolite Part 8: Determination of Calcium Content. 7 pp.
—Drinking Water—Atomic Absorption Spectrometry
DIN ENGL 38406 Pt 3-82. German Standard Methods for the Examination of Water, Waste Water and Sludge; Cations (Group E); Determination of Calcium and Magnesium (E3) (Sept). 7 pp.

INTERNATIONAL AND NON-U.S. NATIONAL STANDARDS
SUBJECT INDEX

Calcium

Calcium Content Analysis (Cont.)

—Fertilizers
EC 89/284/EEC-89. Council Directive Supplementing and Amending Directive 76/116/EEC in Respect of the Calcium, Magnesium, Sodium and Sulphur Content of Fertilizers. 5 pp.
EC 89/519/EEC-89. Commission Directive Supplementing and Amending Directive 77/535/EEC on the Approximation of the Laws of the Member States Relating to Methods of Sampling and Analysis for Fertilizers. 18 pp.

—Food
CNS N5200-82. Method of Test for Calcium and Phosphorus in Food Stuffs (Nov)(9638). 6 pp.

—Fruit Juices
CNS N6227-91. Method of Test for Fruit and Vegetable Juices and Drinks-Determination of Sodium, Potassium, Calcium and Magnesium (Jun)(12638). 4 pp.

—Gypsum Plaster—Volumetric Analysis
MOD UK M 9519/66. Examination of Plaster of Paris (Withdrawn).

—Hard Metals—Atomic Absorption Spectrometry
BSI BS 5600: SUB SEC 4.17.2-86. (WITHDRAWN) 1986 Part 4: Methods of Testing and Chemical Analysis of Hardmetals Section 4.17: Chemical Analysis by Flame Atomic Absorption Spectrometry Subsection 4.17.2: De-termination of Calcium, Potassium, Magnesium and Sodium in Contents from 0.001% to 0.02% (m/m). 4 pp.
BSI BS EN 27627-2-93. 1993 Hardmetals—Chemical Analysis by Flame Atomic Absorption Spectrometry Part 2: Determination of Calcium, Potassium, Magnesium and Sodium, in Contents from 0.001 to 0.02 % (m/m) (ISO 7627-2: 1983) (V). 10 pp.
CEN EN 27627-2-93. Hardmetals—Chemical Analysis by Flame Atomic Absorption Spectrometry—Part 2: Determination of Calcium, Potassium, Magnesium and Sodium in Contents from 0,001 to 0,02 % (m/m) (ISO 7627-2: 1983). 5 pp.
ISO 7627 Pt 2-83. Hardmetals—Chemical Analysis by Flame Atomic Absorption Spectrometry—Part 2: Determination of Calcium, Potassium, Magnesium and Sodium in Contents from 0,001 to 0,02 % (m/m) First Edition (CEN EN 27627-2: 1993). 4 pp.

—Industrial Water—Volumetric Analysis
CPPA H.4P(K)-67. Analysis of Process Waters Calcium and Magnesium. 2 pp.

—Iron Ores—Atomic Absorption Spectrometry
ISO 10203-92. Iron Ores—Determination of Calcium Content—Flame Atomic Absorption Spectrometric Method First Edition. 12 pp.

—Iron Ores—X-Ray Fluorescence Spectrometry
ISO 9516-92. Iron Ores—Determination of Silicon, Calcium, Manganese, Aluminium, Titanium, Magnesium, Phosphorus, Sulfur and Potassium—Wavelength Dispersive X-Ray Fluorescence Spectrometric Method First Edition. 74 pp.

—Lubricating Oils—Atomic Absorption Spectrometry
DIN ENGL 51391 Pt 1-85. Testing of Lubricants; Determination of the Barium, Calcium and Zinc Content of Lubricating Oils; Direct Determination by Atomic Absorption Spectrometry (Dec). 4 pp.

—Magnesium Alloys—Atomic Absorption Spectrometry
JIS H 1341-90. Method for Determination of Calcium in Magnesium Alloys. 6 pp.

—Manganese Ores—Atomic Absorption Spectrometry
ISO 7953-85. Manganese Ores and Concentrates—Determination of Calcium and Magnesium Contents—Flame Atomic Absorption Spectrometric Method First Edition. 6 pp.

—Manganese Ores—Volumetric Analysis
ISO 6233-83. Manganese Ores and Concentrates—Determination of Calcium and Magnesium Contents—EDTA Titrimetric Method First Edition. 7 pp.

—Papers—Atomic Absorption Spectrophotometry
CPPA G.34P-92. Determination of Sodium, Magnesium, Calcium, Manganese, Iron, Copper and Cadmium in Wood, Pulp or Paper by Atomic Absorption Spectrophotometry. 3 pp.

Calcium Content Analysis (Cont.)

—Phosphates—Atomic Absorption Spectrometry
BSI BS 4427: Part 13-82. 1982 Methods of Test for Sodium Tripolyphosphate (Pentasodium Triphosphate) and Sodium Pyro-phosphate (Tetrasodium Pyrophosphate) for Industrial Use Part 13: Condensed Phosphates for Industrial Use (Including Foodstuffs). 7 pp.
ISO 5373-81. Condensed Phosphates for Industrial Use (Including Foodstuffs)—Determination of Calcium Content—Flame Atomic Absorption Spectrometric Method First Edition. 6 pp.

—Phosphoric Acids—Atomic Absorption Spectrophotometry
BSI BS 4258: Part 8-78. 1978 Methods of Test for Phosphoric Acid (Orthophosphoric Acid) for Industrial Use Part 8: Determination of Calcium Content (Flame Atomic Absorption Method). 7 pp.
ISO 3707-76. Phosphoric Acid for Industrial Use (Including Foodstuffs)—Determination of Calcium Content—Flame Atomic Absorption Method First Edition. 6 pp.

—Phosphoric Acids—Volumetric Analysis
ISO 848-81. Phosphoric Acid for Industrial Use—Determination of Calcium Content—Titrimetric Method First Edition. 7 pp.

—Pigments
CGSB 1-GP-71 METH 50.15-81. Methods of Testing Paints and Pigments Pigment Analysis Soluble Calcium Compounds. 1 p.

—Potassium Hydroxide—Atomic Absorption Spectrophotometry
ISO 3698-76. Potassium Hydroxide for Industrial Use—Determination of Calcium and Magnesium Contents—Flame Atomic Absorption Method First Edition. 6 pp.

—Potassium Hydroxide—Complexometric Titrations
ISO 997-76. Potassium Hydroxide for Industrial Use—Determination of Calcium Content—EDTA (Disodium Salt) Complexometric Method First Edition. 5 pp.

—Pulps—Atomic Absorption Spectrometry
BSI BS 4897: Part 1-83. 1983 Trace Metal Contents of Pulps Part 1: Method for Determination of Calcium Content by Edta Titrimetric and Flame Atomic Absorption Spectrometric Methods. 8 pp.
ISO 777-82. Pulps—Determination of Calcium Content—EDTA Titrimetric and Flame Atomic Absorption Spectrometric Methods First Edition. 7 pp.

—Pulps—Atomic Absorption Spectrophotometry
CPPA G.34P-92. Determination of Sodium, Magnesium, Calcium, Manganese, Iron, Copper and Cadmium in Wood, Pulp or Paper by Atomic Absorption Spectrophotometry. 3 pp.

—Pulps—Volumetric Analysis
BSI BS 4897: Part 1-83. 1983 Trace Metal Contents of Pulps Part 1: Method for Determination of Calcium Content by Edta Titrimetric and Flame Atomic Absorption Spectrometric Methods. 8 pp.
ISO 777-82. Pulps—Determination of Calcium Content—EDTA Titrimetric and Flame Atomic Absorption Spectrometric Methods First Edition. 7 pp.

—Silicon—Absorptiometric Analysis
CNS G2142-82. Method of Chemical Analysis for Calcium in Metallic Silicon (GHA Absorptimetric Method) (Oct)(9501).

—Silicon—Atomic Absorption Spectrometry
CNS G2143-82. Method of Chemical Analysis for Calcium in Metallic Silicon (Atomic Absorptimetric Method) (Oct)(9502).

—Silicon—Volumetric Analysis
CNS G2141-82. Method of Chemical Analysis for Calcium in Metallic Silicon (EDTA Titration Method) (Oct)(9500).

—Sodium Borate—Atomic Absorption Spectrometry
BSI BS 5688: Part 28-86. 1986 Methods of Test for Orthoboric Acid (Boric Acid), Diboron Trioxide (Boric Oxide), Disodium Tetraborates, Sodium Perborates and Crude Sodium Borates for Industrial Use Part 28: Determination of Total and AlkaliSoluble Calcium and Magnesium. 7 pp.

Calcium Content Analysis (Cont.)

—Sodium Borate—Atomic Absorption Spectrometry (Cont.)
ISO 6918-84. Crude Sodium Borates for Industrial Use—Determination of Total and Alkali-Soluble Calcium and Magnesium Contents—Flame Atomic Absorption Spectrometric Method First Edition. 6 pp.

—Sodium Borate—Volumetric Analysis
BSI BS 5688: Part 29-86. 1986 Orthoboric Acid, Diboron Trioxide, Disodium Tetraborates, Sodium Perborates and Crude Sodium Borates for Ind. Use Part 29: Detn. of Total and Alkali-Soulble Calcium and Magnesium Contents of Crude Sodium Borates (Titrimetric Method). 17 pp.
ISO 6920-84. Crude Sodium Borates for Industrial Use—Determination of Total and Alkali-Soluble Calcium and Magnesium Contents—Titrimetric Method First Edition. 8 pp.

—Sodium Chloride—Complexometric Titrations
BSI BS 7319: Part 5-90. 1990 Analysis of Sodium Chloride for Industrial Use Part 5: Method for Determination of Calcium and Magnesium Contents. 6 pp.
ISO 2482-73. Sodium Chloride for Industrial Use—Determination of Calcium and Magnesium Contents—EDTA Complexometric Methods First Edition. 5 pp.

—Sodium Hydroxide—Atomic Absorption Spectrophotometry
BSI BS 6075: Part 9-81. 1981 Amd 1 Sampling and Test for Sodium Hydroxide for Industrial Use Part 9: Determination of Calcium and Magnesium Contents. 7 pp.
ISO 3697-76. Sodium Hydroxide for Industrial Use—Determination of Calcium and Magnesium Contents—Flame Atomic Absorption Method First Edition. 6 pp.

—Sodium Hydroxide—Complexometric Titrations
ISO 986-76. Sodium Hydroxide for Industrial Use—Determination of Calcium Content—EDTA (Disodium Salt) Complexometric Method First Edition. 5 pp.

—Sodium Nitrate
MOD UK M 869/91. Examination of Sodium Nitrate, Grade 1.

—Sodium Sulfate—Atomic Absorption Spectrometry
ISO 5994-79. Sodium Sulphate for Industrial Use—Determination of Calcium Content—Flame Atomic Absorption Spectrometric Method First Edition. 6 pp.

—Sodium Sulfate—Complexometric Titrations
ISO 3238-75. Sodium Sulphate for Industrial Use—Determination of Calcium Content—EDTA Complexometric Method First Edition. 5 pp.

—Steels—Atomic Absorption Spectrometry
BSI BS 6200: SUBSEC 3.7.1-87. 1987 Sampling and Analysis of Iron, Steel and Other Ferrous Metals Part 3: Methods of Analysis Section 3.7: Determination of Calcium Subsection 3.7.1: Steel: Flame Atomic Absorption Spectrometric Method. 8 pp.
BSI BS 6200: SUBSEC 3.7.1-01. 1987 Amd 1 Sampling and Analysis of Iron, Steel and Other Ferrous Metals Part 3: Methods of Analysis Sec 3.7: Determination of Calcium Subsec 3.7.1: Steel: Flame Atomic Absorption Spectrometric Method (AMD 7069) February 28, 1992 (V). 13 pp.
BSI BS 6200: Sec 6.1-90. 1990 Sampling and Analysis of Iron, Steel and Other Ferrous Metals: Part 6: Guidelines on Atomic Absorption Spectrometric Techniques: Section 6.1: Recommendations for the Drafting of Standard Methods for the Chemical Anal. of Iron and Steel by Flame Atomic Absorp. 21 pp.
CEN EN 10 177-89. Chemical Analysis of Ferrous Materials Determination of Calcium in Steels Flame Atomic Absorption Spectrometric Method. 6 pp.
DIN ENGL EN 10177-90. Chemical Analysis of Ferrous Materials; Determination of Calcium in Steels; Flame Atomic Absorption Spectrometric Method (Apr). 7 pp.
ISO 10697 Pt 1-92. Steel—Determination of Calcium Content by Flame Atomic Absorption Spectrometry—Part 1: Determination of Acid-Soluble Calcium Content First Edition. 13 pp.

—Titanium Ores
JIS M 8318-67. Determination of Calcium in Titanium Ores.

INDUSTRY STANDARDS

Calcium Content Analysis (Cont.)

—Vegetable Juices
CNS N6227-91. Method of Test for Fruit and Vegetable Juices and Drinks-Determination of Sodium, Potassium, Calcium and Magnesium (Jun)(12638). 4 pp.

—Waste Water—Atomic Absorption Spectrometry
DIN ENGL 38406 Pt 3-82. German Standard Methods for the Examination of Water, Waste Water and Sludge; Cations (Group E); Determination of Calcium and Magnesium (E3) (Sept). 7 pp.

—Waste Water—Volumetric Analysis
CNS K9085-82. Method of Test for Calcium in Industrial Waste Water Titrimetric Method (May)(8841).

—Water—Atomic Absorption Spectrometry
BSI BS 6068: Sec 2.30-87. 1987 Water Quality Part 2: Physical, Chemical and Bio-Chemical Methods Section 2.30: Methods for Determination of Calcium and Magnesium by Atomic Absorption Spectrometric Method. 6 pp.
CNS K9024-80. Method of Test for Calcium and Magnesium in Water (Atomic Absorption, Direct) (Feb)(5222).
DIN ENGL 38406 Pt 3-82. German Standard Methods for the Examination of Water, Waste Water and Sludge; Cations (Group E); Determination of Calcium and Magnesium (E3) (Sept). 7 pp.
ISO 7980-86. Water Quality—Determination of Calcium and Magnesium—Atomic Absorption Spectrometric Method First Edition. 5 pp.
JIS K 0553-90. Testing Methods for Determination of Metallic Elements in Highly Purified Water. 27 pp.

—Water—Volumetric Analysis
BSI BS 6068: Sec 2.8-84. 1984 Water Quality Part 2: Physical, Chemical and Bio-Chemical Methods Section 2.8: Determination of Calcium Content: EDTA Titrimetric Method. 6 pp.
BSI BS 6068: Sec 2.9-84. 1984 Water Quality Part 2: Physical, Chemical and Bio-Chemical Methods Section 2.0: Determination of the Sum of Calcium and Magnesium: EDTA Titrimetric Method. 7 pp.
ISO 6058-84. Water Quality—Determination of Calcium Content—EDTA Titrimetric Method First Edition. 5 pp.
ISO 6059-84. Water Quality—Determination of the Sum of Calcium and Magnesium—EDTA Titrimetric Method First Edition. 6 pp.

—Wood—Atomic Absorption Spectrophotometry
CPPA G.34P-92. Determination of Sodium, Magnesium, Calcium, Manganese, Iron, Copper and Cadmium in Wood, Pulp or Paper by Atomic Absorption Spectrophotometry. 3 pp.

Calcium Cyanamide
See Also: Calcium Nitrate
CNS K6013-71. Method of Analysis for Calcium Cyanamide (Jun)(188).

—Fertilizers
CNS N3005-88. Calcium Cyanamide, Fertilizer Grade (Aug)(266).

Calcium Hydroxides
Use For: Hydrated Lime; Slaked Lime
See Also: Lime Water
CNS K7125-64. Chemical Reagent (Calcium Hydroxide) (Jul)(1625).
JIS K 8575-75. Calcium Hydroxide.
MOD UK DSTAN 68-39-93. Calcium Hydroxide, Technical Issue 1. 14 pp.
MOD UK DEF-97-59. Calcium Hydroxide. 3 pp.

—Chemical Analysis
BSI BS 6463: Part 2-84. 1984 Quicklime, Hydrated Lime and Natural Calcium Carbonate Part 2: Methods of Chemical Analysis. 14 pp.
BSI BS 6463: Part 2-01. 1984 Amd 1 Quicklime, Hydrated Lime and Natural Calcium Carbonate Part 2: Methods of Chemical Analysis (AMD 7586) May 15, 1993 (W). 15 pp.
CPPA J.5-51. Analysis of Lime. 2 pp.

—Fertilizer
CNS N3023-88. Slacked Lime, Fertilizer Grade (Aug)(11846).

—Grit Content
MOD UK M 9514/65. Examination of Calcium Hydroxide.

—Sampling
BSI BS 6463: Part 1-84. 1984 Quicklime, Hydrated Lime and Natural Calcium Carbonate Part 1: Methods of Sampling. 8 pp.

Calcium Hydroxides (Cont.)

—Sieve Analysis
MOD UK M 9514/65. Examination of Calcium Hydroxide.

—Volatile Matter Content
MOD UK M 9514/65. Examination of Calcium Hydroxide.

—Water Treatment
CNS K1147-71. Slaked Lime for Water Purification (Jan)(3167). 1 p.
MOD UK CS 3135. Hydrated Lime.

Calcium Hyperphosphate
CNS K6015-71. Method of Analysis for Calcium Hyperphosphate (Mar)(190).

Calcium Hypochlorite
Use For: Calcium Oxychloride *See Also:* Bleach Liquors; Hazardous Materials
JIS K 1207-64. Calcium Hypochlorite Solution. 4 pp.

—Bleaching Agents
CGSB CAN/CGSB-15.32-93. Calcium Hypochlorite. 8 pp.
CNS K1162-74. Calcium Hypochlorite Bleaching Solution (Sep)(3395).
CNS K6331-74. Method of Test for Calcium Hypochlorite Bleaching Solution (Oct)(3396).
CPPA J.2H-64. Analysis of Calcium Hypochlorite Bleaching Powder, Bleach Liquor, and Bleach Sludge. 4 pp.

—Decontamination
NATO STANAG 4323 ED 1 AMD 0-00. Specification of Calcium Hypochlorite Used for Biological and Chemical Decontamination. 14 pp.
NATO STANAG 4323 ED 1 AMD 0-92. Specification of Calcium Hypochlorite Used for Biological and Chemical Decontamination. 13 pp.

—Decontamination—Military Chemical Agents
MOD UK DSTAN 68-132-89. Calcium Hypochlorite Issue 1. 11 pp.

—Disinfectants
CGSB CAN/CGSB-15.32-93. Calcium Hypochlorite. 8 pp.
MOD UK DSTAN 68-132-89. Calcium Hypochlorite Issue 1. 11 pp.
MOD UK TS 10202A. Calcium Hypochlorite (Superseded by Def Stan 68-132).

—Highway Transportation—Emergency Procedures
SAA AS 1678.5.1. 004-93. Emergency Procedure Guides—Transport—Part 5.1.004: Calcium Hypochlorite (in Professional Package 37A).

—Water Treatment
CEN PREN 900-92. Calcium Hypochlorite Used for Water Intended for Human Consumption. 21 pp.
CGSB CAN/CGSB-15.32-93. Calcium Hypochlorite. 8 pp.

Calcium Inorganic Compounds
Scope Note: Use a more specific term *See:* Bleach Liquors; Calcium Ammonium Nitrate; Calcium Carbide; Calcium Carbonate; Calcium Chloride; Calcium Cyanamide; Calcium Hydroxides; Calcium Hypochlorite; Calcium Nitrate; Calcium Oxide; Calcium Phosphate, Monobasic; Calcium Phosphates; Calcium Silicate; Calcium Silicon; Gypsum; Superphosphates

Calcium Nitrate
See Also: Calcium Cyanamide
CNS K7126-64. Chemical Reagent (Calcium Nitrate) (Jul)(1626).
JIS K 8549-75. Calcium Nitrate.

—Fertilizers
CNS N3065-88. Calcium Nitrate, Fertilizer Grade (Jul)(11952).

Calcium Organic Compounds
Scope Note: Use a more specific term *See:* Calcium Acetate; Calcium Carbonate; Calcium Pantothenate

Calcium Oxalate

—Pyrotechnics
MOD UK DSTAN 68-77-89. Calcium Oxalate Monohydrate Issue 1. 8 pp.

Calcium Oxide
Use For: Lime; Lime Putty; Quicklime
See Also: Calcium Oxide Content Analysis; Calcium Phosphates; Green Liquors; Pesticides

Calcium Oxide (Cont.)
CEN PREN 459 (Part 1)-91. Building Time—Part 1: Definitions, Specifications and Conformity Criteria. 28 pp.
CNS K7127-64. Chemical Reagent (Calcium Oxide) (Jul)(1627).
JIS K 8410-61. Calcium Oxide.
JIS R 9001-81. Industrial Limes. 6 pp.
SAA AS N44-63. Lime, Lime-Sulphur and Sulphur for Use as Agricultural Pesticides. 15 pp.
SAA AS 1672-74. Building Limes Corrig.. 34 pp.
SNZ NZS/AS 1672-74. Building Limes (This is a Joint Standard with SAA AS 1672). 34 pp.

—Carbon Dioxide Content
CGSB 1-GP-71 METH 50.10-80. Methods of Testing Paints and Pigments Pigment Analysis Carbon Dioxide in Carbonates. 1 p.

—Chemical Analysis
BSI BS 6463: Part 2-84. 1984 Quicklime, Hydrated Lime and Natural Calcium Carbonate Part 2: Methods of Chemical Analysis. 14 pp.
BSI BS 6463: Part 2-01. 1984 Amd 1 Quicklime, Hydrated Lime and Natural Calcium Carbonate Part 2: Methods of Chemical Analysis (AMD 7586) May 15, 1993 (W). 15 pp.
CEN PREN 459 (Part 2)-91. Building Time-Part 2: Test Methods. 48 pp.
CPPA J.5-51. Analysis of Lime. 2 pp.
DIN ENGL 1060 Pt 2-82. Building Limes; Methods of Chemical Analysis (Nov). 7 pp.
JIS R 9011-81. Chemical Analysis of Limes. 25 pp.

—Construction Materials
BSI BS 890-72. 1972 Building Limes. 33 pp.
CNS A2002-72. Quicklime for Buildings (Jun)(381). 1 p.

—Fertilizers
CNS N3022-88. Quick Lime, Fertilizer Grade (Aug)(11845).
CNS N3026-88. Mixed Lime, Fertilizer (Jul)(11849).
CNS N3027-88. By-Product Lime Fertilizer (Jul)(11850).

—Fly Ash—Reference Method
CEN PREN 451 (Part 1)-91. Method of Testing Fly Ash—Part 1 Determination of Free Calcium Oxide. 5 pp.

—Furnaces—Heat Balance
CNS R3124-85. Method for Calculating Heat Balance of Kiln and Furnace for Lime (May) (11274).
JIS R 0305-91. Heat Balancing of Kiln and Furnace for Lime. 25 pp.

—Glossaries
CEN PREN 459 (Part 1)-91. Building Time—Part 1: Definitions, Specifications and Conformity Criteria. 28 pp.
DIN ENGL 1060 Pt 1-86. Building Limes; Terminology, Requirements, Supply, Inspection (Jan). 9 pp.
JIS R 9200-86. Glossary of Terms Related to Gypsum, Lime and Magnesia Cement. 32 pp.

—Inspection
DIN ENGL 1060 Pt 1-86. Building Limes; Terminology, Requirements, Supply, Inspection (Jan). 9 pp.

—Kilns—Heat Balance
CNS R3124-85. Method for Calculating Heat Balance of Kiln and Furnace for Lime (May) (11274).
JIS R 0305-91. Heat Balancing of Kiln and Furnace for Lime. 25 pp.

—Physical Testing
BSI BS 6463: Part 4-87. 1987 Quicklime, Hydrated Lime and Natural Calcium Carbonate Part 4: Methods of Test for Physical Properties of Hydrated Lime and Lime Putty. 19 pp.
CEN PREN 459 (Part 2)-91. Building Time-Part 2: Test Methods. 48 pp.
DIN ENGL 1060 Pt 3-82. Building Limes; Methods of Physical Test (Nov). 9 pp.

—Sampling
BSI BS 6463: Part 1-84. 1984 Quicklime, Hydrated Lime and Natural Calcium Carbonate Part 1: Methods of Sampling. 8 pp.
CEN PREN 459 (Part 2)-91. Building Time-Part 2: Test Methods. 48 pp.

—Water Treatment
DIN ENGL 19611-83. High-Calcium Lime for Use in Water Treatment; Technical Delivery Conditions (Apr). 7 pp.

Calcium Oxide Content Analysis
See Also: Calcium Oxide

INTERNATIONAL AND NON-U.S. NATIONAL STANDARDS
SUBJECT INDEX
Calcium

Calcium Oxide Content Analysis (Cont.)

—Blended Cements
SAA AS 2350.10-91. Methods of Testing Portland and Blended Cements—Part 10: Calcium Oxide Content of Blended Cement (in Professional Package 58). 4 pp.

—Dolomite
ISO 10058-92. Magnesites and Dolomites —Chemical Analysis First Edition. 23 pp.

—Dolomite—Volumetric Analysis
CNS M3106-82. Method for Determining Calcium Oxide in Dolomite (EDTA Titration Method) (Sep)(9424).
DIN ENGL 52253 Pt 1-88. Testing the Frost Resistance of Roofing Tiles; Freeze-Thaw Test with Upper Side Freezing After Sprinkling with Water (Dec) (Supersedes DIN 52251 Part 7, January 1981 Edition, Withdrawn in January 1986). 5 pp.

—Glass—Atomic Absorption Spectrometry
ISO 10136 Pt 3-93. Glass and Glassware—Analysis of Extract Solutions—Part 3: Determination of Calcium Oxide and Magnesium Oxide by Flame Atomic Absorption Spectrometry First Edition. 10 pp.

—Glassware—Atomic Absorption Spectrometry
ISO 10136 Pt 3-93. Glass and Glassware—Analysis of Extract Solutions—Part 3: Determination of Calcium Oxide and Magnesium Oxide by Flame Atomic Absorption Spectrometry First Edition. 10 pp.

—Iron Ores—Atomic Absorption Spectrophotometry
JIS M 8221-83. Methods for Determination of Calcium Oxide in Iron Ores. 20 pp.

—Iron Ores—Volumetric Analysis
JIS M 8221-83. Methods for Determination of Calcium Oxide in Iron Ores. 20 pp.

—Limestone
CPPA J.4-92. Analysis of Limestone. 5 pp.

—Limestone—Volumetric Analysis
DIN ENGL 52240 Pt 8-85. Analysis of Raw Materials Used in Glass Production; Chemical Analysis of Limestone Containing Not Less Than 95% of Calcium Carbonate; Determination of Calcium Oxide and Magnesium Oxide (Sept). 4 pp.

—Magnesite
ISO 10058-92. Magnesites and Dolomites —Chemical Analysis First Edition. 23 pp.

—Manganese Ores—Atomic Absorption Spectrophotometry
JIS M 8240-82. Methods for Determination of Calcium Oxide in Manganese Ores. 10 pp.

—Manganese Ores—Photometry
CNS M3072-81. Method for Determination of Calcium Oxide in Manganese Ores (Flame Photometric Method) (Aug)(7837).

—Manganese Ores—Volumetric Analysis
CNS M3070-81. Method for Determination of Calcium Oxide in Manganese Ores Potassium Permanganate (Titration Method) (Aug) (7835).
CNS M3071-81. Method for Determination of Calcium Oxide in Manganese Ores (EDTA Titration Method) (Aug) (7836).
JIS M 8240-82. Methods for Determination of Calcium Oxide in Manganese Ores. 10 pp.

—Stabilized Soil Analysis
BSI BS 1924: Part 2-90. 1990 Stabilized Materials for Civil Engineering Purposes Part 2: Methods of Test for Cement-Stabilized and Lime-Stabilized Materials. 104 pp.

Calcium Oxychloride
Use: Calcium Hypochlorite

Calcium Pantothenate
JIS K 8757-61. Calcium Pantothenate.

Calcium Phosphate, Dibasic
See Also: Calcium Phosphates
CNS K7128-64. Chemical Reagent (Calcium Phosphate, Phosphate, Dibasic) (Jul)(1628).

Calcium Phosphate, Monobasic
See Also: Calcium Phosphates
CNS K7129-64. Chemical Reagent (Calcium Phosphate, Monobasic)(Calcium Biphosphate) (Jul)(1629).

Calcium Phosphates
Use For: Tricalcium Phosphates

Calcium Phosphates (Cont.)
See Also: Animal Feed; Calcium Oxide; Calcium Phosphate, Dibasic; Calcium Phosphate, Monobasic; Calcium Sulfate; Superphosphates

—Animal Feed
CNS N2045-82. Calcium Phosphates (for Feed) (Apr)(4526). 1 p.
CNS N4047-82. Methods of Test for Calcium Phosphates (for Feed)(Apr)(4527).

—Propellants
MOD UK DSTAN 68-91-88. Calcium Phosphate Issue 1. 8 pp.

Calcium Plumbate Paints
See Also: Calcium Plumbate Primers; Paints
BSI BS 3699-64. (WITHDRAWN) 1964 Calcium Plumbate for Paints. 12 pp.

—Steel
CNS K2160-87. Calcium Plumbate Anticorrosive Paint (Oct)(12135). 2 pp.
CNS K6922-87. Method of Test for Calcium Plumbate Anticorrosive Paint (Oct)(12136). 5 pp.
JIS K 5629-92. Calcium Plumbate Anticorrosive Paint. 11 pp.

Calcium Plumbate Primers
See Also: Calcium Plumbate Paints
BSI BS 3698-64. (OBSOLESCENT) 1964 Calcium Plumbate Priming Paints. 14 pp.

Calcium Resinate
CNS K1160-72. Calcium Resinate (Jun)(3342).
CNS K6319-72. Method of Test for Calcium Resinate (Jun)(3343).

Calcium Silicate
See Also: Calcium Silicate Blocks; Calcium Silicate Bricks
CNS K1234-80. Calcium Silicate (Apr)(5449).
CNS K6472-80. Method of Test for Calcium Silicate (Apr)(5453).

—Board—Asbestos Cement—Autoclaved
CNS A2174-84. Autoclave Asbestos Cement Silicate Boards (Apr)(10843).
CNS A2175-84. Decorated Autoclave Asbestos Cement Calcium Silicate Boards (Apr)(10845).
CNS A3205-84. Method of Test for Autoclave Asbestos Cement Silicate Boards (Apr)(10844).
CNS A3206-84. Method of Test for Decorated Autoclave Asbestos Cement Calcium Silicate Boards (Apr)(10846).
JIS A 5418-89. Silica-Asbestos-Cement Boards. 19 pp.

—Flexible Pavements
BSI BS 6677: Part 1-86. 1986 Clay and Calcium Silicate Pavers for Flexible Pavements Part 1: Specification for Pavers. 10 pp.
BSI BS 6677: Part 2-86. 1986 Clay and Calcium Silicate Pavers for Flexible Pavements Part 2: Code of Practice for Design of Lightly Trafficked Pavements. 10 pp.
BSI BS 6677: Part 3-86. 1986 Clay and Calcium Silicate Pavers for Flexible Pavements Part 3: Method for Construction of Pavements. 8 pp.

—Thermal Insulation
BSI BS 3958: Part 2-82. 1982 Thermal Insulating Materials Part 2: Calcium Silicate Preformed Insulation. 8 pp.
CGSB CAN/CGSB-51.2-M88. Thermal Insulation, Calcium Silicate, for Piping, Machinery and Boilers; (Amendment 1 December 1991). 11 pp.
JIS A 9510-89. Thermal Insulation Material Made of Calcium Silicate. 13 pp.
SNZ BS 3958: Part 2-82. Thermal insulating Materials Part 2: Calcium Silicate Preformed Insulation. 8 pp.

—Thermal Insulation—Marine
MOD UK NES 800: Part 2-88. Requirements for Thermal Insulation Material Part 2: Calcium Silicate Porducts Calcium Silicate Filling Compound Issue 2 (11.88). 10 pp.
MOD UK NES 800: Part 1-91. Requirements for Insulation Material Part 1: Calcium Silicate Slab and Section Insulation Material Issue 3 (09.91). 16 pp.
MOD UK NES 800: Part 1-01. Requirements for Thermal Insulation Material Part 1: Calcium Silicate Slab and Section Insulation Material Issue 3 (09.91); Amendment 2. 22 pp.

Calcium Silicate Blocks
See Also: Blocks; Calcium Silicate
DIN ENGL 106 Pt 1-80. Sandlime Bricks and Blocks; Solid Bricks, Perforated Bricks, Solid Blocks, Hollow Blocks (Sept). 6 pp.

—Thermal Insulation
CNS R2043-76. Calcium Silicate Thermal Insulating Plates (Blocks) and Pipes (Mar)(2176).

Calcium Silicate Blocks (Cont.)
—Thermal Insulation (Cont.)
CNS R3045-76. Method of Test for Calcium Silicate for Thermal Insulating Plates (Blocks) and Pipes (Mar)(2177).

Calcium Silicate Bricks
Use For: Sandlime Bricks *See Also:* Bricks; Calcium Silicate
BSI BS 187-78. 1978 Amd 1 Calcium Silicate (Sandlime and Flintlime) Bricks. 15 pp.
BSI BS 6649-85. 1985 Clay and Calcium Silicate Modular Bricks. 4 pp.
CEN PREN 771-2-92. Specification for Masonry Units—Part 2: Calcium Silicate Masonry Units. 18 pp.
CNS A2034-85. Sand-Lime Bricks (Jan)(2220). 2 pp.
CNS A3054-85. Method of Test for Sand-Lime Bricks (Jan)(2221). 2 pp.
CSA A82.3-M1978. Calcium Silicate (Sand-Lime) Building Brick (R 1992). 8 pp.
DIN ENGL 106 Pt 1-80. Sandlime Bricks and Blocks; Solid Bricks, Perforated Bricks, Solid Blocks, Hollow Blocks (Sept). 6 pp.

—Cleaning
BSI BS 6270: Part 1-82. 1982 Amd 2 Cleaning and Surface Repair of Buildings Part 1: Natural Stone, Cast Stone and Clay and Calcium Silicate Brick Masonry (AMD 5605) April 28, 1989. 46 pp.

—Compression Testing
CEN PREN 772-1-92. Methods of Test for Masonry Units—Part 1: Determination of Compressive Strength. 13 pp.

—Facing
DIN ENGL 106 Pt 2-80. Sandlime Bricks and Blocks; Facing Bricks and Hard-Burnt Facing Bricks (Nov). 6 pp.

—Frost Resistance
CEN PREN 772-18-92. Methods of Test for Masonry Units—Part 18: Determination of Frost Resistance of Calcium Silicate Masonry Units. 7 pp.

—Moisture Content
CEN PREN 772-10-92. Methods of Test for Masonry Units—Part 10: Determination of Moisture Content of Calcium Silicate, Aggregate Concrete and Autoclaved Aerated Concrete Masonry Units. 6 pp.

—Repair
BSI BS 6270: Part 1-82. 1982 Amd 2 Cleaning and Surface Repair of Buildings Part 1: Natural Stone, Cast Stone and Clay and Calcium Silicate Brick Masonry (AMD 5605) April 28, 1989. 46 pp.

—Sizes
CEN PREN 772-16-92. Methods of Test for Masonry Units—Part 16: Determination of Size and Dimensions (Excluding Natural Stone Masonry Units). 6 pp.

—Thickness Measurement
CEN PREN 772-16-92. Methods of Test for Masonry Units—Part 16: Determination of Size and Dimensions (Excluding Natural Stone Masonry Units). 6 pp.

—Voids—Sand Filling
CEN PREN 772-9-92. Methods of Test for Masonry Units—Part 9: Determination of Volume of Voids and Net Volume of Clay and Calcium Silicate Masonry Units by Sand Filling. 6 pp.

—Volume
CEN PREN 772-3-92. Methods of Test for Masonry Units—Part 3: Determination of Net Volume of Clay and Calcium Silicate Masonry Units by Weighing. 5 pp.

Calcium Silicide
—Alkalinity
MOD UK M 805/84. Examination of Calcium Silicide.

—Carbide Content Analysis
MOD UK M 805/84. Examination of Calcium Silicide.

—Sieve Analysis
MOD UK M 805/84. Examination of Calcium Silicide.

—Visual Inspection
MOD UK M 805/84. Examination of Calcium Silicide.

Calcium Silicon
CNS G3115-84. Calcium Silicon (Jun) (5124).
DIN ENGL 17580-68. Calcium-Silicon; Technical Conditions of Delivery (Dec). 2 pp.
JIS G 2314-86. Calcium Silicon. 6 pp.

INDUSTRY STANDARDS

Calcium Silicon (Cont.)

—Calcium Content
CNS G2104-81. Method of Chemical Analysis for Calcium and Silica in Calcium-Silicon (Nov)(8123).

—Carbon Content
CNS G2105-81. Method of Chemical Analysis for Carbon in Calcium-Silicon (Nov)(8124).

—Chemical Analysis
JIS G 1324-89. Methods for Chemical Analysis of Calcium-Silicon. 27 pp.
JIS G 1603-85. Methods of Sampling for Chemical Analysis of Ferroalloys (Part 3 Ferrophosphorus, Manganese Metal, Silicon Metal, Chromium Metal, Calcium Silicon and Ferroboron). 10 pp.

—Phosphorus Content
CNS G2117-82. Methods of Chemical Analysis for Phosphorus in Calcium-Silicon (Jan)(8401).

—Sampling
CNS G2215-84. Method of Sampling for Calcium-Silicon (Mar)(10813).
JIS G 1603-85. Methods of Sampling for Chemical Analysis of Ferroalloys (Part 3 Ferrophosphorus, Manganese Metal, Silicon Metal, Chromium Metal, Calcium Silicon and Ferroboron). 10 pp.

—Silica Content
CNS G2104-81. Method of Chemical Analysis for Calcium and Silica in Calcium-Silicon (Nov)(8123).

Calcium Sulfate
See Also: Calcium Phosphates; Calcium Sulfate Content Analysis
CNS K7130-64. Chemical Reagent (Calcium Sulfate) (Jul)(1630).
JIS K 8963-75. Calcium Sulfate, Dihydrate.

—Fillers
MOD UK TS 567B. Calcium Sulphate, Technical.

Calcium Sulfate Content Analysis
See Also: Calcium Sulfate

—Mortars
CNS R3110-82. Method of Test for Calcium Sulfate in Hydrated Portland Cement Mortar (Dec)(9745).

Calcium Sulphate
Use: Calcium Sulfate

Calcium Superphosphates
Use: Superphosphates

Calculating Machines
Use: Calculators

Calculators
Use For: Accounting Machines; Calculating Machines
See Also: Abacus; Adding Machines; Cash Registers; Computers; Office Machines
CNS C6293-87. Method of Test for Calculator (Aug)(12047).

—Glossaries
BSI BS 5478: Part 4-85. 1985 Calculators and Adding Machines Part 4: Glossary of Terms for Calculators. 18 pp.
BSI BS ISO 2382/22-86. 1986 Information Technology—Vocabulary 2382/22: Calculators. 27 pp.
CNS C5136-82. Definition of a Calculator (Jun)(8940).
CNS C5137-82. Definition of Accounting Machines (Jun)(8941).
CNS C5239-90. Information Processing Systems—Vocabulary (Part 22: Calculators) (Aug)(12760).
ISO 2382 Pt 22-86. Information Processing Systems—Vocabulary—Part 22: Calculators First Edition. 23 pp.
JIS X 0022-89. Glossary of Terms Used in Information Processing (Calculators).
JTC1 2382 Pt 22-86. Information Processing Systems—Vocabulary—Part 22: Calculators First Edition. 23 pp.
OSI ISO 2382-22-86. Information Processing Systems—Vocabulary—Part 22: Calculators. 23 pp.
SAA AS 1189.22-87. Data Processing—Vocabulary—Part 22: Calculators. 10 pp.

—Keyboards
BSI BS 5478: Part 1-77. 1977 Calculators and Adding Mahchines Part 1:Numeric Section of Ten-Key Keyboards for Calculators and Adding Machines. 3 pp.
CNS C5119-81. 10-Key Keyboard for Adding and Calculating Machines (Jul)(7649).
CNS C5195-85. Calculating Machines—Numeric Section of Ten-Key Keyboards (Mar)(11219).
CNS C5196-85. Calculating Machines—Keytop and Printed or Displayed Symbols (Mar)(11220).

Calculators (Cont.)

—Keyboards (Cont.)
ISO 1092-74. Adding Machines and Calculating Machines—Numeric Section of Ten-Key Keyboards First Edition. 3 pp.
JIS B 9517-90. Keyboards Arrangement for Calculating Machines with Ten-Key. 6 pp.
JTC1 1092-74. Adding Machines and Calculating Machines—Numeric Section of Ten-Key Keyboards First Edition. 3 pp.
OSI ISO 1092-74. Adding Machines and Calculating Machines—Numeric Sectionb of Ten-Key Keyboards. 3 pp.

—Symbols
BSI BS 5478: Part 1-77. 1977 Calculators and Adding Mahchines Part 1:Numeric Section of Ten-Key Keyboards for Calculators and Adding Machines. 3 pp.
BSI BS 5478: Part 3-82. 1982 Calculators and Adding Machines Part 3: Specification for Keytop and Printed or Displayed Symbols. 4 pp.
CNS C5196-85. Calculating Machines—Keytop and Printed or Displayed Symbols (Mar)(11220).
ISO 1093-81. Adding Machines and Calculating Machines—Keytop and Printed or Displayed Symbols First Edition. 5 pp.
JIS B 9516-90. Keytop and Printed or Displayed Symbols for Adding Machines. 5 pp.
JTC1 1093-81. Adding Machines and Calculating Machines—Keytop and Printed or Displayed Symbols First Edition. 5 pp.
OSI ISO 1093-81. Adding Machines and Calculating Machines—Keytop and Printed or Displayed Symbols. 5 pp.
SAA AS 1412.1-76. Adding and Calculating Machines—Part 1: Basic Informative Symbols. 4 pp.

Calendars
See Also: Julian Date

—Desk
CGSB CAN/CGSB-53.22-M89. Calendar Pads and Stands. 10 pp.

—Stands (Supports)
CGSB CAN/CGSB-53.22-M89. Calendar Pads and Stands. 10 pp.

Calendered Papers
See Also: Papers

—Ammunition
MOD UK DEF-42-A. Paper, Pure, Bleached, Super-Calendered and Paper, Pure, Bleached, Super-Calendered, Lead Free (Superseded by Def Stan 81-97).

Calibers
Use For: Calibres

—Ammunition/Weapons
MOD UK DSTAN 05-15-01. Metric System Measurement of Weapon Calibres and Ammunition Issue 1; Amendment 1. 4 pp.

Calibrating Tanks
Use For: Meter Proving Tanks *See Also:* Tanks (Containers)

—Petroleum Products—Temperature Corrections
BSI BS 6922-88. 1988 Temperature Corrections for Use in the Calibration of Reference Measuring Systems for Petroleum Measurement. 8 pp.
ISO 8222-87. Petroleum Measurement Systems—Calibration—Temperature Corrections for Use with Volumetric Reference Measuring Systems First Edition. 8 pp.

Calibration
Scope Note: For additional listings, see the subheading Calibration under specific products
BSI BS 5781: Part 1-79. 1979 Amd 1 Measurement and Calibration Systems Part 1: Specification of Systems Requirements. 6 pp.
BSI BS 5781: Part 2-81. (WITHDRAWN) 1981 Measurement and Calibration Systems Part 2: Guide to the Use of BS 5781 Part 1 (Superseded by BS 5781: Part 1: 1992). 14 pp.
CSA CAN/CSA-Q419-88. Recommended Practices for Calibration and Use of Devices for Measuring and Testing; (Gen Instr 1). 22 pp.
JIS Z 9090-91. Measurement—General Rules for Calibration System. 49 pp.
MOD UK AQAP-6-76. NATO Measurement and Calibration System Requirements for Industry Edn 2, 7/76. 12 pp.
MOD UK AQAP-7-78. Guide for the Evaluation of a Contractor's Measurement and Calibration System for Compliance with AQAP-6 10/78. 39 pp.
QSS MIL-Std-45662A-88. Calibration Systems Requirements. 9 pp.

Calibration (Cont.)

—Quality Control
ISO 8466 Pt 1-90. Water Quality—Calibration and Evaluation of Analytical Methods and Estimation of Performance Characteristics Part 1: Statistical Evaluation of the Linear Calibration Function First Edition. 11 pp.
ISO 8466 Pt 2-93. Water Quality—Calibration and Evaluation of Analytical Methods and Estimation of Performance Characteristics—Part 2: Calibration Strategy for Non-Linear Second Order Calibration Functions First Edition. 16 pp.

Calibration Gases
See Also: Gas Analysis; Gases
ISO 6143-81. Gas Analysis—Determination of Composition of Calibration Gas Mixtures—Comparison Methods First Edition. 6 pp.

—Preparation
BSI BS 4559: Sec 1.1-83. 1983 Preparation of Calibration Gas Mixtures Part 1: Weighing Methods Section 1.1: Mixtures Containing Components Fully Vaporizable Under Ambient Conditions. 10 pp.
BSI BS 4559: Part 3-83. 1983 Preparation of Calibration Gas Mixtures Part 3: Static Volumetric Methods. 16 pp.
BSI BS 4559: Part 4-88. 1988 Preparation of Calibration Gas Mixtures Part 4: Certificate of Mixture Preparation. 4 pp.
BSI BS 4559: Sec 5.1-87. 1987 Preparation of Calibration Gas Mixtures Part 5: Dynamic Volumetric Methods Section 5.1: Review of Methods of Calibration. 25 pp.
BSI BS 4559: Sec 5.3-87. 1987 Preparation of Calibration Gas Mixtures Part 5: Dynamic Volumetric Methods Section 5.3: Periodic Injections into a Flowing Gas Stream. 9 pp.
BSI BS 4559: Sec 5.4-87. 1987 Preparation of Calibration Gas Mixtures Part 5: Dynamic Volumetric Methods Section 5.4: Continous Injection Method. 8 pp.
BSI BS 4559: Sec 5.6-87. 1987 Preparation of Calibration Gas Mixtures Part 5: Dynamic Volumetric Methods Section 5.6: Sonic Orifices. 9 pp.
BSI BS 4559: Part 6-81. 1981 Preparation of Calibration Gas Mixtures Part 6: Saturation Method. 4 pp.
BSI BS 4559: Part 7-81. 1981 Preparation of Calibration Gas Mixtures Part 7: Permeation Method. 8 pp.
BSI BS 4559: Part 8-85. 1985 Preparation of Calibration Gas Mixtures Part 8: Mass Dynamic Method. 8 pp.
BSI BS 4559: Sec 9.1-88. 1988 Preparation of Calibration Gas Mixtures Part 9: Comparison Methods and Methods for Establishing Traceability Section 9.1: Determination of Comparison Methods. 7 pp.
BSI BS 4559: Sec 9.2-88. 1988 Preparation of Calibration Gas Mixtures Part 9: Comparison Methods and Methods for Establishing Traceability Section 9.2: Checking by a Comparison Method. 4 pp.
ISO 6141-84. Gas Analysis—Calibration Gas Mixtures—Certificate of Mixture Preparation Second Edition. 4 pp.
ISO 6142-81. Gas Analysis—Preparation of Calibration Gas Mixtures—Weighing Methods First Edition; (Addendum 1-1983). 12 pp.
ISO 6144-81. Gas Analysis—Preparation of Calibration Gas Mixtures—Static Volumetric Methods First Edition. 15 pp.
ISO 6145 Pt 1-86. Gas Analysis—Preparation of Calibration Gas Mixtures—Dynamic Volumetric Methods—Part 1: Methods of Calibration First Edition. 23 pp.
ISO 6145 Pt 3-86. Gas Analysis—Preparation of Calibration Gas Mixtures—Dynamic Volumetric Methods—Part 3: Periodic Injections into a Flowing Gas Stream First Edition. 8 pp.
ISO 6145 Pt 4-86. Gas Analysis—Preparation of Calibration Gas Mixtures—Dynamic Volumetric Methods—Part 4: Continuous Injection Method First Edition. 7 pp.
ISO 6145 Pt 6-86. Gas Analysis—Preparation of Calibration Gas Mixtures—Dynamic Volumetric Methods—Part 6: Sonic Orifices First Edition. 8 pp.
ISO 6146-79. Gas Analysis—Preparation of Calibration Gas Mixtures—Manometric Method First Edition. 21 pp.
ISO 6147-79. Gas Analysis—Preparation of Calibration Gas Mixtures—Saturation Method First Edition. 4 pp.
ISO 6349-79. Gas Analysis—Preparation of Calibration Gas Mixtures—Permeation Method First Edition. 8 pp.
ISO 7395-84. Gas Analysis—Preparation of Calibration Gas Mixtures—Mass Dynamic Method First Edition. 7 pp.

INTERNATIONAL AND NON-U.S. NATIONAL STANDARDS
SUBJECT INDEX

Call

Calibration Reference Standards
Use For: Certified Reference Materials; Reference Materials
SAA HB19. Reference Materials.

—Certificates
BSI PD 6532: Part 2-92. 1992 Reference Materials Part 2: Guide to the Contents of Certificates of Reference Materials (ISO Guide 31: 1981) (G). 14 pp.
ISO Guide 31-81. Contents of Certificates of Reference Materials First Edition. 11 pp.
SAA HB19.31-91. Reference Materials—Guide 31: Contents of Certificates of Reference Materials (ISO Guide 31:1981) (SANZ HB19.31—1991). 8 pp.

—Certification
BSI PD 6532: Part 3-92. 1992 Reference Materials Part 3: Guide to the Uses of Certified Reference Materials (ISO Guide 33: 1989) (G). 21 pp.
BSI PD 6532: Part 4-92. 1992 Reference Materials Part 4: Guide to General and Statistical Principles for the Certification of Reference Materials (ISO Guide 35: 1989) (G). 38 pp.
SAA HB19.33-91. Reference Materials—Guide 33: Uses of Certified Reference Materials (ISO Guide 33:1989) (SANZ HB19.33—1991). 14 pp.
SAA HB19.35-91. Reference Materials—Guide 35: Certification of Reference Materials—General and Statistical Principles (ISO Guide 35:1985) (SANZ HB19.35—1991). 28 pp.

—Glossaries
BSI PD 6532: Part 1-93. 1993 Reference Materials Part 1: Guide to Terms and Definitions Used in Connection with Reference Materials (ISO Guide 30: 1992) (G). 16 pp.
ISO Guide 30-92. Terms and Definitions Used in Connection with Reference Materials Second Edition. 13 pp.
SAA HB19.30-91. Reference Materials—Guide 30: Terms and Definitions Used in Connection with Reference Materials (ISO Guide 30:1981) (SANZ HB19.30—1991). 5 pp.

Calibration Systems
—CECC
CECC CECC 00 011 ISSUE 1-82. Basic Specification: Calibration Requirements for the CECC System (En, Fr, Ge). 24 pp.

Calibrators (Sound)
Use: Sound Calibrators

Calibres
Use: Calibers

California Bearing Ratio
—Aggregates
SNZ NZS 4407: Part 3.15-91. Methods of Sampling and Testing Road Aggregates Part 3: Methods of Testing Road Aggregates—Laboratory Tests Test 3.15: The California Bearing Ratio (CBR). 10 pp.

—Soil Analysis
SAA AS 1289.F1.1-77. Methods of Testing Soil for Engineering Purposes—Part F1.1: Soil Strength and Consolidation Tests—Determination of the California Bearing Ratio of a Soil—Standard Laboratory Method for a Remoulded Specimen Corrig. 7 pp.
SAA AS 1289.F1.2-77. Methods of Testing Soil for Engineering Purposes—Part F1.2: Soil Strength and Consolidation Tests—Determination of the California Bearing Ratio of a Soil—Standard Laboratory Method for an Undisturbed Specimen. 3 pp.
SAA AS 1289.F1.3-77. Methods of Testing Soil for Engineering Purposes—Part F1.3: Soil Strength and Consolidation Tests—Determination of the California Bearing Ratio of a Soil—Standard Field-in-Place Method. 3 pp.

Calipers
Scope Note: Use a more specific term
See: Micrometers

Calipers (Measuring Instruments)
Scope Note: For additional listings, use a more specific term *See Also:* Firm Joint Calipers; Measuring Instruments; Micrometers; Spring Calipers; Spring Joint Calipers; Vernier Calipers

—Blanks
BSI BS 1044: Part 1-64. 1964 Gauge Blanks Part 1: Plug, Ring and Caliper Gauges. 55 pp.

—Dial
CGSB 39-GP-19M-79. Calipers and Gages, Vernier and Dial, Standard for. 22 pp.

Call Attempts
Use For: Call Demands

Call Attempts *(Cont.)*
See Also: Call Completion; Completion Ratio (Telephone); Internal Loss Probability; Telecommunication Services; Telephone Services; Unsuccessful Calls

—Global Maritime Distress and Safety Systems
CCIR RECMN 541-3-90. Operational Procedures for the Use of Digital Selective-Calling (DSC) Equipment in the Maritime Mobile Service—Section 8B—Maritime Mobile Service; Telegraphy and Related Subjects. 13 pp.
CCIR RECMN 541-4-92. Operational Procedures for the Use of Digital Selective-Calling (DSC) Equipment in the Maritime Mobile Service—Section 8B—Maritime Mobile Service; Telegraphy and Related Subjects. 14 pp.

—Maritime Mobile Services
CCIR RECMN 541-3-90. Operational Procedures for the Use of Digital Selective-Calling (DSC) Equipment in the Maritime Mobile Service—Section 8B—Maritime Mobile Service; Telegraphy and Related Subjects. 13 pp.
CCIR RECMN 541-4-92. Operational Procedures for the Use of Digital Selective-Calling (DSC) Equipment in the Maritime Mobile Service—Section 8B—Maritime Mobile Service; Telegraphy and Related Subjects. 14 pp.

—Public Land Mobile Networks
CENELEC GSM 02.82-92. Call Offering Supplementary Services. 45 pp.
ETSI GSM 02.82-92. Call Offering Supplementary Services. 45 pp.

—Telecommunication Traffic—CCITT No. 7 Signaling Systems
CCITT RECMN E.713-89. Control Plane Traffic Models—Telephone Network and ISDN—Quality of Service, Network Management and Traffic Engineering (Study Group II) 7 pp. 7 pp.

—Telecommunication Traffic—Integrated Services Digital Networks
CCITT RECMN E.711-92. User Demand Modelling Study Group II) 11 pp. 11 pp.
CCITT RECMN E.712-92. User Plane Traffic Modeling (Study Gruop II) 16 pp. 16 pp.
CCITT RECMN E.713-89. Control Plane Traffic Models—Telephone Network and ISDN—Quality of Service, Network Management and Traffic Engineering (Study Group II) 7 pp. 7 pp.

—Telephone Services
CCITT RECMN E.426-92. General Guide to the Percentage of Effective Attempts Which Should Be Observed for International Telephone Calls—Telephone Network and ISDN Quality of Service, Network Management and Traffic Engineering (Study Group II) 3 pp. 3 pp.

Call Barring
Use For: Incoming Call Barring; Outgoing Call Barring *See Also:* Telecommunication Services; Telephone Services

—Cellular Mobile Radio Equipment
CENELEC PRI-ETS 300 029-90. European Digital Cellular Telecommunications System (Phase 1); Mobile Radio Interface Layer 3 Call Restriction Supplementary Services Specification. 14 pp.
CENELEC PRI-ETS 300 029-91. European Digital Cellular Telecommunications System (Phase 1); Mobile Radio Interface Layer 3 Call Restriction Supplementary Services Specification. 17 pp.
CENELEC I-ETS 300 029-92. European Digital Cellular Telecommunications System (Phase 1); Mobile Radio Interface Layer 3 Call Restriction Supplementary Services Specification (GSM 04.88). 16 pp.
CENELEC GSM 04.88-92. See PRI-ETS 300 029. 15 pp.
ETSI PRI-ETS 300 029-90. European Digital Cellular Telecommunications System (Phase 1) Mobile Radio Interface Layer 3 Call Restriction Supplementary Services Specification (GSM 04. 88). 14 pp.
ETSI I-ETS 300 029-92. European Digital Cellular Telecommunications System (Phase 1); Mobile Radio Interface Layer 3 Call Restriction Supplementary Services Specification (GSM 04.88). 16 pp.
ETSI PRI-ETS 300 029-91. European Digital Cellular Telecommunications System (Phase 1); Mobile Radio Interface Layer 3 Call Restriction Supplementary Services Specification. 17 pp.
ETSI GSM 04.88-92. See PRI-ETS 300 029. 15 pp.

—Integrated Services Digital Networks
CCITT RECMN I.255. 5-92. Outgoing Call Barring (Study Group I) 9 pp. 9 pp.

Call Barring *(Cont.)*
—Public Land Mobile Networks
CENELEC GSM 02.88-92. Call Restriction Supplementary Services. 19 pp.
CENELEC GSM 03.88-92. European Digital Cellular Telecommunication System (Phase 1); Technical Realization of Call Restriction Supplementary Services. 32 pp.
ETSI GSM 02.88-92. Call Restriction Supplementary Services. 19 pp.
ETSI GSM 03.88-92. European Digital Cellular Telecommunication System (Phase 1); Technical Realization of Call Restriction Supplementary Services. 32 pp.

—Public Switched Telephone Networks
BSI BS 6789: Part 7-88. 1988 Apparatus with One or More Particular Functions for Connection to Certain Public Switched Telephone Networks Part 7: Apparatus with Call Bearing Facilities. 11 pp.
BSI PD 6566-92. 1992 Draft Amendment to BS 6789: Part 7: 1988 'Apparatus with One or More Particular Functions for Connection to Public Switched Telephone Networks Run by Certain Public Telecommunications Operators, Part 7: Specification for. 6 pp.
BSI PD 6569: Part 7-92. 1992 Apparatus with One or More Particular Functions for Connection to Public Switched Telephone Networks Run by Certain Public Telecommunications Operators Part 7: Specification for Apparatus with Call Barring Facilities. 20 pp.

Call Clearing Delay
Use: Call Release Time

Call Completion
See Also: Call Attempts; Completion Ratio (Telephone); Telecommunication Services; Telephone Services

—Private Switched Networks
ECMA ECMA 185-92. Private Telecommunication Network (PTN)—Specification, Functional Model and Information Flows—Call Completion Supplementary Services (CCSD). 65 pp.
ECMA ECMA 186-92. Private Telecommunication Network (PTN)—Inter-Exchange Signalling Protocol—Call Completion Supplementary Services (QSIG-CC). 63 pp.

Call Control Procedures
See Also: Call Set Up; Telephone Services

—D Channel—Network Layer—Integrated Services Digital Networks—DSS1
CCITT RECMN Q.931-89. ISDN User-Network Interface Layer 3 Specification for Basic Call Control—Digital Subscriber Signalling System No. 1 (DSS 1), Network Layer, User-Network Management (Study Group XI) 356 pp (Same as Recmn I.451). 356 pp.

—Data Transmission—Public Networks—Interworking
CCITT RECMN X.301-89. Description of the General Arrangements for Call Control Within a Subnetwork and Between Subnetworks for the Provision of Data Transmission Services—Data Communication Networks—Interworking Between Networks, Mobile Data Transmission Systems, Internetwork. 72 pp.

—Interfaces—Network Layer—Integrated Services Digital Networks
CCITT RECMN I.451-89. ISDN User-Network Interface Layer 3 Specification for Basic Call Control—Integrated Services Digital Network (ISDN)—Overall Network Aspects and Functions, ISDN User-Network Interfaces (Study Group XVIII) 1 pp (Same as Recmn Q.931). 1 p.
CENELEC PRETS 300 102-1-90. Integrated Services Digital Network (ISDN); User-Network Interface Layer 3 Specifications for Basic Call Control. 233 pp.
CENELEC ETS 300 102-1-90. Integrated Services Digital Network (ISDN); User-Network Interface Layer 3 Specifications for Basic Call Control. 183 pp.
CENELEC ETS 300 102-1/A1-93. AMD 1 Integrated Services Digital Network (ISDN); User-Network Interface Layer 3 Specifications for Basic Call Control. 6 pp.
CENELEC ETS 300 102-1 PRA1-92. AMD 1 Integrated Services Digital Network (ISDN); User-Network Interface Layer 3 Specifications for Basic Call Control. 6 pp.
CENELEC ETS 300 102-1 PRA2-93. AMD prA2 Integrated Services Digital Network (ISDN); User-Network Interface Layer 3 Specifications for Basic Call Control. 4 pp.
ETSI PRETS 300 102-1-90. Integrated Services Digital Network (ISDN); User Network Interface Layer 3—Specifications for Basic Call Control. 233 pp.

INDUSTRY STANDARDS

Call Control Procedures (Cont.)
—Interfaces—Network Layer—Integrated Services Digital Networks (Cont.)

ETSI ETS 300 102-1/A1-93. AMD 1 Integrated Services Digital Network (ISDN); User-Network Interface Layer 3 Specifications for Basic Call Control. 6 pp.

ETSI ETS 300 102-1 PRA2-93. AMD prA2 Integrated Services Digital Network (ISDN); User-Network Interface Layer 3 Specifications for Basic Call Control. 4 pp.

ETSI ETS 300 102-1 PRA1-92. AMD 1 Integrated Services Digital Network (ISDN); User-Network Interface Layer 3 Specifications for Basic Call Control. 6 pp.

ETSI ETS 300 102-1-90. Integrated Services Digital Network (ISDN); User-Network Interface Layer 3 Specifications for Basic Call Control. 183 pp.

—Phototelegraphy—Telephone Circuits

CCITT RECMN E.320-89. Speeding up the Establishment and Clearing of Phototelegraph Calls—Telephone Network and ISDN—Operation, Numbering, Routing and Mobile Service (Study Group II)1 pp. 1 p.

—Private Integrated Services Networks—DSS1

IEC DIS 11572-92. Information Technology—Telecommunications and Information Exchange Between Systems—Private Integrated Services Network—Circuit Mode Bearer Services—Inter-Exchange Signalling Procedures and Protocol ***CD-ROM ONLY***. 172 pp.

ISO DIS 11572-92. Information Technology—Telecommunications and Information Exchange Between Systems—Private Integrated Services Network—Circuit Mode Bearer Services—Inter-Exchange Signalling Procedures and Protocol ***CD-ROM ONLY***. 172 pp.

—Private Integrated Services Networks—PSS1

JTC1 DIS11572-92. Information Technology—Telecommunications and Information Exchange Between Systems—Private Integrated Services Network—Circuit Mode Bearer Services—Inter-Exchange Signalling Procedures and Protocol ***CD-ROM ONLY***. 172 pp.

—Protocol Implementation Conformance Statement—ISDN—DSS1

CENELEC PRI-ETS 300 314-93. Integrated Services Digital Network (ISDN); Digital Subscriber Signalling System No. One (DSS1) Protocol Implementation Conformance Statement (PICS) Proforma for Basic-Access User for Signalling-Network-Layer Protocol for Circuit-Mode Basic Call Control. 51 pp.

CENELEC PRI-ETS 300 315-93. Integrated Services Digital Network (ISDN); Digital Sub. Signalling Sys. No. One (DSS1) Protocol Implementation Conformance Statement (PICS) Proforma for Primary-Rate-Access User for Signalling-Network-Layer Protocol for Circuit-Mode Basic Call Control. 52 pp.

CENELEC PRI-ETS 300 316-93. Integrated Services Digital Network (ISDN); Digital Subscriber Signalling System No. One (DSS1) Protocol Implementation Conformance Statement (PICS) Proforma for Basic-Access Network for Signalling-Network-Layer Protocol for Circuit-Mode Basic Call Control. 47 pp.

CENELEC PRI-ETS 300 317-93. Integrated Services Digital Network (ISDN); Digital Subscriber Signalling System No. One (DSS1) Protocol Impl. Conformance Statement (PICS) Proforma for Primary-Rate-Access Network for Signalling-Network-Layer Protocol for Circuit-Mode Basic Call Control. 46 pp.

ETSI PRI-ETS 300 314-93. Integrated Services Digital Network (ISDN); Digital Subscriber Signalling System No. One (DSS1) Protocol Implementation Conformance Statement (PICS) Proforma for Basic-Access User for Signalling-Network-Layer Protocol for Circuit-Mode Basic Call Control. 51 pp.

ETSI PRI-ETS 300 315-93. Integrated Services Digital Network (ISDN); Digital Sub. Signalling Sys. No. One (DSS1) Protocol Implementation Conformance Statement (PICS) Proforma for Primary-Rate-Access User for Signalling-Network-Layer Protocol for Circuit-Mode Basic Call Control. 52 pp.

ETSI PRI-ETS 300 316-93. Integrated Services Digital Network (ISDN); Digital Subscriber Signalling System No. One (DSS1) Protocol Implementation Conformance Statement (PICS) Proforma for Basic-Access Network for Signalling-Network-Layer Protocol for Circuit-Mode Basic Call Control. 47 pp.

ETSI PRI-ETS 300 317-93. Integrated Services Digital Network (ISDN); Digital Subscriber Signalling System No. One (DSS1) Protocol Impl. Conformance Statement (PICS) Proforma for Primary-Rate-Access Network for Signalling-Network-Layer Protocol for Circuit-Mode Basic Call Control. 46 pp.

ETSI PRI-ETS 300 318-93. Integrated Services Digital Network (ISDN); Digital Subscriber Signalling System No. One (DSS1) Protocol Implementation eXtra Information for Testing (PIXIT) Proforma for Basic-Access User for Signalling-Network-Layer Protocol for Circuit-Mode Basic Call Control. 13 pp.

ETSI PRI-ETS 300 320-93. Integrated Services Digital Network (ISDN); Digital Subscriber Signalling System No. One (DSS1) Protocol Implementation eXtra Information for Testing (PIXIT) Proforma for Basic-Access Network for Signalling-Network-Layer Protocol for Circuit-Mode Basic Call Control. 13 pp.

ETSI PRI-ETS 300 321-93. Integrated Services Digital Network (ISDN); Digital Sub. Signalling Sys. No. One (DSS1) Protocol Implementation eXtra Information for Testing (PIXIT) Proforma for Primary-Rate-Access Network for Signalling-Network-Layer Protocol for Circuit-Mode Basic Call Control. 13 pp.

—Protocol Implementation Extra Information for Testing—ISDN—DSS1

CENELEC PRI-ETS 300 318-93. Integrated Services Digital Network (ISDN); Digital Subscriber Signalling System No. One (DSS1) Protocol Implementation eXtra Information for Testing (PIXIT) Proforma for Basic-Access User for Signalling-Network-Layer Protocol for Circuit-Mode Basic Call Control. 13 pp.

CENELEC PRI-ETS 300 319-93. Integrated Services Digital Network (ISDN); Digital Subscriber Signalling System No. One (DSS1) Protocol Impl. eXtra Information for Testing (PIXIT) Proforma for Primary-Rate-Access User for Signalling-Network-Layer Protocol for Circuit-Mode Basic Call Control. 13 pp.

CENELEC PRI-ETS 300 320-93. Integrated Services Digital Network (ISDN); Digital Subscriber Signalling System No. One (DSS1) Protocol Implementation eXtra Information for Testing (PIXIT) Proforma for Basic-Access Network for Signalling-Network-Layer Protocol for Circuit-Mode Basic Call Control. 13 pp.

CENELEC PRI-ETS 300 321-93. Integrated Services Digital Network (ISDN); Digital Sub. Signalling Sys. No. One (DSS1) Protocol Implementation eXtra Information for Testing (PIXIT) Proforma for Primary-Rate-Access Network for Signalling-Network-Layer Protocol for Circuit-Mode Basic Call Control. 13 pp.

ETSI PRI-ETS 300 319-93. Integrated Services Digital Network (ISDN); Digital Subscriber Signalling System No. One (DSS1) Protocol Impl. eXtra Information for Testing (PIXIT) Proforma for Primary-Rate-Access User for Signalling-Network-Layer Protocol for Circuit-Mode Basic Call Control. 13 pp.

—Routing

CCITT RECMN E.170-92. Traffic Routing (Study Group II) 10 pp. 10 pp.

—Teleconferencing Services

CCITT RECMN F.701-89. Teleconference Service—Telematic, Data Transmission and Teleconference Services —Operations and Quality of Service (Study Group I) 13 pp (Renumbered from Recmn F.710). 13 pp.

CCITT RECMN F.710-91. General Principles for Audiographic Conference Service—Telematic, Data Transmission and Teleconference Services Operations and Quality of Service (Study Group I) 15 pp. 15 pp.

Call Deflection
Use: Call Forwarding

Call Demands
Use: Call Attempts

Call Diversion Services
Use For: Diversion If Number Busy
See Also: Telephone Services

—Freephone Services

CCITT RECMN E.152-89. International Freephone Service (IFS)—Telephone Network and ISDN—Operation, Numbering, Routing and Mobile Service (Study Group II) 12 pp. 12 pp.

Call Duration
See Also: Tariffs (Telecommunications); Telephone Services; Time

—Automatic Services—Accounting

CCITT RECMN E.260-89. Basic Technical Problems Concerning the Measurement and Recording of Call Durations—Telephone Network and ISDN—Operation, Numbering, Routing and Mobile Service (Study Group II) 4 pp. 4 pp.

—Conference Calling—Tariffs

CCITT RECMN E.151-89. Conditions of Operation and Setting up of Conference Calls—Telephone Network and ISDN—Operation, Numbering, Routing and Mobile Service (Study Group II) 3 pp. 3 pp.

—Data Logging

CCITT RECMN E.261-89. Devices for Measuring and Recording Call Durations—Telephone Network and ISDN—Operation, Numbering, Routing and Mobile Service (Study Group II) 3 pp. 3 pp.

—Data Logging—Automatic Services

CCITT RECMN E.260-89. Basic Technical Problems Concerning the Measurement and Recording of Call Durations—Telephone Network and ISDN—Operation, Numbering, Routing and Mobile Service (Study Group II) 4 pp. 4 pp.

—Data Logging—Telephone Services

CCITT RECMN E.260-89. Basic Technical Problems Concerning the Measurement and Recording of Call Durations—Telephone Network and ISDN—Operation, Numbering, Routing and Mobile Service (Study Group II) 4 pp. 4 pp.

—Maritime Mobile Services—Radiotelephony

CCITT RECMN F.110-89. Operational Provisions for the Maritime Mobile Service—Telegraph and Mobile Services Operations and Quality of Service (Study Group I) 21 pp (Same as Recmn E.200). 21 pp.

—Maritime Mobile Services—Radioteletype Communications

CCITT RECMN F.110-89. Operational Provisions for the Maritime Mobile Service—Telegraph and Mobile Services Operations and Quality of Service (Study Group I) 21 pp (Same as Recmn E.200). 21 pp.

—Measurement

CCITT RECMN E.261-89. Devices for Measuring and Recording Call Durations—Telephone Network and ISDN—Operation, Numbering, Routing and Mobile Service (Study Group II) 3 pp. 3 pp.

—Measurement—Ampere Hour Meters

CCITT RECMN E.261-89. Devices for Measuring and Recording Call Durations—Telephone Network and ISDN—Operation, Numbering, Routing and Mobile Service (Study Group II) 3 pp. 3 pp.

—Measurement—Automatic Services (Telephone)

CCITT RECMN E.260-89. Basic Technical Problems Concerning the Measurement and Recording of Call Durations—Telephone Network and ISDN—Operation, Numbering, Routing and Mobile Service (Study Group II) 4 pp. 4 pp.

—Measurement—Coulombmeters

CCITT RECMN E.261-89. Devices for Measuring and Recording Call Durations—Telephone Network and ISDN—Operation, Numbering, Routing and Mobile Service (Study Group II) 3 pp. 3 pp.

—Measurement—Pulse Counters

CCITT RECMN E.261-89. Devices for Measuring and Recording Call Durations—Telephone Network and ISDN—Operation, Numbering, Routing and Mobile Service (Study Group II) 3 pp. 3 pp.

—Measurement—Telephone Services

CCITT RECMN E.260-89. Basic Technical Problems Concerning the Measurement and Recording of Call Durations—Telephone Network and ISDN—Operation, Numbering, Routing and Mobile Service (Study Group II) 4 pp. 4 pp.

—Teleinformatic Services

CCITT RECMN E.301-89. Impact of Non-Voice Applications on the Telephone Network—Telephone Network and ISDN—Operation, Numbering, Routing and Mobile Service (Study Group II) 5 pp. 5 pp.

—Telephone Services—Accounting

CCITT RECMN E.260-89. Basic Technical Problems Concerning the Measurement and Recording of Call Durations—Telephone Network and ISDN—Operation, Numbering, Routing and Mobile Service (Study Group II) 4 pp. 4 pp.

INTERNATIONAL AND NON-U.S. NATIONAL STANDARDS
SUBJECT INDEX
Call

Call Duration (Cont.)
—Telephone Services—Tariffs
CCITT RECMN E.140-92. Operator-Assisted Telephone Service (Study Group I) 6 pp. 6 pp.
CCITT RECMN E.140-89. Principles for the Operation of International Telephone Services—Telephone Network and ISDN—Operation, Numbering, Routing and Mobile Service (Study Group II) 3 pp. 3 pp.
CCITT RECMN E.230-92. Chargeable Duration of Calls (Study Group I) 4 pp. 4 pp.
CCITT RECMN E.230-89. Charging (Determination of Collection Charges) in the International Telephone Service—Telephone Network and ISDN—Operation, Numbering, Routing and Mobile Service (Study Group II) 1 pp. 1 p.

—Telex Communications
CCITT RECMN F.60-92. Operational Provisions for the International Telex Service (Study Group I) 34 pp. 34 pp.
CCITT RECMN F.60-89. Operational Provisions for the International Telex Service—Telegraph and Mobile Services Operations and Quality of Service (Study Group I) 17 pp. 17 pp.

—Telex Communications—Tariffs
CCITT RECMN D.61-89. Charging and Accounting Provisions Relating to the Measurement of the Chargeable Duration of a Telex Call—General Tariff Principles—Charging and Accounting in International Telecommunications Services (Study Group III) 2 pp. 2 pp.
CCITT RECMN F.61-89. Operational Provisions Relating to the Chargeable Duration of a Telex Call—Telegraph and Mobile Services Operations and Quality of Service (Study Group I) 1 pp. 1 p.

Call Failure
Use: Unsuccessful Calls

Call Forwarding
Use For: Call Deflection See Also: Call Handling; Call Redirection; Telephone Services; Universal Personal Telecommunication Services
CEPT T/CS 20-25-82. Deviation D'Appel. 3 pp.
CEPT T/CS 20-25 E-82. Call Diversion. 3 pp.
CEPT T/CS 21-08-84. Sequence De Traitement D'Appel Pour Le Transfert D'Appels Telephoniques Vers Un Autre Numero. 17 pp.
CEPT T/CS 21-08 E-84. Call Handling Sequences for Diversion of Telephone Calls to Another Number. 17 pp.

—Integrated Services Digital Networks
CCITT RECMN I.252-89. Call Offering Supplementary Services—Integrated Services Digital Network (ISDN)—General Structure and Service Capabilities (Study Group XVIII) 37 pp. 37 pp.
CCITT RECMN I.252. 5-92. Call Deflection (Study Group I) 12 pp. 12 pp.
CENELEC PRETS 300 202-92. Integrated Services Digital Network (ISDN); Call Deflection (CD) Supplementary Service Service Description. 17 pp.
CENELEC PRETS 300 206-92. Integrated Services Digital Network (ISDN); Call Deflection (CD) Supplementary Service Functional Capabilities and Information Flows. 39 pp.
ETSI PRETS 300 202-92. Integrated Services Digital Network (ISDN); Call Deflection (CD) Supplementary Service Service Description. 17 pp.
ETSI PRETS 300 206-92. Integrated Services Digital Network (ISDN); Call Deflection (CD) Supplementary Service Functional Capabilities and Information Flows. 39 pp.

—ISDN—CCITT No. 7 Signaling Systems
CCITT RECMN Q.730-89. ISDN Supplementary Services—Specifications of Signalling System No. 7 (Study Group XI) 59 pp. 59 pp.

—ISDN—Functions/Information Flows
CCITT RECMN Q.82-89. Call Offering Supplementary Services—General Recommendations on Telephone Switching and Signalling—Functions and Information Flows for Services in the ISDN—Supplements (Study Group XI) 25 pp. 25 pp.

Call Forwarding Busy
Use For: CFB See Also: Call Handling; Call Redirection; Telephone Services; Universal Personal Telecommunication Services

—Cellular Mobile Radio Equipment
CENELEC PRI-ETS 300 028-90. European Digital Cellular Telecommunications System (phase 1): Mobile Radio Interface Layer 3 Call Offering Supplementary Services Specification. 38 pp.
CENELEC PRI-ETS 300 028-91. European Digital Cellular Telecommunications System (Phase 1); Mobile Radio Interface Layer 3 Call Offering Supplementary Services Specification. 39 pp.

Call Forwarding Busy (Cont.)
—Cellular Mobile Radio Equipment (Cont.)
CENELEC I-ETS 300 028-92. European Digital Cellular Telecommunications System (Phase 1); Mobile Radio Interface Layer 3 Call Offering Supplementary Services Specification (GSM 04.82). 39 pp.
CENELEC GSM 04.82-92. See PRI-ETS 300 028. 37 pp.
ETSI PRI-ETS 300 028-90. European Digital Cellular Telecommunications System (Phase 1) Mobile Radio Interface Layer 3 Call Offering Supplementary Services Specification (GSM 04. 82). 40 pp.
ETSI I-ETS 300 028-92. European Digital Cellular Telecommunications System (Phase 1); Mobile Radio Interface Layer 3 Call Offering Supplementary Services Specification (GSM 04.82). 39 pp.
ETSI PRI-ETS 300 028-91. European Digital Cellular Telecommunications System (Phase 1); Mobile Radio Interface Layer 3 Call Offering Supplementary Services Specification. 39 pp.
ETSI GSM 04.82-92. See PRI-ETS 300 028. 37 pp.

—Integrated Services Digital Networks
CCITT RECMN I.252. 2-92. Call Forwarding Busy (Study Group I) 17 pp. 17 pp.
CENELEC PRETS 300 199-92. Integrated Services Digital Network (ISDN); Call Forwarding Busy (CFB) Supplementary Service Service Description. 19 pp.
CENELEC PRETS 300 203-92. Integrated Services Digital Network (ISDN); Call Forwarding Busy (CFB) Supplementary Service Functional Capabilities and Information Flows. 45 pp.
ETSI PRETS 300 199-92. Integrated Services Digital Network (ISDN); Call Forwarding Busy (CFB) Supplementary Service Service Description. 19 pp.

—Private Switched Networks
CENELEC PRETS 300 256-92. Private Telecommunication Network (PTN); Specification, Functional Model and Information Flows Diversion Supplementary Services. 71 pp.
CENELEC PRETS 300 257-92. Private Telecommunication Network (PTN); Inter-Exchange Signalling Protocol Diversion Supplementary Services. 62 pp.
ECMA ECMA 173-92. Private Telecommunication Networks (PTN)—Specification, Functional Model and Information Flows—Diversion Supplementary Services (CFSD). 71 pp.
ECMA ECMA 174-92. Private Telecommunication Networks (PTN)—Inter-Exchange Signalling Protocol—Diversion Supplementary Services (QSIG-CF). 62 pp.
ETSI PRETS 300 256-92. Private Telecommunication Network (PTN); Specification, Functional Model and Information Flows Diversion Supplementary Services. 71 pp.
ETSI PRETS 300 257-92. Private Telecommunication Network (PTN); Inter-Exchange Signalling Protocol Diversion Supplementary Services. 62 pp.

—Public Land Mobile Networks
CENELEC GSM 02.82-92. Call Offering Supplementary Services. 45 pp.
CENELEC GSM 03.82-92. European Digital Cellular Telecommunication System (Phase 1); Technical Realization of Call Offering Supplementary Services. 82 pp.
ETSI GSM 02.82-92. Call Offering Supplementary Services. 45 pp.
ETSI GSM 03.82-92. European Digital Cellular Telecommunication System (Phase 1); Technical Realization of Call Offering Supplementary Services. 82 pp.

Call Forwarding No Reply
Use For: CFNR See Also: Call Handling; Call Redirection; Telephone Services; Universal Personal Telecommunication Services

—Cellular Mobile Radio Equipment
CENELEC PRI-ETS 300 028-90. European Digital Cellular Telecommunications System (phase 1): Mobile Radio Interface Layer 3 Call Offering Supplementary Services Specification. 38 pp.
CENELEC PRI-ETS 300 028-91. European Digital Cellular Telecommunications System (Phase 1); Mobile Radio Interface Layer 3 Call Offering Supplementary Services Specification. 39 pp.
CENELEC I-ETS 300 028-92. European Digital Cellular Telecommunications System (Phase 1); Mobile Radio Interface Layer 3 Call Offering Supplementary Services Specification (GSM 04.82). 39 pp.
CENELEC GSM 04.82-92. See PRI-ETS 300 028. 37 pp.
ETSI PRI-ETS 300 028-90. European Digital Cellular Telecommunications System (Phase 1) Mobile Radio Interface Layer 3 Call Offering Supplementary Services Specification (GSM 04. 82). 40 pp.

Call Forwarding No Reply (Cont.)
—Cellular Mobile Radio Equipment (Cont.)
ETSI I-ETS 300 028-92. European Digital Cellular Telecommunications System (Phase 1); Mobile Radio Interface Layer 3 Call Offering Supplementary Services Specification (GSM 04.82). 39 pp.
ETSI PRI-ETS 300 028-91. European Digital Cellular Telecommunications System (Phase 1); Mobile Radio Interface Layer 3 Call Offering Supplementary Services Specification. 39 pp.
ETSI GSM 04.82-92. See PRI-ETS 300 028. 37 pp.

—Integrated Services Digital Networks
CCITT RECMN I.252. 3-92. Call Forwarding No Reply (Study Group I) 12 pp. 12 pp.
CENELEC PRETS 300 201-92. Integrated Services Digital Network (ISDN); Call Forwarding No Reply (CFNR) Supplementary Service Service Description. 20 pp.
CENELEC PRETS 300 205-92. Integrated Services Digital Network (ISDN); Call Forwarding No Reply (CFNR) Supplementary Service Functional Capabilities and Information Flows. 51 pp.
ETSI PRETS 300 201-92. Integrated Services Digital Network (ISDN); Call Forwarding No Reply (CFNR) Supplementary Service Service Description. 20 pp.
ETSI PRETS 300 205-92. Integrated Services Digital Network (ISDN); Call Forwarding No Reply (CFNR) Supplementary Service Functional Capabilities and Information Flows. 51 pp.

—Private Switched Networks
CENELEC PRETS 300 256-92. Private Telecommunication Network (PTN); Specification, Functional Model and Information Flows Diversion Supplementary Services. 71 pp.
CENELEC PRETS 300 257-92. Private Telecommunication Network (PTN); Inter-Exchange Signalling Protocol Diversion Supplementary Services. 62 pp.
ECMA ECMA 173-92. Private Telecommunication Networks (PTN)—Specification, Functional Model and Information Flows—Diversion Supplementary Services (CFSD). 71 pp.
ECMA ECMA 174-92. Private Telecommunication Networks (PTN)—Inter-Exchange Signalling Protocol—Diversion Supplementary Services (QSIG-CF). 62 pp.
ETSI PRETS 300 256-92. Private Telecommunication Network (PTN); Specification, Functional Model and Information Flows Diversion Supplementary Services. 71 pp.
ETSI PRETS 300 257-92. Private Telecommunication Network (PTN); Inter-Exchange Signalling Protocol Diversion Supplementary Services. 62 pp.

—Public Land Mobile Networks
CENELEC GSM 02.82-92. Call Offering Supplementary Services. 45 pp.
CENELEC GSM 03.82-92. European Digital Cellular Telecommunication System (Phase 1); Technical Realization of Call Offering Supplementary Services. 82 pp.
ETSI GSM 02.82-92. Call Offering Supplementary Services. 45 pp.
ETSI GSM 03.82-92. European Digital Cellular Telecommunication System (Phase 1); Technical Realization of Call Offering Supplementary Services. 82 pp.

Call Forwarding Unconditional
Use For: CFU See Also: Call Handling; Call Redirection; Telephone Services; Universal Personal Telecommunication Services

—Cellular Mobile Radio Equipment
CENELEC PRI-ETS 300 028-90. European Digital Cellular Telecommunications System (phase 1): Mobile Radio Interface Layer 3 Call Offering Supplementary Services Specification. 38 pp.
CENELEC PRI-ETS 300 028-91. European Digital Cellular Telecommunications System (Phase 1); Mobile Radio Interface Layer 3 Call Offering Supplementary Services Specification. 39 pp.
CENELEC I-ETS 300 028-92. European Digital Cellular Telecommunications System (Phase 1); Mobile Radio Interface Layer 3 Call Offering Supplementary Services Specification (GSM 04.82). 39 pp.
CENELEC GSM 04.82-92. See PRI-ETS 300 028. 37 pp.
ETSI PRI-ETS 300 028-90. European Digital Cellular Telecommunications System (Phase 1) Mobile Radio Interface Layer 3 Call Offering Supplementary Services Specification (GSM 04. 82). 40 pp.
ETSI I-ETS 300 028-92. European Digital Cellular Telecommunications System (Phase 1); Mobile Radio Interface Layer 3 Call Offering Supplementary Services Specification (GSM 04.82). 39 pp.
ETSI PRI-ETS 300 028-91. European Digital Cellular Telecommunications System (Phase 1); Mobile Radio Interface Layer 3 Call Offering Supplementary Services Specification. 39 pp.

Call Forwarding Unconditional (Cont.)

—**Cellular Mobile Radio Equipment** (Cont.)
ETSI GSM 04.82-92. See PRI-ETS 300 028. 37 pp.

—**Integrated Services Digital Networks**
CCITT RECMN I.252. 4-92. Call Forwarding Unconditional (Study Group I) 13 pp. 13 pp.
CENELEC PRETS 300 200-92. Integrated Services Digital Network (ISDN); Call Forwarding Unconditional (CFU) Supplementary Service Service Description. 19 pp.
CENELEC PRETS 300 204-92. Integrated Services Digital Network (ISDN); Call Forwarding Unconditional (CFU) Supplementary Service Functional Capabilities and Information Flows. 43 pp.
ETSI PRETS 300 200-92. Integrated Services Digital Network (ISDN); Call Forwarding Unconditional (CFU) Supplementary Service Service Description. 19 pp.
ETSI PRETS 300 203-92. Integrated Services Digital Network (ISDN); Call Forwarding Busy (CFB) Supplementary Service Functional Capabilities and Information Flows. 45 pp.
ETSI PRETS 300 204-92. Integrated Services Digital Network (ISDN); Call Forwarding Unconditional (CFU) Supplementary Service Functional Capabilities and Information Flows. 43 pp.

—**Private Switched Networks**
CENELEC PRETS 300 256-92. Private Telecommunication Network (PTN); Specification, Functional Model and Information Flows Diversion Supplementary Services. 71 pp.
CENELEC PRETS 300 257-92. Private Telecommunication Network (PTN); Inter-Exchange Signalling Protocol Diversion Supplementary Services. 62 pp.
ECMA ECMA 173-92. Private Telecommunication Networks (PTN)—Specification, Functional Model and Information Flows—Diversion Supplementary Services (CFSD). 71 pp.
ECMA ECMA 174-92. Private Telecommunication Networks (PTN)—Inter-Exchange Signalling Protocol—Diversion Supplementary Services (QSIG-CF). 62 pp.
ETSI PRETS 300 256-92. Private Telecommunication Network (PTN); Specification, Functional Model and Information Flows Diversion Supplementary Services. 71 pp.
ETSI PRETS 300 257-92. Private Telecommunication Network (PTN); Inter-Exchange Signalling Protocol Diversion Supplementary Services. 62 pp.

—**Public Land Mobile Networks**
CENELEC GSM 02.82-92. Call Offering Supplementary Services. 45 pp.
CENELEC GSM 03.82-92. European Digital Cellular Telecommunication System (Phase 1); Technical Realization of Call Offering Supplementary Services. 82 pp.
ETSI GSM 02.82-92. Call Offering Supplementary Services. 45 pp.
ETSI GSM 03.82-92. European Digital Cellular Telecommunication System (Phase 1); Technical Realization of Call Offering Supplementary Services. 82 pp.

Call Handling
See Also: Alternative Routing; Call Forwarding; Call Forwarding Busy; Call Forwarding No Reply; Call Forwarding Unconditional; Call Holding; Call Redirection; Telephone Services
CEPT T/CS 21-08-84. Sequence De Traitement D'Appel Pour Le Transfert D'Appels Telephoniques Vers Un Autre Numero. 17 pp.
CEPT T/CS 21-08 E-84. Call Handling Sequences for Diversion of Telephone Calls to Another Number. 17 pp.

—**Alarm Call Services**
CEPT T/CS 21-12-84. Traitement D'Appel Pour Les Services Du Reveil. 2 pp.
CEPT T/CS 21-12 E-84. Call Handling for Alarm Call Services. 2 pp.

—**Calling Line Identification—Malicious Calls**
CEPT T/CS 21-05-83. Sequence De Traitement D'Appel Pour La Famille De Services Supplementaires: Identification D'Appels Malveillants. 13 pp.
CEPT T/CS 21-05 E-83. Call Handling Sequences for Supplementary Services Family (MCI) Malicious Call Identification. 13 pp.

—**Calls Terminated Abnormally**
CCITT FASCICLE II.2-88. Telephone Network and ISDN-Operation, Numbering, Routing and Mobile Service. Recommendations E.100—E.333. 362 pp.

Call Handling (Cont.)

—**Land Mobile Services**
CCIR RECMN 687-90. Future Public Land Mobile Telecommunication Systems (FPLMTS)—Section 8A—Land Mobile Service and Related Subjects. 11 pp.
CCIR RECMN 687-1-92. Future Public Land Mobile Telecommunication Systems (FPLMTS)—Section 8A—Land Mobile Service and Related Subjects. 21 pp.

—**Telephone Exchanges**
CEPT T/CS 21-01-83. Flux D'Information Pour Le Traitement D'Appel. 3 pp.
CEPT T/CS 21-01 E-83. Call Handling Information Flow. 3 pp.
CEPT T/CS 62-03-86. Connexions, Signalisation, Commande, Traitement Des Appels Et Fonctions Auxiliaires Pour Commutateurs Numeriques Principaux D'Abonne Ou Mixtes. 2 pp.
CEPT T/CS 62-03 E-86. Connections, Signalling, Control, Call Handling and Ancillary Functions for Digital Local and Combined Exchanges. 2 pp.
CEPT T/CS 68-03-86. Connexions, Signalisation, Commande, Traitement Des Appels Et Fonctions Auxiliaires Pour Commutateurs De Transit Numeriques. 1 p.
CEPT T/CS 68-03 E-86. Connections, Signalling, Control, Call Handling and Ancillary Functions for Digital Transit Exchanges. 1 p.

—**Three Party Services**
CEPT T/STI 21-10-86. Sequence De Traitement D'Appel Pour Le Service a Trois Participants. 17 pp.
CEPT T/STI 21-10 E-86. Call Handling Sequences for Three Party Service. 17 pp.

Call Holding
See Also: Call Handling; Information Interchange; Telephone Services

—**Integrated Services Digital Networks**
CCITT RECMN I.253-89. Call Completion Supplementary Services—Integrated Services Digital Network (ISDN)—General Structure and Service Capabilities (Study Group XVIII) 17 pp. 17 pp.
CCITT RECMN I.253. 2-92. Call Hold (Study Group I) 10 pp. 10 pp.
CENELEC PRETS 300 139-90. Integrated Services Digital Network (ISDN) Call Hold (HOLD) Supplementary Service Service Description (T/NA1(89)27). 14 pp.
CENELEC PRETS 300 139-91. Integrated Services Digital Network (ISDN); Call Hold (HOLD) Supplementary Service Service Description. 16 pp.
CENELEC ETS 300 139-92. Integrated Services Digital Network (ISDN); Call Hold (HOLD) Supplementary Service Service Description. 15 pp.
CENELEC PRETS 300 140-90. Integrated Services Digital Network (ISDN); Call Hold (HOLD) Supplementary Service Functional Capabilities and Information Flows (T/S 22-19). 12 pp.
CENELEC PRETS 300 140-91. Integrated Services Digital Network (ISDN); Call Hold (HOLD) Supplementary Service Functional Capabilities and Information Flows. 27 pp.
CENELEC ETS 300 140-92. Integrated Services Digital Network (ISDN); Call Hold (HOLD) Supplementary Service Functional Capabilities and Information Flows. 26 pp.
CENELEC PRETS 300 141-90. Integrated Services Digital Network (ISDN); Call Hold (HOLD) Supplementary Service Digital Subscriber Signalling One (DSS1) Protocol (T/S 46-335). 11 pp.
CENELEC PRETS 300 141-91. Integrated Services Digital Network (ISDN); Call Hold (HOLD) Supplementary Service Digital Subscriber Signalling System No. One (DSS1) Protocol. 30 pp.
CENELEC ETS 300 141-92. Integrated Services Digital Network (ISDN); Call Hold (HOLD) Supplementary Service Digital Subscriber Signalling System No. One (DSS1) Protocol. 19 pp.
ETSI PRETS 300 139-90. Integrated Services Digital Network (ISDN) Call Hold (HOLD) Supplementary Service Service Description (T/NA1(89)27). 13 pp.
ETSI PRETS 300 140-90. Integrated Services Digital Network (ISDN); Call Hold (HOLD) Supplementary Service Functional Capabilities and Information Flows (T/S 22-19). 12 pp.
ETSI PRETS 300 141-90. Integrated Services Digital Network (ISDN); Call Hold (HOLD) Supplementary Service Digital Subscriber Signalling One (DSS1) Protocol (T/S)46-33S). 11 pp.
ETSI ETS 300 139-92. Integrated Services Digital Network (ISDN); Call Hold (HOLD) Supplementary Service Service Description. 15 pp.
ETSI PRETS 300 139-91. Integrated Services Digital Network (ISDN); Call Hold (HOLD) Supplementary Service Service Description. 16 pp.
ETSI ETS 300 140-92. Integrated Services Digital Network (ISDN); Call Hold (HOLD) Supplementary Service Functional Capabilities and Information Flows. 26 pp.

Call Holding (Cont.)

—**Integrated Services Digital Networks** (Cont.)
ETSI PRETS 300 140-91. Integrated Services Digital Network (ISDN); Call Hold (HOLD) Supplementary Service Functional Capabilities and Information Flows. 27 pp.
ETSI ETS 300 141-92. Integrated Services Digital Network (ISDN); Call Hold (HOLD) Supplementary Service Digital Subscriber Signalling System No. One (DSS1) Protocol. 19 pp.
ETSI PRETS 300 141-91. Integrated Services Digital Network (ISDN); Call Hold (HOLD) Supplementary Service Digital Subscriber Signalling System No. One (DSS1) Protocol. 30 pp.

—**Integrated Services Digital Networks—Functions/Information Flows**
CCITT RECMN Q.83-92. Stage 2 Description for Call Completion Supplementary Services Section 2—Call Hold (REV.1) (Study Group XI) 22 pp. 22 pp.
CCITT RECMN Q.83-91. Stage 2 Description for Call Completion Supplementary Services Section 1—Call Waiting (CW) Section 4—Terminal Portability (Study Group XI) 28 pp. 28 pp.
CCITT RECMN Q.83-89. Call Completion Supplementary Services—General Recommendations on Telephone Switching and Signalling—Functions and Information Flows for Services in the ISDN—Supplements (Study Group XI) 33 pp. 33 pp.

Call Offer Services
See Also: Telephone Services

—**Private Switched Networks**
ECMA ECMA 191-93. Private Telecommunication Networks (PTN)—Specification, Functional Model and Information Flows—Call Offer Supplementary Service (COSD). 42 pp.
ECMA ECMA 192-93. Private Telecommunication Networks (PTN)—Inter-Exchange Signalling Protocol—Call Offer Supplementary Service (QSIG-CO). 38 pp.

Call On Hold
Use: Call Holding

Call Points
Use: Fire Alarm Switches

Call Progress Information
See Also: Telephone Services
CEPT T/CS 21-02-84. Information De Progression D'Appel. 3 pp.
CEPT T/CS 21-02 E-84. Call Progress Information. 3 pp.

—**Packet Switched Data Transmission Services—Public Data Networks**
CCITT RECMN F.601-89. Service and Operational Principles for Packet-Switched Public Data Networks—Telematic, Data Transmission and Teleconference Services—Operations and Quality of Service (Study Group I) 5 pp. 5 pp.

—**Public Data Networks**
CCITT RECMN X.96-89. Call Progress Signals in Public Data Networks—Data Communication Networks—Transmission, Signalling and Switching, Network Aspects, Maintenance and Administrative Arrangements (Study Group VII) 6 pp. 6 pp.

—**Public Land Mobile Networks**
CENELEC GSM 02.40-92. Procedures for Call Progress Indications. 7 pp.
ETSI GSM 02.40-92. Procedures for Call Progress Indications. 7 pp.

Call Redirection
Use For: Call Transfer *See Also:* Alternative Routing; Call Forwarding; Call Forwarding Busy; Call Forwarding No Reply; Call Forwarding Unconditional; Call Handling; Telephone Services

—**Integrated Services Digital Networks**
CCITT RECMN I.252-89. Call Offering Supplementary Services—Integrated Services Digital Network (ISDN)—General Structure and Service Capabilities (Study Group XVIII) 37 pp. 37 pp.
CCITT RECMN I.252. 1-88. Call Transfer—Integrated Services Digital Network (ISDN)—General Structure and Service Capabilities (Study Group XVIII) 9 pp. 9 pp.

—**Private Switched Networks**
CENELEC PRETS 300 260-92. Private Telecommunication Network (PTN); Specification, Functional Model and Information Flows Call Transfer Supplementary Service. 44 pp.

INTERNATIONAL AND NON-U.S. NATIONAL STANDARDS
SUBJECT INDEX

Call

Call Redirection *(Cont.)*
—**Private Switched Networks** *(Cont.)*
CENELEC PRETS 300 261-92. Private Telecommunication Network (PTN); Inter-Exchange Signalling Protocol Call Transfer Supplementary Service. 50 pp.
ECMA ECMA 177-92. Private Telecommunication Networks (PTN)—Specification, Functional Model and Information Flows—Call Transfer Supplementary Service (CTSD). 43 pp.
ECMA ECMA 178-92. Private Telecommunication Networks (PTN)—Inter-Exchange Signalling Protocol—Call Transfer Supplementary Service (QSIG-CT). 49 pp.
ETSI PRETS 300 260-92. Private Telecommunication Network (PTN); Specification, Functional Model and Information Flows Call Transfer Supplementary Service. 44 pp.
ETSI PRETS 300 261-92. Private Telecommunication Network (PTN); Inter-Exchange Signalling Protocol Call Transfer Supplementary Service. 50 pp.

—**Public Land Mobile Networks**
CENELEC GSM 02.82-92. Call Offering Supplementary Services. 45 pp.
ETSI GSM 02.82-92. Call Offering Supplementary Services. 45 pp.

—**Telex Communications**
CCITT RECMN F.63-89. Additional Facilities in the International Telex Service—Telegraph and Mobile Services Operations and Quality of Service (Study Group I) 2 pp. 2 pp.
CCITT RECMN U.41-89. Changed Address Interception and Call Redirection in the Telex Service—Telegraph Switching (Study Group IX) 3 pp. 3 pp.

Call Release Delay
Use: Call Release Time

Call Release Time
See Also: Delay; Release (Telephony); Release Time (Telecommunications); Telecommunication Services; Time

—**Grade of Service—Integrated Services Digital Networks**
CCITT RECMN E.721-91. Network Grade of Service Parameters and Target Values for Circuit-Switched Services in the Evolving ISDN (Study Group II) 8 pp. 8 pp.
CCITT RECMN E.721-89. Network Grade of Service Parameters in ISDN—Telephone Network and ISDN—Quality of Service, Network Management and Traffic Engineering (Study Group II) 2 pp. 2 pp.

—**Quality of Service**
CCITT RECMN E.431-92. Service Quality Assessment for Connection Set-up and Release Delays (Study Group II) 6 pp. 6 pp.

Call Requests
See Also: Call Set Up; Telephone Services
CCITT RECMN E.140-92. Operator-Assisted Telephone Service (Study Group I) 6 pp. 6 pp.
CCITT RECMN E.140-89. Principles for the Operation of International Telephone Services—Telephone Network and ISDN—Operation, Numbering, Routing and Mobile Service (Study Group II) 3 pp. 3 pp.
CCITT RECMN E.230-89. Charging (Determination of Collection Charges) in the International Telephone Service—Telephone Network and ISDN—Operation, Numbering, Routing and Mobile Service (Study Group II) 1 pp. 1 p.

—**Video Telephone Services**
CCITT RECMN F.720-92. Videotelephony Services—General (Study Group I) 11 pp. 11 pp.

Call Routing
Use: Routing (Telecommunications)

Call Set Up
See Also: Automatic Repeat Attempt; Call Control Procedures; Call Requests; Seizing (Telecommunications); Telecommunication Services; Telephone Services

—**CCITT No. 6 Signaling Systems**
CCITT RECMN Q.261-89. 4.1 Normal Call Set-Up—Specifications of Signalling System No. 6 (Study Group XI) 7 pp. 7 pp.

—**CCITT R2 Signaling Systems**
CCITT RECMN Q.460-89. Signalling Procedures—Normal Call Set-Up Procedures for International Working—Specifications of Signalling Sytems R1 and R2 (Study Group XI) 1 pp. 1 p.

Call Set Up *(Cont.)*
—**Cellular Mobile Radio Equipment**
CCIR RECMN 622-86. Technical and Operational Characteristics of Analogue Cellular Systems for Public Land Mobile Telephone Use—Section 8A—Land Mobile Service and Related Subjects. 3 pp.

—**Conference Calling**
CCITT RECMN E.151-92. Telephone Conference Calls (Study Group I) 5 pp. 5 pp.
CCITT RECMN E.151-89. Conditions of Operation and Setting up of Conference Calls—Telephone Network and ISDN—Operation, Numbering, Routing and Mobile Service (Study Group II) 3 pp. 3 pp.

—**DSS1—Integrated Services Digital Networks User Part—Interworking**
CCITT RECMN Q.699-89. Interworking Between the Digital Subscriber Signalling System Layer 3 Protocol and the Signalling System No. 7 ISDN User Part—Interworking of Signalling Systems (Study Group XI) 66 pp. 66 pp.

—**Gentex Networks**
CCITT RECMN F.1-89. Operational Provisions for the International Public Telegram Service—Telegraph and Mobile Services Operations and Quality of Service (Study Group I) 54 pp. 54 pp.

—**Inexperienced Users—Models**
CCITT FASCICLE II.2-88. Telephone Network and ISDN-Operation, Numbering, Routing and Mobile Service. Recommendations E.100—E.333. 362 pp.

—**Maritime Mobile Services—Radiotelephony**
CCITT RECMN F.110-89. Operational Provisions for the Maritime Mobile Service—Telegraph and Mobile Services Operations and Quality of Service (Study Group I) 21 pp (Same as Recmn E.200). 21 pp.

—**Maritime Mobile Services—Radioteletype Communications**
CCITT RECMN F.110-89. Operational Provisions for the Maritime Mobile Service—Telegraph and Mobile Services Operations and Quality of Service (Study Group I) 21 pp (Same as Recmn E.200). 21 pp.

—**Maritime Mobile Services—SELCAL Systems**
CCIR Report 1161-90. Use of MF/HF DSC for Automatic Connection of Calls in the Maritime-Mobile Service MF and HF Bands to the Public Switched Network—Section 8B—Maritime Mobile Service: Telegraphy and Related Subjects. 12 pp.

—**Phototelegraphy—Multiple Address Calls**
CCITT RECMN F.85-89. Message Handling Service; Intercommunication Between the IPM Service and the Telex Service—Telegraph and Mobile Services Operations and Quality of Service (Study Group I) 4 pp (Same as Recmn F.421) (Renumbered from Recmn F.75). 4 pp.

—**Phototelegraphy—Telephone Circuits**
CCITT RECMN E.323-89. Rules for Phototelegraph Communications Set up Over Circuits Normally Used for Telephone Traffic—Telephone Network and ISDN—Operation, Numbering, Routing and Mobile Service (Study Group II) 1 pp (Same as Recmn F.82 and F.107). 1 p.
CCITT RECMN F.82-91. Operational Provisions to Permit Interworking Between the International Telex Service and the Intex Service (Study Group I) 5 pp. 5 pp.
CCITT RECMN F.107-89. Rules for Phototelegraph Calls Established over Circuits Normally Used for Telephone Traffic—Telegraph and Mobile Services Operations and Quality of Service (Study Group I) 4 pp (Renumbered from Recmn F.82) (Same as Recmn E.323). 4 pp.

—**Routing**
CCITT RECMN E.140-92. Operator-Assisted Telephone Service (Study Group I) 6 pp. 6 pp.
CCITT RECMN E.140-89. Principles for the Operation of International Telephone Services—Telephone Network and ISDN—Operation, Numbering, Routing and Mobile Service (Study Group II) 3 pp. 3 pp.

—**Teleconferencing Services**
CCITT RECMN F.701-89. Teleconference Service—Telematic, Data Transmission and Teleconference Services —Operations and Quality of Service (Study Group I) 13 pp (Renumbered from Recmn F.710). 13 pp.
CCITT RECMN F.710-91. General Principles for Audiographic Conference Service—Telematic, Data Transmission and Teleconference Services Operations and Quality of Service (Study Group I) 15 pp. 15 pp.

Call Set Up *(Cont.)*
—**Telex Communications**
CCITT RECMN F.60-92. Operational Provisions for the International Telex Service (Study Group I) 34 pp. 34 pp.
CCITT RECMN F.60-89. Operational Provisions for the International Telex Service—Telegraph and Mobile Services Operations and Quality of Service (Study Group I) 17 pp. 17 pp.

—**Telex Communications—Packet Switched Public Data Networks**
CCITT RECMN F.73-90. Operational Principles for Communication Between Terminals of the International Telex Service and Data Terminal Equipment on Packet Switched Public Data Networks (Study Group I) 9 pp. 9 pp.
CCITT RECMN F.73-89. Operational Principles for Communication Between Terminals on Telex Networks and Data Terminal Equipment on Packet Switched Public Data Networks—Telegraph and Mobile Services Operations and Quality of Service (Study Group I) 5 pp. 5 pp.

—**Video Telephone Services**
CCITT RECMN F.720-92. Videotelephony Services—General (Study Group I) 11 pp. 11 pp.

—**Video Telephone Services—Integrated Services Digital Networks**
CCITT RECMN F.721-92. Videotelephony Teleservice for ISDN (Study Group I) 12 pp. 12 pp.
CCITT RECMN F.721-89. Basic Narrow Band Videophone Service in the ISDN—Telematic, Data Transmission and Teleconference Services—Operations and Quality of Service (Study Group I) 5 pp. 5 pp.

—**Videotex Communications**
CCITT RECMN F.300-89. Videotex Service—Telematic, Data Transmission and Teleconference Services—Operations and Quality of Service (Study Group I) 25 pp. 25 pp.

Call Set Up Delay
Use: Call Set Up Time

Call Set Up Time
See Also: Delay; Telecommunication Services; Telephone Services; Time

—**Quality of Service**
CCITT RECMN E.431-92. Service Quality Assessment for Connection Set-up and Release Delays (Study Group II) 6 pp. 6 pp.

Call Setup
Use: Call Set Up

Call Systems (Medical)
Use For: Luminous Call Systems
See Also: Communication Equipment; Medical Electrical Equipment; Medical Equipment; Signal Devices
CSA C22.2 NO 205-M1983. Signal Equipment (R 1992); (Gen Instr 1). 32 pp.
DIN VDE 0834-82. Luminous Call Systems; Installations in Hospitals, Nursing Homes and Similar Institutions; Installation and Operation (Aug). 9 pp.

Call Testing
Use For: Test Calling; Test Calls
See Also: Telephone Equipment; Telephone Services; Telephone Systems; Testing

—**Integrated Services Digital Networks User Part—CCITT No. 7 Signal**
CCITT RECMN Q.784-91. ISUP Basic Call Test Specification (Study Group XI) 83 pp. 83 pp.
CENELEC PRETS 300 335-93. Integrated Services Digital Network (ISDN); CCITT Signalling System No. 7 Integrated Services User Part (ISUP) Version 1 Test Specification. 120 pp.
ETSI PRETS 300 335-93. Integrated Services Digital Network (ISDN); CCITT Signalling System No. 7 Integrated Services User Part (ISUP) Version 1 Test Specification. 120 pp.

—**Public Telephone Networks**
CCITT RECMN M.1235-89. Use of Automatically Generated Test Calls for Assessment of Network Performance—Maintenance of International Telegraph, Phototelegraph and Leased Circuits—Maintenance of the International Public Telephone Network—Maintenance of Maritime. 2 pp.

—**Telephone Circuits**
CCITT RECMN E.424-92. Test Calls—Telephone Network and ISDN Quality of Service Network Management and Traffic Engineering (Study Group II) 5 pp. 5 pp.

Call Testing (Cont.)

—Telephone Circuits—Tables (Data)
CCITT RECMN E.424-92. Test Calls—Telephone Network and ISDN Quality of Service Network Management and Traffic Engineering (Study Group II) 5 pp. 5 pp.

—Telephone Services—Telephone Relations
CCITT RECMN E.111-89. Extension of International Telephone Services—Telephone Network and ISDN—Operation, Numbering, Routing and Mobile Service (Study Group II) 2 pp. 2 pp.

Call Transfer
Use: Call Redirection

Call Waiting
See Also: Telephone Services

—Digital Subscriber Signaling Systems
CCITT RECMN Q.953-92. Stage 3 Description for Call Completion Supplementary Services Using DSS 1 Section 1—Call Waiting (Study Group XI) 14 pp. 14 pp.

—Integrated Services Digital Networks
CCITT RECMN I.253-89. Call Completion Supplementary Services—Integrated Services Digital Network (ISDN)—General Structure and Service Capabilities (Study Group XVIII) 17 pp. 17 pp.
CCITT RECMN I.253. 1-90. Call Waiting (CW) Supplementary Service (Study Group I) 15 pp. 15 pp.
CENELEC PRETS 300 056-90. Integrated Services Digital Network (ISDN); Call Waiting (CW) Supplementary Service Service Description (T/NAI(89)18). 14 pp.
CENELEC PRETS 300 056-91. Integrated Services Digital Network (ISDN); Call Waiting (CW) Supplementary Service Service Description. 16 pp.
CENELEC ETS 300 056-91. Integrated Services Digital Network (ISDN); Call Waiting (CW) Supplementary Service Service Description. 16 pp.
CENELEC PRETS 300 057-90. Integrated Services Digital Network (ISDN); Call Waiting (CW) Supplementary Service Functional Capabilities and Information Flows (T/S 22.02). 18 pp.
CENELEC PRETS 300 057-91. Integrated Services Digital Network (ISDN); Call Waiting (CW) Supplementary Service Functional Capabilities and Information Flows. 22 pp.
CENELEC ETS 300 057-92. Integrated Services Digital Network (ISDN); Call Waiting (CW) Supplementary Service Functional Capabilities and Information Flows. 21 pp.
ETSI PRETS 300 056-90. Integrated Services Digital Network (ISDN); Call Waiting (CW) Supplementary Service—Service Description (T/NA1 (89)18). 14 pp.
ETSI PRETS 300 057-90. Integrated Services Digital Network (ISDN); Call Waiting (CW) Supplementary Service Functional Capabilities and Information Flows (T/S 22-2). 18 pp.
ETSI ETS 300 056-91. Integrated Services Digital Network (ISDN); Call Waiting (CW) Supplementary Service Service Description. 16 pp.
ETSI PRETS 300 056-91. Integrated Services Digital Network (ISDN); Call Waiting (CW) Supplementary Service Service Description. 16 pp.
ETSI PRETS 300 056-90. Integrated Services Digital Network (ISDN); Call Waiting (CW) Supplementary Service Service Description (T/NAI(89)18). 14 pp.
ETSI ETS 300 057-92. Integrated Services Digital Network (ISDN); Call Waiting (CW) Supplementary Service Functional Capabilities and Information Flows. 21 pp.
ETSI PRETS 300 057-91. Integrated Services Digital Network (ISDN); Call Waiting (CW) Supplementary Service Functional Capabilities and Information Flows. 22 pp.
ETSI PRETS 300 057-90. Integrated Services Digital Network (ISDN); Call Waiting (CW) Supplementary Service Functional Capabilities and Information Flows (T/S 22.02). 18 pp.

—Integrated Services Digital Networks—CCITT No. 7 Signaling
CCITT RECMN Q.733-92. Stage 3 Description for Call Completion Supplementary Services Using No. 7 Signalling System Section 1—Call Waiting (CW)—Specifications of Signalling System No. 7 (Study Group XI) 10 pp. 10 pp.

—Integrated Services Digital Networks—DSS1
CENELEC PRETS 300 058-90. Integrated Services Digital Network (ISDN); Call Waiting (CW) Supplementary Service Digital Subscriber Signalling One (DSS1) Protocol (T/S 46-33f). 14 pp.
CENELEC PRETS 300 058-91. Integrated Services Digital Network (ISDN); Call Waiting (CW) Supplementary Service Digital Subscriber Signalling System No. One (DSS1) Protocol. 18 pp.

Call Waiting (Cont.)

—Integrated Services Digital Networks—DSS1 (Cont.)
CENELEC ETS 300 058-91. Integrated Services Digital Network (ISDN); Call Waiting (CW) Supplementary Service Digital Subscriber Signalling System No. One (DSS1) Protocol. 17 pp.
ETSI PRETS 300 058-90. Integrated Services Digital Network (ISDN); Call Waiting (CW) Supplementary Service Digital Subscriber Signalling One (DSS1) Protocol (T/S 46-33F). 14 pp.
ETSI ETS 300 058-91. Integrated Services Digital Network (ISDN); Call Waiting (CW) Supplementary Service Digital Subscriber Signalling System No. One (DSS1) Protocol. 17 pp.
ETSI PRETS 300 058-90. Integrated Services Digital Network (ISDN); Call Waiting (CW) Supplementary Service Digital Subscriber Signalling One (DSS1) Protocol (T/S 46-33f). 14 pp.

—Integrated Services Digital Networks—Functions/Information Flows
CCITT RECMN Q.83-91. Stage 2 Description for Call Completion Supplementary Services Section 1—Call Waiting (CW) Section 4—Terminal Portability (Study Group XI) 28 pp. 28 pp.
CCITT RECMN Q.83-89. Call Completion Supplementary Services—General Recommendations on Telephone Switching and Signalling—Functions and Information Flows for Services in the ISDN—Supplements (Study Group XI) 33 pp. 33 pp.

—Ringing Tones—Private Automatic Branch Exchanges
CCITT RECMN E.182-89. Application of Tones and Recorded Announcements in Telephone Services—Telephone Network and ISDN—Operation, Numbering, Routing and Mobile Service (Study Group II) 8 pp. 8 pp.

—Telephone Networks
CEPT T/S 22-02-87. Service Supplementaire "Appel En Instance". 19 pp.
CEPT T/S 22-02 E-87. Supplementary Service Call Waiting. 19 pp.

Call Waiting Tones
See Also: Tones (Telephone Services)
CCITT RECMN E.182-89. Application of Tones and Recorded Announcements in Telephone Services—Telephone Network and ISDN—Operation, Numbering, Routing and Mobile Service (Study Group II) 8 pp. 8 pp.
CCITT RECMN Q.35-89. Technical Characteristics of Tones for the Telephone Service—General Recommendations on Telephone Switching and Signalling—Functions and Information Flows for Services in the ISDN —Supplements (Study Group XI) 9 pp. 9 pp.
CENELEC PRETS 300 295-92. Human Factors (HF); Specification of Characteristics of Telephone Services Tones When Locally Generated in Terminals. 9 pp.
ETSI PRETS 300 295-92. Human Factors (HF); Specification of Characteristics of Telephone Services Tones When Locally Generated in Terminals. 9 pp.

—Frequency Bands
CCITT RECMN E.180-89. Technical Characteristics of Tones for the Telephone Service—Telephone Network and ISDN—Operation, Numbering, Routing and Mobile Service (Study Group II) 9 pp. 9 pp.

Called Line Identification
See Also: Address Codes; Identification Systems; Terminal Identification

—Integrated Services Digital Networks
CCITT RECMN E.164-91. Numbering Plan for the ISDN Era (Study Group II) 19 pp (Same as Recmn I.331). 19 pp.
CCITT RECMN E.164-89. Numbering Plan for the ISDN Era—Telephone Network and ISDN—Operation, Numbering, Routing and Mobile Service (Study Group II) 6 pp (Same as Recmn I.331). 6 pp.

Called Party Hold Plugs
Use: Line Traps

Caller Waiting Tones
See Also: Tones (Telephone Services)
CCITT RECMN E.182-89. Application of Tones and Recorded Announcements in Telephone Services—Telephone Network and ISDN—Operation, Numbering, Routing and Mobile Service (Study Group II) 8 pp. 8 pp.
CCITT RECMN Q.35-89. Technical Characteristics of Tones for the Telephone Service—General Recommendations on Telephone Switching and Signalling—Functions and Information Flows for Services in the ISDN —Supplements (Study Group XI) 9 pp. 9 pp.

Caller Waiting Tones (Cont.)

—Frequency Bands
CCITT RECMN E.180-89. Technical Characteristics of Tones for the Telephone Service—Telephone Network and ISDN—Operation, Numbering, Routing and Mobile Service (Study Group II) 9 pp. 9 pp.

Calling Channel Loading

—Maritime Mobile Services—SELCAL Services
CCIR RECMN 822-92. Calling-Channel Loading for Digital Selective-Calling (DSC) for the Maritime Mobile Service. 1 p.

Calling Line Identification
See Also: Address Codes; Calling Line Identification Presentation; Calling Line Identification Restriction; Identification Systems; Terminal Identification

—Integrated Services Digital Networks
CCITT RECMN E.164-91. Numbering Plan for the ISDN Era (Study Group II) 19 pp (Same as Recmn I.331). 19 pp.
CCITT RECMN E.164-89. Numbering Plan for the ISDN Era—Telephone Network and ISDN—Operation, Numbering, Routing and Mobile Service (Study Group II) 6 pp (Same as Recmn I.331). 6 pp.
CEPT T/S 22-01-87. Service Supplementaire Identification. 9 pp.

—Malicious Calls—Integrated Services Digital Networks
CCITT RECMN I.251. 7 (REV 1)-92. Malicious Call Identification (Study Group I) 10 pp. 10 pp.
CENELEC PRETS 300 128-90. Integrated Services Digital Network (ISDN) Malicious Call Identification (MCID) Supplementary Service Service Description (T/NA1(89)03). 11 pp.
CENELEC PRETS 300 128-91. Integrated Services Digital Network (ISDN); Malicious Call Identification (MCID) Supplementary Service Service Description (T/NA1(89)03). 15 pp.
CENELEC ETS 300 128-92. Integrated Services Digital Network (ISDN); Malicious Call Identification (MCID) Supplementary Service Service Description. 14 pp.
CENELEC PRETS 300 129-90. Integrated Services Digital Network (ISDN): Malicious Call Identification (MCID) Supplementary Service Functional Capabilities and Information Flows (T/S 22-10). 24 pp.
CENELEC PRETS 300 129-91. Integrated Services Digital Network (ISDN); Malicious Call Identification (MCID) Supplementary Service Functional Capabilities and Information Flows. 24 pp.
CENELEC ETS 300 129-92. Integrated Services Digital Network (ISDN); Malicious Call Identification (MCID) Supplementary Service Functional Capabilities and Information Flows. 23 pp.
CENELEC PRETS 300 130-90. Integrated Services Digital Network (ISDN); Malicious Call Identification (MCID) Supplementary Service Digital Subscriber Signalling One (DSSI) Protocol (T/S 46-33N). 14 pp.
CENELEC PRETS 300 130-91. Integrated Services Digital Network (ISDN); Malicious Call Identification (MCID) Supplementary Service Digital Subscriber Signalling System No. One (DSS1) Protocol. 16 pp.
CENELEC ETS 300 130-92. Integrated Services Digital Network (ISDN); Malicious Call Identification (MCID) Supplementary Service Digital Subscriber Signalling System No. One (DSS1) Protocol. 15 pp.
ETSI PRETS 300 128-90. Integrated Services Digital Network (ISDN) Malicious Call Identification (MCID) Supplementary Service Service Description (T/NA1(89)03). 11 pp.
ETSI PRETS 300 129-90. Integrated Services Digital Network (ISDN); Malicious Call Identification (MCID) Supplementary Service Functional Capabilities and Information Flows (T/S 22-10). 24 pp.
ETSI PRETS 300 130-90. Intergrated Services Digital Network (ISDN); Malicious Call Identification (MCID) Supplementary Service Digital Subscriber Signalling One (DSS1) Protocol (T/S 46-33N). 14 pp.
ETSI ETS 300 128-92. Integrated Services Digital Network (ISDN); Malicious Call Identification (MCID) Supplementary Service Service Description. 14 pp.
ETSI PRETS 300 128-91. Integrated Services Digital Network (ISDN); Malicious Call Identification (MCID) Supplementary Service Service Description (T/NA1(89)03). 15 pp.

INTERNATIONAL AND NON-U.S. NATIONAL STANDARDS
SUBJECT INDEX
Calling

Calling Line Identification *(Cont.)*
—Malicious Calls—Integrated Services Digital Networks *(Cont.)*
ETSI ETS 300 129-92. Integrated Services Digital Network (ISDN); Malicious Call Identification (MCID) Supplementary Service Functional Capabilities and Information Flows. 23 pp.
ETSI PRETS 300 129-91. Integrated Services Digital Network (ISDN); Malicious Call Identification (MCID) Supplementary Service Functional Capabilities and Information Flows. 24 pp.
ETSI ETS 300 130-92. Integrated Services Digital Network (ISDN); Malicious Call Identification (MCID) Supplementary Service Digital Subscriber Signalling System No. One (DSS1) Protocol. 15 pp.
ETSI PRETS 300 130-91. Integrated Services Digital Network (ISDN); Malicious Call Identification (MCID) Supplementary Service Digital Subscriber Signalling System No. One (DSS1) Protocol. 16 pp.

—Telex Communications
CCITT RECMN F.63-89. Additional Facilities in the International Telex Service—Telegraph and Mobile Services Operations and Quality of Service (Study Group I) 2 pp. 2 pp.

Calling Line Identification Presentation
Use For: CLIP ***See Also:*** Calling Line Identification; Calling Line Identification Restriction; Identification Systems; Telephone Services; Terminal Identification

—Integrated Services Digital Networks
CCITT RECMN I.251-89. Number Identification Supplementary Services—Integrated Services Digital Network (ISDN)—General Structure and Service Capabilities (Study Group XVIII) 28 pp. 28 pp.
CCITT RECMN I.251. 3 (REV 1)-92. Calling Line Identification Presentation (Study Group I) 12 pp. 12 pp.
CENELEC PRETS 300 089-90. Integrated Services Digital Network (ISDN) Calling Line Identification Presentation (CLIP) Supplementary Service Service Description (T/NA1(89)07). 13 pp.
CENELEC PRETS 300 089-91. Integrated Services Digital Network (ISDN); Calling Line Identification Presentation (CLIP) Supplementary Service Service Description. 17 pp.
CENELEC ETS 300 089-92. Integrated Services Digital Network (ISDN); Calling Line Identification Presentation (CLIP) Supplementary Service Service Description. 17 pp.
CENELEC PRETS 300 091-90. Integrated Services Digital Network (ISDN) Calling Line Identification, Presentation and Restriction (CLIP and CLIR) Supplementary Services Functional Capabilities and Information Flows (T/S 22-01). 21 pp.
CENELEC PRETS 300 091-91. Integrated Services Digital Network (ISDN); Calling Line Identification Presentation (CLIP) and Calling Line Identification Restriction (CLIR) Supplementary Services Functional Capabilities and Information Flows. 39 pp.
CENELEC ETS 300 091-92. Integrated Services Digital Network (ISDN); Calling Line Identification Presentation (CLIP) and Calling Line Identification Restriction (CLIR) Supplementary Services Functional Capabilities and Information Flows. 43 pp.
CEPT T/S 22-01 E-87. Supplementary Service Calling Line Identification Presentation. 9 pp.
ETSI PRETS 300 089-90. Integrated Services Digital Network (ISDN) Calling Line Identification Presentation (CLIP) Supplementary Service—Service Description (T/NA1 (89)07). 13 pp.
ETSI PRETS 300 091-90. Integrated Services Digital Network (ISDN) Calling Line Identification, Presentation and Restriction (CLIP and CLIR) Supplementary Services—Functional Capabilities and Information Flows (T/S 22-01). 21 pp.
ETSI ETS 300 089-92. Integrated Services Digital Network (ISDN); Calling Line Identification Presentation (CLIP) Supplementary Service Service Description. 17 pp.
ETSI ETS 300 091-92. Integrated Services Digital Network (ISDN); Calling Line Identification Presentation (CLIP) and Calling Line Identification Restriction (CLIR) Supplementary Services Functional Capabilities and Information Flows. 43 pp.
ETSI PRETS 300 091-91. Integrated Services Digital Network (ISDN); Calling Line Identification Presentation (CLIP) and Calling Line Identification Restriction (CLIR) Supplementary Services Functional Capabilities and Information Flows. 39 pp.

Calling Line Identification Presentation *(Cont.)*
—Integrated Services Digital Networks—DSS1
CENELEC PRETS 300 092-90. Integrated Services Digital Network (ISDN) Calling Line Identification, Presentation (CLIP) Supplementary Service Digital Subscriber Signalling One (DSS1) Protocol (T/S 46-33C). 18 pp.
CENELEC PRETS 300 092-91. Integrated Services Digital Network (ISDN); Calling Line Identification Presentation (CLIP) Supplementary Service Digital Subscriber Signalling System No. One (DSS1) Protocol. 4 pp.
CENELEC ETS 300 092-92. Integrated Services Digital Network (ISDN); Calling Line Identification Presentation (CLIP) Supplementary Service Digital Subscriber Signalling No. One (DSS1) Protocol. 22 pp.
CENELEC ETS 300 092 PRA1-92. AMD 1 Integrated Services Digital Network (ISDN); Calling Line Identification Presentation (CLIP) Supplementary Service Digital Subscriber Signalling System No. One (DSS1) Protocol. 4 pp.
CENELEC ETS 300 092/A1-93. AMD 1 Integrated Services Digital Network (ISDN); Calling Line Identification Presentation (CLIP) Supplementary Service Digital Subscriber Signalling System No. One (DSS1) Protocol. 4 pp.
ETSI PRETS 300 092-90. Integrated Services Digital Network (ISDN) Calling Line Identification Presentation (CLIP) Supplementary Service—Digital Subscriber Signalling One (DSS1) Protocol (T/S 46-33C). 18 pp.
ETSI ETS 300 092/A1-93. AMD 1 Integrated Services Digital Network (ISDN); Calling Line Identification Presentation (CLIP) Supplementary Service Digital Subscriber Signalling System No. One (DSS1) Protocol. 4 pp.
ETSI ETS 300 092 PRA1-92. AMD 1 Integrated Services Digital Network (ISDN); Calling Line Identification Presentation (CLIP) Supplementary Service Digital Subscriber Signalling System No. One (DSS1) Protocol. 4 pp.
ETSI ETS 300 092-92. Integrated Services Digital Network (ISDN); Calling Line Identification Presentation (CLIP) Supplementary Service Digital Subscriber Signalling No. One (DSS1) Protocol. 22 pp.
ETSI PRETS 300 092-91. Integrated Services Digital Network (ISDN); Calling Line Identification Presentation (CLIP) Supplementary Service Digital Subscriber Signalling System No. One (DSS1) Protocol. 4 pp.

—ISDN—CCITT No. 7 Signaling Systems
CCITT RECMN Q.730-89. ISDN Supplementary Services—Specifications of Signalling System No. 7 (Study Group XI) 59 pp. 59 pp.

—ISDN—Functions/Information Flows
CCITT RECMN Q.81-91. Stage 2 Description for Number Identification Supplementary Services (Study Group XI) 46 pp. 46 pp.
CCITT RECMN Q.81-89. Number Identification Supplementary Services—General Recommendations on Telephone Switching and Signalling—Functions and Information Flows for Services in the ISDN—Supplements (Study Group XI) 30 pp. 30 pp.

—Private Switched Networks
CENELEC PRETS 300 173-91. Identification Supplementary Services in Private Telecommunication Networks—Specification, Functional Model and Information Flows-. 50 pp.
CENELEC PRETS 300 173-92. Private Telecommunications Network (PTN); Specification, Functional Models and Information Flows Identification Supplementary Services. 41 pp.
CENELEC ETS 300 173-92. Private Telecommunication Network (PTN); Specification, Functional Models and Information Flows Identification Supplementary Services. 40 pp.
CENELEC PRETS 300 237-92. Private Telecommunications Network (PTN); Specification, Fuctional Models and Information Flows Name Identification Supplementary Services. 44 pp.
CENELEC PRETS 300 237-93. Private Telecommunication Network (PTN); Specification, Functional Models and Information Flows Name Identification Supplementary Services. 42 pp.
CENELEC ETS 300 237-93. Private Telecommunication Network (PTN); Specification, Functional Models and Information Flows Name Identification Supplementary Services. 43 pp.
ECMA ECMA 163-92. Private Telecommunication Networks (PTN)—Specification, Functional Model and Information Flows—Name Identification Supplementary Services (NISD). 41 pp.
ETSI PRETS 300 173-91. Identification Supplementary Services in Private Telecommunication Networks—Specification, Functional Model and Information Flows—(Standard ECMA-148). 49 pp.

Calling Line Identification Presentation *(Cont.)*
—Private Switched Networks *(Cont.)*
ETSI ETS 300 173-92. Private Telecommunication Network (PTN); Specification, Functional Models and Information Flows Identification Supplementary Services. 40 pp.
ETSI PRETS 300 173-92. Private Telecommunications Network (PTN); Specification, Functional Models and Information Flows Identification Supplementary Services. 41 pp.
ETSI PRETS 300 173-91. Identification Supplementary Services in Private Telecommunication Networks—Specification, Functional Model and Information Flows-. 50 pp.
ETSI ETS 300 237-93. Private Telecommunication Network (PTN); Specification, Functional Models and Information Flows Name Identification Supplementary Services. 43 pp.
ETSI PRETS 300 237-93. Private Telecommunication Network (PTN); Specification, Functional Models and Information Flows Name Identification Supplementary Services. 42 pp.
ETSI PRETS 300 237-92. Private Telecommunications Network (PTN); Specification, Fuctional Models and Information Flows Name Identification Supplementary Services. 44 pp.
ETSI ETS 300 238-93. Private Telecommunication Network (PTN); Inter-Exchange Signalling Protocol Name Identificaiton Supplementary Services. 24 pp.

—Private Switched Networks—Signaling Protocols
CENELEC ETS 300 191-92. Private Telecommunication Network (PTN); Signalling Protocol at the S-Reference Point Identification Supplementary Services. 26 pp.
ECMA ECMA 157-91. Protocol for Signalling over the D-Channel of Interfaces at the S Reference Point Between Terminal Equipment and Private Telecommunication Networks for the Support of Identification Supplementary Services. 25 pp.
ECMA ECMA 164-92. Private Telecommunication Networks (PTN)—Signalling Between Private Telecommunication Exchanges—Protocol for the Support of Name Identification Supplementary Services (QSIG-NA). 22 pp.

—Teleinformatic Services
CCITT RECMN F.351-89. General Principles on the Presentation of Terminal Identification to Users of the Telematic Services—Telematic, Data Transmission and Teleconference Services—Operations and Quality of Service (Study Group I) 2 pp. 2 pp.

Calling Line Identification Restriction
Use For: CLIR ***See Also:*** Calling Line Identification; Calling Line Identification Presentation; Telephone Services; Terminal Identification

—ISDN
CCITT RECMN I.251-89. Number Identification Supplementary Services—Integrated Services Digital Network (ISDN)—General Structure and Service Capabilities (Study Group XVIII) 28 pp. 28 pp.
CCITT RECMN I.251. 4 (REV 1)-92. Calling Line Identification Restriction (Study Group I) 8 pp. 8 pp.
CENELEC PRETS 300 090-90. Integrated Services Digital Network (ISDN) Calling Line Identification Restriction (CLIR) Supplementary Service Description (T/NA1(89) 08). 11 pp.
CENELEC PRETS 300 090-91. Integrated Services Digital Network (ISDN); Calling Line Identification Restriction (CLIR) Supplementary Service Service Description. 15 pp.
CENELEC ETS 300 090-92. Integrated Services Digital Network (ISDN); Calling Line Identification Restriction (CLIR) Supplementary Service Service Description. 13 pp.
CENELEC PRETS 300 091-90. Integrated Services Digital Network (ISDN) Calling Line Identification, Presentation and Restriction (CLIP and CLIR) Supplementary Services Functional Capabilities and Information Flows (T/S 22-01). 21 pp.
CENELEC PRETS 300 091-91. Integrated Services Digital Network (ISDN); Calling Line Identification Presentation (CLIP) and Calling Line Identification Restriction (CLIR) Supplementary Services Functional Capabilities and Information Flows. 39 pp.
CENELEC ETS 300 091-92. Integrated Services Digital Network (ISDN); Calling Line Identification Presentation (CLIP) and Calling Line Identification Restriction (CLIR) Supplementary Services Functional Capabilities and Information Flows. 43 pp.
CENELEC PRETS 300 093-90. Integrated Services Digital Network (ISDN) Calling Line Identification Restriction (CLIR) Supplementary Service Digital Subscriber Signalling One (DSSI) Protocol (T/S 46-33D). 9 pp.

INDUSTRY STANDARDS
395

Calling Line Identification Restriction (Cont.)

—ISDN (Cont.)

CENELEC PRETS 300 093-91. Integrated Services Digital Network (ISDN); Calling Line Identification Restriction (CLIR) Supplementary Service Digital Subscriber Signalling System No. One (DSS1) Protocol. 14 pp.

CENELEC ETS 300 093-92. Integrated Services Digital Network (ISDN); Calling Line Identification Restriction (CLIR) Supplementary Service Digital Subscriber Signalling System No. One (DSS1) Protocol. 12 pp.

ETSI PRETS 300 090-90. Integrated Services Digital Network (ISDN) Calling Line Identification Restriction (CLIR) Supplementary Service—Service Description (T/NA1 (89)08). 11 pp.

ETSI PRETS 300 091-90. Integrated Services Digital Network (ISDN) Calling Line Identification, Presentation and Restriction (CLIP and CLIR) Supplementary Services—Functional Capabilities and Information Flows (T/S 22-01). 21 pp.

ETSI PRETS 300 093-90. Integrated Services Digital Network (ISDN) Calling Line Identification Restriction (CLIR) Supplementary Service—Digital Subscriber Signalling One (DSS1) Protocol (T/S 46-33D). 9 pp.

ETSI ETS 300 090-92. Integrated Services Digital Network (ISDN); Calling Line Identification Restriction (CLIR) Supplementary Service Service Description. 13 pp.

ETSI ETS 300 091-92. Integrated Services Digital Network (ISDN); Calling Line Identification Presentation (CLIP) and Calling Line Identification Restriction (CLIR) Supplementary Services Functional Capabilities and Information Flows. 43 pp.

ETSI PRETS 300 091-91. Integrated Services Digital Network (ISDN); Calling Line Identification Presentation (CLIP) and Calling Line Identification Restriction (CLIR) Supplementary Services Functional Capabilities and Information Flows. 39 pp.

ETSI ETS 300 093-92. Integrated Services Digital Network (ISDN); Calling Line Identification Restriction (CLIR) Supplementary Service Digital Subscriber Signalling System No. One (DSS1) Protocol. 12 pp.

ETSI PRETS 300 093-91. Integrated Services Digital Network (ISDN); Calling Line Identification Restriction (CLIR) Supplementary Service Digital Subscriber Signalling System No. One (DSS1) Protocol. 14 pp.

ETSI ETS 300 094-92. Integrated Services Digital Network (ISDN); Connected Line Identification Presentation (COLP) Supplementary Service Service Description. 16 pp.

—ISDN—CCITT No. 7 Signaling Systems

CCITT RECMN Q.730-89. ISDN Supplementary Services—Specifications of Signalling System No. 7 (Study Group XI) 59 pp. 59 pp.

—ISDN—Functions/Information Flows

CCITT RECMN Q.81-91. Stage 2 Description for Number Identification Supplementary Services (Study Group XI) 46 pp. 46 pp.

CCITT RECMN Q.81-89. Number Identification Supplementary Services—General Recommendations on Telephone Switching and Signalling—Functions and Information Flows for Services in the ISDN—Supplements (Study Group XI) 30 pp. 30 pp.

—Private Switched Networks

CENELEC PRETS 300 173-91. Identification Supplementary Services in Private Telecommunication Networks—Specification, Functional Model and Information Flows-. 50 pp.

CENELEC PRETS 300 173-92. Private Telecommunications Network (PTN); Specification, Functional Models and Information Flows Identification Supplementary Services. 41 pp.

CENELEC ETS 300 173-92. Private Telecommunication Network (PTN); Specification, Functional Models and Information Flows Identification Supplementary Services. 40 pp.

CENELEC PRETS 300 237-92. Private Telecommunications Network (PTN); Specification, Fuctional Models and Information Flows Name Identification Supplementary Services. 44 pp.

CENELEC PRETS 300 237-93. Private Telecommunication Network (PTN); Specification, Functional Models and Information Flows Name Identification Supplementary Services. 42 pp.

CENELEC ETS 300 237-93. Private Telecommunication Network (PTN); Specification, Functional Models and Information Flows Name Identification Supplementary Services. 43 pp.

ECMA ECMA 163-92. Private Telecommunication Networks (PTN)—Specification, Functional Model and Information Flows—Name Identification Supplementary Services (NISD). 41 pp.

Calling Line Identification Restriction (Cont.)

—Private Switched Networks (Cont.)

ETSI PRETS 300 173-91. Identification Supplementary Services in Private Telecommunication Networks—Specification, Functional Model and Information Flows—(Standard ECMA-148). 49 pp.

ETSI ETS 300 173-92. Private Telecommunication Network (PTN); Specification, Functional Models and Information Flows Identification Supplementary Services. 40 pp.

ETSI PRETS 300 173-92. Private Telecommunications Network (PTN); Specification, Functional Models and Information Flows Identification Supplementary Services. 41 pp.

ETSI PRETS 300 173-91. Identification Supplementary Services in Private Telecommunication Networks—Specification, Functional Model and Information Flows-. 50 pp.

ETSI PRETS 300 237-93. Private Telecommunication Network (PTN); Specification, Functional Models and Information Flows Name Identification Supplementary Services. 43 pp.

ETSI PRETS 300 237-93. Private Telecommunication Network (PTN); Specification, Functional Models and Information Flows Name Identification Supplementary Services. 42 pp.

ETSI ETS 300 237-93. Private Telecommunication Network (PTN); Specification, Fuctional Models and Information Flows Name Identification Supplementary Services. 44 pp.

ETSI ETS 300 238-93. Private Telecommunication Network (PTN); Inter-Exchange Signalling Protocol Name Identificaiton Supplementary Services. 24 pp.

—Private Switched Networks—Signaling Protocols

CENELEC PRETS 300 191-91. Protocol for Signalling Over the D-Channel of Interfaces at the S Reference Point Between Terminal Equipment and Private Telecommunication Networks for the Support of Identification Supplementary Services. 28 pp.

CENELEC PRETS 300 191-92. Private Telecommunications Network (PTN); Signalling Protocol at the S-Reference Point Identification Supplementary Services. 27 pp.

CENELEC ETS 300 191-92. Private Telecommunication Network (PTN); Signalling Protocol at the S-Reference Point Identification Supplementary Services. 26 pp.

ETSI ETS 300 191-92. Private Telecommunication Network (PTN); Signalling Protocol at the S-Reference Point Identification Supplementary Services. 26 pp.

ETSI PRETS 300 191-92. Private Telecommunications Network (PTN); Signalling Protocol at the S-Reference Point Identification Supplementary Services. 27 pp.

ETSI PRETS 300 191-91. Protocol for Signalling Over the D-Channel of Interfaces at the S Reference Point Between Terminal Equipment and Private Telecommunication Networks for the Support of Identification Supplementary Services. 28 pp.

—Signaling Protocols

ECMA ECMA 157-91. Protocol for Signalling over the D-Channel of Interfaces at the S Reference Point Between Terminal Equipment and Private Telecommunication Networks for the Support of Identification Supplementary Services. 25 pp.

Calling Name Identification Presentation

Use For: CNIP *See Also:* Connected Name Identification Presentation; Telephone Services; Terminal Identification

—Private Switched Networks—Signaling Protocols

CENELEC PRETS 300 238-92. Private Telecommunications Network (PTN); Signalling Between Private Telecommunication Exchanges Protocol for the Support of Name Identification Supplementary Services. 25 pp.

CENELEC PRETS 300 238-93. Private Telecommunication Network (PTN); Inter-Exchange Signalling Protocol Name Identfication Supplementary Services. 24 pp.

CENELEC ETS 300 238-93. Private Telecommunication Network (PTN); Inter-Exchange Signalling Protocol Name Identificaiton Supplementary Services. 24 pp.

ETSI PRETS 300 238-93. Private Telecommunication Network (PTN); Inter-Exchange Signalling Protocol Name Identification Supplementary Services. 24 pp.

ETSI PRETS 300 238-92. Private Telecommunications Network (PTN); Signalling Between Private Telecommunication Exchanges Protocol for the Support of Name Identification Supplementary Services. 25 pp.

Calling Number Indication

Use: Calling Line Identification

Calorific Value

Use For: Heating Value *See Also:* Calorimeters; Calorimetry; Chemical Analysis; Heat of Combustion; High Temperature Testing; Thermodynamic Properties

—Asphalts

CNS K6209-65. Determination of Calorific Value of Asphalt (Sep)(2491)(R 1971).

—Coal

BSI BS 1016: Part 5-77. (WITHDRAWN) 1977 Methods for the Analysis and Testing of Coal and Coke Part 5: Gross Calorific Value of Coal and Coke (Superseded by BS 1016: Part 105: 1992). 12 pp.

BSI BS 1016: Part 105-92. 1992 Methods for Analysis and Testing of Coal and Coke Part 105: Determination of Gross Calorific Value (W). 17 pp.

JIS M 8814-85. Determination of Calorific Value of Coal and Coke; (Erratum). 20 pp.

SAA AS 1038.5.1-88. Methods for the Analysis and Testing of Coal and Coke—Part 5.1: Gross Specific Energy of Coal and Coke—Adiabatic Calorimeters. 9 pp.

SAA AS 1038.5.2-89. Methods for the Analysis and Testing of Coal and Coke—Part 5.2: Gross Specific Energy of Coal and Coke —Automatic Isothermal-Type Calorimeters. 5 pp.

—Coal Gas

CNS K6238-66. Method of Test for Calorific Value of Coal Gas (Sep)(2632)(R 1971).

—Coke

BSI BS 1016: Part 5-77. (WITHDRAWN) 1977 Methods for the Analysis and Testing of Coal and Coke Part 5: Gross Calorific Value of Coal and Coke (Superseded by BS 1016: Part 105: 1992). 12 pp.

BSI BS 1016: Part 105-92. 1992 Methods for Analysis and Testing of Coal and Coke Part 105: Determination of Gross Calorific Value (W). 17 pp.

JIS M 8814-85. Determination of Calorific Value of Coal and Coke; (Erratum). 20 pp.

SAA AS 1038.5.1-88. Methods for the Analysis and Testing of Coal and Coke—Part 5.1: Gross Specific Energy of Coal and Coke—Adiabatic Calorimeters. 9 pp.

SAA AS 1038.5.2-89. Methods for the Analysis and Testing of Coal and Coke—Part 5.2: Gross Specific Energy of Coal and Coke —Automatic Isothermal-Type Calorimeters. 5 pp.

—Construction Materials

ISO 1716-73. Building Materials—Determination of Calorific Potential First Edition. 6 pp.

—Electrical Insulating Liquids

BSI BS EN 61100-93. 1993 Classification of Insulating Liquids According to Fire Point and Net Calorific Value (IEC 1100: 1992) (E). 12 pp.

CENELEC EN 61100-92. Classification of Insulating Liquids According to Fire Point and Net Calorific Value. 5 pp.

IEC 1100-92. Classification of Insulating Liquids According to Fire-Point and Net Calorific Value First Edition. 15 pp.

—Gaseous Fuels

BSI BS 7420-91. 1991 Determination of Calorific Values of Solid, Liquid and Gaseous Fuels (Including Definitions). 11 pp.

—Gases

BSI BS 3156: Part 3-68. (WITHDRAWN) 1968 Methods for the Analysis of Fuel Gases Part 3: Combustion Characteristics. 24 pp.

BSI BS 3804: Part 1-64. (OBSOLESCENT) 1964 Amd 1 Methods for the Determination of Calorific Value of Fuel Gases Part 1: Non-Recording Methods. 61 pp.

DIN ENGL 51858-82. Gaseous Fuels and Other Gases; Calculation of Gross and Nett Calorific Values and of Relative Density of Gas Mixtures (Nov). 7 pp.

JIS K 2301-92. Fuel Gases and Natural Gas—Methods for Chemical Analysis and Testing. 116 pp.

—Glossaries

DIN ENGL 5499-72. Gross and Net Calorific Values; Terms (Jan). 3 pp.

—Liquid Fuels

BSI BS 7420-91. 1991 Determination of Calorific Values of Solid, Liquid and Gaseous Fuels (Including Definitions). 11 pp.

DIN ENGL 51900 Pt 1-89. Determination of Gross Calorific Value of Solid and Liquid Fuels by the Bomb Calorimeter and Calculation of Net Calorific Value; General Information (Nov). 11 pp.

INTERNATIONAL AND NON-U.S. NATIONAL STANDARDS
SUBJECT INDEX

Calorific Value (Cont.)
—Liquid Fuels (Cont.)
DIN ENGL 51900 Pt 2-77. Testing of Solid and Liquid Fuels; Determination of the Gross Calorific Value by the Bomb Calorimeter and Calculation of the Net Calorific Value; Method Using Isothermal Water Jacket (Aug). 8 pp.

—Liquid Hydrocarbons
BSI BS 7577-92. 1992 Calculation Procedures for Static Measurement of Refrigerated Light Hydrocarbon Fluids (ISO 6578: 1991). 28 pp.

ISO 6578-91. Refrigerated Hydrocarbon Liquids—Static Measurement—Calculation Procedure First Edition. 26 pp.

—Natural Gas
DIN ENGL 51858-82. Gaseous Fuels and Other Gases; Calculation of Gross and Nett Calorific Values and of Relative Density of Gas Mixtures (Nov). 7 pp.

ISO 6976-83. Natural Gas—Calculation of Calorific Value, Density and Relative Density First Edition. 16 pp.

—Solid Fuels
BSI BS 7420-91. 1991 Determination of Calorific Values of Solid, Liquid and Gaseous Fuels (Including Definitions). 11 pp.

CNS M3154-84. Method of Test for Gross Calorific Value of Solid Mineral Fuels by the Calorimeter Bomb Method, and Calculation of Net Calorific Value (Mar) (10835).

DIN ENGL 51900 Pt 1-89. Determination of Gross Calorific Value of Solid and Liquid Fuels by the Bomb Calorimeter and Calculation of Net Calorific Value; General Information (Nov). 11 pp.

DIN ENGL 51900 Pt 2-77. Testing of Solid and Liquid Fuels; Determination of the Gross Calorific Value by the Bomb Calorimeter and Calculation of the Net Calorific Value; Method Using Isothermal Water Jacket (Aug). 8 pp.

DIN ENGL 51900 Pt 3-77. Testing of Solid and Liquid Fuels; Determination of the Gross Calorific Value by the Bomb Calorimeter and Calculation of the Net Calorific Value; Method Using Adiabatic Jacket (Aug). 5 pp.

ISO 1928-76. Solid Mineral Fuels—Determination of Gross Calorific Value by the Calorimeter Bomb Method, and Calculation of Net Calorific Value First Edition; (Erratum—June 1979). 18 pp.

Calorifiers
See Also: Heat Exchangers
BSI BS 853-90. 1990 Calorifiers and Storage Vessels for Central Heating and Hot Water Supply. 40 pp.
SNZ NZS/BS 853-81. Specification for Calorifiers and Storage Vessels for Central Heating and Hot Water Supply. 24 pp.

—Hot Water
BSI BS 853-81. 1981 Calorifiers and Storage Vessels for Central Heating and Hot Water Supply. 28 pp.

Calorimeter Bombs
Use: Calorimeters

Calorimeters
Use For: Calorimeter Bombs See Also: Bomb Calorimeters; Calorific Value; Calorimetry; High Temperature Testing; Pyrheliometers; Temperature Measuring Instruments; Thermal Measurement
ISO 651-75. Solid-Stem Calorimeter Thermometers First Edition. 6 pp.
ISO 652-75. Enclosed-Scale Calorimeter Thermometers First Edition. 5 pp.

—Cements
DIN ENGL 1164 Pt 8-78. Portland-, Iron Portland-, Blast-Furnace-, and Trass Cement; Determination of the Heat of Hydration with the Solution Calorimeter (Nov). 6 pp.

—Coal
SAA AS 1038.5.1-88. Methods for the Analysis and Testing of Coal and Coke—Part 5.1: Gross Specific Energy of Coal and Coke—Adiabatic Calorimeters. 9 pp.
SAA AS 1038.5.2-89. Methods for the Analysis and Testing of Coal and Coke—Part 5.2: Gross Specific Energy of Coal and Coke —Automatic Isothermal-Type Calorimeters. 5 pp.

—Coke
SAA AS 1038.5.1-88. Methods for the Analysis and Testing of Coal and Coke—Part 5.1: Gross Specific Energy of Coal and Coke—Adiabatic Calorimeters. 9 pp.
SAA AS 1038.5.2-89. Methods for the Analysis and Testing of Coal and Coke—Part 5.2: Gross Specific Energy of Coal and Coke —Automatic Isothermal-Type Calorimeters. 5 pp.

Calorimeters (Cont.)
—Fuel Gas
BSI BS 3804: Part 1-64. (OBSOLESCENT) 1964 Amd 1 Methods for the Determination of Calorific Value of Fuel Gases Part 1: Non-Recording Methods. 61 pp.

Calorimetry
Use For: Scanning Calorimetry See Also: Calorific Value; Calorimeters; Chemical Analysis; Thermochemical Properties

—Electrical Insulation
BSI BS EN 61074-93. 1993 Method of Test for Determination of Heats and Temperatures of Melting and Crystallization of Electrical Insulating Materials by Differential Scanning Calorimetry (IEC 1074: 1991) (F). 18 pp.
CENELEC EN 61074-93. Determination of Heats and Temperatures of Melting and Crystallization of Electrical Insulating Materials by Differential Scanning Calorimetry (IEC 1074: 1991). 13 pp.
IEC 1074-91. Determination of Heats and Temperatures of Melting and Crystallization of Electrical Insulating Materials by Differential Scanning Calorimetry First Edition; (CENELEC EN 61074:1993). 29 pp.

—Glassy Alloys
JIS H 7151-91. Method of Determining the Crystallization Temperatures of Amorphous Metals. 7 pp.

CAM
Use For: Computer Aided Manufacturing
See Also: CAD; CAD/CAM; CIM; Manufacturing

—Glossaries
CNS B8003-4-89. Glossary of Terms for Computer-Aided Manufacturing (CAM): Terms of Classification and Coding System and Theory for Group Technology (Jul)(11258-4).
CNS B8003-5-89. Glossary of Terms for Computer-Aided Manufacturing (CAM): Terms of Function and Application for Group Technology (Jul)(11258-5).
CNS B8003-6-89. Glossary of Terms for Computer-Aided Manufacturing (CAM): Terms of Component and Layout for Flexible Manufacturing System (Jul)(11258-6).
CNS B8003-7-89. Glossary of Terms for Computer-Aided Manufacturing (CAM): Terms of Application of Automation for Flexible Manufacturing (Jul)(11258-7).

—Information Interchange—Aerospace
EURO CTI/80/10832-87. Recommendation for Geometry Data Exchange. 48 pp.
EURO CTI/85/15617-85. Report of Geometry Data Exchange Study Group. 88 pp.

—Warehouses—Glossaries
CNS B8003-89. Glossary of Terms for Computer-Aided Manufacturing (CAM): Terms of Automated Warehouse System (General) (Jul)(11258).
CNS B8003-1-89. Glossary of Terms for Computer-Aided Manufacturing (CAM): Terms of Equipment for Automated Warehouse System (Jul)(11258-1).
CNS B8003-2-89. Glossary of Terms for Computer-Aided Manufacturing (CAM): Terms of Function for Automated Warehouse System (Jul)(11258-2).
CNS B8003-3-89. Glossary of Terms for Computer-Aided Manufacturing (CAM): Terms of Layout and Scheduling for Automated Warehouse System (Jul)(11258-3).

CAMAC
Use For: Computer Aided Measurement and Control; Computer Automated Measurement and Control
See Also: Computers; Interfaces
BSI BS 5554-78. 1978 Guide to a Modular Instrumentaton Data Handling: CAMAC System. 65 pp.
CENELEC HD 357 S2-87. A modular Instrumentation System for Data Handling; CAMAC System. 3 pp.
CENELEC HD 417 S2-87. Camac-Serial Highway Interface System. 3 pp.
IEC 516-75. Modular Instrumentation System for Data Handling; CAMAC System First Edition; (Amendment 1-1984). 79 pp.
IEC 640-79. CAMAC—Serial Highway Interface System First Edition; (Amendment 1-1984). 265 pp.

—BASIC
IEC 775-83. Real-Time BASIC for CAMAC First Edition. 41 pp.

—Block Transfers
CENELEC HD 431-83. Block Transfer in Camac Systems. 2 pp.
IEC 677-80. Block Transfers in CAMAC Systems First Edition. 43 pp.

CAMAC (Cont.)
—Crate Controllers
BSI BS 5836-80. 1980 Guide to CAMAC: Organization of Multi Crate Systems: Specification of the Branch-Highway and CAMAC Crate Controller Type A1. 43 pp.
CENELEC HD 374 S2-86. CAMac—Organization of Multi-Crate Systems. Specification of the Branch-Highway and CAMAC Crate Controller Type A1. 2 pp.
CENELEC HD 453-84. Multiple Controllers in a Camac Crate. 2 pp.
IEC 552-77. CAMAC—Organization of Multi-Crate Systems Specification of the Branch-Highway and CAMAC Crate Controller Type A1 First Edition; (Amendment 1-1984). 85 pp.
IEC 729-82. Multiple Controllers in a CAMAC Crate First Edition. 69 pp.

—Glossaries
CENELEC HD 432-83. Definitions of Camac Terms Used in IEC Publications. 2 pp.
IEC 678-80. Definitions of CAMAC Terms Used in IEC Publications First Edition. 37 pp.

—Subroutines
CENELEC HD 445-83. Subroutines for Camac. 2 pp.
IEC 713-81. Subroutines for CAMAC First Edition. 61 pp.

Camelback
Use: Retreading—Rubber—Compounded

Camera Shutters
See Also: Cameras
CNS Z9056-81. Shutter Cable Release Tips and Sockets for Cameras (Mar)(7096).
CNS Z9061-81. Focal-Plane Shutters (May)(7474).
ISO 6053-79. Photography—Shutter Cable Release Tip and Socket—Dimensions First Edition. 4 pp.
JIS B 7104-92. Shutter Cable Release Tips and Sockets for Cameras. 7 pp.

—Still Cameras
JIS B 7091-92. Shutters for Still Cameras. 17 pp.

—Timing
CNS Z9002-80. Still Camera Exposure Markings (Jan)(5143).
ISO 516-86. Photography—Camera Shutters—Timing Second Edition. 14 pp.

Camera Tubes
Use For: Pickup Tubes See Also: Electron Tubes; Thermionic Tubes; Vidicons
CECC CECC 13 001 ISSUE 1-80. Blank Detail Specification: Camera Tubes (En, Fr, Ge). 22 pp.
CECC EN 113 000-91. Generic Specification: Camera Tubes (Supersedes CECC 13 000 Issue 1: 1980). 18 pp.
MOD UK DSTAN 59-60: Part 10-01. Valves, Electronic (Electronic Tubes) (Listed on EPIC Database) Part 10: Camera Tubes Issue 1; Amendment 1. 14 pp.

—Electrical Measurement
IEC 151 Pt 26-71. Measurements of the Electrical Properties of Electronic Tubes Part 26: Methods of Measurement for Camera Tubes First Edition. 43 pp.

—Quality Assurance
BSI BS CECC 13000-81. (WITHDRAWN) 1981 Camera Tubes: Generic Specification (Renumbered as BS EN 113000: 1993). 20 pp.
BSI BS CECC 13001-81. 1981 Camera Tubes: Blank Detail Specification. 12 pp.
BSI BS EN 113000-93. 1993 Generic Specification: Camera Tubes (AMD 7573) February 15, 1993 (T). 24 pp.
CENELEC EN 113000-91. Generic Specification: Camera Tubes. 18 pp.

Cameras
See Also: Camera Shutters; Closed Circuit Television Cameras; Focal Length; Microfilm Cameras; Motion Picture Cameras; Ophthalmic Cameras; Optical Equipment; Process Cameras; Scintillation Cameras; Still Cameras; Television Cameras
CSA CAN/CSA-C22. 2 NO 1-M90. Radio, Television, and Electronic Apparatus; (Gen Instr 1 Thru 3). 146 pp.
CSA 1423A Bull. Electrical Bulletin 1423A January 2, 1985 to C22.2 NO 1. 8 pp.
CSA 1434 Bull. Electrical Bulletin 1434 April 1, 1985 to C22.2 NO 1. 2 pp.

—Accessory Shoes
CNS Z9007-80. Camera Accessory Shoes, with and Without Electrical Contacts, for Photoflash Lamps and Electronic Photoflash Units (Jan)(5148).

INDUSTRY STANDARDS

INTERNATIONAL AND NON-U.S. NATIONAL STANDARDS
SUBJECT INDEX

Cameras

Cameras (Cont.)
- **Accessory Shoes (Cont.)**
  ISO 518-77. Photography—Camera Accessory Shoes, with and Without Electrical Contacts, for Photoflash Lamps and Electronic Photoflash Units First Edition. 4 pp.
  JIS B 7101-75. Camera Accessory Shoes and Feet (R 1983). 6 pp.
- **Aerial Films**
  NATO STANAG 3178 ED 5 AMD 1-85. Rolled Air Film and Air Framing Camera Standard Image Format Sizes. 9 pp.
  NATO STANAG 3178 ED 5 AMD 2-85. Rolled Air Film and Air Framing Camera Standard Image Format Sizes. 7 pp.
  NATO STANAG 3179 ED 6 AMD 0-90. 16 mm and 35 mm Roll Film for Aerial Cameras. 6 pp.
- **Automatic Exposure Controls**
  ISO 2721-82. Photography—Cameras—Automatic Controls of Exposure First Edition; (ANSI PH3.301-1990). 10 pp.
- **Back Window Position**
  CNS Z9047-88. Roll Film Camera-Back Window Location and Picture Sizes (Jun)(5641).
  ISO 1203-84. Photography—Roll Film Cameras—Back Window Location and Picture Sizes Third Edition; (ANSI PH3.100-1992). 4 pp.
- **Film Magazines**
  JIS B 7176-69. Dimensions for 16 mm Film Magazines for Still Picture Cameras (R 1983). 6 pp.
- **Lens Aperture Markings**
  CNS Z9003-80. Still Camera Aperture Markings (Jan)(5144).
  ISO 517-73. Photography—Still Cameras—Lens Aperture Markings First Edition. 3 pp.
- **Light Exposure Measurement**
  JIS B 7092-73. Photoelectric Methods for Measuring Light Exposure in Focal Planes of Cameras. 10 pp.
- **Packaging—Export**
  CNS Z5068-80. Packaging and Packing of Cameras for Export (Dec)(6726).
- **Photometric Properties**
  ISO 2721-82. Photography—Cameras—Automatic Controls of Exposure First Edition; (ANSI PH3.301-1990). 10 pp.
- **Picture Apertures**
  CNS Z9028-80. Stereo System Using 35mm Objectives on 35mm Film, Five-Perforation Format — Specification (Jan)(5169).
  ISO 515-73. Photography—Stereo Systems Using 35 mm Objectives on 35 mm Film, Five-Perforation Format—Specifications First Edition. 6 pp.
- **Picture Dimensions**
  CNS Z9038-88. Picture Size of Cameras Using 35mm and 16mm Film (Jun)(5262).
  CNS Z9047-88. Roll Film Camera-Back Window Location and Picture Sizes (Jun)(5641).
  ISO 1203-84. Photography—Roll Film Cameras—Back Window Location and Picture Sizes Third Edition; (ANSI PH3.100-1992). 4 pp.
  ISO 1754-86. Photography—Cameras Using 35 mm and 16 mm Film—Picture Sizes Third Edition. 3 pp.
- **Self Timers**
  CNS Z9057-81. Self-Timers for Cameras (Mar)(7097).
  JIS B 7116-76. Self-Timers for Camera. 4 pp.
- **Tripod—Connectors**
  CNS Z9001-80. Photographic Connections (Jan)(5142).
  ISO 1222-87. Photography—Tripod Connections Second Edition. 4 pp.
  JIS B 7103-75. Tripod Connections for Cameras (R 1983). 5 pp.

Cameron DG 14 Balloons
See Also: Balloons
- **Data Sheets**
  CAA BAS5. Cameron Balloons Ltd DG-14. 2 pp.

Cameron Skystar Balloons
See Also: Balloons
- **Data Sheets**
  CAA BAS3 ISSUE 2. Cameron Balloons Ltd Skystar. 2 pp.

Camouflage
- **Aircraft**
  NATO STANAG 3687 ED 2 AMD 1-85. Camouflage of Aircraft. 5 pp.
  NATO STANAG 3687 ED 3 AMD 1-90. Camouflage of Aircraft. 5 pp.

Camouflage (Cont.)
- **Aircraft (Cont.)**
  NATO STANAG 3687 ED 3 AMD 2-90. Camouflage of Aircraft. 6 pp.
- **Aircraft Landing Areas**
  NATO STANAG 3111 ED 1 AMD 2-86. Airfield Marking Tone Down. 16 pp.
  NATO STANAG 3111 ED 1 AMD 3-86. Airfield Marking Tone Down. 17 pp.

Camouflage Coatings
Use For: Camouflage Paints *See Also:* Coatings
  MOD UK DSTAN 80-122-89. Powder Coating High Durability, Stoving Types: Near Matt Semi-Gloss and Glossy Issue 1. 17 pp.
  MOD UK DSTAN 80-122-01. Powder Coating High Durability, Stoving Types: Near Matt Semi-Gloss and Glossy Issue 1; Amendment 1. 19 pp.
  MOD UK DSTAN 80-128-01. Paint, Finishing, Temporary Camouflage Alkali-Removable Issue 1; Amendment 1. 26 pp.
- **Canvas**
  MOD UK DSTAN 80-125-01. Paint Finishing Emulsion for Canvas Type: Brushing Issue 1; Amendment 1. 21 pp.
- **Marine—Low Luster**
  MOD UK DSTAN 80-111-85. Paint Finishing, Low Sheen, Black Brushing Issue 1. 11 pp.
- **Military Vehicles**
  MOD UK DSTAN 00-23-80. NATO Infra Red Reflective (IRR) Green Colour for Painting Military Equipment Issue 1. 9 pp.
  MOD UK DSTAN 80-42-82. Paint System, Defence Equipment, IRR, Matt, Stoving Types: Spraying Dipping Issue 1. 19 pp.
- **Military Vehicles—Removable**
  NATO STANAG 2836 ED 2 AMD 2-76. Removable Paints for Camouflage. 9 pp.
  NATO STANAG 2836 ED 2 AMD 3-88. Removable Paints for Camouflage. 10 pp.
- **Paints—Temporary**
  MOD UK DSTAN 80-120-87. Paint, Finishing, Temporary Snow Camouflage, Brushing Issue 1. 23 pp.
  MOD UK DSTAN 80-120-01. Paint, Finishing, Temporary Snow Camouflage, Brushing Issue 1; Amendment 1. 25 pp.
- **Submarines**
  MOD UK DSTAN 80-111-85. Paint Finishing, Low Sheen, Black Brushing Issue 1. 11 pp.

Camouflage Paints
Use: Camouflage Coatings

Camp Beds
Use: Cots—Camping

Camp Stoves
See Also: Cooking Appliances; Stoves
  CSA B140.9.2-1975. Portable Pressurized Type Liquid Petroleum Fuelled Camp Stoves (R 1991); (Rev 1-3 November 1977) (Rev 4-5 August 1982). 19 pp.
- **Naphthas**
  MOD UK TS 10184A. Naphtha for Cooking Stoves.

Campgrounds
Use For: Camping Sites *See Also:* Recreational Vehicle Parks
- **Electrical Installations**
  CENELEC HD 384.7.708 S1-92. Electrical Installations of Buildings Part 7: Requirements for Special Installations or Locations Section 708: Electrical Installations in Caravan Parks and Caravans. 12 pp.
  DIN VDE 0100 Pt 708 (D)-90. Erection of Power Installations with Nominal Voltages up to 1000 V; Electrical Installations in Caravan Parks and Caravans; German Version prHD 384.7.708 (1989) (Mar). 27 pp.
  DIN VDE 0100 Pt 721-84. Erection of Power Installations with Rated Voltages up to 1000 V; Caravans, Boats and Yachts, as Well as Power Supply Thereof at Camping Sites and Berths (Apr). 10 pp.
  IEC 364 Pt 7-708-88. Electrical Installations of Buildings Part 7: Requirements for Special Installations or Locations—Section 708: Electrical Installations in Caravan Parks and Caravans First Edition; (Amendment 1-1993). 34 pp.

Camphor
See Also: Ketones
  CNS K1027-57. Camphor (Jul)(375).

Camphor (Cont.)
  CNS K1136-69. Crude Camphor (Tentative) (Jan)(3021)(R 1973).
  CNS K6279-69. Method of Test for Crude Camphor (Jan)(3022)(R 1973).
- **Acidity**
  MOD UK M 9555/72. Examination of Camphor for Celluloid (Superseded by CS 2914).
- **Chloride Content—Volumetric Analysis**
  MOD UK M 9555/72. Examination of Camphor for Celluloid (Superseded by CS 2914).
- **Grit Content**
  MOD UK M 9555/72. Examination of Camphor for Celluloid (Superseded by CS 2914).
- **Halogen Content**
  CNS K6064-57. Method of Test for Natural Camphor (Jul)(818)(R 1973). 2 pp.
- **Melting Points**
  CNS K6064-57. Method of Test for Natural Camphor (Jul)(818)(R 1973). 2 pp.
  MOD UK M 9555/72. Examination of Camphor for Celluloid (Superseded by CS 2914).
- **Moisture Content**
  CNS K6064-57. Method of Test for Natural Camphor (Jul)(818)(R 1973). 2 pp.
- **Nitrocellulose**
  JIS K 6702-77. Testing Method for Camphor Used in Celluloid Industry.
  MOD UK DSTAN 68-110-89. Camphor for Celluloid Issue 1. 14 pp.
- **Nitrocellulose Content**
  MOD UK CS 2914B. Camphor for Celluloid (Superseded by Def Stan 68-110).
- **Nonvolatile Matter Content**
  CNS K6064-57. Method of Test for Natural Camphor (Jul)(818)(R 1973). 2 pp.
- **Solubility**
  CNS K6064-57. Method of Test for Natural Camphor (Jul)(818)(R 1973). 2 pp.
- **Sulfate Content**
  MOD UK M 9555/72. Examination of Camphor for Celluloid (Superseded by CS 2914).
- **Visual Inspection**
  MOD UK M 9555/72. Examination of Camphor for Celluloid (Superseded by CS 2914).

Camphor Oil
See Also: Oils
  CNS K5038-65. Camphor White Oil (Jan)(2405)(R 1971).
  CNS K5058-69. Crude Camphor Oil (Tentative) (Jan)(3023) (R 1973). 1 p.
  CNS K5059-69. Fragrant Camphor Oil (Tentative) (Jan)(3025) (R 1973). 1 p.
  CNS K6192-65. Method of Test for White Camphor Oil (Apr)(2406) (R 1973).
  CNS K6280-69. Method of Test for Camphor Oil (Jul)(3024)(R 1973).

alpha-Camphoric Acid
  CNS K7131-64. Chemical Reagent (Alpha Camphoric Acid) (Jul)(1631).

Camping Sites
Use: Campgrounds

Campylobacter
See Also: Bacteria
- **Food**
  SAA AS 1766.2.13-91. Method for the Microbiological Examination of Food—Part 2: Examination for Specific Organisms—Part 2.13: Campylobacter. 8 pp.

Camshafts
See Also: Engine Valves; Internal Combustion Engines; Shafts (Machine Elements)
- **Internal Combustion Piston Engines—Glossaries**
  BSI BS 7016: Part 3-88. 1988 Components and Systems of Reciprocating Internal Combustion Engines Part 3: Glossary of Terms of Valves, Camshaft Drive and Actuating Mechanisms. 19 pp.
  ISO 7967 Pt 3-87. Reciprocating Internal Combustion Engines—Vocabulary of Components and Systems—Part 3: Valves, Camshaft Drive and Actuating Mechanisms First Edition. 19 pp.

CAN (ICAO)
Use: ICAO Committee on Aircraft Noise

INTERNATIONAL AND NON-U.S. NATIONAL STANDARDS
SUBJECT INDEX

Canned

Canada Balsam
Use: Balsam—Canada

Canadair C 4 Aircraft
See Also: Aircraft
—Accidents
CAA. Canadair C4 and DC4-M-2 (World Airline Accident Summary). 1 p.

Canadair Challenger Aircraft
See Also: Aircraft
—Accidents
CAA. Canadair Challenger. 1 p.

Canadair CL 44 Aircraft
See Also: Aircraft
—Accidents
CAA. Canadair CL 44 Series (World Airline Accident Summary). 1 p.
—Antenna Positions
CAA. Canadair CL44 (Approved Aerial Positions). 1 p.
—Data Sheets
CAA FA3 ISSUE 1. Canadair CL44 Series, Model CL44 D4. 2 pp.

Canadair CL 600 Series Aircraft
See Also: Aircraft
—Foreign Airworthiness Directives
CAA. Canadair CL-600 Series Aircraft. 8 pp.

Canadair CL 600-2B16 Challenger 601-3A Aircraft
See Also: Aircraft
—Certification
CAA. Canadair CL-600-2B16 Challenger 601-3A. 19 pp.
—Data Sheets
CAA FA48 ISSUE 1. Canadair Challenger, Model CL-600-2B16, Variant 601-3A. 2 pp.

Canadair DC 4-M-2 Aircraft
See Also: Aircraft
—Accidents
CAA. Canadair C4 and DC4-M-2 (World Airline Accident Summary). 1 p.

Canadian Content
Use: Procurement—Canadian Content

Canadian Federal Identity Program
—Signs—Aluminum
CGSB 109.1M-89. Signage System, Extruded Aluminum, Federal Identity Program. 19 pp.
—Symbols—Film—Pressure Sensitive Adhesive
CGSB 109.2M-89. Letters and Symbols, Die-Cut Film, Federal Identity Program. 11 pp.

Canadian General Standards Board
Use: CGSB

Canadian Pulp and Paper Association
Use: CPPA

Canadian Standards Association
Use: CSA

Canadian Trade Document Alignment System
Use For: CTDAS
CGSB CAN/CGSB-200.3-M82. Canadian Trade-Document Alignment System—Layout Key and Master. 20 pp.

Cananga Oil
Use: Ilang-Ilang Oil

Candair CL-44 Aircraft
See Also: Aircraft
—Accidents
CAA. Candair CL 44 Series. 1 p.

Candelilla Wax
See Also: Waxes
CNS K1267-81. Candelilla Wax (Oct)(8016).
CNS K6696-81. Method of Test for Candelilla Wax (Oct)(8017).

Candelilla Wax *(Cont.)*
—Propellants
MOD UK DSTAN 91-29-77. Candelilla Wax Issue 1. 8 pp.

Candles
CNS S1017-59. Wax Candle (Sep)(1065)(R 1973).
JIS K 3910-58. Candle.
MOD UK DSTAN 62-3-76. Candles, Illuminating (General Purpose and Arctic) Issue 2. 12 pp.
—Arctic
MOD UK DSTAN 62-3-76. Candles, Illuminating (General Purpose and Arctic) Issue 2. 12 pp.
—Wax
CNS S2017-68. Method of Test for Wax Candles (Apr)(1066) (R 1973).

Candy
See Also: Sugars
CNS N5155-85. Candy (Sep)(4960). 3 pp.
CNS N6206-87. Method of Test for Candy (Mar)(11884). 1 p.

Cane Sugar
See Also: Sugarcane; Sugars
—Ash Content
CNS N6026-79. Method of Test for Cane Sugar (Jan)(1338). 7 pp.
—Glucose Content
CNS N6026-79. Method of Test for Cane Sugar (Jan)(1338). 7 pp.
—Moisture Content
CNS N6026-79. Method of Test for Cane Sugar (Jan)(1338). 7 pp.
—Sucrose Content
CNS N6026-79. Method of Test for Cane Sugar (Jan)(1338). 7 pp.
—Sugar Content
CNS N6026-79. Method of Test for Cane Sugar (Jan)(1338). 7 pp.

Canes
Use For: Tetraped Walking Sticks; Tripod Walking Sticks; Walking Sticks *See Also:* Disability Equipment; Medical Equipment
SNZ NZS 5831: Part 1-89. Walking Aids for the Disabled Part 1: Specification for Walking Sticks. 12 pp.
—Metal
BSI BS 4922-90. 1990 Metal Tripod and Tetrapod Walking Sticks. 9 pp.
BSI BS 5205-90. 1990 Adjustable Metal Walking Sticks. 8 pp.
—Wooden
BSI BS 5181-75. 1975 Wooden Walking Sticks. 8 pp.

Canisters (Containers)
Use: Cans; Cylinders (Containers)

Canned Foods
See Also: Food
CNS N6011-87. Method of Test for Canned Food: General Rules (Jan)(967). 2 pp.
—Baby Foods
CNS N5206-84. Canned Baby Food (Mar)(10838). 4 pp.
—Catsup
CNS N5142-89. Canned Tomato Catsup (Feb)(4681). 4 pp.
—Chinese
CNS N5132-89. Canned Chinese Foods (Dec)(4149). 5 pp.
—Concentrates
CNS N5017-91. Canned Tomato Puree and Paste (Sep)(1251). 6 pp.
—Containers
CNS Z5011-88. Corrugated Paperboard Containers for Canned Food (Dec)(1161). 3 pp.
—Fish
CGSB 32.143M-91. Canned Fish. 11 pp.
CNS N5015-89. Canned Fished (Feb)(1229). 5 pp.
CNS N5138-91. Canned Tuna and Bonito (Feb)(4456). 5 pp.
—Fish—Chemical Analysis
CNS N6104-78. Method of Determination of Bleached Dark Meat in Canned Fish (Jul)(4455). 1 p.

Canned Foods *(Cont.)*
—Fruits
CNS N5011-82. Canned Pineapples (Jan)(822). 12 pp.
CNS N5018-87. Canned Fruits (Sep)(1252). 8 pp.
CNS N5056-82. Canned Mandarin Oranges (for Export)(Tentative) (Jan)(2341). 6 pp.
CNS N5161-85. Canned Apples (May)(5930). 6 pp.
—Fruits—Physical Properties
CNS N6021-59. Method of Test for Canned Pineapple Examination of Slice Dimensions (Apr)(977). 1 p.
—Ham
CGSB 32.65M-90. Canned or Boil-in-Bag Ham. 7 pp.
—Impurities Content
CNS N6023-78. Method of Test for Canned Food (Inspection for Purity) (Jul)(979). 1 p.
—Inspection—Grading/Scoring
CNS N6024-78. Method of Test for Canned Food (Grading and Scoring) (Jul)(980). 1 p.
—Jams
CNS N5154-79. Jams, Packaged (Jun)(4892). 3 pp.
—Meat
CNS N5014-89. Canned Meats (Feb)(1228). 5 pp.
—Mollusks
CNS N5135-85. Canned Snail Meat (Nov)(4337). 3 pp.
—Packaging
CNS N6013-83. Method of Test for Canned Food: Appearance of Can (Nov)(969). 2 pp.
CNS N6015-78. Method of Test for Canned Food (Measurement for Degree of Vacuum) (Jul)(971). 2 pp.
CNS N6017-78. Method of Test for Canned Food (Inspection for Interior of Can) (Jul)(973). 1 p.
—Packaging—Inspection
CNS N6012-82. Method of Test for Canned Food (Packaging Inspection) (May)(968). 1 p.
—Peanut Butter
CNS N5131-85. Canned Peanut Butter (May)(4148). 3 pp.
—Physical Properties
CNS N6018-87. Method of Test for Canned Food: Determination of Net Weight and Drain Weight (Feb)(974). 2 pp.
—Sampling
CNS N6014-87. Method of Test for Canned Food (Sampling) (Sep)(970). 5 pp.
—Sensory Analysis
CNS N6022-87. Method of Test for Canned Food: Test for Flavor and Odor (Feb)(978). 2 pp.
—Shellfish
CGSB 32.143M-91. Canned Fish. 11 pp.
CNS N5109-85. Canned Crab Meat (Jan)(3526). 3 pp.
CNS N5115-85. Canned Shrimp (May)(3756). 2 pp.
—Spices
CNS N5021-82. Canned Ginger (Jan)(1255). 3 pp.
—Tin Content
CNS N6093-75. Method of Test for Metal in Canned Food (Determination of Tin Content) (Jul)(3791). 2 pp.
—Tomatoes
CNS N5017-91. Canned Tomato Puree and Paste (Sep)(1251). 6 pp.
—Vegetables
CGSB 32.253M-89. Canned Fruits or Vegetables and Canned or Refrigerated Juices (Supersedes 32-GP-259A Nov 1967, 32-GP-277A March 1968, 32-GP-280C April 1971). 11 pp.
CNS N5016-85. Canned Mushroom (Nov)(1250). 9 pp.
CNS N5019-89. Canned Bamboo Shoots (Feb)(1253). 6 pp.
CNS N5020-89. Canned Pickled Vegetables (Feb)(1254). 4 pp.
CNS N5022-82. Canned Water Chestnut (Jan)(1256). 3 pp.
CNS N5023-86. Canned Beans (Sep)(1257). 3 pp.
CNS N5046-85. Canned Straw Mushrooms (Nov)(2175). 3 pp.
CNS N5063-89. Canned Asparagus (Jan)(2348). 9 pp.
CNS N5074-83. Canned Vegetables (Dec)(2539). 2 pp.
CNS N5090-80. Canned White Agaric (Aug)(2993). 2 pp.
CNS N5096-87. Canned Baby Corn or Young Corn (Sep)(3184). 4 pp.
CNS N5097-82. Canned Sweet Potatoes (Jan)(3185). 5 pp.
CNS N5100-85. Canned Bean Sprout (Nov)(3314). 2 pp.
CNS N5108-86. Canned Tomatoes (Jan)(3497). 8 pp.

INDUSTRY STANDARDS

Canned Foods (Cont.)

—Vegetables (Cont.)
CNS N5116-87. Canned Oyster Mushroom (Sep)(3757). 3 pp.
CNS N5119-87. Canned Abalone Mushroom (Sep)(3898). 3 pp.
CNS N5127-77. Canned Salsify (Feb)(4092). 4 pp.
CNS N5129-82. Canned Okra (Jan)(4146). 2 pp.
CNS N5220-86. Canned Sweet Corn (Sep)(11712). 3 pp.

Canning
See Also: Cans

—Jars—Caps (Lids)
CGSB CAN/CGSB-143.1-M85. Lids, Home Canning Jar. 8 pp.

Canning Liquids
Use For: Packing Liquids See Also: Liquids

—Acidity—Volumetric Analysis
CNS N6019-87. Method of Test for Canned Food: Determination of Packing Liquid (Feb)(975). 3 pp.

—Oil Content
CNS N6019-87. Method of Test for Canned Food: Determination of Packing Liquid (Feb)(975). 3 pp.

—Salt Content
CNS N6019-87. Method of Test for Canned Food: Determination of Packing Liquid (Feb)(975). 3 pp.

—Sucrose Content—Brix Scale
CNS N6019-87. Method of Test for Canned Food: Determination of Packing Liquid (Feb)(975). 3 pp.

Cannons
Use: Guns

Cannulae
Use: Cannulas

Cannulas
See Also: Catheters; Medical Equipment; Surgical Instruments; Trocars
BSI BS 3348-61. (OBSOLESCENT) 1961 Antral Trocars and Cannulae. 8 pp.

—Intravenous
BSI BS 4843-87. 1987 Sterile Intravenous Cannulae for Single Use. 13 pp.

—Sinus
CNS T1014-81. Cannula for Maxillary Antrum and Frontal Sinus (Jul)(6273).
JIS T 3204-53. Cannula for Maxillary Antrum and Frontal Sinus.

Canoes
See Also: Boats (Marine); Rafts

—Construction
BSI BS MA 91: Part 1-81. 1981 Canoes Part 1: Code of Practice for Canoe Construction. 5 pp.

—Safety
BSI BS MA 91: Part 2-81. 1981 Canoes Part 2: Specification for Safety Features of Canoes. 4 pp.

Canola Meal
See Also: Canola Oil
CGSB CAN/CGSB-32.301-M87. Canola Meal. 8 pp.

Canola Oil
See Also: Canola Meal; Oils
CGSB CAN/CGSB-32.300-M87. Canola Oil, Crude, Degummed and Refined. 11 pp.

Canopies (Aircraft)
See Also: Cockpits; Windshields

—Winding Gear—Manual Controls
SBAC AS 3197 ISSUE 3. Knob (Canopy Winding Gear). 1 p.

Cans
Scope Note: For additional listings, use a more specific term Use For: Canisters (Containers); Tins (Containers) See Also: Bottles; Canning; Containers; Drums (Containers); Shipping Containers; Sliver Cans; Waste Containers

—Aluminum
BSI BS 1133: SUBSEC 10.1-90. 1990 Packaging Code Section 10: Metal Containers Subsection 10.1: Tins and Cans. 20 pp.

—Beverages
JIS Z 1571-85. Hermetically Sealed Metal Cans for Food and Drink. 10 pp.

—Beverages—Capacity Measurement
BSI BS 5774-84. 1984 Internal Diameters of Round Hermetically Sealed Metal Cans for Food and Non-Carbonated Drinks. 4 pp.
BSI BS 6992: Part 3-88. (OBSOLESCENT) 1988 Capacities and Related Cross Sections of Open-Top Cans for Food and Drinks Part 3: Recommendations for Cans for Drinks. 7 pp.

—Capacity Measurement
BSI BS 1764-73. 1973 Nominal Diameters of Round Built-up Tins. 8 pp.
BSI BS 6966: Part 1-88. 1988 Light Gauge Metal Containers Part 1: Glossary of Terms for Open-Top Cans and Methods for Determination of Dimensions and Capacities (Renumbered as BS EN 20090-1: 1993). 18 pp.
BSI BS 6966: Part 2-88. (WITHDRAWN) 1988 Light Gauge Metal Containers Part 2: Glossary of Terms for General Use Containers and Methods for Determination of Dimensions and Capacities (Renumbered as BS EN 20090-2: 1993). 19 pp.
BSI BS EN 20090-1-93. 1993 Amd 1 Light Gauge Metal Containers. Definitions and Determination Methods for Dimensions and Capacities Part 1: Open-Top Cans (ISO 90-1: 1986) (AMD 7478) March 15, 1993 (H). 24 pp.
BSI BS EN 20090-2-93. 1993 Amd 1 Light Gauge Metal Containers. Definitions and Determination Methods for Dimensions and Capacities Part 2: General Use Containers (ISO 90-2: 1986) (AMD 7479) March 15, 1993 (H). 25 pp.
CEN EN 20090-1-92. Light Gauge Metal Containers—Definitions and Determination Methods for Dimensions and Capacities—Part 1: Open-Top Cans (ISO 90-1: 1986). 16 pp.
CEN EN 20090-2-92. Light Gauge Metal Containers—Definitions and Determination Methods for Dimensions and Capacities—Part 2: General Use Containers (ISO 90-2: 1986). 17 pp.
ISO 90 Pt 1-86. Light Gauge Metal Containers—Definitions and Determination Methods for Dimensions and Capacities—Part 1: Open-Top Cans First Edition; (CEN EN 20090-1: 1992). 16 pp.
ISO 90 Pt 2-86. Light Gauge Metal Containers—Definitions and Determination Methods for Dimensions and Capacities—Part 2: General Use Containers First Edition; (CEN EN 20090-2: 1992). 17 pp.
ISO TR10193-89. Round General Use Light Gauge Metal Containers—Nominal Filling Volumes and Nominal Diameters First Edition. 15 pp.
ISO TR10194-89. Non-Round General Use Light Gauge Metal Containers—Nominal Filling Volumes and Nominal Cross-Sections First Edition. 15 pp.
ISO 10653-93. Light-Gauge Metal Containers—Round Open-Top Cans—Cans Defined by Their Nominal Gross Lidded Capacities First Edition; (Replaces 3004 Pt 1 Thru Pt 6). 7 pp.
ISO TR11761-92. Light-Gauge Metal Containers—Round Open-Top Cans—Classification of Can Sizes by Construction Type First Edition; (Replaces 3004 Pt 1 Thru Pt 6). 9 pp.
ISO TR11762-92. Light-Gauge Metal Containers—Round Open-Top Cans for Liquid Products with Added Gas—Classification of Can Sizes by Construction Type First Edition; (Replaces 3004 Pt 1 Thru Pt 6). 7 pp.
SAA AS 2354-91. Open Top Metal Cans—Definitions, and Determination Methods for Dimensions and Capacities (ISO 90-1:1986). 16 pp.

—Carbonated Beverages—Nominal Filling Volumes
ISO 10654-93. Light-Gauge Metal Containers—Round Open-Top Cans—Cans for Liquid Products with Added Gas, Defined by Their Nominal Filling Volumes First Edition; (Replaces 3004 Pt 1 Thru Pt 6). 6 pp.

—Diameters
ISO TR10193-89. Round General Use Light Gauge Metal Containers—Nominal Filling Volumes and Nominal Diameters First Edition. 15 pp.
ISO TR11761-92. Light-Gauge Metal Containers—Round Open-Top Cans—Classification of Can Sizes by Construction Type First Edition; (Replaces 3004 Pt 1 Thru Pt 6). 9 pp.

—Fiber—Spinning
CNS L2076-83. Fiber Cans (Apr)(10174).
JIS L 5133-83. Fibre Cans.

—Fish—Metal—Capacity Measurement
BSI BS 6992: Part 4-89. (OBSOLESCENT) 1989 Capacities and Related Cross Sections of Open-Top Cans for Food and Drinks Part 4: Recommendations for Cans for Fish and Other Fishery Products. 8 pp.

—Food
CNS Z5007-88. Round Metal Cans for Foods (Jan)(827). 7 pp.
CNS Z5018-85. Metal Cans of Irregular Shape for Foods (Dec)(2180). 4 pp.
CNS Z6019-85. Method of Test for Rounded Metal Cans for Foods (Nov)(4060). 6 pp.
ISO 1361-83. Light Gauge Metal Containers—Open-Top Cans—Round Cans—Internal Diameters Third Edition. 4 pp.
JIS Z 1571-85. Hermetically Sealed Metal Cans for Food and Drink. 10 pp.

—Food—Appearance
CNS N6013-83. Method of Test for Canned Food: Appearance of Can (Nov)(969). 2 pp.

—Food—Capacity Measurement
BSI BS 5771-81. 1981 Packages for Certain Pre-Packed Foodstuffs; Capacities of Glass and Metal Containers. 9 pp.
BSI BS 5774-84. 1984 Internal Diameters of Round Hermetically Sealed Metal Cans for Food and Non-Carbonated Drinks. 4 pp.
BSI BS 6992: Part 1-88. (OBSOLESCENT) 1988 Capacities and Related Cross Sections of Open-Top Cans for Food and Drinks Part 1: Recommendations for Cans for General Food. 7 pp.

—Food—Inspection—Grading/Scoring
CNS N6024-78. Method of Test for Canned Food (Grading and Scoring) (Jul)(980). 1 p.

—Food—Interior Inspection
CNS N6017-78. Method of Test for Canned Food (Inspection for Interior of Can) (Jul)(973). 1 p.

—Food—Leakage
CNS N6205-87. Method of Test for Canned Food: Leaking Test (Feb)(11867). 2 pp.

—Food—Metal Coatings
CNS Z6071-83. Method of Test for Coating Materials of Metal Cans for Foods—Determination of the Pinhole of Lacquered Film (Sep)(10587). 1 p.

—Food—Metal Coatings (On Metal)
CNS Z5028-83. Coating Materials for Metal Cans of Foods—General (Sep)(2773). 1 p.

—Food—Metal Coatings (On Metal)—Adhesion Testing
CNS Z6069-83. Method of Test for Coating Materials of Metal Cans for Foods—Adhesive Test for Lacquered Film (Sep)(10585). 2 pp.

—Food—Metal Coatings (On Metal)—Bend Testing
CNS Z6070-83. Method of Test for Coating Materials of Metal Cans for Foods—Impact Bent Test of Lacquered Film (Sep)(10586). 2 pp.

—Food—Metal Coatings (On Metal)—Impact Testing
CNS Z6070-83. Method of Test for Coating Materials of Metal Cans for Foods—Impact Bent Test of Lacquered Film (Sep)(10586). 2 pp.

—Food—Metal Coatings (On Metal)—Sterilization
CNS Z6073-83. Method of Test for Coating Materials of Metal Cans for Foods—Sterilization Resistance Test of Lacquered Film (Sep)(10589). 1 p.

—Food—Metal Coatings (On Metal)—Sulfur Stain Resistance
CNS Z6074-83. Method of Test for Coating Materials of Metal Cans for Foods—Sulfur Stain Resistance Test of Lacquered Film (Sep)(10590). 1 p.

—Food—Metal Coatings (On Metal)—Weight Measurement
CNS Z6072-83. Method of Test for Coating Materials of Metal Cans for Foods—Determination of Coating Weight of Lacquered Film (Sep)(10588). 1 p.

—Food—Vacuum Testing
CNS N6015-78. Method of Test for Canned Food (Measurement for Degree of Vacuum) (Jul)(971). 2 pp.

—Greases
BSI BS 2469-73. (WITHDRAWN) 1973 Round Tins for Lubricating Greases. 8 pp.

—Lids
JIS Z 1607-85. Metal Caps and Nozzles. 7 pp.

—Liquids—Capacity Measurement
BSI BS 5740-89. 1989 Oblong Flat-Top Cans for Liquid Products (H). 11 pp.

INTERNATIONAL AND NON-U.S. NATIONAL STANDARDS
SUBJECT INDEX

Cans *(Cont.)*
—Liquids—Nominal Filling Volumes
ISO 10654-93. Light-Gauge Metal Containers—Round Open-Top Cans—Cans for Liquid Products with Added Gas, Defined by Their Nominal Filling Volumes First Edition; (Replaces 3004 Pt 1 Thru Pt 6). 6 pp.

—Liquids—Steel
DIN ENGL 7274 Pt 1-81. Packaging; Steel Canisters with Nominal Volumes of 5, 10 and 20 l; Dimensions (Sept). 3 pp.

—Liquids—Steel—Safety
DIN ENGL 7274 Pt 2-81. Packaging; Steel Canisters with Nominal Volumes of 5, 10 and 20 l; Safety Requirements and Testing (Sept). 3 pp.

—Lubricating Oils
BSI BS 5614-78. 1978 Round Sealed Metal Cans for Motor Oil and Allied Products. 7 pp.

—Meat—Capacity Measurement
BSI BS 6992: Part 2-91. (OBSOLESENT) 1991 Capacities and Related Cross Sections of Open-Top Cans for Food and Drinks Part 2: Recommendations for Cans for Meat and Meat Products for Human Consumption. 10 pp.

—Metal Coatings (On Metal)
BSI BS 1262-89. 1989 Round Lever Lid Tinplate Cans, Paint Range. 16 pp.

—Milk—Capacity Measurement
BSI BS 6992: Part 5-88. (OBSOLESCENT) 1988 Capacities and Related Cross Sections for Open-Top Cans for Food and Drinks Part 5: Recommendations for Cans for Milk. 7 pp.
ISO TR8610-84. Light Gauge Metal Containers—Round Vent-Hole Cans with Soldered Ends for Milk and Milk Products—Capacities and Related Diameters First Edition. 2 pp.

—Milk—Steel
SAA AS N35-67. Tinned Steel Milk Cans and Lids. 10 pp.

—Motion Picture Film
BSI BS 5550: SUB SEC 3.12.1-79. 1979 Cinematograpy Part 3: 35 mm Film Section 3.12: Miscellaneous Subsection 3.12.1: Metal Cans for Processed 35 mm Motion Picture Film (H). 9 pp.

—Nozzles
JIS Z 1607-85. Metal Caps and Nozzles. 7 pp.

—Plastic—Buttress Threads
DIN ENGL 6063 Pt 1-89. Screw Threads Used Mainly for Plastic Containers; Dimensions of Buttress Threads (May). 2 pp.

—Sheets—Aluminum
DIN ENGL 59606-82. Wrought Aluminium and Aluminium Alloy Sheet and Strip for Cans and Sealing Caps (Nov). 6 pp.

—Sheets—Aluminum Alloys
DIN ENGL 59606-82. Wrought Aluminium and Aluminium Alloy Sheet and Strip for Cans and Sealing Caps (Nov). 6 pp.

—Shipping Containers—Hazardous Materials
CGSB CAN/CGSB-43.124-M89. Inside Metal Containers and Liners (TC-2F and TC-2N). 9 pp.

—Strips—Aluminum
DIN ENGL 59606-82. Wrought Aluminium and Aluminium Alloy Sheet and Strip for Cans and Sealing Caps (Nov). 6 pp.

—Strips—Aluminum Alloys
DIN ENGL 59606-82. Wrought Aluminium and Aluminium Alloy Sheet and Strip for Cans and Sealing Caps (Nov). 6 pp.

—Tin
JIS Z 1602-89. 18 Liter Metal Cans. 8 pp.

—Tinplate
BSI BS 1133: SUBSEC 10.1-90. 1990 Packaging Code Section 10: Metal Containers Subsection 10.1: Tins and Cans. 20 pp.

—Varnishes
BSI BS 1262-89. 1989 Round Lever Lid Tinplate Cans, Paint Range. 16 pp.

Canteens
See Also: Containers

—Aluminum
JIS S 2015-80. Aluminium Sheet Canteens.

Canteens *(Cont.)*
—Metal
CNS S2139-87. Method of Test for Metal Water Canteen (Aug)(12052).

—Plastic
CNS S1132-82. Plastic Water Canteen (Jan)(8327).
CNS S2065-89. Method of Test for Plastic Water Canteen (Aug)(8328). 3 pp.
JIS S 2044-82. Plastics Canteen. 11 pp.

Canteens (Restaurants)
Use: Restaurants

Canvas
See Also: Fabrics
CNS L3060-66. Method of Test for Canvas Band (Nov)(2685) (R 1973).
CNS L4042-66. Canvas Band (Tentative) (Nov)(2684)(R 1973).
MOD UK DSTAN 83-38-01. Cloths, Canvas, Flax (Double-End, Plain Weave) Issue 3; Amendment 1. 17 pp.
MOD UK DTD-642B-61. 14 oz. Flax Canvas. 1 p.
SAA AS 1818-76. Flax Canvas, Including Water-Bag Canvas Reconfirmed 1988. 10 pp.

—Camouflage Coatings
MOD UK DSTAN 80-125-89. Paint Finishing Emulsion for Canvas Type: Brushing Issue 1. 18 pp.
MOD UK DSTAN 80-125-01. Paint Finishing Emulsion for Canvas Type: Brushing Issue 1; Amendment 1. 21 pp.

—Cots
CNS L4054-72. Canvas for Cot (Tentative) (Oct)(3487).

—Cotton
CNS L3064-66. Methods of Test for Cotton Canvas (Nov) (2709)(R 1973). 2 pp.
CNS L4046-66. Cotton Canvas (Tentative) (Nov)(2708) (R 1973). 2 pp.
JIS L 3102-78. Cotton Canvas.
JIS L 3102-61. Cotton Canvas. 7 pp.
MOD UK DSTAN 83-27-01. Cloths, Duck, Cotton and Cloths, Canvas, Cotton Issue 2; Amendment 2. 26 pp.
MOD UK DTD-568-44. 12 1/2 oz. Cotton Canvas (Reprinted November 1955). 1 p.

—Cotton/Polyester—Aerospace
MOD UK DSTAN 83-80-91. Polyester/Cotton Core-Spun Canvas Issue 1. 14 pp.

—Cotton—Shoes (Footwear)
CNS L4055-82. Cotton Canvas for Sport Shoes (Tentative) (Sep) (3488).

—Fire Retardants
CNS Z2016-71. Fire-Proof Agents for Paper, Cloth and Canvas (Tentative) (Jul)(3078).

—Fungus Resistance Testing
MOD UK CS 2877A. Solution, Rot-Proofing and Water-Proofing (Withdrawn).

—Jute—Automotive
CNS L3133-81. Method of Test for Linen or Jute Canvas for Waterproof Roofing of Car (Jul)(7741).
CNS L4136-81. Linen or Tute Canvas for Waterproof Roofing of Car (Jul)(7740).

—Linen
CNS L3128-81. Method of Test for Linen Canvas (Jun)(7504).
CNS L4131-81. Linen Canvas (Jun)(7503).
JIS L 3403-61. Linen Canvas (R 1970). 6 pp.

—Linen—Automotive
CNS L3133-81. Method of Test for Linen or Jute Canvas for Waterproof Roofing of Car (Jul)(7741).
CNS L4136-81. Linen or Tute Canvas for Waterproof Roofing of Car (Jul)(7740).

—Mixed
CNS L3130-81. Method of Test for Mixed Canvas (Jun)(7508).
CNS L4133-81. Mixed Canvas (Jun)(7507).
JIS L 3402-92. Linen and Ramie Canvas. 7 pp.

—Polyester
CNS L4146-82. Polyester Canvas (Nov)(9627).

—Polyvinyl Alcohol—Fibers
CNS L4052-72. Polyvinyl Alcohol Fiber Canvas (Oct)(3393).

—Ramie
CNS L3129-81. Method of Test for Ramie Canvas (Jun)(7506).
CNS L4132-81. Ramie Canvas (Jun)(7505).
JIS L 3404-61. Ramie Canvas. 6 pp.

Canvas *(Cont.)*
—Ships
CNS F4006-80. Application Standard of Ships Canvas (Dec)(6824). 1 p.
JIS F 3440-80. Application of Ships' Canvas.
JIS F 3440-58. Application of Ships' Canvas (R 1964). 3 pp.

—Stretchers
CNS L4053-72. Canvas for Stretcher (Tentative) (Oct)(3486).

—Waterproofing
MOD UK CS 2877A. Solution, Rot-Proofing and Water-Proofing (Withdrawn).

Cap and Pin Insulators
See Also: Electrical Insulators; Insulator Strings; Pin Insulators
IEC 305-78. Characteristics of String Insulator Units of the Cap and Pin Type Third Edition. 12 pp.
IEC 471-77. Dimensions of Clevis and Tongue Couplings of String Insulator Units Second Edition; (Amendment 1-1980). 14 pp.
IEC 797-84. Residual Strength of String Insulator Units of Glass or Ceramic Material for Overhead Lines After Mechanical Damage of the Dielectric First Edition. 18 pp.

Cap Bolts
Use: Cap Screws

Cap Lamps
Use For: Miners' Lamps *See Also:* Helmets; Incandescent Lighting; Lamps; Mining Equipment

—Bulbs
BSI BS 535-87. 1987 Amd 1 Light Sources for Miners' Portable Electric Lamps (AMD 5863) March 31, 1989. 23 pp.
CNS C4108-83. Small Lamps Bulb (for Coal Miner Helmet Use) (Feb)(3375).
JIS C 7502-83. Bulbs for Miners' Electric Cap Lamps. 10 pp.

—Cords
JIS M 7901-87. Cabtyre Cords for Cap Lamp. 7 pp.

—Firedamp
BSI BS EN 50033-91. 1991 Construction and Testing of Miners' Caplamps in Relation to the Risk of Explosion, for Mines Susceptible to Firedamp. 13 pp.
BSI BS EN 50033-01. 1991 Amd 1 Construction and Testing of Miners' Caplamps in Relation to the Risk of Explosion, for Mines Susceptible to Firedamp (AMD 7877) July 15, 1993 (Supersedes BS 6881: 1988) (Q). 14 pp.
CENELEC EN 50 033-86. Electrical Apparatus for Potentially Explosive Atmospheres: Caplamps for Mines Susceptible to Firedamp. 7 pp.
CENELEC EN 50 033-91. Electrical Apparatus for Potentially Explosive Atmospheres Caplights for Mines Susceptible to Firedamp. 17 pp.

—Light Output
BSI BS 4945-89. 1989 Amd 1 Functional and Performance Characteristics of Miners' Caplamps (AMD 6522) December 21, 1990. 13 pp.

—Safety
BSI BS 6881-88. (WITHDRAWN) 1988 Construction and Testing of Miners' Caplamps in Relation to the Risk of Explosion, for Mines Susceptible to Firedamp (Superseded by BS EN 50033: 1991). 9 pp.
CNS M2001-71. Safety Cap Lamp for Miner (Tentative) (Oct)(2415). 7 pp.
JIS M 7607-52. Safety Cap Lamps.

—Safety—Self Service Systems
JIS M 7652-76. Self-Service System Safety Cap Lamps.

Cap Lifters
See Also: Lifting Equipment

—Aircraft—Adapters
SBAC AS 6371 ISSUE 2. Adaptor—Cap Lifter.

Cap Nuts
Use For: Crown Nuts; Dome Nuts
See Also: Hexagonal Nuts; Nuts (Fasteners)
CNS B2262-83. Castle Nuts (Crown Nuts) (Feb)(4469).
CNS B2265-83. Hexagon Cap Nuts (Feb)(4472).
CNS B2266-83. Hexagon Domed Cap Nuts (Feb)(4473).
DIN ENGL 917-87. Hexagon Cap Nuts (June). 5 pp.
DIN ENGL 1587-87. Hexagon Domed Cap Nuts (June). 5 pp.
JIS B 1183-87. Domed Cap Nuts. 13 pp.
MOD UK DSTAN 53-58-73. Nuts, Plain, Cap, Metric (Domed Nuts) Issue 1. 4 pp.

INDUSTRY STANDARDS

Cap Nuts (Cont.)

—Aircraft—Rudder Bars
SBAC AS 2224 ISSUE 4. Cap Nut (Rudder Bar). 1 p.

—Hexagonal—Prevailing Torque
DIN ENGL 986-87. Prevailing Torque Type Hexagon Domed Cap Nuts with Nonmetallic Insert (June). 4 pp.

—Locking
CNS B2194-83. Self-Locking Domed Cap Nuts (Feb)(4315).

—Reduced Shank
CNS B2312-78. Bolted Connections with Reduced Body (Cap Nuts) (Oct)(4582).
DIN ENGL 2510 Pt 6-74. Bolted Connections with Reduced Shank; Cap Nuts (Sept). 2 pp.

Cap Screws
Use For: Cap Bolts *See Also:* Machine Screws; Screws; Socket Head Screws

—Aerospace—Hexagonal Head—Socket
BSI A 287-292-82. 1982 Hexagon Socket Head Cap Screws. Metric Series. 10 pp.
CEN PREN 3303-92. Bolts, Cap Head, Hexagon Socket, Coarse Tolerance Normal Shank, Medium Length Thread, in Alloy Steel, Cadmium Plated Classification: 1 100 MPa (at Ambient Temperature) /235 Degrees C. 6 pp.

—Hexagonal Head
CNS B2358-82. Hexagon Domed Cap Bolts (Oct)(4701).

—Hexagonal Head—Socket
BSI BS 4168: Part 1-81. 1981 Amd 2 Hexagon Socket Screws and Wrench Keys; Metric Series Part 1: Hexagon Socket Head Cap Screws (AMD 5652) July 29, 1987. 11 pp.
CNS B2142-88. Hexagon Socket Head Cap Screws (M3 to M48) (Oct)(3932).
CNS B2289-81. Hexagon Socket Head Cap Screw (M1.4 to M2.5) (Apr)(4555). 3 pp.
CNS B2290-81. Hexagon Socket Head Cap Screws with Reduced Height of Head (Apr)(4556). 4 pp.
CNS B7075-76. Method of Test for Hexagon Socket Head Cap Screws (Metric) (Jun)(3933). 2 pp.
DIN ENGL 912-83. Hexagon Socket Head Cap Screws; (Modified Version of ISO 4762) (Dec). 12 pp.
DIN ENGL 3871-82. Solderless and Soldered Pipe Unions; Male Fittings (Nov). 2 pp.
ISO 4762-89. Hexagon Socket Head Cap Screws—Product Grade A Second Edition. 8 pp.
JIS B 1176-88. Hexagon Socket Head Cap Screws. 19 pp.
MOD UK DSTAN 53-25-01. Screws, Socket Head, Steel—Metric (Course Pitch, Cap Head) Issue 1; Amendment 2. 13 pp.
MOD UK DSTAN 53-67-01. Screws, Socket Head, Unified Coarse Thread (Cap Head) Issue 1; Amendment 2. 22 pp.
MOD UK DSTAN 53-79-01. Screw, Socket Head Unified Fine Thread (Cap Head) Issue 1; Amendment 1. 33 pp.
SAA AS 1420-73. ISO Metric Hexagon Socket Head Cap Screws. 29 pp.
SNZ NZS/BS 4168: Part 1-81. Hexagon Socket Screws and Wrench Keys: Metric Series Part 1: Specification for Hexagon Socket Head Cap Screws Amend: 2. 8 pp.

—Hexagonal Head—Socket—Counterbores
DIN ENGL 74 Pt 2-91. Counterbores (Holes) for Cheese Head, Pan Head and Hexagon Socket Head Cap Screws (May). 5 pp.
DIN ENGL 974 Pt 1-91. Diameters of Counterbores (Holes) for Cheese Head, Pan Head and Hexagon Socket Head Cap Screws (May). 4 pp.

—Hexagonal Head—Socket—Countersunk
DIN ENGL 7991-86. Hexagon Socket Countersunk Head Cap Screws (Jan). 6 pp.

—Hexagonal Head—Socket—Holes
BSI BS 4185: Part 13-85. 1985 Machine Tool Components Part 13: Dimensions of Counterbored Holes. 4 pp.

—Hexagonal Head—Socket—Thin
CNS B2291-81. Hexagon Socket Head Cap Screws (Shallow Head with Pilot Recess for Wrench Key) (Apr)(4557). 4 pp.
DIN ENGL 6912-85. Hexagon Socket Thin Head Cap Screws with Pilot Recess for Wrench Key (May). 7 pp.
DIN ENGL 7984-85. Hexagon Socket Thin Head Cap Screws (May). 6 pp.

CAP 10 Aircraft
See Also: Aircraft

—Antenna Positions
CAA. CAP 10, CAP 20 and CAP 231 Series (Approved Aerial Positions). 1 p.

CAP 20 Aircraft
See Also: Aircraft

—Antenna Positions
CAA. CAP 10, CAP 20 and CAP 231 Series (Approved Aerial Positions). 1 p.

CAP 231 Aircraft
See Also: Aircraft

—Antenna Positions
CAA. CAP 10, CAP 20 and CAP 231 Series (Approved Aerial Positions). 1 p.

Capacitance
See Also: Effective Capacitance; Electrical Properties

—Telephone Cables
CCITT RECMN G.613-89. Characteristics of Symmetric Cable Pairs Usable Wholly for the Transmission of Digital Systems with a Bit Rate of up to 2 Mbits—Transmission Media Characteristics (Study Group XV) 4 pp. 4 pp.

Capacitance Level Indicators
See Also: Level Indicators

—Light Hydrocarbons
ISO 8309-91. Refrigerated Light Hydrocarbon Fluids—Measurement of Liquid Levels in Tanks Containing Liquefied Gases—Electrical Capacitance Gauges First Edition. 15 pp.

—Light Hydrocarbons—Ships
ISO 8309-91. Refrigerated Light Hydrocarbon Fluids—Measurement of Liquid Levels in Tanks Containing Liquefied Gases—Electrical Capacitance Gauges First Edition. 15 pp.

Capacitance Measurement
See Also: Electrical Measurement

—Dry Reed Switches
CNS C6168-87. Method of Test for Dry Reed Switches (Capacitance) (Jul)(9228).

—Electric Switches
CNS C6085-88. Method of Test for Electromechanical Switches (Capacitance) (Dec)(6148).

—Electromechanical Components
BSI BS 5772: Part 9-93. 1993 Electromechanical Components for Electronic Equipment: Basic Testing Procedures and Measuring Methods Part 9: Miscellaneous Tests (IEC 512-9: 1992) (S). 20 pp.
BSI BS 5772: Part 9-79. 1979 Basic Testing Procedures and Measuring Methods for Electromechanical Components for Electronic Equipment Part 9: Cable Clamping Tests, Explosion Hazard Tests, r.f Resistance Test, Capacitance Tests Sheilding and Filtering. 12 pp.
IEC 512 Pt 9-92. Electromechanical Components for Electronic Equipment; Basic Testing Procedures and Measuring Methods Part 9: Miscellaneous Tests Second Edition. 42 pp.

—Quartz Crystal Units
BSI BS 7681: Part 2-93. 1993 Measurement of Quartz Crystal Unit Parameters by Zero Phase Technique in a pi-Network Part 2: Phase Offset Method for Measurement of Motional Capacitance of Quartz Crystal Units (IEC 444-2: 1980) (S). 14 pp.
BSI BS 7681: Part 3-93. 1993 Measurement of Quartz Crystal Unit Parameters by Zero Phase Technique in a Pi-Network Part 3: Basic Method for the Measurement of Two-Terminal Parameters of Quartz Crystal Units in a Pi-Network up to 200 MHz by Phase Technique in a Pi-Network with. 20 pp.
IEC 444 Pt 2-80. Measurement of Quartz Crystal Unit Parameters by Zero Phase Technique in a PI-Network Part 2: Phase Offset Method for Measurement of Motional Capacitance of Quartz Crystal Units First Edition. 21 pp.
IEC 444 Pt 3-86. Measurement of Quartz Crystal Unit Parameters by Zero Phase Technique in a PI-Network Part 3: Basic Method for the Measurement of Two-Terminal Parameters of Quartz Cyrstal Units up to 200 MHz by Phase Technique in a PI-Network with Compensation of the. 37 pp.

Capacitance Operated Intrusion Detectors
See Also: Burglar Alarms; Burglar Detectors

—Burglar
BSI BS 4737: Sec 3.8-78. 1978 Amd 1 Intruder Alarm Systems in Buildings Part 3: Requirements for Detection Devices Section 3.8: Volumetric Capacitive Detectors. 4 pp.
BSI BS 4737: Sec 3.13-78. 1978 Amd 1 Intruder Alarm Systems in Buildings Part 3: Components Section 3.13: Capacitive Proximity Detectors. 4 pp.

Capacitor Antennas
See Also: Antennas

—Antenna Couplers
CNS C7150-82. Across-the-Line Capacitor Antenna-Coupling and Line-by-Pass Components for Radio and Television Type Appliances (Jan)(8382).

Capacitor Dividers
Use: Voltage Dividers

Capacitor Microphones
Use For: Condenser Microphones; Electrostatic Microphones *See Also:* Microphones
CNS C7197-88. Standard Condenser Microphones (Apr)(12252).
JIS C 5515-81. Standard Condenser Microphones. 27 pp.

—Calibration
BSI BS 5677-79. 1979 Method (Precision) for the Pressure Calibration of One-Inch Standard Condenser Microphones by the Reciprocity Technique. 20 pp.
BSI BS 5679-79. 1979 Method (Precision) for Free-Field Calibration of One-Inch Standard Condenser Microphones by the Reciprocity Technique. 20 pp.
IEC 327-71. Precision Method for Pressure Calibration of One-Inch Standard Condenser Microphones by the Reciprocity Technique First Edition. 37 pp.
IEC 402-72. Simplified Method for Pressure Calibration of One-Inch Condenser Microphones by the Reciprocity Technique First Edition. 29 pp.
IEC 486-74. Precision Method for Free-Field Calibration of One-Inch Standard Condenser Microphones by the Reciprocity Technique First Edition. 36 pp.

—Calibration—Telephone Transmission
CCITT RECMN P.61-89. Methods for the Calibration of Condenser Microphones—Telephone Transmission Quality (Study Group XII) 2 pp. 2 pp.

—Sensitivity
BSI BS 5941-80. 1980 Values for the Difference Between Free-Field and Pressure Sensitivity Levels for One-Inch Standard Condenser Microphones. 9 pp.
IEC 655-79. Values for the Difference Between Free-Field and Pressure Sensitivity Levels for One-Inch Standard Condenser Microphones First Edition. 18 pp.

Capacitor Papers
Use For: Condenser Paper; Condenser Tissues
See Also: Capacitors; Dielectrics; Electrical Insulating Papers; Paper Capacitors; Papers
BSI BS 5626: Sec 3.2-85. 1985 Cellulosic Papers for Electrical Purposes Part 3: Specifications for Individual Materials Section 3.2: Capacitor Paper. 10 pp.
CNS C7017-85. High Tension Condenser Paper (Nov)(3067).
IEC 554 Pt 3-2-83. Specification for Cellulosic Papers for Electrical Purposes Part 3: Specifications for Individual Materials Sheet 2: Capacitor Paper First Edition. 21 pp.
JIS C 2306-84. High Tension Capacitor Papers. 7 pp.

—Ammunition
MOD UK DEF-49-A-67. Paper, Tissue, Condenser. 4 pp.

—Electrolytic
BSI BS 5626: Sec 3.4-80. 1980 Cellulosic Papers for Electrical Purposes Part 3: Specifications for Individual Materials Section 3.4: Electrolytic Capacitor Papers. 7 pp.
CNS C4213-85. Electrolytic Capacitor Papers (Sep)(6046).
IEC 554 Pt 3-4-79. Specification for Cellulosic Papers for Electrical Purposes Part 3: Specifications for Individual Materials Sheet 4: Electrolytic Capacitor Paper First Edition. 12 pp.
JIS C 2301-84. Electrolytic Capacitor Papers. 9 pp.

—Metallized
CNS C7018-85. Metalized Papers for Capacitors (Sep)(3068).
JIS C 2316-84. Metalized Papers for Capacitors. 12 pp.

INTERNATIONAL AND NON-U.S. NATIONAL STANDARDS
SUBJECT INDEX
Capacitors

Capacitor Papers *(Cont.)*
—Tissue
CNS C7019-85. Capacitors Tissue Paper (Sep)(3069).
JIS C 2302-84. Capacitor Tissue Papers. 8 pp.

Capacitor Voltage Transformers
See Also: Instrument Transformers; Voltage Transformers
CSA CAN3-C13.1-M79. Capacitor Voltage Transformers. 49 pp.

Capacitors
Scope Note: For additional listings, use a more specific term *Use For:* Condensers (Electric)
See Also: Air Capacitors; Aluminum Capacitors; Ballasts (Electric); Capacitor Papers; Ceramic Capacitors; Ceramic Chip Capacitors; Chip Capacitors; Coupling Capacitors; Dielectrics; Discrete Components; Electrical Components; Film Capacitors; Fixed Capacitors; Metallized Film Capacitors; Metallized Paper Capacitors; Metallizing; Mica Capacitors; Motor Capacitors; Paper Capacitors; Power Capacitors; Pulse Capacitors; Series Capacitors; Shunt Power Capacitors; Tantalum Capacitors; Tantalum Chip Capacitors; Trimmer Capacitors; Tuning Capacitors; Variable Capacitors
BSI BS 5695-80. 1980 Recommendations for Maximum Case Dimensions for Capacitors and Resistors. 4 pp.
BSI BS 6201: Part 1-82. 1982 Fixed Capacitors for Use in Electronic Equipment Part 1: Methods of Test. 29 pp.
BSI BS 6303-82. 1982 Determination of the Space Required by Capacitors and Resistors with Unidirectional Terminations. 6 pp.
BSI BS 9100-83. 1983 Capability Approval for Custom-Built Capacitors and Capacitor Modules: Generic Data and Methods of Test. 19 pp.
BSI BS 9930: Part 0-83. 1983 Fixed Capacitors for Use in Electronic Equipment Part 0: Generic Specification. Specifies Terms, Definitions, Methods of Test and Inspection Requirements in the IECQ System (See BS QC 300000: 1983). 35 pp.
CENELEC HD 242-75. Maximum Case Dimensions for Capacitors and Resistors. 2 pp.
CNS C5020-85. General Rules for Fixed Capacitors for Electronic Equipment (Jan)(4589).
CNS C5021-78. General Rules for Line Capacitors for Electronic Equipment (Oct)(4590).
CNS C7028-90. Capacitors for Electrical Apparatus (Oct)(4327).
DIN VDE 0560 Pt 1-69. Regulations for Capacitors; General Regulations (Dec). 52 pp.
IEC 451-74. Maximum Case Dimensions for Capacitors and Resistors First Edition. 8 pp.
IEC 717-81. Method for the Determination of the Space Required by Capacitors and Resistors with Unidirectional Terminations First Edition. 13 pp.
JIS C 4902-90. High Voltage Power Capacitors. 17 pp.
JIS C 5101-88. General Rules of Fixed Capacitors for Use in Electronic Equipment. 38 pp.
JIS C 5102-86. Test Methods of Fixed Capacitors for Electronic Equipment. 117 pp.
MOD UK DEF-5138-01. General Requirements for Capacitors, Fixed, d.c. Rated (Excluding Electrolytic Types); Corrections. 20 pp.

—Askarels
IEC 588 Pt 1-77. Askarels for Transformers and Capacitors Part 1: General First Edition. 12 pp.
IEC 588 Pt 2-78. Askarels for Transformers and Capacitors Part 2: Test Methods First Edition. 48 pp.
IEC 588 Pt 3-77. Askarels for Transformers and Capacitors Part 3: Specifications for New Askarels First Edition. 22 pp.

—Askarels—Screening Testing
IEC 588 Pt 6-79. Askarels for Transformers and Capacitors Part 6: Screening Test for Effects of Materials on Capacitor Askarels First Edition. 17 pp.

—Automotive
CNS D3041-86. Method of Test for Condensers for Automobiles (Jun)(5120).
JIS D 1604-76. Inspection of Condensers for Automobile Contact Breakers (R 1983). 5 pp.

—Color Coding
BSI BS 5890-80. 1980 Guide for Choice of Colours to Be Used for the Marking of Capacitors and Resistors. 3 pp.
IEC 425-73. Guide for the Choice of Colours to Be Used for the Marking of Capacitors and Resistors First Edition. 7 pp.

—Cylindrical
BSI BS 5692-79. 1979 Measurement of the Dimensions of a Cylindrical Electronic Component Having Two Axial Terminations. 6 pp.

Capacitors *(Cont.)*
—Cylindrical *(Cont.)*
IEC 294-69. Measurement of the Dimensions of a Cylindrical Component Having Two Axial Terminations First Edition. 11 pp.

—Data Processing Equipment—Electrical Insulation
CSA CAN/CSA-C22. 2 NO 220-M91. Information Processing and Business Equipment; (Gen Instr 1). 119 pp.

—Discharge Coils
CNS C4360-82. Discharge Coil for High Voltage Power Capacitor (Jul)(9095).
JIS C 4802-90. Discharge Coils for High Voltage Power Capacitor. 14 pp.

—Discharge Lamps
BSI BS 4017-79. (WITHDRAWN) 1979 Capacitors for Use in Tubular Fluorescent, High Pressure Mercury, and Low Pressure Sodium Vapour Discharge Lamp Circuits (SUPERSEDED BY BS EN 61048 & BS EN 61049). 22 pp.
BSI BS EN 61048-93. 1993 Capacitors for Use in Tubular Fluorescent and Other Discharge Lamp Circuits—General and Safety Requirements (E). 31 pp.
BSI BS EN 61049-93. 1993 Capacitors for Use in Tubular Fluorescent and Other Discharge Lamp Circuits—Performance Requirements (E). 21 pp.
CENELEC EN 61048-93. Capacitors for Use in Tubular Fluorescent and Other Discharge Lamp Circuits—General and Safety Requirements (IEC 1048:1991 + Corrigendum January 1992). 6 pp.
CENELEC EN 61048-93. Capacitors for Use in Tubular Fluorescent and Other Discharge Lamp Circuits—General and Safety Requirements (IEC 1048: 1991, Modified + Corrigendum: 1992). 25 pp.
CENELEC EN 61049-93. Capacitors for Use in Tubular Fluorescent and Other Discharge Lamp Circuits Performance Requirements (IEC 1049: 1991 + Corrigendum 1992, Modified). 14 pp.
DIN VDE 0560 Pt 6-86. Capacitors; Capacitors with Ratings up to 1.5 kvar for Installations with Discharge Lamps, and in Particular Fluorescent Lamps (Jan). 32 pp.
IEC 1048-91. Capacitors for Use in Tubular Fluorescent and Other Discharge Lamp Circuits General and Safety Requirements First Edition; (Replaces 566) (CENELEC EN 61048:1993). 54 pp.
IEC 1049-91. Capacitors for Use in Tubular Fluorescent and Other Discharge Lamp Circuits Performance Requirements First Edition; (Replaces 566) (CENELEC EN 61049: 1993). 30 pp.

—Discharge Lamps—Safety
BSI BS EN 61048-93. 1993 Capacitors for Use in Tubular Fluorescent and Other Discharge Lamp Circuits—General and Safety Requirements (E). 31 pp.
CENELEC EN 61048-93. Capacitors for Use in Tubular Fluorescent and Other Discharge Lamp Circuits—General and Safety Requirements (IEC 1048:1991 + Corrigendum January 1992). 6 pp.
CENELEC EN 61048-93. Capacitors for Use in Tubular Fluorescent and Other Discharge Lamp Circuits—General and Safety Requirements (IEC 1048: 1991, Modified + Corrigendum: 1992). 25 pp.
IEC 1048-91. Capacitors for Use in Tubular Fluorescent and Other Discharge Lamp Circuits General and Safety Requirements First Edition; (Replaces 566) (CENELEC EN 61048:1993). 54 pp.

—Fluorescent Lighting
BSI BS EN 61048-93. 1993 Capacitors for Use in Tubular Fluorescent and Other Discharge Lamp Circuits—General and Safety Requirements (E). 31 pp.
BSI BS EN 61049-93. 1993 Capacitors for Use in Tubular Fluorescent and Other Discharge Lamp Circuits—Performance Requirements (E). 21 pp.
CENELEC EN 61048-93. Capacitors for Use in Tubular Fluorescent and Other Discharge Lamp Circuits—General and Safety Requirements (IEC 1048:1991 + Corrigendum January 1992). 6 pp.
CENELEC EN 61048-93. Capacitors for Use in Tubular Fluorescent and Other Discharge Lamp Circuits—General and Safety Requirements (IEC 1048: 1991, Modified + Corrigendum: 1992). 25 pp.
CENELEC EN 61049-93. Capacitors for Use in Tubular Fluorescent and Other Discharge Lamp Circuits Performance Requirements (IEC 1049: 1991 + Corrigendum 1992, Modified). 14 pp.
DIN VDE 0560 Pt 6-86. Capacitors; Capacitors with Ratings up to 1.5 kvar for Installations with Discharge Lamps, and in Particular Fluorescent Lamps (Jan). 32 pp.
IEC 1048-91. Capacitors for Use in Tubular Fluorescent and Other Discharge Lamp Circuits General and Safety Requirements First Edition; (Replaces 566) (CENELEC EN 61048:1993). 54 pp.

Capacitors *(Cont.)*
—Fluorescent Lighting *(Cont.)*
IEC 1049-91. Capacitors for Use in Tubular Fluorescent and Other Discharge Lamp Circuits Performance Requirements First Edition; (Replaces 566) (CENELEC EN 61049: 1993). 30 pp.

—Fluorescent Lighting—Safety
BSI BS EN 61048-93. 1993 Capacitors for Use in Tubular Fluorescent and Other Discharge Lamp Circuits—General and Safety Requirements (E). 31 pp.
CENELEC EN 61048-93. Capacitors for Use in Tubular Fluorescent and Other Discharge Lamp Circuits—General and Safety Requirements (IEC 1048:1991 + Corrigendum January 1992). 6 pp.
CENELEC EN 61048-93. Capacitors for Use in Tubular Fluorescent and Other Discharge Lamp Circuits—General and Safety Requirements (IEC 1048: 1991, Modified + Corrigendum: 1992). 25 pp.
IEC 1048-91. Capacitors for Use in Tubular Fluorescent and Other Discharge Lamp Circuits General and Safety Requirements First Edition; (Replaces 566) (CENELEC EN 61048:1993). 54 pp.

—Glossaries
BSI BS 4727:Pt2: Group 14-91. 1991 Electrotechnical, Power, Telecommunication, Electronics, Lighting and Colour Terms Part 2: Terms Particular to Power Engineering Group 14: Power Capacitors. 31 pp.

—Holders
MOD UK DSTAN 59-67-01. Retainers, Capacitor Issue 1; Corrigenda. 23 pp.

—Identification Systems
BSI BS 1852-75. 1975 Amd 1 Marking Codes for Resistors and Capacitors. 10 pp.
CENELEC HD 334 S3-91. Marking Codes for Resistors and Capacitors. 4 pp.
IEC 62-92. Marking Codes for Resistors and Capacitors Fourth Edition. 29 pp.
SAA AS 2066-77. Marking Codes for Resistors and Capacitors Corrig. Reconfirmed 1988. 12 pp.

—Induction Heating
CENELEC HD 207-74. Recommendation for Capacitors for Inductive Heat Generating Plants Operating at Frequencies Between 40 and 24000 Hz. 2 pp.
IEC 110-73. Recommendation for Capacitors for Inductive Heat Generating Plants Operating at Frequencies Between 40 and 24000 Hz Second Edition. 40 pp.

—Induction Heating—Disconnecting Switches
BSI BS 7632-93. 1993 Internal Fuses and Internal Overpressure Disconnectors for Capacitors for Inductive Heat Generating Plants (IEC 594: 1977) (E). 16 pp.
IEC 594-77. Internal Fuses and Internal Overpressure Disconnectors for Capacitors for Inductive Heat Generating Plants First Edition; (Amendment 2-1987). 36 pp.

—Induction Heating—Electric Fuses
BSI BS 7632-93. 1993 Internal Fuses and Internal Overpressure Disconnectors for Capacitors for Inductive Heat Generating Plants (IEC 594: 1977) (E). 16 pp.
IEC 594-77. Internal Fuses and Internal Overpressure Disconnectors for Capacitors for Inductive Heat Generating Plants First Edition; (Amendment 2-1987). 36 pp.

—Initiators (Explosives)
CNS M2006-73. Capacitor Type Explosion Initiators (Dec)(2941).
CNS M3008-73. Method of Test for Capacitor Type Explosion Initiators (Dec)(2942).

—Microwave Ovens
DIN VDE 0560 Pt 22-86. Capacitors; Capacitors for Micro-wave Cookers (Nov). 24 pp.

—Office Machines—Electrical Insulation
CSA CAN/CSA-C22. 2 NO 220-M91. Information Processing and Business Equipment; (Gen Instr 1). 119 pp.

—Packaging
MOD UK DSTAN 81-31-90. Packaging of Capacitors Issue 2. 27 pp.

—Preferred Diameters—Wire Terminations
BSI BS 5693-78. 1978 Preferred Diameters of Wire Terminations for Capacitors and Resistors. 2 pp.
CENELEC HD 349-87. Preferred Diameters of Wire Terminations of Capacitors and Resistors. 3 pp.
IEC 301-71. Preferred Diameters of Wire Terminations of Capacitors and Resistors Second Edition; (Amendment 1-1972). 8 pp.

INDUSTRY STANDARDS

INTERNATIONAL AND NON-U.S. NATIONAL STANDARDS
SUBJECT INDEX

Capacitors

Capacitors (Cont.)

—Preferred Diameters—Wire Terminations (Cont.)
SAA AS 1343-77. Preferred Diameters for Wire Terminations for Capacitors and Resistors Reconfirmed 1988. 4 pp.

—Preferred Numbers
BSI BS 2488-66. 1966 Amd 1 Schedule of Preferred Numbers for the Resistance of Resistors and the Capacitance of Capacitors for Telecommunication Equipment. 7 pp.
IEC 63-63. Preferred Number Series for Resistors and Capacitors Second Edition; (Amendment 1-1967) (Amendment 2-1977). 19 pp.
SAA AS 2065-77. Preferred Number Series for Resistors and Capacitors Reconfirmed 1988. 8 pp.

—Preferred Products List
CECC CECC MUAHAG Vol 1 IS 4-90. Preferred Products List; Capacitors (En, Fr, Ge). 64 pp.

—Radio Frequency Interference
DIN VDE 0565 Pt 1-79. Radio Interference Suppression Devices; Radio Interference Suppression Capacitors (Dec). 66 pp.
DIN VDE 0565 Pt 1 A1-84. Radio Interference Suppression Devices; Radio Interference Suppression Capacitors: Amendment 1 (June). 2 pp.
IEC 940-88. Guidance Information on the Application of Capacitors, Resistors, Inductors, and Complete Filter Units for Radio Interference Suppression First Edition. 21 pp.

—Railroad Signaling Systems
BSI BS 3347-61. 1961 Amd 1 Capacitors for Railway Signalling Track-Circuits. 13 pp.

—Reliability Assured
CNS C5030-86. General Rules for Reliability Assured Fixed Capacitors (Jan)(4902).
MOD UK DSTAN 59-44: 90/006-89. Capacitors, Fixed, High Voltage Issue 2. 17 pp.

—Series Reactors
CNS C4359-82. Series Reactors for High Voltage Power Capacitors (Jul)(9094).
JIS C 4801-90. Series Reactors for High Voltage Power Capacitors. 14 pp.

—Symbols
CNS C5111-81. Graphic Symbols for Capacitors (May)(7360).

—Symbols—Identification Systems
CENELEC HD 334 S2-89. Marking Codes for Resistors and Capacitors. 3 pp.

—Tan Delta Measurements
DIN VDE 0560 Pt 1 A1-86. Capacitors; Power Capacitors; Method for Verifying Accuracy of Tan Delta Measurements; Identical with IEC 33 (Central Office) 88: Amendment 1 (Sept). 6 pp.

Capacitors (Power)
Use: Power Capacitors

Capacity Measurement
See Also: Exchange Capacity Measurement

—Aerosol Container—Glossaries
BSI BS 6966: Part 3-88. (WITHDRAWN) 1988 Light Gauge Metal Containers Part 3: Glossary of Terms for Aerosol Cans and Methods for Determination of Dimensions Capacities (Renumbered As BS EN 20090-3: 1993). 11 pp.

—Aerosol Containers
BSI BS 6966: Part 3-88. (WITHDRAWN) 1988 Light Gauge Metal Containers Part 3: Glossary of Terms for Aerosol Cans and Methods for Determination of Dimensions Capacities (Renumbered As BS EN 20090-3: 1993). 11 pp.
BSI BS EN 20090-3-93. 1993 Amd 1 Light Gauge Metal Containers. Definitions and Determination Methods for Dimensions and Capacities Part 3: Aerosol Cans (ISO 90-3: 1986) (AMD 7480) March 15, 1993 (H). 17 pp.
CEN EN 20090-3-92. Light Gauge Metal Containers—Definitions and Determination Methods for Dimensions and Capacities—Part 3: Aerosol Cans (ISO 90-3: 1986). 9 pp.
ISO 90 Pt 3-86. Light Gauge Metal Containers—Definitions and Determination Methods for Dimensions and Capacities—Part 3: Aerosol Cans First Edition; (CEN EN 20090-3: 1992). 9 pp.

—Cans
BSI BS 1764-73. 1973 Nominal Diameters of Round Built-up Tins. 8 pp.
BSI BS 5771-81. 1981 Packages for Certain Pre-Packed Foodstuffs; Capacities of Glass and Metal Containers. 9 pp.

Capacity Measurement (Cont.)

—Cans (Cont.)
BSI BS 5774-84. 1984 Internal Diameters of Round Hermetically Sealed Metal Cans for Food and Non-Carbonated Drinks. 4 pp.
BSI BS 6966: Part 1-88. 1988 Light Gauge Metal Containers Part 1: Glossary of Terms for Open-Top Cans and Methods for Determination of Dimensions and Capacities (Renumbered as BS EN 20090-1: 1993). 18 pp.
BSI BS 6966: Part 2-88. (WITHDRAWN) 1988 Light Gauge Metal Containers Part 2: Glossary of Terms for General Use Containers and Methods for Determination of Dimensions and Capacities (Renumbered as BS EN 20090-2: 1993). 19 pp.
BSI BS 6992: Part 1-88. (OBSOLESCENT) 1988 Capacities and Related Cross Sections of Open-Top Cans for Food and Drinks Part 1: Recommendations for Cans for General Food. 7 pp.
BSI BS 6992: Part 2-91. (OBSOLESENT) 1991 Capacities and Related Cross Sections of Open-Top Cans for Food and Drinks Part 2: Recommendations for Cans for Meat and Meat Products for Human Consumption. 10 pp.
BSI BS 6992: Part 3-88. (OBSOLESCENT) 1988 Capacities and Related Cross Sections of Open-Top Cans for Food and Drinks Part 3: Recommendations for Cans for Drinks. 7 pp.
BSI BS 6992: Part 4-89. (OBSOLESCENT) 1989 Capacities and Related Cross Sections of Open-Top Cans for Food and Drinks Part 4: Recommendations for Cans for Fish and Other Fishery Products. 8 pp.
BSI BS 6992: Part 5-88. (OBSOLESCENT) 1988 Capacities and Related Cross Sections for Open-Top Cans for Food and Drinks Part 5: Recommendations for Cans for Milk. 7 pp.
BSI BS EN 20090-1-93. 1993 Amd 1 Light Gauge Metal Containers. Definitions and Determination Methods for Dimensions and Capacities Part 1: Open-Top Cans (ISO 90-1: 1986) (AMD 7478) March 15, 1993 (H). 24 pp.
BSI BS EN 20090-2-93. 1993 Amd 1 Light Gauge Metal Containers. Definitions and Determination Methods for Dimensions and Capacities Part 2: General Use Containers (ISO 90-2: 1986) (AMD 7479) March 15, 1993 (H). 25 pp.
CEN EN 20090-1-92. Light Gauge Metal Containers—Definitions and Determination Methods for Dimensions and Capacities—Part 1: Open-Top Cans (ISO 90-1: 1986). 16 pp.
CEN EN 20090-2-92. Light Gauge Metal Containers—Definitions and Determination Methods for Dimensions and Capacities—Part 2: General Use Containers (ISO 90-2: 1986). 17 pp.
ISO 90 Pt 1-86. Light Gauge Metal Containers—Definitions and Determination Methods for Dimensions and Capacities—Part 1: Open-Top Cans First Edition; (CEN EN 20090-1: 1992). 16 pp.
ISO 90 Pt 2-86. Light Gauge Metal Containers—Definitions and Determination Methods for Dimensions and Capacities—Part 2: General Use Containers First Edition; (CEN EN 20090-2: 1992). 17 pp.
ISO 1361-83. Light Gauge Metal Containers—Open-Top Cans—Round Cans—Internal Diameters Third Edition. 4 pp.
ISO TR8610-84. Light Gauge Metal Containers—Round Vent-Hole Cans with Soldered Ends for Milk and Milk Products—Capacities and Related Diameters First Edition. 2 pp.
ISO TR10193-89. Round General Use Light Gauge Metal Containers—Nominal Filling Volumes and Nominal Diameters First Edition. 15 pp.
ISO TR10194-89. Non-Round General Use Light Gauge Metal Containers—Nominal Filling Volumes and Nominal Cross-Sections First Edition. 15 pp.
ISO 10653-93. Light-Gauge Metal Containers—Round Open-Top Cans—Cans Defined by Their Nominal Gross Lidded Capacities First Edition; (Replaces 3004 Pt 1 Thru Pt 6). 7 pp.
ISO TR11761-92. Light-Gauge Metal Containers—Round Open-Top Cans—Classification of Can Sizes by Construction Type First Edition; (Replaces 3004 Pt 1 Thru Pt 6). 9 pp.
ISO TR11762-92. Light-Gauge Metal Containers—Round Open-Top Cans for Liquid Products with Added Gas—Classification of Can Sizes by Construction Type First Edition; (Replaces 3004 Pt 1 Thru Pt 6). 7 pp.
SAA AS 2354-91. Open Top Metal Cans—Definitions, and Determination Methods for Dimensions and Capacities (ISO 90-1:1986). 16 pp.

—Cans—Glossaries
BSI BS 6966: Part 2-88. (WITHDRAWN) 1988 Light Gauge Metal Containers Part 2: Glossary of Terms for General Use Containers and Methods for Determination of Dimensions and Capacities (Renumbered as BS EN 20090-2: 1993). 19 pp.

Capacity Measurement (Cont.)

—Clothes Dryers
ISO 9398 Pt 2-93. Specifications for Industrial Laundry Machines—Definitions and Testing of Capacity and Consumption Characteristics—Part 2: Batch Drying Tumblers First Edition. 7 pp.

—Containers
CEN EN 76-78. Packages for Certain Pre-Packed Foodstuffs: Capacities of Glass and Metal Containers. 6 pp.

—Dredges
BSI BS MA 98: Part 1-85. 1985 Multi-Bucket Dredgers Part 1: Scale of Bucket Capacities. 2 pp.
ISO 7607-84. Shipbuilding—Inland Navigation—Multi-Bucket Dredgers—Scale of Bucket Capacities First Edition. 3 pp.

—Dump Trucks
ISO 6483-80. Earth-Moving Machinery—Dumper Bodies—Volumetric Rating First Edition. 7 pp.
JIS A 8804-92. Earth-Moving Machinery—Dumper Bodies—Volumetric Rating.

—Freezers
CSA CAN/CSA-C300-M89. Capacity Measurement and Energy Consumption Test Methods for Refrigerators, Combination Refrigerator-Freezers, and Household Freezers; (Gen Instr 1). 90 pp.
CSA CAN/CSA-C300-M91. Capacity Measurement and Energy Consumption Test Methods for Refrigerators, Combination Refrigerator-Freezers, and Freezers; (Gen Instr 1). 95 pp.

—Front End Loaders
BSI BS 6422-83. 1983 Volumetric Rating of Loader and Front Loading Excavator Buckets Used for Earth-Moving. 12 pp.
ISO 7546-83. Earth-Moving Machinery—Loader and Front Loading Excavator Buckets—Volumetric Ratings First Edition. 10 pp.

—Glass Containers
BSI BS 5771-81. 1981 Packages for Certain Pre-Packed Foodstuffs; Capacities of Glass and Metal Containers. 9 pp.

—Hydraulic Excavating Equipment
BSI BS 6421-83. 1983 Volumetric Rating of Hoe Type Buckets of Hydraulic Excavators Used for Earth-Moving. 8 pp.
ISO 7451-83. Earth-Moving Machinery—Hydraulic Excavators—Hoe Type Buckets—Volumetric Ratings First Edition. 8 pp.

—Hydraulic Motors
BSI BS 7250-89. 1989 Methods of Determining the Derived Capacity of Hydraulic Fluid Power Positive Displacement Pumps and Motors. 12 pp.
ISO 8426-88. Hydraulic Fluid Power—Positive Displacement Pumps and Motors—Determination of Derived Capacity First Edition. 12 pp.

—Hydraulic Power Pumps
BSI BS 7250-89. 1989 Methods of Determining the Derived Capacity of Hydraulic Fluid Power Positive Displacement Pumps and Motors. 12 pp.
ISO 8426-88. Hydraulic Fluid Power—Positive Displacement Pumps and Motors—Determination of Derived Capacity First Edition. 12 pp.

—Irons (Electric)—Flatwork
ISO 9398 Pt 1-93. Specifications for Industrial Laundry Machines—Definitions and Testing of Capacity and Consumption Characteristics—Part 1: Flatwork Ironing Machines First Edition. 8 pp.

—Luggage Compartments—Automobiles
ISO 3832-91. Passenger Cars—Luggage Compartments—Method of Measuring Reference Volume Second Edition; (Corrected and Reprinted -1991). 5 pp.
JIS D 0303-82. Method of Measuring the Reference Volume for the Luggage Compartments of Passenger Cars.

—Metal Containers
CNS Z6014-72. Determination of Capacity for Hermetically Sealed Metal Food Containers (Jun)(3347). 2 pp.

—Packaging
DIN ENGL 32 Pt 4-80. Packaging; Packages for Pre-Packed Goods; Deviations in Capacity (May). 3 pp.

—Refrigerated Display Cabinets
DIN ENGL 8954 Pt 2-74. Open Refrigerated Display Cabinets; Testing of Dimensions, Shelf Areas and Volumes (Feb). 1 p.

INTERNATIONAL AND NON-U.S. NATIONAL STANDARDS
SUBJECT INDEX

Capacity Measurement (Cont.)
—Refrigerator/Freezers
CSA CAN/CSA-C300-M91. Capacity Measurement and Energy Consumption Test Methods for Refrigerators, Combination Refrigerator-Freezers, and Freezers; (Gen Instr 1). 95 pp.

—Refrigerators
CSA CAN/CSA-C300-M89. Capacity Measurement and Energy Consumption Test Methods for Refrigerators, Combination Refrigerator-Freezers, and Household Freezers; (Gen Instr 1). 90 pp.

CSA CAN/CSA-C300-M91. Capacity Measurement and Energy Consumption Test Methods for Refrigerators, Combination Refrigerator-Freezers, and Freezers; (Gen Instr 1). 95 pp.

—Scrapers (Earthmoving Equipment)
BSI BS 6074-81. 1981 Determination of the Volumetric Rating of Tractor-Scrapers for Earth-Moving Machinery. 7 pp.

BSI BS 6114-87. 1987 Method for Measurement of Volumetric Ratings of Elevating Scrapers Used for Earth-Moving. 8 pp.

ISO 6484-86. Earth-Moving Machinery—Elevating Scrapers—Volumetric Ratings Second Edition. 6 pp.

ISO 6485-80. Earth-Moving Machinery—Tractor-Scraper Volumetric Rating First Edition. 6 pp.

JIS D 6508-92. Earth-Moving Machinery—Scraper Volumetric Rating. 7 pp.

—Tractors
BSI BS 6911: Part 1-88. 1988 Testing Earth-Moving Machinery Part 1: Method for Determination of Volumetric Rating of Crawler and Wheel Tractor Dozer Blades. 7 pp.

ISO 9246-88. Earth-Moving Machinery—Crawler and Wheel Tractor Dozer Blades—Volumetric Ratings First Edition. 7 pp.

—Unit Coolers
JIS B 8610-90. Capacity Evaluation for Refrigerating Unit Coolers. 28 pp.

—Unloaders
CNS B4043-80. Calculated Capacity for Unloader (Dec)(6642).

JIS B 8809-68. Calculated Capacity for Unloader (R 1983). 5 pp.

—Washing Machines
ISO 9398 Pt 4-93. Specifications for Industrial Laundry Machines—Definitions and Testing of Capacity and Consumption Characteristics—Part 4: Washers-Extractors First Edition. 9 pp.

—Washing Tunnels
ISO 9398 Pt 3-93. Specifications for Industrial Laundry Machines—Definitions and Testing of Capacity and Consumption Characteristics—Part 3: Washing Tunnels First Edition. 7 pp.

—Water Extractors
ISO 9398 Pt 4-93. Specifications for Industrial Laundry Machines—Definitions and Testing of Capacity and Consumption Characteristics—Part 4: Washers-Extractors First Edition. 9 pp.

Cape Chisels
See Also: Chisels

CGSB 39-GP-43A-74. Chisels, Hand, Cutting, and Swaging (For Metals), Standard for; (Amendment 1 Dec 1978). 15 pp.

CNS B3244-78. Cape Chisels (May)(3976). 1 p.

Capillarity
See Also: Cohesion; Electrical Properties; Porosity; Rheological Properties; Viscosity; Water Absorption

—Abrasives
BSI BS 7425: Part 2-91. 1991 Abrasive Grains Part 2: Method for Determining the Capillarity (ISO 9137: 1990). 8 pp.

CNS R3102-81. Method of Test for Capillarity of Artificial Abrasives (Aug)(7842).

ISO 9137-90. Abrasive Grains—Determination of Capillarity First Edition. 6 pp.

JIS R 6127-69. Testing Method for Capillarity of Artificial Abrasives. 5 pp.

—Cork
ISO 9727-91. Cylindrical Stoppers of Natural Cork—Physical Tests—Reference Methods First Edition. 12 pp.

Capillary Fringe
—Soil Analysis
DIN ENGL 19683 Pt 10-73. Methods of Soil Analysis for Water Management for Agricultural Purposes; Physical Laboratory Tests; Determination of Height of Closed Capillary Fringe (Apr). 2 pp.

Capillary Rheometers
Use: Rheometers—Capillary

Capillary Rise
—Paperboard
ISO 8787-86. Paper and Board—Determination of Capillary Rise—Klemm Method First Edition. 4 pp.

—Papers
ISO 8787-86. Paper and Board—Determination of Capillary Rise—Klemm Method First Edition. 4 pp.

Capillary Viscometers
See Also: Viscometers

ISO 3105-76. Glass Capillary Kinematic Viscometers—Specification and Operating Instructions First Edition; (Erratum—June 1979). 31 pp.

—Pulps
CNS P3051-81. Viscosity of Pulp (Capillary Viscometer Method) (Jul)(7748). 4 pp.

Capital Movements
See Also: Currencies

EC 85/583/EEC-85. Council Directive Amending the Directive of 11 May 1960 on the Implementation of Article 67 of the Treaty. 3 pp.

EC 86/566/EEC-86. Council Directive Amending the First Directive of 11 May 1960 for the Implementation of Article 67 of the Treaty. 7 pp.

EC 88/361/EEC-88. Council Directive for the Implementation of Article 67 of the Treaty. 14 pp.

Capnometers
Use: Carbon Dioxide Analyzers

n-Capric Acid
See Also: Fatty Acids

CNS K7626-82. Chemical Reagent (N-Capric Acid) (Jun)(9009).

Caprines
Use: Goats

n-Caproic Acid
See Also: Fatty Acids

CNS K7627-82. Chemical Reagent (N-Caproic Acid) (Jun)(9010).

Caprolactam
—Absorptance
ISO 7059-82. Caprolactam for Industrial Use—Determination of Absorbance at a Wavelength of 290 nm First Edition. 3 pp.

—Colorimetry
ISO 8112-84. Caprolactam for Industrial Use—Determination of Colour of 50 % Aqueous Caprolactam Solution, Expressed in Hazen Units (Platinum-Cobalt Scale)—Spectrometric Method First Edition. 4 pp.

—Crystallization
ISO 7060-82. Caprolactam for Industrial Use—Determination of Crystallizing Point First Edition. 3 pp.

—Permanganate Number—Spectrometry
ISO 8660-88. Caprolactam for Industrial Use—Determination of Permanganate Index—Spectrometric Method First Edition. 4 pp.

—Volatile Matter Content—Volumetric Analysis
ISO 8661-88. Caprolactam for Industrial Use—Determination of Volatile Bases Content—Titrimetric Method After Distillation First Edition. 6 pp.

Caps (Explosive)
Use: Blasting Caps

Caps (Headgear)
Use: Hats

Caps (Lamps)
Use: Lampholders

Caps (Lids)
Use For: Sealing Caps *See Also:* Blanking Caps; Bottle Caps; Filler Caps; Lids; Pressure Caps; Radiator Caps; Stoppers; Valve Covers

—Aircraft
SBAC AS 1975 ISSUE 4. Cover, Inspection:—Typical Assembly. 1 p.

SBAC AS 2588 ISSUE 3. Cap Assembly. 1 p.

SBAC AS 2945 ISSUE 3. Sealing Cap. 1 p.

SBAC AS 6546 ISSUE 2. Sealing Ca

SBAC AS 6555 ISSUE 2. Sealing Ca

SBAC AS 6559 ISSUE 1. Top, Cover.

Caps (Lids) (Cont.)
—Aircraft (Cont.)
SBAC AS 6631 ISSUE 1. Cover, Inspection (Typical Assembly).

SBAC AS 6746 ISSUE 1. Lid Assembly—Flush Filler Cap Mk.II..

—Aircraft—Aluminum
SBAC AS 41914 (V). Caps Short Protection Metric Thread.

SBAC AS 41915 (V). Caps, Long, Protection, Metric Threads.

—Aircraft—Blanking Caps
SBAC AS 2948 ISSUE 3. Sealing Cap. 1 p.

—Aircraft—Cable Ducts
SBAC AS 8671 ISSUE 3. Capping (Cable Duct) 1.28 Inches Wide, Fire Resistant.

SBAC AS 8672 ISSUE 3. Capping (Cable Duct) 1.09 Inches Wide, Fire Resistant.

SBAC AS 8673 ISSUE 3. Capping (Cable Duct) .88 Inches Wide, Fire Resistant.

SBAC AS 8674 ISSUE 3. Capping (Cable Duct) .72 Inches Wide, Fire Resistant.

—Aircraft—Dipsticks—Assembly
SBAC AS 6268-6269 ISSUE 1(I). Assembly of Cap and Dipstick. 1 p.

—Aircraft—Drain Valves
SBAC AS 2565 ISSUE 5. Screw Cap. 1 p.

SBAC AS 2572 ISSUE 3. Cap Assembly. 1 p.

SBAC AS 2575 ISSUE 4. Screw Cap. 1 p.

SBAC AS 4762 ISSUE 1. Cap 1/4 Inch Bore Drain Valve. 1 p.

SBAC AS 8435 ISSUE 1. Cap 1/4 Bore Drain Valve.

—Aircraft—Fittings
SBAC AGS 3800 ISSUE 2. Cover for 10,000 lb Lashing Fitting Base.

SBAC AGS 3801 ISSUE 2. Cover for 25,000 lb Lashing Fitting Base.

SBAC AGS 3816 ISSUE 2. Cover for 5,000 lb Lashing Fitting Base.

—Aircraft—Flanges/Drives
SBAC AS 4913-4920 ISSUE 1. Protection, Cover for Flanges and Drives. 1 p.

—Aircraft—Fuel Filter Strainers
SBAC AS 3235 ISSUE 2. Top Cap—Strainer—Filter Fuel Type 'A'. 1 p.

SBAC AS 3265 ISSUE 2. Top Cap—Strainer—Filter Fuel Type 'B'. 1 p.

—Aircraft—Handholes
SBAC AS 3179 ISSUE 4. Handhole Cover—Curved. 1 p.

—Aircraft—Handholes—Assembly
SBAC AS 3170 ISSUE 1. Assembly of Handhole Cover. 1 p.

—Aircraft—Handholes—Bridge Pieces
SBAC AS 3175 ISSUE 1. Bridge Piece. 1 p.

—Aircraft—Handholes—Cover Plates
SBAC AS 3171 ISSUE 1. Cover Plate. 1 p.

—Aircraft—Handholes—Finger Hinges
SBAC AS 3174 ISSUE 2. Finger, Hinge Handhole Cover. 1 p.

—Aircraft—Handholes—Lanyard Lugs
SBAC AS 3177 ISSUE 1. Lug for Lanyard. 1 p.

—Aircraft—Handholes—Retaining Rings
SBAC AS 3178 ISSUE 1. Retaining, Ring Handhole Cover. 1 p.

—Aircraft—Handholes—Snap Rings
SBAC AS 3172 ISSUE 1. Circlip. 1 p.

—Aircraft—Handholes—Snap Rings—Housings
SBAC AS 3173 ISSUE 1. Circlip, Housing. 1 p.

—Aircraft—Handholes—Springs (Elastic)
SBAC AS 3176 ISSUE 1. Spring. 1 p.

—Aircraft—Helical Springs
SBAC AS 6276 ISSUE 1. Cap, Spring. 1 p.

SBAC AS 6366 ISSUE 3. Spring—Cap—Oil Drain Valve.

—Aircraft—Pipe Fittings—Metal
SBAC AGS 596 ISSUE 5. Cap Protection—Deep Drawn Metal, B.S.P.F..

SBAC AGS 597 ISSUE 5. Cap Protection—Shallow Drawn Metal, B.S.P.F..

—Aircraft—Pipe Fittings—Steel
SBAC AGS 1236 (V). Pipe Couplings, Cone Caps, (Steel) 3/16" to 1/2".

Caps (Lids) *(Cont.)*

—Aircraft—Pushbuttons
SBAC AS 3051 ISSUE 1. Assembly Push Button Cover. 1 p.
SBAC AS 6603 ISSUE 1. Push Button Cover Assembly.

—Aircraft—Rubber
SBAC AGS 519 ISSUE 3. Protective, Rubber Caps for Heat Exchange Equipment.

—Aircraft—Rudder Bars—Stampings
SBAC AS 2176 ISSUE 4. Stamping for Cap (Rudder Bar). 1 p.

—Aircraft—Rudder Bars—Worms (Mechanical Drives)
SBAC AS 2175 ISSUE 7. Cap for Adjusting Worm (Rudder Bar). 1 p.

—Aircraft—Seals
SBAC AS 6277 ISSUE 3. Cap, Seal. 1 p.
SBAC AS 6301 ISSUE 3. Cap, Seal. 1 p.
SBAC AS 6368 ISSUE 3. Seal—Ca
SBAC AS 41910-912 (V). Seals, for Use with Cap and Plugs Protection, Metric.

—Aircraft—Self Locking Nuts
SBAC AGS 2021 ISSUE 11(I). Stiffnuts—Translucent Ca

—Aircraft—Threads
SBAC AGS 3805-07 ISSUE 1. Caps—Protection, and Sealing, for Inhibited Units, Unified Threads.

—Aircraft—Valves
SBAC AS 6274 ISSUE 3. Valve Cap. 1 p.
SBAC AS 6299 ISSUE 3. Valve Cap. 2 pp.
SBAC AS 6364 ISSUE 4. Cap—Valve.

—Aircraft—Valves—Stainless Steel
SBAC AS 6699 ISSUE 1. Cap—Valve.

—Aircraft—Washers
SBAC AS 2576 ISSUE 4. Jointing Washer (Screw Cap). 1 p.
SBAC AS 3112 ISSUE 1. Sealing Washer for Cap. 1 p.

—Aluminum—Capping Foil/Strip
BSI BS 3313: Part 1-68. 1968 Amd 1 Aluminium Capping Foil and Strip for Dairy Product Containers Part 1: Aluminium Capping Foil for Glass Container. 10 pp.
BSI BS 3313: Part 2-68. 1968 Amd 1 Aluminium Capping Foil and Strip for Dairy Product Containers Part 2: Aluminium Cappint Foil for Skirted Closures for Plastics Containers. 14 pp.
BSI BS EN 28362-3-93. 1993 Injection Containers for Injectables and Accessories Part 3: Aluminium Caps for Injection Vials (ISO 8362-3: 1989) (N). 12 pp.
BSI BS EN 28872-93. 1993 Aluminium Caps for Transfusion, Infusion and Injection Bottles—General Requirements and Test Methods (ISO 8872: 1988) (N). 13 pp.
CEN EN 28362-3-93. Injection Containers for Injectables and Accessories—Part 3: Aluminium Caps for Injection Vials (ISO 8362-3: 1989). 6 pp.
CEN EN 28872-93. Aluminium Caps for Transfusion, Infusion and Injection Bottles—General Requirements and Test Methods (ISO 8872: 1988). 8 pp.
ISO 8362 Pt 3-89. Injection Containers for Injectables and Accessories—Part 3: Aluminium Caps for Injection Vials First Edition (CEN EN 28362-3: 1993). 8 pp.
ISO 8872-88. Aluminium Caps for Transfusion, Infusion and Injection Bottles—General Requirements and Test Methods First Edition; (CEN EN 28872: 1993). 10 pp.

—Automotive
JIS D 2102-76. Sealing Caps for Automobiles (R 1983). 5 pp.

—Blood Transfusion Equipment
ISO 1135 Pt 1-87. Transfusion Equipment for Medical Use—Part 1: Glass Transfusion Bottles, Closures and Caps Second Edition. 13 pp.

—Bottles
CNS Z5005-53. Caps of Cylindrical Glass Bottle for Syrup Storage (Jul)(340)(R 1973).
CNS Z5036-72. Caps for Bottles, Crown Type (Jun)(3365).
CNS Z6015-73. Method of Test for Caps of Bottles of Crown Type (May)(3366).

—Cans
JIS Z 1607-85. Metal Caps and Nozzles. 7 pp.

—Containers
BSI BS 6499-85. 1985 Amd 1 Metal Screw Necks, Screw Caps and Inner Seals for Metal Containers. 14 pp.

Caps (Lids) *(Cont.)*

—Expanding
DIN ENGL 442-70. Sealing Expanding Caps (Nov). 1 p.

—Gas Cylinders—Screw Threads
DIN ENGL 4668-68. Gas Cylinders; Screw Threads for Valve Sockets, Neck Rings and Protecting Caps (Sept). 2 pp.

—Injection Equipment (Medical)—Aluminum/Plastic
ISO 8362 Pt 6-92. Injection Containers for Injectables and Accessories—Part 6: Caps Made of Aluminium-Plastics Combinations for Injection Vials First Edition. 8 pp.
ISO 10985-92. Caps Made of Aluminium-Plastics Combinations for Infusion Bottles and Injection Vials—Requirements and Test Methods First Edition. 8 pp.

—Intravenous Medical Equipment—Aluminum
ISO 8536 Pt 3-92. Infusion Equipment for Medical Use—Part 3: Aluminium Caps for Infusion Bottles First Edition. 10 pp.

—Intravenous Medical Equipment—Aluminum/Plastic
ISO 8536 Pt 7-92. Infusion Equipment for Medical Use—Part 7: Caps Made of Aluminium-Plastics Combinations for Infusion Bottles First Edition. 8 pp.
ISO 10985-92. Caps Made of Aluminium-Plastics Combinations for Infusion Bottles and Injection Vials—Requirements and Test Methods First Edition. 8 pp.

—Jars—Canning—Metal
CGSB CAN/CGSB-143.1-M85. Lids, Home Canning Jar. 8 pp.

—Jars—Canning—Plastic
CGSB CAN/CGSB-143.1-M85. Lids, Home Canning Jar. 8 pp.

—Lamps
BSI BS EN 60309-1-92. 1992 Plugs, Socket-Outlets and Couplers for Industrial Purposes Part 1: General Requirements. 67 pp.
BSI BS EN 60309-1-01. 1992 Amd 1 Plugs, Socket-Outlets and Couplers for Industrial Purposes Part 1: General Requirements (AMD 7889) August 15, 1993 (E). 68 pp.
CENELEC EN 60309-1-92. Plugs, Socket-Outlets and Couplers for Industrial Purposes Part 1: General Requirements. 8 pp.
IEC 309 Pt 1-88. Plugs, Socket-Outlets and Couplers for Industrial Purposes Part 1: General Requirements Second Edition; (Corrigendum—March 1992). 110 pp.
JIS C 7709-89. Lamp Caps and Holders. 126 pp.

—Lighting—Screw Threads
DIN ENGL 40450-70. Electric Light Fittings; Glass Screw Threads for Glass Guards and Caps (Dec). 1 p.

—Markers (Pens)
BSI BS 7272-90. 1990 Amd 1 Safety Caps for Writing and Marking Instruments (L) (AMD 6534) April 30, 1990. 9 pp.
BSI BS 7272-02. 1990 Amd 2 Safety Caps for Writing and Marking Instruments (AMD 6886) October 31, 1991. 12 pp.

—Pens
BSI BS 7272-90. 1990 Amd 1 Safety Caps for Writing and Marking Instruments (L) (AMD 6534) April 30, 1990. 9 pp.
BSI BS 7272-02. 1990 Amd 2 Safety Caps for Writing and Marking Instruments (AMD 6886) October 31, 1991. 12 pp.

—Pens (Writing Implements)—Children's—Safety
ISO 11540-93. Caps for Writing and Marking Instruments Intended for Use by Children Up to 14 Years of Age—Safety Requirements First Edition. 7 pp.

—Push In
DIN ENGL 443-70. Sealing Push-In Caps (Nov). 1 p.

—Rubber
BSI BS 3263-60. (WITHDRAWN) 1960 Amd 1 Rubber Closures for Injectable Products (Superseded by BS EN 28362-2: 1993). 12 pp.

—Sealing—Automotive
CNS D2037-89. Sealing Caps for Automobiles (Jul)(5793).

Caps (Lids) *(Cont.)*

—Sheets—Aluminum and Aluminum Alloys
DIN ENGL 59606-82. Wrought Aluminium and Aluminium Alloy Sheet and Strip for Cans and Sealing Caps (Nov). 6 pp.

—SSMA Coaxial Connectors—Preferred Products List
CECC CECC MUAHAG Vol 3B IS 4-91. Preferred Products List; Connectors; R.F. and Fibre Optics (En, Fr, Ge). 65 pp.

—Strips—Aluminum and Aluminum Alloys
DIN ENGL 59606-82. Wrought Aluminium and Aluminium Alloy Sheet and Strip for Cans and Sealing Caps (Nov). 6 pp.

—Syringes—Dental—Aluminum
ISO 11040 Pt 3-93. Prefilled Syringes—Part 3: Aluminium Caps for Dental Local Anaesthetic Cartridges First Edition. 5 pp.

—Tubes (Packages)
CNS Z5002-53. Caps for Soft Tubes (Jul)(337)(R 1973).

—Valves—Gas Cylinders
CEN PREN 962-92. Gas Cylinders—Valve Protection Caps and Valve Guards for Industrial and Medical Gas Cylinders—Design, Construction and Tests. 12 pp.

Capsaicin Content Analysis

—Chilies—Liquid Chromatography
ISO 7543 Pt 2-93. Chillies and Chilli Oleoresins—Determination of Total Capsaicinoid Content—Part 2: Method Using High-Performance Liquid Chromatography First Edition. 8 pp.

Capsicum

See Also: Spices
CNS N5195-82. Spices and Condiments Hot Pepper—Whole and Ground Specification (Apr)(8734). 4 pp.
ISO 972-85. Chillies and Capsicums, Whole or Ground (Powdered)—Specification First Edition. 7 pp.

Capstan Lathes

Use: Turret Lathes

Capstan Screws

See Also: Screws

—Slotted Head
CNS B2177-80. Slotted Capstan Screws (Jul)(4251).
DIN ENGL 404-86. Slotted Capstan Screws (Sept). 4 pp.

Capstans

Use For: Anchor Capstans

—Ships
BSI BS MA 34-73. 1973 Ships' Deck Machinery; Capstans. 7 pp.
ISO 4568-86. Shipbuilding—Sea-Going Vessels—Windlasses and Anchor Capstans Second Edition. 8 pp.
ISO 6219-83. Shipbuilding—Inland Vessels—Windlasses and Anchor Capstans First Edition. 8 pp.
MOD UK NES 113: Part 3-89. Requirements for Mechanical Handling Part 3: Anchor Capstans, Capstans and Cable Holders Issue 2 (11.89). 14 pp.

—Ships—Acceptance Testing
ISO 4568-86. Shipbuilding—Sea-Going Vessels—Windlasses and Anchor Capstans Second Edition. 8 pp.
ISO 6219-83. Shipbuilding—Inland Vessels—Windlasses and Anchor Capstans First Edition. 8 pp.

—Ships—Safety
ISO 4568-86. Shipbuilding—Sea-Going Vessels—Windlasses and Anchor Capstans Second Edition. 8 pp.
ISO 6219-83. Shipbuilding—Inland Vessels—Windlasses and Anchor Capstans First Edition. 8 pp.

Captafol

—Plants (Botany)—Prohibited Use
EC 90/533/EEC-90. Council Directive Amending the Annex to Directive 79/117/EEC Prohibiting the Placing on the Market and Use of Plant Protection on the Market and Use of Plant Protection Product Containing Certain Active Substances. 1 p.

Captions

Use For: Subtitles

Captions (Cont.)
—Television Program Exchange
CCIR Report 1080-1-90. International Exchange of Television Programmes with Data-Encoded Captions (Sub-Titles)—Section 11B—Ancillary Television Services. 4 pp.

CCIR QUESTION 73/11-90. International Exchange of Captioning Material for Television Programmes—Questions Concerning Study Group 11—Broadcasting Service (Television). 1 p.

Captive Fasteners
Scope Note: For additional listings, use a more specific term Use For: Captive Hardware
See Also: Captive Nuts; Fasteners
—Containers
MOD UK DSTAN 81-73-01. Guide on Use of Captive Fasteners Issue 1; Amendment 1. 34 pp.

Captive Hardware
Use: Captive Fasteners

Captive Nuts
See Also: Captive Fasteners; Nuts (Fasteners)
—Aircraft—Filler Caps
SBAC AS 2408 ISSUE 4. Captive Nut (Filler Cap). 1 p.

SBAC AS 6352 ISSUE 2. Captive, Nut (Filler Cap).

Car Parks
Use: Parking Facilities

Car Ports
Use: Parking Facilities

Car Seats (Infant)
Use: Child Safety Seats

CAR Systems
Use: Computer Assisted Retrieval Systems

Car Top Carriers
Use: Luggage Racks

Car Washes
See Also: Automotive Equipment; Cleaning Equipment and Supplies
—Electrical Equipment—Motor Operated
CSA CAN/CSA-C22. 2 NO 68-92. Motor-Operated Appliances (Household and Commercial); (Gen Instr 1 Thru 2). 115 pp.

CSA 1169 Bull. Electrical Bulletin 1169 June 27, 1978 to C22.2 NO 68. 2 pp.

Carambola
Use: Star Fruit

Caravan Parks
Use: Recreational Vehicle Parks

Caravan Sites
Use: Recreational Vehicle Parks

Caravans
Use: Recreational Vehicles

Caraway
See Also: Caraway Oil; Herbs
ISO 5561-90. Black Caraway and Blond Caraway (Carum Carvi Linnaeus), Whole—Specification Second Edition. 7 pp.

Caraway Oil
See Also: Caraway
ISO 8896-87. Oil of Caraway (Carum carvi Linnaeus) First Edition. 6 pp.

Carbadox Content Analysis
—Animal Feed
CNS N4099-82. Method of Test for Feed Additives: Determination of Carbadox (Aug)(9314).

CNS N4107-83. Method of Test for Feed Additives (Identification of Carbadox) (May)(10289).

Carbamide
Use: Urea

Carbazoles
CNS K7606-82. Chemical Reagent (Carbazole) (May)(8847).

JIS K 2436-90. Naphthalene.Anthracene. Carbazole. 46 pp.

JIS K 8258-61. Carbazole.

Carbide Content Analysis
See Also: Carbides

—Bearing Alloys—Photomicrography
ISO 5949-83. Tool Steels and Bearing Steels—Micrographic Method for Assessing the Distribution of Carbides Using Reference Photomicrographs First Edition. 12 pp.

—Calcium Silicide
MOD UK M 805/84. Examination of Calcium Silicide.

—Tool Steels—Photomicrography
ISO 5949-83. Tool Steels and Bearing Steels—Micrographic Method for Assessing the Distribution of Carbides Using Reference Photomicrographs First Edition. 12 pp.

Carbide Tools
See Also: Cemented Carbides; Cutting Tools; Grinding Wheels; Milling Cutters; Tools
SAA AS 2004-77. Dimensions of Carbide Tips and Tipped Tools Reconfirmed 1987. 36 pp.

—Centers
DIN ENGL 8012-72. Carbide Inserts for Centre Points (May). 1 p.

JIS B 4112-91. Cemented Carbide Centres. 11 pp.

—Dental Instruments
BSI BS 6828: Part 3-87. 1987 Dental Rotary Instruments Part 3: Steel and Carbide Burs (ISO 3823/1: 1986). 16 pp.

BSI BS 6828: Part 3-01. 1987 Amd 1 Dental Rotary Instruments Part 3: Specification for Steel and Carbide Burs (ISO 3823/1: 1986) (AMD 7022) May 1, 1992. 21 pp.

BSI BS 6828: Part 4-87. 1987 Dental Rotary Instruments Part 4: Steel and Carbide Finishing Burs (ISO 3823/2: 1986). 17 pp.

BSI BS 6828: Part 4-01. 1987 Amd 1 Dental Rotary Instruments Part 4: Specification for Steel and Carbide Finishing Burs (ISO 3823/2: 1986) (AMD 7020) May 1, 1992. 23 pp.

BSI BS 6828: Sec 8.2-87. 1987 Dental Rotary Instruments Part 8: Cutters Section 8.2: Carbide Laboratory Cutters. 9 pp.

BSI BS 6828: Sec 8.2-01. 1987 Amd 1 Dental Rotary Instruments Part 8: Cutters Section 8.2: Specification for Carbide Laboratory Cutters (AMD 6346) September 30, 1991. 14 pp.

BSI BS 6828: Sec 8.3-92. 1992 Dental Rotary Instruments Part 8: Cutters Section 8.3: Specification for Carbide Laboratory Cutters for Milling Machines (ISO 7787-3: 1991). 10 pp.

CEN PREN 23 823-1-90. Dental Rotary Instruments—Part 1: Steel and Carbide Burs. 3 pp.

CEN EN 23823-1-91. Dental Rotary Instruments—Part 1: Steel and Carbide Burs. 2 pp.

CEN PREN 23 823-2-90. Dental Rotary Instruments—Part 2: Steel and Carbide Finishing Burs. 3 pp.

CEN EN 23823-2-91. Dental Rotary Instruments—Part 2: Steel and Carbide Finishing Burs. 3 pp.

CEN EN 27 787 (Part 2)-90. Dentistry—Dental Rotary Instruments—Cutters Part 2: Carbide Laboratory Cutters. 4 pp.

CSA CAN/CSA-Z349. 29-35.2-M92. Dental Instruments and Equipment—Volume 1 (CAN/CSA-Z349.29-M92 to CAN/CSA-Z349.35.2 M92); (Gen Instr 1). 183 pp.

ISO 3823 Pt 1-86. Dental Rotary Instruments—Part 1: Steel and Carbide Burs First Edition. 15 pp.

ISO 3823 Pt 2-86. Dental Rotary Instruments—Part 2: Steel and Carbide Finishing Burs First Edition. 16 pp.

ISO 7787 Pt 2-92. Dental Rotary Instruments—Cutters—Part 2: Carbide Laboratory Cutters Second Edition. 14 pp.

ISO 7787 Pt 3-91. Dental Rotary Instruments—Cutters—Part 3: Carbide Laboratory Cutters for Milling Machines First Edition. 7 pp.

—Dies—Cemented
CNS B3194-89. Cemented Carbide Dies for Drawing Wires and Rods (Jun)(3093).

JIS B 4111-91. Cemented Carbide Dies for Drawing Wires and Rods. 9 pp.

—Life (Durability)
CNS B7089-83. Method of Life Test for Single Point Carbide Tools (Jan)(4262).

CNS B7092-78. Method of Life Test for High Speed Steel Single Point Tools (Mar)(4265).

ISO 8688 Pt 1-89. Tool Life Testing in Milling—Part 1: Face Milling First Edition. 32 pp.

JIS B 4011-71. Method of Life Test for Single Point Carbide Tools. 10 pp.

—Machining
BSI BS 7662-93. 1993 Application of Hard Cutting Materials for Machining by Chip Removal-Designation of the Main Groups of Chip Removal and Groups of Application (ISO 513: 1991). 11 pp.

DIN ENGL 4990-72. Groups of Application of Carbides for Machining by Chip Removal (July). 4 pp.

ISO 513-91. Application of Hard Cutting Materials for Machining by Chip Removal—Designation of the Main Groups of Chip Removal and Groups of Application Second Edition. 9 pp.

—Slitting Saws
JIS B 4115-83. Solid Carbide Metal Slitting Saws. 12 pp.

—Tips
CNS B3280-78. Single Point Carbide Tipped Tools (Mar)(4267). 10 pp.

DIN ENGL 8010-63. Carbide Tips for Drills; Point Angle 115 Degree for Heavy Loading (Sept). 1 p.

DIN ENGL 8011-63. Carbide Tips for Reamers, Countersinks, Counterbores and End Mills (Sept). 1 p.

DIN ENGL 8013-63. Carbide Tips for Drills; Point Angle 85 Degree for Light Loading (Sept). 1 p.

DIN ENGL 8037-71. Twist Drills with Parallel Shank with Carbide Tips for Drilling Metal (Aug). 2 pp.

DIN ENGL 8038-71. Twist Drills with Parallel Shank with Carbide Tips for Drilling Plastics (Thermosetting Plastics) (Aug). 2 pp.

DIN ENGL 8041-71. Twist Drills with Morse Taper Shank with Carbide Tips for Drilling Metal (Aug). 2 pp.

ISO 242-75. Carbide Tips for Brazing on Turning Tools First Edition. 4 pp.

ISO 243-75. Turning Tools with Carbide Tips—External Tools First Edition. 6 pp.

ISO 504-75. Turning Tools with Carbide Tips—Designation and Marking First Edition. 3 pp.

ISO 514-75. Turning Tools with Carbide Tips—Internal Tools First Edition. 4 pp.

SAA AS 2004-77. Dimensions of Carbide Tips and Tipped Tools Reconfirmed 1987. 36 pp.

—Tips—Cemented
CNS B7090-78. Method of Test for Cemented Carbide Tips (Mar)(4263).

CNS H3083-89. Cemented Carbide Alloy for Tip (Jun)(5338).

JIS B 4104-89. Hard Tool Material Tips. 13 pp.

JIS B 4105-91. Single Point Tools with Cemented Carbide Tips. 27 pp.

JIS B 4107-83. Side Milling Cutters Brazed with Cemented Carbide Tips. 20 pp.

JIS B 4114-83. Parallel Shank End Mills Brazed with Cemented Carbide Tips. 17 pp.

JIS H 5501-75. Cemented Carbide Alloy of Tip. 7 pp.

—Tips—Identification Systems
ISO 504-75. Turning Tools with Carbide Tips—Designation and Marking First Edition. 3 pp.

—Twist Drills
JIS B 4117-91. Solid Carbide Parallel Shank Stub Drills. 15 pp.

Carbides
Use For: Metal Carbides See Also: Calcium Carbide; Carbide Content Analysis; Cemented Carbides; Hard Metals; Silicon Carbide; Tungsten Carbides

—Carbon Content—Gravimetric Analysis
BSI BS 5600: Sec 4.1-87. (WITHDRAWN) 1987 Powder Metallurgical Materials and Products Part 4: Methods of Testing and Chemical Analysis of Hardmetals Section 4.1: Determination of Total Carbon: Gravimetric Method (Superseded by BS EN 23907: 1993). 5 pp.

BSI BS 5600: Sec 4.2-87. 1987 Powder Metallurgical Materials and Products Part 4: Methods of Testing and Chemical Analysis of Hardmetals Section 4.2: Determination of Insoluble (Free) Carbon: Gravimetric Method. 4 pp.

BSI BS EN 23907-93. 1993 Hardmetals—Determination of Total Carbon Content—Gravimetric Method (ISO 3907: 1985) (V). 10 pp.

BSI BS EN 23908-93. 1993 Hardmetals—Determination of Insoluble (Free) Carbon Content—Gravimetric Method (ISO 3908: 1985) (V). 10 pp.

CEN EN 23908-93. Hardmetals—Determination of Insoluble (Free) Carbon Content—Gravimetric Method (ISO 3908: 1985). 5 pp.

ISO 3907-85. Hardmetals—Determination of Total Carbon Content—Gravimetric Method Second Edition. 5 pp.

ISO 3908-85. Hardmetals—Determination of Insoluble (Free) Carbon Content—Gravimetric Method Second Edition (CEN EN 23908: 1993). 4 pp.

Carbides (Cont.)

—Metallic Element Content—X-Ray Fluorescence Spectrometry
BSI BS 5600: Sec 4.10-80. (WITHDRAWN) 1980 Powder Metallurgical Materials and Products Part 4: Methods of Testing and Chemical Analysis of Hardmetals Section 4.10: Determination of Contents of Metallic Elements by X-ray Fluorescence: Solution Method. 4 pp.

BSI BS 5600: Sec 4.15-80. (WITHDRAWN) 1980 Powder Metallurgical Materials and Products Part 4: Methods of Testing and Chemical Analysis of Hardmetals Section 4.15: Determination of Contents of Metallic Elements by X-ray Fluorescence: Fusion Method. 4 pp.

BSI BS EN 24503-93. 1993 Hardmetals—Determination of Contents of Metallic Elements by X-Ray Fluorescence—Fusion Method (ISO 4503: 1978) (V). 10 pp.

BSI BS EN 24883-93. 1993 Hardmetals—Determination of Contents of Metallic Elements by X-Ray Fluorescence—Solution Method (ISO 4883: 1978) (V). 9 pp.

CEN EN 24503-93. Hardmetals—Determination of Contents of Metallic Elements by X-Ray Fluorescence—Fusion Method (ISO 4503: 1978). 5 pp.

CEN EN 24883-93. Hardmetals—Determination of Contents of Metallic Elements by X-Ray Fluorescence—Solution Method (ISO 4883: 1978). 4 pp.

ISO 4503-78. Hardmetals—Determination of Contents of Metallic Elements by X-Ray Fluorescence—Fusion Method First Edition (CEN EN 24503: 1993). 5 pp.

ISO 4883-78. Hardmetals—Determination of Contents of Metallic Elements by X-Ray Fluorescence—Solution Method First Edition (CEN EN 24883: 1993). 4 pp.

—Titanium Content—Photometry
BSI BS 5600: Sec 4.12-80. 1980 Powder Metallurgical Materials and Products Part 4: Methods of Testing and Chemical Analysis of Hardmetals Section 4.12: Determination of Titanium Content; Photometric Peroxide Method. 4 pp.

ISO 4501-78. Hardmetals—Determination of Titanium—Photometric Peroxide Method First Edition. 5 pp.

Carbohydrates
Scope Note: Use a more specific term
See: Alginates; Cellulose; Dextrins; Galactose; Glucose Syrups; Inulin; Lactose; Lactose Monohydrate; Lithium Chloride; Maltose; Maltose Monohydrate; Maltose Syrups; Starches; Sucrose; Sugars; Xylose

Carbolic Acid
Use: Phenol

Carbominerite Content Analysis

—Anthracite—Petrography
BSI BS 6127: Part 4-90. 1990 Petrographic Analysis of Bituminous Coal and Anthracite Part 4: Method of Determining Microlithotype, Carbominerite and Minerite Composition. 10 pp.

ISO 7404 Pt 4-88. Methods for the Petrographic Analysis of Bituminous Coal and Anthracite—Part 4: Method of Determining Microlithotype, Carbominerite and Minerite Composition First Edition. 8 pp.

—Bituminous Coal—Petrography
BSI BS 6127: Part 4-90. 1990 Petrographic Analysis of Bituminous Coal and Anthracite Part 4: Method of Determining Microlithotype, Carbominerite and Minerite Composition. 10 pp.

ISO 7404 Pt 4-88. Methods for the Petrographic Analysis of Bituminous Coal and Anthracite—Part 4: Method of Determining Microlithotype, Carbominerite and Minerite Composition First Edition. 8 pp.

Carbon
Scope Note: For additional listings, use a more specific term *Use For:* Carbon Materials; Projector Carbons *See Also:* Activated Carbon; Carbon Black; Carbon Black Content Analysis; Carbon Content Analysis; Cast Iron; Charcoal; Coke; Diamonds (Industrial); Graphite

JIS R 7301-60. Projector Carbons (Cinema Carbons).

—Blocks
JIS R 7211-91. Carbon Block. 7 pp.
JIS R 7212-79. Testing Methods for Carbon Blocks (R 1985). 18 pp.

—Modulus of Elasticity—Resonance Method
DIN ENGL 51915-86. Testing of Carbonaceous Materials; Determination of Dynamic Modulus of Elasticity by the Resonance Method; Solids (Apr). 3 pp.

—Sampling—Cathodic Blocks
ISO 8007-85. Carbonaceous Materials Used in the Production of Aluminium—Sampling from Cathodic Blocks and Prebaked Anodes—General First Edition. 4 pp.

Carbon Bisulfide
Use: Carbon Disulfide

Carbon Black
Use For: Lampblack *See Also:* Carbon; Carbon Black Content Analysis; Charcoal; India Ink; Rubber

CNS K1125-67. Carbon Black for Rubber Industry (Aug)(2746).
CNS K6246-67. Method of Test for Carbon Black (Aug)(2747).
JIS K 5107-65. Carbon Black (Pigment). 5 pp.
JIS K 6221-82. Testing Methods of Carbon Black for Rubber Industry. 22 pp.

—Absorption Number
BSI BS 5293: Part 17-93. 1993 Sampling and Testing Carbon Black for Use in the Rubber Industry Part 17: Method for Determination of Dibutyl Phthalate Absorption Number Using an Absorptometer (ISO 4656-1: 1992) (V). 15 pp.

BSI BS 5293: Part 17-88. 1988 Sampling and Testing Carbon Black for Use in the Rubber Industry Part 17: Method for Determination of Dibutylphthalate Absorption Number Using an Absorptometer. 11 pp.

BSI BS 5293: Part 18-92. 1992 Sampling and Testing Carbon Black for Use in the Rubber Industry Part 18: Method for Determination of Dibutyl Phthalate Absorption Number Using Plastograph or Plasticorder (ISO 4656-2: 1991). 11 pp.

DIN ENGL 53601-78. Testing of Carbon Blacks; Determination of the Dibutylphthalate Absorption of Carbon Blacks (Dec). 3 pp.

ISO 4656 Pt 1-92. Rubber Compounding Ingredients—Carbon Black—Determination of Dibutyl Phthalate Absorption Number—Part 1: Method Using Absorptometer Third Edition. 12 pp.

ISO 4656 Pt 2-91. Rubber Compounding Ingredients—Carbon Black—Determination of Dibutyl Phthalate Absorption Number—Part 2: Method Using Plastograph or Plasticorder Second Edition. 9 pp.

—Absorption Number—Sampling
BSI BS 5293: Part 19-91. 1991 Sampling and Testing Carbon Black for Use in the Rubber Industry Part 19: Preparation of Samples for Determination of Dibutyl Phthalate Absorption Number (Compressed Sample) (ISO 6894: 1991). 10 pp.

ISO 6894-91. Rubber Compounding Ingredients—Carbon Black—Preparation of Samples for Determination of Dibutylphthalate Absorption Number (Compressed Sample) Second Edition. 7 pp.

—Area—Adsorption
BSI BS 5293: Part 12-88. 1988 Sampling and Testing Carbon Black for Use in the Rubber Industry Part 12: Methods for Determination of Surface Area by Surfactant Absorption. 15 pp.

ISO 4652-81. Rubber Compounding Ingredients—Carbon Black—Determination of Specific Surface Area—Nitrogen Adsorption Methods First Edition. 17 pp.

ISO 6810-85. Rubber Compounding Ingredients—Carbon Black—Determination of Surface Area—Surfactant Adsorption Methods First Edition. 13 pp.

—Ash Content
BSI BS 5293: Part 7-91. 1991 Sampling and Testing Carbon Black for Use in the Rubber Industry Part 7: Method for Determination of Ash Content (ISO 1125: 1990). 6 pp.

ISO 1125-90. Rubber Compounding Ingredients—Carbon Black—Determination of Ash Content Second Edition. 4 pp.

—Bulk Density
DIN ENGL 53600-76. Testing of Carbon Blacks; Determination of the Bulk Density of Pelletized Carbon Blacks for Use in the Rubber Industry (May). 2 pp.

—Compression Testing
ISO TR8942-88. Rubber Compounding Ingredients—Carbon Black—Determination of Individual Pellet Crushing Strength First Edition. 4 pp.

—Iodine Number
CNS K6514-80. Methods of Test for Iodine Absorption Number of Carbon Black (Jul)(5818).

—Iodine Number—Volumetric Analysis
BSI BS 5293: Part 10-90. 1990 Sampling and Testing Carbon Black for Use in the Rubber Industry Part 10: Method for Determination of Iodine Adsorption Number. 8 pp.

ISO 1304-85. Rubber Compounding Ingredients—Carbon Black—Determination of Iodine Adsorption Number—Titrimetric Method Second Edition. 6 pp.

—Light Transmission
BSI BS 5293: Part 14-91. 1991 Sampling and Testing Carbon Black for Use in the Rubber Industry Part 14: Method for Determination of Light Transmittance of Toluene Extract (Rapid Method) (ISO 3858-1: 1990). 7 pp.

BSI BS 5293: Part 15-91. 1991 Sampling and Testing Carbon Black for Use in the Rubber Industry Part 15: Method for Determination of Light Transmittance of Toluene Extract for Product Evaluation (ISO 3858-2: 1990). 7 pp.

ISO 3858 Pt 1-90. Carbon Black for Use in the Rubber Industry—Determination of Light Transmittance of Toluene Extract—Part 1: Rapid Method Third Edition. 6 pp.

ISO 3858 Pt 2-90. Carbon Black for Use in the Rubber Industry—Determination of Light Transmittance of Toluene Extract—Part 2: Method for Product Evaluation Third Edition. 6 pp.

—Loss of Mass—High Temperature Testing
BSI BS 5293: Part 5-92. 1992 Sampling and Testing Carbon Black for Use in the Rubber Industry Part 5: Method for Determination of Loss of Mass on Heating (ISO 1126: 1992). 7 pp.

BSI BS 5293: Part 5-88. 1988 Sampling and Testing Carbon Black for Use in the Rubber Industry Part 5: Method for Determination of Loss of Mass on Heating. 3 pp.

CNS K6510-80. Methods of Test for Heating Loss of Carbon Black (Jul)(5814).

ISO 1126-92. Rubber Compounding Ingredients—Carbon Black—Determination of Loss on Heating Third Edition. 5 pp.

—Particle Size Distribution
BSI BS 5293: Part 4-90. 1990 Sampling and Testing Carbon Black for Use in the Rubber Industry Part 4: Method for Determination of Pellet Size Distribution (ISO 8511: 1987). 4 pp.

CNS K6512-80. Methods of Test for Fines Content of Pelleted Carbon Black (Jul)(5816).

CNS K6513-80. Methods of Test for Pellet Size Distribution of Carbon Black (Jul)(5817).

ISO 8511-87. Rubber Compounding Ingredients—Carbon Black—Determination of Pellet Size Distribution First Edition. 4 pp.

—PH
CNS K6509-80. Methods of Test for pH Value of Carbon Black (Jul)(5813).

—Pigments
CNS K2016-75. Carbon Black for Paints (Feb)(701). 1 p.
CNS K2103-80. Lamp Black (Jul)(5868). 1 p.

—Pour Density
BSI BS 5293-76. 1976 Amd 4 Sampling and Testing Carbon Black for Use in the Rubber Industry (AMD 6378) May 31, 1991. 22 pp.

BSI BS 5293: Part 2-90. 1990 Sampling and Testing Carbon Black for Use in the Rubber Industry Part 2: Method for Determination of Pour Density (ISO 1306: 1987). 3 pp.

CNS K6511-80. Methods of Test for Poor Density of Pelleted Carbon Black (Jul)(5815).

ISO 1306-87. Rubber Compounding Ingredients—Carbon Black (Pelletized)—Determination of Pour Density Third Edition. 4 pp.

—Propellants
MOD UK TS 10297. Carbon Black (Furnace-Black).

—Residue-on-Ignition Determination
DIN ENGL 53586-88. Testing of Carbon Black; Determination of Residue on Ignition of Carbon Black (Sept). 3 pp.

—Sampling
BSI BS 5293-76. 1976 Amd 4 Sampling and Testing Carbon Black for Use in the Rubber Industry (AMD 6378) May 31, 1991. 22 pp.

BSI BS 5293: Part 1-90. 1990 Sampling and Testing Carbon Black for Use in the Rubber Industry Part 1: Methods of Sampling (ISO 1124: 1988). 4 pp.

ISO 1124-88. Rubber Compounding Ingredients—Carbon Black Shipment Sampling Procedures Third Edition; (Supersedes 1310). 5 pp.

—Sieve Analysis
BSI BS 5293-76. 1976 Amd 4 Sampling and Testing Carbon Black for Use in the Rubber Industry (AMD 6378) May 31, 1991. 22 pp.

INTERNATIONAL AND NON-U.S. NATIONAL STANDARDS
SUBJECT INDEX
Carbon

Carbon Black *(Cont.)*

—Sieve Analysis *(Cont.)*
BSI BS 5293: Part 3-90. 1990 Sampling and Testing Carbon Black for Use in the Rubber Industry Part 3: Method for Determination of Fines Content (ISO 1435: 1988). 4 pp.

BSI BS 5293: Part 6-92. 1992 Sampling and Testing Carbon Black for Use in the Rubber Industry Part 6: Method for Determination of Sieve Residue (ISO 1437: 1992). 8 pp.

BSI BS 5293: Part 6-88. 1988 Sampling and Testing Carbon Black for Use in the Rubber Industry Part 6: Method for Determination of Sieve Residue. 4 pp.

ISO 1435-88. Rubber Compounding Ingredients—Carbon Black (Pelletized)—Determination of Fines Content Third Edition. 4 pp.

ISO 1437-92. Rubber Compounding Ingredients—Carbon Black—Determination of Sieve Residue Third Edition. 5 pp.

—Solvent Extraction
BSI BS 5293: Part 16-90. 1990 Sampling and Testing Carbon Black for Use in the Rubber Industry Part 16: Method for Determination of Solvent Extractable Material (ISO 6209: 1988). 8 pp.

ISO 6209-88. Rubber Compounding Ingredients—Carbon Black—Determination of Solvent Extractable Material Third Edition. 6 pp.

—Standard Reference Blacks
ISO 6809-89. Rubber Compounding Ingredients—Carbon Black—Standard Reference Blacks Second Edition. 4 pp.

—Sulfur Content—Colorimetric Analysis
ISO 1138-81. Rubber Compounding Ingredients—Carbon Black—Determination of Sulphur Content Second Edition. 5 pp.

—Sulfur Content—Combustion Analysis
DIN ENGL 53584-77. Testing of Carbon Black; Determination of the Sulphur Content (Sept). 7 pp.

ISO 1138-81. Rubber Compounding Ingredients—Carbon Black—Determination of Sulphur Content Second Edition. 5 pp.

—Tinting Strength
BSI BS 5293: Part 13-92. 1992 Sampling and Testing Carbon Black for Use in the Rubber Industry Part 13: Method for Determination of Tinting Strength (ISO 5435: 1991). 12 pp.

ISO 5435-91. Rubber Compounding Ingredients—Carbon Black—Determination of Tinting Strength Second Edition. 9 pp.

—Volatile Matter Content
DIN ENGL 53552-77. Testing of Carbon Black; Determination of the Amount of Components of Carbon Black That Are Volatile on Heating (Sept). 2 pp.

Carbon Black Content Analysis
See Also: Carbon; Carbon Black

—Cable Insulation
DIN VDE 0472 Pt 702-89. Testing of Cables, Wires and Flexible Cords; Carbon Black and Filler Content of PE Sheaths (Feb). 8 pp.

—Elastomers—Pyrolysis
DIN ENGL 53585-73. Determination of the Carbon Black Content in Rubber and Elastomers (July). 2 pp.

—Olefin Resins
CNS K6638-81. Method of Test for Carbon Black in Olefin Plastics (Mar)(7049).

—Pipe Fittings—Calcination
ISO 6964-86. Polyolefin Pipes and Fittings—Determination of Carbon Black Content by Calcination and Pyrolysis—Test Method and Basic Specification First Edition. 4 pp.

—Pipe Fittings—Pyrolysis
ISO 6964-86. Polyolefin Pipes and Fittings—Determination of Carbon Black Content by Calcination and Pyrolysis—Test Method and Basic Specification First Edition. 4 pp.

—Plastic Pipes—Calcination
ISO 6964-86. Polyolefin Pipes and Fittings—Determination of Carbon Black Content by Calcination and Pyrolysis—Test Method and Basic Specification First Edition. 4 pp.

—Plastic Pipes—Pyrolysis
ISO 6964-86. Polyolefin Pipes and Fittings—Determination of Carbon Black Content by Calcination and Pyrolysis—Test Method and Basic Specification First Edition. 4 pp.

Carbon Black Content Analysis *(Cont.)*

—Polyethylene—Microscopy Analysis
BSI BS 2782:Pt8: METH 823A-B-78. 1978 Methods of Testing Plastics Part 8: Other Properties Method 823A and 823B: Methods for the Assessment of Carbon Black Dispersion in Polyethylene Using a Microscope. 5 pp.

—Polyolefins
BSI BS 2782:Pt4: METH 452B-93. 1993 Methods of Testing Plastics Part 4: Chemical Properties Method 452B: Determination of Carbon Black Content of Polyolefin Compound. 6 pp.

BSI BS 2782:Pt4: METH 452B-78. 1978 Methods of Testing Plastics Part 4: Chemical Properties Method 452B: Determination of Carbon Black Content of Polyolefin Compound. 2 pp.

—Rubber
SAA AS 1683.17-81. Methods of Test for Elastomers—Part 17: Vulcanized Rubber—Determination of Carbon Black Content—Pyrolytic Method. 6 pp.

—Rubber—Chemical Analysis
BSI BS 903: Part B14-64. (WITHDRAWN) 1964 Methods of Testing Vulcanized Rubber Part B14: Determination of Carbon Black (Superseded by BS 7164: Part 14: 1990). 15 pp.

BSI BS 7164: Part 14-90. 1990 Chemical Tests for Raw and Vulcanized Rubber Part 14: Methods for Determination of Carbon Black Content (ISO 1408: 1987). 15 pp.

ISO 1408-87. Rubber—Determination of Carbon Black Content—Pyrolytic and Chemical Degradation Methods Second Edition. 12 pp.

—Rubber—Pyrolysis
BSI BS 903: Part B14-64. (WITHDRAWN) 1964 Methods of Testing Vulcanized Rubber Part B14: Determination of Carbon Black (Superseded by BS 7164: Part 14: 1990). 15 pp.

BSI BS 7164: Part 14-90. 1990 Chemical Tests for Raw and Vulcanized Rubber Part 14: Methods for Determination of Carbon Black Content (ISO 1408: 1987). 15 pp.

DIN ENGL 53585-73. Determination of the Carbon Black Content in Rubber and Elastomers (July). 2 pp.

ISO 1408-87. Rubber—Determination of Carbon Black Content—Pyrolytic and Chemical Degradation Methods Second Edition. 12 pp.

—Styrene Butadiene Rubber
BSI BS 5293: Part 20-92. 1992 Sampling and Testing Carbon Black for Use in the Rubber Industry Part 20: Method for Evaluation in Styrene-Butadiene Rubbers (ISO 3257: 1992). 8 pp.

ISO 3257-92. Rubber Compounding Ingredients—Carbon Black—Method of Evaluation in Styrene-Butadiene Rubbers Third Edition. 6 pp.

—Vulcanized Rubber—Pyrolysis
CNS K6524-80. Determination of Carbon Black Content of Vulcanized Rubber (Pyrolytic Method) (Jul)(5841).

Carbon Brush Holders
Use: Brush Holders (Machine Components)

Carbon Brushes
Use: Brushes (Machine Components)

Carbon Content Analysis
See Also: Carbon; Fixed Carbon Content Analysis

—Aggregates—Concretes
BSI BS 3681: Part 2-73. (WITHDRAWN) 1973 Methods for the Sampling and Testing of Lightweight Aggregates for Concrete Part 2: Metric Units (Superseded by BS 3797: 1990). 19 pp.

—Asphalts
CNS K6207-65. Determination of Fixed Carbon in Asphalt and Coal Tar (Sep)(2489)(R 1971).

—Calcium Silicide
MOD UK M 805/84. Examination of Calcium Silicide.

—Calcium Silicon
CNS G2105-81. Method of Chemical Analysis for Carbon in Calcium-Silicon (Nov)(8124).

—Carbides—Gravimetric Analysis
BSI BS 5600: Sec 4.1-87. (WITHDRAWN) 1987 Powder Metallurgical Materials and Products Part 4: Methods of Testing and Chemical Analysis of Hardmetals Section 4.1: Determination of Total Carbon: Gravimetric Method (Superseded by BS EN 23907: 1993). 5 pp.

Carbon Content Analysis *(Cont.)*

—Carbides—Gravimetric Analysis *(Cont.)*
BSI BS 5600: Sec 4.2-87. 1987 Powder Metallurgical Materials and Products Part 4: Methods of Testing and Chemical Analysis of Hardmetals Section 4.2: Determination of Insoluble (Free) Carbon: Gravimetric Method. 4 pp.

BSI BS EN 23907-93. 1993 Hardmetals—Determination of Total Carbon Content—Gravimetric Method (ISO 3907: 1985) (V). 10 pp.

BSI BS EN 23908-93. 1993 Hardmetals—Determination of Insoluble (Free) Carbon Content—Gravimetric Method (ISO 3908: 1985) (V). 10 pp.

CEN EN 23908-93. Hardmetals—Determination of Insoluble (Free) Carbon Content—Gravimetric Method (ISO 3908: 1985). 5 pp.

ISO 3907-85. Hardmetals—Determination of Total Carbon Content—Gravimetric Method Second Edition. 5 pp.

ISO 3908-85. Hardmetals—Determination of Insoluble (Free) Carbon Content—Gravimetric Method Second Edition (CEN EN 23908: 1993). 4 pp.

—Cast Iron—Gravimetric Analysis
BSI BS 6200: SUBSEC 3.8.5-91. 1991 Sampling and Analysis of Iron, Steel and Other Ferrous Metals Part 3: Methods of Analysis Section 3.8: Determination of Carbon Subsection 3.8.5: Cast Iron and Pig Iron: Gravimetric Method for the Determination of Non-Combined Carbon. 6 pp.

ISO 437-82. Steel and Cast Iron—Determination of Total Carbon Content—Combustion Gravimetric Method First Edition. 6 pp.

—Chromium
CNS G2119-82. Methods of Chemical Analysis for Chromium and Carbon in Chromium Metal (Jan)(8403).

—Coal
BSI BS 1016: Part 3-73. (WITHDRAWN) 1973 Amd 2 Methods for the Analysis and Testing of Coal and Coke Part 3: Proximate Analysis of Coal (Superseded by BS 1016: Sections 104.1, 104.2, 104.3, and 104.4: 1991). 23 pp.

CNS M3143-84. Method for Calculation of Fixed Carbon of Coal and Coke (Mar)(10824).

SAA AS 1038.6.1-86. Methods for the Analysis and Testing of Coal and Coke—Part 6.1: Ultimate Analysis of Higher Rank Coal—Determination of Carbon and Hydrogen (R 1992). 5 pp.

SAA AS 1038.7-81. Methods for the Analysis and Testing of Coal and Coke—Part 7: Ultimate Analysis of Coke. 21 pp.

SAA AS 1038.23-84. Methods for the Analysis and Testing of Coal and Coke—Part 23: Determination of Carbonate Carbon in Higher Rank Coal. 3 pp.

—Coal—Combustion
CNS M3153-84. Method for Determination of Carbons and Hydrogen of Coal and Coke (Sheffield High Temperature Method) (Mar)(10834).

ISO 609-75. Coal and Coke—Determination of Carbon and Hydrogen—High Temperature Combustion Method First Edition. 9 pp.

—Coal—Liebig Method
CNS M3144-84. Method for Determination of Carbon and Hydrogen of Coal and Coke (Liebig Method) (Mar) (10825).

ISO 625-75. Coal and Coke—Determination of Carbon and Hydrogen—Liebig Method First Edition. 9 pp.

—Coal Tar
CNS K6207-65. Determination of Fixed Carbon in Asphalt and Coal Tar (Sep)(2489)(R 1971).

—Coke
CNS M3143-84. Method for Calculation of Fixed Carbon of Coal and Coke (Mar)(10824).

—Coke—Combustion
CNS M3153-84. Method for Determination of Carbons and Hydrogen of Coal and Coke (Sheffield High Temperature Method) (Mar)(10834).

ISO 609-75. Coal and Coke—Determination of Carbon and Hydrogen—High Temperature Combustion Method First Edition. 9 pp.

—Coke—Liebig Method
CNS M3144-84. Method for Determination of Carbon and Hydrogen of Coal and Coke (Liebig Method) (Mar) (10825).

ISO 625-75. Coal and Coke—Determination of Carbon and Hydrogen—Liebig Method First Edition. 9 pp.

—Copper—Combustion
CEN PREN 723-92. Wrought Copper—Combustion Methods for Determination of Surface Carbon on Tubes. 9 pp.

—Exhaust Gases
CNS D4009-83. Reflection Type Smokemeters for Measuring Smoke Concentration of Exhaust Gas for Diesel Automobiles (Jan)(9845).

INDUSTRY STANDARDS

Carbon Content Analysis (Cont.)

—Exhaust Gases (Cont.)
JIS D 1101-85. Diesel Engine Smoke Measurement. 5 pp.
JIS D 8004-71. Reflection Type Smokemeters for Measuring Carbon Concentration of Exhaust Smoke for Diesel Automobiles. 12 pp.

—Ferrochromium
CNS G2185-83. Method of Chemical Analysis for Carbon in Ferrochromium (Aug)(10505).

—Ferronickel—Infrared Analysis
BSI BS 6783: Part 5-86. 1986 Sampling and Analysis of Nickel, Ferronickel and Nickel Alloys Part 5: Method for Determining of Carbon in Nickel Ferronickel and Nickel Alloys by Infra-Red Absorption After Induction Furnace Combustion. 11 pp.
ISO 7524-85. Nickel, Ferronickel and Nickel Alloys—Determination of Carbon Content—Infra-Red Absorption Method After Induction Furnace Combustion First Edition. 10 pp.

—Ferrosilicochromium
CNS G2175-83. Method of Chemical Analysis for Carbon in Ferrosilico-Chromium (May)(10261).

—Ferrotungsten
CNS G2149-83. Methods of Chemical Analysis for Carbon in Ferrotungsten (Jan)(9860).

—Hard Metals—Gravimetric Analysis
BSI BS 5600: Sec 4.1-87. (WITHDRAWN) 1987 Powder Metallurgical Materials and Products Part 4: Methods of Testing and Chemical Analysis of Hardmetals Section 4.1: Determination of Total Carbon: Gravimetric Method (Superseded by BS EN 23907: 1993). 5 pp.
BSI BS 5600: Sec 4.2-87. 1987 Powder Metallurgical Materials and Products Part 4: Methods of Testing and Chemical Analysis of Hardmetals Section 4.2: Determination of Insoluble (Free) Carbon: Gravimetric Method. 4 pp.
BSI BS EN 23907-93. 1993 Hardmetals—Determination of Total Carbon Content—Gravimetric Method (ISO 3907: 1985) (V). 10 pp.
BSI BS EN 23908-93. 1993 Hardmetals—Determination of Insoluble (Free) Carbon Content—Gravimetric Method (ISO 3908: 1985) (V). 10 pp.
CEN EN 23908-93. Hardmetals—Determination of Insoluble (Free) Carbon Content—Gravimetric Method (ISO 3908: 1985). 5 pp.
ISO 3907-85. Hardmetals—Determination of Total Carbon Content—Gravimetric Method Second Edition. 5 pp.
ISO 3908-85. Hardmetals—Determination of Insoluble (Free) Carbon Content—Gravimetric Method Second Edition (CEN EN 23908: 1993). 4 pp.

—Hard Metals—Metallography
BSI BS 5600: Sec 4.14-80. (WITHDRAWN) 1980 Powder Metallurgical Materials and Products Part 4: Methods of Testing and Chemical Analysis of Hardmetals Section 4.14: Metallographic Determination of Porosity and Uncombined Carbon (Superseded by BS EN 24505: 1993). 7 pp.
BSI BS EN 24505-93. 1993 Hardmetals—Metallographic Determination of Porosity and Uncombined Carbon (ISO 4505: 1978) (V). 12 pp.
CEN EN 24505-93. Hardmetals—Metallographic Determination of Porosity and Uncombined Carbon (ISO 4505: 1978). 7 pp.
ISO 4505-78. Hardmetals—Metallographic Determination of Porosity and Uncombined Carbon First Edition (CEN EN 24505: 1993). 7 pp.

—Iron
CNS G2228-84. Method of Determination for Carbon in Iron and Steel (Oct)(11069).
JIS G 1211-81. Methods for Determination of Carbon in Iron and Steel. 14 pp.

—Iron—Gravimetric Analysis
BSI BS EN 10 036-91. 1991 Chemical Analysis of Ferrous Materials Determination of Total Carbon in Steels and Irons. Gravimetric Method After Combustion in a Stream of Oxygen (V). 10 pp.
CEN EN 10 036-89. Chemical Analysis of Ferous Marterials Determination of Total Carbon in Steels and Irons Gravimetric After Combustion in a Stream of Oxygen. 6 pp.
DIN ENGL EN 10036-90. Chemical Analysis of Ferrous Materials; Determination of Total Carbon in Steel and Iron; Gravimetric Method After Combustion in a Stream of Oxygen (Apr). 7 pp.
SAA AS 1050.2-84. Methods for the Analysis of Iron and Steel—Part 2: Determination of Carbon Content (Gravimetric Method) See Also AS K1. 5 pp.

—Iron—Infrared Analysis
ISO 9686-92. Direct Reduced Iron—Determination of Carbon and/or Sulfur Content—High Frequency Combustion Method with Infrared Measurement First Edition. 15 pp.

Carbon Content Analysis (Cont.)

—Iron—Infrared Analysis (Cont.)
SAA AS 1050.32-84. Methods for the Analysis of Iron and Steel—Part 32: Determination of Carbon Content (Infrared Method). 3 pp.

—Manganese
CNS G2096-81. Method of Chemical Analysis for Carbon in Manganese Metal (Oct)(8004).

—Metals
CNS G2229-84. General Rules of Determination for Carbon in Metallic Materials (Oct)(11070).
JIS Z 2615-79. General Rules for Determination of Carbon in Metallic Materials (R 1984). 63 pp.

—Nickel
JIS H 1427-79. Methods for Determination of Carbon in Nickel Materials for Electron Tubes.

—Nickel Alloys—Infrared Analysis
BSI BS 6783: Part 5-86. 1986 Sampling and Analysis of Nickel, Ferronickel and Nickel Alloys Part 5: Method for Determining of Carbon in Nickel Ferronickel and Nickel Alloys by Infra-Red Absorption After Induction Furnace Combustion. 11 pp.
ISO 7524-85. Nickel, Ferronickel and Nickel Alloys—Determination of Carbon Content—Infra-Red Absorption Method After Induction Furnace Combustion First Edition. 10 pp.

—Nickel Castings
JIS H 1275-88. Methods for Determination of Carbon in Nickel and Nickel Alloy Castings. 11 pp.

—Nickel—Infrared Analysis
BSI BS 6783: Part 5-86. 1986 Sampling and Analysis of Nickel, Ferronickel and Nickel Alloys Part 5: Method for Determining of Carbon in Nickel Ferronickel and Nickel Alloys by Infra-Red Absorption After Induction Furnace Combustion. 11 pp.
ISO 7524-85. Nickel, Ferronickel and Nickel Alloys—Determination of Carbon Content—Infra-Red Absorption Method After Induction Furnace Combustion First Edition. 10 pp.

—Nickel—Photometry
BSI BS 3727: Part 3-66. 1966 Amd 1 Methods for the Analysis of Nickel for Use in Electronic Tubes and Valves Part 3: Determination of Carbon (AMD 6007) July 31, 1989. 20 pp.

—Pig Iron—Gravimetric Analysis
BSI BS 6200: SUBSEC 3.8.5-91. 1991 Sampling and Analysis of Iron, Steel and Other Ferrous Metals Part 3: Methods of Analysis Section 3.8: Determination of Carbon Subsection 3.8.5: Cast Iron and Pig Iron: Gravimetric Method for the Determination of Non-Combined Carbon. 6 pp.

—Pigments
CGSB 1-GP-71 METH 74.1-74. Methods of Testing Paints and Pigments Carbon and Insoluble Matter. 1 p.

—Silicon
CNS G2110-82. Method of Chemical Analysis for Carbon in Metallic Silicon (Electric Method) (Jan)(8284).

—Silicon Carbide
DIN ENGL 51075 Pt 5-82. Testing of Ceramic Materials; Chemical Analysis of Silicon Carbide; Indirect Determination of the Free Carbon Content (Oct). 2 pp.

—Silicon Carbide—Conductometric Analysis
DIN ENGL 51075 Pt 2-84. Testing of Ceramic Materials; Chemical Analysis of Silicon Carbide; Direct Determination of the Free Carbon Content (Mar). 7 pp.
DIN ENGL 51075 Pt 3-82. Testing of Ceramic Materials; Chemical Analysis of Silicon Carbide; Determination of the Total Carbon Content (Oct). 3 pp.

—Silicon Carbide—Coulometric Analysis
DIN ENGL 51075 Pt 2-84. Testing of Ceramic Materials; Chemical Analysis of Silicon Carbide; Direct Determination of the Free Carbon Content (Mar). 7 pp.
DIN ENGL 51075 Pt 3-82. Testing of Ceramic Materials; Chemical Analysis of Silicon Carbide; Determination of the Total Carbon Content (Oct). 3 pp.

—Silicon Carbide—Gravimetric Analysis
DIN ENGL 51075 Pt 2-84. Testing of Ceramic Materials; Chemical Analysis of Silicon Carbide; Direct Determination of the Free Carbon Content (Mar). 7 pp.

Carbon Content Analysis (Cont.)

—Silicon Carbide—Gravimetric Analysis (Cont.)
DIN ENGL 51075 Pt 3-82. Testing of Ceramic Materials; Chemical Analysis of Silicon Carbide; Determination of the Total Carbon Content (Oct). 3 pp.

—Silicon—Conductimetric Analysis
CNS G2109-82. Method of Chemical Analysis for Carbon in Metallic Silicon (Conductivity Method) (Jan)(8283).

—Silicon—Volumetric Analysis
CNS G2108-82. Method of Chemical Analysis for Carbon in Metallic Silicon (Neutralization Titration Method) (Jan)(8282).

—Sintered Materials
BSI BS 5600: Sec 3.3-87. 1987 Powder Metallurgical Materials and Products Part 3: Methods of Test-ing Sintered Metal Materials and Products, Excluding Hardmetals Section 3.3: Preparation of Samples for Chemical Analysis for Determination of Carbon Content. 4 pp.
ISO 7625-83. Sintered Metal Materials, Excluding Hardmetals—Preparation of Samples for Chemical Analysis for Determination of Carbon Content First Edition. 4 pp.

—Steels
CNS G2228-84. Method of Determination for Carbon in Iron and Steel (Oct)(11069).
JIS G 1211-81. Methods for Determination of Carbon in Iron and Steel. 14 pp.

—Steels—Coulometric Analysis
ISO TR4830 Pt IV-78. Steel—Determination of Low Carbon Contents—Part IV: Coulometric Method After Combustion First Edition. 8 pp.

—Steels—Gravimetric Analysis
BSI BS EN 10 036-91. 1991 Chemical Analysis of Ferrous Materials Determination of Total Carbon in Steels and Irons. Gravimetric Method After Combustion in a Stream of Oxygen (V). 10 pp.
CEN EN 10 036-89. Chemical Analysis of Ferrous Marterials Determination of Total Carbon in Steels and Irons Gravimetric After Combustion in a Stream of Oxygen. 6 pp.
DIN ENGL EN 10036-90. Chemical Analysis of Ferrous Materials; Determination of Total Carbon in Steel and Iron; Gravimetric Method After Combustion in a Stream of Oxygen (Apr). 7 pp.
ISO 437-82. Steel and Cast Iron—Determination of Total Carbon Content—Combustion Gravimetric Method First Edition. 6 pp.
SAA AS 1050.2-84. Methods for the Analysis of Iron and Steel—Part 2: Determination of Carbon Content (Gravimetric Method) See Also AS K1. 5 pp.

—Steels—Infrared Analysis
ISO 9556-89. Steel and Iron—Determination of Total Carbon Content—Infrared Absorption Method After Combustion in an Induction Furnace First Edition; (Corrected and Reprinted -1989). 11 pp.
SAA AS 1050.32-84. Methods for the Analysis of Iron and Steel—Part 32: Determination of Carbon Content (Infrared Method). 3 pp.

—Steels—Volumetric Analysis
BSI BS 6200: SUBSEC 3.8.2-91. 1991 Sampling and Analysis of Iron, Steel and Other Ferrous Metals Part 3: Methods of Analysis Section 3.8: Determination of Carbon Subsection 3.8.2: Steel and Cast Iron: Non-Aqueous Titrimetric Method After Combustion. 13 pp.

—Sulfur—Volumetric Analysis
ISO 2866-74. Sulphur for Industrial Use—Determination of Total Carbon Content—Titrimetric Method First Edition. 7 pp.

—Tantalum
JIS H 1681-76. Methods for Determination of Carbon in Tantalum (R 1986). 21 pp.

—Titanium
JIS H 1617-88. Methods for Determination of Carbon in Titanium and Titanium Alloys. 9 pp.

—Waste Water
CNS K9072-81. Method of Test for TC (Total Carbon) and TOC (Total Organic Carbon) in Water and Waste Water (Oct)(8032).

—Water
CNS K9072-81. Method of Test for TC (Total Carbon) and TOC (Total Organic Carbon) in Water and Waste Water (Oct)(8032).
ISO 8245-87. Water Quality—Guidelines for the Determination of Total Organic Carbon (TOC) First Edition. 8 pp.

INTERNATIONAL AND NON-U.S. NATIONAL STANDARDS
SUBJECT INDEX
Carbon

Carbon Content Analysis *(Cont.)*
—Water—Analyzers
JIS K 0805-88. Continuous Total Organic Carbon Analyzer. 13 pp.

—Zirconium
JIS H 1663-88. Methods for Determination of Carbon in Zirconium and Zirconium Alloys. 10 pp.

—Zirconium Alloys
JIS H 1663-88. Methods for Determination of Carbon in Zirconium and Zirconium Alloys. 10 pp.

Carbon Dioxide
See Also: Carbon Dioxide Content Analysis
BSI BS 4105-90. 1990 Specification for Liquid Carbon Dioxide, Industrial. 25 pp.
CNS K1021-81. Liquid Carbon Dioxide (Aug)(195).
CNS K6288-71. Method of Test for Liquefied Carbon Dioxide (Jan)(3145).
JIS K 1106-90. Liquid Carbon Dioxide. 12 pp.

—Fire Extinguishers
SNZ NZS 4508-79. Portable Carbon Dioxide Fire Extinguishers Amend: 1, 1980; 2, 1991. 11 pp.

—Fire Extinguishing Agents
BSI BS 5306: Part 4-86. 1986 Fire Extinguishing Installations and Equipment on Premises Part 4: Carbon Dioxide Systems. 42 pp.
BSI BS 6535: Part 1-90. 1990 Fire Extinguishing Media Part 1: Carbon Dioxide. 18 pp.
ISO 5923-89. Fire Protection—Fire Extinguishing Media—Carbon Dioxide Second Edition. 15 pp.
MOD UK DSTAN 68-65-01. Carbon Dioxide, Technical 74 Gram (Charge, Carbon Dioxide) Issue 1; Amendment 1. 10 pp.

—Standard Gases
JIS K 0003-92. Carbon Dioxide (Standard Reference Gas). 13 pp.

—Water Treatment
CEN PREN 936-92. Carbon Dioxide Used for Water Intended for Human Consumption. 12 pp.

Carbon Dioxide Analyzers
Use For: Capnometers; CO2 Analyzers
See Also: Gas Analyzers

—Medical—Safety
BSI BS 5724: Sec 2.30-93. 1993 Medical Electrical Equipment Part 2: Particular Requirements for Safety Section 2.30: Specification for Capnometers for Use with Humans (ISO 9918: 1993) (N). 32 pp.
CEN PREN 864-92. Medical Electrical Equipment—Capnometers for Use with Humans—Particular Requirements. 50 pp.
ISO 9918-93. Capnometers for Use with Humans—Requirements First Edition. 30 pp.

Carbon Dioxide Content Analysis
See Also: Carbon Dioxide

—Ammonia
BSI BS 4431-89. 1989 Methods of Sampling and Test for Liquefied Anhydrous Ammonia. 16 pp.

—Ammonium Bicarbonate—Volumetric Analysis
ISO 3422-75. Ammonium Hydrogen Carbonate for Industrial Use (Including Foodstuffs—Determination of Total Carbon Dioxide Content—Titrimetric Method First Edition. 5 pp.

—Ammonium Hydroxide
BSI BS 4651: Part 7-88. 1988 Ammonia Solution Part 7: Methods for Determination of Carbon Content. 4 pp.

—Calcium Oxides
CGSB 1-GP-71 METH 50.10-80. Methods of Testing Paints and Pigments Pigment Analysis Carbon Dioxide in Carbonates. 1 p.

—Carbonated Beverages
CNS N6092-82. Method of Test for Carbonated Soft Drink (Jan)(3761). 6 pp.

—Cement Paints (Made From Cement)
CGSB 1-GP-71 METH 50.10-80. Methods of Testing Paints and Pigments Pigment Analysis Carbon Dioxide in Carbonates. 1 p.

—Cements—Volumetric Analysis
BSI BS EN 196-21-92. 1992 Methods of Testing Cement Part 21: Determination of the Chloride, Carbon Dioxide and Alkali Content of Cement. 22 pp.
CEN EN 196 (Part 21)-89. Methods of Testing Cement: Determination of the Chloride, Carbon Dioxide and Alkali Content of Cement. 28 pp.

Carbon Dioxide Content Analysis *(Cont.)*
—Ethylene
CNS K6397-76. Method of Test for Trace Acetylene and Carbon Dioxide in Ethylene (Dec)(4054).

—Ethylene—Gas Chromatography
ISO 6381-81. Ethylene and Propylene for Industrial Use—Determination of Traces of Carbon Monoxide and Carbon Dioxide—Gas Chromatographic Method First Edition. 8 pp.

—Exhaust Gases
JIS D 1030-76. Analytical Procedure for Continuous Measurement of Carbon Monoxide, Carbon Dioxide and Hydrocarbon in Automobile Exhaust Gas.

—Flue Gases
BSI BS 1756: Part 5-71. 1971 Methods for the Sampling and Analysis of Flue Gases Part 5: Semi-Routine Analysis. 25 pp.
BSI BS 1756: Part 5-01. 1971 Amd 1 Methods for the Sampling and Analysis of Flue Gases Part 5: Semi-Routine Analysis (AMD 7756) October 15, 1993 (Q). 26 pp.
CSA CAN/CSA-Z223.2-M86. Method for the Continuous Measurement of Oxygen, Carbon Dioxide, Carbon Monoxide, Sulphur Dioxide, and Oxides of Nitrogen in Enclosed Combustion Flue Gas Streams; (Gen Instr 1). 44 pp.

—Gases
CNS K6230-67. Method of Analysis for Carbon Dioxide, Oxygen, Carbon Monoxide Hydrogen and Methane in Gases (Mar)(2592)(R 1971).
ISO TR6567 Pt 1-81. Gas Analysis—Determination of Carbon Dioxide—Part 1: General Guidance for the Choice of Methods First Edition. 9 pp.

—Manganese Ores—Gravimetric Analysis
ISO 314-81. Manganese Ores—Determination of Carbon Dioxide Content—Gravimetric Method Second Edition. 5 pp.

—Natural Gas—Gas Chromatography
ISO 6568-81. Natural Gas—Simple Analysis by Gas Chromatography First Edition. 6 pp.

—Pigments
CGSB 1-GP-71 METH 50.10-80. Methods of Testing Paints and Pigments Pigment Analysis Carbon Dioxide in Carbonates. 1 p.

—Potassium Hydroxides—Volumetric Analysis
ISO 2900-73. Potassium Hydroxide for Industrial Use—Determination of Carbon Dioxide Content—Titrimetric Method First Edition. 5 pp.

—Propylene—Gas Chromatography
ISO 6381-81. Ethylene and Propylene for Industrial Use—Determination of Traces of Carbon Monoxide and Carbon Dioxide—Gas Chromatographic Method First Edition. 8 pp.

—Slag Cements—Volumetric Analysis
BSI BS EN 196-21-92. 1992 Methods of Testing Cement Part 21: Determination of the Chloride, Carbon Dioxide and Alkali Content of Cement. 22 pp.
CEN EN 196 (Part 21)-89. Methods of Testing Cement: Determination of the Chloride, Carbon Dioxide and Alkali Content of Cement. 28 pp.

—Solid Fuels—Gravimetric Analysis
ISO 925-80. Solid Mineral Fuels—Determination of Carbon Dioxide Content—Gravimetric Method Second Edition. 5 pp.

—Water—Volumetric Analysis
BSI BS 2690: Part 109-84. 1984 Water Used in Industry Part 109: Alkalinity, Acidity, pH Value and Carbon Dioxide. 8 pp.

Carbon Disulfide
Use For: Carbon Bisulfide *See Also:* Carbon Disulfide Content Analysis; Hazardous Materials
BSI BS 662: Part 1-79. 1979 Carbon Disulphide for Industrial Use Part 1: Specification. 4 pp.
BSI BS 662: Part 3-79. 1979 Carbon Disulphide for Industrial Use Part 3: Additional Methods of Test. 4 pp.
CNS K1130-70. Carbon Disulfide (Industrial Grade) (Jan)(2950). 1 p.
CNS K7134-64. Chemical Reagent (Carbon Disulfide) (Jul)(1634).
JIS K 1421-78. Carbon Disulfide. 6 pp.
JIS K 8732-78. Carbon Disulfide.

—Highway Transportation—Emergency Procedures
SAA AS 1678.3.1. 003-89. Emergency Procedure Guide—Transport—Part 3.1.003: Carbon Disulfide (Carbon Bisulfide).

Carbon Disulfide *(Cont.)*
—Insoluble Matter Content
CGSB 1-GP-71 METH 85.1-82. Methods of Testing Paints and Pigments Matter Insoluble in Carbon Disulfide. 1 p.

—Sampling
BSI BS 662: Part 2-78. 1978 Carbon Disulphide for Industrial Use Part 2: Sampling and Methods of Test. 12 pp.
ISO 3144-74. Carbon Disulphide for Industrial Use—Sampling and Methods of Test First Edition; (Erratum—Aug 1978). 11 pp.

Carbon Disulfide Content Analysis
See Also: Carbon Disulfide

—Benzene
CNS K6620-80. Method of Test for Carbon Disulfide in Benzene (Dec)(6835).

—Exhaust Gases
CNS K9110-90. Method of Test for Carbon Disulfide in Exhaust Gas (Dec)(12825).
JIS K 0091-83. Methods for Determination of Carbon Disulfide in Exhaust Gas. 12 pp.

—Waste Water—Photometry
DIN ENGL 38413 Pt 4-86. German Standard Methods for the Examination of Water, Waste Water and Sludge; Individual Constituents (Group P); Determination of Carbon Disulfide (P 4) (Sept). 4 pp.

—Water—Photometry
DIN ENGL 38413 Pt 4-86. German Standard Methods for the Examination of Water, Waste Water and Sludge; Individual Constituents (Group P); Determination of Carbon Disulfide (P 4) (Sept). 4 pp.

Carbon Disulphide
Use: Carbon Disulfide

Carbon Electrodes
Use For: Graphite Electrodes

—Dry Batteries
CNS C4062-85. Carbon Electrode for Dry Batteries (Aug)(2654).

—Electric Arc Furnaces
CENELEC HD 564 S1-90. Nominal Dimensions of Cylindrical Machined Graphite Electrodes with Threaded Sockets and Nipples for Use in Electric Arc Furnaces. 10 pp.
IEC 239-87. Nominal Dimensions of Cylindrical Machined Graphite Electrodes with Threaded Sockets and Nipples for Use in Electric Arc Furnaces Second Edition. 22 pp.

—Graphite
CNS C3032-73. Method of Test for Graphite Electrodes (Nov)(3204).
CNS C4095-71. Graphite Electrode (Jul)(3203).
JIS R 7201-79. Artificial Graphite Electrodes.
JIS R 7202-79. Testing Methods for Artificial Graphite Electrodes (R 1985). 22 pp.

—Pitch—Ash Content
BSI BS 6043: Sec 1.8-86. 1986 Methods of Sampling and Test for Carbonaceous Materials Used in Aluminium Manufacture Part 1: Electrode Pitch Section 1.8: Determination of Ash. 7 pp.
ISO 8006-85. Carbonaceous Materials Used in the Production of Aluminium—Pitch for Electrodes—Determination of Ash First Edition. 5 pp.

—Pitch—Coking Values
BSI BS 6043: Sec 1.6-84. 1984 Methods of Sampling and Test for Carbonaceous Materials Used in Aluminium Manufacture Part 1: Electrode Pitch Section 1.6: Determination of Coking Value. 7 pp.
BSI BS 6043: Sec 1.6-01. 1984 Amd 1 Methods of Sampling and Test for Carbonaceous Materials Used in Aluminium Manufacture Part 1: Electrode Pitch Section 1.6: Determination of Coking Value (AMD 7662) May 15, 1993 (ISO 6998: 1984) (V). 8 pp.
ISO 6998-84. Carbonaceous Materials for the Production of Aluminium—Pitch for Electrodes—Determination of Coking Value First Edition. 5 pp.

—Pitch—Density
BSI BS 6043: Sec 1.7-88. 1988 Methods of Sampling and Test for Carbonaceous Materials Used in Aluminium Manufacture Part 1: Electrode Pitch Section 1.7: Determination of Apparent Density (Bouyancy Method). 7 pp.
ISO 6999-83. Carbonaceous Materials for the Production of Aluminium—Pitch for Electrodes—Determination of Density—Pyknometric Method First Edition. 5 pp.

INDUSTRY STANDARDS

Carbon Electrodes *(Cont.)*

—Pitch—Dynamic Viscosity
ISO 8003-85. Carbonaceous Materials Used in the Production of Aluminium—Pitch for Electrodes—Measurement of Dynamic Viscosity First Edition. 6 pp.

—Pitch—Sampling
BSI BS 6043: Sec 1.1-81. 1981 Methods of Sampling and Test for Carbonaceous Materials Used in Aluminium Manufacture Part 1: Electrode Pitch Section 1.1: Sampling. 19 pp.

ISO 6257-80. Carbonaceous Materials Used in the Production of Aluminium—Pitch for Electrodes—Sampling First Edition. 17 pp.

—Pitch—Softening Point
BSI BS 6043: Sec 1.3-83. 1983 Methods of Sampling and Test for Carbonaceous Materials Used in Aluminium Manufacture Part 1: Electrode Pitch Section 1.3: Determination of Softening Point (Ring and Ball Method). 8 pp.

ISO 5940-81. Carbonaceous Materials for the Production of Aluminium—Pitch for Electrodes—Determination of Softening Point by the Ring-and-Ball Method First Edition. 7 pp.

—Pitch—Water Content
BSI BS 6043: Sec 1.2-81. 1981 Methods of Sampling and Test for Carbonaceous Materials Used in Aluminium Manufacture Part 1: Electrode Pitch Section 1.2: Determination of Water Content (Dean and Stark Method). 8 pp.

ISO 5939-80. Carbonaceous Materials for the Production of Aluminium—Pitch for Electrodes—Determination of Water Content—Azeotropic Distillation (Dean and Stark) Method First Edition. 7 pp.

Carbon Fabrics
See Also: Fabrics

JIS R 7602-89. Testing Methods for Carbon Fibre Woven Fabrics. 9 pp.

Carbon Fiber Content Analysis
See Also: Carbon Fibers; Fiber Content Analysis

—Aerospace
AECMA PREN2559-88. Carbon Fibre Preimpregnates Test Method for the Determination of the Resin and Fibre Content and the Mass of Fibre Per Unit Area. 13 pp.

AECMA PREN2564-88. Carbon Fibre Laminates Test Method for the Determination of the Fibre and Resin Fractions and Porosity Content. 9 pp.

CEN PREN 3783-90. Procedure for the Normalisation of Test Results of Fibre Dominated Composite Mechanical Properties. 6 pp.

—Carbon Fiber Reinforced Plastics
JIS K 7075-91. Testing Methods for Carbon Fiber Content and Void Content of Carbon Fiber Reinforced Plastics. 11 pp.

Carbon Fiber Reinforced Plastics
Use: Reinforced Plastics—Carbon Fiber Reinforced

Carbon Fibers
Use For: Graphite Fibers *See Also:* Carbon Fiber Content Analysis; Fibers; Synthetic Fibers

JIS K 7071-88. Testing Methods for Prepreg, Carbon Fiber and Epoxy Resins. 24 pp.

JIS R 7601-86. Testing Methods for Carbon Fibers. 41 pp.

—Filament Yarns—Aerospace—Sampling
CEN PREN 3675-90. Sampling Plan for Acceptance Testing of Aramid, Carbon Fibre and Textile Glass Filament Yarns. 8 pp.

CEN PREN 3675-92. Sampling Plan for Acceptance Testing of Aramid, Carbon Fibre and Textile Glass Filament Yarns. 7 pp.

—Linear Density
BSI BS 7658: Part 2-93. 1993 Carbon Fibre Part 2: Method for Determination of Linear Density (ISO 10120: 1991) (V). 8 pp.

ISO 10120-91. Carbon Fibre—Determination of Linear Density First Edition. 5 pp.

—Prepreg—Aerospace—Density
AECMA PREN2557-88. Carbon Fibre Preimpregnates Test Method for the Determination of Mass Per Unit Area. 7 pp.

AECMA PREN2559-88. Carbon Fibre Preimpregnates Test Method for the Determination of the Resin and Fibre Content and the Mass of Fibre Per Unit Area. 13 pp.

—Prepreg—Aerospace—Fiber Content
AECMA PREN2559-88. Carbon Fibre Preimpregnates Test Method for the Determination of the Resin and Fibre Content and the Mass of Fibre Per Unit Area. 13 pp.

Carbon Fibers *(Cont.)*

—Prepreg—Aerospace—Fiber Content *(Cont.)*
AECMA PREN2564-88. Carbon Fibre Laminates Test Method for the Determination of the Fibre and Resin Fractions and Porosity Content. 9 pp.

CEN PREN 3783-90. Procedure for the Normalisation of Test Results of Fibre Dominated Composite Mechanical Properties. 6 pp.

—Prepreg—Aerospace—Porosity
AECMA PREN2564-88. Carbon Fibre Laminates Test Method for the Determination of the Fibre and Resin Fractions and Porosity Content. 9 pp.

—Prepreg—Aerospace—Resin Content
AECMA PREN2559-88. Carbon Fibre Preimpregnates Test Method for the Determination of the Resin and Fibre Content and the Mass of Fibre Per Unit Area. 13 pp.

AECMA PREN2564-88. Carbon Fibre Laminates Test Method for the Determination of the Fibre and Resin Fractions and Porosity Content. 9 pp.

—Prepreg—Aerospace—Resin Flow
AECMA PREN2560-88. Carbon Fibre Preimpregnates Test Method for the Determination of the Resin Flow. 9 pp.

—Prepreg—Aerospace—Volatile Matter Content
AECMA PREN2558-88. Carbon Fibre Preimpregnates Test Method for the Determination of the Percentage of Volatile Matter. 8 pp.

—Thermosetting Resins—Aerospace
AECMA PREN2565-87. Preparation of Carbon Fibre Reinforced Resin Panels for Test Purposes. 14 pp.

CEN PREN 2565-93. Aerospace Series Preparation of Carbon Fibre Reinforced Resin Panels for Test Purposes. 12 pp.

—Thermosetting Resins—Aerospace—Compression Testing
CEN PREN 2850-90. Unidirectional Thermosetting Carbon Laminate Compression Test Parallel to the Fibre Direction. 14 pp.

CEN PREN 2850-91. Carbon Thermosetting Resin Unidirectional Laminates Comression Test Parallel to the Fibre Direction. 12 pp.

—Thermosetting Resins—Aerospace—Flexural Strength
AECMA PREN2562-89. Unidirectional Laminates Carbon-Thermosetting Resin Flexural Test. 9 pp.

—Thermosetting Resins—Aerospace—Shear Strength
AECMA PREN2563-89. Unidirectional Laminates Carbon-Thermosetting Resin Test Method—Determination of Apparent Interlaminar Shear Strength. 10 pp.

—Thermosetting Resins—Aerospace—Tensile Testing
AECMA PREN2561-89. Aerospace Series Unidirectional Laminates Carbon-Thermosetting Resin Tensile Test Parallel to the Fibre Direction. 12 pp.

—Yarns—Density
BSI BS 7658: Part 1-93. 1993 Carbon Fibre Part 1: Method for Determination of Density (ISO 10119: 1992) (V). 14 pp.

ISO 10119-92. Carbon Fibre—Determination of Density First Edition. 11 pp.

Carbon Film Resistors
See Also: Carbon Resistors; Fixed Resistors; Resistors; Thin Film Resistors

CNS C7010-84. Fixed Carbon Film Resistors (Oct)(2320).

CNS C7094-85. Insulated Fixed Carbon Film Resistors (Aug)(4972).

JIS C 5212-89. Fixed Carbon Film Chip Resistors Cylindrical Type for Use in Electronic Equipment (Form 27, Characteristic D and G, Grade C). 29 pp.

JIS C 6402-77. Fixed Carbon Film Resistors. 24 pp.

JIS C 6407-72. Insulated Fixed Carbon Film Resistors (R 1983). 13 pp.

JIS C 6408-88. Fixed Carbon Film Resistors for Electronics Equipment (D, G). 30 pp.

—Power
IECQ QC 400100/SG 0001-91. Detail Specification for Electronic Components Fixed Low-Power Non-Wirewound Resistors (Fixed Carbon Film Resistors). 11 pp.

IECQ QC 400101/CN 0002-87. Detail Specification for Electronic Components Fixed Low-Power Non-Wirewound Resistors Fixed Carbon Film Resistors, Type RT14 Assessment Level E. 10 pp.

Carbon Film Resistors *(Cont.)*

—Power *(Cont.)*
IECQ QC 400101/CN 0003-89. Detail Specification for Electronic Components Fixed Low-Power Non-Wirewound Resistors Fixed Carbon Film Resistors, Type RT13 Assessment Level E. 11 pp.

IECQ QC 400101/JP 0002-83. Detail Specification for Electronic Components Fixed Low-Power Non-Wirewound Resistors (Carbon Film Resistors). 13 pp.

IECQ QC 400101/SU 0004-90. Detail Specification for Electronic Components Fixed Low-Power Non-Wirewound Carbon Film a.c. and d.c. Resistors Type CI-4M. 14 pp.

IECQ QC 400101/US 0003-87. Resistors for Use in Electronic Equipment: Detail Specification: Fixed Low-Power Non-Wirewound Resistors Insulated, Carbon Film. 11 pp.

—Power—Printed Circuit Mount
CECC CECC 40 101-801 ISSUE 1-79. Fixed Low Power Non-Wirewound Resistors; Stability 5% (En, Fr, Ge) AMD 1 (En, Fr, Ge) AMD 2 (En, Fr, Ge). 40 pp.

Carbon Manganese Steels
Scope Note: For additional listings, see also specific products made from carbon manganese steels
See Also: Carbon Steels; Low Alloy Steels; Steels

—Bars—Cold Finished
SAA AS 1443-83. Carbon Steels and Carbon-Manganese Steels—Cold-Finished Bars. 15 pp.

—Bars—Hot Rolled
SAA AS 1442-92. Carbon Steels and Carbon-Manganese Steels—Hot-Rolled Bars and Semifinished Products. 13 pp.

—Crystal Structure—Impact Testing
BSI BS 131: Part 5-65. 1965 Methods for Notched Bar Tests Part 5: Determination of Crystallinity. 12 pp.

—Forgings
BSI BS 29-76. 1976 Carbon Steel Forgings Above 150 mm Ruling Section. 12 pp.

SAA AS 1448-81. Carbon Steels and Carbon-Manganese Steels—Forgings (Ruling Section 300 mm Maximum). 12 pp.

—Forgings—Marine
MOD UK NES 848: Part 1-90. Requirements for Carbon Manganese and Low Alloy Steel Forgings Part 1: Carbon Manganese Steel Forgings Issue 1 (08.90). 18 pp.

—Plates
BSI BS 1449: Part 1-83. (WITHDRAWN) 1983 Amd 1 Steel Plate, Sheet and Strip Part 1: Carbon and Carbon-Manganese Plate, Sheet and Strip (Superseded by BS 1449: Sections 1.1 to 1.15: 1991 and BS EN 10130: 1991). 39 pp.

BSI BS 1449: Sec 1.1-91. 1991 Steel Plate, Sheet and Strip Part 1: Carbon and Carbon-Manganese Plate, Sheet and Strip Section 1.1: General Specification. 20 pp.

BSI BS 1501: Part 1-80. 1980 Amd 4 Steels for Fired and Unfired Pressure Vessels: Plates Part 1: Specification for Carbon and Carbon Manganese Steels (AMD 6259) July 31, 1990. 30 pp.

BSI BS 1501: Part 1-05. 1980 Amd 5 Steels for Fired and Unfired Pressure Vessels: Plates Part 1: Specification for Carbon and Carbon Manganese Steels (AMD 6824) October 31, 1991. 34 pp.

BSI BS 1501: Part 1-06. (WITHDRAWN) 1980 Amd 6 Steels for Pressure Purposes Part 1: Specification for Carbon and Carbon Manganese Steels: Plates (AMD 7024) June 15, 1992 (Superseded by BS EN 10028: 1-3: 1993). 35 pp.

SNZ NZS/BS 1449: Pt1:Sec1.1-91. Steel Plate, Sheet and Strip Part 1: Carbon and Carbon-Manganese Plate, Sheet and Strip Section 1.1: General Specification. 16 pp.

SNZ NZS/BS 1501: Part 1-80. Steels for Pressure Purposes: Plates Part 1: Specification for Carbon and Carbon Manganese Steels. 28 pp.

—Plates—Cold Rolled
BSI BS 1449: Sec 1.3-91. 1991 Steel Plate, Sheet and Strip Part 1: Carbon and Carbon-Manganese Plate, Sheet and Strip Section 1.3: Specification for Cold Rolled Steel Plate, Sheet and Wide Strip for Vitreous Enamelling Based on Formability. 10 pp.

BSI BS 1449: Sec 1.5-91. 1991 Steel Plate, Sheet and Strip Part 1: Carbon and Carbon-Manganese Plate, Sheet and Strip Section 1.5: Specification for Cold Rolled Wide Material Based on Specified Minimum Strength. 8 pp.

INTERNATIONAL AND NON-U.S. NATIONAL STANDARDS
SUBJECT INDEX
Carbon

Carbon Manganese Steels *(Cont.)*

—Plates—Cold Rolled *(Cont.)*

BSI BS 1449: Sec 1.7-91. (WITHDRAWN) 1991 Steel Plate, Sheet and Strip Part 1: Carbon and Carbon-Manganese Plate, Sheet and Strip Sec 1.7: Specification for Tolerances on Dimensions and Shape for Cold Rolled Wide Material (Superseded by BS EN 10131: 1991). 11 pp.
BSI BS 1449: Sec 1.9-91. 1991 Steel Plate, Sheet and Strip Part 1: Carbon and Carbon-Manganese Plate, Sheet and Strip Section 1.9: Specification for Cold Rolled Narrow Strip Based on Formability. 11 pp.
BSI BS 1449: Sec 1.11-91. 1991 Steel Plate, Sheet and Strip Part 1: Carbon and Carbon-Manganese Plate, Sheet and Strip Section 1.11: Specification for Cold Rolled Narrow Strip Based on Specified Minimum Strength. 11 pp.
BSI BS 1449: Sec 1.13-91. 1991 Steel Plate, Sheet and Strip Part 1: Carbon and Carbon-Manganese Plate, Sheet and Strip Section 1.13: Specification for Tolerances on Dimensions and Shape for Cold Rolled Narrow Strip. 8 pp.
BSI BS 1449: Sec 1.15-91. 1991 Steel Plate, Sheet and Strip Part 1: Carbon and Carbon-Manganese Plate, Sheet and Strip Section 1.15: Specification for Cold Rolled Narrow Strip Supplied in a Range of Conditions for Heat Treatment and General Engineering Purposes. 12 pp.
SNZ NZS/BS 1449: Pt1:Sec1.3-91. Steel Plate, Sheet and Strip Part 1: Carbon and Carbon-Manganese Plate, Sheet and Strip Section 1.3: Cold Rolled Steel Plate, Sheet and Wide Strip for Vitreous Enamelling Based on Formability. 8 pp.
SNZ NZS/BS 1449: Pt1:Sec1.5-91. Steel Plate, Sheet and Strip Part 1: Carbon and Carbon-Manganese Plate, Sheet and Strip Section 1.5: Cold Rolled Wide Material Based on Specified Minimum Strength. 4 pp.
SNZ NZS/BS 1449: Pt1:Sec1.7-91. Steel Plate, Sheet and Strip Part 1: Carbon and Carbon-Manganese Plate, Sheet and Strip Section 1.7: Tolerances on Dimensions and Shape for Cold Rolled Wide Material. 8 pp.
SNZ NZS/BS 1449: Pt1:Sec1.9-91. Steel Plate, Sheet and Strip Part 1: Carbon and Carbon-Manganese Plate, Sheet and Strip Section 1.9: Cold Rolled Narrow Strip Based on Formability. 8 pp.
SNZ NZS/BS 1449: Pt1:Sec1.11-91. Steel Plate, Sheet and Strip Part 1: Carbon and Carbon-Manganese Plate, Sheet and Strip Section 1.11: Cold Rolled Narrow Strip Based on Specified Minimum Strength. 8 pp.
SNZ NZS/BS 1449: Pt1:Sec1.13-91. Steel Plate, Sheet and Strip Part 1: Carbon and Carbon-Manganese Plate, Sheet and Strip Section 1.13: Tolerances on Dimensions and Shape for Cold Rolled Narrow Strip. 4 pp.
SNZ NZS/BS 1449: Pt1:Sec1.15-91. Steel Plate, Sheet and Strip Part 1: Carbon and Carbon-Manganese Plate, Sheet and Strip Section 1.15: Cold Rolled Narrow Strip Supplied in a Range of Conditions for Heat Treatment and General Engineering Purposes. 8 pp.

—Plates—Hot Rolled

BSI BS 1449: Sec 1.2-91. 1991 Steel Plate, Sheet and Strip Part 1: Carbon and Carbon-Manganese Plate, Sheet and Strip Section 1.2: Specification for Hot Rolled Steel Plate, Sheet and Wide Strip Based on Formability. 10 pp.
BSI BS 1449: Sec 1.4-91. 1991 Steel Plate, Sheet and Strip Part 1: Carbon and Carbon-Manganese Plate, Sheet and Strip Section 1.4: Specification for Hot Rolled Wide Material Based on Specified Minimum Strength. 10 pp.
BSI BS 1449: Sec 1.6-91. (WITHDRAWN) 1991 Steel Plate, Sheet and Strip Part 1: Carbon and Carbon-Manganese Plate, Sheet and Strip Section 1.6: Spec. for Tolerances on Dimensions and Shape for Hot Rolled Wide Material Superseded by BS EN 10051: 1992). 13 pp.
BSI BS 1449: Sec 1.8-91. 1991 Steel Plate, Sheet and Strip Part 1: Carbon and Carbon-Manganese Plate, Sheet and Strip Section 1.8: Specification for Hot Rolled Narrow Strip Based on Formability. 10 pp.
BSI BS 1449: Sec 1.10-91. 1991 Steel Plate, Sheet and Strip Part 1: Carbon and Carbon-Manganese Plate, Sheet and Strip Section 1.10: Specification for Hot Rolled Narrow Strip Based on Specified Minimum Strength. 11 pp.
BSI BS 1449: Sec 1.12-91. 1991 Steel Plate, Sheet and Strip Part 1: Carbon and Carbon-Manganese Plate, Sheet and Strip Section 1.12: Specification for Tolerances on Dimensions and Shape for Hot Rolled Narrow Strip. 8 pp.
BSI BS 1449: Sec 1.14-91. 1991 Steel Plate, Sheet and Strip Part 1: Carbon and Carbon-Manganese Plate, Sheet and Strip Section 1.14: Specification for Hot Rolled Narrow Strip Supplied in a Range of Conditions for Heat Treatment and General Engineering Purposes. 10 pp.
SAA AS 1446-81. Carbon Steels and Carbon-Manganese Steels—Hot-Rolled Slab, Plate and Strip Based on Chemical Composition. 8 pp.

Carbon Manganese Steels *(Cont.)*

—Plates—Hot Rolled *(Cont.)*

SNZ NZS/BS 1449: Pt1:Sec1.2-91. Steel Plate, Sheet and Strip Part 1: Carbon and Carbon-Manganese Plate, Sheet and Strip Section 1.2: Hot Rolled Steel Plate, Sheet and Wide Strip Based on Formability. 8 pp.
SNZ NZS/BS 1449: Pt1:Sec1.4-91. Steel Plate, Sheet and Strip Part 1: Carbon and Carbon-Manganese Plate, Sheet and Strip Section 1.4: Hot Rolled Wide Material Based on Specified Minimum Strength. 8 pp.
SNZ NZS/BS 1449: Pt1:Sec1.6-91. Steel Plate, Sheet and Strip Part 1: Carbon and Carbon-Manganese Plate, Sheet and Strip Section 1.6: Tolerances on Dimensions and Shape for Hot Rolled Wide Material. 12 pp.
SNZ NZS/BS 1449: Pt1:Sec1.8-91. Steel Plate, Sheet and Strip Part 1: Carbon and Carbon-Manganese Plate, Sheet and Strip Section 1.8: Hot Rolled Narrow Strip Based on Formability. 8 pp.
SNZ NZS/BS 1449: Pt1:Sec1.10-91. Steel Plate, Sheet and Strip Part 1: Carbon and Carbon-Manganese Plate, Sheet and Strip Section 1.10: Hot Rolled Narrow Strip Based on Specified Minimum Strength. 8 pp.
SNZ NZS/BS 1449: Pt1:Sec1.12-91. Steel Plate, Sheet and Strip Part 1: Carbon and Carbon-Manganese Plate, Sheet and Strip Section 1.12: Tolerances on Dimensions and Shape for Hot Rolled Narrow Strip. 4 pp.
SNZ NZS/BS 1449: Pt1:Sec1.14-91. Steel Plate, Sheet and Strip Part 1: Carbon and Carbon-Manganese Plate, Sheet and Strip Section 1.14: Hot Rolled Narrow Strip Supplied in a Range of Conditions for Heat Treatment and General Engineering Purposes. 8 pp.

—Semifinished Products—Cold Finished

SAA AS 1443-83. Carbon Steels and Carbon-Manganese Steels—Cold-Finished Bars. 15 pp.

—Semifinished Products—Hot Rolled

SAA AS 1442-92. Carbon Steels and Carbon-Manganese Steels—Hot-Rolled Bars and Semifinished Products. 13 pp.

—Sheets

BSI BS 1449: Part 1-83. (WITHDRAWN) 1983 Amd 1 Steel Plate, Sheet and Strip Part 1: Carbon and Carbon-Manganese Plate, Sheet and Strip (Superseded by BS 1449: Sections 1.1 to 1.15: 1991 and BS EN 10130: 1991). 39 pp.
BSI BS 1449: Sec 1.1-91. 1991 Steel Plate, Sheet and Strip Part 1: Carbon and Carbon-Manganese Plate, Sheet and Strip Section 1.1: General Specification. 20 pp.
SNZ NZS/BS 1449: Pt1:Sec1.1-91. Steel Plate, Sheet and Strip Part 1: Carbon and Carbon-Manganese Plate, Sheet and Strip Section 1.1: General Specification. 16 pp.

—Sheets—Cold Rolled

BSI BS 1449: Sec 1.3-91. 1991 Steel Plate, Sheet and Strip Part 1: Carbon and Carbon-Manganese Plate, Sheet and Strip Section 1.3: Specification for Cold Rolled Steel Plate, Sheet and Wide Strip for Vitreous Enamelling Based on Formability. 10 pp.
BSI BS 1449: Sec 1.5-91. 1991 Steel Plate, Sheet and Strip Part 1: Carbon and Carbon-Manganese Plate, Sheet and Strip Section 1.5: Specification for Cold Rolled Wide Material Based on Specified Minimum Strength. 8 pp.
BSI BS 1449: Sec 1.7-91. (WITHDRAWN) 1991 Steel Plate, Sheet and Strip Part 1: Carbon and Carbon-Manganese Plate, Sheet and Strip Sec 1.7: Specification for Tolerances on Dimensions and Shape for Cold Rolled Wide Material (Superseded by BS EN 10131: 1991). 11 pp.
BSI BS 1449: Sec 1.9-91. 1991 Steel Plate, Sheet and Strip Part 1: Carbon and Carbon-Manganese Plate, Sheet and Strip Section 1.9: Specification for Cold Rolled Narrow Strip Based on Formability. 11 pp.
BSI BS 1449: Sec 1.11-91. 1991 Steel Plate, Sheet and Strip Part 1: Carbon and Carbon-Manganese Plate, Sheet and Strip Section 1.11: Specification for Cold Rolled Narrow Strip Based on Specified Minimum Strength. 11 pp.
BSI BS 1449: Sec 1.13-91. 1991 Steel Plate, Sheet and Strip Part 1: Carbon and Carbon-Manganese Plate, Sheet and Strip Section 1.13: Specification for Tolerances on Dimensions and Shape for Cold Rolled Narrow Strip. 8 pp.
BSI BS 1449: Sec 1.15-91. 1991 Steel Plate, Sheet and Strip Part 1: Carbon and Carbon-Manganese Plate, Sheet and Strip Section 1.15: Specification for Cold Rolled Narrow Strip Supplied in a Range of Conditions for Heat Treatment and General Engineering Purposes. 12 pp.
SNZ NZS/BS 1449: Pt1:Sec1.3-91. Steel Plate, Sheet and Strip Part 1: Carbon and Carbon-Manganese Plate, Sheet and Strip Section 1.3: Cold Rolled Steel Plate, Sheet and Wide Strip for Vitreous Enamelling Based on Formability. 8 pp.

Carbon Manganese Steels *(Cont.)*

—Sheets—Cold Rolled *(Cont.)*

SNZ NZS/BS 1449: Pt1:Sec1.5-91. Steel Plate, Sheet and Strip Part 1: Carbon and Carbon-Manganese Plate, Sheet and Strip Section 1.5: Cold Rolled Wide Material Based on Specified Minimum Strength. 4 pp.
SNZ NZS/BS 1449: Pt1:Sec1.7-91. Steel Plate, Sheet and Strip Part 1: Carbon and Carbon-Manganese Plate, Sheet and Strip Section 1.7: Tolerances on Dimensions and Shape for Cold Rolled Wide Material. 8 pp.
SNZ NZS/BS 1449: Pt1:Sec1.9-91. Steel Plate, Sheet and Strip Part 1: Carbon and Carbon-Manganese Plate, Sheet and Strip Section 1.9: Cold Rolled Narrow Strip Based on Formability. 8 pp.
SNZ NZS/BS 1449: Pt1:Sec1.11-91. Steel Plate, Sheet and Strip Part 1: Carbon and Carbon-Manganese Plate, Sheet and Strip Section 1.11: Cold Rolled Narrow Strip Based on Specified Minimum Strength. 8 pp.
SNZ NZS/BS 1449: Pt1:Sec1.13-91. Steel Plate, Sheet and Strip Part 1: Carbon and Carbon-Manganese Plate, Sheet and Strip Section 1.13: Tolerances on Dimensions and Shape for Cold Rolled Narrow Strip. 4 pp.
SNZ NZS/BS 1449: Pt1:Sec1.15-91. Steel Plate, Sheet and Strip Part 1: Carbon and Carbon-Manganese Plate, Sheet and Strip Section 1.15: Cold Rolled Narrow Strip Supplied in a Range of Conditions for Heat Treatment and General Engineering Purposes. 8 pp.

—Sheets—Hot Rolled

BSI BS 1449: Sec 1.2-91. 1991 Steel Plate, Sheet and Strip Part 1: Carbon and Carbon-Manganese Plate, Sheet and Strip Section 1.2: Specification for Hot Rolled Steel Plate, Sheet and Wide Strip Based on Formability. 10 pp.
BSI BS 1449: Sec 1.4-91. 1991 Steel Plate, Sheet and Strip Part 1: Carbon and Carbon-Manganese Plate, Sheet and Strip Section 1.4: Specification for Hot Rolled Wide Material Based on Specified Minimum Strength. 10 pp.
BSI BS 1449: Sec 1.6-91. (WITHDRAWN) 1991 Steel Plate, Sheet and Strip Part 1: Carbon and Carbon-Manganese Plate, Sheet and Strip Section 1.6: Spec. for Tolerances on Dimensions and Shape for Hot Rolled Wide Material Superseded by BS EN 10051: 1992). 13 pp.
BSI BS 1449: Sec 1.8-91. 1991 Steel Plate, Sheet and Strip Part 1: Carbon and Carbon-Manganese Plate, Sheet and Strip Section 1.8: Specification for Hot Rolled Narrow Strip Based on Formability. 10 pp.
BSI BS 1449: Sec 1.10-91. 1991 Steel Plate, Sheet and Strip Part 1: Carbon and Carbon-Manganese Plate, Sheet and Strip Section 1.10: Specification for Hot Rolled Narrow Strip Based on Specified Minimum Strength. 11 pp.
BSI BS 1449: Sec 1.12-91. 1991 Steel Plate, Sheet and Strip Part 1: Carbon and Carbon-Manganese Plate, Sheet and Strip Section 1.12: Specification for Tolerances on Dimensions and Shape for Hot Rolled Narrow Strip. 8 pp.
BSI BS 1449: Sec 1.14-91. 1991 Steel Plate, Sheet and Strip Part 1: Carbon and Carbon-Manganese Plate, Sheet and Strip Section 1.14: Specification for Hot Rolled Narrow Strip Supplied in a Range of Conditions for Heat Treatment and General Engineering Purposes. 10 pp.
SNZ NZS/BS 1449: Pt1:Sec1.2-91. Steel Plate, Sheet and Strip Part 1: Carbon and Carbon-Manganese Plate, Sheet and Strip Section 1.2: Hot Rolled Steel Plate, Sheet and Wide Strip Based on Formability. 8 pp.
SNZ NZS/BS 1449: Pt1:Sec1.4-91. Steel Plate, Sheet and Strip Part 1: Carbon and Carbon-Manganese Plate, Sheet and Strip Section 1.4: Hot Rolled Wide Material Based on Specified Minimum Strength. 8 pp.
SNZ NZS/BS 1449: Pt1:Sec1.6-91. Steel Plate, Sheet and Strip Part 1: Carbon and Carbon-Manganese Plate, Sheet and Strip Section 1.6: Tolerances on Dimensions and Shape for Hot Rolled Wide Material. 12 pp.
SNZ NZS/BS 1449: Pt1:Sec1.8-91. Steel Plate, Sheet and Strip Part 1: Carbon and Carbon-Manganese Plate, Sheet and Strip Section 1.8: Hot Rolled Narrow Strip Based on Formability. 8 pp.
SNZ NZS/BS 1449: Pt1:Sec1.10-91. Steel Plate, Sheet and Strip Part 1: Carbon and Carbon-Manganese Plate, Sheet and Strip Section 1.10: Hot Rolled Narrow Strip Based on Specified Minimum Strength. 8 pp.
SNZ NZS/BS 1449: Pt1:Sec1.12-91. Steel Plate, Sheet and Strip Part 1: Carbon and Carbon-Manganese Plate, Sheet and Strip Section 1.12: Tolerances on Dimensions and Shape for Hot Rolled Narrow Strip. 4 pp.

INDUSTRY STANDARDS
413

Carbon Manganese Steels (Cont.)
—Sheets—Hot Rolled (Cont.)
SNZ NZS/BS 1449: Pt1:Sec1.14-91. Steel Plate, Sheet and Strip Part 1: Carbon and Carbon-Manganese Plate, Sheet and Strip Section 1.14: Hot Rolled Narrow Strip Supplied in a Range of Conditions for Heat Treatment and General Engineering Purposes. 8 pp.

—Slabs—Hot Rolled
SAA AS 1446-81. Carbon Steels and Carbon-Manganese Steels—Hot-Rolled Slab, Plate and Strip Based on Chemical Composition. 8 pp.

—Strips
BSI BS 1449: Part 1-83. (WITHDRAWN) 1983 Amd 1 Steel Plate, Sheet and Strip Part 1: Carbon and Carbon-Manganese Plate, Sheet and Strip (Superseded by BS 1449: Sections 1.1 to 1.15: 1991 and BS EN 10130: 1991). 39 pp.
BSI BS 1449: Sec 1.1-91. 1991 Steel Plate, Sheet and Strip Part 1: Carbon and Carbon-Manganese Plate, Sheet and Strip Section 1.1: General Specification. 20 pp.
SNZ NZS/BS 1449: Pt1:Sec1.1-91. Steel Plate, Sheet and Strip Part 1: Carbon and Carbon-Manganese Plate, Sheet and Strip Section 1.1: General Specification. 16 pp.

—Strips—Cold Rolled
BSI BS 1449: Sec 1.3-91. 1991 Steel Plate, Sheet and Strip Part 1: Carbon and Carbon-Manganese Plate, Sheet and Strip Section 1.3: Specification for Cold Rolled Steel Plate, Sheet and Wide Strip for Vitreous Enamelling Based on Formability. 10 pp.
BSI BS 1449: Sec 1.5-91. 1991 Steel Plate, Sheet and Strip Part 1: Carbon and Carbon-Manganese Plate, Sheet and Strip Section 1.5: Specification for Cold Rolled Wide Material Based on Specified Minimum Strength. 8 pp.
BSI BS 1449: Sec 1.7-91. (WITHDRAWN) 1991 Steel Plate, Sheet and Strip Part 1: Carbon and Carbon-Manganese Plate, Sheet and Strip Sec 1.7: Specification for Tolerances on Dimensions and Shape for Cold Rolled Wide Material (Superseded by BS EN 10131: 1991). 11 pp.
BSI BS 1449: Sec 1.9-91. 1991 Steel Plate, Sheet and Strip Part 1: Carbon and Carbon-Manganese Plate, Sheet and Strip Section 1.9: Specification for Cold Rolled Narrow Strip Based on Formability. 11 pp.
BSI BS 1449: Sec 1.11-91. 1991 Steel Plate, Sheet and Strip Part 1: Carbon and Carbon-Manganese Plate, Sheet and Strip Section 1.11: Specification for Cold Rolled Narrow Strip Based on Specified Minimum Strength. 11 pp.
BSI BS 1449: Sec 1.13-91. 1991 Steel Plate, Sheet and Strip Part 1: Carbon and Carbon-Manganese Plate, Sheet and Strip Section 1.13: Specification for Tolerances on Dimensions and Shape for Cold Rolled Narrow Strip. 8 pp.
BSI BS 1449: Sec 1.15-91. 1991 Steel Plate, Sheet and Strip Part 1: Carbon and Carbon-Manganese Plate, Sheet and Strip Section 1.15: Specification for Cold Rolled Narrow Strip Supplied in a Range of Conditions for Heat Treatment and General Engineering Purposes. 12 pp.
SNZ NZS/BS 1449: Pt1:Sec1.3-91. Steel Plate, Sheet and Strip Part 1: Carbon and Carbon-Manganese Plate, Sheet and Strip Section 1.3: Cold Rolled Steel Plate, Sheet and Wide Strip for Vitreous Enamelling Based on Formability. 8 pp.
SNZ NZS/BS 1449: Pt1:Sec1.5-91. Steel Plate, Sheet and Strip Part 1: Carbon and Carbon-Manganese Plate, Sheet and Strip Section 1.5: Cold Rolled Wide Material Based on Specified Minimum Strength. 4 pp.
SNZ NZS/BS 1449: Pt1:Sec1.7-91. Steel Plate, Sheet and Strip Part 1: Carbon and Carbon-Manganese Plate, Sheet and Strip Section 1.7: Tolerances on Dimensions and Shape for Cold Rolled Wide Material. 8 pp.
SNZ NZS/BS 1449: Pt1:Sec1.9-91. Steel Plate, Sheet and Strip Part 1: Carbon and Carbon-Manganese Plate, Sheet and Strip Section 1.9: Cold Rolled Narrow Strip Based on Formability. 8 pp.
SNZ NZS/BS 1449: Pt1:Sec1.11-91. Steel Plate, Sheet and Strip Part 1: Carbon and Carbon-Manganese Plate, Sheet and Strip Section 1.11: Cold Rolled Narrow Strip Based on Specified Minimum Strength. 8 pp.
SNZ NZS/BS 1449: Pt1:Sec1.13-91. Steel Plate, Sheet and Strip Part 1: Carbon and Carbon-Manganese Plate, Sheet and Strip Section 1.13: Tolerances on Dimensions and Shape for Cold Rolled Narrow Strip. 4 pp.
SNZ NZS/BS 1449: Pt1:Sec1.15-91. Steel Plate, Sheet and Strip Part 1: Carbon and Carbon-Manganese Plate, Sheet and Strip Section 1.15: Cold Rolled Narrow Strip Supplied in a Range of Conditions for Heat Treatment and General Engineering Purposes. 8 pp.

Carbon Manganese Steels (Cont.)
—Strips—Hot Rolled
BSI BS 1449: Sec 1.2-91. 1991 Steel Plate, Sheet and Strip Part 1: Carbon and Carbon-Manganese Plate, Sheet and Strip Section 1.2: Specification for Hot Rolled Steel Plate, Sheet and Wide Strip Based on Formability. 10 pp.
BSI BS 1449: Sec 1.4-91. 1991 Steel Plate, Sheet and Strip Part 1: Carbon and Carbon-Manganese Plate, Sheet and Strip Section 1.4: Specification for Hot Rolled Wide Material Based on Specified Minimum Strength. 10 pp.
BSI BS 1449: Sec 1.6-91. (WITHDRAWN) 1991 Steel Plate, Sheet and Strip Part 1: Carbon and Carbon-Manganese Plate, Sheet and Strip Section 1.6: Spec. for Tolerances on Dimensions and Shape for Hot Rolled Wide Material Superseded by BS EN 10051: 1992). 13 pp.
BSI BS 1449: Sec 1.8-91. 1991 Steel Plate, Sheet and Strip Part 1: Carbon and Carbon-Manganese Plate, Sheet and Strip Section 1.8: Specification for Hot Rolled Narrow Strip Based on Formability. 10 pp.
BSI BS 1449: Sec 1.10-91. 1991 Steel Plate, Sheet and Strip Part 1: Carbon and Carbon-Manganese Plate, Sheet and Strip Section 1.10: Specification for Hot Rolled Narrow Strip Based on Specified Minimum Strength. 11 pp.
BSI BS 1449: Sec 1.12-91. 1991 Steel Plate, Sheet and Strip Part 1: Carbon and Carbon-Manganese Plate, Sheet and Strip Section 1.12: Specification for Tolerances on Dimensions and Shape for Hot Rolled Narrow Strip. 8 pp.
BSI BS 1449: Sec 1.14-91. 1991 Steel Plate, Sheet and Strip Part 1: Carbon and Carbon-Manganese Plate, Sheet and Strip Section 1.14: Specification for Hot Rolled Narrow Strip Supplied in a Range of Conditions for Heat Treatment and General Engineering Purposes. 10 pp.
SAA AS 1446-81. Carbon Steels and Carbon-Manganese Steels—Hot-Rolled Slab, Plate and Strip Based on Chemical Composition. 8 pp.
SNZ NZS/BS 1449: Pt1:Sec1.2-91. Steel Plate, Sheet and Strip Part 1: Carbon and Carbon-Manganese Plate, Sheet and Wide Strip Based on Formability. 8 pp.
SNZ NZS/BS 1449: Pt1:Sec1.4-91. Steel Plate, Sheet and Strip Part 1: Carbon and Carbon-Manganese Plate, Sheet and Strip Section 1.4: Hot Rolled Wide Material Based on Specified Minimum Strength. 8 pp.
SNZ NZS/BS 1449: Pt1:Sec1.6-91. Steel Plate, Sheet and Strip Part 1: Carbon and Carbon-Manganese Plate, Sheet and Strip Section 1.6: Tolerances on Dimensions and Shape for Hot Rolled Wide Material. 12 pp.
SNZ NZS/BS 1449: Pt1:Sec1.8-91. Steel Plate, Sheet and Strip Part 1: Carbon and Carbon-Manganese Plate, Sheet and Strip Section 1.8: Hot Rolled Narrow Strip Based on Formability. 8 pp.
SNZ NZS/BS 1449: Pt1:Sec1.10-91. Steel Plate, Sheet and Strip Part 1: Carbon and Carbon-Manganese Plate, Sheet and Strip Section 1.10: Hot Rolled Narrow Strip Based on Specified Minimum Strength. 8 pp.
SNZ NZS/BS 1449: Pt1:Sec1.12-91. Steel Plate, Sheet and Strip Part 1: Carbon and Carbon-Manganese Plate, Sheet and Strip Section 1.12: Tolerances on Dimensions and Shape for Hot Rolled Narrow Strip. 4 pp.
SNZ NZS/BS 1449: Pt1:Sec1.14-91. Steel Plate, Sheet and Strip Part 1: Carbon and Carbon-Manganese Plate, Sheet and Strip Section 1.14: Hot Rolled Narrow Strip Supplied in a Range of Conditions for Heat Treatment and General Engineering Purposes. 8 pp.

—Tubes
SAA AS 1450-83. Steel Tubes for Mechanical Purposes. 10 pp.

Carbon Materials
Use: Carbon

Carbon Microphones
See Also: Microphones

—Conditioning—Loudness Ratings—Telephone Transmission
CCITT RECMN P.75-89. Standard Conditioning Method for Handsets with Carbon Microphones—Telephone Transmission Quality (Study Group XII) 1 pp. 1 p.

Carbon Monoxide
See Also: Carbon Monoxide Content Analysis

—Aircraft—Contamination
CAA NOTICE #40 ISSUE 1. Carbon Monoxide Contamination in Aircraft (Airworthiness Notices). 2 pp.
CAA LEAFLET 5-3 07.90. Carbon Monoxide Contamination. 4 pp.

Carbon Monoxide (Cont.)
—Indicator Tubes—Aircraft
MOD UK DTD-922A-48. Manufacture of Carbon Monoxide Indicator Tubes Mark III (Reprinted March 1963). 14 pp.

—Standard Gases
JIS K 0002-92. Carbon Monoxide (Standard Reference Gas). 13 pp.

Carbon Monoxide Analyzers
See Also: Analyzers; Gas Analyzers
ISO 3930-76. Road Vehicles—Carbon Monoxide Analyser Equipment—Technical Specifications First Edition. 4 pp.
JIS B 7951-86. Continuous Analyzer for Carbon Monoxide in Ambient Air. 14 pp.
JIS D 8006-83. Simplified Carbonmonoxide Testers for Automobiles. 9 pp.
SAA AS 2094-77. Carbon Monoxide Analyser Equipment for Road Vehicles Reconfirmed 1988. 8 pp.
SAA AS 2095.1-77. Methods of Test for Pollutants from Road Vehicles—Part 1: Exhaust Carbon Monoxide at Idle Speed Reconfirmed 1988. 4 pp.

Carbon Monoxide Content Analysis
See Also: Carbon Monoxide
—Air
CSA Z223.21-M1978. Method for the Measurement of Carbon Monoxide. 27 pp.
ISO 8760-90. Work-Place Air—Determination of Mass Concentration of Carbon Monoxide—Method Using Detector Tubes for Short-Term Sampling with Direct Indication First Edition. 10 pp.
SAA AS 3580.7.1-92. Methods for Sampling and Analysis of Ambient Air—Part 7: Determination of Carbon Monoxide—Part 7.1: Direct-Reading Instrumental Method. 4 pp.

—Air—Gas Chromatography
ISO 8186-89. Ambient Air—Determination of the Mass Concentration of Carbon Monoxide—Gas Chromatographic Method First Edition. 12 pp.

—Cigarette Smoke
ISO 8454-87. Cigarettes—Determination of Carbon Monoxide in the Vapour Phase of Smoke (NDIR Method) First Edition. 8 pp.

—Ethylene—Gas Chromatography
ISO 6381-81. Ethylene and Propylene for Industrial Use—Determination of Traces of Carbon Monoxide and Carbon Dioxide—Gas Chromatographic Method First Edition. 8 pp.

—Exhaust Gases
CNS D3050-80. Measuring Methods of Carbon Monoxide in Automobile Exhaust Gas (Idling) (May)(5562).
CNS K9109-90. Method of Test for Carbon Monoxide in Exhaust Gas (Dec)(12824).
ISO 3929-76. Road Vehicles—Determination of Exhaust Carbon Monoxide Concentration at Idle Speed First Edition. 4 pp.
JIS D 1028-83. Measuring Method of Carbon Monoxide in Automobile Exhaust Gas (Idling). 5 pp.
JIS D 1030-76. Analytical Procedure for Continuous Measurement of Carbon Monoxide, Carbon Dioxide and Hydrocarbon in Automobile Exhaust Gas.

—Exhaust Gases—Motor Vehicles
SNZ NZS 5429-82. Code of Practice for In-Field Testing of Road Vehicles for Exhaust Carbon Monoxide and Hydrocarbons Concentration at Idle Speed. 8 pp.

—Flue Gases
BSI BS 1756: Part 4-77. 1977 Methods for the Sampling and Analysis of Flue Gases Part 4: Miscellaneous Analyses. 33 pp.
BSI BS 1756: Part 4-01. 1977 Amd 1 Methods for Sampling and Analysis of Flue Gases Part 4: Miscellaneous Analyses (AMD 7755) August 15, 1993. 35 pp.
BSI BS 1756: Part 5-71. 1971 Methods for the Sampling and Analysis of Flue Gases Part 5: Semi-Routine Analysis. 25 pp.
BSI BS 1756: Part 5-01. 1971 Amd 1 Methods for the Sampling and Analysis of Flue Gases Part 5: Semi-Routine Analysis (AMD 7756) October 15, 1993. 26 pp.
CSA CAN/CSA-Z223.2-M86. Method for the Continuous Measurement of Oxygen, Carbon Dioxide, Carbon Monoxide, Sulphur Dioxide, and Oxides of Nitrogen in Enclosed Combustion Flue Gas Streams; (Gen Instr 1). 44 pp.
CSA Z223.21-M1978. Method for the Measurement of Carbon Monoxide. 27 pp.
JIS K 0098-88. Methods for Determination of Carbon Monoxide in Flue Gas. 9 pp.

INTERNATIONAL AND NON-U.S. NATIONAL STANDARDS
SUBJECT INDEX
Carbon

Carbon Monoxide Content Analysis (Cont.)
—Gases
CNS K6230-67. Method of Analysis for Carbon Dioxide, Oxygen, Carbon Monoxide Hydrogen and Methane in Gases (Mar)(2592)(R 1971).

—Propylene—Gas Chromatography
ISO 6381-81. Ethylene and Propylene for Industrial Use—Determination of Traces of Carbon Monoxide and Carbon Dioxide—Gas Chromatographic Method First Edition. 8 pp.

Carbon Monoxide Indicator Tubes
Use: Carbon Monoxide—Indicator Tubes

Carbon Papers
Use For: Carbonizing Papers See Also: Carbonless Papers; Papers; Reproduction Papers
JIS P 3901-76. Carbonizing Paper (R 1984). 6 pp.
JIS P 5103-63. Carbon Paper. 6 pp.

—Double Coated
MOD UK TS 10022. Paper Tissue, Two Side Carbon-Coated (Withdrawn).

—Pencil
CNS S2013-81. Testing Standard for Carbon Paper for Pencil and Typewriting (May)(885).

—Rolls—Teletype
CGSB CAN/CGSB-53.2-M88. Paper, Teletype, Roll. 11 pp.
MOD UK TS 329C. Paper, Tissue, Carbon Coated, Grade I and II (for Telegraph Machines).

—Sets—Duplicating Masters
CGSB CAN/CGSB-53.36-92. Master Duplicating Paper Set. 8 pp.

—Sets—Manifold Papers
CGSB CAN/CGSB-53.113-M88. Paper Set, Manifold and Carbon. 11 pp.

—Solvent Coated
CGSB CAN/CGSB-53.82-M88. Paper, Carbon, Solvent Coated. 10 pp.
CGSB CAN/CGSB-53.140-M88. Carbon Sheet, Film Type, Solvent Coated. 10 pp.

—Typewriter
CGSB CAN/CGSB-53.1-M88. Paper, Carbon, Typewriter. 10 pp.
CNS S2013-81. Testing Standard for Carbon Paper for Pencil and Typewriting (May)(885).

Carbon Potentiometers
See Also: Potentiometers (Resistors)
JIS C 6443-85. Carbon Potentiometers for General Use. 38 pp.

—Rotary—Single Turn—Preferred Products List
CECC CECC MUAHAG Vol 2 IS 4-90. Preferred Products List; Resistors and Potentiometers (En, Fr, Ge). 59 pp.

Carbon Refractories
See Also: Ceramics; Refractory Materials

—Physical Testing
BSI BS 1902: Part 8-88. 1988 Methods for Testing Refractory Materials Part 8: Refractories Containing Carbon. 11 pp.

Carbon Residue Testing
Use For: Conradson Carbon Testing; Ramsbottom Coke Testing See Also: Petroleum Products; Residue Content Analysis
BSI BS 4451-80. 1980 Amd 1 Determination of Carbon Residue of Petroleum Products (Ramsbottom Method) (AMD 6465) November 30, 1990. 14 pp.
CNS C6378-75. Method of Test for Ramsbottom Carbon Residues of Petroleum Products (May)(3776).
DIN ENGL 51551-86. Testing of Lubricants and Liquid Fuels; Determination of Conradson Carbon Residue (Mar). 4 pp.
ISO 4262-93. Petroleum Products—Determination of Carbon Residue—Ramsbottom Method Second Edition. 16 pp.
ISO 6615-93. Petroleum Products—Determination of Carbon Residue—Conradson Method Second Edition. 13 pp.
JIS K 2270-90. Crude Petroleum and Petroleum Products—Determination of Carbon Residue. 21 pp.

—Creosote Oil
CNS K6072-57. Method of Test for Carbon Residue of Creosote Oil (Jul)(915)(R 1971).

Carbon Residue Testing (Cont.)
—Petroleum Products
BSI BS 2000: Part 13-93. 1993 Methods of Test for Petroleum and Its Products Part 13: Determination of Carbon Residue—Conradson Method. 7 pp.
BSI BS 2000: Part 13-90. 1990 Petroleum and Its Products Part 13: Conradson Carbon Residue of Petroleum Products. 8 pp.
CNS K6322-72. Method of Test for Conadson Carbon Residues of Petroleum Products (Oct)(3383).
ISO 10370-93. Petroleum Products—Determination of Carbon Residue—Micro Method First Edition. 12 pp.

Carbon Resistors
Use For: Composition Resistors See Also: Carbon Film Resistors; Fixed Resistors; Resistors; Variable Resistors
CNS C7095-85. Fixed Carbon Composition Resistors (Aug)(4973).

—Power
CECC CECC 40 101-006 ISSUE 1-75. Fixed Low Power Non-Wirewound Resistors—Insulated (En) AMD 1 (En). 7 pp.
CECC CECC 40 101-015. DIN 45 921 TEIL 105; Fixed Low Power Non-Wire-Wound Resistors, Carbon Composition Type, Standard Stability (Ge). 12 pp.

—Power—Printed Circuit Mount
CECC CECC 40 101-034. BS CECC 40 101-034 Issue 1; Fixed Low Power Non-Wirewound Resistors—Insulated (En). 9 pp.

—Quality Assurance
BSI BS E9111 N003-75. 1975 Amd 1 Detail Specification for Fixed Low Power Non-Wirewound Resistors. Typical Construction Carbon Composition. Full Assessment Level. 7 pp.

—Reliability Assured
CNS C5221-88. General Rules for Reliability Assured Fixed Carbon Composition Resistors (May)(12291).
CNS C7205-88. Reliability Assured Fixed Carbon Composition Resistors (Established Failure Rate) (May)(12292).
CNS C7206-88. Reliability Assured Fixed Carbon Composition Resistors (Form 1, Grade X) (May)(12293).
CNS C7207-88. Reliability Assured Fixed Carbon Composition Resistors (Form 2, Grade X) (May)(12294).
JIS C 6406-90. Fixed Carbon Composition Resistors for Use in Electronic Equipment. 22 pp.
MOD UK DSTAN 59-30: Part 4-80. Resistors, Fixed, of Assessed Quality (Listed on EPIC Database) Part 4: Resistors, Fixed, Non-Wirewound, Insulated, Low Power Full Assessment List of Items Conforming to BS CECC 40101-006 Issue 2. 29 pp.

—Variable
CNS C7011-72. Variable Carbon Resistors for General Use (Oct)(2321). 15 pp.
CNS C7097-79. Variable Carbon Resistors for Special Use (Sep)(4975).
JIS C 6444-91. Variable Carbon Composition Resistors for Use in Electronic Equipment—Characteristics Y, W and UC. 56 pp.

Carbon Steel Castings
See Also: Carbon Steels; Castings; Ferroalloy Castings; High Strength Steel Castings; Steel Castings
CNS G2075-86. Method of Test for High Tensile Strength Carbon Steel Castings and Low Alloy Steel Castings for Structural Purpose (Jul)(7146).
CNS G2208-84. Method of Test for Carbon Steel Castings (Feb)(10773).
CNS G3052-84. Carbon Steel Castings (Feb)(2906). 3 pp.
CNS G3136-86. High Tensile Strength Carbon Steel Castings and Low Alloy Steel Castings for Structural Purpose (Jul)(7145).
ISO 3755-91. Cast Carbon Steels for General Engineering Purposes Second Edition. 6 pp.
JIS G 5101-91. Carbon Steel Castings. 8 pp.
JIS G 5111-91. High Tensile Strength Carbon Steel Castings and Low Alloy Steel Castings for Structural Purposes. 9 pp.

—Heat Treated
ISO 9477-92. High Strength Cast Steels for General Engineering and Structural Purposes First Edition. 6 pp.

—Investment
BSI BS 3146: Part 1-74. 1974 Investment Castings in Metal Part 1: Carbon and Low Alloy Steels. 28 pp.

Carbon Steel Coatings
See Also: Carbon Steels; Coatings; Low Carbon Steel Coatings; Metal Coatings (On Metal); Steel Coatings

Carbon Steel Coatings (Cont.)
—Aluminum—Hot Dip
BSI BS 6536-85. 1985 Continuously Hot-Dip Aluminium/Silicon Coated Cold Reduced Carbon Steel Sheet and Strip. 16 pp.
ISO 5000-80. Continuous Hot-Dip Aluminium-Silicon-Coated Cold-Reduced Carbon Steel Sheet of Commercial and Drawing Qualities First Edition; DoD Adopted. 15 pp.

—Cadmium—Electroplated
MOD UK DTD-904C-63. Cadmium Plating (Reprinted December 1966, Incorporating Amendment No. 1) (Superseded by Def Stan 03-19). 5 pp.

—Chromium—Electrodeposited—Aerospace
AECMA PREN2132-82. Electrodeposition of Chromium for Engineering Purposes. 12 pp.

—Lead—Hot Dip
BSI BS 6582-85. 1985 Continuously Hot-Dip Alloy (Terne) Coated Cold Reduced Carbon Steel Flat Rolled Products. 12 pp.
ISO 4999-91. Continuous Hot-Dip Terne (Lead Alloy) Coated Cold-Reduced Carbon Steel Sheet of Commercial and Drawing Qualities Second Edition. 17 pp.

—Silicon—Hot Dip
BSI BS 6536-85. 1985 Continuously Hot-Dip Aluminium/Silicon Coated Cold Reduced Carbon Steel Sheet and Strip. 16 pp.
ISO 5000-80. Continuous Hot-Dip Aluminium-Silicon-Coated Cold-Reduced Carbon Steel Sheet of Commercial and Drawing Qualities First Edition; DoD Adopted. 15 pp.

—Zinc—Electrodeposited
ISO 5002-82. Hot-Rolled and Cold-Reduced Electrolytic Zinc-Coated Carbon Steel Sheet of Commercial and Drawing Qualities First Edition; DoD Adopted. 14 pp.

—Zinc—Hot Dip
ISO 3575-76. Continuous Hot-Dip Zinc-Coated Carbon Steel Sheet of Commercial, Lock-Forming and Drawing Qualities First Edition; DoD Adopted. 17 pp.
ISO 4998-91. Continuous Hot-Dip Zinc-Coated Carbon Steel Sheet of Structural Quality Second Edition. 14 pp.

Carbon Steels
Scope Note: For additional listings, see also products made from carbon steel Use For: Nonalloyed Steels; Unalloyed Steels See Also: Alloy Steels; Carbon Manganese Steels; Carbon Steel Castings; Carbon Steel Coatings; Ferroalloys; Free Machining Steels; High Strength Steels; Low Carbon Steels; Magnetic Materials; Spring Steels; Steels; Structural Steels
BSI BS 1407-70. 1970 Amd 1 High Carbon Bright Steel (Silver Steel). 13 pp.

—Aerospace—Baumann Testing
AECMA PREN2003-13-86. Test Methods for Steel Products—Part 13—Macrographic Examination by Sulphur Print (Baumann Method). 1 p.

—Aircraft
BSI 3S 70-01. (OBSOLESCENT) 1964 Amd 1 '55' Carbon Steel (Normalized) (45/55 tonf/sq in: Limiting Ruling Section 4 in) (PD 6313) January 31, 1968. 4 pp.

—Aluminum Content—Atomic Absorption Spectrometry
BSI BS 6200: SUBSEC 3.1.4-90. 1990 Sampling and Analysis of Iron, Steel and Other Ferrous Metals Part 3: Methods of Analysis Section 3.1: Determination of Aluminium Subsection 3.1.4: Non-Alloyed Steel: Flame Atomic Absorption Spectrometric Method. 13 pp.
BSI BS 6200: SUB SEC3.1.4-01. 1990 Amd 1 Sampling and Analysis of Iron, Steel and Other Ferrous Metals Part 3: Methods of Analysis Section 3.1: Determination of Aluminium Subsection 3.1.4: Non-Alloyed Steel: Flame Atomic Absorption Spectrometric Method. 18 pp.
DIN ENGL EN 29658-92. Determination of Aluminium Content of Steel by Flame Atomic Absorption Spectrometry; (ISO 9658: 1990) (July). 14 pp.
ISO 9658-90. Steel—Determination of Aluminium Content—Flame Atomic Absorption Spectrometric Method First Edition. 14 pp.

—Angles—Rolled
CNS G3160-82. Rerolled Carbon Steel (Jan)(8277). 4 pp.

—Bars
BSI BS 4449-88. 1988 Carbon Steel Bars for the Reinforcement of Concrete. 16 pp.

INDUSTRY STANDARDS

INTERNATIONAL AND NON-U.S. NATIONAL STANDARDS
SUBJECT INDEX — Carbon

Carbon Steels (Cont.)

—Bars (Cont.)
BSI 4S 99-76. 1976 Amd 1 2.5 Per Cent Nickel-Chromium-Molybdenum (High Carbon) Steel Billets, Bars, Forgings and Parts (1230-1420 MPa: Limiting Ruling Section 150 mm). 5 pp.

—Bars—Aircraft
MOD UK DTD-5032-56. Carbon Steel (Suitable for the Manufacture of Tie Rods). 2 pp.

—Bars—Bolts
BSI BS 1506-90. 1990 Carbon, Low Alloy and Stainless Steel Bars and Billets for Bolting Material to be Used in Pressure Retaining Applications. 26 pp.

—Bars—Cold Finished
SAA AS 1443-83. Carbon Steels and Carbon-Manganese Steels—Cold-Finished Bars. 15 pp.

—Bars—Extruded
ISO 4954-93. Steels for Cold Heading and Cold Extruding Second Edition. 42 pp.

—Bars—Hot Rolled
SAA AS 1442-92. Carbon Steels and Carbon-Manganese Steels—Hot-Rolled Bars and Semifinished Products. 13 pp.

—Bars—Rivets—Aerospace
MOD UK DTD-720B-68. 15 Carbon Steel (Suitable for Blind Rivets) (Limiting Ruling Section 20mm). 2 pp.

—Bars—Rolled
CNS G3049-66. Carbon Steel Sash Bar (Dec)(2699)(R 1972). 3 pp.
CNS G3160-82. Rerolled Carbon Steel (Jan)(8277). 4 pp.

—Bars—Seamless
BSI BS 6258-88. 1988 Hollow Steel Bars for Machining. 14 pp.
BSI BS 6258-01. 1988 Amd 1 Hollow Steel Bars for Machining (AMD 7278) September 15, 1992. 15 pp.

—Baumann Testing
BSI BS 6285-82. 1982 Macrographic Examination of Steel by Sulphur Print (Baumann Method). 4 pp.
ISO 4968-79. Steel—Macrographic Examination by Sulphur Print (Baumann Method) First Edition. 5 pp.

—Billets
BSI 4S 99-76. 1976 Amd 1 2.5 Per Cent Nickel-Chromium-Molybdenum (High Carbon) Steel Billets, Bars, Forgings and Parts (1230-1420 MPa: Limiting Ruling Section 150 mm). 5 pp.

—Billets—Aircraft
MOD UK DTD-5032-56. Carbon Steel (Suitable for the Manufacture of Tie Rods). 2 pp.

—Billets—Rolled
CNS G3053-68. Carbon Billet for Forgings and Rollings (Nov)(2907). 3 pp.

—Billets—Semifinished Products
BSI BS 1506-90. 1990 Carbon, Low Alloy and Stainless Steel Bars and Billets for Bolting Material to be Used in Pressure Retaining Applications. 26 pp.

—Boron Content—Spectrophotometry
CEN PREN 10 200-88. Chemical Analysis of Ferrous Materials Determination of Boron in Steel Spectrophotometric Method. 14 pp.

—Case Hardening—Semifinished Products
ISO 683 Pt 11-87. Heat-Treatable Steels, Alloy Steels and Free-Cutting Steels—Part 11: Wrought Case-Hardening Steels First Edition. 21 pp.

—Classification
ISO 4948 Pt 1-82. Steels—Classification—Part 1: Classification of Steels into Unalloyed and Alloy Steels Based on Chemical Composition First Edition. 4 pp.
ISO 4948 Pt 2-81. Steels—Classification—Part 2: Classification of Unalloyed and Alloy Steels According to Main Quality Classes and Main Property or Application Characteristics First Edition. 9 pp.

—Cold Worked
DIN ENGL 1654 Pt 2-89. Cold Heading and Cold Extruding Steels; Technical Delivery Conditions for Killed Unalloyed Steels Not Intended for Heat Treatment (Oct). 4 pp.

—Cold Worked—Bars
CNS G2133-82. Method of Test for Cold Finished Carbon and Alloy Steel Bars (Aug)(9277).

Carbon Steels (Cont.)

—Cold Worked—Bars (Cont.)
CNS G3194-85. Cold Finished Carbon and Alloy Steel Bars (Oct) (9276). 5 pp.
DIN ENGL 1654 Pt 1-89. Cold Heading and Cold Extruding Steels; Technical Delivery Conditions; General (Oct). 13 pp.
ISO 4954-93. Steels for Cold Heading and Cold Extruding Second Edition. 42 pp.
JIS G 3108-87. Rolled Carbon Steel for Cold-Finished Steel Bars.
JIS G 3108-75. Rolled Carbon Steel for Cold-Finished Steel Bars. 7 pp.
JIS G 3123-87. Cold Finished Carbon and Alloy Steel Bars. 11 pp.

—Cold Worked—Sheets
DIN ENGL 1623 Pt 3-87. Steel Flat Products; Cold Reduced Sheet and Strip; Technical Delivery Conditions; Mild Unalloyed Steels for Vitreous Enamelling (Jan). 7 pp.
ISO 3574-86. Cold-Reduced Carbon Steel Sheet of Commercial and Drawing Qualities Second Edition; DoD Adopted. 11 pp.
ISO 5001-80. Cold-Reduced Carbon Steel Sheet for Vitreous Enamelling First Edition; DoD Adopted. 15 pp.
ISO 5002-82. Hot-Rolled and Cold-Reduced Electrolytic Zinc-Coated Carbon Steel Sheet of Commercial and Drawing Qualities First Edition; DoD Adopted. 14 pp.

—Cold Worked—Sheets—Hardness
ISO 5954-84. Cold-Reduced Carbon Steel Sheet to Hardness Requirements First Edition; DoD Adopted. 8 pp.

—Cold Worked—Strips
DIN ENGL 1623 Pt 3-87. Steel Flat Products; Cold Reduced Sheet and Strip; Technical Delivery Conditions; Mild Unalloyed Steels for Vitreous Enamelling (Jan). 7 pp.
ISO 4960-86. Cold-Reduced Carbon Steel Strip with a Carbon Content over 0,25% First Edition; DoD Adopted. 10 pp.
ISO 6932-86. Cold-Reduced Carbon Steel Strip with a Maximum Carbon Content of 0,25% First Edition; DoD Adopted. 11 pp.

—Crystal Structure—Impact Testing
BSI BS 131: Part 5-65. 1965 Methods for Notched Bar Tests Part 5: Determination of Crystallinity. 12 pp.

—Decarburizing
BSI BS 6617: Part 2-87. 1987 Determination of Decarburization in Steel Part 2: Methods for Determining Decarburization by Chemical and Spectrographic Analysis Techniques. 8 pp.
ISO 3887-76. Steel, Non-Alloy and Low-Alloy—Determination of Depth of Decarburization First Edition. 5 pp.
SAA AS 2003-91. Carbon and Low Alloy Steel—Measurement of Decarburization. 8 pp.

—Elongation
BSI BS 3894: Part 1-65. 1965 Conversion for Elongation Values for Steel Part 1: Carbon and Low Alloy Steels. 25 pp.

—Emission Spectroscopy
CNS G2230-84. Emission-Spectroscopic Analysis for Carbon Steel and Low Alloy Steel (Oct)(11071).
JIS G 1252-75. Emission-Spectroscopic Analysis for Carbon Steel and Low Alloy Steel. 8 pp.

—Filler Metals—Welding
DIN ENGL 8573 Pt 1-83. Filler Metals for Welding Unalloyed and Low Alloy Cast Iron Materials; Designation; Technical Delivery Conditions (Jan). 8 pp.

—Flats—Cold Rolled
BSI BS 6830-87. 1987 Continiously Hot-Clip Aluminum/Zinc Alloy Coated Cold Rolled Carbon Steel Flat Products. 15 pp.

—Flats—Rolled
CNS G3160-82. Rerolled Carbon Steel (Jan)(8277). 4 pp.

—Forgings
BSI BS 29-76. 1976 Carbon Steel Forgings Above 150 mm Ruling Section. 12 pp.
BSI BS 1503-89. 1989 Amd 1 Steel Forgings for Pressure Purposes (AMD 6739) September 30, 1991. 35 pp.
BSI BS 1503-02. 1989 Amd 2 Steel Forgings for Pressure Purposes (AMD 7744) July 15, 1993 (Q). 36 pp.
BSI 4S 99-76. 1976 Amd 1 2.5 Per Cent Nickel-Chromium-Molybdenum (High Carbon) Steel Billets, Bars, Forgings and Parts (1230-1420 MPa: Limiting Ruling Section 150 mm). 5 pp.

Carbon Steels (Cont.)

—Forgings (Cont.)
CNS G3048-86. Carbon Steel Forgings for General Use (Jul)(2673).
CNS G3053-68. Carbon Billet for Forgings and Rollings (Nov)(2907). 3 pp.
CNS G3159-82. Carbon Steel Blooms and Billets for Forgings (Jan)(8276).
JIS G 3201-88. Carbon Steel Forgings for General Use. 12 pp.
JIS G 3251-88. Carbon Steel Blooms and Billets for Forgings. 8 pp.
SAA AS 1448-81. Carbon Steels and Carbon-Manganese Steels—Forgings (Ruling Section 300 mm Maximum). 12 pp.
SNZ NZS/BS 1503-89. Steel Forgings for Pressure Purposes Amend: 1, 1991. 32 pp.

—Forgings—Ultrasonic Testing
JIS G 0587-87. Methods for Ultrasonic Examination for Carbon and Low Alloy Steel Forgings. 24 pp.
SAA AS 1065-88. Non-Destructive Testing—Ultrasonic Testing of Carbon and Low Alloy Steel Forgings. 29 pp.

—Gage Lengths
ISO 2566 Pt 1-84. Steel—Conversion of Elongation Values—Part 1: Carbon and Low Alloy Steels Second Edition. 30 pp.

—Heat Treated
BSI 4S 99-76. 1976 Amd 1 2.5 Per Cent Nickel-Chromium-Molybdenum (High Carbon) Steel Billets, Bars, Forgings and Parts (1230-1420 MPa: Limiting Ruling Section 150 mm). 5 pp.

—Hollow Sections—Cold Formed
CEN PREN 10219-2-92. Cold Formed Structural Hollow Sections of Non-Alloy and Fine Grain Structural Steels—Part 2: Tolerances, Dimensions and Sectional Properties. 25 pp.

—Hollow Sections—Hot Finished
CEN PREN 10 210 (Part 1)-90. Hot Finished Structural Hollow Sections of Non-Alloy and Fine Grain Structural Steel—Part 1: Technical Delivery Requirements. 30 pp.
CEN PREN 10210-2-92. Hot Finished Structural Hollow Sections of Non-Alloy and Fine Grain Structural Steels—Part 2: Tolerances, Dimensions and Sectional Properties. 26 pp.

—Hot Rolled
ISO 683 Pt 18-76. Heat-Treated Steels, Alloy Steels and Free-Cutting Steels—Part 18: Wrought Unalloyed Steels in the Normalized, or Normalized and Cold-Drawn, or Hot-Rolled and Cold-Drawn Condition First Edition. 8 pp.

—Lead Content—Gravimetric Analysis
BSI BS 6200: SUB SEC 3.16.1-86. 1986 Sampling and Analysis of Iron, Steel and Other Ferrous Metals Part 3: Methods of Analysis Section 3.16: Determination of Lead Subsection 3.16.1: Carbon Steels and Low Alloy Steels: Gravimetric Method. 4 pp.

—Normalized
ISO 683 Pt 18-76. Heat-Treated Steels, Alloy Steels and Free-Cutting Steels—Part 18: Wrought Unalloyed Steels in the Normalized, or Normalized and Cold-Drawn, or Hot-Rolled and Cold-Drawn Condition First Edition. 8 pp.

—Plates
BSI BS 1449: Part 1-83. (WITHDRAWN) 1983 Amd 1 Steel Plate, Sheet and Strip Part 1: Carbon and Carbon-Manganese Plate, Sheet and Strip (Superseded by BS 1449: Sections 1.1 to 1.15: 1991 and BS EN 10130: 1991). 39 pp.
BSI BS 1449: Sec 1.1-91. 1991 Steel Plate, Sheet and Strip Part 1: Carbon and Carbon-Manganese Plate, Sheet and Strip Section 1.1: General Specification. 20 pp.
BSI BS 1501: Part 1-80. 1980 Amd 4 Steels for Fired and Unfired Pressure Vessels: Plates Part 1: Specification for Carbon and Carbon Manganese Steels (AMD 6259) July 31, 1990. 30 pp.
BSI BS 1501: Part 1-05. 1980 Amd 5 Steels for Fired and Unfired Pressure Vessels: Plates Part 1: Specification for Carbon and Carbon Manganese Steels (AMD 6824) October 31, 1991. 34 pp.
BSI BS 1501: Part 1-06. (WITHDRAWN) 1980 Amd 6 Steels for Pressure Purposes Part 1: Specification for Carbon and Carbon Manganese Steels: Plates (AMD 7024) June 15, 1992 (Superseded by BS EN 10028: 1-3: 1993). 35 pp.
SNZ NZS 1449: Pt1:Sec1.1-91. Steel Plate, Sheet and Strip Part 1: Carbon and Carbon-Manganese Plate, Sheet and Strip Section 1.1: General Specification. 16 pp.
SNZ NZS/BS 1501: Part 1-80. Steels for Pressure Purposes: Plates Part 1: Specification for Carbon and Carbon Manganese Steels. 28 pp.

INTERNATIONAL AND NON-U.S. NATIONAL STANDARDS
SUBJECT INDEX
Carbon

Carbon Steels *(Cont.)*

—Plates—Cold Rolled

BSI BS 1449: Sec 1.3-91. 1991 Steel Plate, Sheet and Strip Part 1: Carbon and Carbon-Manganese Plate, Sheet and Strip Section 1.3: Specification for Cold Rolled Steel Plate, Sheet and Wide Strip for Vitreous Enamelling Based on Formability. 10 pp.

BSI BS 1449: Sec 1.5-91. 1991 Steel Plate, Sheet and Strip Part 1: Carbon and Carbon-Manganese Plate, Sheet and Strip Section 1.5: Specification for Cold Rolled Wide Material Based on Specified Minimum Strength. 8 pp.

BSI BS 1449: Sec 1.7-91. (WITHDRAWN) 1991 Steel Plate, Sheet and Strip Part 1: Carbon and Carbon-Manganese Plate, Sheet and Strip Sec 1.7: Specification for Tolerances on Dimensions and Shape for Cold Rolled Wide Material (Superseded by BS EN 10131: 1991). 11 pp.

BSI BS 1449: Sec 1.9-91. 1991 Steel Plate, Sheet and Strip Part 1: Carbon and Carbon-Manganese Plate, Sheet and Strip Section 1.9: Specification for Cold Rolled Narrow Strip Based on Formability. 11 pp.

BSI BS 1449: Sec 1.11-91. 1991 Steel Plate, Sheet and Strip Part 1: Carbon and Carbon-Manganese Plate, Sheet and Strip Section 1.11: Specification for Cold Rolled Narrow Strip Based on Specified Minimum Strength. 11 pp.

BSI BS 1449: Sec 1.13-91. 1991 Steel Plate, Sheet and Strip Part 1: Carbon and Carbon-Manganese Plate, Sheet and Strip Section 1.13: Specification for Tolerances on Dimensions and Shape for Cold Rolled Narrow Strip. 8 pp.

BSI BS 1449: Sec 1.15-91. 1991 Steel Plate, Sheet and Strip Part 1: Carbon and Carbon-Manganese Plate, Sheet and Strip Section 1.15: Specification for Cold Rolled Narrow Strip Supplied in a Range of Conditions for Heat Treatment and General Engineering Purposes. 12 pp.

SNZ NZS/BS 1449: Pt1:Sec1.3-91. Steel Plate, Sheet and Strip Part 1: Carbon and Carbon-Manganese Plate, Sheet and Strip Section 1.3: Cold Rolled Steel Plate, Sheet and Wide Strip for Vitreous Enamelling Based on Formability. 8 pp.

SNZ NZS/BS 1449: Pt1:Sec1.5-91. Steel Plate, Sheet and Strip Part 1: Carbon and Carbon-Manganese Plate, Sheet and Strip Section 1.5: Cold Rolled Wide Material Based on Specified Minimum Strength. 4 pp.

SNZ NZS/BS 1449: Pt1:Sec1.7-91. Steel Plate, Sheet and Strip Part 1: Carbon and Carbon-Manganese Plate, Sheet and Strip Section 1.7: Tolerances on Dimensions and Shape for Cold Rolled Wide Material. 8 pp.

SNZ NZS/BS 1449: Pt1:Sec1.9-91. Steel Plate, Sheet and Strip Part 1: Carbon and Carbon-Manganese Plate, Sheet and Strip Section 1.9: Cold Rolled Narrow Strip Based on Formability. 8 pp.

SNZ NZS/BS 1449: Pt1:Sec1.11-91. Steel Plate, Sheet and Strip Part 1: Carbon and Carbon-Manganese Plate, Sheet and Strip Section 1.11: Cold Rolled Narrow Strip Based on Specified Minimum Strength. 8 pp.

SNZ NZS/BS 1449: Pt1:Sec1.13-91. Steel Plate, Sheet and Strip Part 1: Carbon and Carbon-Manganese Plate, Sheet and Strip Section 1.13: Tolerances on Dimensions and Shape for Cold Rolled Narrow Strip. 4 pp.

SNZ NZS/BS 1449: Pt1:Sec1.15-91. Steel Plate, Sheet and Strip Part 1: Carbon and Carbon-Manganese Plate, Sheet and Strip Section 1.15: Cold Rolled Narrow Strip Supplied in a Range of Conditions for Heat Treatment and General Engineering Purposes. 8 pp.

—Plates—Hot Rolled

BSI BS 1449: Sec 1.2-91. 1991 Steel Plate, Sheet and Strip Part 1: Carbon and Carbon-Manganese Plate, Sheet and Strip Section 1.2: Specification for Hot Rolled Steel Plate, Sheet and Wide Strip Based on Formability. 10 pp.

BSI BS 1449: Sec 1.4-91. 1991 Steel Plate, Sheet and Strip Part 1: Carbon and Carbon-Manganese Plate, Sheet and Strip Section 1.4: Specification for Hot Rolled Wide Material Based on Specified Minimum Strength. 10 pp.

BSI BS 1449: Sec 1.6-91. (WITHDRAWN) 1991 Steel Plate, Sheet and Strip Part 1: Carbon and Carbon-Manganese Plate, Sheet and Strip Section 1.6: Spec. for Tolerances on Dimensions and Shape for Hot Rolled Wide Material Superseded by BS EN 10051: 1992). 13 pp.

BSI BS 1449: Sec 1.8-91. 1991 Steel Plate, Sheet and Strip Part 1: Carbon and Carbon-Manganese Plate, Sheet and Strip Section 1.8: Specification for Hot Rolled Narrow Strip Based on Formability. 10 pp.

BSI BS 1449: Sec 1.10-91. 1991 Steel Plate, Sheet and Strip Part 1: Carbon and Carbon-Manganese Plate, Sheet and Strip Section 1.10: Specification for Hot Rolled Narrow Strip Based on Specified Minimum Strength. 11 pp.

Carbon Steels *(Cont.)*

—Plates—Hot Rolled *(Cont.)*

BSI BS 1449: Sec 1.12-91. 1991 Steel Plate, Sheet and Strip Part 1: Carbon and Carbon-Manganese Plate, Sheet and Strip Section 1.12: Specification for Tolerances on Dimensions and Shape for Hot Rolled Narrow Strip. 8 pp.

BSI BS 1449: Sec 1.14-91. 1991 Steel Plate, Sheet and Strip Part 1: Carbon and Carbon-Manganese Plate, Sheet and Strip Section 1.14: Specification for Hot Rolled Narrow Strip Supplied in a Range of Conditions for Heat Treatment and General Engineering Purposes. 10 pp.

SAA AS 1446-81. Carbon Steels and Carbon-Manganese Steels—Hot-Rolled Slab, Plate and Strip Based on Chemical Composition. 8 pp.

SNZ NZS/BS 1449: Pt1:Sec1.2-91. Steel Plate, Sheet and Strip Part 1: Carbon and Carbon-Manganese Plate, Sheet and Strip Section 1.2: Hot Rolled Steel Plate, Sheet and Wide Strip Based on Formability. 8 pp.

SNZ NZS/BS 1449: Pt1:Sec1.4-91. Steel Plate, Sheet and Strip Part 1: Carbon and Carbon-Manganese Plate, Sheet and Strip Section 1.4: Hot Rolled Wide Material Based on Specified Minimum Strength. 8 pp.

SNZ NZS/BS 1449: Pt1:Sec1.6-91. Steel Plate, Sheet and Strip Part 1: Carbon and Carbon-Manganese Plate, Sheet and Strip Section 1.6: Tolerances on Dimensions and Shape for Hot Rolled Wide Material. 12 pp.

SNZ NZS/BS 1449: Pt1:Sec1.8-91. Steel Plate, Sheet and Strip Part 1: Carbon and Carbon-Manganese Plate, Sheet and Strip Section 1.8: Hot Rolled Narrow Strip Based on Formability. 8 pp.

SNZ NZS/BS 1449: Pt1:Sec1.10-91. Steel Plate, Sheet and Strip Part 1: Carbon and Carbon-Manganese Plate, Sheet and Strip Section 1.10: Hot Rolled Narrow Strip Based on Specified Minimum Strength. 8 pp.

SNZ NZS/BS 1449: Pt1:Sec1.12-91. Steel Plate, Sheet and Strip Part 1: Carbon and Carbon-Manganese Plate, Sheet and Strip Section 1.12: Tolerances on Dimensions and Shape for Hot Rolled Narrow Strip. 4 pp.

SNZ NZS/BS 1449: Pt1:Sec1.14-91. Steel Plate, Sheet and Strip Part 1: Carbon and Carbon-Manganese Plate, Sheet and Strip Section 1.14: Hot Rolled Narrow Strip Supplied in a Range of Conditions for Heat Treatment and General Engineering Purposes. 8 pp.

—Plates—Ultrasonic Testing

BSI BS 5996-80. 1980 Amd 2 Ultrasonic Testing and Specifying Quality Grades of Ferritic Steel Plate. 10 pp.

SAA AS 1710-86. Non-Destructive Testing of Carbon and Low Alloy Steel Plate—Test Methods and Quality Classification. 8 pp.

—Quench Hardened—Bars

BSI BS EN 10083-2-91. 1991 Quenched and Tempered Steels Part 2: Technical Delivery Conditions for Unalloyed Quality Steels. 23 pp.

CEN EN 10083 (Part 2)-91. Quenched and Tempered Steels—Part 2: Technical Delivery Conditions for Unalloyed Quality Steels. 31 pp.

DIN ENGL EN 10083 Pt 2-91. Quenched and Tempered Steels; Technical Delivery Conditions for Special Steels; English Version of DEN EN 10083 Part 2 (Oct) (This Standard, Together with DIN EN 10083 Part 1, October 1991 Edition, Supersedes DIN 17200, March 1987 Edition). 19 pp.

—Quench Hardened—Flats

BSI BS EN 10083-2-91. 1991 Quenched and Tempered Steels Part 2: Technical Delivery Conditions for Unalloyed Quality Steels. 23 pp.

CEN EN 10083 (Part 2)-91. Quenched and Tempered Steels—Part 2: Technical Delivery Conditions for Unalloyed Quality Steels. 31 pp.

DIN ENGL EN 10083 Pt 2-91. Quenched and Tempered Steels; Technical Delivery Conditions for Special Steels; English Version of DEN EN 10083 Part 2 (Oct) (This Standard, Together with DIN EN 10083 Part 1, October 1991 Edition, Supersedes DIN 17200, March 1987 Edition). 19 pp.

—Quench Hardened—Forgings

BSI BS EN 10083-2-91. 1991 Quenched and Tempered Steels Part 2: Technical Delivery Conditions for Unalloyed Quality Steels. 23 pp.

CEN EN 10083 (Part 2)-91. Quenched and Tempered Steels—Part 2: Technical Delivery Conditions for Unalloyed Quality Steels. 31 pp.

DIN ENGL EN 10083 Pt 2-91. Quenched and Tempered Steels; Technical Delivery Conditions for Special Steels; English Version of DEN EN 10083 Part 2 (Oct) (This Standard, Together with DIN EN 10083 Part 1, October 1991 Edition, Supersedes DIN 17200, March 1987 Edition). 19 pp.

Carbon Steels *(Cont.)*

—Quench Hardened—Plates

BSI BS EN 10083-2-91. 1991 Quenched and Tempered Steels Part 2: Technical Delivery Conditions for Unalloyed Quality Steels. 23 pp.

CEN EN 10083 (Part 2)-91. Quenched and Tempered Steels—Part 2: Technical Delivery Conditions for Unalloyed Quality Steels. 31 pp.

DIN ENGL EN 10083 Pt 2-91. Quenched and Tempered Steels; Technical Delivery Conditions for Special Steels; English Version of DEN EN 10083 Part 2 (Oct) (This Standard, Together with DIN EN 10083 Part 1, October 1991 Edition, Supersedes DIN 17200, March 1987 Edition). 19 pp.

—Quench Hardened—Rods

BSI BS EN 10083-2-91. 1991 Quenched and Tempered Steels Part 2: Technical Delivery Conditions for Unalloyed Quality Steels. 23 pp.

CEN EN 10083 (Part 2)-91. Quenched and Tempered Steels—Part 2: Technical Delivery Conditions for Unalloyed Quality Steels. 31 pp.

DIN ENGL EN 10083 Pt 2-91. Quenched and Tempered Steels; Technical Delivery Conditions for Special Steels; English Version of DEN EN 10083 Part 2 (Oct) (This Standard, Together with DIN EN 10083 Part 1, October 1991 Edition, Supersedes DIN 17200, March 1987 Edition). 19 pp.

—Quench Hardened—Sheets

BSI BS EN 10083-2-91. 1991 Quenched and Tempered Steels Part 2: Technical Delivery Conditions for Unalloyed Quality Steels. 23 pp.

CEN EN 10083 (Part 2)-91. Quenched and Tempered Steels—Part 2: Technical Delivery Conditions for Unalloyed Quality Steels. 31 pp.

DIN ENGL EN 10083 Pt 2-91. Quenched and Tempered Steels; Technical Delivery Conditions for Special Steels; English Version of DEN EN 10083 Part 2 (Oct) (This Standard, Together with DIN EN 10083 Part 1, October 1991 Edition, Supersedes DIN 17200, March 1987 Edition). 19 pp.

—Quench Hardened—Strips

BSI BS EN 10083-2-91. 1991 Quenched and Tempered Steels Part 2: Technical Delivery Conditions for Unalloyed Quality Steels. 23 pp.

CEN EN 10083 (Part 2)-91. Quenched and Tempered Steels—Part 2: Technical Delivery Conditions for Unalloyed Quality Steels. 31 pp.

DIN ENGL EN 10083 Pt 2-91. Quenched and Tempered Steels; Technical Delivery Conditions for Special Steels; English Version of DEN EN 10083 Part 2 (Oct) (This Standard, Together with DIN EN 10083 Part 1, October 1991 Edition, Supersedes DIN 17200, March 1987 Edition). 19 pp.

—Resulfurized

CNS G2037-88. Method of Test for Resulphurized Free Cutting Carbon Steels (Mar)(4005).

CNS G3094-88. Resulphurized Free Cutting Carbon Steels (Mar)(4004). 6 pp.

—Rods—Aircraft

MOD UK DTD-5032-56. Carbon Steel (Suitable for the Manufacture of Tie Rods). 2 pp.

—Rods—Cold Worked

CNS G3167-88. Carbon Steel Wire Rods for Cold Heading and Cold Forgings (May)(8694).

—Scrap

JIS G 3111-87. Rerolled Carbon Steel.

JIS G 3111-76. Rerolled Carbon Steel. 8 pp.

—Semifinished Products

ISO 683 Pt 1-87. Heat-Treatable Steels, Alloy Steels and Free-Cutting Steels—Part 1: Direct-Hardening Unalloyed and Low-Alloyed Wrought Steel in Form of Different Black Products First Edition. 36 pp.

—Semifinished Products—Cold Finished

SAA AS 1443-83. Carbon Steels and Carbon-Manganese Steels—Cold-Finished Bars. 15 pp.

—Semifinished Products—Hot Rolled

SAA AS 1442-92. Carbon Steels and Carbon-Manganese Steels—Hot-Rolled Bars and Semifinished Products. 13 pp.

—Sheets

BSI BS 405-87. 1987 Amd 1 Uncoated Expanded Metal Carbon Steel Sheets for General Purposes (AMD 6660) January 31, 1991. 16 pp.

BSI BS 1449: Part 1-83. (WITHDRAWN) 1983 Amd 1 Steel Plate, Sheet and Strip Part 1: Carbon and Carbon-Manganese Plate, Sheet and Strip (Superseded by BS 1449: Sections 1.1 to 1.15: 1991 and BS EN 10130: 1991). 39 pp.

BSI BS 1449: Sec 1.1-91. 1991 Steel Plate, Sheet and Strip Part 1: Carbon and Carbon-Manganese Plate, Sheet and Strip Section 1.1: General Specification. 20 pp.

INDUSTRY STANDARDS 417

Carbon
INTERNATIONAL AND NON-U.S. NATIONAL STANDARDS
SUBJECT INDEX

Carbon Steels *(Cont.)*

—Sheets *(Cont.)*

SNZ NZS/BS 1449: Pt1:Sec1.1-91. Steel Plate, Sheet and Strip Part 1: Carbon and Carbon-Manganese Plate, Sheet and Strip Section 1.1: General Specification. 16 pp.

—Sheets—Cold Formed

CEN PREN 10249-1-93. Cold Formed Sheet Piling of Non Alloy Steels Part 1: Technical Delivery Conditions. 9 pp.

CEN PREN 10249-2-93. Cold Formed Sheet Piling of Non Alloy Steels Part 2: Tolerances on Shape and Dimensions. 8 pp.

—Sheets—Cold Rolled

BSI BS 1449: Sec 1.3-91. 1991 Steel Plate, Sheet and Strip Part 1: Carbon and Carbon-Manganese Plate, Sheet and Strip Section 1.3: Specification for Cold Rolled Steel Plate, Sheet and Wide Strip for Vitreous Enamelling Based on Formability. 10 pp.

BSI BS 1449: Sec 1.5-91. 1991 Steel Plate, Sheet and Strip Part 1: Carbon and Carbon-Manganese Plate, Sheet and Strip Section 1.5: Specification for Cold Rolled Wide Material Based on Specified Minimum Strength. 8 pp.

BSI BS 1449: Sec 1.7-91. (WITHDRAWN) 1991 Steel Plate, Sheet and Strip Part 1: Carbon and Carbon-Manganese Plate, Sheet and Strip Sec 1.7: Specification for Tolerances on Dimensions and Shape for Cold Rolled Wide Material (Superseded by BS EN 10131: 1991). 11 pp.

BSI BS 1449: Sec 1.9-91. 1991 Steel Plate, Sheet and Strip Part 1: Carbon and Carbon-Manganese Plate, Sheet and Strip Section 1.9: Specification for Cold Rolled Narrow Strip Based on Formability. 11 pp.

BSI BS 1449: Sec 1.11-91. 1991 Steel Plate, Sheet and Strip Part 1: Carbon and Carbon-Manganese Plate, Sheet and Strip Section 1.11: Specification for Cold Rolled Narrow Strip Based on Specified Minimum Strength. 11 pp.

BSI BS 1449: Sec 1.13-91. 1991 Steel Plate, Sheet and Strip Part 1: Carbon and Carbon-Manganese Plate, Sheet and Strip Section 1.13: Specification for Tolerances on Dimensions and Shape for Cold Rolled Narrow Strip. 8 pp.

BSI BS 1449: Sec 1.15-91. 1991 Steel Plate, Sheet and Strip Part 1: Carbon and Carbon-Manganese Plate, Sheet and Strip Section 1.15: Specification for Cold Rolled Narrow Strip Supplied in a Range of Conditions for Heat Treatment and General Engineering Purposes. 12 pp.

CNS G2134-88. Method of Test for Cold Rolled Carbon Steel Sheets and Strips (Jan)(9279). 2 pp.

CNS G3195-88. Cold Rolled Carbon Steel Sheets and Strips (Jan)(9278). 12 pp.

JIS G 3141-90. Cold Rolled Carbon Steel Sheets and Strip. 22 pp.

SNZ NZS/BS 1449: Pt1:Sec1.3-91. Steel Plate, Sheet and Strip Part 1: Carbon and Carbon-Manganese Plate, Sheet and Strip Section 1.3: Cold Rolled Steel Plate, Sheet and Wide Strip for Vitreous Enamelling Based on Formability. 8 pp.

SNZ NZS/BS 1449: Pt1:Sec1.5-91. Steel Plate, Sheet and Strip Part 1: Carbon and Carbon-Manganese Plate, Sheet and Strip Section 1.5: Cold Rolled Wide Material Based on Specified Minimum Strength. 4 pp.

SNZ NZS/BS 1449: Pt1:Sec1.9-91. Steel Plate, Sheet and Strip Part 1: Carbon and Carbon-Manganese Plate, Sheet and Strip Section 1.9: Cold Rolled Narrow Strip Based on Formability. 8 pp.

SNZ NZS/BS 1449: Pt1:Sec1.11-91. Steel Plate, Sheet and Strip Part 1: Carbon and Carbon-Manganese Plate, Sheet and Strip Section 1.11: Cold Rolled Narrow Strip Based on Specified Minimum Strength. 8 pp.

SNZ NZS/BS 1449: Pt1:Sec1.13-91. Steel Plate, Sheet and Strip Part 1: Carbon and Carbon-Manganese Plate, Sheet and Strip Section 1.13: Tolerances on Dimensions and Shape for Cold Rolled Narrow Strip. 4 pp.

SNZ NZS/BS 1449: Pt1:Sec1.15-91. Steel Plate, Sheet and Strip Part 1: Carbon and Carbon-Manganese Plate, Sheet and Strip Section 1.15: Cold Rolled Narrow Strip Supplied in a Range of Conditions for Heat Treatment and General Engineering Purposes. 8 pp.

—Sheets—Expanded

BSI BS 405-87. 1987 Amd 1 Uncoated Expanded Metal Carbon Steel Sheets for General Purposes (AMD 6660) January 31, 1991. 16 pp.

—Sheets—Hot Dip

ISO 5000-80. Continuous Hot-Dip Aluminium-Silicon-Coated Cold-Reduced Carbon Steel Sheet of Commercial and Drawing Qualities First Edition; DoD Adopted. 15 pp.

Carbon Steels *(Cont.)*

—Sheets—Hot Rolled

BSI BS 1449: Sec 1.2-91. 1991 Steel Plate, Sheet and Strip Part 1: Carbon and Carbon-Manganese Plate, Sheet and Strip Section 1.2: Specification for Hot Rolled Steel Plate, Sheet and Wide Strip Based on Formability. 10 pp.

BSI BS 1449: Sec 1.4-91. 1991 Steel Plate, Sheet and Strip Part 1: Carbon and Carbon-Manganese Plate, Sheet and Strip Section 1.4: Specification for Hot Rolled Wide Material Based on Specified Minimum Strength. 10 pp.

BSI BS 1449: Sec 1.6-91. (WITHDRAWN) 1991 Steel Plate, Sheet and Strip Part 1: Carbon and Carbon-Manganese Plate, Sheet and Strip Section 1.6: Spec. for Tolerances on Dimensions and Shape for Hot Rolled Wide Material Superseded by BS EN 10051: 1992). 13 pp.

BSI BS 1449: Sec 1.8-91. 1991 Steel Plate, Sheet and Strip Part 1: Carbon and Carbon-Manganese Plate, Sheet and Strip Section 1.8: Specification for Hot Rolled Narrow Strip Based on Formability. 10 pp.

BSI BS 1449: Sec 1.10-91. 1991 Steel Plate, Sheet and Strip Part 1: Carbon and Carbon-Manganese Plate, Sheet and Strip Section 1.10: Specification for Hot Rolled Narrow Strip Based on Specified Minimum Strength. 11 pp.

BSI BS 1449: Sec 1.12-91. 1991 Steel Plate, Sheet and Strip Part 1: Carbon and Carbon-Manganese Plate, Sheet and Strip Section 1.12: Specification for Tolerances on Dimensions and Shape for Hot Rolled Narrow Strip. 8 pp.

BSI BS 1449: Sec 1.14-91. 1991 Steel Plate, Sheet and Strip Part 1: Carbon and Carbon-Manganese Plate, Sheet and Strip Section 1.14: Specification for Hot Rolled Narrow Strip Supplied in a Range of Conditions for Heat Treatment and General Engineering Purposes. 10 pp.

CEN PREN 10248-1-93. Hot Rolled Sheet Piling of Non Alloy Steels Part 1: Technical Delivery Conditions. 16 pp.

CEN PREN 10248-2-93. Hot Rolled Sheet Piling of Non Alloy Steels Part 2: Tolerances on Shape and Dimensions. 11 pp.

ISO 3573-86. Hot-Rolled Carbon Steel Sheet of Commercial and Drawing Qualities Second Edition; DoD Adopted. 11 pp.

ISO 10384-92. Hot-Rolled Carbon Steel Sheet for Machinery First Edition. 11 pp.

SNZ NZS/BS 1449: Pt1:Sec1.2-91. Steel Plate, Sheet and Strip Part 1: Carbon and Carbon-Manganese Plate, Sheet and Strip Section 1.2: Hot Rolled Steel Plate, Sheet and Wide Strip Based on Formability. 8 pp.

SNZ NZS/BS 1449: Pt1:Sec1.4-91. Steel Plate, Sheet and Strip Part 1: Carbon and Carbon-Manganese Plate, Sheet and Strip Section 1.4: Hot Rolled Wide Material Based on Specified Minimum Strength. 8 pp.

SNZ NZS/BS 1449: Pt1:Sec1.6-91. Steel Plate, Sheet and Strip Part 1: Carbon and Carbon-Manganese Plate, Sheet and Strip Section 1.6: Tolerances on Dimensions and Shape for Hot Rolled Wide Material. 12 pp.

SNZ NZS/BS 1449: Pt1:Sec1.7-91. Steel Plate, Sheet and Strip Part 1: Carbon and Carbon-Manganese Plate, Sheet and Strip Section 1.7: Tolerances on Dimensions and Shape for Cold Rolled Wide Material. 8 pp.

SNZ NZS/BS 1449: Pt1:Sec1.8-91. Steel Plate, Sheet and Strip Part 1: Carbon and Carbon-Manganese Plate, Sheet and Strip Section 1.8: Hot Rolled Narrow Strip Based on Formability. 8 pp.

SNZ NZS/BS 1449: Pt1:Sec1.10-91. Steel Plate, Sheet and Strip Part 1: Carbon and Carbon-Manganese Plate, Sheet and Strip Section 1.10: Hot Rolled Narrow Strip Based on Specified Minimum Strength. 8 pp.

SNZ NZS/BS 1449: Pt1:Sec1.12-91. Steel Plate, Sheet and Strip Part 1: Carbon and Carbon-Manganese Plate, Sheet and Strip Section 1.12: Tolerances on Dimensions and Shape for Hot Rolled Narrow Strip. 4 pp.

SNZ NZS/BS 1449: Pt1:Sec1.14-91. Steel Plate, Sheet and Strip Part 1: Carbon and Carbon-Manganese Plate, Sheet and Strip Section 1.14: Hot Rolled Narrow Strip Supplied in a Range of Conditions for Heat Treatment and General Engineering Purposes. 8 pp.

—Sintered—Structural

CNS G3232-89. Sintered Carbon Steel Structural Parts (Jan)(12488).

—Slabs—Hot Rolled

SAA AS 1446-81. Carbon Steels and Carbon-Manganese Steels—Hot-Rolled Slab, Plate and Strip Based on Chemical Composition. 8 pp.

—Stress Corrosion

DIN ENGL 50915 Pt 1-85. Testing the Unalloyed and Low Alloy Steels Resistance to Intergranular Stress Corrosion Cracking; Unwelded Products (Dec). 3 pp.

Carbon Steels *(Cont.)*

—Strips

BSI BS 1449: Part 1-83. (WITHDRAWN) 1983 Amd 1 Steel Plate, Sheet and Strip Part 1: Carbon and Carbon-Manganese Plate, Sheet and Strip (Superseded by BS 1449: Sections 1.1 to 1.15: 1991 and BS EN 10130: 1991). 39 pp.

BSI BS 1449: Sec 1.1-91. 1991 Steel Plate, Sheet and Strip Part 1: Carbon and Carbon-Manganese Plate, Sheet and Strip Section 1.1: General Specification. 20 pp.

SNZ NZS/BS 1449: Pt1:Sec1.1-91. Steel Plate, Sheet and Strip Part 1: Carbon and Carbon-Manganese Plate, Sheet and Strip Section 1.1: General Specification. 16 pp.

—Strips—Cold Rolled

BSI BS 1449: Sec 1.3-91. 1991 Steel Plate, Sheet and Strip Part 1: Carbon and Carbon-Manganese Plate, Sheet and Strip Section 1.3: Specification for Cold Rolled Steel Plate, Sheet and Wide Strip for Vitreous Enamelling Based on Formability. 10 pp.

BSI BS 1449: Sec 1.5-91. 1991 Steel Plate, Sheet and Strip Part 1: Carbon and Carbon-Manganese Plate, Sheet and Strip Section 1.5: Specification for Cold Rolled Wide Material Based on Specified Minimum Strength. 8 pp.

BSI BS 1449: Sec 1.7-91. (WITHDRAWN) 1991 Steel Plate, Sheet and Strip Part 1: Carbon and Carbon-Manganese Plate, Sheet and Strip Sec 1.7: Specification for Tolerances on Dimensions and Shape for Cold Rolled Wide Material (Superseded by BS EN 10131: 1991). 11 pp.

BSI BS 1449: Sec 1.9-91. 1991 Steel Plate, Sheet and Strip Part 1: Carbon and Carbon-Manganese Plate, Sheet and Strip Section 1.9: Specification for Cold Rolled Narrow Strip Based on Formability. 11 pp.

BSI BS 1449: Sec 1.11-91. 1991 Steel Plate, Sheet and Strip Part 1: Carbon and Carbon-Manganese Plate, Sheet and Strip Section 1.11: Specification for Cold Rolled Narrow Strip Based on Specified Minimum Strength. 11 pp.

BSI BS 1449: Sec 1.13-91. 1991 Steel Plate, Sheet and Strip Part 1: Carbon and Carbon-Manganese Plate, Sheet and Strip Section 1.13: Specification for Tolerances on Dimensions and Shape for Cold Rolled Narrow Strip. 8 pp.

BSI BS 1449: Sec 1.15-91. 1991 Steel Plate, Sheet and Strip Part 1: Carbon and Carbon-Manganese Plate, Sheet and Strip Section 1.15: Specification for Cold Rolled Narrow Strip Supplied in a Range of Conditions for Heat Treatment and General Engineering Purposes. 12 pp.

CNS G2134-88. Method of Test for Cold Rolled Carbon Steel Sheets and Strips (Jan)(9279). 2 pp.

CNS G3195-88. Cold Rolled Carbon Steel Sheets and Strips (Jan)(9278). 12 pp.

JIS G 3141-90. Cold Rolled Carbon Steel Sheets and Strip. 22 pp.

SNZ NZS/BS 1449: Pt1:Sec1.3-91. Steel Plate, Sheet and Strip Part 1: Carbon and Carbon-Manganese Plate, Sheet and Strip Section 1.3: Cold Rolled Steel Plate, Sheet and Wide Strip for Vitreous Enamelling Based on Formability. 8 pp.

SNZ NZS/BS 1449: Pt1:Sec1.5-91. Steel Plate, Sheet and Strip Part 1: Carbon and Carbon-Manganese Plate, Sheet and Strip Section 1.5: Cold Rolled Wide Material Based on Specified Minimum Strength. 4 pp.

SNZ NZS/BS 1449: Pt1:Sec1.7-91. Steel Plate, Sheet and Strip Part 1: Carbon and Carbon-Manganese Plate, Sheet and Strip Section 1.7: Tolerances on Dimensions and Shape for Cold Rolled Wide Material. 8 pp.

SNZ NZS/BS 1449: Pt1:Sec1.9-91. Steel Plate, Sheet and Strip Part 1: Carbon and Carbon-Manganese Plate, Sheet and Strip Section 1.9: Cold Rolled Narrow Strip Based on Formability. 8 pp.

SNZ NZS/BS 1449: Pt1:Sec1.11-91. Steel Plate, Sheet and Strip Part 1: Carbon and Carbon-Manganese Plate, Sheet and Strip Section 1.11: Cold Rolled Narrow Strip Based on Specified Minimum Strength. 8 pp.

SNZ NZS/BS 1449: Pt1:Sec1.13-91. Steel Plate, Sheet and Strip Part 1: Carbon and Carbon-Manganese Plate, Sheet and Strip Section 1.13: Tolerances on Dimensions and Shape for Cold Rolled Narrow Strip. 4 pp.

SNZ NZS/BS 1449: Pt1:Sec1.15-91. Steel Plate, Sheet and Strip Part 1: Carbon and Carbon-Manganese Plate, Sheet and Strip Section 1.15: Cold Rolled Narrow Strip Supplied in a Range of Conditions for Heat Treatment and General Engineering Purposes. 8 pp.

—Strips—Cold Rolled—Magnetic

IEC 404 Pt 8-3-85. Magnetic Materials Part 8: Specifications for Individual Materials Section Three—Specification for Cold-Rolled Magnetic Non-Alloyed Steel Strip Delivered in the Semi-Processed State First Edition. 27 pp.

418
INDEX and DIRECTORY of

INTERNATIONAL AND NON-U.S. NATIONAL STANDARDS
SUBJECT INDEX

Carbon Steels (Cont.)

—Strips—Hot Rolled

BSI BS 1449: Sec 1.2-91. 1991 Steel Plate, Sheet and Strip Part 1: Carbon and Carbon-Manganese Plate, Sheet and Strip Section 1.2: Specification for Hot Rolled Steel Plate, Sheet and Wide Strip Based on Formability. 10 pp.

BSI BS 1449: Sec 1.4-91. 1991 Steel Plate, Sheet and Strip Part 1: Carbon and Carbon-Manganese Plate, Sheet and Strip Section 1.4: Specification for Hot Rolled Wide Material Based on Specified Minimum Strength. 10 pp.

BSI BS 1449: Sec 1.6-91. (WITHDRAWN) 1991 Steel Plate, Sheet and Strip Part 1: Carbon and Carbon-Manganese Plate, Sheet and Strip Section 1.6: Spec. for Tolerances on Dimensions and Shape for Hot Rolled Wide Material Superseded by BS EN 10051: 1992). 13 pp.

BSI BS 1449: Sec 1.8-91. 1991 Steel Plate, Sheet and Strip Part 1: Carbon and Carbon-Manganese Plate, Sheet and Strip Section 1.8: Specification for Hot Rolled Narrow Strip Based on Formability. 10 pp.

BSI BS 1449: Sec 1.10-91. 1991 Steel Plate, Sheet and Strip Part 1: Carbon and Carbon-Manganese Plate, Sheet and Strip Section 1.10: Specification for Hot Rolled Narrow Strip Based on Specified Minimum Strength. 11 pp.

BSI BS 1449: Sec 1.12-91. 1991 Steel Plate, Sheet and Strip Part 1: Carbon and Carbon-Manganese Plate, Sheet and Strip Section 1.12: Specification for Tolerances on Dimensions and Shape for Hot Rolled Narrow Strip. 8 pp.

BSI BS 1449: Sec 1.14-91. 1991 Steel Plate, Sheet and Strip Part 1: Carbon and Carbon-Manganese Plate, Sheet and Strip Section 1.14: Specification for Hot Rolled Narrow Strip Supplied in a Range of Conditions for Heat Treatment and General Engineering Purposes. 10 pp.

CNS G3110-86. Hot-Rolled Carbon Steel Strip for Pipes and Tubes (Jun)(4624). 3 pp.

ISO 6317-82. Hot-Rolled Carbon Steel Strip of Commercial and Drawing Qualities First Edition; DoD Adopted. 8 pp.

SAA AS 1446-81. Carbon Steels and Carbon-Manganese Steels.—Hot-Rolled Slab, Plate and Strip Based on Chemical Composition. 8 pp.

SNZ NZS/BS 1449: Pt1:Sec1.2-91. Steel Plate, Sheet and Strip Part 1: Carbon and Carbon-Manganese Plate, Sheet and Strip Section 1.2: Hot Rolled Steel Plate, Sheet and Wide Strip Based on Formability. 8 pp.

SNZ NZS/BS 1449: Pt1:Sec1.4-91. Steel Plate, Sheet and Strip Part 1: Carbon and Carbon-Manganese Plate, Sheet and Strip Section 1.4: Hot Rolled Wide Material Based on Specified Minimum Strength. 8 pp.

SNZ NZS/BS 1449: Pt1:Sec1.6-91. Steel Plate, Sheet and Strip Part 1: Carbon and Carbon-Manganese Plate, Sheet and Strip Section 1.6: Tolerances on Dimensions and Shape for Hot Rolled Wide Material. 12 pp.

SNZ NZS/BS 1449: Pt1:Sec1.8-91. Steel Plate, Sheet and Strip Part 1: Carbon and Carbon-Manganese Plate, Sheet and Strip Section 1.8: Hot Rolled Narrow Strip Based on Formability. 8 pp.

SNZ NZS/BS 1449: Pt1:Sec1.10-91. Steel Plate, Sheet and Strip Part 1: Carbon and Carbon-Manganese Plate, Sheet and Strip Section 1.10: Hot Rolled Narrow Strip Based on Specified Minimum Strength. 8 pp.

SNZ NZS/BS 1449: Pt1:Sec1.12-91. Steel Plate, Sheet and Strip Part 1: Carbon and Carbon-Manganese Plate, Sheet and Strip Section 1.12: Tolerances on Dimensions and Shape for Hot Rolled Narrow Strip. 4 pp.

SNZ NZS/BS 1449: Pt1:Sec1.14-91. Steel Plate, Sheet and Strip Part 1: Carbon and Carbon-Manganese Plate, Sheet and Strip Section 1.14: Hot Rolled Narrow Strip Supplied in a Range of Conditions for Heat Treatment and General Engineering Purposes. 8 pp.

—Structural

CNS G2028-87. Method of Test for Carbon Steel for Machine Structural Use (May)(3829).

CNS G3086-87. Carbon Steels for Machine Structural Use (May)(3828). 8 pp.

—Surface Finishing—Adhesive Bonding

ISO 4588-89. Adhesives—Preparation of Metal Surfaces for Adhesive Bonding First Edition. 8 pp.

—Tempered—Bars

BSI BS EN 10083-2-91. 1991 Quenched and Tempered Steels Part 2: Technical Delivery Conditions for Unalloyed Quality Steels. 23 pp.

CEN EN 10083 (Part 2)-91. Quenched and Tempered Steels—Part 2: Technical Delivery Conditions for Unalloyed Quality Steels. 31 pp.

Carbon Steels (Cont.)

—Tempered—Bars (Cont.)

DIN ENGL EN 10083 Pt 2-91. Quenched and Tempered Steels; Technical Delivery Conditions for Special Steels; English Version of DEN EN 10083 Part 2 (Oct) (This Standard, Together with DIN EN 10083 Part 1, October 1991 Edition, Supersedes DIN 17200, March 1987 Edition). 19 pp.

—Tempered—Flats

BSI BS EN 10083-2-91. 1991 Quenched and Tempered Steels Part 2: Technical Delivery Conditions for Unalloyed Quality Steels. 23 pp.

CEN EN 10083 (Part 2)-91. Quenched and Tempered Steels—Part 2: Technical Delivery Conditions for Unalloyed Quality Steels. 31 pp.

DIN ENGL EN 10083 Pt 2-91. Quenched and Tempered Steels; Technical Delivery Conditions for Special Steels; English Version of DEN EN 10083 Part 2 (Oct) (This Standard, Together with DIN EN 10083 Part 1, October 1991 Edition, Supersedes DIN 17200, March 1987 Edition). 19 pp.

—Tempered—Forgings

BSI BS EN 10083-2-91. 1991 Quenched and Tempered Steels Part 2: Technical Delivery Conditions for Unalloyed Quality Steels. 23 pp.

CEN EN 10083 (Part 2)-91. Quenched and Tempered Steels—Part 2: Technical Delivery Conditions for Unalloyed Quality Steels. 31 pp.

DIN ENGL EN 10083 Pt 2-91. Quenched and Tempered Steels; Technical Delivery Conditions for Special Steels; English Version of DEN EN 10083 Part 2 (Oct) (This Standard, Together with DIN EN 10083 Part 1, October 1991 Edition, Supersedes DIN 17200, March 1987 Edition). 19 pp.

—Tempered—Plates

BSI BS EN 10083-2-91. 1991 Quenched and Tempered Steels Part 2: Technical Delivery Conditions for Unalloyed Quality Steels. 23 pp.

CEN EN 10083 (Part 2)-91. Quenched and Tempered Steels—Part 2: Technical Delivery Conditions for Unalloyed Quality Steels. 31 pp.

DIN ENGL EN 10083 Pt 2-91. Quenched and Tempered Steels; Technical Delivery Conditions for Special Steels; English Version of DEN EN 10083 Part 2 (Oct) (This Standard, Together with DIN EN 10083 Part 1, October 1991 Edition, Supersedes DIN 17200, March 1987 Edition). 19 pp.

—Tempered—Rods

BSI BS EN 10083-2-91. 1991 Quenched and Tempered Steels Part 2: Technical Delivery Conditions for Unalloyed Quality Steels. 23 pp.

CEN EN 10083 (Part 2)-91. Quenched and Tempered Steels—Part 2: Technical Delivery Conditions for Unalloyed Quality Steels. 31 pp.

DIN ENGL EN 10083 Pt 2-91. Quenched and Tempered Steels; Technical Delivery Conditions for Special Steels; English Version of DEN EN 10083 Part 2 (Oct) (This Standard, Together with DIN EN 10083 Part 1, October 1991 Edition, Supersedes DIN 17200, March 1987 Edition). 19 pp.

—Tempered—Sheets

BSI BS EN 10083-2-91. 1991 Quenched and Tempered Steels Part 2: Technical Delivery Conditions for Unalloyed Quality Steels. 23 pp.

CEN EN 10083 (Part 2)-91. Quenched and Tempered Steels—Part 2: Technical Delivery Conditions for Unalloyed Quality Steels. 31 pp.

DIN ENGL EN 10083 Pt 2-91. Quenched and Tempered Steels; Technical Delivery Conditions for Special Steels; English Version of DEN EN 10083 Part 2 (Oct) (This Standard, Together with DIN EN 10083 Part 1, October 1991 Edition, Supersedes DIN 17200, March 1987 Edition). 19 pp.

—Tempered—Strips

BSI BS EN 10083-2-91. 1991 Quenched and Tempered Steels Part 2: Technical Delivery Conditions for Unalloyed Quality Steels. 23 pp.

CEN EN 10083 (Part 2)-91. Quenched and Tempered Steels—Part 2: Technical Delivery Conditions for Unalloyed Quality Steels. 31 pp.

DIN ENGL EN 10083 Pt 2-91. Quenched and Tempered Steels; Technical Delivery Conditions for Special Steels; English Version of DEN EN 10083 Part 2 (Oct) (This Standard, Together with DIN EN 10083 Part 1, October 1991 Edition, Supersedes DIN 17200, March 1987 Edition). 19 pp.

—Titanium Content—Atomic Absorption Spectrometry

CEN PREN 10211-91. Chemical Analysis of Ferrous Materials—Determination of Titanium in Steel and Iron—Flame Atomic Absorption Spectrometric Method. 16 pp.

—Tubes

BSI BS 3601-87. 1987 Carbon Steel Pipes and Tubes with Specified Room Temperature Properties for Pressure Purposes. 16 pp.

Carbon Steels (Cont.)

—Tubes (Cont.)

SAA AS 1450-83. Steel Tubes for Mechanical Purposes. 10 pp.

—Tubes—Structural

CNS G3102-86. Carbon Steel Tubes for General Structural Purposes (Jul)(4435). 8 pp.

JIS G 3444-88. Carbon Steel Tubes for General Structural Purposes. 18 pp.

—Weld Metal

JIS Z 3183-88. Quality Classification and Test Methods of Submerged Arc Deposited Metal for Carbon Steel and Low Alloy Steel. 18 pp.

SAA AS 1553.1-83. Covered Electrodes for Welding—Part 1: Low Carbon Steel Electrodes for Manual Metal-Arc Welding of Carbon and Carbon-Maganese Steels. 26 pp.

SAA AS 1553.2-87. Covered Electrodes for Welding—Part 2: Low and Intermediate Alloy Steel Electrodes for Manual Metal-Arc Welding of Carbon Steels and Low and Intermediate Alloy Steels. 29 pp.

—Wrought

JIS G 0321-66. Product Analysis and Its Tolerance for Wrought Steel. 8 pp.

—X-Ray Fluorescence Spectrometry

CNS G2182-84. Method for Fluorescent X-Ray Fluorescence Spectrochemical Analysis of Pig Iron, Cast Iron, Carbon Steel and Low Alloy Steel (Oct)(10502).

—Yield Strength

SAA AS 2069-77. Method for Verifying the Minimum Elevated Temperature Lower Yield or Proof Stress Properties of Carbon and Low Alloy Steel Products Reconfirmed 1983. 8 pp.

Carbon Tetrachloride

See Also: Halogenated Hydrocarbons; Hazardous Materials; Methyl Chloride

CNS K7135-64. Chemical Reagent (Carbon Tetrachloride) (Jul)(1635).

ISO 2312-72. Carbon Tetrachloride for Industrial Use—Methods of Test First Edition. 5 pp.

JIS K 1422-85. Carbon Tetrachloride. 15 pp.

JIS K 8459-91. Carbon Tetrachloride.

Carbon Zinc Batteries

Use: Leclanche Batteries

Carbonate Content Analysis

—Detergents

BSI BS 3762: Sec 3.14-85. 1985 Analysis of Formulated Detergents Part 3: Quantitative Test Methods Section 3.14: Method for Determination of Carbonate Content. 4 pp.

CGSB 2-GP-11M METH 26.1-83. Methods of Testing and Analysis of Soaps and Detergents Carbonates. 1 p.

—Fluorite—Volumetric Analysis

BSI BS 5659: Part 4-79. (WITHDRAWN) 1979 Acid-Grade Fluorspar Part 4: Determination of Carbonate Content. 8 pp.

ISO 4283-93. Grades of Fluorspar—Determination of Carbonate Content—Titrimetric Method Third Edition. 8 pp.

—Potassium Silicates—Volumetric Analysis

BSI BS 6092: Part 4-81. 1981 Sampling and Test for Sodium and Potassium Silicates for Industrial Use Part 4: Determination of Carbonate Content. 7 pp.

ISO 1691-76. Sodium and Potassium Silicates for Industrial Use—Determination of Carbonates Content—Gas-Volumetric Method First Edition. 7 pp.

—Soaps

CGSB 2-GP-11M METH 26.1-83. Methods of Testing and Analysis of Soaps and Detergents Carbonates. 1 p.

—Sodium Borates—Gravimetric Analysis

ISO 5936-80. Crude Sodium Borates for Industrial Use—Determination of Carbonate Content—Gravimetric Method First Edition. 6 pp.

—Sodium Fluorides—Gravimetric Analysis

BSI BS 5072: Part 9-80. 1980 Amd 1 Methods of Test for Sodium Fluoride for Industrial Use Part 9: Determination of Carbonate Content. 10 pp.

ISO 4278-77. Sodium Fluoride for Industrial Use—Determination of Carbonate Content—Gravimetric Method First Edition. (Erratum—Sept 1978). 9 pp.

—Sodium Hydroxide—Volumetric Analysis

BSI BS 6075: Part 6-81. 1981 Sampling and Test for Sodium Hydroxide for Industrial Use Part 6: Determination of Carbonate Content. 6 pp.

INTERNATIONAL AND NON-U.S. NATIONAL STANDARDS SUBJECT INDEX

Carbonate Content Analysis (Cont.)

—Sodium Hydroxide—Volumetric Analysis (Cont.)
ISO 3196-75. Sodium Hydroxide for Industrial Use—Determination of Carbonates Content—Titrimetric Method First Edition. 6 pp.

—Sodium Silicates—Volumetric Analysis
BSI BS 6092: Part 4-81. 1981 Sampling and Test for Sodium and Potassium Silicates for Industrial Use Part 4: Determination of Carbonate Content. 7 pp.

ISO 1691-76. Sodium and Potassium Silicates for Industrial Use—Determination of Carbonates Content—Gas-Volumetric Method First Edition. 7 pp.

—Soil Analysis
BSI BS 1377: Part 3-90. 1990 Methods of Test for Soils for Civil Engineering Purposes Part 3: Chemical and Electro-Chemical Tests. 45 pp.

DIN ENGL 19684 Pt 5-77. Methods of Soil Analysis for Water Management for Agricultural Purposes; Chemical Laboratory Tests; Determination of Carbonate Content of Soil (Feb). 2 pp.

Carbonate White Lead
Use: Basic Lead Carbonates

Carbonated Beverage Bottles
See Also: Beverages; Bottles; Carbonated Beverages

BSI BS 7367-91. 1991 Manufacture of Glass Bottles for Carbonated Soft Drinks Including Carbonated Water. 12 pp.

CNS S1099-83. Carbonated Beverage Bottles (Oct)(6527).

JIS S 2351-77. Carbonated Beverage Bottles.

JIS S 2351-74. Carbonated Beverage Bottles. 7 pp.

—Alkali Content
CNS S2111-83. Method of Test for Lead, Arsenic and Alkali Content Extracted from Carbonated Beverage Bottles (Oct)(10633).

—Arsenic Content
CNS S2111-83. Method of Test for Lead, Arsenic and Alkali Content Extracted from Carbonated Beverage Bottles (Oct)(10633).

—Fragment Retention
JIS S 2306-78. Method of Fragment Retention Test for Carbonated Beverage Bottles.

—Impact Testing
CNS S2044-83. Method of Impact Test for Carbonated Beverage Bottles (Jun)(6530).

JIS S 2303-78. Method of Impact Test for Carbonated Beverage Bottles.

JIS S 2303-74. Method of Impact Test for Carbonated Beverage Bottles. 5 pp.

—Lead Content
CNS S2111-83. Method of Test for Lead, Arsenic and Alkali Content Extracted from Carbonated Beverage Bottles (Oct)(10633).

—Pressure Measurement
BSI BS 7367-91. 1991 Manufacture of Glass Bottles for Carbonated Soft Drinks Including Carbonated Water. 12 pp.

CNS S2043-80. Method of Internal Pressure Test for Carbonated Beverage Bottles (Sep)(6529).

JIS S 2302-89. Method of Internal Pressure Test for Carbonated Beverage Bottles. 8 pp.

—Returnable
BSI BS 6119: Part 1-81. 1981 Amd 1 Glass Bottles for Carbonated Soft Drinks Part 1: Specification for 750 ml and 1 Litre Multi-Trip Bottles. 17 pp.

—Strain Testing
JIS S 2305-78. Method of Strain Test for Carbonated Beverage Bottles.

JIS S 2305-74. Method of Strain Test for Carbonated Beverage Bottles. 6 pp.

—Thermal Shock
CNS S2045-80. Method of Thermal Shock Test for Carbonated Beverage Bottles (Sep)(6531).

JIS S 2304-74. Method of Thermal Shock Test for Carbonated Beverage Bottles. 5 pp.

—Thickness Measurement
CNS S2042-80. Method of Measurement of Glass Thickness for Carbonated Beverage Bottles (Sep)(6528).

JIS S 2301-74. Method of Measurement of Glass Thickness for Carbonated Beverage Bottles. 3 pp.

Carbonated Beverages
Use For: Soft Drinks *See Also:* Beverages; Carbonated Beverage Bottles

CNS N5052-82. Carbonated Soft Drinks (Packaged) (Jan)(2270). 2 pp.

Carbonated Beverages (Cont.)

—Brix Testing
CNS N6092-82. Method of Test for Carbonated Soft Drink (Jan)(3761). 6 pp.

—Cans—Nominal Filling Volumes
ISO 10654-93. Light-Gauge Metal Containers—Round Open-Top Cans—Cans for Liquid Products with Added Gas, Defined by Their Nominal Filling Volumes First Edition; (Replaces 3004 Pt 1 Thru Pt 6). 6 pp.

—Carbon Dioxide Content
CNS N6092-82. Method of Test for Carbonated Soft Drink (Jan)(3761). 6 pp.

—Microbiological Analysis
CNS N6092-82. Method of Test for Carbonated Soft Drink (Jan)(3761). 6 pp.

—Sampling
CNS N6092-82. Method of Test for Carbonated Soft Drink (Jan)(3761). 6 pp.

—Soluble Matter Content
CNS N6092-82. Method of Test for Carbonated Soft Drink (Jan)(3761). 6 pp.

—Visual Inspection
CNS N6092-82. Method of Test for Carbonated Soft Drink (Jan)(3761). 6 pp.

Carbonates
Scope Note: Use a more specific term
See: Ammonium Carbonate; Barium Carbonates; Bismuth Subcarbonate; Calcium Carbonate; Carbonate Content Analysis; Copper Carbonate; Esters; Magnesium Carbonate; Magnesium Carbonate, Basic; Manganese Carbonate; Minerals; Nickel Carbonate; Potassium Carbonates; Precipitated Calcium Carbonate; Sodium Bicarbonate; Sodium Carbonate; Sodium Potassium Carbonate; Strontium Carbonate; Zinc Carbonate

Carbonizing Papers
Use: Carbon Papers

Carbonless Papers
See Also: Carbon Papers; Papers; Reproduction Papers

CNS P2072-90. Base Paper for Carbonless Copy Paper (Jul)(12746). 3 pp.

Carbonyl Compound Content Analysis
ISO 1388 Pt 3-81. Ethanol for Industrial Use—Methods of Test—Part 3: Estimation of Content of Carbonyl Compounds Present in Small Amounts—Photometric Method First Edition. 5 pp.

—Acetaldehyde—Volumetric Analysis
ISO 2885-73. Acetaldehyde for Industrial Use—Determination of Total Content of Carbonyl Compounds—Volumetric Method First Edition. 4 pp.

—Alcohols—Potentiometric Analysis
ISO 1843 Pt III-77. Higher Alcohols for Industrial Use—Methods of Test—Part III: Determination of Carbonyl Compounds Content—Potentiometric Method First Edition. 4 pp.

—Essential Oils
CNS K6612-80. Determination of Carbonyl Compounds Content of Essential Oils by Free Hydroxylamine Method (Dec)(6678).

CNS K6613-80. Determination of Carbonyl Compounds Content of Essential Oils by Hydroxylammonium Chloride Method (Dec)(6679).

ISO 1271-83. Essential Oils—Determination of Carbonyl Value—Free Hydroxylamine Method Second Edition; (Corrected and Reprinted -1985). 5 pp.

ISO 1279-84. Essential Oils—Determination of Carbonyl Value—Hydroxylammonium Chloride Method Second Edition. 5 pp.

—Ethanols—Photometry
BSI BS 6392: Part 2-83. 1983 Amd 1 Testing of Ethanol for Industrial Use Part 2: Method for Determination of Carbonyl Compounds Content Present in Small Amounts (Photometric Method). 5 pp.

ISO 1388 Pt 3-81. Ethanol for Industrial Use—Methods of Test—Part 3: Estimation of Content of Carbonyl Compounds Present in Small Amounts—Photometric Method First Edition. 5 pp.

—Ethanols—Volumetric Analysis
BSI BS 6392: Part 3-83. 1983 Amd 1 Testing of Ethanol for Industrial Use Part 3: Method for Determination of Carbonyl Compounds Content Present in Moderate Amounts (Titrimetric Method). 5 pp.

Carbonyl Compound Content Analysis (Cont.)

—Ethanols—Volumetric Analysis (Cont.)
ISO 1388 Pt 4-81. Ethanol for Industrial Use—Methods of Test—Part 4: Estimation of Content of Carbonyl Compounds Present in Moderate Amounts—Titrimetric Method First Edition. 5 pp.

—Furfurals—Volumetric Analysis
ISO 2512-74. Furfural for Industrial Use—Determination of Total Carbonyl Compounds—Volumetric Method First Edition. 4 pp.

Carbonyl J Acid
CNS K2176-87. Carbonyl J-Acid (Disodium Salt) (Dec)(12186). 2 pp.

JIS K 4155-90. 4,4'-Dihydroxy-7,7'-Ureylenedi-2-Naphthalene Sulfonic Acid Disodium Salt (Carbonyl J-Acid Disodium Salt).

Carbonyl Sulfide Content Analysis

—Natural Gas—Potentiometric Analysis
ISO 6326 Pt 3-89. Natural Gas—Determination of Sulfur Compounds—Part 3: Determination of Hydrogen Sulfide, Mercaptan Sulfur and Carbonyl Sulfide Sulfur by Potentiometry First Edition. 9 pp.

Carborundum
Use: Silicon Carbide

2-Carboxy-2'-hydroxy-5'-sulfoformazylbenzene
Use: Zincon (BTN)

3-Carboxy-4-Hydroxybenzenesulfonic Acid
Use: Sulfosalicylic Acid Dihydrate

Carboxyl Content Analysis
Use: Carboxylic Acid Content Analysis

Carboxylic Acid Content Analysis
Use For: Carboxyl Content Analysis

—Pulps
CNS P3037-80. Method of Test for Carboxyl Content of Pulp (Apr)(5469).

Carboxylic Acids
Scope Note: Use a more specific term *See:* Acetic Acid; Carbonyl J Acid; Citric Acid; Fatty Acids; Lactic Acid; Oxalic Acid; Succinic Acid

Carboxylmethyl Cellulose Content Analysis
See Also: Cellulose

—Detergents
BSI BS 3762: Sec 3.15-85. 1985 Analysis of Formulated Detergents Part 3: Quantitative Test Methods Section 3.15: Method for Determination of Carboxymethyl Cellulose Content. 4 pp.

Carboxymethylcellulose Sodium
Use For: Sodium Carboxymethylcellulose

—Initiators (Explosives)
MOD UK DSTAN 68-47-91. Sodium Carboxymethylcellulose Issue 2. 8 pp.

Carburetors
Use For: Carburettors *See Also:* Fuel Lines; Internal Combustion Engines; Spark Ignition Engines

—Aircraft—Control Knobs
SBAC AS 6386 ISSUE 2. Knob (Carburetter Hot/Cold Air Control).

—Aircraft—Deicers
CAA Chapter K5-5 App 04.74. Ice Protection Systems (Light Aeroplanes). 1 p.

CAA Chapter Q5-5 App 12.79. Ice Protection Systems (Non-Rigid Airships). 1 p.

—Automotive
CNS D3083-88. General Rules of Test Code of Carburetors for Automobiles (Dec)(8240).

CNS D3101-89. Operating Parts Test Code of Carburetors for Automobiles (Jan)(8258).

JIS D 1613-91. Carburetors for Automobiles—Test Methods. 23 pp.

—Automotive—Acceleration Testing
CNS D3093-89. Acceleration Test Code of Carburetors for Automobiles (Jan)(8250).

CNS D3099-89. Accelerate Pump Test Code of Carburetors for Automobiles (Jan)(8256).

INTERNATIONAL AND NON-U.S. NATIONAL STANDARDS
SUBJECT INDEX

Carburetors *(Cont.)*

—Automotive—Altitude Testing
CNS D3091-88. Method of Altitude Test of Carburetors for Automobiles (Dec)(8248).

—Automotive—Automatic Control Equipment
CNS D3100-89. Automatic Devices Test Code of Carburetors for Automobiles (Jan)(8257).

—Automotive—Deceleration Testing
CNS D3094-89. Deceleration Test Code of Carburetors for Automobiles (Jan)(8251).

—Automotive—Endurance Testing
CNS D3103-89. Endurance Test Code of Carburetors for Automobiles (Jan)(8260).

—Automotive—Exhaust Gases
CNS D3097-89. Exhaust Gas Measurement of Carburetors for Automobiles (Jan)(8254).

—Automotive—Flanges
CNS D2029-89. Shape and Dimension of Carburetor Flange for Automobiles (Jul)(5785).
JIS D 3701-87. Dimensions of Carburetor Flanges for Automobiles. 5 pp.

—Automotive—Flowbench Testing
CNS D3098-89. Flowbench Test Code of Carburetors for Automobiles (Jan)(8255).

—Automotive—Fuel Leakage
CNS D3102-89. Fuel Leakage Test Code of Carburetors for Automobiles (Jan)(8259).

—Automotive—Loads (Forces)
CNS D3086-88. Method of No-Load Test of Carburetors for Automobiles (Dec)(8243).
CNS D3087-88. Method of Load Test of Carburetors for Automobiles (Dec)(8244).

—Automotive—Low Temperature Testing
CNS D3090-88. Method of Low Temperature Test of Carburetors for Automobiles (Dec)(8247).

—Automotive—Pump Testing
CNS D3099-89. Accelerate Pump Test Code of Carburetors for Automobiles (Jan)(8256).

—Automotive—Road Testing
CNS D3088-88. Method of Incline Test of Carburetors for Automobiles (Dec)(8245).
CNS D3092-88. Method of Urban Driving Test of Carburetors for Automobiles (Dec)(8249).
CNS D3095-89. Cornering Test Code of Carburetors for Automobiles (Jan)(8252).
CNS D3096-89. Rough Road Test Code of Carburetors for Automobiles (Jan)(8253).

—Automotive—Soaking Testing
CNS D3089-88. Method of Hot Soak Test of Carburetors for Automobiles (Dec)(8246).

—Automotive—Starting Testing
CNS D3084-88. Method of Starting Test of Carburetors for Automobiles (Dec)(8241).

—Automotive—Warm-Up
CNS D3085-88. Method of Warm-Up Test of Carburetors for Automobiles (Dec)(8242).

Carburettors
Use: Carburetors

Carburizing
Scope Note: See the subheading Carburizing under the specific metal

Carcinatorn Oscillators
Use: Backward Wave Oscillators

Carcinatrons
Use: Backward Wave Oscillators

Card Cabinets
Use For: Card Cases *See Also:* Cabinets (Furniture); Filing Cabinets

—Plastic
JIS S 6051-85. Plastics Card Cases. 12 pp.

—Steel
CNS S1075-87. Office Steel Card Cabinets (Feb)(3119).

Card Cases
Use: Card Cabinets

Card Connectors
Use: Card Edge Connectors

Card Edge Connectors
Use For: Card Connectors; Edge Connectors; Edge Socket Connectors; Edgeboard Connectors; Edgecard Connectors *See Also:* Connectors; Printed Circuit Connectors
BSI BS 9526-77. (OBSOLESCENT) 1977 Amd 1 Rules for the Preparation of Detail Specifications for Electrical Edge Socket Connectors of Assessed Quality. Full and Basic Assessment Levels. 59 pp.
BSI BS CECC 75101-84. 1984 Two-Part and Edge Socket Connectors for Printed Board Application: Example Detail Specification/Blank Detail Specification. 15 pp.
CECC CECC 75 100 ISSUE 2-91. Sectional Specification: Two Part and Edge Socket Connectors for Printed Board Application (En, Fr, Ge). 125 pp.
CECC CECC 75 101 ISSUE 1-84. Example Detail Specification/Blank Detail Specification: Two Part and Edge Socket Connectors for Printed Board Application (En, Fr). 24 pp.

—.100' Contact Centers
IEC 130 Pt 11-71. Connectors for Frequencies Below 3 MHz Part 11: Edge Socket Connectors with Closed Ends and Having a Contact Spacing of 2.54 mm (0.1 in) Mating Either with Board Mounted Connectors or Printed Wiring Boards with Edge Board Contacts Multi-Row Board Mounted. 50 pp.
IEC 603 Pt 5-87. Connectors for Frequencies Below 3 MHz for Use with Printed Boards Part 5: Edge-Socket Connectors and Two-Part Connectors for Double-Sided Printed Boards with 2.54 mm (0.1 in) Spacing First Edition. 53 pp.
IEC 603 Pt 6-87. Connectors for Frequencies Below 3 MHz for Use with Printed Boards Part 6: Edge-Socket Connectors and Printed-Board Connectors with 2.54 mm (0.1 in) Contact Spacing for Single or Double-Sided Printed Boards of 1.6 mm (0.063 in) Nominal Thickness First Edition. 81 pp.

—Quality Assurance
BSI BS 9526-77. (OBSOLESCENT) 1977 Amd 1 Rules for the Preparation of Detail Specifications for Electrical Edge Socket Connectors of Assessed Quality. Full and Basic Assessment Levels. 59 pp.
BSI BS CECC 75100-84. (OBSOLESCENT) 1984 Two-Part and Edge Socket Connectors for Printed Board Application: Sectional Specification. 38 pp.
BSI BS CECC 75100-01. (OBSOLESCENT) 1984 Amd 1 Sectional Specification: Two-Part and Edge Socket Connectors for Printed Board Application (AMD 7806) June 15, 1993 (T). 41 pp.
BSI BS CECC 75101-84. 1984 Two-Part and Edge Socket Connectors for Printed Board Application: Example Detail Specification/Blank Detail Specification. 15 pp.
BSI BS EN 175100-93. 1993 Sectional Specification: Two Part and Edge Socket Connectors for Printed Board Applications (S). 46 pp.
CENELEC EN 175 100-92. Sectional Specification: Two Part and Edge Socket Connectors for Printed Board Application. 41 pp.

—Removable Contacts—Quality Assurance
BSI BS 9526 N0001-78. (OBSOLESCENT) 1978 Detail Specification for Multi-Contact Edge Socket Electrical Connectors. Single or Double Sided, Closed Ended, with Replaceable Contacts, Through-Board Solder or Wire-Wrap Terminations. Environmental Category. Full Assessment Level. 35 pp.
BSI BS 9526 N0002-79. (OBSOLESCENT) 1979 Detail Spec. for Muit-Contact Edge Socket Electrical Connectors, Single-or Double Sided, Open Ended with Metal Fixing Flanges, Guide Key Loc., Replaceable Contacts and Through-Board Solder or Wire-Warp Terminations. 34 pp.
BSI BS 9526 N0003-79. (OBSOLESCENT) 1979 Multi Contact Edge Socket Electrical Connector of Assessed Quality Single or Double Sided, Open Ended with Guide Key Location Replaceable Contacts, Through Board Solder or Wire Wrap Terminations. Full Assessment Level. 33 pp.

Cardamom
See Also: Spices
CNS N1098-82. Spices and Condiments-Cardamons (Apr)(8732). 3 pp.
ISO 882 Pt 1-93. Cardamom (Elettaria Cardamomum (Linnaeus) Maton Var. Minuscula Burkill)—Specification —Part 1: Whole Capsules First Edition. 12 pp.
ISO 882 Pt 2-93. Cardamom (Elettaria Cardamomum (Linnaeus) Maton Var. Minuscula Burkill)—Specification —Part 2: Seeds First Edition. 7 pp.

Cardamom Oil
See Also: Essential Oils
ISO 4733-81. Oil of Cardamom First Edition. 4 pp.

Cardboard
Use: Paperboard

Cardiac Pacemakers
Use: Pacemakers

Cardiac Valves
Use For: Heart Valve Prostheses
See Also: Prosthetic Devices; Surgical Implants
ISO 5840-89. Cardiovascular Implants—Cardiac Valve Prostheses Second Edition. 14 pp.

—Packaging
BSI BS 6444: Part 1-90. 1990 Cardiovascular Implants Part 1: Methods of Test for Heart Valve Substitutes and Requirements for Their Packaging and Labelling. 15 pp.

—Packaging—Identification Systems
BSI BS 6444: Part 1-90. 1990 Cardiovascular Implants Part 1: Methods of Test for Heart Valve Substitutes and Requirements for Their Packaging and Labelling. 15 pp.

Cardigans
Use: Sweaters

Cards
Scope Note: For additional listings, use a more specific term *See Also:* Aperture Cards; Bank Cards; Cards (Textile Machinery); Credit Cards; Identification Cards; Index Cards; Integrated Circuit Cards; Magnetic Cards; Postcards; Printers' Cards; Punched Cards; Stationery

—Waxed—Test Facilities
MOD UK DEF-134-01. Card, Paraffin-Waxed (for Use at Proof and Experimental Departments); Amendment 1. 4 pp.

Cards (Textile Machinery)
See Also: Flat Cards; Textile Machinery

—Clothings
BSI BS 2936: Part 1-81. 1981 Card Clothing Wire Part 1: Glossary of Terms for Metallic Card Clothing. 24 pp.
ISO 5234-80. Textile Machinery and Accessories—Metallic Card Clothing—Terms and Definitions First Edition; (Erratum—Sept 1981). 25 pp.
JIS L 5135-78. Card Clothing.

—Clothings—Wires
BSI BS ISO 9903-91. 1991 Textile Machinery and Accessories—Wires for Metallic Card Clothing (L). 10 pp.
ISO 4105-78. Textile Machinery and Accessories—Wires for Flexible Card Clothings First Edition. 4 pp.
ISO 6171-82. Textile Machinery and Accessories—Bead Wires and Corresponding Grooves for Cards—Main Types and Dimensions First Edition. 6 pp.
ISO 9903-91. Textile Machinery and Accessories—Wires for Metallic Card Clothing First Edition. 7 pp.

—Condenser Rubber
ISO 6170-83. Spinning Machinery—Condenser Rubbers for Cards First Edition. 5 pp.

—Gages
ISO 2572-82. Textile Machinery and Accessories—Card Gauges Second Edition. 3 pp.

—Glossaries
BSI BS 2936: Part 1-81. 1981 Card Clothing Wire Part 1: Glossary of Terms for Metallic Card Clothing. 24 pp.
ISO 5234-80. Textile Machinery and Accessories—Metallic Card Clothing—Terms and Definitions First Edition; (Erratum—Sept 1981). 25 pp.

—Grooves
ISO 6171-82. Textile Machinery and Accessories—Bead Wires and Corresponding Grooves for Cards—Main Types and Dimensions First Edition. 6 pp.

—Jacquards
JIS L 6311-59. Card for Jacquard.

—Looms—Shuttle Changing
JIS L 6517-61. Cards for Shuttle Change Apparatus (for Silk Loom). 4 pp.

—Woolen
ISO 342-83. Textile Machinery and Accessories—Worsted and Woollen Cards—Width of Cylinder and Width on the Wire Second Edition. 3 pp.
JIS L 5172-84. Working Widths of Worsted and Woolen Cards.

—Worsted
ISO 342-83. Textile Machinery and Accessories—Worsted and Woollen Cards—Width of Cylinder and Width on the Wire Second Edition. 3 pp.
JIS L 5172-84. Working Widths of Worsted and Woolen Cards.

INDUSTRY STANDARDS

Cargo Compartments (Aircraft)
Use: Aircraft Compartments—Cargo

Cargo Containers
Use: Shipping Containers

Cargo Handling Equipment
See Also: Ramps (Loading)

—Helicopters
NATO STANAG 1277 ED 1 AMD 0-85. Vertrep Equipment and Procedures. 13 pp.
NATO STANAG 1277 ED 1 AMD 1-92. Vertrep Equipment and Procedures. 14 pp.
NATO STANAG 2403 ED 1 AMD 1-90. Technical Criteria for External Cargo Carrying Strops/Pendants. 7 pp.
NATO STANAG 3542 ED 2 AMD 0-81. Technical Criteria for the Transport of Cargo by Helicopter. 16 pp.
NATO STANAG 3542 ED 3 AMD 1-90. Technical Criteria for the Transport of Cargo by Helicopter. 14 pp.
NATO STANAG 3542 ED 4 AMD 0-93. Technical Criteria for the Transport of Cargo by Helicopter. 15 pp.

—Shackles
DIN ENGL 82016-71. Cargo Shackles (Feb). 3 pp.

Cargo Holds
Use For: Holds (Cargo) *See Also:* Aircraft Compartments; Materials Handling Equipment

—Ships—Ventilation
ISO 9785-90. Shipbuilding—Ventilation of Cargo Spaces Where Internal Combustion Engine Vehicles May Be Driven—Calculation of Theoretical Total Airflow Required First Edition; (Corrigendum 1-1991). 11 pp.

Cargo Nets
See Also: Materials Handling Equipment; Nets

—Helicopters
MOD UK DSTAN 16-9-69. Nets and Slings, Cargo, Helicopter (Section on Slings is Superseded) Issue 1. 5 pp.
MOD UK DSTAN 16-9: Addendum. Nets and Slings, Helicopter Issue 1. 3 pp.
NATO STANAG 2950 ED 1 AMD 3-83. Technical Criteria for External Cargo Carrying Nets. 8 pp.
NATO STANAG 2950 ED 2 AMD 0-91. Technical Criteria for External Cargo Carrying Nets. 7 pp.

—Pallets—Air Cargo
ISO 4115-87. Air Cargo Equipment—Air/Land Pallet Nets Second Edition. 10 pp.
ISO 4170-87. Air Cargo Equipment—Interline Pallet Nets Second Edition. 8 pp.

—Rail Transportation—Load Restraints
BSI BS 5759-87. 1987 Webbing Load Restraint Assemblies for Use in Surface Transport. 14 pp.
BSI BS 5759-01. 1987 Amd 1 Webbing Load Restraint Assemblies for Use in Surface Transport (AMD 6859) November 29, 1991. 15 pp.
BSI BS 6451-84. 1984 Netting and Fibre Rope Load Restraint Systems in Surface Transport. 8 pp.
SNZ NZS 5445-86. Webbing Load Restraint Assemblies for Use in Surface Transport Amend: A, 1986. 11 pp.

—Stevedoring
BSI BS 6756-86. 1986 Fibre Rope Cargo Nets. 12 pp.

—Unit Loads—Air Cargo
ISO 8097-93. Aircraft—Minimum Airworthiness Requirements and Test Conditions for Certified Air Cargo Unit Load Devices Second Edition. 5 pp.

—Unit Loads—Air Cargo—Environmental Testing
ISO TR8647-90. Environmental Degradation of Textiles Used in Air Cargo Restraint Equipment First Edition. 21 pp.

Cargo Ships
See Also: Barges; Materials Handling Equipment; Merchant Ships; Shipping; Ships; Tanker Ships

—Handrails
ISO 5480-79. Shipbuilding—Guardrails for Cargo Ships First Edition. 8 pp.

—Lighting
CNS F5061-86. Cargo Lights (May)(10256).
CNS F5062-86. Cargo Lights (Simple Type) (May)(10257).
CNS F5063-86. Cargo Lights (Special Type) (May)(10258).
JIS F 8412-83. Cargo Lights.
JIS F 8442-89. Special Type Cargo Lights.

—Stanchions
ISO 5480-79. Shipbuilding—Guardrails for Cargo Ships First Edition. 8 pp.

Cargo Transportation
Use: Shipping

Cargo Vehicles
Scope Note: Use a more specific term *See:* Cargo Ships; Dollies; Electric Trucks; Forklifts; Freight Cars; Ground Vehicles; Hopper Cars; Hospital Carts; Industrial Trucks; Lift Trucks; Materials Handling Equipment; Mine Cars; Tank Cars; Tank Trucks; Trucks

Cargo Winches
See Also: Winches

—Ships
BSI BS MA 31-73. 1973 Ships' Deck Machinery; Cargo Winches. 9 pp.
ISO 3078-87. Shipbuilding—Cargo Winches Second Edition. 7 pp.
JIS F 6708-79. Ships' Cargo Winches.

Carlit
See Also: Explosives
JIS K 4803-68. Carlit (R 1983). 6 pp.

Carminic Acid
CNS K7136-64. Chemical Reagent (Carminic Acid) (Jul)(1636).

Carnauba Wax
See Also: Waxes
MOD UK DSTAN 13-113-90. Wax, Carnauba, LF Quality Issue 1. 10 pp.
MOD UK CS 3078. Wax Carnauba (Superseded by Def Stan 68-126).
MOD UK CS 3079. Wax Carnauba LF Quality (Superseded by Def Stan 13-113).

—Lutes (Material)
MOD UK DSTAN 68-126-90. Wax, Carnauba Issue 1. 10 pp.

Carob
See Also: Food
ISO 7907-87. Carob—Specification First Edition. 7 pp.

Carotene Content Analysis
—Animal Feed
CNS N4062-80. Method of Test for Carotene in Feeds (Dec)(6710).

—Fruit Products
ISO 6558 Pt 2-92. Fruits, Vegetables and Derived Products—Determination of Carotene Content—Part 2: Routine Methods First Edition. 7 pp.

—Fruits
ISO 6558 Pt 2-92. Fruits, Vegetables and Derived Products—Determination of Carotene Content—Part 2: Routine Methods First Edition. 7 pp.

—Vegetable Oils—Spectrophotometry
BSI BS 684: Sec 2.20-77. 1977 Methods of Analysis of Fats and Fatty Oils Part 2: Other Methods Section 2.20: Determination of Carotene in Vegetable Oils. 2 pp.

—Vegetable Products
ISO 6558 Pt 2-92. Fruits, Vegetables and Derived Products—Determination of Carotene Content—Part 2: Routine Methods First Edition. 7 pp.

—Vegetables
ISO 6558 Pt 2-92. Fruits, Vegetables and Derived Products—Determination of Carotene Content—Part 2: Routine Methods First Edition. 7 pp.

Carotenoid Content Analysis
See Also: Pigment Content Analysis

—Orange Oil
ISO 9910-91. Oil of Sweet Orange—Determination of the Total Carotenoids Content First Edition. 6 pp.

Carousels (Playground Equipment)
Use: Merry-Go-Rounds

Carpenters' Squares
Use: Squares (Instruments)

Carpentry
See Also: Construction; Woodworking

—Construction Contracts
DIN ENGL 18334-88. Tendering and Performance Stipulations in Contracts for Construction Works (VOB); Part C: General Technical Specifications in Contracts for Construction Works (ATV); Carpentry (Sept) (This Standard, Together with DIN 18299, September 1988 Edition,. 10 pp.

—Glossaries
BSI BS 6100: Sec 4.4-92. 1992 Glossary of Building and Civil Engineering Terms Part 4: Forest Products Section 4.4: Carpentry and Joinery. 20 pp.
BSI BS 6100: Sec 4.4-85. 1985 Amd 1 Glossary of Building and Civil Engineering Terms Part 4: Forest Products Section 4.4: Carpentry and Joinery (AMD 5552) July 29, 1988. 33 pp.

—Quality Assurance
BSI BS 8000: Part 5-90. 1990 Workmanship on Building Sites Part 5: Code of Practice for Carpentry, Joinery and General Fixings. 33 pp.

Carpet Adhesives
See Also: Adhesives

—Carpet to Concrete
CGSB CAN/CGSB-71.28-M88. Adhesive, for Direct Glue-Down Carpet Installation. 11 pp.

—Carpet to Plywood
CGSB CAN/CGSB-71.28-M88. Adhesive, for Direct Glue-Down Carpet Installation. 11 pp.

Carpet Cleaning Equipment
Use For: Rug Cleaning Equipment; Upholstery Cleaning Equipment *See Also:* Floor Cleaning Equipment; Vacuum Cleaners

—Safety
BSI BS 3456: Sec 102.10-90. (WITHDRAWN) 1990 Safety of Household Electrical Appliances Part 102: Particular Requirements Section 102.10: Floor Treatment Machines and Wet Scrubbing Machines (Superseded by BS EN 60335-2-10: 1992). 18 pp.
BSI BS 3456: Sec 202.2-90. 1990 Safety of Household Electrical Appliances Part 202: Particular Requirements Section 202.2: Vacuum Cleaners and Water Suction Cleaning Appliances. 25 pp.
BSI BS 3456: Sec 202.2-02. 1990 Amd 2 Safety of Household and Similar Electrical Appliances Part 202: Particular Requirements Section 202.2: Vacuum Cleaners and Water Suction Cleaning Appliances (AMD 7296) December 15, 1992 (L). 51 pp.
BSI BS 5415: Sec 2.1-86. 1986 Amd 1 Safety of Electrical Motor-Operated Industrial Cleaning Appliances Part 2: Particular Requirements Section 2.1: Floor Polishing, Scrubbing, Grinding, Scarifying and Carpet Shampooing Appliances (AMD 6450) March 28,1991. 13 pp.
BSI BS EN 60335-2-10-92. 1992 Safety of Household and Similar Electrical Appliances Part 2: Particular Requirements Section 2.10: Floor Treatment Machines and Wet Scrubbing Machines (Supersedes BS 3456: Section 102.10: 1992). 19 pp.
BSI BS EN 60335-2-10-01. 1992 Amd 1 Safety of Household and Similar Electrical Appliances Part 2: Particular Requirements Section 2.10: Floor Treatment Machines and Wet Scrubbing Machines (AMD 7792) July 15, 1993 (L). 22 pp.
CENELEC HD 252-74. Particular Specification for Vacuum Cleaness and Water Suction Cleaning Appliances. 5 pp.
CENELEC HD 252 S3-86. Safety of Household and Similar Electrical Appliances—Part 2: Particular Requirements for Vacuum Cleaners and Water Suction Cleaning Appliances (Superseded by EN 60335-2-2). 4 pp.
CENELEC HD 281 S1-86. Safety of Household and Similar Electrical Appliances—Part 2: Particular Requirements for Floor Treatment Machines and Wet Scrubbing Machines. 10 pp.
CENELEC EN 60 335-2-2-88. Safety of Household and Similar Electrical Appliances Part 2: Particular Requirements for Vacuum Cleaners and Water Suction Cleaning Appliances. 10 pp.
CENELEC EN 60335-2-2 PRAC-93. AMD prAC Safety of Household and Similar Electrical Appliances Part 2: Particular Requirements for Vacuum Cleaners and Water Suction Cleaning Appliances. 3 pp.
CENELEC EN 60335-2-2 /A2-90. AMD 2 Safety of Household and Similar Electrical Appliances Part 2: Particular Requirements for Vacuum Cleaners and Water Suction Cleaning Appliances. 7 pp.
CENELEC EN 60335-2-2 /A52-91. AMD 52 Safety of Household and Similar Electrical Appliances Part 2: Particular Requirements for Vacuum Cleaners and Water Suction Cleaning Appliances. 7 pp.
CENELEC EN 60335-2-1 0-90. Safety of Household and Similar Electrical Appliances Part 2: Particular Requirements for Floor Treatment Machines and Wet Scrubbing Machines. 10 pp.

INTERNATIONAL AND NON-U.S. NATIONAL STANDARDS
SUBJECT INDEX

Carpet Cleaning Equipment (Cont.)
—Safety (Cont.)
CENELEC EN 60335-2-10-90. Safety of Household and Similar Electrical Appliances Part 2: Particular Requirements for Floor Treatment Machines and Wet Scrubbing Machines; (IEC 335-2-10: 1987, Modified). 12 pp.
CENELEC EN 60335-2-10/A1-92. AMD 1 Safety of Household and Similar Electrical Appliances Part 2: Particular Requirements for Floor Treatment Machines and Wet Scrubbing Machines (IEC 335-2-10:1987/A1:1991). 4 pp.
IEC 335 Pt 2-2-93. Safety of Household and Similar Electrical Appliance Part 2: Particular Requirements for Vacuum Cleaners and Water Suction Cleaning Appliances Fourth Edition. 42 pp.
IEC 335 Pt 2-10-92. Safety of Household and Similar Electrical Appliances Part 2: Particular Requirements for Floor Treatment Machines and Wet Scrubbing Machines Fourth Edition; (CENELEC EN 60335-2-10: 1992). 28 pp.

—Spray Extraction
BSI BS 7460-91. 1991 Code of Practice for Use of Spray Extraction Machines on Carpets and Soft Floor Coverings in Hospitals. 9 pp.

—Spray Extraction—Safety
BSI BS 5415: Sec 2.3-86. 1986 Amd 1 Safety of Electrical Motor-Operated Industrial Cleaning Appliances Part 2: Particular Requirements Section 2.3: Spray Extraction Appliances (AMD 6454) December 21, 1990. 12 pp.

Carpet Tiles
See Also: Floor Coverings; Tiles

—Length Measurement
CEN PREN 994-93. Textile Floorcoverings—Determination of the Length of the Edges and Squareness of Tiles. 5 pp.

—Shrinkage
CEN PREN 986-92. Textile Floor Coverings—Tiles—Determination of Dimensional Changes Due to the Effects of Varied Water and Heat Conditions. 7 pp.

—Squareness
CEN PREN 994-93. Textile Floorcoverings—Determination of the Length of the Edges and Squareness of Tiles. 5 pp.

Carpets
Use For: Pile Carpets; Soft Floor Coverings; Velvet Carpets See Also: Fabrics; Floor Coverings; Linoleum; Rugs; Textile Floor Coverings
BSI BS 5229-75. 1975 Amd 1 Determination of Tuft Withdrawal Force of Carpets. 10 pp.
CNS L3122-81. Carpets of Number of Tufts Per Unit Length and Per Unit Area (Jun)(7497).
ISO 1763-86. Carpets—Determination of Number of Tufts and/or Loops per Unit Length and per Unit Area Second Edition. 4 pp.
ISO 4919-78. Carpets—Determination of Tuft Withdrawal Force First Edition. 5 pp.

—Abrasion Testing
CNS L3120-81. Method of Test for Abrasion Resistance of Carpets (Jun)(7495).
JIS L 1023-92. Testing Methods for Several Characteristics of Textile Floor Coverings. 26 pp.

—Backing—Exfoliation
JIS L 1023-92. Testing Methods for Several Characteristics of Textile Floor Coverings. 26 pp.

—Cleaning
JIS L 1023-92. Testing Methods for Several Characteristics of Textile Floor Coverings. 26 pp.
SNZ NZS/AS 3733-90. Textile Floor Coverings—Cleaning Maintenance Techniques for Domestic and Commercial Carpeting (This is a Joint Standard with SAA AS 3733). 26 pp.

—Commercial—Installation
CGSB 4-GP-156-75. Direct Glue-Down Carpet Guide to Selection and Installation. 8 pp.

—Commercial—Selection
CGSB 4-GP-156-75. Direct Glue-Down Carpet Guide to Selection and Installation. 8 pp.

—Cut
BSI BS 4223-89. 1989 Determination of Constructional Details of Textile Floor Coverings with Yarn Pile. 16 pp.

—Electrostatic Properties
JIS L 1023-92. Testing Methods for Several Characteristics of Textile Floor Coverings. 26 pp.

—Felts—Cushions
CGSB 4-GP-36M-78. Carpet Underlay, Fibre Type, Standard for. 9 pp.

Carpets (Cont.)
—Flammability Testing—Sampling
CGSB CAN/CGSB-4.155-M88. Flammability of Soft Floor Coverings—Sampling Plans. 14 pp.

—Flax
CNS L4006-74. Ramie, Flax or Wool Rugs and Carpets (Mar)(1390). 2 pp.

—Hand Knotted
CNS L3101-80. Hand-Made Carpets Determination of Types of Knots (Dec)(6687).
ISO 2549-72. Hand-Knotted Carpets—Determination of Tuft Leg Length Above the Woven Ground First Edition; (Corrigendum 1-1990). 4 pp.
ISO 2550-72. Hand-Made Carpets—Determination of Types of Knots First Edition. 4 pp.
ISO 5086-77. Hand-Knotted Carpets—Sampling and Selection of Areas of Test First Edition. 4 pp.

—Hand Made
CNS L3100-80. Hand-Made Carpets Determination of Pile Length (Dec)(6686).
CNS L3102-80. Hand-Made Carpets Sampling and Selection of Areas of Test (Dec) (6688).

—Knitted—Commercial
CGSB 4-GP-129-72. Carpets, Commercial, Standard for; (Amendment 3 Apr 1977). 16 pp.

—Looped
BSI BS 4223-89. 1989 Determination of Constructional Details of Textile Floor Coverings with Yarn Pile. 16 pp.

—Mass
CNS L3125-81. Carpets for Mass of Pile Above the Substrate (Jun) (7500).

—Needle Punched—Residential
CGSB CAN/CGSB-4.161-M87. Carpet for Residential Use. 42 pp.

—Pile Strength
JIS L 1023-92. Testing Methods for Several Characteristics of Textile Floor Coverings. 26 pp.

—Polymeric—Cushions
CGSB 20-GP-23M-78. Cushion, Carpet, Flexible Polymeric Material, Standard for. 10 pp.

—Ramie
CNS L4006-74. Ramie, Flax or Wool Rugs and Carpets (Mar)(1390). 2 pp.

—Raschel
BSI BS 4223-89. 1989 Determination of Constructional Details of Textile Floor Coverings with Yarn Pile. 16 pp.

—Shampoos
BSI BS 4088-81. 1981 Amd 1 Carpet Shampoos (AMD 6280) September 28, 1990. 12 pp.
CGSB CAN/CGSB-2.180-M87. Carpet and Upholstery Shampoo. 8 pp.
SNZ NZS 8711-86. Specification for Carpet Shampoos Amend: A, 1986. 12 pp.

—Staining
JIS L 1023-92. Testing Methods for Several Characteristics of Textile Floor Coverings. 26 pp.

—Thickness Measurement
CNS L2022-80. Carpet Thickness Gauge (Oct)(6581).
CNS L3123-81. Carpets Thickness of Pile Above the Substrate (Jun) (7498).

—Tufted
BSI BS 4223-89. 1989 Determination of Constructional Details of Textile Floor Coverings with Yarn Pile. 16 pp.
CNS L3124-81. Method of Test for Tuft Bind of Pile Floor Coverings (Jun)(7499).
JIS L 4405-90. Tufted Carpets. 14 pp.
MOD UK NES 851-91. Requirements for Carpets—Woven Tuft Construction Issue 1 (11.91). 24 pp.

—Tufted—Commercial
CGSB 4-GP-129-72. Carpets, Commercial, Standard for; (Amendment 3 Apr 1977). 16 pp.

—Tufted—Residential
CGSB CAN/CGSB-4.161-M87. Carpet for Residential Use. 42 pp.

—Wilton
JIS L 4404-90. Woven Carpets. 12 pp.

—Wool
CNS L4006-74. Ramie, Flax or Wool Rugs and Carpets (Mar)(1390). 2 pp.

—Woven
MOD UK NES 851-91. Requirements for Carpets—Woven Tuft Construction Issue 1 (11.91). 24 pp.

Carpets (Cont.)
—Woven—Commercial
CGSB 4-GP-129-72. Carpets, Commercial, Standard for; (Amendment 3 Apr 1977). 16 pp.

—Woven—Residential
CGSB CAN/CGSB-4.161-M87. Carpet for Residential Use. 42 pp.

Carports
See Also: Buildings

—Components—Metal
JIS A 6604-89. Metal Components for Car Port. 22 pp.

Carriageways
Use: Roads

Carrier Channels
See Also: Carrier Systems; Communication Channels

—Signal to Noise Ratio
CCIR RECMN 454-1-78. Pilot Carrier Level for HF Single-Sideband and Independent-Sideband Reduced-Carrier Systems—Section 3Aa—Technical Characteristics. 2 pp.

Carrier Frequencies
See Also: Carrier Systems; Frequencies; Frequency Deviation; Radio Frequencies; Virtual Carrier Frequencies

—Emergency Position Indicating Radio Beacons
CCIR RECMN 690-90. Transmission Characteristics of Emergency Position-Indicating Radio Beacons (EPIRBs) Operating on Carrier Frequencies of 121.5 MHz and 243 MHz—Section 8D—Radiodetermination, Global Maritime Distress and Safety System and Related Subjects. 1 p.

—Frequency Translation Equipment
CCITT RECMN G.233-89. Recommendations Concerning Translating Equipments—International Analogue Carrier Systems (Study Group XV) 10 pp. 10 pp.

—Groups
CCITT RECMN G.225-89. Recommendations Relating to the Accuracy of Carrier Frequencies—International Analogue Carrier Systems (Study Group XV) 3 pp. 3 pp.

—Master Oscillators—Carrier Systems
CCITT RECMN G.225-89. Recommendations Relating to the Accuracy of Carrier Frequencies—International Analogue Carrier Systems (Study Group XV) 3 pp. 3 pp.

—Mastergroups
CCITT RECMN G.225-89. Recommendations Relating to the Accuracy of Carrier Frequencies—International Analogue Carrier Systems (Study Group XV) 3 pp. 3 pp.

—Mobile Satellite Communications
CCIR Report 1185-90. Technical Aspects of Coordination Among Mobile Satellite Systems Using the Geostationary Satellite Orbit—Section 8H—Efficient Use of the Radio Spectrum Characteristics and Sharing of Frequency Resources. 12 pp.

—Signal Generators—Telephone Circuits
CCITT RECMN G.311-89. General Characteristics of Systems Providing 12 Carrier Telephone Circuits on an Open-Wire Pair—International Analogue Carrier Systems (Study Group XV) 6 pp. 6 pp.

—Supergroups
CCITT RECMN G.225-89. Recommendations Relating to the Accuracy of Carrier Frequencies—International Analogue Carrier Systems (Study Group XV) 3 pp. 3 pp.

—Supermastergroups
CCITT RECMN G.225-89. Recommendations Relating to the Accuracy of Carrier Frequencies—International Analogue Carrier Systems (Study Group XV) 3 pp. 3 pp.

—Telephone Circuits
CCITT RECMN G.225-89. Recommendations Relating to the Accuracy of Carrier Frequencies—International Analogue Carrier Systems (Study Group XV) 3 pp. 3 pp.

Carrier Leaks
See Also: Carrier Systems; Power Levels; Signal Levels

INDUSTRY STANDARDS

Carrier Leaks *(Cont.)*

—Communication Terminal Equipment—Groups

CCITT RECMN G.232-89. 12-Channel Terminal Equipments—International Analogue Carrier Systems (Study Group XV) 13 pp. 13 pp.

CCITT RECMN G.235-89. 16-Channel Terminal Equipments—International Analogue Carrier Systems (Study Group XV) 4 pp. 4 pp.

—Frequency Translation Equipment

CCITT RECMN G.233-89. Recommendations Concerning Translating Equipments—International Analogue Carrier Systems (Study Group XV) 10 pp. 10 pp.

—Open Wire Lines

CCITT RECMN G.361-89. Systems Providing Three Carrier Telephone Circuits on a Pair of Open-Wire Lines—International Analogue Carrier Systems (Study Group XV) 5 pp. 5 pp.

Carrier Power

—Radio Transmitters—Sound Broadcasting—Tropical Regions

CCIR RECMN 215-2-82. Maximum Transmitter Powers for Broadcasting in the Tropical Zone—Section 10A-2—Sound Broadcasting in the Tropical Zone. 1 p.

Carrier Sense Multiple Access/CD

Use For: CSMA/CD

—Local Area Networks

BSI BS 6531: Part 1-84. (WITHDRAWN) 1984 10 Mbps Slotted Ring Local Area Network Part 1: Specification for the Coding of Bits and Structure of Slots and Mini-Packets (Superseded by BS ISO 8802-7: 1991). 9 pp.

BSI BS 7246-90. 1990 Guide for Local Area Networks CSMA/DC 10 Mbits/s Baseband Planning and Installation. 44 pp.

BSI BS ISO 8802/3-89. (WITHDRAWN) 1989 Carrier Sense Multiple Access with Collision Detection (CSMA/CD) Access Method and Physical Layer Specifications (Superseded by BS ISO/IEC 8802-3: 1992). 180 pp.

BSI BS ISO/IEC 8802-3-92. 1992 Information Technology—Local and Metropolitan Area Networks—Part 3: Carrier Sense Multiple Access with Collision Detection (CSMA/CD) Access Method and Physical Layer Specifications. 314 pp.

BSI DD 98: Part 1-84. (WITHDRAWN) 1984 CSMA/CD Local Area Network Part 1: Technical Specification (Superseded by BS ISO 8802-3: 1989). 194 pp.

BSI DD 98: Part 2-84. (WITHDRAWN) 1984 CSMA/CD Local Area Network Part 2: Guidance for Implementors (superseded by BS ISO 8802-3: 1989). 102 pp.

BSI DD ENV 41103-2-93. 1993 Information Systems Interconnection—Provision of the OSI Connection-Mode Transport Service Using the OSI Connection-Mode Network Service in an End System Attached to a LAN Part 2: ISO 8802-3 CSMA/CD Subnetwork Dependent Media Reqmts.. 12 pp.

BSI DD ENV 41801-3-92. 1992 Information Systems Interconnection—Relaying the Connectionless-Mode Network Service—Part 3: ISO 8802-3 CSMA/CD Subnetwork Dependent Media Dependent Requirements. 18 pp.

BSI DD ENV 41802-3-92. 1992 Information Systems Interconnection—X.25 Protocol Relaying Part 3: ISO 8802-3 CSMA/CS Subnetwork Dependent Media Dependent Requirements. 13 pp.

CEN ENV 41 101-86. Information Systems Interconnection: Local Area Networks; Provision of the OSI Connection-Mode Transport Service Using Connectionless-Mode Network Service on a CSMA/CD Single LAN. 16 pp.

CEN ENV 41 102-90. Information Systems Interconnection—Local Area Networks—Provision of the OSI Connection-Mode Transport Service Using the OSI Connectionless-Mode Network Service in an End System Attached to a CSMA/CD LAN. 61 pp.

CEN ENV 41 103-87. Information Systems Interconnection: Local Area Networks; Provision of the OSI Connection-Mode Transport Service and the OSI Connection-Mode Network Service in an End Systems on a CSMA/CD LAN. 23 pp.

CEN ENV 41103-2-92. Information Systems Interconnection—Provision of the OSI Connection-Mode Transport Service Using the OSI Connection-Mode Network Service in an End System Attached to a LAN—Part 2: ISO 8802-3 CSMA/CD Subnetwork Dependent Media Requirements. 10 pp.

CEN ENV 41103-2-92. Information Systems Interconnection—Provision of the OSI Connection-Mode Transport Service Using the OSI Connection-Mode Network Service in an End System Attached to a LAN Part 2: ISO 8802-3 CSMA/CD Subnetwork Dependent Media Requirements. 8 pp.

Carrier Sense Multiple Access/CD *(Cont.)*

—Local Area Networks *(Cont.)*

CEN ENV 41801-3-92. Information Systems Interconnection—Relaying the Connectionless-Mode Network Service—Part 3: ISO 8802-3 CSMA/CD Subnetwork Dependent Media Dependent Requirements. 26 pp.

CEN ENV 41802-3-92. Information Systems Interconnection—X.25 Protocol Relaying—Part 3: ISO 8802-3 CSMA/CD Subnetwork Dependent Media Dependent Requirements. 10 pp.

ECMA ECMA 81-84. Local Area Networks (CSMA/CD Baseband) Physical Layer. 17 pp.

ECMA ECMA 82-84. Local Area Networks (CSMA/CD Baseband) Link Layer. 40 pp.

IEC 907-89. Local Area Networks CSMA/CD 10 Mbit/s Baseband Planning and Installation Guide First Edition. 86 pp.

IEC 8802 Pt 3 Draft AMD 6. Information Processing Systems—Local Area Networks—Part 3: Carrier Sense Multiple Access with Collision Detection (CSMA/CD) Access Method and Physical Layer Specifications Amendment 6; (1991) ***CD-ROM ONLY***. 5 pp.

IEC 8802 Pt 3 Draft AMD 7. Information Processing Systems—Local Area Networks—Part 3: Carrier Sense Multiple Access with Collision Detection (CSMA/CD) Access Method and Physi-cal Layer Specifications Amendment 7: Addition of Layer Management (Section 5); (1991) ***CD-ROM ONLY***. 40 pp.

IEC 8802 Pt 3 Draft AMD 9. Information Processing Systems—Local Area Networks—Part 3: Carrier Sense Multiple Access with Collision Detection (CSMA/CD) Access Method and Physical Layer Specifications Amendment 9: 10Base-T; (1991) ***CD-ROM ONLY***. 54 pp.

IEC 8802 Pt 3 Draft AMD 11. Information Technology—Local and Metropolitan Area Networks—Part 3: Carrier Sense Multiple Access with Collision Detection (CSMA/CD) Access Meth. and Phys. Layer Specs. Amd 11: Layer Mgmt. for 10 Mb/s Baseband Repeaters (Sec. 19); (1993) (IEEE Std. 802.3k) ***CD-ROM ONLY**. 31 pp.

IEC ISP10608 Pt 2-92. Information Technology—International Standard-ized Profile TAnnnn—Connection-Mode Trans-port Service over Con-nectionless-Mode Network Service—Part 2: TA51 Profile Including Sub-network-Dependent Re-quirements for CSMA/CD Local Area Networks (LANs) First Edition. 15 pp.

IEC DISP 10612 Pt 2-92. Information Technology—International Standardized Profile RD5p.5q—Relaying the MAC Service Using Transparent Bridging—Part 2: CSMA/CD LAN Subnetwork-Dependent, Media-Dependent Requirements ***CD-ROM ONLY***. 32 pp.

IEC DISP 10612 Pt 4-92. Information Technology—International Standardized Profile RD5p.5q—Relaying the MAC Service Using Transparent Bridging—Part 4: Profile RD51.51 (CSMA/CD-CSMA/CD) ***CD-ROM ONLY***. 28 pp.

IEC DISP 10614 Pt 3-92. Information Technology—International Standardized Profile RC—X.25 Protocol Relaying —Part 3: ISO 8802-3 CSMA/CD Subnetwork-Dependent, Media-Dependent Requirements ***CD-ROM ONLY***. 37 pp.

ISO 8802 Pt 3 Draft AMD 6. Information Processing Systems—Local Area Networks—Part 3: Carrier Sense Multiple Access with Collision Detection (CSMA/CD) Access Method and Physical Layer Specifications Amendment 6; (1991) ***CD-ROM ONLY***. 5 pp.

ISO 8802 Pt 3 Draft AMD 7. Information Processing Systems—Local Area Networks—Part 3: Carrier Sense Multiple Access with Collision Detection (CSMA/CD) Access Method and Physical Layer Specifications Amendment 7: Addition of Layer Management (Section 5); (1991)***CD-ROM ONLY***. 40 pp.

ISO 8802 Pt 3 Draft AMD 9. Information Processing Systems—Local Area Networks—Part 3: Carrier Sense Multiple Access with Collision Detection (CSMA/CD) Access Method and Physical Layer Specifications Amendment 9: 10Base-T; (1991) ***CD-ROM ONLY***. 54 pp.

ISO 8802 Pt 3 Draft AMD 11. Information Technology—Local and Metropolitan Area Networks—Part 3: Carrier Sense Multiple Access with Collision Detection (CSMA/CD) Access Meth. and Phys. Layer Specs. Amd 11: Layer Mgmt. for 10 Mb/s Baseband Repeaters (Sec. 19); (1993) (IEEE Std. 802.3k) ***CD-ROM ONLY**. 31 pp.

ISO ISP10608 Pt 2-92. Information Technology—International Standard-ized Profile TAnnnn—Connection-Mode Trans-port Service over Con-nectionless-Mode Network Service—Part 2: TA51 Profile Including Sub-network-Dependent Re-quirements for CSMA/CD Local Area Networks (LANs) First Edition. 15 pp.

Carrier Sense Multiple Access/CD *(Cont.)*

—Local Area Networks *(Cont.)*

ISO DISP 10612 Pt 2-92. Information Technology—International Standardized Profile RD5p.5q—Relaying the MAC Service Using Transparent Bridging—Part 2: CSMA/CD LAN Subnetwork-Dependent, Media-Dependent Requirements ***CD-ROM ONLY***. 32 pp.

ISO DISP 10612 Pt 4-92. Information Technology—International Standardized Profile RD5p.5q—Relaying the MAC Service Using Transparent Bridging—Part 4: Profile RD51.51 (CSMA/CD-CSMA/CD) ***CD-ROM ONLY***. 28 pp.

ISO DISP 10614 Pt 3-92. Information Technology—International Standardized Profile RC—X.25 Protocol Relaying —Part 3: ISO 8802-3 CSMA/CD Subnetwork-Dependent, Media-Dependent Requirements ***CD-ROM ONLY***. 37 pp.

JIS X 5252-90. Local Area Networks—Carrier Sense Multiple Access with Collision Detection (CSMA/CD) Access Method and Physical Layer Specifications (ISO 8802/3:1989).

JTC1 907-89. Local Area Networks CSMA/CD 10 Mbit/s Baseband Planning and Installation Guide First Edition. 86 pp.

JTC1 8802 Pt3 Draft AMD 6. Information Processing Systems—Local Area Networks—Part 3: Carrier Sense Multiple Access with Collision Detection (CSMA/CD) Access Method and Physical Layer Specifications Amendment 6; (1991) ***CD-ROM ONLY***. 5 pp.

JTC1 8802 Pt3 Draft AMD 7. Information Processing Systems—Local Area Networks—Part 3: Carrier Sense Multiple Access with Collision Detection (CSMA/CD) Access Method and Physical Layer Specifications Amendment 7: Addition of Layer Management (Section 5); (1991)***CD-ROM ONLY***. 40 pp.

JTC1 8802 Pt3 Draft AMD 9. Information Processing Systems—Local Area Networks—Part 3: Carrier Sense Multiple Access with Collision Detection (CSMA/CD) Access Method and Physical Layer Specifications Amendment 9: 10Base-T; (1991) ***CD-ROM ONLY***. 54 pp.

JTC1 8802 Pt3 Draft AMD 11. Information Technology—Local and Metropolitan Area Networks—Part 3: Carrier Sense Multiple Access with Collision Detection (CSMA/CD) Access Meth. and Phys. Layer Specs. Amd 11: Layer Mgmt. for 10 Mb/s Baseband Repeaters (Sec. 19); (1993) (IEEE Std. 802.3k) ***CD-ROM ONLY**. 31 pp.

JTC1 ISP 10608 Pt 2-92. Information Technology—International Standard-ized Profile TAnnnn—Connection-Mode Trans-port Service over Con-nectionless-Mode Network Service—Part 2: TA51 Profile Including Sub-network-Dependent Re-quirements for CSMA/CD Local Area Networks (LANs) First Edition. 15 pp.

JTC1 DISP10612 Pt 2-92. Information Technology—International Standardized Profile RD5p.5q—Relaying the MAC Service Using Transparent Bridging—Part 2: CSMA/CD LAN Subnetwork-Dependent, Media-Dependent Requirements ***CD-ROM ONLY***. 32 pp.

JTC1 DISP10612 Pt 4-92. Information Technology—International Standardized Profile RD5p.5q—Relaying the MAC Service Using Transparent Bridging—Part 4: Profile RD51.51 (CSMA/CD-CSMA/CD) ***CD-ROM ONLY***. 28 pp.

JTC1 DISP10614 Pt 3-92. Information Technology—International Standardized Profile RC—X.25 Protocol Relaying —Part 3: ISO 8802-3 CSMA/CD Subnetwork-Dependent, Media-Dependent Requirements ***CD-ROM ONLY***. 37 pp.

OSI ISO 8802-3 DAM 4-90. Part 3: Carrier Sense Multiple Access with Collision Detection (CSMA/CD) Access Method and Physical Layer Specifications Amendment 4: Physical Signalling, Medium Attachment and Baseband Medium Specifications, StarLAN, Type 1 Bases. 62 pp.

OSI ISO 8802-3 DAM 3-90. Information Processing Systems—Local Area Networks—Part 3: Carrier Sense Multiple Access with Collision Detection. Amendment 3: Broadband Medium Attachment Unit and Broadband Medium Specification, Type10 BROAD 36. 43 pp.

OSI ISO DIS 8802/3 PDAD-87. Local Area Networks—Part 3: CSMA/CD—Addendum Broadband Medium Attachment Unit and Broadband Medium Specifications, Type IOBROAD36.

OSI ISO 8802-3 DAD 1-87. Information Processing—Local Area Networks PT3: Carrier Sense Multiple Access with Collision Detection Addendum 1: Medium Attachment Unit and Baseband Medium Specifications for Type 10 Base 2. 36 pp.

OSI ISO DIS 8802-3 PDAD-87. Local Area Networks—Part 3: CSMA/CD—Addendum Broadband Medium Attachment Unit and Broadband Medium Specifications, Type IOBROAD36. 230 pp.

424

INDEX and DIRECTORY of

INTERNATIONAL AND NON-U.S. NATIONAL STANDARDS
SUBJECT INDEX

Carrier

Carrier Sense Multiple Access/CD (Cont.)
—Local Area Networks (Cont.)
  OSI ISO/IEC 8802-3-90. Information Processing Systems—Local Area Networks—Part 3: Carrier Sense Multiple Access with Collision Detection (CSMA/CD) Access Method and Physical Layer Specifications (with Correction Sheet 09-90). 217 pp.
  OSI ISO DIS 8802-3-87. Information Processing Systems—Local Area Networks Part 3: Carrier Sense Multiple Access with Collision Detection—Access Method and Physical Layer Specifications. 5 pp.
  OSI ISO 8802-3-89. Information Processing Systems—Local Area Networks Part 3: Carrier Sense Multiple Access with Collision Detection (CSMA/CD) Access Method and Physical Layer Specifications. 178 pp.
  OSI ISO 8802 3 DAD 1-87. Information Processing—Local Area Networks PT3: Carrier Sense Multiple Access with Collision Detection Addendum 1: Medium Attachment Unit and Baseband Medium Specifications for Type 10 Base 2.
  OSI ISO 8802 3 DAD 2-87. Information Processing Systems-Local Area Networks Part 3: Carrier Sense Multiple Access with Collision Detection Addendum 2: Repeater Set and Repeater Unit Specification Foruse with 10 Base 5 and 10 Base 2 Networks.
  OSI ISO/IEC DISP 10608-2-90. Information Technology—International Standardized Profile TAnnnn—Connection-Mode Transport Service over Connectionless-Mode Network Service—Part 2: TA51 Profile Including Subnetwork-Dependent Requirements for CSMS/CD Local Area Networks (LANs). 37 pp.
  SAA AS 4802.3-91. Information Processing Systems—Local Area Networks—Part 3: Carrier Sense Multiple Access with Collision Detection (CSMA/CD) Access Method and Physical Layer Specifications (ISO/IEC 8802-3:1990/IEEE 802.3-1990) (in Prof. Packages 26A, 26C). 191 pp.
  SAA AS 4802.3 Supp 1-91. Information Processing Systems—Local Area Networks—Part 3: Carrier Sense Multiple Access with Collision Detection (CSMA/CD) Access Method and Physical Layer Specifications—Supplement 1: Multisegment 10 Mb/s Baseband Networks:. 50 pp.
  SNZ NZS/ISO 8802. 3-90. Information Processing Systems—Local Area Networks Part 3: Carrier Sense Multiple Access with Collision Detection (CSMA/CD) Access Method and Physical Layer Specifications. 191 pp.

—Packet Switched Networks
  IEC DISP 10614 Pt 3-92. Information Technology—International Standardized Profile RC—X.25 Protocol Relaying —Part 3: ISO 8802-3 CSMA/CD Subnetwork-Dependent, Media-Dependent Requirements ***CD-ROM ONLY***. 37 pp.
  ISO DISP 10614 Pt 3-92. Information Technology—International Standardized Profile RC—X.25 Protocol Relaying —Part 3: ISO 8802-3 CSMA/CD Subnetwork-Dependent, Media-Dependent Requirements ***CD-ROM ONLY***. 37 pp.
  JTC1 DISP10614 Pt 3-92. Information Technology—International Standardized Profile RC—X.25 Protocol Relaying —Part 3: ISO 8802-3 CSMA/CD Subnetwork-Dependent, Media-Dependent Requirements ***CD-ROM ONLY***. 37 pp.

Carrier Sense Multiple Access/CD (Ethernet)
Use: Ethernet (BTN)

Carrier Systems
Use For: Carrier Systems and Equipment; Sub-Carrier Systems; Subcarrier Systems
See Also: Carrier Channels; Carrier Frequencies; Carrier Leaks; Digital Access and Cross Connect Systems; Groups (Communications); Interband Telegraphy; Master Oscillators; Power Line Carriers; Telecommunication Equipment; Transmission Systems; Virtual Carrier Frequencies
  CCIR RECMN 454-1-78. Pilot Carrier Level for HF Single-Sideband and Independent-Sideband Reduced-Carrier Systems—Section 3Aa—Technical Characteristics. 2 pp.

—Fixed Satellite Services—Angle Modulation
  CCIR RECMN 446-2-78. Carrier Energy Dispersal for Systems Employing Angle Modulation by Analogue Signals or Digital Modulation in the Fixed-Satellite Service—Section 4C—Earth Station and Baseband Characteristics —Earth Station Antennas —Maintenance of Earth Stations. 1 p.
  CCIR RECMN 446-3-92. Carrier Energy Dispersal for Systems Employing Angle Modulation by Analogue Signals or Digital Modulation in the Fixed-Satellite Service—Section 4C—Earth Station and Baseband Characteristics —Earth Station Antennas —Maintenance of Earth Stations. 20 pp.

Carrier Systems (Cont.)
—Fixed Satellite Services—Angle Modulation—Power Density
  CCIR Report 792-3-90. Calculation of the Maximum Power Density Averaged over 4 kHz of an Angle-Modulated Carrier—Section 4/9B—Co-Ordination and Interference Calculations. 6 pp.

—Fixed Satellite Services—Digital Modulation
  CCIR RECMN 446-2-78. Carrier Energy Dispersal for Systems Employing Angle Modulation by Analogue Signals or Digital Modulation in the Fixed-Satellite Service—Section 4C—Earth Station and Baseband Characteristics —Earth Station Antennas —Maintenance of Earth Stations. 1 p.
  CCIR RECMN 446-3-92. Carrier Energy Dispersal for Systems Employing Angle Modulation by Analogue Signals or Digital Modulation in the Fixed-Satellite Service—Section 4C—Earth Station and Baseband Characteristics —Earth Station Antennas —Maintenance of Earth Stations. 20 pp.

—Frequency Division Multiplexers—Electrical Impedance
  CCITT RECMN G.371-89. FDM Carrier Systems for Submarine Cable—International Analogue Carrier Systems (Study Group XV) 3 pp. 3 pp.

—Frequency Division Multiplexers—Interfaces
  CCITT RECMN G.371-89. FDM Carrier Systems for Submarine Cable—International Analogue Carrier Systems (Study Group XV) 3 pp. 3 pp.

—Frequency Division Multiplexers—Power Levels
  CCITT RECMN G.371-89. FDM Carrier Systems for Submarine Cable—International Analogue Carrier Systems (Study Group XV) 3 pp. 3 pp.

—Frequency Translation Equipment
  CCITT RECMN G.211-89. Make-up of a Carrier Link—International Analogue Carrier Systems (Study Group XV) 8 pp. 8 pp.

—Frequency Translation Equipment—Noise—Measurement
  CCITT RECMN G.230-89. Measuring Methods for Noise Produced by Modulating Equipment and Through-Connection Filters—International Analogue Carrier Systems (Study Group XV) 4 pp. 4 pp.

—Glossaries
  CCITT RECMN G.211-89. Make-up of a Carrier Link—International Analogue Carrier Systems (Study Group XV) 8 pp. 8 pp.

—Group Links
  CCITT RECMN G.211-89. Make-up of a Carrier Link—International Analogue Carrier Systems (Study Group XV) 8 pp. 8 pp.

—Groups—Attenuation
  CCITT RECMN G.313-89. Open-Wire Lines for Use with 12-Channel Carrier Systems—International Analogue Carrier Systems (Study Group XV) 3 pp. 3 pp.

—Groups—Circuit Noise
  CCITT RECMN G.311-89. General Characteristics of Systems Providing 12 Carrier Telephone Circuits on an Open-Wire Pair—International Analogue Carrier Systems (Study Group XV) 6 pp. 6 pp.
  CCITT RECMN G.322-89. General Characteristics Recommended for Systems on Symmetric Pair Cables —International Analogue Carrier Systems (Study Group XV) 8 pp. 8 pp.
  CCITT RECMN G.325-89. General Characteristics Recommended for Systems Providing 12 Telephone Carrier Circuits on a Symmetric Cable Pair ((12 + 12) Systems)—International Analogue Carrier Systems (Study Group XV) 5 pp. 5 pp.

—Groups—Crosstalk
  CCITT RECMN G.313-89. Open-Wire Lines for Use with 12-Channel Carrier Systems—International Analogue Carrier Systems (Study Group XV) 3 pp. 3 pp.

—Groups—Hypothetical Reference Circuits
  CCITT RECMN G.325-89. General Characteristics Recommended for Systems Providing 12 Telephone Carrier Circuits on a Symmetric Cable Pair ((12 + 12) Systems)—International Analogue Carrier Systems (Study Group XV) 5 pp. 5 pp.

—Groups—Pilot Channels
  CCITT RECMN G.322-89. General Characteristics Recommended for Systems on Symmetric Pair Cables —International Analogue Carrier Systems (Study Group XV) 8 pp. 8 pp.

Carrier Systems (Cont.)
—Groups—Pilot Channels (Cont.)
  CCITT RECMN G.325-89. General Characteristics Recommended for Systems Providing 12 Telephone Carrier Circuits on a Symmetric Cable Pair ((12 + 12) Systems)—International Analogue Carrier Systems (Study Group XV) 5 pp. 5 pp.

—Groups—Surge Protectors
  CCITT RECMN G.313-89. Open-Wire Lines for Use with 12-Channel Carrier Systems—International Analogue Carrier Systems (Study Group XV) 3 pp. 3 pp.

—Groups—Telephone Cables
  CCITT RECMN G.322-89. General Characteristics Recommended for Systems on Symmetric Pair Cables —International Analogue Carrier Systems (Study Group XV) 8 pp. 8 pp.
  CCITT RECMN G.325-89. General Characteristics Recommended for Systems Providing 12 Telephone Carrier Circuits on a Symmetric Cable Pair ((12 + 12) Systems)—International Analogue Carrier Systems (Study Group XV) 5 pp. 5 pp.

—Hypothetical Reference Circuits
  CCITT RECMN G.215-89. Hypothetical Reference Circuit of 5000 km for Analogue Systems—International Analogue Carrier Systems (Study Group XV) 2 pp. 2 pp.

—Line Links—Telephone Networks
  CCITT RECMN G.211-89. Make-up of a Carrier Link—International Analogue Carrier Systems (Study Group XV) 8 pp. 8 pp.

—Maintenance
  CCITT RECMN M.540-89. Routine Maintenance of Carrier and Pilot Generating Equipment—General Maintenance Principles—Maintenance of International Transmission Systems and Telephone Circuits (Study Group IV) 3 pp. 3 pp.

—Master Oscillators—Carrier Frequencies
  CCITT RECMN G.225-89. Recommendations Relating to the Accuracy of Carrier Frequencies—International Analogue Carrier Systems (Study Group XV) 3 pp. 3 pp.

—Mastergroup Links
  CCITT RECMN G.211-89. Make-up of a Carrier Link—International Analogue Carrier Systems (Study Group XV) 8 pp. 8 pp.

—Measuring Instruments—Frequency Shift
  CCITT RECMN O.111-89. Frequency Shift Measuring Equipment for Use on Carrier Channels—Specifications for Measuring Equipment (Study Group IV) 6 pp. 6 pp.

—Open Wire Lines
  CCITT RECMN G.314-89. General Characteristics of Systems Providing Eight Carrier Telephone Circuits on an Open-Wire Pair—International Analogue Carrier Systems (Study Group XV) 1 pp. 1 pp.
  CCITT RECMN G.361-89. Systems Providing Three Carrier Telephone Circuits on a Pair of Open-Wire Lines—International Analogue Carrier Systems (Study Group XV) 5 pp. 5 pp.

—Open Wire Lines—Groups
  CCITT RECMN G.311-89. General Characteristics of Systems Providing 12 Carrier Telephone Circuits on an Open-Wire Pair—International Analogue Carrier Systems (Study Group XV) 6 pp. 6 pp.
  CCITT RECMN G.313-89. Open-Wire Lines for Use with 12-Channel Carrier Systems—International Analogue Carrier Systems (Study Group XV) 3 pp. 3 pp.

—Open Wire Lines—Hypothetical Reference Circuits
  CCITT RECMN G.311-89. General Characteristics of Systems Providing 12 Carrier Telephone Circuits on an Open-Wire Pair—International Analogue Carrier Systems (Study Group XV) 6 pp. 6 pp.

—Open Wire Lines—Interband Telegraphy
  CCITT RECMN G.361-89. Systems Providing Three Carrier Telephone Circuits on a Pair of Open-Wire Lines—International Analogue Carrier Systems (Study Group XV) 5 pp. 5 pp.
  CCITT RECMN R.49-89. Interband Telegraphy over Open-Wire 3-Channel Carrier Systems—Telegraph Transmission (Study Group IX) 2 pp. 2 pp.

—Open Wire Lines—Pilot Channels
  CCITT RECMN G.361-89. Systems Providing Three Carrier Telephone Circuits on a Pair of Open-Wire Lines—International Analogue Carrier Systems (Study Group XV) 5 pp. 5 pp.

INDUSTRY STANDARDS

Carrier Systems (Cont.)

—Open Wire Lines—Signal Levels
CCITT RECMN G.361-89. Systems Providing Three Carrier Telephone Circuits on a Pair of Open-Wire Lines—International Analogue Carrier Systems (Study Group XV) 5 pp. 5 pp.

—Open Wire Lines—Telephone Repeaters
CCITT RECMN G.312-89. Intermediate Repeaters for Open-Wire Carrier Systems Conforming to Recommendation G.311—International Analogue Carrier Systems (Study Group XV) 2 pp. 2 pp.
CCITT RECMN G.361-89. Systems Providing Three Carrier Telephone Circuits on a Pair of Open-Wire Lines—International Analogue Carrier Systems (Study Group XV) 5 pp. 5 pp.

—Sidebands—Signal to Noise Ratio
CCIR RECMN 454-1-78. Pilot Carrier Level for HF Single-Sideband and Independent-Sideband Reduced-Carrier Systems—Section 3Aa—Technical Characteristics. 2 pp.

—Sound-Program Circuits—Crosstalk
CCITT RECMN J.18-89. Crosstalk in Sound-Programme Circuits Set up on Carrier Systems—Line Transmission of Non-Telephone Signals—Transmission of Sound-Programme and Television Signals (Study Group XV) 3 pp. 3 pp.

—Sound-Program Circuits—Equipment Characteristics—10 kHz
CCITT RECMN J.32-89. Characteristics of Equipment and Lines Used for Setting up 10 kHz Type Sound-Programme Circuits—Line Transmission of Non-Telephone Signals—Transmission of Sound-Programme and Television Signals (Study Group XV) 1 pp. 1 p.

—Sound-Program Circuits—Equipment Characteristics—15 kHz
CCITT RECMN J.31-89. Characteristics of Equipment and Lines Used for Setting up 15 kHz Type Sound-Programme Circuits—Line Transmission of Non-Telephone Signals—Transmission of Sound-Programme and Television Signals (Study Group XV) 15 pp. 15 pp.

—Sound-Program Circuits—Equipment Characteristics—6.4 kHz
CCITT RECMN J.33-89. Characteristics of Equipment and Lines Used for Setting up 6.4 kHz Type Sound-Programme Circuits—Line Transmission of Non-Telephone Signals—Transmission of Sound-Programme and Television Signals (Study Group XV) 3 pp. 3 pp.

—Sound-Program Circuits—Equipment Characteristics—7 kHz
CCITT RECMN J.34-89. Characteristics of Equipment Used for Setting up 7 kHz Type Sound-Programme Circuits—Line Transmission of Non-Telephone Signals—Transmission of Sound-Programme and Television Signals (Study Group XV) 3 pp. 3 pp.

—Sound-Program Circuits—Pre-Emphasis
CCITT RECMN J.17-89. Pre-Emphasis Used on Sound-Programme Circuits—Line Transmission of Non-Telephone Signals—Transmission of Sound-Programme and Television Signals (Study Group XV) 2 pp. 2 pp.

—Supergroup Links
CCITT RECMN G.211-89. Make-up of a Carrier Link—International Analogue Carrier Systems (Study Group XV) 8 pp. 8 pp.

—Supergroups—Circuit Noise
CCITT RECMN G.322-89. General Characteristics Recommended for Systems on Symmetric Pair Cables —International Analogue Carrier Systems (Study Group XV) 8 pp. 8 pp.

—Supergroups—Pilot Channels
CCITT RECMN G.322-89. General Characteristics Recommended for Systems on Symmetric Pair Cables —International Analogue Carrier Systems (Study Group XV) 8 pp. 8 pp.

—Supergroups—Telephone Cables
CCITT RECMN G.322-89. General Characteristics Recommended for Systems on Symmetric Pair Cables —International Analogue Carrier Systems (Study Group XV) 8 pp. 8 pp.

—Supermastergroup Links
CCITT RECMN G.211-89. Make-up of a Carrier Link—International Analogue Carrier Systems (Study Group XV) 8 pp. 8 pp.

Carrier Systems (Cont.)

—Telephone Cables
CCITT RECMN G.313-89. Open-Wire Lines for Use with 12-Channel Carrier Systems—International Analogue Carrier Systems (Study Group XV) 3 pp. 3 pp.
CCITT RECMN G.327-89. Valve-Type Systems Offering 12 Carrier Telephone Circuits on a Symmetric Cable Pair ((12 + 12) Systems)—International Analogue Carrier Systems (Study Group XV) 2 pp. 2 pp.
CCITT RECMN G.611-89. Characteristics of Symmetric Cable Pairs for Analogue Transmission—Transmission Media Characteristics (Study Group XV) 5 pp. 5 pp.

—Telephone Cables—Frequency Division Multiplexers
CCITT RECMN G.371-89. FDM Carrier Systems for Submarine Cable—International Analogue Carrier Systems (Study Group XV) 3 pp. 3 pp.

—Telephone Cables—HF—Interfaces
CCITT RECMN G.352-89. Interconnection of Coaxial Carrier Systems of Different Designs—International Analogue Carrier Systems (Study Group XV) 4 pp. 4 pp.

—Telephone Cables—HF—Pilot Channels
CCITT RECMN G.352-89. Interconnection of Coaxial Carrier Systems of Different Designs—International Analogue Carrier Systems (Study Group XV) 4 pp. 4 pp.

—Telephone Cables—HF—Repeater Stations
CCITT RECMN G.352-89. Interconnection of Coaxial Carrier Systems of Different Designs—International Analogue Carrier Systems (Study Group XV) 4 pp. 4 pp.

—Telephone Cables—MF—Interfaces
CCITT RECMN G.352-89. Interconnection of Coaxial Carrier Systems of Different Designs—International Analogue Carrier Systems (Study Group XV) 4 pp. 4 pp.

—Telephone Cables—MF—Pilot Channels
CCITT RECMN G.352-89. Interconnection of Coaxial Carrier Systems of Different Designs—International Analogue Carrier Systems (Study Group XV) 4 pp. 4 pp.

—Telephone Cables—MF—Repeater Stations
CCITT RECMN G.352-89. Interconnection of Coaxial Carrier Systems of Different Designs—International Analogue Carrier Systems (Study Group XV) 4 pp. 4 pp.

—Telephone Cables—VHF—Interfaces
CCITT RECMN G.352-89. Interconnection of Coaxial Carrier Systems of Different Designs—International Analogue Carrier Systems (Study Group XV) 4 pp. 4 pp.

—Telephone Cables—VHF—Pilot Channels
CCITT RECMN G.352-89. Interconnection of Coaxial Carrier Systems of Different Designs—International Analogue Carrier Systems (Study Group XV) 4 pp. 4 pp.

—Telephone Cables—VHF—Repeater Stations
CCITT RECMN G.352-89. Interconnection of Coaxial Carrier Systems of Different Designs—International Analogue Carrier Systems (Study Group XV) 4 pp. 4 pp.

—Telephone Circuits
CCITT RECMN G.125-89. Characteristics of Circuits on Carrier Systems—General Characteristics of International Telephone Connections and Circuits (Study Groups XII and XV) 1 pp. 1 p.
CCITT RECMN G.153-89. Characteristics Appropriate to International Circuits More Than 2500 km in Length—General Characteristics of International Telephone Connections and Circuits (Study Groups XII and XV) 4 pp. 4 pp.

—Telephone Circuits—Attenuation Distortion
CCITT RECMN G.125-89. Characteristics of Circuits on Carrier Systems—General Characteristics of International Telephone Connections and Circuits (Study Groups XII and XV) 1 pp. 1 p.
CCITT RECMN G.141-89. Attenuation Distortion—General Characteristics of International Telephone Connections and Circuits (Study Groups XII and XV) 2 pp. 2 pp.

Carrier Systems (Cont.)

—Telephone Circuits—Circuit Noise
CCITT RECMN G.125-89. Characteristics of Circuits on Carrier Systems—General Characteristics of International Telephone Connections and Circuits (Study Groups XII and XV) 1 pp. 1 p.
CCITT RECMN G.152-89. Characteristics Appropriate to Long-Distance Circuits of a Length Not Exceeding 2500 km—General Characteristics of International Telephone Connections and Circuits (Study Groups XII and XV) 2 pp. 2 pp.

—Telephone Circuits—Frequency Bands
CCITT RECMN G.311-89. General Characteristics of Systems Providing 12 Carrier Telephone Circuits on an Open-Wire Pair—International Analogue Carrier Systems (Study Group XV) 6 pp. 6 pp.

—Telephone Circuits—Pilot Channels
CCITT RECMN G.311-89. General Characteristics of Systems Providing 12 Carrier Telephone Circuits on an Open-Wire Pair—International Analogue Carrier Systems (Study Group XV) 6 pp. 6 pp.

—Telephone Circuits—Signal Levels
CCITT RECMN G.311-89. General Characteristics of Systems Providing 12 Carrier Telephone Circuits on an Open-Wire Pair—International Analogue Carrier Systems (Study Group XV) 6 pp. 6 pp.

—Telephone Equipment—Power Supplies
CCITT RECMN G.231-89. Arrangement of Carrier Equipment—International Analogue Carrier Systems (Study Group XV) 2 pp. 2 pp.

—Telephone Equipment—Tables (Data)
CCITT FASCICLE III.1. General Characteristics of International Telephone Connections and Circuits Recommendations G.101-G.181. 332 pp.

—Telephone Lines—Interfaces—Radio Relay Systems
CCITT RECMN G.421-89. Methods of Interconnection—International Analogue Carrier Systems (Study Group XV) 2 pp. 2 pp.

—Telephone Repeaters—Crosstalk
CCITT RECMN G.312-89. Intermediate Repeaters for Open-Wire Carrier Systems Conforming to Recommendation G.311—International Analogue Carrier Systems (Study Group XV) 2 pp. 2 pp.
CCITT RECMN G.313-89. Open-Wire Lines for Use with 12-Channel Carrier Systems—International Analogue Carrier Systems (Study Group XV) 3 pp. 3 pp.
CCITT RECMN G.322-89. General Characteristics Recommended for Systems on Symmetric Pair Cables —International Analogue Carrier Systems (Study Group XV) 8 pp. 8 pp.

—Telephone Repeaters—Electrical Impedance
CCITT RECMN G.312-89. Intermediate Repeaters for Open-Wire Carrier Systems Conforming to Recommendation G.311—International Analogue Carrier Systems (Study Group XV) 2 pp. 2 pp.
CCITT RECMN G.322-89. General Characteristics Recommended for Systems on Symmetric Pair Cables —International Analogue Carrier Systems (Study Group XV) 8 pp. 8 pp.

—Telephone Repeaters—Gain
CCITT RECMN G.312-89. Intermediate Repeaters for Open-Wire Carrier Systems Conforming to Recommendation G.311—International Analogue Carrier Systems (Study Group XV) 2 pp. 2 pp.

—Telephone Repeaters—Harmonic Distortion
CCITT RECMN G.322-89. General Characteristics Recommended for Systems on Symmetric Pair Cables —International Analogue Carrier Systems (Study Group XV) 8 pp. 8 pp.

—Telephone Repeaters—Noise
CCITT RECMN G.322-89. General Characteristics Recommended for Systems on Symmetric Pair Cables —International Analogue Carrier Systems (Study Group XV) 8 pp. 8 pp.

—Telephone Repeaters—Signal Levels
CCITT RECMN G.322-89. General Characteristics Recommended for Systems on Symmetric Pair Cables —International Analogue Carrier Systems (Study Group XV) 8 pp. 8 pp.

—Telephone Transmission
CCITT FASCICLE III.2-88. International Analogue Carrier Systems Recommendations G.211-G.544. 226 pp.
CCITT RECMN G.221-89. Overall Recommendations Relating to Carrier-Transmission Systems—International Analogue Carrier Systems (Study Group XV) 2 pp. 2 pp.

INTERNATIONAL AND NON-U.S. NATIONAL STANDARDS
SUBJECT INDEX

Cartridge

Carrier Systems *(Cont.)*
—**Telephone Transmission** *(Cont.)*
CCITT RECMN G.222-89. Noise Objectives for Design of Carrier-Transmission Systems of 2500 km—International Analogue Carrier Systems (Study Group XV) 4 pp. 4 pp.

—**Telephone Transmission—Crosstalk**
CCITT RECMN G.221-89. Overall Recommendations Relating to Carrier-Transmission Systems—International Analogue Carrier Systems (Study Group XV) 2 pp. 2 pp.

—**Telephone Transmission—Noise**
CCITT RECMN G.221-89. Overall Recommendations Relating to Carrier-Transmission Systems—International Analogue Carrier Systems (Study Group XV) 2 pp. 2 pp.
CCITT RECMN G.222-89. Noise Objectives for Design of Carrier-Transmission Systems of 2500 km—International Analogue Carrier Systems (Study Group XV) 4 pp. 4 pp.

—**Telephone Transmission—Single Tone Interference**
CCITT RECMN G.221-89. Overall Recommendations Relating to Carrier-Transmission Systems—International Analogue Carrier Systems (Study Group XV) 2 pp. 2 pp.

—**Telephony—Cable Racks**
CCITT RECMN G.231-89. Arrangement of Carrier Equipment—International Analogue Carrier Systems (Study Group XV) 2 pp. 2 pp.

—**Through Connection Filters**
CCITT RECMN G.211-89. Make-up of a Carrier Link—International Analogue Carrier Systems (Study Group XV) 8 pp. 8 pp.

—**Through Connection Filters—Noise—Measurement**
CCITT RECMN G.230-89. Measuring Methods for Noise Produced by Modulating Equipment and Through-Connection Filters—International Analogue Carrier Systems (Study Group XV) 4 pp. 4 pp.

—**Through Group Filters**
CCITT RECMN G.211-89. Make-up of a Carrier Link—International Analogue Carrier Systems (Study Group XV) 8 pp. 8 pp.

—**Through Supergroup Filters**
CCITT RECMN G.211-89. Make-up of a Carrier Link—International Analogue Carrier Systems (Study Group XV) 8 pp. 8 pp.

—**15 Supergroup Assembly Links**
CCITT RECMN G.211-89. Make-up of a Carrier Link—International Analogue Carrier Systems (Study Group XV) 8 pp. 8 pp.

Carrier Systems and Equipment
Use: Carrier Systems

Carrots
See Also: Celery; Food; Vegetables
EC EEC/293/92-92. Commission Regulation Correcting the English, German, Dutch, Danish and Spanish Version of Regulation (EEC) No 920/89 as Regards the Presentation of Carrots. 1 p.

—**Grading**
CNS N1023-82. Grades for Carrots (Jan)(2090). 3 pp.

—**Harvesting Equipment—Rod Links**
BSI BS 4026: Part 4-69. (OBSOLESCENT) 1969 Rod Links for Root Machinery Part 4: Carrot and Sugar Beet Harvesters: Metric Units. 8 pp.

—**Storage**
ISO 2166-81. Carrots—Guide to Storage Second Edition. 4 pp.

Carry Cots
Use: Baby Carriers; Bassinets

Carry Look Ahead Generators
Use: Look Ahead Carry Generators

Carry Lookahead Generators
Use: Look Ahead Carry Generators

Carry Save Adders
See Also: Digital Circuits; Full Adders

—**Dual**
CECC CECC 90 103-079 ISSUE 1-81. BS CECC 90 103-079; Dual Carry-Save Full Adders (En) AMD 1 (En). 19 pp.

Cartons
See Also: Boxes (Containers); Containers; Crates (Shipping Containers); Shipping Containers

—**Chipboard**
MOD UK DSTAN 81-11-79. Cartons, Fibreboard, Fixed Joint, Kraft Lined Chipboard Issue 3. 9 pp.

—**Cleaning Powders—Fiberboard**
BSI BS 5167-78. 1978 Packages for Washing and Cleaning Powders. Dimensions and Volumes of Cartons and Drums from Fibreboard. 9 pp.
CEN EN 23-1-78. Packages for Washing and Cleaning Powders—Part 1: Dimensions and Volumes of Cartons and Drums from Fibreboard. 7 pp.

—**Corrugated**
MOD UK DSTAN 81-9-89. Cartons, Fibreboard, Fixed Joint, Double-Faced Corrugated Issue 3. 10 pp.

—**Creasing**
BSI BS 4818-93. 1993 Method for Determination of the Creasing Quality of Carton Board (Pira Method) (V). 10 pp.
BSI BS 4818-72. 1972 Amd 1 Recommendations for a Method for the Determination of the Creasing Quality of Carton Board as Measured by the PIRA Carton Board Creaser. 12 pp.
BSI BS 6965· Part 1-88. 1988 Creasing Properties of Carton Board Part 1: Method for Determination of Crease Recovery (Spring Back) of 90 Degrees Celsius Foid. 8 pp.

—**Detergents—Fiberboard**
BSI BS 5167-78. 1978 Packages for Washing and Cleaning Powders. Dimensions and Volumes of Cartons and Drums from Fibreboard. 9 pp.
CEN EN 23-1-78. Packages for Washing and Cleaning Powders—Part 1: Dimensions and Volumes of Cartons and Drums from Fibreboard. 7 pp.

—**Fiberboard**
MOD UK DSTAN 81-15-01. Cartons and Boxes, Fibreboard Issue 3; Amendment 3. 21 pp.

Cartop Carriers
Use: Luggage Racks

Cartridge Cases
Use For: Projectile Cases *See Also:* Cartridges (Explosives)

—**Aluminum Sulfate**
MOD UK DSTAN 68-168-93. Aluminium Sulfate (Equivalent to 18% AL2O3) Issue 1. 14 pp.
MOD UK TS 10032. Aluminium Sulphate 18 Per Cent Al2O3 (Superseded by Def Stan 68-168).

—**Brass**
MOD UK DSTAN 95-9-88. Cartridge Brass for Ammunition Components Issue 2. 14 pp.

—**Kraft Pulps**
MOD UK TS 10002B. Pulp Kraft, Bleached for Combustible Charge Components. Type 1 and Type 2.
MOD UK TS 10259. Pulp, Kraft, Unbleached for Combustible Case Components.

—**Papers**
MOD UK DEF-90-A-01. Paper, Cartridge; Amendment 1. 5 pp.

—**Shotguns**
JIS B 9805-77. Cartridge Cases for Shot Guns.

Cartridge Fuse Links
See Also: Cartridge Fuses; Fuse Links
BSI BS 646-58. 1958 Amd 1 Cartridge Fuse-Links (Rated up to 5 Amperes) for a.c. and d.c. Service (AMD 6693) January 31,1991. 18 pp.
BSI BS 2950-58. 1958 Amd 1 Cartridge Fuse-Links for Telecommunication and Light Electrical Apparatus (AMD 6690) January 31, 1991. 18 pp.
MOD UK DSTAN 59-96: Part 1-01. Fuses Links Electrical Part 1: Specifications Issue 1; Amendment 3. 91 pp.
MOD UK DSTAN 59-96: Part 1-01. Fuses Links Electrical Part 1: Specifications Issue 1; Amendment 3 (Reprinted 13 May 1981 Incorporating Amendment 1). 80 pp.
SAA AS C342-65. Cartridge Fuse-Links for Telecommunication and Light Electrical Apparatus Being BS 2950:1958, Endorsed Without Amendment. 16 pp.
SNZ NZS 2035-65. Specification for Cartridge Fuse-Links (Rated up to 5 Amperes) for Alternating Current and Direct Current Service. 16 pp.

—**Fuse Holders**
CENELEC HD 119-74. Fuse-holders for Minature Cartridge Fuse-Links. 2 pp.
CENELEC EN 60 257-90. Fuse-Holders for Minature Cartridge Fuse-Links. 4 pp.

Cartridge Fuse Links *(Cont.)*
—**Fuse Holders** *(Cont.)*
IEC 257-68. Fuse-Holders for Miniature Cartridge Fuse-Links First Edition; (Amendment 2-1989). 51 pp.

—**Miniature**
BSI BS 4265-77. (WITHDRAWN) 1977 Amd 4 Cartridge Fuse Links for Miniature Fuses (AMD 6688) January 31, 1991 (Superseded by BS EN 60127-1, 60127-2, 60127-3 and 60127-5: 1991) (E). 35 pp.
BSI BS EN 60127-2-91. 1991 Miniature Fuses Part 2: Specification for Cartridge Fuse-Links. 25 pp.
CENELEC HD 109.2 S1-90. Miniature Fuses Part 2: Cartridge Fuse-Links. 3 pp.
CENELEC EN 60 127-2-91. Miniature Fuses Part 2: Cartridge Fuse-Links. 6 pp.
IEC 127 Pt 2-89. Miniature Fuses Part 2: Cartridge Fuse-Links First Edition. 33 pp.

—**Miniature—Color Coding**
CENELEC HD 109 S3-83. Cartridge Fuse-Links for Miniature Fuses—First Supplement: Colour Coding. 2 pp.

—**Miniature—Printed Circuit Board Hardware**
BSI BS 6885-88. 1988 Miniature Cartridge Fuse Links for Use on Printed Wiring Boards. 12 pp.

—**Packaging**
MOD UK DEF-175-01. Packaging of Cartridge Fuse-Links; Amendment 1. 9 pp.

—**Subminiature**
BSI BS EN 60127-3-91. 1991 Miniature Fuses Part 3: Specification for Sub-Miniature Fuse-Links. 28 pp.
CENELEC EN 60 127-3-91. Miniature Fuses Part 3: Sub-Miniature Fuse-Links. 6 pp.
IEC 127 Pt 3-88. Miniature Fuses Part 3: Sub-Miniature Fuse-Links Second Edition; (Amendment 1-1991). 42 pp.
SNZ IEC 127: Part 3-88. Miniature Fuses Part 3: Sub-Miniature Fuse-Links. 38 pp.

Cartridge Fuses
See Also: Cartridge Fuse Links; Fuses (Electric)
BSI BS 88: Part 1-88. 1988 Amd 1 Cartridge Fuses for Voltages up to and Including 1000 V a.c. and 1500 V d.c. Part 1: General Requirements (AMD 6448) February 28, 1991. 70 pp.
BSI BS 88: Sec 2.1-88. 1988 Amd 1 Cartridge Fuses for Voltages up to and Including 1000 V a.c. and 1500 V d.c. Part 2: Spec. for Fuses for Use by Authorized Persons (Mainly for Ind. Application) Sec. 2.1: Supp. Requirements (AMD 6684) April 30, 1991. 10 pp.
BSI BS 88: Sec 2.2-88. 1988 Amd 1 Cartridge Fuses for Voltages up to and Including 1000 V a.c. and 1500 V d.c. Part 2: Specification for Fuses for Use by Authorized Persons (Mainly for Industrial Application) Section 2.2: Additional Requirements for Fuses. 17 pp.
BSI BS 88: Part 6-88. 1988 Amd 1 Cartridge Fuses for Voltages up to and Including 1000 V a.c. and 1500 V d.c. Part 6: Spec. of Supplementary Requirements for Fuses of Compact Dimensions for Use in 240/415 V a.c. Industrial and Commercial Electrical Installations. 15 pp.
BSI BS 1361-71. 1971 Amd 3 Cartridge Fuses for a.c. Circuits in Domestic and Similar Premises (AMD 6692) January 31, 1991. 46 pp.
CNS C2044-87. Hard-Drawn Copper Wire Stranded (Non-Insulated) (Jul)(668). 4 pp.
CNS C4073-68. Cartridge Fuse Switch (for Distribution) (Apr)(2803)(R 1973). 5 pp.
JIS C 6575-75. Cartridge Fuses for Electronic Equipment (R 1988). 29 pp.
JIS C 8314-83. Particular Requirements for Low-Voltage Cartridge Fuses. 11 pp.
SNZ NZS 6208: Part 1-80. Specification for Cartridge Fuses for Voltages up to and Including 1000 V Alternating Current and 1500 V Direct Current Part 1: General Requirements Amend: 2A, 1980. 36 pp.
SNZ NZS 6208: Part 2-80. Specification for Cartridge Fuses for Voltages up to and Including 1000 V Alternating Current and 1500 V Direct Current Part 2: Supplementary Requirements for Fuses of Standardized Dimen-sions and Performance for Industrial Purposes Amend: A, 1980. 24 pp.

—**32 V**
CSA C22.2 NO 59.1-M1987. Fuses (Both Plug and Cartridge-Enclosed Types); (Gen Instr 1 Thru 3). 36 pp.
CSA 1137A Bull. Electrical Bulletin 1137A June 9, 1988 to C22.2 NO 59.1. 4 pp.

—**32 V—Recreational Vehicles**
CSA C22.2 NO 59.1-M1987. Fuses (Both Plug and Cartridge-Enclosed Types); (Gen Instr 1 Thru 3). 36 pp.

INDUSTRY STANDARDS

INTERNATIONAL AND NON-U.S. NATIONAL STANDARDS
SUBJECT INDEX

Cartridge

Cartridge Fuses (Cont.)
— 32 V—Recreational Vehicles (Cont.)
  CSA 1137A Bull. Electrical Bulletin 1137A June 9, 1988 to C22.2 NO 59.1. 4 pp.
— 250 V
  CSA C22.2 NO 59.1-M1987. Fuses (Both Plug and Cartridge-Enclosed Types); (Gen Instr 1 Thru 3). 36 pp.
— 250 V—Renewable
  CSA C22.2 NO 59.1-M1987. Fuses (Both Plug and Cartridge-Enclosed Types); (Gen Instr 1 Thru 3). 36 pp.
— 600 V
  CSA C22.2 NO 59.1-M1987. Fuses (Both Plug and Cartridge-Enclosed Types); (Gen Instr 1 Thru 3). 36 pp.
— 600 V—Renewable
  CSA C22.2 NO 59.1-M1987. Fuses (Both Plug and Cartridge-Enclosed Types); (Gen Instr 1 Thru 3). 36 pp.
—Fuse Holders
  CSA C22.2 NO 39-M1987. Fuseholder Assemblies (R 1992); (Gen Instr 1). 38 pp.
  DIN VDE 0820 Pt 2-76. VDE Specification for Miniature Cartridge Fuses; Specification for Fuse-Holders (Feb). 7 pp.
—HRC—600 V—Nonrenewable
  CSA 1437 Bull. Electrical Bulletin 1437 October 23, 1985 to C22.2 NO 106. 2 pp.

Cartridge Tape Drives
Use: Tape Drives

Cartridge Tapes (Magnetic)
Use: Computer Tapes; Magnetic Tapes

Cartridge Valves
See Also: Valves
—Hydraulic—Cavities
  BSI BS 7296: Part 1-90. 1990 Cavities for Hydraulic Fluid Power Cartridge Valves Part 1: Two-Port Slip-in Valves. 23 pp.

Cartridges
See Also: Containers; Data Cartridges; Microfilm Cartridges; Reels; Tape Cartridges
—Motion Picture Film
  ISO 1780-84. Cinematography—Motion-Picture Camera Cartridge, 8 mm Type S Model I—Aperture, Camera Aperture Profile, Film Position, Pressure Pad and Pressure Pad Flatness—Dimensions and Specifications Second Edition; (Corrected and Reprinted -1984). 9 pp.
  ISO 2863-73. Cinematography—Motion-Picture Camera Cartridge, 8 mm Type S Model II—Run Length of Film—Dimensions and Specifications First Edition. 4 pp.
  ISO 3024-83. Cinematography—Motion-Picture Camera Cartridge, 8 mm Type S, Model I—Camera Run Length, Perforation Cut-out and End-of-Run Notch in Film—Specifications Second Edition. 5 pp.
  ISO 3025-74. Cinematography—Motion-Picture Camera Cartridge, 8 mm Type S, Model II—Film Load Position First Edition. 4 pp.
  ISO 3067-83. Cinematography—Motion-Picture Camera Cartridge, 8 mm Type S, Model I—Notches for Film Speed, Film Identification and Colour-Balancing Filter—Dimensions and Positions Second Edition. 9 pp.
  ISO 3641-76. Cinematography—Motion-Picture Camera Cartridge 8 mm Type S Model I—Cartridge Fit and Take-up Core Drive—Dimensions and Specifications First Edition. 7 pp.
  ISO 3646-76. Cinematography—Motion-Picture Camera Cartridge 8 mm Type S Model II—Slots, Projections and Cartridge Hole for Indicating Film Speed, Colour Balance and Film Identification—Dimensions and Positions First Edition. 5 pp.
  ISO 3654-83. Cinematography—Motion-Picture Camera Cartridge, 8 mm Type S, Model I—Cartridge-Camera Interface and Take-up Core Drive—Dimensions and Specifications Second Edition. 8 pp.
  ISO 5759-80. Cinematography—Sound Motion-Picture Camera Cartridge, 8 mm Type S, Model I—Cartridge-Camera Interface and Take-up Core Drive—Dimensions and Specifications First Edition. 7 pp.
  ISO 6903-84. Cinematography—Motion-Picture Camera Cartridge, 8 mm Type S, Model I (Capacity 60 m)—Cartridge-Camera Interface and Sprocket Drive—Dimensions and Specifications First Edition. 9 pp.
  ISO 7453-84. Cinematography—Sound Motion-Picture Camera Cartridge, 8 mm Type S Model II—Cartridge-Camera Fit and Take-up Core Drive—Dimensions and Specifications First Edition. 8 pp.

Cartridges (Cont.)
—Motion Picture Film (Cont.)
  ISO 7454-84. Cinematography—Sound Motion-Picture Camera Cartridge, 8 mm Type S Model II—Camera Run Length and End Notches in Film—Dimensions and Specifications First Edition. 5 pp.
  ISO 7455-84. Cinematography—Sound Motion-Picture Camera Cartridge, 8 mm Type S Model II—Slots and Projection for Film Speed, Cartridge Hole and Projection for Film Identification and Colour-Balancing Filter—Dimensions and Positions First Edition. 7 pp.
  ISO 7456-84. Cinematography—Sound Motion-Picture Camera Cartridge, 8 mm Type S Model II—Film Load Position First Edition; (Corrected and Reprinted -1984). 4 pp.
—Photographic Film
  ISO 3029-83. Photography—126-Size Cartridges—Dimensions of Cartridge, Film and Backing Paper Second Edition. 15 pp.
  ISO 7261-87. Photography—110-Size Cartridges—Dimensions First Edition. 11 pp.
  ISO 7330-88. Photography—110-Size Cartridges—Location and Dimensions of Film Exposure and Film Identification Notches First Edition. 11 pp.
  ISO 7374-87. Photography—110-Size Cartridges—Dimensions and Format of Film and Backing Paper First Edition. 8 pp.
  JIS K 7564-89. Photography-110-Size Cartridges-Dimensions and Format of Film and Backing Paper (ISO 7374—1987). 10 pp.
  JIS K 7565-89. Photography—110-Size Cartridges—Location and Dimensions of Film Exposure and Film Identification Notches (ISO 7330—1988). 14 pp.

Cartridges (Explosives)
See Also: Ammunition; Cartridge Cases; Explosives; Guns; Shot (Ammunition); Small Arms Ammunition
—Aircraft
  NATO STANAG 3821 ED 1 AMD 1-86. Chaff and IR Cartridges for Aircraft. 14 pp.
  NATO STANAG 3821 ED 1 AMD 2-86. Al/1 Chaff and IR Cartridges for Aircraft. 15 pp.
—Shotguns
  JIS B 9806-63. Shots for Shot Gun Cartridges (R 1983). 6 pp.
—Shotguns—Wads
  JIS B 9807-75. Felt Wads and Card Wads for Shot Gun Cartridges.

Carts
Use For: Trolleys (Carts) *See Also:* Cleaning Equipment and Supplies; Hospital Carts; Laundry Carts; Shopping Carts
  BSI BS 6250: Part 2-91. 1991 Domestic and Contract Furniture Part 2: Specification for Performance Requirements for Tables and Trolleys. 16 pp.
  BSI BS 6250: Part 2-01. 1991 Amd 1 Domestic and Contract Furniture Part 2: Specification for Performance Requirements for Tables and Trolleys (AMD 7055) May 1, 1992. 17 pp.
—Aircraft
  CAA NOTICE #99 ISSUE 1. Galley Equipment (Airworthiness Notices). 4 pp.
—Fire Pumps
  CNS Z2048-82. Hand-Cart for Fire Pumps (Jul)(9194).
  CNS Z2049-82. Portable-Cart for Fire Pumps (Jul)(9195).
—Inner Tubes
  CNS K4015-59. Wheel Tubes for Carts (Oct)(1155). 2 pp.
  CNS K6105-59. Method of Test for Tubes for Carts (Oct)(1156). 2 pp.
—Mechanical Testing
  BSI BS 4875: Part 5-85. 1985 Strength and Stability of Furniture Part 5: Methods for Determination of Strength of Tables and Trolleys. 12 pp.
—Sheet Glass—Safety
  BSI BS 7376-90. 1990 Inclusion of Glass in the Construction of Tables or Trolleys. 10 pp.
—Stability Testing
  BSI BS 4875: Part 6-85. 1985 Strength and Stability of Furniture Part 6: Methods for Determination of Stability of Tables and Trolleys. 8 pp.
—Tires
  CNS K4014-59. Tires for Carts (Oct)(1153). 2 pp.
  CNS K6104-59. Method of Test for Tires for Carts (Oct)(1154). 2 pp.

CASA C-212 Aircraft
See Also: Aircraft

CASA C-212 Aircraft (Cont.)
—Accidents
  CAA. CASA 212. 1 p.

CASE
Use: Computer Aided Software Engineering

Case Depth
—Steels
  BSI BS 6479-84. 1984 Determination and Verification of the Effective Depth of Carburized and Hardened Case in Steels. 4 pp.
  CNS G2041-77. Methods of Measuring Case Depth of Steel Hardened by Flame or Induction Hardening Process (Oct)(4179).
  CNS G2168-83. Methods of Measuring Case Depth for Steel (Apr)(10168).
  ISO 2639-82. Steel—Determination and Verification of the Effective Depth of Carburized and Hardened Cases Second Edition; (Corrected and Reprinted -1983). 5 pp.
  JIS G 0557-77. Methods of Measuring Case Depth for Steel.
  JIS G 0557-65. Method of Measuring Case Depth for Steel. 4 pp.
  JIS G 0559-77. Methods of Measuring Case Depth of Steel Hardened by Flame or Induction Hardening Process.
  JIS G 0559-67. Methods of Measuring Case Depth of Steel Hardened by Flame or Induction Hardening Process (R 1969). 5 pp.
  SAA AS 1982-93. Methods for the Measurement of Case Depth in Steels. 6 pp.

Case Hardening
Scope Note: See the subheading Case Hardening or Case Hardened under specific metals

Casein Content Analysis
See Also: Caseins
—Papers
  CNS P3032-79. Method of Test for Casein in Paper (Qualitative) (Jan)(4719). 1 p.

Casein Glue
See Also: Adhesives
  CNS K3091-87. Casein Glue for Wood (Jun)(12003). 2 pp.
  CNS K6908-87. Method of Test for Casein Glue (Jun)(12004). 3 pp.
  CSA O112-M Series 1977. CSA Standards for Wood Adhesives. 91 pp.
  JIS K 6803-79. Casein Glue for Wood. 8 pp.

Caseinates
Use: Caseins

Caseins
See Also: Casein Content Analysis; Dairy Products; Furnish; Sizing (Surface Treatment)
  BSI BS 1416-62. (WITHDRAWN) 1962 Amd 4 Methods for the Sampling and Analysis of Rennet Casein (AMD 5460) December 31, 1986 (Superseded by Relevant Parts of BS 6248 and BS 6394). 21 pp.
  CNS K7137-64. Chemical Reagent (Casein) (Jul)(1637).
  EC 83/417/EEC-83. Council Directive on the Approximation of the Laws of the Member States Relating to Certain Lactoproteins (Caseins and Caseinates) Intended for Human Consumption. 7 pp.
  JIS K 8234-61. Casein (Milk).
—Acidity
  SAA AS 2300.10.6-91. Methods of Chemical and Physical Testing for the Dairying Industry—Part 10: Caseins, Caseinates and Coprecipitates—Part 10.6: Determination of Free Acidity of Caseins (Supersedes AS N60—1970 (in Part)). 3 pp.
—Acidity—Volumetric Analysis
  BSI BS 6248: Part 5-82. 1982 Caseins and Caseinates Part 5: Method for Determination of Free Acidity of Caseins (Reference Method). 4 pp.
  ISO 5547-78. Caseins—Determination of Free Acidity (Reference Method) First Edition. 4 pp.
—Ash Content—Gravimetric Analysis
  BSI BS 6248: Part 2-82. 1982 Caseins and Caseinates Part 2: Method for Determination of 'Fixed Ash' of Caseins (Reference Method). 4 pp.
  BSI BS 6248: Part 3-82. 1982 Caseins and Caseinates Part 3: Method for Determination of Ash of Rennet Caseins and Caseinates (Reference Method). 4 pp.
  ISO 5544-78. Caseins—Determination of "Fixed Ash" (Reference Method) First Edition. 4 pp.
  ISO 5545-78. Rennet Caseins and Caseinates—Determination of Ash (Reference Method) First Edition. 4 pp.

Caseins (Cont.)

—Color
SAA AS 2300.10.3-91. Methods of Chemical and Physical Testing for the Dairying Industry—Part 10: Caseins, Caseinates and Coprecipitates—Part 10.3: Assessment of Colour of Caseins (Supersedes AS N60—1970 (in Part)). 2 pp.
SAA AS 2300.10.3 Supp 1-91. Methods of Chemical and Physical Testing for the Dairying Industry—Part 10: Caseins, Caseinates and Coprecipitates—Part 10.3: Assessment of Colour of Caseins—Supplement 1: Colour Standards for Untreated and Heat-Treated Acid Casein and for Wet-Heat-Treated Rennet Casein.
SAA AS 2300.10.3 Supp 2-91. Methods of Chemical and Physical Testing for the Dairying Industry—Part 10: Caseins, Caseinates and Coprecipitates—Pt. 10.3: Assmt. of Colour of Caseins—Supplement 1: Colour Standards for Untreated Rennet Casein (Supplement to AS 2300. 10.3—1991) (Redesignates AS N60C2).

—Fat Content—Gravimetric Analysis
BSI BS 6248: Part 10-86. 1986 Amd 1 Caseins and Cadeinates Part 10: Method for Determination of Fats (Gravimetric Reference Method) (AMD 5868) July 29, 1988. 11 pp.
ISO 5543-86. Caseins and Caseinates—Determination of Fat Content—Gravimetric Method (Reference Method) First Edition (Corrected and Reprinted -1987). 10 pp.

—Filterable Material Content
SAA AS 2300.10.4 Supp 1-91. Methods of Chemical and Physical Testing for the Dairying Industry—Part 10: Caseins, Caseinates and Coprecipitates—Part 10.4: Determination of Filterable Material—Supplement 1: Photographic Reference Standards (Supplement to AS 2300.10.4—1991) (Redesignates AS N60S).
SAA AS 2300.10.4-91. Methods of Chemical and Physical Testing for the Dairying Industry—Part 10: Caseins, Caseinates and Coprecipitates—Part 10.4: Determination of Filterable Material (Supersedes AS N60—1970 (in Part)). 4 pp.

—Flavor Testing
SAA AS 2300.10.2-91. Methods of Chemical and Physical Testing for the Dairying Industry—Part 10: Caseins, Caseinates and Coprecipitates—Part 10.2: Assessment of Odour and Flavour (Supersedes AS N60—1970 (in Part)). 1 p.

—Lactose Content—Photometry
BSI BS 6248: Part 6-82. 1982 Caseins and Caseinates Part 6: Method for Determination of Lactose Content (Photometric Method). 6 pp.
ISO 5548-80. Caseins and Caseinates—Determination of Lactose Content—Photometric Method First Edition; (Erratum—Jan 1982). 6 pp.
SAA AS 2300.10.7-91. Methods of Chemical and Physical Testing for the Dairying Industry—Part 10: Caseins, Caseinates and Coprecipitates—Part 10.7: Determination of Lactose—Photometric Method (Supersedes AS N60—1970 (in Part)). 4 pp.

—Nitrate Content—Cadmium Reduction
BSI BS 6248: Part 11-87. 1987 Caseins and Caseinates Part 11: Method for the Determination of Nitrate and Nitrite Contents. 8 pp.
ISO 8195-87. Caseins and Caseinates—Determination of Nitrate and Nitrite Contents—Method by Cadmium Reduction and Spectrometry First Edition. 8 pp.

—Nitrite Content—Spectrometry
BSI BS 6248: Part 11-87. 1987 Caseins and Caseinates Part 11: Method for the Determination of Nitrate and Nitrite Contents. 8 pp.
ISO 8195-87. Caseins and Caseinates—Determination of Nitrate and Nitrite Contents—Method by Cadmium Reduction and Spectrometry First Edition. 8 pp.

—Odors
SAA AS 2300.10.2-91. Methods of Chemical and Physical Testing for the Dairying Industry—Part 10: Caseins, Caseinates and Coprecipitates—Part 10.2: Assessment of Odour and Flavour (Supersedes AS N60—1970 (in Part)). 1 p.

—Particle Size Distribution
SAA AS 2300.10.5-91. Methods of Chemical and Physical Testing for the Dairying Industry—Part 10: Caseins, Caseinates and Coprecipitates—Part 10.5: Determination of Particle-Size Distribution in Caseins (Supersedes AS N60—1970 (in Part)). 2 pp.

—PH
BSI BS 6248: Part 4-82. 1982 Caseins and Caseinates Part 4: Method for Determination of pH (Reference Method). 4 pp.
ISO 5546-79. Caseins and Caseinates—Determination of pH (Reference Method) First Edition; (Amendment Slip-1980). 5 pp.

—Protein Content—Volumetric Analysis
BSI BS 6248: Part 7-82. 1982 Caseins and Caseinates Part 7: Method for Determination of Protein Content (Reference Method). 5 pp.
ISO 5549-78. Caseins and Caseinates—Determination of Protein Content (Reference Method) First Edition. 5 pp.

—Sampling
BSI BS 1416-62. (WITHDRAWN) 1962 Amd 4 Methods for the Sampling and Analysis of Rennet Casein (AMD 5460) December 31, 1986 (Superseded by Relevant Parts of BS 6248 and BS 6394). 21 pp.
BSI BS 1417-63. (WITHDRAWN) 1963 Amd 3 Methods for the Sampling and Analysis of Acid Casein (W) (AMD 5459) December 31, 1986 (Solubility Index Method Obsolete, Other Methods Superseded by Relevant Parts of BS 6248 & BS 6394). 19 pp.
BSI BS 6248: Part 1-82. 1982 Caseins and Caseinates Part 1: General Introduction, Including Preparation of Laboratory Samples. 4 pp.
SAA AS 2300.10.1-91. Methods of Chemical and Physical Testing for the Dairying Industry—Part 10: Caseins, Caseinates and Coprecipitates—Part 10.1: General Information and Preparation of Samples (Supersedes AS N60—1970 (in Part)). 2 pp.

—Scorched Particles Content
BSI BS 6248: Part 9-84. 1984 Caseins and Caseinates Part 9: Method for the Determination of Scorched Particles. 10 pp.
ISO 5739-83. Caseins and Caseinates—Determination of Scorched Particles Content First Edition. 10 pp.

—Water Content
BSI BS 6248: Part 8-82. 1982 Caseins and Caseinates Part 8: Method for Determination of Water Content (Reference Method). 4 pp.
ISO 5550-78. Caseins and Caseinates—Determination of Water Content (Reference Method) First Edition. 4 pp.

Casement Fasteners
See Also: Fasteners
BSI BS 6462-85. 1985 Mechanical Performance of Peg-Type Casement Stays and Face-Fixed Wedge-Action Fasteners. 12 pp.

Cases
Scope Note: For additional listings, see the subheading Cases under specific products
CNS Z5039-73. Nailed Board Cases (General Rules) (Jan)(3529).
CNS Z5040-73. Nailed Board Cases (Material) (Jan) (3530).
CNS Z5041-73. Nailed Board Cases (Construction and Style) (Jan)(3531).
CNS Z5042-73. Nailed Board Cases (Requirement for Manufacture) (Jan)(3532).

—Apples—Corrugated
BSI BS 5637-78. (WITHDRAWN) 1978 Non-Returnable Corrugated Cases for Apples. 9 pp.

—Corrugated
BSI BS 1133: Ch 7.5-90. 1990 Packaging Code Section 7: Paper and Board Wrappers, Bags and Containers: Chapter 7.5: Fibreboard Cases. 25 pp.

—Fiberboard
BSI BS 1133: Ch 7.5-90. 1990 Packaging Code Section 7: Paper and Board Wrappers, Bags and Containers: Chapter 7.5: Fibreboard Cases. 25 pp.

—Plywood
CNS Z5116-82. Cleated Plywood Boxes (Dec)(9757).
MOD UK DSTAN 81-23-01. Cases, Plywood, Collapsible Issue 3; Amendment 2. 23 pp.
MOD UK DSTAN 81-23-02. Cases, Plywood, Collapsible Issue 3; Amendment 3. 25 pp.

—Tea—Plywood
CNS Z5008-59. Plywood Cases for Tea (for Export) (Apr)(1128) (R 1972).

Cases (Electrical)
Use: Electrical Enclosures

Cases (Instrument)
Use: Instrument Housings

Cash Registers
See Also: Calculators; Currencies
JIS B 9515-77. Methods of Test for Cash Registers.

—Electronic
JIS B 9510-90. Format of Specification for Electronic Cash Registers. 12 pp.

—Electronic—Glossaries
JIS B 0115-91. Electronic Cash Registers—Vocabulary.

—Glossaries
CNS C5135-82. Definition of a Cash Register (Jun)(8939).
JIS B 0115-83. Glossary of Terms Relating to Cash Registers (R 1988). 21 pp.

—Keyboards
JIS B 9511-77. Registration Symbols for Keyboard Arrangements for Cash Register.

—Printing Papers
JIS B 9518-77. Receipt and Detail Paper for Cash Register.

—Symbols
JIS B 9511-77. Registration Symbols for Keyboard Arrangements for Cash Register.

—Time Switches
BSI BS 7415-91. (WITHDRAWN) 1991 Time Switches for Tariff and Load Control (Renumbered as BS EN 61038: 1993). 35 pp.
BSI BS EN 61038-93. 1993 Amd 1 Time Switches for Tariff and Load Control (AMD 7674) May 15, 1993 (E). 46 pp.
CENELEC EN 61038-92. Time Switches for Tariff and Load Control (IEC 1038:1990). 7 pp.
CENELEC EN 61038-92. Time Switches for Tariff and Load Control (IEC 1038: 1990, Modified). 34 pp.
IEC 1038-90. Time Switches for Tariff and Load Control First Edition; (CENELEC EN 61038:1992). 66 pp.

Cashew Nuts
See Also: Nuts (Food)
ISO 6477-88. Cashew Kernels—Specification First Edition. 7 pp.

Cashew Polymer Sealants

—Screw Threads
CGSB CAN/CGSB-1.168-M90. Cashew Polymer, Lead Free Cement. 17 pp.
MOD UK DSTAN 80-162-93. Sealing Compound (Two-Part) for Screw Threads, Type QX (Cashew Polymer Base, RD 1286) Issue 1. 27 pp.
MOD UK TS 324G. Sealing Compound (Two Part) for Screw Threads, Type QX (Cashew Polymer Base, RD 1286) (Superseded by Def Stan 80-162).

Cashew Resin Enamel Paints
See Also: Cashew Resin Primers; Enamel Paints

—Metal
CNS K2150-86. Cashew Resin Enamel (Dec)(11801). 2 pp.
CNS K6896-89. Method of Test for Cashew Resin Enamel (Jan)(11802). 5 pp.
JIS K 5642-83. Cashew Resin Enamel. 9 pp.

—Wood
CNS K2150-86. Cashew Resin Enamel (Dec)(11801). 2 pp.
CNS K6896-89. Method of Test for Cashew Resin Enamel (Jan)(11802). 5 pp.
JIS K 5642-83. Cashew Resin Enamel. 9 pp.

Cashew Resin Primers
See Also: Cashew Resin Enamel Paints; Primers (Coatings)
CNS K2155-87. Cashew Resin Primer (May)(11946). 1 p.
CNS K6905-87. Method of Test for Cashew Resin Primer (May)(11947). 3 pp.
JIS K 5646-88. Cashew Resin Undercoats. 10 pp.

Cashew Resin Putty
See Also: Putty
CNS K2152-86. Cashew Resin Putty (Dec)(11805). 2 pp.
CNS K6898-86. Method of Test for Cashew Resin Putty (Dec)(11806). 3 pp.

Cashew Resin Varnishes
See Also: Varnishes
CNS K2149-86. Cashew Resin Varnish (Dec)(11799). 2 pp.

—Metal
CNS K2147-86. Cashew Resin Surfacer (Oct)(11729). 2 pp.
CNS K6895-89. Method of Test for Cashew Resin Varnish (Jan)(11800). 5 pp.
JIS K 5641-83. Cashew Resin Varnish. 10 pp.

—Wood
CNS K2147-86. Cashew Resin Surfacer (Oct)(11729). 2 pp.
CNS K6895-89. Method of Test for Cashew Resin Varnish (Jan)(11800). 5 pp.
JIS K 5641-83. Cashew Resin Varnish. 10 pp.

Casing

Casing Pipes
Use: Well Casings

Casings
See Also: Well Casings

—Assembly—Harness Release Gears— Aircraft
SBAC AS 2820 ISSUE 4. Assembly of Front Casing. 1 p.
SBAC AS 2821 ISSUE 5. Assembly of Back Casing. 1 p.

—Core Drills
CNS M2038-80. Casings for Core Drills (Mar)(5365).
JIS M 1411-82. Casing for Core Drills.

—Fuel Filters—Aircraft
SBAC AS 3221 ISSUE 1. Casing—Top Half Filter Fuel Type 'A'. 1 p.
SBAC AS 3223 ISSUE 1. Casing—Bottom Half— Filter Fuel Type 'A'. 1 p.
SBAC AS 3251 ISSUE 1. Casing—Top Half Filter Fuel Type B. 1 p.
SBAC AS 3253 ISSUE 1. Casing—Bottom Half Filter Fuel Type 'B'. 1 p.

—Harness Release Gears—Aircraft
SBAC AS 2819 ISSUE 4. Half Casing for Harness Release Gear. 1 p.

Casings (Meat)
See Also: Meat; Membranes; Sausages
CNS N5034-84. Natural Casings (Apr)(1486). 3 pp.

—Diameters
CNS N6031-81. Method of Test for Sausage Casings (Oct)(1487). 2 pp.

—Length Measurement
CNS N6031-81. Method of Test for Sausage Casings (Oct)(1487). 2 pp.

—Sampling
CNS N6031-81. Method of Test for Sausage Casings (Oct)(1487). 2 pp.

—Visual Inspection
CNS N6031-81. Method of Test for Sausage Casings (Oct)(1487). 2 pp.
CNS N6032-87. Method of Test for Chinese Sausages (Jun)(2103). 2 pp.

CASS Testing
Use: Salt Spray Testing

Casseroles
See Also: Laboratory Equipment

—Aluminum
CNS H3034-86. Aluminium Casserole (Sep)(2701).

—Enameled
CNS R2160-85. Enamel Casserole for Laboratory (Jun)(9035).

—Porcelain
CNS R2113-85. Porcelain Casserols for Chemical Analysis (May)(6612).
JIS R 1305-80. Porcelain Casserols for Chemical Analysis.

Cassette Recorders
Use: Tape Recorders—Cassette

Cassette Tapes
Use: Audio Cassettes

Cassette Tapes (Magnetic)
Use: Magnetic Tapes

Cassia
See Also: Spices
ISO 6538-82. Cassia (Type China, Type Indonesia and Type Viet Nam), Whole or Ground (Powdered)— Specification First Edition. 6 pp.

Cassia Oil
See Also: Essential Oils
CNS K5084-80. Oil of Cassia (Sep)(6455).
ISO 3216-74. Oil of Cassia First Edition. 4 pp.

Cast Iron
Scope Note: For additional listings, see also specific products made from cast iron *See Also:* Austenitic Cast Iron; Carbon; Cast Iron Boilers; Cast Iron Coatings (On Cast Iron); Castings; Ductile Iron; Ductile Iron Castings; Ferroalloys; Gray Iron; Iron; Iron Castings; Malleable Cast Iron; White Iron

Cast Iron (Cont.)

—Aluminum Content—Spectrophotometry
BSI BS 6200: SUBSEC 3.1.2-91. 1991 Sampling and Analysis of Iron, Steel and Other Ferrous Metals Part 3: Methods of Analysis Section 3.1: Determination of Aluminium Subsection 3.1.2: Steel and Cast Iron: Spectrophotometric Method. 10 pp.

—Aluminum Content—Volumetric Analysis
BSI BS 6200: SUBSEC 3.1.1-91. 1991 Sampling and Analysis of Iron, Steel and Other Ferrous Metals Part 3: Methods of Analysis Section 3.1: Determination of Aluminium Subsection 3.1.1: Steel, Cast Iron, Low Carbon Ferro-Chromium Metal: Volumetric Method. 11 pp.

—Carbon Content—Gravimetric Analysis
BSI BS 6200: SUBSEC 3.8.5-91. 1991 Sampling and Analysis of Iron, Steel and Other Ferrous Metals Part 3: Methods of Analysis Section 3.8: Determination of Carbon Subsection 3.8.5: Cast Iron and Pig Iron: Gravimetric Method for the Determination of Non-Combined Carbon. 6 pp.
ISO 437-82. Steel and Cast Iron—Determination of Total Carbon Content—Combustion Gravimetric Method First Edition. 6 pp.

—Chemical Analysis
DIN ENGL 1695-81. Abrasion Resisting Alloy Cast Iron (Sept). 4 pp.

—Chromium Content—Potentiometric Analysis
BSI BS 6200: SUB SEC 3.10.1-85. (WITHDRAWN) 1985 Sampling and Analysis of Iron, Steel and Other Ferrous Metals: Part 3: Methods of Analysis Sec. 3.10: Determ.Chromium: Subsec3.10.1: Steel and Iron: Pptentiometric/Volumetric Method (Superseded by BS EN 24 937: 1991). 8 pp.
BSI BS EN 24937-91. 1991 Steel and Iron Determination of Chromium Content: Potentiometric or Visual Titration Method (V). 14 pp.
BSI BS EN 24937-01. 1991 Amd 1 Steel and Iron Determination of Chromium Content Potentiometric or Visual Titration Method (ISO 4937: 1986) (AMD 7082) July 15, 1992 (V). 15 pp.
CEN EN 24 937-90. Steel and Iron—Determination of Chromium Content—Potentiometric or Visual Method. 2 pp.
DIN ENGL EN 24937-92. Determination of Chromium Content of Steel and Iron by Potentiometric or Visual Titration (ISO 4937: 1986) (Nov). 11 pp.
ISO 4937-86. Steel and Iron—Determination of Chromium Content—Potentiometric or Visual Titration Method First Edition. 10 pp.
SAA AS 1050.9-84. Methods for the Analysis of Iron and Steel—Part 9: Determination of Chromium (Potentiometric Method). 4 pp.

—Chromium Content—Volumetric Analysis
BSI BS 6200: SUB SEC 3.10.1-85. (WITHDRAWN) 1985 Sampling and Analysis of Iron, Steel and Other Ferrous Metals: Part 3: Methods of Analysis Sec. 3.10: Determ.Chromium: Subsec3.10.1: Steel and Iron: Pptentiometric/Volumetric Method (Superseded by BS EN 24 937: 1991). 8 pp.
BSI BS EN 24937-91. 1991 Steel and Iron Determination of Chromium Content: Potentiometric or Visual Titration Method (V). 14 pp.
BSI BS EN 24937-01. 1991 Amd 1 Steel and Iron Determination of Chromium Content Potentiometric or Visual Titration Method (ISO 4937: 1986) (AMD 7082) July 15, 1992 (V). 15 pp.
CEN EN 24 937-90. Steel and Iron—Determination of Chromium Content—Potentiometric or Visual Method. 2 pp.
DIN ENGL EN 24937-92. Determination of Chromium Content of Steel and Iron by Potentiometric or Visual Titration (ISO 4937: 1986) (Nov). 11 pp.
ISO 4937-86. Steel and Iron—Determination of Chromium Content—Potentiometric or Visual Titration Method First Edition. 10 pp.

—Cobalt Content—Spectrophotometry
BSI BS 6200: SUB SEC 3.11.1-91. 1991 Sampling and Analysis of Iron, Steel and Other Ferrous Metals Part 3: Methods of Analysis Section 3.11: Determination of Cobalt Subsection 3.11.1: Steel and Cast Iron: Spectrophotometric Method. 7 pp.

—Copper Content—Atomic Absorption Spectrometry
BSI BS 6200: SUB SEC 3.12.3-86. 1986 Sampling and Analysis of Iron, Steel and Other Ferrous Metals Part 3: Methods of Analysis Section 3.12: Determination of Copper Subsection 3.12.3: Steel and Cast Iron: Flame Atomic Absorption Spectrometric Method. 8 pp.

Cast Iron (Cont.)

—Copper Content—Atomic Absorption Spectrometry (Cont.)
BSI BS 6200: SUB SEC 3.12.3-01. 1986 Amd 1 Sampling and Analysis of Iron, Steel and Other Ferrous Metals Part 3: Methods of Analysis Sec 3.12: Determination of Copper Subsec 3.12.3: Steel and Cast Iron: Flame Atomic Absorption Spectrometric Method (ISO 4943: 1985) (AMD 7071). 13 pp.
CEN EN 24 943-90. Chemical Analysis of Ferrous Metal—Determination of Copper Content—Flame Atomic Absorption Speetrometric Method. 2 pp.
DIN ENGL EN 24943-92. Determination of Copper Content of Steel and Cast Iron by Flame Atomic Absorption Spectrometry; (ISO 4943: 1985) (Oct). 9 pp.
ISO 4943-85. Steel and Cast Iron—Determination of Copper Content—Flame Atomic Absorption Spectrometric Method First Edition. 8 pp.

—Copper Content—Spectrophotometry
BSI BS 6200: SUB SEC 3.12.2-85. 1985 Sampling and Analysis of Iron, Steel and Other Ferrous Metals Part 3: Methods of Analysis Section 3.12: Determination of Copper Subsection 3.12.2: Steel and Cast Iron: Spectrophotometric Method. 8 pp.
BSI BS 6200: SUB SEC 3.12.2-01. 1985 Amd 1 Sampling and Analysis of Iron, Steel and Other Ferrous Metals Part 3: Methods of Analysis Sec 3.12: Determination of Copper Subsec 3.12.2: Steel and Cast Iron: Spectrophotometric Meth (ISO 4946: 1984) (AMD 7070) February 28, 1992. 13 pp.
CEN EN 24 946-90. Determination of Copper Content in Steel and Cast Iron. 2 pp.
DIN ENGL EN 24946-92. Determination of Copper Content of Steel and Cast Iron by the 2,2'-Diquinolyl Spectrophotometric Method (ISO 4946: 1984) (Nov). 9 pp.
ISO 4946-84. Steel and Cast Iron—Determination of Copper Content—2,2'-Diquinolyl Spectrophotometric Method First Edition; (Corrected and Reprinted -1986). 8 pp.
SAA AS 1050.17-88. Methods for the Analysis of Iron and Steel—Part 17: Determination of Copper Content—Spectrophotometric Method. 4 pp.

—Copper Content—Volumetric Analysis
BSI BS 6200: SUB SEC 3.12.1-86. 1986 Sampling and Analysis of Iron, Steel and Other Ferrous Metals Part 3: Methods of Analysis Section 3.12: Determination of Copper Subsection 3.12.1: Steel and Cast Iron: Volumetric Method. 4 pp.
BSI BS 6200: SUB SEC 3.12.4-86. 1986 Sampling and Analysis of Iron, Steel and Other Ferrous Metals Part 3: Methods of Analysis Section 3.12: Determination of Copper Subsection 3.12.4: Cast Iron: Volumetric Method. 4 pp.

—Emission Spectroscopy
CNS G2183-84. Emission-Spectroscopic Analysis for Pig Iron and Cast Iron (Oct)(10503).
JIS G 1251-76. Emission-Spectroscopic Analysis for Pig Iron and Cast Iron (R 1984). 8 pp.

—Graphite
BSI BS 1452-90. 1990 Flake Graphite Cast Iron. 11 pp.
DIN ENGL 1693 Pt 1-73. Cast Iron with Nodular Graphite; Unalloyed and Low Alloy Grades (Oct). 5 pp.

—Graphite—Mechanical Properties
DIN ENGL 1693 Pt 2-77. Cast Iron with Nodular Graphite; Unalloyed and Low Alloy Grades; Properties in Cast-On Test Piece (Oct). 4 pp.

—Graphite—Microstructure
ISO 945-75. Cast Iron—Designation of Microstructure of Graphite First Edition. 10 pp.

—Heat Treatment
DIN ENGL 1695 Suppl 1-81. Wear Resisting Alloyed Cast Iron; Information on Heat Treatment, Mechanical and Physical Properties and Microstructure (Sep). 3 pp.

—Magnesium Content—Volumetric Analysis
BSI BS 6200: SUB SEC 3.17.1-92. 1992 Sampling and Analysis of Iron, Steel and Other Ferrous Metals Part 3: Methods of Analysis Section 3.17: Determination of Magnesium Subsection 3.17.1: Cast Iron: Volumetric Method (V). 12 pp.

—Manganese Content—Spectrophotometry
BSI BS 6200: SUB SEC 3.18.2-85. 1985 Sampling and Analysis of Iron, Steel and Other Ferrous Metals Part 3: Methods of Analysis Section 3.18: Determination of Manganese Subsection 3.18.2: Steel Spectrophotometric Method. 6 pp.
ISO 629-82. Steel and Cast Iron—Determination of Manganese Content—Spectrophotometric Method First Edition. 6 pp.

INTERNATIONAL AND NON-U.S. NATIONAL STANDARDS
SUBJECT INDEX

Cast Iron (Cont.)

—Manganese Content—Volumetric Analysis
BSI BS 6200: SUB SEC 3.18.4-91. 1991 Sampling and Analysis of Iron, Steel and Other Ferrous Metals Part 3: Methods of Analysis Section 3.18: Determination of Manganese Subsection 3.18.4: Steel and Cast Iron: Volumetric Method. 7 pp.

—Mechanical Properties
DIN ENGL 1695 Suppl 1-81. Wear Resisting Alloyed Cast Iron; Information on Heat Treatment, Mechanical and Physical Properties and Microstructure (Sep). 3 pp.

—Microstructures
DIN ENGL 1695 Suppl 1-81. Wear Resisting Alloyed Cast Iron; Information on Heat Treatment, Mechanical and Physical Properties and Microstructure (Sep). 3 pp.

—Molybdenum Content—Photometry
BSI BS 6200: SUB SEC 3.19.1-85. 1985 Sampling and Analysis of Iron, Steel and Other Ferrous Metals Part 3: Methods of Analysis Section 3.19: Determination of Molybdenum Subsection 3.19.1: Steel and Cast Iron: Photometric Method. 6 pp.
ISO 4941-78. Steels and Cast Irons—Determination of Molybdenum Content—Photometric Method First Edition. 6 pp.

—Nickel Content—Atomic Absorption Spectrometry
ISO 4940-85. Steel and Cast Iron—Determination of Nickel Content—Flame Atomic Absorption Spectrometric Method First Edition. 9 pp.
SAA AS 1050.33-88. Methods for the Analysis of Iron and Steel—Part 33: Determination of Nickel Content—Flame Atomic Absorption Spectrometric Method. 4 pp.

—Nickel Content—Atomic Absorption Spectrophotometry
BSI BS 6200: SUB SEC 3.20.4-86. (WITHDRAWN) 1986 Sampling and Analysis of Iron, Steel and Other Ferrous Metals: Part 3: Methods of Anal. Sec. 3.20: Determ. ofNickel: Subsec 3.20.4:Steel & Cast Iron: FlameAtomic Absp.Spectrphotometric (Superseded by BS EN 10 136: 1991). 10 pp.

—Nickel Content—Gravimetric Analysis
BSI BS 6200: SUB SEC 3.20.1-89. 1989 Sampling and Analysis of Iron, Steel and Other Ferrous Metals Part 3: Methods of Analysis Section 3.20: Determination of Nickel Subsection 3.20.1: Steel and Cast Iron Gravimetric or Titrimetric Method (ISO 4938: 1988). 12 pp.
BSI BS 6200: SUB SEC 3.20.1-01. 1989 Amd 1 Sampling and Analysis of Iron, Steel and Other Ferrous Metals Part 3: Methods of Analysis Sec 3.20: Determination of Nickel Subsec 3.20.1: Steel and Cast Iron: Gravimetric or Titrimetric Method (ISO 4938: 1988) (AMD 7074). 17 pp.

—Nickel Content—Spectrophotometry
BSI BS 6200: SUB SEC 3.20.3-85. 1985 Sampling and Analysis of Iron, Steel and Other Ferrous Metals Part 3: Methods of Analysis Section 3.20: Determination of Nickel Subsection 3.20.3: Steel and Cast Iron: Spectrophotometric Method. 8 pp.
ISO 4939-84. Steel and Cast Iron—Determination of Nickel Content—Dimethylglyoxime Spectrophotometric Method First Edition; (Corrected and Reprinted -1986). 8 pp.
SAA AS 1050.19-88. Methods for the Analysis of Iron and Steel—Part 19: Determination of Nickel Content—Spectrophotometric Method. 4 pp.

—Nickel Content—Volumetric Analysis
BSI BS 6200: SUB SEC 3.20.1-89. 1989 Sampling and Analysis of Iron, Steel and Other Ferrous Metals Part 3: Methods of Analysis Section 3.20: Determination of Nickel Subsection 3.20.1: Steel and Cast Iron Gravimetric or Titrimetric Method (ISO 4938: 1988). 12 pp.
BSI BS 6200: SUB SEC 3.20.1-01. 1989 Amd 1 Sampling and Analysis of Iron, Steel and Other Ferrous Metals Part 3: Methods of Analysis Sec 3.20: Determination of Nickel Subsec 3.20.1: Steel and Cast Iron: Gravimetric or Titrimetric Method (ISO 4938: 1988) (AMD 7074). 17 pp.

—Numbering Systems
DIN ENGL 17007 Pt 3-71. Material Numbers; Material Group 3: Pig Iron, Master Alloys and Cast Iron (Jan). 2 pp.

—Phosphorus Content—Spectrophotometry
BSI BS EN 10184-92. 1992 Chemical Analysis of Ferrous Materials—Determination of Phosphorus in Steels and Irons—Spectrophotometric Method (V). 14 pp.

Cast Iron (Cont.)

—Physical Properties
DIN ENGL 1695 Suppl 1-81. Wear Resisting Alloyed Cast Iron; Information on Heat Treatment, Mechanical and Physical Properties and Microstructure (Sep). 3 pp.

—Sampling
BSI BS 6200: Sec 2.1-93. 1993 Sampling and Analysis of Iron, Steel and Other Ferrous Metals Part 2: Sampling and Sample Preparation Section 2.1: Methods for Iron and Steel (Supersedes BS 1837: 1970) (V). 38 pp.

—Silicon Content—Gravimetric Analysis
ISO 439-82. Steel and Cast Iron—Determination of Total Silicon—Gravimetric Method First Edition. 5 pp.

—Silicon Content—Spectrophotometry
BSI BS 6200: SUB SEC 3.26.3-87. 1987 Sampling and Analysis of Iron, Steel and Other Ferrous Metals Part 3: Methods of Analysis Section 3.26: Determination of Silicon Subsection 3.26.3: Determination of Silicon. 10 pp.
BSI BS 6200: SUB SEC 3.26.3-01. 1987 Amd 1 Sampling and Analysis of Iron, Steel and Other Ferrous Metals Part 3: Methods of Analysis Section 3.26: Determination of Silicon Subsec 3.26.3: Steel and Cast Iron: Spectrophotometric Method for Silicon Contents from 0.05%. 16 pp.
BSI BS 6200: SUB SEC 3.26.4-89. 1989 Sampling and Analysis of Iron, Steel and Other Ferrous Metals Part 3: Meth of Analysis Sec 3.26: Detn of Silicon Subsec 3.26.4: Steel and Cast Iron: Spectrophotometric Meth for Si Contents from 0.01% to 0.05% (m/m) (ISO 4829-2: 1988). 8 pp.
BSI BS 6200: SUB SEC 3.26.4-01. 1989 Amd 1 Sampling and Analysis of Iron, Steel and Other Ferrous Metals Part 3: Methods of Analysis Section 3.26: Determination of Silicon Subsection 3.26.4: Steel and Cast Iron: Spectrophotometric Method for Silicon Contents from 0.01%. 13 pp.
CEN EN 24 829-1-90. Steel and Cast Iron—Determination of Total Silicon Content—Reduced Molyoldosilicate Spectrophotometric Method—Part 1: Silicon Content Between 0-05 and 1. 2 pp.
CEN EN 24 829-2-90. Steel and Cast Iron—Determination of Total Silicon Content—Reduced Molyoldosilicate Spectrophotometric Method—Part 2: Silicon Content Between 0-01 and 0-05. 2 pp.
DIN ENGL EN 24829 Pt 1-92. Determination of Total Silicon Content of Steel and Cast Iron by the Reduced Molybdosilicate Spectrophotometric Method; Silicon Contents Between 0,05 and 1,0 %; (ISO 4829-1: 1986) (Oct). 10 pp.
DIN ENGL EN 24829 Pt 2-92. Determination of Total Silicon Content of Steel and Cast Iron by the Reduced Molybdosilicate Spectrophotometric Method; Silicon Contents Between 0,01 and 0,05 %; (ISO 4829-2: 1988) (Oct). 8 pp.
ISO 4829 Pt 1-86. Steel and Cast Iron—Determination of Total Silicon Content—Reduced Molybdosilicate Spectrophotometric Method—Part 1: Silicon Contents Between 0,05 and 1,0 % First Edition. 9 pp.
ISO 4829 Pt 2-88. Steel and Iron—Determination of Total Silicon Content—Reduced Molybdosilicate Spectrophotometric Method—Part 2: Silicon Contents Between 0,01 and 0,05% First Edition. 8 pp.

—Sulfur Content—Gravimetric Analysis
BSI BS 6200: SUB SEC 3.28.1-85. 1985 Sampling and Analysis of Iron, Steel and Other Ferrous Metals Part 3: Methods of Analysis Section 3.28: Determination of Sulphur Subsection 3.28.1: Steel and Cast Iron: Gravimetric Method. 8 pp.
BSI BS 6200: SUB SEC 3.28.1-01. 1985 Amd 1 Sampling and Analysis of Iron, Steel and Other Ferrous Metals Part 3: Methods of Analysis Sec 3.28: Determination of Sulphur Subsec 3.28.1: Steel and Cast Iron: Gravimetric Method (ISO 4934: 1980) (AMD 7079) February 28, 1992 (V). 12 pp.
CEN EN 24 934-89. Steel and Cast Iron Determination of Sulfur Content Gravimetric Method. 2 pp.
DIN ENGL EN 24934-90. Determination of Sulfur Content of Steel and Cast Iron by the Gravimetric Method (ISO 4934: 1980) (Apr). 8 pp.
ISO 4934-80. Steel and Cast Iron—Determination of Sulphur Content—Gravimetric Method First Edition; (Erratum—Dec 1981). 8 pp.

—Sulfur Content—Volumetric Analysis
ISO 671-82. Steel and Cast Iron—Determination of Sulphur Content—Combustion Titrimetric Method First Edition. 7 pp.

—Tin Content—Volumetric Analysis
BSI BS 6200: SUB SEC 3.31.1-92. 1992 Sampling and Analysis of Iron, Steel and Other Ferrous Metals Part 3: Methods of Analysis Section 3.31: Determination of Tin Subsection 3.31.1: Steel and Cast Iron: Volumetric Method. 8 pp.

Cast Iron (Cont.)

—Vanadium Content—Potentiometric Analysis
BSI BS 6200: SUB SEC 3.34.1-87. 1987 Sampling and Analysis of Iron, Steel and Other Ferrous Metals Part 3: Methods of Analysis Section 3.34: Determination of Vanadium Subsection 3.34.1: Steel and Cast Iron: Potentiometric Titration Method. 8 pp.
BSI BS 6200: SUB SEC 3.34.1-01. 1987 Amd 1 Sampling and Analysis of Iron, Steel and Other Ferrous Metals Part 3: Methods of Analysis Sec 3.34: Determin of Vanadium Subsec 3.34.1: Steel and Cast Iron: Potentiometric Titration Method (ISO 4947: 1986) (AMD 7081) February 28, 1992 (V). 13 pp.
DIN ENGL EN 24947-92. Determination of Vanadium Content of Steel and Cast Iron by Potentiometric Titration (ISO 4947: 1986) (Nov). 9 pp.
ISO 4947-86. Steel and Cast Iron—Determination of Vanadium Content—Potentiometric Titration Method First Edition. 8 pp.

—X-Ray Fluorescence Spectrometry
CNS G2182-84. Method for Fluorescent X-Ray Fluorescence Spectrochemical Analysis of Pig Iron, Cast Iron, Carbon Steel and Low Alloy Steel (Oct)(10502).

Cast Iron Boilers
See Also: Boilers; Cast Iron
BSI BS 779-89. 1989 Cast Iron Boilers for Central Heating and Indirect Hot Water Supply (Rated Output 44 kW and Above). 16 pp.
BSI BS 779-01. 1989 Amd 1 Cast Iron Boilers for Central Heating and Indirect Hot Water Supply (Rated Output 44 kW and Above) (AMD 7704) July 15, 1993 (Q). 18 pp.
BSI BS 4433: Part 1-73. 1973 Amd 2 Solid Smokeless Fuel Boilers with Rated Outputs up to 45 kW Part 1: Boilers with Undergrate Ash Removal. 21 pp.
BSI BS 4433: Part 2-69. 1969 Amd 1 Solid Smokeless Fuel Boilers with Rated Outputs up to 45 kW Part 2: Gravity Feed Boilers Designed to Burn Small Anthracite. 28 pp.

—Construction
JIS B 8203-78. Construction of Cast Iron Boilers. 13 pp.

Cast Iron Coatings (On Cast Iron)
See Also: Cast Iron; Metal Coatings (On Metal)

—Vitreous Enamels
BSI BS 1344: Part 13-75. 1975 Methods of Testing Vitreous Enamel Finishes Part 13: Production of Specimens for Testing Coatings on Cast Iron. 7 pp.
ISO 2724-73. Vitreous and Porcelain Enamels for Cast Iron-Production of Specimens for Testing First Edition. 4 pp.

—Zinc—Hot Dip
ISO 1459-73. Metallic Coatings—Protection Against Corrosion by Hot Dip Galvanizing—Guiding Principles First Edition. 4 pp.

Cast Iron Pipes
See Also: Iron Pipes; Pipes
BSI BS 2035-66. 1966 Amd 1 Cast Iron Flanged Pipes and Flanged Fittings. 46 pp.
CEN PREN 877-1-92. Cast Iron Pipes and Fittings, Their Joints and Accessories for the Evacuation of Water from Buildings—Part 1: Technical Specifications. 25 pp.
CNS B7023-67. Method of Test for Cast-Iron Pipes and Fittings (Aug)(788).
SAA AS 1631-74. Cast Iron Non-Pressure Pipes and Pipe Fittings. 19 pp.

—Cement Mortars
CNS A3055-71. Method of Test for Cement Mortar of Cast Iron Pipes (Dec)(2313).

—Drainpipes
BSI BS 437-78. 1978 Amd 1 Cast Iron Spigot and Socket Drain Pipes and Fittings. 42 pp.
ISO 6594-83. Cast Iron Drainage Pipes and Fittings—Spigot Series First Edition; (Corrected and Reprinted -1985). 14 pp.

—Pressure
CNS B5020-82. Cast-Iron Pipes for Pressure Lines with TYTON Joints; Class LA (Mar)(830).
CNS B5021-82. Cast-Iron Pipes for Pressure Lines with TYTON Joints; Class LA (Mar)(831).
CNS B5022-82. Cast-Iron Pipes for Pressure Lines with TYTON Joints; Class B (Mar)(832).
DIN ENGL 28500-77. Cast Iron Pressure Pipes and Special Castings; Technical Conditions of Delivery (Aug). 4 pp.

—Pressure—Flanged
CNS B5023-82. Ductile Cast-Iron Flanged Piper for Pressure Main Lines (Jan)(833).

INTERNATIONAL AND NON-U.S. NATIONAL STANDARDS SUBJECT INDEX

Cast

Cast Iron Pipes (Cont.)

—Pressure—Flanged (Cont.)
CNS B5024-59. Flanged Cast-Iron Pipes, Centrifugal Casting, Class A (Working Pressure Less Than 10 kg/cm2) (Sep)(834).
CNS B5025-59. Flanged Cast-Iron Pipes, Centrifugal Casting, Class B (Working Pressure Less Than 16 kg/cm2) (Sep)(835).

—Pressure—Gas
BSI BS 1211-58. (OBSOLESCENT) 1958 Centrifugally Cast (Spun) Iron Pressure Pipes for Water, Gas and Sewage (Partially Superseded by BS 4772). 26 pp.

—Pressure—Sewer
BSI BS 1211-58. (OBSOLESCENT) 1958 Centrifugally Cast (Spun) Iron Pressure Pipes for Water, Gas and Sewage (Partially Superseded by BS 4772). 26 pp.

—Pressure—Water
BSI BS 1211-58. (OBSOLESCENT) 1958 Centrifugally Cast (Spun) Iron Pressure Pipes for Water, Gas and Sewage (Partially Superseded by BS 4772). 26 pp.

—Quality Assurance
CEN PREN 877-2-92. Cast Iron Pipes and Fittings, Their Joints and Accessories for the Evacuation of Water from Buildings—Part 2: Testing and Quality Control. 25 pp.

—Sewer
BSI BS 416-73. (WITHDRAWN) 1973 Amd 1 Cast Iron Spigot and Socket Soil, Waste and Ventilating Pipes (Sand Cast and Spun) and Fittings (Superseded by BS 416: Parts 1&2: 1990). 38 pp.
BSI BS 416: Part 1-90. 1990 Discharge and Ventilating Pipes and Fittings, Sand-Cast or Spun in Cast Iron Part 1: Specification for Spigot and Socket Systems. 28 pp.
BSI BS 416: Part 2-90. 1990 Discharge and Ventilating Pipes and Fittings, Sand-Cast or Spun in Cast Iron Part 2: Specification for Socketless Systems. 30 pp.

—Soil
BSI BS 416-73. (WITHDRAWN) 1973 Amd 1 Cast Iron Spigot and Socket Soil, Waste and Ventilating Pipes (Sand Cast and Spun) and Fittings (Superseded by BS 416: Parts 1&2: 1990). 38 pp.
BSI BS 416: Part 1-90. 1990 Discharge and Ventilating Pipes and Fittings, Sand-Cast or Spun in Cast Iron Part 1: Specification for Spigot and Socket Systems. 28 pp.
BSI BS 416: Part 2-90. 1990 Discharge and Ventilating Pipes and Fittings, Sand-Cast or Spun in Cast Iron Part 2: Specification for Socketless Systems. 30 pp.
CSA CAN/CSA-B70-M91. Cast Iron Soil Pipe, Fittings, and Means of Joining; (Gen Instr 1). 27 pp.

—Taps (Valves)
DIN ENGL 3543 Pt 1-84. Metal Tapping Valves; Requirements, Testing (Aug). 1 p.

—Vent
BSI BS 416-73. (WITHDRAWN) 1973 Amd 1 Cast Iron Spigot and Socket Soil, Waste and Ventilating Pipes (Sand Cast and Spun) and Fittings (Superseded by BS 416: Parts 1&2: 1990). 38 pp.
BSI BS 416: Part 1-90. 1990 Discharge and Ventilating Pipes and Fittings, Sand-Cast or Spun in Cast Iron Part 1: Specification for Spigot and Socket Systems. 28 pp.
BSI BS 416: Part 2-90. 1990 Discharge and Ventilating Pipes and Fittings, Sand-Cast or Spun in Cast Iron Part 2: Specification for Socketless Systems. 30 pp.

Cast Steels
Use: Steel Castings

Cast Stone
See Also: Concrete Blocks; Concrete Slabs; Rocks
BSI BS 1217-86. 1986 Cast Stone. 8 pp.
BSI BS 1217-75. 1975 Cast Stone. 8 pp.

Castellated Nuts
Use: Castle Nuts

Casters
Use For: Castors

—Chairs (Seats)
JIS S 1038-88. Castors for Office Chairs. 14 pp.

—Furniture—Glossaries
ISO 8555 Pt 8-87. Hardware for Furniture—Terms for Furniture Fittings—Part 8: Castors and Glides First Edition. 8 pp.

—Industrial
JIS B 8923-84. Industrial Castors. 17 pp.

Casters (Cont.)

—Industrial (Cont.)
SAA AS B282-68. Industrial Castors Amdt 1 February 1971. 14 pp.
SAA AS B300-71. Industrial Wheels for Castors and Mobile Equipment. 16 pp.
SAA AS 1961-77. Industrial Wheels and Castors (Dimensions and Capacities) (Based on ISO Recommendations). 8 pp.

—Industrial—Glossaries
ISO 2163-75. Industrial Trucks—Wheels and Castors—Vocabulary First Edition. 12 pp.

—Industrial—Off-Set
ISO 3102-81. Wheels and Castors for Non-Powered Equipment—off-Set for Swivel Castors Second Edition. 3 pp.

—Industrial—Top Plates
ISO 2184-72. Industrial Castors—Dimensions of Top-Plates—Part I: Oblong Top-Plates with 4 Bolt Holes First Edition. 4 pp.
ISO 3101-81. Wheels and Castors—Triangular Top Plates with Three Fixing Holes Second Edition. 4 pp.

—Medical Equipment
BSI BS 2099-89. 1989 Amd 1 Castors for Hospital Equipment (AMD 6252) November 30, 1989. 13 pp.

—Swivel—Chairs (Seats)
CGSB CAN/CGSB-44.119-M89. Chair Swivel Casters for Carpeted or Hard Surface Floors (Supercedes 44-GP-120). 13 pp.
CGSB CAN/CGSB-44.226-M91. Dual Wheel Swivel Casters for Carpeted or Hard-Surfaced Floors; (Amendment 1 September 1992). 14 pp.

Castings
Scope Note: For additional listings, use a more specific term **See Also:** Aluminum Bronze Castings; Aluminum Castings; Austenitic Iron Castings; Brass Castings; Bronze Castings; Carbon Steel Castings; Cast Iron; Centrifugal Castings; Chill Castings; Chromium Molybdenum Steel Castings; Cobalt Castings; Copper Alloy Castings; Copper Castings; Corrosion Resistant Castings; Corrosion Resistant Steel Castings; Die Castings; Ductile Iron Castings; Ferritic Steel Castings; Ferroalloy Castings; Gold Castings; Gunmetal Castings; Heat Resistant Castings; Heat Resistant Steel Castings; Investment Castings; Iron Castings; Lead Castings; Leaded Bronze Castings; Leaded Tin Bronze Castings; Low Alloy Steel Castings; Magnesium Castings; Manganese Steel Castings; Metal Products; Microstructures; Moldings; Molds (Casting); Nickel Castings; Nickel Chromium Molybdenum Niobium Castings; Nickel Chromium Molybdenum Steel Castings; Nickel Steel Castings; Phosphor Bronze Castings; Refractory Metal Castings; Sand Castings; Silicon Bronze Castings; Silicon Castings; Silver Castings; Stainless Steel Castings; Steel Castings; Tin Bronze Castings; Titanium Castings; Zinc Castings
BSI BS 6615-85. 1985 Dimensional Tolerances for Metal and Metal Alloy Castings. 8 pp.
DIN ENGL 1690 Pt 1-85. Technical Delivery Conditions for Castings Made from Metallic Materials; General Conditions (May). 7 pp.
ISO 8062-84. Castings—System of Dimensional Tolerances First Edition. 7 pp.
JIS B 0403-87. Castings-System of Dimensional Tolerances (ISO 8062-1984). 14 pp.

—Aerospace—Radiography
AECMA PREN2002-21-87. Test Methods for Metallic Materials—Part 21—Radiographic Testing of Cast Components. 3 pp.

—Defects—Glossaries
BSI BS 2737-56. 1956 Terminology of Internal Defects in Castings as Revealed by Radiography. 35 pp.

—Impregnating
MOD UK DSTAN 03-1-79. Impregnation of Porous Castings Issue 2. 13 pp.

—Liquid Penetrant Testing
CNS Z8054-85. Liquid Penetrant Test for Castings (Mar)(11225).

—Machine Components
JIS B 0703-87. Roundness of Castings. 14 pp.

—Magnetic Particle Testing
CNS Z8057-85. Magnetic Particle Testing for Castings and Forgings (Sep)(11377).

—Radiography
CNS Z8059-85. Radio Graphic Testing for Castings (Sep)(11379).

Castle Nuts
Use For: Castellated Nuts **See Also:** Hexagonal Nuts; Nuts (Fasteners); Slotted Nuts
CNS B2262-83. Castle Nuts (Crown Nuts) (Feb)(4469).
JIS B 1170-87. Hexagon Slotted and Castle Nuts. 18 pp.
MOD UK DSTAN 53-49-72. Nuts, Slotted and Castellated, Hexagon, Steel, UNF Issue 1. 9 pp.
MOD UK DSTAN 53-50-72. Nuts, Slotted and Castellated, Hexagon, Steel, UNC Issue 1. 8 pp.
SNZ NZS 1069-66. Specification for Precision Hexagon Bolts, Screws and Nuts (B.S.W. and B.S.F. Threads). 32 pp.

—Aerospace
BSI A 242-A 245-74. 1974 Hexagonal Castle Nuts (of Class 3B UNJ Thread). 5 pp.
BSI A 246-A 249-74. 1974 Slotted Hexagonal Thick Nuts (of Class 3B UNJ Thread). 5 pp.
CEN PREN 2868-91. Nuts, Hexagonal, Slotted/Castellated, Normal Height, Normal Across Flats, in Heat Resisting Steel, Silver Plated Classification: 1100 MPa (at Ambient Temperature)/650 Degrees C. 6 pp.
CEN PREN 2868-93. Aerospace Series Nuts, Hexagonal, Slotted/Castellated, Normal Height, Normal Across Flats, in Heat Resisting Steel, Silver Plated Classification: 1 100 MPa (at Ambient Temperature) /650 Degrees C. 5 pp.
CEN PREN 2869-91. Nuts, Hexagonal, Slotted/Castellated, Normal Height, Normal Across Flats, in Heat Resisting Steel, Passivated Classification: 1100 MPa (at Ambient Temperature) /650 Degrees C. 6 pp.
CEN PREN 2869-93. Aerospace Series Nuts, Hexagonal, Slotted/Castellated, Normal Height, Normal Across Flats, in Heat Resisting Steel, Passivated Classification: 1 100 MPa (at Ambient Temperature) /650 Degrees C. 5 pp.
DIN ENGL LN 9345-80. Aerospace; Nuts, Castle (Aug). 3 pp.
ISO 4147-83. Metric Fasteners for Aerospace Construction—Hexagon Slotted (Castellated) Nuts—Strength Classification 1 100 MPa—Maximum Operating Temperature 235 Degrees Celsius Second Edition. 5 pp.

—Aerospace—Locking
CEN PREN 3434-91. Nuts, Hexagon, Slotted, Self-Locking, in Steel, Cadmium Plated, MoS2 Lubricated Classification: 900 MPa (at Ambient Temperature) /235 Degrees C. 6 pp.

—Aerospace—Shear
BSI A 250-A 253-74. 1974 Slotted Hexagonal Shear Nuts (of Class 3B UNJ Thread). 3 pp.

—Aircraft
AECMA PREN2415-82. Nuts, Hexagon, Slotted/Castellated, Thin, in Steel, Cadmium Plated—Classification 900 MPa/235 Degrees Celsius (C3/A49). 6 pp.
AECMA PREN2809-87. Nuts, Hexagon, Slotted/Castellated, Thin Normal Across Flats, in Heat Resisting Steel Silver Plated Classification 600 MPa/650 Degrees Celsius. 5 pp.
AECMA PREN2810-87. Nuts, Hexagon, Slotted/Castellated, Thin Normal Across Flats, in Heat Resisting Steel Passivated Classification 600 MPa/650 Degrees Celsius. 5 pp.
AECMA PREN2811-87. Nuts, Hexagon, Slotted/Castellated in Steel Cadmium Plated Classification 1100 MPa/235 Degrees Celsius. 5 pp.
BSI 3A 107-62. 1962 Amd 1 Aluminium Alloy Nuts (Unified Hexagons and Unified Threads) (Ordinary and Slotted) for Aircraft. 5 pp.
SBAC AS 2113 ISSUE 1. Nut Slotted. 1 p.

—Aircraft—Corrosion Resistant
BSI 2A 24-62. 1962 Amd 2 Corrosion-Resisting Steel Hexagon Nuts (B.A. and B.S.F. Threads) (Ordinary, Thin, Slotted and Castle) for Aircraft (AMD 6066) May 31, 1990. 8 pp.
BSI 3A 105-62. 1962 Corrosion-Resisting Steel Nuts (Unified Hexagons and Unified Threads) (Ordinary, Thin, Slotted and Castle) for Aircraft. 6 pp.

—Aircraft—Thin
BSI 3A 103-62. 1962 Amd 3 Steel Nuts (Unified Hexagons and Unified Threads) (Ordinary, Thin, Slotted and Castle) for Aircraft. 9 pp.

—Automotive—Flat Widths
DIN ENGL 70617-90. Hexagon Slotted Nuts with Small Widths Across Flats for Use in Automotive Engineering (Not Intended for New Designs) (Mar) (This Standard, Together with DIN 70618, March 1990 Edition, Supersedes DIN 70617/DIN 70618, February 1964 Edition). 3 pp.

INTERNATIONAL AND NON-U.S. NATIONAL STANDARDS
SUBJECT INDEX

Castle Nuts (Cont.)
—Automotive—Thin—Flat Widths
DIN ENGL 70618-90. Hexagon Thin Slotted Nuts with Small Widths Across Flats for Use in Automotive Engineering (Not Intended for New Designs) (Mar) (This Standard, Together with DIN 70617, March 1990 Edition, Supersedes DIN 70617/DIN 70618, February 1964 Edition). 3 pp.

—Coarse Pitch Thread
DIN ENGL 935 Pt 1-87. Hexagon Slotted Nuts and Castle Nuts with Metric Coarse and Fine Pitch Thread; Product Grades A and B (Oct). 6 pp.
DIN ENGL 935 Pt 3-87. Hexagon Slotted Nuts with Metric Coarse Pitch Thread; Product Grade C (Oct). 4 pp.
DIN ENGL 979-87. Hexagon Thin Slotted Nuts and Castle Nuts with Metric Coarse and Fine Pitch Thread; Product Grades A and B (Oct). 5 pp.
MOD UK DSTAN 53-26-72. Nuts, Slotted and Castellated, Hexagon, Steel, Metric (Coarse Pitch Precision Nuts) Issue 2. 8 pp.

—Fine Pitch Thread
DIN ENGL 935 Pt 1-87. Hexagon Slotted Nuts and Castle Nuts with Metric Coarse and Fine Pitch Thread; Product Grades A and B (Oct). 6 pp.
DIN ENGL 979-87. Hexagon Thin Slotted Nuts and Castle Nuts with Metric Coarse and Fine Pitch Thread; Product Grades A and B (Oct). 5 pp.

—Flat Widths
CNS B2264-83. Castle Nuts with Small Widths Across Flats (Feb)(4471).

—Round Threads
CNS B2368-79. Hexagon and Castle Nuts with Round Threads (for Motive Power Unit Turnbuckles and Drawbars) (Apr)(4772).

—Thin
CNS B2263-83. Castle Nuts Thin Type (Feb)(4470).
DIN ENGL 937-83. Hexagon Thin Castle Nuts; (Previous Design) (Dec). 4 pp.
DIN ENGL 979-87. Hexagon Thin Slotted Nuts and Castle Nuts with Metric Coarse and Fine Pitch Thread; Product Grades A and B (Oct). 5 pp.

Castor Oil
Use For: Hardened Castor Oils See Also: Vegetable Oils
CNS K5006-82. Castor Oil for Industrial Use (Feb)(118). 2 pp.
CNS K5111-81. Crude Castor Oil (Sep)(7422).
CNS K6011-63. Method of Test for Castor Oil for Industrial Use (Jun)(119). 4 pp.
MOD UK DSTAN 91-41-78. Castor Oil, Dehydrated, for Armament Stores Issue 1. 5 pp.
MOD UK DSTAN 91-58-82. Castor Oil, Hardened Issue 1. 5 pp.

—Aircraft
MOD UK DTD-71B-65. Oil, Castor: Cold Drawn Grade. 1 p.
MOD UK DTD-72A-01. Oil, Castor: Delayed Setting Time Grade Joint Service Designation: OF-300; Amendment 1. 1 p.

—Hydroxyl Value
MOD UK M 9530/66. Examination of Oil, Caster, Hardened.

—Iodine Number
MOD UK M 9530/66. Examination of Oil, Caster, Hardened.

—Melting Points
MOD UK M 9530/66. Examination of Oil, Caster, Hardened.

—Saponification Number
MOD UK M 9530/66. Examination of Oil, Caster, Hardened.

—Visual Inspection
MOD UK M 9530/66. Examination of Oil, Caster, Hardened.

—Weapons
MOD UK DSTAN 68-175-93. Oil, Castor, LF Quality Issue 1. 14 pp.
MOD UK CS 3034. Oil, Caster, LF Quality (Superseded by Def Stan 68-175).

Castor Oil Acid
Use: Ricinoleic Acid

Castor Oil Content Analysis
See Also: Castor Oil Seed Husk Content Analysis; Oil Content Analysis

Castor Oil Content Analysis (Cont.)
—Rubber—Gas Chromatography
ISO 6225 Pt 2-90. Rubber, Raw, Natural—Determination of Castor Oil Content—Part 2: Determination of Total Ricinoleic Acid Content by Gas Chromatography Second Edition. 7 pp.

—Rubber—Thin Layer Chromatography
BSI BS 6603-85. (WITHDRAWN) 1985 Method for Determination of Castor Oil Glycerides Content of Raw, Natural Rubber by Thin Layer Chromatography. 7 pp.
ISO 6225 Pt 1-84. Rubber, Raw, Natural—Determination of Castor Oil Glycerides Content—Thin Layer Chromatographic Method First Edition. 5 pp.

Castor Oil Seed Husk Content Analysis
See Also: Castor Oil Content Analysis

—Animal Feed—Microscopic Analysis
BSI BS 5766: Part 9-83. 1983 Method for Analysis of Animal Feeding Stuffs Part 9: Determination of Castor Oil Seed Husks. 9 pp.
ISO 5061-83. Animal Feeding Stuffs—Determination of Castor Oil Seed Husks—Microscopical Method First Edition. 8 pp.

Castors
Use: Casters

Casualties
See Also: Chemical Agent Casualties; Field Hospitals; Hospitals; Military Personnel; Nuclear Warfare Casualties

—Management—Medical Services
NATO STANAG 2879 ED 2 AMD 2-86. Principles of Medical Policy in the Management of a Mass Casualty Situation. 14 pp.

Casualty Bags
—Inflatable
MOD UK DSTAN 42-18-74. Inflatable Casualty Bag Complete with Stowage Pack Issue 1. 3 pp.

Catalina Aircraft
—Antenna Positions
CAA. Catalina. 1 p.

Cataloging
See Also: Catalogs

—Bibliographic References—Control Characters
ISO 6630-86. Documentation—Bibliographic Control Characters First Edition. 8 pp.
JTC1 6630-86. Documentation—Bibliographic Control Characters First Edition. 8 pp.
OSI ISO 6630-86. Documentation—Bibliographic Control Characters. 8 pp.

Catalogs
See Also: Cataloging

—Aeronautical Navigation Charts
ICAO 7101. Aeronautical Chart Catalogue Twenty-Seventh Edition—January 1992. 292 pp.

—European File
EURO 1989-18. Catalogue 1979-89. 9 pp.
EURO 1990-15. Catalogue 1979-90. 10 pp.

—Interchangeable Ammunition
NATO STANAG 2928 ED 2 AMD 3-81. Land Forces Ammunition Interchangeability Catalogue in Wartime—AOP-6. 9 pp.

—Mine Warfare—Computer Programs
NATO STANAG 1164 ED 2 AMD 2-81. Catalogue of Mine Warfare Computer Programs—AMP-12(Navy). 8 pp.

—Standards
CGSB. 1993 Catalogue. 134 pp.
CNS. CNS Catalog 1993. 358 pp.
DIN ENGL Catalogue-92. English Translations of German Standards Catalogue 1992. 536 pp.
DIN VDE Catalogue-87. List of VDE Specifications, VDE Guidelines Including All DIN VDE Standards and Accompanying Information Sheets and Drafts Thereof Issued by the Association of German Electrical Engineers (VDE) (Dec). 195 pp.
EURO 1989-01 ED 2. CEN Catalogue: Edition 2. 152 pp.
EURO 1990 CEN Catalogue. CEN Catalogue. 123 pp.
EURO 1991 CEN Catalogue. CEN Catalogue. 108 pp.
EURO 1991 LIST. CEN List of Draft Standards. 181 pp.

Catalogs (Cont.)
—Standards (Cont.)
EURO 1992 CEN Catalogue. CEN Catalogue. 104 pp.
EURO 1988 CENE Catalogue. CENELEC Catalogue. 169 pp.
EURO 1989 CENE Catalogue. CENELEC Catalogue. 200 pp.
EURO 1990 CENE Catalogue. CENELEC Catalogue. 241 pp.
EURO 1991 CENE Catalogue. Cenelec Catalogue. 254 pp.
EURO 1992 CENE Catalogue. CENELEC Catalogue. 220 pp.
EURO 1992 CENE Catalogue S1. CENELEC Catalogue Supplement 1. 50 pp.
EURO 1993 CENELEC CATALOGUE. CENELEC Catalogue. 250 pp.
ISO. ISO Catalogue 1993; (Supplement 1-1993). 1004 pp.
OSI ISO TP 1989-89. Technical Programme. 217 pp.
SNZ NZMP 100-93. New Zealand Standards Catalogue 1993. 200 pp.

Catalytic Converters
See Also: Air Pollution; Exhaust Emission Control Equipment

—Automotive
CNS D3059-80. Test Code of Oxidation Catalytic Converters for Automobiles (Aug)(6163).

Cataphoresis
Use: Electrophoresis

Catechol
Use: Pyrocatechol

Catering
See Also: Food Preparation Equipment

—Service Contracts
MOD UK DEFCON 112T-83. Conditions of Contract for Catering Services 9/83. 12 pp.

Catering Equipment
Use: Food Processing Equipment; Food Service Equipment

Caterpillars (Vehicles)
See Also: Crawler Tractors; Earthmoving Equipment; Excavating Equipment; Tanks (Combat Vehicles); Tracked Vehicles

—Crankcase Oils
CNS K6959-89. Method of Test for Evaluating the Performance of Crankcase Lubricants with Single Cylinder Engine-Caterpillar 1H2 Test Method (Jan)(12482).
CNS K6960-89. Method of Test for Evaluating the Performance of Crankcase Lubricants with Single Cylinder Engine-Caterpillar 1G2 Test Method (Jan)(12483).

Catgut
—Sutures
JIS T 4102-76. Surgical Sutures Catgut (Confirmed 1986). 9 pp.

Catheters
See Also: Cannulas; Medical Equipment; Shunts (Medical)

—Anesthesia—Mountings
BSI BS 7143-89. 1989 Catheter Mounts (Flexible Adaptors) for Use with Medical Breathing Systems. 12 pp.
BSI BS 7143-01. 1989 Amd 1 Catheter Mounts (Flexible Adaptors) for Use with Medical Breathing Systems (AMD 7063) June 15, 1992. 13 pp.

—Epidural
BSI BS 6196-89. 1989 Sterile Epidural Catheters and Introducer Needles for Single Use. 16 pp.

—Intravascular
BSI BS 5616-81. (WITHDRAWN) 1981 Sterile Intravascular Catheters for Single Use (Superseded by BS 7174: Parts 1, 2, 3 and 4: 1990). 10 pp.
BSI BS 7174: Part 1-90. 1990 Sterile Intravascular Catheters and Ancillary Devices for Single Use Part 1: General Requirements. 15 pp.
BSI BS 7174: Part 2-90. 1990 Sterile Intravascular Catheters and Ancillary Devices for Single Use Part 2: Specification for Angiographic Catheters. 11 pp.
BSI BS 7174: Part 3-90. 1990 Sterile Intravascular Catheters and Ancillary Devices for Single Use Part 3: Specification for Central Venous Catheters. 10 pp.
BSI BS 7174: Part 4-90. 1990 Sterile Intravascular Catheters and Ancillary Devices for Single Use Part 4: Introducer Needles, Introducer Cannulae, Guide Wires, Dilators, Valves and Connectors. 19 pp.

INTERNATIONAL AND NON-U.S. NATIONAL STANDARDS
SUBJECT INDEX

Catheters

Catheters (Cont.)
—**Intravascular** (Cont.)
BSI BS 7174: Part 5-90. 1990 Sterile Intravascular Catheters and Ancillary Devices for Single Use Part 5: Balloon Dilatation Catheters. 10 pp.

—**Standard Scales**
CNS T1009-80. Standard Scale for Medical Bougie and Catheter (Jul)(5939).
JIS T 4202-53. Standard Scale for Medical Bougie and Catheter.

—**Suction—Respiratory**
BSI BS 7213-89. 1989 Suction Catheters for Use in the Respiratory Tract (ISO 8836: 1988). 12 pp.
ISO 8836-88. Suction Catheters for Use in the Respiratory Tract First Edition; (Corrected and Reprinted -1989). 12 pp.

—**Urethral**
BSI BS 1695: Part 1-90. 1990 Amd 1 Urological Catheters Part 1: Specification for Sterile, Single-Use Urethral Catheters of the Nelaton and Foley Types (AMD 6757) September 30, 1991. 28 pp.
BSI BS 1695: Part 2-90. 1990 Amd 1 Urological Catheters Part 2: Specification for Sterile, Single-Use Urethral Catheters of the Tiemann, Whistle Tip, 3-Way and Haematuria Types (AMD 6758) September 30, 1991. 35 pp.
CNS T1013-81. Urethral Catheters (Jul)(6272).
JIS T 3203-53. Urethral Catheters.

Cathode Copper
Use: Electrolytic Copper

Cathode Ray Oscilloscopes
Use: Oscilloscopes

Cathode Ray Storage Tubes
See Also: Cathode Ray Tubes; Direct View Storage Tubes; Electron Tubes
CECC CECC 11 101 ISSUE 1-84. Blank Detail Specification: Display Storage Tubes (En, Fr). 25 pp.
MOD UK DSTAN 59-60: 09/304-01. Valve, Electronic—Cathode Ray Tube Issue 1; Amendment 2. 72 pp.
MOD UK DSTAN 59-60: 90/076-80. Valve Electronic—Storage Display Cathode Ray Tube Issue 2. 42 pp.

—**Electrical Measurement**
IEC 151 Pt 24-71. Measurements of the Electrical Properties of Electronic Tubes Part 24: Methods of Measurement of Cathode-Ray Charge-Storage Tubes First Edition. 60 pp.

—**Quality Assurance**
BSI BS 9053-73. (WITHDRAWN) 1973 Rules for the Preparation of Detail Specifications for Electrical Charge Storage Display Cathode Ray Tubes of Assessed Quality. 19 pp.

Cathode Ray Tube Displays
Use For: CRT Displays; Video Display Tubes; Video Displays; Visual Display Units; Visual Displays
See Also: Cathode Ray Tubes; Display Devices; Head Up Displays; Multifunction Displays; Radarscopes; Video Display Terminals

—**Emissions**
ECMA ECMA 172-92. Procedure for Measurement of Emissions of Electric and Magnetic Fields from VDU's from 5 Hz to 400 kHz. 17 pp.

—**Monitors**
CECC CECC 11 001-012 ISSUE 2-85. BS CECC 11 001-012; Display Monitor Cathode Ray Tube Assembly (En). 35 pp.

—**Projection**
CECC CECC 11 001-020 ISSUE 1-83. BS CECC 11 001-020; Projection Cathode Ray Tube (En). 20 pp.

Cathode Ray Tube Screens
Use: Cathode Ray Tubes

Cathode Ray Tubes
Use For: CRT See Also: Cathode Ray Storage Tubes; Cathode Ray Tube Displays; Direct View Storage Tubes; Display Devices; Electron Tubes; Image Intensifiers; Oscillators; Picture Tubes; Radarscopes; Thermionic Tubes
CECC CECC 11 000 ISSUE 1-80. Generic Specification: Cathode Ray Tubes (En, Fr, Ge) AMD 1 (En, Fr, Ge). 187 pp.
CECC CECC 11 001 ISSUE 1-86. Blank Detail Specification: Cathode Ray Tubes (En, Fr, Ge). 30 pp.
MOD UK DSTAN 59-60: Part 9-01. Valves, Electronic (Electronic Tubes) (Listed on EPIC Database) Part 9: Cathode Ray Tubes Issue 1; Amendment 7. 29 pp.

Cathode Ray Tubes (Cont.)
MOD UK DSTAN 59-60: Pt 9:Sec E-01. Valves, Electronic (Electronic Tubes) (Listed on EPIC Database) Part 9: Cathode Ray Tubes Section E: Detail Specifications Issue 1; Amendment 1. 4 pp.
MOD UK DSTAN 59-60: 09/305-01. Valve Electronic, Cathode Ray Tube Issue 1; Amendment 1. 24 pp.
MOD UK DSTAN 59-60: 09/305-02. Valve Electronic, Cathode Ray Tube Issue 1; Amendment 3. 28 pp.
MOD UK DSTAN 59-60: 90/002-69. Valve, Electronic, Cathode Ray Tube Issue 1. 14 pp.
MOD UK DSTAN 59-60: 90/003-70. Valve, Electronic, Cathode Ray Tube, Rectangular Face Issue 1. 8 pp.
MOD UK DSTAN 59-60: 90/005-70. Valve Electronic, Cathode Ray Tube Issue 1. 12 pp.
MOD UK DSTAN 59-60: 90/010-70. Valve, Electronic, Cathode Ray Tube Issue 1. 6 pp.
MOD UK DSTAN 59-60: 90/018-70. Electronic Valve, Cathode Ray Tube Issue 1. 17 pp.
MOD UK DSTAN 59-60: 90/023-69. Valve, Electronic, Cathode Ray Tube Issue 1. 12 pp.
MOD UK DSTAN 59-60: 90/026-69. Valve, Electronic, Cathode Ray Tube Issue 1. 20 pp.
MOD UK DSTAN 59-60: 90/036-01. ELectronic Valve, Cathode Ray Tube Issue 1; Amendment 2. 12 pp.
MOD UK DSTAN 59-60: 90/075-01. Valve, Electronic, Cathode Ray Tube Issue 1; Amendment 1. 23 pp.
MOD UK DSTAN 59-60: 90/082-74. Valve, Electronic, Cathode Ray Tube Issue 1. 16 pp.
MOD UK DSTAN 59-60: 90/085-01. Valve, Electronic, Cathode Ray Tube Issue 1; Amendment 1. 21 pp.
SNZ NZSR 15-65. Use of Electronic Valves. 71 pp.

—**Colorimetric Analysis**
IEC 441-74. Photometric and Colorimetric Methods of Measurement of the Light Emitted by a Cathode-Ray Tube Screen First Edition. 26 pp.

—**Construction**
CSA CAN/CSA-C22. 2 NO 228-92. Cathode Ray Tubes (UL 1418); (Gen Instr 1 Thru 2). 48 pp.

—**Deflecting Electrodes**
IEC 236-74. Methods for the Designation of Electrostatic Deflecting Electrodes of Cathode-Ray Tubes Second Edition. 10 pp.

—**Glassware**
CSA CAN/CSA-C22. 2 NO 228-92. Cathode Ray Tubes (UL 1418); (Gen Instr 1 Thru 2). 48 pp.

—**Head Up Displays**
CECC CECC 11 001-021 ISSUE 1-83. BS CECC 11 001-021; Head-Up Display Cathode Ray Tube (En). 27 pp.
CECC CECC 11 001-027 ISSUE 1-84. BS CECC 11 001-027; Radar Display Cathode Ray Tube (En). 18 pp.

—**High Resolution—Miniature**
CECC CECC 11 001-011 ISSUE 2-85. BS CECC 11 001-011; Miniature High Resolution Cathode Ray Tube (En). 47 pp.
CECC CECC 11 001-037 ISSUE 1-85. BS CECC 11 001-037; Miniature High Resolution Cathode Ray Tube Package (En). 32 pp.
CECC CECC 11 001-038 ISSUE 1-86. BS CECC 11 001-038; Miniature High Resolution Cathode Ray Tube (En). 33 pp.
CECC CECC 11 001-041 ISSUE 1-87. BS CECC 11 001-041; Miniature High Resolution Cathode Ray Tube Package (En). 30 pp.
CECC CECC 11 001-044 ISSUE 1-90. BS CECC 11 001-044; Miniature High Resolution Cathode Ray Tube Package (En). 32 pp.

—**Identification Systems**
CSA CAN/CSA-C22. 2 NO 228-92. Cathode Ray Tubes (UL 1418); (Gen Instr 1 Thru 2). 48 pp.

—**Impact Testing**
CSA CAN/CSA-C22. 2 NO 228-92. Cathode Ray Tubes (UL 1418); (Gen Instr 1 Thru 2). 48 pp.

—**Photometry**
IEC 441-74. Photometric and Colorimetric Methods of Measurement of the Light Emitted by a Cathode-Ray Tube Screen First Edition. 26 pp.

—**Quality Assurance**
BSI BS 9050-71. 1971 Amd 3 Cathode Ray Tubes of Assessed Quality. Generic Data and Methods of Test. 81 pp.
BSI BS 9051-71. (WITHDRAWN) 1971 Rules for the Preparation of Detail Specifications for High Precision Professional Instrument Cathode Ray Tubes of Assessed Quality. 17 pp.

Cathode Ray Tubes (Cont.)
—**Quality Assurance** (Cont.)
BSI BS 9052-70. (WITHDRAWN) 1970 Rules for the Preparation of Detail Specifications for General Purposes Professional Instrument Cathode Ray Tubes of Assessed Quality. 13 pp.
BSI BS CECC 11000-80. 1980 Cathode Ray Tubes: Generic Specification. 57 pp.
BSI BS CECC 11000-01. 1980 Amd 1 Generic Specification: Cathode Ray Tubes (AMD 6004) July 15, 1992 (Renumbered as BS EN 111000: 1993) (T). 76 pp.
BSI BS CECC 11001-80. 1980 Cathode Ray Tubes: Blank Detail Specification. 12 pp.
BSI BS EN 111000-93. 1993 Amd 2 Generic Specification: Cathode Ray Tubes (AMD 7600) March 15, 1993 (T). 78 pp.

—**Test Specimens**
CSA CAN/CSA-C22. 2 NO 228-92. Cathode Ray Tubes (UL 1418); (Gen Instr 1 Thru 2). 48 pp.

—**Thermal Shock**
CSA CAN/CSA-C22. 2 NO 228-92. Cathode Ray Tubes (UL 1418); (Gen Instr 1 Thru 2). 48 pp.

Cathodes (Refinery Shapes)
Use: Refinery Shapes

Cathodic Coatings, Electroplated
Use: Electroplated Coatings

Cathodic Protection Equipment
See Also: Corrosion; Corrosion Prevention
BSI CP 1021-73. 1973 Cathodic Protection. 106 pp.

—**Buried Structures**
DIN ENGL 30676-85. Design and Application of Cathodic Protection of External Surfaces (Oct). 14 pp.

—**Cable Joints—Underground**
CCITT RECMN L.7-89. Application of Joint Cathodic Protection—Construction, Installation and Protection of Cable and Other Elements of Outside Plant (Study Group VI) 2 pp. 2 pp.

—**Metal Coatings—Buried Structures**
BSI BS 3900: Part F11-85. 1985 Methods of Test for Paints Group F: Durability Tests on Paint Films Part F11: Determination of Resistance to Cathodic Disbonding of Coatings for Use on Land-Based Buried Structures. 8 pp.

—**Metal Coatings—Marine**
BSI BS 3900: Part F10-85. 1985 Methods of Test for Paints Group F: Durability Tests on Paint Films Part F10: Determination of Resistance to Cathodic Disbonding of Coatings for Use in Marine Environments. 8 pp.

—**Metals**
BSI BS 7361: Part 1-91. 1991 Cathodic Protection Part 1: Code of Practice for Land and Marine Applications (Formerly CP 1021). 115 pp.
SAA AS 2832.2-91. Guide to the Cathodic Protection of Metals—Part 2: Compact Buried Structures Amdt 1 March 1993. 35 pp.
SAA AS 2832.3-92. Guide to the Cathodic Protection of Metals—Part 3: Fixed Immersed Structures Amdt 1 March 1993. 30 pp.

—**Metals—Marine**
BSI BS 7361: Part 1-91. 1991 Cathodic Protection Part 1: Code of Practice for Land and Marine Applications (Formerly CP 1021). 115 pp.

—**Naval Ships**
MOD UK NES 704: Part 1-90. Requirements for Cathodic Protection Part 1: Systems General Issue 3 (11.90). 12 pp.
MOD UK NES 704: Part 2-89. Cathodic Protection Part 2: Impressed Current Cathodic Protection System Issue 3 (07.89). 35 pp.
MOD UK NES 704: Part 3-88. Cathodic Protection Part 3: Sacrificial Anode or Self-Energized System Issue 3 (11.88). 23 pp.
MOD UK NES 704: Part 3-01. Requirements for Cathodic Protection Part 3: Sacrifical Anode or Self-Energized System Issue 4 (12.91); Amendment 1. 27 pp.
MOD UK NES 704: Part 4-01. Requirements for Cathodic Protection Part 4: Ships Fitting—Out, Ships in Refit or Laid—Up Issue 3 (08.90); Amendment 1. 16 pp.
MOD UK NES 704: Part 5-01. Requirements for Cathodic Protection Part 5: General Information on Bi—Metallic Couples Issue 3 (08.90); Amendment 1. 15 pp.

—**Steel Coatings**
CGSB 1-GP-71 METH 151-80. Methods of Testing Paints and Pigments Resistance to Cathodic Protection. 6 pp.

Cathodic Protection Equipment (Cont.)

—Submarines

MOD UK NES 704: Part 1-90. Requirements for Cathodic Protection Part 1: Systems General Issue 3 (11.90). 12 pp.

MOD UK NES 704: Part 2-89. Cathodic Protection Part 2: Impressed Current Cathodic Protection System Issue 3 (07.89). 35 pp.

MOD UK NES 704: Part 3-88. Cathodic Protection Part 3: Sacrificial Anode or Self-Energized System Issue 3 (11.88). 23 pp.

MOD UK NES 704: Part 3-01. Requirements for Cathodic Protection Part 3: Sacrifical Anode or Self-Energized System Issue 4 (12.91); Amendment 1. 27 pp.

MOD UK NES 704: Part 4-01. Requirements for Cathodic Protection Part 4: Ships Fitting—Out, Ships in Refit or Laid—Up Issue 3 (08.90); Amendment 1. 16 pp.

MOD UK NES 704: Part 5-01. Requirements for Cathodic Protection Part 5: General Information on Bi—Metallic Couples Issue 3 (08.90); Amendment 1. 15 pp.

—Water Heaters

DIN ENGL 4753 Pt 6-86. Water Heating Installations for Drinking Water and Service Water; Cathodic Corrosion Protection of Enamelled Steel Containers; Requirements and Testing (Feb). 4 pp.

DIN ENGL 4753 Pt 10-89. Water Heating Installations for Drinking Water and Service Water; Cathodic Corrosion Protection of Uncoated Steel Vessels; Requirements and Testing (May). 3 pp.

Cationic Surfactant Content Analysis

See Also: Anionic Surfactant Content Analysis; Surfactant Content Analysis

—Detergents

BSI BS 3762: Sec 3.4-01. 1991 Amd 1 Analysis of Formulated Detergents Part 3: Quantitative Test Methods Section 3.4: Method for Determination of Lower Molecular Mass Cationic-Active Matter Content (ISO 2871-2: 1990) (AMD 7161) July 15, 1992. 9 pp.

BSI BS 3762: Sec 3.26-89. 1989 Analysis of Formulated Detergents Part 3: Quantitative Test Methods Section 3.26: Method for Determination of High Molecular Mass Cationic-Active Matter Content. 4 pp.

ISO 2871 Pt 1-88. Surface Active Agents—Detergents—Determination of Cationic-Active Matter Content—Part 1: High-Molecular-Mass Cationic-Active Matter First Edition. 5 pp.

ISO 2871 Pt 2-90. Surface Active Agents—Detergents—Determination of Cationic-Active Matter Content—Part 2: Cationic-Active Matter of Low Molecular Mass (Between 200 and 500) First Edition. 7 pp.

—Detergents—Volumetric Analysis

BSI BS 3762: Sec 3.4-91. 1991 Analysis of Formulated Detergents Part 3: Quantitative Test Methods Section 3.4: Method for Determination of Lower Molecular Mass Cationic-Active Matter Content (ISO 2871-2: 1990). 8 pp.

—Drinking Water

DIN ENGL 38409 Pt 20-89. German Standard Methods for the Examination of Water, Waste Water and Sludge; Parameters Characterizing Effects and Substances (Group H); Determination of Substances That React with Disulfine Blue (H 20) (July). 6 pp.

—Surface Waters

DIN ENGL 38409 Pt 20-89. German Standard Methods for the Examination of Water, Waste Water and Sludge; Parameters Characterizing Effects and Substances (Group H); Determination of Substances That React with Disulfine Blue (H 20) (July). 6 pp.

—Waste Water

DIN ENGL 38409 Pt 20-89. German Standard Methods for the Examination of Water, Waste Water and Sludge; Parameters Characterizing Effects and Substances (Group H); Determination of Substances That React with Disulfine Blue (H 20) (July). 6 pp.

Cationic Surfactants

See Also: Surfactants

—Critical Micelle Concentration

ISO 6840-82. Cationic Surface Active Agents (Hydrochlorides and Hydrobromides)—Determination of Critical Micellization Concentration—Method by Measurement of Counter Ion Activity First Edition. 5 pp.

Catsup

Use For: Ketchup; Tomato Catsup See Also: Food
CGSB 32.258M-89. Tomato Catsup. 7 pp.

—Canned

CNS N5142-89. Canned Tomato Catsup (Feb)(4681). 4 pp.

Cattle

Use For: Bovines See Also: Beef Cattle; Cattle Grids; Dairy Cattle

—Breeding Stock

EC 87/328/EEC-87. Council Directive on the Acceptance for Breeding Purposes of Pure-Bred Breeding Animals of the Bovine Species. 2 pp.

—Buildings—Design

BSI BS 5502: Part 40-90. 1990 Design of Buildings and Structures for Agriculture Part 40: Code of Practice for Design and Construction of Cattle Buildings. 18 pp.

BSI BS 5502: Part 40-01. 1990 Amd 1 Buildings and Structures for Agriculture Part 40: Code of Practice for Design and Construction of Cattle Buildings (AMD 7665) June 15, 1993 (R). 20 pp.

—Embryo Collection—Health Certification

EC COM(88) 785-89. Proposal for a Council Regulation (EEC) on Animal Health Conditions Governing Intra-Community Trade in and Importation from Third Countries of Embryos of Domestic Animals of the Bovine Species. 25 pp.

EC 89/556/EEC-89. Council Directive on Animal Health Conditions Governing Intra—Community Trade in and Importation from Third Countries of Embryos of Domestic Animals of the Bovine Species. 11 pp.

—Hair

CNS L3061-80. Method of Test for Cattle Hair Felt (Packing) (Aug)(2687).

CNS L4043-80. Cattle Hair Felt (for Packing) (Aug) (2686).

CNS N1075-85. Cow Tail Hair (May)(2820).

CNS N1077-85. Cow Tail (May)(2822).

—Health and Veterinary Inspection

EC 90/425/EEC-90. Council Directive Concerning Veterinary and Zootechnical Checks Applicable in Intra-Community Trade in Certain Live Animals and Products with a View to the Completion of the Internal Market. 13 pp.

EC 91/628/EEC-91. Council Directive on the Protection of Animals During Transport and Amending Directives 90/425/EEC and 91/496/EEC. 12 pp.

—Health and Veterinary Inspection—Importation

EC 87/64/EEC-86. Council Directive Amending Directive 72/461/EEC on Health Problems Affecting Intra-Community Trade in Fresh Meat and Directive 72/462/EEC on Health and Veterinary Inspection Problems Upon Importation of Bovine Animals and Swine and Fresh Meat from Third. 2 pp.

EC 88/289/EEC-88. Council Directive Amending Directive 72/462/EEC on Health and Veterinary Inspection Problems Upon Importation of Bovine Animals and Swine and Fresh Meat from third Countries. 2 pp.

EC 89/227/EEC-89. Council Directive Amending Directives 72/462/EEC and 77/99/EEC to Take Account of the Introduction of Public Health and Animal Health Rules Which are to Govern Imports of Meat Products from Third Countries. 11 pp.

EC 93/451/EEC-93. Commission Decision Concerning the Animal Health Conditions and Veterinary Certification of Imports of Fresh Meat from Austria. 6 pp.

—Semen—Bacteria Count Methods

ISO TR8607-91. Artificial Insemination of Animals—Frozen Semen of Breeding Bulls—Enumeration of Living Aerobic Micro-Organisms First Edition. 6 pp.

—Semen Collection—Health Certification

EC COM(83) 512-83. Proposal for a Council Directive on Animal Health Problems Affecting Intra-Community Trade and Imports from Third Countries of Semen of Domestic Animals of the Bovine and Porcine Species. 37 pp.

EC COM(86) 657-86. Amendment to the Proposal for a Council Directive on Animal Health Problems Affecting Intra-Community Trade and Imports from Third Countries of Semen of Domestic Animals of the Bovine and Porcine Species. 4 pp.

EC 88/407/EEC-88. Council Directive Laying Down the Animal Health Requirements Applicable to Intra-Community Trade in the Imports of Deep Frozen Semen of Domestic Animals of the Bovine Species. 14 pp.

Cattle (Cont.)

—Semen Collection—Health Certification (Cont.)

EC 92/571/EEC-92. Commission Decision Relating to New Transitional Measures Which are Necessary to Facilitate the Move to the System of Veterinary Checks Provided for in Council Directive 90/675/EEC. 3 pp.

Cattle Grids

BSI BS 4008-91. 1991 Cattle Grids. 14 pp.

BSI BS 4008-01. 1991 Amd 1 Cattle Grids (AMD 7452) January 15, 1993. 15 pp.

BSI BS 4008-73. 1973 Cattle Grids on Private Roads. 9 pp.

—Stalls

ISO 5709-81. Equipment for Internal Farm Work and Husbandry—Metal Grids for Cattle Stalls First Edition. 4 pp.

CATV

Use: Cable Television

CATV Amplifiers

Use: Cable Television Amplifiers

Cauliflowers

See Also: Food; Vegetables

—Grading

CNS N1028-82. Grades and Packaging of Cauliflower (Jan)(2095). 3 pp.

—Packaging

CNS N1028-82. Grades and Packaging of Cauliflower (Jan)(2095). 3 pp.

—Refrigerated Transportation

ISO 949-87. Cauliflowers—Guide to Cold Storage and Refrigerated Transport Second Edition. 5 pp.

—Storage

ISO 949-87. Cauliflowers—Guide to Cold Storage and Refrigerated Transport Second Edition. 5 pp.

Caulking Compounds

See Also: Glazing Compounds; Putty; Sealants

CGSB CAN/CGSB-19.2-M87. Glazing Compound, Nonhardening, Modified Oil Type. 9 pp.

CGSB CAN/CGSB-19.24-M90. Multi-Component, Chemical-Curing Sealing Compound. 9 pp.

CGSB CAN2-19.0-M77. Methods of Testing Putty, Caulking and Sealing Compounds. 7 pp.

CGSB CAN2-19.0-M77METH 7.2-78. Methods of Testing Putty, Caulking and Sealing Compounds Self-Levelling in a Horizontal Test Assembly. 1 p.

CNS A2136-82. Sealing Compounds for Sealing and Glazing in Building (Jun)(8903).

CNS A3154-82. Method of Test for Sealing Compounds for Sealing and Glazing in Building (Jun)(8904).

JIS A 5757-75. Sealing Compounds for Sealing, Glazing and Caulking in Buildings.

JIS A 5758-92. Sealing Compounds for Sealing and Glazing in Building. 36 pp.

—Acrylic

CGSB 19-GP-5M-76. Sealing Compound, One Component, Acrylic Base, Solvent Curing, Standard for (R 1984). 8 pp.

CGSB CAN/CGSB-19.17-M90. One-Component, Acrylic Emulsion Base Sealing Compound. 9 pp.

—Elastomer

CGSB CAN/CGSB-19.13-M87. Sealing Compound, One-Component, Elastomeric, Chemical Curing. 12 pp.

—Glossaries

CGSB CAN/CGSB-19.28-91. Glossary of Terms Related to Sealants. 24 pp.

—Oil Base

CGSB CAN/CGSB-19.6-M87. Caulking Compound, Oil Base. 8 pp.

CNS A2135-82. Oil Based Caulking Compounds for Building (Jun)(8901).

CNS A3153-82. Method of Test for Oil Based Caulking Compounds for Building (Jun)(8902).

JIS A 5751-75. Oil Based Caulking Compounds for Buildings (R 1983). 15 pp.

—Pitch—Marine

MOD UK TS 10198A. Marine Glue Pitch.

—Polyisobutylene

CGSB 19-GP-14M-76. Sealing Compound, One Component, Butyl-Polyisobutylene Polymer Base, Solvent Curing, Standard for (R 1984). 9 pp.

—Polyurethane

CNS A3120-86. Method of Test for Polyurethane for Building Caulking and Flooring (May)(6988).

Caulking Compounds (Cont.)

—Rubber—Ships
MOD UK NES 156-90. Requirements for a Caulking Compound for Wood Decking Issue 1 (08.90). 15 pp.
MOD UK NES 156: Part 2-81. Wood Decking Part 2: Caulking Compound Issue 1 (04.81). 10 pp.

—Slump Testing
CGSB CAN2-19.0-M77METH 7.5-78. Methods of Testing Putty, Caulking and Sealing Compounds Slump of Caulking and Sealing Compounds (Boeing Jig); (Amendment 1 Nov 1978). 3 pp.
CNS K6727-82. Method of Test for Slump of Caulking Compounds and Sealants (Sep)(9403).

—Tensile Testing
CGSB CAN2-19.0-M77METH14.8-78. Methods of Testing Putty, Caulking and Sealing Compounds Extension of Caulking Compounds. 2 pp.

Caustic Potash
Use: Potassium Hydroxides

Caustic Potash Content Analysis
Use: Potassium Hydroxide Content Analysis

Caustic Soda
Use: Sodium Hydroxide

Causticity
Use: Alkalinity

Cautery Equipment
See Also: Medical Equipment
JIS T 3502-80. Cryocauter.

Cavitation Corrosion
See Also: Corrosion; Erosion
IEC 609-78. Cavitation Pitting Evaluation in Hydraulic Turbines, Storage Pumps and Pump-Turbines First Edition. 33 pp.

Cavitation Pitting
Use: Cavitation Corrosion

Cavities (Dental)
Use: Teeth

Cavity Filters
See Also: Electric Filters; Filters; Microwave Filters

—Preferred Products List
CECC CECC MUAHAG Vol 12 IS 1-90. Preferred Products List; Microwave Components (En, Fr, Ge). 76 pp.

Cavity Walls
Use: Walls

CBDS
Use: Connectionless Broadband Data Service

CBPP
Use: Contagious Bovine Pleuropneumonia

CCA
Use: Integrated Circuit Cards

CCE
Use: CEC (Commission of the European Communities)

CCI
Use: Photographic Lenses—Color Contribution Index

CCIR
Use For: Comite Consultatif International des Radiocommunications; International Radio Consultative Committee *See Also:* CCIR Handbooks; CCV; CMTT; IFRB; ITU; Standards

—CCITT—IEC—Organization—Glossaries
CCIR RESOLUTION 113-90. Organisation of Vocabulary Work—Volume XIII—Vocabulary and Related Subjects. 3 pp.

—CCITT—Interface
CCIR OPINION 56-1-86. Location of Interface Between CCIR Study Group 4 and CCITT Responsibilities for Digital Network Recommendations—Volume IV-1—Fixed-Satellite Service. 1 p.

—CCITT—ITU—Organization—Glossaries
CCIR RESOLUTION 113-90. Organisation of Vocabulary Work—Volume XIII—Vocabulary and Related Subjects. 3 pp.

CCIR (Cont.)

—Computer Programs—Access
CCIR RESOLUTION 104-90. Access to and Exchange of Computer Programs—Volume XIV—Administrative Texts of the CCIR. 2 pp.

—Computer Programs—Information Interchange
CCIR RESOLUTION 104-90. Access to and Exchange of Computer Programs—Volume XIV—Administrative Texts of the CCIR. 2 pp.

—Cooperation—BDT—Radio Wave Propagation—Developing Countries
CCIR RESOLUTION 101-90. Cooperation Between the CCIR and BDT in Radio Propagation Measurement Campaigns in the Developing Countries—Volume XIV—Administrative Texts of the CCIR. 2 pp.

—Cooperation—CCITT
CCIR RESOLUTION 82-2-90. Collaboration Between the CCIR and the CCITT—Volume XIV—Administrative Texts of the CCIR. 1 p.
CCITT FASCICLE I.1-88. Minutes and Reports of the Plenary Assembly List of Study Groups and Questions Under Study. 249 pp.

—Cooperation—CISPR
CCIR OPINION 2-2-90. Cooperation with the International Special Committee on Radio Interference—Volume I—Spectrum Utilization and Monitoring. 1 p.

—Cooperation—IEC—Documentation
CCIR RESOLUTION 23-3-90. Collaboration with the International Electrotechnical Commission on Graphical Symbols and Documentation Used in Telecommunications—Volume XIII—Vocabulary and Related Subjects. 1 p.

—Cooperation—IEC—Symbols
CCIR RESOLUTION 23-3-90. Collaboration with the International Electrotechnical Commission on Graphical Symbols and Documentation Used in Telecommunications—Volume XIII—Vocabulary and Related Subjects. 1 p.

—Cooperation—IFRB
CCIR RESOLUTION 83-2-90. Collaboration Between the CCIR and the IFRB—Volume XIV—Administrative Texts of the CCIR. 2 pp.

—Coordination—CCITT—Glossaries
CCITT RECMN A.12-88. Collaboration with the International Electrotechnical Commission on the Subject of Definitions for Telecommunications—Opinions and Resolutions—Recommendations on the Organization and Working Procedures of CCITT (Series A) 1 pp. 1 p.

—Coordination—Glossaries
CCIR RESOLUTION 114-90. Coordination of Vocabulary and Related Subjects—Volume XIII—Vocabulary and Related Subjects. 2 pp.

—Coordination—IEC—Glossaries
CCIR RESOLUTION 114-90. Coordination of Vocabulary and Related Subjects—Volume XIII—Vocabulary and Related Subjects. 2 pp.

—Coordination—IEC—Radio Relay Systems
CCIR OPINION 50-74. Coordination of the Work of the CCIR and the IEC on Measurements for the Adjustment and Maintenance of Radio-Relay Systems—Volume IX-1—Fixed Service Using Radio-Relay Systems. 1 p.

—Coordination—IEC—Symbols
CCIR RESOLUTION 114-90. Coordination of Vocabulary and Related Subjects—Volume XIII—Vocabulary and Related Subjects. 2 pp.

—Coordination—Study Groups
CCIR RESOLUTION 24-7-90. Organization of CCIR Work—Volume XIV—Administrative Texts of the CCIR. 9 pp.

—Coordination—Symbols
CCIR RESOLUTION 114-90. Coordination of Vocabulary and Related Subjects—Volume XIII—Vocabulary and Related Subjects. 2 pp.

—Determination of Coordination Area
CCIR Decision 87-89. Determination of the Coordination Area (Appendix 28 of the Radio Regulations)—Annex to Volumes IV and IX—Part 2—Frequency Sharing and Coordination Between Systems in the Fixed-Satellite Service and Radio-Relay Systems. 2 pp.

—Documentation
CCIR RESOLUTION 24-7-90. Organization of CCIR Work—Volume XIV—Administrative Texts of the CCIR. 9 pp.

CCIR (Cont.)

—Documents
CCIR RESOLUTION 81-2-90. Handbooks and Special Publications—Volume XIV—Administrative Texts of the CCIR. 1 p.

—Documents—Dissemination
CCIR RESOLUTION 105-90. Dissemination of CCIR Texts—Volume XIV—Administrative Texts of the CCIR. 1 p.

—Glossaries
CCIR RESOLUTION 24-7-90. Organization of CCIR Work—Volume XIV—Administrative Texts of the CCIR. 9 pp.

—Guidelines—Glossaries
CCIR RESOLUTION 89-1-90. Guidelines for the Selection of Terms and Preparation of Definitions—Volume XIII—Vocabulary and Related Subjects. 3 pp.

—Information Bulletins
CCIR RESOLUTION 24-7-90. Organization of CCIR Work—Volume XIV—Administrative Texts of the CCIR. 9 pp.

—Information Meetings
CCIR RESOLUTION 108-90. Information Meetings—Volume XIV—Administrative Texts of the CCIR. 2 pp.

—International Technical Cooperation
CCIR RESOLUTION 33-6-90. CCIR and Technical Cooperation—Volume XIV —Administrative Texts of the CCIR. 3 pp.

—ITU—Physical Quantities—Glossaries
CCIR RECMN 663-1-90. Use of Certain Terms Linked with Physical Quantities—Section A—Terminology. 5 pp.

—Liaison—Broadcasting
CCIR RESOLUTION 80-1-90. CCIR Liasion with Organizations Dealing with Radiocommunications Including Broadcasting—Volume XIV—Administrative Texts of the CCIR. 1 p.

—Liaison—Radio Communications
CCIR RESOLUTION 80-1-90. CCIR Liasion with Organizations Dealing with Radiocommunications Including Broadcasting—Volume XIV—Administrative Texts of the CCIR. 1 p.

—Liason—CCITT—Fixed Satellite Services—ISDN
CCIR RESOLUTION 103-90. Liasion and Joint Studies with the CCITT in Mobile and Fixed-Satellite Services and High-Definition Television in ISDNs—Volume XIV—Administrative Texts of the CCIR. 2 pp.

—Liason—CCITT—High Definition Television—ISDN
CCIR RESOLUTION 103-90. Liasion and Joint Studies with the CCITT in Mobile and Fixed-Satellite Services and High-Definition Television in ISDNs—Volume XIV—Administrative Texts of the CCIR. 2 pp.

—Liason—CCITT—Mobile Satellite Communications—ISDN
CCIR RESOLUTION 103-90. Liasion and Joint Studies with the CCITT in Mobile and Fixed-Satellite Services and High-Definition Television in ISDNs—Volume XIV—Administrative Texts of the CCIR. 2 pp.

—Participation—Developing Countries
CCIR RESOLUTION 95-1-90. Participation by the Developing Countries in the Work of the CCIR—Volume XIV—Administrative Texts of the CCIR. 2 pp.

—Planning
CCIR RESOLUTION 106-90. Strategic Review and Planning—Volume XIV—Administrative Texts of the CCIR. 1 p.

—Plenary Assembly
CCIR. Plan of Volumes I to XV, XVIIth Plenary Assembly of the CCIR. 1 p.
CCIR. Distribution of Texts of the XVIIth Plenary Assembly of the CCIR in Volumes I to XV. 5 pp.
CCIR RESOLUTION 24-7-90. Organization of CCIR Work—Volume XIV—Administrative Texts of the CCIR. 9 pp.
CCIR RESOLUTION 97-90. Approval of New and Revised Recommendations Between Plenary Assemblies—Volume XIV—Administrative Texts of the CCIR. 2 pp.

INTERNATIONAL AND NON-U.S. NATIONAL STANDARDS
SUBJECT INDEX

CCITT

CCIR (Cont.)

—Plenary Assembly (Cont.)
CCIR RESOLUTION 109-90. Approval and Categorization of Questions by the XVIITH Plenary Assembly and Subsequent Action by the Study Groups—Volume XIV—Administrative Texts of the CCIR. 47 pp.

—Presentation of Text—Glossaries
CCIR RESOLUTION 78-1-90. Presentation of Texts on Terminology—Volume XIII—Vocabulary and Related Subjects. 1 p.

—Recommendations—Approval
CCIR RESOLUTION 97-90. Approval of New and Revised Recommendations Between Plenary Assemblies—Volume XIV—Administrative Texts of the CCIR. 2 pp.
CCIR OPINION 87-90. CCIR Recommendations Approved Between Plenary Assemblies and Resolution No. 703 of WARC-79—Volume XIV—Administrative Texts of the CCIR. 1 p.

—Recommendations—Radio Regulations
CCIR RESOLUTION 87-2-90. List of Provisions of Radio Regulations Which Include References to Relevant CCIR Recommendations—Volume XIV—Administrative Texts of the CCIR. 42 pp.

—Special Autonomous Groups
CCIR OPINION 77-2-90. CCIR Involvement in the Work of the Special Autonomous Groups (GAS)—Volume XIV—Administrative Texts of the CCIR. 1 p.

—Specification and Description Language—CCITT
CCIR RECMN 664-86. Adoption of the CCITT Specification and Description Language (SDL)—Section B—Graphical Symbols. 2 pp.

—Staff—Technical Assistance
CCIR RESOLUTION 39-3-90. Participation of CCIR Staff in Technical Cooperation and Technical Assistance Activities—Volume XIV—Administrative Texts of the CCIR. 1 p.

—Staff—Technical Cooperation
CCIR RESOLUTION 39-3-90. Participation of CCIR Staff in Technical Cooperation and Technical Assistance Activities—Volume XIV—Administrative Texts of the CCIR. 1 p.

—Standards Preparation—Sound Recording
CCIR OPINION 16-3-86. Organizations Qualified to Set Standards on Sound and Television Recording—Volumes X and XI—Part 3—Sound and Television Recording. 1 p.

—Standards Preparation—Television Recording
CCIR OPINION 16-3-86. Organizations Qualified to Set Standards on Sound and Television Recording—Volumes X and XI—Part 3—Sound and Television Recording. 1 p.

—Strategic Review
CCIR RESOLUTION 106-90. Strategic Review and Planning—Volume XIV—Administrative Texts of the CCIR. 1 p.

—Study Groups
CCIR RESOLUTION 24-7-90. Organization of CCIR Work—Volume XIV—Administrative Texts of the CCIR. 9 pp.
CCIR RESOLUTION 61-4-90. Structure of CCIR Study Groups—Volume XIV—Administrative Texts of the CCIR. 8 pp.
CCIR RESOLUTION 107-90. Restructuring of the CCIR Study Groups—Volume XIV—Administrative Texts of the CCIR. 2 pp.
CCIR RESOLUTION 109-90. Approval and Categorization of Questions by the XVIITH Plenary Assembly and Subsequent Action by the Study Groups—Volume XIV—Administrative Texts of the CCIR. 47 pp.

—Study Questions—Approval
CCIR RESOLUTION 109-90. Approval and Categorization of Questions by the XVIITH Plenary Assembly and Subsequent Action by the Study Groups—Volume XIV—Administrative Texts of the CCIR. 47 pp.

—Study Questions—Classification
CCIR RESOLUTION 109-90. Approval and Categorization of Questions by the XVIITH Plenary Assembly and Subsequent Action by the Study Groups—Volume XIV—Administrative Texts of the CCIR. 47 pp.

—Study Questions—Radio Communications
CCIR RESOLUTION 102-90. Questions for Study by the CCIR—Volume XIV—Administrative Texts of the CCIR. 2 pp.

CCIR (Cont.)

—Terms of Reference
CCIR OPINION 88-90. CCIR Terms of Reference—Volume XIV—Administrative Texts of the CCIR. 1 p.

—World Administrative Radio Conference—Broadcasting
CCIR RESOLUTION 98-90. CCIR Studies to Be Carried out for Submission to the World Administrative Radio Conference (WARC-93) Dealing with Matters Connected with the HF Broadcasting Service—Volume XIV—Administrative Texts of the CCIR. 3 pp.

—World Administrative Radio Conference—Frequency Assignment
CCIR RESOLUTION 100-90. CCIR Studies to Be Carried out for the Preparation of a Report to Be Submitted to the World Administrative Radio Conference (WARC-92) for Frequency Allocations in Certain Parts of the Spectrum—Volume XIV—Administrative Texts of the CCIR. 3 pp.

—World Administrative Radio Conference—Radio Spectra
CCIR RESOLUTION 100-90. CCIR Studies to Be Carried out for the Preparation of a Report to Be Submitted to the World Administrative Radio Conference (WARC-92) for Frequency Allocations in Certain Parts of the Spectrum—Volume XIV—Administrative Texts of the CCIR. 3 pp.

—World Administrative Radio Conference—Study Group 8
CCIR Decision 103-89. Interim Working Party to Develop and Coordinate Contributions from Study Group 8 on the Technical Bases for WARC-92—Annex 3 to Volume VIII—Mobile, Radiodetermination, Amateur and Related Satellite Services. 4 pp.

CCIR Handbooks
See Also: CCIR; CCITT; Handbooks
CCIR RESOLUTION 81-2-90. Handbooks and Special Publications—Volume XIV—Administrative Texts of the CCIR. 1 p.

—Fixed Satellite Services
CCIR Decision 64-1-89. Updating of the Handbook on Satellite Communications (Fixed-Satellite Service)—Annex to Volume IV-1—Fixed-Satellite Service. 1 p.

—Standard Frequency Dissemination—Satellites
CCIR Decision 65-85. Handbook on the Use of Satellite Time and Frequency Dissemination—Annex to Volume VII—Standard Frequencies and Time Signals. 2 pp.

—Time Dissemination—Satellites
CCIR Decision 65-85. Handbook on the Use of Satellite Time and Frequency Dissemination—Annex to Volume VII—Standard Frequencies and Time Signals. 2 pp.

CCITT
Use For: Comite Consultatif International Telegraphique et Telephonique; International Telegraph and Telephone Consultative Committee
See Also: CCIR Handbooks; CCV; CMTT; IFRB; ITU; Network Nodes; Standards

—CCIR—IEC—Organization—Glossaries
CCIR RESOLUTION 113-90. Organisation of Vocabulary Work—Volume XIII—Vocabulary and Related Subjects. 3 pp.

—CCIR—Interface
CCIR OPINION 56-1-86. Location of Interface Between CCIR Study Group 4 and CCITT Responsibilities for Digital Network Recommendations—Volume IV-1—Fixed-Satellite Service. 1 p.

—CCIR—ITU—Organization—Glossaries
CCIR RESOLUTION 113-90. Organisation of Vocabulary Work—Volume XIII—Vocabulary and Related Subjects. 3 pp.

—Cooperation—CCIR
CCIR RESOLUTION 82-2-90. Collaboration Between the CCIR and the CCITT—Volume XIV—Administrative Texts of the CCIR. 1 p.
CCITT FASCICLE I.1-88. Minutes and Reports of the Plenary Assembly List of Study Groups and Questions Under Study. 249 pp.

CCITT (Cont.)

—Cooperation—CCIR—Glossaries
CCITT RECMN A.12-88. Collaboration with the International Electrotechnical Commission on the Subject of Definitions for Telecommunications—Opinions and Resolutions —Recommendations on the Organization and Working Procedures of CCITT (Series A) 1 pp. 1 p.

—Cooperation—Consultative Council for Postal Studies
CCITT FASCICLE I.1-88. Minutes and Reports of the Plenary Assembly List of Study Groups and Questions Under Study. 249 pp.
CCITT FASCICLE I.2-88. Opinions and Resolutions Recommendations on the Organisation and Working Procedures of CCITT (Series A). 64 pp.

—Cooperation—Data Transmission
CCITT RECMN A.20-89. Collaboration with Other International Organizations over Data Transmission—Data Communication over the Telephone Network (Study Group XVII) 2 pp. 2 pp.

—Cooperation—IEC
CCITT RECMN A.20-89. Collaboration with Other International Organizations over Data Transmission—Data Communication over the Telephone Network (Study Group XVII) 2 pp. 2 pp.
CCITT RECMN A.22-89. Collaboration with Other International Organizations on Information Technology—Terminal Equipment and Protocols for Telematic Services (Study Group VIII) 2 pp. 2 pp.
CCITT FASCICLE I.2-88. Opinions and Resolutions Recommendations on the Organisation and Working Procedures of CCITT (Series A). 64 pp.

—Cooperation—IEC—Diagrams
CCITT RECMN A.13-80. Collaboration with the International Electrotechnical Commission on Graphical Symbols and Diagrams Used in Telecommunications—Opinions and Resolutions —Recommendations on the Organization and Working Procedures of CCITT (Series A) 1 pp. 1 p.

—Cooperation—IEC—Glossaries
CCITT RECMN A.12-88. Collaboration with the International Electrotechnical Commission on the Subject of Definitions for Telecommunications—Opinions and Resolutions —Recommendations on the Organization and Working Procedures of CCITT (Series A) 1 pp. 1 p.

—Cooperation—IEC—Symbols
CCITT RECMN A.13-80. Collaboration with the International Electrotechnical Commission on Graphical Symbols and Diagrams Used in Telecommunications—Opinions and Resolutions —Recommendations on the Organization and Working Procedures of CCITT (Series A) 1 pp. 1 p.

—Cooperation—Information Systems
CCITT RECMN A.22-89. Collaboration with Other International Organizations on Information Technology—Terminal Equipment and Protocols for Telematic Services (Study Group VIII) 2 pp. 2 pp.

—Cooperation—International Telecommunication Union
CCITT FASCICLE I.2-88. Opinions and Resolutions Recommendations on the Organisation and Working Procedures of CCITT (Series A). 64 pp.

—Cooperation—ISO
CCITT FASCICLE I.1-88. Minutes and Reports of the Plenary Assembly List of Study Groups and Questions Under Study. 249 pp.
CCITT RECMN A.20-89. Collaboration with Other International Organizations over Data Transmission—Data Communication over the Telephone Network (Study Group XVII) 2 pp. 2 pp.
CCITT RECMN A.21-89. Collaboration with Other International Organizations on CCITT-Defined Telematic Services—Terminal Equipment and Protocols for Telematic Services (Study Group VIII) 2 pp. 2 pp.
CCITT RECMN A.22-89. Collaboration with Other International Organizations on Information Technology—Terminal Equipment and Protocols for Telematic Services (Study Group VIII) 2 pp. 2 pp.
CCITT FASCICLE I.2-88. Opinions and Resolutions Recommendations on the Organisation and Working Procedures of CCITT (Series A). 64 pp.

INDUSTRY STANDARDS

INTERNATIONAL AND NON-U.S. NATIONAL STANDARDS
SUBJECT INDEX

CCITT (Cont.)

—Cooperation—Teleinformatic Services
CCITT RECMN A.21-89. Collaboration with Other International Organizations on CCITT-Defined Telematic Services—Terminal Equipment and Protocols for Telematic Services (Study Group VIII) 2 pp. 2 pp.

—Cooperation—Universal Postal Union
CCITT FASCICLE I.1-88. Minutes and Reports of the Plenary Assembly List of Study Groups and Questions Under Study. 249 pp.
CCITT FASCICLE I.2-88. Opinions and Resolutions Recommendations on the Organisation and Working Procedures of CCITT (Series A). 64 pp.

—Document Formats
CCITT RECMN A.15-80. Presentation of CCITT Texts—Opinions and Resolutions—Recommendations on the Organization and Working Procedures of CCITT (Series A) 9 pp. 9 pp.
CCITT FASCICLE I.2-88. Opinions and Resolutions Recommendations on the Organisation and Working Procedures of CCITT (Series A). 64 pp.

—Document Formats—Machine Readable
CCITT RECMN A.15-80. Presentation of CCITT Texts—Opinions and Resolutions—Recommendations on the Organization and Working Procedures of CCITT (Series A) 9 pp. 9 pp.
CCITT FASCICLE I.2-88. Opinions and Resolutions Recommendations on the Organisation and Working Procedures of CCITT (Series A). 64 pp.

—Documents—Publishing
CCITT FASCICLE I.2-88. Opinions and Resolutions Recommendations on the Organisation and Working Procedures of CCITT (Series A). 64 pp.

—Evolution of Working Procedures
CCITT FASCICLE I.2-88. Opinions and Resolutions Recommendations on the Organisation and Working Procedures of CCITT (Series A). 64 pp.

—Glossaries Preparation
CCITT RECMN A.10-88. Terms and Definitions—Opinions and Resolutions—Recommendations on the Organization and Working Procedures of CCITT (Series A) 1 pp. 1 p.
CCITT FASCICLE I.2-88. Opinions and Resolutions Recommendations on the Organisation and Working Procedures of CCITT (Series A). 64 pp.

—Glossaries—Publication
CCITT RECMN A.14-80. Publication of Definitions—Opinions and Resolutions—Recommendations on the Organization and Working Procedures of CCITT (Series A) 1 pp. 1 p.

—Indexes (Documentation)
CCITT FASCICLE I.4-88. Index of Blue Book. 680 pp.

—Information Services—Developing Countries
CCITT FASCICLE I.1-88. Minutes and Reports of the Plenary Assembly List of Study Groups and Questions Under Study. 249 pp.

—Instructions—Implementation Procedures
CCITT RECMN C.3-89. Instructions for International Communication Services—Terms and Definitions Abbreviations and Acronyms Recommendations on Means of Expression (Series B) General Telecommunications Statistics (Series C) 2 pp. 2 pp.
CCITT RECMN E.141-89. Instructions for the International Telephone Service—Telephone Network and ISDN—Operation, Numbering, Routing and Mobile Service (Study Group X) 1 pp. 1 p.

—Instructions—Legal Aspects
CCITT RECMN C.3-89. Instructions for International Communication Services—Terms and Definitions Abbreviations and Acronyms Recommendations on Means of Expression (Series B) General Telecommunications Statistics (Series C) 2 pp. 2 pp.

—Laboratories—Activities
CCITT FASCICLE I.1-88. Minutes and Reports of the Plenary Assembly List of Study Groups and Questions Under Study. 249 pp.

—Liason—CCIR—Fixed Satellite Services—ISDN
CCIR RESOLUTION 103-90. Liasion and Joint Studies with the CCITT in Mobile and Fixed-Satellite Services and High-Definition Television in ISDNs—Volume XIV—Administrative Texts of the CCIR. 2 pp.

CCITT (Cont.)

—Liason—CCIR—High Definition Television—ISDN
CCIR RESOLUTION 103-90. Liasion and Joint Studies with the CCITT in Mobile and Fixed-Satellite Services and High-Definition Television in ISDNs—Volume XIV—Administrative Texts of the CCIR. 2 pp.

—Liason—CCIR—Mobile Satellite Communications—ISDN
CCIR RESOLUTION 103-90. Liasion and Joint Studies with the CCITT in Mobile and Fixed-Satellite Services and High-Definition Television in ISDNs—Volume XIV—Administrative Texts of the CCIR. 2 pp.

—List of Participants
CCITT FASCICLE I.1-88. Minutes and Reports of the Plenary Assembly List of Study Groups and Questions Under Study. 249 pp.

—Man-Machine Language
CCITT RECMN Z.301-89. Introduction to the CCITT Man-Machine Language—Man-Machine Language (MML) (Study Group X) 4 pp. 4 pp.

—Man-Machine Language—Character Sets
CCITT RECMN Z.314-89. Character Set and Basic Elements—Man-Machine Language (MML) (Study Group X) 9 pp. 9 pp.

—Numbering of Recommendations
CCITT FASCICLE I.2-88. Opinions and Resolutions Recommendations on the Organisation and Working Procedures of CCITT (Series A). 64 pp.

—Open Systems Interconnection
CCITT RECMN X.200-89. Reference Model of Open Systems Interconnection for CCITT Applications—Data Communication Networks—Open Systems Interconnection (OSI) Model and Notation, Service Definition (Study Group VII) 54 pp. 54 pp.
CCITT RECMN X.211-88. Physical Service Definition of Open Systems Interconnection for CCITT Applications—Data Communication Networks—Open Systems Interconnection (OSI) Model and Notation, Service Definition (Study Group VII) 17 pp (Corrigendum-Oct 1992). 19 pp.
CCITT RECMN X.211-88. Physical Service Definition of Open Systems Interconnection for CCITT Applications—Data Communication Networks—Open Systems Interconnection (OSI) Model and Notation, Service Definition (Study Group VII) 17 pp. 17 pp.
CCITT RECMN X.212-88. Data Link Service Definition for Open Systems Interconnection for CCITT Applications—Data Communication Networks—Open Systems Interconnection (OSI) Model and Notation, Service Definition (Study Group VII) 45 pp (Corrigenda-Oct 1992). 45 pp.
CCITT RECMN X.212-88. Data Link Service Definition for Open Systems Interconnection for CCITT Applications—Data Communication Networks—Open Systems Interconnection (OSI) Model and Notation, Service Definition (Study Group VII) 42 pp. 42 pp.
CCITT RECMN X.213-89. Network Service Definition for Open Systems Interconnection for CCITT Applications—Data Communication Networks—Open Systems Interconnection (OSI) Model and Notation, Service Definition (Study Group VII) 60 pp. 60 pp.
CCITT RECMN X.214-89. Transport Service Definition for Open Systems Interconnection for CCITT Applications—Data Communication Networks—Open Systems Interconnection (OSI) Model and Notation, Service Definition (Study Group VII) 25 pp. 25 pp.
CCITT RECMN X.215-89. Session Service Definition for Open Systems Interconnection for CCITT Applications—Data Communication Networks—Open Systems Interconnection (OSI) Model and Notation, Service Definition (Study Group VII) 83 pp. 83 pp.
CCITT RECMN X.216-89. Definicion Del Servicio De Presentacion Para La Interconexion De Sistemas Abiertos Para Aplicaciones Del CCITT—Data Communication Networks—Open Systems Interconnection (OSI) Model and Notation, Service Definition (Study Group VII) 43 pp. 43 pp.
CCITT RECMN X.217-89. Association Contro Service Definition for Open Systems Interconnection for CCITT Applications—Data Communication Networks—Open Systems Interconnection (OSI) Model and Notation, Service Definition (Study Group VII) 21 pp. 21 pp.
CCITT RECMN X.218-89. Reliable Transfer: Model and Service Definition—Data Communication Networks—Open Systems Interconnection (OSI) Model and Notation, Service Definition (Study Group VII) 17 pp. 17 pp.
CCITT RECMN X.219-89. Remote Operations: Model, Notation and Service Definition—Data Communication Networks—Open Systems Interconnection (OSI) Model and Notation, Service Definition (Study Group VII) 38 pp. 38 pp.

CCITT (Cont.)

—Open Systems Interconnection (Cont.)
CCITT RECMN X.220-89. Use of X.200-Series Protocols in CCITT Applications—Data Communication Networks—Open Systems Interconnection (OSI) Protocol Specifications, Conformance Testing (Study Group VII) 3 pp. 3 pp.
CCITT RECMN X.223-89. Use of X.25 to Provide the OSI Connection-Mode Network Service for CCITT Applications—Data Communication Networks—Open Systems Interconnection (OSI) Protocol Specifications, Conformance Testing (Study Group VII) 30 pp. 30 pp.
CCITT RECMN X.224-89. Transport Protocol Specification for Open Systems Interconnection for CCITT Applications—Data Communication Networks—Open Systems Interconnection (OSI) Protocol Specifications, Conformance Testing (Study Group VII) 93 pp. 93 pp.
CCITT RECMN X.225-89. Session Protocol Specification for Open Systems Interconnection for CCITT Applications—Data Communication Networks—Open Systems Interconnection (OSI) Protocol Specifications, Conformance Testing (Study Group VII) 143 pp. 143 pp.
CCITT RECMN X.226-89. Presentation Protocol Specification for Open Systems Interconnection for CCITT Applications—Data Communication Networks—Open Systems Interconnection (OSI) Protocol Specifications, Conformance Testing (Study Group VII) 68 pp. 68 pp.
CCITT RECMN X.227-89. Association Control Protocol Specification for Open Systems Interconnection for CCITT Applications—Data Communication Networks—Open Systems Interconnection (OSI) Protocol Specifications, Conformance Testing (Study Group VII) 41 pp. 41 pp.
CCITT RECMN X.228-89. Reliable Transfer: Protocol Specification—Data Communication Networks—Open Systems Interconnection (OSI) Protocol Specifications, Conformance Testing (Study Group VII) 69 pp. 69 pp.
CCITT RECMN X.229-89. Remote Operations: Protocol Specification—Data Communication Networks—Open Systems Interconnection (OSI) Protocol Specifications, Conformance Testing (Study Group VII) 27 pp. 27 pp.
CCITT RECMN X.290-92. OSI Conformance Testing Methodology and Framework for Protocol Recommendations for CCITT Applications—General Concepts (Study Group VII) 57 pp. 57 pp.
CCITT RECMN X.290-89. OSI Conformance Testing Methodology and Framework for Protocol Recommendations for CCITT Applications—Data Communication Networks—Open Systems Interconnection (OSI) Protocol Specifications, Conformance Testing (Study Group VII) 113 pp. 113 pp.
CCITT RECMN X.291-92. OSI Conformance Testing Methodology and Framework for Protocol Recommendations for CCITT Applications—Abstract Test Suite Specification (Study Group VII) 48 pp. 48 pp.
CCITT RECMN X.293-92. OSI Conformance Testing Methodology and Framework for Protocol Recommendations for CCITT Applications—Test Realization (Study Group VII) 16 pp. 16 pp.
CCITT RECMN X.294-92. OSI Conformance Testing Methodology and Framework for Protocol Recommendations for CCITT Applications—Requirements on Test Laboratories and Clients for the Conformance Assessment Process (Study Group VII) 34 pp. 34 pp.

—Plan Committees
CCITT FASCICLE I.2-88. Opinions and Resolutions Recommendations on the Organisation and Working Procedures of CCITT (Series A). 64 pp.

—Plan Committees—Activities
CCITT FASCICLE I.1-88. Minutes and Reports of the Plenary Assembly List of Study Groups and Questions Under Study. 249 pp.

—Plan Committees—Examination of Study Questions
CCITT FASCICLE I.2-88. Opinions and Resolutions Recommendations on the Organisation and Working Procedures of CCITT (Series A). 64 pp.

—Plenary Assembly—Approval of Recommendations
CCITT FASCICLE I.2-88. Opinions and Resolutions Recommendations on the Organisation and Working Procedures of CCITT (Series A). 64 pp.

—Plenary Assembly—Committee A Report
CCITT FASCICLE I.1-88. Minutes and Reports of the Plenary Assembly List of Study Groups and Questions Under Study. 249 pp.

INTERNATIONAL AND NON-U.S. NATIONAL STANDARDS
SUBJECT INDEX
CCITT

CCITT *(Cont.)*

—**Plenary Assembly—Committee B Report**
CCITT FASCICLE I.1-88. Minutes and Reports of the Plenary Assembly List of Study Groups and Questions Under Study. 249 pp.

—**Plenary Assembly—Committee C Report**
CCITT FASCICLE I.1-88. Minutes and Reports of the Plenary Assembly List of Study Groups and Questions Under Study. 249 pp.

—**Plenary Assembly—Committee D Report**
CCITT FASCICLE I.1-88. Minutes and Reports of the Plenary Assembly List of Study Groups and Questions Under Study. 249 pp.

—**Plenary Assembly—Editorial Committee Report**
CCITT FASCICLE I.1-88. Minutes and Reports of the Plenary Assembly List of Study Groups and Questions Under Study. 249 pp.

—**Plenary Assembly—List of Documents**
CCITT FASCICLE I.1-88. Minutes and Reports of the Plenary Assembly List of Study Groups and Questions Under Study. 249 pp.

—**Plenary Assembly—Minutes**
CCITT FASCICLE I.1-88. Minutes and Reports of the Plenary Assembly List of Study Groups and Questions Under Study. 249 pp.

—**Plenary Assembly—Organization**
CCITT FASCICLE I.1-88. Minutes and Reports of the Plenary Assembly List of Study Groups and Questions Under Study. 249 pp.

—**Plenary Assembly—Report of the Director**
CCITT FASCICLE I.1-88. Minutes and Reports of the Plenary Assembly List of Study Groups and Questions Under Study. 249 pp.

—**Preeminence—Standards**
CCITT FASCICLE I.2-88. Opinions and Resolutions Recommendations on the Organisation and Working Procedures of CCITT (Series A). 64 pp.

—**Recommendations—Formats**
CCITT FASCICLE I.2-88. Opinions and Resolutions Recommendations on the Organisation and Working Procedures of CCITT (Series A). 64 pp.

—**Rules of Procedure**
CCITT FASCICLE I.2-88. Opinions and Resolutions Recommendations on the Organisation and Working Procedures of CCITT (Series A). 64 pp.

—**Rules of Procedure—Contribution of Study Questions**
CCITT RECMN A.1-88. Presentation of Contributions Relative to the Study of Questions Assigned to the CCITT—Opinions and Resolutions—Recommendations on the Organization and Working Procedures of CCITT (Series A) 4 pp. 4 pp.
CCITT FASCICLE I.2-88. Opinions and Resolutions Recommendations on the Organisation and Working Procedures of CCITT (Series A). 64 pp.

—**Rules of Procedure—Drafting of Proposals**
CCITT RECMN A.1-88. Presentation of Contributions Relative to the Study of Questions Assigned to the CCITT—Opinions and Resolutions—Recommendations on the Organization and Working Procedures of CCITT (Series A) 4 pp. 4 pp.
CCITT FASCICLE I.2-88. Opinions and Resolutions Recommendations on the Organisation and Working Procedures of CCITT (Series A). 64 pp.

—**Rules of Procedure—Drafting of Proposals—Forms (Paper)**
CCITT FASCICLE I.2-88. Opinions and Resolutions Recommendations on the Organisation and Working Procedures of CCITT (Series A). 64 pp.

—**Rules of Procedure—Drafting of Questions**
CCITT RECMN A.1-88. Presentation of Contributions Relative to the Study of Questions Assigned to the CCITT—Opinions and Resolutions—Recommendations on the Organization and Working Procedures of CCITT (Series A) 4 pp. 4 pp.
CCITT FASCICLE I.2-88. Opinions and Resolutions Recommendations on the Organisation and Working Procedures of CCITT (Series A). 64 pp.

—**Rules of Procedure—Drafting of Questions—Forms (Paper)**
CCITT FASCICLE I.2-88. Opinions and Resolutions Recommendations on the Organisation and Working Procedures of CCITT (Series A). 64 pp.

CCITT *(Cont.)*

—**Seminars**
CCITT FASCICLE I.2-88. Opinions and Resolutions Recommendations on the Organisation and Working Procedures of CCITT (Series A). 64 pp.

—**Special Autonomous Groups—Activities**
CCITT FASCICLE I.1-88. Minutes and Reports of the Plenary Assembly List of Study Groups and Questions Under Study. 249 pp.

—**Specialized Secretariat—Graphs (Charts)**
CCITT FASCICLE I.1-88. Minutes and Reports of the Plenary Assembly List of Study Groups and Questions Under Study. 249 pp.

—**Specialized Secretariat—Statistical Analysis**
CCITT FASCICLE I.1-88. Minutes and Reports of the Plenary Assembly List of Study Groups and Questions Under Study. 249 pp.

—**Specialized Secretariat—Tables (Data)**
CCITT FASCICLE I.1-88. Minutes and Reports of the Plenary Assembly List of Study Groups and Questions Under Study. 249 pp.

—**Specification and Description Language—CCIR**
CCIR RECMN 664-86. Adoption of the CCITT Specification and Description Language (SDL)—Section B—Graphical Symbols. 2 pp.

—**Study Groups—Activities**
CCITT FASCICLE I.1-88. Minutes and Reports of the Plenary Assembly List of Study Groups and Questions Under Study. 249 pp.

—**Study Groups—Chair Appointments**
CCITT FASCICLE I.2-88. Opinions and Resolutions Recommendations on the Organisation and Working Procedures of CCITT (Series A). 64 pp.

—**Study Groups—Meetings—Invitations**
CCITT FASCICLE I.2-88. Opinions and Resolutions Recommendations on the Organisation and Working Procedures of CCITT (Series A). 64 pp.

—**Study Groups—Meetings—Locations**
CCITT FASCICLE I.2-88. Opinions and Resolutions Recommendations on the Organisation and Working Procedures of CCITT (Series A). 64 pp.

—**Technical Assistance—Developing Nations**
CCITT FASCICLE I.2-88. Opinions and Resolutions Recommendations on the Organisation and Working Procedures of CCITT (Series A). 64 pp.

—**Technical Reports—Dissemination**
CCITT FASCICLE I.2-88. Opinions and Resolutions Recommendations on the Organisation and Working Procedures of CCITT (Series A). 64 pp.

—**Working Parties—Meetings—Invitations**
CCITT FASCICLE I.2-88. Opinions and Resolutions Recommendations on the Organisation and Working Procedures of CCITT (Series A). 64 pp.

—**Working Parties—Meetings—Locations**
CCITT FASCICLE I.2-88. Opinions and Resolutions Recommendations on the Organisation and Working Procedures of CCITT (Series A). 64 pp.

CCITT High Level Language
Use: CHILL

CCITT No. 4 Signaling Systems
Use For: Signaling System No. 4
See Also: Common Channel Signaling Systems; Signaling Systems

—**CCITT No. 5—Interworking**
CCITT RECMN Q.180-89. Interworking of Systems No. 4 and No. 5—Specifications of Signalling Systems Nos. 4 and 5 (Study Group XI) 7 pp. 7 pp.

—**CCITT R2—Interworking—Logic Procedures**
CCITT RECMN Q.634-89. Logic Procedures for Interworking of Signalling System No. 4 to R2—Interworking of Signalling Systems (Study Group XI) 3 pp. 3 pp.

—**Failures**
CCITT RECMN Q.130-89. Special Arrangements in Case of Failures in the Sequence of Signals—Specifications of Signalling Systems Nos. 4 and 5 (Study Group XI) 1 pp. 1 p.

CCITT No. 4 Signaling Systems *(Cont.)*

—**Failures—Blocking Signals**
CCITT RECMN Q.130-89. Special Arrangements in Case of Failures in the Sequence of Signals—Specifications of Signalling Systems Nos. 4 and 5 (Study Group XI) 1 pp. 1 p.

—**Glossaries**
CCITT RECMN Q.120-89. Definition and Function of Signals—Specifications of Signalling Systems Nos. 4 and 5 (Study Group XI) 5 pp. 5 pp.

—**Incoming—Logic Procedures**
CCITT RECMN Q.611-89. Logic Procedures for Incoming Signalling System No. 4—Interworking of Signalling Systems (Study Group XI) 4 pp. 4 pp.

—**International Exchanges—Switching Speed**
CCITT RECMN Q.125-89. Speed of Switching in International Exchanges—Specifications of Signalling Systems Nos. 4 and 5 (Study Group XI) 1 pp. 1 p.

—**Maintenance**
CCITT RECMN Q.134-89. Routine Testing of Equipment (Local Maintenance)—Specifications of Signalling Systems Nos. 4 and 5 (Study Group XI) 2 pp. 2 pp.

—**Measuring Instruments**
CCITT RECMN Q.138-89. Instruments for Checking Equipment and Measuring Signals—Specifications of Signalling Systems Nos. 4 and 5 (Study Group XI) 1 pp. 1 p.

—**Measuring Instruments—Automatic—Numbering**
CCITT RECMN Q.133-89. Numbering for Access to Automatic Measuring and Testing Devices—Specifications of Signalling Systems Nos. 4 and 5 (Study Group XI) 1 pp. 1 p.

—**Outgoing—Logic Procedures**
CCITT RECMN Q.621-89. Logic Procedures for Outgoing Signalling System No. 4—Interworking of Signalling Systems (Study Group XI) 5 pp. 5 pp.

—**Receivers**
CCITT RECMN Q.123-89. Signal Receiver—Specifications of Signalling Systems Nos. 4 and 5 (Study Group XI) 3 pp. 3 pp.

—**Registers (Switching)—Release**
CCITT RECMN Q.127-89. Release of Registers—Specifications of Signalling Systems Nos. 4 and 5 (Study Group XI) 3 pp. 3 pp.
CCITT RECMN Q.131-89. Abnormal Release Conditions of the Outgoing Register Causing Release of the International Circuit—Specifications of Signalling Systems Nos. 4 and 5 (Study Group XI) 1 pp. 1 p.

—**Signal Codes**
CCITT RECMN Q.121-89. Signal Code—Specifications of Signalling Systems Nos. 4 and 5 (Study Group XI) 6 pp. 6 pp.

—**Signal Senders**
CCITT RECMN Q.122-89. Signal Sender—Specifications of Signalling Systems Nos. 4 and 5 (Study Group XI) 1 pp. 1 p.

—**Telephone Circuits—Splitting**
CCITT RECMN Q.124-89. Splitting Arrangements—Specifications of Signalling Systems Nos. 4 and 5 (Study Group XI) 1 pp. 1 p.

—**Telephone Circuits—Switching Systems**
CCITT RECMN Q.128-89. Switching to the Speech Position—Specifications of Signalling Systems Nos. 4 and 5 (Study Group XI) 1 pp. 1 p.

—**Telephone Circuits—Testing Equipment**
CCITT RECMN Q.135-89. Principles of Rapid Transmission Testing Equipment—Specifications of Signalling Systems Nos. 4 and 5 (Study Group XI) 1 pp. 1 p.
CCITT RECMN Q.136-89. Loop Transmission Measurements—Specifications of Signalling Systems Nos. 4 and 5 (Study Group XI) 1 pp. 1 p.
CCITT RECMN Q.137-89. Automatic Testing Equipment—Specifications of Signalling Systems Nos. 4 and 5 (Study Group XI) 2 pp. 2 pp.

—**Testing Equipment**
CCITT RECMN Q.134-89. Routine Testing of Equipment (Local Maintenance)—Specifications of Signalling Systems Nos. 4 and 5 (Study Group XI) 2 pp. 2 pp.

INDUSTRY STANDARDS

CCITT No. 4 Signaling Systems (Cont.)

—Testing Equipment—Automatic—Numbering

CCITT RECMN Q.133-89. Numbering for Access to Automatic Measuring and Testing Devices—Specifications of Signalling Systems Nos. 4 and 5 (Study Group XI) 1 pp. 1 p.

CCITT No. 5 Signaling Systems

Use For: Signaling System No. 5
See Also: Common Channel Signaling Systems; Signaling Systems

CCITT RECMN E.425-92. Internal Automatic Observations Telephone Network and ISDN Service, Network Management and Traffic Engineering (Study Group II) 6 pp. 6 pp.

CCITT FASCICLE VI.2. Specifications of Signalling Systems Nos. 4 and 5. Recommendations Q.120—Q.180 (Study Group XI). 106 pp.

—CCITT No. 6—Interworking—Logic Procedures

CCITT RECMN Q.642-89. Logic Procedures for Interworking of Signalling System No. 5 to No. 6—Interworking of Signalling Systems (Study Group XI) 4 pp. 4 pp.

—CCITT No. 7—Interworking—Logic Procedures

CCITT RECMN Q.643-89. Logic Procedures for Interworking of Signalling System No. 5 to No. 7 (TUP)—Interworking of Signalling Systems (Study Group XI) 4 pp. 4 pp.

—CCITT No. 7—ISDNUP—Interworking

CENELEC PRETS 300 343-93. Integrated Services Digital Network (ISDN); Signalling Interworking Specification for ISDN User Part (ISUP) Version 1. 51 pp.

ETSI PRETS 300 343-93. Integrated Services Digital Network (ISDN); Signalling Interworking Specification for ISDN User Part (ISUP) Version 1. 51 pp.

—CCITT R1—Interworking—Logic Procedures

CCITT RECMN Q.644-89. Logic Procedures for Interworking of Signalling System No. 5 to R1—Interworking of Signalling Systems (Study Group XI) 2 pp. 2 pp.

—CCITT R2—Interworking—Logic Procedures

CCITT RECMN Q.645-89. Logic Procedures for Interworking of Signalling System No. 5 to R2—Interworking of Signalling Systems (Study Group XI) 3 pp. 3 pp.

—End of Pulsing Signals—Register Signaling

CCITT RECMN Q.152-89. End-of-Pulsing Conditions—Register Arrangements Concerning ST (End-of-Pulsing) Signal—Specifications of Signalling Systems Nos. 4 and 5 (Study Group XI) 2 pp. 2 pp.

—Glossaries

CCITT RECMN Q.140-89. Definition and Function of Signals—Specifications of Signalling Systems Nos. 4 and 5 (Study Group XI) 3 pp. 3 pp.

—Incoming—Logic Procedures

CCITT RECMN Q.612-89. Logic Procedures for Incoming Signalling System No. 5—Interworking of Signalling Systems (Study Group XI) 5 pp. 5 pp.

—INMARSAT—Interworking

CCITT RECMN Q.1103-89. Interworking Between Signalling System No.5 and Inmarsat Standard A System—Interworking with Satellite Mobile Systems (Study Group XI) 8 pp. 8 pp.

—Line Signaling—International Exchanges—Switching Speed

CCITT RECMN Q.146-89. Speed of Switching in International Exchanges—Specifications of Signalling Systems Nos. 4 and 5 (Study Group XI) 1 pp. 1 p.

—Line Signaling—Telephone Circuits—Splitting

CCITT RECMN Q.145-89. Splitting Arrangements—Specifications of Signalling Systems Nos. 4 and 5 (Study Group XI) 1 pp. 1 p.

—Maintenance

CCITT RECMN Q.162-89. Routing Testing of Equipment (Local Maintenance)—Specifications of Signalling Systems Nos. 4 and 5 (Study Group XI) 2 pp. 2 pp.

CCITT No. 5 Signaling Systems (Cont.)

—Measuring Instruments

CCITT RECMN Q.164-89. Test Equipment for Checking Equipment and Signals—Specifications of Signalling Systems Nos. 4 and 5 (Study Group XI) 2 pp. 2 pp.

—Outgoing—Logic Procedures

CCITT RECMN Q.622-89. Logic Procedures for Outgoing Signalling System No. 5—Interworking of Signalling Systems (Study Group XI) 4 pp. 4 pp.

—Receivers—Line Signaling

CCITT RECMN Q.144-89. Line Signal Receiver—Specifications of Signalling Systems Nos. 4 and 5 (Study Group XI) 3 pp. 3 pp.

—Receivers—Multifrequency Signaling Systems

CCITT RECMN Q.154-89. Multifrequency Signal Receiver—Specifications of Signalling Systems Nos. 4 and 5 (Study Group XI) 2 pp. 2 pp.

—Registers (Switching)—Release

CCITT RECMN Q.156-89. Release of International Registers—Specifications of Signalling Systems Nos. 4 and 5 (Study Group XI) 2 pp. 2 pp.

—Signal Codes—Line Signaling

CCITT RECMN Q.141-89. Signal Code for Line Signalling—Specifications of Signalling Systems Nos. 4 and 5 (Study Group XI) 5 pp. 5 pp.

—Signal Codes—Register Signaling

CCITT RECMN Q.151-89. Signal Code for Register Signalling—Specifications of Signalling Systems Nos. 4 and 5 (Study Group XI) 2 pp. 2 pp.

—Signal Senders—Line Signaling

CCITT RECMN Q.143-89. Line Signal Sender—Specifications of Signalling Systems Nos. 4 and 5 (Study Group XI) 1 pp. 1 p.

—Signal Senders—Multifrequency Signaling Systems

CCITT RECMN Q.153-89. Multifrequency Signal Sender—Specifications of Signalling Systems Nos. 4 and 5 (Study Group XI) 1 pp. 1 p.

—Telephone Circuits—Both-Way Operation—Double Seizing

CCITT RECMN Q.142-89. Double Seizing with Both-Way Operation—Specifications of Signalling Systems Nos. 4 and 5 (Study Group XI) 2 pp. 2 pp.

—Telephone Circuits—Manual Testing

CCITT RECMN Q.163-89. Manual Testing—Specifications of Signalling Systems Nos. 4 and 5 (Study Group XI) 3 pp. 3 pp.

—Telephone Circuits—Speech Position—Register Signaling

CCITT RECMN Q.157-89. Switching to the Speech Position—Specifications of Signalling Systems Nos. 4 and 5 (Study Group XI) 1 pp. 1 p.

—Testing Equipment

CCITT RECMN Q.162-89. Routing Testing of Equipment (Local Maintenance)—Specifications of Signalling Systems Nos. 4 and 5 (Study Group XI) 2 pp. 2 pp.

CCITT RECMN Q.164-89. Test Equipment for Checking Equipment and Signals—Specifications of Signalling Systems Nos. 4 and 5 (Study Group XI) 2 pp. 2 pp.

CCITT No. 6 Signaling Systems

Use For: Signaling System No. 6
See Also: Common Channel Signaling Systems; Signaling Systems

CCITT FASCICLE VI.3. Specifications of Signalling System No. 6 Recommendations Q.251—Q.300 (Study Group XI). 153 pp.

CCITT RECMN Q.251-89. 1.1 General—Specifications of Signalling System No. 6 (Study Group XI) 4 pp. 4 pp.

—Abbreviations

CCITT FASCICLE VI.3. Specifications of Signalling System No. 6 Recommendations Q.251—Q.300 (Study Group XI). 153 pp.

—Automatic Repeat Attempts

CCITT RECMN Q.264-89. 4.4 Potential for Automatic Repeat Attempt and Re-Routing—Specifications of Signalling System No. 6 (Study Group XI) 1 pp. 1 p.

CCITT No. 6 Signaling Systems (Cont.)

—Blocking Signals

CCITT RECMN Q.266-89. 4.6 Blocking and Unblocking Sequences and Control of Quasi-Associated Signalling—Specifications of Signalling System No. 6 (Study Group XI) 3 pp. 3 pp.

—Call Set Up

CCITT RECMN Q.261-89. 4.1 Normal Call Set-Up—Specifications of Signalling System No. 6 (Study Group XI) 7 pp. 7 pp.

—CCITT No. 5—Interworking—Logic Procedures

CCITT RECMN Q.652-89. Logic Procedures for Interworking of Signalling System No. 6 to No. 5—Interworking of Signalling Systems (Study Group XI) 3 pp. 3 pp.

—CCITT No. 7—Interworking—Logic Procedures

CCITT RECMN Q.653-89. Logic Procedures for Interworking of Signalling System No. 6 to No. 7 (TUP)—Interworking of Signalling Systems (Study Group XI) 4 pp. 4 pp.

—CCITT R1—Interworking—Logic Procedures

CCITT RECMN Q.654-89. Logic Procedures for Interworking of Signalling System No. 6 to R1—Interworking of Signalling Systems (Study Group XI) 2 pp. 2 pp.

—CCITT R2—Interworking—Logic Procedures

CCITT RECMN Q.655-89. Logic Procedures for Interworking of Signalling System No. 6 to R2—Interworking of Signalling Systems (Study Group XI) 3 pp. 3 pp.

—Common Channel—Interworking

CCITT RECMN Q.300-89. Interworking Between CCITT Signalling System No. 6 and National Common Channel Signalling Systems—Specifications of Signalling System No. 6 (Study Group XI) 4 pp. 4 pp.

—Comprehensive Testing

CCITT RECMN Q.295-89. 9.1 Overall Test of Signalling System No. 6—Specifications of Signalling System No. 6 (Study Group XI) 5 pp. 5 pp.

—Continuity Checks—Speech Paths

CCITT RECMN Q.271-89. 1.1 General—Specifications of Signalling System No. 6 (Study Group XI) 3 pp. 3 pp.

—Control Signals—Glossaries

CCITT RECMN Q.255-89. 2.2 Signalling-System-Control Signals—Specifications of Signalling System No. 6 (Study Group XI) 2 pp. 2 pp.

—Control Signals—Signal Units—Formats/Codes

CCITT RECMN Q.259-89. 3.3 Signalling-System-Control Signals—Specifications of Signalling Sytem No. 6 (Study Group XI) 4 pp. 4 pp.

—Glossaries

CCITT FASCICLE VI.3. Specifications of Signalling System No. 6 Recommendations Q.251—Q.300 (Study Group XI). 153 pp.

—Incoming—Logic Procedures

CCITT RECMN Q.613-89. Logic Procedures for Incoming Signalling System No. 6—Interworking of Signalling Systems (Study Group XI) 10 pp. 10 pp.

—International Exchanges—Signal Transfer

CCITT RECMN Q.265-89. 4.5 Speed of Switching and Signal Transfer in International Exchanges—Specifications of Signalling System No. 6 (Study Group XI) 1 pp. 1 p.

—International Exchanges—Switching Speed

CCITT RECMN Q.265-89. 4.5 Speed of Switching and Signal Transfer in International Exchanges—Specifications of Signalling System No. 6 (Study Group XI) 1 pp. 1 p.

—Interworking

CCITT RECMN Q.300-89. Interworking Between CCITT Signalling System No. 6 and National Common Channel Signalling Systems—Specifications of Signalling System No. 6 (Study Group XI) 4 pp. 4 pp.

—Maintenance

CCITT RECMN M.762-89. Maintenance of Common Channel Signalling System No. 6—Maintenance of International Transmission Systems and Telephone Circuits (Study Group IV) 3 pp. 3 pp.

INTERNATIONAL AND NON-U.S. NATIONAL STANDARDS
SUBJECT INDEX — CCITT

CCITT No. 6 Signaling Systems (Cont.)

—Maintenance (Cont.)
CCITT RECMN Q.296-89. 9.6 Monitoring and Maintenance of the Common Signalling Channel—Specifications of Signalling System No. 6 (Study Group XI) 6 pp. 6 pp.

—Management Signals—Glossaries
CCITT RECMN Q.256-89. 2.3 Management Signals—Specifications of Signalling System No. 6 (Study Group XI) 1 pp. 1 p.

—Management Signals—Signal Units—Formats/Codes
CCITT RECMN Q.260-89. 3.4 Management Signals—Specifications of Signalling System No. 6 (Study Group XI) 5 pp. 5 pp.

—Monitoring
CCITT RECMN Q.296-89. 9.6 Monitoring and Maintenance of the Common Signalling Channel—Specifications of Signalling System No. 6 (Study Group XI) 6 pp. 6 pp.

—Network Management Systems
CCITT RECMN Q.297-89. 10. Network Management—Specifications of Signalling System No. 6 (Study Group XI) 13 pp. 13 pp.

—Outgoing—Logic Procedures
CCITT RECMN Q.623-89. Logic Procedures for Outgoing Signalling System No. 6—Interworking of Signalling Systems (Study Group XI) 6 pp. 6 pp.

—Quasiassociated Signaling
CCITT RECMN Q.266-89. 4.6 Blocking and Unblocking Sequences and Control of Quasi-Associated Signalling—Specifications of Signalling System No. 6 (Study Group XI) 3 pp. 3 pp.

—Rerouting
CCITT RECMN Q.264-89. 4.4 Potential for Automatic Repeat Attempt and Re-Routing—Specifications of Signalling System No. 6 (Study Group XI) 1 pp. 1 p.

—Security Arrangements
CCITT RECMN Q.291-89. Security Arrangements—General. Basic Security Arrangements. Types of Failure, Recognition of Failure and Abnormal Error Rates—Specifications of Signalling System No. 6 (Study Group XI) 4 pp. 4 pp.
CCITT RECMN Q.292-89. 8.4 Reserve Facilities Provided—Specifications of Signalling System No. 6 (Study Group XI) 3 pp. 3 pp.
CCITT RECMN Q.293-89. 8.5 Intervals at Which Security Measures are to be Invoked—Specifications of Signalling System No. 6 (Study Group XI) 8 pp. 8 pp.

—Signal Priority
CCITT RECMN Q.285-89. Signal Priority Categories—Specifications of Signalling System No. 6 (Study Group XI) 1 pp. 1 p.

—Signal Transfer Points
CCITT RECMN Q.253-89. 1.3 Association Between Signalling and Speech Networks—Specifications of Signalling System (Study Group XI) 3pp. 3 pp.

—Signal Units—Formats/Codes
CCITT RECMN Q.257-89. 3.1 General—Specifications of Signalling System No. 6 (Study Group XI) 5 pp. 5 pp.

—Signaling Channels
CCITT RECMN Q.286-89. 7.2 Signalling Channel Loading and Queueing Delays—Specifications of Signalling System No. 6 (Study Group XI) 4 pp. 4 pp.

—Signaling Data Links
CCITT RECMN Q.272-89. 6.1 Requirements for the Signalling Data Link—Specifications of Signalling System No. 6 (Study Group XI) 6 pp. 6 pp.

—Signaling Data Links—Maintenance
CCITT RECMN Q.295-89. 9.1 Overall Test of Signalling System No. 6—Specifications of Signalling System No. 6 (Study Group XI) 5 pp. 5 pp.

—Signaling Links—Communication Interfaces
CCITT RECMN Q.274-89. 6.3 Transmission Methods—Specifications of Signalling System No. 6 (Study Group XI) 8 pp. 8 pp.

—Signaling Links—Data Channel Failure Detectors
CCITT RECMN Q.275-89. 6.5 Data Channel Failure Detection—Specifications of Signalling System No. 6 (Study Group XI) 1 pp. 1 p.

CCITT No. 6 Signaling Systems (Cont.)

—Signaling Links—Data Transfer Rates
CCITT RECMN Q.273-89. 6.2 Data Transmission Rate—Specifications of Signalling System No. 6 (Study Group XI) 1 pp. 1 p.

—Signaling Links—Data Transmission
CCITT RECMN Q.274-89. 6.3 Transmission Methods—Specifications of Signalling System No. 6 (Study Group XI) 8 pp. 8 pp.

—Signaling Links—Drift Compensation
CCITT RECMN Q.279-89. 6.9 Drift Compensation—Specifications of Signalling System No. 6 (Study Group XI) 1 pp. 1 p.

—Signaling Links—Error Control
CCITT RECMN Q.277-89. 6.7 Error Control—Specifications of Signalling System No. 6 (Study Group XI) 3 pp. 3 pp.

—Signaling Links—Modems
CCITT RECMN Q.274-89. 6.3 Transmission Methods—Specifications of Signalling System No. 6 (Study Group XI) 8 pp. 8 pp.

—Signaling Links—Reliability
CCITT RECMN Q.276-89. 6.6 Service Dependability—Specifications of Signalling System No. 6 (Study Group XI) 2 pp. 2 pp.

—Signaling Links—Synchronization
CCITT RECMN Q.278-89. 6.8 Synchronization—Specifications of Signalling System No. 6 (Study Group XI) 4 pp. 4 pp.

—Speech Circuits
CCITT RECMN Q.253-89. 1.3 Association Between Signalling and Speech Networks—Specifications of Signalling System (Study Group XI) 3pp. 3 pp.

—Superfluous Messages
CCITT RECMN Q.267-89. 4.7 Unreasonable and Superfluous Messages—Specifications of Signalling System No. 6 (Study Group XI) 4 pp. 4 pp.

—Telecommunication Administrations—Agreements
CCITT RECMN M.750-89. Inter-Administration Agreements on Common Channel Signalling System No. 6—General Maintenance Principles—Maintenance of Transmission Systems and Telephone Circuits (Study Group IV) 4 pp. 4 pp.

—Telephone Circuits—Both-Way Operation—Double Seizing
CCITT RECMN Q.263-89. 4.3 Double Seizing with Both-Way Operation—Specifications of Signalling System No. 6 (Study Group XI) 2 pp. 2 pp.

—Telephone Connections—Release
CCITT RECMN Q.268-89. 4.8 Release of International Connections and Associated Equipment—Specifications of Signalling System No. 6 (Study Group XI) 5 pp. 5 pp.

—Telephone Networks—Maintenance
CCITT RECMN Q.295-89. 9.1 Overall Test of Signalling System No. 6—Specifications of Signalling System No. 6 (Study Group XI) 5 pp. 5 pp.

—Telephone Signals—Glossaries
CCITT RECMN Q.254-89. 2.1 Telephone Signals—Specifications of Signalling System No. 6 (Study Group XI) 4 pp. 4 pp.

—Telephone Signals—Signal Units—Formats/Codes
CCITT RECMN Q.258-89. 3.2 Telephone Signals—Specifications of Signalling System No. 6 (Study Group XI) 8 pp. 8 pp.

—Transfer Links
CCITT RECMN M.760-89. Transfer Link for Common Channel Signalling System No. 6—General Maintenance Principles—Maintenance of Transmission Systems and Telephone Circuits (Study Group IV) 5 pp. 5 pp.
CCITT RECMN M.761-89. Setting up and Lining up a Transfer Link for Common Channel Signalling System No. 6 (Analogue Version)—General Maintenance Principles—Maintenance of International Transmission Systems and Telephone Circuits (Study Group IV) 4 pp. 4 pp.

—Transfer Time
CCITT RECMN Q.287-89. 7.3 Signal Transfer Time Requirements—Specifications of Signalling System No. 6 (Study Group XI) 3 pp. 3 pp.

CCITT No. 6 Signaling Systems (Cont.)

—Transfer Time—Glossaries
CCITT RECMN Q.252-89. 1.2 Signal Transfer Time Definitions—Specifications of Signalling System No. 6 (Study Group XI) 2 pp. 2 pp.

—Unreasonable Messages
CCITT RECMN Q.267-89. 4.7 Unreasonable and Superfluous Messages—Specifications of Signalling System No. 6 (Study Group XI) 4 pp. 4 pp.

CCITT No. 7 Signaling Systems
Use For: Signaling System No. 7
See Also: Common Channel Signaling Systems; Integrated Services Digital Networks User Part; Operations, Maintenance, and Administration Part; Signaling Connection Control Part; Signaling Systems; Transmission Medium Requirements
CCITT RECMN Q.700-89. Introduction to CCITT Signalling System No. 7—Specifications of Signalling System No. 7 (Study Group XI) 24 pp. 24 pp.

—Abbreviations
CCITT FASCICLE VI.7-88. Specifications of Signalling System No. 7 Recommendations Q.700—Q.716. 593 pp.
CCITT FASCICLE VI.8-88. Specifications of Signalling System No. 7 Recommendations Q.721—Q.766. 475 pp.

—CCITT No. 5—Interworking—Logic Procedures
CCITT RECMN Q.662-89. Logic Procedures for Interworking of Signalling System No. 7 (TUP) to No. 5—Interworking of Signalling Systems (Study Group XI) 3 pp. 3 pp.

—CCITT No. 5—R2—Telephone User Part—Interworking
CENELEC PRETS 300 343-93. Integrated Services Digital Network (ISDN); Signalling Interworking Specification for ISDN User Part (ISUP) Version 1. 51 pp.
ETSI PRETS 300 343-93. Integrated Services Digital Network (ISDN); Signalling Interworking Specification for ISDN User Part (ISUP) Version 1. 51 pp.

—CCITT No. 6—Interworking—Logic Procedures
CCITT RECMN Q.663-89. Logic Procedures for Interworking of Signalling System No. 7 (TUP) to No. 6—Interworking of Signalling Systems (Study Group XI) 3 pp. 3 pp.

—CCITT R1—Interworking—Logic Procedures
CCITT RECMN Q.665-89. Logic Procedures for Interworking of Signalling System No. 7 (TUP) to R1—Interworking of Signalling Systems (Study Group XI) 3 pp. 3 pp.

—CCITT R2—Interworking—Logic Procedures
CCITT RECMN Q.666-89. Logic Procedures for Interworking of Signalling System No. 7 (TUP) to R2—Interworking of Signalling Systems (Study Group XI) 3 pp. 3 pp.

—D Channel—Telecommunication Traffic—Call Attempts
CCITT RECMN E.713-89. Control Plane Traffic Models—Telephone Network and ISDN—Quality of Service, Network Management and Traffic Engineering (Study Group II) 7 pp. 7 pp.

—Glossaries
CCITT FASCICLE VI.7-88. Specifications of Signalling System No. 7 Recommendations Q.700—Q.716. 593 pp.
CCITT FASCICLE VI.8-88. Specifications of Signalling System No. 7 Recommendations Q.721—Q.766. 475 pp.

—Grade of Service—Integrated Services Digital Networks
CCITT RECMN E.723-92. Grade-of-Service Parameters for Signalling System No. 7 Networks (Study Group II) 8 pp. 8 pp.

—Incoming—Logic Procedures
CCITT RECMN Q.614-89. Logic Procedures for Incoming Signalling System No. 7 (TUP)—Interworking of Signalling Systems (Study Group XI) 14 pp. 14 pp.

—Integrated Services Digital Networks
CCITT RECMN E.172-89. Call Routing in the ISDN Era—Telephone Network and ISDN—Operation, Numbering, Routing and Mobile Service (Study Group II) 14 pp. 14 pp.

INTERNATIONAL AND NON-U.S. NATIONAL STANDARDS
SUBJECT INDEX

CCITT No. 7 Signaling Systems (Cont.)

—Integrated Services Digital Networks—Supplementary Services
CCITT RECMN Q.730-89. ISDN Supplementary Services—Specifications of Signalling System No. 7 (Study Group XI) 59 pp. 59 pp.
CCITT RECMN Q.731-91. Stage 3 Description for Number Identification Supplementary Services Using Signalling System No. 7 Section 1—Direct Dialling in (DDI) Section 8—Sub-Addressing (SUB)—Specifications of Signalling System No. 7 (Study Group XI) 12 pp. 12 pp.
CCITT RECMN Q.733-92. Stage 3 Description for Call Completion Supplementary Services Using No. 7 Signalling System Section 1—Call Waiting (CW)—Specifications of Signalling System No. 7 (Study Group XI) 10 pp. 10 pp.

—Integrated Services Digital Networks User Part—Call Testing
CCITT RECMN Q.784-91. ISUP Basic Call Test Specification (Study Group XI) 83 pp. 83 pp.
CENELEC PRETS 300 335-93. Integrated Services Digital Network (ISDN); CCITT Signalling System No. 7 Integrated Services User Part (ISUP) Version 1 Test Specification. 120 pp.
ETSI PRETS 300 335-93. Integrated Services Digital Network (ISDN); CCITT Signalling System No. 7 Integrated Services User Part (ISUP) Version 1 Test Specification. 120 pp.

—International Signaling Point Codes—Numbering
CCITT RECMN Q.708-89. Numbering of International Signalling Point Codes—Specifications of Signalling System No. 7 (Study Group XI) 7 pp. 7 pp.

—ISDN—Dimensioning
CCITT RECMN E.730-92. ISDN Dimensioning Methods Overview (Study Group II) 4 pp. 4 pp.
CCITT RECMN E.733-92. Methods for Dimensioning Resources in Signalling System No. 7 Networks (Study Group II) 10 pp. 10 pp.

—ISDNUP—International Exchanges
CENELEC PRETS 300 121-90. Integrated Services Digital Network, (ISDN); Application of the ISDN User part of CCITT Signalling System No.7 for International ISDN Interconnections CCITT Recommendation Q.767 Draft edition 3: 1990-Modified. 239 pp.
CENELEC PRETS 300 121-92. Integrated Services Digital Network (ISDN); Application of the ISDN User Part (ISUP) of CCITT Signalling System No. 7 for International ISDN Interconnections (ISUP Version 1). 7 pp.
CENELEC ETS 300 121-92. Integrated Services Digital Network (ISDN); Application of the ISDN User Part (ISUP) of CCITT Signalling System No. 7 for International ISDN Interconnections (ISUP Version 1). 7 pp.
ETSI PRETS 300 121-90. Integrated Services Digital Network (ISDN); Application of the ISDN User Part of CCITT Signalling System No.7 for International ISDN Interconnections CCITT Recommendation Q.767 Draft Edition 3:1990—Modified (ISUP Version 1—T/S 43-14). 239 pp.
ETSI ETS 300 121-92. Integrated Services Digital Network (ISDN); Application of the ISDN User Part (ISUP) of CCITT Signalling System No. 7 for International ISDN Interconnections (ISUP Version 1). 7 pp.
ETSI PRETS 300 121-92. Integrated Services Digital Network (ISDN); Application of the ISDN User Part (ISUP) of CCITT Signalling System No. 7 for International ISDN Interconnections (ISUP Version 1). 7 pp.
ETSI PRETS 300 121-90. Integrated Services Digital Network, (ISDN); Application of the ISDN User part of CCITT Signalling System No.7 for International ISDN Interconnections CCITT Recommendation Q.767 Draft edition 3: 1990-Modified. 239 pp.

—ISDNUP—Routing (Telecommunications)
CENELEC PRETS 300 334-93. Integrated Services Digital Network (ISDN); Routeing in Support of ISDN User Part (ISUP) Version 2 Services. 18 pp.
ETSI PRETS 300 334-93. Integrated Services Digital Network (ISDN); Routeing in Support of ISDN User Part (ISUP) Version 2 Services. 18 pp.

—ISDNUP—Supplementary Services—Comprehensive Testing
CCITT RECMN Q.785-91. ISUP Protocol Test Specification for Supplementary Services (Study Group XI) 52 pp. 52 pp.

CCITT No. 7 Signaling Systems (Cont.)

—ISDNUP—Telecommunication Connections
CCITT RECMN Q.767-91. Application of the ISDN User Part of CCITT Signalling System No. 7 for International ISDN Interconnections (Study Group XI) 274 pp. 274 pp.

—Land Mobile Services
CCIR Report 1153-90. Future Public Land Mobile Telecommunication Systems—Section 8A—Land Mobile Service and Related Subjects. 62 pp.

—Logic Procedures—Interworking
CCITT RECMN Q.664-89. Logic Procedures for Interworking of Signalling System No. 7 (TUP) to No. 7 (TUP)—Interworking of Signalling Systems (Study Group XI) 5 pp. 5 pp.

—Maintenance
CCITT RECMN M.782-89. Maintenance of Common Signalling System No. 7—General Maintenance Principles—Maintenance of International Transmission Systems and Telephone Circuits (Study Group IV) 6 pp. 6 pp.

—Message Transfer Part
CCITT RECMN Q.701-89. Functional Description of the Message Transfer Part (MTP) of Signalling System No. 7—Specifications of Signalling System No. 7 (Study Group XI) 19 pp. 19 pp.
CCITT RECMN Q.706-89. Message Transfer Part Signalling Performance—Specifications of Signalling System No. 7 (Study Group XI) 18 pp. 18 pp.
CCITT RECMN Q.710-89. Simplified MTP Version for Small Systems—Specifications of Signalling System No. 7 (Study Group XI) 6 pp. 6 pp.
CENELEC PRETS 300 336-93. Integrated Services Digital Network (ISDN); CCITT Signalling System No. 7 Message Transfer Part (MTP) Test Specification. 6 pp.
ETSI PRETS 300 336-93. Integrated Services Digital Network (ISDN); CCITT Signalling System No. 7 Message Transfer Part (MTP) Test Specification. 6 pp.

—Message Transfer Part—Circuit Switched Data Networks
CCITT RECMN X.60-89. Common Channel Signalling for Circuit Switched Data Applications—Data Communication Networks—Transmission, Signalling and Switching, Network Aspects, Maintenance and Administrative Arrangements (Study Group VII) 2 pp. 2 pp.

—Message Transfer Part—Circuit Switched Data Transmission Services
CCITT RECMN X.61-89. Signalling System No. 7—Data User Part—Data Communication Networks—Transmission, Signalling and Switching, Network Aspects, Maintenance and Administrative Arrangements (Study Group VII) 54 pp. 54 pp.

—Message Transfer Part—Measurement
CCITT RECMN Q.791-89. Monitoring and Measurements for Signalling System No. 7 Networks—Specifications of Signalling System No.7 (Study Group XI) 20 pp. 20 pp.

—Message Transfer Part—Mesh Networks
CCITT FASCICLE VI.7-88. Specifications of Signalling System No. 7 Recommendations Q.700—Q.716. 593 pp.

—Message Transfer Part—Monitoring
CCITT RECMN Q.791-89. Monitoring and Measurements for Signalling System No. 7 Networks—Specifications of Signalling System No.7 (Study Group XI) 20 pp. 20 pp.

—Message Transfer Part—Protocol Testing
CCITT RECMN Q.781-89. MTP Level 2 Test Specification—Specifications of Signalling System No. 7 (Study Group XI) 101 pp. 101 pp.
CCITT RECMN Q.782-89. MTP Level 3 Test Specification—Specifications of Signalling System No. 7 (Study Group XI) 141 pp. 141 pp.

—Message Transfer Part—Signaling Data Links
CCITT RECMN Q.702-89. Signalling Data Link—Specifications of Signalling System No. 7 (Study Group XI) 6 pp. 6 pp.

—Message Transfer Part—Signaling Links
CCITT RECMN Q.703-89. Signalling Link—Specifications of Signalling System No. 7 (Study Group XI) 73 pp. 73 pp.

CCITT No. 7 Signaling Systems (Cont.)

—Message Transfer Part—Signaling Networks
CCITT RECMN Q.704-89. Signalling Network Functions and Messages—Specifications of Signalling System No. 7 (Study Group XI) 185 pp. 185 pp.

—Message Transfer Part—Signaling Networks—Design
CCITT RECMN Q.705-89. Signalling Network Structure—Specifications of Signalling System No. 7 (Study Group XI) 20 pp. 20 pp.

—Message Transfer Part—Telecommunication Connections
CCITT RECMN Q.709-89. Hypothetical Signalling Reference Connection—Specifications of Signalling System No. 7 (Study Group XI) 10 pp. 10 pp.

—Network Layer—Signaling Connection Control Part
CCITT FASCICLE VI.7-88. Specifications of Signalling System No. 7 Recommendations Q.700—Q.716. 593 pp.

—Network Management Systems
CCITT RECMN E.415-91. International Network Management Guidance for Common Channel Signalling System No. 7 (Study Group II) 10 pp. 10 pp.

—Outgoing—Logic Procedures
CCITT RECMN Q.624-89. Logic Procedures for Outgoing Signalling System No. 7 (TUP)—Interworking of Signalling Systems (Study Group XI) 11 pp. 11 pp.

—Protocol Testing
CCITT RECMN Q.780-89. Signalling System No. 7 Test Specification General Description—Specifications of Signalling System No. 7 (Study Group XI) 5 pp. 5 pp.

—Public Land Mobile Networks—Mobile Application Part
CCITT RECMN Q.1051-89. Mobile Application Part—Public Land Mobile Network—Mobile Application Part and Interfaces (Study Group XI) 303 pp. 303 pp.

—Signal Transfer Points—Accounting
CCITT RECMN D.211-89. International Accounting for the Use of the Signal Transfer Point (STP) in CCITT Signalling System No. 7—General Tariff Principles—Charging and Accounting in International Telecommunications Services (Study Group III) 1 pp. 1 p.

—Signaling Connection Control Part
CCITT RECMN Q.711-89. Functional Description of the Signalling Connection Control Part—Specifications of Signalling System No. 7 (Study Group XI) 27 pp. 27 pp.
CCITT RECMN Q.712-89. Definition and Function of SCCP Messages—Specifications of Signalling System No. 7 (Study Group XI) 7 pp. 7 pp.
CCITT RECMN Q.713-89. SCCP Formats and Codes—Specifications of Signalling System No. 7 (Study Group XI) 33 pp. 33 pp.
CCITT RECMN Q.714-89. Signalling Connection Control Part Procedures—Specifications of Signalling System No. 7 (Study Group XI) 101 pp. 101 pp.
CCITT RECMN Q.716-89. Signalling Connection Control Part (SCCP) Performances—Specifications of Signalling System No. 7 (Study Group XI) 10 pp. 10 pp.

—Signaling Connection Control Part—Glossaries
CCITT RECMN Q.712-89. Definition and Function of SCCP Messages—Specifications of Signalling System No. 7 (Study Group XI) 7 pp. 7 pp.

—Signaling Connection Control Part—Measurement
CCITT RECMN Q.791-89. Monitoring and Measurements for Signalling System No. 7 Networks—Specifications of Signalling System No.7 (Study Group XI) 20 pp. 20 pp.

—Signaling Connection Control Part—Mobile Global Titles
CCITT RECMN E.214-89. Structure of the Land Mobile Global Title for the Signalling Connection Control Part (SCCP)—Telephone Network and ISDN—Operation, Numbering, Routing and Mobile Service (Study Group II) 3 pp. 3 pp.

INTERNATIONAL AND NON-U.S. NATIONAL STANDARDS
SUBJECT INDEX

CCITT

CCITT No. 7 Signaling Systems (Cont.)

—Signaling Connection Control Part—Monitoring
CCITT RECMN Q.791-89. Monitoring and Measurements for Signalling System No. 7 Networks—Specifications of Signalling System No.7 (Study Group XI) 20 pp. 20 pp.

—Signaling Networks—Comprehensive Testing
CCITT RECMN Q.707-89. Testing and Maintenance—Specifications of Signalling System No. 7 (Study Group XI) 7 pp. 7 pp.

—Signaling Networks—Maintenance
CCITT RECMN Q.707-89. Testing and Maintenance—Specifications of Signalling System No. 7 (Study Group XI) 7 pp. 7 pp.

—Telecommunication Administrations—Agreements
CCITT RECMN M.770-89. Inter-Administration Agreements on Common Channel Signalling System No. 7—General Maintenance Principles—Maintenance of International Transmission Systems and Telephone Circuits (Study Group IV) 4 pp. 4 pp.

—Telecommunication Traffic
CCITT RECMN E.713-89. Control Plane Traffic Models—Telephone Network and ISDN—Quality of Service, Network Management and Traffic Engineering (Study Group II) 7 pp. 7 pp.

—Telecommunication Traffic—Call Attempts
CCITT RECMN E.713-89. Control Plane Traffic Models—Telephone Network and ISDN—Quality of Service, Network Management and Traffic Engineering (Study Group II) 7 pp. 7 pp.

—Telecommunication Traffic—Measurement
CCITT RECMN E.505-92. Measurements of the Performance of Common Channel Signalling Network (Study Group II) 16 pp. 16 pp.

—Telephone User Part
CCITT RECMN Q.721-89. Functional Description of the Signalling System No. 7 Telephone User Part (TUP)—Specifications of Signalling System No. 7 (Study Group XI) 2 pp. 2 pp.
CCITT RECMN Q.723-89. Formats and Codes—Specifications of Signalling System No. 7 (Study Group XI) 29 pp. 29 pp.
CCITT RECMN Q.724-89. Signalling Procedures—Specifications of Signalling System No. 7 (Study Group XI) 75 pp. 75 pp.
CCITT RECMN Q.725-89. Signalling Performance in the Telephone Application—Specifications of System No. 7 (Study Group XI) 7 pp. 7 pp.
CEPT T/S 43-02 E-89. Signalling System Telephone User Part "Plus" (TUP+) (Revised in Edinburgh 1988). 69 pp.
CEPT T/S 43-02-87. Sous-Systeme Utillisateur Telephonique "Plus" (SSUT+). 69 pp.
CEPT T/S 43-02 E-87. Signalling System Telephone User Part "Plus" (TUP+). 69 pp.

—Telephone User Part—Protocol Testing
CCITT RECMN Q.783-89. TUP Test Specification—Specifications of Signalling System No. 7 (Study Group XI) 82 pp. 82 pp.

—Telephone User Part—Telephone Messages
CCITT RECMN Q.722-89. General Function of Telephone Messages and Signals—Specifications of Signalling System No. 7 (Study Group XI) 12 pp. 12 pp.

—Telephone User Part—Telephone Signals
CCITT RECMN Q.722-89. General Function of Telephone Messages and Signals—Specifications of Signalling System No. 7 (Study Group XI) 12 pp. 12 pp.

—Transaction Capabilities Application Part
CCITT RECMN Q.771-89. Functional Description of Transaction Capabilities—Specifications of Signalling System No. 7 (Study Group XI) 30 pp. 30 pp.
CCITT RECMN Q.772-89. Transaction Capabilities Information Element Definitions—Specifications of Signalling System No. 7 (Study Group XI) 5 pp. 5 pp.
CCITT RECMN Q.773-89. Transaction Capabilities Formats and Encoding—Specifications of Signalling System No. 7 (Study Group XI) 23 pp. 23 pp.
CCITT RECMN Q.774-89. Transaction Capabilities Procedures—Specifications of Signalling System No. 7 (Study Group XI) 41 pp. 41 pp.
CCITT RECMN Q.775-89. Guidelines for Using Transaction Capabilities—Specifications of Signalling System No. 7 (Study Group XI) 33 pp. 33 pp.

CCITT No. 7 Signaling Systems (Cont.)

—Transaction Capabilities Application Part (Cont.)
CENELEC PRETS 300 287-92. Integrated Services Digital Network (ISDN); CCITT Signalling System No.7 Transaction Capabilities Application Part (TCAP) (Version 2). 7 pp.
CENELEC PRETS 300 287-93. Integrated Services Digital Network (ISDN); CCITT No. 7 Transaction Capabilities Application Part (TCAP) (Version 2). 8 pp.
CENELEC PRETS 300 344-93. Integrated Services Digital Network (ISDN); CCITT Signalling System No. 7 Transaction Capabilities Application Part (TCAP) Test Specification. 6 pp.
ETSI PRETS 300 287-92. Integrated Services Digital Network (ISDN); CCITT Signalling System No.7 Transaction Capabilities Application Part (TCAP) (Version 2). 7 pp.
ETSI PRETS 300 287-93. Integrated Services Digital Network (ISDN); CCITT Signalling System No.7 Transaction Capabilities Application Part (TCAP) (Version 2). 8 pp.
ETSI PRETS 300 344-93. Integrated Services Digital Network (ISDN); CCITT Signalling System No. 7 Transaction Capabilities Application Part (TCAP) Test Specification. 6 pp.

CCITT R1 Signaling Systems

See Also: Interregister Signaling; Signaling Systems
CCITT FASCICLE VI.4. Specifications of Signalling Systems R1 and R2 Recommendations Q.310—Q.490 (Study Group XI). 181 pp.

—Automatic Transmission Measuring and Signaling Testing Equipment
CCITT RECMN Q.330-89. Automatic Transmission and Signallimg Testing—Specifications of Signalling Systems R1 and R2 (Study Group XI) 1 pp. 1 p.

—CCITT No. 5—Interworking—Logic Procedures
CCITT RECMN Q.671-89. Logic Procedures for Interworking of Signalling System R1 to No. 5—Interworking of Signalling Systems (Study Group XI) 3 pp. 3 pp.

—CCITT No. 6—Interworking—Logic Procedures
CCITT RECMN Q.672-89. Logic Procedures for Interworking of Signalling System R1 to No. 6—Interworking of Signalling Systems (Study Group XI) 3 pp. 3 pp.

—CCITT No. 7—Interworking—Logic Procedures
CCITT RECMN Q.673-89. Logic Procedures for Interworking of Signalling System R1 to No. 7 (TUP)—Interworking of Signalling Systems (Study Group XI) 3 pp. 3 pp.

—CCITT R2—Interworking—Logic Procedures
CCITT RECMN Q.674-89. Logic Procedures for Interworking of Signalling System R1 to R2—Interworking of Signalling Systems (Study Group XI) 3 pp. 3 pp.

—End of Pulsing Signals—Register Signaling
CCITT RECMN Q.321-89. End-of-Pulsing Conditions—Register Arrangements Concerning ST Signal—Specifications of Signalling Systems R1 and R2 (Study Group XI) 1 pp. 1 p.

—Glossaries
CCITT RECMN Q.310-89. Definition and Function of Signals—Specifications of Signalling Systems R1 and R2 (Study Groupo XI) 2 pp. 2 pp.

—Incoming—Logic Procedures
CCITT RECMN Q.615-89. Logic Procedures for Incoming Signalling System R1—Interworking of Signalling Systems (Study Group XI) 3 pp. 3 pp.

—International Exchanges—Switching Speed
CCITT RECMN Q.319-89. Speed of Switching in International Exchanges—Specifications of Signalling Systems R1 and R2 (Study Group XI) 1 pp. 1 p.

—Interworking
CCITT RECMN Q.332-89. Interworking of Signalling System R1 with Other Standardized Systems—Interworking—Specifications of Signalling Systems R1 and R2 (Study Group XI) 2 pp. 2 pp.

—Line Signaling
CCITT RECMN Q.311-89. Line Signalling-2600 Hz Line Signalling—Specifications of Signalling Systems R1 and R2 (Study Group XI) 2 pp. 2 pp.

CCITT R1 Signaling Systems (Cont.)

—Line Signaling (Cont.)
CCITT RECMN Q.317-89. Further Specification Clauses Relative to Line Signalling—Specifications of Signalling Systems R1 and R2 (Study Group XI) 2 pp. 2 pp.

—Line Signaling—Pulse Code Modulation
CCITT FASCICLE VI.4. Specifications of Signalling Systems R1 and R2 Recommendations Q.310—Q.490 (Study Group XI). 181 pp.
CCITT RECMN Q.314-89. PCM Line Signalling Specifications of Signalling Systems R1 and R2 (Study Group XI) 1 pp. 1 p.

—Maintenance
CCITT RECMN Q.328-89. Routine Testing of Equipment (Local Maintenance)—Specifications of Signalling Systems R1 and R2 (Study Group XI) 2 pp. 2 pp.

—Measuring Instruments
CCITT RECMN Q.331-89. Test Equipment for Checking Equipment and Signals—Specifications of Signalling Systems R1 and R2 (Study Group XI) 3 pp. 3 pp.

—Outgoing—Logic Procedures
CCITT RECMN Q.625-89. Logic Procedures for Outgoing Signalling System R1—Interworking of Signalling Systems (Study Group XI) 3 pp. 3 pp.

—Receivers—Line Signaling
CCITT RECMN Q.313-89. 2600 Hz Line Signal Receiving Equipment—Specifications of Signalling Systems R1 and R2 (Study Group XI). 3 pp.

—Receivers—Line Signaling—Pulse Code Modulation
CCITT RECMN Q.316-89. PCM Line Signal Receiver—Specifications of Signalling Systems R1 and R2 (Study Group XI) 1 pp. 1 p.

—Receivers—Multifrequency Signaling Systems
CCITT RECMN Q.323-89. Multifrequency Signal Receiving Equipment—Specifications of Signalling Systems R1 and R2 (Study Group XI) 2 pp. 2 pp.

—Registers (Switching)—Release
CCITT RECMN Q.325-89. Release of Registers—Specifications of Signalling Systems R1 and R2 (Study Group XI) 1 pp. 1 p.

—Routing—Address Codes—Register Signaling
CCITT RECMN Q.324-89. Analysis of Address Information for Routing—Specifications of Signalling Systems R1 and R2 (Study Group XI) 1 pp. 1 p.

—Signal Codes—Register Signaling
CCITT RECMN Q.320-89. Register Signalling-Signal Code for Register Signalling—Specifications of Signalling Systems R1 and R2 (Study Group XI) 2 pp. 2 pp.

—Signal Senders—Line Signaling
CCITT RECMN Q.312-89. Line Signalling-2600 Hz Line Signal Sender (Transmitter)—Specifications of Signalling Systems R1 and R2 (Study Group XI) 1 pp. 1 p.
CCITT RECMN Q.313-89. 2600 Hz Line Signal Receiving Equipment—Specifications of Signalling Systems R1 and R2 (Study Group XI). 3 pp.

—Signal Senders—Line Signaling—Pulse Code Modulation
CCITT RECMN Q.315-89. PCM Line Signal Sender (Transmitter)—Specifications of Signalling Systems R1 and R2 (Study Group XI) 2 pp. 2 pp.

—Signal Senders—Multifrequency Signaling Systems
CCITT RECMN Q.322-89. Multifrequency Signal Sender—Specifications of Signalling Systems R1 and R2 (Study Group XI) 1 pp. 1 p.

—Telephone Circuits—Both-Way Operation—Double Seizing
CCITT RECMN Q.318-89. Double Seizing with Both-Way Operation—Specifications of Signalling Systems R1 and R2 (Study Group XI) 1 pp. 1 p.

—Telephone Circuits—Manual Testing
CCITT RECMN Q.329-89. Manual Testing—Specifications of Signalling Systems R1 and R2 (Study Group XI) 1 pp. 1 p.

—Telephone Circuits—Speech Position—Register Signaling
CCITT RECMN Q.326-89. Switching to the Speech Position—Specifications of Signalling Systems R1 and R2 (Study Group XI) 1 pp. 1 p.

CCITT R1 Signaling Systems (Cont.)

—Test Equipment

CCITT RECMN Q.328-89. Routine Testing of Equipment (Local Maintenance)—Specifications of Signalling Systems R1 and R2 (Study Group XI) 2 pp. 2 pp.

—Testing Equipment

CCITT RECMN Q.331-89. Test Equipment for Checking Equipment and Signals—Specifications of Signalling Systems R1 and R2 (Study Group XI) 3 pp. 3 pp.

CCITT R2 Signaling Systems

See Also: Interregister Signaling; Signaling Systems

CCITT FASCICLE VI.4. Specifications of Signalling Systems R1 and R2 Recommendations Q.310—Q.490 (Study Group XI). 181 pp.

CCITT RECMN Q.480-89. Signalling Procedures—Miscellaneous Procedures—Specifications of Signalling Systems R1 and R2 (Study Group XI) 2 pp. 2 pp.

—Backward Signals—Glossaries

CCITT RECMN Q.400-89. Definitions and Functions of Signals—Forward Line Signals. Backward Line Signals. Forward Register Signals. Backward Register Signals—Specifications of Signalling Systems R1 and R2 (Study Group XI) 4 pp. 4 pp.

—Call Set Up

CCITT RECMN Q.460-89. Signalling Procedures—Normal Call Set-Up Procedures for International Working—Specifications of Signalling Sytems R1 and R2 (Study Group XI) 1 pp. 1 p.

—CCITT No. 4—Interworking—Logic Procedures

CCITT RECMN Q.681-89. Logic Procedures for Interworking of Signalling System R2 to No. 4—Interworking of Signalling Systems (Study Group XI) 2 pp. 2 pp.

—CCITT No. 5—Interworking—Logic Procedures

CCITT RECMN Q.682-89. Logic Procedures for Interworking of Signalling System R2 to No. 5 (TUP)—Interworking of Signalling Systems (Study Group XI) 3 pp. 3 pp.

—CCITT No. 6—Interworking—Logic Procedures

CCITT RECMN Q.683-89. Logic Procedures for Interworking of Signalling System R2 to No. 6—Interworking of Signalling Systems (Study Group XI) 3 pp. 3 pp.

—CCITT No. 7—Interworking—Logic Procedures

CCITT RECMN Q.684-89. Logic Procedures for Interworking of Signalling System R2 to No. 7 (TUP)—Interworking of Signalling Systems (Study Group XI) 4 pp. 4 pp.

—CCITT R1—Interworking—Logic Procedures

CCITT RECMN Q.685-89. Logic Procedures for Interworking of Signalling System R2 to R1—Interworking of Signalling Systems (Study Group XI) 2 pp. 2 pp.

—Comprehensive Testing

CCITT RECMN Q.490-89. Testing and Maintenance—Specifications of Signalling Systems R1 and R2 (Study Group XI) 5 pp. 5 pp.

—Echo Suppressors

CCITT RECMN Q.479-89. Signalling Procedures—Echo-Suppressor Control—Signalling Requirements—Specifications of Signalling Systems R1 and R2 (Study Group XI) 4 pp. 4 pp.

—Forward Signals—Glossaries

CCITT RECMN Q.400-89. Definitions and Functions of Signals—Forward Line Signals. Backward Line Signals. Forward Register Signals. Backward Register Signals—Specifications of Signalling Systems R1 and R2 (Study Group XI) 4 pp. 4 pp.

—Forward Transfer Signals

CCITT FASCICLE VI.4. Specifications of Signalling Systems R1 and R2 Recommendations Q.310—Q.490 (Study Group XI). 181 pp.

CCITT RECMN Q.400-89. Definitions and Functions of Signals—Forward Line Signals. Backward Line Signals. Forward Register Signals. Backward Register Signals—Specifications of Signalling Systems R1 and R2 (Study Group XI) 4 pp. 4 pp.

CCITT RECMN Q.441-89. Interregister Signalling—Signalling Code—Specifications of Signalling Systems R1 and R2 (Study Group XI) 10 pp. 10 pp.

CCITT R2 Signaling Systems (Cont.)

—Forward Transfer Signals—Glossaries

CCITT RECMN Q.400-89. Definitions and Functions of Signals—Forward Line Signals. Backward Line Signals. Forward Register Signals. Backward Register Signals—Specifications of Signalling Systems R1 and R2 (Study Group XI) 4 pp. 4 pp.

—Incoming—Logic Procedures

CCITT RECMN Q.616-89. Logic Procedures for Incoming Signalling System R2—Interworking of Signalling Systems (Study Group XI) 5 pp. 5 pp.

—INMARSAT—Interworking

CCITT RECMN Q.1102-89. Interworking Between Signalling System R2 and Inmarsat Standard A System—Interworking with Satellite Mobile Systems (Study Group XI) 10 pp. 10 pp.

—Interregister Signaling

CCITT RECMN Q.440-89. Interregister Signalling—General—Specifications of Signalling Systems R1 and R2 (Study Group XI) 4 pp. 4 pp.

—Interregister Signaling—Backward Signals—Pulse Transmission

CCITT RECMN Q.442-89. Interregister Signalling—Pulse Transmission of Backward Signals A-3, A-4, A-6 or A-15. Multifrequency Signalling Equipment—Specifications of Signalling Systems R1 and R2 (Study Group XI) 2 pp. 2 pp.

—Interregister Signaling—Range

CCITT RECMN Q.457-89. Interregister Signalling—Range of Interregister Sinalling—Specifications of Signalling Systems R1 and R2 (Study Group XI) 4 pp. 4 pp.

—Interregister Signaling—Range—Transmission Loss—Formulas

CCITT FASCICLE VI.4. Specifications of Signalling Systems R1 and R2 Recommendations Q.310—Q.490 (Study Group XI). 181 pp.

—Interregister Signaling—Regeneration—Transit Exchanges

CCITT RECMN Q.478-89. Signalling Procedures—Relay and Regeneration of R2 Interregister Signals by an Outgoing R2 Register in a Transit Exchange—Specifications of Signalling Systems R1 and R2 (Study Group XI) 2 pp. 2 pp.

—Interregister Signaling—Relay—Transit Exchanges

CCITT RECMN Q.478-89. Signalling Procedures—Relay and Regeneration of R2 Interregister Signals by an Outgoing R2 Register in a Transit Exchange—Specifications of Signalling Systems R1 and R2 (Study Group XI) 2 pp. 2 pp.

—Interregister Signaling—Reliability

CCITT RECMN Q.458-89. Interregister Signalling—Reliability of Interregister Signalling—Specifications of Signalling Systems R1 and R2 (Study Group XI) 5 pp. 5 pp.

—Interregister Signaling—Speed

CCITT RECMN Q.458-89. Interregister Signalling—Reliability of Interregister Signalling—Specifications of Signalling Systems R1 and R2 (Study Group XI) 5 pp. 5 pp.

—Interregister Signaling—Termination—End of Pulsing Signals

CCITT RECMN Q.473-89. Signalling Procedures—Use of End-of-Pulsing Signal I-15 in International Working—Specifications of Signalling Systems R1 and R2 (Study Group XI) 2 pp. 2 pp.

—Interregister Signaling—Termination—Group B Signals

CCITT RECMN Q.474-89. Signalling Procedures—Use of Group B Signals—Specifications of Signalling Systems R1 and R2 (Study Group XI) 3 pp. 3 pp.

—Interregister Signaling—Termination—Telephone Exchanges

CCITT RECMN Q.471-89. Signalling Procedures—at the Last Incoming R2 Register Situated in the Exchange to Which the Called Subscriber is Connected—Specifications of Signalling Systems R1 and R2 (Study Group XI) 2 pp. 2 pp.

—Interregister Signaling—Termination—Transit Exchanges

CCITT RECMN Q.470-89. Signalling Procedures—at an Incoming R2 Register Situated in a Transit Exchange—Specifications of Signalling Systems R1 and R2 (Study Group XI) 2 pp. 2 pp.

CCITT RECMN Q.472-89. Signalling Procedures—at the Last Incoming R2 Register Situated in a Transit Exchange—Specifications of Signalling Systems R1 and R2 (Study Group XI) 2 pp. 2 pp.

CCITT R2 Signaling Systems (Cont.)

—Line Signaling

CCITT FASCICLE VI.4. Specifications of Signalling Systems R1 and R2 Recommendations Q.310—Q.490 (Study Group XI). 181 pp.

—Line Signaling—Both Way Operation

CCITT FASCICLE VI.4. Specifications of Signalling Systems R1 and R2 Recommendations Q.310—Q.490 (Study Group XI). 181 pp.

—Line Signaling—Conversion Equipment

CCITT RECMN Q.430-89. Line Signalling, Digital Version—Conversion Between Analogue and Digital Versions of System R2 Line Signalling—Specifications of Signalling Systems R1 and R2 (Study Group XI) 21 pp. 21 pp.

—Line Signaling—Exchange Lines

CCITT RECMN Q.412-89. Line Signalling, Analogue Version—Clauses for Exchange Line Signalling Equipment. Clauses for Transmission Line Signalling Equipment—Specifications of Signalling Systems R1 and R2 (Study Group XI) 7 pp. 7 pp.

CCITT RECMN Q.422-89. Line Signalling, Digital Version—Clauses for Exchange Line Signalling Equipment—Specifications of Signalling Systems R1 and R2 (Study Group XI) 6 pp. 6 pp.

—Line Signaling—Faulty Transmission

CCITT RECMN Q.424-89. Line Signalling, Digital Version—Protection Against the Effects of Faulty Transmission—Specifications of Signalling Systems R1 and R2 (Study Group XI) 2 pp. 2 pp.

—Line Signaling—Inband

CCITT FASCICLE VI.4. Specifications of Signalling Systems R1 and R2 Recommendations Q.310—Q.490 (Study Group XI). 181 pp.

—Line Signaling—Interruption Control

CCITT RECMN Q.416-89. Line Signalling, Analogue Version—Interruption Control—Specifications of Signalling Systems R1 and R2 (Study Group XI) 5 pp. 5 pp.

—Line Signaling—Metering

CCITT FASCICLE VI.4. Specifications of Signalling Systems R1 and R2 Recommendations Q.310—Q.490 (Study Group XI). 181 pp.

—Line Signaling—Pulse Code Modulation

CCITT FASCICLE VI.4. Specifications of Signalling Systems R1 and R2 Recommendations Q.310—Q.490 (Study Group XI). 181 pp.

—Maintenance

CCITT RECMN Q.490-89. Testing and Maintenance—Specifications of Signalling Systems R1 and R2 (Study Group XI) 5 pp. 5 pp.

—Multifrequency Signaling—Receivers—Interruption Control

CCITT FASCICLE VI.4. Specifications of Signalling Systems R1 and R2 Recommendations Q.310—Q.490 (Study Group XI). 181 pp.

—Multifrequency Signaling Systems

CCITT RECMN Q.450-89. Interregister Signalling—General—Specifications of Signalling Systems R1 and R2 (Study Group XI) 2 pp. 2 pp.

—Multifrequency Signaling Systems—Glossaries

CCITT RECMN Q.451-89. Interregister Signalling—Definitions—Specifications of Signalling Systems R1 and R2 (Study Group XI) 2 pp. 2 pp.

—Multifrequency Signaling Systems—Power Level—Formulas

CCITT FASCICLE VI.4. Specifications of Signalling Systems R1 and R2 Recommendations Q.310—Q.490 (Study Group XI). 181 pp.

CCITT RECMN Q.454-89. Interregister Signalling—the Sending Part of the Multifrequency Signalling Equipment—Specifications of Signalling Systems R1 and R2 (Study Group XI) 2 pp. 2 pp.

—Multifrequency Signaling Systems—Satellite Communications

CCITT FASCICLE VI.4. Specifications of Signalling Systems R1 and R2 Recommendations Q.310—Q.490 (Study Group XI). 181 pp.

—Multifrequency Signaling Systems—Transmission

CCITT RECMN Q.452-89. Interregister Signalling—Requirements Relating to Transmission Conditions—Specifications of Signalling Systems R1 and R2 (Study Group XI) 1 pp. 1 p.

INTERNATIONAL AND NON-U.S. NATIONAL STANDARDS
SUBJECT INDEX

CECC

CCITT R2 Signaling Systems *(Cont.)*
—**Outgoing—Logic Procedures**
CCITT RECMN Q.626-89. Logic Procedures for Outgoing Signalling System R2—Interworking of Signalling Systems (Study Group XI) 5 pp. 5 pp.

—**Receivers—Line Signaling**
CCITT RECMN Q.415-89. Line Signalling, Analogue Version—Signal Receiver—Specifications of Signalling Systems R1 and R2 (Study Group XI) 2 pp. 2 pp.

—**Receivers—Multifrequency Signaling Systems**
CCITT RECMN Q.455-89. Interregister Signalling—the Receiving Part of the Multifrequency Equipment. and Reliability of Interregister Signalling—Specifications of Signalling Systems R1 and R2 (Study Group XI) 4 pp. 4 pp.

—**Registers (Switching)**
CCITT RECMN Q.464-89. Signalling Procedures—Signalling Between the Outgoing International R2 Register and the Last Incoming R2 Register—Specifications of Signalling Systems R1 and R2 (Study Group XI) 1 pp. 1 p.
CCITT RECMN Q.465-89. Signalling Procedures—Particular Cases—Specifications of Signalling Systems R1 and R2 (Study Group XI) 1 pp. 1 p.

—**Registers (Switching)—International Exchanges**
CCITT RECMN Q.462-89. Signalling Procedures—Signalling Between the Outgoing International R2 Register and an Incoming R2 Register in an International Exchange—Specifications of Signalling Systems R1 and R2 (Study Group XI) 3 pp. 3 pp.

—**Registers (Switching)—National Exchanges**
CCITT RECMN Q.463-89. Signalling Procedures—Signalling Between the Outgoing International R2 Register and an Incoming R2 Register in a National Exchange in the Destination Country—Specifications of Signalling Systems R1 and R2 (Study Group XI) 1 pp. 1 p.

—**Registers—Release**
CCITT RECMN Q.475-89. Signalling Procedures—Normal Release of Outgoing and Incoming R2 Registers—Specifications of Signalling Systems R1 and R2 (Study Group XI) 2 PP. 2 pp.
CCITT RECMN Q.476-89. Signalling Procedures—Abnormal Release of Outgoing and Incoming R2 Registers—Specifications of Signalling Systems R1 and R2 (Study Group XI) 2 pp. 2 pp.

—**Signal Codes—Interregister Signaling**
CCITT RECMN Q.441-89. Interregister Signalling—Signalling Code—Specifications of Signalling Systems R1 and R2 (Study Group XI) 10 pp. 10 pp.

—**Signal Codes—Line Signaling**
CCITT RECMN Q.411-89. Line Signalling, Analogue Version-Line Signalling Code—Specifications of Signalling Systems R1 and R2 (Study Group XI) 2 pp. 2 pp.
CCITT RECMN Q.421-89. Line Signalling, Digital Version—Digital Line Signalling Code—Specifications of Signalling Systems R1 and R2 (Study Group XI) 2 pp. 2 pp.

—**Signal Senders—Line Signaling**
CCITT RECMN Q.414-89. Line Signalling, Analogue Version—Signal Sender—Specifications of Signalling Systems R1 and R2 (Study Group XI) 3 PP. 3 pp.

—**Signal Senders—Multifrequency Signaling Systems**
CCITT RECMN Q.454-89. Interregister Signalling—the Sending Part of the Multifrequency Signalling Equipment—Specifications of Signalling Systems R1 and R2 (Study Group XI) 2 pp. 2 pp.

—**Telephone Connections—Release**
CCITT RECMN Q.466-89. Signalling Procedures—Supervision and Release of the Call—Specifications of Signalling Systems R1 and R2 (Study Group XI) 1 pp. 1 p.

—**Telephone Connections—Supervision**
CCITT RECMN Q.466-89. Signalling Procedures—Supervision and Release of the Call—Specifications of Signalling Systems R1 and R2 (Study Group XI) 1 pp. 1 p.

CCPS
Use: Consultative Council for Postal Studies

CCTV Cameras
Use: Closed Circuit Television Cameras

CCTV Equipment
Use: Closed Circuit Television Equipment

CCTV Monitors
Use: Closed Circuit Television Monitors

CCV
Use For: Coordination Commitee For Vocabulary
See Also: CCIR; CCITT
CCIR RESOLUTION 61-4-90. Structure of CCIR Study Groups—Volume XIV—Administrative Texts of the CCIR. 8 pp.

—**Glossaries**
CCIR RESOLUTION 114-90. Coordination of Vocabulary and Related Subjects—Volume XIII—Vocabulary and Related Subjects. 2 pp.

—**Symbols**
CCIR RESOLUTION 114-90. Coordination of Vocabulary and Related Subjects—Volume XIII—Vocabulary and Related Subjects. 2 pp.

CD ROM
Use: CD ROMs

CD ROMs
Use For: Data Compact Disks; Digital Data Compact Disks *See Also:* Compact Disks; Data Storage Devices; Information Interchange; Magnetic Disks; ROMs
BSI BS 4783: Part 7-93. 1993 Storage, Transportation and Maintenance of Media for Use in Data Processing and Information Storage Part 7: Recommendations for Optical Data Disks (CD-ROM). 10 pp.
BSI BS ISO/IEC 10149-89. 1989 Information Technology—Data Interchange on Read-Only 120 mm Optical Data Disks (CD-ROM) (S). 55 pp.
BSI BS ISO/IEC 10149-01. 1989 Amd 1 Information Technology—Data Interchange on Read-Only 120 mm Optical Data Disks (CD-ROM) (AMD 6860) October 31, 1991. 58 pp.
CEN ENV 30 149-89. Information Processing Systems Data Interchange on Read-Only 120 mm Optical Data Disks. 2 pp.
CEN EN 30149-91. Information Processing Systems Interchange on Read-Only 120 mm Optical Data Disks (CD-ROM). 3 pp.
ECMA ECMA 130-88. Data Interchange on Read-Only 120mm Optical Data Disks (CD-ROM). 49 pp.
ECMA ECMA 168-92. Volume and File Structure of Read-Only and Write-Once Compact Disc Media for Information Interchange. 132 pp.
IEC 10149-89. Information Technology—Data Interchange on Read-Only 120 mm Optical Data Disks (CD-ROM) First Edition. 50 pp.
IEC DIS 13490-93. Information Technology—Volume and File Structure of Read-Only and Write-Once Compact Disc Media for Information Interchange ***CD-ROM ONLY***. 143 pp.
ISO 10149-89. Information Technology—Data Interchange on Read-Only 120 mm Optical Data Disks (CD-ROM) First Edition. 50 pp.
ISO DIS 13490-93. Information Technology—Volume and File Structure of Read-Only and Write-Once Compact Disc Media for Information Interchange ***CD-ROM ONLY***. 143 pp.
JTC1 10149-89. Information Technology—Data Interchange on Read-Only 120 mm Optical Data Disks (CD-ROM) First Edition. 50 pp.
JTC1 DIS13490-93. Information Technology—Volume and File Structure of Read-Only and Write-Once Compact Disc Media for Information Interchange ***CD-ROM ONLY***. 143 pp.
OSI ISO IEC 10149-89. Information Technology—Data Interchange on Read-Only 120 mm Optical Data Disks (CD-ROM). 50 pp.

—**File Structure**
BSI BS 7061-89. 1989 Volume and File Structure of Compact Read Only Optical Disks (CD-ROM) for Information Interchange. 35 pp.
CEN EN 29 660-89. Information Processing Volume and File Structure of CD-ROM for Information Onterchange. 2 pp.
ECMA ECMA 119-86. Volume and File Structure of CD Rom for Information Interchange. 67 pp.
ISO 9660-88. Information Processing—Volume and File Structure of CD-ROM for Information Interchange First Edition; (Corrected and Reprinted -1988). 42 pp.
JIS X 0606-90. Volume and File Structure of CD-ROM for Information Interchange.
JTC1 9660-88. Information Processing—Volume and File Structure of CD-ROM for Information Interchange First Edition; (Corrected and Reprinted -1988). 42 pp.
OSI ISO 9660-88. Information Processing—Volume and File Structure of CD-ROM for Information Interchange. 42 pp.
OSI ISO DIS 9660-87. Volume and File Structure of CDROM for Information Interchange. 63 pp.

CD Value
Use: Circular Dichroism Value

CD WOM
Use: CD WOMs

CD WOMs
See Also: Compact Disks; Data Storage Devices; Information Interchange; Optical Disks
ECMA ECMA 168-92. Volume and File Structure of Read-Only and Write-Once Compact Disc Media for Information Interchange. 132 pp.
IEC DIS 13490-93. Information Technology—Volume and File Structure of Read-Only and Write-Once Compact Disc Media for Information Interchange ***CD-ROM ONLY***. 143 pp.
ISO DIS 13490-93. Information Technology—Volume and File Structure of Read-Only and Write-Once Compact Disc Media for Information Interchange ***CD-ROM ONLY***. 143 pp.
JTC1 DIS13490-93. Information Technology—Volume and File Structure of Read-Only and Write-Once Compact Disc Media for Information Interchange ***CD-ROM ONLY***. 143 pp.

—**Hybrid**
ECMA ECMA 168-92. Volume and File Structure of Read-Only and Write-Once Compact Disc Media for Information Interchange. 132 pp.
IEC DIS 13490-93. Information Technology—Volume and File Structure of Read-Only and Write-Once Compact Disc Media for Information Interchange ***CD-ROM ONLY***. 143 pp.
ISO DIS 13490-93. Information Technology—Volume and File Structure of Read-Only and Write-Once Compact Disc Media for Information Interchange ***CD-ROM ONLY***. 143 pp.
JTC1 DIS13490-93. Information Technology—Volume and File Structure of Read-Only and Write-Once Compact Disc Media for Information Interchange ***CD-ROM ONLY***. 143 pp.

CDLR
Use: Connected Line Identification Restriction

CEC (Commission of the European Communities)
Use For: CCE; Commission des Communautes Europeennes; Commission of the European Communities *See Also:* European Communities

—**COM Documents—Indexes (Documentation)**
EURO 1987. Index to 1987 COM Documents of the Commission of the European Communities. 370 pp.

—**Cooperation—CEN**
EURO MEMORANDUM 4-85. General Guidelines. 41 pp.

—**Cooperation—Cenelec**
EURO MEMORANDUM 4-85. General Guidelines. 41 pp.

—**Reports**
EURO 1988 XXIIND. XXIInd General Report on the Activities of the European Communities 1988. 460 pp.

CECC
Use For: Cenelec Electronic Components Committee
See Also: Cenelec; Electronic Components; European Standards; MUAHAG; Standards
CECC CECC 00 500 ISSUE 3. CECC System for Electronic Components of Assessed Quality Introduction to the System (En). 4 pp.

—**Basic Rules**
CECC CECC 00 100 ISSUE 2-88. Basic Rules (En, Fr, Ge). 4 pp.

—**Calibration Systems**
CECC CECC 00 011 ISSUE 1-82. Basic Specification: Calibration Requirements for the CECC System (En, Fr, Ge). 24 pp.

—**Glossaries**
CECC CECC 00 401 ISSUE 1-91. Glossary of Abbreviations, Terms and Definitions of the CECC System (En). 24 pp.

—**Management—Handbooks**
CECC CECC 00 700 ISSUE 1-84. Handbook of Administration (En) AMD 1 & ADDENDA (En) ADDENDA 2 (En) AMD 2 (En) AMD 3 (En) AMD 4 (En) AMD 5 (En) AMD 6 (En) AMD 7 (En) AMD 8 (En). 80 pp.

—**National Documents Register**
CECC CECC 00 300 ISSUE 2-93. Register of National Documents; Implementing CECC Publications and CECC European Standards (En, Fr, Ge). 189 pp.

INDUSTRY STANDARDS

INTERNATIONAL AND NON-U.S. NATIONAL STANDARDS SUBJECT INDEX

CECC (Cont.)

—PPM Approach—Codes of Practice
CECC CECC 00 800 ISSUE 1-86. Code of Practice on the Use of the PPM Approach in Association with the CECC System (En, Fr, Ge). 30 pp.

—Rules of Procedure
BSI BS 9000: Part 2-91. 1991 General Requirements for a System for Electronic Components of Assessed Quality Part 2: Specification for the National Implementation of the CECC System. 8 pp.

BSI BS 9000: Part 2-83. 1983 Amd 1 General Requirements for Electronic Components of Assessed Quality Part 2: National Implementation of CECC Basic Rules and Rules of Procedure. 8 pp.

CECC CECC 00 101 ISSUE 5-91. Rule of Procedure 1: the CENELEC Electronic Components Committee (En, Fr, Ge) AMD 1 (En, Fr, Ge) AMD 2 (En, Fr, Ge). 16 pp.

—Rules of Procedure—Administration
CECC CECC 00 102 ISSUE 2-86. Rule of Procedure 2: Administration Procedures; Part I: Organisation of the CECC General Secretariat Part II: CECC Financial Administration (En, Fr, Ge) AMD 1 (En, Fr, Ge). 7 pp.

—Rules of Procedure—Amendments
CECC CECC 00 106 ISSUE 3-91. Rule of Procedure 6: Amendments to the Rules of Procedure of the System (En, Fr, Ge). 6 pp.

—Rules of Procedure—Capability Approval
CECC CECC 00 107/Pt I ISSUE 3-82. Rule of Procedure 7: Quality Assessment Procedures; Part I: Quality Assessment Procedure for General Usage (En, Fr, Ge). 26 pp.

CECC CECC 00 107/Pt III IS 1-80. Rule of Procedure 7: Quality Assessment Procedures Part III: Procedure for Capability Approval (En, Fr, Ge) AMD 1 (En, Fr, Ge). 11 pp.

—Rules of Procedure—Certification Mark
CECC CECC 00 600 ISSUE 1-82. Rules for the Use and Administration of the CECC Certification Mark (En, Fr, Ge). 57 pp.

—Rules of Procedure—Certified Test Records
BSI BS CECC 00109-91. 1991 Rule of Procedure 9 Certified Test Records. 9 pp.

CECC CECC 00 109 ISSUE 1-74. Rule of Procedure 9: Certified Test Records (En, Fr, Ge). 4 pp.

—Rules of Procedure—Conformance
BSI BS CECC 00108-91. 1991 Rule of Procedure 8 Attestation of Conformity. 20 pp.

CECC CECC 00 108 ISSUE 2-80. Rule of Procedure 8: Attestation of Conformity (En, Fr, Ge). 15 pp.

—Rules of Procedure—ECQAC
CECC CECC 00 103 ISSUE 2-90. Rule of Procedure 3: The Electronic Components Quality Assurance Committee (En, Fr, Ge). 22 pp.

—Rules of Procedure—Membership Register
CECC CECC 00 105 ISSUE 6-89. Rule of Procedure 5: CECC Membership Register (En, Fr, Ge). 18 pp.

—Rules of Procedure—Quality Assurance
CECC CECC 00 107/Pt II IS 1-79. Rule of Procedure 7: Quality Assessment Procedures Part II: Procedure for Enhanced Assessment of Quality (En, Fr, Ge) AMD 1 (En, Fr, Ge). 31 pp.

CECC CECC 00 111/II ISSUE 4-91. Rule of Procedure 11: Specifications Part II: Regulations for CECC Specifications for Components of Enhanced Assessment of Quality (En, Fr, Ge). 7 pp.

CECC CECC 00 114 Pt 0 ISSUE 1-92. Rule of Procedure 14: Quality Assessment Procedures An Introduction to the Types of Approval Available Under the CECC System (En, Fr, Ge). 13 pp.

CECC CECC 00 114 Pt I ISSUE 2-92. Rule of Procedure 14: Quality Assessment Procedures Approval of Manufacturers and Other Organizations (En, Fr, Ge) AMD 1 (En, Fr, Ge). 34 pp.

CECC CECC 00 114 Pt II IS 2-92. Rule of Procedure 14: Quality Assessment Procedures Qualification Approval of Electronic Components (En, Fr, Ge). 31 pp.

CECC CECC 00 114 Pt III IS 2-93. Rule of Procedure 14 Quality Assessment Procedures Part III: Capability Approval of an Electronic Component Manufacturing Activity (En, Fr, Ge) ERRATUM (En, Fr, Ge). 75 pp.

CECC CECC 00 114 Pt IV IS 1-91. Rule of Procedure 14 Quality Assessment Procedures Procedure for Enhanced Assessment of Quality (En, Fr, Ge). 9 pp.

CECC (Cont.)

—Rules of Procedure—Specifications Preparation
BSI BS CECC 00111: Pt 1-91. 1991 Rule of Procedure 11 Specifications Part 1: General Regulations for CECC Specifications. 21 pp.

BSI BS CECC 00111: Pt 1-01. 1991 Amd 1 Rule of Procedure 11 Specifications Part 1: General Regulations for CECC Specifications (AMD 7333) January 15, 1993. 23 pp.

BSI BS CECC 00111: Pt 1-02. 1991 Amd 2 Rule of Procedure 11 Specifications Part 1: General Regulations for CECC Specifications (AMD 7733) May 15, 1993 (T). 41 pp.

BSI BS CECC 00111: Pt 2-91. 1991 Rule of Procedure 11 Specifications Part 2: Regulations for CECC Specifications for Components of Enhanced Assessment of Quality. 10 pp.

BSI BS CECC 00111: Pt 3-91. 1991 Rule of Procedure 11 Specifications Part 3: Regulations for CECC Specifications for Components for General and Professional (Civil and Military) Usage (Excluding Detail Specifications). 31 pp.

BSI BS CECC 00111: Pt 3-01. 1991 Amd 1 Rule of Procedure 11 Specifications Part 3: Regulations for CECC Specifications for Components for General and Professional (Civil and Military) Usage (Excluding Detail Specifications) (AMD 7329) January 15, 1993 (T). 40 pp.

BSI BS CECC 00111: Pt 4-91. 1991 Rule of Procedure 11 Specifications Part 4: Regulations for CECC Detail Specifications. 37 pp.

BSI BS CECC 00111: Pt 4-01. 1991 Amd 1 Rule of Procedure 11 Specifications Part 4: Regulations for CECC Detail Specifications (AMD 7334) January 15, 1993 (T). 44 pp.

BSI BS CECC 00111: Pt 4-02. 1991 Amd 2 Rule of Procedure 11 Specifications Part 4: Regulations for CECC Detail Specifications (AMD 7734) May 15, 1993 (T). 48 pp.

BSI BS CECC 00111: Pt 5-91. 1991 Rule of Procedure 11 Specifications Part 5: Preparation of Specifications Under the Single Originator Procedure. 17 pp.

CECC CECC 00 111/I ISSUE 4-91. Rule of Procedure 11: Specifications Part I: General Regulations for CECC Specifications (En, Fr, Ge) AMD 1 (En, Fr, Ge) AMD 2 (En, Fr, Ge). 24 pp.

CECC CECC 00 111/III ISSUE 4-91. Rule of Procedure 11: Specifications Part III: Regulations for CECC Specifications for Components for General and Professional (Civil and Military) Usage (Excluding Detail Specifications) (En, Fr, Ge) AMD 1 (En, Fr, Ge) AMD 2 (En, Fr, Ge). 29 pp.

CECC CECC 00 111/IV ISSUE 4-91. Rule of Procedure 11: Specifications Part IV: Regulations for CECC Detail Specifications (En, Fr, Ge) AMD 1 (En, Fr, Ge) AMD 2 (En, Fr, Ge). 36 pp.

CECC CECC 00 111/V ISSUE 1-91. Rule of Procedure 11: Specifications: Part V: Preparation of Specifications Under the Single Originator Procedure (En, Fr, Ge). 11 pp.

—Rules of Procedure—Voting
BSI BS CECC 00112-91. 1991 Rule of Procedure 12 Voting Procedures. 17 pp.

CECC CECC 00 112 ISSUE 2-82. Rule of Procedure 12: Voting Procedures (En, Fr, Ge) AMD 1 (En, Fr, Ge) AMD 2 (En, Fr, Ge) AMD 3 (En, Fr, Ge) AMD 4 (En, Fr, Ge). 17 pp.

—Rules of Procedure—Working Groups
BSI BS CECC 00104-91. 1991 Rule of Procedure 4 CECC Working Groups General Rules and Additional Rules. 16 pp.

BSI BS CECC 00104-01. 1991 Amd 1 Rule of Procedure 4 CECC Working Groups General Rules and Additional Rules (AMD 7732) May 15, 1993 (T). 22 pp.

BSI BS CECC 00104-02. 1991 Amd 2 Rule of Procedure 4 CECC Working Groups General Rules and Additional Rules (AMD 7891) September 15, 1993 (T). 24 pp.

—Specifications Preparation—Handbooks
CECC CECC 00 400 ISSUE 3-86. Handbook for the Production of CECC Documents (En) ERRATUM (En, Fr, Ge) AMD 1 (En, Fr, Ge) AMD 2 (En, Fr, Ge) AMD 4 (En, Fr, Ge) AMD 5 (En, Fr, Ge) AMD 6 (En, Fr, Ge) AMD 7 (En, Fr, Ge) AMD 8 (En, Fr, Ge). 103 pp.

—Specifications Register
CECC CECC 00 301 ISSUE 2-93. Register of CECC Specifications and Related Detail Specifications (En, Fr, Ge). 222 pp.

—Standards Preparation—Handbooks
CECC CECC 00 400 ISSUE 3-86. Handbook for the Production of CECC Documents (En) ERRATUM (En, Fr, Ge) AMD 1 (En, Fr, Ge) AMD 2 (En, Fr, Ge) AMD 4 (En, Fr, Ge) AMD 5 (En, Fr, Ge) AMD 6 (En, Fr, Ge) AMD 7 (En, Fr, Ge) AMD 8 (En, Fr, Ge). 103 pp.

Cedarwood Oil
See Also: Essential Oils

ISO 4724-84. Oil of Cedarwood, Virginia (Juniperus Virginiana Linnaeus) First Edition. 4 pp.

ISO 4725-86. Oil of Cedarwood, Texas (Juniperus Mexicana Schiede) First Edition. 4 pp.

—Quality Assurance
ISO 9843-91. Oil of Cedarwood (Cupressus Funebris Endlicher) First Edition. 7 pp.

CEDEX
Use: Information Interchange—Container Equipment Data Exchange (CEDEX)

CEI
Use: IEC

Ceiling Boards
Use: Ceiling Panels *See Also:* Building Board

Ceiling Fans
Use For: Paddle Fans *See Also:* Fans

CNS C4009-86. Ceiling Fan (Jan)(597). 8 pp.

CNS C4055-86. Auto-Rotating Ceiling Fan (Jan)(2450). 6 pp.

CSA C22.2 NO 113-M1984. Fans and Ventilators (R 1993); (Gen Instr 1 Thru 6). 51 pp.

CSA 1169 Bull. Electrical Bulletin 1169 June 27, 1978 to C22.2 NO 113. 2 pp.

—Pendant
CSA C22.2 NO 113-M1984. Fans and Ventilators (R 1993); (Gen Instr 1 Thru 6). 51 pp.

CSA 1169 Bull. Electrical Bulletin 1169 June 27, 1978 to C22.2 NO 113. 2 pp.

—Regulators—Safety
BSI BS 3456: Sec 102.342-88. 1988 Safety of Household Electrical Appliances Part 102: Particular Requirements Section 102.342: Electric Fans and Regulators. 16 pp.

CENELEC HD 280 S1-86. Safety of Household and Similar Electrical Appliances: Particular Requirements for Electric Fans and Regulations. 6 pp.

DIN VDE 0700 Pt 234 (D)-81. Safety of Household and Similar Electrical Appliances; Fans and Associated Regulators (Aug). 25 pp.

IEC 342 Pt 1-81. Safety Requirements for Electric Fans and Regulators Part 1: Fans and Regulators for Household and Similar Purposes Second Edition; (Amendment 1-1982). 30 pp.

Ceiling Heaters
See Also: Heaters

CSA 1219 Bull. Electrical Bulletin 1219 April 2, 1979 to C22.2 NO 46. 2 pp.

Ceiling Panels
Use For: Ceiling Boards *See Also:* Ceilings; Panels

—Adhesives
JIS A 1612-84. Testing Methods for Bonding Strength of Ceiling Boards Adhesives. 11 pp.

JIS A 5539-84. Adhesives for Ceiling Boards. 12 pp.

Ceilings
Use For: Dropped Ceilings; False Celings; Suspended Ceilings *See Also:* Acoustical Insulation; Ceiling Panels; Panels

BSI BS 8290: Part 2-91. 1991 Suspended Ceilings Part 2: Specification for Performance of Components and Assemblies. 19 pp.

BSI BS 8290: Part 2-01. 1991 Amd 1 Suspended Ceilings Part 2: Specification for Performance of Components and Assemblies (AMD 7844) August 15, 1993 (R). 20 pp.

CNS C4033-83. Ceiling Block (Feb)(1267).

JIS A 0030-73. Classification of Performance for Building Elements. 14 pp.

—Acoustics
BSI BS 2750: Part 9-87. 1987 Measurement of Sound Insulation in Buildings and of Building Elements Part 9:Method for Labor-tory Measurement of Room-To-Room Airborne Sound Insulation of a Suspended Ceiling with a Plenum Above it. 10 pp.

ISO 140 Pt 9-85. Acoustics—Measurements of Sound Insulation in Buildings and of Building Elements—Part 9: Laboratory Measurement of Room-to-Room Airborne Sound Insulation of a Suspended Ceiling with a Plenum Above It First Edition. 8 pp.

—Adhesives
BSI BS 5442: Part 2-89. 1989 Classification of Adhesives for Construction Part 2: Adhesives for Use with Interior Wall and Ceiling Coverings (Excluding Decorative Flexible Materials in Roll Form). 8 pp.

INTERNATIONAL AND NON-U.S. NATIONAL STANDARDS
SUBJECT INDEX

Ceilings (Cont.)

—Adhesives (Cont.)
BSI BS 5442: Part 2-78. (WITHDRAWN) 1978 Classification of Adhesives for Construction Part 2: Adhesives for Use with Interior Wall and Ceiling Coverings (Excluding Decorative Flexible Materials in Roll Form). 8 pp.
CNS A2237-89. Adhesives for Ceiling Boards (Sep)(12603).
CNS A3310-89. Method of Test for Adhesives for Ceiling Boards (Sep)(12604).

—Adhesives—Bonding Strength
CNS A3311-89. Method of Test for Bonding Strength of Ceiling Boards Adhesives (Sep)(12605).

—Aggregate Coatings
JIS A 6917-83. Lightweight Aggregate Coating Materials. 16 pp.

—Air Handling Equipment—Compatibility
SAA AS 2946-91. Suspended Ceilings, Recessed Luminaires and Air Diffusers—Interface Requirements for Physical Compatability (In Professional Package 50). 22 pp.

—Binders—Tables (Data)
BSI BS 5268: Sec 7.4-89. 1989 Structural Use of Timber Part 7: Recommendations for the Calculation Basis for Span Tables Section 7.4: Ceiling Binders. 16 pp.

—Construction
BSI CP 290-73. (WITHDRAWN) 1973 Suspended Ceilings and Linings of Dry Construction Using Metal Fixing Systems (Superseded by BS 8290: Part 1: 1991). 52 pp.
DIN ENGL 18168 Pt 1-81. Light Ceiling Linings and Underceilings; Construction Requirements (Oct). 6 pp.

—Design
BSI BS 8290: Part 1-91. 1991 Suspended Ceilings Part 1: Code of Practice for Design. 20 pp.
BSI BS 8290: Part 1-01. 1991 Amd 1 Suspended Ceilings Part 1: Code of Practice for Design (AMD 7843) August 15, 1993 (R). 21 pp.

—Emissivity—Infrared Radiometers
CNS A3269-87. Simplified Test Method for Emissivity by Infrared Radio Meter (Sep)(12066).
JIS A 1423-83. Simplified Test Method for Emissivity by Infrared Radio Meter. 8 pp.

—Fire Testing
ISO TR6167-84. Fire-Resistance Tests—Contribution Made by Suspended Ceilings to the Protection of Steel Beams in Floor and Roof Assemblies First Edition. 13 pp.

—Furrings—Steel
CNS A2206-87. Steel Furrings for Wall and Ceiling in Buildings (Jun)(11984). 6 pp.
CNS A3259-87. Method of Test for Steel Furrings for Wall and Ceiling in Buildings (Jun)(11985).
JIS A 6517-89. Steel Furrings for Wall and Ceiling in Buildings. 16 pp.

—Glossaries
BSI BS 6100: SUBSEC 1.3.3-87. 1987 Glossary of Building and Civil Engineering Terms Part 1: General and Miscellaneous Section 1.3: Parts of Construction Works Subsection 1.3.3: Floors and Ceilings. 12 pp.
BSI BS 6100: SUBSEC 1.3.3-01. 1987 Amd 1 Building and Civil Engineering Terms Part 1: General and Miscellaneous Section 1.3: Parts of Construction Works Subsection 1.3.3: Floors and Ceilings (AMD 7233) August 15, 1992. 14 pp.

—Installation
BSI BS 8290: Part 3-91. 1991 Suspended Ceilings Part 3: Code of Practice for Installation and Maintenace. 12 pp.

—Joists—Tables (Data)
BSI BS 5268: Sec 7.3-89. 1989 Structural Use of Timber Part 7: Recommendations for the Calculation Basis for Span Tables Section 7.3: Ceiling Joists. 16 pp.

—Linings—Construction
DIN ENGL 18168 Pt 1-81. Light Ceiling Linings and Underceilings; Construction Requirements (Oct). 6 pp.

—Linings—Loads (Forces)
DIN ENGL 18168 Pt 2-84. Lightweight Ceiling Linings and Suspended Ceilings; Verification of the Loadbearing Capacity of Metal Substructures and Hangers (Dec). 2 pp.

Ceilings (Cont.)

—Loads (Forces)
DIN ENGL 18168 Pt 2-84. Lightweight Ceiling Linings and Suspended Ceilings; Verification of the Loadbearing Capacity of Metal Substructures and Hangers (Dec). 2 pp.

—Maintenance
BSI BS 8290: Part 3-91. 1991 Suspended Ceilings Part 3: Code of Practice for Installation and Maintenance. 12 pp.

—Rosettes
JIS C 8310-92. Ceiling Rosettes. 25 pp.

—Thermal Insulation—Glass Wool
CNS A2212-87. Glass Wool Loose Fill Thermal Insulation (Sep)(12057).
CNS A3265-87. Method of Test for Glass Wool Loose Fill Thermal Insulation (Sep)(12058).
JIS A 9523-90. Glass Wool Loose Fill Thermal Insulation. 12 pp.

—Thermal Insulation—Rock Wool
CNS A2213-87. Rock Wool Loose Fill Thermal Insulation (Sep)(12059).
CNS A3266-87. Method of Test for Rock Wool Loose Fill Thermal Insulation (Sep)(12060).
JIS A 9524-85. Rock Wool Loose Fill Thermal Insulation. 12 pp.

—Thermal Insulation—Urethane Foam
BSI BS 4841: Part 2-75. 1975 Rigid Urethane Foam for Building Applications Part 2: Laminated Board for Use as a Wall and Ceiling Insulation. 11 pp.

Celery
See Also: Carrots; Vegetables

—Grading
CNS N1107-82. Grades of Celery (Nov)(9635). 2 pp.

Celery Seed
See Also: Spices
ISO 6574-86. Celery Seed (Apium Graveolens Linnaeus)—Specification First Edition. 5 pp.

Celery Seed Oil
See Also: Essential Oils
ISO 3760-79. Oil of Celery Seed First Edition. 4 pp.

Cellophane
Use For: Regenerated Cellulose Film
See Also: Cellulose; Plastic Sheets
CNS K6181-80. Method of Test for Cellophane (Oct)(2269).
CNS S1022-80. Cellophane (Oct)(2268).
JIS Z 1521-76. Cellophane (Regenerated Cellulose Film).
JIS Z 1521-74. Cellophane (Regenerated Cellulose Film). 10 pp.

—Adhesive Tapes
CNS Z5062-78. Pressure Sensitive Adhesive Cellophane (Regenerated Cellulose Film) Tapes (Mar)(4293).
CNS Z6024-78. Method of Test for Pressure Sensitive Adhesive Cellophane (Regenerated Cellulose Film) Tapes (Mar)(4294).
JIS Z 1522-89. Pressure Sensitive Adhesive Cellophane Tapes. 7 pp.

—Food Contact
EC 83/229/EEC-83. Council Directive on the Approximation of the Laws of the Member States Relating to Materials and Articles Made of Regenerated Cellulose Film Intended to Come into Contact with Foodstuffs. 9 pp.

—Polyethylene Coatings
CNS Z5128-83. Polyethylene Extrusion-Coated Cellophane (Jul)(10476).
CNS Z6066-83. Method of Test for Polyethylene Extrusion-Coated Cellophane (Jul)(10477).
JIS Z 1526-76. Polyethylene Extrusion-Coated Cellophane (R 1980). 18 pp.

—Thermal Sealed
CNS K6315-72. Method of Test for Moisture-Proof Heat-Sealing Cellophane (Jun)(3313).
CNS Z5034-71. Moisture-Proof and Thermal-Sealed Cellophane (Jan)(3248).

—Waterproof
CNS K6315-72. Method of Test for Moisture-Proof Heat-Sealing Cellophane (Jun)(3313).
CNS Z5034-71. Moisture-Proof and Thermal-Sealed Cellophane (Jan)(3248).

Cells (Electric)
Scope Note: For additional listings, use a more specific term See Also: Air Cells; Alkaline Cells; Batteries; Cadmium Cells; Lead Acid Cells; Manganese Cells; Nickel Cadmium Cells; Primary Cells

Cells (Electric) (Cont.)
CNS C4202-80. Standard Cells (Aug)(6026).

Cellular Communication Equipment
Use: Cellular Mobile Radio Equipment

Cellular Concretes
Use For: Foamed Concretes See Also: Aerated Concretes; Cellular Materials; Concretes

—Bulk Density
JIS A 1161-73. Testing Methods for Bulk Specific Gravity, Water Content, Absorption and Compressive Strength of Cellular Concrete. 7 pp.

—Compression Testing
JIS A 1161-73. Testing Methods for Bulk Specific Gravity, Water Content, Absorption and Compressive Strength of Cellular Concrete. 7 pp.

—Dimensional Stability
JIS A 1162-73. Testing Methods for Volume Change of Cellular Concrete. 7 pp.

—Water Absorption
JIS A 1161-73. Testing Methods for Bulk Specific Gravity, Water Content, Absorption and Compressive Strength of Cellular Concrete. 7 pp.

—Water Content
JIS A 1161-73. Testing Methods for Bulk Specific Gravity, Water Content, Absorption and Compressive Strength of Cellular Concrete. 7 pp.

Cellular Glass
Use For: Foamed Glass; Porous Glass
See Also: Glass; Thermal Insulation

—Buildings
CGSB CAN/CGSB-51.38-92. Cellular Glass Thermal Insulation. 9 pp.
DIN ENGL 18174-81. Cellular Glass as Insulating Material for Building Construction; Insulating Materials for Thermal Insulation (Jan). 7 pp.

—Equipment
CGSB CAN/CGSB-51.38-92. Cellular Glass Thermal Insulation. 9 pp.

—Pipes
CGSB CAN/CGSB-51.38-92. Cellular Glass Thermal Insulation. 9 pp.

—Tanks
CGSB CAN/CGSB-51.38-92. Cellular Glass Thermal Insulation. 9 pp.

Cellular Materials
See Also: Cellular Concretes; Cellular Glass; Cellular Plastics; Foam Rubber; Latex Foams; Polyethylene Foams; Polystyrene Foams; Polyurethane Foams; Structural Foam; Urea Formaldehyde Foams; Urethane Foams

—Bulk Density
DIN ENGL 53420-78. Testing of Cellular Materials; Determination of Apparent Density (Dec). 2 pp.

—Creep Properties
DIN ENGL 53425-65. Testing of Rigid Foams; Time-Dependent Creep Compression Test Under Heat (Sept). 2 pp.

—Dimensional Stability
DIN ENGL 53424-78. Testing of Rigid Cellular Materials; Determination of Dimensional Stability at Elevated Temperatures with Flexural Load and with Compressive Load (Dec). 3 pp.

—Flexible—Compression Testing
DIN ENGL 53572-86. Testing of Flexible Cellular Materials; Determination of Compression Set After Constant Strain (Nov). 3 pp.
DIN ENGL 53579 Pt 2-85. Testing of Flexible Cellular Materials; Hardness Test on Finished Parts; Compressibility of Profiles (Apr). 3 pp.

—Indentation Hardness Testing
BSI BS 4443: Part 2-88. 1988 Methods of Test for Flexible Cellular Materials Part 2: Method 7, Indentation Hardness Tests. 7 pp.

—Tensile Testing
DIN ENGL 53430-75. Testing of Rigid Cellular Materials; Tensile Test (Sept). 4 pp.

Cellular Mobile Communications
Use: Cellular Mobile Radio Services

Cellular Mobile Radio Equipment
Use For: Cellular Communication Equipment; Cellular Mobile Radio Systems; Cellular Telecommunications Systems

INDUSTRY STANDARDS

Cellular
INTERNATIONAL AND NON-U.S. NATIONAL STANDARDS
SUBJECT INDEX

Cellular Mobile Radio Equipment *(Cont.)*

See Also: Cellular Mobile Radio Services; Communication Equipment; Communication Towers; Mobile Radio Equipment; Mobile Radio Services; Mobile Stations (Communications); Radio Telephones; Radiotelephony

CCIR Report 740-2-86. General Aspects of Cellular Systems—Section 8A—Land Mobile Service and Related Subjects. 9 pp.
CCIR Report 1020-86. Adaptation of System Specification to Ease the Practical Implementation of Radio Equipment—Section 8A—Land Mobile Service and Related Subjects. 5 pp.
CENELEC PRI-ETS 300 020-1-90. European Digital Cellular Telecommunications System (Phase 1); Mobile Station Type Approval Procedure Principles (GSM 11.01) (Candidate NET 10 Part 1). 9 pp.
CENELEC PRETS 300 020-1-91. European Digital Cellular Telecommunications System (Phase 1); Mobile Station Conformity Specifications (GSM 11.10). 569 pp.
CENELEC PRI-ETS 300 020-1-92. European Digital Cellular Telecommunications System (Phase 1); Mobile Station Conformity Specifications (Candidate NET 10, Part One). 531 pp.
CENELEC I-ETS 300 020-1-92. European Digital Cellular Telecommunications System (Phase 1); Mobile Station Conformance Test System System Simulator Specification (Candidate NET 10, Part 1) (GSM 11.10). 530 pp.
CENELEC PRI-ETS 300 020-2-90. European Digital Cellular Telecommunications System (Phase 1); Mobile Station Conformity Specifications. 493 pp.
CENELEC PRETS 300 020-2-91. European Digital Cellular Telecommunications System (Phase 1); Mobile Station Conformance Test System System Simulator Specification (GSM 11.40). 50 pp.
CENELEC PRI-ETS 300 020-2-92. European Digital Cellular Telecommunications System (Phase 1); Mobile Station Conformance Test System System Simulator Specification (Candidate NET 10, Part Two). 51 pp.
CENELEC I-ETS 300 020-2-92. European Digital Cellular Telecommunications System (Phase 1); Mobile Station Conformance Test System System Simulator Specification (Candidate NET 10, Part 2) (GSM 11.40). 50 pp.
CENELEC GSM 11.40-92. See PRI-ETS 300 020-2. 50 pp.
CENELEC GSM 11.40-DCS-92. GSM DCS 1800 System Simulator Conformity Specification. 49 pp.
ETSI PRI-ETS 300 020-1-90. European Digital Cellular Telecommunications System (Phase 1) Mobile Station Type Approval Procedure Principles (GSM 11.01) (Candidate NET 10 Part 1). 9 pp.
ETSI PRI-ETS 300 020-2-90. European Digital Cellular Telecommunications System (Phase 1); Mobile Station Conformity Specifications (GSM 11. 10) (Candidate NET 10 Part 2) (Best Copy Available). 495 pp.
ETSI I-ETS 300 020-1-92. European Digital Cellular Telecommunications System (Phase 1); Mobile Station Conformance Test System System Simulator Specification (Candidate NET 10, Part 1) (GSM 11.10). 530 pp.
ETSI PRI-ETS 300 020-1-92. European Digital Cellular Telecommunications System (Phase 1); Mobile Station Conformity Specifications (Candidate NET 10, Part One). 531 pp.
ETSI PRETS 300 020-1-91. European Digital Cellular Telecommunications System (Phase 1); Mobile Station Conformity Specifications (GSM 11.10). 569 pp.
ETSI I-ETS 300 020-2-92. European Digital Cellular Telecommunications System (Phase 1); Mobile Station Conformance Test System System Simulator Specification (Candidate NET 10, Part 2) (GSM 11.40). 50 pp.
ETSI PRI-ETS 300 020-2-92. European Digital Cellular Telecommunications System (Phase 1); Mobile Station Conformance Test System System Simulator Specification (Candidate NET 10, Part Two). 51 pp.
ETSI PRETS 300 020-2-91. European Digital Cellular Telecommunications System (Phase 1); Mobile Station Conformance Test System System Simulator Specification (GSM 11.40). 50 pp.
ETSI GSM 11.40-92. See PRI-ETS 300 020-2. 50 pp.
ETSI GSM 11.40-DCS-92. GSM DCS 1800 System Simulator Conformity Specification. 49 pp.

—Analog—Call Set Up

CCIR RECMN 622-86. Technical and Operational Characteristics of Analogue Cellular Systems for Public Land Mobile Telephone Use—Section 8A—Land Mobile Service and Related Subjects. 3 pp.

—Analog—Public Land Mobile Networks

CCIR RECMN 622-86. Technical and Operational Characteristics of Analogue Cellular Systems for Public Land Mobile Telephone Use—Section 8A—Land Mobile Service and Related Subjects. 3 pp.

Cellular Mobile Radio Equipment *(Cont.)*

—Analog—Quality of Service

CCIR RECMN 622-86. Technical and Operational Characteristics of Analogue Cellular Systems for Public Land Mobile Telephone Use—Section 8A—Land Mobile Service and Related Subjects. 3 pp.

—Analog—Teleinformatic Services

CCIR RECMN 622-86. Technical and Operational Characteristics of Analogue Cellular Systems for Public Land Mobile Telephone Use—Section 8A—Land Mobile Service and Related Subjects. 3 pp.

—Channel Coding

CENELEC PRI-ETS 300 031-90. European Digital Cellular Telecommunications System (Phase 1): Channel Coding. 23 pp.
CENELEC PRI-ETS 300 031-91. European Digital Cellular Telecommunications System (Phase 1); Channel Coding. 26 pp.
CENELEC I-ETS 300 031-92. European Digital Cellular Telecommunications System (Phase 1); Channel Coding (GSM 05.03). 26 pp.
CENELEC GSM 05.03-92. See PRI-ETS 300 031. 23 pp.
ETSI PRI-ETS 300 031-90. European Digital Cellular Telecommunications System (Phase 1); Channel Coding (GSM 05. 03). 23 pp.
ETSI I-ETS 300 031-92. European Digital Cellular Telecommunications System (Phase 1); Channel Coding (GSM 05.03). 26 pp.
ETSI PRI-ETS 300 031-91. European Digital Cellular Telecommunications System (Phase 1); Channel Coding. 26 pp.
ETSI GSM 05.03-92. See PRI-ETS 300 031. 23 pp.

—Co-Channel Interference

CCIR Report 740-2-86. General Aspects of Cellular Systems—Section 8A—Land Mobile Service and Related Subjects. 9 pp.

—Data Terminal Equipment—Terminal Adaptation Functions

CENELEC PRI-ETS 300 041-90. European Digital Cellular Telecommunications System (phase 1): General on Terminal Adaption Functions for Mobile Stations. 54 pp.
CENELEC PRI-ETS 300 041-91. European Digital Cellular Telecommunications System (Phase 1); General on Terminal Adaptation Functions for Mobile Stations. 57 pp.
CENELEC PRI-ETS 300 041-92. European Digital Cellular Telecommunications System (Phase 1); General on Terminal Adaptation Functions for Mobile Stations. 54 pp.
CENELEC I-ETS 300 041-92. European Digital Cellular Telecommunications System (Phase 1); General on Terminal Adaptation Functions for Mobile Stations (GSM 07.01). 53 pp.
CENELEC PRI-ETS 300 041-93. European Digital Cellular Telecommunications System (Phase 1); General on Terminal Adaptation Functions for Mobile Stations (GSM 07.01). 53 pp.
CENELEC PRI-ETS 300 042-90. European Digital Cellular Telecommunications System (Phase 1): Terminal Adaptation Functions for Services Using Asynchronous Bearer Capabilities. 19 pp.
CENELEC PRI-ETS 300 042-91. European Digital Cellular Telecommunications System (Phase 1); Terminal Adaption Functions for Services Using Asynchronous Bearer Capabilities. 22 pp.
CENELEC PRI-ETS 300 042-92. European Digital Cellular Telecommunications System (Phase 1); Terminal Adaptation Functions for Services Using Asynchronous Bearer Capabilities. 22 pp.
CENELEC I-ETS 300 042-92. European Digital Cellular Telecommunications System (Phase 1); Terminal Adaptation Functions for Services Using Asynchronous Bearer Capabilities (GSM 07.02). 21 pp.
CENELEC PRI-ETS 300 042-93. European Digital Cellular Telecommunications System (Phase 1); Terminal Adaptation Functions for Services Using Asynchronous Bearer Capabilities (GSM 07.02). 22 pp.
CENELEC PRI-ETS 300 043-90. European Digital Cellular Telecommunications System (Phase 1): Terminal Adaptation Functions for Services Using Synchronous Bearer Capabilities. 38 pp.
CENELEC PRI-ETS 300 043-91. European Digital Cellular Telecommunications System (Phase 1); Terminal Adaptation Functions for Services Using Asynchronous Bearer Capabilities. 41 pp.
CENELEC PRI-ETS 300 043-92. European Digital Cellular Telecommunications System (Phase 1); Terminal Adaptation Functions for Services Using Synchronous Bearer Capabilities. 41 pp.
CENELEC I-ETS 300 043-92. European Digital Cellular Telecommunications System (Phase 1); Terminal Adaptation Functions for Services Using Synchronous Bearer Capabilities (GSM 07.03). 40 pp.

Cellular Mobile Radio Equipment *(Cont.)*

—Data Terminal Equipment—Terminal Adaptation Functions *(Cont.)*

CENELEC PRI-ETS 300 043-93. European Digital Cellular Telecommunications System (Phase 1); Terminal Adaption Functions for Services Using Synchronous Bearer Capabilities (GSM 07.03). 40 pp.
CENELEC GSM 07.01-92. See PRI-ETS 300 041. 69 pp.
CENELEC GSM 07.02-92. See PRI-ETS 300 042. 28 pp.
CENELEC GSM 07.03-93. See PRI-ETS 300 043. 48 pp.
ETSI PRI-ETS 300 041-90. European Digital Cellular Telecommunications System (Phase 1); General on Terminal Adaptation Functions for Mobile Stations (GSM 07. 01). 54 pp.
ETSI PRI-ETS 300 042-90. European Digital Cellular Telecommunications System (Phase 1); Terminal Adaptation Functions for Services Using Asynchronous Bearer Capabilities (GSM 07.02). 19 pp.
ETSI PRI-ETS 300 043-90. European Digital Cellular Telecommunications System (Phase 1); Terminal Adaptation Functions for Services Using Synchronous Bearer Capabilities (GSM 07.03). 38 pp.
ETSI I-ETS 300 041-92. European Digital Cellular Telecommunications System (Phase 1); General on Terminal Adaptation Functions for Mobile Stations (GSM 07.01). 53 pp.
ETSI PRI-ETS 300 041-92. European Digital Cellular Telecommunications System (Phase 1); General on Terminal Adaptation Functions for Mobile Stations. 54 pp.
ETSI PRI-ETS 300 041-93. European Digital Cellular Telecommunications System (Phase 1); General on Terminal Adaptation Functions for Mobile Stations (GSM 07.01). 53 pp.
ETSI PRI-ETS 300 041-91. European Digital Cellular Telecommunications System (Phase 1); General on Terminal Adaptation Functions for Mobile Stations. 57 pp.
ETSI I-ETS 300 042-92. European Digital Cellular Telecommunications System (Phase 1); Terminal Adaptation Functions for Services Using Asynchronous Bearer Capabilities (GSM 07.02). 21 pp.
ETSI PRI-ETS 300 042-92. European Digital Cellular Telecommunications System (Phase 1); Terminal Adaptation Functions for Services Using Asynchronous Bearer Capabilities. 22 pp.
ETSI PRI-ETS 300 042-93. European Digital Cellular Telecommunications System (Phase 1); Terminal Adaption Functions for Services Using Asynchronous Bearer Capabilities (GSM 07.02). 22 pp.
ETSI PRI-ETS 300 042-91. European Digital Cellular Telecommunications System (Phase 1); Terminal Adaption Functions for Services Using Asynchronous Bearer Capabilities. 22 pp.
ETSI I-ETS 300 043-92. European Digital Cellular Telecommunications System (Phase 1); Terminal Adaptation Functions for Services Using Synchronous Bearer Capabilities (GSM 07.03). 40 pp.
ETSI PRI-ETS 300 043-92. European Digital Cellular Telecommunications System (Phase 1); Terminal Adaptation Functions for Services Using Synchronous Bearer Capabilities. 41 pp.
ETSI PRI-ETS 300 043-93. European Digital Cellular Telecommunications System (Phase 1); Terminal Adaptation Functions for Services Using Asynchronous Bearer Capabilities (GSM 07.03). 40 pp.
ETSI PRI-ETS 300 043-91. European Digital Cellular Telecommunications System (Phase 1); Terminal Adaptation Functions for Services Using Asynchronous Bearer Capabilities. 41 pp.
ETSI GSM 07.01-92. See PRI-ETS 300 041. 69 pp.
ETSI GSM 07.02-92. See PRI-ETS 300 042. 28 pp.
ETSI GSM 07.03-93. See PRI-ETS 300 043. 48 pp.

—Digital—Bearer Services

CCIR Report 1156-90. Digital Cellular Public Land Mobile Telecommunication Systems (DCPLMTS)—Section 8A—Land Mobile Service and Related Subjects. 28 pp.

—Digital—Public Land Mobile Networks

CCIR Report 1156-90. Digital Cellular Public Land Mobile Telecommunication Systems (DCPLMTS)—Section 8A—Land Mobile Service and Related Subjects. 28 pp.
CENELEC GSM 09.01-92. European Digital Cellular Telecommunication System (Phase 1); General Network Interworking Scenarios. 11 pp.
ETSI GSM 09.01-92. European Digital Cellular Telecommunication System (Phase 1); General Network Interworking Scenarios. 11 pp.
ETSI GSM 09.02-92. See PRI-ETS 300 044. 508 pp.

INTERNATIONAL AND NON-U.S. NATIONAL STANDARDS
SUBJECT INDEX
Cellular

Cellular Mobile Radio Equipment (Cont.)

—Digital—Public Land Mobile Networks (Cont.)
ETSI GSM 09.02-DCS-92. See I-ETS 300 044/A1. 21 pp.

—Digital—Teleservices
CCIR Report 1156-90. Digital Cellular Public Land Mobile Telecommunication Systems (DCPLMTS)—Section 8A—Land Mobile Service and Related Subjects. 28 pp.

—Directional Antennas
CCIR Report 1155-90. Adaptation of Mobile Radiocommunication Technology to the Needs of Developing Countries—Section 8A—Land Mobile Service and Related Subjects. 15 pp.

—Fixed Services—Rural Areas
CCIR Report 1192-90. Application of Cellular Type Mobile Radiocommunication Systems for Use as Fixed Systems—Section 9E1—Line-of-Sight Radio-Relay Systems. 8 pp.

—Frequency Bands
EC 87/372/EEC-87. Council Directive on the Frequency Bands to be Reserved for the Coordinated Introduction of Public Pan-European Cellular Digital Land-Based Mobile Communications in the Community. 2 pp.

—Frequency Modulators
CENELEC PRI-ETS 300 032-90. European Digital Cellular Telecommunications System (Phase 1); Modulation. 5 pp.
CENELEC PRI-ETS 300 032-91. European Digital Cellular Telecommunications System (Phase 1); Modulation. 7 pp.
CENELEC I-ETS 300 032-92. European Digital Cellular Telecommunications System (Phase 1); Modulation (GSM 05.04). 7 pp.
CENELEC GSM 05.04-92. See PRI-ETS 300 032. 4 pp.
ETSI PRI-ETS 300 032-90. European Digital Cellular Telecommunications System (Phase 1); Modulation (GSM 05.04). 5 pp.
ETSI I-ETS 300 032-92. European Digital Cellular Telecommunications System (Phase 1); Modulation (GSM 05.04). 7 pp.
ETSI PRI-ETS 300 032-91. European Digital Cellular Telcommunications System (Phase 1); Modulation. 7 pp.
ETSI GSM 05.04-92. See PRI-ETS 300 032. 4 pp.

—Full Rate Speech Traffic Channels
CENELEC PRI-ETS 300 037-90. European First Generation Digital Cellular Telecommunications System. Substitution and Muting of Lost Frames for Full-Rate Speech Traffic Channels. 8 pp.
CENELEC PRI-ETS 300 037-91. European Digital Cellular Telecommunications System (Phase 1); Substitution and Muting of Lost Frames for Full-Rate Speech Traffic Channels. 10 pp.
CENELEC I-ETS 300 037-92. European Digital Cellular Telecommunications System (Phase 1); Substitution and Muting of Lost Frames for Full-Rate Speech Traffic Channels (GSM 06.11). 9 pp.
CENELEC GSM 06.10-92. See PRI-ETS 300 036. 94 pp.
CENELEC GSM 06.11-92. See PRI-ETS 300 037. 6 pp.
ETSI PRI-ETS 300 037-90. European First Generation Digital Cellular Telecommunications System; Substitution and Muting of Lost Frames for Full-Rate Speech Traffic Channels. 7 pp.
ETSI PRI-ETS 300 037-91. European Digital Cellular Telecommunications System (Phase 1); Substitution and Muting of Lost Frames for Full-Rate Speech Traffic Channels. 10 pp.
ETSI GSM 06.10-92. See PRI-ETS 300 036. 94 pp.
ETSI GSM 06.11-92. See PRI-ETS 300 037. 6 pp.

—Full Rate Speech Traffic Channels—Discontinuous Transmission
CENELEC PRI-ETS 300 039-90. European Digital Cellular Telecommunications System (Phase 1); Discontinuous Transmissions (DTX) for Full-Rate Speech Traffic Channels. 15 pp.
CENELEC PRI-ETS 300 039-91. European Digital Cellular Telecommunications System (Phase 1); Discontinuous Transmission (DTX) for Full-Rate Speech Traffic Channels. 17 pp.
CENELEC I-ETS 300 039-92. European Digital Cellular Telecommunications System (Phase 1); Discontinuous Transmission (DTX) for Full-Rate Speech Traffic Channels (GSM 06.31). 17 pp.
CENELEC GSM 06.31-92. See PRI-ETS 300 039. 14 pp.
ETSI PRI-ETS 300 039-90. European Digital Cellular Telecommunications System (Phase 1); Discontinuous Transmission (DTX) for Full-Rate Speech Traffic Channels (GSM 06.31). 14 pp.

Cellular Mobile Radio Equipment (Cont.)

—Full Rate Speech Traffic Channels—Discontinuous Transmission (Cont.)
ETSI I-ETS 300 039-92. European Digital Cellular Telecommunications System (Phase 1); Discontinuous Transmission (DTX) for Full-Rate Speech Traffic Channels (GSM 06.31). 17 pp.
ETSI PRI-ETS 300 039-91. European Digital Cellular Telecommunications System (Phase 1); Discontinuous Transmission (DTX) for Full-Rate Speech Traffic Channels. 17 pp.
ETSI GSM 06.31-92. See PRI-ETS 300 039. 14 pp.

—Full Rate Speech Traffic Channels—Noise
CENELEC PRI-ETS 300 038-90. European Digital Cellular Telecommunications System (Phase 1); Comfort Noise Aspects for Full-Rate Speech Traffic Channels. 9 pp.
CENELEC PRI-ETS 300 038-91. European Digital Cellular Telecommunications System (Phase 1); Comfort Noise Aspects for Full-Rate Speech Traffic Channels. 11 pp.
CENELEC I-ETS 300 038-92. European Digital Cellular Telecommunications System (Phase 1); Comfort Noise Aspects for Full-Rate Speech Traffic Channels. 10 pp.
CENELEC GSM 06.12-92. Comfort Noise Aspects for Full-Rate Speech Traffic Channels. 7 pp.
ETSI PRI-ETS 300 038-90. European Digital Cellular Telecommunications System (Phase 1); Comfort Noise Aspects for Full-Rate Speech Traffic Channels (GSM 06.12). 7 pp.
ETSI I-ETS 300 037-92. European Digital Cellular Telecommunications System (Phase 1); Substitution and Muting of Lost Frames for Full-Rate Speech Traffic Channels (GSM 06.11). 9 pp.
ETSI I-ETS 300 038-92. European Digital Cellular Telecommunications System (Phase 1); Comfort Noise Aspects for Full-Rate Speech Traffic Channels. 10 pp.
ETSI PRI-ETS 300 038-91. European Digital Cellular Telecommunications System (Phase 1); Comfort Noise Aspects for Full-Rate Speech Traffic Channels. 11 pp.
ETSI GSM 06.12-92. Comfort Noise Aspects for Full-Rate Speech Traffic Channels. 7 pp.

—Full Rate Speech Traffic Channels—Transcoding
CENELEC PRI-ETS 300 036-90. European Digital Cellular Telecommunications System (Phase 1); Full-Rate Speech Transcoding. 95 pp.
CENELEC PRI-ETS 300 036-91. European Digital Cellular Telecommunications System (Phase 1); Full-Rate Speech Transcoding. 96 pp.
CENELEC I-ETS 300 036-92. European Digital Cellular Telecommunications System (Phase 1); Full-Rate Speech Transcoding (GSM 06.10). 97 pp.
ETSI PRI-ETS 300 036-90. European Digital Cellular Telecommunications System (Phase 1); Full Rate Speech Transcoding (GSM 06.10). 94 pp.
ETSI I-ETS 300 036-92. European Digital Cellular Telecommunications System (Phase 1); Full-Rate Speech Transcoding (GSM 06.10). 97 pp.
ETSI PRI-ETS 300 036-91. European Digital Cellular Telecommunications System (Phase 1); Full-Rate Speech Transcoding. 96 pp.

—Full Rate Speech Traffic Channels—Voice Activity Detectors
CENELEC PRI-ETS 300 040-90. European Digital Cellular Telecommunications System (Phase 1): Voice Activity Detection. 39 pp.
CENELEC PRI-ETS 300 040-91. European Digital Cellular Telecommunications System (Phase 1); Voice Activity Detection. 41 pp.
CENELEC I-ETS 300 040-92. European Digital Cellular Telecommunications System (Phase 1); Voice Activity Detection (GSM 06.32). 41 pp.
CENELEC GSM 06.32-92. See PRI-ETS 300 040. 38 pp.
ETSI PRI-ETS 300 040-90. European Digital Cellular Telecommunications System (Phase 1); Voice Activity Detection (GSM 06.32) (Best Copy Available). 39 pp.
ETSI I-ETS 300 040-92. European Digital Cellular Telecommunications System (Phase 1); Voice Activity Detection (GSM 06.32). 41 pp.
ETSI PRI-ETS 300 040-91. European Digital Cellular Telecommunications System (Phase 1); Voice Activity Detection. 41 pp.
ETSI GSM 06.32-92. See PRI-ETS 300 040. 38 pp.

—Inband Control—Rate Adaptors
CENELEC GSM 08.60-92. Inband Control of Remote Transcoders and Rate Adaptors. 31 pp.
ETSI GSM 08.60-92. Inband Control of Remote Transcoders and Rate Adaptors. 31 pp.

Cellular Mobile Radio Equipment (Cont.)

—Inband Control—Remote Transcoders
CENELEC GSM 08.60-92. Inband Control of Remote Transcoders and Rate Adaptors. 31 pp.
ETSI GSM 08.60-92. Inband Control of Remote Transcoders and Rate Adaptors. 31 pp.

—Interfaces
CENELEC PRI-ETS 300 022-90. European Digital Cellular Telecommunications System (Phase 1): Mobile Radio Interface Layer 3 Specification. 449 pp.
CENELEC PRI-ETS 300 022-91. European Digital Cellular Telecommunications System (Phase 1); Mobile Radio Interface Layer 3 Specification. 463 pp.
CENELEC I-ETS 300 022/A1-91. AMD 1 European Digital Cellular Telecommunications System (Phase 1); Mobile Radio Interface Layer 3 Specification. 91 pp.
CENELEC PRI-ETS 300 022-92. European Digital Cellular Telecommunications System (Phase 1); Mobile Radio Interface Layer 3 Specification. 451 pp.
CENELEC I-ETS 300 022-92. European Digital Cellular Telecommunications System (Phase 1); Mobile Radio Interface Layer 3 Specification (GSM 04.08). 451 pp.
CENELEC PRI-ETS 300 022-3-93. European Digital Cellular Telecommunications System (Phase 1); Mobile Radio Interface Layer 3 Specification Part 3: Signalling Support of the Second Ciphering Algorithm (GSM 04.08). 23 pp.
CENELEC GSM 04.08-92. See PRI-ETS 300 022. 452 pp.
CENELEC GSM 04.08-DCS-92. European Digital Cellular Telecommunications System (Phase 1); Mobile Radio Interface—Layer 3 Specification. 85 pp.
CENELEC GSM 08.01-92. European Digital Cellular Telecommunication System (Phase 1); General Aspects on the BSS-MSC Interface. 8 pp.
CENELEC GSM 08.02-92. European Digital Cellular Telecommunication System (Phase 1); BSS-MSC Interface—Interface Principles. 20 pp.
CENELEC GSM 08.08-92. European Digital Cellular Telecommunication System (Phase 1); BSS-MSC Layer 3 Specification. 92 pp.
ETSI PRI-ETS 300 022-90. European Digital Cellular Telecommunications System (Phase 1); Mobile Radio Interface Layer 3 Specification (GSM 04. 08). 439 pp.
ETSI PRI-ETS 300 024-90. European Digital Cellular Telecommunications System (Phase 1) Short Message Service Cell Broadcast (SMSCB) Support on the Mobile Radio Interface (GSM 04.12). 6 pp.
ETSI I-ETS 300 022-92. European Digital Cellular Telecommunications System (Phase 1); Mobile Radio Interface Layer 3 Specification (GSM 04.08). 451 pp.
ETSI PRI-ETS 300 022-92. European Digital Cellular Telecommunications System (Phase 1); Mobile Radio Interface Layer 3 Specification. 451 pp.
ETSI I-ETS 300 022/A1-91. AMD 1 European Digital Cellular Telecommunications System (Phase 1); Mobile Radio Interface Layer 3 Specification. 91 pp.
ETSI PRI-ETS 300 024-91. European Digital Cellular Telecommunications Systems (Phase 1); Short Message Service Cell Broadcast (SMSCB) Support on the Mobile Radio Interface. 9 pp.
ETSI PRI-ETS 300 029-91. European Digital Cellular Telecommunications System (Phase 1); Mobile Radio Interface Layer 3 Call Restriction Supplementary Services Specification. 17 pp.
ETSI PRI-ETS 300 068-90. European Digital Cellular Telecommunications System (Phase 1); Man-Machine Interface of the Mobile Station (GSM 02.30). 22 pp.
ETSI I-ETS 300 068-92. European Digital Cellular Telecommunications System (Phase 1); Man-Machine Interface of the Mobile Station (GSM 02.30). 22 pp.
ETSI PRI-ETS 300 068-92. European Digital Cellular Telecommunications System (Phase 1); Man-Machine Interface of the Mobile Station. 24 pp.
ETSI GSM 04.08-92. See PRI-ETS 300 022. 452 pp.
ETSI GSM 04.08-DCS-92. European Digital Cellular Telecommunications System (Phase 1); Mobile Radio Interface—Layer 3 Specification. 85 pp.
ETSI GSM 08.01-92. European Digital Cellular Telecommunication System (Phase 1); General Aspects on the BSS-MSC Interface. 8 pp.
ETSI GSM 08.02-92. European Digital Cellular Telecommunication System (Phase 1); BSS-MSC Interface—Interface Principles. 20 pp.
ETSI GSM 08.08-92. European Digital Cellular Telecommunication System (Phase 1); BSS-MSC Layer 3 Specification. 92 pp.

—Interfaces—Call Barring
CENELEC PRI-ETS 300 029-90. European Digital Cellular Telecommunications System (Phase 1); Mobile Radio Interface Layer 3 Call Restriction Supplementary Services Specification. 14 pp.

INDUSTRY STANDARDS

Cellular
INTERNATIONAL AND NON-U.S. NATIONAL STANDARDS
SUBJECT INDEX

Cellular Mobile Radio Equipment *(Cont.)*

—Interfaces—Call Barring *(Cont.)*

CENELEC PRI-ETS 300 029-91. European Digital Cellular Telecommunications System (Phase 1); Mobile Radio Interface Layer 3 Call Restriction Supplementary Services Specification. 17 pp.
CENELEC I-ETS 300 029-92. European Digital Cellular Telecommunications System (Phase 1); Mobile Radio Interface Layer 3 Call Restriction Supplementary Services Specification (GSM 04.88). 16 pp.
CENELEC GSM 04.88-92. See PRI-ETS 300 029. 15 pp.
ETSI PRI-ETS 300 029-90. European Digital Cellular Telecommunications System (Phase 1) Mobile Radio Interface Layer 3 Call Restriction Supplementary Services Specification (GSM 04. 88). 14 pp.
ETSI I-ETS 300 029-92. European Digital Cellular Telecommunications System (Phase 1); Mobile Radio Interface Layer 3 Call Restriction Supplementary Services Specification (GSM 04.88). 16 pp.
ETSI PRI-ETS 300 029-91. European Digital Cellular Telecommunications System (Phase 1); Mobile Radio Interface Layer 3 Call Restriction Supplementary Services Specification. 17 pp.
ETSI GSM 04.88-92. See PRI-ETS 300 029. 15 pp.

—Interfaces—Call Forwarding Busy

CENELEC PRI-ETS 300 028-90. European Digital Cellular Telecommunications System (phase 1): Mobile Radio Interface Layer 3 Call Offering Supplementary Services Specification. 38 pp.
CENELEC PRI-ETS 300 028-91. European Digital Cellular Telecommunications System (Phase 1); Mobile Radio Interface Layer 3 Call Offering Supplementary Services Specification. 39 pp.
CENELEC I-ETS 300 028-92. European Digital Cellular Telecommunications System (Phase 1); Mobile Radio Interface Layer 3 Call Offering Supplementary Services Specification (GSM 04.82). 39 pp.
CENELEC GSM 04.82-92. See PRI-ETS 300 028. 37 pp.
ETSI PRI-ETS 300 028-90. European Digital Cellular Telecommunications System (Phase 1) Mobile Radio Interface Layer 3 Call Offering Supplementary Services Specification (GSM 04. 82). 40 pp.
ETSI I-ETS 300 028-92. European Digital Cellular Telecommunications System (Phase 1); Mobile Radio Interface Layer 3 Call Offering Supplementary Services Specification (GSM 04.82). 39 pp.
ETSI PRI-ETS 300 028-91. European Digital Cellular Telecommunications System (Phase 1); Mobile Radio Interface Layer 3 Call Offering Supplementary Services Specification. 39 pp.
ETSI GSM 04.82-92. See PRI-ETS 300 028. 37 pp.

—Interfaces—Call Forwarding No Reply

CENELEC PRI-ETS 300 028-90. European Digital Cellular Telecommunications System (phase 1): Mobile Radio Interface Layer 3 Call Offering Supplementary Services Specification. 38 pp.
CENELEC PRI-ETS 300 028-91. European Digital Cellular Telecommunications System (Phase 1); Mobile Radio Interface Layer 3 Call Offering Supplementary Services Specification. 39 pp.
CENELEC I-ETS 300 028-92. European Digital Cellular Telecommunications System (Phase 1); Mobile Radio Interface Layer 3 Call Offering Supplementary Services Specification (GSM 04.82). 39 pp.
CENELEC GSM 04.82-92. See PRI-ETS 300 028. 37 pp.
ETSI PRI-ETS 300 028-90. European Digital Cellular Telecommunications System (Phase 1) Mobile Radio Interface Layer 3 Call Offering Supplementary Services Specification (GSM 04. 82). 40 pp.
ETSI I-ETS 300 028-92. European Digital Cellular Telecommunications System (Phase 1); Mobile Radio Interface Layer 3 Call Offering Supplementary Services Specification (GSM 04.82). 39 pp.
ETSI PRI-ETS 300 028-91. European Digital Cellular Telecommunications System (Phase 1); Mobile Radio Interface Layer 3 Call Offering Supplementary Services Specification. 39 pp.
ETSI GSM 04.82-92. See PRI-ETS 300 028. 37 pp.

—Interfaces—Call Forwarding Unconditional

CENELEC PRI-ETS 300 028-90. European Digital Cellular Telecommunications System (phase 1): Mobile Radio Interface Layer 3 Call Offering Supplementary Services Specification. 38 pp.
CENELEC PRI-ETS 300 028-91. European Digital Cellular Telecommunications System (Phase 1); Mobile Radio Interface Layer 3 Call Offering Supplementary Services Specification. 39 pp.
CENELEC I-ETS 300 028-92. European Digital Cellular Telecommunications System (Phase 1); Mobile Radio Interface Layer 3 Call Offering Supplementary Services Specification (GSM 04.82). 39 pp.
CENELEC GSM 04.82-92. See PRI-ETS 300 028. 37 pp.

Cellular Mobile Radio Equipment *(Cont.)*

—Interfaces—Call Forwarding Unconditional *(Cont.)*

ETSI PRI-ETS 300 028-90. European Digital Cellular Telecommunications System (Phase 1) Mobile Radio Interface Layer 3 Call Offering Supplementary Services Specification (GSM 04. 82). 40 pp.
ETSI I-ETS 300 028-92. European Digital Cellular Telecommunications System (Phase 1); Mobile Radio Interface Layer 3 Call Offering Supplementary Services Specification (GSM 04.82). 39 pp.
ETSI PRI-ETS 300 028-91. European Digital Cellular Telecommunications System (Phase 1); Mobile Radio Interface Layer 3 Call Offering Supplementary Services Specification. 39 pp.
ETSI GSM 04.82-92. See PRI-ETS 300 028. 37 pp.

—Interfaces—Message Coding

CENELEC PRI-ETS 300 027-90. European Digital Cellular Telecommunications System (Phase 1): Mobile Radio Interface Layer 3 Supplementary Services Specifications Formats and Coding. 39 pp.
CENELEC PRI-ETS 300 027-91. European Digital Cellular Telecommunications System (Phase 1); Mobile Radio Interface Layer 3 Supplementary Services Specification Formats and Coding. 42 pp.
CENELEC I-ETS 300 027-92. European Digital Cellular Telecommunications System (Phase 1); Mobile Radio Interface Layer 3 Supplementary Services Specification Formats and Coding (GSM 04.80). 42 pp.
CENELEC GSM 04.80-92. See PRI-ETS 300 027. 41 pp.
ETSI PRI-ETS 300 027-90. European Digital Cellular Telecommunications System (Phase 1); Mobile Radio Interface Layer 3 Supplementary Services Specification Formats and Coding (GSM 04.80). 39 pp.
ETSI I-ETS 300 027-92. European Digital Cellular Telecommunications System (Phase 1); Mobile Radio Interface Layer 3 Supplementary Services Specification Formats and Coding (GSM 04.80). 42 pp.
ETSI PRI-ETS 300 027-91. European Digital Cellular Telecommunications System (Phase 1); Mobile Radio Interface Layer 3 Supplementary Services Specification Formats and Coding. 42 pp.
ETSI GSM 04.80-92. See PRI-ETS 300 027. 41 pp.

—Interfaces—Protocols

CENELEC PRI-ETS 300 026-90. European Digital Cellular Telecommunications System (Phase 1); Radio Link Protocol (RLP) for Data Telematic Services on Mobile Station-Base Station System (MS-BSS) Interface and Base Station System-Mobile Services Switching Centre (BSS-MSC) Interface. 25 pp.
CENELEC PRI-ETS 300 026-91. European Digital Cellular Telecommunications Sys. (Phase 1); Radio Link Protocol (RLP) for Data and Telematic Services on the Mobile Station—Base Station System (MS-BSS) Interface and the Base Station System —Mobile Services Swg. Centre (BSS-MSC) Interface. 63 pp.
CENELEC PRI-ETS 300 026-92. European Digital Cellular Telecommunications System (Phase 1); Radio Link Protocol (RLP) for Data and Telematic Services on the Mobile Station—Base Station System (MS-BSS) Interface and the Base Station System—Mobile Services Switching Centre (BSS-MSC) Intrfce. 63 pp.
CENELEC I-ETS 300 026-92. European Digital Cellular Telecommunications Sys. (Phase 1); Radio Link Protocol (RLP) for Data and Telematic Servs. on the Mobile Sta.—Base Sta. System (MS-BSS) Interface and the Base Sta. Sys.—Mobile Servs. Switching Centre (BSS-MSC) Interface (GSM 04.22). 63 pp.
CENELEC GSM 04.22-92. See PRI-ETS 300 026. 64 pp.
ETSI PRI-ETS 300 026-90. European Digital Cellular Telecommunications System (Phase 1) Radio Link Protocol (RLP) for Data and Telematic Services on the Mobile Station—Base Station System (MS-BSS) Interface and the Base Station System—Mobile Services Switching Centre (BSS-MSC). 25 pp.
ETSI I-ETS 300 026-92. European Digital Cellular Telecommunications Sys. (Phase 1); Radio Link Protocol (RLP) for Data and Telematic Servs. on the Mobile Sta.—Base Sta. System (MS-BSS) Interface and the Base Sta. Sys.—Mobile Servs. Switching Centre (BSS-MSC) Interface (GSM 04.22). 63 pp.
ETSI PRI-ETS 300 026-92. European Digital Cellular Telecommunications System (Phase 1); Radio Link Protocol (RLP) for Data and Telematic Services on the Mobile Station—Base Station System (MS-BSS) Interface and the Base Station System—Mobile Services Switching Centre (BSS-MSC) Intrfce. 63 pp.
ETSI GSM 04.22-92. See PRI-ETS 300 026. 64 pp.

Cellular Mobile Radio Equipment *(Cont.)*

—Interfaces—Service Messages

CENELEC PRI-ETS 300 023-90. European Digital Cellular Telecommunications System (Phase 1); Point to Point Short Message Service Support on Mobile Radio Interface. 74 pp.
CENELEC PRI-ETS 300 023-91. European Digital Cellular Telecommunications System (Phase 1); Point-to-Point Short Message Service Support on Mobile Radio Interface. 76 pp.
CENELEC PRI-ETS 300 023-92. European Digital Cellular Telecommunications System (Phase 1); Point-to-Point Short Message Service Support on Mobile Radio Interface. 76 pp.
CENELEC I-ETS 300 023-92. European Digital Cellular Telecommunications System (Phase 1); Point-to-Point Short Message Service Support on Mobile Radio Interface (GSM 04.11). 76 pp.
CENELEC PRI-ETS 300 023-93. European Digital Cellular Telecommunications System (Phase 1); Point-to-Point Short Message Service Support on Mobile Radio Interface (GSM 04.11). 78 pp.
CENELEC PRI-ETS 300 024-90. European Digital Cellular Telecommunications System (Phase 1); Short Message Service Cell Broadcast (SMSCB) Support on the Mobile Radio Interface. 6 pp.
CENELEC PRI-ETS 300 024-91. European Digital Cellular Telecommunications Systems (Phase 1); Short Message Service Cell Broadcast (SMSCB) (SMSCB) Support on the Mobile Radio Interface. 9 pp.
CENELEC I-ETS 300 024-92. European Digital Cellular Telecommunications System (Phase 1); Short Message Service Cell Broadcast (SMSCB) Support on the Mobile Radio Interface (GSM 04.12). 8 pp.
CENELEC GSM 04.11-93. See PRI-ETS 300 023. 73 pp.
CENELEC GSM 04.12-92. See PRI-ETS 300 024. 5 pp.
ETSI PRI-ETS 300 023-90. European Digital Cellular Telecommunications System (Phase 1); Point to Point Short Message Service Suppport on Mobile Radio Interface (GSM 04.11). 74 pp.
ETSI PRI-ETS 300 022-3-93. European Digital Cellular Telecommunications System (Phase 1); Mobile Radio Interface Layer 3 Specification Part 3: Signalling Support of the Second Ciphering Algorithm (GSM 04.08). 23 pp.
ETSI I-ETS 300 023-92. European Digital Cellular Telecommunications System (Phase 1); Point-to-Point Short Message Service Support on Mobile Radio Interface (GSM 04.11). 76 pp.
ETSI PRI-ETS 300 023-92. European Digital Cellular Telecommunications System (Phase 1); Point-to-Point Short Message Service Support on Mobile Radio Interface. 76 pp.
ETSI PRI-ETS 300 O23-93. European Digital Cellular Telecommunications System (Phase 1); Point-to-Point Short Message Service Support on Mobile Radio Interface (GSM 04.11). 78 pp.
ETSI GSM 04.11-93. See PRI-ETS 300 023. 73 pp.
ETSI GSM 04.12-92. See PRI-ETS 300 024. 5 pp.

—Intermodulation Interference

CCIR Report 740-2-86. General Aspects of Cellular Systems—Section 8A—Land Mobile Service and Related Subjects. 9 pp.

—Link Control Procedures

CENELEC PRI-ETS 300 034-90. European Digital Cellular Telecommunications System (Phase 1); Radio Sub-System Link Control. 50 pp.
CENELEC PRI-ETS 300 034-91. European Digital Cellular Telecommunications System (Phase 1); Radio Sub-System Link Control. 43 pp.
CENELEC I-ETS 300 034/A1-91. AMD 1 European Digital Cellular Telecommunications System (Phase 1); Radio Sub-System Link Control (GSM 05.08DCS). 25 pp.
CENELEC I-ETS 300 034-92. European Digital Cellular Telecommunications System (Phase 1); Radio Sub-System Link Control (GSM 05.08). 34 pp.
CENELEC GSM 05.08-92. See PRI-ETS 300 034. 39 pp.
CENELEC GSM 05.08-DCS-92. See I-ETS 300 034 A1. 22 pp.
ETSI PRI-ETS 300 034-90. European Digital Cellular Telecommunications System (Phase 1); Radio Sub-System Link Control (GSM 05.08). 50 pp.
ETSI I-ETS 300 034-92. European Digital Cellular Telecommunications System (Phase 1); Radio Sub-System Link Control (GSM 05.08). 34 pp.
ETSI I-ETS 300 034/A1-91. AMD 1 European Digital Cellular Telecommunications System (Phase 1); Radio Sub-System Link Control (GSM 05.08DCS). 25 pp.
ETSI PRI-ETS 300 034-91. European Digital Cellular Telecommunications System (Phase 1); Radio Sub-System Link Control. 43 pp.
ETSI GSM 05.08-92. See PRI-ETS 300 034. 39 pp.

INTERNATIONAL AND NON-U.S. NATIONAL STANDARDS
SUBJECT INDEX
Cellular

Cellular Mobile Radio Equipment *(Cont.)*

—Link Control Procedures *(Cont.)*
ETSI GSM 05.08-DCS-92. See I-ETS 300 034 A1. 22 pp.

—Location Registers
CENELEC GSM 11.31-92. European Digital Cellular Telecommunication System (Phase 1); Home Location Register Specification. 11 pp.
CENELEC GSM 11.32-92. European Digital Cellular Telecommunication System (Phase 1); Visitor Location Register Specification. 15 pp.
ETSI GSM 11.31-92. European Digital Cellular Telecommunication System (Phase 1); Home Location Register Specification. 11 pp.
ETSI GSM 11.32-92. European Digital Cellular Telecommunication System (Phase 1); Visitor Location Register Specification. 15 pp.

—Message Transfer Part (Signaling Systems)
CENELEC GSM 08.06-92. European Digital Cellular Telecommunication System (Phase 1); Signalling Transport Mechanism Specification for the BSS-MSC Interface. 29 pp.
ETSI GSM 08.06-92. European Digital Cellular Telecommunication System (Phase 1); Signalling Transport Mechanism Specification for the BSS-MSC Interface. 29 pp.

—Mobile Application Part—Signaling Systems
CENELEC PRI-ETS 300 044-90. European Digital Cellular Telecommunications System (Phase 1): Mobile Application Part Specification. 559 pp.
CENELEC PRI-ETS 300 044-91. European Digital Cellular Telecommunications System (Phase 1); Mobile Application Part Specification. 528 pp.
CENELEC I-ETS 300 044/A1-91. AMD 1 European Digital Cellular Telecommunications System (Phase 1); Mobile Application Part Specification (GSM 09.02DCS). 24 pp.
CENELEC I-ETS 300 044-92. European Digital Cellular Telecommunications System (Phase 1); Mobile Application Part Specification (GSM 09.02). 512 pp.
CENELEC GSM 09.02-92. See PRI-ETS 300 044. 508 pp.
CENELEC GSM 09.02-DCS-92. See I-ETS 300 044/A1. 21 pp.
ETSI PRI-ETS 300 044-90. European Digital Cellular Telecommunications System (Phase 1); Mobile Application Part Specification (GSM 09. 02). 558 pp.
ETSI I-ETS 300 044-92. European Digital Cellular Telecommunications System (Phase 1); Mobile Application Part Specification (GSM 09.02). 512 pp.
ETSI I-ETS 300 044/A1-91. AMD 1 European Digital Cellular Telecommunications System (Phase 1); Mobile Application Part Specification (GSM 09.02DCS). 24 pp.
ETSI PRI-ETS 300 044-91. European Digital Cellular Telecommunications System (Phase 1); Mobile Application Part Specification. 528 pp.

—Multiplexing
CENELEC PRI-ETS 300 030-90. European Digital Cellular Telecommunications System (Phase 1); Multiplexing and Multiple Access on the Radio Path. 38 pp.
CENELEC PRI-ETS 300 030-91. European Digital Cellular Telecommunications System (Phase 1); Multiplexing and Multiple Access on the Radio Path. 40 pp.
CENELEC PRI-ETS 300 030-92. European Digital Cellular Telecommunications System (Phase 1); Multiplexing and Multiple Access on the Radio Path. 40 pp.
CENELEC I-ETS 300 030-92. European Digital Cellular Telecommunications System (Phase 1); Multiplexing and Multiple Access on the Radio Path (GSM 05.02). 40 pp.
CENELEC GSM 05.02-92. See PRI-ETS 300 030. 41 pp.
ETSI PRI-ETS 300 030-90. European Digital Cellular Telecommunications System (Phase 1) Multiplexing and Multiple Access on the Radio Path (GSM 05.02). 38 pp.
ETSI I-ETS 300 030-92. European Digital Cellular Telecommunications System (Phase 1); Multiplexing and Multiple Access on the Radio Path (GSM 05.02). 40 pp.
ETSI PRI-ETS 300 030-92. European Digital Cellular Telecommunications System (Phase 1); Multiplexing and Multiple Access on the Radio Path. 40 pp.
ETSI GSM 05.02-92. See PRI-ETS 300 030. 41 pp.

—Physical Layer (OSI)
CENELEC GSM 05.01-92. European Digital Cellular Telecommunication System (Phase 1); Physical Layer on the Radio Path: General Description. 15 pp.

Cellular Mobile Radio Equipment *(Cont.)*

—Physical Layer (OSI) *(Cont.)*
CENELEC GSM 05.01-DCS-92. European Digital Cellular Telecommunication System (Phase 1); Physical Layer on the Radio Path: General Description. 7 pp.
ETSI GSM 05.01-92. European Digital Cellular Telecommunication System (Phase 1); Physical Layer on the Radio Path: General Description. 15 pp.
ETSI GSM 05.01-DCS-92. European Digital Cellular Telecommunication System (Phase 1); Physical Layer on the Radio Path: General Description. 7 pp.

—Physical Layer (OSI)—Interfaces
CENELEC GSM 08.04-92. European Digital Cellular Telecommunication System (Phase 1); BSS-MSC Layer 1 Specification. 4 pp.
ETSI GSM 08.04-92. European Digital Cellular Telecommunication System (Phase 1); BSS-MSC Layer 1 Specification. 4 pp.

—Repeaters
CCIR Report 1155-90. Adaptation of Mobile Radiocommunication Technology to the Needs of Developing Countries—Section 8A—Land Mobile Service and Related Subjects. 15 pp.

—Signaling Connection Control Part
CENELEC GSM 08.06-92. European Digital Cellular Telecommunication System (Phase 1); Signalling Transport Mechanism Specification for the BSS-MSC Interface. 29 pp.
ETSI GSM 08.06-92. European Digital Cellular Telecommunication System (Phase 1); Signalling Transport Mechanism Specification for the BSS-MSC Interface. 29 pp.

—Signaling Systems—Interworking
CENELEC GSM 09.11-92. European Digital Cellular Telecommunication System (Phase 1); Signalling Interworking for Supplementary Services. 12 pp.
ETSI GSM 09.11-92. European Digital Cellular Telecommunication System (Phase 1); Signalling Interworking for Supplementary Services. 12 pp.

—Speech Transmission Systems
CENELEC GSM 06.01-92. European Digital Cellular Telecommunication System (Phase 1); Speech Processing Functions: General Description. 11 pp.
ETSI GSM 06.01-92. European Digital Cellular Telecommunication System (Phase 1); Speech Processing Functions: General Description. 11 pp.

—Synchronization (Signaling)
CENELEC PRI-ETS 300 035-90. European Digital Cellular Telecommunications System (Phase 1); Radio Sub-System Synchronization. 8 pp.
CENELEC PRI-ETS 300 035-91. European Digital Cellular Telecommunications System (Phase 1); Radio Sub-System Synchronization. 10 pp.
CENELEC I-ETS 300 035-92. European Digital Cellular Telecommunications System (Phase 1); Radio Sub-System Synchronization (GSM 05.10). 10 pp.
CENELEC PRI-ETS 300 035-93. European Digital Cellular Telecommunications System (Phase 1); Radio Sub-System Synchronisation (GSM 05.10). 11 pp.
CENELEC GSM 05.10-92. See PRI-ETS 300 035. 7 pp.
ETSI PRI-ETS 300 035-90. European Digital Cellular Telecommunications System (Phase 1); Radio Sub-System Synchronization (GSM 05. 10). 8 pp.
ETSI I-ETS 300 035-92. European Digital Cellular Telecommunications System (Phase 1); Radio Sub-System Synchronization (GSM 05.10). 10 pp.
ETSI PRI-ETS 300 035-91. European Digital Cellular Telecommunications System (Phase 1); Radio Sub-System Synchronisation. 10 pp.
ETSI PRI-ETS 300 035-93. European Digital Cellular Telecommunications System (Phase 1); Radio Sub-System Synchronisation (GSM 05.10). 11 pp.
ETSI GSM 05.10-92. See PRI-ETS 300 035. 7 pp.

—System Simulators—Conformance Testing
CENELEC PRI-ETS 300 020-3-90. European Digital Cellular Telecommunications System (Phase 1): Mobile Station Conformance Test System System Simulator Specification (GSM 11.40) (Candidate NET 10 Part 3). 38 pp.
ETSI PRI-ETS 300 020-3-90. European Digital Cellular Telecommunications System (Phase 1) Mobile Station Conformance Test System. System Simulator Specification (GSM 11. 40) (Candidate NET 10 rt 3) (Best Copy Available). 38 pp.
ETSI PRI-ETS 300 020-3-90. European Digital Cellular Telecommunications System (Phase 1): Mobile Station Conformance Test System System Simulator Specification (GSM 11.40) (Candidate NET 10 Part 3). 38 pp.

Cellular Mobile Radio Equipment *(Cont.)*

—Transceivers
CENELEC PRI-ETS 300 033-90. European Digital Cellular Telecommunications System (Phase 1): Radio Transmission and Reception. 22 pp.
CENELEC PRI-ETS 300 033-91. European Digital Cellular Telcommunications System (Phase 1); Radio Transmission and Reception. 23 pp.
CENELEC I-ETS 300 033/A1-91. Text of Amendment 1. 28 pp.
CENELEC PRI-ETS 300 033-92. European Digital Cellular Telecommunications System (Phase 1); Radio Transmission and Reception. 24 pp.
CENELEC I-ETS 300 033-92. European Digital Cellular Telecommunications System (Phase 1); Radio Transmission and Reception (GSM 05.05). 24 pp.
CENELEC PRI-ETS 300 033-1-93. European Digital Cellular Telecommunications System (Phase 1); Radio Transmission and Reception Part 1: Generic (GSM 05.05). 25 pp.
CENELEC GSM 05.05-92. See PRI-ETS 300 033. 21 pp.
CENELEC GSM 05.05-DCS-92. European Digital Cellular Telecommunications System (Phase 1) Radio Transmission and Reception. 31 pp.
ETSI PRI-ETS 300 033-90. European Digital Cellular Telecommunications System (Phase 1); Radio Transmission and Reception (GSM 05.05). 22 pp.
ETSI I-ETS 300 033-92. European Digital Cellular Telecommunications System (Phase 1); Radio Transmission and Reception (GSM 05.05). 24 pp.
ETSI I-ETS 300 033/A1-91. Text of Amendment 1. 28 pp.
ETSI PRI-ETS 300 033-92. European Digital Cellular Telecommunications System (Phase 1); Radio Transmission and Reception. 24 pp.
ETSI PRI-ETS 300 033-1-93. European Digital Cellular Telecommunications System (Phase 1); Radio Transmission and Reception Part 1: Generic (GSM 05.05). 25 pp.
ETSI GSM 05.05-92. See PRI-ETS 300 033. 21 pp.
ETSI GSM 05.05-DCS-92. European Digital Cellular Telecommunications System (Phase 1) Radio Transmission and Reception. 31 pp.

Cellular Mobile Radio Services

Use For: Cellular Mobile Communications
See Also: Cellular Mobile Radio Equipment; Hand Off; Mobile Radio Equipment; Mobile Radio Services; Mobile Satellite Communications; Mobile Stations (Communications); MOBITEX; Public Land Mobile Networks; Radio Telephones; Radiotelephony; Telephone Services

CCIR Report 1155-90. Adaptation of Mobile Radiocommunication Technology to the Needs of Developing Countries—Section 8A—Land Mobile Service and Related Subjects. 15 pp.
EC 87/371/EEC-87. Council Recommendation on the Coordinated Introduction of Public Pan-European Cellular Digital Land-Based Mobile Communications in the Community. 4 pp.

—Accounting
CCITT RECMN D.93-89. Charging and Accounting in the International Land Mobile Telephone Service (Provided Via Cellular Radio Systems)—General Tariff Principles—Charging and Accounting in International Telecommunications Services (Study Group III) 11 pp. 11 pp.

—Automatic Vehicle Location Systems
CCIR Report 904-2-90. Automatic Determination of Location and Guidance in the Land Mobile Service—Section 8A—Land Mobile Service and Related Subjects. 20 pp.

—Developing Countries
CCIR Report 1155-90. Adaptation of Mobile Radiocommunication Technology to the Needs of Developing Countries—Section 8A—Land Mobile Service and Related Subjects. 15 pp.
CCIR QUESTION 77/8-86. Adaptation of Mobile Radiocommunication Technology to the Needs of Developing Countries—Questions Concerning Study Group 8—Mobile, Radiodetermination, Amateur and Related Satellite Services. 1 p.

—Fixed Services (Radio Communications)
CCIR RECMN 757-92. Basic System Requirements and Performance Objectives for Cellular Type Mobile Systems Used as Fixed Systems—Section 9E1—Line-of-Sight Radio Relay Systems. 7 pp.
CCIR QUESTION 140/9-90. Application of Cellular Type Mobile Radiocommunication Systems for Use as Fixed Systems—Questions of Study Group 9 Fixed Service. 1 p.

—References
CCITT RECMN E.201-91. Reference Recommendation for Mobile Services (Study Group II) 12 pp. 12 pp.

INDUSTRY STANDARDS

Cellular Mobile Radio Services (Cont.)

—Routing—Accounting
CCITT RECMN D.93-89. Charging and Accounting in the International Land Mobile Telephone Service (Provided Via Cellular Radio Systems)—General Tariff Principles—Charging and Accounting in International Telecommunications Services (Study Group III) 11 pp. 11 pp.

—Service Messages
CENELEC PRI-ETS 300 069-90. European Digital Cellular Telecommunications System (Phase 1): Technical Realization of the Short Message Service Cell Broadcast. 8 pp.

CENELEC PRI-ETS 300 069-92. European Digital Cellular Telecommunications System (Phase 1): Technical Realization of the Short Message Service Cell Broadcast. 11 pp.

CENELEC I-ETS 300 069-92. European Digital Cellular Telecommunications System (Phase 1); Technical Realization of the Short Message Service Cell Broadcast (GSM 03.41). 11 pp.

CENELEC GSM 03.41-92. See PRI-ETS 300 069. 11 pp.

ETSI PRI-ETS 300 069-90. European Digital Cellular Telecommunications System (Phase 1); Technical Realization of the Short Message Service—Cell Broadcast (GSM 03.41). 8 pp.

ETSI I-ETS 300 069-92. European Digital Cellular Telecommunications System (Phase 1); Technical Realization of the Short Message Service Cell Broadcast (GSM 03.41). 11 pp.

ETSI PRI-ETS 300 069-92. European Digital Cellular Telecommunications System (Phase 1); Technical Realization of the Short Message Service Cell Broadcast. 11 pp.

ETSI GSM 03.41-92. See PRI-ETS 300 069. 11 pp.

—Tariffs
CCIR RECMN 622-86. Technical and Operational Characteristics of Analogue Cellular Systems for Public Land Mobile Telephone Use—Section 8A—Land Mobile Service and Related Subjects. 3 pp.

CCITT RECMN D.93-89. Charging and Accounting in the International Land Mobile Telephone Service (Provided Via Cellular Radio Systems)—General Tariff Principles—Charging and Accounting in International Telecommunications Services (Study Group III) 11 pp. 11 pp.

Cellular Mobile Radio Systems
Use: Cellular Mobile Radio Equipment

Cellular Plastics
Scope Note: For additional listings, see also specific products made from cellular plastic
Use For: Foamed Plastics; Plastic Foam
See Also: Blowing Agents; Foam Rubber; Latex Foams; Plastics; Polyethylene Foams; Polystyrene Foams; Polyurethane Foams; Polyurethane Resins; Sponges (Materials); Structural Foam

CNS K6666-81. Determination of Linear Dimensions of Rigid Cellular Plastics (May)(7406).

ISO 1923-81. Cellular Plastics and Rubbers—Determination of Linear Dimensions Second Edition. 5 pp.

ISO 6453-85. Polymeric Materials, Cellular Flexible—Polyvinylchloride Foam Sheeting—Specification First Edition. 8 pp.

—Acoustical Insulation
DIN ENGL 18164 Pt 2-91. Rigid Cellular Plastics Insulating Building Materials; Polystyrene Foam Impact Sound Insulating Materials (Mar). 9 pp.

—Aerospace
DIN ENGL LN 29898-87. Aerospace; Rigid Cellular Materials; Polyvinyl Chloride and Polymethacrylimide Sheet; Dimensions, Masses (Dec). 6 pp.

—Aging Testing
BSI BS 4443: Part 4-89. 1989 Methods of Test for Flexible Cellular Materials Part 4: Method 10. Determination of Solvent Swelling Method 11. Humidity Ageing at an Elevated Temperature Method 12. Heat Ageing. 12 pp.

BSI BS 4443: Part 4-76. 1976 Methods of Test for Flexible Cellular Materials Part 4: Method 10. Determination of Solvent Swelling Method 11. Humidity Ageing at an Elevated Temperature Method 12. Heat Ageing. 7 pp.

ISO 2440-83. Polymeric Materials, Cellular Flexible—Accelerated Ageing Tests Second Edition. 4 pp.

—Air Flow
BSI BS 4443: Part 6-91. 1991 Methods of Test for Flexible Cellular Materials Part 6: Method 14. Preparation of Water Extract Method 15. Determination of Tear Strength Method 16. Determination of Air Flow Value. 14 pp.

Cellular Plastics (Cont.)

—Air Flow (Cont.)
ISO 7231-84. Polymeric Materials, Cellular Flexible—Method of Assessment of Air Flow Value at Constant Pressure-Drop First Edition. 6 pp.

—Bacteria
ISO 846-78. Plastics—Determination of Behavior Under the Action of Fungi and Bacteria—Evaluation by Visual Examination or Measurement of Change in Mass or Physical Properties First Edition. 12 pp.

—Bend Testing
ISO 1209 Pt 1-90. Cellular Plastics, Rigid —Flexural Tests—Part 1: Bending Test First Edition. 6 pp.

JIS K 7221-84. Testing Method for Flexural Properties of Rigid Cellular Plastics. 12 pp.

—Breaking Load
BSI BS 4370: Part 1-88. 1988 Methods of Test for Rigid Cellular Materials Part 1: Methods 1-5. 12 pp.

—Cell Volume
BSI BS 4370: Part 2-93. 1993 Methods of Test for Rigid Cellular Materials Part 2: Methods 6 to 10 (V). 24 pp.

BSI BS 4370: Part 2-73. 1973 Methods of Test for Rigid Cellular Materials Part 2: Methods 6-10. 29 pp.

BSI BS 4443: Part 1-88. 1988 Amd 1 Methods of Test for Flexible Cellular Materials Part 1: Methods 1 to 6. 16 pp.

ISO 4590-81. Cellular Plastics—Determination of Volume Percentage of Open and Closed Cells of Rigid Materials First Edition. 14 pp.

—Cold Flow
ISO 7616-86. Cellular Plastics, Rigid—Determination of Compressive Creep Under Specified Load and Temperature Conditions First Edition. 4 pp.

ISO 7850-86. Cellular Plastics, Rigid—Determination of Compressive Creep First Edition. 5 pp.

—Compression Testing
BSI BS 4370: Part 1-88. 1988 Methods of Test for Rigid Cellular Materials Part 1: Methods 1-5. 12 pp.

BSI BS 4443: Part 1-88. 1988 Amd 1 Methods of Test for Flexible Cellular Materials Part 1: Methods 1 to 6. 16 pp.

BSI BS 4443: Part 3-88. 1988 Methods of Test for Flexible Cellular Materials Part 3: Method 8, Determination of Creep Method 9, Determination of Dynamic Cushioning Performance. 12 pp.

BSI BS 4443: Part 7-92. 1992 Methods of Test for Flexible Cellular Materials Part 7: Method 17. Determination of Tear Strength of Flexible Cellular Material with an Integral Skin Method 18. Determination of Compression Set Under Humid Conditions. 11 pp.

CNS K6668-81. Compression Test for Rigid Materials of Cellular Plastics (May)(7408).

DIN ENGL 53421-84. Testing of Rigid Cellular Plastics; Compression Test (June). 3 pp.

DIN ENGL 53574-77. Testing of Flexible Cellular Materials; Fatigue Vibration Test by Constant Load Pounding in the Indentation/Pulsation Range (May). 4 pp.

DIN ENGL 53577-88. Determination of Compression Stress Value and Compression Stress-Strain Characteristic for Flexible Cellular Materials (Dec). 3 pp.

ISO 844-78. Cellular Plastics—Compression Test for Rigid Materials First Edition; (Amendment Slip-1979). 6 pp.

ISO 1856-80. Polymeric Materials, Cellular Flexible—Determination of Compression Set Second Edition; (Erratum—Nov 1981). 5 pp.

ISO 3386 Pt 1-86. Polymeric Materials, Cellular Flexible—Determination of Stress-Strain Characteristic in Compression—Part 1: Low-Density Materials Second Edition. 5 pp.

ISO 3386 Pt 2-84. Polymeric Materials, Cellular Flexible—Determination of Stress-Strain Characteristic in Compression—Part 2: High Density Materials First Edition. 4 pp.

ISO 10066-91. Flexible Cellular Polymeric Materials—Determination of Creep in Compression First Edition. 6 pp.

JIS K 7220-83. Testing Method for Compressive Properties of Rigid Cellular Plastics. 10 pp.

—Creep Testing
BSI BS 4443: Part 3-88. 1988 Methods of Test for Flexible Cellular Materials Part 3: Method 8, Determination of Creep Method 9, Determination of Dynamic Cushioning Performance. 12 pp.

ISO 10066-91. Flexible Cellular Polymeric Materials—Determination of Creep in Compression First Edition. 6 pp.

Cellular Plastics (Cont.)

—Density
BSI BS 4370: Part 1-88. 1988 Methods of Test for Rigid Cellular Materials Part 1: Methods 1-5. 12 pp.

BSI BS 4443: Part 1-88. 1988 Amd 1 Methods of Test for Flexible Cellular Materials Part 1: Methods 1 to 6. 16 pp.

CNS K6667-81. Determination of Apparent Density of Cellular Rubbers and Plastics (May)(7407).

ISO 845-88. Cellular Plastics and Rubbers—Determination of Apparent (Bulk) Density Second Edition. 6 pp.

ISO 9054-90. Cellular Plastics, Rigid —Test Methods for Self-Skinned, High-Density Materials First Edition. 8 pp.

JIS K 7222-85. Methods for Determining the Apparent Density of Rigid Cellular Plastics. 7 pp.

—Dimensional Stability
BSI BS 4370: Part 1-88. 1988 Methods of Test for Rigid Cellular Materials Part 1: Methods 1-5. 12 pp.

ISO 2796-86. Cellular Plastics, Rigid—Test for Dimensional Stability Third Edition. 5 pp.

—Elongation
BSI BS 4443: Part 1-88. 1988 Amd 1 Methods of Test for Flexible Cellular Materials Part 1: Methods 1 to 6. 16 pp.

—Fire Testing
BSI BS 4735-74. 1974 Laboratory Methods of Test for Assessment of the Horizontal Burning Characteristics of Specimens No Larger Than 150 mm X 50 mm X 13 mm (Nominal) of Cellular Plastics and Cellular Rubber Materials When Subjected to a Small Flame. 12 pp.

BSI BS 4735-01. 1974 Amd 1 Laboratory Method of Test for Assessment of the Horizontal Burning Char. of Specimens No Larger Than 150 mm X 50 mm X 13 mm (Nominal) of Cellular Plastics and Cellular Rubber Materials When Subjected to a Small Flame (AMD 7687) July 15, 1993 (V). 13 pp.

—Flammability Testing
BSI BS 5111: Part 1-74. 1974 Laboratory Methods of Test for Determination of Smoke Generation Characteristics of Cellular Plastics and Cellular Rubber Materials Part 1: Meth of Testing a 25 mm Cube Test Spec. of Low Density Mat. (up to 130 kg/Metres Cubed) to. 13 pp.

BSI BS 5111: Part 1-01. 1974 Amd 1 Laboratory Methods of Test for Determination of Smoke Generation Characteristics of Cellular Plastics and Cellular Rubber Materials Part 1: Method for Testing a 25 mm Cube Test Specimen of Low Density Material (up to 130 kg/m3) to. 14 pp.

ISO 3582-78. Cellular Plastic and Cellular Rubber Materials—Laboratory Assessment of Horizontal Burning Characteristics of Small Specimens Subjected to a Small Flame First Edition. 9 pp.

—Flexural Strength
BSI BS 4370: Part 4-91. 1991 Methods of Test for Rigid Cellular Materials Part 4: Method 14. Determination of Flexural Properties. 8 pp.

DIN ENGL 53423-75. Testing of Rigid Cellular Materials; Flexural Test (Nov). 3 pp.

ISO 1209 Pt 2-90. Cellular Plastics, Rigid —Flexural Tests—Part 2: Determination of Flexural Properties First Edition. 6 pp.

—Fungi
ISO 846-78. Plastics—Determination of Behavior Under the Action of Fungi and Bacteria—Evaluation by Visual Examination or Measurement of Change in Mass or Physical Properties First Edition. 12 pp.

—Gas Permeability
ISO 4638-84. Polymeric Materials, Cellular Flexible—Determination of Air Flow Permeability First Edition. 9 pp.

—Grindability
BSI BS 4370: Part 3-88. 1988 Methods of Test for Rigid Cellular Materials Part 3: Methods 12 and 13. 12 pp.

—Immersion Testing
BSI BS 4443: Part 4-89. 1989 Methods of Test for Flexible Cellular Materials Part 4: Method 10. Determination of Solvent Swelling Method 11. Humidity Ageing at an Elevated Temperature Method 12. Heat Ageing. 12 pp.

BSI BS 4443: Part 4-76. 1976 Methods of Test for Flexible Cellular Materials Part 4: Method 10. Determination of Solvent Swelling Method 11. Humidity Ageing at an Elevated Temperature Method 12. Heat Ageing. 7 pp.

INTERNATIONAL AND NON-U.S. NATIONAL STANDARDS
SUBJECT INDEX

Cellular Plastics (Cont.)

—**Indentation Hardness Testing**
ISO 2439-80. Polymeric Materials, Cellular Flexible—Determination of Hardness (Indentation Technique) Second Edition; (Erratum—Nov 1981). 6 pp.

—**Loss of Mass**
ISO 846-78. Plastics—Determination of Behavior Under the Action of Fungi and Bacteria—Evaluation by Visual Examination or Measurement of Change in Mass or Physical Properties First Edition. 12 pp.

—**Marine**
MOD UK NES 803: Part 1-01. Requirements for Plastic Foam Thermal Insulation Material Part 1: Plastic Foam Products Ridig Plastic Foam Insulation Slabs, Cut and Moulded Sections Issue 2 (04.89); Amendment 2. 14 pp.

—**Physical Properties**
DIN ENGL 53428-86. Determination of the Behaviour of Cellular Plastics When Exposed to Fluids, Vapours and Solids (Aug). 6 pp.
ISO 846-78. Plastics—Determination of Behavior Under the Action of Fungi and Bacteria—Evaluation by Visual Examination or Measurement of Change in Mass or Physical Properties First Edition. 12 pp.

—**Polyethylene**
ISO 7214-85. Cellular Plastics—Polyethylene—Methods of Test First Edition. 5 pp.

—**Resilience**
ISO 8307-90. Flexible Cellular Polymeric Materials—Determination of Resilience First Edition. 6 pp.

—**Shear Modulus**
BSI BS 4370: Part 2-93. 1993 Methods of Test for Rigid Cellular Materials Part 2: Methods 6 to 10 (V). 24 pp.
BSI BS 4370: Part 2-73. 1973 Methods of Test for Rigid Cellular Materials Part 2: Methods 6-10. 29 pp.

—**Shear Strength**
BSI BS 4370: Part 2-93. 1993 Methods of Test for Rigid Cellular Materials Part 2: Methods 6 to 10 (V). 24 pp.
BSI BS 4370: Part 2-73. 1973 Methods of Test for Rigid Cellular Materials Part 2: Methods 6-10. 29 pp.
CNS K6669-81. Determination of Shear Strength of Rigid Cellular Plastics (May)(7409).
DIN ENGL 53427-86. Determination of Shear Strength of Rigid Cellular Materials Sandwiched Between Metal Plates (Nov). 3 pp.
ISO 1922-81. Cellular Plastics—Determination of Shear Strength of Rigid Materials Second Edition. 6 pp.

—**Tear Strength**
BSI BS 4443: Part 6-91. 1991 Methods of Test for Flexible Cellular Materials Part 6: Method 14. Preparation of Water Extract Method 15. Determination of Tear Strength Method 16. Determination of Air Flow Value. 14 pp.
BSI BS 4443: Part 7-92. 1992 Methods of Test for Flexible Cellular Materials Part 7: Method 17. Determination of Tear Strength of Flexible Cellular Material with an Integral Skin Method 18. Determination of Compression Set Under Humid Conditions. 11 pp.
BSI BS 4443: Part 7-83. 1983 Methods of Test for Flexible Cellular Materials Part 7: Method 17, Tear Strength of Flexible Cellular Material with an Integral Skin. 7 pp.
ISO 8067-89. Flexible Cellular Polymeric Materials—Determination of Tear Strength First Edition. 6 pp.

—**Tensile Testing**
BSI BS 4370: Part 2-93. 1993 Methods of Test for Rigid Cellular Materials Part 2: Methods 6 to 10 (V). 24 pp.
BSI BS 4370: Part 2-73. 1973 Methods of Test for Rigid Cellular Materials Part 2: Methods 6-10. 29 pp.
BSI BS 4443: Part 1-88. 1988 Amd 1 Methods of Test for Flexible Cellular Materials Part 1: Methods 1 to 6. 16 pp.
ISO 1926-79. Cellular Plastics—Determination of Tensile Properties of Rigid Materials Second Edition. 6 pp.

—**Thermal Conductivity**
BSI BS 4370: Part 2-93. 1993 Methods of Test for Rigid Cellular Materials Part 2: Methods 6 to 10 (V). 24 pp.
BSI BS 4370: Part 2-73. 1973 Methods of Test for Rigid Cellular Materials Part 2: Methods 6-10. 29 pp.
ISO 2581-75. Rigid Cellular Plastics—Determination of Apparent Thermal Conductivity by Means of a Heat-Flow Meter First Edition; (Erratum—Dec 1981). 10 pp.

Cellular Plastics (Cont.)

—**Thermal Expansion**
BSI BS 4370: Part 3-88. 1988 Methods of Test for Rigid Cellular Materials Part 3: Methods 12 and 13. 12 pp.

—**Vapor Transmission**
BSI BS 4370: Part 2-93. 1993 Methods of Test for Rigid Cellular Materials Part 2: Methods 6 to 10 (V). 24 pp.
BSI BS 4370: Part 2-73. 1973 Methods of Test for Rigid Cellular Materials Part 2: Methods 6-10. 29 pp.
ISO 1663-81. Cellular Plastics—Determination of Water Vapour Transmission Rate of Rigid Materials First Edition. 6 pp.

—**Water Absorption**
ISO 2896-87. Cellular Plastics, Rigid—Determination of Water Absorption Second Edition. 9 pp.

—**Water Content Analysis**
BSI BS 4443: Part 6-91. 1991 Methods of Test for Flexible Cellular Materials Part 6: Method 14. Preparation of Water Extract Method 15. Determination of Tear Strength Method 16. Determination of Air Flow Value. 14 pp.

Cellular Polymers
Use: Cellular Plastics

Cellular Radio Equipment
Use: Cellular Mobile Radio Equipment

Cellular Rubber
Use: Foam Rubber

Cellular Telecommunications Systems
Use: Cellular Mobile Radio Equipment

Celluloid
Use: Nitrocellulose

Cellulose
See Also: Carboxylmethyl Cellulose Content Analysis; Cellophane; Cellulose Content Analysis; Cellulose Fibers; Chemical Pulps; Dissolving Pulps; Linters; Wood

—**Intrinsic Viscosity**
BSI BS 6306: Part 1-82. 1982 Determination of Limiting Viscosity Number of Celulose in Dilute Solutions Part 1: Cupri-Ethylene-Diamine (CED) Method. 13 pp.
BSI BS 6306: Part 2-82. 1982 Determination of Limiting Viscosity Number of Cellulose in Dilate Solutions Part 2: Iron (III) Sodium Tartrate Complex (EWNN Model NaCl Method). 9 pp.
ISO 5351 Pt 1-81. Cellulose in Dilute Solutions—Determination of Limiting Viscosity Number—Part 1: Method in Cupri-Ethylene-Diamine (CED) Solution First Edition. 14 pp.
ISO 5351 Pt 2-81. Cellulose in Dilute Solutions—Determination of Limiting Viscosity Number—Part 2: Method in Iron(III) Sodium Tartrate Complex (EWNN mod NaCl) Solution First Edition. 8 pp.

—**Metal Finishing—Aeronautical**
BSI X 27-66. (WITHDRAWN) 1966 Cellulose Finishing Scheme for Aeronautical Purposes (Superseded by 2X 27: 1991). 5 pp.
BSI X 30-66. (WITHDRAWN) 1966 Light-Weight Aluminium Finishing Scheme (Cellulose Base) for Aeronautical Purposes (Superseded by 2X 27: 1991). 5 pp.

—**Metal Finishing—Aerospace**
BSI 2X 27-91. 1991 Cellulose Finishing Scheme for Aerospace Purposes. 9 pp.

Cellulose Acetate
Use For: Acetylcellulose; Cellulose Triacetate; Plasticized Cellulose Acetate; Unplasticized Cellulose Acetate *See Also:* Acetates; Plastics

—**Acetic Acid Content**
DIN ENGL 53730-69. Testing of Plastics; Determination of the Acetic Acid Yield of Non-Plasticized Cellulose Acetate Thermoplastics (Oct). 2 pp.
ISO 1597-75. Plastics—Unplasticized Cellulose Acetate—Determination of Acetic Acid Yield First Edition. 5 pp.

—**Acidity—Volumetric Analysis**
BSI BS 2782:Pt4: METH 459A-91. 1991 Methods of Testing Plastics Part 4: Chemical Properties Method 459A: Determination of Free Acidity in Cellulose Acetate (ISO 1061: 1990). 6 pp.
DIN ENGL 53729-68. Testing of Plastics; Determination of Free Acid in Non-Plasticized Cellulose Acetate (Nov). 2 pp.

Cellulose Acetate (Cont.)

—**Acidity—Volumetric Analysis** (Cont.)
ISO 1061-90. Plastics—Unplasticized Cellulose Acetate—Determination of Free Acidity Second Edition. 5 pp.

—**Ash Content**
BSI BS 2782:Pt4: METH 470C-91. 1991 Methods of Testing Plastics Part 4: Chemical Properties Method 470C: Determination of Ash of Unplasticized Cellulose Acetate (ISO 3451/3: 1984). 6 pp.
ISO 3451 Pt 3-84. Plastics—Determination of Ash—Part 3: Unplasticized Cellulose Acetate First Edition. 4 pp.

—**Comprehensive Testing**
BSI BS 2880-91. 1991 Methods of Testing Cellulose Acetate Flake. 14 pp.
BSI BS 2880: Add 1-59. 1959 Amd 1 Methods of Testing Cellulose Acetate Flake Addendum No. 1:. 13 pp.

—**Light Absorption**
BSI BS 2782:Pt5: METH 553A-91. 1991 Methods of Testing Plastics: Part 5: Optical and Colour Properties, Weathering Method 553A: Determination of Light Absorption of Cellulose Acetate, Before and After Further Heating (ISO 1600: 1990). 9 pp.
ISO 1600-90. Plastics—Cellulose Acetate—Determination of Light Absorption on Moulded Specimens Produced Using Different Periods of Heating Second Edition. 8 pp.

—**Moisture Content**
BSI BS 2782:Pt4: METH 431C-91. 1991 Methods of Testing Plastics Part 4: Chemical Properties Method 431C: Determination of Moisture Content of Cellulose Acetate (ISO 585: 1990). 4 pp.
DIN ENGL 53723 Pt 1-68. Testing of Plastics; Determination of Moisture Content of Cellulose Acetate (Non-Plasticized) (Jan). 1 p.
ISO 585-90. Plastics—Unplasticized Cellulose Acetate—Determination of Moisture Content Second Edition. 4 pp.

—**Moldability**
ISO 1599-90. Plastics—Cellulose Acetate—Determination of Viscosity Loss on Moulding Second Edition. 7 pp.

—**Molding Materials**
BSI BS 1524-93. 1993 Cellulose Acetate Moulding Materials. 12 pp.
BSI BS 1524-55. 1955 Amd 2 Cellulose Acetate Moulding Material. 41 pp.
JIS K 6791-76. Cellulose Acetate Molding Materials. 16 pp.
MOD UK TS 391. Cellulose Acetate Moulding Material, Clear, Colourless.

—**Propellants**
MOD UK TS 327C. Cellulose Acetate Flake Type 1 and 2 (Withdrawn).

—**Sheets**
CNS K3075-85. Cellulose Acetate Sheet (Aug)(11339). 3 pp.
CNS K6813-85. Method of Test for Cellulose Acetate Sheet (Aug)(11340). 4 pp.
JIS K 6790-77. Acetyl Cellulose Sheet (R 1980). 9 pp.
MOD UK TS 263C. Cellulose Acetate Film. Types QXA and QXB.
MOD UK TS 383A. Plastic Q Sheet.
MOD UK CS 1943B. Cellulose Acetate Sheet, Type C.
MOD UK CS 3093A. Cellulose Acetate Sheet Type 1 and 2.

—**Solubility**
BSI BS 2782:Pt4: METH 480A-91. 1991 Methods of Testing Plastics Part 4: Chemical Properties Method 480A: Determination of Insoluble Particles in Cellulose Acetate (ISO 1598: 1990). 6 pp.
ISO 1598-90. Plastics—Cellulose Acetate—Determination of Insoluble Particles Second Edition. 6 pp.

—**Solvent Extraction**
ISO 1875-82. Plastics—Plasticized Cellulose Acetate—Determination of Matter Extractable by Diethyl Ether First Edition. 5 pp.

—**Viscosity**
DIN ENGL 53728 Pt 1-70. Testing of Plastics; Determination of Viscosity of Solutions; Cellulose Acetate in Dilute Solution (Jan). 2 pp.
ISO 1599-90. Plastics—Cellulose Acetate—Determination of Viscosity Loss on Moulding Second Edition. 7 pp.

Cellulose Acetate *(Cont.)*
—Viscosity Number
BSI BS 2782:Pt7: METH 733C-91. 1991 Methods of Testing Plastics Part 7: Rheological Properties Method 733C: Determination of Viscosity Number and Viscosity Ratio of Cellulose Acetate (ISO 1157: 1990). 8 pp.
ISO 1157-90. Plastics—Cellulose Acetate in Dilute Solution—Determination of Viscosity Number and Viscosity Ratio Second Edition. 7 pp.

Cellulose Content Analysis
See Also: Cellulose; Holocellulose Content Analysis
—Pulps
CNS P3068-84. Method of Test for Alpha-, Beta-and Gamma-Cellulose in Pulp (Apr) (10865). 5 pp.
CPPA G.21H-65. Alkali-Resistant Cellulose in Pulp (Superseded by G.27P). 2 pp.
CPPA G.27H-88. Alkali-Resistant Cellulose in Pulp (Supersedes G.21). 2 pp.
CPPA G.29P-72. Alpha-, Beta-and Gamma-Cellulose in Bleached Pulp. 3 pp.
—Pulpwood
CNS O2017-84. Method of Test for Holocellulose in Wood (May)(3085). 2 pp.
CNS P3042-81. Method of Test for Cellulose Content of Pulpwood (Jan)(6948). 3 pp.
JIS P 8007-76. Testing Method for Cellulose in Wood for Pulp (R 1984). 7 pp.
JIS P 8008-76. Testing Method for Lignin in Wood for Pulp (R 1984). 6 pp.
JIS P 8012-76. Testing Method for Holocellulose in Pulpwood (R 1984). 7 pp.
—Wood
SAA AS P1.9PM-68. Methods of Test for Pulp and Paper —Part. P9m: Cross and Bevan Cellulose in Wood Being Appita Method P9m—1968 See Also AS 1301. 2 pp.
SAA AS P9M-68. Cross and Bevan Cellulose in Wood. 2 pp.

Cellulose Esters
See Also: Cellulose Acetate; Cellulose Ethers; Nitrocellulose; Thermoplastic Resins; Thermosetting Resins
—Binders (Materials)—Paints
DIN ENGL 55953-81. Binders for Paints and Varnishes; Cellulose Esters of Organic Acids; Testing (May). 6 pp.
—Binders (Materials)—Varnishes
DIN ENGL 55953-81. Binders for Paints and Varnishes; Cellulose Esters of Organic Acids; Testing (May). 6 pp.
—Molding Materials—Test Specimens
DIN ENGL 7742 Pt 2-90. Cellulose Ester (CA,CP, CAB) Moulding Materials; Preparation of Specimens and Determination of Their Properties (Nov). 6 pp.

Cellulose Ethers
See Also: Cellulose Esters; Thermoplastic Resins; Thermosetting Resins
—Binders (Materials)—Paints
DIN ENGL 55952-81. Binders for Paints and Varnishes; Cellulose Ethers; Testing (June). 5 pp.
—Binders (Materials)—Varnishes
DIN ENGL 55952-81. Binders for Paints and Varnishes; Cellulose Ethers; Testing (June). 5 pp.

Cellulose Fibers
See Also: Acetate Fibers; Cellulose; Fibers; Triacetate Fibers
—Cuprammonium Viscosity
BSI BS 2610-78. 1978 Amd 1 Method of Test for the Determination of the Cuprammonium Fluidity of Cotton and Certain Cellulosic Man-Made Fibres (AMD 4965) November 29, 1985. 13 pp.
CGSB CAN/CGSB-4.2 NO.17-M90. Textile Test Methods Cuprammonium Fluidity of Textile Cellulose. 13 pp.
—Sizing (Surface Treatment)
BSI BS 4032-78. 1978 Determination of Certain Water-Or Alkali-Soluble Additives in Cellulosic or Synthetic Fibres, Yarns and Fabrics or Yarns and Fabrics Made from Blends of Such Fibres. 4 pp.
—Thermal Insulation
BSI BS 5803: Part 3-85. 1985 Amd 1 Thermal Insulation for Pitched Roof Spaces in Dwellings Part 3: Cellulose Fibre Thermal Insulation for Application by Blowing (AMD 5829) June 30, 1989. 17 pp.

Cellulose Fibers *(Cont.)*
—Thermal Insulation *(Cont.)*
BSI BS 5803: Part 5-85. 1985 Thermal Insulation for Pitched Roof Spaces in Dwellings Part 5: Installations of Man-Made Mineral Fibre and Cellulose Fibre Insulation. 23 pp.
CGSB 51-GP-44MP-80. Installers of Cellulose Fiber Blown Loose Fill Thermal Insulation; (Amendment 1 April 1982). 50 pp.
CGSB CAN/CGSB-51.60-M90. Cellulose Fibre Loose Fill Thermal Insulation. 17 pp.
CNS A2214-87. Cellulosic Fiber Loose Fill Thermal Insulation (Sep)(12061).
CNS A3267-87. Method of Test for Collulosic Fiber Loose Fill Thermal Insulation (Sep)(12062).
JIS A 9525-91. Cellulosic Fiber Loose Fill Thermal Insulation. 12 pp.

Cellulose Nitrate
Use: Nitrocellulose

Cellulose Nitrate Paints
Use: Nitrocellulose Paints

Cellulose Nitrate Sealants
Use: Nitrocellulose Sealants

Cellulose Triacetate
Use: Cellulose Acetate

Cellulose Wadding
Use: Wadding—Cellulose

CELTIC
Use: Time Assignment Speech Interpolation

Cement Additions
Use: Cement Additives

Cement Additives
Use For: Additions, Cement; Cement Additions
See Also: Admixtures; Cements
—Polymers
CNS A2168-83. Polymer Dispersions for Cement Modifier (Nov)(10639).
JIS A 6203-80. Polymer Dispersions for Cement Modifier. 10 pp.

Cement Coatings (Made From Cement)
See Also: Cement Paints (Made From Cement); Cementitious Primers; Coatings
—Sand Textured
CNS A2180-84. Sand Textured Coating (Cement Type) (Oct)(11054).
CNS A3216-84. Method of Test for Sand Textured Coating (Cement Type) (Oct)(11055).

Cement Content Analysis
—Floor Coverings
CEN PREN 670-92. Identification and Composition of Linoleum—Determination of Cement Content and Ash Residue. 6 pp.
—Stabilized Soil Analysis
BSI BS 1924: Part 2-90. 1990 Stabilized Materials for Civil Engineering Purposes Part 2: Methods of Test for Cement-Stabilized and Lime-Stabilized Materials. 104 pp.

Cement Copper
See Also: Copper
—Moisture Content
JIS M 8081-70. Methods for Sampling and Determination of Moisture Content of Cement Copper.
—Sampling
JIS M 8081-70. Methods for Sampling and Determination of Moisture Content of Cement Copper.

Cement Mortars
See Also: Hydraulic Cement Mortars; Mortars; Portland Cement Mortars
—Aggregates—Impurities Content
CNS A3028-83. Method of Test for Organic Impurities in Fine Aggregate (Jan)(1164).
JIS A 1105-76. Method of Test for Organic Impurities in Fine Aggregate. 5 pp.
—Cast Iron Pipes
CNS A3055-71. Method of Test for Cement Mortar of Cast Iron Pipes (Dec)(2313).

Cement Mortars *(Cont.)*
—Compression Testing
ISO 679-89. Methods of Testing Cements—Determination of Strength First Edition. 18 pp.
—Flexural Strength
ISO 679-89. Methods of Testing Cements—Determination of Strength First Edition. 18 pp.
—Linings—Ductile Iron Pipes
ISO 4179-85. Ductile Iron Pipes for Pressure and Non-Pressure Pipelines—Centrifugal Cement Mortar Lining—General Requirements Second Edition. 6 pp.
ISO 6600-80. Ductile Iron Pipes—Centrifugal Cement Mortar Lining—Composition Controls of Freshly Applied Mortar First Edition. 4 pp.
JIS A 5314-84. Mortar Lining for Ductile Iron Pipes. 10 pp.
—Linings—Pipelines
SAA AS 1516-80. Cement Mortar Lining of Pipelines in Situ. 12 pp.
—Linings—Steel Pipes
SAA AS 1281-93. Cement Mortar Lining of Steel Pipes and Fittings (in Professional Package 58A). 12 pp.
—Plastering
JIS A 1401-63. Method of Test of Cement Mortar for Plastering.
—Plastering—Fire Protection
CNS A3198-83. Method of Plastering of Cement Mortar for Fire Protection (Dec)(10679).
JIS A 7801-63. Method of Plastering of Cement Mortar for Fire Protection.
—Volume
CNS A3049-60. Method of Test for Volume Change of Cement Mortar and Concrete (Apr)(1236)(R 1970).

Cement Paint Powders
Use: Cement Paints (Made From Cement)

Cement Paints (Made From Cement)
Use For: Powder Cement Paints *See Also:* Cement Coatings (Made From Cement); Cementitious Primers; Paints; Portland Cements
BSI BS 4764-86. 1986 Powder Cement Paints. 10 pp.
—Carbon Dioxide Content
CGSB 1-GP-71 METH 50.10-80. Methods of Testing Paints and Pigments Pigment Analysis Carbon Dioxide in Carbonates. 1 p.

Cement Primers (Made From Cement)
Use: Cementitious Primers

Cemented Carbides
See Also: Carbide Tools; Carbides
CNS B7091-78. Method of Test for Cemented Carbide Recommendation (Mar)(4264).
—Machining—Designation
JIS B 4053-89. Hard Tool Materials and Classification of Applicability. 15 pp.
—Tips—Machine Components
CNS B7090-78. Method of Test for Cemented Carbide Tips (Mar)(4263).
CNS H3083-89. Cemented Carbide Alloy for Tip (Jun)(5338).
JIS H 5501-75. Cemented Carbide Alloy of Tip. 7 pp.

Cementitious Hydraulic Slags
Use: Slag Cements

Cementitious Materials
See Also: Cements; Construction Materials; Mortars
—Alkali Content
SAA AS 3583.12-91. Methods of Test for Supplementery Cementitous Materials for Use with Portland Cement—Part 12: Determination of Available Alkali (In Professional Packages 30,58). 4 pp.
—Aluminum Oxide Content Analysis
SAA AS 3583.10-91. Methods of Test for Supplementary Cementitous Materials for Use with Portland Cement—Part 10: Determination of Alumina and Total Iron Content (In Professional Packages 30,58). 4 pp.
—Chloride Ion Content
SAA AS 3583.13-91. Methods of Test for Supplementary Cementitous Materials for Use with Portland Cement—Part 13: Determination of Chloride Ion Content (In Professional Packages 30, 58). 4 pp.

INTERNATIONAL AND NON-U.S. NATIONAL STANDARDS
SUBJECT INDEX
Cements

Cementitious Materials *(Cont.)*

—Insoluble Matter Content
SAA AS 3583.14-91. Methods of Test for Supplementary Cementious Materials for Use with Portland Cement—Part 14: Determination of Insoluble Residue Content (In Professional Packages 30,58). 2 pp.

—Iron Content
SAA AS 3583.10-91. Methods of Test for Supplementary Cementitous Materials for Use with Portland Cement—Part 10: Determination of Alumina and Total Iron Content (In Professional Packages 30,58). 4 pp.

—Magnesium Oxide Content
SAA AS 3583.9-91. Methods of Test for Supplementary Cementitous Materials for Use with Portland Cement—Part 9: Determination of Magnesia Content (In Professional Packages 30,58). 4 pp.

—Manganese Content
SAA AS 3583.11-91. Methods of Test for Supplementary Cementitous Materials for Use with Portland Cement—Part 11: Determination of Manganese Content (In Professional Packages 30,58). 2 pp.

—Sulfide Content
SAA AS 3583.7-91. Methods of Test for Supplementary Cementitous Materials for Use with Portland Cement—Part 7: Determination of Sulfur Content Amdt 1 April 1993 (in Professional Package 58). 4 pp.

—Sulfur Compound Content
SAA AS 3583.8-91. Mehtods or Test for Supplementary Cementitous Materials for Use with Portland Cement—Part 8: Determination of Sulfuric Anhydride Content (In Professional Packages 30,58). 2 pp.

—Water Requirement
SAA AS 3583.6-91. Methods of Test for Suppelementary Cementitous Materials for Use with Portland Cement—Part 6: Determination of Relative Water Requirement and Relative Strength (In Professional Packages 30,58). 3 pp.

Cementitious Primers
See Also: Cement Coatings (Made From Cement); Cement Paints (Made From Cement); Primers (Coatings)
CGSB CAN/CGSB-1.198-92. Cementitious Primer (for Galvanized Surfaces). 8 pp.

Cements
Scope Note: For additional listings, use a more specific term *See Also:* Alkali Aggregate Reactions; Alumina Cements; Aluminate Cements; Asbestos Cements; Asphalt Cements; Bituminous Cements; Blended Cements; Bonding Strength; Cement Additives; Cementitious Materials; Concretes; Construction Materials; Fiberglass Reinforced Cements; Finishing Cements; Grout; Insulating Cements; Latex; Magnesium Oxychloride Cements; Masonry; Masonry Cements; Mortars; Plaster; Portland Cements; Portland Oil Shale Cements; Portland Pozzolan Cements; Portland Slag Cements; Pozzolanic Cements; Pozzolans; Roofing Cements; Screeds; Silicate Cements; Slag Cements; Stucco; Supersulfated Cements; Trass Cements
BSI BS 4550: Part 0-78. 1978 Methods of Testing Cement Part 0: General Introduction. 3 pp.
MOD UK DSTAN 80-181-93. Cement, Celluloid Issue 1. 13 pp.

—Alkali Content—Volumetric Analysis
BSI BS EN 196-21-92. 1992 Methods of Testing Cement Part 21: Determination of the Chloride, Carbon Dioxide and Alkali Content of Cement. 22 pp.
CEN EN 196 (Part 21)-89. Methods of Testing Cement: Determination of the Chloride, Carbon Dioxide and Alkali Content of Cement. 28 pp.

—Bags
CNS Z6003-70. Method of Test for Cement Bags for Domestic Market (Mar)(2109).

—Bags—Paper
CNS P2008-84. Cement Bag Paper (Jan)(1394). 2 pp.
CNS Z5017-77. Paper Bags for Cement (for Domestic Market) (Jun)(2108).
CNS Z5058-77. Paper Sack for Cement (for Export) (Dec)(4210).
CNS Z6020-77. Method of Test for Cement Paper Sack (for Export) (Dec)(4211).
JIS Z 1505-75. Sewn Kraft Paper Sacks for Cement.

Cements *(Cont.)*

—Carbon Dioxide Content—Volumetric Analysis
BSI BS EN 196-21-92. 1992 Methods of Testing Cement Part 21: Determination of the Chloride, Carbon Dioxide and Alkali Content of Cement. 22 pp.
CEN EN 196 (Part 21)-89. Methods of Testing Cement: Determination of the Chloride, Carbon Dioxide and Alkali Content of Cement. 28 pp.

—Chemical Analysis
BSI BS 4550: Part 2-70. 1970 Amd 3 Methods of Testing Cement Part 2: Chemical Tests. 55 pp.
BSI BS 4550: Part 2-04. 1970 Amd 4 Methods of Testing Cement Part 2: Chemical Tests (AMD 7285) July 15, 1992. 48 pp.
CEN EN 196 (Part 2)-87. Methods of Testing Cement; Chemical Analysis of Cement. 48 pp.
CEN EN 196 (Part 2)/A1-89. AMD 1 Methods of Testing Cement; Chemical Analysis of Cement. 2 pp.
CEN ENV 197-1-92. Cement—Composition, Specifications and Conformity Criteria—Part 1: Common Cements. 26 pp.
CEN PREN 197 (Part 2)-87. Cement; Composition, Specifications and Conformity Criteria Part 2: Specifications. 9 pp.
ISO 680-90. Cement—Test Methods—Chemical Analysis First Edition. 24 pp.

—Chloride Content—Volumetric Analysis
BSI BS EN 196-21-92. 1992 Methods of Testing Cement Part 21: Determination of the Chloride, Carbon Dioxide and Alkali Content of Cement. 22 pp.
CEN EN 196 (Part 21)-89. Methods of Testing Cement: Determination of the Chloride, Carbon Dioxide and Alkali Content of Cement. 28 pp.

—Classification
BSI BS 1000: (666-84. 1984 Universal Decimal Classification (UDC). English Full Edition (666): Glass Industry. Ceramics. Cement and Concrete. 60 pp.
SNZ NZS/BS 1000 (666)-84. Universal Decimal Classification Glass Industry. Ceramics. Cement and Concrete. 60 pp.

—Composition
CEN PREN 197-87. Cement; Composition, Specifications and Conformity Criteria Part 1: Definitions and Composition. 11 pp.

—Compression Testing
BSI BS 4550: Sec 3.4-78. 1978 Amd 3 Methods of Testing Cement Part 3: Physical Tests Section 3.4: Strength Test (AMD 5704) July 29, 1988. 13 pp.
CEN EN 196 (Part 1)-87. Methods of Testing Cement; Determination of Strength. 32 pp.
CEN EN 196 (Part 1)/A2-89. AMD 2 Methods of Testing Cement; Determination of Strength. 4 pp.

—Conformance
CEN PREN 197 (Part 3)-87. Cement; Composition, Specifications and Conformity Criteria Part 3: Conformity Criteria. 22 pp.

—Consistency
BSI BS 4550: Sec 3.5-78. 1978 Amd 1 Methods of Testing Cement Part 3: Physical Tests Section 3.5: Determination of Standard Consistence. 5 pp.

—Density
BSI BS 4550: Sec 3.2-78. (WITHDRAWN) 1978 Methods of Testing Cement Part 3: Physical Tests Section 3.2: Density Test (Superseded by BS EN 196-6: 1992). 4 pp.

—Fillers
JIS A 6916-83. Cement Filling Compound for Surface Preparation. 17 pp.

—Fineness
BSI BS EN 196-6-92. 1992 Methods of Testing Cement Part 6: Determination of Fineness. 23 pp.
CEN EN 196 (Part 6)-89. Methods of Testing Cement; Determination of Fineness. 17 pp.

—Flexural Strength
CEN EN 196 (Part 1)-87. Methods of Testing Cement; Determination of Strength. 32 pp.
CEN EN 196 (Part 1)/A2-89. AMD 2 Methods of Testing Cement; Determination of Strength. 4 pp.

—Fly Ash Content
CEN ENV 196 (Part 4)-89. Methods of Testing Cement: Quantitative Determination of Constituents. 31 pp.

Cements *(Cont.)*

—Glossaries
BSI BS 3446: Part 2-90. 1990 Glossary of Terms Associated with Refractory Materials Part 2: Applications in the Coke, Glass, Cement and Other Non-Metallurgical Industries (Supersedes BS 3446: 1962). 49 pp.
CEN PREN 197-87. Cement; Composition, Specifications and Conformity Criteria Part 1: Definitions and Composition. 11 pp.

—Heat of Hydration
BSI BS 4550: Sec 3.8-78. 1978 Methods of Testing Cement Part 3: Physical Tests Section 3.8: Test for Heat Hydration. 8 pp.

—Mechanical Properties
CEN ENV 197-1-92. Cement—Composition, Specifications and Conformity Criteria—Part 1: Common Cements. 26 pp.
CEN PREN 197 (Part 2)-87. Cement; Composition, Specifications and Conformity Criteria Part 2: Specifications. 9 pp.

—Oleic Acid
MOD UK DSTAN 68-170-93. Acid, Oleic, Special, LF Quality Issue 1. 16 pp.

—Particle Size Distribution
BSI BS 4550: Sec 3.3-78. (WITHDRAWN) 1978 Amd 2 Methods of Testing Cement Part 3: Physical Test Section 3.3: Fineness Test (AMD 5703) July 29, 1988 (Superseded by BS EN 196-6: 1992). 9 pp.

—Physical Properties
CEN ENV 197-1-92. Cement—Composition, Specifications and Conformity Criteria—Part 1: Common Cements. 26 pp.
CEN PREN 197 (Part 2)-87. Cement; Composition, Specifications and Conformity Criteria Part 2: Specifications. 9 pp.

—Physical Testing
BSI BS 4550: Sec 3.1-78. 1978 Methods of Testing Cement Part 3: Physical Tests Section 3.1: Introduction. 2 pp.
JIS R 5201-92. Physical Testing Methods for Cement. 26 pp.

—Pozzolana Content
CEN ENV 196 (Part 4)-89. Methods of Testing Cement: Quantitative Determination of Constituents. 31 pp.

—Pozzolanicity
CEN EN 196 (Part 5)-87. Methods of Testing Cement; Pozzolanicity Test for Pozzolanic Cements. 11 pp.

—Sampling
BSI BS 4550: Part 1-78. (WITHDRAWN) 1978 Methods of Testing Cement Part 1: Sampling (Superseded by BS EN 196-7: 1992). 6 pp.
BSI BS EN 196-7-92. 1992 Methods of Testing Cement Part 7: Methods of Taking and Preparing Samples of Cement. 20 pp.
CEN EN 196 (Part 7)-89. Methods of Testing Cement: Methods of Taking and Preparing Samples of Cement. 23 pp.

—Setting Time
BSI BS 4550: Sec 3.6-78. 1978 Amd 1 Methods of Testing Cement Part 3: Physical Tests Section 3.6: Test for Setting Times (AMD 5706) July 29, 1988. 6 pp.
CEN EN 196 (Part 3)-87. Methods of Testing Cement; Determination of Setting Time and Sandness. 12 pp.
ISO 9597-89. Cements—Test Methods—Determination of Setting Time and Soundness First Edition. 10 pp.

—Slag Content
CEN ENV 196 (Part 4)-89. Methods of Testing Cement: Quantitative Determination of Constituents. 31 pp.

—Soundness
BSI BS 4550: Sec 3.7-78. 1978 Amd 1 Methods of Testing Cement Part 3: Physical Tests Section 3.7: Soundness Test. 3 pp.
CEN EN 196 (Part 3)-87. Methods of Testing Cement; Determination of Setting Time and Sandness. 12 pp.
ISO 9597-89. Cements—Test Methods—Determination of Setting Time and Soundness First Edition. 10 pp.

—Spectrophotometry
SAA AS 1378-72. Method for the Spectrophotometric Analysis of Cement (Metric Units). 27 pp.

—Strength—Uniformity
SAA AS 3974-91. Evaluation of Uniformity of Cement Strength from a Single Source (in Professional Package 58). 5 pp.

INDUSTRY STANDARDS

INTERNATIONAL AND NON-U.S. NATIONAL STANDARDS
SUBJECT INDEX

Cements

Cements (Cont.)
—Test Panels
CGSB 1-GP-71 METH 96.1-82. Methods of Testing Paints and Pigments Preparation of Cement Test Panels. 1 p.

Cements, Dental
Use: Dental Materials—Cements

Cemetaries
See Also: Crematoria
—Model Bylaws
SNZ NZS 9201: Chapter 14-72. Model General Bylaws Chapter 14: Cemeteries and Crematoria (Reconfirmed 1980). 27 pp.

CEN
Use For: Comite Europeen de Normalisation; European Committee for Standardization
See Also: European Standards; Standards
EURO 1989 CEN. CEN Makes Sense for Europe. 7 pp.
EURO 1990 CEN. CEN Make Sense for Europe. 8 pp.

—Administration
EURO 1990 Pt 1A. CEN/Cenelec Internal Regulations Introduction. 12 pp.
EURO 1990 Part 1A. CEN/CENELEC Internal Regulations Part 1A: Organization and Administration. 15 pp.
EURO 1991 Part 1. CEN/CENELEC Internal Regulations Part 1: Organization and Administration. 16 pp.

—Bulletins
EURO 1992-05 Bulletin-92. Bulletin of the European Standards Organizations. 19 pp.
EURO 1992-06 Bulletin-92. Bulletin of the European Standards Organizations. 24 pp.
EURO 1992-07 Bulletin-92. Bulletin of the European Standards Organizations. 26 pp.

—Catalogs
EURO 1989-01 ED 2. CEN Catalogue: Edition 2. 152 pp.
EURO 1990 CEN Catalogue. CEN Catalogue. 123 pp.
EURO 1991 CEN Catalogue. CEN Catalogue. 108 pp.
EURO 1991 LIST. CEN List of Draft Standards. 181 pp.
EURO 1992 CEN Catalogue. CEN Catalogue. 104 pp.

—Certification Systems
EURO 1983 Part 3. CEN Internal Regulations Part 3: Directives for Certification Work Edition 1. 30 pp.
EURO 1987 Dec. IT Certification in Europe. 213 pp.
EURO 1989 March. European Framework for Testing and Certification. 27 pp.

—Cooperation—CEC
EURO MEMORANDUM 4-85. General Guidelines. 41 pp.

—Cooperation—EFTA
EURO MEMORANDUM 4-85. General Guidelines. 41 pp.

—Mementos
EURO 1988 ED 1. CEN Memento: Edition 1 Information on Members, Programming Committees, Technical Committees of CEN and Associated Bodies. 165 pp.
EURO 1989 ED 2. CEN Memento: Edition 2. 240 pp.
EURO 1990 ED 3. CEN Memento: Edition 3. 401 pp.
EURO 1991 MEMENTO. CEN Memento. 286 pp.
EURO 1992 MEMENTO. CEN Memento. 364 pp.

—Reviews
EURO 1989-3 REVIEW. Ongoing Activities in European Standards. 37 pp.
EURO 1989-4 REVIEW. Ongoing Activities in European Standards. 31 pp.
EURO 1989-5 REVIEW. Ongoing Activities in European Standards. 37 pp.
EURO 1989-6 REVIEW. Ongoing Activities in European Standards. 39 pp.
EURO 1989-7 REVIEW. Ongoing Activities in European Standards. 30 pp.
EURO 1989-9 REVIEW. Ongoing Activities in European Standards. 29 pp.
EURO 1989-10 REVIEW. Ongoing Activities in European Standards. 26 pp.
EURO 1989-11 REVIEW. Ongoing Activities in European Standards. 30 pp.
EURO 1989-12 REVIEW. Ongoing Activities in European Standards. 26 pp.
EURO 1990-01 REVIEW. Ongoing Activities in European Standards. 16 pp.
EURO 1990-02 REVIEW. Ongoing Activities in European Standards. 28 pp.

CEN (Cont.)
—Reviews (Cont.)
EURO 1990-03 REVIEW. Ongoing Activities in European Standards. 26 pp.
EURO 1990-04 REVIEW. Ongoing Activities in European Standards. 27 pp.
EURO 1990-05 REVIEW. Ongoing Activities in European Standards. 25 pp.
EURO 1990-06 REVIEW. Ongoing Activities in European Standards. 22 pp.
EURO 1990-07 REVIEW. Ongoing Activities in European Standards. 23 pp.
EURO 1990-09 REVIEW. Ongoing Activities in European Standards. 49 pp.
EURO 1990-10 REVIEW. Ongoing Activities in European Standards. 31 pp.
EURO 1990-11 REVIEW. Ongoing Activities in European Standards. 24 pp.
EURO 1990-12 REVIEW. Ongoing Activities in European Standards. 31 pp.
EURO 1991-01 REVIEW. Ongoing Activities in European Standards. 33 pp.
EURO 1991-02 REVIEW. Ongoing Activities in European Standards. 33 pp.
EURO 1991-03 REVIEW. Ongoing Activities in European Standards. 36 pp.
EURO 1991-04 REVIEW. Ongoing Activities in European Standards. 39 pp.
EURO 1991-05 REVIEW. Ongoing Activities in European Standards. 40 pp.
EURO 1991-06 REVIEW. Ongoing Activities in European Standards. 32 pp.
EURO 1991-07 REVIEW. Ongoing Activities in European Standards. 45 pp.
EURO 1991-09 REVIEW. Ongoing Activities in European Standards. 58 pp.
EURO 1991-10 REVIEW. Ongoing Activities in European Standards. 31 pp.
EURO 1991-11 REVIEW. Ongoing Activities in European Standards. 48 pp.
EURO 1991-12 REVIEW. Ongoing Activities in European Standards. 38 pp.
EURO 1992-01 REVIEW. Ongoing Activities in European Standards. 44 pp.
EURO 1992-02 REVIEW. Ongoing Activities in European Standards. 44 pp.
EURO 1992-03 REVIEW. Ongoing Activities in European Standards. 51 pp.

—Standards Preparation
EC SM 5. Public Procurement. 4 pp.
EURO MEMORANDUM 1-88. Status of the European Standard. 8 pp.
EURO MEMORANDUM 2-77. Consumer Interests and the Preparation of Standards. 7 pp.
EURO MEMORANDUM 5-89. Trade Unions and the Preparation of European Standards. 8 pp.
EURO 1988 Part 2. CEN/Cenelec Internal Regulations Part 2 Common Rules for Standards work. 53 pp.
EURO 1989 ED 13. CEN N525: Edition 13. National Implementation of European Standards. 337 pp.
EURO 1990 ED 14. CEN N525: Edition 14. National Implementation of European Standards. 509 pp.
EURO 1990 Part 2. CEN/Cenelec Internal Regulations Part 2: Common Rules for Standard Work. 60 pp.
EURO 1993 CEN STANDARDS. CEN Standards for Access to the European Market. 382 pp.
EURO 1983 Jan. Rules for Presentation of European Standards. 58 pp.
EURO 1988 CS. Common Standards for Enterprises. 74 pp.

—Standards Preparation—Construction Materials
EURO 1989 June. The European Harmonization of Construction Products. 213 pp.

—Standards Preparation—Intellectual Property Rights
EURO MEMORANDUM 8-92. Standardization and Intellectual Property Rights (IPR). 20 pp.

—Standards Preparation—Medical Equipment
EURO 1988 Dec 12-14. Workshop of the European Standardization of Medical Devices. 611 pp.

—Technical Committees—Secretaries
EURO 1989 Feb. Guidelines for New TC Secretaries. 18 pp.

—Technical Reports
EURO 1990 Report. General Technical Report. 439 pp.

Cenelec
Use For: Comite Europeen de Normalisation Electrotechnique; European Committee for Electrotechnical Standardization Use: Cenelec
See Also: CECC; European Standards; Standards

Cenelec (Cont.)
—Administration
EURO 1990 Pt 1A. CEN/Cenelec Internal Regulations Introduction. 12 pp.
EURO 1990 Part 1A. CEN/CENELEC Internal Regulations Part 1A: Organization and Administration. 15 pp.
EURO 1991 Part 1. CEN/CENELEC Internal Regulations Part 1: Organization and Administration. 16 pp.

—Affiliation
EURO MEMORANDUM 17-91. Affiliation with CENELEC. 8 pp.

—Annual Reports
EURO 1992 ANNUAL Report. CENELEC Annual Report 1992. 16 pp.

—Articles of Association
EURO Articles OF ASSOCIATION-81. Memorandum and Articles of Association. 28 pp.

—Bulletins
EURO 1992-05 Bulletin-92. Bulletin of the European Standards Organizations. 19 pp.
EURO 1992-06 Bulletin-92. Bulletin of the European Standards Organizations. 24 pp.
EURO 1992-07 Bulletin-92. Bulletin of the European Standards Organizations. 26 pp.

—Catalogs
EURO 1988 CENE Catalogue. CENELEC Catalogue. 169 pp.
EURO 1989 CENE Catalogue. CENELEC Catalogue. 200 pp.
EURO 1990 CENE Catalogue. CENELEC Catalogue. 241 pp.
EURO 1991 CENE Catalogue. Cenelec Catalogue. 254 pp.
EURO 1992 CENE Catalogue. CENELEC Catalogue. 220 pp.
EURO 1992 CENE Catalogue S1. CENELEC Catalogue Supplement 1. 50 pp.
EURO 1993 CENELEC CATALOGUE. CENELEC Catalogue. 250 pp.

—Certification Systems
EURO 1987 Dec. IT Certification in Europe. 213 pp.
EURO 1989 March. European Framework for Testing and Certification. 27 pp.
EURO MEMORANDUM 6-88. Adoption of New Standards as a Basis for the Approval of Products in the Cenelec Member Countries. 13 pp.
EURO MEMORANDUM 11-82. Elements for the Assessment of Reciprocity in the Field of Certification. 16 pp.
EURO MEMORANDUM 13-84. Cenelec Certification Agreement (CCA). 48 pp.

—Cooperation—CEC
EURO MEMORANDUM 4-85. General Guidelines. 41 pp.

—Cooperation—EFTA
EURO MEMORANDUM 4-85. General Guidelines. 41 pp.

—Current Activities
EURO 1988-09 Report. CENELEC Report on Current Activities as of September 1988. 194 pp.
EURO 1989-03 Report. CENELEC Report on Current Activities as of March 1989. 233 pp.
EURO 1989-09 Report. CENELEC Report on Current Activities as of September 1989. 293 pp.
EURO 1990-09 Report. CENELEC Report on Current Activities as of September 1990. 380 pp.
EURO 1992-03 Report. CENELEC Report on Current Activities as of March 1992. 374 pp.
EURO 1992-09 Report. CENELEC Report on Current Activities as of September 1992. 352 pp.

—Mementos
EURO 1988 ISSUE 14. Cenelec Memento: Issue 14. 102 pp.
EURO 1989 ISSUE 16. Cenelec Memento: Issue 16. 103 pp.
EURO 1989 ISSUE 17. Cenelec Memento: Issue 17. 106 pp.
EURO 1990 ISSUE 19. Cenelec Memento: Issue 19. 111 pp.
EURO 1990 ISSUE 20. Cenelec Memento: Issue 20. 119 pp.
EURO 1990 ISSUE 21. Cenelec Memento: Issue 21. 120 pp.
EURO 1993 ISSUE 93/1. CENELEC Memento: Issue 93/1. 116 pp.

—Newsletters
EURO 1988 NL 41. Cenelec Newsletter 41. 26 pp.
EURO 1989 NL 42. Cenelec Newsletter 42. 30 pp.
EURO 1989 NL 43. Cenelec Newsletter 43. 18 pp.
EURO 1989 NL 44. Cenelec Newsletter 44. 37 pp.

INTERNATIONAL AND NON-U.S. NATIONAL STANDARDS
SUBJECT INDEX

Centrifugal

Cenelec (Cont.)
—Reviews
EURO 1989-3 REVIEW. Ongoing Activities in European Standards. 37 pp.
EURO 1989-4 REVIEW. Ongoing Activities in European Standards. 31 pp.
EURO 1989-5 REVIEW. Ongoing Activities in European Standards. 37 pp.
EURO 1989-6 REVIEW. Ongoing Activities in European Standards. 39 pp.
EURO 1989-7 REVIEW. Ongoing Activities in European Standards. 30 pp.
EURO 1989-9 REVIEW. Ongoing Activities in European Standards. 29 pp.
EURO 1989-10 REVIEW. Ongoing Activities in European Standards. 26 pp.
EURO 1989-11 REVIEW. Ongoing Activities in European Standards. 30 pp.
EURO 1989-12 REVIEW. Ongoing Activities in European Standards. 26 pp.
EURO 1990-01 REVIEW. Ongoing Activities in European Standards. 16 pp.
EURO 1990-02 REVIEW. Ongoing Activities in European Standards. 28 pp.
EURO 1990-03 REVIEW. Ongoing Activities in European Standards. 26 pp.
EURO 1990-04 REVIEW. Ongoing Activities in European Standards. 27 pp.
EURO 1990-05 REVIEW. Ongoing Activities in European Standards. 25 pp.
EURO 1990-06 REVIEW. Ongoing Activities in European Standards. 22 pp.
EURO 1990-07 REVIEW. Ongoing Activities in European Standards. 23 pp.
EURO 1990-09 REVIEW. Ongoing Activities in European Standards. 49 pp.
EURO 1990-10 REVIEW. Ongoing Activities in European Standards. 31 pp.
EURO 1990-11 REVIEW. Ongoing Activities in European Standards. 24 pp.
EURO 1990-12 REVIEW. Ongoing Activities in European Standards. 31 pp.
EURO 1991-01 REVIEW. Ongoing Activities in European Standards. 33 pp.
EURO 1991-02 REVIEW. Ongoing Activities in European Standards. 33 pp.
EURO 1991-03 REVIEW. Ongoing Activities in European Standards. 36 pp.
EURO 1991-04 REVIEW. Ongoing Activities in European Standards. 39 pp.
EURO 1991-05 REVIEW. Ongoing Activities in European Standards. 40 pp.
EURO 1991-06 REVIEW. Ongoing Activities in European Standards. 32 pp.
EURO 1991-07 REVIEW. Ongoing Activities in European Standards. 45 pp.
EURO 1991-09 REVIEW. Ongoing Activities in European Standards. 58 pp.
EURO 1991-10 REVIEW. Ongoing Activities in European Standards. 31 pp.
EURO 1991-11 REVIEW. Ongoing Activities in European Standards. 48 pp.
EURO 1991-12 REVIEW. Ongoing Activities in European Standards. 38 pp.
EURO 1992-01 REVIEW. Ongoing Activities in European Standards. 44 pp.
EURO 1992-02 REVIEW. Ongoing Activities in European Standards. 44 pp.
EURO 1992-03 REVIEW. Ongoing Activities in European Standards. 51 pp.

—Standards Preparation
EURO MEMORANDUM 1-88. Status of the European Standard. 8 pp.
EURO MEMORANDUM 2-77. Consumer Interests and the Preparation of Standards. 7 pp.
EURO MEMORANDUM 5-89. Trade Unions and the Preparation of European Standards. 8 pp.
EURO 1988 Part 2. CEN/Cenelec Internal Regulations Part 2 Common Rules for Standards work. 53 pp.
EURO 1990 Part 2. CEN/Cenelec Internal Regulations Part 2: Common Rules for Standard Work. 60 pp.
EURO MEMORANDUM 1-74. Interrelation Between Regulations and Standards. 13 pp.
EURO MEMORANDUM 15-89. Establishment of Technical Specification. 11 pp.
EURO 1983 Jan. Rules for Presentation of European Standards. 58 pp.
EURO 1984 Dec. Cenelec EN/HD Rules. 55 pp.
EURO 1989-01-30. National Implementation of Approved Documents. 636 pp.
EURO 1988 CS. Common Standards for Enterprises. 74 pp.
EURO 1989. Electrotechnical Standards for the European Economic Space. 34 pp.

—Standards Preparation—Electrical Equipment—Safety
EURO MEMORANDUM 2-74. Preparation of Standards For Safety in the Design and Construction of Electrical Equipment. 5 pp.

—Standards Preparation—Intellectual Property Rights
EURO MEMORANDUM 8-92. Standardization and Intellectual Property Rights (IPR). 20 pp.

—Standards Preparation—Low Voltage Directive
EURO MEMORANDUM 3-87. Implementation of the EEC Low Voltage Direction of 19 February 1973 with Respect to Reports, Certificates and Manufacturer's Declaration of Conformity. 50 pp.
EURO MEMORANDUM 4-84. Certification Organizations for Electrical Equipment Within Cenelec Countries Granting Marks of Conformity Related to Safety. 21 pp.
EURO MEMORANDUM 5-78. Policy Paper on National Deviations to Harmonisation Documents with Particular Reference to the Low Voltage Directive 73/23/ EEC of the European Economic Community. 13 pp.
EURO MEMORANDUM 10-82. Publication of Cenelec Results in the Field of the Low Voltage Directive in the Form of European Standards. 13 pp.

—Standards Preparation—Medical Equipment
EURO 1988 Dec 12-14. Workshop of the European Standardization of Medical Devices. 611 pp.

—Third Countries
EURO MEMORANDUM 16-91. Relationship with Third Countries. 14 pp.

Cenelec Electronic Components Committee
Use: CECC

Center Drills
Use For: Centre Drills See Also: Center Holes; Cutting Tools; Drills
BSI BS 328: Part 2-72. 1972 Drills and Reamers Part 2: Combined Drills and Countersinks (Centre Drills). 15 pp.
CNS B3053-79. 60 Degree Center Drill (Radius Form) (Jan)(226). 2 pp.
CNS B3054-79. 60 Degree Center Drills (Straight Form) (Jan)(227). 2 pp.
CNS B3055-79. 60 Degree Center Drills (Straight Form with Conical Protective Chamfer) (Jan)(228). 2 pp.
DIN ENGL 333-86. 60 Deg. Centre Drills; Types R, A, and B (Apr). 4 pp.
ISO 866-75. Centre Drills for Centre Holes Without Protecting Chamfers—Type A First Edition. 4 pp.
ISO 2540-73. Centre Drills for Centre Holes with Protecting Chamfer—Type B First Edition. 4 pp.
ISO 2541-72. Centre Drills for Centre Holes with Radius Form—Type R First Edition. 4 pp.
JIS B 4304-88. Centre Drills for Centre Holes. 13 pp.
MOD UK DSTAN 51-5-74. Drills, Twist. Drills, Pivot. Drills, Masonry, Rotary Driven. Countersink and Drill (Centre Drills) Issue 2. 42 pp.
SAA AS 1913-76. Centre Drills (Reconfirmed 1987). 12 pp.
SNZ NZS/BS 328: Part 2-72. Drills and Reamers Part 2: Combined Drills and Countersinks (Centre Drills). 16 pp.

Center Finders
Use: Center Gages

Center Gages
Use For: Wigglers
CGSB CAN/CGSB-39.27-M90. Scribers. 11 pp.
CGSB CAN/CGSB-39.58-M90. Centre-Finders. 8 pp.

Center Holes
Scope Note: For center holes of specific products, see the subheading Center Holes under the specific product See Also: Center Drills; Machining Centers
DIN ENGL 332 Pt 1-86. 60 Degree Centre Holes; Types R, A, B, and C (Apr). 5 pp.
JIS B 0041-83. Representation of Centre Holes. 5 pp.
JIS B 1011-87. Centre Holes. 8 pp.

—Engineering Drawings
ISO 6411-82. Technical Drawings—Simplified Representation of Centre Holes First Edition. 7 pp.

Center of Gravity
See Also: Mass

—Air Cushion Vehicles
CAA Chapter A5-1 08.81. Weight and Balance Data. 2 pp.
CAA Chapter A5-1 App 08.81. Weight and Balance Data. 2 pp.

—Aircraft
CAA Chapter A6-4 App 1 07.89. Weight and Balance of Aircraft—Fleet Mean Weight and Fleet Mean Centre-of-Gravity. 2 pp.
CAA Chapter B6-4 App 1 7.89. Weight and Balance of Aircraft—Fleet Mean Weight and Fleet Mean Centre-Of-Gravity.
CAA Chapter B7-1 0 App1 07.89. Weight and Centre-Of-Gravity Schedules for Aircraft Exceeding 2740 kg. 3 pp.
CAA Chapter B7-1 0 App2 07.89. Weight and Centre-Of-Gravity and Loading and Distribution Schedules—Aircraft Not Exceeding 2730 kg. 3 pp.

—Aircraft—Scheduling
CAA Chapter A7-10 App 1. Weight and Centre-of-Gravity Schedules for Aircraft Exceeding 2730 kg. 3 pp.
CAA Chapter A7-1 0 App2 09.91. Weight and Centre-of-Gravity and Loading and Distribution Schedules—Aircraft Not Exceeding 2730 kg. 5 pp.

—Earthmoving Equipment
BSI BS 3318-78. 1978 Earth-Moving Machinery. Method for Locating the Centre of Gravity. 9 pp.
CNS A4022-88. Earth-Moving Machinery Method for Locating the Center of Gravity (Dec)(9953).
ISO 5005-77. Earth-Moving Machinery—Method for Locating the Centre of Gravity First Edition. 8 pp.
JIS A 8915-82. Earth-Moving Machinery-Method for Locating the Centre of Gravity. 12 pp.

—Motor Vehicles
BSI BS AU 246-92. 1992 Method for Determination of the Centre of Gravity of Road Vehicles with Two Axles (ISO 10392: 1992) (E). 11 pp.
ISO 10392-92. Road Vehicles with Two Axles—Determination of Centre of Gravity First Edition. 9 pp.

—Motorcycles
ISO 9130-89. Motorcycles—Measurement Method for Location of Centre of Gravity First Edition. 10 pp.

—Tractors
BSI BS 6347: Part 6-83. 1983 Performance Assessment of Agricultural Tractors Part 6: Method for Determination of Centre of Gravity. 9 pp.
ISO 789 Pt 6-82. Agricultural Tractors—Test Procedures—Part 6: Centre of Gravity First Edition. 8 pp.

Center Punches
See Also: Hand Tools; Punches (Tools); Tools
CNS B3196-85. Center Punches (9.5mm, Solid) (Dec)(3095).

Center Testers
Use: Center Gages

Centers (Machine Tool Components)
Use: Machining Centers

Central Air Conditioners
Use: Air Conditioners

Central Furnaces
Use: Furnaces

Central Processing Units
Use For: CPU; Main Frames See Also: Computers; Data Processing Equipment

—Glossaries
JIS X 0011-89. Glossary of Terms Used in Information Processing (Processing Units).
OSI ISO 2382-11-87. Information Processing Systems—Vocabulary—Part 11: Processing Units. 23 pp.

—Pascal
BSI BS ISO/IEC 10206-91. 1991 Information Technology—Programming Languages—Extended Pascal. 229 pp.
IEC 10206-91. Information Technology—Programming Languages—Extended Pascal First Edition. 226 pp.
ISO 10206-91. Information Technology—Programming Languages—Extended Pascal First Edition. 226 pp.
JTC1 10206-91. Information Technology—Programming Languages—Extended Pascal First Edition. 226 pp.
SAA AS 3981-91. Information Technology—Programming Languages—Extended Pascal (ISO/IEC 10206:1991) (in Professional Package 26A). 214 pp.

Centre Drills
Use: Center Drills

Centre of Gravity
Use: Center of Gravity

Centrifugal Castings
See Also: Castings

INDUSTRY STANDARDS

INTERNATIONAL AND NON-U.S. NATIONAL STANDARDS
SUBJECT INDEX

Centrifugal Castings (Cont.)

—Cobalt Alloy—Aerospace
BSI HC 100-72. 1972 Amd 3 Inspection and Testing Procedure for Iron, Nickel, Copper, Cobalt and Refractory Metal Base Alloy Castings (AMD 6574) May 28, 1991. 17 pp.

—Copper Alloy—Aerospace
BSI HC 100-72. 1972 Amd 3 Inspection and Testing Procedure for Iron, Nickel, Copper, Cobalt and Refractory Metal Base Alloy Castings (AMD 6574) May 28, 1991. 17 pp.

—Iron Alloy—Aerospace
BSI HC 100-72. 1972 Amd 3 Inspection and Testing Procedure for Iron, Nickel, Copper, Cobalt and Refractory Metal Base Alloy Castings (AMD 6574) May 28, 1991. 17 pp.

—Nickel Alloy—Aerospace
BSI HC 100-72. 1972 Amd 3 Inspection and Testing Procedure for Iron, Nickel, Copper, Cobalt and Refractory Metal Base Alloy Castings (AMD 6574) May 28, 1991. 17 pp.

—Nickel Alloys—Marine
MOD UK NES 822: Part 1-86. Requirements for Nickel Chromium Molybdenum Niobium Alloy 625 Part 1: Centrifugal Castings Issue 1 (04.86). 21 pp.

—Nickel Aluminum Bronze—Marine
MOD UK NES 747: Part 1-90. Nickel Aluminium Bronze Castings and Ingots Part 1: Requirements for Nickel Aluminium Bronze (Naval Alloy) Centrifugal Castings and Ingots Issue 2 (11.90). 20 pp.

—Refractory Metal Alloy—Aerospace
BSI HC 100-72. 1972 Amd 3 Inspection and Testing Procedure for Iron, Nickel, Copper, Cobalt and Refractory Metal Base Alloy Castings (AMD 6574) May 28, 1991. 17 pp.

Centrifugal Fans
See Also: Blowers (Ventilation); Fans

—Forward Curved Blades
JIS B 8331-83. Forward-Curved Bladed Fans. 33 pp.

—Mining Equipment
CNS M2069-80. Centrifugal Type Local Fans (Oct)(6593).
JIS M 7613-80. Centrifugal Type Local Fans for Underground.

Centrifugal Pumps
See Also: Axial Flow Pumps; Deep Well Pumps; Feed Pumps; Fire Pumps; Gear Pumps; Pumps; Rotary Pumps; Turbine Pumps; Volute Pumps

BSI BS 6836-87. (WITHDRAWN) 1987 Centrifugal Pumps: Class 11 (Renumbered as BS EN 25199: 1992). 41 pp.
BSI BS EN 25199-92. 1992 Amd 1 Technical Specifications for Centrifugal Pumps—Class II (ISO 5199: 1986) (AMD 7527) December 15, 1992. 48 pp.
CEN EN 25199-92. Technical Specifications for Centrifugal Pumps—Class II; (ISO 5199: 1986). 40 pp.
ISO 5198-87. Centrifugal, Mixed Flow and Axial Pumps—Code for Hydraulic Performance Tests—Precision Class First Edition. 86 pp.
ISO 5199-86. Technical Specifications for Centrifugal Pumps—Class II First Edition; (CEN EN 25199:1992). 40 pp.
JIS B 8301-90. Testing Methods for Centrifugal Pumps, Mixed Flow Pumps and Axial Flow Pumps. 33 pp.
JIS B 8327-89. Testing Methods for Performance of Pump, Using Model Pump. 38 pp.

—Acceptance Testing
BSI BS 5316: Part 1-76. 1976 Acceptance Tests for Centifugal Mixed Flow and Axial Pumps Part 1: Class C Tests. 36 pp.
BSI BS 5316: Part 2-77. 1977 Acceptance Tests for Centifugal Mixed Flow and Axial Pumps Part 2: Class B Tests. 40 pp.
BSI BS 5316: Part 3-88. 1988 Acceptance Tests for Centifugal Mixed Flow and Axial Pumps Part 3: Precision Class Tests. 87 pp.
DIN ENGL 1944-68. Acceptance Tests on Centrifugal Pumps; (VDI Rules for Centrifugal Pumps) (Oct). 44 pp.
ISO 2548-73. Centrifugal, Mixed Flow and Axial Pumps—Code for Acceptance Tests—Class C First Edition; (Amendment Slip-1980). 39 pp.
ISO 3555-77. Centrifugal, Mixed Flow and Axial Pumps—Code for Acceptance Tests—Class B First Edition. 40 pp.

Centrifugal Pumps (Cont.)

—Fire
CEN PREN 1028-1-93. Fire Fighting Pumps—Part 1: Requirements of Fire Fighting Centrifugal Pumps with Primer. 13 pp.

—Fire—Classification
CEN PREN 1028-1-93. Fire Fighting Pumps—Part 1: Requirements of Fire Fighting Centrifugal Pumps with Primer. 13 pp.

—Fire—Stationary
CNS B4052-86. Stationary Centrifugal Fire Pump (Mar)(8917). 4 pp.

—Machine Tools
BSI BS 3766-79. 1979 Coolant Pumps for Machine Tools. 10 pp.

—Multistage
CNS B4063-85. Small Size Multi-Stage Centrifugal Pumps (Mar)(10847).
JIS B 8319-87. Small Size Multi-Stage Centrifugal Pumps. 31 pp.

—Oil
JIS B 8306-90. Testing Methods for Centrifugal Type Oil Pumps. 12 pp.

—Self Priming
CEN PREN 1028-2-93. Fire Fighting Pumps—Part 2: Testing of Fire Fighting Centrifugal Pumps with Primer. 15 pp.
JIS B 8305-90. Testing Methods for Self-Priming Centrifugal Pumps. 8 pp.

—Single Stage
BSI BS 4082: Part 1-69. 1969 External Dimensions for Vertical In-Line Centrifugal Pumps Part 1: I Type. 16 pp.
BSI BS 4082: Part 2-69. 1969 External Dimensions for Vertical in-Line Centrifugal Pumps Part 2: U Type. 16 pp.

—Steam Turbines
JIS B 8304-90. Testing Methods for Condensate Pumps. 9 pp.

—Suction
BSI BS 5257-75. 1975 Horizontal End-Section Centrifugal Pumps (16 Bar). 14 pp.
BSI BS EN 22858-93. 1993 End-Suction Centrifugal Pumps (Rating 16 Bar)—Designation, Nominal Duty Point and Dimensions (ISO 2858: 1975) (Q). 9 pp.
BSI BS EN 23661-93. 1993 End-Suction Centrifugal Pumps—Baseplate and Installation Dimensions (ISO 3661: 1977) (Q). 11 pp.
CEN PREN 733-92. End-Suction Centrifugal Pumps, Rating with 10 Bar with Bearing Bracket —Nominal Duty Point, Main Dimensions, Designation System. 5 pp.
CEN EN 22858-93. End-Suction Centrifugal Pumps (Rating 16 Bar)—Designation, Nominal Duty Point and Dimensions (ISO 2858: 1975). 4 pp.
CEN EN 23661-93. End-Suction Centrifugal Pumps—Baseplate and Installation Dimensions (ISO 3661: 1977). 6 pp.
ISO 2858-75. End-Suction Centrifugal Pumps (Rating 16 Bar)—Designation, Nominal Duty Point and Dimensions Second Edition; (CEN EN 22858: 1993). 4 pp.
ISO 3661-77. End-Suction Centrifugal Pumps—Baseplate and Installation Dimensions First Edition (CEN EN 23661: 1993). 5 pp.
JIS B 8313-91. End Suction Centrifugal Pumps. 37 pp.

—Suction—Cavities
ISO 3069-74. End Suction Centrifugal Pumps—Dimensions of Cavities for Mechanical Seals and for Soft Packing First Edition. 3 pp.

—Suction—Single Stage
CEN PREN 735-92. Overall Dimensions of Rotodynamic Pumps—Tolerances. 5 pp.

Centrifuge Tubes
See Also: Laboratory Ware; Tubes

—Glass
BSI BS 6898-87. 1987 Glass Centrifuge Tubes for General Laboratory Use. 8 pp.

Centrifuges
See Also: Agitators; Centrifuging; Classifiers; Separators (Mechanical)
BSI BS 767-83. 1983 Amd 1 Centrifuges of the Basket and Bowl Type for Use in Industrial and Commercial Applications. 27 pp.
DIN ENGL 24400 Pt 1-71. Three-Column Suspended Basket Centrifuges; Connecting Dimensions to Foundation (Oct). 1 p.

Centrifuges (Cont.)

DIN ENGL 24400 Pt 2-79. Three-Column Suspended Basket Centrifuges; Baskets; Housings; Dimensions and Requirements (Dec). 7 pp.
SNZ NZS/BS 767-83. Specification for Centrifuges of the Basket and Bowl Type for Use in Industrial and Commercial Applications Amend: 1. 22 pp.

—Blood—Hematocrit Value
JIS T 7326-89. Hematocrit Centrifuge. 9 pp.

—Electrical Safety
IEC 1010 Pt 2-020-92. Safety Requirements for Electrical Equipment for Measurement, Control, and Laboratory Use Part 2-020: Particular Requirements for Laboratory Centrifuges First Edition. 60 pp.

—Medical
JIS T 1701-83. Centrifuge for Medical Use.

—Rotors (Machine Elements)—Safety
ISO 6178-83. Centrifuges—Construction and Safety Rules—Method for the Calculation of the Tangential Stress in the Shell of a Cylindrical Centrifuge Rotor First Edition. 8 pp.

—Rotors (Machine Elements)—Sheer Stress
ISO 6178-83. Centrifuges—Construction and Safety Rules—Method for the Calculation of the Tangential Stress in the Shell of a Cylindrical Centrifuge Rotor First Edition. 8 pp.

—Safety
BSI BS 4402-82. 1982 Amd 1 Safety Requirements for Laboratory Centrifuges. 17 pp.

—Ships
MOD UK NES 323-86. Centrifuges Issue 3 (11.86). 27 pp.
MOD UK NES 323-91. Requirements for Centrifuges Issue 4 (10.91). 29 pp.

—Submarines
MOD UK NES 323-86. Centrifuges Issue 3 (11.86). 27 pp.
MOD UK NES 323-91. Requirements for Centrifuges Issue 4 (10.91). 29 pp.

Centrifuging
See Also: Centrifuges; Resource Recovery

—Bunker Fuel Oils
CPPA J.24(F)-77. Bunker "C" Oil—Water and Sediment by Centrifuge. 2 pp.

—Crude Oils
BSI BS 2882-80. (WITHDRAWN) 1980 Determination of Water and Sediment in Crude Petroleum and Fuel Oils (Centrifuge Method). 8 pp.
CNS K6577-80. Method of Test for Water and Sediment in Crude Oils and Fuel Oils by Centrifuge (Aug)(6358).
DIN ENGL 51793-71. Testing of Liquid Fuels; Determination of Water Content and Sediments in Fuel Oils and Crude Oils; Centrifuge Method (July). 3 pp.
ISO 3734-76. Crude Petroleum and Fuel Oils—Determination of Water and Sediment—Centrifuge Method First Edition. 6 pp.
ISO 9030-90. Crude Petroleum—Determination of Water and Sediment—Centrifuge Method First Edition. 12 pp.

—Extenders
BSI BS 3483: Part B9-80. 1980 Methods for Testing Pigments in Paints Part B9: Determination of Density Relative to Water (Using a Centrifuge). 4 pp.

—Fuel Oils
BSI BS 2882-80. (WITHDRAWN) 1980 Determination of Water and Sediment in Crude Petroleum and Fuel Oils (Centrifuge Method). 8 pp.
CNS K6577-80. Method of Test for Water and Sediment in Crude Oils and Fuel Oils by Centrifuge (Aug)(6358).
DIN ENGL 51793-71. Testing of Liquid Fuels; Determination of Water Content and Sediments in Fuel Oils and Crude Oils; Centrifuge Method (July). 3 pp.
ISO 3734-76. Crude Petroleum and Fuel Oils—Determination of Water and Sediment—Centrifuge Method First Edition. 6 pp.

—Paints
CGSB 1-GP-71 METH 21.1-81. Methods of Testing Paints and Pigments Pigment Content Ordinary Centrifuge Method. 5 pp.
CNS K6716-82. Determination of the Pigment Content of Solvent-Type Paints by High Speed Centrifuging (Mar)(8595).

INTERNATIONAL AND NON-U.S. NATIONAL STANDARDS
SUBJECT INDEX
Ceramic

Centrifuging *(Cont.)*
—**Pigments**
BSI BS 3483: Part B9-80. 1980 Methods for Testing Pigments in Paints Part B9: Determination of Density Relative to Water (Using a Centrifuge). 4 pp.

CEPT
See Also: European Standards; Standards
—**Certification Systems**
EURO 1987 Dec. IT Certification in Europe. 213 pp.

Ceramal
Use: Cermet

Ceramet
Use: Cermet

Ceramic Capacitors
See Also: Capacitors; Ceramic Chip Capacitors; Fixed Capacitors; Glass Capacitors; Porcelain Capacitors; Variable Capacitors
BSI BS 6201: Part 2-82. (WITHDRAWN) 1982 Fixed Capacitors for Use in Electronic Equipment Part 2: Ceramic Dielectric, Class 2. Selection of Methods of Test and General Requirements (Superseded by BS QC 300700: 1992). 23 pp.
CECC CECC 31 400 ISSUE 1-82. Sectional Specification: Fixed Ceramic Capacitors of Dielectric Class 1, for Electrical Shock Hazard Protection (En, Fr, Ge) AMD 1 (En, Fr, Ge) AMD 2 (En, Fr, Ge). 72 pp.
CNS C7088-79. Fixed Ceramic Capacitor for Electronic Equipment (High Dielectric Constant) (May)(4860). 15 pp.
CNS C7146-90. High Voltage Ceramic Dielectric Capacitor, Class 2 (Oct)(7227).
CNS C7194-87. Fixed Ceramic Capacitors for Electronic Equipment (Semiconductor) (Apr)(11900).
MOD UK DEF-5133-1-01. Capacitors, Fixed, Ceramic Dielectric, with Specified Capacitance Temperature Coefficients (Normal Permittivity) (Reprinted April 1965, Incorporating Amendments Nos. 1, 2 and 3); Amendment 6. 65 pp.
MOD UK DEF-5133-1-02. Capacitors, Fixed, Ceramic Dielectric, with Specified Capacitance Temperature Coefficients (Normal Permittivity) (Reprinted November 1965 Incorporating Amendments Nos. 1 to 4); Amendment 6. 63 pp.
MOD UK DEF-5133-2-57. Capacitors, Fixed, Ceramic Dielectric (High Permittivity) (Reprinted February 1972, Incorporating Amendments Nos. 1, 2 and 3). 20 pp.
SAA AS C343-64. Fixed Ceramic Dielectric Capacitors (Temperature Compensating) Type 1 Corrig.. 23 pp.

—**Class I**
CECC CECC 30 600 ISSUE 2-89. Sectional Specification: Fixed Capacitors of Ceramic Dielectric, Class 1 (En, Fr, Ge) AMD 2 (En, Fr, Ge) AMD 3 (En, Fr, Ge). 154 pp.
CECC CECC 31 400 ISSUE 1-82. Sectional Specification: Fixed Ceramic Capacitors of Dielectric Class 1, for Electrical Shock Hazard Protection (En, Fr, Ge) AMD 1 (En, Fr, Ge) AMD 2 (En, Fr, Ge). 72 pp.
IECQ QC 300600-88. Fixed Capacitors for Use in Electronic Equipment Part 8: Sectional Specification: Fixed Capacitors of Ceramic Dielectric, Class 1 (Amendment 1-1993) (IEC 384-8 ED 2). 91 pp.
IECQ QC 300601-88. Fixed Capacitors for Use in Electronic Equipment Part 8: Blank Detail Specification: Fixed Capacitors of Ceramic Dielectric, Class 1 Assessment Level E (Amendment 1-1993) (IEC 384-8-1). 44 pp.
IECQ QC 300601/CN 0001-92. Detail Specification for Electronic Components Fixed Capacitors of Ceramic Dielectric, Class 1 Type CC1 Assessment Level E. 17 pp.
IECQ QC 300601/JP 0001-89. Fixed Capacitors for Use in Equipment Part 8: Detail Specification: Fixed Capacitors of Ceramic Dielectric, Class 1 Assessment Level E. 14 pp.
IECQ QC 300601/JP 0002-89. Fixed Capacitors for Use in Electronic Equipment Part 8: Detail Specification: Fixed Capacitors of Ceramic Dielectric, Class 1 Assessment Level E. 14 pp.
IECQ QC 300601/JP 0003-88. Fixed Capacitors for Use in Electronic Equipment Detail Specification: Fixed Ceramic Dielectric Capacitors, Class 1 Assessment Level E. 17 pp.
IECQ QC 300601/JP 0004-91. Fixed Capacitors for Use in Electronic Equipment Part 8: Detail Specification: Fixed Capacitors of Ceramic Dielectric, Class 1 Assessment Level E. 15 pp.
IECQ QC 300601/SG 0001-89. Fixed Capacitors for Use in Electronic Equipment Part 8: Detail Specification: Fixed Capacitors of Ceramic Dielectric, Class 1 Assessment Level E (IEC 384-8-1 ED 1). 14 pp.

Ceramic Capacitors *(Cont.)*
—**Class I** *(Cont.)*
IECQ QC 300601/US 0001-87. Fixed Capacitors for Use in Electronic Equipment: Detail Specification: Fixed Capacitors, Multilayer Ceramic Dielectric, Class 1B Conformal Insulated Coating, Axial Leads. 15 pp.
IECQ QC 300601/US 0002-88. Fixed Capacitors for Use in Electronic Equipment: Detail Specification: Fixed Capacitors, Multilayer Ceramic Dielectric, Class 1B Conformal Insulated Coating, Radial Leads. 15 pp.
IECQ QC 300601/US 0003-88. Fixed Capacitors for Use in Electronic Equipment: Detail Specification: Fixed Capacitors of Ceramic Dielectric, Class 1, Assessment Level E. 23 pp.

—**Class I—Printed Circuit Mount**
CECC CECC 30 601-004. CEI CECC 30 601-004; Fixed Ceramic Dielectric Capacitors (Class 1) (En). 11 pp.

—**Class II**
CECC CECC 30 700 ISSUE 2-89. Sectional Specification: Fixed Capacitors of Ceramic Dielectric, Class 2 (En, Fr, Ge) AMD 2 (En, Fr, Ge). 118 pp.
CECC CECC 30 701 ISSUE 2-89. Blank Detail Specification: Fixed Capacitors of Ceramic Dielectric, Class 2 (En, Fr, Ge). 48 pp.
CECC CECC 30 702 ISSUE 1-89. Blank Detail Specification: Fixed Capacitors of Ceramic Dielectric, Class 2 (En, Fr, Ge). 46 pp.
CECC CECC 31 500 ISSUE 1-82. Sectional Specification: Fixed Ceramic Capacitors of Dielectric Class 2, for Electrical Shock Hazard Protection (En, Fr, Ge) AMD 1 (En, Fr, Ge). 72 pp.
IECQ QC 300700-88. Fixed Capacitors for Use in Electronic Equipment Part 9: Sectional Specification: Fixed Capacitors of Ceramic Dielectric, Class 2 (IEC 384-9 ED 2). 67 pp.
IECQ QC 300701-88. Fixed Capacitors for Use in Electronic Equipment Part 9: Blank Detail Specification: Fixed Capacitors of Ceramic Dielectric, Class 2 Assessment Level E (IEC 384-9-1 ED 1). 33 pp.
IECQ QC 300701/CN 0001-92. Detail Specification for Electronic Components Fixed Capacitors of Ceramic Dielectric, Class 2 Type CT1 Assessment Level E (IEC 384-9-1 ED 1). 16 pp.
IECQ QC 300701/JP 0001-89. Fixed Capacitors for Use in Equipment Part 9: Detail Specification: Fixed Capacitors of Ceramic Dielectric, Class 2 Assessment Level E. 14 pp.
IECQ QC 300701/JP 0002-89. Fixed Capacitors for Use in Electronic Equipment Part 9: Detail Specification: Fixed Capacitors of Ceramic Dielectric, Class 2 Assessment Level E. 13 pp.
IECQ QC 300701/JP 0003-88. Fixed Capacitors for Use in Electronic Equipment Detail Specification: Fixed Ceramic Dielectric Capacitors, Class 2 Assessment Level E. 16 pp.
IECQ QC 300701/JP 0004-91. Fixed Capacitors for Use in Electronic Equipment Part 9: Detail Specification: Fixed Capacitors of Ceramic Dielectric, Class 2 Assessment Level E. 12 pp.
IECQ QC 300701/KR 0001-91. Fixed Capacitors for Use in Electronic Equipment Part 9: Detail Specification: Fixed Capacitors of Ceramic Dielectric, Class 2 Assessment Level E. 13 pp.
IECQ QC 300701/SG 0001-89. Fixed Capacitors for Use in Electronic Equipment Part 9: Detail Specification: Fixed Capacitors of Ceramic Dielectric, Class 2 Assessment Level E. 13 pp.
IECQ QC 300701/US 0001 ISSUE 2-91. Fixed Capacitors for Use in Electronic Equipment: Detail Specification: Fixed Capacitors, Multilayer Ceramic Dielectric, Class 2E6 Conformal Insulated Coating, Axial Leads. 16 pp.
IECQ QC 300701/US 0002 ISSUE 2-91. Fixed Capacitors for Use in Electrotechnical Equipment: Detail Specification: Fixed Capacitors, Multilayer Ceramic Dielectric, Class 2R1 Conformal Insulated Coating, Radial Leads. 16 pp.
IECQ QC 300701/US 0003 ISSUE 2-91. Fixed Capacitors for Use in Electrotechnical Equipment: Detail Specification: Fixed Capacitors, Multilayer Ceramic Dielectric, Class 2R1 Conformal Insulated Coating, Axial Leads. 15 pp.

—**Class III**
CECC CECC 31 100 ISSUE 1-81. Sectional Specification: Fixed Ceramic Dielectric Capacitors of Barrier Layer Type (Dielectric Class 3) (En, Fr, Ge) AMD 1 (En, Fr, Ge). 80 pp.
CECC CECC 31 101 ISSUE 1-81. Blank Detail Specification: Fixed Ceramic Dielectric Capacitors of Barrier Layer Type (Dielectric Class 3) (En, Fr, Ge) AMD 1 (En, Fr, Ge) AMD 2 (En, Fr, Ge). 47 pp.

—**Cylindrical—Class II—Preferred Products List**
CECC CECC MUAHAG Vol 1 IS 4-90. Preferred Products List; Capacitors (En, Fr, Ge). 64 pp.

Ceramic Capacitors *(Cont.)*
—**Dipped—Class I**
CECC CECC 30 601-801 ISSUE 1-88. Fixed Ceramic Dielectric Capacitors, Class 1B (En). 14 pp.

—**Dipped—Class I—Preferred Products List**
CECC CECC MUAHAG Vol 1 IS 4-90. Preferred Products List; Capacitors (En, Fr, Ge). 64 pp.

—**Dipped—Class II**
CECC CECC 30 701-801 ISSUE 1-88. Fixed Ceramic Dielectric Capacitors Class 2C1 (En, Fr, Ge). 14 pp.

—**Dipped—Class II—Preferred Products List**
CECC CECC MUAHAG Vol 1 IS 4-90. Preferred Products List; Capacitors (En, Fr, Ge). 64 pp.

—**Dipped—Class II—Printed Circuit Mount**
CECC CECC 30 701-002. CEI CECC 30 701-002; Fixed Ceramic Dielectric Capacitors (Class 2) (En). 9 pp.
CECC CECC 30 701-003. CEI CECC 30 701-003; Fixed Ceramic Dielectric Capacitors (Class 2) (En). 10 pp.

—**Electric Shock Protection—Preferred Products List**
CECC CECC MUAHAG Vol 1 IS 4-90. Preferred Products List; Capacitors (En, Fr, Ge). 64 pp.

—**Molded—Class I**
IECQ QC 300601/US 0004 ISSUE 1-91. Specifications for Use in Electrotechnical Equipment: Detail Specification: Fixed Capacitors of Ceramic Dielectric Class 1B Molded Case Multilayer Ceramic—Axial Leads. 13 pp.
IECQ QC 300601/US 0005 ISSUE 1-91. Specifications for Use in Electrotechnical Equipment: Detail Specification: Fixed Capacitors of Ceramic Dielectric Class 1B Molded Case Multilayer Ceramic—Radial Leads. 13 pp.

—**Molded—Class I—Preferred Products List**
CECC CECC MUAHAG Vol 1 IS 4-90. Preferred Products List; Capacitors (En, Fr, Ge). 64 pp.

—**Molded—Class II**
CECC CECC 30 701-801 ISSUE 1-88. Fixed Ceramic Dielectric Capacitors Class 2C1 (En, Fr, Ge). 14 pp.

—**Molded—Class II—Preferred Products List**
CECC CECC MUAHAG Vol 1 IS 4-90. Preferred Products List; Capacitors (En, Fr, Ge). 64 pp.

—**Molded—Rectangular—Class I—Preferred Products List**
CECC CECC MUAHAG Vol 1 IS 4-90. Preferred Products List; Capacitors (En, Fr, Ge). 64 pp.

—**Molded—Rectangular—Class II—Preferred Products List**
CECC CECC MUAHAG Vol 1 IS 4-90. Preferred Products List; Capacitors (En, Fr, Ge). 64 pp.

—**Monolithic—Cylindrical—Class I**
CECC CECC 30 601-010. BS CECC 30 601-010 Issue 1; Fixed Capacitor Multilayer Ceramic Dielectric Class 1B; Cylindrical Insulated Non-Metalic Case Axial Centred Leads (En). 10 pp.

—**Monolithic—Cylindrical—Class II**
CECC CECC 30 701-015. BS CECC 30 701-015 Issue 1; Fixed Capacitor Multilayer Ceramic Dielectric Class 2F4; Cylindrical Insulated Non-Metallic Case Axial Centred Leads (En). 10 pp.

—**Monolithic—Dipped—Class II**
IECQ QC 300701/US 0004 ISSUE 2-91. Fixed Capacitors for Use in Electrotechnical Equipment: Detail Specification: Fixed Capacitors, Multilayer Ceramic Dielectric, Class 2E6 Conformal Insulated Coating, Radial Leads. 16 pp.

—**Monolithic—Dipped—Class II—Printed Circuit Mount**
CECC CECC 30 701-001. NBN C 83-170 Feuille 1; Fixed Ceramic Dielectric Capacitors (Class 2) (En) SUPP 1 (En). 10 pp.
CECC CECC 30 701-006 ISSUE 2-84. BS CECC 30 701-006; Fixed Capacitor Multilayer Ceramic Dielectric Class 2F4; Conformal Insulated Coating Non-Metallic Case Radial Leads (En) AMD 1 (En). 11 pp.

INTERNATIONAL AND NON-U.S. NATIONAL STANDARDS SUBJECT INDEX

Ceramic

Ceramic Capacitors (Cont.)

—Monolithic—Dipped—Cylindrical—Class I

CECC CECC 30 601-008 ISSUE 2-84. BS CECC 30 601-008; Fixed Capacitor Multilayer Ceramic Dielectric Class 1B; Conformal Insulated Coating Non-Metalic Case Radial Leads (En). 10 pp.

—Monolithic—Disk—Class II

IECQ QC 300701/US 0005-88. Fixed Capacitors for Use in Electronic Equipment: Detail Specification: Fixed Capacitors of Ceramic Dielectric, Class 2, Assessment Level E. 21 pp.

—Monolithic—Hexagonal—Class II—Printed Circuit Mount

CECC CECC 30 701-024. BS CECC 30 701-024 Issue 1; Fixed Capacitor Monolithic Discoidal Ceramic Dielectric Class 2 (En). 17 pp.

—Monolithic—Molded—Class II

IECQ QC 300701/US 0006 ISSUE 1-91. Specifications for Use in Electrotechnical Equipment: Detail Specification: Fixed Capacitors of Ceramic Dielectric Class 2 X 1 Molded Case Multilayer Ceramic—Axial Leads. 15 pp.

IECQ QC 300701/US 0007 ISSUE 1-91. Specifications for Use in Electrotechnical Equipment: Detail Specification: Fixed Capacitors of Ceramic Dielectric Class 2 X 1 Molded Case Multilayer Ceramic—Radial Leads. 15 pp.

—Monolithic—Plate—Dipped—Rectangular—Class II—Printed Circuit

CECC CECC 30 701-017. NBN C 83-170-3 Feuille 3; Fixed Capacitors with Ceramic Dielectric, Class 2 (En, Fr, Ne). 11 pp.

—Monolithic—Plate—Molded—Rectangular—Class II—Printed Circuit

CECC CECC 30 701-029. NBN C 83-170-4; Capacitors with Ceramic Dielectric, Class 2 (En). 11 pp.

—Monolithic—Plate—Molded—Tubular—Class II

CECC CECC 30 701-022 ISSUE 3-85. I.S. 642; Fixed Ceramic Dielectric Capacitor, Class 2 (En). 12 pp.

CECC CECC 30 701-034 ISSUE 1-85. Fixed Capacitor Dielectric Capacitor, Class 2 (En). 10 pp.

—Monolithic—Rectangular—Class I

CECC CECC 30 601-009. BS CECC 30 601-009 Issue 1; Fixed Capacitor Multilayer Ceramic Dielectric Class 1B; Rectangular Insulated Non-Metallic Case Radial Leads (En) AMD 1 (En). 11 pp.

—Monolithic—Rectangular—Class II—Printed Circuit Mount

CECC CECC 30 701-007. BS CECC 30 701-007 Issue 1; Fixed Capacitor Multilayer Ceramic Dielectric Class 2C1; Rectangular Insulated Non-Metallic Case Radial Leads (En) AMD 2 (En). 11 pp.

CECC CECC 30 701-008. BS CECC 30 701-008 Issue 1; Fixed Capacitor Multilayer Ceramic Dielectric Class 2F4; Rectangular Insulated Non-Metallic Case Radial Leads (En) AMD 2 (En). 11 pp.

—Monolithic—Tubular—Class I

CECC CECC 30 601-017 ISSUE 1-85. I.S. CECC 30 601-017; 1985; Fixed Ceramic Dielectric Capacitor, Class 1 (En). 10 pp.

—Preferred Products List

CECC CECC MUAHAG Vol 1 IS 4-90. Preferred Products List; Capacitors (En, Fr, Ge). 64 pp.

—Quality Assurance

BSI BS 9070: Sec 5-71. 1971 Amd 6 Fixed Capacitors of Assessed Quality: Generic Data and Methods of Test Section 5: Ceramic Dielectric Capacitors. 21 pp.

BSI BS 9070: Sec 5-07. 1971 Amd 7 Fixed Capacitors of Assessed Quality: Generic Data and Methods of Test Section 5: Ceramic Dielectric Capacitors (AMD 7396) March 15, 1993. 24 pp.

BSI BS 9075 N001-81. 1981 Amd 2 Detail Specification for Fixed Ceramic Dielectric Capacitors, Type 1 and Type 2; Rectangular Monolithic Chips. Full Assessment Level. 20 pp.

BSI BS 9075 N001-03. 1981 Amd 3 Detail Specification for Fixed Ceramic Dielectric Capacitors, Type 1 and Type 2; Rectangular Monolithic Chips. Full Assessment Level (AMD 7511) June 15, 1993 (T). 24 pp.

BSI BS 9075 N023-78. 1978 Amd 3 Detail Specification for Fixed Capacitors of Assessed Quality: Monolithic Ceramic Dielectric Type 1B. Full Assessment Level (AMD 3835) December 31, 1931. 12 pp.

Ceramic Capacitors (Cont.)

—Quality Assurance (Cont.)

BSI BS 9075 N023-04. 1978 Amd 4 Detail Specification for Fixed Monolithic Ceramic Dielectric Capacitors (Type 1B). Rectangular Non-Metallic Case, Centred Wires on One Face. Full Assessment Level (AMD 7413) January 15, 1993. 13 pp.

BSI BS 9075 N024-78. 1978 Amd 2 Detail Specification for Fixed Capacitors of Assessed Quality: Monolithic Ceramic Dielectric Type 2C1. Full Assessment. 12 pp.

BSI BS 9075 N025-78. 1978 Amd 3 Detail Specification for Fixed Capacitors of Assessed Quality: Monolithic Ceramic Dielectric Type 2F4. Full Assessment. 12 pp.

BSI BS QC 300600-91. 1991 Fixed Capacitors for Use in Electronic Equipment. Sectional Specification for Fixed Capacitors of Ceramic Dielectric, Class 1 (IEC 384-8: 1988). 48 pp.

BSI BS QC 300601-93. 1993 Fixed Capacitors for Use in Electronic Equipment Blank Detail Specification Fixed Capacitors of Ceramic Dielectric, Class 1 Assessment Level E (IEC 384-8-1: 1988). 15 pp.

BSI BS QC 300700-92. 1992 Fixed Capacitors for Use in Electronic Equipment. Sectional Specification for Fixed Capacitors of Ceramic Dielectric, Class 2 (IEC 384-9: 1988) (T). 35 pp.

BSI BS QC 300701-93. 1993 Fixed Capacitors for Use in Electronic Equipment Blank Detail Specification Fixed Capacitors of Ceramic Dielectric, Class 2 Assessment Level E (IEC 384-9-1: 1988). 16 pp.

BSI BS CECC 30600-79. 1979 Amd 2 Fixed Ceramic Capacitors, Type 1: Sectional Specification. 33 pp.

BSI BS CECC 30600-03. 1979 Amd 3 Sectional Specification: Fixed Ceramic Capacitors, Type 1 (AMD 5983) July 15, 1992. 34 pp.

BSI BS CECC 30601-80. 1980 Amd 1 Fixed Ceramic Dielectric Capacitors. Class 1: Blank Detail Specification (AMD 4551) March 30, 1984. 15 pp.

BSI BS CECC 30700-91. 1991 Fixed Capacitors of Ceramic Dielectric Class 2: Sectional Specification. 39 pp.

BSI BS CECC 30700-01. 1991 Amd 1 Sectional Specification: Fixed Capacitors of Ceramic Dielectric, Class 2 (AMD 7401) February 15, 1993. 45 pp.

BSI BS CECC 30701-93. 1993 Blank Detail Specification: Fixed Capacitors of Ceramic Dielectric, Class 2 (Assessment Level E). 16 pp.

BSI BS CECC 30701-80. 1980 Amd 1 Fixed Ceramic Capacitors. Class 2: Sectional Specification. 15 pp.

BSI BS CECC 30702-93. 1993 Blank Detail Specification: Fixed Capacitors of Ceramic Dielectric, Class 2 (Assessment Level P. Telecom Level). 17 pp.

BSI BS CECC 31100-81. 1981 Fixed Ceramic Dielectric Capacitors of Barrier Layer Type (Dielectric Class 3): Sectional Specification. 28 pp.

BSI BS CECC 31100-01. 1981 Amd 1 Sectional Specification: Fixed Ceramic Dielectric Capacitors of Barrier Layer Type (Dielectric Class 3) (AMD March 15, 1993. 31 pp.

BSI BS CECC 31101-81. 1981 Amd 1 Fixed Ceramic Dielectric Capacitors of Barrier Layer Type (Dielectric Class 3): Blank Detail Specification. 14 pp.

BSI BS CECC 31400-82. 1982 Fixed Ceramic Capacitors of Dielectric Class 1, for Electrical Shock Hazard Protection: Sectional Specification. 25 pp.

BSI BS CECC 31400-01. 1982 Amd 1 Sectional Specification: Fixed Ceramic Capacitors of Dielectric Class 1, for Electrical Shock Hazard Protection (AMD 7404) February 15, 1993. 28 pp.

BSI BS CECC 31401-88. 1988 Fixed Ceramic Dielectric Capacitors, Dielectric Class 1, for Electrical Shock Hazard Protection. 14 pp.

BSI BS CECC 31500-82. 1982 Fixed Ceramic Capacitors of Dielectric Class 2, for Electrical Shock Hazard Protection: Sectional Specification. 25 pp.

BSI BS CECC 31501-88. 1988 Fixed Ceramic Dielectric Capacitors, Dielectric Class 2, for Electrical Shock Hazard Protection. 14 pp.

IEC 384 Pt 8-88. Fixed Capacitors for Use in Electronic Equipment Part 8: Sectional Specification: Fixed Capacitors of Ceramic Dielectric Second Edition; (Amendment 1-1993) (IECQ QC 300600). 91 pp.

IEC 384 Pt 9-88. Fixed Capacitors for Use in Electronic Equipment Part 9: Sectional Specification: Fixed Capacitors of Ceramic Dielectric, Class 2 Second Edition (IECQ QC 300700). 67 pp.

IEC 384 Pt 9-1-88. Fixed Capacitors for Use in Electronic Equipment Part 9: Blank Detail Specification: Fixed Capacitors of Ceramic Dielectric, Class 2 Assessment Level E First Edition (IECQ QC 300701). 33 pp.

MOD UK DSTAN 59-44: Pt 5:Sec 3-01. Capacitors, Fixed, of Assessed Quality (Listed on EPIC Database) Part 5: Capacitors, Fixed, Ceramic Dielectric Section 3: List of Items Conforming to BS 9075 N023 Issue 1; Corrigenda. 24 pp.

Ceramic Capacitors (Cont.)

—Quality Assurance (Cont.)

MOD UK DSTAN 59-44: Pt 5:Sec 4-01. Capacitors, Fixed, of Assessed Quality (Listed on EPIC Database) Part 5: Capacitors, Fixed, Ceramic Dielectric Section 4: List of Items Conforming to BS 9075 N024 Issue 1; Amendment 1. 26 pp.

MOD UK DSTAN 59-44: Pt 5:Sec 5-80. Capacitors, Fixed, of Assessed Quality (Listed on EPIC Database) Part 5: Capacitors, Fixed, Ceramic Dielectric Section 5: List of Items Conforming to BS 9075 N025 Issue 1. 17 pp.

MOD UK DSTAN 59-44: Pt 5:Sec 7-81. Capacitors, Fixed, of Assessed Quality (Listed on EPIC Database) Part 5: Capacitors, Fixed, Ceramic Dielectric Section 7: List of Items Conforming to BS 9075 F0031 Issue 1. 11 pp.

SAA AS 1455-73. Fixed Ceramic Dielectric Capacitors—Type 2. 24 pp.

—Reliability Assured

CNS C5028-85. General Rules for Fixed Ceramic Capacitor for Electronic Equipment (Jun)(4857).

CNS C5032-86. General Rules for Reliability Assured Fixed Ceramic Capacitors (Jan)(4904).

CNS C7066-85. AC Mains Supply Ceramic Capacitors (Jan)(4594).

IEC 234-67. Dimensions of Ceramic Dielectric Capacitors of the Plate Type First Edition; (Supplement A-1970). 12 pp.

IEC 324-70. Ceramic Dielectric Capacitors Type 3 First Edition. 33 pp.

IEC 384 Pt 8-1-88. Fixed Capacitors for Use in Electronic Equipment Part 8: Blank Detail Specification: Fixed Capacitors of Ceramic Dielectric Class 1 Assessment Level E First Edition; (Amendment 1-1993) (IECQ QC 300601). 42 pp.

JIS C 5130-89. General Rules of Fixed Ceramic Capacitors for Use in Electronic Equipment. 42 pp.

JIS C 5154-80. AC Mains Supply Ceramic Capacitors (R 1985). 17 pp.

JIS C 6422-91. Fixed Ceramic Capacitors Class 2 for Use in Electronic Equipment. 39 pp.

JIS C 6423-91. Fixed Ceramic Capacitors Class 1 for Use in Electronic Equipment. 58 pp.

—Shock Protection

BSI BS CECC 31400-82. 1982 Fixed Ceramic Capacitors of Dielectric Class 1, for Electrical Shock Hazard Protection: Sectional Specification. 25 pp.

BSI BS CECC 31400-01. 1982 Amd 1 Sectional Specification: Fixed Ceramic Capacitors of Dielectric Class 1, for Electrical Shock Hazard Protection (AMD 7404) February 15, 1993. 28 pp.

BSI BS CECC 31401-88. 1988 Fixed Ceramic Dielectric Capacitors, Dielectric Class 1, for Electrical Shock Hazard Protection. 14 pp.

BSI BS CECC 31500-82. 1982 Fixed Ceramic Capacitors of Dielectric Class 2, for Electrical Shock Hazard Protection: Sectional Specification. 25 pp.

BSI BS CECC 31501-88. 1988 Fixed Ceramic Dielectric Capacitors, Dielectric Class 2, for Electrical Shock Hazard Protection. 14 pp.

—Temperature Compensating

CNS C7023-85. Temperature Compensating Fixed Ceramic Capacitors for Electronic Devices (Jan)(3516). 20 pp.

—Temperature Compensating—Reliability Assured

CNS C7089-79. Reliability Assured Fixed Ceramic Capacitors (Temperature Compensation) (Jul)(4905).

—Trimmer—Circular—Printed Circuit Mount—Preferred Products List

CECC CECC MUAHAG Vol 1 IS 4-90. Preferred Products List; Capacitors (En, Fr, Ge). 64 pp.

—Variable

CNS C7062-85. Variable Ceramic Capacitors for Electronic Equipment (Jan)(4588). 19 pp.

IEC 499 Pt 1-74. Ceramic Dielectric Disc-Style Rotary Variable Pre-Set Capacitors: Grade 2 Part 1: General Requirements for Tests and Measuring Methods First Edition. 32 pp.

JIS C 6447-84. Pre-Set Ceramic Capacitors (Type C) for Electronic Equipment. 23 pp.

Ceramic Chip Capacitors

See Also: Capacitors; Ceramic Capacitors; Fixed Capacitors

IEC 384 Pt 10-89. Fixed Capacitors for Use in Electronic Equipment Part 10: Sectional Specification: Fixed Multilayer Ceramic Chips Capacitors Second Edition; (Amendment 1-1993) (IECQ QC 301900). 100 pp.

IEC 384 Pt 10-1-89. Fixed Capacitors for Use in Electronic Equipment Part 10: Blank Detail Specification: Fixed Multilayer Ceramic Chips Capacitors Assessment Level E First Edition; (Amendment 1-1993) (IECQ QC 301901). 36 pp.

INTERNATIONAL AND NON-U.S. NATIONAL STANDARDS
SUBJECT INDEX

Ceramic

Ceramic Chip Capacitors *(Cont.)*
IECQ QC 301900-89. Fixed Capacitors for Use in Electronic Equipment Part 10: Sectional Specification: Fixed Multilayer Ceramic Chip Capacitors (Amendment 1-1993) (IEC 384-10 ED 2). 100 pp.

IECQ QC 301901-89. Fixed Capacitors for Use in Electronic Equipment Part 10: Blank Detail Specification: Fixed Multilayer Ceramic Chip Capacitors Assessment Level E (Amendment 1-1993) (IEC 384-10-1 ED 1). 36 pp.

IECQ QC 301901/JP 0001-87. Detail Specification for Electronic Components Fixed Multilayer Ceramic Chip Capacitors Assessment Level E. 17 pp.

IECQ QC 301901/JP 0002-87. Detail Specification for Electronic Components Fixed Multilayer Ceramic Chip Capacitors Assessment Level E. 20 pp.

IECQ QC 301901/SG 000-90. Detail Specification for Electronic Components Fixed Multilayer Ceramic Chip Capacitors Assessment Level E. 17 pp.

IECQ QC 301901/US 0001 ISSUE 1-90. Capacitors for Use in Electrotechnical Equipment Detail Specification: Fixed Multilayer Ceramic Chip Capacitors Subclass 2R1. 14 pp.

IECQ QC 301901/US 0002 ISSUE 1-90. Capacitors for Use in Electrotechnical Equipment Detail Specification: Fixed Multilayer Ceramic Chip Capacitors Subclass 1B. 14 pp.

IECQ QC 301901/US 0003 ISSUE 1-91. Capacitors for Use in Electrotechnical Equipment: Detail Specificaton: Fixed Multilayer Ceramic Chip Capacitors for Mounting on Rigid and Non-Rigid Substrates Subclass 2R1 and 2X1 Rated Voltages: 50 Volts & 100 Volts. 18 pp.

IECQ QC 301901/US 0004 ISSUE 1-91. Capacitors for Use in Electrotechnical Equipment: Detail Specification: Fixed Multilayer Ceramic Chip Capacitors for Mounting on Rigid and Non-Rigid Substrates Subclass 1B Rated Voltages: 50 Volts & 100 Volts. 17 pp.

IECQ PQC 86-89. Fixed Capacitors for Use in Electronic Equipment: Blank Detail Specification: Fixed Chip Capacitors with Thin Film Dielectric Assessment Level E. 10 pp.

IECQ PQC 86/IL 0001 ISSUE C-89. Fixed Chip Capacitors Thin Film Dielectric. 13 pp.

JIS C 6429-89. Fixed Multilayer Ceramic Chip Capacitors for Use in Electronic Equipment. 48 pp.

—**Class I—Hybrid Circuits**
CECC CECC 32 100 ISSUE 1-88. Sectional Specification: Fixed Multilayer Ceramic Chip Capacitors (En, Fr, Ge) Erratum (En, Fr, Ge). 155 pp.

—**Class I—Printed Circuit Mount**
CECC CECC 32 100 ISSUE 1-88. Sectional Specification: Fixed Multilayer Ceramic Chip Capacitors (En, Fr, Ge) Erratum (En, Fr, Ge). 155 pp.

CECC CECC 32 101 ISSUE 1-88. Blank Detail Specification: Fixed Multilayer Ceramic Chip Capacitors (En, Fr, Ge). 38 pp.

—**Class I—Surface Mount**
CECC CECC 32 101-801 ISSUE 1-88. Fixed Multilayer Ceramic Chip Capacitors (En). 21 pp.

—**Class I—Surface Mount—Preferred Products List**
CECC CECC MUAHAG Vol 1 IS 4-90. Preferred Products List; Capacitors (En, Fr, Ge). 64 pp.

—**Class II—Hybrid Circuits**
CECC CECC 32 100 ISSUE 1-88. Sectional Specification: Fixed Multilayer Ceramic Chip Capacitors (En, Fr, Ge) Erratum (En, Fr, Ge). 155 pp.

—**Class II—Printed Circuit Mount**
CECC CECC 32 100 ISSUE 1-88. Sectional Specification: Fixed Multilayer Ceramic Chip Capacitors (En, Fr, Ge) Erratum (En, Fr, Ge). 155 pp.

CECC CECC 32 101 ISSUE 1-88. Blank Detail Specification: Fixed Multilayer Ceramic Chip Capacitors (En, Fr, Ge). 38 pp.

—**Class II—Surface Mount**
CECC CECC 32 101-801 ISSUE 1-88. Fixed Multilayer Ceramic Chip Capacitors (En). 21 pp.

—**Class II—Surface Mount—Preferred Products List**
CECC CECC MUAHAG Vol 1 IS 4-90. Preferred Products List; Capacitors (En, Fr, Ge). 64 pp.

—**Hybrid Circuits**
IECQ PQC 85-89. Fixed Capacitors for Use in Electronic Equipment: Sectional Specification: Fixed Chip Capacitors with Thin Film Dielectric; Selection of Methods of Test and General Requirements. 32 pp.

Ceramic Chip Capacitors *(Cont.)*
—**Microwave—Preferred Products List**
CECC CECC MUAHAG Vol 1 IS 4-90. Preferred Products List; Capacitors (En, Fr, Ge). 64 pp.

—**Preferred Products List**
CECC CECC MUAHAG Vol 1 IS 4-90. Preferred Products List; Capacitors (En, Fr, Ge). 64 pp.

—**Quality Assurance**
BSI BS QC 301900-92. 1992 Fixed Capacitors for Use in Electronic Equipment Sectional Specification for Fixed Multilayer Ceramic Chip Capacitors (IEC 384-10: 1989) (T). 47 pp.

BSI BS QC 301901-92. 1992 Fixed Capacitors for Use in Electronic Equipment. Part 10: Blank Detail Specification: Fixed Multilayer Ceramic Chip Capacitors Assessment Level E. 12 pp.

IECQ PQC 85-89. Fixed Capacitors for Use in Electronic Equipment: Sectional Specification: Fixed Chip Capacitors with Thin Film Dielectric; Selection of Methods of Test and General Requirements. 32 pp.

—**Reliability Assured**
BSI BS CECC 32100-92. 1992 Sectional Specification: Fixed Multilayer Ceramic Chip Capacitors. 55 pp.

Ceramic Clay
See Also: Ceramics; Clays

—**Moisture Content**
CNS R3050-69. Method of Test for Free Moisture Content in Fine Ceramic and Porcelain Clay (Jan) (2886).

—**Shrinkage**
CNS R3051-69. Determination of Drying and Firing Shrinkage of Fine Ceramic and Porcelain Clay (Jan) (2887).

Ceramic Coatings (Made From Ceramics)
See Also: Glazes; Nonmetallic Coatings; Oxide Coatings; Sprayed Coatings; Vitreous Enamels

JIS R 4204-63. Method of Testing for Ceramic Coating.

—**Sprayed**
JIS H 8304-90. Thermal Sprayed Ceramic Coatings. 9 pp.

JIS H 8666-90. Testing Methods for Thermal Sprayed Ceramic Coatings. 14 pp.

MOD UK DSTAN 03-6: Part 3-72. Guide to Flame Spraying Processes Part 3: Ceramic and Cermet Coatings Issue 1. 16 pp.

Ceramic Fiber Blankets
Use: Ceramic Fiber Insulation—Blankets

Ceramic Fiber Felts
Use: Ceramic Fiber Mats

Ceramic Fiber Insulation
See Also: Ceramic Fiber Mats; Ceramic Fiber Papers; Ceramic Fibers

—**Alumina Silica—Blankets**
JIS R 3311-91. Ceramic Fiber Blanket. 10 pp.

—**Blankets**
BSI BS 7225: Sec 3.1-90. 1990 Classification of Refractories Part 3: Section 3.1: Blankets, Mats, Felts and Paper. 8 pp.

Ceramic Fiber Mats
Use For: Ceramic Fiber Felts *See Also:* Ceramic Fiber Insulation; Ceramic Fibers; Fiberglass Mats; Mats

BSI BS 7225: Sec 3.1-90. 1990 Classification of Refractories Part 3: Section 3.1: Blankets, Mats, Felts and Paper. 8 pp.

Ceramic Fiber Papers
Use For: Ceramic Papers *See Also:* Ceramic Fiber Insulation; Ceramic Fibers; Papers

BSI BS 7225: Sec 3.1-90. 1990 Classification of Refractories Part 3: Section 3.1: Blankets, Mats, Felts and Paper. 8 pp.

Ceramic Fibers
Use For: Ceramic Fibres *See Also:* Ceramic Fiber Insulation; Ceramic Fiber Mats; Ceramic Fiber Papers; Fibers

—**Comprehensive Testing**
BSI BS 1902: Part 6-86. 1986 Methods for Testing Refractory Materials Part 6: Ceramic Fibre Products. 24 pp.

Ceramic Fibers *(Cont.)*
—**Felts**
BSI BS 7225: Sec 3.1-90. 1990 Classification of Refractories Part 3: Section 3.1: Blankets, Mats, Felts and Paper. 8 pp.

Ceramic Fibres
Use: Ceramic Fibers

Ceramic Filters
Use For: Piezoelectric Ceramic Filters
See Also: Dielectric Filters; Electric Filters; Piezoelectric Devices

—**Communication Equipment**
IECQ PQC 62/JP 0001-87. Detail Specification for: Piezoelectric Ceramic Filters for Sound IF Circuit of TV Assessment Level E. 10 pp.

IECQ PQC 63-85. Sectional Specification for: Piezoelectric Ceramic Filters for Communication Equipment. 17 pp.

IECQ PQC 64-85. Blank Detail Specification for: Piezoelectric Ceramic Filters for Use in Communication Equipment Assessment Level E. 8 pp.

IECQ PQC 64/JP 0002-91. Detail Specification for: Piezoelectric Ceramic Filters for Use in Communication Equipment Assessment Level E. 10 pp.

—**Electromagnetic Interference Filters—Military—PPL**
CECC CECC MUAHAG Vol 11 IS 2-92. Preferred Products List; Filters (En, Fr, Ge). 88 pp.

—**Radio Receivers**
IECQ PQC 58-85. Blank Detail Specification for: Piezoelectric Ceramic Filters for Use in FM Radios Assessment Level E. 8 pp.

IECQ PQC 58/JP 0001-87. Detail Specification for: Piezoelectric Ceramic Filters for Use in FM Radios Assessment Level E. 10 pp.

IECQ PQC 58/SG 0001-90. Detail Specification for: Piezoelectric Ceramic Filters for Use in FM Radios Assessment Level E. 10 pp.

IECQ PQC 59-85. Sectional Specification for: Piezoelectric Ceramic Filters for AM Radios. 18 pp.

IECQ PQC 60-85. Blank Detail Specification for: Piezoelectric Ceramic Filters for Use in AM Radios Assessment Level E. 8 pp.

IECQ PQC 60/JP 0001-87. Detail Specification for: Piezoelectric Ceramic Filters for Use in AM Radios Assessment Level E. 10 pp.

IECQ PQC 60/SG 0001-90. Detail Specification for: Piezoelectric Ceramic Filters for Use in AM Radios Assessment Level E. 10 pp.

IECQ PQC 64/JP 0001-87. Blank Detail Specification for: Piezoelectric Ceramic Filters for Use in Communication Equipment Assessment Level E. 10 pp.

—**Television Receivers**
IECQ PQC 61-85. Sectional Specification for: Piezoelectric Ceramic Filters for Sound IF Circuit of Television. 18 pp.

—**Television Receivers—Reliability Assured**
IECQ PQC 62-85. Blank Detail Specification for: Piezoelectric Ceramic Filters for Sound IF Circuit of TV Assessment Level E. 8 pp.

Ceramic Magnet Cores
Use: Magnetic Cores

Ceramic Materials
Use: Ceramics

Ceramic Matrix Composites
See Also: Ceramics; Composite Materials

—**Flexural Strength**
BSI DD ENV 658-3-93. 1993 Advanced Technical Ceramics—Mechanical Properties of Ceramic Composites at Room Temperature Part 3: Determination of Flexural Strength (V). 11 pp.

—**Tensile Testing**
BSI DD ENV 658-1-93. 1993 Advanced Technical Ceramics—Mechanical Properties of Ceramic Composites at Room Temperature Part 1: Determination of Tensile Strength (V). 15 pp.

Ceramic Papers
Use: Ceramic Fiber Papers

Ceramic Pipes
See Also: Pipes

CNS R1005-56. Shapes and Dimensions of Ceramic Pipes (May)(481)(R 1973). 1 p.

CNS R3006-56. Testing Standard for Ceramic Pipes (May) (482)(R 1973). 3 pp.

INDUSTRY STANDARDS

Ceramic

Ceramic Pipes (Cont.)

—Drainpipes

SNZ NZS 3302-83. Specification for Ceramic Pipes, Fittings and Joints (Reconfirmed 1991). 27 pp.

—Sewer Pipes

SNZ NZS 3302-83. Specification for Ceramic Pipes, Fittings and Joints (Reconfirmed 1991). 27 pp.

Ceramic Powders

See Also: Ceramics

—Area—Nitrogen Absorption

CEN PREN 725-6-92. Advanced Technical Ceramics—Methods of Test for Ceramic Powders—Part 6: Determination of the Specific Surface Area. 17 pp.

—Bulk Density

CEN PREN 725-8-93. Advanced Technical Ceramics—Methods of Test for Ceramic Powders—Part 8: Determination of Tapped Bulk Density. 7 pp.

CEN PREN 725-9-93. Advanced Technical Ceramics—Methods of Test for Ceramic Powders—Part 9: Determination of Untamped Bulk Density. 7 pp.

—Compactibility

CEN PREN 725-10-93. Advanced Technical Ceramics—Method of Test for Ceramic Powders—Part 10: Determination of Compaction Properties. 10 pp.

—Oxygen Content—Extraction Analysis

CEN PREN 725-3-92. Advanced Technical Ceramics—Methods of Test for Ceramic Powders—Part 3: Determination of the Oxygen Content of Non-Oxides by Thermal Extraction. 8 pp.

—Particle Size Distribution

CEN PREN 725-5-92. Advanced Technical Ceramics—Methods of Test for Ceramic Powders—Part 5: Determination of the Particle Size Distribution. 26 pp.

Ceramic Resistors

See Also: Fixed Resistors; Resistors

—Power—Wirewound

CECC CECC 40 201-010. CEI—CECC 40 201-010 Edition 1; Fixed Power Resistors, Wirewound, Ceramic Encased, Insulated, Stability—Plus or Minus (5% + 0,1 Ohm). 6 pp.

Ceramic Tiles

Use For: Dust Pressed Tiles See Also: Ceramics; Construction Materials; Floor Tiles; Tile Adhesives; Tiles; Wall Tiles

CGSB CAN/CGSB-75.1-M88. Tile, Ceramic. 17 pp.

CNS R1018-86. General Rule for Ceramic Tiles (Aug)(9737). 6 pp.

CNS R2172-87. Ceramic Extrusion Tiles (May)(10631).

CNS R3071-86. Method of Test for Ceramic Tiles (Aug)(3299). 5 pp.

JIS A 5209-87. Ceramic Tiles. 32 pp.

—Acceptance Testing

CEN EN 163-85. Ceramic Tiles: Sampling and Basis for Acceptance. 7 pp.

DIN ENGL EN 163-92. Ceramic Tiles; Sampling and Acceptance Inspection (Jan). 5 pp.

—Floor

BSI BS 6431: Part 10-84. 1984 Ceramic Floor and Wall Tiles Part 10: Method for Determination of Dimensions and Surface Quality. 11 pp.

BSI BS 6431: Part 10-01. 1984 Amd 1 Ceramic Floor and Wall Tiles Part 10: Method for Determination of Dimensions and Surface Quality (AMD 7099) July 15, 1992. 14 pp.

CEN EN 98-84. Ceramic Tiles: Determination of Dimensions and Surface Quality. 9 pp.

CEN EN 98-91. Ceramic Tiles—Determination of Dimensions and Surface Quality. 9 pp.

CNS R2162-86. Ceramic Floor Tile-Earthenware (Aug)(9738). 6 pp.

CNS R2163-86. Ceramic Floor Tile-Stoneware (Aug)(9739). 6 pp.

CNS R2164-86. Ceramic Floor Tile-Vitreous Ware (Aug) (9740). 6 pp.

DIN ENGL 18166 (S)-86. Ceramic Split Tiles and Split Tile Accessories (Oct) (Superseded by DIN EN 186 Parts 1 and 2, and DIN EN 188, December 1991 Editions). 5 pp.

—Floor—Acceptance Testing

BSI BS 6431: Part 23-86. 1986 Ceramic Floor and Wall Tiles Part 23: Sampling and Basis for Acceptance. 9 pp.

BSI BS 6431: Part 23-01. 1986 Amd 1 Ceramic Floor and Wall Tiles Part 23: Specification for Sampling and Basis for Acceptance (AMD 7112) July 15, 1992. 12 pp.

Ceramic Tiles (Cont.)

—Floor—Classification

BSI BS 6431: Part 1-83. 1983 Ceramic Floor and Wall Tiles Part 1: Classification and Marking, Including Definitions and Characteristics. 9 pp.

BSI BS 6431: Part 1-01. 1983 Amd 1 Ceramic Floor and Wall Tiles Part 1: Specification for Classification and Marking, Including Definitions and Characteristics (AMD 7088) July 15, 1992. 12 pp.

CEN EN 87-82. Ceramic Floor and Wall Tiles Definitions, Classification, Characteristics and Marking. 7 pp.

CEN EN 87-91. Ceramic Floor and Wall Tiles—Definitions, Classification, Characteristics and Marking. 12 pp.

CEN EN 87-1-83. Ceramic Floor and Wall Tiles—Part 1: Specification for Classification and Marking, Including Definitions and Characteristics. 7 pp.

—Floor—Design

BSI BS 5385: Part 3-89. 1989 Wall and Floor Tiling Part 3: Code of Practice for the Design and Installation of Ceramic Floor Tiles and Mosaics. 49 pp.

BSI BS 5385: Part 3-01. 1989 Amd 1 Wall and Floor Tiling Part 3: Code of Practice for the Design and Installation of Ceramic Floor Tiles and Mosaics (AMD 7059) February 28, 1992. 50 pp.

—Floor—Frost Resistance

BSI BS 6431: Part 22-86. 1986 Ceramic Floor and Wall Tiles Part 22: Methods of Determination of Frost Resistance. 10 pp.

BSI BS 6431: Part 22-01. 1986 Amd 1 Ceramic Floor and Wall Tiles Part 22: Method for Determination of Frost Resistance (AMD 7111) July 15, 1992. 13 pp.

—Floor—Glazed—Abrasion Testing

BSI BS 6431: Part 20-84. 1984 Ceramic Floor and Wall Tiles Part 20: Method of Determination of Resistance to Surface Abrasion. Glazed Tiles. 10 pp.

BSI BS 6431: Part 20-01. 1984 Amd 1 Ceramic Floor and Wall Tiles Part 20: Method for Determination of Resistance to Surface Abrasion. Glazed Tiles (AMD 7109) July 15, 1992. 13 pp.

CEN EN 154-84. Ceramic Tiles: Determination of Resistance to Surface Abrasion; Glazed Tiles. 8 pp.

CEN EN 154-91. Ceramic Tiles—Determination of Resistance to Surface Abrasion—Glazed Tiles. 7 pp.

—Floor—Glazed—Chemical Resistance

BSI BS 6431: Part 19-01. 1984 Amd 1 Ceramic Floor and Wall Tiles Part 19: Method for Determination of Chemical Resistance. Glazed Tiles (AMD 7108) July 15, 1992 (R). 13 pp.

CEN EN 122-84. Ceramic Tiles: Determination of Chemical Resistance; Glazed Tiles. 8 pp.

CEN EN 122-91. Ceramic Tiles—Determination of Chemical Resistance—Glazed Tiles. 14 pp.

—Floor—Glazed—Cracking (Fracturing)

BSI BS 6431: Part 17-83. 1983 Ceramic Floor and Wall Tiles Part 17: Method for Determination of Glazing Resistance. Glazed Tiles. 8 pp.

BSI BS 6431: Part 17-01. 1983 Amd 1 Ceramic Floor and Wall Tiles Part 17: Method for Determination of Glazing Resistance. Glazed Tiles (AMD 7106) July 15, 1992. 11 pp.

CEN EN 105-81. Ceramic Tiles: Determination of Crazing Resistance; Glazed Tiles. 6 pp.

CEN EN 105-91. Ceramic Tiles—Determination of Crazing Resistance—Glazed Tiles. 5 pp.

—Floor—Glossaries

BSI BS 6431: Part 1-83. 1983 Ceramic Floor and Wall Tiles Part 1: Classification and Marking, Including Definitions and Characteristics. 9 pp.

BSI BS 6431: Part 1-01. 1983 Amd 1 Ceramic Floor and Wall Tiles Part 1: Specification for Classification and Marking, Including Definitions and Characteristics (AMD 7088) July 15, 1992. 12 pp.

CEN EN 87-82. Ceramic Floor and Wall Tiles Definitions, Classification, Characteristics and Marking. 7 pp.

CEN EN 87-91. Ceramic Floor and Wall Tiles—Definitions, Classification, Characteristics and Marking. 12 pp.

CEN EN 87-1-83. Ceramic Floor and Wall Tiles—Part 1: Specification for Classification and Marking, Including Definitions and Characteristics. 7 pp.

—Floor—Identification Systems

BSI BS 6431: Part 1-83. 1983 Ceramic Floor and Wall Tiles Part 1: Classification and Marking, Including Definitions and Characteristics. 9 pp.

BSI BS 6431: Part 1-01. 1983 Amd 1 Ceramic Floor and Wall Tiles Part 1: Specification for Classification and Marking, Including Definitions and Characteristics (AMD 7088) July 15, 1992. 12 pp.

Ceramic Tiles (Cont.)

—Floor—Identification Systems (Cont.)

CEN EN 87-82. Ceramic Floor and Wall Tiles Definitions, Classification, Characteristics and Marking. 7 pp.

CEN EN 87-91. Ceramic Floor and Wall Tiles—Definitions, Classification, Characteristics and Marking. 12 pp.

CEN EN 87-1-83. Ceramic Floor and Wall Tiles—Part 1: Specification for Classification and Marking, Including Definitions and Characteristics. 7 pp.

—Floor—Installation

BSI BS 5385: Part 3-89. 1989 Wall and Floor Tiling Part 3: Code of Practice for the Design and Installation of Ceramic Floor Tiles and Mosaics. 49 pp.

BSI BS 5385: Part 3-01. 1989 Amd 1 Wall and Floor Tiling Part 3: Code of Practice for the Design and Installation of Ceramic Floor Tiles and Mosaics (AMD 7059) February 28, 1992. 50 pp.

BSI BS 5385: Part 4-92. 1992 Wall and Floor Tiling Part 4: Code of Practice for Tiling and Mosaics in Specific Conditions. 32 pp.

BSI BS 5385: Part 4-86. 1986 Wall and Floor Tiling Part 4: Code of Practice for Ceramic Tiling and Mosaics in Specific Conditions. 35 pp.

BSI BS 8000: Sec 11.1-89. 1989 Workmanship on Building Sites Part 11: Code of Practice for Wall and Floor Tiling Section 11.1: Ceramic Tiles, Terrazzo Tiles and Mosaics. 22 pp.

—Floor—Interior/Exterior

BSI BS 6431: Part 2-84. 1984 Ceramic Floor and Wall Tiles Part 2: Specification for Extruded Ceramic Tiles with a Low Water Absorption (EMV3%). Group A1. 11 pp.

BSI BS 6431: Part 2-01. 1984 Amd 1 Ceramic Floor and Wall Tiles Part 2: Specification for Extruded Ceramic Tiles with a Low Water Absorption (E less then or Equal to 3%) Group AI (AMD 7089) July 15, 1992 (R). 14 pp.

BSI BS 6431: Sec 3.1-86. 1986 Ceramic Floor and Wall Tiles Part 3: Extruded Ceramic Tiles with a Water Absorption of 3% Less EMV 6% Section 3.1: General Products. 11 pp.

BSI BS 6431: Sec 3.1-01. 1986 Amd 1 Ceramic Floor and Wall Tiles Part 3: Extruded Ceramic Tiles with a Water Absorption of 3% Less Than E Less Than or Equal to 6% Group AIIa. Section 3.1: Specification for General Products (AMD 7090) July 15, 1992. 14 pp.

BSI BS 6431: Sec 3.2-86. 1986 Ceramic Floor and Wall Tiles Part 3: Extruded Ceramic Tiles with a Water Absorption of 3% Less Than EMV 6% Section 3.2: Specific Products (Terre Cuite, Cotto, Baldosin Catalan). 11 pp.

BSI BS 6431: Sec 3.2-01. 1986 Amd 1 Ceramic Floor and Wall Tiles Part 3: Extruded Ceramic Tiles with a Water Absorption of 3% Less Than E Less Than or Equal to 6% Group AIIa. Section 3.2: Spec. for Specific Products (Terre Cuite, Cotto, Baldosin Catalan) (AMD 7091) July 15, 1992. 14 pp.

BSI BS 6431: Sec 4.1-86. 1986 Ceramic Floor and Wall Tiles Part 4: Extruded Ceramic Tiles with a Water Absorption of 6% Less Than EMV 10% Section 4.1: General Products. 11 pp.

BSI BS 6431: Sec 4.1-01. 1986 Amd 1 Ceramic Floor and Wall Tiles Part 4: Extruded Ceramic Tiles with a Water Absorption of 6% Less Than E Less Than or Equal to 10% Group AIIb. Sect. 4.1: Specification for General Products (AMD 7092) July 15, 1992. 14 pp.

BSI BS 6431: Sec 4.2-86. 1986 Ceramic Floor and Wall Tiles Part 4: Extruded Ceramic Tiles with a Water Absorption of 6% Less Than EMV 10% Section 4.2: Specific Products (Terre Cuite, Cotto, Baldosin Catalan). 11 pp.

BSI BS 6431: Sec 4.2-01. 1986 Amd 1 Ceramic Floor and Wall Tiles Part 4: Extruded Ceramic Tiles with a Water Absorption of 6% Less Than E Less Than or Equal to 10% Group AIIb. Sec 4.2: Spec. for Specific Products (Terre Cuite, Cotto, Baldosin Catalan) (AMD 7093) July 15, 1992. 14 pp.

BSI BS 6431: Part 5-86. 1986 Ceramic Floor and Wall Tiles Part 5: Extruded Ceramic Tiles with a Water of Absorption E Greater Than 10%. 11 pp.

BSI BS 6431: Part 5-01. 1986 Amd 1 Ceramic Floor and Wall Tiles Part 5: Specification for Extruded Ceramic Tiles with a Water of Absorption of E Greater Than 10% Group AIII (AMD 7094) July 15, 1992. 14 pp.

BSI BS 6431: Part 6-84. 1984 Ceramic Floor and Wall Tiles Part 6: Specification for Dust-Pressed Ceramic Tiles with a Low Water Absorption (EMN3%). Group BI. 11 pp.

BSI BS 6431: Part 6-01. 1984 Amd 1 Ceramic Floor and Wall Tiles Part 6: Specification for Dust-Pressed Ceramic Tiles with a Low Water Absorption (E Less Than or Equal to 3%) Group BI (AMD 7095) July 15, 1992. 14 pp.

INTERNATIONAL AND NON-U.S. NATIONAL STANDARDS
SUBJECT INDEX
Ceramic

Ceramic Tiles (Cont.)

—Floor—Interior/Exterior (Cont.)

BSI BS 6431: Part 7-86. 1986 Ceramic Floor and Wall Tiles Part 7: Dust-Pressed Ceramic Tiles with a Water Absorption of 3% Less Than E Less Than 10%. 11 pp.

BSI BS 6431: Part 7-01. 1986 Amd 1 Ceramic Floor and Wall Tiles Part 7: Specification for Dust-Pressed Ceramic Tiles with a Water Absorption of 3% Less Than E Less Than or Equal to 6% Group BIIa (AMD 7096) July 15, 1992. 14 pp.

BSI BS 6431: Part 8-86. 1986 Ceramic Floor and Wall Tiles Part 8: Dust-Pressed Ceramic Tiles with a Water Absorption of 6% Less Than E Less Than 10%. 11 pp.

BSI BS 6431: Part 8-01. 1986 Amd 1 Ceramic Floor and Wall Tiles Part 8: Specification for Dust-Pressed Ceramic Tiles with a Water Absorption of 6% Less Than E Less Than or Equal to 10% Group BIIb (AMD 7097) July 15, 1992. 14 pp.

BSI BS 6431: Part 9-84. 1984 Ceramic Floor and Wall Tiles Part 9: Specification for Dust-Pressed Ceramic Tiles with a Water Absorption of (EMN10%). Group BIII. 11 pp.

BSI BS 6431: Part 9-01. 1984 Amd 1 Ceramic Floor and Wall Tiles Part 9: Specification for Dust-Pressed Ceramic Tiles with a Water Absorption of E Greater Than 10% Group BIII (AMD 7098) July 15, 1992. 14 pp.

CEN EN 121-84. Extruded Ceramic Tiles with Low Water Absorption (E Less Than or Equal to 3%); Group A1. 9 pp.

CEN EN 121-91. Extruded Ceramic Tiles with Low Water Absorption (E<=3%)—Group A 1. 8 pp.

CEN EN 159-84. Dust-Pressed Ceramic Tiles with Water Absorption E Greater Than 10%; Group BIII. 9 pp.

CEN EN 159-91. Dust-Pressed Ceramic Tiles with Water Absorption E>10%—Group BIII. 8 pp.

CEN EN 176-84. Dust-Pressed Ceramic Tiles with a Low Water Absorption (E Less Than or Equal to 3%) Group BI. 9 pp.

CEN EN 176-91. Dust-Pressed Ceramic Tiles with a Low Water Absorption (E<=3%)—Group BI. 8 pp.

CEN EN 177-84. Ceramic Tiles: Dust-Pressed Ceramic Tiles with a Water Absorption of 3% Less Than E Less Than or Equal to 6% (Group B11a). 9 pp.

CEN EN 177-91. Dust-Pressed Ceramic Tiles with a Water Absorption of 3%<E<=6% (Group BIIa). 8 pp.

CEN EN 178-84. Ceramic Tiles: Dust-Pressed Ceramic Tiles with a Water Absorption 6% Less than E Less Than or Equal to 10% (Group B11b). 9 pp.

CEN EN 178-91. Dust-Pressed Ceramic Tiles with a Water Absorption of 6%<E<=10% (Group BIIb). 8 pp.

CEN EN 186-1-85. Ceramic Tiles: Extruded Ceramic Tiles with a Water Absorption of 3% Less Than E Less Than or Equal to 6% (Group A11a) —Part 1. 9 pp.

CEN EN 186-1-91. Ceramic Tiles—Extruded Ceramic Tiles with a Water Absorption of 3%<E<=6% (Group AIIa) Part 1. 8 pp.

CEN EN 186-2-85. Ceramic Tiles: Extruded Ceramic Tiles with a Water Absorption of 3% Less Than E Less Than or Equal to 6% (Group A11a) —Part 2. 9 pp.

CEN EN 186-2-91. Ceramic Tiles—Extruded Ceramic Tiles with a Water Absorption of 3%<E<=6% (Group AIIa) Part 2. 8 pp.

CEN EN 187-1-85. Ceramic Tiles: Extruded Ceramic Tiles with a Water Absorption of 6% Less Than E Less Than or Equal to 10% (Group A11b)-Part 1. 9 pp.

CEN EN 187-1-91. Ceramic Tiles—Extruded Ceramic Tiles with a Water Absorption of 6%<E<=10% (Group AIIb) Part 1. 8 pp.

CEN EN 187-2-85. Ceramic Tiles: Extruded Ceramic Tiles with a Water Absorption of 6% Less Than E Less Than or Equal to 10% (Group A11b)-Part 2. 9 pp.

CEN EN 187-2-91. Ceramic Tiles—Extruded Ceramic Tiles with a Water Absorption of 6%<E<=10% (Group AIIb) Part 2. 8 pp.

CEN EN 188-85. Ceramic Tiles: Extruded Ceramic Tiles with a Water Absorption of E Greater Than 10% (Group A111). 9 pp.

CEN EN 188-91. Ceramic Tiles—Extruded Ceramic Tiles with a Water Absorption of E>10% (Group AIII). 8 pp.

DIN ENGL EN 121-91. Extruded Ceramic Tiles with a Low Water Absorption (E Less Than or Equal To 3%) (Group AI) (Dec)(This Standard, Together with DIN EN 186 Parts 1 and 2, DIN EN 187 Parts 1 and 2, and DIN EN 188, December 1991 Editions, Supersedes DIN 18 166 October 1986 Edition). 8 pp.

DIN ENGL EN 159-91. Dust-Pressed Ceramic Tiles with a High Water Absorption (E Greater Than 10%)—Group B III (Dec). 7 pp.

DIN ENGL EN 176-92. Dust-Pressed Ceramic Tiles with a Low Water Absorption (E Less Than or Equal To 3%) (Group B I) (Jan). 6 pp.

DIN ENGL EN 177-91. Dust-Pressed Ceramic Tiles with a Water Absorption of 3% Less Than E Less Than or Equal To 6% (Group B IIa) (Dec). 6 pp.

DIN ENGL EN 178-91. Dust-Pressed Ceramic Tiles with a Water Absorption of 6% Less Than or Equal To 10% (Group B IIb) (Dec). 6 pp.

DIN ENGL EN 186 Pt 1-91. Extruded Ceramic Tiles with a Water Absorption of 3% Less Than or Equal To 6% (Group AIIa)—Part 1 (Dec) (This Standard, Together with DIN EN 121, DIN EN 186 Part 2, DIN EN 187 Parts 1 and 2 and DIN EN 188 Dec 1991 Editions, Supersedes DIN 18166, Oct 1986 Edition). 7 pp.

DIN ENGL EN 186 Pt 2-91. Extruded Ceramic Tiles with a Water Absorption of 3% Less Than or Equal To 6% (Group AIIa)—Part 2 (Dec) (This Standard, Together with DIN EN 121, DIN EN 186 Part 1, DIN EN 187 Parts 1 and 2 and DIN EN 188 Dec 1991 Editions, Supersedes DIN 18166, Oct 1986 Edition). 7 pp.

DIN ENGL EN 187 Pt 1-91. Extruded Ceramic Tiles with a Water Absorption of 6% Less Than E Less Than or Equal To 10% (Group AIIb)—Part 1 (Dec) (This Standard, Together with DIN EN 121, DIN EN 186 Parts 1 and 2, DIN EN 187 Part 2 and DIN EN 188 Dec 1991 Editions, Supersedes DIN 18166, Oct 1986 Edition). 7 pp.

DIN ENGL EN 187 Pt 2-91. Extruded Ceramic Tiles with a Water Absorption of 6% Less Than E Less Than or Equal To 10% (Group AIIb)—Part 2 (Dec) (This Standard, Together with DIN EN 121, DIN EN 186 Parts 1 and 2, DIN EN 187 Part 1 and DIN EN 188 Dec 1991 Editions, Supersedes DIN 18166, Oct 1986 Edition). 7 pp.

DIN ENGL EN 188-91. Extruded Ceramic Tiles with a Water Absorption of E Greater Than 10% (Group AIII) (Dec) (This Standard Together with DIN EN 121, DIN EN 186 Parts 1 and 2, December 1991 Editions, Supersedes DIN 18166, October 1986 Edition). 7 pp.

—Floor—Maintenance

BSI BS 5385: Part 3-89. 1989 Wall and Floor Tiling Part 3: Code of Practice for the Design and Installation of Ceramic Floor Tiles and Mosaics. 49 pp.

BSI BS 5385: Part 3-01. 1989 Amd 1 Wall and Floor Tiling Part 3: Code of Practice for the Design and Installation of Ceramic Floor Tiles and Mosaics (AMD 7059) February 28, 1992. 50 pp.

—Floor—Modulus of Rupture

BSI BS 6431: Part 12-83. 1983 Ceramic Floor and Wall Tiles Part 12: Method for Determination of Modulus of Rupture. 8 pp.

BSI BS 6431: Part 12-01. 1983 Amd 1 Ceramic Floor and Wall Tiles Part 12: Method for Determination of Modulus of Rupture (AMD 7101) July 15, 1992. 11 pp.

CEN EN 100-82. Ceramic Tiles: Determination of Modulus of Rupture. 6 pp.

CEN EN 100-91. Ceramic Tiles—Determination of Modulus of Rupture. 5 pp.

—Floor—Sampling

BSI BS 6431: Part 23-86. 1986 Ceramic Floor and Wall Tiles Part 23: Sampling and Basis for Acceptance. 9 pp.

BSI BS 6431: Part 23-01. 1986 Amd 1 Ceramic Floor and Wall Tiles Part 23: Specification for Sampling and Basis for Acceptance (AMD 7112) July 15, 1992. 12 pp.

—Floor—Scratch Hardness Testing

BSI BS 6431: Part 13-86. 1986 Ceramic Floor and Wall Tiles Part 13: Method for Determination of Scratch Harness of Surface According to Mohs. 7 pp.

BSI BS 6431: Part 13-01. 1986 Amd 1 Ceramic Floor and Wall Tiles Part 13: Method for Determination of Scratch Hardness of Surface According to Mohs (AMD 7102) July 15, 1992. 10 pp.

CEN EN 101-84. Ceramic Tiles: Determination of Scratch Hardness of Surface According to Mohs. 5 pp.

CEN EN 101-91. Ceramic Tiles—Determination of Scratch Hardness of Surface According to Mohs. 4 pp.

—Floor—Thermal Expansion

BSI BS 6431: Part 15-83. 1983 Ceramic Floor and Wall Tiles Part 15: Method for Determination of Linear of Thermal Expansion. 7 pp.

BSI BS 6431: Part 15-01. 1983 Amd 1 Ceramic Floor and Wall Tiles Part 15: Method for Determination of Linear Thermal Expansion (AMD 7104) July 15, 1992. 10 pp.

—Floor—Thermal Shock

BSI BS 6431: Part 16-83. 1983 Ceramic Floor and Wall Tiles Part 16: Method for Determination of Resistance to Thermal Shock. 7 pp.

—Floor—Thermal Shock (Cont.)

BSI BS 6431: Part 16-01. 1983 Amd 1 Ceramic Floor and Wall Tiles Part 16: Method for Determination of Resistance to Thermal Shock (AMD 7105) July 15, 1992. 10 pp.

CEN EN 104-82. Ceramic Tiles: Determination of Resistance to Thermal Shock. 5 pp.

CEN EN 104-91. Ceramic Tiles—Determination of Resistance to Thermal Shock. 4 pp.

—Floor—Unglazed—Abrasion Testing

BSI BS 6431: Part 14-83. 1983 Ceramic Floor and Wall Tiles Part 14: Method for Determination of Resistance to Deep Abrasion. Unglazed Tiles. 10 pp.

BSI BS 6431: Part 14-01. 1983 Amd 1 Ceramic Floor and Wall Tiles Part 14: Method for Determination of Resistance to Deep Abrasion. Unglazed Tiles (AMD 7103) July 15, 1992. 13 pp.

CEN EN 102-82. Ceramic Tiles: Determination of Resistance to Deep Abrasion; Unglazed Tiles. 8 pp.

CEN EN 102-91. Ceramic Tiles—Determination of Resistance to Deep Abrasion—Unglazed Tiles. 7 pp.

—Floor—Unglazed—Chemical Resistance

BSI BS 6431: Part 18-83. 1983 Ceramic Floor and Wall Tiles Part 18: Method for Determination of Chemical Resistance. Unglazed Tiles. 7 pp.

BSI BS 6431: Part 18-01. 1983 Amd 1 Ceramic Floor and Wall Tiles Part 18: Method for Determination of Chemical Resistance. Unglazed Tiles (AMD 7107) July 15, 1992. 10 pp.

BSI BS 6431: Part 19-84. 1984 Ceramic Floor and Wall Tiles Part 19: Method for Determination of Chemical Resistance. Glazed Tiles. 10 pp.

CEN EN 106-82. Ceramic Tiles: Determination of Chemical Resistance; Unglazed Tiles. 5 pp.

CEN EN 106-91. Ceramic Tiles—Determination of Chemical Resistance—Unglazed Tiles. 4 pp.

—Floor—Unglazed—Expansion

BSI BS 6431: Part 21-84. 1984 Ceramic Floor and Tiles Part 21: Method for Determination of Moisture Expansion Using Boiling Water. Unglazed Tiles. 7 pp.

BSI BS 6431: Part 21-01. 1984 Amd 1 Ceramic Floor and Tiles Part 21: Method for Determination of Moisture Expansion Using Boiling Water. Unglazed Tiles (AMD 7110) July 15, 1992. 10 pp.

CEN EN 155-84. Ceramic Tiles: Determination of Moisture Expansion Using Boiling Water; Unglazed Tiles. 5 pp.

CEN EN 155-91. Ceramic Tiles—Determination of Moisture Expansion Using Boiling Water—Unglazed Tiles. 4 pp.

—Floor—Water Absorption

BSI BS 6431: Part 11-83. 1983 Ceramic Floor and Wall Tiles Part 11: Method for Determination of Water Absorption. 7 pp.

BSI BS 6431: Part 11-01. 1983 Amd 1 Ceramic Floor and Wall Tiles Part 11: Method for Determination of Water Absorption (AMD 7100) July 15, 1992. 10 pp.

CEN EN 99-82. Ceramic Tiles: Determination of Water Absorption. 5 pp.

CEN EN 99-91. Ceramic Tiles—Determination of Water Absorption. 4 pp.

—Freeze Thaw Testing

CEN EN 202-85. Ceramic Tiles: Determination of Frost Resistance. 8 pp.

CEN EN 202-91. Ceramic Tiles—Determination of Frost Resistance. 7 pp.

—Inspection

CEN EN 163-85. Ceramic Tiles: Sampling and Basis for Acceptance. 7 pp.

—Installation

SAA AS 3958.1-91. Ceramic Tiles—Part 1: Guide to the Installation of Ceramic Tiles Amdt 1 January/February 1992 (in Professional Package 30). 67 pp.

—Sampling

CEN EN 163-85. Ceramic Tiles: Sampling and Basis for Acceptance. 7 pp.

CEN EN 163-91. Ceramic Tiles—Sampling and Basis for Acceptance. 6 pp.

DIN ENGL EN 163-92. Ceramic Tiles; Sampling and Acceptance Inspection (Jan). 5 pp.

—Selection

SAA AS 3958.2-92. Ceramic Tiles—Part 2: Guide to the Selection of a Ceramic Tiling System (in Professional Package 30). 38 pp.

—Terrazzo

CNS R2167-86. Ceramic Mosaic Tile-Vitreous Ware (Aug) (9743). 4 pp.

CNS R2168-86. Ceramic Mosaic Tile—Stoneware (Aug)(9744).

INDUSTRY STANDARDS

Ceramic
INTERNATIONAL AND NON-U.S. NATIONAL STANDARDS
SUBJECT INDEX

Ceramic Tiles *(Cont.)*

—Thermal Expansion

CEN EN 103-82. Ceramic Tiles: Determination of Linear Thermal Expansion. 5 pp.
CEN EN 103-91. Ceramic Tiles—Determination of Linear Thermal Expansion. 4 pp.

—Thermal Shock

CNS R3068-85. Method of Test for Thermal Shock Resistance of Glazed Ceramic Tile (May) (3245).

—Wall

BSI BS 6431: Part 10-84. 1984 Ceramic Floor and Wall Tiles Part 10: Method for Determination of Dimensions and Surface Quality. 11 pp.
BSI BS 6431: Part 10-01. 1984 Amd 1 Ceramic Floor and Wall Tiles Part 10: Method for Determination of Dimensions and Surface Quality (AMD 7099) July 15, 1992. 14 pp.
CEN EN 98-84. Ceramic Tiles: Determination of Dimensions and Surface Quality. 9 pp.
CEN EN 98-91. Ceramic Tiles—Determination of Dimensions and Surface Quality. 8 pp.
CNS R2064-86. Ceramic Wall Tile-Earthenware (Aug) (3298). 7 pp.
CNS R2165-86. Ceramic Wall Tile-Stoneware (Aug)(9741). 7 pp.
CNS R2166-86. Ceramic Wall Tile-Vitreous Ware (Aug) (9742). 8 pp.
DIN ENGL 18166 (S)-86. Ceramic Split Tiles and Split Tile Accessories (Oct) (Superseded by DIN EN 186 Parts 1 and 2, and DIN EN 188, December 1991 Editions). 5 pp.

—Wall—Acceptance Testing

BSI BS 6431: Part 23-86. 1986 Ceramic Floor and Wall Tiles Part 23: Sampling and Basis for Acceptance. 9 pp.
BSI BS 6431: Part 23-01. 1986 Amd 1 Ceramic Floor and Wall Tiles Part 23: Specification for Sampling and Basis for Acceptance (AMD 7112) July 15, 1992. 12 pp.

—Wall—Classification

BSI BS 6431: Part 1-83. 1983 Ceramic Floor and Wall Tiles Part 1: Classification and Marking, Including Definitions and Characteristics. 9 pp.
BSI BS 6431: Part 1-01. 1983 Amd 1 Ceramic Floor and Wall Tiles Part 1: Specification for Classification and Marking, Including Definitions and Characteristics (AMD 7088) July 15, 1992. 12 pp.
CEN EN 87-82. Ceramic Floor and Wall Tiles Definitions, Classification, Characteristics and Marking. 7 pp.
CEN EN 87-91. Ceramic Floor and Wall Tiles—Definitions, Classification, Characteristics and Marking. 12 pp.
CEN EN 87-1-83. Ceramic Floor and Wall Tiles—Part 1: Specification for Classification and Marking, Including Definitions and Characteristics. 7 pp.

—Wall—Design

BSI BS 5385: Part 2-91. 1991 Wall and Floor Tiling Part 2: Code of Practice for the Design and Installation of External Ceramic Wall Tiling and Mosaics (Including Terra Cotta and Faience Tiles). 32 pp.
BSI BS 5385: Part 2-78. 1978 Amd 1 Wall and Floor Tiling Part 2: Code of Practice for External Ceramic Wall Tiling and Mosaics. 25 pp.

—Wall—Frost Resistance

BSI BS 6431: Part 22-86. 1986 Ceramic Floor and Wall Tiles Part 22: Methods of Determination of Frost Resistance. 10 pp.
BSI BS 6431: Part 22-01. 1986 Amd 1 Ceramic Floor and Wall Tiles Part 22: Method for Determination of Frost Resistance (AMD 7111) July 15, 1992. 13 pp.

—Wall—Glazed—Abrasion Testing

BSI BS 6431: Part 20-84. 1984 Ceramic Floor and Wall Tiles Part 20: Method of Determination of Resistance to Surface Abrasion. Glazed Tiles. 10 pp.
BSI BS 6431: Part 20-01. 1984 Amd 1 Ceramic Floor and Wall Tiles Part 20: Method for Determination of Resistance to Surface Abrasion. Glazed Tiles (AMD 7109) July 15, 1992. 13 pp.
CEN EN 154-84. Ceramic Tiles: Determination of Resistance to Surface Abrasion; Glazed Tiles. 8 pp.
CEN EN 154-91. Ceramic Tiles—Determination of Resistance to Surface Abrasion—Glazed Tiles. 7 pp.

—Wall—Glazed—Chemical Resistance

BSI BS 6431: Part 19-01. 1984 Amd 1 Ceramic Floor and Wall Tiles Part 19: Method for Determination of Chemical Resistance. Glazed Tiles (AMD 7108) July 15, 1992 (R). 13 pp.
CEN EN 122-84. Ceramic Tiles: Determination of Chemical Resistance; Glazed Tiles. 8 pp.
CEN EN 122-91. Ceramic Tiles—Determination of Chemical Resistance—Glazed Tiles. 14 pp.

Ceramic Tiles *(Cont.)*

—Wall—Glazed—Cracking (Fracturing)

BSI BS 6431: Part 17-83. 1983 Ceramic Floor and Wall Tiles Part 17: Method for Determination of Glazing Resistance. Glazed Tiles. 8 pp.
BSI BS 6431: Part 17-01. 1983 Amd 1 Ceramic Floor and Wall Tiles Part 17: Method for Determination of Glazing Resistance. Glazed Tiles (AMD 7106) July 15, 1992. 11 pp.
CEN EN 105-81. Ceramic Tiles: Determination of Crazing Resistance; Glazed Tiles. 6 pp.
CEN EN 105-91. Ceramic Tiles—Determination of Crazing Resistance—Glazed Tiles. 5 pp.

—Wall—Glossaries

BSI BS 6431: Part 1-83. 1983 Ceramic Floor and Wall Tiles Part 1: Classification and Marking, Including Definitions and Characteristics. 9 pp.
BSI BS 6431: Part 1-01. 1983 Amd 1 Ceramic Floor and Wall Tiles Part 1: Specification for Classification and Marking, Including Definitions and Characteristics (AMD 7088) July 15, 1992. 12 pp.
CEN EN 87-82. Ceramic Floor and Wall Tiles Definitions, Classification, Characteristics and Marking. 7 pp.
CEN EN 87-91. Ceramic Floor and Wall Tiles—Definitions, Classification, Characteristics and Marking. 12 pp.
CEN EN 87-1-83. Ceramic Floor and Wall Tiles—Part 1: Specification for Classification and Marking, Including Definitions and Characteristics. 7 pp.

—Wall—Identification Systems

BSI BS 6431: Part 1-83. 1983 Ceramic Floor and Wall Tiles Part 1: Classification and Marking, Including Definitions and Characteristics. 9 pp.
BSI BS 6431: Part 1-01. 1983 Amd 1 Ceramic Floor and Wall Tiles Part 1: Specification for Classification and Marking, Including Definitions and Characteristics (AMD 7088) July 15, 1992. 12 pp.
CEN EN 87-82. Ceramic Floor and Wall Tiles Definitions, Classification, Characteristics and Marking. 7 pp.
CEN EN 87-91. Ceramic Floor and Wall Tiles—Definitions, Classification, Characteristics and Marking. 12 pp.
CEN EN 87-1-83. Ceramic Floor and Wall Tiles—Part 1: Specification for Classification and Marking, Including Definitions and Characteristics. 7 pp.

—Wall—Installation

BSI BS 5385: Part 1-90. 1990 Wall and Floor Tiling Part 1: Code of Practice for the Design and Installation of Internal Ceramic Wall Tiling and Mosaics in Normal Conditions. 36 pp.
BSI BS 5385: Part 1-01. 1990 Amd 1 Wall and Floor Tiling Part 1: Code of Practice for the Design and Installation of Internal Ceramic Wall Tiling and Mosaics in Normal Conditions (AMD 7058) February 28, 1992. 37 pp.
BSI BS 5385: Part 2-91. 1991 Wall and Floor Tiling Part 2: Code of Practice for the Design and Installation of External Ceramic Wall Tiling and Mosaics (Including Terra Cotta and Faience Tiles). 32 pp.
BSI BS 5385: Part 2-78. 1978 Amd 1 Wall and Floor Tiling Part 2: Code of Practice for External Ceramic Wall Tiling and Mosaics. 25 pp.
BSI BS 5385: Part 4-92. 1992 Wall and Floor Tiling Part 4: Code of Practice for Tiling and Mosaics in Specific Conditions. 32 pp.
BSI BS 5385: Part 4-86. 1986 Wall and Floor Tiling Part 4: Code of Practice for Ceramic Tiling and Mosaics in Specific Conditions. 35 pp.
BSI BS 8000: Sec 11.1-89. 1989 Workmanship on Building Sites Part 11: Code of Practice for Wall and Floor Tiling Section 11.1: Ceramic Tiles, Terrazzo Tiles and Mosaics. 22 pp.

—Wall—Interior/Exterior

BSI BS 6431: Part 2-84. 1984 Ceramic Floor and Wall Tiles Part 2: Specification for Extruded Ceramic Tiles with a Low Water Absorption (EMV3%). Group A1. 11 pp.
BSI BS 6431: Part 2-01. 1984 Amd 1 Ceramic Floor and Wall Tiles Part 2: Specification for Extruded Ceramic Tiles with a Low Water Absorption (E less then or Equal to 3%) Group AI (AMD 7089) July 15, 1992 (R). 14 pp.
BSI BS 6431: Sec 3.1-86. 1986 Ceramic Floor and Wall Tiles Part 3: Extruded Ceramic Tiles with a Water Absorption of 3% Less Than EMV 6% Section 3.1: General Products. 11 pp.
BSI BS 6431: Sec 3.1-01. 1986 Amd 1 Ceramic Floor and Wall Tiles Part 3: Extruded Ceramic Tiles with a Water Absorption of 3% Less Than E Less Than or Equal to 6% Group AIIa. Section 3.1: Specification for General Products (AMD 7090) July 15, 1992. 14 pp.

Ceramic Tiles *(Cont.)*

—Wall—Interior/Exterior *(Cont.)*

BSI BS 6431: Sec 3.2-86. 1986 Ceramic Floor and Wall Tiles Part 3: Extruded Ceramic Tiles with a Water Absorption of 3% Less Than EMV 6% Section 3.2: Specific Products (Terre Cuite, Cotto, Baldosin Catalan). 11 pp.
BSI BS 6431: Sec 3.2-01. 1986 Amd 1 Ceramic Floor and Wall Tiles Part 3: Extruded Ceramic Tiles with a Water Absorption of 3% Less Than E Less Than or Equal to 6% Group AIIa. Section 3.2: Spec. for Specific Products (Terre Cuite, Cotto, Baldosin Catalan) (AMD 7091) July 15, 1992. 14 pp.
BSI BS 6431: Sec 4.1-86. 1986 Ceramic Floor and Wall Tiles Part 4: Extruded Ceramic Tiles with a Water Absorption of 6% Less Than EMV 10% Section 4.1: General Products. 11 pp.
BSI BS 6431: Sec 4.1-01. 1986 Amd 1 Ceramic Floor and Wall Tiles Part 4: Extruded Ceramic Tiles with a Water Absorption of 6% Less Than E Less Than or Equal to 10% Group AIIb. Sect. 4.1: Specification for General Products (AMD 7092) July 15, 1992. 14 pp.
BSI BS 6431: Sec 4.2-86. 1986 Ceramic Floor and Wall Tiles Part 4: Extruded Ceramic Tiles with a Water Absorption of 6% Less Than EMV 10% Section 4.2: Specific Products (Terre Cuite, Cotto, Baldosin Catalan). 11 pp.
BSI BS 6431: Sec 4.2-01. 1986 Amd 1 Ceramic Floor and Wall Tiles Part 4: Extruded Ceramic Tiles with a Water Absorption of 6% Less Than E Less Than or Equal to 10% Group AIIb. Sec 4.2: Spec. for Specific Products (Terre Cuite, Cotto, Baldosin Catalan) (AMD 7093) July 15, 1992. 14 pp.
BSI BS 6431: Part 5-86. 1986 Ceramic Floor and Wall Tiles Part 5: Extruded Ceramic Tiles with a Water of Absorption E Greater Than 10%. 11 pp.
BSI BS 6431: Part 5-01. 1986 Amd 1 Ceramic Floor and Wall Tiles Part 5: Specification for Extruded Ceramic Tiles with a Water Absorption of E Greater Than 10% Group AIII (AMD 7094) July 15, 1992. 14 pp.
BSI BS 6431: Part 6-84. 1984 Ceramic Floor and Wall Tiles Part 6: Specification for Dust-Pressed Ceramic Tiles with a Low Water Absorption (EMN3%). Group BI. 11 pp.
BSI BS 6431: Part 6-01. 1984 Amd 1 Ceramic Floor and Wall Tiles Part 6: Specification for Dust-Pressed Ceramic Tiles with a Low Water Absorption (E Less Than or Equal to 3%) Group BI (AMD 7095) July 15, 1992. 14 pp.
BSI BS 6431: Part 7-86. 1986 Ceramic Floor and Wall Tiles Part 7: Dust-Pressed Ceramic Tiles with a Water Absorption of 3% Less Than E Less Than 10%. 11 pp.
BSI BS 6431: Part 7-01. 1986 Amd 1 Ceramic Floor and Wall Tiles Part 7: Specification for Dust-Pressed Ceramic Tiles with a Water Absorption of 3% Less Than E Less Than or Equal to 6% Group BIIa (AMD 7096) July 15, 1992. 14 pp.
BSI BS 6431: Part 8-86. 1986 Ceramic Floor and Wall Tiles Part 8: Dust-Pressed Ceramic Tiles with a Water Absorption of 6% Less Than E Less Than 10%. 11 pp.
BSI BS 6431: Part 8-01. 1986 Amd 1 Ceramic Floor and Wall Tiles Part 8: Specification for Dust-Pressed Ceramic Tiles with a Water Absorption of 6% Less Than E Less Than or Equal to 10% Group BIIb (AMD 7097) July 15, 1992. 14 pp.
BSI BS 6431: Part 9-84. 1984 Ceramic Floor and Wall Tiles Part 9: Specification for Dust-Pressed Ceramic Tiles with a Water Absorption of (EMN10%). Group BIII. 11 pp.
BSI BS 6431: Part 9-01. 1984 Amd 1 Ceramic Floor and Wall Tiles Part 9: Specification for Dust-Pressed Ceramic Tiles with a Water Absorption of E Greater Than 10% Group BIII (AMD 7098) July 15, 1992. 14 pp.
CEN EN 121-84. Extruded Ceramic Tiles with Low Water Absorption (E Less Than or Equal to 3%); Group A1. 9 pp.
CEN EN 121-91. Extruded Ceramic Tiles with Low Water Absorption (E< =3%)—Group A 1. 8 pp.
CEN EN 159-84. Dust-Pressed Ceramic Tiles with Water Absorption E Greater Than 10%; Group B111. 9 pp.
CEN EN 159-91. Dust-Pressed Ceramic Tiles with Water Absorption E>10%—Group BIII. 8 pp.
CEN EN 176-84. Dust-Pressed Ceramic Tiles with a Low Water Absorption (E Less Than or Equal to 3%) Group B1. 9 pp.
CEN EN 176-91. Dust-Pressed Ceramic Tiles with a Low Water Absorption (E< =3%)—Group BI. 8 pp.
CEN EN 177-84. Ceramic Tiles: Dust-Pressed Ceramic Tiles with a Water Absorption of 3% Less Than E Less Than or Equal to 6% (Group B11a). 9 pp.
CEN EN 177-91. Dust-Pressed Ceramic Tiles with a Water Absorption of 3%<E< =6% (Group BIIa). 8 pp.
CEN EN 178-84. Ceramic Tiles: Dust-Pressed Ceramic Tiles with a Water Absorption 6% Less than E Less Than or Equal to 10% (Group B11b). 9 pp.

INTERNATIONAL AND NON-U.S. NATIONAL STANDARDS
SUBJECT INDEX

Ceramics

Ceramic Tiles (Cont.)

—Wall—Interior/Exterior (Cont.)

CEN EN 178-91. Dust-Pressed Ceramic Tiles with a Water Absorption of 6%<E<=10% (Group BIIb). 8 pp.

CEN EN 186-1-85. Ceramic Tiles: Extruded Ceramic Tiles with a Water Absorption of 3% Less Than E Less Than or Equal to 6% (Group A11a) —Part 1. 9 pp.

CEN EN 186-1-91. Ceramic Tiles—Extruded Ceramic Tiles with a Water Absorption of 3%<E<=6% (Group AIIa) Part 1. 8 pp.

CEN EN 186-2-85. Ceramic Tiles: Extruded Ceramic Tiles with a Water Absorption of 3% Less Than E Less Than or Equal to 6% (Group A11a) —Part 2. 9 pp.

CEN EN 186-2-91. Ceramic Tiles—Extruded Ceramic Tiles with a Water Absorption of 3%<E<=6% (Group AIIa) Part 2. 8 pp.

CEN EN 187-1-85. Ceramic Tiles: Extruded Ceramic Tiles with a Water Absorption of 6% Less Than E Less Than or Equal to 10% (Group A11b)-Part 1. 9 pp.

CEN EN 187-1-91. Ceramic Tiles—Extruded Ceramic Tiles with a Water Absorption of 6%<E<=10% (Group AIIb) Part 1. 8 pp.

CEN EN 187-2-85. Ceramic Tiles: Extruded Ceramic Tiles with a Water Absorption of 6% Less Than E Less Than or Equal to 10% (Group A11b)-Part 2. 9 pp.

CEN EN 187-2-91. Ceramic Tiles—Extruded Ceramic Tiles with a Water Absorption of 6%<E<=10% (Group AIIb) Part 2. 8 pp.

CEN EN 188-85. Ceramic Tiles: Extruded Ceramic Tiles with a Water Absorption of E Greater Than 10% (Group A111). 9 pp.

CEN EN 188-91. Ceramic Tiles—Extruded Ceramic Tiles with a Water Absorption of E>10% (Group AIII). 8 pp.

DIN ENGL EN 121-91. Extruded Ceramic Tiles with a Low Water Absorption (E Less Than or Equal To 3%) (Group AI) (Dec)(This Standard, Together with DIN EN 186 Parts 1 and 2, DIN EN 187 Parts 1 and 2, and DIN EN 188, December 1991 Editions, Supersedes DIN 18 166 October 1986 Edition). 8 pp.

DIN ENGL EN 159-91. Dust-Pressed Ceramic Tiles with a High Water Absorption (E Greater Than 10%)—Group B III (Dec). 7 pp.

DIN ENGL EN 176-92. Dust-Pressed Ceramic Tiles with a Low Water Absorption (E Less Than or Equal To 3%) (Group B I) (Jan). 6 pp.

DIN ENGL EN 177-91. Dust-Pressed Ceramic Tiles with a Water Absorption of 3% Less Than E Less Than or Equal To 6% (Group B IIa) (Dec). 6 pp.

DIN ENGL EN 178-91. Dust-Pressed Ceramic Tiles with a Water Absorption of 6% Less Than E Less Than or Equal To 10% (Group B IIb) (Dec). 6 pp.

DIN ENGL EN 186 Pt 1-91. Extruded Ceramic Tiles with a Water Absorption of 3% Less Than E Less Than or Equal To 6% (Group AIIa)—Part 1 (Dec) (This Standard, Together with DIN EN 121, DIN EN 186 Part 2, DIN EN 187 Parts 1 and 2 and DIN EN 188 Dec 1991 Editions, Supersedes DIN 18166, Oct 1986 Edition). 7 pp.

DIN ENGL EN 186 Pt 2-91. Extruded Ceramic Tiles with a Water Absorption of 3% Less Than E Less Than or Equal To 6% (Group AIIa)—Part 2 (Dec) (This Standard, Together with DIN EN 121, DIN EN 186 Part 1, DIN EN 187 Parts 1 and 2 and DIN EN 188 Dec 1991 Editions, Supersedes DIN 18166, Oct 1986 Edition). 7 pp.

DIN ENGL EN 187 Pt 1-91. Extruded Ceramic Tiles with a Water Absorption of 6% Less Than E Less Than or Equal To 10% (Group AIIb)—Part 1 (Dec) (This Standard, Together with DIN EN 121, DIN EN 186 Parts 1 and 2, DIN EN 187 Part 2 and DIN EN 188 Dec 1991 Editions, Supersedes DIN 18166, Oct 1986 Edition). 7 pp.

DIN ENGL EN 187 Pt 2-91. Extruded Ceramic Tiles with a Water Absorption of 6% Less Than E Less Than or Equal To 10% (Group AIIb)—Part 2 (Dec) (This Standard, Together with DIN EN 121, DIN EN 186 Parts 1 and 2, DIN EN 187 Part 1 and DIN EN 188 Dec 1991 Editions, Supersedes DIN 18166, Oct 1986 Edition). 7 pp.

DIN ENGL EN 188-91. Extruded Ceramic Tiles with a Water Absorption of E Greater Than 10% (Group AIII) (Dec) (This Standard Together with DIN EN 121, DIN EN 186 Parts 1 and 2, December 1991 Editions, Supersedes DIN 18166, October 1986 Edition). 7 pp.

—Wall—Modulus of Rupture

BSI BS 6431: Part 12-83. 1983 Ceramic Floor and Wall Tiles Part 12: Method for Determination of Modulus of Rupture. 8 pp.

BSI BS 6431: Part 12-01. 1983 Amd 1 Ceramic Floor and Wall Tiles Part 12: Method for Determination of Modulus of Rupture (AMD 7101) July 15, 1992. 11 pp.

CEN EN 100-82. Ceramic Tiles: Determination of Modulus of Rupture. 6 pp.

Ceramic Tiles (Cont.)

—Wall—Modulus of Rupture (Cont.)

CEN EN 100-91. Ceramic Tiles—Determination of Modulus of Rupture. 5 pp.

—Wall—Sampling

BSI BS 6431: Part 23-86. 1986 Ceramic Floor and Wall Tiles Part 23: Sampling and Basis for Acceptance. 9 pp.

BSI BS 6431: Part 23-01. 1986 Amd 1 Ceramic Floor and Wall Tiles Part 23: Specification for Sampling and Basis for Acceptance (AMD 7112) July 15, 1992. 12 pp.

—Wall—Scratch Hardness Testing

BSI BS 6431: Part 13-86. 1986 Ceramic Floor and Wall Tiles Part 13: Method for Determination of Scratch Harness of Surface According to Mohs. 7 pp.

BSI BS 6431: Part 13-01. 1986 Amd 1 Ceramic Floor and Wall Tiles Part 13: Method for Determination of Scratch Hardness of Surface According to Mohs (AMD 7102) July 15, 1992. 10 pp.

CEN EN 101-84. Ceramic Tiles: Determination of Scratch Hardness of Surface According to Mohs. 5 pp.

CEN EN 101-91. Ceramic Tiles—Determination of Scratch Hardness of Surface According to Mohs. 4 pp.

—Wall—Thermal Expansion

BSI BS 6431: Part 15-83. 1983 Ceramic Floor and Wall Tiles Part 15: Method for Determination of Linear of Thermal Expansion. 7 pp.

BSI BS 6431: Part 15-01. 1983 Amd 1 Ceramic Floor and Wall Tiles Part 15: Method for Determination of Linear Thermal Expansion (AMD 7104) July 15, 1992. 10 pp.

—Wall—Thermal Shock

BSI BS 6431: Part 16-83. 1983 Ceramic Floor and Wall Tiles Part 16: Method for Determination of Resistance to Thermal Shock. 7 pp.

BSI BS 6431: Part 16-01. 1983 Amd 1 Ceramic Floor and Wall Tiles Part 16: Method for Determination of Resistance to Thermal Shock (AMD 7105) July 15, 1992. 10 pp.

CEN EN 104-82. Ceramic Tiles: Determination of Resistance to Thermal Shock. 5 pp.

CEN EN 104-91. Ceramic Tiles—Determination of Resistance to Thermal Shock. 4 pp.

—Wall—Unglazed—Abrasion Testing

BSI BS 6431: Part 14-83. 1983 Ceramic Floor and Wall Tiles Part 14: Method for Determination of Resistance to Deep Abrasion. Unglazed Tiles. 10 pp.

BSI BS 6431: Part 14-01. 1983 Amd 1 Ceramic Floor and Wall Tiles Part 14: Method for Determination of Resistance to Deep Abrasion. Unglazed Tiles (AMD 7103) July 15, 1992. 13 pp.

CEN EN 102-82. Ceramic Tiles: Determination of Resistance to Deep Abrasion; Unglazed Tiles. 8 pp.

CEN EN 102-91. Ceramic Tiles—Determination of Resistance to Deep Abrasion—Unglazed Tiles. 7 pp.

—Wall—Unglazed—Chemical Resistance

BSI BS 6431: Part 18-83. 1983 Ceramic Floor and Wall Tiles Part 18: Method for Determination of Chemical Resistance. Unglazed Tiles. 7 pp.

BSI BS 6431: Part 18-01. 1983 Amd 1 Ceramic Floor and Wall Tiles Part 18: Method for Determination of Chemical Resistance. Unglazed Tiles (AMD 7107) July 15, 1992. 10 pp.

BSI BS 6431: Part 19-84. 1984 Ceramic Floor and Wall Tiles Part 19: Method for Determination of Chemical Resistance. Glazed Tiles. 10 pp.

CEN EN 106-82. Ceramic Tiles: Determination of Chemical Resistance; Unglazed Tiles. 5 pp.

CEN EN 106-91. Ceramic Tiles—Determination of Chemical Resistance—Unglazed Tiles. 4 pp.

—Wall—Unglazed—Expansion

BSI BS 6431: Part 21-84. 1984 Ceramic Floor and Tiles Part 21: Method for Determination of Moisture Expansion Using Boiling Water. Unglazed Tiles. 7 pp.

BSI BS 6431: Part 21-01. 1984 Amd 1 Ceramic Floor and Wall Tiles Part 21: Method for Determination of Moisture Expansion Using Boiling Water. Unglazed Tiles (AMD 7110) July 15, 1992. 10 pp.

CEN EN 155-84. Ceramic Tiles: Determination of Moisture Expansion Using Boiling Water; Unglazed Tiles. 5 pp.

CEN EN 155-91. Ceramic Tiles—Determination of Moisture Expansion Using Boiling Water—Unglazed Tiles. 4 pp.

—Wall—Water Absorption

BSI BS 6431: Part 11-83. 1983 Ceramic Floor and Wall Tiles Part 11: Method for Determination of Water Absorption. 7 pp.

BSI BS 6431: Part 11-01. 1983 Amd 1 Ceramic Floor and Wall Tiles Part 11: Method for Determination of Water Absorption (AMD 7100) July 15, 1992. 10 pp.

Ceramic Tiles (Cont.)

—Wall—Water Absorption (Cont.)

CEN EN 99-82. Ceramic Tiles: Determination of Water Absorption. 5 pp.

CEN EN 99-91. Ceramic Tiles—Determination of Water Absorption. 4 pp.

Ceramic Tubes

See Also: Ceramics

CNS R1004-84. General Rules of Ceramic Tubes (Jan)(480).

Ceramics

Scope Note: For additional listings, see also specific products and materials made from ceramics
Use For: Ceramic Materials *See Also:* Abrasives; Bricks; Carbon Refractories; Ceramic Clay; Ceramic Matrix Composites; Ceramic Powders; Ceramic Tubes; Cermet; Clays; Firebricks; Fireclay Refractories; Glass; Glass Ceramics; Masonry; Refractory Materials; Refractory Mortars; Stoneware; Tableware; Tiles; Vitreous Ceramic Materials; Vitreous China; Vitreous Clay Pipes; Wall Tiles

—Bulk Density

DIN ENGL 51065 Pt 1-85. Testing of Ceramic Materials; Determination of Bulk Density of Shaped Products and Broken Pieces (Aug). 2 pp.

—Cadmium Content—Atomic Absorption Spectrometry

CNS R3073-86. Method of Test for Lead and Cadmium Extracted from Glazed Ceramic Surfaces (Apr)(3503). 4 pp.

—Classification

BSI BS 1000: (666)-84. 1984 Universal Decimal Classification (UDC). English Full Edition (666): Glass Industry. Ceramics. Cement and Concrete. 60 pp.

SNZ NZS/BS 1000 (666)-84. Universal Decimal Classification Glass Industry. Ceramics. Cement and Concrete. 60 pp.

—Compression Testing

CEN ENV 658-2-93. Advanced Technical Ceramics—Mechanical Properties of Ceramic Composites at Room Temperature—Part 2: Determination of Compressive Strength. 17 pp.

DIN ENGL 51067 Pt 1-77. Testing of Ceramic Raw and Finished Materials; Determination of the Compressive Strength at Room Temperature (KDF) of Refractories with a Total Porosity of up to 45% (May). 5 pp.

DIN ENGL 51067 Pt 2-77. Testing of Ceramic Raw and Finished Materials; Determination of the Compressive Strength at Room Temperature (KDF) of Refractories with a Total Porosity Exceeding 45% (May). 3 pp.

JIS R 1608-90. Testing Methods for Compressive Strength of High Performance Ceramics. 8 pp.

—Cracking (Fracturing)

CNS R3067-71. Method of Test for Crazing Resistance of Fired and Glazed Fine Ceramic and Porcelain Products (by Autoclave Method) (Jun)(3244).

—Deformation—High Temperature Testing

BSI BS 7134: Sec 3.5-89. 1989 Testing of Engineering Ceramics Part 3: Thermo-Mechanical Properties Section 3.5: Methods of Determination of Pyroplastic Deformation (Sagging). 10 pp.

CEN ENV 820-2-92. Advanced Technical Ceramics—Methods of Test for Monolithic Ceramics—Thermo-Mechanical Properties—Part 2: Determination of Selfloaded Deformation. 10 pp.

—Density

CEN PREN 623-2-91. Methods of Testing Advanced Technical Ceramics—General and Textural Properties—Part 2: Determination of Density and Porosity. 15 pp.

CNS R3057-69. Method of Test for Specific Gravity of Fired Fine Ceramic and Porcelain Material (Jan)(2893)(R 1975).

—Diffusivity

CEN PREN 821-2-92. Advanced Technical Ceramics—Thermo-Physical Properties of Monolithic Ceramics—Part 2: Determination of Thermal Diffusivity by the Laser Flash (or Heat Pulse) Method. 20 pp.

JIS R 1611-91. Testing Methods of Thermal Diffusivity, Specific Heat Capacity, and Thermal Conductivity for High Performance Ceramics by Laser Flash Method. 16 pp.

INDUSTRY STANDARDS

INTERNATIONAL AND NON-U.S. NATIONAL STANDARDS
SUBJECT INDEX

Ceramics

Ceramics (Cont.)

—**Electrical Insulation—Cracking (Fracturing)**
BSI BS 7134: Sec 1.1-89. 1989 Testing of Engineering Ceramics Part 1: General and Textural Properties Section 1.1: Method for Determination of the Presence of Cracks and Other Defects by Dye Penetration Tests. 4 pp.

—**Electrical Insulation—Density**
BSI BS 7134: Sec 1.2-89. 1989 Testing of Engineering Ceramics Part 1: General and Textural Properties Section 1.2: Methods of Determination of Density and Porosity. 10 pp.

—**Electrical Insulation—Porosity**
BSI BS 7134: Sec 1.2-89. 1989 Testing of Engineering Ceramics Part 1: General and Textural Properties Section 1.2: Methods of Determination of Density and Porosity. 10 pp.

—**Electrical Insulation—Sampling**
BSI BS 7134: Part 0-89. 1989 Testing of Engineering Ceramics Part 0: Introduction and Guide to Sampling. 7 pp.

—**Electronic Equipment—Bulk Density**
CNS C6068-80. Method of Test for Apparent Density of Ceramics for Electron Device and Semiconductor Application (Jul)(5751). 3 pp.

—**Expansion**
CNS R3056-69. Method of Test for Moisture Expansion of Filed Fine Ceramic and Porcelain Products (Jan)(2892)(R 1975).

—**Fire Testing**
CNS R3054-69. Method of Test for Fire-Resistance of Dry Fired and Pressed Fine Ceramic and Porcelain Specimens at Room Temperature (Jan)(2890) (R 1975).

—**Flame Spraying**
BSI BS 4495-69. 1969 Recommendations for the Flame Spraying of Ceramic and Cement Coatings. 12 pp.

—**Flexural Strength**
CEN ENV 658-3-92. Advanced Technical Ceramics—Mechanical Properties of Ceramic Composites at Room Temperature—Part 3: Determination of Flexural Strength. 10 pp.
CEN PREN 843-1-92. Advanced Technical Ceramics—Mechanical Properties of Monolithic Ceramics at Room Temperature—Part 1: Determination of Flexural Strength. 19 pp.
JIS R 1601-81. Testing Method for Flexural Strength (Modulus of Rupture) of High Performance Ceramics. 8 pp.

—**Flexural Strength—High Temperature Testing**
JIS R 1604-87. Testing Method for Flexural Strength (Modulus of Rupture) of High Performance Ceramics at Elevated Temperature. 9 pp.

—**Food Contact**
EC 84/500/EEC-84. Council Directive on the Approximation of the Laws of the Member States Relating to Ceramic Articles Intended to Come Into Contact with Foodstuffs. 5 pp.

—**Fracture Testing**
JIS R 1607-90. Testing Methods for Fracture Toughness of High Performance Ceramics. 16 pp.

—**Grain Size**
BSI DD ENV 623-3-93. 1993 Advanced Technical Ceramics Monolithic Ceramics—General and Textural Properties Part 3: Determination of Grain Size (V). 15 pp.
CEN ENV 623-3-93. Advanced Technical Ceramics—Monolithic Ceramics—General and Textural Properties—Part 3: Determination of Grain Size. 15 pp.
CEN ENV 623-3-93. Advanced Technical Ceramics—Monolithic Ceramics—General and Textural Properties—Part 3: Determination of Grain Size. 10 pp.

—**High Temperature Testing**
JIS R 1609-90. Testing Methods for Oxidation Resistance of Non-Oxide High Performance Ceramics. 8 pp.

—**Lead Content—Atomic Absorption Spectrometry**
CNS R3073-86. Method of Test for Lead and Cadmium Extracted from Glazed Ceramic Surfaces (Apr)(3503). 4 pp.

—**Modulus of Elasticity**
JIS R 1602-86. Testing Methods for Elastic Modulus of High Performance Ceramics. 14 pp.

Ceramics (Cont.)

—**Modulus of Elasticity** (Cont.)
JIS R 1605-89. Testing Methods for Elastic Modulus of High Performance Ceramics at Elevated Temperature. 10 pp.

—**Modulus of Rupture**
CNS R3055-69. Method of Test for Rupture Modulus of Fired Cast or Extruded Fine Ceramic and Porcelain Products (Jan)(2891)(R 1975).

—**Oxidation Resistance**
JIS R 1609-90. Testing Methods for Oxidation Resistance of Non-Oxide High Performance Ceramics. 8 pp.

—**Porosity**
CEN PREN 623-2-91. Methods of Testing Advanced Technical Ceramics—General and Textural Properties—Part 2: Determination of Density and Porosity. 15 pp.
DIN ENGL 51056-85. Testing of Ceramic Materials; Determination of Water Absorption and Apparent Porosity (Aug). 3 pp.

—**Semiconductors—Bulk Density**
CNS C6068-80. Method of Test for Apparent Density of Ceramics for Electron Device and Semiconductor Application (Jul)(5751). 3 pp.

—**Shear Strength**
CEN ENV 658-4-92. Advanced Technical Ceramics—Mechanical Properties of Ceramic Composites at Room Temperature—Part 4: Determination of Shear Strength by Compression Loading of Notched Specimens. 9 pp.

—**Sieve Analysis**
CNS R3066-71. Sieve Analysis for Non-Plasticible Pulverized Ceramic Material (Apr)(3243).

—**Specific Heat**
JIS R 1611-91. Testing Methods of Thermal Diffusivity, Specific Heat Capacity, and Thermal Conductivity for High Performance Ceramics by Laser Flash Method. 16 pp.

—**Surface Defects—Liquid Penetrant Testing**
CEN PREN 623-1-92. Advanced Technical Ceramics—General and Textural Properties of Monolithic Ceramics—Part 1: Determination of the Presence of Defects by Dye Penetration Tests. 8 pp.

—**Tensile Testing**
CEN ENV 658-1-93. Advanced Technical Ceramics—Mechanical Properties of Ceramic Composites at Room Temperature—Part 1: Determination of Tensile Strength. 10 pp.
JIS R 1606-90. Testing Methods for Tensile Strength of High Performance Ceramics at Room and Elevated Temperature. 10 pp.

—**Test Specimens—Chemical Analysis**
DIN ENGL 51062-61. Testing of Ceramic Raw Materials and Process Materials; Preparation and Drying of Samples for Chemical Analysis (Nov). 2 pp.

—**Thermal Conductivity**
BSI BS 7134: Sec 4.2-90. 1990 Testing of Engineering Ceramics Part 4: Thermo-Physical Properties Section 4.2: Method for the Determination of Thermal Diffusivity, by the Laser Flash (or Heat Pulse) Method. 14 pp.
JIS R 1611-91. Testing Methods of Thermal Diffusivity, Specific Heat Capacity, and Thermal Conductivity for High Performance Ceramics by Laser Flash Method. 16 pp.

—**Thermal Expansion**
BSI BS 7134: Sec 4.1-90. 1990 Testing of Engineering Ceramics Part 4: Thermo-Physical Properties Section 4.1: Method for Determination of Thermal Expansion. 15 pp.
CEN PREN 821-1-92. Advanced Technical Ceramics—Thermo-Physical Properties of Monolithic Ceramics—Part 1: Determination of Thermal Expansion. 21 pp.
CNS R3069-71. Determination of Linean Thermal Expansion of Fired Fine Ceramic and Porcelain Products (Jun)(3246).

—**Vickers Hardness Testing**
JIS R 1610-91. Testing Method for Vickers Hardness of High Performance Ceramics. 7 pp.

—**Water Absorption**
DIN ENGL 51056-85. Testing of Ceramic Materials; Determination of Water Absorption and Apparent Porosity (Aug). 3 pp.

Cereal Products

See Also: Bakery Products; Cereals; Flours (Food); Food; Pasta; Semolina

Cereal Products (Cont.)

—**Amylase Activity—Amylographs**
BSI BS 4317: Part 28-93. 1993 Methods of Test for Cereals and Pulses Part 28: Determination of Viscosity of Flour Using an Amylograph (ISO 7973: 1992). 12 pp.
ISO 7973-92. Cereals and Milled Cereal Products—Determination of the Viscosity of Flour—Method Using an Amylograph First Edition. 11 pp.

—**Amylase Activity—Colorimetric Analysis**
BSI BS 4317: Part 14-80. 1980 Methods of Test for Cereals and Pulses Part 14: Determination of Alpha-Amylase Activity in Cereals and Cereal Products (Colorimetric Method). 8 pp.
ISO 3983-77. Cereals and Cereal Products—Determination of Alpha-Amylase Activity—Colorimetric Method First Edition. 7 pp.

—**Ash Content**
BSI BS 4317: Part 10-93. 1993 Methods of Test for Cereals and Pulses Part 10: Determination of Ash of Cereals and Milled Cereal Products (ISO 2171: 1993) (W). 11 pp.
BSI BS 4317: Part 10-81. 1981 Methods of Test for Cereals and Pulses Part 10: Determination of Ash in Cereals, Pulses and Their Derived Products. 7 pp.
ISO 2171-93. Cereals and Milled Cereal Products—Determination of Total Ash Third Edition. 8 pp.

—**Baby Foods**
CNS N5201-83. Cereal Based Foods for Infant and Children (Jan)(9906). 4 pp.

—**Bacteria Count Methods**
ISO 7698-90. Cereals, Pulses and Derived Products—Enumeration of Bacteria, Yeasts and Moulds First Edition. 10 pp.

—**Fat Content**
ISO 7302-82. Cereals and Cereal Products—Determination of Total Fat Content First Edition. 5 pp.

—**Fungi—Count Methods**
ISO 7698-90. Cereals, Pulses and Derived Products—Enumeration of Bacteria, Yeasts and Moulds First Edition. 10 pp.

—**Moisture Content**
BSI BS 4317: Part 2-87. 1987 Cereals and Pulses Part 2: Determination of Moisture Content of Cereals and Cereal Products (Reference Method). 8 pp.
BSI BS 4317: Part 3-87. 1987 Cereals and Pulses Part 3: Determination of Moisture Content of Cereals and Cereal Products (Routine Method). 6 pp.
ISO 711-85. Cereals and Cereal Products—Determination of Moisture Content (Basic Reference Method) Second Edition. 7 pp.
ISO 712-85. Cereals and Cereal Products—Determination of Moisture Content (Routine Reference Method) Second Edition. 5 pp.

—**Sampling**
BSI BS 6298-82. 1982 Automatic Sampling of Cereals and Milled Cereal Products by Mechanical Means. 8 pp.
ISO 6644-81. Cereals and Milled Cereal Products—Automatic Sampling by Mechanical Means First Edition. 7 pp.

—**Yeasts—Count Methods**
ISO 7698-90. Cereals, Pulses and Derived Products—Enumeration of Bacteria, Yeasts and Moulds First Edition. 10 pp.

Cereals

Use For: Grains (Food) *See Also:* Agricultural Products; Barley; Cereal Products; Congee; Corn; Durum Wheat; Legumes; Oats; Rice; Rye; Sorghum; Wheat

—**Air Flow—Pressure Measurement**
ISO 4174-80. Cereals and Pulses—Measurement of Unit Pressure Losses Due to Single-Dimension Air Flow Through a Batch of Grain First Edition. 13 pp.

—**Amylase Activity—Amylographs**
BSI BS 4317: Part 28-93. 1993 Methods of Test for Cereals and Pulses Part 28: Determination of Viscosity of Flour Using an Amylograph (ISO 7973: 1992). 12 pp.
ISO 7973-92. Cereals and Milled Cereal Products—Determination of the Viscosity of Flour—Method Using an Amylograph First Edition. 11 pp.

—**Amylase Activity—Colorimetric Analysis**
BSI BS 4317: Part 14-80. 1980 Methods of Test for Cereals and Pulses Part 14: Determination of Alpha-Amylase Activity in Cereals and Cereal Products (Colorimetric Method). 8 pp.

INTERNATIONAL AND NON-U.S. NATIONAL STANDARDS
SUBJECT INDEX

Cereals (Cont.)
—Amylase Activity—Colorimetric Analysis (Cont.)
ISO 3983-77. Cereals and Cereal Products—Determination of Alpha-Amylase Activity—Colorimetric Method First Edition. 7 pp.

—Ash Content
BSI BS 4317: Part 10-93. 1993 Methods of Test for Cereals and Pulses Part 10: Determination of Ash of Cereals and Milled Cereal Products (ISO 2171: 1993) (W). 11 pp.
BSI BS 4317: Part 10-81. 1981 Methods of Test for Cereals and Pulses Part 10: Determination of Ash in Cereals, Pulses and Their Derived Products. 7 pp.
ISO 2171-93. Cereals and Milled Cereal Products—Determination of Total Ash Third Edition. 8 pp.

—Bacteria Count Methods
ISO 7698-90. Cereals, Pulses and Derived Products—Enumeration of Bacteria, Yeasts and Moulds First Edition. 10 pp.

—Bulk Density
BSI BS 4317: Part 23-90. 1990 Methods of Test for Cereals and Pulses Part 23: Determination of Bulk Density of Cereals, Called 'Mass per Hectolitre'. 8 pp.
ISO 7971-86. Cereals—Determination of Bulk Density, Called "Mass per Hectolitre" (Reference Method) First Edition. 7 pp.
ISO 8840-87. Refractory Materials—Determination of Bulk Density of Granular Materials (Grain Density) First Edition. 8 pp.

—Conveyors
CNS B4020-81. Mobile Pnematic Grain Conveyor (Jan)(4116).

—Fat Content
ISO 7302-82. Cereals and Cereal Products—Determination of Total Fat Content First Edition. 5 pp.

—Fiber Content
BSI BS 6215: Part 2-81. 1981 Amd 1 Agricultural Food Products Part 2: Determination of Crude Fibre Content (Modified Scharrer Method). 8 pp.
ISO 6541-81. Agricultural Food Products—Determination of Crude Fibre Content—Modified Scharrer Method First Edition. 6 pp.

—Fungi—Count Methods
ISO 7698-90. Cereals, Pulses and Derived Products—Enumeration of Bacteria, Yeasts and Moulds First Edition. 10 pp.

—Glossaries
BSI BS 5939-80. 1980 Cereals and Cereal Products. 9 pp.
BSI BS 6860-87. 1987 Nomenclature for Cereals, Pulses and Other Food Grains. 15 pp.
ISO 5526-86. Cereals, Pulses and Other Food Grains—Nomenclature First Edition. 19 pp.
ISO 5527 Pt 1-79. Cereals—Vocabulary—Part 1 First Edition. 9 pp.

—Insect Contamination
BSI BS 4317: Part 18-88. 1988 Cereals and Pulses Part 18: Determination of Hidden Insect Infestation. 12 pp.
ISO 6639 Pt 1-86. Cereals and Pulses—Determination of Hidden Insect Infestation—Part 1: General Principles First Edition. 4 pp.
ISO 6639 Pt 3-86. Cereals and Pulses—Determination of Hidden Insect Infestation—Part 3: Reference Method First Edition. 6 pp.
ISO 6639 Pt 4-87. Cereals and Pulses—Determination of Hidden Insect Infestation—Part 4: Rapid Methods First Edition; (Supersedes 1162). 20 pp.

—Insect Contamination—Sampling
ISO 6639 Pt 2-86. Cereals and Pulses—Determination of Hidden Insect Infestation—Part 2: Sampling First Edition. 6 pp.

—Mass
BSI BS 4317: Part 1-80. 1980 Methods of Test for Cereals and Pulses Part 1: Determination of the Mass of 1000 Grains. 4 pp.
ISO 520-77. Cereals and Pulses—Determination of the Mass of 1 000 Grains First Edition. 4 pp.

—Milled—Sampling
BSI BS 5333-81. 1981 Sampling Cereals and Pulses (As Milled Products). 11 pp.
ISO 2170-80. Cereals and Pulses—Sampling of Milled Products Second Edition. 10 pp.

—Moisture Content
BSI BS 4317: Part 2-87. 1987 Cereals and Pulses Part 2: Determination of Moisture Content of Cereals and Cereal Products (Reference Method). 8 pp.

Cereals (Cont.)
—Moisture Content (Cont.)
BSI BS 4317: Part 3-87. 1987 Cereals and Pulses Part 3: Determination of Moisture Content of Cereals and Cereal Products (Routine Method). 6 pp.
CNS N4030-89. Methods of Test for Cereals (Nov)(3287). 10 pp.
ISO 711-85. Cereals and Cereal Products—Determination of Moisture Content (Basic Reference Method) Second Edition. 7 pp.
ISO 712-85. Cereals and Cereal Products—Determination of Moisture Content (Routine Reference Method) Second Edition. 5 pp.

—Moisture Meters—Calibration
BSI BS 4317: Part 24-90. 1990 Methods of Test for Cereals and Pulses Part 24: Method of Checking the Calibration of Moisture Meters for Cereals. 10 pp.
ISO 7700 Pt 1-84. Check of the Calibration of Moisture Meters—Part 1: Moisture Meters for Cereals First Edition. 8 pp.

—Oil Content
CNS N4030-89. Methods of Test for Cereals (Nov)(3287). 10 pp.

—Pesticide Residues
EC 86/362/EEC-86. Council Directive on the Fixing of Maximum Levels for Pesticide Residues in and on Cereals. 6 pp.
EC 93/57/EEC-93. Council Directive Amending the Annexes to Directives 86/362/EEC and 86/363/EEC on the Fixing of Maximum Levels for Pesticide Residues in and on Cereals and Foodstuffs of Animal Origin Respectively. 5 pp.

—Sampling
BSI BS 4510-80. 1980 Methods of Sampling Cereals (As Grain). 9 pp.
BSI BS 6298-82. 1982 Automatic Sampling of Cereals and Milled Cereal Products by Mechanical Means. 8 pp.
CNS N4030-89. Methods of Test for Cereals (Nov)(3287). 10 pp.
ISO 950-79. Cereals—Sampling (as Grain) First Edition. 8 pp.
ISO 6644-81. Cereals and Milled Cereal Products—Automatic Sampling by Mechanical Means First Edition. 7 pp.

—Seeds—Marketing
EC 88/380/EEC-88. Council Directive Amending Directives 66/400/EEC, 66/401/EEC, 66/ 402/EEC, 66/403/EEC, 69/ 208/EEC, 70/457/EEC AND 70/458/EEC on the Marketing of Beet Seed, Fodder Plant Seed, Cereal Seed, Seed Potatoes, Seed of Oil and Fibre Plants and Vegetable Seed. 18 pp.

—Sieve Analysis
BSI BS 6219-88. 1988 Test Sieves for Cereals. 6 pp.
CNS N4030-89. Methods of Test for Cereals (Nov)(3287). 10 pp.

—Storage
BSI BS 6279: Part 1-82. 1982 Storage of Cereals and Pulses Part 1: Guide to Particular Problems Encountered in the Storage of Cereals and Pulses. 11 pp.
BSI BS 6279: Part 2-82. 1982 Storage of Cereals and Pulses Part 2: Code of Practice for the Storage of Cereals and Pulses. 9 pp.
BSI BS 6279: Part 2-01. 1982 Amd 1 Storage of Cereals and Pulses Part 2: Code of Practice for the Storage of Cereals and Pulses (ISO 6322/2-1981) (AMD October 15, 1993 (W). 11 pp.
ISO 6322 Pt 1-81. Storage of Cereals and Pulses—Part 1: General Considerations in Keeping Cereals First Edition. 10 pp.
ISO 6322 Pt 2-81. Storage of Cereals and Pulses—Part 2: Essential Requirements First Edition. 8 pp.

—Storage—Pest Control
BSI BS 6279: Part 3-90. 1990 Storage of Cereals and Pulses Part 3: Guide to Control of Attack by Pests. 10 pp.
ISO 6322 Pt 3-89. Storage of Cereals and Pulses—Part 3: Control of Attack by Pests Second Edition. 8 pp.

—Temperature Measurement—Silos
BSI BS 4317: Part 26-91. 1991 Methods of Test for Cereals and Pulses Part 26: Measurement of Temperature of Grain During Bulk Storage (ISO 4112: 1990). 8 pp.
ISO 4112-90. Cereals and Pulses—Guidance on Measurement of the Temperature of Grain Stored in Bulk Second Edition. 7 pp.

—Yeasts—Count Methods
ISO 7698-90. Cereals, Pulses and Derived Products—Enumeration of Bacteria, Yeasts and Moulds First Edition. 10 pp.

Ceric Ammonium Nitrate
Use: Ammonium Ceric Nitrate

Ceric Ammonium Sulfate
CNS K7138-64. Chemical Reagent (Cermic Ammonium Sulfate (Jul)(1638).
JIS K 8977-88. Cerium (IV) Diammonium Sulfate (Dihydrate, Tetrahydrate).

Ceric Sulfate
JIS K 8976-61. Ceric Sulfate.

Cerium Diammonium Sulfate
Use: Ceric Ammonium Sulfate

Cermet
Use For: Metal Ceramics See Also: Ceramics; Metals

—Flame Spraying
BSI BS 4495-69. 1969 Recommendations for the Flame Spraying of Ceramic and Cement Coatings. 12 pp.

—Thermal Spraying—Marine
MOD UK NES 828-89. Thermal Spray Deposition of Metals and Ceramics for Engineering Purposes Issue 1 (11.89). 47 pp.

Cermet Potentiometers
Use For: Metal Ceramic Potentiometers
See Also: Potentiometers (Resistors); Variable Resistors

—Multiturn—Preferred Products List
CECC CECC MUAHAG Vol 2 IS 4-90. Preferred Products List; Resistors and Potentiometers (En, Fr, Ge). 59 pp.

—Multiturn—Sealed—Preferred Products List
CECC CECC MUAHAG Vol 2 IS 4-90. Preferred Products List; Resistors and Potentiometers (En, Fr, Ge). 59 pp.

—Preferred Products List
CECC CECC MUAHAG Vol 2 IS 4-90. Preferred Products List; Resistors and Potentiometers (En, Fr, Ge). 59 pp.

—Preset—Multiturn—Surface Mount—Preferred Products List
CECC CECC MUAHAG Vol 2 IS 4-90. Preferred Products List; Resistors and Potentiometers (En, Fr, Ge). 59 pp.

—Preset—Single Turn—Surface Mount—Preferred Products List
CECC CECC MUAHAG Vol 2 IS 4-90. Preferred Products List; Resistors and Potentiometers (En, Fr, Ge). 59 pp.

—Rotary—Single Turn—Sealed—Preferred Products List
CECC CECC MUAHAG Vol 2 IS 4-90. Preferred Products List; Resistors and Potentiometers (En, Fr, Ge). 59 pp.

—Single Turn—Preferred Products List
CECC CECC MUAHAG Vol 2 IS 4-90. Preferred Products List; Resistors and Potentiometers (En, Fr, Ge). 59 pp.

—Single Turn—Sealed—Preferred Products List
CECC CECC MUAHAG Vol 2 IS 4-90. Preferred Products List; Resistors and Potentiometers (En, Fr, Ge). 59 pp.

Certificate of Airworthiness
See Also: Type Certificates

—Aircraft
CAA Part 2 CAP 396. Obtaining Certification (Civil Air Publications: Air Navigation—Order and Regulations). 20 pp.
CAA NOTICE #32 ISSUE 2. Overhauls, Modifications, Repairs and Replacements to Aircraft Not Exceeding 2730 kg with Certificates of Airworthiness in the Special Category (Airworthiness Notices). 2 pp.
CAA NOTICE #63 ISSUE 9. Certification and Maintenance of Aircraft Not Exceeding 2730 kg (Airworthiness Notices). 4 pp.
CAA. Contents—Section A 07.91. 5 pp.
CAA Chapter A3-2 07.89. Issue of Certificate of Airworthiness. 4 pp.
CAA Chapter A3-4 09.91. Renewal of Certifcate of Airworthiness. 4 pp.
CAA Chapter A3-8 07.89. 'A' Conditions. 2 pp.
CAA Chapter A3-9. 'B' Conditions. 2 pp.
CAA Chapter B1-2 07.89. Categories of Aircraft. 3 pp.
CAA Chapter B3-2 07.89. Issue of Certificate of Airworthiness. 3 pp.

INTERNATIONAL AND NON-U.S. NATIONAL STANDARDS
SUBJECT INDEX

Certificate

Certificate of Airworthiness (Cont.)
—Aircraft (Cont.)
- CAA Chapter B3-4 07.89. Renewal of Certificate of Airworthiness. 5 pp.
- CAA Chapter B3-6 09.91. Certificates of Airworthiness for Export from the UK. 2 pp.
- CAA Chapter B3-8 09.91. 'A' Conditions. 1 p.
- CAA Chapter B3-9 06.90. 'B' Conditions. 2 pp.
- CAA Chapter B5-1 06.90. General. 2 pp.
- CAA Supp TO Sec B 09.91. Sub-Section A8- Approvals. 110 pp.

—Aircraft Engines
- CAA NOTICE #16 ISSUE 9. Aircraft Engines, Propellers and Related Equipment Obtained from Sources Not Under the Airworthiness Control of the CAA (Airworthiness Notices). 6 pp.
- CAA Chapter B4-2 06.90. Type Certification or Validation of Engines.

—Aircraft Equipment
- CAA NOTICE #16 ISSUE 9. Aircraft Engines, Propellers and Related Equipment Obtained from Sources Not Under the Airworthiness Control of the CAA (Airworthiness Notices). 6 pp.
- CAA NOTICE #17 ISSUE 2. Acceptance of Aircraft Components (Appendices Included). 12 pp.
- CAA NOTICE #18 ISSUE 6. Acceptance Standards for the Maintenance Overhaul and Repair of Second-Hand Imported Aircraft for Which a UK C of A is Sought (Airworthiness Notices). 2 pp.
- CAA NOTICE #36 ISSUE 10. Mandatory Modifications and Inspections (Airworthiness Notices). 6 pp.
- CAA Chapter B5-7 09.91. Master Minimum Equipment List. 4 pp.
- CAA Chapter B6-5 09.91. Minimum Equipment Lists. 1 p.
- CAA Chapter B7-6 09.91. Minimum Equipment Lists. 1 p.
- CAA Chapter 6 03.92. Procedures for the Grant of JAR-145 Approval by the JAA National Aviation Authority (Airworthiness Notices). 4 pp.

—Aircraft—Export
- CAA Chapter A3-6 06.90. Certificates of Airworthiness for Export. 2 pp.

—Aircraft—Fees
- CAA NOTICE #25 ISSUE 19. Charges for the CAA's Airworthiness and Noise Certification (Airworthiness Notices). 13 pp.

—Aircraft—Flight Testing
- CAA Chapter A3-3 07.89. Flight Testing for Issue of Certificate of Airworthiness or a Permit to Fly. 4 pp.
- CAA Chapter A3-5 06.90. Flight Testing for Renewal of Certificates of Airworthiness or Permits to Fly. 6 pp.

—Aircraft—Imported
- CAA LEAFLET 11-4 12.90. UK Certification of Imported Aircraft Not Exceeding 2730 kg MTWA Which Are Eligible for the Issue of Airworthiness. 6 pp.

—Aircraft—Inspection and Repair
- CAA Chapter A6-7. Certification of Inspections, Overhauls, Modifications, Repairs and Replacements. 6 pp.

—Aircraft—Permit to Fly
- CAA Chapter A3-7 07.89. Issue and Renewal of Permits to Fly. 4 pp.
- CAA Chapter A3-7 App 1 09.91. Evidence to Substantiate Applications. 1 p.
- CAA Chapter A3-7 App 2 07.89. Flight Release Certificate. 1 p.

—Aircraft Propellers
- CAA JAR-P Section 3. Approval of Propellers and Associated Equipment (Joint Airworthiness Requirements). 4 pp.

—Aircraft—Radio
- CAA Chapter A3-11 07.89. Aircraft Radio Installations. 4 pp.
- CAA Chapter B3-11 07.89. Radio Installations. 3 pp.

—Auxiliary Power Units
- CAA. Contents (Joint Airworthiness Requirements). 1 p.
- CAA. Foreword. 1 p.
- CAA. Check List of Pages (Joint Airworthiness Requirements). 1 p.
- CAA JAR-APU Section 1. Regulations (Joint Airworthiness Requirements). 12 pp.
- CAA JAR-APU Section 2. Acceptable Means of Compliance and Interpretations (Joint Airworthiness Requirements). 4 pp.
- CAA JAR-APU Section 3. Advisory Material Joint—AMJ (Joint Airworthiness Requirements). 5 pp.

Certificate of Airworthiness (Cont.)
—Auxiliary Power Units (Cont.)
- CAA JAR-APU Section 4. Approval of Gas Turbine Auxiliary Power Units and Associated Equipment. 5 pp.

—Instrument Landing Systems
- CAA JAR-AWO SUBPART 3. Airworthiness Certification of Aeroplanes for Operations with Decision Heights Below 30M (100ft) or No Decision Height—Category 3 Operations (Joint Airworthiness Requirements). 10 pp.

Certificates
Scope Note: For specific types of certificates, see specific topics **See Also:** Certificate of Airworthiness; Experimental Certificates; Flight Testing; Type Certificates

—Bids
- MOD UK DEFCON 48-81. Bona-Fide Tendering Certificate (See Leaflet N3 of DCH) 8/81. 1 p.

—Contractor Employee Attendance
- MOD UK DEFCON 12-87. Certificate of Attendance of Contractor's Personnel 11/87. 2 pp.

—Testing Equipment
- MOD UK DEFCON 103-75. Special Test Equipment Certificate 10/75. 2 pp.

Certification
Scope Note: For specific types of certification, see specific topics **See Also:** Standards

- IEC Guide 7-82. Requirements for Standards Suitable for Product Certification First Edition. 7 pp.
- IEC Guide 16-78. Code of Principles on Third Party Certification Systems and Related Standards First Edition. 5 pp.
- IEC Guide 22-82. Information on Manufacturer's Declaration of Conformity with Standards or Other Technical Specifications First Edition. 7 pp.
- IEC Guide 23-82. Methods of Indicating Conformity with Standards for Third-Party Certification Systems First Edition. 7 pp.
- IEC Guide 28-82. General Rules for a Model Third-Party Certification System for Products First Edition. 19 pp.
- IEC Guide 40-83. General Requirements for the Acceptance of Certification Bodies First Edition. 6 pp.
- ISO Guide 7-82. Requirements for Standards Suitable for Product Certification First Edition. 7 pp.
- ISO Guide 16-78. Code of Principles on Third Party Certification Systems and Related Standards First Edition. 5 pp.
- ISO Guide 22-82. Information on Manufacturer's Declaration of Conformity with Standards or Other Technical Specifications First Edition. 7 pp.
- ISO Guide 23-82. Methods of Indicating Conformity with Standards for Third-Party Certification Systems First Edition. 7 pp.
- ISO Guide 27-83. Guidelines for Corrective Action to Be Taken by a Certification Body in the Event of Misuse of its Mark of Conformity First Edition; (Corrected and Reprinted -1983). 8 pp.
- ISO Guide 28-82. General Rules for a Model Third-Party Certification System for Products First Edition. 19 pp.
- ISO Guide 40-83. General Requirements for the Acceptance of Certification Bodies First Edition. 6 pp.
- ISO Guide 48-86. Guidelines for Third-Party Assessment and Registration of a Supplier's Quality System First Edition. 12 pp.
- ISO Guide 53-88. Approach to the Utilization of a Supplier's Quality System in Third Party Product Certification First Edition. 17 pp.
- SAA HB18. Guidelines for Third-Party Certification and Accreditation.
- SAA HB18.7-91. Guidelines for Third-Party Certification and Accreditation—Guide 7: Requirements for Standards Suitable for Product Certification (ISO/IEC Guide 7:1982) (SANZ HB18.7—1991). 4 pp.
- SAA HB18.16-91. Guidelines for Third-Party Certification and Accreditation—Guide 16: Code of Principles on Third Party Certification Systems and Related Standards (ISO/IEC Guide 16:1978) (SANZ HB18.16—1991). 2 pp.
- SAA HB18.22-91. Guidelines for Third-Party Certification and Accreditation—Guide 22: Information on Manufacturer's Declaration of Conformity with Standards or Other Technical Specifications (ISO/IEC Guide 22:1982) (SANZ HB18.22—1991). 4 pp.
- SAA HB18.23-91. Guidelines for Third-Party Certification and Accreditation—Guide 23: Methods of Indicating Conformity with Standards for Third-Party Certification Systems (ISO/IEC Guide 23:1982) (SANZ HB18.23—1991). 4 pp.

Certification (Cont.)
- SAA HB18.25-91. Guidelines for Third-Party Certification and Accreditation—Guide 25: General Requirements for the Competence of Calibration and Testing Laboratories (ISO/IEC Guide 25:1990) (SANZ HB18.25—1991). 7 pp.
- SAA HB18.27-91. Guidelines for Third-Party Certification and Accreditation—Guide 27: Guidelines for Corrective Action to be Taken by a Certification Body in the Event of Either Misapplication of its Mark of Conformity to a Product, or Products Which Bear the Mark of the. 5 pp.
- SAA HB18.27 CONT-91. Certification Body Being Found to Subject Persons or Property to Risk (ISO/IEC Guide 27:1983) (SANZ HB18.27—1991).
- SAA HB18.28-91. Guidelines for Third-Party Certification and Accreditation—Guide 28: General Rules for a Model Third-Party Certification System for Products (ISO/IEC Guide 28:1982) (SANZ HB18.28—1991). 16 pp.
- SAA HB18.40-91. Guidelines for Third-Party Certification and Accreditation—Guide 40: General Requirements for the Acceptance of Certification Bodies (ISO/IEC Guide 40:1983) (SANZ HB18.40—1991). 3 pp.
- SAA HB18.43-91. Guidelines for Third-Party Certification and Accreditation—Guide 43: Development and Operation of Laboratory Proficiency Testing (ISO/IEC Guide 43:1984) (SANZ HB18.43—1991). 6 pp.
- SAA HB18.48-91. Guidelines for Third-Party Certification and Accreditation—Guide 48: Guidelines for Third-Party Assessment and Registration of a Supplier's Quality System (ISO/IEC Guide 48:1986) (SANZ HB18.48—1991). 9 pp.
- SAA HB18.53-91. Guidelines for Third-Party Certification and Accreditation—Guide 53: An Approach to the Utilization of a Supplier's Quality System in Third Party Product Certification (ISO/IEC Guide 53:1988) (SANZ HB18.53—1991). 13 pp.
- SAA HB18.54-91. Guidelines for Third-Party Certification and Accreditation—Guide 54: Testing Laboratory Accreditation Systems—General Requirements for the Acceptance of Accreditation Bodies (ISO/IEC Guide 54:1988) (SANZ HB18.54—1991). 3 pp.
- SAA HB18.55-91. Guidelines for Third-Party Certification and Accreditation—Guide 55: Testing Laboratory Accreditation Systems—General Recommendations for Operation (ISO/IEC Guide 55:1988) (SANZ HB18.55—1991). 4 pp.
- SNZ SANZ/SAA HB 18.7-91. Guidelines for Third-Party Certification and Accreditation Guide 7 Requirements for Standards Suitable for Product Certification (ISO/IEC Guide 7-1982). 4 pp.
- SNZ SANZ/SAA HB 18.16-91. Guidelines for Third-Party Certification and Accreditation Guide 16 Code of Principles on Third-Party Certification Systems and Related Standards (ISO/IEC Guide 16-1978). 2 pp.
- SNZ SANZ/SAA HB 18.22-91. Guidelines for Third-Party Certification and Accreditation Guide 22 Information on Manufacturer's Declaration of Conformity with Standards or Other Technical Specifications (ISO/IEC Guide 22-1982). 4 pp.
- SNZ SANZ/SAA HB 18.23-91. Guidelines for Third-Party Certification and Accreditation Guide 23 Methods of Indicating Conformity with Standards for Third-Party Certification Systems (ISO/IEC Guide 23-1982). 4 pp.
- SNZ SANZ/SAA HB 18.27-91. Guidelines for Third-Party Certification and Accreditation Guide 27 Guidelines for Corrective Action to Be Taken by a Certification Body in the Event of Either Misapplication of Its Mark of Conformity to a Product, or Products Which Bear the Mark of the. 5 pp.
- SNZ SANZ/SAA HB 18.27 CONT-91. Certification Body Being Found to Subject Persons or Property to Risk (ISO/IEC Guide 27-1983).
- SNZ SANZ/SAA HB 18.28-91. Guidelines for Third-Party Certification and Accreditation Guide 28 General Rules for a Model Third-Party Certification System for Products (ISO/IEC Guide 28-1982). 16 pp.
- SNZ SANZ/SAA HB 18.40-91. Guidelines for Third-Party Certification and Accreditation Guide 40 General Requirements for the Acceptance of Certification Bodies (ISO/IEC Guide 40-1983). 3 pp.
- SNZ SANZ/SAA HB 18.48-91. Guidelines for Third-Party Certification and Accreditation Guide 48 Guidelines for Third-Party Assessment and Registration of a Supplier's Quality System (ISO/IEC Guide 48-1986). 9 pp.
- SNZ SANZ/SAA HB 18.53-91. Guidelines for Third-Party Certification and Accreditation Guide 53 an Approach to the Utilisation of a Supplier's Quality System in Third-Party Product Certification (ISO/IEC Guide 53-1988). 13 pp.

INDEX and DIRECTORY of

Certification (Cont.)

SNZ SANZ/SAA HB 18.54-88. Guidelines for Third-Party Certification and Accreditation Guide 54 Testing Laboratory Accreditation Systems—General Recommendations for the Acceptance of Accreditation Bodies (ISO/IEC Guide 54-1988). 3 pp.

—AECMA
EURO Publication LIST-92. Main A.E.C.M.A. Publications Revision 8. 12 pp.

—Binders (Files)
BSI BS 5641-82. 1982 Loose-Leaf Publications. 7 pp.

—CCIR
CCIR RESOLUTION 81-2-90. Handbooks and Special Publications—Volume XIV—Administrative Texts of the CCIR. 1 p.

—CECC
CECC CECC 00 600 ISSUE 1-82. Rules for the Use and Administration of the CECC Certification Mark (En, Fr, Ge). 57 pp.

—International Systems
IEC Guide 42-84. Guidelines for a Step-by-Step Approach to an International Certification System First Edition. 9 pp.
IEC Guide 44-85. General Rules for ISO or IEC International Third-Party Certification Schemes for Products First Edition. 16 pp.
ISO Guide 42-84. Guidelines for a Step-by-Step Approach to an International Certification System First Edition. 9 pp.
ISO Guide 44-85. General Rules for ISO or IEC International Third-Party Certification Schemes for Products First Edition. 16 pp.
SAA HB18.42-91. Guidelines for Third-Party Certification and Accreditation—Guide 42: Guidelines for a Step-by-Step Approach to an International Certification System (ISO/IEC Guide 42:1984) (SANZ HB18.42—1991). 6 pp.
SAA HB18.44-91. Guidelines for Third-Party Certification and Accreditation—Guide 44: General Rules for ISO or IEC International Third-Party Certification Schemes for Products (ISO/IEC Guide 44:1985) (SANZ HB18.44—1991). 13 pp.
SNZ SANZ/SAA HB 18.42-91. Guidelines for Third-Party Certification and Accreditation Guide 42 Guidelines for a Step-by-Step Approach to an International Certification System (ISO/IEC Guide 42-1984). 6 pp.
SNZ SANZ/SAA HB 18.44-85. Guidelines for Third-Party Certification and Accreditation Guide 44 General Rules for ISO or IEC International Third-Party Certification Schemes for Products (ISO/IEC Guide 44-1985). 13 pp.

—Quality Control
ISO Guide 56-89. Approach to the Review by a Certification Body of Its Own Internal Quality System First Edition. 7 pp.
SAA HB18.56-91. Guidelines for Third-Party Certification and Accreditation—Guide 56: An Approach to the Review by a Certification Body of Its Own Internal Quality System (ISO/IEC Guide 56:1989) (SANZ HB18.56—1991). 4 pp.
SNZ SANZ/SAA HB 18.56-88. Guidelines for Third-Party Certification and Accreditation Guide 56 an Approach to the Review by a Certification Body of Its Own Internal Quality System (ISO/IEC Guide 56-1989). 4 pp.

Certified Reference Materials
Use: Calibration Reference Standards

Cesium Chloride
See Also: Radioactive Cesium Content Analysis
CNS K7550-88. Chemical Reagent (Cesium Chloride) (Oct)(7708).

Cessna F 406 Caravan II Aircraft
See Also: Aircraft
—Certification
CAA. Reims Cessna F406—Caravan II. 10 pp.

Cessna S 550 Aircraft
See Also: Aircraft
—Certification
CAA. Cessna S550—Citation SII. 13 pp.

Cessna Series Aircraft
See Also: Aircraft
—Foreign Airworthiness Directives
CAA. Cessna Series Aircraft (Foreign Airworthiness Directives). 18 pp.

Cessna T 303 Aircraft
See Also: Aircraft

Cessna T 303 Aircraft (Cont.)
—Antenna Positions
CAA. Cessna T303 (Approved Aerial Positions). 1 p.

Cessna 120 Aircraft
See Also: Aircraft
—Antenna Positions
CAA. Cessna 120 and 140 (Approved Aerial Positions). 1 p.

Cessna 140 Aircraft
See Also: Aircraft
—Antenna Positions
CAA. Cessna 120 and 140 (Approved Aerial Positions). 1 p.

Cessna 150 Aircraft
See Also: Aircraft
—Antenna Positions
CAA. Cessna 150 and 152 (Approved Aerial Positions). 1 p.

Cessna 152 Aircraft
See Also: Aircraft
—Antenna Positions
CAA. Cessna 150 and 152 (Approved Aerial Positions). 1 p.

Cessna 170 Aircraft
See Also: Aircraft
—Antenna Positions
CAA. Cessna 170/172/175/177/180/182/185/205/206/210 Series (Approved Aerial Positions). 1 p.

Cessna 172 Aircraft
See Also: Aircraft
—Antenna Positions
CAA. Cessna 170/172/175/177/180/182/185/205/206/210 Series (Approved Aerial Positions). 1 p.

Cessna 175 Aircraft
See Also: Aircraft
—Antenna Positions
CAA. Cessna 170/172/175/177/180/182/185/205/206/210 Series (Approved Aerial Positions). 1 p.

Cessna 177 Aircraft
See Also: Aircraft
—Antenna Positions
CAA. Cessna 170/172/175/177/180/182/185/205/206/210 Series (Approved Aerial Positions). 1 p.

Cessna 180 Aircraft
See Also: Aircraft
—Antenna Positions
CAA. Cessna 170/172/175/177/180/182/185/205/206/210 Series (Approved Aerial Positions). 1 p.

Cessna 182 Aircraft
See Also: Aircraft
—Antenna Positions
CAA. Cessna 170/172/175/177/180/182/185/205/206/210 Series (Approved Aerial Positions). 1 p.

Cessna 185 Aircraft
See Also: Aircraft
—Antenna Positions
CAA. Cessna 170/172/175/177/180/182/185/205/206/210 Series (Approved Aerial Positions). 1 p.

Cessna 205 Aircraft
See Also: Aircraft
—Antenna Positions
CAA. Cessna 170/172/175/177/180/182/185/205/206/210 Series (Approved Aerial Positions). 1 p.

Cessna 206 Aircraft
See Also: Aircraft
—Antenna Positions
CAA. Cessna 170/172/175/177/180/182/185/205/206/210 Series (Approved Aerial Positions). 1 p.

Cessna 207 Aircraft
See Also: Aircraft
—Antenna Positions
CAA. Cessna 207 (Approved Aerial Positions). 1 p.

Cessna 208 Caravan 1
See Also: Aircraft
—Certification
CAA. Cessna Model 208 Caravan 1. 16 pp.

Cessna 210 Aircraft
See Also: Aircraft
—Antenna Positions
CAA. Cessna 170/172/175/177/180/182/185/205/206/210 Series (Approved Aerial Positions). 1 p.

Cessna 300 Aircraft
See Also: Aircraft
—Airworthiness Notices
CAA Chapter 7 03.92. Procedures for the Renewal of the JAR-145 Approval by the JAA National Aviation Authority (Airworthiness Notices). 1 p.

Cessna 310 Aircraft
See Also: Aircraft
—Antenna Positions
CAA. Cessna 310 and 320 (Approved Aerial Positions). 1 p.

Cessna 320 Aircraft
See Also: Aircraft
—Antenna Positions
CAA. Cessna 310 and 320 (Approved Aerial Positions). 1 p.

Cessna 335 Aircraft
See Also: Aircraft
—Antenna Positions
CAA. Cessna 335 and 340 (Approved Aerial Positions). 1 p.

Cessna 336 Aircraft
See Also: Aircraft
—Antenna Positions
CAA. Cessna 336/337 (Approved Aerial Positions). 1 p.

Cessna 337 Aircraft
See Also: Aircraft
—Antenna Positions
CAA. Cessna 336/337 (Approved Aerial Positions). 1 p.

Cessna 340 Aircraft
See Also: Aircraft
—Antenna Positions
CAA. Cessna 335 and 340 (Approved Aerial Positions). 1 p.

Cessna 400 Aircraft
See Also: Aircraft
—Airworthiness Notices
CAA Chapter 7 03.92. Procedures for the Renewal of the JAR-145 Approval by the JAA National Aviation Authority (Airworthiness Notices). 1 p.

Cessna 401 Aircraft
See Also: Aircraft
—Antenna Positions
CAA. Cessna 401/402/404/411/414/421/425/441 (Approved Aerial Positions). 1 p.
—Certification
CAA. Cessna 401, 402, 402A, 402B and 411. 7 pp.

Cessna 402 Aircraft
See Also: Aircraft
—Antenna Positions
CAA. Cessna 401/402/404/411/414/421/425/441 (Approved Aerial Positions). 1 p.
—Certification
CAA. Cessna 401, 402, 402A, 402B and 411. 7 pp.

Cessna 402A Aircraft
See Also: Aircraft
—Certification
CAA. Cessna 401, 402, 402A, 402B and 411. 7 pp.

Cessna 402B Aircraft
See Also: Aircraft
—Certification
CAA. Cessna 401, 402, 402A, 402B and 411. 7 pp.

Cessna

Cessna 402C Aircraft
See Also: Aircraft
—Certification
CAA. Cessna Model 402C. 7 pp.

Cessna 404 Aircraft
See Also: Aircraft
—Antenna Positions
CAA. Cessna 401/402/404/411/414/421/425/441 (Approved Aerial Positions). 1 p.
—Certification
CAA. Cessna 404. 8 pp.

Cessna 406 Aircraft
See Also: Aircraft
—Antenna Positions
CAA. Cessna 406. 1 p.

Cessna 411 Aircraft
See Also: Aircraft
—Antenna Positions
CAA. Cessna 401/402/404/411/414/421/425/441 (Approved Aerial Positions). 1 p.
—Certification
CAA. Cessna 401, 402, 402A, 402B and 411. 7 pp.

Cessna 414 Aircraft
See Also: Aircraft
—Antenna Positions
CAA. Cessna 401/402/404/411/414/421/425/441 (Approved Aerial Positions). 1 p.

Cessna 414A Chancellor Aircraft
See Also: Aircraft
—Certification
CAA. Cessna 414A—Chancellor. 8 pp.

Cessna 421 Aircraft
See Also: Aircraft
—Antenna Positions
CAA. Cessna 401/402/404/411/414/421/425/441 (Approved Aerial Positions). 1 p.

Cessna 421B Aircraft
See Also: Aircraft
—Certification
CAA. Cessna 421B. 8 pp.

Cessna 425 Aircraft
See Also: Aircraft
—Antenna Positions
CAA. Cessna 401/402/404/411/414/421/425/441 (Approved Aerial Positions). 1 p.
—Certification
CAA. Cessna 425—Corsair. 11 pp.

Cessna 441 Aircraft
See Also: Aircraft
—Antenna Positions
CAA. Cessna 401/402/404/411/414/421/425/441 (Approved Aerial Positions). 1 p.

Cessna 500 Aircraft
See Also: Aircraft
—Antenna Positions
CAA. Cessna 500 (Approved Aerial Positions). 1 p.
—Certification
CAA. Cessna 500—Citation. 10 pp.
—Data Sheets
CAA FA13 ISSUE 2. Cessna Models 500 and 550. 4 pp.

Cessna 500 Citation I Aircraft
See Also: Aircraft
—Accidents
CAA. Cessna 500 and 600 Series Citation I, II and III (World Airline Accident Summary). 1 p.

Cessna 500 Citation II Aircraft
See Also: Aircraft
—Accidents
CAA. Cessna 500 and 600 Series Citation I, II and III (World Airline Accident Summary). 1 p.

Cessna 500 Citation III Aircraft
See Also: Aircraft
—Accidents
CAA. Cessna 500 and 600 Series Citation I, II and III (World Airline Accident Summary). 1 p.

Cessna 550 Aircraft
See Also: Aircraft
—Antenna Positions
CAA. Cessna 550 (Approved Aerial Positions). 1 p.
—Data Sheets
CAA FA13 ISSUE 2. Cessna Models 500 and 550. 4 pp.

Cessna 550 Citation II Aircraft
See Also: Aircraft
—Certification
CAA. Cessna 550—Citation II. 8 pp.

Cessna 560 Citation V Aircraft
See Also: Aircraft
—Certification
CAA. Cessna 560—Citation V. 12 pp.

Cessna 600 Citation I Aircraft
See Also: Aircraft
—Accidents
CAA. Cessna 500 and 600 Series Citation I, II and III (World Airline Accident Summary). 1 p.

Cessna 600 Citation II Aircraft
See Also: Aircraft
—Accidents
CAA. Cessna 500 and 600 Series Citation I, II and III (World Airline Accident Summary). 1 p.

Cessna 600 Citation III Aircraft
See Also: Aircraft
—Accidents
CAA. Cessna 500 and 600 Series Citation I, II and III (World Airline Accident Summary). 1 p.

Cessna 650 Aircraft
See Also: Aircraft
—Antenna Positions
CAA. Cessna 650 Serial No. 650-0179 and on. 1 p.

Cessna 650 Citation III Aircraft
See Also: Aircraft
—Certification
CAA. Cessna 650—Citation III. 13 pp.

Cesspools
See Also: Sewage Systems; Sewage Treatment
—Design
BSI BS 6297-83. 1983 Amd 1 Design and Installation of Small Sewage Treatment Works and Cesspools (AMD 6150) December 21, 1990. 38 pp.
—Installation
BSI BS 6297-83. 1983 Amd 1 Design and Installation of Small Sewage Treatment Works and Cesspools (AMD 6150) December 21, 1990. 38 pp.

Cetane Number
See Also: Diesel Fuels
JIS K 2280-86. Testing Methods for Octane Number and Cetane Number. 60 pp.
—Diesel Fuels
BSI BS 2000: Part 218-82. (OBSOLESCENT) 1982 Amd 1 Petroleum and Its Products Part 218: Calculation of Cetane Index of Diesel Fuels (Range 55 and Above). 5 pp.
BSI BS 2000: Part 364-85. 1985 Petroleum and Its Products Part 364: Calculated Cetane Index of Diesel Fuels. 6 pp.
BSI BS 5580-78. 1978 Diesel Fuels. Determination of Ignition Quality. Cetane Method. 4 pp.
DIN ENGL 51773-71. Testing of Liquid Fuels; Determination of Ignition Quality (Cetane Number) of Diesel Fuels (July). 4 pp.
ISO 5165-92. Diesel Fuels—Determination of Ignition Quality—Cetane Method Second Edition. 11 pp.
—Distillate Fuels
CNS K6915-87. Method of Calculated Cetane Index of Distillate Fuels (Jul)(12016).
CNS K6972-90. Method for Calculated Cetane Index of Distillate Fuels (Four Variable Equation Method) (Aug)(12761).

CFB
Use: Call Forwarding Busy

CFC
Use: Chlorofluorocarbons

CFNR
Use: Call Forwarding No Reply

CFU
Use: Call Forwarding Unconditional

CGI
Use: Computer Graphics Interface

CGSB
Use For: Canadian General Standards Board
See Also: Standards
—Catalogs
CGSB. 1993 Catalogue. 134 pp.

Chaffcutter Saws
Use: Combines—Chaffcutter Saws

Chain Assemblies
Use: Chains

Chain Cases
Use For: Chain Guards *See Also:* Guards (Protective)
—Aircraft—Sprockets
SBAC RS 407 ISSUE 1. Relation of Sprocket to Guard Non Reversible Chain.
—Aircraft—Trim Controls
SBAC AS 884 ISSUE 2. Chain Guard—Trim Control. 1 p.
SBAC AS 6259 ISSUE 1. Chain Guard—Trim Control. 1 p.
—Bicycles
JIS D 9454-83. Chain Cases for Bicycles.

Chain Conveyors
See Also: Chains; Conveyors; Materials Handling Equipment; Slat Conveyors
BSI BS 6849-87. 1987 Drop-Forged Rivetless Chains for Conveyors. 7 pp.
CNS B2156-77. Conveyer Chains, Attachments and Chain Wheels (Oct)(4173).
ISO 6973-86. Drop-Forged Rivetless Chains for Conveyors First Edition. 6 pp.
—Attachments
BSI BS 4116: Part 3-92. 1992 Conveyor Chains, Their Attachments and Associated Chain Wheels Part 3: Specification of Attachments (Metric Series) (ISO 1977/3: 1974) (Q). 7 pp.
BSI BS 4116-71. 1971 Steel Roller Chains, Chainwheels and Attachments for Conveyors. 20 pp.
DIN ENGL 8167 Pt 2-86. ISO-Type M Conveyor Chains with Solid Pins; Attachments; Material and Connecting Dimensions (Mar). 3 pp.
DIN ENGL 8168 Pt 2-86. ISO-Type MC Conveyor Chains with Hollow Pins; Attachments; Material and Connecting Dimensions (Mar). 2 pp.
ISO 1977 Pt III-74. Conveyor Chains, Attachments and Chain Wheels—Part III: Attachments—Metric Series First Edition. 5 pp.
SNZ NZS/BS 4116-71. Specification for Steel Roller Chains, Chainwheels and Attachments for Conveyors. 20 pp.
—Link Chains
BSI BS 2969-80. 1980 High-Tensile Steel Chains (Round Link) for Chain Conveyors and Coal Ploughs. 22 pp.
CNS B2607-81. Round Steel Link Chains for Conveyer, Long Link (Oct)(7940).
CNS M2043-80. Round Link Chain for Mine Double Chain Conveyor (Mar)(7939).
DIN ENGL 762 Pt 1-82. Calibrated and Tested Round Steel Link Chains for Continuous Conveyors; Grade 2, Pitch 5d (Jan). 4 pp.
DIN ENGL 762 Pt 2-82. Calibrated and Tested Round Steel Link Chains for Continuous Conveyors; Grade 3, Pitch 5d (Jan). 4 pp.
DIN ENGL 764 Pt 1-82. Calibrated and Tested Round Steel Link Chains for Continuous Conveyors; Grade 2, Pitch 3,5d (Jan). 3 pp.
DIN ENGL 764 Pt 2-82. Calibrated and Tested Round Steel Link Chains for Continuous Conveyors; Grade 3, Pitch 3,5d (Jan). 3 pp.
ISO 610-90. High-Tensile Steel Chains (Round Link) for Chain Conveyors and Coal Ploughs Second Edition. 21 pp.
JIS M 6504-76. Round Link Chain for Mine Double Chain Conveyors.

INTERNATIONAL AND NON-U.S. NATIONAL STANDARDS
SUBJECT INDEX
Chain

Chain Conveyors (Cont.)
—Roller Chains
BSI BS 4116: Part 1-92. 1992 Conveyor Chains, Their Attachments and Associated Chain Wheels Part 1: Specification for Chains (Metric Series) (ISO 1977/1: 1976) (SUPERSEDES BS 4116: 1971). 11 pp.
BSI BS 4116-71. 1971 Steel Roller Chains, Chainwheels and Attachments for Conveyors. 20 pp.
ISO 1977 Pt I-76. Conveyor Chains, Attachments and Chain Wheels—Part I: Chains—Metric Series First Edition. 7 pp.
SNZ NZS/BS 4116-71. Specification for Steel Roller Chains, Chainwheels and Attachments for Conveyors. 20 pp.

—Rollers
DIN ENGL 8169-86. Rollers for ISO-Types M, MT, MC and MCT Conveyor Chains with Solid and Hollow Pins (Mar). 2 pp.

—Safety
BSI BS 5667: Part 15-79. 1979 Continuous Mechanical Handling Equipment. Safety Requirements Part 15: Unit Loads: Crate Carrying Chain Conveyors Having Biplanar Chains for Flat-Bottomed Unit Loads. 3 pp.
ISO 5041-77. Continuous Mechanical Handling Equipment for Unit Loads—Crate-Carrying Chain Conveyors Having Biplanar Chains for Flat-Bottomed Unit Loads—Safety Code First Edition. 3 pp.
ISO TR5046-77. Continuous Mechanical Handling Equipment—Safety Code for Conveyors and Elevators with Chain-Elements—Examples for Guarding of Nip Points First Edition. 23 pp.
ISO TR5047-82. Continuous Mechanical Handling Equipment—Chain Conveyors with Bearing Devices or Load Carriers—Examples of Protection Against Injuries by Load Carriers First Edition. 20 pp.

—Scraper Bars—Mining Equipment
BSI BS 6591-85. 1985 Scraper Bars for Chain Conveyors for Mining. 8 pp.
ISO 5612-90. Mining—Scraper Bars for Chain Conveyors Second Edition. 8 pp.

—Scrapers
CNS M2041-80. Scraper for Mine Double Chain Conveyor (Mar) (5371).
JIS M 6552-75. Scraper for Mine Double Chain Conveyor.

—Shackles
BSI BS 4831-85. 1985 Shackle Type Connector Units for Chain Conveyors for Mining. 9 pp.
CNS M2040-82. Connector Shackle for Double Chains Conveyor and High-Tensile Round-Link Steel Chains for Mining (Sep)(5370).
CNS M3100-82. Method of Test for Connector Shackle for Double Chains Conveyor and High-Tensile Round-Link Steel Chains for Mining (Sep)(9418).
ISO 1082-90. Mining—Shackle Type Connector Units for Chain Conveyors Second Edition. 8 pp.
JIS M 6551-75. Shackle for Mine Double Chain Conveyor.

—Sprockets
BSI BS 4116-71. 1971 Steel Roller Chains, Chainwheels and Attachments for Conveyors. 20 pp.
BSI BS 5801-79. 1979 Flat-Top Chains and Associated Chainwheels for Conveyors. 12 pp.
BSI BS 6592-85. 1985 Drive Sprocket Assemblies for Chain Conveyors for Mining. 10 pp.
ISO 4348-83. Flat-Top Chains and Associated Chain Wheels for Conveyors Second Edition. 11 pp.
ISO 5613-84. Mining—Drive Sprocket Assemblies for Chain Conveyors First Edition. 9 pp.
ISO TR8865-90. Mining—Guidance on Methods of Verifying Dimensions of Sprocket Assemblies for Chain Conveyors First Edition. 6 pp.
SNZ NZS/BS 4116-71. Specification for Steel Roller Chains, Chainwheels and Attachments for Conveyors. 20 pp.

—Troughs
CNS M2042-80. Trough for Mine Double Chain Conveyor (Mar) (5372).
CNS M2044-80. V Type Trough Conveyor Chain for Mine (Mar) (5374).
CNS M2045-80. H Type Trough Conveyor Chain for Mine (Mar)(5375).
JIS M 6553-75. Trough for Mine Double Chain Conveyor.

Chain Guards
Use: Chain Cases

Chain Hoists
Use: Hoists

Chain Link Fences
BSI BS 1722: Part 1-86. 1986 Fences Part 1: Chain Link Fences. 30 pp.

Chain Link Fences (Cont.)
BSI BS 1722: Part 10-90. 1990 Fences Part 10: Anti-Intruder Fences in Chain Link and Welded Mesh. 29 pp.
BSI BS 1722: Part 10-72. (WITHDRAWN) 1972 Amd 2 Fences Part 10: Anti-Intruder Chain Link Fences. 24 pp.
SAA AS 1725-75. Galvanized Rail-Less Chainwire Security Fences and Gates. 16 pp.

—Framework—Galvanized Steel
CGSB CAN/CGSB-138.2-M80. Fence, Chain Link, Framework, Zinc-Coated, Steel; (Amendment 1 June 1982). 16 pp.

—Gates
CGSB CAN/CGSB-138.4-M82. Fence, Chain Link, Gates. 12 pp.

—Installation
CGSB CAN/CGSB-138.3-M80. Fence, Chain Link—Installation. 12 pp.

—Tennis Courts
BSI BS 1722: Part 13-78. 1978 Fences Part 13: Chain Link Fences for Tennis Court Surrounds. 7 pp.

—Wire Cloth
CGSB CAN/CGSB-138.1-M80. Fence, Chain Link, Fabric. 14 pp.

—Wire Cloth—Low Carbon Steel
CNS G3176-86. Chain Link Wire Netting (Jun)(8826).
JIS G 3552-84. Chain Link Wire Netting. 9 pp.

Chain Pulley Blocks
See Also: Materials Handling Equipment; Pulley Blocks
BSI BS 3243-90. 1990 Hand-Operated Chain Blocks. 14 pp.
BSI BS 3243-01. 1990 Amd 1 Hand-Operated Chain Blocks (Q) (AMD 6778) October 31, 1991. 15 pp.

—Naval Ships
MOD UK NES 113: Part 10-92. Requirements for Mechanical Handling Part 10: Non-Traversing Hand Operated Chain Blocks and Their Associated Non-Geared Overhead Runway Trolleys Issue 1 (06.92). 34 pp.

—Steering Gear
CNS F3104-81. Leading Block for Chain Type Hand Steering Gear (May)(7369).

—Submarines
MOD UK NES 113: Part 10-92. Requirements for Mechanical Handling Part 10: Non-Traversing Hand Operated Chain Blocks and Their Associated Non-Geared Overhead Runway Trolleys Issue 1 (06.92). 34 pp.

Chain Saws
Use For: Chainsaws See Also: Forestry Equipment; Saws; Woodworking Equipment
BSI BS 2769: Sec 2.12-91. 1991 Hand-Held Electric-Motor-Operated Tools Part 2: Particular Requirements Section 2.12: Chain Saws. 12 pp.
BSI BS 6916: Part 2-88. 1988 Chain Saws Part 2: Method for Specifying Requirements (ISO 6532: 1982). 4 pp.
CENELEC HD 400.3 S2-89. Hand Held Motor Operated Tools Part II: Particular Specifications Section L: Chain Saws. 5 pp.
EC 89/392/EEC-89. Council Directive on the Approximation of the Laws of the Member States Relating to Machinery (NEW APP) (for Rel Stds See SM 1) (for Rel Stds See TEST/89/392, TEST/89/392 A1, TEST/89/392 A2). 24 pp.
EC TEST/89/392-89. Commission Communication in the Framework of the Implementation of Council Directive 89/392/EEC of 14 June 1989 in Relation to Machinery, as Amended by Directive 91/368/EEC of 20 June 1991. 3 pp.
EC TEST/89/392 A1-89. AMD 1 Commission Communication in the Framework of the Implementation of Council Directive 89/392/EEC of 14 June 1989 in Relation to Machinery, as Amended by Council Directive 91/368/EEC of 20 June 1991. 1 p.
EC TEST/89/392 A2-89. AMD 2 Commission Communication in the Framework of the Implementation of Council Directive 89/392/EEC of 14 June 1989 in Relation to Machinery, as Amended by Council Directive 91/368/EEC of 20 June 1991. 1 p.
EC TEST/89/398-89. List of Competent Authorities of the Member States Within the Meaning of Article 9 of Council Directive 89/398/EEC on the Approximation of the Laws of the Member States Relating to Foodstuffs Intended for Foods for Particular Nutritional Uses. 2 pp.
EC 91/368/EEC-91. Council Directive Amending Directive 89/392/EEC on the Approximation of the Laws of the Member States Relating to Machinery (NEW APP) (for Rel Stds See SM 1). 18 pp.

Chain Saws (Cont.)
EC 93/68/EEC-93. Council Directive Amending Directives 87/404/EEC (Simple Pressure Vessels), 88/378/EEC (Safety of Toys), 89/106/EEC (Construction Products), 89/336/EEC (Electromagnetic Compatibility), 89/392/EEC (Machinery), 89/686/EEC (Personal Protective). 22 pp.
ISO 6532-82. Machinery for Forestry—Portable Chain Saws—Technical Data First Edition. 4 pp.

—Balancing
BSI BS 6916: Part 10-88. 1988 Chain Saws Part 10: Specification for Longitudinal Balance (ISO 8334: 1985). 3 pp.
ISO 8334-85. Forestry Machinery—Portable Chain-Saws—Determination of Balance First Edition. 3 pp.

—Chain Catchers
ISO 10726-92. Portable Chain-Saws—Chain Catcher—Dimensions and Mechanical Strength First Edition. 5 pp.

—Chain Catchers—Mechanical Properties
ISO 10726-92. Portable Chain-Saws—Chain Catcher—Dimensions and Mechanical Strength First Edition. 5 pp.

—Electric
CSA CAN/CSA-C22. 2 NO 147-M90. Motor-Operated Gardening Appliances; (Gen Instr 1 Thru 3). 85 pp.

—Glossaries
BSI BS 6916: Part 1-88. 1988 Chain Saws Part 1: Glossary of Terms. 11 pp.
ISO 6531-89. Machinery for Forestry—Portable Chain Saws—Vocabulary First Edition. 13 pp.

—Guards (Protective)
BSI BS 6916: Part 5-88. 1988 Chain Saws Part 5: Methods of Evaluating Chain Brake Performance (ISO 6535: 1983). 4 pp.
ISO 6535-91. Portable Chain-Saws—Chain Brake Performance Second Edition. 5 pp.

—Guards (Protective)—Impact Testing
BSI BS 6916: Part 4-88. 1988 Chain Saws Part 4: Specification for Strength of Front Hand Guards (ISO 6534: 1985). 4 pp.
ISO 6534-92. Portable Chain-Saws—Hand-Guards—Mechanical Strength Second Edition. 6 pp.

—Hand Guards
BSI BS 6916: Part 3-88. 1988 Chain Saws Part 3: Specification for Dimensions of Front Hand Guards (ISO 6533: 1983). 4 pp.
ISO 6533-93. Forestry Machinery—Portable Chain-Saw Front Hand-Guard—Dimensions Second Edition. 6 pp.

—Handles
ISO 7914-86. Forestry Machinery—Portable Chain Saws—Minimum Handle Clearance and Sizes First Edition. 5 pp.

—Handles—Static Loads
BSI BS 6916: Part 9-88. 1988 Chain Saws Part 9: Specification for Strength of Handles (ISO 7915: 1985). 4 pp.
ISO 7915-91. Forestry Machinery—Portable Chain-Saws—Determination of Handle Strength Second Edition. 4 pp.

—Internal Combustion Engines
BSI BS 6916: Part 7-88. 1988 Chain Saws Part 7: Method of Evaluating Engine Performance Including Fuel Consumption (ISO 7293: 1983). 6 pp.
BSI BS 6916: Part 11-89. 1989 Chain Saws Part 11: Specification for General Design, Performance, Safety and Construction Requirements for Internal Combustion Engined Machines. 10 pp.
ISO 7293-83. Forestry Machinery—Portable Chain Saws—Engine Performance and Fuel Consumption First Edition. 5 pp.

—Kickback
CSA Z62.3-M1990. Chain Saw Kickback; (Gen Instr 1). 48 pp.
ISO TR9412-91. Portable Chain-Saws—Automatic Chain Brake and Cutting Equipment—Operator's Safety Test First Edition. 11 pp.
ISO 9518-92. Forestry Machinery—Portable Chain-Saws—Kickback Test First Edition. 40 pp.

—Noise
BSI BS 6916: Part 6-88. 1988 Chain Saws Part 6: Method of Measurement of Airborne Noise at the Operator's Position (ISO 7182: 1984). 8 pp.
BSI BS 6916: Part 6-01. 1988 Amd 1 Chain Saws Part 6: Method of Measurement of Airborne Noise at the Operator's Position (ISO 7182: 1984) (AMD 6872) January 31, 1992. 12 pp.
CEN EN 27182-91. Acoustics—Measurement at the Operator's Position of Airborne Noise Emitted by Chain Saws. 3 pp.

INTERNATIONAL AND NON-U.S. NATIONAL STANDARDS
SUBJECT INDEX

Chain

Chain Saws (Cont.)
—Noise (Cont.)
DIN ENGL EN 27182-91. Acoustics; Measurement, at Operator's Position, of Airborne Noise Emitted by Hand-Held Chain Saws (ISO 7182: 1984) (Oct). 6 pp.
ISO 7182-84. Acoustics—Measurement at the Operator's Position of Airborne Noise Emitted By Chain Saws First Edition. 5 pp.

—Protective Clothing
BSI BS EN 381-1-93. 1993 Protective Clothing for Users of Hand-Held Chain Saws Part 1: Test Rig for Testing Resistance to Cutting by a Chain Saw (N). 16 pp.
CEN PREN 381 (Part 1)-90. Protective Clothing for Users of Hand Held Chain Saws—Part 1: Test Rig for Testing Resistance to Cutting by a Chain Saw. 12 pp.
CEN EN 381-1-93. Protective Clothing for Users of Hand-Held Chain Saws—Part 1: Test Rig for Testing Resistance to Cutting by a Chain Saw. 10 pp.
CEN PREN 381 (Part 2)-90. Protective Clothing for Users of Hand Held Chain Saws—Part 2: Test Rig for Leg Protection. 9 pp.
CEN PREN 381 (Part 3)-90. Protective Clothing for Users of Hand Held Chain Saws—Part 3: Test Method for Boots. 14 pp.
CEN PREN 381-4-92. Protective Clothing for Users of Hand-Held Chainsaws—Part 4: Test Method for Chainsaw Protective Gloves. 13 pp.
CEN PREN 381 (Part 5)-90. Protective Clothing for Users of Hand Held Chain Saws—Part 5: Requirements for Leg Protection. 8 pp.
CEN PREN 381 (Part 6)-90. Protective Clothing for Users of Hand Held Chain Saws—Part 6: Requirements for Boots. 8 pp.
CEN PREN 381-7-92. Protective Clothing for Users of Hand-Held Chainsaws—Part 7: Requirements for Chainsaw Protective Gloves. 7 pp.
CEN PREN 381-8-91. Protective Clothing for Users of Hand-Held Chainsaws: Part 8—Test Method for Chainsaw Protective Gaiters. 15 pp.
CEN PREN 381-9-91. Protective Clothing for Users of Hand-Held Chainsaws: Part 9—Requirements for Chainsaw Protective Gaiters. 9 pp.
SNZ NZS 5840-88. Specification for Protective Legwear for Chain Saw Users. 14 pp.

—Safety
BSI BS 2769: Sec 2.12-91. 1991 Hand-Held Electric-Motor-Operated Tools Part 2: Particular Requirements Section 2.12: Chain Saws. 12 pp.
CEN PREN 608-91. Safety Requirements for Agricultural and Forestry Machinery—Portable Chain Saws. 23 pp.
CENELEC HD 400.3-81. Hand-Held Motor Operated Tools—Part 2: Particular Specifications. 28 pp.
CSA CAN3-Z62.1-M85. Chain Saws; (Gen Instr 2). 41 pp.
DIN ENGL 38822-81. Woodworking Machines; Portable Cutter Bar Chain Sawing Machines for One-Man Operation; Safety Requirements and Testing (Aug). 8 pp.
IEC 745 Pt 2-13-89. Safety of Hand-Held Motor-Operated Electric Tools Part 2: Particular Requirements for Chain Saws First Edition; (Amendment 1-1992). 32 pp.
ISO TR9412-91. Portable Chain-Saws—Automatic Chain Brake and Cutting Equipment—Operator's Safety Test First Edition. 11 pp.
SNZ NZS 5819-82. Chain Saw Safety (Reconfirmed 1989). 12 pp.

—Sound Pressure Levels
DIN ENGL EN 27182-91. Acoustics; Measurement, at Operator's Position, of Airborne Noise Emitted by Hand-Held Chain Saws (ISO 7182: 1984) (Oct). 6 pp.

—Vibration
BSI BS 6916: Part 8-88. 1988 Chain Saws Part 8: Method of Measurement of Hand-Transmitted Vibration (ISO 7505: 1986). 12 pp.
DIN ENGL 45675 Pt 2-87. Exposure to Mechanical Vibration Transmitted to the Hand-Arm System; Measurement of Vibration Induced by Portable Chain Saws (Sept). 3 pp.
ISO 7505-86. Forestry Machinery—Chain Saws—Measurement of Hand-Transmitted Vibration First Edition. 10 pp.

Chain Slings
Use: Hoisting Slings

Chain Stoppers
Use: Cable Stoppers

Chain Wheels
Use: Sprockets

Chain Wrenches
See Also: Adjustable Wrenches; Pipe Wrenches; Wrenches
CGSB 39-GP-48B-82. Wrenches, Pipe, Standard for. 24 pp.
CNS B3268-85. Chain Pipe Wrenches (Feb)(4103).
CNS B3268-84. Chain Pipe Wrenches (Oct)(4103). 3 pp.

Chains
Use For: Chain Assemblies *See Also:* Bush Chains; Chain Conveyors; Conveyor Chains; Cranked Link Chains; Door Chains; Drag Chains; Fasteners; Lifting Chains; Link Chains; Roller Chains; Tire Chains; Towing Chains
CNS B2136-72. 16mm x 4267mm Chain (Tentative) (Jun)(3369).

—Agricultural Equipment
CNS B4032-82. Feed Chains for Agricultural Machinery (Jan)(6007).
JIS B 9204-81. Feed Chains for Agricultural Machinery (R 1987). 11 pp.

—Aircraft
CAA LEAFLET 5-4 07.90. Control Chains, Chain Wheels and Pulleys. 8 pp.
MOD UK DTD-1090A-54. Testing of Aircraft Chain Assemblies. 2 pp.
NATO STANAG 3774 ED 3 AMD 0-87. Control Procedures for Pallets and Associated Restraint Equipment Used in Combined Air Transport Operations. 5 pp.
NATO STANAG 3774 ED 3 AMD 3-87. Control Procedures for Pallets and Associated Restraint Equipment Used in Combined Air Transport Operations. 10 pp.
SBAC RS 413 ISSUE 1. Non-Reversible Chains.

—Aircraft—Assembly
SBAC RS 404 ISSUE 1. Typical Assembly Showing Incorrect and Correct Method Non Reversible, Chain.
SBAC RS 405 ISSUE 1. Typical Assembly Showing Incorrect and Correct Method Non-Reversible Chain.
SBAC RS 406 ISSUE 1. Typical Assembly Showing Incorrect and Correct Method Non-Reversible Chain.

—Aircraft—Bolt Assemblies
SBAC AS 168 ISSUE 4. Nut and Bolt Assembly for Chains. 1 p.
SBAC AS 46792 ISSUE 1. Nut and Bolt Assembly (for Chains).

—Aircraft—Clips
SBAC AS 8437 ISSUE 1. Chain, Clip 1/4 Inch Bore Drain Valve.

—Aircraft—Clips—Drain Valves
SBAC AS 3339 ISSUE 2. Chain Clip 1/4 Bore Drain Hole. 1 p.

—Aircraft—Connectors
SBAC AS 7770-7777 ISSUE 2. Connector, Chain, Internal Thread.
SBAC AS 7779-7780 ISSUE 2(I). Connector—6 UNC (8m/m Chain).
SBAC AS 8505-8506 ISSUE 2. Connector Chain Internal Thread (8 UNC).
SBAC AS 8507-8508 ISSUE 1. Connector 8 UNC (8m/m Chain).

—Aircraft—Fittings
SBAC AS 8509-8552 ISSUE 3. B.S.I. Chain (8 m/m Pitch) and 8 UNC Fittings Assembly Code Non-Reversible Control.
SBAC AS 8553-8570 ISSUE 3. B.S.I. Chain (8 m/m Pitch) and 8 UNC Fittings Assembly Code Non-Reversible Control.
SBAC AS 8571-8588 ISSUE 3. B.S.I. Chain (8 m/m Pitch) 8 UNC Fittings Assembly Code Non-Reversible Control.

—Aircraft—Nut Assemblies
SBAC AS 168 ISSUE 4. Nut and Bolt Assembly for Chains. 1 p.
SBAC AS 46792 ISSUE 1. Nut and Bolt Assembly (for Chains).

—Aircraft—Projecting Plates
SBAC RS 408 ISSUE 1. Projecting Plates Non Reversible Chain.

—Aircraft—Pulleys
CAA LEAFLET 5-4 07.90. Control Chains, Chain Wheels and Pulleys. 8 pp.
SBAC AS 1868 ISSUE 5. Pulley—(B.S.1 Chain—8 m/m Pitch). 1 p.
SBAC AS 1871 ISSUE 7. Pulley—(B.S.6. Chain—1/2 Inch Pitch). 1 p.

Chains (Cont.)
—Aircraft—Sprockets
CAA LEAFLET 5-4 07.90. Control Chains, Chain Wheels and Pulleys. 8 pp.
SBAC AS 4830-4853 ISSUE 1. Sprocket (Blanks) for Chains Non-Reversible Control. 1 p.
SBAC AS 4854-4877 ISSUE 1. Sprocket (Blanks) for Chains Non-Reversible Control. 1 p.

—Aircraft—Tags
SBAC AS 2570 ISSUE 3. Chain Tag. 1 p.

—Aircraft—Tags—Drain Valves
SBAC AS 3338 ISSUE 2. Chain Tag, Washer—1/4 Inch Bore Drain Valve Drain Valve. 1 p.

—Anchor
BSI BS 7160-90. 1990 Anchor Chains for Small Craft. 9 pp.
ISO 4565-86. Small Craft—Anchor Chains First Edition. 5 pp.

—Anchor—Steel
JIS F 3302-75. Cast Steel Anchor Chain Cables. 13 pp.

—Bicycles
CNS B2047-75. Chain for Bicycles (Dec)(357). 2 pp.
CNS B7062-91. Method of Test for Chain of Bicycles (Jun)(3875). 1 p.
JIS D 9417-76. Chains for Bicycles.

—Blocks
CNS B4048-81. Chain Blocks (Sep)(7878).

—Buoys
CNS B2753-83. Buoy Chains (Oct)(10592).

—Cycles
ISO 9633-92. Cycle Chains—Characteristics and Test Methods First Edition. 13 pp.

—Flat
CNS L2080-83. Flat Chains for Card (Apr)(10178).
JIS L 5142-92. Flat Chains.
JIS L 5142-61. Flat Chains. 4 pp.

—Lifting Hooks
BSI BS 2903-80. 1980 Amd 1 Higher Tensile Steel Hooks for Chains, Slings, Blocks and General Engineering Purposes. 28 pp.
ISO 4779-86. Forged Steel Lifting Hooks with Point and Eye for Use with Steel Chains of Grade M(4) First Edition. 7 pp.
ISO 7597-87. Forged Steel Lifting Hooks with Point and Eye for Use with Steel Chains of Grade T(8) First Edition. 8 pp.
SNZ NZS/BS 2903-80. Specification for Higher Tensile Steel Hooks for Chains, Slings, Blocks and General Engineering Purposes Amend: 1. 28 pp.

—Lumber
JIS F 2102-90. Chains for Lumber Lashing.

—Mining Equipment
SAA AS 3637.6-91. Underground Mining—Winding Suspension Equipment—Part 6: Shackles and Chains (in Professional Package 32). 8 pp.

—Motorcycles—Adjusters
CNS D2138-88. Drive Chain Adjuster for Motorcycle (Jul)(8955).

—Recreational Vehicles
SAA AS 1872-76. Safety Chains for Trailers and Caravans. 8 pp.

—Rings
CNS B5072-70. Chain Ring (1.6mm) (Tentative) (Dec)(3041).
CNS G3070-86. Steel Rods for Chains Rings (Jun)(3291).

—Round—Steel
CNS B1273-81. Round Steel Chains (General Rules) (Oct)(7939).

—Safety—Recreational Vehicles
SAA AS 1872-76. Safety Chains for Trailers and Caravans. 8 pp.

—Safety—Trailers
SAA AS 1872-76. Safety Chains for Trailers and Caravans. 8 pp.

—Set—Components
CNS B2532-80. 125mm Empty Set Chain Component (Dec)(6736).

—Ships
CNS F3079-80. Ships' Chainlets (Dec)(6831).
CNS F3128-81. Chain for Lumber Lashing (Aug)(7790).
CNS F3130-81. Ships' Chain for General Use (Aug)(7792).
JIS F 2102-75. Lumber Lashing Chains. 6 pp.

INDEX and DIRECTORY of

Chains (Cont.)
—Ships (Cont.)
JIS F 2106-88. Ships' Chains for General Use.
JIS F 2106-76. Ships' Chains for General Use. 6 pp.
JIS F 3906-87. Ships' Chainlets.
JIS F 3906-75. Ships' Chainlets. 4 pp.

—Ships—Eyeplates
CNS F3081-80. Ships' Eye Plates for Chainlet (Dec)(6833).
JIS F 3908-87. Ships' Eye Plates for Chainlet.
JIS F 3908-75. Ships' Eye Plates for Chainlet. 4 pp.

—Ships—S Rings
CNS F3080-80. Ships' Ring of Chainlet (Dec)(6832).
JIS F 3907-87. Ships' S Rings of Chainlet.
JIS F 3907-75. Ships' S Rings of Chainlet. 4 pp.

—Steel
CNS B2755-83. Round Steel Chains Without Quality Requirements (Dec) (10683).
ISO 4779-86. Forged Steel Lifting Hooks with Point and Eye for Use with Steel Chains of Grade M(4) First Edition. 7 pp.
ISO 7597-87. Forged Steel Lifting Hooks with Point and Eye for Use with Steel Chains of Grade T(8) First Edition. 8 pp.

—Trailers
SAA AS 1872-76. Safety Chains for Trailers and Caravans. 8 pp.

Chainsaws
Use: Chain Saws

Chairs (Seats)
See Also: Armchairs; Dental Chairs; Furniture; High Chairs; Office Furniture; Recliners (Chairs); Rocking Chairs; Seats; Swivel Chairs
CNS S1052-68. One-Man Sofa Chair (Apr)(2843).

—Back Support
CNS S1046-68. Chairs with Full Back Support (Apr)(2837).

—Casters
CGSB CAN/CGSB-44.119-M89. Chair Swivel Casters for Carpeted or Hard Surface Floors (Supercedes 44-GP-120). 13 pp.
CGSB CAN/CGSB-44.226-M91. Dual Wheel Swivel Casters for Carpeted or Hard-Surfaced Floors; (Amendment 1 September 1992). 14 pp.

—Children's—Safety
BSI BS 7495-91. 1991 Safety Requirements for Children's Table-Mounted Chairs for Domestic Use (L). 15 pp.
BSI BS 7495-01. 1991 Amd 1 Safety Requirements for Children's Table-Mounted Chairs for Domestic Use (AMD 7194) July 15, 1992. 16 pp.
SNZ NZS 5856-93. Safety Requirements for Children's Table-Mounted Chairs Amend: A, 1993. 16 pp.

—Fatigue Testing
BSI BS 4875: Part 1-85. 1985 Strength and Stability of Furniture Part 1: Methods for Determination of Strength of Chairs and Stools. 20 pp.
ISO 7173-89. Furniture—Chairs and Stools—Determination of Strength and Durability First Edition. 23 pp.

—Folding—Wooden
CNS S1049-68. Folding Chairs, Wooden (Apr)(2840).

—Household—Student
CNS S1200-86. Student Chair for Domestic (Aug)(11675).
JIS S 1062-84. Student Chairs for Domestic Use.

—Impact Testing
BSI BS 4875: Part 1-85. 1985 Strength and Stability of Furniture Part 1: Methods for Determination of Strength of Chairs and Stools. 20 pp.
ISO 7173-89. Furniture—Chairs and Stools—Determination of Strength and Durability First Edition. 23 pp.

—Medical
BSI BS 3516-62. 1962 Chairs for Chiropodists' Patients. 8 pp.
BSI BS 3622-75. 1975 General Purpose Stools and Anaesthetists' Chairs for Hospital Use. 10 pp.

—Motor Operated
CSA CAN/CSA-C22. 2 NO 68-92. Motor-Operated Appliances (Household and Commercial); (Gen Instr 1 Thru 2). 115 pp.
CSA 1169 Bull. Electrical Bulletin 1169 June 27, 1978 to C22.2 NO 68. 2 pp.

—Office
BSI BS 5459: Part 1-77. 1977 Performance Requirements and Tests for Office Furniture Part 1: Desks and Tables. 12 pp.

Chairs (Seats) (Cont.)
—Office (Cont.)
BSI BS 5940: Part 1-80. 1980 Office Furniture Part 1: Design and Dimensions of Office Workstations, Desks, Tables and Chairs. 14 pp.
JIS S 1011-78. Standard Size of Chairs for Office.

—Office—Adjustable
BSI BS 5459: Part 2-90. 1990 Performance Requirements and Tests for Office Furniture Part 2: Office Seating. 27 pp.
BSI BS 5459: Part 2-01. 1990 Amd 1 Performance Requirements and Tests for Office Furniture Part 2: Office Seating (AMD 7169) July 15, 1992. 29 pp.

—Office—Casters
JIS S 1038-88. Castors for Office Chairs. 14 pp.

—Office—Conference Rooms
JIS S 1042-75. Chairs for Conference (Office Furniture).

—Office—Side
CGSB 44.183M-86. Chair, Side, Arm, Metal Frame. 11 pp.
CGSB 44.184M-86. Chair, Side, Armless, Metal Frame. 9 pp.

—Office—Steel
CNS S1072-70. Steel Office chair (Sep)(3089).
JIS S 1032-91. Office Furniture—Steel Chairs. 13 pp.

—Office—Swivel
BSI BS 5459: Part 2-90. 1990 Performance Requirements and Tests for Office Furniture Part 2: Office Seating. 27 pp.
BSI BS 5459: Part 2-01. 1990 Amd 1 Performance Requirements and Tests for Office Furniture Part 2: Office Seating (AMD 7169) July 15, 1992. 29 pp.

—Office—Wooden
JIS S 1023-89. Office Wooden Furniture (Desk, Table and Chair).

—Office—Wooden—Upholstery
JIS S 1028-89. Standard Upholstering for Office Wooden Chair.

—Rattan
CNS S1048-68. Chairs, Rattan Seat, Straight Back (Apr)(2839).
CNS S1050-68. Chairs, Rattan Seat, Curved Back (Apr)(2841).
CNS S1054-68. One-Man Rattan Chair (Apr)(2845).

—Reception Areas
JIS S 1052-78. Chairs for Reception (Office Furniture).

—Schools
ISO 5970-79. Furniture—Chairs and Tables for Educational Institutions—Functional Sizes First Edition. 7 pp.

—Schools—Finishes
BSI BS 5873: Part 1-80. 1980 Amd 1 Educational Furniture Part 1: Functional Dimensions, Identification and Finish of Chairs and Tables for Educational Institutions. 13 pp.

—Schools—Identification Systems
BSI BS 5873: Part 1-80. 1980 Amd 1 Educational Furniture Part 1: Functional Dimensions, Identification and Finish of Chairs and Tables for Educational Institutions. 13 pp.

—Schools—Lecture Rooms
JIS S 1015-74. Sizes and Dimensions of Fixed Desk and Chair for Lecture Room.
JIS S 1016-76. Fixed Desk and Chair for Lecture Room.
JIS S 1021-91. School Furniture (Desks and Chairs for Class Room). 23 pp.

—Schools—Library
JIS S 1073-80. School Furniture (Tables and Chairs for Library).

—Schools—Stability
BSI BS 5873: Part 2-91. 1991 Amd 1 Educational Furniture Part 2: Specification for Strength and Stability of Chairs for Educational Institutions (AMD 6809) August 30, 1991. 21 pp.

—Shells—Plastic
DIN ENGL 68872-76. Plastic Chair Shells for Indoor Use; Requirements; Testing (Jan). 3 pp.

—Stability Testing
BSI BS 4875: Part 2-85. 1985 Strength and Stability of Furniture Part 2: Methods for Determination of Stability of Chairs and Stools. 14 pp.
ISO 7174 Pt 1-88. Furniture—Chairs—Determination of Stability Part 1: Upright Chairs and Stools First Edition. 10 pp.

Chairs (Seats) (Cont.)
—Stacking—Steel
CGSB 44.15M-89. Straight Stacking Chair, Steel. 8 pp.

—Stacking—Upholstered
CGSB 44.157M-91. Stacking Chairs with Metal Frame. 11 pp.

—Static Loads
BSI BS 4875: Part 1-85. 1985 Strength and Stability of Furniture Part 1: Methods for Determination of Strength of Chairs and Stools. 20 pp.
ISO 7173-89. Furniture—Chairs and Stools—Determination of Strength and Durability First Edition. 23 pp.

—Workstations—Posture
CGSB CAN/CGSB-44.200-M90. Posture Chairs. 40 pp.

Chalk
Use For: Precipitated Chalk; Prepared Chalk
See Also: Calcium Carbonate; Chalkboards
BSI BS 3060-58. (WITHDRAWN) 1958 School Chalks. 8 pp.
CGSB 53-GP-59M-78. Chalk, Marking, White and Colored. 7 pp.
CNS S1122-82. Writing Chalks (Feb)(7446).
CNS S2066-82. Method of Test for Writing Chalks (Feb)(8543).
JIS S 6009-84. White Chalks Made of Calcined Plaster.
JIS S 6010-84. White Chalks Made of Calcium Carbonate.

Chalk Boards
Use: Chalkboards

Chalkboards
Use For: Chalk Boards *See Also:* Chalk
BSI PD 6482-83. 1983 Preparation of Visual Aids for Lectures. 15 pp.
CNS S1206-87. General Rules for Chalkboards (Jul)(12035).
CNS S2138-87. Method of Test for Chalkboards (Jul)(12036).
JIS S 6101-84. General Rules for Chalkboards.

—Baking Finished
CNS S1208-87. Porcelain-Enameled and Baking Finished Chalkboards (Jul)(12038).
JIS S 6045-86. Porcelain-Enameled and Baking Finished Chalkboards.

—Grinding Finished
CNS S1207-87. Grinding Finished Chalkboards (Jul)(12037).
JIS S 6007-86. Grinding Finished Chalkboards.

—Paints
MOD UK DEF-1442-64. Paint, Finishing, Matt, Black, for Blackboards. 4 pp.

—Vitreous Enameled
CNS S1208-87. Porcelain-Enameled and Baking Finished Chalkboards (Jul)(12038).
CNS S1209-87. Surface Material for Porcelain Enamel Chalkboards (Jul)(12039).
JIS S 6044-89. Surface Material for Porcelain Enamel Chalkboards. 6 pp.
JIS S 6045-86. Porcelain-Enameled and Baking Finished Chalkboards.
JIS S 6052-87. Porcelain-Enameled Marker Board.

Chalking Testing
—Paints
DIN ENGL 53159-77. Testing of Paints, Varnishes and Similar Coating Materials; Determination of the Degree of Chalking of Paint Coatings and Similar Coatings by Kempf's Method (Sept). 3 pp.
SAA AS 1580.481. 1.11-91. Paints and Related Materials—Methods of Test—Part 481: Coatings—Part 481.1: Assessment of Individual Defects of Exposed Films —Part 481.1.11: Exposed to Weathering—Degree of Chalking (Supersedes AS 1580.481.1—1975 (in Part)) (in Professional Packages 30, 39). 3 pp.
SNZ NZS/AS 1580. 481.1.11-91. Methods of Test for Paints and Related Materials Part 481.1.11: Coatings—Exposed to Weathering—Degree of Chalking (This is a Joint Standard with SAA AS 1580.481.11). 3 pp.

—Varnishes
DIN ENGL 53159-77. Testing of Paints, Varnishes and Similar Coating Materials; Determination of the Degree of Chalking of Paint Coatings and Similar Coatings by Kempf's Method (Sept). 3 pp.

Chamfers
See Also: Machining
CNS B3025-47. Fillet and Chamfer (Mar)(149)(R1973).

Chamois

Chamois Leather
See Also: Leather
BSI BS 6715-91. 1991 Chamois Leather. 9 pp.

Change Gears
See Also: Gears

—Gear Teeth—Machine Tools
CNS B2025-84. Number of Teeth of Change Gears for Geared Lathe, Milling Machine and Gear Generating Equipment (Aug)(186).

—Machine Tools
CNS B2024-84. Dimensions of Change Gear for Machine Tools (Aug)(185).

Change of Temperature Testing
Use: Temperature Change Testing

Channel Flow
Use: Open Channel Flow

Channel Induction Furnaces
Use: Induction Furnaces—Channel

Channel Loading
Scope Note: Use a more specific term *See:* Calling Channel Loading; Communication Channels; Communication Circuits

Channel Nuts
Use For: Gang Channel Nuts *See Also:* Nuts (Fasteners)

—Aerospace—Locking—Counterbore
DIN ENGL LN 29993-74. Gang Channel Nuts; Deep Counterbore, Self-Locking, Floating for Temperatures up to 120 Degrees C (Oct). 2 pp.

Channel Sequence Numbers (Telegraphy)
See Also: Numbers; Telegraph Channels; Telegraphy

—Message Switching
CCITT RECMN F.35-89. Provisions Applying to the Operation of an International Public Automatic Message Switching Service for Equipments Utilizing the International Telegraph Alphabet No. 2—Telegraph and Mobile Services Operations and Quality of Service (Study Group I) 7 pp. 7 pp.

—Telegraph Repeaters
CCITT RECMN F.31-89. Telegram Retransmission System—Telegraph and Mobile Services Operations and Quality of Service (Study Group I) 12 pp. 12 pp.

Channel Serial Numbers (Telegraphy)
Use: Channel Sequence Numbers (Telegraphy)

Channel Spacing
See Also: Communication Channels; Television Channels

—Sound Broadcasting
CCIR RECMN 597-1-86. Channel Spacing for Sound Broadcasting in Band 7 (HF)—Section 10A-1—Amplitude-Modulation Sound Broadcasting in Bands 5 (LF), 6 (MF) and 7 (HF). 1 p.

—Sound Broadcasting—FM—VHF
CCIR RECMN 412-5-90. Planning Standards for FM Sound Broadcasting at VHF—Section 10B—Frequency-Modulation Sound Broadcasting in Bands 8 (VHF) and 9 (UHF). 6 pp.

Channels (Communication)
Use: Communication Channels

Channels (Hydraulic)
Use: Conduits (Channels)

Channels (Structural Shapes)
See Also: Angle Iron; Steels; Structural Members

—Aircraft—Clips
SBAC AS 8646 ISSUE 1. Channel.

—Aircraft—Electrical Grounding
SBAC AS 3114 ISSUE 1. Channel, Earthing Point 19 Amp. 1 p.
SBAC AS 3115 ISSUE 1. Channel, Earthing Point 37 Amp. 1 p.

—Cables (Electric)—Naval Ships
MOD UK NES 2046-90. Specification for Channel Perforated Steel for Electric Cable Issue 2 (06.90). 10 pp.
MOD UK NES 2047-90. Specification for Channel Perforated Aluminium Alloy for Electric Cable Issue 2 (06.90). 10 pp.

—Cables (Electric)—Submarines
MOD UK NES 2046-90. Specification for Channel Perforated Steel for Electric Cable Issue 2 (06.90). 10 pp.
MOD UK NES 2047-90. Specification for Channel Perforated Aluminium Alloy for Electric Cable Issue 2 (06.90). 10 pp.

—Cold Formed
CSA CAN/CSA-G40.20-M92. General Requirements for Rolled or Welded Structural Quality Steel; (Gen Instr 1 Thru 3). 81 pp.

—Cold Formed—Rolled
CSA CAN/CSA-G40.20-M92. General Requirements for Rolled or Welded Structural Quality Steel; (Gen Instr 1 Thru 3). 81 pp.

—Extruded—Aircraft
SBAC RS 91-100, 106-115 IS 1. Symmetric Square Channel Extrusions.

—Hot Rolled
ISO 657 Pt 11-80. Hot-Rolled Steel Sections—Part 11: Sloping Flange Channel Sections (Metric Series)—Dimensions and Sectional Properties First Edition. 4 pp.
ISO 657 Pt 13-81. Hot-Rolled Steel Sections—Part 13: Tolerances on Sloping Flange Beam, Column and Channel Sections First Edition. 6 pp.

—Linings
BSI BS 7364-90. 1990 Galvanized Steel Studs and Channels for Stud and Sheet Partitions and Linings Using Screw Fixed Gypsum Wallboards. 10 pp.

—Partitions
BSI BS 7364-90. 1990 Galvanized Steel Studs and Channels for Stud and Sheet Partitions and Linings Using Screw Fixed Gypsum Wallboards. 10 pp.

—Washers
CNS B2016-78. Square Taper Washers for Channels (Nov)(156).
CNS B2325-83. Square Taper Washers for High Prestressed Channels (Mar)(4647).
DIN ENGL 434-90. Square Taper Washers for Use with Channel Sections (Apr). 3 pp.
DIN ENGL 6918-90. Square Taper Washers for High-Strength Structural Bolting of Steel Channel Sections (Apr). 3 pp.

Character Error Rate
See Also: Error Analysis; Error Control

—Radiotelegraphy—Start-Stop Transmission
CCITT RECMN F.112-89. Quality Objectives for 50-Baud Start-Stop Telegraph Transmission in the Maritime Mobile-Satellite Service—Telegraph and Mobile Services Operations and Quality of Service (Study Group I) 1 pp. 1 p.

—Telegraph Equipment—Start-Stop Transmission
CCITT RECMN F.10-89. Character Error Rate Objective for Telegraph Communication Using 5-Unit Start-Stop Equipment—Telegraph and Mobile Services Operations and Quality of Service (Study Group I) 1 pp. 1 p.

—Telegraphy
CCITT RECMN F.10-89. Character Error Rate Objective for Telegraph Communication Using 5-Unit Start-Stop Equipment—Telegraph and Mobile Services Operations and Quality of Service (Study Group I) 1 pp. 1 p.

Character Imaging Devices
See Also: Coded Character Sets; Data Processing Equipment; Page Readers
BSI BS 7204-89. (WITHDRAWN) 1989 Control Functions for ISO 7-Bit and 8-Bit Coded Character Sets (Superseded by BS ISO/IEC 6429: 1992). 55 pp.
BSI BS ISO/IEC 6429-92. 1992 Information Technology—Control Functions for Coded Character Sets (S). 100 pp.
ECMA ECMA 48-91. Control Functions for Coded Character Sets. 108 pp.
ECMA ECMA TR 53-92. Handling of Bi-Directional Texts. 24 pp.
ECMA ECMA-TR 53-90. Handling of Bi-Directional Texts. 29 pp.
IEC 6429-92. Information Technology—Control Functions for Coded Character Sets Third Edition. 98 pp.
ISO 6429-92. Information Technology—Control Functions for Coded Character Sets Third Edition. 98 pp.
JIS X 0211-86. Additional Control Functions for Character-Imaging Devices. 77 pp.
JTC1 6429-92. Information Technology—Control Functions for Coded Character Sets Third Edition. 98 pp.
OSI ISO 6429-88. Information Processing—Control Functions for 7-Bit and 8-Bit Coded Character Sets. 64 pp.

—FAX Machines
CCITT RECMN T.351-89. Imaging Process of Character Information on Facsimile Apparatus—Terminal Equipment and Protocols for Telematic Services (Study Group VIII) 4 pp. 4 pp.

Character Sets
Scope Note: For additional listings, use a more specific term *Use For:* Alphabets
See Also: Alphanumeric Character Sets; Check Characters; Coded Character Sets; Control Characters; Graphic Characters; International Reference Alphabet; International Telegraph Alphabet No. 2; International Telegraph Alphabet No. 5; Numeric Character Sets
SAA AS 1070-71. Definition of 4-Bit Character Sets Derived from the Australian Standard 7-Bit Coded Character Set Corrig. Reconfirmed 1985. 11 pp.

—International System of Units
BSI BS 6430-83. 1983 Representing SI and Other Units in Information Processing Systems with Limited Character Sets. 8 pp.
BSI PD 6462-72. 1972 Modification of Keyboards to Include Symbols for SI Units. 4 pp.
ISO 2955-83. Information Processing—Representation of SI and Other Units in Systems with Limited Character Sets Second Edition. 7 pp.
JTC1 2955-83. Information Processing—Representation of SI and Other Units in Systems with Limited Character Sets Second Edition. 7 pp.
OSI ISO 2955-83. Information Processing—Representation of SI and Other Unit in Systems with Limited Character Sets. 7 pp.
SAA AS 1340-75. Symbols for SI Units for Systems with Limited Character Set. 11 pp.

—Man-Machine Language—CCITT
CCITT RECMN Z.314-89. Character Set and Basic Elements—Man-Machine Language (MML) (Study Group X) 9 pp. 9 pp.

—Telegraphy
CCITT RECMN F.1-89. Operational Provisions for the International Public Telegram Service—Telegraph and Mobile Services Operations and Quality of Service (Study Group I) 54 pp. 54 pp.

—Video Display Terminals
DIN ENGL 66234 Pt 1-80. VDU Work Stations; Geometrical Design of Characters (Mar). 3 pp.

Characteristic Impedance
Use For: Surge Impedance *See Also:* Electrical Impedance; Transmission Loss

—Coaxial Cables
CENELEC HD 120-74. Characteristic Impedances and Dimensions of Radio Frequency Coaxial Cables. 2 pp.
IEC 78-67. Characteristic Impedances and Dimensions of Radio-Frequency Coaxial Cables Third Edition. 11 pp.

—Coaxial Connectors
CENELEC HD 134.7-76. Radio-Frequency Connector Part 7: R.F. Coxial Connectors with Inner Diameter of Outer Conductor 9.5mm (0.374 in) with Bayonet Lock-Characteristic Impedance 50 Ohms (Type C). 2 pp.
IEC 169 Pt 7-75. Radio-Frequency Connectors Part 7: R.F. Coaxial Connectors with Inner Diameter of Outer Conductor 9,5 mm (0.374 in) with Bayonet Lock—Characteristic Impedance 50 ohms (Type C) First Edition; (Amendment 1-1993). 51 pp.
IEC 169 Pt 8-78. Radio-Frequency Connectors Part 8: R.F. Coaxial Connectors with Inner Diameter of Outer Conductor 6.5 mm (0.256 in) with Bayonet Lock—Characteristic Impedance 50 Ohms (Type BNC) First Edition. 51 pp.
IEC 169 Pt 9-78. Radio-Frequency Connectors Part 9: R.F. Coaxial Connectors with Inner Diameter of Outer Conductor 3 mm (0.12 in) with Screw Coupling—Characteristic Impedance 50 Ohms (Type SMC) First Edition. 41 pp.
IEC 169 Pt 10-83. Radio-Frequency Connectors Part 10: R.F. Coaxial Connectors with Inner Diameter of Outer Conductor 3 mm (0.12 in) with Snap-on Coupling—Characteristic Impedance 50 Ohms (Type SMB) First Edition; (Amendment 1-1986). 55 pp.
IEC 169 Pt 11-77. Radio-Frequency Connectors Part 11: R.F. Coaxial Connectors with Inner Diameter of Outer Conductor 9.5 mm (0.374 in) with Screw Coupling-Characteristic Impedance 50 Ohms (Type 4.1/9.5) First Edition. 32 pp.

INTERNATIONAL AND NON-U.S. NATIONAL STANDARDS
SUBJECT INDEX

Characteristic Impedance *(Cont.)*
—**Coaxial Connectors** *(Cont.)*
IEC 169 Pt 12-79. Radio-Frequency Connectors Part 12: R.F. Coaxial Connectors with Screw Coupling, Unmatched (Type UHF) First Edition. 15 pp.
IEC 169 Pt 13-76. Radio Frequency Connectors Part 13: R.F. Coaxial Connectors with Inner Diameter of Outer Conductor 5.6 mm (0.22 in)—Characteristic Impedance 75 Ohms (Type 1.6/5.6)-Characteristic Impedance 50 Ohms (Type 1.8/5.6) with Similar Mating Dimensions First Edition. 39 pp.
IEC 169 Pt 14-77. Radio-Frequency Connectors Part 14: R.F. Coaxial Connectors with Inner Diameter of Outer Conductor 12 mm (0.472 in) with Screw Coupling-Characteristic Impedance 75 Ohms (Type 3.5/12) First Edition. 39 pp.
IEC 169 Pt 15-79. Radio-Frequency Connectors Part 15: R.F. Coaxial Connectors with Inner Diameter of Outer Conductor 4.13 mm (0.163 in) with Screw Coupling-Characteristic Impedance 50 Ohms (Type SMA) First Edition. 16 pp.
IEC 169 Pt 16-82. Radio-Frequency Connectors Part 16: R.F. Coaxial Connectors with Inner Diameter of Outer Conductor 7 mm (0.276 in) with Screw Coupling-Characteristic Impedance 50 Ohms (75 Ohms) (Type N) First Edition. 27 pp.
IEC 169 Pt 17-80. Radio-Frequency Connectors Part 17: R.F. Coaxial Connectors with Inner Diameter of Outer Conductor 6,5 mm (0,256 in) with Screw Coupling—Characteristic Impedance 50 ohms (Type TNC) First Edition; (Amendment 1-1993). 92 pp.
IEC 169 Pt 18-85. Radio-Frequency Connectors Part 18: R.F. Coaxial Connectors with Inner Diameter of Outer Conductor 2.79 mm (0.110 in) with Screw Coupling—Characteristic Impedance 50 Ohms (Type SSMA) First Edition. 20 pp.
IEC 169 Pt 19-85. Radio-Frequency Connectors Part 19: R.F. Coaxial Connectors with Inner Diameter of Outer Conductor 2.08 mm (0.082 in) with Snap Coupling—Characteristic Impedance 50 Ohms (Type SSMB) First Edition. 22 pp.
IEC 169 Pt 20-85. Radio-Frequency Connectors Part 20: R.F. Coaxial Connectors with Inner Diameter of Outer Conductor 2.08 mm (0.082 in) with Screw Coupling—Characteristic Impedance 50 Ohms (Type SSMC) First Edition. 20 pp.
IEC 169 Pt 26-93. Radio-Frequency Connectors—Part 26: R.F. Coaxial Connectors with Screw Coupling—Characteristic Impedance 50 ohms—Frequency Range 0 to 18 GHz (Type TNC 18 GHz) First Edition. 27 pp.

Characters
Use: Symbols

Charcoal
Use For: Grill Charcoal *See Also:* Activated Carbon; Carbon; Carbon Black; Coke; Solid Fuels
CNS Z7019-72. Charcoal (for Fuel) (Tentative) (Oct)(2598).
CNS Z8009-72. Method of Test for Charcoal for Fuel (Oct)(3512).

—**Explosives**
MOD UK DSTAN 91-31-89. Charcoal, Wood for Explosives and Pyrotechnics Types: A, D, F, G and J Issue 2. 9 pp.

—**Powdered**
CNS K6374-75. Method of Test for Charcoal Powder (Feb)(3754).
MOD UK TS 10253A. Charcoal Powered Type: 2000.

—**Pyrotechnics**
MOD UK DSTAN 91-31-89. Charcoal, Wood for Explosives and Pyrotechnics Types: A, D, F, G and J Issue 2. 9 pp.

—**Triethylenediamine**
MOD UK TS 10237. Triethylenediamine (TEDA).

Charcoal Cloth
Use: Charcoal Fabrics

Charcoal Fabrics
—**Copper/Chromium Impregnated**
MOD UK DSTAN 83-84-93. Charcoal Cloth Impregnated with Copper and Chromium, Type 1303 Issue 1. 28 pp.

—**Silver Nitrate Impregnated**
MOD UK TS 10206. Charcoal Cloth Impregnated with Silver Nitrate, Type 1301 (Withdrawn).

Charcoal Tube/Gas Chromatographic Method
Use: Gas Chromatography

Charge Card Calling Services
Use: Credit Card Calling Services

Charge Coupled Devices
See Also: Semiconductor Devices
—**Delay Lines—Video**
MOD UK DSTAN 59-62: 90/174-82. Charge Coupled Device Issue 1. 35 pp.

—**Image Sensing**
CECC CECC 20 000 ISSUE 1-82. Generic Specification: Semiconductor Optoelectronic and Liquid Crystal Devices (En, Fr, Ge) AMD 3 (En, Fr, Ge) SUPP 1 (En, Fr, Ge) SUPP 2 (En, Fr, Ge). 401 pp.

Charge Storage Cathode Ray Tubes
Use: Cathode Ray Storage Tubes

Charge Transfer Devices
See Also: Digital Circuits
BSI BS 6493: Sec 2.2-86. 1986 Semiconductor Devices Part 2: Integrated Circuits Section 2.2: Recommendations for Digital Integrated Circuits. 143 pp.
BSI BS 6493: Sec 2.2-01. 1986 Amd 1 Semiconductor Devices Part 2: Integrated Circuits Section 2.2: Recommendations for Digital Integrated Circuits (IEC 748-2: 1985) (AMD 7003) July 15, 1992. 166 pp.
IEC 748 Pt 2-85. Semiconductor Devices Integrated Circuits Part 2: Digital Integrated Circuits First Edition; (Amendment 1-1991). 312 pp.

Charged Particles
—**Emulsified Asphalts**
CNS K6781-83. Method of Test for Particle Charge of Emulsified Asphalts (Jun)(10364).

Charging (Billing), Telecommunications
Use: Tariffs (Telecommunications)

Charging Valves
Use: Pressure Control Valves

Charpy Impact Testing
Use: Impact Testing

Charpy Impact Testing Equipment
Use: Impact Testing Equipment—Charpy

Chart Display Systems, Electronic
Use: Electronic Chart Display Systems

Chart Recorders
Use For: Drum Chart Recorders; Electric Chart Recorders; Pneumatic Chart Recorders; X-t Recorders; XT Recorders *See Also:* Electric Measuring Instruments; Recording Instruments; XY Recorders
BSI BS 7610: Part 1-92. 1992 Electrical Measuring Instruments—X-t Recorders Part 1: Definitions and Requirements (IEC 1143-1: 1992). 34 pp.
BSI BS 7610: Part 2-93. 1993 Electrical Measuring Instruments—X-t Recorders Part 2: Recommended Additional Test Methods (IEC 1143-2: 1992) (E). 12 pp.
IEC 1143 Pt 1-92. Electrical Measuring Instruments—X-t Recorders Part 1: Definitions and Requirements First Edition. 64 pp.
IEC 1143 Pt 2-92. Electrical Measuring Instruments—X-t Recorders Part 2: Recommended Additional Test Methods First Edition. 20 pp.

—**Alarm Metering Systems**
BSI BS 2740-69. 1969 Amd 1 Simple Smoke Alarms and Alarm Metering Devices. 16 pp.

—**Digital**
BSI BS 7528: Part 1-91. 1991 Digital Recorders for Measurements in High-Voltage Impulse Tests Part 1: Requirements for Digital Recorders (Renumbered as BS EN 61083-1: 1993). 32 pp.

—**Glossaries**
BSI BS 7610: Part 1-92. 1992 Electrical Measuring Instruments—X-t Recorders Part 1: Definitions and Requirements (IEC 1143-1: 1992). 34 pp.
IEC 1143 Pt 1-92. Electrical Measuring Instruments—X-t Recorders Part 1: Definitions and Requirements First Edition. 64 pp.

—**Meteorological—Drums**
BSI BS 3883-65. 1965 Drum Chart Mechanisms for Meteorological Recorders. 8 pp.
SNZ NZS 2110-66. Specification for Drum Chart Mechanisms for Meteorological Recorders. 8 pp.

—**Process Control Equipment**
BSI BS 7623-93. 1993 Electrical and Pneumatic Analogue Chart Recorders for Use in Industrial-Process Control Systems. Guidance for Inspection and Routine Testing (IEC 1153: 1992). 10 pp.

Chart Recorders *(Cont.)*
—**Process Control Equipment** *(Cont.)*
BSI BS EN 60873-93. 1993 Amd 2 Methods of Evaluating the Performance of Electrical and Pneumatic Analogue Chart Recorders for Use in Industrial-Process Control Systems (Q). 44 pp.
CENELEC HD 504 S1-88. Methods of Evaluating the Performance of Electrical and Pneumatic Analogue Chart Recorders for Use in Industrial-Process Control Systems. 4 pp.
CENELEC EN 60873-93. Methods of Evaluating the Performance of Electrical and Pneumatic Analogue Chart Recorders for Use in Industrial-Process Control Systems (IEC 873: 1986, Modified). 35 pp.
IEC 873-86. Methods of Evaluating the Performance of Electrical and Pneumatic Analogue Chart Recorders for Use in Industrial-Process Control Systems First Edition; (CENELEC EN 60873:1993). 71 pp.
IEC 1153-92. Electrical and Pneumatic Analogue Chart Recorders for Use in Industrial-Process Control Systems—Guidance for Inspection and Routine Testing First Edition. 17 pp.

Charts
Scope Note: For additional listings, use a more specific term *See Also:* Aeronautical Navigation Charts; Amphibious Charts; Audiovisual Equipment; Color Charts; Combat Charts; Control Charts; Diagrams; Drawings; Electronic Chart Display Systems; Flow Charts; Graphic Methods; Graphs (Charts); Maps; Meteorological Charts; Navigational Charts; Process Charts; Scale (Ratio); Symbols; Visual Aids
BSI PD 6482-83. 1983 Preparation of Visual Aids for Lectures. 15 pp.
BSI DD 52-77. (WITHDRAWN) 1977 Recommendations for the Presentation of Tables, Graphs and Charts (Superseded by BS 7581: 1992). 16 pp.

—**Classification**
CENELEC HD 246.1-75. Diagrams, charts, tables Part 1: Definitions and classification. 2 pp.
DIN ENGL 40719 Pt 1-73. Circuit Representations; Definitions; Classification (June). 4 pp.

—**Designations**
CENELEC HD 246.2-75. Diagrams, Charts and Tables Part 2: Item Designation. 2 pp.

—**Electrical Components—Glossaries**
SAA AS 1103.1-73. Diagrams, Charts and Tables for Electrotechnology—Part 1: Definitions and Classifications Reconfirmed 1982. 5 pp.
SNZ NZS/AS 1103. 1-73. Diagrams, Charts and Tables for Electrotechnology Part 1: Definitions and Classifications (This is a Joint Standard with SAA AS 1103.1). 5 pp.

—**Glossaries**
CENELEC HD 246.1-75. Diagrams, charts, tables Part 1: Definitions and classification. 2 pp.
DIN ENGL 40719 Pt 1-73. Circuit Representations; Definitions; Classification (June). 4 pp.

—**Military Operations**
NATO STANAG 2205 ED 4 AMD 3-88. Use of Identical Maps and Charts (Excluding Nautical Charts). 7 pp.

—**Spelling**
NATO STANAG 3689 ED 2 AMD 2-84. Place Name Spelling on Maps and Charts. 25 pp.
NATO STANAG 3689 ED 2 AMD 5-84. Place Name Spelling on Maps and Charts. 24 pp.

—**Symbols**
JIS X 0121-86. Documentation Symbols and Conventions for Data Program and System Flowcharts, Program Networks Charts and System Resources Charts.

—**Varnishes**
MOD UK DSTAN 80-103-85. Varnish, Chart Issue 1. 8 pp.

Chassis
See Also: Electronic Equipment; Running Gear
CNS C5214-87. Dimensions of Racks and Unit Chassis of Electronic Equipment for General Use (Apr)(11901).
JIS C 6010-69. Dimensions of Racks and Unit Chassis of Electronic Equipment for General Use (R 1988). 10 pp.

—**Buses (Vehicles)**
BSI BS AU 116-65. (OBSOLESCENT) 1965 Dimensions of Chassis for 27 ft Public Service Vehicle (Double Deck). 2 pp.
BSI BS AU 117-65. (OBSOLESCENT) 1965 Dimensions of Chassis for 30 ft Public Service Vehicle (Double Deck). 2 pp.

INTERNATIONAL AND NON-U.S. NATIONAL STANDARDS
SUBJECT INDEX

Chassis (Cont.)
—Electric Wire—Color Coding
CNS C5072-80. Color Coding of Chassis Wiring (Aug)(6321).
—Glossaries
IEC 916-88. Mechanical Structures for Electronic Equipment Terminology First Edition. 16 pp.

Chassis Dynamometers
—Mopeds—Air Pollution
ISO TR6970-81. Road Vehicles—Pollution Tests for Motorcycles and Mopeds—Bench (Chassis Dynamometer) First Edition. 10 pp.
—Motorcycles—Air Pollution
ISO TR6970-81. Road Vehicles—Pollution Tests for Motorcycles and Mopeds—Bench (Chassis Dynamometer) First Edition. 10 pp.

Chatter
Use For: Contact Chatter See Also: Electric Contacts; Vibration
—Electric Switches
CNS C6094-88. Method of Test for Electromechanical Switches (Monitoring Contact Chatter) (Dec)(6157).

Chayotes
See Also: Cucumbers; Food; Vegetables
—Grading
CNS N1039-82. Grades of Chayote (for Export) (Sep)(2124). 2 pp.

Check Characters
See Also: Character Sets
—Information Interchange
BSI BS 6541-85. 1985 Check Character Systems for Use in Information Interchange, and Guidance on Choice and Methods of Application. 16 pp.
CNS C5200-85. Data Processing (Check Character Systems) (Nov)(11402).
ISO 7064-83. Data Processing—Check Character Systems First Edition. 15 pp.
JTC1 7064-83. Data Processing—Check Character Systems First Edition. 15 pp.
OSI ISO 7064-83. Data Processing Character Systems. 15 pp.

Check Gages
DIN ENGL 7163-66. Workshop Gap Gauges and Check Gauges for ISO Fit Sizes from 1 to 500 mm Nominal Dimension; Gauge Dimensions and Manufacturing Tolerances (Aug). 15 pp.

Check Valves
Use For: Nonreturn Valves See Also: Angle Check Valves; Backflow Preventers; Backwater Valves; Ball Valves; Flap Valves; Globe Check Valves; Hydraulic Valves; Lift Check Valves; Pneumatic Valves; Stop Valves; Swing Check Valves; Valves
BSI BS 1868-75. 1975 Amd 1 Steel Check Valves (Flanged and Butt-Welding Ends) for the Petroleum, Petrochemical and Allied Industries (AMD 6563) July 31, 1990 July 31, 1990. 27 pp.
BSI BS 5153-74. 1974 Amd 3 Cast Iron Check Valves for General Purposes (AMD 6067) July 31, 1989. 20 pp.
BSI BS 5352-81. 1981 Amd 2 Steel Wedge Gate, Globe and Check Valves 50 mm and Smaller for the Petroleum, Petrochemical and Allied Industries (AMD 6560) August 31, 1990 (Supersedes BS 2995: 1966). 25 pp.
BSI BS 6282: Part 1-82. 1982 Devices with Moving Parts for the Prevention of Contamination of Water by Backflow Part 1: Check Valves of Nominal Size up to and Including DN 54. 8 pp.
BSI BS 6282: Part 4-82. 1982 Devices with Moving Parts for the Prevention of Contamination of Water by Backflow Part 4: Combined Check and Anti-Vacuum Valves of Nominal Size up to and Including DN 42. 4 pp.
CEN PREN 1074-3-93. Valves for Water Supply—Specification for Use and Appropriate Verification Tests—Part 3: Check Valves. 10 pp.
CNS B2497-80. Face-to-Face and End-to-End Dimensions of Check Valves (Jul)(5713).
DIN ENGL 86502-73. Gunmetal Screwed Bonnet Non-Return Valves with Non-Soldered Pipe Unions with Cutting Ring (July). 6 pp.
DIN ENGL 86512-73. Shut-Off Check Valves, Screwed-Bonnet Type of Gun Metal with 25 Deg. Taper-Bushing Pipe Unions for Brazing (July). 6 pp.
JIS B 2002-87. Face-to-Face and End-to-End Dimensions of Valves.
JIS B 2002-68. Face-to-Face and End-to-End Dimensions of Valves. 14 pp.
JIS B 2011-88. Bronze Gate, Globe, Angle and Check Valves. 41 pp.

Check Valves (Cont.)
SNZ NZS/BS 5153-74. Specification for Cast Iron Check Valves for General Purposes Amend: 3. 20 pp.
SNZ NZS/BS 5352-81. Specification for Steel Wedge Gate, Globe and Check Valves 50 mm and Smaller for the Petroleum, Petrochemical and Allied Industries Amend: 1. 24 pp.
—Aircraft
SBAC AS 3376 (V). Non-Return Valve—Hydraulic. 2 pp.
SBAC AS 3377 ISSUE 1. Body—Non-Return Valve. 1 p.
SBAC AS 3379 ISSUE 1. Valve. 1 p.
SBAC AS 3382 ISSUE 3. 3/8 Inch Non-Return, Valve (Hydraulic). 1 p.
SBAC AS 3383 ISSUE 1. Body. 1 p.
SBAC AS 3385 ISSUE 1. Valve. 1 p.
SBAC AS 3388 ISSUE 3. 1/2 Inch Non Return, Valve—Hydraulic. 1 p.
SBAC AS 3389 ISSUE 1. Body. 1 p.
SBAC AS 3391 ISSUE 1. Valve. 1 p.
—Aircraft—Guards (Protective)—Pipe Threads
SBAC AGS 3415 ISSUE 1. Non Return Valve, Safety Guard for Use with 1/4 Inch B.S.P.F. Thread.
—Aircraft—Sealing Rings
SBAC AS 3381 ISSUE 1. Sealing, Ring. 1 p.
SBAC AS 3387 ISSUE 1. Sealing, Ring. 1 p.
SBAC AS 3393 ISSUE 1. Sealing, Ring. 1 p.
—Aircraft—Springs (Elastic)
SBAC AS 3380 ISSUE 1. Spring. 1 p.
SBAC AS 3386 ISSUE 1. Spring. 1 p.
SBAC AS 3392 ISSUE 1. Spring. 1 p.
—Aircraft—Stoppers
SBAC AS 3378 ISSUE 1. Plug. 1 p.
SBAC AS 3384 ISSUE 1. Plug. 1 p.
SBAC AS 3390 ISSUE 1. Plug. 1 p.
—Gas Industry
CSA CAN/CSA-Z245.15-M91. Steel Valves; (Gen Instr 1). 81 pp.
—Globe
BSI BS 5154-91. 1991 Copper Alloy Globe, Globe Stop and Check, Check and Gate Valves. 22 pp.
SNZ NZS/BS 5154-89. Specification for Copper Alloy Globe, Globe Stop and Check, Check and Gate Valves. 20 pp.
SNZ NZS/BS 5154-91. Specification for Copper Alloy Globe, Globe Stop and Check, Check and Gate Valves. 24 pp.
—Hydraulic—Aircraft
SBAC AS 3382 ISSUE 3. 3/8 Inch Non-Return, Valve (Hydraulic). 1 p.
SBAC AS 3388 ISSUE 3. 1/2 Inch Non Return, Valve—Hydraulic. 1 p.
—Hydraulic—Mounting Surfaces
BSI BS 6494: Part 6-89. 1989 Hydraulic Fluid Power Valve Mounting Surfaces Part 6: Pressure-Control Valves (Excluding Pressure-Relief Valves), Sequence Valves, Unloading Valves, Throttle Valves and Check Valves. 22 pp.
ISO 5781-87. Hydraulic Fluid Power—Pressure-Control Valves (Excluding Pressure-Relief Valves), Sequence Valves, Unloading Valves, Throttle Valves and Check Valves—Mounting Surfaces First Edition. 20 pp.
—Petroleum Industry
CSA CAN/CSA-Z245.15-M91. Steel Valves; (Gen Instr 1). 81 pp.
—Wafer
BSI BS 7438-91. 1991 Steel and Copper Alloy Wafer Check Valves, Single Disk, Spring-Loaded Type. 14 pp.
—Water—Irrigation Equipment (Agricultural)
ISO 9952-93. Agricultural Irrigation Equipment—Check Valves First Edition. 11 pp.

Checking Resistance
See Also: Coatings; Cracking (Fracturing)
—Lacquers—Temperature Change Testing
CGSB 1-GP-71 METH 121.2-79. Methods of Testing Paints and Pigments Resistance to Temperature Changes Cold-Check Resistance. 1 p.

Checking Resistance (Cont.)
—Paints
SAA AS 1580.481.1.7-91. Paints and Related Materials—Methods of Test—Part 481: Coatings—Part 481.1: Assessment of Individual Defects of Exposed Films —Part 481.1.7: Exposed to Weathering—Degree of Checking (Supersedes AS 1580.481.1—1975 (in Part)) (in Professional Packages 30, 39). 3 pp.
SNZ NZS/AS 1580. 481.1.7-91. Methods of Test for Paints and Related Materials Part 481.1.7: Coatings—Exposed to Weathering—Degree of Checking (This is a Joint Standard with SAA AS 1580.481.1.7). 3 pp.

Checks
Use For: Cheques See Also: Banking Documents
—Data Transmission
ISO 6680-87. International Cheque Remittance First Edition. 8 pp.
OSI ISO 6680-87. International Cheque Remittance. 8 pp.
—Printing Papers—Bond—Magnetic Ink Character Recognition
CGSB 9-GP-47M-79. Paper, Bond, for Magnetic Ink Character Recognition Cheque Printing, Standard for; (Amendment 1 Mar 1980). 9 pp.

Checkweighing Scales
See Also: Scales (Weight)
EC 78/1031/EEC-78. Council Directive on the Approximation of the Laws of the Member States Relating to Automatic Checkweighing and Weight Grading Machines. 21 pp.

Cheese Head Screws
See Also: Screws
—Counterbores
DIN ENGL 74 Pt 2-91. Counterbores (Holes) for Cheese Head, Pan Head and Hexagon Socket Head Cap Screws (May). 5 pp.
DIN ENGL 974 Pt 1-91. Diameters of Counterbores (Holes) for Cheese Head, Pan Head and Hexagon Socket Head Cap Screws (May). 4 pp.
—Countersunk—Electric Contacts
CNS B2233-81. Cheese Head Countersunk Screws for Electric Contact Set (Apr)(4413). 3 pp.
—Gage Handles
DIN ENGL 2240 Pt 2-89. Gauge Handles for Gauging Members over 40 mm Nominal Diameter; Handles, Fixing Screws and Locking Prongs (Nov). 4 pp.
—Phillips
CNS B2729-88. Cross-Recessed Binding (or Cheese) Head Machine Screws (Apr)(9674).
—Phillips—Oval
DIN ENGL 7985-90. Cross Recessed Raised Cheese Head Screws (Aug). 5 pp.
—Safety Cups
DIN ENGL 526-73. Safety Cups for Cheese Head Screws According to DIN 84 (May). 1 p.
—Slotted
BSI BS 7635-93. 1993 Slotted Cheese Head Screws: Product Grade A (ISO 1207: 1992) (E). 10 pp.
CNS B2200-80. Slotted Cheese Head Screws (Jul)(4355). 4 pp.
CNS B2207-80. Slotted Cheese Head Screws for Fine Mechanics (M 0.3 to M 1.4) (Jul)(4362). 4 pp.
DIN ENGL 84-90. Product Grade A Slotted Cheese Head Screws (Aug). 4 pp.
DIN ENGL 8243-69. Slotted Cheese Head Screws for Fine Mechanics M 0.3 to M 1.4 (Nov). 2 pp.
ISO 1207-92. Slotted Cheese Head Screws—Product Grade A Third Edition. 7 pp.
—Slotted—Counterbores
CNS B1130-83. Counterbores for Hexagon Socket Head Screws and Slotted Cheese Head Screws (Mar)(4807).
—Slotted—Oval
CNS B2728-88. Slotted Binding (or Oval Cheese) Head Machine Screws (Apr)(9673).
—Spring Lock Washers
DIN ENGL 7980-87. Spring Lock Washers with Square Ends for Cheese Head Screws (Oct). 3 pp.
—Washers
CNS B2329-78. Washers for Cheese Head Screws (Nov)(4651). 2 pp.
DIN ENGL 433 Pt 1-90. Product Grade A Washers with a Hardness up to 250 HV Designed for Use with Cheese Head Screws (Mar). 3 pp.

INTERNATIONAL AND NON-U.S. NATIONAL STANDARDS
SUBJECT INDEX
Chemical

Cheese Head Screws *(Cont.)*
—Washers *(Cont.)*
DIN ENGL 433 Pt 2-90. Product Grade A Washers with a Hardness from 300 HV Designed for Use with Cheese Head Screws (Mar). 3 pp.

Cheese Molds
Use For: Cheese Moulds; Molds (Cheese)
See Also: Dairy Equipment
BSI BS 3633: Part 1-69. 1969 Cheese-Pressing Moulds Part 1: Rectangular Moulds. 12 pp.
BSI BS 3633: Part 2-63. 1963 Amd 1 Cheese-Pressing Moulds Part 2: Circular Moulds (AMD PD 6380) April 19, 1968. 11 pp.

Cheese Moulds
Use: Cheese Molds

Cheese Products
Use: Cheeses

Cheesecloth
See Also: Cotton Fabrics
CGSB 4-GP-81MB-90. Cotton Cheesecloth. 8 pp.

Cheeses
Use For: Cheese Products; Processed Cheese Food; Whey Cheese *See Also:* Animal Products; Dairy Products; Milk
CGSB 32.172M-88. Cheese and Cheese Products (Supersedes 32-GP-162M and 32-GP-173M). 10 pp.
CNS N5087-90. Processed Cheese (Edible) (Jun)(2879).
CNS N5087-86. Processed Cheese (Edible) (Sep)(2879). 2 pp.

—Aluminum Foil
BSI BS 1683-87. 1987 Coated Aluminium Foil for Wrapping Processed Cheese. 8 pp.

—Ash Content
CNS N6089-74. Methods of Test for Milk and Milk Products (Determination of Ash of Edible Cheese) (Apr)(3722). 1 p.

—Boxes (Containers)
CGSB CAN/CGSB-43.43-M86. Boxes, Bulk Cheese, and Liners. 13 pp.

—Chemical Analysis
BSI BS 770: Part 1-86. 1986 Chemical Analysis of Cheese Part 1: General Introduction (Including the Method for Preparation of the Sample). 4 pp.
SAA AS N75-70. Methods for the Sampling and Chemical Analysis of Cheese. 19 pp.

—Chloride Content—Potentiometric Analysis
BSI BS 770: Part 4-89. 1989 Chemical Analysis of Cheese Part 4: Determination of Chloride Content. 4 pp.
ISO 5943-88. Cheese and Processed Cheese Products—Determination of Chloride Content—Potentiometric Titration Method Second Edition. 5 pp.

—Citric Acid Content
BSI BS 770: Part 7-76. 1976 Chemical Analysis of Cheese Part 7: Determination of Citric Acid Content (Reference Method). 3 pp.
ISO 2963-74. Cheese and Processed Cheese Products—Determination of Citric Acid Content (Reference Method) First Edition; (Erratum—Dec 1981). 5 pp.

—Dry Matter Content
ISO 2920-74. Whey Cheese—Determination of Dry Matter Content (Reference Method) First Edition. 4 pp.

—Fat Content
CNS N6066-72. Methods of Test for Milk and Milk Products (Detection of Iso-Fat) (Oct)(3450). 1 p.
SNZ NZS 501: Part 2-71. Gerber Method for the Determination of Fat in Milk and Milk Products Part 2: Methods Amend: A, 1982. 28 pp.

—Fat Content—Gravimetric Analysis
BSI BS 770: Part 3-89. 1989 Chemical Analysis of Cheese Part 3: Determination of Fat Content (Reference Method). 12 pp.
ISO 1735-87. Cheese and Processed Cheese Products—Determination of Fat Content—Gravimetric Method (Reference Method) Second Edition. 11 pp.
ISO 1854-87. Whey Cheese—Determination of Fat Content—Gravimetric Method (Reference Method) Second Edition. 11 pp.

—Fat Content—Van Gulik Method
ISO 3432-75. Cheese—Determination of Fat Content—Butyrometer for Van Gulik Method First Edition. 6 pp.
ISO 3433-75. Cheese—Determination of Fat Content—Van Gulik Method First Edition. 6 pp.

Cheeses *(Cont.)*
—Hoops
SAA AS N47-63. Rectangular Cheese Hoops. 11 pp.

—Microbiological Analysis
SAA AS 1095.2.4-81. Microbiological Methods for the Dairy Industry—Part 2: Methods for the Examination of Specific Dairy Products—Part 2. 4: Cheese. 3 pp.

—Natamycin Content—Liquid Chromatography
ISO 9233-91. Cheese and Cheese Rind—Determination of Natamycin Content—Method by Molecular Absorption Spectrometry and by High-Performance Liquid Chromatography First Edition. 13 pp.

—Natamycin Content—Molecular Absorption Spectrophotometry
ISO 9233-91. Cheese and Cheese Rind—Determination of Natamycin Content—Method by Molecular Absorption Spectrometry and by High-Performance Liquid Chromatography First Edition. 13 pp.

—Nisin Content
BSI BS 4020-74. 1974 Amd 1 Estimation and Differentiation of Nisin in Processed Cheese. 16 pp.

—Nitrate Content—Cadmium Reduction
BSI BS 770: Part 9-85. 1985 Chemical Analysis of Cheese Part 9: Determination of Nitrate and Nitrate Contents. 8 pp.
ISO 4099-84. Cheese—Determination of Nitrate and Nitrite Contents—Method by Cadmium Reduction and Photometry Second Edition. 8 pp.
ISO 6739-88. Whey Cheese—Determination of Nitrate and Nitrite Contents—Method by Cadmium Reduction and Spectrometry Second Edition. 8 pp.

—Nitrate Content—Photometry
BSI BS 770: Part 9-85. 1985 Chemical Analysis of Cheese Part 9: Determination of Nitrate and Nitrate Contents. 8 pp.
ISO 4099-84. Cheese—Determination of Nitrate and Nitrite Contents—Method by Cadmium Reduction and Photometry Second Edition. 8 pp.
ISO 6739-88. Whey Cheese—Determination of Nitrate and Nitrite Contents—Method by Cadmium Reduction and Spectrometry Second Edition. 8 pp.

—Nitrite Content—Cadmium Reduction
ISO 4099-84. Cheese—Determination of Nitrate and Nitrite Contents—Method by Cadmium Reduction and Photometry Second Edition. 8 pp.
ISO 6739-88. Whey Cheese—Determination of Nitrate and Nitrite Contents—Method by Cadmium Reduction and Spectrometry Second Edition. 8 pp.

—Nitrite Content—Photometry
ISO 4099-84. Cheese—Determination of Nitrate and Nitrite Contents—Method by Cadmium Reduction and Photometry Second Edition. 8 pp.
ISO 6739-88. Whey Cheese—Determination of Nitrate and Nitrite Contents—Method by Cadmium Reduction and Spectrometry Second Edition. 8 pp.

—Nitrogen Content
BSI BS 770: Part 8-87. 1987 Chemical Analysis of Cheese Part 8: Determination of Nitrogen Content (Reference Method). 4 pp.

—PH—Chemical Analysis
BSI BS 770: Part 5-76. 1976 Chemical Analysis of Cheese Part 5: Determination of pH Value. 2 pp.

—Phosphorus Content—Molecular Absorption Spectrophotometry
BSI BS 770: Part 6-84. 1984 Chemical Analysis of Cheese Part 6: Determination of Total Phosphorus Content (Molecular Absorption Spectrometric Method). 4 pp.
ISO 2962-84. Cheese and Processed Cheese Products—Determination of Total Phosphorus Content—Molecular Absorption Spectrometric Method Second Edition. 5 pp.

—Protective Coverings
SAA AS 2183-91. Flexible Wrappings (Including Laminates) for Rindless Cheese. 10 pp.

—Rennet
BSI BS 3624-63. 1963 Determination of the Milk Coagulating Power of Rennet. 8 pp.

—Salt Content
CNS N6090-74. Methods of Test for Milk and Milk Products (Determination of Salt of Edible Cheese) (Apr)(3723). 1 p.

—Sampling
SAA AS N75-70. Methods for the Sampling and Chemical Analysis of Cheese. 19 pp.

Cheeses *(Cont.)*
—Sodium Chloride
BSI BS 998-69. (WITHDRAWN) 1969 Amd 1 Vacuum Salt for Butter and Cheese Making and Other Food Uses (Superseded by BS 998: 1990 & BS 7319: Parts 1 to 12: 1990). 22 pp.
BSI BS 998-90. 1990 Vacuum Salt for Food Use. 9 pp.

—Solids Content
BSI BS 770: Part 10-86. 1986 Chemical Analysis of Cheese Part 10: Determination of Total Solids Content (Reference Method). 5 pp.
ISO 5534-85. Cheese and Processed Cheese—Determination of Total Solids Content (Reference Method) First Edition. 4 pp.

—Test Specimens
CNS N6056-80. Method of Test for Milk and Milk Products (General Rules) (Aug)(3440). 2 pp.

—Test Specimens—Microbiological Analysis
DIN ENGL 10191 Pt 4-86. Microbiological Examinations; Preparation of Samples of Cheese and Cheese Products (Oct). 2 pp.

—Water Content
BSI BS 770: Part 2-76. 1976 Chemical Analysis of Cheese Part 2: Determination of Water Content (Reference and Routine Methods). 2 pp.

Chelate Cleaning
Use: Chemical Cleaning

Chelating Agents Content Analysis
—Waste Water
DIN ENGL 38409 Pt 26-89. German Standard Methods for the Examination of Water, Waste Water and Sludge; Parameters Characterizing Effects and Substances (Group H); Determination of the Bismuth-Chelating Index, IBiK (H 26) (May). 4 pp.

Chemical Admixtures
Use: Admixtures

Chemical Agent Casualties
See Also: Casualties
—Documents
NATO STANAG 2917 ED 1 AMD 0-85. Chemical Casualty Assessment Exercise Publication (AXP-7). 5 pp.

Chemical Analysis
Scope Note: For chemical analysis of specific products or materials, see the subheading Chemical Analysis under the specific material or product. For additional references, consult the following list
Use For: Analytical Chemistry; Chemical Testing
See Also: Absorptiometric Analysis; Assaying; Atomic Absorption Analysis; Calorific Value; Calorimetry; Colorimetric Analysis; Electrolytic Analysis; Epoxy Equivalent; Fluorometric Analysis; Gas Analysis; Gas Sampling; Hydrometers; Photometry; Physical Testing; Polarographic Analysis; Qualitative Analysis; Quantitative Analysis; Radioactivation Analysis; Reagents; Solubility; Spectrochemical Analysis; Spectrography; Turbidimeters; Volumetric Analysis; Water Analysis
CNS K0022-90. General Considerations for Chemical Analysis (Dec)(9179).
CNS Z4032-86. General Rules for Allowance of Chemical Analysis and Physical Test (Nov)(11771).
ISO 78 Pt 2-82. Layouts for Standards—Part 2: Standard for Chemical Analysis First Edition. 14 pp.
JIS K 0050-91. General Rules for Chemical Analysis. 29 pp.
JIS Z 8402-91. General Rules for Permissible Tolerance of Chemical Analyses and Physical Tests. 110 pp.

—Glossaries
CNS K0009-80. Technical Terms for Analytical Chemistry (Optic Parts) (Sep) (6490).
CNS K0044-89. Technical Terms for Analytical Chemistry (General Part) (Aug)(12586).
CNS K0045-89. Technical Terms for Analytical Chemistry (Electro Chemistry Division) (Aug)(12587).
CNS K0046-89. Technical Terms for Analytical Chemistry (Chromatography Part) (Aug)(12588).
JIS K 0211-87. Technical Terms for Analytical Chemistry (General Part). 90 pp.
JIS K 0215-89. Technical Terms for Analytical Chemistry (Analytical Instrument Part).

—Laboratories—Quality Assurance
CSA Q641-1992SP. Guide for Quality Assurance Programs for Physical and Chemical Testing Laboratories; (Gen Instr 1). 35 pp.

INDUSTRY STANDARDS

Chemical Analysis (Cont.)
—Water
SAA AS 2031.1-86. Selection of Containers and Preservation of Water Samples for Chemical and Micro-biological Analysis—Part 1: Chemical. 7 pp.

Chemical Apparatus
Use: Chemical Equipment

Chemical Cleaning
Use For: Acid Pickling; Chelate Cleaning; Pickling
See Also: Cleaning; Metal Cleaning
—Steels
SAA AS 1627.5-75. Metal Finishing—Preparation and Pretreatment of Surfaces—Part 5: Pickling Steel Surfaces (This is a Joint Standard with SANZ NZS 1627). 12 pp.
SNZ NZS/AS 1627. 5-75. Metal Finishing-Preparation and Pretreatment of Surfaces Part 5: Pickling Steel Surfaces (This is a Joint Standard with SAA AS 1627.5). 12 pp.

Chemical Coatings
Use: Conversion Coatings

Chemical Conversion Coatings
Use: Conversion Coatings

Chemical Equipment
Use For: Chemical Apparatus See Also: Equipment
—Organic Coatings
DIN ENGL 28054-90. Organic Coatings for Application to Metallic Components of Chemical Apparatus; Requirements and Testing (Apr). 6 pp.
—Stands (Supports)
DIN ENGL 28081 Pt 1-85. Tubular Vessel Supports; Dimensions (June). 4 pp.

Chemical Hazards
Use: Hazardous Materials

Chemical Hoses
See Also: Hoses
CNS K4070-82. Chemical Resistant Hoses (Nov)(9619).
—Rubber
JIS K 6342-82. Rubber Hoses for Chemical Use.

Chemical Indexes
Scope Note: Use a more specific term See: Acetyl Value; Base Number; Black Smoke Index; Bomer Number; Bromine Number; Chemical Analysis; Chlorine Number; Copper Number; Cresol Red Index; Dichromate Number; Ester Value; Hydroxyl Value; Hypo Number; Iodine Number; Kappa Number; Neutralization Number; Permanganate Number; Peroxide Number; Rancidity Index; Saponification Number; Saprobic Index; Scoville Index; Sedimentation Number; Viscosity Number

Chemical Industry
Scope Note: For additional listings, see the subheading Chemical Industry under the specific types of equipment
—Classification
BSI BS 1000: (66/66.0)-83. 1983 Universal Decimal Classification (UDC). English Full Edition (66/66.0): Chemical Technology. Chemical Engineering, Production and Processing, Operations and Plant. 36 pp.
SNZ NZS/BS 1000 (6/60)-79. Universal Decimal Classification Applied Sciences in General. Inventions and Discoveries. 8 pp.
SNZ NZS/BS 1000 (66/66.0)-83. Universal Decimal Classification Chemical Technology. Chemcial Engineering, Production and Processing, Operations and Plant. 36 pp.
—Quality Assurance
SAA AS 3905.1-91. Quality System Guidelines—Part 1: Guidelines to AS 3901/NZS 9001/ISO 9001 for Chemical and Allied Industries (in Professional Package 46). 15 pp.
—Stoneware Pipes—Acid Resistant
CNS R2106-80. Conical Flanged Straight Pipe of Acid-Proof Pottery for Chemical Industry (Sep)(6520).
CNS R2109-80. Conical Flanged T Pipe of Acid-Proof Pottery for Chemical Industry (Sep)(6523).
CNS R2110-80. Straight Pipe with Socket of Acid-Proof Pottery for Chemical Industry (Sep)(6524).
JIS R 1511-91. Conical Flanged Straight Pipe of Acid Proof Porcelains for Chemical Industry. 5 pp.
JIS R 1514-91. Conical Flanged Tee Pipes of Acid Proof Porcelains for Chemical Industry. 5 pp.

Chemical Neutrality
—Aromatic Hydrocarbons
ISO 5276-79. Aromatic Hydrocarbons—Test for Neutrality First Edition. 3 pp.

Chemical Oxygen Demand
Use: Oxygen Demand

Chemical Plants
See Also: Industrial Plants
—Fire Protection
BSI BS 5908-90. 1990 Code of Practice for Fire Precautions in the Chemical and Allied Industries. 86 pp.
—Glass
CNS R2194-88. Glass Components for Chemical Plants (Oct)(12445).
—Pipelines—Symbols
JIS Z 8209-88. Symbols for Piping Drawings of Chemical Plants. 28 pp.

Chemical Plating
Use: Electrodeposited Coatings; Electroplated Coatings

Chemical Process Equipment
See Also: Equipment
—Covers
DIN ENGL 28125 Pt 1-89. DN 150 to DN 600 Round Hinged Closures (Aug). 20 pp.
—Flanges
DIN ENGL 28030 Pt 2-89. Flanged Joints for Chemical Apparatus; Tolerances on Flange Dimensions (Feb). 3 pp.
DIN ENGL 28031-89. Flanges for Welding for Use on Unalloyed and Stainless Steel Vessels Not Subject to Pressure (Feb). 3 pp.
DIN ENGL 28032-89. Flanges for Welding for Use on Unalloyed Steel Pressure Vessels (June). 4 pp.
DIN ENGL 28034-89. Weld-Neck Flanges for Use on Unalloyed Steel Pressure Vessels (Feb). 4 pp.
DIN ENGL 28036-89. Flanges for Welding for Use on Stainless Steel Pressure Vessels (Feb). 3 pp.
DIN ENGL 28038-89. Flanges for Welding, with Cylindrical Hub, for Use on Stainless Steel Pressure Vessels (Feb). 5 pp.
—Joints—Flanged
DIN ENGL 28030 Pt 1-89. Flanged Joints for Chemical Apparatus; Design and Construction (Feb). 7 pp.
—Joints—Flanged—Gaskets
DIN ENGL 28040-89. Gaskets for Use with Flanged Joints (Feb). 3 pp.
—Organic Coatings
DIN ENGL 28053-88. Chemical Apparatus; Metallic Substrates and Semi-Finished Products to Receive Organic Coatings and Linings (Nov). 8 pp.
—Organic Linings
DIN ENGL 28053-88. Chemical Apparatus; Metallic Substrates and Semi-Finished Products to Receive Organic Coatings and Linings (Nov). 8 pp.
DIN ENGL 28055 Pt 1-90. Organic Linings for Application to Metallic Components of Chemical Apparatus; Requirements (Sept) (Supersedes Parts of VDI-Richtlinie (VDI Code of Practice) 2534 Suppl. 1, Aug. 1966 Ed., Suppl. 2, June 1972 Ed., Suppl. 3, Jan. 1967 Ed., and VDI-Richtlinie 2537, Jan. 1976 Ed.). 4 pp.
DIN ENGL 28055 Pt 2-91. Organic Linings for Application to Metallic Components of Chemical Apparatus; Testing (Feb) (Supersedes VDI-Richtlinie (VDI Code of Practice) 2539, August 1967 Edition). 4 pp.

Chemical Products
—Chloride Content—Mercurimetric Analysis
ISO 5790-79. Inorganic Chemical Products for Industrial Use—General Method for Determination of Chloride Content—Mercurimetric Method First Edition. 11 pp.
—Chloride Ion Content—Potentiometric Analysis
BSI BS 6337: Part 4-84. 1984 General Methods of Chemical Analysis Part 4: Method for Determination of Chloride Ions by Potentiometry. 11 pp.
ISO 6227-82. Chemical Products for Industrial Use—General Method for Determination of Chloride Ions—Potentiometric Method First Edition. 10 pp.

Chemical Products (Cont.)
—Classification
BSI BS 1000: (661)-83. 1983 Universal Decimal Classification (UDC). English Full Edition (661): Chemical Products. 44 pp.
SNZ NZS/BS 1000 (661)-83. Universal Decimal Classification Chemical Products. 44 pp.
—Colorimetry
BSI BS 5339-76. 1976 Measurement of Colour in Hazen Units (Platinum-Cobalt Scale) of Liquid Chemical Products. 5 pp.
ISO 2211-73. Liquid Chemical Products—Measurement of Colour in Hazen Units (Platinum-Cobalt Scale) First Edition. 4 pp.
—Crystallization
BSI BS 4633 & 4634-70. 1970 Determination of Crystallizing Point. Determination of Melting Point and/or Melting Range. 12 pp.
BSI BS 4633 & 4634-01. 1970 Amd 1 Method for the Determination of Crystallizing Point Method for the Determination of Melting Point and/or Melting Range (AMD 7715) July 15, 1993 (L). 14 pp.
—Defense Contracts
MOD UK DEFCON 112S-90. Basic Set of Conditions of Contract—Paints and Chemicals 10/90. 3 pp.
—Density
CNS K0015-82. Method for Determining Specific Gravity of Chemical Products (May)(8834).
CNS K6560-80. Determination of Density of Liquid Chemical Products (at 20 Degrees Celsius) (Aug)(6205).
ISO 758-76. Liquid Chemical Products for Industrial Use—Determination of Density at 20 Degrees Celsius First Edition. 4 pp.
JIS K 0061-92. Test Methods for Density and Relative Density of Chemical Products. 60 pp.
—Density—Pycnometric Analysis
CNS K6560-80. Determination of Density of Liquid Chemical Products (at 20 Degrees Celsius) (Aug)(6205).
ISO 758-76. Liquid Chemical Products for Industrial Use—Determination of Density at 20 Degrees Celsius First Edition. 4 pp.
—Distillation Methods
CNS K0023-82. Method for Determining Distillation of Chemical Products (Jul)(9180).
JIS K 0066-92. Test Methods for Distillation of Chemical Products. 22 pp.
—Ester Value
JIS K 0070-92. Test Methods for Acid Value, Saponification Value, Ester Value, Iodine Value, Hydroxyl Value and Unsaponifiable Matter of Chemical Products. 20 pp.
—Freezing Points
CNS K0017-82. Method for Determining Freezing Point of Chemical Products (May)(8836).
CNS K7007-82. Chemical Reagent (Freezing Point Determination) (Jun)(1507).
JIS K 0065-92. Test Method for Freezing Point of Chemical Products. 7 pp.
—Glossaries
BSI BS 2474-93. 1993 Names for Chemicals Used in Industry (W). 37 pp.
BSI BS 2474-83. 1983 Names for Chemicals Used in Industry. 36 pp.
ISO 6206-79. Chemical Products for Industrial Use—Sampling—Vocabulary First Edition. 10 pp.
—Hydroxyl Value
JIS K 0070-92. Test Methods for Acid Value, Saponification Value, Ester Value, Iodine Value, Hydroxyl Value and Unsaponifiable Matter of Chemical Products. 20 pp.
—Iodine Number
JIS K 0070-92. Test Methods for Acid Value, Saponification Value, Ester Value, Iodine Value, Hydroxyl Value and Unsaponifiable Matter of Chemical Products. 20 pp.
—Iron Content—Spectrophotometry
BSI BS 6337: Part 3-83. 1983 General Methods of Chemical Analysis Part 3: Method for Determination of Iron Content (1.10-Phenanthroline Spectrophotometric Method). 7 pp.
ISO 6685-82. Chemical Products for Industrial Use—General Method for Determination of Iron Content—1,10-Phenanthroline Spectrophotometric Method First Edition. 6 pp.
—Losses
CNS K0019-82. Method for Determining Loss and Residue of Chemical Products (May)(8838).
JIS K 0067-92. Test Methods for Loss and Residue of Chemical Products. 18 pp.

INTERNATIONAL AND NON-U.S. NATIONAL STANDARDS
SUBJECT INDEX
Chemistry

Chemical Products (Cont.)

—Melting Points
BSI BS 4633 & 4634-70. 1970 Determination of Crystallizing Point. Determination of Melting Point and/or Melting Range. 12 pp.
BSI BS 4633 & 4634-01. 1970 Amd 1 Method for the Determination of Crystallizing Point Method for the Determination of Melting Point and/or Melting Range (AMD 7715) July 15, 1993 (L). 14 pp.
CNS K0016-82. Method for Determining Melting Point of Chemical Products (May)(8835).
CNS K7008-82. Chemical Reagent (Melting Point Determination) (May)(1508).
JIS K 0064-92. Test Methods for Melting Point and Melting Range of Chemical Products. 11 pp.

—Refractivity
CNS K0020-88. Method of Test for Refractive Index of Chemicals (Sep)(8839).
JIS K 0062-92. Test Methods for Refractive Index of Chemical Products. 15 pp.

—Residue Content
CNS K0019-82. Method for Determining Loss and Residue of Chemical Products (May)(8838).
JIS K 0067-92. Test Methods for Loss and Residue of Chemical Products. 18 pp.

—Safety Data Sheets
DIN ENGL 52900-83. DIN Safety Data Sheet for Chemical Substances and Preparations; Form and Instructions on How to Fill in the Form (Feb). 10 pp.

—Sampling
BSI BS 5309: Part 1-76. 1976 Methods for Sampling Chemical Products Part 1: Introduction and General Principles. 24 pp.
ISO 3165-76. Sampling of Chemical Products for Industrial Use—Safety in Sampling First Edition. 9 pp.
ISO 6206-79. Chemical Products for Industrial Use—Sampling—Vocabulary First Edition. 10 pp.
ISO 8213-86. Chemical Products for Industrial Use—Sampling Techniques—Solid Chemical Products in the Form of Particles Varying from Powders to Coarse Lumps First Edition. 33 pp.

—Saponification Number
JIS K 0070-92. Test Methods for Acid Value, Saponification Value, Ester Value, Iodine Value, Hydroxyl Value and Unsaponifiable Matter of Chemical Products. 20 pp.

—Sieve Analysis
JIS K 0069-92. Test Methods for Sieving of Chemical Products. 13 pp.

—Specific Rotation
CNS K0018-82. Method for Determining Specific Rotation of Chemical Products (May)(8837).
CNS K7013-82. Chemical Reagent (Specific Rotation) (Jun)(1513).
JIS K 0063-92. Test Methods for Optical Rotation of Chemical Products. 13 pp.

—Storage
BSI BS 5502: Part 81-89. 1989 Design of Buildings and Structures for Agriculture Part 81: Code of Practice for Design and Construction of Chemical Stores (R). 12 pp.

—Sulfur Compound Content—Reduction
BSI BS 6337: Part 1-83. 1983 General Methods of Chemical Analysis Part 1: Method for Determination of Traces of Sulphur Compounds by Reduction and Titrimetry. 10 pp.
ISO 6228-80. Chemical Products for Industrial Use—General Method for Determination of Traces of Sulphur Compounds, as Sulphate, by Reduction and Titrimetry First Edition; (Erratum—Dec 1981). 10 pp.

—Sulfur Compound Content—Volumetric Analysis
BSI BS 6337: Part 1-83. 1983 General Methods of Chemical Analysis Part 1: Method for Determination of Traces of Sulphur Compounds by Reduction and Titrimetry. 10 pp.
ISO 6228-80. Chemical Products for Industrial Use—General Method for Determination of Traces of Sulphur Compounds, as Sulphate, by Reduction and Titrimetry First Edition; (Erratum—Dec 1981). 10 pp.

—Unsaponifiable Matter
JIS K 0070-92. Test Methods for Acid Value, Saponification Value, Ester Value, Iodine Value, Hydroxyl Value and Unsaponifiable Matter of Chemical Products. 20 pp.

—Water Content
JIS K 0068-92. Test Methods for Water Content of Chemical Products. 35 pp.

Chemical Properties
Scope Note: Use a more specific term *See:* Acidity; Alkalinity; Aniline Point; Buffer Value; Calorific Value; Chemical Neutrality; Chromaticity; Heat of Combustion; Heat of Crystallization; Heat of Fusion; Heat of Hydration; Molecular Weight; Optical Activity; Passivation; pH; Solubility; Thermodynamic Properties; Toxicity

Chemical Pulps
See Also: Cellulose; Dissolving Pulps; Kraft Pulps; Pulps; Sulfite Pulps

—Bleaching Agents
CPPA J.16P-92. Analysis of Peroxides. 3 pp.

—Fiber Length
CPPA D.11U-77. Determination of Fibre Length Index (De Montigny Grid). 2 pp.

—Impurities Content
ISO 5350 Pt 1-82. Pulps—Estimation of Dirt and Shives—Part 1: Unbleached Chemical Pulps First Edition. 7 pp.

—Kappa Number
CPPA G.8U-77. Rapid Determination of Kappa Number. 2 pp.

Chemical Reagents
Use: Reagents

Chemical Resistant Clothing
Use For: Splash Resistant Clothing
See Also: Exposure Suits

—Gloves
CEN PREN 374 (Part 1)-90. Protective Gloves Against Chemicals and Micro-Organisms—Part 1: Terminology and Performance Requirements. 6 pp.
CEN PREN 374 (Part 5)-90. Protective Gloves Against Chemicals and Micro-Organisms—Part 5; Determination of Resistance to Permeation by Chemicals. 12 pp.
DIN ENGL 4841 Pt 5-87. Protective Gloves; Grade 2 Gloves Affording Protection Against Chemicals; Safety Requirements, Testing (Apr). 2 pp.

—Liquid Permeability
BSI BS 7182-89. 1989 Air-Impermeable Chemical Protective Clothing. 15 pp.
BSI BS 7184-89. 1989 Selection, Use and Maintenance of Chemical Protective Clothing. 25 pp.
BSI BS EN 368-93. 1993 Protective Clothing—Protection Against Liquid Chemicals—Test Method: Resistance of Materials to Penetration by Liquids (N). 11 pp.
BSI BS EN 369-93. 1993 Protective Clothing—Protection Against Liquid Chemicals—Test Method: Resistance of Materials to Permeation by Liquids (N). 14 pp.
CEN PREN 368-90. Protective Clothing: Protection Against Liquid Chemicals: Resistance of Materials to Penetration by Liquids. 10 pp.
CEN PREN 369-90. Protective Clothing: Protection Against Liquid Chemicals: Resistance of Air-Impermeable Materials to Permeation by Liquids. 15 pp.
CEN EN 369-93. Protective Clothing—Protection Against Liquid Chemicals—Test Method: Resistance of Materials to Permeation by Liquids. 12 pp.
CEN PREN 464-91. Chemical Protective Clothing—Protective Against Gases and Vapours—Method of Test—Determination of Leak—Tightness (Internal Pressure Test). 6 pp.
CEN PREN 465-91. Protective Clothing—Protection Against Liquid Chemicals—Performance Requirements —Type 4 Equipment—Protective Suits with Spray—Tight Connections Between Different Parts of the Protective Suit. 12 pp.
CEN PREN 466-91. Chemical Protection Clothing—Protection Against Liquid Chemicals (Including Liquid Aerosols)—Performance Requirements—Type 3 Equipment—Chemical Protective Clothing with Liquid Tight—Connections Between Different Parts of the Clothing. 12 pp.
CEN PREN 467-91. Protective Clothing—Protective Against Liquid Chemicals—Performance Requirements —Type 5 Equipment—Garments Providing Chemical Protection to Parts of the Body. 11 pp.
CEN PREN 945-92. Protective Clothing for Use Against Liquid and Gaseous Chemicals, Including Liquid Aerosols and Solid Particles—Performance Requirements for Air Fed Protective Clothing with Non Gas Tight Connections (Type 2 Equipment). 9 pp.
DIN ENGL EN 368-93. Protective Clothing for Use Against Liquid Chemicals; Method of Determining the Resistance of Materials to Penetration by Liquids (Jan) (Supersedes Parts of DIN 32763, September 1986 Edition). 7 pp.

Chemical Resistant Clothing (Cont.)

—Liquid Permeability (Cont.)
ISO 6529-90. Protective Clothing—Protection Against Liquid Chemicals—Determination of Resistance of Air-Impermeable Materials to Permeation by Liquids First Edition. 11 pp.

—Penetration Resistance
CEN PREN 463-91. Chemical Protective Clothing—Protection Against Liquid Chemicals—Method of Test—Determination of Resistance to Penetration by Liquids (Jet Test). 6 pp.
CEN PREN 468-91. Chemical Protective Clothing—Protection Against Liquid Chemicals—Method of Test—Determination of Resistance to Penetration by Spray. 9 pp.
CEN PREN 943-92. Protective Clothing for Use Against Liquid and Gaseous Chemicals, Including Liquid and Aerosols and Solid Particles—Performance Requirements for Unventilated Protective Clothing with Gas Tight Connections (Type 1B Equipment). 12 pp.
CEN PREN 944-92. Protective Clothing for Use Against Liquid and Gaseous Chemicals, Including Liquid Aerosols and Solid Particles—Performance Requirements for Air Fed Protective Clothing with Gas Tight Connections (Type 1C Equipment). 26 pp.
CEN PREN 946-92. Protective Clothing for Use Against Liquid and Gaseous Chemicals, Including Liquid Aerosols and Solid Particles—Performance Requirements for Unventilated Encapsulating Protective Clothing with Gas Tight Connections (Type 1A Equipment). 14 pp.
DIN ENGL 32763-86. Grade 2 Clothing for Protection Against Chemicals; Safety Requirements, Testing (Sept) (Partially Supersedes DIN 4846 Part 1, December 1965 Edition and DIN 4847, September 1969 Edition) (Partially Superseded by DIN EN 368). 7 pp.
ISO 6530-90. Protective Clothing—Protection Against Liquid Chemicals—Determination of Resistance of Materials to Penetration by Liquids Second Edition. 7 pp.

Chemical Testing
Use: Chemical Analysis

Chemical Toilets
Use For: Water Closets, Chemical *See Also:* Toilet Facilities; Toilets
BSI BS 2081: Part 1-80. 1980 Closets for Use with Chemicals Part 1: Portable and Transportable Types. 6 pp.
BSI BS 2081: Part 2-81. 1981 Closets for Use with Chemicals Part 2: Permanent Installations. 7 pp.

Chemical Warfare
Use: Military Chemical Operations

Chemical Warfare Agents
Use: Military Chemical Agents

Chemicals
Scope Note: Use a more specific term
See: Agricultural Chemicals; Cyclohexane; Molybdate; Monomethylamine Hydrochloride; Monosodium Glutamate; Papermaking Chemicals; o-Phenanthroline Hydrochloride; 1,10-Phenanthroline Hydrochloride; p-Phenetidine; Phenicarbazide; Phenolphthalein; Phenolsulfonphthalein; L-Phenylalanine; m-Phenylenediamine Hydrochloride; m-Phenylenediamine; Phenylfluorone; Phenylmercuric Compounds; Phosphorus Pentachloride; Phosphorus-32; Rubber Chemicals

Chemiluminescence

—Air
BSI BS 1747: Part 9-87. 1987 Methods for the Measurement of Air Pollution Part 9: Determination of the Mass Concentration of Nitrogen Oxides in Ambient Air Chemiluminescence Method. 12 pp.
BSI BS 1747: Part 12-93. 1993 Methods for Measurement of Air Pollution Part 12: Determination of Mass the Concentration of Ozone in Ambient Air: Chemiluminescence Method (ISO 10313: 1993) (N). 14 pp.
CSA Z223.23-M1981. Method for the Measurement of Ozone in Air; (Gen Instr 1). 17 pp.
CSA Z223.24-M1983. Method for the Measurement of Nitric Oxide and Nitrogen Dioxide in Air; (Gen Instr 1). 24 pp.
ISO 7996-85. Ambient Air—Determination of the Mass Concentration of Nitrogen Oxides—Chemiluminescence Method First Edition. 11 pp.
ISO 10313-93. Ambient Air—Determination of the Mass Concentration of Ozone—Chemiluminescence Method First Edition. 12 pp.

Chemistry
See Also: Physical Chemistry

Chemistry (Cont.)

—Classification
- BSI BS 1000: (54)-72. 1972 Amd 2 Universal Decimal Classification (UDC). English Full Edition (54): Chemistry. Crystallography. Mineralogy. 117 pp.
- BSI BS 1000: (66/66.0)-83. 1983 Universal Decimal Classification (UDC). English Full Edition (66/66.0): Chemical Technology. Chemical Engineering, Production and Processing, Operations and Plant. 36 pp.
- SNZ NZS/BS 1000 (54)-72. Universal Decimal Classification Chemistry. Crystallography. Mineralogy. 120 pp.
- SNZ NZS/BS 1000 (6/60)-79. Universal Decimal Classification Applied Sciences in General. Inventions and Discoveries. 8 pp.
- SNZ NZS/BS 1000 (66/66.0)-83. Universal Decimal Classification Chemical Technology. Chemcial Engineering, Production and Processing, Operations and Plant. 36 pp.

—Glossaries
- IEC. Vocabulary of Fundamental Concepts. 237 pp.
- IEC 50 Chap 111-01-82. Advance Edition of the International Electrotechnical Vocabulary Chapter 111: Physics and Chemistry Section 111-01—Physical Concepts. 47 pp.
- IEC 50 Chap 111-03-77. Advance Edition of the International Electrotechnical Vocabulary Chapter 111: Physics and Chemistry Section 111-03—Concepts Related to Quantities and Units. 22 pp.
- SAA AS 1852.111. 01-83. International Electrotechnical Vocabulary—Part 111.01: Physics and Chemistry—Physical Concepts. 16 pp.
- SAA AS 1852.111. 03-78. International Electrotechnical Vocabulary—Part 111.03: Physics and Chemistry—Concepts Relating to Quantities and Units Being IEC 50(111-03). 22 pp.
- SNZ IEC 50: 50(111-01)-82. International Electrotechnical Vocabulary 50(111-01): Physics and Chemistry Section 111-01: Physical Concepts. 37 pp.
- SNZ IEC 50: 50(111-03)-77. International Electrotechnical Vocabulary 50(111-03): Physics and Chemistry Section 111-03: Concepts Related to Quantities and Units. 18 pp.

—Symbols
- JIS Z 8202-85. Quantity Symbols, Unit Symbols and Chemical Symbols. 173 pp.

Cheques
Use: Checks

Cherries
See Also: Food; Fruits

—Dried
- ISO 6755-84. Dried Sour Cherries—Specification First Edition. 6 pp.
- ISO 7908-91. Dried Sweet Cherries—Specification First Edition. 9 pp.

—Kernels
- ISO 6757-84. Decorticated Kernels of Mahaleb Cherries—Specification First Edition. 6 pp.

—Refrigerated Transportation
- ISO 7920-84. Sweet Cherries and Sour Cherries—Guide to Cold Storage and Refrigerated Transport First Edition. 4 pp.

—Storage
- ISO 7920-84. Sweet Cherries and Sour Cherries—Guide to Cold Storage and Refrigerated Transport First Edition. 4 pp.

Chestnuts

—Puree
- EC 79/693/EEC-79. Council Directive on the Approximation of the Laws of the Member States Relating to Fruit Jams, Jellies and Marmalades and Chestnut Puree. 12 pp.
- EC 88/593/EEC-88. Council Directive Amending Directive 79/693/EEC on the Approximation of the Laws of the Member States Relating to Fruit Jams, Jellies and Marmalades and Chestnut Puree. 4 pp.

Chicken (Meat)
Use: Poultry Meat—Chicken

Chicory
See Also: Vegetables

—Extracts
- EC 77/436/EEC-77. Council Directive on the Approximation of the Laws of the Member States Relating to Coffee Extracts and Chicory Extracts. 5 pp.
- EC 85/573/EEC-85. Council Directive Amending Directive 77/436/EEC on the Approximation of the Laws of the Member States Relating to Coffee Extracts and Chicory Extracts. 2 pp.

Chicory (Cont.)

—Extracts—Dry Matter Content
- EC 79/1066/EEC-79. First Commission Directive Laying Down Community Methods of Analysis for Testing Coffee Extracts and Chicory Extracts. 12 pp.

Child Resistant Packaging
See Also: Consumer Safety; Packaging
- BSI BS 6652-89. (WITHDRAWN) 1989 Packagings Resistant to Opening by Children (Superseded by BS EN 28317: 1993). 16 pp.
- BSI BS EN 28317-93. 1993 Child-Resistant Packaging—Requirements and Testing Procedures for Reclosable Packages (ISO 8317: 1989) (H). 19 pp.
- CEN EN 28317-92. Child-Resistant Packaging—Requirements and Testing Procedures for Reclosable Packages (ISO 8317: 1989). 14 pp.
- CSA CAN/CSA-Z76.1-M90. Recloseable Child-Resistant Packages; (Gen Instr 1). 33 pp.
- DIN ENGL 55559-80. Packaging; Child Resistant Packages; Requirements Testing (Oct). 6 pp.
- ISO 8317-89. Child-Resistant Packaging—Requirements and Testing Procedures for Reclosable Packages First Edition; (CEN EN 28317: 1992). 15 pp.
- SAA AS 1928-82. Child-Resistant Packaging. 12 pp.
- SNZ NZS 5825-91. Child-Resistant Packages. 18 pp.

—Classification
- CNS Z5126-83. Classification of Child-Resistant Packages (Mar)(10135).

—Drugs
- BSI BS 7236-89. 1989 Code of Practice for Non-Reclosable Packaging for Solid Dose Units of Medicinal Products. 8 pp.
- BSI DD 30: Part 2-73. 1973 Resistance of Pharamaceutical Packages to Opening by Children Part 2: Non-Reclosable Unit Packages. 9 pp.

Child Restraining Devices
See Also: Child Safety Seats; Leading Reins (Babies); Safety Belts

—Adjustment Devices
- SAA AS 3629.4-91. Methods of Testing Child Restraints—Part 4: Determination of Adjustment Device Forces (AS 3629: Part 4:1991/NZS 5466: Part 4:1991). 2 pp.
- SNZ NZS 5466.4-91. Methods of Testing Child Restraints Part 4: Determination of Adjustment Device Forces Amend: 1, 1992 (NZS 5466.4:1991 AS 3629.4-1991). 4 pp.

—Dynamic Testing
- SNZ NZS 5466.1-91. Methods of Testing Child Restraints Part 1: Dynamic Testing Amend: 1, 1992 (NZS 5466.1:1991/AS 3629.1-1991). 10 pp.

—Throat Contact—Hazardous
- SNZ NZS/AS 3629. 2-90. Methods of Testing Child Restraints Part 2: Determination of Hazardous Throat Contact in Abnormal Situations Amend: 1, 1992 (This is a Joint Standard with SAA AS 3629.2). 1 p.

—Upper Anchorages—Dynamic Testing
- SAA AS 3629.3-91. Methods of Testing Child Restraints—Part 3: Dynamic Testing of Upper Anchorage Components (AS 3629: Part 3:1991/NZS 5466: Part 3:1991) Amdt 1 October 1992 Amdt 2 April 1993. 3 pp.
- SNZ NZS 5466.3-91. Methods of Testing Child Restraints Part 3: Dynamic Testing of Upper Anchorage Components Amend: 1, 1992; 2, 1993 (NZS 5466.3:1991/AS 3629.3-1991). 4 pp.

Child Safety Barriers
See Also: Barriers
- BSI BS 4125-91. 1991 Safety Requirements for Child Safety Barriers for Domestic Use. 15 pp.
- SAA AS 1879-76. Child Barriers for Domestic Premises. 8 pp.
- SNZ NZS 5843-88. Specification for Safety Requirements for Child Safety Barriers for Domestic Use Amend: 1, 1982; A, 1988. 8 pp.

Child Safety Seats
See Also: Child Restraining Devices; Safety Belts
- BSI BS 3254: Part 2-88. 1988 Seat Belt Assemblies for Road Vehicles Part 2: Specification for Restraining Devices for Children (Supersedes BS 3254: 1960) (E). 16 pp.
- BSI BS AU 186A-83. 1983 Amd 1 Carry Cot Restraints. 13 pp.
- CNS D2188-86. Child Restraint Devices for Automobiles (Feb)(11497). 7 pp.
- CNS D3167-86. Method of Test for Child Restraint Devices for Automobiles (Feb)(11498). 4 pp.
- JIS D 0401-90. Child Restraints for Automobiles. 25 pp.

Child Safety Seats (Cont.)

—Automotive
- BSI BS AU 202A-85. 1985 Amd 1 Rearward-Facing Restraining Devices for Infants for Use in Road Vehicles. 24 pp.
- SAA AS 1754-75. Child Restraint Systems for Use in Motor Vehicles (AS 1754:1991/NZS 5411:1991) Amdt 1 October 1992 Amdt 2 January/February 1993 Amdt 3 October 1993 (Superseded by 1754.1—1989, AS 1754.2 1989, AS 1754.4—1989, AS 3629.1—1989, AS 3629.2—1989). 41 pp.
- SNZ NZS 5411-91. Child Restraint Systems Amend: 1, 1992 (NZS 5411:1991/AS 1754-1991). 28 pp.

—Cushions
- BSI BS AU 185-83. 1983 Seat Belt Booster Cushions. 18 pp.

Children's Lamps
Use: Lamps—Children's

Chilies
Use For: Chillies *See Also:* Spices
- BSI BS 7087: Part 17-93. 1993 Herbs and Spices Ready for Food Use Part 17: Specification for Chillies (Whole and Ground). 11 pp.
- ISO 972-85. Chillies and Capsicums, Whole or Ground (Powdered)—Specification First Edition. 7 pp.

—Capsaicin Content—Liquid Chromatography
- ISO 7543 Pt 2-93. Chillies and Chilli Oleoresins—Determination of Total Capsaicinoid Content—Part 2: Method Using High-Performance Liquid Chromatography First Edition. 8 pp.

—Scoville Index
- BSI BS 4585: Part 7-89. 1989 Methods of Test for Spices and Condiments Part 7: Determination of Scoville Index of Chillies. 8 pp.
- CNS N6157-81. Method of Test for Spices and Condiments—Chillies—Determination of Scoville Index (Oct)(8050). 2 pp.
- ISO 3513-77. Spices and Condiments—Chillies—Determination of Scoville Index First Edition. 4 pp.

—Sensory Analysis
- CNS N6157-81. Method of Test for Spices and Condiments—Chillies—Determination of Scoville Index (Oct)(8050). 2 pp.

CHILL
Use For: CCITT High Level Language
See Also: Programming Languages
- BSI BS ISO/IEC 9496-89. 1989 CCITT High Level Language (Chill). 262 pp.
- CCITT FASCICLE VI.12. Public Land Mobile Network Interworking with ISDN and PSTN. Recommendations Q.1000—Q.1032 (Study Group XI). 95 pp.
- CCITT FASCICLE X.6. CCITT High Level Language (CHILL). Recommendation Z.200 (Study Group X). 260 pp.
- CCITT RECMN Z.200-89. CCITT High Level Language (CHILL)(Study Group X) 260 pp. 260 pp.
- OSI ISO IEC DIS 9496-2-88. Information Processing—Programming Languages—CCITT High Level Languages (CHILL). 242 pp.

—Design
- CCITT FASCICLE X.6. CCITT High Level Language (CHILL). Recommendation Z.200 (Study Group X). 260 pp.

—Design—Composite Modes
- CCITT FASCICLE X.6. CCITT High Level Language (CHILL). Recommendation Z.200 (Study Group X). 260 pp.

—Design—Concurrent Execution
- CCITT FASCICLE X.6. CCITT High Level Language (CHILL). Recommendation Z.200 (Study Group X). 260 pp.

—Design—Discrete Modes
- CCITT FASCICLE X.6. CCITT High Level Language (CHILL). Recommendation Z.200 (Study Group X). 260 pp.

—Design—Dynamic Modes
- CCITT FASCICLE X.6. CCITT High Level Language (CHILL). Recommendation Z.200 (Study Group X). 260 pp.

—Design—Implementation
- CCITT FASCICLE X.6. CCITT High Level Language (CHILL). Recommendation Z.200 (Study Group X). 260 pp.

INTERNATIONAL AND NON-U.S. NATIONAL STANDARDS
SUBJECT INDEX

CHILL (Cont.)
—Design—Input/Output Modes
CCITT FASCICLE X.6. CCITT High Level Language (CHILL). Recommendation Z.200 (Study Group X). 260 pp.

—Design—Input/Output Reference Models
CCITT FASCICLE X.6. CCITT High Level Language (CHILL). Recommendation Z.200 (Study Group X). 260 pp.

—Design—Instance Modes
CCITT FASCICLE X.6. CCITT High Level Language (CHILL). Recommendation Z.200 (Study Group X). 260 pp.

—Design—Location/Access
CCITT FASCICLE X.6. CCITT High Level Language (CHILL). Recommendation Z.200 (Study Group X). 260 pp.

—Design—Mode Classifications
CCITT FASCICLE X.6. CCITT High Level Language (CHILL). Recommendation Z.200 (Study Group X). 260 pp.

—Design—Mode Definitions
CCITT FASCICLE X.6. CCITT High Level Language (CHILL). Recommendation Z.200 (Study Group X). 260 pp.

—Design—Powerset Modes
CCITT FASCICLE X.6. CCITT High Level Language (CHILL). Recommendation Z.200 (Study Group X). 260 pp.

—Design—Procedure Modes
CCITT FASCICLE X.6. CCITT High Level Language (CHILL). Recommendation Z.200 (Study Group X). 260 pp.

—Design—Program Structure
CCITT FASCICLE X.6. CCITT High Level Language (CHILL). Recommendation Z.200 (Study Group X). 260 pp.

—Design—Reference Modes
CCITT FASCICLE X.6. CCITT High Level Language (CHILL). Recommendation Z.200 (Study Group X). 260 pp.

—Design—Synchronization Modes
CCITT FASCICLE X.6. CCITT High Level Language (CHILL). Recommendation Z.200 (Study Group X). 260 pp.

—Design—Timing Modes
CCITT FASCICLE X.6. CCITT High Level Language (CHILL). Recommendation Z.200 (Study Group X). 260 pp.

—Design—Values/Operations
CCITT FASCICLE X.6. CCITT High Level Language (CHILL). Recommendation Z.200 (Study Group X). 260 pp.

Chill Castings
See Also: Castings; Metal Products

—Aluminum Alloy
ISO 2378-72. Aluminium Alloy Chill Castings—Reference Test Bar First Edition. 4 pp.

—Aluminum Alloy—Aerospace
AECMA PREN2722-86. Aluminium Alloy A1-C12-T4 Chill Castings. 3 pp.
AECMA PREN2724-86. Aluminium Alloy A1-C12-T6 Chill Castings. 3 pp.
AECMA PREN2727-86. Aluminium Alloy A1-C26-T6 Chill Castings. 3 pp.
AECMA PREN2729-86. Aluminium Alloy A1-C27-T6 Chill Castings. 3 pp.
BSI L 173-91. 1991 Castings of Aluminium-Silicon-Magnesium Alloy, Chill Cast (Solution Treated and Precipitation Treated to an Overaged (T7) Condition). 6 pp.
CEN EN 3549-92. Aluminium Alloy AL-C21002 T7 Chill Castings. 4 pp.

—Copper and Copper Alloy
DIN ENGL 17655-81. Unalloyed and Low Alloy Copper Materials for Casting; Castings (Nov). 4 pp.

—Magnesium Alloy—Aerospace
AECMA PREN2732-86. Magnesium Alloy Mg-C51-T6 Chill Castings. 3 pp.
AECMA PREN2734-86. Magnesium Alloy Mg-C81-T5 Chill Castings. 3 pp.
AECMA PREN2736-86. Magnesium Alloy Mg-C91-T5 Chill Castings. 3 pp.
AECMA PREN2739-86. Magnesium Alloy Mg-C43-T5 Chill Castings. 3 pp.

Chill Castings (Cont.)
—Magnesium Alloy—Aerospace (Cont.)
BSI L 173-91. 1991 Castings of Aluminium-Silicon-Magnesium Alloy, Chill Cast (Solution Treated and Precipitation Treated to an Overaged (T7) Condition). 6 pp.

—Silicon Alloy—Aerospace
BSI L 173-91. 1991 Castings of Aluminium-Silicon-Magnesium Alloy, Chill Cast (Solution Treated and Precipitation Treated to an Overaged (T7) Condition). 6 pp.

Chillers
Use For: Liquid Chillers *See Also:* Condensing Units; Refrigeration Equipment; Water Coolers
CSA CAN/CSA-C22. 2 NO 236-M90. Heating and Cooling Equipment; (Gen Instr 1) (ANSI/UL 1995). 136 pp.

—Centrifugal—Packaged—Air Conditioning Equipment—Buildings
CSA C743-93. Performance Standard for Rating Packaged Water Chillers; (Gen Instr 1). 38 pp.

—Reciprocating—Packaged—Air Conditioning Equipment—Buildings
CSA C743-93. Performance Standard for Rating Packaged Water Chillers; (Gen Instr 1). 38 pp.

—Screw—Packaged—Air Conditioning Equipment—Buildings
CSA C743-93. Performance Standard for Rating Packaged Water Chillers; (Gen Instr 1). 38 pp.

Chillies
Use: Chilies

Chimney Sweeping
Use: Chimneys—Cleaning

Chimney Terminals
Use: Flue Terminals

Chimneys
See Also: Boilers; Dust Collectors; Exhaust Systems; Firebacks; Flare Stacks; Flue Blocks; Flue Exhausters; Flue Terminals; Flues; Smoke Viewers; Solid Fuel Burning Equipment; Vents
BSI BS 4543: Part 1-90. 1990 Factory-Made Insulated Chimneys Part 1: Methods of Test. 20 pp.
CSA CAN/CSA-B415.1-92. Performance Testing of Solid-Fuel-Burning Stoves, Inserts, and Low-Burn-Rate Factory-Built Fireplaces; (Gen Instr 1 Thru 2). 89 pp.

—Access Fittings
BSI BS 3572-86. 1986 Amd 1 Access Fittings for Chimneys and Other High Structures in Concrete or Brickwork. 20 pp.
BSI BS 3678-86. 1986 Access Hooks for Chimneys and Other High Structures. 8 pp.

—Asbestos Cement
CNS A1039-83. Standard Practice for Fitting Asbestos Cement Chimney (Apr)(10140).
JIS A 5405-82. Asbestos Cement Pipes. 10 pp.
JIS A 7802-62. Standard Practice for Fitting Asbestos Cement Chimney.

—Clay
CSA CAN/CSA-A324-M88. Clay Flue Liners; (Gen Instr 1). 35 pp.

—Cleaning—Service Contracts
MOD UK DEFCON 112BC-87. Conditions of Contract Chimney Sweeping 3/87. 6 pp.

—Deposits—Sampling
BSI BS 2455: Part 2-83. 1983 Methods of Sampling and Examining Deposits from Boilers and Associated Industrial Plant Part 2: Methods for Sampling and Examining Free-Side Deposits. 41 pp.
BSI BS 2455: Part 2-01. 1983 Amd 1 Sampling and Examining Deposits from Boilers and Associated Industrial Plant Part 2: Methods for Sampling and Examining Fire-Side Deposits (AMD 7751) October 15, 1993 (Q). 42 pp.

—Fire Testing
ISO 4736-79. Fire Tests—Small Chimneys—Testing at Elevated Temperatures First Edition. 8 pp.

—Free Standing—Bricks
DIN ENGL 1057 Pt 1-85. Building Materials for Free-Standing Chimneys; Compass Bricks; Requirements, Testing, Inspection (July). 3 pp.

—Free Standing—Cements
DIN ENGL 1057 Pt 3-85. Building Materials for Free-Standing Chimneys; Acid-Resistant Waterglass Cements; Requirements, Testing, Inspection (July). 2 pp.

Chimneys (Cont.)
—Free Standing—Construction
DIN ENGL 1056-84. Solid Construction, Free-Standing Chimneys; Design and Construction (Oct). 29 pp.

—Free Standing—Linings—Blocks
DIN ENGL 1057 Pt 2-85. Building Materials for Free-Standing Chimneys; Lining Blocks; Requirements, Testing, Inspection (July). 6 pp.

—Installation
BSI BS 6461: Part 2-84. (WITHDRAWN) 1984 Installation of Chimneys and Flues for Domestic Appliances Burning Solid Solid-Fuel (Including Wood and Peat) Part 2: Code of Practice for Factory-Made Insulated Chimneys for Internal Applications (Superseded by BS 7566: Parts 1, 2, 3 & 4: 1992). 15 pp.
BSI BS 7566: Part 1-92. 1992 Installation of Factory-Made Chimneys to BS 4543 for Domestic Appliances Part 1: Method of Specifying Installation Design Information. 8 pp.
BSI BS 7566: Part 2-92. 1992 Installation of Factory-Made Chimneys to BS 4543 for Domestic Appliances Part 2: Specification for Installation Design. 15 pp.
BSI BS 7566: Part 3-92. 1992 Installation of Factory-Made Chimneys to BS 4543 for Domestic Appliances Part 3: Specification for Site Installation. 8 pp.
BSI BS 7566: Part 4-92. 1992 Installation of Factory-Made Chimneys to BS 4543 for Domestic Appliances Part 4: Recommendations for Installation Design and Installation. 11 pp.

—Ladders
BSI BS 4211-87. 1987 Ladders for Permanent Access to Chimneys, Other High Structures, Silos and Bins. 16 pp.
BSI BS 4211-01. 1987 Amd 1 Ladders for Permanent Access to Chimneys, Other High Structures, Silos and Bins (AMD 7064) June 15, 1992 (R). 18 pp.

—Liners—Refractory Materials—Installation
BSI BS 4207-89. 1989 Installation of Monolithic Linings for Steel Chimneys and Flues. 8 pp.

—Liners—Stainless Steel
BSI BS 4543: Part 2-90. 1990 Factory-Made Insulated Chimneys Part 2: Chimneys with Stainless Steel Flue Linings for Use with Solid Fuel Fired Appliances. 12 pp.
BSI BS 4543: Part 3-90. 1990 Factory-Made Chimneys Part 3: Chimneys with Stainless Steel Fluelining for Use with Oil Fired Appliances. 12 pp.
BSI BS 4543: Part 3-76. 1976 Amd 2 Factory-Made Insulated Chimneys Part 3: Specification for Chimneys for Oil fired Appliances (AMD 3476) November 28, 1980. 10 pp.

—Masonry
CSA CAN/CSA-A405-M87. Design and Construction of Masonry Chimneys and Fireplaces; (Gen Instr 1). 40 pp.

—Masonry—Installation
BSI BS 6461: Part 1-84. 1984 Installation of Chimneys and Flues for Domestic Appliances Burning Solid-Fuel (Including Wood and Peat) Part 1: Code of Practice for Masonry Chimneys and Flue Pipes. 23 pp.

—Noise
DIN ENGL 45635 Pt 47-85. Measurement of Noise Emitted by Machines; Airborne Noise Emission; Enveloping Surface Method; Chimneys (June). 4 pp.

—Steel—Design
BSI BS 4076-89. 1989 Specification for Steel Chimneys. 24 pp.
BSI BS 4076-78. (WITHDRAWN) 1978 Amd 1 Specification for Steel Chimneys. 18 pp.

China
—Aircraft—Sales Financing
EURO AFS/HS/4650/435-79. China-Sales Financing of Civil Aircraft. 10 pp.

China Clay
Use: Kaolin

Chinese (ROC) National Standards
Use: CNS

Chinese Apricots
Use: Apricots

Chinese Backing Papers
Use: Beaubey Papers

Chinese Cabbages
See Also: Food; Vegetables

INTERNATIONAL AND NON-U.S. NATIONAL STANDARDS SUBJECT INDEX

Chinese

Chinese Cabbages (Cont.)
—Grading
CNS N1027-82. Grades and Packing of Chinese Cabbage (Jan) (2094). 3 pp.

—Packaging
CNS N1027-82. Grades and Packing of Chinese Cabbage (Jan) (2094). 3 pp.

Chinese Characters
See Also: Graphic Characters

—Coded Character Sets
CNS C5209-86. Standard Interchange Code for Generally-Used Chinese Characters (Aug)(11643).

—Engraving
JIS Z 8903-84. Standard Type of Letters Used in Mechanical Engraving (Joyo Kanji, Common-Use Chinese Characters) (R 1989). 43 pp.

—Romanization
BSI BS 7014-89. 1989 Guide to the Romanization of Chinese. 11 pp.
ISO 7098-91. Information and Documentation Romanization of Chinese Second Edition. 8 pp.

Chinese Chives
Use: Chives

Chinese Ham
Use: Ham

Chinese Lacquers
See Also: Lacquers
CNS K2048-86. Chinese Lacquer (Apr)(2809). 1 p.
CNS K6265-86. Method of Test for Chinese Lacquer (Apr)(2810).

Chinese Noodles
Use: Dry Noodles

Chinese Plums
Use: Plums

Chinese Sausages
Use: Sausages

Chip Capacitors
Scope Note: For additional listings, use a more specific term See Also: Aluminum Chip Capacitors; Capacitors; Ceramic Chip Capacitors; Metallized Film Chip Capacitors; Tantalum Chip Capacitors

—Metallized Film
IECQ QC 302200-93. Fixed Capacitors for Use in Electronic Equipment Part 19: Sectional Specification: Fixed Metallized Polyethylene-Terephthalate Film Dielectric Chip d.c. Capacitors (IEC 384-19 ED 1). 63 pp.
IECQ QC 302201-93. Fixed Capacitors for Use in Electronic Equipment Part 19: Blank Detail Specification: Fixed Metallized Polyethylene-Terephthalate Film Dielectric Chip d.c. Capacitors Assessment Level E (IEC 384-19-1 ED 1). 33 pp.

Chip Detectors
Use For: Metal Chip Dectectors

—Aircraft—Housings
MOD UK DSTAN 47-3-82. Metal Chip Detectors (Housings) Issue 4. 6 pp.

Chip Resistors
See Also: Fixed Resistors; Resistors
BSI BS QC 400600-90. 1990 Fixed Resistors for Use in Electronic Equipment: Sectional Specification: Fixed Surface Mounting (Chip) Resistors. 20 pp.
IEC 115 Pt 8-89. Fixed Resistors for Use in Electronic Equipment Part 8: Sectional Specification: Fixed Chip Resistors First Edition (IECQ QC 400600). 39 pp.
IEC 115 Pt 8-1-89. Fixed Resistors for Use in Electronic Equipment Part 8: Blank Detail Specification: Fixed Chip Resistors. Assessment Level E First Edition (IECQ QC 400601). 25 pp.
IECQ QC 400600-89. Fixed Resistors for Use in Electronic Equipment Part 8: Sectional Specification: Fixed Chip Resistors (IEC 115-8 ED 1). 39 pp.
IECQ QC 400601-89. Fixed Resistors for Use in Electronic Equipment Part 8: Blank Detail Specification: Fixed Chip Resistors Assessment Level E (IEC 115-8-1 ED 1). 25 pp.
IECQ QC 400601/JP 0001-88. Detail Specification for Electronic Components Fixed Chip Resistors. 10 pp.
IECQ QC 400601/US 0001 ISSUE 1-90. Resistors for Use in Electrotechnical Equipment Detail Specification: Fixed Chip Resistors Style RR 0805. 13 pp.

Chip Resistors (Cont.)
IECQ QC 400601/US 0002 ISSUE 1-90. Resistors for Use in Electrotechnical Equipment Detail Specification: Fixed Chip Resistors Style RR 1206. 13 pp.
JIS C 5212-89. Fixed Carbon Film Chip Resistors Cylindrical Type for Use in Electronic Equipment (Form 27, Characteristic D and G, Grade C). 29 pp.
JIS C 5213-89. Fixed Metal Film Chip Resistors Cylindrical Type for Use in Electronic Equipment (Form 27 and 27S, Characteristic G, Grade C). 28 pp.
JIS C 5222-89. Fixed Metal Film Chip Resistors Rectangular Type for Use in Electronic Equipment (Form 72 and 73, Characteristic F, G and H, Grade C). 33 pp.
JIS C 5223-89. Fixed Thick Film Chip Resistors Rectangular Type for Use in Electronic Equipment (Form 72 and 73, Characteristic H, K and M, Grade C). 34 pp.

—Ceramic—Printed Circuit Mount
CECC CECC 40 401-002. NL-CECC 40 401-002 Issue 4; Fixed Low Power Non-Wirewound Chip Resistors with Rectangular Base Without Leads (En). 7 pp.

—Preferred Products List
CECC CECC MUAHAG Vol 2 IS 4-90. Preferred Products List; Resistors and Potentiometers (En, Fr, Ge). 59 pp.

—Quality Assurance
BSI BS CECC 40400-90. 1990 Fixed Low Power Surface Mounting (Chip) Resistors. 19 pp.
BSI BS CECC 40401-90. 1990 Blank Detail Specification Fixed Low Power Non-Wirewound Surface Mounting (Chip) Resistors Assessment Level 'S'. 13 pp.

—Surface Mount
CECC CECC 40 400 ISSUE 1-89. Sectional Specification: Fixed Low Power Surface Mounting (Chip) Resistors (En, Fr, Ge). 44 pp.
CECC CECC 40 401 ISSUE 1-89. Blank Detail Specification: Fixed Low Power Non-Wirewound Surface Mounting (Chip) Resistors (En, Fr, Ge). 34 pp.

—Surface Mount—Preferred Products List
CECC CECC MUAHAG Vol 2 IS 4-90. Preferred Products List; Resistors and Potentiometers (En, Fr, Ge). 59 pp.

—Thin Film—Preferred Products List
CECC CECC MUAHAG Vol 2 IS 4-90. Preferred Products List; Resistors and Potentiometers (En, Fr, Ge). 59 pp.

Chipboard
See Also: Leatherboards; Millboards; Packaging Materials; Paperboard; Particle Board
BSI BS 5669: Part 2-89. 1989 Amd 1 Particleboard Part 2: Wood Chipboard (AMD 6611) March 29, 1991. 21 pp.
BSI BS 5669: Part 2-02. 1989 Amd 2 Particleboard Part 2: Specification for Wood Chipboard (AMD 6947) September 15, 1992 (SUPERSEDES BS 5669: 1979) (R). 23 pp.
BSI BS 5669: Part 2-03. 1989 Amd 3 Particleboard Part 2: Specification for Wood Chipboard (AMD 7778) August 15, 1993 (R). 24 pp.
BSI BS 7331-90. 1990 Direct Surfaced Wood Chipboard Based on Thermosetting Resins. 17 pp.
CNS P2049-84. Chip Paper Board (Jan)(3498). 2 pp.
DIN ENGL 68761 Pt 1-86. Chipboard; General Purpose; Flat Pressed Particle Board; FPY Board (Nov). 2 pp.
DIN ENGL 68761 Pt 4-82. Chipboard; Flat Pressed Board for General Purposes; FPO Board (Feb). 4 pp.
DIN ENGL 68762-82. Chipboard for Special Purposes in Building Construction; Concepts, Requirements, Testing (Mar). 4 pp.
DIN ENGL 68763-90. Flat Pressed Particleboard for Use in Building Construction; Concepts, Requirements, Testing and Inspection (Sept). 8 pp.
SNZ NZS 3608-75. Specification for Resin-Bonded Wood Chipboard Amend: 1, 1975; 2, 1975; 3, 1975; 3A, 1975. 24 pp.

—Ammunition
MOD UK DSTAN 81-40-80. Chipboard, Lined, and Chipboard, Unlined Issue 1. 10 pp.
MOD UK DEF-47-A-01. Paper, Chip, Patch and/or Bitumen Sized; Amendment 1. 7 pp.
MOD UK DEF-48-A-01. Board, Pitch and/or Bitumen Sized; Amendment 1. 6 pp.

—Boxes
MOD UK DSTAN 81-11-79. Cartons, Fibreboard, Fixed Joint, Kraft Lined Chipboard Issue 3. 9 pp.

Chipboard (Cont.)
—Kraft Lined
MOD UK DSTAN 81-1-01. Fibreboard, Solid, Kraft-Lined Chipboard Issue 2; Amendment 1. 9 pp.
MOD UK DSTAN 81-11-79. Cartons, Fibreboard, Fixed Joint, Kraft Lined Chipboard Issue 3. 9 pp.
MOD UK DEF-1250-01. Chipboard, Kraft Lined, High Wet-Strength; Amendment 2. 8 pp.

—Kraft Lined—Waterproof
MOD UK DSTAN 81-109-93. Chipboard, Kraft-Lined, Waterproof Issue 1. 17 pp.
MOD UK TS 416C. Chipboard, Kraft-Lined, Waterproof.

—Test Facilities
MOD UK DSTAN 93-59-89. Chipboard (for Use at Proof and Experimental Establishments) Issue 1. 9 pp.

—Tubes
MOD UK DSTAN 81-18-70. Tubes, Packing, Open End (Metric Units) Issue 1. 6 pp.

Chipmunk Aircraft
See Also: Aircraft

—Antenna Positions
CAA. Chipmunk (Approved Aerial Positions). 1 p.

Chipping
See Also: Cutting; Machining; Pulping

—Carbide Tools
BSI BS 7662-93. 1993 Application of Hard Cutting Materials for Machining by Chip Removal-Designation of the Main Groups of Chip Removal and Groups of Application (ISO 513: 1991). 11 pp.
ISO 513-91. Application of Hard Cutting Materials for Machining by Chip Removal—Designation of the Main Groups of Chip Removal and Groups of Application Second Edition. 9 pp.

Chipping Hammers
Use For: Scaling Hammers See Also: Chisels; Hammers; Hand Tools

—Power—Vibration
ISO 8662 Pt 2-92. Hand-Held Portable Power Tools—Measurement of Vibrations at the Handle—Part 2: Chipping Hammers and Riveting Hammers First Edition. 15 pp.

—Safety
CEN PREN 792-4-92. Handheld Non-Electric Power Tools—Safety Requirements—Part 4: Percussive Non-Rotary Power Tools. 8 pp.

Chips (Wood)
Use For: Wood Chips See Also: Pulpwood; Wood

—Density
JIS P 8014-76. Testing Method for Density of Pulpwood and Woodchips (R 1985). 8 pp.
SAA AS 1301.P1S-79. Methods of Test for Pulp and Paper (Metric Units)—Part P1s: Basic Density of Wood Chips (This is a Joint Standard with SANZ NZS 1301).
SNZ NZS/AS 1301. P1S-79. Methods of Test for Pulp and Paper Basic Density of Wood Chips (This is a Joint Standard with SAA AS 1301.P1S). 2 pp.

—Moisture Content
SAA AS 1301.P10R P-86. Methods of Test for Pulp and Paper (Metric Units)—Part P10rp: Determination of Moisture in Wood Chips—Distillation Method (This is a Joint Standard with SANZ NZS 1301).
SNZ NZS/AS 1301. P10RP-86. Methods of Test for Pulp and Paper Determination of Moisture in Wood Chips—Distillation Method (This is a Joint Standard with SAA AS 1301.P10RP). 2 pp.

—Moisture Content—Toluene Method
JIS P 8015-76. Testing Method for Moisture in Woodchips and Sawdusts by Toluene Method (R 1985). 7 pp.

Chisel Bars
Use: Chisels

Chisels
Use For: Gouges See Also: Cape Chisels; Chipping Hammers; Cold Chisels; Dental Instruments; Diamond Chisels; Flat Chisels; Fuller Chisels; Hand Tools; Osteotomes; Rivet Busters; Roundnose Chisels; Tools
CGSB 39-GP-45A-74. Bars, Steel, Standard for; (Amendment 1 Mar 1975). 18 pp.
CNS B3198-70. Chisel (Tentative) (May)(3097).
CNS B3204-70. Hand Chisel (Tentative) (May)(3152).
CNS B3207-83. Chisels of Stone Work (Mar)(3155).

INTERNATIONAL AND NON-U.S. NATIONAL STANDARDS
SUBJECT INDEX

Chloride

Chisels (Cont.)
—Grooving
CNS B3246-78. Grooving Chisels (May)(3978). 1 p.
—Nonsparking
CNS M2057-86. Non-Sparking Beryllium Copper Alloy Tools Chisels (Nov)(5905).
—Shanks
CNS B3245-78. Pneumatic Tools, Shank Chisels (May)(3977). 3 pp.
—Woodworking
BSI BS 1943-89. 1989 Woodworking Chisels and Gouges. 14 pp.
ISO 2729-73. Woodworking Tools—Chisels and Gouges First Edition. 9 pp.
MOD UK DSTAN 51-11: Part 27-01. Hand Tools, General Purpose Part 27: Chisels and Gouges, Woodworking Issue 1; Corrigendum. 17 pp.

Chisels (Surgical)
Use: Osteotomes

Chives
See Also: Onions
—Grading
CNS N1104-82. Grades of Flower Stalk of Chinese Chive (Nov) (9632). 2 pp.

Chloral Hydrate
CNS K7139-64. Chemical Reagent (Chloral Hydrate) (Jul)(1639).
JIS K 8869-61. Chloral Hydrate.

Chloramine-T
See Also: Chloramine-T Content Analysis
CNS K7140-64. Chemical Reagent (Chloramine-T)(Sodium P-Toluenesulfonch-loramide) (Jul)(1640).

Chloramine-T Content Analysis
Use For: Tosylchloramide Sodium Content Analysis
See Also: Chloramine-T
—Cosmetics
EC 83/514/EEC-83. Third Commission Directive on the Approximation of the Laws of the Member States Relating to Methods of Analysis Necessary for Checking the Composition of Cosmetic Products. 38 pp.

Chloramine-T Trihydrate
Use For: Sodium p-Toluenesulfonchloramide Trihydrate
JIS K 8318-80. Sodium p-Toluenesulfon-chloramide Trihydrate.

Chlorate Content Analysis
—Cosmetics
EC 85/490/EEC-85. Fourth Commission Directive on the Approximation of the Laws of the Member States Relating to Methods of Analysis Necessary for Checking the Composition of Cosmetic Products. 16 pp.
—Magnesium Carbonate
MOD UK M 870/67. Examination of Magnesium Carbonate, Heavy.
—Milk
BSI BS 1741: Part 11-89. 1989 Methods for Chemical Analysis of Liquid Milk and Cream Part 11: Detection of Detergent/Disinfectant Residues. 4 pp.
—Sodium Chlorate—Volumetric Analysis
ISO 3199-75. Sodium Chlorate for Industrial Use—Determination of Chlorate Content—Dichromate Titrimetric Method First Edition. 4 pp.
—Strontium Nitrate
MOD UK M 867/92. Examination of Strontium Nitrate (Anhydrous).

Chlorella
Use: Algae

Chlorfenvinphos
See Also: Pesticides
SAA AS 1870.11D-79. Standard for Development—Pesticides for Agricultural Use—Part 11D: Chlorfenvinphos. 15 pp.

Chloric Acid
—Packaging
CNS Z5012-72. Packing of Sulfuric Acid, Chloric Acid and Nitric Acid (Oct)(1259).

Chloride Content Analysis
—Acetic Acid—Visual Limit Testing
ISO 753 Pt 8-81. Acetic Acid for Industrial Use—Methods of Test—Part 8: Visual Limit Test for Inorganic Chlorides First Edition. 4 pp.
—Admixtures
CEN PREN 480 (Part 10)-91. Admixtures for Concrete, Mortar and Grout—Test Methods—Part 10: Determination of the Chloride Content. 7 pp.
—Aggregates—Concretes
SAA AS 1012.20-92. Methods of Testing Concrete—Part 20: Determination of Chloride and Sulfate in Hardened Concrete and Concrete Aggregates (in Professional Packages 30, 58). 4 pp.
—Ammonium Dihydrogen Orthophosphate
MOD UK M 9515/65. Examination of Ammonium Dihydrogen Orthophosphate (Withdrawn).
—Ammonium Hydroxide
BSI BS 4651: Part 6-88. 1988 Ammonia Solution Part 6: Methods for Determination of Chloride Content. 4 pp.
—Ammonium Nitrate—Colorimetric Analysis
BSI BS 4267: Part 8-87. 1987 Ammonium Nitrate Part 8: Methods for Determination of Chloride Content. 4 pp.
—Ammonium Nitrate—Potentiometric Analysis
BSI BS 4267: Part 8-87. 1987 Ammonium Nitrate Part 8: Methods for Determination of Chloride Content. 4 pp.
—Animal Feed
BSI BS 5766: Part 5-81. 1981 Amd 1 Methods for Analysis of Animal Feeding Stuffs Part 5: Determination of Water-Soluble Chlorides Content. 7 pp.
CNS N4024-13-86. Method of Test for Feeds: Determination of Water Soluble Chloride (Aug)(2770-13).
ISO 6495-80. Animal Feeding Stuffs—Determination of Water-Soluble Chlorides Content First Edition. 5 pp.
—Barium Peroxide
MOD UK M 819/72. Examination of Barium Peroxide.
—Boric Acids—Mercurimetric Analysis
BSI BS 5688: Part 16-79. 1979 Orthoboric Acid (Boric Acid), Diboron Trioxide (Boric Oxide), Disodium Tetraborates, Sodium Perborates and Crude Sodium Borates for Industrial Use Part 16: Determin. of Chloride Content of Boric Acid, Boric Oxide and Disodium Tetraborates. 4 pp.
ISO 3121-76. Boric Acid, Boric Oxide and DiSodium Tetraborates for Industrial Use—Determination of Chloride Content—Mercurimetric Method First Edition. 5 pp.
—Boron Oxides—Mercurimetric Analysis
BSI BS 5688: Part 16-79. 1979 Orthoboric Acid (Boric Acid), Diboron Trioxide (Boric Oxide), Disodium Tetraborates, Sodium Perborates and Crude Sodium Borates for Industrial Use Part 16: Determin. of Chloride Content of Boric Acid, Boric Oxide and Disodium Tetraborates. 4 pp.
ISO 3121-76. Boric Acid, Boric Oxide and DiSodium Tetraborates for Industrial Use—Determination of Chloride Content—Mercurimetric Method First Edition. 5 pp.
—Camphor—Volumetric Analysis
MOD UK M 9555/72. Examination of Camphor for Celluloid (Superseded by CS 2914).
—Cements—Volumetric Analysis
BSI BS EN 196-21-92. 1992 Methods of Testing Cement Part 21: Determination of the Chloride, Carbon Dioxide and Alkali Content of Cement. 22 pp.
CEN EN 196 (Part 21)-89. Methods of Testing Cement: Determination of the Chloride, Carbon Dioxide and Alkali Content of Cement. 28 pp.
—Cheeses—Potentiometric Analysis
BSI BS 770: Part 4-89. 1989 Chemical Analysis of Cheese Part 4: Determination of Chloride Content. 4 pp.
ISO 5943-88. Cheese and Processed Cheese Products—Determination of Chloride Content—Potentiometric Titration Method Second Edition. 5 pp.
—Chromic Acid
MOD UK M 9526/66. Examination of Chromic Acid Crystals.

Chloride Content Analysis (Cont.)
—Cleaning Agents
CGSB 31-GP-0A METH 32.3-62. Methods of Testing Corrosion-Prevention Materials and Processes Chloride Content; (Re-Edited March 1982). 2 pp.
—Concretes
SAA AS 1012.20-92. Methods of Testing Concrete—Part 20: Determination of Chloride and Sulfate in Hardened Concrete and Concrete Aggregates (in Professional Packages 30, 58). 4 pp.
—Detergents—Volumetric Analysis
BSI BS 3762: Sec 3.16-85. 1985 Analysis of Formulated Detergents Part 3: Quantitative Test Methods Section 3.16: Method for Determination of Chloride Content. 4 pp.
—Dicyandiamide
MOD UK M 812/77. Examination of Dicyandiamide.
—Drying Agents
BSI BS 3482: Part 4-91. 1991 Methods of Test for Desiccants Part 4: Determination of Water-Soluble Chlorides Content. 5 pp.
—Dyes
MOD UK DEF-1053: METH NO. 79. Standard Methods of Testing Paint, Varnish, Lacquer and Related Products. Indices Method 79: Determination of Water-Soluble Chloride Content in Dyestuffs (Withdrawn).
MOD UK M 818/63. Examination of Dyestuffs for Use in Pyrotechnic Compositions and HE Substitutes.
—Fabrics—Gravimetric Analysis
BSI BS 3266-81. 1981 Determination of Conductivity, pH, Water Soluble Matter, Chloride and Sulphate in Aqueous Extracts of Textile Materials. 9 pp.
—Fabrics—Volumetric Analysis
BSI BS 3266-81. 1981 Determination of Conductivity, pH, Water Soluble Matter, Chloride and Sulphate in Aqueous Extracts of Textile Materials. 9 pp.
—Formaldehyde—Visual Limit Testing
ISO 2221-72. Formaldehyde Solutions for Industrial Use—Limit Test for Inorganic Chlorides First Edition. 4 pp.
—Formic Acid—Turbidity
BSI BS 4341-68. 1968 Methods of Test for Formic Acid. 13 pp.
—Formic Acid—Visual Limit Testing
ISO 731 Pt IV-77. Formic Acid for Industrial Use—Methods of Test—Part IV: Visual Limit Test for Inorganic Chlorides First Edition. 4 pp.
—Glucose
CNS N6055-72. Method of Test for Powdered Glucose (Jun)(3351). 3 pp.
—Glycerol—Turbidimetry
BSI BS 5711: Part 13-79. 1979 Methods of Sampling and Test for Glycerol Part 13: Limit Test for Chloride. 2 pp.
BSI BS 5711: Part 14-79. 1979 Methods of Sampling and Test for Glycerol Part 14: Limit Test for Organic Chloride. 2 pp.
—Glycerol—Volumetric Analysis
BSI BS 5711: Part 12-79. 1979 Methods of Sampling and Test for Glycerol Part 12: Determination of Chloride Content. 2 pp.
—Industrial Water
CPPA H.4P(N)-67. Analysis of Process Waters Chloride. 2 pp.
—Inorganic Chemical Products—Mercurimetric Analysis
ISO 5790-79. Inorganic Chemical Products for Industrial Use—General Method for Determination of Chloride Content—Mercurimetric Method First Edition. 11 pp.
—Iron Ores—Ion Selective Electrode Method
BSI BS 7020: Part 20-90. 1990 Analysis of Iron Ores Part 20: Method for the Determination of Water Soluble Chloride Content: Ion-Selective Electrode Method. 21 pp.
ISO 9517-89. Iron Ores—Determination of Water Soluble Chloride Content—Ion-Selective Electrode Method First Edition. 11 pp.
—Magnesium Carbonate
MOD UK M 870/67. Examination of Magnesium Carbonate, Heavy.

INDUSTRY STANDARDS

Chloride Content Analysis (Cont.)

—Magnesium Oxides
MOD UK M 9541/69. Examination of Magnesium Oxide and Magnesium Oxide, Special, Lead Free (Withdrawn).

—Meat
BSI BS 4401: Part 6-81. 1981 Methods of Test for Meat and Meat Products Part 6: Determination of Chloride Content. 5 pp.
CNS N6146-80. Method of Test for Meat and Meat Products—Determination of Chloride Content (Sep)(6512). 3 pp.
ISO 1841-81. Meat and Meat Products—Determination of Chloride Content (Reference Method) First Edition. 5 pp.

—Milk
BSI BS 1741: Part 8-88. 1988 Amd 2 The Chemical Analysis of Liquid Milk and Cream Part 8: Determination of Chloride Content of Liquid Milk (AMD 1234) December 31, 1998. 8 pp.

—Mortars
CEN PREN 1015-17-93. Methods of Test for Mortar for Masonry—Part 17: Determination of Soluble Chloride Content of Fresh and Hardened Mortars. 9 pp.

—Olefin Resins
BSI BS 2782:Pt4: METH 452D-F-78. 1978 Methods of Testing Plastics: Part 4: Chemical Properties: 452D-F: Determination of: (D) pH of Water Extract of Polyolefin Compound: (E) Water-Soluble Sulphates in Polyolefin Compound: (F) Water-Soluble Chlorides in Polyolefin Compound. 2 pp.

—Oleic Acid
MOD UK M 9545/68. Examination of Oleic Acid, Special, Lead Free.

—Paperboard
BSI BS 2924: Part 4-90. 1990 Aqueous Extracts of Paper, Board and Pulp Part 4: General Method for Determination of Water-Soluble Chlorides (V) (ISO 9197-1: 1989). 7 pp.
BSI BS 2924: Part 4-01. 1990 Amd 1 Aqueous Extracts of Paper, Board and Pulp Part 4: General Method for Determination of Water-Soluble Chlorides (ISO 9197-1: 1989) (AMD 6360) June 28, 1991 (Supersedes BS 2924: 1968). 8 pp.
DIN ENGL 53125-85. Determination of Chloride Content in Aqueous Extracts of Paper and Board (Aug). 4 pp.
ISO 9197 Pt 1-89. Paper, Board and Pulps—Determination of Water-Soluble Chlorides—Part 1: General Method First Edition. 6 pp.
ISO 9197 Pt 2-90. Paper, Board and Pulps—Determination of Water-Soluble Chlorides—Part 2: Method for High Purity Products First Edition. 6 pp.

—Paperboard—Potentiometric Analysis
CNS P3050-81. Analytical Methods for Water Soluble Chlorides in Paper and Paperboard (Jul)(7747).
JIS P 8144-76. Analytical Method for Water Soluble Chlorides in Paper and Paperboard (R 1984). 10 pp.

—Paperboard—Volumetric Analysis
CNS P3050-81. Analytical Methods for Water Soluble Chlorides in Paper and Paperboard (Jul)(7747).
JIS P 8144-76. Analytical Method for Water Soluble Chlorides in Paper and Paperboard (R 1984). 10 pp.

—Papers
BSI BS 2924: Part 4-90. 1990 Aqueous Extracts of Paper, Board and Pulp Part 4: General Method for Determination of Water-Soluble Chlorides (V) (ISO 9197-1: 1989). 7 pp.
BSI BS 2924: Part 4-01. 1990 Amd 1 Aqueous Extracts of Paper, Board and Pulp Part 4: General Method for Determination of Water-Soluble Chlorides (ISO 9197-1: 1989) (AMD 6360) June 28, 1991 (Supersedes BS 2924: 1968). 8 pp.
DIN ENGL 53125-85. Determination of Chloride Content in Aqueous Extracts of Paper and Board (Aug). 4 pp.
ISO 9197 Pt 1-89. Paper, Board and Pulps—Determination of Water-Soluble Chlorides—Part 1: General Method First Edition. 6 pp.
ISO 9197 Pt 2-90. Paper, Board and Pulps—Determination of Water-Soluble Chlorides—Part 2: Method for High Purity Products First Edition. 6 pp.

—Papers—Potentiometric Analysis
CNS P3050-81. Analytical Methods for Water Soluble Chlorides in Paper and Paperboard (Jul)(7747).
JIS P 8144-76. Analytical Method for Water Soluble Chlorides in Paper and Paperboard (R 1984). 10 pp.

—Papers—Volumetric Analysis
CNS P3050-81. Analytical Methods for Water Soluble Chlorides in Paper and Paperboard (Jul)(7747).

Chloride Content Analysis (Cont.)

—Papers—Volumetric Analysis (Cont.)
JIS P 8144-76. Analytical Method for Water Soluble Chlorides in Paper and Paperboard (R 1984). 10 pp.
MOD UK DSTAN 13-10-68. Paper, Wrapping, Unglazed, and Paper, Wrapping, Unglazed, Lead Free Issue 1. 9 pp.
MOD UK M 6512/61. Examination of Paper Wrapping, Unglazed and Paper, Wrapping, Unglazed, LF Quality (Superseded by Def Stan 13-10).

—Phosphates—Potentiometric Analysis
BSI BS 4427: Part 14-89. 1989 Methods of Test for Sodium Tripolyphosphate (Pentasodium Triphosphate) and Sodium Pyro-phosphate (Tetrasodium Pyrophosphate) for Industrial Use Part 14: Condensed Phosphates for Indus Use (Including Foodstuffs): Determin of Chloride Content. 4 pp.
ISO 5374-78. Condensed Phosphates for Industrial Use (Including Foodstuffs)—Determination of Chloride Content—Potentiometric Method First Edition. 6 pp.

—Phosphoric Acids—Potentiometric Analysis
BSI BS 4258: Part 9-89. 1989 Methods of Test for Phosphoric Acid (Orthophosphoric Acid) for Industrial Use Part 9: Determination of Chloride Content. 4 pp.
ISO 3708-76. Phosphoric Acid for Industrial Use (Including Foodstuffs)—Determination of Chloride Content—Potentiometric Method First Edition. 6 pp.

—Potassium Cryolite
MOD UK M 828/68. Examination of Potassium Cryolite.

—Potassium Hydroxides—Mercurimetric Analysis
ISO 992-75. Potassium Hydroxide for Industrial Use—Determination of Chlorides Content-Mercurimetric Method First Edition. 4 pp.

—Potassium Hydroxides—Photometry
ISO 3177-75. Potassium Hydroxide for Industrial Use—Determination of Chlorides Content—Photometric Method First Edition. 5 pp.

—Potassium Sulfate—Mercurimetric Analysis
ISO 2488-73. Potassium Sulphate for Industrial Use—Determination of Chloride Content—Mercurimetric Method First Edition. 4 pp.

—Pulps
BSI BS 2924: Part 4-90. 1990 Aqueous Extracts of Paper, Board and Pulp Part 4: General Method for Determination of Water-Soluble Chlorides (V) (ISO 9197-1: 1989). 7 pp.
BSI BS 2924: Part 4-01. 1990 Amd 1 Aqueous Extracts of Paper, Board and Pulp Part 4: General Method for Determination of Water-Soluble Chlorides (ISO 9197-1: 1989) (AMD 6360) June 28, 1991 (Supersedes BS 2924: 1968). 8 pp.
ISO 9197 Pt 1-89. Paper, Board and Pulps—Determination of Water-Soluble Chlorides—Part 1: General Method First Edition. 6 pp.
ISO 9197 Pt 2-90. Paper, Board and Pulps—Determination of Water-Soluble Chlorides—Part 2: Method for High Purity Products First Edition. 6 pp.

—Selenious Acid—Volumetric Analysis
MOD UK M 9525/67. Examination of Selenious Acid Crystals.

—Slag Cements—Volumetric Analysis
BSI BS EN 196-21-92. 1992 Methods of Testing Cement Part 21: Determination of the Chloride, Carbon Dioxide and Alkali Content of Cement. 22 pp.
CEN EN 196 (Part 21)-89. Methods of Testing Cement: Determination of the Chloride, Carbon Dioxide and Alkali Content of Cement. 28 pp.

—Soaps
CGSB 2-GP-11M METH 11.1-83. Methods of Testing and Analysis of Soaps and Detergents Chlorides. 1 p.

—Soaps—Potentiometric Analysis
BSI BS 1715: Sec 2.7-89. 1989 Analysis of Soaps Part 2: Quantitative Test Methods Section 2.7: Method for Determination of Chloride Content (ISO 4323: 1977). 7 pp.
ISO 4323-77. Soaps—Determination of Chloride Content—Potentiometric Method First Edition. 5 pp.

—Soaps—Volumetric Analysis
ISO 457-83. Soaps—Determination of Chloride Content—Titrimetric Method Second Edition. 4 pp.

—Sodium Azide
MOD UK M 807/91. Examination of Sodium Azide.

Chloride Content Analysis (Cont.)

—Sodium Bicarbonate—Mercurimetric Analysis
ISO 2201-72. Sodium Hydrogen Carbonate for Industrial Use—Determination of Chloride Content—Mercurimetric Method First Edition. 4 pp.

—Sodium Borates—Mercurimetric Analysis
BSI BS 5688: Part 16-79. 1979 Orthoboric Acid (Boric Acid), Diboron Trioxide (Boric Oxide), Disodium Tetraborates, Sodium Perborates and Crude Sodium Borates for Industrial Use Part 16: Determin. of Chloride Content of Boric Acid, Boric Oxide and Disodium Tetraborates. 4 pp.
ISO 3121-76. Boric Acid, Boric Oxide and DiSodium Tetraborates for Industrial Use—Determination of Chloride Content—Mercurimetric Method First Edition. 5 pp.

—Sodium Carbonates—Mercurimetric Analysis
BSI BS 6070: Part 2-81. 1981 Methods of Sampling and Test for Sodium Carbonate for Industrial Use Part 2: Determination of Chloride Content. 4 pp.
ISO 742-73. Sodium Carbonate for Industrial Use—Determination of Chloride Content—Mercurimetric Method First Edition. 4 pp.

—Sodium Chlorate—Mercurimetric Analysis
ISO 2463-73. Sodium Chlorate for Industrial Use—Determination of Chloride Content—Mercurimetric Method First Edition. 4 pp.

—Sodium Fluorides—Turbidimetry
BSI BS 5072: Part 8-80. 1980 Methods of Test for Sodium Fluoride for Industrial Use Part 8: Determination of Chlorides Content. 4 pp.
ISO 3566-76. Sodium Fluoride Primarily Used for the Production of Aluminium—Determination of Chlorides Content—Turbidimetric Method First Edition. 4 pp.

—Sodium Hydroxide
MOD UK M 808/73. Examination of Sodium Hydroxide Pure (Withdrawn).

—Sodium Hydroxide—Mercurimetric Analysis
BSI BS 6075: Part 2-81. 1981 Sampling and Test for Sodium Hydroxide for Industrial Use Part 2: Determination of Chloride Content (Mercurimetric Method). 4 pp.
ISO 981-73. Sodium Hydroxide for Industrial Use—Determination of Chloride Content—Mercurimetric Method First Edition. 4 pp.

—Sodium Hydroxide—Photometry
BSI BS 6075: Part 7-81. 1981 Sampling and Test for Sodium Hydroxide for Industrial Use Part 7: Determination of Chloride Content (Photometric Method). 8 pp.
ISO 3197-75. Sodium Hydroxide for Industrial Use—Determination of Chlorides Content—Photometric Method First Edition. 5 pp.

—Sodium Metasilicate
BSI BS 6092: Part 11-81. 1981 Sampling and Test for Sodium and Potassium Silicates for Industrial Use Part 11: Determination of Chloride Content of Sodium Metasilicate. 2 pp.

—Sodium Nitrate
MOD UK M 869/91. Examination of Sodium Nitrate, Grade 1.

—Sodium Sulfates
CNS K6085-74. Methods of Test for Sodium Sulfate of Industrial Grade (Oct)(1004). 5 pp.

—Sodium Sulfates—Mercurimetric Analysis
ISO 3236-75. Sodium Sulphate for Industrial Use—Determination of Chlorides Content—Mercurimetric Method First Edition. 6 pp.

—Soil Analysis
BSI BS 1377: Part 3-90. 1990 Methods of Test for Soils for Civil Engineering Purposes Part 3: Chemical and Electro-Chemical Tests. 45 pp.

—Soldering Fluxes
MOD UK M 9524/66. Examination of Flux Soldering Solution (Withdrawn).
MOD UK M 9542/70. Examination of Flux, Soldering (Zinc Chloride Solution).

—Spectrophotometry
CNS J2041-82. Chloride by Thiocyanate Spectrophotometric Method (Feb)(8511).

—Starches—Potentiometric Analysis
ISO 5810-82. Starches and Derived Products—Determination of Chloride Content—Potentiometric Method First Edition. 4 pp.

INTERNATIONAL AND NON-U.S. NATIONAL STANDARDS
SUBJECT INDEX

Chloride Content Analysis (Cont.)

—Steels—Surface Analysis
BSI BS 7079: Part B2-92. 1992 Preparation of Steel Substrates Before Application of Paints and Related Products Group B: Methods for the Assessment of Surface Cleanliness Part B2: Method for Detn. of Chlorides on Cleaned Surfaces (ISO 8502-2: 1992). 9 pp.
ISO 8502 Pt 2-92. Preparation of Steel Substrates Before Application of Paints and Related Products—Tests for the Assessment of Surface Cleanliness—Part 2: Laboratory Determination of Chloride on Cleaned Surfaces First Edition. 9 pp.

—Strontium Nitrate
MOD UK M 867/92. Examination of Strontium Nitrate (Anhydrous).

—Strontium Peroxide
MOD UK M 884/70. Examination of Strontium Peroxide (Anhydrous).

—Sulfur—Photometry
ISO 5793-78. Sulphur for Industrial Use—Determination of Chloride Content—Photometric Method First Edition. 5 pp.

—Sulfuric Acid—Potentiometric Analysis
ISO 2877-74. Sulphuric Acid for Industrial Use—Determination of Chlorides Content—Potentiometric Method First Edition; (Amendment 1-1976). 6 pp.

—Tetranitro Oxanilide
MOD UK M 887/72. Examination of Tetranitro-Oxanilide.

—Trichlorotrifluoroethane
CNS K6486-80. Method of Test for Chloride in Trichlorotrifluoroethane (May)(5590).

—Vegetables
ISO 3634-79. Vegetable Products—Determination of Chloride Content First Edition. 5 pp.

—Water
BSI BS 2690: Part 6-68. (WITHDRAWN) 1968 Water Used in Industry Part 6: Chloride and Sulphate (SUPERSEDED BY BS 6068: SECTIONS 2.37 AND 2.39). 24 pp.

—Water—Volumetric Analysis
BSI BS 6068: Sec 2.37-90. 1990 Water Quality Part 2: Physical, Chemical and Bio-Chemical Methods Section 2.37: Method for the Determination of Chloride Via a Silver Nitrate Titration with Chromate Indicator (Mohr's Method) (SUP'S BS 2690: PART 6: 1968). 8 pp.
ISO 9297-89. Water Quality—Determination of Chloride—Silver Nitrate Titration with Chromate Indicator (Mohr's Method) First Edition. 7 pp.

—Zinc Oxides
MOD UK M 806/91. Examination of Zinc Oxide.

Chloride Ion Content Analysis

—Aggregates
BSI BS 812: Part 117-88. 1988 Testing Aggregates Part 117: Method for Determination of Water-Soluble Chloride Salts. 11 pp.

—Ammonium Nitrate—Potentiometric Analysis
ISO 3695-77. Ammonium Nitrate for Industrial Use—Determination of Chloride Ions Content—Potentiometric Method First Edition. 6 pp.

—Ammonium Sulfate—Potentiometric Analysis
ISO 3694-77. Ammonium Sulphate for Industrial Use—Determination of Chloride Ions Content—Potentiometric Method First Edition. 6 pp.

—Cementitious Materials
SAA AS 3583.13-91. Methods of Test for Supplementary Cementitous Materials for Use with Portland Cement—Part 13: Determination of Chloride Ion Content (In Professional Packages 30, 58). 4 pp.

—Chemical Products—Potentiometric Analysis
BSI BS 6337: Part 4-84. 1984 General Methods of Chemical Analysis Part 4: Method for Determination of Chloride Ions by Potentiometry. 11 pp.
ISO 6227-82. Chemical Products for Industrial Use—General Method for Determination of Chloride Ions—Potentiometric Method First Edition. 10 pp.

—Nitric Acid—Potentiometric Analysis
ISO 3693-77. Nitric Acid for Industrial Use—Determination of Chloride Ions Content—Potentiometric Method First Edition. 6 pp.

Chloride Ion Content Analysis (Cont.)

—Waste Water
CNS K9039-81. Method of Test for Chloride Ion in Waste Water (Oct)(5858).

—Waste Water—Ion Exchange Chromatography
DIN ENGL 38405 Pt 20-91. German Standard Methods for the Examination of Water, Waste Water and Sludge; Anions (Group D); Determination of Dissolved Bromide, Chloride, Nitrate, Nitrite, (Ortho) Phosphate and Sulfate Anions in Waste Water by Ion Chromatography (D 20) (Sept). 11 pp.

—Waste Water—Volumetric Analysis
DIN ENGL 38405 Pt 1-85. German Standard Methods for the Examination of Water, Waste Water and Sludge; Anions (Group D); Determination of Chloride Ions (D 1) (Dec). 7 pp.

—Water—Ion Exchange Chromatography
JIS K 0556-90. Testing Methods for Determination of Anions in Highly Purified Water. 18 pp.

—Water—Liquid Chromatography
DIN ENGL 38405 Pt 19-88. German Standard Methods for the Examination of Water, Waste Water and Sludge; Anions (Group D) Determination of Fluo-ride, Chloride, Nitrite, (Ortho)Phosphate, Brom-ide, Nitrate and Sulfate Anions in Water with a Low Pollution Level by Ion Exchange Chromatography (D 19) (Feb). 11 pp.
ISO 10304 Pt 1-92. Water Quality—Determination of Dissolved Fluoride, Chloride, Nitrite, Orthophosphate, Bromide, Nitrate and Sulfate Ions, Using Liquid Chromatography of Ions—Part 1: Method for Water with Low Contamination First Edition. 16 pp.

—Water—Volumetric Analysis
DIN ENGL 38405 Pt 1-85. German Standard Methods for the Examination of Water, Waste Water and Sludge; Anions (Group D); Determination of Chloride Ions (D 1) (Dec). 7 pp.

Chlorides
Scope Note: Use a more specific term *See:* Acetyl Chloride; Aluminum Chloride; Ammonium Chloride; Ammonium Cupric Chloride; Barium Chlorides; Benzidine Dihydrochloride; Benzoyl Chloride; Bismuth Chloride; Cadmium Chloride; Calcium Chloride; Cesium Chloride; Chlorobenzene; Choline Chloride; Chromium Chlorides; Cobaltous Chloride; Cupric Chloride; Cupric Potassium Chloride; Cuprous Chloride; 3,5-Dinitrobenzoyl Chloride; Ferric Chloride; Ferrous Chloride; Hydrochloric Acid; Hydrogen Chloride; Lanthanum Chloride; Lead Chloride; Lithium Chloride; L-Lysine Monohydrochloride; Magnesium Chloride; Manganese Chloride Tetrahydrate; Mercuric Chloride; Mercurous Chloride; Methylene Chloride; Methylmercuric Chloride; Nickel Chloride; p-Nitrobenzoyl Chloride; Palladium Chloride; Pararosaniline Hydrochloride; 1, 10-Phenanthroline Hydrochloride; m-Phenylenediamine Hydrochloride; Phenylhydrazine Hydrochloride; Phenylhydrazinium Chloride; Phosphorus Oxychloride; Phosphorus Pentachloride; Phosphorus Trichloride; Platinic Chloride; Platinic Chloride Hexahydrate; Plumbous Chloride; Quaternary Ammonium Chloride; Silver Chloride; Sodium Chloride; Stannic Chloride; Stannous Chloride; Strontium Chloride; Sulfur Chloride; Sulfuryl Chloride; Thionyl Chloride; Thorium Chloride; 1,1,1-Trichloroethane; Triphenyltetrazolium Chloride; Zinc Chloride; Zinc Hydrosulfite; Zirconium Oxychloride

Chlorinated Hydrocarbon Content Analysis
Use: Halogenated Hydrocarbon Content Analysis

Chlorinated Lime
Use: Bleaching Powder

Chlorinated Polyethylene
Use: Polyethylene

Chlorinated Rubber Coatings
Use: Chlorinated Rubber Paints

Chlorinated Rubber Paints
Use For: Chlorinated Rubber Coatings
See Also: Paints
CNS K2070-89. Chlorinated Rubber Finish Paint (Jan)(4919). 2 pp.
CNS K6849-86. Method of Test for Chlorinated Rubber Finish Paint (May)(11579).

Chlorinated Rubber Paints (Cont.)

—Acid Resistant—Alkali Resistant
MOD UK DSTAN 80-72-91. Paint, System, Acid and Alkali Resisting Issue 2. 16 pp.
MOD UK DSTAN 80-72-01. Paint, System, Acid and Alkali Resisting Issue 2; Amendment 1. 19 pp.

—Antifouling—Steel—Ships
CNS K2072-86. Chlorinated Rubber Antifouling Paint for Steel Ship (May)(4921). 2 pp.
CNS K6864-86. Method of Test for Chlorinated Rubber Antifouling Paint for Steel Ship (May)(11594).

—Steel
JIS K 5639-89. Chlorinated Rubber Paints. 17 pp.

—Steel—Ships
CNS K2073-89. Chlorinated Rubber Boottopping Paint for Steel Ship (Jan)(4922). 2 pp.
CNS K2074-86. Chlorinated Rubber Topside Paint for Steel Ship (May)(4923). 2 pp.
CNS K2075-86. Chlorinated Rubber Deck Paint for Steel Ship (May)(4924). 2 pp.
CNS K6865-86. Method of Test for Chlorinated Rubber Boottopping Paint for Steel Ship (May)(11595).
CNS K6866-86. Method of Test for Chlorinated Rubber Topside Paint for Steel Ship (May)(11596).
CNS K6867-86. Method of Test for Chlorinated Rubber Deck Paint for Steel Ship (May)(11597).

—Steel—Ships—Corrosion Inhibitive
CNS K2071-86. Chlorinated Rubber Anticorrosive Paint for Steel Ship (May)(4920). 2 pp.
CNS K6863-86. Method of Test for Chlorinated Rubber Anticorrosive Paint for Steel Ship (May)(11593).

Chlorinated Rubber Primers
See Also: Primers (Coatings)
CNS K2066-86. Chlorinated Rubber Red Lead Anticorrosive Primer (Apr)(4915). 2 pp.
CNS K2067-86. Chlorinated Rubber Zinc Chromate Anticorrosive Primer (Apr)(4916). 2 pp.
CNS K2068-86. Chlorinated Rubber LZI Anti-Corrosive Primer (May)(4917). 2 pp.
CNS K2069-86. Chlorinated Rubber Red Iron Oxide Anticorrosive Primer (May)(4918). 2 pp.
CNS K6845-86. Method of Test for Chlorinated Rubber Red Lead Anticorrosive Primer (Apr)(11565).
CNS K6846-86. Method of Test for Chlorinated Rubber Zinc Chromate Anticorrosive Primer (Apr)(11566).
CNS K6847-86. Method of Test for Chlorinated Rubber LZI Anticorrosive Primer (May)(11577).
CNS K6848-86. Method of Test for Chlorinated Rubber Red Iron Oxide Anticorrosive Primer (May)(11578).

—Aluminum
CNS K2203-88. Chlorinated Rubber Aluminum Tri-Polyphosphate Anti-Corrosive Primer (Apr)(12270). 2 pp.

—Aluminum—Tripolyphosphate Content
CNS K6954-86. Method of Test for Chlorinated Rubber Aluminum Tri-Polyphosphate Anti-Corrosive Primer (Apr)(12271). 3 pp.

Chlorinated Solvents
Use: Solvents—Chlorinated

Chlorination
See Also: Biochemical Oxygen Demand; Bleaching; Chlorine Number; Sewage Treatment; Water Treatment

—Fabrics—Colorfastness Testing
CGSB CAN/CGSB-4.2 NO.52.1-M90. Textile Test Methods Colourfastness to Chlorinated Water. 9 pp.
CGSB CAN/CGSB-4.2 NO.52.2-M91. Textile Test Methods Textiles—Tests for Colourfastness—Part E03: Colourfastness to Chlorinated Water (Swimming-Pool Water) (ISO 105-E03:1987). 7 pp.
CGSB CAN/CGSB-4.2 NO.55-M90. Textile Test Methods Loss in Strength and Colour Change of Fabrics Due to Retained Chlorine. 13 pp.
CNS L3183-82. Method of Test for Colour Fastness to Chlorination (Aug)(9306).
ISO 105 Pt E03-87. Textiles—Tests for Colour Fastness—Part E03: Colour Fastness to Chlorinated Water (Swimming-Bath Water) Second Edition. 4 pp.
ISO 105 Pt X14-87. Textiles—Tests for Colour Fastness—Part X14: Colour Fastness to Acid Chlorination of Wool: Sodium Dichloroisocyanurate Second Edition.
JIS L 0873-74. Testing Method for Colour Fastness to Chlorination.

Chlorinators
See Also: Water Treatment

Chlorinators

Chlorinators (Cont.)

—Water Treatment
DIN ENGL 19606-83. Chlorinators for Water Treatment; Equipment, Installation and Operation (Feb). 3 pp.

Chlorine

See Also: Chlorine Content Analysis; Hazardous Materials

BSI BS 3947-76. 1976 Specification for Liquid Chlorine. 25 pp.
BSI BS 3947-01. 1976 Amd 1 Liquid Chloring (AMD 7587) April 15, 1993 (W). 26 pp.
CNS K1034-81. Liquid Chlorine (Nov)(432). 1 p.
CNS K6079-72. Method of Test for Anhydrous Chlorine (Jun)(963).
CNS K6096-75. Method of Test for Liquid Chlorine (Jan)(1059).
CNS K6146-69. Method of Test for Argon (Jun)(1375)(R 1973).
JIS K 1102-59. Liquid Chlorine. 4 pp.

—Highway Transportation—Emergency Procedures
SAA AS 1678.2.2. 002-91. Emergency Procedure Guide—Transport—Part 2.2.002: Chlorine (in Professional Package 37).

—Sampling
ISO 1552-76. Liquid Chlorine for Industrial Use—Method of Sampling (for Determining Only the Volumetric Chlorine Content) First Edition. 8 pp.

—Shipping—Safety
CNS Z1010-87. Safety Code for Storage, Transporting and Use of Liquefied Chlorine (Oct) (2571).

—Storage—Safety
CNS Z1010-87. Safety Code for Storage, Transporting and Use of Liquefied Chlorine (Oct) (2571).

—Tablets—Water Purification
MOD UK DSTAN 68-136-90. Chlorine Release Tablets Issue 1. 12 pp.

—Vaporizing
ISO 2120-72. Liquid Chlorine for Industrial Use—Determination of the Content of Chlorine by Volume in the Vaporized Product First Edition. 5 pp.

—Water Content—Electrolytic Analysis
ISO 2202-72. Liquid Chlorine for Industrial Use—Determination of Water Content Using an Electrolytic Analyser First Edition; (Addendum 1-1975). 12 pp.

—Water Content—Gravimetric Analysis
ISO 2121-72. Liquid Chlorine for Industrial Use—Determination of Water Content—Gravimetric Method First Edition. 8 pp.

—Water Treatment
CEN PREN 937-92. Chlorine Used for Treatment of Water Intended for Human Consumption. 15 pp.

Chlorine Analyzers

See Also: Analyzers

JIS B 7955-79. Continuous Analyzers for Chlorine in Ambient Air.

Chlorine Content Analysis

See Also: Chlorine

—Bleach Liquors
CPPA J.22P-86. Analysis of Chlorine Solutions, Hypochlorite Bleach Liquors, and Spent Bleach Liquors. 3 pp.

—Coal
CNS M3151-84. Method for Determination of Chlorine in Coal (Mar)(10832).
SAA AS 1038.8-80. Methods for the Analysis and Testing of Coal and Coke—Part 8: Chlorine in Coal and Coke (Superseded (in Part) by AS 1038.8.1—1992 Will Remain Current). 12 pp.

—Coal—Combustion
BSI BS 1016: Part 8-77. 1977 Amd 1 Methods for Analysis and Testing of Coal and Coke Part 8: Chlorine in Coal and Coke (AMD 3472) December 31, 1980. 9 pp.
BSI BS 1016: Part 8-02. 1977 Amd 2 Methods for Analysis and Testing of Coal and Coke Part 8: Chlorine in Coal and Coke (AMD 6984) May 1, 1992. 12 pp.

—Coal—Eschka Method
BSI BS 1016: Part 8-77. 1977 Amd 1 Methods for Analysis and Testing of Coal and Coke Part 8: Chlorine in Coal and Coke (AMD 3472) December 31, 1980. 9 pp.

Chlorine Content Analysis (Cont.)

—Coal—Eschka Method (Cont.)
BSI BS 1016: Part 8-02. 1977 Amd 2 Methods for Analysis and Testing of Coal and Coke Part 8: Chlorine in Coal and Coke (AMD 6984) May 1, 1992. 12 pp.
SAA AS 1038.8.1-92. Coal and Coke—Analysis and Testing—Part 8.1: Coal and Coke—Chlorine —Eschka Method (Supersedes AS 1038.8—1980 (in Part)) Which Will Remain Current) (in Professional Package 32). 2 pp.

—Coke
SAA AS 1038.8-80. Methods for the Analysis and Testing of Coal and Coke—Part 8: Chlorine in Coal and Coke (Superseded (in Part) by AS 1038.8.1—1992 Will Remain Current). 12 pp.

—Coke—Combustion
BSI BS 1016: Part 8-77. 1977 Amd 1 Methods for Analysis and Testing of Coal and Coke Part 8: Chlorine in Coal and Coke (AMD 3472) December 31, 1980. 9 pp.
BSI BS 1016: Part 8-02. 1977 Amd 2 Methods for Analysis and Testing of Coal and Coke Part 8: Chlorine in Coal and Coke (AMD 6984) May 1, 1992. 12 pp.

—Coke—Eschka Method
BSI BS 1016: Part 8-77. 1977 Amd 1 Methods for Analysis and Testing of Coal and Coke Part 8: Chlorine in Coal and Coke (AMD 3472) December 31, 1980. 9 pp.
BSI BS 1016: Part 8-02. 1977 Amd 2 Methods for Analysis and Testing of Coal and Coke Part 8: Chlorine in Coal and Coke (AMD 6984) May 1, 1992. 12 pp.
SAA AS 1038.8.1-92. Coal and Coke—Analysis and Testing—Part 8.1: Coal and Coke—Chlorine —Eschka Method (Supersedes AS 1038.8—1980 (in Part)) Which Will Remain Current) (in Professional Package 32). 2 pp.

—Copolymers
JIS K 7229-87. Determination of Chlorine in Chlorine-Containing Polymers, Copolymers and Their Compounds.

—Detergents—Volumetric Analysis
BSI BS 3762: Sec 3.18-86. 1986 Analysis of Formulated Detergents Part 3: Quantitative Test Methods Section 3.18: Method for Determination of Chlorine Oxidizing Agents Content. 4 pp.

—Drinking Water
CNS N6188-84. Method of Test for Bactericides in Water Test of Residual Chlorine (May)(10892). 3 pp.

—Epoxy Resins
BSI BS 2782:Pt4: METH 433A-79. 1979 Methods of Testing Plastics Part 4: Chemical Properties Method 433A: Determination of Inorganic Chlorine in Epoxide Resins and Glycidyl Esters. 4 pp.
BSI BS 2782:Pt4: METH 433B-79. 1979 Methods of Testing Plastics Part 4: Chemical Properties Method 433B: Determination of Easily Saponifiable Chlorine in Epoxide Resins and Related Materials. 4 pp.
ISO 4573-78. Plastics—Epoxide Resins and Glycidyl Esters—Determination of Inorganic Chlorine First Edition. 4 pp.
ISO 4583-78. Plastics—Epoxide Resins and Related Materials—Determination of Easily Saponifiable Chlorine First Edition. 5 pp.

—Epoxy Resins—Combustion Analysis
ISO 4615-79. Plastics—Unsaturated Polyesters and Epoxide Resins—Determination of Total Chlorine Content First Edition. 7 pp.

—Exhaust Gases
CNS K9062-81. Methods of Test for Chlorine in Exhaust Gas (O-Tolidine Method) (Mar)(7054).
JIS K 0106-82. Methods for Determination of Chlorine in Exhaust Gas. 14 pp.

—Glycidyl Esters
BSI BS 2782:Pt4: METH 433A-79. 1979 Methods of Testing Plastics Part 4: Chemical Properties Method 433A: Determination of Inorganic Chlorine in Epoxide Resins and Glycidyl Esters. 4 pp.
BSI BS 2782:Pt4: METH 433B-79. 1979 Methods of Testing Plastics Part 4: Chemical Properties Method 433B: Determination of Easily Saponifiable Chlorine in Epoxide Resins and Related Materials. 4 pp.
ISO 4573-78. Plastics—Epoxide Resins and Glycidyl Esters—Determination of Inorganic Chlorine First Edition. 4 pp.
ISO 4583-78. Plastics—Epoxide Resins and Related Materials—Determination of Easily Saponifiable Chlorine First Edition. 5 pp.

Chlorine Content Analysis (Cont.)

—Hydrochloric Acid—Volumetric Analysis
ISO 908-80. Hydrochloric Acid for Industrial Use—Determination of Oxidizing or Reducing Substances Content—Titrimetric Method First Edition. 5 pp.

—Industrial Water
CPPA H.4P(C)-67. Determination of Free Chlorine in Process Waters. 1 p.

—Lubricant Additives—X-Ray Fluorescence Spectrometry
DIN ENGL 51577 Pt 2-85. Testing of Petroleum Products; Determination of the Chlorine Content of Lubricating Oils and Lubricating Oil Additives by X-Ray Fluorescence Analysis (Sept). 3 pp.

—Lubricants
CNS K6760-83. Method of Test for Chlorine in New and Used Lubricants (Sodium Alcoholate Method) (Mar)(10095).

—Lubricating Oils—X-Ray Fluorescence Spectrometry
DIN ENGL 51577 Pt 2-85. Testing of Petroleum Products; Determination of the Chlorine Content of Lubricating Oils and Lubricating Oil Additives by X-Ray Fluorescence Analysis (Sept). 3 pp.

—Mineral Oils—Combustion Analysis
DIN ENGL 51408 Pt 1-83. Testing of Liquid Petroleum Hydrocarbons; Determination of Chlorine Content; Wickbold Combustion Method (June). 4 pp.

—Olefins—Combustion Analysis
ISO 8915-87. Light Olefins for Industrial Use—Determination of Traces of Chlorine—Wickbold Combustion Method First Edition. 6 pp.

—Organic Coatings
CNS K6804-16-84. Method of Test for Organic Coating (Chemical Analysis) — Qualitative Test of Chlorine in Resin Content (Jul)(10880-16).

—Perchloroethylene
BSI BS 1593-63. (WITHDRAWN) 1963 Amd 1 Perchloroethylene (Tetrachloroethylene). 13 pp.
SAA AS K105-66. Perchloroethylene (Tetrachloroethylene) Being BS 1593:1963, Endorsed Without Amendment. 12 pp.

—Plastics
DIN ENGL 53474-76. Testing of Plastics, Rubber and Elastomers; Determination of the Chlorine Content (Aug). 11 pp.

—Polyester Resins—Combustion Analysis
ISO 4615-79. Plastics—Unsaturated Polyesters and Epoxide Resins—Determination of Total Chlorine Content First Edition. 7 pp.

—Polymers
JIS K 7229-87. Determination of Chlorine in Chlorine-Containing Polymers, Copolymers and Their Compounds.

—PVC—Combustion Analysis
ISO 1158-84. Plastics—Vinyl Chloride Homopolymers and Copolymers—Determination of Chlorine Second Edition. 6 pp.

—Rubber
DIN ENGL 53474-76. Testing of Plastics, Rubber and Elastomers; Determination of the Chlorine Content (Aug). 11 pp.

—Rubber—Combustion
BSI BS 7164: Sec 22.2-92. 1992 Chemical Tests for Raw and Vulcanized Rubber Part 22: Methods for Determination of Bromine and Chlorine Content Section 22.2: Oxygen Flask Combustion Technique for Determination of Bromine and Chlorine (ISO 7725: 1991) (V). 14 pp.
ISO 7725-91. Rubber and Rubber Products—Determination of Bromine and Chlorine Content—Oxygen Flask Combustion Technique First Edition. 11 pp.

—Sodium Chloride—Mercurimetric Analysis
BSI BS 7319: Part 1-90. 1990 Analysis of Sodium Chloride for Industrial Use Part 1: Method for Determination of Sodium Chloride Content. 8 pp.
ISO 2481-73. Sodium Chloride for Industrial Use—Determination of Halogens, Expressed as Chlorine—Mercurimetric Method First Edition. 4 pp.

—Solid Fuels—Combustion
ISO 352-81. Solid Mineral Fuels—Determination of Chlorine—High Temperature Combustion Method Second Edition. 6 pp.

INTERNATIONAL AND NON-U.S. NATIONAL STANDARDS
SUBJECT INDEX
Chlorofluorocarbons

Chlorine Content Analysis *(Cont.)*
—Solid Fuels—Eschka Method
ISO 587-81. Solid Mineral Fuels—Determination of Chlorine Using Eschka Mixture Second Edition. 5 pp.

—Spent Liquors
CPPA J.22P-86. Analysis of Chlorine Solutions, Hypochlorite Bleach Liquors, and Spent Bleach Liquors. 3 pp.

—Titanium
JIS H 1615-73. Methods for Determination of Chlorine in Titanium.

—Titanium Tetrachloride
MOD UK M 888/72. Examination of Titanium Tetrachloride (Withdrawn).

—Uranium Hexafluoride—Volumetric Analysis
CNS J2064-82. Method for Determining Chlorine in UF6 Titrimetric Method (Aug)(9282).

—Vulcanized Rubber
BSI BS 903: Part B11-B12-60. 1960 Amd 1 Methods of Testing Vulcanized Rubber Part B11 to B12: Rubber (Polymer) Determinations (AMD 6627) April 30, 1991. 17 pp.
BSI BS 903: Part B11-B12-02. 1960 Amd 2 Methods of Testing Vulcanized Rubber Parts B11 & B12 Rubber (Polymer) Determinations (AMD 7134) July 15, 1992. 19 pp.

—Waste Water
CNS K9071-81. Method of Test for Residual Chlorine in Waste Water (Oct)(8031).

—Water—Colorimetric Analysis
BSI BS 6068: Sec 2.26-86. 1986 Water Quality Part 2: Physical, Chemical and Bio-Chemical Methods Section 2.26: Method for Determination of Free Chlorine and Total Chlorine: Colorimetric Method Using N, N-Diethy-1, 4-Phenylenediamine, for Routine Control Purposes. 10 pp.
ISO 7393 Pt 2-85. Water Quality—Determination of Free Chlorine and Total Chlorine—Part 2: Colorimetric Method Using N,N-Diethyl-1,4-Phenylenediamine, for Routine Control Purposes First Edition. 10 pp.

—Water—Iodometry
BSI BS 6068: Sec 2.27-90. 1990 Water Quality Part 2: Physical, Chemical and Bio-Chemical Methods Section 2.27: Method for Determination of Total Chlorine: Iodometric Titration Method. 12 pp.
ISO 7393 Pt 3-90. Water Quality—Determination of Free Chlorine and Total Chlorine—Part 3: Iodometric Titration Method for the Determination of Total Chlorine Second Edition. 10 pp.

—Water—Volumetric Analysis
BSI BS 6068: Sec 2.25-86. 1986 Water Quality Part 2: Physical, Chemical and Bio-Chemical Methods Section 2.25: Method for Determination of Free Chlorine and Total Chlorine: Titrimetric Method Using N, N-Diethy-1, 4-Phenylenediamine. 10 pp.
ISO 7393 Pt 1-85. Water Quality—Determination of Free Chlorine and Total Chlorine—Part 1: Titrimetric Method Using N,N-Diethyl-1,4-Phenylenediamine First Edition. 10 pp.

Chlorine Dioxide
—Bleaching Agents
CPPA J.14P-84. Chlorine Dioxide Plant Analyses. 5 pp.

Chlorine Dioxide Content Analysis
—Water—Volumetric Analysis
DIN ENGL 38408 Pt 5-90. German Standard Methods for the Examination of Water, Waste Water and Sludge; Gaseous Components (Group G); Determination of Chlorine Dioxide (G 5) (June). 4 pp.

Chlorine Inorganic Compounds
Scope Note: Use a more specific term *See:* Chlorine Dioxide; Hypochlorites; Thionyl Chloride

Chlorine Number
See Also: Chlorination; Lignin Content Analysis
—Pulps
CPPA G.16H-88. Chlorine Number of Pulp. 5 pp.
CPPA G.1U-77. Modified Test for Roe Chlorine Number. 1 p.

Chlorine Organic Compounds
Scope Note: Use a more specific term *See:* Acetyl Chloride; Benzyl Chloride; Carbon Tetrachloride; Chlorobenzene; Chloroform; Chloronitrobenzenes; Ethyl Chloride; Methyl Chloride

Chlorine Sulfonic Acid
CNS K6285-72. Method of Test for Chlorine Sulfonic Acid (Jun)(3116).

Chlorites
Scope Note: Use a more specific term *See:* Sodium Hypochlorite

1-Chloro-2,4-dinitrobenzene
Use For: 2,4-Dinitrochlorobenzene
CNS K7197-65. Chemical Reagent (2,4-Dinitrochlorobenzene) (Jan)(1697).
JIS K 4103-82. Chloronitrobenzenes (o-Chloronitrobenzene-p-Chloronitrobenzene. 1,4-Dichloro-2-Nitrobenzene. 1-Chloro-1,4-Dinitobenzene).
JIS K 8478-78. 2,4-Dinitrobenzene.

1-Chloro-2,3-Epoxy Propane
Use: Epichlorohydrin

Chloro-IPC
Use: Chlorpropham

p-Chloro-o-nitroaniline
Use: 4-Chloro-2-nitroaniline

4-Chloro-2-nitroaniline
Use For: p-Chloro-o-nitroaniline
CNS K2205-88. p-Chloro-o-Nitroaniline (Apr)(12273). 6 pp.
JIS K 4168-86. p-Chloro-o-Nitroanline.

Chloroacetic Acid
Use For: Monochloroacetic Acid
CNS K7323-70. Chemical Reagent (Monochloroacetic Acid) (Jan)(1822).
CNS K7630-82. Chemical Reagent (Chloroacetic Acid) (Aug)(9287).
JIS K 8899-80. Chloroacetic Acid.

2'-Chloroacetoacetanilide
JIS K 4173-81. Acetoacetanilides (Acetoacetanilide 2-Methylaceto-acetanilide 2',4'-Dimethylaceto-acetanilide.2'-Chloroacetoacetanilide).

Chloroalkane
—Pyrotechnics
MOD UK TS 10173. Chloralkane No. 1.

o-Chloroaniline
See Also: Aniline
CNS K2233-88. Chloroanilines (Aug)(12401).
JIS K 4142-81. Chloroanilines (o-Chloroaniline, p-Chloroaniline).

p-Chloroaniline
See Also: Aniline
CNS K2233-88. Chloroanilines (Aug)(12401).
JIS K 4142-81. Chloroanilines (o-Chloroaniline, p-Chloroaniline).

Chloroauric Acid
Use: Gold Trichloride

Chlorobenzene
Use For: Benzene Chloride; Monochlorobenzene
See Also: Benzene; o-Dichlorobenzene; p-Dichlorobenzene; Halogenated Hydrocarbons; Hazardous Materials; Insecticides
CNS K2171-87. Chlorobenzenes (Dec)(12181).
ISO 1697-77. Chlorobenzene for Industrial Use—List of Methods of Test First Edition. 3 pp.
JIS K 4102-83. Chlorobenzenes (Chlorobenzene.o-Dichlorobene.p-Dichlorobenzen).
SAA AS N28-58. Benzene Hexachloride for Insecticidal Preparations. 11 pp.

Chlorobutanol Content Analysis
—Cosmetics
EC 85/490/EEC-85. Fourth Commission Directive on the Approximation of the Laws of the Member States Relating to Methods of Analysis Necessary for Checking the Composition of Cosmetic Products. 16 pp.

Chlorobutyl Rubber
See Also: Butyl Rubber; Elastomers; Rubber
MOD UK TS 10252. Chlorobutyl Rubber, Black, Hardness 45 IRHD.

—Test Specimens
MOD UK DSTAN 93-52: Part 6-88. Vulcanized Rubbers for Use as Standard Test Materials Part 6: Vulcanized Chlorobutyl Rubber Sheet, Standard CIIR/1 Issue 1. 10 pp.

Chlorobutyl Rubber Content Analysis
—Elastomers
DIN ENGL 53621 Pt 5-76. Testing of Rubber and Elastomers; Quantitative Determination of Polymers; Determination of the Chlorobutyl Rubber Content (July). 2 pp.

—Rubber
DIN ENGL 53621 Pt 5-76. Testing of Rubber and Elastomers; Quantitative Determination of Polymers; Determination of the Chlorobutyl Rubber Content (July). 2 pp.

Chlorocarbon Content Analysis
—Uranium Hexafluoride
CNS J2066-82. Method for Determining Hydrocarbons, Chlorocarbons and Partially Substituted Halohydrocarbons in UF6 (Sept)(9399).

Chlorodifluoromethane
Use For: FLON 22; Monochlorodifluoromethane
See Also: Chlorofluorocarbons
JIS K 1517-82. Fluoromethanes Trichloromonofluoromethane (FLON 11) Dichlorodifluoromethane (FLON 12) Monochlorodifluoro-methane (FLON 22).
JIS K 1517-73. Dichlorodifluoromethane (FLON 12) (R 1978). 4 pp.

Chloroethane
Use: Ethyl Chloride

Chloroethanes
Use: Ethyl Chloride

Chloroethene
Use: Vinyl Chloride

Chloroethylene
Use: Vinyl Chloride

Chlorofibers
See Also: Fibers
—Binary Mixtures—Quantitative Analysis
EC 87/185/EEC-87. Commission Recommendation on Quantitative Methods of Analysis for the Identification of Acrylic and Modacrylic Fibres, Chlorofibres and Trivinyl Fibres. 6 pp.

Chlorofluorinated Hydrocarbons
Use: Chlorofluorocarbons

Chlorofluorocarbons
Use For: CFC; Chlorofluorinated Hydrocarbons; Fluorochlorinated Hydrocarbons
See Also: Chlorodifluoromethane; Dichlorotetrafluoroethane; Fluorocarbons; Halogenated Hydrocarbons; Halon; Hydrocarbons; 1,1,2-Trichloro-1,2,2-Trifluoroethane; Trichlorofluoromethane

—Acidity
BSI BS 5598: Part 6-79. 1979 Methods of Sampling and Test for Halogenated Hydrocarbons Part 6: Determination of Acidity of Fluorochlorinated Hydrocarbons. 6 pp.
ISO 3363-76. Fluorochlorinated Hydrocarbons for Industrial Use—Determination of Acidity—Titrimetric Method First Edition. 5 pp.

—Boiling Points
BSI BS 5598: Part 9-81. 1981 Methods of Sampling and Test for Halogenated Hydrocarbons Part 9: Determination of Boiling Range of Chlorofluorinated Hydrocarbons. 6 pp.
ISO 5917-80. Chlorofluorinated Hydrocarbons—Determination of Boiling Range—Test for Product Characterization First Edition; (Amendment Slip-1981). 6 pp.

—Gas Chromatography
ISO 5921-82. Chlorofluorohydrocarbons for Industrial Use—Analysis by Gas Chromatography—General Principles First Edition. 9 pp.

—Inert Gases Content—Gas Chromatography
BSI BS 5598: Part 11-83. 1983 Methods of Sampling and Test for Halogenated Hydrocarbons Part 11: Determination of 'Inert' Gas Content of Chlorofluorohydrocarbons Gas Chromatographic Method, General Purposes. 7 pp.
ISO 5918-82. Chlorofluorohydrocarbons for Industrial Use—Determination of Inert Gas Content—Gas Chromatographic Method—General Principles First Edition. 6 pp.

—Water Content—Gravimetric Analysis
ISO 5920-83. Chlorofluorinated Hydrocarbons—Determination of Water Content—Gravimetric Method First Edition. 9 pp.

INDUSTRY STANDARDS

Chlorofluorohydrocarbons
Use: Chlorofluorocarbons

Chloroform
Use For: Trichloromethane *See Also:* Chloroform Content Analysis; Halogenated Hydrocarbons; Methyl Chloride
BSI BS 4774-72. 1972 Methods of Test for Chloroform for Industrial Use. 12 pp.
CNS K7141-64. Chemical Reagent (Chloroform) (Jul)(1641).
ISO 1870-77. Chloroform for Industrial Use—List of Methods of Test First Edition. 4 pp.
JIS K 8322-91. Chloroform.

—Insoluble Matter Content
CNS K6657-81. Method of Test for Matter Insoluble in Chloroform in Oiticica Oils (Apr)(7258).

Chloroform Content Analysis
See Also: Chloroform

—Toothpastes
EC 80/1335/EEC-80. First Commission Directive on the Approximation of the Laws of the Member States Relating to Methods of Analysis Neccessary for Checking the Composition of Cosmetic Products. 20 pp.

Chloromethanes
Use: Methyl Chloride

(Chloromethyl) Benzene
Use: Benzyl Chloride

Chloromethyloxirane
Use: Epichlorohydrin

o-Chloronitrobenzene
JIS K 4103-82. Chloronitrobenzenes (o-Chloronitrobenzene-p-Chloronitrobenzene. 1,4-Dichloro-2-Nitrobenzene. 1-Chloro-1,4-Dinitobenzene).

p-Chloronitrobenzene
JIS K 4103-82. Chloronitrobenzenes (o-Chloronitrobenzene-p-Chloronitrobenzene. 1,4-Dichloro-2-Nitrobenzene. 1-Chloro-1,4-Dinitobenzene).

Chloronitrobenzenes
See Also: Benzene
CNS K2172-87. Chloronitrobenzenes (Dec)(12182).

Chlorophyll Content Analysis
—Algae
CNS N6106-78. Method of Test for Edible Chlorella (Oct)(4597). 9 pp.

—Bioassay—Spectrometry
ISO 10260-92. Water Quality—Measurement of Biochemical Parameters—Spectrometric Determination of the Chlorophyll-a Concentration First Edition. 10 pp.

—Rapeseeds—Spectrometry
BSI BS 4289: Part 10-92. 1992 Methods for Analysis of Oilseeds Part 10: Determination of Chlorophyll Content of Rapeseed (ISO 10519: 1992). 14 pp.
ISO 10519-92. Rapeseed—Determination of Chlorophyll Content—Spectrometric Method First Edition. 11 pp.

—Surface Waters—Spectrometry
ISO 10260-92. Water Quality—Measurement of Biochemical Parameters—Spectrometric Determination of the Chlorophyll-a Concentration First Edition. 10 pp.

—Surface Waters—Spectrophotometry
DIN ENGL 38412 Pt 16-85. German Standard Methods for the Examination of Water, Waste Water and Sludge; Test Methods Using Water Organisms (Group L); Determination of Chlorophyll a in Surface Water (L 16) (Dec). 4 pp.

—Tobacco—Residues
ISO TR8452-92. Raw Tobacco—Determination of Chlorophyll Residues Content (Green Index) First Edition. 7 pp.

Chlorophyta
Use: Algae

Chloropicrin
Use For: Nitrochloroform; Trichloronitromethane
See Also: Hazardous Materials

—Highway Transportation—Emergency Procedures
SAA AS 1678.6.0. 013-84. Emergency Procedure Guides—Transport—Part 6.0.013: Chlorpicrin (Trichloronitromethane).

Chloroplatinic
Use: Platinic Chloride

Chloroprene Rubber
Use For: CR; Polychloroprene Rubber
See Also: Rubber
BSI BS 2752-90. 1990 Amd 1 Chloroprene Rubber Compounds. 9 pp.
CNS K6740-83. Method of Test for Chloroprene Rubber (CR) (Feb)(10011).
JIS K 6388-77. Testing Methods for Synthetic Rubber CR (R 1989). 16 pp.
MOD UK DSTAN 93-3-01. Chloroprene Rubber Type QX Issue 2; Amendment 1. 12 pp.
MOD UK DSTAN 93-66-92. Polychloroprene Rubber with Defined Burning Characteristics (80 IRHD) Issue 1. 10 pp.
MOD UK TS 10189. Polychloroprene Rubber with Defined Burning Characteristics (80 IRHD) (Superseded by Def Stan 93-66).

—Aerospace
AECMA PREN2109-78. Chloroprene Rubber (CR) Hardness 40—Characteristics (C7/SC1/01). 4 pp.
AECMA PREN2110-78. Chloroprene Rubber (CR) Hardness 50—Characteristics (SC7/SC1/01). 4 pp.
AECMA PREN2111-78. Chloroprene Rubber (CR) Hardness 60—Characteristics (C7/SC1/01). 4 pp.
AECMA PREN2112-78. Chloroprene Rubber (CR) Hardness 70—Characteristics (C7/SC1/01). 4 pp.
AECMA PREN2113-78. Chloroprene Rubber (CR) Hardness 80—Characteristics (C7/SC1/01). 4 pp.
CEN PREN 2109-89. Chloroprene—Rubber (CR) Hardness 40 1 RHD. 5 pp.
CEN PREN 2110-89. Chloroprene—Rubber (CR) Hardness 50 1 RHD. 5 pp.
CEN PREN 2111-89. Chloroprene—Rubber (CR) Hardness 60 1 RHD. 3 pp.
CEN PREN 2112-89. Chloroprene—Rubber (CR) Hardness 70 1 RHD. 5 pp.
CEN PREN 2113-89. Chloroprene—Rubber (CR) Hardness 80 1 RHD. 5 pp.

—Gaskets
MOD UK DSTAN 93-21-91. Chloroprene Rubber, Cellular, Closed Cell, Minimum Tarnishing Issue 3. 16 pp.

—Molding Materials
MOD UK DSTAN 93-25-84. Chloroprene Rubber, Black, Type MD Issue 1. 8 pp.

—Test Specimens
MOD UK DSTAN 93-52: Part 9-90. Vulcanized Rubbers for Use as Standard Test Materials Part 9: Vulcanized Polychloroprene Rubber Sheet, Standard CR/1 Issue 1. 9 pp.

—Vulcanization
BSI BS 5375-92. 1992 Method for Evaluation of General Purpose Chloroprene Rubber (CR) (ISO 2475: 1990) (V). 10 pp.
BSI BS 5375-76. 1976 Methods of Test for Raw General Purpose Chloroprene Rubbers. 7 pp.
ISO 2475-90. Rubber, Chloroprene (CR)—General Purpose Types—Evaluation Procedures Third Edition; (Corrigendum 1-1993). 8 pp.

Chloroprene Rubber Adhesives
See Also: Adhesives
CNS K3026-81. Polychloroprene Rubber Adhesives (Jul)(3295). 2 pp.
CNS K6311-81. Method of Test for Chloroprene Rubber Adhesive (Jul)(3296).

Chloroprene Rubber Coatings (Made From Chloroprene Rubber)
Use For: Rubber Coatings, Chloroprene
See Also: Coatings

—Roofing
CNS A2130-88. Roof Coating for Waterproofing (Chloroprene Rubber Type) (Jan)(8642).

Chloroprene Rubber Content Analysis
—Elastomers
DIN ENGL 53621 Pt 2-74. Testing of Rubber and Elastomers; Quantitative Determination of Polymers; Determination of the Chloroprene Rubber Content (Nov). 2 pp.

—Rubber
DIN ENGL 53621 Pt 2-74. Testing of Rubber and Elastomers; Quantitative Determination of Polymers; Determination of the Chloroprene Rubber Content (Nov). 2 pp.

Chloroprene Rubber Latex
See Also: Latex
CNS K6741-83. Method of Test for Chloroprene Rubber (CR) Latex (Feb)(10012).
JIS K 6393-76. Testing Methods for Synthetic Rubber CR Latex (R 1989). 7 pp.

Chlorosulfonic Acid
Use For: Chlorosulfuric Acid
CNS K1141-72. Chlorosulfonic Acid (Jun)(3115).
JIS K 1411-60. Chlorosulfonic Acid.

Chlorosulfuric Acid
Use: Chlorosulfonic Acid

Chlorothene
Use For: 1,1,1-Trichloroethane

a-Chlorotoluene
Use: Benzyl Chloride

alpha-Chlorotoluene
Use: Benzyl Chloride

o-Chlorotoluene
CNS K2212-88. Chlorotoluenes (May)(12300).
ISO 1695-77. o-Chlorotoluene for Industrial Use—List of Methods of Test First Edition. 6 pp.
JIS K 4183-82. Chlorotoluenes (o-Chlorotoluene, p-Chlorotoluene).

p-Chlorotoluene
CNS K2212-88. Chlorotoluenes (May)(12300).
ISO 1696-77. p-Chlorotoluene for Industrial Use—List of Methods of Test First Edition. 6 pp.
JIS K 4183-82. Chlorotoluenes (o-Chlorotoluene, p-Chlorotoluene).

Chlorotrimethyl Silane
Use: Trimethylchlorosilane

Chlorpropham
Use For: Chloro-IPC *See Also:* Pesticides; Propham
SAA AS 1870.15D-79. Standard for Development—Pesticides for Agricultural Use—Part 15D: Chlorpropham and Propham. 8 pp.

Chlortetracycline Content Analysis
—Animal Feed
CNS N4101-82. Method of Test for Feed Additives: Determination of Chloratetracycline (Aug)(9316).

Chocks
See Also: Automotive Equipment; Marine Equipment

—Closed
CNS F3034-80. Closed Chocks (Apr)(5444).
JIS F 2005-75. Closed Chocks. 5 pp.

—Open
CNS F3033-80. Open Chocks (Apr)(5443).
JIS F 2006-76. Open Chocks.
JIS F 2006-68. Open Chocks (R 1971). 5 pp.

—Panama
CNS F3032-80. Panama Chocks (Apr)(5442).
JIS F 2017-82. Panama Chocks.
JIS F 2017-75. Panama Chocks. 5 pp.

Choke Coils
Use For: Chokes (Electric); Suppression Chokes
See Also: Electric Coils; Inductors; Radio Frequency Chokes
DIN VDE 0565 Pt 2-78. Radio Interference Suppression Devices; Radio Interference Suppressions Chokes up to 16 A and Protective Conductor Chokes from 16 to 36 A (Sept). 49 pp.

—Fluorescent Lighting—Ships
MOD UK NES 2002-88. Specification for Chokes and Transformers Used in Fluorescent Fittings Issue 2 (12.88). 25 pp.
MOD UK NES 2002-92. Specification for Chokes and Transformers Used in Fluorescent Fittings Issue 3 (01.92). 25 pp.

—Fluorescent Lighting—Submarines
MOD UK NES 2002-88. Specification for Chokes and Transformers Used in Fluorescent Fittings Issue 2 (12.88). 25 pp.
MOD UK NES 2002-92. Specification for Chokes and Transformers Used in Fluorescent Fittings Issue 3 (01.92). 25 pp.

—Laminates
BSI BS 2857-76. 1976 Nickel-Iron Transformer and Choke Laminations. 30 pp.

—Magnetic Cores—Telecommunications
BSI BS 9925: Part 03.0-88. 1988 Inductor and Transformer Cores for Telecommunications Part 03.0: Magnetic Oxide Cores for Transformers and Chokes for Power Applications. 10 pp.
BSI BS 9925: Part 03.01-88. 1988 Inductor and Transformer Cores for Telecommunications Part 03.01: Magnetic Oxide Cores for Transformers and Chokes for Power Applications. Assessment Level A. 8 pp.

INTERNATIONAL AND NON-U.S. NATIONAL STANDARDS
SUBJECT INDEX
Chromatography

Choke Coils (Cont.)
—Magnetic Cores—Telecommunications (Cont.)
IECQ QC 250300-87. Inductor and Transformer Cores for Telecommunications Part 4: Sectional Specification: Magnetic Oxide Cores for Transformers and Chokes for Power Applications (IEC 723-4 ED 1). 21 pp.
IECQ QC 250301-87. Inductor and Transformer Cores for Telecommunications Part 4: Blank Detail Specification: Magnetic Oxide Cores for Transformers and Chokes for Power Applications; Assessment Level A (IEC 723-4-1 ED 1). 21 pp.

Chokes (Electric)
Use: Choke Coils

Cholesterol
CNS K7142-66. Chemical Reagent (Cholesterol) (Mar)(1642).
JIS K 8350-80. Cholesterol.

Choline Chloride
CNS K7143-66. Chemical Reagent (Choline Chloride) (Mar)(1643).

Cholinesterase
MOD UK TS 684. Horse-Serum Cholinesterase.

Chopper Transistors
See Also: Transistors
—Amplifiers—Preferred Products List
CECC CECC MUAHAG Vol 9 IS 3-90. Preferred Products List; Semiconductors (En, Fr, Ge) AMD 1 (En, Fr, Ge). 51 pp.
—FET
MOD UK DSTAN 59-61: Part 6-01. Semiconductor Devices (Listed on EPIC Database) Part 6: Field Effect Transistors Issue 1; Amendment 1. 28 pp.
—Switching—Preferred Products List
CECC CECC MUAHAG Vol 9 IS 3-90. Preferred Products List; Semiconductors (En, Fr, Ge) AMD 1 (En, Fr, Ge). 51 pp.

Choppers
See Also: Food Processing Equipment
—Electric—Safety
DIN VDE 0700 Pt 251-86. Safety of Household and Similar Electrical Appliances; Food Processors and Food Choppers (Oct). 21 pp.
—Meat
CSA C22.2 NO 195-M1987. Motor Operated Food Processing Appliances (Household and Commercial); (Gen Instr 1 Thru 2). 55 pp.

Choppers (Circuits)
See Also: Switching Circuits
CNS C7122-80. Choppers (Aug)(6132).
—Rolling Stock
JIS E 6201-91. General Rules of Choppers for Electric Rolling Stock.

Chow Mein Noodles
Use: Dry Noodles

Christen Pitts S-1 Aircraft
See Also: Aircraft
—Foreign Airworthiness Directives
CAA. Christen Industies (Pitts) S-1 and S-2 Series Aircraft. 1 p.

Christen Pitts S-2 Aircraft
See Also: Aircraft
—Foreign Airworthiness Directives
CAA. Christen Industies (Pitts) S-1 and S-2 Series Aircraft. 1 p.

Christmas Tree Lights
Use: Lighting Chains

Chromate Coatings
See Also: Coatings; Conversion Coatings; Nonmetallic Coatings; Paints; Phosphate Coatings; Undercoatings
—Aluminum
DIN ENGL 50939-88. Corrosion Protection; Chromating of Aluminium; Principles and Methods of Test (Apr). 5 pp.
ISO 10546-93. Chemical Conversion Coatings—Rinsed and Non-Rinsed Chromate Conversion Coatings on Aluminium and Aluminium Alloys First Edition. 10 pp.

Chromate Coatings (Cont.)
—Aluminum (Cont.)
MOD UK DSTAN 03-18-79. Chromate Conversion Coatings (Chromate Filming Treatments) for Aluminium and Aluminium Alloys Issue 1. 12 pp.
—Aluminum—Aerospace
CEN PREN2437-90. Chromate Conversion Coatings (Yellow) for Aluminium and Aluminium Alloys. 10 pp.
—Aluminum Alloys—Aerospace
CEN PREN2437-90. Chromate Conversion Coatings (Yellow) for Aluminium and Aluminium Alloys. 10 pp.
—Aluminum—Surface Density
BSI BS 5411: Part 14-82. 1982 Method of Test for Metallic and Related Coatings Part 14: Gravimetric Method for Determination of Coating Mass per Unit Area of Conversion Coatings on Metallic Materials. 7 pp.
ISO 3892-80. Conversion Coatings on Metallic Materials—Determination of Coating Mass per Unit Area—Gravimetric Methods First Edition. 5 pp.
—Cadmium
BSI BS 5411: Part 13-82. 1982 Amd 1 Method of Test for Metallic and Related Coatings Part 13: Chromate Conversion Coatings on Zinc and Cadmium. 9 pp.
CGSB 31-GP-115MA-87. Coating, Conversion, Chromate and Phosphate, for Zinc and Cadmium Surfaces. 8 pp.
ISO 3613-80. Chromate Conversion Coatings on Zinc and Cadmium—Test Methods First Edition; (Erratum—Aug 1982). 7 pp.
—Cadmium—Electroplated
BSI BS 6338-82. 1982 Chromate Conversion Coatings on Electroplated Zinc and Cadmium Coatings. 4 pp.
ISO 4520-81. Chromate Conversion Coatings on Electroplated Zinc and Cadmium Coatings First Edition. 4 pp.
JIS H 8625-93. Chromate Conversion Coatings on Electroplated Zinc and Cadmium Coatings. 14 pp.
SAA AS 1791-86. Chromate Conversion Coatings—Zinc and Cadmium (R 1993). 5 pp.
—Cadmium—Passivation
MOD UK DEF-130-61. Chromate Passivation of Cadmium and Zinc Surfaces (Reprinted March 1977, Incorporating Amendment No. 1). 4 pp.
—Cadmium—Surface Density
BSI BS 5411: Part 14-82. 1982 Method of Test for Metallic and Related Coatings Part 14: Gravimetric Method for Determination of Coating Mass per Unit Area of Conversion Coatings on Metallic Materials. 7 pp.
ISO 3892-80. Conversion Coatings on Metallic Materials—Determination of Coating Mass per Unit Area—Gravimetric Methods First Edition. 5 pp.
—Zinc
BSI BS 5411: Part 13-82. 1982 Amd 1 Method of Test for Metallic and Related Coatings Part 13: Chromate Conversion Coatings on Zinc and Cadmium. 9 pp.
CGSB 31-GP-115MA-87. Coating, Conversion, Chromate and Phosphate, for Zinc and Cadmium Surfaces. 8 pp.
ISO 3613-80. Chromate Conversion Coatings on Zinc and Cadmium—Test Methods First Edition; (Erratum—Aug 1982). 7 pp.
—Zinc—Electroplated
BSI BS 6338-82. 1982 Chromate Conversion Coatings on Electroplated Zinc and Cadmium Coatings. 4 pp.
ISO 4520-81. Chromate Conversion Coatings on Electroplated Zinc and Cadmium Coatings First Edition. 4 pp.
JIS H 8625-93. Chromate Conversion Coatings on Electroplated Zinc and Cadmium Coatings. 14 pp.
SAA AS 1791-86. Chromate Conversion Coatings—Zinc and Cadmium (R 1993). 5 pp.
—Zinc—Passivation
MOD UK DEF-130-61. Chromate Passivation of Cadmium and Zinc Surfaces (Reprinted March 1977, Incorporating Amendment No. 1). 4 pp.
—Zinc—Surface Density
BSI BS 5411: Part 13-82. 1982 Amd 1 Method of Test for Metallic and Related Coatings Part 13: Chromate Conversion Coatings on Zinc and Cadmium. 9 pp.
ISO 3892-80. Conversion Coatings on Metallic Materials—Determination of Coating Mass per Unit Area—Gravimetric Methods First Edition. 5 pp.

Chromated Copper Arsenate
Use: Copper Arsenate

Chromated Zinc Chloride
Use: Zinc Chloride

Chromates
Scope Note: Use a more specific term
See: Ammonium Chromate; Ammonium Dichromate; Barium Chromates; Chromate Coatings; Cupric Chromate; Lead Chromate; Lead Chrome Green; Lead Silicochromate; Molybdate; Pigments; Potassium Chromates; Potassium Dichromate; Silver Chromate; Sodium Chromate; Sodium Dichromate; Strontium Chromate; Zinc Chromate

Chromatic Dispersion
See Also: Dispersion; Fiber Optic Equipment; Optical Waveguides; Wavelength
—Fiber Optic Cables
CCITT RECMN G.651-89. Characteristics of a 50/125 Micrometer Multimode Graded Index Optical Fibre Cable—Transmission Media Characteristics (Study Group XV) 30 pp. 30 pp.
CCITT RECMN G.652-89. Characteristics of a Single-Mode Optical Fibre Cable—Transmission Media Characteristics (Study Group XV) 34 pp. 34 pp.
CCITT RECMN G.653-89. Characteristics of a Dispersion-Shifted Single-Mode Optical Fibre Cable—Transmission Media Characteristics (Study Group XV) 5 pp. 5 pp.
CCITT RECMN G.654-89. Characteristics of a 1550 mm Wavelength Loss-Minimized Single-Mode Optical Fibre Cable—Transmission Media Characteristics (Study Group XV) 3 pp. 3 pp.
—Optical Fibers
CNS C6332-89. Method of Test for Fiber Optic Devices (FOTP-169 Chromatic Dispersion Measurement of Optical Fiber by the Phase—Shift Method) (Jul)(12563).

Chromaticity
See Also: Color; Electromagnetic Properties; Optical Properties
—Classification
CNS Z1041-84. Chromaticity Classification (Mar)(10839).
—Colorimetry
BSI BS 950: Part 1-67. 1967 Artificial Daylight for the Assessment of Colour Part 1: Illuminant for Colour Matching and Colour Appraisal. 13 pp.
CNS Z7192-85. Specification of Colours According to the CIE 1931 Standard Colorimetric System and the CIE 1964 Supplementary Standard Colormetric System (Apr)(11256).
CNS Z7197-85. Methods of Measurement for Colour of Reflecting or Transmitting Objects (Aug)(11351).
CNS Z7199-85. Method of Measurement for Light Source Colour (Aug)(11353).
JIS Z 8701-82. Specification of Colours According to the CIE 1931 Standard Colorimetric System and the CIE 1964 Supplementary Standard Colorimetric System. 24 pp.
JIS Z 8722-82. Methods of Measurement for Colour of Reflecting or Transmitting Objects. 22 pp.
JIS Z 8724-83. Methods of Measurement for Light Source Colour. 25 pp.
JIS Z 8729-80. Specification of Colour of Materials According to the CIE 1976 (L*A*B) Space and the CIE 1976 (L*U*V) Space (R 1985). 15 pp.
—Discharge Lamps
CNS Z1042-84. Chromaticity Domain of Discharge Lamps (Mar)(10840).
JIS Z 9113-73. Chromaticity Domain of Discharge Lamps.
—Electric Switches
CNS C6237-84. Method of Test for Electromechanical Switches (Chromaticity) (Nov)(11100).
—Fabrics
CGSB CAN/CGSB-4.2 NO.64-M91. Textile Test Methods Chromatic Transference Scale. 9 pp.
—Fluorescent Lighting
JIS Z 9112-90. Classification of Fluorescent Lamps by Chromaticity and Colour Rendering Property. 10 pp.

Chromatography
See Also: Adsorption; Chemical Analysis; Colorimetric Analysis; Colorimetry; Column Chromatography; Gas Chromatography; Ion Chromatography; Ion Exchange Chromatography; Liquid Chromatography; Paper Chromatography; Spectrophotometry; Thin Layer Chromatography; Volumetric Analysis
CNS K7036-61. Chemical Reagent (Aluminum Oxide) (for Chomatography) (Dec)(1536).

Chromatography (Cont.)

—Glossaries
CNS K0046-89. Technical Terms for Analytical Chemistry (Chromatography Part) (Aug)(12588).

Chrome
Use: Chromium

Chrome Bricks
Use: Chrome Refractories

Chrome Orange
See Also: Pigments

—Pigments—Lead Chromate Content
CGSB 1-GP-71 METH 50.19-81. Methods of Testing Paints and Pigments Pigment Analysis Lead Chromate. 2 pp.

Chrome Ores
Use: Chromium Ores

Chrome Oxide Green
See Also: Pigments

—Pigments
BSI BS 318-88. 1988 Chromic Oxide Green Pigments for Paints. 9 pp.
CNS K2106-86. Chrome Oxide Green Pigment (Feb)(5871). 1 p.
CNS K6825-86. Method of Test for Chrome Oxide Green Pigment (Feb)(11507).
ISO 4621-86. Chrome Oxide Green Pigments—Specifications and Methods of Test First Edition. 7 pp.

Chrome Refractories
Use For: Chrome Bricks *See Also:* Refractory Materials
CNS R2068-71. Chrome Brick (Jul)(3322). 2 pp.
JIS R 2301-76. Chrome Brick. 6 pp.

—Chemical Analysis
BSI BS 1902: Sec 2.2-74. 1974 Amd 1 Methods for Testing Refractory Materials Part 2: Chemical Analysis (Wet Methods) Section 2.2: Chemical Analysis of Chrome-Bearing Materials (AMD 4960) February 28, 1986. 14 pp.
JIS R 2901-71. Chemical Analysis of Chrome Ore for Refractories. 12 pp.

—Magnesia
CNS R2069-71. Chrome-Magnesite Bricks and Magnesite-Chrome Bricks (Sep)(3323). 2 pp.
JIS R 2306-76. Chrome-Magnesia Brick. 7 pp.

Chrome Yellow
See Also: Pigments

—Pigments
CNS K2056-89. Chrome Yellow (Pigment) (Jul)(3945). 2 pp.
CNS K6387-89. Method of Test for Chrome Yellow (Pigment) (Jul)(3946). 2 pp.
JIS K 5110-65. Chrome Yellow (Pigment). 5 pp.

—Pigments—Lead Chromate Content
CGSB 1-GP-71 METH 50.19-81. Methods of Testing Paints and Pigments Pigment Analysis Lead Chromate. 2 pp.

Chromic Acid
Use: Chromium Trioxide

Chromic Anhydride
Use: Chromium Trioxide

Chromic Potassium Sulfate
CNS K7145-66. Chemical Reagent (Chromium Potassium Sulfate)(Chrome Alum) (Mar)(1645).

Chromium
Scope Note: For additional listings, see also specific products made from chromium *Use For:* Chrome
See Also: Chromium Coatings (Made From Chromium); Chromium Content Analysis; Metals
CNS G2119-82. Methods of Chemical Analysis for Chromium and Carbon in Chromium Metal (Jan)(8403).
CNS G3116-84. Chromium Metal (Jun) (5125).
DIN ENGL 17565-68. Ferro-Chromium, Ferro-Chromium-Silicon and Chromium; Technical Conditions of Delivery (Dec). 2 pp.
JIS G 2313-86. Chromium Metal. 5 pp.

—Aluminum Content—Volumetric Analysis
BSI BS 6200: SUBSEC 3.1.1-91. 1991 Sampling and Analysis of Iron, Steel and Other Ferrous Metals Part 3: Methods of Analysis Section 3.1: Determination of Aluminium Subsection 3.1.1: Steel, Cast Iron, Low Carbon Ferro-Chromium Metal: Volumetric Method. 11 pp.

Chromium (Cont.)

—Aluminum Content—Volumetric Analysis (Cont.)
CNS G2120-82. Methods of Chemical Analysis for Silicon, Phosphorus, Sulphur Ferrite and Aluminum in Chromium Metal (Jan)(8404).

—Carbon Content
CNS G2119-82. Methods of Chemical Analysis for Chromium and Carbon in Chromium Metal (Jan)(8403).

—Charcoal Fabrics
MOD UK DSTAN 83-84-93. Charcoal Cloth Impregnated with Copper and Chromium, Type 1303 Issue 1. 28 pp.

—Chemical Analysis
JIS G 1323-89. Methods for Chemical Analysis of Chromium Metal. 28 pp.

—Cosmetics
CNS S2102-83. Method of Hygienic Test for Cosmetics Chromium (Jun)(10380).

—Phosphorus Content
CNS G2120-82. Methods of Chemical Analysis for Silicon, Phosphorus, Sulphur Ferrite and Aluminum in Chromium Metal (Jan)(8404).

—Salts—Fabrics—Colorfastness Testing
CNS L3189-82. Method of Test for Colour Fastness to Metals in the Dyebath: Chromium Salt (Aug) (9312).
JIS L 0870-75. Testing Method for Colour Fastness to Metals in the Dyebath: Chromium Salt.

—Sampling
CNS G2214-84. Method of Sampling for Metallic Chromium (Mar)(10812).
JIS G 1603-85. Methods of Sampling for Chemical Analysis of Ferroalloys (Part 3 Ferrophosphorus, Manganese Metal, Silicon Metal, Chromium Metal, Calcium Silicon and Ferroboron). 10 pp.

—Silicon Content
CNS G2120-82. Methods of Chemical Analysis for Silicon, Phosphorus, Sulphur Ferrite and Aluminum in Chromium Metal (Jan)(8404).

—Standard Solution
CNS K0037-88. Chromium Standard Solution (Nov)(12467).
JIS K 0024-83. Chromium Standard Solution. 14 pp.

—Sulfur Ferrite Content
CNS G2120-82. Methods of Chemical Analysis for Silicon, Phosphorus, Sulphur Ferrite and Aluminum in Chromium Metal (Jan)(8404).

—Tanning Materials
CNS K6817-85. Method of Test for Chrome Tanning Agent (Dec)(11454).
JIS K 6506-77. Analytical Method for Chrome Tanning Agents.

—Toys—Migration Testing
BSI BS 5665: Part 3-89. 1989 Safety of Toys Part 3: Migration of Certain Elements (Supersedes BS 3443: 1968). 11 pp.
CEN EN 71-3-88. Safety of Toys Part 3 Migration of Certain Elements. 14 pp.
CEN PREN 71-3-92. Safety of Toys—Part 3: Migration of Certain Elements. 24 pp.

—Wood Preservatives
BSI BS 4072: Part 1-87. 1987 Amd 1 Wood Preservation by Means of Copper/Chromium /Arsenic Compositions Part 1: Preservatives (AMD 6200) June 30, 1989. 5 pp.

Chromium Carbide Coatings (Made From Chromium Carbide)
See Also: Coatings

—Electrodeposited—Aerospace
MOD UK DTD-943-74. Electrodeposited Cobalt/ Chromium Carbide Composite Coatings. 2 pp.

Chromium Chlorides
CNS K7144-66. Chemical Reagent (Chromium Chloride) (Mar)(1644).
JIS K 8128-61. Chromium Chloride.

Chromium Coatings (Made From Chromium)
Scope Note: Includes coatings made from chromium alloys *See Also:* Chromium; Coatings; Metal Coatings (Made From Metal)

—Aircraft—Electrodeposited
JIS W 1113-84. Chromium Plating (Electrodeposited for Aircraft).

Chromium Coatings (Made From Chromium) (Cont.)

—Alloy Steel—Electrodeposited—Aerospace
AECMA PREN2132-82. Electrodeposition of Chromium for Engineering Purposes. 12 pp.

—Aluminum—Electroplated
BSI BS 1224-70. 1970 Amd 1 Electroplated Coatings of Nickel and Chromium. 29 pp.

—Brasses—Electroplated
BSI BS 3382: Parts 3 & 4-65. 1965 Amd 1 Electroplated Coatings on Threaded Components Part 3: Nickel or Nickel Plus Chromium on Steel Components Part 4: Nickel or Nickel Plus Chromium on Copper and Copper Alloy (Including Brass) Components. 33 pp.
JIS H 9121-53. Standard Operation of Nickel Plating and Chromium Plating. 9 pp.

—Carbon Steel—Electrodeposited—Aerospace
AECMA PREN2132-82. Electrodeposition of Chromium for Engineering Purposes. 12 pp.

—Copper—Electroplated
BSI BS 1224-70. 1970 Amd 1 Electroplated Coatings of Nickel and Chromium. 29 pp.
BSI BS 3382: Parts 3 & 4-65. 1965 Amd 1 Electroplated Coatings on Threaded Components Part 3: Nickel or Nickel Plus Chromium on Steel Components Part 4: Nickel or Nickel Plus Chromium on Copper and Copper Alloy (Including Brass) Components. 33 pp.

—Corrosion Testing
ISO 4539-80. Electrodeposited Chromium Coatings—Electrolytic Corrosion Testing (EC Test) First Edition. 8 pp.

—Electrodeposited
MOD UK DSTAN 03-31-93. Guide to the Use of Chromium Plating for Engineering Purposes Issue 1. 14 pp.
MOD UK DG-13. Design Aspects of Chromium Plating for Engineering Purposes (Superseded by Def Stan 03-31).

—Electroplated
BSI BS 4641-86. 1986 Electroplated Coatings of Chromium for Engineering Purposes. 14 pp.
CNS H3060-87. Electroplated Coatings of Nickel and Chromium (Sep)(4157).
CNS H3109-82. Recommended Practice for Copper-Nickel-Chromium Plating (Jun)(8977).
CNS H3154-90. Electroplated Coatings of Chromium for Engineering Use (Mar)(12685).
ISO 1456-88. Metallic Coatings—Electrodeposited Coatings of Nickel Plus Chromium and of Copper Plus Nickel Plus Chromium Second Edition; (Supersedes 1457). 19 pp.
ISO 6158-84. Metallic Coatings—Electroplated Coatings of Chromium for Engineering Purposes First Edition. 9 pp.
JIS H 8615-93. Electroplated Coatings of Chromium for Engineering Purposes. 12 pp.
JIS H 8617-91. Electroplated Coatings of Nickel and Chromium. 17 pp.

—Iron—Electroplated
BSI BS 1224-70. 1970 Amd 1 Electroplated Coatings of Nickel and Chromium. 29 pp.

—Low Carbon Steel—Electrodeposited
CEN PREN 10 171-87. Double Cold Reduced Electrolytic Chromium/Chromium Oxide Coated Steel: Sheet. 29 pp.
CEN PREN 10 172-87. Single Cold Reduced Electrolytic Chromium/Chromium Oxide Coated Steel: Coil for Subsequent Cutting into Sheets. 29 pp.
CEN PREN 10 173-87. Double Cold Reduced Electrolytic Chromium/Chromium Oxide Coated Steel: Coil for Subsequent Cutting into Sheets. 29 pp.

—Metal—Electroplated
SAA AS 1192-82. Electroplated Coatings—Nickel and Chromium (R 1993). 9 pp.

—Plastic—Electroplated
BSI BS 4601-70. 1970 Electroplated Coatings of Nickel Plus Chromium on Plastics Materials. 20 pp.
CNS H3136-87. Electroplated Coatings of Copper Nickel Chromium on Plastics (Oct)(11391).
ISO 4525-85. Metallic Coatings—Electroplated Coatings of Nickel Plus Chromium on Plastics Materials First Edition. 7 pp.

—Steel—Electrodeposited
BSI BS EN 10 202-90. 1990 Cold Reduced Electrolytic Chromium/Chromium Steel Coated Steel. 24 pp.

INTERNATIONAL AND NON-U.S. NATIONAL STANDARDS
SUBJECT INDEX
Chromium

Chromium Coatings (Made From Chromium) *(Cont.)*

—**Steel—Electrodeposited** *(Cont.)*
CEN EN 10 202-89. Cold Reduced Electrolytic Chromium/Chromium Oxide Coated Steel. 34 pp.
CNS G2240-85. Method of Test for Chromium Plated Tin Free Steel (Apr)(11242).
CNS G3226-85. Chromium Plated Tin Free Steel (Apr)(11241).
DIN ENGL EN 10202-90. Cold Reduced Electrolytic Chromium/Chromium Oxide Coated Steel; English Version of DIN EN 10202 (Mar). 17 pp.
ISO 8110 Pt 1-88. Single Cold-Reduced Electrolytic Chromium/Chromium Oxide-Coated Steel—Part 1: Sheets First Edition. 18 pp.
ISO 8111 Pt 1-88. Double Cold-Reduced Electrolytic Chromium/Chromium Oxide-Coated Steel—Part 1: Sheets First Edition. 20 pp.
JIS G 3315-87. Chromium Plated Tin Free Steel. 29 pp.

—**Steel—Electroplated**
BSI BS 1224-70. 1970 Amd 1 Electroplated Coatings of Nickel and Chromium. 29 pp.
BSI BS 3382: Parts 3 & 4-65. 1965 Amd 1 Electroplated Coatings on Threaded Components Part 3: Nickel or Nickel Plus Chromium on Steel Components Part 4: Nickel or Nickel Plus Chromium on Copper and Copper Alloy (Including Brass) Components. 33 pp.
JIS H 9121-53. Standard Operation of Nickel Plating and Chromium Plating. 9 pp.

—**Zinc—Electroplated**
BSI BS 1224-70. 1970 Amd 1 Electroplated Coatings of Nickel and Chromium. 29 pp.
CNS H3106-86. Operation Standard for Chromium Plating on Zinc Alloys (Feb)(8508).
JIS H 9125-65. Operation Standard for Chromium Plating on Zinc Alloys (R 1983). 9 pp.

Chromium Content Analysis
See Also: Chromium

—**Alloy Steels—Atomic Absorption Spectrometry**
ISO 10138-91. Steel and Iron—Determination of Chromium Content—Flame Atomic Absorption Spectrometric Method First Edition. 12 pp.

—**Aluminum Alloys**
JIS H 1358-72. Methods for Determination of Chromium in Aluminium Alloy.

—**Aluminum Alloys—Atomic Absorption Spectrometry**
ISO 4193-81. Aluminium and Aluminium Alloys—Determination of Chromium Content—Flame Atomic Absorption Spectrometric Method First Edition. 5 pp.

—**Aluminum Alloys—Photometry**
BSI BS 1728: Part 16-68. 1968 Methods for the Analysis of Aluminium and Aluminium Alloys: Part 16: Chromium (Photometric Method). 10 pp.

—**Aluminum Alloys—Spectrophotometry**
ISO 3978-76. Aluminium and Aluminium Alloys—Determination of Chromium—Spectrophotometric Method Using Diphenylcarbazide, After Extraction First Edition. 6 pp.

—**Aluminum Alloys—Volumetric Analysis**
BSI BS 1728: Part 17-68. 1968 Methods for the Analysis of Aluminium and Aluminium Alloys: Part 17: Chromium (Volumetric Method). 8 pp.

—**Aluminum—Atomic Absorption Spectrometry**
ISO 4193-81. Aluminium and Aluminium Alloys—Determination of Chromium Content—Flame Atomic Absorption Spectrometric Method First Edition. 5 pp.

—**Aluminum—Photometry**
BSI BS 1728: Part 16-68. 1968 Methods for the Analysis of Aluminium and Aluminium Alloys: Part 16: Chromium (Photometric Method). 10 pp.

—**Aluminum—Spectrophotometry**
ISO 3978-76. Aluminium and Aluminium Alloys—Determination of Chromium—Spectrophotometric Method Using Diphenylcarbazide, After Extraction First Edition. 6 pp.

—**Aluminum—Volumetric Analysis**
BSI BS 1728: Part 17-68. 1968 Methods for the Analysis of Aluminium and Aluminium Alloys: Part 17: Chromium (Volumetric Method). 8 pp.

—**Animal Feed**
CNS N4024-18-86. Method of Test for Feeds: Determination of Chromium (Aug)(2770-18).

Chromium Content Analysis *(Cont.)*

—**Boric Acids—Photometry**
BSI BS 5688: Part 14-79. 1979 Orthoboric Acid (Boric Acid), Diboron Trioxide (Boric Oxide), Disodium Tetraborates, Sodium Perborates and Crude Sodium Borates for Industrial Use Part 14: Determ. of Chromium Content of Boric Acid, Boric Oxide and Disodium Tetraborates. 4 pp.
ISO 3119-76. Boric Acid, Boric Oxide and diSodium Tetraborates for Industrial Use—Determination of Chromium Content—Diphenylcarbazide Photometric Method First Edition. 5 pp.

—**Boron Oxides—Photometry**
BSI BS 5688: Part 14-79. 1979 Orthoboric Acid (Boric Acid), Diboron Trioxide (Boric Oxide), Disodium Tetraborates, Sodium Perborates and Crude Sodium Borates for Industrial Use Part 14: Determ. of Chromium Content of Boric Acid, Boric Oxide and Disodium Tetraborates. 4 pp.
ISO 3119-76. Boric Acid, Boric Oxide and diSodium Tetraborates for Industrial Use—Determination of Chromium Content—Diphenylcarbazide Photometric Method First Edition. 5 pp.

—**Cast Iron—Potentiometric Analysis**
BSI BS 6200: SUB SEC 3.10.1-85. (WITHDRAWN) 1985 Sampling and Analysis of Iron, Steel and Other Ferrous Metals: Part 3: Methods of Analysis Sec. 3.10: Determ.Chromium: Subsec3.10.1: Steel and Iron: Pptentiometric/Volumetric Method (Superseded by BS EN 24 937: 1991). 8 pp.
BSI BS EN 24937-91. 1991 Steel and Iron Determination of Chromium Content: Potentiometric or Visual Titration Method (V). 14 pp.
BSI BS EN 24937-01. 1991 Amd 1 Steel and Iron Determination of Chromium Content Potentiometric or Visual Titration Method (ISO 4937: 1986) (AMD 7082) July 15, 1992 (V). 15 pp.
CEN EN 24 937-90. Steel and Iron—Determination of Chromium Content—Potentiometric or Visual Method. 2 pp.
DIN ENGL EN 24937-92. Determination of Chromium Content of Steel and Iron by Potentiometric or Visual Titration (ISO 4937: 1986) (Nov). 11 pp.
ISO 4937-86. Steel and Iron—Determination of Chromium Content—Potentiometric or Visual Titration Method First Edition. 10 pp.
SAA AS 1050.9-84. Methods for the Analysis of Iron and Steel—Part 9: Determination of Chromium (Potentiometric Method). 4 pp.

—**Cast Iron—Volumetric Analysis**
BSI BS 6200: SUB SEC 3.10.1-85. (WITHDRAWN) 1985 Sampling and Analysis of Iron, Steel and Other Ferrous Metals: Part 3: Methods of Analysis Sec. 3.10: Determ.Chromium: Subsec3.10.1: Steel and Iron: Pptentiometric/Volumetric Method (Superseded by BS EN 24 937: 1991). 8 pp.
BSI BS EN 24937-91. 1991 Steel and Iron Determination of Chromium Content: Potentiometric or Visual Titration Method (V). 14 pp.
BSI BS EN 24937-01. 1991 Amd 1 Steel and Iron Determination of Chromium Content Potentiometric or Visual Titration Method (ISO 4937: 1986) (AMD 7082) July 15, 1992 (V). 15 pp.
CEN EN 24 937-90. Steel and Iron—Determination of Chromium Content—Potentiometric or Visual Method. 2 pp.
DIN ENGL EN 24937-92. Determination of Chromium Content of Steel and Iron by Potentiometric or Visual Titration (ISO 4937: 1986) (Nov). 11 pp.
ISO 4937-86. Steel and Iron—Determination of Chromium Content—Potentiometric or Visual Titration Method First Edition. 10 pp.

—**Chromium Ores—Volumetric Analysis**
ISO 6331-83. Chromium Ores and Concentrates—Determination of Chromium Content—Titrimetric Method First Edition. 6 pp.

—**Coatings**
CGSB CAN2-1.500-75 METH 8-73. Methods of Test for Toxic Trace Elements in Protective Coatings Determination of Leachable Chromium (Cr) in Low Concentration. 1 p.

—**Copper Alloys—Atomic Absorption Spectrometry**
ISO 4744-84. Copper and Copper Alloys—Determination of Chromium Content—Flame Atomic Absorption Spectrometric Method First Edition. 5 pp.

—**Copper Alloys—Atomic Absorption Spectrophotometry**
BSI BS 6721: Part 4-86. 1986 Sampling and Analysis of Copper and Copper Alloys Part 4: Method for Determination of Chromium in Copper and Copper Alloys by Flame Atomic Absorption Spectrophotometry. 7 pp.

Chromium Content Analysis *(Cont.)*

—**Copper Alloys—Atomic Absorption Spectrophotometry** *(Cont.)*
BSI BS 6721: Part 5-86. 1986 Sampling and Analysis of Copper and Copper Alloys Part 5: Method for Determination of Zinc in Copper and Copper Alloys by Flame Atomic Absorption Spectrophotometry. 7 pp.

—**Copper Alloys—Volumetric Analysis**
ISO 6437-84. Copper Alloys—Determination of Chromium Content—Titrimetric Method First Edition. 5 pp.

—**Copper—Atomic Absorption Spectrometry**
ISO 4744-84. Copper and Copper Alloys—Determination of Chromium Content—Flame Atomic Absorption Spectrometric Method First Edition. 5 pp.

—**Copper—Atomic Absorption Spectrophotometry**
BSI BS 6721: Part 4-86. 1986 Sampling and Analysis of Copper and Copper Alloys Part 4: Method for Determination of Chromium in Copper and Copper Alloys by Flame Atomic Absorption Spectrophotometry. 7 pp.

—**Electrolytic Copper—Atomic Absorption Spectrophotometry**
BSI BS 7317: Part 1-90. 1990 Methods for Analysis of High Purity Copper Cathode Cu-CATH-1 Part 1: Method for Determination of Cadmium Manganese and Silver (Screening Procedure for Chromium, Cobalt, Iron, Nickel and Zinc) by Atomic Absorption. 9 pp.
BSI BS 7317: Part 2-90. 1990 Methods for Analysis of High Purity Copper Cathode Cu-CATH-1 Part 2: Method for Determination of Chromium, Cobalt, Iron, Nickel and Zinc by Discrete Volume Nebulization Atomic Absorption Spectrophotometry. 9 pp.
BSI DD 95: Part 2-84. (WITHDRAWN) 1984 Amd 1 Analysis of Higher Purity Copper Cathode Cu-CATH-1 Part 2: Method for Determination of Chromium, Cobalt, Iron, Nickel, and Zinc by Discrete Volume Nebulization Atomic Absorption Spectrophotmetry. 10 pp.

—**Ferrochromium**
CNS G2184-83. Method of Chemical Analysis for Chromium in Ferrochromium (Aug)(10504).
SAA AS 3587.3-91. Ferroalloys—Chemical Analysis—Part 3: Determination of Chromium Content of Ferrochromium and Ferrosilicochromium. 4 pp.

—**Ferrochromium—Potentiometric Analysis**
BSI BS 6200: SUB SEC 3.10.4-85. 1985 Sampling and Analysis of Iron, Steel and Other Ferrous Metals Part 3: Methods of Analysis Section 3.10: Determination of Chromium Subsection 3.10.4: Ferrochromium and Ferrosilicochromium: Potentiometric Method. 6 pp.
ISO 4140-79. Ferrochromium and Ferrosilicochromium—Determination of Chromium Content—Potentiometric Method First Edition. 5 pp.

—**Ferrochromium—Volumetric Analysis**
BSI BS 6200: SUB SEC 3.10.5-92. 1992 Sampling and Analysis of Iron, Steel and Other Ferrous Metals Part 3: Methods of Analysis Section 3.10: Determination of Chromium Subsection 3.10.5: Ferrochromium: Volumetric Method. 8 pp.

—**Ferrosilicochromium**
CNS G2174-83. Method of Chemical Analysis for Chromium in Ferrosilico-Chromium (May)(10260).
SAA AS 3587.3-91. Ferroalloys—Chemical Analysis—Part 3: Determination of Chromium Content of Ferrochromium and Ferrosilicochromium. 4 pp.

—**Ferrosilicochromium—Potentiometric Analysis**
BSI BS 6200: SUB SEC 3.10.4-85. 1985 Sampling and Analysis of Iron, Steel and Other Ferrous Metals Part 3: Methods of Analysis Section 3.10: Determination of Chromium Subsection 3.10.4: Ferrochromium and Ferrosilicochromium: Potentiometric Method. 6 pp.
ISO 4140-79. Ferrochromium and Ferrosilicochromium—Determination of Chromium Content—Potentiometric Method First Edition. 5 pp.

—**Flue Gases**
JIS K 0096-75. Methods for Determination of Chromium and Manganese in Stack Gas (R 1989). 16 pp.

—**Gallium Arsenide**
JIS H 1191-91. Methods for Chemical Analysis of Gallium Arsenide. 57 pp.

INDUSTRY STANDARDS

Chromium Content Analysis (Cont.)

—Hard Metals—Atomic Absorption Spectrometry

BSI BS 5600: SUB SEC 4.17.6-86. (WITHDRAWN) 1986 Part 4: Methods of Testing and Chemical Analysis of Hardmetals Section 4.17: Chemical Analysis by Flame Atomic Absorption Spectrometry Subsection 4.17.6: Determination of Chromium in Contents from 0.01% to 2% (m/m). 4 pp.

BSI BS EN 27627-6-93. 1993 Hardmetals—Chemical Analysis by Flame Atomic Absorption Spectrometry Part 6: Determination of Chromium in Contents from 0.01 to 2 % (m/m) (ISO 7627-6: 1985) (V). 10 pp.

CEN EN 27627-6-93. Hardmetals—Chemical Analysis by Flame Atomic Absorption Spectrometry—Part 6: Determination of Chromium in Contents from 0,01 to 2 % (m/m) (ISO 7627-6: 1985). 5 pp.

ISO 7627 Pt 6-85. Hardmetals—Chemical Analysis by Flame Atomic Absorption Spectrometry—Part 6: Determination of Chromium in Contents from 0,01 to 2 % (m/m) First Edition (CEN EN 27627-6: 1993). 4 pp.

—Hydrofluoric Acid—Emission Spectroscopy

DIN ENGL 50451 Pt 2-90. Determination of Cobalt, Chromium, Copper, Iron and Nickel as Impurities in Hydrofluoric Acid for Use in Semiconductor Technology by Plasma-Induced Emission Spectroscopy (Oct). 2 pp.

—Iron

CNS G2245-85. Method of Determination for Chromium in Iron and Steel (Jul)(11302).

JIS G 1217-81. Methods for Determination of Chromium in Iron and Steel. 20 pp.

—Iron—Atomic Absorption Spectrometry

BSI BS 6200: SUB SEC 3.10.2-89. 1989 Sampling and Analysis of Iron, Steel and Other Ferrous Metals Part 3: Methods of Analysis Section 3.10 Determination of Chromium Subsection 3.10.2: Steel and Cast Iron: Flame Atomic Absorption Spectrometric Method. 12 pp.

CEN EN 10 188-89. Chemical Analysis of Ferrous Materials Determination of Chromium in Steels and Irons Flame Atomic Absorption Spectrometic Method. 9 pp.

DIN ENGL EN 10188-90. Chemical Analysis of Ferrous Materials; Determination of Chromium in Steels and Iron; Flame Atomic Absorption Spectrometic Method (Apr). 10 pp.

ISO 10138-91. Steel and Iron—Determination of Chromium Content—Flame Atomic Absorption Spectrometric Method First Edition. 12 pp.

—Iron Ores—Atomic Absorption Spectrometry

BSI BS 7020: Part 16-93. 1993 Analysis of Iron Ores Part 16: Method for the Determination of Nickel and/or Chromium Contents: Flame Atomic Absorption Spectrometric Method (ISO 9685: 1991) (Q). 16 pp.

ISO 9685-91. Iron Ores—Determination of Nickel and/or Chromium Contents—Flame Atomic Absorption Spectrometric Method First Edition. 14 pp.

—Iron Ores—Spectrophotometry

JIS M 8224-83. Methods for Determination of Chromium in Iron Ores.

JIS M 8224-71. Methods for Determination of Chromium in Iron Ores. 11 pp.

—Iron Ores—Volumetric Analysis

JIS M 8224-83. Methods for Determination of Chromium in Iron Ores.

JIS M 8224-71. Methods for Determination of Chromium in Iron Ores. 11 pp.

—Iron—Potentiometric Analysis

JIS G 1238-92. Steel and Iron—Determination of Chromium Content—Potentiometric on Visual Titration Method.

—Manganese Ores—Photometry

ISO 619-81. Manganese Ores—Determination of Chromium Content—Diphenylcarbazide Photometric Method and Silver Persulphate Titrimetric Method Second Edition. 6 pp.

—Manganese Ores—Volumetric Analysis

ISO 619-81. Manganese Ores—Determination of Chromium Content—Diphenylcarbazide Photometric Method and Silver Persulphate Titrimetric Method Second Edition. 6 pp.

—Nickel Alloys—Atomic Absorption Spectrometry

BSI BS 7455: Part 3-91. 1991 Analysis of Nickel Alloys by Flame Atomic Absorption Spectrometry Part 3: Method for the Determination of Chromium (ISO 7530-3: 1990). 8 pp.

Chromium Content Analysis (Cont.)

—Nickel Alloys—Atomic Absorption Spectrometry (Cont.)

ISO 7530 Pt 3-90. Nickel Alloys—Flame Atomic Absorption Spectrometric Analysis—Part 3: Determination of Chromium Content First Edition. 6 pp.

—Nickel Alloys—Potentiometric Analysis

ISO 7529-89. Nickel Alloys—Determination of Chromium Content—Potentiometric Titration Method with Ammonium Iron(II) Sulfate First Edition. 8 pp.

—Nickel Castings—Atomic Absorption Spectrophotometry

JIS H 1279-88. Methods for Determination of Chromium in Nickel Alloy Castings. 12 pp.

—Nickel Castings—Volumetric Analysis

JIS H 1279-88. Methods for Determination of Chromium in Nickel Alloy Castings. 12 pp.

—Nickel—Photometry

BSI BS 3727: Part 4-64. 1964 Methods for the Analysis of Nickel for Use in Electronic Tubes and Valves Part 4: Determination of Chromium (Photometric Method). 8 pp.

—Paints—Atomic Absorption Spectrometry

BSI BS 3900: Part B11-86. 1986 Methods of Test for Paints Group B: Tests Involving Chemical Examination of Liquid Paints and Dried Paint Films Part B11: Determination of Total Chromium Content of Liquid Matter. 6 pp.

ISO 3856 Pt 6-84. Paints and Varnishes—Determination of "Soluble" Metal Content—Part 6: Determination of Total Chromium Content of the Liquid Portion of the Paint—Flame Atomic Absorption Spectrometric Method Second Edition. 5 pp.

—Paints—Spectrophotometry

BSI BS 3900: Part B10-86. 1986 Methods of Test for Paints Group B: Tests Involving Chemical Examination of Liquid Paints and Dried Paint Films Part B10: Determination of Hexavalent Chromium Content of Solid Matter. 6 pp.

ISO 3856 Pt 5-84. Paints and Varnishes—Determination of "Soluble" Metal Content-Part 5: Determination of Hexavalent Chromium Content of the Pigment Portion of the Liquid Paint or the Paint in Powder Form—Diphenyl-carbazide Spectrophotometric Method Second Edition. 5 pp.

—Pigments

CGSB 1-GP-71 METH 50.18-81. Methods of Testing Paints and Pigments Pigment Analysis Zinc and Chromium in Basic Zinc Chromate. 2 pp.

—Sodium Borates—Photometry

BSI BS 5688: Part 14-79. 1979 Orthoboric Acid (Boric Acid), Diboron Trioxide (Boric Oxide), Disodium Tetraborates, Sodium Perborates and Crude Sodium Borates for Industrial Use Part 14: Determ. of Chromium Content of Boric Acid, Boric Oxide and Disodium Tetraborates. 4 pp.

ISO 3119-76. Boric Acid, Boric Oxide and diSodium Tetraborates for Industrial Use—Determination of Chromium Content—Diphenylcarbazide Photometric Method First Edition. 5 pp.

—Steels

CNS G2245-85. Method of Determination for Chromium in Iron and Steel (Jul)(11302).

JIS G 1217-81. Methods for Determination of Chromium in Iron and Steel. 20 pp.

—Steels—Atomic Absorption Spectrometry

BSI BS 6200: SUB SEC 3.10.2-89. 1989 Sampling and Analysis of Iron, Steel and Other Ferrous Metals Part 3: Methods of Analysis Section 3.10 Determination of Chromium Subsection 3.10.2: Steel and Cast Iron: Flame Atomic Absorption Spectrometric Method. 12 pp.

CEN EN 10 188-89. Chemical Analysis of Ferrous Materials Determination of Chromium in Steels and Irons Flame Atomic Absorption Spectrometic Method. 9 pp.

DIN ENGL EN 10188-90. Chemical Analysis of Ferrous Materials; Determination of Chromium in Steels and Iron; Flame Atomic Absorption Spectrometric Method (Apr). 10 pp.

ISO 10138-91. Steel and Iron—Determination of Chromium Content—Flame Atomic Absorption Spectrometric Method First Edition. 12 pp.

—Steels—Potentiometric Analysis

BSI BS 6200: SUB SEC 3.10.1-85. (WITHDRAWN) 1985 Sampling and Analysis of Iron, Steel and Other Ferrous Metals: Part 3: Methods of Analysis Sec. 3.10: Determ.Chromium: Subsec3.10.1: Steel and Iron: Pptentiometric/Volumetric Method (Superseded by BS EN 24 937: 1991). 8 pp.

Chromium Content Analysis (Cont.)

—Steels—Potentiometric Analysis (Cont.)

BSI BS EN 24937-91. 1991 Steel and Iron Determination of Chromium Content: Potentiometric or Visual Titration Method (V). 14 pp.

BSI BS EN 24937-01. 1991 Amd 1 Steel and Iron Determination of Chromium Content Potentiometric or Visual Titration Method (ISO 4937: 1986) (AMD 7082) July 15, 1992 (V). 15 pp.

CEN EN 24 937-90. Steel and Iron—Determination of Chromium Content—Potentiometric or Visual Method. 2 pp.

DIN ENGL EN 24937-92. Determination of Chromium Content of Steel and Iron by Potentiometric or Visual Titration (ISO 4937: 1986) (Nov). 11 pp.

ISO 4937-86. Steel and Iron—Determination of Chromium Content—Potentiometric or Visual Titration Method First Edition. 10 pp.

JIS G 1238-92. Steel and Iron—Determination of Chromium Content—Potentiometric on Visual Titration Method.

SAA AS 1050.9-84. Methods for the Analysis of Iron and Steel—Part 9: Determination of Chromium (Potentiometric Method). 4 pp.

—Steels—Volumetric Analysis

BSI BS 6200: SUB SEC 3.10.1-85. (WITHDRAWN) 1985 Sampling and Analysis of Iron, Steel and Other Ferrous Metals: Part 3: Methods of Analysis Sec. 3.10: Determ.Chromium: Subsec3.10.1: Steel and Iron: Pptentiometric/Volumetric Method (Superseded by BS EN 24 937: 1991). 8 pp.

BSI BS EN 24937-91. 1991 Steel and Iron Determination of Chromium Content: Potentiometric or Visual Titration Method (V). 14 pp.

BSI BS EN 24937-01. 1991 Amd 1 Steel and Iron Determination of Chromium Content Potentiometric or Visual Titration Method (ISO 4937: 1986) (AMD 7082) July 15, 1992 (V). 15 pp.

CEN EN 24 937-90. Steel and Iron—Determination of Chromium Content—Potentiometric or Visual Method. 2 pp.

DIN ENGL EN 24937-92. Determination of Chromium Content of Steel and Iron by Potentiometric or Visual Titration (ISO 4937: 1986) (Nov). 11 pp.

ISO 4937-86. Steel and Iron—Determination of Chromium Content—Potentiometric or Visual Titration Method First Edition. 10 pp.

—Tantalum

JIS H 1687-76. Methods for Determination of Chromium in Tantalum (R 1986). 9 pp.

—Titanium Ores

JIS M 8316-76. Determination of Chromium in Titanium Ores.

—Uranium Hexafluoride—Atomic Absorption Analysis

CNS J2087-83. Method for Determining Chromium Soluble in UF6 Method (Jan)(9886).

CNS J2088-83. Method for Determining Chromium Insoluble in UF6 Method (Jan)(9887).

—Waste Water—Atomic Absorption Spectrometry

DIN ENGL 38406 Pt 10-85. German Standard Methods for the Examination of Water, Waste Water and Sludge; Cations (Group E); Determination of Chromium (E 10) (June). 6 pp.

—Waste Water—Photometry

DIN ENGL 38405 Pt 24-87. German Standard Methods for the Examination of Water, Waste Water and Sludge; Anions (Group D); Photometric Determination of Chromium(VI) Using 1,5-Diphenylcarbonohydrazide (D 24) (May). 5 pp.

—Water—Atomic Absorption Spectrometry

BSI BS 6068: Sec 2.38-90. 1990 Water Quality Part 2: Physical, Chemical and Bio-Chemical Methods Section 2.38: Methods for the Determination of Total Chromium by Atomic Absorption Spectrometry. 11 pp.

DIN ENGL 38406 Pt 10-85. German Standard Methods for the Examination of Water, Waste Water and Sludge; Cations (Group E); Determination of Chromium (E 10) (June). 6 pp.

ISO 9174-90. Water Quality—Determination of Total Chromium—Atomic Absorption Spectrometric Methods First Edition. 10 pp.

JIS K 0553-90. Testing Methods for Determination of Metallic Elements in Highly Purified Water. 27 pp.

—Water—Photometry

DIN ENGL 38405 Pt 24-87. German Standard Methods for the Examination of Water, Waste Water and Sludge; Anions (Group D); Photometric Determination of Chromium(VI) Using 1,5-Diphenylcarbonohydrazide (D 24) (May). 5 pp.

INTERNATIONAL AND NON-U.S. NATIONAL STANDARDS
SUBJECT INDEX
Chromium

Chromium Content Analysis (Cont.)

—**Wood Preservatives—Atomic Absorption Spectrophotometry**

BSI BS 5666: Part 3-91. 1991 Methods of Analysis of Wood Preservatives and Treated Timber Part 3: Quantitative Analysis of Preservatives and Treated Timber Containing Copper/Chromium/Arsenic Formulations. 14 pp.

—**Wood Preservatives—Colorimetric Analysis**

BSI BS 5666: Part 3-91. 1991 Methods of Analysis of Wood Preservatives and Treated Timber Part 3: Quantitative Analysis of Preservatives and Treated Timber Containing Copper/Chromium/Arsenic Formulations. 14 pp.

—**Zirconium**

JIS H 1656-89. Methods for Determination of Chromium in Zirconium and Zirconium Alloys. 10 pp.

—**Zirconium Alloys**

JIS H 1656-89. Methods for Determination of Chromium in Zirconium and Zirconium Alloys. 10 pp.

Chromium Cupronickel

See Also: Copper Alloys; Copper Nickel Alloys

—**Ingots—Marine**

MOD UK NES 824: Part 1-01. Copper Nickel Chromium Sand Castings and Ingots Part 1: Production Requirements Issue 2 (05.89); Amendment 2. 27 pp.

MOD UK NES 824: Part 1-93. Copper Nickel Chromium Sand Castings and Ingots Part 1: Production Requirements Issue 3 (07.93). 23 pp.

MOD UK NES 824: Part 2-01. Copper Nickel Chromium Sand Castings Part 2: Guide to Production Methods Issue 1 (09.87); Amendment 1. 125 pp.

Chromium Cupronickel Castings

—**Sand—Marine**

MOD UK NES 824: Part 1-01. Copper Nickel Chromium Sand Castings and Ingots Part 1: Production Requirements Issue 2 (05.89); Amendment 2. 27 pp.

MOD UK NES 824: Part 1-93. Copper Nickel Chromium Sand Castings and Ingots Part 1: Production Requirements Issue 3 (07.93). 23 pp.

MOD UK NES 824: Part 2-01. Copper Nickel Chromium Sand Castings Part 2: Guide to Production Methods Issue 1 (09.87); Amendment 1. 125 pp.

—**Welding—Marine**

MOD UK NES 825-01. Requirements for the Welding Copper Nickel Chromium Sand Castings (Including Welding Wire) Issue 2 (06.89); Amendment 1. 31 pp.

Chromium Inorganic Compounds

Scope Note: Use a more specific term
See: Chromium Chlorides; Chromium Oxides; Chromium Trioxide

Chromium Iron Alloys

See Also: Chromium Steels; Ferroalloys

—**Aircraft**

MOD UK DTD-462-45. High Chromium Alloy Cast Iron (Suitable for Piston Ring Pots) (Centrifugally Cast) (Reprinted October 1961). 4 pp.

MOD UK DTD-614-44. Medium-Chromium Alloy Cast Iron (Suitable for Piston Ring Pots) (Centrifugally Cast) (Reprinted October 1952). 4 pp.

Chromium Molybdenum Steel Castings

See Also: Castings; Chromium Molybdenum Steels; Ferroalloy Castings; Steel Castings

MOD UK DTD-5229-67. 3 Per Cent Chromium—Molybdenum Steel Investment Castings (Tensile Strength 63-80 kgf/mm2) (Withdrawn). 2 pp.

—**Aerospace**

BSI HC 3-73. 1973 1% Chromium-Molybdenum Low Alloy Steel Castings (700 N/Square mm). 2 pp.

BSI HC 4-73. 1973 3% Chromium-Molybdenum Steel Castings (620-770 N/Square mm). 3 pp.

BSI HC 7-74. 1974 Amd 1 3% Chromium-Molybdenum Steel Castings (880-1080 Mpa). 3 pp.

BSI HC 8-74. 1974 Amd 1 3% Chromium-Molybdenum Steel Castings (1150-1300 MPa). 3 pp.

—**Nitriding—Aerospace**

BSI HC 6-73. 1973 3% Chromium-Molybdenum Nitriding Steel Castings (850-1000 N/Square mm). 2 pp.

Chromium Molybdenum Steels

Scope Note: For additional listings, see also specific products made from chromium molybdenum steels
Use For: Molybdenum Chromium Steels
See Also: Alloy Steels; Aluminum Chromium Molybdenum Steels; Chromium Molybdenum Steel Castings; Chromium Molybdenum Vanadium Steels; Chromium Steels; Corrosion Resistant Steels; Ferroalloys; Heat Resistant Steels; Low Alloy Steels; Steels

—**Aerospace**

MOD UK DTD-5082A-68. 1 Per Cent Chromium-Molybdenum Steel (115 Hbar) (Limiting Ruling Section 16mm) (Suitable for Welding). 3 pp.

MOD UK DTD-5122A-68. 1 Per Cent Chromium-Molybdenum Steel (115 Hbar) (Limiting Ruling Section 12.5mm) (Suitable for Welding by Specialised Processes). 2 pp.

—**Bars—Aerospace**

BSI 2S 142-77. 1977 Amd 1 1 Per Cent Chromium-Molybdenum Steel Billets, Bars, Forgings and Parts (900-1100 MPa: Limiting Ruling Section 40 mm). 5 pp.

BSI S 158-77. 1977 1 Per Cent Chromium-Molybdenum Steel Bars for the Manufacture of Forged Bolts and Forged Nuts. 4 pp.

—**Billets—Aerospace**

BSI 2S 142-77. 1977 Amd 1 1 Per Cent Chromium-Molybdenum Steel Billets, Bars, Forgings and Parts (900-1100 MPa: Limiting Ruling Section 40 mm). 5 pp.

—**Forgings**

CNS G2038-76. Method of Test for Chromium Molybdenum Steel Forgings for General Use (Dec)(4042).

CNS G3095-76. Chromium Molybdenum Steel Forgings for General Use (Dec)(4041).

JIS G 3221-88. Chromium Molybdenum Steel Forgings for General Use. 17 pp.

—**Forgings—Aerospace**

BSI 2S 142-77. 1977 Amd 1 1 Per Cent Chromium-Molybdenum Steel Billets, Bars, Forgings and Parts (900-1100 MPa: Limiting Ruling Section 40 mm). 5 pp.

—**Nitriding—Bars—Aerospace**

BSI 4S 106-76. 1976 Amd 1 3 Per Cent Chromium-Molybdenum Steel Billets, Bars, Forgings and Parts (930-1080 MPa: Limiting Ruling Section 150 mm) (Suitable for Nitriding). 3 pp.

BSI 3S 132-76. 1976 Amd 2 3 Per Cent Chromium-Molybdenum-Vanadium Steel Billets, Bars, Forgings and Parts (1320-1470 MPa: Limiting Ruling Section 70 mm) (Suitable for Nitriding). 4 pp.

—**Nitriding—Billets—Aerospace**

BSI 4S 106-76. 1976 Amd 1 3 Per Cent Chromium-Molybdenum Steel Billets, Bars, Forgings and Parts (930-1080 MPa: Limiting Ruling Section 150 mm) (Suitable for Nitriding). 3 pp.

BSI 3S 132-76. 1976 Amd 2 3 Per Cent Chromium-Molybdenum-Vanadium Steel Billets, Bars, Forgings and Parts (1320-1470 MPa: Limiting Ruling Section 70 mm) (Suitable for Nitriding). 4 pp.

—**Nitriding—Forgings—Aerospace**

BSI 4S 106-76. 1976 Amd 1 3 Per Cent Chromium-Molybdenum Steel Billets, Bars, Forgings and Parts (930-1080 MPa: Limiting Ruling Section 150 mm) (Suitable for Nitriding). 3 pp.

BSI 3S 132-76. 1976 Amd 2 3 Per Cent Chromium-Molybdenum-Vanadium Steel Billets, Bars, Forgings and Parts (1320-1470 MPa: Limiting Ruling Section 70 mm) (Suitable for Nitriding). 4 pp.

—**Plates**

CNS G2196-83. Method of Test for Chromium-Molybdenum Alloy Steel Plates for Pressure Vessels (Dec)(10717).

—**Sheets—Aircraft**

MOD UK DTD-5112-59. 80 Ton 1 Per Cent Chromium-Molybdenum Steel Sheets (Suitable for Welding by Specialised Processes). 2 pp.

—**Sheets—Welding—Aerospace**

BSI S 534-69. 1969 Chromium-Molybdenum Steel Sheet and Strip (88/108 Hbar) (Suitable for Welding). 3 pp.

BSI S 535-69. 1969 Chromium-Molybdenum Steel Sheet and Strip (115/130 Hbar) (Suitable for Welding). 3 pp.

—**Strips—Welding—Aerospace**

BSI S 534-69. 1969 Chromium-Molybdenum Steel Sheet and Strip (88/108 Hbar) (Suitable for Welding). 3 pp.

BSI S 535-69. 1969 Chromium-Molybdenum Steel Sheet and Strip (115/130 Hbar) (Suitable for Welding). 3 pp.

Chromium Molybdenum Steels (Cont.)

—**Structural**

CNS G3063-87. Chromium Molybdenum Steels for Machine Structural Use (May)(3229).

—**Structural—Hot Worked**

JIS G 4105-79. Chromium Molybdenum Steels. 14 pp.

—**Tubes—Aerospace**

BSI 3T 53-80. 1980 Chromium-Molybdenum Steel Tube (700 MPa) (Tube 12.5 mm Outside Diameter and Greater) (Weldable). 2 pp.

BSI 3T 60-80. 1980 Chromium-Molybdenum Steel Tube (1150 MPa) (Maximum Wall Thickness 8 mm). 2 pp.

BSI T 77-80. 1980 Chromium-Molybdenum Steel Tube (900 MPa) (Weldable). 2 pp.

—**Tubes—Aircraft**

MOD UK DTD-167A-50. 45-Ton Chrome-Molybdendum Steel Tubes (Not Suitable for Welding) (Reprinted December 1950). 2 pp.

MOD UK DTD-5132-59. 80 Ton 1 Per Cent Chromium-Molybdenum Steel Tubes (Suitable for Welding by Specialised Processes). 2 pp.

—**Tubes—Hot Rolled—Aircraft**

MOD UK DTD-5142-59. 80 Ton 1 Per Cent Chromium-Molybdenum Steel Tubes (Hot Rolled) (Suitable for Welding by Specialised Processes). 2 pp.

Chromium Molybdenum Vanadium Steels

Scope Note: For additional listings, see also specific products made from chromium molybdenum vanadium steels *See Also:* Alloy Steels; Chromium Molybdenum Steels; Chromium Steels; Corrosion Resistant Steels; Heat Resistant Steels; Low Alloy Steels; Steels

—**Bolts—Aerospace**

MOD UK DTD-5222-69. 5 Per Cent Chromium-Molybdenum—Vanadium Steel Suitable for Forged Bolts (180 Hbar) (Vacuum Re-Melted—Limiting Ruling Section 25 mm). 5 pp.

Chromium Nickel Molybdenum Steel Castings

Use: Nickel Chromium Molybdenum Steel Castings

Chromium Nickel Molybdenum Steels

Use: Nickel Chromium Molybdenum Steels

Chromium Nickel Steel Castings

Use: Nickel Chromium Steel Castings

Chromium Nickel Steels

Use: Nickel Chromium Steels

Chromium Ores

Use For: Chrome Ores *See Also:* Ores

—**Aluminum Content—Complexometric Titrations**

ISO 8889-88. Chromium Ores and Concentrates—Determination of Aluminium Content—Complexometric Method First Edition. 6 pp.

—**Aluminum Oxide Content**

CNS M3202-88. Method of Determination for Aluminium Oxide in Chrome Ores (May)(12319).

JIS M 8266-82. Method for Determination of Aluminium Oxide in Chrome Ores.

—**Calcium Content—Volumetric Analysis**

ISO 5975-83. Chromium Ores—Determination of Calcium and Magnesium Contents—EDTA Titrimetric Method First Edition. 6 pp.

—**Chemical Analysis**

CNS M3197-88. General Rules for Chemical Analysis of Chromium Ores (May)(12314).

ISO 6629-81. Chromium Ores and Concentrates—Methods of Chemical Analysis—General Instructions First Edition. 4 pp.

JIS M 8261-82. General Rules for Chemical Analysis of Chrome Ores. 6 pp.

—**Chromium Content—Volumetric Analysis**

ISO 6331-83. Chromium Ores and Concentrates—Determination of Chromium Content—Titrimetric Method First Edition. 6 pp.

—**Chromium Oxide Content**

CNS M3198-88. Method of Determination for Chromium Oxide in Chrome Ores (May)(12315).

JIS M 8262-82. Method for Determination of Chromium Oxide in Chrome Ores.

INDUSTRY STANDARDS

Chromium

Chromium Ores (Cont.)

—Iron Content
CNS M3199-88. Method of Determination for Iron in Chrome Ores (May)(12316).
JIS M 8263-82. Methods for Determination of Iron in Chrome Ores.

—Iron Content—Volumetric Analysis
ISO 6130-85. Chromium Ores—Determination of Total Iron Content—Titrimetric Method After Reduction First Edition. 5 pp.

—Magnesium Content—Volumetric Analysis
ISO 5975-83. Chromium Ores—Determination of Calcium and Magnesium Contents—EDTA Titrimetric Method First Edition. 6 pp.

—Magnesium Oxide Content
CNS M3201-88. Method of Determination for Magnesium Oxide in Chrome Ores (May)(12318).

—Magnesium Peroxide Content
JIS M 8265-82. Method for Determination of Magnesium Dioxide in Chrome Ores.

—Moisture Content
CNS M3194-88. Method of Sampling, Moisture Determination and Size Determination for Chromium Ores, Manganese Ores and Ferruginous Manganese Ores (Apr)(12280).
ISO 8531-86. Manganese and Chromium Ores—Experimental Methods for Checking the Precision of Moisture Determination First Edition. 7 pp.
JIS M 8108-92. Chromium, Manganese and Ferruginous Manganese Ores—Sampling and Determination of Moisture Content and Size Distribution. 89 pp.

—Moisture Content—Gravimetric Analysis
ISO 6129-81. Chromium Ores—Determination of Hygroscopic Moisture Content in Analytical Samples—Gravimetric Method First Edition. 3 pp.

—Particle Size Distribution
CNS M3194-88. Method of Sampling, Moisture Determination and Size Determination for Chromium Ores, Manganese Ores and Ferruginous Manganese Ores (Apr)(12280).
JIS M 8108-92. Chromium, Manganese and Ferruginous Manganese Ores—Sampling and Determination of Moisture Content and Size Distribution. 89 pp.

—Phosphorus Content
CNS M3203-88. Method of Determination for Phosphorus in Chrome Ores (May)(12320).
JIS M 8267-88. Method for Determination of Phosphorus in Chromium Ores. 6 pp.

—Phosphorus Content—Photometry
ISO 6127-81. Chromium Ores—Determination of Phosporus Content—Reduced Molybdophosphate Photometric Method First Edition. 6 pp.

—Refractories—Chemical Analysis
BSI BS 1902: Sec 2.2-74. 1974 Amd 1 Methods for Testing Refractory Materials Part 2: Chemical Analysis (Wet Methods) Section 2.2: Chemical Analysis of Chrome-Bearing Materials (AMD 4960) February 28, 1986. 14 pp.
JIS R 2901-71. Chemical Analysis of Chrome Ore for Refractories. 12 pp.

—Sampling
CNS M3194-88. Method of Sampling, Moisture Determination and Size Determination for Chromium Ores, Manganese Ores and Ferruginous Manganese Ores (Apr)(12280).
ISO 6153-89. Chromium Ores—Increment Sampling First Edition. 14 pp.
ISO 6154-89. Chromium Ores—Preparation of Samples First Edition. 15 pp.
ISO 8530-86. Manganese and Chromium Ores—Experimental Methods for Checking the Precision of Sample Division First Edition. 7 pp.
ISO 8541-86. Manganese and Chromium Ores—Experimental Methods for Checking the Bias of Sampling and Sample Preparation First Edition. 8 pp.
ISO 8542-86. Manganese and Chromium Ores—Experimental Methods for Evaluation of Quality Variation and Methods for Checking the Precision of Sampling First Edition. 11 pp.
JIS M 8108-92. Chromium, Manganese and Ferruginous Manganese Ores—Sampling and Determination of Moisture Content and Size Distribution. 89 pp.

—Silica Content
CNS M3200-88. Method of Determination for Silicon Dioxide in Chrome Ores (May)(12317).

Chromium Ores (Cont.)

—Silicon Content—Gravimetric Analysis
ISO 5997-84. Chromium Ores and Concentrates—Determination of Silicon Content—Molecular Absorption Spectrometric Method and Gravimetric Method First Edition. 7 pp.

—Silicon Content—Molecular Absorption Spectrophotometry
ISO 5997-84. Chromium Ores and Concentrates—Determination of Silicon Content—Molecular Absorption Spectrometric Method and Gravimetric Method First Edition. 7 pp.

—Silicon Dioxide Content
JIS M 8264-82. Method for Determination of Silicon Dioxide in Chrome Ores.

—Sulfur Content
CNS M3204-88. Method of Determination for Sulfur in Chrome Ores (May)(12321).
JIS M 8268-90. Chromium Ores—Determination of Sulfur Content. 29 pp.

Chromium Oxide Coatings
See Also: Chromium Oxides; Coatings; Oxide Coatings

—Low Carbon Steel—Electrodeposited
CEN PREN 10 171-87. Double Cold Reduced Electrolytic Chromium/Chromium Oxide Coated Steel: Sheet. 29 pp.
CEN PREN 10 172-87. Single Cold Reduced Electrolytic Chromium/Chromium Oxide Coated Steel: Coil for Subsequent Cutting into Sheets. 29 pp.
CEN PREN 10 173-87. Double Cold Reduced Electrolytic Chromium/Chromium Oxide Coated Steel: Coil for Subsequent Cutting into Sheets. 29 pp.

—Steel—Electrodeposited
ISO 8110 Pt 1-88. Single Cold-Reduced Electrolytic Chromium/Chromium Oxide-Coated Steel—Part 1: Sheets First Edition. 18 pp.
ISO 8111 Pt 1-88. Double Cold-Reduced Electrolytic Chromium/Chromium Oxide-Coated Steel—Part 1: Sheets First Edition. 20 pp.

Chromium Oxide Content Analysis

—Chromium Ores
CNS M3198-88. Method of Determination for Chromium Oxide in Chrome Ores (May)(12315).
JIS M 8262-82. Method for Determination of Chromium Oxide in Chrome Ores.

—Dolomite
ISO 10058-92. Magnesites and Dolomites—Chemical Analysis First Edition. 23 pp.

—Leather
CNS K6135-81. Method of Test for Chromium Oxide Content of Leather (Aug)(1291).

—Magnesite
ISO 10058-92. Magnesites and Dolomites—Chemical Analysis First Edition. 23 pp.

—Sodium Sulfate
CNS K6085-74. Methods of Test for Sodium Sulfate of Industrial Grade (Oct)(1004). 5 pp.

Chromium Oxides
See Also: Chromium Oxide Coatings
JIS K 1401-92. Dichromium Trioxide. 16 pp.
JIS K 8434-80. Chromium (VI) Oxide.

—Pyrotechnics
MOD UK DSTAN 13-71-91. Chromium (III) Oxide Issue 2. 8 pp.

Chromium Steel Bearing Alloys
Use: Bearing Alloys—Chromium Steel

Chromium Steels
Scope Note: For additional listings, see also specific products made from chromium steels
See Also: Alloy Steels; Chromium Iron Alloys; Chromium Molybdenum Steels; Chromium Molybdenum Vanadium Steels; Corrosion Resistant Steels; Ferroalloys; Heat Resistant Steels; Low Alloy Steels; Nickel Chromium Steels; Stainless Steels; Steels

—Bars
BSI 2S 135-77. 1977 1 Per Cent Carbon-Chromium Steel Billets, Bars, Forgings and Parts (Limiting Ruling Section 25 mm). 4 pp.
BSI 2S 136-77. 1977 1 Per Cent Carbon-Chromium Steel (Vacuum Arc Remelted) Billets, Bars, Forgings and Parts (Limiting Ruling Section 25 mm). 4 pp.

Chromium Steels (Cont.)

—Billets
BSI 2S 135-77. 1977 1 Per Cent Carbon-Chromium Steel Billets, Bars, Forgings and Parts (Limiting Ruling Section 25 mm). 4 pp.
BSI 2S 136-77. 1977 1 Per Cent Carbon-Chromium Steel (Vacuum Arc Remelted) Billets, Bars, Forgings and Parts (Limiting Ruling Section 25 mm). 4 pp.

—Bolts—Aerospace
MOD UK DTD-5066-01. 12 Per Cent Chromium Heat Resisting Steel Suitable for Bolts, Studs, Set Screws and Nuts (Limiting Ruling Section 50 mm; 110 Hbar); Amendment 1. 4 pp.

—Forgings
BSI 2S 135-77. 1977 1 Per Cent Carbon-Chromium Steel Billets, Bars, Forgings and Parts (Limiting Ruling Section 25 mm). 4 pp.
BSI 2S 136-77. 1977 1 Per Cent Carbon-Chromium Steel (Vacuum Arc Remelted) Billets, Bars, Forgings and Parts (Limiting Ruling Section 25 mm). 4 pp.

—Hot Worked
JIS G 4104-79. Chromium Steels. 12 pp.

—Mechanical Properties
ISO 10446-90. Welding—All-Weld Metal Test Assembly for the Classification of Corrosion-Resisting Chromium and Chromium-Nickel Steel Covered Arc Welding Electrodes First Edition; (Corrected and Reprinted -1991). 6 pp.

—Nuts—Aerospace
MOD UK DTD-5066-01. 12 Per Cent Chromium Heat Resisting Steel Suitable for Bolts, Studs, Set Screws and Nuts (Limiting Ruling Section 50 mm; 110 Hbar); Amendment 1. 4 pp.

—Structural
CNS G3065-87. Chromium Steels for Machine Structural Use (May)(3231).

—Studs—Aerospace
MOD UK DTD-5066-01. 12 Per Cent Chromium Heat Resisting Steel Suitable for Bolts, Studs, Set Screws and Nuts (Limiting Ruling Section 50 mm; 110 Hbar); Amendment 1. 4 pp.

—Tubes—Aircraft
MOD UK DTD-97B-50. 28-Ton 12 Per Cent Chromium Corrosion-Resistant Steel Tubes. 1 p.

Chromium Trioxide
Use For: Chromic Anhydride
CNS K1100-65. Chromium Trioxide (Industrial Grade) (Aug)(2479)(R 1971).
CNS K6200-73. Method of Test for Chromium Trioxide of Industrial Grade (Nov)(2480).
CNS K7146-66. Chemical Reagent (Chromium Trioxide) (Chromic Alum) (Mar)(1646).
JIS K 1402-92. Chromium Trioxide. 15 pp.
MOD UK DSTAN 68-109-89. Chromium Trioxide (Chromic Acid) Issue 1. 12 pp.
MOD UK CS 3119. Chromic Acid (Superseded by Def Stan 68-109).

—Chloride Content
MOD UK M 9526/66. Examination of Chromic Acid Crystals.

—Concentration
MOD UK M 9526/66. Examination of Chromic Acid Crystals.

—Sulfate Content
MOD UK M 9526/66. Examination of Chromic Acid Crystals.

Chromium Trioxide Content Analysis

—Basic Lead Silicochromate
CNS K6536-80. Method of Test for Chromium Trioxide in Basic Lead Silica-Chromate (Jul)(5865).

—Organic Coatings—Insoluble Matter
CNS K6804-24-89. Method of Test for Organic Coating (Chemical Analysis) — Quantative Test of Chromic Anhydride in Solvent Insolubles (Jan)(10880-24).

Chromotropic Acid
CNS K7147-66. Chemical Reagent (Chromotropic Acid) (1,8-Dihydroxynaphthalein-3,6-Disulfonic Acid) (Mar)(1647).

Chromotropic Acid Disodium Salt
JIS K 8316-78. Chromotropic Acid Disodium Salt, Dihydrate 4,5-Dihydroxy-2,7-Naphthalene-disulfonic Acid Disodium Salt, Dihydrate.

Chronometers
See Also: Measuring Instruments; Stop Watches

Chronometers (Cont.)
—Electronic
MOD UK DSTAN 66-30-81. Chronometer. Chronometer, Electronic, Quartz Issue 1. 9 pp.

Chronometers (Wrist)
Use: Wrist Watches—Precision

Chronometric Tachometers
Use For: Jaquet Chronometric Tachometers
See Also: Tachometers

—Portable
CNS B6057-81. Portable Chronometric Tachometers (Oct)(7960).
CNS B7176-81. Method of Test for Portable Chronometric Tachometers (Oct)(7961).
JIS B 7521-79. Portable Chronometric Tachometers.

Chrysoidine
JIS K 8296-72. Chrysoidine.

Chrysotile
Use: Serpentine

Chrysotile Asbestos
Use: Serpentine

Chuck Bushings
See Also: Bushings; Tool Holders

—Drill Steels
ISO 723-91. Rock Drilling Equipment—Forged Collared Shanks and Corresponding Chuck Bushings for Hollow Hexagonal Drill Steels Second Edition. 6 pp.

—Pneumatic Tools
BSI BS 673-84. 1984 Amd 1 Shanks for Pneumatic Tools and Fitting Dimensions of Chuck Bushings. 32 pp.
ISO 1180-83. Shanks for Pneumatic Tools and Dimensions of Chuck Bushings First Edition; (Addendum 1-1985). 24 pp.
SNZ NZS/ISO 1180-83. Shanks for Pneumatic Tools and Fitting Dimensions of Chuck Bushings Amend: 1. 12 pp.

—Rock Drills
CNS M2046-80. Chuck Bushing for Rock Drill (May)(5619).
JIS M 3904-74. Chack Bushing for Rock Drill.

Chucks
Scope Note: For additional listings, use a more specific term. See Also: Clamps; Collets; Drill Chucks; Fasteners; Holders; Lathe Chucks; Magnetic Chucks; Mechanical Guides; Scroll Chucks; Self Centering Chucks

—Broaching Machines
BSI BS 5326-76. 1976 Internal Broaching Tools and Pull Chuck Locations. 4 pp.

—Collet Pads
BSI BS 1983: Part 3-87. 1987 Chucks for Machine Tools and Portable Power Tools Part 3: Dimensions of Collet Pads for Base Jaws of Power Operated Workholding Chucks. 10 pp.

—Independent
BSI BS 1983: Part 2-72. 1972 Chucks for Machine Tools and Portable Power Tools Part 2: Work-Holding Chucks. 17 pp.
JIS B 6154-77. Independent Chucks (R 1980). 9 pp.

—Jaws—Mountings
ISO 9401-91. Machine Tools—Jaw Mountings on Power Chucks First Edition. 10 pp.

—Jaws—Mountings—T Slot Nuts
ISO 9401-91. Machine Tools—Jaw Mountings on Power Chucks First Edition. 10 pp.

—Tool
BSI BS 1983: Part 1-69. 1969 Amd 1 Chucks for Machine Tools and Portable Power Tools Part 1: Tool Holding Chucks. 32 pp.
BSI BS 1983: Part 4-87. 1987 Chucks for Machine Tools and Portable Power Tools Part 4: Criteria to be Stated Affecting Performance of Power Operated Workholding Chucks at Speed. 10 pp.
BSI BS 6386: Part 1-86. 1986 Tool Chucks with Clamp Screws for Flatted Parallel Shanks Part 1: Dimensions of Chuck Nose. 4 pp.
BSI BS 6386: Part 2-83. 1983 Tool Chucks with Clamp Screws for Flatted Parallel Shanks Part 2: Dimensions of Taper Shanks. 8 pp.
ISO 5414 Pt 1-85. Tool Chucks (End Mill Holders) with Clamp Screws for Flatted Parallel Shank Tools—Part 1: Dimensions of the Driving System of Tool Shanks Second Edition. 4 pp.

Chucks (Cont.)
—Tool (Cont.)
ISO 5414 Pt 2-82. Tool Chucks (End Mill Holders) with Clamp Screws for Flatted Parallel Shank Tools—Part 2: Connecting Dimensions of Chucks First Edition. 6 pp.

CI Mordant Black 11
Use For: Eriochrome Black T (TM)
CNS K7209-66. Chemical Reagent (Eriochrome Black T) (Jun)(1709).
JIS K 8736-72. Eriochrome Black T (1-(1-Hydroxy-2-Naphthylazo)-6-Nitro-2-Naphthol-4-Sulfonic Acid Sodium Salt).

CI Mordant Blue 29
JIS K 9557-76. C.I. Mordant Blue 29.

Cider
See Also: Apple Juice; Beverages

—Bottles
BSI BS 6118-81. 1981 Amd 3 Multi-Trip Glass Bottles for Beer and Cider. 19 pp.

CIE
Use: Commision International de l'Eclairage

Cigarette Filters
See Also: Cigarettes; Filters

—Alkaloid Retention—Spectrometry
BSI BS 5202: Part 9-92. 1992 Methods for Chemical Analysis of Tobacco and Tobacco Products Part 9: Determination of Alkaloid Retention by Filters of Cigarettes (Spectrometric Method) (ISO 3401: 1991) (W). 15 pp.
BSI BS 5202: Part 9-77. 1977 Methods for Chemical Analysis of Tobacco and Tobacco Products Part 9: Determination of Alkaloid Retention by Filters of Cigarettes. 10 pp.
CNS N4051-80. Methods of Test for Tobacco and Tobacco Products—Determination of Alkaloid Retention by Filters of Cigarettes (Jul)(5920). 7 pp.
ISO 3401-91. Cigarettes—Determination of Alkaloid Retention by the Filters—Spectrometric Method Second Edition. 12 pp.

—Draw Resistance
BSI BS 5381: Part 5-83. 1983 Determination of Physical Properties of Tobacco and Tobacco Products Part 5: Draw Resistance of Cigarettes and Filter Rods. 7 pp.
ISO 6565-83. Tobacco and Tobacco Products—Draw Resistance of Cigarettes and Filter Rods—Definitions, Standard Conditions and General Aspects First Edition. 6 pp.

—Nominal Diameters
BSI BS 5381: Part 2-88. 1988 Determination of Physical Properties of Tobacco and Tobacco Products Part 2: Nominal Diameter of Cigarettes and Filters (Pneumatic Method). 8 pp.
ISO 2971-87. Cigarettes and Filters—Determination of Nominal Diameter—Pneumatic Method Second Edition. 8 pp.

—Smoke Condensate Retention—Spectrophotometry
CNS N4053-80. Methods of Test for Tobacco and Tobacco Products—Determination of Retention of Coloured Part of Smoke Condensate by Cigarette Filters—Direct Spectrophotometric Method (Jul)(5922). 3 pp.
ISO 4388-91. Cigarettes—Determination of the Smoke Condensate Retention Index of a Filter—Direct Spectrometric Method Second Edition. 6 pp.

—Standard Atmospheres
BSI BS 5743-91. 1991 Atmosphere for Conditioning and Testing Tobacco and Tobacco Products (ISO 3402: 1991). 10 pp.
BSI BS 5743-79. 1979 Atmospheres for Conditioning and Testing Tobacco and Tobacco Products. 3 pp.
CNS N4048-80. Method of Test for Tobacco and Tobacco Products Atmospheres for Conditioning and Testing (Jul)(5917). 1 p.
ISO 3402-91. Tobacco and Tobacco Products—Atmosphere for Conditioning and Testing Third Edition. 6 pp.

Cigarette Lighters
Use: Lighters

Cigarette Papers
See Also: Cigarettes; Papers

—Bobbins
CNS P2024-84. Cigarette Paper Bobbins (Feb)(2304). 2 pp.

Cigarette Papers (Cont.)
—Combustion
CNS P3065-84. Combustion Test for Cigarette Paper (Feb)(10778). 2 pp.

—Combustion Rate
CNS P3015-84. Method of Test for Combustion Rate of Cigarette Paper (Feb)(2388). 2 pp.

—Gas Permeability
BSI BS 5381: Part 4-80. (WITHDRAWN) 1980 Determination of Physical Properties of Tobacco and Tobacco Products Part 4: Determination of Air Permeability of Material Used as Cigarette Papers. 6 pp.
ISO 2965-79. Material Used as Cigarette Papers—Determination of Air Permeability First Edition. 5 pp.

—Standard Atmospheres
BSI BS 5743-91. 1991 Atmosphere for Conditioning and Testing Tobacco and Tobacco Products (ISO 3402: 1991). 10 pp.
BSI BS 5743-79. 1979 Atmospheres for Conditioning and Testing Tobacco and Tobacco Products. 3 pp.
CNS N4048-80. Method of Test for Tobacco and Tobacco Products Atmospheres for Conditioning and Testing (Jul)(5917). 1 p.
ISO 3402-91. Tobacco and Tobacco Products—Atmosphere for Conditioning and Testing Third Edition. 6 pp.

Cigarette Smoke
See Also: Cigarettes; Smoke

—Alkaloid Content—Spectrophotometry
BSI BS 5202: Part 7-90. 1990 Methods for Chemical Analysis of Tobacco and Tobacco Products Part 7: Determination of Alkaloids in Smoke Condensate of Cigarettes (Spectrophotometric Method). 8 pp.
CNS N4050-85. Method of Test for Tobacco and Tobacco Products-Determination of Alkaloids in Cigarette Smoke Condensates-Spectrophotometric Method (Dec)(5919). 3 pp.
ISO 3400-89. Cigarettes—Determination of Alkaloids in Smoke Condensates—Spectrometric Method Second Edition. 6 pp.

—Carbon Monoxide Content
ISO 8454-87. Cigarettes—Determination of Carbon Monoxide in the Vapour Phase of Smoke (NDIR Method) First Edition. 8 pp.

—Condensates—Nicotine Content—Gas Chromatography
BSI BS 5202: Part 12-91. 1991 Methods for Chemical Analysis of Tobacco and Tobacco Products Part 12: Determination of Nicotine in Smoke Condensate of Cigarettes (Gas-Chromatographic Method) (ISO 10315: 1991). 11 pp.
BSI BS 5202: Part 12-01. 1991 Amd 1 Methods for Chemical Analysis of Tobacco and Tobacco Products Part 12: Determination of Nicotine in Smoke Condensate of Cigarettes (Gas-Chromatographic Method) (ISO 10315: 1991) (AMD 7155) July 15, 1992. 12 pp.
ISO 10315-91. Cigarettes—Determination of Nicotine in Smoke Condensates—Gas-Chromatographic Method First Edition. 9 pp.

—Condensates—Water Content—Gas Chromatography
BSI BS 5202: Part 13-91. 1991 Methods for Chemical Analysis of Tobacco and Tobacco Products Part 13: Determination of Water in Smoke Condensate of Cigarettes (Gas-Chromatographic Method) (W) (ISO 10362-1: 1991). 11 pp.
BSI BS 5202: Part 13-01. 1991 Amd 1 Methods for Chemical Analysis of Tobacco and Tobacco Products Part 13: Determination of Water in Smoke Condensate of Cigarettes (Gas-Chromatographic Method) (ISO 10362-1: 1991) (AMD 7156) July 15, 1992. 12 pp.
ISO 10362 Pt 1-91. Cigarettes—Determination of Water in Smoke Condensates—Part 1: Gas-Chromatographic Method First Edition. 8 pp.

—Discoloration—Coated Fabrics
ISO 6449-82. Rubber or Plastics Coated Fabrics—Determination of Discolouration by Cigarette Smoke First Edition. 4 pp.

—Particulates Content
BSI BS 5202: Part 14-92. 1992 Methods for Chemical Analysis of Tobacco and Tobacco Products Part 14: Determination of Total and Nicotine-Free Dry Particulate Matter Using a Routine Analytical Smoking Machine (ISO 4387: 1991) (W). 21 pp.
ISO 4387-91. Cigarettes—Determination of Total and Nicotine-Free Dry Particulate Matter Using a Routine Analytical Smoking Machine Second Edition. 18 pp.

Cigarette Smoke (Cont.)
—Particulates Content (Cont.)
ISO 8453-87. Cigarettes—Determination of Total and Dry Particulate Matter Using a Routine Analytical Cigarette-Smoking Machine—Electrostatic Smoke Trap Method First Edition. 12 pp.

Cigarettes
See Also: Cigarette Filters; Cigarette Papers; Cigarette Smoke; Smoking Machines; Tobacco

—Combustion Rate
BSI BS 5381: Part 3-79. 1979 Determination of Physical Properties of Tobacco and Tobacco Products Part 3: Determination of Free Combustion Rate of Cigarettes. 11 pp.
ISO 3612-77. Tobacco and Tobacco Products—Cigarettes—Determination of Rate of Free Combustion First Edition. 10 pp.

—Draw Resistance
BSI BS 5381: Part 5-83. 1983 Determination of Physical Properties of Tobacco and Tobacco Products Part 5: Draw Resistance of Cigarettes and Filter Rods. 7 pp.
ISO 6565-83. Tobacco and Tobacco Products—Draw Resistance of Cigarettes and Filter Rods—Definitions, Standard Conditions and General Aspects First Edition. 6 pp.

—Excise Taxes
EC COM(87) 325/2-87. Proposal for a Council Directive on the Approximation of Taxes on Cigarettes. 23 pp.

—Glossaries
BSI BS 7680-93. 1993 Glossary of Terms for Tobacco and Tobacco Products (ISO 10185: 1993) (W). 32 pp.
ISO 10185-93. Tobacco and Tobacco Products—Vocabulary First Editions. 31 pp.

—Nominal Diameters
BSI BS 5381: Part 2-88. 1988 Determination of Physical Properties of Tobacco and Tobacco Products Part 2: Nominal Diameter of Cigarettes and Filters (Pneumatic Method). 8 pp.
ISO 2971-87. Cigarettes and Filters—Determination of Nominal Diameter—Pneumatic Method Second Edition. 8 pp.

—Sampling
BSI BS 6245: Part 2-83. 1983 Sampling Tobacco and Tobacco Products Part 2: Method of Preparing Samples for Collaborative Studies. 8 pp.
BSI BS 6245: Part 3-92. 1992 Sampling Tobacco and Tobacco Products Part 3: Methods of Sampling Cigarettes (ISO 8243: 1991) (W). 17 pp.
BSI BS 6245: Part 3-89. 1989 Sampling Tobacco and Tobacco Products Part 3: Methods of Sampling Cigarettes. 11 pp.
ISO 8243-91. Cigarettes—Sampling Second Edition. 14 pp.

—Standard Atmospheres
BSI BS 5743-91. 1991 Atmosphere for Conditioning and Testing Tobacco and Tobacco Products (ISO 3402: 1991). 10 pp.
BSI BS 5743-79. 1979 Atmospheres for Conditioning and Testing Tobacco and Tobacco Products. 3 pp.
CNS N4048-80. Method of Test for Tobacco and Tobacco Products Atmospheres for Conditioning and Testing (Jul)(5917). 1 p.
ISO 3402-91. Tobacco and Tobacco Products—Atmosphere for Conditioning and Testing Third Edition. 6 pp.

—Test Specimens
CGSB CAN/CGSB-176.1-92. Preparation of Cigarettes from Cigarette Tobacco for Testing. 9 pp.

—Tobacco Loss—Vibration Testing
BSI BS 5381: Part 1-86. (WITHDRAWN) 1986 Determination of Physical Properties of Tobacco and Tobacco Products Part 1: Loss of Tobacco from the Ends of Cigarettes. 6 pp.
CNS N4054-85. Method of Test for Tobacco and Tobacco Products-Determination of Loss of Tobacco from the Ends (Dec)(5923). 3 pp.
ISO 3550-85. Cigarettes—Determination of Loss of Tobacco from the Ends Second Edition. 5 pp.

—Ventilation
BSI BS 5381: Part 6-93. 1993 Methods for Determination of Physical Properties of Tobacco and Tobacco Products Part 6: Ventilation of Cigarettes (ISO 9512: 1993) (W). 12 pp.
ISO 9512-93. Cigarettes—Determination of Ventilation—Definitions and Measurement Principles First Edition. 9 pp.

CIM
Use For: Computer Integrated Manufacturing
See Also: CAD/CAM; CAM; Computers; Manufacturing
BSI DD 194-90. 1990 Computer Integrated Manufacturing (CIM): CIM Systems Architecture Framework for Modelling. 29 pp.

—Open Systems Interconnection
BSI BS ISO/IEC 9506-1-90. 1990 Industrial Automation Systems—Manufacturing Message Specification—Part 1: Service Definition (S). 329 pp.
BSI BS ISO/IEC 9506-1-01. 1990 Amd 1 Industrial Automation Systems—Manufacturing Message Specification—Part 1: Service Definition (AMD 7991) November 15, 1993 (S). 339 pp.
BSI BS ISO/IEC 9506-2-90. 1990 Industrial Automation Systems—Manufacturing Message Specification—Part 2: Protocol Specification (S). 142 pp.
BSI BS ISO/IEC 9506-4-92. 1992 Industrial Automation Systems—Manufacturing Message Specification—Part 4: Companion Standard for Numerical Control (S). 142 pp.
BSI PD 6526-90. 1990 Evaluation Report on CIM Architectures. 172 pp.
CEN ENV 40 003-90. Computer Integrated Manufacturing—Systems Architecture—Framework for Enterprise Modelling. 24 pp.
IEC 9506 Pt 1-90. Industrial Automation Systems—Manufacturing Message Specification—Part 1: Service Definition First Edition. 334 pp.
IEC 9506 Pt 1 Draft AMD 1. Industrial Automation Systems—Manufacturing Message Specification—Part 1: Service Definition Amendment 1: Data Exchange; (1992) ***CD-ROM ONLY***. 11 pp.
IEC 9506 Pt 2-90. Industrial Automation Systems—Manufacturing Message Specification—Part 2: Protocol Specification AMMENDMENT 1: Data Exchange First Edition; (Ammendment 1-1993). 147 pp.
IEC 9506 Pt 2 Draft AMD 1. Industrial Automation Systems—Manufacturing Message Specification—Part 2: Protocol Specification Amendment 1: Data Exchange; (1992) ***CD-ROM ONLY***. 7 pp.
IEC 9506 Pt 4-92. Industrial Automation Systems—Manufacturing Message Specification—Part 4: Companion Standard for Numerical Control First Edition. 139 pp.
ISO 9506 Pt 1-90. Industrial Automation Systems—Manufacturing Message Specification—Part 1: Service Definition First Edition; (Amendment 1-1993). 334 pp.
ISO 9506 Pt 1 Draft AMD 1. Industrial Automation Systems—Manufacturing Message Specification—Part 1: Service Definition Amendment 1: Data Exchange; (1992) ***CD-ROM ONLY***. 11 pp.
ISO 9506 Pt 2-90. Industrial Automation Systems—Manufacturing Message Specification—Part 2: Protocol Specification AMENDMENT 1: Data Exchange First Edition; (Amendment 1-1993). 147 pp.
ISO 9506 Pt 2 Draft AMD 1. Industrial Automation Systems—Manufacturing Message Specification—Part 2: Protocol Specification Amendment 1: Data Exchange; (1992) ***CD-ROM ONLY***. 7 pp.
ISO 9506 Pt 4-92. Industrial Automation Systems—Manufacturing Message Specification—Part 4: Companion Standard for Numerical Control First Edition. 139 pp.
JTC1 9506-90. Industrial Automation Systems—Manufacturing Message Specification—Part 1: Service Definition First Edition. 326 pp.
JTC1 9506 Pt1 Draft AMD 1. Industrial Automation Systems—Manufacturing Message Specification—Part 1: Service Definition Amendment 1: Data Exchange; (1992) ***CD-ROM ONLY***. 11 pp.
JTC1 9506-90. Industrial Automation Systems—Manufacturing Message Specification—Part 2: Protocol Specification First Edition. 139 pp.
JTC1 9506 Pt2 Draft AMD 1. Industrial Automation Systems—Manufacturing Message Specification—Part 2: Protocol Specification Amendment 1: Data Exchange; (1992) ***CD-ROM ONLY***. 7 pp.
JTC1 9506-92. Industrial Automation Systems—Manufacturing Message Specification—Part 4: Companion Standard for Numerical Control First Edition. 139 pp.
OSI ISO/IEC 9506-1-90. Industrial Automation Systems—Manufacturing Message Specification—Part 1: Service Definition. 325 pp.
OSI ISO IEC DIS 9506-1-88. Manufacturing Message Specification—Part 1: Service Definition. 224 pp.
OSI ISO/IEC 9506-2-90. Industrial Automation Systems—Manufacturing Message Specification—Part 2: Protocol Specification. 139 pp.
OSI ISO IEC DIS 9506-2-88. Manufacturing Message Specification—Part 2: Protocol Specification. 90 pp.
SAA AS 4038.1-92. Industrial Automation Systems—Manufacturing Message Specification—Part 1: Service Definition (ISO/IEC 9506-1:1990). 316 pp.
SAA AS 4038.2-92. Industrial Automation Systems—Manufacturing Message Specification—Part 2: Protocol Specification (ISO/IEC 9506-2:1990). 128 pp.
SAA AS 4038.3-92. Industrial Automation Systems—Manufacturing Message Specification—Part 3: Companion Standard for Robotics (ISO/IEC 9506-3:1991). 119 pp.

Cinchonine
CNS K7148-66. Chemical Reagent (Cinchonine) (Mar)(1648).
JIS K 8571-79. Cinchonine.

Cinchonine Hydrochloride
CNS K7600-82. Chemical Reagent (Cinchonine Hydrochloride) (Feb)(8519).
JIS K 8194-61. Cinchonine Hydrochloride.

Cinder Blocks
See Also: Concrete Blocks; Concretes

—Fillers
CGSB CAN/CGSB-1.188-M90. Emulsion Type Filler Masonry Block. 10 pp.

Cinema Carbons
Use: Carbon

Cinema Screens
Use: Screens (Projection)

Cinemagraphic Cameras
Use: Motion Picture Cameras

Cinemas
Use: Theaters

Cinematographic Films
Use: Motion Picture Films

Cinematography
Use For: Motion Picture Photography
See Also: Motion Picture Films

—Background Noise—Sound Pressure Levels
ISO 9568-93. Cinematography—Background Acoustic Noise Levels in Theatres, Review Rooms and Dubbing Rooms First Edition. 7 pp.

—Glossaries
BSI BS 5550: Sec 8.1-80. 1980 Amd 1 Cinematography Part 8: Glossaries Section 8.1: Glossary of Terms Used in the Motion-Picture Industry. 47 pp.
ISO 4246-84. Cinematography—Vocabulary First Edition. 85 pp.

—Location Lighting—Electric Power Distribution
BSI BS 5550: SUBSEC 7.5.1-81. 1981 Cinematography Part 7: Production and Presentation Section 7.5: Film and Television Location Lighting Subsection 7.5.1: Code of Practice for Distribution of A.C. Electricity for Location Lighting. 15 pp.
BSI BS 5550: SUBSEC 7.5.2-81. 1981 Cinematography Part 7: Production and Presentation Section 7.5: Film and Television Location Lighting Subsection 7.5.2: Location Lighting Power Distribution Units. 18 pp.

Cineol Content Analysis
Use: Eucalyptol Content Analysis

Cinnamon
See Also: Spices
BSI BS 7087: Part 15-93. 1993 Herbs and Spices Ready for Food Use Part 15: Specification for Cinnamon (Whole and Ground). 9 pp.
ISO 6539-83. Cinnamon (Type Sri Lanka (Ceylon), Type Seychelles and Type Madagascar), Whole or Ground (Powdered)—Specification First Edition. 8 pp.

Cinnamon Leaf Oil
CNS K5091-80. Oil of Cinnamon Leaf (Sep)(6462).
ISO 3524-77. Oil of Cinnamon Leaf First Edition. 4 pp.

Circlips
Use: Snap Rings

Circuit Access Points
Use: Testing Points

Circuit Analyzers
Use: Multimeters

Circuit Boxes
See Also: Connectors; Electrical Enclosures; Junction Boxes

Circuit Boxes (Cont.)

—Coaxial—Microwave—Preferred Products List

CECC CECC MUAHAG Vol 12 IS 1-90. Preferred Products List; Microwave Components (En, Fr, Ge). 75 pp.

Circuit Breakers

See Also: Air Circuit Breakers; Air Switches; Circuit Protectors; Control Systems Equipment; Current Limiters; Current Limiting Circuit Breakers; Disconnecting Switches; Earth Leakage Circuit Breakers; Electric Switches; Electrical Protection Equipment; Ground Fault Circuit Interrupters; Molded Case Circuit Breakers; Overcurrent Circuit Breakers; Relays; Switching Circuits

BSI BS 7395-91. 1991 Circuit-Breakers for Equipment (CBE) (EN 60934: 1990). 60 pp.

BSI BS 7395-01. 1991 Amd 1 Circuit-Breakers for Equipment (CBE) (AMD 7291) February 15, 1993. 80 pp.

CENELEC EN 60 934-90. Circuit-Breakers for Equipment (CBE). 5 pp.

CENELEC EN 60934/A1-92. AMD 1 Circuit-Breakers for Equipment (CBE). 6 pp.

IEC 934-93. Circuit-Breakers for Equipment (CBE) Second Edition. 142 pp.

JIS C 4610-90. Circuit-Breakers for Equipment. 34 pp.

—

BSI BS 4752: Part 1-77. (WITHDRAWN) 1977 Amd 1 Switchgear and Controlgear for Voltages up to and Including 1000V a.c. And 1200V d.c. Part 1: Circuit-Breakers (Superseded by BS EN 60947-2: 1992). 80 pp.

DIN VDE 0660 Pt 108-82. Switchgear and Control Gear; Circuit Breakers; Additional Requirements for d.c. Circuitbreakers Rated Between 1200 V and 3000 V (Sept). 4 pp.

JIS C 4603-90. AC Circuit Breakers for 3.3 kV or 6.6 kV. 52 pp.

SAA AS 1930-76. Circuit-Breakers for Distribution Circuits (up to and Including 1 000 V a.c. and 1 200 V d.c.) Corrig.. 81 pp.

SAA AS 2006-86. High Voltage a.c. Switchgear and Controlgear—Circuit Breakers for Rated Voltages Above 1 000 V. 117 pp.

— 250-500 V AC—Miniature

DIN VDE 0641-78. Miniature Circuit Breakers up to 63A Rated Current 415 V ac (June). 71 pp.

DIN VDE 0641 A2 (D)-84. Miniature Circuit Breakers up to 63 A Rated Current and up to 415 a.c.: Amendment 2 (VDE Specification) (May). 3 pp.

DIN VDE 0641 A3 (D)-86. Miniature Circuit Breakers up to 63 A Rated Current and up to 415 C ac: Amendment 3 (Oct). 4 pp.

DIN VDE 0641 A4-88. Miniature Circuit Breakers up to 63A Nominal Current and up to a.c. 415 V; Tripping Characteristics B and C: Amendment 4 (Nov). 20 pp.

— 250-500 V DC—Miniature

DIN VDE 0641 Pt 2-84. Miniature Circuit Breakers up to 63A Rated Current, up to 440 V d.c. (Apr). 56 pp.

— 550-1000 V AC

SAA AS 1930-76. Circuit-Breakers for Distribution Circuits (up to and Including 1 000 V a.c. and 1 200 V d.c.) Corrig.. 81 pp.

—Aerospace

AECMA PREN 2350-89. Circuit Breakers Technical Specification. 25 pp.

BSI BS EN 2350-91. 1991 Circuit Breakers. Technical Specification. 30 pp.

CEN EN 2350-90. Circuit Breakers Technical Specification. 25 pp.

—Failure (Quality Control)—Aircraft

CAA NOTICE #87 ISSUE 1. (This Notice Gives Details of a Mandatory Action) Failure of Mechanical Products Inc. Circuit Breakers (Airworthiness Notices). 2 pp.

—High Speed—Rolling Stock

JIS E 5005-87. Test Methods of High Speed Circuit Breakers for Railway Rolling Stock.

—High Voltage

BSI BS 5311-88. 1988 Amd 2 High-Voltage Alternating-Current Circuit-Breakers (AMD 6737) September 30, 1991. 186 pp.

BSI BS 7426-91. 1991 High-Voltage Alternating Current Switch-Fuse Combinations. 55 pp.

CENELEC HD 348 S2-79. High-voltage Alternating-Current Circuit-Breakers. 5 pp.

CENELEC HD 348 S3-89. High Voltage Alternating-Current Circuit-Breakers. 4 pp.

CENELEC HD 348 S4-91. High-Voltage Alternating-Current Circuit-Breakers. 5 pp.

Circuit Breakers (Cont.)

—High Voltage (Cont.)

CNS C4142-79. High-Voltage Alternating-Current Circuit-Breakers (Jan)(4734).

CNS C4356-82. Cubicle Type High Voltage Power Receiving Unit (Jul)(9090).

IEC 56-87. High-Voltage Alternating-Current Circuit-Breakers Fourth Edition; (Corrigendum—April 1989) (Amendment 1-1992). 360 pp.

IEC 420-90. High-Voltage Alternating Current Switch-Fuse Combinations Second Edition. 101 pp.

JIS C 4620-92. Cubicle Type High Voltage Power Receiving Units. 43 pp.

SNZ IEC 56-87. High-Voltage Alternating Current Circuit-Breakers. 330 pp.

—High Voltage—Maintenance

IEC 1208-92. High-Voltage Alternating Current Circuit-Breakers —Guide for Maintenance First Edition. 27 pp.

—High Voltage—Seismic Design

BSI BS EN 61166-93. 1993 High-Voltage Alternating Current Circuit-Breakers Guide for Seismic Qualification of High-Voltage Alternating Current Circuit-Breakers (IEC 1166: 1993) (E). 24 pp.

IEC 1166-93. High-Voltage Alternating Current Circuit-Breakers —Guide for Seismic Qualification of High-Voltage Alternating Current Circuit-Breakers First Edition. 35 pp.

—Low Voltage

BSI BS EN 60947-2-92. 1992 Low-Voltage Switchgear and Controlgear Part 2: Circuit-Breakers (E). 74 pp.

BSI BS EN 60947-2-01. 1992 Amd 1 Low-Voltage Switchgear and Controlgear Part 2: Circuit-Breakers (AMD 7853) September 15, 1993 (E). 104 pp.

CENELEC HD 418-82. Low-Voltage Switchgear and Controlgear Circuit-Breakers. 9 pp.

CENELEC PREN 60 898/A1-89. AMD 1 Circuit-Breakers for Overcurrent Protection for Household and Similar Installations. 9 pp.

CENELEC PREN 60 898/A2-89. AMD 2 Circuit-Breakers for Overcurrent Protection for Household and Similar Installations. 4 pp.

CENELEC PREN 60 898/A3-89. AMD 3 Circuit-Breakers for Overcurrent Protection for Household and Similar Installations. 8 pp.

CENELEC PREN 60 898/A4-89. AMD 4 Circuit-Breakers for Overcurrent Protection for Household and Similar Installations. 4 pp.

CENELEC EN 60898-91. Circuit-Breakers for Overcurrent Protection for Household and Similar Installations. 33 pp.

CENELEC EN 60947-2-91. Low-Voltage Switchgear and Controlgear Part 2: Circuit-Breakers. 7 pp.

DIN VDE 0660 Pt 101-82. Switchgear and Controlgear; Low Voltage Switchgear and Controlgear, Circuit-Breaker (Sept). 9 pp.

DIN VDE 0660 Pt 101 A6 (D)-92. Low-Voltage Switchgear and Controlgear; Part 2: Circuit-Breaker; Amendment 6 to Draft DIN VDE 0660 Part 1-1; Identical with IEC 17B(Sec)410 (Mar). 9 pp.

IEC 947 Pt 2-89. Low-Voltage Switchgear and Controlgear Part 2: Circuit-Breakers First Edition; (Corrigendum—June 1989) (Corrigendum—April 1990) (Amendment 1-1992). 197 pp.

JIS C 8372-91. Low-Voltage Circuit Breakers. 43 pp.

—Mining Equipment

BSI BS 5126: Part 2-76. 1976 Mining Type Flameproof Supply and Control Units for Use on Systems up to 1100V Part 2: 300A Circuit Breaker Units. 18 pp.

BSI BS 7202-89. 1989 Non-Incendive Low Voltage Control/Interlock and Low Voltage Earth Fault Monitoring Circuits for Use in Mines. 8 pp.

—Mining Equipment—600-999 V AC

BSI BS 787: Part 2-68. 1968 Amd 1 Mining Type Flameproof Gate-End Boxes Part 2: Gate-End Boxes with Air-Break Circuit Breakers (For Use on 3-Phase A.C. Circuits up to 650 V). 26 pp.

—Out of Phase Switching—Three Phase Electrical Systems

CENELEC HD 409-76. Guide to the Testing of Circuit-Breakers with Respect to out-of-Phase Switching. 2 pp.

CENELEC HD 409 S1-88. Guide to the Testing of Circuit-Breakers with Respect to Out-of-Phase Switching. 3 pp.

—Railroad Traction Equipment

CENELEC PREN 50123-2-93. Railway Applications—D.C. Switchgear for Stationary Installations in Traction Systems—Part Two: D.C. Circuit Breakers. 37 pp.

—Residual Current

BSI BS 4293-83. 1983 Amd 2 Residual Current-Operated Circuit-Breakers (AMD 6279) December 21, 1990. 19 pp.

Circuit Breakers (Cont.)

—Residual Current (Cont.)

CNS C3077-80. Method of Test for Residual Current Protective Device (Apr)(5423). 11 pp.

CNS C4176-80. Residual Current Protective Device (Apr)(5422). 12 pp.

DIN VDE 0661-88. Portable Protective Devices to Increase the Protection Level for 230 V a.c. Rated Voltage 16 A Rated Current and Rated Residual Current I Delta n Less Than or Equal to 30 mA (Apr). 38 pp.

IEC 1008 Pt 1-90. Residual Current Operated Circuit-Breakers Without Integral Overcurrent Protection for Household and Similar Uses (RCCB'S) Part 1: General Rules First Edition; (Corrigendum—Dec 1990) (Amendment 1-1992). 226 pp.

IEC 1008 Pt 2-1-90. Residual Current Operated Circuit-Breakers Without Integral Overcurrent Protection for Household and Similar Uses (RCCB'S) Part 2-1: Applicability of the General Rules to RCCB's Functionally Independent of Line Voltage First Edition. 11 pp.

IEC 1008 Pt 2-2-90. Residual Current Operated Circuit-Breakers Without Integral Overcurrent Protection for Household and Similar Uses (RCCB'S) Part 2-2: Applicability of the General Rules to RCCB's Functionally Dependent on Line Voltage First Edition. 9 pp.

IEC 1009 Pt 1-91. Residual Current Operated Circuit-Breakers with Integral Overcurrent Protection for Household and Similar Uses (RCBO's) Part 1: General Rules First Edition. 238 pp.

IEC 1009 Pt 2-1-91. Residual Current Operated Circuit-Breakers with Integral Overcurrent Protection for Household and Similar Uses (RCBO's) Part 2-1: Applicability of the General Rules to RCBO's Functionally Independent of Line Voltage First Edition. 12 pp.

IEC 1009 Pt 2-2-91. Residual Current Operated Circuit-Breakers with Integral Overcurrent Protection for Household and Similar Uses (RCBO's) Part 2-2: Applicability of the General Rules to RCBO's Functionally Dependent on Line Voltage First Edition. 12 pp.

JIS C 8371-92. Residual Current Operated Circuit Breakers. 94 pp.

—Residual Current—Household

IEC 1009 Pt 1-91. Residual Current Operated Circuit-Breakers with Integral Overcurrent Protection for Household and Similar Uses (RCBO's) Part 1: General Rules First Edition. 238 pp.

IEC 1009 Pt 2-1-91. Residual Current Operated Circuit-Breakers with Integral Overcurrent Protection for Household and Similar Uses (RCBO's) Part 2-1: Applicability of the General Rules to RCBO's Functionally Independent of Line Voltage First Edition. 12 pp.

IEC 1009 Pt 2-2-91. Residual Current Operated Circuit-Breakers with Integral Overcurrent Protection for Household and Similar Uses (RCBO's) Part 2-2: Applicability of the General Rules to RCBO's Functionally Dependent on Line Voltage First Edition. 12 pp.

—Residual Current—Short Circuit Current Testing

CENELEC PRHD 536 S1-89. Test Requirements for the Evaluation of the Emission of Ionized Gases During Short-Circuit Test of Residual Current Operated Circuit-Breakers for Household and Similar Use. 12 pp.

CENELEC HD 536 S1-91. Test Requirements for the Evaluation of the Emission of Ionized Gases During Short-Circuit Tests of Residual Current Operated Circuit-Breakers for Household and Similar Use. 5 pp.

CENELEC PRHD 537 S1-89. Test Requirements for the Evaluation of the Emission of Ionized Gases During Short-Circuit Tests of Residual Current Operated Circuit-Breakers with Integral Overcurrent Protection for Household and Similar Use. 14 pp.

CENELEC HD 537 S1-91. Test Requirements for the Evaluation of the Emission of Ionized Gases During Short-Circuit Tests of Residual Current Operated Circuit-Breakers with Integrated Overcurrent Protection for Household and Similar Use. 5 pp.

—Residual Current—250-500 V AC

DIN VDE 0664 Pt 1-85. Residual Current-Operated Protective Devices; Residual Current-Operated Circuit Breakers Rated up to 500 V a.c. and up to 63 A (Oct). 40 pp.

DIN VDE 0664 Pt 2-88. Residual Current Operated Devices; Residual Current Operated Circuit Breakers with Overcurrent Protection up to 415 V ac and up to 63 A (Aug). 19 pp.

—Section Boards—Ships

CNS F5055-87. Marine Section Boards (Circuit Breaker Type, Below 250V) (May)(10080).

JIS F 8829-84. Marine Section Boards with Circuit Breakers.

Circuit Breakers (Cont.)

—Short Circuit Current—Ships
IEC 363-72. Short-Circuit Current Evaluation with Special Regard to Rated Short-Circuit Capacity of Circuit-Breakers in Installations in Ships First Edition. 65 pp.

—Short Circuit Current Testing
BSI BS 7441-91. (WITHDRAWN) 1991 Synthetic Testing of High-Voltage Alternating Current Circuit-Breakers (Implementation of CENELEC HD 580 S1) (IEC 427: 1989) (Renumbered as BS EN 60427: 1992). 73 pp.
BSI BS EN 60427-92. 1992 Synthetic Testing of High-Voltage Alternating Current Circuit-Breakers (IEC 427: 1989) (AMD 7451) December 15, 1992. 82 pp.
BSI BS EN 60427-01. 1992 Amd 1 Synthetic Testing of High-Voltage Alternating Current Circuit-Breakers (IEC 427: 1989) (AMD 7802) September 15, 1993 (E). 86 pp.
CENELEC HD 580 S1-90. Synthetic Testing of High-Voltage Alternating Current Circuit-Breakers. 3 pp.
CENELEC EN 60427-92. Report on Synthetic Testing of High Voltage Alternating Current Circuit-Breakers. 5 pp.
CNS C4002-71. A.C. Circuit Breakers (for Short Circuit Test) (Jan)(318).
IEC 427-89. Synthetic Testing of High-Voltage Alternating Current Circuit-Breakers Second Edition; (Amendment 1-1992). 126 pp.

—Single Pole—Aerospace
AECMA PREN2495-89. Single-Pole Circuit Breakers Temperature Compensated Rated Currents up to a 25 A Product Standard. 9 pp.
BSI BS EN 2495-91. 1991 Single-Pole Circuit Breakers, Temperature Compensated, Rated Currents up to 25 A. Product Standard. 14 pp.
CEN EN 2495-90. Single-Pole Circuit Breakers Temperatures Compensated Rated Currents up to 25 A Product Standard. 9 pp.
CEN PREN 2666-001-91. Single-Pole Circuit Breakers Rated Currents 50 to 100 A Technical Requirements. 12 pp.
CEN PREN 2666-003-91. Single-Pole Circuit Breakers Rated Currents 50 to 100 A Distance Between Terminals 36.5 mm Product Standard. 11 pp.
CEN PREN 2794-001-91. Single-Pole Circuit Breakers Temperature Compensated Rated Currents 20 to 50 A Technical Specification. 12 pp.
CEN PREN 2794-003-91. Single-Pole Circuit Breakers Temperature Compensated Rated Currents 20 to 50 A Product Standard—Metric Thread. 8 pp.
CEN PREN 2794-004-91. Single-Pole Circuit Breakers Temperature Compensated Rated Currents 20 to 50 A Product Standard—U.N. Thread. 8 pp.
CEN PREN 3773-001-91. Temperature Compensated Single-Pole Circuit Breakers Rated Currents up to 25A Switching Capacity 65In/ 1000A Max Technical Specification. 12 pp.
CEN PREN 3773-003-92. Single-Pole Circuit Breakers Temperature Compensated Rated Currents 1A to 25A Switching Capacity 65In/1000A Max Product Standard—Metric Thread Distance Between Terminal Centres 12mm. 7 pp.
CEN PREN 3773-004-91. Temperature Compensated Single-Pole Circuit Breakers with Signal Contact Rated Currents up to 25A Switching Capacity 65 In/1000A Max Product Standard UN Fasteners (ASME). 8 pp.

—Single Pole—Aircraft
ISO 530-75. Aircraft—General Purpose Push-Pull Single-Pole Circuit-Breakers—Dimensions First Edition. 4 pp.
ISO 1467-73. General Purpose Push-Pull Single-Pole Circuit-Breakers for Aircraft—Performance Requirements First Edition. 11 pp.

—Single Pole—Miniature—Aerospace
BSI G 216-75. 1975 Hand Operated, Thermally Compensated, Miniature, Single-and Triple-Pole Circuit-Breakers. 32 pp.

—Switchboards—Ships
CNS F5054-87. Marine Distribution Boards (Circuit Breaker Type, Below 250V) (May)(10079).
JIS F 8828-84. Marine Distribution Boards with Circuit Breakers.

—Symbols
CNS C1060-83. Symbols for Sequential Control (Breakers and Switches) (Nov)(5528).

—Testing Equipment
DIN VDE 0413 Pt 6-87. Appliance for Testing the Protective Devices in Power Installations; Measuring Instruments for Testing the Efficiency of Fault-Current and Fault-Voltage Operated Protective Devices in TN and TT Systems (Aug). 18 pp.

—Three Pole—Aerospace
AECMA PREN2592-89. Three-Pole Circuit Breakers Temperature Compensated Rated Currents up to 25 A Product Standard. 10 pp.
BSI BS EN 2592-91. 1991 Three-Pole Circuit Breakers, Temperature Compensated, Rated Currents up to 25 A. Product Standard. 15 pp.
CEN EN 2592-90. Three-Pole Circuit Breakers Temperature Compensated Rated Currents up to 25 A Product Standard. 10 pp.
CEN PREN 2665-001-91. Three-Pole Circuit Breakers Temperature Compensated Rated Currents 20 to 50 A Technical Specification. 12 pp.
CEN PREN 2665-003-91. Three-Pole Circuit Breakers Temperature Compensated Rated Currents 20 to 50 A Product Standard—Metric Thread. 8 pp.
CEN PREN 2665-004-91. Three-Pole Circuit Breakers Temperature Compensated Rated Currents 20 to 50 A Product Standard—U.N Thread. 8 pp.
CEN PREN 3774-001-91. Temperature Compensated Three-Pole Circuit Breakers Rated Currents up to 25A Switching Capacity 65In Technical Specification. 11 pp.
CEN PREN 3774-003-92. Three-Pole Circuit Breakers Temperature Compensated Rated Currents 2A to 25A Switching Capacity 65In/1000A Max Product Standard—Metric Fastners Distance Between Terminal Centres 12.7mm. 7 pp.
CEN PREN 3774-004-91. Temperature Compensated Three-Pole Circuit Breakers Rated Currents up to 25A Switching Capacity 65 In Product Standard UN Fasteners (ASME). 8 pp.

—Three Pole—Aircraft
ISO 1033-75. Aircraft—Dimensions for General Purpose Push-Pull Three-Pole Circuit-Breakers First Edition. 3 pp.
ISO 1509-73. General Purpose Push-Pull Three-Pole Circuit-Breakers for Aircraft—Performance Requirements First Edition. 11 pp.

—Three Pole—Miniature—Aerospace
BSI G 216-75. 1975 Hand Operated, Thermally Compensated, Miniature, Single-and Triple-Pole Circuit-Breakers. 32 pp.

Circuit Card Assemblies
Use: Integrated Circuit Cards

Circuit Control Stations (Telecommunications)
Use: Telecommunication Circuit Control Stations

Circuit Diagrams
Use For: Wiring Diagrams *See Also:* Diagrams; Schematic Diagrams

CENELEC HD 246.6 S2-88. Diagrams, Charts, Tables Part 6: Preparation of Unit Wiring Diagrams and Tables. 3 pp.
CNS C1057-83. Circuit Diagram (Schematic Diagram) for Sequential Control (Nov)(5525).
CSA B78.4-1979. Electrical and Electronics Diagrams. 7 pp.
IEC 113 Pt 4-75. Diagrams, Charts, Tables Part 4: Recommendations for the Preparation of Circuit Diagrams First Edition. 70 pp.
IEC 113 Pt 6-76. Diagrams, Charts, Tables Part 6: Preparation of Unit Wiring Diagrams and Tables First Edition; (Amendment 1-1983). 23 pp.
JIS C 0401-82. Circuit Diagram (Schematic Diagram) for Sequential Control. 30 pp.

—Aircraft
BSI G 207-69. 1969 Aircraft Electrical Circuit Diagrams. 5 pp.
ISO 2042-73. Aircraft Electrical Circuit Diagrams First Edition. 6 pp.

—Aircraft—Automation
EURO CTI/79/9343-79. Recommendation for the Automation of Wiring Diagrams. 50 pp.

—Classification
DIN ENGL 40719 Pt 1-73. Circuit Representations; Definitions; Classification (June). 4 pp.

—Electrical Components
SAA AS 1103.4-92. Diagrams, Charts and Tables for Electrotechnology—Part 4: Guiding Principles for the Preparation of Circuit Diagrams (in Professional Package 56). 46 pp.
SAA AS 1103.6-81. Diagrams, Charts and Tables for Electrotechnology—Part 6: Preparation of Unit Wiring Diagrams and Tables. 14 pp.
SNZ NZS/AS 1103. 4-78. Diagrams, Charts and Tables for Electrotechnology Part 4: Guiding Principles for the Preparation of Circuit Diagrams (This is a Joint Standard with SAA AS 1103.4). 44 pp.
SNZ NZS/AS 1103. 6-81. Diagrams, Charts and Tables for Electrotechnology Part 6: Preparation of Unit Wiring Diagrams and Tables (This is a Joint Standard with SAA AS 1103.6). 14 pp.

—Glossaries
DIN ENGL 40719 Pt 1-73. Circuit Representations; Definitions; Classification (June). 4 pp.

—Marine—Identification Systems
CNS F5114-87. Markings and Arrangements of Main Connections and Small-Wiring for Marine Use (Jan)(11816).

—Railroad Signals—Symbols
BSI BS 376: Part 2-54. 1954 Railway Signalling Symbols Part 2: Wiring Symbols and Written Circuits. 79 pp.

—Residential Buildings
DIN ENGL 18015 Pt 3-90. Electrical Installations in Dwellings; Circuit Arrangement and Location of Electrical Equipment (July). 3 pp.

—Welding Equipment
CNS C1048-79. Electric Circuit Diagram for Welding Machine (Nov)(5018).
JIS C 9310-89. Electric Circuit Diagrams for Welding Machines. 10 pp.

Circuit Groups (Telecommunications)
See Also: Groups (Communications); Telecommunication Circuits; Telephone Circuits; Trunk Groups

—Automatic Services
CCITT RECMN E.520-89. Number of Circuits to be Provided in Automatic and/or Semiautomatic Operation, Without Overflow Facilities—Telephone Network and ISDN—Quality of Service, Network Management and Traffic Engineering (Study Group II) 3 pp. 3 pp.

—Grade of Service
CCITT RECMN E.540-89. Overall Grade of Service of the International Part of an International Connection —Telephone Network and ISDN—Quality of Service, Network Management and Traffic Engineering (Study Group II) 1 pp. 1 p.

—Grade of Service—Busy Hour
CCITT RECMN E.521-89. Calculation of the Number of Circuits in a Group Carrying Overflow Traffic—Telephone Network and ISDN—Quality of Service, Network Management and Traffic Engineering (Study Group II) 11 pp. 11 pp.

—High Usage
CCITT RECMN E.522-89. Number of Circuits in a High-Usage Group—Telephone Network and ISDN—Quality of Service, Network Management and Traffic Engineering (Study Group II) 10 pp. 10 pp.

—High Usage—Alternative Routing—Expenses
CCITT RECMN E.522-89. Number of Circuits in a High-Usage Group—Telephone Network and ISDN—Quality of Service, Network Management and Traffic Engineering (Study Group II) 10 pp. 10 pp.

—High Usage—Alternative Routing—Traffic Units
CCITT RECMN E.522-89. Number of Circuits in a High-Usage Group—Telephone Network and ISDN—Quality of Service, Network Management and Traffic Engineering (Study Group II) 10 pp. 10 pp.

—High Usage—Expenses
CCITT RECMN E.522-89. Number of Circuits in a High-Usage Group—Telephone Network and ISDN—Quality of Service, Network Management and Traffic Engineering (Study Group II) 10 pp. 10 pp.

—High Usage—Tables (Data)
CCITT RECMN E.522-89. Number of Circuits in a High-Usage Group—Telephone Network and ISDN—Quality of Service, Network Management and Traffic Engineering (Study Group II) 10 pp. 10 pp.

—High Usage—Traffic Peakedness Factor
CCITT RECMN E.521-89. Calculation of the Number of Circuits in a Group Carrying Overflow Traffic—Telephone Network and ISDN—Quality of Service, Network Management and Traffic Engineering (Study Group II) 11 pp. 11 pp.

—Manual Demand Operating
CCITT RECMN E.510-89. Determination of the Number of Circuits in Manual Operation—Telephone Network and ISDN—Quality of Service, Network Management and Traffic Engineering (Study Group II) 5 pp. 5 pp.

INTERNATIONAL AND NON-U.S. NATIONAL STANDARDS
SUBJECT INDEX

Circuit

Circuit Groups (Telecommunications) (Cont.)

—Overflow Traffic
CCITT RECMN E.521-89. Calculation of the Number of Circuits in a Group Carrying Overflow Traffic—Telephone Network and ISDN—Quality of Service, Network Management and Traffic Engineering (Study Group II) 11 pp. 11 pp.

—Overflow Traffic—Approximation
CCITT RECMN E.524 (REV 1)-92. Overflow Approximations for Non-Random Inputs (Study Group II) 14 pp. 14 pp.
CCITT RECMN E.524-89. Overflow Approximations for Non-Random Inputs —Telephone Network and ISDN—Quality of Service, Network Management and Traffic Engineering (Study Group II) 9 pp. 9 pp.

—Overflow Traffic—Formulas
CCITT RECMN E.524 (REV 1)-92. Overflow Approximations for Non-Random Inputs (Study Group II) 14 pp. 14 pp.
CCITT RECMN E.524-89. Overflow Approximations for Non-Random Inputs —Telephone Network and ISDN—Quality of Service, Network Management and Traffic Engineering (Study Group II) 9 pp. 9 pp.

—Overflow Traffic—Grade of Service
CCITT RECMN E.525 (REV 1)-92. Designing Networks to Control Grade of Service (Study Group II) 10 pp. 10 pp.

—Overflow Traffic—Statistical Analysis
CCITT RECMN E.521-89. Calculation of the Number of Circuits in a Group Carrying Overflow Traffic—Telephone Network and ISDN—Quality of Service, Network Management and Traffic Engineering (Study Group II) 11 pp. 11 pp.

—Overflow Traffic—Tables (Data)
CCITT RECMN E.524 (REV 1)-92. Overflow Approximations for Non-Random Inputs (Study Group II) 14 pp. 14 pp.
CCITT RECMN E.524-89. Overflow Approximations for Non-Random Inputs —Telephone Network and ISDN—Quality of Service, Network Management and Traffic Engineering (Study Group II) 9 pp. 9 pp.

—Overflow Traffic—Traffic Peakedness Factor
CCITT RECMN E.525-89. Service Protection Methods—Telephone Network and ISDN—Quality of Service, Network Management and Traffic Engineering (Study Group II) 5 pp. 5 pp.

—Planning
CCITT FASCICLE II.3-88. Telephone Network and ISDN Quality of Service, Network Management and Traffic Engineering. Recommendations E.401-E.880. 368 pp.

—Semiautomatic Demand Operating
CCITT RECMN E.520-89. Number of Circuits to be Provided in Automatic and/or Semiautomatic Operation, Without Overflow Facilities—Telephone Network and ISDN—Quality of Service, Network Management and Traffic Engineering (Study Group II) 3 pp. 3 pp.

—Telecommunication Traffic—Estimation
CCITT RECMN E.501 (REV 1)-92. Estimation of Traffic Offered in the Network (Study Group II) 19 pp. 19 pp.
CCITT RECMN E.501-89. Estimation of Traffic Offered in the International Network —Telephone Network and ISDN—Quality of Service, Network Management and Traffic Engineering (Study Group II) 8 pp. 8 pp.

—Telecommunication Traffic—Measurement
CCITT RECMN E.500 (REV 1)-92. Traffic Intensity Measurement Principles (Study Group II) 17 pp. 17 pp.
CCITT RECMN E.500-89. Traffic Intensity Measurement Principles —Telephone Network and ISDN—Quality of Service, Network Management and Traffic Engineering (Study Group II) 12 pp. 12 pp.
CCITT RECMN E.502 (REV 1)-92. Traffic Measurement Requirements for Digital Telecommunication Exchanges (Study Group II) 23 pp. 23 pp.
CCITT RECMN E.502-89. Traffic Measurement Requirements for SPC (Especially Digital) Telecommunication Exchanges—Telephone Network and ISDN—Quality of Service, Network Management and Traffic Engineering (Study Group II) 15 pp. 15 pp.

Circuit Groups (Telecommunications) (Cont.)

—Telex Communications—Traffic Units
CCITT RECMN F.64-89. Determination of the Number of International Telex Circuits Required to Carry a Given Volume of Traffic—Telegraph and Mobile Services Operations and Quality of Service (Study Group I) 5 pp. 5 pp.

—Telex Communications—Traffic Units—Tables (Data)
CCITT RECMN F.64-89. Determination of the Number of International Telex Circuits Required to Carry a Given Volume of Traffic—Telegraph and Mobile Services Operations and Quality of Service (Study Group I) 5 pp. 5 pp.

Circuit Interrupters
Use: Interrupter Switches

Circuit Mode Bearer Services
Use: Bearer Services—Circuit Mode

Circuit Monitors
See Also: Monitors

—Telecommunication Circuits
CCITT RECMN E.261-89. Devices for Measuring and Recording Call Durations—Telephone Network and ISDN—Operation, Numbering, Routing and Mobile Service (Study Group II) 3 pp. 3 pp.

—Telecommunication Equipment
CCITT RECMN E.261-89. Devices for Measuring and Recording Call Durations—Telephone Network and ISDN—Operation, Numbering, Routing and Mobile Service (Study Group II) 3 pp. 3 pp.

Circuit Multiplication Equipment
See Also: Digital Circuit Multiplication Equipment; Packet Circuit Multiplication Equipment; Telecommunication Equipment

—International Switching Centers—Signaling Systems
CCITT RECMN Q.50-89. Signalling Between Circuit Multiplication Equipments (CME) and International Switching Centres (ISC)—General Recommendations on Telephone Switching and Signalling—Functions and Information Flows for Services in the ISDN—Supplements. 17 pp.

Circuit Noise
See Also: Circuit Noise Levels; Circuits; Noise (Spurious Signals); Psophometric Power; Telecommunication Circuits

—Automatic Exchanges
CCITT RECMN G.123-89. Circuit Noise in National Networks—General Characteristics of International Telephone Connections and Circuits (Study Groups XII and XV) 6 pp. 6 pp.

—Band Reject Filters—Measurement
CCITT RECMN G.228-89. Measurement of Circuit Noise in Cable Systems Using a Uniform-Spectrum Random Noise Loading—International Analogue Carrier Systems (Study Group XV) 11 pp. 11 pp.

—Bandpass Filters—Measurement
CCITT RECMN G.228-89. Measurement of Circuit Noise in Cable Systems Using a Uniform-Spectrum Random Noise Loading—International Analogue Carrier Systems (Study Group XV) 11 pp. 11 pp.

—Data Circuits
CCITT RECMN G.143-89. Circuit Noise and the Use of Compandors—General Characteristics of International Telephone Connections and Circuits (Study Groups XII and XV) 5 pp. 5 pp.

—Echo Suppressors
CCITT RECMN G.164-89. Echo Suppressors—General Characteristics of International Telephone Connections and Circuits (Study Groups XII and XV) 36 pp. 36 pp.

—Electrostatic Induction—Power Lines—Telephone Networks
CCITT RECMN G.123-89. Circuit Noise in National Networks—General Characteristics of International Telephone Connections and Circuits (Study Groups XII and XV) 6 pp. 6 pp.

—Frequency Division Multiplexers—Radio Relay Systems
CCITT RECMN G.441-89. Permissable Circuit Noise on Frequency-Division Multiplex Radio-Relay Systems—International Analogue Carrier Systems (Study Group XV) 1 pp. 1 p.

Circuit Noise (Cont.)

—Groups—Carrier Systems
CCITT RECMN G.311-89. General Characteristics of Systems Providing 12 Carrier Telephone Circuits on an Open-Wire Pair—International Analogue Carrier Systems (Study Group XV) 6 pp. 6 pp.
CCITT RECMN G.325-89. General Characteristics Recommended for Systems Providing 12 Telephone Carrier Circuits on a Symmetric Cable Pair ((12 + 12) Systems)—International Analogue Carrier Systems (Study Group XV) 5 pp. 5 pp.

—Hypothetical Reference Circuits
CCITT RECMN G.215-89. Hypothetical Reference Circuit of 5000 km for Analogue Systems—International Analogue Carrier Systems (Study Group XV) 2 pp. 2 pp.

—Leased Circuits—Data Transmission
CCITT RECMN G.143-89. Circuit Noise and the Use of Compandors—General Characteristics of International Telephone Connections and Circuits (Study Groups XII and XV) 5 pp. 5 pp.

—Telephone Circuits
CCITT RECMN G.143-89. Circuit Noise and the Use of Compandors—General Characteristics of International Telephone Connections and Circuits (Study Groups XII and XV) 5 pp. 5 pp.
CCITT RECMN G.153-89. Characteristics Appropriate to International Circuits More Than 2500 km in Length—General Characteristics of International Telephone Connections and Circuits (Study Groups XII and XV) 4 pp. 4 pp.

—Telephone Circuits—Carrier Systems
CCITT RECMN G.125-89. Characteristics of Circuits on Carrier Systems—General Characteristics of International Telephone Connections and Circuits (Study Groups XII and XV) 1 pp. 1 p.
CCITT RECMN G.152-89. Characteristics Appropriate to Long-Distance Circuits of a Length Not Exceeding 2500 km—General Characteristics of International Telephone Connections and Circuits (Study Groups XII and XV) 2 pp. 2 pp.

—Telephone Circuits—Radio Relay Systems
CCITT RECMN G.153-89. Characteristics Appropriate to International Circuits More Than 2500 km in Length—General Characteristics of International Telephone Connections and Circuits (Study Groups XII and XV) 4 pp. 4 pp.

—Telephone Equipment
CCITT RECMN G.143-89. Circuit Noise and the Use of Compandors—General Characteristics of International Telephone Connections and Circuits (Study Groups XII and XV) 5 pp. 5 pp.

—Telephone Lines—Measurement
CCITT RECMN G.228-89. Measurement of Circuit Noise in Cable Systems Using a Uniform-Spectrum Random Noise Loading—International Analogue Carrier Systems (Study Group XV) 11 pp. 11 pp.

—Telephone Networks
CCITT RECMN G.123-89. Circuit Noise in National Networks—General Characteristics of International Telephone Connections and Circuits (Study Groups XII and XV) 6 pp. 6 pp.

—Telephone Systems—HF
CCITT RECMN G.334-89. 18 MHz Systems on Standardized 2.6/9.5 mm Coaxial Cable Pairs—International Analogue Carrier Systems (Study Group XV) 9 pp. 9 pp.
CCITT RECMN G.345-89. 12 MHz Systems on Standardized 1.2/4.4 mm Coaxial Cable Pairs—International Analogue Carrier Systems (Study Group XV) 1 pp. 1 p.
CCITT RECMN G.346-89. 18 MHz Systems on Standardized 1.2/4.4 mm Coaxial Cable Pairs—International Analogue Carrier Systems (Study Group XV) 1 pp. 1 p.

—Telephone Systems—HF/VHF
CCITT RECMN G.333-89. 60 MHz Systems on Standardized 2.6/9.5 mm Coaxial Cable Pairs—International Analogue Carrier Systems (Study Group XV) 10 pp. 10 pp.

—Telephone Systems—MF/HF
CCITT RECMN G.332-89. 12 MHz Systems on Standardized 2.6/9.5 mm Coaxial Cable Pairs—International Analogue Carrier Systems (Study Group XV) 11 pp. 11 pp.

—Underwater Telephone Cables
CCITT RECMN G.153-89. Characteristics Appropriate to International Circuits More Than 2500 km in Length—General Characteristics of International Telephone Connections and Circuits (Study Groups XII and XV) 4 pp. 4 pp.

INDUSTRY STANDARDS

INTERNATIONAL AND NON-U.S. NATIONAL STANDARDS
SUBJECT INDEX

Circuit Noise (Cont.)
—Voice Band Data Transmission
CCITT RECMN G.113-89. Transmission Impairments—General Characteristics of International Telephone Connections and Circuits (Study Groups XII and XV) 22 pp. 22 pp.

Circuit Noise Levels
See Also: Circuit Noise; Circuits; Telecommunication Circuits

—Telephone Circuits
CCITT RECMN G.113-89. Transmission Impairments—General Characteristics of International Telephone Connections and Circuits (Study Groups XII and XV) 22 pp. 22 pp.

—Telephone Connections
CCITT RECMN G.113-89. Transmission Impairments—General Characteristics of International Telephone Connections and Circuits (Study Groups XII and XV) 22 pp. 22 pp.

Circuit Noise Meters
Use For: Psophometers See Also: Acoustic Measuring Instruments; Noise Meters

—Telephone Circuits
CCITT RECMN O.41-89. Psophometer for Use on Telephone-Type Circuits—Specifications for Measuring Equipment (Study Group IV) 8 pp (Same as Recmn P.53). 8 pp.

Circuit Protectors
See Also: Circuit Breakers; Electrical Protection Equipment; Fuses (Electric); Fusible Switches

—Spacecraft Charging
CCIR Report 674-2-90. Spacecraft Charging—Section 2A—Research in Space Technology. 7 pp.

Circuit Selector Switches
Use: Selector Switches

Circuit Switched Connections
Use For: International Digital Connections

—Integrated Services Digital Networks
CCITT RECMN I.120-89. Integrated Services Digital Networks (ISDNs) —Integrated Services Digital Network (ISDN)—General Structure and Service Capabilities (Study Group XVIII) 2 pp. 2 pp.
IEC 8473 Draft AMD 5. Information Processing Systems—Data Commun-ications—Protocol for Providing the Connec-tionless-Mode Network Service Amendment 5: Provision of the Underlying Service for Operation over ISDN Circuit-Switched B-Channel; (1992) ***CD-ROM ONLY***. 9 pp.
ISO 8473 Draft AMD 5. Information Processing Systems—Data Commun-ications—Protocol for Providing the Connec-tionless-Mode Network Service Amendment 5: Provision of the Underlying Service for Operation over ISDN Circuit-Switched B-Channel; (1992) ***CD-ROM ONLY***. 9 pp.
JTC1 8473 Draft AMD 5. Information Processing Systems—Data Commun-ications—Protocol for Providing the Connec-tionless-Mode Network Service Amendment 5: Provision of the Underlying Service for Operation over ISDN Circuit-Switched B-Channel; (1992) ***CD-ROM ONLY***. 9 pp.

—Integrated Services Digital Networks—Controlled Slip Rate
CCITT RECMN G.822-89. Controlled Slip Rate Objectives on an International Digital Connection—Digital Networks, Digital Sections and Digital Line Systems (Study Groups XV and XVIII) 3 pp. 3 pp.

—Integrated Services Digital Networks—Error Performance
CCITT RECMN G.821-89. Error Performance of an International Digital Connection Forming Part of an Integrated Services Digital Network—Digital Networks, Digital Sections and Digital Line Systems (Study Groups XV and XVIII) 10 pp. 10 pp.

Circuit Switched Data Networks
See Also: Communication Networks; Packet Switched Networks; Public Data Networks

—Message Transfer Part—CCITT No. 7 Signaling Systems
CCITT RECMN X.60-89. Common Channel Signalling for Circuit Switched Data Applications—Data Communication Networks—Transmission, Signalling and Switching, Network Aspects, Maintenance and Administrative Arrangements (Study Group VII) 2 pp. 2 pp.

Circuit Switched Data Networks (Cont.)
—Network Services
IEC 8880 Pt 2 Draft AMD 2. Info. Technology—Tele-communications and Info. Exchange Between Systems —Protocol Combinations to Provide and Support the OSI Network Service—Part 2: Provision and Support of the Connec-tion-Mode Network Service Amd 2: Addition of PSTN & CSDN Environments (1991) ***CD-ROM ONLY***. 6 pp.
ISO 8880 Pt 2 Draft AMD 2. Information Technology—Telecommunications and Info. Exchange Between Systems—Protocol Com-binations to Provide and Support the OSI Network Service—Part 2: Provi-sion and Support of the Connection-Mode Network Service Amd 2: Addition of PSTN & CSDN; Environ-ments;(1991)*CD-ROM ONLY. 6 pp.
JTC1 8880 Pt2 Draft AMD 2. Info. Technology—Tele-communications and Info. Exchange Between Systems —Protocol Combinations to Provide and Support the OSI Network Service—Part 2: Provision and Support of the Connec-tion-Mode Network Service Amd 2: Addition of PSTN & CSDN Environments (1991) ***CD-ROM ONLY***. 6 pp.

—Public—Bearer Services—Directory Databases
CCITT RECMN E.115-91. Computerized Information Service for Telephone Subscriber Numbers in Foreign Countries (Directory Assistance), Reserved for Operators (Study Group I) 19 pp. 19 pp.
CCITT RECMN E.115-89. Computerized Information Service for Telephone Subscriber Numbers in Foreign Countries (Directory Assistance), Reserved for Operators—Telephone Network and ISDN—Operation, Numbering, Routing and Mobile Service (Study Group II) 11 pp. 11 pp.

—Public—Call Blocking
CCITT RECMN X.131-89. Call Blocking in Public Data Networks when Providing International Synchronous Circuit-Switched Data Services—Data Communication Networks—Transmission, Signalling and Switching, Network Aspects, Maintenance and Administrative Arrangements. 4 pp.

—Public—Call Processing Delays
CCITT RECMN X.130-89. Call Processing Delays in Public Data Networks when Providing International Synchronous Circuit-Switched Data Services—Data Communication Networks—Transmission, Signalling and Switching, Network Aspects, Maintenance and Administrative. 13 pp.

—Public—FAX Communications—CCITT Group 4
CCITT RECMN F.184-89. Operational Provisions for the International Public Facsimile Service Between Subscriber Stations with Group 4 Facsimile Machines (Telefax 4)—Telematic, Data Transmission and Teleconference Services—Operations and Quality of Service (Study Group I) 9 pp. 9 pp.

—Public—Hypothetical Reference Connections
CCITT RECMN X.92-89. Hypothetical Reference Connections for Public Synchronous Data Networks—Data Communication Networks—Transmission, Signalling and Switching, Network Aspects, Maintenance and Administrative Arrangements (Study Group VII) 4 pp. 4 pp.

—Public—ISDN—Data Transmission—Interworking
CCITT RECMN I.540-89. General Arrangements For Interworking Between Circuit Switched Public Data Networks (CSPDNs) and Integrated Services Digital Networks (ISDNs) for the Provision of Data Transmission—Integrated Services Digital Network (ISDN)—Internetwork Interfaces and Maintenance. 1 p.
CCITT RECMN X.321-89. General Arrangements for Interworking Between Circuit Switched Public Data Networks (CSPDNs) and Integrated Service Digital Networks (ISDNs) for the Provision of Data Transmission Services—Data Communication Networks—Interworking Between Networks, Mobile Data. 7 pp.

—Public—ISDN—Interworking
CCITT RECMN E.166-89. Numbering Plan Interworking in the ISDN Era—Telephone Network and ISDN—Operation, Numbering, Routing and Mobile Service (Study Group II) 14 pp (Same as Recmn X.122). 14 pp.
CCITT RECMN X.81-89. Interworking Between an ISDN Circuit-Switched and a Circuit-Switched Public Data Network (CSPDN)—Data Communication Networks—Transmission, Signalling and Switching, Network Aspects, Maintenance and Administrative Arrangements (Study Group VII) 18 pp. 18 pp.

Circuit Switched Data Networks (Cont.)
—Public Land Mobile Networks—Interworking
CENELEC GSM 09.04-92. Interworking Between the PLMN and the CSPDN. 13 pp.
ETSI GSM 09.04-92. Interworking Between the PLMN and the CSPDN. 13 pp.

—Public—Message Conversion Systems
CCITT RECMN F.202-89. Interworking Between the Telex Service and the Teletex Service—General Procedures and Operational Requirements for the International Interconnection of Telex/Teletex Conversion Facilities—Telematic, Data Transmission and Teleconference Services—Operations and. 4 pp.

—Public—Numbering Plans
CCITT RECMN E.166-89. Numbering Plan Interworking in the ISDN Era—Telephone Network and ISDN—Operation, Numbering, Routing and Mobile Service (Study Group II) 14 pp (Same as Recmn X.122). 14 pp.

—Public—Packet Switched Networks—Data Transmission—Interworking
CCITT RECMN X.322-89. General Arrangements for Interworking Between Packet Switched Public Data Networks (PSPDNs) and Circuit Switched Public Data Networks (CSPDNs) for the Provision of Data Transmission Services—Data Communication Networks—Interworking Between Networks, Mobile. 7 pp.

—Public—Packet Switched Networks—Interworking
CCITT RECMN X.82-89. Detailed Arrangements for Interworking Between CSPDNs and PSPDNs Based on Recommendation T.70—Data Communication Networks—Transmission, Signalling and Switching, Network Aspects, Maintenance and Administrative Arrangements (Study Group VII) 20 pp. 20 pp.

—Public—Teleinformatic Services—Terminal Identification
CCITT RECMN F.351-89. General Principles on the Presentation of Terminal Identification to Users of the Telematic Services—Telematic, Data Transmission and Teleconference Services—Operations and Quality of Service (Study Group I) 2 pp. 2 pp.

—Public—Teletex Communications
CCITT RECMN F.200-92. Teletex Service (Study Group I) 28 pp. 28 pp.
CCITT RECMN F.200-89. Teletex Service—Telematic, Data Transmission and Teleconference Services —Operations and Quality of Service (Study Group I) 20 pp. 20 pp.

—Signaling Systems—Interworking
CCITT RECMN X.80-89. Interworking of Interexchange Signalling Systems for Circuit Switched Data Services—Data Communication Networks—Transmission, Signalling and Switching, Network Aspects, Maintenance and Administrative Arrangements (Study Group VII) 11 pp. 11 pp.

Circuit Switched Data Transmission Services
See Also: Communication Networks; Data Communication; Data Transmission

—Grade of Service—Integrated Services Digital Networks
CCITT RECMN E.701-92. Reference Connections for Traffic Engineering—Telephone Network and ISDN—Quality of Service, Network Management and Traffic Engineering (Study Group II) 4 pp. 4 pp.
CCITT RECMN E.721-91. Network Grade of Service Parameters and Target Values for Circuit-Switched Services in the Evolving ISDN (Study Group II) 8 pp. 8 pp.
CCITT RECMN E.721-89. Network Grade of Service Parameters in ISDN—Telephone Network and ISDN—Quality of Service, Network Management and Traffic Engineering (Study Group II) 2 pp. 2 pp.
CCITT RECMN E.723-92. Grade-of-Service Parameters for Signalling System No. 7 Networks (Study Group II) 8 pp. 8 pp.

—Message Transfer Part—CCITT No. 7 Signaling Systems
CCITT RECMN X.61-89. Signalling System No. 7—Data User Part—Data Communication Networks—Transmission, Signalling and Switching, Network Aspects, Maintenance and Administrative Arrangements (Study Group VII) 54 pp. 54 pp.

Circuit Switched Data Transmission Services *(Cont.)*

—Open Systems Interconnection—Network Service
CEN ENV 41 107-88. Teleprocessing, Open System Interconnection, Data Transmission, Connection Oriented Transmission Digital Circuit, Services, Protocols, Physical Layer Network Layer, Data Link Layer, Transport Layer, Implementation. 42 pp.

—Open Systems Interconnection—Transport Service
CEN ENV 41 106-88. Information Systems Interconnection, Digital Date, Circuit (Circuit Switched Data Networks); Provision of the OSI Connection-Mode Transport Service in the T-70 Case for Telematic End Systems. 29 pp.
CEN ENV 41 107-88. Teleprocessing, Open System Interconnection, Data Transmission, Connection Oriented Transmission Digital Circuit, Services, Protocols, Physical Layer Network Layer, Data Link Layer, Transport Layer, Implementation. 42 pp.

Circuit Switched Networks

—ISDN—Dimensioning
CCITT RECMN E.730-92. ISDN Dimensioning Methods Overview (Study Group II) 4 pp. 4 pp.

—Network Performance
CENELEC ETR 011-90. Network Aspects (NA); the Relationship Between Network Component Performance and the Overall Network Performance. 8 pp.
ETSI ETR 011-90. Network Aspects (NA); the Relationship Between Network Component Performance and the Overall Network Performance. 8 pp.

—Private—Telex Communications
CCITT RECMN F.71-89. Interconnection of Private Teleprinter Networks with the Telex Network—Telegraph and Mobile Services Operations and Quality of Service (Study Group I) 2 pp. 2 pp.

—Public Data Communication—Reverse Charging
CCITT RECMN D.30-89. Implementation of Reverse Charging on International Public Data Communication Services—General Tariff Principles—Charging and Accounting in International Telecommunications Services (Study Group III) 5 pp. 5 pp.

—Routing (Telecommunications)—Management Model
CENELEC PRI-ETS 300 292-93. Network Aspects (NA); Functional Specification of Call Routeing Information Management on the Operations System/Network Element (OS/NE) Interface. 71 pp.
ETSI PRI-ETS 300 292-93. Network Aspects (NA); Functional Specification of Call Routeing Information Management on the Operations System/Network Element (OS/NE) Interface. 71 pp.

Circuit Switching Systems

See Also: Digital Switching Systems; Switching Systems

—Analog
CECC CECC 90 203 ISSUE 1-85. Blank Detail Specification: Integrated Analogue Switching Circuits (En, Fr, Ge). 46 pp.

Circuit Testers
JIS C 1202-86. Circuit Testers. 17 pp.

—Telegraph Equipment
CCITT RECMN R.51 BIS-89. Standardized Text for Testing the Elements of a Complete Circuit—Telegraph Transmission (Study Group IX) 1 pp. 1 p.

Circuits

Scope Note: For additional listings, use a more specific term *Use For:* Electric Circuits
See Also: Amplifiers; Automatic Frequency Control; Branch Circuits; Circuit Noise; Circuit Noise Levels; Communication Circuits; Control Circuits; Data Networks; Digital Circuits; Electric Wire; Electrical Grounding; Electron Tubes; Flip Flops; Heating Circuits; Hypothetical Reference Circuits; Integrated Circuit Cards; Integrated Circuits; Interchange Circuits; Interlock Circuits; Lighting Circuits; Linear Circuits; Magnetic Circuits; Metallic Circuits; Multivibrators; Noise Suppressors; Passive Elements; Plug In Units; Polyphase Circuits; Power Circuits; Printed Circuit Boards; Private Leased Circuits; Pulse Generators; Radiotelephone Circuits; Remote Control Circuits; Service Circuits; Short Circuits; Signal Circuits; Signal Generators; Switching Circuits;

Circuits *(Cont.)*

See Also: (Cont.)
Tandem Data Circuits; Telecommunication Circuits; Transmission Lines
CAA LEAFLET 9-1 07.90. Bonding and Circuit Testing. 12 pp.
IEC 375-72. Conventions Concerning Electric and Magnetic Circuits First Edition. 55 pp.

—Aircraft
CAA 706 (K). Section K Light Aeroplanes Electrical Supply, Systems and Equipment (Blue Papers). 54 pp.
CAA 707 (G). Section G Rotorcraft Electrical Supply, Systems and Equipment (Blue Papers). 47 pp.
CAA Chapter J2-3 06.48. Circuit Control and Protection. 2 pp.
CAA Chapter G6-14 11.85. Utilisation and Installation of Electrically Operated Systems and Equipment (Rotocraft). 15 pp.

—Aircraft—Identification Systems
NATO STANAG 3347 ED 4 AMD 0-85. Aircraft Electrical Circuit Identification. 11 pp.
NATO STANAG 3347 ED 4 AMD 1-85. Aircraft Electrical Circuit Identification. 9 pp.
NATO STANAG 3347 ED 5 AMD 1-92. Identification of Aircraft Electrical Circuits. 9 pp.
NATO STANAG 3347 ED 5 AMD 2-92. Identification of Aircraft Electrical Circuits. 10 pp.

—Aircraft—Symbols
NATO STANAG 3455 ED 2 AMD 2-78. Basic Symbols for Aircraft Electrical Circuits. 2 pp.
NATO STANAG 3455 ED 3 AMD 0-90. Basic Symbols for Aircraft Electrical Circuits. 5 pp.

—Glossaries
BSI BS 4727:Pt1: Group 01-83. 1983 Amd 1 Electrotechnical, Power, Telecommunication, Electronics, Lighting and Colour Terms Part 1: Terms Common to Power, Telecommunications and Electronics Group 01: Fundamental Terminology. 56 pp.
IEC. Vocabulary of Fundamental Concepts. 237 pp.
IEC 50 Chap 131-78. International Electrotechnical Vocabulary Chapter 131: Electric and Magnetic Circuits; (Supplement A-1982) (Amendment 1-1984). 87 pp.
SAA AS 1852.131-88. International Electrotechnical Vocabulary—Part 131: Electric and Magnetic Circuits. 37 pp.
SNZ IEC 50: 50(131)-78. International Electrotechnical Vocabulary 50(131): Electric and Magnetic Circuits Amend: 1. 43 pp.

—Solid Fuel Burning Equipment
CSA 1343 Bull. Electrical Bulletin 1343 November 3, 1981 to C22.2 NO 3. 4 pp.

—Symbols
CNS C1019-82. Symbols of Electric Networks (Jul)(403).

—Telecommunication Systems
BSI BS 6328: Part 1-85. 1985 Amd 2 Apparatus for Connection to Private Circuits Run by Certain Public Telecommunication Operators Part 1: Apparatus for Connection to Speechband Circuits (AMD 5977) December 30, 1988. 39 pp.
BSI BS 6328: Part 2-85. 1985 Amd 1 Apparatus for Connection to Private Circuits Run by Certain Public Telecommunication Operators Part 2: Apparatus for Connection to Baseband Circuits. 23 pp.
BSI BS 6328: Part 3-83. 1983 Amd 1 Apparatus for Connection to Private Circuits Run by Certain Public Telecommunication Operators Part 3: Apparatus for Connection to Direct Current Circuits. 18 pp.
BSI BS 6328: Part 7-93. 1993 Apparatus for Connection to Private Circuits Run by Certain Public Telecommunication Operators Part 7: Specification for Apparatus for Connection to Digital Circuits According to CCITT Recommendation X.21 bis (S). 12 pp.
BSI BS 6328: Part 7-90. 1990 Apparatus for Connection to Private Circuits Run by Certain Public Tele-communication Operators Part 7: Apparatus for Connection to Digital Circuits According to CCITT Recommendation X.21 bis. 13 pp.
BSI BS 6328: Sec 8.1-93. 1993 Apparatus for Connection to Private Circuits Run by Certain Public Telecommunication Operators Part 8: Specification for Apparatus for Connection to Digital Circuits with Interfaces According to CCITT G-Series Recommendations. 29 pp.
BSI BS 6328: Sec 8.1-90. 1990 Apparatus for Connection to Private Circuits Run by Certain Public Tele-communication Operators Part 8: Apparatus for Connection to Digital Circuits with Interfaces According to CCITT G-Series Recm. Sec. 8.1: G.703 Services at 2048 kbit/s and 8448 kbit/s. 30 pp.

Circuits *(Cont.)*

—Telecommunication Systems *(Cont.)*
BSI BS 6328: Sec 8.2-93. 1993 Apparatus for Connection to Private Circuits Run by Certain Public Tele-communication Operators Part 8: Apparatus for Connection to Digital Circuits with Interfaces According to CCITT G-Series Recm. Sec 8.2: G. 703 64 Kbit/S Serv. Using Codirectional Opt.. 18 pp.
BSI BS 6328: Sec 8.2-90. 1990 Apparatus for Connection to Private Circuits Run by Certain Public Tele-Communication Operators Part 8: Apparatus for Connectionto Digital Circuits with Interfaces According to CCITT G-Series Recm. Sec 8.2: G. 703 64 Kbit/S Serv. Using Codirectional Opt.. 20 pp.

—Telephone Networks
CCITT RECMN H.11-89. Characteristics of Circuits in the Switched Telephone Network—Line Transmission of Non-Telephone Signals—Transmission of Sound-Programme and Television Signals (Study Group XV) 1 pp. 1 p.

Circular Connectors

Use For: Circular Multipin Connectors; Concentric Connectors; Cylindrical Connectors *See Also:* Audio Connectors; Coaxial Connectors; Connectors
CECC CECC MUAHAG Vol 3A IS 4-91. Preferred Products List; Connectors; L. F. (En, Fr, Ge). 46 pp.
CECC EN 175 200-91. Sectional Specification: Circular Connectors (Supersedes CECC 75 200 Issue 1: 1985). 57 pp.
CNS C7083-85. Cylindrical Connectors for Electronic Equipment (Jun)(4854).
IEC 149 Pt 2-65. Sockets and Accessories for Electronic Plug-in Devices Part 2: Specification Sheets for Sockets and Dimensions of Wiring Jigs and Pin Straighteners First Edition; (Supplement A-H:1972) (Supplement J-L:1976). 257 pp.
JIS C 5432-76. Cylindrical Connector for Electronic Equipment (R 1985). 17 pp.
MOD UK DSTAN 59-35: Pt 1:Sec 8-70. Plugs and Sockets, Electrical Pattern 121 for d.c. and Low Frequency Applications Issue 1. 75 pp.
MOD UK DSTAN 59-35: Pt 3:Sec 5-01. Connectors, Electrical Part 3: Connectors for d.c. and Low Frequency Applications (See Also EPIC Database) Section 5: Pattern 9522 F0038 List of Items Conforming to BS 9522 F0038 Issue 1; Amendment 1. 39 pp.
MOD UK DSTAN 59-35: Pt 3:Sec 7-01. Connectors, Electrical Part 3: Connectors for d.c. and Low Frequency Applications (See Also EPIC Database) Section 7: Pattern 9522 F0017 List of Items Conforming to BS 9522 F0017 Issue 1; Amendment 1. 70 pp.
MOD UK DSTAN 59-56: Part 1-01. Plugs and Sockets, Electrical, Pattern 602 Part 1: Standard and Maintenance Ranges Issue 1; Amendment 1. 40 pp.

—Accessories—Quality Assurance
MOD UK DSTAN 59-35: Part 6-93. Connectors Electrical Part 6: Circular Connector Accessories General Requirements and Test Methods for Capability Approval Issue 1. 41 pp.

—Aerospace
CEN PREN 3372-008-89. Circular Electrical Connectors Medium and High Density, Scoop-Proof with Bayonet Coupling, 175 Degrees C and 200 Degrees C (150 Degrees in the Case of Solder Contacts). 7 pp.

—Aerospace—Accessories
CEN PREN 3660-001-91. Cable Outlet Accessories for Circular and Rectangular Electrical and Optical Connectors Part 001—Technical Specification. 22 pp.
CEN PREN 3660-001-93. Aerospace Series Cable Outlet Accessories for Circular and Rectangular Electrical and Optical Connectors Part 001—Technical Specification. 12 pp.

—Aerospace—Accessories—Indexes (Documentation)
CEN PREN 3660-002-93. Aerospace Series Cable Outlet Accessories for Circular and Rectangular Electrical and Optical Connectors Part 002—Index of Product Standards. 3 pp.

—Aerospace—Plugs
CEN PREN 2997 (Part 008)-90. Connectors, Electrical, Circular, Coupled by Threaded Ring, Operating Temperatures 175 Degrees Celsius Continuous, 200 Degrees Celsius Continuous, 260 Degrees Celsius Peak and Fire Resistant Part 008-Plug Product Standard. 6 pp.
CEN PREN 2997 (Part 009)-90. Connectors, Electrical, Circular, Coupled by Threaded Ring, Operating Temperatures 175 Degrees Celsius Continuous, 200 Degrees Celsius Continuous, 260 Degrees Celsius Peak and Fire Resistant Part 009-Flight Flank for Receptacle Product Standard. 5 pp.

Circular Connectors (Cont.)

—Aerospace—Quality Assurance
CEN PREN 2997 (Part 002)-90. Connectors, Electrical, Circular, Coupled by Threaded Ring, Operating Temperatures 175 Degrees Celsius Continuous, 200 Degrees Celsius Continuous, 260 Degrees Celsius Peak and Fire Resistant Part 002- Specification of Performance and Contact Arrangements. 14 pp.

—Aerospace—Quick Disconnect
CEN PREN 3372-001-89. Circular Electrical Connectors Medium and High Density, Scoop-Proof with Bayonet Coupling, 175 Degrees C and 200 Degrees C Technical Specification. 71 pp.

CEN PREN 3372-002-89. Circular Electrical Connectors Medium and High Density, Scoop-Proof with Bayonet Coupling 175 Degrees C and 200 Degrees C Data for Selection and Contact Arrangement. 12 pp.

CEN PREN 3372-003-89. Circular Electrical Connectors Medium and High Density, Scoop-Proof with Bayonet Coupling, 175 Degrees C and 200 Degrees C. 7 pp.

CEN PREN 3372-004-89. Circular Electrical Connectors Medium and High Density, Scoop-Proof with Bayonet Coupling, 175 Degrees C and 200 Degrees C Type B, Fixed Connector, Single-Hole Assembly Individual Sheet. 7 pp.

CEN PREN 3372-005-89. Circular Electrical Connectors Medium and High Density, Scoop-Proof with Bayonet Coupling, 175 Degrees C Type C, Fixed Connector, Hermetic, Solder Flange, Male Contacts, Solder Connection Individual Sheet. 7 pp.

CEN PREN 3372-006-89. Circular Electrical Connectors Medium and High Density, Scoop-Proof with Bayonet Coupling, 175 Degrees C and 200 Degrees C, Type D, Free Connector with Earth Spring Ring Individual Sheet. 6 pp.

CEN PREN 3372-007-89. Circular Electrical Connectors Medium and High Density, Scoop Proof with Bayonet Coupling Type H, Fixed Connector, Hermetic, Single-Hole Mounting, Male contact, Solder Connection Individual Sheet. 7 pp.

CEN PREN 3372-009-89. Circular Electrical Connectors Medium and High Density, Scoop-Proof with Bayonet Coupling, 175 Degrees C and 200 Degrees C Type N Fixed Connector, Data Bus, Earth Connection Single-Hole Mounting Individual Sheet. 7 pp.

CEN PREN 3372-010-89. Circular Electrical Connectors Medium and High Density, Scoop-Proof with Bayonet Coupling, 175 Degrees C and 200 Degrees C Type V1, V2, Protective Cover for Fixed Connector Individual Sheet. 7 pp.

CEN PREN 3372-011-89. Circular Electrical Connectors Medium and High Density, Scoop-Proof with Bayonet Coupling, 175 Degrees C and 200 Degrees C Type V3, Protective Cover for the Connector Individual Sheet. 6 pp.

CEN PREN 3372-012-89. Circular Electrical Connectors Medium and High Density, Scoop-Proof with Bayonet Coupling, 175 Degrees C and 200 Degrees C Type W, Dummy Individual Sheet. 7 pp.

CEN PREN 3372-013-89. Circular Electrical Connectors Medium and High Density, Scoop-Proof with Bayonet Coupling, 175 Degrees C and 200 Degrees C Type P, Fixed Connector, Square Flange Without Rear Accessory. 7 pp.

—Aircraft
SBAC TS 300 ISSUE 2. SBAC Manufacturing Specification TS 300 for ESC Series of Circular Electrical Connectors.

—Aircraft—Accessories
SBAC TS 302 ISSUE 2. SBAC Manufacturing Specification TS 302 for ESC Series of Circular Electrical Connector Accessories.

—Crimp—Preferred Products List
CECC CECC MUAHAG Vol 3A IS 4-91. Preferred Products List; Connectors; L. F. (En, Fr, Ge). 46 pp.

—Data Buses—Crimp—Preferred Products List
CECC CECC MUAHAG Vol 3A IS 4-91. Preferred Products List; Connectors; L. F. (En, Fr, Ge). 46 pp.

—Data Buses—Quick Disconnect—Crimp—Preferred Products List
CECC CECC MUAHAG Vol 3A IS 4-91. Preferred Products List; Connectors; L. F. (En, Fr, Ge). 46 pp.

—Electronic Equipment
JIS C 5432-76. Cylindrical Connector for Electronic Equipment (R 1985). 17 pp.

—Fire Resistant—Preferred Products List
CECC CECC MUAHAG Vol 3A IS 4-91. Preferred Products List; Connectors; L. F. (En, Fr, Ge). 46 pp.

—Hermaphroditic—Preferred Products List
CECC CECC MUAHAG Vol 3A IS 4-91. Preferred Products List; Connectors; L. F. (En, Fr, Ge). 46 pp.

—Plugs
MOD UK DSTAN 59-4: Part 1-01. Plugs and Sockets, Electrical Part 1: Breeze Type Issue 2; Amendment 2. 18 pp.

MOD UK DSTAN 59-4: Part 2-67. Plugs and Sockets, Electrical Part 2: Miscellaneous, Non-Insulated Types Issue 1. 7 pp.

MOD UK DSTAN 59-4: Part 3-67. Plugs and Sockets, Electrical Part 3: All-Insulated Sealed Types Issue 1. 20 pp.

MOD UK DSTAN 59-4: Part 4-01. Plugs and Sockets, Electrical Part 4: Pattern 104 and Intermateable Types Issue 1; Amendment 2 (Parts of This Document Have Been Superseded). 245 pp.

MOD UK DSTAN 59-4: Part 6-71. Plugs and Sockets, Electrical Part 6: Plugs and Sockets for Power Connection Issue 2. 18 pp.

MOD UK DSTAN 59-35: Pt 1:Sec 1-01. Plugs and Sockets, Electrical Pattern 103 Issue 1; Amendment 1. 30 pp.

MOD UK DSTAN 59-35: Pt 1:Sec 2-01. Plugs and Sockets, Electrical Pattern 104, Sealed, with Soldered Terminations and Threaded Coupling, for d.c. and Low Frequency Applications Issue 1; Amendment 1. 105 pp.

MOD UK DSTAN 59-35: Pt 1:Sec 2-02. Plugs and Sockets, Electrical Pattern 104, Sealed, with Soldered Terminations and Threaded Coupling, for d.c. and Low Frequency Applications Issue 1; Amendment 2. 108 pp.

MOD UK DSTAN 59-35: Pt 1:Sec 8B-01. Plugs and Sockets, Electrical Pattern 121B for d.c. and Low Frequency Applications Issue 1; Amendment 1. 89 pp.

MOD UK DSTAN 59-35: Pt 3:Sec 5-01. Connectors, Electrical Part 3: Connectors for d.c. and Low Frequency Applications (See Also EPIC Database) Section 5: Pattern 9522 F0038 List of Items Conforming to BS 9522 F0038 Issue 1; Amendment 1. 39 pp.

MOD UK DSTAN 59-35: Pt 3:Sec 7-01. Connectors, Electrical Part 3: Connectors for d.c. and Low Frequency Applications (See Also EPIC Database) Section 7: Pattern 9522 F0017 List of Items Conforming to BS 9522 F0017 Issue 1; Amendment 1. 70 pp.

MOD UK DSTAN 59-35: Part 5-01. Plugs and Sockets, Electrical, Pattern 608, Sealed, with Soldered Terminations: Non-Sealed, with Crimped Terminations: Coarse Threaded Coupling, for d.c. and Low Frequency Applications Issue 2; Amendment 6. 108 pp.

MOD UK DSTAN 59-56: Part 2-01. Plugs and Sockets, Electrical, Pattern 602 Part 2: Items No Longer to Be Provisioned Issue 1; Amendment 1. 49 pp.

—Plugs—Batteries
MOD UK DSTAN 59-35: Pt 1:Sec 7-70. Plugs and Sockets, Electrical Pattern 112 for Use with Primary Batteries Issue 1. 17 pp.

—Preferred Products List
CECC CECC MUAHAG Vol 3A IS 4-91. Preferred Products List; Connectors; L. F. (En, Fr, Ge). 46 pp.

—Quality Assurance
BSI BS 9522-90. (OBSOLESCENT) 1990 Circular Electrical Connectors of Assessed Quality for d.c. and Low Frequency Applications Full Basic and Full Plus Airframe Assessment Levels. 58 pp.

BSI BS CECC 75200-86. (WITHDRAWN) 1986 Circular Connectors: Sectional Specification (Renumbered as BS EN 175200: 1992). 59 pp.

BSI BS CECC 75201-92. 1992 Blank Detail Specification Interim Example Detail Specification/Blank Detail Specification Circular Connectors for Frequencies Below 3 MHz. 4 pp.

BSI BS EN 175200-92. 1992 Amd 1 Sectional Specification: Circular Connectors (AMD 7483) January 15, 1993. 64 pp.

CENELEC EN 175 200-91. Sectional Specification: Circular Connectors. 62 pp.

—Quick Disconnect
IEC 130 Pt 7-71. Connectors for Frequencies Below 3 MHz Part 7: Circular Multipole Connectors with Bayonet or Push-Pull Coupling First Edition. 57 pp.

—Quick Disconnect—Crimp—Preferred Products List
CECC CECC MUAHAG Vol 3A IS 4-91. Preferred Products List; Connectors; L. F. (En, Fr, Ge). 46 pp.

—Quick Disconnect—Hermetically Sealed—Crimp
CECC CECC 75 201-001 ISSUE 1-91. BS CECC 75 201-001; Multi Contact Circular Electrical Connectors for Frequencies Below 3 MHz, Bayonet Coupling Rear Released Crimp Contacts, Barrier and Hermetically Sealed Styles (En). 94 pp.

—Quick Disconnect—Preferred Products List
CECC CECC MUAHAG Vol 3A IS 4-91. Preferred Products List; Connectors; L. F. (En, Fr, Ge). 46 pp.

—Quick Disconnect—Quality Assurance
BSI BS 9522 N0001-82. 1982 Amd 1 Detail Specification for MultiContact Circular Electrical Connectors. Bayonet Coupling with Front Release Rear Removable Crimp Contacts. Full Assessment (AMD 6821) September 30, 1991. 48 pp.

BSI BS 9522 N0003-90. 1990 Multi-Contact Circular Electrical Connectors Bayonet Coupling Non-Barrier Sealed, Environment Resistant with Rear Insertable, Rear Release, Rear Removable Crimp Contacts Also Barrier Sealed with Non-Removable Solder Contact Styles. 142 pp.

—Receptacles
MOD UK DSTAN 59-35: Pt 3:Sec 5-01. Connectors, Electrical Part 3: Connectors for d.c. and Low Frequency Applications (See Also EPIC Database) Section 5: Pattern 9522 F0038 List of Items Conforming to BS 9522 F0038 Issue 1; Amendment 1. 39 pp.

MOD UK DSTAN 59-35: Pt 3:Sec 7-01. Connectors, Electrical Part 3: Connectors for d.c. and Low Frequency Applications (See Also EPIC Database) Section 7: Pattern 9522 F0017 List of Items Conforming to BS 9522 F0017 Issue 1; Amendment 1. 70 pp.

MOD UK DSTAN 59-35: Part 5-01. Plugs and Sockets, Electrical, Pattern 608, Sealed, with Soldered Terminations: Non-Sealed, with Crimped Terminations: Coarse Threaded Coupling, for d.c. and Low Frequency Applications Issue 2; Amendment 6. 108 pp.

—Sockets
MOD UK DSTAN 59-4: Part 1-01. Plugs and Sockets, Electrical Part 1: Breeze Type Issue 2; Amendment 2. 18 pp.

MOD UK DSTAN 59-4: Part 2-67. Plugs and Sockets, Electrical Part 2: Miscellaneous, Non-Insulated Types Issue 1. 7 pp.

MOD UK DSTAN 59-4: Part 3-67. Plugs and Sockets, Electrical Part 3: All-Insulated Sealed Types Issue 1. 20 pp.

MOD UK DSTAN 59-4: Part 4-01. Plugs and Sockets, Electrical Part 4: Pattern 104 and Intermateable Types Issue 1; Amendment 2 (Parts of This Document Have Been Superseded). 245 pp.

MOD UK DSTAN 59-4: Part 6-71. Plugs and Sockets, Electrical Part 6: Plugs and Sockets for Power Connection Issue 2. 18 pp.

MOD UK DSTAN 59-35: Pt 1:Sec 1-01. Plugs and Sockets, Electrical Pattern 103 Issue 1; Amendment 1. 30 pp.

MOD UK DSTAN 59-35: Pt 1:Sec 2-01. Plugs and Sockets, Electrical Pattern 104, Sealed, with Soldered Terminations and Threaded Coupling, for d.c. and Low Frequency Applications Issue 1; Amendment 1. 105 pp.

MOD UK DSTAN 59-35: Pt 1:Sec 2-02. Plugs and Sockets, Electrical Pattern 104, Sealed, with Soldered Terminations and Threaded Coupling, for d.c. and Low Frequency Applications Issue 1; Amendment 2. 108 pp.

MOD UK DSTAN 59-35: Pt 1:Sec 8B-01. Plugs and Sockets, Electrical Pattern 121B for d.c. and Low Frequency Applications Issue 1; Amendment 1. 89 pp.

MOD UK DSTAN 59-56: Part 2-01. Plugs and Sockets, Electrical, Pattern 602 Part 2: Items No Longer to Be Provisioned Issue 1; Amendment 1. 49 pp.

—Sockets—Batteries
MOD UK DSTAN 59-35: Pt 1:Sec 7-70. Plugs and Sockets, Electrical Pattern 112 for Use with Primary Batteries Issue 1. 17 pp.

—Threaded
IEC 130 Pt 4-66. Connectors for Frequencies Below 3 MHz Part 4: Circular Multipole Connectors with Threaded Coupling First Edition; (Supplement A-1970). 51 pp.

Circular Connectors (Cont.)
—Threaded—Aerospace
CEN PREN 2997 (Part 001)-90. Connectors, Electrical, Circular, Coupled by Threaded Ring, Operating Temperatures 175 Degrees Celsius Continuous, 200 Degrees Celsius Continuous, 260 Degrees Celsius Peak and Fire Resistant Part 001- Technical Specification. 58 pp.

CEN PREN 2997 (Part 002)-90. Connectors, Electrical, Circular, Coupled by Threaded Ring, Operating Temperatures 175 Degrees Celsius Continuous, 200 Degrees Celsius Continuous, 260 Degrees Celsius Peak and Fire Resistant Part 002- Specification of Performance and Contact Arrangements. 14 pp.

CEN PREN 2997 (Part 003)-90. Connectors, Electrical, Circular, Coupled by Threaded Ring, Operating Temperatures 175 Degrees Celsius Continuous, 200 Degrees Celsius Continuous, 260 Degrees Celsius Peak and Fire Resistant Part 003- Square Flange Receptacle Product Standard. 6 pp.

CEN PREN 2997 (Part 004)-90. Connectors, Electrical, Circular, Coupled by Threaded Ring, Operating Temperatures 175 Degrees Celsius Continuous, 200 Degrees Celsius Continuous, 260 Degrees Celsius Peak and Fire Resistant Part 004- Jam-Nut Mounted Receptacle Product Standard. 6 pp.

CEN PREN 2997 (Part 005)-90. Connectors, Electrical, Circular, Coupled by Threaded Ring, Operating Temperatures 175 Degrees Celsius Continuous, 200 Degrees Celsius Continuous, 260 Degrees Celsius Peak and Fire Resistant Part 005- Hermetic Square Flange Receptacle Product Standard. 6 pp.

CEN PREN 2997 (Part 006)-90. Connectors, Electrical, Circular, Coupled by Threaded Ring, Operating Temperatures 175 Degrees C Continuous, 200 Degrees C Continuous, 260 Degrees C Peak and Fire Resistant Part 006—Hermetic Jam-Nut Mounted Receptacle Product Standard. 6 pp.

CEN PREN 2997 (Part 007)-90. Connectors, Electrical, Circular, Coupled by Threaded Ring, Operating Temperatures 175 Degrees Celsius Continuous, 200 Degrees Celsius Continuous, 260 Degrees Celsius Peak and Fire Resistant Part 007- Hermetic Receptacle with Round Flange Attached by Soldering or Brazing. 6 pp.

CEN PREN 2997 (Part 010)-90. Connectors, Electrical, Circular, Coupled by Threaded Ring, Operating Temperatures 175 Degrees C Continuous, 200 Degrees C Continuous, 260 Degrees C Peak and Fire Resistant Part 010—Flight Blank for Plug Product Standard. 6 pp.

CEN PREN 2997 (Part 011)-90. Connectors, Electrical, Circular, Coupled by Threaded Ring, Operating Temperatures 175 Degrees C Continuous, 200 Degrees C Continous, 260 Degrees C Peak and Fire Resistant Part 011-Dummy Receptacle Product Standard. 7 pp.

Circular Dichroism Value
Use For: CD Value
—Citrus Oils—Ultraviolet Spectrophotometry
ISO 4735-81. Oils of Citrus—Determination of CD Value by Ultraviolet Spectrophotometric Analysis First Edition. 5 pp.

Circular Knives
See Also: Knives
—Electric—Hand Held—Safety
BSI BS 2769: Sec 2.5-91. 1991 Hand-Held Electric Motor-Operated Tools Part 2: Particular Requirements Section 2.5: Specification for Circular Saws and Circular Knives. 16 pp.

BSI BS 2769: Sec 2.5-01. 1991 Amd 1 Hand-Held Electric Motor-Operated Tools Part 2: Particular Requirements Section 2.5: Specification for Circular Saws and Circular Knives (AMD 6920) April 1, 1992. 17 pp.

CENELEC HD 400.2 S2-79. Hand-Held Motor Operated Tools Part II: Particular Specifications Sections A to G. 34 pp.

CENELEC HD 400.2 S2-89. Hand-Held Motor Operated Tools—Part II: Particular Specifications Section E: Circular Saws and Circular Knives. 29 pp.

CENELEC HD 400.2E/A1-91. AMD 1 Hand-Held Motor Operated Tools Part II: Particular Specifications Section E: Circular Saws and Circular Knives. 3 pp.

CENELEC HD 400.2E S2 /A1-91. AMD 1 Hand-Held Motor Operated Tools Part II: Particular Specifications Section E: Circular Saws and Circular Knives. 4 pp.

—Safety
IEC 745 Pt 2-5-93. Safety of Hand-Held Motor-Operated Electric Tools Part 2: Particular Requirements for Circular Saws and Circular Knives Second Edition. 44 pp.

Circular Letters
Use: Newsletters

Circular Multipin Connectors
Use: Circular Connectors

Circular Saw Blades
See Also: Circular Saw Blades
—Metal
CNS B3214-72. Circular Saw Blade (for Metal Cutting) (Jun)(3409).
DIN ENGL 1837-70. Circular Metal Slitting Saw Blades; Fine-Toothed (Aug). 2 pp.
DIN ENGL 1838-70. Circular Metal Slitting Saw Blades; Coarse-Toothed (Aug). 2 pp.
DIN ENGL 1840-70. Circular Metal Slitting Saw Blades; Tooth Forms; Side Clearance; Manufacturing Tolerances (Aug). 3 pp.

—Safety
CEN PREN 847-1-92. Tools for Woodworking—Safety Requirements—Part 1: Milling Tools, Circular Saw Blades. 37 pp.

—Woodworking
CNS B2696-82. Saw Blades for Woodworking; Technical Conditions of Delivery for Circular Saw Blades of Tool Steel (Aug)(9225).
CNS B3248-76. Woodworking Circular Saw Blades (Sep)(3980).
DIN ENGL 5134 Pt 4-75. Saw Blades for Woodworking; Technical Conditions of Delivery for Circular Saw Blades of Tool Steel (Sept). 2 pp.
DIN ENGL 8809-75. Circular Saw Blades of Tool Steel for Woodworking (Aug). 3 pp.
ISO 2935-74. Circular Saw Blades for Woodworking—Dimensions First Edition. 4 pp.

Circular Saws
See Also: Cutting Tools; Diamond Saws; Machine Tools; Radial Saws; Saws; Scroll Saws; Woodworking Equipment

BSI BS 2064: Part 2-76. 1976 Diamond Abrasive Products Part 2: Diamond Abrasive Circular Saws and Frame Saws. 12 pp.
CNS B1299-83. Nominal Dimension of Circular Saw Machines (Mar)(10051).
EC 89/392/EEC-89. Council Directive on the Approximation of the Laws of the Member States Relating to Machinery (NEW APP) (for Rel Stds See SM 1) (for Rel Stds See TEST/89/392, TEST/89/392 A1, TEST/89/392 A2). 24 pp.
EC TEST/89/392-89. Commission Communication in the Framework of the Implementation of Council Directive 89/392/EEC of 14 June 1989 in Relation to Machinery, as Amended by Directive 91/368/EEC of 20 June 1991. 3 pp.
EC TEST/89/392 A1-89. AMD 1 Commission Communication in the Framework of the Implementation of Council Directive 89/392/EEC of 14 June 1989 in Relation to Machinery, as Amended by Council Directive 91/368/EEC of 20 June 1991. 1 p.
EC TEST/89/392 A2-89. AMD 2 Commission Communication in the Framework of the Implementation of Council Directive 89/392/EEC of 14 June 1989 in Relation to Machinery, as Amended by Council Directive 91/368/EEC of 20 June 1991. 1 p.
EC TEST/89/398-89. List of Competent Authorities of the Member States Within the Meaning of Article 9 of Council Directive 89/ 398/EEC on the Approximation of the Laws of the Member States Relating to Foodstuffs Intended for Foods for Particular Nutritional Uses. 2 pp.
EC 91/368/EEC-91. Council Directive Amending Directive 89/392/EEC on the Approximation of the Laws of the Member States Relating to Machinery (NEW APP) (for Rel Stds See SM 1). 18 pp.
EC 93/68/EEC-93. Council Directive Amending Directives 87/404/EEC (Simple Pressure Vessels), 88/378/EEC (Safety of Toys), 89/106/EEC (Construction Products), 89/336/EEC (Electromagnetic Compatibility), 89/392/EEC (Machinery), 89/686/ EEC (Personal Protective. 22 pp.

—Electric—Hand Held—Performance Measurement
IEC 1176-93. Hand-Held Electric Mains Voltage Operated Circular Saws—Methods for Measuring the Performance First Edition. 31 pp.

—Electric—Portable
MOD UK DSTAN 51-19: Part 2-92. Hand Tools, Powered Part 2: Electric Issue 1. 31 pp.

—Electric—Portable—Safety
BSI BS 2769: Sec 2.5-91. 1991 Hand-Held Electric Motor-Operated Tools Part 2: Particular Requirements Section 2.5: Specification for Circular Saws and Circular Knives. 16 pp.

BSI BS 2769: Sec 2.5-01. 1991 Amd 1 Hand-Held Electric Motor-Operated Tools Part 2: Particular Requirements Section 2.5: Specification for Circular Saws and Circular Knives (AMD 6920) April 1, 1992. 17 pp.
CENELEC HD 400.2 S2-79. Hand-Held Motor Operated Tools Part II: Particular Specifications Sections A to G. 34 pp.
CENELEC HD 400.2 S2-89. Hand-Held Motor Operated Tools—Part II: Particular Specifications Section E: Circular Saws and Circular Knives. 29 pp.
CENELEC HD 400.2E/A1-91. AMD 1 Hand-Held Motor Operated Tools Part II: Particular Specifications Section E: Circular Saws and Circular Knives. 3 pp.
CENELEC HD 400.2E S2 /A1-91. AMD 1 Hand-Held Motor Operated Tools Part II: Particular Specifications Section E: Circular Saws and Circular Knives. 4 pp.

—Metal Cutting—Drives
BSI BS 387-79. 1979 Interchangeability Dimensions of the Drive of Solid and Segmental Circular Saws for Cold Cutting of Metals: Saw Diameter Range 224mm to 2240mm. 4 pp.
ISO 2924-73. Solid and Segmental Circular Saws for Cold Cutting of Metals—Interchangeability Dimensions of the Drive—Saw Diameter Range 224 to 2 240 mm First Edition. 4 pp.

—Safety
IEC 745 Pt 2-5-93. Safety of Hand-Held Motor-Operated Electric Tools Part 2: Particular Requirements for Circular Saws and Circular Knives Second Edition. 44 pp.
IEC 1029 Pt 2-1-93. Safety of Transportable Motor-Operated Electric Tools Part 2: Particular Requirements for Circular Saws First Edition. 32 pp.

—Woodworking
BSI BS 411-69. 1969 Circular Saws for Woodworking and Their Attachment. 27 pp.
CGSB 45-GP-10C-72. Saw, Circular, Electric, Portable, Heavy Duty, Standard for. 9 pp.
CNS B7270-84. Test Code for Accuracy of Circular Sawing Machines (Jul)(10956).
CNS B7271-84. Test Code for Performance of Circular Sawing Machines (Jul)(10957).
CNS C3169-83. Method of Test for Portable Electric Circular Saws (Jan)(9812).
CNS C4383-83. Portable Electric Circular Saws (Jan)(9811).
JIS B 4802-73. Wood Circular Saws.
JIS B 4805-89. Circular Saws with Cemented Carbide Tips. 11 pp.
JIS B 6508-90. Test Methods for Performance and Accuracy of Circular Saw Machines. 17 pp.
JIS B 6596-91. Double Sizers—Test and Inspection Methods. 21 pp.
JIS C 9626-92. Portable Electric Circular Saws.

—Woodworking—Acceptance Testing
BSI BS 4361: Part 18-89. 1989 Woodworking Machines Part 18: Single Blade Stroke Circular Sawing Machines for Lengthwise Cutting of Solid Woods and Panels. 18 pp.
BSI BS 4361: Part 19-89. 1989 Woodworking Machines Part 19: Double Edging Precision Circular Sawing Machines. 14 pp.
BSI BS 4361: Part 20-90. 1990 Woodworking Machines Part 20: Single Blade Circular Sawing Machines with Travelling Table. 15 pp.
ISO 7958-87. Woodworking Machines—Single Blade Stroke Circular Sawing Machines for Lengthwise Cutting of Solid Woods and Panels—Nomenclature and Acceptance Conditions First Edition. 17 pp.
ISO 7959-87. Woodworking Machines—Double Edging Precision Circular Sawing Machines—Nomenclature and Acceptance Conditions First Edition. 13 pp.
ISO 7983-88. Woodworking Machines—Single Blade Circular Sawing Machines with Travelling Table—Nomenclature and Acceptance Conditions First Edition. 14 pp.

—Woodworking—Benches
BSI BS 4361: Part 5-88. 1988 Woodworking Machines Part 5: Single Blade Circular Saw Benches with or Without Travelling Table. 16 pp.
ISO 7008-83. Woodworking Machines—Single Blade Circular Saw Benches with or Without Travelling Table—Nomenclature and Acceptance Conditions First Edition. 15 pp.

—Woodworking—Glossaries
BSI BS 4361: Part 18-89. 1989 Woodworking Machines Part 18: Single Blade Stroke Circular Sawing Machines for Lengthwise Cutting of Solid Woods and Panels. 18 pp.
BSI BS 4361: Part 19-89. 1989 Woodworking Machines Part 19: Double Edging Precision Circular Sawing Machines. 14 pp.

Circular Saws (Cont.)
—Woodworking—Glossaries (Cont.)
BSI BS 4361: Part 20-90. 1990 Woodworking Machines Part 20: Single Blade Circular Sawing Machines with Travelling Table. 15 pp.
BSI BS 4361: Part 39-90. 1990 Woodworking Machines Part 39: Nomenclature for Circular Sawing Machines for Building Sites. 10 pp.
ISO 7958-87. Woodworking Machines—Single Blade Stroke Circular Sawing Machines for Lengthwise Cutting of Solid Woods and Panels—Nomenclature and Acceptance Conditions First Edition. 17 pp.
ISO 7959-87. Woodworking Machines—Double Edging Precision Circular Sawing Machines—Nomenclature and Acceptance Conditions First Edition. 13 pp.
ISO 7983-88. Woodworking Machines—Single Blade Circular Sawing Machines with Travelling Table—Nomenclature and Acceptance Conditions First Edition. 14 pp.
ISO 9616-89. Woodworking Machines—Circular Sawing Machines for Building Sites—Nomenclature First Edition. 7 pp.

—Woodworking—Pneumatic
CGSB 45-GP-9-54. Specification for Saws; Circular, Pneumatic, Portable. 8 pp.

—Woodworking—Safety
BSI BS 6854: Part 2-88. 1988 Code of Practice for Safeguarding Woodworking Machines Part 2: Circular Sawing Machines (Sawing Machines with Rotating Tools). 42 pp.

—Woodworking—Spindles
ISO 7006-81. Woodworking Machines—Diameters of Spindles for Receiving Circular Sawblades First Edition. 3 pp.

Circular Screwing Dies
Use: Threading Dies

Circular Waveguides
See Also: Waveguides
CENELEC HD 123.4-75. Hollow Metallic Waveguides Part 4: Relevant Specifications for Circular Waveguides. 2 pp.
IEC 153 Pt 4-73. Hollow Metallic Waveguides Part 4: Relevant Specifications for Circular Waveguides Second Edition. 25 pp.
MOD UK DSTAN 59-64: Part 4-75. Waveguide, Tubing, Rigid Part 4: Waveguide—Circular, Standard Wall Thickness, Full Assessment Issue 1. 5 pp.

—Flanges
CENELEC HD 129.4-77. Flanges for Waveguides-Part 4: Relevant Specifications for Flanges for Circular Waveguides. 2 pp.
CENELEC HD 129.7-77. Flanges for Waveguides Part 7: Relevant Specifications for Flanges for Circular Waveguides. 2 pp.
IEC 154 Pt 4-69. Flanges for Waveguides Part 4: Relevant Specifications for Flanges for Circular Waveguides First Edition. 25 pp.

—Quality Assurance
BSI BS 9220 N004-71. 1971 Detail Specification for Circular Waveguide Tubing of Assessed Quality. General Application Category. 5 pp.

Circulating Fans
Use: Fans

Circulating Pumps
Use For: Circulators (Pumps) *See Also:* Pumps; Purging Equipment
BSI BS 1394: Part 2-87. 1987 Stationary Circulation Pumps for Heating and Hot Water Service Systems Part 2: Physical and Performance Requirements. 12 pp.

—Boilers—Safety
BSI BS 5258: Part 1-86. 1986 Safety of Domestic Gas Appliances Part 1: Central Heating Boilers and Circulators. 43 pp.
BSI BS 5258: Pt 1: Supp 1-83. 1983 Safety of Domestic Gas Appliances Part 1: Central Heating Boilers and Circulators Supplement 1: Fan-Powered Appliances. 13 pp.
BSI DD 189-90. 1990 Safety of Condensing Boilers (2nd and 3rd Family Gases). 9 pp.

—Boilers—Thermal Efficiency
BSI BS 6332: Part 1-88. 1988 Thermal Performance of Domestic Gas Appliances Part 1: Thermal Performance of Central Heating Boilers and Circulators. 12 pp.

—Construction
CEN PREN 1151-93. Pumps—Rotodynamic Pumps—Circulation Pumps Having an Electrical Effect not Exceeding 200 W for Heating Installations and Domestic Hot Water Installations—Requirements, Testing, Marking. 15 pp.

Circulating Pumps (Cont.)
—Safety
BSI BS 1394: Part 1-87. (WITHDRAWN) 1987 Stationary Circulation Pumps for Heating and Hot Water Service Systems Part 1: Safety Requirements (Superseded by BS EN 60335-2-51: 1991). 55 pp.
DIN VDE 0700 Pt 237-84. Safety of Household and Similar Electrical Appliances; Stationary Circulation Pumps for Heating and Service Water Plants (VDE Specifications) (Feb) (Replaces DIN 57730 Part 2 ZL/VDE 0730 Part 2 ZL/05.79). 21 pp.

Circulators (Laboratory)
See Also: Laboratory Equipment
—Liquid—Constant Temperature—Safety
DIN ENGL 12879 Pt 1-79. Electrical Laboratory Equipment; Constant-Temperature Liquid Circulators; General and Safety Requirements, Testing (May). 6 pp.

Circulators, Microwave
Use: Microwave Circulators

Circulators (Pumps)
Use: Circulating Pumps

Circulators, Waveguide
Use: Microwave Circulators—Waveguide

CISPR
Use For: International Special Committee on Radio Interference *See Also:* Radio Frequency Interference
IEC CISPR 10-92. Organization, Rules and Procedures of the CISPR Fourth Edition. 50 pp.

—Cooperation—CCIR
CCIR OPINION 2-2-90. Cooperation with the International Special Committee on Radio Interference—Volume I—Spectrum Utilization and Monitoring. 1 p.

Cisterns
See Also: Tanks (Containers)
—Boilers
JIS A 4006-76. Cisterns for Hot-Water Boiler.

—Valves
BSI BS 1212: Part 4-91. 1991 Float Operated Valves Part 4: Specification for Compact Type Float Operated Valves for WC Flushing Cisterns (Including Floats). 20 pp.

—Water Storage
BSI BS 7491: Part 1-01. 1991 Amd 1 Glass Fibre Reinforced Plastics Cisterns for Cold Water Storage Part 1: Specification for One-Piece Cisterns of Capacity up to 500 L (AMD 7382) December 15, 1992. 21 pp.
BSI BS 7491: Part 1-91. 1991 Glass Fibre Reinforced Plastics Cisterns for Cold Water Storage Part 1: Specification for One-Piece Cisterns of Capacity up to 500 L. 19 pp.
BSI BS 7491: Part 2-92. 1992 Glass Fibre Reinforced Plastics Cisterns for Cold Water Storage Part 2: Specification for One-Piece Cisterns of Nominal Capacity from 500 1 to 25 000 1. 17 pp.

—Water Supply
BSI BS 4213-91. 1991 Cold Water Storage and Combined Feed and Expansion Cisterns (Polyolefin or Olefin Copolymer) up to 500 L Capacity Used for Domestic Purposes (R). 26 pp.
BSI BS 4213-01. 1991 Amd 1 Cold Water Storage and Combined Feed and Expansion Cisterns (Polyolefin or Olefin Copolymer) up to 500 L Capacity Used for Domestic Purposes (AMD 6797) November 29, 1991. 27 pp.
BSI BS 7181-89. 1989 Storage Cisterns up to 500L Actual Capacity for Water Supply for Domestic Purposes. 11 pp.

CIT/A1 Aircraft
See Also: Aircraft
—Antenna Positions
CAA. CIT/A1 (Approved Aerial Positions). 1 p.

Citations
Use: Bibliographic References

Cities
Use: Urban Areas

Citral Content Analysis
—Lemon Oil—Gas Chromatography
ISO 7611-85. Oils of Lemon and Petitgrain Citronnier, and Oil of Lime Obtained by a Mechanical Process—Determination of Citral (Neral + Geranial) Content—Gas Chromatographic Method on Capillary Columns First Edition. 7 pp.

Citral Content Analysis (Cont.)
—Lime Oil—Gas Chromatography
ISO 7611-85. Oils of Lemon and Petitgrain Citronnier, and Oil of Lime Obtained by a Mechanical Process—Determination of Citral (Neral + Geranial) Content—Gas Chromatographic Method on Capillary Columns First Edition. 7 pp.

—Petitgrain Oil—Gas Chromatography
ISO 7611-85. Oils of Lemon and Petitgrain Citronnier, and Oil of Lime Obtained by a Mechanical Process—Determination of Citral (Neral + Geranial) Content—Gas Chromatographic Method on Capillary Columns First Edition. 7 pp.

Citrates
Scope Note: Use a more specific term
See: Ammonium Citrate, Dibasic; Ammonium Ferric Citrate; Ferric Citrate; Potassium Citrate; Potassium Citrate Monohydrate; Sodium Citrate

Citric Acid
See Also: Polybasic Organic Acids
CNS K1069-62. Citric Acid (Industrial Grade) (Dec)(2076)(R 1973). 1 p.
CNS K6164-74. Method of Test for Citric Acid of Industrial Grade (Oct)(2077).
CNS K7149-66. Chemical Reagent (Citric Acid) (Mar)(1649).
JIS K 1355-54. Citric Acid.
JIS K 8283-86. Citric Acid Monohydrate.

—Acid Resistance Testing—Enamels
BSI BS 1344: Part 2-75. 1975 Methods of Testing Vitreous Enamel Finishes Part 2: Resistance to Citric Acid at Room Temperature. 8 pp.
BSI BS 1344: Part 8-84. 1984 Methods of Testing Vitreous Enamel Finishes Part 8: Resistance to Boiling Citric Acid Solution. 6 pp.
CNS R3096-80. Method of Test for Resistance of Porcelain Enameled Utensils to Boiling Acid (Dec)(6852). 5 pp.
ISO 2722-73. Vitreous and Porcelain Enamels—Determination of Resistance to Citric Acid at Room Temperature First Edition. 5 pp.
ISO 2742-83. Vitreous and Porcelain Enamels—Determination of Resistance to Boiling Citric Acid Second Edition. 4 pp.

Citric Acid Content Analysis
See Also: Isocitric Acid Content Analysis; Organic Acid Content Analysis
—Cheeses
BSI BS 770: Part 7-76. 1976 Chemical Analysis of Cheese Part 7: Determination of Citric Acid Content (Reference Method). 3 pp.
ISO 2963-74. Cheese and Processed Cheese Products—Determination of Citric Acid Content (Reference Method) First Edition; (Erratum—Dec 1981). 5 pp.

—Fruit Juices—Enzymatic Method
CNS N6225-91. Method of Test for Fruit and Vegetable Juices and Drinks-Determination of Citric Acid by Enzymatic Analysis (Jun)(12636). 2 pp.

—Vegetable Juices—Enzymatic Method
CNS N6225-91. Method of Test for Fruit and Vegetable Juices and Drinks-Determination of Citric Acid by Enzymatic Analysis (Jun)(12636). 2 pp.

Citronella Oil
See Also: Essential Oils
CNS K5008-68. Citronella Oil (Jan)(207) (R 1973). 1 p.
CNS K5098-80. Oil of Java Citronella (Sep)(6469).
CNS K5121-81. Oil of Ceylon Citronella (Nov)(8133).
ISO 3848-76. Oil of Java Citronella First Edition. 4 pp.
ISO 3849-81. Oil of Ceylon Citronella First Edition. 4 pp.

—Alcohol Content
CNS K6063-65. Method of Test for Citronella Oil (Aug)(817)(R 1971). 3 pp.

—Citronellal Content
CNS K6063-65. Method of Test for Citronella Oil (Aug)(817)(R 1971). 3 pp.

—Distillation Methods
CNS K6063-65. Method of Test for Citronella Oil (Aug)(817)(R 1971). 3 pp.

—Physical Properties
CNS K6063-65. Method of Test for Citronella Oil (Aug)(817)(R 1971). 3 pp.

Citronellal Content Analysis
—Citronella Oil
CNS K6063-65. Method of Test for Citronella Oil (Aug)(817)(R 1971). 3 pp.

Citrus Fruit Products
Use: Citrus Fruits

Citrus Fruits
Use For: Citrus Fruit Products *See Also:* Food; Fruits; Grapefruits; Lemons; Limes; Mandarin Oranges; Oranges

—Essential Oil Content
ISO 1955-82. Citrus Fruits and Derived Products—Determination of Essential Oils Content (Reference Method) First Edition. 5 pp.

—Storage
ISO 3631-78. Citrus Fruits—Guide to Storage First Edition. 11 pp.

Citrus Grandis
Use: Grapefruits

Citrus Oil
See Also: Essential Oils

—Circular Dichroism Value—Ultraviolet Spectrophotometry
ISO 4735-81. Oils of Citrus—Determination of CD Value by Ultraviolet Spectrophotometric Analysis First Edition. 5 pp.

Citrus Powders

—Beverages
CNS N5104-72. Citrus Essence Powder (Tentative) (Jun)(3438). 1 p.
CNS N5105-72. Citrus Juice Powder (Tentative) (Jun)(3439). 1 p.

Civil Aviation
Use: Air Transportation

Civil Aviation Authority
Use: CAA

Civil Engineering
Scope Note: For additional listings, use a more specific term *See Also:* Air Transportation; Architecture; Construction; Highways; Land Use Planning; Mapping; Mechanical Engineering; Military Engineering; Sanitary Engineering; Snow Loads; Stress Analysis; Structural Design; Structural Engineering; Structural Forms; Transportation; Water Supply Engineering

—Classification
BSI BS 1000: (624)-81. 1981 Universal Decimal Classification (UDC). English Full Edition (624): Civil and Structural Engineering in General. 32 pp.
SNZ NZS/BS 1000 (624)-81. Universal Decimal Classification Civil and Structural Engineering in General. 32 pp.

—Designations
DIN ENGL 1080 Pt 1-76. Terms, Symbols and Units Used in Civil Engineering; Principles (June). 16 pp.

—Earthworks
BSI BS 6031-81. 1981 Amd 1 Code of Practice for Earthworks. 90 pp.

—Fabrics
CNS A2183-85. Nonwoven Fabric Used in Civil Engineering (Apr)(11228).

—Glossaries
BSI BS 6100: Part 0-92. 1992 Building and Civil Engineering Terms Part 0: Introduction. 101 pp.
BSI BS 6100: Part 0-90. 1990 Building and Civil Engineering Terms Part 0: Introduction. 58 pp.
BSI BS 6100: Part 0-86. (WITHDRAWN) 1986 Glossary of Building and Civil Engineering Terms Part 0: Introduction. 34 pp.
BSI BS 6100: Sec 1.0-92. 1992 Building and Civil Engineering Terms Part 1: General and Miscellaneous Section 1.0: General. 40 pp.
BSI BS 6100: Sec 1.0-84. 1984 Glossary of Building and Civil Engineering Terms Part 1: General and Miscellaneous Section 1.0: General and Miscellaneous. 18 pp.
BSI BS 6100: Sec 1.2-85. 1985 Glossary of Building and Civil Engineering Terms Part 1: General and Miscellaneous Section 1.2: Spaces. 8 pp.
BSI BS 6100: Sec 1.2-92. 1992 Glossary of Building and Civil Engineering Terms Part 1: General and Miscellaneous Section 1.2: Spaces. 11 pp.
BSI BS 6100: SUBSEC 1.3.0-91. 1991 Building and Civil Engineering Terms Part 1: General and Miscellaneous Section 1.3: Parts of Construction Works Subsection 1.3.0: External Works. 16 pp.
BSI BS 6100: SUBSEC 1.3.0-01. 1991 Amd 1 Glossary of Building and Civil Engineering Terms Part 1: General and Miscellaneous Section 1.3: Parts of Construction Works Subsection 1.3.0: External Works (AMD 7231) August 15, 1992. 21 pp.
BSI BS 6100: SUBSEC 1.3.6-91. 1991 Building and Civil Engineering Terms Part 1: General and Miscellaneous Section 1.3: Parts of Construction Works Subsection 1.3.6: Jointing Products, Builders' Hardware and Accessories. 32 pp.
BSI BS 6100: SUBSEC 1.3.6-01. 1991 Amd 1 Glossary of Building and Civil Engineering Terms Part 1: General and Miscellaneous Section 1.3: Parts of Construction Works Subsection 1.3.6: Jointing Products, Builders' Hardware and Accessories (AMD 7236) August 15, 1992. 35 pp.
BSI BS 6100: SUBSEC 1.5.5-91. 1991 Building and Civil Engineering Terms Part 1: General and Miscellaneous Section 1.5: Operations; Associated Plant and Equipment Subsection 1.5.5: Plant and Equipment. 27 pp.
BSI BS 6100: SUBSEC 1.5.5-01. 1991 Amd 1 Glossary of Building and Civil Engineering Terms Part 1: General and Miscellaneous Section 1.5: Operations; Associated Plant and Equipment Subsection 1.5.5: Plant and Equipment (AMD 7243) August 15, 1992. 30 pp.
BSI BS 6100: SUBSEC 1.5.6-88. 1988 Glossary of Building and Civil Engineering Terms Part 1: General and Miscellaneous Section 1.5: Operations: Associated Plant and Equipment Subsection 1.5.6: Documentation Excluding Drawings. 13 pp.
BSI BS 6100: SUBSEC 1.5.6-01. 1988 Amd 1 Glossary of Building and Civil Engineering Terms Part 1: General Section 1.5: Operations; Associated Plant and Equipment Subsection 1.5.6: Documentation Excluding Drawings (AMD 7244) August 15, 1992. 14 pp.
BSI BS 6100: SUBSEC 1.7.1-86. 1986 Glossary of Building and Civil Engineering Terms Part 1: General and Miscellaneous Section 1.7: Characteristics and Performance Subsection 1.7.1: Performance. 5 pp.
BSI BS 6100: SUBSEC 1.7.1-01. 1986 Amd 1 Building and Civil Engineering Terms Part 1: General and Miscellaneous Section 1.7: Characteristics and Performance Subsection 1.7.1: Performance (AMD 7247) August 15, 1992. 6 pp.
BSI BS 6100: SUBSEC 1.7.2-89. 1989 Glossary of Building and Civil Engineering Terms Part 1: General and Miscellaneous Section 1.7: Characteristics and Performance Subsection 1.7.2: Characteristics. 15 pp.
BSI BS 6100: SUBSEC 1.7.2-01. 1989 Amd 1 Glossary of Building and Civil Engineering Terms Part 1: General and Miscellaneous Section 1.7: Characteristics and Performance Subsec. 1.7.2: Characteristics (AMD 7248) Augsut 15, 1992. 18 pp.
BSI BS 6100: Sec 2.5-91. 1991 Building and Civil Engineering Terms Part 2: Civil Engineering Section 2.5: Hydraulic Engineering and Construction Work. 36 pp.
BSI BS 6100: Sec 2.5-01. 1991 Amd 1 Glossary of Building and Civil Engineering Terms Part 2: Civil Engineering Section 2.5: Water Engineering (AMD 7253) August 15, 1992. 40 pp.
BSI BS 6100: Sec 2.6-91. 1991 Building and Civil Engineering Terms Part 2: Civil Engineering Section 2.6: Natural Waters. Inland. Coastal. Marine. 45 pp.
BSI BS 6100: Sec 2.6-01. 1991 Amd 1 Glossary of Building and Civil Engineering Terms Part 2: Civil Engineering Section 2.6: Inland Waters, Coastal and Maritime Engineering (AMD 7254) August 15, 1992. 48 pp.
BSI BS 6100: Sec 3.3-92. 1992 Glossary of Building and Civil Engineering Terms Part 3: Services Section 3.3: Sanitation. 39 pp.
BSI BS 6100: Sec 3.3-91. 1991 Building and Civil Engineering Terms Part 3: Services Section 3.3: Sanitation. 42 pp.
DIN ENGL 1080 Pt 1-76. Terms, Symbols and Units Used in Civil Engineering; Principles (June). 16 pp.
DIN ENGL 1080 Pt 2-80. Terms, Symbols and Units Used in Civil Engineering; Statics (Mar). 10 pp.
ISO 6707 Pt 1-89. Building and Civil Engineering—Vocabulary—Part 1: General Terms Second Edition; (Corrected and Reprinted -1989). 37 pp.

—Industrial Wastes
BSI BS 6543-85. 1985 Guide to Use of Industrial By-Products and Waste Materials in Building and Civil Engineering. 40 pp.

—Measurement
SNZ NZS 4224-83. Code of Practice for Measurement of Civil Engineering Quantities. 64 pp.

—Personnel—Glossaries
BSI BS 6100: Sec 1.6-85. 1985 Glossary of Building and Civil Engineering Terms Part 1: General and Miscellaneous Section 1.6: Persons. 6 pp.
BSI BS 6100: Sec 1.6-01. 1985 Amd 1 Glossary of Building and Civil Engineering Terms Part 1: General and Miscellaneous Section 1.6: Persons (AMD 7246) August 15, 1992. 7 pp.

—Schedules
SAA AS 1181-82. Method of Measurement of Civil Engineering Works and Associated Building Works Reconfirmed 1990. 50 pp.

—Soil Analysis
BSI BS 1377: Part 1-90. 1990 Methods of Test for Soils for Civil Engineering Purposes Part 1: General Requirements and Sample Preparation. 30 pp.

—Soils—Classification
DIN ENGL 18196-88. Soil Classification for Civil Engineering Purposes (Oct). 7 pp.

—Standards Preparation
BSI PD 6501: Part 1-82. 1982 Amd 1 Preparation of British Standards for Building and Civil Engineering Part 1: Guide to the Types of British Standard, Their Aims, Relationship, Content and Application. 20 pp.
BSI PD 6501: Part 2-84. 1984 Preparation of British Standards for Building and Civil Engineering Part 2: Guide to Presentation. 34 pp.
BSI PD 6501: Part 3-85. 1985 Preparation of British Standards for Building and Civil Engineering Part 3: Guidance to BSI Committees on the Preparation of British Standards for Use with the Building Act 1984. 47 pp.
BSI PD 6501: Part 5-90. 1990 Preparation of British Standards for Building and Civil Engineering Part 5: Guide to Technical Committees on the Preparation of Performance Specifications in British Standards for Construction Products. 24 pp.

—Symbols
DIN ENGL 1080 Pt 1-76. Terms, Symbols and Units Used in Civil Engineering; Principles (June). 16 pp.
DIN ENGL 1080 Pt 2-80. Terms, Symbols and Units Used in Civil Engineering; Statics (Mar). 10 pp.
DIN ENGL 1080 Pt 5-80. Terms, Symbols and Units Used in Civil Engineering; Timber Construction (Mar). 4 pp.

Civil Engineering Drawings
Use: Architectural Drawings

Clad Metals
Use For: Explosive Bonded Clad Metals
See Also: Coatings; Laminated Metals; Laminates; Metals

—Welded Joints—Steel
DIN ENGL 8553-91. Joining of Clad Steel Plate by Welding; Requirements and Design (Feb). 12 pp.

Cladding (Roofs)
Use: Roofing

Cladding (Walls)
Use: Siding

Cladding (Waveguides)
See Also: Cores (Waveguides); Dielectrics; Optical Fibers; Optical Waveguides; Waveguides

—Optical Fibers
DIN VDE 0472 Pt 212-86. Testing of Cables, Wires and Flexible Cords; Core and Cladding Dimensions or Optical Fibres (Aug). 5 pp.

—Optical Fibers—Diameters
CCITT RECMN G.651-89. Characteristics of a 50/125 Micrometer Multimode Graded Index Optical Fibre Cable—Transmission Media Characteristics (Study Group XV) 30 pp. 30 pp.
CCITT RECMN G.652-89. Characteristics of a Single-Mode Optical Fibre Cable—Transmission Media Characteristics (Study Group XV) 34 pp. 34 pp.
CCITT RECMN G.653-89. Characteristics of a Dispersion-Shifted Single-Mode Optical Fibre Cable—Transmission Media Characteristics (Study Group XV) 5 pp. 5 pp.
CCITT RECMN G.654-89. Characteristics of a 1550 mm Wavelength Loss-Minimized Single-Mode Optical Fibre Cable—Transmission Media Characteristics (Study Group XV) 3 pp. 3 pp.

Claims
Scope Note: Written demands by contracting parties for payment, adjustment or interpretation of terms, or other relief relating to contracts
See Also: Contractors
MOD UK DEFCON 44-76. Contractor's Financial Claim 10/76. 4 pp.

Clamp Straps
See Also: Clamps; Straps (Fasteners)

—Aircraft
SBAC AS 6702 ISSUE 3. Bonding Clip Assembly (Brass Strap).

Clamp Straps (Cont.)
—Aircraft (Cont.)
SBAC AS 6703 ISSUE 3. Bonding Clip Assembly (Non-Corrodible Steel Strap).

Clamped Fittings
Use For: V Band Couplings *See Also:* Pipe Fittings
—Aircraft—Flanges
ISO 2563-74. Aircraft Ducting and Piping—Profile Dimensions for Flanges of V-Band Couplings First Edition. 3 pp.
MOD UK DSTAN 53-68-01. V-Flange Couplings for Aircraft Piping and Ducting (Metric) Issue 2; Amendment 2. 8 pp.
—Polyethylene
JIS B 2354-91. Crosslinked Polyethylene (XPE) Pipe Clamp Type Fittings. 11 pp.

Clamping Bolts
See Also: Bolts
CNS B2336-84. Tommy Screws with Fixed Clamp Bolt (Jul)(4658).
CNS B2337-84. Tommy Screws with Moveable Clamping Bolt (Jul)(4659).
CNS B2338-84. Tommy Nuts with Fixed Clamping Bolt (Jul)(4660).
CNS B2339-84. Tommy Nuts with Moveable Clamping Bolt (Jul)(4661).
—Wheels
MOD UK DSTAN 53-5-01. Bolts and Screws for Clamping Divided Wheels Issue 1; Amendment 1. 9 pp.

Clamping Nuts
See Also: Fasteners; Nuts (Fasteners)
CNS B2338-84. Tommy Nuts with Fixed Clamping Bolt (Jul)(4660).
CNS B2339-84. Tommy Nuts with Moveable Clamping Bolt (Jul)(4661).
DIN ENGL 6305-71. Tommy Nuts with Fixed Bar (Sept). 1 p.
DIN ENGL 6307-67. Clamping Nuts with Loose Bar (Aug). 2 pp.

Clamping Screws
See Also: Screws
—Aircraft—Fuel Filters—Stirrups
SBAC AS 3325 ISSUE 2. Clamping, Screw—Filter Stirrup. 1 p.
—Aircraft—Pins
SBAC AS 3326 ISSUE 2. Centre, Pin (Clamping Screw). 1 p.
—Aircraft—Tank Straps
SBAC AS 6686 ISSUE 1. Clamping Screw-No.10 UNF Tank Straps.
—Chucks
BSI BS 6386: Part 1-86. 1986 Tool Chucks with Clamp Screws for Flatted Parallel Shanks Part 1: Dimensions of Chuck Nose. 4 pp.
BSI BS 6386: Part 2-83. 1983 Tool Chucks with Clamp Screws for Flatted Parallel Shanks Part 2: Dimensions of of Taper Shanks. 8 pp.
ISO 5414 Pt 1-85. Tool Chucks (End Mill Holders) with Clamp Screws for Flatted Parallel Shank Tools—Part 1: Dimensions of the Driving System of Tool Shanks Second Edition. 4 pp.
ISO 5414 Pt 2-82. Tool Chucks (End Mill Holders) with Clamp Screws for Flatted Parallel Shank Tools—Part 2: Connecting Dimensions of Chucks First Edition. 6 pp.
—Drawing Equipment
CNS B2381-83. Clamping Screws for Drawing Instruments (Feb)(4946).
—Wheels
MOD UK DSTAN 53-5-01. Bolts and Screws for Clamping Divided Wheels Issue 1; Amendment 1. 9 pp.

Clamping Units
Use: Clamps

Clamps
Scope Note: For additional listings, use a more specific term *Use For:* P Clamps
See Also: Aerospace Clamps; Band Clamps; Bands; Bonding Clamps; Cable Clamps; Chucks; Clamp Straps; Clips; Duct Joints; Fasteners; Grounding Clamps; Holders; Hose Clamps; Laboratory Clamps; Loop Clamps; Mounting Clamps; Omega Clamps; Pipe Clamps; Quick Release Fasteners; Straps (Fasteners)
—Aircraft
SBAC AS 41900 (V). Clamp, Loop, Bonding, Metric.
SBAC AS 41901 (V). Clamp, Loop, Bonding, Metric.
—Automotive—Battery Cables
CNS C4350-82. Cables Clamps for Automobiles Lead Storage Batteries (Jul)(9077).
—Components
DIN ENGL 6324-71. Operating Elements for Clamping Devices; Survey (Jan). 2 pp.
—Double—Fixed—Laboratory
CNS Z7101-81. Metal Laboratory Ware; Bossheads Fixed Double Clamp (Nov)(8178).
—Double—Rotary—Laboratory
CNS Z7102-81. Metal Laboratory Ware; Bossheads Rotary Double Clamp (Nov)(8179).
—Glass Joints—Spherical
CNS Z7096-81. Clamps for Spherical-Ground Joints (Nov)(8173).
—Jig (Positioners)
BSI BS 5078-74. 1974 Jig and Fixture Components. 58 pp.
CNS B3379-80. Clamps for Jigs (Sep)(6405).
JIS B 5227-89. Clamps for Jigs and Fixtures. 12 pp.
—Mountings—Spindle Noses
DIN ENGL 55028-80. Machine Tools; Mountings for Clamping Devices; Connecting Dimensions for Spindle Noses According to DIN 55026 and DIN 55027 (Mar). 4 pp.
—Multispindle Heads
DIN ENGL 69001 Pt 45-81. Machine Tools; Multi-Spindle Heads; Clamping Pieces; Type A (Oct). 1 p.
—Operating Elements
CNS B1156-80. Operating Elements for Clamping Devices Survey (Mar)(5271).
—Ropes—Mountaineering Equipment—Safety
BSI BS EN 567-93. 1993 Mountaineering Equipment Rope Clamp Safety Requirements and Test Method (N). 9 pp.
CEN PREN 567-91. Mountaineering Equipment —Rope Clamp—Safety Requirements, Testing, Marking. 6 pp.
CEN EN 567-92. Mountaineering Equipment —Rope Clamps—Safety Requirements and Test Method. 4 pp.
—Ships—Derricks
BSI BS MA 17-72. 1972 Derrick Boom Clamp. 7 pp.
—Transmission Lines
CNS C4310-81. Malleable Iron Casting Clamp and Aluminum Alloy Casting Clamp for Transmission Line Use (Jun)(7597).
—V Blocks
CGSB 39-GP-38A-75. "V" Blocks and Clamps, Standard for. 16 pp.

Clapboard Siding
See Also: Siding
—Asbestos Cement
CGSB CAN/CGSB-34.4-M89. Siding, Asbestos-Cement, Shingles and Clapboards. 12 pp.

Clarifiers
See Also: Absorbers (Equipment); Precipitators; Separators (Mechanical)
JIS B 9942-87. Testing Methods of Filtration Equipments for Clarifying. 13 pp.

Clarinets
See Also: Recorders (Musical)
CNS S1173-83. Clarinet (Nov)(10665).
JIS S 8514-88. Clarinets. 10 pp.

Clarke Orbits
Use: Geostationary Orbits

Classifications
Scope Note: See specific products

Classifiers
Use For: Refiners (Classifiers) *See Also:* Agitators; Centrifuges; Concentrators
—Pulp Fibers
CPPA C.3U-77. Fibre Classification—Clark Method. 2 pp.
CPPA D.7U-77. Fibre Size Classification of Pulp. 2 pp.

Claw Bars
Use: Crowbars (Tools)

Clay Bricks
Use: Bricks

Clay Building Blocks
Use: Building Blocks—Clay

Clay Content Analysis
—Aggregates
CNS A3035-86. Method of Test for Clay Lumps and Friable Particles in Aggregate (Aug)(1171).
JIS A 1137-89. Method of Test for Clay Contained in Aggregates. 6 pp.
SNZ NZS 4407: Part 3.5-91. Methods of Sampling and Testing Road Aggregates Part 3: Methods of Testing Road Aggregates—Laboratory Tests Test 3.5: The Clay Index. 6 pp.
—Foundry Sands
CNS Z8027-82. Method for Determining Clay Content of Foundry Sand (May)(8894).
JIS Z 2601-76. Method for Determining Clay Content of Foundry Sands.

Clay Pipes
See Also: Pipes; Vitreous Clay Pipes
JIS R 1201-91. Clay Pipes. 25 pp.
—Drainpipes
BSI BS 1196-89. 1989 Clayware Field Drain Pipes and Juntions. 8 pp.
BSI BS 1196-71. 1971 Clayware Field Drain Pipes. 8 pp.
SNZ BS 1196-89. Specification for Clayware Field Drain Pipes and Junctions. 4 pp.
—Sewer
DIN ENGL 1230 Pt 6-83. Clayware for Sewerage; Pipes and Fittings Without Sockets; Dimensions (Aug). 7 pp.
DIN ENGL 1230 Pt 7-83. Clayware for Sewerage; Pipes and Fittings Without Sockets; Technical Delivery Conditions (Aug). 18 pp.

Clays
Scope Note: For products made from clay, see specific products. For additional references, consult the following list *See Also:* Binders (Materials); Bricks; Ceramic Clay; Ceramics; Furnish; Grout; Masonry; Refractory Clay; Refractory Materials; Sediments; Silicates; Sizing (Surface Treatment)
—Modulus of Rupture
CNS R3052-76. Determination of Rupture Modulus of Unfired Clay Specimens (Jun)(2888).
—Paper Coatings
CNS M3030-81. Methods of Test for Filler and Coating Clay for Paper (Jan) (6938).
JIS M 8016-91. Testing Methods of Filler and Coating Clay for Paper.
—Supporting Liquids—Retaining Walls
DIN ENGL 4127-86. Earthworks and Foundation Engineering; Diaphragm Wall Clays for Supporting Liquids; Requirements, Testing, Supply, Inspection (Aug). 11 pp.

CLDATA
Use: Numerical Control—Cutter Location

Clean Rooms
Use For: Dust Free Rooms *See Also:* Controlled Atmospheres
BSI BS 5295: Part 1-89. 1989 Environmental Cleanliness in Enclosed Spaces Part 1: Specification for Clean Rooms and Clean Air Devices (N). 15 pp.
BSI BS 5295: Part 1-01. 1989 Amd 1 Environmental Cleanliness in Enclosed Spaces Part 1: Clean Rooms and Clean Air Devices (AMD 6602) December 21, 1990. 17 pp.
BSI BS 5295: Part 2-89. 1989 Environmental Cleanliness in Enclosed Spaces Part 2: Method for Specifying the Design, Construction and Commissioning of Clean Rooms and Clean Air Devices (N). 14 pp.
BSI BS 5295: Part 2-01. 1989 Amd 1 Environmental Cleanliness in Enclosed Spaces Part 2: Method for Specifying the Design, Construction and Commissioning of Clean Rooms and Clean Air Devices (AMD 6603) December 21, 1990. 15 pp.
BSI BS 5295: Part 3-89. 1989 Environmental Cleanliness in Enclosed Spaces Part 3: Guide to Operational Procedures and Disciplines Applicable to Clean Rooms and Clean Air Devices. 8 pp.
BSI BS 5295: Part 4-89. 1989 Environmental Cleanliness in Enclosed Spaces Part 4: Specification for Monitoring Clean Rooms and Clean Air Devices to Prove Continued Compliance with BS 5295: Part 1. 10 pp.

INTERNATIONAL AND NON-U.S. NATIONAL STANDARDS
SUBJECT INDEX

Cleaning

Clean Rooms (Cont.)

SAA AS 1386.1-89. Cleanrooms and Clean Workstations—Part 1: Principles of Clean Space Control. 22 pp.

SAA AS 1386.6-89. Cleanrooms and Clean Workstations—Part 6: Operation and Inspection of Cleanrooms. 3 pp.

SAA AS 1807.0-89. Cleanrooms, Workstations and Safety Cabinets—Methods of Test—Part 0: Apparatus. 7 pp.

SAA AS 1807.5-89. Cleanrooms, Workstations and Safety Cabinets—Methods of Test—Part 5: Determination of Work Zone Integrity. 4 pp.

—Air Cleaners

BSI BS 5295: Part 1-01. 1989 Amd 1 Environmental Cleanliness in Enclosed Spaces Part 1: Clean Rooms and Clean Air Devices (AMD 6602) December 21, 1990. 17 pp.

BSI BS 5295: Part 2-89. 1989 Environmental Cleanliness in Enclosed Spaces Part 2: Method for Specifying the Design, Construction and Commissioning of Clean Rooms and Clean Air Devices (N). 14 pp.

BSI BS 5295: Part 2-01. 1989 Amd 1 Environmental Cleanliness in Enclosed Spaces Part 2: Method for Specifying the Design, Construction and Commissioning of Clean Rooms and Clean Air Devices (AMD 6603) December 21, 1990. 15 pp.

BSI BS 5295: Part 3-89. 1989 Environmental Cleanliness in Enclosed Spaces Part 3: Guide to Operational Procedures and Disciplines Applicable to Clean Rooms and Clean Air Devices. 8 pp.

—Air Filters—HEPA

SAA AS 1807.6-89. Cleanrooms, Workstations and Safety Cabinets—Methods of Test—Part 6: Determination of Integrity of Terminally Mounted HEPA Filter Installations. 2 pp.

SAA AS 1807.7-89. Cleanrooms, Workstations and Safety Cabinets—Methods of Test—Part 7: Determination of Integrity of HEPA Filter Installations Not Terminally Mounted. 2 pp.

—Air Flow

SAA AS 1807.1-89. Cleanrooms, Workstations and Safety Cabinets—Methods of Test—Part 1: Determination of Air Velocity and Uniformity of Air Velocity in Clean Workstations and Laminar Flow Safety Cabinets. 1 p.

SAA AS 1807.2-89. Cleanrooms, Workstations and Safety Cabinets—Methods of Test—Part 2: Determination of Performance of Clean Workstations and Laminar Flow Safety Cabinets Under Loaded Filter Conditions. 2 pp.

SAA AS 1807.3-89. Cleanrooms, Workstations and Safety Cabinets—Methods of Test—Part 3: Determination of Air Velocity and Uniformity of Air Velocity in Laminar Flow Cleanrooms. 2 pp.

SAA AS 1807.4-89. Cleanrooms, Workstations and Safety Cabinets—Methods of Test—Part 4: Determination of Performance of Laminar Flow Cleanrooms Under Loaded Filter Conditions. 2 pp.

SAA AS 1807.11-89. Cleanrooms, Workstations and Safety Cabinets—Methods of Test—Part 11: Determination of Airflow Parallelism in Laminar Flow Cleanrooms. 1 p.

—Air Pressure Measurement

SAA AS 1807.10-89. Cleanrooms, Workstations and Safety Cabinets—Methods of Test—Part 10: Determination of Air Pressure of Cleanrooms. 1 p.

—Aircraft Equipment

CAA LEAFLET 2-2 07.90. Clean Rooms. 19 pp.

—Clothing

SAA AS 2013.1-89. Cleanroom Garments—Part 1: Product Requirements. 2 pp.

SAA AS 2013.2-89. Cleanroom Garments—Part 2: Processing and Use. 2 pp.

—Contamination

JIS B 9920-89. Measuring Methods for Airborne Particles in Clean Room and Evaluating Methods for Air Cleanliness of Clean Room. 33 pp.

—Contamination—Clothing

JIS B 9923-86. Methods for Sizing and Counting Particle Contaminants in and on Clean Room Garments.

SAA AS 1807.19-89. Cleanrooms, Workstations and Safety Cabinets—Methods of Test—Part 19: Sizing and Counting of Particulate Contaminants in and on Cleanroom Garments. 5 pp.

—Dust

JIS B 9926-91. Clean Room—Test Methods for Dust Generation from Moving Mechanisms. 11 pp.

—Glossaries

BSI BS 5295: Part 0-89. 1989 Environmental Cleanliness in Enclosed Spaces Part 0: General Introduction, Terms and Definitions for Clean Rooms and Clean Air Devices. 4 pp.

Clean Rooms (Cont.)

—Humidity

SAA AS 1807.13-89. Cleanrooms, Workstations and Safety Cabinets—Methods of Test—Part 13: Determination of Relative Humidity in Cleanrooms. 2 pp.

—Illuminance

SAA AS 1807.15-89. Cleanrooms, Workstations and Safety Cabinets—Methods of Test—Part 15: Determination of Illuminance. 1 p.

—Inspection

SAA AS 1386.6-89. Cleanrooms and Clean Workstations—Part 6: Operation and Inspection of Cleanrooms. 3 pp.

—Laminar Flow

SAA AS 1386.2-89. Cleanrooms and Clean Workstations—Part 2: Laminar Flow Cleanrooms. 6 pp.

—Noise

SAA AS 1807.16-89. Cleanrooms, Workstations and Safety Cabinets—Methods of Test—Part 16: Determination of Sound Level in Cleanrooms. 2 pp.

—Nonlaminar Flow

SAA AS 1386.3-89. Cleanrooms and Clean Workstations—Part 3: Non-Laminar Flow Cleanrooms—Class 350 and Cleaner. 4 pp.

SAA AS 1386.4-89. Cleanrooms and Clean Workstations—Part 4: Non-Laminar Flow Cleanrooms—Class 3500. 3 pp.

—Particle Counting

SAA AS 1807.8-89. Cleanrooms, Workstations and Safety Cabinets—Methods of Test—Part 8: Particle Counting in Work Zone by Automatic Particle Counter. 4 pp.

SAA AS 1807.9-89. Cleanrooms, Workstations and Safety Cabinets—Methods of Test—Part 9: Particle Counting in Cleanrooms by Microscopic Sizing and Counting. 6 pp.

—Recovery Time

SAA AS 1807.24-89. Cleanrooms, Workstations and Safety Cabinets—Methods of Test—Part 24: Determination of Recovery Times of Cleanrooms. 2 pp.

—Temperature Measurement

SAA AS 1807.12-89. Cleanrooms, Workstations and Safety Cabinets—Methods of Test—Part 12: Determination of Temperature in Work Zones. 1 p.

—Vibration

SAA AS 1807.17-89. Cleanrooms, Workstations and Safety Cabinets—Methods of Test—Part 17: Determination of Vibration in Cleanrooms. 1 p.

SAA AS 1807.18-89. Cleanrooms, Workststions and Safety Cabinets—Methods of Test—Part 18: Determination of Vibration in Workstations and Safety Cabinets. 1 p.

Cleaners (Equipment)

Use: Cleaning Equipment and Supplies

Cleaning

See Also: Chemical Cleaning; Corrosion Prevention; Laundering; Shot Blasting; Window Cleaning

—Carpets

JIS L 1023-92. Testing Methods for Several Characteristics of Textile Floor Coverings. 26 pp.

SNZ NZS/AS 3733-90. Textile Floor Coverings—Cleaning Maintenance Techniques for Domestic and Commercial Carpeting (This is a Joint Standard with SAA AS 3733). 26 pp.

—Medical Electrical Equipment

CSA CAN/CSA-C22. 2NO601.1-M90. Medical Electrical Equipment, Part 1: General Requirements for Safety; (Gen Instr 1). 240 pp.

—Service Contracts

MOD UK DEFCON 112AV-85. Conditions of Contract for Accommodation Cleaning 4/85. 15 pp.

Cleaning Agents

Use For: Cleaning Materials *See Also:* Abrasives; Alkaline Cleaners; Baths; Bleaching Agents; Cleaning Equipment and Supplies; Degreasers (Cleaning Agents); Descaling Compounds; Detergents; Dishwashing Compounds; Floor Stripping Agents; Hand Cleaners; Metal Cleaners; Paint Removers; Polishes; Scouring Powders; Shampoos; Soaps; Solvents; Stain Removers; Steam Cleaning Compounds; Surfactants; Sweeping Compounds; Wetting Agents; Windshields

Cleaning Agents (Cont.)

—Abrasion Testing

CGSB 2-GP-11M METH 15.1-83. Methods of Testing and Analysis of Soaps and Detergents Abrasion Test. 1 p.

—Aircraft—Exterior

MOD UK DTD-5507B-01. Foaming and General Purpose Cleaning Material for Exterior Surfaces of Aircraft; Amendment 1. 6 pp.

MOD UK DTD-5507B-02. Foaming and General Purpose Cleaning Material for Exterior Surfaces of Aircraft; Amendment 2. 7 pp.

MOD UK DTD-5600-01. Heavy Duty Cleaner for the Exterior Surfaces of Aircraft; Amendment 1. 5 pp.

MOD UK TS 10281A. Compound, Cleaning, Foaming, for Aircraft Surfaces.

MOD UK TS 10304A. Compound, Cleaning Gel, for Aircraft Surfaces.

—Ammonia Content

CGSB 2-GP-11M METH 6.4-88. Methods of Testing and Analysis of Soaps and Detergents Free Ammonia; (Amendment 1 Nov 1988). 2 pp.

CGSB 31-GP-0A METH 68.2-62. Methods of Testing Corrosion-Prevention Materials and Processes Ammonia (Quantitative Determination); (Re-Edited April 1982). 2 pp.

—Building Maintenance—Rinsing Properties

CGSB 2-GP-11M METH 31.2-83. Methods of Testing and Analysis of Soaps and Detergents Rinsing Properties. 1 p.

—Chloride Content

CGSB 31-GP-0A METH 32.3-62. Methods of Testing Corrosion-Prevention Materials and Processes Chloride Content; (Re-Edited March 1982). 2 pp.

—Compressors—Gas Turbine Engines

MOD UK TS 10268A. Cleaning Fluid for Compressors for Gas Turbine Engines.

—Containers—Fiberboard

BSI BS 5167-78. 1978 Packages for Washing and Cleaning Powders. Dimensions and Volumes of Cartons and Drums from Fibreboard. 9 pp.

CEN EN 23-1-78. Packages for Washing and Cleaning Powders—Part 1: Dimensions and Volumes of Cartons and Drums from Fibreboard. 7 pp.

—Corrosion Testing

CGSB 31-GP-0A METH 11.4A-62. Methods of Testing Corrosion-Prevention Materials and Processes Corrosion (Cleaning Compounds) (Supersedes 11.4 15 March 1957). 1 p.

—Dairy Equipment

BSI BS 5305-84. 1984 Cleaning and Disinfecting of Plant and Equipment Used in the Dairying Industry. 62 pp.

—Dairy Equipment—Elastomeric Sealant

DIN ENGL 11483 Pt 2-84. Dairy Installations; Cleaning and Disinfection; Consideration of the Action on Sealing Materials (Feb). 2 pp.

—Dairy Equipment—Stainless Steel

DIN ENGL 11483 Pt 1-83. Dairy Installations; Cleaning and Disinfection; Consideration of the Action on Stainless Steel (Jan). 3 pp.

—Dental

JIS T 5205-88. Dental Pulp Canal Cleansers. 7 pp.

—Drainpipes

CGSB CAN/CGSB-2.49-M87. Alkali Drain Cleaner. 10 pp.

—Dry—Sieve Analysis

CGSB 2-GP-11M METH 16.2-83. Methods of Testing and Analysis of Soaps and Detergents Sieve Analysis. 1 p.

—Dust

CGSB 31-GP-0A METH 45.1-57. Methods of Testing Corrosion-Prevention Materials and Processes Dusting Characteristics. 1 p.

—Emulsions—Acidity

CGSB 31-GP-0A METH 51.1-57. Methods of Testing Corrosion-Prevention Materials and Processes Free Alkali or Free Acid. 1 p.

—Emulsions—Alkalinity

CGSB 31-GP-0A METH 51.1-57. Methods of Testing Corrosion-Prevention Materials and Processes Free Alkali or Free Acid. 1 p.

—Emulsions—Corrosion Testing

CGSB 31-GP-0A METH 11.5-57. Methods of Testing Corrosion-Prevention Materials and Processes Corrosion (Emulsion Cleaners); (Amended September 1966) (Re-Edited March 1982). 1 p.

INDUSTRY STANDARDS

Cleaning Agents (Cont.)

—Emulsions—Sodium Chromate Content
CGSB 31-GP-0A METH 52.1-57. Methods of Testing Corrosion-Prevention Materials and Processes Determination of Sodium Chromate. 2 pp.

—Floors
CGSB 2-GP-11M METH 44.2-61. Methods of Testing and Analysis of Soaps and Detergents Deleterious Action on Floors. 1 p.

—Glass
CGSB CAN/CGSB-2.55-92. Glass Cleaner. 7 pp.
MOD UK TS 10306. Liquid Cleaner for Use on Aircraft Transparencies.

—Granular—Caking
CGSB 31-GP-0A METH 47.1-62. Methods of Testing Corrosion-Prevention Materials and Processes Caking Characteristics. 1 p.

—Granular—Water Content
CGSB 31-GP-0A METH 48.4-57. Methods of Testing Corrosion-Prevention Materials and Processes Water Content. 1 p.

—Hydrocarbons—Bilges
MOD UK DSTAN 79-13-89. Quick-Break Type Cleaners for Ships' Bilges Issue 1. 19 pp.

—Liquid—Surface Tension
CGSB 2-GP-11M METH 46.1-83. Methods of Testing and Analysis of Soaps and Detergents Determination of Surface Tension. 1 p.

—Paste—Sampling
CGSB 2-GP-11M METH 1.4-83. Methods of Testing and Analysis of Soaps and Detergents Sampling-Pastes. 1 p.

—Phosphate Content
CGSB 31-GP-0A METH 36.2-57. Methods of Testing Corrosion-Prevention Materials and Processes Determination of Phosphates. 2 pp.

—Sodium Carbonates
CGSB CAN/CGSB-15.5-93. Soda Ash. 7 pp.

—Sodium Hydroxide
CGSB CAN/CGSB-15.7-92. Caustic Soda (Lye). 7 pp.

—Sodium Tripolyphosphate
CGSB CAN/CGSB-15.40-93. Sodium Tripolyphosphate. 8 pp.

—Solubility—Kerosene
CGSB 31-GP-0A METH 54.1-57. Methods of Testing Corrosion-Prevention Materials and Processes Solubility (in Kerosine). 1 p.

—Toilet Facilities
CGSB CAN/CGSB-2.46-M87. Toilet Bowl and Urinal Cleaning Compound. 8 pp.
CGSB CAN/CGSB-2.47-M87. Toilet Bowl Cleaning Compound; (Amendment 1 September 1992). 9 pp.

—Water Content—Distillation Methods
CGSB 2-GP-11M METH 13.2-83. Methods of Testing and Analysis of Soaps and Detergents Water Content (Distillation Method). 3 pp.

—Windshields
CGSB CAN2-3.532-M85. Antifreeze, Windshield Washer, Methanol Base. 10 pp.
JIS K 2398-89. Windshield Washer Fluids for Automobiles. 12 pp.
MOD UK DSTAN 68-199-93. Windscreen Wash Liquid Concentrate Issue 1. 10 pp.
MOD UK TS 10072A. Windscreen Wash Liquid Concentrate (Superseded by Def Stan 68-199).
MOD UK TS 10238A. Wash Fluid for Thermal Image Sight.

Cleaning Appliances
Use: Cleaning Equipment and Supplies

Cleaning Cloths
Use For: Lubricating Cloths; Polishing Cloths
See Also: Abrasive Cloth; Wipes

BSI BS 7033: Part 1-89. 1989 Cleaning and Polishing Cloths Part 1: Optical Polishing Cloths for Use on Transparent Plastics Panels. 4 pp.
BSI BS 7033: Part 2-89. 1989 Cleaning and Polishing Cloths Part 2: Industrial Wipers from Reclaimed Materials, and Waste Yarn for Cleaning and Lubricating Purposes. 8 pp.
BSI BS 7033: Part 3-89. 1989 Amd 1 Cleaning and Polishing Cloths Part 3: Specification for Floorcloths and Dishcloths (AMD 6361) June 28, 1991. 8 pp.
BSI BS 7033: Part 4-90. 1990 Cleaning and Polishing Cloths Part 4: Window Cleaning Cloths. 8 pp.
BSI BS 7033: Part 5-90. 1990 Cleaning and Polishing Cloths Part 5: General Polishing Cloths. 8 pp.

Cleaning Cloths (Cont.)

—Aircraft
BSI F 137-89. 1989 Warp Knitted Cotton Polishing and Polish Application Cloths for Aircraft Transparencies. 6 pp.

—Glass
CNS L4062-74. Glass Cleaning Cloth (Tentative) (Mar)(3713).

Cleaning Compounds
Use: Cleaning Agents

Cleaning Equipment
Scope Note: Use a more specific term *See:* Air Cleaners; Carpet Cleaning Equipment; Carts; Gas Cleaning Equipment; Vacuum Blowers; Vacuum Cleaners; Washing Tunnels; Water Suction Cleaning Equipment

Cleaning Equipment and Supplies
Scope Note: For additional listings, use a more specific term *Use For:* Janitorial Supplies
See Also: Air Cleaners; Baths; Brooms; Brushes (Cleaning/Polishing); Buckets (Pails); Car Washes; Carpet Cleaning Equipment; Carts; Cleaning Agents; Cleaning Cloths; Dry Cleaning Equipment; Dusters; Dusting Brushes; Floor Cleaning Equipment; Gas Cleaning Equipment; High Pressure Water Cleaning Equipment; Mops; Parts Cleaning Equipment; Scrub Brushes; Steam Cleaning Equipment; Sterilizers; Toilet Bowl Brushes; Vacuum Cleaners; Washers (Cleaners); Water Suction Cleaning Equipment; Window Cleaning Equipment; Wipes

—Safety
CENELEC EN 60 335-2-54-91. Safety of Household and Similar Electrical Appliances Part 2: Particular Requirements for General Purpose Cleaning Appliances. 6 pp.
CENELEC EN 60335-2-54-92. Safety of Household and Similar Electrical Appliances Part 2: Particular Requirements for General Purpose Cleaning Appliances. 14 pp.
IEC 335 Pt 2-54-88. Safety of Household and Similar Electrical Appliances Part 2: Particular Requirements for General Purpose Cleaning Appliances First Edition. 25 pp.

Cleaning Materials
Use: Cleaning Agents

Clear Coatings
See Also: Coatings; Lacquers

—Colorfastness Testing
CGSB 1-GP-71 METH 120.3-79. Methods of Testing Paints and Pigments Color Stability Clear Films. 1 p.

—Mercury Content
CGSB 1-GP-71 METH 89.1-82. Methods of Testing Paints and Pigments Fungicide Content Dithizone Method for Determination of Total Mercury in Clears and Paints. 2 pp.

Clear Point
See Also: Cloud Point; Thermodynamic Properties

—Detergents
BSI BS 3762: Sec 4.1-89. 1989 Analysis of Formulated Detergents Part 4: Physical Test Methods Section 4.1: Method for Determination of Clear Point of Liquid Detergents. 3 pp.

Cleats

—Aircraft—Wiring
SBAC AGS 1672 (V). Cleat, Wiring No.14.

—Hatches
CNS F3017-80. Hatch Cleats (Mar)(5325).
CNS F3018-80. Hatch Cleats (Mar)(5326).
CNS F3172-82. Hatch Cleats (Simple Type) (Apr)(8686).

—Horn
BSI BS MA 11-71. 1971 Horncleats. 3 pp.
CNS F3075-80. Horn Cleats (Dec)(6827).
JIS F 1010-84. Yacht's Horn Cleats.
JIS F 3414-74. Horn Cleats. 5 pp.
JIS F 3416-74. Ships' Derrick Guy Cleats. 5 pp.

—Horn—Ships
CNS F3077-80. Ships' Derrick Guy Cleats (Dec)(6829).
JIS F 3414-89. Ships' Horn Cleats.
JIS F 3416-90. Guy Cleats for Ships.
JIS F 3416-74. Ships' Derrick Guy Cleats. 5 pp.

Clevis
See Also: Clevis Pins; Couplings; Fasteners; Holders; Pins; Quick Release Fasteners

BSI BS 5594-93. 1993 Leaf Chains, Clevises and Sheaves (ISO 4347: 1992). 22 pp.
BSI BS 5594-86. 1986 Leaf Chains, Clevises and Sheaves. 16 pp.
DIN ENGL 8152 Pt 4-89. Leaf Chains; Clevises and Sheaves for Heavy Series LH Chains; Connecting Dimensions (Feb). 3 pp.
ISO 4347-92. Leaf Chains, Clevises and Sheaves Third Edition. 19 pp.

—Automotive
CNS D2047-87. Adjustable Clevis Joints for Automobiles (Mar)(6165).

—Automotive—Yokes
CNS D2048-87. Yokes of Adjustable Clevis Joints for Automobiles (Mar)(6166).

—Hitches
ISO 6489 Pt 2-80. Agricultural Vehicles—Mechanical Connections on Towing Vehicles—Part 2: Clevis Type—Dimensions First Edition. 4 pp.

—Pneumatic Cylinders
ISO 8140-91. Pneumatic Fluid Power—Cylinders, 1 000 kPa (10 Bar) Series—Rod End Clevis-Mounting Dimensions Second Edition. 8 pp.

—Shipboard Handling Equipment
ISO 6043-85. Shipbuilding and Marine Structures—Eye and Fork Assemblies Under Tension Load—Main Dimensions First Edition. 5 pp.

Clevis Pins
See Also: Clevis; Pins

BSI BS 5893-80. (WITHDRAWN) 1980 Clevis Pins. Metric Series (Superseded by BS EN 22340: 1992). 4 pp.
BSI BS 5894-80. (WITHDRAWN) 1980 Clevis Pins with Heads. Metric Series (Superseded by BS EN 22341: 1992). 5 pp.
BSI BS EN 22340-92. 1992 Clevis Pins Without Head (ISO 2340: 1986) (E). 11 pp.
BSI BS EN 22341-92. 1992 Clevis Pins with Head (ISO 2341: 1986) (E). 11 pp.
CEN EN 22340-92. Clevis Pins Without Head. 6 pp.
CEN EN 22341-92. Clevis Pins with Head. 6 pp.
CNS B2190-83. Clevis Pins (Mar)(4311).
CNS B2191-83. Clevis Pins with Heads (Mar)(4312).
DIN ENGL EN 22341-92. Clevis Pins with Head; (ISO 2341: 1986) (Oct) (Supersedes DIN 1444, March 1974 Edition). 9 pp.
ISO 2340-86. Clevis Pins Without Head Second Edition. 6 pp.
ISO 2341-86. Clevis Pins with Head Second Edition. 6 pp.

—Automotive
CNS D2049-87. Pins of Adjustable Clevis Joints for Automobiles (Mar)(6167).

—Headless
DIN ENGL EN 22340-92. Clevis Pins Without Head; (ISO 2340: 1986) (Oct) (Supersedes DIN 1443, March 1974 Edition). 9 pp.

—Plain Washers
BSI BS EN 28738-92. 1992 Plain Washers for Clevis Pins—Product Grade A (ISO 8738: 1986) (E). 10 pp.
CEN EN 28738-92. Plain Washers for Clevis Pins—Product Grade A. 5 pp.
DIN ENGL EN 28738-92. Plain Washers for Use with Clevis Pins, Product Grade A; (ISO 8738: 1986) (Oct) (Supersedes DIN 1440, July 1974 Edition). 7 pp.
ISO 8738-86. Plain Washers for Clevis Pins—Product Grade A First Edition. 5 pp.

—Rolling Stock
CNS E1044-85. Clevis Pins with Head for Railway Rolling Stock (Jun)(11281).
JIS E 4111-84. Clevis Pins with Head for Railway Rolling Stock.

—Shipboard Handling Equipment
ISO 6043-85. Shipbuilding and Marine Structures—Eye and Fork Assemblies Under Tension Load—Main Dimensions First Edition. 5 pp.

Climatic Testing
Use: Environmental Testing

Climatology
See Also: Humidity; Polar Regions; Sea Conditions; Temperature; Tropical Regions; Water Waves; Wind (Meteorology)

—Glossaries
DIN ENGL 50010 Pt 1-77. Climates and Their Technical Application; Climatic Concepts; General Climatic Concepts (Oct). 4 pp.

Climatology (Cont.)

—Glossaries (Cont.)
DIN ENGL 50010 Pt 2-81. Climates and Their Technical Application; Climatological Terms and Definitions; Physical Terms and Definitions (Aug). 5 pp.

—Maps
DIN ENGL 50019 Pt 1-79. Climates and Their Technical Application; Technical Climatology; Characterization and Cartographic Representation of Open-Air Climates (Nov). 14 pp.

—Statistical Models
DIN ENGL 50019 Pt 3-79. Climates and Their Technical Application; Technical Climatology; Statistical Models of Climates (Nov). 28 pp.
DIN ENGL 50019 Pt 3 Suppl. 1-79. Climates and Their Technical Application; Technical Climatology; Geographical Summary of the Statistical Models of Open-Air Climates (Nov). 7 pp.

Climbers' Helmets
Use: Helmets—Climbers'

Climbing Equipment, Technical
Use: Mountaineering Equipment

Climbing Ropes
Use For: Ropes (Gymnastic)   *See Also:* Ropes
BSI BS 1892: Sec 2.2-72. (WITHDRAWN) 1972 Gymnasium Equipment Part 2: Particular Requirements Section 2.2: Ropes. 12 pp.

Clinch Nuts
Use: Anchor Nuts

Clinical Investigation Instruments
Use: Diagnostic Instruments (Medical)

Clinical Pathology
See Also: Pathology

—Glossaries
CEN PREN 406-90. In-Vitro Diagnostic Systems—Definitions of Disciplines. 3 pp.

Clinical Thermometers
Use For: Fever Thermometers   *See Also:* Medical Equipment; Temperature Measuring Instruments; Thermometers
BSI BS 691-87. 1987 Amd 1 Solid-Stem Clinical Maximum Thermometers (Mercury-in-Glass) (AMD 6834) September 2, 1991. 14 pp.
CGSB CAN/CGSB-14.1-M87. Mercury-in-Glass Clinical Thermo-Meters. 14 pp.
CNS T1061-86. Clinical Thermometers (Aug)(11690).
JIS T 1306-89. Continuous Measuring Clinical Electrical Thermometers. 13 pp.
JIS T 4206-89. Clinical Thermometers, Mercury-in-Glass, with Maximum Device. 8 pp.

Clinkers
See Also: Aggregates; Ashes; Portland Cements
BSI BS 1165-85. (WITHDRAWN) 1985 Clinker and Furnace Bottom Ash Aggregates for Concrete (R) (Superseded by BS 3797: 1990). 4 pp.

Clinometers
See Also: Surveying Instruments

—Ships
CNS F3087-81. Ships' Clinometers (Jan)(6917).
JIS F 3613-82. Ships' Clinometers.
JIS F 3613-58. Ships' Clinometers (R 1976). 5 pp.

CLIP
Use: Calling Line Identification Presentation

Clip Bolts
Use: Bolts—Clip

Clip Nuts
See Also: Nuts (Fasteners)

—Aerospace
CEN PREN 3240-92. Nuts, Self-Locking, Clip, in Heat Resisting Steel FE-PA92HT (A286), Uncoated Classification: 1 100 MPa (at Ambient Temperature) /425 Degrees C. 4 pp.
CEN PREN 3241-92. Nuts, Self-Locking, Clip, in Heat Resisting Steel FE-PA92HT (A286), Silver Coated Classification: 1 100 MPa (at Ambient Temperature) /425 Degrees C. 4 pp.

—Aerospace—Hole Size
CEN PREN 3741-91. Clip Nuts, Metric Installation Holes and Assembly. 6 pp.
CEN PREN 3741-93. Aerospace Series Nuts, Clip, Metric Installation Holes and Assembly. 5 pp.

Clipboards
See Also: Arch Clipboard Files
CGSB CAN/CGSB-53.28-M88. Arch Clip Board Files. 8 pp.

Clipper Machines
Use: Cutting Machines

Clippers (Hair)
Use: Hair Clippers

Clipping Machines
Use: Cutting Machines

Clips
See Also: Anchors (Fasteners); Bands; Cable Clips; Clamps; Connectors; Couplings; Fasteners; Holders; Quick Release Fasteners; Retaining Rings; Spring Clips; Truss Clips; Wire Rope Clips

—Aircraft
SBAC AS 486 ISSUE 7. Clips, 360 Degree, with Pressed Ends (M.S.). 1 p.
SBAC AS 487 ISSUE 6. Clips, 180 Degree, with Pressed Ends (M.S.). 1 p.
SBAC AS 490 ISSUE 3(I). Clips, 180 Degree, with Pressed Ends (Aluminium Alloy). 1 p.
SBAC AS 491 ISSUE 4. Clips, Aluminium, 1/2 Inch Wide. 1 p.
SBAC AS 492 ISSUE 5. Clips, aluminium, 5/8 Inch Wide. 1 p.
SBAC AS 493 ISSUE 5. Clips, Aluminium, 3/4 Inch Wide. 1 p.
SBAC AS 1157 ISSUE 5. Clips, 360 Degree, with Pressed Ends M.S.. 1 p.
SBAC AS 1158 ISSUE 4. Clips, 180 Degree, with Pressed Ends M.S.. 1 p.
SBAC AS 1161 ISSUE 2. Clips, 180 Degree, with Pressed Ends (Aluminium Alloy). 1 p.
SBAC AS 1162 ISSUE 2. Clips, 360 Degree with Pressed Ends Aluminium Alloy. 1 p.
SBAC AS 1163 ISSUE 2. Clips, 360 Degree with Pressed Ends Aluminium Alloy. 1 p.
SBAC AS 3181 ISSUE 3. "P" Clips. 1 p.
SBAC AS 4580 ISSUE 6(I). Clip. 1 p.
SBAC AS 4581 ISSUE 4. Clip with Earthing Tongue. 1 p.
SBAC AS 4697 ISSUE 1. Clips. 1 p.
SBAC AS 5418 ISSUE 8(I). Support Clips Rubber Covered. 1 p.
SBAC AS 5419 ISSUE 7(I). Support Clips Rubber Covered. 1 p.
SBAC AS 12810-839 ISSUE 3. Clip, Spring, Tension, Inner.
SBAC AS 12840-869 ISSUE 3. Clip, Spring, Tension, Outer.

—Aircraft—Pipe Fittings
SBAC AGS 3431 ISSUE 4. Pipe, Clip Rubber Covered for 1/4 Inch Diameter Fixing (For Pipes 1 3/4 Inch Diameter and Over).
SBAC AGS 3432 ISSUE 4. Pipe, Clip Rubber Covered for 1/4 Inch Diameter Fixing with Earthing Tongue (For Pipes 1 3/4 Inch Diameter and Over).

—Aircraft—Pipelines
SBAC AGS 584 ISSUE 9. Clip for Detachable Pipe Line.

—Alignment—Leaf Springs
CNS B2600-81. Alignment Clips for Laminated (Leaf) Spring (Sep)(7861). 3 pp.

—Bushed—Aircraft
SBAC AS 4676-4693 ISSUE 2(I). Clips Nylastic, Bushed. 1 p.

—Chains—Aircraft
SBAC AS 8437 ISSUE 1. Chain, Clip 1/4 Inch Bore Drain Valve.

—Chains—Drain Valves—Aircraft
SBAC AS 3339 ISSUE 2. Chain Clip 1/4 Bore Drain Hole. 1 p.

—Channels—Aircraft
SBAC AS 8646 ISSUE 1. Channel.

—Doors—Ships
CNS F3050-80. Ships' Steel Watertight Doors (Aug)(6182). 9 pp.
CNS F3211-83. Fitting for Weathertight Steel Doors of Small Ships (Mar)(10072).
JIS F 2316-87. Fittings for Weather-Tight Steel Doors.
JIS F 2316-70. Fittings for Ships' Weathertight Steel Doors. 9 pp.
JIS F 2330-80. Fittings for Small Ships' Weather-Tight Steel Doors.
JIS F 2330-75. Fittings for Small Ships' Weathertight Steel Doors. 13 pp.

Clips (Cont.)

—Drain Valves—Aircraft
SBAC AS 8437 ISSUE 1. Chain, Clip 1/4 Inch Bore Drain Valve.

—Identification Bands—Aircraft
SBAC AGS 2089 ISSUE 4. Clip Identification Band.
SBAC AGS 2091 ISSUE 4. Clip Identification Band.
SBAC AGS 2093 ISSUE 4. Clip—Identification Band.
SBAC AGS 2095 ISSUE 4. Clip—Identification Band.
SBAC AGS 2114-17 ISSUE 4. Clip Identification Band.

—Micrometers
CNS Z7070-81. Fit of Clips (Jan)(6984).
JIS B 7146-67. Fit of Clips (R 1973). 3 pp.

—Nut Assemblies—Locking—Aircraft
SBAC AS 41100-109 (V). Nut, Self-Locking, Clip, Assembly of.

—Pressed Ends—Aircraft
SBAC AS 485 ISSUE 5. Pressed End for Clips—to Suit 2 BA Bolts. 1 p.
SBAC AS 1156 ISSUE 4. Pressed End for Clips to Suit 1/4 Diameter Bolts. 1 p.

—Rubber Covered—Aircraft
SBAC AS 3180 ISSUE 12. P. Clips Rubber Covered. 1 p.

—Saddle—Aircraft
SBAC AS 488 ISSUE 6. Clips, Saddle, with Pressed Ends (M.S.). 1 p.
SBAC AS 489 ISSUE 4. Clips, Saddle, with Pressed Ends (Aluminium Alloy). 1 p.
SBAC AS 1159 ISSUE 5. Clips, Saddle, with Pressed Ends M.S.. 1 p.
SBAC AS 1160 ISSUE 3. Clips, Saddle, with Pressed Ends (Aluminium Alloy). 1 p.

—Straps
CNS Z7171-84. Clips for Strap End (Aug)(10988).

—Support—Cushions—Rubber—Aircraft
SBAC AS 4525 ISSUE 8. Support Clips Rubber Covered with Earthing Tongue. 1 p.
SBAC AS 4526 ISSUE 8. Support Clips Rubber Covered. 1 p.

—Wire
CNS B2443-80. Clip for Wire Rope (Apr)(5398).
JIS B 2809-77. Wire Clips for Wire Strand or Wire Rope (R 1980). 16 pp.

—Wire—Automotive
CNS D2149-82. Piping and Wiring Clips for Automobiles (Sep)(9375).

—Wire—Electric Railroads
JIS E 2209-89. Wire Clips for Overhead Contact System of Electric Railway. 12 pp.

CLIR
Use: Calling Line Identification Restriction

Clock Movements
See Also: Clocks; Time Measuring Instruments
JIS B 7009-82. Determination of Size for Watch and Clock Movements.
JIS B 7009-57. Determination of Size of Movement of Watch and Clock (R 1965). 3 pp.

Clock Operated Switches
Use: Time Switches

Clocks
See Also: Clock Movements; Electric Clocks; Plesiochronous Signaling; Reference Clocks; Slave Clocks; Time Clocks; Watches
CNS B7032-66. General Rules for Testing of Watches and Clocks (Jan)(2467)(R 1973).
CNS B7033-66. Testing Standard for Hanging and Table Clocks (Ordinary Grade) (Jan)(2469)(R 1973).
CNS B7034-66. Testing Standard for Hanging and Table Clocks (Middle and High Grade) (Jan)(2470)(R 1973).
JIS B 7001-83. Testing Methods for Watches and Clocks. 34 pp.
JIS B 7301-89. Clocks with Chart Drum. 10 pp.

—Collets
DIN ENGL 8255 Pt 1-72. Collets for Watches and Clocks; Slotted, Round (Nov). 3 pp.

—Quartz
CNS C6205-83. Method of Test for Main Part of Quartz Clock (Jun)(10329).

—Safety
BSI BS 3456: Sec 102.26-88. (WITHDRAWN) 1988 Safety of Household Electrical Appliances Part 102: Particular Requirements Section 102.26: Clocks (Superseded by BS EN 60335-2-26: 1991). 16 pp.

Clocks (Cont.)

—Safety (Cont.)

BSI BS EN 60335-2-26-91. 1991 Safety of Household and Similar Electrical Appliances Part 2: Particular Requirements Section 2.26: Clocks (Supersedes BS 3456: Section 102.26: 1988). 19 pp.

CENELEC HD 255 S2-83. Safety of Household and Similar Electric Appliances Particular Requirements for Clocks. 7 pp.

CENELEC EN 60 335-2-26-90. Safety of Household and Similar Electrical Appliances Part 2: Particular Requirements for Clocks. 6 pp.

IEC 335 Pt 2-26-87. Safety of Household and Similar Electrical Appliances Part 2: Particular Requirements for Clocks Second Edition. 25 pp.

—Size Notation

CNS B1029-66. Notation of Size for Watches and Clocks (Jan)(2468)(R 1973).

—Symbols

CNS C5130-82. Graphic Symbols for Clocks and Electrical Time Service Devices (Apr)(8662).

CNS C5131-82. Graphic Symbols for Components for Precision Engineering Devices, Especially Clocks (Apr)(8663).

DIN ENGL 40700 Pt 23-76. Graphical Symbols; Clocks and Electrical Time Service Devices (June). 6 pp.

DIN ENGL 40700 Pt 24-76. Graphical Symbols; Components for Precision Engineering Devices, Especially Clocks (June). 5 pp.

Clogs (Footwear)

See Also: Footwear

—Antistatic

ISO 7232-86. Rubber or Plastics Footwear—Antistatic Sandals, Sabots and Clogs First Edition. 6 pp.

—Safety

BSI BS 1870: Part 1-88. 1988 Amd 1 Safety Footwear Part 1: Safety Footwear Other Than All-Rubber and All-Plastics Moulded Types (AMD 6273) February 28, 1990. 24 pp.

Closed Circuit Television Cameras

Use For: CCTV Cameras See Also: Cameras; Closed Circuit Television Equipment; Television Cameras; Television Equipment

CNS C5090-82. Electrical Performance Standards for Closed Circuit Television Camera 525/60 Interlaced 2:1 (May)(6900).

—Black and White

CNS C5067-80. Electrical Performance Standards for Monochrome Closed Circuit Television Cameras 525/60 Random Interlace (Aug)(6315).

CNS C5091-82. Engineering Specifications Outline for Monochrome CCTV Camera Equipment (May)(6901).

CNS C5120-83. Electrical Performance Standards for High Resolution Monochrome Close Circuit Television Camera (Jan)(7651).

—Color

CNS C5066-80. Engineering Specifications Format for Color CCTV Camera Equipment (Aug)(6314).

Closed Circuit Television Equipment

Use For: CCTV Equipment See Also: Closed Circuit Television Cameras; Closed Circuit Television Monitors; Television Equipment

—Design—Ships

MOD UK NES 572-91. Guide to Design of Training and Recreational Television Systems Issue 2 (08.91). 73 pp.

—Design—Submarines

MOD UK NES 572-91. Guide to Design of Training and Recreational Television Systems Issue 2 (08.91). 73 pp.

Closed Circuit Television Monitors

Use For: CCTV Monitors See Also: Closed Circuit Television Equipment; Monitors; Television Equipment; Television Receivers; Video Monitors

CNS C5121-83. Electrical Performance Standard for Direct View Monochrome Closed Circuit Television Monitors 525/60 Interlaced 2:1 (Jan)(7652).

CNS C5122-83. Electrical Performance Standards for Direct View High Resolution Monochrome Closed Circuit Television Monitors (Jan)(7653).

Closed Conduit Flow Measurement

Use: Flow Measurement

Closed Die Forging

Use: Die Forgings

Closed User Groups

Use For: CUG See Also: Telephone Services

Closed User Groups (Cont.)

—Connectionless Broadband Data Service

CENELEC PRETS 300 217-3-91. Network Aspects (NA); Connectionless Broadband Data Service (CBDS) Part 3: Definition of Supplementary Services (DE/NA-53201-3). 9 pp.

CENELEC PRETS 300 217-3-92. Network Aspects (NA); Connectionless Broadband Data Service (CBDS) Part 3: Definition of Supplementary Services. 8 pp.

CENELEC ETS 300 217-3-92. Network Aspects (NA); Connectionless Broadband Data Service (CBDS) Part 3: Definition of Supplementary Services. 8 pp.

ETSI ETS 300 217-3-92. Network Aspects (NA); Connectionless Broadband Data Service (CBDS) Part 3: Definition of Supplementary Services. 8 pp.

ETSI PRETS 300 217-3-92. Network Aspects (NA); Connectionless Broadband Data Service (CBDS) Part 3: Definition of Supplementary Services. 8 pp.

ETSI PRETS 300 217-3-91. Network Aspects (NA); Connectionless Broadband Data Service (CBDS) Part 3: Definition of Supplementary Services (DE/NA-53201-3). 9 pp.

—Data Network Identification Codes—ISDN

CCITT RECMN E.167-89. ISDN Network Identification Codes—Telephone Network and ISDN—Operation, Numbering, Routing and Mobile Service (Study Group II) 2 pp. 2 pp.

—Digital Subscriber Signaling Systems

CCITT RECMN Q.955-92. Stage 3 Description for Community of Interest Supplementary Services Using DSS 1 Section 1—Closed User Group (Study Group XI) 24 pp. 24 pp.

—FAX Communications—Store and Forward Mode

CCITT RECMN F.162-89. Operational Requirements of an International Store-and-Forward Facsimile Switching Service (Comfax)—Telematic, Data Transmission and Teleconference Services—Operations and Quality of Service—(Study Group I) 7 pp. 7 pp.

—ISDN

CCITT RECMN I.255-89. Community of Interest Supplementary Services—Integrated Services Digital Network (ISDN)—General Structure and Service Capabilities (Study Group XVIII) 10 pp. 10 pp.

CCITT RECMN I.255. 1-92. Closed User Group (Study Group I) 17 pp. 17 pp.

CENELEC PRETS 300 136-90. Integrated Services Digital Network (ISDN) Closed User Group (CUG) Supplementary Service Service Description (T/NA1(89)21). 19 pp.

CENELEC PRETS 300 136-91. Integrated Services Digital Network (ISDN); Closed User Group (CUG) Supplementary Service Service Description. 18 pp.

CENELEC ETS 300 136-92. Integrated Services Digital Network (ISDN); Closed User Group (CUG) Supplementary Service Service Description. 17 pp.

CENELEC PRETS 300 137-90. Integrated Services Digital Network (ISDN); Closed User Group (CUG) Supplementary Service Functional Capabilities and Information Flows (T/S 22-03). 34 pp.

CENELEC PRETS 300 137-91. Integrated Services Digital Network (ISDN); Closed User Group (CUG) Supplementary Service Functional Capabilities and Information Flows. 33 pp.

CENELEC ETS 300 137-92. Integrated Services Digital Network (ISDN); Closed User Group (CUG) Supplementary Service Functional Capabilities and Information Flows. 32 pp.

CENELEC PRETS 300 138-90. Integrated Services Digital Network (ISDN): Closed User Group (CUG) Supplementary Service Digital Subscriber Signalling One (DSS1) Protocol (T/S 46-33H). 29 pp.

CENELEC PRETS 300 138-91. Integrated Services Digital Network (ISDN); Closed User Group (CUG) Supplementary Service Digital Subscriber Signalling No. One (DSS1) Protocol. 30 pp.

CENELEC ETS 300 138-92. Integrated Services Digital Network (ISDN); Closed User Group (CUG) Supplementary Service Digital Subscriber Signalling System No. One (DSS1) Protocol. 29 pp.

CEPT T/S 22-03-87. Service Supplementaire "Groupe Ferme D'Usagers". 8 pp.

CEPT T/S 22-03 E-87. Supplementary Service "Closed User Group". 8 pp.

ETSI PRETS 300 136-90. Intergrated Services Digital Network (ISDN) Closed User Group (CUG) Supplementary Service Service Description (T/NA1(89)21). 19 pp.

ETSI PRETS 300 137-90. Integrated Services Digital Network (ISDN); Closed User Group Supplementary Service Functional Capabilities and Information Flows (T/S 22-03). 34 pp.

ETSI PRETS 300 138-90. Integrated Services Digital Network (ISDN); Closed User Group (CUG) Supplementary Service Digital Subscriber Signalling One (DSS1) Protocol (T/S 46-33H). 29 pp.

Closed User Groups (Cont.)

—ISDN (Cont.)

ETSI ETS 300 136-92. Integrated Services Digital Network (ISDN); Closed User Group (CUG) Supplementary Service Service Description. 17 pp.

ETSI PRETS 300 136-91. Integrated Services Digital Network (ISDN); Closed User Group (CUG) Supplementary Service Service Description. 18 pp.

ETSI ETS 300 137-92. Integrated Services Digital Network (ISDN); Closed User Group (CUG) Supplementary Service Functional Capabilities and Information Flows. 32 pp.

ETSI PRETS 300 137-91. Integrated Services Digital Network (ISDN); Closed User Group (CUG) Supplementary Service Functional Capabilities and Information Flows. 33 pp.

ETSI ETS 300 138-92. Integrated Services Digital Network (ISDN); Closed User Group (CUG) Supplementary Service Digital Subscriber Signalling System No. One (DSS1) Protocol. 29 pp.

ETSI PRETS 300 138-91. Integrated Services Digital Network (ISDN); Closed User Group (CUG) Supplementary Service Digital Subscriber Signalling No. One (DSS1) Protocol. 30 pp.

—ISDN—CCITT No. 7 Signaling Systems

CCITT RECMN Q.730-89. ISDN Supplementary Services—Specifications of Signalling System No. 7 (Study Group XI) 59 pp. 59 pp.

—ISDN—Functions/Information Flows

CCITT RECMN Q.85-92. Stage 2 Description for Community of Interest Supplementary Services Section 1—Closed User Group (CUG) Section 3—Multi-Level Precedence and Preemption (MLPP) (Rev.1) (Study Group XI) 59 pp. 59 pp.

CCITT RECMN Q.85-89. Community of Interest Supplementary Services—General Recommendations on Telephone Switching and Signalling—Functions and Information Flows for Services in the ISDN—Supplements (Study Group XI) 14 pp. 14 pp.

—Management

CCITT RECMN X.180-89. Administrative Arrangements for International Closed User Groups (CUGs)—Data Communication Networks—Transmission, Signalling and Switching, Network Aspects, Maintenance and Administrative Arrangements (Study Group VII) 3 pp. 3 pp.

—Videotex Communications

CCITT RECMN F.300-89. Videotex Service—Telematic, Data Transmission and Teleconference Services—Operations and Quality of Service (Study Group I) 25 pp. 25 pp.

Clostridium

See Also: Bacteria

—Water

BSI BS 6068: Sec 4.8-93. 1993 Water Quality Detection and Enumeration of the Spores of Sulfite-Reducing Anaerobes (Clostridia) Part 1: Method by Enrichment in a Liquid Medium (ISO 6461/1: 1986) (N). 10 pp.

BSI BS 6068: Sec 4.9-93. 1993 Water Quality Detection and Enumeration of the Spores of Sulfite-Reducing Anaerobes (Clostridia) Part 2: Method by Membrane Filtration (ISO 6461/2: 1986) (N). 10 pp.

BSI BS EN 26461-1-93. 1993 Water Quality Detection and Enumeration of the Spores of Sulfite-Reducing Anaerobes (Clostridia) Part 1: Method by Enrichment in a Liquid Medium (ISO 6461/1: 1986) (N). 10 pp.

BSI BS EN 26461-2-93. 1993 Water Quality Detection and Enumeration of the Spores of Sulfite-Reducing Anaerobes (Clostridia) Part 2: Method by Membrane Filtration (ISO 6461/2: 1986) (N). 10 pp.

CEN EN 26461-1-93. Water Quality—Detection and Enumeration of the Spores of Sulfite-Reducing Anaerobes (Clostridia)—Part 1: Method by Enrichment in a Liquid Medium (ISO 6461-1: 1986). 5 pp.

CEN EN 26461-2-93. Water Quality—Detection and Enumeration of the Spores of Sulfite-Reducing Anaerobes (Clostridia)—Part 2: Method by Membrane Filtration (ISO 6461/2: 1986). 5 pp.

ISO 6461 Pt 1-86. Water Quality—Detection and Enumeration of the Spores of Sulfite-Reducing Anaerobes (Clostridia)—Part 1: Method by Enrichment in a Liquid Medium First Edition; (CEN EN 26461-1: 1993). 5 pp.

ISO 6461 Pt 2-86. Water Quality—Detection and Enumeration of the Spores of Sulfite-Reducing Anaerobes (Clostridia)—Part 2: Method by Membrane Filtration First Edition; (CEN EN 26461-2: 1993). 5 pp.

Clostridium Botulinum

See Also: Bacteria

INTERNATIONAL AND NON-U.S. NATIONAL STANDARDS SUBJECT INDEX

Clostridium Botulinum (Cont.)

—Food
SAA AS 1766.2.7-91. Methods for the Microbiological Examination of food—Part 2.7: Examination for Specific Organisms—Clostridium Botulinum and Clostridium Botulinum Toxin (Supersedes AS 1766.2.1.7 Addendum 1—1976). 5 pp.

—Meat
DIN ENGL 10102-88. Microbiological Analysis of Meat and Meat Products; Detection of Clostridium Botulinum and Botulinum Toxin (June). 5 pp.

Clostridium Perfringens
See Also: Bacteria

—Dairy Products
BSI BS 4285: Sec 3.13-91. 1991 Microbiological Examination for Dairy Purposes Part 3: Methods for Detection and/or Enumeration of Specific Groups of Microorganisms Section 3.13: Enumeration of Clostridium Perfringens. 6 pp.

—Food
BSI BS 5763: Part 9-86. 1986 Microbiological Examination of Food and Animal Feeding Stuffs Part 9: Enumeration of Clostridium Perfringens. 10 pp.

ISO 7937-85. Microbiology—General Guidance for Enumeration of Clostridium Perfringens—Colony Count Technique First Edition. 9 pp.

SAA AS 1766.2.8-91. Methods for the Microbiological Examination of Food—Part 2: Examination for Specific Organisms—Part 2.8: Clostridium Perfringens. 6 pp.

Closures
Use: Stoppers

Cloth
Use: Fabrics

Clothes Driers
Use: Clothes Dryers

Clothes Dryers
Use For: Drying Cabinets See Also: Appliances; Dryers; Washing Machines

BSI BS 3999: Part 4-67. 1967 Amd 1 Methods of Measuring the Performance of Household Electrical Appliances Part 4: Electric Clothes Drying Cabinets and Racks. 14 pp.

BSI BS 6246: Part 4-83. 1983 Industrial Laundry Machinery Part 4: Methods for the Assessment of the Effect of Batch Drying Tumblers on Textiles. 11 pp.

CNS C3059-87. Method of Test for Electrical Wash Dryer (Jan)(4674). 2 pp.

CNS C4141-87. Electrical Wash Dryers (Jan)(4673). 3 pp.

JIS C 9608-86. Tumbler Type Electric Clothes Dryers. 32 pp.

SNZ NZS 6311-93. Approval and Test Specification—Particular Requirements for Tumble Dryers (NZS 6311:1993/AS 3317-1993). 20 pp.

—Electric—Commercial
CSA C22.2 NO 112-M1990. Clothes Drying Machines; (Gen Instr 1 Thru 4). 54 pp.

CSA 1169 Bull. Electrical Bulletin 1169 June 27, 1978 to C22.2 NO 112. 2 pp.

—Electric—Commercial—Capacity Measurement
ISO 9398 Pt 2-93. Specifications for Industrial Laundry Machines—Definitions and Testing of Capacity and Consumption Characteristics—Part 2: Batch Drying Tumblers First Edition. 7 pp.

—Electric—Commercial—Energy Consumption
ISO 9398 Pt 2-93. Specifications for Industrial Laundry Machines—Definitions and Testing of Capacity and Consumption Characteristics—Part 2: Batch Drying Tumblers First Edition. 7 pp.

—Electric—Household
CENELEC PREN 61121-93. Method for Measuring the Performance of Tumbler Dryers for Household Use. 27 pp.

CSA C22.2 NO 112-M1990. Clothes Drying Machines; (Gen Instr 1 Thru 4). 54 pp.

CSA 1169 Bull. Electrical Bulletin 1169 June 27, 1978 to C22.2 NO 112. 2 pp.

IEC 1121-91. Method for Measuring the Performance of Tumbler Dryers for Household Use First Edition. 37 pp.

SAA AS 3196-86. Approval and Test Specifications—Electric Clothes Drying Cabinets for Household Use (This is a Joint Standard with SANZ NZS 3196). 5 pp.

Clothes Dryers (Cont.)

—Electric—Household (Cont.)
SNZ NZS/AS 3196-86. Approval and Test Specification—Electric Clothes Drying Cabinets for Household Use (This is a Joint Standard with SAA AS 3196). 5 pp.

—Electric—Household—Energy Consumption
CSA CAN/CSA-C361-92. Test Method for Measuring Energy Consumption and Drum Volume of Electrically Heated Household Tumble-Type Clothes Dryers; (Gen Instr 1). 25 pp.

—Electric—Household—Safety
CENELEC EN 60 335-2-11-89. Safety of Household and Similar Electrical Appliances—Part 2: Particular Requirements for Tumbler Dryers. 10 pp.

CENELEC EN 60 335-2-11/PRAB-89. AMD prAB Safety of Household and Similar Electrical Appliances—Part 2: Particular Requirements for Tumbler Dryers. 2 pp.

CENELEC EN 60335-2-11/A1-90. AMD 1 Safety of Household and Similar Electrical Appliances—Part 2: Particular Requirements for Tumbler Dryers. 7 pp.

CENELEC EN 60335-2-11/A2-92. AMD 2 Safety of Household and Similar Electrical Appliances Part 2: Particular Requirements for Tumbler Dryers (IEC 335-2-11:1984/A2:1991). 4 pp.

IEC 335 Pt 2-11-93. Safety of Household and Similar Electrical Appliances Part 2: Particular Requirements for Tumbler Dryers Fourth Edition. 37 pp.

—Gas—Commercial
CSA C22.2 NO 112-M1990. Clothes Drying Machines; (Gen Instr 1 Thru 4). 54 pp.

CSA 1169 Bull. Electrical Bulletin 1169 June 27, 1978 to C22.2 NO 112. 2 pp.

—Gas—Household
CSA C22.2 NO 112-M1990. Clothes Drying Machines; (Gen Instr 1 Thru 4). 54 pp.

CSA 1169 Bull. Electrical Bulletin 1169 June 27, 1978 to C22.2 NO 112. 2 pp.

JIS S 2130-91. Gas Burning Clothes Dryers for Domestic Use. 24 pp.

—Gas—Household—Installation
BSI BS 7624-93. 1993 Installation of Domestic Direct Gas-Fired Tumble Dryers of up to 3 kW Heat Input (2nd and 3rd Family Gases) (L). 14 pp.

—Installation
BSI CP 335: Part 1-73. 1973 Selection and Installation of Miscellaneous Town Gas Appliances Part 1: Domestic Laundering and Miscellaneous Appliances. 17 pp.

—Safety
BSI BS 3456: Sec 102.11-88. (WITHDRAWN) 1988 Safety of Household Electrical Appliances Part 102: Particular Requirements Section 102.11: Tumbler Dryers (Superseded by BS 3456: Section 202.11: 1990). 20 pp.

BSI BS 3456: Sec 202.11-90. 1990 Safety of Household Electrical Appliances Part 202: Particular Requirements Section 202.11: Tumbler Dryers. 25 pp.

BSI BS 3456: Sec 202.11-01. 1990 Amd 1 Safety of Household and Similar Electrical Appliances Part 202: Particular Requirements Section 202.11: Tumbler Dryers (AMD 6913) July 15, 1992. 36 pp.

BSI BS 3456: Sec 202.11-02. 1990 Amd 2 Safety of Household and Similar Electrical Appliances Part 202: Particular Requirements Section 202.11: Tumbler Dryers (AMD 7664) June 15, 1992 (L). 42 pp.

BSI BS 3456: Sec 202.11-03. 1990 Amd 3 Safety of Household and Similar Electrical Appliances Part 202: Particular Requirements Section 202.11: Tumbler Dryers (AMD 7869) July 15, 1993 (SUPERSEDES BS 3456: SECTION 102. 11: 1988). 51 pp.

BSI BS 3456: Sec 202.43-90. 1990 Safety of Household Electrical Appliances Part 202: Particular Requirements Section 202.43: Clothes Dryers and Towel Rails. 20 pp.

BSI BS 3456: Sec 202.43-01. 1990 Amd 1 Safety of Household and Similar Electrical Appliances Part 202: Particular Requirements Section 202.43: Clothes Dryers and Towel Rails (AMD 6901) April 1, 1992 (SUPERSEDES BS 3456 SECTION 2.9: 1970 & 3. 6: 1979). 26 pp.

BSI BS 3456: Sec 202.43-02. 1990 Amd 2 Safety of Household and Similar Electrical Appliances Part 202: Particular Requirements Section 202.43: Clothes Dryers and Towel Rails (AMD 7550) JAN 15, 1993 SUPERSEDES BS 3456 SECTION 2. 9: 1970 & 3. 6: 1979. 31 pp.

Clothes Dryers (Cont.)

—Safety (Cont.)
BSI BS 5258: Part 17-92. 1992 Safety of Domestic Gas Appliances Part 17: Specification for Direct Gas-Fired Tumble Dryers (2nd and 3rd Family Gases). 32 pp.

BSI BS EN 60335-2-45-91. 1991 Safety of Household and Similar Electrical Appliances Part 2: Particular Requirements Section 2.45: Portable Electric Heating Tools and Similar Appliances. 23 pp.

BSI BS EN 60335-2-45-01. 1991 Amd 1 Safety of Household and Similar Electrical Appliances Part 2: Particular Requirements Section 2.45: Portable Electric Heating Tools And Similar Appliances. 44 pp.

CENELEC HD 263 S1-84. Particular Specification for Clothes Dryers and Towel Dryers. 5 pp.

CENELEC HD 268-76. Particular Specification for Clothes Dryers of the Tumbler Type (Superseded by EN 60335-2-11). 9 pp.

CENELEC HD 268 S2-84. Safety of Household and Similar Electrical Appliances-Particular Requirements for Tumble Dryers. 14 pp.

CENELEC EN 60 335-2-43-89. Safety of Household and Similar Electrical Appliances—Part 2: Particular Requirements for Clothes Dryers and Towel Rails. 7 pp.

CENELEC EN 60 335-2-43/A1-90. AMD 1 Safety of Household and Similar Electrical Appliances—Part 2: Particular Requirements for Clothes Dryers and Towel Rails. 6 pp.

CENELEC EN 60335-2-43/A51-92. AMD 51 Safety of Household and Similar Electrical Appliances—Part 2: Particular Requirements for Clothes Dryers and Towel Rails. 4 pp.

CENELEC EN 60 335-2-45-90. Safety of Household and Similar Electrical Appliances Part 2: Particular Requirements for Portable Electric Heating Tools and Similar Appliances. 11 pp.

CENELEC EN 60335-2-45/PRAB-93. AMD AB Safety of Household and Similar Electric Appliances. Part 2: Particular Requirements for Portable Electric Heating Tools and Similar Appliances—Sub-Clauses 8.1 and 22.107. 2 pp.

CENELEC EN 60335-2-45/A1-92. AMD 1 Safety of Household and Similar Electrical Appliances Part 2: Particular Requirements for Portable Electric Heating Tools and Similar Appliances (IEC 335-2-45:1986/A1:1990). 5 pp.

DIN VDE 0700 Pt 600 (D)-88. Safety of Household and Similar Electrical Appliances; Connection of Washing Machines, Dishwashers and Tumble Dryers to the Supply Mains; German Version prHD 274 S2: 1987 (Oct). 37 pp.

IEC 335 Pt 2-43-84. Safety of Household and Similar Electrical Appliances Part 2: Particular Requirements for Clothes Dryers and Towel Rails First Edition; (Amendment 1-1988). 35 pp.

IEC 335 Pt 2-45-86. Safety of Household and Similar Electrical Appliances Part 2: Particular Requirements for Portable Electric Heating Tools and Similar Appliances First Edition; (Amendment 1-1990) (CENELEC EN 60335-2-45/A1:1992). 48 pp.

—Steam—Commercial
CSA C22.2 NO 112-M1990. Clothes Drying Machines; (Gen Instr 1 Thru 4). 54 pp.

CSA 1169 Bull. Electrical Bulletin 1169 June 27, 1978 to C22.2 NO 112. 2 pp.

—Water Pipelines—Connection
BSI BS 6614-91. (WITHDRAWN) 1991 Safety Devices and Water Supply Connections for Washing Machines, Dishwashers and Tumbler Dryers Connected to the Water Supply Mains (Superseded by BS EN 50084: 1992). 18 pp.

BSI BS 6614-85. 1985 Safety Devices and Water Supply Connections for Washing Machines and Dishwashers Connected to the Water Supply Mains. 25 pp.

BSI BS EN 50084-92. 1992 Safety of Household and Similar Electrical Appliances. Requirements for the Connection of Washing Machines, Dishwashers and Tumbler Dryers to the Water Mains (L). 24 pp.

CENELEC HD 274 S2-90. Safety of Household and Similar Electrical Appliances Requirements for the Connection of Washing Machines, Dishwashers and Tumble Dryers to the Water Supply Mains. 24 pp.

CENELEC EN 50084-92. Safety of Household and Similar Electrical Appliances—Requirements for the Connection of Washing Machines, Dishwashers and Tumbler Dryers to the Water Mains. 21 pp.

—Water Pipelines—Safety
BSI BS EN 50084-92. 1992 Safety of Household and Similar Electrical Appliances. Requirements for the Connection of Washing Machines, Dishwashers and Tumbler Dryers to the Water Mains (L). 24 pp.

CENELEC EN 50084-92. Safety of Household and Similar Electrical Appliances—Requirements for the Connection of Washing Machines, Dishwashers and Tumbler Dryers to the Water Mains. 21 pp.

Clothes

Clothes Drying Cabinets
Use: Clothes Dryers

Clothes Drying Machines
Use: Clothes Dryers

Clothes Lockers
Use: Lockers

Clothes Washing Machines
Use: Washing Machines

Clothing

Scope Note: For additional listings, use a more specific term *See Also:* Australian Garment Marks; Bathrobes; Belts (Clothing); Blazers; Blouses; Boots (Footwear); Buttons (Fasteners); Clothing Racks; Coats; Combat Uniforms; Coveralls; Dress Uniforms; Dresses; Exposure Suits; Fire Fighting Clothing; Footwear; Foundation Garments; Gaiters; Gloves; Gowns (Protective); Hangers (Clothing); Hats; Headgear; Helmets; Hosiery; Jackets; Life Preservers; Neckties; Nightgowns; Nightwear; Outerwear; Overalls; Pajamas; Pants; Pressure Suits; Protective Clothing; Raincoats; Rompers; Shirts; Shoes (Footwear); Shorts (Clothing); Skirts (Clothing); Slips (Underwear); Smocks; Snow Suits; Socks; Suits; Sunsuits; Surgical Gloves; Sweat Suits; Sweaters; Swim Suits; Underwear; Uniforms; Vests; Wardrobes; Work Clothes

—**Boys'—Mannequins**
CGSB CAN/CGSB-49.8-M89. Girls' and Boys' Canada Standard Sizes Model Forms, Regular Range—Dimensions. 21 pp.

—**Boys'—School**
SAA AS 1994.1-77. School Wear for Boys and Girls—Part 1: Manufacturing Requirements. 8 pp.

—**Classification**
BSI BS 1000: (687)-84. 1984 Universal Decimal Classification (UDC). English Full Edition (687): Clothing Industry. Beauty Culture Industries. 30 pp.
EC EEC/548/89. Commission Regulation Concerning the Classification of Certain Goods in the Combined Nomenclature. 3 pp.
SNZ NZS/BS 1000 (687)-84. Universal Decimal Classification Clothing Industry. Beauty Culture Industries. 32 pp.

—**Clean Rooms**
SAA AS 2013.1-89. Cleanroom Garments—Part 1: Product Requirements. 2 pp.
SAA AS 2013.2-89. Cleanroom Garments—Part 2: Processing and Use. 2 pp.

—**Defense Contracts**
MOD UK DEFCON 112B-91. Conditions of Contract for Clothing and Textiles 2/91. 7 pp.

—**Design—Fire Hazards**
ISO TR9240-92. Textiles—Design of Apparel for Reduced Fire Hazard First Edition. 12 pp.

—**Dimensional Stability**
BSI BS 4931-86. 1986 Preparation, Marking and Measuring of Textile Fabrics, Garments and Fabric Assemblies in Tests for Assessing Dimensional Change. 8 pp.

—**Dry Cleaning—Dimensional Stability**
ISO 3175-79. Textiles—Determination of Dimensional Change on Dry Cleaning in Perchlorethylene—Machine Method Second Edition. 6 pp.

—**Fasteners—Tapes**
JIS L 3416-88. Touch and Close Fastener. 13 pp.

—**Flammability Testing**
CNS L3197-83. Method of Test for Flammability of Clothes — Test for Burning Speed (May)(10286).
CNS L3198-83. Method of Test for Flammability of Clothes — Number Test of Ignition (May)(10287).

—**Girls'—Mannequins**
CGSB CAN/CGSB-49.8-M89. Girls' and Boys' Canada Standard Sizes Model Forms, Regular Range—Dimensions. 21 pp.

—**Girls'—School**
SAA AS 1994.1-77. School Wear for Boys and Girls—Part 1: Manufacturing Requirements. 8 pp.

—**Glossaries**
BSI BS 1903-81. 1981 Glossary of Terms Used by the Light Clothing Industry. 22 pp.
CNS L1022-90. Glossary of Terms Used for Apparel (Measuring Part) (Jan)(12658).
ISO 3635-81. Size Designation of Clothes—Definitions and Body Measurement Procedure Third Edition. 7 pp.
JIS L 0111-83. Glossary of Terms Used in Body Measurements for Clothes.
JIS L 0112-86. Glossary of Terms on Parts and Measurements of Clothes.
JIS L 0215-84. Glossary of Terms on Clothes.

—**Government Supplied Property—Forms**
MOD UK DEFCON 198-91. Clothing and Textile Contracts Issues of Government Property on Embodiment Loan Terms 2/91. 2 pp.
MOD UK DEFCON 198B-80. Materials to Be Issued to the Contractor 9/80. 1 p.

—**Hooks**
CNS S1110-77. Clothes Pegs (Single Arm Light Duty) (Mar)(7071). 1 p.
CNS S1111-81. Clothes-Pegs (Single Arm, Heavy Duty) (Mar)(7072). 1 p.
CNS S1112-81. Hanger for Clothes-Pegs (Mar)(7073). 1 p.
CNS S1115-81. Clothes-Pegs (Double Arm) (Mar)(7076). 1 p.

—**Infants'—Mannequins**
CGSB CAN/CGSB-49.9-M89. Infants' Canada Standard Sizes Model Forms—Dimensions. 8 pp.

—**Invitation for Bids**
MOD UK DEFCON 47D-86. Invitation to Tender for Clothing and Textile Contracts 8/86. 2 pp.
MOD UK DEFCON 47D-91. Invitation to Tender for Clothing and Textile Contracts 9/91. 2 pp.

—**Labels**
SAA AS 1957-87. Care Labelling of Clothing, Household Textiles, Furnishings, Upholstered Furniture, Bedding, Piece Goods and Yarns. 7 pp.

—**Linings—Quality Assurance**
CNS L4035-66. Colored Lining Fabrics (for Winter Clothes) (Tentative) (Sep)(2564) (R 1973). 1 p.
CNS L4036-66. Colored Lining Fabrics, for Clothes and Raincoats (Tentative) (Sep)(2565)(R1973). 2 pp.
CNS L4040-66. Lining Fabrics for Woolen and Worsted Clothes (Tentative) (Sep)(2639) (R 1973). 2 pp.

—**Liquid Resistance**
BSI BS 4724: Part 1-86. 1986 Resistance of Clothing Materials to Permeation by Liquids Part 1: Method for Assessment of Breakthrough Time. 8 pp.
BSI BS 4724: Part 2-88. (WITHDRAWN) 1988 Resistance of Clothing Material to Permeation by Liquids Part 2: Method for Determination of Liquid Permeating After Breakthrough (Superseded by BS EN 369: 1993). 11 pp.

—**Nylon Fabrics—Waterproof—Visual Inspection**
CNS L4039-84. Polymethyl Acrylate Resin Treated Waterproof Nylon Cloth (Jul)(2637). 2 pp.

—**Patterns**
JIS L 0110-90. Symbol Marks for Paper Pattern. 15 pp.

—**Polyurethane Foams**
CNS K3038-80. Flexible Urethane Foam for Garments (Jul)(5844). 2 pp.
CNS K6526-80. Method of Test for Flexible Urethane Foam (for Garments) (Jul)(5845).
JIS K 6402-76. Flexible Urethane Foam for Garments (R 1979). 9 pp.

—**Quality Assurance**
BSI BS 6476-84. 1984 Garment Quality and Relevant British Standards. 7 pp.

—**Seam Strength**
CNS L3142-81. Method of Test for Seam Strength of Clothes (Nov)(8150).
JIS L 1093-78. Testing Methods for Seam Strength of Clothes (R 1988). 11 pp.

—**Sizes**
BSI BS 5511-77. 1977 Amd 1 Size Designation of Clothes. Definitions and Body Measurement Procedure. 9 pp.
ISO TR10652-91. Standard Sizing Systems for Clothes First Edition. 29 pp.
JIS L 0103-90. General Rule on Sizing Systems and Designation for Clothes. 26 pp.

—**Sizes—Boys'**
BSI BS 3728-82. 1982 Size Designation of Childrens and Infants Wear. 12 pp.
CGSB CAN/CGSB-49.5-M85. Canada Standard System for Sizing Girls' and Boys' Apparel. 100 pp.
CGSB CAN/CGSB-49.6-M78. Application of the Canada Standard System for the Sizing of Girls' and Boys' Apparel; (Amendment 1 Nov 1983) (Supplement No. 1 June 1991). 98 pp.
JIS L 4002-84. Sizing Systems for Boys' Garments. 21 pp.
SAA AS 1182-80. Size Coding Scheme for Infants' and Childrens' Clothing (Underwear and Outerwear). 16 pp.
SNZ NZS 8774-73. Size Designations and Body Measurements for the Sizing of Boy's Ready-To-Wear Clothing (Reconfirmed 1983). 8 pp.

—**Sizes—Children's**
BSI BS 3728-82. 1982 Size Designation of Childrens and Infants Wear. 12 pp.
SAA AS 1182-80. Size Coding Scheme for Infants' and Childrens' Clothing (Underwear and Outerwear). 16 pp.
SNZ NZS 8771-73. Size Designations and Body Measurements for the Sizing of Ready-To-Wear Clothing for Babies and Infants Aged 3 Months to 5 Years (Reconfirmed 1983). 8 pp.

—**Sizes—Color Coding**
CGSB 38-GP-38-74. Color Code, for Size Identification of Clothing and Linen, Standard for. 5 pp.

—**Sizes—Girls'**
BSI BS 3728-82. 1982 Size Designation of Childrens and Infants Wear. 12 pp.
CGSB CAN/CGSB-49.5-M85. Canada Standard System for Sizing Girls' and Boys' Apparel. 100 pp.
CGSB CAN/CGSB-49.6-M78. Application of the Canada Standard System for the Sizing of Girls' and Boys' Apparel; (Amendment 1 Nov 1983) (Supplement No. 1 June 1991). 98 pp.
JIS L 4003-84. Sizing Systems for Girls' Garments. 22 pp.
SAA AS 1182-80. Size Coding Scheme for Infants' and Childrens' Clothing (Underwear and Outerwear). 16 pp.
SNZ NZS 8772-73. Size Designations and Body Measurements for the Sizing of Girls' Ready-To-Wear Clothing (Reconfirmed 1983). 16 pp.

—**Sizes—Infants'**
BSI BS 3728-82. 1982 Size Designation of Childrens and Infants Wear. 12 pp.
CGSB CAN/CGSB-49.7-M78. Canada Standard System for Sizing Infants' Apparel; (Amendment 2 Apr 1985) (Supplement No. 1 June 1991). 19 pp.
ISO 3638-77. Size Designation of Clothes—Infants' Garments First Edition. 5 pp.
JIS L 4001-83. Sizing Systems for Infants' Garments.
SNZ NZS 8771-73. Size Designations and Body Measurements for the Sizing of Ready-To-Wear Clothing for Babies and Infants Aged 3 Months to 5 Years (Reconfirmed 1983). 8 pp.

—**Sizes—Men's**
BSI BS 6185-82. 1982 Amd 1 Size Designation of Men's Wear. 9 pp.
JIS L 4004-80. Sizing Systems for Men's Garments.
SAA AS 1954-76. Size Designation Scheme for Men's Clothing (Including Multiple Fitting Outerwear and Industrial Wear) Amdt 1 March 1988. 16 pp.
SNZ NZS 8775-74. Size Designations and Body Measurements for the Sizing of Men's Ready-To-Wear Clothing Other Than Shirts (Reconfirmed 1983). 12 pp.

—**Sizes—Women's**
BSI BS 3666-82. 1982 Size Designation of Women's Wear. 8 pp.
CGSB CAN/CGSB-49.201-92. Canada Standard System for Sizing Women's Apparel. 105 pp.
CGSB CAN/CGSB-49.202-92. Application of the Canada Standard System for the Sizing of Women's Wearing Apparel. 19 pp.
CGSB CAN/CGSB-49.203-M87. Canada Standard Sizes for Women's Apparel—Trade Sizes; (Supplement No. 1 June 1991). 99 pp.
JIS L 4005-85. Sizing Systems for Women's Garments. 27 pp.
SAA AS 1344-75. Size Coding Scheme for Women's Clothing (Underwear, Outerwear and Foundation Garments) Amdt 1 October 1984. 22 pp.
SNZ NZS 8773-73. Size Designations and Body Measurements for the Sizing of Women's Ready-To-Wear Clothing (Reconfirmed 1983). 12 pp.

—**Water Resistance Testing**
CNS L3201-83. Method of Test for Water Resistance of Clothes Pressure Test (Jul) (10460).
CNS L3202-83. Method of Test for Water Resistance of Clothes (Jul)(10461).
CNS L3203-83. Method of Test for Water Resistance of Clothes (Jul)(10462).
CNS L3223-88. Method of Test for Water Vapor Permeability of Clothes (Jan)(12222).
JIS L 1092-92. Testing Methods for Water Resistance of Clothes. 36 pp.
JIS L 1099-85. Testing Methods for Water Vapour Permeability of Clothes. 9 pp.

—**White**
JIS L 4107-92. White Garments.

INTERNATIONAL AND NON-U.S. NATIONAL STANDARDS
SUBJECT INDEX

Clothing (Cont.)
—White (Cont.)
JIS L 4107-85. White Garments. 11 pp.
—Women's—Mannequins
CGSB CAN/CGSB-49.204-M89. Junior, Misses and Women's Canada Standard Sizes Model Forms—Dimensions. 13 pp.

Clothing Hangers
Use: Hangers (Clothing)

Clothing Racks
Use For: Garment Racks *See Also:* Clothing; Racks (Storage)
—Metal
CGSB 44.160A-93. Metal Costumer. 7 pp.
—Steel
CGSB 44.14A-93. Metal Rack for Clothing. 7 pp.

Cloths (Cleaning)
Use: Cleaning Cloths

Cloud Point
See Also: Clear Point; Thermodynamic Properties
—Beeswax
MOD UK M 2012/71. Examination of Beeswax GS and Beeswax, Lead Free.
—Crude Oil
CNS K6914-87. Method of Test for Cloud Point of Petroleum Oils (Jul)(12015).
JIS K 2269-87. Testing Methods for Pour Point and Cloud Point of Crude Oil and Petroleum Products. 19 pp.
—Fats
BSI BS 684: Sec 1.5-87. 1987 Amd 1 Methods of Analysis of Fats and Fatty Oils Part 1: Physical Methods Section 1.5: Determination of Cloud Point. 7 pp.
CNS N6086-82. Methods of Test for Edible Oils and Fats (Cloud Point Test) (Jan)(3651). 1 p.
—Halogenated Hydrocarbons
BSI BS 5598: Part 5-79. 1979 Methods of Sampling and Test for Halogenated Hydrocarbons Part 5: Determination of Cloud Point. 4 pp.
ISO 1394-77. Liquid Halogenated Hydrocarbons for Industrial Use—Determination of Cloud Point First Edition. 4 pp.
—Nonionic Surfactants
BSI BS 6829: Sec 4.1-91. 1991 Analysis of Surface Active Agents (Raw Materials) Part 4: Ethylene Oxide Adducts Section 4.1: Methods for Determination of Cloud Temperature (Cloud Point). 11 pp.
BSI BS 6829: Sec 4.1-88. 1988 Analysis of Surface Active Agents (Raw Materials) Part 4: Ethylene Oxide Adducts Section 4.1: Methods for Determination of Cloud Temperature (Cloud Point). 7 pp.
ISO 1065-91. Non-Ionic Surface-Active Agents Obtained from Ethylene Oxide and Mixed Non-Ionic Surface-Active Agents—Determination of Cloud Point Second Edition. 9 pp.
—Nonionic Surfactants—Volumetric Analysis
ISO 4320-77. Non-Ionic Surface Active Agents—Determination of Cloud Point Index—Volumetric Method First Edition; (Erratum—June 1980). 6 pp.
—Petroleum Products
BSI BS 2000: Part 219-93. 1993 Methods of Test for Petroleum and Its Products Part 219: Petroleum Products—Determination of Cloud Point (ISO 3015: 1992) (W). 6 pp.
BSI BS 2000: Part 219-83. 1983 Petroleum and Its Products Part 219: Cloud Point of Petroleum Oils. 6 pp.
ISO 3015-92. Petroleum Products—Determination of Cloud Point Second Edition. 7 pp.
JIS K 2269-87. Testing Methods for Pour Point and Cloud Point of Crude Oil and Petroleum Products. 19 pp.
—Surfactants
BSI BS 6829: Sec 4.4-89. 1989 Analysis of Surface Active Agents (Raw Materials) Section 4.4: Method for Determination of Cloud Point Index. 7 pp.
—Vegetable Oils
CNS N6086-82. Methods of Test for Edible Oils and Fats (Cloud Point Test) (Jan)(3651). 1 p.

Clove Oil
See Also: Eugenol
CNS K5080-80. Oil of Clove Leaf (Sep)(6451).
CNS K5081-80. Oil of Clove Bud (Sep)(6452).

Clove Oil (Cont.)
CNS K5082-80. Oil of Clove Stem (Sep)(6453).
ISO 3141-86. Oil of Clove Leaf (Syzygium Aromaticum (Linnaeus) Merrill Et Perry Syn. Eugenia Caryophyllus (C.Sprengel) Bullock Et Harrison) Second Edition. 5 pp.
ISO 3142-74. Oil of Clove Bud First Edition. 4 pp.
ISO 3143-75. Oil of Clove Stem First Edition. 3 pp.

Cloves
See Also: Spices
BSI BS 7087: Part 4-90. 1990 Herbs and Spices Ready for Food Use Part 4: Cloves (Whole and Ground). 10 pp.
ISO 2254-80. Cloves, Whole and Ground (Powdered)—Specification First Edition. 6 pp.

Clutch Facings
Use: Clutches—Facings

Clutch Housings
Use: Bell Housings

Clutch Linings
Use: Clutches—Linings

Clutches
See Also: Couplings; Disk Clutches; Electromagnetic Clutches; Friction Clutches; Hydraulic Clutches; Master Cylinders; Slip Friction Clutches
—Automotive
CNS D3113-82. Method of Bench Performance Test for Clutch for Automobiles (Apr)(8677).
—Ball Bearings—Automotive
CNS D2085-81. Clutch Release Ball Bearings for Automobiles (Mar)(7133).
—Covers—Automotive
CNS D2126-87. Clutch Cover Assemblies for Automobiles (Jun)(8673).
CNS D3111-87. Method of Test for Clutch Cover Assemblies for Automobiles (Jun)(8674).
—Facings—Automotive
BSI BS AU 208-86. 1986 Assessment of Sticking of Automobile Clutch Facings. 6 pp.
CNS D2005-85. Clutch Facings for Automobiles (Jun)(3011). 7 pp.
CNS D2006-85. Rivets for Brake Linings and Clutch Facings for Automobiles (Jun)(3012).
JIS D 4311-90. Clutch Facings for Automobiles. 15 pp.
JIS D 4312-90. Rivets of Brake Linings and Clutch Facings for Automobiles. 8 pp.
—Facings—Automotive—Hardness Testing
JIS D 4421-87. Method of Hardness Test for Brake Linings, Pads and Clutch Facings of Automobiles. 11 pp.
—Glossaries
JIS B 0152-73. Glossary of Terms Relating to Clutches and Brakes.
—Linings—Asbestos—Industrial Equipment
CNS R2179-85. Asbestos Brake Linings for Industrial Machines (May)(11268).
JIS R 3455-79. Asbestos Brake Linings for Industrial Machines. 15 pp.
—Linings—Rivets
BSI BS 3575-63. 1963 Rivets for the Attachment of Friction Linings. 8 pp.
CNS B2582-81. Rivets for Brake and Clutch Lining (Jun)(7570).
DIN ENGL 7338-83. Rivets for Brake Linings and Clutch Linings (Dec). 5 pp.
JIS D 4312-90. Rivets of Brake Linings and Clutch Facings for Automobiles. 8 pp.
—Lubricants
BSI BS 6413: Part 2-83. 1983 Lubricants, Industrial Oils and Related Products (Class L) Part 2: Classification for Family F (Spindle Bearings, Bearings and Associated Clutches). 3 pp.
ISO 6743 Pt 2-81. Lubricants, Industrial Oils and Related Products (Class L)—Classification—Part 2: Family F (Spindle Bearings, Bearings and Associated Clutches) First Edition. 3 pp.

CMOS Digital Circuits
Use: Digital Circuits—CMOS

CMTT
See Also: CCIR; CCITT
CCIR RESOLUTION 61-4-90. Structure of CCIR Study Groups—Volume XIV—Administrative Texts of the CCIR. 8 pp.

CNC Controllers
Use: Computer Numerical Control

CNC Units
Use: Computer Numerical Control

CNIP
Use: Calling Name Identification Presentation

CNS
Use For: Chinese (ROC) National Standards
See Also: Standards
—Catalogs
CNS. CNS Catalog 1993. 358 pp.
—Classification
CNS Z7005-78. Classification of Chinese National Standards (Jul)(88).
—Indexes (Documentation)
CNS. Alphabetical Subject Index.
—Standards Preparation
CNS Z7043-86. Style for Chinese National Standard (Sep)(3689).

Co-Channel Interference
See Also: Radio Frequency Interference
CCIR Report 485-1-82. Contribution to the Planning of Broadcasting Services—Section 11E—Planning of Television Networks, Protection Ratios, Television Receivers and Antennas. 5 pp.
—Cellular Mobile Radio Equipment
CCIR Report 740-2-86. General Aspects of Cellular Systems—Section 8A—Land Mobile Service and Related Subjects. 9 pp.
—Mobile Radio Equipment—Adaptive Antennas
CCIR Report 973-86. Adaptive Antennas for Land Mobile Applications in Reducing Co-Channel Interference—Section 1D—Spectrum Utilization and Applications. 5 pp.
—Multiple Phase Shift Keying
CCIR Report 528-1-78. Co-Channel Interference Effects on Multiple Phase Shift Keying (MPSK) System Performance—Section 1B—Spectrum Sharing and Planning Principles and Techniques. 9 pp.
—Single Sideband Communication Equipment
CCIR Report 1018-1-90. Co-Channel and Adjacent-Channel Coordination Criteria for Simultaneous Use of Different Modulation Techniques in the Mobile Service—Section 8A—Land Mobile Service and Related Subjects. 6 pp.
—Television Broadcasting
CCIR Report 481-70. Ratio of Wanted-to-Unwanted Signal in Television Subjective Assessment of Multiple Co-Channel Interference—Section 11E—Planning of Television Networks, Protection Ratios, Television Receivers and Antennas. 2 pp.
CCIR QUESTION 75/11-90. Methods of Reducing Interference to the Broadcasting Service (Television) from Other Services Operating in the Same or Adjacent Bands—Questions Concerning Study Group 11—Broadcasting Service (Television). 1 p.
—Television Channels—Radio Relay Systems
CCIR Report 931-82. Study of the Effects of Co-Channel Interference in the TV Channels of FM Radio-Relay Systems—Section 9A—Performance Objectives, Propagation and Interference Effects. 4 pp.

Coach Screws
Use: Lag Bolts

Coagulant Content Analysis
—Latex
SAA AS 1683.5-74. Methods of Test for Elastomers—Part 5: Coagulum Content of Rubber Latices Reconfirmed 1989.
—Latex—Sieve Analysis
BSI BS 6057: Sec 3.8-87. 1987 Rubber Latices Part 3: Methods of Test Section 3.8: Determination of Coagulum Content (Sieve Residue). 4 pp.
ISO 706-85. Rubber Latex—Determination of Coagulum Content (Sieve Residue) Third Edition. 4 pp.

Coagulation
Scope Note: Use a more specific term
See: Sedimentation

Coal
Scope Note: For additional listings, use a more specific term

Coal (Cont.)

See Also: Anthracite; Bituminous Coal; Coal Ash; Coal Preparation Plants; Fossil Fuels; Lignite; Magnetite; Pulverized Fuels; Rocks; Solid Mineral Fuels; Sub-Bituminous Coal; Vitrinite

JIS M 8815-76. Methods for Analysis of Coal Ash and Coke Ash.

SAA TR2.13-92. Certified Reference Materials—Part 13: Coal—Preparation and Certification of ASCRM-013 (Supersedes TR2.9—1985). 5 pp.

—Abrasion Testing

BSI BS 1016: Part 19-80. (WITHDRAWN) 1980 Methods for the Analysis and Testing of Coal and Coke Part 19: Determination of the Index of Abrasion of Coal (Superseded by BS 1016: Part 111: 1990). 7 pp.

BSI BS 1016: Part 111-93. 1993 Methods for Analysis and Testing of Coal and Coke Part 111: Determination of Abrasion Index of Coal (W). 12 pp.

BSI BS 1016: Part 111-90. 1990 Methods for the Analysis and Testing of Coal and Coke Part 111: Determination of Abrasion Index of Coal. 11 pp.

SAA AS 1038.19-89. Methods for the Analysis and Testing of Coal and Coke—Part 19: Determination of the Abrasion Index of Higher Rank Coal. 6 pp.

—Arsenic Content

SAA AS 1038.10-80. Methods for the Analysis and Testing of Coal and Coke—Part 10: Arsenic in Coal and Cokes. 16 pp.

—Arsenic Content—Oxidation Method

BSI BS 1016: Part 10-77. 1977 Methods for the Analysis and Testing of Coal and Coke Part 10: Arsenic in Coal and Coke. 8 pp.

SAA AS 1038.10-80. Methods for the Analysis and Testing of Coal and Coke—Part 10: Arsenic in Coal and Cokes. 16 pp.

—Ash Content

BSI BS 1016: Part 3-73. (WITHDRAWN) 1973 Amd 2 Methods for the Analysis and Testing of Coal and Coke Part 3: Proximate Analysis of Coal (Superseded by BS 1016: Sections 104.1, 104.2, 104.3, and 104.4: 1991). 23 pp.

BSI BS 1016: Part 14-63. 1963 Methods for the Analysis and Testing of Coal and Coke Part 14: Analysis of Coal Ash and Coke Ash. 37 pp.

CNS M3141-84. Method for Determination of Ash of Coal and Coke (Mar)(10822).

CNS M3155-84. Method for Analysis of Coal Ash and Coke Ash (Mar)(10836).

JIS M 8812-84. Methods for Proximate Analysis of Coal and Coke. 28 pp.

SAA AS 1038.3-89. Methods for the Analysis and Testing of Coal and Coke—Part 3: Proximate Analysis of Higher Rank Coal. 12 pp.

—Ash Content—Atomic Absorption Spectrometry—Acid Digestion

SAA AS 1038.14.2-85. Methods for the Analysis and Testing of Coal and Coke—Part 14.2: Analysis of Higher Rank Coal Ash, Coke Ash (Acid Digestion)—Flame Atomic Absorption Spectrometric Method). 11 pp.

—Ash Content—Atomic Absorption Spectrometry—Borate Fusion

SAA AS 1038.14.1-81. Methods for the Analysis and Testing of Coal and Coke—Part 14.1: Analysis of Coal Ash, Coke Ash and Mineral Matter (Borate Fusion—Flame Atomic Absorption Spectrometric Method). 18 pp.

—Ash Content—Fusibility

SAA AS 1038.15-87. Methods for the Analysis and Testing of Coal and Coke—Part 15: Fusibility of Higher Rank Coal Ash and Coke Ash. 6 pp.

—Ash Content—Reference Method

BSI BS 1016: Sec 104.4-91. 1991 Analysis and Testing of Coal and Coke Part 104: Proximate Analysis Section 104.4: Determination of Ash. 9 pp.

—Bin Flow

SAA AS 3880-91. Bin Flow Properties of Coal (in Professional Package 32). 14 pp.

—Boron Content—Spectrophotometry

SAA AS 1038.10.3-88. Methods for the Analysis and Testing of Coal and Coke—Part 10.3: Determination of Trace Elements—Coal, Coke and Fly-Ash—Determination of Boron Content—Spectrophotometric Method. 5 pp.

—Bulk Density

SAA AS 3899-91. Higher Rank Coal and Coke—Bulk Density (in Professional Package 32). 7 pp.

Coal (Cont.)

—Caking—Gray-King Coke Test

BSI BS 1016: Sec 107.2-91. 1991 Analysis and Testing of Coal and Coke Part 107: Caking and Swelling Properties of Coal Section 107.2: Assessment of Caking Power by Gray-King Coke Test. 17 pp.

CNS M3169-84. Method for Determination of Caking Power of Coal-Gray-King Coke Test (Sep)(11022).

ISO 502-82. Coal—Determination of Caking Power—Gray-King Coke Test Second Edition. 11 pp.

—Caking—Roga Test

CNS M3168-84. Method for Determination of Caking Power of Hard Coal-Roga Test (Sep) (11021).

ISO 335-74. Hard Coal—Determination of Caking Power—Roga Test First Edition. 6 pp.

—Calorific Value

BSI BS 1016: Part 5-77. (WITHDRAWN) 1977 Methods for the Analysis and Testing of Coal and Coke Part 5: Gross Calorific Value of Coal and Coke (Superseded by BS 1016: Part 105: 1992). 12 pp.

BSI BS 1016: Part 105-92. 1992 Methods for Analysis and Testing of Coal and Coke Part 105: Determination of Gross Calorific Value (W). 17 pp.

JIS M 8814-85. Determination of Calorific Value of Coal and Coke; (Erratum). 20 pp.

SAA AS 1038.5.1-88. Methods for the Analysis and Testing of Coal and Coke—Part 5.1: Gross Specific Energy of Coal and Coke—Adiabatic Calorimeters. 9 pp.

SAA AS 1038.5.2-89. Methods for the Analysis and Testing of Coal and Coke—Part 5.2: Gross Specific Energy of Coal and Coke —Automatic Isothermal-Type Calorimeters. 5 pp.

—Carbon Content

BSI BS 1016: Part 3-73. (WITHDRAWN) 1973 Amd 2 Methods for the Analysis and Testing of Coal and Coke Part 3: Proximate Analysis of Coal (Superseded by BS 1016: Sections 104.1, 104.2, 104.3, and 104.4: 1991). 23 pp.

CNS M3143-84. Method for Calculation of Fixed Carbon of Coal and Coke (Mar)(10824).

JIS M 8812-84. Methods for Proximate Analysis of Coal and Coke. 28 pp.

SAA AS 1038.6.1-86. Methods for the Analysis and Testing of Coal and Coke—Part 6.1: Ultimate Analysis of Higher Rank Coal—Determination of Carbon and Hydrogen (R 1992). 5 pp.

SAA AS 1038.7-81. Methods for the Analysis and Testing of Coal and Coke—Part 7: Ultimate Analysis of Coke. 21 pp.

—Carbon Content—Combustion

CNS M3153-84. Method for Determination of Carbons and Hydrogen of Coal and Coke (Sheffield High Temperature Method) (Mar)(10834).

ISO 609-75. Coal and Coke—Determination of Carbon and Hydrogen—High Temperature Combustion Method First Edition. 9 pp.

—Carbon Content—Gravimetric Analysis

SAA AS 1038.23-84. Methods for the Analysis and Testing of Coal and Coke—Part 23: Determination of Carbonate Carbon in Higher Rank Coal. 3 pp.

—Carbon Content—Liebig Method

CNS M3144-84. Method for Determination of Carbon and Hydrogen of Coal and Coke (Liebig Method) (Mar) (10825).

ISO 625-75. Coal and Coke—Determination of Carbon and Hydrogen—Liebig Method First Edition. 9 pp.

—Chemical Analysis

CNS M3139-84. General Rules for Testing of Coal and Coke (Mar) (10820).

CPPA J.7H-68. Sampling and Analysis of Coal. 10 pp.

JIS M 8810-89. General Rules for Sampling, Analysis and Testing of Coal and Coke. 26 pp.

JIS M 8813-88. Methods for Ultimate Analysis of Coal and Coke. 100 pp.

—Chemical Analysis—Formulas

ISO 1170-77. Coal and Coke—Calculation of Analyses to Different Bases First Edition. 6 pp.

—Chemical Analysis—Glossaries

ISO 1213 Pt 2-92. Solid Mineral Fuels—Vocabulary—Part 2: Terms Relating to Sampling, Testing and Analysis First Edition. 16 pp.

—Chlorine Content

CNS M3151-84. Method for Determination of Chlorine in Coal (Mar)(10832).

SAA AS 1038.8-80. Methods for the Analysis and Testing of Coal and Coke—Part 8: Chlorine in Coal and Coke (Superseded (in Part) by AS 1038.8.1—1992 Will Remain Current). 12 pp.

Coal (Cont.)

—Chlorine Content—Combustion

BSI BS 1016: Part 8-77. 1977 Amd 1 Methods for Analysis and Testing of Coal and Coke Part 8: Chlorine in Coal and Coke (AMD 3472) December 31, 1980. 9 pp.

BSI BS 1016: Part 8-02. 1977 Amd 2 Methods for Analysis and Testing of Coal and Coke Part 8: Chlorine in Coal and Coke (AMD 6984) May 1, 1992. 12 pp.

—Chlorine Content—Eschka Method

BSI BS 1016: Part 8-77. 1977 Amd 1 Methods for Analysis and Testing of Coal and Coke Part 8: Chlorine in Coal and Coke (AMD 3472) December 31, 1980. 9 pp.

BSI BS 1016: Part 8-02. 1977 Amd 2 Methods for Analysis and Testing of Coal and Coke Part 8: Chlorine in Coal and Coke (AMD 6984) May 1, 1992. 12 pp.

SAA AS 1038.8.1-92. Coal and Coke—Analysis and Testing—Part 8.1: Coal and Coke—Chlorine—Eschka Method (Supersedes AS 1038.8—1980 (in Part)) Which Will Remain Current) (in Professional Package 32). 2 pp.

—Classification

SAA AS 1634-74. Recommended Procedures for the Expression and Presentation of Results of Tests of Coal Size Classifying Equipment Corrig. Reconfirmed 1989. 12 pp.

SAA AS 2096-87. Classification and Coding Systems for Australian Coals. 5 pp.

—Coking

CNS M1010-82. Coking Coal (Sep)(2508).

—Comprehensive Testing—Reputing

SAA AS 1038.16-86. Methods for the Analysis and Testing of Coal and Coke—Part 16: Acceptance and Reporting of Results. 15 pp.

—Defense Contracts

MOD UK DEFCON 112AA-83. Basic Set of Conditions of Contract for the Supply of Coal and Coke 9/83. 4 pp.

—Density

SAA AS 1038.21-83. Methods for the Analysis and Testing of Coal and Coke—Part 21: Determination of the Relative Density and Apparent Relative Density of Hard Coal (Superseded (in Part) by AS 1038.21.2—1992). 10 pp.

—Extraction Analysis

SAA AS 1038.9.2-92. Coal and Coke—Analysis and Testing—Part 9.2: Higher Rank Coal—Phosphorus—Coal Extraction Method (in Professional Package 32). 3 pp.

—Float and Sink Analysis

CNS M3172-84. Method of Test for Float and Sink of Coal (Sep)(11025).

—Fluorine Content—Pyrohydrolysis

SAA AS 1038.10.4-89. Methods for the Analysis and Testing of Coal and Coke—Part 10.4: Determinaton of Trace Elements—Coal, Coke and Fly-Ash—Determination of Fluorine Content—Phrohydrolysis Method Amdt 1 October 1990. 11 pp.

—Fragmentation

CNS M3161-84. Method of Drop Shatter Test for Coal (Jun) (10946).

—Froth Flotation Testing

BSI BS 7530: Part 1-92. 1992 Methods for Froth Flotation Testing of Hard Coal Part 1: Laboratory Procedure. 12 pp.

—Fusibility

BSI BS 1016: Part 15-70. 1970 Methods for the Analysis and Testing of Coal and Coke Part 15: Fusibility of Coal Ash and Coke Ash. 16 pp.

CNS M3163-84. Method of Test for Fusibility of Coal and Coke Ash (Jun)(10948).

—Glossaries

BSI BS 3323-78. 1978 Glossary of Coal Terms. 12 pp.

BSI BS 3552-82. 1982 Coal Preparation Terms. 31 pp.

JIS M 0104-84. Technical Terms Used in Coal Utilization.

—Hardgrove Grindability Index

BSI BS 1016: Part 20-81. 1981 Methods for the Analysis and Testing of Coal and Coke Part 20: Determination of Hardgrove Grindability Index of Hard Coal. 12 pp.

CNS M3164-84. Method of Test for Grindability of Coal by the Hardgrove-Machine Method (Sep)(11017).

ISO 5074-80. Hard Coal—Determination of Hardgrove Grindability Index First Edition. 10 pp.

INTERNATIONAL AND NON-U.S. NATIONAL STANDARDS
SUBJECT INDEX

Coal

Coal (Cont.)

—Hardgrove Grindability Index (Cont.)

SAA AS 1038.20-92. Coal and Coke—Analysis and Testing—Part 20: Higher Rank Coal—Hardgrove Grindability Index (in Professional Package 32). 6 pp.

SAA CRM 011E-91. Certified Reference Materials—Certified Reference Coal (Hardgrove Grindability Index)—Part E: Hardgrove Grindability Index: 36 (Superseded by CRM 011-3A—1992).

SAA CRM 011F-91. Certified Reference Materials—Certified Reference Coal (Hardgrove Grindability Index)—Part F: Hardgrove Grindability Index: 48 (Superseded by CRM 011-3B—1992).

SAA CRM 011G-91. Certified Reference Materials—Certified Reference Coal (Hardgrove Grindability Index)—Part G: Hardgrove Grindability Index: 68 (Superseded by CRM 011-3C—1992).

SAA CRM 011H-91. Certified Reference Materials—Certified Reference Coal (Hardgrove Grindability Index)—Part H: Hardgrove Grindability Index: 96 (Superseded by CRM 011-3D—1992).

SAA TR2.11-92. Certified Reference Materials—Part 11: Coal—Preparation and Certification of ASCRM-011-3 (Hardgrove Grindability Index) (in Professional Package 32). 2 pp.

—Hydrogen Content

SAA AS 1038.6.1-86. Methods for the Analysis and Testing of Coal and Coke—Part 6.1: Ultimate Analysis of Higher Rank Coal—Determination of Carbon and Hydrogen (R 1992). 5 pp.

SAA AS 1038.7-81. Methods for the Analysis and Testing of Coal and Coke—Part 7: Ultimate Analysis of Coke. 21 pp.

—Hydrogen Content—Combustion

CNS M3153-84. Method for Determination of Carbons and Hydrogen of Coal and Coke (Sheffield High Temperature Method) (Mar)(10834).

ISO 609-75. Coal and Coke—Determination of Carbon and Hydrogen—High Temperature Combustion Method First Edition. 9 pp.

—Hydrogen Content—Liebig Method

CNS M3144-84. Method for Determination of Carbon and Hydrogen of Coal and Coke (Liebig Method) (Mar) (10825).

ISO 625-75. Coal and Coke—Determination of Carbon and Hydrogen—Liebig Method First Edition. 9 pp.

—Inorganic Content Analysis

SAA AS 2434.9-91. Methods for the Analysis and Testing of Lower Rank Coal and its Chars—Part 9: Determination of Four Acid-Extractable Inorganic Ions in Lower Rank Coal (In Professional Package 32). 4 pp.

—Invitation for Bids

MOD UK DEFCON 47C-89. Invitation to Tender for the Supply and Delivery of Coal and/or Coke 11/89. 2 pp.

—Mineral Content

ISO 602-83. Coal—Determination of Mineral Matter Second Edition. 6 pp.

JIS M 8818-86. Method for Determination of Mineral Matter in Coal. 12 pp.

SAA AS 1038.22-92. Coal and Coke—Analysis and Testing—Part 22: Higher Rank Coal—Mineral Matter and Water of Constitution (in Professional Package 32). 12 pp.

—Moisture Content

BSI BS 1016: Part 1-73. 1973 Methods for the Analysis and Testing of Coal and Coke Part 1: Total Moisture of Coal. 11 pp.

BSI BS 1016: Part 3-73. (WITHDRAWN) 1973 Amd 2 Methods for the Analysis and Testing of Coal and Coke Part 3: Proximate Analysis of Coal (Superseded by BS 1016: Sections 104.1, 104.2, 104.3, and 104.4: 1991. 23 pp.

BSI BS 1016: Part 21-81. 1981 Methods for the Analysis and Testing of Coal and Coke Part 21: Determination of Moisture-Holding Capacity of Hard Coal. 10 pp.

CNS M3140-84. Method for Determination of Inherent Moisture of Coal and Coke (Mar) (10821).

JIS M 8803-76. Methods for Test of Equilibrium Moisture of Coal at 96 to 97 Percent Relative Humidity and 30 Degrees C.

JIS M 8811-76. Method for Sampling and Determination of Total Moisture and Adherent Moisture of Coal and Coke.

JIS M 8812-84. Methods for Proximate Analysis of Coal and Coke. 28 pp.

SAA AS 1038.1-92. Coal and Coke—Analysis and Testing—Part 1: Higher Rank Coal—Total Moisture (in Professional Package 32) Amdt 1 October—1993. 9 pp.

SAA AS 1038.3-89. Methods for the Analysis and Testing of Coal and Coke—Part 3: Proximate Analysis of Higher Rank Coal. 12 pp.

Coal (Cont.)

—Moisture Content (Cont.)

SAA AS 1038.17-89. Methods for the Analysis and Testing of Coal and Coke Part 17: Determination of Moisture-Holding Capacity (Equilibrium Moisture) of Higher Rank Coal. 5 pp.

SAA AS 2434.1-91. Methods for the Analysis and Testing of Lower Rank Coal and its Chars—Part 1: Determination of the Total Moisture Content of Lower Rank Coal (In Professional Package 32). 7 pp.

—Moisture Content—Gravimetric Analysis

ISO 331-83. Coal—Determination of Moisture in the Analysis Sample—Direct Gravimetric Method Second Edition. 5 pp.

—Moisture Content—Liebig Method

CNS M3144-84. Method for Determination of Carbon and Hydrogen of Coal and Coke (Liebig Method) (Mar) (10825).

ISO 625-75. Coal and Coke—Determination of Carbon and Hydrogen—Liebig Method First Edition. 9 pp.

—Moisture Content—Reference Method

BSI BS 1016: Sec 104.1-91. 1991 Analysis and Testing of Coal and Coke Part 104: Proximate Analysis Section 104.1: Determination of Moisture Content of the General Analysis Sample of Coal. 9 pp.

—Nitrogen Content

SAA AS 1038.6.2-86. Methods for the Analysis and Testing of Coal and Coke—Part 6.2: Ultimate Analysis of Higher Rank Coal—Determination of Nitrogen (R 1992). 4 pp.

SAA AS 1038.7-81. Methods for the Analysis and Testing of Coal and Coke—Part 7: Ultimate Analysis of Coke. 21 pp.

—Nitrogen Content—Kjeldahl Method

CNS M3147-84. Method for Determination of Nitrogen of Coal and Coke (Kjeldahl Method) (Mar)(10828).

ISO 333-83. Coal—Determination of Nitrogen—Semi-Micro Kjeldahl Method Third Edition. 6 pp.

—Oxygen Content

CNS M3148-84. Method for Calculation of Oxygen Content of Coal and Coke (Mar)(10829).

ISO 1994-76. Hard Coal—Determination of Oxygen Content First Edition. 14 pp.

—Petrography

CNS M3167-84. Method for Microscopical Determination of Volume Percent of Petrographic Components of Coal (Sep)(11020).

—Phosphorus Content—Absorptiometric Analysis

CNS M3149-84. Method for Determination of Phosphorus of Coal and Coke (Molybdenum Blue Absorptiometric Method) (Mar)(10830).

—Phosphorus Content—Ash Digestion Method

SAA AS 1038.9.1-92. Coal and Coke—Analysis and Testing—Part 9.1: Coal and Coke—Phosphorus—Ash Digestion/Molybdenum Blue Method (Supersedes AS 1038.9—1977) Amdt 1 April 1993 (in Professional Package 32). 4 pp.

SAA AS 1038.9.3-91. Coal and Coke—Analysis and Testing—Part .9.3: Coal and Coke—Phosphorus—Ash Digestion Method (In Professional Package 32). 3 pp.

—Phosphorus Content—Coal Extraction Method

SAA AS 1038.9.2-92. Coal and Coke—Analysis and Testing—Part 9.2: Higher Rank Coal—Phosphorus—Coal Extraction Method (in Professional Package 32). 3 pp.

—Phosphorus Content—Molybdenum Blue Method

SAA AS 1038.9.1-92. Coal and Coke—Analysis and Testing—Part 9.1: Coal and Coke—Phosphorus—Ash Digestion/Molybdenum Blue Method (Supersedes AS 1038.9—1977) Amdt 1 April 1993 (in Professional Package 32). 4 pp.

—Phosphorus Content—Oxidation Method

BSI BS 1016: Part 9-77. 1977 Methods for the Analysis and Testing of Coal and Coke Part 9: Phosphorus in Coal and Coke. 8 pp.

—Physical Properties

JIS M 8801-79. Methods for Testing of Coal. 71 pp.

—Plastic Properties

CNS M3170-84. Method of Test for Plastic Properties of Coal by the Arnu-Audibert Dilatometer (Sep)(11023).

CNS M3171-84. Method of Test for Plastic Properties of Coal by the Gieseler Plastometer (Sep) (11024).

Coal (Cont.)

—Reflectance

CNS M3165-84. Method of Preparing Coal Samples for Microscopical Analysis by Reflected Light (Sep)(11018).

CNS M3166-84. Method for Microscopical Determination of the Reflectance of the Organic Components in a Polished Specimen of Coal (Sep)(11019).

JIS M 8816-86. Methods of Microscopical Measurement for the Macerals and Reflectance of Coal. 18 pp.

—Reserves

JIS M 1002-78. Calculation of Coal Reserves.

—Samples—Microscopy

SAA AS 2061-89. Preparation of Coal Samples for Incident Light Microscopy. 2 pp.

—Sampling

BSI BS 1017: Part 1-89. 1989 Methods for Sampling of Coal and Coke Part 1: Sampling of Coal. 66 pp.

BSI BS 4845: Part 2-79. 1979 Sampling Manufactured Domestic Solid Smokeless Fuels in Small Consignments of Mass 50Kg to 5000Kg Either in Bulk or in Bags Part 2: Sampling of Solid Smokeless Fuels Other Than Coke. 4 pp.

BSI BS 7067-90. 1990 Guide to Determination and Presentation of Float and Sink Characteristics of Raw Coal and of Products from Coal Preparation Plants. 20 pp.

CNS M3137-83. Method of Sampling for Coal (Nov)(10661).

CNS M3139-84. General Rules for Testing of Coal and Coke (Mar) (10820).

CNS M3140-84. Method for Determination of Inherent Moisture of Coal and Coke (Mar) (10821).

CNS M3156-84. Method for Sampling and Fineness Test of Pulverized Coal (Jun) (10941).

CPPA J.7H-68. Sampling and Analysis of Coal. 10 pp.

JIS M 8810-89. General Rules for Sampling, Analysis and Testing of Coal and Coke. 26 pp.

JIS M 8811-76. Method for Sampling and Determination of Total Moisture and Adherent Moisture of Coal and Coke.

—Seams—Desorbable Gas Content

SAA AS 3980-91. Guide to the Determination of Desorbable Gas Content of Coal Seams—Direct Method (in Professional Package 32). 11 pp.

—Sieve Analysis

BSI BS 1016: Part 17-79. 1979 Methods for the Analysis and Testing of Coal and Coke Part 17: Size Analysis of Coal. 11 pp.

BSI BS 1293 & 2074-65. 1965 Amd 1 Methods for the Size Analysis of Coal and Coke. 52 pp.

CNS M3156-84. Method for Sampling and Fineness Test of Pulverized Coal (Jun) (10941).

CNS M3159-84. Method of Sieve Analysis of Coal (Jun)(10944).

SAA AS 1634-74. Recommended Procedures for the Expression and Presentation of Results of Tests of Coal Size Classifying Equipment Corrig. Reconfirmed 1989. 12 pp.

SAA AS 3881-91. Higher Rank Coal—Size Analysis (in Professional Package 32). 14 pp.

SNZ NZS 535, NZS 2175-67. Methods for the Size Analysis of Coal and Coke Amend: 1; 1A, 1988. 52 pp.

—Sulfur Content

BSI BS 1016: Part 11-77. 1977 Methods for the Analysis and Testing of Coal and Coke Part 11: Forms of Sulphur in Coal. 8 pp.

CNS M3150-84. Method for Determination of Uncombustible Sulfur in Coal Ash and Coke Ash (Mar)(10831).

CNS M3152-84. Method for Determination of Forms of Sulfur in Coal (Mar)(10833).

ISO 157-75. Hard Coal—Determination of Forms of Sulphur First Edition. 12 pp.

JIS M 8817-84. Methods for Determination of Forms of Sulfur in Coal.

SAA AS 1038.7-81. Methods for the Analysis and Testing of Coal and Coke—Part 7: Ultimate Analysis of Coke. 21 pp.

SAA AS 1038.11-93. Coal and Coke—Analysis and Testing—Part 11: Coal and Coke—Forms of Sulfur (in Professional Package 32). 14 pp.

—Sulfur Content—Combustion

CNS M3146-84. Method for Determination of Total Sulfur of Coal and Coke (High Temperature Combustion Method) (Mar)(10827).

ISO 351-84. Solid Mineral Fuels—Determination of Total Sulfur—High Temperature Combustion Method Second Edition. 6 pp.

SAA AS 1038.6.3. 2-86. Methods for the Analysis and Testing of Coal and Coke—Part 6.3.2: Ultimate Analysis of Higher Rank Coal—Determination of Total Sulphur (High Temperature Combustion Method) Amdt 1 September 1986. 4 pp.

INDUSTRY STANDARDS

Coal (Cont.)

—Sulfur Content—Eschka Method
BSI BS 1016: SUB SEC 106.4.1-93. 1993 Methods for Analysis and Testing of Coal and Coke Part 106: Ultimate Analysis of Coal and Coke Section 106.4: Determination of Total Sulfur Content Subsection 106.4.1: Eschka Method (W). 11 pp.
CNS M3145-84. Method for Determination of Total Sulfur of Coal and Coke (Eschka Method) (Mar)(10826).
ISO 334-92. Solid Mineral Fuels—Determination of Total Sulfur—Eschka Method Second Edition. 9 pp.
SAA AS 1038.6.3. 1-86. Methods for the Analysis and Testing of Coal and Coke—Part 6.3.1: Ultimate Analysis of Higher Rank Coal—Determination of Total Sulphur (Eschka Method) (R 1992). 2 pp.

—Sulfur Content—Infrared Analysis
SAA AS 1038.6.3. 3-86. Methods for the Analysis and Testing of Coal and Coke—Part 6.3.3: Ultimate Analysis of Higher Rank Coal—Determination of Total Sulphur (Infrared Method) (R 1992). 3 pp.

—Swelling
SAA AS 1038.12.1-93. Coal and Coke—Analysis and Testing—Part 12.1: Higher Rank Coal—Caking And Coking Properties—Crucible Swelling Number (in Professional Package 32) Amdt 1—October 1993. 7 pp.

—Swelling—Dilatometry
BSI BS 1016: Part 12-80. (WITHDRAWN) 1980 Methods for the Analysis and Testing of Coal and Coke Part 12: Caking and Swelling Properties of Coal (Superseded by BS 1016: Sections 107.1, 107.2: 1991 and 107.3: 1990). 23 pp.
ISO 349-75. Hard Coal—Audibert-Arnu Dilatometer Test First Edition. 13 pp.
SAA AS 1038.12.3-84. Methods for the Analysis and Testing of Coal and Coke—Part 12.3: Determination of the Dilatometer Characteristics of Higher Rank Coal. 9 pp.

—Swelling—Grey-King Coke Test
SAA AS 1038.12.2-90. Methods for the Analysis and Testing of Coal and Coke—Part 12.2: Carbonization Properties of Higher Rank Coal—Determinaiton of Gray-King Coke Type. 6 pp.

—Swelling Index
CNS M3006-73. Free Swelling Index of Coal (Aug)(2507).
CNS M3162-84. Method of Test for Free-Swelling Index of Coal (Jun)(10947).
ISO 501-81. Coal—Determination of the Crucible Swelling Number Second Edition. 7 pp.

—Swelling—Reference Method
BSI BS 1016: Sec 107.1-91. 1991 Analysis and Testing of Coal and Coke Part 107: Caking and Swelling Properties of Coal Section 107.1: Determination of Crucible Swelling Number. 13 pp.

—Test Reports
BSI BS 1016: Part 16-81. 1981 Amd 1 Methods for the Analysis and Testing of Coal and Coke Part 16: Methods for Reporting Results. 19 pp.
BSI BS 3620-63. 1963 Expression and Presentation of Results of Coal Cleaning Tests. 39 pp.

—Trace Element Content
SAA TR2.9 Supp 1-92. Certified Reference Materials—Part 9: Coal—Preparation and Certification of ASCRM-009—Supplement 1: Trace Element Data (Supplement to SAA TR2.9—1985). 1 p.

—Trace Element Content—Atomic Absorption Spectrometry
SAA AS 1038.10.1-86. Methods for the Analysis and Testing of Coal and Coke—Part 10.1: Determination of Trace Elements—Determination of Eleven Trace Elements in Coal, Coke and Fly-Ash—Flame Atomic Absorption Spectrometric Method (R 1992). 10 pp.

—Ultimate Analysis
BSI BS 1016: Part 6-77. 1977 Methods for Analysis and Testing of Coal and Coke Part 6: Ultimate Analysis of Coal. 15 pp.
BSI BS 1016: Part 6-01. 1977 Amd 1 Methods for Analysis and Testing of Coal and Coke Part 6: Ultimate Analysis of Coal (AMD 6982) May 1, 1992 (W). 22 pp.
BSI BS 1016: Part 6-02. 1977 Amd 2 Methods for Analysis and Testing of Coal and Coke Part 6: Ultimate Analysis of Coal (AMD 7689) May 15, 1993 (W). 23 pp.

—Volatile Matter Content
BSI BS 1016: Part 3-73. (WITHDRAWN) 1973 Amd 2 Methods for the Analysis and Testing of Coal and Coke Part 3: Proximate Analysis of Coal (Superseded by BS 1016: Sections 104.1, 104.2, 104.3, and 104.4: 1991). 23 pp.

Coal (Cont.)

—Volatile Matter Content (Cont.)
CNS M3142-84. Method for Determination of Volatile Matter of Coal and Coke (Mar) (10823).
CNS M3144-84. Method for Determination of Carbon and Hydrogen of Coal and Coke (Liebig Method) (Mar) (10825).
ISO 562-81. Hard Coal and Coke—Determination of Volatile Matter Content Second Edition. 7 pp.
JIS M 8812-84. Methods for Proximate Analysis of Coal and Coke. 28 pp.
SAA AS 1038.3-89. Methods for the Analysis and Testing of Coal and Coke—Part 3: Proximate Analysis of Higher Rank Coal. 12 pp.

—Volatile Matter Content—Reference Method
BSI BS 1016: Sec 104.3-91. 1991 Analysis and Testing of Coal and Coke Part 104: Proximate Analysis Section 104.3: Determination of Volatile Matter Content. 11 pp.

—Washing—Bucket Elevators
CNS M2082-80. Bucket Elevator for Coal Washing (Dec)(6843).
JIS M 4601-52. Bucket Elevator for Coal Washing. 17 pp.

—Water of Hydration Content
SAA AS 1038.22-92. Coal and Coke—Analysis and Testing—Part 22: Higher Rank Coal—Mineral Matter and Water of Constitution (in Professional Package 32). 12 pp.

Coal Ash
See Also: Coal

—Trace Element Content
SAA TR2.10 Supp 1-92. Certified Reference Materials—Part 10: Coal Ash—Preparation and Certification of ASCRM-010—Supplement 1: Trace Element Data (Supplement to SAA TR2.10—1985). 1 p.

Coal Cleaning Plants
Use: Coal Preparation Plants

Coal Content Analysis

—Sand
CNS A3036-59. Method of Test for Coal and Lignite in Sand (Sep)(1172)(R 1970).

Coal Gas
Use For: Coke Oven Gas; Illuminating Gas
See Also: Gases
CNS K1122-66. Coal Gas Distilled for Industrial Fuel (Jun)(2631)(R 1971).

—Ammonia Content
CNS K6240-66. Method of Test for Ammonia in Coal Gas (Sep)(2634)(R 1971).

—Calorific Value
CNS K6238-66. Method of Test for Calorific Value of Coal Gas (Sep)(2632)(R 1971).

—Distilled
CNS S1038-71. Distilled Coal Gas (for Household Fuel) (Jul)(2630).

—Hydrogen Sulfide Content
CNS K6239-66. Method of Test for Hydrogen Sulfide in Coal Gas (Sep)(2633)(R 1971).

—Sulfur Content
CNS K6241-66. Method of Test for Total Sulfur in Coal Gas (Sep)(2635)(R 1971).

Coal Gas Content Analysis

—Lignite—Distillation Methods
ISO 647-74. Brown Coals and Lignites—Determination of the Yields of Tar, Water, Gas and Coke Residue by Low Temperature Distillation First Edition. 8 pp.

Coal Mining Equipment
Use: Mining Equipment

Coal Plows
Use: Plows—Coal

Coal Preparation Plants
See Also: Coal

—Efficiency
ISO 923-75. Coal Cleaning Tests—Expression and Presentation of Results First Edition. 23 pp.

—Effluents
JIS M 0201-74. Testing Method for Effluents from Coal Preparation Plant.

Coal Preparation Plants (Cont.)

—Flow Charts
BSI BS 3567-85. 1985 Preparation of Flowsheets for the Design of Coal Preparation Plant. 12 pp.
BSI BS 3567-01. 1985 Amd 1 Preparation of Flowsheets for the Design of Coal Preparation Plant (AMD 6905) May 15, 1992 (Q). 13 pp.
ISO 924-89. Coal Preparation Plant—Principles and Conventions for Flowsheets Second Edition. 9 pp.
JIS M 4001-85. Methods of Presentation of Flowsheet for Mineral, Coal and Rock Processing.
SAA AS 1414-90. Flowsheets and Symbols Relating to Coal Preparation Plant. 30 pp.

—Flow Charts—Symbols
BSI BS 3553-92. 1992 Graphical Symbols for Coal Preparation Plant Flowsheets (Q). 21 pp.
BSI BS 3553-86. 1986 Graphical Symbols for Coal Preparation Plant Flowsheets. 20 pp.
ISO 561-89. Coal Preparation Plant—Graphical Symbols Second Edition; (Corrected and Reprinted -1992). 17 pp.

—Glossaries
ISO 1213 Pt 1-82. Solid Mineral Fuels—Vocabulary—Part 1: Terms Relating to Coal Preparation First Edition; (Addendum 1-1989). 31 pp.

—Separation
BSI BS 7067-90. 1990 Guide to Determination and Presentation of Float and Sink Characteristics of Raw Coal and of Products from Coal Preparation Plants. 20 pp.

Coal Tar
See Also: Bitumens; Coal Tar Roofing; Creosote; Tars
CNS K5037-72. Coal Tar (Tentative) (Jun)(2325).

—Ash Content
CNS K6205-65. Method of Test for Ash of Asphalt and Coal Tar (Sep)(2487)(R 1973).

—Carbon Content
CNS K6207-65. Determination of Fixed Carbon in Asphalt and Coal Tar (Sep)(2489)(R 1971).

—Creosote
CNS K5062-77. Coal-Tar for Blending with Creosote (Jun)(4128).

—Density
CNS K6211-65. Method for Determination of Specific Gravity of Coal Tar (Sep)(2493)(R 1971).

—Distillation Methods
CNS K6213-65. Method of Fractional Distillation Test for Coal Tar (Sep)(2495)(R 1971).
CNS K6214-65. Method of Fractional Distillation Test for Crude Methyl Phenol, M-Methyl Phenol and Coal Tar (High Boiling Point) (Sep)(2496)(R 1971).

—Insoluble Matter Content
CNS K6206-65. Determination of Benzene-Insoluble Substance in Asphalt and Coal Tar (Sep)(2488)(R 1973).

—Moisture Content
CNS K6208-65. Determination of Moisture Contents in Asphalt, Coal Tar and Phenols (Sep)(2490)(R 1971).

—Paving Materials
DIN ENGL 1995 Pt 5-89. Bituminous Binders; Road Tars, Pitch-Bitumen Mixtures and Cold Pitch Solutions; Requirements (Oct). 5 pp.

—Plastic Properties
DIN ENGL 52012-85. Testing of Bitumen; Determination of the Fraass Breaking Point (Aug). 6 pp.

—Viscosity
CNS K6212-65. Method for Determination of Viscosity of Coal Tar (Sep)(2494)(R 1971).

Coal Tar Coatings
Use For: Coal Tar Paints *See Also:* Asphalt Coatings (Made From Asphalt); Bituminous Coatings; Coatings; Nonmetallic Coatings; Waterproof Coatings

—Adhesion Testing
MOD UK M 3004/67. Protective PX-2 and Composition Rust Preventative Type B.

—Bend Testing
MOD UK M 3004/67. Protective PX-2 and Composition Rust Preventative Type B.

—Concrete
CGSB 1-GP-184MA-83. Coating, Coal Tar-Epoxy, Standard for. 17 pp.

Coal Tar Coatings (Cont.)
—Concrete (Cont.)
CGSB CAN/CGSB-37.33-M90. Fibrated, Mineral Colloid Type, Emulsified Coal Tar Pitch for Roof Coatings and for Damp-Proofing and Water-Proofing. 12 pp.
—Construction Materials
BSI BS 1310-84. (WITHDRAWN) 1984 Coal Tar Pitches for Building Purposes. 16 pp.
—Content Analysis
MOD UK M 3004/67. Protective PX-2 and Composition Rust Preventative Type B.
—Flexible Pavements
CGSB 37-GP-39M-77. Pitch, Emulsified Coal Tar, Mineral Colloid Type Filled, for Coating Bituminous Pavements, Standard for. 12 pp.
CGSB 37-GP-40M-77. Application of Coal Tar Pitch Emulsion as a Bituminous Pavement Coating, Standard for. 7 pp.
—Iron
BSI BS 4164-87. 1987 Amd 2 Coal Tar Based Hot Applied Coating Materials for Protecting Iron and Steel, Including a Suitable Primer (AMD 6216) August 31, 1989. 26 pp.
—Masonry
CGSB 1-GP-184MA-83. Coating, Coal Tar-Epoxy, Standard for. 17 pp.
CGSB CAN/CGSB-37.33-M90. Fibrated, Mineral Colloid Type, Emulsified Coal Tar Pitch for Roof Coatings and for Damp-Proofing and Water-Proofing. 12 pp.
—Metal
CGSB 1-GP-184MA-83. Coating, Coal Tar-Epoxy, Standard for. 17 pp.
—Roofing
CGSB CAN/CGSB-37.35-M89. Application of Emulsified, Coal Tar Pitch as a Roof Coating. 9 pp.
—Solums
BSI BS 2832-57. 1957 Amd 1 Hot Applied Damp Resisting Coatings for Solums (AMD 5956) July 31, 1989. 14 pp.
—Steel
BSI BS 4164-87. 1987 Amd 2 Coal Tar Based Hot Applied Coating Materials for Protecting Iron and Steel, Including a Suitable Primer (AMD 6216) August 31, 1989. 26 pp.
CGSB CAN/CGSB-1.161-M91. Cold-Application Coal Tar Base Coating. 6 pp.
—Steel Pipes
ISO 5256-85. Steel Pipes and Fittings for Buried or Submerged Pipelines—External and Internal Coating by Bitumen or Coal Tar Derived Materials First Edition. 57 pp.
—Steel Pipes—Water
JIS G 3492-77. Coal-Tar Enamel Protective Coatings for Steel Water Pipe (R 1983). 28 pp.
—Steel Structures
SAA AS 3887-91. Paints for Steel Structures—Coal Tar Epoxy (Two-Pack) (Supersedes AS K172—1970). 21 pp.
—Wood
CGSB 1-GP-184MA-83. Coating, Coal Tar-Epoxy, Standard for. 17 pp.

Coal Tar Creosote
Use: Creosote Oil

Coal Tar Epoxy Coatings
Use: Coal Tar Coatings

Coal Tar Paints
Use: Coal Tar Coatings

Coal Tar Pitch Coatings
Use: Coal Tar Coatings

Coal Tar Primers
—Concrete
CGSB CAN/CGSB-37.32-M89. Coal Tar Primer for Coal Tar Roofing, Damp-Proofing and Water-Proofing. 8 pp.
CGSB CAN/CGSB-37.34-M89. Application of Coal Tar Primer for Coal Tar Roofing, Dampproofing and Waterproofing. 8 pp.
—Felt
CGSB CAN/CGSB-37.32-M89. Coal Tar Primer for Coal Tar Roofing, Damp-Proofing and Water-Proofing. 8 pp.
CGSB CAN/CGSB-37.34-M89. Application of Coal Tar Primer for Coal Tar Roofing, Dampproofing and Waterproofing. 8 pp.
—Iron
BSI BS 4164-87. 1987 Amd 2 Coal Tar Based Hot Applied Coating Materials for Protecting Iron and Steel, Including a Suitable Primer (AMD 6216) August 31, 1989. 26 pp.
—Metal
CGSB CAN/CGSB-37.32-M89. Coal Tar Primer for Coal Tar Roofing, Damp-Proofing and Water-Proofing. 8 pp.
CGSB CAN/CGSB-37.34-M89. Application of Coal Tar Primer for Coal Tar Roofing, Dampproofing and Waterproofing. 8 pp.
—Steel
BSI BS 4164-87. 1987 Amd 2 Coal Tar Based Hot Applied Coating Materials for Protecting Iron and Steel, Including a Suitable Primer (AMD 6216) August 31, 1989. 26 pp.
—Steel Pipes
SAA AS 2043-77. Coal-Tar and Synthetic (Fast Dry) Primers for Steel Pipes Bound Together with Standards AS 2043-2046.
SAA AS 2044-77. Coal-Tar Enamel for Steel Pipes Bound Together with Standards AS 2043-2046.
SAA AS 2045-77. Materials Associated with the Coating and Lining of Steel Pipes with Coal-Tar Primer/Enamel Systems Bound Together with Standards AS 2043-2046.
SAA AS 2046-77. Code of Practice for the Coating and Lining of Steel Pipes with Coal-Tar Primer/Enamel Systems Bound Together with Standards AS 2043-2046.
—Wood
CGSB CAN/CGSB-37.32-M89. Coal Tar Primer for Coal Tar Roofing, Damp-Proofing and Water-Proofing. 8 pp.
CGSB CAN/CGSB-37.34-M89. Application of Coal Tar Primer for Coal Tar Roofing, Dampproofing and Waterproofing. 8 pp.

Coal Tar Roofing
See Also: Coal Tar; Roofing
—Coal Tar Coatings
CGSB CAN/CGSB-37.35-M89. Application of Emulsified, Coal Tar Pitch as a Roof Coating. 9 pp.
—Tar Coatings
CGSB CAN/CGSB-37.25-M89. Application of Fibrated, Cutback Tar Roof Coating. 7 pp.

Coalescence
—Paints—Low Temperature
SAA AS 1580.409. 2-92. Paints and Related Materials—Methods of Test—Part 409.2: Low Temperature Coalescence (in Professional Packages 30, 39). 2 pp.
SNZ NZS/AS 1580. 409.2-92. Methods of Test for Paints and Related Materials Part 409.2: Low Temperature Coalescence (This is a Joint Standard with SAA AS 1580.409.2). 2 pp.

Coalescence Separators
Use: Coalescers

Coalescers
Use For: Coalescence Separators
See Also: Separators (Mechanical)
—Fuel Filters
MOD UK DSTAN 49-4-01. Filter Elements, Fluid Pressure (Coalescer Elements), Gasoline and Kerosine Fuels Issue 1; Amendment 1. 13 pp.
—Fuel Separators/Filters
NATO STANAG 3967 ED 1 AMD 0-93. Design and Performance Requirements for Aviation Fuel Filter Separator Vessels and Coalescer and Separator Elements. 33 pp.
—Fuel Separators/Filters—Design
NATO STANAG 3967 ED 1 AMD 0-93. Design and Performance Requirements for Aviation Fuel Filter Separator Vessels and Coalescer and Separator Elements. 33 pp.

Coast Earth Stations
Use: Earth Stations

Coast Protection
Use: Shore Protection

Coast Station Identity
See Also: Identification Systems; Numbers; Ship Station Identity
—Maritime Mobile Satellite Communications—VHF/UHF
CCITT RECMN F.120-89. Ship Station Identification for VHF/UHF and Maritime Mobile-Satellite Services—Telegraph and Mobile Services Operations and Quality of Service (Study Group I) 6 pp (Same as Recmn E.210). 6 pp.
—Maritime Mobile Services—Radiotelephony
CCIR RECMN 587-1-86. Coast Station Identities and Initiation of Location Registration in an Automated VHF/UHF Maritime Mobile Telephone System—Section 8C—Maritime Mobile Service; Telephony and Related Subjects. 1 p.
—Maritime Mobile Services—Telegraph Printers
CCIR RECMN 625-1-90. Direct-Printing Telegraph Equipment Employing Automatic Identification in the Maritime Mobile Service—Section 8B—Maritime Mobile Service; Telegraphy and Related Subjects. 55 pp.
CCIR RECMN 625-2-92. Direct-Printing Telegraph Equipment Employing Automatic Identification in the Maritime Mobile Service—Section 8B—Maritime Mobile Service; Telegraphy and Related Subjects. 59 pp.

Coastal Dunes
Use: Dunes—Coastal

Coasts
See Also: Shore Protection
—Military Geographic Documentation
NATO STANAG 2263 ED 4 AMD 4-76. Military Geograghic Documentation—Coastal Areas and Landing Beaches. 18 pp.
NATO STANAG 2263 ED 4 AMD 5-76. MGD—Coastal Areas and Landing Beaches. 28 pp.

Coated Abrasive Sleeves
Use: Abrasive Sleeves

Coated Abrasives
See Also: Abrasives; Coatings
BSI BS 5367-76. 1976 Dimensions of Coated Abrasives. 8 pp.
CNS R3159-87. Standard Recommended Practice for Selection of Coated Abrasives (Nov)(12166).
JIS R 6260-80. Standard Recommended Practice for Selection of Coated Abrasives. 11 pp.
SAA AS 1863-76. Coated Abrasives (Technical Products). 12 pp.
—Grain Size Analysis
JIS R 6010-91. Coated Abrasive Grain Sizes. 7 pp.
JIS R 6012-91. Testing Method for Grain Size of Coated Abrasive Microgrits (P240—P1200). 14 pp.

Coated Board
See Also: Coatings; Paperboard
CNS P2066-84. One-Side Coated White Board (Feb)(10776). 2 pp.

Coated Electrodes
Use For: Covered Electrodes See Also: Electrodes; Filler Metal; Weld Metal; Welding Electrodes; Welding Rods
DIN ENGL 8555 Pt 1-83. Filler Metals Used for Surfacing; Filler Wires, Filler Rods, Wire Electrodes, Covered Electrodes; Designation; Technical Delivery Conditions (Nov). 8 pp.
SAA AS 1553.3-92. Covered Electrodes for Welding—Part 3: Corrosion-Resisting Chromium and Chromium-Nickel Steel Electrodes (in Professional Package 36). 23 pp.
—Austenitic Ferrite Number
BSI BS 6787-87. 1987 Determination of Ferrite Number in Austenitic Weld Metal Deposited by Covered Cr-Ni Steel Electrodes. 16 pp.
ISO 8249-85. Welding—Determination of Ferrite Number in Austenitic Weld Metal Deposited by Covered Cr-Ni Steel Electrodes First Edition. 15 pp.
—Chemical Analysis
DIN ENGL 8556 Pt 2-65. Filler Metals for Welding Stainless and Heat Resisting Steels; Testing of Covered Rod Electrodes; Weld Metal Specimen (Apr). 2 pp.
—Deposition Rates
ISO 2401-72. Covered Electrodes—Determination of the Efficiency, Metal Recovery and Deposition Coefficient First Edition. 5 pp.
JIS Z 3182-91. Method of Deposition Rate Measurement for Covered Electrodes. 6 pp.

Coated Electrodes (Cont.)

—Designations
DIN ENGL 8555 Pt 1-83. Filler Metals Used for Surfacing; Filler Wires, Filler Rods, Wire Electrodes, Covered Electrodes; Designation; Technical Delivery Conditions (Nov). 8 pp.

—Efficiency
DIN ENGL 32523-84. Determination of the Electrode Efficiency When Welding with Covered Electrodes (Apr.). 2 pp.
ISO 2401-72. Covered Electrodes—Determination of the Efficiency, Metal Recovery and Deposition Coefficient First Edition. 5 pp.

—Fumes
JIS Z 3930-79. Method of Measuring Total Amount of Weld Fumes Generated by Covered Electrode. 8 pp.

—Hard Surfacing
JIS Z 3251-91. Covered Electrodes for Hardfacing. 14 pp.

—Metal Arc Welding
BSI BS 639-86. 1986 Covered Carbon and Carbon Manganese Steel Electrodes for Manual Metal-Arc Welding. 21 pp.
BSI BS 2493-85. 1985 Low Alloy Steel Electrodes for Manual Metal-Arc Welding. 20 pp.
BSI BS 2926-84. 1984 Chromium and Chromium-Nickel Steel Electrodes for Manual Metal-Arc Welding. 16 pp.
CEN PREN 499-91. Classification of Covered Electrodes for Manual Metal Arc Welding of Carbon Steels, Carbon-Manganese Steels and Micro Alloyed Steels. 12 pp.
CEN PREN 757-92. Classification of Covered Electrodes for Manual Metal Arc Welding of High Strength Steels. 11 pp.
CNS C4031-85. Covered Electrodes for Mild Steel (Dec)(1215).
CSA W48.3-M1982. Low-Alloy Steel Covered Arc Welding Electrodes; (Gen Instr 1 Thru 2). 44 pp.
DIN ENGL 1913 Pt 1-84. Covered Electrodes for the Joint Welding of Unalloyed and Low Alloy Steel; Classification, Designation, Technical Delivery Conditions (June). 10 pp.
DIN ENGL 8529 Pt 1-81. Covered Electrodes for Joint Welding of High Tensile Fine-Grained Structural Steels; Basic Covered Electrodes; Classification, Designation, Technical Delivery Conditions (Apr). 11 pp.
ISO 6847-85. Covered Electrodes for Manual Metal Arc Welding—Deposition of a Weld Metal Pad for Chemical Analysis First Edition. 4 pp.
ISO 10446-90. Welding—All-Weld Metal Test Assembly for the Classification of Corrosion-Resisting Chromium and Chromium-Nickel Steel Covered Arc Welding Electrodes First Edition; (Corrected and Reprinted -1991). 6 pp.
JIS Z 3211-91. Covered Electrodes for Mild Steel. 11 pp.
JIS Z 3212-90. Covered Electrodes for High Tensile Strength Steel. 14 pp.
JIS Z 3214-87. Covered Electrodes for Atmospheric Corrosion Resisting Steel. 18 pp.
JIS Z 3221-89. Stainless Steel Covered Electrodes. 20 pp.
JIS Z 3223-87. Molybdenum Steel and Chromium Molybdenum Steel Covered Electrodes. 15 pp.
JIS Z 3224-91. Nickel and Nickel-Alloy Covered Electrodes. 15 pp.
JIS Z 3225-90. Covered Electrodes for 9% Nickel Steel. 9 pp.
JIS Z 3231-89. Copper and Copper Alloy Covered Electrodes. 15 pp.
JIS Z 3241-88. Covered Electrodes for Low Temperature Service Steel. 13 pp.
JIS Z 3252-92. Covered Electrodes for Cast Iron. 9 pp.

—Metal Arc Welding—Core Wire
CNS G3033-88. Wire Rods for Core Wire of Covered Electrode (May)(2067). 3 pp.
CNS G3181-88. Core Wires for Covered Electrode (May)(8968).
JIS G 3503-80. Wire Rods for Core Wire of Covered Electrode. 6 pp.
JIS G 3523-80. Core Wires for Covered Electrode. 7 pp.

—Metal Arc Welding—Designations
DIN ENGL 1913 Pt 1-84. Covered Electrodes for the Joint Welding of Unalloyed and Low Alloy Steel; Classification, Designation, Technical Delivery Conditions (June). 10 pp.
DIN ENGL 8529 Pt 1-81. Covered Electrodes for Joint Welding of High Tensile Fine-Grained Structural Steels; Basic Covered Electrodes; Classification, Designation, Technical Delivery Conditions (Apr). 11 pp.

Coated Electrodes (Cont.)

—Metal Arc Welding—Symbols
ISO 1071-83. Covered Electrodes for Manual Arc Welding of Cast Iron—Symbolization First Edition. 7 pp.
ISO 2560-73. Covered Electrodes for Manual Arc Welding of Mild Steel and Low Alloy Steel—Code of Symbols for Identification First Edition; (Addendum 1-1974). 10 pp.
ISO 3580-75. Covered Electrodes for Manual Arc Welding of Creep-Resisting Steels—Code of Symbols for Identification First Edition. 5 pp.
ISO 3581-76. Covered Electrodes for Manual Arc Welding of Stainless and Other Similar High Alloy Steels—Code of Symbols for Identification First Edition. 7 pp.

—Shielded Metal Arc Welding
CSA W48.1-M1991. Carbon Steel Covered Electrodes for Shielded Metal Arc Welding; (Gen Instr 1 Thru 2). 56 pp.
CSA W48.2-M1992. Chromium and Chromium-Nickel Steel Covered Electrodes for Shielded Metal Arc Welding; (Gen Instr 1). 44 pp.

Coated Fabrics

Use For: Silicon Carbide Coated Fabrics
See Also: Coatings; Fabrics; Laminated Fabrics; Leathercloth

BSI BS 3424: Part 0-82. 1982 Testing Coated Fabrics Part 0: Foreword and General Introduction. 4 pp.
BSI BS 3424: Part 0-01. 1982 Amd 1 Testing Coated Fabrics Part 0: Foreword and General Introduction (AMD 6845) January 31, 1992. 5 pp.
BSI BS 3424: Part 2-92. 1992 Testing Coated Fabrics Part 2: Method 4. Pre-Conditioning and Conditioning of Coated Fabrics for Testing Purposes (ISO 2231: 1989). 6 pp.
BSI BS 3424: Part 2-82. 1982 Testing Coated Fabrics Part 2: Method 4. Conditioning and Selection of Test Specimens. 4 pp.
BSI BS 3424-73. (WITHDRAWN) 1973 Amd 2 Methods of Test for Coated Fabrics. 49 pp.
MOD UK DSTAN 53-77-82. Rubber Sheet, Solid, Cloth Insert Issue 2. 9 pp.
SAA AS 1440-73. Vinyl (PVC) Coated Fabrics for Upholstery and Other Purposes. 11 pp.
SAA AS 1441. Methods of Test for Coated Fabrics Bound Together. 37 pp.

—Abrasion Resistance Testing
BSI BS 3424: Part 11-82. 1982 Testing Coated Fabrics Part 11: Method 13. Method for Determination of Resistance to Blocking. 4 pp.
BSI BS 3424: Part 24-90. 1990 Testing Coated Fabrics Part 24: Methods 27A and 27B. Determinaton of Abrasion Resistance. 8 pp.
BSI BS 3424: Part 31-90. 1990 Testing Coated Fabrics Part 31: Method 34 Method for Determination of Resistance to Scuffing and Snagging. 8 pp.

—Adhesion Testing
BSI BS 3424: Part 7-82. 1982 Testing Coated Fabrics Part 7: Method 9. Methods for Determination of Coating Adhesion Strength. 6 pp.
SAA AS 1441.5-73. Methods of Test for Coated Fabrics—Part 5: Method for Determination of Coating and Ply Adhesion.

—Aging Testing
BSI BS 3424: Part 12-90. 1990 Testing Coated Fabrics Part 12: Methods 14A, 14B, 14C, 14D and 14E. Accelerated Ageing Tests. 6 pp.

—Aircraft
CAA NOTICE #20 ISSUE 6. Cotton, Linen and Synthetic Fabric Covered Aircraft (Airworthiness Notices). 3 pp.
MOD UK DTD-436B-44. Proofed Cotton Fabric (Reprinted November 1960). 1 p.
MOD UK DTD-891B-62. Single Ply Rubber Proofed Fabric and Tape. 5 pp.
MOD UK DTD-5551-60. Two Ply Rubber Proofed Silk Fabric. 3 pp.

—Bend Testing
SAA AS 1441.6-73. Methods of Test for Coated Fabrics—Part 6: Method for Determination of Resistance to Flex Cracking.

—Breaking Load
BSI BS 3424: Part 4-82. 1982 Testing Coated Fabrics Part 4: Method 6. Method for Determination of Breaking Strength and Elongation at Break. 4 pp.

—Colorfastness Testing
SAA AS 1441.10-73. Methods of Test for Coated Fabrics—Part 10: Method for Determination of Colour Bleeding.
SAA AS 1441.11-73. Methods of Test for Coated Fabrics—Part 11: Method for Determination of Colour Fastness to Light.

Coated Fabrics (Cont.)

—Colorfastness Testing—Chemical Resistance
DIN ENGL 53378-65. Testing of Plastic Films; Determination of Colour Fastness to Hydrogen Sulphide (June). 2 pp.

—Colorfastness Testing—Rubbing
BSI BS 3424: Part 14-85. 1985 Testing Coated Fabrics Part 14: Method 16. Methods for Determination of Colour Fastness to Wet and Dry Rubbing and Determination of Resistance to Printwear Using a Common Testing Apparatus. 6 pp.

—Cracking (Fracturing)
SAA AS 1441.14-73. Methods of Test for Coated Fabrics—Part 14: Method for Determination of Resistance to Cold Cracking.

—Dimensional Stability
BSI BS 3424: Part 17-87. 1987 Testing Coated Fabrics Part 17: Methods 20. Method for Determination of Dimersional Stability to Water Immersion. 4 pp.
BSI BS 3424: Part 20-87. 1987 Testing Coated Fabrics Part 20: Methods for Determination of Dimensional Changes on Mechanical Relaxation at Zero Tension. 4 pp.
BSI BS 3424: Part 36-93. 1993 Testing Coated Fabrics Part 36: Method 39. Method for Determination of the Dimensional Stability of Coated Fabrics to Domestic Washing. 7 pp.
CGSB CAN/CGSB-4.2 NO.67-M90. Textile Test Methods Dimensional Change and Appearance After Laundering of Coated, Bonded, Laminated and Fused Fabrics. 11 pp.

—Dry Cleaning—Dimensional Stability
CGSB CAN/CGSB-4.2 NO.66-M91. Textile Test Methods Dimensional Change and Appearance After Dry Cleaning of Coated, Bonded, Laminated and Fused Fabrics. 9 pp.

—Elastomers—Aerospace
BSI F 127-83. (WITHDRAWN) 1983 Amd 1 Nylon Fabrics Suitable for Coating with Natural or Synthetic Elastomers (Superseded by BS 2F 127: 1991). 6 pp.
BSI 2F 127-91. 1991 Nylon Fabrics Suitable for Coating with Natural or Synthetic Elastomers for Aerospace Purposes. 8 pp.
BSI 2F 127-01. 1991 Amd 1 Nylon Fabrics Suitable for Coating with Natural or Synthetic Elastomers for Aerospace Purposes (AMD 7967) October 15, 1993 (S). 9 pp.

—Elastomers—Test Specimens
DIN ENGL 53502-72. Testing of Elastomers and of Fabrics Coated with Elastomers; Test Specimens; Directions for Preparing (Aug.). 3 pp.

—Elongation
BSI BS 3424: Part 4-82. 1982 Testing Coated Fabrics Part 4: Method 6. Method for Determination of Breaking Strength and Elongation at Break. 4 pp.
BSI BS 3424: Part 21-93. 1993 Testing Coated Fabrics Part Method for Determination of Elongation and Tension Set. 10 pp.
BSI BS 3424: Part 21-87. 1987 Testing Coated Fabrics Part Method for Determination of Elongation and Tension Set. 6 pp.
SAA AS 1441.9-73. Methods of Test for Coated Fabrics—Part 9: Method for Determination of Elongation and Stretch-Set.

—Flammability Testing
SAA AS 1441.13-73. Methods of Test for Coated Fabrics—Part 13: Method for Determination of Flammability.

—Flexural Strength
BSI BS 3424: Part 9-90. 1990 Testing Coated Fabrics Part 9: Methods 11A, 11B, 11C and 11D. Methods for Determination of Resistance to Damage by Flexing. 12 pp.

—Friction
BSI BS 3424: Part 10-87. 1987 Testing Coated Fabrics Part 10: Methods 12A and 12B. Determination of Surface Drag. 12 pp.

—Fungus Resistance Testing
SAA AS 1157.4-78. Methods of Testing Materials for Resistance to Fungal Growth—Part 4: Resistance of Coated Fabrics to Fungal Growth. 10 pp.

—Glass—Waterproofing Membranes
CGSB 37-GP-63M-77. Cloth, Glass, Coated, for Membrane Waterproofing Systems and Built-Up Roofing, Standard for. 9 pp.

—Heat Resistance
SAA AS 1441.7-73. Methods of Test for Coated Fabrics—Part 7: Method for Determination of Resistance to Heat.

INTERNATIONAL AND NON-U.S. NATIONAL STANDARDS
SUBJECT INDEX

Coated Fabrics (Cont.)

—Life Rafts—Aircraft
MOD UK DTD-537E-61. Proofed Fabric and Tape for Inflatable Liferaft Equipment (Reprinted January 1966, Incorporating Amendment No. 1). 7 pp.

—Low Temperature Testing
BSI BS 3424: Part 8-83. 1983 Testing Coated Fabrics Part 8: Methods 10A, 10B and 10C. Methods for Determination of Low Temperature Performance. 10 pp.

—Mass
BSI BS 3424: Part 3-82. 1982 Amd 1 Testing Coated Fabrics Part 3: Methods 5A, 5B and 5C. Methods for Determination of Mass Per Unit Area. 5 pp.
SAA AS 1441.2-73. Methods of Test for Coated Fabrics—Part 2: Method for Determination of Mass per Unit Area and Coating Mass per Unit Area.

—Plastic
ISO 2231-89. Rubber-or Plastics-Coated Fabrics—Standard Atmospheres for Conditioning and Testing Second Edition. 6 pp.

—Plastic—Abrasion Resistance Testing
ISO 5470-80. Rubber or Plastics Coated Fabrics—Determination of Abrasion Resistance First Edition. 5 pp.
ISO 5978-90. Rubber-or Plastics-Coated Fabrics—Determination of Blocking Resistance Second Edition. 6 pp.
ISO 5981-82. Rubber or Plastics Coated Fabrics—Determination of Flex Abrasion First Edition. 6 pp.

—Plastic—Adhesion Testing
ISO 2411-91. Rubber-or Plastics-Coated Fabrics—Determination of Coating Adhesion Second Edition. 10 pp.

—Plastic—Adhesive Tear Strength
ISO 6133-81. Rubber and Plastics—Analysis of Multi-Peak Traces Obtained in Determinations of Tear Strength and Adhesion Strength First Edition. 5 pp.

—Plastic—Aging Testing
ISO 1419-77. Fabrics Coated with Rubber or Plastics—Accelerated Ageing and Simulated Service Tests First Edition. 4 pp.

—Plastic—Bend Testing
ISO 4675-90. Rubber-or Plastics-Coated Fabrics—Low-Temperature Bend Test Second Edition. 8 pp.

—Plastic—Breaking Load
ISO 1421-77. Fabrics Coated with Rubber or Plastics—Determination of Breaking Strength and Elongation at Break First Edition. 4 pp.

—Plastic—Bursting Strength
ISO 3303-90. Rubber-or Plastics-Coated Fabrics—Determination of Bursting Strength Second Edition. 7 pp.

—Plastic—Comprehension Testing
ISO 5473-79. Rubber-or Plastics-Coated Fabrics—Determination of Crush Resistance First Edition. 5 pp.

—Plastic—Cracking (Fracturing)
ISO 3011-81. Rubber or Plastics Coated Fabrics—Determination of Resistance to Ozone Cracking Under Static Conditions Second Edition. 4 pp.

—Plastic—Discoloration—Cigarette Smoke
ISO 6449-82. Rubber or Plastics Coated Fabrics—Determination of Discolouration by Cigarette Smoke First Edition. 4 pp.

—Plastic—Elongation
ISO 1421-77. Fabrics Coated with Rubber or Plastics—Determination of Breaking Strength and Elongation at Break First Edition. 4 pp.

—Plastic—Flexibility
ISO 5979-82. Rubber or Plastics Coated Fabrics—Determination of Flexibility—Flat Loop Method First Edition. 5 pp.

—Plastic—Flexural Strength
ISO 7854-84. Rubber-or Plastics-Coated Fabrics—Determination of Resistance to Damage by Flexing (Dynamic Method) First Edition. 6 pp.

—Plastic—Impact Testing
ISO 4646-89. Rubber-or Plastics-Coated Fabrics—Low-Temperature Impact Test Second Edition. 9 pp.

—Plastic—Low Temperature Testing
ISO 4675-90. Rubber-or Plastics-Coated Fabrics—Low-Temperature Bend Test Second Edition. 8 pp.

Coated Fabrics (Cont.)

—Plastic—Protective Clothing
ISO 8096 Pt 1-89. Rubber-or Plastics-Coated Fabrics for Water-Resistant Clothing—Specification—Part 1: PVC-Coated Fabrics First Edition; (Corrigendum 1-1991). 9 pp.
ISO 8096 Pt 2-89. Rubber-or Plastics-Coated Fabrics for Water-Resistant Clothing—Specification—Part 2: Polyurethane-and Silicone Elastomer-Coated Fabrics First Edition. 16 pp.
ISO 8096 Pt 3-88. Rubber-or Plastics-Coated Fabrics for Water-Resistant Clothing—Specification—Part 3: Natural Rubber-and Synthetic Rubber-Coated Fabrics First Edition. 20 pp.

—Plastic—Roll
CEN PREN 22286-92. Rubber-or Plastics-Coated Fabrics—Determination of Roll Characteristics (ISO 2286:1986). 7 pp.
ISO 2286-86. Rubber-or Plastics-Coated Fabrics—Determination of Roll Characteristics Second Edition; (CEN PREN 22286:1992). 6 pp.

—Plastic—Test Strength
ISO 4674-77. Fabrics Coated with Rubber or Plastics—Determination of Tear Resistance First Edition. 6 pp.

—Plastic—Upholstery
ISO 7617 Pt 1-88. Plastics-Coated Fabrics for Upholstery—Part 1: Specification for PVC-Coated Knitted Fabrics First Edition. 12 pp.
ISO 7617 Pt 2-88. Plastics-Coated Fabrics for Upholstery—Part 2: Specification for PVC-Coated Woven Fabrics First Edition. 8 pp.
ISO 7617 Pt 3-88. Plastics-Coated Fabrics for Upholstery—Part 3: Specification for Polyurethane-Coated Woven Fabrics First Edition. 8 pp.

—Plastic—Water Resistance Testing
ISO 1420-87. Rubber-or Plastics-Coated Fabrics—Determination of Resistance to Penetration by Water Second Edition. 8 pp.

—Plasticizers—Evaporation
SAA AS 1441.8-73. Methods of Test for Coated Fabrics—Part 8: Method for Determination of Plasticizer Evaporation.

—Polyurethane—Protective Clothing
BSI BS 3546: Part 1-81. (WITHDRAWN) 1981 Coated Fabrics for Water Resistant Clothing Part 1: Specifications for Polyurethane and Silicone Elastomer Coated Fabrics (Superseded by BS 3546: Part 2: 1993). 12 pp.

—Polyurethane—Upholstery
BSI DD 37-74. (WITHDRAWN) 1974 Polyurethane Coated Fabrics for Upholstered Furniture. 12 pp.

—Protective Clothing
BSI BS 3546: Part 4-91. 1991 Coated Fabrics for Use in the Manufacturer of Water Penetration Resistant Clothing Part 4: Specification for Water Vapour Permeable Coated Fabrics. 17 pp.
BSI BS 6408-83. 1983 Clothing Made from Coated Fabrics for Protection Against Wet Weather. 8 pp.
ISO 8096 Pt 1-89. Rubber-or Plastics-Coated Fabrics for Water-Resistant Clothing—Specification—Part 1: PVC-Coated Fabrics First Edition; (Corrigendum 1-1991). 9 pp.
MOD UK TS 10312. Fabric. Non-Woven, Water-Repellent.

—PVC
CNS K3061-82. Polyvinylchloride Coated Fabric (Oct)(9519). 3 pp.
CNS K6730-82. Method of Test for Polyvinyl Chloride Coated Fabric (Oct)(9520).
JIS K 6772-76. Polyvinylchloride Coated Fabric (R 1979). 13 pp.

—PVC—Fusion
BSI BS 3424: Part 22-83. 1983 Testing Coated Fabrics Part 22: Methods for Determination of Fusion of PVC Coatings and the State of Cure of Vulcanized Rubber Coatings. 4 pp.
ISO 6451-82. Plastics Coated Fabrics—Polyvinyl Chloride Coatings—Rapid Method for Checking Fusion First Edition. 3 pp.

—PVC—Tarpaulins
ISO 8095-90. PVC-Coated Fabrics for Tarpaulins—Specification First Edition. 8 pp.

—PVC—Upholstery
BSI BS 5790: Part 1-79. 1979 Amd 1 Coated Fabrics for Upholstery Part 1: PVC Coated Knitted Fabrics. 9 pp.
BSI BS 5790: Part 2-79. 1979 Amd 1 Coated Fabrics for Upholstery Part 2: PVC Coated Woven Fabrics. 9 pp.

Coated Fabrics (Cont.)

—Raincoats
CNS L4029-82. Silicone Treated Fabrics for Raincoat Cloth (Tentative) (Sep)(2414).

—Rolls
BSI BS 3424: Part 1-82. 1982 Testing Coated Fabrics Part 1: Method 1. Method for Determination of Roll Characteristics. 4 pp.

—Rubber
ISO 2231-89. Rubber-or Plastics-Coated Fabrics—Standard Atmospheres for Conditioning and Testing Second Edition. 6 pp.
JIS K 6328-81. Rubber Coated Fabrics. 42 pp.
JIS S 4007-76. Hand Sticked Rubberised Fabric Products.
JIS S 4008-86. Rubberised Fabric Products.

—Rubber—Abrasion Resistance Testing
ISO 5470-80. Rubber or Plastics Coated Fabrics—Determination of Abrasion Resistance First Edition. 5 pp.
ISO 5978-90. Rubber or Plastics Coated Fabrics—Determination of Blocking Resistance Second Edition. 6 pp.
ISO 5981-82. Rubber or Plastics Coated Fabrics—Determination of Flex Abrasion First Edition. 6 pp.

—Rubber—Adhesion Testing
ISO 2411-91. Rubber-or Plastics-Coated Fabrics—Determination of Coating Adhesion Second Edition. 10 pp.

—Rubber—Adhesive Tear Strength
BSI BS 903: Part A47-82. 1982 Methods of Testing Vulcanized Rubber Part A47: Analysis of Multi-Peak Traces Obtained in Determinations of Tear Strength and Adhesion Strength. 7 pp.
ISO 6133-81. Rubber and Plastics—Analysis of Multi-Peak Traces Obtained in Determinations of Tear Strength and Adhesion Strength First Edition. 5 pp.

—Rubber—Aerospace
BSI F 131-86. 1986 Amd 1 Chlorosulphonated Polyethylene Rubber Coated Nylon Fabric (300 G/M2) for Aerospace Purposes (AMD 5795) December 30, 1988. 5 pp.
BSI F 136-89. 1989 Pigmented Chlorosulphonated Polyethylene Rubber-Coated Nylon Fabric (190 g/m Squared) for Aerospace Purposes. 4 pp.
BSI F 139-89. 1989 Aluminized Polychloroprene Rubber-Coated Nylon Fabric (350 g/m Squared) for Aerospace Purposes. 4 pp.

—Rubber—Aging Testing
ISO 1419-77. Fabrics Coated with Rubber or Plastics—Accelerated Ageing and Simulated Service Tests First Edition. 4 pp.

—Rubber—Bend Testing
ISO 4675-90. Rubber-or Plastics-Coated Fabrics—Low-Temperature Bend Test Second Edition. 8 pp.

—Rubber—Breaking Load
ISO 1421-77. Fabrics Coated with Rubber or Plastics—Determination of Breaking Strength and Elongation at Break First Edition. 4 pp.

—Rubber—Bursting Strength
ISO 3303-90. Rubber-or Plastics-Coated Fabrics—Determination of Bursting Strength Second Edition. 7 pp.

—Rubber—Comprehension Testing
ISO 5473-79. Rubber-or Plastics-Coated Fabrics—Determination of Crush Resistance First Edition. 5 pp.

—Rubber—Cracking (Fracturing)
ISO 3011-81. Rubber or Plastics Coated Fabrics—Determination of Resistance to Ozone Cracking Under Static Conditions Second Edition. 4 pp.

—Rubber—Discoloration—Cigarette Smoke
ISO 6449-82. Rubber or Plastics Coated Fabrics—Determination of Discolouration by Cigarette Smoke First Edition. 4 pp.

—Rubber—Elongation
ISO 1421-77. Fabrics Coated with Rubber or Plastics—Determination of Breaking Strength and Elongation at Break First Edition. 4 pp.

—Rubber—Flexibility
ISO 5979-82. Rubber or Plastics Coated Fabrics—Determination of Flexibility—Flat Loop Method First Edition. 5 pp.

—Rubber—Flexural Strength
ISO 7854-84. Rubber-or Plastics-Coated Fabrics—Determination of Resistance to Damage by Flexing (Dynamic Method) First Edition. 6 pp.

Coated Fabrics (Cont.)

—Rubber—Fusion
BSI BS 3424: Part 22-83. 1983 Testing Coated Fabrics Part 22: Methods for Determination of Fusion of PVC Coatings and the State of Cure of Vulcanized Rubber Coatings. 4 pp.
ISO 6451-82. Plastics Coated Fabrics—Polyvinyl Chloride Coatings—Rapid Method for Checking Fusion First Edition. 3 pp.

—Rubber—Impact Testing
ISO 4646-89. Rubber-or Plastics-Coated Fabrics—Low-Temperature Impact Test Second Edition. 9 pp.

—Rubber—Low Temperature Testing
ISO 4675-90. Rubber-or Plastics-Coated Fabrics—Low-Temperature Bend Test Second Edition. 8 pp.

—Rubber—Protective Clothing
BSI BS 3546: Part 3-83. (WITHDRAWN) 1983 Coated Fabrics for Water Resistant Clothing Part 3: Specification for Natural Rubber and Synthetic Rubber Polymer Coated Fabrics (Superseded by BS 3546: Part 2: 1993). 16 pp.
ISO 8096 Pt 1-89. Rubber-or Plastics-Coated Fabrics for Water-Resistant Clothing—Specification—Part 1: PVC-Coated Fabrics First Edition; (Corrigendum 1-1991). 9 pp.
ISO 8096 Pt 2-89. Rubber-or Plastics-Coated Fabrics for Water-Resistant Clothing—Specification—Part 2: Polyurethane-and Silicone Elastomer-Coated Fabrics First Edition. 16 pp.
ISO 8096 Pt 3-88. Rubber-or Plastics-Coated Fabrics for Water-Resistant Clothing—Specification—Part 3: Natural Rubber-and Synthetic Rubber-Coated Fabrics First Edition. 20 pp.

—Rubber—Roll
CEN PREN 22286-92. Rubber-or Plastics-Coated Fabrics—Determination of Roll Characteristics (ISO 2286:1986). 7 pp.
ISO 2286-86. Rubber-or Plastics-Coated Fabrics—Determination of Roll Characteristics Second Edition; (CEN PREN 22286:1992). 6 pp.

—Rubber—Test Strength
ISO 4674-77. Fabrics Coated with Rubber or Plastics—Determination of Tear Resistance First Edition. 6 pp.

—Rubber—Vapor Transmission
ISO 6179-89. Vulcanized Rubber Sheet, and Fabrics Coated with Vulcanized Rubber—Determination of Transmission Rate of Volatile Liquids (Gravimetric Technique) Second Edition. 7 pp.

—Rubber—Water Resistance Testing
ISO 1420-87. Rubber-or Plastics-Coated Fabrics—Determination of Resistance to Penetration by Water Second Edition. 8 pp.

—Silicone Rubber—Protective Clothing
BSI BS 3546: Part 1-81. (WITHDRAWN) 1981 Coated Fabrics for Water Resistant Clothing Part 1: Specifications for Polyurethane and Silicone Elastomer Coated Fabrics (Superseded by BS 3546: Part 2: 1993). 12 pp.

—Staining
BSI BS 3424: Part 19-89. 1989 Testing Coated Fabrics Part 19: Methods 22A, 22B and 22C Determination of Sulphur Staining. 4 pp.

—Targets—Aircraft
MOD UK DTD-778A-59. Proofed Fabric for Targets. 2 pp.

—Tear Strength
BSI BS 3424: Part 5-82. 1982 Testing Coated Fabrics Part 5: Methods 7A, 7B and 7C. Methods for Determination of Tear Strength. 6 pp.

—Tensile Testing
BSI BS 3424: Part 21-93. 1993 Testing Coated Fabrics Part Method for Determination of Elongation and Tension Set. 10 pp.
BSI BS 3424: Part 21-87. 1987 Testing Coated Fabrics Part Method for Determination of Elongation and Tension Set. 6 pp.

—Test Specimens
BSI BS 3424: Part 2-82. 1982 Testing Coated Fabrics Part 2: Method 4. Conditioning and Selection of Test Specimens. 4 pp.
SAA AS 1441.1-73. Methods of Test for Coated Fabrics—Part 1: Method for Conditioning of Test Specimens.

—Thickness Measurement
BSI BS 3424: Part 23-87. 1987 Testing Coated Fabrics Part 23: Method for Determination of the Thickness of Coated Fabrics. 4 pp.

Coated Fabrics (Cont.)

—Thickness Measurement (Cont.)
BSI BS 3424: Part 25-93. 1993 Testing Coated Fabrics Part 25: Method 28. Method for Determination of the Coating Thickness and Thickness of Any Expanded Layer. 6 pp.
DIN ENGL 53353-71. Testing of Artificial Leather and Similar Sheet Materials; Determination of Thickness with Mechanical Feelers (June). 2 pp.
SAA AS 1441.3-73. Methods of Test for Coated Fabrics—Part 3: Method for Determination of Thickness.

—Vapor Transmission
BSI BS 903: Part A46-91. 1991 Physical Testing of Rubber Part A46: Method for Determination of the Transmission Rate of Volatile Liquids (ISO 6179: 1989). 9 pp.
BSI BS 2782:Pt8: METH 820A-92. 1992 Methods of Testing Plastics Part 8: Other Properties Method 820A: Determination of Water Vapour Transmission Rate (Dish Method) (ISO 2528: 1974) (V). 12 pp.
BSI BS 3424: Part 34-92. 1992 Testing Coated Fabrics Part 34: Method 37 Method for Determination of Water Vapour Permeability Index (WVPI). 10 pp.
ISO 2528-74. Sheet Materials—Determination of Water Vapour Transmission Rate—Dish Method First Edition. 11 pp.

—Vinyl—Upholstery
CGSB CAN/CGSB-4.124-M90. Vinyl-Coated Fabric for Upholstery Covering. 11 pp.
CGSB CAN/CGSB-4.125-M90. Expanded Vinyl-Coated Fabric for Upholstery Covering. 11 pp.
CGSB CAN/CGSB-4.149-M90. Vinyl-Coated Fabric for Public Seating. 11 pp.

—Vinyl—Wallpaper
CGSB 41-GP-30M-77. Wallcoverings, Vinyl-Coated Fabrics, Standard for (R 1982). 14 pp.

—Wallpaper
CEN EN 233-89. Wallcoverings in Roll Form Specification for Finished Wallpapers, Wall Vinyls and Plastic Wallcoverings. 22 pp.

—Water Resistance Testing
BSI BS 3424: Part 26-86. 1986 Testing Coated Fabrics Part 26: Methods 29A, 29B and 29C. Methods for Determination of Resistance to Penetration by Water. 7 pp.
BSI BS 3424: Part 26-90. 1990 Testing Coated Fabrics Part 26: Methods 29A, 29B, 29C, and 29D Methods for Determination of Resistance to Water Penetration and Surface Wetting. 11 pp.
BSI BS 3546: Part 2-93. 1993 Coated Fabrics for Use in the Manufacture of Water Penetration Resistant Clothing Part 2: Specification for Non-Water Vapour Permeable Coated Fabrics (Supersedes BS 3546: Part 1: 1981 & Part 3: 1983). 16 pp.
CGSB CAN/CGSB-4.2 NO.26.5-M89. Textile Test Methods Water Resistance—High-Pressure Penetration Test. 9 pp.

—Waterproofing Membranes
CGSB 37-GP-63M-77. Cloth, Glass, Coated, for Membrane Waterproofing Systems and Built-Up Roofing, Standard for. 9 pp.

—Wear Testing
SAA AS 1441.12-73. Methods of Test for Coated Fabrics—Part 12: Method for Determination of Resistance to Wear.

—Wicking Resistance Testing
BSI BS 3424: Part 18-86. 1986 Testing Coated Fabrics Part 18: Methods for 21A and 21B. Methods for Determination of Resistance to Wicking and Lateral Leakage. 6 pp.

Coated Fabrics (Electrical Insulation)
Use: Electrical Insulating Fabrics

Coated Glass
See Also: Glass; Insulating Glass

—Buildings—Chemical Properties
CEN PREN 1096-1-93. Coated Glass for Use in Buildings—Part 1: Characteristics and Properties. 12 pp.

Coated Macadam
Use: Macadam Aggregates—Coated

Coating Processes
Use: Coatings

Coatings
Scope Note: For additional listings, use a more specific term *Use For:* Coating Processes; Decorative Coatings; Protective Coatings *See Also:* Additives; Aggregate Coatings; Aluminum Coatings (Made From Aluminum); Aluminum Coatings (On Aluminum);

Coatings (Cont.)

See Also: (Cont.)
Anodic Coatings; Antifouling Coatings; Antireflection Coatings; Asphalt Coatings (Made From Asphalt); Baking Finishes; Baking Primers; Bituminous Coatings; Black Oxide Coatings; Bonded Coatings; Brass Coatings (On Brasses); Brush Coatings; Cadmium Coatings (Made From Cadmium); Cadmium Coatings (On Cadmium); Camouflage Coatings; Carbon Steel Coatings; Cast Iron Coatings (On Cast Iron); Cement Coatings (Made From Cement); Ceramic Coatings (Made From Ceramics); Checking Resistance; Chloroprene Rubber Coatings (Made From Chloroprene Rubber); Chromate Coatings; Chromium Carbide Coatings (Made From Chromium Carbide); Chromium Coatings (Made From Chromium); Chromium Oxide Coatings; Clad Metals; Clear Coatings; Coal Tar Coatings; Coated Abrasives; Coated Board; Coated Fabrics; Concrete Coatings; Conformal Coatings; Conversion Coatings; Copper Coatings (Made From Copper); Copper Coatings (On Copper); Corrosion; Corrosion Prevention; Corrosion Resistant Steel Coatings; Corrosion Resistant Steels; Dopes; Elastomer Coatings; Electrodeposited Coatings; Electroless Coatings; Epoxy Coatings; Ethyl Cellulose Coatings; Ethyl Cellulose Lacquers; Ethylene Copolymer Bitumen Coatings; Fabric Finishes; Fabrics; Ferroalloy Coatings (On Ferroalloys); Flame Spraying; Flat Finishes; Fluorescent Coatings; Galvanized Steel Coatings; Glass Coatings (On Glass); Glazes; Gold Coatings (Made From Gold); Hard Surfacing; Hot Dip Coatings; Immersion Coatings; Infrared Coatings; Intumescent Coatings; Iron Coatings (On Iron); Iron Phosphate Coatings (Made From Iron Phosphate); Lacquers; Laminates; Lead Coatings (Made From Lead); Light Metal Alloy Coatings (On Light Metal Alloys); Linings; Low Alloy Steel Coatings; Low Carbon Steel Coatings; Marine Coatings; Masonry Coatings; Matte Finishes; Metal Cleaning; Metal Coatings (Made From Metal); Metal Coatings (On Metal); Metal Finishing; Metallizing; Mineral Coatings; Neoprene Coatings (Made From Neoprene); Nickel Coatings (Made From Nickel); Nickel Coatings (On Nickel); Nonmagnetic Coatings; Nonmetallic Coatings; Orange Shellac; Organic Coatings; Oxide Coatings; Paint Driers; Paints; Palladium Coatings (Made From Palladium); Paper Coatings; Passivation; Phosphate Coatings; Pipe Coatings *See Also:* Plastic Coatings (On Plastic); Plasticizers; Polyethylene Coatings (Made From Polyethylene); Polymer Coatings (Made From Polymers); Polyurethane Coatings (Made From Polyurethane); Powder Coatings; Propylene Rubber Coatings (Made From Propylene Rubber); Reflective Coatings; Release Agents; Rhodium Coatings (Made From Rhodium); Roof Coatings; Sealants; Semigloss; Shellac Varnishes; Shellacs; Sherardized Coatings; Silica Coatings (Made From Silica); Silicon Coatings (Made From Silicon); Silver Coatings (Made From Silver); Solder Masks; Solvents; Sprayed Coatings; Steel Coatings; Stucco Coatings; Surface Finishing; Surfacers; Synthetic Resin Emulsion Coatings; Tar Coatings; Tin Coatings (Made From Tin); Titanium Coatings (On Titanium); Titanium Nitride Coatings; Undercoatings; Vacuum Deposited Coatings; Vacuum Evaporated Coatings; Varnishes; Veneers; Vinyl Coatings (Made From Vinyl); Vitreous Enamels; Waterproof Coatings; Waxes; Wood Coatings; Yellowing; Zinc Coatings (Made From Zinc); Zinc Coatings (On Zinc)
CGSB 31-GP-1M-88. Corrosion Preventive Compound, Cold Application, Hard Film. 11 pp.
CGSB 31-GP-3M-88. Corrosion Preventive Compound, Cold Application, Soft Film. 11 pp.
CNS A2181-85. Finish Coatings and Wall Coverings for Decorative Use (Feb)(11194).
CNS A3076-83. Method of Test for Multi-Layer Coatings with Decorative Pattern (Nov)(4684).
CNS A3221-85. Method of Test for Finish Coatings and Wall Coverings for Decorative Use (Feb)(11195).
JIS K 5400-90. Testing Methods for Paints. 198 pp.
MOD UK DSTAN 80-85-01. Corrosion Preventive Compound: Soft Film, Hot Application NATO Code No: C-628 Joint Service Designation: PX-11 Issue 1; Amendment 1. 7 pp.

—Abrasion Testing
CGSB 1-GP-71 METH 104.1-74. Methods of Testing Paints and Pigments Abrasion Resistance. 1 p.
JIS A 1436-91. Test Methods for Movement Capability of Coatings and Sheets Fully Adhered on Substrate. 15 pp.

—Accelerated Testing
CGSB 1-GP-71 METH 122.2-74. Methods of Testing Paints and Pigments Accelerated Weathering General Procedure. 2 pp.

INTERNATIONAL AND NON-U.S. NATIONAL STANDARDS
SUBJECT INDEX
Coatings

Coatings (Cont.)

—**Accelerated Testing—Test Panels**
CGSB 1-GP-71 METH 122.1-74. Methods of Testing Paints and Pigments Accelerated Weathering Selection, Exposure and Examination of Panels. 1 p.

—**Adhesion Testing**
CGSB 1-GP-71 METH 135.1-74. Methods of Testing Paints and Pigments Adhesion Knife Test. 1 p.
CGSB 1-GP-71 METH 135.2-78. Methods of Testing Paints and Pigments Adhesion Toughness and Adhesion. 1 p.
CGSB 1-GP-71 METH 135.8-74. Methods of Testing Paints and Pigments Adhesion Cross-Cut Adhesion Test. 2 pp.

—**Aerosol—Solids Content**
CNS K6737-83. Method of Test for Solid Contents of Aerosol Coatings (Jan)(9895).

—**Aerospace**
JIS W 2014-89. Finishes and Coatings for Protection of Aerospace Systems, Structures and Parts.

—**Aftertack**
CGSB 1-GP-71 METH 5.10-74. Methods of Testing Paints and Pigments Drying Time After-Tack. 1 p.

—**Aircraft**
MOD UK DSTAN 80-83-82. Corrosion Preventive Compound: Aircraft Structures Joint Service Designation: PX-32 Issue 1. 21 pp.

—**Alcohol Resistance Testing**
CGSB 1-GP-71 METH 146.1-78. Methods of Testing Paints and Pigments Alcohol Resistance. 1 p.

—**Antimony Content**
CGSB CAN2-1.500-75 METH 4-73. Methods of Test for Toxic Trace Elements in Protective Coatings Determination of Leachable Antimony (Sb) in Low Concentration. 2 pp.

—**Arsenic Content**
CGSB CAN2-1.500-75 METH 7-73. Methods of Test for Toxic Trace Elements in Protective Coatings Determination of Leachable Arsenic (As) in Low Concentration. 2 pp.

—**Ash Content**
CGSB 1-GP-71 METH 69.3-82. Methods of Testing Paints and Pigments Ash Content of Varnish, Tar-Base and Other Coatings. 1 p.

—**Automotive**
CNS D1010-81. General Rules of Coating Films for Automobile Parts (Jul)(7670).
JIS D 0202-88. General Rules of Coating Films for Automobile Parts. 17 pp.
MOD UK TS 10131. PX 28 Preservative for Hollow Sections and Automotive Underbodies.

—**Automotive—Underbodies**
MOD UK DSTAN 80-168-93. Compound Protective for Vehicle Underbodies Water-Based Issue 1. 18 pp.

—**Barium Content**
CGSB CAN2-1.500-75 METH 3-73. Methods of Test for Toxic Trace Elements in Protective Coatings Determination of Leachable Barium (Ba) in Low Concentration. 2 pp.

—**Benzene Content**
CGSB 1-GP-71 METH 65.1-82. Methods of Testing Paints and Pigments Detection of Benzene (Benzol). 2 pp.

—**Bicycles**
CNS B7069-75. Standard for Coating of Bicycle Parts (Dec)(3882). 1 p.

—**Bicycles—Impact Testing**
CNS B7070-75. Method of Test for Coating of Bicycle Parts (Dec)(3883). 1 p.

—**Bicycles—Salt Spray Testing**
CNS B7070-75. Method of Test for Coating of Bicycle Parts (Dec)(3883). 1 p.

—**Binders (Materials)**
DIN ENGL 55928 Pt 9-91. Corrosion Protection of Steel Structures by the Application of Organic or Metallic Coatings; Composition of Binders and Pigments for Coating Materials (May). 6 pp.

—**Cadmium Content**
CGSB CAN2-1.500-75 METH 2-73. Methods of Test for Toxic Trace Elements in Protective Coatings Determination of Leachable Cadmium (Cd) in Low Concentration. 2 pp.

—**Cans—Metal—Food**
CNS Z5028-83. Coating Materials for Metal Cans of Foods—General (Sep)(2773). 1 p.

Coatings (Cont.)

—**Chromium Content**
CGSB CAN2-1.500-75 METH 8-73. Methods of Test for Toxic Trace Elements in Protective Coatings Determination of Leachable Chromium (Cr) in Low Concentration. 1 p.

—**Classification**
EC 83/265/EEC-83. Council Directive Amending Directive 77/728/EEC on the Approximation of the Laws, Regulations and Administrative Provisions of the Member States Relating to the Classification, Packaging and Labelling of Paints, Varnishes, Printing Inks, Adhesives. 7 pp.

—**Colorfastness Testing**
CGSB 1-GP-71 METH 120.1-74. Methods of Testing Paints and Pigments Color Stability Fading by Light. 2 pp.

—**Containers**
BSI BS 1262-89. 1989 Round Lever Lid Tinplate Cans, Paint Range. 16 pp.

—**Corrosion Testing**
DIN ENGL 50928-85. Corrosion of Metals; Testing and Assessment of the Corrosion Protection of Coated Metallic Materials in Contact with Aqueous Corrosive Agents (Sept). 8 pp.

—**Density**
DIN ENGL 53217 Pt 1-91. Determination of Density of Paints, Varnishes and Similar Coating Materials; Survey of Test Methods (Mar). 2 pp.
DIN ENGL 53217 Pt 3-91. Determination of Density of Paints, Varnishes and Similar Coating Materials by the Displacement Float Method (Mar). 4 pp.
DIN ENGL 53217 Pt 5-91. Determination of Density of Paints, Varnishes and Similar Coating Materials by the Vibration Method (Mar). 4 pp.

—**Density—Hydrometry**
DIN ENGL 53217 Pt 4-91. Determination of Density of Paints, Varnishes and Similar Coating Materials by the Hydrometer Method (Mar). 3 pp.

—**Density—Pycnometric Analysis**
DIN ENGL 53217 Pt 2-91. Determination of Density of Paints, Varnishes and Similar Coating Materials by the Pyknometer Method (Mar). 5 pp.

—**Drying**
CGSB 1-GP-71 METH 103.1-79. Methods of Testing Paints and Pigments Drying Conditions Air Drying. 1 p.
CGSB 1-GP-71 METH 103.2-79. Methods of Testing Paints and Pigments Drying Conditions Oven Drying. 1 p.
CGSB 1-GP-71 METH 103.3-79. Methods of Testing Paints and Pigments Drying Conditions Low-Temperature Conditioning in Air. 1 p.

—**Environmental Testing**
CGSB 1-GP-71 METH 122.2-74. Methods of Testing Paints and Pigments Accelerated Weathering General Procedure. 2 pp.
CGSB 1-GP-71 METH 143.1-80. Methods of Testing Paints and Pigments Long-Term Outdoor Performance Selection, Exposure and Examination of Panels. 2 pp.

—**Environmental Testing—Test Panels**
CGSB 1-GP-71 METH 122.1-74. Methods of Testing Paints and Pigments Accelerated Weathering Selection, Exposure and Examination of Panels. 1 p.

—**Evaporation Residue Analysis**
CGSB 1-GP-71 METH 20.1-78. Methods of Testing Paints and Pigments Residue on Evaporation. 1 p.

—**Explosives**
MOD UK CS 2724. Ethyl Cellulose, High Viscosity.

—**Fatty Acid Content**
CNS K6630-81. Method of Test for Fatty Acids Used in Protective Coatings (Mar)(7039).

—**Faucets**
BSI BS 5412 & 5413: Part 5-76. 1976 Amd 3 Performance of Draw-Off Taps with Metal Bodies for Water Services and Taps with Plastic Bodies for Water Services Part 5: Physio-Chemical Characteristics: Materials, Coatings. 15 pp.

—**Fire Testing**
SNZ NZMP 9-89. Fire Properties of Building Materials and Elements of Structure. 207 pp.

—**Furniture—Adhesion Testing**
CNS S2135-86. Method of Test for Adhesion of Furniture Coatings (Aug)(11684).

—**Furniture—Corrosion Inhibitors**
CNS S2136-86. Method of Test for Antirust Property of Furniture Coating (Aug)(11685).

Coatings (Cont.)

—**Gloss**
CGSB 1-GP-71 METH 13.1-75. Methods of Testing Paints and Pigments Gloss After 48 Hours. 1 p.
CGSB 1-GP-71 METH 13.2-75. Methods of Testing Paints and Pigments Gloss After 7 Days. 1 p.
CGSB 1-GP-71 METH 13.3-75. Methods of Testing Paints and Pigments Gloss Gloss After One Hour Baking. 1 p.

—**Hiding Power**
CGSB 1-GP-71 METH 14.1-81. Methods of Testing Paints and Pigments Hiding Power. 6 pp.
CGSB 1-GP-71 METH 14.2-81. Methods of Testing Paints and Pigments Hiding Power Quick-Drying Materials. 1 p.

—**Humidity**
CGSB 1-GP-71 METH 113.5-74. Methods of Testing Paints and Pigments Corrosion Humidity Cabinet Test. 5 pp.

—**Identification Systems**
EC 83/265/EEC-83. Council Directive Amending Directive 77/728/EEC on the Approximation of the Laws, Regulations and Administrative Provisions of the Member States Relating to the Classification, Packaging and Labelling of Paints, Varnishes, Printing Inks, Adhesives. 7 pp.

—**Impact Testing**
CGSB 1-GP-71 METH 147.1-74. Methods of Testing Paints and Pigments Impact Resistance. 1 p.
CGSB 1-GP-71 METH 147.2-78. Methods of Testing Paints and Pigments Impact Resistance Falling Ball Method. 2 pp.

—**Inks**
CGSB 1-GP-71 METH 130.9-79. Methods of Testing Paints and Pigments Behavior Towards Topcoats Resistance to Stencil Inks. 1 p.

—**Knife Testing**
CGSB 1-GP-71 METH 135.1-74. Methods of Testing Paints and Pigments Adhesion Knife Test. 1 p.

—**Lead Content**
CGSB 1-GP-71 METH 52.3-81. Methods of Testing Paints and Pigments Metal Content. 2 pp.
CGSB CAN2-1.500-75 METH 1-73. Methods of Test for Toxic Trace Elements in Protective Coatings Determination of Lead (Pb) in Low Concentration. 2 pp.

—**Lead Free—Toys**
CNS K2097-86. Lead-Free Coatings for Toy (Apr)(5211). 2 pp.
CNS K6836-86. Method of Test for Lead-Free Coating for Toy (Apr)(11549).

—**Leather—Thickness Measurement**
BSI BS 3144: Part 28-87. 1987 Methods of Sampling and Physical Testing of Leather Part 28: Method for Measurement of Thickness of Surface Coatings on Leather. 4 pp.

—**Light Testing**
CNS K6708-82. Method of Test for Light Stability of Clear Coatings (Feb)(8515).

—**Manganese Content**
CGSB 1-GP-71 METH 52.3-81. Methods of Testing Paints and Pigments Metal Content. 2 pp.

—**Mercury Content**
CGSB CAN2-1.500-75 METH 6-73. Methods of Test for Toxic Trace Elements in Protective Coatings Determination of Leachable Mercury (Hg) in Low Concentration. 2 pp.

—**Methanol Content**
CGSB 1-GP-71 METH 84.1-82. Methods of Testing Paints and Pigments Methyl Alcohol Content Method for Determining Methyl Alcohol Content. 1 p.

—**Multilayer**
CNS A2064-83. Multi-Layer Coatings with Decorative Pattern (Nov)(4683). 3 pp.

—**Nonconductive—Continuity Testing**
SAA AS 3894.1-91. Site Testing of Protective Coatings—Part 1: Non-Conductive Coatings—Continuity Testing—High Voltage ('Brush') Method (in Professional Package 39). 16 pp.
SAA AS 3894.2-91. Site Testing of Protective Coatings—Part 2: Non-Conductive Coatings—Continuity Testing—Wet Sponge Method (in Professional Package 39). 5 pp.

—**Nonvolatile Matter Content**
CGSB 1-GP-71 METH 19.1-80. Methods of Testing Paints and Pigments Nonvolatile Vehicle Nonvolatile Vehicle by Difference. 1 p.

INDUSTRY STANDARDS

INTERNATIONAL AND NON-U.S. NATIONAL STANDARDS
SUBJECT INDEX

Coatings

Coatings (Cont.)

—**Packaging**
EC 83/265/EEC-83. Council Directive Amending Directive 77/728/EEC on the Approximation of the Laws, Regulations and Administrative Provisions of the Member States Relating to the Classification, Packaging and Labelling of Paints, Varnishes, Printing Inks, Adhesives. 7 p.

—**Pencil Scratch Testers**
CNS Z7202-86. Pencil Scratch Tester for Coated Film (Oct)(11723).
JIS K 5401-69. Pencil Scratch Tester for Coated Film (R 1984). 7 pp.

—**Phenol Formaldehyde Resins—Qualitative Analysis**
CGSB 1-GP-71 METH 75.2-82. Methods of Testing Paints and Pigments Detection of Resins Qualitative Test for Phenol-Formaldehyde Resins. 1 p.

—**Pigments**
DIN ENGL 55928 Pt 9-91. Corrosion Protection of Steel Structures by the Application of Organic or Metallic Coatings; Composition of Binders and Pigments for Coating Materials (May). 6 pp.

—**Pressurized Containers**
BSI BS 1101-77. 1977 Amd 1 Pressure Containers for Paint and Other Similar Substances. 17 pp.

—**Rocker Hardness Testing**
CGSB 1-GP-71 METH 116.1-74. Methods of Testing Paints and Pigments Hardness Sward Rocker Hardness. 1 p.

—**Rosin Acid Content**
CGSB 1-GP-71 METH 60.2-74. Methods of Testing Paints and Pigments Total Rosin Acids Content of Coating Vehicles. 3 pp.

—**Rosin Content**
CGSB 1-GP-71 METH 60.1-74. Methods of Testing Paints and Pigments Rosin Qualitative Determination. 1 p.

—**Salt Spray Testing**
CGSB 1-GP-71 METH 129.1-79. Methods of Testing Paints and Pigments Salt-Spray Corrosion Resistance General Procedure. 1 p.

—**Sanding**
CGSB 1-GP-71 METH 127.1-79. Methods of Testing Paints and Pigments Sanding Properties Wet Sanding. 1 p.
CGSB 1-GP-71 METH 127.3-79. Methods of Testing Paints and Pigments Sanding Properties Dry Sanding. 1 p.

—**Selenium Content**
CGSB CAN2-1.500-75 METH 5-73. Methods of Test for Toxic Trace Elements in Protective Coatings Determination of Leachable Selenium (Se) in Low Concentration. 2 pp.

—**Slick Resistance**
CGSB 1-GP-71 METH 148.1-78. Methods of Testing Paints and Pigments Slick Resistance. 1 p.

—**Staining**
CGSB 1-GP-71 METH 114.3-79. Methods of Testing Paints and Pigments Cleansability Asphalt Stain Test. 1 p.

—**Tape Testing**
CGSB 1-GP-71 METH 137.2-78. Methods of Testing Paints and Pigments Tape Test over Bend. 1 p.

—**Test Panels**
CGSB 1-GP-71 METH 13.4-75. Methods of Testing Paints and Pigments Gloss Test Panels. 1 p.

—**Thickness Measurement**
BSI BS 5868-80. 1980 Guide to Ionizing Radiation Thickness Meters for Materials in the Form of Sheets, Coatings or Laminates. 27 pp.
CGSB 1-GP-71 METH 128.1-79. Methods of Testing Paints and Pigments Measurement of Film Thickness Dry Film Thickness. 1 p.
CGSB 1-GP-71 METH 128.2-79. Methods of Testing Paints and Pigments Measurement of Film Thickness Dry Film—Soft Thick Film. 1 p.
DIN ENGL 50933-87. Measurement of Coating Thickness by Differential Measurement Using a Stylus Instrument (Aug). 4 pp.
DIN ENGL 50982 Pt 3-87. Principles of Coating Thickness Measurement; Selection Criteria and Basic Measurement Procedures (Aug). 6 pp.

—**Thickness Measurement—Glossaries**
DIN ENGL 50982 Pt 1-87. Principles of Coating Thickness Measurement; Terminology Associated with Coating Thickness and Measuring Areas (Aug). 3 pp.

Coatings (Cont.)

—**Thinners—Aromatic Hydrocarbon Content**
CGSB 1-GP-71 METH 66.3-82. Methods of Testing Paints and Pigments Hydrocarbons and Esters Aromatic Hydrocarbons. 1 p.

—**Toughness**
CGSB 1-GP-71 METH 135.2-78. Methods of Testing Paints and Pigments Adhesion Toughness and Adhesion. 1 p.

—**Vehicles**
MOD UK TS 310B. Compound, Protective for Vehicle Under-Bodies Solvent-Based.

—**Viscosity—Ford Cup**
JIS K 5402-71. Ford Cup for Determining Consistency of Coatings.

—**Visual Inspection**
CGSB 1-GP-71 METH 134.19-78. Methods of Testing Paints and Pigments Applicability and Appearance Examination of Applied Films (Viewing Standard). 1 p.

—**Water Contact—Odors**
CGSB 1-GP-71 METH 144.1-78. Methods of Testing Paints and Pigments Odor and Taste Test. 2 pp.

—**Water Contact—Taste**
CGSB 1-GP-71 METH 144.1-78. Methods of Testing Paints and Pigments Odor and Taste Test. 2 pp.

—**Water Resistance Testing**
CGSB 1-GP-71 METH 110.1-78. Methods of Testing Paints and Pigments Water Resistance General Method. 1 p.
CGSB 1-GP-71 METH 138.1-78. Methods of Testing Paints and Pigments Resistance to Chlorinated Water. 1 p.

—**Yellowing**
CGSB 1-GP-71 METH 120.2-74. Methods of Testing Paints and Pigments Color Stability Accelerated Yellowing in Ammonia Atmosphere. 1 p.
CGSB 1-GP-71 METH 120.4-74. Methods of Testing Paints and Pigments Color Stability Accelerated Yellowing of Coatings Exposed to the Dark, in a Warm, Moist Atmosphere. 1 p.

Coats
See Also: Clothing; Jackets; Outerwear; Raincoats

—**Boys'**
CGSB CAN/CGSB-49.30-M91. Canada Standard Children's Sizes 2 to 6X, Girls' Sizes 7 to 16 and Boys' Sizes 7 to 20, Unlined or Lightweight-Lined Outerwear Coats and Jackets, Regular Range—Dimensions. 19 pp.
CGSB CAN/CGSB-49.72-M81. Coats, Winter, Little Girls', and Little Boys', Regular Range—Dimensions; (Amendment 1 Nov 1983). 11 pp.
CGSB CAN/CGSB-49.77-M81. Coats, Winter, Boys', Regular Range—Dimensions; (Amendment 1 Nov 1983). 11 pp.

—**Children's**
CGSB CAN/CGSB-49.30-M91. Canada Standard Children's Sizes 2 to 6X, Girls' Sizes 7 to 16 and Boys' Sizes 7 to 20, Unlined or Lightweight-Lined Outerwear Coats and Jackets, Regular Range—Dimensions. 19 pp.

—**Girls'**
CGSB CAN/CGSB-49.30-M91. Canada Standard Children's Sizes 2 to 6X, Girls' Sizes 7 to 16 and Boys' Sizes 7 to 20, Unlined or Lightweight-Lined Outerwear Coats and Jackets, Regular Range—Dimensions. 19 pp.
CGSB CAN/CGSB-49.72-M81. Coats, Winter, Little Girls', and Little Boys', Regular Range—Dimensions; (Amendment 1 Nov 1983). 11 pp.
CGSB CAN/CGSB-49.73-M81. Coats, Winter, Girls', Regular Range—Dimensions. 10 pp.

—**Juniors'—Sizes**
CGSB CAN/CGSB-49.216-92. Juniors, Misses and Women's Canada Standard Sizes Coats—Dimensions. 16 pp.

—**Women's—Sizes**
CGSB CAN/CGSB-49.216-92. Juniors, Misses and Women's Canada Standard Sizes Coats—Dimensions. 16 pp.

Coaxial Adapters
Use For: Radio Frequency Coaxial Adapters
See Also: Adapters (Electric); Coaxial Connectors

—**BNC—50 Ohms**
CECC CECC 22 121-813 ISSUE 1-85. Detail Specification: Radio Frequency Coaxial Connectors; Series BNC (En, Fr, Ge). 21 pp.

Coaxial Adapters (Cont.)

—**BNC—50 Ohms** (Cont.)
CECC CECC 22 121-814 ISSUE 1-85. Detail Specification: Radio Frequency Coaxial Connectors; Series BNC (En, Fr, Ge). 21 pp.
CECC CECC 22 121-815 ISSUE 1-85. Detail Specification: Radio Frequency Coaxial Connectors; Series BNC (En, Fr, Ge). 21 pp.

—**BNC—50 Ohms—Right Angle**
CECC CECC 22 121-812 ISSUE 1-85. Detail Specification: Radio Frequency Coaxial Connectors; Series BNC (En, Fr, Ge) AMD 1 (En, Fr, Ge). 24 pp.

—**BNC—50 Ohms—Right Angle—Preferred Products List**
CECC CECC MUAHAG Vol 3B IS 4-91. Preferred Products List; Connectors; R.F. and Fibre Optics (En, Fr, Ge). 65 pp.

—**BNC—50 Ohms—Tee**
CECC CECC 22 121-816 ISSUE 1-85. Detail Specification: Radio Frequency Coaxial Connectors; Series BNC (En, Fr, Ge). 21 pp.

—**BNC—50 Ohms—Tee—Preferred Products List**
CECC CECC MUAHAG Vol 3B IS 4-91. Preferred Products List; Connectors; R.F. and Fibre Optics (En, Fr, Ge). 65 pp.

—**N—50 Ohms**
CECC CECC 22 211-815 ISSUE 1-88. Detail Specificatio: Radio Frequency Coaxial Connectors; Series N (En, Fr, Ge). 41 pp.
CECC CECC 22 211-816 ISSUE 1-88. Detail Specification: Radio Frequency Coaxial Connectors; Series N (En, Fr, Ge). 41 pp.
CECC CECC 22 211-817 ISSUE 1-88. Detail Specification: Radio Frequency Coaxial Connectors; Series N (En, Fr, Ge). 41 pp.

—**N—50 Ohms—Preferred Products List**
CECC CECC MUAHAG Vol 3B IS 4-91. Preferred Products List; Connectors; R.F. and Fibre Optics (En, Fr, Ge). 65 pp.

—**N—50 Ohms—Right Angle—Preferred Products List**
CECC CECC MUAHAG Vol 3B IS 4-91. Preferred Products List; Connectors; R.F. and Fibre Optics (En, Fr, Ge). 65 pp.

—**N—50 Ohms—Tee—Preferred Products List**
CECC CECC MUAHAG Vol 3B IS 4-91. Preferred Products List; Connectors; R.F. and Fibre Optics (En, Fr, Ge). 65 pp.

—**SMA—50 Ohms—Preferred Products List**
CECC CECC MUAHAG Vol 3B IS 4-91. Preferred Products List; Connectors; R.F. and Fibre Optics (En, Fr, Ge). 65 pp.

—**SMA—50 Ohms—Right Angle—Preferred Products List**
CECC CECC MUAHAG Vol 3B IS 4-91. Preferred Products List; Connectors; R.F. and Fibre Optics (En, Fr, Ge). 65 pp.

—**SMB—75 Ohms—Preferred Products List**
CECC CECC MUAHAG Vol 3B IS 4-91. Preferred Products List; Connectors; R.F. and Fibre Optics (En, Fr, Ge). 65 pp.

—**SSMA—50 Ohms**
CECC CECC 22 161-813 ISSUE 1-87. Detail Specification: Radio Frequency Coaxial Connectors Series SSMA (En, Fr, Ge). 33 pp.
CECC CECC 22 161-814 ISSUE 1-87. Detail Specification: Radio Frequency Coaxial Connectors Series SSMA (En, Fr, Ge). 33 pp.
CECC CECC 22 161-815 ISSUE 1-87. Detail Specification: Radio Frequency Coaxial Connectors Series SSMA (En, Fr, Ge). 33 pp.

—**SSMA—50 Ohms—Preferred Products List**
CECC CECC MUAHAG Vol 3B IS 4-91. Preferred Products List; Connectors; R.F. and Fibre Optics (En, Fr, Ge). 65 pp.

—**SSMB—50 Ohms—Preferred Products List**
CECC CECC MUAHAG Vol 3B IS 4-91. Preferred Products List; Connectors; R.F. and Fibre Optics (En, Fr, Ge). 65 pp.

Coaxial Adapters (Cont.)

—SSMC—50 Ohms—Preferred Products List
CECC CECC MUAHAG Vol 3B IS 4-91. Preferred Products List; Connectors; R.F. and Fibre Optics (En, Fr, Ge). 65 pp.

—Waveguide—Preferred Products List
CECC CECC MUAHAG Vol 12 IS 1-90. Preferred Products List; Microwave Components (En, Fr, Ge). 76 pp.

Coaxial Attenuators

—Microwave—50 Ohms—Preferred Products List
CECC CECC MUAHAG Vol 12 IS 1-90. Preferred Products List; Microwave Components (En, Fr, Ge). 76 pp.

—Variable—Microwave—50 Ohms—Preferred Products List
CECC CECC MUAHAG Vol 12 IS 1-90. Preferred Products List; Microwave Components (En, Fr, Ge). 76 pp.

Coaxial Cable Assemblies

Use For: Radio Frequency Coaxial Cable Assemblies
See Also: Cable Assemblies (Electric); Radio Frequency Equipment
IEC 966 Pt 1-88. Generic Specification for Radio Frequency and Coaxial Cable Assemblies Part 1: General Requirements and Test Methods First Edition; (Amendment 1-1990). 76 pp.

—Flexible
IEC 966 Pt 2-1-91. Radio Frequency and Coaxial Cable Assemblies Part 2-1: Sectional Specification for Flexible Coaxial Cable Assemblies First Edition. 31 pp.

—Flexible—Quality Assurance
IEC 966 Pt 2-2-92. Radio Frequency and Coaxial Cable Assemblies Part 2-2: Blank Detail Specification for Flexible Coaxial Cable Assemblies First Edition. 19 pp.

—Semiflexible
IEC 966 Pt 3-92. Radio Frequency and Coaxial Cable Assemblies Part 3: Sectional Specification for Semi-Flexible Coaxial Cable Assemblies First Edition. 30 pp.

—Semiflexible—Quality Assurance
IEC 966 Pt 3-1-92. Radio Frequency and Coaxial Cable Assemblies Part 3-1: Blank Detail Specification for Semi-Flexible Coaxial Cable Assemblies First Edition. 19 pp.

—Semirigid
IEC 966 Pt 4-92. Radio Frequency and Coaxial Cable Assemblies Part 4: Sectional Specification for Semi-Rigid Coaxial Cable Assemblies First Edition. 36 pp.

—Semirigid—Quality Assurance
IEC 966 Pt 4-1-92. Radio Frequency and Coaxial Cable Assemblies Part 4-1: Blank Detail Specification for Semi-Rigid Coaxial Cable Assemblies First Edition. 19 pp.

Coaxial Cables

Use For: Coaxial Lines; Radio Frequency Coaxial Cables *See Also:* Cable Television Cables; Cables (Electric); Coaxial Connectors; Coaxial Transmission Lines (Waveguide); Communication Transmission Lines; Leaky Cables; Microwave Coaxial Cables; Power Lines; Radio Frequency Equipment; Telephone Cables; Transmission Lines
BSI BS 2316: Parts 1 & 2-68. 1968 Amd 3 Radio-Frequency Cables Part 1: General Requirements and Tests Part 2: British Government Services Requirements. 52 pp.
BSI BS 2316: Part 3-69. 1969 Amd 3 Radio-Frequency Cables Part 3: Cable Data Sheets. 93 pp.
BSI BS 5425: Part 1-86. 1986 Coaxial Cables for Wired Distribution Systems Part 1: Single Unit, Semi-Airspaced Cables for Wideband Distribution Systems. 22 pp.
CENELEC HD 120-74. Characteristic Impedances and Dimensions of Radio Frequency Coaxial Cables. 2 pp.
CEPT T/TTT 5-75. Relative a L'Utilisation De Paires Coaxiales Du Type 0.7/2.9mm. 3 pp.
CNS C2063-89. 50 (Omega Particals) and 75 (Omega Particals) Radio Frequency Coaxial Cables (May)(4735).
IEC 78-67. Characteristic Impedances and Dimensions of Radio-Frequency Coaxial Cables Third Edition. 11 pp.
IEC 96 Pt 0-1-90. Radio-Frequency Cables Part 0: Guide to the Design of Detail Specifications Section 1 —Coaxial Cables Second Edition. 35 pp.

Coaxial Cables (Cont.)

IEC 96 Pt 3-82. Radio-Frequency Cables Part 3: General Requirements and Tests for Single-Unit Coaxial Cables for Use in Cabled Distribution Systems First Edition. 13 pp.
IECQ PQC 71-87. Radio Frequency Cables for Use in Electrotechnical Equipment: Generic Specification: Radio Frequency Cables of Assessed Quality. 36 pp.
IECQ PQC 72-87. Radio Frequency Cables for Use in Electrotechnical Equipment: Sectional Specification: Radio Frequency Cables, Flexible, for Operation at a Maximum Center Conductor Temperature of 85 Degrees C. 12 pp.
JIS C 3501-87. Radio-Frequency Coaxial Cables. 11 pp.
MOD UK DSTAN 59-6-67. Lines, Radio Frequency Transmission (Rigid Coaxial) Issue 1. 4 pp.

— 50 Ohms
CNS C2063-89. 50 (Omega Particals) and 75 (Omega Particals) Radio Frequency Coaxial Cables (May)(4735).

— 75 Ohms
CNS C2063-89. 50 (Omega Particals) and 75 (Omega Particals) Radio Frequency Coaxial Cables (May)(4735).
DIN VDE 0887 Pt 1-80. Radio Frequency Cables; Coaxial, Z=75 Ohms for Television Antenna Systems; General Requirements; (VDE Specifications) (Dec). 17 pp.

—Aircraft
ISO 3389-75. Aircraft—Radio Frequency Flexible Coaxial Cables—Dimensions and Electrical Characteristics First Edition. 5 pp.

—Cable Television—75 Ohms
DIN VDE 0887 Pt 3-84. Coaxial Radio Frequency Cable, Z = 75 Ohm for Receiving and Transmission Systems Including CATV Systems; Outdoor Cable with Sealed Outer Conductor (Dec). 10 pp.

—Communication—Construction
CSA CAN/CSA-C22. 2 NO 214-M90. Communications Cables; (Gen Instr 1 Thru 3). 60 pp.

—Communication—Multiconductor—Construction
CSA CAN/CSA-C22. 2 NO 214-M90. Communications Cables; (Gen Instr 1 Thru 3). 60 pp.

—Communication—Multiconductor—Testing
CSA CAN/CSA-C22. 2 NO 214-M90. Communications Cables; (Gen Instr 1 Thru 3). 60 pp.

—Communication—Testing
CSA CAN/CSA-C22. 2 NO 214-M90. Communications Cables; (Gen Instr 1 Thru 3). 60 pp.

—Crimp Barrels
CENELEC HD 489 S1-87. Recommended Dimensions for Hexagonal and Square Crimping-Die Cavities, Indentors, Gauges, Outer Conductor Crimp Sleeves and Centre Contact Crimp Barrels for r.f. Cables and Connectors. 4 pp.
IEC 803-84. Recommended Dimensions for Hexagonal and Square Crimping-Die Cavities, Indentors, Gauges, Outer Conductor Crimp Sleeves and Centre Contact Crimp Barrels for R.F. Cables and Connectors First Edition. 40 pp.

—Crimp Sleeves
CENELEC HD 489 S1-87. Recommended Dimensions for Hexagonal and Square Crimping-Die Cavities, Indentors, Gauges, Outer Conductor Crimp Sleeves and Centre Contact Crimp Barrels for r.f. Cables and Connectors. 4 pp.
IEC 803-84. Recommended Dimensions for Hexagonal and Square Crimping-Die Cavities, Indentors, Gauges, Outer Conductor Crimp Sleeves and Centre Contact Crimp Barrels for R.F. Cables and Connectors First Edition. 40 pp.

—Die Cavities—Hexagonal/Square Crimping
CENELEC HD 489 S1-87. Recommended Dimensions for Hexagonal and Square Crimping-Die Cavities, Indentors, Gauges, Outer Conductor Crimp Sleeves and Centre Contact Crimp Barrels for r.f. Cables and Connectors. 4 pp.
IEC 803-84. Recommended Dimensions for Hexagonal and Square Crimping-Die Cavities, Indentors, Gauges, Outer Conductor Crimp Sleeves and Centre Contact Crimp Barrels for R.F. Cables and Connectors First Edition. 40 pp.

—Ethernet
ECMA ECMA 80-84. Local Area Networks (CSMA/CD Baseband) Coaxial Cable System. 18 pp.

Coaxial Cables (Cont.)

—Ethernet (Cont.)
ECMA ECMA-TR 26-90. Planning and Installation Guide for CSMA/CD 10 Mbit/s Baseband LAN Coaxial Cable Systems. 64 pp.

—Ferrules
MOD UK DSTAN 59-23-70. Ferrules, Screened Cable Issue 1. 18 pp.

—Flat
CSA CAN/CSA-C22. 2 NO 214-M90. Communications Cables; (Gen Instr 1 Thru 3). 60 pp.

—Flat—Construction
CSA CAN/CSA-C22. 2 NO 214-M90. Communications Cables; (Gen Instr 1 Thru 3). 60 pp.

—Flat—Multiconductor
CSA CAN/CSA-C22. 2 NO 214-M90. Communications Cables; (Gen Instr 1 Thru 3). 60 pp.

—Flat—Multiconductor—Construction
CSA CAN/CSA-C22. 2 NO 214-M90. Communications Cables; (Gen Instr 1 Thru 3). 60 pp.

—Gages
CENELEC HD 489 S1-87. Recommended Dimensions for Hexagonal and Square Crimping-Die Cavities, Indentors, Gauges, Outer Conductor Crimp Sleeves and Centre Contact Crimp Barrels for r.f. Cables and Connectors. 4 pp.
IEC 803-84. Recommended Dimensions for Hexagonal and Square Crimping-Die Cavities, Indentors, Gauges, Outer Conductor Crimp Sleeves and Centre Contact Crimp Barrels for R.F. Cables and Connectors First Edition. 40 pp.

—Indentors
CENELEC HD 489 S1-87. Recommended Dimensions for Hexagonal and Square Crimping-Die Cavities, Indentors, Gauges, Outer Conductor Crimp Sleeves and Centre Contact Crimp Barrels for r.f. Cables and Connectors. 4 pp.
IEC 803-84. Recommended Dimensions for Hexagonal and Square Crimping-Die Cavities, Indentors, Gauges, Outer Conductor Crimp Sleeves and Centre Contact Crimp Barrels for R.F. Cables and Connectors First Edition. 40 pp.

—Leaky—Land Mobile Services
CCIR Report 902-1-90. Leaky-Feeder Systems in the Land Mobile Service—Section 8A—Land Mobile Service and Related Subjects. 10 pp.

—Local Area Networks
ECMA ECMA 80-84. Local Area Networks (CSMA/CD Baseband) Coaxial Cable System. 18 pp.
ECMA ECMA-TR 26-90. Planning and Installation Guide for CSMA/CD 10 Mbit/s Baseband LAN Coaxial Cable Systems. 64 pp.

—Military
IECQ PQC 73-87. Radio Frequency Cables for Use in Electrotechnical Equipment: Blank Detail Specification: Flexible Radio Frequency Cables for Operation at a Maximum Center Conductor Temperature of 85 Degrees C Assessment Level H. 16 pp.
IECQ PQC 73/US 0001-88. Radio Frequency Cables for Use in Electrotechnical Equipment: Detail Specification: Radio Frequency Cables, Flexible, for Operation at a Maximum Center Conductor Temperature of 85 Degrees C Assessment Level H. 16 pp.
IECQ PQC 74-87. Radio Frequency Cables for Use in Electrotechnical Equipment: Blank Detail Specification: Flexible Radio Frequency Cables for Operation at a Maximum Center Conductor Temperature of 85 Degrees C Assessment Level U. 16 pp.
IECQ PQC 74/US 0001-88. Radio Frequency Cables for Use in Electrotechnical Equipment: Detail Specification: Radio Frequency Cables, Flexible, for Operation at a Maximum Center Conductor Temperature of 85 Degrees C Assessment Level U. 16 pp.

—Naval Ships
MOD UK NES 1051-86. Requirements for the Installation of Corrugated Copper Air-Spaced RF Cable Issue 1 (09.86). 138 pp.
MOD UK NES 1053-91. Requirements for the Installation of Corrugated Copper Foam Dielectric R F Cable Issue 1 (12.91). 76 pp.
MOD UK NES 512: Part 5-91. Guide to Cables, Electrical and Associated Items Part 5: Cables, Electrical, Radio Frequency Issue 2 (04.91). 82 pp.

Coaxial Cables (Cont.)

—Naval Ships (Cont.)

MOD UK NES 512: Part 5-01. Guide to Cables, Electrical and Associated Items Part 5: Cables, Electrical, Radio Frequency Issue 2 (04.91); Amendment 1. 85 pp.

MOD UK NES 517: Part 3-92. Requirements for Cables, Electric Part 3: Co-Axial Issue 1 (06.92). 61 pp.

—Quality Assurance

BSI BS 9215-91. 1991 Capability Approval of Radio Frequency Connector Cable Assemblies and Radio Frequency Cable Assemblies: Generic Specification, Test Methods, Procedures, Customer Blank Detail Specification and Guidance. 39 pp.

IECQ PQC 71-87. Radio Frequency Cables for Use in Electrotechnical Equipment: Generic Specification: Radio Frequency Cables of Assessed Quality. 36 pp.

—Semirigid—Flanges

MOD UK DSTAN 59-18-01. Flanges for Rigid and Semi-Rigid Coaxial Cable; Amendment 1. 6 pp.

—Semirigid—Polytetrafluoroethylene Insulated

IEC 1196 Pt 2-93. Radio-Frequency Cables—Specifications Part 2: Semi-Rigid Radio-Frequency and Coaxial Cables with Polytetrafluoroethylene (PTFE) Insulation—Sectional Specification First Edition. 56 pp.

—Submarines

MOD UK NES 1051-86. Requirements for the Installation of Corrugated Copper Air-Spaced RF Cable Issue 1 (09.86). 138 pp.

MOD UK NES 1053-91. Requirements for the Installation of Corrugated Copper Foam Dielectric R F Cable Issue 1 (12.91). 76 pp.

MOD UK NES 512: Part 5-91. Guide to Cables, Electrical and Associated Items Part 5: Cables, Electrical, Radio Frequency Issue 2 (04.91). 82 pp.

MOD UK NES 512: Part 5-01. Guide to Cables, Electrical and Associated Items Part 5: Cables, Electrical, Radio Frequency Issue 2 (04.91); Amendment 1. 85 pp.

MOD UK NES 517: Part 3-92. Requirements for Cables, Electric Part 3: Co-Axial Issue 1 (06.92). 61 pp.

—Television Antennas—75 Ohms

DIN VDE 0887 Pt 1-80. Radio Frequency Cables; Coaxial, Z=75 Ohms for Television Antenna Systems; General Requirements; (VDE Specifications) (Dec). 17 pp.

—Television Receivers

CNS C2083-84. Coaxial Cable for Television Receive (Jun)(6077).

CNS C3100-84. Method of Test for Coaxial Cable for Television Receive (Jun)(6078).

JIS C 3502-87. Coaxial Cables for Television Receivers. 9 pp.

—Television Transmission—Interfaces

CCITT RECMN J.75-89. Interconnection of Systems for Television Transmission on Coaxial Pairs and on Radio-Relay Links—Line Transmission of Non-Telephone Signals—Transmission of Sound-Programme and Television Signals (Study Group XV) 2 pp. 2 pp.

—Television—75 Ohms

DIN VDE 0887 Pt 2-86. Radio Frequency Cables, Coaxial, Z=75 Ohms for Television Systems, Indoor Cables (July). 11 pp.

Coaxial Circulators

Use: Microwave Circulators—Coaxial

Coaxial Connectors

Use For: Radio Frequency Coaxial Connectors; RF Connectors *See Also:* APC Coaxial Connectors; BNC Coaxial Connectors; C Coaxial Connectors; Circular Connectors; Coaxial Adapters; Coaxial Cables; Connectors; F Coaxial Connectors; Hybrid Junctions; MCX Coaxial Connectors; Microwave Coaxial Connectors; N Coaxial Connectors; Radio Frequency Equipment; SMA Coaxial Connectors; SMA Coaxial Microwave Connectors; SMB Coaxial Connectors; SMC Coaxial Connectors; SSMA Coaxial Connectors; SSMB Coaxial Connectors; SSMC Coaxial Connectors; TNC Coaxial Connectors

BSI BS 3041: Part 1-77. 1977 Radio-Frequency Connectors Part 1: General Requirements and Measuring Methods. 29 pp.

BSI BS 3041: Part 2-77. 1977 Amd 1 Radio-Frequency Connectors Part 2: Coaxial Unmatched Connector. 14 pp.

BSI BS 3041: Part 12-81. 1981 Radio-Frequency Connectors Part 12: Specification for R.F. Coaxial Connectors with Screw Coupling, Unmatched (Type UHF). 8 pp.

Coaxial Connectors (Cont.)

BSI BS 7587-92. 1992 Interconnection of Radio and TV Receivers to Feeder System Outlets (IEC 1022: 1989). 7 pp.

BSI BS 9210 N006: Pt 3-85. 1985 Detail Specification for Radio Frequency Connectors of Assessed Quality Part 3: Full Assessment Level. Detail Specification, Parameters. 70 pp.

CECC CECC 22 000 ISSUE 3-92. Generic Specification: Radio Frequency Coaxial Connectors; (Parts I, II and III) (En, Fr, Ge). 270 pp.

CECC CECC 22 002 ISSUE 1-91. Blank Detail Specification: Radio Frequency Coaxial Connectors; Series (En, Fr, Ge). 22 pp.

CECC CECC 22 190 ISSUE 2-93. Sectional Specification: Radio Frequency Coaxial Connectors; Series 7-16 (En, Fr, Ge). 100 pp.

CENELEC HD 134.1-74. Radio-Frequency Connectors Part 1: General Requirements and Measuring Methods. 2 pp.

CENELEC HD 134.2 S2-84. Radio-Frequency Connectors—Part 2: Coaxial Unmatched Connector. 2 pp.

CENELEC HD 134.3-74. Radio Frequency Connectors Part 3: Two Pin Connector for Twin Balanced Aerial Feeders. 2 pp.

CENELEC HD 134.4 S2-82. Radio-Frequency Connectors—Part 4: R.F. Coaxial Connectors with Inner Diameter of Outer Conductor 16 mm (0.63 in) with Screw Lock. Characteristic Impedance 50 Ohms (Type 7-16). 2 pp.

CENELEC HD 134.5-76. Radio-Frequency Connectors Part 5: R.F. Coaxial Connectors for Cables 96 IEC 50-17 and Larger. 2 pp.

CENELEC HD 134.6-76. Radio Frequency Connectors Part 6: R.F. Coaxial Connectors for Cables 96 IEC 75-17 and Larger. 2 pp.

CNS C5025-86. General Rules of Connectors for Radio Frequency Coaxial Cables (Jun)(4738).

CNS C7080-86. C11 Type Connectors for Radio Frequency Coaxial Cables (Jun)(4744).

IEC 169 Pt 1-87. Radio-Frequency Connectors Part 1: General Requirements and Measuring Methods Second Edition. 71 pp.

IEC 169 Pt 1-1-87. Radio-Frequency Connectors Part 1: General Requirements and Measuring Methods Section One—Electrical Tests and Measuring Procedures: Reflection Factor First Edition. 35 pp.

IEC 169 Pt 1-3-88. Radio-Frequency Connectors Part 1: General Requirements and Measuring Methods Section Three—Electrical Tests and Measuring Procedures: Screening Effectiveness First Edition. 31 pp.

IEC 169 Pt 2-65. Radio-Frequency Connectors Part 2: Coaxial Unmatched Connector First Edition; (Corrigendum—Oct 1975) (Amendment 1-1982). 21 pp.

IEC 169 Pt 3-65. Radio-Frequency Connectors Part 3: Two Pin Connector for Twin Balanced Aerial Feeders First Edition. 13 pp.

IEC 169 Pt 5-70. Radio-Frequency Connectors Part 5: R.F. Coaxial Connectors for Cables 96 IEC 50-17 and Larger First Edition. 27 pp.

IEC 169 Pt 6-71. Radio-Frequency Connectors Part 6: R.F. Coaxial Connectors for Cables 96 IEC 75-17 and Larger First Edition. 21 pp.

IEC 169 Pt 21-85. Radio-Frequency Connectors Part 21: Two Types of Radio-Frequency Connectors with Inner Diameter of Outer Conductor 9.5 mm (0.374 in) with Different Versions of Screw Coupling—Characteristic Impedance 50 Ohms (Types SC-A and SC-B) First Edition. 39 pp.

IEC 169 Pt 22-85. Radio-Frequency Connectors Part 22: R.F. Two-Pole Bayonet Coupled Connectors for Use with Shielded Balanced Cables Having Twin Inner Conductors (Type BNO) First Edition. 25 pp.

IEC 169 Pt 25-92. Radio-Frequency Connectors Part 25: Two-Pole Screw (3/4-20 UNEF) Coupled Connectors for Use with Shielded Balanced Cables Having Twin Inner Conductors with Inner Diameter of Outer Conductor 13,56 mm (0,534 in) (Type TWHN) First Edition. 20 pp.

IEC 1022-89. Interconnection of Radio and TV Receivers to Feeder System Outlets First Edition. 12 pp.

IEC 1169 Pt 1-92. Radio-Frequency Connectors Part 1: Generic Specification—General Requirements and Measuring Methods First Edition; (IECQ QC 220000). 170 pp.

JIS C 5410-91. General Rules of Connectors for Radio Frequency Coaxial Cables. 46 pp.

— 50 Ohms

CECC CECC 22 191 ISSUE 1-84. Blank Detail Specification: Radio Frequency Coaxial Connectors; Series 7-16 (En, Fr, Ge) AMD 1 (En, Fr, Ge). 45 pp.

CECC CECC 22 250 ISSUE 2-93. Sectional Specification: Radio Frequency Coaxial Connectors; Series 1,4/4,4 (En, Fr, Ge). 89 pp.

CECC CECC 22 251 ISSUE 1-89. Blank Detail Specification: Radio Frequency Coaxial Connectors; Series 1,4/4,4 (En, Fr, Ge). 45 pp.

Coaxial Connectors (Cont.)

— 50 Ohms (Cont.)

CENELEC HD 134.7-76. Radio-Frequency Connector Part 7: R.F. Coxial Connectors with Inner Diameter of Outer Conductor 9.5mm (0.374 in) with Bayonet Lock-Characteristic Impedance 50 Ohms (Type C). 2 pp.

IEC 169 Pt 4-75. Radio-Frequency Connectors Part 4: R.F. Coaxial Connectors with Inner Diameter of Outer Conductor 16 mm (0.63 in) with Screw Lock Characteristic Impedance 50 Ohms (Type 7-16) Second Edition. 35 pp.

IEC 169 Pt 26-93. Radio-Frequency Connectors—Part 26: R.F. Coaxial Connectors with Screw Coupling—Characteristic Impedance 50 ohms—Frequency Range 0 to 18 GHz (Type TNC 18 GHz) First Edition. 27 pp.

— 50 Ohms—Flanges

CECC CECC 22 150 ISSUE 1-86. Sectional Specification: Radio Frequency Coaxial Connectors; Series EIA Flange (En, Fr, Ge) AMD 1 (En, Fr, Ge) ERRATA (En, Fr, Ge). 133 pp.

CECC CECC 22 151 ISSUE 1-86. Blank Detail Specification: Radio Frequency Coaxial Connectors; Series EIA Flange (En, Fr, Ge). 42 pp.

— 50 Ohms—Plugs

CECC CECC 22 201-801 ISSUE 1-88. Detail Specification: Radio Frequency Coaxial Connectors; Series TNC (En, Fr, Ge). 38 pp.

CECC CECC 22 231-802 ISSUE 1-91. Detail Specification: Radio Frequency Coaxial Connectors; Series 1,0/2,3 (En, Fr, Ge). 18 pp.

CECC CECC 22 231-806 ISSUE 1-91. Detail Specification: Radio Frequency Coaxial Connectors; Series 1,0/2,3 (En, Fr, Ge). 18 pp.

CECC CECC 22 231-808 ISSUE 1-91. Detail Specification: Radio Frequency Coaxial Connectors; Series 1,0/2,3 (En, Fr, Ge). 18 pp.

— 50 Ohms—Receptacles

CECC CECC 22 231-804 ISSUE 1-91. Detail Specification: Radio Frequency Coaxial Connectors; Series 1,0/2,3 (En, Fr, Ge). 18 pp.

CECC CECC 22 231-805 ISSUE 1-91. Detail Specification: Radio Frequency Coaxial Connectors; Series 1,0/2,3 (En, Fr, Ge). 18 pp.

CECC CECC 22 231-807 ISSUE 1-91. Detail Specification: Radio Frequency Coaxial Connectors; Series 1,0/2,3 (En, Fr, Ge). 18 pp.

— 50 Ohms—Right Angle—Plugs

CECC CECC 22 231-803 ISSUE 1-91. Detail Specification: Radio Frequency Coaxial Connectors; Series 1,0/2,3 (En, Fr, Ge). 18 pp.

— 75 Ohms

CECC CECC 22 240 ISSUE 1-89. Sectional Specification: Radio Frequency Coaxial Connectors; Series 1,6/5,6 (En, Fr, Ge). 93 pp.

CECC CECC 22 241 ISSUE 1-89. Blank Detail Specification: Radio Frequency Coaxial Connectors; Series 1,6/5,6 (En, Fr, Ge). 48 pp.

— 75 Ohms—Flanges

CECC CECC 22 150 ISSUE 1-86. Sectional Specification: Radio Frequency Coaxial Connectors; Series EIA Flange (En, Fr, Ge) AMD 1 (En, Fr, Ge) ERRATA (En, Fr, Ge). 133 pp.

CECC CECC 22 151 ISSUE 1-86. Blank Detail Specification: Radio Frequency Coaxial Connectors; Series EIA Flange (En, Fr, Ge). 42 pp.

—Crimp Barrels

CENELEC HD 489 S1-87. Recommended Dimensions for Hexagonal and Square Crimping-Die Cavities, Indentors, Gauges, Outer Conductor Crimp Sleeves and Centre Contact Crimp Barrels for r.f. Cables and Connectors. 4 pp.

IEC 803-84. Recommended Dimensions for Hexagonal and Square Crimping-Die Cavities, Indentors, Gauges, Outer Conductor Crimp Sleeves and Centre Contact Crimp Barrels for R.F. Cables and Connectors First Edition. 40 pp.

—Crimp Sleeves

CENELEC HD 489 S1-87. Recommended Dimensions for Hexagonal and Square Crimping-Die Cavities, Indentors, Gauges, Outer Conductor Crimp Sleeves and Centre Contact Crimp Barrels for r.f. Cables and Connectors. 4 pp.

IEC 803-84. Recommended Dimensions for Hexagonal and Square Crimping-Die Cavities, Indentors, Gauges, Outer Conductor Crimp Sleeves and Centre Contact Crimp Barrels for R.F. Cables and Connectors First Edition. 40 pp.

INTERNATIONAL AND NON-U.S. NATIONAL STANDARDS
SUBJECT INDEX

Coaxial

Coaxial Connectors (Cont.)
—Crimping Tools
BSI BS 5310: Part 2A-76. 1976 Amd 1 Hand Crimping Tools for the Termination of Electrical Cables and Wires for Low Frequency and Radio Frequency Applications Part 2: Hand Crimping Tools (Fixed Die, Sizes A to E), for Radio Frequency Connectors and Concentric Contacts. 9 pp.
BSI BS 5310: Part 2B-81. 1981 Hand Crimping Tools for the Termination of Electrical Cables and Wires for Low Frequency and Radio Frequency Applications Part 2B: Hand Crimping Tools (Removable and Inter-Changeable Dies, Sizes A to G & Q to S) for Radio Freq Conn & Conc Contact. 8 pp.

—Die Cavities—Hexagonal/Square Crimping
CENELEC HD 489 S1-87. Recommended Dimensions for Hexagonal and Square Crimping-Die Cavities, Indentors, Gauges, Outer Conductor Crimp Sleeves and Centre Contact Crimp Barrels for r.f. Cables and Connectors. 4 pp.
IEC 803-84. Recommended Dimensions for Hexagonal and Square Crimping-Die Cavities, Indentors, Gauges, Outer Conductor Crimp Sleeves and Centre Contact Crimp Barrels for R.F. Cables and Connectors First Edition. 40 pp.

—Engineering Drawings—Quality Assurance
BSI BS 9210 N009: Pt 2-78. 1978 Detail Specification for Electronic Components of Assessed Quality, Soldered Part 2: Control Drawings, Mating Face Details and Gauge Information. 6 pp.

—Engineering Drawings—50 Ohms
BSI BS 9210 N006: Pt 2-78. 1978 Detail Specification for Radio Frequency Connect-ors (Series SMA), 50 ohms, Unsealed, Solder-ed, Centre Contact, Clamp Outer for Flexible Cables, Solder Outer for Semi-Rigid Cables: Part 2: Control Drawings, Mating Face Details and Gauge Information. 6 pp.

—Flanges
MOD UK DSTAN 59-35: Pt 4:Sec 12-80. Connectors, Electrical Part 4: Radio Frequency Coaxial Connectors (See Also EPIC Database) Section 12: Detail Specification-Series EIA Flange Issue 1 (Obsolescent). 24 pp.

—Frequency Limits
IEC 1141-92. Upper Frequency Limit of r.f. Coaxial Connectors First Edition. 50 pp.

—Gages
CENELEC HD 489 S1-87. Recommended Dimensions for Hexagonal and Square Crimping-Die Cavities, Indentors, Gauges, Outer Conductor Crimp Sleeves and Centre Contact Crimp Barrels for r.f. Cables and Connectors. 4 pp.
IEC 803-84. Recommended Dimensions for Hexagonal and Square Crimping-Die Cavities, Indentors, Gauges, Outer Conductor Crimp Sleeves and Centre Contact Crimp Barrels for R.F. Cables and Connectors First Edition. 40 pp.

—Indentors
CENELEC HD 489 S1-87. Recommended Dimensions for Hexagonal and Square Crimping-Die Cavities, Indentors, Gauges, Outer Conductor Crimp Sleeves and Centre Contact Crimp Barrels for r.f. Cables and Connectors. 4 pp.
IEC 803-84. Recommended Dimensions for Hexagonal and Square Crimping-Die Cavities, Indentors, Gauges, Outer Conductor Crimp Sleeves and Centre Contact Crimp Barrels for R.F. Cables and Connectors First Edition. 40 pp.

—Military
CECC CECC 22 001 ISSUE 2-92. Blank Detail Specification: CECC Military Specification for RF Connectors; (Type MIL-C-39012) (En, Fr, Ge). 23 pp.

—Nuclear Instruments
IEC 313-83. Coaxial Cable Connectors Used in Nuclear Instrumentation Second Edition. 9 pp.
IEC 498-75. High-Voltage Coaxial Connectors Used in Nuclear Instrumentation First Edition. 20 pp.

—Pin and Socket
IEC 169 Pt 23-91. Radio-Frequency Connectors Part 23: Pin and Socket Connector for Use with 3,5 mm Rigid Precision Coaxial Lines with Inner Diameter of Outer Conductor 3,5 mm (0,1378 in) First Edition. 20 pp.

—Plugs
MOD UK DSTAN 59-35: Part 4-01. Connectors Electrical (See Also EPIC Database) Part 4: Radio Frequency Coaxial Connectors Issue 2; Amendment 1. 24 pp.

Coaxial Connectors (Cont.)
—Printed Circuit—Crimp—Preferred Products List
CECC CECC MUAHAG Vol 3A IS 4-91. Preferred Products List; Connectors; L. F. (En, Fr, Ge). 46 pp.

—Printed Circuit—Preferred Products List
CECC CECC MUAHAG Vol 3A IS 4-91. Preferred Products List; Connectors; L. F. (En, Fr, Ge). 46 pp.

—Printed Circuit—50 Ohms—Right Angle—Receptacles
CECC CECC 22 231-809 ISSUE 1-91. Detail Specification: Radio Frequency Coaxial Connectors; Series 1,0/2,3 (En, Fr, Ge). 18 pp.

—Quality Assurance
BSI BS 9210-04. 1984 Amd 4 Radio Frequency Connectors of Assessed Quality: Generic Data and Methods of Test (AMD 7654) July 15, 1993. 94 pp.
BSI BS 9210 N004-81. 1981 Detail Specification for Radio Frequency Coaxial Connector (Series BNC), Unsealed, Soldered, Captive Contact 50 ohms, Basic Assessement Level. 17 pp.
BSI BS 9215-91. 1991 Capability Approval of Radio Frequency Connector Cable Assemblies and Radio Frequency Cable Assemblies: Generic Specification, Test Methods, Procedures, Customer Blank Detail Specification and Guidance. 39 pp.
BSI BS CECC 22000-93. 1993 Generic Specification: Radio Frequency Coaxial Connectors (Parts I, II and III) (T). 95 pp.
BSI BS CECC 22000-80. 1980 Radio Frequency Coaxial Connectors: Generic Specification. 43 pp.

—Radio Frequency—Telecommunication Equipment
IECQ QC 220000-92. Radio-Frequency Connectors Part 1: Generic Specification—General Requirements and Measuring Methods (IEC 1169-1 ED 1). 170 pp.

—Radio Receivers—Automotive
ISO 10599 Pt 1-92. Car Radios—Coaxial Aerial Connectors—Part 1: Dimensions First Edition. 5 pp.

—Receptacles
MOD UK DSTAN 59-35: Part 4-01. Connectors Electrical (See Also EPIC Database) Part 4: Radio Frequency Coaxial Connectors Issue 2; Amendment 1. 24 pp.

—Stripline—50 Ohms—Receptacles
CECC CECC 22 231-810 ISSUE 1-91. Detail Specification: Radio Frequency Coaxial Connectors; Series 1,0/2,3 (En, Fr, Ge). 18 pp.

Coaxial Diodes
—Quality Assurance
BSI BS 9300 C377-378-71. 1971 Detail Requirements for Silicon Coaxial Mixer Diodes. 10 pp.
BSI BS 9300 C534-71. (OBSOLESCENT) 1971 Detail Requirements for a Silicon Coaxial Resistive Switching Diode. 13 pp.
BSI BS 9300 C771-772-71. 1971 Detail Requirements for Coaxial Mixer Diodes. 11 pp.
BSI BS 9300 C776-777-71. (OBSOLESCENT) 1971 Detail Requirements for Germanium Coaxial Mixer Diodes. 12 pp.
BSI BS 9300 C778-71. (OBSOLESCENT) 1971 Detail Requirements for a Matched Pair of Germanium Coaxial Mixer Diodes. 3 pp.

Coaxial Filters
See Also: Electric Filters

—Microwave—50 Ohms—Preferred Products List
CECC CECC MUAHAG Vol 12 IS 1-90. Preferred Products List; Microwave Components (En, Fr, Ge). 76 pp.

Coaxial Isolators
—Radio Frequency
MOD UK DSTAN 59-85-01. Isolators, Radio Frequency (Reflection) and Circulators, Radio Frequency Issue 1; Amendment 1. 37 pp.
MOD UK DSTAN 59-85: Part 90-81. Isolators, Radio Frequency (Reflection) and Circulators, Radio Frequency Part 90: Detail Specifications Issue 1. 3 pp.
MOD UK DSTAN 59-85: 90/007-80. Isolators/Circulators, Radio Frequency, Reflection Issue 1. 12 pp.

Coaxial Lines
Use: Coaxial Cables

Coaxial Microwave Connectors
Use: Microwave Coaxial Connectors

Coaxial Phase Shifters
See Also: Microwave Phase Shifters

—50 Ohms—Preferred Products List
CECC CECC MUAHAG Vol 12 IS 1-90. Preferred Products List; Microwave Components (En, Fr, Ge). 76 pp.

Coaxial Relays
See Also: Coaxial Transmission Lines (Waveguide); Relays

—All-Or-Nothing—0-49 V DC—Crystal Can—Severe Environments—PPL
CECC CECC MUAHAG Vol 5 IS 3-88. Preferred Products List; Relays (En, Fr, Ge). 45 pp.

—Microwave—50 Ohms—Preferred Products List
CECC CECC MUAHAG Vol 12 IS 1-90. Preferred Products List; Microwave Components (En, Fr, Ge). 76 pp.

Coaxial Short Circuits
Use For: Coaxial Shorts

—SSMA
CECC CECC MUAHAG Vol 3B IS 4-91. Preferred Products List; Connectors; R.F. and Fibre Optics (En, Fr, Ge). 65 pp.

—SSMA—Plugs
CECC CECC MUAHAG Vol 3B IS 4-91. Preferred Products List; Connectors; R.F. and Fibre Optics (En, Fr, Ge). 65 pp.

Coaxial Shorts
Use: Coaxial Short Circuits

Coaxial Transmission Lines (Waveguide)
Use For: Rigid Coaxial Cables *See Also:* Coaxial Cables; Coaxial Relays; Communication Transmission Lines; Waveguides

CENELEC HD 350.1-76. General Purpose Rigid Coaxial Transmission Lines and Their Associated Flange Connectors Part 1: General Requirements and Measuring Methods. 2 pp.
CENELEC HD 350.2-76. General Purpose Rigid Coaxial Transmission Lines and Their Associated Flange Connectors Part 1: Detail Specifications. 2 pp.
IEC 339 Pt 1-71. General Purpose Rigid Coaxial Transmission Lines and Their Associated Flange Connectors Part 1: General Requirements and Measuring Methods First Edition. 23 pp.
IEC 339 Pt 2-72. General Purpose Rigid Coaxial Transmission Lines and Their Associated Flange Connectors Part 2: Detail Specifications First Edition; (Corrigendum—July 1976). 24 pp.
MOD UK DSTAN 59-6-67. Lines, Radio Frequency Transmission (Rigid Coaxial) Issue 1. 4 pp.

—Flanges
CENELEC HD 350.1-76. General Purpose Rigid Coaxial Transmission Lines and Their Associated Flange Connectors Part 1: General Requirements and Measuring Methods. 2 pp.
CENELEC HD 350.2-76. General Purpose Rigid Coaxial Transmission Lines and Their Associated Flange Connectors Part 1: Detail Specifications. 2 pp.
IEC 339 Pt 1-71. General Purpose Rigid Coaxial Transmission Lines and Their Associated Flange Connectors Part 1: General Requirements and Measuring Methods First Edition. 23 pp.
IEC 339 Pt 2-72. General Purpose Rigid Coaxial Transmission Lines and Their Associated Flange Connectors Part 2: Detail Specifications First Edition; (Corrigendum—July 1976). 24 pp.

—Precision
CENELEC HD 351.1-76. Rigid Precision Coaxial Lines and Their Associated Precision Connectors Part 1: General Requirements and Measuring Methods. 2 pp.
CENELEC HD 351.2-76. Rigid Precision Coaxial Lines and Their Associated Precision Connectors Part 2: 50ohm 7mm Rigid Precision Coaxial Line and Associated Hermaphroditic Precision Coaxial Connector. 2 pp.
CENELEC HD 351.3 S2-81. Rigid Precision Coaxial Lines and Their Associated Precision Connectors—Part 3: 14 mm Rigid Precision Coaxial Line and Associated Hermaphroditic Precision Coaxial Connector—Characteristic Impedances 50 Ohms and 75 Ohms. 2 pp.

Coaxial Transmission Lines (Waveguide) (Cont.)
—Precision (Cont.)
CENELEC HD 351.4 S2-84. Rigid Precision Coaxial Lines and Their Associated Precision Connectors—Part 4: 21 mm Rigid Precision Coaxial line and Associated Hermaphroditic Precision Coaxial Connector Characteristic Impedance 50 OHMs (Type 9/21) Characteristic Impedance. 2 pp.

CENELEC HD 351.5-86. Rigid Precision Coaxial Lines and Their Associated Precision Connectors Part 5: 50 Ohms 3.5mm Rigid Precision Coaxial Line with Provision for Mounting Connectors. 2 pp.

IEC 457 Pt 1-74. Rigid Precision Coaxial Lines and Their Associated Precision Connectors Part 1: General Requirements and Measuring Methods First Edition. 30 pp.

IEC 457 Pt 2-74. Rigid Precision Coaxial Lines and Their Associated Precision Connectors Part 2: 50 Ohm 7 mm Rigid Precision Coaxial Line and Associated Hermaphroditic Precision Coaxial Connector First Edition. 15 pp.

IEC 457 Pt 3-80. Rigid Precision Coaxial Lines and Their Associated Precision Connectors Part 3: 14 mm Rigid Precision Coaxial Line and Associated Hermaphroditic Precision Coaxial Connector—Characteristic Impedances 50 Ohms and 75 Ohms Second Edition. 17 pp.

IEC 457 Pt 4-78. Rigid Precision Coaxial Lines and Their Associ-ated Precision Connec-tors Part 4: 21 mm Rigid Precision Coaxial Line and Associated Hermaphr-oditic Precision Coaxial Connector Characteristic Impedance 50 Ohms (Type 9/21)—Characteristic Impedance 75 Ohms (Type 6/21) Second Edition. 16 pp.

IEC 457 Pt 5-84. Rigid Precision Coaxial Lines and Their Associated Precision Connectors Part 5: 50 Ohms 3.5 mm Rigid Precision Coaxial Line with Provision for Mounting Connectors First Edition. 9 pp.

Coba
See Also: Agricultural Products
—Grading
CNS N1100-82. Grades of Coba (Nov)(9628). 2 pp.

Cobalt
Scope Note: For additional listings, see also specific products made from cobalt *See Also:* Cobalt Alloys; Cobalt Castings; Cobalt Coatings (Made From Cobalt); Cobalt Content Analysis; Cobalt Wire; Metals
—Standard Solutions
JIS K 0014-83. Cobalt Standard Solution. 14 pp.

Cobalt Alloy Castings
Use: Cobalt Castings

Cobalt Alloys
Scope Note: For additional listings, see also specific products made from cobalt alloys *See Also:* Alloys; Cobalt; Cobalt Castings; Heat Resistant Alloys
—Annealed—Bars—Aerospace
AECMA PREN2162-80. Heat Resisting Cobalt Base Alloy CO-P92HT-Annealed—Bars (C5/48). 3 pp.

AECMA PREN2166-80. Heat Resisting Cobalt Base Alloy CO-P92HT-Annealed—Bars and Wires D Less Than or Equal to 4 mm (C5/48). 3 pp.
—Annealed—Bars—Rings—Aerospace
AECMA PREN2164-80. Heat Resisting Cobalt Base Alloy CO-P92HT-Annealed—Bars and Sections for Welded Rings (C5/48). 3 pp.
—Annealed—Forgings—Aerospace
AECMA PREN2163-80. Heat Resisting Cobalt Base Alloy CO-P92HT-Annealed—Forgings (C5/48). 3 pp.
—Annealed—Sections—Rings—Aerospace
AECMA PREN2164-80. Heat Resisting Cobalt Base Alloy CO-P92HT-Annealed—Bars and Sections for Welded Rings (C5/48). 3 pp.
—Annealed—Sheets—Aerospace
AECMA PREN2165-80. Heat Resisting Cobalt Base Alloy CO-P92HT-Annealed—Sheets and Strips a Less Than or Equal to 3 mm (C5/48). 3 pp.
—Annealed—Strips—Aerospace
AECMA PREN2165-80. Heat Resisting Cobalt Base Alloy CO-P92HT-Annealed—Sheets and Strips a Less Than or Equal to 3 mm (C5/48). 3 pp.
—Bars—Aerospace
BSI HR 40-72. 1972 Amd 1 Cobalt-Chromium-Tungsten-Nickel-Manganese Heat-Resisting Alloy Billets, Bars and Forgings(Cobalt Base Cr 20, W 15, Ni 10, Mn 1.5). 3 pp.

Cobalt Alloys (Cont.)
—Billets—Aerospace
BSI HR 40-72. 1972 Amd 1 Cobalt-Chromium-Tungsten-Nickel-Manganese Heat-Resisting Alloy Billets, Bars and Forgings(Cobalt Base Cr 20, W 15, Ni 10, Mn 1.5). 3 pp.
—Forgings—Aerospace
BSI HR 40-72. 1972 Amd 1 Cobalt-Chromium-Tungsten-Nickel-Manganese Heat-Resisting Alloy Billets, Bars and Forgings(Cobalt Base Cr 20, W 15, Ni 10, Mn 1.5). 3 pp.
—Heat Treated—Bars—Aerospace
AECMA PREN2657-88. Heat Resisting Cobalt Base Alloy Co-P93HT Solution Treated Rm Greater Than or Equal to 860 MPa Bar for Machining De Less Than or Equal to 150 mm. 5 pp.
—Heat Treated—Forgings—Aerospace
AECMA PREN2658-88. Heat Resisting Cobalt Base Alloy Co-P93HT Not Heat Treated Reference Heat Treatment: Solution Treated Forging Stock De Less Than or Equal to 360 mm. 5 pp.

AECMA PREN2659-88. Heat Resisting Cobalt Base Alloy Co-P93HT Solution Treated Rm Greater Than or Equal to 860 MPa Forgings. 5 pp.
—Heat Treated—Sheets—Aerospace
AECMA PREN2661-88. Heat Resisting Cobalt Base Alloy Co-P93HT Solution Treated Rm Greater Than or Equal to 850 MPa Sheet and Strip A Less Than or Equal to 3 mm. 5 pp.
—Heat Treated—Strips—Aerospace
AECMA PREN2661-88. Heat Resisting Cobalt Base Alloy Co-P93HT Solution Treated Rm Greater Than or Equal to 850 MPa Sheet and Strip A Less Than or Equal to 3 mm. 5 pp.
—Plates—Aerospace
BSI HR 240-72. 1972 Amd 1 Cobalt-Chromium-Tungsten Nickel-Manganese Heat-Resisting Alloy Plate, Sheet and Strip (Cobalt Base W 15, Ni 10, Mn 1.5). 3 pp.
—Remelting Stock—Aerospace
AECMA PREN2103-88. Steel Nickel Base and Cobolt Base Alloy Remelting Stock and Castings—Technical Specification—Part 1—General Requirements. 11 pp.

AECMA PREN2103-02-88. Steel, Nickel Base and Cobalt Base Alloy Remelting Stock and Castings—Technical Specification—Part 2—Remelting Stock. 6 pp.

BSI BS EN 2103: Part 1-92. 1992 Steel, Nickel Base and Cobalt Base Alloy Remelting Stock and Castings. Technical Specification Part 1: General Requirements (V). 16 pp.

BSI BS EN 2103: Part 2-92. 1992 Steel, Nickel Base and Cobalt Base Alloy Remelting Stock and Castings. Technical Specification Part 2: Remelting Stock (V). 11 pp.

CEN EN 2103-1-91. Steel, Nickel Base and Cobalt Base Alloy Remelting Stock and Castings—Technical Specification—Part 1—General Requirements. 12 pp.

CEN EN 2103-2-91. Steel, Nickel Base and Cobalt Base Alloy Remelting Stock and Castings—Technical Specification—Part 2—Remelting Stock. 7 pp.
—Sheets—Aerospace
BSI HR 240-72. 1972 Amd 1 Cobalt-Chromium-Tungsten Nickel-Manganese Heat-Resisting Alloy Plate, Sheet and Strip (Cobalt Base W 15, Ni 10, Mn 1.5). 3 pp.
—Strips—Aerospace
BSI HR 240-72. 1972 Amd 1 Cobalt-Chromium-Tungsten Nickel-Manganese Heat-Resisting Alloy Plate, Sheet and Strip (Cobalt Base W 15, Ni 10, Mn 1.5). 3 pp.

Cobalt Castings
Use For: Cobalt Alloy Castings *See Also:* Castings; Cobalt; Cobalt Alloys; Metal Products
—Aerospace
AECMA PREN2103-88. Steel Nickel Base and Cobolt Base Alloy Remelting Stock and Castings—Technical Specification—Part 1—General Requirements. 11 pp.

AECMA PREN2103-03-88. Steel, Nickel Base and Cobalt Base Alloy Remelting Stock and Castings—Technical Specification—Part 3—Pre-Production and Production Castings. 17 pp.

BSI BS EN 2103: Part 1-92. 1992 Steel, Nickel Base and Cobalt Base Alloy Remelting Stock and Castings. Technical Specification Part 1: General Requirements (V). 16 pp.

BSI BS EN 2103: Part 3-92. 1992 Steel, Nickel Base and Cobalt Base Alloy Remelting Stock and Castings. Technical Specification Part 3: Pre-Production and Production Castings (V). 22 pp.

BSI HC 301-73. 1973 Cobalt Base Chromium-Nickel-Tungsten Alloy Castings (Cr 25.5, Ni 105, W 7.5). 2 pp.

Cobalt Castings (Cont.)
—Aerospace (Cont.)
CEN EN 2103-1-91. Steel, Nickel Base and Cobalt Base Alloy Remelting Stock and Castings—Technical Specification—Part 1—General Requirements. 12 pp.

CEN EN 2103-3-91. Steel, Nickel Base and Cobalt Base Alloy Remelting Stock and Castings—Technical Specification—Part 3—Pre-Production and Production Castings. 18 pp.
—Centrifugal—Aerospace
BSI HC 100-72. 1972 Amd 3 Inspection and Testing Procedure for Iron, Nickel, Copper, Cobalt and Refractory Metal Base Alloy Castings (AMD 6574) May 28, 1991. 17 pp.
—Investment
BSI BS 3146: Part 2-75. 1975 Amd 1 Investment Castings in Metal Part 2: Corrosion and Heat Resisting Steels, Nickel and Cobalt Base Alloys. 28 pp.
—Investment—Aerospace
AECMA PREN2161-78. Heat Resisting Cobalt Base Alloy CO-C91-HT—as Cast—Precision Castings (C5/48). 3 pp.

BSI HC 100-72. 1972 Amd 3 Inspection and Testing Procedure for Iron, Nickel, Copper, Cobalt and Refractory Metal Base Alloy Castings (AMD 6574) May 28, 1991. 17 pp.
—Sand—Aerospace
BSI HC 100-72. 1972 Amd 3 Inspection and Testing Procedure for Iron, Nickel, Copper, Cobalt and Refractory Metal Base Alloy Castings (AMD 6574) May 28, 1991. 17 pp.
—Surgical Implants
BSI BS 7254: Part 5-90. 1990 Orthopaedic Implants Part 5: Production of Castings Made of Cobalt-Chromium-Molybdenum Alloy. 9 pp.

Cobalt Chloride
Use: Cobaltous Chloride

Cobalt Coatings (Made From Cobalt)
Scope Note: Includes coatings made from cobalt alloys *See Also:* Cobalt; Metal Coatings (Made From Metal)
—Electrodeposited—Aerospace
MOD UK DTD-943-74. Electrodeposited Cobalt/Chromium Carbide Composite Coatings. 2 pp.
—Sprayed—Self Fluxing
CNS H3097-82. Spray Fused Deposits of Self-Fluxing Alloys (Jan)(8292).

JIS H 8303-89. Spray Fused Deposits of Self-Fluxing Alloys. 9 pp.

Cobalt Content Analysis
See Also: Cobalt
—Boric Acids—Photometry
ISO 5932-80. Boric Acid, Boric Oxide and DiSodium Tetraborates for Industrial Use—Determination of Cobalt Content—2-Nitroso-1-Naphthol Photometric Method First Edition. 6 pp.
—Boron Oxides—Photometry
ISO 5932-80. Boric Acid, Boric Oxide and DiSodium Tetraborates for Industrial Use—Determination of Cobalt Content—2-Nitroso-1-Naphthol Photometric Method First Edition. 6 pp.
—Cast Iron—Spectrophotometry
BSI BS 6200: SUB SEC 3.11.1-91. 1991 Sampling and Analysis of Iron, Steel and Other Ferrous Metals Part 3: Methods of Analysis Section 3.11: Determination of Cobalt Subsection 3.11.1: Steel and Cast Iron: Spectrophotometric Method. 7 pp.
—Copper—Absorptiometric Analysis
JIS H 1060-89. Methods for Determination of Cobalt in Copper and Copper Alloys. 12 pp.
—Copper Alloys—Absorptiometric Analysis
JIS H 1060-89. Methods for Determination of Cobalt in Copper and Copper Alloys. 12 pp.
—Copper Alloys—Atomic Absorption Spectrometry
JIS H 1060-89. Methods for Determination of Cobalt in Copper and Copper Alloys. 12 pp.
—Copper—Atomic Absorption Spectrometry
JIS H 1060-89. Methods for Determination of Cobalt in Copper and Copper Alloys. 12 pp.

INTERNATIONAL AND NON-U.S. NATIONAL STANDARDS
SUBJECT INDEX
Cobalt

Cobalt Content Analysis *(Cont.)*

—**Drinking Water—Voltametry**

DIN ENGL 38406 Pt 16-90. German Standard Methods for the Examination of Water, Waste Water and Sludge; Cations (Group E); Determination of Zinc, Cadmium, Lead, Copper, Thallium, Nickel, Cobalt by Voltammetry (E16) (Mar). 8 pp.

—**Electrolytic Copper—Atomic Absorption Spectrophotometry**

BSI BS 7317: Part 1-90. 1990 Methods for Analysis of High Purity Copper Cathode Cu-CATH-1 Part 1: Method for Determination of Cadmium Manganese and Silver (Screening Procedure for Chromium, Cobalt, Iron, Nickel and Zinc) by Atomic Absorption. 9 pp.

BSI BS 7317: Part 2-90. 1990 Methods for Analysis of High Purity Copper Cathode Cu-CATH-1 Part 2: Method for Determination of Chromium, Cobalt, Iron, Nickel and Zinc by Discrete Volume Nebulization Atomic Absorption Spectrophotometry. 9 pp.

BSI DD 95: Part 2-84. (WITHDRAWN) 1984 Amd 1 Analysis of Higher Purity Copper Cathode Cu-CATH-1 Part 2: Method for Determination of Chromium, Cobalt, Iron, Nickel, and Zinc by Discrete Volume Nebulization Atomic Absorption Spectrophotmetry. 10 pp.

—**Ferronickel—Atomic Absorption Spectrometry**

BSI BS 6783: Part 3-86. 1986 Sampling and Analysis of Nickel, Ferronickel and Nickel Alloys Part 3: Method for Determination of Cobalt in Ferronickel by Flame Atomic Absorption Spectrometry. 6 pp.

BSI BS 6783: Part 3-01. 1986 Amd 1 Sampling and Analysis of Nickel, Ferronickel and Nickel Alloys Part 3: Method for Determination of Cobalt in Ferronickel by Flame Atomic Absorption Spectrometry (ISO 7520: 1985) (AMD 6992) February 28, 1992 (V). 10 pp.

CEN EN 27520-91. Ferronickel—Determination of Cobalt Content—Flame Atomic Absorption Spectrometric Method. 3 pp.

DIN ENGL EN 27520-92. Determination of Cobalt Content of Ferronickel; Flame Atomic Absorption Spectrometric Method (ISO 7520: 1985) (Feb). 6 pp.

ISO 7520-85. Ferronickel—Determination of Cobalt Content—Flame Atomic Absorption Spectrometric Method First Edition. 5 pp.

—**Fertilizers**

EC COM(88) 562-88. Proposal for a Council Directive on the Approximation of the Laws of the Member States in Respect of the Trace Elements Boron, Cobalt, Copper, Iron, Manganese, Molybdenum and Zinc Contained in Fertilizers. 20 pp.

EC 89/530/EEC-89. Council Directive Supplementing and Amending Directive 76/116/EEC in Respect of the Trace Elements Boron, Cobalt, Copper Iron, Manganese, Molybdenum and Zinc Contained in Fertilizers. 9 pp.

—**Ground Water—Voltametry**

DIN ENGL 38406 Pt 16-90. German Standard Methods for the Examination of Water, Waste Water and Sludge; Determination of Zinc, Cadmium, Lead, Copper, Thallium, Nickel, Cobalt by Voltammetry (E16) (Mar). 8 pp.

—**Hard Metals—Atomic Absorption Spectrometry**

BSI BS 5600: SUB SEC 4.17.3-86. (WITHDRAWN) 1986 Part 4: Methods of Testing and Chemical Analysis of Hardmetals Section 4.17: Chemical Analysis by Flame Atomic Absorption Spectrometry Subsection 4.17.3: De-termination of Cobalt, Iron, Manganese and Nickel in Contents from 0.01% to 0.05% (m/m). 4 pp.

BSI BS 5600: SUB SEC 4.17.5-86. (WITHDRAWN) 1986 Part 4: Methods of Testing and Chemical Analysis of Hardmetals Section 4.17: Chemical Analysis by Flame Atomic Absorption Spectrometry Sub 4.17.5:Determination of Cobalt, Iron, Manganese, Molybdenum, Nickel Titanium and Vanadium in Cont from 0.5% to 2% m/m. 4 pp.

BSI BS EN 27627-3-93. 1993 Hardmetals—Chemical Analysis by Flame Atomic Absorption Spectrometry Part 3: Determination of Cobalt, Iron, Manganese and Nickel in Contents from 0.01 to 0.5 % (m/m) (ISO 7627-3: 1983) (V). 10 pp.

BSI BS EN 27627-5-93. 1993 Hardmetals—Chemical Analysis by Flame Atomic Absorption Spectrometry Part 5: Determination of Cobalt, Iron, Manganese, Molybdenum, Nickel, Titanium and Vanadium in Contents from 0.5 to 2 % (m/m) (ISO 7627-5: 1985) (V). 10 pp.

CEN EN 27627-3-93. Hardmetals—Chemical Analysis by Flame Atomic Absorption Spectrometry—Part 3: Determination of Cobalt, Iron, Manganese and Nickel in Contents from 0,01 to 0,5 % (m/m) (ISO 7627-3: 1983). 5 pp.

Cobalt Content Analysis *(Cont.)*

—**Hard Metals—Atomic Absorption Spectrometry** *(Cont.)*

CEN EN 27627-5-93. Hardmetals—Chemical Analysis by Flame Atomic Absorption Spectrometry—Part 5: Determination of Cobalt, Iron, Manganese, Molybdenum, Nickel, Titanium and Vanadium in Contents from 0,5 to 2 % (m/m) (ISO 7627-5: 1983). 5 pp.

ISO 7627 Pt 3-83. Hardmetals—Chemical Analysis by Flame Atomic Absorption Spectrometry—Part 3: Determination of Cobalt, Iron, Manganese and Nickel in Contents from 0,01 to 0,5 % (m/m) First Edition (CEN EN 27627-3: 1993). 4 pp.

ISO 7627 Pt 5-83. Hardmetals—Chemical Analysis by Flame Atomic Absorption Spectrometry—Part 5: Determination of Cobalt, Iron, Manganese, Molybdenum, Nickel, Titanium and Vanadium in Contents from 0,5 to 2 % (m/m) First Edition (CEN EN 27627-5: 1993). 4 pp.

—**Hard Metals—Potentiometric Analysis**

BSI BS 5600: Sec 4.3-79. 1979 Powder Metallurgical Materials and Products Part 4: Methods of Testing and Chemical Analysis of Hardmetals Section 4.3: Determination of Cobalt: Potentiometric Method. 4 pp.

ISO 3909-76. Hardmetals—Determination of Cobalt—Potentiometric Method First Edition. 5 pp.

—**Hydrofluoric Acid—Emission Spectroscopy**

DIN ENGL 50451 Pt 2-90. Determination of Cobalt, Chromium, Copper, Iron and Nickel as Impurities in Hydrofluoric Acid for Use in Semiconductor Technology by Plasma-Induced Emission Spectrometry (Oct). 2 pp.

—**Iron**

CNS G2237-84. Method of Determination for Cobalt in Iron and Steel (Dec)(11168).

JIS G 1222-81. Methods for Determination of Cobalt in Iron and Steel.

—**Iron—Atomic Absorption Spectrometry**

BSI BS 6200: Sec 6.1-90. 1990 Sampling and Analysis of Iron, Steel and Other Ferrous Metals: Part 6: Guidelines on Atomic Absorption Spectrometric Techniques: Section 6.1: Recommendations for the Drafting of Standard Methods for the Chemical Anal. of Iron and Steel by Flame Atomic Absorp. 21 pp.

SAA AS 1050.29-89. Methods for the Analysis of Iron and Steel—Part 29: Determination of Cobalt Content—Flame Atomic Absorption Spectrometric Method. 3 pp.

—**Iron Ores—Absorptiometric Analysis**

CNS M3016-80. Method for Determination of Cobalt in Iron Ores (Mar)(5378).

JIS M 8210-83. Methods for Determination of Cobalt in Iron Ores. 11 pp.

—**Iron Ores—Atomic Absorption Spectrophotometry**

CNS M3016-80. Method for Determination of Cobalt in Iron Ores (Mar)(5378).

JIS M 8210-83. Methods for Determination of Cobalt in Iron Ores. 11 pp.

—**Iron—Photometry**

SAA AS K1.21-65. Methods for the Sampling and Analysis of Iron and Steel Cobalt in Iron and Steel (Photometric Method) Corrig. Reconfirmed 1987. 6 pp.

—**Iron—Spectrophotometry**

BSI BS 6200: SUB SEC 3.11.2-91. 1991 Sampling and Analysis of Iron, Steel and Other Ferrous Metals Part 3: Methods of Analysis Section 3.11: Determination of Cobalt Subsec 3.11.2: Steel, Irons and Steelmaking Materials: Spectrophotometric Method for Trace Amounts. 10 pp.

—**Manganese Ores—Photometry**

ISO 316-82. Manganese Ores—Determination of Cobalt Content—Nitroso-R-Salt Photometric Method Second Edition. 4 pp.

—**Nickel**

JIS H 1431-79. Methods for Determination of Cobalt in Nickel Materials for Electron Tubes.

—**Nickel Alloys—Atomic Absorption Spectrometry**

BSI BS 7455: Part 2-91. 1991 Analysis of Nickel Alloys by Flame Atomic Absorption Spectrometry Part 2: Method for the Determination of Cobalt (ISO 7530-2: 1990). 8 pp.

ISO 7530 Pt 2-90. Nickel Alloys—Flame Atomic Absorption Spectrometry Analysis—Part 2: Determination of Cobalt Content First Edition. 6 pp.

Cobalt Content Analysis *(Cont.)*

—**Nickel—Atomic Absorption Spectrometry**

BSI BS 6783: Part 1-86. 1986 Sampling and Analysis of Nickel, Ferronickel and Nickel Alloys Part 1: Method for Determination of Silver, Bismuth, Cadmium, Cobalt, Copper, Iron, Manganese, Lead and Zinc in Nickel by Flame Atomic Absorption Spectrometry. 15 pp.

ISO 6351-85. Nickel—Determination of Silver, Bismuth, Cadmium Cobalt, Copper, Iron, Manganese, Lead and Zinc Contents—Flame Atomic Absorption Spectrometric Method First Edition. 13 pp.

—**Nickel—Photometry**

BSI BS 3727: Part 5-64. 1964 Methods for the Analysis of Nickel for Use in Electronic Tubes and Valves Part 5: Determination of Cobalt (Photometric Method). 9 pp.

—**Nickel—Potentiometric Analysis**

BSI BS 6783: Part 11-90. 1990 Sampling and Analysis of Nickel, Ferronickel and Nickel Alloys Part 11: Method for the Determination of Cobalt in Nickel Alloys (Potentiometric Titration Method Using Potassium Hexacyanoferrate) (III). 11 pp.

ISO 9389-89. Nickel Alloys—Determination of Cobalt Content—Potentiometric Titration Method with Potassium Hexacyanoferrate(III) First Edition. 8 pp.

—**Ores**

JIS M 8129-81. Methods for Determination of Cobalt in Ores. 20 pp.

—**Ores—Absorptiometric Analysis**

CNS M3057-81. Method for Determination of Cobalt in Ores (Absorptiometric Method) (Jun)(7519).

—**Ores—Gravimetric Analysis**

CNS M3054-81. Method for Determination of Cobalt in Ores (Gravimetric Method) (Jun)(7516).

—**Ores—Polarographic Analysis**

CNS M3056-81. Method for Determination of Cobalt in Ores (Polarographic Method) (Jun)(7518).

—**Ores—Potentiometric Analysis**

CNS M3055-81. Method for Determination of Cobalt in Ores (Potentiometric Method) (Jun)(7517).

—**Paints**

CNS K6628-81. Method of Test for Cobalt in Paint Drier by EDTA Method (Jan)(6932).

—**Precipitation—Voltametry**

DIN ENGL 38406 Pt 16-90. German Standard Methods for the Examination of Water, Waste Water and Sludge; Cations (Group E); Determination of Zinc, Cadmium, Lead, Copper, Thallium, Nickel, Cobalt by Voltammetry (E16) (Mar). 8 pp.

—**Sodium Borates—Photometry**

ISO 5932-80. Boric Acid, Boric Oxide and DiSodium Tetraborates for Industrial Use—Determination of Cobalt Content—2-Nitroso-1-Naphthol Photometric Method First Edition. 6 pp.

—**Steels**

CNS G2237-84. Method of Determination for Cobalt in Iron and Steel (Dec)(11168).

JIS G 1222-81. Methods for Determination of Cobalt in Iron and Steel.

—**Steels—Atomic Absorption Spectrometry**

BSI BS 6200: Sec 6.1-90. 1990 Sampling and Analysis of Iron, Steel and Other Ferrous Metals: Part 6: Guidelines on Atomic Absorption Spectrometric Techniques: Section 6.1: Recommendations for the Drafting of Standard Methods for the Chemical Anal. of Iron and Steel by Flame Atomic Absorp. 21 pp.

SAA AS 1050.29-89. Methods for the Analysis of Iron and Steel—Part 29: Determination of Cobalt Content—Flame Atomic Absorption Spectrometric Method. 3 pp.

—**Steels—Photometry**

MOD UK M 9005/63. Determination of Cobalt in Steel: Photometric Method (No Information).

SAA AS K1.21-65. Methods for the Sampling and Analysis of Iron and Steel Cobalt in Iron and Steel (Photometric Method) Corrig. Reconfirmed 1987. 6 pp.

—**Steels—Spectrophotometry**

BSI BS 6200: SUB SEC 3.11.1-91. 1991 Sampling and Analysis of Iron, Steel and Other Ferrous Metals Part 3: Methods of Analysis Section 3.11: Determination of Cobalt Subsection 3.11.1: Steel and Cast Iron: Spectrophotometric Method. 7 pp.

INDUSTRY STANDARDS

Cobalt

Cobalt Content Analysis (Cont.)
—Steels—Spectrophotometry (Cont.)
  BSI BS 6200: SUB SEC 3.11.2-91. 1991 Sampling and Analysis of Iron, Steel and Other Ferrous Metals Part 3: Methods of Analysis Section 3.11: Determination of Cobalt Subsec 3.11.2: Steel, Irons and Steelmaking Materials: Spectrophotometric Method for Trace Amounts. 10 pp.

—Surface Waters—Voltametry
  DIN ENGL 38406 Pt 16-90. German Standard Methods for the Examination of Water, Waste Water and Sludge; Cations (Group E); Determination of Zinc, Cadmium, Lead, Copper, Thallium, Nickel, Cobalt by Voltammetry (E16) (Mar). 8 pp.

—Water—Atomic Absorption Spectrometry
  BSI BS 6068: Sec 2.29-87. 1987 Water Quality Part 2: Physical, Chemical and Bio-Chemical Methods Section 2.29: Determination of Cobalt, Nickel. Copper, Zinc, Cadmium and Lead: Flame Atomic Absorption Spectrometric Methods. 13 pp.
  CNS K9031-80. Method of Test for Cobalt in Water (Atomic Absorption, Direct) (Apr)(5465).
  ISO 8288-86. Water Quality—Determination of Cobalt, Nickel, Copper, Zinc, Cadmium and Lead—Flame Atomic Absorption Spectrometric Methods First Edition. 13 pp.

—Zirconium
  JIS H 1658-85. Methods for Determination of Cobalt in Zirconium and Zirconium Alloys. 9 pp.

—Zirconium Alloys
  JIS H 1658-85. Methods for Determination of Cobalt in Zirconium and Zirconium Alloys. 9 pp.

Cobalt Nitrate
Use: Cobaltous Nitrate

Cobalt Sulfate
Use: Cobaltous Sulfate

Cobalt Wire
Scope Note: Includes wire made from cobalt alloys
See Also: Cobalt; Wire

—Aerospace
  AECMA PREN2166-80. Heat Resisting Cobalt Base Alloy CO-P92HT-Annealed—Bars and Wires D Less Than or Equal to 4 mm (C5/48). 3 pp.

Cobaltous Acetate
Use For: Coblatic Acetate *See Also:* Acetates
  CNS K7151-66. Chemical Reagent (Cobalt Acetate) (Mar)(1651).

Cobaltous Chloride
Use For: Cobalt Chloride
  CNS K7152-66. Chemical Reagent (Cobalt Chloride) (Mar)(1652).
  JIS K 8129-88. Cobalt (II) Chloride, Hexahydrate.

—Drying Agents
  BSI BS 3482: Part 9-91. 1991 Methods of Test for Desiccants Part 9: Determination of Cobalt Chloride Content. 5 pp.

—Test Papers—Humidity
  MOD UK DEF-1296-58. Paper, Humidity Indicator, Cobalt Chloride (Reprinted April 1970). 3 pp.

Cobaltous Diammonium Sulfate
  JIS K 8969-80. Cobalt (II) Diammonium Sulfate Hexahydrate.

Cobaltous Nitrate
Use For: Cobalt Nitrate
  CNS K7153-66. Chemical Reagent (Cobalt Nitrate) (Mar)(1653).
  JIS K 8552-88. Cobalt (II) Nitrate Hexahydrate.

Cobaltous Sulfate
Use For: Cobalt Sulfate
  CNS K7154-66. Chemical Reagent (Cobalt Sulfate) (Mar)(1654).

Coblatic Acetate
Use: Cobaltous Acetate

COBOL
Use For: Common Business-Oriented Language
See Also: FORTRAN; PL/1; Programming Languages
  BSI BS 7147-90. 1990 Programming Language COBOL. 4 pp.
  CEN EN 21 989-89. Programming Languages Cobal. 2 pp.
  CSA CAN/CSA-Z243.39-90. Programming Language—COBOL (IDT ISO 1989-1985). 12 pp.

COBOL (Cont.)
  IEC 1989 AMD 1. Amendment 1—Programming Languages—COBOL Amendment 1: Intrinsic Function Module; (1992). 4 pp.
  IEC 1989 Draft AMD 2. Programming Languages—COBOL Amendment 2: Correction and Clarification Amendment for COBOL; (1993) ***CD-ROM ONLY***. 66 pp.
  ISO 1989-85. Programming Languages—COBOL Amendment 1: Intrinsic Function Module Second Edition; (Amendment 1-1992). 7 pp.
  ISO 1989 Draft AMD 2. Programming Languages—COBOL Amendment 2: Correction and Clarification Amendment for COBOL; (1993) ***CD-ROM ONLY***. 66 pp.
  JIS X 3002-88. Programming Language COBOL.
  JTC1 1989-85. Programming Languages—COBOL Amendment 1: Intrinsic Function Module Second Edition; (Amendment 1-1992). 3 pp.
  JTC1 1989 Draft AMD 2. Programming Languages—COBOL Amendment 2: Correction and Clarification Amendment for COBOL; (1993) ***CD-ROM ONLY***. 66 pp.
  OSI ISO 1989 DAD 1-85. Programming Languages—COBAL—Addendum 1: Intrinsic Function Module. 79 pp.
  OSI ISO 1989-85. Programming Languages—COBOL. 719 pp.
  SAA AS 1209-78. Programming Language COBOL. 541 pp.
  SNZ NZS/ISO 1989-78. Programming Languages—COBOL. 541 pp.

—Open Systems Interconnection—File Transfer, Access and Management
  IEC DISP 10607 Pt1 DR AMD 1. Information Technology—International Standardized Profiles AFTnn—File Transfer, Access and Management—Part 1: Specification of ASCE, Presentation and Session Protocols for the Use by FTAM Amd. 1: Additional Specs. for COBOL Doc. Types; (1993) ***CD-ROM ONLY***. 10 pp.
  IEC DISP 10607 Pt2 DR AMD 2. Information Technology—International Standardized Profiles AFTnn—File Transfer, Access and Management—Part 2: Definition of Document Types, Contraint Sets and Syntaxes Amendment 2: Additional Defs. for COBOL Doc. Types; (1993) ***CD-ROM ONLY***. 54 pp.
  IEC DISP 10607 Pt4 DR AMD 1. Information Technology—International Standardized Profiles AFTnn—File Transfer, Access and Management—Part 4: AFT12—Positional File Transfer Service (Flat) Amendment 1: Additional Specifications for COBOL Document Types; (1993) ***CD-ROM ONLY***. 15 pp.
  IEC DISP 10607 Pt5 DR AMD 1. Information Technology—International Standardized Profiles AFTnn—File Transfer, Access and Management—Part 5: AFT22—Positional File Access Service (Flat) Amendment 1: Additional Specifications for COBOL Document Types; (1993) ***CD-ROM ONLY***. 15 pp.
  ISO DISP 10607 Pt1 DR AMD 1. Information Technology—International Standardized Profiles AFTnn—File Transfer, Access and Management—Part 1: Specification of ASCE, Presentation and Session Protocols for the Use by FTAM Amd. 1: Additional Specs. for COBOL Doc. Types; (1993) ***CD-ROM ONLY***. 10 pp.
  ISO DISP 10607 Pt2 DR AMD 2. Information Technology—International Standardized Profiles AFTnn—File Transfer, Access and Management—Part 2: Definition of Document Types, Contraint Sets and Syntaxes Amendment 2: Additional Defs. for COBOL Doc. Types; (1993) ***CD-ROM ONLY***. 54 pp.
  ISO DISP 10607 Pt4 DR AMD 1. Information Technology—International Standardized Profiles AFTnn—File Transfer, Access and Management—Part 4: AFT12—Positional File Transfer Service (Flat) Amendment 1: Additional Specifications for COBOL Document Types; (1993) ***CD-ROM ONLY***. 15 pp.
  ISO DISP 10607 Pt5 DR AMD 1. Information Technology—International Standardized Profiles AFTnn—File Transfer, Access and Management—Part 5: AFT22—Positional File Access Service (Flat) Amendment 1: Additional Specifications for COBOL Document Types; (1993) ***CD-ROM ONLY***. 15 pp.
  JTC1 DISP10607 Pt1 DR AMD 1. Information Technology—International Standardized Profiles AFTnn—File Transfer, Access and Management—Part 1: Specification of ASCE, Presentation and Session Protocols for the Use by FTAM Amd. 1: Additional Specs. for COBOL Doc. Types; (1993) ***CD-ROM ONLY***. 10 pp.

COBOL (Cont.)
—Open Systems Interconnection—File Transfer, Access and Management (Cont.)
  JTC1 DISP10607 Pt2 DR AMD 2. Information Technology—International Standardized Profiles AFTnn—File Transfer, Access and Management—Part 2: Definition of Document Types, Contraint Sets and Syntaxes Amendment 2: Additional Defs. for COBOL Doc. Types; (1993) ***CD-ROM ONLY***. 54 pp.
  JTC1 DISP10607 Pt4 DR AMD 1. Information Technology—International Standardized Profiles AFTnn—File Transfer, Access and Management—Part 4: AFT12—Positional File Transfer Service (Flat) Amendment 1: Additional Specifications for COBOL Document Types; (1993) ***CD-ROM ONLY***. 15 pp.
  JTC1 DISP10607 Pt5 DR AMD 1. Information Technology—International Standardized Profiles AFTnn—File Transfer, Access and Management—Part 5: AFT22—Positional File Access Service (Flat) Amendment 1: Additional Specifications for COBOL Document Types; (1993) ***CD-ROM ONLY***. 15 pp.

—Source Programs
  ECMA ECMA 53-78. Representation of Source Program for Program Interchange-APL, COBOL, FORTRAN, Minimal BASIC and PL/1. 16 pp.

Cockpit Voice Recorders
Use For: Voice Recorders *See Also:* Aircraft Equipment; Flight Recorders; Recording Instruments

—Aircraft
  CAA LEAFLET 14-14 06.91. Cockpit Voice Recorder System Fairchild A100 Series. 2 pp.
  CAA LEAFLET 14-16 06.92. Cockpit Voice Recorder System Sundstrand AV557 Series. 2 pp.
  EUROCAE ED-56 02.88. MOPR for Cockpit Voice Recorder System. 121 pp.

Cockpits
See Also: Aircraft Cabins; Aircraft Compartments; Aircraft Equipment; Canopies (Aircraft); Pressurized Cabins

—Aircraft—Control Knobs
  SBAC AS 6397 ISSUE 2. Knob (For Cockpit Heating Control).

—Gliders—Signs
  CAA JAR-22 Appendix G. Cockpit Placards (Joint Airworthiness Requirements). 1 p.

Cockroaches
Use For: Roaches

—DDT
  MOD UK CS 3099A. DDT Cockroach Powder (Withdrawn).

Cocks
See Also: Ball Valves; Drain Cocks; Faucets; Gas Valves; Hydraulic Valves; Plumbing Fixtures; Stopcocks; Valves
  CNS B2498-80. Face-to-Face and End-to-End Dimensions of Cocks (Jul)(5714).
  JIS B 2002-87. Face-to-Face and End-to-End Dimensions of Valves.
  JIS B 2002-68. Face-to-Face and End-to-End Dimensions of Valves. 14 pp.

—Aircraft—Fuel—Manual Controls
  SBAC AS 3196 ISSUE 1. Fuel Cock, Knob (Low Pressure). 1 p.

—Chemical Analysis
  CNS R2133-86. Glass Cock for Chemical Analysis (Mar)(7309).

—Gas
  JIS S 2120-92. Gas Valves. 51 pp.

—Gland
  CNS B2554-83. Screwed Bronze Gland Cocks (Jan)(7117).
  JIS B 2191-84. Screwed Bronze Plug Cocks and Gland Cocks. 14 pp.

—Motorcycles—Fuel Tank
  CNS D2028-88. Dimensions of Cock Connectors of Fuel Tanks for Motor Cycles (Jul)(5784). 1 p.
  JIS D 3903-66. Dimensions of Cock Connections of Fuel Tanks for Motorcycles (R 1983). 4 pp.

—Rolling Stock
  CNS E1050-85. Three-Way Cocks for Railway Rolling Stock (Nov)(11408).
  CNS E1052-85. Cut-Out Cocks for Railway Rolling Stock (Nov)(11410).
  CNS E1054-85. Cut-Out Cocks with Side Vent for Railway Rolling Stock (Nov)(11412).

INTERNATIONAL AND NON-U.S. NATIONAL STANDARDS
SUBJECT INDEX

Cocks *(Cont.)*
—Rolling Stock *(Cont.)*
JIS E 4103-88. Cut-out Cocks for Railway Rolling Stock.
JIS E 4105-91. Three-Way Cocks for Railway Rolling Stock.

—Ships
CNS F1006-83. General Rules for Inspection of Marine Valves and Cocks (Mar)(3914).
CNS F3188-82. Marine Bronze 5 kgf/cm2 Flanged Cocks (Jun)(8964).
CNS F3189-82. Marine Bronze 16 kgf/cm2 Cocks (Jun)(8965).
JIS F 7300-89. Application for Marine Valves and Cocks.
JIS F 7381-88. Bronze 5 K Flanged Cocks for Marine Use.
JIS F 7387-88. Bronze 16 K Cocks for Marine Use.
JIS F 7400-85. General Rules for Inspection of Marine Valves and Cocks.

—Ships—Locks
CNS F3192-82. Marine Cocks with Lock (Aug)(9254).
JIS F 7390-89. Marine Cocks with Locks.

—Ships—Pressure Gages
CNS F3156-82. Marine Bronze 20 kgf/cm2 Pressure Gauge Cocks (Jan)(8271).
JIS F 7343-88. Bronze 20 K Pressure Gauge Cocks for Marine Use.

—Ships—Water Gages—Boilers
JIS F 5609-88. Forged Steel 20 K Reflex Type Water Gauges with Cocks for Marine Boilers.

—Sprayers
CNS B4033-80. Cocks for Sprayer (Aug)(6008).
JIS B 9121-89. Cocks for Sprayer.
JIS B 9121-75. Cocks for Sprayer (R 1984). 12 pp.

—Water Gages—Boilers
CNS B2746-83. Water Gages for Boilers — with 10 kgf/cm2 Cocks Tubular Type (Feb)(9973).
JIS B 8216-77. 10 kgf/cm2 Tubular Type Water Gauges with Cocks for Boilers (R 1982). 8 pp.

Cocoa
See Also: Cocoa Beans; Food
CNS N5203-83. Cocoa (Edible) (Feb)(10028). 1 p.

Cocoa Beans
See Also: Cocoa; Cocoa Butter
CNS N1096-81. Cocoa Beans—Specification (Apr) (7292). 2 pp.
ISO 2451-73. Cocoa Beans—Specification First Edition. 5 pp.

—Cut Testing
CNS N4073-81. Method of Test for Cocoa Beans-Cut Test (Apr)(7294). 1 p.
ISO 1114-77. Cocoa Beans—Cut Test First Edition. 3 pp.

—Moisture Content
CNS N4074-81. Method of Test for Cocoa Beans—Determination of Moisture Content (Apr)(7295). 2 pp.
ISO 2291-80. Cocoa Beans—Determination of Moisture Content (Routine Method) Second Edition. 4 pp.

—Sampling
CNS N4072-81. Sampling of Cocoa Beans (Apr)(7293). 5 pp.
ISO 2292-73. Cocoa Beans—Sampling First Edition. 7 pp.

Cocoa Butter
See Also: Cocoa Beans; Coconut Oil; Fats
CNS N5182-88. Cocoa Butter (Edible) (May)(7526). 2 pp.

Coconut Oil
See Also: Cocoa Butter; Vegetable Oils
CNS K5028-82. Coconut Oil for Industrial Use (Feb)(2240). 1 p.
CNS K5115-81. Crude Coconut Oil (Sep)(7426).
CNS N5146-89. Edible Coconut Oil (Sep)(4834). 2 pp.

—Colorimetric Analysis
CNS K6180-74. Method of Test for Coconut Oil (Technical Grade) (Oct)(2241). 5 pp.

—Iodine Number
CNS K6180-74. Method of Test for Coconut Oil (Technical Grade) (Oct)(2241). 5 pp.

—Physical Properties
CNS K6180-74. Method of Test for Coconut Oil (Technical Grade) (Oct)(2241). 5 pp.

—Residue Content
CNS K6180-74. Method of Test for Coconut Oil (Technical Grade) (Oct)(2241). 5 pp.

Coconut Oil *(Cont.)*
—Saponification Number
CNS K6180-74. Method of Test for Coconut Oil (Technical Grade) (Oct)(2241). 5 pp.

—Unsaponifiable Matter Content
CNS K6180-74. Method of Test for Coconut Oil (Technical Grade) (Oct)(2241). 5 pp.

—Water Content
CNS K6180-74. Method of Test for Coconut Oil (Technical Grade) (Oct)(2241). 5 pp.

Coconut Oil Meal
Use: Oil Meal

Coconuts
See Also: Food

—Dried
CNS N5202-83. Coconut Dried Prepared (Edible) (Feb)(10027). 2 pp.

—Fatty Acids
CNS K1261-81. Distilled Coconut Fatty Acids (May)(7429).

COD
Use: Oxygen Demand

CODEC
Use: Encoder/Decoders

Codecs
Use: Encoder/Decoders

Coded Character Sets
Use For: Picture Representations (Coded)
See Also: Character Imaging Devices; Character Sets; Codes; Computer Graphics; Information Interchange
BSI BS 6692: Part 1-86. 1986 Coded Character Sets for Text Communication Part 1: General Introduction. 15 pp.
CENELEC HD 40 001-85. Section 2: Requirements for Information Technology Equipment. 10 pp.
CENELEC HD 40 001-89. Information Systems Interconnection; Combinations of Reference Documents for the Elaboration of Functional Standards. 4 pp.
ISO 6937 Pt 1-83. Information Processing—Coded Character Sets for Text Communication—Part 1: General Introduction First Edition. 15 pp.
JIS X 5001-82. Character Structure on the Transmission Circuits and Horizontal Parity Method.
JTC1 6937 Pt 1-83. Information Processing—Coded Character Sets for Text Communication—Part 1: General Introduction First Edition. 15 pp.
OSI ISO/IEC 6937-91. Information Technology—Coded Graphic Character Set for Text Communication—Latin Alphabet. 47 pp.
OSI ISO 6937-1-83. Coded Character Sets for Text Communication—Part 1: General Introduction. 16 pp.

—4-Bit—Punched Tapes
BSI BS 3880: Part 3-71. 1971 Amd 1 Paper Tape for Data Processing Part 3: Representation of Codes on Paper Tape. 23 pp.

—5-Bit—Punched Tapes
BSI BS 3880: Part 3-71. 1971 Amd 1 Paper Tape for Data Processing Part 3: Representation of Codes on Paper Tape. 23 pp.

—6-Bit—Punched Tapes
BSI BS 3880: Part 3-71. 1971 Amd 1 Paper Tape for Data Processing Part 3: Representation of Codes on Paper Tape. 23 pp.

—7-Bit
BSI BS 4730-93. 1993 UK 7-Bit Coded Character Set. 8 pp.
BSI BS 4730-85. (WITHDRAWN) 1985 Amd 1 UK 7 Bit Coded Character Set (AMD 5436) March 31, 1987 (Superseded by BS ISO/IEC 646: 1991). 22 pp.
BSI BS 6727-87. 1987 Representation of Numerical Values in Character Strings for Information Interchange. 15 pp.
BSI BS ISO/IEC 646-91. 1991 Information Technology—ISO 7-Bit Coded Character Set for Information Interchange. 23 pp.
BSI BS ISO/IEC 646-01. 1991 Amd 1 Information Technology—ISO 7-Bit Coded Character Set for Information Interchange (AMD 7619) February 15, 1993. 24 pp.
CNS C5036-89. 7-Bit Coded Character Set for Information Processing Interchange (Jun)(5205).
CNS C5123-89. Code Extension Techniques for Use with 7-Bit Coded Character Set (Jul)(7654).
CNS C5125-89. 8-Bit Coded Character Set for Information Interchange (Jul)(7656).

Coded Character Sets *(Cont.)*
—7-Bit *(Cont.)*
CNS C5126-84. Numerical Control of Machines Character Set (Jan)(8380).
CSA CAN/CSA-Z243.4-87. 7-Bit and 8-Bit Coded Character Sets for Information Processing and Interchange; (Gen Instr 1). 59 pp.
IEC 646-91. Information Technology—ISO 7-Bit Coded Character Set for Information Interchange Third Edition. 20 pp.
ISO 646-91. Information Technology—ISO 7-Bit Coded Character Set for Information Interchange Third Edition. 20 pp.
ISO 6093-85. Information Processing—Representation of Numerical Values in Character Strings for Information Interchange First Edition. 15 pp.
JIS X 0210-86. Representation of Numerical Values in Character Strings for Information Interchange. 21 pp.
JTC1 646-91. Information Technology—ISO 7-Bit Coded Character Set for Information Interchange Third Edition. 20 pp.
JTC1 6093-85. Information Processing—Representation of Numerical Values in Character Strings for Information Interchange First Edition. 15 pp.
NATO STANAG 5036 ED 3 AMD 1-86. Parameters and Practices for the Use of the NATO 7-Bit Code. 31 pp.
OSI ISO 646-91. Information Technology—ISO 7-Bit Coded Character Set for Information Interchange. 20 pp.
OSI ISO 6093-85. Information Processing-Representation of Numerical Values in Character Strings for Information Interchange. 15 pp.
SAA AS 1776-80. Information Processing—7-Bit Coded Character Set for Information Interchange. 16 pp.
SAA AS 1953-87. Information Processing—ISO 7-Bit and 8-Bit Coded Character Sets—Code Extension Techniques. 25 pp.

—7-Bit—Basic Mode Control
BSI BS 4505: Part 1-81. 1981 Digital Data Transimission Part 1: Basic Mode Control Procedures. 22 pp.
ISO 1745-75. Information Processing—Basic Mode Control Procedures for Data Communication Systems First Edition. 22 pp.
JTC1 1745-75. Information Processing—Basic Mode Control Procedures for Data Communication Systems First Edition. 22 pp.
OSI ISO 1745-75. Information Processing—Basic Mode Control Procedures for Data Communication Systems. 22 pp.
SNZ NZS/ISO 1745-75. Information Processing—Basic Mode Control Procedures for Data Communication Systems. 19 pp.

—7-Bit—Character Conversion
BSI BS 6429-89. 1989 Method for Conversion Between the Two Coded Character Sets of BS 4730 (ISO 646) and BS 6692: Part 2: (ISO 6937-2) and the CCITT International Telegraph Alphabet No.2 (ITA 2). 8 pp.
ISO 6936-88. Information Processing—Conversion Between the Two Coded Character Sets of ISO 646 and ISO 6937-2 and the CCITT International Telegraph Alphabet No. 2 (ITA 2) Second Edition. 8 pp.
JTC1 6936-88. Information Processing—Conversion Between the Two Coded Character Sets of ISO 646 and ISO 6937-2 and the CCITT International Telegraph Alphabet No. 2 (ITA 2) Second Edition. 8 pp.
OSI ISO DIS 6936-88. Information Processing—Conversion Between the Two Coded Character Sets of ISO 646 and ISO 6937-2 and the CCITT International Telegraph Alphabet No.2 (ITA)2. 9 pp.

—7-Bit—Character Structure
BSI BS 4505: Part 2-90. 1990 Amd 1 Digital Data Transmission Part 2: Character Structure for Start/Stop and Synchronous Transmission. 7 pp.
CNS C5106-89. Information Processing — Character Structure for Start/Stop and Synchronous Transmission (Mar)(7222).
ISO 1177-85. Information Processing—Character Structure for Start/Stop and Synchronous Character Oriented Transmission Second Edition. 4 pp.
JTC1 1177-85. Information Processing—Character Structure for Start/Stop and Synchronous Character Oriented Transmission Second Edition. 4 pp.
OSI ISO 1177-85. Information Processing—Character Structure for Start/Stop and Synchronous Character Oriented Transmission. 4 pp.

—7-Bit—Code Extensions
ECMA ECMA 35-93. Character Code Structure and Extension Techniques. 55 pp.
ISO 2022-86. Information Processing—ISO 7-Bit and 8-Bit Coded Character Sets—Code Extension Techniques Third Edition. 30 pp.

Coded Character Sets (Cont.)
— 7-Bit—Code Extensions (Cont.)
JTC1 2022-86. Information Processing—ISO 7-Bit and 8-Bit Coded Character Sets—Code Extension Techniques Third Edition. 30 pp.
OSI ISO 2022-86. Information Processing-ISO 7-Bit and 8-Bit Coded Character Sets-Code Extension Techniques. 28 pp.
OSI ISO/IEC DIS 6429-91. Information Technology—Control Functions for Coded Character Sets. 110 pp.

— 7-Bit—Control Functions
BSI BS 7204-89. (WITHDRAWN) 1989 Control Functions for ISO 7-Bit and 8-Bit Coded Character Sets (Superseded by BS ISO/IEC 6429: 1992). 55 pp.
BSI BS ISO/IEC 6429-92. 1992 Information Technology—Control Functions for Coded Character Sets (S). 100 pp.
ECMA ECMA 48-91. Control Functions for Coded Character Sets. 108 pp.
IEC 6429-92. Information Technology—Control Functions for Coded Character Sets Third Edition. 98 pp.
ISO 6429-92. Information Technology—Control Functions for Coded Character Sets Third Edition. 98 pp.
JIS X 0211-91. Information Processing—Control Functions for 7-Bit and 8-Bit Coded Character Sets.
JTC1 6429-92. Information Technology—Control Functions for Coded Character Sets Third Edition. 98 pp.
OSI ISO 6429-88. Information Processing—Control Functions for 7-Bit and 8-Bit Coded Character Sets. 64 pp.
OSI ISO/IEC DIS 6429-91. Information Technology—Control Functions for Coded Character Sets. 110 pp.
OSI ISO DP 9429-86. ISO/DP 6429—ISO 7-Bit and 8-Bit Coded Character Sets—Control Functions.

— 7-Bit—Graphic Characters
BSI BS 6856-87. 1987 Code Extension Techniques for United Kingdom 7-Bit and 8-Bit Coded Character Sets. 28 pp.
CNS C5107-83. Information Processing — Graphical Representations for the Control Characters of the 7 Bit Coded Character Set (Feb)(7223).
ECMA ECMA 17-68. Graphic Representation of the Control Characters of the ECMA 7-Bit Coded Character Set for Information Interchange. 8 pp.
ISO 2022-86. Information Processing—ISO 7-Bit and 8-Bit Coded Character Sets—Code Extension Techniques Third Edition. 30 pp.
ISO 2047-75. Information Processing—Graphical Representations for the Control Characters of the 7-Bit Coded Character Set First Edition. 8 pp.
ISO 9036-87. Information Processing—Arabic 7-Bit Coded Character Set for Information Interchange First Edition. 16 pp.
JTC1 2022-86. Information Processing—ISO 7-Bit and 8-Bit Coded Character Sets—Code Extension Techniques Third Edition. 30 pp.
JTC1 2047-75. Information Processing—Graphical Representations for the Control Characters of the 7-Bit Coded Character Set First Edition. 8 pp.
JTC1 9036-87. Information Processing—Arabic 7-Bit Coded Character Set for Information Interchange First Edition. 16 pp.
OSI ISO 2047-75. Information Processing—Graphical Representations for the Control Characters of the 7-Bit Coded Character Set. 8 pp.
OSI ISO 9036-87. Information Processing—Arabic 7-Bit Coded Character Set for Information Interchange. 22 pp.
OSI ISO DIS 9036-86. Information Processing—Arabic 7-Bit Coded Character Set for Information Interchange. 22 pp.

— 7-Bit—International Reference Alphabet
CCITT RECMN T.50-92. International Reference Alphabet (IRA) (Formerly International Alphabet No. 5 or IA5) Information Technology—7-Bit Coded Character Set for Information Interchange (Study Group VIII) 22 pp. 22 pp.
CCITT RECMN T.50-89. International Alphabet No. 5—Terminal Equipment and Protocols for Telematic Services (Study Group VIII) 16 pp. 16 pp.

— 7-Bit—Keyboards
BSI BS 4822-80. 1980 Keyboard Arrangements of the Graphic Characters of the United Kingdom 7-Bit Data Code, for Data Processing. 8 pp.
CNS C5150-83. Keyboard for International Information Processing Interchange Using 7-Bit Coded Character Set (Alphanumeric Area) (Apr)(10153).
ISO 2530-75. Keyboard for International Information Processing Interchange Using the ISO 7-Bit Coded Character Set—Alphanumeric Area First Edition. 9 pp.

Coded Character Sets (Cont.)
— 7-Bit—Keyboards (Cont.)
JTC1 2530-75. Keyboard for International Information Processing Interchange Using the ISO 7-Bit Coded Character Set—Alphanumeric Area First Edition. 9 pp.
OSI ISO 2530-75. Keyboard for International Information Processing Interchange Using the ISO 7-Bit Coded Character Set-Alphanumeric Area. 8 pp.

— 7-Bit—Magnetic Tapes
ISO 962-74. Information Processing—Implementation of the 7-Bit Coded Character Set and Its 7-Bit and 8-Bit Extensions on 9-Track 12,7 mm (0.5 in) Magnetic Tape First Edition. 5 pp.
ISO 3275-74. Information Processing—Implementation of the 7-Bit Coded Character Set and Its 7-Bit and 8-Bit Extensions on 3,81 mm Magnetic Tape Cassette for Data Interchange First Edition. 3 pp.
JTC1 962-74. Information Processing—Implementation of the 7-Bit Coded Character Set and Its 7-Bit and 8-Bit Extensions on 9-Track 12,7 mm (0.5 in) Magnetic Tape First Edition. 5 pp.
JTC1 3275-74. Information Processing—Implementation of the 7-Bit Coded Character Set and Its 7-Bit and 8-Bit Extensions on 3,81 mm Magnetic Tape Cassette for Data Interchange First Edition. 3 pp.
OSI ISO 962-74. Information Processing—Implementation of the 7-Bit Coded Character Set and Its 7-Bit and 8-Bit Extensions on 9-Track 12,7mm (0.5 in) Magnetic Tape. 5 pp.
OSI ISO 3275-74. Information Processing—Implementation of the 7-bit Coded Character Set and Its 7-bit and 8-bit Extension on 3.81mm Magnetic Tape Cassette for Data Interchange. 3 pp.

— 7-Bit—Parity
BSI BS 4505: Part 3-81. (OBSOLESCENT) 1981 Digital Data Transmission Part 3: The Use of Longitudinal Parity to Detect Errors in Information Messages. 3 pp.
CNS C5105-83. Information Processing — Use of Longitudinal Parity to Detect Error in Information Messages (Feb)(7221).
ISO 1155-78. Information Processing—Use of Longitudinal Parity to Detect Errors in Information Messages Second Edition. 3 pp.
JTC1 1155-78. Information Processing—Use of Longitudinal Parity to Detect Errors in Information Messages Second Edition. 3 pp.
OSI ISO 1155-78. Information Processing—Use of Longitudinal Parity to Detect Errors in Information Messages. 3 pp.
SNZ NZS/ISO 1155-78. Information Processing—Use of Longitudinal Parity to Detect Errors in Information Messages. 1 p.

— 7-Bit—Pictures
BSI BS ISO 9282-1-88. 1988 Information Processing—Coded Representation of Pictures—Part 1: Encoding Principles for Picture Representation in a 7-Bit or 8-Bit Environment (S). 31 pp.
BSI BS ISO/IEC 9282-2-92. 1992 Information Processing—Coded Representation of Pictures—Part 2: Incremental Encoding of Point Lists in a 7-Bit or 8-Bit Environment. 31 pp.
IEC 9282 Pt 2-92. Information Processing—Coded Representation of Pictures—Part 2: Incremental Encoding of Point Lists in a 7-Bit or 8-Bit Environment First Edition. 28 pp.
ISO 9282 Pt 1-88. Information Processing—Coded Representation of Pictures—Part 1: Encoding Principles for Picture Representation in a 7-Bit or 8-Bit Environment First Edition. 29 pp.
ISO 9282 Pt 2-92. Information Processing—Coded Representation of Pictures—Part 2: Incremental Encoding of Point Lists in a 7-Bit or 8-Bit Environment First Edition. 28 pp.
JTC1 9282 Pt 1-88. Information Processing—Coded Representation of Pictures—Part 1: Encoding Principles for Picture Representation in a 7-Bit or 8-Bit Environment First Edition. 29 pp.
JTC1 9282-92. Information Processing—Coded Representation of Pictures—Part 2: Incremental Encoding of Point Lists in a 7-Bit or 8-Bit Environment First Edition. 28 pp.
OSI ISO/IEC 9281-2-90. Information Technology—Picture Coding Methods—Part 2: Procedure for Registration. 7 pp.
OSI ISO DIS 9282-1-87. Information Processing—Coded Representation of Pictures—Part 1: Encoding Principles for Picture Representation in a 7 or 8 Bit Environment. 34 pp.
OSI ISO/IEC DIS 9282-2-90. Information Technology—Coded Representation of Pictures—Part 2: Incremental Encoding of Point Lists in a 7-Bit Environment. 28 pp.
OSI ISO DIS 9282-2-90. Information Technology—Coded Representation of Pictures—Part 2: Incremental Encoding of Point Lists in a 7-Bit or 8-bit Environment. 28 pp.

Coded Character Sets (Cont.)
— 7-Bit—Pictures (Cont.)
SNZ NZS/ISO 9282. 1-88. Computer Graphics—Coded Representation of Pictures Part 1: Encoding Principles for Picture Representation in a 7-Bit or 8-Bit Environment. 23 pp.

— 7-Bit—Punched Cards
BSI BS 4636: Part 3-86. 1986 Punched Cards Part 3: Representation of 7-Bit and 8-Bit Coded Character Sets on 12-Row Punched Cards. 8 pp.
CNS C5149-83. Data Processing Implementation of 7-Bit and 8-Bit Coded Character Sets on Punched Cards (Feb)(9982).
ECMA ECMA 44-75. Implementation of the ECMA 7-Bit and 8-Bit Coded Character Sets on Punched Cards. 12 pp.
ISO 6586-80. Data Processing—Implementation of the ISO 7-Bit and 8-Bit Coded Character Sets on Punched Cards First Edition. 7 pp.
JTC1 6586-80. Data Processing—Implementation of the ISO 7-Bit and 8-Bit Coded Character Sets on Punched Cards First Edition. 7 pp.
OSI ISO 6586-80. Data Processing—Implementation of the ISO 7-Bit and 8-Bit Coded Character Sets on Punched Cards. 7 pp.

— 7-Bit—Punched Tapes
BSI BS 3880: Part 3-71. 1971 Amd 1 Paper Tape for Data Processing Part 3: Representation of Codes on Paper Tape. 23 pp.
CNS C5103-83. Information Processing — Representation of 6 and 7 Bit Coded Character Sets on Punched Tape (Feb)(7219).
CNS C5158-83. Subset of Standard Code for Information Interchange for Numerical Machine Control Perforated Tape (May)(10245).
ISO 1113-79. Information Processing—Representation of the 7-Bit Coded Character Set on Punched Tape Second Edition. 4 pp.
JTC1 1113-79. Information Processing—Representation of the 7-Bit Coded Character Set on Punched Tape Second Edition. 4 pp.
OSI ISO 1113-79. Information Processing—Representation of the 7-Bit Coded Character Set on Punched Tape. 4 pp.

— 7-Bit—Teleinformatic Services
CCITT RECMN T.51-92. Latin Based Coded Character Sets for Telematic Services (Study Group VIII) 28 pp. 28 pp.

— 7-Bit—Teleprinters
NATO STANAG 5045 ED 2 AMD 1-86. Interoperability Characteristics for Teleprinters Using the NATO 7-Bit Code. 10 pp.

— 7-Bit—Teletex Communications—Character Repertoire
CCITT RECMN T.61-89. Character Repertoire and Coded Character Sets for the International Teletex Service—Terminal Equipment and Protocols for Telematic Services (Study Group VIII) 50 pp. 50 pp.

— 8-Bit
BSI BS 6006-87. (WITHDRAWN) 1987 Structure and Rules for Implementation of United Kingdom 8-bit Coded Character Set (Superseded by BS ISO/IEC 4873: 1991). 20 pp.
BSI BS 7204-89. (WITHDRAWN) 1989 Control Functions for ISO 7-Bit and 8-Bit Coded Character Sets (Superseded by BS ISO/IEC 6429: 1992). 55 pp.
BSI BS ISO/IEC 4873-91. 1991 Information Technology—ISO 8-Bit Code for Information Interchange—Structure and Rules for Implementation. 27 pp.
CSA CAN/CSA-Z243.4-87. 7-Bit and 8-Bit Coded Character Sets for Information Processing and Interchange; (Gen Instr 1). 59 pp.
ECMA ECMA 6-91. 8-Bit Coded Character Set. 21 pp.
ECMA ECMA 43-91. 8-Bit Coded Character Set Structure and Rules. 25 pp.
IEC 4873-91. Information Technology—ISO 8-Bit Code for Information Interchange—Structure and Rules for Implementation Third Edition. 25 pp.
ISO 2022-86. Information Processing—ISO 7-Bit and 8-Bit Coded Character Sets—Code Extension Techniques Third Edition. 30 pp.
ISO 4873-91. Information Technology—ISO 8-Bit Code for Information Interchange—Structure and Rules for Implementation Third Edition. 25 pp.
JTC1 2022-86. Information Processing—ISO 7-Bit and 8-Bit Coded Character Sets—Code Extension Techniques Third Edition. 30 pp.
JTC1 4873-91. Information Technology—ISO 8-Bit Code for Information Interchange—Structure and Rules for Implementation Third Edition. 25 pp.
OSI ISO 4873-86. Information Processing-ISO 8-Bit Code for Information Interchange-Structure and Rules for Implementation. 20 pp.

INTERNATIONAL AND NON-U.S. NATIONAL STANDARDS
SUBJECT INDEX
Coded

Coded Character Sets *(Cont.)*

— 8-Bit *(Cont.)*

OSI ISO/IEC 4873-91. Information Technology—ISO 8-bit Code for Information Interchange—Structure and Rules for Implementation. 25 pp.
OSI ISO 6429-88. Information Processing—Control Functions for 7-Bit and 8-Bit Coded Character Sets. 64 pp.
SAA AS 1953-87. Information Processing—ISO 7-Bit and 8-Bit Coded Character Sets—Code Extension Techniques. 25 pp.

— 8-Bit—Character Structure

BSI BS 4505: Part 2-90. 1990 Amd 1 Digital Data Transmission Part 2: Character Structure for Start/Stop and Synchronous Transmission. 7 pp.
CNS C5106-89. Information Processing — Character Structure for Start/Stop and Synchronous Transmission (Mar)(7222).
ISO 1177-85. Information Processing—Character Structure for Start/Stop and Synchronous Character Oriented Transmission Second Edition. 4 pp.
JTC1 1177-85. Information Processing—Character Structure for Start/Stop and Synchronous Character Oriented Transmission Second Edition. 4 pp.
OSI ISO 1177-85. Information Processing—Character Structure for Start/Stop and Synchronous Character Oriented Transmission. 4 pp.

— 8-Bit—Code Extensions

BSI BS 6856-87. 1987 Code Extension Techniques for United Kingdom 7-Bit and 8-Bit Coded Character Sets. 28 pp.
ISO 2022-86. Information Processing—ISO 7-Bit and 8-Bit Coded Character Sets—Code Extension Techniques Third Edition. 30 pp.
JTC1 2022-86. Information Processing—ISO 7-Bit and 8-Bit Coded Character Sets—Code Extension Techniques Third Edition. 30 pp.
OSI ISO 2022-86. Information Processing-ISO 7-Bit and 8-Bit Coded Character Sets-Code Extension Techniques. 28 pp.
OSI ISO/IEC DIS 6429-91. Information Technology—Control Functions for Coded Character Sets. 110 pp.

— 8-Bit—Control Functions

BSI BS 7204-89. (WITHDRAWN) 1989 Control Functions for ISO 7-Bit and 8-Bit Coded Character Sets (Superseded by BS ISO/IEC 6429: 1992). 55 pp.
BSI BS ISO/IEC 6429-92. 1992 Information Technology—Control Functions for Coded Character Sets (S). 100 pp.
ECMA ECMA 48-91. Control Functions for Coded Character Sets. 108 pp.
IEC 6429-92. Information Technology—Control Functions for Coded Character Sets Third Edition. 98 pp.
ISO 6429-92. Information Technology—Control Functions for Coded Character Sets Third Edition. 98 pp.
JIS X 0211-91. Information Processing—Control Functions for 7-Bit and 8-Bit Coded Character Sets.
JTC1 6429-92. Information Technology—Control Functions for Coded Character Sets Third Edition. 98 pp.
OSI ISO 6429-88. Information Processing—Control Functions for 7-Bit and 8-Bit Coded Character Sets. 64 pp.
OSI ISO/IEC DIS 6429-91. Information Technology—Control Functions for Coded Character Sets. 110 pp.
OSI ISO DP 9429-86. ISO/DP 6429—ISO 7-Bit and 8-Bit Coded Character Sets—Control Functions.

— 8-Bit—Graphic Characters

BSI BS 7203: Part 1-89. 1989 8-Bit Single-Byte Coded Graphic Character Sets: Part 1: Latin Alphabet No. 1. 11 pp.
BSI BS 7203: Part 2-89. 1989 8-Bit Single-Byte Coded Graphic Character Sets: Part 2: Latin Alphabet No. 2 (S). 10 pp.
BSI BS ISO/IEC 8859-10-92. 1992 Information Technology—8-Bit Single-Byte Coded Graphic Character Sets—Part 10: Latin Alphabet No. 6 (S). 22 pp.
BSI BS ISO/IEC 10367-91. 1991 Information Technology—Standardized Coded Graphic Character Sets for Use in 8-Bit Codes. 58 pp.
ECMA ECMA 94-86. 8-Bit Single-Byte Coded Graphic Character Sets—Latin Alphabets No.1 to No. 4. 37 pp.
ECMA ECMA 113-88. 8-Bit Single-Byte Coded Graphic Character Sets, Latin/Cyrillic Alphabet. 15 pp.
ECMA ECMA 114-86. 8-Bit Single Byte Coded Graphic Character Sets, Latin/Arabic Alphabet. 17 pp.
ECMA ECMA 118-86. 8-Bit Single-Byte Coded Graphic Character Sets, Latin/Greek Alphabet. 15 pp.
ECMA ECMA 121-87. 8-Bit Single-Byte Coded Graphic Character Sets. 14 pp.

Coded Character Sets *(Cont.)*

— 8-Bit—Graphic Characters *(Cont.)*

ECMA ECMA 128-88. 8-Bit Single Byte Coded Graphic Character Sets Latin Alphabet No.5. 15 pp.
ECMA ECMA 144-92. 8-Bit Single Byte Coded Graphic Character Set Latin Alphabet No. 6. 16 pp.
ECMA ECMA 144-90. 8-Bit Single-Byte Coded Graphic Character Sets—Latin Alphabet No. 6. 16 pp.
IEC 8859 Pt 5-88. Information Processing—8-Bit Single-Byte Coded Graphic Character Sets—Part 5: Latin/Cyrillic Alphabet First Edition. 10 pp.
IEC 8859 Pt 9-89. Information Processing—8-Bit Single-Byte Coded Graphic Character Sets—Part 9: Latin Alphabet No. 5 First Edition. 9 pp.
IEC 8859 Pt 10-92. Information Technology—8-Bit Single-Byte Coded Graphic Character Sets—Part 10: Latin Alphabet No. 6 First Edition. 19 pp.
IEC 10367-91. Information Technology—Standardized Coded Graphic Character Sets for Use in 8-Bit Codes First Edition. 55 pp.
ISO 8859 Pt 1-87. Information Processing—8-Bit Single-Byte Coded Graphic Character Sets—Part 1: Latin Alphabet No. 1 First Edition. 11 pp.
ISO 8859 Pt 2-87. Information Processing—8-Bit Single-Byte Coded Graphic Character Sets—Part 2: Latin Alphabet No. 2 First Edition. 10 pp.
ISO 8859 Pt 3-88. Information Processing—8-Bit Single-Byte Coded Graphic Character Sets Part 3: Latin Alphabet No. 3 First Edition. 10 pp.
ISO 8859 Pt 4-88. Information Processing—8-Bit Single-Byte Coded Graphic Character Sets Part 4: Latin Alphabet No. 4 First Edition. 10 pp.
ISO 8859 Pt 5-88. Information Processing—8-Bit Single-Byte Coded Graphic Character Sets—Part 5: Latin/Cyrillic Alphabet First Edition. 10 pp.
ISO 8859 Pt 6-87. Information Processing—8-Bit Single-Byte Coded Graphic Character Sets—Part 6: Latin/Arabic Alphabet First Edition. 9 pp.
ISO 8859 Pt 7-87. Information Processing—8-Bit Single-Byte Coded Graphic Character Sets—Part 7: Latin/Greek Alphabet First Edition. 9 pp.
ISO 8859 Pt 8-88. Information Processing—8-Bit Single-Byte Coded Graphic Character Sets—Part 8: Latin/Hebrew Alphabet First Edition. 10 pp.
ISO 8859 Pt 9-89. Information Processing—8-Bit Single-Byte Coded Graphic Character Sets—Part 9: Latin Alphabet No. 5 First Edition. 9 pp.
ISO 8859 Pt 10-92. Information Technology—8-Bit Single-Byte Coded Graphic Character Sets—Part 10: Latin Alphabet No. 6 First Edition. 19 pp.
ISO 10367-91. Information Technology—Standardized Coded Graphic Character Sets for Use in 8-Bit Codes First Edition. 55 pp.
JTC1 8859 Pt 1-87. Information Processing—8-Bit Single-Byte Coded Graphic Character Sets—Part 1: Latin Alphabet No. 1 First Edition. 11 pp.
JTC1 8859 Pt 2-87. Information Processing—8-Bit Single-Byte Coded Graphic Character Sets—Part 2: Latin Alphabet No. 2 First Edition. 10 pp.
JTC1 8859 Pt 3-88. Information Processing—8-Bit Single-Byte Coded Graphic Character Sets Part 3: Latin Alphabet No. 3 First Edition. 10 pp.
JTC1 8859 Pt 4-88. Information Processing—8-Bit Single-Byte Coded Graphic Character Sets Part 4: Latin Alphabet No. 4 First Edition. 10 pp.
JTC1 8859-88. Information Processing—8-Bit Single-Byte Coded Graphic Character Sets—Part 5: Latin/Cyrillic Alphabet First Edition. 10 pp.
JTC1 8859 Pt 6-87. Information Processing—8-Bit Single-Byte Coded Graphic Character Sets—Part 6: Latin/Arabic Alphabet First Edition. 9 pp.
JTC1 8859 Pt 7-87. Information Processing—8-Bit Single-Byte Coded Graphic Character Sets—Part 7: Latin/Greek Alphabet First Edition. 9 pp.
JTC1 8859 Pt 8-88. Information Processing—8-Bit Single-Byte Coded Graphic Character Sets—Part 8: Latin/Hebrew Alphabet First Edition. 10 pp.
JTC1 8859-89. Information Processing—8-Bit Single-Byte Coded Graphic Character Sets—Part 9: Latin Alphabet No. 5 First Edition. 9 pp.
JTC1 8859-92. Information Technology—8-Bit Single-Byte Coded Graphic Character Sets—Part 10: Latin Alphabet No. 6 First Edition. 19 pp.
JTC1 10367-91. Information Technology—Standardized Coded Graphic Character Sets for Use in 8-Bit Codes First Edition. 55 pp.
OSI ISO/IEC DIS 7350.2. Information Technology—Registration of Subrepertoires of the Graphic Character Repertoire of ISO 10367. 23 pp.
OSI ISO 8859-1-87. Information Processing—8-Bit Single-Byte Coded Graphic Character Sets—Part 1: Latin Alphabet No. 1. 11 pp.
OSI ISO DP 8859-2-87. Information Processing—8-Bit Single-Byte Coded Graphic Character Sets—Part 2: Latin Alphabet No. 2. 10 pp.
OSI ISO 8859-3-88. Information Processing—8-Bit Single-Byte Coded Graphic Character Sets—Part 3: Latin Alphabet No. 3. 10 pp.
OSI ISO DP 8859-3-86. Information Processing—8-Bit Single Byte Coded Graphic Character Sets—Part 3: Latin Alphabet No. 3. 14 pp.

Coded Character Sets *(Cont.)*

— 8-Bit—Graphic Characters *(Cont.)*

OSI ISO 8859-4-88. Information Processing—8-Bit Single-Byte Coded Graphic Character Sets—Part 4: Latin Alphabet No. 4. 10 pp.
OSI ISO DP 8859-4-86. Information Processing—8-Bit Single Byte Coded Graphic Character Sets—Part 4: Latin Alphabet No. 4. 14 pp.
OSI ISO IEC 8859-5-88. Information Processing—8-Bit Single-Byte Coded Graphic Character Sets—Part 5: Latin/Cyrillic Alphabet. 10 pp.
OSI ISO DIS 8859-5-86. Information Processing—8-Bit Single-Byte Coded Graphic Character Sets—Part 5: Latin/Cyrillic Alphabet. 18 pp.
OSI ISO 8859-6-87. Information Processing—8-Bit Single-Byte Coded Graphic Character Sets—Part 6: Latin/Arabic Alphabet. 9 pp.
OSI ISO DIS 8859-7-87. Information Processing—8-Bit Single-Byte Coded Graphic Character Sets—Part 7: Latin/Greek Alphabet. 9 pp.
OSI ISO 8859-8-88. Information Processing—8-Bit Single-Byte Coded Graphic Character Sets—Part 8: Latin/Hebrew Alphabet. 10 pp.
OSI ISO DIS 8859-8-87. Information Processing—8-Bit Single-Byte Coded Graphic Character Sets—Part 8: Latin/Hebrew Alphabet. 16 pp.
OSI ISO 8859 2-87. Information Processing-8-Bit Single-Byte Coded Graphic Character Sets-Part 2: Latin Alphabet No. 2.
OSI ISO DP 8859 3-86. Information Processing-8-Bit Single Byte Coded Graphic Character Sets-Part 3: Latin Alphabet No. 3.
OSI ISO DP 8859 4-86. Information Processing-8-Bit Single Byte Coded Graphic Character Sets-Part 4: Latin Alphabet No. 4.
OSI ISO DIS 8859 5-86. Information Processing-8-Bit Single-Byte Coded Graphic Characters Sets-Part 5: Latin/Cyrillic Alphabet.
OSI ISO 8859 7-87. Information Processing-8-Bit Single-Byte Coded Graphic Character Sets-Part 7: Latin/Greek Alphabet.
OSI ISO DIS 8859 8-87. Information Processing-8-Bit Single-Byte Coded Graphic Character Sets-Part 8: Latin/Hebrew Alphabet.
OSI ISO/IEC DIS 8859-10-91. Information Technology—8-Bit Single-Byte Coded Graphic Character Sets—Part 10: Latin Alphabet No.6. 18 pp.
OSI ISO IEC DIS 10367-90. Information Processing—Repertoire of Standardized Coded Graphic Character Sets for Use in 8-Bit Codes. 55 pp.

— 8-Bit—Magnetic Tapes

CNS C5149-83. Data Processing Implementation of 7-Bit and 8-Bit Coded Character Sets on Punched Cards (Feb)(9982).
ISO 962-74. Information Processing—Implementation of the 7-Bit Coded Character Set and Its 7-Bit and 8-Bit Extensions on 9-Track 12,7 mm (0.5 in) Magnetic Tape First Edition. 5 pp.
ISO 3275-74. Information Processing—Implementation of the 7-Bit Coded Character Set and Its 7-Bit and 8-Bit Extensions on 3,81 mm Magnetic Tape Cassette for Data Interchange First Edition. 3 pp.
JTC1 962-74. Information Processing—Implementation of the 7-Bit Coded Character Set and Its 7-Bit and 8-Bit Extensions on 9-Track 12,7 mm (0.5 in) Magnetic Tape First Edition. 5 pp.
JTC1 3275-74. Information Processing—Implementation of the 7-Bit Coded Character Set and Its 7-Bit and 8-Bit Extensions on 3,81 mm Magnetic Tape Cassette for Data Interchange First Edition. 3 pp.
OSI ISO 962-74. Information Processing—Implementation of the 7-Bit Coded Character Set and Its 7-Bit and 8-Bit Extensions on 9-Track 12, 7mm (0.5 in) Magnetic Tape. 5 pp.
OSI ISO 3275-74. Information Processing—Implementation of the 7-bit Coded Character Set and Its 7-bit and 8-bit Extension on 3.81mm Magnetic Tape Cassette for Data Interchange. 3 pp.

— 8-Bit—Pictures

BSI BS ISO 9282-1-88. 1988 Information Processing—Coded Representation of Pictures—Part 1: Encoding Principles for Picture Representation in a 7-Bit or 8-Bit Environment (S). 31 pp.
BSI BS ISO/IEC 9282-2-92. 1992 Information Processing—Coded Representation of Pictures—Part 2: Incremental Encoding of Point Lists in a 7-Bit or 8-Bit Environment. 31 pp.
IEC 9282 Pt 2-92. Information Processing—Coded Representation of Pictures—Part 2: Incremental Encoding of Point Lists in a 7-Bit or 8-Bit Environment First Edition. 28 pp.
ISO 9282 Pt 1-88. Information Processing—Coded Representation of Pictures—Part 1: Encoding Principles for Picture Representation in a 7-Bit or 8-Bit Environment First Edition. 29 pp.
ISO 9282 Pt 2-92. Information Processing—Coded Representation of Pictures—Part 2: Incremental Encoding of Point Lists in a 7-Bit or 8-Bit Environment First Edition. 28 pp.

INDUSTRY STANDARDS
531

Coded Character Sets *(Cont.)*
— 8-Bit—Pictures *(Cont.)*
JTC1 9282 Pt 1-88. Information Processing—Coded Representation of Pictures—Part 1: Encoding Principles for Picture Representation in a 7-Bit or 8-Bit Environment First Edition. 29 pp.

JTC1 9282-92. Information Processing—Coded Representation of Pictures—Part 2: Incremental Encoding of Point Lists in a 7-Bit or 8-Bit Environment First Edition. 28 pp.

OSI ISO/IEC 9281-2-90. Information Technology—Picture Coding Methods—Part 2: Procedure for Registration. 7 pp.

OSI ISO DIS 9282-1-87. Information Processing—Coded Representation of Pictures—Part 1: Encoding Principles for Picture Representation in a 7 or 8 Bit Environment. 34 pp.

OSI ISO/IEC DIS 9282-2-90. Information Technology—Coded Representation of Pictures—Part 2: Incremental Encoding of Point Lists in a 7-Bit Environment. 28 pp.

OSI ISO DIS 9282-2-90. Information Technology—Coded Representation of Pictures—Part 2: Incremental Encoding of Point Lists in a 7-Bit or 8-bit Environment. 28 pp.

SNZ NZS/ISO 9282. 1-88. Computer Graphics—Coded Representation of Pictures Part 1: Encoding Principles for Picture Representation in a 7-Bit or 8-Bit Environment. 23 pp.

— 8-Bit—Punched Cards
BSI BS 4636: Part 3-86. 1986 Punched Cards Part 3: Representation of 7-Bit and 8-Bit Coded Character Sets on 12-Row Punched Cards. 8 pp.

ECMA ECMA 44-75. Implementation of the ECMA 7-Bit and 8-Bit Coded Character Sets on Punched Cards. 12 pp.

ISO 6586-80. Data Processing—Implementation of the ISO 7-Bit and 8-Bit Coded Character Sets on Punched Cards First Edition. 7 pp.

JTC1 6586-80. Data Processing—Implementation of the ISO 7-Bit and 8-Bit Coded Character Sets on Punched Cards First Edition. 7 pp.

OSI ISO 6586-80. Data Processing—Implementation of the ISO 7-Bit and 8-Bit Coded Character Sets on Punched Cards. 7 pp.

—Bibliographic—Graphic Characters
OSI ISO DIS 6862.2-90. Documentation—Mathematical Coded Character Set for Bibliographic Information Interchange. 18 pp.

—Bit Sequencing
CNS C5160-89. Information Interchange — the Bit Sequencing for Serial-by-Bit Data Transmission (Jul)(10331).

—Character Structure
CNS C5161-89. Information Interchange — Character Structure and Character Parity Sense for Serial-by-Bit Data Communication (Jul)(10426).

CNS C5162-89. Information Interchange — Character Structure and Character Parity Sense for Parallel-by-Bit Data Communication (Jul)(10427).

—Control Functions
OSI ISO IEC DIS 10538-90. Information Technology—Control Functions for Text Communication. 64 pp.

—Graphic Characters
BSI BS 6474: Part 1-84. 1984 Coded Character Sets for Bibliographic Information Interchange Part 1: Specification for Extension of the Latin Alphabet Coded Character Set. 9 pp.

BSI BS 6474: Part 2-85. 1985 Coded Character Sets for Bibliographic Information Interchange Part 2: Greek Alphabet Coded Character Set. 8 pp.

BSI BS 6474: Part 4-86. 1986 Coded Character Sets for Bibliographic Information Interchange Part 4: Extension of the Cyrillic Alphabet Coded Character Set. 7 pp.

BSI BS 6692: Part 2-90. 1990 Coded Character Sets for Text Communication Part 2: Latin Alphabetic and Non-Alphabetic Graphic Characters. 48 pp.

BSI BS ISO/IEC 7350-91. 1991 Information Technology—Registration of Repertoires of Graphic Characters from ISO/IEC 10367. 19 pp.

BSI DD ENV 41501-91. 1991 Information Systems Interconnection—Graphic Character Repertoire for Information Received from or Transmitted to CEPT Videotex Services or Private Videotex Systems. 14 pp.

BSI DD ENV 41502-91. 1991 Information Systems Interconnection—Character Repertoires for Information Received from or Transmitted to the Teletex Service or Private Information Processing Systems Using Teletex Technology. 12 pp.

BSI DD ENV 41503-91. 1991 Information Systems Interconnection—European Character Repertoires and Their Coding (Incorporating Corrigendum June 1991). 15 pp.

BSI DD ENV 41504-91. 1991 Information Systems Interconnection—Character Repertoire and Coding for Interworking with Telex Services (Incorporating Corrigendum June 1991). 14 pp.

BSI DD ENV 41505-91. 1991 Information Systems Interconnection—Graphic Character Repertoire and Coding for Line Drawing. 9 pp.

BSI DD ENV 41506-91. 1991 Information Systems Interconnection—Data Stream Formats for Information Received from or Transmitted to the Teletex Service or Private Information Processing Systems Using Teletex Technology. 11 pp.

BSI DD ENV 41507-91. 1991 Information Systems Interconnection—Data String Formats for Information Received from or Transmitted to CEPT Videotex Services or Private Videotex Systems. 11 pp.

BSI DD ENV 41508-91. 1991 Information Systems Interconnection—East European Graphic Character Repertoires and Their Coding (Incorporating Corrigendum June 1991). 13 pp.

CEN ENV 41 501-87. Information Systems Interconnection: Graphic Chartacter Repertoire and Coding for Interworking with CEPT Videotex Services. 7 pp.

CEN ENV 41501-91. Information Systems Interconnection—Graphic Character Repertoire for Information Received from or Transmitted to CEPT Videotex Services or Private Videotex Systems. 12 pp.

CEN ENV 41 502-87. Information Systems Interconnection: Graphic Character Repertoire and Coding for Interworking with CEPT Teletex Services. 7 pp.

CEN ENV 41502-91. Information Systems Interconnection Character Repertoires for Information Received from or Transmitted to the Teletex Service or Private Information Processing Systems Using Teletex Technology. 10 pp.

CEN ENV 41 502/A1-92. AMD 1 Information Systems Interconnection—Character Repertoires for Information Received from or Transmitted to the Teletex Service or Private Information Processing Systems Using Teletex Technology. 5 pp.

CEN ENV 41 503-87. Information Systems Interconnection: European Graphic Character Repertoires and Their Coding. 9 pp.

CEN ENV 41503-90. Information Systems Interconnection—European Character Repertoires and Their Coding. 14 pp.

CEN ENV 41504-90. Information Systems Interconnection—Character Repertoire and Coding for Interworking with Telex Services. 12 pp.

CEN PRENV 41505-91. Information Systems Interconnection—Graphic Character Repertoire and Coding for Line Drawing. 7 pp.

CEN ENV 41505-91. Information Systems Interconnection—Graphic Character Repertoire and Coding for Line Drawing. 7 pp.

CEN ENV 41507-91. Information Systems Interconnection—Data String Formats for Information Received from or Transmitted to CEPT Videotex Services or Private Videotex Systems. 9 pp.

CEN ENV 41508-90. Information Systems Interconnection—East European Graphic Character Repertoires and Their Coding. 11 pp.

CNS C5209-86. Standard Interchange Code for Generally-Used Chinese Characters (Aug)(11643).

IEC DIS 6937-91. Information Technology—Coded Graphic Character Set for Text Communication—Latin Alphabet ***CD-ROM ONLY***. 50 pp.

IEC 7350-91. Information Technology—Registration of Repertoires of Graphic Characters from ISO/IEC 10367 Second Edition. 17 pp.

ISO 5426-83. Extension of the Latin Alphabet Coded Character Set for Bibliographic Information Interchange Second Edition. 8 pp.

ISO 5427-84. Extension of the Cyrillic Alphabet Coded Character Set for Bibliographic Information Interchange First Edition. 6 pp.

ISO 5428-84. Greek Alphabet Coded Character Set for Bibliographic Information Interchange Second Edition. 7 pp.

ISO 6438-83. Documentation—African Coded Character Set for Bibliographic Information Interchange First Edition. 8 pp.

ISO DIS 6937-91. Information Technology—Coded Graphic Character Set for Text Communication—Latin Alphabet ***CD-ROM ONLY***. 50 pp.

ISO 6937 Pt 2-83. Information Processing—Coded Character Sets for Text Communication—Part 2: Latin Alphabetic and Non-Alphabetic Graphic Characters First Edition; (Addendum 1-1989). 49 pp.

ISO 7350-91. Information Technology—Registration of Repertoires of Graphic Characters from ISO/IEC 10367 Second Edition. 17 pp.

JIS X 0212-90. Code of the Supplementary Japanese Graphic Character Set for Information Interchange. 86 pp.

JTC1 DIS6937-91. Information Technology—Coded Graphic Character Set for Text Communication—Latin Alphabet ***CD-ROM ONLY***. 50 pp.

JTC1 6937 Pt 2-83. Information Processing—Coded Character Sets for Text Communication—Part 2: Latin Alphabetic and Non-Alphabetic Graphic Characters First Edition; (Addendum 1-1989). 49 pp.

JTC1 7350-91. Information Technology—Registration of Repertoires of Graphic Characters from ISO/IEC 10367 Second Edition. 17 pp.

OSI ISO 5426-83. Extension of the Latin Alphabet Coded Character Set for Bibliographic Information Interchange. 8 pp.

OSI ISO 5428-84. Greek Alphabet Coded CharacterSet for Bibliographic Information Interchange. 7 pp.

OSI ISO 6438-83. Documentation—African Coded Character Set for Bibliographic Information Interchange. 8 pp.

OSI ISO 6937-2-83. Information Processing—Coded Character Sets for Text Communication—Part 2: Latin Alphabet and Non-Alphabetic Graphic Characters. 40 pp.

OSI ISO DP 6937-2 DAD 1-86. Information Processing—Coded Character Sets for Text Communication—Part 2: Latin Alphabetic and Non-Alphabetic Graphic Characters. 9 pp.

OSI ISO 6937-2 AD1-89. Information Processing—Coded Character Sets for Text Communication—Part 2: Latin Alphabetic and Non-Alphabetic Graphic Characters. 9 pp.

OSI ISO DIS 6937-3-84. Information Processing—Coded Character Sets for Text Communication. 54 pp.

OSI ISO DIS 6937-8-87. Coded Character Sets for Text Communication—Part 8: Cyrillic Graphic Characters. 13 pp.

OSI ISO 7350-84. Text Communication—Registration of Graphic Character Subrepertoires. 13 pp.

OSI ISO/IEC DIS 10646-90. Information Technology—Universal Coded Character Set (UCS). 162 pp.

—Parity
CNS C5161-89. Information Interchange — Character Structure and Character Parity Sense for Serial-by-Bit Data Communication (Jul)(10426).

CNS C5162-89. Information Interchange — Character Structure and Character Parity Sense for Parallel-by-Bit Data Communication (Jul)(10427).

—Pictures
IEC 9281 Pt 1-90. Information Technology—Picture Coding Methods—Part 1: Identification First Edition. 11 pp.

ISO 9281 Pt 1-90. Information Technology—Picture Coding Methods—Part 1: Identification First Edition. 11 pp.

JTC1 9281-90. Information Technology—Picture Coding Methods—Part 1: Identification First Edition. 11 pp.

OSI ISO DIS 9281-88. Information Processing—Identification of Picture Coding Methods. 13 pp.

SAA AS 3874.1-91. Information Processing —Picture Coding Methods—Part 1: Identification (ISO/IEC 9581.1:1990). 8 pp.

SNZ NZS/ISO-IEC 9281.1-90. Information Processing—Picture Coding Methods Part 1: Identification. 8 pp.

—Pictures—Registration Procedures
BSI BS ISO/IEC 9281/1-90. 1990 Information Technology—Picture Coding Methods—Part 1: Identification (S). 14 pp.

BSI BS ISO/IEC 9281-2-90. 1990 Information Technology—Picture Coding Methods—Part 2: Procedure for Registration. 9 pp.

IEC 9281 Pt 2-90. Information Technology—Picture Coding Methods—Part 2: Procedure for Registration First Edition. 7 pp.

ISO 9281 Pt 2-90. Information Technology—Picture Coding Methods—Part 2: Procedure for Registration First Edition. 7 pp.

JTC1 9281-90. Information Technology—Picture Coding Methods—Part 2: Procedure for Registration First Edition. 7 pp.

—SGML
BSI BS ISO/IEC TR 9573-13-91. 1991 Information Technology—SGML Support Facilities—Techniques for Using SGML—Part 13: Public Entity Sets for Mathematics and Science. 91 pp.

IEC TR9573 Pt 13-91. Information Technology—SGML Support Facilities—Techniques for Using SGML—Part 13: Public Entity Sets for Mathematics and Science First Edition. 89 pp.

ISO TR9573 Pt 13-91. Information Technology—SGML Support Facilities—Techniques for Using SGML—Part 13: Public Entity Sets for Mathematics and Science First Edition; (Replaces ISO 8879 Annex D (in part)). 89 pp.

INTERNATIONAL AND NON-U.S. NATIONAL STANDARDS
SUBJECT INDEX

Coded Character Sets (Cont.)
—SGML (Cont.)
OSI ISO IEC TR 9573-13-91. Information Technology—SGML Support Facilities—for Using SGML—Part 13: Public Entity Sets for Mathematicsand Science.

—Source Programs
ECMA ECMA 53-78. Representation of Source Program for Program Interchange-APL, COBOL, FORTRAN, Minimal BASIC and PL/1. 16 pp.

—Universal Multiple-Octet—Architecture
BSI BS ISO/IEC 10646-1-93. 1993 Information Technology—Universal Multiple-Octet Coded Character Set (UCS)—Part 1: Architecture and Basic Multilingual Plane (S). 762 pp.
BSI PD 6581-93. 1993 Information Technology—Universal Multiple-Octet Coded Character Set (UCS) Part 1: Architecture and Basic Multilingual Plane (S). 212 pp.
IEC 10646 Pt 1-93. Information Technology—Universal Multiple-Octet Coded Character Set (UCS)—Part 1: Architecture and Basic Multilingual Plane First Edition. 761 pp.
ISO 10646 Pt 1-93. Information Technology—Universal Multiple-Octet Coded Character Set (UCS)—Part 1: Architecture and Basic Multilingual Plane First Edition. 761 pp.
JTC1 10646 Pt 1-93. Information Technology—Universal Multiple-Octet Coded Character Set (UCS)—Part 1: Architecture and Basic Multilingual Plane First Edition. 761 pp.

—Universal Multiple-Octet—Basic Multilingual Plane
BSI BS ISO/IEC 10646-1-93. 1993 Information Technology—Universal Multiple-Octet Coded Character Set (UCS)—Part 1: Architecture and Basic Multilingual Plane (S). 762 pp.
BSI PD 6581-93. 1993 Information Technology—Universal Multiple-Octet Coded Character Set (UCS) Part 1: Architecture and Basic Multilingual Plane (S). 212 pp.
IEC 10646 Pt 1-93. Information Technology—Universal Multiple-Octet Coded Character Set (UCS)—Part 1: Architecture and Basic Multilingual Plane First Edition. 761 pp.
ISO 10646 Pt 1-93. Information Technology—Universal Multiple-Octet Coded Character Set (UCS)—Part 1: Architecture and Basic Multilingual Plane First Edition. 761 pp.
JTC1 10646 Pt 1-93. Information Technology—Universal Multiple-Octet Coded Character Set (UCS)—Part 1: Architecture and Basic Multilingual Plane First Edition. 761 pp.

Coder/Decoders
Use: Encoder/Decoders

Codes
Scope Note: For additional listings, use a more specific term Use For: Numerical Codes
See Also: Access Codes; Accounting Authority Identification Codes; Address Codes; Answerback; Bar Codes; Boiler Codes; Building Codes; Coded Character Sets; Codewords; Control Codes; Country Codes; Data Codes; Data Network Identification Codes; Dimensional Codes; DVGW Codes; Electrical Codes; Encoded Data Transmission; Encoded Information Types; End of Address; End of Message Codes; End of Transmission Codes; Fire Codes; Gentex Code; Information Interchange; International Securities Identification Numbering System; International Signaling Point Codes; International Standard Recording Code; International Telegraph Alphabet No. 2; International Telegraph Alphabet No. 5; Mechanical Codes; Message Authentication Codes; Mobile Country Codes; Morse Code; Morse Code Equipment; Numerical Control Codes; Phonetic Alphabet and Figure Code; Provider Codes; Provider Oriented Codes; Signal Codes; Speech Encoding; Start of Message Codes; Subscriber Control Codes; Telex Message Identifiers; Telex Network Identification Codes; Telex Service Codes; Time and Address Codes; Time Codes; Uniform Commodity Codes; Vehicle Identification Numbers

—Aircraft—Designations
NATO STANAG 3236 ED 8 AMD 0-91. Designation System for Aircraft and Designation of Role Codes of Aircraft. 10 pp.

—Classification—Military Equipment
NATO STANAG 3150 ED 6 AMD 1-85. Codification of Equipment—Uniform System of Supply Classification. 5 pp.

Codes (Cont.)
—Common User Item Lists—Military Equipment
NATO STANAG 2386 ED 1 AMD 0-90. (DRAFT) Codification of Equipment—Establishing Common User Item List and Matrices. 22 pp.
NATO STANAG 2386 ED 1 AMD 0-92. Codification of Equipment—Establishing Common User Item Lists and Matrices. 6 pp.

—Data Acquisition Systems
NATO STANAG 4177 ED 1 AMD 1-86. Codification of Items of Supply—Uniform System of Data Acquisition. 10 pp.
NATO STANAG 4177 ED 2 AMD 0-90. Codification of Items of Supply—Uniform System of Data Acquisition. 8 pp.

—Identification
CSA CAN/CSA-Z234.4-89. All-Numeric Dates and Times; (Gen Instr 1). 19 pp.

—Identification—Accounting
JIS X 0406-84. Accounts Code.

—Identification—Boats (Marine)
BSI BS 7490-91. 1991 Coding System for Hull Identification of Small Craft (ISO 10087: 1990). 7 pp.
ISO 10087-90. Small Craft—Hull Identification—Coding System First Edition. 4 pp.

—Identification—Commodities
JIS X 0405-77. Commodity Classification Code.

—Identification—Currencies
BSI BS 5716-81. (WITHDRAWN) 1981 Codes for the Representation of Currencies and Funds (Superseded by ISO 4217: 1990 (Ed 4) Which Will Be Dual-Numbered as BS 7095). 47 pp.
CGSB CAN/CGSB-200.6-91. Codes for the Representation of Currencies and Funds (ISO 4217:1990); (Amendment 1 September 1992) (Amendment 2 October 1992) (Amendment 3 December 1992) (Amendment 4 December 1992) (Amendment 5 December 1992) (Amendment 6 April 1993). 54 pp.
ISO 4217-90. Codes for the Representation of Currencies and Funds Fourth Edition. 36 pp.

—Identification—Educational Institutions
JIS X 0408-89. Identification Code for Universities and Colleges.

—Identification—Geographical Locations
JIS X 0401-73. To-Do-Fu-Ken (Prefecture) Identification Code.
JIS X 0402-88. Identification Code for Cities, Towns and Villages.

—Identification—Geographical Types
JIS X 0411-76. Code for Categories of Land.

—Identification—Grid Squares
JIS X 0410-76. Grid Square Code.

—Identification—Industries
JIS X 0403-86. Industry Classification Code.

—Identification—Languages
ISO 639-88. Code for the Representation of Names of Languages First Edition. 21 pp.
ISO 3166-88. Code for the Representation of Names of Countries Third Edition; (Revision and Redesignation of ANSI Z39.27-84). 59 pp.
OSI ISO 3166-88. Codes for the Representation of Names of Countries. 59 pp.

—Identification—Military Equipment
NATO STANAG 3151 ED 6 AMD 1-85. Codification of Equipment—Uniform System of Item Identification. 6 pp.
NATO STANAG 3151 ED 7 AMD 0-90. Codification of Equipment—Uniform System of Item Identification. 7 pp.
NATO STANAG 4199 ED 2 AMD 0-91. Codification of Equipment—Uniform System of Exchange of Material Management Data. 6 pp.

—Identification—Occupations
JIS X 0404-87. Occupation Classification Code.

—Identification—Trade Documents
CGSB CAN/CGSB-200.8-87. Location of Codes in Trade Documents (ISO 8440-1986). 12 pp.
ISO 8440-86. Location of Codes in Trade Documents First Edition. 5 pp.
JTC1 8440-86. Location of Codes in Trade Documents First Edition. 5 pp.

—Radio Paging Systems
CCIR RECMN 584-1-86. Standard Codes and Formats for International Radio Paging—Section 8A—Land Mobile Service and Related Subjects. 5 pp.

Codes (Cont.)
—Radio Paging Systems (Cont.)
CCIR Report 900-2-90. Radio-Paging Systems Standardization of Code and Format—Section 8A—Land Mobile Service and Related Subjects. 22 pp.

—Telecommunication Services—Books
CCITT RECMN F.92-89. Service Codes—Telegraph and Mobile Services Operations and Quality of Service (Study Group I) 3 pp. 3 pp.

—Warsaw Pact Aircraft—Designations
NATO STANAG 3236 ED 7 AMD 0-88. Designation System for Warsaw Pact Aircraft and Guided Missiles and Designation of Role Codes of Aircraft. 10 pp.

—Warsaw Pact Guided Missiles—Designations
NATO STANAG 3236 ED 7 AMD 0-88. Designation System for Warsaw Pact Aircraft and Guided Missiles and Designation of Role Codes of Aircraft. 10 pp.

Codewords
See Also: Codes

—Maritime
NATO STANAG 1401 ED 1 AMD 0-91. Allied Maritime Codewords—APP-7. 6 pp.

Coding (Data Conversion)
Use: Encoding

Codirectional Interfaces
See Also: Communication Interfaces

—Bit Rates—Digital Hierarchy—Electrical Characteristics
CCITT RECMN G.703-91. Physical/Electrical Characteristics of Hierarchical Digital Interfaces (Study Group XVIII) 42 pp. 42 pp.
CCITT RECMN G.703-89. Physical/Electrical Characteristics of Hierarchical Digital Interfaces—General Aspects of Digital Transmission Systems; Terminal Equipments (Study Groups XV and XVIII) 31 pp. 31 pp.

—Bit Rates—Digital Hierarchy—Physical Characteristics
CCITT RECMN G.703-91. Physical/Electrical Characteristics of Hierarchical Digital Interfaces (Study Group XVIII) 42 pp. 42 pp.
CCITT RECMN G.703-89. Physical/Electrical Characteristics of Hierarchical Digital Interfaces—General Aspects of Digital Transmission Systems; Terminal Equipments (Study Groups XV and XVIII) 31 pp. 31 pp.

Coefficient of Discharge
Use: Discharge Coefficient

Coefficient of Thermal Expansion
Use: Thermal Expansion

Coercive Force
Use For: Coercivity

—Hard Metals
BSI BS 5600: Sec 4.9-79. (WITHDRAWN) 1979 Powder Metallurgical Materials and Products Part 4: Methods of Testing and Chemical Analysis of Hardmetals Section 4.9: Determination of (the Magnetization) Coercivity (Superseded by BS EN 23326: 1993). 4 pp.
BSI BS EN 23326-93. 1993 Hardmetals—Determination of (the Magnetization) Coercivity (ISO 3326: 1975) (V). 9 pp.
CEN EN 23326-93. Hardmetals—Determination of (the Magnetization) Coercivity (ISO 3326: 1975). 4 pp.
ISO 3326-75. Hardmetals—Determination of (the Magnetization) Coercivity First Edition (CEN EN 23326: 1993). 4 pp.

—Magnetic Materials
BSI BS 6404: Part 7-86. 1986 Magnetic Materials Part 7: Methods of Measurement of the Coercuity of Magnetic Materials in an Open Magnetic Circuit. 12 pp.
IEC 404 Pt 7-82. Magnetic Materials Part 7: Method of Measurement of the Coercivity of Magnetic Materials in an Open Magnetic Circuit First Edition. 20 pp.

Coercivity
Use: Coercive Force

Coffee
Use For: Coffee Products; Green Coffee
See Also: Beverages; Tea

INDUSTRY STANDARDS

Coffee (Cont.)
BSI BS 7601-92. 1992 Guide to Specifying Green Coffee (ISO 9116: 1992). 7 pp.
ISO 9116-92. Green Coffee—Guidance on Methods of Specification First Edition. 6 pp.

—Caffeine Content
BSI BS 5752: Part 3-83. 1983 Coffee and Coffee Products Part 3: Coffee: Determination of Caffein Content (Reference Method). 9 pp.
ISO 4052-83. Coffee—Determination of Caffeine Content (Reference Method) First Edition. 8 pp.

—Caffeine Content—Liquid Chromatography
BSI BS 5752: Part 12-92. 1992 Methods of Test for Coffee and Coffee Products Part 12: Coffee: Determination of Caffeine Content (Routine Method by HPLC) (ISO 10095: 1992). 12 pp.
ISO 10095-92. Coffee—Determination of Caffeine Content—Method Using High-Performance Liquid Chromatography First Edition. 9 pp.

—Decaffeinated—Caffeine Content—Liquid Chromatography
BSI BS 5752: Part 12-92. 1992 Methods of Test for Coffee and Coffee Products Part 12: Coffee: Determination of Caffeine Content (Routine Method by HPLC) (ISO 10095: 1992). 12 pp.
ISO 10095-92. Coffee—Determination of Caffeine Content—Method Using High-Performance Liquid Chromatography First Edition. 9 pp.

—Defects
BSI BS 7683-93. 1993 Guide to Defects of Green Coffee (ISO 10470: 1993) (W). 29 pp.
ISO 10470-93. Green Coffee—Defect Reference Chart First Edition. 29 pp.

—Extracts
EC 77/436/EEC-77. Council Directive on the Approximation of the Laws of the Member States Relating to Coffee Extracts and Chicory Extracts. 5 pp.
EC 85/573/EEC-85. Council Directive Amending Directive 77/436/EEC on the Approximation of the Laws of the Member States Relating to Coffee Extracts and Chicory Extracts. 2 pp.

—Extracts—Caffeine Content
EC 79/1066/EEC-79. First Commission Directive Laying Down Community Methods of Analysis for Testing Coffee Extracts and Chicory Extracts. 12 pp.

—Extracts—Dry Matter Content
EC 79/1066/EEC-79. First Commission Directive Laying Down Community Methods of Analysis for Testing Coffee Extracts and Chicory Extracts. 12 pp.

—Glossaries
BSI BS 5456-89. 1989 Glossary of Terms Relating to Coffee and Its Products. 15 pp.
CNS N1097-81. Coffee and Its Products—Vocabulary (Nov)(8151). 5 pp.
ISO 3509-89. Coffee and Its Products—Vocabulary Third Edition. 18 pp.

—Ground
CGSB 32-GP-110M-83. Coffee, Roasted and Ground, Standard for. 9 pp.

—Identification Systems
BSI BS 7601-92. 1992 Guide to Specifying Green Coffee (ISO 9116: 1992). 7 pp.
ISO 9116-92. Green Coffee—Guidance on Methods of Specification First Edition. 6 pp.

—Insect Damage
BSI BS 5752: Part 8-86. 1986 Coffee and Coffee Products Part 8: Green Coffee: Determination of Proportion of Insect-Damaged Beans. 25 pp.
ISO 6667-85. Green Coffee—Determination of Proportion of Insect-Damaged Beans First Edition. 14 pp.

—Inspection
BSI BS 5752: Part 4-80. 1980 Coffee and Coffee Products Part 4: Green Coffee: Olfactory and Visual Examination and Determination of Foreign Matter and Defects. 5 pp.
ISO 4149-80. Green Coffee—Olfactory and Visual Examination and Determination of Foreign Matter and Defects First Edition. 4 pp.

—Instant
CGSB 32.113M-91. Instant Coffee. 7 pp.
CNS N5204-85. Coffee, Instant (Jun)(10029). 1 p.

—Instant—Bulk Density
BSI BS 5752: Part 11-87. 1987 Coffee and Coffee Products Part 11: Instant Coffee: Determination of Free-Flow and Compacted Bulk Densities. 9 pp.
ISO 8460-87. Instant Coffee—Determination of Free-Flow and Compacted Bulk Densities First Edition. 9 pp.

—Instant—Caffeine Content
CNS N6180-83. Method of Test for Beverage: Determination of Methanol Content (May)(10291). 1 p.

—Instant—Insoluble Matter Content
BSI BS 5752: Part 9-86. 1986 Coffee and Coffee Products Part 9: Instant Coffee: Determination of Insoluble Matter Content. 6 pp.
CNS N6180-83. Method of Test for Beverage: Determination of Methanol Content (May)(10291). 1 p.
CNS N6199-85. Instant Coffee—Determination of Insoluble Matter (Jun)(11289). 3 pp.
ISO 7534-85. Instant Coffee—Determination of Insoluble Matter Content First Edition. 5 pp.

—Instant—Loss of Mass
BSI BS 5752: Part 6-83. 1983 Coffee and Coffee Products Part 6: Instant Coffee: Determination of Loss in Mass at 70 Degrees Celsius Under Reduced Pressure. 4 pp.
ISO 3726-83. Instant Coffee—Determination of Loss in Mass at 70 Degrees Celsius Under Reduced Pressure First Edition. 4 pp.

—Instant—Moisture Content
CNS N6179-85. Method of Test for Instant Coffee (Jun)(10030). 1 p.

—Instant—Sampling
BSI BS 6379: Part 2-84. 1984 Sampling of Coffee and Coffee Products Part 2: Methods of Sampling Instant Coffee in Cases with Liners. 8 pp.
ISO 6670-83. Instant Coffee in Cases with Liners—Sampling First Edition. 8 pp.

—Instant—Sieve Analysis
BSI BS 5752: Part 10-86. 1986 Coffee and Coffee Products Part 10: Instant Coffee: Size Analysis. 4 pp.
ISO 7532-85. Instant Coffee—Size Analysis First Edition. 4 pp.

—Loss of Mass—High Temperature Testing
BSI BS 5752: Part 7-84. 1984 Coffee and Coffee Products Part 7: Determination of Loss in Mass at 105 Degrees Celsius. 4 pp.
ISO 6673-83. Green Coffee—Determination of Loss in Mass at 105 Degrees C First Edition. 4 pp.

—Moisture Content
BSI BS 5752: Part 1-79. 1979 Coffee and Coffee Products Part 1: Green Coffee: Determination of Moisture Content (Basic Reference Method). 8 pp.
BSI BS 5752: Part 2-84. 1984 Coffee and Coffee Products Part 2: Green Coffee; Determination of Moisture Content (Routine Method). 4 pp.
CNS N4076-81. Green Coffee-Determination of Moisture Content (Basic Reference Method) (Nov)(8152). 5 pp.
CNS N4077-81. Green Coffee-Determination of Moisture Content (Routine Method) (Nov)(8153). 2 pp.
ISO 1446-78. Green Coffee—Determination of Moisture Content (Basic Reference Method) First Edition. 7 pp.
ISO 1447-78. Green Coffee—Determination of Moisture Content (Routine Method) First Edition. 4 pp.

—Packaging
BSI BS 7601-92. 1992 Guide to Specifying Green Coffee (ISO 9116: 1992). 7 pp.
ISO 9116-92. Green Coffee—Guidance on Methods of Specification First Edition. 6 pp.

—Sampling
BSI BS 6379: Part 1-83. 1983 Sampling of Coffee and Coffee Products Part 1: Method of Sampling Green Coffee in Bags. 5 pp.
BSI BS 6379: Part 3-83. 1983 Sampling of Coffee and Coffee Products Part 3: Coffee Trier for Sampling Green Coffee. 4 pp.
BSI BS 7601-92. 1992 Guide to Specifying Green Coffee (ISO 9116: 1992). 7 pp.
ISO 4072-82. Green Coffee in Bags—Sampling First Edition. 5 pp.
ISO 6666-83. Coffee Triers First Edition. 4 pp.
ISO 9116-92. Green Coffee—Guidance on Methods of Specification First Edition. 6 pp.

—Sensory Analysis—Sampling
BSI BS 6379: Part 4-91. 1991 Sampling of Coffee and Coffee Products Part 4: Method for Preparation of Samples of Green Coffee for Use in Sensory Analysis (ISO 6668: 1991). 7 pp.
ISO 6668-91. Green Coffee—Preparation of Samples for Use in Sensory Analysis First Edition. 6 pp.

—Shipping
BSI BS 6827-87. 1987 Storage and Transport of Green Coffee in Bags. 4 pp.
BSI BS 7601-92. 1992 Guide to Specifying Green Coffee (ISO 9116: 1992). 7 pp.
ISO 8455-86. Green Coffee in Bags—Guide to Storage and Transport First Edition. 5 pp.
ISO 9116-92. Green Coffee—Guidance on Methods of Specification First Edition. 6 pp.

—Sieve Analysis
BSI BS 5752: Part 5-91. 1991 Methods of Test for Coffee and Coffee Products Part 5: Green Coffee: Size Analysis by Manual Sieving (ISO 4150: 1991). 12 pp.
BSI BS 5752: Part 5-81. 1981 Coffee and Coffee Products Part 5: Green Coffee: Size Analysis by Manual Sieving (W). 7 pp.
ISO 4150-91. Green Coffee—Size Analysis—Manual Sieving Second Edition. 9 pp.

—Storage
BSI BS 6827-87. 1987 Storage and Transport of Green Coffee in Bags. 4 pp.
BSI BS 7601-92. 1992 Guide to Specifying Green Coffee (ISO 9116: 1992). 7 pp.
ISO 8455-86. Green Coffee in Bags—Guide to Storage and Transport First Edition. 5 pp.
ISO 9116-92. Green Coffee—Guidance on Methods of Specification First Edition. 6 pp.

Coffee Beans
Use: Coffee

Coffee Grinders
See Also: Appliances; Food Processing Equipment; Grinders

—Commercial
CSA C22.2 NO 195-M1987. Motor Operated Food Processing Appliances (Household and Commercial); (Gen Instr 1 Thru 2). 55 pp.

—Household
CSA C22.2 NO 195-M1987. Motor Operated Food Processing Appliances (Household and Commercial); (Gen Instr 1 Thru 2). 55 pp.

—Safety
BSI BS 3456: Sec 102.33-88. (WITHDRAWN) 1988 Safety of Household Electrical Appliances Part 102: Particular Requirements Section 102.33: Coffee Mills and Coffee Grinders (Superseded by BS EN 60335-2-33: 1992). 18 pp.
BSI BS EN 60335-2-33-92. 1992 Safety of Household and Similar Electrical Appliances Part 2: Particular Requirements Section 2.33: Coffee Mills and Coffee Grinders. 19 pp.
CENELEC HD 260-75. Particular Specification for Coffee Grinders and Mills. 5 pp.
CENELEC HD 260 S3-87. Safety of Household and Similar Electrical Appliances Part 2: Particular Requirements for Coffee Mills and Coffee Grinders (Superseded by EN 60335-2-33). 8 pp.
CENELEC EN 60 335-2-33-90. Safety of Household and Similar Electrical Appliances Part 2: Particular Requirements for Coffee Mills and Coffee Grinders (Supersedes HD 260 S3). 6 pp.
DIN VDE 0700 Pt 33 A2 (D)-85. Safety of Household and Similar Electrical Appliances; Coffee and Grain Mills: Amendment 2 (Sept). 4 pp.
IEC 335 Pt 2-33-87. Safety of Household and Similar Electrical Appliances Part 2: Particular Requirements for Coffee Mills and Coffee Grinders Second Edition; (Amendment 1-1990). 36 pp.

Coffee Makers
See Also: Appliances; Coffee Urns; Cooking Appliances
BSI BS 3999: Part 8-84. 1984 Methods of Measuring the Performance of Household Electrical Appliances Part 8: Coffee Makers. 8 pp.
BSI BS 4167: Part 6-69. 1969 Amd 1 Electrically-Heated Catering Equipment Part 6: Bulk Liquid Heaters. 46 pp.
CNS C4067-87. Electric Percolators (Jan)(2662). 3 pp.
CSA CAN/CSA-C22. 2 NO 64-M91. Household Cooking and Liquid-Heating Appliances; (Gen Instr 1 Thru 2). 91 pp.
IEC 661-80. Methods for Measuring the Performance of Electric Household Coffee Makers First Edition; (Amendment 1-1992). 17 pp.

—Commercial—Electric
CSA C22.2 NO 109-M1981. Commercial Cooking Appliances; (Gen Instr 1 Thru 3). 36 pp.
CSA 1169 Bull. Electrical Bulletin 1169 June 27, 1978 to C22.2 NO 109. 2 pp.

Coffee Mills
Use: Coffee Grinders

Coffee Products
Use: Coffee

Coffee Urns
See Also: Coffee Makers
CGSB CAN/CGSB-52.8-M86. Urns, Beverage, Food Service. 12 pp.

Cofferdams
See Also: Dams

—Steel Piles—Hot Rolled—Sheet
JIS A 5528-88. Hot Rolled Steel Sheet Piles. 11 pp.

—Steel Piles—Pipe—Sheet
JIS A 5530-88. Steel Pipe Sheet Piles. 20 pp.

—Steel Piles—Sheet
CNS A2109-81. Steel Sheet Piles (Sep)(7851). 4 pp.

Cohesion
See Also: Adhesives; Bearing Capacity; Binders (Materials); Capillarity; Plastic Properties

—Aggregates
SAA AS 1141.52-87. Methods of Sampling and Testing Aggregates—Part 52: Unconfined Cohesion of Compacted Pavement Materials. 6 pp.

—Joint Sealants
BSI BS EN 29046-91. 1991 Building Construction—Sealants—Determination of Adhesion/Cohesion Properties at Constant Temperature (ISO 9046: 1987). 10 pp.
CEN EN 29 046-90. Building Construction—Jointing Products—Determination of Adhesion Properties at Constant Temperatures. 9 pp.
DIN ENGL 52455 Pt 1-87. Testing of Building Construction Sealants; Determination of Adhesion/Cohesion After Conditioning in Standard Atmosphere, Water, or at Elevated Temperature (Apr). 4 pp.
DIN ENGL 52455 Pt 4-87. Testing of Building Construction Sealants; Determination of Adhesion/Cohesion When Subjected to an Extension/Compression Cycle at Alternating Temperature (Apr). 3 pp.
DIN ENGL EN 29046-91. Sealants in Building Construction; Determination of Adhesion/Cohesion Properties at Constant Temperature (ISO 9046: 1987) (May). 6 pp.
ISO 9046-87. Building Construction—Sealants—Determination of Adhesion/Cohesion Properties at Constant Temperature First Edition. 6 pp.
ISO 9047-89. Building Construction—Sealants—Determination of Adhesion/Cohesion Properties at Variable Temperatures First Edition. 6 pp.
ISO 10590-91. Building Construction—Sealants—Determination of Adhesion/Cohesion Properties at Maintained Extension After Immersion in Water First Edition. 6 pp.
ISO 10591-91. Building Construction—Sealants—Determination of Adhesion/Cohesion Properties After Immersion in Water First Edition. 6 pp.

—Laminated Plastics—Tubes
BSI BS 2782:Pt3: METH 346A-84. 1984 Methods of Testing Plastics Part 3: Mechanical Properties Method 346A: Determination of Cohesion Between Layers of Laminated Tube. 4 pp.

—Sealants
ISO 11431-93. Building Construction—Sealants—Determination of Adhesion/Cohesion Properties After Exposure to Artificial Light Through Glass First Edition. 7 pp.

—Wood Fibers
CPPA D.27U-92. Zero-Span Breaking Length of Pulp (Pulmac Zero-Span Method). 4 pp.

Cohesive Strength
Use: Cohesion

Cohesive Testing
Use: Cohesion

Cohesiveness
Use: Cohesion

Coil Loader Reels
Use: Cable Reels

Coil Springs
See Also: Helical Springs; Springs (Elastic)

—Aircraft
SBAC AS 2816 ISSUE 3. Drum Spring. 1 p.

—Automotive—Suspension Systems
CNS D2161-82. Suspension Coil Springs for Automobiles (Nov)(9597).

—Shape Memory Alloy—Loads (Forces)
JIS H 7104-91. Method of Fixed Temperature Load Test for Coil Spring of Shape Memory Alloys. 10 pp.

—Ships
JIS F 0503-87. Coiled Springs for Marine Machinery.
JIS F 0503-60. Coil Springs for Ship Machinery (R 1976). 19 pp.

—Steel Wire
BSI BS 4637-70. 1970 Amd 1 Carbon Steel Wire for Coiled Springs (Bedding and Seating). 12 pp.
BSI BS 4637-03. 1970 Amd 3 Carbon Steel Wire for Coiled Springs (Bedding and Seating) (AMD 7609) June 15, 1993. 14 pp.

Coils (Electric)
Use: Electric Coils

Coin Operated Vending Machines
Use: Vending Machines

Coke
See Also: Calcined Coke; Carbon; Charcoal; Coke Content Analysis; Coking; Solid Mineral Fuels
CNS M1003-82. Coke for Blast Furnace (Sep)(1007).
CNS M1004-82. Foundry Coke (Sep)(1008).
CNS M1005-65. Coal Fuel (for Taiwan Area) (Sep)(1118).
CNS M1014-82. Coke for General Purpose (Sep)(3345).
CNS M1015-82. Coke for Producer (Sep)(3346).
JIS K 2161-61. Cokes (R 1969). 13 pp.
JIS K 2151-77. Methods for Testing of Coke (R 1980). 48 pp.
JIS M 8815-76. Methods for Analysis of Coal Ash and Coke Ash.

—Arsenic Content
SAA AS 1038.10-80. Methods for the Analysis and Testing of Coal and Coke—Part 10: Arsenic in Coal and Cokes. 16 pp.

—Arsenic Content—Oxidation Method
BSI BS 1016: Part 10-77. 1977 Methods for the Analysis and Testing of Coal and Coke Part 10: Arsenic in Coal and Coke. 8 pp.
SAA AS 1038.10-80. Methods for the Analysis and Testing of Coal and Coke—Part 10: Arsenic in Coal and Cokes. 16 pp.

—Ash Content
BSI BS 1016: Part 4-73. (WITHDRAWN) 1973 Amd 2 Methods for the Analysis and Testing of Coal and Coke Part 4: Moisture, Volatile Matter and Ash in the Analysis Sample of Coke (Superseded by BS 1016: Sections 104.1, 104.2, 104.3 and 104.4: 1991). 18 pp.
BSI BS 1016: Part 14-63. 1963 Methods for the Analysis and Testing of Coal and Coke Part 14: Analysis of Coal Ash and Coke Ash. 37 pp.
BSI BS 6043: Sec 2.2-85. 1985 Methods of Sampling and Test for Carbonaceous Materials Used in Aluminium Manufacture Part 2: Electrode Coke Section 2.2: Determination of Ash Content of Green and Calcined Cokes. 6 pp.
CNS M3141-84. Method for Determination of Ash of Coal and Coke (Mar)(10822).
CNS M3155-84. Method for Analysis of Coal Ash and Coke Ash (Mar)(10836).
ISO 8005-84. Carbonaceous Materials Used in the Production of Aluminium—Green and Calcined Coke—Determination of Ash Content First Edition. 4 pp.
JIS M 8812-84. Methods for Proximate Analysis of Coal and Coke. 28 pp.
SAA AS 1038.4-79. Methods for the Analysis and Testing of Coal and Coke—Part 4: Proximate Analysis of Coke. 16 pp.

—Ash Content—Atomic Absorption Spectrometry—Acid Digestion
SAA AS 1038.14.2-85. Methods for the Analysis and Testing of Coal and Coke—Part 14.2: Analysis of Higher Rank Coal Ash, Coke Ash (Acid Digestion—Flame Atomic Absorption Spectrometric Method). 11 pp.

—Ash Content—Atomic Absorption Spectrometry—Borate Fusion
SAA AS 1038.14.1-81. Methods for the Analysis and Testing of Coal and Coke—Part 14.1: Analysis of Coal Ash, Coke Ash and Mineral Matter (Borate Fusion—Flame Atomic Absorption Spectrometric Method). 18 pp.

—Ash Content—Fusibility
SAA AS 1038.15-87. Methods for the Analysis and Testing of Coal and Coke—Part 15: Fusibility of Higher Rank Coal Ash and Coke Ash. 6 pp.

—Ash Content—Reference Method
BSI BS 1016: Sec 104.4-91. 1991 Analysis and Testing of Coal and Coke Part 104: Proximate Analysis Section 104.4: Determination of Ash. 9 pp.

—Boron Content—Spectrophotometry
SAA AS 1038.10.3-88. Methods for the Analysis of Coal and Coke—Part 10.3: Determination of Trace Elements—Coal, Coke and Fly-Ash—Determination of Boron Content—Spectrophotometric Method. 5 pp.

—Bulk Density
BSI BS 1016: Sec 108.3-91. 1991 Analysis and Testing of Coal and Coke Part 108: Tests Special to Coke Section 108.3: Determination of Bulk Density (Small Container). 7 pp.
BSI BS 1016: Sec 108.4-91. 1991 Analysis and Testing of Coal and Coke Part 108: Tests Special to Coke Section 108.4: Determination of Bulk Density (Large Container). 6 pp.
CNS M3173-84. Method of Test for Bulk Density of Coke in a Larger Container (Nov) (11115).
CNS M3174-84. Method of Test for Bulk Density of Coke in a Small Container (Nov) (11116).
ISO 567-74. Coke—Determination of the Bulk Density in a Small Container First Edition. 4 pp.
ISO 1013-75. Coke—Determination of Bulk Density in a Large Container First Edition. 4 pp.
SAA AS 3899-91. Higher Rank Coal and Coke—Bulk Density (in Professional Package 32). 7 pp.

—Calorific Value
BSI BS 1016: Part 5-77. (WITHDRAWN) 1977 Methods for the Analysis and Testing of Coal and Coke Part 5: Gross Calorific Value of Coal and Coke (Superseded by BS 1016: Part 105: 1992). 12 pp.
BSI BS 1016: Part 105-92. 1992 Methods for Analysis and Testing of Coal and Coke Part 105: Determination of Gross Calorific Value (W). 17 pp.
JIS M 8814-85. Determination of Calorific Value of Coal and Coke; (Erratum). 20 pp.
SAA AS 1038.5.1-88. Methods for the Analysis and Testing of Coal and Coke—Part 5.1: Gross Specific Energy of Coal and Coke—Adiabatic Calorimeters. 9 pp.
SAA AS 1038.5.2-89. Methods for the Analysis and Testing of Coal and Coke—Part 5.2: Gross Specific Energy of Coal and Coke —Automatic Isothermal-Type Calorimeters. 5 pp.

—Carbon Content
CNS M3143-84. Method for Calculation of Fixed Carbon of Coal and Coke (Mar)(10824).
JIS M 8812-84. Methods Proximate Analysis of Coal and Coke. 28 pp.

—Carbon Content—Combustion
CNS M3153-84. Method for Determination of Carbons and Hydrogen of Coal and Coke (Sheffield High Temperature Method) (Mar)(10834).
ISO 609-75. Coal and Coke—Determination of Carbon and Hydrogen—High Temperature Combustion Method First Edition. 9 pp.

—Carbon Content—Liebig Method
CNS M3144-84. Method for Determination of Carbon and Hydrogen of Coal and Coke (Liebig Method) (Mar) (10825).
ISO 625-75. Coal and Coke—Determination of Carbon and Hydrogen—Liebig Method First Edition. 9 pp.

—Chemical Analysis
CNS M3139-84. General Rules for Testing of Coal and Coke (Mar) (10820).
JIS M 8810-89. General Rules for Sampling, Analysis and Testing of Coal and Coke. 26 pp.
JIS M 8813-88. Methods for Ultimate Analysis of Coal and Coke. 100 pp.

—Chemical Analysis—Formulas
ISO 1170-77. Coal and Coke—Calculation of Analyses to Different Bases First Edition. 6 pp.

—Chlorine Content
SAA AS 1038.8-80. Methods for the Analysis and Testing of Coal and Coke—Part 8: Chlorine in Coal and Coke (Superseded (in Part) by AS 1038.8.1—1992 Will Remain Current). 12 pp.

—Chlorine Content—Combustion
BSI BS 1016: Part 8-77. 1977 Amd 1 Methods for Analysis and Testing of Coal and Coke Part 8: Chlorine in Coal and Coke (AMD 3472) December 31, 1980. 9 pp.
BSI BS 1016: Part 8-02. 1977 Amd 2 Methods for Analysis and Testing of Coal and Coke Part 8: Chlorine in Coal and Coke (AMD 6984) May 1, 1992. 12 pp.

Coke (Cont.)

—Chlorine Content—Eschka Method
BSI BS 1016: Part 8-77. 1977 Amd 1 Methods for Analysis and Testing of Coal and Coke Part 8: Chlorine in Coal and Coke (AMD 3472) December 31, 1980. 9 pp.

BSI BS 1016: Part 8-02. 1977 Amd 2 Methods for Analysis and Testing of Coal and Coke Part 8: Chlorine in Coal and Coke (AMD 6984) May 1, 1992. 12 pp.

SAA AS 1038.8.1-92. Coal and Coke—Analysis and Testing—Part 8.1: Coal and Coke—Chlorine—Eschka Method (Supersedes AS 1038.8—1980 (in Part)) Which Will Remain Current) (in Professional Package 32). 2 pp.

—Comprehensive Testing
BSI BS 1016: Part 13-80. 1980 Amd 1 Methods for Analysis and Testing of Coal and Coke Part 13: Tests Special to Coke. 15 pp.

BSI BS 1016: Part 13-02. (WITHDRAWN) 1980 Amd 2 Methods for Analysis and Testing of Coal and Coke Part 13: Tests Special to Coke (AMD 6894) January 31, 1992 (Superseded by BS 1016: Various Sections) (W). 16 pp.

SAA AS 1038.13-90. Methods for the Analysis and Testing of Coal and Coke—Part 13: Tests Specific to Coke. 17 pp.

—Defense Contracts
MOD UK DEFCON 112AA-83. Basic Set of Conditions of Contract for the Supply of Coal and Coke 9/83. 4 pp.

—Density
BSI BS 1016: Sec 108.5-92. 1992 Methods for Analysis and Testing of Coal and Coke Part 108: Tests Special to Coke Section 108.5: Determination of Density and Porosity. 10 pp.

CNS M3179-84. Method for Determination of True Relative Density Apparent Relative Density and Porosity of Coke (Nov)(11121).

ISO 1014-85. Coke—Determination of True Relative Density, Apparent Relative Density and Porosity Second Edition. 5 pp.

—Electrodes—Sampling
BSI BS 6043: Sec 2.1-85. 1985 Methods of Sampling and Test for Carbonaceous Materials Used in Aluminium Manufacture Part 2: Electrode Coke Section 2.1: Sampling. 8 pp.

ISO 6375-80. Carbonaceous Materials for the Production of Aluminium—Cokes for Electrodes—Sampling First Edition. 7 pp.

—Fluorine Content—Pyrohydrolosis
SAA AS 1038.10.4-89. Methods for the Analysis and Testing of Coal and Coke—Part 10.4: Determinaton of Trace Elements—Coal, Coke and Fly-Ash—Determination of Fluorine Content—Prohydrolysis Method Amdt 1 October 1990. 11 pp.

—Fragmentation
BSI BS 1016: Sec 108.1-92. 1992 Methods for Analysis and Testing of Coal and Coke Part 108: Tests Special to Coke Section 108.1: Determination of Shatter Indices (W). 11 pp.

CNS M3177-84. Method of Test for Shatter of Coke (Nov) (11119).

ISO 616-77. Coke—Determination of Shatter Indices First Edition. 8 pp.

—Fusibility
BSI BS 1016: Part 15-70. 1970 Methods for the Analysis and Testing of Coal and Coke Part 15: Fusibility of Coal Ash and Coke Ash. 16 pp.

CNS M3163-84. Method of Test for Fusibility of Coal and Coke Ash (Jun)(10948).

—Glossaries
BSI BS 3446: Part 2-90. 1990 Glossary of Terms Associated with Refractory Materials Part 2: Applications in the Coke, Glass, Cement and Other Non-Metallurgical Industries (Supersedes BS 3446: 1962). 49 pp.

—Hydrogen Content—Combustion
CNS M3153-84. Method for Determination of Carbons and Hydrogen of Coal and Coke (Sheffield High Temperature Method) (Mar)(10834).

ISO 609-75. Coal and Coke—Determination of Carbon and Hydrogen—High Temperature Combustion Method First Edition. 9 pp.

—Hydrogen Content—Liebig Method
CNS M3144-84. Method for Determination of Carbon and Hydrogen of Coal and Coke (Liebig Method) (Mar) (10825).

ISO 625-75. Coal and Coke—Determination of Carbon and Hydrogen—Liebig Method First Edition. 9 pp.

—Invitation for Bids
MOD UK DEFCON 47C-89. Invitation to Tender for the Supply and Delivery of Coal and/or Coke 11/89. 2 pp.

Coke (Cont.)

—Mechanical Properties
BSI BS 1016: Sec 108.2-92. 1992 Methods for Analysis and Testing of Coal and Coke Part 108: Tests Special to Coke Section 108.2: Determination of Micum and Irsid Indices (W). 11 pp.

CNS M3178-84. Method of Test for Mechanical Strength of Coke (Greater Than 20 mm in Size) (Nov)(11120).

ISO 556-80. Coke (Greater Than 20 mm in Size)—Determination of Mechanical Strength First Edition. 7 pp.

—Moisture Content
BSI BS 1016: Part 2-73. 1973 Methods for the Analysis and Testing of Coal and Coke Part 2: Total Moisture of Coke. 8 pp.

BSI BS 1016: Part 4-73. (WITHDRAWN) 1973 Amd 2 Methods for the Analysis and Testing of Coal and Coke Part 4: Moisture, Volatile Matter and Ash in the Analysis Sample of Coke (Superseded by BS 1016: Sections 104.1, 104.2, 104.3 and 104.4: 1991). 18 pp.

CNS M3140-84. Method for Determination of Inherent Moisture of Coal and Coke (Mar) (10821).

ISO 579-81. Coke—Determination of Total Moisture Content Second Edition. 4 pp.

JIS M 8811-76. Method for Sampling and Determination of Total Moisture and Adherent Moisture of Coal and Coke. 28 pp.

JIS M 8812-84. Methods for Proximate Analysis of Coal and Coke. 28 pp.

SAA AS 1038.2-79. Methods for the Analysis and Testing of Coal and Coke—Part 2: Total Moisture in Coke. 8 pp.

SAA AS 1038.4-79. Methods for the Analysis and Testing of Coal and Coke—Part 4: Proximate Analysis of Coke. 16 pp.

—Moisture Content—Gravimetric Analysis
ISO 687-74. Coke—Determination of Moisture in the Analysis Sample First Edition. 4 pp.

—Moisture Content—Liebig Method
CNS M3144-84. Method for Determination of Carbon and Hydrogen of Coal and Coke (Liebig Method) (Mar) (10825).

ISO 625-75. Coal and Coke—Determination of Carbon and Hydrogen—Liebig Method First Edition. 9 pp.

—Moisture Content—Reference Method
BSI BS 1016: Sec 104.2-91. 1991 Analysis and Testing of Coal and Coke Part 104: Proximate Analysis Section 104.2: Determination of Moisture Content of the General Analysis Sample of Coke. 9 pp.

—Nitrogen Content—Kjeldahl Method
CNS M3147-84. Method for Determination of Nitrogen of Coal and Coke (Kjeldahl Method) (Mar)(10828).

—Phosphorus Content—Absorptiometric Analysis
CNS M3149-84. Method for Determination of Phosphorus of Coal and Coke (Molybdenum Blue Absorptiometric Method) (Mar)(10830).

—Phosphorus Content—Ash Digestion Method
SAA AS 1038.9.1-92. Coal and Coke—Analysis and Testing—Part 9.1: Coal and Coke—Phosphorus—Ash Digestion/Molybdenum Blue Method (Supersedes AS 1038.9—1977) Amdt 1 April 1993 (in Professional Package 32). 4 pp.

SAA AS 1038.9.3-91. Coal and Coke—Analysis and Testing—Part .9.3: Coal and Coke—Phosphorus—Ash Digestion Method (In Professional Package 32). 3 pp.

—Phosphorus Content—Molybdenum Blue Method
SAA AS 1038.9.1-92. Coal and Coke—Analysis and Testing—Part 9.1: Coal and Coke—Phosphorus—Ash Digestion/Molybdenum Blue Method (Supersedes AS 1038.9—1977) Amdt 1 April 1993 (in Professional Package 32). 4 pp.

—Phosphorus Content—Oxidation Method
BSI BS 1016: Part 9-77. 1977 Methods for the Analysis and Testing of Coal and Coke Part 9: Phosphorus in Coal and Coke. 8 pp.

—Physical Properties
ISO TR7517-83. Coke—Comparison of Different Tests Used to Assess the Physical Strength First Edition. 14 pp.

—Porosity
BSI BS 1016: Sec 108.5-92. 1992 Methods for Analysis and Testing of Coal and Coke Part 108: Tests Special to Coke Section 108.5: Determination of Density and Porosity. 10 pp.

Coke (Cont.)

—Porosity (Cont.)
CNS M3179-84. Method for Determination of True Relative Density Apparent Relative Density and Porosity of Coke (Nov)(11121).

ISO 1014-85. Coke—Determination of True Relative Density, Apparent Relative Density and Porosity Second Edition. 5 pp.

—Reactivity
BSI BS 1016: Sec 108.6-92. 1992 Methods for Analysis and Testing of Coal and Coke Part 108: Tests Special to Coke Section 108.6: Determination of Critical Air Blast Value. 11 pp.

CNS M3180-84. Method of Test for Reactivity of Coke (Nov)(11122).

—Sampling
BSI BS 1017: Part 2-60. 1960 Amd 2 Methods for Sampling of Coal and Coke Part 2: Sampling of Coke. 107 pp.

BSI BS 4845: Part 1-72. 1972 Amd 1 Sampling Manufactured Domestic Solid Smokeless Fuels in Small Consignments of Mass 50Kg to 5000Kg Either in Bulk or in Bags Part 1: Sampling of Coke. 10 pp.

CNS M3138-83. Method of Sampling for Coke (Nov)(10662).

CNS M3139-84. General Rules for Testing of Coal and Coke (Mar) (10820).

CNS M3140-84. Method for Determination of Inherent Moisture of Coal and Coke (Mar) (10821).

ISO 2309-80. Coke—Sampling Second Edition. 27 pp.

JIS M 8810-89. General Rules for Sampling, Analysis and Testing of Coal and Coke. 26 pp.

JIS M 8811-76. Method for Sampling and Determination of Total Moisture and Adherent Moisture of Coal and Coke.

—Screening Plants
BSI BS 3407-82. 1982 Guide for Assessing Coke Screening Plant. 10 pp.

—Sieve Analysis
BSI BS 1016: Part 18-81. 1981 Methods for Analysis and Testing of Coal and Coke Part 18: Size Analysis of Coke. 8 pp.

BSI BS 1293 & 2074-65. 1965 Amd 1 Methods for the Size Analysis of Coal and Coke. 52 pp.

CNS M3175-84. Method for Size Analysis of Coke (Nominal Top Size Greater Than 20mm) (Nov)(11117).

CNS M3176-84. Method for Size Analysis of Coke (Nominal Top Size 20 mm or Less) (Nov)(11118).

ISO 728-81. Coke (Nominal Top Size Greater Than 20 mm)—Size Analysis Second Edition. 5 pp.

ISO 2325-86. Coke—Size Analysis (Nominal Top Size 20 mm or Less) Third Edition. 5 pp.

SAA AS 1038.18-81. Methods for the Analysis and Testing of Coal and Coke—Part 18: Size Analysis of Coke. 8 pp.

SNZ NZS 535, NZS 2175-67. Methods for the Size Analysis of Coal and Coke Amend: 1; 1A, 1988. 52 pp.

—Sulfur Content
CNS M3150-84. Method for Determination of Uncombustible Sulfur in Coal Ash and Coke Ash (Mar)(10831).

—Sulfur Content—Combustion
CNS M3146-84. Method for Determination of Total Sulfur of Coal and Coke (High Temperature Combustion Method) (Mar)(10827).

ISO 351-84. Solid Mineral Fuels—Determination of Total Sulfur—High Temperature Combustion Method Second Edition. 6 pp.

—Sulfur Content—Eschka Method
BSI BS 1016: SUB SEC 106.4.1-93. 1993 Methods for Analysis and Testing of Coal and Coke Part 106: Ultimate Analysis of Coal and Coke Section 106.4: Determination of Total Sulfur Content Subsection 106.4.1: Eschka Method (W). 11 pp.

CNS M3145-84. Method for Determination of Total Sulfur of Coal and Coke (Eschka Method) (Mar)(10826).

ISO 334-92. Solid Mineral Fuels—Determination of Total Sulfur—Eschka Method Second Edition. 9 pp.

—Technical Quality
BSI BS 4262-84. 1984 Method for Specifying the Technical Quality of Coke for Use in Blast Furnaces. 11 pp.

—Test Reports
BSI BS 1016: Part 16-81. 1981 Amd 1 Methods for the Analysis and Testing of Coal and Coke Part 16: Methods for Reporting Results. 19 pp.

INTERNATIONAL AND NON-U.S. NATIONAL STANDARDS
SUBJECT INDEX

Coke *(Cont.)*

—Trace Element Content—Atomic Absorption Spectrometry
BSI BS 6043: Sec 2.3-89. 1989 Methods of Sampling and Test for Carbonaceous Materials Used in Aluminium Manufacture Part 2: Electrode Coke Section 2.3: Determination of Trace Elements by Flame Atomic Absorption Spectrometry. 10 pp.
SAA AS 1038.10.1-86. Methods for the Analysis and Testing of Coal and Coke—Part 10.1: Determination of Trace Elements—Determination of Eleven Trace Elements in Coal, Coke and Fly-Ash—Flame Atomic Absorption Spectrometric Method (R 1992). 10 pp.

—Ultimate Analysis
BSI BS 1016: Part 7-77. 1977 Methods for Analysis and Testing of Coal and Coke Part 7: Ultimate Analysis of Coke. 12 pp.
BSI BS 1016: Part 7-01. 1977 Amd 1 Methods for Analysis and Testing of Coal and Coke Part 7: Ultimate Analysis of Coke (AMD 6983) May 1, 1992 (W). 17 pp.
BSI BS 1016: Part 7-02. 1977 Amd 2 Methods for Analysis and Testing of Coal and Coke Part 7: Ultimate Analysis of Coke (AMD 7690) May 15, 1993 (W). 18 pp.

—Volatile Matter Content
BSI BS 1016: Part 4-73. (WITHDRAWN) 1973 Amd 2 Methods for the Analysis and Testing of Coal and Coke Part 4: Moisture, Volatile Matter and Ash in the Analysis Sample of Coke (Superseded by BS 1016: Sections 104.1, 104.2, 104.3 and 104.4: 1991). 18 pp.
ISO 562-81. Hard Coal and Coke—Determination of Volatile Matter Content Second Edition. 7 pp.
JIS M 8812-84. Methods for Proximate Analysis of Coal and Coke. 28 pp.
SAA AS 1038.4-79. Methods for the Analysis and Testing of Coal and Coke—Part 4: Proximate Analysis of Coke. 16 pp.

—Volatile Matter Content—Gravimetric Analysis
BSI BS 6043: Sec 2.11-89. 1989 Methods of Sampling and Test for Carbonaceous Materials Used in Aluminium Manufacture Part 2: Electrode Coke Section 2.11: Determination of Volatile Matter Content (ISO 9406: 1988). 7 pp.
ISO 9406-88. Carbonaceous Materials for the Production of Aluminium—Green Coke—Determination of Volatile Matter Content by Gravimetric Analysis First Edition. 7 pp.

—Volatile Matter Content—Reference Method
BSI BS 1016: Sec 104.3-91. 1991 Analysis and Testing of Coal and Coke Part 104: Proximate Analysis Section 104.3: Determination of Volatile Matter Content. 11 pp.

—Xylene—Density
BSI BS 6043: Sec 2.6-86. 1986 Methods of Sampling and Test for Carbonaceous Materials Used in Aluminium Manufacture Part 2: Electrode Coke Section 2.6: Determination of the Density in Xylene of Calcined Coke. 8 pp.
ISO 8004-85. Carbonaceous Materials for the Production of Aluminium—Calcined Coke and Calcined Carbon Products—Determination of the Density in Xylene—Pyknometric Method First Edition. 6 pp.

Coke Content Analysis
See Also: Lignite

—Lignite—Distillation Methods
ISO 647-74. Brown Coals and Lignites—Determination of the Yields of Tar, Water, Gas and Coke Residue by Low Temperature Distillation First Edition. 8 pp.

Coke Oven Gas
Use: Coal Gas

Coke Ovens
Use: Coking

Coking
Use For: Coke Ovens *See Also:* Coke

—Aluminosilicate Refractories
BSI BS 6886-87. 1987 Alumino-Silicate Refractories for Use in Coke Ovens. 12 pp.

—Bituminous Coal
CNS M1001-65. Bituminous Coal for Coking (for Export from Taiwan Area) (Sep)(204) (R 1972). 1 p.

—Coal
CNS M1010-82. Coking Coal (Sep)(2508).

Coking *(Cont.)*

—Coal *(Cont.)*
JIS M 8801-79. Methods for Testing of Coal. 71 pp.

—Comprehensive Testing
BSI BS 999-62. 1962 Schedule of Tests for Coke Ovens. 43 pp.

—Pitch
BSI BS 6043: Sec 1.6-84. 1984 Methods of Sampling and Test for Carbonaceous Materials Used in Aluminium Manufacture Part 1: Electrode Pitch Section 1.6: Determination of Coking Value. 7 pp.
BSI BS 6043: Sec 1.6-01. 1984 Amd 1 Methods of Sampling and Test for Carbonaceous Materials Used in Aluminium Manufacture Part 1: Electrode Pitch Section 1.6: Determination of Coking Value (AMD 7662) May 15, 1993 (ISO 6998: 1984) (V). 8 pp.
ISO 6998-84. Carbonaceous Materials for the Production of Aluminium—Pitch for Electrodes—Determination of Coking Value First Edition. 5 pp.

—Silica Bricks
BSI BS 4966: Part 1-82. (WITHDRAWN) 1982 Silica Refractories Part 1: Refractories for Use in Coke Ovens (Superseded by BS 4966: 1991). 11 pp.
JIS R 2401-76. Silica Brick for Coke Oven (R 1986). 7 pp.

—Silicate Refractories
BSI BS 4966-91. 1991 Silica Refractories for Use in Coke Ovens. 13 pp.

Cold Bending
Scope Note: See the subheading Cold Worked under the specific metal or alloy

Cold Cathode Counter Tubes
See Also: Electron Tubes; Photomultiplier Tubes; Phototubes
BSI BS CECC 46000-78. 1978 Cold Cathode Indicator Tubes: Generic Specification. 12 pp.
BSI BS CECC 46001-77. 1977 Cold Cathode Indicator Tubes: Blank Detail Specification. 8 pp.

—Electrical Measurement
IEC 151 Pt 22-70. Measurements of the Electrical Properties of Electronic Tubes and Valves Part 22: Methods of Measurement for Cold Cathode Counting and Indicator Tubes First Edition. 28 pp.

Cold Cathode Tubes
See Also: Electron Tubes; Photomultiplier Tubes; Phototubes; Protector Tubes; Thermionic Tubes; Trigger Tubes
SNZ NZSR 15-65. Use of Electronic Valves. 71 pp.

—Electrical Measurement
IEC 151 Pt 17-69. Measurements of the Electrical Properties of Electronic Tubes and Valves Part 17: Methods of Measurement of Gasfilled Tubes and Valves Second Edition. 51 pp.

—Indicator
CECC CECC 46 000 ISSUE 1-76. Generic Specification: Cold Cathode Indicator Tubes (En, Fr, Ge). 27 pp.
CECC CECC 46 001 ISSUE 1-76. Blank Detail Specification: Cold Cathode Indicator Tubes (En, Fr, Ge). 18 pp.

—Lighting—Transformers
CSA C22.2 NO 13-1962. Transformers for Luminous-Tube Signs, Oil-or Gas-Burner Ignition Equipment, Cold-Cathode Interior Lighting (R 1992). 46 pp.
CSA 1169 Bull. Electrical Bulletin 1169 June 27, 1978 to C22.2 NO 13. 2 pp.
SAA AS 3143-82. Approval and Test Specification—for Transformers for Cold-Cathode Electric Discharge Lamps and Lighting Systems (This is a Joint Standard with SANZ NZS 3143). 4 pp.
SNZ NZS/AS 3143-82. Approval and Test Specification for Transformers for Cold-Cathode Electric Discharge Lamps and Lighting Systems (This is a Joint Standard with SAA AS 3143). 4 pp.

Cold Chisels
See Also: Chisels; Hand Tools
BSI BS 3066-81. 1981 Amd 1 Engineers' Cold Chisels and Allied Tools. 17 pp.
CGSB 39-GP-43A-74. Chisels, Hand, Cutting, and Swaging (For Metals), Standard for; (Amendment 1 Dec 1978). 15 pp.
CNS B3242-78. Cold Chisels (May)(3974). 2 pp.
DIN ENGL 5107-76. Blacksmith's Chisels; Cold Chisels (Dec). 2 pp.
MOD UK DSTAN 51-11: Part 29-01. Hand Tools, General Purpose Part 29: Chisels, Cold, Hand Issue 1; Amendment 2. 17 pp.

Cold Curing

—Polyurethane Adhesives
MOD UK TS 10041A. Polyurethane Adhesive, Mineral Filled, Two Part, Type QX.

Cold Drawn
Scope Note: See the subheading Drawn under the specific metal or alloy

Cold Extruding
Scope Note: See the subheading Extruded under the specific metal or alloy

Cold Finishing
Scope Note: See the subheading Cold Worked under the specific metal or alloy

Cold Flow
Use For: Compressive Creep *See Also:* Creep Properties; Plastic Properties

—Cellular Plastics
ISO 7616-86. Cellular Plastics, Rigid—Determination of Compressive Creep Under Specified Load and Temperature Conditions First Edition. 4 pp.
ISO 7850-86. Cellular Plastics, Rigid—Determination of Compressive Creep First Edition. 5 pp.

Cold Forming
Scope Note: See the subheading Cold Worked under the specific metal or alloy

Cold Heading
Scope Note: See the subheading Cold Worked under the specific metal or alloy

Cold Injuries
See Also: Injuries

—Military Personnel—Prevention
NATO STANAG 2981 ED 1 AMD 3. Prevention of Cold Injury. 14 pp.

Cold Reduction
Scope Note: See the subheading Cold Worked under the specific metal or alloy

Cold Rolled Products
Scope Note: See the subheading Cold Rolled under the specific metal or alloy

Cold Rooms
See Also: Refrigeration Systems

—Construction—Ships
MOD UK NES 111-91. Requirements for the Insulation and Fittings for the Refrigeration Spaces in Surface Ships and Submarines Issue 2 (06.91). 35 pp.

—Fittings—Ships
MOD UK NES 111-91. Requirements for the Insulation and Fittings for the Refrigeration Spaces in Surface Ships and Submarines Issue 2 (06.91). 35 pp.

—Insulation—Ships
MOD UK NES 803: Part 2-89. Requirements for Plastic Foam Thermal Insulation Material Part 2: Plastic Foam Products Phenolic Foam Slabs for Cold and Cool Rooms Issue 1 (06.89). 13 pp.
MOD UK NES 111-91. Requirements for the Insulation and Fittings for the Refrigeration Spaces in Surface Ships and Submarines Issue 2 (06.91). 35 pp.

—Insulation—Submarines
MOD UK NES 803: Part 2-89. Requirements for Plastic Foam Thermal Insulation Material Part 2: Plastic Foam Products Phenolic Foam Slabs for Cold and Cool Rooms Issue 1 (06.89). 13 pp.

—Walk-In
BSI BS 2502-79. 1979 Manufacture of Sectional Cold Rooms (Walk-In Type). 4 pp.

Cold Storage
See Also: Controlled Atmospheres; Frozen Foods

—Apples
ISO 1212-76. Apples—Guide to Cold Storage First Edition. 8 pp.

—Apricots
ISO 2826-74. Apricots—Guide to Cold Storage First Edition. 5 pp.

—Avocados
ISO 2295-74. Avocados—Guide for Storage and Transport First Edition. 6 pp.

Cold Storage (Cont.)

—Berries
ISO 6664-83. Bilberries and Blueberries—Guide to Cold Storage First Edition. 5 pp.
ISO 6665-83. Strawberries—Guide to Cold Storage First Edition. 5 pp.

—Cabbages
ISO 2167-91. Round-Headed Cabbage—Guide to Cold Storage and Refrigerated Transport Second Edition. 7 pp.

—Cauliflowers
ISO 949-87. Cauliflowers—Guide to Cold Storage and Refrigerated Transport Second Edition. 5 pp.

—Cherries
ISO 7920-84. Sweet Cherries and Sour Cherries—Guide to Cold Storage and Refrigerated Transport First Edition. 4 pp.

—Fish
CGSB 32.72M-91. Handling, Packaging and Labelling of Meat, Poultry and Fish for Food Services. 10 pp.

—Fruits
ISO 2169-81. Fruits and Vegetables—Physical Conditions in Cold Stores—Definitions and Measurement Second Edition. 7 pp.
ISO 3659-77. Fruit and Vegetables—Ripening After Cold Storage First Edition. 6 pp.

—Fuel Oils
BSI BS 2000: Part 230-93. 1993 Methods of Test for Petroleum and Its Products Part 230: Determination of Minimum Handling and Storage Temperatures of Fuel Oil—Ferranti Viscometer Method (W). 5 pp.

—Garlic
ISO 6663-83. Garlic—Guide to Cold Storage First Edition. 4 pp.

—Grapes
ISO 2168-74. Table Grapes—Guide to Cold Storage First Edition. 7 pp.

—Green Beans
ISO 9930-93. Green Beans—Storage and Refrigerated Transport First Edition. 5 pp.

—Leeks
ISO 7922-85. Leeks—Guide to Cold Storage and Refrigerated Transport First Edition. 4 pp.

—Meat
CGSB 32.72M-91. Handling, Packaging and Labelling of Meat, Poultry and Fish for Food Services. 10 pp.

—Mushrooms
ISO 7561-84. Cultivated Mushrooms—Guide to Cold Storage and Refrigerated Transport First Edition. 4 pp.

—Peaches
ISO 873-80. Peaches—Guide to Cold Storage First Edition. 7 pp.

—Pears
ISO 1134-80. Pears—Guide to Cold Storage First Edition. 9 pp.

—Peppers
ISO 6659-81. Sweet Pepper—Guide to Refrigerated Storage and Transport First Edition. 6 pp.

—Plums
ISO 6662-83. Plums—Guide to Cold Storage First Edition. 6 pp.

—Tomatoes
ISO 5524-91. Tomatoes—Guide to Cold Storage and Refrigerated Transport Second Edition. 7 pp.

—Vegetables
ISO 2169-81. Fruits and Vegetables—Physical Conditions in Cold Stores—Definitions and Measurement Second Edition. 7 pp.
ISO 3659-77. Fruit and Vegetables—Ripening After Cold Storage First Edition. 6 pp.

Cold Testing
Use: Low Temperature Testing

Cold Work Tool Steels
Use: Tool Steels

Cold Working
Scope Note: See the subheading Cold Worked under the specific metal or alloy.

Coleoptera
Use For: Anobiidae; Hylotrupes; Lyctus
See Also: Insect Contamination

Coleoptera (Cont.)

—Damage—Wood—Preservatives
BSI BS 5217-75. (WITHDRAWN) 1975 Amd 1 Wood Preservatives. Determination of the Preventive Action Against Lyctus Brunneus (Stephens) (Laboratory Method) (Superseded by BS EN 20-1: 1992). 16 pp.
BSI BS 5218-89. 1989 Wood Preservatives. Determination of the Toxic Values Against Anobium Punctatum (De Geer) by Larval Transfer (Laboratory Method). 15 pp.
BSI BS 5219-75. 1975 Wood Preservatives. Determination of Eradicant Action Against Hylotrupes Bajulus (Linnaeus) Larvae (Laboratory Method). 15 pp.
BSI BS 5434-89. 1989 Wood Preservatives. Determination of the Preventive Action Against Recently Hatched Larvae of Hylotrupes Bajulus (Linnaeus) (Laboratory Method). 14 pp.
BSI BS 5435-89. 1989 Wood Preservatives. Determination of the Toxic Values Against Larvae of Hylotrupes Bajulus (Linnaeus) Larvae (Laboratory Method). 14 pp.
BSI BS 5436-89. 1989 Wood Preservatives. Determination of Eradicant Action Against Larvae of Anobium Punctatum (De Geer) (Laboratory Method). 16 pp.
BSI BS 5437-77. (WITHDRAWN) 1977 Amd 1 Wood Preservatives. Determination of Toxic Values Against Anobium Punctatum (De Geer) by Egg-Laying and Larval Survival (Laboratory Method) (Superseded by BS EN 49-2: 1992). 13 pp.

Coliform Bacteria
See Also: Bacteria; Enterobacteriaceae; Escherichia Coli

—Animal Feed
CNS N4046-78. Methods of Test for Bacteria in Feeds (Mar) (3682). 3 pp.

—Dairy Products
BSI BS 4285: Sec 3.7-87. 1987 Microbiological Examination for Dairy Purposes Part 3: Methods for Detection and/or Enumeration of Specific Groups of Microorganisms Section 3.7: Enumeration of Coliform Bacteria. 4 pp.
ISO 5541 Pt 1-86. Milk and Milk Products—Enumeration of Coliforms—Part 1: Colony Count Technique at 30 Degrees Celsius First Edition. 9 pp.
ISO 5541 Pt 2-86. Milk and Milk Products—Enumeration of Coliforms—Part 2: Most Probable Number Technique at 30 Degrees Celsius First Edition. 10 pp.

—Food
BSI BS 5763: Part 2-91. 1991 Microbiological Examination of Food and Animal Feeding Stuffs Part 2: Enumeration of Coliforms—Colony Count Technique (ISO 4832: 1991). 11 pp.
BSI BS 5763: Part 3-91. 1991 Microbiological Examination of Food and Animal Feeding Stuffs Part 3: Enumeration of Coliforms—Most Probable Number Technique (ISO 4831: 1991). 19 pp.
CNS N6194-88. Methods of Test for Food Microbiology-Test of Coliform Bacteria (May)(10984). 3 pp.
ISO 4831-91. Microbiology—General Guidance for the Enumeration of Coliforms—Most Probable Number Technique Second Edition. 16 pp.
ISO 4832-91. Microbiology—General Guidance for the Enumeration of Coliforms —Colony Count Technique Second Edition. 9 pp.
SAA AS 1766.2.3-92. Food Microbiology—Part 2: Examination for Specific Organisms—Part 2.3: Coliforms and Escherichia Coli (Supersedes AS 1095.3.1—1987). 7 pp.

—Meat
ISO 3811-79. Meat and Meat Products—Detection and Enumeration of Presumptive Coliform Bacteria and Presumptive Escherichia Coli (Reference Method) First Edition. 7 pp.

—Water
DIN ENGL 38411 Pt 6-91. German Standard Methods for the Examination of Water, Waste Water and Sludge; Microbiological Methods (Group K); Determination of Escherichia Coli and Coliform Organisms (K 6) (June). 6 pp.
ISO 9308 Pt 1-90. Water Quality—Detection and Enumeration of Coliform Organisms, Thermotolerant Coliform Organisms and Presumptive Escherichia Coli—Part 1: Membrane Filtration Method First Edition. 14 pp.
ISO 9308 Pt 2-90. Water Quality—Detection and Enumeration of Coliform Organisms, Thermotolerant Coliform Organisms and Presumptive Escherichia Coli—Part 2: Multiple Tube (Most Probable Number) Method First Edition. 13 pp.

Coliform Bacteria (Cont.)

—Water (Cont.)
SAA AS 1095.4.1. 3-81. Microbiological Methods for the Dairy Industry—Part 4: Methods for the Examination of Water and Air—Part 4.1.3: Microbiological Examination of Water—Coliforms by Multiple Tube Dilution. 3 pp.
SAA AS 1095.4.1. 5-81. Microbiological Methods for the Dairy Industry—Part 4: Methods for the Examination of Water and Air—Part 4.1.5: Microbiological Examination of Water—Coliforms by Membrane Filtration. 4 pp.

Collapsible Plastic Containers
Use: Plastic Containers—Collapsible

Collapsible Tubes
Use: Tubes (Packages)

Collars (Mechanical Components)
See Also: Mechanical Components

—Aircraft
SBAC AS 2393-2395 ISSUE 2. Collar. 1 p.
SBAC AS 5369 ISSUE 3. Loose Collar. 1 p.
SBAC AS 44408 ISSUE 1. Collar, Aluminium Alloy, Anodised, Metric.

—Arbors
CNS B3355-80. Arbor Collars for Milling Machines (Jul)(5670).
CNS B3356-80. Arbor Bearing Collars for Milling Machine (Jul)(5671).
CNS B3357-80. Arbor End Collars for Milling Machine (Jul)(5672).
CNS B3412-81. Bearing Collars for Milling Machine Arbors for Milling Cutters with Axial Keyway (Apr)(7196).
DIN ENGL 2084 Pt 2-64. Spacer Collars for Hob Arbors for Hobs with Longitudinal Keyway (July). 2 pp.
DIN ENGL 2084 Pt 3-64. Spacer Collars for Hob Arbors for Hobs with Transverse Slot (July). 1 p.

—Bolts—Aerospace
DIN ENGL LN 65047 Pt 1-78. Aerospace; Bolts, Close Tolerance, in Titanium Alloy, with Self-Locking Collars in Aluminium Alloy; Procurement Specification; Bolts (May). 17 pp.
DIN ENGL LN 65047 Pt 2-78. Aerospace; Bolts, Close Tolerance, in Titanium Alloy, with Self-Locking Collars in Aluminium Alloy; Procurement Specification; Collars (May). 10 pp.

—Butt Joints
DIN ENGL 3867-82. Solderless Pipe Unions with Cutting Ring; Thrust Collars for Butt Joints (Nov). 2 pp.

—Cotter Pins
CNS B2007-72. Collar (Using Cotter Pin or Taper Pin) (Jun)(122).

—Cylindrical Roller Bearings
BSI BS 5646: Part 6-80. 1980 Rolling Bearings—Accessories Part 6: Dimensions of Separate Thrust Collars for Cylindrical Roller Bearings. 3 pp.
CNS B2613-81. Cylindrical Roller Bearings; Separate Thrust Collar; E Types (Oct)(7950).
ISO 246-78. Cylindrical Roller Bearings—Separate Thrust Collars—Boundary Dimensions First Edition. 3 pp.

—Flanges
DIN ENGL 2641-75. Lapped Flanges; Welding Neck Flanges; Plain Collars; Nominal Pressure 6 (Mar). 2 pp.
DIN ENGL 2642-75. Slip-On Flanges; Upturned Welding Flanges; Plain Collars; Nominal Pressure 10 (Mar). 2 pp.
DIN ENGL 2655-75. Lapped Flanges; Plain Collars; Nominal Pressure 25 (Mar). 2 pp.
DIN ENGL 2656-75. Lapped Flanges; Plain Collars; Nominal Pressure 40 (Mar). 2 pp.

—Locking—Radial Bearings
BSI BS 6010: Part 4-93. 1993 Rolling Bearings—Bearings with Spherical Outside Surface and Extended Inner Ring Width Part 4: Specification for Insert Bearings and Eccentric Locking Collars (ISO 9628: 1992) (SUPERSEDES BS 6010: PARTS 1 & 2: 1980) (E). 18 pp.
ISO 9628-92. Rolling Bearings—Insert Bearings and Eccentric Locking Collars First Edition. 15 pp.

—Machine Tools
BSI BS 4185: Part 2-67. 1967 Amd 1 Machine Tool Components Part 2: Collars. 17 pp.

—Manholes—Storage Tanks
DIN ENGL 6627-89. Collars for Masonry Manhole Shafts for Underground Tanks Designed for the Storage of Flammable and Non-Flammable Water Polluting Liquids (Sept). 2 pp.

INTERNATIONAL AND NON-U.S. NATIONAL STANDARDS
SUBJECT INDEX

Collars (Mechanical Components) (Cont.)

—**Multispindle Heads**
DIN ENGL 69001 Pt 40-81. Machine Tools; Multi-Spindle Heads; Thrust Collars; Types A to C (Oct). 4 pp.

—**Pilot Seats—Gears—Aircraft**
SBAC AS 2112 ISSUE 1. Stop Collar, Pilots Seat-Raising Gear. 1 p.

—**Pilot Seatslot Seats**
SBAC AS 2124 ISSUE 6 (I). Stop Collar. 1 p.

—**Pins—Aerospace**
AECMA PREN2365-82. Collars, Aluminium Alloy.
BSI BS EN 2365-89. 1989 Collars, Aluminium Alloy. 8 pp.
CEN EN 2365-89. Collars, Aluminium Alloy. 5 pp.

—**Pipe Fittings**
CNS B5028-82. Ductile Cast-Iron Special Casting for Pressure Main Lines, Collars (Mar)(838).
DIN ENGL 16966 Pt 6-82. Glass Fibre Reinforced Polyester Resin (UP-GF) Pipe Fittings and Joint Assemblies; Collars, Flanges, Joint Rings, Dimensions (July). 11 pp.
DIN ENGL 28624-90. Ductile Iron Collars for Use with Gas and Water Pipes; Dimensions and Mass (Jan). 3 pp.

—**Pipe Fittings—Aircraft**
SBAC AGS 952 ISSUE 18. Pipe Couplings Collar.

—**Pipe Joints**
DIN ENGL 16966 Pt 6-82. Glass Fibre Reinforced Polyester Resin (UP-GF) Pipe Fittings and Joint Assemblies; Collars, Flanges, Joint Rings, Dimensions (July). 11 pp.

—**Plain Bearings—Surface Roughness**
DIN ENGL 31670 Pt 8-86. Plain Bearings; Quality Assurance for Plain Bearings; Checking the Geometrical Tolerances and Surface Roughness of Shafts, Collars, and Thrust Collars (July). 4 pp.
DIN ENGL 31699-86. Plain Bearings; Shafts, Collars, Thrust Collars; Geometrical Tolerances and Surface Roughness (July). 3 pp.

—**Press Tools—Springs (Elastic)**
BSI BS 7568: Part 13-92. 1992 Press Tools Part 13: Elastomer Pressure Springs—Specification of Accessories (ISO 10069-2: 1991). 7 pp.
ISO 10069 Pt 2-91. Tools for Pressing—Elastomer Pressure Springs—Part 2: Specification of Accessories First Edition. 6 pp.

—**Rudder Bars—Worms (Mechanical Drives)—Aircraft**
SBAC AS 2178 ISSUE 6. Collar for Adjusting Worm (Rudder Bar). 1 p.

—**Set Screws**
CNS B2005-47. Collar (Using Light Duty Set Screw) (Mar)(120)(R 1973).
CNS B2006-47. Collar (Using Heavy Duty Set Screw) (Mar)(121)(R 1973).

—**Shear Pins—Aerospace**
DIN ENGL LN 29798-80. Aerospace; Collars, Threaded; Self-Locking for Head Pins of Titanium Alloy (Sept). 3 pp.

—**Shear Pins—Aircraft**
BSI SP 121-58. 1958 Collars for Shear Pins for Aircraft. 2 pp.

—**Stop Valves—Connecting Rods—Aircraft**
SBAC AS 4893 ISSUE 1. Collar—Float Needle Connecting Rod. 1 p.

—**Taper Pins**
CNS B2007-72. Collar (Using Cotter Pin or Taper Pin) (Jun)(122).

Collect Calls
See Also: Operator Assisted Calls; Reverse Charging; Telephone Services; Telephones
CCITT RECMN E.140-89. Principles for the Operation of International Telephone Services—Telephone Network and ISDN—Operation, Numbering, Routing and Mobile Service (Study Group II) 3 pp. 3 pp.

—**Pay Telephones**
CEPT T/TPH 38 E-83. Prevention of Fraudulent Collect to Public Call Offices. 1 p.

Collect Calls (Cont.)

—**Tariffs**
CCITT RECMN D.174-89. Conventional Transmission of Information Necessary for Billing and Accounting Regarding Collect and Credit Card Calls—General Tariff Principles—Charging and Accounting in International Telecommunications Services (Study Group III) 1 pp. 1 p.
CCITT RECMN E.277-89. Conventional Transmission of Information Necessary for the Collection of Charges and the Accounting Regarding Collect and Credit Card Calls—Telephone Network and ISDN—Operation, Numbering, Routing and Mobile Service (Study Group II) 1pp(Same as Recmn D.174). 1 p.

Collective Investments (Securities)
Use: Securities—Collective Investments

Collective Nuclear, Biological, and Chemical Protection
Use: Nuclear, Biological, and Chemical Warfare—Collective Protection

Collectors' Items

—**Value-Added Tax**
EC COM(88) 846-89. Proposal for a Council Directive Supplementing the Common System of Value-Added Tax and Amending Articles 32 and 28 of Directive 77/388/EEC—Special Arrangements for Second-Hand Goods, Works of Art, Antiques and Collectors' Items. 24 pp.

Collets
Use For: Draw Back Collets; Pushout Collets
See Also: Chucks; Sleeves (Fittings)
DIN ENGL 6341 Pt 1-57. Drawback Collets and Taper Sleeves for Collets (Nov). 2 pp.
DIN ENGL 6343-61. Push-Out Collets (Apr). 2 pp.

—**Aircraft**
SBAC AGS 1657 ISSUE 8. Collet.

—**Machine Tools**
BSI BS 4185: Part 7-73. 1973 Machine Tool Components Part 7: Wide Range Collets. 12 pp.

—**Packaging**
MOD UK DSTAN 81-57-90. Packaging of Knobs, Finger and Ancillary Items Issue 1. 9 pp.

—**Pads—Chucks**
BSI BS 1983: Part 3-87. 1987 Chucks for Machine Tools and Portable Power Tools Part 3: Dimensions of Collet Pads for Base Jaws of Power Operated Workholding Chucks. 10 pp.

—**Threading Dies**
CGSB 46-GP-6-60. Dies; Threadcutting and Rethreading, Collets and Diestocks. 13 pp.

—**Threads**
DIN ENGL 6341 Pt 2-59. Drawback Collets; Collet Thread; Nominal Dimensions; Tolerances; Limits (June). 2 pp.

—**Watches**
DIN ENGL 8255 Pt 1-72. Collets for Watches and Clocks; Slotted, Round (Nov). 3 pp.

Colliery Equipment
Use: Mining Equipment

Collimators
See Also: Autocollimators; Optical Equipment

—**Headlamps**
BSI BS AU 162A-76. 1976 Amd 1 Headlamp Optical Aiming Devices. 8 pp.

Collision Avoidance
See Also: Accident Prevention; Air Traffic Control; Air Traffic Control Equipment; Near Misses (Aircraft)

—**Aircraft**
ICAO Circular 174. Secondary Surveillance Radar Mode S Advisory Circular—1983. 42 pp.
ICAO Circular 213. Pilot Skills to Make "Look-Out" More Effective in Visual Collision Avoidance—1989. 16 pp.
ICAO 9274. Manual on the Use of the Collision Risk Model (CRM) for ILS Operations First Edition—1980. 307 pp.

Collision Research
Use For: Automobile Accidents; Motor Vehicle Accidents; Traffic Accidents *See Also:* Acceleration Testing; Accident Investigations; Aircraft Accidents; Impact Testing; Motor Vehicle Operators; Road Testing

Collision Research (Cont.)

—**Barriers**
ISO 3560-75. Road Vehicles—Frontal Fixed Barrier Collision Test Method First Edition. 4 pp.
ISO 3984-82. Road Vehicles—Passenger Cars—Moving Barrier Rear Collision Test Method Second Edition. 6 pp.
JIS D 1060-82. Frontal and Rear Vehicle Collision Test Procedure. 7 pp.

—**Bumpers**
ISO 2958-73. Road Vehicles—Exterior Protection for Passenger Cars First Edition. 6 pp.

—**Fuel Leakage**
ISO 3437-75. Road Vehicles—Determination of Fuel Leakage in the Event of a Collision First Edition. 3 pp.
JIS D 1042-84. Determination of Fuel Leakage in the Event of a Collision for Passenger Cars. 5 pp.

—**Glossaries**
BSI BS AU 191-83. 1983 Road Vehicle Collisions. 10 pp.
ISO 6813-81. Road Vehicles—Collisions—Terminology First Edition. 10 pp.

—**Impact Velocity**
BSI BS AU 164-77. 1977 Measurement of Impact Velocity in Collision Tests on Road Vehicles. 4 pp.
ISO 3784-76. Road Vehicles—Measurement of Impact Velocity in Collision Tests First Edition. 5 pp.

—**Safety Belts**
ISO TR6546-79. Road Vehicles—Information Core Appropriate to the Field Study of Accidents in Which Seat Belts Are Used First Edition. 4 pp.

—**Safety Belts—Sled Testing**
ISO 7862-92. Passenger Cars—Sled Test Procedure for Evaluating Adult Restraint Systems in Simulated Frontal Collisions First Edition. 12 pp.

—**Steering Gear**
EC 74/297/EEC-74. Council Directive on the Approximation of the Laws of the Member States Relating to the Interior Fittings of Motor Vehicles (the Behaviour of the Steering Mechanism in the Event of an Impact). 10 pp.
JIS D 1061-85. Steering Control System Laboratory Impact Test Procedure for Passenger Cars.

Collision Risk Model
Use: Collision Avoidance—Aircraft

Collodion Cotton
Use: Nitrocellulose

Colony Count Technique
Use: Bacteria Count Methods

Colophonium
Use: Rosins

Colophonium Content Analysis
Use: Rosin Content Analysis

Color
Scope Note: For additional listings, see the subheading Colors under specific products or materials *Use Use:* Colour; Hue
See Also: Chromaticity; Color Charts; Color Coding; Color Matching; Color Temperature; Colorimetry; Contrast; Discoloration; Electromagnetic Properties; Light (Visible Radiation); Optical Properties; Printing Inks; Symbols; Tristimulus Values
CNS Z7195-85. Specification of Colours According to Their Three Attributes (Jun)(11295).
JIS Z 8721-77. Specification of Colours According to Their Three Attributes. 45 pp.

—**Aircraft Cabins**
NATO STANAG 3701 ED 2 AMD 4-78. Aircraft Interior Colour Schemes. 14 pp.
NATO STANAG 3701 ED 3 AMD 0-92. Aircraft Interior Colour Schemes. 6 pp.
NATO STANAG 3701 ED 3 AMD 1-92. Aircraft Interior Colour Schemes. 7 pp.

—**Classification**
BSI BS 1000: (667)-79. 1979 Universal Decimal Classification (UDC). English Full Edition (667): Colour Industries (Dyes, Inks, Paints Etc.). 20 pp.
SNZ NZS/BS 1000 (667)-79. Universal Decimal Classification Colour Industries (Dyes, Inks, Paints, Etc.). 20 pp.

INDUSTRY STANDARDS

INTERNATIONAL AND NON-U.S. NATIONAL STANDARDS
SUBJECT INDEX

Color (Cont.)

—Glossaries
BSI BS 4727:Pt4: Group 02-71. (WITHDRAWN) 1971 Amd 1 Electrotechnical, Power Telecommunications, Electronics, Lighting and Colour Terms Part 4: Terms Particular to Lighting and Colour Group 02: Vision and and Colour Terminology. 27 pp.
JIS Z 8102-85. Names of Non-Luminous Object Colours. 34 pp.
JIS Z 8105-82. Glossary of Colour Terms. 66 pp.
JIS Z 8110-91. Names of Light-Source Colours. 8 pp.

—Lighting Booths—Visual Inspection
SAA AS 4004-92. Lighting Booths for Visual Assessment of Colour and Colour Matching (in Professional Package 39). 5 pp.

—Open Document Architecture
CCITT RECMN T.410 Series(REV2)-92. Revision (February, 1992) of the T-410 Series (1988) of Recommendations Contained in the CCITT Blue Book, Fascicle VII.6 on the Subject of "Colour" (Study Group VIII) 64 pp. 64 pp.

—Paints—Visual Inspection
SAA AS 1580.601. 1-92. Paints and Related Materials—Methods of Test—Part 601.1: Colour—Visual Comparison (in Professional Packages 30, 39). 8 pp.
SAA AS 1580.601. 2-75. Paints and Related Materials—Methods of Test—Part 601.2: Colour—Instrumental Measurement of Colour Difference Using the 'Colormaster' Differential Colorimeter. 4 pp.
SNZ NZS/AS 1580. 601.1-92. Methods of Test for Paints and Related Materials Part 601.1: Colour—Visual Comparison (This is a Joint Standard with SAA AS 1580.601.1). 8 pp.
SNZ NZS/AS 1580. 601.2-75. Methods of Test for Paints and Related Materials Part 601.2: Colour-Instrumental Measurement of Colour Difference Using the Colourmaster Differential Colorimeter (This is a Joint Standard with SAA AS 1580.601.2). 4 pp.

Color Bar Signals
See Also: Test Signals
CCIR RECMN 471-1-86. Nomenclature and Description of Colour Bar Signals—Section 11A—Characteristics of Systems for Monochrome and Colour Television. 2 pp.

Color Bar Test Patterns
Use: Test Patterns—Color Bar

Color Charts
Use For: Color Matching Fans *See Also:* Charts; Color; Colorimetry
BSI BS 5252F-76. 1976 Framework for Colour Co-Ordination for Building Purposes: Colour Matching Fan. 1 p.

—Colorimetry
DIN ENGL 6164 Pt 1-80. DIN Colour Chart; System Based on the 2 Deg. Standard Colorimetric Observer (Feb). 13 pp.
DIN ENGL 6164 Pt 2-80. DIN Colour Chart; Specification of Colour Samples (Feb). 2 pp.
DIN ENGL 6164 Pt 3-81. DIN Colour Chart; System Based on the 10 Deg. Standard Colorimetric Observer (July). 13 pp.

—Toasters
SAA AS 1907P-76. Reference Print for Evaluating Toast Colour Reconfirmed 1988.

Color Coding
Use For: Colour Codes *See Also:* Color; Electric Wire; Identification Systems; Symbols
BSI BS 381C-88. 1988 Colours for Identification, Coding and Special Purposes. 22 pp.
SNZ NZS 5807C-80. Code of Practice for Industrial Identification by Colour, Wording, or Other Coding. 1 p.
SNZ NZS 7702-89. Specification for Colours for Identification, Coding and Special Purposes Amend: A, 1989. 28 pp.

—Aircraft Instruments
MOD UK DSTAN 66-26: Part 7-86. General Requirements for Aircraft Instruments and Displays Part 7: Colours and Markings Used to Denote Operating Ranges of Instruments Issue 1. 11 pp.
NATO STANAG 3436 ED 4 AMD 3-85. Colours and Markings Used to Denote Operating Ranges of Aircraft Instruments. 15 pp.
NATO STANAG 3436 ED 4 AMD 4-85. Colours and Markings Used to Denote Operating Ranges of Aircraft Instruments. 16 pp.

—Aircraft Instruments—Emergency
NATO STANAG 3341 ED 3 AMD 6-72. Emergency Control Colour Schemes (Withdrawn 93-06). 6 pp.

Color Coding (Cont.)

—Aircraft Landing Areas—Markings
NATO STANAG 3711 ED 1 AMD 6-76. Airfield Marking and Lighting Colour Standards. 16 pp.
NATO STANAG 3711 ED 2 AMD 0-92. Airfield Marking and Lighting Colour Standards. 15 pp.
NATO STANAG 3711 ED 2 AMD 1-92. Airfield Marking and Lighting Colour Standards. 16 pp.

—Airport Lighting
NATO STANAG 3711 ED 1 AMD 6-76. Airfield Marking and Lighting Colour Standards. 16 pp.
NATO STANAG 3711 ED 2 AMD 0-92. Airfield Marking and Lighting Colour Standards. 15 pp.
NATO STANAG 3711 ED 2 AMD 1-92. Airfield Marking and Lighting Colour Standards. 16 pp.

—Ammunition
NATO STANAG 2321 ED 4 AMD 1-87. The NATO Code of Colours for the Indentification of Ammunition (Except Ammunition of a Calibre Below 20mm. 13 pp.
NATO STANAG 2953 ED 1 AMD 0-90. Method of Application of the NATO Code of Colours for the Identification of Ammunition (Except Ammunition of a Calibre Below 20 mm)—AOP-2(A). 4 pp.

—Annunciators
BSI BS 4099· Part 2-77. (WITHDRAWN) 1977 Amd 1 Colours of Indicator Lights, Push Buttons, Annunciators and Digital Readouts: Flashing Lights, Annunciators and Digital Readouts (Superseded by BS EN 60073: 1993). 8 pp.

—Cable Insulation
DIN VDE 0206-64. Recommendations on Colours for the Overall Sheaths and Coverings of Plastics or Rubber for Cables and Flexible Cords (June). 5 pp.

—Cables (Electric)
BSI BS 1843-52. 1952 Amd 1 Colour Code for Twin Compensating Cables for Thermocouples. 8 pp.
BSI BS AU 7A-83. 1983 Colour Code for Road Vehicle Electrical Cables. 11 pp.
BSI PD 2379-90. 1990 Register of Colours of Manufacturers' Identification Threads for Electric Cables and Cords. 25 pp.
DIN VDE 0293-90. Identification of Cores and Cables and Flexible Cords Used in Power Installations with Voltage Ratings up to 1000 V (Jan). 14 pp.
SNZ NZS/BSPD 2379-90. Register of Colours of Manufacturers' Identification Threads for Electric Cables and Cords. 26 pp.

—Capacitors
BSI BS 5890-80. 1980 Guide for Choice of Colours to Be Used for the Marking of Capacitors and Resistors. 3 pp.
IEC 425-73. Guide for the Choice of Colours to Be Used for the Marking of Capacitors and Resistors First Edition. 7 pp.

—Clothing
CGSB 38-GP-38-74. Color Code, for Size Identification of Clothing and Linen, Standard for. 5 pp.

—Compression Springs
BSI BS 7568: Part 4-92. 1992 Press Tools Part 4: Compression Springs with Rectangular Section—Housing Dimensions and Colour Coding (ISO 10243: 1991). 22 pp.
ISO 10243-91. Tools for Pressing—Compression Springs with Rectangular Section—Housing Dimensions and Colour Coding First Edition. 20 pp.

—Cordage
BSI BS 6033-81. 1981 Colour Code for Identification of the Constituent Material of Ropes and Cordage. 4 pp.

—Cords (Electric)
BSI PD 2379-90. 1990 Register of Colours of Manufacturers' Identification Threads for Electric Cables and Cords. 25 pp.
DIN VDE 0293-90. Identification of Cores and Cables and Flexible Cords Used in Power Installations with Voltage Ratings up to 1000 V (Jan). 14 pp.
SNZ NZS/BSPD 2379-90. Register of Colours of Manufacturers' Identification Threads for Electric Cables and Cords. 26 pp.

—Cranes—Manual Controls
ISO 7296 Pt 1-91. Cranes—Graphic Symbols —Part 1: General First Edition. 18 pp.

—Dental Instruments
BSI BS 6828: Sec 6.3-93. 1993 Dental Rotary Instruments Part 6: Diamond Instruments Section 6.3: Specification for Grit Sizes, Designation and Colour Code (ISO 7711-3: 1992). 7 pp.
ISO 7711 Pt 3-92. Dental Rotary Instruments—Diamond Instruments—Part 3: Grit sizes, Designation and Colour Code First Edition. 5 pp.

Color Coding (Cont.)

—Digital Readouts
BSI BS 4099: Part 2-77. (WITHDRAWN) 1977 Amd 1 Colours of Indicator Lights, Push Buttons, Annunciators and Digital Readouts: Flashing Lights, Annunciators and Digital Readouts (Superseded by BS EN 60073: 1993). 8 pp.

—Electric Conductors
CENELEC HD 27-78. Colours of the Cores of Flexible Cables and Cords. 7 pp.
CENELEC HD 308-76. Identification and Use of Cores of Flexible Cables. 4 pp.
CENELEC HD 324-76. Identification of Insulated and Bare Conductors by Colours. 2 pp.
CNS C1051-84. General Rules of Color Identification for Protective Conductors and Neutral Conductor and Terminal Marking for Apparatus (Dec)(5202).
IEC 173-64. Colours of the Cores of Flexible Cables and Cords First Edition. 5 pp.
IEC 446-89. Identification of Conductors by Colours or Numerals Second Edition. 18 pp.
JIS C 0602-84. General Rules of Colour Identification for Protective Conductor and Neutral Conductor and Terminal Marking for Apparatus. 5 pp.

—Electric Wire
CNS C5072-80. Color Coding of Chassis Wiring (Aug)(6321).
MOD UK DSTAN 61-7: Part 1-90. Identification of Electrical and Electronic Systems. Wiring and Components Part 1: Colour Coding and Marking for Polarity and Phase Distinction in Electrical Systems Issue 6. 11 pp.

—Electrical Equipment
BSI BS 7645-93. 1993 Code for Designation of Colours (IEC 757: 1983) (H). 9 pp.
CENELEC HD 457-85. Code for Designation of Colours. 2 pp.
CNS C5076-80. Colors for Color Indentification and Coding (Dec)(6661).
IEC 757-83. Code for Designation of Colours First Edition. 10 pp.
SNZ IEC 757-83. Code for Designation of Colours. 7 pp.

—Electronic Components
CNS C5138-82. Color Code for Electronic Parts (Jul)(9079).
JIS C 0801-64. Color Code for Electronic Parts (R 1970). 4 pp.

—Engineering Drawings—Ventilation Equipment—Ships
ISO 5571-81. Shipbuilding—Identification Colours for Schemes for Ventilation Systems First Edition. 3 pp.

—Flashing Lights
BSI BS 4099: Part 2-77. (WITHDRAWN) 1977 Amd 1 Colours of Indicator Lights, Push Buttons, Annunciators and Digital Readouts: Flashing Lights, Annunciators and Digital Readouts (Superseded by BS EN 60073: 1993). 8 pp.

—Forklifts—Operator Controls—Symbols
JIS D 6022-85. Control Symbols for Fork Lift Trucks. 20 pp.

—Foundry Patterns
DIN ENGL 1511-78. Pattern Equipment for Foundries; Production and Quality (Apr). 10 pp.
SAA AS B8-59. Methods for the Colouring and Marking of Foundry Patterns (Withdrawn).

—Gas Cylinders
CEN PREN 1089-3-93. Cylinder Identification—Part 3: Colour Coding System for Gas Cylinders for Use in Europe. 8 pp.

—Gas Cylinders—Medical
ISO 32-77. Gas Cylinders for Medical Use—Marking for Identification of Content First Edition; (Erratum—Aug 1979). 5 pp.

—Grounding Conductors—Medical Electrical Equipment
CSA CAN/CSA-C22. 2NO601.1-M90. Medical Electrical Equipment, Part 1: General Requirements for Safety; (Gen Instr 1). 240 pp.

—Harnesses (Electric)—Automotive
CNS D1052-83. Colour Code for Wiring Harness for Automobiles (Apr)(10159).

—Hypodermic Needles
BSI BS 7128-93. 1993 Colour Coding of Hypodermic Needles for Single Use (ISO 6009: 1982). 15 pp.
BSI BS 7128-89. 1989 Colour Coding of Hypodermic Needles for Single Use (ISO 6009: 1988). 7 pp.
ISO 6009-92. Hypodermic Needles for Single Use—Colour Coding for Identification Third Edition. 12 pp.

Color Coding (Cont.)

—Indicator Lights
BSI BS 4099: Part 1-86. (WITHDRAWN) 1986 Colours of Indicator Lights, Push Buttons, Annunciators and Digital Readouts Part 1: Colours of Indicator Lights and Push Buttons (Superseded by BS EN 60073: 1993). 12 pp.

BSI BS EN 60073-93. 1993 Coding of Indicating Devices and Actuators by Colours and Supplementary Means (IEC 73: 1991) (H). 29 pp.

CENELEC HD 354 S2-87. Colours of Indicator Lights and Push-Buttons. 4 pp.

CENELEC EN 60073-93. Coding of Indicating Devices and Actuators by Colours and Supplementary Means (IEC 73:1991) (Supersedes HD 354 S2:1987). 5 pp.

CNS C1067-80. Colors of Indicator Lights and Push Buttons (Jul)(5741).

CSA CAN/CSA-Z431-M89. Colours of Indicator Lights and Push-Buttons (IEC 73-1984). 23 pp.

DIN VDE 0199 (D)-88. Coding of Indicating Devices and Actuators by Colours and Supplementary Means; Requirements for Safety; Identical with IEC 16 (Secretariat) 302 (Dec). 36 pp.

IEC 73-91. Coding of Indicating Devices and Actuators by Colours and Supplementary Means Fourth Edition; (CENELEC EN 60073: 1993). 48 pp.

SNZ BS 4099: Part 1-86. Colours of Indicator Lights, Push-Buttons, Annunciators and Digital Readouts Part 1: Specification for Colours of Indicator Lights and Push-Buttons. 12 pp.

—Indicator Lights—Ships
ISO 2412-82. Shipbuilding—Colours of Indicator Lights Second Edition. 3 pp.

—Industrial Equipment
SAA AS 1318-85. Use of Colour for the Marking of Physical Hazards and the Identification of Certain Equipment in Industry (Known as the SAA Industrial Safety Colour Code) (Incorporating Amdt 1). 6 pp.

—Linen
CGSB 38-GP-38-74. Color Code, for Size Identification of Clothing and Linen, Standard for. 5 pp.

—Lumber
SAA AS 1613-74. Colours for Marking Stress Graded Timber. 3 pp.

—Materiel—Air Drop Operations
NATO STANAG 3427 ED 4 AMD 1-77. Colours for the Identification of Airdropped Supplies. 6 pp.

NATO STANAG 3427 ED 4 AMD 2-77. Colours for the Identification of Airdropped Supplies. 7 pp.

—Medical Electrical Equipment
CSA CAN/CSA-C22. 2NO601.1-M90. Medical Electrical Equipment, Part 1: General Requirements for Safety; (Gen Instr 1). 240 pp.

—Metals—Aerospace
MOD UK DSTAN 05-69-93. Standard Colour Scheme of Metallic Materials for Aerospace Applications Issue 1. 66 pp.

MOD UK QTR 7/AQD. Standard Colour Scheme for Metallic Materials for Aerospace Applications Issue 1.

—Pajamas
CGSB CAN/CGSB-38.104-93. Men's Hospital Pajamas—Dimensions and Colour Size Coding. 8 pp.

—Pipelines
DIN ENGL 2403-84. Identification of Pipelines According to the Fluid Conveyed (Mar). 8 pp.

DIN ENGL 2404-42. Identification Colour Code for Heating System Pipelines (Dec). 1 p.

—Pipettes
BSI BS 3996-78. 1978 Colour Coding for One-Mark and Graduated Pipettes (Including Requirements for the Service Performance of the Colour Coding Enamels). 7 pp.

ISO 1769-75. Laboratory Glassware—Pipettes—Colour Coding First Edition. 4 pp.

—Power Cables
CSA 1369 Bull. Electrical Bulletin 1369 May 4, 1982 to C22.2 NO 96. 3 pp.

—Protective Clothing—Flight Decks
NATO STANAG 1050 ED 3 AMD 1-83. Distinctive Colour of Clothing for Flight Deck Personnel. 10 pp.

NATO STANAG 1050 ED 3 AMD 2-83. Distinctive Colour of Clothing for Flight Deck Personnel. 12 pp.

Color Coding (Cont.)

—Pushbutton Switches
BSI BS 4099: Part 1-86. (WITHDRAWN) 1986 Colours of Indicator Lights, Push Buttons, Annunciators and Digital Readouts Part 1: Colours of Indicator Lights and Push Buttons (Superseded by BS EN 60073: 1993). 12 pp.

BSI BS EN 60073-93. 1993 Coding of Indicating Devices and Actuators by Colours and Supplementary Means (IEC 73: 1991) (H). 29 pp.

CENELEC HD 354 S2-87. Colours of Indicator Lights and Push-Buttons. 4 pp.

CENELEC EN 60073-93. Coding of Indicating Devices and Actuators by Colours and Supplementary Means (IEC 73:1991) (Supersedes HD 354 S2:1987). 5 pp.

CNS C1067-80. Colors of Indicator Lights and Push Buttons (Jul)(5741).

CSA CAN/CSA-Z431-M89. Colours of Indicator Lights and Push-Buttons (IEC 73-1984). 23 pp.

DIN VDE 0199 (D)-88. Coding of Indicating Devices and Actuators by Colours and Supplementary Means; Requirements for Safety; Identical with IEC 16 (Secretariat) 302 (Dec). 36 pp.

IEC 73-91. Coding of Indicating Devices and Actuators by Colours and Supplementary Means Fourth Edition; (CENELEC EN 60073: 1993). 48 pp.

SNZ BS 4099: Part 1-86. Colours of Indicator Lights, Push-Buttons, Annunciators and Digital Readouts Part 1: Specification for Colours of Indicator Lights and Push-Buttons. 12 pp.

—Resistors
BSI BS 5890-80. 1980 Guide for Choice of Colours to Be Used for the Marking of Capacitors and Resistors. 3 pp.

CNS C1087-82. Colour Code for Miniature Fixed Resistor (Jul)(9080).

IEC 425-73. Guide for the Choice of Colours to Be Used for the Marking of Capacitors and Resistors First Edition. 7 pp.

JIS C 0802-90. Colour Code for Fixed Resistors for Electronic Equipment. 8 pp.

—Rivets
MOD UK DTD-913A-61. Identification Colouring of Rivets in Aluminium and Aluminium Alloys (Reprinted January 1962). 2 pp.

—Robots—Controllers—Symbols
JIS B 8434-91. Industrial Robots—Identification Symbols for Operator Controls. 26 pp.

—Ropes
BSI BS 6033-81. 1981 Colour Code for Identification of the Constituent Material of Ropes and Cordage. 4 pp.

—Rubber—Aircraft
SBAC RS 614 (V)(I). Recommended Colour Code for the Identification of Rubbers in Aircraft.

—Safety
CNS Z1024-87. General Rule of Safety Colour (May)(9328).

CNS Z1027-87. General Rules of Coloured Light for Safety (9331).

CNS Z1033-87. General Code of Fluorescent Safety Colour (Mar)(9646).

ISO 3864-84. Safety Colours and Safety Signs First Edition. 16 pp.

JIS Z 9101-86. General Code of Safety Colour.

JIS Z 9104-87. General Rules of Coloured Light for Safety. 9 pp.

JIS Z 9106-90. Fluorescent Safety Colours—General Rules for Application.

SAA AS 1318-85. Use of Colour for the Marking of Physical Hazards and the Identification of Certain Equipment in Industry (Known as the SAA Industrial Safety Colour Code) (Incorporating Amdt 1). 6 pp.

SNZ NZS 5807-80. Code of Practice for Industrial Identification by Colour, Wording or Other Coding Amend: 1, 1983; 2, 1988. 20 pp.

—Safety—Lifting Equipment
DIN ENGL 15026-78. Lifting Appliances; Marking of Points of Hazard (Jan). 2 pp.

—Safety—Railroad Crossings
JIS E 3701-84. Safety Colour Code for Railway Crossing Equipment. 12 pp.

—Safety—Signs
CNS Z1034-87. Fluorescent Safety Signs Board (Mar)(9647).

—Semiconductor Devices
CNS C5059-88. Color Coding of Discrete Semiconductor Devices (Jan)(6137).

—Signal Devices—Automotive
BSI BS AU 143C-84. 1984 Symbols for Controls, Indicators and Tell-Tales for Road Vehicles. 8 pp.

Color Coding (Cont.)

—Signal Devices—Automotive (Cont.)
ISO 2575-82. Road Vehicles—Symbols for Controls, Indicators and Tell-Tales Fourth Edition. 9 pp.

JIS D 0032-82. Symbols for Controls, Indicators and Tell-Tales for Automobiles. 17 pp.

—Signal Lights
BSI BS 1376-74. 1974 Amd 1 Colours of Light Signals. 49 pp.

—Steel Tube Fittings
ISO 9095-90. Steel Tubes—Continuous Character Marking and Colour Coding for Material Identification First Edition. 9 pp.

—Steel Tubes
BSI BS 5383-86. 1986 Material Identification of Steel, Nickel Alloy and Titanium Alloy Tubes by Continuous Character Marking and Colour Coding of Steel Tubes. 10 pp.

ISO 9095-90. Steel Tubes—Continuous Character Marking and Colour Coding for Material Identification First Edition. 9 pp.

—Tactical Data Systems—Marine
NATO STANAG 4420 ED 1 AMD 1-90. (DRAFT) Display Symbology and Colours for NATO Maritime Units (Withdrawn 93-06). 100 pp.

Color Contribution Index, Photographic Lenses
Use: Photographic Lenses—Color Contribution Index

Color Filters
See Also: Filters; Optical Filters

BSI BS 3944-65. (WITHDRAWN) 1965 Amd 1 Colour Filters for Theatre Lighting and Other Purposes (Superseded by BS 3944: Part 1: 1992). 23 pp.

—Dimensional Stability
BSI BS 3944: Part 1-92. 1992 Colour and Diffusion Filter Material for Theatre, Television and Similar Entertainment Purposes Part 1: Specification for Flammability and Dimensional Stability. 8 pp.

—Flammability Testing
BSI BS 3944: Part 1-92. 1992 Colour and Diffusion Filter Material for Theatre, Television and Similar Entertainment Purposes Part 1: Specification for Flammability and Dimensional Stability. 8 pp.

—Theaters
BSI BS 3944-65. (WITHDRAWN) 1965 Amd 1 Colour Filters for Theatre Lighting and Other Purposes (Superseded by BS 3944: Part 1: 1992). 23 pp.

Color Index Testing
Use: Colorimetry

Color Matching
See Also: Color; Colorimetry; Dyeing

—Standard Colorimetric Observers
ISO CIE10527-91. CIE Standard Colorimetric Observers First Edition. 31 pp.

Color Matching Fans
Use: Color Charts

Color Photographic Films
Use: Photographic Films

Color Stability
Use: Colorfastness Testing

Color Temperature
See Also: Color; Temperature

—Light Sources
JIS Z 8725-87. Methods for Determining Distribution Temperature and Colour Temperature or Correlated Colour Temperature of Light Sources. 28 pp.

Color Testing
See Also: Colorimetry; Volumetric Analysis

—Adipates
ISO 2524-74. Adipate Esters for Industrial Use—Measurement of Colour After Heat Treatment First Edition. 4 pp.

—Alcohols
CNS K6414-78. Test for Colour with Sulphuric Acid of Higher Alcohols for Industrial Use (Mar)(4283).

ISO 1843 Pt 8-82. Higher Alcohols for Industrial Use—Methods of Test—Part 8: Sulphuric Acid Colour Test First Edition. 4 pp.

Color Testing (Cont.)

—Aromatic Hydrocarbons
CNS K6254-74. Method of Test for Colors of Aromatic Hydrocarbons (Jun)(2755).
CNS K6256-74. Method of Test for Acid-Wash Colors of Aromatic Hydrocarbons (Oct)(2757).
ISO 5274-79. Aromatic Hydrocarbons—Acid-Wash Test First Edition. 4 pp.
ISO 5276-79. Aromatic Hydrocarbons—Test for Neutrality First Edition. 3 pp.

—Butanols
ISO 755 Pt 3-81. Butan-1-ol for Industrial Use—Methods of Test—Part 3: Sulphuric Acid Colour Test First Edition; (Amendment Slip-1982). 6 pp.

—Cresylic Acids
ISO 1897 Pt VII-77. Phenol, o-Cresol, m-Cresol, p-Cresol, Cresylic Acid and Xylenols for Industrial Use—Methods of Test—Part VII: Measurement of Colour (Cresylic Acid and Xylenols Only) First Edition. 4 pp.

—Drinking Water
BSI BS 6920: Sec 2.3-90. 1990 Suitability of Non-Metallic Products for Use in Contact with Water Intended for Human Consumption with Regard to Their Effect on the Quality of the Water Part 2: Methods of Test Section 2.3: Appearance of Water. 7 pp.
SNZ NZS/BS 6920: Pt 2:Sec 2.3-90. Suitability of Non-Metallic Products for Use in Contact with Water Intended for Human Consumption with Regard to Their Effect on the Quality of Water Part 2: Methods of Test Section 2: Taste of Water Section 2.3: Appearance of Water. 8 pp.

—Fats—Lovibond Tintometer
CNS K6764-83. Method of Test for Colour of Fats and Oils by Means of Lovibond Tintometer (Mar)(10099).
CNS N6076-82. Methods of Test for Edible Oils and Fats (Color Test) (Jan)(3641). 1 p.
JIS K 3503-62. Testing Method of Colour of Fats and Oils by Means of Lovibond Tintometer.

—Fructose Syrups
CNS N6200-85. Method of Test for High Fructose Syrup (Sep)(11370). 3 pp.

—Glucose
CNS N6053-73. Method of Test for Dried Glucose and Glucose Syrup (Jan)(3218). 9 pp.

—Glucose Syrups
CNS N6053-73. Method of Test for Dried Glucose and Glucose Syrup (Jan)(3218). 9 pp.

—Gray Scale
CNS L1005-82. Grey Scale for Assessing Change in Color (Feb)(3839).
JIS L 0804-83. Grey Scale for Assessing Change in Colour. 6 pp.

—Maleic Anhydride
CNS K6566-80. Measurement of Colour of the Molten Material of Maleic Anhydride for Industrial Use (Aug)(6211).
ISO 1390 Pt II-77. Maleic Anhydride for Industrial Use—Methods of Test—Part II: Measurement of Colour of the Molten Material First Edition. 3 pp.

—Oils—Lovibond Tintometer
CNS K6764-83. Method of Test for Colour of Fats and Oils by Means of Lovibond Tintometer (Mar)(10099).
JIS K 3503-62. Testing Method of Colour of Fats and Oils by Means of Lovibond Tintometer.

—Petroleum Products
CNS K6330-72. Method of Test for Colors of Petroleum Products (Oct)(3391).
JIS K 2580-80. Testing Methods for Color of Petroleum Products (R 1985). 21 pp.

—Phthalate Esters
ISO 1385 Pt II-77. Phthalate Esters for Industrial Use—Methods of Test—Part II: Measurement of Colour After Heat Treatment (Diallyl Phthalate Excluded) First Edition. 3 pp.

—Phthalic Anhydride
ISO 1389 Pt II-77. Phthalic Anhydride for Industrial Use—Methods of Test—Part II: Measurement of Colour of the Molten Material First Edition. 4 pp.
ISO 1389 Pt III-77. Phthalic Anhydride for Industrial Use—Methods of Test—Part III: Measurement of Colour Stability First Edition. 4 pp.
ISO 1389 Pt IV-77. Phthalic Anhydride for Industrial Use—Methods of Test—Part IV: Measurement of Colour After Treatment with Sulphuric Acid First Edition. 4 pp.

Color Testing (Cont.)

—Polyester Resins
CNS K6731-82. Method of Test for Liquid Unsaturated Polyester Resin (Dec)(9716). 11 pp.

—Reagents
CNS K7621-82. Chemical Reagent (Determination of Colored Material of Sulfuric Acid) (Jun)(8993).

—Sulfur Dioxide—Air
CSA Z223.22-M1980. Method for the Measurement of Sulphur Dioxide in Air; (Gen Instr 1). 27 pp.

—Vegetable Oils—Lovibond Tintometer
CNS N6076-82. Methods of Test for Edible Oils and Fats (Color Test) (Jan)(3641). 1 p.

—Water
BSI BS 2690: Part 101-84. (WITHDRAWN) 1984 Water Used in Industry Part 101: Dissolved Oxygen (Superseded by BS EN 25813). 10 pp.

—Xylenols
ISO 1897 Pt VII-77. Phenol, o-Cresol, m-Cresol, p-Cresol, Cresylic Acid and Xylenols for Industrial Use—Methods of Test—Part VII: Measurement of Colour (Cresylic Acid and Xylenols Only) First Edition. 4 pp.

Colorants

Use For: Colors (Materials) *See Also:* Dyes; Food Colors; Paints; Pigments

—Medicinal Products
EC 78/25/EEC-78. Council Directive on the Approximation of the Laws of the Member States Relating to the Colouring Matters Which May Be Added to Medicinal Products. 3 pp.

Colorfastness Testing

See Also: Bleaching; Colorfastness Testing Equipment; Yellowing

—Anodic Coatings
BSI BS 6161: Part 8-81. 1981 Methods of Test for Anodic Oxidation Coatings on Aluminium and Its Alloys Part 8: Determination of the Fastness to Ultraviolet Light to Coloured Anodic Oxide Coatings. 4 pp.
CNS H2061-82. Methods of Accelerated Test for Light Fastness of Colored Anodic Oxidation Coatings on Aluminium and Aluminium Alloys (Jan)(8408). 3 pp.
ISO 6581-80. Anodizing of Aluminium and Its Alloys—Determination of the Fastness to Ultra-Violet Light of Coloured Anodic Oxide Coatings First Edition. 4 pp.
JIS H 8685-88. Accelerated Test Methods for Lightfastness of Coloured Anodic Oxidation Coatings on Aluminium and Aluminium Alloys. 11 pp.

—Coated Fabrics
SAA AS 1441.10-73. Methods of Test for Coated Fabrics—Part 10: Method for Determination of Colour Bleeding.
SAA AS 1441.11-73. Methods of Test for Coated Fabrics—Part 11: Method for Determination of Colour Fastness to Light.

—Coatings
CGSB 1-GP-71 METH 120.3-79. Methods of Testing Paints and Pigments Color Stability Clear Films. 1 p.

—Coatings—Light
CGSB 1-GP-71 METH 120.1-74. Methods of Testing Paints and Pigments Color Stability Fading by Light. 2 pp.

—Condoms
CEN PREN 600-4-91. Latex Rubber Condoms—Part 4: Determination of Colour Fastness. 4 pp.
ISO 4074 Pt 4-80. Rubber Condoms—Part 4: Determination of Colour Fastness First Edition. 3 pp.

—Dental Materials
BSI BS 6790-87. 1987 Determination of Colour Stability of Dental Polymeric Materials (ISO 7491: 1985). 4 pp.
BSI BS 6790-01. 1987 Amd 1 Determination of Colour Stability of Dental Polymeric Materials (ISO 7491: 1985) (AMD 7021) May 1, 1992. 9 pp.
CEN PREN 27 491-90. Dentistry—Dental Materials—Determination of Colour Stability of Dental Polymeric Materials. 3 pp.
CEN EN 27491-91. Dentistry—Dental Materials—Determination of Colour Stability of Dental Polymeric Materials. 3 pp.
ISO 7491-85. Dental Materials—Determination of Colour Stability of Dental Polymeric Materials First Edition. 4 pp.

Colorfastness Testing (Cont.)

—Fabrics
BSI BS 1006-90. 1990 Methods of Test for Colour Fastness of Textiles and Leather (L). 234 pp.
BSI BS 1006-01. 1990 Amd 1 Methods of Test for Colour Fastness of Textiles and Leather (AMD 7284) October 15, 1992 (L). 247 pp.
BSI BS 1006-02. 1990 Amd 2 Methods of Test for Colour Fastness of Textiles and Leather (AMD 7201) January 15, 1993 (L). 248 pp.
BSI BS EN 20105-A01-92. 1993 Textiles—Tests for Colour Fastness Part A01: General Principles of Testing (ISO 105-A01: 1989) (L). 15 pp.
BSI BS EN 20105-A02-93. 1993 Textiles—Tests for Colour Fastness Part A02: Grey Scale for Assessing Change in Colour (ISO 105-A02: 1987) (L). 9 pp.
BSI BS EN 20105-A03-93. 1993 Textiles—Tests for Colour Fastness Part A03: Grey Scale for Assessing Staining (ISO 105-A03: 1987) (L). 9 pp.
CEN EN 20105-A01-92. Textiles—Tests for Colour Fastness Part A01: General Principles of Testing (ISO 105-A01: 1989). 7 pp.
CEN EN 20105-A03-92. Textiles—Tests for Colour Fastness Part A03: Grey Scale for Assessing Staining (ISO 105-A03: 1987). 4 pp.
CGSB CAN/CGSB-4.2 NO.46-M90. Textile Test Methods Textiles—Tests for Colourfastness—Part A02: Grey Scale for Assessing Change in Colour (ISO 105-A02:1987). 10 pp.
CGSB CAN/CGSB-4.2 NO.47-M90. Textile Test Methods Textiles—Tests for Colourfastness—Part A03: Grey Scale for Assessing Staining (ISO 105-A03:1987). 10 pp.
CGSB CAN/CGSB-4.2 NO.68-92. Textile Test Methods Textiles—Tests for Colourfastness—Part J01: Measurement of Colour and Colour Differences (ISO 105-J01:1989). 8 pp.
CNS L3150-82. General Principles of Testing for Colour Fastness (Jan)(8429).
CNS L4152-84. Weft Knitted Fabrics (Finished) (Jul)(10970). 2 pp.
ISO 105 Pt A01-89. Textiles—Tests for Colour Fastness—Part A01: General Principles of Testing Fourth Edition; (Corrected and Reprinted -1990) (CEN EN 20105-A01:1992). 10 pp.
ISO 105 Pt A02-93. Textiles—Tests for Colour Fastness—Part A02: Grey Scale for Assessing Change in Colour Fourth Edition. 5 pp.
ISO 105 Pt A03-93. Textiles—Tests for Colour Fastness—Part A03: Grey Scale for Assessing Staining Fourth Edition. 5 pp.
ISO 105 Pt A04-89. Textiles—Tests for Colour Fastness—Part A04: Method for the Instrumental Assessment of the Degree of Staining of Adjacent Fabrics First Edition. 4 pp.
ISO 105 Pt F10-89. Textiles—Tests for Colour Fastness—Part F10: Specification for Adjacent Fabric: Multifibre First Edition. 7 pp.
ISO 105 Pt G-78. Textiles—Tests for Colour Fastness Part G: Colour Fastness to Atmospheric Contaminants First Edition. 20 pp.
ISO 105 Pt J01-89. Textiles—Tests for Colour Fastness—Part J01: Measurement of Colour and Colour Differences Third Edition; (CAN/CGSB-4.2 No.68-92). 4 pp.
ISO 105 Pt J02-87. Textiles—Tests for Colour Fastness—Part J02: Method for the Instrumental Assessment of Whiteness First Edition. 4 pp.
ISO 105 Pt Z-78. Textiles—Tests for Colour Fastness Part Z: Colorant Characteristics First Edition. 8 pp.
JIS L 0801-78. General Principles of Testing Methods for Colour Fastness (R 1983). 14 pp.
JIS Z 8715-91. Whiteness of Near-White Opaque Materials—Specifying Method. 8 pp.
MOD UK DSTAN 83-63-01. Cloth, Cellular, Cotton Issue 1; Amendment 2. 17 pp.
SAA AS 2001.4. Methods of Test for Textiles—Part 4: Colourfastness Tests Subset in Binder.
SAA AS 2001.4.1-80. Methods of Test for Textiles—Part 4: Colourfastness Tests—Part 4.1: Definitions and General Requirements. 4 pp.
SNZ NZS/BS 1006-78. Methods of Test for Colour Fastness of Textiles and Leather Amend: 7. 212 pp.

—Fabrics—Acid
SAA AS 2001.4.12-81. Methods of Test for Textiles—Part 4: Colourfastness Tests—Part 4.12: Determination of Colourfastness to Acid Spotting. 2 pp.
SAA AS 2001.4.13-87. Methods of Test for Textiles—Part 4: Colourfastness Tests—Part 4.13: Determination of Colourfastness to Acid Milling. 2 pp.

—Fabrics—Alkali
SAA AS 2001.4.9-81. Methods of Test for Textiles—Part 4: Colourfastness Tests—Part 4.9: Determination of Colourfastness to Alkali Spotting. 1 p.

INTERNATIONAL AND NON-U.S. NATIONAL STANDARDS
SUBJECT INDEX
Colorfastness

Colorfastness Testing *(Cont.)*

—Fabrics—Alkaline

SAA AS 2001.4.10-87. Methods of Test for Textiles—Part 4: Colourfastness Tests—Part 4.10: Determination of Colourfastness to Alkaline Milling. 2 pp.

—Fabrics—Artificial Light

ISO 105 Pt B06-92. Textiles—Tests for Colour Fastness—Part B06: Colour Fastness to Artificial Light at High Temperatures: Xenon Arc Fading Lamp Test First Edition. 15 pp.
SAA AS 2001.4.21-79. Methods of Test for Textiles—Part 4: Colourfastness Tests—Part 4.21: Determination of Colourfastness to Light Using an Artificial Light Source (Mercury Vapour—Tungsten Filament—Internally Phosphor-Coated Lamp) Reconfirmed 1986. 4 pp.

—Fabrics—Bleaching

CNS L3031-83. Method of Test for Colour Fastness to Bleaching with Hypochlorite (Mar)(1498).
CNS L3161-82. Method of Test for Colour Fastness to Bleaching with Peroxide (Mar)(8619).
CNS L3163-82. Method of Test for Colour Fastness to Bleaching Sodium Chlorite (Mar)(8621).
ISO 105 Pt N-78. Textiles—Tests for Colour Fastness Part N: Colour Fastness to Bleaching Agencies First Edition. 15 pp.
JIS L 0856-83. Testing Methods for Colour Fastness to Bleaching with Hypochlorite. 7 pp.
JIS L 0857-75. Testing Method for Colour Fastness to Bleaching with Peroxide. 5 pp.
JIS L 0859-75. Testing Method for Colour Fastness to Bleaching with Sodium Chlorite (R 1988). 7 pp.
JIS L 0889-87. Testing Methods for Colour Fastness to Bleaching with Sodium Percarbonate. 10 pp.

—Fabrics—Carbonizing

CNS L3187-82. Method of Test for Colour Fastness to Carbonizing with Aluminium Chloride (Aug)(9310).
CNS L3188-82. Method of Test for Colour Fastness to Carbonizing with Sulphuric Acid (Aug) (9311).
ISO 105 Pt X01-87. Textiles—Tests for Colour Fastness—Part X01: Colour Fastness to Carbonizing: Aluminium Chloride Third Edition. 4 pp.
ISO 105 Pt X02-87. Textiles—Tests for Colour Fastness—Part X02: Colour Fastness to Carbonizing: Sulfuric Acid Third Edition. 4 pp.
JIS L 0866-76. Testing Method for Colour Fastness to Carbonizing with Aluminium Chloride.
JIS L 0867-76. Testing Method for Colour Fastness to Carbonizing with Sulphuric Acid.

—Fabrics—Chemical Resistance

CGSB CAN/CGSB-4.2 NO.35.1-M90. Textile Test Methods Colourfastness to Burnt Gas Fumes. 11 pp.
CGSB CAN/CGSB-4.2 NO.52.1-M90. Textile Test Methods Colourfastness to Chlorinated Water. 9 pp.
CGSB CAN/CGSB-4.2 NO.55-M90. Textile Test Methods Loss in Strength and Colour Change of Fabrics Due to Retained Chlorine. 13 pp.
CNS L3152-82. Method of Test for Colour Fastness to Organic Solvents (Jan) (8431).
CNS L3153-82. Method of Test for Colour Fastness to Rubbing with Organic Solvents (Jan)(8432).
CNS L3159-82. Method of Test for Colour Fastness to Soda Boiling (Mar)(8617).
CNS L3172-82. Method of Test for Colour Fastness to Formaldehyde (Jun)(9015).
CNS L3173-82. Method of Test for Colour Fastness to Nitrogen Oxide (Jun) (9016).
CNS L3174-82. Method of Test for Colour Fastness to Chlorinated Water (Jun) (9017).
CNS L3176-82. Method of Test for Colour Fastness to Acid Spotting (Jun) (9019).
CNS L3177-82. Method of Test for Colour Fastness to Alkali Spotting (Jun) (9020).
CNS L3183-82. Method of Test for Colour Fastness to Chlorination (Aug)(9306).
CNS L3186-82. Method of Test for Colour Fastness to Alkali Milling (Aug) (9309).
DIN ENGL 53378-65. Testing of Plastic Films; Determination of Colour Fastness to Hydrogen Sulphide (June). 2 pp.
ISO 105 Pt D02-87. Textiles—Tests for Colour Fastness—Part D02: Colour Fastness to Rubbing: Organic Solvents Third Edition. 4 pp.
ISO 105 Pt E03-87. Textiles—Tests for Colour Fastness—Part E03: Colour Fastness to Chlorinated Water (Swimming-Bath Water) Second Edition. 4 pp.
ISO 105 Pt E05-89. Textiles—Tests for Colour Fastness—Part E05: Colour Fastness to Spotting: Acid Third Edition. 4 pp.
ISO 105 Pt E06-89. Textiles—Tests for Colour Fastness—Part E06: Colour Fastness to Spotting: Alkali Third Edition. 4 pp.
ISO 105 Pt E12-89. Textiles—Tests for Colour Fastness—Part E12: Colour Fastness to Milling: Alkaline Milling Third Edition. 6 pp.

Colorfastness Testing *(Cont.)*

—Fabrics—Chemical Resistance *(Cont.)*

ISO 105 Pt E13-87. Textiles—Tests for Colour Fastness—Part E13: Colour Fastness to Acid-Felting: Severe Second Edition. 4 pp.
ISO 105 Pt E14-87. Textiles—Tests for Colour Fastness—Part E14: Colour Fastness to Acid-Felting: Mild Second Edition. 4 pp.
ISO 105 Pt G04-89. Textiles—Tests for Colour Fastness Part G04: Colour Fastness to Oxides of Nitrogen in the Atmosphere at High Humidities First Edition. 7 pp.
ISO 105 Pt X05-87. Textiles—Tests for Colour Fastness—Part X05: Colour Fastness to Organic Solvents Third Edition. 4 pp.
ISO 105 Pt X06-87. Textiles—Tests for Colour Fastness—Part X06: Colour Fastness to Soda Boiling Third Edition. 4 pp.
ISO 105 Pt X09-87. Textiles—Tests for Colour Fastness—Part X09: Colour Fastness to Formaldehyde Third Edition. 4 pp.
JIS L 0851-75. Testing Method for Colour Fastness to Acid Spotting. 4 pp.
JIS L 0852-67. Testing Method for Colour Fastness to Alkali Spotting. 4 pp.
JIS L 0855-92. Testing Methods for Colour Fastness to Nitrogen Oxides. 10 pp.
JIS L 0861-75. Testing Method for Colour Fastness to Organic Solvents (R 1983). 5 pp.
JIS L 0862-75. Testing Method for Colour Fastness to Rubbing with Organic Solvents (R 1983). 7 pp.
JIS L 0864-76. Testing Method for Colour Fastness to Soda Boiling.
JIS L 0868-75. Testing Method for Colour Fastness to Formaldehyde. 5 pp.
JIS L 0873-74. Testing Method for Colour Fastness to Chlorination.
JIS L 0876-70. Testing Method for Colour Fastness to Alkali Milling.
JIS L 0884-83. Testing Methods for Colour Fastness to Chlorinated Water. 6 pp.

—Fabrics—Chlorinated Water

CGSB CAN/CGSB-4.2 NO.52.1-M90. Textile Test Methods Colourfastness to Chlorinated Water. 9 pp.
CGSB CAN/CGSB-4.2 NO.52.2-M91. Textile Test Methods Textiles—Tests for Colourfastness—Part E03: Colourfastness to Chlorinated Water (Swimming-Pool Water) (ISO 105-E03:1987). 7 pp.
CGSB CAN/CGSB-4.2 NO.55-M90. Textile Test Methods Loss in Strength and Colour Change of Fabrics Due to Retained Chlorine. 13 pp.
CNS L3183-82. Method of Test for Colour Fastness to Chlorination (Aug)(9306).
ISO 105 Pt E03-87. Textiles—Tests for Colour Fastness—Part E03: Colour Fastness to Chlorinated Water (Swimming-Bath Water) Second Edition. 4 pp.
ISO 105 Pt X14-87. Textiles—Tests for Colour Fastness—Part X14: Colour Fastness to Acid Chlorination of Wool: Sodium Dichloroisocyanurate Second Edition. 6 pp.
JIS L 0873-74. Testing Method for Colour Fastness to Chlorination.
SAA AS 2001.4.5-81. Methods of Test for Textiles—Part 4: Colourfastness Tests—Part 4.5: Determination of Colourfastness to Chlorinated Swimming Pool Water. 4 pp.

—Fabrics—Daylight

SAA AS 2001.4.2-82. Methods of Test for Textiles—Part 4: Colourfastness Tests—Part 4.2: Determination of Colourfastness to Daylight. 3 pp.

—Fabrics—Decating

CNS L3184-82. Method of Test for Colour Fastness to Decatizing (Aug)(9307).
ISO 105 Pt E10-87. Textiles—Tests for Colour Fastness—Part E10: Colour Fastness to Decatizing Second Edition. 4 pp.
JIS L 0874-75. Testing Method for Colour Fastness to Decatizing (R 1983). 8 pp.

—Fabrics—Degumming

ISO 105 Pt X08-87. Textiles—Tests for Colour Fastness—Part X08: Colour Fastness to Degumming Third Edition. 4 pp.

—Fabrics—Dry Cleaning

CGSB CAN/CGSB-4.2 NO.29.1-M89. Textile Test Methods Colourfastness to Dry Cleaning Solvent. 9 pp.
CNS L3155-82. Method of Test for Colour Fastness to Dry Cleaning (Jan)(8434).
ISO 105 Pt D01-87. Textiles—Tests for Colour Fastness—Part D01: Colour Fastness to Dry Cleaning Third Edition. 4 pp.
JIS L 0860-74. Testing Method for Colour Fastness to Dry Cleaning. 4 pp.
SAA AS 2001.4.16-81. Methods of Test for Textiles—Part 4: Colourfastness Tests—Part 4.16: Determination of Colourfastness to Dry Cleaning Solvents. 2 pp.

Colorfastness Testing *(Cont.)*

—Fabrics—Dry Heat

CNS L3178-82. Method of Test for Colour Fastness to Dry Heating (Jun)(9021).
ISO 105 Pt P-78. Textiles—Tests for Colour Fastness Part P: Colour Fastness to Heat Treatments First Edition. 11 pp.
JIS L 0879-75. Testing Method for Colour Fastness to Dry Heating (R 1983). 6 pp.

—Fabrics—Flat Abrasion

SAA AS 2001.4.22-91. Methods of Test for Textiles—Part 4: Colourfastness Tests—Part 4.22: Determination of Colour Change Due to Flat Abrasion (Frosting) of Textile Fabrics (Screen Wire Method). 5 pp.

—Fabrics—Hot Pressing

SAA AS 2001.4.6-90. Methods of Test for Textiles—Part 4: Colourfastness Tests—Part 4.6: Determination of Colourfastness to Hot Pressing. 3 pp.

—Fabrics—Ironing

CGSB CAN/CGSB-4.2 NO.31-M89. Textile Test Methods Textiles—Tests for Colourfastness—Part X11: Colourfastness to Hot Pressing (ISO 105/X11-1987). 11 pp.
CNS L3156-83. Method of Test for Colour Fastness to Hot Pressing (Jun)(8532).
ISO 105 Pt X11-87. Textiles—Tests for Colour Fastness—Part X11: Colour Fastness to Hot Pressing Third Edition. 4 pp.
JIS L 0850-75. Testing Method for Colour Fastness to Hot Pressing (R 1978). 9 pp.

—Fabrics—Laundering

BSI BS EN 20105-C01-93. 1993 Textiles—Tests for Colour Fastness Part C01: Colour Fastness to Washing: Test 1 (ISO 105-C01: 1989) (L). 10 pp.
BSI BS EN 20105-C02-93. 1993 Textiles—Tests for Colour Fastness Part C02: Colour Fastness to Washing: Test 2 (ISO 105-C02: 1989) (L). 10 pp.
BSI BS EN 20105-C03-93. 1993 Textiles—Tests for Colour Fastness Part C03: Colour Fastness to Washing: Test 3 (ISO 105-C03: 1989) (L). 10 pp.
BSI BS EN 20105-C04-93. 1993 Textiles—Tests for Colour Fastness Part C04: Colour Fastness to Washing: Test 4 (ISO 105-C04: 1989) (L). 10 pp.
BSI BS EN 20105-C05-93. 1993 Textiles—Tests for Colour Fastness Part C05: Colour Fastness to Washing: Test 5 (ISO 105-C05: 1989) (L). 10 pp.
BSI BS EN 20105-C06-93. 1993 Textiles—Tests for Colour Fastness Part C06: Colour Fastness to Domestic and Commercial Laundering (ISO 105-C06: 1987) (L). 11 pp.
CEN EN 20105-C01-92. Textiles—Tests for Colour Fastness Part C01: Colour Fastness to Washing: Test 1 (ISO 105-C01: 1989). 7 pp.
CEN EN 20105-C03-92. Textiles—Tests for Colour Fastness Part C03: Colour Fastness to Washing: Test 3 (ISO 105-C03: 1989). 7 pp.
CEN EN 20105-C04-92. Textiles—Tests for Colour Fastness—Part C04: Colour Fastness to Washing—Test 4 (ISO 105-C04: 1989). 7 pp.
CEN EN 20105-C05-92. Textiles—Tests for Colour Fastness—Part C05: Colour Fastness to Washing—Test 5 (ISO 105-C05: 1989). 5 pp.
CEN EN 20105-C06-92. Textiles—Tests for Colour Fastness—Part C06: Colour Fastness to Domestic and Commercial Laundering (ISO 105-C06: 1987). 8 pp.
CGSB CAN/CGSB-4.2 NO.19.1-M90. Textile Test Methods Colourfastness to Washing—Accelerated Test—Launder-Ometer; (Amendment 1 June 1992). 12 pp.
CGSB CAN/CGSB-4.2 NO.24-M91. Textile Test Methods Colourfastness and Dimensional Change in Commercial Laundering. 7 pp.
CGSB CAN/CGSB-4.2 NO.55-M90. Textile Test Methods Loss in Strength and Colour Change of Fabrics Due to Retained Chlorine. 13 pp.
CNS L3027-83. Method of Test for Colour Fastness to Washing (Mar)(1494). 6 pp.
DIN ENGL EN 20105 Pt C01-93. Tests for Colour Fastness of Textiles; Colour Fastness to Washing: Test 1 (ISO 105-C01: 1989) (Mar). 6 pp.
DIN ENGL EN 20105 Pt C02-93. Tests for Colour Fastness of Textiles; Colour Fastness to Washing: Test 2 (ISO 105-C02: 1989) (Mar). 6 pp.
DIN ENGL EN 20105 Pt C03-93. Tests for Colour Fastness of Textiles; Colour Fastness to Washing: Test 3 (ISO 105-C03: 1989) (Mar). 6 pp.
DIN ENGL EN 20105 Pt C04-93. Tests for Colour Fastness of Textiles; Colour Fastness to Washing: Test 4 (ISO 105-C04: 1989) (Mar). 6 pp.
DIN ENGL EN 20105 Pt C05-93. Tests for Colour Fastness of Textiles; Colour Fastness to Washing: Test 5 (ISO 105-C05: 1989) (Mar). 6 pp.
ISO 105 Pt C01-89. Textiles—Tests for Colour Fastness—Part C01: Colour Fastness to Washing: Test 1 Fourth Edition; (CEN EN 20105-C01:1992). 6 pp.

INDUSTRY STANDARDS
543

Colorfastness

Colorfastness Testing (Cont.)

—Fabrics—Laundering (Cont.)

ISO 105 Pt C02-89. Textiles—Tests for Colour Fastness—Part C02: Colour Fastness to Washing: Test 2 Fourth Edition. 6 pp.

ISO 105 Pt C03-89. Textiles—Tests for Colour Fastness—Part C03: Colour Fastness to Washing: Test 3 Fourth Edition; (CEN EN 20105-C03:1992). 6 pp.

ISO 105 Pt C04-89. Textiles—Tests for Colour Fastness—Part C04: Colour Fastness to Washing: Test 4 Fourth Edition; (CEN EN 20105-C04:1992). 6 pp.

ISO 105 Pt C05-89. Textiles—Tests for Colour Fastness—Part C05: Colour Fastness to Washing: Test 5 Fourth Edition; (CEN EN 20105-C05:1992). 6 pp.

ISO 105 Pt C06-87. Textiles—Tests for Colour Fastness—Part C06: Colour Fastness to Domestic and Commercial Laundering Second Edition; (CEN EN 20105-C06:1992). 7 pp.

JIS L 0844-86. Testing Methods for Colour Fastness to Washing and Laundering. 14 pp.

—Fabrics—Light

BSI BS EN 20105-B02-93. 1993 Textiles—Tests for Colour Fastness Part B02: Colour Fastness to Artificial Light (Xenon Arc Fading Lamp Test) (ISO 105-B02: 1988) (L). 18 pp.

CEN EN 20105-B02-92. Textiles—Tests for Colour Fastness Part B02. Colour Fastness to Artificial Light: Xenon Arc Fading Lamp Test (ISO 105-B02: 1988). 13 pp.

CGSB CAN/CGSB-4.2 NO.18.1-M90. Textile Test Methods Colourfastness to Artificial Light: Carbon-Arc Radiation. 20 pp.

CGSB CAN/CGSB-4.2 NO.18.2-M90. Textile Test Methods Textiles—Tests for Colourfastness—Part B01: Colourfastness to Light: Daylight (ISO 105-B01:1988). 20 pp.

CGSB CAN/CGSB-4.2 NO.18.3-M90. Textile Test Methods Textiles—Tests for Colourfastness—Part B02: Colourfastness to Artificial Light: Xenon Arc Fading Lamp Test (ISO 105-B02:1988). 28 pp.

CNS L3026-82. Method of Test for Colour Fastness to Sunlight and Daylight (Jun)(1493).

CNS L3074-82. Method of Test for Colour Fastness to Carbon Arc Lamp Light (Jun)(3845). 4 pp.

CNS L3075-82. Method of Test for Colour Fastness to Xenon Arc Lamp Light (Jun)(3846).

CNS L3193-83. Method of Test for Colour Fastness to Light and Perspiration (Jan)(9902).

CNS L3194-83. Method of Test for Detection and Colour Fastness to Photochromism (Jan) (9903).

ISO 105 Pt B01-89. Textiles—Tests for Colour Fastness—Part B01: Colour Fastness to Light: Daylight Fourth Edition. 10 pp.

ISO 105 Pt B02-88. Textiles—Tests for Colour Fastness—Part B02: Colour Fastness to Artifical Light: Xenon Arc Fading Lamp Test Third Edition; (CEN EN 20105-B02:1992). 14 pp.

ISO 105 Pt B03-88. Textiles—Tests for Colour Fastness—Part B03: Colour Fastness to Weathering: Outdoor Exposure Third Edition. 8 pp.

ISO 105 Pt B04-88. Textiles—Tests for Colour Fastness—Part B04: Colour Fastness to Weathering: Xenon Arc Third Edition. 8 pp.

ISO 105 Pt B05-88. Textiles—Tests for Colour Fastness—Part B05: Detection and Assessment of Photochromism Third Edition. 7 pp.

JIS L 0841-92. Testing Methods for Colour Fastness to Daylight. 12 pp.

JIS L 0842-88. Testing Methods for Colour Fastness to Carbon Arc Lamp Light. 6 pp.

JIS L 0843-88. Testing Methods for Colour Fastness to Xenon Arc Lamp Light. 9 pp.

JIS L 0886-92. Testing Methods for the Detection and Assessment of Photochromism. 7 pp.

JIS L 0888-88. Testing Method for Colour Fastness to Light and Perspiration.

—Fabrics—Mercerizing

CNS L3158-82. Method of Test for Colour Fastness to Mercerizing (Mar)(8616).

ISO 105 Pt X04-87. Textiles—Tests for Colour Fastness—Part X04: Colour Fastness to Mercerizing Third Edition. 4 pp.

JIS L 0863-92. Testing Method for Colour Fastness to Mercerizing. 6 pp.

—Fabrics—Metals

CNS L3189-82. Method of Test for Colour Fastness to Metals in the Dyebath: Chromium Salt (Aug)(9312).

CNS L3190-82. Method of Test for Colour Fastness to Metals in the Dyebath: Iron and Copper (Aug)(9313).

JIS L 0870-75. Testing Method for Colour Fastness to Metals in the Dyebath: Chromium Salt.

JIS L 0871-75. Testing Method for Colour Fastness to Metals in the Dyebath: Iron and Copper.

Colorfastness Testing (Cont.)

—Fabrics—Perspiration

CGSB CAN/CGSB-4.2 NO.23-M90. Textile Test Methods Colourfastness to Perspiration. 9 pp.

CNS L3029-83. Method of Test for Colour Fastness to Perspiration (Mar)(1496).

CNS L3193-83. Method of Test for Colour Fastness to Light and Perspiration (Jan)(9902).

ISO 105 Pt E04-89. Textiles—Tests for Colour Fastness—Part E04: Colour Fastness to Perspiration Third Edition. 6 pp.

JIS L 0848-78. Testing Method for Colour Fastness to Perspiration. 10 pp.

JIS L 0888-88. Testing Method for Colour Fastness to Light and Perspiration.

SAA AS 2001.4.17-80. Methods of Test for Textiles—Part 4: Colourfastness Tests—Part 4.17: Determination of Colourfastness to Perspiration. 4 pp.

—Fabrics—Pleating

CNS L3180-82. Method of Test for Colour Fastness to Pleating (Jun)(9023).

JIS L 0880-77. Testing Method for Colour Fastness to Pleating.

—Fabrics—Resin Finish

CNS L3168-82. Method of Test for Colour Fastness to Resin Finish (Apr)(8711).

JIS L 0885-92. Testing Method for Colour Fastness to Resin Finish. 7 pp.

—Fabrics—Rubbing

BSI BS 3424: Part 14-85. 1985 Testing Coated Fabrics Part 14: Method 16. Methods for Determination of Colour Fastness to Wet and Dry Rubbing and Determination of Resistance to Printwear Using a Common Testing Apparatus. 6 pp.

CGSB CAN/CGSB-4.2 NO.22-M90. Textile Test Methods Colourfastness to Rubbing (Crocking). 11 pp.

CNS L3032-83. Method of Test for Colour Fastness to Rubbing (Mar)(1499). 2 pp.

CNS L3153-82. Method of Test for Colour Fastness to Rubbing with Organic Solvents (Jan)(8432).

ISO 105 Pt D02-87. Textiles—Tests for Colour Fastness—Part D02: Colour Fastness to Rubbing: Organic Solvents Third Edition. 4 pp.

ISO 105 Pt X12-93. Textiles—Tests for Colour Fastness—Part X12: Colour Fastness to Rubbing Fourth Edition. 5 pp.

JIS L 0849-71. Testing Method for Colour Fastness to Rubbing (R 1978). 6 pp.

JIS L 0862-75. Testing Method for Colour Fastness to Rubbing with Organic Solvents (R 1983). 7 pp.

SAA AS 2001.4.3-81. Methods of Test for Textiles—Part 4: Colourfastness Tests—Part 4.3: Determination of Colourfastness to Rubbing. 2 pp.

—Fabrics—Seawater

SAA AS 2001.4.14-80. Methods of Test for Textiles—Part 4: Colourfastness Tests—Part 4.14: Determination of Colourfastness to Seawater. 4 pp.

—Fabrics—Soda Boiling

SAA AS 2001.4.7-81. Methods of Test for Textiles—Part 4: Colourfastness Tests—Part 4.7: Determination of Colourfastness to Soda Boiling. 2 pp.

—Fabrics—Sodium Hypochlorite

SAA AS 2001.4.11-82. Methods of Test for Textiles—Part 4: Colourfastness Tests—Part 4.11: Determination of Colourfastness to Bleaching with Sodium Hypochlorite. 3 pp.

—Fabrics—Standard Adjacent Fabrics

ISO 105 Pt F-85. Textiles—Tests for Colour Fastness—Part F: Standard Adjacent Fabrics Third Edition. 24 pp.

—Fabrics—Steaming

CNS L3154-82. Method of Test for Colour Fastness to Steaming (Jan)(8433).

CNS L3179-82. Method of Test for Colour Fastness to High Temperature Steaming (Jun)(9022).

ISO 105 Pt E11-87. Textiles—Tests for Colour Fastness—Part E11: Colour Fastness to Steaming Second Edition. 4 pp.

ISO 105 Pt P-78. Textiles—Tests for Colour Fastness Part P: Colour Fastness to Heat Treatments First Edition. 11 pp.

JIS L 0869-77. Testing Method for Colour Fastness to Steaming.

JIS L 0878-75. Testing Method for Colour Fastness to High Temperature Steaming (R 1983). 9 pp.

—Fabrics—Storage

CNS L3157-82. Method of Test for Colour Fastness to Sublimation in Storage (Feb)(8533).

JIS L 0854-75. Testing Method for Colour Fastness to Sublimation in Storage (R 1983). 5 pp.

Colorfastness Testing (Cont.)

—Fabrics—Stoving

CNS L3162-82. Method of Test for Colour Fastness to Stoving (Mar)(8620).

JIS L 0858-75. Testing Method for Colour Fastness to Stoving (R 1983). 6 pp.

—Fabrics—Vulcanization

CNS L3169-82. Method of Test for Colour Fastness to Vulcanizing with Sulphur Monochloride (Apr)(8712).

CNS L3170-82. Method of Test for Colour Fastness to Steam Vulcanizing (Apr) (8713).

CNS L3171-82. Method of Test for Colour Fastness to Vulcanizing with Hot Air (Apr)(8714).

ISO 105 Pt S-78. Textiles—Tests for Colour Fastness Part S: Colour Fastness to Vulcanizing First Edition. 11 pp.

JIS L 0881-71. Testing Method for Colour Fastness to Vulcanizing with Sulphur Monochloride.

JIS L 0882-75. Testing Method for Colour Fastness to Steam Vulcanizing.

JIS L 0883-71. Testing Method for Colour Fastness to Vulcanizing with Hot Air.

—Fabrics—Washing/Laundering

SAA AS 2001.4.15-87. Methods of Test for Textiles—Part 4: Colourfastness Tests—Part 4.15: Determination of Colourfastness to Washing. 5 pp.

—Fabrics—Water

CGSB CAN/CGSB-4.2 NO.20-M89. Textile Test Methods Colourfastness to Water. 9 pp.

CGSB CAN/CGSB-4.2 NO.21-M90. Textile Test Methods Colourfastness to Sea Water. 9 pp.

CGSB CAN/CGSB-4.2 NO.52.1-M90. Textile Test Methods Colourfastness to Chlorinated Water. 9 pp.

CNS L3028-83. Method of Test for Colour Fastness to Hot Water (Mar)(1495).

CNS L3030-83. Method of Test for Colour Fastness to Water (Mar)(1497).

CNS L3151-82. Method of Test for Colour Fastness to Sea Water (Jan)(8430).

CNS L3174-82. Method of Test for Colour Fastness to Chlorinated Water (Jun) (9017).

CNS L3175-82. Method of Test for Colour Fastness to Water Spotting (Jun) (9018).

ISO 105 Pt E01-89. Textiles—Tests for Colour Fastness—Part E01: Colour Fastness to Water Third Edition. 5 pp.

ISO 105 Pt E02-89. Textiles—Tests for Colour Fastness—Part E02: Colour Fastness to Sea Water Third Edition. 5 pp.

ISO 105 Pt E03-87. Textiles—Tests for Colour Fastness—Part E03: Colour Fastness to Chlorinated Water (Swimming-Bath Water) Second Edition. 4 pp.

ISO 105 Pt E07-89. Textiles—Tests for Colour Fastness—Part E07: Colour Fastness to Spotting: Water Third Edition. 4 pp.

ISO 105 Pt E08-87. Textiles—Tests for Colour Fastness—Part E08: Colour Fastness to Water: Hot Water Second Edition. 4 pp.

ISO 105 Pt E09-89. Textiles—Tests for Colour Fastness—Part E09: Colour Fastness to Potting Third Edition. 4 pp.

JIS L 0845-75. Testing Method for Colour Fastness to Hot Water (R 1983). 6 pp.

JIS L 0846-92. Testing Method for Colour Fastness to Water. 6 pp.

JIS L 0847-75. Testing Method for Colour Fastness to Sea Water (R 1983). 6 pp.

JIS L 0853-75. Testing Method for Colour Fastness to Water Spotting. 4 pp.

JIS L 0884-83. Testing Methods for Colour Fastness to Chlorinated Water. 6 pp.

SAA AS 2001.4.4-81. Methods of Test for Textiles—Part 4: Colourfastness Tests—Part 4.4: Determination of Colourfastness to Water Spotting. 1 p.

SAA AS 2001.4.8-79. Methods of Test for Textiles—Part 4: Colourfastness Tests—Part 4.8: Determination of Colourfastness to Water Reconfirmed 1986. 4 pp.

—Fibers—Water

CGSB CAN/CGSB-4.2 NO.20-M89. Textile Test Methods Colourfastness to Water. 9 pp.

Floor Tiles

CNS K6312-77. Method of Test for Polyvinyl Chloride Asbestos Tiles for Flooring (Dec)(3309). 4 pp.

—Fluorescent Whitening Agents

CNS L3192-83. Method of Test for Colour Fastness to Light of Fluorescent Whitening Agents and Fluorescent Whitened Textiles (Jan)(9901).

JIS L 0887-75. Testing Method for Colour Fastness to Light of Fluorescent Whitening Agents and Fluorescent Whitened Textiles.

INTERNATIONAL AND NON-U.S. NATIONAL STANDARDS
SUBJECT INDEX

Colorfastness Testing (Cont.)

—Footwear—Paperboard
BSI BS 5131: Sec 4.12-90. 1990 Methods of Test for Footwear and Footwear Materials Part 4: Other Components Section 4.12: Fastness of Fibreboard Finishes to Rubbing in the Presence of Water and Perspiration. 10 pp.

—Hosiery
CNS L3045-64. Inspection Standard of Knitted Hoses and Socks (for Export) (Jul)(2280) (R 1970). 10 pp.

—Leather
BSI BS 1006-90. 1990 Methods of Test for Colour Fastness of Textiles and Leather (L). 234 pp.
BSI BS 1006-01. 1990 Amd 1 Methods of Test for Colour Fastness of Textiles and Leather (AMD 7284) October 15, 1992 (L). 247 pp.
BSI BS 1006-02. 1990 Amd 2 Methods of Test for Colour Fastness of Textiles and Leather (AMD 7201) January 15, 1993 (L). 248 pp.
CNS K6117-60. Method of Test for Leather (Decoloration Test) (May)(1273).
JIS K 6547-76. Testing Method for Colour Fastness to Rubbing of Leathers.
JIS K 6547-67. Testing Method for Color Fastness to Rubbing of Leather. 6 pp.
SNZ NZS/BS 1006-78. Methods of Test for Colour Fastness of Textiles and Leather Amend: 7. 212 pp.

—Oxide Coatings
BSI BS 6161: Part 8-81. 1981 Methods of Test for Anodic Oxidation Coatings on Aluminium and Its Alloys Part 8: Determination of the Fastness to Ultraviolet Light to Coloured Anodic Oxide Coatings. 4 pp.
CNS H2061-82. Methods of Accelerated Test for Light Fastness of Colored Anodic Oxidation Coatings on Aluminium and Aluminium Alloys (Jan)(8408). 3 pp.
ISO 6581-80. Anodizing of Aluminium and Its Alloys—Determination of the Fastness to Ultra-Violet Light of Coloured Anodic Oxide Coatings First Edition. 4 pp.
JIS H 8685-88. Accelerated Test Methods for Lightfastness of Coloured Anodic Oxidation Coatings on Aluminium and Aluminium Alloys. 11 pp.

—Plastic Sheets
CNS K6160-62. Method of Test for Soft Polyvinyl Chloride Sheet and Film (Apr)(1441) (R 1973). 6 pp.
DIN ENGL 53378-65. Testing of Plastic Films; Determination of Colour Fastness to Hydrogen Sulphide (June). 2 pp.

—Plastics
BSI BS 2782:Pt5: METH 541A-78. 1978 Methods of Testing Plastics Part 5: Optical and Colour Properties, Weathering Method 541A: Determination of Colour Fastness to Water. 2 pp.
BSI BS 2782:Pt5: METH 542A-79. 1979 Methods of Testing Plastics Part 5: Optical and Colour Properties, Weathering Method 542A: Qualitative Evaluation of Bleeding in Colorants. 4 pp.
BSI BS 2782:Pt5: METH 552A-81. 1981 Methods of Testing Plastics: Part 5: Optical and Colour Properties, Weathering: Method 552A: Determination of Changes in Colour Variations in Prop After Exposure to Daylight Under Glass, Natural Weathering or Artificial Light. 10 pp.
BSI BS 4618: Sec 4.3-74. 1974 Recommendations for the Presentation of Plastics Design Data Part 4: Enviromental and Chemical Effects Section 4.3: Resistance to Colour Change Produced by Exposure to Light. 8 pp.
CNS K6516-80. Determination of Resistance of Plastics to Colour Change upon Exposure to Light of the Enclosed Carbon Arc (Jul)(5821).
ISO 183-76. Plastics—Qualitative Evaluation of the Bleeding of Colorants First Edition. 4 pp.
ISO 4582-80. Plastics—Determination of Changes in Colour and Variations in Properties After Exposure to Daylight Under Glass, Natural Weathering or Artificial Light First Edition. 8 pp.
JIS K 7101-81. Testing Method for Colour Fastness of Plastics Under Window Glass upon Exposure to Daylight. 10 pp.
JIS K 7102-81. Testing Method for Colour Fastness of Plastics upon Exposure to Light of the Carbon Arc. 8 pp.

—Plywood
CNS O2046-82. Method of Test for Color Changing of Special Plywood (Mar)(8630).

—Polyester Fabrics
CNS L4082-80. Polyester Fabrics (Finished) (Jul)(5894). 2 pp.
CNS L4083-80. Polyester Filament Fabrics (Unfinished) (Jul)(5895). 2 pp.
CNS L4084-80. Polyester Filament Fabrics (Finished) (Jul) (5896). 3 pp.

Colorfastness Testing (Cont.)

—Polyurethane Resins
CNS K6771-83. Method of Test for Polyurethane Leather (May)(10270). 3 pp.

—PVC
DIN ENGL 53775 Pt 3-84. Testing of Colorants in Plastics; Testing of Colorants in Plasticized Polyvinyl Chloride (PVC-P); Determination of Bleeding of Colorants (June). 2 pp.

—PVC Coatings
ISO 105 Pt X10-87. Textiles—Tests for Colour Fastness—Part X10: Assessment of Migration of Textile Colours into Polyvinyl Chloride Coatings Third Edition. 4 pp.

—Sealing Materials
CGSB CAN2-19.0-M77METH 9.1-78. Methods of Testing Putty, Caulking and Sealing Compounds Staining of Concrete and Masonry, and Color Stability of Sealing Compounds. 1 p.

—Silk
CNS L4067-77. Silk Fabrics (Finished) (Jun)(3856). 2 pp.

—Silk—Degumming
CNS L3160-82. Method of Test for Colour Fastness to Degumming (Mar)(8618).
JIS L 0865-75. Testing Method for Colour Fastness to Degumming (R 1983). 6 pp.

—Socks
CNS L3045-64. Inspection Standard of Knitted Hoses and Socks (for Export) (Jul)(2280) (R 1970). 10 pp.

—Textile Floor Coverings
BSI BS 6659: Part 1-86. 1986 Amd 1 Producing and Assessing Changes in Surface Structure and Colour of Textile Floor Coverings Part 1: Method for Assessment of the Changes. 10 pp.
BSI BS 6659: Part 2-86. 1986 Producing and Assessing Changes in Surface Structure and Colour of Textile Floor Coverings Part 2: Method for Fatiguing Using the Hexapod Tumbler Tester. 7 pp.

—Towel Fabrics
CNS L4086-89. Woven Towel Fabrics (Finished) (Jul)(5898). 2 pp.

—Toys
DIN ENGL 53160-74. Testing of Coloured Toys for Resistance to Saliva and Perspiration (June). 2 pp.

—Toys—Fabrics—Perspiration
CNS Z8016-9-86. Method of Test for Toy Safety (Testing for Color Fastness of Textile Materials to Perspiration) (Jun)(4798-9). 2 pp.

—Vitreous Enamels
CNS R3094-80. Method of Test for Color Retention of Red, Orange and Yellow Porcelain Enamels (Dec)(6850). 2 pp.

—Wool
CNS L3182-82. Method of Test for Colour Fastness to Cross-Dyeing of Wool (Aug)(9305).
CNS L3185-82. Method of Test for Colour Fastness to Potting (Aug)(9308).
ISO 105 Pt X07-87. Textiles—Tests for Colour Fastness—Part X07: Colour Fastness to Cross-Dyeing: Wool Third Edition. 4 pp.
ISO 105 Pt X13-87. Textiles—Tests for Colour Fastness—Part X13: Colour Fastness of Wool Dyes to Processes Using Chemical Means for Creasing, Pleating and Setting Third Edition. 6 pp.
ISO 105 Pt X14-87. Textiles—Tests for Colour Fastness—Part X14: Colour Fastness to Acid Chlorination of Wool: Sodium Dichloroisocyanurate Second Edition. 6 pp.
JIS L 0872-75. Testing Method for Colour Fastness to Cross-Dyeing of Wool.
JIS L 0875-75. Testing Method for Colour Fastness to Potting. 4 pp.

—Wool Fabrics—Monoethanolamine Sulfite
SAA AS 2001.4.18-81. Methods of Test for Textiles—Part 4: Colourfastness Tests—Part 4.18: Determination of Colourfastness of Wool Textiles to Chemical Processes Using Monoethanolamine Sulphite. 2 pp.

—Yarns—Dry Cleaning
CGSB CAN/CGSB-4.2 NO.29.1-M89. Textile Test Methods Colourfastness to Dry Cleaning Solvent. 9 pp.

—Yarns—Water
CGSB CAN/CGSB-4.2 NO.20-M89. Textile Test Methods Colourfastness to Water. 9 pp.

—Yarns—Worsted
CNS L4123-86. Blended Worsted Yarns, Finished (Jun)(7065). 5 pp.

Colorfastness Testing Equipment
See Also: Colorfastness Testing

—Fabrics—Laundering
CNS L2027-81. Apparatus for Testing of Color Fastness to Washing (Nov)(8149).
JIS L 0821-83. Apparatus for Testing of Colour Fastness to Washing. 5 pp.

—Fabrics—Light
CNS L2053-82. Apparatus for Testing of Colour Fastness to Carbon Arc Lamp Light (Jun)(9024).
JIS L 0824-88. Apparatus for Testing of Colour Fastness to Carbon Arc Lamp Light. 9 pp.

—Fabrics—Perspiration
CNS L2020-75. Perspiration Testing Machine for Color Fastness Test (Aug)(3842).
JIS L 0822-92. Apparatus for Testing of Colour Fastness to Perspiration. 5 pp.

—Fabrics—Rubbing
CNS L2021-75. Rubbing Testing Machine for Color Fastness Test (Aug)(3843).
JIS L 0823-71. Apparatus for Testing of Colour Fastness to Rubbing.

Colorimetric Analysis
See Also: Colorimetry; Photometry; Quantitative Analysis; Spectrochemical Analysis; Volumetric Analysis; Water Analysis

—Ammonium Nitrate
BSI BS 4267: Part 8-87. 1987 Ammonium Nitrate Part 8: Methods for Determination of Chloride Content. 4 pp.

—Animal Fats
BSI BS 684: Sec 2.16-76. 1976 Methods of Analysis of Fats and Fatty Oils Part 2: Other Methods Section 2.16: Determination of Copper. Colorimetric Method. 2 pp.
BSI BS 684: Sec 2.17-76. 1976 Methods of Analysis of Fats and Fatty Oils Part 2: Other Methods Section 2.17: Determination of Iron. Colorimetric Method. 2 pp.

—Carbon Black
ISO 1138-81. Rubber Compounding Ingredients—Carbon Black—Determination of Sulphur Content Second Edition. 5 pp.

—Cathode Ray Tubes
IEC 441-74. Photometric and Colorimetric Methods of Measurement of the Light Emitted by a Cathode-Ray Tube Screen First Edition. 26 pp.

—Cereal Products
BSI BS 4317: Part 14-80. 1980 Methods of Test for Cereals and Pulses Part 14: Determination of Alpha-Amylase Activity in Cereals and Cereal Products (Colorimetric Method). 8 pp.
ISO 3983-77. Cereals and Cereal Products—Determination of Alpha-Amylase Activity—Colorimetric Method First Edition. 7 pp.

—Cereals
BSI BS 4317: Part 14-80. 1980 Methods of Test for Cereals and Pulses Part 14: Determination of Alpha-Amylase Activity in Cereals and Cereal Products (Colorimetric Method). 8 pp.
ISO 3983-77. Cereals and Cereal Products—Determination of Alpha-Amylase Activity—Colorimetric Method First Edition. 7 pp.

—Chloroform
BSI BS 4774-72. 1972 Methods of Test for Chloroform for Industrial Use. 12 pp.

—Coconut Oil
CNS K6180-74. Method of Test for Coconut Oil (Technical Grade) (Oct)(2241). 5 pp.

—Enamels
CNS K6031-87. Method of Test for Lacquer Enamel (May). 6 pp.

—Ethanols
BSI BS 6392: Part 4-83. 1983 Amd 1 Testing of Ethanol for Industrial Use Part 4: Method for Determination of Aldehydes Content. 5 pp.
BSI BS 6392: Part 7-83. 1983 Amd 1 Testing of Ethanol for Industrial Use Part 7: Method for Determination of Methanol Content (0.10% (V/V) to 1.50% (V/V)) (Visual Colorimetric Method). 5 pp.
ISO 1388 Pt 5-81. Ethanol for Industrial Use—Methods of Test—Part 5: Determination of Aldehydes Content—Visual Colorimetric Method First Edition. 5 pp.
ISO 1388 Pt 8-81. Ethanol for Industrial Use—Methods of Test—Part 8: Determination of Methanol Content (Methanol Contents Between 0,10 and 1,50 % (V/V))—Visual Colorimetric Method First Edition. 5 pp.

INDUSTRY STANDARDS

Colorimetric Analysis (Cont.)

—Glycerols
BSI BS 5711: Part 18-79. 1979 Methods of Sampling and Test for Glycerol Part 18: Detection of Sugars. 2 pp.

—Lacquers
CNS K6386-76. Method of Test for Lacquer in Surface of Leather Shoes (Mar) (3922). 2 pp.
MOD UK DEF-1053: METH NO. 81. Standard Methods of Testing Paint, Varnish, Lacquer and Related Products. Indices Method No. 81: Determination of Lead in 'Lead Free' Paints, Varnishes and Allied Products and Their Containers. (Withdrawn).

—Linseed Oil
CNS K6055-75. Method of Test for Boiled Linseed Oil (Feb)(769). 4 pp.

—Methanols
CNS K6263-80. Method of Test for Methyl Alcohol (Methanol) (Feb)(2790). 7 pp.

—Paints
MOD UK DEF-1053: METH NO. 81. Standard Methods of Testing Paint, Varnish, Lacquer and Related Products. Indices Method 81: Determination of Lead in 'Lead Free' Paints, Varnishes and Allied Products and Their Containers. (Withdrawn).

—Paraffin Wax
CNS K6035-75. Method of Test for Paraffin Wax (Mar)(632). 2 pp.

—Phenol Formaldehyde Resins
BSI BS 2782:Pt4: METH 451D-78. 1978 Methods of Testing Plastics Part 4: Chemical Properties Method 451D: Determination of Free Ammonia and Ammonium Compounds in Phenol-Formaldehyde Mouldings (Colormetric Comparison Method). 4 pp.
ISO 120-77. Plastics—Phenol-Formaldehyde Mouldings—Determination of Free Ammonia and Ammonium Compounds—Colorimetric Comparison Method First Edition. 4 pp.

—Phenolic Resins
BSI BS 2782:Pt4: METH 451F-J-78. 1978 Methods of Testing Plastics: Part 4: Chemical Properties: Method 451F; Determination of Formaldehyde in Phenolic Mouldings (Colorimetric Method) Method 451G; Deter of Formaldehyde in Phenolic Mouldings (Gravimetric Method). 4 pp.

—Phosphoric Acids
CNS K6235-72. Method of Test for Phosphoric Acid (Industrial Use) (Jun)(2620). 2 pp.

—Phthalic Anhydride
ISO 1389 Pt X-77. Phthalic Anhydride for Industrial Use—Methods of Test—Part X: Determination of 1, 4-Naphthaquinone Content—Colorimetric Method First Edition. 4 pp.

—Pigments
CNS K6094-89. Method of Test for Red Lead (Pigment) (Jul)(1043). 2 pp.

—Soils
SNZ NZS 4402: Pt 3:TEST 3.3.2-86. Methods of Testing Soils for Civil Engineering Purposes Part 3: Soil Chemical Tests. Section 3.3: Determination of the pH Value. Test 3.3.2: Subsidiary Method (Colorimetric). 1 p.

—Varnishes
CNS K6057-70. Method of Test for Clear Varnish for Baking Varnish (Jan)(773). 3 pp.
MOD UK DEF-1053: METH NO. 81. Standard Methods of Testing Paint, Varnish, Lacquer and Related Products. Indices Method 81: Determination of Lead in 'Lead Free' Paints, Varnishes and Allied Products and Their Containers. (Withdrawn).

—Vegetable Fats
BSI BS 684: Sec 2.16-76. 1976 Methods of Analysis of Fats and Fatty Oils Part 2: Other Methods Section 2.16: Determination of Copper. Colorimetric Method. 2 pp.
BSI BS 684: Sec 2.17-76. 1976 Methods of Analysis of Fats and Fatty Oils Part 2: Other Methods Section 2.17: Determination of Iron. Colorimetric Method. 2 pp.

—Water
BSI BS 6068: Sec 2.12-90. 1990 Water Quality Part 2: Physical, Chemical and Bio-Chemical Methods Section 2.1: Determination of Phenol Index: 4-Aminoantipyrine (4-Aminophenazone) Spectrometric Methods After Distillation. 12 pp.
BSI BS 6068: Sec 2.26-86. 1986 Water Quality Part 2: Physical, Chemical and Bio-Chemical Methods Section 2.26: Method for Determination of Free Chlorine and Total Chlorine: Colorimetric Method Using N, N-Diethy-1, 4-Phenylenediamine, for Routine Control Purposes. 10 pp.

Colorimetric Analysis (Cont.)

—Water (Cont.)
CNS K9036-80. Method of Test for Boron in Water (Carmine Colorimetric Method) (May)(5577).
CNS K9046-80. Method of Test for Bromide in Water (Colorimetric Method) (Aug)(6228).
CNS K9048-80. Method of Test for Iodide in Water (Colorimetric Method) (Aug)(6230).
CNS K9049-80. Method of Test for Potassium in Water (Colorimetric Method) (Aug)(6231).
ISO 6439-90. Water Quality—Determination of Phenol Index—4-Aminoantipyrine Spectrometric Methods After Distillation Second Edition. 11 pp.
ISO 7393 Pt 2-85. Water Quality—Determination of Free Chlorine and Total Chlorine—Part 2: Colorimetric Method Using N,N-Diethyl-1,4-Phenylenediamine, for Routine Control Purposes First Edition. 10 pp.

—Wood Preservatives
BSI BS 5666: Part 3-91. 1991 Methods of Analysis of Wood Preservatives and Treated Timber Part 3: Quantitative Analysis of Preservatives and Treated Timber Containing Copper/Chromium/Arsenic Formulations. 14 pp.
BSI BS 5666: Part 4-79. 1979 Amd 1 Wood Preservatives and Treated Timber Part 4: Quantitative Analysis of Preservatives and Treated Timber Containing Copper Naphthenate. 5 pp.
BSI BS 5666: Part 5-86. 1986 Wood Preservatives and Treated Timber Part 5: Determination of Zinc Naphthenate in Preservative Solutions and Treated Timber. 8 pp.
BSI BS 5666: Part 6-83. 1983 Amd 1 Wood Preservatives and Treated Timber: Part 6: Quantitative Analysis or Preservative Solution and Treated Timber Containing Pentachloro-phenol, Phenachloro-phenyl Laurate, GS Hexachlorocyclohexane and Dieldrin (AMD 6224). 11 pp.
BSI BS 5666: Part 7-91. 1991 Methods of Analysis of Wood Preservatives and Treated Timber Part 7: Quantitative Analysis of Preservatives Containing Bis(tri-n-butyltin)oxide: Determination of Total Tin. 12 pp.

—Wool Fibers
ISO 2913-75. Wool—Colorimetric Determination of Cystine Plus Cysteine in Hydrolysates First Edition. 6 pp.
ISO 2915-75. Wool—Determination of Cysteic Acid Content of Wool Hydrolysates by Paper Electrophoresis and Colorimetry First Edition. 7 pp.

Colorimetry

Use For: Color Index Testing *See Also:* Color; Color Charts; Color Matching; Color Testing; Colorimetric Analysis; Commision International de l'Eclairage; Gray Scale; Materials Testing; Spectrophotometry

BSI BS 6923-88. 1988 Calculation of Small Colour Differences. 10 pp.
CNS Z7198-85. Method of Comparison for Surface Colour (Aug)(11352).
DIN ENGL 6174-79. Colorimetric Evaluation of Colour Differences of Surface Colours According to the CIELAB Formula (Jan). 3 pp.
JIS Z 8723-88. Methods of Visual Comparison for Surface Colours. 11 pp.

—Aircraft Fuels
CNS K6369-73. Method of Test for Mercaptan Sulfur in Aviation Fuels (Color Indicator Method) (Nov)(3478).

—Anodic Coatings
ISO TR8125-84. Anodizing of Aluminium and Its Alloys—Determination of Colour and Colour Difference of Coloured Anodic Coatings First Edition. 33 pp.

—Binders (Materials)
BSI BS 6782: Part 5-87. 1987 Binders for Paints Part 5: Method for Estimation of Colour of Clear Liquids by the Gardner Colour Scale. 8 pp.
ISO 4630-81. Binders for Paints and Varnishes—Estimation of Colour of Clear Liquids by the Gardner Colour Scale First Edition. 7 pp.

—Chemical Products
BSI BS 5339-76. 1976 Measurement of Colour in Hazen Units (Platinum-Cobalt Scale) of Liquid Chemical Products. 5 pp.
ISO 2211-73. Liquid Chemical Products—Measurement of Colour in Hazen Units (Platinum-Cobalt Scale) First Edition. 4 pp.

—Chromaticity
BSI BS 950: Part 1-67. 1967 Artificial Daylight for the Assessment of Colour Part 1: Illuminant for Colour Matching and Colour Appraisal. 13 pp.

Colorimetry (Cont.)

—Chromaticity (Cont.)
CNS Z7192-85. Specification of Colours According to the CIE 1931 Standard Colorimetric System and the CIE 1964 Supplementary Standard Colormetric System (Apr)(11256).
CNS Z7197-85. Methods of Measurement for Colour of Reflecting or Transmitting Objects (Aug)(11351).
CNS Z7199-85. Method of Measurement for Light Source Colour (Aug)(11353).
JIS Z 8701-82. Specification of Colours According to the CIE 1931 Standard Colorimetric System and the CIE 1964 Supplementary Standard Colorimetric System. 24 pp.
JIS Z 8722-82. Methods of Measurement for Colour of Reflecting or Transmitting Objects. 22 pp.
JIS Z 8724-83. Methods of Measurement for Light Source Colour. 25 pp.
JIS Z 8729-80. Specification of Colour of Materials According to the CIE 1976 (L*A*B) Space and the CIE 1976 (L*U*V) Space (R 1985). 15 pp.

—Color Charts
DIN ENGL 6164 Pt 1-80. DIN Colour Chart; System Based on the 2 Deg. Standard Colorimetric Observer (Feb). 13 pp.
DIN ENGL 6164 Pt 2-80. DIN Colour Chart; Specification of Colour Samples (Feb). 2 pp.
DIN ENGL 6164 Pt 3-81. DIN Colour Chart; System Based on the 10 Deg. Standard Colorimetric Observer (July). 13 pp.

—Color Coding
BSI BS 381C-88. 1988 Colours for Identification, Coding and Special Purposes. 22 pp.
SNZ NZS 7702-89. Specification for Colours for Identification, Coding and Special Purposes Amend: A, 1989. 28 pp.

—Cotton Fibers—Lighting
ISO 4911-80. Textiles—Cotton Fibres—Equipment and Artificial Lighting for Cotton Classing Rooms First Edition. 9 pp.

—Drying Oils
CNS K6655-81. Method of Test for Color After Heating of Drying Oils (Apr)(7256).

—Fabrics
CGSB CAN/CGSB-4.2 NO.41-M91. Textile Test Methods Standard Light Sources for Colour Matching of Textiles. 6 pp.
CGSB CAN/CGSB-4.2 NO.64-M91. Textile Test Methods Chromatic Transference Scale. 9 pp.

—Fats
BSI BS 684: Sec 1.14-87. 1987 Methods of Analysis of Fats and Fatty Oils Part 1: Physical Methods Section 1.14: Determination of Colour. 4 pp.

—Fatty Acids
CNS K6647-81. Method of Test for Color After Heating of Fatty Acid (Mar)(7166).

—Fillers (Papermaking)
CPPA J.3U-77. Colour of Mineral Fillers and Pigments. 1 p.

—Fluorescent Materials
JIS Z 8717-89. Methods of Measurement for Colour of Fluorescent Objects. 39 pp.

—Light Sources
BSI BS 950: Part 1-67. 1967 Artificial Daylight for the Assessment of Colour Part 1: Illuminant for Colour Matching and Colour Appraisal. 13 pp.
CGSB CAN/CGSB-4.2 NO.41-M91. Textile Test Methods Standard Light Sources for Colour Matching of Textiles. 6 pp.
CNS Z7194-85. Standard Illuminants and Sources for Colorimetry (Jun)(11294).
CNS Z7199-85. Method of Measurement for Light Source Colour (Aug)(11353).
ISO CIE10526-91. CIE Standard Colorimetric Illuminants First Edition. 15 pp.
JIS Z 8110-91. Names of Light-Source Colours. 8 pp.
JIS Z 8720-83. Standard Illuminants and Sources for Colorimetry. 25 pp.
JIS Z 8724-83. Methods of Measurement for Light Source Colour. 25 pp.
JIS Z 8726-90. Method of Specifying Colour Rendering Properties of Light Sources. 21 pp.

—Light Sources—Fluorescent Lighting
JIS Z 8716-91. Fluorescent Lamp as a Simulator of CIE Standard Illuminant D65 for a Visual Comparison of Surface Colours—Type and Characteristics. 8 pp.

—Light Sources—Metamerism
JIS Z 8718-90. Evaluation Method of Degree of Metamerism for Change in Observers. 22 pp.
JIS Z 8719-84. Evaluation Method of Degree of Metamerism for Change in Illuminants. 15 pp.

INTERNATIONAL AND NON-U.S. NATIONAL STANDARDS
SUBJECT INDEX

Colorimetry (Cont.)

—Liquids
- CNS K6644-81. Method of Test for Color of Transparent Liquids (Gardner Color Scale) (Mar)(7163).
- CNS K6663-81. Method of Test for Color of Clear Liquids (Platinum Scale) (May)(7403).
- ISO 6271-81. Clear Liquids—Estimation of Colour by the Platinum-Cobalt Scale First Edition. 5 pp.

—Mineral Coatings
- CPPA J.3U-77. Colour of Mineral Fillers and Pigments. 1 p.

—Naphthalene
- CNS K6182-82. Method of Test for Coloration of Naphthalene by Sulfuric Acid (Aug)(2331).
- CNS K6649-81. Method of Test for Acid Wash Color for Refined Grade Naphthalene (Mar)(7168).

—Nitrocellulose Lacquers
- CGSB 1-GP-71 METH 12.7-74. Methods of Testing Paints and Pigments Color Color of Nitrocellulose Sealers and Lacquers. 1 p.

—Nitrocellulose Sealants
- CGSB 1-GP-71 METH 12.7-74. Methods of Testing Paints and Pigments Color Color of Nitrocellulose Sealers and Lacquers. 1 p.

—Oils
- BSI BS 684: Sec 1.14-87. 1987 Methods of Analysis of Fats and Fatty Oils Part 1: Physical Methods Section 1.14: Determination of Colour. 4 pp.

—Opaque Objects
- CNS Z7160-83. Method for Specification of Colour Differences for Opaque Materials (Mar)(10136).
- JIS Z 8730-80. Method for Specification of Colour Differences for Opaque Materials (R 1985). 24 pp.

—Paints
- BSI BS 3900: Part D1-78. 1978 Methods of Test for Paints Group D: Optical Tests on Paint Films Part D1: Visual Comparison of the Colour of Paints. 6 pp.
- BSI BS 3900: Part D8-86. 1986 Methods of Test for Paints Group D: Optical Tests on Paint Films Part D8: Determination of Colour and Colour Difference Principles. 7 pp.
- BSI BS 3900: Part D9-86. 1986 Methods of Test for Paints Group D: Optical Tests on Paint Films Part D9: Determination of Colour and Colour Difference Measurement. 8 pp.
- BSI BS 3900: Part D10-86. 1986 Methods of Test for Paints Group D: Optical Tests on Paint Films Part D10: Determination of Colour and Colour Difference: Calculation. 4 pp.
- CGSB 1-GP-71 METH 12.1-74. Methods of Testing Paints and Pigments Color Color Difference by Instrumental Evaluation. 1 p.
- CGSB 1-GP-71 METH 12.8-74. Methods of Testing Paints and Pigments Color Color Uniformity. 1 p.
- CGSB 1-GP-71 METH 12.9-74. Methods of Testing Paints and Pigments Color Visual Comparison. 2 pp.
- ISO 3668-76. Paints and Varnishes—Visual Comparison of the Colour of Paints First Edition. 6 pp.
- ISO 7724 Pt 1-84. Paints and Varnishes—Colorimetry—Part 1: Principles First Edition. 8 pp.
- ISO 7724 Pt 2-84. Paints and Varnishes—Colorimetry—Part 2: Colour Measurement First Edition. 9 pp.
- ISO 7724 Pt 3-84. Paints and Varnishes—Colorimetry—Part 3: Calculation of Colour Differences First Edition. 6 pp.
- SAA AS 1580.481. 1.12-91. Paints and Related Materials—Methods of Test—Part 481: Coatings—Part 481.1: Assessment of Individual Defects of Exposed Films —Part 481.1.12: Exposed to Weathering—Degree of Colour Change (Supersedes AS 1580.481. 1—1975 (in Part)) (in Prof. Packages 30, 39). 2 pp.
- SAA AS 1580.601. 2-75. Paints and Related Materials—Methods of Test—Part 601.2: Colour-Instrumental Measurement of Colour Difference Using the 'Colormaster' Differential Colorimeter. 4 pp.
- SAA AS 1580.601. 3-75. Paints and Related Materials—Methods of Test—Part 601.3: Colour-Instrumental Measurement of Colour Difference Using the Colour Eye. 4 pp.
- SNZ NZS/AS 1580. 481.1.12-91. Methods of Test for Paints and Related Materials Part 481.1.12: Coatings—Exposed to Weathering—Degree of Colour Change (This is a Joint Standard with SAA AS 1580.481.12). 2 pp.
- SNZ NZS/AS 1580. 601.2-75. Methods of Test for Paints and Related Materials Part 601.2: Colour-Instrumental Measurement of Colour Difference Using the Colourmaster Differential Colorimeter (This is a Joint Standard with SAA AS 1580.601.2). 4 pp.

Colorimetry (Cont.)

—Paints (Cont.)
- SNZ NZS/AS 1580. 601.3-75. Methods of Test for Paints and Related Materials Part 601.3: Colour-Instrumental Measurement of Colour Difference Using the Colour Eye (This is a Joint Standard with SAA AS 1580.601.3). 4 pp.

—Paperboard
- CPPA E.5P-86. Pulp, Paper and Paperboard—Colour Measurement with a Diffuse/Zero Geometry Tristimulus Reflectometer. 3 pp.

—Papers
- CPPA E.5P-86. Pulp, Paper and Paperboard—Colour Measurement with a Diffuse/Zero Geometry Tristimulus Reflectometer. 3 pp.
- SAA AS/NZS 1301. 455S-93. Methods of Test for Pulp and Paper—Part 455S: Colour Measurement with a Diff/0 Degrees Celeius Geometry Tristimulus Reflectometer. 5 pp.
- SNZ NZS/AS 1301. 455S-91. Methods of Test for Pulp and Paper Colour Measurement with a Diff/0 Degrees Geometry Tristimulus Reflectometer (This is a Joint Standard with SAA AS 1301.455S). 3 pp.

—Petroleum Products
- BSI BS 5859-80. 1980 Determination of Colour of Petroleum Products. 4 pp.
- ISO 2049-72. Petroleum Products—Determination of Colour First Edition. 4 pp.

—Pigments
- BSI BS 3483: Part A1-83. 1983 Methods for Testing Pigments in Paints Part A1: Comparison of Colour. 6 pp.
- CGSB 1-GP-71 METH 12.1-74. Methods of Testing Paints and Pigments Color Color Difference by Instrumental Evaluation. 1 p.
- DIN ENGL 55987-81. Testing of Pigments; Determination of Hiding Power Value of Pigmented Media; Colorimetric Method (Feb). 4 pp.

—Plastics
- BSI BS 2782:Pt5: METH 530A-B-76. 1976 Methods of Testing Plastics Part 5: Optical and Colour Properties, Weathering Method 530A: Determination of the Colour of Near-White or Near-Colourless Materials. 6 pp.

—Printing Inks
- ISO 2845-75. Set of Printing Inks for Letterpress Printing—Colorimetric Characteristics First Edition. 7 pp.
- ISO 2846-75. Set of Printing Inks for Offset Printing—Colorimetric Characteristics First Edition. 7 pp.

—Pulps
- CPPA E.5P-86. Pulp, Paper and Paperboard—Colour Measurement with a Diffuse/Zero Geometry Tristimulus Reflectometer. 3 pp.
- SAA AS/NZS 1301. 455S-93. Methods of Test for Pulp and Paper—Part 455S: Colour Measurement with a Diff/0 Degrees Celeius Geometry Tristimulus Reflectometer. 5 pp.
- SNZ NZS/AS 1301. 455S-91. Methods of Test for Pulp and Paper Colour Measurement with a Diff/0 Degrees Geometry Tristimulus Reflectometer (This is a Joint Standard with SAA AS 1301.455S). 3 pp.

—Rubber
- BSI BS 7596-92. 1992 Method for Determination of Colour of Raw Natural Rubber (Colour Index Test) (ISO 4660: 1991). 9 pp.
- ISO 4660-91. Rubber, Raw Natural—Colour Index Test Second Edition. 7 pp.

—Television Systems
- CCIR Report 476-1-74. Colorimetric Standards in Colour Television—Section 11A—Characteristics of Systems for Monochrome and Colour Television. 2 pp.

—Urea
- ISO 2750-74. Urea for Industrial Use—Measurement of Colour in Hazen Units (Platinum-Cobalt Scale) of a Urea-Formaldehyde Solution First Edition. 4 pp.

—Varnishes
- BSI BS 3900: Part D1-78. 1978 Methods of Test for Paints Group D: Optical Tests on Paint Films Part D1: Visual Comparison of the Colour of Paints. 6 pp.
- ISO 3668-76. Paints and Varnishes—Visual Comparison of the Colour of Paints First Edition. 6 pp.
- ISO 7724 Pt 1-84. Paints and Varnishes—Colorimetry—Part 1: Principles First Edition. 8 pp.
- ISO 7724 Pt 2-84. Paints and Varnishes—Colorimetry—Part 2: Colour Measurement First Edition. 9 pp.
- ISO 7724 Pt 3-84. Paints and Varnishes—Colorimetry—Part 3: Calculation of Colour Differences First Edition. 6 pp.

Colorimetry (Cont.)

—Waste Water
- DIN ENGL 38404 Pt 1-76. German Standard Methods for Analysing of Water, Waste Water and Sludge; Physical and Physical-Chemical Parameters (Group C); Determination of Colour (C1) (Dec). 4 pp.

—Water
- BSI BS 6068: Sec 2.22-86. 1986 Water Quality Part 2: Physical, Chemical and Bio-Chemical Methods Section 2.22: Examination and Determination of Colour. 10 pp.
- DIN ENGL 38404 Pt 1-76. German Standard Methods for Analysing of Water, Waste Water and Sludge; Physical and Physical-Chemical Parameters (Group C); Determination of Colour (C1) (Dec). 4 pp.
- ISO 7887-85. Water Quality—Examination and Determination of Colour First Edition. 10 pp.

—Wool Fibers
- SNZ NZS 8707-84. Method for the Measurement of the Colour of Wool. 17 pp.

Coloring Power
Use For: Tinting Strength

—Carbon Black
- BSI BS 5293: Part 13-92. 1992 Sampling and Testing Carbon Black for Use in the Rubber Industry Part 13: Method for Determination of Tinting Strength (ISO 5435: 1991). 12 pp.
- ISO 5435-91. Rubber Compounding Ingredients—Carbon Black—Determination of Tinting Strength Second Edition. 9 pp.

—Pigments
- BSI BS 3483: Part E1-91. 1991 Testing Pigments for Paints Part E1: Assessment of Dispersion Characteristics from the Change in Tinting Strength of Coloured Pigments (ISO 8781-1: 1990). 10 pp.
- CNS K6705-82. Method of Test for Tinting Strength of White Pigments (Jan)(8421).
- CNS K6707-82. Mass Color and Tinting Strength of Color Pigments (Feb)(8514).
- ISO 8781 Pt 1-90. Pigments and Extenders—Methods of Assessment of Dispersion Characteristics—Part 1: Assessment from the Change in Tinting Strength of Coloured Pigments First Edition. 9 pp.

—Turmeric—Spectrophotometry
- BSI BS 4585: Part 13-83. 1983 Methods of Test for Spices and Condiments Part 13: Determination of Colouring Power of Tumeric. 4 pp.
- ISO 5566-82. Turmeric—Determination of Colouring Power—Spectrophotometric Method First Edition. 4 pp.

Colors (Materials)
Use: Colorants

Colour
Use: Color

Colour Codes
Use: Color Coding

Colour-Fastness Tests
Use: Colorfastness Testing

COLP
Use: Connected Line Identification Presentation

Colt GA 42 Airships
Use: Thunder and Colt GA 42 Airships

Column Chromatography

—Sodium Tripolyphosphate
- BSI BS 4427: Part 11-83. 1983 Methods of Test for Sodium Tripolyphosphate (Pentasodium Triphosphate) and Sodium Pyro-phosphate (Tetrasodium Pyrophosphate) for Industrial Use Part 11: Separation by Column Chromatography and Determination of the Diff Phosphate Forms. 12 pp.
- ISO 3358-79. Sodium Tripolyphosphate and Sodium Pyrophosphate for Industrial Use—Separation by Column Chromatography and Determination of the Different Phosphate Forms Second Edition; (Erratum—Nov 1979). 13 pp.

—Tetrasodium Pyrophosphate
- BSI BS 4427: Part 11-83. 1983 Methods of Test for Sodium Tripolyphosphate (Pentasodium Triphosphate) and Sodium Pyro-phosphate (Tetrasodium Pyrophosphate) for Industrial Use Part 11: Separation by Column Chromatography and Determination of the Diff Phosphate Forms. 12 pp.

Column Chromatography *(Cont.)*
—Tetrasodium Pyrophosphate *(Cont.)*
ISO 3358-79. Sodium Tripolyphosphate and Sodium Pyrophosphate for Industrial Use—Separation by Column Chromatography and Determination of the Different Phosphate Forms Second Edition; (Erratum—Nov 1979). 13 pp.

Columns (Supports)
Use For: Composite Columns; Pillars (Supports); Tubes (Supports) *See Also:* Beams (Supports); Composite Structures; Control Columns; Footings; Structural Members
DIN ENGL 18806 Pt 1-84. Composite Steel and Concrete Structures; Composite Columns (Mar). 10 pp.
ISO 2891-77. Modular Units for Machine Tool Construction—Centre Bases and Columns Second Edition. 5 pp.
ISO 3589-75. Modular Units for Machine Tool Construction—Integral Way Columns First Edition; (Amendment 1-1982). 5 pp.
ISO 3970-77. Modular Units for Machine Tool Construction—Integral Way Columns—Floor-Mounted Type First Edition; (Amendment 1-1982). 5 pp.

—Aircraft—Rudder Bars
SBAC AS 2191 ISSUE 5. Rudder Bar Pedestal (Upper). 1 p.
SBAC AS 2193 ISSUE 5. Rudder Bar Pedestal (Lower). 1 p.
SBAC AS 6331 ISSUE 1. Rudder Bar, Pedestal (Upper). 1 p.

—Aircraft—Rudder Bars—Castings
SBAC AS 2192 ISSUE 6. Casting for Rudder Bar Pedestal (Upper). 1 p.
SBAC AS 2194 ISSUE 5. Casting for Rudder Bar Pedestal (Lower). 1 p.

—Containers—Loads (Forces)
DIN ENGL 28081 Pt 3-85. Type B Tubular Vessel Supports; Maximum Permissible Loads on Tubular Supports for Dished Ends (Sept). 7 pp.

—Drilling
BSI BS 1937-53. 1953 Engineers' Ratchet Braces and Drilling Pillars. 24 pp.
BSI BS 7191-89. 1989 Amd 1 Weldable Structural Steels for Fixed Offshore Structures (R) (AMD 6750) September 30, 1991. 48 pp.
BSI BS 7191-02. 1989 Amd 2 Weldable Structural Steels for Fixed Offshore Structures (AMD 6885) December 24, 1991. 52 pp.

—Hot Rolled
BSI BS 4: Part 1-80. 1980 Amd 5 Structural Steel Sections Part 1: Hot-Rolled Sections. 30 pp.
ISO 657 Pt 13-81. Hot-Rolled Steel Sections—Part 13: Tolerances on Sloping Flange Beam, Column and Channel Sections First Edition. 6 pp.
ISO 657 Pt 16-80. Hot-Rolled Steel Sections—Part 16: Sloping Flange Column Sections (Metric Series)—Dimensions and Sectional Properties First Edition. 4 pp.
SNZ NZS/BS 4: Part 1-80. Structural Steel Sections Part 1: Specification for Hot-Rolled Sections. 27 pp.

—Metal—Adjustable
CGSB CAN/CGSB-7.2-M88. Adjustable Metal Columns (Replaces 115-GP-1). 9 pp.

—Steel
BSI BS 7191-02. 1989 Amd 2 Weldable Structural Steels for Fixed Offshore Structures (AMD 6885) December 24, 1991. 52 pp.

—Steel—Adjustable
CEN PREN 1065-93. Adjustable Telescopic Steel Props—Classification, Configuration, Requirements, Design and Assessment by Calculation and Tests. 36 pp.

—Wing Bases
ISO 2934-73. Modular Units for Machine Tool Construction—Wing Base for Columns First Edition. 4 pp.
SNZ NZS 5220-74. Modular Units for Machine Tool Construction. Wing Base for Columns Amend: A, 1974. 2 pp.

Colza Seeds
Use: Rapeseeds

COM
Use: Computer Output Microfilm

Combat
Use: Warfare

Combat Charts
See Also: Amphibious Charts; Amphibious Operations; Charts
NATO STANAG 1022 ED 5 AMD 8-78. Combat Charts, Amphibious Charts and Combat/Landing Charts. 24 pp.
NATO STANAG 1022 ED 5 AMD 9-78. Combat Charts, Amhibious Charts and Combat/Landing Charts. 29 pp.
NATO STANAG 1022 ED 5 AMD 10-78. Combat Charts, Amphibious Charts and Combat/Landing Charts. 31 pp.
NATO STANAG 1022 ED 5 AMD 11-78. Combat Charts, Amphibious Charts and Combat/Landing Charts. 32 pp.

Combat Clothing
Use: Combat Uniforms

Combat Net Radios
See Also: Radio Equipment
—Interfaces
NATO STANAG 4448 ED 1 AMD 0. (Draft) Stanag 4448 on Technical Standards for Non-Secure Voice Interoperability of Very High Frequency (VHF) Combat Net Radios (CNR) by Use of Common Interface Radio Adapter Devices. 13 pp.

Combat Uniforms
Use For: Combat Clothing *See Also:* Clothing; Dress Uniforms; Footwear; Helmets; Munitions; Uniforms
NATO STANAG 2333 ED 3 AMD 2-74. Performance and Protective Properties of Combat Clothing. 7 pp.
NATO STANAG 2333 ED 4 AMD 0-91. Performance and Protective Properties of Combat Clothing. 8 pp.
NATO STANAG 2333 ED 4 AMD 0-92. Performance and Protective Properties of Combat Clothing. 6 pp.

—Anthropometric Characteristics
NATO STANAG 2177 ED 1 AMD 0-00. Methodology for Anthropometric Data. 6 pp.

—Load Carrying Equipment
NATO STANAG 2311 ED 3 AMD 1-83. Principles Governing the Design of the Individual Load-Carrying Equipment of the Combat Soldier. 8 pp.

—Sizes—Interchangeable
NATO STANAG 2335 ED 2 AMD 2-76. Interchangeability of Combat Clothing Sizes. 15 pp.
NATO STANAG 2335 ED 2 AMD 3-76. Interchangeability of Combat Clothing Sizes. 24 pp.

—Troop Trial
NATO STANAG 2138 ED 3 AMD 1-78. Troop Trial Procedures—Combat Clothing and Equipment. 5 pp.

Combat Vehicles
Use For: Fighting Vehicles; Tactical Vehicles
See Also: Military Vehicles; Motor Vehicles; Munitions; Trailers; Trucks; Weapons
NATO STANAG 4381 ED 1 AMD 0-00. (Draft) Blackout Lighting Systems for Tactical Land Vehicles. 18 pp.

—Auxiliary Power Units—Connectors
NATO STANAG 4074 ED 1 AMD 3-69. Auxiliary Power Unit Connections for Starting Combat and Tactical Vehicles. 11 pp.
NATO STANAG 4074 ED 2 AMD 0-00. (Draft) Auxilliary Power Unit Connections for Starting Tactical Land Vehicles. 15 pp.

—Brakes—Antilock Devices—Connectors
NATO STANAG 4395 ED 1 AMD 0-00. Connector for Tactical Land Vehicles with Anti-Lock Braking Systems. 4 pp.

—Connectors
NATO STANAG 4040 ED 1 AMD 1-69. Waterproof Electrical Connectors. 19 pp.

—Cooling Systems—Hoses
NATO STANAG 4043 ED 1 AMD 4-58. Cooling System Hoses for Wheeled Transport Tactical Vehicles. 7 pp.

—Diesel Fuels—Low Temperature
NATO STANAG 2415 ED 1 AMD 2-90. Procedures for Operation of Mechanical Ground Equipment to Minimize Diesel Fuel Problems at Low Ambient Temperature. 9 pp.

—Electrical Installations—Standard Voltages
NATO STANAG 2601 ED 2 AMD 0-79. Standardization of Voltage of Electrical Systems in Tactical Vehicles. 9 pp.
NATO STANAG 2601 ED 3 AMD 0-00. Standardization of Electrical Systems in Tactical Vehicles. 5 pp.

—Flotation
NATO STANAG 2805 ED 4 AMD 0-90. Fording and Flotation Requirements for Combat and Support Ground Vehicles. 11 pp.

—Hydraulic Equipment—Pipe Fittings
MOD UK DSTAN 47-20-78. Brazed Nipple Pipe Fittings-Metric-High and Low Pressure for Fighting Vehicle Applications Issue 1. 20 pp.

—Ignition Cables
NATO STANAG 4006 Pt2 ED1 AMD1-65. Shielded Ignition Cables for Wheeled Tactical Vehicles. 11 pp.

—Name Plates
MOD UK DSTAN 99-4-01. Plates, Identification for Vehicles, Trailers, and Transportable Containers Issue 1; Amendment 2. 9 pp.

—Spark Plugs
NATO STANAG 4006 Pt1 AMD 1-65. Shielded Sparking Plugs (for 5 mm Lead) for Wheeled Tactical Vehicles. 9 pp.

—Starter Batteries
NATO STANAG 4015 Pt1 ED2 AMD1-65. Starter Battery Spaces for Wheeled Tactical Vehicles. 8 pp.
NATO STANAG 4015 ED 3 AMD 0-00. Stater Battery Spaces for Tactical Land Vehicules. 7 pp.

—Starter Batteries—Electric Terminals
NATO STANAG 4015 Pt2 ED1 AMD1-65. Sizes and Positions of Starter Battery Terminal Posts for Wheeled Tactical Vehicles. 8 pp.
NATO STANAG 4015 Pt2 ED2 AMD1-00. Sizes and Positions of Starter Battery Terminal Posts for Wheeled Tactical Vehicles. 11 pp.

—Towing Attachments—Interoperability
NATO STANAG 4019 ED 1 AMD 3-57. Emergency Towing Facilities. 5 pp.
NATO STANAG 4019 ED 2 AMD 0-92. Emergency Towing Facilities. 9 pp.

Combat Zones
See Also: Warfare
—Air Traffic Control—Allied Tactical Publications
NATO STANAG 3805 ED 2 AMD 0-80. Doctrine and Procedures for Airspace Control in the Combat Zone. 5 pp.
NATO STANAG 3805 ED 3 AMD 0-89. Doctrine and Procedures for Airspace Control in the Combat Zone—ATP-40. 5 pp.

Comber Boards
See Also: Looms
JIS L 6114-61. Comber Board for Silk Loom. 4 pp.

Combination Doors
See Also: Doors
—Aluminum
CGSB 82-GP-3M-79. Doors, Aluminum, Combination Storm and Screen, Standard for; (Amendment 1 Aug 1981). 24 pp.

—Steel
CGSB 82-GP-4M-79. Doors, Steel, Combination Storm and Screen, Standard for. 21 pp.

Combination Wrenches
See Also: Adjustable Wrenches; Wrenches
BSI BS 7586-92. 1992 Combination Wrenches (F). 18 pp.
CGSB 39-GP-11C-86. Wrenches, Box, Open End, and Combination, (Inch Series). 32 pp.
CGSB 39-GP-11M-79. Wrenches, Box, Open End, Combination and Flare Nut, Standard for; (Amendment 1 Apr 1984). 29 pp.
CNS B3282-78. Combination Wrenches (Long Series) (Mar)(4296). 2 pp.
CNS B3283-78. Combination Wrenches (Short Series) (Mar)(4297). 3 pp.
CNS B3465-88. Multi-Wrenches (Jun)(12336).
ISO 3318-90. Assembly Tools for Screws and Nuts—Double-Headed Open-Ended Wrenches, Double-Headed Ring Wrenches and Combination Wrenches—Maximum Widths of Heads Third Edition. 4 pp.
ISO 7738-90. Spanners and Wrenches—Combination Wrenches—Minimum Length and Thickness of Heads Second Edition. 6 pp.
JIS B 4651-89. Combination Wrenches. 11 pp.

INTERNATIONAL AND NON-U.S. NATIONAL STANDARDS
SUBJECT INDEX

Combination Wrenches (Cont.)
SAA AS 1354-84. Wrenches—Combination—Ring and Open Ended. 8 pp.

Combined Exchanges
See Also: Telephone Exchanges

—Design
CCITT RECMN Q.541-89. Digital Exchange Design Objectives—General—Digital Local, Transit, Combined and International Exchanges in Integrated Digital Networks and Mixed Analogue-Digital Networks—Supplements (Study Group XI) 6 pp. 6 pp.

CCITT RECMN Q.542-89. Digital Exchange Design Objectives—Operations and Maintenance—Digital Local, Transit, Combined and International Exchanges in Integrated Digital Networks and Mixed Analogue-Digital Networks—Supplements (Study Group XI) 21 pp. 21 pp.

CCITT RECMN Q.543-89. Digital Exchange Performance Design Objectives—Digital Local, Transit, Combined and International Exchanges in Integrated Digital Networks and Mixed Analogue-Digital Networks—Supplements (Study Group XI) 40 pp. 40 pp.

—Integrated Digital Networks
CCITT RECMN Q.500-89. Digital Local, Combined, Transit and International Exchanges Introduction and Field of Application—Digital Local, Transit, Combined and International Exchanges in Integrated Digital Networks and Mixed Analogue-Digital Networks—Supplements (Study Group XI) 4 pp. 4 pp.

CCITT RECMN Q.521-89. Exchange Functions—Digital Local, Transit, Combined and International Exchanges in Integrated Digital Networks and Mixed Analogue-Digital Networks—Supplements (Study Group XI) 3 pp. 3 pp.

—Integrated Services Digital Networks
CCITT RECMN Q.500-89. Digital Local, Combined, Transit and International Exchanges Introduction and Field of Application—Digital Local, Transit, Combined and International Exchanges in Integrated Digital Networks and Mixed Analogue-Digital Networks—Supplements (Study Group XI) 4 pp. 4 pp.

CCITT RECMN Q.521-89. Exchange Functions—Digital Local, Transit, Combined and International Exchanges in Integrated Digital Networks and Mixed Analogue-Digital Networks—Supplements (Study Group XI) 3 pp. 3 pp.

CCITT RECMN Q.522-89. Digital Exchange Connections, Signalling and Ancillary Functions—Digital Local, Transit, Combined and International Exchanges in Integrated Digital Networks and Mixed Analogue-Digital Networks—Supplements (Study Group XI) 16 pp. 16 pp.

—Interfaces
CCITT RECMN Q.511-89. Exchange Interfaces Towards Other Exchanges—Digital Local, Transit, Combined and International Exchanges in Integrated Digital Networks and Mixed Analogue-Digital Networks—Supplements (Study Group XI) 4 pp. 4 pp.

—Interfaces—Operations, Administration and Maintenance
CCITT RECMN Q.513-89. Exchange Interfaces for Operations, Administration and Maintenance—Digital Local, Transit, Combined and International Exchanges in Integrated Digital Networks and Mixed Analogue-Digital Networks—Supplements (Study Group XI) 4 pp. 4 pp.

—Interfaces—Subscriber Access
CCITT RECMN Q.512-89. Exchange Interfaces for Subscriber Access—Digital Local, Transit, Combined and International Exchanges in Integrated Digital Networks and Mixed Analogue-Digital Networks—Supplements (Study Group XI) 9 pp. 9 pp.

—Interfaces—Transmission
CCITT RECMN Q.552-89. Transmission Characteristics at 2-Wire Analogue Interfaces of Digital Exchange—Digital Local, Transit, Combined and International Exchanges in Integrated Digital Networks and Mixed Analogue-Digital Networks—Supplements (Study Group XI) 25 pp. 25 pp.

CCITT RECMN Q.553-89. Transmission Characteristics at 4-Wire Analogue Interfaces of a Digital Exchange—Digital Local, Transit, Combined and International Exchanges in Integrated Digital Networks and Mixed Analogue-Digital Networks—Supplements (Study Group XI) 12 pp. 12 pp.

CCITT RECMN Q.554-89. Transmission Characteristics at Digital Interfaces of a Digital Exchange—Digital Local, Transit, Combined and International Exchanges in Integrated Digital Networks and Mixed Analogue-Digital Networks—Supplements (Study Group XI) 3 pp. 3 pp.

Combined Exchanges (Cont.)
—Maintenance
CCITT RECMN Q.542-89. Digital Exchange Design Objectives—Operations and Maintenance—Digital Local, Transit, Combined and International Exchanges in Integrated Digital Networks and Mixed Analogue-Digital Networks—Supplements (Study Group XI) 21 pp. 21 pp.

—Measurement
CCITT RECMN Q.544-89. Digital Exchange Measurements—Digital Local, Transit, Combined and International Exchanges in Integrated Digital Networks and Mixed Analogue-Digital Networks—Supplements (Study Group XI) 15 pp. 15 pp.

—Mixed Analog/Digital Networks
CCITT RECMN Q.500-89. Digital Local, Combined, Transit and International Exchanges Introduction and Field of Application—Digital Local, Transit, Combined and International Exchanges in Integrated Digital Networks and Mixed Analogue-Digital Networks—Supplements (Study Group XI) 4 pp. 4 pp.

CCITT RECMN Q.521-89. Exchange Functions—Digital Local, Transit, Combined and International Exchanges in Integrated Digital Networks and Mixed Analogue-Digital Networks—Supplements (Study Group XI) 3 pp. 3 pp.

—Operation
CCITT RECMN Q.542-89. Digital Exchange Design Objectives—Operations and Maintenance—Digital Local, Transit, Combined and International Exchanges in Integrated Digital Networks and Mixed Analogue-Digital Networks—Supplements (Study Group XI) 21 pp. 21 pp.

—Signaling Systems—Interworking
CCITT RECMN Q.522-89. Digital Exchange Connections, Signalling and Ancillary Functions—Digital Local, Transit, Combined and International Exchanges in Integrated Digital Networks and Mixed Analogue-Digital Networks—Supplements (Study Group XI) 16 pp. 16 pp.

—Traffic Units
CCITT RECMN Q.544-89. Digital Exchange Measurements—Digital Local, Transit, Combined and International Exchanges in Integrated Digital Networks and Mixed Analogue-Digital Networks—Supplements (Study Group XI) 15 pp. 15 pp.

—Transmission
CCITT RECMN Q.551-89. Transmission Characteristics of Digital Exchanges—Digital Local, Transit, Combined and International Exchanges in Integrated Digital Networks and Mixed Analogue-Digital Networks—Supplements (Study Group XI) 14 pp. 14 pp.

Combines
See Also: Agricultural Equipment
JIS B 9218-86. Standard Form of Specifications of Head-Feeding Type Combines. 20 pp.

—Beater Bars
DIN ENGL 11701-80. Agricultural Machinery; Angular Beater Bars for Combine-Harvesters; Profile (Apr). 1 p.

—Capacity Measurement
BSI BS 7560-92. 1992 Method of Assessing, Testing and Reporting Characteristics and Performance of Combine Harvesters (ISO 8210: 1989). 12 pp.
ISO 8210-89. Equipment for Harvesting —Combine Harvesters—Test Procedure First Edition. 10 pp.

—Chaffcutter Saws
JIS B 9216-83. Chaff-Cutter Saws of Combines.

—Components
BSI BS 1562: Part 2-73. (OBSOLESCENT) 1973 Agricultural Mower and Combine-Harvester Parts Part 2: Metric Units. 15 pp.

—Components—Glossaries
ISO 5702-83. Equipment for Harvesting—Combine Harvester Component Parts—Equivalent Terms First Edition. 8 pp.
ISO 6689-81. Equipment for Harvesting—Combines and Functional Components—Definitions, Characteristics and Performance First Edition. 11 pp.

—Finger Liners
CNS B2632-82. Finger Liners for Agricultural Binders and Rice Combines (Jan)(8202).

—Functional Testing
BSI BS 7560-92. 1992 Method of Assessing, Testing and Reporting Characteristics and Performance of Combine Harvesters (ISO 8210: 1989). 12 pp.
ISO 8210-89. Equipment for Harvesting —Combine Harvesters—Test Procedure First Edition. 10 pp.

Combines (Cont.)
—Hoppers—Capacity Measurement
BSI BS 6621-85. 1985 Determination of Combine Harvester Capacity and Unloading Rate. 4 pp.
ISO 5687-81. Equipment for Harvesting—Combine Harvester—Determination and Designation of Grain Tank Capacity and Unloading Device Performance First Edition. 3 pp.

—Hoppers—Unloading Rate
BSI BS 6621-85. 1985 Determination of Combine Harvester Capacity and Unloading Rate. 4 pp.
ISO 5687-81. Equipment for Harvesting—Combine Harvester—Determination and Designation of Grain Tank Capacity and Unloading Device Performance First Edition. 3 pp.

—Knife Sections
CNS B2563-81. Knife Section for Agricultural Binders and Rice Combines (Jun)(7551).
JIS B 9213-89. Knife Section and Ledger Plate of Agricultural Binders and Head-Feeding Combines. 7 pp.

—Rice
CNS B4022-86. Rice Combine (Apr)(4601). 3 pp.
CNS B7104-86. Method of Test for Rice Combine (Apr)(4602). 7 pp.

—Safety
DIN ENGL 11001 Pt 2-80. Agricultural Machines and Tractors; Combine-Harvesters; Special Technical Safety Requirements and Testing (Dec). 3 pp.

—Self Propelled—Safety
CEN PREN 632-92. Safety Requirements for Agricultural and Forestry Machinery—Combine Harvesters and Forage Harvesters. 26 pp.

—Tractor Drawn—Safety
CEN PREN 632-92. Safety Requirements for Agricultural and Forestry Machinery—Combine Harvesters and Forage Harvesters. 26 pp.

Combustibility Testing
Use: Flammability Testing

Combustible Gas Detectors
See Also: Gas Detectors
BSI BS 6020: Part 1-81. (WITHDRAWN) 1981 Amd 2 Instruments for the Detection of Combustible Gases Part 1: Specification for General Requirements and Test Methods (Superseded by BS EN 50054: 1991). 22 pp.

BSI BS 6020: Part 4-81. (WITHDRAWN) 1981 Amd 1 Instruments for the Detection of Combustible Gases Part 4: Specification for Performance Requirements for Group II Instruments Reading up to 100% Lower Explosive Limit (Superseded by BS EN 50057: 1991). 9 pp.

BSI BS 6020: Part 5-82. (WITHDRAWN) 1982 Amd 1 Instruments for the Detection of Combustible Gases Part 5: Performance Requirements for Group II Instruments Reading up to 100% Gas (Superseded by BS EN 50058: 1991). 7 pp.

BSI BS 6959-89. 1989 Code of Practice for Selection, Installation, Use and Maintenance of Apparatus for the Detection and Measurement of Combustible Gases (Other Than for Mining Applications or Explosives Processing and Manufacture). 16 pp.

BSI BS EN 50057-91. 1991 Electrical Apparatus for the Detection and Measurement of Combustible Gases Performance Requirements for Group II Apparatus Indicating up to 100 Percent Lower Explosive Limit (Q). 10 pp.

BSI BS EN 50058-91. 1991 Electrical Apparatus for the Detection and Measurement of Combustible Gases Performance Requirements for Group II Apparatus Indicating up to 100 Percent (V/V) Gas. 10 pp.

—Hazardous Locations
CSA C22.2 NO 152-M1984. Combustible Gas Detection Instruments (R 1989); (Gen Instr 1 Thru 3). 37 pp.

—Methane
BSI BS 6020: Part 2-81. (WITHDRAWN) 1981 Amd 2 Instruments for the Detection of Combustible Gases Part 2: Specification for Safety and Performance Requirements for Group I Instruments Reading up to 5% Methane in Air (Superseded by BS EN 50055: 1991). 9 pp.

BSI BS 6020: Part 3-82. (WITHDRAWN) 1982 Amd 1 Instruments for the Detection of Combustible Gases Part 3: Safety and Performance Requirements for Group I Instruments Reading up to 100% Methane (Superseded by BS EN 50056: 1991). 5 pp.

Combustible Gas Detectors *(Cont.)*

—Methane—Portable
BSI BS EN 50055-91. 1991 Electrical Apparatus for the Detection and Measurement of Combustible Gases Performance Requirements for Group I Apparatus Indicating up to 5 % (V/V) Methane in Air. 10 pp.
BSI BS EN 50056-91. 1991 Electrical Apparatus for the Detection and Measurement of Combustible Gases Performance Requirements for Group I Apparatus Indicating up to 100 Percent (V/V) Methane. 10 pp.

—Portable
BSI BS EN 50054-91. 1991 Electrical Apparatus for the Detection and Measurement of Combustible Gases General Requirements and Test Methods. 36 pp.
BSI BS EN 50057-91. 1991 Electrical Apparatus for the Detection and Measurement of Combustible Gases Performance Requirements for Group II Apparatus Indicating up to 100 Percent Lower Explosive Limit (Q). 10 pp.
BSI BS EN 50058-91. 1991 Electrical Apparatus for the Detection and Measurement of Combustible Gases Performance Requirements for Group II Apparatus Indicating up to 100 Percent (V/V) Gas. 10 pp.
CENELEC EN 50054-91. Electrical Apparatus for the Detection and Measurement of Combustible Gases General Requirements and Test Methods. 55 pp.
CENELEC EN 50057-91. Electrical Apparatus for the Detection and Measurement of Combustible Gases Performance Requirements for Group II Apparatus Indicating up to 100 % Lower Explosive Limit. 13 pp.
CENELEC EN 50058-91. Electrical Apparatus for the Detection and Measurement of Combustible Gases Performance Requirements for Group II Apparatus Indicating up to 100 % (v/v) Gas. 13 pp.
CSA C22.2 NO 152-M1984. Combustible Gas Detection Instruments (R 1989); (Gen Instr 1 Thru 3). 37 pp.

—Portable—Hazardous Locations
CSA C22.2 NO 152-M1984. Combustible Gas Detection Instruments (R 1989); (Gen Instr 1 Thru 3). 37 pp.

Combustible Liquids
Use: Flammable Liquids

Combustible Materials
Use: Flammable Materials

Combustible Matter Content Analysis

—Glass Fabrics
ISO 1887-80. Textile Glass—Determination of Combustible Matter Content Second Edition. 5 pp.

—Thermal Insulation
DIN ENGL 52273-84. Testing of Mineral Fibre Insulating Materials; Determination of Combustible Matter Content (Dec). 3 pp.

—Wood
CNS Z3009-74. Method of Test for Combustile Properties of Treated Wood by the Fire-Tube Apparatus (Aug)(3514).
CNS Z3010-74. Method of Test for Combustile Properties of Treated Wood by the Crib Test (Aug)(3580).

Combustion
See Also: Burning Quality; Burning Rate; Combustion Rate; Explosions; Fires; Flame Propagation; Flammability Testing; Flammable Liquids; Flammable Materials; Heat of Combustion; Ignition Loss; Ignition Temperature; Punking Behavior; Scorching

—Cables (Electric)
BSI BS 6425: Part 1-90. 1990 Gases Evolved During Combustion of Electric Cables Part 1: Method for Determination of Amount of Halogen Acid Gas Evolved During Combustion of Polymeric Materials Taken from Cables. 4 pp.
BSI BS 6425: Part 2-93. 1993 Test on Gases Evolved During the Combustion of Materials from Cables Part 2: Determination of Degree of Acidity (Corrosivity) of Gases by Measuring pH and Conductivity (E). 19 pp.
CENELEC HD 602 S1-92. Test on Gases Evolved During Combustion of Electric Cables Part 2: Determination of Degree of Acidity of Gases Evolved During the Combustion of Materials Taken from Electric Cables by Measuring pH and Conductivity. 4 pp.
CENELEC HD 602 S1-92. Test on Gases Evolved During Combustion of Materials from Cables Determination of Degree of Acidity (Corrosivity) of Gases by Measuring pH and Conductivity. 4 pp.
IEC 754 Pt 1-82. Test on Gases Evolved During Combustion of Electric Cables Part 1: Determination of the Amount of Halogen Acid Gas Evolved During the Combustion of Polymeric Materials Taken from Cables First Edition. 11 pp.
IEC 754 Pt 2-91. Test on Gases Evolved During Combustion of Electric Cables Part 2: Determination of Degree of Acidity of Gases Evolved During the Combustion of Materials Taken from Electric Cables by Measuring pH and Conductivity First Edition. 33 pp.
SAA AS 1660.5.3-88. Methods of Test for Electrical Cables, Cords and Conductors—Part 5: Fire Tests—Part 5.3: Determination of the Amount of Halogen Acid Gas Evolved During Combustion of Polymeric Materials Taken from Cables. 2 pp.

—Carbon Black
DIN ENGL 53584-77. Testing of Carbon Black; Determination of the Sulphur Content (Sept). 7 pp.
ISO 1138-81. Rubber Compounding Ingredients—Carbon Black—Determination of Sulphur Content Second Edition. 5 pp.

—Cigarette Papers
CNS P3065-84. Combustion Test for Cigarette Paper (Feb)(10778). 2 pp.

—Coal
BSI BS 1016: Part 8-77. 1977 Amd 1 Methods for Analysis and Testing of Coal and Coke Part 8: Chlorine in Coal and Coke (AMD 3472) December 31, 1980. 9 pp.
BSI BS 1016: Part 8-02. 1977 Amd 2 Methods for Analysis and Testing of Coal and Coke Part 8: Chlorine in Coal and Coke (AMD 6984) May 1, 1992. 12 pp.
CNS M3146-84. Method for Determination of Total Sulfur of Coal and Coke (High Temperature Combustion Method) (Mar)(10827).
CNS M3153-84. Method for Determination of Carbons and Hydrogen of Coal and Coke (Sheffield High Temperature Method) (Mar)(10834).
ISO 351-84. Solid Mineral Fuels—Determination of Total Sulfur—High Temperature Combustion Method Second Edition. 6 pp.
ISO 609-75. Coal and Coke—Determination of Carbon and Hydrogen—High Temperature Combustion Method First Edition. 9 pp.
SAA AS 1038.6.3. 2-86. Methods for the Analysis and Testing of Coal and Coke—Part 6.3.2: Ultimate Analysis of Higher Rank Coal—Determination of Total Sulphur (High Temperature Combustion Method) Amdt 1 September 1986. 4 pp.

—Coke
BSI BS 1016: Part 8-77. 1977 Amd 1 Methods for Analysis and Testing of Coal and Coke Part 8: Chlorine in Coal and Coke (AMD 3472) December 31, 1980. 9 pp.
BSI BS 1016: Part 8-02. 1977 Amd 2 Methods for Analysis and Testing of Coal and Coke Part 8: Chlorine in Coal and Coke (AMD 6984) May 1, 1992. 12 pp.
CNS M3153-84. Method for Determination of Carbons and Hydrogen of Coal and Coke (Sheffield High Temperature Method) (Mar)(10834).
ISO 351-84. Solid Mineral Fuels—Determination of Total Sulfur—High Temperature Combustion Method Second Edition. 6 pp.
ISO 609-75. Coal and Coke—Determination of Carbon and Hydrogen—High Temperature Combustion Method First Edition. 9 pp.

—Copper
CEN PREN 723-92. Wrought Copper—Combustion Methods for Determination of Surface Carbon on Tubes. 9 pp.

—Epoxy Resins
ISO 4615-79. Plastics—Unsaturated Polyesters and Epoxide Resins—Determination of Total Chlorine Content First Edition. 7 pp.

—Iron
ISO 9556-89. Steel and Iron—Determination of Total Carbon Content—Infrared Absorption Method After Combustion in an Induction Furnace First Edition; (Corrected and Reprinted -1989). 11 pp.

—Iron Ores
BSI BS 7020: Sec 7.2-88. 1988 Analysis of Iron Ores Part 7: Methods for the Determination of Sulphur Content Section 7.2: Combustion Method. 12 pp.
ISO 4690-86. Iron Ores—Determination of Sulfur Content—Combustion Method First Edition. 10 pp.

—Mineral Oils—Chlorine Content
DIN ENGL 51408 Pt 1-83. Testing of Liquid Petroleum Hydrocarbons; Determination of Chlorine Content; Wickbold Combustion Method (June). 4 pp.

—Natural Gas
ISO 4260-87. Petroleum Products and Hydrocarbons—Determination of Sulfur Content—Wickbold Combustion Method First Edition. 20 pp.
ISO 6326 Pt 5-89. Natural Gas—Determination of Sulfur Compounds—Part 5: Lingener Combustion Method First Edition. 13 pp.

—Nickel
BSI BS 3727: Part 12-70. 1970 Amd 1 Methods for the Analysis of Nickel for Use in Electronic Tubes and Valves Part 12: Determination of Sulphur (Combustion Method) (AMD 6005) July 31, 1970. 13 pp.

—Olefins
ISO 4260-87. Petroleum Products and Hydrocarbons—Determination of Sulfur Content—Wickbold Combustion Method First Edition. 20 pp.
ISO 8915-87. Light Olefins for Industrial Use—Determination of Traces of Chlorine—Wickbold Combustion Method First Edition. 6 pp.

—Oxygen—Ships
MOD UK NES 714-81. Determination of the Oxygen Index of Small Specimens of Material Issue 2 (02.81) (Withdrawn). 21 pp.

—Petroleum Products
BSI BS 2000: Part 61-93. 1993 Methods of Test for Petroleum and Its Products Part 61: Determination of Sulphur—Bomb Method. 5 pp.
BSI BS 2000: Part 61-82. 1982 Petroleum and Its Products Part 61: Sulphur in Petroleum Products (Bomb Method). 7 pp.
BSI BS 2000: Part 107-93. 1993 Methods of Test for Petroleum and Its Products Part 107: Determination of Sulphur—Lamp Combustion Method. 12 pp.
BSI BS 2000: Part 107-91. 1991 Petroleum and Its Products Part 107: Sulphur in Petroleum Products (Lamp Method) (Identical with IP 107/86). 16 pp.
BSI BS 5379-76. 1976 Determination of the Sulphur Content of Petroleum Products by the Wickbold Combustion Method. 18 pp.
CEN EN 41-75. Determination of the Sulphur Content of Petroleum Products by the Wickbold Combustion Method. 16 pp.
ISO 4260-87. Petroleum Products and Hydrocarbons—Determination of Sulfur Content—Wickbold Combustion Method First Edition. 20 pp.

—Plastics—Gas Analysis
JIS K 7217-83. Analytical Method for Determining Gases Evolved from Burning Plastics. 15 pp.

—Plastics—Gas Concentration
JIS K 7228-87. Method for Measuring Smoke Density and Concentration of Gases Evolved by Incineration or Decomposition of Plastics. 18 pp.

—Plastics—Smoke Density
JIS K 7228-87. Method for Measuring Smoke Density and Concentration of Gases Evolved by Incineration or Decomposition of Plastics. 18 pp.

—Polyester Resins
ISO 4615-79. Plastics—Unsaturated Polyesters and Epoxide Resins—Determination of Total Chlorine Content First Edition. 7 pp.

—PVC
ISO 1158-84. Plastics—Vinyl Chloride Homopolymers and Copolymers—Determination of Chlorine Second Edition. 6 pp.

—Rubber
BSI BS 7164: Sec 22.2-92. 1992 Chemical Tests for Raw and Vulcanized Rubber Part 22: Methods for Determination of Bromine and Chlorine Content Section 22.2: Oxygen Flask Combustion Technique for Determination of Bromine and Chlorine (ISO 7725: 1991) (V). 14 pp.
BSI BS 7164: Sec 23.1-93. 1993 Chemical Tests for Raw and Vulcanized Rubber Part 23: Methods for Determination of Total Sulfur Content Section 23.1: Oxygen Combustion Flask Method (ISO 6528-1: 1992). 14 pp.
ISO 6528 Pt 1-92. Rubber—Determination of Total Sulfur Content—Part 1: Oxygen Combustion Flask Method Second Edition. 12 pp.
ISO 6528 Pt 3-88. Rubber—Determination of Total Sulfur Content—Part 3: Furnace Combustion Method First Edition. 11 pp.
ISO 7725-91. Rubber and Rubber Products—Determination of Bromine and Chlorine Content—Oxygen Flask Combustion Technique First Edition. 11 pp.

—Solid Fuels
ISO 352-81. Solid Mineral Fuels—Determination of Chlorine—High Temperature Combustion Method Second Edition. 6 pp.

INTERNATIONAL AND NON-U.S. NATIONAL STANDARDS
SUBJECT INDEX
Commission

Combustion (Cont.)
—Steels
ISO 9556-89. Steel and Iron—Determination of Total Carbon Content—Infrared Absorption Method After Combustion in an Induction Furnace First Edition; (Corrected and Reprinted -1989). 11 pp.

—Temperature Testing—Ships
MOD UK NES 715-81. Determination of the Temperature Index of Small Specimens of Materials Issue 2 (03.81) (Withdrawn). 19 pp.

Combustion Boats
See Also: Laboratory Equipment
—Platinum
CNS K0025-82. Platinum Boats (Nov)(9623).
JIS H 6203-86. Platinum Boats for Chemical Analysis.
JIS H 6203-62. Platinum Boat for Chemical Analysis (R 1968). 4 pp.

—Porcelain
CNS R2062-85. Porcelain Combustion Boats for Chemical Analysis (May)(3242).
JIS R 1306-87. Porcelain Combustion Boats for Chemical Analysis.
JIS R 1306-80. Porcelain Combustion Boats for Chemical Analysis (R 1985). 8 pp.

Combustion Chambers
See Also: Engine Valves; Firebricks; Internal Combustion Engines; Pistons; Spark Plugs

—Aircraft—Numbering
BSI 2M 41-77. 1977 Methods of Numbering Propulsion Units and Components and Describing Their Direction of Rotation. 6 pp.
ISO 482-77. Aircraft—Propulsion Units and Components—Methods of Numbering and Describing Direction of Rotation First Edition. 6 pp.

—Deposits—Sampling
BSI BS 2455: Part 2-83. 1983 Methods of Sampling and Examining Deposits from Boilers and Associated Industrial Plant Part 2: Methods for Sampling and Examining Free-Side Deposits. 41 pp.
BSI BS 2455: Part 2-01. 1983 Amd 1 Sampling and Examining Deposits from Boilers and Associated Industrial Plant Part 2: Methods for Sampling and Examining Fire-Side Deposits (AMD 7751) October 15, 1993 (Q). 42 pp.

Combustion Control Equipment
See Also: Process Control Equipment; Remote Control Equipment
CSA CAN/CSA-C22. 2 NO 74-92. Equipment for Use with Electric Discharge Lamps (Incorporating Electrical Bulletin Nos. 523F, 753, 846, 1124A, 1125A, 1325, and 1326); (Gen Instr 1). 82 pp.

—Gas Burners
BSI BS 6505-84. 1984 Amd 2 Rubber-Type Materials Used for Controls Components for Use with 1st, 2nd and 3rd Family Gases. 13 pp.
BSI BS 6505-03. 1984 Amd 3 Rubber-Type Controls Components for Use with 1st, 2nd and 3rd Family Gases (AMD 7178) July 15, 1992. 15 pp.
CSA CAN/CSA-C22. 2 NO 199-M89. Combustion Safety Controls and Solid-State Igniters for Gas-and Oil-Burning Equipment; (Gen Instr 1). 94 pp.

—Gas Burners—Appliances
BSI BS 6047: Part 1-81. 1981 Flame Supervision Devices for Domestic, Commercial and Catering Gas Appliances Part 1: Heat Sensitive Types. 11 pp.
BSI BS 6067-81. 1981 Amd 1 Multifunctional Gas Controls for Domestic, Commercial and Catering Appliances. 12 pp.

—Oil Burners
CSA CAN/CSA-C22. 2 NO 199-M89. Combustion Safety Controls and Solid-State Igniters for Gas-and Oil-Burning Equipment; (Gen Instr 1). 94 pp.
JIS B 8412-81. Combustion Safety Controllers for Guntype Oil Burners. 28 pp.

Combustion Furnaces
Use: Furnaces

Combustion Heaters
Use: Heaters

Combustion Rate
See Also: Combustion

—Cigarette Papers
CNS P3015-84. Method of Test for Combustion Rate of Cigarette Paper (Feb)(2388). 2 pp.

Combustion Rate (Cont.)
—Cigarettes
BSI BS 5381: Part 3-79. 1979 Determination of Physical Properties of Tobacco and Tobacco Products Part 3: Determination of Free Combustion Rate of Cigarettes. 11 pp.
ISO 3612-77. Tobacco and Tobacco Products—Cigarettes—Determination of Rate of Free Combustion First Edition. 10 pp.

Combustion Safety Control Equipment
Use: Combustion Control Equipment

Combustion Tubes
See Also: Glass Tubes; Laboratory Equipment
—Porcelain
CNS R2104-80. Porcelain Combustion Tubes for Chemical Analysis (Sep)(6518).
JIS R 1307-87. Porcelain Combustion Tubes for Chemical Analysis. 8 pp.

Comedo Extractors
See Also: Medical Electrical Equipment; Medical Equipment
CNS T1032-80. Comedo Extractor (Aug)(6291).
JIS T 2611-88. Comedo Extractors. 4 pp.

Comet 4 Aircraft
See Also: Aircraft
—Antenna Positions
CAA. Comet 4 (Approved Aerial Positions). 1 p.

Comet 4B Aircraft
See Also: Aircraft
—Antenna Positions
CAA. Comet 4B and 4C (Approved Aerial Positions). 1 p.

Comet 4C Aircraft
See Also: Aircraft
—Antenna Positions
CAA. Comet 4B and 4C (Approved Aerial Positions). 1 p.

Comfort Tones
See Also: Tones (Telephone Services)
CCITT RECMN E.182-89. Application of Tones and Recorded Announcements in Telephone Services—Telephone Network and ISDN—Operation, Numbering, Routing and Mobile Service (Study Group II) 8 pp. 8 pp.

Comite Consultatif International des Radiocommunications
Use: CCIR

Comite Consultatif International Telegraphique et Telephonique
Use: CCITT

Comite Europeen de Normalisation
Use: CEN

Comite Europeen de Normalisation Electrotechnique
Use: Cenelec

Command and Control
See Also: Command and Control Systems; Military Operations; Military Personnel; Tactical Warfare; Warfare

—Amphibious Operations
NATO STANAG 1356 ED 1 AMD 1-87. Guidance for the Command and Control/Coordination of Forces in the Amphibious Objective Area (AOA) Not Assigned to the Commander Amphibious Task Force (CATF) (Withdrawn). 10 pp.

—Military Communications—Interoperability
NATO STANAG 4312 Pt2 ED1 AMD0-00. Interoperability of Army Short Range Air Defence Surveillance, Command and Control Systems Part II: Interface Requirements and Bit-Oriented Messages. 60 pp.
NATO STANAG 4312 ED 2 AMD 0-89. (DRAFT) Interoperability of Army Short Range Air Defence Surveillance, Command and Control Systems Part II: Common Interface Requirements and Bit-Oriented Messages. 12 pp.

—Signal Devices
NATO STANAG 2100 ED 6 AMD 0-88. Signs, Signals and Markings to Be Used in Controlling Combined Exercises. 6 pp.

Command and Control Systems
Use For: Telecommand Systems See Also: Command and Control; Control Systems Equipment; Telecommunication Systems

—Ship Movement
CCIR QUESTION 55-2/8-90. Development and Future Implementation of Data Exchange Systems and Ship Movement Telemetry and Telecommand Systems—Questions Concerning Study Group 8—Mobile, Radiodetermination, Amateur and Related Satellite Services. 1 p.

—Ships
CCIR Report 1044-86. Development and Future Implementation of Exchange Systems and Ship Movement Telemetry and Telecommand Systems—Section 8D—Radiodetermination, Global Maritime Distress and Safety System and Related Subjects. 1 p.

Command Guidance
See Also: Ground Support Equipment; Radio Guidance Systems

—Meteorological Satellites
CCIR QUESTION 141/7-90. Command and Data Communication Systems for Meteorological Satellites—Questions Concerning Study Group 7—Science Services. 1 p.

Commerce
Use For: Trade See Also: Balance of Payments; Balance of Trade; INTRASTAT System; Wholesaling

—European Communities—Commercial Agents
EC 86/653/EEC-86. Council Directive on the Coordination of the Laws of the Member States Relating to Self-Employed Commercial Agents. 5 pp.

—European Communities—Statistics
EC EEC/3330/91-91. Council Regulation on the Statistics Relating to the Trading of Goods Between Member States. 11 pp.

—European Communities—Transit
EC EEC 1901/85-85. Council Regulation Amending Regulation (EEC) No 222/77 on Community Transit. 3 pp.
EC EEC/1674/87-87. Council Regulations Amending Regulation 222/ 77 on Community Transit. 2 pp.
EC EEC/474/90-90. Council Regulation Amending, with a View to Abolishing Lodgement of the Transit Advice Note on Crossing an Internal Frontier of the Community, Regulation (EEC) No 222/77 on Community Transit. 2 pp.
EC EEC/2726/90-90. Council Regulation on Community Transit. 10 pp.

Commercial Land Use
Use: Land Use Planning

Commercial Vehicles (Trucks)
Use: Trucks

Comminution
Use For: Particle Size Reduction (Crushing); Size Reduction (Comminution) See Also: Disintegrators

—Glossaries
DIN ENGL 24100 Pt 1-78. Machines for the Building and Building Material Industry; Mechanical Communication; Concepts Relating to Materials Processing (May). 5 pp.

Commision International de l'Eclairage
Use For: CIE; International Commission on Illumination See Also: Colorimetry

—CIELAB Formula—Color Differences
DIN ENGL 6174-79. Colorimetric Evaluation of Colour Differences of Surface Colours According to the CIELAB Formula (Jan). 3 pp.

Commission des Communautes Europeennes
Use: CEC (Commission of the European Communities)

Commission Electrotechnique Internationale
Use: IEC

Commission of the European Communities
Use: CEC (Commission of the European Communities)

INDUSTRY STANDARDS

INTERNATIONAL AND NON-U.S. NATIONAL STANDARDS
SUBJECT INDEX

Committee on Aircraft Engine Emissions (ICAO)
Use: ICAO Committee on Aircraft Engine Emissions

Committee on Aircraft Noise (ICAO)
Use: ICAO Committee on Aircraft Noise

Committee on Aviation Environmental Protection (ICAO)
Use: ICAO Committee on Aviation Environmental Protection

Common Business-Oriented Language
Use: COBOL

Common Carriers
See Also: Recognized Private Operating Agencies; Telecommunication Administrations; Telecommunication Services; Telecommunication Systems

—**Statistical Analysis—Yearbooks**
CCITT RECMN C.1-89. Yearbook of Common Carrier Telecommunication Statistics—Terms and Definitions Abbreviations and Acronyms Recommendations on Means of Expression (Series B) General Telecommunications Statistics (Series C) 9 pp. 9 pp.

Common Channel Signaling
Use: Common Channel Signaling Systems

Common Channel Signaling Networks
Use: Common Channel Signaling Systems

Common Channel Signaling Systems
See Also: CCITT No. 4 Signaling Systems; CCITT No. 5 Signaling Systems; CCITT No. 6 Signaling Systems; CCITT No. 7 Signaling Systems; Continuity Checks; Drift Compensation; Integrated Services Digital Networks User Part; Message Transfer Part (Signaling Systems); Quasiassociated Signaling; Security Arrangements (Signaling Systems); Signaling Channels; Signaling Systems; Superfluous Messages; Telephone Signals; Unreasonable Messages

CEPT T/CS 14-05-80. Systeme De Signalisation Sur Voie Commune. 1 p.
CEPT T/CS 14-05 E-80. Use of Common Channel Signalling. 1 p.

—**Planning—Failure (Quality Control)**
CCITT RECMN E.413-89. International Network Management—Planning —Telephone Network and ISDN—Quality of Service, Network Management and Traffic Engineering (Study Group II) 5 pp. 5 pp.

—**Public Packet Switched Networks—Interworking**
CCITT RECMN X.326-89. General Arrangements for Interworking Between Packet Switched Public Data Networks (PSPDNs) and Common Channel Signalling Network (CCSN)—Data Communication Networks—Interworking Between Networks, Mobile Data Transmission Systems, Internetwork Management. 9 pp.

—**Routing**
CCITT RECMN E.170-92. Traffic Routing (Study Group II) 10 pp. 10 pp.

—**Telecommunication Traffic—Measurement**
CCITT RECMN E.502 (REV 1)-92. Traffic Measurement Requirements for Digital Telecommunication Exchanges (Study Group II) 23 pp. 23 pp.
CCITT RECMN E.502-89. Traffic Measurement Requirements for SPC (Especially Digital) Telecommunication Exchanges—Telephone Network and ISDN—Quality of Service, Network Management and Traffic Engineering (Study Group II) 15 pp. 15 pp.
CCITT RECMN E.505-92. Measurements of the Performance of Common Channel Signalling Network (Study Group II) 16 pp. 16 pp.

Common Market
Use: European Communities

Common Mode Rejection
See Also: Differential Amplifiers

—**Cable Couplers**
CEN PREN 2591 (Part G3)-92. Elements of Electrical and Optical Connection Test Methods Part G3—Common Mode Rejection. 4 pp.

Common Salt
Use: Sodium Chloride

Communication Cables
Use For: Telecommunication Cables *See Also:* Cable Television Cables; Cables (Electric); Communication Equipment; Communication Transmission Lines; Electric Wire; Fiber Optic Cables; Local Area Network Cables; Local Cables; Microphone Cables; Radiating Cables; Stranded Conductors; Telephone Cables; Telephone Lines; Television Cables; Transmission Lines; Waveguides

CCITT FASCICLE III.3. Transmission Media Characteristics Recommendations G.601—G.654. 134 pp.
SAA AS 3080-92. Telecommunications Installations—Integrated Communications Cabling Systems for Commercial Premises Amdt 1 November 1992. 83 pp.

—**Aluminum Tape—Adhesive Strength**
DIN VDE 0472 Pt 618-83. Testing of Cables, Wires and Flexible Cords; Adhesive Strength (Jan). 6 pp.

—**Capacitance Measurement**
DIN VDE 0472 Pt 504-83. Testing of Cables, Wires and Flexible Cords; Effective Capacitance (Apr). 5 pp.

—**Capacitive Unbalance**
DIN VDE 0472 Pt 506-83. Testing of Cables, Wires and Flexible Cords Capacitative Unbalance (VDE Specification) (Apr). 5 pp.

—**Construction**
CCITT Volume 1X. Protection Against Interference—Series K Recommendations Construction, Installation and Protection of Cable and Other Elements of Outside Plant—Series L Recommendations. 174 pp.
CSA CAN/CSA-C22. 2 NO 214-M90. Communications Cables; (Gen Instr 1 Thru 3). 60 pp.

—**Copper—Electrical Resistance**
IEC 344-80. Guide to the Calculation of Resistance of Plain and Coated Copper Conductors of Low-Frequency Cables and Wires Second Edition; (Amendment 1-1985). 16 pp.

—**Data Transmission Equipment**
MOD UK DSTAN 61-12: Part 23-01. Wires, Cords, and Cables, Electrical—Metric Units Part 23: General Requirements for Cables, Special Purpose, Electrical (Digital Data Transmission) Issue 1; Amendment 1. 25 pp.
MOD UK DSTAN 61-12: 90/002-01. Cables, Special Purpose, Electrical (Digital Data Transmission) Issue 2; Amendment 1. 14 pp.

—**Effective Capacitance**
DIN VDE 0472 Pt 504-83. Testing of Cables, Wires and Flexible Cords; Effective Capacitance (Apr). 5 pp.

—**Flat—Construction**
CSA CAN/CSA-C22. 2 NO 214-M90. Communications Cables; (Gen Instr 1 Thru 3). 60 pp.

—**Flat—Multiconductor**
CSA CAN/CSA-C22. 2 NO 214-M90. Communications Cables; (Gen Instr 1 Thru 3). 60 pp.

—**Flat—Multiconductor—Construction**
CSA CAN/CSA-C22. 2 NO 214-M90. Communications Cables; (Gen Instr 1 Thru 3). 60 pp.

—**Installation—Safety**
DIN VDE 0800 Pt 4-86. Telecommunications; Erection of Telecommunication Lines (Mar) (Partially Supersedes DIN 57800 Part1/VDE 0800 Part 1/04.84). 75 pp.

—**Multiconductor—Construction**
CSA CAN/CSA-C22. 2 NO 214-M90. Communications Cables; (Gen Instr 1 Thru 3). 60 pp.

—**Multiconductor—Testing**
CSA CAN/CSA-C22. 2 NO 214-M90. Communications Cables; (Gen Instr 1 Thru 3). 60 pp.

—**Overhead—Power Lines—Electrical Protection**
CSA C22.3 NO 5.1-93. Recommended Practices for Electrical Protection—Electric Contact Between Overhead Supply and Communication Lines; (Gen Instr 1). 49 pp.

Communication Cables (Cont.)
—**Paper Insulated**
DIN VDE 0816 Pt 3-88. Outdoor Cables for Telecommunications and Information Processing Systems; Cables with Paper Insulation (Feb) (Together with DIN VDE 0816 Part 1/02.88 and DIN VDE 0816 Part 2/02.88 Replaces DIN 57816/02.79). 58 pp.

—**Polyethylene Insulated**
SAA AS 1049-86. Telecommunication Cables—Insulation and Sheath—Polyethylene. 64 pp.

—**Polyethylene Insulated—Polyethylene Sheathed**
BSI BS 3573-90. 1990 Polyolefin Copper-Conductor Telecommunication Cables. 22 pp.
DIN VDE 0816 Pt 1-88. Outdoor Cables for Telecommunications and Information Processing Systems; Cables with Insulation and Sheath of PE of Unit Construction (Feb) (Together with DIN VDE 0816 Part 2/02.88 and DIN VDE 0816 Part 3/02.88 Replaces DIN 57816/VDE 0816/02.79). 39 pp.
DIN VDE 0891 Pt 6-90. Use of Cables and Insulated Cords for Telecommunications and Information Processing Installations; Special Requirements for Outside Cables According to DIN VDE 0816 Part 1 to Part 3 (May). 31 pp.

—**Polyethylene Insulated—PVC Sheathed**
DIN VDE 0816 Pt 2-88. Outdoor Cables for Telecommunication and Information Processing Systems; Signalling and Measuring Cables, Mine Cables (Feb) (Together with DIN VDE 0816 Part 1/02.88 and DIN VDE 0816 Part 3/02.88, Replaces DIN 57816/VDE 0816/02.79). 37 pp.
DIN VDE 0891 Pt 7-84. Use of Cables and Insulated Cords for Telecommunications Systems and Data Processing Systems; Special Guidelines for Cables with Stranded Conductors for Increased Mechanical Stress for Telecommunications Systems and Data Processing Systems. 8 pp.

—**Protection**
CCITT Volume 1X. Protection Against Interference—Series K Recommendations Construction, Installation and Protection of Cable and Other Elements of Outside Plant—Series L Recommendations. 174 pp.

—**PVC Insulated—PVC Sheathed**
BSI BS 4808: Part 1-72. 1972 L.F. Cables and Wires with PVC Insulation and PVC Sheath for Telecommunication Part 1: General Requirements and Test. 9 pp.
BSI BS 4808: Part 3-72. 1972 L.F. Cables and Wires with PVC Insulation and PVC Sheath for Telecommunication Part 3: Cables and Equipment Wires, with Solid or Stranded Conductors, Screened, Single. 9 pp.
BSI BS 4808: Part 5-73. 1973 L.F. Cables and Wires with PVC Insulation and PVC Sheath for Telecommunication Part 5: Cables with Solid or Stranded Conductors, Screened and Sheathed, One Pair. 9 pp.

—**PVC Insulated—PVC Sheathed—Flameproof**
DIN VDE 0891 Pt 7-84. Use of Cables and Insulated Cords for Telecommunications Systems and Data Processing Systems; Special Guidelines for Cables with Stranded Conductors for Increased Mechanical Stress for Telecommunications Systems and Data Processing Systems. 8 pp.

—**Receivers**
BSI BS EN 50080-91. 1991 RF Characteristics of MAC AM-VSB Cable Receivers. 12 pp.
CENELEC EN 50080-91. RF Characteristics of MAC AM-VSB Cable Receivers. 10 pp.

—**Reduction Factor**
DIN VDE 0472 Pt 507-83. Testing of Cables, Wires and Flexible Cords; Reduction Factor (VDE Specification) (Apr) (Partly Replaces 0472 9.71). 6 pp.

—**Ships**
IEC 92 Pt 376-83. Electrical Installations in Ships Part 376: Shipboard Multicore Cables for Control Circuits First Edition. 22 pp.

—**Switchboards**
DIN VDE 0891 Pt 3-90. Use of Cables and Insulated Cords for Telecommunications and Information Processing Installations; Particular Requirements for Switchboard Cables According to DIN VDE 0813 (May). 5 pp.

—**Telecommunication Systems—Standards**
CCITT FASCICLE I.2-88. Opinions and Resolutions Recommendations on the Organisation and Working Procedures of CCITT (Series A). 64 pp.

INTERNATIONAL AND NON-U.S. NATIONAL STANDARDS
SUBJECT INDEX
Communication

Communication Cables *(Cont.)*

—Testing
  CSA CAN/CSA-C22. 2 NO 214-M90. Communications Cables; (Gen Instr 1 Thru 3). 60 pp.

—Underground—Corrosion Prevention
  CCITT RECMN K.29-92. Coordinated Protection Schemes for Telecommunication Cables Below Ground (Study Group V) 7 pp. 7 pp.

—Underground—Electromagnetic Interference
  CCITT RECMN K.29-92. Coordinated Protection Schemes for Telecommunication Cables Below Ground (Study Group V) 7 pp. 7 pp.

—Underground—Lighting Protection
  CCITT RECMN K.29-92. Coordinated Protection Schemes for Telecommunication Cables Below Ground (Study Group V) 7 pp. 7 pp.

—Varnishes—Elastic Properties
  DIN VDE 0472 Pt 624-83. Testing of Cables, Wires and Flexible Cords; Testing of the Elasticity of the Varnished Coating (Jan). 4 pp.

—Water Resistance Testing
  DIN VDE 0472 Pt 811-87. Testing of Cables, Wires and Flexible Cords; Longitudinal Watertightness (Oct). 8 pp.

Communication Channel Assignment
Use: Frequency Assignment

Communication Channels
Use For: Channels (Communication)
See Also: Carrier Channels; Channel Spacing; Data Channels; Digital Blocks; Frequency Assignment; Frequency Separation; Groups (Communications); Mastergroups (Communications); Narrowband Channels; Pilot Channels; Radiotelegraph Channels; Service Channels; Single Channel Per Carrier; Standby Channels; Supergroups (Communications); Supermastergroups (Communications); Telephone Channels; Television Channels; Voice Frequency Telegraph Channels

—
  CCIR QUESTION 7-3/8-90. Characteristics of Equipment and Principles Governing the Allocation of Frequency Channels in the Land Mobile Services Between 25 and 3000 MHz—Questions Concerning Study Group 8—Mobile, Radiodetermination, Amateur and Related Satellite Services. 2 pp.

—
  CCIR RECMN 478-4-90. Technical Characteristics of Equipment and Principles Governing the Allocation of Frequency Channels Between 25 and 1000 MHz for the Land Mobile Service—Section 8A—Land Mobile Service and Related Subjects. 4 pp.

—
  CCIR QUESTION 58/CMTT-90. Standards for the Digital Transmission of Sound-Programme Signals on One, Two or Three 64 kbit/s Channels—Questions Concerning the CMTT CCIR/CCITT Joint Study Group for Television and Sound Transmission. 2 pp.

— 300—3400 Hz—Independent Sideband
  CCIR Report 703-2-90. Use of Channels with Bandwidth 300-3400 Hz in SSB and ISB Systems—Section 3Cb—Data Transmission. 3 pp.

— 300—3400 Hz—Single Sideband
  CCIR Report 703-2-90. Use of Channels with Bandwidth 300-3400 Hz in SSB and ISB Systems—Section 3Cb—Data Transmission. 3 pp.

— 320 kbit/s—Sound Transmission
  CCIR RECMN 719-90. Transmission of High-Quality Sound-Programme Analogue Signals over Mixed Analogue/Digital Circuits at 320 kbit/s—Section CMTT C—Transmission Standards and Performance Objectives for Sound-Programme Channels. 6 pp.

— 384 kbit/s—Sound Transmisison
  CCIR RECMN 660-86. Transmission of Analogue High-Quality Sound-Programme Signals on Mixed Analogue-and-Digital Circuits Using 384 kbit/s Channels—Section CMTT C—Transmission Standards and Performance Objectives for Sound-Programme Channels. 5 pp.

— 480 kbit/s—Sound Transmission
  CCIR RECMN 718-90. Digital Transmission of High-Quality Sound-Programme Signals on Distribution Circuits Using 480 kbit/s (496 kbit/s) per Audio Channel—Section CMTT C—Transmission Standards and Performance Objectives for Sound-Programme Channels. 9 pp.

Communication Channels *(Cont.)*

—Analog—Data Transmission
  CCIR Report 903-2-90. Digital Transmission in the Land Mobile Service—Section 8A—Land Mobile Service and Related Subjects. 34 pp.

—Analog—Line Up
  CCITT RECMN M.470-89. Setting up and Lining up Analogue Channels for International Telecommunication Services—General Maintenance Principles—Maintenance of International Transmission Systems and Telephone Circuits (Study Group IV) 1 pp. 1 p.
  CCITT RECMN M.475-89. Setting up and Lining up Mixed Analogue/Digital Channels for International Telecommunication Services—General Maintenance Principles—Maintenance of International Transmission Systems and Telephone Circuits (Study Group IV) 3 pp. 3 pp.

—Analog—Radio Frequencies
  CEPT T/R 12-01 E-91. Harmonized Radio Frequency Channel Arrangements for Analogue and Digital Terrestrial Fixed Systems Operating in the Band 37 GHz-39.5 GHz. 5 pp.

—Analog—Set Up
  CCITT RECMN M.470-89. Setting up and Lining up Analogue Channels for International Telecommunication Services—General Maintenance Principles—Maintenance of International Transmission Systems and Telephone Circuits (Study Group IV) 1 pp. 1 p.
  CCITT RECMN M.475-89. Setting up and Lining up Mixed Analogue/Digital Channels for International Telecommunication Services—General Maintenance Principles—Maintenance of International Transmission Systems and Telephone Circuits (Study Group IV) 3 pp. 3 pp.

—Broadcast Transmitters—Sound Broadcasting
  CCIR Report 1201-90. Number of HF Sound Broadcasting Transmitters Using a Single Channel—Section 10A-1—Amplitude-Modulation Sound Broadcasting in Bands 5 (LF), 6 (MF) and 7 (HF). 2 pp.

—Data Networks
  CCITT RECMN X.57-89. Method of Transmitting a Single Lower Speed Data Channel on a 64 kBit/s Data Stream—Data Communication Networks—Transmission, Signalling and Switching, Network Aspects, Maintenance and Administrative Arrangements (Study Group VII) 2 pp. 2 pp.

—Digital—Audiovisual Equipment
  CCITT RECMN H.242-90. System for Establishing Communication Between Audiovisual Terminals Using Digital Channels up to 2 Mbit/s (Study Group XV) 35 pp. 35 pp.

—Digital—Line Up
  CCITT RECMN M.475-89. Setting up and Lining up Mixed Analogue/Digital Channels for International Telecommunication Services—General Maintenance Principles—Maintenance of International Transmission Systems and Telephone Circuits (Study Group IV) 3 pp. 3 pp.

—Digital—Maritime Mobile Services
  CCIR RECMN 626-86. Evaluation of the Quality of Digital Channels in the Maritime Mobile Service—Section 8B—Maritime Mobile Service; Telegraphy and Related Subjects. 1 p.

—Digital—Maritime Mobile Services—Quality of Service
  CCIR RECMN 626-86. Evaluation of the Quality of Digital Channels in the Maritime Mobile Service—Section 8B—Maritime Mobile Service; Telegraphy and Related Subjects. 1 p.
  CCIR Report 743-1-82. Transmission Quality Assessment of Digital Channels in Maritime Mobile Services—Section 8B—Maritime Mobile Service: Telegraphy and Related Subjects. 13 pp.
  CCIR QUESTION 42-1/8-90. Characteristics of Digital Channels in the Maritime Mobile Service—Questions Concerning Study Group 8—Mobile, Radiodetermination, Amateur and Related Satellite Services. 1 p.

—Digital—Private—Leasing
  CCITT RECMN D.3-92. Principles for the Lease of Analogue International Circuits for Private Service (Study Group III) 5 pp (Replaces Recmn D2 and D3). 5 pp.
  CCITT RECMN D.3-89. Special Conditions for the Lease of Intercontinental Telecommunication Circuits for Private Service—General Tariff Principles—Charging and Accounting in International Telecommunications Services (Study Group III) 4 pp. 4 pp.

Communication Channels *(Cont.)*

—Digital—Private—Tariffs
  CCITT RECMN D.3-92. Principles for the Lease of Analogue International Circuits for Private Service (Study Group III) 5 pp (Replaces Recmn D2 and D3). 5 pp.
  CCITT RECMN D.3-89. Special Conditions for the Lease of Intercontinental Telecommunication Circuits for Private Service—General Tariff Principles—Charging and Accounting in International Telecommunications Services (Study Group III) 4 pp. 4 pp.

—Digital—Radio Frequencies
  CEPT T/R 12-01 E-91. Harmonized Radio Frequency Channel Arrangements for Analogue and Digital Terrestrial Fixed Systems Operating in the Band 37 GHz-39.5 GHz. 5 pp.

—Digital—Set Up
  CCITT RECMN M.475-89. Setting up and Lining up Mixed Analogue/Digital Channels for International Telecommunication Services—General Maintenance Principles—Maintenance of International Transmission Systems and Telephone Circuits (Study Group IV) 3 pp. 3 pp.
  CCITT RECMN M.556-89. Setting up and Initial Testing of Digital Channels on an International Digital Path or Block—General Maintenance Principles—Maintenance of International Transmission Systems and Telephone Circuits (Study Group IV) 1 pp. 1 p.

—Digital—Telecommunication Services—Tariffs—Europe
  CCITT RECMN D.307R-91. Remuneration of Digital Systems and Channels Used in Telecommunication Relations Between the Countries of Europe and the Mediterranean Basin (Study Group III) 7 pp. 7 pp.
  CCITT RECMN D.307R-89. Remuneration of Digital Systems and Channels Used in Telecommunication Relations Between the Countries of Europe and the Mediterranean Basin—General Tariff Principles—Charging and Accounting in Intl Telecom. Serv. (Study Group III) 3 pp. 3 pp.

—Digital—Telecommunication Services—Tariffs—Mediterranean Basin
  CCITT RECMN D.307R-91. Remuneration of Digital Systems and Channels Used in Telecommunication Relations Between the Countries of Europe and the Mediterranean Basin (Study Group III) 7 pp. 7 pp.
  CCITT RECMN D.307R-89. Remuneration of Digital Systems and Channels Used in Telecommunication Relations Between the Countries of Europe and the Mediterranean Basin—General Tariff Principles—Charging and Accounting in Intl Telecom. Serv. (Study Group III) 3 pp. 3 pp.

—Glossaries
  CCIR RECMN 662-1-90. Terms and Definitions—Section A—Terminology. 19 pp.

—Integrated Services Digital Networks—Interfaces
  CCITT RECMN I.412-89. ISDN User-Network Interfaces Interface Structures and Access Capabilities—Integrated Services Digital Network (ISDN)—Overall Network Aspects and Functions, ISDN User-Network Interfaces (Study Group XVIII) 6 pp. 6 pp.
  CCITT RECMN I.431-89. Primary Rate User-Network Interface—Layer 1 Specification—Integrated Services Digital Network (ISDN)—Overall Network Aspects and Functions, ISDN User-Network Interfaces (Study Group XVIII) 29 pp. 29 pp.

—Integrated Services Digital Networks—Interfaces—Glossaries
  CCITT RECMN I.431-89. Primary Rate User-Network Interface—Layer 1 Specification—Integrated Services Digital Network (ISDN)—Overall Network Aspects and Functions, ISDN User-Network Interfaces (Study Group XVIII) 29 pp. 29 pp.

—Integrated Services Digital Networks—Interfaces—Network Layer
  CCITT RECMN I.430-89. Basic User-Network Interface—Layer 1 Specification—Integrated Services Digital Network (ISDN)—Overall Network Aspects and Functions, ISDN User-Network Interfaces (Study Group XVIII) 70 pp. 70 pp.

—Mobile Stations (Communications)
  CENELEC GSM 04.03-92. MS—BSS Interface: Channel Structures and Access Capabilities. 12 pp.
  ETSI GSM 04.03-92. MS—BSS Interface: Channel Structures and Access Capabilities. 12 pp.

INDUSTRY STANDARDS

INTERNATIONAL AND NON-U.S. NATIONAL STANDARDS
SUBJECT INDEX

Communication

Communication Channels (Cont.)

—Multiplexed—Allocations

CCITT RECMN X.54-89. Allocation of Channels on International Multiplex Links at 64 kBit/s—Data Communication Networks—Transmission, Signalling and Switching, Network Aspects, Maintenance and Administrative Arrangements (Study Group VII) 2 pp. 2 pp.

—Multiplexed—Numbering Systems

CCITT RECMN X.53-89. Numbering of Channels on International Multiplex Links at 64 kBit/s—Data Communication Networks—Transmission, Signalling and Switching, Network Aspects, Maintenance and Administrative Arrangements (Study Group VII) 2 pp. 2 pp.

—Multiplexed—Satellite Communications—10 Bit Envelope

CCITT RECMN X.56-89. Interface Between Synchronous Data Networks Using an 8 + 2 Envelope Structure and Single Channel Per Carrier (SCPC) Satellite Channels—Data Communication Networks—Transmission, Signalling and Switching, Network Aspects, Maintenance and Administrative. 5 pp.

—Multiplexed—Satellite Communications—8 Bit Envelope

CCITT RECMN X.55-89. Interface Between Synchronous Data Networks Using a 6 + 2 Envelope Structure and Single Channel Per Carrier (SCPC) Satellite Channels—Data Communication Networks—Transmission, Signalling and Switching, Network Aspects, Maintenance and Administrative. 3 pp.

—Multiplexed—Signaling Systems

CCITT RECMN U.20-89. Telex and Gentex Signalling on Radio Channels (Synchronous 7-Unit Systems Affording Error Correction by Automatic Repetition)—Telegraph Switching (Study Group IX) 6 pp. 6 pp.

—Numbering—Data Transmission Equipment

CCITT RECMN M.1320-89. Numbering of Channels in Data Transmission Systems—Maintenance of International Telegraph, Phototelegraph and Leased Circuits—Maintenance of the International Public Telephone Network—Maintenance of Maritime Satellite and Data. 2 pp.

—Numbering—Groups

CCITT RECMN M.320-89. Numbering of the Channels in a Group—General Maintenance Principles—Maintenance of International Transmission Systems and Telephone Circuits (Study Group IV) 2 pp. 2 pp.

—Numbering—Transmission Systems

CCITT RECMN M.390-89. Numbering in Systems on Symmetric Pair Cable—General Maintenance Principles—Maintenance of International Transmission Systems and Telephone Circuits (Study Group IV) 4 pp. 4 pp.

CCITT RECMN M.400-89. Numbering in Radio-Relay Links or Open-Wire Line Systems—General Maintenance Principles—Maintenance of International Transmission Systems and Telephone Circuits (Study Group IV) 1 pp. 1 p.

—Radio Communication—Radio Relay Systems—Supervision

CCIR Report 787-3-90. Preferred Methods and Characteristics for the Supervision and Protection of Digital Radio-Relay Systems—Section 9D—Maintenance. 8 pp.

—Radio Communications

CCIR Report 1020-86. Adaptation of System Specification to Ease the Practical Implementation of Radio Equipment—Section 8A—Land Mobile Service and Related Subjects. 5 pp.

—Radio Communications—Delay—Error Correction

CCITT RECMN U.22-89. Signals Indicating Delay in Transmission on Calls Set up by Means of Synchronous Systems with Automatic Error Correction by Repetition—Telegraph Switching (Study Group IX) 2 pp. 2 pp.

—Radio Communications—Maritime Mobile Services—SELCAL Systems

CCIR Report 908-1-86. Channel Requirements for a Digital Selective-Calling System—Section 8B—Maritime Mobile Service: Telegraphy and Related Subjects. 15 pp.

Communication Channels (Cont.)

—Radio Communications—Radio Relay Systems

CCIR RECMN 595-2-90. Radio-Frequency Channel Arrangements for Digital Radio-Relay Systems in the 17.7 to 19.7 GHz Frequency Band—Section 9B1—Radio-Frequency Channel Arrangements. 3 pp.

CCIR RECMN 595-3-92. Radio-Frequency Channel Arrangements for Radio-Relay Systems Operating in the 18 GHz Band—Section 9B1—Radio-Frequency Channel Arrangements. 8 pp.

CCIR RECMN 635-1-90. Radio-Frequency Channel Arrangements Based on a Homogeneous Pattern for High Capacity Digital Radio-Relay Systems Operating in the 4 GHz Band—Section 9B1—Radio-Frequency Channel Arrangements. 2 pp.

CCIR RECMN 635-2-92. Radio-Frequency Channel Arrangements Based on a Homogeneous Pattern for Radio-Relay Systems Operating in the 4 GHz Band—Section 9B1—Radio-Frequency Channel Arrangements. 5 pp.

CCIR RECMN 636-1-90. Radio-Frequency Channel Arrangements for Radio-Relay Systems Operating in the 15 GHz Band—Section 9B1—Radio-Frequency Channel Arrangements. 3 pp.

CCIR RECMN 636-2-92. Radio-Frequency Channel Arrangements for Radio-Relay Systems Operating in the 15 GHz Band—Section 9B1—Radio-Frequency Channel Arrangements. 4 pp.

CCIR RECMN 637-86. Radio-Frequency Channel Arrangements for Analogue and Digital Radio-Relay Systems in the 21.2 to 23.6 GHz Frequency Band—Section 9B1—Radio-Frequency Channel Arrangements. 1 p.

CCIR RECMN 637-1-92. Radio-Frequency Channel Arrangements for Radio-Relay Systems Operating in the 23 GHz Band—Section 9B1—Radio-Frequency Channel Arrangements. 6 pp.

CCIR RECMN 746-91. Radio-Frequency Channel Arrangements for Radio-Relay Systems—Section 9B1—Radio-Frequency Channel Arrangements. 14 pp.

CCIR RECMN 747-92. Radio-Frequency Channel Arrangements for Radio-Relay Systems Operating in the 10 GHz Band—Section 9B1—Radio-Frequency Channel Arrangements. 3 pp.

CCIR RECMN 748-92. Radio-Frequency Channel Arrangements for Radio-Relay Systems Operating in the 25.25 to 27.5 GHz and 27.5 to 29.5 GHz Bands—Section 9B1—Radio-Frequency Channel Arrangements. 1 p.

CCIR RECMN 749-92. Radio-Frequency Channel Arrangements for Radio-Relay Systems Operating in the 36.0 to 40.5 GHz Band—Section 9B1—Radio-Frequency Channel Arrangements. 4 pp.

CCIR RECMN 388-63. Radio-Frequency Channel Arrangements for Trans-Horizon Radio-Relay Systems—Section 9E2—Trans-Horizon Radio-Relay Systems. 1 p.

CCIR Report 607-4-90. Radio-Frequency Channel Arrangements for Radio-Relay Systems in the Ranges 10.5 to 10.68 GHz and 11.7 to 15.35 GHz—Section 9B1—Radio-Frequency Channel Arrangements. 7 pp.

CCIR Report 782-3-90. Radio-Frequency Channel Arrangements for High-Capacity Digital Radio-Relay Systems in the 11 GHz Frequency Band—Section 9B1—Radio-Frequency Channel Arrangements. 6 pp.

CCIR Report 934-2-90. Radio-Frequency Channel Arrangements for High and Medium-High Capacity Digital Radio-Relay Systems Operating in the Frequency Bands Below About 10 GHz—Section 9B1—Radio-Frequency Channel Arrangements. 15 pp.

CCIR Report 936-2-90. Radio-Frequency Channel Arrangements in the Bands Above About 17 GHz—Section 9B1—Radio-Frequency Channel Arrangements. 11 pp.

CCIR Report 1055-1-90. Radio-Frequency Channel Arrangements for Analogue and Small and Medium Capacity Digital Radio-Relay Systems Operating in the Bands Below About 10 GHz—Section 9B1—Radio-Frequency Channel Arrangements. 8 pp.

CCIR QUESTION 103/9-90. Trans-Horizon Radio-Relay Systems—Questions of Study Group 9 Fixed Service. 1 p.

CCIR QUESTION 108/9-90. Channel Spacings and Arrangements for Radio-Relay Systems Operating in Frequency Bands Above About 17 GHz—Questions of Study Group 9 Fixed Service. 1 p.

CCIR QUESTION 136/9-90. Radio-Frequency Channel Arrangements for Digital Radio-Relay Systems Below About 17 GHz—Questions of Study Group 9 Fixed Service. 1 p.

—Radio Communications—Radio Relay Systems—Radio Links

CCIR OPINION 14-6-90. Preferred Radio-Frequency Channel Arrangements for Radio-Relay Links for International Connections—Volume IX-1—Fixed Service Using Radio-Relay Systems. 2 pp.

Communication Channels (Cont.)

—Radio Communications—Signaling Systems

CCITT RECMN U.20-89. Telex and Gentex Signalling on Radio Channels (Synchronous 7-Unit Systems Affording Error Correction by Automatic Repetition)—Telegraph Switching (Study Group IX) 6 pp. 6 pp.

—Radio Communications—Supervision

CCIR RECMN 622-86. Technical and Operational Characteristics of Analogue Cellular Systems for Public Land Mobile Telephone Use—Section 8A—Land Mobile Service and Related Subjects. 3 pp.

—Radio Frequency Interference—Test Signals

CCIR RECMN 571-2-90. Conventional Test Signal Simulating Sound-Programme Signals for Measuring Interference in Other Channels—Section CMTT D—Methods of Operation and Assessment of Performance of Sound-Programme Transmission Channels. 2 pp.

CCIR Report 497-3-82. Conventional Test Signal Simulating Sound-Programme Signals for Measuring Interference in Other Channels—Section CMTT D—Methods of Operation and Assessment of Performance of Sound-Programme Transmission Channels. 4 pp.

—Radiotelephony

CCIR Report 1029-86. Future Use of the Band 2170-2194 kHz—Section 8B—Maritime Mobile Service: Telegraphy and Related Subjects. 7 pp.

—Radiotelephony—Maritime Mobile Services

CCIR Report 748-1-86. Improved Use of the HF Radiotelephone Channels for Coast Stations in the Bands Allocated Exclusively to the Maritime Mobile Service—Section 8C—Maritime Mobile Service; Telephony and Related Subjects. 2 pp.

CCIR QUESTION 30-3/8-90. Improved Use of the HF Radiotelephone Channels for Coast Stations in the Bands Allocated Exclusively to the Maritime Mobile Service—Questions Concerning Study Group 8—Mobile, Radiodetermination, Amateur and Related Satellite Services. 1 p.

—Radiotelephony—Mobile Satellite Communications

CCIR Report 760-2-90. Link Power Budgets for a Maritime Mobile-Satellite Service—Section 8I—Technical and Operating Characteristics of Mobile Satellite Services. 11 pp.

—Radiotelephony—Performance Measurement

CCIR Report 920-2-90. Maritime Satellite System Performance at Low Elevation Angles—Section 8I—Technical and Operating Characteristics of Mobile Satellite Services. 14 pp.

—Radiotelephony—Radio Frequency Interference

CCIR Report 917-2-90. Permissible Levels of Interference into Telephone Channels in the Maritime Mobile-Satellite Service—Section 8G—Availability, Performance Objectives and Interworking with Terrestrial Networks. 6 pp.

—Radiotelephony—Radio Relay Systems

CCIR RECMN 283-5-90. Radio-Frequency Channel Arrangements for Low and Medium Capacity Analogue or Digital Radio-Relay Systems Operating in the 2 GHz Band—Section 9B1—Radio-Frequency Channel Arrangements. 3 pp.

CCIR RECMN 382-5-90. Radio-Frequency Channel Arrangements for Medium and High Capacity Analogue Radio-Relay Systems Operating in the 2 and 4 GHz Bands, or for Medium and High Capacity Digital Radio-Relay Systems Operating in the 4 GHz Band—Section 9B1—Radio-Frequency Channel Argmts. 4 pp.

CCIR RECMN 382-6-91. Radio-Frequency Channel Arrangements for Radio-Relay Systems Operating in the 2 and 4 GHz Bands—Section 9B1—Radio-Frequency Channel Arrangements. 3 pp.

CCIR RECMN 383-4-90. Radio-Frequency Channel Arrangements, for High Capacity Analogue or Digital Radio-Relay Systems Operating in the Lower 6 GHz Band—Section 9B1—Radio-Frequency Channel Arrangements. 3 pp.

CCIR RECMN 383-5-92. Radio-Frequency Channel Arrangements for High Capacity Radio-Relay Systems Operating in the Lower 6 GHz Band—Section 9B1—Radio-Frequency Channel Arrangements. 4 pp.

INTERNATIONAL AND NON-U.S. NATIONAL STANDARDS
SUBJECT INDEX
Communication

Communication Channels *(Cont.)*

—**Radiotelephony—Radio Relay Systems** *(Cont.)*

CCIR RECMN 384-5-90. Radio-Frequency Channel Arrangements for Medium and High Capacity Analogue or High Capacity Digital Radio-Relay Systems Operating in the Upper 6 GHz Band—Section 9B1—Radio-Frequency Channel Arrangements. 3 pp.

CCIR RECMN 385-4-90. Radio-Frequency Channel Arrangements for Low Capacity Analogue Radio-Relay Systems Operating in the 7 GHz Band—Section 9B1—Radio-Frequency Channel Arrangements. 2 pp.

CCIR RECMN 385-5-92. Radio-Frequency Channel Arrangements for Radio-Relay Systems Operating in the 7 GHz Band—Section 9B1—Radio-Hz Frequency Channel Arrangements. 6 pp.

CCIR RECMN 386-3-86. Radio-Frequency Channel Arrangements for Systems with a Capacity of 960 Telephone Channels, or the Equivalent, Operating in the 8 GHz Band—Section 9B1—Radio-Frequency Channel Arrangements. 3 pp.

CCIR RECMN 386-4-92. Radio-Frequency Channel Arrangements for Radio-Relay Systems Operating in the 8 GHz Band—Section 9B1—Radio-Frequency Channel Arrangements. 6 pp.

CCIR RECMN 387-5-90. Radio-Frequency Channel Arrangements for Medium and High Capacity Analogue or Digital Radio-Relay Systems Operating in the 11 GHz Band—Section 9B1—Radio-Frequency Channel Arrangements. 5 pp.

CCIR RECMN 387-6-92. Radio-Frequency Channel Arrangements for Radio-Relay Systems Operating in the 11 GHz Band—Section 9B1—Radio-Frequency Channel Arrangements. 6 pp.

CCIR RECMN 389-2-74. Preferred Characteristics of Auxiliary Radio-Relay Systems Operating in the 2, 4, 6 or 11 GHz Bands—Section 9B1—Radio-Frequency Channel Arrangements. 3 pp.

CCIR RECMN 497-3-90. Radio-Frequency Channel Arrangements for Low and Medium Capacity Analogue or Medium and High Capacity Digital Radio-Relay Systems Operating in the 13 GHz Band—Section 9B1—Radio-Frequency Channel Arrangements. 4 pp.

CCIR RECMN 497-4-92. Radio-Frequency Channel Arrangements for Radio-Relay Systems Operating in the 13 GHz Frequency Band—Section 9B1—Radio-Frequency Channel Arrangements. 5 pp.

CCIR RECMN 701-90. Radio-Frequency Channel Arrangements for Analogue and Digital Point-to-Multipoint Radio Systems Operating in Frequency Bands in the Range 1.427 to 2.690 GHz (1.5, 1.8, 2.0, 2.2, 2.4 and 2.6 GHz)—Section 9E1—Line-of-Sight Radio-Relay Systems. 3 pp.

CCIR Report 607-4-90. Radio-Frequency Channel Arrangements for Radio-Relay Systems in the Ranges 10.5 to 10.68 GHz and 11.7 to 15.35 GHz—Section 9B1—Radio-Frequency Channel Arrangements. 7 pp.

—**Radiotelephony—Supervision**

CCIR RECMN 586-1-86. Automated VHF/UHF Maritime Mobile Telephone System—Section 8C—Maritime Mobile Service; Telephony and Related Subjects. 44 pp.

—**Radiotelephony—Telegraph Printers—Maritime Mobile Services**

CCIR Report 584-1-86. Direct-Printing and Other Data Signals Using Audio-Frequency Techniques in the VHF Radiotelephony Channels in the Maritime Mobile Service—Section 8C—Maritime Mobile Service; Telephony and Related Subjects. 2 pp.

—**Radioteletype Communications—Performance Measurement**

CCIR Report 920-2-90. Maritime Satellite System Performance at Low Elevation Angles—Section 8I—Technical and Operating Characteristics of Mobile Satellite Services. 14 pp.

—**Single Frequency—Telegraph Printers**

CCIR Report 1026-1-90. Use of Narrow-Band Direct-Printing Telegraph Equipment on a Single-Frequency Radio Channel—Section 8B—Maritime Mobile Service: Telegraphy and Related Subjects. 9 pp.

—**Single Frequency—Telegraph Printers—Maritime Mobile Services**

CCIR RECMN 692-90. Narrow-Band Direct-Printing Telegraph Equipment Using a Single-Frequency Channel—Section 8B—Maritime Mobile Service; Telegraphy and Related Subjects. 1 p.

—**Sound Transmission**

CCIR RECMN 724-90. Transmission of Digital Studio Quality Sound Signals over H1 Channels—Section CMTT C—Transmission Standards and Performance Objectives for Sound-Programme Channels. 9 pp.

Communication Channels *(Cont.)*

—**Sound Transmission** *(Cont.)*

CCIR Decision 77-1-89. Transmission of Digital Sound Programme Signals of Digital Studio Quality on Circuits Using the H1 Channel—Annex to Volume XII—Television and Sound Transmission (CMTT). 3 pp.

—**Synchronization—Voice Frequency Telegraph Systems—Radio Circuits**

CCIR Report 863-82. Synchronization of Channels in Multi-Channel Voice-Frequency Telegraph Systems Using ARQ on Long-Range Radio Circuits—Section 3Ca—Radiotelegraph Circuits. 4 pp.

—**Telex Communications—Delay—Error Correction**

CCITT RECMN U.22-89. Signals Indicating Delay in Transmission on Calls Set up by Means of Synchronous Systems with Automatic Error Correction by Repetition—Telegraph Switching (Study Group IX) 2 pp. 2 pp.

—**Telex Communications—Signaling Systems**

CCITT RECMN U.20-89. Telex and Gentex Signalling on Radio Channels (Synchronous 7-Unit Systems Affording Error Correction by Automatic Repetition)—Telegraph Switching (Study Group IX) 6 pp. 6 pp.

—**Time Division Multiplexers—Numbering Systems**

CCITT RECMN R.114-89. Numbering of International TDM Channels—Telegraph Transmission (Study Group IX) 3 pp. 3 pp.

—**Transmitters—Communication Circuits**

CCIR RECMN 348-4-90. Arrangement of Channels in Multi-Channel Single-Sideband and Independent-Sideband Transmitters for Long-Range Circuits Operating at Frequencies Below About 30 MHz—Section 3Aa—Technical Characteristics. 2 pp.

Communication Circuits

See Also: Attenuation Distortion; Circuits; Half Duplex Transmission; Private Leased Circuits; Propagation Time; Radio Circuits; Signal Circuits; Telecommunication Circuits

—**Below 30 MHz—Transmitters—Communication Channels**

CCIR RECMN 348-4-90. Arrangement of Channels in Multi-Channel Single-Sideband and Independent-Sideband Transmitters for Long-Range Circuits Operating at Frequencies Below About 30 MHz—Section 3Aa—Technical Characteristics. 2 pp.

—**Mobile Satellite Communications—Availability**

CCIR RECMN 828-92. Definition of Availability for Communication Circuits in the Mobile-Satellite Services. 2 pp.

—**Satellite News Gathering—Earth Stations**

CCIR RECMN 771-92. Auxiliary Coordination Satellite Circuits for SNG Terminals. 2 pp.

Communication Controllers

Scope Note: Use a more specific term
See: Controllers; Data Communication Equipment; Gateways; Interface Circuits; Modems

Communication Equipment

Scope Note: For additional listings, see specific types of equipment *Use For:* Communication Systems and Equipment; Communications Systems
See Also: Antennas; Audio Amplifiers; Call Systems (Medical); Cellular Mobile Radio Equipment; Communication Cables; Communication Transmission Line Hardware; Communication Transmission Lines; Conference Communication Systems; Electrical Equipment; Equipment; Fiber Optic Equipment; Gateways; Inductive Coordination; Mobile Radio Equipment; Morse Code Equipment; Protection Ratio; Radio Equipment; Radio Paging Systems; Single Sideband Communication Equipment; Speech Transmission Systems; Telecommunication Equipment
JIS C 6020-70. General Rules and Testing Methods for Intercommunication Equipments.

—**Aluminum Capacitors**

CNS C7009-84. Electrolytic Capacitor for Special Use (Oct)(2319). 15 pp.

—**Classification**

BSI BS 1000: (383/389)-71. 1971 Amd 1 Universal Decimal Classification (UDC). English Full Edition (383/389): Communications. Transport. Metrology. Weights and Measures. 13 pp.

Communication Equipment *(Cont.)*

—**Classification** *(Cont.)*

SNZ NZS/BS 1000 (383/389)-71. Universal Decimal Classification Communications. Transport. Metrology. Weights and Measures Amend: 1. 12 pp.

—**Connectors**

CSA CAN/CSA-C22. 2NO182.4-M90. Plugs, Receptacles, and Connectors for Communication Systems; (Gen Instr 1 Thru 3). 33 pp.

—**Crystal Filters—Ceramic**

IECQ PQC 62/JP 0001-87. Detail Specification for: Piezoelectric Ceramic Filters for Sound IF Circuit of TV Assessment Level E. 10 pp.

IECQ PQC 63-85. Sectional Specification for: Piezoelectric Ceramic Filters for Communication Equipment. 17 pp.

IECQ PQC 64-85. Blank Detail Specification for: Piezoelectric Ceramic Filters for Use in Communication Equipment Assessment Level E. 8 pp.

IECQ PQC 64/JP 0002-91. Detail Specification for: Piezoelectric Ceramic Filters for Use in Communication Equipment Assessment Level E. 10 pp.

—**Diagnostic Equipment**

CCIR Report 993-86. Automatic Diagnosis of HF Equipment—Section 3Ad—Operational Questions. 4 pp.

—**Electric Contacts**

CNS C4340-82. Electric Contact Material for Electric Communication Equipment and Apparatus (Jan)(8369).

JIS C 2509-65. Electric Contact Materials for Electric Communication Equipment and Apparatus. 7 pp.

—**Glossaries**

IEC. Electricity, Electronics and Telecommunications Multilingual Dictionary Volume 1: English-French-Russian-German-Spanish-Dutch-Italian-Swedish-Polish Second Edition. 1944 pp.

IEC. Electricity, Electronics and Telecommunications Multilingual Dictionary Volume 2: English-French-Russian-German-Spanish-Dutch-Italian-Swedish-Polish Second Edition. 964 pp.

IEC. Electricity, Electronics and Telecommunications Multilingual Dictionary Volume 3: German, Dutch and Swedish Indexes Second Edition. 793 pp.

IEC. Electricity, Electronics and Telecommunications Multilingual Dictionary Volume 4: Spanish and Italian Indexes Second Edition. 540 pp.

IEC. Electricity, Electronics and Telecommunications Multilingual Dictionary Volume 5: Russian and Polish Indexes Second Edition. 578 pp.

—**Hookup Wire**

DIN VDE 0881-86. Jumper Wires and Stranded Hook-Up Wires with Extended Temperature Range for Telecommunications Systems and Data Processing Systems (Mar). 34 pp.

—**Inductive Coordination**

CSA C22.3 NO 3-1954. Inductive Co-Ordination Definitions, Principles and Practices (R 1970). 22 pp.

—**Inductive Coordination—Glossaries**

CSA C22.3 NO 3-1954. Inductive Co-Ordination Definitions, Principles and Practices (R 1970). 22 pp.

—**Jumpers (Connectors)**

DIN VDE 0881-86. Jumper Wires and Stranded Hook-Up Wires with Extended Temperature Range for Telecommunications Systems and Data Processing Systems (Mar). 34 pp.

—**Phone Jacks**

CSA CAN/CSA-C22. 2NO182.4-M90. Plugs, Receptacles, and Connectors for Communication Systems; (Gen Instr 1 Thru 3). 33 pp.

—**Phone Plugs**

CSA CAN/CSA-C22. 2NO182.4-M90. Plugs, Receptacles, and Connectors for Communication Systems; (Gen Instr 1 Thru 3). 33 pp.

—**Power Transformers**

CNS C7004-83. Power Transformers for Communications (Dec)(1264).

—**Radio Transceivers**

CENELEC PRI-ETS 300 033-90. European Digital Cellular Telecommunications System (Phase 1): Radio Transmission and Reception. 22 pp.

CENELEC PRI-ETS 300 033-91. European Digital Cellular Telecommunications System (Phase 1); Radio Transmission and Reception. 23 pp.

CENELEC I-ETS 300 033/A1-91. Text of Amendment 1. 28 pp.

INDUSTRY STANDARDS

INTERNATIONAL AND NON-U.S. NATIONAL STANDARDS
SUBJECT INDEX

Communication

Communication Equipment *(Cont.)*

—Radio Transceivers *(Cont.)*
CENELEC PRI-ETS 300 033-92. European Digital Cellular Telecommunications System (Phase 1); Radio Transmission and Reception. 24 pp.
CENELEC I-ETS 300 033-92. European Digital Cellular Telecommunications System (Phase 1); Radio Transmission and Reception (GSM 05.05). 24 pp.
CENELEC PRI-ETS 300 033-1-93. European Digital Cellular Telecommunications System (Phase 1); Radio Transmission and Reception Part 1: Generic (GSM 05.05). 25 pp.
CENELEC GSM 05.05-92. See PRI-ETS 300 033. 21 pp.
CENELEC GSM 05.05-DCS-92. European Digital Cellular Telecommunications System (Phase 1) Radio Transmission and Reception. 31 pp.
CEPT T/R 20-07-84. Libre Circulation, En Vue De L'Utilisation D'un Pays a L'Autre, Des Emetteurs-Recepteurs De Faible Puissance Mobiles Et Protatifs Dans La Bande Des 27 MHz (Appareils PR 27, Recommandation T/R 20-02). 1 p.
CEPT T/R 20-07 E-88. Free Circulation, for Use in Different Countries, of Low-Power Mobile and Portable Transmitter-Receivers in the 27 MHz Band (PR 27 Equipment, Recommendation T/R 20-02). 2 pp.
CEPT T/R 21-07 E-89. Border Crossing and Use of Mobile Transmitter-Receivers in CEPT Member Countries. 9 pp.
ETSI PRI-ETS 300 033-90. European Digital Cellular Telecommunications System (Phase 1); Radio Transmission and Reception (GSM 05.05). 22 pp.
ETSI I-ETS 300 033-92. European Digital Cellular Telecommunications System (Phase 1); Radio Transmission and Reception (GSM 05.05). 24 pp.
ETSI I-ETS 300 033/A1-91. Text of Amendment 1. 28 pp.
ETSI PRI-ETS 300 033-92. European Digital Cellular Telecommunications System (Phase 1); Radio Transmission and Reception. 24 pp.
ETSI PRI-ETS 300 033-1-93. European Digital Cellular Telecommunications System (Phase 1); Radio Transmission and Reception Part 1: Generic (GSM 05.05). 25 pp.
ETSI GSM 05.05-92. See PRI-ETS 300 033. 21 pp.
ETSI GSM 05.05-DCS-92. European Digital Cellular Telecommunications System (Phase 1) Radio Transmission and Reception. 31 pp.

—Ships
MOD UK NES 544-01. Guide to System Design and Description of Rationalised Internal Communications Equipment MK1 and MK1 Variations Issue 3 (02.91); Amendment 2. 54 pp.
MOD UK NES 550-01. Guide to Design of AF 250 Series Main Broadcast Systems Issue 3 (07.89); Amendment 1. 54 pp.
MOD UK NES 551-89. Guide to Internal Communication Equipment Main Broadcast and Associated Items Issue 3 (08.89). 86 pp.
MOD UK NES 555-81. Guide to Internal Communication Equipment. Flight Deck Communications Issue 2 (01.81). 31 pp.

—Submarines
MOD UK NES 544-01. Guide to System Design and Description of Rationalised Internal Communications Equipment MK1 and MK1 Variations Issue 3 (02.91); Amendment 2. 54 pp.
MOD UK NES 550-01. Guide to Design of AF 250 Series Main Broadcast Systems Issue 3 (07.89); Amendment 1. 54 pp.
MOD UK NES 551-89. Guide to Internal Communication Equipment Main Broadcast and Associated Items Issue 3 (08.89). 86 pp.
MOD UK NES 555-81. Guide to Internal Communication Equipment. Flight Deck Communications Issue 2 (01.81). 31 pp.

—Symbols
CNS C5001-82. Fundamental Symbols for Electric Communications Engineering (Jul)(413).

—Symbols—Engineering Drawings—Ships
CNS F5003-84. Graphical Symbols for Electrical Apparatus (Communication) for Marine Engineering Drawings (May)(7786).
JIS F 8013-91. Graphical Symbols for Electrical Apparatus for Marine Use Engineering Drawings—Communication, Instrumentation, Navigation and Radio.

Communication Interfaces
Use For: Computer Interfaces; Input/Output Interfaces **See Also:** Application Program Interfaces; Asynchronous Communications Interfaces; Codirectional Interfaces; Computer Graphics Interface; Counter/Timer Modules; Data Buses; Data Circuit Terminating Equipment; Data Communication Equipment; Data Links; Interface Boards; Man-Machine Interfaces; Military Communications; Network Interfaces; Parallel Interfaces; Peripheral

Communication Interfaces *(Cont.)*
See Also: *(Cont.)*
Interfaces; Q Interfaces; RS-232; Serial Interfaces; Small Computer System Interfaces; Subscriber Line Interface Circuits; User Network Interfaces
BSI BS ISO 2110-89. 1989 Information Technology—Data Communication—25-Pole DTE/DCE Interface Connector and Contact Number Assignments. 18 pp.
CEPT T/L 03-01-81. Jonctions. 1 p.
CEPT T/L 03-01 E-81. Interfaces. 1 p.
ISO 8480-87. Information Processing—Data Communication—DTE/DCE Interface Back-up Control Operation Using the 25-Pole Connector First Edition. 9 pp.
JTC1 8480-87. Information Processing—Data Communication—DTE/DCE Interface Back-up Control Operation Using the 25-Pole Connector First Edition. 9 pp.
OSI ISO 8480-87. Information Processing—Data Communication—DTE/DCE Interface Back-Up Control Operation Using the 25-Pole Connector.

—Aeronautical Passenger Communications
CENELEC PRETS 300 326-2-93. Radio Equipment and Systems (RES); Terrestrial Flight Telephone System (TFTS) Part 2: Speech Services, Radio Interface. 316 pp.
ETSI PRETS 300 326-2-93. Radio Equipment and Systems (RES); Terrestrial Flight Telephone System (TFTS) Part 2: Speech Services, Radio Interface. 316 pp.

—Analog Output—Private Automatic Branch Exchanges
CENELEC PRETS 300 004-90. Transmission Characteristics at 2-Wire Analogue Interface of a Digital Private Automatic Branch Exchange (PABX) (TE 10-02). 44 pp.
CENELEC PRI-ETS 300 004-91. Business Telecommunications (BT); Transmission Characteristics at 2-Wire Analogue Interfaces of a Digital Private Automatic Branch Exchange (PABX). 44 pp.
CENELEC I-ETS 300 004-91. Business Telecommunications (BT); Transmission Characteristics at 2-Wire Analogue Interfaces of a Digital Private Automatic Branch Exchange (PABX). 44 pp.
CENELEC PRETS 300 005-90. Transmission Characteristics at 4-Wire Analogue Interface of a Digital Private Automatic Branch Exchange (PABX) (TE 10-03). 28 pp.
CENELEC PRI-ETS 300 005-91. Business Telecommunications (BT); Transmission Characteristics at 4-Wire Analog Interfaces of a Digital Private Automatic Branch Exchange (PABX). 28 pp.
CENELEC I-ETS 300 005-91. Business Telecommunications (BT); Transmission Characteristics at 4-Wire Analogue Interfaces of a Digital Private Automatic Branch Exchange (PABX). 28 pp.
ETSI PRETS 300 004-90. Transmission Characteristics at 2-Wire Analogue Interface of a Digital Private Automatic Branch Exchange (PABX)—Also Known as TE 10-02. 43 pp.
ETSI PRETS 300 005-90. Transmission Characteristics at 4-Wire Analogue Interface of a Digital Private Automatic Branch Exchange (PABX)—Also Known as TE 10-03. 28 pp.
ETSI I-ETS 300 003-91. Business Telecommunications (BT); Transmission Characteristics of Digital Private Automatic Branch Exchanges (PABXs). 26 pp.
ETSI I-ETS 300 004-91. Business Telecommunications (BT); Transmission Characteristics at 2-Wire Analogue Interfaces of a Digital Private Automatic Branch Exchange (PABX). 44 pp.
ETSI PRETS 300 004-90. Transmission Characteristics at 2-Wire Analogue Interface of a Digital Private Automatic Branch Exchange (PABX) (TE 10-02). 44 pp.
ETSI I-ETS 300 005-91. Business Telecommunications (BT); Transmission Characteristics at 4-Wire Analogue Interface of a Digital Private Automatic Branch Exchange (PABX). 28 pp.
ETSI PRETS 300 005-90. Transmission Characteristics at 4-Wire Analogue Interface of a Digital Private Automatic Branch Exchange (PABX) (TE 10-03). 28 pp.

—Audio Equipment
BSI BS 7239-01. 1989 Amd 1 Digital Audio Interface (IEC 958: 1989) (AMD 6413) October 31, 1991. 31 pp.

—Base Station Systems
CENELEC PRI-ETS 300 078-90. European Digital Cellular Telecommunications System (Phase 1): Layer 1—General Requirements. 21 pp.
CENELEC PRI-ETS 300 078-92. European Digital Cellular Telecommunications System (Phase 1); MS-BSS Layer 1—General Requirements. 24 pp.

Communication Interfaces *(Cont.)*
—Base Station Systems *(Cont.)*
CENELEC I-ETS 300 078-92. European Digital Cellular Telecommunications System (Phase 1); MS-BSS Layer 1—General Requirements (GSM 04.04). 24 pp.
CENELEC GSM 04.04-92. See PRI-ETS 300 078. 25 pp.
CENELEC GSM 04.05-92. MS—BSS Data Link Layer General Aspects. 27 pp.
CENELEC GSM 08.51-92. BSC-BTS Interface General Aspects. 7 pp.
CENELEC GSM 08.52-92. BSC-BTS Interface Principles. 18 pp.
CENELEC GSM 08.54-92. BSC-BTS: Layer 1 Structure of Physical Circuits. 6 pp.
CENELEC GSM 08.56-92. BSC-BTS Specification. 14 pp.
CENELEC GSM 08.58-92. BSC-BTS Specification. 83 pp.
CENELEC GSM 08.58-DCS-92. European Digital Cellular Telecommunication System (Phase 1); Base Station Controller (BSC) to Base Station Tranceiver (BTS) Interface Layer 3 Specification. 5 pp.
CENELEC GSM 08.59-92. BSC—BTS O&M Signalling Transport. 9 pp.
ETSI PRI-ETS 300 078-90. European Digital Cellular Telecommunications System (Phase 1); Layer 1—General Requirements (GSM 04.04). 21 pp.
ETSI I-ETS 300 078-92. European Digital Cellular Telecommunications System (Phase 1); MS-BSS Layer 1—General Requirements (GSM 04.04). 24 pp.
ETSI PRI-ETS 300 078-92. European Digital Cellular Telecommunications System (Phase 1); MS-BSS Layer 1—General Requirements. 24 pp.
ETSI GSM 04.04-92. See PRI-ETS 300 078. 25 pp.
ETSI GSM 04.05-92. MS—BSS Data Link Layer General Aspects. 27 pp.
ETSI GSM 08.52-92. BSC-BTS Interface Principles. 18 pp.
ETSI GSM 08.54-92. BSC-BTS: Layer 1 Structure of Physical Circuits. 6 pp.
ETSI GSM 08.56-92. BSC-BTS Specification. 14 pp.
ETSI GSM 08.58-92. BSC-BTS Specification. 83 pp.
ETSI GSM 08.58-DCS-92. European Digital Cellular Telecommunication System (Phase 1); Base Station Controller (BSC) to Base Station Tranceiver (BTS) Interface Layer 3 Specification. 5 pp.
ETSI GSM 08.59-92. BSC—BTS O&M Signalling Transport. 9 pp.

—Cellular Mobile Radio Equipment
CENELEC PRI-ETS 300 022-90. European Digital Cellular Telecommunications System (Phase 1): Mobile Radio Interface Layer 3 Specification. 449 pp.
CENELEC PRI-ETS 300 022-91. European Digital Cellular Telecommunications System (Phase 1); Mobile Radio Interface Layer 3 Specification. 463 pp.
CENELEC I-ETS 300 022/A1-91. AMD 1 European Digital Cellular Telecommunications System (Phase 1); Mobile Radio Interface Layer 3 Specification. 91 pp.
CENELEC PRI-ETS 300 022-92. European Digital Cellular Telecommunications System (Phase 1); Mobile Radio Interface Layer 3 Specification. 451 pp.
CENELEC I-ETS 300 022-92. European Digital Cellular Telecommunications System (Phase 1); Mobile Radio Interface Layer 3 Specification (GSM 04.08). 451 pp.
CENELEC PRI-ETS 300 022-3-93. European Digital Cellular Telecommunications System (Phase 1); Mobile Radio Interface Layer 3 Specification Part 3: Signalling Support of the Second Ciphering Algorithm (GSM 04.08). 23 pp.
CENELEC PRI-ETS 300 023-90. European Digital Cellular Telecommunications System (Phase 1); Point to Point Short Message Service Support on Mobile Radio Interface. 74 pp.
CENELEC PRI-ETS 300 023-91. European Digital Cellular Telecommunications System (Phase 1); Point-to-Point Short Message Service Support on Mobile Radio Interface. 76 pp.
CENELEC PRI-ETS 300 023-92. European Digital Cellular Telecommunications System (Phase 1); Point-to-Point Short Message Service Support on Mobile Radio Interface. 76 pp.
CENELEC I-ETS 300 023-92. European Digital Cellular Telecommunications System (Phase 1); Point-to-Point Short Message Service Support on Mobile Radio Interface (GSM 04.11). 76 pp.
CENELEC PRI-ETS 300 023-93. European Digital Cellular Telecommunications System (Phase 1); Point-to-Point Short Message Service Support on Mobile Radio Interface (GSM 04.11). 78 pp.
CENELEC PRI-ETS 300 024-90. European Digital Cellular Telecommunications System (Phase 1); Short Message Service Cell Broadcast (SMSCB) Support on the Mobile Radio Interface. 6 pp.

INTERNATIONAL AND NON-U.S. NATIONAL STANDARDS
SUBJECT INDEX
Communication

Communication Interfaces *(Cont.)*

—Cellular Mobile Radio Equipment *(Cont.)*

CENELEC PRI-ETS 300 024-91. European Digital Cellular Telecommunications Systems (Phase 1); Short Message Service Cell Broadcast (SMSCB) (SMSCB) Support on the Mobile Radio Interface. 9 pp.

CENELEC I-ETS 300 024-92. European Digital Cellular Telecommunications System (Phase 1); Short Message Service Cell Broadcast (SMSCB) Support on the Mobile Radio Interface (GSM 04.12). 8 pp.

CENELEC PRI-ETS 300 027-90. European Digital Cellular Telecommunications System (Phase 1): Mobile Radio Interface Layer 3 Supplementary Services Specifications Formats and Coding. 39 pp.

CENELEC PRI-ETS 300 027-91. European Digital Cellular Telecommunications System (Phase 1); Mobile Radio Interface Layer 3 Supplementary Services Specification Formats and Coding. 42 pp.

CENELEC I-ETS 300 027-92. European Digital Cellular Telecommunications System (Phase 1); Mobile Radio Interface Layer 3 Supplementary Services Specification Formats and Coding (GSM 04.80). 42 pp.

CENELEC PRI-ETS 300 028-90. European Digital Cellular Telecommunications System (phase 1): Mobile Radio Interface Layer 3 Call Offering Supplementary Services Specification. 38 pp.

CENELEC PRI-ETS 300 028-91. European Digital Cellular Telecommunications System (Phase 1); Mobile Radio Interface Layer 3 Call Offering Supplementary Services Specification. 39 pp.

CENELEC I-ETS 300 028-92. European Digital Cellular Telecommunications System (Phase 1); Mobile Radio Interface Layer 3 Call Offering Supplementary Services Specification (GSM 04.82). 39 pp.

CENELEC PRI-ETS 300 029-90. European Digital Cellular Telecommunications System (Phase 1); Mobile Radio Interface Layer 3 Call Restriction Supplementary Services Specification. 14 pp.

CENELEC PRI-ETS 300 029-91. European Digital Cellular Telecommunications System (Phase 1); Mobile Radio Interface Layer 3 Call Restriction Supplementary Services Specification. 17 pp.

CENELEC I-ETS 300 029-92. European Digital Cellular Telecommunications System (Phase 1); Mobile Radio Interface Layer 3 Call Restriction Supplementary Services Specification (GSM 04.88). 16 pp.

CENELEC GSM 04.08-92. See PRI-ETS 300 022. 452 pp.

CENELEC GSM 04.08-DCS-92. European Digital Cellular Telecommunication System (Phase 1); Mobile Radio Interface—Layer 3 Specification. 85 pp.

CENELEC GSM 04.11-93. See PRI-ETS 300 023. 73 pp.

CENELEC GSM 04.12-92. See PRI-ETS 300 024. 5 pp.

CENELEC GSM 04.80-92. See PRI-ETS 300 027. 41 pp.

CENELEC GSM 04.82-92. See PRI-ETS 300 028. 37 pp.

CENELEC GSM 04.88-92. See PRI-ETS 300 029. 15 pp.

ETSI PRI-ETS 300 022-90. European Digital Cellular Telecommunications System (Phase 1); Mobile Radio Interface Layer 3 Specification (GSM 04. 08). 439 pp.

ETSI PRI-ETS 300 023-90. European Digital Cellular Telecommunications System (Phase 1); Point to Point Short Message Service Suppport on Mobile Radio Interface (GSM 04.11). 74 pp.

ETSI PRI-ETS 300 024-90. European Digital Cellular Telecommunications System (Phase 1) Short Message Service Cell Broadcast (SMSCB) Support on the Mobile Radio Interface (GSM 04.12). 6 pp.

ETSI PRI-ETS 300 027-90. European Digital Cellular Telecommunications System (Phase 1); Mobile Radio Interface Layer 3 Supplementary Services Specification Formats and Coding (GSM 04.80). 39 pp.

ETSI PRI-ETS 300 028-90. European Digital Cellular Telecommunications System (Phase 1) Mobile Radio Interface Layer 3 Call Offering Supplementary Services Specification (GSM 04. 82). 40 pp.

ETSI PRI-ETS 300 029-90. European Digital Cellular Telecommunications System (Phase 1) Mobile Radio Interface Layer 3 Call Restriction Supplementary Services Specification (GSM 04. 88). 14 pp.

ETSI I-ETS 300 022-92. European Digital Cellular Telecommunications System (Phase 1); Mobile Radio Interface Layer 3 Specification (GSM 04.08). 451 pp.

ETSI PRI-ETS 300 022-92. European Digital Cellular Telecommunications System (Phase 1); Mobile Radio Interface Layer 3 Specification. 451 pp.

ETSI I-ETS 300 022/A1-91. AMD 1 European Digital Cellular Telecommunications System (Phase 1); Mobile Radio Interface Layer 3 Specification. 91 pp.

ETSI PRI-ETS 300 022-3-93. European Digital Cellular Telecommunications System (Phase 1); Mobile Radio Interface Layer 3 Specification Part 3: Signalling Support of the Second Ciphering Algorithm (GSM 04.08). 23 pp.

ETSI I-ETS 300 023-92. European Digital Cellular Telecommunications System (Phase 1); Point-to-Point Short Message Service Support on Mobile Radio Interface (GSM 04.11). 76 pp.

ETSI PRI-ETS 300 023-92. European Digital Cellular Telecommunications System (Phase 1); Point-to-Point Short Message Service Support on Mobile Radio Interface. 76 pp.

ETSI PRI-ETS 300 O23-93. European Digital Cellular Telecommunications System (Phase 1); Point-to-Point Short Message Service Support on Mobile Radio Interface (GSM 04.11). 78 pp.

ETSI I-ETS 300 024-92. European Digital Cellular Telecommunications System (Phase 1); Short Message Service Cell Broadcast (SMSCB) Support on the Mobile Radio Interface (GSM 04.12). 8 pp.

ETSI PRI-ETS 300 024-91. European Digital Cellular Telecommunications Systems (Phase 1); Short Message Service Cell Broadcast (SMSCB) (SMSCB) Support on the Mobile Radio Interface. 9 pp.

ETSI I-ETS 300 027-92. European Digital Cellular Telecommunications System (Phase 1); Mobile Radio Interface Layer 3 Supplementary Services Specification Formats and Coding (GSM 04.80). 42 pp.

ETSI PRI-ETS 300 027-91. European Digital Cellular Telecommunications System (Phase 1); Mobile Radio Interface Layer 3 Supplementary Services Specification Formats and Coding. 42 pp.

ETSI I-ETS 300 028-92. European Digital Cellular Telecommunications System (Phase 1); Mobile Radio Interface Layer 3 Call Offering Supplementary Services Specification (GSM 04.82). 39 pp.

ETSI PRI-ETS 300 028-91. European Digital Cellular Telecommunications System (Phase 1); Mobile Radio Interface Layer 3 Call Offering Supplementary Services Specification. 39 pp.

ETSI I-ETS 300 029-92. European Digital Cellular Telecommunications System (Phase 1); Mobile Radio Interface Layer 3 Call Restriction Supplementary Services Specification (GSM 04.88). 16 pp.

ETSI PRI-ETS 300 029-91. European Digital Cellular Telecommunications System (Phase 1); Mobile Radio Interface Layer 3 Call Restriction Supplementary Services Specification. 17 pp.

ETSI I-ETS 300 068-92. European Digital Cellular Telecommunications System (Phase 1); Man-Machine Interface of the Mobile Station (GSM 02.30). 22 pp.

ETSI GSM 04.08-92. See PRI-ETS 300 022. 452 pp.

ETSI GSM 04.08-DCS-92. European Digital Cellular Telecommunication System (Phase 1); Mobile Radio Interface—Layer 3 Specification. 85 pp.

ETSI GSM 04.11-93. See PRI-ETS 300 023. 73 pp.

ETSI GSM 04.12-92. See PRI-ETS 300 024. 5 pp.

ETSI GSM 04.80-92. See PRI-ETS 300 027. 41 pp.

ETSI GSM 04.82-92. See PRI-ETS 300 028. 37 pp.

ETSI GSM 04.88-92. See PRI-ETS 300 029. 15 pp.

—Connectors

BSI BS ISO 2110-01. 1989 Amd 1 Information Technology—Data Communication—25-Pole DTE/DCE Interface Connector and Contact Number Assignments (AMD 7165) July 15, 1992. 26 pp.

BSI BS ISO 4902-89. 1989 Information Technology—Data Communication—37 Pole DTE/DCE Interface Connector and Contact Number Assignments. 24 pp.

BSI BS ISO 4903-89. 1989 Information Technology—Data Communication—15-Pole DTE/DCE Interface Connector and Contact Number Assignments. 22 pp.

BSI BS ISO/IEC 2593-93. 1993 Information Technology—Telecommunications and Information Exchange Between Systems—34-Pole DTE/DCE Interface Connector Mateability Dimensions and Contact Number Assignments (S). 23 pp.

CSA Z243.12.2-1981. Data Communication—15-Pin DTE/DCE Interface Connector and Pin Assignments (ISO 4903-1980 has Been Adopted as a CSA Standard) (R 1988). 25 pp.

IEC 2593-93. Information Technology—Telecommunications and Information Exchange Between Systems—34-Pole DTE/DCE Interface Connector Mateability Dimensions and Contact Number Assignments Third Edition. 21 pp.

IEC 11569-93. Information Technology—Telecommunications and Information Exchange Between Systems—26-Pole Interface Connector Mateability Dimensions and Contact Number Assignments First Edition. 11 pp.

ISO 2110-89. Information Technology—Data Communication—25-Pole DTE/DCE Inter-face Connector and Con-tact Number Assignments Amendment 1: Interface Connector and Contact Assignments for a DTE/DCE Interface for Data Signalling Rates Above 20 000 Bit/s Third Ed.; (Amendment 1-1991). 22 pp.

Communication Interfaces *(Cont.)*

—Connectors *(Cont.)*

ISO 2593-93. Information Technology—Telecommunications and Information Exchange Between Systems—34-Pole DTE/DCE Interface Connector Mateability Dimensions and Contact Number Assignments Third Edition. 21 pp.

ISO 4902-89. Information Technology—Data Communication—37-Pole DTE/DCE Interface Connector and Contact Number Assignments Second Edition. 20 pp.

ISO 4903-89. Information Technology—Data Communication—15-Pole DTE/DCE Interface Connector and Contact Number Assignments Second Edition. 20 pp.

ISO 11569-93. Information Technology—Telecommunications and Information Exchange Between Systems—26-Pole Interface Connector Mateability Dimensions and Contact Number Assignments First Edition. 11 pp.

JTC1 2110-89. Information Technology—Data Communication—25-Pole DTE/DCE Inter-face Connector and Con-tact Number Assignments Amendment 1: Interface Connector and Contact Assignments for a DTE/DCE Interface for Data Signalling Rates Above 20 000 Bit/s Third Ed.; (Amendment 1-1991). 22 pp.

JTC1 2593-93. Information Technology—Telecommunications and Information Exchange Between Systems—34-Pole DTE/DCE Interface Connector Mateability Dimensions and Contact Number Assignments Third Edition. 21 pp.

JTC1 4902-89. Information Technology—Data Communication—37-Pole DTE/DCE Interface Connector and Contact Number Assignments Second Edition. 20 pp.

JTC1 4903-89. Information Technology—Data Communication—15-Pole DTE/DCE Interface Connector and Contact Number Assignments Second Edition. 20 pp.

OSI ISO 2110-89. Information Technology—Date Communication—25-Pole DTE/DCE Interface Connector and Contact Number Assignment—Amd 1: Interface Connector and Contact Assignments for a DTE/DCE Interface for Data Signalling Rates Above 20 000 bits/s. 23 pp.

OSI ISO 2110 DAM 1-89. Interface Connector and Contact Number Assignments for a DTE/DCE Interface for Data Signalling Rates Above 20 000 bits per Second. 5 pp.

OSI ISO/IEC DIS 2593-89. Information Technology—Data Communication—34-Pole DTE/DCE Interface Connector and Contact Number Assignments. 14 pp.

OSI ISO 2593-84. Data Communication—34 Pin DTE/DCE Interface Connector and Pin Assignments. 7 pp.

OSI ISO 4902-89. Information Technology—Date Communication—37 Pole DTE/DCE Interface Connector and Contact Number Assignments. 20 pp.

OSI ISO 4903-89. Information Technology—Data Communication—15-Pole DTE/DCE Interface Connector and Contact Number Assignments. 20 pp.

SAA AS 2748-91. Information Technology—Data Communication—25-Pole DTE/DCE Interface Connector and Contact Number Assignments (ISO/IEC 2110:1989) (in Professional Packages 26A, 26C). 13 pp.

SAA AS 3613-91. Information Technology—Data Communications—15-Pole DTE/DCE Interface Connector and Contact Number Assignments (ISO 4903:1989) (in Professional Packages 26A, 26C). 17 pp.

SNZ NZS/ISO 4902-80. Data Communication—37-Pin and 9-Pin DTE/DCE Interface Connectors and Pin Assignments. 24 pp.

SNZ NZS/ISO 4903-89. Information Technology—Data Communications—15-Pole DTE/DCE Interface Connector and Contact Number Assignments. 17 pp.

—Data Processing Equipment—Glossaries

JIS X 0021-87. Glossary of Terms Used in Information Processing (Interfaces Between Process Computer Systems and Technical Processes).

—Data Processing Equipment—Naval

NATO STANAG 4146 ED 1 AMD 2-76. Interim Standard Specifications for Input /Output Interfaces in NATO Naval Data Handling Equipments (Withdrawn). 157 pp.

—Data Processing Equipment—Private Automatic Branch Exchanges

ECMA ECMA-TR 24-85. Interfaces Between Data Processing Equipment and Private Automatic Bench Exchange. Circuit Switching Application. 63 pp.

—Data Terminal Equipment—Interchange Circuits—Electrical Measure

CCITT RECMN V.10-89. Electrical Characteristics for Unbalanced Double-Current Interchange Circuits for General Use with Integrated Circuit Equipment in the Field of Data Communications—Data Communication Over the Telephone Network (Study Group XVII) 17 pp. 17 pp.

INDUSTRY STANDARDS 557

Communication
INTERNATIONAL AND NON-U.S. NATIONAL STANDARDS
SUBJECT INDEX

Communication Interfaces *(Cont.)*

—Data Terminal Equipment—Interchange Circuits—Electrical Measure *(Cont.)*

CCITT RECMN V.11-89. Electrical Characteristics for Balanced Double-Current Interchange Circuits for General Use with Integrated Circuit Equipment in the Field of Data Communications—Data Communication Over the Telephone Network (Study Group XVII) 12 pp. 12 pp.

—Data Transmission Equipment—Avionics

MOD UK DSTAN 00-18: Pt 2:Supp A-90. Avionic Data Transmission Interface Systems Part 2: Serial, Time Division, Command/Response Multiplex Data Bus Supplement A: Fibre Optic Supplement for a Point to Point Link Issue 1. 15 pp.

MOD UK DSTAN 00-18: Pt 2:Supp B-90. Avionic Data Transmission Interface Systems Part 2: Serial, Time Division, Command/Response Multiplex Data Bus Supplement B: Fibre Optic Supplement for a Single Transmissive Star Issue 1. 15 pp.

MOD UK DSTAN 00-18: Pt 2:Supp C-90. Avionic Data Transmission Interface Systems Part 2: Serial, Time Division, Command/Response Multiplex Data Bus Supplement C: Fibre Optic Supplement for a Single Reflective Star Issue 1. 15 pp.

MOD UK DSTAN 00-18: Pt 2:Supp D-90. Avionic Data Transmission Interface Systems Part 2: Serial, Time Division, Command/Response Multiplex Data Bus Supplement D: Fibre Optic Supplement for a Multi-Local Transmissive Star Issue 1. 20 pp.

—Digital—Audio Equipment

BSI BS 7239-89. 1989 Digital Audio Interface (S) (IEC 958: 1989). 26 pp.

BSI BS 7239-01. 1989 Amd 1 Digital Audio Interface (IEC 958: 1989) (AMD 6413) October 31, 1991. 31 pp.

CENELEC EN 60 958-90. Digital Audio Interface. 4 pp.

IEC 958-89. Digital Audio Interface First Edition; (Amendment 1-1993). 91 pp.

—Digital—Audio Equipment—Data Channels—Formats

CCIR RECMN 776-92. Format for User Data Channel of the Digital Audio Interface—Section 10C—Audio-Frequency Characteristics of Sound-Broadcasting Signals. 18 pp.

—Digital—Cellular Mobile Radio Equipment

CENELEC GSM 08.01-92. European Digital Cellular Telecommunication System (Phase 1); General Aspects on the BSS-MSC Interface. 8 pp.

CENELEC GSM 08.02-92. European Digital Cellular Telecommunication System (Phase 1); BSS-MSC Interface—Interface Principles. 20 pp.

CENELEC GSM 08.08-92. European Digital Cellular Telecommunication System (Phase 1); BSS-MSC Layer 3 Specification. 92 pp.

ETSI GSM 08.01-92. European Digital Cellular Telecommunication System (Phase 1); General Aspects on the BSS-MSC Interface. 8 pp.

ETSI GSM 08.02-92. European Digital Cellular Telecommunication System (Phase 1); BSS-MSC Interface—Interface Principles. 20 pp.

ETSI GSM 08.08-92. European Digital Cellular Telecommunication System (Phase 1); BSS-MSC Layer 3 Specification. 92 pp.

—Digital Exchanges

CCITT RECMN Q.511-89. Exchange Interfaces Towards Other Exchanges—Digital Local, Transit, Combined and International Exchanges in Integrated Digital Networks and Mixed Analogue-Digital Networks—Supplements (Study Group XI) 4 pp. 4 pp.

—Digital Exchanges—Operations, Administration and Maintenance

CCITT RECMN Q.513-89. Exchange Interfaces for Operations, Administration and Maintenance—Digital Local, Transit, Combined and International Exchanges in Integrated Digital Networks and Mixed Analogue-Digital Networks—Supplements (Study Group XI) 4 pp. 4 pp.

—Digital Exchanges—Subscriber Access

CCITT RECMN Q.512-89. Exchange Interfaces for Subscriber Access—Digital Local, Transit, Combined and International Exchanges in Integrated Digital Networks and Mixed Analogue-Digital Networks—Supplements (Study Group XI) 9 pp. 9 pp.

—Digital Exchanges—Transmission

CCITT RECMN Q.552-89. Transmission Characteristics at 2-Wire Analogue Interfaces of Digital Exchange—Digital Local, Transit, Combined and International Exchanges in Integrated Digital Networks and Mixed Analogue-Digital Networks—Supplements (Study Group XI) 25 pp. 25 pp.

Communication Interfaces *(Cont.)*

—Digital Exchanges—Transmission *(Cont.)*

CCITT RECMN Q.553-89. Transmission Characteristics at 4-Wire Analogue Interfaces of a Digital Exchange—Digital Local, Transit, Combined and International Exchanges in Integrated Digital Networks and Mixed Analogue-Digital Networks—Supplements (Study Group XI) 12 pp. 12 pp.

CCITT RECMN Q.554-89. Transmission Characteristics at Digital Interfaces of a Digital Exchange—Digital Local, Transit, Combined and International Exchanges in Integrated Digital Networks and Mixed Analogue-Digital Networks—Supplements (Study Group XI) 3 pp. 3 pp.

—Digital—Private Automatic Branch Exchanges

CENELEC PRETS 300 006-90. Transmission Characteristics at Digital Interfaces of a Digital Private Automatic Branch Exchange (PABX) (TE 10-04). 7 pp.

CENELEC PRI-ETS 300 006-91. Business Telecommunications (BT) Transmission Characteristics at Digital Interfaces of a Digital Private Automatic Branch Exchange (PABX). 8 pp.

CENELEC I-ETS 300 006-91. Business Telecommunications (BT); Transmission Characteristics at Digital Interfaces of a Digital Private Automatic Branch Exchange (PABX). 9 pp.

ETSI PRETS 300 003-90. Transmission Characteristics of Digital Private Automatic Branch Exchanges (PABXs) Also known as TE 10-01. 25 pp.

ETSI PRETS 300 006-90. Transmission Characteristics at Digital Interfaces of a Digital Private Automatic Branch Exchange (PABX)—Also Known as TE 10-04. 7 pp.

ETSI I-ETS 300 006-91. Business Telecommunications (BT); Transmission Characteristics at Digital Interfaces of a Digital Private Automatic Branch Exchange (PABX). 9 pp.

ETSI PRETS 300 006-90. Transmission Characteristics at Digital Interfaces of a Digital Private Automatic Branch Exchange (PABX) (TE 10-04). 7 pp.

—Digital—Protocols—Cellular Mobile Radio Equipment

CENELEC PRI-ETS 300 026-90. European Digital Cellular Telecommunications System (Phase 1); Radio Link Protocol (RLP) for Data Telematic Services on Mobile Station-Base Station System (MS-BSS) Interface and Base Station System-Mobile Services Switching Centre (BSS-MSC) Interface. 25 pp.

CENELEC PRI-ETS 300 026-91. European Digital Cellular Telecommunications Sys. (Phase 1); Radio Link Protocol (RLP) for Data and Telematic Services on the Mobile Station—Base Station System (MS-BSS) Interface and the Base Station System —Mobile Services Swg. Centre (BSS-MSC) Interface. 63 pp.

CENELEC PRI-ETS 300 026-92. European Digital Cellular Telecommunications System (Phase 1); Radio Link Protocol (RLP) for Data and Telematic Services on the Mobile Station—Base Station System (MS-BSS) Interface and the Base Station System—Mobile Services Switching Centre (BSS-MSC) Intrfce. 63 pp.

CENELEC I-ETS 300 026-92. European Digital Cellular Telecommunications Sys. (Phase 1); Radio Link Protocol (RLP) for Data and Telematic Servs. on the Mobile Sta.—Base Sta. System (MS-BSS) Interface and the Base Sta. Sys.—Mobile Servs. Switching Centre (BSS-MSC) Interface (GSM 04.22). 63 pp.

CENELEC GSM 04.22-92. See PRI-ETS 300 026. 64 pp.

ETSI PRI-ETS 300 026-90. European Digital Cellular Telecommunications System (Phase 1) Radio Link Protocol (RLP) for Data and Telematic Services on the Mobile Station—Base Station System (MS-BSS) Interface and the Base Station System—Mobile Services Switching Centre (BSS-MSC). 25 pp.

ETSI I-ETS 300 026-92. European Digital Cellular Telecommunications Sys. (Phase 1); Radio Link Protocol (RLP) for Data and Telematic Servs. on the Mobile Sta.—Base Sta. System (MS-BSS) Interface and the Base Sta. Sys.—Mobile Servs. Switching Centre (BSS-MSC) Interface (GSM 04.22). 63 pp.

ETSI PRI-ETS 300 026-92. European Digital Cellular Telecommunications System (Phase 1); Radio Link Protocol (RLP) for Data and Telematic Services on the Mobile Station—Base Station System (MS-BSS) Interface and the Base Station System—Mobile Services Switching Centre (BSS-MSC) Intrfce. 63 pp.

ETSI GSM 04.22-92. See PRI-ETS 300 026. 64 pp.

—Digital—Testing Points

CCITT RECMN G.772-89. Digital Protected Monitoring Points—General Aspects of Digital Transmission Systems; Terminal Equipments (Study Groups XV and XVIII) 4 pp. 4 pp.

Communication Interfaces *(Cont.)*

—Digital—Voice/Data Systems—Naval

NATO STANAG 4186 ED 1 AMD 1-88. Interim Standard Specification for a Parallel Digital Interface Between User Units and a Ship Distribution Network for Voice Communication (Withdrawn). 82 pp.

—Electronic Measuring Instruments

BSI BS 6146: Part 1-81. 1981 Interface System for Programmable Measuring Instruments (Byte Serial, Bit Parallel) Part 1: Specification for Functional, Elect-rical and Mechanical Requirements, System Applications and Require -ments for Designer and User. 112 pp.

BSI BS 6146: Part 2-81. (WITHDRAWN) 1981 Interface System for Programmable Measuring Instruments (Byte Serial, Bit Parallel) Part 2: Guide for Code and Format Conventions (Superseded by Revision of IEC 625-2). 27 pp.

CENELEC HD 414.1-80. An Interface System for Programmable Measuring Instruments (Byte Serial, Bit Parallel) Part 1: Functional Specifications, Electrical Specifications System Application and Requirements for the Designer and User. 2 pp.

CENELEC HD 414.2-83. An Interface System for Programmable Measuring Instruments (Byte Serial, Bit Parrallel) Part 2: Code and Format Conventions. 2 pp.

IEC 625 Pt 1-79. Interface System for Programmable Measuring Instruments (Byte, Serial, Bit Parallel) Part 1: Functional Specifications, Electri-cal, Specifications, Mechanical Specifica-tions, System Applica-tions and Requirements for the Designer and User First Edition. 226 pp.

IEC 625 Pt 2-80. An Interface System for Programmable Measuring Instruments (Byte Serial, Bit Parallel) Part 2: Code and Format Conventions First Edition. 54 pp.

JIS C 1901-87. Interface System for Programmable Measuring Instruments. 146 pp.

—Glossaries

BSI BS ISO 2382/21-85. 1985 Information Technology—Vocabulary 2382/21: Interfaces Between Process Computer Systems and Technical Processes. 24 pp.

CNS C5238-90. Information Processing Systems—Vocabulary (Part 21: Interfaces Between Process Computer Systems and Technical Processes) (Aug)(12759).

ISO 2382 Pt 21-85. Data Processing—Vocabulary—Part 21: Interfaces Between Process Computer Systems and Technical Processes First Edition. 21 pp.

JIS X 0021-87. Glossary of Terms Used in Information Processing (Interfaces Between Process Computer Systems and Technical Processes).

JTC1 2382 Pt 21-85. Data Processing—Vocabulary—Part 21: Interfaces Between Process Computer Systems and Technical Processes First Edition. 21 pp.

OSI ISO 2382-21-85. Data Processing-Vocabulary—Part 21: Interfaces Between Process Computer Systems and Technical Processes. 21 pp.

SAA AS 1189.21-87. Data Processing—Vocabulary—Part 21: Interfaces Between Process Computers Systems and Technical Processes. 9 pp.

—Information Resource Dictionary System

BSI BS ISO/IEC 10728-93. 1993 Information Technology—Information Resource Dictionary System (IRDS) Services Interface (S). 119 pp.

ISO 10728-93. Information Technology—Information Resource Dictionary System (IRDS) Services Interface First Edition. 117 pp.

JTC1 10728-93. Information Technology—Information Resource Dictionary System (IRDS) Services Interface First Edition. 117 pp.

—INMARSAT—Integrated Services Digital Networks

CCITT RECMN Q.1111-89. Interfaces Between the Inmarsat Standard B System and the International Public Switched Telephone Network/ISDN—Interworking with Satellite Mobile Systems (Study Group XI) 22 pp. 22 pp.

CCITT RECMN Q.1151-89. Interfaces Between the Inmarsat Aeronautical Mobile-Satellite System and the International Public Switched Telephone Network/ISDN—Interworking with Satellite Mobile Systems (Study Group XI) 19 pp. 19 pp.

—INMARSAT—Public Switched Telephone Networks

CCITT RECMN Q.1111-89. Interfaces Between the Inmarsat Standard B System and the International Public Switched Telephone Network/ISDN—Interworking with Satellite Mobile Systems (Study Group XI) 22 pp. 22 pp.

INTERNATIONAL AND NON-U.S. NATIONAL STANDARDS
SUBJECT INDEX
Communication

Communication Interfaces *(Cont.)*

—**INMARSAT—Public Switched Telephone Networks** *(Cont.)*
CCITT RECMN Q.1151-89. Interfaces Between the Inmarsat Aeronautical Mobile-Satellite System and the International Public Switched Telephone Network/ISDN—Interworking with Satellite Mobile Systems (Study Group XI) 19 pp. 19 pp.

—**INMARSAT—Telex Communications**
CCITT RECMN U.61-89. Detailed Requirements to be Met in Interfacing the International Telex Network with Maritime Satellite Systems—Telegraph Switching (Study Group IX) 10 pp. 10 pp.

—**Local Exchanges**
CEPT T/CS 62-02-86. Interfaces Pour Commutateurs Numeriques Principaux D'Abonne Ou Mixtes. 2 pp.
CEPT T/CS 62-02 E-86. Interfaces for Digital Local and Combined Exchanges. 2 pp.

—**Maritime Mobile Satellite Communications—Radiotelegraphy**
CCIR RECMN 553-78. Interface Requirements for 50-Baud Start-Stop Telegraph Transmission in the Maritime Mobile-Satellite Service—Section 8I—Technical and Operating Characteristics of Mobile Satellite Services. 2 pp.

—**Metropolitan Area Networks**
CENELEC PRETS 300 275-92. Network Aspects (NA); Metropolitan Area Network (MAN) Interconnection of MANs. 15 pp.
ETSI PRETS 300 275-92. Network Aspects (NA); Metropolitan Area Network (MAN) Interconnection of MANs. 15 pp.

—**Mobile Radio Services**
CENELEC GSM 04.10-92. European Digital Cellular Telecommunication System (Phase 1); Mobile Radio Interface Layer 3 Supplementary Services Specification General Aspects. 25 pp.
ETSI GSM 04.10-92. European Digital Cellular Telecommunications System (Phase 1); Mobile Radio Interface Layer 3 Supplementary Services Specification General Aspects. 25 pp.

—**Mobile Satellite Communications—Telex Communications**
CCITT RECMN U.60-89. General Requirements to be Met in Interfacing the International Telex Network with Maritime Satellite Systems—Telegraph Switching (Study Group IX) 2 pp. 2 pp.

—**Mobile Stations (Communications)**
CENELEC PRI-ETS 300 045-90. European Digital Cellular Telecommunications System (Phase 1): Subscriber Identity Module—Mobile Equipment (SIM-ME) Interface Specification. 139 pp.
CENELEC PRI-ETS 300 045-91. European Digital Cellular Telecommunications System (Phase 1); Subscriber Identity Module—Mobile Equipment (SIM-ME) Interface Specification. 139 pp.
CENELEC I-ETS 300 045-91. AMD 1 European Digital Cellular Telecommunications System (Phase 1); Subscriber Identity Module—Mobile Equipment (SIM-ME) Interface Specification (GSM 11.11). 22 pp.
CENELEC PRI-ETS 300 045-92. European Digital Cellular Telecommunications System (Phase 1); Subscriber Identity Module—Mobile Equipment (SIM-ME) Interface Specification. 135 pp.
CENELEC I-ETS 300 045-92. European Digital Cellular Telecommunications System (Phase 1); Subscriber Identity Module—Mobile Equipment (SIM-ME) Interface Specification (GSM 11.11). 135 pp.
CENELEC PRI-ETS 300 045-1-93. European Digital Cellular Telecommunications System (Phase 1); Subscriber Identity Module—Mobile Equipment (SIM-ME) Interface Specification Part 1: Generic (GSM 11.11). 136 pp.
CENELEC GSM 11.11-92. See PRI-ETS 300 045. 155 pp.
CENELEC GSM 11.11-DCS-92. European Digital Cellular Telecommunication System (Phase 1); Specification of the SIM-ME Interface. 19 pp.
ETSI PRI-ETS 300 045-90. European Digital Cellular Telecommunications System (Phase 1); Subscriber Identity Module—Mobile Equipment (SIM-ME) Interface Specification (GSM 11.11). 138 pp.
ETSI PRI-ETS 300 045-92. European Digital Cellular Telecommunications System (Phase 1); Subscriber Identity Module—Mobile Equipment (SIM-ME) Interface Specification. 135 pp.
ETSI PRI-ETS 300 045-1. European Digital Cellular Telecommunications System (Phase 1); Subscriber Identity Module—Mobile Equipment (SIM-ME) Interface Specification Part 1: Generic (GSM 11.11). 136 pp.
ETSI GSM 11.11-92. See PRI-ETS 300 045. 155 pp.
ETSI GSM 11.11-DCS-92. European Digital Cellular Telecommunication System (Phase 1); Specification of the SIM-ME Interface. 19 pp.

Communication Interfaces *(Cont.)*

—**Monitors—Remote—Thermal Containers**
ISO 10368-92. Freight Thermal Containers—Remote Condition Monitoring First Edition. 95 pp.

—**NMOS—Programmable—Microprocessors—Preferred Products List**
CECC CECC MUAHAG Vol 7 IS 8-92. Preferred Products List; Active Microcircuits (En, Fe, Ge). 89 pp.

—**Private Automatic Branch Exchanges**
CENELEC PRETS 300 003-90. Transmission Characteristics of Digital Private Automatic Branch Exchanges (PABXs) (TE 10-01). 25 pp.
CENELEC PRI-ETS 300 003-91. Business Telecommunications (BT) Transmission Characteristics of Digital Private Automatic Branch Exchanges (PABXs). 25 pp.
CENELEC I-ETS 300 003-91. Business Telecommunications (BT); Transmission Characteristics of Digital Private Automatic Branch Exchanges (PABXs). 26 pp.
ETSI PRETS 300 003-90. Transmission Characteristics of Digital Private Automatic Branch Exchanges (PABXs) (TE 10-01). 25 pp.

—**Protocols**
MOD UK NES 1028: Part 2-85. Standard for Inter-System Communications Protocols Part 2: Rules and Protocols for Inter—Communication Across Digital Highways Issue 1 (03.85). 47 pp.

—**Public Land Mobile Networks**
CENELEC PRI-ETS 300 025-90. European Digital Cellular Telecommunications System (Phase 1); Rate Adaptation on the Mobile Station—Base Station System (MS-B55) Interface. 13 pp.
CENELEC PRI-ETS 300 025-91. European Digital Cellular Telecommunications System (Phase 1); Rate Adaption on the Mobile Station—Base Station System (MS-BSS) Interface. 15 pp.
CENELEC PRI-ETS 300 025-92. European Digital Cellular Telecommunications System (Phase 1); Rate Adaptation on the Mobile Station—Base Station System (MS-BSS) Interface. 16 pp.
CENELEC I-ETS 300 025-92. European Digital Cellular Telecommunications System (Phase 1); Rate Adaptation on the Mobile Station—Base Station System (MS-BSS) Interface (GSM 04.21). 16 pp.
CENELEC GSM 04.21-92. See PRI-ETS 300 025. 17 pp.
CENELEC GSM 08.20-92. Rate Adaptation on the BSS-MSC Interface. 6 pp.
ETSI PRI-ETS 300 025-90. European Digital Cellular Telecommunications System (Phase 1) Rate Adaptation on the Mobile Station—Base Station System (MS-BSS) Interface (GSM 04.21). 13 pp.
ETSI I-ETS 300 025-92. European Digital Cellular Telecommunications System (Phase 1); Rate Adaptation on the Mobile Station—Base Station System (MS-BSS) Interface (GSM 04.21). 16 pp.
ETSI PRI-ETS 300 025-92. European Digital Cellular Telecommunications System (Phase 1); Rate Adaptation on the Mobile Station—Base Station System (MS-BSS) Interface. 16 pp.
ETSI GSM 04.21-92. See PRI-ETS 300 025. 17 pp.
ETSI GSM 08.20-92. Rate Adaptation on the BSS-MSC Interface. 6 pp.

—**Radio Equipment—Telex Communications**
CCITT RECMN U.62-89. General Requirements to be Met in Interfacing the International Telex Network with the Fully Automated Maritime VHF/UHF Radio System—Telegraph Switching (Study Group IX) 5 pp. 5 pp.

—**Remote Control Equipment**
BSI BS 7404: Part 3-91. 1991 Telecontrol Equipment and Systems Part 3: Specification for Interfaces (Electrical Characteristics) (IEC 870-3: 1989). 37 pp.
CENELEC HD 546.3 S1-91. Telecontrol Equipment and Systems Part 3: Interfaces (Electrical Characteristics). 5 pp.
IEC 870 Pt 3-89. Telecontrol Equipment and Systems Part 3: Interfaces (Electrical Characteristics) First Edition. 68 pp.

—**Signal Generators—Interchange Circuits—Electrical Measurement**
CCITT RECMN V.10-89. Electrical Characteristics for Unbalanced Double-Current Interchange Circuits for General Use with Integrated Circuit Equipment in the Field of Data Communications—Data Communication Over the Telephone Network (Study Group XVII) 17 pp. 17 pp.
CCITT RECMN V.11-89. Electrical Characteristics for Balanced Double-Current Interchange Circuits for General Use with Integrated Circuit Equipment in the Field of Data Communications—Data Communication Over the Telephone Network (Study Group XVII) 12 pp. 12 pp.

Communication Interfaces *(Cont.)*

—**Signal Quality**
BSI BS 6638: Part 1-87. (WITHDRAWN) 1987 Guide to Transmission Signal Quality at DTE/DCE Interfaces Part 1: Start-Stop Signal Quality (Superseded by BS ISO/IEC 7480: 1991). 13 pp.
BSI BS 6638: Part 2-89. 1989 Guide to Transmission Signal Quality at DTE/DCE Interfaces Part 2: Synchronous Transmission. 15 pp.
BSI BS ISO/IEC 7480-91. 1991 Information Technology—Telecommunications and Information Exchange Between Systems—Start-Stop Transmission Signal Quality at DTE/DCE Interfaces. 17 pp.
IEC 7480-91. Information Technology—Telecommunications and Information Exchange Between Systems—Start-Stop Transmission Signal Quality at DTE/DCE Interfaces Second Edition. 14 pp.
ISO 7480-91. Information Technology—Telecommunications and Information Exchange Between Systems—Start-Stop Transmission Signal Quality at DTE/DCE Interfaces Second Edition. 14 pp.
JTC1 7480-91. Information Technology—Telecommunications and Information Exchange Between Systems—Start-Stop Transmission Signal Quality at DTE/DCE Interfaces Second Edition. 14 pp.
SNZ NZS/ISO 7480-84. Information Processing—Start-Stop Transmission Signal Quality at DTE/DCE Interfaces. 10 pp.

—**Signaling Links—CCITT No. 6 Signaling Systems**
CCITT RECMN Q.274-89. 6.3 Transmission Methods—Specifications of Signalling System No. 6 (Study Group XI) 8 pp. 8 pp.

—**Store and Forward Mode—Telex Communications—Interworking**
CCITT RECMN U.82-89. International Telex Store and Forward—Interconnection of Telex Store and Forward Units—Telegraph Switching (Study Group IX) 39 pp. 39 pp.

—**Store and Forward Units—FAX Machines**
CCITT RECMN F.163-92. Operational Requirements of the Interconnection of Facsimile Store-and-Forward Units (Study Group I) 10 pp. 10 pp.

—**Synchronous Digital Hierarchy**
CCITT RECMN G.957-90. Optical Interfaces for Equipments and Systems Relating to the Synchronous Digital Hierarchy (Study Group XV) 32 pp. 32 pp.
CENELEC PRETS 300 232-92. Transmission and Multiplexing (TM); Optical Parameter for Interfaces for the Synchronous Digital Hierarchy (SDH). 9 pp.
CENELEC PRETS 300 232-93. Transmission and Multiplexing (TM); Optical Interfaces for Equipments and Systems Relating to the Synchronous Digital Hierarchy. 36 pp.
CENELEC ETS 300 232-93. Transmission and Multiplexing (TM); Optical Interfaces for Equipments and Systems Relating to the Synchronous Digital Hierarchy (ITU-T Recommendation G.957 (1993) Modified). 6 pp.
ETSI ETS 300 232-93. Transmission and Multiplexing (TM); Optical Interfaces for Equipments and Systems Relating to the Synchronous Digital Hierarchy (ITU-T Recommendation G.957 (1993) Modified). 6 pp.
ETSI PRETS 300 232-93. Transmission and Multiplexing (TM); Optical Interfaces for Equipments and Systems Relating to the Synchronous Digital Hierarchy. 36 pp.
ETSI PRETS 300 232-92. Transmission and Multiplexing (TM); Optical Parameter for Interfaces for the Synchronous Digital Hierarchy (SDH). 9 pp.

—**Synchronous Digital Hierarchy—Digital Multiplexers**
CCITT RECMN G.783-90. Characteristics of Synchronous Digital Hierarchy (SDH) Multiplexing Equipment Functional Blocks (Study Group XV) 65 pp. 65 pp.

—**Telecommunication Systems—Military Communications**
NATO STANAG 5040 ED 3 AMD 2-85. NATO Automatic and Semi-Automatic Interfaces Between the National Switched Telecommunications Systems of the Combat Zone and Between These Systems and the NATO Integrated Communications System (NICS)—Period From 1979 to the 1990s. 129 pp.

—**Telegraph Printers—Telex Communications**
CCITT RECMN U.63-89. General Requirements to be Met in Interfacing the International Telex Network with the Maritime "Direct Printing" System—Telegraph Switching (Study Group IX) 2 pp. 2 pp.

INDUSTRY STANDARDS

INTERNATIONAL AND NON-U.S. NATIONAL STANDARDS
SUBJECT INDEX

Communication

Communication Interfaces (Cont.)
—**Telephone Exchanges**
CEPT T/S 61-10 E-88. Exchange Interfaces Towards Other Exchanges. 1 p.
CEPT T/S 61-20 E-88. Exchange Interfaces to Subscribers. 9 pp.

—**Telephone Networks—Military Communications**
NATO STANAG 5018 ED 1 AMD 4-78. NATO Manual Interface Between the Manual Switched Telecommunications Systems of the Combat Zone. 7 pp.

—**Television Transmission—Coaxial Cables**
CCITT RECMN J.75-89. Interconnection of Systems for Television Transmission on Coaxial Pairs and on Radio-Relay Links—Line Transmission of Non-Telephone Signals—Transmission of Sound-Programme and Television Signals (Study Group XV) 2 pp. 2 pp.

—**Television Transmission—Radio Relay Systems**
CCITT RECMN J.75-89. Interconnection of Systems for Television Transmission on Coaxial Pairs and on Radio-Relay Links—Line Transmission of Non-Telephone Signals—Transmission of Sound-Programme and Television Signals (Study Group XV) 2 pp. 2 pp.

—**Transit Exchanges**
CEPT T/CS 68-02-86. Interfaces Pour Commutateurs De Transit Numeriques. 1 p.
CEPT T/CS 68-02 E-86. Interfaces for Digital Transit Exchanges. 1 p.

—**Video Signals**
CCIR RECMN 799-92. Interfaces for Digital Component Video Signals in 525-Line and 625-Line Television Systems Operating at the 4:4:4 Level of Recommendation 601. 17 pp.

—**Videotex Equipment**
CCITT RECMN F.300-89. Videotex Service—Telematic, Data Transmission and Teleconference Services—Operations and Quality of Service (Study Group I) 25 pp. 25 pp.

—**V5.1—Access Networks—Local Exchanges**
CENELEC PRETS 300 324-1-93. Signalling Protocols and Switching (SPS); V5.1 Interface Specification for the Support of Access Network. 270 pp.
CENELEC PRETS 300 324-2-93. Signalling Protocols and Switching (SPS); V5.1 Interface Specification for the Suport of Access Network Protocol Implementation Conformance Statement (PICS) Proforma. 31 pp.
ETSI PRETS 300 324-1-93. Signalling Protocols and Switching (SPS); V5.1 Interface Specification for the Support of Access Network. 270 pp.
ETSI PRETS 300 324-2-93. Signalling Protocols and Switching (SPS); V5.1 Interface Specification for the Suport of Access Network Protocol Implementation Conformance Statement (PICS) Proforma. 31 pp.

—**V5.1—Protocol Implementation Conformance Statement**
CENELEC PRETS 300 324-2-93. Signalling Protocols and Switching (SPS); V5.1 Interface Specification for the Suport of Access Network Protocol Implementation Conformance Statement (PICS) Proforma. 31 pp.
ETSI PRETS 300 324-2-93. Signalling Protocols and Switching (SPS); V5.1 Interface Specification for the Suport of Access Network Protocol Implementation Conformance Statement (PICS) Proforma. 31 pp.

Communication Links
Scope Note: Use a more specific term *See:* Data Links; Downlinks; Feeder Links; Group Links; Line Links; Mastergroup Links; Optical Communication Links; Radio Links; Satellite Links; Sound-Program Links; Supergroup Links; Supermastergroup Links; Telecommunication Equipment; Television Links; Terrestrial Link Antennas; Uplinks; 15 Supergroup Assembly Links

Communication Networks
Scope Note: For additional listings, use a more specific term *Use For:* Data Communication Networks; Networks (Communication)
See Also: Access Codes; Answering Time; Branching Networks; Cable Networks; Circuit Switched Data Networks; Circuit Switched Data Transmission Services; Data Communication Equipment; Data Highways; Data Network Identification Codes; Data Transmission; Deemphasis Networks; Digital Transmission Networks; European Telephone Networks; Gateways; Gentex Networks; Grade of Service; Information Interchange; Integrated Digital Networks; Integrated Services Digital Networks; Intelligent

Communication Networks (Cont.)
See Also: (Cont.)
Networks; Local Area Networks; Mixed Analog/Digital Networks; Network Congestion; Network Management Systems; Network Nodes; Network Performance; Open Systems Interconnection; Packet Switched Networks; Private Integrated Services Networks; Public Data Networks; Public Land Mobile Networks; Public Switched Networks; Public Switched Telephone Networks; Radio Local Area Networks; Satellite Networks; Shaping Networks; Signaling Networks; Switched Networks; Telecommunication Equipment; Telecommunication Networks; Telecommunication Systems; Telegraph Networks; Telemessage Services; Telephone Networks; Time Out; Transmitter Networks; Virtual Private Networks; Weighting Networks
ECMA ECMA TR/58-92. Databases and Networking. 26 pp.

—**Data Transmission Equipment—Interfaces—Interworking**
CCITT FASCICLE VIII.6. Data Communication Networks Interworking Between Networks, Mobile Data Transmission Systems, Internetwork Management Recommendations X.300-X.370 (Study Group VII). 240 pp.

—**Directory Services**
CCITT FASCICLE VIII.8. Data Communication Networks Directory Recommendation X.500—X.521 (Study Group VII). 231 pp.

—**Integrated Services Digital Networks—Interfaces**
CCITT FASCICLE VIII.2. Data Communications Networks: Services and Facilities, Interfaces Recommendations X.1—X.32 (Study Group VII). 563 pp.

—**Interworking**
CENELEC ETR 030-92. Network Aspects (NA); Interworking Aspects of MoU-ISDN Priority 1 + 2 Services. 20 pp.
ETSI ETR 030-92. Network Aspects (NA); Interworking Aspects of MoU-ISDN Priority 1 + 2 Services. 20 pp.

—**Message Handling Systems**
CCITT RECMN F.400-92. Message Handling Services: Message Handling System and Service Overview (Study Group I) 82 pp (Same as Recmn X.400). 82 pp.
CCITT FASCICLE VIII.7. Data Communication Networks: Message Handling Systems Recommendation X.400—X.420 (Study Group VII). 634 pp.
CCITT RECMN X.400-93. Message Handling Services: Message Handling System and Service Overview (Study Group VII) 82 pp (Same as Recmn F.400). 82 pp.

—**Message Handling Systems—Glossaries**
CCITT RECMN X.400-89. Message Handling System and Service Overview—Data Communication Networks—Message Handling Systems (Study Group VII) 73 pp. 73 pp.

—**Open Systems Interconnection—Layer—Protocol—Service**
CCITT FASCICLE VIII.4-88. Data Communication Networks. Open Systems Interconnection (OSI). Model and Notation, Service Definition. 508 pp.

—**Open Systems Interconnection—Protocols—Conformance**
CCITT FASCICLE VIII.5. Data Communication Networks Open Systems Interconnection (OSI) Protocol Specifications, Conformance Testing Recommendations X.200—X.290 (Study Group VII). 598 pp.

—**Public Data Networks—Signal Systems—Maintenance—Management**
CCITT FASCICLE VIII.3-89. Data Communications Networks Transmission, Signalling and Switching, Network Aspects, Maintenance and Administrative Arrangements Recommendations X.40—X.181 (Study Group VII). 481 pp.

—**Radio Paging Systems**
CENELEC PRETS 300 133-3-91. Paging Systems; European Radio Message System (ERMES) Part 3: Network Aspects. 272 pp.
CENELEC ETS 300 133-3-92. Paging Systems (PS); European Radio Message System (ERMES) Part 3: Network Aspects (Without Annexes). 294 pp.
CENELEC ETS 300 133 PRA1-93. AMD prA1 Paging Systems (PS); European Radio Message System (ERMES) Part 3: Network Aspects. 19 pp.
ETSI PRETS 300 133-3-91. Paging Systems; European Radio Message System (ERMES) Part 3: Network Aspects. 270 pp.

Communication Networks (Cont.)
—**Radio Paging Systems (Cont.)**
ETSI PRI-ETS 300 133-3 PRA1-93. AMD prA1 Paging Systems (PS); European Radio Message System (ERMES) Part 3: Network Aspects. 19 pp.
ETSI ETS 300 133-3-92. Paging Systems (PS); European Radio Message System (ERMES) Part 3: Network Aspects (Without Annexes). 294 pp.
ETSI PRETS 300 133-3-91. Paging Systems; European Radio Message System (ERMES) Part 3: Network Aspects. 272 pp.

—**Transport Protocol**
BSI BS ISO/IEC 11570-92. 1992 Information Technology—Telecommunications and Information Exchange Between Systems—Open Systems Interconnection—Transport Protocol Identification Mechanism (S). 12 pp.
IEC 11570-92. Information Technology—Telecommunications and Information Exchange Between Systems—Open Systems Interconnection—Transport Protocol Identification Mechanism First Edition. 10 pp.
ISO 11570-92. Information Technology—Telecommunications and Information Exchange Between Systems—Open Systems Interconnection—Transport Protocol Identification Mechanism First Edition. 10 pp.
JTC1 11570-92. Information Technology—Telecommunications and Information Exchange Between Systems—Open Systems Interconnection—Transport Protocol Identification Mechanism First Edition. 10 pp.

Communication Protocols
Use: Protocols

Communication Receivers (Radio)
Use: Radio Receivers

Communication Satellites
Use: Communications Satellites

Communication Systems and Equipment
Use: Communication Equipment

Communication Terminal Equipment
Use For: Computerized Communication Terminals; Telecommunication Terminal Equipment
See Also: Customer Premises Equipment; Data Terminal Equipment; Multipoint Control Units (Telecommunications); Signal Senders; Telecommunication Equipment; Terminal Identification
BSI BS 6403-83. 1983 Amd 2 Simple Telex Terminals Using Single Channel Voice Frequency Signal-ling for Connection with British Telecommunicat-ions Telex Network Inco-rporating the Designa-tions of the Secretary of State for Trade and Industry in Relation to Telecommunications Stds.. 50 pp.
BSI BS 7378: Part 1-91. 1991 Apparatus for Connection to Public Telecommunications Systems Using the Digital Access Signalling System No. 2 (DASS 2) Via a 2048 kbit/s CCITT Recommendation G.703 Interface Part 1: General Requirements. 44 pp.
BSI BS 7378: Part 2-91. 1991 Apparatus for Connection to Public Telecommunications Systems Using the Digital Access Signalling System No. 2 (DASS 2) Via a 2048 kbit/s CCITT Recommendation G.703 Interface Part 2: Specification. 12 pp.
CENELEC PRETS 300 015-90. Terminal Equipment Requirements for Teletex Terminal Equipment Participating in the Teletex Service (T/TE 07-01). 40 pp.
CENELEC PRETS 300 015-92. Terminal Equipment (TE); Basic and Recommended Additional Requirements for Terminal Equipment Supporting Teletex Application. 32 pp.
CENELEC PRETS 300 016-90. Terminal Equipment Service Intercommunication Requirements for Teletex Terminal Equipment Participating in a Regulated Teletex Service (T/TE 07-04). 59 pp.
CENELEC PRETS 300 017-90. Terminal Equipment Test Procedures for Teletex (T/TE 07-05). 482 pp.
CENELEC PRETS 300 017-93. Terminal Equipment (TE); Test Procedures for Teletex. 415 pp.
CENELEC PRETS 300 018-90. Terminal Equipment Attachment Requirements for Teletex Terminal Equipment Participating in a Regulated Teletex Service (T/TE 07-07) (Candidate NET 32). 14 pp.
CENELEC PRETS 300 148-91. Terminal Equipment (TE); Requirements for Teletex Systems Participating in the Service (T/TE 07-10). 14 pp.
CENELEC PRETS 300 154-91. Terminal Equipment (TE); Optional Applications Between Teletex Equipments Transparent Mode and Local Dispatching at the Receiving Side (T/TE 07-09). 22 pp.

INTERNATIONAL AND NON-U.S. NATIONAL STANDARDS
SUBJECT INDEX
Communication

Communication Terminal Equipment (Cont.)

CEPT T/A 07-01-86. Harmonisation des Terminaux de Teletex.

CEPT T/SF 41 E-85. Conditions of Use of Private Terminal Equipment for Connection to the Administrations Telecommunications System. 3 pp.

CEPT T/SF 47-86. Procedures Formelles D'Agrement D'Equipements Terminaux Fournis a Titre Prive Pour Le Raccordement Aux Reseaux Publics De Telecommunication Et/Ou Pour L'Acces Aux Services Publics. 8 pp.

CEPT T/SF 47 E-86. Formal Approval Procedures for Privately Supplied Terminal Equipment for Connection to Public Telecommunications Networks and/or Access to Public Services. 8 pp.

CEPT T/SF 48-87. Procedures De Delivrance Et D'Utilisation De Certificats De Conformite Aux Normes Pour Les Equipements Terminaux De Telecommunications. 5 pp.

CEPT T/SF 48E-87. Procedures for Issue and Use of Certificates of Conformity to Standards for Telecommunications Terminal Equipment. 5 pp.

CEPT T/TE 07-01 BIS E-88. Harmonization of Teletex Terminal Equipment. 23 pp.

CEPT T/TE 07-04 E-88. Service Intercommunication Requirements for Teletex Terminals. 28 pp.

EC 86/361/EEC-86. Council Directive on the Initial Stage of the Mutual Recognition of Type Approval for Telecommunications Terminal Equipment. 5 pp.

ETSI PRETS 300 015-90. Terminal Equipment—Requirements for Telex Terminal Equipment Participating in the Telex Service (T/TE 07-01). 38 pp.

ETSI PRETS 300 016-90. Terminal Equipment Service Intercommunication Requirements for Teletex Terminal Equipment Participating in a Regulated Service (T/TE 07-04). 57 pp.

ETSI PRETS 300 017-90. Terminal Equipment—Test Procedures for Teletex (T/TE 07-05). 478 pp.

ETSI PRETS 300 018-90. Terminal Equipment Attachment Requirements for Teletex Terminal Equipment Participating in a Regulated Teletex Service (T/TE 07-07) (Candidate NET 32). 12 pp.

ETSI PRETS 300 015-92. Terminal Equipment (TE); Basic and Recommended Additional Requirements for Terminal Equipment Supporting Teletex Application. 32 pp.

ETSI PRETS 300 015-90. Terminal Equipment Requirements for Teletex Terminal Equipment Participating in the Teletex Service (T/TE 07-01). 40 pp.

ETSI PRETS 300 016-90. Terminal Equipment Service Intercommunication Requirements for Teletex Terminal Equipment Participating in a Regulated Teletex Service (T/TE 07-04). 59 pp.

ETSI PRETS 300 017-93. Terminal Equipment (TE); Test Procedures for Teletex. 415 pp.

ETSI PRETS 300 017-90. Terminal Equipment Test Procedures for Teletex (T/TE 07-05). 482 pp.

ETSI PRETS 300 018-90. Terminal Equipment Attachment Requirements for Teletex Terminal Equipment Participating in a Regulated Teletex Service (T/TE 07-07) (Candidate NET 32). 14 pp.

ETSI PRETS 300 148-91. Terminal Equipment (TE); Requirements for Teletex Systems Participating in the Teletex Service (T/TE 07-10).

ETSI PRETS 300 154-91. Terminal Equipment (TG); Optional Applications Between Teletex Equipments—Transparent Mode and Local Dispatching at the Receiving 07-09).

ETSI PRETS 300 148-91. Terminal Equipment (TE); Requirements for Teletex Systems Participating in the Service (T/TE 07-10). 14 pp.

ETSI PRETS 300 154-92. Terminal Equipment (TE); Optional Application for Teletex and Facsimile Group 4 Equipments Telematic File Transfer. 12 pp.

ETSI PRETS 300 154-91. Terminal Equipment (TE); Optional Applications Between Teletex Equipments Transparent Mode and Local Dispatching at the Receiving Side (T/TE 07-09). 22 pp.

—**8 Channel—Telephone Channel Units**

CCITT RECMN G.234-89. 8-Channel Terminal Equipments—International Analogue Carrier Systems (Study Group XV) 1 pp. 1 p.

—**Certification**

CEPT YEAR Book-92. Information on Certification, Approval and Testing Laboratories for Telecommunications Terminal Equipment. 100 pp.

—**Groups**

CCITT RECMN G.232-89. 12-Channel Terminal Equipments—International Analogue Carrier Systems (Study Group XV) 13 pp. 13 pp.

—**Groups—Attenuation Distortion**

CCITT RECMN G.232-89. 12-Channel Terminal Equipments—International Analogue Carrier Systems (Study Group XV) 13 pp. 13 pp.

Communication Terminal Equipment (Cont.)

—**Groups—Carrier Leaks**

CCITT RECMN G.232-89. 12-Channel Terminal Equipments—International Analogue Carrier Systems (Study Group XV) 13 pp. 13 pp.

—**Groups—Crosstalk**

CCITT RECMN G.232-89. 12-Channel Terminal Equipments—International Analogue Carrier Systems (Study Group XV) 13 pp. 13 pp.

—**Groups—Electrical Impedance**

CCITT RECMN G.232-89. 12-Channel Terminal Equipments—International Analogue Carrier Systems (Study Group XV) 13 pp. 13 pp.

—**Groups—Envelope Delay Distortion**

CCITT RECMN G.232-89. 12-Channel Terminal Equipments—International Analogue Carrier Systems (Study Group XV) 13 pp. 13 pp.

—**Groups—Linearity**

CCITT RECMN G.232-89. 12-Channel Terminal Equipments—International Analogue Carrier Systems (Study Group XV) 13 pp. 13 pp.

—**Groups—Pilot Channels**

CCITT RECMN G.232-89. 12-Channel Terminal Equipments—International Analogue Carrier Systems (Study Group XV) 13 pp. 13 pp.

—**Groups—Return Loss**

CCITT RECMN G.232-89. 12-Channel Terminal Equipments—International Analogue Carrier Systems (Study Group XV) 13 pp. 13 pp.

—**Groups—Signal Levels**

CCITT RECMN G.232-89. 12-Channel Terminal Equipments—International Analogue Carrier Systems (Study Group XV) 13 pp. 13 pp.

—**Groups—Signal Limiters**

CCITT RECMN G.232-89. 12-Channel Terminal Equipments—International Analogue Carrier Systems (Study Group XV) 13 pp. 13 pp.

—**IA5—Interpersonal Messaging Services**

CCITT RECMN F.420-92. Message Handling Services: Public Interpersonal Messaging Service (Study Group I) 16 pp. 16 pp.

CCITT RECMN F.420-89. Message Handling Services: the Public Interpersonal Messaging Service—Message Handling and Directory Services—Operations and Definition of Service (Study Group I) 15 pp. 15 pp.

—**Integrated Services Digital Networks**

CCITT RECMN I.333-89. Terminal Selection in ISDN—Integrated Services Digital Network (ISDN)—Overall Network Aspects and Functions, ISDN User-Network Interfaces (Study Group XVIII) 18 pp. 18 pp.

CENELEC ETR 026-92. Network Aspects (NA); Terminal Selection Principles for Priority 1 and 2 Services of MoU—ISDN Applicable in Multi-Terminal Environments at Customer Premises. 21 pp.

ETSI ETR 026-92. Network Aspects (NA); Terminal Selection Principles for Priority 1 and 2 Services of MoU—ISDN Applicable in Multi-Terminal Environments at Customer Premises. 21 pp.

—**Integrated Services Digital Networks—Bearer Services**

CCITT RECMN I.333-89. Terminal Selection in ISDN—Integrated Services Digital Network (ISDN)—Overall Network Aspects and Functions, ISDN User-Network Interfaces (Study Group XVIII) 18 pp. 18 pp.

CENELEC ETR 026-92. Network Aspects (NA); Terminal Selection Principles for Priority 1 and 2 Services of MoU—ISDN Applicable in Multi-Terminal Environments at Customer Premises. 21 pp.

ETSI ETR 026-92. Network Aspects (NA); Terminal Selection Principles for Priority 1 and 2 Services of MoU—ISDN Applicable in Multi-Terminal Environments at Customer Premises. 21 pp.

—**Integrated Services Digital Networks—Interfaces**

CCITT RECMN I.410-89. General Aspects and Principles Relating to Recommendations on ISDN User-Network Interfaces—Integrated Services Digital Network (ISDN)—Overall Network Aspects and Functions, ISDN User-Network Interfaces (Study Group XVIII) 3 pp. 3 pp.

Communication Terminal Equipment (Cont.)

—**Integrated Services Digital Networks—Network Capabilities**

CCITT RECMN I.333-89. Terminal Selection in ISDN—Integrated Services Digital Network (ISDN)—Overall Network Aspects and Functions, ISDN User-Network Interfaces (Study Group XVIII) 18 pp. 18 pp.

CENELEC ETR 026-92. Network Aspects (NA); Terminal Selection Principles for Priority 1 and 2 Services of MoU—ISDN Applicable in Multi-Terminal Environments at Customer Premises. 21 pp.

ETSI ETR 026-92. Network Aspects (NA); Terminal Selection Principles for Priority 1 and 2 Services of MoU—ISDN Applicable in Multi-Terminal Environments at Customer Premises. 21 pp.

—**Integrated Services Digital Networks—Teleservices**

CCITT RECMN I.333-89. Terminal Selection in ISDN—Integrated Services Digital Network (ISDN)—Overall Network Aspects and Functions, ISDN User-Network Interfaces (Study Group XVIII) 18 pp. 18 pp.

CENELEC ETR 026-92. Network Aspects (NA); Terminal Selection Principles for Priority 1 and 2 Services of MoU—ISDN Applicable in Multi-Terminal Environments at Customer Premises. 21 pp.

ETSI ETR 026-92. Network Aspects (NA); Terminal Selection Principles for Priority 1 and 2 Services of MoU—ISDN Applicable in Multi-Terminal Environments at Customer Premises. 21 pp.

—**Interfaces**

CSA CAN/CSA-T538-M92. Carrier-to-Customer Installation—DS1 Metallic Interface; (Gen Instr 1). 58 pp.

—**Laboratories**

CEPT YEAR Book-92. Information on Certification, Approval and Testing Laboratories for Telecommunications Terminal Equipment. 100 pp.

CEPT T/G 01-01-87. Criteres D'Accreditation Des Laboratoires D'Essais D'Equipements Terminaux De Telecommunication. 4 pp.

CEPT T/G 01-01 E-87. Criteria for Accreditation of Testing Laboratories for Telecommunications Terminal Equipment. 4 pp.

—**Protocols—Teleinformatic Services**

CCITT FASCICLE VII.3. Terminal Equipment and Protocols for Telematic Services—Recommendations T.0-T.63. 494 pp.

—**Radio Relay Systems—Telephone Networks**

CCITT RECMN G.412-89. Terminal Equipments of Radio-Relay Systems Forming Part of a General Telecommunication Network—International Analogue Carrier Systems (Study Group XV) 1 pp. 1 p.

—**Repeater Stations**

CCITT RECMN G.544-89. Specifications for Terminal Equipment and Intermediate Repeater Stations—International Analogue Carrier Systems (Study Group XV) 1 pp. 1 p.

—**Teleconferencing Equipment**

CCITT RECMN F.701-89. Teleconference Service—Telematic, Data Transmission and Teleconference Services—Operations and Quality of Service (Study Group I) 13 pp (Renumbered from Recmn F.710). 13 pp.

CCITT RECMN F.710-91. General Principles for Audiographic Conference Service—Telematic, Data Transmission and Teleconference Services Operations and Quality of Service (Study Group I) 15 pp. 15 pp.

CENELEC PRETS 300 101-90. Integrated Services Digital Network (ISDN) International Digital Audiographic Teleconference (T/N 33-01). 185 pp.

CENELEC PRI-ETS 300 101-92. Integrated Services Digital Network (ISDN); International Digital Audiographic Teleconference. 179 pp.

CENELEC I-ETS 300 101-93. Integrated Services Digital Network (ISDN); International Digital Audiographic Teleconference. 179 pp.

ETSI PRETS 300 101-90. Integrated Services Digital Network (ISDN) International Digital Audiographic Teleconference (T/N 33-01). 181 pp.

ETSI I-ETS 300 101-93. Integrated Services Digital Network (ISDN); International Digital Audiographic Teleconference. 179 pp.

ETSI PRI-ETS 300 101-92. Integrated Services Digital Network (ISDN); International Digital Audiographic Teleconference. 179 pp.

ETSI PRETS 300 101-90. Integrated Services Digital Network (ISDN) International Digital Audiographic Teleconference (T/N 33-01). 185 pp.

INDUSTRY STANDARDS

INTERNATIONAL AND NON-U.S. NATIONAL STANDARDS
SUBJECT INDEX

Communication

Communication Terminal Equipment (Cont.)

—Teleinformatic Services
- CCITT RECMN F.350-89. Application of Series T Recommendations—Telematic, Data Transmission and Teleconference Services—Operations and Quality of Service (Study Group I) 2 pp. 2 pp.
- CCITT FASCICLE VII.3. Terminal Equipment and Protocols for Telematic Services—Recommendations T.0-T.63. 494 pp.

—Teleinformatic Services—ISDN
- CCITT RECMN T.90-92. Characteristics and Protocols for Terminals for Telematic Services in ISDN (Study Group VIII) 61 pp. 61 pp.

—Teleinformatic Services—Protocols
- CCITT FASCICLE VII.5-89. Terminal Equipment and Protocols for Telematic Services—Recommendations T.65-T.101, T.150-T.390. 388 pp.
- CCITT RECMN T.65-89. Applicability of Telematic Protocols and Terminal Characteristics to Computerized Communication Terminals (CCTs)—Terminal Equipment and Protocols for Telematic Services (Study Group VIII) 9 pp. 9 pp.

—Teleinformatic Services—Protocols—ISDN
- CCITT RECMN T.90-92. Characteristics and Protocols for Terminals for Telematic Services in ISDN (Study Group VIII) 61 pp. 61 pp.

—Teleinformatic Services—Raster Graphics
- CCITT RECMN T.410 Series-91. First Extension (January 1991) to the T.410 Series (1988) of Recommendations Contained in the CCITT Blue Book, Fascicle VII.6 (Study Group VIII) 58 pp. 58 pp.

—Telephony—Integrated Services Digital Networks
- CCITT RECMN I.330-89. ISDN Numbering and Addressing Principles—Integrated Services Digital Network (ISDN)—Overall Network Aspects and Functions, ISDN User-Network Interfaces (Study Group XVIII) 8 pp. 8 pp.
- CENELEC PRETS 300 281-93. Integrated Services Digital Network (ISDN); Telephony 7 kHz Teleservice Terminal Requirements Necessary for End-to-End Compatibility. 53 pp.
- ETSI PRETS 300 281-93. Integrated Services Digital Network (ISDN); Telephony 7 kHz Teleservice Terminal Requirements Necessary for End-to-End Compatibility. 53 pp.

—16 Channel—Groups
- CCITT RECMN G.235-89. 16-Channel Terminal Equipments—International Analogue Carrier Systems (Study Group XV) 4 pp. 4 pp.

—16 Channel—Groups—Attenuation Distortion
- CCITT RECMN G.235-89. 16-Channel Terminal Equipments—International Analogue Carrier Systems (Study Group XV) 4 pp. 4 pp.

—16 Channel—Groups—Carrier Leaks
- CCITT RECMN G.235-89. 16-Channel Terminal Equipments—International Analogue Carrier Systems (Study Group XV) 4 pp. 4 pp.

—16 Channel—Groups—Crosstalk
- CCITT RECMN G.235-89. 16-Channel Terminal Equipments—International Analogue Carrier Systems (Study Group XV) 4 pp. 4 pp.

—16 Channel—Groups—Group Delay
- CCITT RECMN G.235-89. 16-Channel Terminal Equipments—International Analogue Carrier Systems (Study Group XV) 4 pp. 4 pp.

—16 Channel—Groups—Linearity
- CCITT RECMN G.235-89. 16-Channel Terminal Equipments—International Analogue Carrier Systems (Study Group XV) 4 pp. 4 pp.

—16 Channel—Groups—Pilot Channels
- CCITT RECMN G.235-89. 16-Channel Terminal Equipments—International Analogue Carrier Systems (Study Group XV) 4 pp. 4 pp.

—16 Channel—Groups—Psophometric Power
- CCITT RECMN G.235-89. 16-Channel Terminal Equipments—International Analogue Carrier Systems (Study Group XV) 4 pp. 4 pp.

—16 Channel—Groups—Signal Limiters
- CCITT RECMN G.235-89. 16-Channel Terminal Equipments—International Analogue Carrier Systems (Study Group XV) 4 pp. 4 pp.

Communication Terminal Equipment (Cont.)

—16 Channel—Groups—Virtual Carrier Frequencies
- CCITT RECMN G.235-89. 16-Channel Terminal Equipments—International Analogue Carrier Systems (Study Group XV) 4 pp. 4 pp.

Communication Towers
Use For: Antenna Towers; Radio Towers
See Also: Antenna Masts; Antennas; Cellular Mobile Radio Equipment; Microwave Equipment; Radio Equipment; Radio Stations; Telecommunication Equipment

—Safety
- MOD UK DSTAN 05-39-01. Safety Requirements for the Design, Erection and Maintenance of Land Based Masts, Towers, Turntables, Antenna Supporting, and Related Structures Issue 1; Amendment 2. 22 pp.

Communication Transmission Line Hardware
See Also: Cable Supports; Communication Equipment; Power Line Hardware
- CSA CAN/CSA-C83-M87. Communication and Power Line Hardware; (Gen Instr 2). 313 pp.

Communication Transmission Lines
See Also: Coaxial Cables; Coaxial Transmission Lines (Waveguide); Communication Cables; Communication Equipment; Echo Cancellers; Echo Suppressors; Line Links; Network Nodes; Private Leased Circuits; Regulated Line Sections; Telecommunication Equipment; Telephone Cables; Telephone Lines; Transmission Lines; Transmission Loss; Waveguides

—Access Points—Maintenance
- CCITT RECMN M.120-89. Access Points for Maintenance—General Maintenance Principles—Maintenance of International Transmission Systems and Telephone Circuits (Study Group IV) 2 pp. 2 pp.

—Commercial Buildings
- CSA CAN/CSA-T529-M91. Design Guidelines for Telecommunications Wiring Systems in Commercial Buildings; (Gen Instr 1). 76 pp.

—Construction—Safety
- DIN VDE 0800 Pt 4-86. Telecommunications; Erection of Telecommunication Lines (Mar) (Partially Supersedes DIN 57800 Part1/VDE 0800 Part 1/04.84). 75 pp.

—Crossbars
- CNS C4076-67. Concrete Crossbar (for Communications Line) (Tentative) (Oct)(2806).

—Insulators—Ceramic
- CNS C3016-70. Method of Test for Ceramic Insulators for Communication Line (Jan)(907).

—Line Up
- CCITT RECMN M.35-89. Principles Concerning Line-up and Maintenance Limits—General Maintenance Principles—Maintenance of International Transmission Systems and Telephone Circuits (Study Group IV) 1 pp. 1 pp.

—Maintenance
- CCITT RECMN M.35-89. Principles Concerning Line-up and Maintenance Limits—General Maintenance Principles—Maintenance of International Transmission Systems and Telephone Circuits (Study Group IV) 1 pp. 1 pp.

—Overhead
- CSA CAN/CSA-C22. 3 NO 1-M87. Overhead Systems; (Gen Instr 1). 131 pp.

—Poles (Supports)
- CNS C4075-88. Centrifugal Pre-Stressed Concrete Poles (for Communications Line) (Apr)(2805).

—PVC Insulated
- BSI PD 6455-70. 1970 Metric Dimensions for l.f. Cables and Wire for Telecommunication. 4 pp.

—Underground
- CSA CAN3-C22.3 NO 7-M86. Underground Systems; (Gen Instr 1). 39 pp.

Communications Interfaces
Use: Communication Interfaces

Communications Satellites
Use For: Communication Satellites; Radio Communications Satellites; Radio Relay Satellites; Radiocommunication Satellites; Radiocommunications Satellites

Communications Satellites (Cont.)
See Also: Broadcasting Satellite Systems; Data Relay Satellites; Fixed Satellite Services; Intersatellite Services; Satellites

—Amateur Radio Services
- CCIR Report 1154-90. Techniques and Frequency Usage in the Amateur and Amateur-Satellite Services—Section 8L—Amateur Service; Amateur-Satellite Service. 24 pp.

—Coordination
- CCIR Report 870-2-90. Technical Coordination Methods for Communication-Satellite Systems—Section 4D2—Coordination Methods. 13 pp.
- CCIR QUESTION 49/4-90. Technical Coordination Methods for Systems in the Fixed-Satellite Service—Questions of Study Group 4 Fixed-Satellite Service. 1 p.

—Frequency Band Sharing
- CCIR Decision 2-7-89. Frequency Sharing Between Radiocommunication Satellites Technical Considerations Affecting the Efficient Use of the Geostationary-Satellite Orbit—Annex to Volume IV-1—Fixed-Satellite Service. 4 pp.

—Geostationary
- CCIR Report 771-3-90. Considerations for the Mobile-Satellite Service —Section 8F—Frequencies, Orbits and Systems. 7 pp.

—Geostationary Satellite Orbit—Frequency Band Sharing—Efficiency
- CCIR Report 453-5-90. Technical Factors Influencing the Efficiency of Use of the Geostationary-Satellite Orbit by Radiocommunication Satellites Sharing the Same Frequency Bands General Summary—Section 4D2—Coordination Methods. 36 pp.
- CCIR QUESTION 48/4-90. Technical Factors Influencing the Efficiency of Use of the Geostationary-Satellite Orbit by Radiocommunication Sat. Networks Sharing Frequency Bands Allocated to the Fixed-Sat. Service—Questions of Study Group 4 Fixed-Satellite Service. 2 pp.

—Television Program Exchange
- CCIR OPINION 38-70. Exchange of Monochrome and Colour Television Programmes Via Satellites—Volume XI-1—Broadcasting Service (Television). 1 p.

Communications Systems
Use: Communication Equipment

Community Antenna Television
Use: Cable Television

Commutators
See Also: Brushes (Machine Components); Distributors (Electrical); Rotating Machines
- IEC 356-71. Dimensions for Commutators and Slip-Rings First Edition. 23 pp.

—Bars
- BSI BS 1434-85. 1985 Copper for Electrical Purposes: Copper Sections in Bars, Blanks and Segments for Commutators. 14 pp.
- CNS C4411-84. Commutator Bar (Jun)(10916).
- JIS C 2801-91. Commutator Bars. 11 pp.

—Glossaries
- BSI BS 4999: Part 146-88. 1988 General Requirements for Rotating Electrical Machines Part 146: Definitions and Nomenclature for Carbon Brushes, Brush Holders Commutators and Slip-Rings. 36 pp.
- BSI BS 4999: Part 146-01. 1988 Amd 1 General Requirements for Rotating Electrical Machines Part 146: Definitions and Nomenclature for Carbon Brushes, Brush-Holders, Commutators and Slip-Rings (IEC 276: 1968) (AMD 7014) September 15, 1992. 40 pp.
- CENELEC HD 56 S2-91. Definitions and Nomenclature for Carbon Brushes, Brush-Holders, Commutators and Slip-Rings. 4 pp.
- IEC 276-68. Definitions and Nomenclature for Carbon Brushes, Brush-Holders, Commutators and Slip-Rings First Edition; (Amendment 1-1987). 41 pp.

—Mica
- DIN VDE 0332-68. Specifications for Mica Products (Nov). 49 pp.
- DIN VDE 0332A-71. Specifications for Mica Products: Amendment A (Sept). 14 pp.
- DIN VDE 0332 Pt 5-81. Mica Products for Electrical Insulation; Formed Parts of Rigid Mica for Commutators (July). 13 pp.

—Rotating Machines—Glossaries
- SAA AS 1359.46-88. Rotating Electrical Machines—General Requirements—Part 46: Brushes, Brush-Holders, Commutators, and Slip-Rings—Glossary of Terms. 19 pp.

Commutators (Cont.)

—Rotating Machines—Railroad Traction Equipment

IEC 638-79. Criteria for Assessing and Coding of the Commutation of Rotating Electrical Machines for Traction First Edition. 34 pp.

—Separators—Mica

CENELEC HD 352.3.1 S2-87. Specification for Insulating Materials Based on Mica Part 3: Specifications for Individual Materials Sheet 1: Commutator Separators and Materials. 4 pp.

CENELEC HD 352.3.1 S2-89. Specification for Insulating Materials Based on Mica-Part 3: Specifications for Individual Materials. Sheet One: Commutator Separators and Materials. 3 pp.

IEC 371 Pt 3-1-84. Specification for Insulating Materials Based on Mica Part 3: Specifications for Individual Materials Sheet 1: Commutator Separators and Materials Second Edition. 19 pp.

JIS C 2252-92. Rigid Mica Materials for Commutator Separators. 7 pp.

Commuter Cars
See Also: Electric Rolling Stock; Rolling Stock

—Electric

JIS E 6002-89. General Rules for Performance of Electric Railcars for Commuter Use.

JIS E 6003-85. General Rules for Design of Driving Cabs of Electric Commuter Cars.

JIS E 7103-89. General Rules for Design of Bodies of Electric Railcars for Commute Use.

Compact Disk Players
Use For: Laser Audio Disk Players *See Also:* Audio Equipment; Disk Drives (Computer); Laser Equipment

BSI BS 7064-89. (WITHDRAWN) 1989 Amd 1 Compact Disc Digital Audio System (AMD 6401) September 30, 1991 (Superseded by BS EN 60908: 1993). 42 pp.

BSI BS EN 60908-93. 1993 Compact Disc Digital Audio System (IEC 908: 1987) (S). 51 pp.

BSI BS EN 61096-93. 1993 Methods of Measuring the Characteristics of Reproducing Equipment for Digital Audio Compact Discs (IEC 1096: 1992) (S). 28 pp.

CENELEC HD 529 S1-89. Compact Disc Digital Audio System. 3 pp.

CENELEC EN 60908-92. Compact Disc Digital Audio System (Includes Amendment A1: 1993) (IEC 908: 1987/A1: 1992). 46 pp.

CENELEC EN 61096-93. Methods of Measuring the Characteristics of Reproducing Equipment for Digital Audio Compact Discs (IEC 1096:1992). 23 pp.

CNS C6266-86. Method of Test for Compact Disc Players (Oct)(11715).

CSA CAN/CSA-C22. 2 NO 1-M90. Radio, Television, and Electronic Apparatus; (Gen Instr 1 Thru 3). 146 pp.

IEC 908-87. Compact Disc Digital Audio System First Edition; (Amendment 1-1992) (Corrigendum—March 1993, to Amendment 1) (CENELEC EN 60908/A1: 1993). 89 pp.

IEC 1096-92. Methods of Measuring the Characteristics of Reproducing Equipment for Digital Audio Compact Discs First Edition; (CENELEC EN 61096:1993). 48 pp.

Compact Disks
See Also: Audio Equipment; CD ROMs; CD WOMs; Sound Recording

BSI BS 7064-89. (WITHDRAWN) 1989 Amd 1 Compact Disc Digital Audio System (AMD 6401) September 30, 1991 (Superseded by BS EN 60908: 1993). 42 pp.

BSI BS EN 60908-93. 1993 Compact Disc Digital Audio System (IEC 908: 1987) (S). 51 pp.

BSI BS EN 61096-93. 1993 Methods of Measuring the Characteristics of Reproducing Equipment for Digital Audio Compact Discs (IEC 1096: 1992) (S). 28 pp.

CENELEC HD 529 S1-89. Compact Disc Digital Audio System. 3 pp.

CENELEC EN 60908-92. Compact Disc Digital Audio System (Includes Amendment A1: 1993) (IEC 908: 1987/A1: 1992). 46 pp.

CENELEC EN 61096-93. Methods of Measuring the Characteristics of Reproducing Equipment for Digital Audio Compact Discs (IEC 1096:1992). 23 pp.

IEC 908-87. Compact Disc Digital Audio System First Edition; (Amendment 1-1992) (Corrigendum—March 1993, to Amendment 1) (CENELEC EN 60908/A1: 1993). 89 pp.

IEC 1096-92. Methods of Measuring the Characteristics of Reproducing Equipment for Digital Audio Compact Discs First Edition; (CENELEC EN 61096:1993). 48 pp.

Compact Disks (Cont.)

—Address Codes

BSI BS 6288: Part 11-89. 1989 Magnetic Tape Sound Recording and Reproducing Systems Part 11: Guide to an Address Code for Compact Cassettes. 11 pp.

Compatibility
See Also: Bulk Density; Compacting; Compressibility Factors; Green Strength; Moldability; Porosity

—Aggregates

BSI BS 5835: Part 1-80. 1980 Amd 3 Recommendations for Testing of Aggregates Part 1: Compactibility Test for Graded Aggregates. 21 pp.

—Asphalts

DIN ENGL 1996 Pt 7-83. Testing of Bituminous Materials for Road Building and Related Purposes; Determination of Density and Voids (Jan). 8 pp.

—Ceramic Powders

CEN PREN 725-10-93. Advanced Technical Ceramics—Method of Test for Ceramic Powders—Part 10: Determination of Compaction Properties. 10 pp.

—Concretes

ISO 4111-79. Fresh Concrete—Determination of Consistency—Degree of Compactibility (Compaction Index) First Edition. 4 pp.

—Metal Powders

BSI BS 5600: Sec 2.4-88. (WITHDRAWN) 1988 Powder Metallurgical Materials and Products Part 2: Methods of Sampling and Testing Metallic Powders Section 2.4: Determination of Compressibility in Uniaxial Compression (Superseded by BS EN 23927: 1993). 6 pp.

BSI BS EN 23927-93. 1993 Metallic Powders, Excluding Powders for Hardmetals—Determination of Compactibility (Compressibility) in Uniaxial Compression (ISO 3927: 1985) (V). 11 pp.

CEN EN 23927-93. Metallic Powders, Excluding Powders for Hardmetals—Determination of Compactibility (Compressibility) in Uniaxial Compression (ISO 3927: 1985). 6 pp.

ISO 3927-85. Metallic Powders, Excluding Powders for Hardmetals—Determination of Compactibility (Compressibility) in Uniaxial Compression Second Edition (CEN EN 23927: 1993). 6 pp.

—Mineral Aggregates

DIN ENGL 52102-88. Determination of Absolute Density, Dry Density, Compactness and Porosity of Natural Stone and Mineral Aggregates (Aug). 10 pp.

—Refractory Materials

BSI BS 1902: Sec 7.2-87. 1987 Refractory Materials Part 7: Unshaped Refractories Used in Monolithic Construction Section 7.2: Testing of Material as Supplied and Received (Method 1902-702). 8 pp.

—Rocks

DIN ENGL 52102-88. Determination of Absolute Density, Dry Density, Compactness and Porosity of Natural Stone and Mineral Aggregates (Aug). 10 pp.

—Soil Analysis

BSI BS 1377: Part 4-90. 1990 Methods of Test for Soils for Civil Engineering Purposes Part 4: Compaction-Related Tests. 52 pp.

SNZ NZS 4402: Pt 4:TEST 4.1.1-86. Methods of Testing Soils for Civil Engineering Purposes Part 4: Soil Compaction Tests. Section 4.1: Determination of the Dry Density/Water Content Relationship. Test 4.1.1: New Zealand Standard Compaction Test. 6 pp.

SNZ NZS 4402: Pt 4:TEST 4.1.2-86. Methods of Testing Soils for Civil Engineering Purposes Part 4: Soil Compaction Tests. Section 4.1: Determination of the Dry Density/Water Content Relationship. Test 4.1.2: New Zealand Heavy Compaction Test. 6 pp.

SNZ NZS 4402: Pt 4:TEST 4.1.3-86. Methods of Testing Soils for Civil Engineering Purposes Part 4: Soil Compaction Tests. Section 4.1: Determination of the Dry Density/Water Content Relationship. Test 4.1.3: New Zealand Vibrating Hammer Compaction Test. 8 pp.

SNZ NZS 4402: Pt 4:TEST 4.2.1-88. Methods of Testing Soils for Civil Engineering Purposes Part 4: Soil Compaction Tests. Section 4.2: Determination of the Minimum and Maximum Dry Densities and Relative Density of a Cohesionless Soil. Test 4.2.1: Minimum Dry Density. 3 pp.

SNZ NZS 4402: Pt 4:TEST 4.2.2-88. Methods of Testing Soils for Civil Engineering Purposes Part 4: Soil Compaction Tests. Section 4.2: Determination of the Minimum and Maximum Dry Densities and Relative Density of a Cohesion-less Soil. Test 4.2.2: Maximum Dry Density. 6 pp.

Compatibility (Cont.)

—Soil Analysis (Cont.)

SNZ NZS 4402: Pt 4:TEST 4.2.3-88. Methods of Testing Soils for Civil Engineering Purposes Part 4: Soil Compaction Tests. Section 4.2: Determination of the Minimum and Maximum Dry Densities and Relative Density of a Cohesion-less Soil. Test 4.2.3: Relative Density. 2 pp.

Compacting
See Also: Caking; Compactibility; Metal Powders; Moldability; Packaging; Powder Metallurgy; Presses; Vibration

—Concretes

BSI BS 1881: Part 103-83. 1983 Amd 2 Testing Concrete Part 103: Method for Determination of Compacting Factor (AMD 6726) July 31, 1991. 12 pp.

—Concretes—Volumetric Analysis

BSI DD 90-83. 1983 Volumetric Method for Determination of Compacting Factor of Fresh Concrete. 7 pp.

—Flexible Pavements

CNS A3288-88. Method of Test for Determining Degree of Pavement Compaction of Bituminous Aggregate Mixtures (Aug)(12390).

—Metal Powders

BSI BS 5600: Sec 2.7-84. (WITHDRAWN) 1984 Powder Metallurgical Materials Products Part 2: Methods of Sampling and Testing Metallic Powders Section 2.7: Determination of Dimens-ional Changes Associated With Compacting and Sintering, Excluding Powders for Hardmetals. 7 pp.

BSI BS EN 24492-93. 1993 Metallic Powders, Excluding Powders for Hardmetals—Determination of Dimensional Changes Associated with Compacting and Sintering (ISO 4492: 1985) (V). 13 pp.

CEN EN 24492-93. Metallic Powders, Excluding Powders for Hardmetals—Determination of Dimensional Changes Associated with Compacting and Sintering (ISO 4492: 1985). 8 pp.

ISO 4492-85. Metallic Powders, Excluding Powders for Hardmetals—Determination of Dimensional Changes Associated with Compacting and Sintering Second Edition (CEN EN 24492: 1993). 7 pp.

Compaction Equipment
Scope Note: Use a more specific term
See: Construction Equipment; Earthmoving Equipment

Compactors
Scope Note: For additional listings use a more specific term *See Also:* Construction Equipment; Earthmoving Equipment; Plate Compactors; Rollers (Compactors); Trash Compactors

—Safety

CEN PREN 500-4-91. Mobile Road Construction Machines—Safety—Part 4: Specific Requirements for Compaction Machines. 29 pp.

Compandors
Use For: Lincompex Systems; Syncompex Systems *See Also:* Electronic Equipment; Integrated Circuits; Signal to Noise Ratio

—Land Mobile Services—Radiotelephony

CCIR Report 899-1-90. Systems of Modulation with High Spectrum Efficiency for the Land Mobile Service—Section 8A—Land Mobile Service and Related Subjects. 11 pp.

—Radiotelephone Circuits—Maritime Mobile Services

CCIR Report 500-3-82. Improvements in the Performance of Radiotelephone Circuits in the MF and HF Bands Linked Compressor and Expandor Systems—Section 8C—Maritime Mobile Service; Telephony and Related Subjects. 4 pp.

—Single Sideband Radio Telephones—Maritime Mobile Services

CCIR RECMN 475-1-74. Improvements in the Performance of Radiotelephone Circuits in the MF and HF Maritime Mobile Bands—Section 8C—Maritime Mobile Service; Telephony and Related Subjects. 17 pp.

—Single Sideband Radio Telephones—Mobile Radio Services

CCIR RECMN 489-1-78. Technical Characteristics of VHF Radiotelephone Equipment Operating in the Maritime Mobile Service in Channels Spaced by 25 kHz—Section 8C—Maritime Mobile Service; Telephony and Related Subjects. 2 pp.

INTERNATIONAL AND NON-U.S. NATIONAL STANDARDS SUBJECT INDEX

Compandors

Compandors (Cont.)

—**Sound Broadcasting—FM**
CCIR QUESTION 97/10-90. Use of Compandor Systems in Frequency-Modulation Sound Broadcasting—Questions Concerning Study Group 10—Broadcasting Service (Sound). 1 p.

—**Sound-Program Circuits**
CCIR Report 493-3-82. Compandors for Sound Programme Circuits—Section CMTT C—Transmission Standards and Performance Objectives for Sound-Programme Channels. 7 pp.

—**Syllabic—Frequency Division Multiplexers**
CCITT RECMN G.166-89. Characteristics of Syllabic Compandors for Telephony on High Capacity Long Distance Systems—General Characteristics of International Telephone Connections and Circuits (Study Groups XII and XV) 7 pp. 7 pp.

—**Syllabic—Tandem Data Circuits**
CCITT RECMN G.143-89. Circuit Noise and the Use of Compandors—General Characteristics of International Telephone Connections and Circuits (Study Groups XII and XV) 5 pp. 5 pp.

—**Syllabic—Telephone Circuits**
CCITT RECMN G.143-89. Circuit Noise and the Use of Compandors—General Characteristics of International Telephone Connections and Circuits (Study Groups XII and XV) 5 pp. 5 pp.
CCITT RECMN G.166-89. Characteristics of Syllabic Compandors for Telephony on High Capacity Long Distance Systems—General Characteristics of International Telephone Connections and Circuits (Study Groups XII and XV) 7 pp. 7 pp.

—**Syllabic—Telephone Networks**
CCITT RECMN G.166-89. Characteristics of Syllabic Compandors for Telephony on High Capacity Long Distance Systems—General Characteristics of International Telephone Connections and Circuits (Study Groups XII and XV) 7 pp. 7 pp.

—**Telephone Circuits**
CCITT RECMN G.162-89. Characteristics of Compandors for Telephony—General Characteristics of International Telephone Connections and Circuits (Study Groups XII and XV) 8 pp. 8 pp.
CCITT RECMN G.166-89. Characteristics of Syllabic Compandors for Telephony on High Capacity Long Distance Systems—General Characteristics of International Telephone Connections and Circuits (Study Groups XII and XV) 7 pp. 7 pp.
CCITT RECMN M.590-89. Setting up and Lining up a Circuit Fitted with a Compandor—General Maintenance Principles—Maintenance of International Transmission Systems and Telephone Circuits (Study Group IV) 3 pp. 3 pp.
CCITT RECMN M.670-89. Maintenance of a Circuit Fitted with a Compandor—General Maintenance Principles—Maintenance of International Transmission Systems and Telephone Circuits (Study Group IV) 1 pp. 1 p.

—**Telephone Circuits—Compression Ratio**
CCITT RECMN G.162-89. Characteristics of Compandors for Telephony—General Characteristics of International Telephone Connections and Circuits (Study Groups XII and XV) 8 pp. 8 pp.
CCITT RECMN G.166-89. Characteristics of Syllabic Compandors for Telephony on High Capacity Long Distance Systems—General Characteristics of International Telephone Connections and Circuits (Study Groups XII and XV) 7 pp. 7 pp.

—**Telephone Circuits—Connection Stability**
CCITT RECMN G.162-89. Characteristics of Compandors for Telephony—General Characteristics of International Telephone Connections and Circuits (Study Groups XII and XV) 8 pp. 8 pp.
CCITT RECMN G.166-89. Characteristics of Syllabic Compandors for Telephony on High Capacity Long Distance Systems—General Characteristics of International Telephone Connections and Circuits (Study Groups XII and XV) 7 pp. 7 pp.

—**Telephone Circuits—Electrical Impedance**
CCITT RECMN G.162-89. Characteristics of Compandors for Telephony—General Characteristics of International Telephone Connections and Circuits (Study Groups XII and XV) 8 pp. 8 pp.
CCITT RECMN G.166-89. Characteristics of Syllabic Compandors for Telephony on High Capacity Long Distance Systems—General Characteristics of International Telephone Connections and Circuits (Study Groups XII and XV) 7 pp. 7 pp.

—**Telephone Circuits—Expansion Ratio**
CCITT RECMN G.162-89. Characteristics of Compandors for Telephony—General Characteristics of International Telephone Connections and Circuits (Study Groups XII and XV) 8 pp. 8 pp.

Compandors (Cont.)

—**Telephone Circuits—Expansion Ratio (Cont.)**
CCITT RECMN G.166-89. Characteristics of Syllabic Compandors for Telephony on High Capacity Long Distance Systems—General Characteristics of International Telephone Connections and Circuits (Study Groups XII and XV) 7 pp. 7 pp.

—**Telephone Circuits—Harmonic Distortion**
CCITT RECMN G.162-89. Characteristics of Compandors for Telephony—General Characteristics of International Telephone Connections and Circuits (Study Groups XII and XV) 8 pp. 8 pp.
CCITT RECMN G.166-89. Characteristics of Syllabic Compandors for Telephony on High Capacity Long Distance Systems—General Characteristics of International Telephone Connections and Circuits (Study Groups XII and XV) 7 pp. 7 pp.

—**Telephone Circuits—Intermodulation**
CCITT RECMN G.162-89. Characteristics of Compandors for Telephony—General Characteristics of International Telephone Connections and Circuits (Study Groups XII and XV) 8 pp. 8 pp.
CCITT RECMN G.166-89. Characteristics of Syllabic Compandors for Telephony on High Capacity Long Distance Systems—General Characteristics of International Telephone Connections and Circuits (Study Groups XII and XV) 7 pp. 7 pp.

—**Telephone Circuits—Noise**
CCITT RECMN G.162-89. Characteristics of Compandors for Telephony—General Characteristics of International Telephone Connections and Circuits (Study Groups XII and XV) 8 pp. 8 pp.
CCITT RECMN G.166-89. Characteristics of Syllabic Compandors for Telephony on High Capacity Long Distance Systems—General Characteristics of International Telephone Connections and Circuits (Study Groups XII and XV) 7 pp. 7 pp.

—**Telephone Circuits—Return Loss**
CCITT RECMN G.162-89. Characteristics of Compandors for Telephony—General Characteristics of International Telephone Connections and Circuits (Study Groups XII and XV) 8 pp. 8 pp.
CCITT RECMN G.166-89. Characteristics of Syllabic Compandors for Telephony on High Capacity Long Distance Systems—General Characteristics of International Telephone Connections and Circuits (Study Groups XII and XV) 7 pp. 7 pp.

—**Telephone Circuits—Signal Levels**
CCITT RECMN G.162-89. Characteristics of Compandors for Telephony—General Characteristics of International Telephone Connections and Circuits (Study Groups XII and XV) 8 pp. 8 pp.

—**Telephone Circuits—Transient Response**
CCITT RECMN G.162-89. Characteristics of Compandors for Telephony—General Characteristics of International Telephone Connections and Circuits (Study Groups XII and XV) 8 pp. 8 pp.
CCITT RECMN G.166-89. Characteristics of Syllabic Compandors for Telephony on High Capacity Long Distance Systems—General Characteristics of International Telephone Connections and Circuits (Study Groups XII and XV) 7 pp. 7 pp.

—**Underwater Telephone Cables**
CCITT RECMN G.153-89. Characteristics Appropriate to International Circuits More Than 2500 km in Length—General Characteristics of International Telephone Connections and Circuits (Study Groups XII and XV) 4 pp. 4 pp.

Companies
See: Limited Liability Companies; Private Limited Companies; Public Limited Companies

Comparative Analysis
Use: Comparative Testing

Comparative Testing
Use For: Comparative Analysis
DIN ENGL 66054-82. Comparative Testing; Principles Regarding Procedure (Nov). 3 pp.
IEC Guide 46-85. Comparative Testing of Consumer Products and Related Services—General Principles First Edition. 5 pp.
ISO Guide 46-85. Comparative Testing of Consumer Products and Related Services—General Principles First Edition. 5 pp.

—**Food—Sensory Analysis**
BSI BS 5929: Part 2-82. 1982 Methods for Sensory Analysis of Food Part 2: Paired Comparison Test. 10 pp.
BSI BS 5929: Part 5-88. 1988 Methods for Sensory Analysis of Food Part 5: 'A'-'Not A' Test. 9 pp.

Comparative Testing (Cont.)

—**Ionospheric Channel Simulators**
CCIR RECMN 520-1-82. Use of High Frequency Ionospheric Channel Simulators—Section 3Ac—Influence of the Ionosphere. 2 pp.
CCIR RECMN 520-2-92. Use of High Frequency Ionospheric Channel Simulators—Section 3Ac—Influence of the Ionosphere. 4 pp.

—**Modems—Telephone Circuits**
CCITT RECMN V.56-89. Comparative Tests of Modems for Use over Telephone-Type Circuits—Data Communication over the Telephone Network (Study Group XVII) 8 pp. 8 pp.

—**Sensory Analysis**
ISO 5495-83. Sensory Analysis—Methodology—Paired Comparison Test Second Edition. 8 pp.
ISO 8588-87. Sensory Analysis—Methodology—"A"—"Not A" Test First Edition. 8 pp.

Comparators
See Also: Analyzers; Differential Comparators; Equality Comparators; Magnitude Comparators; Measuring Instruments; Voltage Comparators
BSI BS 1054-75. 1975 Engineers' Comparators for External Measurement. 9 pp.

—**Electronic**
CNS B6054-81. Electrical Comparators (Oct)(7956).
JIS B 7536-82. Electrical Comparators. 19 pp.

Compass Base Surveying
Use: Compasses (Indicating Instruments)—Aircraft

Compass Bricks
See Also: Bricks

—**Chimneys**
DIN ENGL 1057 Pt 1-85. Building Materials for Free-Standing Chimneys; Compass Bricks; Requirements, Testing, Inspection (July). 3 pp.

Compass Saws
See Also: Hand Tools; Keyhole Saws; Saws
MOD UK DSTAN 51-11: Part 11-75. Hand Tools, General Purpose Part 11: Saw, Compass Issue 1. 6 pp.

Compasses (Drawing)
See Also: Bow Compasses; Drawing Equipment
CNS Z7107-82. Compass (Half Set) with Interchangeable Point for Precision Drawing Instruments (P) (May)(8888).

—**Adapters (Fittings)**
CNS Z7157-83. Indian Ink Writing and Drawing Instruments, Adapters for Compasses for Ink Writing Pen for Precision Drawing Instruments (P) (Jan)(9933).
ISO 9176-88. Tubular Technical Pens—Adaptor for Compasses First Edition. 7 pp.

—**School**
CNS Z7129-82. Compass with Interchangeable Point for School Drawing Instruments (Dec)(9771).
JIS S 6011-77. Pair of Compasses for School Children.

Compasses (Indicating Instruments)
Use For: Lensatic Compasses; Magnetic Compasses
See Also: Gyrocompasses; Indicating Instruments; Navigational Aids; Surveying Instruments
CNS B6042-81. Pocket Compasses (Apr)(7202).
MOD UK DSTAN 66-5-78. Compass, Magnetic, Unmounted Issue 2. 5 pp.

—**Aircraft**
CAA Chapter P6-8. Compasses (Provisional Airworthiness Requirements for Civil Powered Lift Aircraft). 2 pp.
CAA LEAFLET 8-1 07.90. Compass Base Surveying. 8 pp.

—**Aircraft—Direct Reading**
BSI G 118-49. 1949 Aircraft Magnetic Compass, Direct-Reading. 3 pp.

—**Aircraft—Electromagnetic Interference**
CAA 706 (K). Section K Light Aeroplanes Electrical Supply, Systems and Equipment (Blue Papers). 54 pp.
CAA 707 (G). Section G Rotorcraft Electrical Supply, Systems and Equipment (Blue Papers). 47 pp.
CAA Chapter J4-1 App 09.66. Interference Avoidance of Magnetic Interference with the Compass. 1 p.
CAA Chap G6-14 App #2 11.85. Interference (Rotorcraft). 2 pp.

—**Aircraft—Remote Reading**
BSI G 123-49. 1949 Aircraft Magnetic Compass, Remote Reading (Non-Stabilized). 3 pp.

Compasses (Indicating Instruments) (Cont.)

—Alarms—Ships
BSI BS MA 2: Part 8-85. 1985 Magnetic Compasses and Binnacles Part 8: Transmitting Systems and Off-Course Alarms for Class A Magnetic Compasses. 4 pp.

—Certification
ISO 2269-92. Shipbuilding—Class A Magnetic Compasses, Azimuth Reading Devices and Binnacles—Tests and Certification Second Edition. 26 pp.

—Marine
CNS F5098-86. Marine Magnetic Compasses (Apr)(11535).
JIS F 9101-89. Marine Magnetic Compasses.

—Military
CNS Z7204-87. Lensatic Compass (Military Type) (Mar)(11889).

—Ships
BSI BS MA 2: Part 1-85. 1985 Magnetic Compasses and Binnacles Part 1: Class A Magnetic Compasses, Their Binnacles and Azimuth Reading Devices (Including Test Procedures). 19 pp.
BSI BS MA 2: Part 2-69. 1969 Magnetic Compasses and Binnacles Part 2: Class B Instruments. 6 pp.
BSI BS MA 2: Part 8-85. 1985 Magnetic Compasses and Binnacles Part 8: Transmitting Systems and Off-Course Alarms for Class A Magnetic Compasses. 4 pp.
ISO 449-79. Shipbuilding—Magnetic Compasses and Binnacles, Class A First Edition. 8 pp.
ISO 613-82. Shipbuilding—Magnetic Compasses, Binnacles and Azimuth Reading Devices—Class B First Edition; (Corrigendum 1-1991). 7 pp.
ISO 10316-90. Shipbuilding—Class B Magnetic Compasses—Tests and Certification First Edition. 11 pp.

—Ships—Glossaries
BSI BS MA 2: Part 4-85. 1985 Magnetic Compasses and Binnacles Part 4: Glossary of Terms. 10 pp.
ISO 1069-73. Magnetic Compasses and Binnacles for Sea Navigation—Vocabulary First Edition. 40 pp.

—Ships—Positioning
BSI BS MA 2: Part 3-69. 1969 Amd 1 Magnetic Compasses and Binnacles Part 3: Recommendations for the Positioning of Compasses (AMD 5029) March 31, 1986. 8 pp.
ISO R694-68. Positioning of Magnetic Compasses in Ships First Edition; (Erratum—Aug 1969). 9 pp.

—Type Testing
ISO 2269-92. Shipbuilding—Class A Magnetic Compasses, Azimuth Reading Devices and Binnacles—Tests and Certification Second Edition. 26 pp.

Compatibility (Communications)

—Broadcast Stations—Aeronautical Mobile Services—Radio Navigation
CCIR Decision 71-1-89. Continuation of Studies on Compatibility Between the Aeronautical Radionavigation Service in the Band 108-117.975 MHz, the Aeronautical Mobile (R) Service in the Band 117.975-137 MHz and the FM Sound Broadcasting Stations in the Band About 87-108 MHz—Annex 3 to Volume. 1 p.

—High Definition Television Systems—Television Standards
CCIR QUESTION 69/11-90. Compatibility of the HDTV Standard with Existing and Future Standards—Questions Concerning Study Group 11—Broadcasting Service (Television). 1 p.

—Radio Spectra Utilization
CCIR QUESTION 44-1/1-86. System Models for the Evaluation of Compatibility in Spectrum Use—Questions Concerning Study Group 1—Spectrum Management Techniques (Spectrum Engineering, Planning, Sharing, Monitoring and Utilization). 1 p.

—Sound Broadcasting—VHF—Television Broadcasting
CCIR QUESTION 72/10-90. Compatibility Between VHF Sound Broadcasting and Television Broadcasting Using the Same Band or Adjacent Bands—Questions Concerning Study Group 10—Broadcasting Service (Sound). 1 p.

—Television Broadcasting
CCIR QUESTION 39-1/11-90. Compatibility of the Broadcasting Service (Television) with Other Services—Questions Concerning Study Group 11—Broadcasting Service (Television). 1 p.

Compensation
See Also: Linearity; Radio Transmission; Transient Response

—Doppler Shift—Fixed Satellite Services
CCIR RECMN 730-92. Compensation of the Effects of Switching Discontinuities for Voice Band Data and of Doppler Frequency-Shifts in the Fixed-Satellite Service—Section 4B2—Performance and Availability. 12 pp.

—Military Support
NATO STANAG 3113 ED 5 AMD 5-82. Provision of Support to Visiting Personnel, Aircraft and Vehicles. 18 pp.
NATO STANAG 3113 ED 6 AMD 0-93. Provision of Support to Visiting Personnel, Aircraft and Vehicles. 12 pp.
NATO STANAG 3381 ED 1 AMD 0-89. NATO Standard Procedures for Compensation and Form for Request and Receipt of Support in the Form of Supplies and Services. 17 pp.
NATO STANAG 3381 ED 1 AMD 1-89. NATO Standard Procedures for Compensation and Form for Request and Receipt of Support in the Form of Supplies and Services. 19 pp.

—Multipath Fading—Forward Error Correction
CCIR Report 921-2-90. System Aspects of Digital Ship Earth Stations—Section 8F—Frequencies, Orbits and Systems. 46 pp.

—Switching Discontinuities—Fixed Satellite Services
CCIR RECMN 730-92. Compensation of the Effects of Switching Discontinuities for Voice Band Data and of Doppler Frequency-Shifts in the Fixed-Satellite Service—Section 4B2—Performance and Availability. 12 pp.

Competitive Bidding
See Also: Bids
MOD UK DEFCON Guide no. 10-01. Guidelines on Competitive Tendering 9/86; Addendum 1. 10 pp.

—International
NATO AC/4-D/2261 AMD 0-74. Procedures for International Competitive Bidding for Commonly Financed NATO Infrastructure Works. 43 pp.
NATO AC/4-D/2261-87. Procedures for International Competitive Bidding for Commonly-Financed NATO Infrastructure Works. 55 pp.

Competitive Business Practices
Use: Business Practices

Compilers

—CORAL
MOD UK DSTAN 05-47-83. Computer On-Line Real-Time Applications Language CORAL 66 Specification for Compilers Issue 2. 6 pp.

Completion Ratio (Telephone)
See Also: Call Attempts; Call Completion; Telecommunication Networks; Telephone Networks; Telephone Services

—International Switching Centers
CCITT RECMN E.426-92. General Guide to the Percentage of Effective Attempts Which Should Be Observed for International Telephone Calls—Telephone Network and ISDN Quality of Service, Network Management and Traffic Engineering (Study Group II) 3 pp. 3 pp.

Complexometric Titrations
See Also: Volumetric Analysis

—Aluminum
ISO 2297-73. Chemical Analysis of Aluminium and Its Alloys—Complexometric Determination of Magnesium First Edition. 7 pp.

—Aluminum Alloys
ISO 2297-73. Chemical Analysis of Aluminium and Its Alloys—Complexometric Determination of Magnesium First Edition. 7 pp.

—Chromium Ores
ISO 8889-88. Chromium Ores and Concentrates—Determination of Aluminium Content—Complexometric Method First Edition. 6 pp.

—Elastomers
DIN ENGL 53581-74. Testing of Rubber and Elastomers; Determination of the Zinc Content (Mar). 4 pp.

Complexometric Titrations (Cont.)

—Fuels
DIN ENGL 51769 Pt 5-84. Testing of Petroleum Products; Determination of Lead Content (Total Lead) of Gasolines; Complexometric Method (June). 7 pp.

—Latex
DIN ENGL 53581-74. Testing of Rubber and Elastomers; Determination of the Zinc Content (Mar). 4 pp.

—Potassium Hydroxides
ISO 997-76. Potassium Hydroxide for Industrial Use—Determination of Calcium Content—EDTA (Disodium Salt) Complexometric Method First Edition. 5 pp.

—Rubber
DIN ENGL 53581-74. Testing of Rubber and Elastomers; Determination of the Zinc Content (Mar). 4 pp.

—Sodium Chloride
BSI BS 7319: Part 5-90. 1990 Analysis of Sodium Chloride for Industrial Use Part 5: Method for Determination of Calcium and Magnesium Contents. 6 pp.
ISO 2482-73. Sodium Chloride for Industrial Use—Determination of Calcium and Magnesium Contents—EDTA Complexometric Methods First Edition. 5 pp.

—Sodium Hydroxide
ISO 986-76. Sodium Hydroxide for Industrial Use—Determination of Calcium Content—EDTA (Disodium Salt) Complexometric Method First Edition. 5 pp.

—Sodium Sulfates
ISO 3238-75. Sodium Sulphate for Industrial Use—Determination of Calcium Content—EDTA Complexometric Method First Edition. 5 pp.

—Zinc Alloys
ISO 2741-73. Zinc Alloys—Complexometric Determination of Magnesium First Edition. 5 pp.

Complexometry
Use: Complexometric Titrations

Composite Beams
Use: Beams (Supports)

Composite Bridges
Use: Bridges (Structures)

Composite Columns
Use: Columns (Supports)

Composite Construction
Use: Composite Structures

Composite Materials
Scope Note: For additional listings, use a more specific term See Also: Aerospace Composite Materials; Ceramic Matrix Composites; Cermet; Coatings; Composite Structures; Copolymers; Epoxy Laminates; Fiber Composites; Fiberglass Reinforced Cements; Fiberglass Reinforced Plastics; Fibers; Foil Papers; Laminated Glass; Laminated Plastics; Laminated Wood; Laminates; Metals; Phenolic Laminates; Plastics; Plywood; Polymers; Reinforced Concrete; Sheet Laminates; Thermosetting Resins

—Boards
CNS O1009-59. Dimension of Composite Boards (May)(889). 1 p.

—Electric Conduits
IEC 614 Pt 2-6-92. Specification for Conduits for Electrical Installations Part 2: Particular Specifications for Conduits Section 6: Pliable Conduits of Metal or Composite Materials First Edition. 37 pp.

—Metals—Glossaries
JIS H 7006-91. Glossary of Terms Used in Metal Matrix Composites. 13 pp.

Composite Piles
See Also: Composite Structures; Concrete Piles

—Construction
DIN ENGL 4128-83. Small Diameter Injection Piles (Cast-In-Place Concrete Piles and Composite Piles); Construction Procedure, Design and Permissible Loading (Apr). 7 pp.

Composite Piles (Cont.)
—Loads (Forces)
DIN ENGL 4128-83. Small Diameter Injection Piles (Cast-In-Place Concrete Piles and Composite Piles); Construction Procedure, Design and Permissible Loading (Apr). 7 pp.

Composite Structures
Use For: Composite Construction
See Also: Columns (Supports); Composite Piles; Concrete Construction

—Glossaries
DIN ENGL 1080 Pt 4-80. Terms, Symbols and Units Used in Civil Engineering; Concrete Construction, Composite Steel Construction and Steel Girders in Concrete (Mar). 10 pp.

—Steel Construction—Glossaries
DIN ENGL 1080 Pt 4-80. Terms, Symbols and Units Used in Civil Engineering; Concrete Construction, Composite Steel Construction and Steel Girders in Concrete (Mar). 10 pp.

—Steel Construction—Symbols
DIN ENGL 1080 Pt 4-80. Terms, Symbols and Units Used in Civil Engineering; Concrete Construction, Composite Steel Construction and Steel Girders in Concrete (Mar). 10 pp.

Composition Resistors
Use: Carbon Resistors

Composts
—Fertilizer
CNS N3020-85. Garbage Compost, Fertilizer Grade (Dec)(3960).

Compressed Air
See Also: Air; Compressors (Pressure Equipment); Pneumatic Equipment

—Aircraft
ISO 2434-73. Compressed Non-Breathing Air for Use in Aircraft First Edition. 3 pp.
MOD UK DSTAN 16-7-90. Compressed Air, Characteristics, Supply, Pressure and Hoses for Technical Purposes Issue 3. 7 pp.
MOD UK DSTAN 16-8-01. Compressed Breathing Air for Aircraft Systems Issue 3; Amendment 1. 8 pp.
MOD UK DSTAN 16-8-93. Air, Compressed (Breathing) for Aircraft Systems Issue 4. 7 pp.

—Aircraft—Interdepartmental Procurement
NATO STANAG 3054 ED 4 AMD 4-73. Characteristics of Compressed Air for Technical Purposes, Supply Pressure and Hoses. 6 pp.
NATO STANAG 3054 ED 5 AMD 0-89. Characteristics of Compressed Air for Technical Purposes, Supply Pressure and Hoses. 6 pp.
NATO STANAG 3054 ED 5 AMD 1-89. Characteristics of Compressed Air for Technical Purposes, Supply Pressure and Hoses. 6 pp.

—Breathing
CSA CAN3-Z180.1-M85. Compressed Breathing Air and Systems; (Gen Instr 1). 34 pp.
NATO STANAG 3611 ED 2 AMD 4-73. Compressed Breathing Air Characteristics. 6 pp.

—Breathing—Purity
SAA AS 2568-91. Medical Gases—Purity of Compressed Medical Breathing Air. 4 pp.

—Breathing—Scuba Diving Equipment
JIS S 7306-89. Compressed Air for SCUBA Diving. 9 pp.

—Contamination—Classification
ISO 8573 Pt 1-91. Compressed Air for General Use—Part 1: Contaminants and Quality Classes First Edition. 13 pp.

—Safety
BSI BS 6158-82. 1982 Safety Devices for Fuel Gases and Oxygen or Compressed Air for Welding, Cutting and Related Processes. 12 pp.
CEN PREN 730-92. Equipment Used in Gas Welding, Cutting and Allied Processes—Safety Devices for Fuel Gases and Oxygen or Compressed Air—General Specifications, Requirements and Tests. 15 pp.
CSA CAN/CSA-Z275.3-M86. Occupational Safety Code for Construction Work in Compressed Air. 46 pp.
SAA AS CA12-70. Rules for Work in Compressed air (Known as the SAA Compressed Air Code). 58 pp.

Compressed Air Cylinders
Use For: Compressed Air Storage Containers
See Also: Compressed Air Equipment

BSI BS 6005-93. 1993 Moulded Transparent Polycarbonate Bowls Used in Compressed Air Filters and Lubricators (E). 10 pp.
BSI BS 6005-81. 1981 Amd 1 Moulded Transparent Polycarbonate Bowls Used in Compressed Air Filters and Lubricators. 8 pp.

—Internal Combustion Engines
DIN ENGL 6274-82. Internal Combustion Engines for General Purposes; Compressed Air Containers with Valve Block; 38 mm Bore; Assembly (Apr). 6 pp.
DIN ENGL 6275-82. Internal Combustion Engines for General Purposes; Compressed Air Containers for Permissible Working Overpressures up to 30 Bar (Apr). 4 pp.

Compressed Air Dryers
See Also: Compressed Air Equipment; Dryers
BSI BS 6754-86. 1986 Specifications and Testing of Compressed Air Dryers. 18 pp.
ISO 7183-86. Compressed Air Dryers—Specifications and Testing First Edition. 16 pp.

Compressed Air Equipment
See Also: Compressed Air Cylinders; Compressed Air Dryers; Pneumatic Equipment

—Cleanliness—Naval Ships
MOD UK NES 373-87. Code of Practice for Compressed Air System Cleanliness Issue 1 (07.87). 46 pp.

—Cleanliness—Submarines
MOD UK NES 373-87. Code of Practice for Compressed Air System Cleanliness Issue 1 (07.87). 46 pp.

—Cleanliness—Weapons Systems
MOD UK NES 373-87. Code of Practice for Compressed Air System Cleanliness Issue 1 (07.87). 46 pp.

Compressed Air Hoses
Use: Pneumatic Hoses

Compressed Air Storage Containers
Use: Compressed Air Cylinders

Compressed Gas Containers
Use: Gas Cylinders

Compressed Gases
—Highway Transportation—Emergency Procedures
SAA AS 1678. Emergency Procedure Guide—Transport Group Text EPGs for Class 2 Substances—Compressed and Liquefied Gases Pads of 20 Forms.
SAA AS 1678.2A1-87. Emergency Procedure Guide—Transport Group Text EPGs for Class 2 Substances—Compressed and Liquefied Gases—Part 2A1: Flammable, Compressed Gas (Supersedes AS 1678.2.1.007—1983 and AS 1678.2.1.009—1987).
SAA AS 1678.2A4-87. Emergency Precedure Guide—Transport Group Text EPGs for Class 2 Substances—Compressed and Liquefied Gases—Part 2A4: Flammable, Poisonous Gas.
SAA AS 1678.2A5-87. Emergency Precedure Guide—Transport Group Text EPGs for Class 2 Substances—Compressed and Liquefied Gases—Part 2A5: Flammable, Poisonous—Special Low-Temperature Control Gas.
SAA AS 1678.2B1-87. Emergency Precedure Guide—Transport Group Text EPGs for Class 2 Substances—Compressed and Liquefied Gases—Part 2B1: Poisonous (Non-Flammable) Gas.
SAA AS 1678.2B2-87. Emergency Procedure Guide—Transport Group Text EPGs for Class 2 Substances—Compressed and Liquefied Gases—Part 2B2: Poisonous, Flammable Gas.
SAA AS 1678.2B3-87. Emergency Procedure Guide—Transport Group Text EPGs for Class 2 Substances—Compressed and Liquefied Gases—Part 2B3: Poisonous Gas (Will Burn, Corrode) (Supersedes AS 1678.8.0.010—1984 (in Part)).
SAA AS 1678.2B4-87. Emergency Procedure Guide—Transport Group Text EPGs for Class 2 Substances—Compressed and Liquefied Gases—Part 2B4: Poisonous, Flammable (Spontaneously Combustible) Gas.
SAA AS 1678.2B5-87. Emergency Procedure Guide—Transport Group Text EPGs for Class 2 Substances—Compressed and Liquefied Gases—Part 2B5: Poisonous, Oxidizing Gas.
SAA AS 1678.2B6-87. Emergency Precedure Guide—Transport Group Text EPGs for Class 2 Substances—Compressed and Liquefied Gases—Part 2B6: Poisonous, Oxidizing Gas (Flourine-Type).
SAA AS 1678.2B7-87. Emergency Procedure Guide—Transport Group Text EPGs for Class 2 Substances—Compressed and Liquefied Gases—Part 2B7: Poisonous, Oxiding Gas, Corrosive Gas.
SAA AS 1678.2B8-87. Emergency Procedure Guide—Transport Group Text EPGs for Class 2 Substances—Compressed and Liquefied Gases—Part 2B8: Poisonous and Corrosive Gas.
SAA AS 1678.2C1-87. Emergency Procedure Guide—Transport Group Text EPGs for Class 2 Substances—Compressed and Liquefied Gases—Part 2C1: Non-Flammable, Compressed Gas.
SAA AS 1678.2C5-87. Emergency Procedure Guide—Transport Group Text EPGs for Class 2 Substances—Compressed and Liquefied Gases—Part 2C5: Oxidizing, Compressed Gas—Breathable, Non-Flammable.
SAA AS 1678.2C6-87. Emergency Procedure Guide—Transport Group Text EPGs for Class 2 Substances—Compressed and Liquefied Gases—Part 2C6: Non-Flammable, Oxidizing, Compressed Gas—Breathable.
SAA AS 1678.2C8-87. Emergency Procedure Guide—Transport Group Text EPGs for Class 2 Substances—Compressed and Liquefied Gases—Part 2C8: Non-Flammable, Oxidizing, Compressed Gas—Asphyxiating.
SAA AS 1678.2D1-87. Emergency Procedure Guide—Transport Group Text EPGs for Class 2 Substances—Compressed and Liquefied Gases—Part 2D1: Aerosol Dispensers.

Compressibility Factors
See Also: Compactibility; Thermodynamic Properties

—Interlaminar Insulation—Magnetic Materials
BSI BS 6404: Part 12-93. 1993 Magnetic Materials Part 12: Guide to Methods of Assessment of Temperature Capability of Interlaminar Insulation Coatings (IEC 404-12: 1992) (V). 16 pp.
IEC 404 Pt 12-92. Magnetic Materials Part 12: Guide to Methods of Assessment of Temperature Capability of Interlaminar Insulation Coatings First Edition. 28 pp.

Compression Connectors
Use For: Press On Connectors
See Also: Compression Couplings; Connectors; Solderless Connections

—Power Cables
BSI BS 7609-92. 1992 Installation and Inspection of Uninsulated Compression and Mechanical Connectors for Power Cables with Copper or Aluminium Conductors (E). 17 pp.
IEC 1238 Pt 1-93. Compression and Mechanical Connectors for Power Cables with Copper or Aluminium Conductors Part 1: Test Methods and Requirements First Edition. 65 pp.

—Power Line Hardware
BSI BS 4579: Part 3-76. 1976 Amd 1 Performance of Mechanical and Compression Joints in Electric Cable and Wire Connectors Part 3: Mechanical and Compression Joints in Aluminium Conductors. 21 pp.

Compression Couplings
See Also: Compression Connectors; Couplings
BSI BS 5114-75. 1975 Amd 2 Performance Requirements for Joints and Compression Fittings for Use with Polyethylene Pipes (AMD 5336) July 31, 1987. 17 pp.
DIN ENGL 2353-91. Compression Fittings and Couplings (June). 16 pp.
DIN ENGL 3859-84. Compression Couplings; Technical Delivery Conditions (Mar). 6 pp.
DIN ENGL 3900-84. Solderless Compression Couplings with Olive; Compression Couplings with Tapered Stud End (Apr). 3 pp.
DIN ENGL 3901-87. Olive Type Compression Couplings; Male Stud Compression Couplings (Sept). 4 pp.
DIN ENGL 3903-84. Solderless Compression Couplings with Olive; Taper Stud Elbows (Apr). 3 pp.
DIN ENGL 3906-84. Solderless Compression Couplings with Olive; Taper Stud Tees (Stud Branch) (Apr). 3 pp.
DIN ENGL 3910-84. Solderless Compression Couplings with Olive; Male Stud Bulkhead Compression Couplings (Apr). 3 pp.
DIN ENGL 3911-84. Solderless Compression Couplings with Olive; Elbow Bulkhead Compression Couplings (Apr). 3 pp.
DIN ENGL 3912-84. Solderless Compression Couplings with Olive; Straight Bulkhead Compression Couplings (Apr). 2 pp.
DIN ENGL 3913-84. Solderless Compression Couplings with Olive; Taper Stud Tees (Stud Run) (Apr). 3 pp.
DIN ENGL 7632-85. Compression Couplings with Spherical Liner; Reducers (Jan). 3 pp.

INTERNATIONAL AND NON-U.S. NATIONAL STANDARDS
SUBJECT INDEX
Compression

Compression Couplings (Cont.)
—Bolts
DIN ENGL 7644-85. Compression Couplings with Spherical Liner; Male Studed Banjo Bolts for Ring Type Nipples (Jan). 4 pp.

—Straight
DIN ENGL 3902-84. Solderless Compression Couplings with Olive; Straight Compression Couplings (Apr). 3 pp.

DIN ENGL 3912-84. Solderless Compression Couplings with Olive; Straight Bulkhead Compression Couplings (Apr). 2 pp.

DIN ENGL 7631-85. Compression Couplings with Spherical Liner; Straight Compression Couplings (Jan). 3 pp.

—Tubes
BSI BS 4368: Part 1-72. 1972 Amd 4 Carbon and Stainless Steel Compression Part 1: Couplings for Tubes Heavy Series. 43 pp.

BSI BS 4368: Part 3-74. 1974 Amd 3 Carbon and Stainless Steel Compression Couplings for Tubes Part 3: Compression Couplings for Tubes. 22 pp.

BSI BS 4368: Part 4-84. 1984 Amd 1 Carbon and Stainless Steel Compression Couplings for Tubes Part 4: Type Test Requirements. 9 pp.

—Valves
DIN ENGL 3202 Pt 5-84. Face-to-Face and Centre-to-Face Dimensions of Valves; Valves for Connection with Compression Couplings (Sept). 2 pp.

—Weld On
DIN ENGL 3909-84. Solderless Compression Couplings with Olive; Weld-On Compression Couplings (Apr). 3 pp.

Compression Fittings
Scope Note: Includes components such as couplings, unions, nipples, tees, bends, and elbows for connecting pipes **See Also:** Compression Joints; Pipe Fittings

DIN ENGL 2353-91. Compression Fittings and Couplings (June). 16 pp.

—Bolts
DIN ENGL 7644-85. Compression Couplings with Spherical Liner; Male Studed Banjo Bolts for Ring Type Nipples (Jan). 4 pp.

—Bulkhead
DIN ENGL 3910-84. Solderless Compression Couplings with Olive; Male Stud Bulkhead Compression Couplings (Apr). 3 pp.

DIN ENGL 3911-84. Solderless Compression Couplings with Olive; Elbow Bulkhead Compression Couplings (Apr). 3 pp.

DIN ENGL 3912-84. Solderless Compression Couplings with Olive; Straight Bulkhead Compression Couplings (Apr). 2 pp.

—Crosses
DIN ENGL 7640-85. Compression Couplings with Spherical Liner; Equal Crosses (Jan). 3 pp.

—Elbows
DIN ENGL 3904-87. Olive Type Compression Couplings; Male Stud Elbows (Sept). 4 pp.

DIN ENGL 3905-84. Solderless Compression Couplings with Olive; Equal Elbows (Apr). 3 pp.

DIN ENGL 3911-84. Solderless Compression Couplings with Olive; Elbow Bulkhead Compression Couplings (Apr). 3 pp.

DIN ENGL 3914-87. Olive Type Compression Couplings; Male Stud Tees (Stud Run) (Sept). 4 pp.

DIN ENGL 3951-84. Solderless Compression Couplings with Olive; Equal Crosses (Apr). 3 pp.

DIN ENGL 3952-84. Solderless Compression Couplings with Olive; Male Stud Elbow Assemblies (Apr). 3 pp.

DIN ENGL 3954-84. Solderless Compression Couplings with Olive; Male Stud Tee (Stud Run) Assemblies (Apr). 3 pp.

DIN ENGL 7638-85. Compression Couplings with Spherical Liner; Equal Elbows (Jan). 3 pp.

—Male Stud
DIN ENGL 3955-87. Olive Type Compression Couplings; Male Stud Coupling Assemblies (Sept). 4 pp.

DIN ENGL 7647-85. Compression Couplings with Spherical Liner; Male Stud Compression Couplings (Jan). 3 pp.

—Nipples (Fittings)
DIN ENGL 7633-85. Compression Couplings with Spherical Liner; Brazing Nipples (Jan). 2 pp.

—Pipes—Copper
BSI BS 864: Part 5-90. 1990 Capillary and Compression Tube Fittings of Copper and Copper Alloy Part 5: Compression Fittings for Polyethylene Pipes with Outside Diameters to BS 5556. 12 pp.

Compression Fittings (Cont.)
—Pipes—Copper (Cont.)
BSI BS 2051: Part 1-73. (OBSOLESCENT) 1973 Amd 2 Tube and Pipe Fittings for Engineering Purposes Part 1: Copper and Copper Alloy Capillary and Compression Tube Fittings for Engineering Purposes. 19 pp.

BSI BS 2051: Part 2-84. (OBSOLESCENT) 1984 Tube and Pipe Fittings for Engineering Purposes Part 2: Specification for Olive Type Copper Alloy Compression Tube Fittings. 15 pp.

SNZ NZS/BS 2051: Part 1-73. Specification for Tube and Pipe Fittings for Engineering Purposes Part 1: Copper and Copper Alloy Capillary and Compression Tube Fittings for Engineering Purposes. 20 pp.

—Pipes—Metal
BSI BS 5114-75. 1975 Amd 2 Performance Requirements for Joints and Compression Fittings for Use with Polyethylene Pipes (AMD 5336) July 31, 1987. 17 pp.

DIN ENGL 8076 Pt 1-84. Thermoplastics Pressure Pipelines; Metal Compression Fittings for Polyethylene (PE) Pipes; General Quality Requirements; Testing (Mar). 5 pp.

—Plastic
BSI BS 5114-75. 1975 Amd 2 Performance Requirements for Joints and Compression Fittings for Use with Polyethylene Pipes (AMD 5336) July 31, 1987. 17 pp.

—Reducers
DIN ENGL 7632-85. Compression Couplings with Spherical Liner; Reducers (Jan). 3 pp.

—Refrigeration Equipment
DIN ENGL 8906-90. PN 40 Flared Flange Solderless Compression Couplings for Use in Refrigerating Systems (June). 15 pp.

—Steel
SNZ NZS 202-66. Specification for Steel Pipes and Joints for Hydraulic Purposes Amend: 3A, 1977. 16 pp.

—Stoppers—Threaded
DIN ENGL 3852 Pt 1-92. Stud Ends and Tapped Holes with Metric Fine Pitch Thread, for Use with Compression Couplings, Valves and Screw Plugs; Dimensions (Feb). 10 pp.

DIN ENGL 3852 Pt 2-91. Stud Ends and Tapped Holes with Pipe Thread, for Use with Compression Couplings, Valves and Screw Plugs; Dimensions (Nov). 7 pp.

DIN ENGL 3852 Pt 3-91. Type F Stud Ends and Type W Tapped Holes with Metric Fine Pitch Thread, for Use with Compression Couplings, Valves and Screw Plugs, Sealed by O-Ring; Dimensions (Nov). 6 pp.

—Tapped Holes
DIN ENGL 3852 Pt 1-92. Stud Ends and Tapped Holes with Metric Fine Pitch Thread, for Use with Compression Couplings, Valves and Screw Plugs; Dimensions (Feb). 10 pp.

—Tees
DIN ENGL 3907-87. Olive Type Compression Couplings; Male Stud Tees (Stud Branch) (Sept). 4 pp.

DIN ENGL 3908-84. Solderless Compression Couplings with Olive; Equal Tees (Apr). 3 pp.

DIN ENGL 3953-84. Solderless Compression Couplings with Olive; Male Stud Tee (Stud Branch) Assemblies (Apr). 3 pp.

DIN ENGL 7639-85. Compression Couplings with Spherical Liner; Equal Tees (Jan). 3 pp.

—Tubes—Copper
BSI BS 864: Part 2-83. 1983 Amd 2 Capillary and Compression Tube Fittings of Copper and Copper Alloy Part 2: Capilliary and Compression Fittings for Copper Tubes (AMD 5651) April 30, 1987. 22 pp.

BSI BS 864: Part 2-03. 1983 Amd 3 Capillary and Compression Tube Fittings of Copper and Copper Alloy Part 2: Specification for Capillary and Compression Fittings for Copper Tubes (AMD 7067) February 28, 1992. 25 pp.

BSI BS 864: Part 5-90. 1990 Capillary and Compression Tube Fittings of Copper and Copper Alloy Part 5: Compression Fittings for Polyethylene Pipes with Outside Diameters to BS 5556. 12 pp.

BSI BS 4368: Part 1-72. 1972 Amd 4 Carbon and Stainless Steel Compression Part 1: Couplings for Tubes Heavy Series. 43 pp.

BSI BS 4368: Part 3-74. 1974 Amd 3 Carbon and Stainless Steel Compression Couplings for Tubes Part 3: Compression Couplings for Tubes. 22 pp.

Compression Fittings (Cont.)
—Tubes—Copper Alloy
BSI BS 864: Part 2-83. 1983 Amd 2 Capillary and Compression Tube Fittings of Copper and Copper Alloy Part 2: Capilliary and Compression Fittings for Copper Tubes (AMD 5651) April 30, 1987. 22 pp.

BSI BS 864: Part 2-03. 1983 Amd 3 Capillary and Compression Tube Fittings of Copper and Copper Alloy Part 2: Specification for Capillary and Compression Fittings for Copper Tubes (AMD 7067) February 28, 1992. 25 pp.

—Tubes—Stainless Steel
BSI BS 4368: Part 1-72. 1972 Amd 4 Carbon and Stainless Steel Compression Part 1: Couplings for Tubes Heavy Series. 43 pp.

BSI BS 4368: Part 3-74. 1974 Amd 3 Carbon and Stainless Steel Compression Couplings for Tubes Part 3: Compression Couplings for Tubes. 22 pp.

Compression Ignition Engines
Use: Diesel Engines

Compression Joints
See Also: Compression Fittings; Joints

—Copper Conductors
BSI BS 4579: Part 1-70. 1970 Amd 3 Performance of Mechanical and Compression Joints in Electric Cable and Wire Connectors Part 1: Compression Joints in Copper Conductors (AMD 4687) March 29, 1985. 12 pp.

BSI BS 4579: Part 2-73. 1973 Amd 1 Performance of Mechanical and Compression Joints in Electric Cable and Wire Connectors Part 2: Compression Joints in Nickel, Iron and Platted Copper Conductors. 11 pp.

Compression Molding Machines
See Also: Molding Machines

—Plastics
CNS B1298-86. Dimensions Relating to Molds for Plastics Compression Molding and Plastics Transfer Molding Machines (Feb)(9966).

JIS B 6702-92. Dimensions Relating to Molds for Plastics Compression Molding Machines and Plastics Transfer Molding Machines. 13 pp.

—Plastics—Safety
CEN PREN 289-89. Technical Safety Requirements for the Design and Construction of Compression and Transfer Moulding Machines for Plastics and Rubber. 22 pp.

—Rubber—Safety
CEN PREN 289-89. Technical Safety Requirements for the Design and Construction of Compression and Transfer Moulding Machines for Plastics and Rubber. 22 pp.

Compression Ratio (Transmission Gain)
See Also: Expansion Ratio (Transmission Gain); Signal Levels

—Compandors—Telephone Circuits
CCITT RECMN G.162-89. Characteristics of Compandors for Telephony—General Characteristics of International Telephone Connections and Circuits (Study Groups XII and XV) 8 pp. 8 pp.

CCITT RECMN G.166-89. Characteristics of Syllabic Compandors for Telephony on High Capacity Long Distance Systems—General Characteristics of International Telephone Connections and Circuits (Study Groups XII and XV) 7 pp. 7 pp.

Compression Rings
Use: Retaining Rings

Compression Rivets
See Also: Rivets

CNS B2581-81. Compression Rivets (Jun)(7569).

Compression Springs
See Also: Springs (Elastic)

BSI BS 7568: Part 4-92. 1992 Press Tools Part 4: Compression Springs with Rectangular Section—Housing Dimensions and Colour Coding (ISO 10243: 1991). 22 pp.

ISO 10243-91. Tools for Pressing—Compression Springs with Rectangular Section—Housing Dimensions and Colour Coding First Edition. 20 pp.

—Brake Blocks
DIN ENGL 37082-75. Adjusting Equipment for Brake Blocks; Compression Springs; Spring Bolt (Sept). 1 p.

Compression Springs (Cont.)

—Color Coding
BSI BS 7568: Part 4-92. 1992 Press Tools Part 4: Compression Springs with Rectangular Section—Housing Dimensions and Colour Coding (ISO 10243: 1991). 22 pp.

ISO 10243-91. Tools for Pressing—Compression Springs with Rectangular Section—Housing Dimensions and Colour Coding First Edition. 20 pp.

—Helical
CNS B1258-81. Helical Springs Made of Round Wire; Information to be Given in Order to Obtain the Tension Springs Required (Jun)(7590).

CNS B1259-81. Helical Springs Made of Round Wire and Rod; Information to be Given in Order to Obtain the Compression Springs Required (Jun)(7591).

CNS B1260-81. Helical Compression Springs Made of Flat Barsteel Calculation (Jun)(7592).

CNS B1269-81. Calculation and Design for Helical Compression Spring Circular Wire or Bar (Sep)(7857).

CNS B2593-81. Helical Springs Made of Round Wire; Dimensions for Cold Coiled Compression Springs of Less Than 9.5mm Wire Diameter (Jun)(7593).

CNS B2594-81. Helical Springs Made of Round Wire; Dimensions for Cold Coiled Compression Springs with Wire Diameter from 0.5 Upwards (Jun)(7594).

CNS B2596-81. Quality Specifications for Cold Coiled Compression Springs Made of Round Wire (Jun)(7596). 9 pp.

CNS B2598-82. Helical Compression Spring Made of Round Rod; Quality Requirements for Mass Productions (Nov)(7855).

DIN ENGL 2089 Pt 1-84. Helical Compression Springs Made from Round Wire or Rod; Calculation and Design (Dec). 28 pp.

DIN ENGL 2090-71. Helical Compression Springs Made of Flat Bar Steel; Calculation (Jan). 2 pp.

DIN ENGL 2098 Pt 1-68. Helical Springs Made of Round Wire; Dimensions for Cold Coiled Compression Springs with Wire Diameter from 0.5 mm Upwards (Oct). 4 pp.

DIN ENGL 2098 Pt 2-70. Helical Springs Made of Round Wire; Dimensions for Cold-Coiled Compression Springs of Less Than 0.5 mm Wire Diameter (Aug). 4 pp.

DIN ENGL 2099 Pt 1-73. Helical Springs Made of Round Wire and Rod; Data for Compression Springs; Printed Form (Nov). 3 pp.

JIS B 2704-87. Design of Helical Compression and Extension Springs.

JIS B 2704-78. Design of Helical Compression and Extension Springs. 30 pp.

JIS B 2707-87. Cold Coiled Helical Compression Springs. 11 pp.

—Helical—Design
BSI BS 1726: Part 1-87. 1987 Amd 2 Coil Springs Part 1: Guide for the Design of Helical Compression Springs (AMD 6218) April 30, 1990. 32 pp.

—Helical—Harness Release Gears—Aircraft
SBAC AS 2817 ISSUE 4. Catch Spring for Harness Release Gear. 1 p.

—Helical—Quality Assurance
DIN ENGL 2095-73. Helical Springs Made of Round Wire; Quality Specifications for Cold Coiled Compression Springs (May). 8 pp.

DIN ENGL 2096 Pt 1-81. Helical Compression Springs Made of Round Wire and Rod; Quality Requirements for Hot Formed Compression Springs (Nov). 7 pp.

DIN ENGL 2096 Pt 2-79. Cylindrical Coil Compression Springs Made from Round Rods; Quality Requirements for Mass Production (Jan). 7 pp.

—Testing Equipment
DIN ENGL 51232-86. Materials Testing Machines; Spring Testing Machines for the Static Testing of Springs; General Requirements (May). 4 pp.

JIS B 7738-84. Testing Machines for Helical Compression and Extension Springs. 11 pp.

Compression Testers
See Also: Stress (Testing); Testing Equipment

CNS B6073-82. Compression Testing Machines (Aug)(9211).

CNS B7197-82. Method of Test for Compression Testing Machines (Aug)(9212).

DIN ENGL 51223-87. Materials Testing Machines; Compression Testing Machines; Requirements (Oct). 6 pp.

DIN ENGL 51302 Pt 1-85. Materials Testing Machines; Verification of Tensile, Compression and Bend Testing Machines; Principles (Mar). 6 pp.

Compression Testers (Cont.)
DIN ENGL 51302 Pt 2-86. Materials Testing Machines; Verification of Tensile, Compression and Bend Testing Machines; Supplementary Verification of Compression Testing Machines for Building Materials (Aug). 6 pp.

JIS B 7733-92. Compression Testing Machines. 14 pp.

—Concretes
BSI BS 1881: Part 115-86. 1986 Amd 1 Testing Concrete Part 115: Specification for Compression Testing Machines for Concrete (AMD 6536) December 21, 1990. 12 pp.

—Packaging
BSI BS 4826: Part 9-86. 1986 Complete, Filled Transport Packages Part 9: Method of Test for Stacking Using Compression Tester (Renumbered as BS EN 22874: 1993). 4 pp.

BSI BS EN 22874-93. 1993 Amd 1 Complete, Filled Transport Packages. Method of Test for Stacking Using Compression Tester (ISO 2874: 1985) (AMD 7460) March 15, 1993 (H). 11 pp.

CEN EN 22874-92. Packaging—Complete, Filled Transport Packages—Stacking Test Using Compression Tester (ISO 2874: 1985). 5 pp.

ISO 2874-85. Packaging—Complete, Filled Transport Packages—Stacking Test Using Compression Tester Second Edition; (CEN EN 22874: 1992). 4 pp.

—Papers
SAA AS 1301.449S-92. Methods of Test for Pulp and Paper—Part 449s: Description of Crush Testing Equipment. 3 pp.

SNZ NZS/AS 1301. 449S-92. Methods of Test for Pulp and Paper Description of Crush Testing Equipment (This is a Joint Standard with SAA AS 1301.449S). 3 pp.

Compression Testing
Use For: Biaxial Compression Testing
See Also: Creep Properties; Ductility; Dynamic Testing; Fatigue Testing; Flattening Testing; Green Strength; Hardness Testing; High Temperature Testing; Hydrostatic Testing; Impact Testing; Indentation Hardness Testing; Low Temperature Testing; Shear Testing; Stacking Testing; Static Testing

—Adhesives
BSI BS 5350: Part C15-90. 1990 Methods of Test for Adhesives Group C: Adhesively Bonded Joints: Mechanical Tests Part C15: Determination of Bond Strength in Compressive Shear. 5 pp.

CNS K6505-80. Methods of Test for Strength Properties of Adhesives in Shear by Compression Loading (Jul)(5809).

JIS K 6852-76. Testing Methods for Strength Properties of Adhesives in Shear by Compression Loading. 5 pp.

—Aggregates
BSI BS 812: Part 110-90. 1990 Testing Aggregates Part 110: Method for Determination of Aggregate Crushing Value (ACV). 12 pp.

BSI BS 812: Part 111-90. 1990 Testing Aggregates Part 111: Methods for Determination of Ten per Cent Fines Value (TFV). 13 pp.

SAA AS 1141.51-85. Methods for Sampling and Testing Aggregates—Part 51: Unconfined Compressive Strength of Compacted Bound Materials. 4 pp.

SNZ NZS 4407: Part 3.10-91. Methods of Sampling and Testing Road Aggregates Part 3: Methods of Testing Road Aggregates—Laboratory Tests Test 3.10: The Crushing Resistance of Coarse Aggregate Under a Specific Load. 4 pp.

—Asphalts
DIN ENGL 1996 Pt 12-85. Testing of Asphalt; Compression Testing of Mastic Asphalt (Feb). 3 pp.

—Bearing Alloys
ISO 4385-81. Plain Bearings—Compression Testing of Metallic Bearing Materials First Edition. 6 pp.

—Bearings
CNS Z8043-82. Method for Determining Radial Crushing Strength Constant of Metal Powder Sintered Bearing, Oil Impregnated (Jul)(9207).

JIS Z 2507-89. Method for Determination of Radial Crushing Strength Constant of Metal Powder Sintered Bearing, Oil Impregnated. 7 pp.

—Bitumens
BSI BS 598: Part 104-89. 1989 Amd 1 Sampling and Examination of Bituminous Mixtures for Roads and Other Paved Areas Part 104: Methods of Test for the Determination of Density and Compaction (AMD 6738) September 30, 1991 (Supersedes BS 598: Part 3: 1985). 17 pp.

DIN ENGL 1996 Pt 11-81. Testing of Bituminous Materials for Road Building and Related Purposes; Determination of Marshall Stability and of Marshall Flow Value (July). 4 pp.

Compression Testing (Cont.)

—Boxes (Containers)
SAA AS 1301.800S-87. Methods of Test for Pulp and Paper (Metric Units)—Part 800s: Compression Resistance of Fibreboard Boxes (Cases) (This is a Joint Standard with SANZ NZS 1301). 2 pp.

SNZ NZS/AS 1301. 800S-87. Methods of Test for Pulp and Paper Compression Resistance of Fibreboard Boxes (Cases) (This is a Joint Standard with SAA AS 1301.800S). 2 pp.

—Brake Linings
BSI BS AU 180: Part 1-83. 1983 Brake Linings Part 1: Method for Measurement of Compressibility of Brake Lining Material. 5 pp.

ISO 6310-81. Road Vehicles—Brake Linings—Compressibility—Test Procedure First Edition. 6 pp.

—Brake Pads
BSI BS AU 180: Part 1-83. 1983 Brake Linings Part 1: Method for Measurement of Compressibility of Brake Lining Material. 5 pp.

CNS D3177-87. Method of Test of Compressibility for Disc Brake Pads of Automobiles (Dec)(12173).

JIS D 4413-86. Test Procedure of Compressibility for Disc Brake Pads of Automobiles. 9 pp.

—Brake Shoes
BSI BS AU 180: Part 1-83. 1983 Brake Linings Part 1: Method for Measurement of Compressibility of Brake Lining Material. 5 pp.

—Bricks
CEN PREN 772-1-92. Methods of Test for Masonry Units—Part 1: Determination of Compressive Strength. 13 pp.

SAA AS 1226.4-84. Methods of Sampling and Testing Clay Building Bricks—Part 4: Method for Determining Compressive Strength. 3 pp.

SNZ NZS 366-63. Clay Building Bricks. 11 pp.

—Building Stones
CNS A3226-85. Method of Test for Compressive Strength of Natural Building Stone (Aug)(11319).

—Bushings
BSI BS 5600: Sec 3.9-79. 1979 Powder Metallurgical Materials and Products Part 3: Methods of Testing Sintered Metal Materials and Products, Excluding Hardmetals Section 3.9: Determination of Radial Crushing Strength for Sintered Metal Bushes. 4 pp.

ISO 2739-73. Sintered Metal Bushes—Determination of Radial Crushing Strength First Edition. 4 pp.

—Calcium Silicate Bricks
CEN PREN 772-1-92. Methods of Test for Masonry Units—Part 1: Determination of Compressive Strength. 13 pp.

—Carbon Black
ISO TR8942-88. Rubber Compounding Ingredients—Carbon Black—Determination of Individual Pellet Crushing Strength First Edition. 4 pp.

—Cellular Materials
DIN ENGL 53425-65. Testing of Rigid Foams; Time-Dependent Creep Compression Test Under Heat (Sept). 2 pp.

DIN ENGL 53572-86. Testing of Flexible Cellular Materials; Determination of Compression Set After Constant Strain (Nov). 3 pp.

DIN ENGL 53579 Pt 2-85. Testing of Flexible Cellular Materials; Hardness Test on Finished Parts; Compressibility of Profiles (Apr). 3 pp.

—Cellular Plastics
BSI BS 4370: Part 1-88. 1988 Methods of Test for Rigid Cellular Materials Part 1: Methods 1-5. 12 pp.

BSI BS 4443: Part 1-88. 1988 Amd 1 Methods of Test for Flexible Cellular Materials Part 1: Methods 1 to 6. 16 pp.

BSI BS 4443: Part 3-88. 1988 Methods of Test for Flexible Cellular Materials Part 3: Method 8, Determination of Creep Method 9, Determination of Dynamic Cushioning Performance. 12 pp.

BSI BS 4443: Part 7-92. 1992 Methods of Test for Flexible Cellular Materials Part 7: Method 17. Determination of Tear Strength of Flexible Cellular Material with an Integral Skin Method 18. Determination of Compression Set Under Humid Conditions. 11 pp.

CNS K6668-81. Compression Test for Rigid Materials of Cellular Plastics (May)(7408).

DIN ENGL 53421-84. Testing of Rigid Cellular Plastics; Compression Test (June). 3 pp.

DIN ENGL 53574-77. Testing of Flexible Cellular Materials; Fatigue Vibration Test by Constant Load Pounding in the Indentation/Pulsation Range (May). 4 pp.

DIN ENGL 53577-88. Determination of Compression Stress Value and Compression Stress-Strain Characteristic for Flexible Cellular Materials (Dec). 3 pp.

INTERNATIONAL AND NON-U.S. NATIONAL STANDARDS
SUBJECT INDEX — Compression

Compression Testing (Cont.)

—Cellular Plastics (Cont.)
ISO 844-78. Cellular Plastics—Compression Test for Rigid Materials First Edition; (Amendment Slip-1979). 6 pp.
ISO 1856-80. Polymeric Materials, Cellular Flexible—Determination of Compression Set Second Edition; (Erratum—Nov 1981). 5 pp.
ISO 3386 Pt 1-86. Polymeric Materials, Cellular Flexible—Determination of Stress-Strain Characteristic in Compression—Part 1: Low-Density Materials Second Edition. 5 pp.
ISO 3386 Pt 2-84. Polymeric Materials, Cellular Flexible—Determination of Stress-Strain Characteristic in Compression—Part 2: High Density Materials First Edition. 4 pp.
ISO 10066-91. Flexible Cellular Polymeric Materials—Determination of Creep in Compression First Edition. 6 pp.
JIS K 7220-83. Testing Method for Compressive Properties of Rigid Cellular Plastics. 10 pp.

—Cement Mortars
CNS R3032-83. Method of Test for Compressive Strength of Hydraulic Cement Mortar (Using 50mm Cube Specimens) (Jul)(1010).
ISO 679-89. Methods of Testing Cements—Determination of Strength First Edition. 18 pp.

—Cements
BSI BS 4550: Sec 3.4-78. 1978 Amd 3 Methods of Testing Cement Part 3: Physical Tests Section 3.4: Strength Test (AMD 5704) July 29, 1988. 13 pp.
CEN EN 196 (Part 1)-87. Methods of Testing Cement; Determination of Strength. 32 pp.
CEN EN 196 (Part 1)/A2-89. AMD 2 Methods of Testing Cement; Determination of Strength. 4 pp.

—Ceramics
CEN ENV 658-2-93. Advanced Technical Ceramics—Mechanical Properties of Ceramic Composites at Room Temperature—Part 2: Determination of Compressive Strength. 17 pp.
CNS R3060-76. Method of Test for Compressive Strength of Chemical Stoneware (Jun) (3187).
DIN ENGL 51067 Pt 1-77. Testing of Ceramic Raw and Finished Materials; Determination of the Compressive Strength at Room Temperature (KDF) of Refractories with a Total Porosity of up to 45% (May). 5 pp.
DIN ENGL 51067 Pt 2-77. Testing of Ceramic Raw and Finished Materials; Determination of the Compressive Strength at Room Temperature (KDF) of Refractories with a Total Porosity Exceeding 45% (May). 3 pp.
JIS R 1608-90. Testing Methods for Compressive Strength of High Performance Ceramics. 8 pp.

—Coated Fabrics
ISO 5473-79. Rubber-or Plastics-Coated Fabrics—Determination of Crush Resistance First Edition. 5 pp.

—Concrete Blocks
CEN PREN 772-1-92. Methods of Test for Masonry Units—Part 1: Determination of Compressive Strength. 13 pp.

—Concretes
BSI BS 1881: Part 116-83. 1983 Amd 2 Testing Concrete Part 116: Method for Determination of Compressive Strength of Concrete Cubes (AMD 6720) July 31, 1991. 11 pp.
BSI BS 1881: Part 119-83. 1983 Methods of Testing Concrete Part 119: Method for Determination of Compressive Strength Using Portions of Beams Broken in Flexure (Equivalent Cube Method). 5 pp.
BSI BS 1881: Part 120-83. 1983 Amd 1 Methods of Testing Concrete Part 120: Method for Determination of the Compressive Strength of Concrete Cores (AMD 6109) July 31, 1989. 10 pp.
BSI BS 1881: Part 127-90. 1990 Methods of Testing Concrete Part 127: Method of Verifying the Performance of a Concrete Cube Compression Machine Using the Comparative Cube Test. 16 pp.
CEN PREN 679-92. Determination of the Compressive Strength of Autoclaved Aerated Concrete. 5 pp.
CNS A3045-82. Method of Test for Compressive Strength of Cylindrical Concrete Specimens (Jan)(1232). 4 pp.
CNS A3051-84. Method of Test for Compressive and Flexural Strength of Drilled Cores and Sawed Beams of Concrete (Apr)(1238). 3 pp.
CNS A3208-84. Method of Test for Compressive Strength of Spun Concrete (Aug)(10979).
DIN ENGL 1048 Pt 2-91. Testing Concrete; Testing of Hardened Concrete (Specimens Taken In Situ) (June). 6 pp.
ISO 3893-77. Concrete—Classification by Compressive Strength First Edition. 3 pp.
ISO 4012-78. Concrete—Determination of Compressive Strength of Test Specimens First Edition. 6 pp.
JIS A 1107-78. Method of Obtaining and Testing Drilled Cores and Sawed Beams of Concrete. 7 pp.
JIS A 1108-76. Method of Test for Compressive Strength of Concrete. 5 pp.
JIS A 1114-76. Method of Test for Compressive Strength of Concrete Using Portions of Beams Broken in Flexure. 7 pp.
JIS A 1136-78. Method of Test for Compressive Strength of Spun Concrete. 9 pp.
JIS A 1139-78. Method of Test for Concrete Under Biaxial Compressive Loads. 7 pp.
JIS A 1140-78. Method of Test for Compressive Strength of Reinforced Concrete Street and House Inlet. 5 pp.
JIS A 1161-73. Testing Methods for Bulk Specific Gravity, Water Content, Absorption and Compressive Strength of Cellular Concrete. 7 pp.
JIS A 1182-78. Method of Test for Compressive Strength of Polyester Resin Concrete. 6 pp.
JIS A 1183-78. Method of Test for Compressive Strength of Polyester Resin Concrete Using Portions of Beams Broken in Flexure. 7 pp.
SAA AS 1012.9-86. Methods of Testing Concrete—Part 9: Method for the Determination of the Compressive Strength of Concrete Specimens. 11 pp.
SAA AS 1012.14-91. Methods of Testing Concrete—Part 14: Method for Securing and Testing Cores from Hardened Concrete for Compressive Strength (in Professional Package 58). 7 pp.
SNZ NZS 3112: Part 2-86. Methods of Test for Concrete Part 2: Tests Relating to the Determination of Strength of Concrete Amend: 1. 22 pp.

—Connectors
CEN PREN 2591-F12-93. Aerospace Series Elements of Electrical and Optical Connection Test Methods Part F12—Optical Elements Effectiveness of Cable Attachment Cable Axial Compression. 3 pp.
CNS C6144-87. Method of Test for Low Frequency (Below 3 MHz) Electrical Connectors (TP-40 Crush Test) (Oct)(8222).

—Containers
CNS Z6016-72. Method of Test for Compressive Strength of Corrugated Fiberboard Container (Oct)(3511).

—Contraceptive Diaphragms
ISO 8009 Pt 7-85. Reusable Rubber Contraceptive Diaphragms—Part 7: Determination of Compression Resistance of Coil Spring and Flat Spring Diaphragms First Edition. 4 pp.
ISO 8009 Pt 8-85. Reusable Rubber Contraceptive Diaphragms—Part 8: Determination of Twisting During Compression of Coil Spring and Flat Spring Diaphragms First Edition. 4 pp.

—Cork
ISO 9727-91. Cylindrical Stoppers of Natural Cork—Physical Tests—Reference Methods First Edition. 12 pp.

—Corrugated Board
BSI BS 4686-71. 1971 Determination of the Flat Crush Resistance of Single Faced and Single Wall Corrugated Fibreboard. 8 pp.
BSI BS 6036-85. 1985 Method for Determination of the Edgewise Crush Resistance of Corrugated Fibre Board. 4 pp.
BSI BS 6844-87. 1987 Determination of the Flat Crush Resistance of Corrugating Medium After Laboratory Fluting (CMT Concora Medium Test). 8 pp.
CNS P3024-85. Method of Test for Ring Crush of Corrugating Medium and Paperboard (Aug)(2956). 4 pp.
CNS P3088-87. Method of Test for Flat Crush of Corrugating Medium (CMT Test) (Sep)(12107).
CNS Z6081-86. Method of Test for Edgewise Compressive Strength of Corrugated Fiberboard-Rectangle Form Specimen (Jan)(11485).
CNS Z6082-86. Method of Test for Edgewise Compressive Strength of Corrugated Fiberboard-Butterfly Form Specimen (Jan)(11486).
CNS Z6083-86. Method of Test for Edgewise Compressive Strength of Corrugated Fiberboard-Butterfly Form Specimen (Jan)(11487).
CNS Z6084-86. Method of Test for Flat Crush of Corrugated Fiberboard (Jan)(11488).
CPPA D.20P-80. Flat Crush Test of Corrugated Board. 2 pp.
CPPA D.24P-88. Flat Crush Test of Corrugating Medium. 1 pp.
CPPA D.30P-77. Edgewise Compression Resistance of Fluted Corrugating Medium. 2 pp.
ISO 3035-82. Single-Faced and Single-Wall Corrugated Fibreboard—Determination of Flat Crush Resistance Second Edition. 4 pp.
ISO 3037-82. Corrugated Fibreboard—Determination of Edgewise Crush Resistance Second Edition. 4 pp.
ISO 7263-85. Corrugating Medium—Determination of the Flat Crush Resistance After Laboratory Fluting First Edition. 7 pp.
JIS Z 0401-85. Method of Compression Test for Corrugated Fibreboard.
SAA AS 1301.429S-89. Methods of Test for Pulp and Paper (Metric Units)—Part 429s: Flat Crush Resistance of Corrugated Board (This is a Joint Standard with SANZ NZS 1301). 3 pp.
SAA AS 1301.434S-91. Methods of Test for Pulp and Paper—Part 434S: Crush Resistance of Corrugating Medium Amdt 1 August 1991 (This is a Joint Standard with SANZ NZS 1301.434S).
SAA AS 1301.439S-91. Methods of Test for Pulp and Paper —Part 439s: Bendsten Roughness of Paper and Paperboard (This is a Joint Standard with SANZ NZS 1301). 8 pp.
SAA AS 1301.444S-92. Methods of Test for Pulp and Paper—Part 444s: Edgewise Compression Resistance of Corrugated Fibreboard. 3 pp.
SNZ NZS/AS 1301. 429S-89. Methods of Test for Pulp and Paper Flat Crush Resistance of Corrugated Board (This is a Joint Standard with SAA AS 1301.429S). 3 pp.
SNZ NZS/AS 1301. 434S-91. Methods of Test for Pulp and Paper Crush Resistance of Corrugating Medium (This is a Joint Standard with SAA AS 1301.434S). 4 pp.
SNZ NZS/AS 1301. 439S-91. Methods of Test for Pulp and Paper Bendsten Roughness of Paper and Paperboard (This is a Joint Standard with SAA AS 1301.439S). 8 pp.
SNZ NZS/AS 1301. 444S-92. Methods of Test for Pulp and Paper Edgewise Compression Resistance of Corrugated Fibreboard (This is a Joint Standard with SAA AS 1301.444S). 3 pp.

—Ebonite
CNS K6449-80. Method of Test for Crushing Strength of Ebonite (Jan)(5139).
ISO 2474-72. Ebonite—Determination of Crushing Strength First Edition. 4 pp.

—Elastomers
DIN ENGL 53517-87. Testing of Rubber and Elastomers; Determination of Compression Set After Constant Strain (Apr) (Supersedes January 1972 Editions of DIN 53517 Parts 1 and 2). 4 pp.

—Fasteners—Masonry
CEN PREN 846-4-92. Methods of Test for Ancillary Components for Masonry—Part 4: Determination of Tensile and Compressive Load Capacity and Stiffness of Wall Ties (Wall Test). 14 pp.
CEN PREN 846-5-92. Methods of Test for Ancillary Components for Masonry—Part 5: Determination of Tensile and Compressive Load Capacity and Stiffness of Wall Ties (Couplet Test). 13 pp.
CEN PREN 846-6-92. Methods of Test for Ancillary Components for Masonry—Part 6: Determination of Tensile and Compressive Load Capacity and Stiffness of Wall Ties (Single End Test). 11 pp.

—Fiber Optic Cables
CNS C6318-88. Method of Test for Fiber Optic Devices (FOTP-41 Compressive Loading Resistance of Fiber Optic Cable) (Jul)(12366).

—Fiber Optic Cables—Stuffing Tubes
CNS C6328-89. Method of Test for Fiber Optic Devices (FOTP-94 Fiber Cable Stuffing Tube Compression) (Jul)(12559).

—Fiberglass Reinforced Plastics
CNS K6980-90. Method of Test for Compressive Properties of Glass Fiber Reinforced Plastics (Sep)(12781).
ISO 8515-91. Textile-Glass-Reinforced Plastics—Determination of Compressive Properties in the Direction Parallel to the Plane of Lamination First Edition. 11 pp.
JIS K 7056-87. Testing Method for Compressive Properties of Glass Fiber Reinforced Plastics. 11 pp.

—Fire Extinguishers
CEN EN 3-2-78. Portable Fire Extinguishers—Part 2. 7 pp.
CEN PREN 3-2-93. Fight Against Fire—Portable Fire Extinguishers—Part 2: Retention of Charge, Dielectric Test, Compaction, Special Provision. 9 pp.
CEN EN 3-5-84. Portable Fire Extinguishers—Part 5: Complementary Requirements and Tests. 17 pp.
CEN PREN 3-5-93. Fight Against Fire—Portable Fire Extinguishers—Part 5: Complementary Requirements and Tests. 23 pp.

—Firebricks
CNS R3010-78. Method of Test for Cold Crushing Strength and Modulus of Rupture of Refractory Brick and Shapes at Room Temperature (Nov)(616).
CNS R3143-86. Method of Test for Crushing Strength of Insulating Fire Bricks (Oct)(11740).
JIS R 2206-91. Test Method for Cold Crushing Strength of Refractory Bricks. 6 pp.
JIS R 2615-85. Testing Method for Crushing Strength of Insulating Fire Bricks. 6 pp.

INDUSTRY STANDARDS

Compression

Compression Testing (Cont.)

—Foundry Sands
CNS Z8030-82. Method for Determining Strength of Foundry Sand (May)(8897).
JIS Z 2604-76. Method for Determining Strength of Foundry Sands. 11 pp.

—Gaskets
CNS O2014-73. Method of Test for Cork Binder Discs (May)(2303). 1 p.

—Glass Containers
DIN ENGL 52346-82. Testing of Glass; Thrust Load Test on Containers; Testing by Attributes and by Variables (June). 2 pp.
ISO 8113-85. Glass Containers—Resistance to Vertical Load—Test Method First Edition. 4 pp.

—Glass Fabrics
ISO 3605-87. Textile Glass—Rovings—Determination of Compressive Strength of Rod Composites Second Edition. 8 pp.
ISO 3616-77. Textile Glass—Mats—Determination of Average Thickness, Thickness Under Load and Recovery After Compression First Edition. 6 pp.

—Grout
SNZ NZS 3112: Part 4-86. Methods of Test for Concrete Part 4: Tests Relating to Grout. 10 pp.

—Hard Metals
BSI BS 5600: Sec 4.16-81. (WITHDRAWN) 1981 Powder Metallurgical Materials and Products Part 4: Methods of Testing and Chemical Analysis of Hardmetals Section 4.16: Compression Test (Superseded by BS EN 24506: 1993). 6 pp.
BSI BS EN 24506-93. 1993 Hardmetals—Compression Test (ISO 4506: 1979) (V). 11 pp.
CEN EN 24506-93. Hardmetals—Compression Test (ISO 4506: 1979). 6 pp.
ISO 4506-79. Hardmetals—Compression Test First Edition (CEN EN 24506: 1993). 6 pp.

—Helical Springs
CNS B2596-81. Quality Specifications for Cold Coiled Compression Springs Made of Round Wire (Jun)(7596). 9 pp.
JIS B 2704-87. Design of Helical Compression and Extension Springs.
JIS B 2704-78. Design of Helical Compression and Extension Springs. 30 pp.
JIS B 2707-87. Cold Coiled Helical Compression Springs. 11 pp.

—Hoses
BSI BS 5173: Sec 103.2-90. 1990 Rubber and Plastics Hoses and Hose Assemblies Part 103: Physical Tests Section 103.2: Determination of Crush Resistance of Hoses. 8 pp.

—Hosiery
BSI BS 6612-01. 1985 Amd 1 Graduated Compression Hoisery (AMD 7728) June 15, 1993. 15 pp.
BSI BS 7563-92. 1992 Method for Determination of Compression Values and Stiffness of Non-Prescriptive Graduated Support Hosiery (N). 13 pp.
BSI BS 7672-93. 1993 Compression, Stiffness and Labelling of Anti-Embolism Hosiery (L). 16 pp.

—Insulation Board
CNS K6224-65. Method of Test for Foam Polystyrene Heat Insulating Material (Sep)(2536) (R 1971). 3 pp.

—Iron Ores
BSI BS 6599-85. 1985 Method for Determination of Crushing Strength of Iron Ore Pellets. 4 pp.
ISO 4700-83. Iron Ore Pellets—Determination of Crushing Strength First Edition. 4 pp.
JIS M 8718-88. Test Method for Determination of Crushing Strength of Iron Ore Pellets. 6 pp.

—Iron Pipes
CSA CAN/CSA-B70-M91. Cast Iron Soil Pipe, Fittings, and Means of Joining; (Gen Instr 1). 27 pp.

—Joints
ISO 8969-90. Timber Structures—Testing of Unilateral Punched Metal Plate Fasteners and Joints First Edition. 12 pp.

—Laminated Plastics
CEN PREN 2850-90. Unidirectional Thermosetting Carbon Laminate Compression Test Parallel to the Fibre Direction. 14 pp.
CEN PREN 2850-91. Carbon Thermosetting Resin Unidirectional Laminates Comression Test Parallel to the Fibre Direction. 12 pp.

—Leather
CNS K6121-60. Method of Test for Leather (Compression Test) (May)(1277) (R 1973).

Compression Testing (Cont.)

—Lumber
ISO 8906-88. Sawn Timber—Test Methods—Determination of Resistance to Local Transverse Compression First Edition. 6 pp.

—Masonry
CEN PREN 1052-1-93. Methods of Test for Masonry—Part 1: Determination of Compressive Strength. 13 pp.
DIN ENGL 18554 Pt 1-85. Testing of Masonry; Determination of Compressive Strength and of Elastic Modulus (Dec). 4 pp.

—Masonry Prisms
CSA CAN/CSA-A369.1-M90. Method of Test for Compressive Strength of Masonry Prisms; (Gen Instr 1). 22 pp.

—Masonry—Stone
CEN PREN 772-1-92. Methods of Test for Masonry Units—Part 1: Determination of Compressive Strength. 13 pp.

—Metals
DIN ENGL 50106-78. Testing of Metallic Materials; Compression Test (Dec). 4 pp.

—Mortars
CEN PREN 1015-11-93. Methods of Test for Mortar for Masonry—Part 11: Determination of Flexural and Compressive Strength of Hardened Mortar. 12 pp.
CNS A3220-84. Method of Test for Effect of Organic Impurities in Fine Aggregate on Strength of Mortar (Dec)(11153). 4 pp.
DIN ENGL 18555 Pt 3-82. Testing of Mortars Containing Mineral Binders; Hardened Mortars; Determination of Flexural Strength, Compressive Strength and Bulk Density (Sept). 4 pp.
DIN ENGL 18555 Pt 4-86. Testing of Mortars Containing Mineral Binders; Hardened Mortars; Determination of Linear and Transverse Strain and of Deformation Characteristics of Masonry Mortars by the Static Pressure Test (Mar). 4 pp.

—O Rings
MOD UK DTD-5613A-82. Elastomeric Toroidal Sealing Rings, (O Rings) Low Compression Set Fluorocarbon Type (Incorporating Amendment No. 1). 5 pp.

—Optical Waveguide Connectors
CNS C6275-86. Method of Test for Fiber Optic Devices (FOTP-26 Crush Resistance of Fiber Optic Cable Interconnecting Devices) (Dec)(11787).
CNS C6283-87. Method of Test for Fiber Optic Devices (FOTP-83 Cable to Interconnecting Device Axial Compressive Loading) (Apr)(11905).

—Packaging
BSI BS 4826: Part 7-86. (WITHDRAWN) 1986 Complete, Filled Transport Packages Part 7: Methods for Determination of Resistance to Compression (Superseded by BS EN 22872: 1993). 4 pp.
BSI BS EN 22872-93. 1993 Complete, Filled Transport Packages Method for Determination of Resistance to Compression (ISO 2872: 1985) (Supersedes BS 4826: Part 7: 1986) (H). 9 pp.
CEN EN 22872-92. Packaging—Complete, Filled Transport Packages—Compression Test (ISO 2872: 1985). 4 pp.
ISO 2872-85. Packaging—Complete, Filled Transport Packages—Compression Test Second Edition; (CEN EN 22872:1992). 4 pp.

—Packaging Materials
CNS Z6068-83. Method of Test for Static Compression for Package Cushioning Materials (Jul)(10482).
JIS Z 0200-87. General Rules of Performance Testing for Packaged Freights. 12 pp.
JIS Z 0212-87. Method of Compression Test for Packaged Freights and Containers. 8 pp.
JIS Z 0234-76. Testing Methods of Static Compression for Package Cushioning Materials (R 1984). 7 pp.
JIS Z 0235-76. Testing Methods of Dynamic Compression for Package Cushioning Materials (R 1984). 7 pp.

—Paints
CNS K6143-87. Method of Test for Traffic Paints (May)(1334). 19 pp.

—Pallets
CNS Z6041-81. Method of Test for Flat Pallet (Nov)(8170).
JIS Z 0602-88. Test Methods for Flat Pallets. 10 pp.

—Paperboard
BSI BS 7325-90. 1990 Method for Determination of the Compressive Strength of Paper and Board by the Short Span Test. 11 pp.

Compression Testing (Cont.)

—Paperboard (Cont.)
CNS P3024-85. Method of Test for Ring Crush of Corrugating Medium and Paperboard (Aug)(2956). 4 pp.
CPPA D.33P-77. Ring Crush of Paperboard. 2 pp.
ISO 9895-89. Paper and Board—Compressive Strength—Short Span Test First Edition. 8 pp.
JIS P 8126-87. Method of Test for Ring Crush of Paperboard. 7 pp.
SAA AS 1301.450R P-89. Methods of Test for Pulp and Paper (Metric Units)—Part 450rp: Compression Strength of Paper and Board—Short Span Test Amdt 1 November 1992 (This is a Joint Standard with SANZ NZS 1301). 5 pp.
SNZ NZS/AS 1301. 450RP-89. Methods of Test for Pulp and Paper Compression Strength of Paper and Board—Short Span Test (This is a Joint Standard with SAA AS 1301.450RP). 4 pp.

—Papers
BSI BS 7325-90. 1990 Method for Determination of the Compressive Strength of Paper and Board by the Short Span Test. 11 pp.
BSI BS 7387-91. 1991 Method for Determination of the Bulking Thickness, Apparent Bulk Density, Compressibility and Compressibility Index of Soft Creped Tissue Paper. 12 pp.
ISO 9895-89. Paper and Board—Compressive Strength—Short Span Test First Edition. 8 pp.
SAA AS 1301.450R P-89. Methods of Test for Pulp and Paper (Metric Units)—Part 450rp: Compression Strength of Paper and Board—Short Span Test Amdt 1 November 1992 (This is a Joint Standard with SANZ NZS 1301). 5 pp.
SNZ NZS/AS 1301. 450RP-89. Methods of Test for Pulp and Paper Compression Strength of Paper and Board—Short Span Test (This is a Joint Standard with SAA AS 1301.450RP). 4 pp.

—Pipe Fittings—Iron
CSA CAN/CSA-B70-M91. Cast Iron Soil Pipe, Fittings, and Means of Joining; (Gen Instr 1). 27 pp.

—Pipe Fittings—PVC
ISO 9853-91. Injection-Moulded Unplasticized Poly(Vinyl Chloride) (PVC-U) Fittings for Pressure Pipe Systems—Crushing Test First Edition. 8 pp.

—Plastic Pipes
CNS K6140-87. Method of Test for Unplasticized Polyvinyl Chloride Pipes (Feb)(1299). 4 pp.
CNS K6641-81. Method of Test for Rigid Polyvinyl Chloride Corrugated Pipes (Mar)(7057). 4 pp.
DIN ENGL 53769 Pt 3-88. Testing of Glass Fibre Reinforced Plastics Pipes; Determination of Initial and Long-Term Ring Stiffness (Nov). 5 pp.
DIN ENGL 53769 Pt 5-89. Testing of Glass Fibre Reinforced Plastics Pipes; Determination of Chemical Resistance of Pipes Under Compression (Apr). 4 pp.

—Plastics
BSI BS 2782:Pt1: METH 131C-D-78. 1978 Methods of Testing Plastics Part 1: Thermal Properties Method 131C-D: Crushing Strength After Heating (Heat Resistance) of Thermosetting (C) Moulding Material. (D) Laminated Sheet or Mouldings.. 2 pp.
BSI BS 2782:Pt3: METH 345A-79. 1979 Methods of Testing Plastics Part 3: Mechanical Properties Method 345A: Determination of Compressive Properties by Deformation at Constant Rate. 8 pp.
DIN ENGL 53454-71. Testing of Plastics; Compression Test (Apr). 6 pp.
DIN ENGL 53457-87. Testing of Plastics; Determination of the Elastic Modulus by Tensile, Compression and Bend Testing (Oct). 7 pp.
ISO 604-93. Plastics—Determination of Compressive Properties Second Edition. 12 pp.
JIS K 7208-75. Testing Method for Compressive Properties of Plastics. 12 pp.

—Plywood
CNS O2039-87. Method of Test for Compression Strength of Plywood (May)(8537). 2 pp.

—Polyester Resins
CNS K3062-82. Liquid Unsaturated Polyester Resin for Reinforced Plastics (Dec)(9715). 15 pp.

—Polyurethane Foams
SAA AS 2282.9-91. Methods for Testing Flexible Cellualar Polyurethane—Part 9: Determination of Compression Set. 2 pp.
SAA AS 2282.12-91. Methods for Testing Flexible Cellular Polyurethane—Part 12: Determination of Compression Fatigue. 3 pp.

—Porcelain
CNS R3053-84. Method of Test for Compressive (Crushing) Strength of Fired Whiteware Clay Materials (Jan)(2889).

INTERNATIONAL AND NON-U.S. NATIONAL STANDARDS
SUBJECT INDEX
Compressor

Compression Testing (Cont.)

—Pressure Pipe Fittings
CEN PREN 802-92. Plastics Piping and Ducting Systems—Injection-Moulded Thermoplastics Fittings for Pressure Piping Systems—Test Methods for Maximum Deformation by Crushing. 7 pp.

—Pressure Vessels
BSI BS 5500: ENQ CASE 34-81. 1981 Enquiry Case Case 34: Longitudinal Compressive Stress Limit for Pressure Vessels. 1 p.

—Refractory Materials
BSI BS 1902: Sec 4.0-85. 1985 Methods for Testing Refractory Materials Part 4: Properties Measured Under an Applied Stress Section 4.0: Introduction. 4 pp.

BSI BS 1902: Sec 4.1-84. 1984 Methods for Testing Refractory Materials Part 4: Properties Measured Under an Applied Stress: Section 4.1: Determination of the Cold Crushing Strength of Shaped Insulating Refractory Products (Method 1902-401). 4 pp.

BSI BS 1902: Sec 4.3-85. 1985 Amd 1 Methods for Testing Refractory Materials Part 4: Properties Measured Under an Applied Stress Sec. 4.3: Detrm. of Cold Crushing Strength of Dense Shaped Products (Method 1902-403) (AMD 6063) July 31, 1989. 9 pp.

BSI BS 1902: Sec 4.10-90. 1990 Methods for Testing Refractory Materials Part 4: Properties Measured Under an Applied Stress Section 4.10: Determination of Creep in Compression (Method 1902-410). 16 pp.

BSI BS 6043: Sec 4.9-91. 1991 Methods of Sampling and Test for Carbonaceous Materials Used in Aluminium Manufacture Part 4: Cold Ramming Pastes Section 4.9: Determination of the Cold Crushing Strength of Baked Rammed Paste. 8 pp.

CEN PREN 1094-5-93. Insulating Refractory Products—Part 5: Determination of Cold Crushing Strength. 7 pp.

CNS R3148-87. Method of Test for Crushing Strength and Modulus of Rupture of Castable Refractories (May)(11974).

CNS R3151-87. Method of Test for Crushing Strength and Modulus of Rupture of Light Weight Castable Refractories (May) (11977).

CNS R3153-87. Method of Test for Crushing Strength and Modulus of Rupture of High Alumina and Fireclay Plastic Refractories (May) (11973).

ISO 8895-86. Shaped Insulating Refractory Products—Determination of Cold Crushing Strength First Edition. 4 pp.

ISO 10059 Pt 1-92. Dense, Shaped Refractory Products—Determination of Cold Compressive Strength—Part 1: Referee Test Without Packing First Edition. 9 pp.

JIS R 2553-92. Testing Method for Crushing Strength and Modulus of Rupture of Castable Refractories. 14 pp.

JIS R 2575-92. Testing Method for Crushing Strength and Modulus of Rupture of High Alumina and Fireclay Plastic Refractories. 10 pp.

SAA AS 1774.1-91. Refractories and Refractory Materials—Physical Test Methods—Part 1: Determination of Cold Compressive Strength. 3 pp.

—Reinforced Plastics
JIS K 7076-91. Testing Methods for Compressive Properties of Carbon Fibre Reinforced Plastics. 17 pp.

—Resins
BSI BS 6319: Part 2-83. 1983 Testing of Resin Composition for Use in Construction Part 2: Method for Measurement of Compressive Strength. 4 pp.

—Rocks
CNS M3012-80. Method of Test for Compressive Strength of Rock (Mar)(5367).

CNS M3110-82. Method of Test for Triaxial Compressive Strength of Undrained Rock Core Specimens Without Pore Pressure Measurements (Oct) (9531).

DIN ENGL 52105-88. Testing the Compressive Strength of Natural Stone (Aug). 2 pp.

JIS M 0302-75. Method of Test of Compressive Strength of Rock.

—Rubber
DIN ENGL 53517-87. Testing of Rubber and Elastomers; Determination of Compression Set After Constant Strain (Apr) (Supersedes January 1972 Editions of DIN 53517 Parts 1 and 2). 4 pp.

ISO 7743-89. Rubber, Vulcanized or Thermoplastic—Determination of Compression Stress-Strain Properties First Edition; (Corrigendum 1-1990). 9 pp.

—Sandwich Structures
DIN ENGL 53291-82. Testing of Sandwiches; Compression Test Perpendicular to the Faces (Feb). 4 pp.

Compression Testing (Cont.)

—Sealants
CGSB CAN2-19.0-M77METH14.4-78. Methods of Testing Putty, Caulking and Sealing Compounds Tensile Compression Cycling—Hockman Procedure. 2 pp.

ISO 11432-93. Building Construction—Sealants—Determination of Resistance to Compression First Edition. 6 pp.

—Shipping Containers
CNS Z6006-76. Method of Compression Test for Packing and Shipping Containers (Dec)(2543).

JIS Z 0212-87. Method of Compression Test for Packaged Freights and Containers. 8 pp.

—Soil Analysis
SAA AS 1289.F4.1-77. Methods of Testing Soil for Engineering Purposes—Part F4.1: Soil Strength and Consolidation Tests—Detn. of the Compressive Str. of a Soil—Compressive Str. of a Spec. Tested in Undrained Triaxial Compression Without Mst. of Pore Pressure. 3 pp.

SNZ NZS 4402: Pt 6:TEST 6.2.1-86. Methods of Testing Soils for Civil Engineering Purposes Part 6: Soil Strength Tests. Section 6.2: Determination of the Compressive Strength of Specimens Tested in Undrained Triaxial Compression. Test 6.2.1: Method Without Measurement of Pore. 9 pp.

SNZ NZS 4402: Pt 6:TEST 6.3.1-86. Methods of Testing Soils for Civil Engineering Purposes Part 6: Soil Strength Tests. Section 6.3: Determination of the Unconfined Compressive Strength of Cohesive Soil. Test 6.3.1: Standard Method Using Laboratory Apparatus. 6 pp.

SNZ NZS 4402: Pt 6:TEST 6.3.2-86. Methods of Testing Soils for Civil Engineering Purposes Part 6: Soil Strength Tests. Section 6.3: Determination of the Unconfined Compressive Strength of Cohesive Soil. Test 6.3.2: Alternative Method Using Autographic Apparatus. 6 pp.

—Soils
BSI BS 1377: Part 5-90. 1990 Methods of Test for Soils for Civil Engineering Purposes Part 5: Compressibility, Permeability and Durability Tests. 37 pp.

CNS A3282-88. Method of Test for Unconfined Compressive Strength of Cohesive Soil (Aug)(12384).

DIN ENGL 18136-87. Soil Analysis; Determination of Unconfined Compressive Strength; Unconfined Compression Test (Mar). 3 pp.

JIS A 1216-90. Method for Unconfined Compression Test of Soils. 9 pp.

—Stabilized Soils
BSI BS 1924-75. (WITHDRAWN) 1975 Methods of Test for Stabilized Soils (R) (Superseded by BS 1924: Parts 1 & 2: 1990). 98 pp.

BSI BS 1924: Part 2-90. 1990 Stabilized Materials for Civil Engineering Purposes Part 2: Methods of Test for Cement-Stabilized and Lime-Stabilized Materials. 104 pp.

—Textile Floor Coverings
DIN ENGL 54327-83. Testing of Textiles; Determination of the Tread Comfort Value of Textile Floor Coverings by Measuring the Strain Energy Under Compression (July). 5 pp.

—Thermal Insulation
CEN PREN 826-92. Thermal Insulating Products for Building Applications—Determination of Compression Behaviour. 12 pp.

—Toys
CNS Z8016-23-87. Method of Test for Toy Safety (Compression Test) (Dec)(4798-23). 2 pp.

CNS Z8016-28-87. Method of Test for Toy Safety (Compression Test for Wheeled Toys Having an Axle Retained by a Snap-in Fixture) (Dec)(4798-28). 2 pp.

—Uppers (Footwear)
BSI BS 5131: Sec 3.4-78. 1978 Methods of Test for Footwear and Footwear Materials Part 3: Uppers, Textiles and Threads Section 3.4: Edgewise Compressibility, Upper Materials. 2 pp.

—Vulcanized Rubber
BSI BS 903: Part A4-90. 1990 Physical Testing of Rubber Part A4: Determination of Compression Stress-Strain Properties (V) (ISO 7743: 1989). 12 pp.

BSI BS 903: Part A4-01. 1990 Amd 1 Physical Testing of Rubber Part A4: Determination of Compression Stress-Strain Properties (V) (ISO 7743: 1989) (AMD 6741) September 30, 1991. 14 pp.

BSI BS 903: Part A6-92. 1992 Physical Testing of Rubber Part A6: Method for Determination of Compression Set at Ambient, Elevated or Low Temperatures (ISO 815: 1991) (V). 16 pp.

BSI BS 903: Part A6-69. 1969 Amd 2 Methods of Testing Vulcanized Rubber Part A6: Determination of Compression Set After Constant Strain. 12 pp.

Compression Testing (Cont.)

—Vulcanized Rubber (Cont.)
BSI BS 903: Part A15-90. 1990 Physical Testing of Rubber Part A15: Method for Determination of Creep in Compression or Shear (ISO 8013: 1988). 18 pp.

BSI BS 903: Part A39-80. (WITHDRAWN) 1980 Methods of Testing Vulcanized Rubber Part A39: Determination of Compression Set Under Constant Deflection at Low Temperatures (Superseded by BS 903: Part A6: 1992). 6 pp.

CNS K6351-85. Method of Test for Permanent Compression Set Test for Vulcanized Rubber (Dec)(3560).

CNS K6747-83. Method of Test for Compression of Vulcanized Rubber (Feb)(10019).

CNS K6750-83. Method of Test for Permanent Compression Set Under Low Temperature of Vulcanized Rubber (Feb)(10022).

ISO 815-91. Rubber, Vulcanized or Thermoplastic—Determination of Compression Set at Ambient, Elevated or Low Temperatures Second Edition; (Corrigendum 1-1993). 15 pp.

ISO 6471-83. Rubber, Vulcanized—Determination of Crystallization Effects Under Compression First Edition. 8 pp.

ISO 8013-88. Rubber, Vulcanized—Determination of Creep in Compression or Shear First Edition. 15 pp.

SAA AS 1683.13-92. Methods of Test for Elastomers—Part 13: Compression Set of Vulcanized Rubber Under Constant Deflection (Supersedes AS 1683.13B—1976). 3 pp.

—Welded Joints—Windows
CEN PREN 514-91. Unplasticised Polyvinylchloride (PVC-U) Profiles for the Construction of Windows—Determination of the Strength of Welded Corners and T-Joints. 16 pp.

CEN PREN 514-93. Unplasticised Polyvinylchloride (PVC-U) Profiles for the Construction of Windows—Determination of the Strength of Welded Corners and T-Joints. 15 pp.

—Wood
CNS O2004-81. Method of Test for Compression of Wood (Mar)(453). 3 pp.

ISO 3132-75. Wood—Testing in Compression Perpendicular to Grain First Edition. 4 pp.

ISO 3787-76. Wood—Test Methods—Determination of Ultimate Stress in Compression Parallel to Grain First Edition. 4 pp.

JIS Z 2111-77. Method of Compression Test for Wood.

JIS Z 2111-63. Method of Compression Test for Wood (R 1970). 6 pp.

Compressive Creep
Use: Cold Flow

Compressive Properties
Use: Compression Testing

Compressive Strength Testing
Use: Compression Testing

Compressor Blades
See Also: Compressors (Pressure Equipment); Turbine Blades

—Aircraft
SBAC RS 414 (V). Standard Method of Dimensioning Compressor Blades.

—Forgings—Aluminum Alloys—Aerospace
MOD UK DTD-745A-71. Forging Stock and Compressor Blade Forgings of Aluminium—Copper—Magnesium—Nickel—Iron Alloy (Solution Treated and Precipitation Treated) (Cu 2.5, Mg 1.5, Ni 1.2, Fe 1.0). 2 pp.

Compressor/Evaporators
See Also: Air Conditioning Equipment; Compressors (Pressure Equipment)
CSA CAN/CSA-C22. 2 NO 236-M90. Heating and Cooling Equipment; (Gen Instr 1) (ANSI/UL 1995). 136 pp.

—Condensing Units
CSA CAN/CSA-C22. 2 NO 236-M90. Heating and Cooling Equipment; (Gen Instr 1) (ANSI/UL 1995). 136 pp.

Compressor Lubricants
See Also: Compressors (Pressure Equipment); Lubricants

—Air Compressors
BSI BS 6413:Pt3: Sec 3.1-88. 1988 Lubricants, Industrial Oils and Related Products (Class L) Part 3: Classification for Family D (Compressors) Section 3.1: Air Compressors. 7 pp.

INDUSTRY STANDARDS

INTERNATIONAL AND NON-U.S. NATIONAL STANDARDS
SUBJECT INDEX

Compressor

Compressor Lubricants *(Cont.)*
—**Air Compressors** *(Cont.)*
BSI BS 6413:Pt3: Sec 3.2-89. 1989 Lubricants, Industrial Oils and Related Products (Class L) Part 3: Classification for Family D (Compressors) Section 3.2: Gas and Refrigeration Compressors. 7 pp.
CNS K5049-88. Compressor Oil (Jan)(2974). 2 pp.
DIN ENGL 51506-85. Lubricants; VB and VC Lubricating Oils with and Without Additives and VDL Lubricating Oils; Classification and Requirements (Sept). 3 pp.
ISO 6743 Pt 3A-87. Lubricants, Industrial Oils and Related Products (Class L)—Classification—Part 3A: Family D (Compressors) First Edition. 7 pp.
ISO 6743 Pt 3B-88. Lubricants, Industrial Oils and Related Products (Class L)—Classification—Part 3B: Family D (Gas and Refrigeration Compressors) First Edition. 6 pp.
MOD UK DSTAN 91-42-78. Lubricating Oil, Petroleum: Compressor, Light Joint Service Designation: OM-58 Lubricating Oil, Petroleum: Compressor, Medium Joint Service Designation: OM-160 Issue 1. 6 pp.

Compressors (HVAC)
Use: Refrigerant Compressors

Compressors (Pressure Equipment)
See Also: Air Compressors; Air Conditioning Equipment; Compressed Air; Compressor Blades; Compressor/Evaporators; Compressor Lubricants; Condensing Units; Coolers; Cooling Systems; Displacement Compressors; Motor Compressors; Pneumatic Conveyors; Reciprocating Compressors; Refrigerant Compressors; Refrigeration Equipment; Refrigeration Systems; Rotary Compressors; Screw Compressors; Vacuum Pumps

—**Air Filters**
BSI BS 1701-70. (WITHDRAWN) 1970 Air Filters for Air Supply to Internal Combustion Engines and Compressors Other Than for Aircraft (Superseded by BS 7226: 1989). 40 pp.
BSI BS 2806-56. (OBSOLESCENT) 1956 Limiting Dimensions of Air Filters for Internal Combustion Engines and Compressors, Other Than for Aircraft. 17 pp.
BSI BS 7226-89. 1989 Methods of Test for Performance of Inlet Air Cleaning Equipment for Internal Combustion Engines and Compressors. 28 pp.
DIN ENGL 24189-86. Testing of Air Cleaners for Internal Combustion Engines and Compressors; Test Methods (Jan). 10 pp.
DIN ENGL 71459-87. Air Filter Elements for Commercial Vehicles; Dimensions (May). 6 pp.
ISO 5011-88. Inlet Air Cleaning Equipment for Internal Combustion Engines and Compressors—Performance Testing First Edition. 28 pp.

—**Gas Cylinders—Refueling**
SNZ NZS 5425: Part 1-80. Code of Practice for CNG Compressor and Refuelling Stations Part 1: On Site Storage and Location of Equipment Amend: 1, 1986. 27 pp.
SNZ NZS 5425: Part 2-82. Code of Practice for CNG Compressor and Refuelling Stations Part 2: Compressor Equipment. 16 pp.

—**Gas Turbine Engines—Cleaning Agents**
MOD UK TS 10268A. Cleaning Fluid for Compressors of Gas Turbine Engines.

—**Gas Turbine Engines—Corrosion Inhibitors**
MOD UK DSTAN 68-10-89. Corrosion Preventive Compound: Water Displacing NATO Code: C-634 Joint Service Designation: PX-24 Issue 3. 28 pp.
MOD UK DSTAN 68-10-01. Corrosion Preventive Compound: Water Displacing NATO Code: C-634 Joint Service Designation: PX-24 Issue 3; Amendment 1. 29 pp.

—**Glossaries**
BSI BS 5791: Part 1-79. 1979 Glossary of Terms for Compressors, Pneumatic Tools and Machines Part 1: General. 8 pp.
ISO 3857 Pt I-77. Compressors, Pneumatic Tools and Machines—Vocabulary—Part I: General First Edition. 9 pp.
ISO 3857 Pt II-77. Compressors Pneumatic Tools and Machines—Vocabulary—Part II: Compressors First Edition. 7 pp.
ISO 5390-77. Compressors—Classification First Edition. 9 pp.
JIS B 0132-84. Glossary of Terms for Fans, Blowers and Compressors.

—**Liquefied Petroleum Gas**
CNS Z2067-88. Compressor for Liquefied Petroleum Gas Use (Aug)(12406).

Compressors (Pressure Equipment) *(Cont.)*
—**Lubricating Oils**
CNS K5049-88. Compressor Oil (Jan)(2974). 2 pp.

—**Noise**
CNS Z8024-82. Method of Noise Level Measurement for Fans Blowers and Compressors (Apr)(8753).
CNS Z8025-82. Method of Noise Spectrum Measurement for Fans Blowers and Compressors (Apr)(8754).
ISO 2151-72. Measurement of Airborne Noise Emitted by Compressor/Primemover-Units Intended for Outdoor Use First Edition. 9 pp.

—**Power Levels**
EC 84/533/EEC-84. Council Directive on the Approximation of the Laws of the Member States Relating to the Permissible Sound Power Level of Compressors. 7 pp.

—**Preferred Pressures**
ISO 5941-79. Compressors, Pneumatic Tools and Machines—Preferred Pressures First Edition. 4 pp.

—**Safety**
CEN PREN 1012-1-93. Compressors and Vacuum Pumps—Safety Requirements—Part 1: Compressors. 50 pp.

—**Sound Pressure**
JIS B 8346-91. Fans, Blowers and Compressors—Determination of A-Weighted Sound Pressure Level. 60 pp.

—**Steam Jet**
DIN ENGL 28430-84. Vacuum Technology; Rules for the Measurement of Steam Jet Vacuum Pumps and Steam Jet Compressors Using Steam as the Working Fluid (Nov). 26 pp.

—**Symbols**
BSI BS 1553: Part 3-50. 1950 Specification for Graphical Symbols for General Engineering Part 3: Graphical Symbols for Compressing Plant. 14 pp.

—**Turbo**
ISO 5389-92. Turbocompressors—Performance Test Code First Edition. 177 pp.

Computer Aided Design
Use: CAD

Computer Aided Design/Computer Aided Manufacturing
Use: CAD/CAM

Computer Aided Manufacturing
Use: CAM

Computer Aided Measurement and Control
Use: CAMAC

Computer Aided Software Engineering
Use For: CASE *See Also:* Software

—**Reference Models**
ECMA ECMA-TR 55-90. Reference Model for Frameworks of Computer Assisted Software Engineering Environments. 102 pp.

Computer Assisted Publishing
Use: Desktop Publishing

Computer Assisted Retrieval Systems
Use For: CAR Systems; Information Retrieval Systems, Computer; Search and Retrieve Systems, Computer *See Also:* Computer Systems; Document Storage and Retrieval Systems

—**Hypermedia—Architecture**
CENELEC ETR 084-93. Terminal Equipment (TE); Multimedia & Hypermedia Information Retrieval Services (M&HIRS) Investigation of Candidate Architectures for M&HIRS. 83 pp.
ETSI ETR 084-93. Terminal Equipment (TE); Multimedia & Hypermedia Information Retrieval Services (M&HIRS) Investigation of Candidate Architectures for M&HIRS. 83 pp.

—**Multimedia Services—Architecture**
CENELEC ETR 084-93. Terminal Equipment (TE); Multimedia & Hypermedia Information Retrieval Services (M&HIRS) Investigation of Candidate Architectures for M&HIRS. 83 pp.
ETSI ETR 084-93. Terminal Equipment (TE); Multimedia & Hypermedia Information Retrieval Services (M&HIRS) Investigation of Candidate Architectures for M&HIRS. 83 pp.

Computer Assisted Retrieval Systems *(Cont.)*
—**Open Systems Interconnection**
BSI BS ISO 10162-93. 1993 Information and Documentation—Open Systems Interconnection—Search and Retrieve Application Service Definition (S). 28 pp.
BSI BS ISO 10163-1-93. 1993 Information and Documentation—Open Systems Interconnection—Search and Retrieve Application Protocol Specification—Part 1: Protocol Specification (S). 50 pp.
ISO 10162-93. Information and Documentation—Open Systems Interconnection—Search and Retrieve Application Service Definition First Ediiton. 26 pp.
ISO 10163 Pt 1-93. Information and Documentation—Open Systems Interconnection—Search and Retrieve Application Protocol Specification—Part 1: Protocol Specification First Edition. 41 pp.
OSI ISO/DIS 10162-90. Documentation—Search and Retrieve Service Definition. 36 pp.
OSI ISO/DIS 10163-90. Documentation—Search and Retrieve Protocol Specification. 52 pp.

—**Search Commands**
ISO 8777-93. Information and Documentation—Commands for Interactive Text Searching First Edition. 28 pp.

Computer Automated Measurement and Control
Use: CAMAC

Computer Control
Use: Control Systems Equipment

Computer Control Units
Use: Control Units

Computer Controllers
Use: Control Units

Computer Controls
Use: Control Units

Computer Data Safes
Use: Data Safes

Computer Equipment
Scope Note: For additional listings, use a more specific term *See Also:* Acoustic Couplers; Computer Numerical Control; Computer Systems; Computer Tapes; Computers; Control Units; Data Cartridges; Disk Cartridges; Disk Drives (Computer); Disk Packs; Display Drivers; Electronic Equipment; Firmware; Floppy Disk Controllers; Information Systems; Magnetic Disks; Magnetic Heads; Magnetic Tapes; Modems; Peripheral Controllers; Peripheral Equipment; Tape Drives; Video Display Terminals; Video Monitors

—**Engineering Drawings—Circuit Diagrams—Aircraft**
EURO CTI/79/9343-79. Recommendation for the Automation of Wiring Diagrams. 50 pp.

—**Environmental Conditions**
BSI BS 7083-89. 1989 Amd 1 Recommendations for the Accommodation and Operating Environment of Computer Equipment (AMD 6624) December 21, 1990. 17 pp.

—**Environmental Conditions—Ships**
JIS F 0416-91. Commercial Computing Apparatus and Their Peripherals—Environmental Conditions in Ships.

—**Noise**
BSI BS 7135: Part 1-89. 1989 Noise Emitted by Computer and Business Equipment Part 1: Method of Measurement of Airborne Noise. 42 pp.
BSI BS 7135: Part 1-01. 1989 Amd 1 Noise Emitted by Computer and Business Equipment Part 1: Method of Measurement of Airborne Noise (ISO 7779: 1988) (AMD 6977) April 1, 1992. 46 pp.
BSI BS 7135: Part 2-89. 1989 Noise Emitted by Computer and Business Equipment Part 2: Method of Measurement of High-Frequency Noise. 15 pp.
BSI BS 7135: Part 2-01. 1989 Amd 1 Noise Emitted by Computer and Business Equipment Part 2: Method of Measurement of High-Frequency Noise (ISO 9295: 1988) (AMD 6978) April 1, 1992. 19 pp.
BSI BS 7135: Part 3-89. 1989 Noise Emitted by Computer and Business Equipment Part 3: Method for Determining and Verifying Declared Noise Emission Values. 11 pp.
CEN EN 29295-91. Acoustics—Measurement of High-Frequency Noise Emitted by Computer and Business Equipment. 3 pp.

INTERNATIONAL AND NON-U.S. NATIONAL STANDARDS
SUBJECT INDEX
Computer

Computer Equipment *(Cont.)*

—Noise *(Cont.)*

DIN ENGL EN 27779-91. Acoustics; Measurement of Airborne Noise Emitted by Computer and Business Equipment (ISO 7779: 1988) (Nov). 42 pp.
DIN ENGL EN 29295-91. Acoustics; Measurement of High-Frequency Noise Emitted by Computer and Business Equipment (ISO 9295: 1988) (Nov). 15 pp.
ECMA ECMA 74-87. Measurement of Airborne Noise Emitted by Computers and Business Equipment. 87 pp.
ECMA ECMA 74-92. Measurement of Airborne Noise Emitted by Computer and Business Equipment. 90 pp.
ECMA ECMA 108-89. Measurement of High-Frequency Noise Emitted by Computer and Business Equipment. 24 pp.
ECMA ECMA 109-87. Declared Noise Emission Values of Computer and Business Equipment. 16 pp.
ECMA ECMA 109-92. Declared Noise Emission Values of Computer and Business Equipment. 15 pp.
ECMA ECMA 160-91. Determination of Sound Power Levels of Computer and Business Equipment Using Sound Intensity Measurements; Scanning Method in Controlled Rooms. 18 pp.
ECMA ECMA 160-92. Determination of Sound Power Levels of Computer and Business Equipment Using Sound Intensity Measurements; Scanning Method in Controlled Rooms. 33 pp.
ISO 7779-88. Acoustics—Measurement of Airborne Noise Emitted by Computer and Business Equipment First Edition. 41 pp.
ISO 9295-88. Acoustics—Measurement of High-Frequency Noise Emitted by Computer and Business Equipment First Edition; (Corrected and Reprinted -1989). 14 pp.
ISO 9296-88. Acoustics—Declared Noise Emission Values of Computer and Business Equipment First Edition. 11 pp.
JTC1 9295-88. Acoustics—Measurement of High-Frequency Noise Emitted by Computer and Business Equipment First Edition; (Corrected and Reprinted -1989). 14 pp.
JTC1 9296-88. Acoustics—Declared Noise Emission Values of Computer and Business Equipment First Edition. 11 pp.

Computer Graphics

See Also: Coded Character Sets; Computer Graphics Interface; Computer Graphics Software; Computer Programs; Computers; Raster Graphics; Software
ECMA ECMA 116-86. BASIC: ECMA Basic 1, ECMA Basic 2, ECMA Graphics Module. 202 pp.

—Architectural Drawings—Information Interchange

SAA AS 3643.2-92. Computer Graphics—Initial Graphics Exchange Specification (IGES) for Digital Exchange of Product Definition Data—Part 2: Subset of AS 3643.1—Two-Dimensional Drawings for Architectural, Engineering and Construction (AEC) Industries. 37 pp.

—Conformance Testing

BSI BS ISO/IEC 10641-93. 1993 Information Technology—Computer Graphics and Image Processing—Conformance Testing of Implementations of Graphics Standards (S). 34 pp.
IEC 10641-93. Information Technology—Computer Graphics and Image Processing—Conformance Testing of Implementations of Graphics Standards First Edition. 31 pp.
ISO 10641-93. Information Technology—Computer Graphics and Image Processing—Conformance Testing of Implementations of Graphics Standards First Edition. 31 pp.
JTC1 10641-93. Information Technology—Computer Graphics and Image Processing—Conformance Testing of Implementations of Graphics Standards First Edition. 32 pp.

—Glossaries

BSI BS ISO 2382/13-84. 1984 Information Technology—Vocabulary 2382/13: Computer Graphics. 31 pp.
CNS C5203-85. Data Processing Vocabulary (Part 13: Computer Graphics) (Nov)(11405).
ISO 2382 Pt 13-84. Data Processing Vocabulary—Part 13: Computer Graphics First Edition. 27 pp.
JIS X 0013-87. Glossary of Terms Used in Information Processing (Computer Graphics).
JTC1 2382 Pt 13-84. Data Processing Vocabulary—Part 13: Computer Graphics First Edition. 27 pp.
OSI ISO 2382-13-84. Data Processing Vocabulary—Part 13: Computer Graphics. 27 pp.

—Information Interchange

SNZ NZS/AS 3643. 1-89. Computer Graphics—Initial Graphics Exchange Specification (IGES) for Digital Exchange of Product Definition Data Part 1: General (This is a Joint Standard with SAA AS 3643.1). 516 pp.

Computer Graphics *(Cont.)*

—Reference Models

BSI BS ISO/IEC 11072-92. 1992 Information Technology—Computer Graphics—Computer Graphics Reference Model. 43 pp.
IEC 11072-92. Information Technology—Computer Graphics—Computer Graphics Reference Model First Edition. 41 pp.
ISO 11072-92. Information Technology—Computer Graphics—Computer Graphics Reference Model First Edition. 41 pp.
JTC1 11072-92. Information Technology—Computer Graphics—Computer Graphics Reference Model First Edition. 41 pp.

—Registration Procedures

BSI DD 182-90. 1990 Computer Graphics: Procedures for Registration of Graphical Items. 69 pp.
IEC TR9973-88. Information Processing—Procedures for Registration of Graphical Items First Edition. 69 pp.
IEC DIS 9973-93. Information Technology—Computer Graphics and Image Processing—Procedures for Registration of Graphical Items ***CD-ROM ONLY***. 36 pp.
ISO TR9973-88. Information Processing—Procedures for Registration of Graphical Items First Edition. 69 pp.
ISO DIS 9973-93. Information Technology—Computer Graphics and Image Processing—Procedures for Registration of Graphical Items ***CD-ROM ONLY***. 36 pp.
JTC1 TR9973-88. Information Processing—Procedures for Registration of Graphical Items First Edition. 69 pp.
JTC1 DIS9973-93. Information Technology—Computer Graphics and Image Processing—Procedures for Registration of Graphical Items ***CD-ROM ONLY***. 36 pp.
OSI ISO IEC TR 9973-88. Information Processing—Procedures for Registration of Graphical Items. 69 pp.

Computer Graphics Interface

Use For: CGI ***See Also:*** Communication Interfaces; Computer Graphics
BSI BS ISO/IEC 9636-1-91. 1991 Information Technology—Computer Graphics—Interfacing Techniques for Dialogues with Graphical Devices (CGI)—Functional Specification—Part 1: Overview, Profiles, and Conformance. 91 pp.
BSI BS ISO/IEC 9636-2-91. 1991 Information Technology—Computer Graphics—Interfacing Techniques for Dialogues with Graphical Devices (CGI)—Functional Specification—Part 2: Control. 53 pp.
IEC 9636 Pt 1-91. Information Technology—Computer Graphics—Interfacing Techniques for Dialogues with Graphical Devices (CGI)—Functional Specification—Part 1: Overview, Profiles, and Conformance First Edition. 89 pp.
ISO 9636 Pt 1-91. Information Technology—Computer Graphics—Interfacing Techniques for Dialogues with Graphical Devices (CGI)—Functional Specification—Part 1: Overview, Profiles, and Conformance First Edition. 89 pp.
ISO 9636 Pt 2-91. Information Technology—Computer Graphics—Interfacing Techniques for Dialogues with Graphical Devices (CGI)—Functional Specification—Part 2: Control First Edition. 50 pp.
JTC1 9636-91. Information Technology—Computer Graphics—Interfacing Techniques for Dialogues with Graphical Devices (CGI)—Functional Specification—Part 1: Overview, Profiles, and Conformance First Edition. 89 pp.
JTC1 9636-91. Information Technology—Computer Graphics—Interfacing Techniques for Dialogues with Graphical Devices (CGI)—Functional Specification—Part 2: Control First Edition. 50 pp.
OSI ISO IEC DIS 9636-1-90. Information Technology—Computer Graphics—Interfacing Techniques for Dialogues with Graphical Devices—Functional Specification Part 1: Overview, Profiles and Conformance. 101 pp.

—Conformance

BSI BS ISO/IEC 9636-1-91. 1991 Information Technology—Computer Graphics—Interfacing Techniques for Dialogues with Graphical Devices (CGI)—Functional Specification—Part 1: Overview, Profiles, and Conformance. 91 pp.
BSI BS ISO/IEC 9636-2-91. 1991 Information Technology—Computer Graphics—Interfacing Techniques for Dialogues with Graphical Devices (CGI)—Functional Specification—Part 2: Control. 53 pp.
IEC 9636 Pt 1-91. Information Technology—Computer Graphics—Interfacing Techniques for Dialogues with Graphical Devices (CGI)—Functional Specification—Part 1: Overview, Profiles, and Conformance First Edition. 89 pp.

Computer Graphics Interface *(Cont.)*

—Conformance *(Cont.)*

ISO 9636 Pt 1-91. Information Technology—Computer Graphics—Interfacing Techniques for Dialogues with Graphical Devices (CGI)—Functional Specification—Part 1: Overview, Profiles, and Conformance First Edition. 89 pp.
ISO 9636 Pt 2-91. Information Technology—Computer Graphics—Interfacing Techniques for Dialogues with Graphical Devices (CGI)—Functional Specification—Part 2: Control First Edition. 50 pp.
JTC1 9636-91. Information Technology—Computer Graphics—Interfacing Techniques for Dialogues with Graphical Devices (CGI)—Functional Specification—Part 1: Overview, Profiles, and Conformance First Edition. 89 pp.
JTC1 9636-91. Information Technology—Computer Graphics—Interfacing Techniques for Dialogues with Graphical Devices (CGI)—Functional Specification—Part 2: Control First Edition. 50 pp.

—Coordinate Space Control

BSI BS ISO/IEC 9636-2-91. 1991 Information Technology—Computer Graphics—Interfacing Techniques for Dialogues with Graphical Devices (CGI)—Functional Specification—Part 2: Control. 53 pp.
IEC 9636 Pt 2-91. Information Technology—Computer Graphics—Interfacing Techniques for Dialogues with Graphical Devices (CGI)—Functional Specification—Part 2: Control First Edition. 50 pp.
ISO 9636 Pt 2-91. Information Technology—Computer Graphics—Interfacing Techniques for Dialogues with Graphical Devices (CGI)—Functional Specification—Part 2: Control First Edition. 50 pp.
JTC1 9636-91. Information Technology—Computer Graphics—Interfacing Techniques for Dialogues with Graphical Devices (CGI)—Functional Specification—Part 2: Control First Edition. 50 pp.

—Encoding

BSI BS ISO/IEC 9636-5-91. 1991 Information Technology—Computer Graphics—Interfacing Techniques for Dialogues with Graphical Devices (CGI)—Functional Specification—Part 5: Input and Echoing. 103 pp.
BSI BS ISO/IEC 9637-2-92. 1992 Information Technology—Computer Graphics—Interfacing Techniques for Dialogues with Graphical Devices (CGI)—Data Stream Binding—Part 2: Binary Encoding (S). 80 pp.
IEC 9636 Pt 5-91. Information Technology—Computer Graphics—Interfacing Techniques for Dialogues with Graphical Devices (CGI)—Functional Specification—Part 5: Input and Echoing First Edition. 101 pp.
IEC DIS 9637 Pt 1-92. Information Technology—Interface Techniques for Dialogues with Graphical Devices—CGI Data Stream Encoding—Part 1: Character Encoding ***CD-ROM ONLY***. 69 pp.
IEC 9637 Pt 2-92. Information Technology—Computer Graphics—Interfacing Techniques for Dialogues with Graphical Devices (CGI)—Data Stream Binding—Part 2: Binary Encoding First Edition. 78 pp.
ISO 9636 Pt 5-91. Information Technology—Computer Graphics—Interfacing Techniques for Dialogues with Graphical Devices (CGI)—Functional Specification—Part 5: Input and Echoing First Edition. 101 pp.
ISO DIS 9637 Pt 1-92. Information Technology—Interface Techniques for Dialogues with Graphical Devices—CGI Data Stream Encoding—Part 1: Character Encoding ***CD-ROM ONLY***. 69 pp.
ISO 9637 Pt 2-92. Information Technology—Computer Graphics—Interfacing Techniques for Dialogues with Graphical Devices (CGI)—Data Stream Binding—Part 2: Binary Encoding First Edition. 78 pp.
JTC1 9636-91. Information Technology—Computer Graphics—Interfacing Techniques for Dialogues with Graphical Devices (CGI)—Functional Specification—Part 5: Input and Echoing First Edition. 101 pp.
JTC1 9637-92. Information Technology—Computer Graphics—Interfacing Techniques for Dialogues with Graphical Devices (CGI)—Data Stream Binding—Part 2: Binary Encoding First Edition. 78 pp.
OSI ISO/IEC DIS 9636-5-90. Information Technology—Computer Graphics—Interfacing Techniques for Dialogues with Graphical Devices—Functional Specification Part 5: Input and Echoing. 141 pp.
OSI ISO IEC DIS 9637-2-91. Information Technology—Computer Graphics—Interfacing Techniques for Dialogues with Graphical Devices (CGI)—Data Stream Binding—Part 2: Binary Encoding.

—Error Control

BSI BS ISO/IEC 9636-2-91. 1991 Information Technology—Computer Graphics—Interfacing Techniques for Dialogues with Graphical Devices (CGI)—Functional Specification—Part 2: Control. 53 pp.
IEC 9636 Pt 2-91. Information Technology—Computer Graphics—Interfacing Techniques for Dialogues with Graphical Devices (CGI)—Functional Specification—Part 2: Control First Edition. 50 pp.

INDUSTRY STANDARDS
573

Computer Graphics Interface *(Cont.)*

—Error Control *(Cont.)*

ISO 9636 Pt 2-91. Information Technology—Computer Graphics—Interfacing Techniques for Dialogues with Graphical Devices (CGI)—Functional Specification—Part 2: Control First Edition. 50 pp.

JTC1 9636-91. Information Technology—Computer Graphics—Interfacing Techniques for Dialogues with Graphical Devices (CGI)—Functional Specification—Part 2: Control First Edition. 50 pp.

—Input—Echoing

BSI BS ISO/IEC 9636-5-91. 1991 Information Technology—Computer Graphics—Interfacing Techniques for Dialogues with Graphical Devices (CGI)—Functional Specification—Part 5: Input and Echoing. 103 pp.

IEC 9636 Pt 5-91. Information Technology—Computer Graphics—Interfacing Techniques for Dialogues with Graphical Devices (CGI)—Functional Specification—Part 5: Input and Echoing First Edition. 101 pp.

ISO 9636 Pt 5-91. Information Technology—Computer Graphics—Interfacing Techniques for Dialogues with Graphical Devices (CGI)—Functional Specification—Part 5: Input and Echoing First Edition. 101 pp.

JTC1 9636-91. Information Technology—Computer Graphics—Interfacing Techniques for Dialogues with Graphical Devices (CGI)—Functional Specification—Part 5: Input and Echoing First Edition. 101 pp.

—Output

BSI BS ISO/IEC 9636-3-91. 1991 Information Technology—Computer Graphics—Interfacing Techniques for Dialogues with Graphical Devices (CGI)—Functional Specification—Part 3: Output. 160 pp.

IEC 9636 Pt 3-91. Information Technology—Computer Graphics—Interfacing Techniques for Dialogues with Graphical Devices (CGI)—Functional Specification—Part 3: Output First Edition. 158 pp.

ISO 9636 Pt 3-91. Information Technology—Computer Graphics—Interfacing Techniques for Dialogues with Graphical Devices (CGI)—Functional Specification—Part 3: Output First Edition. 158 pp.

JTC1 9636-91. Information Technology—Computer Graphics—Interfacing Techniques for Dialogues with Graphical Devices (CGI)—Functional Specification—Part 3: Output First Edition. 158 pp.

OSI ISO IEC DIS 9636-3-90. Information Technology—Computer Graphics—Interfacing Techniques for Dialogues with Graphical Devices—Functional Specification Part 3: Output. 203 pp.

—Profiles

BSI BS ISO/IEC 9636-1-91. 1991 Information Technology—Computer Graphics—Interfacing Techniques for Dialogues with Graphical Devices (CGI)—Functional Specification—Part 1: Overview, Profiles, and Conformance. 91 pp.

BSI BS ISO/IEC 9636-2-91. 1991 Information Technology—Computer Graphics—Interfacing Techniques for Dialogues with Graphical Devices (CGI)—Functional Specification—Part 2: Control. 53 pp.

IEC 9636 Pt 1-91. Information Technology—Computer Graphics—Interfacing Techniques for Dialogues with Graphical Devices (CGI)—Functional Specification—Part 1: Overview, Profiles, and Conformance First Edition. 89 pp.

ISO 9636 Pt 1-91. Information Technology—Computer Graphics—Interfacing Techniques for Dialogues with Graphical Devices (CGI)—Functional Specification—Part 1: Overview, Profiles, and Conformance First Edition. 89 pp.

ISO 9636 Pt 2-91. Information Technology—Computer Graphics—Interfacing Techniques for Dialogues with Graphical Devices (CGI)—Functional Specification—Part 2: Control First Edition. 50 pp.

JTC1 9636-91. Information Technology—Computer Graphics—Interfacing Techniques for Dialogues with Graphical Devices (CGI)—Functional Specification—Part 1: Overview, Profiles, and Conformance First Edition. 89 pp.

JTC1 9636-91. Information Technology—Computer Graphics—Interfacing Techniques for Dialogues with Graphical Devices (CGI)—Functional Specification—Part 2: Control First Edition. 50 pp.

—Raster Graphics

BSI BS ISO/IEC 9636-6-91. 1991 Information Technology—Computer Graphics—Interfacing Techniques for Dialogues with Graphical Devices (CGI)—Functional Specification—Part 6: Raster. 58 pp.

IEC 9636 Pt 6-91. Information Technology—Computer Graphics—Interfacing Techniques for Dialogues with Graphical Devices (CGI)—Functional Specification—Part 6: Raster First Edition. 55 pp.

ISO 9636 Pt 6-91. Information Technology—Computer Graphics—Interfacing Techniques for Dialogues with Graphical Devices (CGI)—Functional Specification—Part 6: Raster First Edition. 55 pp.

Computer Graphics Interface *(Cont.)*

—Raster Graphics *(Cont.)*

JTC1 9636-91. Information Technology—Computer Graphics—Interfacing Techniques for Dialogues with Graphical Devices (CGI)—Functional Specification—Part 6: Raster First Edition. 55 pp.

OSI ISO IEC DIS 9636-6-90. Information Technology—Computer Graphics—Interfacing Techniques for Dialogues with Graphical Devices-Functional Specification Part 6: Raster. 66 pp.

—Segments

BSI BS ISO/IEC 9636-4-91. 1991 Information Technology—Computer Graphics—Interfacing Techniques for Dialogues with Graphical Devices (CGI)—Functional Specification—Part 4: Segments (S). 50 pp.

IEC 9636 Pt 4-91. Information Technology—Computer Graphics—Interfacing Techniques for Dialogues with Graphical Devices (CGI)—Functional Specification—Part 4: Segments First Edition. 47 pp.

ISO 9636 Pt 4-91. Information Technology—Computer Graphics—Interfacing Techniques for Dialogues with Graphical Devices (CGI)—Functional Specification—Part 4: Segments First Edition. 47 pp.

JTC1 9636-91. Information Technology—Computer Graphics—Interfacing Techniques for Dialogues with Graphical Devices (CGI)—Functional Specification—Part 4: Segments First Edition. 47 pp.

OSI ISO IEC DIS 9636-4-90. Information Technology—Computer Graphics—Interfacing Techniques for Dialogues with Graphical Devices—Functional Specification Part 4: Segments. 60 pp.

—Virtual Device Control

BSI BS ISO/IEC 9636-2-91. 1991 Information Technology—Computer Graphics—Interfacing Techniques for Dialogues with Graphical Devices (CGI)—Functional Specification—Part 2: Control. 53 pp.

IEC 9636 Pt 2-91. Information Technology—Computer Graphics—Interfacing Techniques for Dialogues with Graphical Devices (CGI)—Functional Specification—Part 2: Control First Edition. 50 pp.

ISO 9636 Pt 2-91. Information Technology—Computer Graphics—Interfacing Techniques for Dialogues with Graphical Devices (CGI)—Functional Specification—Part 2: Control First Edition. 50 pp.

JTC1 9636-91. Information Technology—Computer Graphics—Interfacing Techniques for Dialogues with Graphical Devices (CGI)—Functional Specification—Part 2: Control First Edition. 50 pp.

OSI ISO IEC DIS 9636-2-90. Information Technology—Computer Graphics—Interfacing Techniques for Dialogues with Graphical Devices—Functional Specification Part 2: Control. 58 pp.

Computer Graphics Software

Use For: Graphics Software ***See Also:*** Computer Graphics; Software

JTC1 9592-89. Information Processing Systems—Computer Graphics—Programmer's Hierarchical Interactive Graphics System (PHIGS) —Part 2: Archive File Format First Edition; (Amendment 1-1992). 5 pp.

—Graphical Kernel System

BSI BS 6390-85. 1985 Amd 1 Set of Functions for Computer Graphics Programming, the Graphical Kernel System (GKS) (AMD 5341) June 28, 1991. 260 pp.

BSI BS 7040: Part 3-89. (WITHDRAWN) 1989 Computer Graphics: Graphical Kernel System (GKS) Language Bindings Part 3: GKS Language Binding for Ada (Renumbered as BS EN 28651-3: 1992). 188 pp.

BSI BS 7196-89. (WITHDRAWN) 1989 Set of Functions for Computer Graphics Programming: Graphical Kernel System for Three Dimensions (GKS-3D) (Renumbered BS EN 28805: 1992). 387 pp.

BSI BS EN 28805-92. 1992 Set of Functions for Computer Graphics Programming: Graphical Kernel System for Three Dimensions (GKS-3D) (ISO 8805: 1988) (AMD 7371) November 15, 1992. 389 pp.

BSI DD 128-86. (WITHDRAWN) 1986 Computer Graphics: Graphical Kernel System for Three Dimensions (GKS-3D) Functional Description (Superseded by BS 7196: 1989). 186 pp.

CEN EN 27942-86. Information Processing Systems Computer Graphics Graphical Kernel System (GKS)—Functional Description. 249 pp.

CEN EN 28805-92. Information Processing Systems—Computer Graphics—Graphical Kernel System for Three Dimensions (GKS-3D) Functional Description. 389 pp.

CEN EN 28805-92. Information Processing Systems—Computer Graphics—Graphical Kernel System for Three Dimensions (GKS-3D) Functional Description. 381 pp.

CENELEC PRETS 300 073-90. Videotex Presentation Layer Data Syntax Geometric Display (CEPT Recommendation T/TE 06-02, Edinburgh 1988). 125 pp.

Computer Graphics Software *(Cont.)*

—Graphical Kernel System *(Cont.)*

CENELEC ETS 300 073-90. Videotex Presentation Layer Data Syntax Geometric Display (CEPT Recommendation T/TE 06. 20, Edinburgh 1988). 126 pp.

CEPT T/TE 06-02 E-88. Videotex Presentation Layer Data Syntax Geometric Display. 122 pp.

CEPT T/TE 06-04 E-88. Videotex Presentation Layer Data Syntax Processable Data. 88 pp.

ECMA ECMA 96-85. Syntax of Graphical Data for Multiple-Workstation Interface (GDS). 273 pp.

ETSI PRETS 300 073-90. Videotex Presentation Layer Data Syntax Geometric Display (CEPT Recommendation T/TE 06-02, Ediburgh 1988). 124 pp.

ETSI ETS 300 072-90. Terminal Equipment (TE); Videotex Presentation Layer Protocol Videotex Presentation Layer Data Syntax. 159 pp.

ETSI ETS 300 073-90. Videotex Presentation Layer Data Syntax Geometric Display (CEPT Recommendation T/TE 06. 20, Edinburgh 1988). 126 pp.

IEC 7942 AMD 1. Amendment 1—Information Processing Systems—Computer Graphics—Graphical Kernel System (GKS) Functional Description; (1991). 36 pp.

ISO 7942-85. Information Processing Systems—Computer Graphics—Graphical Kernel System (GKS) Functional Description First Edition; (Amendment 1-1991). 287 pp.

ISO 8805-88. Information Processing Systems—Computer Graphics—Graphical Kernel System for Three Dimensions (GKS-3D) Functional Description First Edition. 387 pp.

JIS X 4201-90. Information Processing Systems—Computer Graphics—Graphical Kernel System (GKS) Functional Description.

JTC1 7942-85. Information Processing Systems—Computer Graphics—Graphical Kernel System (GKS) Functional Description First Edition; (Amendment 1-1991). 287 pp.

JTC1 8805-88. Information Processing Systems—Computer Graphics—Graphical Kernel System for Three Dimensions (GKS-3D) Functional Description First Edition. 387 pp.

OSI ISO 7942-85. Information Processing Systems—Computer Graphics—Graphical Kernel System (GKS) Functional Description (with Amendment 1). 288 pp.

OSI ISO 8805-88. Information Processing Systems—Computer Graphics—Graphical Kernal System for the Three Dimensions (GKS-3D) Functional Description. 387 pp.

OSI ISO DIS 8805-88. Information Processing Systems—Computer Graphics—Graphical Kernel System for Three Dimensions (GKS-3D) Functional Description. 383 pp.

OSI ISO DP 8805-87. Computer Graphics—Graphical Kernel System for Three Dimensions (GKS-3D). 388 pp.

OSI ISO DIS 8805-88. Information Processing Systems—Computer Graphics—Graphical Kernel System for Three Dimensions (GKS-3D) Functional Description.

OSI ISO DP 8805-87. Computer Graphics-Graphical Kernel System for Three Dimensions (GKS-3D).

SNZ NZS/ISO 7942-85. Information Processing Systems—Computer Graphics—Graphical Kernel System (GKS) Functional Description. 245 pp.

SNZ NZS/ISO 8805-88. Computer Graphics—Graphical Kernel System for Three Dimensions (GKS-3D) Functional Description. 379 pp.

—Graphical Kernel System—Image Processing

IEC DIS 12087 Pt 2-92. Information Technology—Computer Graphics and Image Processing—Image Processing and Interchange (IPI)—Functional Specification —Part 2: Programmer's Imaging Kernel System Application Programme Interface ***CD-ROM ONLY***. 915 pp.

ISO DIS 12087 Pt 2-92. Information Technology—Computer Graphics and Image Processing—Image Processing and Interchange (IPI)—Functional Specification —Part 2: Programmer's Imaging Kernel System Application Programme Interface ***CD-ROM ONLY***. 915 pp.

JTC1 DIS12087 Pt 2-92. Information Technology—Computer Graphics and Image Processing—Image Processing and Interchange (IPI)—Functional Specification —Part 2: Programmer's Imaging Kernel System Application Programme Interface ***CD-ROM ONLY***. 915 pp.

—Graphical Kernel System—Language Binding—ADA

BSI BS EN 28651-3-92. 1992 Information Processing Systems—Computer Graphics—Graphical Kernel System (GKS) Language Bindings Part 3: Ada (ISO 8651-3: 1988) (AMD 7539) December 15, 1992. 194 pp.

INTERNATIONAL AND NON-U.S. NATIONAL STANDARDS
SUBJECT INDEX
Computer

Computer Graphics Software *(Cont.)*

—Graphical Kernel System—Language Binding—ADA *(Cont.)*

CEN EN 28651-3-92. Information Processing Systems—Computer Graphics—Graphical Kernel System (GKS) Language Bindings Part 3 Ada; (ISO 8651-3: 1988). 186 pp.

ISO 8651 Pt 3-88. Information Processing Systems—Computer Graphics—Graphical Kernel System (GKS) Language Bindings—Part 3: ADA First Edition; (CEN EN 28651-3:1992). 188 pp.

JTC1 8651 Pt 3-88. Information Processing Systems—Computer Graphics—Graphical Kernel System (GKS) Language Bindings—Part 3: ADA First Edition. 188 pp.

OSI ISO 8651-3-88. Information Processing Systems—Computer Graphics—Graphical Kernal System (GKS) Language Bindings Part 3: Ada. 188 pp.

OSI ISO DP 8651-3.2-87. Computer Graphics—GKS—Language Bindings—Part 3: Ada. 185 pp.

SNZ NZS/ISO 8651. 3-88. Computer Graphics—Graphical Kernel System (GKS) Language Bindings Part 3: Ada. 184 pp.

—Graphical Kernel System—Language Binding—C Language

BSI BS ISO/IEC 8651-4-91. 1991 Information Technology—Computer Graphics—Graphical Kernel System (GKS) Language Bindings—Part 4: C. 205 pp.

IEC 8651 Pt 4-91. Information Technology—Computer Graphics—Graphical Kernel System (GKS) Language Bindings—Part 4: C First Edition. 202 pp.

ISO 8651 Pt 4-91. Information Technology—Computer Graphics—Graphical Kernel System (GKS) Language Bindings—Part 4: C First Edition. 202 pp.

JTC1 8651-91. Information Technology—Computer Graphics—Graphical Kernel System (GKS) Language Bindings—Part 4: C First Edition. 202 pp.

OSI ISO/IEC DIS 8651-4-90. Information Technology—Computer Graphics—Graphic Kernel System (GKS) Language Bindings—Part 4: C. 189 pp.

OSI ISO/IEC DIS 8806-4-90. Information Technology—Computer Graphics—Graphical Kernel System for Three Dimensions (GKS-3D) Language Bindings—Part 4: C. 253 pp.

—Graphical Kernel System—Language Binding—FORTRAN

BSI BS 7040: Part 1-89. (WITHDRAWN) 1989 Computer Graphics: Graphical Kernel System (GKS) Language Bindings Part 1: GKS Language Binding for FORTRAN (Renumbered as BS EN 28651-1: 1992). 122 pp.

BSI BS EN 28651-1-92. 1992 Information Processing Systems—Computer Graphics—Graphical Kernel System (GKS) Language Bindings Part 1: FORTRAN (ISO 8651-1: 1988) (AMD 7537) December 15, 1992. 126 pp.

BSI DD 150: Part 1-87. (WITHDRAWN) 1987 Computer Graphics: Graphical Kernel System for Three Dimensions (GKS-3D) Language Bindings Part 1: Fortran (Superseded by ISO/DIS 8806-1). 48 pp.

CEN EN 28651-1-92. Information Processing Systems—Computer Graphics—Graphical Kernel System (GKS) Language Bindings Part 1 FORTRAN; (ISO 8651-1: 1988). 118 pp.

ISO 8651 Pt 1-88. Information Processing Systems—Computer Graphics—Graphical Kernel System (GKS) Language Bindings—Part 1: FORTRAN First Edition; (CEN EN 28651-1:1992). 119 pp.

JTC1 8651 Pt 1-88. Information Processing Systems—Computer Graphics—Graphical Kernel System (GKS) Language Bindings—Part 1: FORTRAN First Edition. 119 pp.

OSI ISO DP 8651-1-86. Computer Graphics-Graphical Kernel System (GKS) Language Bindings: Part 1: FORTRAN. 115 pp.

OSI ISO 8651-1-88. Information Processing Systems—Computer Graphics—Graphical Kernel Systems (GKS) Language Bindings—Part 1: FORTRAN. 119 pp.

OSI ISO/IEC DIS 8806-1-88. Information Processing Systems—Computer Graphics—Graphical Kernal System for Three Dimensions (GKS-3D) Language Bindings—Part 1: Fortran. 137 pp.

SNZ NZS/ISO 8651. 1-88. Computer Graphics—Graphical Kernel System (GKS) Language Bindings Part 1: FORTRAN. 116 pp.

—Graphical Kernel System—Language Binding—Pascal

BSI BS 7040: Part 2-89. (WITHDRAWN) 1989 Computer Graphics: Graphical Kernel System (GKS) Language Bindings Part 2: GKS Language Binding for Pascal (Renumbered as BS EN 28651-2: 1992). 173 pp.

Computer Graphics Software *(Cont.)*

—Graphical Kernel System—Language Binding—Pascal *(Cont.)*

BSI BS EN 28651-2-92. 1992 Information Processing Systems—Computer Graphics—Graphical Kernel System (GKS) Language Bindings Part 2: Pascal (ISO 8651-2: 1988) (AMD 7538) December 15, 1992. 178 pp.

CEN EN 28651-2-92. Information Processing Systems—Computer Graphics—Graphical Kernel System (GKS) Language Bindings Part 2 Pascal; (ISO 8651-2: 1988). 171 pp.

ISO 8651 Pt 2-88. Information Processing Systems—Computer Graphics—Graphical Kernel System (GKS) Language Bindings—Part 2: Pascal First Edition; (CEN EN 28651-2:1992). 174 pp.

JTC1 8651 Pt 2-88. Information Processing Systems—Computer Graphics—Graphical Kernel System (GKS) Language Bindings—Part 2: Pascal First Edition. 174 pp.

OSI ISO 8651-2-88. Information Processing Systems-Computer Graphics-Graphical Kernel System (GKS) Language Bindings—Part 2: Pascal. 175 pp.

SNZ NZS/ISO 8651. 2-88. Computer Graphics—Graphical Kernel System (GKS) Language Bindings Part 2: Pascal. 168 pp.

—Metafiles

BSI BS 6945: Part 1-88. (WITHDRAWN) 1988 Amd 1 Computer Graphics: Metafile for the Storage and Transfer of Picture Description Information (CGM) Part 1: Functional Specification (ISO 8632-1: 1987) (AMD 6353) August 30, 1991 (Renumbered as BS EN 28632-1: 1992). 332 pp.

BSI BS EN 28632-1-92. 1992 Computer Graphics: Metafile for the Storage and Transfer of Picture Description Information (CGM) Part 1: Functional Specification (AMD 7372) November 15, 1992. 336 pp.

CEN EN 28632-1-92. Information Processing Systems—Computer Graphics—Metafile for the Storage and Transfer of Picture Description Information—Part 1: Functional Specification. 296 pp.

IEC 8632 Pt 1-92. Information Technology—Computer Graphics—Metafile for the Storage and Transfer of Picture Description Information—Part 1: Functional Specification Second Edition. 346 pp.

IEC 8632 Pt 1 Draft AMD 1. Information Technology—Computer Graphics—Metafile for the Storage and Transfer of Picture Description Information—Part 1: Functional Specification; (1993) ***CD-ROM ONLY***. 159 pp.

ISO 8632 Pt 1-92. Information Technology—Computer Graphics—Metafile for the Storage and Transfer of Picture Description Information—Part 1: Functional Specification Second Edition. 346 pp.

ISO 8632 Pt 1 Draft AMD 1. Information Technology—Computer Graphics—Metafile for the Storage and Transfer of Picture Description Information—Part 1: Functional Specification; (1993) ***CD-ROM ONLY***. 159 pp.

JTC1 8632-92. Information Technology—Computer Graphics—Metafile for the Storage and Transfer of Picture Description Information—Part 1: Functional Specification Second Edition. 346 pp.

JTC1 8632 Pt1 Draft AMD 1. Information Technology—Computer Graphics—Metafile for the Storage and Transfer of Picture Description Information—Part 1: Functional Specification; (1993) ***CD-ROM ONLY***. 159 pp.

OSI ISO 8632-1/DAM 3-90. Information Processing Systems—Computer Graphics—Metafile for the Storge and Transfer of Picture Description Information—Part 1: Functional Specification AMENDMENT 3. 124 pp.

OSI ISO 8632-1-87. Information Processing Systems—Computer Graphics—Metafile for the Storage and Transfer of Picture Description Information—Part 1: Functional Specification with Amendment 1. 295 pp.

OSI ISO 8632-1-90. Information Processing Systems—Computer Graphics—Metafile for the Storage and Transfer of Picture Description Information—Part 1: Functional Specification—Amendment 1. 108 pp.

OSI ISO 8632-1. Information Processing Systems—Computer Graphics—Metafile for the Storage and Transfer of Picture Description Information.

SNZ NZS/ISO 8632. 1-87. Computer Graphics—Metafiles for the Storage and Transfer of Picture Description Information Part 1: Functional Specifications. 144 pp.

—Metafiles—Encoding

BSI BS 6945: Part 2-88. (WITHDRAWN) 1988 Amd 1 Computer Graphics: Metafile for the Storage and Transfer of Picture Description Information (CGM) Part 2: Character Encoding (ISO 8632-2: 1987) (AMD 6354) August 30, 1992 (Renumbered as BS EN 28632-2: 1992). 88 pp.

BSI BS 6945: Part 3-88. 1988 Computer Graphics: Metafile for the Storage and Transfer of Picture Description Information Part 3: Binary Encoding. 54 pp.

Computer Graphics Software *(Cont.)*

—Metafiles—Encoding *(Cont.)*

BSI BS 6945: Part 3-01. (WITHDRAWN) 1988 Amd 1 Computer Graphics: Metafile for the Storage and Transfer of Picture Description Information (CGM) Part 3: Binary Encoding (AMD 6355) November 29, 1991 (Renumbered as BS EN 28632-3: 1992). 71 pp.

BSI BS 6945: Part 4-88. (WITHDRAWN) 1988 Amd 1 Computer Graphics: Metafile for the Storage and Transfer of Picture Description Information (CGM) Part 4: Clear Text Encoding (ISO 8632-4: 1987) (AMD 6356) August 30, 1991 (Renumbered as BS EN 28632-4: 1992). 36 pp.

BSI BS EN 28632-2-92. 1992 Computer Graphics: Metafile for the Storage and Transfer of Picture Description Information (CGM) Part 2: Character Encoding (AMD 7373) November 15, 1992. 93 pp.

BSI BS EN 28632-3-92. 1992 Computer Graphics: Metafile for the Storage and Transfer of Picture Description Information (CGM) Part 3: Binary Encoding (AMD 7374) November 15, 1992. 77 pp.

BSI BS EN 28632-4-92. 1992 Computer Graphics: Metafile for the Storage and Transfer of Picture Description Information (CGM) Part 4: Clear Text Encoding (AMD 7375) November 15, 1992. 61 pp.

CEN EN 28632-2-92. Information Processing Systems—Computer Graphics—Metafile for the Storage and Transfer of Picture Description Information—Part 2: Character Encoding. 89 pp.

CEN EN 28632-3-92. Information Processing Systems—Computer Graphics—Metafile for the Storage and Transfer of Picture Description Information—Part 3: Binary Encoding. 56 pp.

CEN EN 28632-4-92. Information Processing Systems—Computer Graphics—Metafile for the Storage and Transfer of Picture Description Information—Part 4: Clear Text Encoding. 60 pp.

CEN EN 28632-4-92. Information Processing Systems—Computer Graphics—Metafile for the Storage and Transfer of Picture Description Information Part 4: Clear Text Encoding. 56 pp.

IEC 8632 Pt 2-92. Information Technology—Computer Graphics—Metafile for the Storage and Transfer of Picture Description Information—Part 2: Character Encoding Second Edition. 98 pp.

IEC 8632 Pt 2 Draft AMD 1. Information Technology—Computer Graphics—Metafile for the Storage and Transfer of Picture Description Information—Part 2: Character Encoding; (1993) ***CD-ROM ONLY***. 6 pp.

IEC 8632 Pt 3-92. Information Technology—Computer Graphics—Metafile for the Storage and Transfer of Picture Description Information—Part 3: Binary Encoding Second Edition. 79 pp.

IEC 8632 Pt 3 Draft AMD 1. Information Technology—Computer Graphics—Metafile for the Storage and Transfer of Picture Description Information—Part 3: Binary Encoding; (1993) ***CD-ROM ONLY***. 6 pp.

IEC 8632 Pt 4-92. Information Technology—Computer Graphics—Metafile for the Storage and Transfer of Picture Description Information—Part 4: Clear Text Encoding Second Edition. 63 pp.

IEC 8632 Pt 4 Draft AMD 1. Information Technology—Computer Graphics—Metafile for the Storage and Transfer of Picture Description Information—Part 4: Clear Text Encoding; (1993) ***CD-ROM ONLY***. 6 pp.

ISO 8632 Pt 2-92. Information Technology—Computer Graphics—Metafile for the Storage and Transfer of Picture Description Information—Part 2: Character Encoding Second Edition; (CEN EN 28632-2:1992). 98 pp.

ISO 8632 Pt 2 Draft AMD 1. Information Technology—Computer Graphics—Metafile for the Storage and Transfer of Picture Description Information—Part 2: Character Encoding; (1993) ***CD-ROM ONLY***. 6 pp.

ISO 8632 Pt 3-92. Information Technology—Computer Graphics—Metafile for the Storage and Transfer of Picture Description Information—Part 3: Binary Encoding Second Edition. 79 pp.

ISO 8632 Pt 3 Draft AMD 1. Information Technology—Computer Graphics—Metafile for the Storage and Transfer of Picture Description Information—Part 3: Binary Encoding; (1993) ***CD-ROM ONLY***. 6 pp.

ISO 8632 Pt 4-92. Information Technology—Computer Graphics—Metafile for the Storage and Transfer of Picture Description Information—Part 4: Clear Text Encoding Second Edition. 63 pp.

ISO 8632 Pt 4 Draft AMD 1. Information Technology—Computer Graphics—Metafile for the Storage and Transfer of Picture Description Information—Part 4: Clear Text Encoding; (1993) ***CD-ROM ONLY***. 6 pp.

JTC1 8632-92. Information Technology—Computer Graphics—Metafile for the Storage and Transfer of Picture Description Information—Part 2: Character Encoding Second Edition. 98 pp.

INDUSTRY STANDARDS
575

Computer Graphics Software *(Cont.)*
—Metafiles—Encoding *(Cont.)*

JTC1 8632 Pt2 Draft AMD 1. Information Technology—Computer Graphics—Metafile for the Storage and Transfer of Picture Description Information—Part 2: Character Encoding; (1993) ***CD-ROM ONLY***. 6 pp.

JTC1 8632-92. Information Technology—Computer Graphics—Metafile for the Storage and Transfer of Picture Description Information—Part 3: Binary Encoding Second Edition. 79 pp.

JTC1 8632 Pt3 Draft AMD 1. Information Technology—Computer Graphics—Metafile for the Storage and Transfer of Picture Description Information—Part 3: Binary Encoding; (1993) ***CD-ROM ONLY***. 6 pp.

JTC1 8632-92. Information Technology—Computer Graphics—Metafile for the Storage and Transfer of Picture Description Information—Part 4: Clear Text Encoding Second Edition. 63 pp.

JTC1 8632 Pt4 Draft AMD 1. Information Technology—Computer Graphics—Metafile for the Storage and Transfer of Picture Description Information—Part 4: Clear Text Encoding; (1993) ***CD-ROM ONLY***. 6 pp.

OSI ISO 8632-2-90. Information Processing Systems—for Computer Graphics—Metafile for the Storage and Transfer of Picture Description Information—Part 2: Character Encoding—Amendment 1. 14 pp.

OSI ISO 8632-2 DAM 3-90. Information Processing Systems—Computer Graphics—Metafile for the Storage and Transfer of Picture Description Information—Part 2: Character Encoding AMENDMENT 3. 28 pp.

OSI ISO 8632-2 PDAD 1-87. Information Processing Systems; Computer Graphics; Metafile for the Storage and Transfer of Picture Descriptioninformation-Part 2: Character Encoding, Proposed Draft Addendum 1 to ISO 8632-2.

OSI ISO 8632-3-87. Information Processing Systems—Computer Graphics—Metafile for the Storage and Transfer of Picture Description Information—Part 3—Binary Encoding (with Amendment 1). 74 pp.

OSI ISO 8632-3/DAM 3-91. Information Processing Systems—Computer Graphics—Metafile for the Storage and Transfer of Picture Description Information—Part 3—Binary Encoding AMENDMENT 3. 22 pp.

OSI ISO 8632-3 PDAD 1-87. Information Processing Systems; Computer Graphics; Metafile for the Storage and Transfer of Picture Description Information Part 3: Binary Encoding, Proposed Draft Addendum 1 to ISO 8632-3.

OSI ISO 8632-4-87. Information Processing Systems—Computer Graphics—Metafile for the Storage and Transfer of Picture Description Information—Part 4: Clear Text Encoding. 59 pp.

OSI ISO 8632-4 DAM 3-91. Information Processing Systems—Computer Graphics—Metafile for the Storage and Transfer of Picture Description Information—Part 4: Clear Text Encoding—Amendment 3. 18 pp.

OSI ISO 8632-4 PDAD 1-87. Information Processing Systems; Computer Graphics; Metafile for the Storage and Transfer of Picture Description Information Part 4: Clear Text Encoding, Proposed Draft Addendum 1 to ISO 8632—4.

SNZ NZS/ISO 8632. 2-87. Computer Graphics—Metafiles for the Storage and Transfer of Picture Description Information Part 2: Character Encoding. 60 pp.

SNZ NZS/ISO 8632. 3-87. Computer Graphics—Metafiles for the Storage and Transfer of Picture Description Information Part 3: Binary Encoding. 50 pp.

SNZ NZS/ISO 8632. 4-87. Computer Graphics—Metafiles for the Storage and Transfer of Picture Description Information Part 4: Clear Text Encoding. 33 pp.

—Programmers' Hierarchical Interactive Graphics Systems

BSI BS 7217: Part 1-89. 1989 Amd 1 Computer Graphics: Programmer's Hierarchical Interactive Graphics System (PHIGS) Part 1: Specification for a Set of Functions (ISO/IEC 9592-1: 1989) (AMD 6385) September 30, 1991. 392 pp.

BSI BS 7217: Part 2-89. 1989 Amd 1 Computer Graphics: Programmer's Hierarchical Interactive Graphics System (PHIGS) Part 2: Specification for Archive File Format (AMD 6387) September 30, 1991. 16 pp.

BSI BS 7217: Part 3-89. 1989 Amd 1 Computer Graphics: Programmer's Hierarchical Interactive Graphics System (PHIGS) Part 3: Specification for Clear-Text Encoding of Archive File (AMD 6388) September 30, 1991. 41 pp.

BSI BS ISO/IEC 9592-4-92. 1992 Information Processing Systems—Computer Graphics—Programmer's Hierarchical Interactive Graphics System (PHIGS) Part 4: Plus Lumiere und Surfaces, PHIGS PLUS. 179 pp.

Computer Graphics Software *(Cont.)*
—Programmers' Hierarchical Interactive Graphics Systems *(Cont.)*

CEN EN 29 592 (Part 1)-91. Information Processing Systems Graphics Hierarchical Interactive Graphics System (PHIGS) — Part 1: Functional Description. 4 pp.

CEN EN 29 592 (Part 2)-91. Information Processing Systems Graphics Hierarchical Interactive Graphics System (PHIGS) — Part 2: Archive File Format. 2 pp.

CEN EN 29 592 (Part 3)-91. Information Processing Systems Graphics Hierarchical Interactive Graphics System (PHIGS) — Part 3: Clear-Text Encoding of Archive File. 2 pp.

IEC 9592 Pt 1-89. Information Processing Systems—Computer Graphics—Programmer's Hierarchical Interactive Graphics System (PHIGS) —Part 1: Functional Description First Edition; (Amendment 1-1992). 389 pp.

IEC 9592 Pt 2-89. Information Processing Systems—Computer Graphics—Programmer's Hierarchical Interactive Graphics System (PHIGS) —Part 2: Archive File Format First Edition; (Amendment 1-1992). 17 pp.

IEC 9592 Pt 3-89. Information Processing Systems—Computer Graphics—Programmer's Hierarchical Interactive Graphics System (PHIGS) —Part 3: Clear-Text Encoding of Archive File First Edition; (Amendment 1-1992). 61 pp.

IEC 9592 Pt 4-92. Information Processing Systems—Computer Graphics—Programmer's Hierarchical Interactive Graphics System (PHIGS)—Part 4: Plus Lumiere und Surfaces, PHIGS PLUS First Edition. 181 pp.

ISO 9592 Pt 1-89. Information Processing Systems—Computer Graphics—Programmer's Hierarchical Interactive Graphics System (PHIGS) —Part 1: Functional Description First Edition; (Amendment 1-1992). 389 pp.

ISO 9592 Pt 2-89. Information Processing Systems—Computer Graphics—Programmer's Hierarchical Interactive Graphics System (PHIGS) —Part 2: Archive File Format First Edition; (Amendment 1-1992). 17 pp.

ISO 9592 Pt 3-89. Information Processing Systems—Computer Graphics—Programmer's Hierarchical Interactive Graphics System (PHIGS) —Part 3: Clear-Text Encoding of Archive File First Edition; (Amendment 1-1992). 61 pp.

ISO 9592 Pt 4-92. Information Processing Systems—Computer Graphics—Programmer's Hierarchical Interactive Graphics System (PHIGS)—Part 4: Plus Lumiere und Surfaces, PHIGS PLUS First Edition. 181 pp.

JTC1 9592-89. Information Processing Systems—Computer Graphics—Programmer's Hierarchical Interactive Graphics System (PHIGS) —Part 1: Functional Description First Edition; (Amendment 1-1992). 5 pp.

JTC1 9592-89. Information Processing Systems—Computer Graphics—Programmer's Hierarchical Interactive Graphics System (PHIGS) —Part 3: Clear-Text Encoding of Archive File First Edition; (Amendment 1-1992). 25 pp.

JTC1 9592-92. Information Processing Systems—Computer Graphics—Programmer's Hierarchical Interactive Graphics System (PHIGS)—Part 4: Plus Lumiere und Surfaces, PHIGS PLUS ***CD-ROM ONLY***. 190 pp.

OSI ISO IEC 9592-1-89. Information Processing Systems—Computer Graphics—Programmer's Hierarchical Interactive Graphics System (PHIGS)—Part 1: Functional Description. 384 pp.

OSI ISO IEC 9592-1 DAM 1-91. Information Processing Systems—Computer Graphics—Programmer's Hierarchical Interactive Graphics System (PHIGS)—Part 1: Functional Description—Amendment 1.

OSI ISO IEC DIS 9592-1-87. Information Processing Systems-Computer Graphics-Programmers Hierarchical Interactive Graphic System (PHIGS) Part 1: Functional Description. 355 pp.

OSI ISO IEC 9592-2-89. Information Processing Systems-Computer Graphics-Programmers Hierarchical Interactive Graphics System (PHIGS) Part 2: Archive File Format. 12 pp.

OSI ISO IEC 9592-2 DAM 1-91. Information Processing Systems—Computer Graphics—Programmer's Hierarchical Interactive Graphics System (PHIGS)—Part 2: Archive File Format—Amendment 1.

OSI ISO DIS 9592-2-87. Information Processing Systems—Computer Graphics—Programmers Hierarchical Interactive Graphic System (PHIGS) Part 2: Archive File Format. 14 pp.

OSI ISO IEC 9592-3-89. Information Processing Systems—Computer Graphics—Programmer's Hierarchical Interactive Graphics System (PHIGS)—Part 3: Clear-Text Encoding of Archive File. 36 pp.

Computer Graphics Software *(Cont.)*
—Programmers' Hierarchical Interactive Graphics Systems *(Cont.)*

OSI ISO IEC 9592-3 DAM 1-91. Information Processing Systems—Computer Graphics—Programmer's Hierarchical Interactive Graphics System (PHIGS)—Part 3: Clear-Text Encoding of Archive File—Amendment 1.

OSI ISO IEC DIS 9592-4-91. Information Processing Systems—Computer Graphics—Programmer's Hierarchical Interactive Graphics System (PHIGS)—Part 4: PHIGS Plus.

OSI ISO IEC DIS 9592-3-87. Information Processing Systems—Computer Graphics—Programmers Hierarchical Interactive Graphic System (PHIGS) Part 3: Clear-Text Encoding of Archive File. 34 pp.

OSI ISO IEC 9593-1-90. Information Processing Systems—Computer Graphics—Programmers Hierarchical Interactive Graphics System (PHIGS) Language Bindings—Part 1: FORTRAN. 219 pp.

SNZ NZS/ISO-IEC 9592.1-89. Computer Graphics—Programmer's Hierarchical Interactive Graphics System (PHIGS) Part 1: Functional Description. 337 pp.

SNZ NZS/ISO-IEC 9592.2-89. Computer Graphics—Programmer's Hierarchical Interactive Graphics System (PHIGS) Part 2: Archive File Format. 32 pp.

SNZ NZS/ISO-IEC 9592.3-89. Computer Graphics—Programmer's Hierarchical Interactive Graphics System (PHIGS) Part 3: Clear-Text Encoding of Archive File. 8 pp.

SNZ NZS/ISO-IEC 9593.1-90. Computer Graphics—Programmer's Hierarchical Interactive Graphics System (PHIGS) Language Bindings Part 1: FORTRAN. 213 pp.

—Programmers' Hierarchical Interactive Graphics Systems—ADA

BSI BS ISO/IEC 9593/3-90. 1990 Information Technology—Computer Graphics—Programmers Hierarchical Interactive Graphics System (PHIGS) Language Bindings 9593/3: Ada. 294 pp.

IEC 9593 Pt 3-90. Information Technology—Computer Graphics—Programmer's Hierarchical Interactive Graphics System (PHIGS) Language Bindings—Part 3: Ada First Edition; (Technical Corrigendum 1 —1993) (ANSI 9593.3-1990) (FIPS PUB 153). 292 pp.

IEC 9593 Pt 3 Draft AMD 1. Information Technology—Computer Graphics—Programmer's Hierarchical Interactive Graphics System (PHIGS) Language Bindings—Part 3: Ada Amendment 1: Incorporation of PHIGS Plus; (1992) ***CD-ROM ONLY***. 339 pp.

ISO 9593 Pt 3-90. Information Technology—Computer Graphics—Programmer's Hierarchical Interactive Graphics System (PHIGS) Language Bindings—Part 3: Ada First Edition; (Technical Corrigendum 1 —1993) (ANSI 9593.3-1990) (FIPS PUB 153). 292 pp.

ISO 9593 Pt 3 Draft AMD 1. Information Technology—Computer Graphics—Programmer's Hierarchical Interactive Graphics System (PHIGS) Language Bindings—Part 3: Ada Amendment 1: Incorporation of PHIGS Plus; (1992) ***CD-ROM ONLY***. 339 pp.

JTC1 9593 Pt 3-90. Information Technology—Computer Graphics—Programmer's Hierarchical Interactive Graphics System (PHIGS) Language Bindings—Part 3: Ada First Edition; (Technical Corrigendum 1 —1993) (ANSI 9593.3-1990) (FIPS PUB 153). 292 pp.

JTC1 9593 Pt3 Draft AMD 1. Information Technology—Computer Graphics—Programmer's Hierarchical Interactive Graphics System (PHIGS) Language Bindings—Part 3: Ada Amendment 1: Incorporation of PHIGS Plus; (1992) ***CD-ROM ONLY***. 339 pp.

OSI ISO/IEC 9593-3-90. Information Technology—Computer Graphics—Programmer's Hierarchical Interactive Graphics System (PHIGS) Language Bindings—Part 3: Ada. 291 pp.

SAA AS 3794.3-91. Computer Graphics—Programmer's Hierarchical Interactive Graphics System (PHIGS) Language Bindings—Part 3: Ada (ISO/IEC 9593-3) (In Professional Packages 26A, 26D). 285 pp.

SNZ NZS/ISO-IEC 9593.3-90. Computer Graphics—Programmer's Hierarchical Interactive Graphics System (PHIGS) Language Bindings Part 3: Ada. 285 pp.

—Programmers' Hierarchical Interactive Graphics Systems—C Language

BSI BS EN 28806-4-93. 1993 Amd 0 Information Technology—Computer Graphics—Graphical Kernel System for Three Dimensions (GKS-3D) Language Bindings Part 4: C (ISO/IEC 8806-4: 1991) (AMD 7932) September 15, 1993 (S). 276 pp.

Computer Graphics Software (Cont.)
—Programmers' Hierarchical Interactive Graphics Systems—C Language (Cont.)
BSI BS ISO/IEC 8806-4-91. 1991 Information Technology—Computer Graphics—Graphical Kernel System for Three Dimensions (GKS-3D) Language Bindings Part 4: C (Renumbered as BS EN 28806-4: 1993). 273 pp.

BSI BS ISO/IEC 9593-4-91. 1991 Information Technology—Computer Graphics—Programmer's Hierarchical Interactive Graphics System (PHIGS) Language Bindings—Part 4: C. 319 pp.

CEN EN 28806-4-93. Information Technology—Computer Graphics—Graphical Kernel System for Three Dimensions (GKS-3D) Language Bindings—Part 4: C (ISO/IEC 8806-4: 1991). 271 pp.

IEC 8806 Pt 4-91. Information Technology—Computer Graphics—Graphical Kernel System for Three Dimensions (GKS-3D) Language Bindings—Part 4: C First Edition (CEN EN 28806-4: 1993). 270 pp.

IEC 9593 Pt 4-91. Information Technology—Computer Graphics—Programmer's Hierarchical Interactive Graphics System (PHIGS) Language Bindings—Part 4: C First Edition. 316 pp.

IEC 9593 Pt 4 Draft AMD 1. Information Technology—Computer Graphics—Programmer's Hierarchical Interactive Graphics System (PHIGS) Language Bindings—Part 4: C; (1992) ***CD-ROM ONLY***. 167 pp.

ISO 8806 Pt 4-91. Information Technology—Computer Graphics—Graphical Kernel System for Three Dimensions (GKS-3D) Language Bindings—Part 4: C First Edition (CEN EN 28806-4: 1993). 270 pp.

ISO 9593 Pt 4-91. Information Technology—Computer Graphics—Programmer's Hierarchical Interactive Graphics System (PHIGS) Language Bindings—Part 4: C First Edition; (ANSI/ISO 9593.4-1991). 316 pp.

ISO 9593 Pt 4 Draft AMD 1. Information Technology—Computer Graphics—Programmer's Hierarchical Interactive Graphics System (PHIGS) Language Bindings—Part 4: C; (1992) ***CD-ROM ONLY***. 167 pp.

JTC1 8806-91. Information Technology—Computer Graphics—Graphical Kernel System for Three Dimensions (GKS-3D) Language Bindings—Part 4: C First Edition. 270 pp.

JTC1 9593-91. Information Technology—Computer Graphics—Programmer's Hierarchical Interactive Graphics System (PHIGS) Language Bindings—Part 4: C First Edition. 316 pp.

JTC1 9593 Pt4 Draft AMD 1. Information Technology—Computer Graphics—Programmer's Hierarchical Interactive Graphics System (PHIGS) Language Bindings—Part 4: C; (1992) ***CD-ROM ONLY***. 167 pp.

OSI ISO/IEC DIS 9593-4-90. Information Technology—Computer Graphics—Programmer's Hierarchical Interactive Graphics System (PHIGS) Language Bindings—Part 4: C. 298 pp.

—Programmers' Hierarchical Interactive Graphics Systems—FORTRAN
BSI BS ISO/IEC 9593/1-90. 1990 Information Technology—Computer Graphics—Programmers Hierarchical Interactive Graphics System (PHIGS) Language Bindings Part 1: FORTRAN. 223 pp.

IEC 9593 Pt 1-90. Information Processing Systems—Computer Graphics—Programmer's Hierarchical Interactive Graphics System (PHIGS) Language Bindings—Part 1: FORTRAN First Edition; (Technical Corrigendum 1 —1993). 221 pp.

IEC 9593 Pt 1 Draft AMD 1. Information Processing Systems—Computer Graphics—Programmer's Hierarchical Interactive Graphics System (PHIGS) Language Bindings—Part 1: FORTRAN; (1992) ***CD-ROM ONLY***. 162 pp.

ISO 9593 Pt 1-90. Information Processing Systems—Computer Graphics—Programmer's Hierarchical Interactive Graphics System (PHIGS) Language Bindings—Part 1: FORTRAN First Edition; (Technical Corrigendum 1 —1993). 221 pp.

ISO 9593 Pt 1 Draft AMD 1. Information Processing Systems—Computer Graphics—Programmer's Hierarchical Interactive Graphics System (PHIGS) Language Bindings—Part 1: FORTRAN; (1992) ***CD-ROM ONLY***. 162 pp.

JTC1 9593 Pt 1-90. Information Processing Systems—Computer Graphics—Programmer's Hierarchical Interactive Graphics System (PHIGS) Language Bindings—Part 1: FORTRAN First Edition; (Technical Corrigendum 1 —1993). 221 pp.

JTC1 9593 Pt1 Draft AMD 1. Information Processing Systems—Computer Graphics—Programmer's Hierarchical Interactive Graphics System (PHIGS) Language Bindings—Part 1: FORTRAN; (1992) ***CD-ROM ONLY***. 162 pp.

OSI ISO DIS 9593-1-88. Information Processing Systems—Computer Graphics—Programmers Hierarchical Interactive Graphics System (PHIGS) Language Bindings Part 1: Fortran Binding. 205 pp.

SAA AS 3794.1-91. Computer Graphics—Programmer's Hierarchical Interactive Graphics System (PHIGS) Language Bindings—Part 1: FORTRAN (ISO/IEC 9593-1) (In Professional Packages 26A, 26D). 213 pp.

—Programmers' Hierarchical Interactive Graphics Systems—Pascal
OSI ISO IEC DP 9593-2-88. Information Processing Systems—Computer Graphics—Programmer's Hierarchical Interactive Graphics System (PHIGS) Language Bindings Part 2: Extended Pascal. 146 pp.

Computer Integrated Manufacturing
Use: CIM

Computer Interfaces
Use: Communication Interfaces; Interfaces

Computer Micrographics
Use For: Micrographics (Computer)
See Also: Computer Output Microfilm; Microfilming; Photomicrography

—Glossaries
BSI BS ISO 6196: Part 7-92. 1992 Micrographics—Vocabulary Part 7: Computer Micrographics. 27 pp.

ISO 6196 Pt 7-92. Micrographics—Vocabulary—Part 07: Computer Micrographics First Edition. 24 pp.

JTC1 6196 Pt 7-92. Micrographics—Vocabulary—Part 07: Computer Micrographics First Edition. 24 pp.

Computer Numerical Control
Use For: CNC Controllers; CNC Units
See Also: Computer Equipment; Control Systems Equipment; Controllers; Numerical Control; Programmable Controllers

—Interface Boards—Lathes
CNS B3462-87. Interface Board for CNC Lathe (Aug)(12044).

Computer On-Line Real-Time Applications Language
Use: CORAL

Computer Output Microfiche
Use: Computer Output Microfilm

Computer Output Microfilm
Use For: COM; Computer Output Microfiche; Computer Output Microform   See Also: Computer Micrographics; Microfilm
BSI BS 5644-78. 1978 Amd 1 Computer Output Microfiche (COM), A6 Size. 22 pp.

CGSB CAN/CGSB-72.15-M83. Computer Output Microfilm (COM), Microfiche. 25 pp.

CGSB CAN/CGSB-72.16-M83. Computer Output Microfilm (COM) 16 mm Roll. 17 pp.

ISO 5126-80. Micrographics—Computer Output Microfiche (COM)—Microfiche A6 Second Edition. 16 pp.

JIS Z 6007-82. Computer Output Microfiche.

JTC1 5126-80. Micrographics—Computer Output Microfiche (COM)—Microfiche A6 Second Edition. 16 pp.

—Aerospace
AECMA PREN2499-83. Computer Output Microfiche (COM) A6 Microfiche. 4 pp.

BSI BS EN 2499-90. 1990 Computer Output Microfiche (COM) A6 Microfiche. 7 pp.

CEN EN 2499-87. Computers Output Microfiche (COM) A6 Microfiche. 4 pp.

—Quality Assurance
BSI BS 7353-91. 1991 Methods of Measuring Quality of Output of Graphic COM Recorders. 15 pp.

BSI BS DD 27-73. (WITHDRAWN) 1973 Quality Requirements for Computer Output on Microfilm (COM) (Superseded by BS ISO 8514-1 & 8514-2: 1992). 10 pp.

—Quality Control
BSI BS ISO 8514-1-92. 1992 Micrographics—Alphanumeric Computer Output Microforms—Quality Control—Part 1: Characteristics of the Test Slide and Test Data (S). 11 pp.

BSI BS ISO 8514-2-92. 1992 Micrographics—Alphanumeric Computer Output Microforms—Quality Control—Part 2: Method (S). 13 pp.

ISO 8514 Pt 1-92. Micrographics—Alphanumeric Computer Output Microforms—Quality Control—Part 1: Characteristics of the Test Slide and Test Data First Edition. 9 pp.

ISO 8514 Pt 2-92. Micrographics—Alphanumeric Computer Output Microforms—Quality Control—Part 2: Method First Edition. 10 pp.

JTC1 8514 Pt 1-92. Micrographics—Alphanumeric Computer Output Microforms—Quality Control—Part 1: Characteristics of the Test Slide and Test Data First Edition. 9 pp.

JTC1 8514 Pt 2-92. Micrographics—Alphanumeric Computer Output Microforms—Quality Control—Part 2: Method First Edition. 10 pp.

—Test Slides
BSI BS ISO 8514-1-92. 1992 Micrographics—Alphanumeric Computer Output Microforms—Quality Control—Part 1: Characteristics of the Test Slide and Test Data (S). 11 pp.

ISO 8514 Pt 1-92. Micrographics—Alphanumeric Computer Output Microforms—Quality Control—Part 1: Characteristics of the Test Slide and Test Data First Edition. 9 pp.

JTC1 8514 Pt 1-92. Micrographics—Alphanumeric Computer Output Microforms—Quality Control—Part 1: Characteristics of the Test Slide and Test Data First Edition. 9 pp.

Computer Output Microform
Use: Computer Output Microfilm

Computer Printouts
Use: Printouts

Computer Programming
Use For: Programming (Computers)
See Also: Algorithms; Computer Programs; Data Processing; Flow Charts; Programming Languages; Systems Manuals

—Files—Record Groups
CNS C5227-89. Information Processing Program Flow for Processing Sequential Files in Terms of Record Groups (Dec)(12644).

—Glossaries
BSI BS ISO/IEC 2382-7-89. 1989 Computer Programming. 34 pp.

CNS C5145-85. Data Processing Vocabulary (Part 7: Digital Computer Programming) (Jul)(9693).

IEC 2382 Pt 7-89. Information Technology—Vocabulary—Part 07: Computer Programming Second Edition. 31 pp.

ISO 2382 Pt 7-89. Information Technology—Vocabulary—Part 07: Computer Programming Second Edition. 31 pp.

JIS X 0007-87. Glossary of Terms Used in Information Processing (Digital Computer Programming).

JTC1 2382 Pt 7-89. Information Technology—Vocabulary—Part 07: Computer Programming Second Edition. 31 pp.

OSI ISO 2382-7-89. Information Technology—Vocabulary—Part 7: Computer Programming. 31 pp.

SAA AS 1189.7-91. Data Processing—Vocabulary—Part 7: Computer Programming (ISO/IEC 2382-7:1989) (In Professional Package 26). 12 pp.

—Representations
CNS C5229-89. Information Processing Program Constructs and Conventions for Their Representation (Dec)(12646).

Computer Programming Languages
Use: Programming Languages

Computer Programs
See Also: Algorithms; Computer Graphics; Computer Programming; Computers; Object Modules; Programming Languages; Software; Systems Manuals
BSI BS 6976-90. 1990 Program Constructs and Conventions for Their Representation (Renumbered as BS EN 28631: 1993). 11 pp.

BSI BS EN 28631-93. 1993 Amd 1 Information Technology—Program Constructs and Conventions for Their Respresentation (ISO/IEC 8631: 1989) (AMD 7942) September 15, 1993 (S). 16 pp.

BSI BS ISO 6593-90. 1990 Information Processing—Program Flow for Processing Sequential Files in Terms of Record Groups. 12 pp.

IEC 8631-89. Information Technology—Program Constructs and Conventions for Their Representation Second Edition. 12 pp.

ISO 6593-85. Information Processing—Program Flow for Processing Sequential Files in Terms of Record Groups First Edition. 9 pp.

ISO 8631-89. Information Technology—Program Constructs and Conventions for Their Representation Second Edition. 12 pp.

JIS X 0128-88. Program Constructs and Conventions for Their Representation. 9 pp.

JTC1 6593-85. Information Processing—Program Flow for Processing Sequential Files in Terms of Record Groups First Edition. 9 pp.

JTC1 8631-89. Information Technology—Program Constructs and Conventions for Their Representation Second Edition. 12 pp.

INTERNATIONAL AND NON-U.S. NATIONAL STANDARDS SUBJECT INDEX

Computer Programs (Cont.)

OSI ISO 6593-85. Information Processing—Program Flow for Processing Sequential Files in Terms of Record Groups. 9 pp.

OSI ISO IEC 8631-89. Information Technology—Program Constructs and Conventions for Their Representation. 12 pp.

SAA AS 3875-91. Information Technology—Program Constructs and Conventions for Their Representation (ISO/IEC 8631:1989). 7 pp.

SNZ NZS/ISO-IEC 8631-89. Information Technology—Program Constructs and Conventions for Their Representation. 7 pp.

—Abstracts of Standards/Specifications

CNS C5147-83. Computer Program Abstracts (Feb)(9980).

—Broadcasting Satellite Services—Planning

CCIR Report 812-3-90. Computer Programs for Planning Broadcasting-Satellite Services in the 12 GHz Band—Section 10/11D—Planning. 7 pp.

—CCIR—Access

CCIR RESOLUTION 104-90. Access to and Exchange of Computer Programs—Volume XIV—Administrative Texts of the CCIR. 2 pp.

—CCIR—Information Interchange

CCIR RESOLUTION 104-90. Access to and Exchange of Computer Programs—Volume XIV—Administrative Texts of the CCIR. 2 pp.

—Diagrams—Program Design

BSI BS 6224-87. 1987 Design Structure Diagrams for Use in Program Design and Other Logical Applications. 36 pp.

—Ionosphere—Forecasting

CCIR RESOLUTION 63-3-90. Computer Programs for the Prediction of Ionospheric Characteristics, Sky-Wave Transmission Loss and Noise—Volume VI—Propagation in Ionized Media. 7 pp.

—Legal Protection

EC 91/250/EEC-91. Council Directive on the Legal Protection of Computer Programs. 7 pp.

—Mine Warfare—Catalogs

NATO STANAG 1164 ED 2 AMD 2-81. Catalogue of Mine Warfare Computer Programs—AMP-12(Navy). 8 pp.

—Radio Spectra

CCIR RESOLUTION 88-1-90. Computer Programs for Radio-Frequency Management—Volume I—Spectrum Utilization and Monitoring. 2 pp.

—Radio Spectra—Information Interchange

CCIR RECMN 668-90. Methods of Exchanging Computer Programs and Data for Spectrum Management Purposes—Section 1A—Spectrum Engineering and Computer-Aided Principles and Techniques. 1 p.

—Reliability

BSI BS 5760: Part 6-91. 1991 Reliability of Systems, Equipment and Components Part 6: Guide to Programmes for Reliability Growth (IEC 1014: 1989) (G). 34 pp.

IEC 1014-89. Programmes for Reliability Growth First Edition. 64 pp.

—Shipbuilding—Glossaries

ISO 7462-85. Shipbuilding—Principal Ship Dimensions—Terminology and Definitions for Computer Applications First Edition. 11 pp.

JTC1 7462-85. Shipbuilding—Principal Ship Dimensions—Terminology and Definitions for Computer Applications First Edition. 11 pp.

—Shipbuilding—Symbols

BSI BS 7467-91. 1991 Symbols for Computer Applications in Ship Building and Marine Technology (ISO 7463: 1990). 34 pp.

ISO 7463-90. Shipbuilding and Marine Structures—Symbols for Computer Applications First Edition. 33 pp.

JTC1 7463-90. Shipbuilding and Marine Structures—Symbols for Computer Applications First Edition. 33 pp.

OSI ISO 7463-90. Shipbuilding and Marine Structures—Symbols for Computer Applications. 33 pp.

Computer Security
Use: Information Security

Computer Software
Use: Software

Computer Support System Software
Use: System Software

Computer Systems

Use For: Computer Systems, Industrial
See Also: Computer Assisted Retrieval Systems; Computer Equipment; Computers; Distributed Processing Systems; Global Positioning Systems; Information Systems; Point of Sale Systems

BSI BS 5887-80. 1980 Code of Practice for Testing of Computer-Based Systems. 10 pp.

BSI BS 6238-82. 1982 Performance Monitoring of Computer-Based Systems. 12 pp.

—Configuration Management

BSI BS 6488-84. 1984 Configuration Management of Computer-Based Systems. 7 pp.

—Documentation

BSI BS 5515-84. 1984 Code of Practice for Documentation of Computer Based Systems. 8 pp.

—Symbols

BSI BS 7153-89. 1989 Guide for Computer System Configuration Diagram (Symbols and Conventions). 18 pp.

CNS C5230-89. Information Processing Systems System Configuration Diagram Symbols and Conventions (Dec)(12647).

ISO 8790-87. Information Processing Systems—Computer System Configuration Diagram Symbols and Conventions First Edition. 18 pp.

JIS X 0127-88. Computer System Configuration Diagram Symbols and Conventions. 21 pp.

JTC1 8790-87. Information Processing Systems—Computer System Configuration Diagram Symbols and Conventions First Edition. 18 pp.

—User Requirements

BSI BS 6719-86. 1986 Guide to Specifying User Requirements for a Computer-Based System. 16 pp.

Computer Systems, Industrial
Use: Computer Systems

Computer Tapes

Use For: Reel Tapes (Magnetic); Reel to Reel Tapes
See Also: Computer Equipment; Data Cartridges; Data Processing Equipment; Disk Cartridges; Information Interchange; Magnet Wire; Magnetic Cards; Magnetic Devices; Magnetic Disks; Magnetic Heads; Magnetic Tapes; Memory (Data Storage); Reperforators; Tape Cartridges; Tape Drives; Tape Recorders; Tape Recording; Tapes

BSI BS 5079-81. 1981 Amd 1 Information Interchange on 3.81 mm (0.150 in) Magnetic Tape Cassette at 4 cpmm (100 cpi), Phase Encoded at 63 ftpmm (1600 ftpi). 28 pp.

BSI BS 7062-02. 1988 Amd 2 Construction and Use of 12.7 mm (0.5 in) Wide Magnetic Tape Cartridges for Data Interchange, Recording at 1491 Data Bytes per Millimetre (37 871 Data Bytes per Inch), on 18 Tracks (ISO 9661: 1988) (AMD 6861) January 31, 1992. 53 pp.

BSI BS 7080: Part 1-89. 1989 Guide to Construction and Use of 8.30mm (0.25 in) Magnetic Tape Cartridge for Data Interchange, Using GCR Recording at 394 ftpmm (100000 ftpi), 39 cpmm (1000 cip) Part 1: Mechanical, Physical an Magnetic Properties. 36 pp.

BSI BS 7080: Part 2-89. 1989 Guide to Construction and Use of 8.30mm (0.25 in) Magnetic Tape Cartridge for Data Interchange, Using GCR Recording at 394 ftpmm (100000 ftpi), 39 cpmm (1000 cip) Part 2: Streaming Mode. 15 pp.

BSI BS 7081: Part 1-89. 1989 Guide to Construction and Use of 6.30 mm (0.25 in) Magnetic Tape Cartridge for Data Interchange, Using IMFM Recording at 252 ft pmm (6400 ftpi) Part 1: Mechanical, Physical and Magnetic Properties. 38 pp.

BSI BS EN 21864-91. 1991 Information Processing—Unrecorded 12,7 mm (0.5 in) Wide Magnetic Tape for Information Interchange—32 ftpmm (800 ftpi) NRZ1, 126 ftpmm (3 200 ftpi) Phase Encoded and 356 ftpmm (9 042 ftpi) NRZ1 (ISO 1864: 1985). 19 pp.

BSI BS ISO 4057-86. 1986 Information Processing—Data Interchange on 6,30 mm (0.25 in) Magnetic Tape Cartridge, 63 bpmm (1 600 bpi) Phase-Encoded. 37 pp.

BSI BS ISO 5652-84. 1984 Information Processing—9-Track, 12.7mm (0.5in) Wide Magnetic Tape for Information Interchange—Format and Recording, Using Group Coding at 246 cpmm (6 250 cpi). 22 pp.

BSI BS ISO 5652-01. 1984 Amd 1 Information Processing—9-Track, 12,7mm (0.5in) Wide Magnetic Tape for Information Interchange—Format and Recording, Using Group Coding at 246 cpmm (6 250 cpi) (AMD 7136) July 15, 1992. 26 pp.

BSI BS ISO 6098-84. 1984 Information Processing—Self-Loading Cartridges for 12,7 mm (0.5 in) Wide Magnetic Tape. 17 pp.

Computer Tapes (Cont.)

BSI BS ISO/IEC 1863-90. 1990 Information Processing—9-Track, 12,7 mm (0,5 in) Wide Magnetic Tape for Information Interchange Using NRZI at 32 ftpmm (800 ftpi)—32 cpmm (800 cpi). 14 pp.

BSI BS ISO/IEC 3788-90. 1990 Information Processing—9-Track, 12,7mm (0, 5in) Wide Magnetic Tape for Information Interchange Using Phase Encoding at 126 ftpmm (3 200 ftpi)—63 cpmm (1 600 cpi). 19 pp.

BSI BS ISO/IEC 8441-1-91. 1991 Information Technology—High Density Digital Recording (HDDR)—Part 1: Unrecorded Magnetic Tape for (HDDR) Applications. 38 pp.

CNS C5044-86. Precision Reel for Magnetic Tape (Oct)(5558).

CNS C5083-88. Information Processing — General Purpose Hubs and Reels, with 76mm (3in) Centre Hole for Magnetic Tape Used in Interchange Instrumentation Applications (Dec)(6892).

CNS C5084-88. Information Processing — Unrecorded Magnetic Tapes for Interchange Instrumentation Applications Dimensional Requirements (Dec)(6893).

CNS C5085-88. Information Processing — Precision Reels for Magnetic Tape Used in Interchange Instrumentation Applications (Dec)(6894).

CNS C5087-88. Information Processing — General Purpose Reels with 8mm (5/16in) Centre Hole for Magnetic Tape for Interchange Instrumentation Applications (Dec)(6896).

CNS C5088-81. One Half Inch (12.7mm) Magnetic Tape Reel for Computer Use (Jan)(6897).

CNS C7110-80. 1/2 Inch Type B Plastic Reel for Magnetic Tape (Requirements for Interchange) (Jul)(5767).

CNS C7111-80. Type A Hubs and Reels for Magnetic Tape (Requirements for Interchange) (Jul)(5768).

ECMA ECMA 34-76. Data Interchanges on 3.81 mm Magnetic Tape Cassette (32 bpmm, Phase Encoded). 35 pp.

ECMA ECMA 46-76. Data Interchange on 6.30 mm Magnetic Tape Cartridge (63 bpmm, Phase Encoded). 38 pp.

ECMA ECMA 56-78. Self-Loading Cartridges for 12.7mm Wide Magnetic Tapes. 23 pp.

ECMA ECMA 62-85. Data Interchange on 12.7 mm 9-Track Magnetic Tape. 76 pp.

ECMA ECMA 79-85. Data Interchange on 6.30 mm Magnetic Tape Cartridge Using IMFM Recording at 252 ftpmm. 57 pp.

ECMA ECMA 98-85. Data Interchange on 6.30 mm Magnetic Tape Cartridge Using NRZ1 Recording at 394 ftpmm Streaming Mode. 64 pp.

ECMA ECMA 120-87. Data Interchange on 12.7mm 18-Track Magnetic Tape Cartridges. 78 pp.

ECMA ECMA-139-90. 3,81mm Wide Magnetic Tape Cartridge for Information Interchange—Helical Scan Recording—DDS Format. 112 pp.

ECMA ECMA 145-90. 8mm Wide Magnetic Tape Cartridge for Information Interchange—Helical Scan Recording. 74 pp.

ECMA ECMA 146-90. 3,81mm Wide Magnetic Tape Cartridge for Information Interchange—Helical Scan Recording —Data/Dat Format. 139 pp.

ECMA ECMA 152-91. Data Interchange on 12.7 mm 18-Track Magnetic Tape Cartridges Extended Format. 82 pp.

IEC 3788-90. Information Processing—9-Track, 12,7 mm (0,5 in) Wide Magnetic Tape for Information Interchange Using Phase Encoding at 126 ftpmm (3 200 ftpi)—63 cpmm (1 600 cpi) Second Edition. 16 pp.

IEC 8441 Pt 1-91. Information Technology—High Density Digital Recording (HDDR)—Part 1: Unrecorded Magnetic Tape for (HDDR) Applications First Edition. 35 pp.

IEC 9661 COR 1. Corrigendum 1 Information Processing—Data Interchange on 12,7 mm (0.5 in) Wide Magnetic Tape Cartridges —18 Tracks, 1 491 Data Bytes per Millimetre (37 871 Data Bytes per Inch); (1990). 1 p.

IEC 9661 COR 2. Corrigendum 2 Information Processing—Data Interchange on 12,7 mm (0.5 in) Wide Magnetic Tape Cartridges —18 Tracks, 1 491 Data Bytes per Millimetre (37 871 Data Bytes per Inch); (1992). 7 pp.

ISO 962-74. Information Processing—Implementation of the 7-Bit Coded Character Set and Its 7-Bit and 8-Bit Extensions on 9-Track 12,7 mm (0.5 in) Magnetic Tape First Edition. 5 pp.

ISO 1861-75. Information Processing—7-Track, 12,7 mm (0.5 in) Wide Magnetic Tape for Information Interchange Recorded at 8 rpmm (200 rpi) First Edition. 9 pp.

ISO 1862-75. Information Processing—9-Track, 12,7 mm (0.5 in) Wide Magnetic Tape for Information Interchange Recorded at 8 rpmm (200 rpi) First Edition. 9 pp.

ISO 1863-90. Information Processing—9-Track, 12,7 mm (0,5 in) Wide Magnetic Tape for Information Interchange Using NRZI at 32 ftpmm (800 ftpi)—32 cpmm (800 cpi) Second Edition. 11 pp.

Computer Tapes *(Cont.)*

ISO 3275-74. Information Processing—Implementation of the 7-Bit Coded Character Set and Its 7-Bit and 8-Bit Extensions on 3,81 mm Magnetic Tape Cassette for Data Interchange First Edition. 3 pp.
ISO 3407-83. Information Processing—Information Interchange on 3,81 mm (0.150 in) Magnetic Tape Cassette at 4 cpmm (100 cpi), Phase Encoded at 63 ftpmm (1 600 ftpi) Second Edition. 24 pp.
ISO 3788-90. Information Processing—9-Track, 12,7 mm (0,5 in) Wide Magnetic Tape for Information Interchange Using Phase Encoding at 126 ftpmm (3 200 ftpi)—63 cpmm (1 600 cpi) Second Edition. 16 pp.
ISO 4057-86. Information Processing—Data Interchange on 6,30 mm (0.25 in) Magnetic Tape Cartridge, 63 bpmm (1 600 bpi) Phase-Encoded Second Edition. 38 pp.
ISO 5652-84. Information Processing—9-Track, 12,7 mm (0.5 in) Wide Magnetic Tape for Information Interchange—Format and Recording, Using Group Coding at 246 cpmm (6 250 cpi) Second Edition. 18 pp.
ISO 6098-84. Information Processing—Self-Loading Cartridges for 12,7 mm (0.5 in) Wide Magnetic Tape Second Edition. 14 pp.
ISO 6156-87. Magnetic Tape Exchange Format for Terminological/Lexicographical Records (MATER) First Edition. 28 pp.
ISO 8063 Pt 1-86. Information Processing—Data Interchange on 6,30 mm (0.25 in) Wide Magnetic Tape Cartridge Using IMFM Recording at 252 ftpmm (6 400 ftpi)—Part 1: Mechanical, Physical and Magnetic Properties First Edition. 37 pp.
ISO 8441 Pt 1-91. Information Technology—High Density Digital Recording (HDDR)—Part 1: Unrecorded Magnetic Tape for (HDDR) Applications First Edition. 35 pp.
ISO 8462 Pt 1-86. Information Processing—Data Interchange on 6,30 mm (0.25 in) Magnetic Tape Cartridge Using GCR Recording at 394 ftpmm (10 000 ftpi), 39 cpmm (1 000 cpi)—Part 1: Mechanical, Physical and Magnetic Properties First Edition. 38 pp.
ISO 8462 Pt 2-86. Information Processing—Data Interchange on 6,30 mm (0.25 in) Magnetic Tape Cartridge Using GCR Recording at 394 ftpmm (10 000 ftpi), 39 cpmm (1 000 cpi)—Part 2: Streaming Mode First Edition. 15 pp.
ISO 9661-88. Information Processing—Data Interchange on 12,7 mm (0.5 in) Wide Magnetic Tape Cartridges—18 Tracks, 1 491 Data Bytes Per Millimetre (37 871 Data Bytes per Inch) First Edition; (Corrigendum 2-1992). 55 pp.
JIS X 0204-69. Implementation of Code for Information Interchange on Magnetic Tape.
JIS X 0206-87. Implementation of Code for Information Interchange on Magnetic Tape Cassette.
JIS X 6101-91. Magnetic Tape for Information Interchange. 19 pp.
JIS X 6103-88. 9-Track, 12.7 mm Wide Magnetic Tape for Information Interchange-Format and Recording, Using NRZ-1 at 32 cpmm. 14 pp.
JIS X 6104-88. 9-Track, 12.7 mm Wide Magnetic Tape for Information Interchange-Format and Recording, Using Phase Encoding at 63 cpmm. 17 pp.
JIS X 6121-75. Magnetic Tape Cassette for Information Interchange.
JTC1 962-74. Information Processing—Implementation of the 7-Bit Coded Character Set and Its 7-Bit and 8-Bit Extensions on 9-Track 12,7 mm (0.5 in) Magnetic Tape First Edition. 5 pp.
JTC1 1861-75. Information Processing—7-Track, 12,7 mm (0.5 in) Wide Magnetic Tape for Information Interchange Recorded at 8 rpmm (200 rpi) First Edition. 9 pp.
JTC1 1862-75. Information Processing—9-Track, 12,7 mm (0.5 in) Wide Magnetic Tape for Information Interchange Recorded at 8 rpmm (200 rpi) First Edition. 9 pp.
JTC1 1863-90. Information Processing—9-Track, 12,7 mm (0,5 in) Wide Magnetic Tape for Information Interchange Using NRZ1 at 32 ftpmm (800 ftpi)—32 cpmm (800 cpi) Second Edition. 11 pp.
JTC1 3275-74. Information Processing—Implementation of the 7-Bit Coded Character Set and Its 7-Bit and 8-Bit Extensions on 3,81 mm Magnetic Tape Cassette for Data Interchange First Edition. 3 pp.
JTC1 3407-83. Information Processing—Information Interchange on 3,81 mm (0.150 in) Magnetic Tape Cassette at 4 cpmm (100 cpi), Phase Encoded at 63 ftpmm (1 600 ftpi) Second Edition. 24 pp.
JTC1 3788-90. Information Processing—9-Track, 12,7 mm (0,5 in) Wide Magnetic Tape for Information Interchange Using Phase Encoding at 126 ftpmm (3 200 ftpi)—63 cpmm (1 600 cpi) Second Edition. 16 pp.
JTC1 4057-86. Information Processing—Data Interchange on 6,30 mm (0.25 in) Magnetic Tape Cartridge, 63 bpmm (1 600 bpi) Phase-Encoded Second Edition. 38 pp.

Computer Tapes *(Cont.)*

JTC1 5652-84. Information Processing—9-Track, 12,7 mm (0.5 in) Wide Magnetic Tape for Information Interchange—Format and Recording, Using Group Coding at 246 cpmm (6 250 cpi) Second Edition. 18 pp.
JTC1 6098-84. Information Processing—Self-Loading Cartridges for 12,7 mm (0.5 in) Wide Magnetic Tape Second Edition. 14 pp.
JTC1 6156-87. Magnetic Tape Exchange Format for Terminological/Lexicographical Records (MATER) First Edition. 28 pp.
JTC1 8063 Pt 1-86. Information Processing—Data Interchange on 6,30 mm (0.25 in) Wide Magnetic Tape Cartridge Using IMFM Recording at 252 ftpmm (6 400 ftpi)—Part 1: Mechanical, Physical and Magnetic Properties First Edition. 37 pp.
JTC1 8441 Pt 1-91. Information Technology—High Density Digital Recording (HDDR)—Part 1: Unrecorded Magnetic Tape for (HDDR) Applications First Edition. 35 pp.
JTC1 8462 Pt 1-86. Information Processing—Data Interchange on 6,30 mm (0.25 in) Magnetic Tape Cartridge Using GCR Recording at 394 ftpmm (10 000 ftpi), 39 cpmm (1 000 cpi)—Part 1: Mechanical, Physical and Magnetic Properties First Edition. 38 pp.
JTC1 8462 Pt 2-86. Information Processing—Data Interchange on 6,30 mm (0.25 in) Magnetic Tape Cartridge Using GCR Recording at 394 ftpmm (10 000 ftpi), 39 cpmm (1 000 cpi)—Part 2: Streaming Mode First Edition. 15 pp.
JTC1 9661-88. Information Processing—Data Interchange on 12,7 mm (0.5 in) Wide Magnetic Tape Cartridges—18 Tracks, 1 491 Data Bytes Per Millimetre (37 871 Data Bytes per Inch) First Edition; (Corrigendum 2-1992). 55 pp.
OSI ISO 962-74. Information Processing—Implementation of the 7-Bit Coded Character Set and Its 7-Bit and 8-Bit Extensions on 9-Track 12, 7mm (0.5 in) Magnetic Tape. 5 pp.
OSI ISO 1861-75. Information Processing—7-Track, 12,7mm (0.5in) Wide Magnetic Tape for Information Interchange Recorded at 8 rpmm (200rpi). 9 pp.
OSI ISO 1862-75. Information Processing—9-Track, 12,7mm (0.5in) Wide Magnetic Tape for Information Interchange Recorded at 8 rpmm (200rpi). 9 pp.
OSI ISO/IEC 1863-90. Information Processing—9-Track, 12,7mm (0.5in) Wide Magnetic Tape for Information Interchange Using NRZI at 32ftpmm (800 ftpi)-32cpmm (800cpi). 11 pp.
OSI ISO 3275-74. Information Processing—Implementation of the 7-bit Coded Character Set and Its 7-bit and 8-bit Extension on 3.81mm Magnetic Tape Cassette for Data Interchange. 3 pp.
OSI ISO 3407-83. Information Processing—Information Interchange on 3,81 mm (0.150 in) Magnetic Tape Cassette at 4 cpmm (100 cpi), Phase Encoded at 63 ft pmm (1 600 ft pi). 24 pp.
OSI ISO/IEC 3788-90. Information Processing—9-Track 12,7mm (0,5in) Wide Magnetic Tape for Information Interchange Using Phase Encoding at 126 ftpmm (3 200 ftpi)—63 cpmm (1 600 cpi). 27 pp.
OSI ISO 4057-86. Information Processing—Data Interchange on 6,30mm (0.25in)Magnetic Tape Cartridge. 63bpmm (1600 bpi) Phase-Coded. 36 pp.
OSI ISO 5652-84. Information Processing—9 Track, 12,7mm (0.5in) Wide Magnetic Tape for Information Interchange—Format and Recording, Using Group Coding at 246 cpmm (6 250 cpi). 18 pp.
OSI ISO 6098-84. Information Processing-Self Loading Cartridges for 12,7mm (0.5in) Wide Magnetic Tape. 14 pp.
OSI ISO 8063-1-86. Information Processing—Data Interchange on 6,30mm (0.25in)Wide Magnetic Tape Cartridge Using IMFM Recording at 252 ftpmm (6400 ftpi)—Part 1: Mechanical, Physical and Magnetic Properties. 37 pp.
OSI ISO IEC DIS 8441-1-89. Information Technology—High Density Digital Recording (HDDR)—Part 1: Unrecorded Magnetic Tape for (HDDR) Applications. 53 pp.
OSI ISO 8462-1-86. Information Processing—Data Interchange on 6, 30mm (0.25in) Magnetic Tape Cartridge Using GCR Recording at 394 ftpmm (10 000 ftpi) 39cpmm (1 000 cpi) Part 1: Mechanical, Physical and Magnetic Properties. 38 pp.
OSI ISO 8462-2-86. Information Processing—Data Interchange on 6, 30mm (0.25in) Magnetic Tape Cartridge Using GCR Recording at 394 ftpmm (10.000 ftpi), 39 cpmm (1.000 cpi)—Part 2: Streaming Mode. 15 pp.
OSI ISO 9661-88. Information Processing—Data Interchange on 12, 7mm (0.5 in) Wide Magnetic Tape Cartridges—18 Tracks, 1 491 Data Bytes Per Millimetre (37 871 Data Bytes per Inch) (with Technical Corrigendum 1). 49 pp.
OSI ISO/IEC/DIS 10777-90. Information Technology—3,81mm Wide Magnetic Tape Cartridge for Information Interchange—Helical Scan Recording—DDS Format. 115 pp.

Computer Tapes *(Cont.)*

OSI ISO/IEC DIS 11319-91. Information Technology—8mm Wide Magnetic Tape Cartridge for Information Interchange—Helical Scan Recording. 76 pp.
OSI ISO/IEC DIS 11321-91. Information Technology—3,81 mm Wide Magnetic Tape Cartridge for Information Interchange—Helical Scan Recording—Data/Dat Format. 141 pp.
SAA AS 1009-83. Information Processing—9-Track, 12.7 mm Wide Magnetic Tape for Information Interchange Recorded at 32 rpmm. 12 pp.
SAA AS 1011-86. Information Processing—Unrecorded 12.7 mm (0.5 in) Wide Magnetic Tape for Information Interchange—32 ftpmm (800 ftpi) NRZ-1,126 ftpmm (3 200 ftpi) Phase Encoded and 356 ftpmm (9 042 ftpi) NRZ-1. 11 pp.
SAA AS 3977.1-91. Informaiton Processing Systems—High Density Digital Recording (HDDR) —Part 1: Unrecorded Magnetic Tape for HDDR Applications (ISO/IEC 8441-1:1991) (in Professional Package 26A). 27 pp.

—Audio

CENELEC HD 369.10 S4-91. Audiovisual, Video and Television Equipment and Systems Part 10: Audio Cassette Systems. 5 pp.
CNS C7195-88. Magnetic Sound Recording Tapes (Apr)(12250).

—Audio—Reels

CNS C7196-88. Reels for Magnetic Recording Tape (Apr)(12251).

—Calibration

JIS C 5563-89. Magnetic Tape Sound Recording and Reproducing Systems Part 2: Calibration Tapes; (IEC 94-2-1986).

—Character Position—OCR

ECMA ECMA 21-69. Character Positioning on OCR Journal Tape. 9 pp.

—Electronic Prepress Systems—User Exchange Format

BSI BS 7607-92. 1992 Graphic Technology—Prepress Digital Data Exchange—Colour Picture Data on Magnetic Tape (ISO 10755: 1992) (H). 15 pp.
ISO 10755-92. Graphic Technology—Prepress Digital Data Exchange—Colour Picture Data on Magnetic Tape First Edition; (ANSI IT8.1-1988). 14 pp.
JIS X 0651-91. User Exchange Format for the Exchange of Color Picture Data Between Electronic Prepress Systems Via Magnetic Tape. 24 pp.
JTC1 10755-92. Graphic Technology—Prepress Digital Data Exchange—Colour Picture Data on Magnetic Tape First Edition; (ANSI IT8.1-1988). 14 pp.
OSI ISO/DIS 10755-90. User Exchange Format (UEF00) for the Exchange of Colour Picture Data Between Electronic Prepress Systems Via Magnetic Tape (DDE500). 18 pp.
OSI ISO/DIS 10756-90. User Exchange Format (UEF01) for the Exchange of Line of Data Between Electronic Prepress Systems Via Magnetic Tape (DDE500). 21 pp.
OSI ISO/DIS 10757-90. User Exchange Format (UEF02) for the Exchange of Geometric Information Between Electronic Prepress Systems Via Magnetic Tape (DDE500). 48 pp.
OSI ISO/DIS 10759-90. User Exchange Format (UEF03) for the Exchange of Monochrome Image Data Between Electronic Prepress Systems Via Magnetic Tape (DDE500). 21 pp.

—File Structure

BSI BS 4732-88. 1988 File Structure and Labelling of Magnetic Tape for Information Interchange. 20 pp.
ECMA ECMA 13-85. File Structure and Labelling of Magnetic Tapes for Information Interchange. 44 pp.
ECMA ECMA 41-73. Magnetic Tape Cassette Labelling and File Structure for Information Exchange. 26 pp.
ISO 1001-86. Information Processing—File Structure and Labelling of Magnetic Tapes for Information Interchange Second Edition. 23 pp.
ISO 4341-78. Information Processing—Magnetic Tape Cassette and Cartridge Labelling and File Structure for Information Interchange First Edition; (ANSI 4341-1978). 14 pp.
JIS X 0601-90. Information Processing—File Structure and Labeling of Magnetic Tapes for Information Interchange.
JIS X 0602-82. Magnetic Tape Cassette Labelling and File Structure for Information Interchange.
JTC1 1001-86. Information Processing—File Structure and Labelling of Magnetic Tapes for Information Interchange Second Edition. 23 pp.
JTC1 4341-78. Information Processing—Magnetic Tape Cassette and Cartridge Labelling and File Structure for Information Interchange First Edition; (ANSI 4341-1978). 14 pp.
OSI ISO 1001-86. Information Processing—File Structure and Labelling of Magnetic Tapes for Information Interchange. 23 pp.

Computer — INTERNATIONAL AND NON-U.S. NATIONAL STANDARDS SUBJECT INDEX

Computer Tapes (Cont.)

—File Structure (Cont.)
OSI ISO 4341-78. Information Processing—Magnetic Tape Cassette and Cartridge Labelling and File Structure for Information Interchange. 14 pp.
SAA AS 1068-87. Information Processing—File Structure and Labelling of Magnetic Tapes for Information Change. 17 pp.

—Glossaries
BSI BS ISO 2382/12-88. 1988 Information Technology—Vocabulary 2382/12: Peripheral Equipment. 55 pp.
CNS C5154-90. Data Processing Vocabulary (Part 12: Data Media, Storage and Related Equipment) (Nov)(10241).
ISO 2382 Pt 12-88. Information Processing Systems—Vocabulary—Part 12: Peripheral Equipment Second Edition. 51 pp.
JTC1 2382 Pt 12-88. Information Processing Systems—Vocabulary—Part 12: Peripheral Equipment Second Edition. 51 pp.
OSI ISO 2382-12-87. Information Processing Systems—Vocabulary—Part 12: Peripheral Equipment. 51 pp.
SAA AS 1189.12-91. Data Processing—Vocabulary—Part 12: Peripheral Equipment (ISO 2382-12:1988) (In Professional Package 26). 24 pp.

—Handling
BSI BS 4783: Part 5-91. 1991 Storage, Transportation and Maintenance of Media for Use in Data Processing and Information Storage Part 5: Recommendations for 12.7 mm Magnetic Tape Cartridges for Data Interchange, Recording at 1491 Data Bytes Per Millimetre on 18 Tracks. 13 pp.
ECMA ECMA-TR 11-81. Guidelines for Magnetic Tape Handling and Storage. 9 pp.

—Identification Systems
BSI BS 4732-88. 1988 File Structure and Labelling of Magnetic Tape for Information Interchange. 20 pp.
BSI BS 5769: Part 1-79. 1979 Magnetic Tape Cassette and Cartridge Labelling and File Structure, for Information Interchange Part 1: Label Standard Version 1. 14 pp.
ECMA ECMA 13-85. File Structure and Labelling of Magnetic Tapes for Information Interchange. 44 pp.
ECMA ECMA 41-73. Magnetic Tape Cassette Labelling and File Structure for Information Exchange. 26 pp.
ISO 1001-86. Information Processing—File Structure and Labelling of Magnetic Tapes for Information Interchange Second Edition. 23 pp.
ISO 4341-78. Information Processing—Magnetic Tape Cassette and Cartridge Labelling and File Structure for Information Interchange First Edition; (ANSI 4341-1978). 14 pp.
JIS X 0601-90. Information Processing—File Structure and Labeling of Magnetic Tapes for Information Interchange.
JIS X 0602-82. Magnetic Tape Cassette Labelling and File Structure for Information Interchange.
JTC1 1001-86. Information Processing—File Structure and Labelling of Magnetic Tapes for Information Interchange Second Edition. 23 pp.
JTC1 4341-78. Information Processing—Magnetic Tape Cassette and Cartridge Labelling and File Structure for Information Interchange First Edition; (ANSI 4341-1978). 14 pp.
OSI ISO 1001-86. Information Processing—File Structure and Labelling of Magnetic Tapes for Information Interchange. 23 pp.
OSI ISO 4341-78. Information Processing—Magnetic Tape Cassette and Cartridge Labelling and File Structure for Information Interchange. 14 pp.
SAA AS 1068-87. Information Processing—File Structure and Labelling of Magnetic Tapes for Information Change. 17 pp.

—Information Interchange
BSI BS EN 21864-91. 1991 Information Processing—Unrecorded 12,7 mm (0.5 in) Wide Magnetic Tape for Information Interchange—32 ftpmm (800 ftpi) NRZ1, 126 ftpmm (3 200 ftpi) Phase Encoded and 356 ftpmm (9 042 ftpi) NRZ1 (ISO 1864: 1985). 19 pp.
BSI BS ISO/IEC 8441-2-91. 1991 Information Technology—High Density Digital Recording (HDDR)—Part 2: Guide for Interchange Practice. 41 pp.
CEN EN 21864-91. Information Processing—Unrecorded 12,7 mm (0.5 in) Wide Magnetic Tape for Information Interchange—32 ftpmm (800 ftpi) NRZ1, 126 ftpmm (3 200 ftpi) Phase Encoded and 356 ftpm (9 042 ftpi) NRZ1. 3 pp.
IEC 1864-92. Information Technology—Unrecorded 12,7 mm (0,5 in) Wide Magnetic Tape for Information Interchange—32 ftpmm (800 ftpi), NRZ1, 126 ftpmm (3 200 ftpi) Phase Encoded and 356 ftpmm (9 042 ftpi), NRZ1 Fourth Edition. 18 pp.
IEC 8441 Pt 2-91. Information Technology—High Density Digital Recording (HDDR)—Part 2: Guide for Interchange Practice First Edition. 38 pp.
ISO 1864-92. Information Technology—Unrecorded 12,7 mm (0,5 in) Wide Magnetic Tape for Information Interchange—32 ftpmm (800 ftpi), NRZ1, 126 ftpmm (3 200 ftpi) Phase Encoded and 356 ftpmm (9 042 ftpi), NRZ1 Fourth Edition. 18 pp.
ISO 8441 Pt 2-91. Information Technology—High Density Digital Recording (HDDR)—Part 2: Guide for Interchange Practice First Edition. 38 pp.
JTC1 1864-92. Information Technology—Unrecorded 12,7 mm (0,5 in) Wide Magnetic Tape for Information Interchange—32 ftpmm (800 ftpi), NRZ1, 126 ftpmm (3 200 ftpi) Phase Encoded and 356 ftpmm (9 042 ftpi), NRZ1 Fourth Edition. 18 pp.
JTC1 8441 Pt 2-91. Information Technology—High Density Digital Recording (HDDR)—Part 2: Guide for Interchange Practice First Edition. 38 pp.
OSI ISO/IEC DIS 1864-91. Information Processing—Unrecorded 12,7mm (0.5in) Wide Magnetic Tape for Information Interchange—32ftpmm (800 fti) NRZI, 126ftpmm (3200 ftpi) Phase Encoded and 356ftpmm (9 042 ftpi) NRZI. 18 pp.
SAA AS 3977.2-92. Information Processing Systems—High Density Digital Recording (HDDR) —Part 2: Guide for Interchange Practice (ISO/IEC 8441-2:1991) (in Professional Package 26A). 31 pp.
SNZ NZS/AS 1011-93. Information Processing Systems—Unrecorded 12.7 mm (0.5 in) Wide Magnetic Tape for Info. Interchange—32 ftpmm (800 ftpi), NRZ1, 126 ftpmm (3 200 ftpi) Phase Encoded and 356 ftpmm (9 042 ftpi), NRZ1, (ISO/IEC 1864:1992) 1992) (This is a Joint Stand. with SAA AS 1011). 15 pp.

—Mechanical Properties
JIS C 5565-89. Magnetic Tape Sound Recording and Reproducing Systems Part 4: Mechanical Magnetic Tape Properties; (IEC 94-4-1986).

—Punched
BSI BS 3880: Part 2-71. 1971 Paper Tape for Data Processing Part 2: Dimensions and Locations of Punched Holes in Paper Tape. 12 pp.
BSI BS 3880: Part 4-72. 1972 Amd 1 Paper Tape for Data Processing Part 4: General Requirements for Data Interchange on Punched Paper Tape. 17 pp.
CNS C5104-83. Information Processing — Punched Paper Tape — Dimensions and Location of Feed Holes and Code Holes (Feb)(7220).
CNS C5108-83. Data Interchange on Rolled-Up Punched Paper Tape Requirement (Apr)(7224).
ECMA ECMA 10-70. Data Interchange on Punched Tape. 22 pp.
ISO 1154-75. Information Processing—Punched Paper Tape—Dimensions and Location of Feed Holes and Code Holes First Edition. 4 pp.
ISO 2195-72. Data Interchange on Rolled-up Punched Paper Tape—General Requirements First Edition. 3 pp.
JIS X 6192-71. Dimensions for Punched Paper Tape for Information Interchange.
JTC1 1154-75. Information Processing—Punched Paper Tape—Dimensions and Location of Feed Holes and Code Holes First Edition. 4 pp.
JTC1 2195-72. Data Interchange on Rolled-Up Punched Paper Tape—General Requirements First Edition. 3 pp.
OSI ISO 1154-75. Information Processing—Punched Paper Tape—Dimensions and Locations of Feed Holes and Code Holes. 4 pp.
OSI ISO 2195-72. Data Interchange on Rolled-up Punched Paper Tape—General Requirements. 3 pp.
SAA AS 1069-71. Dimensions for Punched Paper Tape for Data Interchange Reconfirmed 1985. 4 pp.
SAA AS 1305.1-74. Punched Tape for Data Interchange—Part 1: General Requirements Reconfirmed 1989. 4 pp.

—Punched—Coded Character Sets
BSI BS 3880: Part 3-71. 1971 Amd 1 Paper Tape for Data Processing Part 3: Representation of Codes on Paper Tape. 23 pp.
CNS C5103-83. Information Processing — Representation of 6 and 7 Bit Coded Character Sets on Punched Tape (Feb)(7219).
CNS C5158-83. Subset of Standard Code for Information Interchange for Numerical Machine Control Perforated Tape (May)(10245).
ISO 1113-79. Information Processing—Representation of the 7-Bit Coded Character Set on Punched Tape Second Edition. 4 pp.
JIS X 0203-73. Implementation of Code for Information Interchange and Numerical Control of Machines on Paper Tape.
JTC1 1113-79. Information Processing—Representation of the 7-Bit Coded Character Set on Punched Tape Second Edition. 4 pp.
OSI ISO 1113-79. Information Processing—Representation of the 7-Bit Coded Character Set on Punched Tape. 4 pp.

—Punched—Cores
CNS C5086-88. Information Processing — Reels and Cores for 25.4mm (1in) Perforated Paper Tape for Information Interchange — Dimensions (Dec)(6895).
ISO 3692-76. Information Processing—Reels and Cores for 25,4 mm (1 in) Perforated Paper Tape for Information Interchange—Dimensions First Edition. 5 pp.
JTC1 3692-76. Information Processing—Reels and Cores for 25,4 mm (1 in) Perforated Paper Tape for Information Interchange—Dimensions First Edition. 5 pp.
OSI ISO 3692-76. Information Processing—Reels and Cores for 25,4mm (1 in) Perforated Paper Tape for Information Interchange—Dimensions. 5 pp.

—Punched—Greaseproof Papers
MOD UK TS 288A. Paper, Vegetable Parchment (for Telegraph Machines).

—Punched—Reels
CNS C5086-88. Information Processing — Reels and Cores for 25.4mm (1in) Perforated Paper Tape for Information Interchange — Dimensions (Dec)(6895).
ISO 3692-76. Information Processing—Reels and Cores for 25,4 mm (1 in) Perforated Paper Tape for Information Interchange—Dimensions First Edition. 5 pp.
JTC1 3692-76. Information Processing—Reels and Cores for 25,4 mm (1 in) Perforated Paper Tape for Information Interchange—Dimensions First Edition. 5 pp.
OSI ISO 3692-76. Information Processing—Reels and Cores for 25,4mm (1 in) Perforated Paper Tape for Information Interchange—Dimensions. 5 pp.

—Punched—Teletypewriters
CGSB CAN/CGSB-53.3-M87. Tape, Teletype, Perforator, Oiled. 9 pp.

—Punched—Variable Block Formats—Numerical Control
CNS B1266-84. Punched Tape Variable Block Format for Contouring and Contouring/Positioning Numerically Controlled Machine Tools (Jan)(7776).
JIS B 6312-76. Punched Tape Variable Block Format for Positioning and Straight-Cut Numerically Controlled Machine Tools (R 1984). 20 pp.
JIS B 6313-74. Punched Tape Variable Block Format for Contouring and Contouring/Positioning Numerically Controlled Machine Tools (R 1986). 24 pp.

—Storage
BSI BS 4783: Part 5-91. 1991 Storage, Transportation and Maintenance of Media for Use in Data Processing and Information Storage Part 5: Recommendations for 12.7 mm Magnetic Tape Cartridges for Data Interchange, Recording at 1491 Data Bytes Per Millimetre on 18 Tracks. 13 pp.
ECMA ECMA-TR 11-81. Guidelines for Magnetic Tape Handling and Storage. 9 pp.

—Time and Address Codes
BSI BS 6288: Part 10-89. 1989 Amd 1 Magnetic Tape Sound Recording and Reproducing Systems Part 10: Specification for Time and Address Codes (IEC 94-10: 1988) (AMD 6405) August 30, 1991. 11 pp.

—Track Format
BSI BS 4748-82. 1982 Bibliographic Information Interchange Format for Magnetic Tape. 8 pp.
BSI BS 7081: Part 2-89. 1989 Guide to Construction and Use of 6.30 mm (0.25 in) Magnetic Tape Cartridge for Data Interchange, Using IMFM Recording at 252 ft pmm (6400 ftpi) Part 2: Track Format and Method of Recording for Data Interchange in Start/Stop Mode. 8 pp.
IEC 1863-90. Information Processing—9-Track, 12,7 mm (0,5 in) Wide Magnetic Tape for Information Interchange Using NRZ1 at 32 ftpmm (800 ftpi)—32 cpmm (800 cpi) Second Edition. 11 pp.
ISO 2709-81. Documentation—Format for Bibliographic Information Interchange on Magnetic Tape Second Edition. 7 pp.
ISO 8063 Pt 2-86. Information Processing—Data Interchange on 6,30 mm (0.25 in) Wide Magnetic Tape Cartridge Using IMFM Recording at 252 ftpmm (6 400 ftpi)—Part 2: Track Format and Method of Recording for Data Interchange in Start/Stop Mode First Edition. 7 pp.
JTC1 2709-81. Documentation—Format for Bibliographic Information Interchange on Magnetic Tape Second Edition. 7 pp.
JTC1 8063 Pt 2-86. Information Processing—Data Interchange on 6,30 mm (0.25 in) Wide Magnetic Tape Cartridge Using IMFM Recording at 252 ftpmm (6 400 ftpi)—Part 2: Track Format and Method of Recording for Data Interchange in Start/Stop Mode First Edition. 7 pp.

INTERNATIONAL AND NON-U.S. NATIONAL STANDARDS
SUBJECT INDEX

Concrete

Computer Tapes *(Cont.)*
—Track Format *(Cont.)*
OSI ISO 8063-2-86. Information Processing—Data Interchange on 6,30mm (0.25in)Wide Magnetic Tape Cartridge Using IMFM Recording at 252 ftpmm (6400 ftpi)—Part 2: Track Format and Method of Recording for Data Interchange in Start-Stop Mode. 7 pp.

—Unpunched
BSI BS 3880: Part 1-71. 1971 Amd 1 Paper Tape for Data Processing Part 1: Unpunched Paper Tape. 31 pp.
ISO 1729-73. Information Processing—Unpunched Paper Tape—Specification First Edition. 17 pp.
JIS X 6191-76. Unpunched Paper Tape for Information Interchange (R 1987). 14 pp.
JTC1 1729-73. Information Processing—Unpunched Paper Tape—Specification First Edition. 17 pp.
OSI ISO 1729-73. Information Processing—Unpunched Paper Tape—Specification. 17 pp.

—Video
CENELEC HD 369.10 S4-91. Audiovisual, Video and Television Equipment and Systems Part 10: Audio Cassette Systems. 5 pp.
CNS C7195-88. Magnetic Sound Recording Tapes (Apr)(12250).

—Video—Reels
CNS C7196-88. Reels for Magnetic Recording Tape (Apr)(12251).

Computer Terminals
Use: Video Display Terminals

Computerized Communication Terminals
Use: Communication Terminal Equipment

Computers
See Also: Analog Computers; Artillery Computers; Avionic Computers; Bus Controllers; Calculators; CAMAC; Central Processing Units; CIM; Computer Graphics; Computer Programs; Computer Systems; Data Processing; Data Processing Equipment; Dot Matrix Printers; Editing Systems; Image Processing Systems; Information Interchange; Information Systems; Memory (Data Storage); Personal Computers; Programming Languages; Software; Word Processors
BSI PP 7315-87. (WITHDRAWN) 1987 Standardization for Information Technology. 50 pp.
MOD UK DSTAN 00-21: Part 1-83. M 700 Computers Part 1: Basic Specification Issue 2. 25 pp.
MOD UK DSTAN 00-21: Part 2-83. M 700 Computers Part 2: Class 1 Issue 2. 11 pp.

—Environments
BSI BS 7083-89. 1989 Amd 1 Recommendations for the Accommodation and Operating Environment of Computer Equipment (AMD 6624) December 21, 1990. 17 pp.

—Glossaries
BSI BS ISO 2382/19-89. 1989 Information Technology—Vocabulary 2382/19: Analog Computing. 21 pp.
ISO 2382 Pt 19-89. Information Processing Systems—Vocabulary—Part 19: Analog Computing Second Edition. 18 pp.
JTC1 2382 Pt 19-89. Information Processing Systems—Vocabulary—Part 19: Analog Computing Second Edition. 18 pp.
OSI ISO 2382-19-89. Information Processing Systems—Part 19: Analog Computing. 18 pp.
SAA AS 1189.19-91. Data Processing—Vocabulary—Part 19: Analog Computing (ISO 2382-19:1989) (In Professional Package 26). 5 pp.

—Noise
CEN EN 27779-91. Acoustics—Measurement of Airborne Noise Emitted by Computer and Business Equipment. 3 pp.
ECMA ECMA-TR 27-85. Method for the Prediction of Installation Noise Levels. 38 pp.

—Nuclear Reactors
BSI BS 6078-81. 1981 Application of Digital Computers to Nuclear Reactor Instrumentation and Control. 18 pp.
IEC 643-79. Application of Digital Computers to Nuclear Reactor Instrumentation and Control First Edition. 36 pp.

—Nuclear Safety
IEC 987-89. Programmed Digital Computers Important to Safety for Nuclear Power Stations First Edition. 51 pp.

Computers *(Cont.)*
—Operation
BSI BS 6650-85. (WITHDRAWN) 1985 Control of the Operation of a Computer. 11 pp.

—Project Management
BSI BS 6046: Part 3-92. 1992 Use of Network Techniques in Project Management Part 3: Guide to the Use of Computers (G). 28 pp.
BSI BS 6046: Part 3-81. 1981 Use of Network Techniques in Project Management Part 3: Guide to the Use of Computers. 22 pp.

—Quality Assurance
BSI DD 198-91. 1991 Assessment of Reliability of Systems Containing Software. 85 pp.

—Safety
BSI BS 7002-89. 1989 Amd 1 Safety of Information Technology Equipment Including Electrical Business Equipment (AMD 6650) June 28, 1991. 140 pp.
BSI BS 7002-02. (WITHDRAWN) 1989 Amd 2 Safety of Information Technology Equipment Including Electrical Business Equipment (AMD 6974) January 31, 1992 (Superseded by BS EN 60950: 1992). 180 pp.

—Standards Preparation—Certification/Conformance
ECMA ECMA-TR 18-83. The Meaning of Conformance to Standards. 16 pp.

Computing Machines
Use: Adding Machines

Concave Cutters
Use: Milling Cutters—Concave

Concentration (Composition)
See Also: Solubility

—Chromic Acid
MOD UK M 9526/66. Examination of Chromic Acid Crystals.

—Disinfectants
BSI BS 6905-87. 1987 Method for Estimation of Concentration of Disinfectants Used in 'Dirty' Conditions in Hospitals by the Modified Kelsey-Sykes Test. 11 pp.

—Exhaust Gases
CNS D3185-89. Method of Test for Exhausts Concentration for Gasoline Engine Automobiles (Jul)(12567).

—Fumes—Welding
CNS Z8020-81. Method for Determining Average Concentration of Weld Fume in Welding Environment (Apr)(7250).
CNS Z8021-81. Method of Measuring Weld Fume Concentration (Apr)(7330).
JIS Z 3950-86. Methods of Measurement for Weld Fume Concentration.
JIS Z 3950-75. Methods of Measuring Weld Fume Concentration (R 1978). 6 pp.
JIS Z 3951-86. Methods for Determining Average Concentration of Weld Fume in Welding Environment.
JIS Z 3951-77. Methods for Determining Average Concentration of Weld Fume in Welding Environment (R 1980). 5 pp.
JIS Z 3952-90. Methods of Measurement for Gas Concentration in Welding Environment. 16 pp.

—Gas Mixtures
ISO 6711-81. Gas Analysis—Checking of Calibration Gas Mixtures by a Comparison Method First Edition. 4 pp.

—Glossaries
DIN ENGL 1310-84. Composition of (Gaseous, Liquid and Solid) Mixtures; Concepts, Symbols (Feb). 3 pp.

—Hydrochloric Acid—Density
ISO 905-76. Hydrochloric Acid for Industrial Use—Evaluation of Hydrochloric Acid Concentration by Measurement of Density First Edition. 4 pp.

—Particulates—Ducts
BSI BS 893-78. 1978 Amd 1 Measurement of the Concentration of Particulate Material in Ducts Carrying Gases. 16 pp.
BSI BS 6069: Sec 4.3-92. 1992 Characterization of Air Quality Part 4: Stationary Source Emissions Section 4.3: Method for the Manual Gravimetric Detn. of Concentration and Mass Flow Rate of Particulate Material in Gas-Carrying Ducts (ISO 9096: 1992). 37 pp.
ISO 9096-92. Stationary Source Emissions—Determination of Concentration and Mass Flow Rate of Particulate Material in Gas-Carrying Ducts—Manual Gravimetric Method First Edition. 36 pp.

Concentration (Composition) *(Cont.)*
—Pigments
CGSB 1-GP-71 METH 21.2-81. Methods of Testing Paints and Pigments Pigment Content Pigment Volume Concentration (PVC). 2 pp.

—Pulps
DIN ENGL 54359-75. Testing of Pulp; Determination of Stock Concentration of Fibre Suspensions (Oct). 2 pp.
SAA AS 1301.207S-89. Methods of Test for Pulp and Paper (Metric Units)—Part 207s: Determination of Stock Concentration (This is a Joint Standard with SANZ NZS 1301). 3 pp.
SNZ NZS/AS 1301. 207S-89. Methods of Test for Pulp and Paper Determination of Stock Concentration (This is a Joint Standard with SAA AS 1301.207S). 3 pp.

—Sulfuric Acid—Density
ISO 911-77. Sulphuric Acid for Industrial Use—Evaluation of Sulphuric Acid Concentration by Measurement of Density First Edition. 5 pp.

—Symbols
DIN ENGL 1310-84. Composition of (Gaseous, Liquid and Solid) Mixtures; Concepts, Symbols (Feb). 3 pp.

—Tables (Data)
SAA AS 1377.4-73. Conversion Tables—Part 4: Mass, Density, Concentration and Flow Rate (Incorporating Amdt 1). 28 pp.

Concentration Gradient Density Columns
Use: Density Columns

Concentrator Exploiting the Idle Time of Circuits
Use: Time Assignment Speech Interpolation

Concentrators
See Also: Accumulators; Classifiers; Filtration; Separators (Mechanical)

—Aircraft—Oxygen—Molecular Sieves
NATO STANAG 3865 ED 2 AMD 2-86. Physiological Requirements for Aircraft Molecular Sieve Oxygen Concentrating Systems. 11 pp.
NATO STANAG 3865 ED 2 AMD 3-86. Physiological Requirements for Aircraft Molecular Sieve Oxygen Concentrating Systems. 13 pp.

—Oxygen—Medical—Gas Pipelines
BSI BS 7634-93. 1993 Oxygen Concentrators for Use with Medical Gas Pipeline Systems (ISO 10083: 1992) (N). 15 pp.
CSA CAN/CSA-Z305.6-92. Medical Oxygen Concentrator Central Supply System: with Nonflammable Medical Gas Piping Systems; (Gen Instr 1). 47 pp.
ISO 10083-92. Oxygen Concentrators for Use with Medical Gas Pipeline Systems First Edition. 14 pp.

—Oxygen—Safety
BSI BS 5724: Sec 2.23-89. 1989 Medical Electrical Equipment Part 2: Particular Requirements for Safety Section 2.23: Specification for Oxygen Concentrators (ISO 8359: 1988). 20 pp.
ISO 8359-88. Oxygen Concentrators for Medical Use—Safety Requirements First Edition. 19 pp.
SAA AS 3200.2.20 0-92. Approval and Test Specification—Medical Electrical Equipment—Part 2: Particular Requirements for Safety—Part 2.200: Oxygen Concentrators for Individual Patient Use (in Professional Package 17C, 28) (This is a Joint Standard with SNZ NZS 3200.2.200). 17 pp.

Concentric Connectors
Use: Circular Connectors

Concorde Aircraft
See Also: Aircraft

—Antenna Positions
CAA. Concorde (Approved Aerial Positions). 1 p.

Concrete
Use: Concretes

Concrete Anchors
See Also: Anchor Bolts; Anchors (Fasteners); Concrete Construction

—Stud Welding
CNS B2346-83. Studs for Drawn Arc Stud Welding; Concrete Anchor and Shear Connectors (Jan)(4689).
DIN ENGL 32500 Pt 3-79. Studs for Drawn Arc Stud Welding; Concrete Anchors and Shear Connectors (Jan). 3 pp.

INDUSTRY STANDARDS

INTERNATIONAL AND NON-U.S. NATIONAL STANDARDS
SUBJECT INDEX — Concrete

Concrete Anchors *(Cont.)*
—Tensile Testing
BSI BS 5080: Part 1-93. 1993 Structural Fixings in Concrete and Masonry Part 1: Method of Test for Tensile Loading. 15 pp.
BSI BS 5080: Part 1-74. 1974 Amd 1 Methods of Test for Structural Fixings in Concrete and Masonry Part 1: Tensile Loading. 13 pp.
CSA CAN3-A370-M84. Connectors for Masonry; (Gen Instr 1 Thru 2). 59 pp.

Concrete Beams
See Also: Beams (Supports); Structural Members

—Compression Testing
BSI BS 1881: Part 119-83. 1983 Methods of Testing Concrete Part 119: Method for Determination of Compressive Strength Using Portions of Beams Broken in Flexure (Equivalent Cube Method). 5 pp.
BSI BS 1881: Part 120-83. 1983 Amd 1 Methods of Testing Concrete Part 120: Method for Determination of the Compressive Strength of Concrete Cores (AMD 6109) July 31, 1989. 10 pp.
CNS A3051-84. Method of Test for Compressive and Flexural Strength of Drilled Cores and Sawed Beams of Concrete (Apr)(1238). 3 pp.
JIS A 1107-78. Method of Obtaining and Testing Drilled Cores and Sawed Beams of Concrete. 7 pp.
JIS A 1114-76. Method of Test for Compressive Strength of Concrete Using Portions of Beams Broken in Flexure. 7 pp.

—Flexural Strength
CNS A3051-84. Method of Test for Compressive and Flexural Strength of Drilled Cores and Sawed Beams of Concrete (Apr)(1238). 3 pp.
JIS A 1107-78. Method of Obtaining and Testing Drilled Cores and Sawed Beams of Concrete. 7 pp.

—Polyester Resin—Compression Testing
JIS A 1183-78. Method of Test for Compressive Strength of Polyester Resin Concrete Using Portions of Beams Broken in Flexure. 7 pp.

—Prestressed—Beam Bridges
JIS A 5316-91. Prestressed Concrete Beams for Beam Bridges. 44 pp.

—Prestressed—Slab Bridges
JIS A 5313-91. Prestressed Concrete Beams for Slab Bridges. 83 pp.
JIS A 5319-92. Prestressed Concrete Beams for Light Load Slab Bridges. 31 pp.

—Test Specimens
BSI BS 1881: Part 109-83. 1983 Amd 1 Methods of Testing Concrete Part 109: Method for Making Test Beams from Fresh Concrete (AMD 6104) July 31, 1989. 7 pp.

—Test Specimens—Curing
BSI BS 1881: Part 112-83. 1983 Amd 1 Methods of Testing Concrete Part 112: Methods of Accelerated Curing of Test Cubes (AMD 6100) July 31, 1989. 8 pp.

Concrete Blocks
Use For: Aerated Concrete Blocks; Concrete Bricks; Concrete Masonry Units; Masonry Units, Concrete; Precast Concrete Blocks; Reconstructed Stone Masonry Units; Reinforced Concrete Blocks; Unit Masonry, Concrete *See Also:* Bricks; Building Blocks; Cast Stone; Cinder Blocks; Construction Materials; Masonry
BSI BS 6073: Part 1-81. 1981 Amd 2 Precast Concrete Masonry Units Part 1: Specification for Precast Concrete Masonry Units. 20 pp.
BSI BS 6073: Part 2-81. 1981 Amd 1 Precast Concrete Masonry Units Part 2: Method for Specifying Precast Concrete Masonry Units (AMD 4508) March 30, 1984. 11 pp.
BSI BS 6457-84. 1984 Reconstructed Stone Masonry Units. 7 pp.
CEN PREN 771-3-92. Specification for Masonry Units—Part 3: Aggregate Concrete Masonry Units (Dense and Light-Weight Aggregates). 26 pp.
CEN PREN 771-4-92. Specification for Masonry Units—Part 4: Autoclaved Aerated Concrete Masonry Units. 17 pp.
CNS A2137-87. Hollow Concrete Blocks (Jun)(8905). 5 pp.
CNS A3042-87. Method of Test for Hollow Concrete Blocks (Jun)(1178). 3 pp.
CSA CAN3-A165 Series-M85. CSA Standards on Concrete Masonry Units; (Gen Instr 1 Thru 2). 49 pp.
DIN ENGL 4165-86. Autoclaved Aerated Concrete Blocks and Flat Elements (Dec). 4 pp.
DIN ENGL 18151-87. Lightweight Concrete Hollow Blocks (Sept). 8 pp.
DIN ENGL 18152-87. Lightweight Concrete Solid Bricks and Blocks (Apr). 7 pp.
DIN ENGL 18153-89. Normal-Weight Concrete Masonry Units (Sept). 12 pp.

Concrete Blocks *(Cont.)*
JIS A 5406-87. Hollow Concrete Blocks. 11 pp.
JIS A 5408-87. Concrete Blocks for Fully Grouted Masonry. 9 pp.
SNZ NZS 3102-83. Concrete Masonry Units. 18 pp.

—Aggregates
BSI BS 3797-90. 1990 Amd 1 Lightweight Aggregates for Masonry Units and Structural Concrete (AMD 6796) August 30, 1991. 12 pp.

—Buildings—Cleaning
BSI BS 6270: Part 2-85. 1985 Cleaning and Surface Repair of Buildings Part 2: Concrete and Precast Concrete Masonry. 17 pp.

—Buildings—Repair
BSI BS 6270: Part 2-85. 1985 Cleaning and Surface Repair of Buildings Part 2: Concrete and Precast Concrete Masonry. 17 pp.

—Compression Testing
CEN PREN 772-1-92. Methods of Test for Masonry Units—Part 1: Determination of Compressive Strength. 13 pp.

—Culverts
JIS A 5328-90. Reinforced Concrete Built-up Culvert Blocks. 10 pp.

—Decorative
JIS A 5407-87. Decorated Concrete Blocks. 12 pp.

—Designations
DIN ENGL 18153-89. Normal-Weight Concrete Masonry Units (Sept). 12 pp.

—Dimensional Stability
CEN PREN 772-14-92. Methods of Test for Masonry Units—Part 14: Determination of Moisture Movement of Aggregate Concrete Masonry Units. 9 pp.
CEN PREN 772-19-93. Methods of Test for Masonry Units—Part 19: Determination of Dimensional Stability of Large Clay Masonry Units. 7 pp.

—Fillers
CGSB CAN/CGSB-1.188-M90. Emulsion Type Filler Masonry Block. 10 pp.

—Flatness
CEN PREN 772-20-93. Methods of Test for Masonry Units—Part 20: Determination of Flatness of Faces and Squareness of Angles and Edges. 7 pp.

—Flexural Strength
CEN PREN 1052-2-93. Methods of Test for Masonry—Part 2: Determination of Flexural Strength. 13 pp.

—Grout
CSA A179-1975. Mortar and Grout for Unit Masonry. 20 pp.
CSA A179-M1976. Mortar and Grout for Unit Masonry; (Erratum September 1977). 20 pp.

—Inspection
DIN ENGL 18153-89. Normal-Weight Concrete Masonry Units (Sept). 12 pp.

—Manhole—Sewers
CNS A2072-88. Reinforced Concrete Manhole Blocks for Sewerage Work (Dec) (4995).
JIS A 5317-90. Reinforced Concrete Manhole Blocks for Sewerage Work. 9 pp.

—Markings
DIN ENGL 18153-89. Normal-Weight Concrete Masonry Units (Sept). 12 pp.

—Moisture Content
CEN PREN 772-10-92. Methods of Test for Masonry Units—Part 10: Determination of Moisture Content of Calcium Silicate, Aggregate Concrete and Autoclaved Aerated Masonry Units. 6 pp.

—Mortars
CSA A179-1975. Mortar and Grout for Unit Masonry. 20 pp.
CSA A179-M1976. Mortar and Grout for Unit Masonry; (Erratum September 1977). 20 pp.

—Paving Materials
BSI BS 6717: Part 1-86. 1986 Precast Concrete Paving Blocks Part 1: Paving Blocks. 8 pp.
DIN ENGL 18501-82. Concrete Pavement Setts (Nov). 4 pp.
SNZ NZS 3116-91. Interlocking Concrete Block Paving. 36 pp.

—Paving Materials—Installation
BSI BS 6717: Part 3-89. 1989 Precast Concrete Paving Blocks Part 3: Code of Practice for Laying. 14 pp.

Concrete Blocks *(Cont.)*
—Retaining Walls
CNS A2221-88. Concrete Blocks for Retaining Wall and Revetment (Jun)(12326).
CNS A3274-88. Method of Test for Concrete Blocks for Retaining Wall and Revetment (Jun)(12327).
JIS A 5323-90. Concrete Blocks for Retaining Wall and Revetment. 9 pp.

—Sewers
CNS A3083-88. Method of Test for Reinforced Concrete Blocks for Sewerage Work (Dec)(4996).

—Sizes
CEN PREN 772-16-92. Methods of Test for Masonry Units—Part 16: Determination of Size and Dimensions (Excluding Natural Stone Masonry Units). 6 pp.

—Squareness
CEN PREN 772-20-93. Methods of Test for Masonry Units—Part 20: Determination of Flatness of Faces and Squareness of Angles and Edges. 7 pp.

—Tensile Testing
CEN PREN 772-6-92. Methods of Test for Masonry Units—Part 6: Determination of Bending Tensile Strength of Concrete Masonry Units. 6 pp.

—Thickness Measurement
CEN PREN 772-16-92. Methods of Test for Masonry Units—Part 16: Determination of Size and Dimensions (Excluding Natural Stone Masonry Units). 6 pp.

—Water Absorption
CEN PREN 772-8-92. Methods of Test for Masonry Units—Part 8: Determination of Water Absorption of Aggregate Concrete Masonry Units by Soaking. 6 pp.
CEN PREN 772-11-92. Methods of Test for Masonry Units—Part 11: Determination of Water Absorption of Clay and Aggregate Concrete Masonry Units Due to Capillary Action. 7 pp.

Concrete Breakers
See Also: Construction Equipment

—Shanks—Steel
CNS A1035-86. Shapes and Dimensions of Steel Shank for Concrete Breakers (Dec)(9553).

—Sound Power
EC 84/537/EEC-84. Council Directive on the Approximation of the Laws of the Member States Relating to the Permissible Sound Power Level of Powered Hand-Held Concrete-Breakers and Picks. 15 pp.

Concrete Bricks
Use: Concrete Blocks

Concrete Bridges
See Also: Bridges (Structures); Concrete Construction
BSI BS 5400: Part 1-88. 1988 Steel, Concrete and Composite Bridges Part 1: General Statement. 10 pp.

—Construction
DIN ENGL 1075-81. Concrete Bridges; Dimensioning and Construction (Apr). 15 pp.

—Design
BSI BS 5400: Part 4-90. 1990 Steel, Concrete and Composite Bridges Part 4: Code of Practice for Design of Concrete Bridges. 65 pp.
BSI BS 5400: Part 4-84. (WITHDRAWN) 1984 Steel, Concrete and Composite Bridges Part 4: Code of Practice for Design of Concrete Bridges. 65 pp.

—Fatigue (Materials)
BSI BS 5400: Part 10-80. 1980 Steel, Concrete and Composite Bridges Part 10: Code of Practice for Fatigue. 55 pp.
BSI BS 5400: Part 10C-80. 1980 Steel, Concrete and Composite Bridges Part 10C: Charts for Classification of Details for Fatigue. 2 pp.

—Loads (Forces)
BSI BS 5400: Part 2-78. 1978 Amd 1 Steel, Concrete and Composite Bridges Part 2: Specification for Loads. 45 pp.

Concrete Coatings
See Also: Coatings; Concrete Sealants; Concretes; Masonry Coatings

—Asphalt
CGSB CAN/CGSB-37.2-M88. Emulsified Asphalt, Mineral—Colloid Type, Unfilled, for Dampproofing and Waterproofing and for Roof Coatings. 13 pp.
CGSB CAN/CGSB-37.28-M89. Reinforced, Mineral Colloid Type, Emulsified Asphalt for Roof Coatings and Waterproofing. 14 pp.

INTERNATIONAL AND NON-U.S. NATIONAL STANDARDS
SUBJECT INDEX
Concrete

Concrete Coatings (Cont.)
—Asphalt (Cont.)
CGSB CAN/CGSB-37.65-M88. Mastic Asphalt (Hot Process) for Flooring. 19 pp.

—Coal Tar
CGSB CAN/CGSB-37.33-M90. Fibrated, Mineral Colloid Type, Emulsified Coal Tar Pitch for Roof Coatings and for Damp-Proofing and Water-Proofing. 12 pp.

—Coal Tar—Interior/Exterior
CGSB 1-GP-184MA-83. Coating, Coal Tar-Epoxy, Standard for. 17 pp.

—Enamels—Floors—Brushed
CGSB 1-GP-71 METH 134.6-79. Methods of Testing Paints and Pigments Applicability and Appearance Enamels for Concrete Floors. 1 p.

—Enamels—Floors—Roller Applied
CGSB 1-GP-71 METH 134.6-79. Methods of Testing Paints and Pigments Applicability and Appearance Enamels for Concrete Floors. 1 p.

—Epoxy
CGSB 1-GP-146M-79. Coating, Epoxy, Cold Cured, Gloss, Standard for; (Amendment 1 Feb 1981) (QPL Apr 1986). 14 pp.
CGSB CAN/CGSB-1.153-M90. High Build, Gloss, Epoxy Coating. 13 pp.

—Epoxy—Corrosion Inhibitive
CGSB CAN/CGSB-1.207-M91. Low Temperature Curing Epoxy Coating. 11 pp.

—Nonmetallic
DIN ENGL 28052 Pt 1-92. Nonmetallic Protective Coatings and Linings for Concrete Structural Elements in Process Plants; Concepts and Selection Criteria (Feb) (Supersedes June 1972 Edition of VDI 2533). 12 pp.

—Nuclear Power Plants
DIN ENGL 55991 Pt 1-83. Coating Materials; Coatings for Nuclear Installations; Requirements, Testing (Aug). 5 pp.

Concrete Column Slabs
Use: Concrete Slabs

Concrete Construction
See Also: Composite Structures; Concrete Anchors; Concrete Bridges; Concrete Floors; Concrete Pavements; Concrete Structures; Construction; Construction Joints; Earthquake Resistant Structures; Formwork (Construction); Precast Concrete; Prestressed Concrete; Strands (Wire)

BSI BS 8110: Part 1-85. 1985 Structural Use of Concrete Part 1: Code of Practice for Design and Construction. 126 pp.
BSI BS 8110: Part 1-02. 1985 Amd 2 Structural Use of Concrete Part 1: Code of Practice for Design and Construction (AMD 6276) December 22, 1989. 127 pp.
BSI BS 8110: Part 1-03. 1985 Amd 3 Structural Use of Concrete Part 1: Code of Practice for Design and Construction (AMD 7583) March 15, 1993 (R). 142 pp.
BSI BS 8110: Part 1-04. 1985 Structural Use of Concrete Part 1: Code of Practice for Design and Construction (AMD 7973) September 15, 1993 (R). 143 pp.
BSI BS 8110: Part 2-85. 1985 Amd 1 Structural Use of Concrete Part 2: Code of Practice for Special Circumstances. 50 pp.
CSA CAN/CSA-A23. 1&2-M90. CAN/CSA-A23.1-M90, Concrete Materials and Methods of Concrete Construction; CAN/CSA-A23.2-M90, Methods of Test for Concrete; (Gen Instr 1 Thru 2). 276 pp.
CSA CAN3-A438-M84. Concrete Construction for Housing and Small Buildings; (Gen Instr 1). 35 pp.
DIN ENGL 4126-86. Cast-In-Situ Concrete Diaphragm Walls; Design and Construction (Aug). 26 pp.
SNZ NZS 3124-87. Specification for Concrete Construction for Minor Works. 14 pp.

—Cements—Waterproof Coatings
CNS A2047-78. Waterproof Agent of Cement for Concrete Construction (Mar)(3763). 1 p.
CNS A3060-78. Method of Test for Waterproof Agent of Cement for Concrete Construction (Mar)(3764). 4 pp.
JIS A 1404-77. Method of Test for Waterproof Agent of Cement for Concrete Construction. 8 pp.

Concrete Construction (Cont.)
—Construction Contracts
DIN ENGL 18331-88. Tendering and Performance Stipulations in Contracts for Construc-tion Works (VOB); Pt C: General Technical Speci-fications in Contracts for Construction Works (ATV); Concrete Work (Sept) (This Standard, Together with DIN 18299, Sept 1988 Ed., Supsds. October 1979 Edition). 6 pp.

—Design
BSI PP 7312-88. 1988 Extracts from British Standards for Students of Structural Design. 156 pp.
DIN ENGL 1045-88. Structural Use of Concrete; Design and Construction (July). 92 pp.

—Designations
DIN ENGL 1080 Pt 3-80. Terms, Symbols and Units Used in Civil Engineering; Concrete Construction and Reinforced Concrete Construction, Prestressed Concrete Construction, Masonry Construction (Mar). 4 pp.
DIN ENGL 1080 Pt 4-80. Terms, Symbols and Units Used in Civil Engineering; Concrete Construction, Composite Steel Construction and Steel Girders in Concrete (Mar). 10 pp.

—Forms—Metal
CNS A2104-81. Metal Panels for Concrete Form (May)(7334). 4 pp.
JIS A 8652-79. Metal Panels for Concrete Form. 8 pp.

—Reinforced
CSA CAN/CSA-G30.18-M92. Billet-Steel Bars for Concrete Reinforcement; (Gen Instr 1 Thru 2). 26 pp.
CSA W186-M1990. Welding of Reinforcing Bars in Reinforced Concrete Construction; (Gen Instr 1). 75 pp.

—Symbols
DIN ENGL 1080 Pt 3-80. Terms, Symbols and Units Used in Civil Engineering; Concrete Construction and Reinforced Concrete Construction, Prestressed Concrete Construction, Masonry Construction (Mar). 4 pp.
DIN ENGL 1080 Pt 4-80. Terms, Symbols and Units Used in Civil Engineering; Concrete Construction, Composite Steel Construction and Steel Girders in Concrete (Mar). 10 pp.

Concrete Flooring Systems
Use: Concrete Floors

Concrete Floors
Use For: Concrete Flooring Systems
See Also: Concrete Construction; Floors
BSI BS 8204: Part 1-87. 1987 In-Situ Floorings Part 1: Code of Practice for Concrete Bases and Screeds to Receive In-Situ Floorings. 19 pp.
BSI BS 8204: Part 2-87. 1987 In-Situ Floorings Part 2: Code of Practice for Concrete Wearing Surfaces. 20 pp.

—Asphalt Coatings—Waterproof
CGSB 37-GP-12MA-84. Application of Unfilled Cutback Asphalt for Dampproofing, Standard for. 6 pp.
CGSB CAN/CGSB-37.16-M89. Filled, Cutback Asphalt for Dampproofing and Waterproofing. 10 pp.

—Bituminous Coatings
DIN ENGL 18195 Pt 4-83. Waterproofing of Buildings and Structures; Damp-Proofing Against Moisture from the Ground; Design and Workmanship (Aug). 8 pp.

—Bricks
DIN ENGL 4159-78. Structurally Cooperating Bricks for Floors and Wall Panels (Apr). 9 pp.
DIN ENGL 4160-78. Structurally Non-Cooperating Bricks for Floors (Aug). 6 pp.

—Finishes
BSI CP 204: Part 2-70. 1970 Amd 3 In-Situ Floor Finishes Part 2: Metric Units. 40 pp.
BSI CP 204: Part 2-04. 1970 Amd 4 In-Situ Floor Finishes Part 2: Metric Units (AMD 6968) July 15, 1992. 24 pp.

—Painting—Interior/Exterior
CGSB 85-GP-32M-79. Painting Concrete Floors. 14 pp.

—Paints (Enamel)
CGSB 1-GP-71 METH 134.6-79. Methods of Testing Paints and Pigments Applicability and Appearance Enamels for Concrete Floors. 1 p.

—Paints (Enamel)—Interior
CGSB CAN/CGSB-1.66-M89. Interior Enamel for Concrete Floors. 10 pp.

Concrete Floors (Cont.)
—Paints (Enamel)—Interior/Exterior
CGSB CAN/CGSB-1.73-M91. Exterior and Interior Enamel for Floors. 8 pp.

—Paints (Latex)
CGSB CAN/CGSB-1.154-M89. Latex Type Paint for Concrete Floors. 10 pp.

—Paints—Gasoline Resistant
MOD UK DSTAN 80-84-82. Paint, Floor, Petrol and Oil Resisting Types: Brushing Spraying Issue 1. 9 pp.

—Paints—Oil Resistant
MOD UK DSTAN 80-84-82. Paint, Floor, Petrol and Oil Resisting Types: Brushing Spraying Issue 1. 9 pp.

—Polymer Finishes—Design
BSI BS 8204: Part 3-93. 1993 In-Situ Floorings Part 3: Code of Practice for Polymer Modified Cementitious Wearing Surfaces (R). 26 pp.

—Prestressed—Filler Joists
DIN ENGL 4158-78. Filler Concrete Joists for Reinforced and Prestressed Concrete Floors (May). 7 pp.

—Reinforced—Filler Joists
DIN ENGL 4158-78. Filler Concrete Joists for Reinforced and Prestressed Concrete Floors (May). 7 pp.

—Reinforced—Wood Wool
BSI BS 3809-93. 1993 Wood Wool Permanent Formwork and Infill Units for Reinforced Concrete Floors and Roofs (R). 10 pp.
BSI BS 3809-71. 1971 Wood Wool Permanent Formwork and Infill Units for Reinforced Concrete Floors and Roofs. 8 pp.

—Waterproofing Membranes—Plastic Sheets
DIN ENGL 18195 Pt 4-83. Waterproofing of Buildings and Structures; Damp-Proofing Against Moisture from the Ground; Design and Workmanship (Aug). 8 pp.

Concrete Formwork
Use: Formwork (Construction)

Concrete Masonry Units
Use: Concrete Blocks

Concrete Mix
Use: Concretes

Concrete Mixers
See Also: Tilting Mixers
BSI BS 1305-74. 1974 Batch Type Concrete Mixers. 17 pp.
BSI BS 3963-74. (OBSOLESCENT) 1974 Method for Testing the Mixing Performance of Concrete Mixers. 13 pp.
CNS A4002-86. Drum-Type Concrete Mixers (Dec)(7102).
JIS A 8601-78. Drum Type Mixers. 7 pp.
SNZ NZS 3105-86. Specification for Concrete Mixers (Batch Type and Truck Type). 23 pp.

—Forced Mixing
CNS A4003-86. Forced Mixing-Type Concrete Mixers (Dec)(7103).
JIS A 8603-77. Forced Mixing Type Mixers. 6 pp.

—Stationary—Glossaries
DIN ENGL 459-72. Concrete Mixers; Definitions; Sizes; Requirements (Aug). 4 pp.

Concrete Pavements
Use For: Rigid Pavements *See Also:* Concrete Construction; Paving Materials

—Construction Contracts
DIN ENGL 18316-88. Tendering and Performance Stipulations in Contracts for Construction Works (VOB); Pt C: General Technical Specifications in Contracts for Construction Works (ATV); Pavements—Surfacings with Hydraulic Binders (Sept) (This Standard,. 6 pp.

—Joint Sealants
BSI BS 2499: Part 1-93. 1993 Hot-Applied Joint Sealant Systems for Concrete Pavements Part 1: Specification for Joint Sealants (R). 10 pp.
BSI BS 2499: Part 2-92. 1992 Hot-Applied Joint Sealant Systems for Concrete Pavements Part 2: Code of Practice for the Application and Use of Joint Sealants (Superseded BS 2499: 1973). 14 pp.
BSI BS 2499: Part 3-93. 1993 Hot-Applied Joint Sealant Systems for Concrete Pavements Part 3: Methods of Test (Superseded BS 2499: 1973) (R). 24 pp.

Concrete Pavements *(Cont.)*
—Joint Sealants *(Cont.)*
BSI BS 2499-73. (WITHDRAWN) 1973 Hot Applied Joint Sealants for Concrete Pavements (Superseded by BS 2499: Parts 1 & 3: 1992 & Part 2: 1992). 28 pp.
BSI BS 5212-75. (WITHDRAWN) 1975 Cold Poured Joint Sealants for Concrete Pavements (Superseded by BS 5212: Parts 1, 2, & 3: 1990). 19 pp.
BSI BS 5212: Part 1-90. 1990 Cold Applied Joint Sealant Systems for Concrete Pavements Part 1: Joint Sealants. 11 pp.
BSI BS 5212: Part 2-90. 1990 Cold Applied Joint Sealant Systems for Concrete Pavements Part 2: Code of Practice for the Application and Use of Joint Sealants. 14 pp.

—Joint Sealants—Comprehensive Testing
BSI BS 5212: Part 3-90. 1990 Cold Applied Joint Sealant Systems for Concrete Pavements Part 3: Methods of Test. 22 pp.

Concrete Paving Slabs
Use: Concrete Slabs

Concrete Piles
See Also: Composite Piles; Steel Piles; Wooden Piles
CNS A1038-83. Standard Practice for Execution of Spun Concrete Piles (Apr)(10137).
CNS A2031-82. Concrete Base Piles by Centrifugal Process (Dec)(1260). 7 pp.
CNS A2037-89. Pretensioned Spun Concrete Piles (Oct)(2602). 11 pp.
JIS A 5310-87. Reinforced Spun Concrete Piles. 17 pp.
JIS A 5329-91. Pressed Concrete Sheet Piles. 17 pp.
JIS A 5335-87. Pretensioned Spun Concrete Piles. 13 pp.
JIS A 5337-87. Pretensioned Spun High Strength Concrete Piles. 18 pp.
JIS A 7201-87. Standard Practice for Execution of Spun Concrete Piles. 23 pp.

—Cast In Place—Construction
DIN ENGL 4128-83. Small Diameter Injection Piles (Cast-In-Place Concrete Piles and Composite Piles); Construction Procedure, Design and Permissible Loading (Apr). 7 pp.

—Cast In Place—Loads (Forces)
DIN ENGL 4128-83. Small Diameter Injection Piles (Cast-In-Place Concrete Piles and Composite Piles); Construction Procedure, Design and Permissible Loading (Apr). 7 pp.

—Driven—Loads (Forces)
DIN ENGL 4026-75. Driven Piles; Manufacture, Dimensioning and Permissible Loading (Aug). 11 pp.
DIN ENGL 4026 Suppl. 1-75. Driven Piles; Manufacture, Dimensioning and Permissible Loading; Explanations (Aug). 6 pp.

—Sheet
CNS A2113-81. Reinforced Concrete Sheet Piles (Oct)(7933).
CNS A2115-86. Prestressed Concrete Sheet Piles (Dec)(7935).
JIS A 5326-88. Prestressed Concrete Sheet Piles. 20 pp.

Concrete Pipes
Use For: Reinforced Concrete Pipes; Spun Concrete Pipes *See Also:* Drains; Pipes
BSI BS 1194-69. (WITHDRAWN) 1969 Concrete Porous Pipes for Under-Drainage (Superseded by BS 5911: Part 114: 1992). 24 pp.
BSI BS 5911: Part 3-82. (WITHDRAWN) 1982 Precast Concrete Pipes and Fittings for Drainage and Sewerage Part 3: Ogee Jointed Concrete Pipes, Bends and Junctions, Unreinforced or Reinforced with Steel Cages or Hoops (Superseded by BS 5911: Part 110: 1992). 19 pp.
CNS A1001-79. Concrete Pipes (May)(483). 3 pp.
CNS A1008-86. Dimensions of Concrete Pipes (Sep)(1086).
CNS A2189-88. Prestressed Concrete Non-Cylinder Pipes (Nov)(11691). 13 pp.
CNS A2220-88. Prestressed Concrete Cylinder Pipes (May)(12285).
CNS A3003-75. Method of Test for Concrete Pipes (Dec)(484).
CNS A3238-88. Method of Test for Prestressed Concrete Non-Cylinder Pipes (May)(11692).
CNS A3273-88. Method of Test for Prestressed Concrete Cylinder Pipes (May)(12286).
DIN ENGL 2410 Pt 3-78. Pipes; Survey of Standards for Pipes from Concrete, Reinforced Concrete and Prestressed Concrete (Mar). 1 p.
DIN ENGL 4032-81. Concrete Pipes and Fittings; Dimensions; Technical Conditions of Delivery (Jan). 19 pp.
JIS A 5302-90. Reinforced Concrete Pipes. 13 pp.

Concrete Pipes *(Cont.)*
JIS A 5303-90. Centrifugal Reinforced Concrete Pipes. 25 pp.
JIS A 5333-88. Core Type Prestressed Concrete Pipes. 16 pp.
SAA AS 4058-92. Precast Concrete Pipes (Pressure and Non-Pressure) (in Professional Package 61B) (Supersedes AS 1342—1973 and AS 1392—1974). 30 pp.

—Drainpipes
BSI BS 5178-75. 1975 Prestressed Concrete Pipes for Drainage and Sewage. 17 pp.
BSI BS 5911: Part 1-81. 1981 Amd 1 Precast Concrete Pipes and Fittings for Drainage and Sewerage Part 1: Specification for Pipes and Fittings with Flexible Joints and Manholes (AMD 4035) September 30, 1982. 42 pp.
BSI BS 5911: Part 100-88. 1988 Amd 1 Precast Concrete Pipes and Fittings for Drainage and Sewerage Part 100: Unreinforced and Reinforced Pipes and Fittings with Flexible Joints (AMD 6269) December 22, 1989. 39 pp.
BSI BS 5911: Part 100-02. 1988 Amd 2 Precast Concrete Pipes, Fittings and Ancillary Products Part 100: Specification for Unreinforced and Reinforced Pipes and Fittings with Flexible Joints (AMD 7588) April 15, 1993 (R). 40 pp.
BSI BS 5911: Part 101-88. 1988 Precast Concrete Pipes and Fittings for Drainage and Sewerage Part 101: Glass Composite Concrete (GCC) Pipes and Fittings with Flexible Joints. 32 pp.
BSI BS 5911: Part 110-92. 1992 Precast Concrete Pipes, Fittings and Ancillary Products Part 110: Specification for Ogee Pipes and Fittings (Including Perforated). 34 pp.
BSI BS 5911: Part 120-89. 1989 Precast Concrete Pipes and Fittings for Drainage and Sewerage Part 120: Reinforced Jacking Pipes with Flexible Joints. 25 pp.

—Irrigation
JIS A 5322-90. Reinforced Spun-Concrete Pipes with Socket. 9 pp.
JIS A 5330-90. Concrete Pipes with Socket. 9 pp.

—Ogee
BSI BS 5911: Part 110-92. 1992 Precast Concrete Pipes, Fittings and Ancillary Products Part 110: Specification for Ogee Pipes and Fittings (Including Perforated). 34 pp.

—Porous
BSI BS 5911: Part 114-92. 1992 Precast Concrete Pipes, Fittings and Ancillary Products Part 114: Specification for Porous Pipes (Supersedes BS 1194: 1969). 17 pp.

—Pressure
BSI BS 4625-70. 1970 Amd 2 Prestressed Concrete Pressure Pipes (Including Fittings). 26 pp.
BSI BS 8010: Sec 2.4-88. 1988 Pipelines Part 2: Pipelines on Land: Design, Construction and Installation Section 2.4: Prestressed Concrete Pressure Pipelines. 19 pp.
CEN PREN 642-92. Prestressed Concrete Pressure Pipes, Cylinder and Non-Cylinder, Including Joints, Fittings and Specific Requirement for Prestressing Steel for Pipes. 26 pp.
SAA AS 4058-92. Precast Concrete Pipes (Pressure and Non-Pressure) (in Professional Package 61B) (Supersedes AS 1342—1973 and AS 1392—1974). 30 pp.

—Pressure—Earth
SAA AS 3725-89. Loads on Buried Concrete Pipes (This is a Joint Standard with SANZ NZS 3725). 16 pp.
SNZ NZS/AS 3725-89. Loads on Buried Concrete Pipes (This is a Joint Standard with SAA AS 3725). 16 pp.
SNZ NZS/AS 3725 Supplement 1-89. Loads on Buried Concrete Pipes. Commentary (This is a Joint Standard with SAA AS 3725). 25 pp.

—Pressure—Nominal Pressures
DIN ENGL 2401 Pt 3-68. Pipelines; Pressure Ratings; Permissible Working Pressures for Pipeline Components of Reinforced Concrete and Prestressed Concrete (Sept). 1 p.

—Pressure—Water—Pressure Measurement
DIN ENGL 4279 Pt 5-75. Testing of Pressure Pipelines for Water by Internal Pressure; Reinforced Concrete Pressure Pipes and Prestressed Concrete Pressure Pipes (Nov). 2 pp.

—Sewer
BSI BS 5178-75. 1975 Prestressed Concrete Pipes for Drainage and Sewage. 17 pp.
BSI BS 5911: Part 1-81. 1981 Amd 1 Precast Concrete Pipes and Fittings for Drainage and Sewerage Part 1: Specification for Pipes and Fittings with Flexible Joints and Manholes (AMD 4035) September 30, 1982. 42 pp.

Concrete Pipes *(Cont.)*
—Sewer *(Cont.)*
BSI BS 5911: Part 100-88. 1988 Amd 1 Precast Concrete Pipes and Fittings for Drainage and Sewerage Part 100: Unreinforced and Reinforced Pipes and Fittings with Flexible Joints (AMD 6269) December 22, 1989. 39 pp.
BSI BS 5911: Part 100-02. 1988 Amd 2 Precast Concrete Pipes, Fittings and Ancillary Products Part 100: Specification for Unreinforced and Reinforced Pipes and Fittings with Flexible Joints (AMD 7588) April 15, 1993 (R). 40 pp.
BSI BS 5911: Part 101-88. 1988 Precast Concrete Pipes and Fittings for Drainage and Sewerage Part 101: Glass Composite Concrete (GCC) Pipes and Fittings with Flexible Joints. 32 pp.
BSI BS 5911: Part 120-89. 1989 Precast Concrete Pipes and Fittings for Drainage and Sewerage Part 120: Reinforced Jacking Pipes with Flexible Joints. 25 pp.
BSI DD 76: Part 2-83. 1983 Precast Concrete Pipes of Composite Construction Part 2: Precast Concrete Pipes Strengthened by Chopped Zinc-Coated Steel Fibres. 20 pp.
CNS A2050-84. Concrete Pipes for Sewerage (Use for Jacking Method) (Apr)(3905). 4 pp.
JIS A 5322-90. Reinforced Spun-Concrete Pipes with Socket. 9 pp.
JIS A 5330-90. Concrete Pipes with Socket. 9 pp.

—Water
BSI DD 76: Part 2-83. 1983 Precast Concrete Pipes of Composite Construction Part 2: Precast Concrete Pipes Strengthened by Chopped Zinc-Coated Steel Fibres. 20 pp.

—Water Absorption
CSA CAN/CSA-A257 Series-M92. Standards for Concrete Pipe (A257.0 thru A257. 4); (Gen Instr 1). 81 pp.

Concrete Primers
Scope Note: Primer coatings for concrete surfaces
See Also: Concretes

—Asphalt
CGSB 37-GP-9MA-83. Primer, Asphalt, Unfilled, for Asphalt Roofing, Dampproofing and Waterproofing, Standard for. 8 pp.
CGSB 37-GP-15M-76. Application of Asphalt Primer for Asphalt Roofing, Dampproofing and Waterproofing, Standard for (R 1984). 7 pp.

—Coal Tar
CGSB CAN/CGSB-37.32-M89. Coal Tar Primer for Coal Tar Roofing, Damp-Proofing and Water-Proofing. 8 pp.
CGSB CAN/CGSB-37.34-M89. Application of Coal Tar Primer for Coal Tar Roofing, Dampproofing and Waterproofing. 8 pp.

Concrete Reinforcing Steels
Use: Reinforcing Steels

Concrete Sealants
See Also: Concrete Coatings; Sealants

—Alkali Resistant
CGSB CAN/CGSB-1.142-M89. Clear Sealer, Alkali Resistant. 9 pp.

Concrete Slabs
Use For: Concrete Paving Slabs; Reinforced Concrete Slabs; Slabs, Concrete *See Also:* Cast Stone; Construction Materials; Paving Materials; Structural Members
DIN ENGL 4028-82. Reinforced Concrete Slabs Made of Lightweight Concrete with Internally Porous Texture; Requirements, Testing, Design, Construction, Installation (Jan). 11 pp.
DIN ENGL 18162-76. Light Weight Concrete Wall Slabs; Non-Reinforced (Aug). 3 pp.

—Aerated
CNS A2133-88. Autoclaved Lightweight Aerated Concrete Panels (Mar)(8646).
CNS A3146-88. Method of Test for Autoclaved Lightweight Aerated Concrete Panels (Mar)(8647).
DIN ENGL 4166-86. Autoclaved Aerated Concrete Slabs and Panels (Dec). 3 pp.

—Precast
CSA A231.1-1972. Precast Concrete Paving Slabs. 16 pp.
CSA CAN3-A231.2-M85. Precast Concrete Pavers; (Gen Instr 1) (Errata April 1985). 18 pp.

—Precast—Lightweight
SNZ NZS 3151-74. Specification for Precast Lightweight Concrete Panels and Slabs Amend: 1, 1976; 2, 1985. 11 pp.

INTERNATIONAL AND NON-U.S. NATIONAL STANDARDS
SUBJECT INDEX

Concretes

Concrete Slabs *(Cont.)*

—Prestressed
DIN ENGL 51231 Pt 1-72. Testing Installations for Building Components; Floor Slabs or Beams (July). 4 pp.

—Prestressed—Double T
JIS A 5412-90. Prestressed Concrete Slab (Double-T Type). 19 pp.

—Seals
ISO 4635-82. Rubber, Vulcanized—Preformed Compression Seals for Use Between Concrete Motorway Paving Sections—Specification for Material First Edition. 5 pp.

—Thermal Insulation
BSI BS 3958: Part 1-82. 1982 Thermal Insulating Materials Part 1: Magnesia Preformed Insulation. 7 pp.
BSI BS 3958: Part 2-82. 1982 Thermal Insulating Materials Part 2: Calcium Silicate Preformed Insulation. 8 pp.
SNZ BS 3958: Part 1-82. Thermal insulating Materials Part 1: Magnesia Preformed Insulation. 8 pp.
SNZ BS 3958: Part 2-82. Thermal insulating Materials Part 2: Calcium Silicate Preformed Insulation. 8 pp.

Concrete Steel
Use: Reinforcing Steels

Concrete Structures
See Also: Concrete Construction; Locks (Waterways); Structures

CSA CAN3-A23.3-M84. Design of Concrete Structures for Buildings; (Erratums March 1988). 290 pp.
SAA AS 3600 Supp 3-91. Supplement 3:—Concrete Structures—Extracts from AS 3600 Concrete Structures—Concrete Construction Requirements (Supplement to AS 3600—1988)(In Professional Packages 20,21,30,58,62-69). 34 pp.
SNZ NZS 3109-87. Specification for Concrete Construction Amend: 1, 1992; 2, 1993. 51 pp.

—Design
BSI DD ENV 1992-1-1-92. 1992 Eurocode 2: Design of Concrete Structures Part 1: General Rules and Rules for Buildings (Together with United Kingdom National Application Document) (R). 272 pp.
CEN ENV 1992-1-1-91. Eurocode 2: Design of Concrete Structures—Part 1: General Rules and Rules for Buildings. 254 pp.
CEN ENV 1994-1-1-92. Design of Composite Steel and Concrete Structures—Part 1-1: General Rules and Rules for Buildings. 180 pp.
SAA AS MP28 Sect. C9-75. Commentary on AS 1480 SAA Concrete Strctures Code—Sect.C9: General Design Requirement. 4 pp.
SNZ NZS 3101: Part 1-82. Design of Concrete Structures Part 1: Code of Practice for the Design of Concrete Structures Amend: 1, 1989; 2, 1992. 127 pp.
SNZ NZS 3101: Part 2-82. Design of Concrete Structures Part 2: Commentary on the Design of Concrete Structures Amend: 1, 1989; 2, 1992. 156 pp.

—Design—Waterproof
BSI BS 8007-87. 1987 Design of Concrete Structures for Retaining Aqueous Liquids. 31 pp.

—Nuclear Containment
CSA N287.1-93. General Requirements for Concrete Containment Structures for CANDU Nuclear Power Plants; (Gen Instr 1). 47 pp.
CSA N287.3-93. Design Requirements for Concrete Containment Structures for CANDU Nuclear Power Plants; (Gen Instr 1). 65 pp.
CSA CAN/CSA-N287.4-92. Construction, Fabrication, and Installation Requirements for Concrete Containment Structures for CANDU Nuclear Power Plants; (Gen Instr 1). 47 pp.
CSA N287.5-93. Examination and Testing Requirements for Concrete Containment Structures for CANDU Nuclear Power Plants; (Gen Instr 1). 37 pp.
CSA CAN3-N287.6-M80. Pre-Operational Proof and Leakage Rate Testing Requirements for Concrete Containment Structures for CANDU Nuclear Power Plants; (Gen Instr 1 Thru 2). 15 pp.
CSA CAN3-N287.7-M80. In-Service Examination and Testing Requirements for Concrete Containment Structures for CANDU Nuclear Power Plants; (Gen Instr 1). 16 pp.

—Reinforced—Design
CNS A1011-72. Design Standards for Reinforced Concrete Buildings (Jun)(3035).

—Reinforcing Steels
SAA AS MP28 Sect. C6-77. Commentary on AS 1480 SAA Concrete Structures Code—Sect.C6: Details of Reinforcement. 6 pp.

Concrete Structures *(Cont.)*

—Reinforcing Steels—Stresses
SAA AS MP28 Sect. C11-77. Commentary on AS 1480 SAA Concrete Structures Code—Sect.C11: Developement of Stress in Reinforcement. 11 pp.

—Roofing—Plastic Sheets
CNS A3182-83. Method of Test for Synthetic Polymer Roofing Sheets for Waterproof (Apr)(10144).
CNS A3183-83. Method of Test for Synthetic Polymer Roofing Sheets Laminated with Cloth and Others (Apr)(10146).
JIS A 6008-92. Roofing Sheets of Synthetic Polymer. 31 pp.

—Roofing—Waterproof Coatings
JIS A 6021-89. Liquid-Applied Compounds for Waterproofing Membrane Coating of Roof. 19 pp.

—Serviceability
SAA AS MP28 Sect. C10-75. Commentary on AS 1480 SAA Concrete Structures Code—Sect.C10: Serviceability. 7 pp.

—Storage Tanks—Liquids
SAA AS 3735-91. Concrete Structures for Retaining Liquids (In Professional Packages 20,21,30,58, 62-69). 14 pp.
SAA AS 3735 Supp 1-91. Concrete Structures for Retaining Liquids —Supplement 1: Concrete Structures for Retaining Liquids—Commentary (Supplement to AS 3735—1991) (In Professional Packages 20,21,30,58, 62-69). 44 pp.

—Strength
BSI BS 6089-81. 1981 Assessment of Concrete Strength in Existing Structures. 15 pp.

Concrete Vibrators
See Also: Construction Equipment

BSI BS 2769: Sec 2.11-84. 1984 Hand-Held Electric-Motor-Operated Tools Part 2: Particular Requirements Section 2.11: Concrete Vibrators. 6 pp.
DIN ENGL 4235 Pt 1-78. Compacting of Concrete by Vibrating; Vibrators and Vibration Mechanics (Dec). 6 pp.
DIN ENGL 4235 Pt 2-78. Compacting of Concrete by Vibrating; Compacting by Internal Vibrators (Dec). 4 pp.
DIN ENGL 4235 Pt 3-78. Compacting of Concrete by Vibrating; Compacting by External Vibrators During the Manufacture of Precast Components (Dec). 4 pp.
DIN ENGL 4235 Pt 4-78. Compacting of Concrete by Vibrating; Compacting of In-Situ by Formwork Vibrators (Dec). 3 pp.
DIN ENGL 4235 Pt 5-78. Compacting of Concrete by Vibrating; Compacting by Surface Vibrators (Dec). 2 pp.

—Electric—Portable
CSA CAN/CSA-C22. 2 NO71.1-M89. Portable Electric Tools; (Gen Instr 1 Thru 4). 89 pp.

—External
DIN ENGL 4235 Pt 3-78. Compacting of Concrete by Vibrating; Compacting by External Vibrators During the Manufacture of Precast Components (Dec). 4 pp.

—Form
CNS A2080-87. Form Vibrators for Concrete (Mar)(5648). 3 pp.
CNS A3097-87. Method of Test for Form Vibrator for Concrete (Mar)(5649). 2 pp.
DIN ENGL 4235 Pt 4-78. Compacting of Concrete by Vibrating; Compacting of In-Situ by Formwork Vibrators (Dec). 3 pp.
JIS A 8611-77. Form Vibrators for Concrete. 9 pp.

—Internal
CNS A2079-87. Internal Vibrators for Concrete (Mar)(5646). 4 pp.
CNS A3096-87. Method of Test for Internal Vibrators for Concrete (Mar)(5647). 1 p.
DIN ENGL 4235 Pt 2-78. Compacting of Concrete by Vibrating; Compacting by Internal Vibrators (Dec). 4 pp.
IEC 745 Pt 2-12-82. Safety of Hand-Held Motor-Operated Electric Tools Part 2: Particular Requirements for Concrete Vibrators (Internal Vibrators) First Edition; (Amendment 1-1991). 30 pp.
JIS A 8610-78. Internal Vibrators for Concrete. 8 pp.

—Safety
BSI BS 2769: Sec 2.11-84. 1984 Hand-Held Electric-Motor-Operated Tools Part 2: Particular Requirements Section 2.11: Concrete Vibrators. 6 pp.
IEC 745 Pt 2-12-82. Safety of Hand-Held Motor-Operated Electric Tools Part 2: Particular Requirements for Concrete Vibrators (Internal Vibrators) First Edition; (Amendment 1-1991). 30 pp.

Concrete Vibrators *(Cont.)*

—Surface
DIN ENGL 4235 Pt 5-78. Compacting of Concrete by Vibrating; Compacting by Surface Vibrators (Dec). 2 pp.

Concrete Wall Slabs
Use: Concrete Slabs

Concretes
Scope Note: For additional listings, see also specific products made from concrete *Use For:* Concrete; Concrete Mix; Fresh Concrete; Hardened Concrete; Portland Cement Concretes *See Also:* Admixtures; Aerated Concretes; Alkali Aggregate Reactions; Architectural Concrete; Cellular Concretes; Cements; Cinder Blocks; Concrete Coatings; Concrete Primers; Construction Materials; Curing Agents; Grout; Lightweight Concrete; Masonry; Mortars; Polymer Concretes; Portland Cements; Precast Concrete; Prestressed Concrete; Reinforced Concrete; Screeds; Shotcrete; Spun Concrete; Structural Forms; Structural Members; Terrazzo

BSI BS 1881: Part 124-88. 1988 Methods of Testing Concrete Part 124: Methods for Analysis of Hardened Concrete. 24 pp.
BSI BS 5328-81. (WITHDRAWN) 1981 Amd 2 Methods for Specifying Concrete, Including Ready-Mixed Concrete (AMD 4970) October 31, 1985 (Superseded by BS 5328: Parts 1, 2, 3, and 4: 1990). 21 pp.
BSI BS 5328: Part 1-91. 1991 Concrete Part 1: Guide to Specifying Concrete. 26 pp.
BSI BS 5328: Part 1-01. 1991 Amd 1 Concrete Part 1: Guide to Specifying Concrete (AMD 7174) July 15, 1992. 30 pp.
BSI BS 5328: Part 1-90. 1990 Concrete Part 1: Guide to Specifying Concrete (R). 24 pp.
BSI BS 5328: Part 2-91. 1991 Concrete Part 2: Methods for Specifying Concrete Mixes. 23 pp.
BSI BS 5328: Part 2-01. 1991 Amd 1 Concrete Part 2: Methods for Specifying Concrete Mixes (AMD 7175) July 15, 1992. 26 pp.
BSI BS 5328: Part 2-90. 1990 Concrete Part 2: Methods for Specifying Concrete Mixes (R). 18 pp.
BSI BS 5838: Part 1-80. 1980 Dry Packaged Cementitious Mixes Part 1: Prepacked Concrete Mixes. 4 pp.
BSI BS 8110: Part 1-85. 1985 Structural Use of Concrete Part 1: Code of Practice for Design and Construction. 126 pp.
BSI BS 8110: Part 1-02. 1985 Amd 2 Structural Use of Concrete Part 1: Code of Practice for Design and Construction (AMD 6276) December 22, 1989. 127 pp.
BSI BS 8110: Part 1-03. 1985 Amd 3 Structural Use of Concrete Part 1: Code of Practice for Design and Construction (AMD 7583) March 15, 1993 (R). 142 pp.
BSI BS 8110: Part 1-04. 1985 Structural Use of Concrete Part 1: Code of Practice for Design and Construction (AMD 7973) September 15, 1993 (R). 143 pp.
BSI BS 8110: Part 2-85. 1985 Amd 1 Structural Use of Concrete Part 2: Code of Practice for Special Circumstances. 50 pp.
BSI PD 6534-93. 1993 Guide to the use in the UK of DD ENV 206: 1992 Concrete. Production, Placing and Compliance Criteria (R). 55 pp.
BSI DD ENV 206-92. 1992 Concrete Performance, Production, Placing and Compliance Criteria (R). 41 pp.
CAA JAR-25 Section 1. Requirements. 295 pp.
CEN ENV 206-90. Concrete—Performance, Production, Placing and Compliance Criteria. 36 pp.
CSA CAN/CSA-A23. 1&2-M90. CAN/CSA-A23.1-M90, Concrete Materials and Methods of Concrete Construction; CAN/CSA-A23.2-M90, Methods of Test for Concrete; (Gen Instr 1 Thru 2). 276 pp.
DIN ENGL 1048 Pt 1-91. Testing Concrete; Testing of Fresh Concrete (June) (This Standard, Together with DIN 1048 Part 5, June 1991 Edition, Supersedes DIN 1048 Part 1, December 1978 Edition). 5 pp.
SAA AS 1379-91. Specification and Manufacture of Concrete (in Professional Packages 30, 55, 58). 27 pp.
SNZ NZS 3104-91. Specification for Concrete Production—High Grade and Special Grade. 36 pp.
SNZ NZS 3108-83. Specification for Concrete Production—Ordinary Grade Amend: 1; 2. 10 pp.

—Admixtures
BSI BS 5075: Part 1-82. 1982 Amd 2 Concrete Admixtures Part 1: Accelerating Admixtures, Retarding Admixtures, and Water Reducing Admixtures. 21 pp.
CEN PREN 104.301 (Part 1)-91. Admixtures for Concrete, Mortar and Grout Test Methods Reference Concrete and Reference Mortar for Testing. 3 pp.

INTERNATIONAL AND NON-U.S. NATIONAL STANDARDS
SUBJECT INDEX

Concretes

Concretes (Cont.)

—Admixtures (Cont.)
CEN PREN 934-2-92. Admixtures for Concrete, Mortars and Grouts—Part 2: Concrete Admixtures—Definitions, Specifications and Conformity Criteria. 12 pp.
CNS A2170-83. Expansive Additive for Concrete (Nov)(10641).
CNS A2219-88. Chemical Admixtures for Concrete (May)(12283).
CNS A3167-82. Method of Test for Air-Entraining Admixtures for Concrete (Oct)(9208).
CNS A3272-88. Method of Test for Chemical Admixture for Concrete (May)(12284).
CSA CAN3-A266.2-M78. Chemical Admixtures for Concrete; (Amd 1 September 1978). 26 pp.
CSA CAN3-A266.4-M78. Guidelines for the Use of Admixtures in Concrete; (Amd 1-11 November 1983). 45 pp.
JIS A 6202-80. Expansive Additive for Concrete. 31 pp.
JIS A 6204-87. Chemical Admixtures for Concrete. 31 pp.
SAA AS 1478-92. Chemical Admixtures for Concrete (in Professional Packages 30, 58A) (Supersedes AS 1479—1973). 21 pp.
SAA AS 2073-77. Methods for the Testing of Expanding Admixtures for Concrete, Mortar and Grout. 28 pp.
SAA MP20.2-75. Part 2: Thickening Admixtures for Use in Concrete and Mortar. 9 pp.
SAA MP20.3-77. Part 3: Expanding Admixtures for Use in Concrete, Mortar and Grout. 12 pp.
SNZ NZS 3113-79. Specification for Chemical Admixtures for Concrete. 15 pp.

—Admixtures—Chloride Content
CEN PREN 480 (Part 10)-91. Admixtures for Concrete, Mortar and Grout—Test Methods—Part 10: Determination of the Chloride Content. 7 pp.

—Admixtures—Density
CEN PREN 480 (Part 7)-91. Admixtures for Concrete, Mortar and Grout—Test Methods—Part 7: Determination of the Density of Liquid Admixtures. 4 pp.

—Admixtures—Dry Matter Content Analysis
CEN PREN 480 (Part 8)-91. Admixtures for Concrete, Mortar and Grout—Test Methods—Part 8: Determination of the Conventional Dry Material Content. 5 pp.

—Admixtures—Infrared Analysis
CEN PREN 480 (Part 6)-91. Admixtures for Concrete, Mortar and Grout—Test Methods—Part 6: Infrared Analysis. 5 pp.

—Admixtures—PH
CEN PREN 480 (Part 9)-91. Admixtures for Concrete, Mortar and Grout—Test Methods—Part 9: Determination of the Ph Value. 6 pp.

—Admixtures—Sampling
SAA AS 2072-77. Methods for the Sampling of Expanding Admixtures for Concrete, Mortar and Grout. 4 pp.

—Aggregates
BSI BS 877: Part 2-73. (WITHDRAWN) 1973 Amd 1 Foamed or Expanded Blast-Furnace Slag Light Weight Aggregate for Concrete Part 2: Metric Units (Superseded by BS 3797: 1990). 9 pp.
BSI BS 882-92. 1992 Aggregates from Natural Sources for Concrete. 12 pp.
BSI BS 882-83. 1983 Amd 1 Aggregates from Natural Sources for Concrete. 11 pp.
BSI BS 3797-90. 1990 Amd 1 Lightweight Aggregates for Masonry Units and Structural Concrete (AMD 6796) August 30, 1991. 12 pp.
BSI BS 4550: Part 4-78. 1978 Methods of Testing Cement Part 4: Standard Coarse Aggregate for Concrete Cubes. 2 pp.
CNS A2029-86. Concrete Aggregates (Aug)(1240).
CNS A2046-74. Lightweight Aggregate for Structural Concrete (Jan)(3691). 5 pp.
CNS A2202-87. Air-Cooled Iron-Blast-Furnace Slag Coarse Aggregate for Concrete (Feb)(11824).
CNS A3254-87. Method of Test for Air-Cooled Iron-Blast-Furnace Slag Coarse Aggregate for Concrete (Feb)(11825).
CNS A3257-87. Method of Test for Granulated Blast Furnace Slag Fine Aggregate for Concrete (Apr)(11891).
JIS A 5002-78. Light Weight Aggregates for Structural Concrete. 12 pp.
JIS A 5011-77. Air-Cooled Iron-Blast-Furnace Slag Aggregate for Concrete. 16 pp.
JIS A 5012-81. Granulated Blast Furnace Slag Fine Aggregate for Concrete. 10 pp.
SNZ NZS 3111-86. Methods of Test for Water and Aggregate for Concrete Amend: 1, 1988. 50 pp.

Concretes (Cont.)

—Aggregates (Cont.)
SNZ NZS 3121-86. Specification for Water and Aggregate for Concrete. 8 pp.

—Aggregates—Alkali Reactions
JIS A 1804-92. Mehtods of Test for Production Control of Concrete—Method of Rapid Test for Identification of the Alkali Reactivity of Aggregates. 9 pp.

—Aggregates—Bleeding
CEN PREN 480 (Part 4)-91. Admixtures for Concrete, Mortar and Grout—Test Methods—Part 4: Determination of Bleeding. 5 pp.

—Aggregates—Chloride Content
SAA AS 1012.20-92. Methods of Testing Concrete—Part 20: Determination of Chloride and Sulfate in Hardened Concrete and Concrete Aggregates (in Professional Packages 30, 58). 4 pp.

—Aggregates—Density
CEN PREN 992-92. Determination of the Dry Density of Lightweight Aggregate Concrete with Open Structure. 3 pp.
JIS A 1134-89. Methods of Test for Bulk Specific Gravity and Absorption of Light Weight Fine Aggregates for Structural Concrete. 9 pp.
JIS A 1135-89. Methods of Test for Bulk Specific Gravity and Absorption of Light Weight Coarse Aggregates for Structural Concrete. 6 pp.

—Aggregates—Impurities Content
CNS A3028-83. Method of Test for Organic Impurities in Fine Aggregate (Jan)(1164).
JIS A 1105-76. Method of Test for Organic Impurities in Fine Aggregate. 5 pp.

—Aggregates—Mass
JIS A 1119-89. Method of Test for Variability of Constituents in Freshly Mixed Concrete. 9 pp.

—Aggregates—Moisture Content
CNS A3008-84. Method of Test for Surface Moisture in Fine Aggregate (Apr)(489). 2 pp.
JIS A 1111-76. Method of Test for Surface Moisture in Fine Aggregate. 7 pp.
JIS A 1802-89. Methods of Test for Production Control of Concrete (Method of Test for Surface Moisture in Fine Aggregate by Centrifugal Force). 7 pp.
JIS A 1803-91. Method of Test for Production Control of Concrete—Method of Test for Surface Moisture in Coarse Aggregate. 6 pp.

—Aggregates—Sand Content
JIS A 1801-89. Methods of Test for Production Control of Concrete (Method of Test for Sand Equivalent Value of Fine Aggregates for Concrete). 7 pp.

—Aggregates—Shrinkage
BSI BS 812: Part 120-89. 1989 Testing Aggregates Part 120: Method for Testing and Classifying Drying Shrinkage of Aggregates in Concrete. 8 pp.

—Aggregates—Sieve Analysis
CNS A3005-59. Method of Sieve Analysis for Coarse and Fine Aggregates (Oct)(486) (R 1970).
CNS A3010-84. Method of Test for Substance in Aggregates Finer Than 75 Micrometer CNS 386 Test Sieve (Dec)(491). 2 pp.
ISO 6274-82. Concrete—Sieve Analysis of Aggregates First Edition. 4 pp.
JIS A 1102-89. Method of Test for Sieve Analysis of Aggregates. 6 pp.
JIS A 1103-89. Method of Test for Amount of Material Passing Standard Sieve 75 Micrometer in Aggregates. 6 pp.

—Aggregates—Solids Content
CNS A3027-86. Method of Test for Unit Weight and Voids in Aggregate (Aug)(1163).
JIS A 1104-76. Method of Test for Unit Weight of Aggregate and Solid Content in Aggregate. 7 pp.

—Aggregates—Sulfate Content
BSI BS 812: Part 118-88. 1988 Testing Aggregates Part 118: Method for Determination of Sulphate Content. 15 pp.
SAA AS 1012.20-92. Methods of Testing Concrete—Part 20: Determination of Chloride and Sulfate in Hardened Concrete and Concrete Aggregates (in Professional Packages 30, 58). 4 pp.

—Air Content
BSI BS 1881: Part 106-83. 1983 Amd 2 Testing Concrete Part 106: Methods for Determination of Air Content of Fresh Concrete (AMD 6723) July 31, 1991. 14 pp.
ISO 4848-80. Concrete—Determination of Air Content of Freshly Mixed Concrete—Pressure Method First Edition. 12 pp.

Concretes (Cont.)

—Air Content (Cont.)
SAA AS 1012.4-83. Methods of Testing Concrete—Part 4: Methods for the Determination of Air Content of Freshly Mixed Concrete. 27 pp.
SNZ NZS 3112: Part 1-86. Methods of Test for Concrete Part 1: Tests Relating to Fresh Concrete. 20 pp.

—Air Content—Gravimetric Analysis
CNS A3218-84. Method of Test for Unit Weight, Yield and Air Content (Gravimetric) of Concrete (Dec)(11151). 5 pp.
JIS A 1116-75. Method of Test for Unit Weight and Air Content (Gravimetric) of Fresh Concrete. 5 pp.

—Air Content—Pressure Measurement
CNS A3174-87. Method of Test for Air Content of Freshly Mixed Concrete by the Pressure Method (Dec)(9661).
JIS A 1128-75. Method of Test for Air Content of Fresh Concrete by Pressure Method. 7 pp.

—Air Content—Volumetric Analysis
CNS A3175-87. Method of Test for Air Content of Freshly Mixed Concrete by the Volumetric Method (Dec)(9662).
JIS A 1118-75. Method of Test for Air Content of Fresh Concrete by Volumetric Method. 8 pp.

—Air Entraining Agents
BSI BS 5075: Part 2-82. 1982 Amd 2 Concrete Admixtures Part 2: Air-Entraining Admixtures. 12 pp.
CNS A2043-86. Air-Entraining Admixtures for Concrete (Dec)(3091). 5 pp.
CNS A3167-82. Method of Test for Air-Entraining Admixtures for Concrete (Oct)(9208).
CSA CAN3-A266.1-M78. Air-Entraining Admixtures for Concrete; (Amd 1 September 1978) (Amd 2 November 1981). 23 pp.

—Anchors (Fasteners)—Tensile Testing
BSI BS 5080: Part 1-93. 1993 Structural Fixings in Concrete and Masonry Part 1: Method of Test for Tensile Loading. 15 pp.
BSI BS 5080: Part 1-74. 1974 Amd 1 Methods of Test for Structural Fixings in Concrete and Masonry Part 1: Tensile Loading. 13 pp.

—Binders (Materials)
SAA AS 1012. Methods of Testing Concrete Complete Set in Binder.

—Bleeding
CNS A3048-86. Method of Test for Bleeding Water from Concrete (Dec)(1235).
JIS A 1123-75. Method of Test for Bleeding of Concrete. 5 pp.
SAA AS 1012.6-83. Methods of Testing Concrete—Part 6: Method for the Determination of Bleeding of Concrete. 9 pp.
SNZ NZS 3112: Part 1-86. Methods of Test for Concrete Part 1: Tests Relating to Fresh Concrete. 20 pp.

—Building Board
CNS A2033-63. Bituminous Bagasse Boards for Concrete (Sep)(2179) (R 1973). 1 p.

—Building Sites
BSI BS 8000: Sec 2.2-90. 1990 Workmanship on Building Sites Part 2: Code of Practice for Concrete Work Section 2.2: Sitework with In Situ and Precast Concrete. 22 pp.

—Building Sites—Mixing
BSI BS 8000: Sec 2.1-90. 1990 Workmanship on Building Sites Part 2: Code of Practice for Concrete Work Section 2.1: Mixing and Transporting Concrete. 14 pp.

—Building Sites—Transportation
BSI BS 8000: Sec 2.1-90. 1990 Workmanship on Building Sites Part 2: Code of Practice for Concrete Work Section 2.1: Mixing and Transporting Concrete. 14 pp.

—Cement Content
CNS A3039-86. Method of Test for Cement Content of Hardened Portland Cement Concrete (Aug)(1175). 5 pp.
CNS A3218-84. Method of Test for Unit Weight, Yield and Air Content (Gravimetric) of Concrete (Dec)(11151). 5 pp.

—Chemical Resistance
DIN ENGL 4030 Pt 1-91. Assessment of Water, Soil and Gases for Their Aggressiveness to Concrete; Principles and Limiting Values (June). 6 pp.

INTERNATIONAL AND NON-U.S. NATIONAL STANDARDS SUBJECT INDEX

Concretes

Concretes (Cont.)

—Chemical Resistance—Sampling
DIN ENGL 4030 Pt 2-91. Assessment of Water, Soil and Gases for Their Aggressiveness to Concrete; Collection and Examination of Water and Soil Samples (June). 12 pp.

—Chloride Content
SAA AS 1012.20-92. Methods of Testing Concrete—Part 20: Determination of Chloride and Sulfate in Hardened Concrete and Concrete Aggregates (in Professional Packages 30, 58). 4 pp.

—Classification
BSI BS 1000: (666)-84. 1984 Universal Decimal Classification (UDC). English Full Edition (666): Glass Industry. Ceramics. Cement and Concrete. 60 pp.
ISO 3893-77. Concrete—Classification by Compressive Strength First Edition. 3 pp.
SNZ NZS/BS 1000 (666)-84. Universal Decimal Classification Glass Industry. Ceramics. Cement and Concrete. 60 pp.

—Compacting
BSI BS 1881: Part 103-83. 1983 Amd 2 Testing Concrete Part 103: Method for Determination of Compacting Factor (AMD 6726) July 31, 1991. 12 pp.

—Compacting—Volumetric Analysis
BSI DD 90-83. 1983 Volumetric Method for Determination of Compacting Factor of Fresh Concrete. 7 pp.

—Composition
BSI DD 83-83. 1983 Assessment of the Composition of Fresh Concrete. 46 pp.

—Compression Testers
BSI BS 1881: Part 115-86. 1986 Amd 1 Testing Concrete Part 115: Specification for Compression Testing Machines for Concrete (AMD 6536) December 21, 1990. 12 pp.

—Compression Testing
BSI BS 1881: Part 116-83. 1983 Amd 2 Testing Concrete Part 116: Method for Determination of Compressive Strength of Concrete Cubes (AMD 6720) July 31, 1991. 11 pp.
BSI BS 1881: Part 119-83. 1983 Methods of Testing Concrete Part 119: Method for Determination of Compressive Strength Using Portions of Beams Broken in Flexure (Equivalent Cube Method). 5 pp.
BSI BS 1881: Part 120-83. 1983 Amd 1 Methods of Testing Concrete Part 120: Method for Determination of the Compressive Strength of Concrete Cores (AMD 6109) July 31, 1989. 10 pp.
BSI BS 1881: Part 127-90. 1990 Methods of Testing Concrete Part 127: Method of Verifying the Performance of a Concrete Cube Compression Machine Using the Comparative Cube Test. 16 pp.
DIN ENGL 1048 Pt 2-91. Testing Concrete; Testing of Hardened Concrete (Specimens Taken In Situ) (June). 6 pp.
ISO 3893-77. Concrete—Classification by Compressive Strength First Edition. 3 pp.
ISO 4012-78. Concrete—Determination of Compressive Strength of Test Specimens First Edition. 6 pp.
ISO 4013-78. Concrete—Determination of Flexural Strength of Test Specimens First Edition. 6 pp.
JIS A 1108-76. Method of Test for Compressive Strength of Concrete. 6 pp.
JIS A 1139-78. Method of Test for Concrete Under Biaxial Compressive Loads. 7 pp.

—Compressive Testing
SNZ NZS 3112: Part 2-86. Methods of Test for Concrete Part 2: Tests Relating to the Determination of Strength of Concrete Amend: 1. 22 pp.

—Consistency
ISO 4103-79. Concrete—Classification of Consistency First Edition. 4 pp.
ISO 4109-80. Fresh Concrete—Determination of the Consistency—Slump Test First Edition. 4 pp.
ISO 4110-79. Fresh Concrete—Determination of the Consistency—Vebe Test First Edition. 4 pp.
ISO 4111-79. Fresh Concrete—Determination of Consistency—Degree of Compactibility (Compaction Index) First Edition. 4 pp.
SAA AS 1012.3-83. Methods of Testing Concrete—Part 3: Methods for the Determination of Properties Related to the Consistence of Concrete. 23 pp.
SNZ NZS 3112: Part 1-86. Methods of Test for Concrete Part 1: Tests Relating to Fresh Concrete. 20 pp.

—Content Analysis
DIN ENGL 52171-42. Proportion of the Constituents and the Mixing Ratio of Fresh Mortar and Fresh Concrete (July). 2 pp.

Concretes (Cont.)

—Cores—Compression Testing
CNS A3045-82. Method of Test for Compressive Strength of Cylindrical Concrete Specimens (Jan)(1232). 4 pp.
CNS A3051-84. Method of Test for Compressive and Flexural Strength of Drilled Cores and Sawed Beams of Concrete (Apr)(1238). 3 pp.
JIS A 1107-78. Method of Obtaining and Testing Drilled Cores and Sawed Beams of Concrete. 7 pp.
SAA AS 1012.14-91. Methods of Testing Concrete—Part 14: Method for Securing and Testing Cores from Hardened Concrete for Compressive Strength (in Professional Package 58). 7 pp.

—Cores—Length
CNS A3053-86. Method of Test for Measuring Length of Drilled Concrete Cores (Dec)(1241).

—Creep Properties
SAA AS 1012.16-74. Methods of Testing Concrete—Part 16: Determination of Creep of Concrete Cylinders in Compression. 9 pp.

—Crushed Stone
CNS A1027-80. Crushed Stone for Concrete (Aug)(6299). 6 pp.
JIS A 5005-87. Crushed Stone for Concrete. 9 pp.

—Density
BSI BS 1881: Part 107-83. 1983 Amd 2 Testing Concrete Part 107: Method for Determination of Density of Compacted Fresh Concrete (AMD 6722) July 31, 1991. 8 pp.
BSI BS 1881: Part 114-83. 1983 Amd 2 Testing Concrete Part 114: Methods for Determination of Density of Hardened Concrete (AMD 6721) July 31, 1991. 11 pp.
BSI BS 1881: Part 129-92. 1992 Testing Concrete Part 129: Method for Determination of Density of Partially Compacted Semi-Dry Fresh Concrete. 10 pp.
ISO 6275-82. Concrete, Hardened—Determination of Density First Edition. 4 pp.
ISO 6276-82. Concrete, Compacted Fresh—Determination of Density First Edition. 3 pp.
SNZ NZS 3112: Part 3-86. Methods of Test for Concrete Part 3: Tests on Hardened Concrete Other Than for Strength. 11 pp.

—Dimensional Stability
JIS A 1129-75. Methods of Test for Length Change of Mortar and Concrete. 12 pp.

—Expansion
BSI BS 1881: Part 5-70. 1970 Amd 2 Methods of Testing Concrete Part 5: Methods of Testing Hardened Concrete for Other Than Strength (AMD 6267) March 30, 1990. 27 pp.
SNZ NZS 3112: Part 3-86. Methods of Test for Concrete Part 3: Tests on Hardened Concrete Other Than for Strength. 11 pp.

—Fillers
CGSB CAN/CGSB-1.188-M90. Emulsion Type Filler Masonry Block. 10 pp.
CNS A3212-84. Method of Testing Preformed Expansion Joint Fillers for Concrete (Nonextruding and Resilient Types) (Sep)(10993).

—Flexural Strength
BSI BS 1881: Part 118-83. 1983 Amd 1 Methods of Testing Concrete Part 118: Method for Determination of Flexural Strength (AMD 6095) July 31, 1989. 8 pp.
CNS A3046-84. Method of Test for Flexural Strength of Concrete (Using Simple Beam with Third-Point Loading) (Apr)(1233). 3 pp.
CNS A3047-84. Method of Test for Flexural Strength of Concrete (Using Simple Beam with Center-Point Loading) (Apr)(1234). 3 pp.
JIS A 1106-76. Method of Test for Flexural Strength of Concrete. 8 pp.
SNZ NZS 3112: Part 2-86. Methods of Test for Concrete Part 2: Tests Relating to the Determination of Strength of Concrete Amend: 1. 22 pp.

—Flow Testing
BSI BS 1881: Part 105-84. 1984 Amd 2 Testing Concrete Part 105: Method for Determination of Flow (AMD 6724) July 31, 1991. 10 pp.

—Fly Ash
CEN PREN 450-91. Fly Ash for Concrete—Definitions, Requirements and Quality Control. 7 pp.
SAA AS 1129-72. Fly Ash for Use in Concrete Withdrawn.

—Formboards—Plywood
CNS O1022-81. Plywood for Concrete-Form (Oct)(8057). 4 pp.

Concretes (Cont.)

—Glossaries
BSI BS 6100: Sec 6.2-86. 1986 Glossary of Building and Civil Engineering Terms Part 6: Concrete and Plaster Section 6.2: Concrete. 7 pp.
BSI BS 6100: Sec 6.2-01. 1986 Amd 1 Glossary of Building and Civil Engineering Terms Part 6: Concrete and Plaster Section 6.2: Concrete (AMD 7264) August 15, 1992. 8 pp.
BSI BS 6100: SUBSEC 6.6.1-92. 1992 Glossary of Building and Civil Engineering Terms Part 6: Concrete and Plaster Section 6.6: Products, Applications and Operations Subsection 6.6.1: Concrete and Mortar. 27 pp.
BSI BS 6100: SUBSEC 6.6.1-91. 1991 Building and Civil Engineering Terms Part 6: Concrete and Plaster Section 6.6: Products, Applications and Operations Subsection 6.6.1: Concrete and Mortar. 27 pp.
JIS A 0203-80. Concrete Terminology. 31 pp.

—Hardness Testing
BSI BS 1881: Part 202-86. 1986 Methods of Testing Concrete Part 202: Recommendations for Surface Hardness Testing by Rebound Hammer. 8 pp.

—Laboratories
CSA A283-1980. Qualification Code for Concrete Testing Laboratories; (Gen Instr 2 Thru 6) (Supplement 1 1987). 27 pp.

—Mass
CNS A3218-84. Method of Test for Unit Weight, Yield and Air Content (Gravimetric) of Concrete (Dec)(11151). 5 pp.
JIS A 1116-75. Method of Test for Unit Weight and Air Content (Gravimetric) of Fresh Concrete. 5 pp.
SAA AS 1012.5-83. Methods of Testing Concrete—Part 5: Method for Determination of Mass Per Unit Volume of Freshly Mixed Concrete. 7 pp.
SAA AS 1012.12-86. Methods of Testing Concrete—Part 12: Methods for the Determination of Mass Per Unit of Volume Hardened Concrete. 6 pp.
SNZ NZS 3112: Part 1-86. Methods of Test for Concrete Part 1: Tests Relating to Fresh Concrete. 20 pp.

—Modulus of Elasticity
BSI BS 1881: Part 121-83. 1983 Methods of Testing Concrete Part 121: Method of Determination of Static Modulus of Elasticity in Compression. 7 pp.
BSI BS 1881: Part 209-90. 1990 Methods of Testing Concrete Part 209: Recommendations for the Measurement of Dynamic Modulus of Elasticity. 8 pp.
ISO 6784-82. Concrete—Determination of Static Modulus of Elasticity in Compression First Edition. 5 pp.

—Modulus of Elasticity—Resonance Vibration
JIS A 1127-76. Method of Test for Dynamic Modulus of Elasticity, Rigidity and Dynamic Poisson's Ratio of Concrete Specimens by Resonance Vibration. 11 pp.

—Nondestructive Testing
BSI BS 1881: Part 201-86. 1986 Methods of Testing Concrete Part 201: Guide to the Use of Non-Destructive Methods of Test for Hardened Concrete. 26 pp.

—Penetration Resistance
CNS A3200-84. Method of Test for Penetration Resistance of Hardened Concrete (Jan)(10733).

—Plasticizers
BSI BS 5075: Part 3-85. 1985 Concrete Admixtures Part 3: Super Plasticizing Admixtures. 11 pp.
CSA A266.5-M1981P. Guidelines for the Use of Superplasticizing Admixtures in Concrete. 14 pp.
CSA CAN3-A266.6-M85. Superplasticizing Admixtures for Concrete; (Gen Instr 1). 21 pp.

—Poisson Ratio—Resonance Vibration
JIS A 1127-76. Method of Test for Dynamic Modulus of Elasticity, Rigidity and Dynamic Poisson's Ratio of Concrete Specimens by Resonance Vibration. 11 pp.

—Polyester Resin—Compression Testing
JIS A 1182-78. Method of Test for Compressive Strength of Polyester Resin Concrete. 6 pp.

—Polyester Resin—Flexural Strength
JIS A 1184-78. Method of Test for Flexural Strength of Polyester Resin Concrete. 7 pp.

—Polyester Resin—Life Testing
JIS A 1186-78. Measuring Methods for Working Life of Polyester Resin Concrete. 8 pp.

—Polyester Resin—Tensile Strength
JIS A 1185-78. Method of Test for Splitting Tensile Strength of Polyester Resin Concrete. 6 pp.

INDUSTRY STANDARDS

Concretes (Cont.)

—Polyester Resin—Test Specimens—Production
JIS A 1181-78. Method of Making Polyester Resin Concrete Specimens. 11 pp.

—Portland Cement Content
SAA AS 1012.15-79. Methods of Testing Concrete—Part 15: Method for the Estimation of Portland Cement Content of Hardened Concrete. 11 pp.

—Production Methods
BSI BS 5328: Part 3-90. 1990 Concrete Part 3: Procedures to Be Used in Producing and Transporting Concrete (R). 12 pp.
BSI BS 5328: Part 3-01. 1990 Amd 1 Concrete Part 3: Specification for the Procedures to Be used in Producing and Transporting Concrete (AMD 6927) November 29, 1991 (R). 13 pp.
BSI BS 5328: Part 3-02. 1990 Amd 2 Concrete Part 3: Specification for the Procedures to Be Used in Producing and Transporting Concrete (AMD 7176) July 15, 1992. 13 pp.

—Quality Assurance
BSI BS 5328: Part 4-90. 1990 Concrete Part 4: Procedures to Be Used in Sampling, Testing and Assessing Compliance of Concrete (R). 12 pp.
BSI BS 5328: Part 4-01. 1990 Amd 1 Concrete Part 4: Specification for the Procedures to Be Used in Sampling, Testing and Assessing Compliance of Concrete (AMD 6928) November 29, 1991. 13 pp.
DIN ENGL 1084 Pt 1-78. Control (Quality Control) of Concrete Structures and Reinforced Concrete Structures; Concrete B II on Building Sites (Dec). 11 pp.
SAA AS MP28 Sect. C4-75. Commentary on AS 1480 SAA Concrete Structures Code—Sect. C4: Quality of Concrete. 4 pp.

—Radiography
BSI BS 1881: Part 205-86. 1986 Methods of Testing Concrete Part 205: Recommendations for Radiography of Concrete. 16 pp.

—Ready Mixed
BSI BS 5328-81. (WITHDRAWN) 1981 Amd 2 Methods for Specifying Concrete, Including Ready-Mixed Concrete (AMD 4970) October 31, 1985 (Superseded by BS 5328: Parts 1, 2, 3, and 4: 1990). 21 pp.
CNS A2042-91. Ready-Mixed Concrete (Jun)(3090). 15 pp.
JIS A 5308-89. Ready-Mixed Concrete. 71 pp.

—Ready Mixed—Quality Assurance
DIN ENGL 1084 Pt 3-78. Control (Quality Control) of Concrete Structures and Reinforced Concrete Structures; Ready-Mixed Concrete (Dec). 11 pp.

—Reinforcing Steels—Bonding Strength
CEN PREN 989-92. Determination of the Bond Behaviour Between Reinforcing Steel and Autoclaved Aerated Concrete by the "Push-Out" Test. 5 pp.
CNS A3219-84. Method of Test for Comparing Concrete on the Basis of the Bond Developed with Reinforcing Steel (Dec)(11152).

—Resilience
CNS A3199-84. Method of Test for Rebound Number of Hardened Concrete (Jan)(10732).

—Sampling
BSI BS 1881: Part 101-83. 1983 Amd 2 Testing Concrete Part 101: Method of Sampling Fresh Concrete on Site (AMD 6728) July 31, 1991. 10 pp.
BSI BS 1881: Part 125-86. 1986 Amd 1 Methods of Testing Concrete Part 125: Methods for Mixing and Sampling Fresh Concrete in the Laboratory (AMD 6107) July 31, 1989. 9 pp.
BSI BS 5328: Part 4-90. 1990 Concrete Part 4: Procedures to Be Used in Sampling, Testing and Assessing Compliance of Concrete (R). 12 pp.
BSI BS 5328: Part 4-01. 1990 Amd 1 Concrete Part 4: Specification for the Procedures to Be Used in Sampling, Testing and Assessing Compliance of Concrete (AMD 6928) November 29, 1991. 13 pp.
CNS A3038-86. Method of Sampling Fresh Concrete (Aug)(1174). 3 pp.
ISO 2736 Pt 1-86. Concrete Tests—Test Specimens—Part 1: Sampling of Fresh Concrete First Edition. 3 pp.
JIS A 1115-75. Method of Sampling Fresh Concrete.
SAA AS 1012.1-93. Methods of Testing Concrete—Part 1: Sampling of Fresh Concrete (in Professional Packages 30A, 58A). 6 pp.
SNZ NZS 3112: Part 1-86. Methods of Test for Concrete Part 1: Tests Relating to Fresh Concrete. 20 pp.

—Sands
BSI BS 4550: Part 5-78. 1978 Methods of Testing Cement Part 5: Standard Sand for Concrete Cubes. 6 pp.

Concretes (Cont.)

—Sands (Cont.)
JIS A 5004-87. Manufactured Sand for Concrete. 9 pp.

—Setting Time
SAA AS 1012.18-75. Methods of Testing Concrete—Part 18: Determination of Setting Time of Fresh Concrete, Mortar and Grout by Penetration Resistance. 10 pp.
SNZ NZS 3112: Part 1-86. Methods of Test for Concrete Part 1: Tests Relating to Fresh Concrete. 20 pp.

—Sewers
DIN ENGL 4281-85. Concrete for Drainage Units; Manufacture, Requirements and Testing (Mar). 4 pp.

—Shear Modulus—Resonance Vibration
JIS A 1127-76. Method of Test for Dynamic Modulus of Elasticity, Rigidity and Dynamic Poisson's Ratio of Concrete Specimens by Resonance Vibration. 11 pp.

—Shipping
BSI BS 5328: Part 3-90. 1990 Concrete Part 3: Procedures to Be Used in Producing and Transporting Concrete (R). 12 pp.
BSI BS 5328: Part 3-01. 1990 Amd 1 Concrete Part 3: Specification for the Procedures to Be used in Producing and Transporting Concrete (AMD 6927) November 29, 1991 (R). 13 pp.
BSI BS 5328: Part 3-02. 1990 Amd 2 Concrete Part 3: Specification for the Procedures to Be Used in Producing and Transporting Concrete (AMD 7176) July 15, 1992. 13 pp.

—Shrinkage
BSI BS 1881: Part 5-70. 1970 Amd 2 Methods of Testing Concrete Part 5: Methods of Testing Hardened Concrete for Other Than Strength (AMD 6267) March 30, 1990. 27 pp.
SAA AS 1012.13-92. Methods of Testing Concrete—Part 13: Determination of the Drying Shrinkage of Concrete for Samples Prepared in the Field or in the Laboratory (in Professional Packages 30, 58) Amdt October—1993. 12 pp.
SNZ NZS 3112: Part 3-86. Methods of Test for Concrete Part 3: Tests on Hardened Concrete Other Than for Strength. 11 pp.

—Slags
CNS A2233-90. Ground Granulated Blast Furnace Slag for Use in Concrete and Mortars (Sep)(12549).

—Slump Testing
BSI BS 1881: Part 102-83. 1983 Amd 2 Testing Concrete Part 102: Method for Determination of Slump (AMD 6727) July 31, 1991. 8 pp.
CNS A3040-84. Method of Slump Test for Concrete (Dec)(1176). 3 pp.
JIS A 1101-75. Method of Test for Slump of Concrete. 6 pp.

—Strain Testing
BSI BS 1881: Part 206-86. 1986 Methods of Testing Concrete Part 206: Recommendations for Determination of Strain in Concrete. 9 pp.

—Strength Testing
BSI BS 1881: Part 207-92. 1992 Testing Concrete Part 207: Recommendations for the Assessment of Concrete Strength by Near-to-Surface Tests (R). 17 pp.
ISO 2736 Pt 2-86. Concrete Tests—Test Specimens—Part 2: Making and Curing of Test Specimens for Strength Tests First Edition. 4 pp.
SNZ NZS 3112: Part 2-86. Methods of Test for Concrete Part 2: Tests Relating to the Determination of Strength of Concrete Amend: 1. 22 pp.

—Sulfate Content
SAA AS 1012.20-92. Methods of Testing Concrete—Part 20: Determination of Chloride and Sulfate in Hardened Concrete and Concrete Aggregates (in Professional Packages 30, 58). 4 pp.

—Supports (Construction)—Shear Testing
BSI BS 5080: Part 2-86. 1986 Methods of Test for Structural Fixings in Concrete and Masonry Part 2: Method for Determination of Resistance to Loading in Shear. 15 pp.
BSI BS 5080: Part 2-01. 1986 Amd 1 Methods of Test for Structural Fixings in Concrete and Masonry Part 2: Method for Determination of Resistance to Loading in Shear (AMD 7602) July 15, 1993 (R). 19 pp.

—Surfaces
DIN ENGL 18217-81. Concrete Surfaces and Formwork Surface (Dec). 2 pp.
SAA HB34-92. Near-to-Surface Testing of Hardened Concrete (in Professional Package 58A). 15 pp.

Concretes (Cont.)

—Surfaces (Cont.)
SNZ NZS 3114-87. Specification for Concrete Surface Finishes. 44 pp.

—Tensile Testing
BSI BS 1881: Part 117-83. 1983 Amd 1 Methods of Testing Concrete Part 117: Method for Determination of Tensile Splitting Strength (AMD 6096) July 31, 1989. 9 pp.
CNS A3061-85. Method of Test for Splitting Tensile Strength of Cylindrical Concrete Specimens (Feb)(3801). 5 pp.
ISO 4108-80. Concrete—Determination of Tensile Splitting Strength of Test Specimens First Edition. 5 pp.
JIS A 1113-76. Method of Test for Splitting Tensile Strength of Concrete. 6 pp.
SAA AS 1012.10-85. Methods of Testing Concrete—Part 10: Method for the Determination of Indirect Tensile Strength of Concrete Cylinders ('Brazil' or Splitting Test) Amdt 1 March 1987. 7 pp.
SNZ NZS 3112: Part 2-86. Methods of Test for Concrete Part 2: Tests Relating to the Determination of Strength of Concrete Amend: 1. 22 pp.

—Test Specimens
BSI BS 1881: Part 108-83. 1983 Amd 1 Methods of Testing Concrete Part 108: Method for Making Test Cubes from Fresh Concrete (AMD 6105) July 31, 1989. 7 pp.
BSI BS 1881: Part 110-83. 1983 Amd 1 Methods of Testing Concrete Part 110: Methods for Making Test Cylinders from Fresh Concretes (AMD 6103) July 31, 1989. 11 pp.
BSI BS 1881: Part 113-83. 1983 Amd 1 Methods of Testing Concrete Part 113: Method for Making and Curing No-Fines Test Cubes (AMD 6099) July 31, 1989. 6 pp.
CNS A3043-85. Laboratory Method for Making and Curing Concrete Test Specimens (Jun)(1230).
CNS A3044-82. Method of Marking and Curing Concrete Test Specimens in the Field (Oct)(1231). 7 pp.
DIN ENGL 1048 Pt 5-91. Testing Concrete; Testing of Hardened Concrete (Specimens Prepared in Mould) (June) (This Standard, Together with DIN 1048 Part 1, June 1991 Edition, Supersedes DIN 1048 Part 1, December 1978 Edition). 8 pp.
ISO 1920-76. Concrete Tests—Dimensions, Tolerances and Applicability of Test Specimens First Edition. 4 pp.
JIS A 1132-76. Method of Making and Curing Concrete Specimens. 12 pp.
JIS A 1138-75. Method of Making Test Sample of Concrete in the Laboratory. 6 pp.
SAA AS 1012.2-83. Methods of Testing Concrete—Part 2: Method for Preparation of Concrete Mixes in the Laboratory. 13 pp.
SAA AS 1012.8-86. Methods of Testing Concrete—Part 8: Method for Making and Curing Concrete Compression, Indirect Tensile and Flexure Test Specimens in the Laboratory or in the Field Amdt 1 July 1989. 14 pp.

—Test Specimens—Capping
CNS A3224-85. Method of Capping Cylindrical Concrete Specimens (Jul)(11297).

—Test Specimens—Compression Testing
CNS A3045-82. Method of Test for Compressive Strength of Cylindrical Concrete Specimens (Jan)(1232). 4 pp.
SAA AS 1012.9-86. Methods of Testing Concrete—Part 9: Method for the Determination of the Compressive Strength of Concrete Specimens. 11 pp.

—Test Specimens—Curing
BSI BS 1881: Part 111-83. 1983 Amd 1 Methods of Testing Concrete Part 111: Method of Normal Curing of Test Specimens (20 Degrees Celsius Method) (AMD 6102) July 31, 1989. 4 pp.
BSI BS 1881: Part 112-83. 1983 Amd 1 Methods of Testing Concrete Part 112: Methods of Accelerated Curing of Test Cubes (AMD 6100) July 31, 1989. 8 pp.
BSI BS 1881: Part 113-83. 1983 Amd 1 Methods of Testing Concrete Part 113: Method for Making and Curing No-Fines Test Cubes (AMD 6099) July 31, 1989. 6 pp.
BSI DD 92-84. 1984 Amd 1 A Method for Temperature-Matched Curing of Concrete Specimens. 9 pp.
SAA AS 1012.8-86. Methods of Testing Concrete—Part 8: Method for Making and Curing Concrete Compression, Indirect Tensile and Flexure Test Specimens in the Laboratory or in the Field Amdt 1 July 1989. 14 pp.
SAA AS 1012.19-88. Methods of Testing Concrete—Part 19: Accelerated Curing of Concrete Compression Test Specimens (Laboratory or Field)—Hot Water and Warm Water Methods. 6 pp.

INTERNATIONAL AND NON-U.S. NATIONAL STANDARDS
SUBJECT INDEX

Concretes *(Cont.)*

—Test Specimens—Flexural Strength
SAA AS 1012.11-85. Methods of Testing Concrete—Part 11: Method for the Determination of the Flexural Strength of Concrete Specimens. 5 pp.

—Test Specimens—Freeze Thaw Testing
CNS A3032-84. Method of Test for Resistance of Concrete Specimens to Rapid Freezing and Thawing in Water (Apr)(1168).

CNS A3033-84. Method of Test for Resistance of Concrete Specimens to Rapid Freezing in Air and Rapid Thawing in Water (Apr)(1169).

CNS A3034-84. Method of Test for Resistance of Concrete Specimens to Slow Freezing and Thawing in Water (Apr)(1170).

—Test Specimens—Modulus of Elasticity
SAA AS 1012.17-76. Methods of Testing Concrete—Part 17: Methods for the Determination of the Static Chord Modulus of Elasticity and Poisson's Ratio of Concrete Specimens. 15 pp.

—Test Specimens—Molds (Casting)
CNS A1040-84. Molds for Forming Concrete Test Cylinders Vertically (Sep)(10991).

CNS A3211-84. Method of Test for Molds for Forming Concrete Test Cylinders Vertically (Sep)(10992).

—Test Specimens—Poisson Ratio
SAA AS 1012.17-76. Methods of Testing Concrete—Part 17: Methods for the Determination of the Static Chord Modulus of Elasticity and Poisson's Ratio of Concrete Specimens. 15 pp.

—Test Specimens—Resonance Frequency
CNS A3052-87. Method of Test for Fundamental Transverse, Longitudinal and Torsional Frequencies of Concrete Specimens (Dec)(1239). 6 pp.

—Ultrasonic Testing
BSI BS 1881: Part 203-86. 1986 Amd 2 Testing Concrete Part 203: Recommendations for Measurement of Velocity of Ultrasonic Pulses in Concrete (AMD 6766) August 30, 1991. 22 pp.

—Vebe Time
BSI BS 1881: Part 104-83. 1983 Amd 2 Testing Concrete Part 104: Method for Determination of Vebe Time (AMD 6725) July 31, 1991. 10 pp.

—Volume
CNS A3049-60. Method of Test for Volume Change of Cement Mortar and Concrete (Apr)(1236)(R 1970).

—Washing Testing
JIS A 1112-89. Method of Test for Washing Analysis of Fresh Concrete. 9 pp.

—Water
SNZ NZS 3121-86. Specification for Water and Aggregate for Concrete. 8 pp.

—Water Absorption
BSI BS 1881: Part 5-70. 1970 Amd 2 Methods of Testing Concrete Part 5: Methods of Testing Hardened Concrete for Other Than Strength (AMD 6267) March 30, 1990. 27 pp.

BSI BS 1881: Part 122-83. 1983 Amd 1 Methods of Testing Concrete Part 122: Method for Determination of Water Absorption (AMD 6108) July 31, 1989. 4 pp.

—Water Analysis
CNS A3050-84. Method of Test for Quality of Water in Concrete (Dec)(1237).

SNZ NZS 3111-86. Methods of Test for Water and Aggregate for Concrete Amend: 1, 1988. 50 pp.

—Workability
SNZ NZS 3112: Part 1-86. Methods of Test for Concrete Part 1: Tests Relating to Fresh Concrete. 20 pp.

—Yield
CNS A3218-84. Method of Test for Unit Weight, Yield and Air Content (Gravimetric) of Concrete (Dec)(11151). 5 pp.

Condensate Chambers
Use: Condensate Wells

Condensate Content Analysis

—Natural Gas
BSI BS 3156: SUB SEC 11.2.1-86. 1986 Methods for the Analysis of Fuel Gases Part 11: Methods for Non-Manufactured Gases Section 11.2: Determination of Potential Hydrocarbon Liquid Content Subsection 11.2.1: General Introduction. 12 pp.

—Natural Gas *(Cont.)*
ISO 6570 Pt 1-83. Natural Gas—Determination of Potential Hydrocarbon Liquid Content—Part 1: Principles and General Requirements First Edition. 11 pp.

—Natural Gas—Volumetric Analysis
BSI BS 3156: SUB SEC 11.2.3-86. 1986 Methods for the Analysis of Fuel Gases Part 11: Methods for Non-Manufactured Gases Section 11.2: Determination of Potential Hydrocarbon Liquid Content Subsection 11.2.3: Volumetric Method. 11 pp.

ISO 6570 Pt 3-84. Natural Gas—Determination of Potential Hydrocarbon Liquid Content—Part 3: Volumetric Method First Edition. 9 pp.

—Natural Gas—Weight Measurement
BSI BS 3156: SUB SEC 11.2.2-86. 1986 Methods for the Analysis of Fuel Gases Part 11: Methods for Non-Manufactured Gases Section 11.2: Determination of Potential Hydrocarbon Liquid Content Subsection 11.2.2: Weighing Method. 11 pp.

ISO 6570 Pt 2-84. Natural Gas—Determination of Potential Hydrocarbon Liquid Content—Part 2: Weighing Method First Edition. 9 pp.

Condensate Gas
Use: Natural Gas

Condensate Wells
Use For: Condensate Chambers

—Flowmeters
DIN ENGL 19211-80. Methods for Measurement of Fluid Flow; Rectifying Vessels for Throttling Appliances in Flow Measurement Devices (Sept). 3 pp.

Condensation Resins
Scope Note: Use a more specific term *See:* Alkyd Resins; Amino Resins; Phenol Formaldehyde Resins; Phenolic Resins; Polyamide Resins; Polycarbonate Resins; Polyester Resins; Polyethylene Terephthalate; Polyurethane Resins; Silicone Resins; Urea Formaldehyde Resins; Urea Resins

Condensed Milk
See Also: Beverages; Dairy Products; Evaporated Milk; Milk

CGSB 32.166M-89. Milk, Condensed or Evaporated (Supercedes 32-GP-167M October 1978, 32-GP-174M October 1978). 8 pp.

CNS N5029-84. Sweetened Condensed Whole Milk (Dec)(1347). 2 pp.

CNS N5059-80. Sweetened Condensed Milk (Non-Fat) (Aug)(2344). 2 pp.

EC 76/118/EEC-75. Council Directive on the Approximation of the Laws of the Member States Relating to Certain Partly or Wholly Dehydrated Preserved Milk for Human Consumption. 7 pp.

EC 78/630/EEC-78. Council Directive Amending for the First Time Directive 76/118/EEC on the Approximation of the Laws of the Member States Relating to Certain Partly or Wholly Dehydrated Preserved Milk for Human Consumption. 1 p.

EC 83/635/EEC-83. Council Directive Amending for the Second Time Directive 76/118/EEC on the Approximation of the Laws of the Member States Relating to Certain Partly or Wholly Dehydrated Preserved Milk for Human Consumption. 3 pp.

SAA AS 2300.5.1-91. Methods of Chemical and Physical Testing for the Dairying Industry—Part 5: Condensed Milk—Part 5.1: General Information and Preparation of Samples (Supersedes AS N48—1965 (in Part)). 2 pp.

—Ash Content
BSI BS 1742: Part 4-91. 1991 Chemical Analysis of Condensed Milks Part 4: Determination of Ash from Condensed Milks. 8 pp.

—Chemical Analysis
BSI BS 1742-51. 1951 Amd 1 Methods for the Chemical Analysis of Condensed Milk (AMD 6313) December 21, 1990. 17 pp.

—Dry Matter Content
EC 79/1067/EEC-79. First Commission Directive Laying Down Community Methods of Analysis for Testing Certain Partly or Wholly Dehydrated Preserved Milk for Human Consumption. 24 pp.

—Fat Content
CNS N6066-72. Methods of Test for Milk and Milk Products (Detection of Iso-Fat) (Oct)(3450). 1 p.

EC 79/1067/EEC-79. First Commission Directive Laying Down Community Methods of Analysis for Testing Certain Partly or Wholly Dehydrated Preserved Milk for Human Consumption. 24 pp.

Condensed Milk *(Cont.)*

—Fat Content—Gravimetric Analysis
BSI BS 1742: Part 3-90. 1990 Methods for Chemical Analysis of Condensed Milks Part 3: Determination of Fat Content. 12 pp.

ISO 1737-85. Evaporated Milk and Sweetened Condensed Milk—Determination of Fat Content—Gravimetric Method (Reference Method) Second Edition. 10 pp.

—Microbiological Analysis
SAA AS 1095.2.9-77. Microbiological Methods for the Dairy Industry—Part 2: Methods for the Examination of Specific Dairy Products—Part 2. 9: Sweetened Condensed Milk. 5 pp.

—Nitrogen Content—Reference Method
BSI BS 1742: Sec 5.1-91. 1991 Chemical Analysis of Condensed Milks Part 5: Determination of Nitrogen Content Section 5.1: Reference Method. 9 pp.

—Nitrogen Content—Routine Method
BSI BS 1742: Sec 5.2-91. 1991 Chemical Analysis of Condensed Milks Part 5: Determination of Nitrogen Content Section 5.2: Routine Method. 9 pp.

—Sampling
EC 87/524/EEC-87. First Commission Directive Laying down Community Methods of Sampling for Chemical Analysis for the Monitoring of Preserved Milk Products. 8 pp.

SAA AS 2300.5.1-91. Methods of Chemical and Physical Testing for the Dairying Industry—Part 5: Condensed Milk—Part 5.1: General Information and Preparation of Samples (Supersedes AS N48—1965 (in Part)). 2 pp.

—Solids Content
BSI BS 1742: Part 2-90. 1990 Methods for Chemical Analysis of Condensed Milks Part 2: Determination of Total Solids Content of Sweetened Condensed Milks. 7 pp.

ISO 6734-89. Sweetened Condensed Milk—Determination of Total Solids Content (Reference Method) First Edition. 6 pp.

SAA AS 2300.5.3-91. Methods of Chemical and Physical Testing for the Dairying Industry—Part 5: Condensed Milk—Part 5.3: Determination of Total Milk Solids (Supersedes AS N48—1965 (in Part)). 2 pp.

—Sucrose Content
EC 79/1067/EEC-79. First Commission Directive Laying Down Community Methods of Analysis for Testing Certain Partly or Wholly Dehydrated Preserved Milk for Human Consumption. 24 pp.

—Sucrose Content—Polarimetric Analysis
BSI BS 1742: Part 6-90. 1990 Methods for Chemical Analysis of Condensed Milks Part 6: Determination of Sucrose Content of Sweetened Condensed Milk. 7 pp.

ISO 2911-76. Sweetened Condensed Milk—Determination of Sucrose Content—Polarimetric Method First Edition. 5 pp.

SAA AS 2300.5.2-91. Methods of Chemical and Physical Testing for the Dairying Industry—Part 5: Condensed Milk—Part 5.2: Determination of Sucrose—Polarimetric Method (Supersedes N48—1965 (in Part)).

—Test Specimens
CNS N6056-80. Method of Test for Milk and Milk Products (General Rules) (Aug)(3440). 2 pp.

—Viscosity
CNS N6071-72. Methods of Test for Milk and Milk Products (Determination of Viscosity of Condensed Milk) (Oct)(3455). 2 pp.

Condenser Microphones
Use: Capacitor Microphones

Condenser Paper
Use: Capacitor Papers

Condenser Tissues
Use: Capacitor Papers

Condenser Tubes
See Also: Condensers (Laboratory); Condensing Units; Tubes

—Aluminum Alloy
CNS H2032-81. Method of Test for Aluminium-Alloy Drawn Seamless Tubes for Condenser and Heat Exchangers Use (Jun) (2967).

CNS H3037-81. Aluminium Alloy Drawn Seamless Tubes for Condensers and Heat Exchangers (Jun)(2966).

Condenser

Condenser Tubes (Cont.)

—Copper
DIN ENGL 1785-83. Wrought Copper and Copper Alloy Tubes for Condensers and Heat Exchangers (Oct). 7 pp.
ISO 1635 Pt 2-87. Seamless Wrought Copper and Copper Alloy Tube—Part 2: Technical Conditions of Delivery for Condenser and Heat-Exchanger Tubes First Edition. 7 pp.

—Copper Alloy
DIN ENGL 1785-83. Wrought Copper and Copper Alloy Tubes for Condensers and Heat Exchangers (Oct). 7 pp.
ISO 1635 Pt 2-87. Seamless Wrought Copper and Copper Alloy Tube—Part 2: Technical Conditions of Delivery for Condenser and Heat-Exchanger Tubes First Edition. 7 pp.

—Nickel
ISO 6207-92. Seamless Nickel and Nickel Alloy Tube First Edition. 17 pp.

—Nickel Alloy
ISO 6207-92. Seamless Nickel and Nickel Alloy Tube First Edition. 17 pp.

Condensers (Electric)
Use: Capacitors

Condensers (Laboratory)
See Also: Condenser Tubes; Laboratory Equipment

—Glass
SAA AS K137-62. Glass Condensers Being BS 1848:1952, (Including the Amdts of October 1954, September 1957 and March 1960) Endorsed Subject to Australian Amendment Amdt 1 February 1962. 14 pp.

Condensers (Liquefiers)
Use: Condensing Units

Condensers (Optics)
Use: Optical Condensers

Condensing Units
Use For: Condensers (Liquefiers) *See Also:* Chillers; Compressors (Pressure Equipment); Condenser Tubes; Cooling Systems; Dehumidifiers; Demisters; Heat Exchangers; Refrigerant Condensers

—Air Conditioners
CSA CAN/CSA-C746-93. Performance Standard for Rating Large Air Conditioners and Heat Pumps; (Gen Instr 1). 31 pp.

—Compressor/Evaporators
CSA CAN/CSA-C22. 2 NO 236-M90. Heating and Cooling Equipment; (Gen Instr 1) (ANSI/UL 1995). 136 pp.

—Laboratory
BSI BS 5922-80. 1980 Glass Condensers for Laboratory Use. 10 pp.
CNS R2142-86. Glass Condensors for Chemical Analysis (Mar)(7318).
ISO 4799-78. Laboratory Glassware—Condensers First Edition. 8 pp.

—Plates—Copper
DIN ENGL 17675 Pt 2-80. Plates of Copper and Wrought Copper Alloys for Condensers and Heat Exchangers; Technical Conditions of Delivery (July). 3 pp.
DIN ENGL 17675 Pt 3-80. Plates of Copper and Wrought Copper Alloys for Condensers and Heat Exchangers; Dimensions (July). 3 pp.

—Plates—Copper Alloy
DIN ENGL 17675 Pt 2-80. Plates of Copper and Wrought Copper Alloys for Condensers and Heat Exchangers; Technical Conditions of Delivery (July). 3 pp.
DIN ENGL 17675 Pt 3-80. Plates of Copper and Wrought Copper Alloys for Condensers and Heat Exchangers; Dimensions (July). 3 pp.

—Plates—Copper Alloy—Mechanical Properties
DIN ENGL 17675 Pt 1-80. Plates of Copper and Wrought Copper Alloys for Condensers and Heat Exchangers; Strength Properties (July). 4 pp.

—Plates—Copper—Mechanical Properties
DIN ENGL 17675 Pt 1-80. Plates of Copper and Wrought Copper Alloys for Condensers and Heat Exchangers; Strength Properties (July). 4 pp.

—Refrigeration Equipment
CSA C22.2 NO 140.3-M1987. Refrigerant-Containing Components for Use in Electrical Equipment (R 1993). 19 pp.

Condensing Units (Cont.)

—Water Formed Deposits—Sampling
BSI BS 2455: Part 1-73. 1973 Methods of Sampling and Examining Deposits from Boilers and Associated Industrial Plant Part 1: Water-Side Deposits. 31 pp.
BSI BS 2455: Part 1-01. 1973 Amd 1 Methods of Sampling and Examining Deposits from Boilers and Associated Industrial Plant Part 1: Water-Side Deposits (AMD 7753) October 15, 1993 (Q). 32 pp.

Condiments
See Also: Food; Horseradish; Miso; Mustard; Soy Sauce; Spices; Vinegars
CNS N5190-81. Flavored Condiments (Oct)(8055). 3 pp.

—Ash Content
BSI BS 4585: Part 3-81. 1981 Methods of Test for Spices and Condiments Part 3: Determination of Total Total Ash. 4 pp.
BSI BS 4585: Part 9-81. 1981 Methods of Test for Spices and Condiments Part 9: Determination of Acid-Insoluble Ash. 4 pp.
BSI BS 4585: Part 10-81. (WITHDRAWN) 1981 Methods of Test for Spices and Condiments Part 10: Determination of Water-Soluble Ash. 4 pp.
CNS N6153-81. Spices and Condiments—Determination of Total Ash (Sep)(7919). 3 pp.
CNS N6154-81. Spices and Condiments—Determination of Water-Insoluble Ash (Sep)(7920). 2 pp.
CNS N6155-81. Spices and Condiments—Determination of Acid-Insoluble Ash (Sep)(7921). 2 pp.
ISO 928-80. Spices and Condiments—Determination of Total Ash First Edition. 4 pp.
ISO 929-80. Spices and Condiments—Determination of Water-Insoluble Ash First Edition; (Erratum—Dec 1981). 5 pp.
ISO 930-80. Spices and Condiments—Determination of Acid-Insoluble Ash First Edition. 4 pp.

—Glossaries
BSI BS 4488-83. 1983 Nomenclature for Spice and Condiments. 14 pp.
CNS N5187-81. Nomenclature of Spices and Condiments (Oct)(8048). 3 pp.
ISO 676-82. Spices and Condiments—Nomenclature—First List First Edition. 18 pp.

—Grain Size Analysis
BSI BS 4585: Part 8-77. 1977 Methods of Test for Spices and Condiments Part 8: Determination of Degree of Fineness of Grinding; Hand Sieving Method (Reference Method). 4 pp.
CNS N6158-81. Spices and Condiments—Determination of Degree of Fineness of Grinding—Hand Sieving Method (Oct)(8051). 2 pp.
ISO 3588-77. Spices and Condiments—Determination of Degree of Fineness of Grinding—Hand Sieving Method (Reference Method) First Edition. 4 pp.

—Impurities Content
BSI BS 4585: Part 1-83. 1983 Methods of Test for Spices and Condiments Part 1: Determination of Extraneous Matter. 3 pp.
BSI BS 4585: Part 14-83. 1983 Methods of Test for Spices and Condiments Part 14: Determination of Filth. 8 pp.
CNS N6125-80. Spices and Condiments—Determination of Extraneous Matter (Jul)(5929). 2 pp.
ISO 927-82. Spices and Condiments—Determination of Extraneous Matter Content Second Edition. 3 pp.
ISO 1208-82. Spices and Condiments—Determination of Filth First Edition. 7 pp.

—Moisture Content
CNS N6161-81. Method of Test for Flavored Condiments (Oct)(8056). 2 pp.

—Moisture Content—Entrainment Method
BSI BS 4585: Part 2-82. 1982 Methods of Test for Spices and Condiments Part 2: Determination of Moisture Content (Entrainment Method). 8 pp.
CNS N6151-81. Method of Test for Spices and Condiments—Determination of Moisture Content—Entrainment Method (Feb)(7070). 3 pp.
ISO 939-80. Spices and Condiments—Determination of Moisture Content—Entrainment Method First Edition. 6 pp.

—Nitrogen Content
CNS N6161-81. Method of Test for Flavored Condiments (Oct)(8056). 2 pp.

—Nonvolatile Matter Content
BSI BS 4585: Part 6-92. 1992 Methods of Test for Spices and Condiments Part 6: Determination of Non-Volatile Ether Extract. 7 pp.
BSI BS 4585: Part 6-81. 1981 Methods of Test for Spices and Condiments Part 6: Determination of Non-Volatile Ether Extract. 4 pp.

Condiments (Cont.)

—Nonvolatile Matter Content (Cont.)
CNS N6156-81. Spices and Condiments—Determination of Non-Volatile Ether Extract (Sep)(7922). 2 pp.
ISO 1108-92. Spices and Condiments—Determination of Non-Volatile Ether Extract Second Edition. 5 pp.

—Salt Content
CNS N6161-81. Method of Test for Flavored Condiments (Oct)(8056). 2 pp.

—Sampling
BSI BS 4540: Part 1-81. 1981 Sampling of Spices and Condiments Part 1: Methods of Sampling. 4 pp.
CNS N6122-80. Spices and Condiments—Sampling (Jul)(5926). 3 pp.
CNS N6123-80. Spices and Condiments—Preparation of a Ground Sample for Analysis Jul)(5927). 1 p.
CNS N6161-81. Method of Test for Flavored Condiments (Oct)(8056). 2 pp.
ISO 948-80. Spices and Condiments—Sampling First Edition. 4 pp.

—Soluble Matter Content
BSI BS 4585: Part 4-88. 1988 Methods of Test for Spices and Condiments Part 4: Determination of Alcohol-Soluble Extract. 4 pp.
BSI BS 4585: Part 5-81. 1981 Methods of Test for Spices and Condiments Part 5: Determination of Cold Water Soluble Extract. 4 pp.
CNS N6124-80. Spices and Condiments—Determination of Alcohol-Soluble Extract (Jul)(5928). 2 pp.
CNS N6140-80. Spices and Condiments—Determination of Cold Water Soluble Extract (Sep)(6502). 2 pp.
ISO 940-79. Spices and Condiments—Determination of Alcohol-Soluble Extract First Edition. 4 pp.

—Starch Content
CNS N6161-81. Method of Test for Flavored Condiments (Oct)(8056). 2 pp.

—Sugar Content
CNS N6161-81. Method of Test for Flavored Condiments (Oct)(8056). 2 pp.

—Test Specimens
BSI BS 4540: Part 2-82. 1982 Sampling of Spices and Condiments Part 2: Preparation of a Ground Sample for Analysis. 3 pp.
ISO 2825-81. Spices and Condiments—Preparation of a Ground Sample for Analysis Second Edition. 3 pp.

—Volatile Oil Content
BSI BS 4585: Part 15-85. 1985 Methods of Test for Spices and Condiments Part 15: Determination of Volatile Oil Content. 7 pp.
ISO 6571-84. Spices, Condiments and Herbs—Determination of Volatile Oil Content First Edition. 6 pp.

Conditioning (Treating)
Use: Treatment

Condoms
BSI BS 3704-89. 1989 Natural Rubber Latex Condoms (N). 16 pp.
BSI BS 3704-01. 1989 Amd 1 Natural Rubber Latex Condoms (AMD 6644) November 30, 1990. 19 pp.
CEN PREN 600-1-91. Latex Rubber Condoms—Part 1: Condoms for Consumer Use. 62 pp.
CEN PREN 600-7-91. Latex Rubber Condoms—Part 7: Oven Treatment of Condoms. 3 pp.
CNS T2008-80. Condoms (Oct)(6629).
JIS T 9111-85. Rubber Condoms. 5 pp.
JIS T 9111-66. Condoms. 5 pp.
SNZ NZS/BS 3704-89. Specification for Natural Rubber Latex Condoms Amend: 1. 16 pp.

—Bursting Strength
CEN PREN 600-6-91. Latex Rubber Condoms—Part 6: Determination Bursting Volume and Pressure. 13 pp.
ISO 4074 Pt 6-84. Rubber Condoms—Part 6: Determination of Bursting Volume and Pressure Second Edition. 4 pp.

—Colorfastness Testing
CEN PREN 600-4-91. Latex Rubber Condoms—Part 4: Determination of Colour Fastness. 4 pp.
ISO 4074 Pt 4-80. Rubber Condoms—Part 4: Determination of Colour Fastness First Edition. 3 pp.

—Elongation
CEN PREN 600-8-91. Latex Rubber Condoms—Part 8: Determination of Force and Elongation at Break. 7 pp.

—Identification Systems
CEN PREN 600-9-91. Latex Rubber Condoms—Part 9: Packaging and Labelling. 14 pp.

INTERNATIONAL AND NON-U.S. NATIONAL STANDARDS
SUBJECT INDEX

Condoms (Cont.)
—Identification Systems (Cont.)
ISO 4074 Pt 10-90. Rubber Condoms—Part 10: Packaging and Labelling—Condoms in Consumer Packages Second Edition. 6 pp.

—Leakage
CEN PREN 600-5-91. Latex Rubber Condoms—Part 5: Testing for Perforations. 10 pp.
ISO 4074 Pt 5-84. Rubber Condoms—Part 5: Testing for Holes Second Edition. 3 pp.

—Length Measurement
CEN PREN 600-2-91. Latex Rubber Condoms—Part 2: Determination of Length. 5 pp.
ISO 4074 Pt 2-80. Rubber Condoms—Part 2: Determination of Length First Edition. 4 pp.

—Packaging
CEN PREN 600-9-91. Latex Rubber Condoms—Part 9: Packaging and Labelling. 14 pp.
ISO 4074 Pt 1-90. Rubber Condoms—Part 1: Requirements—Condoms in Consumer Packages First Edition. 14 pp.
ISO 4074 Pt 10-90. Rubber Condoms—Part 10: Packaging and Labelling—Condoms in Consumer Packages Second Edition. 6 pp.

—Storage
ISO 4074 Pt 7-86. Rubber Condoms—Part 7: Determination of Resistance to Deterioration During Storage First Edition. 3 pp.

—Tensile Testing
ISO 4074 Pt 9-80. Rubber Condoms—Part 9: Determination of Tensile Properties First Edition. 4 pp.

—Width Measurement
CEN PREN 600-3-91. Latex Rubber Condoms—Part 3: Determination of Width. 4 pp.
ISO 4074 Pt 3-80. Rubber Condoms—Part 3: Determination of Width First Edition. 3 pp.

Conductance Meters
BSI BS 89: Part 6-90. 1990 Direct Acting Indicating Analogue Electrical Measuring Instruments and Their Accessories Part 6: Special Requirements for Ohmmeters (Impedance Meters) and Conductance Meters. 14 pp.
CENELEC EN 60 051-6-89. Direct Acting Indicating Analogue Electrical Measuring Instruments and Their Accessories Part 6: Special Requirements for Shinmeters (Impedance Meters) and Conductance Meters (Supersedes HD 233). 4 pp.
IEC 51 Pt 6-84. Direct Acting Indicating Analogue Electrical Measuring Instruments and Their Accessories Part 6: Special Requirements for Ohmmeters (Impedance Meters) and Conductance Meters Fourth Edition. 20 pp.

Conducted Interference
Use: Electromagnetic Interference

Conductimetric Method
See Also: Karl Fischer Method; Water Content Analysis

—Anhydrous Hydrogen Fluoride
BSI BS 5365: Part 6-81. (WITHDRAWN) 1981 Methods of Sampling and Test for Anhydrous Fluoride for Industrial Use Part 6: Determination of Water Content (Conductimetric Method). 11 pp.
ISO 3700-80. Anhydrous Hydrogen Fluoride for Industrial Use—Determination of Water Content—Conductimetric Method First Edition. 10 pp.

—Silicon
CNS G2109-82. Method of Chemical Analysis for Carbon in Metallic Silicon (Conductivity Method) (Jan)(8283).

Conduction Heating
See Also: Heaters; Heating Equipment

—Safety
BSI BS 7699: Part 3-93. 1993 Safety in Electroheat Installations Part 3: Particular Requirements for Induction and Conduction Heating and Induction Melting Installations (F). 26 pp.
CENELEC HD 491.3 S1-90. Safety in Electroheat Installations Part 3: Particular Requirements for Induction and Conduction Heating and Induction Melting Installations. 4 pp.
IEC 519 Pt 3-88. Safety in Electroheat Installations Part 3: Particular Requirements for Induction and Conduction Heating and Induction Melting Installations Second Edition. 46 pp.

Conductive Bridge Telemetering Systems
Use: Telemetry

Conductive Floors
Use: Floors—Conductive

Conductive Plastic Potentiometers
See Also: Potentiometers (Resistors)

—Precision—Single Turn—Preferred Products List
CECC CECC MUAHAG Vol 2 IS 4-90. Preferred Products List; Resistors and Potentiometers (En, Fr, Ge). 59 pp.

Conductivity
Scope Note: Use a more specific term
See: Electrical Properties; Electrical Resistivity; Ground Conductivity; Photoconductivity

Conductometric Analysis
Use For: Conductometric Titration

—Nickel Castings
JIS H 1277-88. Methods for Determination of Sulfur in Nickel and Nickel Alloy Castings. 10 pp.

—Silicon Carbide
DIN ENGL 51075 Pt 2-84. Testing of Ceramic Materials; Chemical Analysis of Silicon Carbide; Direct Determination of the Free Carbon Content (Mar). 7 pp.
DIN ENGL 51075 Pt 3-82. Testing of Ceramic Materials; Chemical Analysis of Silicon Carbide; Determination of the Total Carbon Content (Oct). 3 pp.

Conductometric Titration
Use: Conductometric Analysis

Conductor Busways
Use: Busways

Conductors (Electric)
Use: Electric Conductors

Conduit Boxes
Use: Electric Conduit Boxes

Conduit Fittings (Electric)
Use: Electric Conduit Fittings

Conduits (Channels)
Use For: Channels (Hydraulic) See Also: Gutters; Pipes

—Construction Contracts
DIN ENGL 18305-88. Tendering and Performance Stipulations in Contracts for Construction Works (VOB) Part C: General Technical Specifications in Contracts for Construction Works (ATV); Groundwater Lowering (Sept) (This Standard, Together with DIN 18299, September. 4 pp.

—Identification Systems
SAA AS 1345-82. Identification of the Contents of Piping, Conduits and Ducts (Incorporating Amdt 1) Amdt 2 March 1986. 18 pp.

—Precast Concrete
BSI BS 340-79. (WITHDRAWN) 1979 Precast Concrete Kerbs, Channels, Edgings and Quadrants (Superseded by BS 7263: Part 1: 1990). 12 pp.
BSI BS 7263: Part 1-90. 1990 Precast Concrete Flags, Kerbs, Channels Edgings and Quadrants Part 1: Specification. 19 pp.

—Pressure Measurement
ISO 9110 Pt 2-90. Hydraulic Fluid Power—Measurement Techniques—Part 2: Measurement of Average Steady-State Pressure in a Closed Conduit First Edition. 10 pp.

—Stone
BSI BS 435-75. 1975 Amd 1 Dressed Natural Stone Kerbs, Channels, Quadrants and Setts. 11 pp.

Conduits (Electric)
Use: Electric Conduits

Cone Proof Load Testing
Use: Loads (Forces)

Cones
See Also: Cones (Bobbins); Machine Tapers; Seger Cones

—Medical Equipment
SAA AS 1600.1-88. Medical Equipment—Conical Fittings with a 6 Percent(Luer) Taper for Syringes, Needles and Certain Other Medical Equipment—Part 1: General Requirements. 7 pp.

Cones (Bobbins)
See Also: Bobbins (Thread); Cones; Packaging; Textile Machinery
BSI BS 2547: Part 6-78. 1978 Basic Dimensions of Cones and Tubes for Winding Textile Yarns Part 6: Transfer Cones. Half Angle of the Cone of the 4 Degrees 20. 4 pp.

—Cross Winding
BSI BS 2547: Part 10-86. 1986 Basic Dimensions of Cones and Tubes for Winding Textile Yarns Part 10: Cones for Cross-Wound Winding: Values of Half Angles, Lengths and Large Inner Diameters. 4 pp.
ISO 110-78. Textile Machinery and Accessories—Cones for Yarn Winding (Cross Wound)—Half Angle of the Cone 9 Degrees 15 Feet First Edition. 4 pp.
ISO 111-78. Textile Machinery and Accessories—Cones for Yarn Winding (Cross Wound)—Half Angle of the Cone 4 Degrees 20 Feet First Edition. 4 pp.
ISO 112-83. Textile Machinery and Accessories—Cones for Yarn Winding (Cross Wound)—Half Angle of the Cone 3 Degrees 30 Feet Second Edition. 5 pp.
ISO 324-78. Textile Machinery and Accessories—Cones for Cross Winding for Dyeing Purposes—Half Angle of the Cone 4 Degrees 20 Feet First Edition. 4 pp.
ISO 575-78. Textile Machinery and Accessories—Transfer Cones—Half Angle of the Cone 4 Degrees 20 Feet First Edition. 4 pp.
ISO 5237-78. Textile Machinery and Accessories—Cones for Yarn Winding (Cross Wound)—Half Angle of the Cone 5 Degrees 57 Feet First Edition. 4 pp.
ISO 8489 Pt 1-85. Textile Machinery and Accessories—Cones for Cross-Wound Winding—Part 1: Values of Half-Angles, Lengths and Large Inner Diameters First Edition. 4 pp.

—Paper
JIS L 6417-82. Paper Cones for Yarn Winding (Cross Wound).

Cones (Tapers)
Use: Machine Tapers

Conference Call Services
Use: Conference Calling

Conference Calling
See Also: Add On Conference; Conference Communication Systems; Meet Me Conference; Telephone Services; Three Party Services
CCITT RECMN E.140-89. Principles for the Operation of International Telephone Services—Telephone Network and ISDN—Operation, Numbering, Routing and Mobile Service (Study Group II) 3 pp. 3 pp.
CCITT RECMN E.151-92. Telephone Conference Calls (Study Group I) 5 pp. 5 pp.
CCITT RECMN E.151-89. Conditions of Operation and Setting up of Conference Calls—Telephone Network and ISDN—Operation, Numbering, Routing and Mobile Service (Study Group II) 3 pp. 3 pp.

—Accounting
CCITT RECMN D.110 (REV 1)-92. Charging and Accounting for Conference Calls (Study Group III) 4 pp. 4 pp.
CCITT RECMN D.110-89. Charging and Accounting for Conference Calls—General Tariff Principles—Charging and Accounting in International Telecommunications Services (Study Group III) 2 pp. 2 pp.

—Bridges (Electrical)
CCITT RECMN G.172-89. Transmission Plan Aspects of International Conference Calls—General Characteristics of International Telephone Connections and Circuits (Study Groups XII and XV) 3 pp. 3 pp.

—Bridges (Electrical)—Automatic Gain Control
CCITT RECMN G.172-89. Transmission Plan Aspects of International Conference Calls—General Characteristics of International Telephone Connections and Circuits (Study Groups XII and XV) 3 pp. 3 pp.

—Bridges (Electrical)—Echo
CCITT RECMN G.172-89. Transmission Plan Aspects of International Conference Calls—General Characteristics of International Telephone Connections and Circuits (Study Groups XII and XV) 3 pp. 3 pp.

INDUSTRY STANDARDS

Conference

Conference Calling *(Cont.)*

—Bridges (Electrical)—Noise
- CCITT RECMN G.172-89. Transmission Plan Aspects of International Conference Calls—General Characteristics of International Telephone Connections and Circuits (Study Groups XII and XV) 3 pp. 3 pp.

—Bridges (Electrical)—Transmission Loss
- CCITT RECMN G.172-89. Transmission Plan Aspects of International Conference Calls—General Characteristics of International Telephone Connections and Circuits (Study Groups XII and XV) 3 pp. 3 pp.

—Call Set Up
- CCITT RECMN E.151-92. Telephone Conference Calls (Study Group I) 5 pp. 5 pp.
- CCITT RECMN E.151-89. Conditions of Operation and Setting up of Conference Calls—Telephone Network and ISDN—Operation, Numbering, Routing and Mobile Service (Study Group II) 3 pp. 3 pp.

—Integrated Services Digital Networks
- CCITT RECMN I.254-89. Multiparty Supplementary Services—Integrated Services Digital Network (ISDN)—General Structure and Service Capabilities (Study Group XVIII) 30 pp. 30 pp.
- CCITT RECMN I.254. 1-88. Conference Calling Service Description—Integrated Services Digital Network (ISDN)—General Structure and Service Capabilities (Study Group XVIII) 19 pp. 19 pp.

—Loudness Ratings
- CCITT RECMN G.172-89. Transmission Plan Aspects of International Conference Calls—General Characteristics of International Telephone Connections and Circuits (Study Groups XII and XV) 3 pp. 3 pp.

—Tariffs
- CCITT RECMN D.110 (REV 1)-92. Charging and Accounting for Conference Calls (Study Group III) 4 pp. 4 pp.
- CCITT RECMN D.110-89. Charging and Accounting for Conference Calls—General Tariff Principles—Charging and Accounting in International Telecommunications Services (Study Group III) 2 pp. 2 pp.
- CCITT RECMN E.151-92. Telephone Conference Calls (Study Group I) 5 pp. 5 pp.
- CCITT RECMN E.151-89. Conditions of Operation and Setting up of Conference Calls—Telephone Network and ISDN—Operation, Numbering, Routing and Mobile Service (Study Group II) 3 pp. 3 pp.

—Tariffs—Call Duration
- CCITT RECMN E.151-89. Conditions of Operation and Setting up of Conference Calls—Telephone Network and ISDN—Operation, Numbering, Routing and Mobile Service (Study Group II) 3 pp. 3 pp.

—Telephone Credit Cards
- CCITT RECMN E.116-89. International Telephone Credit Cards for Use in a Non-Automated Environment—Telephone Network and ISDN—Operation, Numbering, Routing and Mobile Service (Study Group II) 3 pp. 3 pp.

—Video Telephone Services
- CCITT RECMN F.720-92. Videotelephony Services—General (Study Group I) 11 pp. 11 pp.

—Videotex Communications
- CCITT RECMN F.300-89. Videotex Service—Telematic, Data Transmission and Teleconference Services—Operations and Quality of Service (Study Group I) 25 pp. 25 pp.

Conference Communication Systems
See Also: Audio Equipment; Communication Equipment; Conference Calling

—Circuits
- CEPT T/TTVS 1-61. Utilisation Des Circuits De Conversation Comme Circuit De Conference. 1 p.

—Interpretation Systems—Audio Equipment
- BSI BS 7154-89. 1989 Amd 1 Conference Systems: Electrical and Audio Requirements (AMD 6408) September 30, 1991. 47 pp.
- CENELEC HD 549 S1-89. Conference Systems—Electrical and Audio Requirements. 3 pp.
- IEC 914-88. Conference Systems—Electrical and Audio Requirements First Edition. 91 pp.

—Language Distribution Systems—Audio Equipment
- BSI BS 7154-89. 1989 Amd 1 Conference Systems: Electrical and Audio Requirements (AMD 6408) September 30, 1991. 47 pp.

Conference Communication Systems *(Cont.)*

—Language Distribution Systems—Audio Equipment *(Cont.)*
- CENELEC HD 549 S1-89. Conference Systems—Electrical and Audio Requirements. 3 pp.
- IEC 914-88. Conference Systems—Electrical and Audio Requirements First Edition. 91 pp.

—Voting Systems—Audio Equipment
- BSI BS 7154-89. 1989 Amd 1 Conference Systems: Electrical and Audio Requirements (AMD 6408) September 30, 1991. 47 pp.
- CENELEC HD 549 S1-89. Conference Systems—Electrical and Audio Requirements. 3 pp.
- IEC 914-88. Conference Systems—Electrical and Audio Requirements First Edition. 91 pp.

Conference Room Tables
Use For: Boardroom Tables; Conference Tables
See Also: Furniture; Office Furniture; Tables (Stands)
- CGSB 44.178-85. Table, Office, Round Conference and Reception. 9 pp.
- JIS S 1041-92. Tables for Conference (Office Furniture). 16 pp.

—Modular
- CGSB 44.154-85. Tables, Modular, Office and Conference. 20 pp.

Conference Tables
Use: Conference Room Tables

Confidence Interval
Use: Confidence Limits

Confidence Limits
Use For: Confidence Interval *See Also:* Statistical Analysis
- BSI BS 2846: Part 2-81. 1981 Guide to Statistical Interpretation of Data Part 2: Estimation of the Mean: Confidence Limit. 8 pp.
- ISO 2602-80. Statistical Interpretation of Test Results—Estimation of the Mean—Confidence Interval Second Edition. 7 pp.
- MOD UK DSTAN 00-26-88. Guide to the Evaluation and Expression of the Uncertainties Associated with the Results of Electrical Measurements Issue 2. 36 pp.

Configuration Management

—MOD (UK)—Materiel
- MOD UK DSTAN 05-57-01. Configuration Management Policy and Procedures for Defence Materiel Issue 2; Corrigendum. 35 pp.
- MOD UK DSTAN 05-57-93. Configuration Management Issue 3. 44 pp.

—NATO—Multinational Joint Projects
- NATO STANAG 4159 ED 2 AMD 0-91. NATO Material Configuration Management Policy and Procedures for Multinational Joint Projects. 50 pp.
- NATO STANAG 4159 ED 2 AMD 1-91. NATO Materiel Configuration Management Policy and Procedures for Multinational Joint Projects. 55 pp.
- NATO STANAG 4427 ED 1 AMD 0-93. (DRAFT) Introduction of Allied Configuration Management Publications (ACMP's). 4 pp.

—NES (Naval Engineering Standards)
- MOD UK NES 41-89. Requirements for the Configuration Management and Ship Fit Definitions Issue 3 (05.89). 45 pp.
- MOD UK NES 41-92. Requirements for Configuration Management and Ship Fit Definitions Issue 4 (01.92). 42 pp.

Conformal Coatings
See Also: Coatings; Plastic Coatings (Made From Plastic)

—Printed Circuit Boards
- IEC 1086 Pt 1-92. Specification for Coatings for Loaded Printed Wire Boards (Conformal Coatings) Part 1: Definitions, Classification and General Requirements First Edition. 18 pp.
- IEC 1086 Pt 2-92. Specification for Coatings for Loaded Printed Wire Boards (Conformal Coatings) Part 2: Methods of Test First Edition. 47 pp.

Congee
See Also: Cereals; Rice

—Canned
- CNS N5224-89. Canned Mixed Congee (Dec)(12650). 5 pp.

Congestion Tones
See Also: Tones (Telephone Services)

Congestion Tones *(Cont.)*
- CCITT RECMN Q.35-89. Technical Characteristics of Tones for the Telephone Service—General Recommendations on Telephone Switching and Signalling—Functions and Information Flows for Services in the ISDN —Supplements (Study Group XI) 9 pp. 9 pp.
- CENELEC PRETS 300 295-92. Human Factors (HF); Specification of Characteristics of Telephone Services Tones When Locally Generated in Terminals. 9 pp.
- ETSI PRETS 300 295-92. Human Factors (HF); Specification of Characteristics of Telephone Services Tones When Locally Generated in Terminals. 9 pp.

—Frequency Bands
- CCITT RECMN E.180-89. Technical Characteristics of Tones for the Telephone Service—Telephone Network and ISDN—Operation, Numbering, Routing and Mobile Service (Study Group II) 9 pp. 9 pp.

—Periodicity
- CCITT RECMN E.180-89. Technical Characteristics of Tones for the Telephone Service—Telephone Network and ISDN—Operation, Numbering, Routing and Mobile Service (Study Group II) 9 pp. 9 pp.

—Transit Exchanges
- CCITT RECMN Q.118-89. Special Release Arrangements—General Recommendations on Telephone Switching and Signalling—Functions and Information Flows for Services in the ISDN—Supplements (Study Group XI) 1 pp. 1 p.
- CCITT RECMN Q.118 BIS-89. Indication of Congestion Conditions at Transit Exchanges—General Recommendations on Telephone Switching and Signalling—Functions and Information Flows for Services in the ISDN—Supplements (Study Group XV) 1 pp. 1 p.

Congo Paper
- CNS K7157-66. Chemical Reagent (Congo Paper) (Mar)(1657).

Congo Red
- CNS K7156-66. Chemical Reagent (Congo Red) (Mar)(1656).
- JIS K 8352-72. Congo Red.

Conical Seats
See Also: Fasteners
- CNS B2392-79. Spherical Washers and Conical Seats (Dec)(5054).
- DIN ENGL 6319-91. Spherical Washers and Conical Seats (Sept). 4 pp.

Conical Spring Washers
See Also: Washers (Fasteners)
- CNS B2188-78. Conical Spring Washers (Mar)(4309).
- DIN ENGL 137-87. Curved and Wave Spring Washers (Oct). 3 pp.
- JIS B 1252-77. Conical Spring Washers. 9 pp.
- SAA AS 1970-76. Conical Spring Washers (Metric Series). 12 pp.

—Bolt Assemblies
- DIN ENGL 6796-87. Conical Spring Washers for Bolt/Nut Assemblies (Oct). 2 pp.

—Screw Assemblies
- CNS B2389-79. Conical Spring Washers for Screw Assemblies (Dec)(5051).
- DIN ENGL 6900 Pt 5-90. Screw and Washer Assemblies; Coarse Threaded Screws with Captive Conical Spring Washer (Dec). 3 pp.
- DIN ENGL 6908-90. Conical Spring Washers for Screw and Washer Assemblies (Dec). 2 pp.

—Seating Bolts—Clamping
- CNS B2241-81. Seating Bolts and Conical Spring Washers for Clamping (Apr)(4421).

Conical Tapers
Use: Machine Tapers

Coniferous Woods
Use: Softwoods

Connected Line Identification Presentation
Use For: COLP *See Also:* Connected Line Identification Restriction; Telephone Services; Terminal Identification

—Integrated Services Digital Networks
- CCITT RECMN I.251. 5-88. Connected Line Identification Presentation (Study Group I) 5 pp. 5 pp.
- CENELEC PRETS 300 094-90. Integrated Services Digital Network Connected Line Identification Presentation (COLP) Supplementary Service Service Description (T/NA1 (89)09). 12 pp.

INTERNATIONAL AND NON-U.S. NATIONAL STANDARDS
SUBJECT INDEX
Connected

Connected Line Identification Presentation *(Cont.)*

—Integrated Services Digital Networks *(Cont.)*

CENELEC PRETS 300 094-91. Integrated Services Digital Network (ISDN); Connected Line Identification Presentation (COLP) Supplementary Service Service Description. 17 pp.
CENELEC ETS 300 094-92. Integrated Services Digital Network (ISDN); Connected Line Identification Presentation (COLP) Supplementary Service Service Description. 16 pp.
CENELEC PRETS 300 096-90. Integrated Services Digital Network (ISDN) Connected Line Identification, Presentation and Restriction (COLP and COLR) Supplementary Services Functional Capabillities and Information Flows (T/S 22-05). 23 pp.
CENELEC PRETS 300 096-91. Integrated Services Digital Network (ISDN); Connected Line Identification Presentation (COLP) and Connected Line Identification Restriction (COLR) Supplementary Services Functional Capabilities and Information Flows. 34 pp.
CENELEC ETS 300 096-92. Integrated Services Digital Network (ISDN); Connected Line Identification Presentation (COLP) and Connected Line Identification Restriction (COLR) Supplementary Services Functional Capabilities and Information Flows. 39 pp.
CENELEC PRETS 300 097-90. Integrated Services Digital Network (ISDN) Connected Line Identification Presentation (COLP) Supplementary Service Digital Subscriber One (DSS1) Protocol (T/S 46-33L). 17 pp.
CENELEC PRETS 300 097-91. Integrated Services Digital Network (ISDN); Connected Line Identification Presentation (COLP) Supplementary Service Digital Subscriber Signalling System No. One (DSS1) Protocol. 24 pp.
CENELEC ETS 300 097-92. Integrated Services Digital Network (ISDN); Connected Line Identification Presentation (COLP) Supplementary Service Digital Subscriber Signalling System No. One (DSS1) Protocol. 21 pp.
CEPT T/S 22-05 E-89. Connected Line Identification Presentation Service. 11 pp.
ETSI PRETS 300 094-90. Integrated Services Digital Network Connected Line Identification Presentation (COLP) Supplementary Service—Service Description (T/NA1 (89)09). 12 pp.
ETSI PRETS 300 096-90. Integrated Services Digital Network (ISDN) Connected Line Identification, Presentation and Restriction (COLP and COLR) Supplementary Services—Functional Capabilities and Information Flows (T/S 22-05). 23 pp.
ETSI PRETS 300 097-90. Integrated Services Digital Network (ISDN) Connected Line Identification, Presentation (COLP) Supplementary Service—Digital Subscriber Signalling One (DSSI) Protocol (T/S 46-33L). 17 pp.
ETSI ETS 300 094-92. Integrated Services Digital Network (ISDN); Connected Line Identification Presentation (COLP) Supplementary Service Service Description. 16 pp.
ETSI PRETS 300 094-91. Integrated Services Digital Network (ISDN); Connected Line Identification Presentation (COLP) Supplementary Service Service Description. 17 pp.
ETSI ETS 300 096-92. Integrated Services Digital Network (ISDN); Connected Line Identification Presentation (COLP) and Connected Line Identification Restriction (COLR) Supplementary Services Functional Capabilities and Information Flows. 39 pp.
ETSI PRETS 300 096-91. Integrated Services Digital Network (ISDN); Connected Line Identification Presentation (COLP) and Connected Line Identification Restriction (COLR) Supplementary Services Functional Capabilities and Information Flows. 34 pp.
ETSI ETS 300 097-92. Integrated Services Digital Network (ISDN); Connected Line Identification Presentation (COLP) Supplementary Service Digital Subscriber Signalling System No. One (DSS1) Protocol. 21 pp.
ETSI PRETS 300 097-91. Integrated Services Digital Network (ISDN); Connected Line Identification Presentation (COLP) Supplementary Service Digital Subscriber Signalling System No. One (DSS1) Protocol. 24 pp.
ETSI ETS 300 098-92. Integrated Services Digital Network (ISDN); Connected Line Identification Restriction (COLR) Supplementary Service Digital Subscriber Signalling System No. One (DSS1) Protocol. 12 pp.

—Integrated Services Digital Networks—Functions/Information Flows

CCITT RECMN Q.81-91. Stage 2 Description for Number Identification Supplementary Services (Study Group XI) 46 pp. 46 pp.

Connected Line Identification Presentation *(Cont.)*

—Integrated Services Digital Networks—Functions/Information Flows *(Cont.)*

CCITT RECMN Q.81-89. Number Identification Supplementary Services—General Recommendations on Telephone Switching and Signalling—Functions and Information Flows for Services in the ISDN—Supplements (Study Group XI) 30 pp. 30 pp.

—Private Switched Networks

CENELEC PRETS 300 173-91. Identification Supplementary Services in Private Telecommunication Networks—Specification, Functional Model and Information Flows-. 50 pp.
CENELEC PRETS 300 173-92. Private Telecommunications Network (PTN); Specification, Functional Models and Information Flows Identification Supplementary Services. 41 pp.
CENELEC ETS 300 173-92. Private Telecommunication Network (PTN); Specification, Functional Models and Information Flows Identification Supplementary Services. 40 pp.
CENELEC PRETS 300 237-92. Private Telecommunications Network (PTN); Specification, Fuctional Models and Information Flows Name Identification Supplementary Services. 44 pp.
CENELEC PRETS 300 237-93. Private Telecommunication Network (PTN); Specification, Functional Models and Information Flows Name Identification Supplementary Services. 42 pp.
CENELEC ETS 300 237-93. Private Telecommunication Network (PTN); Specification, Functional Models and Information Flows Name Identification Supplementary Services. 43 pp.
ECMA ECMA 163-92. Private Telecommunication Networks (PTN)—Specification, Functional Model and Information Flows—Name Identification Supplementary Services (NISD). 41 pp.
ETSI PRETS 300 173-91. Identification Supplementary Services in Private Telecommunication Networks—Specification, Functional Model and Information Flows—(Standard ECMA-148). 49 pp.
ETSI ETS 300 173-92. Private Telecommunication Network (PTN); Specification, Functional Models and Information Flows Identification Supplementary Services. 40 pp.
ETSI PRETS 300 173-92. Private Telecommunications Network (PTN); Specification, Functional Models and Information Flows Identification Supplementary Services. 41 pp.
ETSI PRETS 300 173-91. Identification Supplementary Services in Private Telecommunication Networks—Specification, Functional Model and Information Flows-. 50 pp.
ETSI ETS 300 237-93. Private Telecommunication Network (PTN); Specification, Functional Models and Information Flows Name Identification Supplementary Services. 43 pp.
ETSI PRETS 300 237-93. Private Telecommunication Network (PTN); Specification, Functional Models and Information Flows Name Identification Supplementary Services. 42 pp.
ETSI PRETS 300 237-92. Private Telecommunications Network (PTN); Specification, Fuctional Models and Information Flows Name Identification Supplementary Services. 44 pp.
ETSI ETS 300 238-93. Private Telecommunication Network (PTN); Inter-Exchange Signalling Protocol Name Identificaiton Supplementary Services. 24 pp.

—Private Switched Networks—Signaling Protocols

CENELEC PRETS 300 191-91. Protocol for Signalling Over the D-Channel of Interfaces at the S Reference Point Between Terminal Equipment and Private Telecommunication Networks for the Support of Identification Supplementary Services. 28 pp.
CENELEC PRETS 300 191-92. Private Telecommunications Network (PTN); Signalling Protocol at the S-Reference Point Identification Supplementary Services. 27 pp.
CENELEC ETS 300 191-92. Private Telecommunication Network (PTN); Signalling Protocol at the S-Reference Point Identification Supplementary Services. 26 pp.
ECMA ECMA 157-91. Protocol for Signalling over the D-Channel of Interfaces at the S Reference Point Between Terminal Equipment and Private Telecommunication Networks for the Support of Identification Supplementary Services. 25 pp.
ECMA ECMA 164-92. Private Telecommunication Networks (PTN)—Signalling Between Private Telecommunication Exchanges—Protocol for the Support of Name Identification Supplementary Services (QSIG-NA). 22 pp.
ETSI ETS 300 191-92. Private Telecommunication Network (PTN); Signalling Protocol at the S-Reference Point Identification Supplementary Services. 26 pp.

Connected Line Identification Presentation *(Cont.)*

—Private Switched Networks—Signaling Protocols *(Cont.)*

ETSI PRETS 300 191-92. Private Telecommunications Network (PTN); Signalling Protocol at the S-Reference Point Identification Supplementary Services. 27 pp.
ETSI PRETS 300 191-91. Protocol for Signalling Over the D-Channel of Interfaces at the S Reference Point Between Terminal Equipment and Private Telecommunication Networks for the Support of Identification Supplementary Services. 28 pp.

Connected Line Identification Restriction

Use For: CDLR ***See Also:*** Connected Line Identification Presentation; Telephone Services

—Integrated Services Digital Networks

CCITT RECMN I.251. 6-88. Connected Line Identification Restriction (Study Group I) 7 pp. 7 pp.
CENELEC PRETS 300 095-90. Integrated Services Digital Network (ISDN) Connected Line Identification Restriction (COLR) Supplementary Service Service Description (T/NA1(89)10). 12 pp.
CENELEC PRETS 300 095-91. Integrated Services Digital Network (ISDN); Connected Line Identification Restriction (COLR) Supplementary Service Service Description. 15 pp.
CENELEC ETS 300 095-92. Integrated Services Digital Network (ISDN); Connected Line Identification Restriction (COLR) Supplementary Service Service Description. 14 pp.
CENELEC PRETS 300 096-90. Integrated Services Digital Network (ISDN) Connected Line Identification, Presentation and Restriction (COLP and COLR) Supplementary Services Functional Capabillities and Information Flows (T/S 22-05). 23 pp.
CENELEC PRETS 300 096-91. Integrated Services Digital Network (ISDN); Connected Line Identification Presentation (COLP) and Connected Line Identification Restriction (COLR) Supplementary Services Functional Capabilities and Information Flows. 34 pp.
CENELEC ETS 300 096-92. Integrated Services Digital Network (ISDN); Connected Line Identification Presentation (COLP) and Connected Line Identification Restriction (COLR) Supplementary Services Functional Capabilities and Information Flows. 39 pp.
CENELEC PRETS 300 098-90. Integrated Services Digital Network (ISDN) Connected Line Identification Restriction (COLR) Supplementary Service Digital Subscriber Signalling One (DSS1) Protocol (T/S 46-33M). 9 pp.
CENELEC PRETS 300 098-91. Integrated Services Digital Network (ISDN); Connected Line Identification Restriction (COLR) Supplementary Service Digital Subscriber Signalling System No. One (DSS1) Protocol. 15 pp.
CENELEC ETS 300 098-92. Integrated Services Digital Network (ISDN); Connected Line Identification Restriction (COLR) Supplementary Service Digital Subscriber Signalling System No. One (DSS1) Protocol. 12 pp.
ETSI PRETS 300 095-90. Integrated Services Digital Network (ISDN) Connected Line Identification Restriction (COLR) Supplementary Service—Service Description (T/NA1 (89)10). 12 pp.
ETSI PRETS 300 096-90. Integrated Services Digital Network (ISDN) Connected Line Identification, Presentation and Restriction (COLP and COLR) Supplementary Services—Functional Capabilities and Information Flows (T/S 22-05). 23 pp.
ETSI PRETS 300 098-90. Integrated Services Digital Network (ISDN)—Connected Line Identification Restriction (COLR) Supplementary Service—Digital Subscriber Signalling One (DSS1) (T/S 46-33M). 9 pp.
ETSI ETS 300 095-92. Integrated Services Digital Network (ISDN); Connected Line Identification Restriction (COLR) Supplementary Service Service Description. 14 pp.
ETSI PRETS 300 095-91. Integrated Services Digital Network (ISDN); Connected Line Identification Restriction (COLR) Supplementary Service Service Description. 15 pp.
ETSI ETS 300 096-92. Integrated Services Digital Network (ISDN); Connected Line Identification Presentation (COLP) and Connected Line Identification Restriction (COLR) Supplementary Services Functional Capabilities and Information Flows. 39 pp.
ETSI PRETS 300 096-91. Integrated Services Digital Network (ISDN); Connected Line Identification Presentation (COLP) and Connected Line Identification Restriction (COLR) Supplementary Services Functional Capabilities and Information Flows. 34 pp.

INDUSTRY STANDARDS
593

Connected Line Identification Restriction *(Cont.)*

—Integrated Services Digital Networks *(Cont.)*

ETSI PRETS 300 098-91. Integrated Services Digital Network (ISDN); Connected Line Identification Restriction (COLR) Supplementary Service Digital Subscriber Signalling System No. One (DSS1) Protocol. 15 pp.

—Integrated Services Digital Networks—Functions/Information Flows

CCITT RECMN Q.81-91. Stage 2 Description for Number Identification Supplementary Services (Study Group XI) 46 pp. 46 pp.

CCITT RECMN Q.81-89. Number Identification Supplementary Services—General Recommendations on Telephone Switching and Signalling—Functions and Information Flows for Services in the ISDN—Supplements (Study Group XI) 30 pp. 30 pp.

—Private Switched Networks

CENELEC PRETS 300 173-91. Identification Supplementary Services in Private Telecommunication Networks—Specification, Functional Model and Information Flows-. 50 pp.

CENELEC PRETS 300 173-92. Private Telecommunications Network (PTN); Specification, Functional Models and Information Flows Identification Supplementary Services. 41 pp.

CENELEC ETS 300 173-92. Private Telecommunication Network (PTN); Specification, Functional Models and Information Flows Identification Supplementary Services. 40 pp.

CENELEC PRETS 300 237-92. Private Telecommunications Network (PTN); Specification, Fuctional Models and Information Flows Name Identification Supplementary Services. 44 pp.

CENELEC PRETS 300 237-93. Private Telecommunication Network (PTN); Specification, Functional Models and Information Flows Name Identification Supplementary Services. 42 pp.

CENELEC ETS 300 237-93. Private Telecommunication Network (PTN); Specification, Functional Models and Information Flows Name Identification Supplementary Services. 43 pp.

ECMA ECMA 163-92. Private Telecommunication Networks (PTN)—Specification, Functional Model and Information Flows—Name Identification Supplementary Services (NISD). 41 pp.

ETSI PRETS 300 173-91. Identification Supplementary Services in Private Telecommunication Networks—Specification, Functional Model and Information Flows—(Standard ECMA-148). 49 pp.

ETSI ETS 300 173-92. Private Telecommunication Network (PTN); Specification, Functional Models and Information Flows Identification Supplementary Services. 40 pp.

ETSI PRETS 300 173-92. Private Telecommunications Network (PTN); Specification, Functional Models and Information Flows Identification Supplementary Services. 41 pp.

ETSI PRETS 300 173-91. Identification Supplementary Services in Private Telecommunication Networks—Specification, Functional Model and Information Flows-. 50 pp.

ETSI ETS 300 237-93. Private Telecommunication Network (PTN); Specification, Functional Models and Information Flows Name Identification Supplementary Services. 43 pp.

ETSI PRETS 300 237-93. Private Telecommunication Network (PTN); Specification, Functional Models and Information Flows Name Identification Supplementary Services. 42 pp.

ETSI PRETS 300 237-92. Private Telecommunications Network (PTN); Specification, Fuctional Models and Information Flows Name Identification Supplementary Services. 44 pp.

ETSI ETS 300 238-93. Private Telecommunication Network (PTN); Inter-Exchange Signalling Protocol Name Identificaiton Supplementary Services. 24 pp.

—Private Switched Networks—Signaling Protocols

CENELEC PRETS 300 191-91. Protocol for Signalling Over the D-Channel of Interfaces at the S Reference Point Between Terminal Equipment and Private Telecommunication Networks for the Support of Identification Supplementary Services. 28 pp.

CENELEC PRETS 300 191-92. Private Telecommunications Network (PTN); Signalling Protocol at the S-Reference Point Identification Supplementary Services. 27 pp.

CENELEC ETS 300 191-92. Private Telecommunication Network (PTN); Signalling Protocol at the S-Reference Point Identification Supplementary Services. 26 pp.

ETSI ETS 300 191-92. Private Telecommunication Network (PTN); Signalling Protocol at the S-Reference Point Identification Supplementary Services. 26 pp.

Connected Line Identification Restriction *(Cont.)*

—Private Switched Networks—Signaling Protocols *(Cont.)*

ETSI PRETS 300 191-92. Private Telecommunications Network (PTN); Signalling Protocol at the S-Reference Point Identification Supplementary Services. 27 pp.

ETSI PRETS 300 191-91. Protocol for Signalling Over the D-Channel of Interfaces at the S Reference Point Between Terminal Equipment and Private Telecommunication Networks for the Support of Identification Supplementary Services. 28 pp.

—Signaling Protocols

ECMA ECMA 157-91. Protocol for Signalling over the D-Channel of Interfaces at the S Reference Point Between Terminal Equipment and Private Telecommunication Networks for the Support of Identification Supplementary Services. 25 pp.

Connected Name Identification Presentation

Use For: CONP *See Also:* Calling Name Identification Presentation; Telephone Services; Terminal Identification

—Private Switched Networks—Signaling Protocols

CENELEC PRETS 300 238-92. Private Telecommunications Network (PTN); Signalling Between Private Telecommunication Exchanges Protocol for the Support of Name Identification Supplementary Services. 25 pp.

CENELEC PRETS 300 238-93. Private Telecommunication Network (PTN); Inter-Exchange Signalling Protocol Name Identification Supplementary Services. 24 pp.

CENELEC ETS 300 238-93. Private Telecommunication Network (PTN); Inter-Exchange Signalling Protocol Name Identificaiton Supplementary Services. 24 pp.

ETSI PRETS 300 238-93. Private Telecommunication Network (PTN); Inter-Exchange Signalling Protocol Name Identification Supplementary Services. 24 pp.

ETSI PRETS 300 238-92. Private Telecommunications Network (PTN); Signalling Between Private Telecommunication Exchanges Protocol for the Support of Name Identification Supplementary Services. 25 pp.

Connecting Rods

Use: Piston Rods

Connection Accessibility

See Also: Quality of Service; Serveability Performance; Telecommunication Services; Telephone Connections; Telephone Services

—Telecommunication Networks

CCITT RECMN E.845-89. Connection Accessibility Objective for the International Telephone Service—Telephone Network and ISDN—Quality of Service, Network Management and Traffic Engineering (Study Group II) 7 pp. 7 pp.

—Telecommunication Networks—Busy Hour

CCITT RECMN E.845-89. Connection Accessibility Objective for the International Telephone Service—Telephone Network and ISDN—Quality of Service, Network Management and Traffic Engineering (Study Group II) 7 pp. 7 pp.

—Telephone Circuits—Measurement

CCITT RECMN E.845-89. Connection Accessibility Objective for the International Telephone Service—Telephone Network and ISDN—Quality of Service, Network Management and Traffic Engineering (Study Group II) 7 pp. 7 pp.

—Telephone Exchanges—Measurement

CCITT RECMN E.845-89. Connection Accessibility Objective for the International Telephone Service—Telephone Network and ISDN—Quality of Service, Network Management and Traffic Engineering (Study Group II) 7 pp. 7 pp.

—Telephone Services—Measurement

CCITT RECMN E.845-89. Connection Accessibility Objective for the International Telephone Service—Telephone Network and ISDN—Quality of Service, Network Management and Traffic Engineering (Study Group II) 7 pp. 7 pp.

—Telephone Services—Models

CCITT RECMN E.820-92. Call Models for Serveability and Service Integrity Performance (Study Group II) 11 pp. 11 pp.

CCITT RECMN E.830-92. Models for the Specification, Evaluation and Allocation of Serveability and Service Integrity (Study Group II) 5 pp. 5 pp.

Connection Accessibility *(Cont.)*

—Telephone Services—Quality of Service

CCITT RECMN E.845-89. Connection Accessibility Objective for the International Telephone Service—Telephone Network and ISDN—Quality of Service, Network Management and Traffic Engineering (Study Group II) 7 pp. 7 pp.

Connection Integrity

Use: Connection Stability

Connection Retainability

See Also: Quality of Service; Serveability Performance; Telecommunication Services; Telephone Connections; Telephone Services

—Quality of Service—Telephone Services

CCITT RECMN E.428-92. Connection Retention—Telephone Network and ISDN of Service, Network Management and Traffic Engineering (Study Group II) 4 pp. 4 pp.

—Telephone Services

CCITT RECMN E.850-92. Connection Retainability Objective for the International Telephone Service—Telephone Network and ISDN—Quality of Service, Network Management and Traffic Engineering (Study Group II) 7 pp. 7 pp.

—Telephone Services—Measurement

CCITT RECMN E.850-92. Connection Retainability Objective for the International Telephone Service—Telephone Network and ISDN—Quality of Service, Network Management and Traffic Engineering (Study Group II) 7 pp. 7 pp.

—Telephone Services—Models

CCITT RECMN E.820-92. Call Models for Serveability and Service Integrity Performance (Study Group II) 11 pp. 11 pp.

CCITT RECMN E.830-92. Models for the Specification, Evaluation and Allocation of Serveability and Service Integrity (Study Group II) 5 pp. 5 pp.

Connection Splitting

See Also: Telephone Services

CEPT T/CS 20-20-82. Double Appel. 2 pp.

CEPT T/CS 20-20 E-86. Connection Splitting. 2 pp.

Connection Stability

Use For: Stability (Telecommunications)
See Also: Frequency Error; Quality of Service; Stability Loss (Transmission); Telecommunication Services; Telephone Connections; Telephone Services

CCITT RECMN G.122-89. Influence of National Systems on Stability, Talker Echo, and Listener Echo in International Connections—General Characteristics of International Telephone Connections and Circuits (Study Groups XII and XV) 13 pp. 13 pp.

CCITT RECMN G.131-89. Stability and Echo—General Characteristics of International Telephone Connections and Circuits (Study Groups XII and XV) 13 pp. 13 pp.

CCITT RECMN Q.42-89. Stability and Echo (Echo Suppressors)—General Recommendations on Telephone Switching and Signalling—Functions and Information Flows for Services in the ISDN—Supplements (Study Group XI) 1 pp. 1 p.

—Compandors—Telephone Circuits

CCITT RECMN G.162-89. Characteristics of Compandors for Telephony—General Characteristics of International Telephone Connections and Circuits (Study Groups XII and XV) 8 pp. 8 pp.

CCITT RECMN G.166-89. Characteristics of Syllabic Compandors for Telephony on High Capacity Long Distance Systems—General Characteristics of International Telephone Connections and Circuits (Study Groups XII and XV) 7 pp. 7 pp.

—Telephone Circuits—Private Switched Networks

CCITT RECMN G.171-89. Transmission Plan Aspects of Privately Operated Networks—General Characteristics of International Telephone Connections and Circuits (Study Groups XII and XV) 19 pp. 19 pp.

—Telephone Lines

CCITT RECMN G.214-89. Line Stability of Cable Systems—International Analogue Carrier Systems (Study Group XV) 1 pp. 1 p.

—Telephone Services

CCITT RECMN E.855-89. Connection Integrity Objective for International Telephone Service—Telephone Network and ISDN—Quality of Service, Network Management and Traffic Engineering (Study Group II) 5 pp. 5 pp.

Connection Stability *(Cont.)*

—Telephone Services—Measurement

CCITT RECMN E.855-89. Connection Integrity Objective for International Telephone Service—Telephone Network and ISDN—Quality of Service, Network Management and Traffic Engineering (Study Group II) 5 pp. 5 pp.

—Telephone Services—Models

CCITT RECMN E.820-92. Call Models for Serveability and Service Integrity Performance (Study Group II) 11 pp. 11 pp.
CCITT RECMN E.830-92. Models for the Specification, Evaluation and Allocation of Serveability and Service Integrity (Study Group II) 5 pp. 5 pp.

—Virtual Analog Switching Points

CCITT RECMN G.122-89. Influence of National Systems on Stability, Talker Echo, and Listener Echo in International Connections—General Characteristics of International Telephone Connections and Circuits (Study Groups XII and XV) 13 pp. 13 pp.

—Virtual Carrier Frequencies—Groups

CCITT RECMN G.232-89. 12-Channel Terminal Equipments—International Analogue Carrier Systems (Study Group XV) 13 pp. 13 pp.
CCITT RECMN G.235-89. 16-Channel Terminal Equipments—International Analogue Carrier Systems (Study Group XV) 4 pp. 4 pp.

Connectionless Broadband Data Service

Use For: CBDS ***See Also:*** Bearer Services; Data Communication; Telecommunication Equipment; Telecommunication Services

CENELEC PRETS 300 217-1-91. Network Aspects (NA); Connectionless Broadband Data Service (CBDS) Part 1: Overview (DE/NA-53201-1). 16 pp.
CENELEC PRETS 300 217-1-92. Network Aspects (NA); Connectionless Broadband Data Service (CBDS) Part 1: Overview. 16 pp.
CENELEC ETS 300 217-1-92. Network Aspects (NA); Connectionless Broadband Data Service (CBDS) Part 1: Overview. 16 pp.
CENELEC ETR 082-93. Network Aspects (NA); Connectionless Broadband Data Service (CBDS) Complementary Information to ETS 300 217. 13 pp.
CEPT T/SF 43 E-86. Broadband Services. 3 pp.
ETSI ETS 300 217-1-92. Network Aspects (NA); Connectionless Broadband Data Service (CBDS) Part 1: Overview. 16 pp.
ETSI PRETS 300 217-1-92. Network Aspects (NA); Connectionless Broadband Data Service (CBDS) Part 1: Overview. 16 pp.
ETSI PRETS 300 217-1-91. Network Aspects (NA); Connectionless Broadband Data Service (CBDS) Part 1: Overview (DE/NA-53201-1). 16 pp.
ETSI ETR 082-93. Network Aspects (NA); Connectionless Broadband Data Service (CBDS) Complementary Information to ETS 300 217. 13 pp.

—Addressing Systems

CENELEC PRETS 300 217-3-91. Network Aspects (NA); Connectionless Broadband Data Service (CBDS) Part 3: Definition of Supplementary Services (DE/NA-53201-3). 9 pp.
CENELEC PRETS 300 217-3-92. Network Aspects (NA); Connectionless Broadband Data Service (CBDS) Part 3: Definition of Supplementary Services. 8 pp.
CENELEC ETS 300 217-3-92. Network Aspects (NA); Connectionless Broadband Data Service (CBDS) Part 3: Definition of Supplementary Services. 8 pp.
CENELEC PRETS 300 217-4-91. Network Aspects (NA); Connectionless Broadband Data Service (CBDS) Part 4: Address Screening Supplementary Services (DE/NA-53201-4). 11 pp.
CENELEC PRETS 300 217-4-92. Network Aspects (NA); Connectionless Broadband Data Service (CBDS) Part 4: Address Screening Supplementary Service. 10 pp.
CENELEC ETS 300 217-4-92. Network Aspects (NA); Connectionless Broadband Data Service (CBDS) Part 4: Address Screening Supplementary Service. 10 pp.
ETSI ETS 300 217-3-92. Network Aspects (NA); Connectionless Broadband Data Service (CBDS) Part 3: Definition of Supplementary Services. 8 pp.
ETSI PRETS 300 217-3-92. Network Aspects (NA); Connectionless Broadband Data Service (CBDS) Part 3: Definition of Supplementary Services. 8 pp.
ETSI PRETS 300 217-3-91. Network Aspects (NA); Connectionless Broadband Data Service (CBDS) Part 3: Definition of Supplementary Services (DE/NA-53201-3). 9 pp.
ETSI ETS 300 217-4-92. Network Aspects (NA); Connectionless Broadband Data Service (CBDS) Part 4: Address Screening Supplementary Service. 10 pp.
ETSI PRETS 300 217-4-92. Network Aspects (NA); Connectionless Broadband Data Service (CBDS) Part 4: Address Screening Supplementary Service. 10 pp.

Connectionless Broadband Data Service *(Cont.)*

—Addressing Systems *(Cont.)*

ETSI PRETS 300 217-4-91. Network Aspects (NA); Connectionless Broadband Data Service (CBDS) Part 4: Address Screening Supplementary Services (DE/NA-53201-4). 11 pp.

—Bearer Services

CCITT RECMN F.812-92. Broadband Connectionless Data Bearer Service (Study Group I) 8 pp. 8 pp.
CENELEC PRETS 300 217-2-91. Network Aspects (NA); Connectionless Broadband Data Service (CBDS) Part 2: Basic Bearer Service Definition (DE/NA-53201-2). 22 pp.
CENELEC PRETS 300 217-2-92. Network Aspects (NA); Connectionless Broadband Data Service (CBDS) Part 2: Basic Bearer Service Definition. 20 pp.
CENELEC ETS 300 217-2-92. Network Aspects (NA); Connectionless Broadband Data Service (CBDS) Part 2: Basic Bearer Service Definition. 20 pp.
ETSI ETS 300 217-2-92. Network Aspects (NA); Connectionless Broadband Data Service (CBDS) Part 2: Basic Bearer Service Definition. 20 pp.
ETSI PRETS 300 217-2-92. Network Aspects (NA); Connectionless Broadband Data Service (CBDS) Part 2: Basic Bearer Service Definition. 20 pp.
ETSI PRETS 300 217-2-91. Network Aspects (NA); Connectionless Broadband Data Service (CBDS) Part 2: Basic Bearer Service Definition (DE/NA-53201-2). 22 pp.

—Closed User Groups

CENELEC PRETS 300 217-3-91. Network Aspects (NA); Connectionless Broadband Data Service (CBDS) Part 3: Definition of Supplementary Services (DE/NA-53201-3). 9 pp.
CENELEC PRETS 300 217-3-92. Network Aspects (NA); Connectionless Broadband Data Service (CBDS) Part 3: Definition of Supplementary Services. 8 pp.
CENELEC ETS 300 217-3-92. Network Aspects (NA); Connectionless Broadband Data Service (CBDS) Part 3: Definition of Supplementary Services. 8 pp.
ETSI ETS 300 217-3-92. Network Aspects (NA); Connectionless Broadband Data Service (CBDS) Part 3: Definition of Supplementary Services. 8 pp.
ETSI PRETS 300 217-3-92. Network Aspects (NA); Connectionless Broadband Data Service (CBDS) Part 3: Definition of Supplementary Services. 8 pp.
ETSI PRETS 300 217-3-91. Network Aspects (NA); Connectionless Broadband Data Service (CBDS) Part 3: Definition of Supplementary Services (DE/NA-53201-3). 9 pp.

—International Private Leased Circuits

CCITT RECMN D.3-92. Principles for the Lease of Analogue International Circuits for Private Service (Study Group III) 5 pp (Replaces Recmn D2 and D3). 5 pp.

Connector Housings

See Also: Electrical Enclosures; Housings; Shells (Electric)

—Power Lines

SAA AS 3124-82. Approval and Test Specification—for Overhead Line Connector Boxes Amdt 1 April 1985 (This is a Joint Standard with SANZ NZS 3124). 5 pp.
SNZ NZS/AS 3124-82. Approval and Test Specification for Overhead Line Connector Boxes Amend: 1, 1985 (This is a Joint Standard with SAA AS 3124). 5 pp.

Connectors

Use For: Electric Connections; Electric Connectors
See Also: Adapters (Fittings); Antenna Connectors; Audio Connectors; Battery Connectors; Binding Posts; Cable Couplers; Cable Joints; Card Edge Connectors; Circuit Boxes; Circular Connectors; Clips; Coaxial Connectors; Compression Connectors; Couplings; D Subminiature Connectors; Data Bus Couplers; DIN Connectors; Electric Contacts; Electric Inlets; Electric Outlets; Electric Plug Adapters; Electric Plug Receptacles; Electric Plugs; Electric Terminals; Electrical Components; Fasteners; Feedthrough Connectors; Fiber Optic Connectors; Flat Cable Connectors; Grounding Connectors; Hermaphroditic Connectors; Interconnection Systems; Joints; Jumpers (Electric); Lighting Tracks; MCX Coaxial Connectors; Microwave Coaxial Connectors; Peritelevision Connectors; Pneumatic Connectors; Printed Circuit Connectors; Rack and Panel Connectors; Rectangular Connectors; SMB Coaxial Connectors; SMC Coaxial Connectors; Spacers (Electric); SSMA Coaxial Connectors; Tap Changers; Taps (Electric); Telephone Connectors; Testing Points; Thermocouple Connectors; Ties (Fasteners); Video Connectors; Waveguide Connectors

Connectors *(Cont.)*

BSI BS 196-61. 1961 Amd 4 Protected-Type Non-Reversible Plugs, Socket-Outlets, Cable-Couplers and Appliance Couplers, with Earthing Contacts for Single Phase a.c. Circuits up to 250 Volts. 64 pp.
BSI BS 7215-90. 1990 Separable Insulated Cable Connector System Above 1 kV and up to 36 kV. 21 pp.
CNS C5212-86. General Rule for Low Frequency (Below 3MHz) Electrical Connectors (Dec)(11786).
CNS C6127-86. General Rule for Low Frequency (Below 3MHz) Electrical Connectors (Dec)(7657).
CNS C6132-81. Test Procedure for Low Frequency (Below 3 MHz) Electrical Connectors (TP-23 Low Level Circuit Test) (Jul)(7662).
CNS C6308-88. Method of Test for Low Frequency (Below 3 MHz) Electrical Connectors (TP-15 Contact Strength) (May)(12297).
CNS C6309-88. Method of Test for Low Frequency (Below 3 MHz) Electrical Connectors (TP-53 Nitric Acid Vapor Test, Gold Finish) (May)(12298).
CSA 655 Bull. Electrical Bulletin 655 October 12, 1966 to C22.2 NO 4. 4 pp.
CSA CAN/CSA-C22. 2 NO 18-92. Outlet Boxes, Conduit Boxes, and Fittings; (Gen Instr 1 Thru 2). 118 pp.
CSA C22.2 NO 65-93. Wire Connectors; (Gen Instr 1). 67 pp.
CSA 655 Bull. Electrical Bulletin 655 October 12, 1966 to C22.2 NO 65. 4 pp.
CSA 812 Bull. Electrical Bulletin 812 February 4, 1971 to C22.2 NO 65. 5 pp.
CSA 837 Bull. Electrical Bulletin 837 August 18, 1971 to C22.2 NO 65. 1 p.
CSA 892 Bull. Electrical Bulletin 892 August 30, 1972 to C22.2 NO 65. 3 pp.
CSA 1122A Bull. Electrical Bulletin 1122A January 10, 1979 to C22.2 NO 65. 3 pp.
CSA 812A Bull. Electrical Bulletin 812A April 21, 1980 to C22.2 NO 65. 1 p.
CSA 1287 Bull. Electrical Bulletin 1287 September 30, 1980 to C22.2 NO 65. 1 p.
CSA 1165A Bull. Electrical Bulletin 1165A October 31, 1980 to C22.2 NO 65. 22 pp.
CSA 1324 Bull. Electrical Bulletin 1324 June 15, 1981 to C22.2 NO 65. 1 p.
CSA 1165B Bull. Electrical Bulletin 1165B June 25, 1984 to C22.2 NO 65. 5 pp.
CSA 1422 Bull. Electrical Bulletin 1422 July 6, 1984 to C22.2 NO 65. 3 pp.
CSA 1165C Bull. Electrical Bulletin 1165C December 4, 1984 to C22.2 NO 65. 2 pp.
CSA 1438 Bull. Electrical Bulletin 1438 October 23, 1985 to C22.2 NO 65. 20 pp.
CSA 1165D Bull. Electrical Bulletin 1165D April 23, 1986 to C22.2 NO 65. 2 pp.
DIN VDE 0220 Pt 2-71. Pressure-Connectors in Heavy-Current Cable Installations (Nov). 20 pp.
DIN VDE 0220 Pt 2A-78. Regulations for Press Connectors in Heavy-Current Cable Installations (Apr). 7 pp.
DIN VDE 0627-86. Connectors and Plug-And-Socket Devices for Rated Voltages up to 1000 V a.c. and 1200 V d.c. and Rated Currents up to 500 A for Each Pole (June). 36 pp.
IEC 130 Pt 1-88. Connectors for Frequencies Below 3 MHz Part 1: General Requirements and Measuring Methods Second Edition. 27 pp.
IEC 998 Pt 1-90. Connecting Devices for Low Voltage Circuits for Household and Similar Purposes Part 1: General Requirements First Edition. 54 pp.
MOD UK DSTAN 59-35: Part 0-90. Connectors Electrical Part 0: Guide to Connectors Electrical, Direct Current, Low Frequency and Radio Frequency, and Their Application Issue 3. 153 pp.
MOD UK DSTAN 59-35: Pt 1:Sec 14-84. Connectors, Electrical: Connectors, Butting, Pattern 100 (10 PR Coupler), Sealed and Non-Sealed, with Soldered Terminations for d.c. and Low Frequency Applications Issue 2 (Obsolescent). 17 pp.
MOD UK DSTAN 59-35: Part 3-72. Connectors for d.c and Low Frequency Applications (See Also EPIC Database) Issue 1. 6 pp.
MOD UK DSTAN 59-35: Pt 3:Sec 4-78. Plugs and Sockets Electrical Part 3: Electrical Connectors Below 3MHz (See Also EPIC Database) Section 4: Pattern 9525 F0003 Issue 1. 7 pp.
MOD UK DSTAN 59-35: Part 5-01. Plugs and Sockets, Electrical, Pattern 608, Sealed, with Soldered Terminations: Non-Sealed, with Crimped Terminations: Coarse Threaded Coupling, for d.c. and Low Frequency Applications Issue 2; Amendment 6. 108 pp.
MOD UK DSTAN 59-35: Part 90-01. Connectors, Electrical Part 90: Detail Specifications Issue 1; Amendment 1. 8 pp.
MOD UK DSTAN 61-12: Part 22-01. Wires, Cords, and Cables, Electrical—Metric Units Part 22: Cables, Special Purpose, Electrical (Equipment Interconnecting Cables) Issue 1; Amendment 1. 29 pp.

INDUSTRY STANDARDS
595

INTERNATIONAL AND NON-U.S. NATIONAL STANDARDS
SUBJECT INDEX
Connectors

Connectors *(Cont.)*
SNZ NZS 1989-65. Specification for Protected-Type Non-Reversible Plugs, Socket-Outlets, Cable-Couplers and Appliance-Couplers, with Earthing Contacts for Single Phase Alternating Current Circuits up to 250 Volts. 64 pp.

—**Acceleration Testing**
CNS C6184-88. Method of Test for Low Frequency (Below 3 MHz) Electrical Connectors (TP-1 Acceleration) (Apr)(9363).

—**Aerospace**
BSI BS EN 2591-92. 1992 Elements of Electrical and Optical Connection—Test Methods—General. 13 pp.
CEN PREN 2591-91. Elements of Electrical and Optical Connection Test Methods—General. 8 pp.
CEN EN 2591-92. Aerospace Series—Elements of Electrical and Optical Connection—Test Methods—General. 8 pp.
CEN PREN 3660-002-91. Index of Detail Specification Sheets. 8 pp.

—**Aerospace—Acceleration Testing**
BSI BS EN 2591-D1-92. 1992 Elements of Electrical and Optical Connection—Test Methods Part D1: Acceleration, Steady State. 9 pp.
CEN PREN 2591 (Part D1)-91. Elements of Electrical and Optical Connection Test Methods Part D1—Acceleration, Steady State. 4 pp.
CEN EN 2591-D1-92. Aerospace Series—Elements of Electrical and Optical Connection—Test Methods—Part D1: Acceleration, Steady State. 4 pp.

—**Aerospace—Air Leakage**
CEN PREN 2591 (Part C12)-92. Elements of Electrical and Optical Connection Test Methods Part C12—Air Leakage. 5 pp.

—**Aerospace—Attenuation**
CEN PREN 2591-F2-93. Aerospace Series Elements of Electrical and Optical Connection Test Methods Part F2—Optical Elements Variation of Attenuation and Optical Discontinuity. 3 pp.

—**Aerospace—Axial Loads**
CEN PREN 2591 (Part D5)-92. Elements of Electrical and Optical Connection Test Methods Part D5—Axial Load. 4 pp.

—**Aerospace—Bend Testing**
CEN PREN 2591 (Part D4)-92. Elements of Electrical and Optical Connection Test Methods Part D4—Transverse Load (External Bending Moment). 4 pp.
CEN PREN 2591 (Part D20)-92. Elements of Electrical and Optical Connection Test Methods Part D20—Mechanical Strength of Rear Accessories. 5 pp.

—**Aerospace—Compression Testing**
CEN PREN 2591-F12-93. Aerospace Series Elements of Electrical and Optical Connection Test Methods Part F12—Optical Elements Effectiveness of Cable Attachment Cable Axial Compression. 3 pp.

—**Aerospace—Contact Force**
CEN PREN 2591 (Part D8)-92. Elements of Electrical and Optical Connection Test Methods Part D8—Mating and Unmating Forces. 4 pp.
CEN PREN 2591 (Part D9)-92. Elements of Electrical and Optical Connection Test Methods Part D9—Contact Retention in Insert. 4 pp.
CEN PREN 2591 (Part D10)-92. Elements of Electrical and Optical Connection Test Methods Part D10—Insert Retention in Housing (Axial). 4 pp.
CEN PREN 2591 (Part D11)-92. Elements of Electrical and Optical Connection Test Methods Part D11—Insert Retention in Housing (Torsional). 4 pp.
CEN PREN 2591 (Part D12)-92. Elements of Electrical and Optical Connection Test Methods Part D12—Contact Insertion and Extraction Forces. 4 pp.
CEN PREN 2591-F10-93. Aerospace Series Elements of Electrical and Optical Connection Test Methods Part F10—Optical Elements Effectiveness of Cable Attachment Cable Pulling. 3 pp.

—**Aerospace—Contact Resistance**
BSI BS EN 2591-B1-92. 1992 Elements of Electrical and Optical Connection—Test Methods Part B1: Contact Resistance—Low Level. 10 pp.
CEN PREN 2591 (Part B1)-91. Elements of Electrical and Optical Connection Test Methods Part B1—Contact Resistance—Low Level. 4 pp.
CEN EN 2591-B1-92. Aerospace Series—Elements of Electrical and Optical Connection—Test Methods—Part B 1: Contact Resistance—Low Level. 5 pp.
CEN PREN 2591 (Part B2)-92. Elements of Electrical and Optical Connection Test Methods Part B2—Contact Resistance at Rated Current. 4 pp.
CEN PREN 2591 (Part B3)-92. Elements of Electrical and Optical Connection Test Methods Part B3—Electrical Continuity at Microvolt Level. 3 pp.

Connectors *(Cont.)*
—**Aerospace—Contact Resistance** *(Cont.)*
CEN PREN 2591 (Part B4)-92. Elements of Electrical and Optical Connection Test Methods Part B4—Discontinuity of Contacts in the Microsecond Range. 4 pp.
CEN PREN 2591-F11-93. Aerospace Series Elements of Electrical and Optical Connection Test Methods Part F11—Optical Elements Effectiveness of Cable Attachment Cable Torsion. 4 pp.

—**Aerospace—Cyclic Flexing**
CEN PREN 2591-F9-93. Aerospace Series Elements of Electrical and Optical Connection Test Methods Part F9—Optical Elements Effectiveness of Cable Attachment Cable Cyclic Flexing. 3 pp.

—**Aerospace—Damp Heat Testing**
CEN PREN 2591 (Part C3)-92. Elements of Electrical and Optical Connection Test Methods Part C3—Cold/Low Pressure and Damp Heat. 6 pp.
CEN PREN 2591 (Part C21)-92. Elements of Electrical and Optical Connection Test Methods Part C21—Damp Heat, Cyclic Test. 4 pp.
CEN PREN 2591-FC3-93. Aerospace Series Elements of Electrical and Optical Connection Test Methods Part FC3—Optical Elements Cold/Low Pressure and Damp Heat. 3 pp.

—**Aerospace—Data Links**
CEN PREN 2591 (Part G)-92. Elements of Electrical and Optical Connection Test Methods Part G-. 4 pp.

—**Aerospace—Dry Heat Testing**
CEN PREN 2591 (Part C9)-92. Elements of Electrical and Optical Connection Test Methods Part C9—Dry Heat. 4 pp.

—**Aerospace—Dust**
CEN PREN 2591 (Part C8)-92. Elements of Electrical and Optical Connection Test Methods Part C8—Sand and Dust. 5 pp.

—**Aerospace—Endurance Testing**
CEN PREN 2591 (Part D6)-92. Elements of Electrical and Optical Connection Test Methods Part D6—Mechanical Endurance. 4 pp.

—**Aerospace—Environmental Testing**
CEN PREN 2591 (Part C2)-92. Elements of Electrical and Optical Connection Test Methods Part C2—Climatic Sequence. 6 pp.

—**Aerospace—Fire Testing**
CEN PREN 2591 (Part C18)-92. Elements of Electrical and Optical Connection Test Methods Part C18—Fire-Resistance. 6 pp.
CEN PREN 2591-FC18-93. Aerospace Series Elements of Electrical and Optical Connection Test Methods Part FC18—Optical Elements Fire Resistance. 3 pp.

—**Aerospace—Flammability Testing**
CEN PREN 2591 (Part C17)-92. Elements of Electrical and Optical Connection Test Methods Part C17—Flammability. 4 pp.
CEN PREN 2591-FC17-93. Aerospace Series Elements of Electrical and Optical Connection Test Methods Part FC17—Optical Elements Flammability. 3 pp.

—**Aerospace—Fluid Resistance**
CEN PREN 2591 (Part C15)-92. Elements of Electrical and Optical Connection Test Methods Part C15—Fluid Resistance. 6 pp.
CEN PREN 2591-FC15-93. Aerospace Series Elements of Electrical and Optical Connection Test Methods Part FC15—Optical Elements Fluid Resistance. 3 pp.

—**Aerospace—Fungus Resistance Testing**
CEN PREN 2591 (Part C6)-92. Elements of Electrical and Optical Connection Test Methods Part C6—Mould Growth. 11 pp.
CEN PREN 2591-FC6-93. Aerospace Series Elements of Electrical and Optical Connection Test Methods Part FC6—Optical Elements Mould Growth. 3 pp.

—**Aerospace—High Temperature Testing**
BSI BS EN 2591-C1-92. 1992 Elements of Electrical and Optical Connection—Test Methods Part C1: Endurance at Temperature. 10 pp.
CEN PREN 2591 (Part C1)-91. Elements of Electrical and Optical Connection Test Methods Part C1—Endurance at Temperature. 6 pp.
CEN EN 2591-C1-92. Aerospace Series—Elements of Electrical and Optical Connection—Test Methods—Part C1: Endurance at Temperature. 5 pp.
CEN PREN 2591-FC1-93. Aerospace Series Elements of Electrical and Optical Connection Test Methods Part FC1—Optical Elements Endurance at Temperature. 3 pp.

Connectors *(Cont.)*
—**Aerospace—Humidity**
CEN PREN 2591 (Part C4)-92. Elements of Electrical and Optical Connection Test Methods Part C4—Damp Heat Steady State. 6 pp.

—**Aerospace—Insertion Force**
CEN PREN 2591 (Part D8)-92. Elements of Electrical and Optical Connection Test Methods Part D8—Mating and Unmating Forces. 4 pp.
CEN PREN 2591 (Part D12)-92. Elements of Electrical and Optical Connection Test Methods Part D12—Contact Insertion and Extraction Forces. 4 pp.

—**Aerospace—Insertion Loss**
CEN PREN 2591-F1-93. Aerospace Series Elements of Electrical and Optical Connection Test Methods Part F1—Optical Elements Insertion Loss. 7 pp.

—**Aerospace—Insulation Resistance**
CEN PREN 2591 (Part B6)-92. Elements of Electrical and Optical Connection Test Methods Part B6—Measurement of Insulation Resistance. 3 pp.

—**Aerospace—Low Pressure Testing**
CEN PREN 2591 (Part C3)-92. Elements of Electrical and Optical Connection Test Methods Part C3—Cold/Low Pressure and Damp Heat. 6 pp.
CEN PREN 2591 (Part C11)-92. Elements of Electrical and Optical Connection Test Methods Part C11—Low Air Pressure. 4 pp.
CEN PREN 2591 (Part C14)-92. Elements of Electrical and Optical Connection Test Methods Part C14—Immersion at Low Air Pressure. 5 pp.
CEN PREN 2591-FC3-93. Aerospace Series Elements of Electrical and Optical Connection Test Methods Part FC3—Optical Elements Cold/Low Pressure and Damp Heat. 3 pp.
CEN PREN 2591-FC14-93. Aerospace Series Elements of Electrical and Optical Connection Test Methods Part FC14—Optical Elements Immersion at Low Air Pressure. 3 pp.

—**Aerospace—Low Temperature Testing**
CEN PREN 2591 (Part C10)-92. Elements of Electrical and Optical Connection Test Methods Part C10—Cold. 4 pp.

—**Aerospace—Lubricant Resistance**
CEN PREN 2591 (Part C15)-92. Elements of Electrical and Optical Connection Test Methods Part C15—Fluid Resistance. 6 pp.
CEN PREN 2591-FC15-93. Aerospace Series Elements of Electrical and Optical Connection Test Methods Part FC15—Optical Elements Fluid Resistance. 3 pp.

—**Aerospace—Mass**
CEN PREN 2591 (Part A2)-92. Elements of Electrical and Optical Connection Test Methods Part A2—Examination of Dimensions and Mass. 2 pp.

—**Aerospace—Mechanical Shock**
CEN PREN 2591 (Part D2)-92. Elements of Electrical and Optical Connection Test Methods Part D2—Shock. 6 pp.

—**Aerospace—Optical Discontinuity**
CEN PREN 2591-F2-93. Aerospace Series Elements of Electrical and Optical Connection Test Methods Part F2—Optical Elements Variation of Attenuation and Optical Discontinuity. 3 pp.

—**Aerospace—Optical Face—Cleaning Capability**
CEN PREN 2591-F4-93. Aerospace Series Elements of Electrical and Optical Connection Test Methods Part F4—Optical Elements Cleaning Capability of Optical Face. 3 pp.

—**Aerospace—Overload Testing**
CEN PREN 2591 (Part B10)-92. Elements of Electrical and Optical Connection Test Methods Part B10—Electrical Overload. 4 pp.

—**Aerospace—Ozone Resistance**
CEN PREN 2591 (Part C16)-92. Elements of Electrical and Optical Connection Test Methods Part C16—Ozone Resistance. 3 pp.
CEN PREN 2591-FC16-93. Aerospace Series Elements of Electrical and Optical Connection Test Methods Part FC16—Optical Elements Ozone Resistance. 2 pp.

—**Aerospace—Random Vibration**
CEN PREN 2591 (Part D3)-92. Elements of Electrical and Optical Connection Test Methods Part D3—Sinusoidal and Random Vibrations. 10 pp.

—**Aerospace—Residual Flux Density**
CEN PREN 2591 (Part E4)-92. Elements of Electrical and Optical Connection Test Methods Part E4—Residual Magnetism. 4 pp.

INTERNATIONAL AND NON-U.S. NATIONAL STANDARDS
SUBJECT INDEX
Connectors

Connectors *(Cont.)*

—**Aerospace—Restricted Entry Testing**
CEN PREN 2591 (Part E2)-92. Elements of Electrical and Optical Connection Test Methods Part E2—Restricted Entry. 3 pp.

—**Aerospace—Salt Spray Testing**
CEN PREN 2591 (Part C7)-92. Elements of Electrical and Optical Connection Test Methods Part C7—Salt Mist. 5 pp.
CEN PREN 2591-FC7-93. Aerospace Series Elements of Electrical and Optical Connection Test Methods Part FC7—Optical Elements Salt Mist. 3 pp.

—**Aerospace—Sand**
CEN PREN 2591 (Part C8)-92. Elements of Electrical and Optical Connection Test Methods Part C8—Sand and Dust. 5 pp.

—**Aerospace—Sealing**
CEN PREN 2591 (Part C24)-92. Elements of Electrical and Optical Connection Test Methods Part C24—Interfacial Sealing. 4 pp.

—**Aerospace—Sealing—Wear Testing**
CEN PREN 2591 (Part D7)-92. Elements of Electrical and Optical Connection Test Methods Part D7—Durability of Contact Retention System and Seals (Maintenance Ageing). 5 pp.

—**Aerospace—Solar Radiation**
CEN PREN 2591 (Part C20)-92. Elements of Electrical and Optical Connection Test Methods Part C20—Simulated Solar Radiation at Ground Level. 5 pp.

—**Aerospace—Temperature Change Testing**
CEN PREN 2591 (Part B8)-92. Elements of Electrical and Optical Connection Test Methods Part B8—Temperature Rise Due to Rated Current. 3 pp.
CEN PREN 2591 (Part B9)-92. Elements of Electrical and Optical Connection Test Methods Part B9—Current Temperature Derating. 5 pp.
CEN PREN 2591 (Part C5)-92. Elements of Electrical and Optical Connection Test Methods Part C5—Rapid Change of Temperature. 5 pp.
CEN PREN 2591-FC5-93. Aerospace Series Elements of Electrical and Optical Connection Test Methods Part FC5—Optical Elements Rapid Change of Temperature. 3 pp.

—**Aerospace—Tensile Testing**
CEN PREN 2591 (Part D20)-92. Elements of Electrical and Optical Connection Test Methods Part D20—Mechanical Strength of Rear Accessories. 5 pp.

—**Aerospace—Torque**
CEN PREN 2591 (Part D8)-92. Elements of Electrical and Optical Connection Test Methods Part D8—Mating and Unmating Forces. 4 pp.
CEN PREN 2591 (Part D20)-92. Elements of Electrical and Optical Connection Test Methods Part D20—Mechanical Strength of Rear Accessories. 5 pp.

—**Aerospace—Torsion**
CEN PREN 2591 (Part D11)-92. Elements of Electrical and Optical Connection Test Methods Part D11—Insert Retention in Housing (Torsional). 4 pp.

—**Aerospace—Vibration**
CEN PREN 2591 (Part D3)-92. Elements of Electrical and Optical Connection Test Methods Part D3—Sinusoidal and Random Vibrations. 10 pp.

—**Aerospace—Visual Inspection**
CEN PREN 2591 (Part A1)-92. Elements of Electrical and Optical Connection Test Methods Part A1—Visual Examination. 2 pp.
CEN PREN 2591-FA1-93. Aerospace Series Elements of Electrical and Optical Connection Test Methods Part FA1—Optical Elements Visual Examination. 2 pp.

—**Aerospace—Voltage Measurement**
CEN PREN 2591 (Part B7)-92. Elements of Electrical and Optical Connection Test Methods Part B7—Voltage Proof Test. 3 pp.

—**Aerospace—Water Resistance Testing**
CEN PREN 2591 (Part C13)-92. Elements of Electrical and Optical Connection Test Methods Part C13—Driving Rain (Artificial). 5 pp.

—**Aerospace—Wear Testing**
CEN PREN 2591 (Part D7)-92. Elements of Electrical and Optical Connection Test Methods Part D7—Durability of Contact Retention System and Seals (Maintenance Ageing). 5 pp.

Connectors *(Cont.)*

—**Aging Testing**
CNS C6133-87. Method of Test for Low Frequency (Below 3 MHz) Electrical Connectors (TP-24 Maintenance Aging Test) (Oct)(7663).

—**Agricultural Equipment**
BSI BS 6777: Part 1-86. (WITHDRAWN) 1986 Electrical Connections for Agricultural Machinery Part 1: Double-Pole Connectors (Superseded by BS EN 24165: 1992). 3 pp.

—**Agricultural Equipment—Towing Attachments**
CSA CAN/CSA-M663-92. Seven-Pin Electrical Connector and Cable for Agricultural Towing/Towed Equipment; (Gen Instr 1). 21 pp.

—**Air Conditioners—Aircraft**
NATO STANAG 3208 ED 4 AMD 4-82. Air Conditioning Connections. 9 pp.
NATO STANAG 3208 ED 5 AMD 0-92. Air Conditioning Connections. 9 pp.
NATO STANAG 3208 ED 5 AMD 1-92. Air Conditioning Connections. 10 pp.

—**Aircraft**
BSI 2G 202-88. 1988 Design and Performance Requirements for Airframe-Fit Electrical Connecotors for D.C. and Low Frequency A.C. Applications. 11 pp.
CAA. Contents (Civil Air Publications: Approved Aerial Positions). 1 p.
CAA. Foreword (Civil Air Publications: Approved Aerial Positions). 3 pp.
ISO 1949-87. Aircraft—Electrical Connectors—Design Requirements Second Edition. 16 pp.
ISO 2100-87. Aircraft—Electrical Connectors—Tests Second Edition. 15 pp.
SBAC TS 350 ISSUE 1. Test for Electrical Connectors Contact Engaging and Separating Forces.
SBAC TS 357 ISSUE 1. Tests for Electrical Connectors Restricted Entry.
SBAC TS 365 ISSUE 1. Test for Electrical Connectors Holding Force of Grounding Spring System.
SBAC TS 366 ISSUE 1. Test for Electrical Connectors Stability of Male Contacts in the Insert.
SBAC TS 368 ISSUE 1. Test for Electrical Connectors Robustness of Protective Cover Attachment.
SBAC ESC 10 ISSUE 3. High Temperature Electrical Connectors (Generally to Mil-C-83723 Series III K). 7 pp.

—**Aircraft—Air/Nitrogen—Replenishment**
NATO STANAG 3806 ED 1 AMD 5-79. Aircraft Gaseous Air/Nitrogen Systems Replenishment Connectors. 6 pp.

—**Aircraft—Bend Testing**
SBAC TS 356 ISSUE 1. Test for Electrical Connectors Contact Bending Strength.

—**Aircraft—Contact Force**
SBAC TS 364 ISSUE 1. Test for Electrical Connectors Contact Insertion and Extraction Forces.

—**Aircraft—Contact Resistance**
SBAC TS 340 ISSUE 1. Test for Electrical Connectors Contact Resistance—Low Current.
SBAC TS 341 ISSUE 1. Test for Electrical Connectors Contact Resistance at Rated Current.

—**Aircraft—Contact Retention**
SBAC TS 320 ISSUE 2. Test for Electrical Connectors Contact Retention.
SBAC TS 321 ISSUE 1. Test for Electrical Connectors Insert Retention.

—**Aircraft—Continuity Testing**
SBAC TS 342 ISSUE 1. Test for Electrical Connectors Housing (Shell) Electrical Continuity.

—**Aircraft—Corrosion Testing**
SBAC TS 333 ISSUE 1. Test for Electrical Connectors Corrosion.

—**Aircraft—Coupling Torque**
SBAC TS 322 ISSUE 2. Test for Electrical Connectors Coupling Torque.

—**Aircraft—Damp Heat Testing**
SBAC TS 346 ISSUE 1. Test for Electrical Connectors Damp Heat—Cyclic Test.

—**Aircraft—Dielectric Strength**
SBAC TS 323 ISSUE 1. Test for Electrical Connectors Dielectric Withstand.

—**Aircraft—Dust and Sand Contamination**
SBAC TS 330 ISSUE 1. Test for Electrical Connectors Dust and Sand Contamination.

Connectors *(Cont.)*

—**Aircraft—Electrical Grounding**
NATO STANAG 3632 ED 3 AMD 2-86. Electrical Safety Connections for Aircraft and Ground Support Equipment. 12 pp.
NATO STANAG 3632 ED 4 AMD 0-91. Aircraft and Ground Support Equipment Electrical Connections for Static Grounding. 9 pp.
NATO STANAG 3632 ED 4 AMD 1-91. Aircraft and Ground Support Equipment Electrical Connections for Static Grounding. 10 pp.

—**Aircraft—Electromagnetic Interference**
SBAC ESC 65 ISSUE 3. Backshell, Assembly, Straight, RFI/EMI Cone Grounding, Anti-Decoupling. (MS 3155 Connector Interface MoD). 7 pp.
SBAC ESC 66 ISSUE 3. Backshell, Assembly, 90 Degrees RFI/EMI Cone Grounding, Anti Decoupling. (MS 3155 Connector Interface MoD). 7 pp.
SBAC ESC 67 ISSUE 3. Backshell, Assembly, 45 Degrees RFI/EMI Grounding, Anti—Decoupling. (MS 3155 Connector Interface MoD). 7 pp.
SBAC ESC 72 ISSUE 1. Backshell, Assembly, Straight, RFI/EMI Cone Grounding, Anti-Decoupling. (MS 3155 Connector Interface MoD). 7 pp.
SBAC ESC 73 ISSUE 1. Backshell, Assembly, 90 Degrees RFI/EMI Cone Grounding, Anti Decoupling. (MS 3155 Connector Interface MoD). 7 pp.
SBAC ESC 74 ISSUE 1. Backshell, Assembly, 45 Degrees RFI/EMI Grounding, Anti—Decoupling. (MS 3155 Connector Interface MoD). 7 pp.

—**Aircraft—Endurance Testing**
SBAC TS 363 ISSUE 1. Test for Electrical Connectors Mechanical Endurance.

—**Aircraft Engines—Starting**
NATO STANAG 3947 ED 1 AMD 5-83. Ground Half Air and Electrical Connectors for Low Pressure Air Starting of Aircraft Engines. 8 pp.
NATO STANAG 3947 ED 1 AMD 6-83. Ground Half Air and Electrical Connectors for Low Pressure Air Starting of Aircraft Engines. 9 pp.

—**Aircraft—Environmental Testing**
SBAC TS 331 ISSUE 2. Test for Electrical Connectors Climatic Tests.
SBAC TS 337 ISSUE 1. Test for Electrical Connectors Environmental Sealing.

—**Aircraft—Fire Testing**
SBAC TS 370 ISSUE 1. Test for Electrical Connectors Fire Resistance.

—**Aircraft—Fluid Resistance Testing**
SBAC TS 334 ISSUE 2. Test for Electrical Connectors Fluid Resistance Test.
SBAC TS 377 ISSUE 1. High Temperature Endurance and Fluid Resistance.

—**Aircraft—High Temperature Testing**
SBAC TS 332 ISSUE 2. Test for Electrical Connectors High Temperature Cyclic Endurance.
SBAC TS 377 ISSUE 1. High Temperature Endurance and Fluid Resistance.

—**Aircraft—Identification Systems**
SBAC RS 743 ISSUE 2. Connector Specifications Identification System.

—**Aircraft—Immersion Testing**
SBAC TS 344 ISSUE 1. Test for Electrical Connectors Immersion at Low Air Pressure.

—**Aircraft—Insulation Resistance**
SBAC TS 324 ISSUE 2. Test for Electrical Connectors Insulation Resistance.

—**Aircraft—Low Temperature Testing**
SBAC TS 326 ISSUE 1. Test for Electrical Connectors Low Temperature Handling.

—**Aircraft—Mass**
SBAC TS 339 ISSUE 1. Test for Electrical Connectors Examination of Dimensions and Mass.

—**Aircraft—Mechanical Shock**
SBAC TS 329 ISSUE 2. Test for Electrical Connectors Physical Shock.

—**Aircraft—Mechanical Testing**
SBAC TS 372 ISSUE 1. Mechanical Strength of Rear Accessories.

—**Aircraft—Overload Testing**
SBAC TS 371 ISSUE 1. Test for Electrical Connectors Electrical Overload.

—**Aircraft—Ozone Resistance**
SBAC TS 345 ISSUE 1. Test for Electrical Connectors Ozone Resistance.

INDUSTRY STANDARDS

INTERNATIONAL AND NON-U.S. NATIONAL STANDARDS
SUBJECT INDEX — Connectors

Connectors (Cont.)

—Aircraft Refueling Equipment
NATO STANAG 3681 ED 1 AMD 6-74. Criteria for Pressure Fuelling of Aircraft. 6 pp.
NATO STANAG 3681 ED 2 AMD 0-93. Criteria for Pressure Fuelling/Defuelling of Aircraft. 7 pp.

—Aircraft—Residual Magnetism Testing
SBAC TS 373 ISSUE 1. Residual Magnetism.

—Aircraft—Sealing
SBAC TS 349 ISSUE 1. Test for Electrical Connectors Interfacial Sealing.

—Aircraft—Shielding Testing
SBAC TS 343 ISSUE 1. Test for Electrical Connectors Shielding Effectiveness from 100 MHz to 1 GHz.

—Aircraft Stores—Location
NATO STANAG 3558 ED 3 AMD 0-86. Locations for Aircraft Electrical Control Connections for Aircraft Stores. 2 pp.
NATO STANAG 3558 ED 3 AMD 1-86. Locations for Aircraft Electrical Control Connections for Aircraft Stores. 9 pp.
NATO STANAG 3558 ED 3 AMD 4-86. Locations for Aircraft Electrical Control Connections for Aircraft Stores. 9 pp.
NATO STANAG 3558 ED 3 AMD 5-86. Locations for Aircraft Electrical Control Connections for Aircraft Stores. 10 pp.

—Aircraft—Tensile Testing
SBAC TS 367 ISSUE 1. Test for Electrical Connectors Contact Retention System Effectiveness (Removable Contact Walkout).

—Aircraft—Test Probes
SBAC TS 351 ISSUE 1. Test for Electrical Connectors Test Probe Damage.

—Aircraft—Testing Equipment
SBAC TS 369 ISSUE 1. Test for Electrical Connectors Use of Tools.

—Aircraft—Thermal Shock
SBAC TS 325 ISSUE 1. Test for Electrical Connectors Thermal Shock.
SBAC TS 348 ISSUE 1. Test for Electrical Connectors Thermal Shock.

—Aircraft—Transverse Strain
SBAC TS 362 ISSUE 1. Test for Electrical Connectors Transverse Load (External Bending Moment).

—Aircraft—Vibration
SBAC TS 360 ISSUE 1. Test for Electrical Connectors Random Vibration.

—Aircraft—Visual Inspection
SBAC TS 338 ISSUE 1. Test for Electrical Connectors Visual Examination.

—Aircraft—Wear Testing
SBAC TS 361 ISSUE 1. Test for Electrical Connectors Durability of Contact Retention Systems and Seals (Maintenance Aging).

—Aluminum Conductors
CSA C22.2 NO 65-93. Wire Connectors; (Gen Instr 1). 67 pp.

—Antilock Devices—Brakes—Combat Vehicles
NATO STANAG 4395 ED 1 AMD 0-00. Connector for Tactical Land Vehicles with Anti-Lock Braking Systems. 4 pp.

—Appliances
CSA 655 Bull. Electrical Bulletin 655 October 12, 1966 to C22.2 NO 4. 4 pp.
CSA C22.2 NO 65-93. Wire Connectors; (Gen Instr 1). 67 pp.
CSA 655 Bull. Electrical Bulletin 655 October 12, 1966 to C22.2 NO 65. 4 pp.
CSA 812 Bull. Electrical Bulletin 812 February 4, 1971 to C22.2 NO 65. 5 pp.
CSA 1122A Bull. Electrical Bulletin 1122A January 10, 1979 to C22.2 NO 65. 3 pp.
CSA 1165B Bull. Electrical Bulletin 1165B June 25, 1984 to C22.2 NO 65. 5 pp.
CSA 1422 Bull. Electrical Bulletin 1422 July 6, 1984 to C22.2 NO 65. 3 pp.
CSA 1165C Bull. Electrical Bulletin 1165C December 4, 1984 to C22.2 NO 65. 2 pp.
CSA 1438 Bull. Electrical Bulletin 1438 October 23, 1985 to C22.2 NO 65. 20 pp.
CSA 1165D Bull. Electrical Bulletin 1165D April 23, 1986 to C22.2 NO 65. 2 pp.

Connectors (Cont.)

—Arc Welding Cables
IEC 501-75. Safety Requirements for Arc Welding Equipment—Plugs, Socket-Outlets and Couplers for Welding Cables First Edition. 13 pp.

—Armored Cables
CSA CAN/CSA-C22. 2 NO 18-92. Outlet Boxes, Conduit Boxes, and Fittings; (Gen Instr 1 Thru 2). 118 pp.

—Artillery
NATO STANAG 4007 ED 2 AMD 0-92. (Draft) Electrical Connectors Between Prime Movers, Trailers and Towed Artillery. 17 pp.
NATO STANAG 4007 ED 2 AMD 0-00. Electrical Connectors Between Prime Movers, Trailers and Towed Artillery. 21 pp.

—Atmospheric Corrosion Testing
BSI BS 2011: Part 2.1KC-91. 1991 Environmental Testing Part 2.1: Tests Test Kc and Guidance. Sulphur Dioxide Test for Contacts and Connections. 21 pp.
BSI BS 2011: Part 2.1KD-77. 1977 Basic Environmental Testing Procedures Part 2.1: Tests Part 2.1KD: Test Kd. Hydrogen Sulfide Test for Contacts and Connections. 8 pp.
BSI BS 2011: Part 2.2KD-84. 1984 Basic Environmental Testing Procedures Part 2.2: Guidance Part 2.2KD: Test Kd. Guidance on Test Kd: Hydrogen Sulfide Test for Contacts and Connections. 12 pp.
CENELEC HD 323.2.46 S1-88. Basic Environmental Testing Procedures—Part 2: Tests. Guidance to Test Kd: Hydrogen Sulphide Test for Contacts and Connections. 3 pp.
IEC 68 Pt 2-42-82. Basic Environmental Testing Procedures Part 2: Tests Test Kc: Sulphur Dioxide Test for Contacts and Connections Second Edition. 18 pp.
IEC 68 Pt 2-43-76. Basic Environmental Testing Procedures Part 2: Tests Test Kd: Hydrogen Sulphide Test for Contacts and Connections First Edition. 16 pp.
IEC 68 Pt 2-46-82. Basic Environmental Testing Procedures Part 2: Tests Guidance to Test Kd: Hydrogen Sulphide Test for Contacts and Connections First Edition. 27 pp.
IEC 68 Pt 2-49-83. Basic Environmental Testing Procedures Part 2: Tests Guidance to Test Kc: Sulphur Dioxide Test for Contacts and Connections First Edition. 23 pp.
IEC 68 Pt 2-60 TTD-89. Environmental Testing Part 2: Tests—Test Ke: Corrosion Tests in Artificial Atmosphere at Very Low Concentration of Polluting Gas(es). 25 pp.
SAA AS 1099.2KC-81. Basic Environmental Testing Procedures for Electrotechnology—Part 2: Tests—Part 2Kc: Sulphur Dioxide Test for Contacts and Connections. 9 pp.
SAA AS 1099.2KD-81. Basic Environmental Testing Procedures for Electrotechnology—Part 2: Tests—Part 2Kd: Hydrogen Sulphide Test for Contacts and Connections. 6 pp.
SNZ IEC 68: Part 2-42-82. Basic Environmental Testing Procedures Part 2-42: Test Kc: Sulphur Dioxide Test for Controls and Connections. 14 pp.
SNZ IEC 68: Part 2-43-76. Basic Environmental Testing Procedures Part 2-43: Test Kd: Hydrogen Sulphide Test for Contacts and Connections. 14 pp.
SNZ IEC 68: Part 2-46-82. Basic Environmental Testing Procedures Part 2-46: Guidance to Test Kd: Hydrogen Sulphide Test for Contacts and Connections. 23 pp.
SNZ IEC 68: Part 2-49-83. Basic Environmental Testing Procedures Part 2-49: Guidance to Test Kc: Sulphur Dioxide for Contacts and Connections. 19 pp.

—Automotive
BSI BS EN 24165-92. 1992 Road Vehicles—Electrical Connections—Double-Pole Connector (ISO 4165: 1979). 8 pp.
CEN EN 24165-91. Road Vehicles—Electrical Connections—Double Pole Connector. 3 pp.
CNS D2154-88. Multi-Connectors for Automotives (Jan)(9381).
CNS D3141-88. Method of Test for Multi-Connectors for Automobiles (Jan)(9382).
ISO 4165-79. Road Vehicles—Electrical Connections—Double-Pole Connector First Edition. 3 pp.

—Automotive—Radio Receivers
ISO 10487 Pt 1-92. Passenger Car Radio Connections—Part 1: Dimensions and General Requirements First Edition. 6 pp.

—Autotransformers
BSI BS 171: Part 4-78. 1978 Power Transformers Part 4: Tappings and Connections. 18 pp.

Connectors (Cont.)

—Auxiliary Power Units—Combat Vehicles
NATO STANAG 4074 ED 1 AMD 3-69. Auxiliary Power Unit Connections for Starting Combat and Tactical Vehicles. 11 pp.
NATO STANAG 4074 ED 2 AMD 0-00. (Draft) Auxilliary Power Unit Connections for Starting Tactical Land Vehicles. 15 pp.

—Batteries
IEC 130 Pt 3-65. Connectors for Frequencies Below 3 MHz (Mc/s) Part 3: Battery Connectors First Edition; (Corrigendum—Jan 1993). 27 pp.

—Battery Terminals—Automotive
BSI BS AU 16A-89. 1989 Electrical Cable Connectors (17.5 A Maximum Continuous Rating). 8 pp.
BSI BS AU 91-65. 1965 Dimensions of Electrical Cable Connectors for Starter Battery Taper Terminal Posts (Cars and Light Commercial Vehicles Using Positive Earth System). 2 pp.

—Bend Testing
CNS C6145-87. Method of Test for Low Frequency (Below 3 MHz) Electrical Connectors (TP-41 Circular Jacket Cable Flexing Test) (Oct)(8223).
CNS C6183-88. Method of Test for Low Frequency (Below 3 MHz) Electrical Connectors (TP-43 Bending Moment Test Connector) (May)(9243).

—Brakes
CNS D1018-82. Mounting Method of Brake Line and Electrical Connectors Between Truck Tractors and Trailers (Jan)(8237).
ISO 7638-85. Road Vehicles—Brake Anti-Lock Device Connector First Edition. 13 pp.
JIS D 6604-73. Mounting Method of Brake Line and Electrical Connectors Between Truck Tractors and Trailers (R 1983). 5 pp.

—Burglar Alarms
BSI BS 4166-67. (WITHDRAWN) 1967 Amd 2 Automatic Intruder Alarm Terminating Equipment in Police Stations (AMD 5802) May 31, 1989. 28 pp.

—Busbars
BSI BS 159-92. 1992 High-Voltage Busbars and Busbar Connections. 16 pp.
BSI BS 159-57. 1957 Amd 3 Busbars and Busbar Connections. 37 pp.

—Bushings—Aircraft
SBAC ESC 53 ISSUE 3. Connector Accessory: Bush (Cable Support). 3 pp.

—Classification
CNS C6189-82. Method of Test for Low Frequency (Below 3MHz) Electrical Connectors (TP-34 Classification) (Sep)(9368).

—Closed End
CNS C4267-80. Insulated Closed-End Connectors (Dec)(6766).
JIS C 2807-92. Insulated Closed-End Connectors. 18 pp.

—Communication Equipment
CSA CAN/CSA-C22. 2NO182.4-M90. Plugs, Receptacles, and Connectors for Communication Systems; (Gen Instr 1 Thru 3). 33 pp.

—Compression Testing
CNS C6144-87. Method of Test for Low Frequency (Below 3 MHz) Electrical Connectors (TP-40 Crush Test) (Oct)(8222).

—Contact Force
CNS C6141-86. Method of Test for Low Frequency (Below 3 MHz) Electrical Connectors (TP-29 Contact Retention Test) (Sep)(8219).
CNS C6181-82. Method of Test for Low Frequency (Below 3MHz) Electrical Connectors (TP-37 Contact Separation Force) (Aug)(9241).
CNS C6185-88. Method of Test for Low Frequency (Below 3 MHz) Electrical Connectors (TP-5 Contact Insertion and Removal Forces) (Apr)(9364).
CNS C6187-88. Method of Test for Low Frequency (Below 3 MHz) Electrical Connectors (TP-13 Mating and Unmating Forces) (Apr)(9366).
CNS C6190-88. Method of Test for Low Frequency (Below 3 MHz) Electrical Connectors (TP-35 Insert Retention) (Apr)(9369).

—Contact Resistance
CNS C6186-88. Method of Test for Low Frequency (Below 3 MHz) Electrical Connectors (TP-6 Contact Resistance) (Apr)(9365).

—Continuity Testing
CNS C6255-86. Method of Test for Low Frequency (Below 3MHz) Electrical Connectors (TP-46 Continuity Test) (May)(11572).

INTERNATIONAL AND NON-U.S. NATIONAL STANDARDS
SUBJECT INDEX
Connectors

Connectors *(Cont.)*

—**Controlgear—Low Voltage—Identification Systems**
CENELEC EN 50 043-85. Low-Voltage Switchgear and Controlgear for Industrial Use: Size Numbers and Gauges for Flat Connections. 6 pp.

—**Copper Conductors**
CSA C22.2 NO 65-93. Wire Connectors; (Gen Instr 1). 67 pp.

—**Couplers—Aerospace—Endurance Testing**
CEN PREN 2591 (Part CG1)-92. Elements of Electrical and Optical Connection Test Methods Part CG1—Endurance Test. 4 pp.

—**Covers—Aerospace—Mechanical Testing**
CEN PREN 2591 (Part D27)-92. Elements of Electrical and Optical Connection Test Methods Part D27—Robustness of Protective Cover Attachment. 4 pp.

—**Crimp Contacts**
CNS C4185-90. Non-Insulated Crimp-Style Connecting Sleeves for Copper Conductors (Dec)(5518).
MOD UK DSTAN 59-71-01. Crimped Electrical Connections for Copper Conductors Generic Specification Issue 1; Amendment 1. 138 pp.
MOD UK DSTAN 59-71: Part 0-85. Crimped Electrical Connections Part 0: Guide to the Crimping of Electrical Connections for Low Frequency and Radio Frequency Applications Issue 1. 15 pp.

—**Crimp Contacts—Aerospace—Deformation**
CEN PREN 2591 (Part E3)-92. Elements of Electrical and Optical Connection Test Methods Part E3—Contact Deformation After Crimping. 4 pp.

—**Crimp Contacts—Aerospace—Tensile Testing**
CEN PREN 2591 (Part D17)-92. Elements of Electrical and Optical Connection Test Methods Part D17—Tensile Strength (Crimped Connection). 4 pp.

—**Crimp Contacts—Aircraft**
SBAC TS 355 ISSUE 1. Test for Electrical Connectors Contact Deformation After Crimping.

—**Crimp Contacts—Aircraft—Identification Systems**
ISO 8843-91. Aircraft—Crimp-Removable Contacts for Electrical Connectors—Identification System First Edition. 5 pp.

—**Crimping Tools**
BSI BS 5310: Part 1-76. 1976 Hand Crimping Tools for the Termination of Electrical Cables and Wires for Low Frequency and Radio Frequency Applications Part 1: General Requirements and Tests. 7 pp.

—**Current Cycling Testing**
CNS C6258-86. Method of Test for Low Frequency (Below 3MHz) Electrical Connectors (TP-55 Current Cycling Test) (May)(11575).

—**Data Processing Equipment**
DIN VDE 0627-86. Connectors and Plug-And-Socket Devices for Rated Voltages up to 1000 V a.c. and 1200 V d.c. and Rated Currents up to 500 A for Each Pole (June). 36 pp.

—**Defibrillators**
CSA CAN/CSA-C22. 2NO601.1-M90. Medical Electrical Equipment, Part 1: General Requirements for Safety; (Gen Instr 1). 240 pp.

—**Drawing**
IEC 130 Pt 0-70. Connectors for Frequencies Below 3 MHz Part 0: Guide to Drawing Information in Detail Specifications First Edition; (Errata—June 1971). 18 pp.

—**Dust**
CNS C6256-86. Method of Test for Low Frequency (Below 3MHz) Electrical Connectors (TP-50 Sand and Dust Test) (May)(11573).

—**Electric Conductors**
CSA 1122A Bull. Electrical Bulletin 1122A January 10, 1979 to C22.2 NO 65. 3 pp.
CSA C22.2 NO 188-M1983. Splicing Wire and Cable Connectors (R 1989); (Gen Instr 1 Thru 2). 38 pp.
CSA 1122A Bull. Electrical Bulletin 1122A January 10, 1979 to C22.2 NO 188. 3 pp.
CSA 1122B Bull. Electrical Bulletin 1122B September 30, 1985 to C22.2 NO 188. 1 p.

—**Electric Corona**
CNS C6254-86. Method of Test for Low Frequency (Below 3MHz) Electrical Connectors (TP-44 Corona Test) (May)(11571).

Connectors *(Cont.)*

—**Electric Vehicles**
BSI BS 3214-74. 1974 Amd 3 Locking Connectors for Battery Operated Vehicles, Excluding Industrial Trucks, (320 Ampere Rating) (AMD 6125) July 31, 1989. 14 pp.

—**Electric Wiring**
CNS C1075-85. General Rules on Wire Connectors for Interior Wiring (Jun)(6768).
JIS C 2810-84. General Rules on Wire Connectors for Interior Wiring. 21 pp.

—**Electrical Codes**
CSA 1165 Bull. Electrical Bulletin 1165 April 12, 1978 to C22.2 NO 4. 2 pp.
CSA 932 Bull. Electrical Bulletin 932 January 21, 1974 to C22.2 NO 65. 2 pp.
CSA 1008 Bull. Electrical Bulletin 1008 June 24, 1975 to C22.2 NO 65. 1 p.
CSA 1165 Bull. Electrical Bulletin 1165 April 12, 1978 to C22.2 NO 65. 2 pp.

—**Electrical Flexible Tubing**
CSA CAN/CSA-C22. 2 NO 18-92. Outlet Boxes, Conduit Boxes, and Fittings; (Gen Instr 1 Thru 2). 118 pp.

—**Electronic Equipment**
CNS C5024-86. General Rules of Connectors for Electronic Equipment (Jun)(4736).
CNS C6032-86. Method of Test for Connectors for Electronic Equipment (Jun)(4737).
JIS C 5401-91. General Rules of Connectors for Use in Electronic Equipment. 18 pp.
JIS C 5402-92. Method for Test of Connectors for Use in Electronic Equipment. 124 pp.

—**EMI Shielded—Aerospace—Shielding Testing**
CEN PREN 2591 (Part B13)-92. Elements of Electrical and Optical Connection Test Methods Part B13—Shielding Effectiveness From 100 MHz to 1 GHz. 6 pp.

—**Engine Starters—Aircraft**
NATO STANAG 3372 ED 5 AMD 2-88. Low Pressure Air and Associated Electrical Connections for Aircraft Engine Starting. 10 pp.
NATO STANAG 3372 ED 5 AMD 3-88. Low Pressure Air and Associated Electrical Connections for Aircraft Engine Starting. 11 pp.

—**Entertainment Equipment—Portable**
IEC 130 Pt 10-71. Connectors for Frequencies Below 3 MHz Part 10: Connectors for Coupling an External Low-Voltage Power Supply to Portable Entertainment Equipment First Edition. 21 pp.

—**Fall Arresting Devices**
BSI BS EN 362-93. 1993 Personal Protective Equipment Against Falls from a Height—Connectors (Supersedes BS 1397: 1979 & BS 5062: Part 1: 1985) (N). 10 pp.
CEN PREN 362-90. Personal Fall Arresting Systems: Connectors. 4 pp.
CEN EN 362-92. Personal Protective Equipment Against Falls from a Height—Connectors. 4 pp.

—**Flexible Metal Conduits**
CSA CAN/CSA-C22. 2 NO 18-92. Outlet Boxes, Conduit Boxes, and Fittings; (Gen Instr 1 Thru 2). 118 pp.

—**Flexural Strength**
CSA C22.2 NO 188-M1983. Splicing Wire and Cable Connectors (R 1989); (Gen Instr 1 Thru 2). 38 pp.

—**Fuel Pumps—Automotive**
ISO 9534-89. Road Vehicles—Fuel Pump Electric Connections First Edition. 7 pp.

—**Gas Discharge Tubes**
CSA C22.2 NO 34-M1987. Electrode Receptacles, Fittings, and Connectors for Gas Tubes (R 1993); (Gen Instr 1 Thru 2). 23 pp.

—**Gas Inlet—Medical Electrical—Construction**
CSA CAN/CSA-C22. 2NO601.1-M90. Medical Electrical Equipment, Part 1: General Requirements for Safety; (Gen Instr 1). 240 pp.

—**Ground Support Equipment—Aircraft**
BSI 4G 173: Part 1-85. 1985 Connectors for Ground Electrical Supplies for Aircraft Part 1: Design, Performance and Test Requirements. 6 pp.
BSI 4G 173: Part 2-85. 1985 Connectors for Ground Electrical Supplies for Aircraft Part 2: Dimensions. 9 pp.
ISO 461 Pt 1-85. Aircraft—Connectors for Ground Electrical Supplies—Part 1: Design, Performance and Test Requirements First Edition. 6 pp.

Connectors *(Cont.)*

—**Ground Support Equipment—Aircraft** *(Cont.)*
ISO 461 Pt 2-85. Aircraft—Connectors for Ground Electrical Supplies—Part 2: Dimensions First Edition. 9 pp.
NATO STANAG 3917 ED 2 AMD 2-84. Air Conditioning Connections, Ground Half Connectors. 9 pp.
NATO STANAG 3917 ED 2 AMD 3-84. Air Conditioning Connections, Ground Half Connectors. 10 pp.

—**Ground Support Equipment—Electrical Grounding**
NATO STANAG 3632 ED 3 AMD 2-86. Electrical Safety Connections for Aircraft and Ground Support Equipment. 12 pp.
NATO STANAG 3632 ED 4 AMD 0-91. Aircraft and Ground Support Equipment Electrical Connections for Static Grounding. 9 pp.
NATO STANAG 3632 ED 4 AMD 1-91. Aircraft and Ground Support Equipment Electrical Connections for Static Grounding. 10 pp.

—**Grounding Devices—Aerospace—Contact Force**
CEN PREN 2591 (Part D13)-92. Elements of Electrical and Optical Connection Test Methods Part D13—Holding Force of Grounding Spring System. 4 pp.

—**Hazardous Environments**
JIS C 4501-77. Explosion-Proof Type Plug Connection Devices for Coal Mines (R 1985). 18 pp.

—**Hermetic Seals—Aerospace**
CEN PREN 2591 (Part C22)-92. Elements of Electrical and Optical Connection Test Methods Part C22—Hermeticity. 4 pp.

—**Hermetic Seals—Aerospace—Thermal Shock**
CEN PREN 2591 (Part C23)-92. Elements of Electrical and Optical Connection Test Methods Part C23—Thermal Shock. 5 pp.

—**Hermetic Seals—Aircraft**
SBAC TS 347 ISSUE 1. Test for Electrical Connectors Hermeticity.

—**Humidity**
CNS C6142-86. Method of Test for Low Frequency (Below 3 MHz) Electrical Connectors (TP-31 Humidity Test) (Sep)(8220).

—**Hydraulic Equipment**
BSI BS 5630-78. 1978 Interface Dimensions of Snap-On Connections for Use with Electrically Controlled Hydraulic Equipment. 4 pp.

—**Hydrostatic Testing**
CNS C6192-88. Method of Test for Low Frequency (Below 3 MHz) Electrical Connectors (TP-39 Hydrostatic Test) (Apr)(9371).

—**Ice Resistance**
CNS C6257-86. Method of Test for Low Frequency (Below 3MHz) Electrical Connectors (TP-51 Ice Resistance of Mated Connectors Test) (May)(11574).

—**Identification Systems**
CSA C22.2 NO 65-93. Wire Connectors; (Gen Instr 1). 67 pp.
CSA 1122A Bull. Electrical Bulletin 1122A January 10, 1979 to C22.2 NO 65. 3 pp.
CSA C22.2 NO 188-M1983. Splicing Wire and Cable Connectors (R 1989); (Gen Instr 1 Thru 2). 38 pp.
SNZ NZS 73-62. Specification for the Marking and Arrangement of Switchgear, Busbars, Main Connections and Small Wiring (Reconfirmed 1979) Amend: 1, 1971. 18 pp.

—**Immersion Testing**
CNS C6177-88. Method of Test for Low Frequency (Below 3 MHz) Electrical Connector (TP-3 Altitude Immersion) (May)(9237).

—**Impact Testing**
CNS C6182-88. Method of Test for Low Frequency (Below 3 MHz) Electrical Connectors (TP-42 Impact) (May)(9242).

—**Indexes (Documentation)**
SBAC AS (V). AS Index for High Temperature Electrical Connectors. 32 pp.
SBAC AG (V). AGS Index for High Temperature Electrical Connectors. 13 pp.
SBAC TS (V). TS Index for High Temperature Electrical Connectors. 6 pp.
SBAC ESC (V). ESC Index for High Temperature Electrical Connectors. 1 p.

INDUSTRY STANDARDS

Connectors

Connectors (Cont.)

—Industrial

BSI BS 4343-92. 1992 Plugs, Socket-Outlets and Couplers for Industrial Purposes Part 2: Dimensional Interchangeability Requirements for Pin and Contact-Tube Accessories of Harmonized Configurations (E). 74 pp.

BSI BS 4343-01. 1992 Amd 1 Plugs, Socket-Outlets and Couplers for Industrial Purposes Part 2: Dimensional Interchangeability Requirements for Pin and Contact-Tube Accessories of Harmonized Configurations (AMD 7890) August 15, 1993 (E). 75 pp.

BSI BS 4343-68. 1968 Amd 1 Industrial Plugs, Socket-Outlets and Couplers for a.c and d.c. Supplies. 139 pp.

BSI BS EN 60309-1-92. 1992 Plugs, Socket-Outlets and Couplers for Industrial Purposes Part 1: General Requirements. 67 pp.

BSI BS EN 60309-1-01. 1992 Amd 1 Plugs, Socket-Outlets and Couplers for Industrial Purposes Part 1: General Requirements (AMD 7889) August 15, 1993 (E). 68 pp.

BSI BS EN 60309-2-92. 1992 Plugs, Socket-Outlets and Couplers for Industrial Purposes Part 2: Dimensional Interchangeability Requirements for Pin and Contact-Tube Accessories of Harmonized Configurations (E). 74 pp.

BSI BS EN 60309-2-01. 1992 Amd 1 Plugs, Socket-Outlets and Couplers for Industrial Purposes Part 2: Dimensional Interchangeability Requirements for Pin and Contact-Tube Accessories of Harmonized Configurations (AMD 7890) August 15, 1993 (E). 75 pp.

CENELEC HD 196-75. Plugs, Sockets—Outlets and Coupless for Industrial Purposes. 7 pp.

CENELEC EN 60309-1-92. Plugs, Socket-Outlets and Couplers for Industrial Purposes Part 1: General Requirements. 8 pp.

CENELEC EN 60309-2-92. Plugs, Socket-Outlets and Couplers for Industrial Purposes Part 2: Dimensional Interchangeability Requirements for Pin and Contact-Tube Accessories. 7 pp.

DIN VDE 0623 & 0623A-77. Regulations for Plugs, Socket-Outlets, Couplers and Connectors for Industrial Purposes up to 200 A and up to 750 V (Mar). 68 pp.

IEC 309 Pt 1-88. Plugs, Socket-Outlets and Couplers for Industrial Purposes Part 1: General Requirements Second Edition; (Corrigendum—March 1992). 110 pp.

IEC 309 Pt 2-89. Plugs, Socket-Outlets and Couplers for Industrial Purposes Part 2: Dimensional Interchangeability Requirements for Pin and Contact-Tube Accessories Second Edition; (Corrigendum—April 1992). 95 pp.

—Inspection

CNS C6180-88. Method of Test for Low Frequency (Below 3 MHz) Electrical Connectors (TP-18 Visual and Dimensional Inspection) (May)(9240).

—Insulation Resistance

CNS C6130-86. Method of Test for Low Frequency (Below 3 MHz) Electrical Connectors (TP-21 Insulation Resistance) (Dec)(7660).

—Integrated Services Digital Networks

CENELEC ENV 41 001-87. ISDN Connector up to 8 Pins and up to 2,048 M bit/s. 59 pp.

IEC 10173-91. Information Technology—Integrated Services Digital Network (ISDN) Primary Access Connector at Reference Points S and T First Edition. 15 pp.

ISO 8877-92. Information Technology—Telecommunications and Information Exchange Between Systems—Interface Connector and Contact Assignments for ISDN Basic Access Interface Located at Reference Points S and T Second Edition. 15 pp.

SNZ NZS/AS 4102-93. Information Technology—Integrated Services Digital Network Primary Access Connector at Reference Points S and T (This is a Joint Standard with SAA AS 4102). 12 pp.

SNZ NZS/ISO 8877-87. Information Processing Systems—Interface Connector and Contact Assignments for ISDN Basic Access Interface Located at Reference Points S and T. 9 pp.

—Intervalometers—Rocket Launchers

NATO STANAG 3576 ED 3 AMD 0-91. Electrical Connector for Dispensers and Internal Intervalometer Type Rocket Launchers for Aircraft. 6 pp.

—Leakage

CNS C6176-88. Method of Test for Low Frequency (Below 3 MHz) Electrical Connector (TP-2 Air Leakage) (May)(9236).

—Life (Durability)

CNS C6131-87. Method of Test for Low Frequency (Below 3 MHz) Electrical Connectors (TP-22 Life Test) (Oct)(7661).

Connectors (Cont.)

—Life (Durability) (Cont.)

CNS C6178-88. Method of Test for Low Frequency (Below 3 MHz) Electrical Connectors (TP-9 Durability) (May)(9238).

CNS C6179-82. Method of Test for Low Frequency (Below 3MHz) Electrical Connectors (TP-17 Temperature Life) (Aug)(9239).

—Link

IEC 130 Pt 12-76. Connectors for Frequencies Below 3 MHz Part 12: Link and Test Connectors Second Edition. 42 pp.

—Local Area Networks

BSI BS ISO/IEC TR 9578-90. 1990 Information Technology—Communication Interface Connectors Used in Local Area Networks. 49 pp.

IEC TR9578-90. Information Technology—Communication Interface Connectors Used in Local Area Networks First Edition. 47 pp.

ISO TR9578-90. Information Technology—Communication Interface Connectors Used in Local Area Networks First Edition. 47 pp.

JTC1 TR9578-90. Information Technology—Communication Interface Connectors Used in Local Area Networks First Edition. 47 pp.

OSI ISO/IEC TR 9578-90. Information Technology—Communication Interface Connectors Used in Local Area Networks. 47 pp.

SAA AS 4028-92. Information Technology—Communication Interface Connectors Used in Local Area Networks (ISO/IEC/TR 9578:1990) (in Professional Package 26A). 42 pp.

—Lockwire Holes—Aircraft

SBAC TS 375 ISSUE 1. Locking Wire Hole Strength.

—Low Profile—Pacemakers

ISO 5841 Pt 3-92. Cardiac Pacemakers—Part 3: Low-Profile Connectors (IS-1) for Implantable Pacemakers First Edition. 14 pp.

—Marine

CNS F5111-86. Minimum Length Between the Cable Entrances and the Terminals of Electric Machinery and Equipment for Marine Use (Dec)(11794).

—Marine—Identification Systems

CNS F5114-87. Markings and Arrangements of Main Connections and Small-Wiring for Marine Use (Jan)(11816).

—Mechanical Shock

CNS C6139-86. Method of Test for Low Frequency (Below 3MHz) Electrical Connectors (TP-27 Mechanical Shock Pulse) (Sep)(8217).

—Mechanical Testing

BSI BS 5772: Part 8-85. 1985 Amd 1 Basic Testing Procedures and Measuring Methods for Electromechanical Components for Electronic Equip. Part 8: Connector Tests (Mechanical) and Mech. Tests on Contacts and Terminations (AMD 5076) May 31, 1990. 32 pp.

IEC 512 Pt 8-93. Electromechanical Components for Electronic Equipment; Basic Testing Procedures and Measuring Methods Part 8: Connector Tests (Mechanical) and Mechanical Tests on Contacts and Terminations Third Edition. 72 pp.

—Microscope Tubes

ISO 8040-86. Optics and Optical Instruments—Microscopes—Connecting Dimensions of Tube Slides and Tube Slots First Edition. 4 pp.

—Motor Starters

BSI BS AU 233-89. 1989 Starter Motor Electrical Connections for Passenger Cars. 4 pp.

ISO 9458-88. Passenger Cars—Starter Motor Electrical Connections First Edition. 4 pp.

—Mounting—Automotive

BSI BS AU 201: Part 1-85. 1985 Relays and Flashers Part 1: Mounting and Positioning Dimensions of Male Tabs and Socket Apertures for Relays and Flashers for Road Vehicles. 5 pp.

ISO 7588-83. Road Vehicles—Relays and Flashers—Mounting and Positioning Dimensions of Male Tabs and Socket Apertures for Relays and Flashers First Edition. 5 pp.

—Overhead Line Conductors

CSA C57-1966. Electric Power Connectors for Use in Overhead Line Conductors. 27 pp.

—Ozone Resistance

CNS C6188-88. Method of Test for Low Frequency (Below 3 MHz) Electrical Connectors (TP-14 Ozone Exposure) (Apr)(9367).

Connectors (Cont.)

—Power Cables

IEC 1238 Pt 1-93. Compression and Mechanical Connectors for Power Cables with Copper or Aluminium Conductors Part 1: Test Methods and Requirements First Edition. 65 pp.

—Power Line Carriers

DIN VDE 0850-80. Coupling Devices for Power Line Carrier Systems (PLC Systems) (VDE Specification) (Mar). 33 pp.

IEC 481-74. Coupling Devices for Power Line Carrier Systems First Edition. 28 pp.

—Power Lines

DIN VDE 0220 Pt 2-71. Pressure-Connectors in Heavy-Current Cable Installations (Nov). 20 pp.

DIN VDE 0220 Pt 2A-78. Regulations for Press Connectors in Heavy-Current Cable Installations (Apr). 7 pp.

—Power Supplies—Aircraft

NATO STANAG 3302 ED 5 AMD 4-81. Connectors for 28 Volt "DC" Servicing Power. 14 pp.

NATO STANAG 3302 ED 5 AMD 5-81. Connectors for 28 Volt "DC" Servicing Power. 16 pp.

NATO STANAG 3303 ED 3 AMD 2-86. Connectors for 115/200 Volts, 400 Hertz, 3 Phase, AC Servicing Power. 14 pp.

NATO STANAG 3303 ED 3 AMD 3-86. Connectors for 115/200 Volts, 400 Hertz, 3 Phase, AC Servicing Power. 15 pp.

NATO STANAG 3303 ED 3 AMD 4-86. Connectors for 115/200 Volts, 400 Hertz, 3 Phase, AC Servicing Power. 16 pp.

—Power Transformers

BSI BS 171: Part 4-78. 1978 Power Transformers Part 4: Tappings and Connections. 18 pp.

CENELEC HD 398.4-79. Power Transformers Part 4: Tappings and connections. 5 pp.

—Preferred Products List

CECC CECC MUAHAG Vol 3A IS 4-91. Preferred Products List; Connectors; L. F. (En, Fr, Ge). 46 pp.

CECC CECC MUAHAG Vol 3B IS 4-91. Preferred Products List; Connectors; R.F. and Fibre Optics (En, Fr, Ge). 65 pp.

—Prime Movers (Vehicles)

NATO STANAG 4007 ED 2 AMD 0-92. (Draft) Electrical Connectors Between Prime Movers, Trailers and Towed Artillery. 17 pp.

NATO STANAG 4007 ED 2 AMD 0-00. Electrical Connectors Between Prime Movers, Trailers and Towed Artillery. 21 pp.

—Probes—Aerospace—Damage Testing

CEN PREN 2591 (Part D15)-92. Elements of Electrical and Optical Connection Test Methods Part D15—Test Probe Damage (Female Contact). 5 pp.

—Probes—Damage Testing

CNS C6134-87. Method of Test for Low Frequency (Below 3 MHz) Electrical Connectors (TP-25 Probe Damage Test) (Oct)(7664).

—Proximity Switches—Identification Systems

BSI BS 6519-84. 1984 Low Voltage Switchgear and Controlgear for Industrial Use. Inductive Proximity Switches. Identification of Connections. 5 pp.

—Proximity Switches—Low Voltage—Identification Systems

CENELEC EN 50 044-81. Low Voltage Switchgear and Controlgear for Industrial Use: Inductive Proximity Switches: Identification of Connections. 3 pp.

—Pullout Testing

CNS C6191-82. Method of Test for Low Frequency (Below 3 MHz) Electrical Connectors (TP-38 Cable Pull Out) (Sep)(9370).

—Quality Assurance

BSI BS 9520-83. 1983 Amd 3 Electrical Connectors of Assessed Quality for d.c. and Low Frequency Application: Generic Data, Methods of Test and Capability Approval Procedures (AMD 6555) April 30, 1991. 68 pp.

BSI BS 9520-04. 1983 Amd 4 Electrical Connectors of Assessed Quality for d.c. and Low Frequency Application: Generic Data, Methods of Test and Capability Approval Procedures (AMD 6329) April 1, 1992 (T). 88 pp.

BSI BS 9520-05. 1983 Amd 5 Electrical Connectors of Assessed Quality for d.c. and Low Frequency Application: Generic Data, Methods of Test and Capability Approval Procedures (AMD 7418) February 15, 1993. 89 pp.

Connectors (Cont.)

—Quality Assurance (Cont.)
BSI BS 9530-88. (OBSOLESCENT) 1988 Cable Fitting Accessories of Assessed Quality for Circular Electrical Connectors: Generic Data, Methods of Test and Rules for the Preparation of Detail Specifications. 29 pp.

BSI BS 9550-89. (OBSOLESCENT) 1989 Capability Approval Procedures for d.c. and Low Frequency Connector Cable Assemblies and Wiring Harnesses: Generic Data. 24 pp.

—Quick Disconnect
MOD UK DSTAN 59-35: Pt 3:Sec 5-01. Connectors, Electrical Part 3: Connectors for d.c. and Low Frequency Applications (See Also EPIC Database) Section 5: Pattern 9522 F0038 List of Items Conforming to BS 9522 F0038 Issue 1; Amendment 1. 39 pp.

MOD UK DSTAN 59-35: Pt 3:Sec 7-01. Connectors, Electrical Part 3: Connectors for d.c. and Low Frequency Applications (See Also EPIC Database) Section 7: Pattern 9522 F0017 List of Items Conforming to BS 9522 F0017 Issue 1; Amendment 1. 70 pp.

—Quick Disconnect—Aerospace—Contact Force
CEN PREN 2591 (Part D14)-92. Elements of Electrical and Optical Connection Test Methods Part D14—Unmating of Lanyard Release Connectors. 4 pp.

—Quick Disconnect—Aircraft
MOD UK DSTAN 59-13-01. Butting Connectors Issue 1; Amendment 2. 7 pp.

—Radio Equipment
IEC 130 Pt 2-65. Connectors for Frequencies Below 3 MHz Part 2: Connectors for Radio Receivers and Associated Sound Equipment Second Edition; (Amendment 1-1969). 38 pp.

IEC 130 Pt 8-76. Connectors for Frequencies Below 3 MHz Part 8: Concentric Connectors for Audio Circuits in Radio Receivers Second Edition. 13 pp.

—Radio Receivers
IEC 130 Pt 2-65. Connectors for Frequencies Below 3 MHz Part 2: Connectors for Radio Receivers and Associated Sound Equipment Second Edition; (Amendment 1-1969). 38 pp.

—Receptacles
BSI BS 91-73. 1973 Electric Cable Soldering Sockets. 13 pp.

—Safety Belts
BSI BS EN 362-93. 1993 Personal Protective Equipment Against Falls from a Height—Connectors (Supersedes BS 1397: 1979 & BS 5062: Part 1: 1985) (N). 10 pp.

CEN PREN 362-90. Personal Fall Arresting Systems: Connectors. 4 pp.

CEN EN 362-92. Personal Protective Equipment Against Falls from a Height—Connectors. 4 pp.

—Salt Spray Testing
CNS C6138-86. Method of Test for Low Frequency (Below 3MHz) Electrical Connectors (TP-26 Salt Spray Corrosion Test) (Sep)(8216).

—Sand
CNS C6256-86. Method of Test for Low Frequency (Below 3MHz) Electrical Connectors (TP-50 Sand and Dust Test) (May)(11573).

—Sealing
SBAC TS 349 ISSUE 1. Test for Electrical Connectors Interfacial Sealing.

—Service Entrance Cables
CSA CAN/CSA-C22. 2 NO 18-92. Outlet Boxes, Conduit Boxes, and Fittings; (Gen Instr 1 Thru 2). 118 pp.

—Ships
MOD UK NES 512: Part 10-01. Guide to Cables, Electrical and Associated Items Part 10: Terminations and Connectors Issue 2 (03.91); Amendment 1. 51 pp.

MOD UK NES 514-81. Guide to Cable Entry, Termination and Junction Components for Equipment Issue 2 (01.81). 101 pp.

MOD UK NES 514-92. Guide to Cable Entry, Termination and Junction Components for Equipment Issue 3 (04.92). 79 pp.

—Solder Contacts—Quality Assurance
BSI BS 9522 N0003-90. 1990 Multi-Contact Circular Electrical Connectors Bayonet Coupling Non-Barrier Sealed, Environment Resistant with Rear Insertable, Rear Release, Rear Removable Crimp Contacts Also Barrier Sealed with Non-Removable Solder Contact Styles. 142 pp.

—Soldering—Inspection
CNS C5039-80. Criteria for Inspection for Highly Reliable Soldered Connection in Electronic and Electrical Applications (Apr)(5429).

—Spark Plugs
ISO 3412-92. Road Vehicles—Screened and Waterproof Spark-Plugs and Their Connections—Types 1A and 1B Third Edition. 6 pp.

ISO 3895-86. Road Vehicles—Screened and Waterproof Spark-Plug and Its Connection—Type 2 Third Edition. 5 pp.

ISO 3896-86. Road Vehicles—Screened and Waterproof Spark-Plug and Its Connection—Type 3 Third Edition. 5 pp.

—Straight—Aerospace
CEN PREN 3660-004-91. Part 004—Detail Specification Sheet Cable Outlet, Style A, Straight, Unsealed, with Clamp Strain Relief. 11 pp.

CEN PREN 3660-005-91. Part 005—Detail Specification Sheet Cable Outlet, Style A, 90 Degree, Unsealed, with Clamp Strain Relief. 11 pp.

—Submarines
MOD UK NES 512: Part 10-01. Guide to Cables, Electrical and Associated Items Part 10: Terminations and Connectors Issue 2 (03.91); Amendment 1. 51 pp.

—Switchgear
BSI BS 6904-87. 1987 Guide for Cable Connections for Gas-Insulated Metal-Enclosed Switchgear for Rated Voltages of 72.5kV and Above. 12 pp.

IEC 859-86. Cable Connections for Gas-Insulated Metal-Enclosed Switchgear for Rated Voltages of 72.5 kV and Above First Edition. 25 pp.

—Switchgear—Low Voltage—Identification Systems
CENELEC EN 50 043-85. Low-Voltage Switchgear and Controlgear for Industrial Use: Size Numbers and Gauges for Flat Connections. 6 pp.

—Symbols—Diagrams
BSI BS 3939: Part 3-85. 1985 Graphical Symbols for Electrical Power, Telecommunications and Electronics Diagrams Part 3: Conductors and Connecting Devices (G). 11 pp.

IEC 617 Pt 3-83. Graphical Symbols for Diagrams Part 3: Conductors and Connecting Devices First Edition. 16 pp.

SAA AS 1102.103-89. Graphical Symbols for Electrotechnology—Part 103: Conductors and Connecting Devices. 10 pp.

SNZ IEC 617: Part 3-83. Graphical Symbols for Diagrams Part 3: Conductors and Connecting Devices. 13 pp.

—Television Equipment
BSI BS 5550: SUBSEC 7.5.3-81. 1981 Cinematography Part 7: Production and Presentation Section 7.5: Film and Television Locating Lighting Subsection 7.5.3: Plugs and Socket Connectors. 35 pp.

BSI BS 5550: SUBSEC 7.5.4-81. 1981 Cinematography Part 7: Production and Presentation Section 7.5: Film and Television Location Lighting Subsection 7.5.4: Single Pole High Current Plugs and Socket Connectors. 23 pp.

—Television Equipment—Identification Systems—Voltage Ratings
IEC 1062-91. Audiovisual Equipment and Systems—Rating Plates—Marking of Electricity Supply First Edition (ANSI IT7.105-1992). 12 pp.

—Television Receivers
IEC 246-67. Connecting Wires Having a Rated Voltage of 20 kV and 25 kV D.C. and a Maximum Working Temperature of 105 Degrees C for Use in Television Receivers First Edition. 17 pp.

—Tensile Testing
CNS C6128-86. Method of Test for Low Frequency (Below 3MHz) Electrical Connectors (TP-8 Crimp Tensile Strength) (Dec)(7658).

—Test
IEC 130 Pt 12-76. Connectors for Frequencies Below 3 MHz Part 12: Link and Test Connectors Second Edition. 42 pp.

—Thermocouples—Aircraft
ISO 8056 Pt 4-87. Aircraft—Nickel-Chromium and Nickel-Aluminium Thermocouple Extension Cables—Part 4: Crimp-Type Butt Connectors—Dimensions First Edition. 4 pp.

SBAC TS 353 ISSUE 1. Test for Electrical Connectors Thermocouple Contacts.

—Threaded
MOD UK DSTAN 59-35: Part 5-01. Plugs and Sockets, Electrical, Pattern 608, Sealed, with Soldered Terminations: Non-Sealed, with Crimped Terminations: Coarse Threaded Coupling, for d.c. and Low Frequency Applications Issue 2; Amendment 6. 108 pp.

—Torque
CSA C22.2 NO 188-M1983. Splicing Wire and Cable Connectors (R 1989); (Gen Instr 1 Thru 2). 38 pp.

—Towed Vehicles
BSI BS AU 194-84. 1984 Performance of Electrical Connections Between Towing Vehicles and Trailers. 4 pp.

BSI BS AU 195-84. (WITHDRAWN) 1984 Mounting Electrical Connections on Rear Cross Members of Towing Vehicles (Superseded by BS AU 195A: 1991). 4 pp.

BSI BS AU 195A-91. 1991 Mounting Electrical Connections on Rear Cross Members of Towing Vehicles (ISO 4009: 1989). 6 pp.

BSI BS AU 197-84. 1984 Electrical Connections Between Towing Vehicles and Trailers with 24V Equipment: Type 24N (Normal). 7 pp.

BSI BS AU 198-84. 1984 Electrical Connections Between Towing and Trailers with 24V Electrical Equipment: Type 24S (Supplementary). 7 pp.

ISO 4009-89. Towing Vehicles—Mounting of Electrical Connections on Rear Cross-Members Second Edition. 4 pp.

ISO 4091-92. Road Vehicles—Connectors for Electrical Connections Between Towing Vehicles and Trailers—Test Methods and Performance Requirements Second Edition. 10 pp.

—Towed Vehicles—Interchangeability
BSI BS AU 149A-80. 1980 Electrical Connections Between Towing Vehicles and Trailers with 6V or 12V Electrical Equipment: Type 12N (Normal). 6 pp.

BSI BS AU 177A-80. 1980 Amd 1 Electrical Connections Between Towing Vehicles and Trailers with 6V or 12V Electrical Equipment: Type 12S (Supplementary). 7 pp.

ISO 1185-75. Road Vehicles—Electrical Connections Between Towing Vehicles and Towed Vehicles with 24 V Electrical Equipment—Type 24 N (Normal) First Edition. 7 pp.

ISO 1724-80. Road Vehicles—Electrical Connections Between Towing Vehicles and Towed Vehicles with 6 or 12 V Electrical Equipment—Type 12 N (Normal) Second Edition. 6 pp.

ISO 3731-80. Road Vehicles—Electrical Connections Between Towing Vehicles and Trailers with 24 V Electrical Equipment—Type 24 S (Supplementary) Second Edition. 6 pp.

ISO 3732-82. Road Vehicles—Electrical Connections Between Towing Vehicles and Trailers with 6 or 12 V Electrical Equipment—Type 12 S (Supplementary) Second Edition. 6 pp.

—Tractor Trucks
NATO STANAG 4007 ED 1 AMD 7-59. Electrical Connectors Between Tractors and Trailers. 18 pp.

—Tractors—Towed Vehicles
EC 75/323/EEC-75. Council Directive on the Approximation of the Laws of the Member States Relating to the Power Connection Fitted on Wheeled Agricultural or Forestry Tractors for Lighting and Light-Signalling Devices on Tools, Machinery or Trailers Intended for Agriculture or Forestry. 2 pp.

—Trailers
NATO STANAG 4007 ED 1 AMD 7-59. Electrical Connectors Between Tractors and Trailers. 18 pp.

NATO STANAG 4007 ED 2 AMD 0-92. (Draft) Electrical Connectors Between Prime Movers, Trailers and Towed Artillery. 17 pp.

NATO STANAG 4007 ED 2 AMD 0-00. Electrical Connectors Between Prime Movers, Trailers and Towed Artillery. 21 pp.

SAA AS 2513-82. Electrical Connectors for Trailer Vehicles (This is a Joint Standard with SANZ NZS 2513). 16 pp.

SNZ NZS/AS 2513-82. Electrical Connectors for Trailer Vehicles (This is a Joint Standard with SAA AS 2513). 16 pp.

—Trucks
CNS D2118-86. Seven Conductor Electrical Connectors Between Truck Tractors and Trailers (Jun)(8239).

JIS D 6606-76. Seven Conductor Electrical Connectors Between Truck Tractors and Trailers (R 1983). 13 pp.

—Vehicles
CNS D1018-82. Mounting Method of Brake Line and Electrical Connectors Between Truck Tractors and Trailers (Jan)(8237).

INTERNATIONAL AND NON-U.S. NATIONAL STANDARDS
SUBJECT INDEX

Connectors

Connectors (Cont.)

—Vehicles (Cont.)
CNS D2118-86. Seven Conductor Electrical Connectors Between Truck Tractors and Trailers (Jun)(8239).
IEC 783-84. Wiring and Connectors for Electric Road Vehicles First Edition. 22 pp.
JIS D 6604-73. Mounting Method of Brake Line and Electrical Connectors Between Truck Tractors and Trailers (R 1983). 5 pp.
JIS D 6606-76. Seven Conductor Electrical Connectors Between Truck Tractors and Trailers (R 1983). 13 pp.

—Vibration
CNS C6140-86. Method of Test for Low Frequency (Below 3 MHz) Electrical Connectors (TP-28 Vibration Test Procedure) (Sep)(8218).

—Waterproof—Automotive
CNS D2155-88. Multi-Connectors Use in Electronics and Waterproof for Automotives (Jan)(9383).
CNS D3142-88. Method of Test for Multi-Connectors Use in Electronics and Waterproof for Automotives (Jan)(9384).

—Waterproof—Combat Vehicles
NATO STANAG 4040 ED 1 AMD 1-69. Waterproof Electrical Connectors. 19 pp.

—Withstand Voltage
CNS C6129-86. Method of Test for Low Frequency (Below 3 MHz) Electrical Connectors (TP-20 Withstanding Voltage Test) (Dec)(7659).

CONP
Use: Connected Name Identification Presentation

Conradson Carbon Testing
Use: Carbon Residue Testing

Conservancy

—Service Contracts
MOD UK DEFCON 112BF-88. Conditions of Contract Conservancy Services 4/88. 3 pp.

Consistency
See Also: Mobilometers; Reliability; Tolerances

—Concretes
ISO 4103-79. Concrete—Classification of Consistency First Edition. 4 pp.
ISO 4109-80. Fresh Concrete—Determination of the Consistency—Slump Test First Edition. 4 pp.
ISO 4110-79. Fresh Concrete—Determination of the Consistency—Vebe Test First Edition. 4 pp.
ISO 4111-79. Fresh Concrete—Determination of Consistency—Degree of Compactibility (Compaction Index) First Edition. 4 pp.
SAA AS 1012.3-83. Methods of Testing Concrete—Part 3: Methods for the Determination of Properties Related to the Consistence of Concrete. 23 pp.

—Enamels
CNS K6030-72. Method of Test for Enamel (Jun)(627). 7 pp.

—Hydraulic Cements
CNS R3075-88. Method of Test for Normal Consistency of Hydraulic Cement (Jun)((3590).

—Lacquers
SAA AS 1580.214. 1-90. Paints and Related Materials—Methods of Test—Part 214.1: Consistency—Stormer Viscometer Amdt 1 January/February 1992 Amdt 2 January/February 1993 (in Professional Packages 30, 39). 1 p.
SAA AS 1580.214. 2-90. Paints and Related Materials—Methods of Test—Part 214.2: Consistency—Flow Cup. 6 pp.
SAA AS 1580.214. 4-90. Paints and Related Materials—Methods of Test—Part 214.4: Consistency—Rotothinner. 4 pp.
SAA AS 1580.214. 5-90. Paints and Related Materials—Methods of Test—Part 214.5: Consistency—Ratational Viscometer. 10 pp.
SNZ NZS/AS 1580. 214.1-90. Methods of Test for Paints and Related Materials Part 214.1: Consistency—Stormer Viscometer Amend: 1, 1992 (This is a Joint Standard with SAA AS 1580.214.1). 6 pp.
SNZ NZS/AS 1580. 214.2-90. Methods of Test for Paints and Related Materials Part 214.2: Consistency—Flow Cup (This is a Joint Standard with SAA AS 1580.214.2). 6 pp.
SNZ NZS/AS 1580. 214.4-90. Methods of Test for Paints and Related Materials Part 214.4: Consistency—Rotothinner (This is a Joint Standard with SAA AS 1580.214.4). 4 pp.
SNZ NZS/AS 1580. 214.5-90. Methods of Test for Paints and Related Materials Part 214.5: Consistency—Rotational Viscometer (This is a Joint Standard with SAA AS 1580.214.5). 10 pp.

Consistency (Cont.)

—Mortars
CEN PREN 1015-3-93. Methods of Test for Mortar for Masonry—Part 3: Determination of Consistence of Fresh Mortar (by Flow Table) (Reference Method). 10 pp.
CEN PREN 1015-4-93. Methods of Test for Mortar for Masonry—Part 4: Determination of Consistence of Fresh Mortar (by Plunger Penetration) (Alternative Method). 8 pp.

—Paints
CGSB 1-GP-71 METH 4.5-74. Methods of Testing Paints and Pigments Consistency Stormer Consistency. 2 pp.
CNS K6494-80. Method of Test for Consistency of Paints (Stormer Viscometer) (May)(5598).
SAA AS 1580.214. 1-90. Paints and Related Materials—Methods of Test—Part 214.1: Consistency—Stormer Viscometer Amdt 1 January/February 1992 Amdt 2 January/February 1993 (in Professional Packages 30, 39). 1 p.
SAA AS 1580.214. 2-90. Paints and Related Materials—Methods of Test—Part 214.2: Consistency—Flow Cup. 6 pp.
SAA AS 1580.214. 4-90. Paints and Related Materials—Methods of Test—Part 214.4: Consistency—Rotothinner. 4 pp.
SAA AS 1580.214. 5-90. Paints and Related Materials—Methods of Test—Part 214.5: Consistency—Ratational Viscometer. 10 pp.
SNZ NZS/AS 1580. 214.1-90. Methods of Test for Paints and Related Materials Part 214.1: Consistency—Stormer Viscometer Amend: 1, 1992 (This is a Joint Standard with SAA AS 1580.214.1). 6 pp.
SNZ NZS/AS 1580. 214.2-90. Methods of Test for Paints and Related Materials Part 214.2: Consistency—Flow Cup (This is a Joint Standard with SAA AS 1580.214.2). 6 pp.
SNZ NZS/AS 1580. 214.4-90. Methods of Test for Paints and Related Materials Part 214.4: Consistency—Rotothinner (This is a Joint Standard with SAA AS 1580.214.4). 4 pp.
SNZ NZS/AS 1580. 214.5-90. Methods of Test for Paints and Related Materials Part 214.5: Consistency—Rotational Viscometer (This is a Joint Standard with SAA AS 1580.214.5). 10 pp.

—Pastes
MOD UK M 9523/66. Examination of Paste, Water Detecting (Withdrawn).

—Pulps
CNS P3066-84. Method of Test for Consistency of Pulp Suspensions (Apr)(10863). 3 pp.
CPPA D.6U-77. Rapid Consistency Determinations. 1 p.

—Putty
CGSB 1-GP-71 METH 4.7-74. Methods of Testing Paints and Pigments Consistency Consistency of Putty by Means of Gardner Mobilometer. 1 p.

—Sealants
SAA AS 1937.6-77. Methods of Test for Sealers and Adhesives for Automotive Purposes—Part 6: Determination by Pressure Extrudiometer of the Consistency of Sealers Corrig. Reconfirmed 1989. 3 pp.

—Sealants—Aging Testing
SAA AS 1937.7-77. Methods of Test for Sealers and Adhesives for Automotive Purposes—Part 7: Determination by Pressure Extrudiometer of the Consistency of Sealers After Accelerated Ageing Reconfirmed 1989. 2 pp.

—Slush Pulps
CPPA D.16-84. Consistency of Stocks; Reprinted—1986. 2 pp.

—Varnishes
SAA AS 1580.214. 1-90. Paints and Related Materials—Methods of Test—Part 214.1: Consistency—Stormer Viscometer Amdt 1 January/February 1992 Amdt 2 January/February 1993 (in Professional Packages 30, 39). 1 p.
SAA AS 1580.214. 2-90. Paints and Related Materials—Methods of Test—Part 214.2: Consistency—Flow Cup. 6 pp.
SAA AS 1580.214. 4-90. Paints and Related Materials—Methods of Test—Part 214.4: Consistency—Rotothinner. 4 pp.
SAA AS 1580.214. 5-90. Paints and Related Materials—Methods of Test—Part 214.5: Consistency—Ratational Viscometer. 10 pp.
SNZ NZS/AS 1580. 214.1-90. Methods of Test for Paints and Related Materials Part 214.1: Consistency—Stormer Viscometer Amend: 1, 1992 (This is a Joint Standard with SAA AS 1580.214.1). 6 pp.

Consistency (Cont.)

—Varnishes (Cont.)
SNZ NZS/AS 1580. 214.2-90. Methods of Test for Paints and Related Materials Part 214.2: Consistency—Flow Cup (This is a Joint Standard with SAA AS 1580.214.2). 6 pp.
SNZ NZS/AS 1580. 214.4-90. Methods of Test for Paints and Related Materials Part 214.4: Consistency—Rotothinner (This is a Joint Standard with SAA AS 1580.214.4). 4 pp.
SNZ NZS/AS 1580. 214.5-90. Methods of Test for Paints and Related Materials Part 214.5: Consistency—Rotational Viscometer (This is a Joint Standard with SAA AS 1580.214.5). 10 pp.

Consoles
See Also: Control Consoles; Data Processing Equipment; Keyboards

—Design—Ships
MOD UK NES 520-89. Guide to Design Procedures for Versatile Console System Issue 3 (07.89). 23 pp.
MOD UK NES 521-90. Guide to Versatile Console System Units Issue 3 (01.90). 177 pp.

—Design—Submarines
MOD UK NES 520-89. Guide to Design Procedures for Versatile Console System Issue 3 (07.89). 23 pp.
MOD UK NES 521-90. Guide to Versatile Console System Units Issue 3 (01.90). 177 pp.

—Glossaries
IEC 916-88. Mechanical Structures for Electronic Equipment Terminology First Edition. 16 pp.

Construction
Use For: Building Construction *See Also:* Aluminum Structures; Architecture; Building Services; Carpentry; Civil Engineering; Concrete Construction; Construction Contracts; Construction Costs; Construction Equipment; Construction Joints; Construction Materials; Construction Sites; Demolition; Fabrication Services; Masonry; Pile Foundations; Quantity Surveying; Specifications; Stonework; Structural Engineering; Structural Forms; Structural Members; Structures; Tunneling (Excavations); Underground Construction; Welding; Wood Construction

BSI BS 2654-89. 1989 Manufacture of Vertical Steel Welded Non-Refrigerated Storage Tanks with Butt-Welded Shells for the Petroleum Industry. 92 pp.
BSI BS 5964-80. (WITHDRAWN) 1980 Amd 1 Setting Out and Measurement of Buildings: Permissible Measuring Deviations (Superseded by BS 5964: Part 1: 1990). 19 pp.
BSI BS 5964: Part 1-90. 1990 Building Setting Out and Measurement Part 1: Methods of Measuring, Planning and Organization and Acceptance Criteria. 29 pp.
BSI BS 6019-80. 1980 Recommendations for Performance Standards in Buildings: Contents and Presentation. 4 pp.
BSI BS 7308-90. 1990 Method for Presentation of Dimensional Accuracy Data in Building Constructions. 16 pp.
BSI BS 7334: Part 1-90. 1990 Measuring Instruments for Building Construction Part 1: Methods for Determining Accuracy in Use: Theory. 10 pp.
BSI BS 7334: Part 2-90. 1990 Measuring Instruments for Building Construction Part 2: Methods for Determining Accuracy in Use: Measuring Tapes. 11 pp.
BSI BS 7334: Part 3-90. 1990 Measuring Instruments for Building Construction Part 3: Methods for Determining Accuracy in Use: Optical Levelling Instruments. 15 pp.
BSI HANDBOOK NO. 3-85. 1992 Ad 41 Summaries of British Standards for Building (R). 2371 pp.
BSI HANDBOOK NO. 20-85. (WITHDRAWN) 1985 Building Regulations 1985 British Standards Summarized. 239 pp.
ISO TR6116-81. Actions on Structures First Edition. 14 pp.
ISO 6240-80. Performance Standards in Building—Contents and Presentation First Edition; (Erratum—July 1981). 5 pp.
ISO 6241-84. Performance Standards in Building—Principles for Their Preparation and Factors to Be Considered First Edition. 12 pp.
ISO 7077-81. Measuring Methods for Building—General Principles and Procedures for the Verification of Dimensional Compliance First Edition. 4 pp.
ISO 8322 Pt 1-89. Building Construction—Measuring Instruments—Procedures for Determining Accuracy in Use—Part 1: Theory First Edition. 8 pp.
ISO 8322 Pt 2-89. Building Construction—Measuring Instruments—Procedures for Determining Accuracy in Use—Part 2: Measuring Tapes First Edition. 8 pp.

INTERNATIONAL AND NON-U.S. NATIONAL STANDARDS
SUBJECT INDEX
Construction

Construction (Cont.)

ISO 8322 Pt 3-89. Building Construction—Measuring Instruments—Procedures for Determining Accuracy in Use—Part 3: Optical Levelling Instruments First Edition. 12 pp.

SNZ NZS/BS 2654-89. Specification for Manufacture of Vertical Steel Welded Non-Refrigerated Storage Tanks with Butt-Welded Shells for the Petroleum Industry. 88 pp.

—Abbreviations

SAA HB24-92. Symbols and Abbreviations for Building and Construction. 111 pp.

—Accuracy Testing

BSI BS 5606-90. 1990 Accuracy in Building. 57 pp.

—Classification

BSI BS 1000: (69)-81. 1981 Universal Decimal Classification (UDC). English Full Edition (69): Building. 51 pp.

SNZ NZS/BS 1000 (69)-81. Universal Decimal Classification Building. 52 pp.

—Documentation

BSI BS 4940-73. 1973 Recommendations for the Presentation of Technical Information About Products and Services in the Construction Industry. 29 pp.

SAA AS 1388-74. Recommendations for Trade and Technical Literature for the Building Industry. 10 pp.

SNZ BS 4940-73. Recommendations for the Presentation of Technical Information About Products and Services in the Construction Industry. 32 pp.

—Drawings

BSI BS 1192: Part 1-84. 1984 Construction Drawing Practice Part 1: Recommendations for General Principles. 34 pp.

BSI BS 1192: Part 1-01. 1984 Amd 1 Construction Drawing Practice Part 1: Recommendations for General Principles (AMD 7031) June 15, 1992. 35 pp.

—Drawings—CAD

BSI BS 1192: Part 5-90. 1990 Construction Drawing Practice Part 5: Guide to Structuring of Computer Graphic Information. 22 pp.

SAA AS 3883-91. Computer Graphics—Computer Aided Design (CAD)—Guide for Structuring of Computer Graphic Information (BS 1192:5:1990). 16 pp.

—Electrical Installations—Controlgear—Fire Protection

IEC 364 Pt 5-53-86. Electrical Installations of Buildings Part 5: Selection and Erection of Electrical Equipment—Chapter 53: Switchgear and Controlgear First Edition; (Amendment 2-1992, Includes Amendment 1). 30 pp.

—Electrical Installations—Switchgear—Fire Protection

IEC 364 Pt 5-53-86. Electrical Installations of Buildings Part 5: Selection and Erection of Electrical Equipment—Chapter 53: Switchgear and Controlgear First Edition; (Amendment 2-1992, Includes Amendment 1). 30 pp.

—Explosives—Safety

BSI BS 5607-88. 1988 Safe Use of Explosives in the Construction Industry. 36 pp.

BSI BS 5607-78. 1978 Code of Practice for the Safe Use of Exolosives in the Construction Industry. 30 pp.

—Fire Hazards

SAA AS 3959-91. Construction of Buildings in Bushfire-Prone Areas (in Professional Package 44). 10 pp.

—Fire Testing

SNZ NZS/BS 476: Part 8-72. Fire Tests on Building Materials and Structures Part 8: Test Methods and Criteria for the Fire Resistance of Elements of Building Construction Amend: 3. 17 pp.

—Generators—Gaseous Mines

CSA C22.2 NO 145-M1986. Motors and Generators for Use in Hazardous Locations (R 1992); (Gen Instr 1 Thru 2). 47 pp.

—Generators—Hazardous Environments

CSA C22.2 NO 145-M1986. Motors and Generators for Use in Hazardous Locations (R 1992); (Gen Instr 1 Thru 2). 47 pp.

—Glossaries

BSI BS 6100: Part 0-92. 1992 Building and Civil Engineering Terms Part 0: Introduction. 101 pp.

BSI BS 6100: Part 0-90. 1990 Building and Civil Engineering Terms Part 0: Introduction. 58 pp.

Construction (Cont.)
—Glossaries (Cont.)

BSI BS 6100: Part 0-86. (WITHDRAWN) 1986 Glossary of Building and Civil Engineering Terms Part 0: Introduction. 34 pp.

BSI BS 6100: Sec 1.0-92. 1992 Building and Civil Engineering Terms Part 1: General and Miscellaneous Section 1.0: General. 40 pp.

BSI BS 6100: Sec 1.0-84. 1984 Glossary of Building and Civil Engineering Terms Part 1: General and Miscellaneous Section 1.0: General and Miscellaneous. 18 pp.

BSI BS 6100: SUBSEC 1.3.0-91. 1991 Building and Civil Engineering Terms Part 1: General and Miscellaneous Section 1.3: Parts of Construction Works Subsection 1.3.0: External Works. 18 pp.

BSI BS 6100: SUBSEC 1.3.0-01. 1991 Amd 1 Glossary of Building and Civil Engineering Terms Part 1: General and Miscellaneous Section 1.3: Parts of Construction Works Subsection 1.3.0: External Works (AMD 7231) August 15, 1992. 21 pp.

BSI BS 6100: SUBSEC 1.3.4-89. 1989 Glossary of Building and Civil Engineering Terms Part 1: General and Miscellaneous Section 1.3: Parts of Construction Works Subsection 1.3.4: Stairs and Circulation Elements. 8 pp.

BSI BS 6100: SUBSEC 1.3.4-01. 1989 Amd 1 Glossary of Building and Civil Engineering Terms Part 1: General and Miscellaneous Section 1.3: Parts of Construction Works Subsection 1.3.4: Stairs and Circulation Elements (AMD 7234) August 15, 1992. 10 pp.

BSI BS 6100: SUBSEC 1.3.6-91. 1991 Building and Civil Engineering Terms Part 1: General and Miscellaneous Section 1.3: Parts of Construction Works Subsection 1.3.6: Jointing Products, Builders' Hardware and Accessories. 32 pp.

BSI BS 6100: SUBSEC 1.3.6-01. 1991 Amd 1 Glossary of Building and Civil Engineering Terms Part 1: General and Miscellaneous Section 1.3: Parts of Construction Works Subsection 1.3.6: Jointing Products, Builders' Hardware and Accessories (AMD 7236) August 15, 1992. 35 pp.

BSI BS 6100: SUBSEC 1.5.5-91. 1991 Building and Civil Engineering Terms Part 1: General and Miscellaneous Section 1.5: Operations; Associated Plant and Equipment Subsection 1.5.5: Plant and Equipment. 27 pp.

BSI BS 6100: SUBSEC 1.5.5-01. 1991 Amd 1 Glossary of Building and Civil Engineering Terms Part 1: General and Miscellaneous Section 1.5: Operations; Associated Plant and Equipment Subsection 1.5.5: Plant and Equipment (AMD 7243) August 15, 1992. 30 pp.

BSI BS 6100: SUBSEC 1.5.6-88. 1988 Glossary of Building and Civil Engineering Terms Part 1: General and Miscellaneous Section 1.5: Operations: Associated Plant and Equipment Subsection 1.5.6: Documentation Excluding Drawings. 13 pp.

BSI BS 6100: SUBSEC 1.5.6-01. 1988 Amd 1 Glossary of Building and Civil Engineering Terms Part 1: General Section 1.5: Operations; Associated Plant and Equipment Subsection 1.5.6: Documentation Excluding Drawings (AMD 7244) August 15, 1992. 14 pp.

BSI BS 6100: SUBSEC 1.7.1-86. 1986 Glossary of Building and Civil Engineering Terms Part 1: General and Miscellaneous Section 1.7: Characteristics and Performance Subsection 1.7.1: Performance. 5 pp.

BSI BS 6100: SUBSEC 1.7.1-01. 1986 Amd 1 Building and Civil Engineering Terms Part 1: General and Miscellaneous Section 1.7: Characteristics and Performance Subsection 1.7.1: Performance (AMD 7247) August 15, 1992. 6 pp.

BSI BS 6100: SUBSEC 1.7.2-89. 1989 Glossary of Building and Civil Engineering Terms Part 1: General and Miscellaneous Section 1.7: Characteristics and Performance Subsection 1.7.2: Characteristics. 15 pp.

BSI BS 6100: SUBSEC 1.7.2-01. 1989 Amd 1 Glossary of Building and Civil Engineering Terms Part 1: General and Miscellaneous Section 1.7: Characteristics and Performance Subsec. 1.7.2: Characteristics (AMD 7248) Augsut 15, 1992. 18 pp.

BSI BS 6100: Sec 2.5-91. 1991 Building and Civil Engineering Terms Part 2: Civil Engineering Section 2.5: Hydraulic Engineering and Construction Work. 36 pp.

BSI BS 6953-88. 1988 Terms for Procedures for Setting out, Measurement and Surveying in Building Construction (Including Guidance Notes). 27 pp.

BSI BS EN 26927-91. 1991 Building Construction—Jointing Products—Sealants—Vocabulary (ISO 6927: 1981). 10 pp.

CEN EN 26927-90. Building Construction—Jointing Products—Sealants—Vocabulary. 6 pp.

DIN ENGL 18201-84. Tolerances in Building; Terminology, Principles, Application, Verification (Dec). 3 pp.

Construction (Cont.)
—Glossaries (Cont.)

DIN ENGL EN 26927-91. Jointing Products in Building Construction; Sealants; Vocabulary (ISO 6927:1981) (May) (Supersedes DIN 52460, August 1979 Edition). 6 pp.

ISO 1803 Pt 1-85. Building Construction—Tolerances—Vocabulary—Part 1: General Terms Second Edition. 6 pp.

ISO 1803 Pt 2-86. Building Construction—Tolerances—Vocabulary—Part 2: Derived Terms First Edition. 5 pp.

ISO 1804-72. Doors—Terminology First Edition. 10 pp.

ISO 3880 Pt 1-77. Building Construction—Stairs—Vocabulary—Part 1: First Edition. 7 pp.

ISO 6707 Pt 1-89. Building and Civil Engineering—Vocabulary—Part 1: General Terms Second Edition; (Corrected and Reprinted -1989). 37 pp.

ISO 6927-81. Building Construction—Jointing Products—Sealants—Vocabulary First Edition. 6 pp.

ISO 7078-85. Building Construction—Procedures for Setting out, Measurement and Surveying—Vocabulary and Guidance Notes First Edition. 51 pp.

SAA HB23-92. Thesaurus of Australian Construction Terms. 277 pp.

SAA HB25-92. Australian Building and Construction Definitions. 539 pp.

SNZ NZMP 4212-79. Glossary of Building Terminology. 160 pp.

—Industrial Wastes

BSI BS 6543-85. 1985 Guide to Use of Industrial By-Products and Waste Materials in Building and Civil Engineering. 40 pp.

—Measuring Instruments

SAA AS 1290-80. General Requirements for Linear Measuring Instruments for Use in Construction. 12 pp.

—Metric System

SAA AS 1155-74. Metric Units for Use in the Construction Industry Amdt 1 June 1979. 16 pp.

—Modular

BSI BS 6750-86. 1986 Modular Co-Ordination in Building. 23 pp.

BSI PD 6446-70. 1970 Recommendations for the Co-Ordination of Dimensions in Building. Combinations of Sizes. 26 pp.

CNS A1010-87. Basis for Coordination of Dimensions of Building (Jun)(2927).

CNS A1012-87. Planning Module for Buildings (Jun)(3537).

CNS A1015-78. Building Module (Mar)(4112).

CNS A1018-78. Principle of Modular Coordination in Buildings (Mar)(4115).

CNS A1030-81. Modular Dimension for Housing Components (Nov)(8084).

CSA CAN3-A31-M75. Series of Standards for Metric Dimensional Co-Ordination in Building (R 1981). 35 pp.

DIN ENGL 4172-55. Modular Co-Ordination in Building Construction (July). 2 pp.

DIN ENGL 18000-84. Modular Coordination in Building (May). 4 pp.

DIN ENGL 30798 Pt 2-82. Modular Systems; Modular Coordination; Principles (Sept). 2 pp.

DIN ENGL 30798 Pt 3-82. Modular Systems; Modular Coordination; Principles for the Application (Sept). 4 pp.

ISO 1006-83. Building Construction—Modular Coordination—Basic Module Second Edition. 4 pp.

ISO 1040-83. Building Construction—Modular Coordination—Multimodules for Horizontal Coordinating Dimensions Second Edition. 3 pp.

ISO 2848-84. Building Construction—Modular Coordination—Principles and Rules Second Edition. 7 pp.

ISO 6511-82. Building Construction—Modular Coordination—Modular Floor Plane for Vertical Dimensions First Edition. 5 pp.

ISO 6512-82. Building Construction—Modular Coordination—Storey Heights and Room Heights First Edition. 4 pp.

ISO 6513-82. Building Construction—Modular Coordination—Series of Preferred Multimodular Sizes for Horizontal Dimensions First Edition. 4 pp.

ISO 6514-82. Building Construction—Modular Coordination—Sub-Modular Increments First Edition. 3 pp.

ISO TR8389-84. Building Construction—Modular Coordination—System of Preferred Numbers Defining Multimodular Sizes First Edition. 10 pp.

ISO TR8390-84. Building Construction—Modular Coordination—Application of Horizontal Multimodules First Edition. 7 pp.

JIS A 0001-63. Building Module (R 1966). 5 pp.

JIS A 0004-64. Principle of Modular Co-Ordination in Buildings.

JIS A 0018-92. Modular Co-Ordination of Equipment Unit for Dwellings. 16 pp.

INDUSTRY STANDARDS

Construction

Construction (Cont.)

—Modular—Agricultural Buildings
BSI BS 5502: Sec 3.8-78. (WITHDRAWN) 1978 Amd 1 Design of Buildings and Structures for Agriculture Part 3: Appendices: Legislation, Technical Data and References Section 3.8: Dimensional Co-Ordination (AMD 5003) April 30, 1986 (Superseded by BS 5502:. 5 pp.

—Modular—Air Conditioners
CNS A1023-79. Modular Coordinating Sizes of Air Conditioning Unit for Dwellings (Apr) (4769).
JIS A 0014-76. Modular Co-Ordinating Sizes of Air-Conditioning Unit for Dwellings.

—Modular—Appliances—Built In
JIS A 0016-79. Modular Coordination-Coordinating Size of Opening for Built-in Appliances in Storage Furniture.

—Modular—Architectural Drawings
ISO 8560-86. Technical Drawings—Construction Drawings—Representation of Modular Sizes, Lines and Grids First Edition. 8 pp.

—Modular—Bricks
CNS R2065-72. General Type Bricks for Building (Modulus Coordination) (Jan) (3319). 1 p.

—Modular—Building Board
ISO 2777-74. Modular Co-Ordination—Co-Ordinating Sizes for Rigid Flat Sheet Boards Used in Building First Edition. 3 pp.
SNZ NZS 4207-75. Modular Co-Ordination. Preferred Co-Ordinating Sizes for Rigid Flat Sheet Boards Used in Building. 1 p.

—Modular—Doors
ISO 2776-74. Modular Co-Ordination—Co-Ordinating Sizes for Doorsets—External and Internal First Edition. 3 pp.

—Modular—Glossaries
CNS A1016-77. Glossary of Terms Used in Building Module (Jun)(4113).
DIN ENGL 30798 Pt 1-82. Modular Systems; Modular Coordination; Terminology (Sept). 2 pp.
ISO 1791-83. Building Construction—Modular Coordination—Vocabulary Second Edition. 9 pp.
JIS A 0002-66. Glossary of Terms Used in Building Module.

—Modular—Kitchens
BSI BS 5957-80. 1980 Recommendations for Modules for the Coordinating Dimensions of Catering Equipment Using Containers to BS 4874. 11 pp.
CNS A1022-78. Modular Coordinating Sizes of Kitchen Unit for Dwellings (Jul) (4440).
JIS A 0017-92. Kitchen Equipment—Coordinating Sizes. 11 pp.

—Modular—Packaging—Pallets
DIN ENGL 55510-82. Packaging; Modular Coordination in Packaging; Modular Sub-Multiples of the 600 mm x 400 mm Area Module (Mar). 4 pp.

—Modular—Pipelines
CNS A1024-79. Modular Coordinating Sizes of Piping Unit for Dwellings (Apr)(4770).
JIS A 0015-76. Modular Co-Ordinating Sizes of Piping Unit for Dwellings.

—Modular—Ships
BSI BS MA 44-74. 1974 Recommendations for the Coordination of Dimensions in Shipbuilding: Principles of Dimensional Coordination. 23 pp.
BSI BS MA 45-74. 1974 Recommendations for the Coordination of Dimensions in Shipbuilding: Controlling Dimensions. 4 pp.
BSI BS MA 71-77. 1977 Coordination of Dimensions in Shipbuilding: Coordinating Spaces for Internal Subdivision. 18 pp.
BSI BS MA 72-76. 1976 Coordination of Dimensions in Shipbuilding: Coordinating Spaces for Furniture. 15 pp.
BSI BS MA 73-76. 1976 Coordination of Dimensions in Shipbuilding: Coordinating Sizes for Fixtures, Fittings and Equipment. 11 pp.
BSI BS MA 77-76. 1976 Recommendations for the Coordination of Dimensions in Shipbuilding: Coordinating Sizes for Services. 10 pp.
ISO 3827 Pt I-77. Shipbuilding—Co-Ordination of Dimensions in Ships' Accommodation—Part I: Principles of Dimensional Co-Ordination First Edition. 24 pp.
ISO 3827 Pt III-77. Shipbuilding—Co-Ordination of Dimensions in Ships' Accommodation—Part III: Co-Ordinating Sizes for Components and Assemblies First Edition. 7 pp.
ISO 3827 Pt IV-77. Shipbuilding—Co-Ordination of Dimensions in Ships' Accommodation—Part IV: Controlling Dimensions First Edition. 5 pp.

Construction (Cont.)

—Modular—Ships (Cont.)
ISO 3827 Pt 5-79. Shipbuilding—Co-Ordination of Dimensions in Ships' Accommodation—Part 5: Co-Ordinating Sizes for Key Components First Edition. 3 pp.

—Modular—Ships—Glossaries
ISO 3827 Pt II-77. Shipbuilding—Co-Ordination of Dimensions in Ships' Accommodation—Part II: Glossary of Terms First Edition. 6 pp.

—Modular—Stairways
BSI BS 5578: Part 2-78. 1978 Building Construction. Stairs Part 2: Modular Coordination. Specification for Coordinating Dimensions for Stairs and Stair Openings. 6 pp.
ISO 3881-77. Building Construction—Modular Co-Ordination—Stairs and Stair Openings—Co-Ordinating Dimensions First Edition. 5 pp.

—Modular—Toilet Facilities
CNS A1021-78. Modular Coordinating Sizes of Sanitary Unit for Dwellings (Jul) (4439).
JIS A 0012-80. Modular Co-Ordination—Sizes of Sanitary Unit for Dwellings.

—Motors—Gaseous Mines
CSA C22.2 NO 145-M1986. Motors and Generators for Use in Hazardous Locations (R 1992); (Gen Instr 1 Thru 2). 47 pp.

—Motors—Hazardous Environments
CSA C22.2 NO 145-M1986. Motors and Generators for Use in Hazardous Locations (R 1992); (Gen Instr 1 Thru 2). 47 pp.

—Noise
SNZ NZS 6803P-84. Measurement and Assessment of Noise from Construction, Maintenance and Demolition Work. 32 pp.

—Personnel—Glossaries
BSI BS 6100: Sec 1.6-85. 1985 Glossary of Building and Civil Engineering Terms Part 1: General and Miscellaneous Section 1.6: Persons. 6 pp.
BSI BS 6100: Sec 1.6-01. 1985 Amd 1 Glossary of Building and Civil Engineering Terms Part 1: General and Miscellaneous Section 1.6: Persons (AMD 7246) August 15, 1992. 7 pp.

—Protection—Buildings
DIN ENGL 4123-72. Protection of Buildings in the Area of Excavations, Foundations and Underpinnings (May). 5 pp.

—Reliability
ISO 2394-86. General Principles on Reliability for Structures Second Edition; (Addendum 1-1988). 30 pp.

—Reliability—Glossaries
ISO 8930-87. General Principles on Reliability for Structures—List of Equivalent Terms First Edition. 23 pp.

—Safety
CNS A1009-75. Safety Codes for Construction Contractors (Dec)(2857). 14 pp.

—Schedules
SAA AS 1181-82. Method of Measurement of Civil Engineering Works and Associated Building Works Reconfirmed 1990. 50 pp.

—Scheduling
BSI BS 1192: Part 1-84. 1984 Construction Drawing Practice Part 1: Recommendations for General Principles. 34 pp.
BSI BS 1192: Part 1-01. 1984 Amd 1 Construction Drawing Practice Part 1: Recommendations for General Principles (AMD 7031) June 15, 1992. 35 pp.

—Sites—Roads—Traffic Control
SAA AS 1742.3-85. Manual of Uniform Traffic Control Devices—Part 3: Traffic Control Devices for Works on Roads. 111 pp.

—Spaces—Glossaries
BSI BS 6100: Sec 1.2-85. 1985 Glossary of Building and Civil Engineering Terms Part 1: General and Miscellaneous Section 1.2: Spaces. 8 pp.
BSI BS 6100: Sec 1.2-92. 1992 Glossary of Building and Civil Engineering Terms Part 1: General and Miscellaneous Section 1.2: Spaces. 11 pp.

—Standards Preparation
BSI BS 7642-93. 1993 Performance Standards in Building—Contents and Format of Standards for Evaluation of Performance (ISO 7162: 1992). 11 pp.

Construction (Cont.)

—Standards Preparation (Cont.)
BSI PD 6501: Part 1-82. 1982 Amd 1 Preparation of British Standards for Building and Civil Engineering Part 1: Guide to the Types of British Standard, Their Aims, Relationship, Content and Application. 20 pp.
BSI PD 6501: Part 2-84. 1984 Preparation of British Standards for Building and Civil Engineering Part 2: Guide to Presentation. 34 pp.
BSI PD 6501: Part 3-85. 1985 Preparation of British Standards for Building and Civil Engineering Part 3: Guidance to BSI Committees on the Preparation of British Standards for Use with the Building Act 1984. 47 pp.
BSI PD 6501: Part 5-90. 1990 Preparation of British Standards for Building and Civil Engineering Part 5: Guide to Technical Committees on the Preparation of Performance Specifications in British Standards for Construction Products. 24 pp.
ISO 7162-92. Performance Standards in Building—Contents and Format of Standards for Evaluation of Performance First Edition. 9 pp.

—Statistical Analysis—Tolerances
BSI BS 6954: Part 2-88. 1988 Tolerances for Building Part 2: Recommendations for Statistical Basis for Predicting Fit Between Components Having a Normal Distribution of Sizes. 8 pp.

—Symbols
SAA HB24-92. Symbols and Abbreviations for Building and Construction. 111 pp.

—Thesauri
SAA HB23-92. Thesaurus of Australian Construction Terms. 277 pp.

—Tolerances
BSI BS 6954: Part 1-88. 1988 Tolerances for Building Part 1: Recommendations for Basic Principles for Evaluation and Specification. 4 pp.
BSI BS 6954: Part 3-88. 1988 Tolerances for Building Part 3: Recommendations for Selecting Target Size and Predicting Fit. 16 pp.
BSI BS 7307: Part 1-90. 1990 Building Tolerances: Measurement of Buildings and Building Products Part 1: Methods and Instruments. 87 pp.
BSI BS 7307: Part 2-90. 1990 Building Tolerances: Measurement of Buildings and Building Products Part 2: Position of Measuring Points. 30 pp.
DIN ENGL 18202-86. Tolerances in Building; Buildings (May) (This Standard Supersedes DIN 18202 Part 1, March 1969 Edition, DIN 18202 Part 4, June 1974 Edition, Supplement 1 to DIN 18202 Part 4, August 1977 Edition, and DIN 18202 Part 5, October 1979 Edition). 7 pp.
ISO 3443 Pt 1-79. Tolerances for Building—Part 1: Basic Principles for Evaluation and Specification First Edition. 4 pp.
ISO 3443 Pt 2-79. Tolerances for Building—Part II: Statistical Basis for Predicting Fit Between Components Having a Normal Distribution of Sizes First Edition. 8 pp.
ISO 3443 Pt 3-87. Tolerances for Building—Part 3: Procedures for Selecting Target Size and Predicting Fit First Edition. 16 pp.
ISO 3443 Pt 4-86. Tolerances for Building—Part 4: Method for Predicting Deviations of Assemblies and for Allocation of Tolerances First Edition. 14 pp.
ISO 3443 Pt 5-82. Building Construction—Tolerances for Building—Part 5: Series of Values to Be Used for Specification of Tolerances First Edition. 3 pp.
ISO 3443 Pt 6-86. Tolerances for Building—Part 6: General Principles for Approval Criteria, Control of Conformity with Dimensional Tolerance Specifications and Statistical Control—Method 1 First Edition. 6 pp.
ISO 3443 Pt 7-88. Tolerances for Building—Part 7: General Principles for Approval Criteria, Control of Conformity with Dimensional Tolerance Specifications and Statistical Control—Method 2 (Statistical Control Method) First Edition. 22 pp.
ISO 3443 Pt 8-89. Tolerances for Building—Part 8: Dimensional Inspection and Control of Construction Work First Edition. 7 pp.
ISO 4463 Pt 1-89. Measurement Methods for Building—Setting-out and Measurement—Part 1: Planning and Organization, Measuring Procedures, Acceptance Criteria First Edition. 27 pp.
ISO 4464-80. Tolerances for Building—Relationship Between the Different Types of Deviations and Tolerances Used for Specification First Edition. 11 pp.
ISO 7737-86. Tolerances for Building—Presentation of Dimensional Accuracy Data First Edition. 14 pp.
ISO 7976 Pt 1-89. Tolerances for Building—Methods of Measurement of Buildings and Building Products—Part 1: Methods and Instruments First Edition. 86 pp.
ISO 7976 Pt 2-89. Tolerances for Building—Methods of Measurement of Buildings and Building Products—Part 2: Position of Measuring Points First Edition. 28 pp.

INTERNATIONAL AND NON-U.S. NATIONAL STANDARDS
SUBJECT INDEX
Construction

Construction *(Cont.)*
—Tolerances *(Cont.)*
SAA HB31-92. Handbook of Building Construction Tolerances—Extracts from Building Products and Structures Standards (in Professional Packages 30, 56). 52 pp.

Construction Contracts
Scope Note: For additional listings, see the subheading Construction Contracts under the specific product or material *Use For:* Building Contracts *See Also:* Construction; Cost Reimbursement Contracts; Service Contracts
 DIN ENGL 1960-90. Tendering and Performance Stipulations in Contracts for Construction Works (VOB) Part A: General Provisions Relating to the Award of Contracts for Construction Works (July). 26 pp.
 DIN ENGL 1961-92. Construction Contract Procedures (VOB); Part B: General Conditions of Contract Relating to the Execution of Construction Work (Dec). 8 pp.
 DIN ENGL 18299-92. Construction Contract Procedures (VOB); Part C: General Technical Specifications in Construction Contracts (ATV); General Rules Applying to All Types of Construction Work (Dec). 6 pp.
 MOD UK DEFCON 112 LWS-91. Basic Set of Conditions of Contract for Building, Civil Engineering, Mechanical and Electrical Small Works 5/91. 9 pp.
 SNZ NZS 3902-74. Agreement for Small Building Contracts (Including Supply of Land) (Reconfirmed 1991). 20 pp.

—Models
 SNZ NZS 3901-74. Agreement for Small Building Contracts (Excluding Supply of Land) (Reconfirmed 1991). 15 pp.
 SNZ NZS 3910-87. Conditions of Contract for Building and Civil Engineering Construction. 84 pp.

Construction Costs
Use For: Building Costs *See Also:* Construction
 DIN ENGL 276 Pt 3-81. Building Costs; Ascertainment of Costs (Apr). 35 pp.

—Classification
 DIN ENGL 276 Pt 2-81. Building Costs; Classification of Costs (Apr). 24 pp.

—Glossaries
 DIN ENGL 276 Pt 1-81. Building Costs; Terminology (Apr). 2 pp.

Construction Equipment
Scope Note: For additional listings, use a more specific term *Use For:* Construction Plant *See Also:* Backup Alarms; Bulldozers; Concrete Breakers; Concrete Vibrators; Construction; Construction Costs; Construction Sites; Cradles (Hoists); Cranes; Dredging Equipment; Earthmoving Equipment; Excavating Equipment; Falsework; Fascia; Finishing Machinery; Forestry Equipment; Gabions; Hoists; Industrial Equipment; Ladders; Monorails; Off Road Equipment; Pile Drivers; Plate Compactors; Platforms; Prime Movers (Vehicles); Rock Drills; Rollers (Compactors); Scaffolds; Scarifiers; Stagings; Tampers; Tractors; Trench Excavators; Trestles (Supports)
 EC 84/532/EEC-84. Council Directive on the Approximation of the Laws of the Member States Relating to Common Provisions for Construction Plant and Equipment. 12 pp.

—Access Fittings
 JIS A 8302-86. Construction Machinery—Access Systems. 10 pp.

—Air Compressors
 MOD UK DEF-1017-A-65. Air Compressors for Construction Purposes. 22 pp.

—Air Filters
 DIN ENGL 71459-87. Air Filter Elements for Commercial Vehicles; Dimensions (May). 6 pp.

—Ammeters
 JIS A 8106-71. Ammeters for Construction Machinery.

—Diesel Engines
 CNS D1044-83. Standard Form of Specification of Diesel Engines for Construction Machineries (Jan)(9851).
 JIS D 0006-86. Standard Form of Specifications of Diesel Engines for Construction Machinery. 16 pp.
 JIS D 1005-86. Testing Methods of Diesel Engines for Construction Machinery. 22 pp.

—Ergonomics
 JIS A 8301-86. Construction Machinery-Minimum Access Dimensions. 9 pp.

Construction Equipment *(Cont.)*
—Falling Object Protective Structures
 EC 86/296/EEC-86. Council Directive on the Approximation of the Laws of the Member States Relating to Falling-Object Protective Structures (FOPS) for Certain Construction Plant. 9 pp.

—Gages—Mechanical Shock
 CNS A3164-82. Testing Method of Vibration and Shock for Construction Machinery Gauges (Jul)(9054).
 JIS A 8101-78. Vibration and Shock Testing Method for Construction Machinery Gauges.

—Gages—Vibration
 CNS A3164-82. Testing Method of Vibration and Shock for Construction Machinery Gauges (Jul)(9054).
 JIS A 8101-78. Vibration and Shock Testing Method for Construction Machinery Gauges.

—Glossaries
 BSI BS 6100: SUBSEC 1.5.5-91. 1991 Building and Civil Engineering Terms Part 1: General and Miscellaneous Section 1.5: Operations; Associated Plant and Equipment Subsection 1.5.5: Plant and Equipment. 27 pp.
 BSI BS 6100: SUBSEC 1.5.5-01. 1991 Amd 1 Glossary of Building and Civil Engineering Terms Part 1: General and Miscellaneous Section 1.5: Operations; Associated Plant and Equipment Subsection 1.5.5: Plant and Equipment (AMD 7243) August 15, 1992. 30 pp.

—Hoists—Safety
 BSI BS 7212-89. 1989 Code of Practice for Safe Use of Construction Hoists. 27 pp.

—Hydraulic Torque Converters
 JIS D 1007-76. Test Code of Hydraulic Torque Converters for Construction Machineries and Industrial Vehicles (R 1984). 22 pp.

—Laser Instruments (Distance Measurement)—Accuracy Testing
 BSI BS 7334: Part 6-92. 1992 Measuring Instruments for Building Construction Part 6: Methods for Determining Accuracy in Use of Laser Instruments (ISO 8322-6: 1991). 19 pp.
 ISO 8322 Pt 6-91. Building Construction—Measuring Instruments—Procedures for Determining Accuracy in Use—Part 6: Laser Instruments First Edition. 17 pp.

—Maintenance—Hand Tools
 JIS A 8905-86. Construction Machinery—Service Tools—Hand Tools for Maintenance and Adjustment Works. 11 pp.

—Measuring Instruments
 BSI BS 4035-66. 1966 Amd 1 Linear Measuring Instruments for Use on Building and Civil Engineering Constructional Works. Steel Measuring Tapes, Steel Bands and Retractable Steel Pocket Rules. 21 pp.

—Measuring Tapes—Accuracy Testing
 BSI BS 7334: Part 7-92. 1992 Measuring Instruments for Building Construction Part 7: Methods for Determining Accuracy in Use of Instruments When Used for Setting Out (ISO 8322-7: 1991). 16 pp.
 ISO 8322 Pt 7-91. Building Construction—Measuring Instruments—Procedures for Determining Accuracy in Use—Part 7: Instruments When Used for Setting out First Edition. 15 pp.

—Meters—Flexible Shafts
 CNS A4009-88. Meter Flexible Shafts for Construction Machinery (Oct)(9457).
 JIS A 8104-78. Meter Flexible Shafts for Construction Machinery.

—Mobile—Soil Stabilization
 CEN PREN 500-3-91. Mobile Road Construction Machines—Safety—Part 3: Specific Requirements for Soil Stabilization Machines. 8 pp.

—Noise
 CSA CAN/CSA-Z107.32-M86. Test Procedure for the Measurement of Sound Emitted from Construction, Forestry, and Mining Machines to the Operator Station and Exterior of the Machine; (Gen Instr 1 Thru 2). 36 pp.
 ISO 4872-78. Acoustics—Measurement of Airborne Noise Emitted by Construction Equipment Intended for Outdoor Use—Method for Determining Compliance with Noise Limits First Edition; (Erratum—June 1979) (SAE J 2101). 14 pp.
 JIS A 8305-88. Method for the Measurement of Airborne Noise Emitted by Construction Equipment Intended for Outdoor Use. 20 pp.

—Oil Pressure Gages
 JIS A 8107-92. Oil Pressure Gauges for Engine of Construction Machinery. 14 pp.

Construction Equipment *(Cont.)*
—Optical Plummets—Accuracy Testing
 BSI BS 7334: Part 5-92. 1992 Measuring Instruments for Building Construction Part 5: Methods for Determining Accuracy in Use of Optical Plumbing Instruments (ISO 8322-5: 1991). 16 pp.
 ISO 8322 Pt 5-91. Building Construction—Measuring Instruments—Procedures for Determining Accuracy in Use—Part 5: Optical Plumbing Instruments First Edition. 14 pp.

—Pavement Finishers—Safety
 CEN PREN 500-6-91. Mobile Road Construction Machines—Safety—Part 6: Specific Requirements for Paver-Finishers. 6 pp.

—Piles—Safety
 CEN PREN 996-93. Piling Equipment—Safety Requirements. 62 pp.

—Piles—Specification Forms
 CNS A4027-89. Standard Form of Specification of Piling Equipment (Jan)(10139).

—Recording Instruments
 JIS A 8108-75. Working Recorders for Construction Machinery.

—Road—Mobile—Safety
 CEN PREN 500-1-91. Mobile Road Construction Machines—Safety—Part 1: General Requirements. 42 pp.

—Roll Over Protective Structures
 CSA B352-M1980. Rollover Protective Structures (ROPS) for Agricultural, Construction, Earthmoving, Forestry, Industrial, and Mining Machines; (Gen Instr 1). 76 pp.
 EC 86/295/EEC-86. Council Directive on the Approximation of the Laws of the Member States Relating to Roll-Over Protective Structures (ROPS) for Certain Construction Plant. 9 pp.

—Rotary Compressors
 JIS A 8109-84. Testing Methods of Rotary Compressors for Construction Machinery. 10 pp.

—Safety Belts
 CSA Z259.1-1976. Fall Arresting Safety Belts and Lanyards for the Construction and Mining Industries; (Amd 1-9 May 1979) (Amd 10-11 July 1993) (Gen Instr 2). 29 pp.

—Safety Lines
 CSA Z259.2-M1979. Fall-Arresting Devices, Personnel Lowering Devices, and Life Lines; (Amd 1-2 July 1993) (Gen Instr 2). 23 pp.

—Safety Nets
 BSI BS 8093-91. 1991 Code of Practice for the Use of Safety Nets, Containment Nets and Sheets on Constructional Works. 39 pp.
 BSI CP 93-72. (WITHDRAWN) 1972 Amd 1 Use of Safety Nets on Constructional Works (AMD 6203) November 30, 1989 (Superseded by BS 8093: 1991). 30 pp.

—Seats—Vibration
 JIS A 8304-87. Testing Method of Operator Seat-Transmitted Vibration for Construction Machinery. 17 pp.

—Speedometers
 CNS A4010-88. Speedometers for Construction Machinery (Oct)(9458).
 JIS A 8103-78. Speedometers for Construction Machinery.

—Suspended Safety Chairs
 BSI BS 2830-73. 1973 Suspended Safety Chairs and Cradles for Use in the Construction Industry. 12 pp.

—Tachometers
 CNS A4008-88. Engine Tachometers for Construction Machinery (Oct)(9055).
 JIS A 8102-78. Engine Tachometers for Construction Machinery.

—Temperature Gages
 JIS A 8105-86. Temperature Gauges for Construction Machinery. 15 pp.

—Temperature Measuring Instruments
 CNS A4011-82. Temperature Indicators for Construction Machinery (Oct)(9459).

—Theodolites—Accuracy Testing
 BSI BS 7334: Part 4-92. 1992 Measuring Instruments for Building Construction Part 4: Methods for Determining Accuracy in Use of Theodolites (ISO 8322-4: 1991). 19 pp.

Construction

Construction Equipment (Cont.)
—Theodolites—Accuracy Testing (Cont.)
BSI BS 7334: Part 7-92. 1992 Measuring Instruments for Building Construction Part 7: Methods for Determining Accuracy in Use of Instruments When Used for Setting Out (ISO 8322-7: 1991). 16 pp.
ISO 8322 Pt 4-91. Building Construction—Measuring Instruments—Procedures for Determining Accuracy in Use—Part 4: Theodolites First Edition. 17 pp.
ISO 8322 Pt 7-91. Building Construction—Measuring Instruments—Procedures for Determining Accuracy in Use—Part 7: Instruments When Used for Setting out First Edition. 15 pp.

—Tires
CGSB 20-GP-5D-80. Tires, Pneumatic, Low Speed, Off Highway, Standard for. 23 pp.

—Turning Radius
JIS A 8303-87. Measurement of Turning Dimensions of Wheeled Construction Machinery. 12 pp.

—Vibration
BSI BS 6794-86. 1986 Reporting Measured Vibration Data for Land Vehicles. 14 pp.
ISO 8002-86. Mechanical Vibrations—Land Vehicles—Method for Reporting Measured Data First Edition. 12 pp.

Construction Joints
Scope Note: Includes standards on joints between concrete structures; for standards on joints used in building construction, see Joints—Construction (Excludes Concrete Structures) *See Also:* Bonded Joints; Concrete Construction; Construction; Joints
DIN ENGL 4227 Pt 3 (P)-83. Prestressed Concrete; Segmental Type Structural Components; Design and Workmanship of Joints (Dec). 9 pp.

—Horizontal
ISO 7728-85. Typical Horizontal Joints Between an External Wall of Prefabricated Ordinary Concrete Components and a Concrete Floor—Properties, Characteristics and Classification Criteria First Edition. 8 pp.
ISO 7845-85. Horizontal Joints Between Load-Bearing Walls and Concrete Floors—Laboratory Mechanical Tests—Effect of Vertical Loading and of Moments Transmitted by the Floors First Edition. 8 pp.

—Joint Sealants
CNS K3031-90. Flexible Polyvinylchloride Water-Stops (Jun)(3895).
CNS K6384-90. Method of Test for Flexible PVC Water-Stops (Jun)(3896).
DIN ENGL 7865 Pt 1-82. Elastomeric Joint Sealing Strip for Sealing Joints in Concrete; Shape and Dimensions (Feb). 12 pp.
DIN ENGL 7865 Pt 2-82. Elastomeric Joint Sealing Strip for Sealing Joints in Concrete; Material Requirements and Testing (Feb). 3 pp.
JIS K 6773-77. Flexible Polyvinylchloride Water-Stops. 16 pp.

—Mechanical Testing
ISO 7844-85. Grooved Vertical Joints with Connecting Bars and Concrete Infill Between Large Reinforced Concrete Panels—Laboratory Mechanical Tests—Effect of Tangential Loading First Edition. 6 pp.
ISO 7845-85. Horizontal Joints Between Load-Bearing Walls and Concrete Floors—Laboratory Mechanical Tests—Effect of Vertical Loading and of Moments Transmitted by the Floors First Edition. 8 pp.

—Vertical
ISO 7729-85. Typical Vertical Joints Between Two Prefabricated Ordinary Concrete External Wall Components—Properties, Characteristics and Classification Criteria First Edition. 9 pp.
ISO 7844-85. Grooved Vertical Joints with Connecting Bars and Concrete Infill Between Large Reinforced Concrete Panels—Laboratory Mechanical Tests—Effect of Tangential Loading First Edition. 6 pp.

Construction Machinery
Use: Construction Equipment

Construction Materials
Scope Note: For additional listings, use a more specific term *Use For:* Building Materials; Construction Products *See Also:* Asbestos Cements; Baseboards; Bricks; Building Blocks; Building Board; Building Papers; Building Stones; Cementitious Materials; Cements; Ceramic Tiles; Concrete Blocks; Concrete Slabs; Concretes; Construction; Construction Sites; Floor Tiles; Framing; Furring; Glass Blocks; Glazing Systems; Gypsum Wallboard; Hardware; Joint Sealants; Lathing; Lumber; Moldings; Mortars; Paints; Panels; Partitions (Walls); Portland Cements; Portland Slag Cements; Roofing; Sealants; Sheathing; Sheeting; Siding; Silicate Cements; Slag Cements; Slates; Soffits; Stonework; Strands (Wire); Structural Members; Structural Steels; Tiles; Trass Cements;

Construction Materials (Cont.)
See Also: (Cont.)
Veneers; Wall Tiles; Wallboard; Wood
CNS A1017-77. Basic Tolerances for Building Components (Jun)(4114).
JIS A 0003-63. Basic Tolerances for Building Components (R 1966). 4 pp.

—Abrasion Resistance Testing
CNS A3160-82. Method of Abrasion Test for Building Material and Part of Building Construction (Method of Abrasion Test for Flooring Materials Method with Rotating Disk Fitted Friction and Impact) (Jun)(8912).
CNS A3202-84. Method of Abrasion Test for Building Materials and Building Construction Parts (Falling Sand Method) (Mar)(10784).
CNS A3203-84. Method of Abrasion Test for Building Materials and Building Construction Parts (Abrasive-Paper Method) (Mar)(10785).
JIS A 1451-70. Method of Abrasion Test for Building Materials and Part of Building Construction (Method of Abrasion Test for Flooring Materials Method with Rotating Disk Fitted Friction and Impact).
JIS A 1452-72. Method of Abrasion Test for Building Materials and Part of Building Construction (Falling Sand Method).
JIS A 1453-73. Method of Abrasion Test for Building Materials and Part of Building Construction (Abrasive-Paper Method).

—Agricultural Buildings
BSI BS 5502: Part 21-90. 1990 Design of Buildings and Structures for Agriculture Part 21: Code of Practice for Selection and Use of Construction Materials. 23 pp.
BSI BS 5502: Part 21-01. 1990 Amd 1 Buildings and Structures for Agriculture Part 21: Code of Practice for Selection and Use of Construction Materials (AMD 7676) June 15, 1993 (R). 28 pp.

—Aluminum
BSI BS 4868-72. 1972 Profiled Aluminium Sheet for Building. 8 pp.
BSI CP 118-69. 1969 Amd 1 Structural Use of Aluminium. 184 pp.
DIN ENGL 4113-58. Aluminium in Building Construction; Directions for Calculation and Construction of Aluminium Building Components (Feb). 7 pp.
SAA AS 1664-79. Rules for the Use of Aluminium in Structures (Known as the SAA Aluminium Structures Code) (Incorporating Corrig.) Interpretation 1 June 1985. 104 pp.

—Aluminum Alloy
BSI BS 1161-77. 1977 Amd 1 Aluminium Alloy Sections for Structural Purposes. 16 pp.

—Angle of Repose
DIN ENGL 1055 Pt 1-78. Design Loads for Buildings; Stored Materials, Building Materials and Structural Members; Dead Load and Angle of Friction (July). 15 pp.

—Cabling Systems (Electric Power)—Fire Testing
DIN ENGL 4102 Pt 12-91. Fire Behaviour of Building Materials and Elements; Fire Resistance of Electrical Cable Systems; Requirements and Testing (Jan). 6 pp.

—Calcium Oxide
BSI BS 890-72. 1972 Building Limes. 33 pp.
CNS A2002-72. Quicklime for Buildings (Jun)(381). 1 p.

—Calorific Value
ISO 1716-73. Building Materials—Determination of Calorific Potential First Edition. 6 pp.

—Ceramics—Flexural Strength
CEN PREN 843-1-92. Advanced Technical Ceramics—Mechanical Properties of Monolithic Ceramics at Room Temperature—Part 1: Determination of Flexural Strength. 19 pp.
JIS R 1601-81. Testing Method for Flexural Strength (Modulus of Rupture) of High Performance Ceramics. 8 pp.

—Ceramics—Modulus of Elasticity
JIS R 1602-86. Testing Methods for Elastic Modulus of High Performance Ceramics. 14 pp.

—Coal Tar Coatings
BSI BS 1310-84. (WITHDRAWN) 1984 Coal Tar Pitches for Building Purposes. 16 pp.

—Color
BSI BS 5252-76. 1976 Framework for Colour Co-Ordination for Building Purposes: Colour Matching Fan. 11 pp.
SNZ NZS/BS 5252-76. Framework for Colour Co-Ordination for Building Purposes Amend: 1. 4 pp.

Construction Materials (Cont.)
—Durability
BSI BS 7543-92. 1992 Durability of Buildings and Building Elements, Products and Components. 45 pp.

—Fasteners
BSI BS 1494: Part 1-64. 1964 Amd 1 Fixing Accessories for Building Purposes Part 1: Fixings for Sheet, Roof and Wall Coverings. 27 pp.

—Fiberglass Reinforced Plastic Sheets—Corrugated
BSI BS 4154: Part 1-85. 1985 Corrugated Plastics Translucent Sheets Made from Thermo-Setting Polyester Resin (Glass Fibre Reinforced) Part 1: Material and Performance Requirements. 16 pp.
BSI BS 4154: Part 2-85. 1985 Corrugated Plastics Translucent Sheets Made from Thermo-Setting Polyester Resin (Glass Fibre Reinforced) Part 2: Profiles and Dimensions. 4 pp.

—Fire Testing
BSI BS 476: Part 6-89. 1989 Amd 1 Fire Tests on Building Materials and Structures Part 6: Method of Test for Fire Propagation for Products. 16 pp.
BSI BS 476: Part 6-81. 1981 Amd 1 Fire Tests on Building Materials and Structures Part 6: Method of Test for Fire Propagation for Products (AMD 4329) August 31, 1983. 20 pp.
BSI BS 476: Part 7-01. 1987 Amd 1 Fire Tests on Building Materials and Structures Part 7: Method for Classification of the Surface Spread of Flame of Products (AMD 6249) January 31, 1990. 27 pp.
BSI BS 476: Part 7-02. 1987 Amd 2 Fire Tests on Building Materials and Structures Part 7: Method for Classification of the Surface Spread of Flame of Products (AMD 7030) January 30, 1992. 28 pp.
BSI BS 476: Part 7-03. 1987 Amd 3 Fire Tests on Building Materials and Structures Part 7: Method for Classification of the Surface Spread of Flame of Products (AMD 7612) April 15, 1993 (R). 28 pp.
BSI BS 476: Part 7-71. 1971 Fire Tests on Building Materials and Structures Part 7: Surface Spread of Flame Tests for Materials. 18 pp.
BSI BS 476: Part 10-83. 1983 Fire Tests on Building Materials and Structures Part 10: Guide to the Principles and Application of Fire Testing. 12 pp.
BSI BS 476: Part 11-82. 1982 Fire Tests on Building Materials and Structures Part 11: Method for Assessing the Heat Emission from Building Materials. 18 pp.
BSI BS 476: Part 12-91. 1991 Fire Tests on Building Materials and Structures Part 12: Method of Test for Ignitability of Products by Direct Flame Impingement. 20 pp.
BSI BS 476: Part 13-87. 1987 Amd 1 Fire Tests on Building Materials and Structures Part 13: Method of Measuring the Ignitability of Products Subjected to Thermal Irradiance. 37 pp.
BSI BS 476: Part 15-93. 1993 Fire Tests on Building Materials and Structures Part 15: Method of Measuring the Rate of Heat Release of Products (ISO 5660-1: 1993) (R). 38 pp.
BSI BS 476: Part 20-87. 1987 Amd 1 Fire Tests on Building Materials and Structures Part 20: Method for Determination of the Fire Resistance of Elements of Construction (General Principles) (AMD 6487) April 30, 1990. 44 pp.
BSI BS 476: Part 24-87. 1987 Fire Tests on Building Materials and Structures Part 24: Method for Determination of the Fire Resistance of Ventilation Ducts. 24 pp.
BSI BS 476: Part 32-89. 1989 Fire Tests on Building Materials and Structures Part 32: Guide to Full Scale Fire Tests Within Buildings. 16 pp.
BSI BS 6401-83. 1983 Measurement, in the Laboratory, of the Specific Optical Density of Smoke Generated by Materials. 32 pp.
BSI PD 6496-81. 1981 Amd 1 Comparison Between the Technical Requirements of BS 576 Part 8: 1972 with Other Relevant International Standards and Documents on Fire Resistance Tests. 38 pp.
BSI PD 6520-88. 1988 Guide to Fire test Methods for Building Materials and Elements of Construction. 14 pp.
CNS A3113-80. Method of Test for Incombustibility of Internal Finish Material of Buildings (Oct)(6532).
CNS A3125-81. Method of Test for Incombustibility of Thin Materials for Buildings (Jul)(7614).
CNS Z3011-73. Fire Test of Building Construction and Materials (Nov)(3581).
CNS Z3013-74. Method of Test for Surface Burning Characteristics of Building Materials (Mar)(3690).
DIN ENGL 4102 Suppl. 1-81. Fire Behaviour of Building Materials and Building Components; Tables of Contents (May). 5 pp.
DIN ENGL 4102 Pt 1-81. Fire Behaviour of Building Materials and Building Components; Building Materials; Concepts, Requirements and Tests (May) (Superseded in Parts by DIN 4102 Part 15). 16 pp.

INTERNATIONAL AND NON-U.S. NATIONAL STANDARDS
SUBJECT INDEX
Construction

Construction Materials *(Cont.)*
—**Fire Testing** *(Cont.)*
DIN ENGL 4102 Pt 4-81. Fire Behaviour of Building Materials and Building Components; Summary and Use of Classified Building Materials, Building Components and Special Building Components (Mar). 108 pp.
DIN ENGL 4102 Pt 11-85. Fire Behaviour of Building Materials and Building Components; Pipe Encasements, Pipe Bushings, Service Shafts and Ducts, and Barriers Across Inspection Openings; Terminology, Requirements and Testing (Dec). 12 pp.
DIN ENGL 4102 Pt 15-90. Fire Behaviour of Building Materials and Elements; 'Brandschacht' (May) (Supersedes Parts of DIN 4102 Part 1, May 1981 Edition). 15 pp.
ISO 1182-90. Fire Tests—Building Materials—Non-Combustibility Test Third Edition. 22 pp.
ISO TR3814-89. Tests for Measuring "Reaction-to-Fire" of Building Materials—Their Development and Application Second Edition; (Replaces TR6585). 17 pp.
ISO 5657-86. Fire Tests—Reaction to Fire—Ignitability of Building Products First Edition. 36 pp.
ISO 5660 Pt 1-93. Fire Tests—Reaction to Fire—Part 1: Rate of Heat Release from Building Products (Cone Calorimeter Method) First Edition. 36 pp.
ISO TR5924-89. Fire Tests—Reaction to Fire—Smoke Generated by Building Products (Dual-Chamber Test) First Edition. 38 pp.
ISO TR6167-84. Fire-Resistance Tests—Contribution Made by Suspended Ceilings to the Protection of Steel Beams in Floor and Roof Assemblies First Edition. 13 pp.
JIS A 1321-75. Testing Method for Incombustibility of Internal Finish Material and Procedure of Buildings. 11 pp.
JIS A 1322-66. Testing Method for Incombustibility of Thin Materials for Buildings (R 1982). 8 pp.
SAA AS 1530.4-90. Methods for Fire Tests on Building Materials, Components and Structures—Part 4: Fire-Resistance Test of Elements of Building Construction (This is a Joint Standard with SNZ NZS 1530.4). 27 pp.
SNZ NZMP 9-89. Fire Properties of Building Materials and Elements of Structure. 207 pp.
SNZ NZS/BS 476: Part 20-87. Fire Tests on Building Materials and Structures Part 20: Method for Determination of the Fire Resistance of Elements of Construction (General Principles). 40 pp.
SNZ NZS/AS 1530. 4-90. Methods for Fire Tests on Building Materials, Components and Structures Part 4: Fire-Resistance Test of Elements of Building Construction (This is a Joint Standard with SAA AS 1530.4). 27 pp.

—**Fire Testing—Furnaces**
DIN ENGL 4102 Pt 8-86. Fire Behaviour of Building Materials and Components; Small Scale Test Furnace (May). 4 pp.

—**Flammability Testing**
BSI BS 476: Part 4-70. 1970 Amd 2 Fire Tests on Building Materials and Structures Part 4: Non-Combustibility Test for Materials (AMD 4390) September 30, 1983. 13 pp.
BSI BS 476: Part 5-79. (WITHDRAWN) 1979 Amd 1 Fire Tests on Building Materials and Structures Part 5: Method of Test for Ignitability (Superseded by BS 476: Part 12 1991). 8 pp.
SAA AS 1530.1-84. Methods for Fire Tests on Building Materials, Components and Structures—Part 1: Combustibility Test for Materials Amdt 1 July 1993. 15 pp.
SAA AS 1530.2-93. Methods for Fire Tests on Building Materials, Components and Structures—Part 2: Test for Flammability of Materials (in Professional Package 63). 8 pp.
SAA AS 1530.3-89. Methods for Fire Tests on Building Materials, Components and Struc-tures—Part 3: Simul-taneous Determination of Ignitability, Flame Propagation, Heat Release and Smoke Release Amdt 1 April 1992 (in Professional Packages 20, 21, 30, 44, 55, 62-69). 1 p.
SAA AS 1530.5-89. Methods for Fire Tests on Building Materials, Components and Structures—Part 5: Test for Piloted Ignitability. 29 pp.
SNZ NZS/BS 476: Part 4-70. Fire Tests on Building Materials and Structures Part 4: Non-Combustibility Test for Materials. 12 pp.
SNZ NZS/AS 1530. 1-84. Methods for Fire Tests on Building Materials, Components and Structures Part 1: Combustibility Test for Materials (This is a Joint Standard with SAA AS 1530.1). 20 pp.
SNZ NZS/AS 1530. 2-73. Methods for Fire Tests on Building Materials, Components and Structures Part 2: Test for Flammability of Materials (This is a Joint Standard with SAA AS 1530.2). 8 pp.
SNZ NZS/AS 1530. 3-89. Methods for Fire Tests on Building Materials, Components and Structures Part 3: Simultaneous Determination of Ignitability, Flame Propagation, Heat Release and Smoke Release (This is a Joint Standard with SAA AS 1530.3). 13 pp.

Construction Materials *(Cont.)*
—**Glossaries**
CNS A1033-87. Glossary of Terms Used in Building Materials for Performing Uses (Mar)(9453). 1 p.
CNS A1034-87. Glossary of Terms Used in Building Materials for Finishing Surface (Mar)(9454).
CNS A1036-87. Glossary of Terms Used in Building Materials for Shape (Mar)(9653).
CNS A1037-87. Glossary of Terms Used in Building Materials (Mar)(9654).
JIS A 0201-71. Glossary of Terms Used in Finishing Building Materials for Interior and Exterior Uses.

—**Inspection**
DIN ENGL 18200-86. Inspection of Construction Materials, Structural Members and Types of Construction; General Principles (Dec). 3 pp.

—**Lead—Sheets**
BSI BS 1178-82. 1982 Milled Lead Sheet for Building Purposes. 7 pp.
BSI CP 143: Part 11-70. (WITHDRAWN) 1970 Sheet Roof and Wall Coverings Part 11: Lead: Metric Units (Superseded by BS 6915). 39 pp.

—**Lead—Sheets—Milled**
BSI BS 1178-82. 1982 Milled Lead Sheet for Building Purposes. 7 pp.
BSI BS 6915-88. 1988 Design and Construction of Fully Supported Lead Sheet Roof and Wall Coverings. 28 pp.

—**Organic—Fungus Resistance Testing**
BSI BS 1982-68. (WITHDRAWN) 1968 Methods of Test for Fungal Resistance of Manufactured Building Materials Made of or Containing Materials of Organic Origin (R) (Superseded by BS 1982: Parts 0, 1, 2 and 3: 1990). 16 pp.
BSI BS 1982: Part 0-90. 1990 Fungal Resistance of Panel Products Made of or Containing Materials of Organic Origin Part 0: Guide to Methods for Determination. 10 pp.
BSI BS 1982: Part 1-90. 1990 Fungal Resistance of Panel Products Made of or Containing Materials of Organic Origin Part 1: Method for Determination of Resistance to Wood-Rotting Basidiomycetes. 19 pp.
BSI BS 1982: Part 1-01. 1990 Amd 1 Fungal Resistance of Panel Products Made of or Containing Materials of Organic Origin Part 1: Method for Determination of Resistance to Wood-Rotting Basidiomycetes (AMD 7780) June 15, 1993 (R). 20 pp.
BSI BS 1982: Part 2-90. 1990 Fungal Resistance of Panel Products Made of or Containing Materials of Organic Origin Part 2: Method for Determination of Resistance to Cellulose-Decomposing Microfungi. 16 pp.
BSI BS 1982: Part 3-90. 1990 Fungal Resistance of Panel Products Made of or Containing Materials of Organic Origin Part 3: Methods for Determination of Resistance to Mould or Mildew. 12 pp.

—**Plastic—Environmental Testing**
CNS A3156-82. Method of Test for Change in Properties of Plastics Building Materials Resulting from Out-Door Exposure (Jun)(8908).
CNS A3157-82. Method of Test for Out-Door Exposure of Plastics Building Materials (Jun)(8909).
CNS A3158-82. Method of Test for Accelerated Artificial Exposure of Plastics Building Materials (Jun)(8910).
JIS A 1410-68. Recommended Practice for Out-Door Exposure of Plastics Building Materials (R 1982). 7 pp.
JIS A 1411-68. Standard Method of Test for Change in Properties of Plastics Building Materials Resulting from Out-Door Exposure (R 1982). 8 pp.
JIS A 1415-77. Recommended Practice for Accelerated Artificial Exposure of Plastics Building Materials (R 1982). 18 pp.

—**PVC**
BSI BS 3869-65. 1965 Amd 2 Rigid Expanded Polyvinyl Chloride for Thermal Insulation Purposes and Building Applications. 17 pp.

—**Resins**
BSI BS 6319: Part 1-83. 1983 Testing of Resin Compositions for Use in Construction Part 1: Method for Preparation of Test Specimens. 7 pp.

—**Resins—Bonding Strength—Slant Shear Method**
BSI BS 6319: Part 4-84. 1984 Testing of Resin Compositions for Use in Construction Part 4: Method for Measurement of Bond Strength (Slant Shear Method). 12 pp.

—**Resins—Compression Testing**
BSI BS 6319: Part 2-83. 1983 Testing of Resin Composition for Use in Construction Part 2: Method for Measurement of Compressive Strength. 4 pp.

Construction Materials *(Cont.)*
—**Resins—Exothermic Peak Temperature**
BSI BS 6319: Part 9-87. 1987 Testing of Resin Compositions for Use in Construction Part 9: Method for Measurement and Classification of Peak Exotherm Temperature. 4 pp.

—**Resins—Flexural Strength**
BSI BS 6319: Part 3-90. 1990 Testing of Resin Compositions for Use in Construction Part 3: Methods for Measurement of Modulus of Elasticity in Flexure and Flexural Strength. 9 pp.

—**Safety**
EC COM(86) 756-87. Commission Proposal for a Council Directive on the Approximation of the Laws, Regulations and Administrative Provisions of the Member States Relating to Construction Products. 22 pp.
EC 89/106/EEC-88. Council Directive on the Approximation of Laws, Regulations and Administrative Provisions of the Member States Relating to Construction Products (NEW APP). 15 pp.
EC 93/68/EEC-93. Council Directive Amending Directives 87/404/EEC (Simple Pressure Vessels), 88/378/EEC (Safety of Toys), 89/106/EEC (Construction Products), 89/336/EEC (Electromagnetic Compatibility), 89/392/EEC (Machinery), 89/686/ EEC (Personal Protective. 22 pp.

—**Sealants—Adhesion Testing**
ISO 11431-93. Building Construction—Sealants—Determination of Adhesion/Cohesion Properties After Exposure to Artificial Light Through Glass First Edition. 7 pp.

—**Sealants—Classification**
ISO 11600-93. Building Construction—Sealants—Classification and Requirements First Edition; (Corrected and Reprinted -1993). 8 pp.

—**Sealants—Cohesion**
ISO 11431-93. Building Construction—Sealants—Determination of Adhesion/Cohesion Properties After Exposure to Artificial Light Through Glass First Edition. 7 pp.

—**Sealants—Compatibility Testing**
DIN ENGL 52452 Pt 1-89. Testing of Building Sealants for Compatibility with Construction Materials (Oct) (Supersedes September 1978 Edition of DIN 52453 Part 1 and April 1977 Edition of DIN 53453 Part 3.). 2 pp.

—**Sealants—Compression Testing**
ISO 11432-93. Building Construction—Sealants—Determination of Resistance to Compression First Edition. 6 pp.

—**Sealants—Environmental Testing**
ISO 11431-93. Building Construction—Sealants—Determination of Adhesion/Cohesion Properties After Exposure to Artificial Light Through Glass First Edition. 7 pp.

—**Sealants—Light Testing**
ISO 11431-93. Building Construction—Sealants—Determination of Adhesion/Cohesion Properties After Exposure to Artificial Light Through Glass First Edition. 7 pp.

—**Sizes**
SAA AS 1224-72. Preferred Sizes of Building Components (Metric Units). 12 pp.

—**Slags**
DIN ENGL 4301-81. Ferrous and Non-Ferrous Metallurgical Slag for Civil Engineering and Building Construction Use (Apr). 4 pp.

—**Smoke Density**
BSI BS 6401-83. 1983 Measurement, in the Laboratory, of the Specific Optical Density of Smoke Generated by Materials. 32 pp.
JIS A 1306-83. Measuring Method of Smoke Density Using Light Extinction Method. 7 pp.

—**Sprayed Coatings—Fire Protection**
BSI BS 8202: Part 1-87. 1987 Coatings for Fire Protection of Building Elements Part 1: Code of Practice for the Selection and Installation of Sprayed Mineral Coatings. 20 pp.

—**Standards Preparation**
EURO 1989 June. The European Harmonization of Construction Products. 213 pp.

—**Static Loads**
DIN ENGL 1055 Pt 1-78. Design Loads for Buildings; Stored Materials, Building Materials and Structural Members; Dead Load and Angle of Friction (July). 15 pp.

INTERNATIONAL AND NON-U.S. NATIONAL STANDARDS SUBJECT INDEX

Construction

Construction Materials (Cont.)

—**Steel—Vitreous Enamelled**
BSI BS 3830-73. 1973 Amd 1 Vitreous Enamelled Steel Building Components. 8 pp.

—**Straw Slabs**
BSI BS 4046: Part 2-71. (WITHDRAWN) 1971 Compressed Straw Building Slabs: Metric Units (Superseded by BS 4046: 1991). 13 pp.

—**Thermal Conductivity**
DIN ENGL 52611 Pt 1-91. Determination of Thermal Resistance of Building Elements; Laboratory Method (Jan). 6 pp.
ISO TR9165-88. Practical Thermal Properties of Building Materials and Products First Edition. 22 pp.

—**Thermal Insulation—Calculation Methods**
ISO 6946 Pt 1-86. Thermal Insulation—Calculation Methods—Part 1: Steady State Thermal Properties of Building Components and Building Elements First Edition. 6 pp.

—**Thermal Resistance**
DIN ENGL 52611 Pt 1-91. Determination of Thermal Resistance of Building Elements; Laboratory Method (Jan). 6 pp.
DIN ENGL 52611 Pt 2-90. Determination of Thermal Resistance of Building Elements; Use of Measured Values in Building Applications (Apr). 5 pp.
ISO TR9165-88. Practical Thermal Properties of Building Materials and Products First Edition. 22 pp.

—**Thermal Resistance—Calculation Methods**
CEN PREN 30211-92. Building Components and Building Elements—Thermal Resistance and Thermal Transmittance—Calculation Method. 35 pp.

—**Thermal Transmittance—Calculation Methods**
CEN PREN 30211-92. Building Components and Building Elements—Thermal Resistance and Thermal Transmittance—Calculation Method. 35 pp.

—**Vapor Transmission**
BSI BS 7374-90. 1990 Methods of Test for Water Vapour Transmission Resistance of Board Materials Used in Buildings. 15 pp.
DIN ENGL 52615-87. Testing of Thermal Insulating Materials; Determination of Water Vapour (Moisture) Permeability of Construction and Insulating Materials (Nov) (Supersedes DIN 52615 Part 1, June 1973 Edition and DIN 53429, February 1971 Edition). 5 pp.

—**Water Absorption**
DIN ENGL 52617-87. Determination of the Water Absorption Coefficient of Construction Materials (May). 4 pp.

—**Weight Measurement**
BSI BS 648-64. 1964 Amd 2 Schedule of Weights of Building Materials. 50 pp.

Construction Plant
Use: Construction Equipment

Construction Products
Use: Construction Materials

Construction Safety
Use: Occupational Safety and Health

Construction Sites
Use For: Building Sites *See Also:* Construction; Construction Equipment; Construction Materials; Earthwork

—**Earthwork**
BSI BS 8000: Part 1-89. 1989 Workmanship on Building Sites Part 1: Code of Practice for Excavation and Filling. 15 pp.

—**Electric Power Distribution**
BSI BS 4363-91. 1991 Distribution Assemblies for Electricity Supplies for Construction and Building Sites. 20 pp.
BSI BS 4363-68. 1968 Amd 2 Distribution Units for Electricity Supplies for Construction and Building Sites. 26 pp.
BSI BS 7375-91. 1991 Code of Practice for Distribution of Electricity on Construction and Building Sites. 23 pp.
BSI CP 1017-69. (WITHDRAWN) 1969 Amd 2 Distribution of Electricity on Construction and Building Sites (Superseded by BS 7375: 1991). 24 pp.
DIN VDE 0612-74. VDE-Specifications for Electricity Distribution Units for Use on Construction and Building Sites at Rated Voltages up to 380 V a.c. and Rated Currents up to 630 A (May). 24 pp.

Construction Sites (Cont.)

—**Electrical Installations**
BSI BS 4363-68. 1968 Amd 2 Distribution Units for Electricity Supplies for Construction and Building Sites. 26 pp.
IEC 364 Pt 7-704-89. Electrical Installations of Buildings Part 7: Requirements for Special Installations or Locations Section 704—Construction and Demolition Site Installations First Edition. 15 pp.

—**Noise Reduction**
BSI BS 5228: Part 1-84. 1984 Code of Practice for Noise Control on Construction and Demolition Sites Part 1: Code of Practice for Basic Information and Procedures for Noise Control. 63 pp.

—**Noise Reduction—Legislation**
BSI BS 5228: Part 2-84. 1984 Code of Practice for Noise Control on Construction and Demolition Sites Part 2: Guide to Noise Control Legislation for Construction and Demolition, Including Road Construction and Maintenance. 8 pp.

Construction Timber
Use: Structural Timber

Consultants
See Also: Personnel

—**Service Contracts**
MOD UK DEFCON 112Z-86. Conditions of Contract for Consultancy Services 8/86. 3 pp.

Consultative Committee on International Telegraphy and Telephony
Use: CCITT

Consultative Council for Postal Studies
Use For: CCPS

—**Cooperation—CCITT**
CCITT FASCICLE I.1-88. Minutes and Reports of the Plenary Assembly List of Study Groups and Questions Under Study. 249 pp.
CCITT FASCICLE I.2-88. Opinions and Resolutions Recommendations on the Organisation and Working Procedures of CCITT (Series A). 64 pp.

Consumable Electrodes

—**Structural Steels—Submarines**
MOD UK NES 769-01. Welding Consumables for Structural Steel Approval System Issue 3 (03.82); Amendment 8. 65 pp.
MOD UK NES 769-92. Approval System for Welding Consumables for Structural Steels Issue 4 (12.92). 62 pp.

Consumer Information Systems
Use For: Product Information for Consumers
See Also: Information Systems
IEC Guide 14-77. Product Information for Consumers First Edition. 5 pp.
ISO Guide 14-77. Product Information for Consumers First Edition. 5 pp.

—**Product Identification**
IEC Guide 14-77. Product Information for Consumers First Edition. 5 pp.
ISO Guide 14-77. Product Information for Consumers First Edition. 5 pp.

—**Product Instructions**
IEC Guide 37-83. Instructions for Use of Products of Consumer Interest First Edition. 8 pp.
ISO Guide 37-83. Instructions for Use of Products of Consumer Interest First Edition. 8 pp.

Consumer Safety
See Also: Child Resistant Packaging; Electrical Safety; Marine Safety; Medical Safety; Skid Resistance
DIN ENGL 31001 Pt 1-83. Safety Design of Technical Products; Safety Devices; Concepts, Safety Distances for Adults and Children (Apr) (Superseded in Part by EN 294, August 1992 Edition). 7 pp.
EC 92/59/EEC-92. Council Directive on General Product Safety. 9 pp.
EC 88/41/EEC-87. Council Recommendation on the Involvement and Improvement of Consumer Participation in Standardization. 1 p.
EURO MEMORANDUM 2-77. Consumer Interests and the Preparation of Standards. 7 pp.

—**Aerosol Containers**
CNS Z5083-81. Safe Fill of Aerosol Containers (Mar)(7182). 1 p.

Consumer Safety (Cont.)

—**Air Cleaners**
DIN VDE 0700 Pt 259-88. Safety of Household and Similar Electrical Appliances; Air Cleaning Appliances (July). 16 pp.
IEC 335 Pt 2-65-93. Safety of Household and Similar Electrical Appliances Part 2: Particular Requirements for Air-Cleaning Appliances First Edition. 29 pp.

—**Air Conditioners**
BSI BS 3456: Sec 2.34-76. 1976 Amd 2 Safety of Household Electrical Appliances Part 2: Particular Requirements Section 2.34: Room Air Conditioners (AMD 6077) April 30, 1990. 17 pp.
DIN ENGL 1946 Pt 2-83. Air Conditioning; Health Requirements (VDI Ventilation Rules) (Jan). 9 pp.

—**Air Preheaters**
CEN PREN 1020-93. Requirements for Non-Domestic Gas-Fired Forced Convection Air Heaters for Space Heating Incorporating a Fan to Assist Transportation of Combustion Air and/or Combustion Products. 137 pp.

—**Amusement Rides**
CSA Z267-M1983. Safety Code for Amusement Rides; (Gen Instr 1). 29 pp.

—**Anesthesia Equipment—Electric**
CENELEC HD 395.2.13 S1-89. Medical Electrical Equipment Part 2: Particular Requirements for the Safety of Anaesthetic Machines. 3 pp.

—**Appliances**
BSI BS 3456: Part 101-87. 1987 Amd 1 Safety of Household Electrical Appliances Part 101: General Requirements (AMD 6055) August 31, 1990. 85 pp.
BSI BS 3456: Sec 102.15-87. 1987 Safety of Household Electrical Appliances Part 102: Particular Requirements Section 102.15: Appliances for Heating Liquids. 28 pp.
BSI BS 3456: Sec 102.27-87. (WITHDRAWN) 1987 Safety of Household and Similar Electrical Appliances Part 102: Particular Requirements Section 102.27: Ultra-Violet and Infra-Red Radiation Skin Treatment Appliances for Household Use (Superseded by BS 3456: Sec. 202.27: 1991). 20 pp.
BSI BS 3456: Part 201-90. 1990 Amd 3 Safety of Household and Similar Electrical Appliances Part 201: General Requirements (AMD 6494) November 30, 1990 (L). 196 pp.
BSI BS 3456: Part 201-06. 1990 Amd 6 Safety of Household and Similar Electrical Appliances Part 201: General Requirements (AMD 7347) November 15, 1992 (L). 216 pp.
BSI BS 3456: Part 201-07. 1990 Amd 7 Safety of Household and Similar Electrical Appliances Part 201: General Requirements (AMD 7564) March 15, 1993 (L). 225 pp.
BSI BS 3456: Part 201-08. 1990 Amd 8 Safety of Household and Similar Electrical Appliances Part 201: General Requirements (AMD 7887) August 15, 1993 (L). 230 pp.
BSI BS 3456: Part 1-69. 1969 Amd 5 Safety of Household Elecrical Appliances Part 1: General Requirements. 110 pp.
BSI BS 5258: Part 12-01. 1990 Amd 1 Safety of Domestic Gas Appliances Part 12: Specification for Decorative Fuel Effect Gas Appliances (2nd and 3rd Family Gases) (AMD 7507) January 15, 1993. 34 pp.
BSI BS 5415: Part 1-85. 1985 Amd 1 Safety of Electrical Motor-Operated Industrial Cleaning Appliances Part 1: General Requirements (AMD 6452) November 30, 1990. 73 pp.
BSI BS 7462-91. 1991 Electrical Safety of Domestic Gas Appliances. 52 pp.
BSI BS EN 60335-2-54-92. 1992 Safety of Household and Similar Electrical Appliances Part 2: Particular Requirements Section 2.54: General Purpose Cleaning Appliances. 18 pp.
BSI PP 888-82. (WITHDRAWN) 1982 Safety and Performance of Domestic Electrical Appliances. 33 pp.
CEN EN 30-79. Domestic Cooking Appliances Burning Gas. 50 pp.
CEN PREN 30 (Part 1)-90. Domestic Cooking Appliances Burning Gas—Part 1: Safety. 122 pp.
CENELEC HD 250-74. General Specification for Electric Motor Operated Appliances for Household and Similar Purposes. 12 pp.
CENELEC HD 250.2-77. General Specification for Electric Motor Operated Appliances for Household and Similar Purposes. 2 pp.
CENELEC HD 251 S3-82. Safety of Household and Similar Electrical Appliances: General Requirements. 2 pp.
CENELEC HD 251 S3-87. Safety of Household and Similar Electrical Appliances: General Requirements. 38 pp.
CENELEC HD 251 S3/A1-85. AMD 1 Safety of Household and Similar Electrical Appliances: General Requirements. 18 pp.

INTERNATIONAL AND NON-U.S. NATIONAL STANDARDS
SUBJECT INDEX

Consumer

Consumer Safety (Cont.)
—Appliances (Cont.)

CENELEC HD 251 S3/A2-87. AMD 2 Safety of Household and Similar Electrical Appliances: General Requirements. 8 pp.

CENELEC HD 264 S2-84. Safety of Household and Similar Electrical Appliances—Particular Requirements for Appliances for Heating Liquids. 8 pp.

CENELEC HD 289 S1-90. Safety of Household and Similar Appliances Particular Rules for Routine Tests Referring to Appliances Under the Scope of EN 60 335-1. 7 pp.

CENELEC HD 289 S1/A1-92. AMD 1 Safety of Household and Similar Appliances Particular Rules for Routine Tests Referring to Appliances Under the Scope of EN 60 335-1. 6 pp.

CENELEC EN 60 335-1-88. Safety of Household and Similar Electrical Appliances Part 1: General Requirements. 32 pp.

CENELEC EN 60 335-1/A2-88. AMD 2 Safety of Household and Similar Electrical Appliances Part 1: General Requirements. 8 pp.

CENELEC EN 60 335-1/A5-89. AMD 5 Safety of Household and Similar Electrical Appliances Part 1: General Requirements. 6 pp.

CENELEC EN 60 335-1/A6-89. AMD 6 Safety of Household and Similar Electrical Appliances Part 1: General Requirements. 7 pp.

CENELEC EN 60335-1/A51-91. AMD 51 Safety of Household and Similar Electrical Appliances Part 1: General Requirements. 7 pp.

CENELEC EN 60335-1/A52-92. AMD 52 Safety of Household and Similar Electrical Appliances Part 1: General Requirements. 7 pp.

CENELEC EN 60335-1/A53-92. AMD 53 Safety of Household and Similar Electrical Appliances Part 1: General Requirements. 6 pp.

CENELEC EN 60335-1/A54-92. AMD 54 Safety of Household and Similar Electrical Appliances Part 1: General Requirements. 5 pp.

CNS C4126-90. Safety Testing Standards for Household Electrical Appliances (Oct)(3766).

CSA C22.2 NO 1335.1-93. Portable Electrical Motor-Operated and Heating Appliances: General Requirements; (Gen Instr 1). 208 pp.

DIN VDE 0700 Pt 1 A2-86. Safety of Household and Similar Electrical Appliances; General Requirements; Amendment 2 (Sept). 8 pp.

DIN VDE 0700 Pt 1 A3 (D)-81. Safety of Household and Similar Electrical Appliances; General Requirements (VDE Specification): Amendment 3 (July). 4 pp.

DIN VDE 0700 Pt 1 A7 (D)-82. Safety of Household and Similar Electrical Appliances; Part 1: General Requirements—Calibration of Spring-Operated Impact Test Apparatus: Amendment 7 (VDE Specification) (Feb). 10 pp.

DIN VDE 0700 Pt 1 A8 (D)-82. Safety of Household and Similar Electrical Appliances; General Requirements: Amendment 8 (VDE Specification) (Mar). 6 pp.

DIN VDE 0700 Pt 1 A11 (D)-83. (Amendment to IEC 335-1) Safety of Household and Similar Electrical Appliances; Part 1: General Requirements: Amendment 11 (VDE Specification) (DIN IEC 61(CO)330) (May). 9 pp.

DIN VDE 0700 Pt 1 A13 (D)-84. (Amendment to IEC 335-1) Safety of Household and Similar Electrical Appliances; Part 1: General Requirements: Amendment 13 (VDE Specification) (DIN IEC 61(CO)342) (Mar). 7 pp.

DIN VDE 0700 Pt 1 A14 (D)-84. (Amendment to IEC 335-1) Safety of Household and Similar Electrical Appliances; Part 1: General Requirements: Amendment 14 (VDE Specification) (DIN IEC 61(CO)346) (Mar). 5 pp.

DIN VDE 0700 Pt 1 A15 (D)-84. Safety of Household and Similar Electrical Appliances; General Requirements: Amendment 15 (VDE Specification) (Mar). 4 pp.

DIN VDE 0700 Pt 1 A17 (D)-84. Safety of Household and Similar Electrical Appliances; General Requirements: Amendment 17 (VDE Specification) (Sept). 6 pp.

DIN VDE 0700 Pt 1 A18 (D)-84. Safety of Household and Similar Electrical Appliances; General Requirements: Amendment 18 (VDE Specification) (Dec). 12 pp.

DIN VDE 0700 Pt 1 A19 (D)-85. Safety of Household and Similar Electrical Appliances; General Requirements: Amendment 19 (June). 8 pp.

DIN VDE 0700 Pt 1 A21 (D)-86. Safety of Household and Similar Electrical Appliances; General Requirements: Amendment 21; Identical to IEC 61(CO)458 (Sept). 7 pp.

DIN VDE 0700 Pt 1 A24 (D)-86. Safety of Household and Similar Electrical Appliances; General Requirements: Amendment 24; Identical to IEC 61(CO)468 (Sept). 7 pp.

DIN VDE 0700 Pt 1 A25 (D)-87. Safety of Household and Similar Electrical Appliances; General Requirements: Amendment 25 (Mar). 6 pp.

Consumer Safety (Cont.)
—Appliances (Cont.)

DIN VDE 0700 Pt 1 A26 (D)-87. Safety of Household and Similar Electrical Appliances; General Requirements: Amendment 26; Identical to IEC 61(CO)484 (Sept) (Supersedes Draft 0700 Part 1 A20/5.86). 18 pp.

DIN VDE 0700 Pt 1 A27 (D)-87. Safety of Household and Similar Electrical Appliances; General Requirements: Amendment 27 (Sept). 6 pp.

DIN VDE 0700 Pt 1 A28 (D)-87. Safety of Household and Similar Electrical Appliances; General Requirements: Amendment 28 (Oct). 5 pp.

DIN VDE 0700 Pt 1 A29 (D)-88. Safety of Household and Similar Electrical Appliances; General Requirements: Amendment 29 (Oct). 4 pp.

DIN VDE 0700 Pt 1 A30 (D)-89. Safety of Household and Similar Electrical Appliances; General Requirements: Amendment 30 (Mar). 5 pp.

IEC 335 Pt 1-91. Safety of Household and Similar Electrical Appliances Part 1: General Requirements Third Edition; (I-SH 01—Feb 1993). 246 pp.

IEC 335 Pt 2-18-84. Safety of Household and Similar Electrical Appliances Part 2: Guide for Preparing Safety Requirements for Battery-Powered Motor-Operated Appliances and Their Charging and Battery Assemblies Second Edition. 39 pp.

IEC 730 Pt 2-9-92. Automatic Electrical Controls for Household and Similar Use Part 2: Particular Requirements for Temperature Sensing Controls First Edition. 52 pp.

ISO 7421-91. Leisure Accommodation Vehicles—Liquefied Petroleum Gas Systems First Edition. 7 pp.

—Awnings

BSI BS 5576-85. 1985 Safety Features of Camping Tents, Awnings, Trailer Tents and Caravan Awnings. 4 pp.

—Baby Carriages

BSI BS 4139-86. (WITHDRAWN) 1986 Amd 1 Safety Requirements for Perambulators (Baby Carriages) (AMD 6474) March 30, 1990 (Superseded by BS 7409: 1991). 22 pp.

BSI BS 7409-91. 1991 Safety Requirements for Wheeled Child Conveyances. 32 pp.

BSI BS 7409-01. 1991 Amd 1 Safety Requirements for Wheeled Child Conveyances (AMD 7177) July 15, 1992. 34 pp.

SAA AS 2088-89. Prams and Strollers—Safety Requirements Amdt 1 May 1990 Amdt 2 January 1991 Amdt 3 June 1991 (Superseded by AS/NZS 2088:1993).

SNZ NZS/AS 2088-93. Prams and Strollers—Safety Requirements (This is a Joint Standard with SAA AS 2088). 43 pp.

SNZ NZS 5804-86. Safety Requirements for Baby Carriages Amend: 1, 1987. 26 pp.

—Baby Walkers

BSI BS 4648-89. 1989 Amd 1 Safety Requirements for Baby Walking Frames (AMD 6680) April 30, 1991. 13 pp.

BSI BS 4648-02. 1989 Amd 2 Safety Requirements for Baby Walking Frames (AMD 6948) February 28, 1992. 14 pp.

SNZ NZS 5846-88. Specification for Safety Requirements for Baby Walking Frames Amend: 1, 1976; 2, 1980; A, 1988. 8 pp.

—Band Saws

IEC 1029 Pt 2-5-93. Safety of Transportable Motor-Operated Electric Tools Part 2: Particular Requirements for Band Saws First Edition. 24 pp.

—Barbecues

BSI BS 5258: Part 14-84. 1984 Safety of Domestic Gas Appliances Part 14: Specification for Barbecues (3rd Family Gas). 19 pp.

—Bassinets

BSI BS 3881-86. (WITHDRAWN) 1986 Amd 1 Safety Requirements for Carry Cots. Carry Cot Stands and Carry Cot Transporters (AMD 6473) April 30, 1990 (Superseded by BS 7409: 1991 & BS 7551: 1992). 15 pp.

BSI BS 7551-92. 1992 Safety Requirements for Carry Cots and Similar Handled Products and Stands. 13 pp.

—Batteries

BSI BS 6132-83. 1983 Safe Operation of Alkaline Secondary Cells and Batteries. 13 pp.

BSI BS 6133-85. 1985 Amd 1 Code of Practice for Safe Operation of Lead-Acid Stationary Cells and Batteries. 19 pp.

BSI BS 6604-85. 1985 Code of Practice for Safe Operation of Starter Batteries. 8 pp.

Consumer Safety (Cont.)
—Battery Chargers

BSI BS 3456: Sec 3.25-81. (WITHDRAWN) 1981 Amd 1 1972-1981 Edition. Specification for Safety of Household Electrical Appliances Part 3: Complete Particular Specifications Section 3.25: Battery Chargers (Superseded by BS EN 60335-2-29: 1991). 56 pp.

BSI BS EN 60335-2-29-91. 1991 Safety of Household and Similar Electrical Appliances Part 2: Particular Requirements Section 2.29: Battery Chargers. 23 pp.

CENELEC EN 60 335-2-29-91. Safety of Household and Similar Electrical Appliances Part 2: Particular Requirements for Battery Chargers. 9 pp.

DIN VDE 0700 Pt 207-82. Safety of Household and Similar Electrical Appliances; Particular Requirements for Power Supply Appliances up to 20 VA (Oct). 21 pp.

IEC 335 Pt 2-29-87. Safety of Household and Similar Electrical Appliances Part 2: Particular Requirements for Battery Chargers Second Edition; (Amendment 1-1989) (Amendment 2-1991). 62 pp.

SNZ NZS 6329-90. Safety of Household and Similar Electrical Appliances. Particular Requirements for Battery Chargers. 20 pp.

—Beds

BSI BS EN 747-1-93. 1993 Furniture—Bunk Beds for Domestic Use Part 1: Safety Requirements (L). 10 pp.

BSI BS EN 747-2-93. 1993 Furniture—Bunk Beds for Domestic Use Part 2: Test Methods (L). 15 pp.

CEN EN 747-1-93. Furniture—Bunk Beds for Domestic Use—Part 1: Safety Requirements. 7 pp.

CEN EN 747-2-93. Furniture—Bunk Beds for Domestic Use—Part 2: Test Methods. 12 pp.

—Bench Grinders

IEC 1029 Pt 2-4-93. Safety of Transportable Motor-Operated Electric Tools Part 2: Particular Requirements for Bench Grinders First Edition. 38 pp.

—Bicycles

BSI BS 6101: Part 1-81. 1981 Machine Tool Ball Screws Part 1: Methods of Calculating Dynamic Load and Life Ratings. 6 pp.

BSI BS 6102: Part 1-92. 1992 Cycles Part 1: Specification for Safety Requirements for Bicycles. 37 pp.

BSI BS 6102: Part 1-81. 1981 Amd 4 Cycles Part 1: Safety Requirements for Bicycles (AMD 5540) May 29, 1987. 45 pp.

CSA CAN3-D113.1-M80. Bicycles. 34 pp.

ISO 4210-89. Cycles—Safety Requirements of Bicycles Third Edition. 34 pp.

ISO 8098-89. Cycles—Safety Requirements for Bicycles for Young Children First Edition; (Amendment 1-1992). 24 pp.

JIS D 9102-90. Bicycles for General Use —Safety Requirements. 19 pp.

SAA AS 1927-89. Pedal Bicycles for Normal Road Use—Safety Requirements (This is a Joint Standard with SANZ NZS 1927:1989). 31 pp.

SNZ NZMP 205-91. Guide to New Zealand Product Safety Standard: Pedal Bicycles for Normal Use—Safety Requirements. 36 pp.

SNZ NZS/AS 1927-89. Pedal Bicycles for Normal Road Use—Safety Requirements (This is a Joint Standard with SAA AS 1927). 31 pp.

—Blankets

SNZ NZS 6317-88. Specification for Safety of Electric Blankets. 60 pp.

—Boilers

BSI BS 5258: Part 1-86. 1986 Safety of Domestic Gas Appliances Part 1: Central Heating Boilers and Circulators. 43 pp.

BSI BS 5258: Pt 1: Supp 1-83. 1983 Safety of Domestic Gas Appliances Part 1: Central Heating Boilers and Circulators Supplement 1: Fan-Powered Appliances. 13 pp.

BSI BS 5258: Part 8-80. 1980 Safety of Domestic Gas Appliances Part 8: Combined Appliances: Gas Fire/Back Boiler. 24 pp.

BSI BS 5258: Part 15-90. 1990 Safety of Domestic Gas Appliances Part 15: Combination Boilers. 52 pp.

BSI DD 189-90. 1990 Safety of Condensing Boilers (2nd and 3rd Family Gases). 9 pp.

—Brush Saws

CEN PREN 31806-92. Safety Requirements for Agricultural and Forestry Machinery—Brush Cutters and Grass Trimmers. 28 pp.

—Buntings (Clothing)

BSI BS 6595-85. 1985 Safety Requirements for Baby Nests. 4 pp.

INDUSTRY STANDARDS

Consumer Safety *(Cont.)*

—Cable Cutters

CEN PREN 792-10-92. Handheld Non-Electric Power Tools—Safety Requirements—Part 10: Compression Power Tools, Squeeze Riveters, Presses, Punches. 6 pp.

—Carpet Cleaning Equipment

BSI BS 3456: Sec 202.2-90. 1990 Safety of Household Electrical Appliances Part 202: Particular Requirements Section 202.2: Vacuum Cleaners and Water Suction Cleaning Appliances. 25 pp.

BSI BS 3456: Sec 202.2-02. 1990 Amd 2 Safety of Household and Similar Electrical Appliances Part 202: Particular Requirements Section 202.2: Vacuum Cleaners and Water Suction Cleaning Appliances (AMD 7296) December 15, 1992 (L). 51 pp.

BSI BS EN 60335-2-10-92. 1992 Safety of Household and Similar Electrical Appliances Part 2: Particular Requirements Section 2.10: Floor Treatment Machines and Wet Scrubbing Machines (Supersedes BS 3456: Section 102.10: 1992). 19 pp.

BSI BS EN 60335-2-10-01. 1992 Amd 1 Safety of Household and Similar Electrical Appliances Part 2: Particular Requirements Section 2.10: Floor Treatment Machines and Wet Scrubbing Machines (AMD 7792) July 15, 1993 (L). 22 pp.

CENELEC HD 252-74. Particular Specification for Vacuum Cleaness and Water Suction Cleaning Appliances. 5 pp.

CENELEC HD 252 S3-86. Safety of Household and Similar Electrical Appliances—Part 2: Particular Requirements for Vacuum Cleaners and Water Suction Cleaning Appliances (Superseded by EN 60335-2-2). 4 pp.

CENELEC HD 281 S1-86. Safety of Household and Similar Electrical Appliances—Part 2: Particular Requirements for Floor Treatment Machines and Wet Scrubbing Machines. 10 pp.

CENELEC EN 60 335-2-2-88. Safety of Household and Similar Electrical Appliances Part 2: Particular Requirements for Vacuum Cleaners and Water Suction Cleaning Appliances. 10 pp.

CENELEC EN 60335-2-2 PRAC-93. AMD prAC Safety of Household and Similar Electrical Appliances Part 2: Particular Requirements for Vacuum Cleaners and Water Suction Cleaning Appliances. 3 pp.

CENELEC EN 60335-2-2 /A2-90. AMD 2 Safety of Household and Similar Electrical Appliances Part 2: Particular Requirements for Vacuum Cleaners and Water Suction Cleaning Appliances. 7 pp.

CENELEC EN 60335-2-2 /A52-91. AMD 52 Safety of Household and Similar Electrical Appliances Part 2: Particular Requirements for Vacuum Cleaners and Water Suction Cleaning Appliances. 7 pp.

CENELEC EN 60335-2-1 0-90. Safety of Household and Similar Electrical Appliances Part 2: Particular Requirements for Floor Treatment Machines and Wet Scrubbing Machines. 10 pp.

CENELEC EN 60335-2-10-90. Safety of Household and Similar Electrical Appliances Part 2: Particular Requirements for Floor Treatment Machines and Wet Scrubbing Machines; (IEC 335-2-10: 1987, Modified). 12 pp.

CENELEC EN 60335-2-10/A1-92. AMD 1 Safety of Household and Similar Electrical Appliances Part 2: Particular Requirements for Floor Treatment Machines and Wet Scrubbing Machines (IEC 335-2-10:1987/A1:1991). 4 pp.

IEC 335 Pt 2-2-93. Safety of Household and Similar Electrical Appliance Part 2: Particular Requirements for Vacuum Cleaners and Water Suction Cleaning Appliances Fourth Edition. 42 pp.

IEC 335 Pt 2-10-92. Safety of Household and Similar Electrical Appliances Part 2: Particular Requirements for Floor Treatment Machines and Wet Scrubbing Machines Fourth Edition; (CENELEC EN 60335-2-10: 1992). 28 pp.

—Cells

BSI BS 6132-83. 1983 Safe Operation of Alkaline Secondary Cells and Batteries. 13 pp.

BSI BS 6133-85. 1985 Amd 1 Code of Practice for Safe Operation of Lead-Acid Stationary Cells and Batteries. 19 pp.

—Chain Saws

CEN PREN 608-91. Safety Requirements for Agricultural and Forestry Machinery—Portable Chain Saws. 23 pp.

IEC 745 Pt 2-13-89. Safety of Hand-Held Motor-Operated Electric Tools Part 2: Particular Requirements for Chain Saws First Edition; (Amendment 1-1992). 32 pp.

SNZ NZS 5819-82. Chain Saw Safety (Reconfirmed 1989). 12 pp.

—Chairs—Children's

BSI BS 7495-01. 1991 Amd 1 Safety Requirements for Children's Table-Mounted Chairs for Domestic Use (AMD 7194) July 15, 1992. 16 pp.

SNZ NZS 5856-93. Safety Requirements for Children's Table-Mounted Chairs Amend: A, 1993. 16 pp.

Consumer Safety *(Cont.)*

—Children

ISO Guide 50-87. Child Safety and Standards—General Guidelines First Edition. 16 pp.

—Cleaning Equipment and Supplies

CENELEC EN 60 335-2-54-91. Safety of Household and Similar Electrical Appliances Part 2: Particular Requirements for General Purpose Cleaning Appliances. 6 pp.

CENELEC EN 60335-2-54-92. Safety of Household and Similar Electrical Appliances Part 2: Particular Requirements for General Purpose Cleaning Appliances. 14 pp.

IEC 335 Pt 2-54-88. Safety of Household and Similar Electrical Appliances Part 2: Particular Requirements for General Purpose Cleaning Appliances First Edition. 25 pp.

—Clocks

BSI BS 3456: Sec 102.26-88. (WITHDRAWN) 1988 Safety of Household Electrical Appliances Part 102: Particular Requirements Section 102.26: Clocks (Superseded by BS EN 60335-2-26: 1991). 16 pp.

BSI BS EN 60335-2-26-91. 1991 Safety of Household and Similar Electrical Appliances Part 2: Particular Requirements Section 2.26: Clocks (Supersedes BS 3456: Section 102.26: 1988). 19 pp.

CENELEC HD 255 S2-83. Safety of Household and Similar Electric Appliances Particular Requirements for Clocks. 7 pp.

CENELEC EN 60 335-2-26-90. Safety of Household and Similar Electrical Appliances Part 2: Particular Requirements for Clocks. 6 pp.

IEC 335 Pt 2-26-87. Safety of Household and Similar Electrical Appliances Part 2: Particular Requirements for Clocks Second Edition. 25 pp.

—Clothes Dryers

BSI BS 3456: Sec 102.11-88. (WITHDRAWN) 1988 Safety of Household Electrical Appliances Part 102: Particular Requirements Section 102.11: Tumbler Dryers (Superseded by BS 3456: Section 202.11: 1990). 20 pp.

BSI BS 3456: Sec 202.11-90. 1990 Safety of Household Electrical Appliances Part 202: Particular Requirements Section 202.11: Tumbler Dryers. 25 pp.

BSI BS 3456: Sec 202.11-01. 1990 Amd 1 Safety of Household and Similar Electrical Appliances Part 202: Particular Requirements Section 202.11: Tumbler Dryers (AMD 6913) July 15, 1992. 36 pp.

BSI BS 3456: Sec 202.11-02. 1990 Amd 2 Safety of Household and Similar Electrical Appliances Part 202: Particular Requirements Section 202.11: Tumbler Dryers (AMD 7664) June 15, 1992 (L). 42 pp.

BSI BS 3456: Sec 202.11-03. 1990 Amd 3 Safety of Household and Similar Electrical Appliances Part 202: Particular Requirements Section 202.11: Tumbler Dryers (AMD 7869) July 15, 1993 (SUPERSEDES BS 3456: SECTION 102. 11: 1988). 51 pp.

BSI BS 3456: Sec 202.43-90. 1990 Safety of Household Electrical Appliances Part 202: Particular Requirements Section 202.43: Clothes Dryers and Towel Rails. 20 pp.

BSI BS 3456: Sec 202.43-01. 1990 Amd 1 Safety of Household and Similar Electrical Appliances Part 202: Particular Requirements Section 202.43: Clothes Dryers and Towel Rails (AMD 6901) April 1, 1992 (SUPERSEDES BS 3456 SECTION 2.9: 1970 & 3. 6: 1979). 26 pp.

BSI BS 3456: Sec 202.43-02. 1990 Amd 2 Safety of Household and Similar Electrical Appliances Part 202: Particular Requirements Section 202.43: Clothes Dryers and Towel Rails (AMD 7550) JAN 15, 1993 SUPERSEDES BS 3456 SECTION 2. 9: 1970 & 3. 6: 1979. 31 pp.

BSI BS 5258: Part 17-92. 1992 Safety of Domestic Gas Appliances Part 17: Specification for Direct Gas-Fired Tumble Dryers (2nd and 3rd Family Gases). 32 pp.

BSI BS EN 50084-92. 1992 Safety of Household and Similar Electrical Appliances. Requirements for the Connection of Washing Machines, Dishwashers and Tumbler Dryers to the Water Mains (L). 24 pp.

BSI BS EN 60335-2-45-91. 1991 Safety of Household and Similar Electrical Appliances Part 2: Particular Requirements Section 2.45: Portable Electric Heating Tools and Similar Appliances. 23 pp.

BSI BS EN 60335-2-45-01. 1991 Amd 1 Safety of Household and Similar Electrical Appliances Part 2: Particular Requirements Section 2.45: Portable Electric Heating Tools And Similar Appliances. 44 pp.

CENELEC HD 263 S1-84. Particular Specification for Clothes Dryers and Towel Dryers. 5 pp.

CENELEC HD 268-76. Particular Specification for Clothes Dryers of the Tumbler Type (Superseded by EN 60335-2-11). 9 pp.

CENELEC HD 268 S2-84. Safety of Household and Similar Electrical Appliances-Particular Requirements for Tumble Dryers. 14 pp.

Consumer Safety *(Cont.)*

—Clothes Dryers *(Cont.)*

CENELEC EN 50084-92. Safety of Household and Similar Electrical Appliances—Requirements for the Connection of Washing Machines, Dishwashers and Tumbler Dryers to the Water Mains. 21 pp.

CENELEC EN 60 335-2-11-89. Safety of Household and Similar Electrical Appliances—Part 2: Particular Requirements for Tumbler Dryers. 10 pp.

CENELEC EN 60 335-2-11/PRAB-89. AMD prAB Safety of Household and Similar Electrical Appliances—Part 2: Particular Requirements for Tumbler Dryers. 2 pp.

CENELEC EN 60335-2-11/A1-90. AMD 1 Safety of Household and Similar Electrical Appliances—Part 2: Particular Requirements for Tumbler Dryers. 7 pp.

CENELEC EN 60335-2-11/A2-92. AMD 2 Safety of Household and Similar Electrical Appliances Part 2: Particular Requirements for Tumbler Dryers (IEC 335-2-11:1984/A2:1991). 4 pp.

CENELEC EN 60 335-2-43-89. Safety of Household and Similar Electrical Appliances—Part 2: Particular Requirements for Clothes Dryers and Towel Rails. 7 pp.

CENELEC EN 60 335-2-43/A1-90. AMD 1 Safety of Household and Similar Electrical Appliances—Part 2: Particular Requirements for Clothes Dryers and Towel Rails. 6 pp.

CENELEC EN 60335-2-43/A51-92. AMD 51 Safety of Household and Similar Electrical Appliances—Part 2: Particular Requirements for Clothes Dryers and Towel Rails. 4 pp.

CENELEC EN 60 335-2-45-90. Safety of Household and Similar Electrical Appliances Part 2: Particular Requirements for Portable Electric Heating Tools and Similar Appliances. 11 pp.

CENELEC EN 60335-2-45/PRAB-93. AMD AB Safety of Household and Similar Electric Appliances. Part 2: Particular Requirements for Portable Electric Heating Tools and Similar Appliances—Sub-Clauses 8.1 and 22.107. 2 pp.

CENELEC EN 60335-2-45/A1-92. AMD 1 Safety of Household and Similar Electrical Appliances Part 2: Particular Requirements for Portable Electric Heating Tools and Similar Appliances (IEC 335-2-45:1986/A1:1990). 5 pp.

DIN VDE 0700 Pt 600 (D)-88. Safety of Household and Similar Electrical Appliances; Connection of Washing Machines, Dishwashers and Tumble Dryers to the Supply Mains; German Version prHD 274 S2: 1987 (Oct). 37 pp.

IEC 335 Pt 2-11-93. Safety of Household and Similar Electrical Appliances Part 2: Particular Requirements for Tumbler Dryers Fourth Edition. 37 pp.

IEC 335 Pt 2-43-84. Safety of Household and Similar Electrical Appliances Part 2: Particular Requirements for Clothes Dryers and Towel Rails First Edition; (Amendment 1-1988). 35 pp.

IEC 335 Pt 2-45-86. Safety of Household and Similar Electrical Appliances Part 2: Particular Requirements for Portable Electric Heating Tools and Similar Appliances First Edition; (Amendment 1-1990) (CENELEC EN 60335-2-45/A1:1992). 48 pp.

—Coffee Grinders

BSI BS 3456: Sec 102.33-88. (WITHDRAWN) 1988 Safety of Household Electrical Appliances Part 102: Particular Requirements Section 102.33: Coffee Mills and Coffee Grinders (Superseded by BS EN 60335-2-33: 1992). 18 pp.

BSI BS EN 60335-2-33-92. 1992 Safety of Household and Similar Electrical Appliances Part 2: Particular Requirements Section 2.33: Coffee Mills and Coffee Grinders. 19 pp.

CENELEC HD 260-75. Particular Specification for Coffee Grinders and Mills. 5 pp.

CENELEC HD 260 S3-87. Safety of Household and Similar Electrical Appliances Part 2: Particular Requirements for Coffee Mills and Coffee Grinders (Superseded by EN 60335-2-33). 8 pp.

CENELEC EN 60 335-2-33-90. Safety of Household and Similar Electrical Appliances Part 2: Particular Requirements for Coffee Mills and Coffee Grinders (Supersedes HD 260 S3). 6 pp.

DIN VDE 0700 Pt 33 A2 (D)-85. Safety of Household and Similar Electrical Appliances; Coffee and Grain Mills: Amendment 2 (Sept). 4 pp.

IEC 335 Pt 2-33-87. Safety of Household and Similar Electrical Appliances Part 2: Particular Requirements for Coffee Mills and Coffee Grinders Second Edition; (Amendment 1-1990). 36 pp.

—Concrete Vibrators

IEC 745 Pt 2-12-82. Safety of Hand-Held Motor-Operated Electric Tools Part 2: Particular Requirements for Concrete Vibrators (Internal Vibrators) First Edition; (Amendment 1-1991). 30 pp.

INTERNATIONAL AND NON-U.S. NATIONAL STANDARDS
SUBJECT INDEX
Consumer

Consumer Safety *(Cont.)*

—Cooking Appliances

BSI BS 3456: Sec 102.9-91. (WITHDRAWN) 1991 Safety of Household and Similar Electrical Appliances Part 102: Particular Requirements Section 102.9: Toasters, Grills, Roasters and Similar Appliances (Superseded by BS EN 60335-2-9: 1991). 26 pp.

BSI BS 3456: Sec 102.13-91. (WITHDRAWN) 1991 Safety of Household and Similar Electrical Appliances Part 102: Particular Requirements Section 102.13: Frying Pans, Deep Fat Fryers and Similar Appliances (SUPERSEDED BY BS EN 60335-2-13: 1991). 24 pp.

BSI BS 3456: Sec 102.25-88. (WITHDRAWN) 1988 Safety of Household Electrical Appliances Part 102: Particular Requirements Section 102.25: Appliances for Heating Food by Means of Microwave Energy (Superseded by BS EN 60335-2-25: 1991). 39 pp.

BSI BS 3456: Sec 202.14-90. 1990 Safety of Household Electrical Appliances Part 202: Particular Requirements Section 202.14: Electric Kitchen Machines. 31 pp.

BSI BS 3456: Sec 202.14-02. 1990 Amd 2 Safety of Household and Similar Electrical Appliances Part 202: Particular Requirements Section 202.14: Electric Kitchen Machines (AMD 7532) January, 15, 1993 (L). 50 pp.

BSI BS 3456: Sec 202.14-03. 1990 Amd 3 Safety of Household and Similar Electrical Appliances Part 202: Particular Requirements Section 202.14: Electric Kitchen Machines (AMD 7800) June 15, 1993 (SUPERSEDES BS 3456: SECTION 3. 12: 1979). 57 pp.

BSI BS 3456: Sec A1-66. (WITHDRAWN) 1966 Amd 4 Safety of Household Electrical Appliances Part A: Heating and Cooking Appliances Section A1: General Requirements. 57 pp.

BSI BS 3456: Sec A4-71. (WITHDRAWN) 1971 Amd 9 Safety of Household Electrical Appliances Part A: Heating and Cooking Appliances Section A4: Electrically Heated Blankets (Withdrawn, superseded by BS En 60967: 1991). 67 pp.

BSI BS 5258: Part 2-75. (OBSOLESCENT) 1975 Amd 2 Safety of Domestic Gas Appliances Part 2: Cooking Appliances. 34 pp.

BSI BS EN 60335-2-6-91. 1991 Safety of Household and Similar Electrical Appliances Part 2: Particular Requirements Section 2.6: Cooking Ranges, Cooking Tables, Ovens and Similar Appliances for Household Use. 52 pp.

BSI BS EN 60335-2-6-01. 1991 Amd 1 Safety of Household and Similar Electrical Appliances Part 2: Particular Requirements Section 2.6: Cooking Ranges, Cooking Tables, Ovens and Similar Appliances for Household Use (AMD 7677) July 15, 1993 (L). 70 pp.

BSI BS EN 60335-2-6-02. 1991 Amd 2 Safety of Household and Similar Electrical Appliances Part 2: Particular Requirements Section 2.6: Cooking Ranges, Cooking Tables, Ovens and Similar Appliances for Household Use (AMD 7915) September 15, 1993 (L). 81 pp.

BSI BS EN 60335-2-6-03. 1991 Amd 3 Safety of Household and Similar Electrical Appliances Part 2: Particular Requirements Section 2.6: Cooking Ranges, Cooking Tables, Ovens and Similar Appliances for Household Use (AMD 7916) October 15, 1993 (L). 84 pp.

BSI BS EN 60335-2-15-91. 1991 Safety of Household and Similar Electrical Appliances Part 2: Particular Requirements Section 2.15: Appliances for Heating Liquids (L). 29 pp.

BSI BS EN 60335-2-15-02. 1991 Amd 2 Safety of Household and Similar Electrical Appliances Part 2: Particular Requirements Section 2.15: Appliances for Heating Liquids (AMD 7281) October 15, 1992 (L). 52 pp.

BSI BS EN 60335-2-15-03. 1991 Amd 3 Safety of Household and Similar Electrical Appliances Part 2: Particular Requirements Section 2.15: Appliances for Heating Liquids (AMD 7695) July 15, 1993 (L). 58 pp.

BSI BS EN 60335-2-50-92. 1992 Safety of Household and Similar Electrical Appliances Part 2: Particular Requirements Section 2.50: Commercial Electric Bains-Marie. 22 pp.

BSI BS EN 60335-2-50-01. 1992 Amd 1 Safety of Household and Similar Electrical Appliances Part 2: Particular Requirements Section 2.50: Commercial Electric Bains-Marie (AMD 7833) July 15, 1993 (L). 35 pp.

CENELEC HD 261-80. Particular Specification for Kitchen Machines. 7 pp.

CENELEC HD 261 S1-84. Particular Specification for Kitchen Machines (Replaced by EN 60 335-2-14-1988). 3 pp.

CENELEC HD 261.2-77. Particular Specification for Kitchen Machines. 5 pp.

CENELEC HD 261.3-78. Particular Specification for Kitchen Machines. 5 pp.

CENELEC HD 261.4 S1-88. Particular Specification for Kitchen Machines. 5 pp.

Consumer Safety *(Cont.)*

—Cooking Appliances *(Cont.)*

CENELEC HD 265 S2-84. Safety of Household and Similar Electrical Appliances—Particular Requirements for Appliances for Toasters, Grills, Roasters and Similar Appliances. 8 pp.

CENELEC HD 265 S2/A1-88. AMD 1 Safety of Household and Similar Electrical Appliances—Particular Requirements for Appliances for Toasters, Grills, Roasters and Similar Appliances. 4 pp.

CENELEC HD 270 S1-83. Safety of Household and Similar Electrical Appliances—Particular Requirements for Microwave Cooking Appliances (Superseded by EN 41003). 17 pp.

CENELEC HD 275 S1-86. Safety of Household and Similar Electrical Appliances—Part 2: Particular Requirements for Cooking Ranges, Cooking Tables, Ovens and Similar Appliances for Household Use. 30 pp.

CENELEC HD 275 S1/A1-89. AMD 1 Safety of Household and Similar Electrical Appliances—Part 2: Particular Requirements for Cooking Ranges, Cooking Tables, Ovens and Similar Appliances for Household Use. 9 pp.

CENELEC PRHD 1003-88. Heating in Contact with the Front of the Domestic Cooking Appliances Burning Gas. 10 pp.

CENELEC HD 1003-90. Heating in Contact with the Front of the Domestic Cooking Appliances Burning Gas. 10 pp.

CENELEC EN 60335-2-6-90. Safety of Household and Similar Electrical Appliances Part 2: Particular Requirements for Cooking Ranges, Cooking Tables, Ovens and Similar Appliances for Household Use (Supercedes HD 275 S1: 1986 and Its Amendment 1: 1988). 38 pp.

CENELEC EN 60335-2-6-91. CORRIGENDUM Safety of Household and Similar Electrical Appliances Part 2: Particular Requirements for Cooking Ranges, Cooking Tables, Ovens and Similar Appliances for Household Use. 2 pp.

CENELEC EN 60335-2-6 /A2-92. AMD 2 Safety of Household and Similar Electrical Appliances Part 2: Particular Requirements for Cooking Ranges, Cooking Tables, Ovens and Similar Appliances for Household Use; (IEC 335-2-6:1986/A2: 1990). 6 pp.

CENELEC EN 60 335-2-9/A51-91. AMD 51 Safety of Household and Similar Electrical Appliances Part 2: Particular Requirements for Toasters, Grills, Roasters and Similar Appliances. 6 pp.

CENELEC EN 60 335-2-14-88. Safety of Household and Similar Electrical Appliances Part 2: Particular Requirements for Electric Kitchen Machines (Replaces HD 261 S1-1984). 11 pp.

CENELEC EN 60335-2-14/A1-90. AMD 1 Safety of Household and Similar Electrical Appliances Part 2: Particular Requirements for Electric Kitchen Machines; (Replaces HD 261 S1:1984). 3 pp.

CENELEC EN 60335-2-14/A51-91. AMD 51 Safety of Household and Similar Electrical Appliances Part 2: Particular Requirements for Electric Kitchen Machines; (Replaces HD 261 S1:1984). 6 pp.

CENELEC EN 60335-2-14/A52-92. AMD 52 Safety of Household and Similar Electrical Appliances Part 2: Particular Requirements for Electric Kitchen Machines. 4 pp.

CENELEC EN 60335-2-15-90. Safety of Household and Similar Electrical Appliances Part 2: Particular Requirements for Heating Liquids. 10 pp.

CENELEC EN 60335-2-15/A1-91. AMD 1 Safety of Household and Similar Electrical Appliances Part 2: Particular Requirements for Heating Liquids. 10 pp.

CENELEC EN 60335-2-15/A2-92. AMD 2 Safety of Household and Similar Electrical Appliances Part 2: Particular Requirements for Heating Liquids. 7 pp.

CENELEC EN 60335-2-15/A52-92. AMD 52 Safety of Household and Similar Electrical Appliances Part 2: Particular Requirements for Appliances for Heating Liquids. 4 pp.

CENELEC EN 60335-2-50-91. Safety of Household and Similar Electrical Appliances Part 2: Particular Requirements for Commercial Electric Bains-Marie. 8 pp.

CENELEC EN 60335-2-50/A1-92. AMD 1 Safety of Household and Similar Electrical Appliances Part 2: Particular Requirements for Commercial Electric Bains-Marie. 5 pp.

CSA C22.2 NO 1335.2.9-93. Portable Electrical Motor-Operated and Heating Appliances: Particular Requirements for Portable Electric Cooking Appliances; (Gen Instr 1). 40 pp.

DIN VDE 0700 Pt 6 A8 (D)-89. Safety of Household and Similar Electrical Appliances; Cooking Ranges, Cooking Tables, Ovens and Similar Appliances for Household Use; Amendment 8 to Draft DIN 57700 Part 6/VDE 0700 Part 6/02.84 (Apr). 15 pp.

IEC 335 Pt 2-6-86. Safety of Household and Similar Electrical Appliances Part 2: Part-icular Requirements for Cooking Ranges, Cooking Tables, Ovens and Sim-ilar Appliances for Household Use Third Edition; (Amendment 2-1990, Incorporating Amendment 1) (Amendment 3-1992). 106 pp.

Consumer Safety *(Cont.)*

—Cooking Appliances *(Cont.)*

IEC 335 Pt 2-14-84. Safety of Household and Similar Electrical Appliances Part 2: Particular Requirements for Electric Kitchen Machines Second Edition; (Amendment 1-1989) (Amendment 2-1990). 69 pp.

IEC 335 Pt 2-15-86. Safety of Household and Similar Electrical Appliances Part 2: Particular Requirements for Appliances for Heating Liquids Third Edition; (Corrigendum—Dec 1986) (Amendment 1-1988) (Amendment 2-1990) (Amendment 3-1992). 62 pp.

IEC 335 Pt 2-50-89. Safety of Household and Similar Electrical Appliances Part 2: Particular Requirements for Commercial Electric Bains-Marie First Edition; (Amendment 1-1990). 48 pp.

IEC 335 Pt 2-64-91. Safety of Household and Similar Electrical Appliances Part 2: Particular Requirements for Commercial Electric Kitchen Machines First Edition. 44 pp.

—Cooking Utensils

BSI BS 3456: Sec 102.39-89. (WITHDRAWN) 1989 Safety of Household Electrical Appliances Part 102: Particular Requirements Section 102.39: Commercial Electric Multi-Purpose Cooking Pans (Superseded by BS 3456: Section 202.39: 1990). 16 pp.

BSI BS 3456: Sec 202.39-90. (WITHDRAWN) 1990 Safety of Household Electrical Appliances Part 202: Particular Requirements Section 202.39: Commercial Electric Multi-Purpose Cooking Pans (Supersedes BS 3456: Section 102.39: 1989). 20 pp.

BSI BS 3456: Sec 202.39-01. 1990 Amd 1 Safety of Household and Similar Electrical Appliances Part 202: Particular Requirements Section 202.39: Commercial Electric Multi-Purpose Cooking Pans (AMD 7616) March 15, 1993 (L). 28 pp.

BSI BS 3456: Sec 202.39-02. 1990 Amd 2 Safety of Household and Similar Electrical Appliances Part 202: Particular Requirements Section 202.39: Commercial Electric Multi-Purpose Cooking Pans (AMD 7830) July 15, 1993 (L). 44 pp.

BSI BS EN 60335-2-47-03. 1992 Amd 3 Safety of Household and Similar Electrical Appliances Part 2: Particular Requirements Section 2.47: Commercial Electric Boiling Pans (AMD 7425) November 15, 1992. 46 pp.

BSI BS EN 60335-2-47-92. 1992 Safety of Household and Similar Electrical Appliances Part 2: Particular Requirements Section 2.47: Commercial Electric Boiling Pans (L). 23 pp.

CENELEC HD 287 S1-87. Safety of Household and Similar Electrical Appliances—Part 2: Particular Requirements for Commercial Electric Multi-Purpose Cooking Pans (Replaced by EN 335-2-39-1989). 9 pp.

CENELEC EN 60 335-2-39-89. Safety of Household and Similar Electrical Appliances Part 2: Particular Requirements for Commercial Electric Multi-Purpose Cooking Pans (Replaces HD 287 S1-1987). 10 pp.

CENELEC EN 60335-2-39/A1-92. AMD 1 Safety of Household and Similar Electrical Appliances Part 2: Particular Requirements for Commercial Electric Multi-Purpose Cooking Pans; (Replaces HD 287 S1:1987). 5 pp.

CENELEC EN 60 335-2-39/A51-91. AMD 51 Safety of Household and Similar Electrical Appliances Part 2: Particular Requirements for Commercial Electric Multi-Purpose Cooking Pans; (Replaces HD 287 S1:1987). 7 pp.

CENELEC EN 60 335-2-47-90. Safety of Household and Similar Electrical Appliances Part 2 Particular Requirements for Commercial Electric Boiling Pans. 10 pp.

CENELEC EN 60335-2-47/A1-92. AMD 1 Safety of Household and Similar Electrical Appliances Part 2 Particular Requirements for Commercial Electric Boiling Pans. 5 pp.

CENELEC EN 60335-2-47/A51-91. AMD 51 Safety of Household and Similar Electrical Appliances Part 2 Particular Requirements for Commercial Electric Boiling Pans. 7 pp.

CENELEC EN 60335-2-47/A52-92. AMD 52 Safety of Household and Similar Electrical Appliances Part 2 Particular Requirements for Commercial Electric Boiling Pans. 7 pp.

CENELEC EN 60335-2-47/A52-92. AMD 52 Safety of Household and Similar Electrical Appliances Part 2: Particular Requirements for Commercial Electric Boiling Pans. 2 pp.

IEC 335 Pt 2-39-86. Safety of Household and Similar Electrical Appliances Part 2: Particular Requirements for Commercial Electric Multi-Purpose Cooking Pans Second Edition; (Amendment 1-1990). 48 pp.

IEC 335 Pt 2-47-87. Safety of Household and Similar Electrical Appliances Part 2: Particular Requirements for Commercial Electric Boiling Pans First Edition; (Amendment 1-1990). 46 pp.

INDUSTRY STANDARDS
611

Consumer Safety (Cont.)

—Corn Grinders
DIN VDE 0700 Pt 33 A2 (D)-85. Safety of Household and Similar Electrical Appliances; Coffee and Grain Mills: Amendment 2 (Sept). 4 pp.

—Cradles (Furniture)
CEN PREN 1130-1-93. Furniture—Cribs and Cradles for Domestic Use—Part 1: Safety Requirements. 6 pp.

—Cribs (Furniture)
BSI BS 1753-87. 1987 Amd 1 Safety Requirements for Children's Cots for Domestic Use (AMD 6265) January 31, 1990. 18 pp.
BSI BS 1753-02. 1987 Amd 2 Safety Requirements for Children's Cots for Domestic Use (AMD 7657) June 15, 1993. 21 pp.
BSI BS 7423-91. 1991 Safety Requirements for Children's Travel Cots of Internal Base Length Not Less Than 900 mm. 16 pp.
BSI BS 7423-01. 1991 Amd 1 Safety Requirements for Children's Travel Cots of Internal Base Length Not Less Than 900 mm (AMD 7656) June 15, 1993. 17 pp.
CEN PREN 716-1-92. Furniture—Children's and Nursery Furniture—Children's Cots and Folding Cots for Domestic Use—Part 1: Safety Requirements. 9 pp.
CEN PREN 716-2-92. Furniture—Children's and Nursery Furniture—Children's Cots and Folding Cots for Domestic Use—Part 2: Test Methods. 18 pp.
CEN PREN 1130-1-93. Furniture—Cribs and Cradles for Domestic Use—Part 1: Safety Requirements. 6 pp.
SAA AS 2172-91. Cots for Household Use (Safety Requirements). 15 pp.
SNZ NZS 5810-92. Cots for Household Use (Safety Requirements) Amend: 1, 1992. 15 pp.

—Cutting Tools
DIN VDE 0700 Pt 30-83. Test Certificate, Safety Test of Household & Similar Electrical Appliances. Portable Electric Heating Tools (Jan). 26 pp.
DIN VDE 0700 Pt 30 A1 (D)-84. Safety of Household and Similar Electrical Appliances; Particular Specifications for Portable Electric Heating Tools (VDE Specification): Amendment 1 (Nov). 19 pp.

—Deep Fat Fryers
BSI BS EN 60335-2-13-91. 1991 Safety of Household and Similar Electrical Appliances Part 2: Particular Requirements Section 2.13: Frying Pans, Deep Fat Fryers and Similar Appliances (L). 24 pp.
BSI BS EN 60335-2-13-01. 1991 Amd 1 Safety of Household and Similar Electrical Appliances Part 2: Particular Requirements Section 2.13: Frying Pans, Deep Fat Fryers and Similar Appliances (AMD 7585) March 15, 1993 (L). 30 pp.
CENELEC EN 60 335-2-13-90. Safety of Household and Similar Electrical Appliances Part 2: Particular Requirements For Frying Pans, Deep Fat Fryers and Similar Appliances. 7 pp.
CENELEC EN 60335-2-13/A1-92. AMD 1 Safety of Household and Similar Electrical Appliances Part 2: Particular Requirements for Frying Pans, Deep Fat Fryers and Similar Appliances. 4 pp.

—Dehumidifiers
DIN VDE 0700 Pt 258-88. Safety of Household and Similar Electrical Appliances; Dehumidifiers (May). 24 pp.

—Diamond Drills
IEC 1029 Pt 2-6-93. Safety of Transportable Motor-Operated Electric Tools Part 2: Particular Requirements for Diamond Drills with Water Supply First Edition. 24 pp.

—Diamond Saws
IEC 1029 Pt 2-7-93. Safety of Transportable Motor-Operated Electric Tools Part 2: Particular Requirements for Diamond Saws with Water Supply First Edition. 24 pp.

—Dishwashers
BSI BS 3456: Sec 102.5-88. (WITHDRAWN) 1988 Safety of Household Electrical Appliances Part 102: Particular Requirements Section 102.5: Dishwashers (Superseded by BS 3456: Section 202.5: 1990). 24 pp.
BSI BS 3456: Sec 202.5-90. 1990 Safety of Household Electrical Appliances Part 202: Particular Requirements Section 202.5: Dishwashers. 28 pp.
BSI BS 3456: Sec 202.5-01. 1990 Amd 1 Safety of Household and Similar Electrical Appliances Part 202: Particular Requirements Section 202.5: Dishwashers (AMD 6899) April 1, 1992 (L). 36 pp.
BSI BS 3456: Sec 202.5-03. 1990 Amd 3 Safety of Household and Similar Electrical Appliances Part 202: Particular Requirements Section 202.5: Dishwashers (AMD 7663) June 15, 1993 (SUPERSEDES BS 3456: SECTION 102. 5: 1988). 56 pp.
BSI BS 3456: Sec 202.5-02. 1990 Amd 2 Safety of Household and Similar Electrical Appliances Part 202: Particular Requirements Section 202.5: Dishwashers (AMD 7348) November 15, 1992 (Supersedes BS 3456: Section 102.5: 1988). 50 pp.
BSI BS 6614-85. 1985 Safety Devices and Water Supply Connections for Washing Machines and Dishwashers Connected to the Water Supply Mains. 25 pp.
BSI BS EN 50084-92. 1992 Safety of Household and Similar Electrical Appliances. Requirements for the Connection of Washing Machines, Dishwashers and Tumbler Dryers to the Water Mains (L). 24 pp.
BSI BS EN 60335-2-58-93. 1993 Safety of Household and Similar Electrical Appliances Part 2: Particular Requirements Section 2.58: Commercial Electric Dishwashing Machines (L). 33 pp.
CENELEC HD 257 S2-84. Safety of Household and Similar Electrical Apliances Particular Requirements for Dishwashers. 17 pp.
CENELEC EN 50084-92. Safety of Household and Similar Electrical Appliances—Requirements for the Connection of Washing Machines, Dishwashers and Tumbler Dryers to the Water Mains. 21 pp.
CENELEC EN 60 335-2-5-89. Safety of Household and Similar Electrical Appliances—Part 2: Particular Requirements for Dishwashers. 12 pp.
CENELEC EN 60 335-2-5/A1-90. AMD 1 Safety of Household and Similar Electrical Appliances—Part 2: Particular Requirements for Dishwashers. 7 pp.
CENELEC EN 60 335-2-5/A2-90. AMD 2 Safety of Household and Similar Electrical Appliances—Part 2: Particular Requirements for Dishwashers. 5 pp.
CENELEC EN 60335-2-5/A3-92. AMD 3 Safety of Household and Similar Electrical Appliances Part 2: Particular Requirements for Dishwashers (IEC 335-2-5:1984/A3:1990). 5 pp.
CENELEC EN 60335-2-58-93. Safety of Household and Similar Electrical Appliances Part 2: Particular Requirements for Commercial Electric Dishwashing Machines (IEC 335-2-58: 1990, Modified). 25 pp.
DIN VDE 0700 Pt 231-84. Safety of Household & Similar Electrical Appliances; Commercial Dishwashers (VDE Specification) (Aug) (Supplements or Replaces Corresponding Sections in DIN 57700 Part 1/VDE 0700 Part 1). 34 pp.
DIN VDE 0700 Pt 600 (D)-88. Safety of Household and Similar Electrical Appliances; Connection of Washing Machines, Dishwashers and Tumble Dryers to the Supply Mains; German Version prHD 274 S2: 1987 (Oct). 37 pp.
IEC 335 Pt 2-5-92. Safety of Household and Similar Electrical Appliances Part 2: Particular Requirements for Dishwashers Fourth Edition; (CENELEC EN 60335-2-5/A3:1992). 44 pp.
IEC 335 Pt 2-58-90. Safety of Household and Similar Electrical Appliances Part 2: Particular Requirements for Commercial Electric Dishwashing Machines First Edition (CENELEC EN 60335-2-58: 1993). 53 pp.

—Disk Sanders
BSI BS 2769: Sec 2.3-84. 1984 Hand-Held Electric-Motor-Operated Tools Part 2: Particular Requirements Section 2.3: Grinders, Polishers and Disc-Type Sanders. 16 pp.
BSI BS 2769: Sec 2.3-01. 1984 Amd 1 Hand-Held Electric Motor-Operated Tools Part 2: Particular Requirements Section 2.3: Specification for Grinders, Polishers and Disc-Type Sanders (AMD 6924) May 1, 1992. 17 pp.

—Drills
CEN PREN 792-3-92. Handheld Non-Electric Power Tools—Safety Requirements—Part 3: Rotary Drilling Tools, Drills, Tappers. 8 pp.
CEN PREN 792-5-92. Handheld Non-Electric Power Tools—Safety Requirements—Part 5: Rotary Percussive Drills. 8 pp.
CENELEC HD 400.2A S1 PRA1-93. AMD 1 Hand-Held Motor Operated Tools Part II: Particular Specifications Section A: Drills. 4 pp.

—Ear Muffs
CEN PREN 352 (Part 1)-90. Hearing Protectors—Safety Requirements and Testing—Part 1: Ear Muffs. 33 pp.

—Ear Plugs (Hearing Protection)
CEN PREN 352 (Part 2)-90. Hearing Protectors—Safety Requirements and Testing—Part 2: Ear Plugs. 20 pp.

—Electric Blankets
BSI BS 3456: Sec A4-71. (WITHDRAWN) 1971 Amd 9 Safety of Household Electrical Appliances Part A: Heating and Cooking Appliances Section A4: Electrically Heated Blankets (Withdrawn, superseded by BS En 60967: 1991). 67 pp.
IEC 335 Pt 17-74. Safety of Household and Similar Electrical Appliances Particular Requirements for Electrically Heated Blankets, Pads and Mattresses, Recm. First Edition. 95 pp.

—Electric Mattresses
IEC 335 Pt 17-74. Safety of Household and Similar Electrical Appliances Particular Requirements for Electrically Heated Blankets, Pads and Mattresses, Recm. First Edition. 95 pp.

—Electronic Equipment
BSI BS 415-90. 1990 Safety Requirements for Mains-Operated Electronic and Related Apparatus for Household and Similar General Use. 80 pp.
CENELEC HD 195 S4-85. Safety Requirements for Mains Operated Electronic and Related Apparatus for Household and Similar General Use. 4 pp.
CENELEC HD 195 S5-88. Safety Requirements for Mains Operated Electronic and Related Apparatus for Household and Similar General Use. 4 pp.
CENELEC HD 195 S6-89. Safety Requirements for Mains Operated Electronic and Related Apparatus for Household and Similar General Use. 19 pp.
CENELEC HD 401-80. Safety Requirements for Electronic Measuring Apparatus. 2 pp.

—European Communities
EURO 1990-14. European Community and Consumer Protection. 12 pp.

—Exercise Equipment
CEN PREN 957-1-92. Stationary Training Equipment—Part 1: General Safety Requirements and Test Methods. 9 pp.
CEN PREN 957-2-92. Stationary Training Equipment—Part 2: Additional Specific Safety Requirements and Test Methods for Domestic Strength Training Equipment (Type H). 5 pp.
CEN PREN 957-3-92. Stationary Training Equipment—Part 3: Additional Specific Safety Requirements and Test Methods for Institutional Strength Training Equipment (Type S). 7 pp.
CEN PREN 957-4-92. Stationary Training Equipment—Part 4: Additional Specific Safety Requirements and Test Methods for Strength Training Benches (Type B). 5 pp.
CEN PREN 957-5-92. Stationary Training Equipment—Part 5: Additional Specific Safety Requirements and Test Methods for Pedal Crank Training Equipment (Type P). 10 pp.

—Eyeglasses
SAA AS 1067.1-90. Sunglasses and Fashion Spectacles—Part 1: Safety Requirements Amdt 1 December 1990 Amdt 2 July 1993. 15 pp.

—Fall Arresting Devices
CEN PREN 958-92. Mountaineering Equipment—Fall Arrestor System—Safety Requirements and Test Method. 5 pp.

—Fans—Regulators
BSI BS 3456: Sec 102.342-88. 1988 Safety of Household Electrical Appliances Part 102: Particular Requirements Section 102.342: Electric Fans and Regulators. 16 pp.
CENELEC HD 280 S1-86. Safety of Household and Similar Electrical Appliances: Particular Requirements for Electric Fans and Regulations. 6 pp.
DIN VDE 0700 Pt 234 (D)-81. Safety of Household and Similar Electrical Appliances; Fans and Associated Regulators (Aug). 25 pp.
IEC 342 Pt 1-81. Safety Requirements for Electric Fans and Regulators Part 1: Fans and Regulators for Household and Similar Purposes Second Edition; (Amendment 1-1982). 30 pp.

—Files (Tools)
CEN PREN 792-9-92. Handheld Non-Electric Power Tools—Safety Requirements—Part 9: Die Grinders. 9 pp.

—Floor Cleaning Equipment
BSI BS 3456: Sec 102.10-90. (WITHDRAWN) 1990 Safety of Household Electrical Appliances Part 102: Particular Requirements Section 102.10: Floor Treatment Machines and Wet Scrubbing Machines (Superseded by BS EN 60335-2-10: 1992). 18 pp.
BSI BS EN 60335-2-10-92. 1992 Safety of Household and Similar Electrical Appliances Part 2: Particular Requirements Section 2.10: Floor Treatment Machines and Wet Scrubbing Machines (Supersedes BS 3456: Section 102.10: 1992). 19 pp.

INTERNATIONAL AND NON-U.S. NATIONAL STANDARDS
SUBJECT INDEX
Consumer

Consumer Safety *(Cont.)*

—Floor Cleaning Equipment *(Cont.)*

BSI BS EN 60335-2-10-01. 1992 Amd 1 Safety of Household and Similar Electrical Appliances Part 2: Particular Requirements Section 2.10: Floor Treatment Machines and Wet Scrubbing Machines (AMD 7792) July 15, 1993 (L). 22 pp.

CENELEC HD 281 S1-86. Safety of Household and Similar Electrical Appliances—Part 2: Particular Requirements for Floor Treatment Machines and Wet Scrubbing Machines. 10 pp.

CENELEC EN 60335-2-1 0-90. Safety of Household and Similar Electrical Appliances Part 2: Particular Requirements for Floor Treatment Machines and Wet Scrubbing Machines. 10 pp.

CENELEC EN 60335-2-10-90. Safety of Household and Similar Electrical Appliances Part 2: Particular Requirements for Floor Treatment Machines and Wet Scrubbing Machines; (IEC 335-2-10: 1987, Modified). 12 pp.

CENELEC EN 60335-2-10/A1-92. AMD 1 Safety of Household and Similar Electrical Appliances Part 2: Particular Requirements for Floor Treatment Machines and Wet Scrubbing Machines (IEC 335-2-10:1987/A1:1991). 4 pp.

DIN VDE 0700 Pt 205-87. Safety of Household and Similar Electrical Appliances; Floor Treatment Appliances for Commercial Use (Feb). 42 pp.

IEC 335 Pt 2-10-92. Safety of Household and Similar Electrical Appliances Part 2: Particular Requirements for Floor Treatment Machines and Wet Scrubbing Machines Fourth Edition; (CENELEC EN 60335-2-10: 1992). 28 pp.

IEC 335 Pt 2-67-92. Safety of Household and Similar Electrical Appliances Part 2: Particular Requirements for Floor Treatment and Floor Cleaning Machines, for Industrial and Commercial Use First Edition. 34 pp.

IEC 335 Pt 2-68-92. Safety of Household and Similar Electrical Appliances Part 2: Particular Requirements for Spray Extraction Appliances, for Industrial and Commercial Use First Edition. 42 pp.

—Floor Polishers

BSI BS 3456: Sec 2.20-71. (WITHDRAWN) 1971 Amd 1 1972-1981 Edition. Specification for Safety of Household Electrical Appliances Part 2: Particular Requirements Section 2.20: Electric Floor Polishers (AMD 2500) March 31, 1978 (Superseded by BS 3456: Section 102.10: 1990). 15 pp.

—Food Additives

EC 80/1089/EEC-80. Commission Recommendation to the Member States Concerning Tests Relating to the Safety Evaluation of Food Additives. 1 p.

—Food Processing Equipment

DIN VDE 0700 Pt 251-86. Safety of Household and Similar Electrical Appliances; Food Processors and Food Choppers (Oct). 21 pp.

—Food Warmers

BSI BS 3456: Sec 2.6-70. (WITHDRAWN) 1970 Amd 2 1972-1981 Edition. Specification for Safety of Household Electrical Appliances: Part 2: Particular Requirements: Section 2.6: Frying Pans, Grills, Plate Warmers, and Other Dry Cooking Appliances (Superseded by BS 3456: Sec 3.16 & 3.17). 18 pp.

BSI BS EN 60335-2-12-91. 1991 Safety of Household and Similar Electrical Appliances Part 2: Particular Requirements Section 2.12: Warming Plates and Similar Appliances. 20 pp.

BSI BS EN 60335-2-49-92. 1992 Safety of Household and Similar Electrical Appliances Part 2: Particular Requirements Section 2.49: Commercial Catering Electric Hot Cupboards (L). 22 pp.

BSI BS EN 60335-2-49-01. 1992 Amd 1 Safety of Household and Similar Electrical Appliances Part 2: Particular Requirements Section 2.49: Commercial Catering Electric Hot Cupboards (AMD 7516) January 15, 1993 (L). 32 pp.

CENELEC EN 60 335-2-12-90. Safety of Household and Similar Electrical Appliances Part 2: Particular Requirements for Warming Plates and Similar Appliances. 7 pp.

CENELEC EN 60335-2-49-90. Safety of Household and Similar Electrical Appliances Part 2 Particular Requirements for Commercial Catering Electric Hot Cupboards; (IEC 335-2-49:1988, Modified). 9 pp.

CENELEC EN 60335-2-49-90. Safety of Household and Similar Electrical Appliances Part 2: Particular Requirements for Commercial Catering Electric Hot Cupboards; (IEC 335-2-49:1988, Modified). 17 pp.

CENELEC EN 60335-2-49/A1-92. AMD 1 Safety of Household and Similar Electrical Appliances Part 2: Particular Requirements for Commercial Catering Electric Hot Cupboards; (IEC 335-2-49:1988, Modified). 5 pp.

CENELEC EN 60335-2-49/A1-92. AMD 1 Safety of Household and Similar Electrical Appliances Part 2: Particular Requirements for Commercial Catering Electric Hot Cupboards; (IEC 335-2-49:1988, Modified). 7 pp.

Consumer Safety *(Cont.)*

—Food Warmers *(Cont.)*

CENELEC EN 60335-2-49/A1-92. CORRIGENDUM Safety of Household and Similar Electric Appliances Part 2: Particular Requirements for Commercial Catering Electric Hot Cupboards (IEC 335-2-49:1988/A1: 1990). 1 p.

IEC 335 Pt 2-12-92. Safety of Household and Similar Electrical Appliances Part 2: Particular Requirements for Warming Plates and Similar Appliances Fourth Edition. 30 pp.

IEC 335 Pt 2-49-88. Safety of Household and Similar Electrical Appliances Part 2: Particular Requirements for Commercial Electric Hot Cupboards First Edition; (Amendment 1-1990) (CENELEC EN 60335-2-49: 1990). 44 pp.

—Foods—Imitation

EC 87/357/EEC-87. Council Directive on the Approximation of the Laws of the Member States Concerning Products Which, Appearing to be Other Than They are, Endanger the Health of Safety of Consumers. 2 pp.

—Freezers

BSI BS 3456: Sec 102.24-88. (WITHDRAWN) 1988 Safety of Household Electrical Appliances Part 102: Particular Requirements Section 102.24: Refrigerators and Food Freezers (Superseded by BS 3456: Section 202.24: 1990). 31 pp.

BSI BS 3456: Sec 202.24-90. 1990 Safety of Household Electrical Appliances Part 202: Particular Requirements Section 202.24: Refrigerators and Food Freezers (L). 42 pp.

BSI BS 5258: Part 6-88. 1988 Safety of Domestic Gas Appliances Part 6: Refrigerators and Food Freezers. 16 pp.

BSI BS 5258: Part 6-75. 1975 Safety of Domestic Gas Appliances Part 6: Refrigerators and Food Freezers. 15 pp.

CENELEC HD 269 S2-86. Safety of Household and Similar Electrical Appliances-Part: Particular Requirements for Refrigerators and Food Freezers. 7 pp.

CENELEC HD 269 S2-86. Safety of Household and Similar Electrical Appliances Part: 2 Particular Requirements for Refrigerators and Food Freezers (IEC 335-2-24 (1976) Ed 1 Modified). 18 pp.

CENELEC HD 269 S2-85. Amd 1 Safety of Household Similar Electrical Appliances Particular Requirements for Refrigerators and Food Freezers. 4 pp.

CENELEC EN 60 335-2-24-89. Safety of Household and Similar Electrical Appliances-Part 2: Particular Requirements for Refrigerators and Food Freezers. 13 pp.

DIN VDE 0700 Pt 24-86. Safety of Household and Similar Electrical Appliances; Refrigerators and Food Freezers (Mar). 33 pp.

DIN VDE 0700 Pt 240-83. Safety of Household and Similar Electrical Appliances; Particular Requirements for Refrigerators and Freezers for Special Purposes and Ice Makers (Feb) (Supersedes VDE 0730 Part 2 ZK/10.78). 61 pp.

IEC 335 Pt 2-24-92. Safety of Household and Similar Electrical Appliances Part 2: Particular Requirements for Refrigerators, Food-Freezers and Ice-Makers Third Edition: (CENELEC HD 269 S2:1986). 78 pp.

SNZ NZS 6324-90. Approval and Test Specification—Particular Requirements for Refrigerators and Food Freezers Amend: 1, 1991; 2, 1992 (NZS 6324:1990/ AS 3303:1990). 31 pp.

—Frying Utensils

BSI BS 3456: Sec 102.13-91. (WITHDRAWN) 1991 Safety of Household and Similar Electrical Appliances Part 102: Particular Requirements Section 102.13: Frying Pans, Deep Fat Fryers and Similar Appliances (SUPERSEDED BY BS EN 60335-2-13: 1991). 24 pp.

BSI BS 3456: Sec 102.37-89. (WITHDRAWN) 1989 Safety of Household Electrical Appliances Part 102: Particular Requirements Section 102.37: Commercial Electric Deep Fat Fryers (Superseded by BS 3456: Section 202.37: 1990). 18 pp.

BSI BS 3456: Sec 202.37-90. (WITHDRAWN) 1990 Safety of Household Electrical Appliances Part 202: Particular Requirements Section 202.37: Commercial Electric Deep Fat Fryers (Supersedes BS 3456: Section 102.37: 1989). 22 pp.

BSI BS 3456: Sec 202.37-01. 1990 Amd 1 Safety of Household and Similar Electrical Appliances Part 202: Particular Requirements Section 202.37: Commercial Electric Deep Fat Fryers (AMD 7811) June 15, 1993 (L). 41 pp.

BSI BS 3456: Sec 2.6-70. (WITHDRAWN) 1970 Amd 2 1972-1981 Edition. Specification for Safety of Household Electrical Appliances: Part 2: Particular Requirements: Section 2.6: Frying Pans, Grills, Plate Warmers, and Other Dry Cooking Appliances (Superseded by BS 3456: Sec 3.16 & 3.17). 18 pp.

Consumer Safety *(Cont.)*

—Frying Utensils *(Cont.)*

BSI BS 3456: Sec 3.16-80. (WITHDRAWN) 1980 1972-1981 Edition. Specification for Safety of Household Electrical Appliances: Part 3: Complete Particular Sepcifications: Section 3.16: Frying Pans, Deep Fat Fryers and Similar Appliances (Superseded by BS 3456: Section 102.13: 1991). 46 pp.

BSI BS EN 60335-2-13-91. 1991 Safety of Household and Similar Electrical Appliances Part 2: Particular Requirements Section 2.13: Frying Pans, Deep Fat Fryers and Similar Appliances (L). 24 pp.

BSI BS EN 60335-2-13-01. 1991 Amd 1 Safety of Household and Similar Electrical Appliances Part 2: Particular Requirements Section 2.13: Frying Pans, Deep Fat Fryers and Similar Appliances (AMD 7585) March 15, 1993 (L). 30 pp.

CENELEC HD 279 S1-86. Safety of Household and Similar Electrical Appliances-Part 2: Particular Requirements for Frying Pans, Deep Fat Fryers and Similar Appliances. 10 pp.

CENELEC HD 285 S1-87. Safety of Household and Similar Electrical Appliances—Part 2: Particular Requirements for Commercial Electric Deep Fat Fryers. 10 pp.

CENELEC EN 60 335-2-13-90. Safety of Household and Similar Electrical Appliances Part 2: Particular Requirements For Frying Pans, Deep Fat Fryers and Similar Appliances. 7 pp.

CENELEC EN 60335-2-13/A1-92. AMD 1 Safety of Household and Similar Electrical Appliances Part 2: Particular Requirements for Frying Pans, Deep Fat Fryers and Similar Appliances. 4 pp.

CENELEC EN 60 335-2-37-89. Safety of Household and Similar Electrical Appliances Part 2: Particular Requirements for Commercial Electric Deep Fat Fryers. 10 pp.

CENELEC EN 60335-2-37/A1-92. AMD 1 Safety of Household and Similar Electrical Appliances Part 2: Particular Requirements for Commercial Electric Deep Fat Fryers. 5 pp.

IEC 335 Pt 2-37-86. Safety of Household and Similar Electrical Appliances Part 2: Particular Requirements for Commercial Electric Deep Fat Fryers Second Edition; (Amendment 1-1990). 48 pp.

—Gages

BSI BS 5101: Part 3-75. (WITHDRAWN) 1975 Amd 5 Lamp Caps and Holders Together with Gauges for the Control of Interchangeability and Safety Part 3: Gauges (AMD 6520) January 31, 1991 (Superseded by BS EN 60061-3: 1993). 338 pp.

BSI BS 5101: Part 4-80. 1980 Amd 1 Lamp Caps and Holders Together with Gauges for the Control of Interchangeability and Safety Part 4: Lamp Caps, Lampholders and Gauges Used in the United Kingdom but Not Specified in Parts 1, 2 or 3. 18 pp.

BSI BS 5101: Part 5-90. (WITHDRAWN) 1990 Lamp Caps and Holders Together with Gauges for the Control of Interchangeability and Safety Part 5: Guidelines and General Information (Renumbered as BS EN 60061-4: 1992). 22 pp.

BSI BS EN 60061-3-93. 1993 Lamp Caps and Holders Together with Gauges for the Control of Interchangeability and Safety Part 3: Gauges (E). 425 pp.

BSI BS EN 60061-4-92. 1992 Amd 1 Lamp Caps and Holders Together with Gauges for the Control of Interchangeability and Safety Part 4: Guidelines and General Information (AMD 7556) December 15, 1992. 24 pp.

CENELEC HD 65-82. Lamp Caps and Holders Together with Gauges for the Control of Interchangeability and Safety. 4 pp.

CENELEC EN 60061-3-93. Lamp Caps and Holders Together with Gauges for the Control of Interchangeability and Safety Part 3: Gauges (IEC 61-3: 1969 + Supplements A: 1970 to M: 1992, Modified). 420 pp.

CENELEC EN 60061-4-92. Lamp Caps and Holders Together with Gauges for the Control of Interchangeability and Safety Part 4: Guidelines and General Information; (IEC 61-4: 1990, Mod). 4 pp.

CENELEC EN 60061-4-92. Lamp Caps and Holders Together with Gauges for the Control of Interchangeability and Safety Part 4: Guidelines and General Information. 13 pp.

IEC 61 Pt 1-69. Lamp Caps and Holders Together with Gauges for the Control of Interchangeability and Safety Part 1: Lamp Caps Third Edition; (Supplements A-J-1970-1980)(Supplement K-1983) (Supplement L-1987) (Supplement M-1989) (Supplement N-1992); (CENELEC EN60061-1:1993). 279 pp.

IEC 61 Pt 2-69. Lamp Caps and Holders Together with Gauges for the Control of Inter-changeability and Safety Part 2: Lampholders Third Edition; (Supplements A-F-1970-1980)(Supplement G-1983) (Supplement H-1987) (Supplement J-1989) (Supplement K-1992) (CENE EN 60061-2: 1993). 202 pp.

Consumer
INTERNATIONAL AND NON-U.S. NATIONAL STANDARDS
SUBJECT INDEX

Consumer Safety *(Cont.)*

—Gages *(Cont.)*

IEC 61 Pt 3-69. Lamp Caps and Holders Together with Gauges for the Control of Interchangeability and Safety Part 3: Gauges Third Edition; (Errata—Dec 1984) (Supplement K-1987) (Supplement L-1989) (Supplement M-1992) (CENE EN 60061-3: 1993). 506 pp.

IEC 61 Pt 4-90. Lamp Caps and Holders Together with Gauges for the Control of Interchangeability and Safety Part 4: Guidelines and General Information First Edition; (Supplement A-1992) (CENELEC EN 60061-4: 1992). 85 pp.

—Game (Meat)

EC 92/45/EEC-92. Council Directive on Public Health and Animal Health Problems Relating to the Killing of Wild Game and the Placing on the Market of Wild-Game Meat. 19 pp.

—Garbage Disposals

BSI BS 3456: Sec 202.16-90. 1990 Safety of Household Electrical Appliances Part 202: Particular Requirements Section 202.16: Food Waste Disposers. 24 pp.

BSI BS 3456: Sec 2.30-71. 1971 Testing and Approval of Household Electrical Appliances Part 2: Particular Requirements Section 2.30: Food Waste Disposal Units. 16 pp.

BSI BS 3456: Sec 3.8-79. (WITHDRAWN) 1979 1972-1981 Edition. Specification for Safety of Household Electrical Appliances Part 3: Complete Particular Specifications Section 3.8: Food Waste Disposal Units. 46 pp.

CENELEC HD 259-76. Particular Specification for Food Waste Disposal Units. 6 pp.

CENELEC HD 259.2-77. Particular Specification for Food Waste Disposal Units. 5 pp.

CENELEC EN 60 335-2-16-89. Safety of Household and Similar Electrical Appliances Part 2: Particular Requirements for Food Waste Disposers. 9 pp.

IEC 335 Pt 2-16-86. Safety of Household and Similar Electrical Appliances Part 2: Particular Requirements for Food Waste Disposers Third Edition. 37 pp.

—Gas Logs

BSI BS 5258: Part 12-90. 1990 Safety of Domestic Gas Appliances Part 12: Decorative Fuel Effect Gas Appliances (2nd and 3rd Family Gases). 34 pp.

BSI BS 5258: Part 12-80. 1980 Amd 1 Safety of Domestic Gas Appliances Part 12: Decorative Gas Log and Other Fuel Effect Appliances (2nd and 3rd Family Gases) (AMD 5434) May 29, 1987. 20 pp.

—Griddles

BSI BS 3456: Sec 102.38-89. (WITHDRAWN) 1989 Safety of Household Electrical Appliances Part 102: Particular Requirements Section 102.38: Commercial Electric Griddles and Griddle Grills (Superseded by BS 3456: Section 202.38: 1990). 18 pp.

BSI BS 3456: Sec 202.38-90. (WITHDRAWN) 1990 Safety of Household Electrical Appliances Part 202: Particular Requirements Section 202.38: Commercial Electric Griddles and Griddle Grills (Supersedes BS 3456: Section 102.38: 1989). 22 pp.

BSI BS 3456: Sec 202.38-01. 1990 Amd 1 Safety of Household and Similar Electrical Appliances Part 202: Particular Requirements Section 202.38: Commercial Electric Griddles and Griddle Grills (AMD 7821) July 15, 1993 (L). 37 pp.

BSI BS 3456: Sec 202.38-02. 1990 Amd 2 Safety of Household and Similar Electrical Appliances Part 202: Particular Requirements Section 202.38: Commercial Electric Griddles and Griddle Grills (AMD 7822) JULY 15, 1993 SUPERSEDES BS 3456: SECTION 102. 38: 1989 (L). 45 pp.

CENELEC HD 286 S1-87. Safety of Household and Similar Electrical Appliances—Part 2: Particular Requirements for Commercial Electric Griddles and Griddle Grills (Replaced by EN 60 335-2-38-1989). 10 pp.

CENELEC EN 60 335-2-38-89. Safety of Household and Similar Electrical Appliances Part 2: Particular Requirements for Commercial Electric Griddles and Griddle Grills (Replaces HD 286 S1-1987). 11 pp.

CENELEC EN 60335-2-38/A1-92. AMD 1 Safety of Household and Similar Electrical Appliances Part 2: Particular Requirements for Commercial Electric Griddles and Griddle Grills; (Replaces HD 286 S1:1987). 5 pp.

CENELEC EN 60335-2-38/A51-92. AMD 51 Safety of Household and Similar Electrical Appliances Part 2: Particular Requirements for Commercial Electric Griddles and Griddle Grills; (Replaces HD 286 S1:1987). 6 pp.

DIN VDE 0700 Pt 6 A8 (D)-89. Safety of Household and Similar Electrical Appliances; Cooking Ranges, Cooking Tables, Ovens and Similar Appliances for Household Use; Amendment 8 to Draft DIN 57700 Part 6/VDE 0700 Part 6/02.84 (Apr). 15 pp.

Consumer Safety *(Cont.)*

—Griddles *(Cont.)*

IEC 335 Pt 2-38-86. Safety of Household and Similar Electrical Appliances Part 2: Particular Requirements for Commercial Electric Griddles and Griddle Grills Second Edition; (Amendment 1-1990). 46 pp.

—Grills (Appliances)

BSI BS 3456: Sec 2.6-70. (WITHDRAWN) 1970 Amd 2 1972-1981 Edition. Specification for Safety of Household Electrical Appliances: Part 2: Particular Requirements: Section 2.6: Frying Pans, Grills, Plate Warmers, and Other Dry Cooking Appliances (Superseded by BS 3456: Sec 3.16 & 3.17). 18 pp.

BSI BS EN 60335-2-6-91. 1991 Safety of Household and Similar Electrical Appliances Part 2: Particular Requirements Section 2.6: Cooking Ranges, Cooking Tables, Ovens and Similar Appliances for Household Use. 52 pp.

BSI BS EN 60335-2-6-01. 1991 Amd 1 Safety of Household and Similar Electrical Appliances Part 2: Particular Requirements Section 2.6: Cooking Ranges, Cooking Tables, Ovens and Similar Appliances for Household Use (AMD 7677) July 15, 1993 (L). 70 pp.

BSI BS EN 60335-2-6-02. 1991 Amd 2 Safety of Household and Similar Electrical Appliances Part 2: Particular Requirements Section 2.6: Cooking Ranges, Cooking Tables, Ovens and Similar Appliances for Household Use (AMD 7915) September 15, 1993 (L). 81 pp.

BSI BS EN 60335-2-6-03. 1991 Amd 3 Safety of Household and Similar Electrical Appliances Part 2: Particular Requirements Section 2.6: Cooking Ranges, Cooking Tables, Ovens and Similar Appliances for Household Use (AMD 7916) October 15, 1993 (L). 84 pp.

BSI BS EN 60335-2-9-91. 1991 Safety of Household and Similar Electrical Appliances Part 2: Particular Requirements Section 2.9: Toasters, Grills, Roasters and Similar Appliances. 26 pp.

BSI BS EN 60335-2-9-01. 1991 Amd 1 Safety of Household and Similar Electrical Appliances Part 2: Particular Requirements Section 2.9: Toasters, Grills, Roasters and Similar Appliances (AMD 6897) April 1, 1992. 32 pp.

BSI BS EN 60335-2-9-02. 1991 Amd 2 Safety of Household and Similar Electrical Appliances Part 2: Particular Requirements Section 2.9: Toasters, Grills, Roasters and Similar Appliances (AMD 7584) March 15, 1993 (L). 42 pp.

BSI BS EN 60335-2-48-92. 1992 Safety of Household and Similar Electrical Appliances Part 2: Particular Requirements Section 2.48: Commercial Electric Grillers and Toasters. 22 pp.

BSI BS EN 60335-2-48-01. 1992 Amd 1 Safety of Household and Similar Electrical Appliances Part 2: Particular Requirements Section 2.48: Commercial Electric Grillers and Toasters (AMD 7832) August 15, 1993 (L). 33 pp.

CENELEC HD 265 S2-84. Safety of Household and Similar Electrical Appliances—Particular Requirements for Appliances for Toasters, Grills, Roasters and Similar Appliances. 8 pp.

CENELEC EN 60335-2-6-90. Safety of Household and Similar Electrical Appliances Part 2: Particular Requirements for Cooking Ranges, Cooking Tables, Ovens and Similar Appliances for Household Use (Supercedes HD 275 S1: 1986 and Its Amendment 1: 1988). 38 pp.

CENELEC EN 60335-2-6-91. CORRIGENDUM Safety of Household and Similar Electrical Appliances Part 2: Particular Requirements for Cooking Ranges, Cooking Tables, Ovens and Similar Appliances for Household Use. 2 pp.

CENELEC EN 60335-2-6 /A2-92. AMD 2 Safety of Household and Similar Electrical Appliances Part 2: Particular Requirements for Cooking Ranges, Cooking Tables, Ovens and Similar Appliances for Household Use; (IEC 335-2-6:1986/A2: 1990). 6 pp.

CENELEC EN 60 335-2-9-90. Safety of Household and Similar Electrical Appliances Part 2: Particular Requirements for Toasters, Grills, Roasters and Similar Appliances. 8 pp.

CENELEC EN 60335-2-9 /A2-92. AMD 2 Safety of Household and Similar Electrical Appliances Part 2: Particular Requirments for Toasters, Grills, Roasters and Similar Appliances. 5 pp.

CENELEC EN 60335-2-48-90. Safety of Household and Similar Electrical Appliances Part 2: Particular Requirements for Commercial Electric Grillers and Toasters. 10 pp.

CENELEC EN 60335-2-48/A1-92. AMD 1 Safety of Household and Similar Electrical Appliances Part 2: Particular Requirements for Commercial Electric Grillers and Toasters. 5 pp.

DIN VDE 0700 Pt 6 A8 (D)-89. Safety of Household and Similar Electrical Appliances; Cooking Ranges, Cooking Tables, Ovens and Similar Appliances for Household Use; Amendment 8 to Draft DIN 57700 Part 6/VDE 0700 Part 6/02.84 (Apr). 15 pp.

Consumer Safety *(Cont.)*

—Grills (Appliances) *(Cont.)*

IEC 335 Pt 2-9-93. Safety of Household and Similar Electrical Appliances Part 2: Particular Requirements for Toasters, Grills, Roasters and Similar Appliances Fourth Edition. 46 pp.

IEC 335 Pt 2-48-88. Safety of Household and Similar Electrical Appliances Part 2: Particular Requirements for Commercial Electric Grillers and Toasters First Edition; (Amendment 1-1990). 44 pp.

—Grinders

BSI BS 2769: Sec 2.3-84. 1984 Hand-Held Electric-Motor-Operated Tools Part 2: Particular Requirements Section 2.3: Grinders, Polishers and Disc-Type Sanders. 16 pp.

BSI BS 2769: Sec 2.3-01. 1984 Amd 1 Hand-Held Electric Motor-Operated Tools Part 2: Particular Requirements Section 2.3: Specification for Grinders, Polishers and Disc-Type Sanders (AMD 6924) May 1, 1992. 17 pp.

CENELEC HD 400.2C/A1-91. AMD 1 Hand-Held Motor Operated Tools Part II: Particular Specifications (Section C). 5 pp.

CENELEC HD 400.2C S1 PRA2-93. AMD 2 Hand-Held Motor Operated Tools Part II: Particular Specifications Section C: Grinders, Polishers and Disc Type Sanders. 2 pp.

IEC 745 Pt 2-3-84. Safety of Hand-Held Motor-Operated Electric Tools Part 2: Particular Requirements for Grinders, Polishers and Disk-Type Sanders First Edition. 21 pp.

—Grooming Appliances

BSI BS 3456: Sec 102.8-88. (WITHDRAWN) 1988 Safety of Household Electrical Appliances Part 102: Particular Requirements Section 102.8: Electric Shavers, Hair Clippers and Similar Appliances (Superseded by BS EN 60335-2-8: 1991). 19 pp.

BSI BS 3456: Sec 202.19-90. 1990 Safety of Household Electrical Appliances Part 202: Particular Requirements Section 202.19: Battery-Powered Shavers, Hair Clippers and Similar Appliances and Their Charging and Battery Assemblies. 28 pp.

BSI BS EN 60335-2-8-91. 1991 Safety of Household and Similar Electrical Appliances Part 2: Particular Requirements Section 2.8: Electric Shavers, Hair Clippers and Similar Appliances (Supersedes BS 3456: Section 102.8: 1988) (L). 19 pp.

CENELEC HD 254 S2-84. Safety of Household and Similar Electric Appliances Particular Requirements for Electric Shavers, Hair Clippers and Similar Appliances (Superseded by EN 60335-2-8). 10 pp.

CENELEC EN 60335-2-8-90. Safety of Household and Similar Electrical Appliances Part 2: Particular Requirements for Electric Shavers, Hair Clippers and Similar Appliances (Supersedes HD 254 S2). 7 pp.

CENELEC EN 60 335-2-19-89. Safety of Household and Similar Electrical Appliances Part 2: Particular Requirements for Battery—Powered Shavers, Hair Clippers and Similar Appliances and Their Charging and Battery Assemblies. 7 pp.

—Guillotine Shears

SAA AS 1893-77. Code of Practice for the Guarding and Safe Use of Metal and Paper Cutting Guillotines. 32 pp.

—Gymnastic Equipment

CEN PREN 913-92. Gymnastic Equipment—General Safety Requirements and Test Methods. 11 pp.

CEN PREN 914-92. Gymnastic Equipment—Parallel Bars and Combination Asymmetric/Parallel Bars. 11 pp.

—Hair Care Appliances

BSI BS 3456: Sec 3.13-79. (WITHDRAWN) 1979 Amd 2 1972-1981 Edition. Specification for Safety of Household Electrical Appliances Part 3: Complete Particular Specifications Section 3.13: Appliances for Skin or Hair Treatment (Superseded by BS EN 60335-2-23: 1991). 60 pp.

BSI BS EN 60335-2-23-91. 1991 Safety of Household and Similar Electrical Appliances Part 2: Particular Requirements Section 2.23: Appliances for Skin or Hair Care (Supersedes BS 3456: Section 3.13: 1979). 32 pp.

BSI BS EN 60335-2-23-01. 1991 Amd 1 Safety of Household and Similar Electrical Appliances Part 2: Particular Requirements Section 2.23: Appliances for Skin or Hair Care (AMD 7793) July 15, 1993 (L). 44 pp.

BSI BS EN 60335-2-23-02. 1991 Amd 2 Safety of Household and Similar Electrical Appliances Part 2: Particular Requirements Section 2.23: Appliances for Skin or Hair Care (AMD 7901) August 15, 1993 (L). 50 pp.

CENELEC HD 266-75. Particular Specification for Appliances for Skin or Hair Treatment. 6 pp.

CENELEC HD 266.2-77. Particular Specification for Appliances for Skin and Hair Treatment. 9 pp.

CENELEC HD 266.3-83. Particular Specification for Appliances for Skin or Hair Treatment. 7 pp.

614
INDEX and DIRECTORY of

INTERNATIONAL AND NON-U.S. NATIONAL STANDARDS
SUBJECT INDEX Consumer

Consumer Safety *(Cont.)*

—Hair Care Appliances *(Cont.)*

CENELEC HD 266.4-83. Particular Specification for Appliances for Skin or Hair Treatment. 6 pp.
CENELEC EN 60 335-2-23-90. Safety of Household and Similar Electrical Appliances Part 2: Particular Requirements for Appliances for Skin or Hair Care. 11 pp.
CENELEC EN 60335-2-23/A1-92. AMD 1 Safety of Household and Similar Electrical Appliances Part 2: Particular Requirements for Appliances for Skin or Hair Care. 5 pp.
CENELEC EN 60335-2-23/A51-92. AMD 51 Safety of Household and Similar Electrical Appliances Part 2: Particular Requirements for Appliances for Skin or Hair Care. 5 pp.
IEC 335 Pt 2-23-86. Safety of Household and Similar Electrical Appliances Part 2: Particular Requirements for Appliances for Skin or Hair Care Third Edition; (Amendment 1-1990). 50 pp.

—Hair Clippers

BSI BS 3456: Sec 102.8-88. (WITHDRAWN) 1988 Safety of Household Electrical Appliances Part 102: Particular Requirements Section 102.8: Electric Shavers, Hair Clippers and Similar Appliances (Superseded by BS EN 60335-2-8: 1991). 19 pp.
BSI BS 3456: Sec 202.19-90. 1990 Safety of Household Electrical Appliances Part 202: Particular Requirements Section 202.19: Battery-Powered Shavers, Hair Clippers and Similar Appliances and Their Charging and Battery Assemblies. 28 pp.
CENELEC HD 254 S2-84. Safety of Household and Similar Electric Appliances Particular Requirements for Electric Shavers, Hair Clippers and Similar Appliances (Superseded by EN 60335-2-8). 10 pp.
CENELEC EN 60335-2-8-90. Safety of Household and Similar Electrical Appliances Part 2: Particular Requirements for Electric Shavers, Hair Clippers and Similar Appliances (Supersedes HD 254 S2). 7 pp.
CENELEC EN 60 335-2-19-89. Safety of Household and Similar Electrical Appliances Part 2: Particular Requirements for Battery—Powered Shavers, Hair Clippers and Similar Appliances and Their Charging and Battery Assemblies. 7 pp.
IEC 335 Pt 2-8-92. Safety of Household and Similar Electrical Appliances Part 2: Particular Requirements for Shavers, Hair Clippers and Similar Appliances Fourth Edition. 29 pp.
IEC 335 Pt 2-19-84. Safety of Household and Similar Electrical Appliances Part 2: Particular Requirements for Battery-Powered Shavers, Hair Clippers and Similar Appliances and Their Charging and Battery Assemblies Second Edition. 45 pp.

—Hammers

BSI BS 2769: Sec 2.6-84. 1984 Hand-Held Electric-Motor-Operated Tools Part 2: Particular Requirements Section 2.6: Hammers. 6 pp.
CENELEC HD 400.2F S1 PRA1-93. AMD 1 Hand-Held Motor Operated Tools Part II: Particular Specifications Section F: Hammers. 4 pp.
CENELEC PREN 50XXX-2-6-92. Safety of Hand-Held Electric Motor Operated Tools Part 2: Particular Requirements for Hammers. 15 pp.
IEC 745 Pt 2-6-89. Safety of Hand-Held Motor-Operated Electric Tools Part 2: Particular Requirements for Hammers First Edition; (Corrigendum—Dec 1989) (Amendment 1-1992). 36 pp.

—Hand Tools

CEN PREN 792-1-92. Handheld Non-Electric Power Tools—Safety Requirements—Part 1: General Safety Requirements for all Types of Handheld Non-Electric Power Tools. 17 pp.
CEN PREN 792-2-92. Handheld Non-Electric Power Tools—Safety Requirements—Part 2: Safety Requirements Related to the Energy Supply of Power Tools. 9 pp.
CEN PREN 792-4-92. Handheld Non-Electric Power Tools—Safety Requirements—Part 4: Percussive Non-Rotary Power Tools. 8 pp.
CEN PREN 792-9-92. Handheld Non-Electric Power Tools—Safety Requirements—Part 9: Die Grinders. 9 pp.
CEN PREN 792-10-92. Handheld Non-Electric Power Tools—Safety Requirements—Part 10: Compression Power Tools, Squeeze Riveters, Presses, Punches. 6 pp.
CENELEC HD 400.1-79. Hand-Held Motor Operated Tools-Part 1: General Specifications. 59 pp.
CENELEC HD 400.1-80. Hand-Held Motor Operated Tools Part I: General Specifications. 60 pp.
CENELEC HD 400.1/A1-81. AMD 1 Hand-Held Motor Operated Tools Part I: General Specifications. 8 pp.
CENELEC HD 400.1 S1/A1-91. AMD 1 Hand-Held Motor Operated Tools Part I: General Specifications. 6 pp.
CENELEC HD 400.2 S2-79. Hand-Held Motor Operated Tools Part II: Particular Specifications Sections A to G. 34 pp.

Consumer Safety *(Cont.)*

—Hand Tools *(Cont.)*

CENELEC HD 400.2 S2-89. Hand-Held Motor Operated Tools—Part II: Particular Specifications Section E: Circular Saws and Circular Knives. 29 pp.
CENELEC HD 400.2E/A1-91. AMD 1 Hand-Held Motor Operated Tools Part II: Particular Specifications Section E: Circular Saws and Circular Knives. 3 pp.
CENELEC HD 400.2E S2 /A1-91. AMD 1 Hand-Held Motor Operated Tools Part II: Particular Specifications Section E: Circular Saws and Circular Knives. 4 pp.
DIN VDE 0740 Pt 1-81. Hand-Held Motor Operated Tools; General Requirements (Apr). 97 pp.
DIN VDE 0740 Pt 1 A1-82. Hand-Held Motor Operated Electric Tools; General Specifications; Amendment 1; (VDE Specification) (July). 8 pp.
DIN VDE 0740 Pt 22-91. Hand-Held Motor-Operated Tools; Additional Specified Requirements (Apr). 41 pp.

—Heat Pumps

DIN VDE 0700 Pt 222-84. Safety of Household and Similar Electrical Appliances; Heat Pumps for Room Heating Purposes (VDE Specification) (Sept). 40 pp.
IEC 335 Pt 2-40-92. Safety of Household and Similar Electrical Appliances Part 2: Particular Requirements for Electrical Heat Pumps, Air-Conditioners and Dehumidifiers Second Edition; (Replaces 378). 74 pp.

—Heaters

CNS C4125-90. Safety Testing Code for Household Electric Heaters (Jan)(3765).
CSA C22.2 NO 1335.1-93. Portable Electrical Motor-Operated and Heating Appliances: General Requirements; (Gen Instr 1). 208 pp.

—Heaters—Waterbeds

IEC 335 Pt 2-66-93. Safety of Household and Similar Electrical Appliances Part 2: Particular Requirements for Water-Bed Heaters First Edition. 35 pp.

—Heating Equipment

BSI BS 3456: Sec 102.30-88. 1988 Amd 1 Safety of Household and Similar Electrical Appliances Part 102: Particular Requirements Section 102.30: Room Heaters (AMD 6670) May 31, 1991. 37 pp.
BSI BS 3456: Sec 202.43-90. 1990 Safety of Household Electrical Appliances Part 202: Particular Requirements Section 202.43: Clothes Dryers and Towel Rails. 20 pp.
BSI BS 3456: Sec 202.43-01. 1990 Amd 1 Safety of Household and Similar Electrical Appliances Part 202: Particular Requirements Section 202.43: Clothes Dryers and Towel Rails (AMD 6901) April 1, 1992 (SUPERSEDES BS 3456 SECTION 2.9: 1970 & 3. 6: 1979). 26 pp.
BSI BS 3456: Sec 202.43-02. 1990 Amd 2 Safety of Household and Similar Electrical Appliances Part 202: Particular Requirements Section 202.43: Clothes Dryers and Towel Rails (AMD 7550) JAN 15, 1993 SUPERSEDES BS 3456 SECTION 2. 9: 1970 & 3. 6: 1979. 31 pp.
BSI BS 3456: Sec A1-66. (WITHDRAWN) 1966 Amd 4 Safety of Household Electrical Appliances Part A: Heating and Cooking Appliances Section A1: General Requirements. 57 pp.
BSI BS 3456: Sec A4-71. (WITHDRAWN) 1971 Amd 9 Safety of Household Electrical Appliances Part A: Heating and Cooking Appliances Section A4: Electrically Heated Blankets (Withdrawn, superseded by BS En 60967: 1991). 67 pp.
BSI BS 3456: Sec 2.21-72. 1972 Amd 3 Safety of Household and Similar Electrical Appliances Part 2: Particular Requirements Section 2.21: Electric Immersion Heaters (AMD 6345) September 30, 1991. 21 pp.
BSI BS 3456: Sec 2.22-72. 1972 Amd 4 Safety of Household Electrical Appliances Part 2: Particular Requirements Section 2.22: Safety of Household Electrical Appliances (AMD 6226) February 28, 1990. 38 pp.
BSI BS 3456: Sec 3.21-81. (OBSOLESCENT) 1981 Amd 2 1972-1981 Edition. Specification for Safety of Household Electrical Appliances Part 3: Complete Particular Specifications Section 3.21: Portable Immersion Heater (AMD 5895) August 31, 1988. 45 pp.
BSI BS 5258: Part 4-87. 1987 Safety of Domestic Gas Appliances Part 4: Fanned-Circulation Ducted-Air Heaters. 28 pp.
BSI BS 5258: Part 5-89. 1989 Safety of Domestic Gas Appliances Part 5: Gas Fires. 27 pp.
BSI BS 5258: Part 5-75. 1975 Amd 2 Safety of Domestic Gas Appliances Part 5: Gas Fires (AMD 4745) March 29, 1985. 37 pp.
BSI BS 5258: Part 8-80. 1980 Safety of Domestic Gas Appliances Part 8: Combined Appliances: Gas Fire/ Back Boiler. 24 pp.

Consumer Safety *(Cont.)*

—Heating Equipment *(Cont.)*

BSI BS 5258: Part 9-89. 1989 Safety of Domestic Gas Appliances Part 9: Combined Appliances: Fanned-Circulation Ducted-Air Heaters/Circulators. 32 pp.
BSI BS 5258: Part 10-80. 1980 Amd 1 Safety of Domestic Gas Appliances Part 10: FluelessSpace Heaters (Excluding Catalytic Combustion Heaters) (3rd Family Gases). 18 pp.
BSI BS 5258: Part 11-80. 1980 Amd 1 Safety of Domestic Gas Appliances Part 11: Flueless Catalytic Combustion Heaters (3rd Family Gases). 17 pp.
BSI BS 5258: Part 13-86. 1986 Safety of Domestic Gas Appliances Part 13: Convector Heaters. 33 pp.
BSI BS 5258: Part 16-91. 1991 Safety of Domestic Gas Appliances Part 16: Specification for Inset Live Fuel Effect Gas Fires (2nd and 3rd Family Gases). 38 pp.
BSI BS 5986-80. 1980 Electrical Safety and Performance of Gas Fired Space Heating Appliances with Inputs 60 kW to 2 mW. 18 pp.
BSI BS 7261-90. 1990 Safety of Small Non-Domestic Flueless Space Heaters Burning 3rd Family Gases. 25 pp.
BSI BS EN 60335-2-45-91. 1991 Safety of Household and Similar Electrical Appliances Part 2: Particular Requirements Section 2.45: Portable Electric Heating Tools and Similar Appliances. 23 pp.
BSI BS EN 60335-2-45-01. 1991 Amd 1 Safety of Household and Similar Electrical Appliances Part 2: Particular Requirements Section 2.45: Portable Electric Heating Tools And Similar Appliances. 44 pp.
BSI BS EN 60967-91. 1991 Safety of Electrically Heated Blankets, Pads and Similar Flexible Heating Appliances for Household Use. 123 pp.
CEN PREN 416 (Part 1)-90. Single Burner Gas-Fired Overhead Radiant Tube Heaters for Non-Domestic Use—Part 1: Safety. 69 pp.
CEN PREN 419 (Part 1)-90. Non-Domestic Gas-Fired Overhead Luminous Radiant Heaters—Part 1: Safety. 60 pp.
CEN PREN 525-91. Requirements for Non-Domestic Direct Gas-Fired Forced Convection Air Heaters for Space Heating. 74 pp.
CEN PREN 624-91. LPG Space Heating Equipment in Vehicles and Boats. 60 pp.
CEN PREN 777-1-92. Multi-Burner Gas-Fired Overhead Radiant Tube Heater Systems for Non-Domestic Use—Part 1: System D—Safety. 69 pp.
CEN PREN 777-2-92. Multi-Burner Gas-Fired Overhead Radiant Tube Heater Systems for Non-Domestic Use—Part 2: System E—Safety. 69 pp.
CEN PREN 777-3-92. Multi-Burner Gas-Fired Overhead Radiant Tube Heater Systems for Non-Domestic Use—Part 3: System F—Safety. 71 pp.
CEN PREN 777-4-92. Multi-Burner Gas-Fired Overhead Radiant Tube Heater Systems for Non-Domestic Use—Part 4: System H—Safety. 71 pp.
CENELEC HD 278 S1-87. Safety of Household and Similar Electrical Appliances—Part 2: Particular Requirements for Room Heaters; (Superseded by EN 60335-2-30:1992). 32 pp.
CENELEC EN 60 335-2-43-89. Safety of Household and Similar Electrical Appliances—Part 2: Particular Requirements for Clothes Dryers and Towel Rails. 7 pp.
CENELEC EN 60 335-2-43/A1-90. AMD 1 Safety of Household and Similar Electrical Appliances—Part 2: Particular Requirements for Clothes Dryers and Towel Rails. 6 pp.
CENELEC EN 60335-2-43/A51-92. AMD 51 Safety of Household and Similar Electrical Appliances—Part 2: Particular Requirements for Clothes Dryers and Towel Rails. 4 pp.
CENELEC EN 60 335-2-45-90. Safety of Household and Similar Electrical Appliances Part 2: Particular Requirements for Portable Electric Heating Tools and Similar Appliances. 11 pp.
CENELEC EN 60335-2-45/PRAB-93. AMD AB Safety of Household and Similar Electric Appliances. Part 2: Particular Requirements for Portable Electric Heating Tools and Similar Appliances—Sub-Clauses 8.1 and 22.107. 2 pp.
CENELEC EN 60335-2-45/A1-92. AMD 1 Safety of Household and Similar Electrical Appliances Part 2: Particular Requirements for Portable Electric Heating Tools and Similar Appliances (IEC 335-2-45:1986/A1:1990). 5 pp.
CENELEC EN 60 967-90. Safety of Electrically Heated Blankets, Pads and Similar Flexible Heating Appliances for Household Use. 13 pp.
CENELEC EN 60967-92. Safety of Electrically Heated Blankets, Pads and Similar Flexible Heating Appliances for Household Use. 3 pp.
DIN VDE 0700 Pt 30-83. Test Certificate, Safety Test of Household & Similar Electrical Appliances. Portable Electric Heating Tools (Jan). 26 pp.
DIN VDE 0700 Pt 30 A1 (D)-84. Safety of Household and Similar Electrical Appliances; Particular Specifications for Portable Electric Heating Tools (VDE Specification): Amendment 1 (Nov). 19 pp.

INDUSTRY STANDARDS 615

INTERNATIONAL AND NON-U.S. NATIONAL STANDARDS SUBJECT INDEX

Consumer Safety (Cont.)

Heating Equipment (Cont.)

IEC 335 Pt 2-43-84. Safety of Household and Similar Electrical Appliances Part 2: Particular Requirements for Clothes Dryers and Towel Rails First Edition; (Amendment 1-1988). 35 pp.

IEC 335 Pt 2-45-86. Safety of Household and Similar Electrical Appliances Part 2: Particular Requirements for Portable Electric Heating Tools and Similar Appliances First Edition; (Amendment 1-1990) (CENELEC EN 60335-2-45/A1:1992). 48 pp.

IEC 335 Pt 17-74. Safety of Household and Similar Electrical Appliances Particular Requirements for Electrically Heated Blankets, Pads and Mattresses, Recm. First Edition. 95 pp.

Hedge Trimmers

CEN PREN 774-92. Safety Requirements for Agricultural and Forestry Machinery—Portable Powered Hedge Trimmers. 22 pp.

CENELEC HD 400.3N S2-92. Hand-Held Motor Operated Tools Part II: Particular Specifications Section N: Hedge Trimmers and Scissor-Type Grass Shears. 13 pp.

IEC 745 Pt 2-15-84. Safety of Hand-Held Motor-Operated Electric Tools Part 2: Particular Requirements for Hedge Trimmers and Grass Shears First Edition. 15 pp.

ISO 10517-93. Portable Powered Hedge Trimmers—Definitions, Mechanical Safety Requirements and Tests First Edition. 12 pp.

High Chairs

BSI BS 5799-86. 1986 Amd 1 Safety Requirements for Children's High Chairs and Multi-Purpose High Chairs for Domestic Use (AMD 6681) April 30, 1991 April 30, 1991. 20 pp.

ISO 9221 Pt 1-92. Furniture—Children's High Chairs—Part 1: Safety Requirements First Edition. 6 pp.

SNZ NZS/BS 5799-86. Specification for Safety Requirements for Children's High Chairs and Multi-Purpose High Chairs for Domestic Use Amend: 1. 20 pp.

Humidifiers

BSI BS 3456: Sec 202.139-91. 1991 Safety of Household and Similar Electrical Appliances Part 202: Particular Requirements Section 202.139: Room Humidifiers (SUPERSEDES BS 3456 SECTION 2. 39: 1973). 9 pp.

BSI BS 3456: Sec 2.39-73. (WITHDRAWN) 1973 Amd 1 1972-1981 Edition. Specification for Safety of Household Electrical Appliances Part 2: Section 2.39: Room Humidifiers (Superseded by BS 3456: Section 202.139: 1991). 12 pp.

Ice Cream Makers

BSI BS EN 60335-2-57-92. 1992 Safety of Household and Similar Electrical Appliances Part 2: Particular Requirements Section 2.57: Ice-Cream Appliances with Incorporated Motor-Compressors. 23 pp.

CENELEC EN 60335-2-57-92. Safety of Household and Similar Electrical Appliances Part 2: Particular Requirements for Ice-Cream Appliances with Incorporated Motor-Compressors. 8 pp.

IEC 335 Pt 2-57-89. Safety of Household and Similar Electrical Appliances Part 2: Particular Requirements for Ice-Cream Appliances with Incorporated Motor-Compressors First Edition. 36 pp.

Ice Makers

DIN VDE 0700 Pt 240-83. Safety of Household and Similar Electrical Appliances; Particular Requirements for Refrigerators and Freezers for Special Purposes and Ice Makers (Feb) (Supersedes VDE 0730 Part 2 ZK/10.78). 61 pp.

IEC 335 Pt 2-24-92. Safety of Household and Similar Electrical Appliances Part 2: Particular Requirements for Refrigerators, Food-Freezers and Ice-Makers Third Edition: (CENELEC HD 269 S2:1986). 78 pp.

Impact Wrenches

CEN PREN 792-6-92. Handheld Non-Electric Power Tools—Safety Requirements—Part 6: Assembly Power Tools for Threaded Fasteners. 9 pp.

IEC 745 Pt 2-2-82. Safety of Hand-Held Motor-Operated Electric Tools Part 2: Particular Requirements for Screwdrivers and Impact Wrenches First Edition; (Amendment 1-1991). 20 pp.

Infrared Lamps

BSI BS 3456: Sec 102.27-87. (WITHDRAWN) 1987 Safety of Household and Similar Electrical Appliances Part 102: Particular Requirements Section 102.27: Ultra-Violet and Infra-Red Radiation Skin Treatment Appliances for Household Use (Superseded by BS 3456: Sec. 202.27: 1991). 20 pp.

Consumer Safety (Cont.)

Infrared Lamps (Cont.)

BSI BS 3456: Sec 202.27-91. 1991 Safety of Household Electrical Appliances Part 202: Particular Requirements Section 202.27: Ultra-Violet and Infra-Red Radiation Skin Treatment Appliances for Household and Similar Use. 34 pp.

BSI BS 3456: Sec 202.27-01. 1991 Amd 1 Safety of Household and Similar Electrical Appliances Part 202: Particular Requirements Section 202.27: Ultra-Violet and Infra-Red Radiation Skin Treatment Appliances for Household and Similar Use (AMD 7963) October 15, 1993 (L). 35 pp.

BSI BS EN 60335-2-27-92. 1992 Safety of Household and Similar Electrical Appliances Part 2: Particular Requirements Section 2.27: Ultra-Violet and Infra-Red Radiation Skin Treatment Appliances for Household and Similar Use. 28 pp.

BSI BS EN 60335-2-27-01. 1992 Amd 1 Safety of Household and Similar Electrical Appliances Part 2: Particular Requirements Section 2.27: Ultra-Violet and Infra-Red Radiation Skin Treatment Appliances for Household and Similar Use (AMD 7794) June 15, 1993 (L). 35 pp.

CENELEC HD 272 S3-87. Safety of Household and Similar Electrical Appliances—Part 2: Particular Requirements for Ultra-Violet and Infra-Red Radiation Skin Treatment Appliances for Household Use. 8 pp.

CENELEC EN 60 335-2-27-89. Safety of Household and Similar Electrical Appliance Part 2: Particular Requirements for Ultra-Violet and Infra-Red Radiation Skin Treatment Appliances for Household. 17 pp.

CENELEC EN 60335-2-27-92. Safety of Household and Similar Electrical Appliances Part 2: Particular Requirements for Ultra-Violet and Infra-Red Radiation Skin Treatment Appliances for Household Use; (IEC 335-2-27:1987 + A1:1989). 7 pp.

CENELEC EN 60335-2-27-92. Safety of Household and Similar Electrical Appliances Part 2: Particular Requirements for Ultra-Violet and Infra-Red Radiation Skin Treatment Appliances for Household and Similar Use; (IEC 335-2-27: 1987 + Amendment 1: 1989, Modified). 20 pp.

CENELEC EN 60335-2-27-92. Safety of Household and Similar Electrical Appliances Part 2: Particular Requirements for Ultra-Violet and Infra-Red Radiation Skin Treatment Appliances for Household Use; (IEC 335-2-27:1987 + A1:1989). 9 pp.

CENELEC EN 60335-2-27/A2-92. AMD 2 Safety of Household and Similar Electrical Appliances Part 2: Particular Requirements for Ultra-Violet and Infra-Red Radiation Skin Treatment Appliances for Household and Similar Use (IEC 335-2-27:1987/A2: 1991). 4 pp.

CENELEC EN 60335-2-37/A51-92. AMD 51 Safety of Household and Similar Electrical Appliances Part 2: Particular Requirements for Commercial Electric Deep Fat Fryers. 6 pp.

IEC 335 Pt 2-27-87. Safety of Household and Similar Electrical Appliances Part 2: Particular Requirements for Ultra-Violet and Infra-Red Radiation Skin Treatment Appliances for Household Use Second Edition; (Amendment 1-1989) (Amendment 2-1991); (CENELEC EN 60335-2-27:/A2:1992). 60 pp.

Insect Electrocusion Devices

IEC 335 Pt 2-59-90. Safety of Household and Similar Electrical Appliances Part 2: Particular Requirements for Insect Killers First Edition. 33 pp.

Irons (Electric)

BSI BS 3456: Sec 102.3-90. (WITHDRAWN) 1990 Safety of Household Electrical Appliances Part 102: Particular Requirements Section 102.3: Electric Irons (Superseded by BS EN 60335-2-3: 1991). 20 pp.

BSI BS EN 60335-2-3-91. 1991 Safety of Household and Similar Electrical Appliances Part 2: Particular Requirements Section 2.3: Electric Irons (L). 25 pp.

BSI BS EN 60335-2-3-01. 1991 Amd 1 Safety of Household and Similar Electrical Appliances Part 2: Particular Requirements Section 2.3: Electric Irons (AMD 7280) October 15, 1992. 32 pp.

BSI BS EN 60335-2-3-02. 1991 Amd 2 Safety of Household and Similar Electrical Appliances Part 2: Particular Requirements Section 2.3: Electric Irons (AMD 7694) June 15, 1993 (L). 38 pp.

BSI BS EN 60335-2-44-91. 1991 Safety of Household and Similar Electrical Appliances Part 2: Particular Requirements Section 2.44: Electric Ironers. 22 pp.

CENELEC HD 253-74. Particular Specification for Electric Irons, Ironess and Pressing Machines. 6 pp.

CENELEC HD 253 S3-86. Safety of Household and Similar Electrical Appliances Part 2: Particular Requirements for Electric Irons. 3 pp.

CENELEC HD 253 S3/A1-87. AMD 1 Safety of Household and Similar Electrical Appliances Part 2: Particular Requirements for Electric Irons. 5 pp.

CENELEC EN 60 335-2-3-90. Safety of Household and Similar Electrical Appliances Part 2: Particular Requirements for Electric Irons. 18 pp.

Consumer Safety (Cont.)

Irons (Electric) (Cont.)

CENELEC EN 60335-2-3 /PRAB-91. AMD prAB Safety of Household and Similar Electrical Appliances Part 2: Particular Requirements for Electric Irons. 1 p.

CENELEC EN 60335-2-3 /A1-92. AMD 1 Safety of Household and Similar Electrical Appliances Part 2: Particular Requirements for Electric Irons. 6 pp.

CENELEC EN 60335-2-3 /A52-92. AMD 52 Safety of Household and Similar Electrical Appliances Part 2: Particular Requirements for Electric Irons. 4 pp.

CENELEC EN 60335-2-3-92. CORRIGENDUM Safety of Household and Similar Electrical Appliances Part 2: Particular Requirements for Electric Irons. 1 p.

CENELEC EN 60 335-2-44-91. Safety of Household and Similar Electrical Appliances Part 2 Particular Requirements for Electric Ironers. 9 pp.

DIN VDE 0700 Pt 212-81. Safety of Household and Similar Electrical Appliances; Particular Requirements for Rotary Ironers and Flat Bed Ironers (July). 16 pp.

IEC 335 Pt 2-3-93. Safety of Household and Similar Electrical Appliances Part 2: Particular Requirements for Electric Irons Fourth Edition. 38 pp.

IEC 335 Pt 2-44-87. Safety of Household and Similar Electrical Appliances Part 2: Particular Requirements for Electric Ironers First Edition. 27 pp.

Jigsaws

IEC 745 Pt 2-11-84. Safety of Hand-Held Motor-Operated Electric Tools Part 2: Particular Requirements for Reciprocating Saws (Jig and Sabre Saws) First Edition. 19 pp.

Kitchen Furniture

CEN PREN 1153-93. Kitchen Furniture—Safety Requirements and Test Methods for Built-in and Free Standing Kitchen Cabinets and Work Tops. 16 pp.

Knives

BSI BS 2769: Sec 2.5-91. 1991 Hand-Held Electric Motor-Operated Tools Part 2: Particular Requirements Section 2.5: Specification for Circular Saws and Circular Knives. 16 pp.

BSI BS 2769: Sec 2.5-01. 1991 Amd 1 Hand-Held Electric Motor-Operated Tools Part 2: Particular Requirements Section 2.5: Specification for Circular Saws and Circular Knives (AMD 6920) April 1, 1992. 17 pp.

IEC 745 Pt 2-5-93. Safety of Hand-Held Motor-Operated Electric Tools Part 2: Particular Requirements for Circular Saws and Circular Knives Second Edition. 44 pp.

Lampholders

BSI BS 5101: Part 3-75. (WITHDRAWN) 1975 Amd 5 Lamp Caps and Holders Together with Gauges for the Control of Interchangeability and Safety Part 3: Gauges (AMD 6520) January 31, 1991 (Superseded by BS EN 60061-3: 1993). 338 pp.

BSI BS 5101: Part 4-80. 1980 Amd 1 Lamp Caps and Holders Together with Gauges for the Control of Interchangeability and Safety Part 4: Lamp Caps, Lampholders and Gauges Used in the United Kingdom but Not Specified in Parts 1, 2 or 3. 18 pp.

BSI BS 5101: Part 5-90. (WITHDRAWN) 1990 Lamp Caps and Holders Together with Gauges for the Control of Interchangeability and Safety Part 5: Guidelines and General Information (Renumbered as BS EN 60061-4: 1992). 22 pp.

BSI BS EN 60061-1-93. 1993 Lamp Caps and Holders Together with Gauges for the Control of Interchangeability and Safety Part 1: Lamp Caps (Supersedes BS 5101: part 1: 1975) (E). 202 pp.

BSI BS EN 60061-2-93. 1993 Lamp Caps and Holders Together with Gauges for the Control of Interchangebility and Safety Part 2: Lampholders (Supersedes BS 5101: part 2: 1975) (E). 158 pp.

BSI BS EN 60061-3-93. 1993 Lamp Caps and Holders Together with Gauges for the Control of Interchangeability and Safety Part 3: Gauges (E). 425 pp.

BSI BS EN 60061-4-92. 1992 Amd 1 Lamp Caps and Holders Together with Gauges for the Control of Interchangeability and Safety Part 4: Guidelines and General Information (AMD 7556) December 15, 1992. 24 pp.

CENELEC HD 65-82. Lamp Caps and Holders Together with Gauges for the Control of Interchangeability and Safety. 4 pp.

CENELEC EN 60061-1-93. Lamp Caps and Holders Together with Gauges for the Control of Interchangeability and Safety Part 1: Lamp Caps (IEC 61-1: 1969 + Supplements A: 1970 to N: 1992, Modified) (Supersedes HD 65.1 S1: 1978). 197 pp.

INTERNATIONAL AND NON-U.S. NATIONAL STANDARDS
SUBJECT INDEX

Consumer Safety (Cont.)
—Lampholders (Cont.)
CENELEC EN 60061-2-93. Lamp Caps and Holders Together with Gauges for the Control of Interchangeability and Safety Part 2: Lampholders (IEC 61-2: 1969 + Supplements A: 1970 to K: 1992, Modified). 152 pp.

CENELEC EN 60061-3-93. Lamp Caps and Holders Together with Gauges for the Control of Interchangeability and Safety Part 3: Gauges (IEC 61-3: 1969 + Supplements A: 1970 to M: 1992, Modified). 420 pp.

CENELEC EN 60061-4-92. Lamp Caps and Holders Together with Gauges for the Control of Interchangeability and Safety Part 4: Guidelines and General Information. (IEC 61-4: 1990, Mod). 4 pp.

CENELEC EN 60061-4-92. Lamp Caps and Holders Together with Gauges for the Control of Interchangeability and Safety Part 4: Guidelines and General Information. 13 pp.

IEC 61 Pt 1-69. Lamp Caps and Holders Together with Gauges for the Control of Interchangeability and Safety Part 1: Lamp Caps Third Edition; (Supplements A-J-1970-1980)(Supplement K-1983) (Supplement L-1987) (Supplement M-1989) (Supplement N-1992); (CENELEC EN60061-1:1993). 279 pp.

IEC 61 Pt 2-69. Lamp Caps and Holders Together with Gauges for the Control of Inter-changeability and Safety Part 2: Lampholders Third Edition; (Supplements A-F-1970-1980)(Supplement G-1983) (Supplement H-1987) (Supplement J-1989) (Supplement K-1992) (CENE EN 60061-2: 1993). 202 pp.

IEC 61 Pt 3-69. Lamp Caps and Holders Together with Gauges for the Control of Interchangeability and Safety Part 3: Gauges Third Edition; (Errata—Dec 1984) (Supplement K-1987) (Supplement L-1989) (Supplement M-1992) (CENE EN 60061-3: 1993). 506 pp.

IEC 61 Pt 4-90. Lamp Caps and Holders Together with Gauges for the Control of Interchangeability and Safety Part 4: Guidelines and General Information First Edition; (Supplement A-1992) (CENELEC EN 60061-4: 1992). 85 pp.

—Lamps
BSI BS 5101: Part 1-75. (WITHDRAWN) 1975 Amd 5 Lamp Caps and Holders Together with Gauges for the Control of Interchangeability and Safety Part 1: Lamp Caps (IEC 61-1) (AMD 6519) January 31, 1991 (Superseded by BS EN 60061-2: 1993). 184 pp.

BSI BS 5101: Part 2-75. (WITHDRAWN) 1975 Amd 5 Lamp Caps and Holders Together with Gauges for the Control of Interchangeability and Safety Part 2: Lamp Holders (IEC 61-2) (AMD 6521) January 31, 1991 (Superseded by BS EN 60061-2: 1993). 137 pp.

BSI BS 5971-88. 1988 Amd 1 Safety of Tungsten Filament Lamps for Domestic and Similar General Lighting Purposes (EN 60 432: 1988) (AMD 6369) March 28, 1991. 55 pp.

BSI BS 6982-88. 1988 Amd 1 Single-Capped Fluorescent Lamps—Safety and Performance Requirements (IEC 901: 1987) (AMD 6572) February 28, 1991. 40 pp.

BSI BS 6982-02. 1988 Amd 2 Single-Capped Fluorscent Lamps—Safety and Performance Requirements (IEC 901: 1987) (AMD 6708) April 30, 1991. 48 pp.

BSI BS 7173-89. 1989 Amd 1 Self-Ballasted Lamps for General Lighting Services Safety Requirements (AMD 6707) September 30, 1991. 24 pp.

BSI BS 7173-02. 1989 Amd 2 Self-Ballasted Lamps for General Lighting Services Safety Requirements (AMD 7716) May 15, 1993. 30 pp.

CENELEC HD 65.1 S1-88. Lamp Caps and Holders Together with Gauges for the Control of Interchangeability and Safety Part 1: Lamp Caps. 3 pp.

CENELEC HD 65.2-87. Lamp Caps and Holders Together with Gauges for the Control of Interchangeability and Safety Part 2: Lampholders. 3 pp.

CENELEC HD 65.3-87. Lamp Caps and Holders Together with Gauges for the Control of Interchangeability and Safety Part 3: Gauges. 3 pp.

CENELEC EN 60 432-88. Safety Requirements for Tungsten Filament Lamps for Domestic and Similar General Lighting Purposes. 30 pp.

CENELEC EN 60 432/A1-89. AMD 1 Safety Requirements for Tungsten Filament Lamps for Domestic and Similar General Lighting Purposes. 4 pp.

CENELEC EN 60 901-90. Single-Capped Fluorescent Lamps—Safety and Performance Requirements. 6 pp.

CENELEC EN 60901/A1-90. Single-Capped Fluorescent Lamps Safety and Performance Requirements. 4 pp.

CENELEC EN 60968-90. Self-Ballasted Lamps for General Lighting Services Safety Requirements. 5 pp.

Consumer Safety (Cont.)
—Lamps (Cont.)
CENELEC EN 60968/A1-93. AMD 1 Self-Ballasted Lamps for General Lighting Services Safety Requirements (IEC 968:1988/A1: 1991). 4 pp.

IEC 901-87. Single-Capped Fluorescent Lamps—Safety and Performance Requirements First Edition; (Amendment 1-1989) (Amendment 2-1992) (Corrigendum—June 1992). 211 pp.

IEC 968-88. Self-Ballasted Lamps for General Lighting Services Safety Requirements First Edition; (Amendment 1-1991) (CENELEC EN 60968/A1: 1993). 34 pp.

SNZ NZS 6708: Part 2-89. Specification for Tungsten Filament Lamps for Domestic and Similar General Lighting Purposes Part 2: Safety Requirements Amend: A, 1 (Incorporated in Amend. No 2), 1A, 2, 2A, 3. 55 pp.

—Lawn Mowers
BSI BS 3456: Sec 2.32-74. 1974 Amd 3 1972-1981 Edition. Specification for Safety of Household Electrical Appliances Part 2: Section 2.32: Mains-Operated Electric Lawnmowers. 23 pp.

BSI BS 3456: Sec 2.42-77. 1977 1972-1981 Edition. Specification for Safety of Household Electrical Appliances Part 2: Section 2.42: Battery-Operated Lawnmowers. 10 pp.

DIN VDE 0700 Pt 206 (D)-88. Safety of Household and Similar Electrical Appliances Particular Requirements for Pedestrian Controlled Mains Operated Electrical Lawn-Mowers and Similar Gardening Appliances; Identical to IEC 61F(Central Office) 58 (Jan). 37 pp.

—Lighters
BSI BS 3456: Sec 2.19-70. (WITHDRAWN) 1970 Amd 1 1972-1981 Edition. Specification for Safety of Household Electrical Appliances Part 2: Section 2.19: Electric Firelighters (Superseded by BS EN 60335-2-45: 1991). 13 pp.

BSI BS 6908-90. 1990 Safety Requirements for Gas Fuelled Smokers' Lighters. 19 pp.

CEN EN 123-80. Gas Fuelled Smokers' Lighters: Safety Requirements. 10 pp.

CEN EN 29 994-90. Lighters—Safety Specification. 16 pp.

DIN ENGL EN 29994-91. Lighters; Safety Requirements (ISO 9994: 1989) (Feb). 15 pp.

ISO 9994-89. Lighters—Safety Specification First Edition. 16 pp.

—Lighting Chains
BSI BS 4647-70. (WITHDRAWN) 1970 Amd 5 Lighting Sets for Christmas Trees and Decorative Purposes for Indoor Use (AMD 6397) May 31, 1991 (Superseded by BS EN 60598-2-20: 1992). 22 pp.

—Liquefied Petroleum Gas
CNS Z1004-86. Safety Code for Liquefied Petroleum Gas (Sep)(1332). 9 pp.

ISO 7421-91. Leisure Accommodation Vehicles—Liquefied Petroleum Gas Systems First Edition. 7 pp.

—Log Splitters
CEN PREN 609-91. Safety Requirements for Agricultural and Forestry Machinery—Log Splitters. 16 pp.

—Massage Equipment
BSI BS 3456: Sec 102.32-88. (WITHDRAWN) 1988 Safety of Household Electrical Appliances Part 102: Particular Requirements Section 102.32: Massage Appliances (Superseded by BS EN 60335-2-32: 1991). 15 pp.

BSI BS EN 60335-2-32-91. 1991 Safety of Household and Similar Electrical Appliances Part 2: Particular Requirements Section 2.32: Massage Appliances (Supersedes BS 3456: Section 102.32: 1988). 21 pp.

CENELEC HD 258 S2-83. Safety of Household and Similar Electrical Appliances—Particular Requirements for Massage Appliances. 7 pp.

CENELEC EN 60 335-2-32-90. Safety of Household and Similar Electrical Appliances Part 2: Particular Requirements for Massage Appliances. 7 pp.

CENELEC EN 60335-2-32-92. Safety of Household and Similar Electrical Appliances Part: 2 Particular Requirements for Massage Appliances. 17 pp.

DIN VDE 0700 Pt 32-84. Safety of Household and Similar Electrical Appliances; Particular Requirements for Massage Appliances (VDE Specification) Amended IEC 335-2-32 (Feb). 25 pp.

DIN VDE 0700 Pt 32 A1 (D)-84. Safety of Household and Similar Electrical Appliances; Particular Requirements for Massage: Amendment 1 (VDE Specification) (July). 5 pp.

IEC 335 Pt 2-32-87. Safety of Household and Similar Electrical Appliances Part 2: Particular Requirements for Massage Appliances Second Edition. 27 pp.

Consumer Safety (Cont.)
—Meat
SAA AS 2056-77. Code of Practice for Safety in the Meat Industry. 24 pp.

—Microwave Ovens
BSI BS 3456: Sec 102.25-88. (WITHDRAWN) 1988 Safety of Household Electrical Appliances Part 102: Particular Requirements Section 102.25: Appliances for Heating Food by Means of Microwave Energy (Superseded by BS EN 60335-2-25: 1991). 39 pp.

BSI BS 5175-76. 1976 Amd 4 Safety of Commercial Electrical Appliances Using Microwave Energy for Heating Foodstuffs. 51 pp.

BSI BS EN 60335-2-25-91. 1991 Safety of Household and Similar Electrical Appliances Part 2: Particular Requirements Section 2.25: Microwave Ovens. 34 pp.

BSI BS EN 60335-2-25-01. 1991 Amd 1 Safety of Household and Similar Electrical Appliances Part 2: Particular Requirements Section 2.25: Microwave Ovens (AMD 7217) September 15, 1992. 41 pp.

BSI BS EN 60335-2-25-02. 1991 Amd 2 Safety of Household and Similar Electrical Appliances Part 2: Particular Requirements Section 2.25: Microwave Ovens (AMD 7880) September 15, 1993 (L). 47 pp.

CENELEC HD 270 S1-83. Safety of Household and Similar Electrical Appliances—Particular Requirements for Microwave Cooking Appliances (Superseded by EN 41003). 17 pp.

CENELEC EN 60 335-2-25-90. Safety of Household and Similar Electrical Appliances Part 2: Particular Requirements for Micro Wave Ovens. 9 pp.

CENELEC EN 60335-2-25/A51-92. AMD 1 Safety of Household and Similar Electrical Appliances Part 2: Particular Requirements for Micro Wave Ovens. 7 pp.

IEC 335 Pt 2-25-93. Safety of Household and Similar Electrical Appliances Part 2: Particular Requirements for Microwave Ovens Third Edition. 70 pp.

IEC 335 Pt 25-76. Safety of Household and Similar Electrical Appliances Part 2: Particular Requirements for Microwave Cooking Appliances First Edition. 55 pp.

—Milk Coolers
BSI BS EN 50087-93. 1993 Safety of Household and Similar Electrical Appliances—Particular Requirements for Bulk-Milk Coolers (L). 18 pp.

CENELEC PREN 50 087-90. Safety of Household and Similar Electrical Appliances Part 2: Particular Requirements for Bulk Milk Coolers. 17 pp.

CENELEC EN 50087-93. Safety of Household and Similar Electrical Appliances Particular Requirements for Bulk-Milk Coolers. 16 pp.

CENELEC EN 50087-93. Safety of Household and Similar Electrical Appliances Particular Requirements for Bulk-Milk Coolers. 13 pp.

—Milking Machines
IEC 335 Pt 2-70-93. Safety of Household and Similar Electrical Appliances Part 2: Particular Requirements for Milking Machines First Edition. 35 pp.

—Motor Compressors
BSI BS 3456: Sec 102.34-87. 1987 Safety of Household and Similar Electrical Appliances Part 102: Particular Requirements Section 102.34: Motor-Compressors. 24 pp.

BSI BS 3456: Sec 102.34-01. 1987 Amd 1 Safety of Household and Similar Electrical Appliances Part 102: Particular Requirements Section 102.34: Motor-Compressors (AMD 6876) April 1, 1992. 25 pp.

CENELEC HD 277 S1-85. Safety of Household and Similar Electrical Appliances—Particular Requirements for Motor-Compressors. 7 pp.

IEC 335 Pt 2-34-80. Safety of Household and Similar Electrical Appliances Part 2: Particular Requirements for Motor-Compressors First Edition; (Corrigendum—Jan 1989) (Amendment 3-1992, Incorporating Amendment 2). 61 pp.

—Motorcycles
CNS Z3002-79. Testing Standard for Safety of 50 cc Motorcycles (Tentative) (Nov)(2652).

—Mountaineering Equipment
BSI BS EN 564-93. 1993 Mountaineering Equipment Accessory Cord Safety Requirements and Test Method. 9 pp.

BSI BS EN 565-93. 1993 Mountaineering Equipment Tape Safety Requirements and Test Method. 8 pp.

BSI BS EN 566-93. 1993 Mountaineering Equipment Slings Safety Requirements and Test Method (N). 8 pp.

BSI BS EN 567-93. 1993 Mountaineering Equipment Rope Clamp Safety Requirements and Test Method (N). 9 pp.

BSI BS EN 568-93. 1993 Mountaineering Equipment Ice Anchors Safety Requirements and Test Method (N). 10 pp.

Consumer

INTERNATIONAL AND NON-U.S. NATIONAL STANDARDS
SUBJECT INDEX

Consumer Safety *(Cont.)*

—Mountaineering Equipment *(Cont.)*

BSI BS EN 569-93. 1993 Mountaineering Equipment Pitons Safety Requirements and Test Method (N). 10 pp.
CEN PREN 564-91. Mountaineering Equipment —Accessory Cord—Safety Requirements, Testing, Marking. 5 pp.
CEN EN 564-92. Mountaineering Equipment —Accessory Cord—Safety Requirements and Test Method. 4 pp.
CEN PREN 565-91. Mountaineering Equipment —Tape Material—Safety Requirements, Testing, Marking. 5 pp.
CEN EN 565-92. Mountaineering Equipment —Tape— Safety Requirements and Test Method. 3 pp.
CEN PREN 566-91. Mountaineering Equipment —Slings—Safety Requirements, Testing, Marking. 5 pp.
CEN EN 566-92. Mountaineering Equipment —Slings—Safety Requirements and Test Method. 3 pp.
CEN PREN 567-91. Mountaineering Equipment —Rope Clamp—Safety Requirements, Testing, Marking. 6 pp.
CEN EN 567-92. Mountaineering Equipment —Rope Clamps—Safety Requirements and Test Method. 4 pp.
CEN PREN 568-91. Mountaineering Equipment —Ice Anchors—Safety Requirements, Testing, Marking. 8 pp.
CEN EN 568-92. Mountaineering Equipment —Ice Anchors—Safety Requirements and Test Method. 5 pp.
CEN PREN 569-91. Mountaineering Equipment —Pitons—Safety Requirements, Testing, Marking. 9 pp.
CEN EN 569-92. Mountaineering Equipment —Pitons—Safety Requirements and Test Method. 5 pp.
CEN PREN 892-1-92. Mountaineering Equipment —Ropes—Part 1: Safety Requirements, Testing, Marking. 12 pp.
CEN PREN 959-92. Mountaineering Equipment —Rock Anchors—Safety Requirements and Test Method. 7 pp.

—Nibblers

BSI BS 2769: Sec 2.8-84. 1984 Hand-Held Electric-Motor-Operated Tools Part 2: Particular Requirements Section 2.8: Sheet Metal Shears and Nibblers. 5 pp.
CEN PREN 792-11-92. Handheld Non-Electric Power Tools—Safety Requirements—Part 11: Cutting Power Tools. 6 pp.
CENELEC PREN 50XXX-2-8-92. Safety of Hand-Held Electric Motor Operated Tools Part 2: Particular Requirements for Sheet Metal Shears and Nibblers. 12 pp.

—Nutdrivers

CEN PREN 792-6-92. Handheld Non-Electric Power Tools—Safety Requirements—Part 6: Assembly Power Tools for Threaded Fasteners. 9 pp.

—Oral Irrigators

BSI BS EN 60335-2-52-91. 1991 Safety of Household and Similar Electrical Appliances Part 2: Particular Requirements Section 2.52: Oral Hygiene Appliances Connected to the Mains Supply Through a Safety Isolating Transformer. 18 pp.
CENELEC EN 60 335-2-52-91. Safety of Household and Similar Electrical Appliances Part 2: Particular Requirements for Oral Hygiene Appliances Connected to the Mains Supply Through a Safety Isolating Transformer. 7 pp.
DIN VDE 0700 Pt 223-84. Safety of Household and Similar Electrical Appliances; Oral Hygiene Appliances Connected to the Mains Supply Through a Safety Isolating Transformer (Dec) (Partially Supersedes VDE 0730 Part 2 ZC/05.76). 12 pp.
DIN VDE 0700 Pt 228-86. Safety of Household and Similar Electrical Appliances; Mains Operated Oral Hygiene Appliances (Mar) (Supersedes VDE 0730 Part 2. 11 pp.
DIN VDE 0730 Pt 2 ZCB-78. Requirements for Electrical Equipment for Household and Similar Purposes; Part 2 ZCB: Specific Requirements for Battery-Powered Tooth and Oral Hygiene Equipment (Dec). 23 pp.
IEC 335 Pt 2-52-88. Safety of Household and Similar Electrical Appliances Part 2: Particular Requirements for Oral Hygiene Appliances Connected to the Mains Supply Through a Safety Isolating Transformer First Edition. 23 pp.

—Ovens

BSI BS 3456: Sec 102.36-89. (WITHDRAWN) 1989 Safety of Household Electrical Appliances Part 102: Particular Requirements Section 102.36: Commercial Electric Ranges, Ovens and Hob Elements (Superseded by BS 3456: Section 202.36: 1990). 20 pp.

Consumer Safety *(Cont.)*

—Ovens *(Cont.)*

BSI BS 3456: Sec 102.42-89. (WITHDRAWN) 1989 Safety of Household Electrical Appliances Part 102: Particular Requirements Section 102.42: Commercial Electric Forced Convention Ovens (Superseded by BS 3456: Section 202.42: 1990). 19 pp.
BSI BS 3456: Sec 202.36-01. 1990 Amd 1 Safety of Household and Similar Electrical Appliances Part 202: Particular Requirements Section 202.36: Commercial Electric Ranges, Ovens and Hob Elements (AMD 7624) March 15, 1993 (L). 32 pp.
BSI BS 3456: Sec 202.36-02. 1990 Amd 2 Safety of Household and Similar Electrical Appliances Part 202: Particular Requirements Section 202.36: Commercial Electric Ranges, Ovens and Hob Elements (AMD 7820) July 15, 1993 (L). 46 pp.
BSI BS 3456: Sec 202.42-90. (WITHDRAWN) 1990 Safety of Household Electrical Appliances Part 202: Particular Requirements Section 202.42: Commercial Electric Forced Convection Ovens (Supersedes BS 3456: Section 102.42: 1989). 22 pp.
BSI BS 3456: Sec 202.42-01. 1990 Amd 1 Safety of Household and Similar Electrical Appliances Part 202: Particular Requirements Section 202.42: Commercial Electric Forced Convection Ovens (AMD 7831) July 15, 1993 (L). 38 pp.
BSI BS 5386: Part 6-91. 1991 Gas Burning Appliances Part 6: Specification for Domestic Gas Cooking Appliances with Forced-Convection Ovens. 9 pp.
BSI BS 5386: Part 6-01. 1991 Amd 1 Gas Burning Appliances Part 6: Specification for Domestic Gas Cooking Appliances with Forced-Convection Ovens (AMD 6926) February 28, 1992. 10 pp.
BSI BS EN 60335-2-6-91. 1991 Safety of Household and Similar Electrical Appliances Part 2: Particular Requirements Section 2.6: Cooking Ranges, Cooking Tables, Ovens and Similar Appliances for Household Use. 52 pp.
BSI BS EN 60335-2-6-01. 1991 Amd 1 Safety of Household and Similar Electrical Appliances Part 2: Particular Requirements Section 2.6: Cooking Ranges, Cooking Tables, Ovens and Similar Appliances for Household Use (AMD 7677) July 15, 1993 (L). 70 pp.
BSI BS EN 60335-2-6-02. 1991 Amd 2 Safety of Household and Similar Electrical Appliances Part 2: Particular Requirements Section 2.6: Cooking Ranges, Cooking Tables, Ovens and Similar Appliances for Household Use (AMD 7915) September 15, 1993 (L). 81 pp.
BSI BS EN 60335-2-6-03. 1991 Amd 3 Safety of Household and Similar Electrical Appliances Part 2: Particular Requirements Section 2.6: Cooking Ranges, Cooking Tables, Ovens and Similar Appliances for Household Use (AMD 7916) October 15, 1993 (L). 84 pp.
BSI BS EN 60335-2-9-91. 1991 Safety of Household and Similar Electrical Appliances Part 2: Particular Requirements Section 2.9: Toasters, Grills, Roasters and Similar Appliances. 26 pp.
BSI BS EN 60335-2-9-01. 1991 Amd 1 Safety of Household and Similar Electrical Appliances Part 2: Particular Requirements Section 2.9: Toasters, Grills, Roasters and Similar Appliances (AMD 6897) April 1, 1992. 32 pp.
BSI BS EN 60335-2-9-02. 1991 Amd 2 Safety of Household and Similar Electrical Appliances Part 2: Particular Requirements Section 2.9: Toasters, Grills, Roasters and Similar Appliances (AMD 7584) March 15, 1993 (L). 42 pp.
CENELEC HD 275 S1-86. Safety of Household and Similar Electrical Appliances—Part 2: Particular Requirements for Cooking Ranges, Cooking Tables, Ovens and Similar Appliances for Household Use. 30 pp.
CENELEC HD 275 S1/A1-89. AMD 1 Safety of Household and Similar Electrical Appliances—Part 2: Particular Requirements for Cooking Ranges, Cooking Tables, Ovens and Similar Appliances for Household Use. 9 pp.
CENELEC HD 284 S1-87. Safety of Household and Similar Electrical Appliances—Part 2: Particular Requirements for Commercial Electric Ranges, Ovens and Hob Elements. 10 pp.
CENELEC HD 288 S1-87. Safety of Household and Similar Electric Appliances—Part 2: Particular Requirements for Commercial Electric Forced Convection Ovens. 9 pp.
CENELEC EN 60335-2-6-90. Safety of Household and Similar Electrical Appliances Part 2: Particular Requirements for Cooking Ranges, Cooking Tables, Ovens and Similar Appliances for Household Use (Supercedes HD 275 S1: 1986 and Its Amendment 1: 1988). 38 pp.
CENELEC EN 60335-2-6-91. CORRIGENDUM Safety of Household and Similar Electrical Appliances Part 2: Particular Requirements for Cooking Ranges, Cooking Tables, Ovens and Similar Appliances for Household Use. 2 pp.

Consumer Safety *(Cont.)*

—Ovens *(Cont.)*

CENELEC EN 60335-2-6 /A2-92. AMD 2 Safety of Household and Similar Electrical Appliances Part 2: Particular Requirements for Cooking Ranges, Cooking Tables, Ovens and Similar Appliances for Household Use; (IEC 335-2-6:1986/A2: 1990). 6 pp.
CENELEC EN 60 335-2-9-90. Safety of Household and Similar Electrical Appliances Part 2: Particular Requirements for Toasters, Grills, Roasters and Similar Appliances. 8 pp.
CENELEC EN 60335-2-9 /A2-92. AMD 2 Safety of Household and Similar Electrical Appliances Part 2: Particular Requirments for Toasters, Grills, Roasters and Similar Appliances. 5 pp.
CENELEC EN 60 335-2-36-89. Safety of Household and Similar Electrical Appliances Part 2: Particular Requirements for Commercial Electric Ranges, Ovens and Hob Elements. 10 pp.
CENELEC EN 60335-2-36/A1-92. AMD 1 Safety of Household and Similar Electrical Appliances Part 2: Particular Requirements for Commercial Electric Ranges, Ovens and Hob Elements. 5 pp.
CENELEC EN 60 335-2-42-89. Safety of Household and Similar Electrical Appliances Part 2 Particular Requirements for Commercial Electric Forced Convection Ovens. 9 pp.
CENELEC EN 60335-2-42/A1-92. AMD 1 Safety of Household and Similar Electrical Appliances Part 2 Particular Requirements for Commercial Electric Forced Convection Ovens. 6 pp.
DIN VDE 0700 Pt 6 A8 (D)-89. Safety of Household and Similar Electrical Appliances; Cooking Ranges, Cooking Tables, Ovens and Similar Appliances for Household Use; Amendment 8 to Draft DIN 57700 Part 6/VDE 0700 Part 6/02.84 (Apr). 15 pp.
IEC 335 Pt 2-6-86. Safety of Household and Similar Electrical Appliances Part 2: Part-icular Requirements for Cooking Ranges, Cooking Tables, Ovens and Sim-ilar Appliances for Household Use Third Edition; (Amendment 2-1990, Incorporating Amendment 1) (Amendment 3-1992). 106 pp.
IEC 335 Pt 2-9-93. Safety of Household and Similar Electrical Appliances Part 2: Particular Requirements for Toasters, Grills, Roasters and Similar Appliances Fourth Edition. 46 pp.
IEC 335 Pt 2-25-93. Safety of Household and Similar Electrical Appliances Part 2: Particular Requirements for Microwave Ovens Third Edition. 70 pp.
IEC 335 Pt 2-36-93. Safety of Household and Similar Electrical Appliances Part 2: Particular Requirements for Commercial Electric Cooking Ranges, Ovens, Hobs and Hob Elements Third Edition. 65 pp.
IEC 335 Pt 2-42-87. Safety of Household and Similar Electrical Appliances Part 2: Particular Requirements for Commercial Electric Forced Convection Ovens Second Edition; (Amendment 1-1990). 48 pp.

—Packaging

ISO Guide 41-84. Standards for Packaging—Consumer Requirements First Edition. 6 pp.

—Panel Boards (Electrical)—Plug Fuses

CSA 1077A Bull. Electrical Bulletin 1077A February 14, 1977 to C22.2 NO 29. 2 pp.

—Paper Cutters

SAA AS 1893-77. Code of Practice for the Guarding and Safe Use of Metal and Paper Cutting Guillotines. 32 pp.

—Pens (Writing Implements)—Children's

ISO 11540-93. Caps for Writing and Marking Instruments Intended for Use by Children Up to 14 Years of Age—Safety Requirements First Edition. 7 pp.

—Photographic Flash Equipment

CENELEC HD 327 S2-88. Safety Requirements for Electronic Flash Apparatus for Photographic Purposes. 5 pp.
CENELEC HD 327 S2/A1-91. AMD 1 Safety Requirements for Electronic Flash Apparatus for Photographic Purposes. 5 pp.
IEC 491-84. Safety Requirements for Electronic Flash Apparatus for Photographic Purposes Second Edition. 105 pp.

—Planers (Tools)

BSI BS 2769: Sec 2.13-92. 1992 Hand-Held Electric Motor-Operated Tools Part 2: Particular Requirements Section 2.13: Specification for Planers. 13 pp.
BSI BS 2769: Sec 2.13-84. 1984 Hand-Held Electric-Motor-Operated Tools Part 2: Particular Requirements Section 2.13: Planers. 6 pp.
CENELEC HD 400.3-81. Hand-Held Motor Operated Tools—Part 2: Particular Specifications. 28 pp.
CENELEC HD 400.3M S2-92. Hand-Held Motor Operated Tools Part II: Particular Specifications Section M: Planers. 11 pp.

618

INDEX and DIRECTORY of

INTERNATIONAL AND NON-U.S. NATIONAL STANDARDS
SUBJECT INDEX

Consumer

Consumer Safety (Cont.)

—**Planers (Tools)** (Cont.)

IEC 745 Pt 2-14-84. Safety of Hand-Held Motor-Operated Electric Tools Part 2: Particular Requirements for Planers First Edition. 19 pp.

IEC 1029 Pt 2-3-93. Safety of Transportable Motor-Operated Electric Tools Part 2: Particular Requirements for Planers and Thicknessers First Edition. 36 pp.

JIS B 6601-83. Safety Standards for Construction of Single Surface Planers. 9 pp.

—**Playground Equipment**

CNS A1043-89. Safety Aspects of Playground Equipment Design & Installation (Dec)(12642).

CNS A1044-89. Safety Aspects of Playground Equipment Inspection & Maintenance (Dec)(12643).

SAA AS 1924.2-81. Playground Equipment for Parks, Schools and Domestic Use—Part 2: Design and Construction—Safety Aspects (Incorporating Amdt 1). 19 pp.

—**Playpens**

BSI BS 4863-91. 1991 Safety Requirements for Playpens for Domestic Use. 21 pp.

—**Polishers**

BSI BS 2769: Sec 2.3-84. 1984 Hand-Held Electric-Motor-Operated Tools Part 2: Particular Requirements Section 2.3: Grinders, Polishers and Disc-Type Sanders. 16 pp.

BSI BS 2769: Sec 2.3-01. 1984 Amd 1 Hand-Held Electric Motor-Operated Tools Part 2: Particular Requirements Section 2.3: Specification for Grinders, Polishers and Disc-Type Sanders (AMD 6924) May 1, 1992. 17 pp.

CEN PREN 792-8-92. Handheld Non-Electric Power Tools—Safety Requirements—Part 8: Polishers and Sanders. 8 pp.

CENELEC HD 400.2C/A1-91. AMD 1 Hand-Held Motor Operated Tools Part II: Particular Specifications (Section C). 5 pp.

CENELEC HD 400.2C S1 PRA2-93. AMD 2 Hand-Held Motor Operated Tools Part II: Particular Specifications Section C: Grinders, Polishers and Disc Type Sanders. 2 pp.

IEC 745 Pt 2-3-84. Safety of Hand-Held Motor-Operated Electric Tools Part 2: Particular Requirements for Grinders, Polishers and Disk-Type Sanders First Edition. 21 pp.

—**Power Supplies**

BSI BS 4435: Part 1-69. (WITHDRAWN) 1969 Amd 1 Power Supply Units Part 1: Units for Indoor Use (Superseded by BS 3535: Part 1: 1990). 17 pp.

DIN VDE 0700 Pt 207-82. Safety of Household and Similar Electrical Appliances; Particular Requirements for Power Supply Appliances up to 20 VA (Oct). 21 pp.

—**Presses**

CEN PREN 792-10-92. Handheld Non-Electric Power Tools—Safety Requirements—Part 10: Compression Power Tools, Squeeze Riveters, Presses, Punches. 6 pp.

—**Pressure Cookers**

BSI BS 1746-87. 1987 Domestic Pressure Cookers. 8 pp.

—**Projectors**

BSI BS 3456: Sec 3.14-80. 1980 1972-1981 Edition. Specification for Safety of Household Electrical Appliances Part 3: Particular Specifications Section 3.14: Projectors. 52 pp.

BSI BS EN 60335-2-56-91. 1991 Safety of Household and Similar Electrical Appliances Part 2: Particular Requirements Section 2.56: Projectors and Similar Appliances. 24 pp.

CENELEC EN 60335-2-56-91. Safety of Household and Similar Electrical Appliances Part 2: Particular Requirements for Projectors and Similar Appliances. 6 pp.

IEC 335 Pt 2-56-90. Safety of Household and Similar Electrical Appliances Part 2: Particular Requirements for Projectors and Similar Appliances First Edition. 37 pp.

—**Pumps**

BSI BS 3456: Sec 202.41-91. 1991 Safety of Household and Similar Electrical Appliances Part 202: Particular Requirements Section 202.41: Electric Pumps for Liquids Having a Temperature Not Exceeding 35 Degrees C (EN 60335-2-41: 1990). 23 pp.

BSI BS 3456: Sec 202.41-01. 1991 Amd 1 Safety of Household and Similar Electrical Appliances Part 202: Particular Requirements Sec 202.41:Electric Pumps for Liquids Having a Temperature Not Exceeding 35 Degrees C (AMD 6898) May 1, 1992. 29 pp.

Consumer Safety (Cont.)

—**Pumps** (Cont.)

BSI BS 3456: Sec 2.35-75. (WITHDRAWN) 1975 1972-1981 Edition. Specification for Safety of Household Electrical Appliances Part 2: Section 2.35: Electrical Pumps (Superseded by BS 3456: Section 202.41: 1991). 12 pp.

BSI BS 5258: Part 1-86. 1986 Safety of Domestic Gas Appliances Part 1: Central Heating Boilers and Circulators. 43 pp.

BSI BS 5258: Pt 1: Supp 1-83. 1983 Safety of Domestic Gas Appliances Part 1: Central Heating Boilers and Circulators Supplement 1: Fan-Powered Appliances. 13 pp.

BSI BS EN 60335-2-51-91. 1991 Safety of Household and Similar Electrical Appliances Part 2: Particular Requirements Section 2.51: Stationary Circulation Pumps for Heating and Service Water Installations (Supersedes BS 1394: Part 1: 1987). 18 pp.

BSI DD 189-90. 1990 Safety of Condensing Boilers (2nd and 3rd Family Gases). 9 pp.

CEN PREN 809-92. Pumps and Pump Units for Liquids—Safety Requirements. 15 pp.

CENELEC EN 60 335-2-41-90. Safety of Household and Similar Electrical Appliances Part 2: Particular Requirements for Electric Pumps for Liquids Having Temperature not Exceeding 35 degrees Celsius. 9 pp.

CENELEC EN 60335-2-41/PRAC-93. AMD prAC Safety of Household and Similar Electrical Appliances Part 2: Particular Requirements for Electric Pumps for Liquids Having a Temperature Not Exceeding 35 Degrees C. 5 pp.

CENELEC EN 60335-2-41/A51-91. AMD 51 Safety of Household and Similar Electrical Appliances Part 2: Particular Requirements for Electric Pumps for Liquids Having Temperature Not Exceeding 35 Degrees Celsius. 6 pp.

CENELEC EN 60 335-2-51-91. Safety of Household and Similar Electrical Appliances Part 2: Particular Requirements for Stationary Circulation Pumps for Heating and Service Water Installations. 7 pp.

CENELEC EN 60335-2-51-92. Safety of Household and Similar Electrical Appliances Part 2: Particular Requirements for Stationary Circulation Pumps for Heating and Service Water Installations. 14 pp.

DIN VDE 0700 Pt 237-84. Safety of Household and Similar Electrical Appliances; Stationary Circulation Pumps for Heating and Service Water Plants (VDE Specifications) (Feb) (Replaces DIN 57730 Part 2 ZL/VDE 0730 Part 2 ZL/05.79). 21 pp.

IEC 335 Pt 2-40-92. Safety of Household and Similar Electrical Appliances Part 2: Particular Requirements for Electrical Heat Pumps, Air-Conditioners and Dehumidifiers Second Edition; (Replaces 378). 74 pp.

IEC 335 Pt 2-41-84. Safety of Household and Similar Electrical Appliances Part 2: Particular Requirements for Electric Pumps for Liquids Having a Temperature Not Exceeding 35 Degrees Celsius First Edition; (Amendment 1-1990). 40 pp.

IEC 335 Pt 2-51-88. Safety of Household and Similar Electrical Appliances Part 2: Particular Requirements for Stationary Circulation Pumps for Heating and Service Water Installations First Edition. 27 pp.

—**Punches (Tools)**

CEN PREN 792-10-92. Handheld Non-Electric Power Tools—Safety Requirements—Part 10: Compression Power Tools, Squeeze Riveters, Presses, Punches. 6 pp.

—**Radial Saws**

IEC 1029 Pt 2-2-93. Safety of Transportable Motor-Operated Electric Tools Part 2: Particular Requirements for Radial Arm Saws First Edition. 28 pp.

—**Radio Equipment**

BSI BS 3192-90. 1990 Safety Requirements for Radio (Including Television) Transmitting Equipment. 30 pp.

CENELEC HD 220 S2-80. Safety Requirements for Radio Transmitting Equipment (Superseded by EN 60215). 3 pp.

CENELEC EN 60 215-89. Safety Requirements for Radio Transmitting Equipment (Supersedes HD 220 S2). 5 pp.

DIN VDE 0868 (D)-90. Guide for Safe Handling and Operation of Mobile Radio Equipment; Identical to IEC 12F(CO) 154 (Sept). 39 pp.

—**Radio Transmitters**

BSI BS 3192-90. 1990 Safety Requirements for Radio (Including Television) Transmitting Equipment. 30 pp.

CENELEC HD 220 S2-80. Safety Requirements for Radio Transmitting Equipment (Superseded by EN 60215). 3 pp.

CENELEC EN 60 215-89. Safety Requirements for Radio Transmitting Equipment (Supersedes HD 220 S2). 5 pp.

Consumer Safety (Cont.)

—**Radio Transmitters** (Cont.)

CENELEC EN 60215/A1-92. AMD 1 Safety Requirements for Radio Transmitting Equipment; (Supersedes HD 220 S2). 4 pp.

IEC 215-87. Safety Requirements for Radio Transmitting Equipment Third Edition; (Amendment 2-1993, Incorporating Amendment 1). 61 pp.

SNZ IEC 215-87. Safety Requirements for Radio Transmitting Equipment. 49 pp.

—**Range Hoods**

BSI BS 3456: Sec 102.31-87. (WITHDRAWN) 1987 Safety of Household Electrical Appliances Part 102: Particular Requirements Section 102.31: Range Hoods (Superseded by BS EN 60335-2-31: 1991). 19 pp.

BSI BS EN 60335-2-31-91. 1991 Safety of Household and Similar Electrical Appliances Part 2: Particular Requirements Section 2.31: Range Hoods (Supersedes BS 3456: Section 102.31: 1987) (L). 25 pp.

BSI BS EN 60335-2-31-01. 1991 Amd 1 Safety of Household and Similar Electrical Appliances Part 2: Particular Requirements Section 2.31: Range Hoods (AMD 7190) October 15, 1992 (Supersedes BS 3456: Section 102.31: 1987). 34 pp.

CENELEC HD 276 S1-85. Safety of Household and Similar Electrical Appliances Particular Requirements for Range Hoods. 9 pp.

CENELEC HD 276 S1/A1-86. AMD 1 Safety of Household and Similar Electrical Appliances Particular Requirements for Range Hoods. 4 pp.

CENELEC EN 60 335-2-31-90. Safety of Household and Similar Electrical Appliances Part 2: Particular Requirements for Range Loads. 10 pp.

CENELEC EN 60335-2-31/A1-91. AMD 1 Safety of Household and Similar Electrical Appliances Part 2: Particular Requirements for Range Loads. 7 pp.

IEC 335 Pt 2-31-88. Safety of Household and Similar Electrical Appliances Part 2: Particular Requirements for Range Hoods Second Edition; (Amendment 1-1990). 34 pp.

—**Ranges**

BSI BS 3456: Sec 102.31-87. (WITHDRAWN) 1987 Safety of Household Electrical Appliances Part 102: Particular Requirements Section 102.31: Range Hoods (Superseded by BS EN 60335-2-31: 1991). 19 pp.

BSI BS 3456: Sec 102.36-89. (WITHDRAWN) 1989 Safety of Household Electrical Appliances Part 102: Particular Requirements Section 102.36: Commercial Electric Ranges, Ovens and Hob Elements (Superseded by BS 3456: Section 202.36: 1990). 20 pp.

BSI BS 3456: Sec 202.36-01. 1990 Amd 1 Safety of Household and Similar Electrical Appliances Part 202: Particular Requirements Section 202.36: Commercial Electric Ranges, Ovens and Hob Elements (AMD 7624) March 15, 1993 (L). 32 pp.

BSI BS 3456: Sec 202.36-02. 1990 Amd 2 Safety of Household and Similar Electrical Appliances Part 202: Particular Requirements Section 202.36: Commercial Electric Ranges, Ovens and Hob Elements (AMD 7820) July 15, 1993 (L). 46 pp.

BSI BS EN 60335-2-6-91. 1991 Safety of Household and Similar Electrical Appliances Part 2: Particular Requirements Section 2.6: Cooking Ranges, Cooking Tables, Ovens and Similar Appliances for Household Use. 52 pp.

BSI BS EN 60335-2-6-01. 1991 Amd 1 Safety of Household and Similar Electrical Appliances Part 2: Particular Requirements Section 2.6: Cooking Ranges, Cooking Tables, Ovens and Similar Appliances for Household Use (AMD 7677) July 15, 1993 (L). 70 pp.

BSI BS EN 60335-2-6-02. 1991 Amd 2 Safety of Household and Similar Electrical Appliances Part 2: Particular Requirements Section 2.6: Cooking Ranges, Cooking Tables, Ovens and Similar Appliances for Household Use (AMD 7915) September 15, 1993 (L). 81 pp.

BSI BS EN 60335-2-6-03. 1991 Amd 3 Safety of Household and Similar Electrical Appliances Part 2: Particular Requirements Section 2.6: Cooking Ranges, Cooking Tables, Ovens and Similar Appliances for Household Use (AMD 7916) October 15, 1993 (L). 84 pp.

CENELEC HD 275 S1-86. Safety of Household and Similar Electrical Appliances—Part 2: Particular Requirements for Cooking Ranges, Cooking Tables, Ovens and Similar Appliances for Household Use. 30 pp.

CENELEC HD 275 S1/A1-89. AMD 1 Safety of Household and Similar Electrical Appliances—Part 2: Particular Requirements for Cooking Ranges, Cooking Tables, Ovens and Similar Appliances for Household Use. 9 pp.

CENELEC HD 284 S1-87. Safety of Household and Similar Electrical Appliances—Part 2: Particular Requirements for Commercial Electric Ranges, Ovens and Hob Elements. 10 pp.

Consumer
INTERNATIONAL AND NON-U.S. NATIONAL STANDARDS
SUBJECT INDEX

Consumer Safety *(Cont.)*

—Ranges *(Cont.)*

CENELEC EN 60335-2-6-90. Safety of Household and Similar Electrical Appliances Part 2: Particular Requirements for Cooking Ranges, Cooking Tables, Ovens and Similar Appliances for Household Use (Supercedes HD 275 S1: 1986 and Its Amendment 1: 1988). 38 pp.

CENELEC EN 60335-2-6-91. CORRIGENDUM Safety of Household and Similar Electrical Appliances Part 2: Particular Requirements for Cooking Ranges, Cooking Tables, Ovens and Similar Appliances for Household Use. 2 pp.

CENELEC EN 60335-2-6 /A2-92. AMD 2 Safety of Household and Similar Electrical Appliances Part 2: Particular Requirements for Cooking Ranges, Cooking Tables, Ovens and Similar Appliances for Household Use; (IEC 335-2-6:1986/A2: 1990). 6 pp.

CENELEC EN 60 335-2-36-89. Safety of Household and Similar Electrical Appliances Part 2: Particular Requirements for Commercial Electric Ranges, Ovens and Hob Elements. 10 pp.

CENELEC EN 60335-2-36/A1-92. AMD 1 Safety of Household and Similar Electrical Appliances Part 2: Particular Requirements for Commercial Electric Ranges, Ovens and Hob Elements. 5 pp.

DIN VDE 0700 Pt 6 A8 (D)-89. Safety of Household and Similar Electrical Appliances; Cooking Ranges, Cooking Tables, Ovens and Similar Appliances for Household Use; Amendment 8 to Draft DIN 57700 Part 6/VDE 0700 Part 6/02.84 (Apr). 15 pp.

IEC 335 Pt 2-6-86. Safety of Household and Similar Electrical Appliances Part 2: Part-icular Requirements for Cooking Ranges, Cooking Tables, Ovens and Sim-ilar Appliances for Household Use Third Edition; (Amendment 2-1990, Incorporating Amendment 1) (Amendment 3-1992). 106 pp.

IEC 335 Pt 2-36-93. Safety of Household and Similar Electrical Appliances Part 2: Particular Requirements for Commercial Electric Cooking Ranges, Ovens, Hobs and Hob Elements Third Edition. 65 pp.

—Recreational Vehicles

BSI BS 4989-84. (WITHDRAWN) 1984 Amd 1 Holiday Caravans (Superseded by BS 6794: 1991). 10 pp.

BSI BS 6764-91. 1991 Habitation and Stability Requirements for Leisure Accommodation Vehicles: Leisure Homes (ISO 8817: 1990). 18 pp.

CSA CAN/CSA-Z241 Series-92. Park Model Trailers; (Gen Instr 1). 126 pp.

ISO 7422-91. Leisure Accommodation Vehicles—Caravans—Habitation Requirements First Edition. 17 pp.

ISO 8817-90. Leisure Accommodation Vehicles—Leisure Homes—Habitation and Stability Requirements First Edition. 16 pp.

—Refrigerators

BSI BS 3456: Sec 102.24-88. (WITHDRAWN) 1988 Safety of Household Electrical Appliances Part 102: Particular Requirements Section 102.24: Refrigerators and Food Freezers (Superseded by BS 3456: Section 202.24: 1990). 31 pp.

BSI BS 3456: Sec 202.24-90. 1990 Safety of Household Electrical Appliances Part 202: Particular Requirements Section 202.24: Refrigerators and Food Freezers (L). 42 pp.

BSI BS 5258: Part 6-88. 1988 Safety of Domestic Gas Appliances Part 6: Refrigerators and Food Freezers. 16 pp.

BSI BS 5258: Part 6-75. 1975 Safety of Domestic Gas Appliances Part 6: Refrigerators and Food Freezers. 15 pp.

CENELEC HD 269 S2-86. Safety of Household and Similar Electrical Appliances-Part: Particular Requirements for Refrigerators and Food Freezers. 7 pp.

CENELEC HD 269 S2-86. Safety of Household and Similar Electrical Appliances Part: 2 Particular Requirements for Refrigerators and Food Freezers (IEC 335-2-24 (1976) Ed 1 Modified). 18 pp.

CENELEC HD 269 S2-85. Amd 1 Safety of Household Similar Electrical Appliances Particular Requirements for Refrigerators and Food Freezers. 4 pp.

CENELEC HD 270 S1-83. Safety of Household Similar Electrical Appliances Particular Requirements for Microwave Cooking Appliances. 17 pp.

CENELEC EN 60 335-2-24-89. Safety of Household and Similar Electrical Appliances-Part 2: Particular Requirements for Refrigerators and Food Freezers. 13 pp.

DIN VDE 0700 Pt 24-86. Safety of Household and Similar Electrical Appliances; Refrigerators and Food Freezers (Mar). 33 pp.

DIN VDE 0700 Pt 240-83. Safety of Household and Similar Electrical Appliances; Particular Requirements for Refrigerators and Freezers for Special Purposes and Ice Makers (Feb) (Supersedes VDE 0730 Part 2 ZK/10.78). 61 pp.

Consumer Safety *(Cont.)*

—Refrigerators *(Cont.)*

IEC 335 Pt 2-24-92. Safety of Household and Similar Electrical Appliances Part 2: Particular Requirements for Refrigerators, Food-Freezers and Ice-Makers Third Edition: (CENELEC HD 269 S2:1986). 78 pp.

SNZ NZS 6324-90. Approval and Test Specification—Particular Requirements for Refrigerators and Food Freezers Amend: 1, 1991; 2, 1992 (NZS 6324:1990/ AS 3303:1990). 31 pp.

—Riveters

CEN PREN 792-10-92. Handheld Non-Electric Power Tools—Safety Requirements—Part 10: Compression Power Tools, Squeeze Riveters, Presses, Punches. 6 pp.

—Roasters

BSI BS 3456: Sec 3.17-80. (WITHDRAWN) 1980 Amd 2 1972-1981 Edition. Specification for Safety of Household Electrical Appliances: Part 3: Complete Particular Specif-ications: Section 3.17: Toasters. Grills, Waffle Irons, Roasters & Other Dry Cooking Appliances (Superseded by BS 3456: Section 102.9: 1991). 48 pp.

BSI BS EN 60335-2-9-91. 1991 Safety of Household and Similar Electrical Appliances Part 2: Particular Requirements Section 2.9: Toasters, Grills, Roasters and Similar Appliances. 26 pp.

BSI BS EN 60335-2-9-01. 1991 Amd 1 Safety of Household and Similar Electrical Appliances Part 2: Particular Requirements Section 2.9: Toasters, Grills, Roasters and Similar Appliances (AMD 6897) April 1, 1992. 32 pp.

BSI BS EN 60335-2-9-02. 1991 Amd 2 Safety of Household and Similar Electrical Appliances Part 2: Particular Requirements Section 2.9: Toasters, Grills, Roasters and Similar Appliances (AMD 7584) March 15, 1993 (L). 42 pp.

CENELEC HD 265 S2-84. Safety of Household and Similar Electrical Appliances—Particular Requirements for Appliances for Toasters, Grills, Roasters and Similar Appliances. 8 pp.

CENELEC EN 60 335-2-9-90. Safety of Household and Similar Electrical Appliances Part 2: Particular Requirements for Toasters, Grills, Roasters and Similar Appliances. 8 pp.

CENELEC EN 60335-2-9 /A2-92. AMD 2 Safety of Household and Similar Electrical Appliances Part 2: Particular Requirments for Toasters, Grills, Roasters and Similar Appliances. 5 pp.

IEC 335 Pt 2-9-93. Safety of Household and Similar Electrical Appliances Part 2: Particular Requirements for Toasters, Grills, Roasters and Similar Appliances Fourth Edition. 46 pp.

—Room Heaters

BSI BS EN 60335-2-30-93. 1993 Safety of Household and Similar Electrical Appliances Part 2: Particular Requirements Section 2.30: Room Heaters (L). 45 pp.

CENELEC EN 60335-2-30-92. Safety of Household and Similar Electrical Appliances Part 2: Particular Requirements for Room Heaters (Supersedes HD 278 S1: 1987 + Its Amendments); (IEC 335-2-30:1990 + A1). 13 pp.

IEC 335 Pt 2-30-90. Safety of Household and Similar Electrical Appliances Part 2: Particular Requirements for Room Heaters Second Edition; (Amendment 1-1990) (CENELEC EN 60335-2-30: 1992). 64 pp.

SNZ NZS 1999-65. Safety Requirements for Electric Room Heaters Amend: 1, 1968; 2, 1971; 3, 1973. 15 pp.

SNZ NZS 6331-72. Specification for Thermal Storage Electric Room Heaters. 12 pp.

—Rotisseries

BSI BS EN 60335-2-9-91. 1991 Safety of Household and Similar Electrical Appliances Part 2: Particular Requirements Section 2.9: Toasters, Grills, Roasters and Similar Appliances. 26 pp.

BSI BS EN 60335-2-9-01. 1991 Amd 1 Safety of Household and Similar Electrical Appliances Part 2: Particular Requirements Section 2.9: Toasters, Grills, Roasters and Similar Appliances (AMD 6897) April 1, 1992. 32 pp.

BSI BS EN 60335-2-9-02. 1991 Amd 2 Safety of Household and Similar Electrical Appliances Part 2: Particular Requirements Section 2.9: Toasters, Grills, Roasters and Similar Appliances (AMD 7584) March 15, 1993 (L). 42 pp.

CENELEC EN 60 335-2-9-90. Safety of Household and Similar Electrical Appliances Part 2: Particular Requirements for Toasters, Grills, Roasters and Similar Appliances. 8 pp.

CENELEC EN 60335-2-9 /A2-92. AMD 2 Safety of Household and Similar Electrical Appliances Part 2: Particular Requirments for Toasters, Grills, Roasters and Similar Appliances. 5 pp.

Consumer Safety *(Cont.)*

—Rotisseries *(Cont.)*

IEC 335 Pt 2-9-93. Safety of Household and Similar Electrical Appliances Part 2: Particular Requirements for Toasters, Grills, Roasters and Similar Appliances Fourth Edition. 46 pp.

—Routers (Tools)

BSI BS 2769: Sec 2.15-92. 1992 Hand-Held Electric Motor-Operated Tools Part 2: Particular Requirements Section 2.15: Specification for Routers. 11 pp.

CENELEC HD 400.3O S1-92. Hand-Held Motor Operated Tools Part II: Particular Specifications Section O: Routers. 9 pp.

IEC 745 Pt 2-17-89. Safety of Hand-Held Motor-Operated Electric Tools Part 2: Particular Requirements for Routers and Trimmers First Edition. 21 pp.

—Sanders (Tools)

BSI BS 2769: Sec 2.4-84. 1984 Hand-Held Electric-Motor-Operated Tools Part 2: Particular Requirements Section 2.4: Sanders. 5 pp.

BSI BS 2769: Sec 2.4-01. 1984 Amd 1 Hand-Held Electric Motor-Operated Tools Part 2: Particular Requirements Section 2.4: Specification for Sanders (AMD 6925) May 1, 1992. 6 pp.

CEN PREN 792-8-92. Handheld Non-Electric Power Tools—Safety Requirements—Part 8: Polishers and Sanders. 8 pp.

CENELEC HD 400.2C/A1-91. AMD 1 Hand-Held Motor Operated Tools Part II: Particular Specifications (Section C). 5 pp.

CENELEC HD 400.2C S1 PRA2-93. AMD 2 Hand-Held Motor Operated Tools Part II: Particular Specifications Section C: Grinders, Polishers and Disc Type Sanders. 2 pp.

CENELEC HD 400.2D/A1-91. AMD 1 Hand-Held Motor Operated Tools Part II: Particular Specifications (Section D). 5 pp.

CENELEC HD 400.2D S1 PRA2-93. AMD 2 Hand-Held Motor Operated Tools Part II: Particular Specifications Section D: Sanders. 3 pp.

IEC 745 Pt 2-3-84. Safety of Hand-Held Motor-Operated Electric Tools Part 2: Particular Requirements for Grinders, Polishers and Disk-Type Sanders First Edition. 21 pp.

IEC 745 Pt 2-4-83. Safety of Hand-Held Motor-Operated Electric Tools Part 2: Particular Requirements for Sanders First Edition; (Amendment 1-1992). 24 pp.

—Saunas

BSI BS EN 60335-2-53-91. 1991 Safety of Household and Similar Electrical Appliances Part 2: Particular Requirements Section 2.53: Electric Sauna Heating Appliances. 30 pp.

CENELEC EN 60 335-2-53-91. Safety of Household and Similar Electrical Appliances Part 2: Particular Requirements for Electric Sauna Heating Appliances. 7 pp.

IEC 335 Pt 2-53-88. Safety of Household and Similar Electical Appliances Part 2: Particular Requirements for Electric Sauna Heating Appliances First Edition. 44 pp.

—Saws

BSI BS 2769: Sec 2.5-91. 1991 Hand-Held Electric Motor-Operated Tools Part 2: Particular Requirements Section 2.5: Specification for Circular Saws and Circular Knives. 16 pp.

BSI BS 2769: Sec 2.5-01. 1991 Amd 1 Hand-Held Electric Motor-Operated Tools Part 2: Particular Requirements Section 2.5: Specification for Circular Saws and Circular Knives (AMD 6920) April 1, 1992. 17 pp.

IEC 745 Pt 2-5-93. Safety of Hand-Held Motor-Operated Electric Tools Part 2: Particular Requirements for Circular Saws and Circular Knives Second Edition. 44 pp.

IEC 745 Pt 2-11-84. Safety of Hand-Held Motor-Operated Electric Tools Part 2: Particular Requirements for Reciprocating Saws (Jig and Sabre Saws) First Edition. 19 pp.

IEC 1029 Pt 2-1-93. Safety of Transportable Motor-Operated Electric Tools Part 2: Particular Requirements for Circular Saws First Edition. 32 pp.

—Screwdrivers

CEN PREN 792-6-92. Handheld Non-Electric Power Tools—Safety Requirements—Part 6: Assembly Power Tools for Threaded Fasteners. 9 pp.

IEC 745 Pt 2-2-82. Safety of Hand-Held Motor-Operated Electric Tools Part 2: Particular Requirements for Screwdrivers and Impact Wrenches First Edition; (Amendment 1-1991). 20 pp.

—Sewing Machines

BSI BS 3456: Sec 102.28-88. (WITHDRAWN) 1988 Safety of Household Electrical Appliances Part 102: Particular Requirements Section 102.28: Sewing Machines (Superseded by BS EN 60335-2-28: 1991). 16 pp.

INTERNATIONAL AND NON-U.S. NATIONAL STANDARDS
SUBJECT INDEX

Consumer

Consumer Safety *(Cont.)*

—Sewing Machines *(Cont.)*

BSI BS EN 60335-2-28-91. 1991 Safety of Household and Similar Electrical Appliances Part 2: Particular Requirements Section 2.28: Sewing Machines (Supersedes BS 3456: Section 102.28: 1987). 17 pp.

CENELEC HD 273 S2-87. Safety of Household and Similar Electrical Appliances—Part 2: Particular Requirements for Sewing Machines. 8 pp.

CENELEC HD 273 S2/A1-89. AMD 1 Safety of Household and Similar Electrical Appliances—Part 2: Particular Requirements for Sewing Machines. 5 pp.

CENELEC EN 60335-2-28-90. Safety of Household and Similar Electrical Appliances Part 2: Particular Requirements for Sewing Machines. 7 pp.

IEC 335 Pt 2-28-87. Safety of Household and Similar Electrical Appliances Part 2: Particular Requirements for Sewing Machines Second Edition. 27 pp.

—Shavers

BSI BS 3456: Sec 102.8-88. (WITHDRAWN) 1988 Safety of Household Electrical Appliances Part 102: Particular Requirements Section 102.8: Electric Shavers, Hair Clippers and Similar Appliances (Superseded by BS EN 60335-2-8: 1991). 19 pp.

BSI BS 3456: Sec 202.19-90. 1990 Safety of Household Electrical Appliances Part 202: Particular Requirements Section 202.19: Battery-Powered Shavers, Hair Clippers and Similar Appliances and Their Charging and Battery Assemblies. 28 pp.

CENELEC HD 254 S2-84. Safety of Household and Similar Electric Appliances Particular Requirements for Electric Shavers, Hair Clippers and Similar Appliances (Superseded by EN 60335-2-8). 10 pp.

CENELEC EN 60335-2-8-90. Safety of Household and Similar Electrical Appliances Part 2: Particular Requirements for Electric Shavers, Hair Clippers and Similar Appliances (Supersedes HD 254 S2). 7 pp.

CENELEC EN 60 335-2-19-89. Safety of Household and Similar Electrical Appliances Part 2: Particular Requirements for Battery—Powered Shavers, Hair Clippers and Similar Appliances and Their Charging and Battery Assemblies. 7 pp.

IEC 335 Pt 2-8-92. Safety of Household and Similar Electrical Appliances Part 2: Particular Requirements for Shavers, Hair Clippers and Similar Appliances Fourth Edition. 29 pp.

IEC 335 Pt 2-19-84. Safety of Household and Similar Electrical Appliances Part 2: Particular Requirements for Battery-Powered Shavers, Hair Clippers and Similar Appliances and Their Charging and Battery Assemblies Second Edition. 45 pp.

—Shears—Grass

BSI BS 2769: Sec 2.14-92. 1992 Hand-Held Electric Motor-Operated Tools Part 2: Particular Requirements Section 2.14: Specification for Hedge Trimmers and Scissor-Type Grass Shears (F). 13 pp.

BSI BS 2769: Sec 2.14-01. 1992 Hand-Held Electric Motor-Operated Tools Part 2: Particular Requirements Section 2.14: Specification for Hedge Trimmers and Scissor-Type Grass Shears (AMD 7505) November 15, 1992. 14 pp.

BSI BS 2769: Sec 2.14-84. 1984 Hand-Held Electric-Motor-Operated Tools Part 2: Particular Requirements Section 2.14: Hedge Trimmers and Scissor-Type Grass Shears. 6 pp.

CENELEC HD 400.3N S2-92. Hand-Held Motor Operated Tools Part II: Particular Specifications Section N: Hedge Trimmers and Scissor-Type Grass Shears. 13 pp.

IEC 745 Pt 2-15-84. Safety of Hand-Held Motor-Operated Electric Tools Part 2: Particular Requirements for Hedge Trimmers and Grass Shears First Edition. 15 pp.

—Shears—Sheet Metal

BSI BS 2769: Sec 2.8-84. 1984 Hand-Held Electric-Motor-Operated Tools Part 2: Particular Requirements Section 2.8: Sheet Metal Shears and Nibblers. 5 pp.

CEN PREN 792-11-92. Handheld Non-Electric Power Tools—Safety Requirements—Part 11: Cutting Power Tools. 6 pp.

CENELEC PREN 50XXX-2-8-92. Safety of Hand-Held Electric Motor Operated Tools Part 2: Particular Requirements for Sheet Metal Shears and Nibblers. 12 pp.

IEC 745 Pt 2-8-82. Safety of Hand-Held Motor-Operated Electric Tools Part 2: Particular Requirements for Sheet Metal Shears First Edition; (Amendment 1-1992). 20 pp.

—Sheet Glass

BSI BS 7376-90. 1990 Inclusion of Glass in the Construction of Tables or Trolleys. 10 pp.

Consumer Safety *(Cont.)*

—Sinks

BSI BS EN 60335-2-62-93. 1993 Safety of Household and Similar Electrical Appliances Part 2: Particular Requirements Section 2.62: Commercial Electric Rinsing Sinks (L). 22 pp.

CENELEC EN 60335-2-62-92. Safety of Household and Similar Electrical Appliances Part 2: Particular Requirements for Commercial Electric Rinsing Sinks. 8 pp.

CENELEC EN 60335-2-62-92. Safety of Household and Similar Electrical Appliances Part 2: Particular Requirements for Commercial Electric Rinsing Sinks (IEC 335-2-62: 1990, Modified). 16 pp.

IEC 335 Pt 2-62-90. Safety of Household and Similar Electrical Appliances Part 2: Particular Requirements for Commercial Electric Rinsing Sinks First Edition; (CENELEC EN 60335-2-62: 1992). 38 pp.

—Ski Bindings

DIN ENGL 7881 Pt 1-82. Winter Sports Equipment; Release Bindings for Alpine Downhill Skiing; Terms and Definitions, Safety Requirements, Testing (Feb). 11 pp.

—Skin Care Appliances

BSI BS 3456: Sec 3.13-79. (WITHDRAWN) 1979 Amd 2 1972-1981 Edition. Specification for Safety of Household Electrical Appliances Part 3: Complete Particular Specifications Section 3.13: Appliances for Skin or Hair Treatment (Superseded by BS EN 60335-2-23: 1991). 60 pp.

BSI BS EN 60335-2-23-91. 1991 Safety of Household and Similar Electrical Appliances Part 2: Particular Requirements Section 2.23: Appliances for Skin or Hair Care (Supersedes BS 3456: Section 3.13: 1979). 32 pp.

BSI BS EN 60335-2-23-01. 1991 Amd 1 Safety of Household and Similar Electrical Appliances Part 2: Particular Requirements Section 2.23: Appliances for Skin or Hair Care (AMD 7793) July 15, 1993 (L). 44 pp.

BSI BS EN 60335-2-23-02. 1991 Amd 2 Safety of Household and Similar Electrical Appliances Part 2: Particular Requirements Section 2.23: Appliances for Skin or Hair Care (AMD 7901) August 15, 1993 (L). 50 pp.

CENELEC HD 266-75. Particular Specification for Appliances for Skin or Hair Treatment. 6 pp.

CENELEC HD 266.2-77. Particular Specification for Appliances for Skin and Hair Treatment. 9 pp.

CENELEC HD 266.3-83. Particular Specification for Appliances for Skin or Hair Treatment. 7 pp.

CENELEC HD 266.4-83. Particular Specification for Appliances for Skin or Hair Treatment. 6 pp.

CENELEC EN 60 335-2-23-90. Safety of Household and Similar Electrical Appliances Part 2: Particular Requirements for Appliances for Skin or Hair Care. 11 pp.

CENELEC EN 60335-2-23/A1-92. AMD 1 Safety of Household and Similar Electrical Appliances Part 2: Particular Requirements for Appliances for Skin or Hair Care. 5 pp.

CENELEC EN 60335-2-23/A51-92. AMD 51 Safety of Household and Similar Electrical Appliances Part 2: Particular Requirements for Appliances for Skin or Hair Care. 5 pp.

IEC 335 Pt 2-23-86. Safety of Household and Similar Electrical Appliances Part 2: Particular Requirements for Appliances for Skin or Hair Care Third Edition; (Amendment 1-1990). 50 pp.

—Soldering Equipment

DIN VDE 0700 Pt 30-83. Test Certificate, Safety Test of Household & Similar Electrical Appliances. Portable Electric Heating Tools (Jan). 26 pp.

DIN VDE 0700 Pt 30 A1 (D)-84. Safety of Household and Similar Electrical Appliances; Particular Specifications for Portable Electric Heating Tools (VDE Specification): Amendment 1 (Nov). 19 pp.

—Soldering Irons

BSI BS 3456: Sec 2.14-71. (WITHDRAWN) 1971 1972-1981 Edition. Specification for Safety of Household Electrical Appliances Part 2: Section 2.14: Electric Soldering Irons (Superseded by BS EN 60335-2-45: 1991). 15 pp.

—Spray Guns

IEC 745 Pt 2-7-89. Safety of Hand-Held Motor-Operated Electric Tools Part 2: Particular Requirements for Spray Guns for Non-Flammable Liquids First Edition. 19 pp.

—Starters

BSI BS EN 60926-91. 1991 General and Safety Requirements for Starting Devices (Other Than Glow Starters). 36 pp.

CENELEC EN 60 926-90. Starting Devices (Other Than Glow Starters) General and Safety Requirements. 6 pp.

Consumer Safety *(Cont.)*

—Steamers

BSI BS 3456: Sec 202.46-90. 1990 Safety of Household Electrical Appliances Part 202: Particular Requirements Section 202.46: Commercial Electric Steam Cookers. 23 pp.

BSI BS 3456: Sec 202.46-01. 1990 Amd 1 Safety of Household and Similar Electrical Appliances Part 202: Particular Requirements Section 202.46: Commercial Electric Steam Cookers (AMD 7823) August 15, 1993 (L). 36 pp.

CENELEC EN 60 335-2-46-89. Safety of Household and Similar Electrical Appliances Part 2: Particular Requirements for Commercial Electric Steam Cookers. 10 pp.

CENELEC EN 60335-2-46/A1-92. AMD 1 Safety of Household and Similar Electrical Appliances Part 2: Particular Requirements for Commercial Electric Steam Cookers. 6 pp.

IEC 335 Pt 2-46-86. Safety of Household and Similar Electrical Appliances Part 2: Particular Requirements for Commercial Electric Steam Cookers First Edition; (Amendment 1-1990). 46 pp.

—Storage Heaters

CENELEC HD 283 S1-92. Safety of Household and Similar Electrical Appliances Particular Requirement for the Maximum Temperature Allowed for the Surfaces of Air-Outlet Grilles of Thermal Storage Room Heating Appliances. 9 pp.

IEC 335 Pt 2-61-92. Safety of Household and Similar Electrical Appliances Part 2: Particular Requirements for Thermal Storage Room Heaters First Edition. 44 pp.

—Stoves

CENELEC HD 275 S1-86. Safety of Household and Similar Electrical Appliances—Part 2: Particular Requirements for Cooking Ranges, Cooking Tables, Ovens and Similar Appliances for Household Use. 30 pp.

CENELEC HD 275 S1/A1-89. AMD 1 Safety of Household and Similar Electrical Appliances—Part 2: Particular Requirements for Cooking Ranges, Cooking Tables, Ovens and Similar Appliances for Household Use. 9 pp.

IEC 335 Pt 2-6-86. Safety of Household and Similar Electrical Appliances Part 2: Part-icular Requirements for Cooking Ranges, Cooking Tables, Ovens and Sim-ilar Appliances for Household Use Third Edition; (Amendment 2-1990, Incorporating Amendment 1) (Amendment 3-1992). 106 pp.

—Strollers

BSI BS 4792-84. 1984 Amd 2 Safety Requirements for Pushchairs (AMD 6303) February 28, 1990. 18 pp.

SAA AS 2088-89. Prams and Strollers—Safety Requirements Amdt 1 May 1990 Amdt 2 January 1991 Amdt 3 June 1991 (Superseded by AS/NZS 2088:1993).

SNZ NZS/AS 2088-93. Prams and Strollers—Safety Requirements (This is a Joint Standard with SAA AS 2088). 43 pp.

—Sunglasses

SAA AS 1067.1-90. Sunglasses and Fashion Spectacles—Part 1: Safety Requirements Amdt 1 December 1990 Amdt 2 July 1993. 15 pp.

—Swimming Pools

SAA AS 2020-77. Safety Covers for Private Swimming Pools and Wading Pools (for the Protection of Children 5 Years of Age and Under) Corrig.. 12 pp.

—Tanks (Containers)—Anhydrous Ammonia

CNS Z1022-86. Safety Code for Liquefied Ammonia Tank on Automobiles (Feb)(7249).

—Taps (Threading Tools)

CEN PREN 792-3-92. Handheld Non-Electric Power Tools—Safety Requirements—Part 3: Rotary Drilling Tools, Drills, Tappers. 8 pp.

IEC 745 Pt 2-9-84. Safety of Hand-Held Motor-Operated Electric Tools Part 2: Particular Requirements for Tappers First Edition. 19 pp.

—Television Equipment

BSI BS 3192-90. 1990 Safety Requirements for Radio (Including Television) Transmitting Equipment. 30 pp.

—Television Transmitters

BSI BS 3192-90. 1990 Safety Requirements for Radio (Including Television) Transmitting Equipment. 30 pp.

—Temperature Controllers

IEC 730 Pt 2-9-92. Automatic Electrical Controls for Household and Similar Use Part 2: Particular Requirements for Temperature Sensing Controls First Edition. 52 pp.

INDUSTRY STANDARDS

Consumer
INTERNATIONAL AND NON-U.S. NATIONAL STANDARDS
SUBJECT INDEX

Consumer Safety *(Cont.)*

—Tents

BSI BS 5576-85. 1985 Safety Features of Camping Tents, Awnings, Trailer Tents and Caravan Awnings. 4 pp.

—Toasters

BSI BS 3456: Sec 3.17-80. (WITHDRAWN) 1980 Amd 2 1972-1981 Edition. Specification for Safety of Household Electrical Appliances: Part 3: Complete Particular Specif-ications: Section 3.17: Toasters. Grills, Waffle Irons, Roasters & Other Dry Cooking Appliances (Superseded by BS 3456: Section 102.9: 1991). 48 pp.

BSI BS EN 60335-2-9-91. 1991 Safety of Household and Similar Electrical Appliances Part 2: Particular Requirements Section 2.9: Toasters, Grills, Roasters and Similar Appliances. 26 pp.

BSI BS EN 60335-2-9-01. 1991 Amd 1 Safety of Household and Similar Electrical Appliances Part 2: Particular Requirements Section 2.9: Toasters, Grills, Roasters and Similar Appliances (AMD 6897) April 1, 1992. 32 pp.

BSI BS EN 60335-2-9-02. 1991 Amd 2 Safety of Household and Similar Electrical Appliances Part 2: Particular Requirements Section 2.9: Toasters, Grills, Roasters and Similar Appliances (AMD 7584) March 15, 1993 (L). 42 pp.

BSI BS EN 60335-2-48-92. 1992 Safety of Household and Similar Electrical Appliances Part 2: Particular Requirements Section 2.48: Commercial Electric Grillers and Toasters. 22 pp.

BSI BS EN 60335-2-48-01. 1992 Amd 1 'Safety of Household and Similar Electrical Appliances Part 2: Particular Requirements Section 2.48: Commercial Electric Grillers and Toasters (AMD 7832) August 15, 1993 (L). 33 pp.

CENELEC HD 265 S2-84. Safety of Household and Similar Electrical Appliances—Particular Requirements for Appliances for Toasters, Grills, Roasters and Similar Appliances. 8 pp.

CENELEC EN 60 335-2-9-90. Safety of Household and Similar Electrical Appliances Part 2: Particular Requirements for Toasters, Grills, Roasters and Similar Appliances. 8 pp.

CENELEC EN 60335-2-9 /A2-92. AMD 2 Safety of Household and Similar Electrical Appliances Part 2: Particular Requirments for Toasters, Grills, Roasters and Similar Appliances. 5 pp.

CENELEC EN 60335-2-48-90. Safety of Household and Similar Electrical Appliances Part 2: Particular Requirements for Commercial Electric Grillers and Toasters. 10 pp.

CENELEC EN 60335-2-48/A1-92. AMD 1 Safety of Household and Similar Electrical Appliances Part 2: Particular Requirements for Commercial Electric Grillers and Toasters. 5 pp.

IEC 335 Pt 2-9-93. Safety of Household and Similar Electrical Appliances Part 2: Particular Requirements for Toasters, Grills, Roasters and Similar Appliances Fourth Edition. 46 pp.

IEC 335 Pt 2-48-88. Safety of Household and Similar Electrical Appliances Part 2: Particular Requirements for Commercial Electric Grillers and Toasters First Edition; (Amendment 1-1990). 44 pp.

—Tools

BSI BS EN 50 059-91. 1991 Electrostatic Hand-Held Spraying Equipment for Non-Flammable Material for Painting and Finishing. 15 pp.

CENELEC EN 50 059-90. Specification for Electrostatic Hand-Held Spraying Equipment for Non-Flammable Material for Painting and Finishing. 24 pp.

IEC 745 Pt 1-82. Safety of Hand-Held Motor-Operated Electric Tools Part 1: General Requirements First Edition. 155 pp.

IEC 745 Pt 2-1-89. Safety of Hand-Held Motor-Operated Electric Tools Part 2: Particular Requirements for Drills First Edition; (Amendment 1-1992). 37 pp.

IEC 1029 Pt 1-90. Safety of Transportable Motor-Operated Electric Tools Part 1: General Requirements First Edition. 187 pp.

SAA AS 1895-77. Code of Practice for Guarding and Safe Use of Portable Electric Tools for Domestic Use. 16 pp.

—Toothbrushes

BSI BS 3456: Sec 202.20-90. 1990 Safety of Household and Similar Electrical Appliances Part 202: Particular Requirements Section 202.20: Battery-Powered Tooth-Brushes and Their Charging and Battery Assemblies. 23 pp.

BSI BS 3456: Sec 3.11-79. (WITHDRAWN) 1979 Amd 2 1972-1981 Edition. Specification for Safety of Household Electrical Appliances Part 3: Complete Particular Specifications Section 3.11: Mains-Operated Toothbrushes (Superseded by BS EN 60335-2-52: 1991). 37 pp.

Consumer Safety *(Cont.)*

—Toothbrushes *(Cont.)*

BSI BS EN 60335-2-52-91. 1991 Safety of Household and Similar Electrical Appliances Part 2: Particular Requirements Section 2.52: Oral Hygiene Appliances Connected to the Mains Supply Through a Safety Isolating Transformer. 18 pp.

CENELEC EN 60 335-2-20-89. Safety of Household and Similar Electrical Appliances Part 2: Particular Requirements for Battery—Powered Tooth—Brushes and Their Charging and Battery Assemblies. 8 pp.

CENELEC EN 60 335-2-52-91. Safety of Household and Similar Electrical Appliances Part 2: Particular Requirements for Oral Hygiene Appliances Connected to the Mains Supply Through a Safety Isolating Transformer. 7 pp.

DIN VDE 0700 Pt 223-84. Safety of Household and Similar Electrical Appliances; Oral Hygiene Appliances Connected to the Mains Supply Through a Safety Isolating Transformer (Dec) (Partially Supersedes VDE 0730 Part 2 ZC/05.76). 12 pp.

DIN VDE 0700 Pt 228-86. Safety of Household and Similar Electrical Appliances; Mains Operated Oral Hygiene Appliances (Mar) (Supersedes VDE 0730 Part 2. 11 pp.

DIN VDE 0730 Pt 2 ZCB-78. Requirements for Electrical Equipment for Household and Similar Purposes; Part 2 ZCB: Specific Requirements for Battery-Powered Tooth and Oral Hygiene Equipment (Dec). 23 pp.

IEC 335 Pt 2-20-84. Safety of Household and Similar Electrical Appliances Part 2: Particular Requirements for Battery-Powered Tooth-Brushes and Their Charging and Battery Assemblies Second Edition. 35 pp.

IEC 335 Pt 2-52-88. Safety of Household and Similar Electrical Appliances Part 2: Particular Requirements for Oral Hygiene Appliances Connected to the Mains Supply Through a Safety Isolating Transformer First Edition. 23 pp.

—Towel Rails (Heated)

BSI BS 3456: Sec 202.43-90. 1990 Safety of Household Electrical Appliances Part 202: Particular Requirements Section 202.43: Clothes Dryers and Towel Rails. 20 pp.

BSI BS 3456: Sec 202.43-01. 1990 Amd 1 Safety of Household and Similar Electrical Appliances Part 202: Particular Requirements Section 202.43: Clothes Dryers and Towel Rails (AMD 6901) April 1, 1992 (SUPERSEDES BS 3456 SECTION 2.9: 1970 & 3. 6: 1979). 26 pp.

BSI BS 3456: Sec 202.43-02. 1990 Amd 2 Safety of Household and Similar Electrical Appliances Part 202: Particular Requirements Section 202.43: Clothes Dryers and Towel Rails (AMD 7550) JAN 15, 1993 SUPERSEDES BS 3456 SECTION 2. 9: 1970 & 3. 6: 1979. 31 pp.

CENELEC HD 263 S1-84. Particular Specification for Clothes Dryers and Towel Dryers. 5 pp.

CENELEC EN 60 335-2-43-89. Safety of Household and Similar Electrical Appliances—Part 2: Particular Requirements for Clothes Dryers and Towel Rails. 7 pp.

CENELEC EN 60 335-2-43/A1-90. AMD 1 Safety of Household and Similar Electrical Appliances—Part 2: Particular Requirements for Clothes Dryers and Towel Rails. 6 pp.

CENELEC EN 60335-2-43/A51-92. AMD 51 Safety of Household and Similar Electrical Appliances—Part 2: Particular Requirements for Clothes Dryers and Towel Rails. 4 pp.

IEC 335 Pt 2-43-84. Safety of Household and Similar Electrical Appliances Part 2: Particular Requirements for Clothes Dryers and Towel Rails First Edition; (Amendment 1-1988). 35 pp.

—Toys

BSI BS 5665: Part 1-89. 1989 Safety of Toys Part 1: Mechanical and Physical Properties (Supersedes BS 3443: 1968). 19 pp.

BSI BS 5665: Part 2-89. 1989 Safety of Toys Part 2: Flammability Requirements (Supersedes BS 3443: 1968). 8 pp.

BSI BS 5665: Part 3-89. 1989 Safety of Toys Part 3: Migration of Certain Elements (Supersedes BS 3443: 1968). 11 pp.

BSI BS 5665: Part 4-90. 1990 Safety of Toys Part 4: Experimental Sets for Chemistry and Related Activities. 13 pp.

BSI BS 5665: Part 5-93. 1993 Safety of Toys Part 5: Chemical Toys (Sets) Other Than Experimental Sets (L). 24 pp.

BSI BS EN 71-5-93. 1993 Safety of Toys Part 5: Chemical Toys (Sets) Other Than Experimental Sets (L). 24 pp.

BSI PD 6530-92. 1992 Explanatory Supplement to BS 5665 'Safety of Toys' Part 1: 1989 'Specification for Mechanical and Physical Properties'. 20 pp.

CEN EN 71-1-78. Safety of Toys—Part 1: Mechanical and Physical Properties. 11 pp.

CEN EN 71-1-88. Safety of Toys Part 1: Mechanical and Physical Properties. 35 pp.

Consumer Safety *(Cont.)*

—Toys *(Cont.)*

CEN EN 71-2-88. Safety of Toys—Part 2: Flammability. 6 pp.

CEN PREN 71-2-92. Safety of Toys—Part 2: Flammability. 10 pp.

CEN EN 71 (Part 4)-90. Safety of Toys—Part 4: Experimental Sets for Chemistry and Metal Related Activities. 14 pp.

CEN EN 71-5-93. Safety of Toys Part 5: Chemical Toys (Sets) Other Than Experimental Sets. 18 pp.

CENELEC HD 271 S1-86. Safety of Household and Similar Electrical Appliances Particular Requirements for Electric Toys Supplied at Safety Extra-Low Voltage. 2 pp.

CENELEC HD 271 S1/A1-86. AMD 1 Safety of Household and Similar Electrical Appliances Particular Requirements for Electric Toys Supplied at Safety Extra-Low Voltage. 5 pp.

CNS Z7066-86. Toy Safety (General Requirements) (Jun)(4797). 5 pp.

CNS Z7066-1-86. Toy Safety (Flammability Requirements) (Jun)(4797-1). 2 pp.

CNS Z7066-2-86. Toy Safety Toxicological Requirements (Jun)(4797-2). 4 pp.

CNS Z7066-3-87. Toy Safety (Constructional Requirements) (Dec) (4797-3). 19 pp.

CNS Z8016-86. Method of Test for Toy Safety (General Rule) (Jun)(4798). 1 p.

CNS Z8016-1-86. Method of Test for Toy Safety (Testing for Flammability of Textile Materials) (Jun)(4798-1). 3 pp.

CNS Z8016-2-86. Method of Test for Toy Safety (Testing for Flammability of Rigid and Pliable Solids) (Jun)(4798-2). 3 pp.

CNS Z8016-3-86. Method of Test for Toy Safety (Testing for Flammability of Self-Pressurized Container) (Jun)(4798-3). 2 pp.

CNS Z8016-4-86. Method of Test for Toy Safety (Testing for Durability and Legibility of Labels) (Jun)(4798-4). 2 pp.

CNS Z8016-5-86. Method of Test for Toy Safety (Testing for Toxic Substances Content) (Jun)(4798-5). 2 pp.

CNS Z8016-6-86. Method of Test for Toy Safety (Testing for Irritant Substances Content) (Jun)(4798-6). 4 pp.

CNS Z8016-7-86. Method of Test for Toy Safety (Testing for Corrosive Substance Content) (Jun)(4798-7). 1 p.

CNS Z8016-8-86. Method of Test for Toy Safety (Testing for Wet/Dry Crocking of Textile Materials) (Jun)(4798-8). 2 pp.

CNS Z8016-9-86. Method of Test for Toy Safety (Testing for Color Fastness of Textile Materials to Perspiration) (Jun)(4798-9). 2 pp.

CNS Z8016-10-86. Method of Test for Toy Safety (Testing for Lead and Other Hazardous Heavy Metal Contents) (Nov)(4798-10). 4 pp.

CNS Z8016-11-87. Method of Test for Toy Safety (Testing for Accessibility) (Dec)(4798-11). 2 pp.

CNS Z8016-12-87. Method of Test for Toy Safety (Testing for Hazardous Sharp Edge) (Dec)(4798-12). 2 pp.

CNS Z8016-13-87. Method of Test for Safety (Testing for Hazardous Sharp Point) (Dec)(4798-13). 2 pp.

CNS Z8016-14-87. Method of Test for Toy Safety (Testing for Ingestion or Inhalation Hazards) (Dec)(4798-14). 2 pp.

CNS Z8016-15-87. Method of Test for Toy Safety (Testing for Moving Components) (Dec)(4798-15). 2 pp.

CNS Z8016-16-87. Method of Test for Toy Safety (Testing for Toys Which Are Labelled as Being Washable) (Dec)(4798-16). 2 pp.

CNS Z8016-17-87. Method of Test for Toy Safety (Drop Test) (Dec)(4798-17). 2 pp.

CNS Z8016-18-87. Method of Test for Toy Safety (Tip-Over Test for Toy of Large Area or Volume) (Oct)(4798-18). 1 p.

CNS Z8016-19-87. Method of Test for Toy Safety (Tumble Test for Wheeled Toys) (Dec)(4798-19). 2 pp.

CNS Z8016-20-87. Method of Test for Toy Safety (Torque Test) (Dec)(4798-20). 2 pp.

CNS Z8016-21-87. Method of Test for Toy Safety (Tension Test) (Dec)(4798-21). 2 pp.

CNS Z8016-22-87. Method of Test for Toy Safety (Bite Test) (Dec)(4798-22). 2 pp.

CNS Z8016-23-87. Method of Test for Toy Safety (Compression Test) (Dec)(4798-23). 2 pp.

CNS Z8016-24-87. Method of Test for Toy Safety (Impact Test for Projectiles Propelled by a Discharge Mechanism) (Dec) (4798-24). 1 p.

CNS Z8016-25-87. Method of Test for Toy Safety (Seam Strength Test) (Dec)(4798-25). 3 pp.

CNS Z8016-26-87. Method of Test for Safety (Tension Test for Tyre Removal) (Dec)(4798-26). 2 pp.

CNS Z8016-27-87. Method of Test for Toy Safety (Tension Test for Wheeled Toys Having an Axle Retained by a Snap-in Fixture) (Dec)(4798-27). 3 pp.

622
INDEX and DIRECTORY of

INTERNATIONAL AND NON-U.S. NATIONAL STANDARDS
SUBJECT INDEX
Consumer

Consumer Safety *(Cont.)*

—Toys *(Cont.)*

CNS Z8016-28-87. Method of Test for Toy Safety (Compression Test for Wheeled Toys Having an Axle Retained by a Snap-in Fixture) (Dec)(4798-28). 2 pp.
CNS Z8016-29-87. Method of Test for Safety (Flexure Test for Toys with Stiffening Means for Retention of Form) (Dec)(4798-29). 2 pp.
CNS Z8016-30-87. Method of Test for Toy Safety (Testing for Mouth-Actuated Toys) (Dec)(4798-30). 1 p.
CNS Z8016-31-87. Method of Test for Toy Safety (Impact Test for Toys That Cover the Eyes) (Dec)(4798-31). 2 pp.
CNS Z8016-32-87. Method of Test for Toy Safety (Measurement of Noise Level) (Dec)(4798-32). 2 pp.
CNS Z8016-33-87. Method of Test for Toy Safety (Testing for Handhold Rattles or Other Handhold Toy Which Are Designed to Produce Sound When Shaken) (Dec)(4798-33). 2 pp.
CNS Z8016-34-87. Method of Test for Toy Safety (Testing for Penetration of Projectiles) (Dec)(4798-34). 4 pp.
CNS Z8016-35-87. Method of Test for Toy Safety (Testing for Stability of Ride-on or Sit-on Toys) (Dec)(4798-35). 5 pp.
CNS Z8016-36-87. Method of Test for Toy Safety (Testing for Toys Containing a Heat Source) (Dec)(4798-36). 2 pp.
CNS Z8016-37-87. Method of Test for Toy Safety (Testing for Toys Containing a Heat Engine) (Dec)(4798-37). 2 pp.
DIN ENGL 53160-74. Testing of Coloured Toys for Resistance to Saliva and Perspiration (June). 2 pp.
DIN VDE 0700 Pt 209-84. Safety of Household and Similar Electrical Appliances; Particular Requirements for Appliances for Playing and Learning; (VDE Specification) (Mar). 25 pp.
EC 88/378/EEC-88. Council Directive on the Approximation of the Laws of the Member States Concerning the Safety of Toys (NEW APP) (for Rel Stds See COMMUN/88/378, TEST/88/378). 13 pp.
EC COMMUN/88/378-88. Commission Communication Pursuant to Article 9 (2) of Council Directive 88/378/EEC Regarding the List of Bodies Approved by the Member States Responsible for Carrying out the EC Type-Examination Referred to in Article 8 (2) and 10 of That Directive (Safety of Toys). 2 pp.
EC TEST/88/378-88. Commission Communication Pursuant to Article 9 of Council Directive 88/378/EEC Regarding the List of Bodies Approved by the Member States Responsible for Carrying out the EC Type -Examination Referred to in Articles 8 and 10 of That Directive (Safety of Toys). 1 p.
EC 93/68/EEC-93. Council Directive Amending Directives 87/404/EEC (Simple Pressure Vessels), 88/378/EEC (Safety of Toys), 89/106/EEC (Construction Products), 89/336/EEC (Electromagnetic Compatibility), 89/392/EEC (Machinery), 89/686/ EEC (Personal Protective. 22 pp.
EC 93/C 237/02-93. Commission Communication in the Framework of the Implementation of Council Directive No 88/378/EEC in Relation to Safety of Toys. 1 p.
IEC 335 Pt 22-75. Safety of Household and Similar Electrical Appliances Particular Requirements for Electric Toys (Mains-Operated) First Edition. 40 pp.
SAA AS 1647.1-90. Children's Toys (Safety Requirements)—Part 1: General Requirements. 4 pp.
SAA AS 1647.2-92. Children's Toys (Safety Requirements)—Part 2: Constructional Requirements. 73 pp.
SNZ NZS 5820-82. Specification for the Safety of Toys Amend: 1, 1990. 95 pp.
SNZ NZS 5822-92. Prevention of Ingestion and Inhalation Hazards in Toys Intended for Use by Children Under Three Years of Age. 48 pp.

—Traction Batteries

BSI BS 6287-82. 1982 Safe Operation of Traction Batteries. 4 pp.

—Trimmers

BSI BS 2769: Sec 2.14-92. 1992 Hand-Held Electric Motor-Operated Tools Part 2: Particular Requirements Section 2.14: Specification for Hedge Trimmers and Scissor-Type Grass Shears (F). 13 pp.
BSI BS 2769: Sec 2.14-01. 1992 Hand-Held Electric Motor-Operated Tools Part 2: Particular Requirements Section 2.14: Specification for Hedge Trimmers and Scissor-Type Grass Shears (AMD 7505) November 15, 1992. 14 pp.
BSI BS 2769: Sec 2.14-84. 1984 Hand-Held Electric-Motor-Operated Tools Part 2: Particular Requirements Section 2.14: Hedge Trimmers and Scissor-Type Grass Shears. 6 pp.
BSI BS 2769: Sec 2.16-92. 1992 Hand-Held Electric Motor-Operated Tools Part 2: Particular Requirements Section 2.16: Specification for Trimmers. 11 pp.

Consumer Safety *(Cont.)*

—Trimmers *(Cont.)*

CEN PREN 786-92. Safety Requirements for Agricultural and Forestry Machinery—Electrically Powered Walk-Behind and Hand-Held Lawn Trimmers and Lawn Edge Trimmers. 21 pp.
CEN PREN 31806-92. Safety Requirements for Agricultural and Forestry Machinery—Brush Cutters and Grass Trimmers. 28 pp.
CENELEC HD 400.3-81. Hand-Held Motor Operated Tools—Part 2: Particular Specifications. 28 pp.
CENELEC PRHD 400.3 Section R S1-91. Hand-Held Motor Operated Tools Particular Specifications Section R: Trimmers. 8 pp.
CENELEC HD 400.3R S1-92. Hand-Held Motor Operated Tools Part 3: Particular Specifications Section R: Trimmers. 10 pp.
IEC 745 Pt 2-17-89. Safety of Hand-Held Motor-Operated Electric Tools Part 2: Particular Requirements for Routers and Trimmers First Edition. 21 pp.
ISO 10518-91. Powered Walk-Behind and Hand-Held Lawn Trimmers and Lawn Edge Trimmers—Mechanical Safety Requirements and Test Methods First Edition; (Corrected and Reprinted -1991). 11 pp.
SAA AS 4057-92. Powered Walk-Behind and Hand-Held Lawn Trimmers and Lawn Edge Trimmers—Mechanical Safety Requirements and Test Methods (ISO 10518:1991). 8 pp.

—Ultraviolet Lamps

BSI BS 3456: Sec 102.27-87. (WITHDRAWN) 1987 Safety of Household and Similar Electrical Appliances Part 102: Particular Requirements Section 102.27: Ultra-Violet and Infra-Red Radiation Skin Treatment Appliances for Household Use (Superseded by BS 3456: Sec. 202.27: 1991). 20 pp.
BSI BS 3456: Sec 202.27-91. 1991 Safety of Household Electrical Appliances Part 202: Particular Requirements Section 202.27: Ultra-Violet and Infra-Red Radiation Skin Treatment Appliances for Household and Similar Use. 34 pp.
BSI BS 3456: Sec 202.27-01. 1991 Amd 1 Safety of Household and Similar Electrical Appliances Part 202: Particular Requirements Section 202.27: Ultra-Violet and Infra-Red Radiation Skin Treatment Appliances for Household and Similar Use (AMD 7963) October 15, 1993 (L). 35 pp.
BSI BS EN 60335-2-27-92. 1992 Safety of Household and Similar Electrical Appliances Part 2: Particular Requirements Section 2.27: Ultra-Violet and Infra-Red Radiation Skin Treatment Appliances for Household and Similar Use. 28 pp.
BSI BS EN 60335-2-27-01. 1992 Amd 1 Safety of Household and Similar Electrical Appliances Part 2: Particular Requirements Section 2.27: Ultra-Violet and Infra-Red Radiation Skin Treatment Appliances for Household and Similar Use (AMD 7794) June 15, 1993 (L). 35 pp.
CENELEC HD 272 S3-87. Safety of Household and Similar Electrical Appliances—Part 2: Particular Requirements for Ultra-Violet and Infra-Red Radiation Skin Treatment Appliances for Household Use. 8 pp.
CENELEC EN 60 335-2-27-89. Safety of Household and Similar Electrical Appliance Part 2: Particular Requirements for Ultra-Violet and Infra-Red Radiation Skin Treatment Appliances for Household. 17 pp.
CENELEC EN 60335-2-27-92. Safety of Household and Similar Electrical Appliances Part 2: Particular Requirements for Ultra-Violet and Infra-Red Radiation Skin Treatment Appliances for Household Use; (IEC 335-2-27:1987 + A1:1989). 7 pp.
CENELEC EN 60335-2-27-92. Safety of Household and Similar Electrical Appliances Part 2: Particular Requirements for Ultra-Violet and Infra-Red Radiation Skin Treatment Appliances for Household and Similar Use; (IEC 335-2-27: 1987 + Amendment 1: 1989, Modified). 20 pp.
CENELEC EN 60335-2-27-92. Safety of Household and Similar Electrical Appliances Part 2: Particular Requirements for Ultra-Violet and Infra-Red Radiation Skin Treatment Appliances for Household Use; (IEC 335-2-27:1987 + A1:1989). 9 pp.
CENELEC EN 60335-2-27/A2-92. AMD 2 Safety of Household and Similar Electrical Appliances Part 2: Particular Requirements for Ultra-Violet and Infra-Red Radiation Skin Treatment Appliances for Household and Similar Use (IEC 335-2-27:1987/A2: 1991). 4 pp.
CENELEC EN 60335-2-37/A51-92. AMD 51 Safety of Household and Similar Electrical Appliances Part 2: Particular Requirements for Commercial Electric Deep Fat Fryers. 6 pp.
IEC 335 Pt 2-27-87. Safety of Household and Similar Electrical Appliances Part 2: Particular Requirements for Ultra-Violet and Infra-Red Radiation Skin Treatment Appliances for Household Use Second Edition; (Amendment 1-1989) (Amendment 2-1991); (CENELEC EN 60335-2-27:/ A2:1992). 60 pp.

Consumer Safety *(Cont.)*

—Ultraviolet Lamps *(Cont.)*

IEC 1228-93. Method of Measuring and Specifying the UV-Radiation of Ultraviolet Lamps Used for Sun-Tanning First Edition. 23 pp.

—Vacuum Cleaners

BSI BS 3456: Sec 102.2-87. (WITHDRAWN) 1987 Safety of Household Electrical Appliances Part 102: Particular Requirements Section 102.2: Vacuum Cleaners (Superseded by BS 3456: Section 202.2: 1990). 17 pp.
BSI BS 3456: Sec 202.2-90. 1990 Safety of Household Electrical Appliances Part 202: Particular Requirements Section 202.2: Vacuum Cleaners and Water Suction Cleaning Appliances. 25 pp.
BSI BS 3456: Sec 202.2-02. 1990 Amd 2 Safety of Household and Similar Electrical Appliances Part 202: Particular Requirements Section 202.2: Vacuum Cleaners and Water Suction Cleaning Appliances (AMD 7296) December 15, 1992 (L). 51 pp.
CENELEC HD 252-74. Particular Specification for Vacuum Cleaness and Water Suction Cleaning Appliances. 5 pp.
CENELEC HD 252 S3-86. Safety of Household and Similar Electrical Appliances—Part 2: Particular Requirements for Vacuum Cleaners and Water Suction Cleaning Appliances (Superseded by EN 60335-2-2). 4 pp.
CENELEC EN 60 335-2-2-88. Safety of Household and Similar Electrical Appliances Part 2: Particular Requirements for Vacuum Cleaners and Water Suction Cleaning Appliances. 10 pp.
CENELEC EN 60335-2-2 PRAC-93. AMD prAC Safety of Household and Similar Electrical Appliances Part 2: Particular Requirements for Vacuum Cleaners and Water Suction Cleaning Appliances. 3 pp.
CENELEC EN 60335-2-2 /A2-90. AMD 2 Safety of Household and Similar Electrical Appliances Part 2: Particular Requirements for Vacuum Cleaners and Water Suction Cleaning Appliances. 7 pp.
CENELEC EN 60335-2-2 /A52-91. AMD 52 Safety of Household and Similar Electrical Appliances Part 2: Particular Requirements for Vacuum Cleaners and Water Suction Cleaning Appliances. 7 pp.
IEC 335 Pt 2-2-93. Safety of Household and Similar Electrical Appliance Part 2: Particular Requirements for Vacuum Cleaners and Water Suction Cleaning Appliances Fourth Edition. 42 pp.
IEC 335 Pt 2-67-92. Safety of Household and Similar Electrical Appliances Part 2: Particular Requirements for Floor Treatment and Floor Cleaning Machines, for Industrial and Commercial Use First Edition. 34 pp.

—Valves

IEC 730 Pt 2-8-92. Automatic Electrical Controls for Household and Similar Use Part 2: Particular Requirements for Electrically Operated Water Valves, Including Mechanical Requirements First Edition. 70 pp.

—Viewers

BSI BS 3456: Sec 3.14-80. 1980 1972-1981 Edition. Specification for Safety of Household Electrical Appliances Part 3: Particular Specifications Section 3.14: Projectors. 52 pp.

—Waffle Irons

BSI BS 3456: Sec 3.17-80. (WITHDRAWN) 1980 Amd 2 1972-1981 Edition. Specification for Safety of Household Electrical Appliances: Part 3: Complete Particular Specif-ications: Section 3.17: Toasters. Grills, Waffle Irons, Roasters & Other Dry Cooking Appliances (Superseded by BS 3456: Section 102.9: 1991). 48 pp.
BSI BS EN 60335-2-9-91. 1991 Safety of Household and Similar Electrical Appliances Part 2: Particular Requirements Section 2.9: Toasters, Grills, Roasters and Similar Appliances. 26 pp.
BSI BS EN 60335-2-9-01. 1991 Amd 1 Safety of Household and Similar Electrical Appliances Part 2: Particular Requirements Section 2.9: Toasters, Grills, Roasters and Similar Appliances (AMD 6897) April 1, 1992. 32 pp.
BSI BS EN 60335-2-9-02. 1991 Amd 2 Safety of Household and Similar Electrical Appliances Part 2: Particular Requirements Section 2.9: Toasters, Grills, Roasters and Similar Appliances (AMD 7584) March 15, 1993 (L). 42 pp.
CENELEC EN 60 335-2-9-90. Safety of Household and Similar Electrical Appliances Part 2: Particular Requirements for Toasters, Grills, Roasters and Similar Appliances. 8 pp.
CENELEC EN 60335-2-9 /A2-92. AMD 2 Safety of Household and Similar Electrical Appliances Part 2: Particular Requirments for Toasters, Grills, Roasters and Similar Appliances. 5 pp.
IEC 335 Pt 2-9-93. Safety of Household and Similar Electrical Appliances Part 2: Particular Requirements for Toasters, Grills, Roasters and Similar Appliances Fourth Edition. 46 pp.

INDUSTRY STANDARDS
623

Consumer

INTERNATIONAL AND NON-U.S. NATIONAL STANDARDS
SUBJECT INDEX

Consumer Safety *(Cont.)*

—Washing Machines

BSI BS 3456: Sec 102.7-88. (WITHDRAWN) 1988 Safety of Household Electrical Appliances Part 102: Particular Requirements Section 102.7: Washing Machines (Superseded by BS 3456: Section 202.7: 1991). 32 pp.

BSI BS 3456: Sec 202.7-91. 1991 Safety of Household Electrical Appliances Part 202: Particular Requirements Section 202.7: Washing Machines (L). 35 pp.

BSI BS 3456: Sec 202.7-01. 1991 Amd 1 Safety of Household and Similar Electrical Appliances Part 202: Particular Requirements Section 202.7: Washing Machines (AMD 6900) April 1, 1992, (L). 43 pp.

BSI BS 3456: Sec 202.7-02. 1991 Amd 2 Safety of Household and Similar Electrical Appliances Part 202: Particular Requirements Section 202.7: Washing Machines (AMD 7312) October 15, 1992 (L). 55 pp.

BSI BS 3456: Sec 202.7-03. 1991 Amd 3 Safety of Household and Similar Electrical Appliances Part 202: Particular Requirements Section 202.7: Washing Machines (AMD 7549) January 15, 1993 (Supersedes BS 3456: Section 102.7: 1988). 60 pp.

BSI BS 6614-85. 1985 Safety Devices and Water Supply Connections for Washing Machines and Dishwashers Connected to the Water Supply Mains. 25 pp.

BSI BS EN 50084-92. 1992 Safety of Household and Similar Electrical Appliances. Requirements for the Connection of Washing Machines, Dishwashers and Tumbler Dryers to the Water Mains (L). 24 pp.

CENELEC HD 256 S2-84. Safety of Household and Similar Electrical Appliances Particular Requirements for Washing Machines. 21 pp.

CENELEC HD 256 S2-85. Safety of Household and Similar Electrical Appliances. Particular Requirements for Washing Machines. 19 pp.

CENELEC EN 50084-92. Safety of Household and Similar Electrical Appliances—Requirements for the Connection of Washing Machines, Dishwashers and Tumbler Dryers to the Water Mains. 21 pp.

CENELEC EN 60 335-2-7-90. Safety of Household and Similar Electrical Appliances Part 2: Particular Requirements for Washing Machines. 14 pp.

CENELEC EN 60335-2-7 /A1-90. AMD 1 Safety of Household and Similar Electrical Appliances Part 2: Particular Requirements for Washing Machines. 6 pp.

CENELEC EN 60335-2-7 /A2-92. AMD 2 Safety of Household and Similar Electrical Appliances Part 2: Particular Requirements for Washing Machines. 6 pp.

CENELEC EN 60335-2-7 /A51-92. AMD 51 Safety of Household and Similar Electrical Appliances Part 2: Particular Requirements for Washing Machines. 5 pp.

DIN VDE 0700 Pt 600 (D)-88. Safety of Household and Similar Electrical Appliances; Connection of Washing Machines, Dishwashers and Tumble Dryers to the Supply Mains; German Version prHD 274 S2: 1987 (Oct). 37 pp.

IEC 335 Pt 2-7-93. Safety of Household and Similar Electrical Appliances Part 2: Particular Requirements for Washing Machines Fourth Edition. 49 pp.

—Water Extractors

BSI BS 3456: Sec 102.4-87. (WITHDRAWN) 1987 Safety of Household Electrical Appliances Part 102: Particular Requirements Section 102.4: Spin Extractors (Superseded by BS 3456: Section 202.4: 1990). 19 pp.

BSI BS 3456: Sec 202.4-90. 1990 Safety of Household Electrical Appliances Part 202: Particular Requirements Section 202.4: Spin Extractors. 21 pp.

BSI BS 3456: Sec 202.4-01. 1990 Amd 1 Safety of Household and Similar Electrical Appliances Part 202: Particular Requirements Section 202.4: Spin Extractors (AMD 6904) April 1, 1992 (L) (L). 27 pp.

BSI BS 3456: Sec 202.4-02. 1990 Amd 2 Safety of Household and Similar Electrical Appliances Part 202: Particular Requirements Section 202.4: Spin Extractors (AMD 7279) October 15, 1992 (Supersedes BS 3456: Section 102.4: 1987). 37 pp.

CENELEC HD 267 S1-82. Safety of Household and Similar Electrical Appliances—Particular Requirements for Spin Extractors. 11 pp.

CENELEC EN 60 335-2-4-89. Safety of Household and Similar Electrical Appliances Part 2: Particular Requirements for Spin Extractors. 10 pp.

CENELEC EN 60 335-2-4-89. Safety of Household and Similar Electrical Appliances Part 2: Particular Requirements for Spin Extractors (IEC 335-2-4:1984 3rd Edition, Modified). 7 pp.

CENELEC EN 60 335-2-4/AA-90. AMD A Safety of Household and Similar Electrical Appliances Part 2: Particular Requirements for Spin Extractors; (IEC 335-2-4:1984 3rd Edition, Modified). 3 pp.

Consumer Safety *(Cont.)*

—Water Extractors *(Cont.)*

CENELEC EN 60 335-2-4/AB-90. AMD B Safety of Household and Similar Electrical Appliances Part 2: Particular Requirements for Spin Extractors; (IEC 335-2-4:1984 3rd Edition, Modified). 3 pp.

CENELEC EN 60335-2-4 /A2-92. AMD 2 Safety of Household and Similar Electrical Appliances Part 2: Particular Requirements for Spin Extractors; (IEC 335-2-4:1984 3rd Edition, Modified). 6 pp.

CENELEC EN 60335-2-4 /A51-91. AMD 51 Safety of Household and Similar Electrical Appliances Part 2: Particular Requirements for Spin Extractors; (IEC 335-2-4:1984 3rd Edition, Modified). 6 pp.

IEC 335 Pt 2-4-93. Safety of Household and Similar Electrical Appliances Part 2: Particular Requirements for Spin Extractors Fourth Edition. 36 pp.

—Water Heaters

BSI BS 3456: Sec 102.21-88. 1988 Safety of Household Electrical Appliances Part 102: Particular Requirement Section 102.21: Storage Water Heaters. 27 pp.

BSI BS 3456: Sec 3.9-79. (WITHDRAWN) 1979 Amd 4 1972-1981 Edition. Specification for Safety of Household Electrical Appliances Part 3: Complete Particular Specif Section 3.9: Stationary Instantaneous Water Heaters (AMD 5598) July 29, 1988. 54 pp.

BSI BS 5258: Part 7-77. 1977 Safety of Domestic Gas Appliances Part 7: Storage Water Heaters. 19 pp.

BSI BS EN 60335-2-21-92. 1992 Safety of Household and Similar Electrical Appliances Part 2: Particular Requirements Section 2-21: Storage Water Heaters (L). 33 pp.

BSI BS EN 60335-2-51-91. 1991 Safety of Household and Similar Electrical Appliances Part 2: Particular Requirements Section 2.51: Stationary Circulation Pumps for Heating and Service Water Installations (Supersedes BS 1394: Part 1: 1987). 18 pp.

BSI BS EN 60335-2-63-93. 1993 Safety of Household and Similar Electrical Appliances Part 2: Particular Requirements Section 2.63: Commercial Electric Water Boilers and Liquid Heaters (L). 26 pp.

CENELEC HD 282 S1-90. Safety of Household and Similar Electrical Appliances Part 2: Particular Requirements for Instantaneous Water Heaters. 12 pp.

CENELEC HD 282 S1/A1-92. AMD 1 Safety of Household and Similar Electrical Appliances Part 2: Particular Requirements for Instantaneous Water Heaters. 8 pp.

CENELEC EN 60335-2-21-92. Safety of Household and Similar Electrical Appliances Part 2: Particular Requirements for Storage Water Heaters. 24 pp.

CENELEC PREN 60335-2-35 PRAA-93. AMD prAA Safety of Household and Similar Electrical Appliances Part 2: Particular Requirements for Instantaneous Water Heaters. 2 pp.

CENELEC EN 60 335-2-51-91. Safety of Household and Similar Electrical Appliances Part 2: Particular Requirements for Stationary Circulation Pumps for Heating and Service Water Installations. 7 pp.

CENELEC EN 60335-2-51-92. Safety of Household and Similar Electrical Appliances Part 2: Particular Requirements for Stationary Circulation Pumps for Heating and Service Water Installations. 14 pp.

CENELEC EN 60335-2-63-93. Safety of Household and Similar Electrical Appliances Part 2: Particular Requirements for Commercial Electric Water Boilers and Liquid Heaters (IEC 335-2-63: 1990, Modified). 18 pp.

DIN VDE 0700 Pt 21 A1 (D)-83. Safety of Household and Similar Electrical Appliances; Part 2: Particular Requirements for Water Heaters (Storage Heaters and Boilers); Amendment 1 (Apr). 5 pp.

DIN VDE 0700 Pt 21 A3 (D)-87. Safety of Household and Similar Electrical Appliances; Water Heaters (Storage Water Heaters and Hot Water Boilers); Amendment 3 to Draft DIN VDE 0700 Part 21 (Apr). 7 pp.

IEC 335 Pt 2-21-89. Safety of Household and Similar Electrical Appliances Part 2: Particular Requirements for Storage Water Heaters Third Edition; (Amendment 1-1990) (Amendment 2-1990) (Amendment 3-1992). 65 pp.

IEC 335 Pt 2-35-91. Safety of Household and Similar Electrical Appliances Part 2: Particular Requirements for Instantaneous Water Heaters Second Edition; (I-SH 02—Nov 1991). 47 pp.

IEC 335 Pt 2-51-88. Safety of Household and Similar Electrical Appliances Part 2: Particular Requirements for Stationary Circulation Pumps for Heating and Service Water Installations First Edition. 27 pp.

IEC 335 Pt 2-63-90. Safety of Household and Similar Electrical Appliances Part 2: Particular Requirements for Commercial Electric Water Boilers and Liquid Heaters First Edition; (CENELEC EN 60035-2-63: 1993). 40 pp.

Consumer Safety *(Cont.)*

—Water Heaters *(Cont.)*

SAA AS 3888(INT)-91. Water Heaters (All Types)—Basic Safety and Public Health Requirements (Expires 31 January 1993) Amdt 1 August 1991 (Superseded by AS 3498(Int)—1993).

SNZ NZS 6335-90. Safety of Household and Similar Electrical Appliances. Particular Requirements for Instantaneous Water Heaters Amend: 1, 1991. 28 pp.

—Water Pumps

BSI BS EN 60335-2-51-91. 1991 Safety of Household and Similar Electrical Appliances Part 2: Particular Requirements Section 2.51: Stationary Circulation Pumps for Heating and Service Water Installations (Supersedes BS 1394: Part 1: 1987). 18 pp.

CENELEC EN 60 335-2-51-91. Safety of Household and Similar Electrical Appliances Part 2: Particular Requirements for Stationary Circulation Pumps for Heating and Service Water Installations. 7 pp.

CENELEC EN 60335-2-51-92. Safety of Household and Similar Electrical Appliances Part 2: Particular Requirements for Stationary Circulation Pumps for Heating and Service Water Installations. 14 pp.

IEC 335 Pt 2-51-88. Safety of Household and Similar Electrical Appliances Part 2: Particular Requirements for Stationary Circulation Pumps for Heating and Service Water Installations First Edition. 27 pp.

—Welding Equipment

DIN VDE 0700 Pt 30-83. Test Certificate, Safety Test of Household & Similar Electrical Appliances. Portable Electric Heating Tools (Jan). 26 pp.

DIN VDE 0700 Pt 30 A1 (D)-84. Safety of Household and Similar Electrical Appliances; Particular Specifications for Portable Electric Heating Tools (VDE Specification): Amendment 1 (Nov). 19 pp.

—Whirlpool Baths

BSI BS EN 60335-2-60-91. 1991 Safety of Household and Similar Electrical Appliances Part 2: Particular Requirements Section 2.60: Whirlpool Baths and Similar Equipment. 19 pp.

CENELEC EN 60335-2-60-91. Safety of Household and Similar Electrical Appliances Part 2: Particular Requirements for Whirlpool Baths and Similar Equipment. 7 pp.

CENELEC EN 60335-2-60/PRAB-93. AMD prAb Safety of Household and Similar Electrical Appliances Part 2: Particular Requirements for Whirlpool Baths and Similar Equipment. 2 pp.

CENELEC EN 60335-2-60/A51-93. AMD 51 Safety of Household and Similar Electrical Appliances Part 2: Particular Requirements for Whirlpool Baths and Similar Equipment. 2 pp.

IEC 335 Pt 2-60-90. Safety of Household and Similar Electrical Appliances Part 2: Particular Requirements for Whirlpool Baths and Similar Equipment First Edition. 29 pp.

—Woodburning Tools

DIN VDE 0700 Pt 30-83. Test Certificate, Safety Test of Household & Similar Electrical Appliances. Portable Electric Heating Tools (Jan). 26 pp.

DIN VDE 0700 Pt 30 A1 (D)-84. Safety of Household and Similar Electrical Appliances; Particular Specifications for Portable Electric Heating Tools (VDE Specification): Amendment 1 (Nov). 19 pp.

Consumer Service Conductors

Use: Service Entrance Cables

Consumer Service Equipment

Use: Service Entrance Equipment

Consumption Rate

—Motor Oils

ISO 3046 Pt 1-86. Reciprocating Internal Combustion Engines—Performance—Part 1: Standard Reference Conditions and Declarations of Power, Fuel Consumption and Lubricating Oil Consumption Third Edition; (Amendment 1-1987). 19 pp.

Contact Adhesives

See Also: Adhesives

MOD UK DSTAN 80-121-01. General Purpose Contact Adhesive, Non-Flammable Issue 1; Amendment 1. 16 pp.

MOD UK DSTAN 80-121-92. General Purpose Contact Adhesive Non-Flammable Issue 2. 11 pp.

—Plastic Laminates to Wood

CGSB CAN/CGSB-71.19-M88. Adhesive, Contact, Sprayable. 12 pp.

CGSB CAN/CGSB-71.20-M88. Adhesive, Contact, Brushable. 12 pp.

Contact Adhesives *(Cont.)*

—Wood

BSI BS 1204: Part 2-79. (WITHDRAWN) 1979 Amd 1 Synthetic Resin Adhesives (Phenolic and Amino-Plastic) for Wood Part 2: Close-Contact Adhesives (AMD 6503) December 21, 1990 (Superseded by BS EN 301: 1992 & BS EN 302: Parts 1-4: 1992). 13 pp.

SNZ NZS 7202: Part 2-86. Synthetic Resin Adhesives (Phenolic and Aminoplastic) for Wood Part 2: Specification for Close-Contact Adhesives Amend: A, 1986. 12 pp.

Contact Bounce

See Also: Electric Contacts; Relay Contacts

—Dry Reed Switches

CNS C6151-87. Method of Test for Dry Reed Switches (Operate, Bounce, Release and Transfer (SPDT) Time) (Jul)(8229).

—Electric Switches

CNS C6235-84. Method of Test for Electromechanical Switches (Contact Bounce) (Nov)(11098).

Contact Breakers

Use: Breaker Points

Contact Chatter

Use: Chatter

Contact Force

Use For: Contact Retention; Retention Force; Withdrawal Force *See Also:* Force; Insertion Force

—Connectors

CEN PREN 2591 (Part D8)-92. Elements of Electrical and Optical Connection Test Methods Part D8—Mating and Unmating Forces. 4 pp.

CEN PREN 2591 (Part D9)-92. Elements of Electrical and Optical Connection Test Methods Part D9—Contact Retention in Insert. 4 pp.

CEN PREN 2591 (Part D10)-92. Elements of Electrical and Optical Connection Test Methods Part D10—Insert Retention in Housing (Axial). 4 pp.

CEN PREN 2591 (Part D11)-92. Elements of Electrical and Optical Connection Test Methods Part D11—Insert Retention in Housing (Torsional). 4 pp.

CEN PREN 2591 (Part D12)-92. Elements of Electrical and Optical Connection Test Methods Part D12—Contact Insertion and Extraction Forces. 4 pp.

CEN PREN 2591 (Part D14)-92. Elements of Electrical and Optical Connection Test Methods Part D14—Unmating of Lanyard Release Connectors. 4 pp.

CEN PREN 2591-F10-93. Aerospace Series Elements of Electrical and Optical Connection Test Methods Part F10—Optical Elements Effectiveness of Cable Attachment Cable Pulling. 3 pp.

CNS C6141-86. Method of Test for Low Frequency (Below 3 MHz) Electrical Connectors (TP-29 Contact Retention Test) (Sep)(8219).

CNS C6181-82. Method of Test for Low Frequency (Below 3MHz) Electrical Connectors (TP-37 Contact Separation Force) (Aug)(9241).

CNS C6185-88. Method of Test for Low Frequency (Below 3 MHz) Electrical Connectors (TP-5 Contact Insertion and Removal Forces) (Apr)(9364).

CNS C6187-88. Method of Test for Low Frequency (Below 3 MHz) Electrical Connectors (TP-13 Mating and Unmating Forces) (Apr)(9366).

—Electric Contacts

CEN PREN 2591 (Part D26)-92. Elements of Electrical and Optical Connection Test Methods Part D26—Contact Retention System Effectiveness (Removable Contact Walkout). 4 pp.

—Electric Contacts—Gages

CEN PREN 2591 (Part D18)-92. Elements of Electrical and Optical Connection Test Methods Part D18—Gauge Insertion and Extraction Forces in a Female Contact. 4 pp.

—Fiberboard—Screws

BSI BS EN 320-93. 1993 Fibreboards—Determination of Resistance to Axial Withdrawal of Screws (R). 11 pp.

CEN PREN 320-90. Fibreboards—Determination of Resistance to Axial Withdrawal of Screws. 6 pp.

CEN EN 320-93. Fibreboards—Determination of Resistance to Axial Withdrawal of Screws. 6 pp.

—Grounding Connectors

CEN PREN 2591 (Part D13)-92. Elements of Electrical and Optical Connection Test Methods Part D13—Holding Force of Grounding Spring System. 4 pp.

Contact Lenses

See Also: Eyeglasses; Lenses; Optical Lenses

Contact Lenses *(Cont.)*

BSI BS 7208: Part 3-92. 1992 Contact Lenses Part 3: Methods of Test for Contact Lenses. 11 pp.

SAA AS 1887-76. Contact Lenses Corrig.. 28 pp.

—Glossaries

BSI BS 3521: Part 3-88. 1988 Glossary of Terms Relating to Ophthalmic Lenses and Spectacle Frames Part 3: Glossary of Terms Relating to Contact Lenses. 17 pp.

ISO 8320-86. Optics and Optical Instruments—Contact Lenses—Vocabulary and Symbols First Edition. 15 pp.

—Hard

BSI BS 7208: Part 1-92. 1992 Contact Lenses Part 1: Specification for Rigid Corneal and Scleral Contact Lenses (ISO 8321-1: 1991). 19 pp.

BSI BS 7208: Part 1-89. (WITHDRAWN) 1989 Contact Lenses Part 1: Specification for Rigid Contact Lenses (Superseded by BS 7208: Part 1 & 3: 1992). 12 pp.

CNS T2036-88. Hard Contact Lenses (Oct)(12447).

ISO 8321 Pt 1-91. Optics and Optical Instruments—Contact Lenses—Part 1: Specification for Rigid Corneal and Scleral Contact Lenses First Edition. 15 pp.

—Materials—Classification

BSI BS 7208: Part 2-91. 1991 Contact Lenses Part 2: Method of Classifying Contact Lens Materials. 14 pp.

—Soft

CNS T2035-88. Soft Contact Lenses (Oct)(12446).

Contact Resistance

See Also: Electrical Properties; Electrical Resistance

—Connectors

BSI BS EN 2591-B1-92. 1992 Elements of Electrical and Optical Connection—Test Methods Part B1: Contact Resistance—Low Level. 10 pp.

CEN PREN 2591 (Part B1)-91. Elements of Electrical and Optical Connection Test Methods Part B1—Contact Resistance—Low Level. 4 pp.

CEN EN 2591-B1-92. Aerospace Series—Elements of Electrical and Optical Connection—Test Methods—Part B 1: Contact Resistance—Low Level. 5 pp.

CEN PREN 2591 (Part B2)-92. Elements of Electrical and Optical Connection Test Methods Part B2—Contact Resistance at Rated Current. 4 pp.

CEN PREN 2591 (Part B3)-92. Elements of Electrical and Optical Connection Test Methods Part B3—Electrical Continuity at Microvolt Level. 3 pp.

CEN PREN 2591 (Part B4)-92. Elements of Electrical and Optical Connection Test Methods Part B4—Discontinuity of Contacts in the Microsecond Range. 4 pp.

CEN PREN 2591-F11-93. Aerospace Series Elements of Electrical and Optical Connection Test Methods Part F11—Optical Elements Effectiveness of Cable Attachment Cable Torsion. 4 pp.

CNS C6186-88. Method of Test for Low Frequency (Below 3 MHz) Electrical Connectors (TP-6 Contact Resistance) (Apr)(9365).

SBAC TS 340 ISSUE 1. Test for Electrical Connectors Contact Resistance—Low Current.

SBAC TS 341 ISSUE 1. Test for Electrical Connectors Contact Resistance at Rated Current.

—Electromechanical Components

BSI BS 5772: Part 2-79. 1979 Basic Testing Procedures and Measuring Methods for Electromechanical Components for Electronic Equipment Part 2: General Examination, Electrical Continuity and Contact Resistance Tests, Insulation Tests and Voltage Stress Tests. 16 pp.

BSI BS 5772: Pt 2: Supp 1-81. 1981 Electromechanical Components for Electronic Equipment: Basic Testing Procedures Part 2: General Examination, Electrical Continuity and Contact Resistance Tests, Insulation Tests and Voltage Stress Supp 1: 1981 Test 4c. 4 pp.

IEC 512 Pt 2-85. Electromechanical Components for Electronic Equipment; Basic Testing Procedures and Measuring Methods Part 2: General Examination, Electrical Continuity and Contact Resistance Tests, Insulation Tests and Voltage Stress Tests Second Edition. 39 pp.

—Shells (Electric)

CEN PREN 2591 (Part B5)-92. Elements of Electrical and Optical Connection Test Methods Part B5—Housing (Shell) Electrical Continuity. 2 pp.

Contact Resistance Testing

Use: Contact Resistance

Contact Retention

Use: Contact Force

Contact Wire (Trolley Wire)

Use: Trolley Wire

Contactor Relays

Use: Contactors

Contactors

Use For: Electric Contactors *See Also:* Agitators; Control Systems Equipment; Controlgear; Magnetic Contactors; Mercury Wetted Contact Relays; Motor Starters; Relays; Semiconductor Contactors; Sprayers; Switchgear

BSI BS 4794: Sec 2.3 & 2.4-77. (WITHDRAWN) 1977 Control Switches (Switching Devices, Including Contactor Relays, for Controland Auxiliary Circuits for Voltages up to and Including 1000 V a.c and 1200 V d.c. Part 2: Special Requirements for Specific Types of. 8 pp.

—Aircraft

ISO 4084-77. Aircraft—Repairable Contactors (Not Hermetically Sealed)—Performance Requirements First Edition. 10 pp.

—High Voltage—AC

DIN VDE 0660 Pt 103-84. Switchgear and Controlgear; High Voltage Alternating Current Contactors (Mar). 47 pp.

IEC 470-74. High-Voltage Alternating Current Contactors First Edition; (Amendment 1-1975). 56 pp.

SAA AS 1864-76. High-Voltage Alternating Current Contactors Reconfirmed 1988. 48 pp.

—Household

BSI BS EN 61095-93. 1993 Electromechanical Contactors for Household and Similar Purposes (IEC 1095: 1992) (L). 124 pp.

BSI BS EN 61095-01. 1993 Amd 1 Electromechanical Contactors for Household and Similar Purposes (IEC 1095: 1992) (AMD 7858) July 15, 1993 (L). 125 pp.

CENELEC EN 61095-93. Electromechanical Contactors for Household and Similar Purposes; (IEC 1095:1992). 7 pp.

IEC 1095-92. Electromechanical Contactors for Household and Similar Purposes First Edition; (CENELEC EN 61095:1993). 215 pp.

—Low Voltage

BSI BS 4794: Part 1-79. (WITHDRAWN) 1979 Amd 1 Control Switches (Switching Devices, Including Contactor Relays, for Control and Auxiliary Circuits for Voltages up to and Including 1000 V a.c. and 1200 V d.c.) Part 1: General Requirements. 28 pp.

BSI BS 4794: Sec 2.1-77. (WITHDRAWN) 1977 Control Switches (Switching Devices, Including Contactor Relays, for Control and Auxiliary Circuits for Voltages up to and Including 1000 V a.c. and 1200 V d.c.) Part 2: Special Requirements for Specific Types of. 12 pp.

BSI BS 5424: Part 1-77. (WITHDRAWN) 1977 Amd 1 Controlgear for Voltages up to and Including 1000 V a.c. and 1200 V d.c. Part 1: Contactors (AMD 5103) March 31, 1988 (Superseded by BS EN 60947-4-1: 1992). 59 pp.

BSI BS 5424: Part 3-88. 1988 Controlgear for Voltages up to and Including 1000 V a.c. and 1200 V d.c. Part 3: Additional Requirements for Contactors Subject to Certification. 21 pp.

BSI BS EN 60947-4-1-01. 1992 Amd 1 Low-Voltage Switchgear and Controlgear Part 4: Contactors and Motor-Starters Section 1: Electromechanical Contactors and Motor-Starters (AMD 7337) December 15, 1992 (E). 103 pp.

BSI BS EN 60947-4-1-02. 1992 Amd 2 Low-Voltage Switchgear and Controlgear Part 4: Contactors and Motor-Starters Section 1: Electromechanical Contactors and Motor-Starters (AMD 7702) May 15, 1993 (E). 103 pp.

BSI BS EN 60947-4-1-03. 1992 Amd 3 Low-Voltage Switchgear and Controlgear Part 4: Contactors and Motor-Starters Section 1: Electromechanical Contactors and Motor-Starters (AMD 7849) July 15, 1993 (S). 105 pp.

BSI BS EN 60947-4-1-92. 1992 Low-Voltage Switchgear and Controlgear Part 4: Contactors and Motor-Starters Section 1: Electromechanical Contactors and Motor-Starters. 102 pp.

CENELEC HD 419-82. Low-Voltage Switchgear and Controlgear Contactors. 7 pp.

CENELEC HD 419.3 S1-88. Low-Voltage Controlgear Part 3: Additional Requirements for Contactors Subject to Certification. 4 pp.

CENELEC HD 420-82. Control Switches (Low Voltage-Devices for Control and Auxiliary Circuits, Including Contactor Relays). 5 pp.

CENELEC HD 420 S2-88. Control Switches (Low Voltage Switching—Devices for Control and Auxiliary Circuit, Including Contactor Relays). 6 pp.

CENELEC EN 60947-4-1-92. Low-Voltage Switchgear and Controlgear Part 4: Contactors and Motor-Starters Section One—Electromechanical Contactors and Motor-Starters (IEC 947-4-1:1990). 6 pp.

INTERNATIONAL AND NON-U.S. NATIONAL STANDARDS
SUBJECT INDEX

Contactors

Contactors (Cont.)
—Low Voltage (Cont.)
CENELEC EN60947-4-1-92. Low-Voltage Switchgear and Controlgear Part 4: Contactors and Motor-Starters Section One—Electromechanical Contactors and Motor-Starters (IEC 947-4-1:1990). 6 pp.
CENELEC EN 60947-4-1-92. CORRIGENDUM Low-Voltage Switchgear and Controlgear Part 4: Contactors and Motor-Starters Section One—Electromechanical Contactors and Motor-Starters (IEC 947-4-1:1990). 1 p.
CENELEC EN 60947-4-1-92. CORRIGENDUM Low-Voltage Switchgear and Controlgear Part 4: Contactors and Motor-Starters Section One—Electromechanical Contactors and Motor-Starters (IEC 947-4-1:1990). 1 p.
DIN VDE 0660 Pt 102-82. Switchgear; Low-Voltage Switchgear; Contractors (Sept). 7 pp.
DIN VDE 0660 Pt 102 A2 (D)-92. Switchgear & Controlgear; Low-Voltage Switchgear & Controlgear; Contactors & Motor-Starters; Amendment 2 to Draft DIN VDE 0660 Part 102; Identical with IEC 17B (Secretariat) 409 (Jan). 10 pp.
DIN VDE 0660 Pt 201-82. Switchgear and Control Gear; Low Voltage Switchgear and Control Gear; Control Switches; Additional Specifications for Pushbuttons and Related Control Switches (Sept). 18 pp.
DIN VDE 0660 Pt 203-82. Switchgear and Control Gear; Low Voltage Switchgear and Control Gear; Control Switches; Additional Specification for Contractor Relays (Sept). 9 pp.
IEC 158 Pt 3-85. Low-Voltage Controlgear Part 3: Additional Requirements for Contactors Subject to Certification First Edition. 39 pp.
IEC 947 Pt 4-1-90. Low-Voltage Switchgear and Controlgear Part 4: Contactors and Motor-Starters Section One—Electromechanical Contactors and Motor-Starters First Edition (CENELEC EN 60947-4-1: 1992). 186 pp.
JIS C 4531-82. Contactor Relays. 25 pp.

—Low Voltage—Emergency Lighting
BSI BS 764-90. 1990 Automatic Change-Over Contactors for Emergency Lighting Systems. 12 pp.

—Low Voltage—Terminals—Identification Systems
BSI BS 5582-78. 1978 Low-Voltage Switch Gear and Controlgear for Industrial Use. Terminal Marking and Distinctive Number for Auxiliary Contacts of Particular Contractors. 6 pp.
BSI BS 5583-78. 1978 Low-Voltage Switch-Gear and Controlgear for Industrial Use. Terminal Marking, Distinctive Number and Distinctive Letter for Particular Contractor Relays. 8 pp.
CENELEC EN 50 011-77. Low Voltage Switchgear and Controlgear for Industrial Use Terminal Marking, Distinctive Number and Distinctive Letter for Particular Contactor Relays. 17 pp.
CENELEC EN 50 011/PRAB-90. AMD prAB Low Voltage and Controlgear for Industrial Use Terminal Marking, Distinctive Number and Distinctive Letter for Particular Contactor Relays. 3 pp.
CENELEC EN 50 012-77. Low Voltage Switchgear and Controlgear for Industrial use Terminal Marking and Distinctive Number for Auxiliary Contacts of Particular Contactors. 4 pp.

—Numbering
BSI BS 5582-78. 1978 Low-Voltage Switch Gear and Controlgear for Industrial Use. Terminal Marking and Distinctive Number for Auxiliary Contacts of Particular Contractors. 6 pp.
BSI BS 5583-78. 1978 Low-Voltage Switch-Gear and Controlgear for Industrial Use. Terminal Marking, Distinctive Number and Distinctive Letter for Particular Contractor Relays. 8 pp.

—Reed—Quality Assurance
BSI BS 9200-72. 1972 Amd 3 Reed Contact Units of Assessed Quality: Generic Data and Methods of Test. 41 pp.

—Symbols
CNS C5116-81. Graphic Symbols for Switches and Contractors (May)(7365).

—1 kV—AC
BSI BS 775: Part 2-74. 1974 Amd 1 Contactors Part 2: A.C. Contactors for Voltages Above 1 kV and up to and Including 12 kV. 23 pp.

Contacts (Electric)
Use: Electric Contacts

Contagious Bovine Pleuropneumonia
Use For: CBPP See Also: Animal Diseases
—Portugal
EC 89/145/EEC-89. Council Decision Introducing a Community Financial Measure for the Eradication of Contagious Bovine Pleuropneumonia (CBPP) in Portugal. 3 pp.

Container Equipment Data Exchange (CEDEX)
Use: Information Interchange—Container Equipment Data Exchange (CEDEX)

Containers
Scope Note: For additional listings, use a more specific term Use For: Inland Containers; Receptacles (Containers); Vessels See Also: Aerosol Containers; Autoclaves; Bags; Bales; Barrels (Containers); Baskets; Beakers; Bottles; Boxes (Containers); Buckets (Pails); Bulk Storage; Bulk Transporters; Burets; Cans; Canteens; Cartons; Cartridges; Crates (Shipping Containers); Crucibles; Cryogenic Containers; Drums (Containers); Flasks; Food Containers; Fuel Tanks; Gas Cylinders; Glass Containers; Humidity Cabinets; Jars; Jerry Cans; Jugs; Laboratory Containers; Livestock Crates; Luggage; Luggage Compartments; Lunch Boxes; Materials Handling Equipment; Measuring Containers; Medicine Measures; Metal Containers; Open Top Containers; Packaging; Packaging Materials; Plastic Containers; Radioactive Materials; Reels; Refrigerated Containers; Septic Tanks; Sewage Holding Tanks; Shipping Containers; Storage Tanks; Tanks (Containers); Thermal Containers; Trays; Troughs; Tubes (Packages); Waste Containers; Waste Disposal; Water Storage Tanks
DIN ENGL 28005 Pt 1-88. General Tolerances for Vessels; General-Purpose Vessels (Nov). 7 pp.
MOD UK DSTAN 00-4-01. Pallets, Packages, Containers, Unit Loads and Associated Materials Handling Equipment Issue 2; Amendment 2. 62 pp.

—Agitator
DIN ENGL 28136 Pt 1-88. Mixing Vessels; Principal Dimensions (May). 5 pp.

—Agitator—Jackets—Strength
DIN ENGL 28145 Pt 7 Suppl. 1-87. Welded Attachments for Steel, Glass-Lined Agitator Vessels; Bracket Supports Welded to Jackets; Dimensions, Maximum Permissible Loads; Stresses in the Jacket (Nov). 7 pp.

—Agitator—Manholes
DIN ENGL 28136 Pt 2-88. Unalloyed and Stainless Steel Mixing Vessels; Arrangement and Size of Cover Nozzles and Manholes (May). 12 pp.
DIN ENGL 28136 Pt 3-89. Glass-Lined Steel Mixing Vessels; Arrangement and Size of Cover Nozzles and Manholes (Oct) (This Standard, Together with the October 1989 Edition of DIN 28136 Part 11, Supersedes April 1984 Edition). 15 pp.

—Agitator—Nozzles
DIN ENGL 28136 Pt 2-88. Unalloyed and Stainless Steel Mixing Vessels; Arrangement and Size of Cover Nozzles and Manholes (May). 12 pp.
DIN ENGL 28136 Pt 3-89. Glass-Lined Steel Mixing Vessels; Arrangement and Size of Cover Nozzles and Manholes (Oct) (This Standard, Together with the October 1989 Edition of DIN 28136 Part 11, Supersedes April 1984 Edition). 15 pp.

—Agitator—Supports—Brackets—Loads (Forces)
DIN ENGL 28145 Pt 7-87. Welded Attachments for Steel, Glass-Lined Agitator Vessels; Bracket Supports Welded to Jackets; Dimensions, Maximum Permissible Loads (Nov). 9 pp.

—Agitator—Thickness Measurement
DIN ENGL 28136 Pt 4-87. Mixing Vessels; Wall Thicknesses for Unalloyed and Stainless Steel Mixing Vessels (May). 6 pp.

—Aluminum—Coatings
BSI BS 1101-77. 1977 Amd 1 Pressure Containers for Paint and Other Similar Substances. 17 pp.

—Automotive—Lead Acid Batteries
SAA AS D23-70. Containers and Cell Covers for Lead-Acid Batteries of the Automobile Type. 20 pp.

—Batteries
JIS C 2335-91. Containers for Lead-Acid Batteries. 10 pp.

—Blood Specimen
BSI BS 4851-82. 1982 Single Use Labelled Medical Specimen Containers for Haematology and Biochemistry. 7 pp.
BSI BS 6242-82. 1982 Single Use Unlabelled Medical Specimen Containers for Haematology and Biochemistry. 4 pp.

—Blood Transfusion
BSI BS 2463: Part 1-90. 1990 Transfusion Equipment for Medical Use Part 1: Collapsible Containers for Blood and Blood Components. 15 pp.
ISO 4822-81. Single Use Blood Specimen Containers up to 25 ml Capacity First Edition. 6 pp.

Containers (Cont.)
—Brackets—Loads (Forces)
DIN ENGL 28083 Pt 1-87. Bracket Supports; Dimensions and Maximum Loads (Jan). 8 pp.
DIN ENGL 28083 Pt 2-87. Bracket Supports; Maximum Bending Moments Imposed on the Vessel Wall Resulting from Type A Bracket Support Loads (Jan). 23 pp.

—Brake Fluids—Identification Systems
ISO 3871-80. Road Vehicles—Labelling of Containers for Petroleum or Non-Petroleum Base Brake Fluid Third Edition. 3 pp.

—Canned Foods—Corrugated Board
CNS Z5011-88. Corrugated Paperboard Containers for Canned Food (Dec)(1161). 3 pp.

—Chemical Industry—Steel—Welded Joints
DIN ENGL 8558 Pt 2-83. Design and Workmanship of Welded Joints; Steel Vessels and Apparatus Used in the Chemical Industry (Sept). 34 pp.

—Components—Designation
CNS Z5122-83. Method of Designating Component Parts and Points of Containers for Test (Feb)(10034).
JIS Z 0201-89. Method of Designating Component Parts and Points of Containers for Test.

—Composite Materials
BSI BS 1133: SUBSEC 7.7-90. 1990 Packaging Code Section 7: Paper and Board Wrappers, Bags and Containers Subsection 7.7: Composite Materials. 10 pp.

—Compression Testing
JIS Z 0212-87. Method of Compression Test for Packaged Freights and Containers. 8 pp.

—Copper Linings
BSI BS 5624-78. 1978 The Lining of Vessels and Equipment for Chemical Processes: Copper and Copper Alloys. 21 pp.

—Corrosion Prevention
DIN ENGL 50927-85. Planning and Application of Electrochemical Corrosion Protection of Internal Surfaces of Apparatus, Containers and Tubes (Internal Protection) (Aug). 12 pp.

—Corrugated Board
CNS Z5020-88. Corrugated Paperboard Containers for Outer Packaging (Dec)(2354). 11 pp.

—Corrugated Board—Compression Testing
CNS Z6016-72. Method of Test for Compressive Strength of Corrugated Fiberboard Container (Oct)(3511).

—Corrugated Board—Waterproof Adhesives—Adhesion Testing
CPPA D.10U-77. Ply Separation of Combined Container Board. 1 p.

—Fasteners
MOD UK DSTAN 81-73-01. Guide on Use of Captive Fasteners Issue 1; Amendment 1. 34 pp.

—Fiberboard
BSI BS 1133: Sec 7-67. (WITHDRAWN) 1967 Amd 5 Packaging Code Section 7: Paper and Board Wrappers, Bags and Containers Section 7: NB Clauses 1 to 9, 12 to 14, 16 and 17 are Sup'd by 1133 Subsec 7.1, 7.2, 7.3, 7.6: 1986 and 1133 Ch 7.5: 1981 The Remaining Clauses are Obsolete. 47 pp.

—Film Strips
BSI BS 2698-71. 1971 Containers and Notes for Filmstrips. 12 pp.

—Fish—Metal
JIS Z 1608-63. Metal Fish Containers and Freezing Pans.

—Flanges
DIN ENGL 28031-89. Flanges for Welding for Use on Unalloyed and Stainless Steel Vessels Not Subject to Pressure (Feb). 3 pp.

—Food—Aluminum Foil
BSI BS 5439-77. 1977 Aluminium Foil Catering Containers. 8 pp.

—Food—Glass
CNS Z5026-75. Glass Container for Foods (General) (Feb)(2574). 5 pp.

—Food—Metal
CNS S1213-88. Metal Food Container (May)(12325).

INTERNATIONAL AND NON-U.S. NATIONAL STANDARDS
SUBJECT INDEX
Containment

Containers (Cont.)

—Food—Metal—Capacity Measurement
CEN EN 76-78. Packages for Certain Pre-Packed Foodstuffs: Capacities of Glass and Metal Containers. 6 pp.

—Food Preparation Equipment
BSI BS 4874-72. 1972 Catering Container Dimensions. 11 pp.

CEN PREN 631-1-92. Materials and Articles in Contact with Foodstuffs—Catering Containers—Part 1: Dimensions of Containers. 10 pp.

—Food Preparation Equipment—Aluminum
BSI BS 5313-76. 1976 Aluminium Catering Containers and Lids. 8 pp.

—Food Preparation Equipment—Stainless Steel
BSI BS 5312-76. 1976 Stainless Steel Catering Containers and Lids. 8 pp.

—Food—Pressurized—Product Retention
CNS Z6051-82. Determination of Percent Product Retention in Pressurized Food Containers (Aug)(9333).

—Food—PVC
SNZ NZS 7608-87. Plastics Materials for Food Contact Use Amend: A. 20 pp.

—Food Service Equipment
SAA AS 4027-92. Food-Service Container Dimensions. 6 pp.

—Fruits—Corrugated Board
CNS Z5033-88. Corrugated Paperboard Containers for Fresh Fruits (Dec)(3247). 4 pp.

—Fuels—Portable—Metal
SAA AS 2906-91. Fuel Containers—Portable—Plastics and Metal (in Professional Package 47). 13 pp.

—Fuels—Portable—Plastic
SAA AS 2906-91. Fuel Containers—Portable—Plastics and Metal (in Professional Package 47). 13 pp.

—Gasoline
JIS Z 1580-86. Fuels Cans for Handy Type.

—Gasoline—Paints—Exterior
MOD UK DSTAN 80-30-73. Paint, Finishing, Gasoline-Resistant Types \ Brushing Spraying Dipping Issue 1. 9 pp.

—Gasoline—Polyethylene
CEN PREN 227-84. Jerrycans of Polyethylene (PE) with up to 20 Litres Capacity Requirements Test Methods. 27 pp.

—Gasoline—Portable
CSA B376-M1980. Portable Containers for Gasoline and Other Petroleum Fuels (R 1992); (Gen Instr 1 Thru 3). 20 pp.

—Gasoline—Steel
CNS Z5134-86. Steel Gasoline Can (18 Type) (Feb)(11514).

—Glossaries
CNS Z5095-81. Terminology of Various Packages & Containers (Oct)(8070).

—Handles
MOD UK DSTAN 81-29-01. Handles, Webbing or Rope, with Metal Brackets or Wood Blocks for Use on Containers Issue 2; Amendment 1. 16 pp.

—Ice Cream—Waxed Paperboard
BSI BS 4879-73. 1973 Waxed Board for Packaging Ice Cream and Frozen Confectionery. 9 pp.

—Lacquers
MOD UK DEF-1053: METH NO. 81. Standard Methods of Testing Paint, Varnish, Lacquer and Related Products. Indices Method 81: Determination of Lead in 'Lead Free' Paints, Varnishes and Allied Products and Their Containers. (Withdrawn).

—Magnetic Tapes
BSI BS 5550: SUBSEC 5.1.2-93. 1993 Cinematography Part 5: Common to More Than One Film Gauge Section 5.1: Raw Stock Subsection 5.1.2: Specification for the Minimum Information for the Labelling of Containers for Raw Stock Motion-Picture Films and Magnetic Films (ISO 3042: 1992). 12 pp.

ISO 3042-92. Cinematography—Labelling of Containers for Raw-Stock Motion-Picture Films and Magnetic Films—Minimum Information Specifications Third Edition. 9 pp.

Containers (Cont.)

—Manhole Covers
DIN ENGL 28124 Pt 1-89. Manhole Closures for Use on Unalloyed and Stainless Steel Vessels Not Subject to Pressure (Mar). 6 pp.

—Motion Picture Film
BSI BS 5550: SUBSEC 5.1.2-93. 1993 Cinematography Part 5: Common to More Than One Film Gauge Section 5.1: Raw Stock Subsection 5.1.2: Specification for the Minimum Information for the Labelling of Containers for Raw Stock Motion-Picture Films and Magnetic Films (ISO 3042: 1992). 12 pp.

ISO 3042-92. Cinematography—Labelling of Containers for Raw-Stock Motion-Picture Films and Magnetic Films—Minimum Information Specifications Third Edition. 9 pp.

—Paints
BSI BS 3900: Part A15-76. (WITHDRAWN) 1976 Methods of Test for Paints Group A: Tests on Liquid Paints (Excluding Chemical Tests) Part A15: Determination of Quantity of Material in a Container. 1 p.

ISO 3232-74. Paints and Varnishes—Determination of Quantity of Material in a Container First Edition. 3 pp.

MOD UK DEF-1053: METH NO. 81. Standard Methods of Testing Paint, Varnish, Lacquer and Related Products. Indices Method 81: Determination of Lead in 'Lead Free' Paints, Varnishes and Allied Products and Their Containers. (Withdrawn).

—Paperboard
BSI BS 1133: Sec 7-67. (WITHDRAWN) 1967 Amd 5 Packaging Code Section 7: Paper and Board Wrappers, Bags and Containers Section 7: NB Clauses 1 to 9, 12 to 14, 16 and 17 are Sup'd by 1133 Subsec 7.1, 7.2, 7.3, 7.6: 1986 and 1133 Ch 7.5: 1981 The Remaining Clauses are Obsolete. 47 pp.

—Photographic Films
ISO 10214-91. Photography—Processed Photographic Materials—Filing Enclosures for Storage First Edition; (Corrigendum 1-1992). 14 pp.

—Platform
JIS Z 1625-79. Platform Containers for International Trade.

—Radioactive Materials
BSI BS 5288-76. 1976 Sealed Radioactive Sources. 15 pp.

BSI BS 5915-80. 1980 Equipment for Minehead Assay and Sorting Radioactive Ores in Containers. 16 pp.

IEC 600-79. Equipment for Minehead Assay and Shorting Radioactive Ores in Containers First Edition. 23 pp.

ISO 1677-77. Sealed Radioactive Sources—General First Edition. 5 pp.

ISO 2919-80. Sealed Radioactive Sources—Classification First Edition. 12 pp.

—Radioactive Materials—Leakage
ISO 2855-76. Radioactive Materials—Packagings—Tests for Contents Leakage and Radiation Leakage First Edition; (Amendment Slip-1976). 7 pp.

ISO 9978-92. Radiation Protection—Sealed Radioactive Sources—Leakage Test Methods First Edition. 15 pp.

—Repeaters—Temperature—Installation
CCITT FASCICLE III.5-89. Digital Networks, Digital Sections and Digital Line Systems—Recommendations G.801—G.961. 292 pp.

—Rings
DIN ENGL 28084-82. Ring Supports; Dimensions, Maximum Operating Mass (Sept). 10 pp.

DIN ENGL 28084 Suppl. 1-82. Ring Supports; Examples of Calculation (Sept). 4 pp.

—Rubber
MOD UK DSTAN 93-53-88. Butyl Rubber, Calendered Sheet for Flexible Containers Issue 1. 13 pp.

MOD UK TS 10161A. Butyl Rubber, Calendered Sheet for Flexible Containers (Superseded by Def Stan 93-53).

—Saddles (Supports)
DIN ENGL 28080-86. Saddle Supports for Horizontal Vessels; Dimensions (Jan) (Supersedes DIN 28080 Part 1, November 1975 Edition, DIN 28080 Part 2, March 1977 Edition, Supplement 1 to DIN 28080, December 1978 Edition and DIN 28188, March 1971 Edition). 12 pp.

—Safety Matches
SNZ NZS 7204-90. Safety Matches and Containers. Safety Requirements Amend: A, 1990. 6 pp.

Containers (Cont.)

—Stands (Supports)
DIN ENGL 28081 Pt 2-88. Vessel Supports Made of Steel Sections; Dimensions, Material and Design (Jan). 4 pp.

—Stands (Supports)—Chemical Equipment
DIN ENGL 28081 Pt 1-85. Tubular Vessel Supports; Dimensions (June). 4 pp.

—Stands (Supports)—Loads (Forces)
DIN ENGL 28081 Pt 3-85. Type B Tubular Vessel Supports; Maximum Permissible Loads on Tubular Supports for Dished Ends (Sept). 7 pp.

—Steel—Bituminous Coatings
DIN ENGL 30673-86. Bitumen Coatings and Linings for Steel Pipes, Fittings and Vessels (Dec). 6 pp.

—Steel—Coatings
BSI BS 1101-77. 1977 Amd 1 Pressure Containers for Paint and Other Similar Substances. 17 pp.

—Steel—Linings
DIN ENGL 30673-86. Bitumen Coatings and Linings for Steel Pipes, Fittings and Vessels (Dec). 6 pp.

—Supports—Loads (Forces)
DIN ENGL 28081 Pt 4-88. Steel Profile Supports for Vessels; Maximum Permissible Moments on the Vessel Wall Due to Profile Support Loads (Jan). 8 pp.

—Test Specimens—Medical
BSI BS 5213-75. 1975 Medical Specimen Containers for Microbiology. 10 pp.

—Tinplate—Lids—Plastic—Apertures
BSI BS 5567-78. 1978 Apertures in Tinplate Containers to Receive Plug-In Plastics Closures (Supersedes BS 2878: 1976). 2 pp.

—Varnishes
BSI BS 3900: Part A15-76. (WITHDRAWN) 1976 Methods of Test for Paints Group A: Tests on Liquid Paints (Excluding Chemical Tests) Part A15: Determination of Quantity of Material in a Container. 1 p.

ISO 3232-74. Paints and Varnishes—Determination of Quantity of Material in a Container First Edition. 3 pp.

MOD UK DEF-1053: METH NO. 81. Standard Methods of Testing Paint, Varnish, Lacquer and Related Products. Indices Method 81: Determination of Lead in 'Lead Free' Paints, Varnishes and Allied Products and Their Containers. (Withdrawn).

—Vibration
JIS Z 0232-87. Method of Vibration Test for Packaged Freights and Containers.

—Waste Disposal
BSI BS 7320-90. 1990 Sharps Containers. 11 pp.

—Water Resistance Testing
JIS Z 0216-91. Water Spray Test for Packages and Containers. 6 pp.

—Wooden—Fittings
MOD UK DSTAN 81-86-89. Metal Fitting for Wooden Containers Issue 1. 30 pp.

Containment
Scope Note: Use a more specific term
See: Containment Cabinets; Nuclear Reactor Containment Structures; Sealing

Containment Cabinets
Use For: Biological Containment Cabinets
See Also: Cabinets (Furniture); Laboratory Equipment

BSI BS 5726: Part 2-92. 1992 Microbiological Safety Cabinets Part 2: Recommendations for Information to Be Exchanged Between Purchaser, Vendor and Installer and Recommendations for Installation. 15 pp.

BSI BS 5726: Part 3-92. 1992 Microbiological Safety Cabinets Part 3: Specification for Performance After Installation. 8 pp.

BSI BS 5726: Part 4-92. 1992 Microbiological Safety Cabinets Part 4: Recommendations for Selection, Use and Maintenance. 10 pp.

BSI BS 5726: Part 4-01. 1992 Amd 1 Microbiological Safety Cabinets Part 4: Recommendations for Selection, Use and Maintenance (AMD 7784) June 15, 1993 SUPERSEDES BS 5726: 1979. 11 pp.

BSI BS 5726-79. (WITHDRAWN) 1979 Amd 3 Microbiological Safety Cabinets (AMD 6119) March 30, 1990 (Superseded by BS 5726: Parts 1,2,3, & 4: 1992). 21 pp.

CSA CAN/CSA-Z316.3-M87. Biological Containment Cabinets: Installation and Field Testing. 23 pp.

Containment Cabinets (Cont.)

SAA AS 1807.22-89. Cleanrooms, Workstations and Safety Cabinets—Methods of Test—Part 22: Determination of Air Barrier Containment of Laminar Flow Safety Cabinets. 2 pp.

—Air Flow
SAA AS 1807.1-89. Cleanrooms, Workstations and Safety Cabinets—Methods of Test—Part 1: Determination of Air Velocity and Uniformity of Air Velocity in Clean Workstations and Laminar Flow Safety Cabinets. 1 p.

SAA AS 1807.2-89. Cleanrooms, Workstations and Safety Cabinets—Methods of Test—Part 2: Determination of Performance of Clean Workstations and Laminar Flow Safety Cabinets Under Loaded Filter Conditions. 2 pp.

SAA AS 1807.21-89. Cleanrooms, Workstations and Safety Cabinets—Methods of Test—Part 21: Determination of Inward Air Velocity of Class 1 Biological Safety Cabinets. 2 pp.

—Air Permeability
SAA AS 1807.25-90. Cleanrooms, Workstations and Safety Cabinets—Methods of Test—Part 25: Determination of Gas Tightness of Outer Shell of Biological Safety Cabinets. 1 p.

—Construction
BSI BS 5726: Part 1-92. 1992 Microbiological Safety Cabinets Part 1: Specification for Design, Construction and Performance Prior to Installation. 20 pp.

—Design
BSI BS 5726: Part 1-92. 1992 Microbiological Safety Cabinets Part 1: Specification for Design, Construction and Performance Prior to Installation. 20 pp.

—Noise
SAA AS 1807.20-89. Cleanrooms, Workstations and Safety Cabinets—Methods of Test—Part 20: Determination of Sound Level at Installed Workstations and Safety Cabinets. 2 pp.

—Vibration
SAA AS 1807.18-89. Cleanrooms, Workststions and Safety Cabinets—Methods of Test—Part 18: Determination of Vibration in Workstations and Safety Cabinets. 1 p.

Contamination

See Also: Decontamination; Effluents; Impurities Content Analysis; Purity; Radioactive Decontamination Index; Sand

—Air Handling Equipment—Ground Effect Machines
CAA 42.
CAA Chapter B6-6 01.91. Heating and Ventilation Systems. 1 p.

—Aircraft Fuel Systems
CAA NOTICE #21 ISSUE 3. Microbiological Contamination of Fuel Tanks of Turbine Engined Aircraft (Airworthiness Notices). 3 pp.

—Clean Rooms
JIS B 9920-89. Measuring Methods for Airborne Particles in Clean Room and Evaluating Methods for Air Cleanliness of Clean Room. 33 pp.

—Clean Rooms—Clothing
JIS B 9923-86. Methods for Sizing and Counting Particle Contaminants in and on Clean Room Garments.
SAA AS 1807.19-89. Cleanrooms, Workstations and Safety Cabinets—Methods of Test—Part 19: Sizing and Counting of Particulate Contaminants in and on Cleanroom Garments. 5 pp.

—Compressed Air—Classification
ISO 8573 Pt 1-91. Compressed Air for General Use—Part 1: Contaminants and Quality Classes First Edition. 13 pp.

—Connectors
SBAC TS 330 ISSUE 1. Test for Electrical Connectors Dust and Sand Contamination.

—Control—Glossaries
JIS Z 8122-88. Glossary of Terms for Contamination Control. 59 pp.

—Drinking Water
SAA AS 3855(INT)-91. Products for Use in Contact with Water Intended for Human Consumption with Regard to Their Effect on the Quality of Water Amdt 1 May 1992 (Redesignated AS 4020(Int)—1992 by This Amendment) (in Professional Packages 61A, 61B). 1 p.

Contamination (Cont.)

—Drinking Water (Cont.)
SAA AS 4020(INT)-92. Products for Use in Contact with Water Intended for Human Consumption with Regard to Their Effect on the Quality of Water (Replaces AS 3855(Int)-1991).

—Food
CNS N6120-80. Methods of Analysis for Foodstuffs—Detection of Extraneous Matter (May)(5629). 3 pp.
EC EEC/315/93-93. Council Regulation Laying Down Community Procedures for Contaminants in Food. 3 pp.

—Fuel Systems—Aircraft
MOD UK DERD 2153-78. Fuel and Control Systems for Gas Turbine and Ramjet Engines: Rig Tests for Assessing Sensitivity to Abrasion and Blockage by Fuel Borne Solids Issue 4. 13 pp.

—Hydraulic Equipment
BSI BS 7265-90. 1990 Code of Practice for Assessing the Cleanliness of Hydraulic Fluid Power Components. 11 pp.

—Hydraulic Fluids
BSI BS 5540: Part 1-78. 1978 Evaluating Particulate Contamination of Hydraulic Fluids Part 1: Qualifying and Controlling of Cleaning Methods for Sample Containers. 4 pp.
BSI BS 5540: Part 3-78. 1978 Evaluating Particulate Contamination of Hydraulic Fluids Part 3: Method of Bottling Fluid Samples. 4 pp.
BSI BS 5540: Part 4-88. 1988 Evaluating Particulate Contamination of Hydraulic Fluids Part 4: Method of Defining Levels of Contamination (Solid Contaminant Code). 7 pp.
BSI BS 5540: Part 5-87. 1987 Evaluating Particulate Contamination of Hydraulic Fluids Part 5: Method of Reporting Contamination Analysis Data. 10 pp.
BSI BS 5540: Part 8-91. 1991 Evaluating Particulate Contamination of Hydraulic Fluids Part 8: Method of Determining the Level of Contamination (by the Counting Method Using a Microscope) (W) (ISO 4407: 1991). 13 pp.
ISO 3722-76. Hydraulic Fluid Power—Fluid Sample Containers—Qualifying and Controlling Cleaning Methods First Edition. 4 pp.
ISO 3938-86. Hydraulic Fluid Power—Contamination Analysis—Method for Reporting Analysis Data First Edition. 9 pp.
ISO 4406-87. Hydraulic Fluid Power—Fluids—Method for Coding Level of Contamination by Solid Particles First Edition. 7 pp.
ISO 4407-91. Hydraulic Fluid Power—Fluid Contamination—Determination of Particulate Contamination by the Counting Method Using a Microscope First Edition. 10 pp.
ISO 5884-87. Aerospace—Fluid Systems and Components—Methods for System Sampling and Measuring the Solid Particle Contamination of Hydraulic Fluids First Edition. 19 pp.
JIS B 9930-77. Determination of Particulate Contamination in Hydraulic Fluids by the Particle Count Method (R 1986). 22 pp.
JIS B 9931-90. Fluid Contamination-Determination of Contaminants by the Gravimetric Method. 10 pp.

—Hydraulic Fluids—Sampling
ISO 4021-92. Hydraulic Fluid Power—Particulate Contamination Analysis—Extraction of Fluid Samples from Lines of an Operating System Second Edition. 10 pp.
ISO 11217-93. Aerospace—Hydraulic System Fluid Contamination—Location of Sampling Points and Criteria for Sampling First Edition. 33 pp.

—Hydraulic Power Pumps
ISO 9632-92. Hydraulic Fluid Power—Fixed Displacement Pumps—Flow Degradation Due to Classified AC Fine Test Dust Contaminant—Test Method First Edition. 12 pp.

—Hydraulic Systems—Aircraft
SBAC TS 74 ISSUE 1. Code of Practice for Controlling Contamination Introduced into Aircraft Hydraulic Components and Systems During Manufacture and Assembly.

—Land
BSI DD 175-88. 1988 Code of Practice for the Identification of Potentially Contaminated Land and Its Investigation. 28 pp.

—Petroleum Products
DIN ENGL 51419-83. Testing of Liquid Fuels; Determination of Total Contamination in Highly Fluid Petroleum Products (June). 3 pp.

Contamination Meters (Radiation)
Use: Radiation Meters

Contamination Monitors (Radiation)
Use: Radiation Monitors

Continental Quilts
Use: Quilts

Continuity Checks
See Also: Common Channel Signaling Systems

—Speech Paths—CCITT No. 6 Signaling Systems
CCITT RECMN Q.271-89. 1.1 General—Specifications of Signalling System No. 6 (Study Group XI) 3 pp.

Continuity Pilots
See Also: Pilot Channels

—Frequencies—Radio Relay Systems
CCIR RECMN 401-2-70. Frequencies and Deviations of Continuity Pilots for Frequency Modulation Radio-Relay Systems for Television and Telephony—Section 9D—Maintenance. 2 pp.

—Frequency Deviations—Radio Relay Systems
CCIR RECMN 401-2-70. Frequencies and Deviations of Continuity Pilots for Frequency Modulation Radio-Relay Systems for Television and Telephony—Section 9D—Maintenance. 2 pp.

Continuity Testing
See Also: Physical Testing; Testing Equipment

—Coatings—Nonconductive
SAA AS 3894.1-91. Site Testing of Protective Coatings—Part 1: Non-Conductive Coatings—Continuity Testing—High Voltage ('Brush') Method (in Professional Package 39). 16 pp.
SAA AS 3894.2-91. Site Testing of Protective Coatings—Part 2: Non-Conductive Coatings—Continuity Testing—Wet Sponge Method (in Professional Package 39). 5 pp.

—Connectors
CNS C6255-86. Method of Test for Low Frequency (Below 3MHz) Electrical Connectors (TP-46 Continuity Test) (May)(11572).
SBAC TS 342 ISSUE 1. Test for Electrical Connectors Housing (Shell) Electrical Continuity.

—Electric Wiring—Aircraft
AECMA PREN2283-88. Testing of Aircraft Wiring. 9 pp.

—Electromechanical Components
BSI BS 5772: Part 2-79. 1979 Basic Testing Procedures and Measuring Methods for Electromechanical Components for Electronic Equipment Part 2: General Examination, Electrical Continuity and Contact Resistance Tests, Insulation Tests and Voltage Stress Tests. 16 pp.
BSI BS 5772: Pt 2: Supp 1-81. 1981 Electromechanical Components for Electronic Equipment: Basic Testing Procedures Part 2: General Examination, Electrical Continuity and Contract Resistance Tests, Insulation Tests and Voltage Stress Supp 1: 1981 Test 4c. 4 pp.
IEC 512 Pt 2-85. Electromechanical Components for Electronic Equipment; Basic Testing Procedures and Measuring Methods Part 2: General Examination, Electrical Continuity and Contact Resistance Tests, Insulation Tests and Voltage Stress Tests Second Edition. 39 pp.

—Insulated Wire
CNS C3006-80. Testing Standard for Electric Continuance of Insulated Wires (Jul)(684). 1 p.

—Oxide Coatings—Anodic
ISO 2085-76. Anodizing of Aluminium and Its Alloys—Check of Continuity of Thin Anodic Oxide Coatings—Copper Sulphate Test First Edition. 3 pp.

—Plating—Stranded Conductors—Aircraft
CEN PREN 3475 (Part 506)-92. Cables, Electrical, Aircraft Use Test Methods Part 506—Plating Continuity. 2 pp.

Continuous Blenders
Use: Blenders

Continuous Broaching Machines
Use: Broaching Machines

Continuous Current Ratings
Use: Current Ratings

Continuous Dryers
Use: Dryers

Continuous Flow Inhalation Anesthetic Apparatus
Use: Anesthesia Gas Machines

INTERNATIONAL AND NON-U.S. NATIONAL STANDARDS
SUBJECT INDEX

Continuous Forms
See Also: Business Forms; Data Processing Equipment; Forms (Paper); Paper Products
- BSI BS 5537-91. 1991 Forms Design Sheet and Layout Chart. 10 pp.
- CNS P1002-47. Format and Dimensions for Standard Paper (Mar)(90).
- CNS P2075-90. Continuous Computer Forms (Continuous Business Forms) (Dec)(12827). 4 pp.
- DIN ENGL 6720 Pt 1-78. Papers for Teleprinters; Page Printer Papers in Rolls and as Continuous Forms (Feb). 7 pp.
- DIN ENGL 9771-74. Papers for Data Processing; Papers for Continuous Forms; Dimensions (July). 4 pp.
- ISO 3535-77. Forms Design Sheet and Layout Chart First Edition. 6 pp.
- JIS P 4502-78. Basic Paper of Continuous Business Forms.
- JIS X 6195-75. Continuous Forms for Information Processing.
- JTC1 2784-74. Continuous Forms Used for Information Processing—Sizes and Sprocket Feed Holes First Edition; (Amendment-1974). 6 pp.
- JTC1 3535-77. Forms Design Sheet and Layout Chart First Edition. 6 pp.
- MOD UK TS 10139. Paper for Continuous Stationery.
- OSI ISO 3535-77. Forms Design Sheet and Layout Chart. 8 pp.

—Folded—Impact Printers
- BSI BS 4623-89. 1989 Folded Continuous Stationery for Impact Printers. 11 pp.

—Paper Punches
- JIS B 9614-92. Printing Machine—Punching Tools and Perforating for Continuous Business Forms—Classification and Dimensions. 10 pp.

—Printing Papers
- CNS P2073-90. Base Paper for Continuous Business Forms (Jul)(12747). 2 pp.

—Sizes
- CGSB CAN/CGSB-9.62-81. Paper Sizes for Single Part Continuous Business Forms. 13 pp.
- ISO 2784-74. Continuous Forms Used for Information Processing—Sizes and Sprocket Feed Holes First Edition; (Amendment-1974). 6 pp.
- JIS P 0201-66. Sizes of Continuous Business Forms.
- OSI ISO 2784-74. Continuous Forms Used for Information Processing—Sizes and Sprocket Feed Holes. 6 pp.

—Sprocket Punched
- ECMA ECMA-TR 3-72. Continuous Sprocket-Punched Stationery Part II (Physical Properties, Fastenings, Packaging and Storage). 8 pp.
- ECMA ECMA-TR 7-73. Continuous Sprocket Punched Stationery Part I (Recommended Sizes). 10 pp.
- ISO 2784-74. Continuous Forms Used for Information Processing—Sizes and Sprocket Feed Holes First Edition; (Amendment-1974). 6 pp.
- OSI ISO 2784-74. Continuous Forms Used for Information Processing—Sizes and Sprocket Feed Holes. 6 pp.

Continuous Handling Equipment
Use: Conveyors

Continuous Stationery
Use: Continuous Forms

Contraceptive Diaphragms
Use For: Diaphragms (Contraceptive)
- BSI BS 4028-89. 1989 Reusable Contraceptive Diaphragms. 16 pp.
- ISO 8009 Pt 2-85. Reusable Rubber Contraceptive Diaphragms—Part 2: Determination of Size First Edition. 3 pp.
- SAA AS 1808-84. Contraceptive Devices—Diaphragms. 6 pp.

—Aging Testing
- ISO 8009 Pt 6-85. Reusable Rubber Contraceptive Diaphragms—Part 6: Determination of Deterioration After Accelerated Ageing First Edition. 3 pp.

—Compression Testing
- ISO 8009 Pt 7-85. Reusable Rubber Contraceptive Diaphragms—Part 7: Determination of Compression Resistance of Coil Spring and Flat Spring Diaphragms First Edition. 4 pp.
- ISO 8009 Pt 8-85. Reusable Rubber Contraceptive Diaphragms—Part 8: Determination of Twisting During Compression of Coil Spring and Flat Spring Diaphragms First Edition. 4 pp.

—Identification Systems
- ISO 8009 Pt 9-85. Reusable Rubber Contraceptive Diaphragms—Part 9: Packaging and Labelling First Edition. 4 pp.

Contraceptive Diaphragms (Cont.)
—Packaging
- ISO 8009 Pt 9-85. Reusable Rubber Contraceptive Diaphragms—Part 9: Packaging and Labelling First Edition. 4 pp.

—Storage
- ISO 8009 Pt 10-85. Reusable Rubber Contraceptive Diaphragms—Part 10: Recommendations for Storage First Edition. 3 pp.

—Tensile Testing
- ISO 8009 Pt 5-85. Reusable Rubber Contraceptive Diaphragms—Part 5: Determination of Tensile Properties First Edition. 4 pp.

—Thickness Measurement
- ISO 8009 Pt 3-85. Reusable Rubber Contraceptive Diaphragms—Part 3: Determination of Dome Thickness First Edition. 3 pp.

—Visual Inspection
- ISO 8009 Pt 4-85. Reusable Rubber Contraceptive Diaphragms—Part 4: Freedom from Visible Defects First Edition. 5 pp.

Contraceptives
Scope Note: Use a more specific term
See: Condoms; Contraceptive Diaphragms; Drugs; Intrauterine Devices

Contract Alterations
Use: Contract Modifications

Contract Costs
See Also: Contract Pricing; Contracts; Research and Development Costs

—Production Contracts
- MOD UK DEFCON Guide no.4 (PROV)-82. Cost Management of Unpriced Production Contracts 9/82. 4 pp.

Contract Financing
Use For: Contract Payments See Also: Contracts

—Defense Contracts
- MOD UK DEFCON Guide no. 9-86. Financing Terms for Defence Contracts 10/86. 5 pp.

Contract Modifications
Use For: Contract Alterations See Also: Contracts

—Defense Contracts
- MOD UK DEFCON 56-78. Authorisation of Alterations Navy Contracts D of C (SE) 3/78. 2 pp.

Contract Payments
Use: Contract Financing

Contract Pricing
See Also: Contract Costs; Contracts

—Sole Source Contracts
- MOD UK DEFCON Guide no. 3-77. Working Guidelines for the Pricing of Non-Competitive Risk Contracts 1/77. 12 pp.

—Sole Source Contracts—Research and Development Costs
- MOD UK Addendum NO. 1-70. Guidance Document Agreed Between the Government and the Confederation of British Industry on Admission in the Pricing of Non-Competitive Government Contracts of Private Venture Research and Development Expenditure. 5 pp.
- MOD UK Addendum NO. 2-70. Interim Revision of Profit Formula Applicable from 1 March 1975 to Non-Competitive Government Contracts. 6 pp.
- MOD UK Addendum NO. 3-70. Profit Formula Arrangements After 29 February 1976. 1 p.
- MOD UK Addendum NO. 4-70. Revision of Profit Formula Applicable from 1 October 1977 to Non-Competitive Government Contracts. 6 pp.

Contractor Employees
Use For: Temporary Employees
See Also: Contractors; Personnel
- MOD UK DEFCON 76-91. General Conditions of Contract Applicable to Work Performed by Contractor's Personnel at Government Establishments 6/91. 4 pp.

—Administrative Support—Classification
- CGSB CAN/CGSB-168.1-91. Classification and Descriptions for Administrative Support Temporary Help. 27 pp.

—Attendance Certificates
- MOD UK DEFCON 12-87. Certificate of Attendance of Contractor's Personnel 11/87. 2 pp.

Contractor Liability
See Also: Contractors

—Defense Contracts
- MOD UK DEFCON 6-76. Liabilities During Running Trials 2/76. 2 pp.

Contractor Liability, Government Supplied Property
Use: Government Supplied Property—Contractor Liability

Contractors
Use For: Subcontractors See Also: Claims; Contractor Employees; Contractor Liability; Defense Contracts; Services

—Collaboration—Classification
- EURO CTI/89/20337-89. Recommendations for the Operation of a Concession/Production Permit System for Collaborative Projects. 9 pp.

—Collaboration—Permits
- EURO CTI/89/20337-89. Recommendations for the Operation of a Concession/Production Permit System for Collaborative Projects. 9 pp.

—Competitive Business Practices
- MOD UK DEFCON 176B-90. MOD Requirements for Competition in Sub-Contracting (Competitive Main Contract) 9/90. 1 p.
- MOD UK DEFCON Guide no. 8-86. Code of Practice for Competitive Sub-Contracting 5/86. 8 pp.

—Failure of Performance—Defense Contracts
- MOD UK DEFCON 92-90. Failure of Performance 8/90. 1 p.

—Quality Assurance
- MOD UK DSTAN 05-61: Part 3-76. Quality Assurance Procedural Requirements Part 3: Quality Assurance of Sub-Contract Work Issue 1. 4 pp.
- MOD UK. Defence Sub-Contract Arrangements.
- MOD UK DSTAN 05-62: Part 2-76. Guidance on Quality Assurance Procedures Part 2: Guide to Quality Assurance Arrangements for Sub-Contracted Work Issue 1. 3 pp.
- MOD UK. Contractor Assessment.
- MOD UK AQAP-1-84. NATO Requirements for an Industrial Quality Control System Edn 3, 5/84. 21 pp.
- MOD UK AQAP-2-84. Guide for the Evaluation of a Contractor's Quality Control System for Compliance with AQAP-1 Edn 3, 5/84. 45 pp.
- MOD UK AQAP-4-76. NATO Inspection System Requirements for Industry Edn 2, 6/76. 11 pp.
- MOD UK AQAP-5-76. Guide for the Evaluation of a Contractor's Inspection System for Compliance with AQAP-4 Edn 2, 3/76. 33 pp.
- MOD UK AQAP-7-78. Guide for the Evaluation of a Contractor's Measurement and Calibration System for Compliance with AQAP-6 10/78. 39 pp.
- MOD UK AQAP-9-76. NATO Basic Inspection Requirements for Industry Edn 2, 3/76. 8 pp.
- MOD UK AQAP-10-01. NATO Guide for a Government Quality Assurance Programme Edn 2, 9/87; Amendment 1. 22 pp.
- MOD UK AQAP-11-79. NATO Guideline for the Specification of Technical Publications 1/79. 50 pp.
- MOD UK DSTAN 05-126-87. Contractors Repair Supply Procedure Issue 1. 42 pp.

—Software—Quality Assurance
- MOD UK AQAP-13-81. NATO Software Quality Control System Requirements 8/81. 12 pp.
- MOD UK AQAP-14-84. Guide for the Evaluation of a Contractor's Software Quality Control System for Compliance with AQAP-13 5/84. 32 pp.

—Surplus Materiel
- MOD UK DEFCON 44A-77. Schedule of Surpluses with Main Contractor or Sub-Contractors 3/77. 2 pp.
- MOD UK DEFCON 44A CONT.-75. Schedule of Surpluses (Continuation Sheet) 11/75. 1 p.
- MOD UK DEFCON 44C-80. Instructions Relating to Redundant Items/Materials 11/80. 2 pp.

—Surplus Materiel—Disposal
- MOD UK DEFCON 43-78. Memorandum to Contractor on Disposal of Materials, Etc Made Surplus by the Termination, Amendment or Reduction of Ministry of Defence Contracts 9/78. 2 pp.

—Training—Small Value Orders
- MOD UK DEFCON 53F (TU)-84. Small Value Order for Courses of Instruction at Contractors Works 6/84. 2 pp.

Contracts
Scope Note: For additional listing, use a more specific term

INDUSTRY STANDARDS

INTERNATIONAL AND NON-U.S. NATIONAL STANDARDS
SUBJECT INDEX

Contracts (Cont.)
See Also: Construction Contracts; Contract Costs; Contract Financing; Contract Modifications; Contract Pricing; Cost Reimbursement Contracts; Defense Contracts; Fixed Price Contracts; Incentive Contracts; Production Contracts; Repair Contracts; Repair Services Contracts; Research and Development Contracts; Service Contracts; Supply Contracts; Works Contracts

SAA HB42-92. General Conditions of Contract (AS 2124—1992)—User Guide. 146 pp.
SAA MP65-92. Lump Sum Contract with Administration by Proprietor (RAIA ABP-1—1992) Amdt 1 August 1993. 137 pp.

Contrast
See Also: Color; Printing Inks; Resolution Guides

—Microfilm Readers—Screens (Projection)
BSI BS 6354-83. 1983 Measuring the Screen Luminance, Contrast and Reflectance of Microform Readers. 9 pp.

—Projector Lenses—Test Charts
JIS B 7182-70. Brilliancy Test Charts for Projection Lenses (R 1986). 6 pp.

Contrast Ratio
See Also: Covering Power

—Emulsion Paints
DIN ENGL 53778 Pt 3-83. Emulsion Paints; Determination of Contrast Ratio and Lightness of Coatings (Aug). 3 pp.

—Paints
BSI BS 3900: Part D4-74. 1974 Methods of Test for Paints Group D: Optical Tests on Paint Films Part D4: Comparison of Contrast Ratio (Hiding Power) of Paints of the Same Type and Colour. 3 pp.
CGSB 1-GP-71 METH 14.7-82. Methods of Testing Paints and Pigments Hiding Power Contrast Ratio Method. 2 pp.

—Television Pictures—Displays
CCIR RECMN 814-92. Specifications and Alignment Procedures for Setting of Brightness and Contrast of Displays. 4 pp.
CCIR RECMN 815-92. Specification of a Signal for Measurement of the Contrast Ratio of Displays. 2 pp.

Control and Indication Signals
See Also: Control Signals; Signaling Systems

—Audiovisual Services—Transmission Systems
CCITT RECMN H.230-90. Frame-Synchronous Control and Indication Signals for Audiovisual Systems (Study Group XV) 8 pp. 8 pp.

Control and Status Registers

—Architecture
IEC DIS 13213-92. Information Technology—Standard Control and Status Register (CSR) Architecture for Microcomputer Buses ***CD-ROM ONLY***. 134 pp.
ISO DIS 13213-92. Information Technology—Standard Control and Status Register (CSR) Architecture for Microcomputer Buses ***CD-ROM ONLY***. 134 pp.
JTC1 DIS13213-92. Information Technology—Standard Control and Status Register (CSR) Architecture for Microcomputer Buses ***CD-ROM ONLY***. 134 pp.

—Data Buses
IEC DIS 13213-92. Information Technology—Standard Control and Status Register (CSR) Architecture for Microcomputer Buses ***CD-ROM ONLY***. 134 pp.
ISO DIS 13213-92. Information Technology—Standard Control and Status Register (CSR) Architecture for Microcomputer Buses ***CD-ROM ONLY***. 134 pp.
JTC1 DIS13213-92. Information Technology—Standard Control and Status Register (CSR) Architecture for Microcomputer Buses ***CD-ROM ONLY***. 134 pp.

—Glossaries
IEC DIS 13213-92. Information Technology—Standard Control and Status Register (CSR) Architecture for Microcomputer Buses ***CD-ROM ONLY***. 134 pp.
ISO DIS 13213-92. Information Technology—Standard Control and Status Register (CSR) Architecture for Microcomputer Buses ***CD-ROM ONLY***. 134 pp.
JTC1 DIS13213-92. Information Technology—Standard Control and Status Register (CSR) Architecture for Microcomputer Buses ***CD-ROM ONLY***. 134 pp.

Control and Status Registers (Cont.)

—ROM
IEC DIS 13213-92. Information Technology—Standard Control and Status Register (CSR) Architecture for Microcomputer Buses ***CD-ROM ONLY***. 134 pp.
ISO DIS 13213-92. Information Technology—Standard Control and Status Register (CSR) Architecture for Microcomputer Buses ***CD-ROM ONLY***. 134 pp.
JTC1 DIS13213-92. Information Technology—Standard Control and Status Register (CSR) Architecture for Microcomputer Buses ***CD-ROM ONLY***. 134 pp.

Control Boards
Use: Control Panels

Control Cables (Electric)
See Also: Brake Cables; Cables (Electric); Marine Cables (Electric); Tray Cables

JIS C 3401-87. Control Cables. 17 pp.
JIS T 9221-92. Control Cable System Units. 7 pp.

—Dielectric Strength
CSA CAN/CSA-C22. 2 NO 239-M91. Control and Instrumentation Cables; (Gen Instr 1 Thru 3). 50 pp.

—Electrical Codes
CSA 916 Bull. Electrical Bulletin 916 July 3, 1973 to C22.2 NO 16. 2 pp.
CSA 916 Bull. Electrical Bulletin 916 July 3, 1973 to C22.2 NO 35. 2 pp.

—Elevator—300 V
CSA CAN/CSA-C22. 2 NO 49-92. Flexible Cords and Cables; (Gen Instr 1). 121 pp.

—Elevator—600 V
CSA CAN/CSA-C22. 2 NO 49-92. Flexible Cords and Cables; (Gen Instr 1). 121 pp.

—Identification Systems
CSA 1335 Bull. Electrical Bulletin 1335 August 31, 1981 to C22.2 NO 35. 1 p.

—Impact Testing
CSA CAN/CSA-C22. 2 NO 239-M91. Control and Instrumentation Cables; (Gen Instr 1 Thru 3). 50 pp.

—Instrumentation—Multiconductor—150 V
CSA CAN/CSA-C22. 2 NO 239-M91. Control and Instrumentation Cables; (Gen Instr 1 Thru 3). 50 pp.

—Instrumentation—Multiconductor—300 V
CSA CAN/CSA-C22. 2 NO 239-M91. Control and Instrumentation Cables; (Gen Instr 1 Thru 3). 50 pp.

—Instrumentation—Multiconductor—600 V
CSA CAN/CSA-C22. 2 NO 239-M91. Control and Instrumentation Cables; (Gen Instr 1 Thru 3). 50 pp.

—Insulation Resistance
CSA CAN/CSA-C22. 2 NO 239-M91. Control and Instrumentation Cables; (Gen Instr 1 Thru 3). 50 pp.

—Low Temperature Testing
CSA CAN/CSA-C22. 2 NO 239-M91. Control and Instrumentation Cables; (Gen Instr 1 Thru 3). 50 pp.

—Low Voltage
CSA 1334 Bull. Electrical Bulletin 1334 August 28, 1981 to C22.2 NO 16. 1 p.
CSA C22.2 NO 35-M1987. Extra-Low-Voltage Control Circuit Cables, Low-Energy Control Cable, and Extra-Low-Voltage Control Cable (R 1993); (Gen Instr 1). 28 pp.
CSA 1334 Bull. Electrical Bulletin 1334 August 28, 1981 to C22.2 NO 35. 1 p.
CSA 1335 Bull. Electrical Bulletin 1335 August 31, 1981 to C22.2 NO 35. 1 p.

—PVC Insulated—PVC Sheathed
CNS C2064-87. Polyvinyl Chloride Insulated and Sheathed Control Cables (Nov)(4898).

—Sheathed
CSA 1334 Bull. Electrical Bulletin 1334 August 28, 1981 to C22.2 NO 16. 1 p.

—Shielded
CNS C2172-90. Control Cables with Shields (Jun)(12726).
CNS C3208-90. Method of Test for Control Cables with Shields (Jun)(12727).

Control Cables (Electric) (Cont.)

—Spark Testing
CSA CAN/CSA-C22. 2 NO 239-M91. Control and Instrumentation Cables; (Gen Instr 1 Thru 3). 50 pp.

—Thermocouple Extension
CSA CAN/CSA-C22. 2 NO 239-M91. Control and Instrumentation Cables; (Gen Instr 1 Thru 3). 50 pp.

—Thermoset Insulated
CSA 1334 Bull. Electrical Bulletin 1334 August 28, 1981 to C22.2 NO 16. 1 p.

Control Cables (Mechanical)
See Also: Cables (Mechanical); Control Rods; Control Systems Equipment; Power Takeoffs

—Aerospace—End Fittings—Swaged
AECMA PREN2569-89. Control Cable Fittings and Twinbarrels Technical Specification. 13 pp.
BSI BS EN 2569-91. 1991 Aerospace Series—Control Cable Fittings and Turnbarrels—Technical Specification. 18 pp.
CEN EN 2569-91. Control Cable Fittings and Turnbarrels Technical Specification. 15 pp.

—Aerospace—Turnbarrels
AECMA PREN2569-89. Control Cable Fittings and Twinbarrels Technical Specification. 13 pp.
BSI BS EN 2569-91. 1991 Aerospace Series—Control Cable Fittings and Turnbarrels—Technical Specification. 18 pp.
CEN EN 2569-91. Control Cable Fittings and Turnbarrels Technical Specification. 15 pp.

—Aerospace—Wire Rope
ISO 2020-84. Aerospace—Mechanical System Parts—Preformed Flexible Steel Wire Rope for Aircraft Controls—Technical Specification Second Edition. 10 pp.

—Aircraft
BSI SP 33-39-51. 1951 Amd 3 Turnbarrels, Tension Rods and Swaged Cable-End Connections for Aircraft. 23 pp.
BSI SP 101-104-56. 1956 Amd 1 Turnbarrels, Tension Rods and Swaged Cable-End Connections (Unified Threads) for Aircraft. 12 pp.

—Aircraft—Assemblies
AECMA PREN2348-86. Control Cable Assemblies —Technical Specification. 10 pp.
AECMA PREN2641-86. Control Cable Assemblies —Combinations and Dimensions. 4 pp.
BSI BS EN 2348-89. 1989 Control Cable Assemblies. Technical Specification. 13 pp.
BSI BS EN 2641-89. 1989 Control Cable Assemblies. Combinations and Dimensions. 8 pp.
CEN PREN 2348-87. Control Cable Assemblies Technical Specification. 10 pp.
CEN EN 2348-88. Control Cable Assemblies Technical Specification. 10 pp.
CEN EN 2641-88. Control Cable Assemblies Combinations and Dimensions. 9 pp.

—Aircraft—Ball Ends
AECMA PREN2361-85. Ball-Ends, Double Shank in Corrosion Resisting Steel Swaged on Type, Control Cable—Dimensions and Loads. 4 pp.
AECMA PREN2362-85. Ball-Ends in Corrosion Resisting Steel Swaged on Type, Control Cable—Dimensions and Loads (C1/20-04). 4 pp.
BSI BS EN 2361-89. 1989 Ball-Ends, Double Shank in Corrosion Resisting Steel Swaged on Type, Control Cable. Dimensions and Loads. 7 pp.
BSI BS EN 2362-89. 1989 Ball-Ends in Corrosion Resisting Steel Swaged on Type, Control Cable. Dimensions and Loads. 7 pp.
CEN EN 2361-88. Ball-Ends, Double Shank in Corrosion Resisting Steel Swayed on Type, Control Cable Dimensions and Loads. 4 pp.
CEN EN 2362-88. Ball-Ends in Corrosion Resisting Steel Swaged on Type, Control Cable Dimensions and Loads. 6 pp.

—Aircraft—Eye Fasteners
AECMA PREN2358-85. Eye-Ends in Corrosion Resisting Steel Swaged on Type, Control Cable—Dimensions and Loads. 6 pp.
BSI BS EN 2358-89. 1989 Eye-Ends in Corrosion Resisting Steel Swaged on Type, Control Cable. Dimensions and Loads. 9 pp.
CEN EN 2358-88. Eye-Ends in Corrosion Resisting Steel Swaged on Type, Control Cable Dimensions and Loads. 6 pp.

—Aircraft—Fork Ends
AECMA PREN2360-85. Fork-Ends for Rolling Bearings in Corrosion Resisting Steel Swaged on Type, Control Cable—Dimensions and Loads. 6 pp.
BSI BS EN 2360-89. 1989 Fork-Ends for Rolling Bearings in Corrosion Resisting Steel Swaged on Type, Control Cable. Dimensions and Loads. 9 pp.

INTERNATIONAL AND NON-U.S. NATIONAL STANDARDS
SUBJECT INDEX

Control

Control Cables (Mechanical) (Cont.)

—Aircraft—Fork Ends (Cont.)
CEN EN 2360-88. Fork-Ends for Rolling Bearings in Corrosion Resisting Steel Swaged on Type, Control Cable Dimensions and Loads. 8 pp.

—Aircraft—Fork Ends—Fittings
AECMA PREN2359-85. Fork-Ends in Corrosion Resisting Steel Swaged on Type, Control Cable—Dimensions and Loads. 6 pp.
BSI BS EN 2359-89. 1989 Fork-Ends in Corrosion Resisting Steel Swaged on Type, Control Cable. Dimensions and Loads. 9 pp.
CEN EN 2359-88. Fork-Ends in Corrosion Resisting Steel Swaged on Type, Control Cable Dimensions and Loads. 6 pp.

—Aircraft—Pulleys
AECMA PREN2062-79. Fully Non-Metallic Body Pulleys with Bearing for Control Cables—Technical Specification (C1/22-01). 19 pp.
AECMA PREN2081-77. Non-Metallic Bodied Pulleys with Ball Bearing—for Control Cables—Dimensions and Loads (C1/22-02). 5 pp.
AECMA PREN2496-80. Bearing, Ball for Control Cable pulleys—Dimensions and Load (C1/22). 5 pp.
AECMA PREN3182-89. Ball Bearings, Rigid in Corrosion Resisting Steel Cadmium Plated, for Control Cable Pulleys Dimensions and Loads. 5 pp.
AECMA PREN3629-89. Ball Bearings, Rigid in Corrosion Resisting Steel for Control Cable Pulleys Dimensions and Loads. 5 pp.
CNS C2064-87. Polyvinyl Chloride Insulated and Sheathed Control Cables (Nov)(4898).
ISO 7938-86. Aircraft—Ball Bearings for Control Cable Pulleys—Dimensions and Loads First Edition. 6 pp.
ISO 7939-88. Aircraft—Non-Metallic Pulleys with Ball Bearings for Control Cables—Dimensions and Loads First Edition. 6 pp.

—Aircraft—Steel Wire Rope
AECMA PREN2080-85. Flexible Wire Ropes for Aircraft Controls in Corrosion Resisting Steel—Construction—Dimensions—Loads (C1/19-03). 4 pp.
AECMA PREN2160-85. Flexible Wire Ropes for Aircraft Controls in Zinc Coated Carbon Steel—Construction—Dimensions—Loads (C1/19-03). 4 pp.

—Aircraft—Stud Ends
AECMA PREN2357-86. Stud-Ends in Corrosion Resisting Steel Swaged on Type, Control Cable; Dimensions and Loads. 6 pp.
BSI BS EN 2357-89. 1989 Stud-Ends in Corrosion Resisting Steel Swaged on Type, Control Cable. 9 pp.
CEN EN 2357-88. Stud-Ends in Corrosion Resisting Steel Swaged on Type, Control Cable Dimensions and Loads. 8 pp.

—Aircraft—Tie Rods
BSI SP 33-39-51. 1951 Amd 3 Turnbarrels, Tension Rods and Swaged Cable-End Connections for Aircraft. 23 pp.
BSI SP 101-104-56. 1956 Amd 1 Turnbarrels, Tension Rods and Swaged Cable-End Connections (Unified Threads) for Aircraft. 12 pp.

—Aircraft—Turnbarrels
AECMA PREN2353-85. Turnbarrels, Control Cable in Corrosion Resisting Steel—Dimensions and Loads (C1/20-02). 6 pp.
AECMA PREN2609-85. Turnbarrels, Control Cable in Copper-Zinc Alloys—Dimensions and Loads (C1/20-02). 6 pp.
BSI BS EN 2353-89. 1989 Turnbarrels, Control Cable in Corrosion Resisting Steel. Dimensions and Loads. 9 pp.
BSI BS EN 2609-89. 1989 Turnbarrels, Control Cable in Copper-Zinc Alloys Dimensions and Loads. 9 pp.
BSI SP 33-39-51. 1951 Amd 3 Turnbarrels, Tension Rods and Swaged Cable-End Connections for Aircraft. 23 pp.
BSI SP 101-104-56. 1956 Amd 1 Turnbarrels, Tension Rods and Swaged Cable-End Connections (Unified Threads) for Aircraft. 12 pp.
CEN EN 2353-88. Turnbarrels, Control Cable in Corrosion Resisting Steel Dimensions and Loads. 6 pp.
CEN EN 2609-88. Turnbarrels, Control Cable in Copper—Zinc Alloys Dimensions and Loads. 8 pp.

—Aircraft—Turnbuckles—Eye Fasteners
AECMA PREN2354-85. Eye-Ends, Threaded, Control Cable in Corrosion Resisting Steel—Dimensions and Loads (C1/20-02). 6 pp.
AECMA PREN2355-86. Fork-Ends, Threaded, Control Cable in Corrosion Resisting Steel; Dimensions and Loads. 6 pp.
BSI BS EN 2354-89. 1989 Eye-Ends, Threaded, Control Cable in Corrosion Resisting Steel. Dimensions and Loads. 9 pp.

Control Cables (Mechanical) (Cont.)

—Aircraft—Turnbuckles—Eye Fasteners (Cont.)
BSI BS EN 2355-89. 1989 Fork-Ends, Threaded, Control Cable in Corrosion Resisting Steel. Dimensions and Loads. 9 pp.
CEN EN 2354-88. Eye-Ends, Threaded, Control Cable in Corrosion Resisting Steel Dimensions and Loads. 8 pp.
CEN EN 2355-88. Fork-Ends, Threaded, Control Cable in Corrosion Resisting Steel Dimensions and Loads. 8 pp.

—Aircraft—Turnbuckles—Fork Ends
AECMA PREN2356-86. Fork-Ends Threaded, Control Cable for Rolling Bearings in Corrosion Resisting Steel; Dimensions and Loads. 6 pp.
BSI BS EN 2356-89. 1989 Fork-Ends Threaded, Control Cable for Rolling Bearings in Corrosion Resisting Steel. Dimensions and Loads. 9 pp.
CEN EN 2356-88. Fork-Ends Threaded, Control Cable for Rolling Bearings in Corrosion Resisting Steel Dimensions and Loads. 8 pp.

—Aircraft—Turnbuckles—Locking Clips
AECMA PREN2363-81. Locking Clips for Turnbuckles of Control Cables—Dimensions. 4 pp.
BSI BS EN 2363-89. 1989 Locking Clips for Turnbuckles of Control Cables. Dimensions. 7 pp.
CEN EN 2363-89. Locking Clips for Turnbuckles of Control Cables Dimensions. 7 pp.

—Aircraft—Wire Rope
BSI 2W 11-67. 1967 Preformed Corrosion-Resisting Steel Wire Rope. 6 pp.

—Automobiles
CNS D2124-85. Control Cables for Automobiles (Feb)(8482).
CNS D3104-85. Method of Test for Control Cables for Automobiles (Feb)(8484).

—Motorcycles
CNS D2125-85. Control Cables for Motorcycle (Feb)(8483).
CNS D3105-85. Method of Test Control Cables for Motorcycles (Feb)(8485).

—Wire Rope
CNS G3207-83. Wire Ropes for Mechanical Control (Feb)(10000).
JIS G 3540-88. Wire Ropes for Mechanical Control. 14 pp.

Control Characters

See Also: Character Sets; Escape Characters; Information Interchange; Keyboards; Symbols

—Bibliographic References
ISO 6630-86. Documentation—Bibliographic Control Characters First Edition. 8 pp.
JTC1 6630-86. Documentation—Bibliographic Control Characters First Edition. 8 pp.
OSI ISO 6630-86. Documentation—Bibliographic Control Characters. 8 pp.

—Japanese Character Sets
JIS X 0207-79. Code of the Control Character Set for Japanese Graphic Characters for Information Interchange.
JIS X 0208-90. Code of the Japanese Graphic Character Set for Information Interchange. 83 pp.

—Symbols
JIS X 0209-76. Graphical Representations of Control Characters for Information Interchange.

—7-Bit Character Sets
CNS C5107-83. Information Processing — Graphical Representations for the Control Characters of the 7 Bit Coded Character Set (Feb)(7223).
ISO 2047-75. Information Processing—Graphical Representations for the Control Characters of the 7-Bit Coded Character Set First Edition. 8 pp.
JTC1 2047-75. Information Processing—Graphical Representations for the Control Characters of the 7-Bit Coded Character Set First Edition. 8 pp.
OSI ISO 2047-75. Information Processing—Graphical Representations for the Control Characters of the 7-Bit Coded Character Set. 8 pp.

Control Charts

See Also: Charts; Cusum Charts; Inventory Control Systems; Process Charts; Quality Assurance; Quality Control

BSI BS 2564-55. 1955 Amd 1 Control Chart Technique When Manufacturing to a Specification, with Special Reference to Articles Machined to Dimensional Tolerances. 81 pp.
BSI BS 5701-80. 1980 Amd 1 Guide to Number-Defective Charts for Quality Control. 23 pp.
BSI BS 6002-79. 1979 Sampling Procedures and Charts for Inspection by Variables for Percent Defective. 95 pp.

Control Charts (Cont.)

CNS Z4003-74. Control Chart Method of Analyzing Data (Mar)(2312). 30 pp.
CNS Z4005-74. Control Chart Method of Controlling Quality During Production (Mar)(2580). 44 pp.
CNS Z4024-83. X Control Chart (May)(10301).
JIS Z 9021-54. Control Chart Method.
JIS Z 9022-59. Median Control Chart.
JIS Z 9023-63. X Control Chart. 17 pp.

—Shewhart—Statistical Process Control
ISO 8258-91. Shewhart Control Charts First Edition; (Corrigendum 1-1993). 37 pp.
SNZ NZS/AS 3944-93. Shewhart Control Charts (ISO 8258:1991) (This is a Joint Standard with SAA AS 3944). 29 pp.

Control Circuit Cables
Use: Control Cables (Electric)

Control Circuits

See Also: Circuits; Control Systems Equipment; Dimmers

—Delay Timers
JIS C 4551-81. Motor Driven Timer. 27 pp.

—Electromagnetic Relays
JIS C 5440-80. General Rules for Reliability Assured Lowpower Electromagnetic Relays for Industrial Control Circuits.
JIS C 5442-86. Test Methods of Low Power Electromagnetic Relays for Industrial Control Circuits. 70 pp.

—Lighting
CSA C22.2 NO 184-M1988. Solid-State Lighting Controls; (Gen Instr 1 Thru 3). 33 pp.

—Low Voltage
SAA AS 1431.1-89. Low Voltage Switchgear and Controlgear—Control Circuit Devices and Switching Elements —Part 1: General Requirements (This is a Joint Standard with SNZ NZS 1431.1). 27 pp.
SNZ NZS/AS 1431. 1-89. Low Voltage Switchgear and Controlgear—Control Circuit Devices and Switching Elements Part 1: General Requirements (This is a Joint Standard with SAA AS 1431.1). 27 pp.

—Mining Equipment
BSI BS 3101-86. 1986 Control and Interlock Circuits Primarily Associated with Flameproof Restrained Plugs and Sockets for Use in Coal Mines. 10 pp.

—Photoswitches
JIS C 4525-92. Photoelectric Switches. 33 pp.

—1000 V AC
BSI BS EN 60947-5-1-01. 1992 Amd 1 Low-Voltage Switchgear and Controlgear Part 5: Control Circuit Devices and Switching Elements Section 1: Electromechanical Control Circuit Devices (AMD 7338) December 15, 1992 (E). 75 pp.
BSI BS EN 60947-5-1-02. 1992 Amd 2 Low-Voltage Switchgear and Controlgear Part 5: Control Circuit Devices and Switching Elements Section 1: Electromechanical Control Circuit Devices (AMD 7850) July 15, 1993 (E). 76 pp.
BSI BS EN 60947-5-1-92. 1992 Low-Voltage Switchgear and Controlgear Part 5: Control Circuit Devices and Switching Elements Section 1: Electromechanical Control Circuit Devices. 74 pp.
CENELEC EN 60947-5-1-91. Low-Voltage Switchgear and Controlgear Part 5: Control Circuit Devices and Switching Elements Section One—Electromechanical Control Circuit Devices. 6 pp.
CENELEC EN 60947-5-1-92. CORRIGENDUM Low-Voltage Switchgear and Controlgear Part 5: Control Circuit Devices and Switching Elements Section One—Electromechanical Control Circuit Devices. 1 p.
DIN VDE 0660 Pt 200 A1 (D)-87. Switchgear and Controlgear; Low-Voltage Switchgear and Controlgear; Control Circuit Devices and Switching Elements (Aug). 15 pp.
DIN VDE 0660 Pt 200 A2 (D)-89. Switchgear and Controlgear; Low-Voltage Switchgear and Controlgear; Control Circuit Devices and Switching Elements; Amendment 2 (Aug). 9 pp.
IEC 947 Pt 5-1-90. Low-Voltage Switchgear and Controlgear Part 5: Control Circuit Devices and Switching Elements Section One—Electromechanical Control Circuit Devices First Edition. 131 pp.

Control Codes

See Also: Codes; Control Functions; End of Address; Numerical Control Codes

—Audiovisual Equipment
BSI BS 5817: Sec 5.1-81. 1981 Amd 1 Audio-Visual, Video and Television Equipment and Systems Part 5: Control, Synchronization and Address Codes Section 5.1: Synchronize Tape/Visual Operating Practice (AMD 6286) December 21, 1990. 10 pp.

Control Codes (Cont.)

—Audiovisual Equipment (Cont.)
BSI BS 5817: Sec 5.2-84. 1984 Amd 1 Audio-Visual, Video and Television Equipment and Systems Part 5: Control, Synchronization and Address Codes Sec 5.2: Method of Operation for Pulsed Cue Tone Control Systems for Two Still Projectors (AMD 6492) January 31, 1991. 12 pp.

CENELEC HD 369.5-87. Audio-Visual, Video and Television Equipment and Systems Part 5: Control, Synchronization and Address Codes Chapter 1: Synchronized Tape/Visual Operating Practice. 2 pp.

IEC 574 Pt 5-80. Audio-Visual, Video and Television Equipment and Systems Part 5: Control, Synchronization and Address Codes Chapter I: Synchronized Tape/Visual Operating Practice First Edition. 18 pp.

IEC 574 Pt 5-2-83. Audio-Visual, Video and Television Equipment and Systems Part 5: Control, Synchronization and Address Codes Chapter II: Control Systems for Two Still Projectors Operating Practice First Edition. 21 pp.

—Magnetic Tapes—Television Recording
CCIR Report 963-1-86. Time and Control Code for Television Recordings on Magnetic Tape—Section 10/11I—Utilization and Synchronization of Different Programme Supports. 1 p.

CCIR QUESTION 112/11-90. Recording of Time and Control Code Information on Magnetic Tapes for Television—Questions Concerning Study Group 11—Broadcasting Service (Television). 1 p.

—Television Equipment
BSI BS 5817: Sec 5.1-81. 1981 Amd 1 Audio-Visual, Video and Television Equipment and Systems Part 5: Control, Synchronization and Address Codes Section 5.1: Synchronize Tape/Visual Operating Practice (AMD 6286) December 21, 1990. 10 pp.

BSI BS 5817: Sec 5.2-84. 1984 Amd 1 Audio-Visual, Video and Television Equipment and Systems Part 5: Control, Synchronization and Address Codes Sec 5.2: Method of Operation for Pulsed Cue Tone Control Systems for Two Still Projectors (AMD 6492) January 31, 1991. 12 pp.

CENELEC HD 369.5-87. Audio-Visual, Video and Television Equipment and Systems Part 5: Control, Synchronization and Address Codes Chapter 1: Synchronized Tape/Visual Operating Practice. 2 pp.

CENELEC HD 369.5.2-86. Audio-Visual, Video and Television Equipment and Systems Part 5: Control Synchronization and Address Codes Chapter II: Control Systems for Two Still Projectors Operating Practice. 2 pp.

IEC 574 Pt 5-80. Audio-Visual, Video and Television Equipment and Systems Part 5: Control, Synchronization and Address Codes Chapter I: Synchronized Tape/Visual Operating Practice First Edition. 18 pp.

IEC 574 Pt 5-2-83. Audio-Visual, Video and Television Equipment and Systems Part 5: Control, Synchronization and Address Codes Chapter II: Control Systems for Two Still Projectors Operating Practice First Edition. 21 pp.

—Text Communication
BSI BS ISO/IEC 10538-91. 1991 Information Technology—Control Functions for Text Communication. 63 pp.

IEC 10538-91. Information Technology—Control Functions for Text Communication First Edition. 60 pp.

ISO 10538-91. Information Technology—Control Functions for Text Communication First Edition. 60 pp.

JTC1 10538-91. Information Technology—Control Functions for Text Communication First Edition. 60 pp.

—Video Equipment
BSI BS 5817: Sec 5.1-81. 1981 Amd 1 Audio-Visual, Video and Television Equipment and Systems Part 5: Control, Synchronization and Address Codes Section 5.1: Synchronize Tape/Visual Operating Practice (AMD 6286) December 21, 1990. 10 pp.

BSI BS 5817: Sec 5.2-84. 1984 Amd 1 Audio-Visual, Video and Television Equipment and Systems Part 5: Control, Synchronization and Address Codes Sec 5.2: Method of Operation for Pulsed Cue Tone Control Systems for Two Still Projectors (AMD 6492) January 31, 1991. 12 pp.

CENELEC HD 369.5-87. Audio-Visual, Video and Television Equipment and Systems Part 5: Control, Synchronization and Address Codes Chapter 1: Synchronized Tape/Visual Operating Practice. 2 pp.

CENELEC HD 369.5.2-86. Audio-Visual, Video and Television Equipment and Systems Part 5: Control, Synchronization and Address Codes Chapter II: Control Systems for Two Still Projectors Operating Practice. 2 pp.

Control Codes (Cont.)

—Video Equipment (Cont.)
IEC 574 Pt 5-80. Audio-Visual, Video and Television Equipment and Systems Part 5: Control, Synchronization and Address Codes Chapter I: Synchronized Tape/Visual Operating Practice First Edition. 18 pp.

IEC 574 Pt 5-2-83. Audio-Visual, Video and Television Equipment and Systems Part 5: Control, Synchronization and Address Codes Chapter II: Control Systems for Two Still Projectors Operating Practice First Edition. 21 pp.

—Video Tape Recorders
BSI BS 6865-87. 1987 Amd 1 Time and Control Codes for Video Tape Recorders (AMD 6418) August 31, 1990. 33 pp.

CENELEC HD 507 S1-88. Time and Control Code for Video Tape Recorders. 3 pp.

IEC 461-86. Time and Control Code for Video Tape Recorders Second Edition. 65 pp.

—Video Tape Recording—Television Program Exchange
CCIR RECMN 780-92. Time and Control Code Standards for the International Exchange of Television Programmes on Magnetic Tapes. 1 p.

Control Columns
See Also: Columns (Supports)

—Aircraft—Assembly
SBAC AS 2254 ISSUE 9. Control Column Assembly. 1 p.

SBAC AS 3293 ISSUE 2. G.A. and Assembly of Control Column. 1 p.

SBAC AS 6248 ISSUE 1. GA and Assembly of Control Column. 1 p.

SBAC AS 6249 ISSUE 1. Control Column Assembly. 1 p.

—Aircraft—Ball Race Spigots
SBAC AS 6251 ISSUE 1. Front Ball Race, Spigot Control Column. 1 p.

—Aircraft—Ball Races
SBAC AS 867 ISSUE 3. Ball Race and Housing—Control Column. 1 p.

—Aircraft—Bolts
SBAC AS 2292 ISSUE 6. Bolt 3/8 Inch Dia-Control Column. 1 p.

SBAC AS 2293 ISSUE 6. Bolt 5/16 Inch Dia-Control Column. 1 p.

SBAC AS 6252 ISSUE 2. Adjustable Stop, Control Column. 1 p.

SBAC AS 6255 ISSUE 2. Bolt 5/16 Inch Diameter (Unified) Control Column. 1 p.

—Aircraft—Bushings
SBAC AS 860 ISSUE 6. Combined Dust Washer and Bush Control Column. 1 p.

SBAC AS 861 ISSUE 4. Distance Piece—Control Column. 1 p.

—Aircraft—Distance Pieces
SBAC AS 862 ISSUE 7. Distance Piece—Control Column. 1 p.

—Aircraft—Housings
SBAC AS 852 ISSUE 11. Housing—Control Column. 1 p.

SBAC AS 867 ISSUE 3. Ball Race and Housing—Control Column. 1 p.

SBAC AS 5365 ISSUE 1. Forging for Control Column Housing. 1 p.

SBAC AS 5366 ISSUE 1. Forging for Control Column Housing. 1 p.

SBAC AS 6250 ISSUE 1. Housing—Control Column. 1 p.

—Aircraft—Housings—Covers
SBAC AS 866 ISSUE 4. Cover for Control Column Housing. 1 p.

—Aircraft—Plug Ends
SBAC AS 2256 ISSUE 5. Bottom Plug End—Control Column. 1 p.

—Aircraft—Shims
SBAC AS 869 ISSUE 4. Shim Column Control. 1 p.

—Aircraft—Stops
SBAC AS 2261 ISSUE 8. Adjustable Stop, Control Column. 1 p.

SBAC AS 6253 ISSUE 2. Sleeve for Adjustable Stop Control Column. 1 p.

—Aircraft—Stops—Rubber Pads
SBAC AS 4960 ISSUE 1. Rubber, Pad—Control Stops. 1 p.

—Aircraft—Stops—Sleeves
SBAC AS 2260 ISSUE 3. Sleeve for Adjustable Stops Control Column. 1 p.

Control Columns (Cont.)

—Aircraft—Stops—Sleeves (Cont.)
SBAC AS 6254 ISSUE 1. Sleeve for Adjustable Stop, Control Column. 1 p.

—Aircraft—Tubes
SBAC AS 2255 ISSUE 8. Control Column Tube. 1 p.

—Aircraft—Tubes—Plug Ends
SBAC AS 2257 ISSUE 4. Top Plug End for Control Column: Tube. 1 p.

—Aircraft—Tubes—Sleeves
SBAC AS 2239 ISSUE 5. Sleeve for Control Column Tube. 1 p.

—Aircraft—Washers
SBAC AS 860 ISSUE 6. Combined Dust Washer and Bush Control Column. 1 p.

SBAC AS 5362 ISSUE 2. Washer—Front Spigot—Control Column. 1 p.

Control Consoles
See Also: Consoles; Control Panels; Control Systems Equipment

—Marine Engines
CNS F5025-82. Outline Dimensions of Marine Engine Remote Control Consoles (Mar)(8580).

CNS F5026-82. Internal Wirings of Marine Engine Remote Control Consoles (Mar)(8581).

JIS F 0411-87. Outline Dimensions of Marine Engine Control Consoles.

JIS F 0414-87. Internal Wirings and Pipings of Marine Engine Control Console.

—Marine Engines—Pipes
CNS F5027-82. Internal Piping of Marine Engine Remote Control Consoles (Mar)(8582).

JIS F 0414-87. Internal Wirings and Pipings of Marine Engine Control Console.

—Marine Engines—Positioning
JIS F 0415-89. Arrangement of Main Engine Control Stand in Wheelhouse and Local Alarm Panel for Small Shi

—Wheelhouses—Positioning
JIS F 0415-89. Arrangement of Main Engine Control Stand in Wheelhouse and Local Alarm Panel for Small Shi

Control Desks
Use: Control Consoles

Control Devices
Use: Control Systems Equipment

Control Engineering
See Also: Control Systems Equipment

—Glossaries
DIN ENGL 19226-68. Control Engineering; Definitions and Terms (May). 33 pp.

Control Functions
See Also: Control Codes; Data Processing
JIS X 0211-91. Information Processing—Control Functions for 7-Bit and 8-Bit Coded Character Sets.

—Text Communication
BSI BS ISO/IEC 10538-91. 1991 Information Technology—Control Functions for Text Communication. 63 pp.

IEC 10538-91. Information Technology—Control Functions for Text Communication First Edition. 60 pp.

ISO 10538-91. Information Technology—Control Functions for Text Communication First Edition. 60 pp.

JTC1 10538-91. Information Technology—Control Functions for Text Communication First Edition. 60 pp.

Control Keys
Use: Function Keys

Control Knobs
Use: Knobs; Manual Controls

Control Panels
Use For: Control Boards; Monitor Panels; Regulating Panels *See Also:* Control Consoles; Control Systems Equipment; Electrical Enclosures; Panel Boards (Electrical)

—Ships
MOD UK NES 533-81. Requirements for Control Boards for Main Electrical Supply Systems Issue 2 (01.81) (Superseded by NES 530). 35 pp.

Control Pedals
Use: Manual Controls

INTERNATIONAL AND NON-U.S. NATIONAL STANDARDS
SUBJECT INDEX
Control

Control Procedures
See Also: Basic Mode Control Procedures; Information Interchange; Link Control Procedures; Telephone Exchanges

—FAX Communications—Teletex
CCITT RECMN T.62-89. Control Procedures for Teletex and Group 4 Facsimile Services—Terminal Equipment and Protocols for Telematic Services (Study Group VIII) 180 pp. 180 pp.

—FAX Communications—Teletex—X.215
CCITT RECMN T.62 BIS-89. Control Procedures for Teletex and G4 Facsimile Services Based on Recommendations X.215 and X.225—Terminal Equipment and Protocols for Telematic Services (Study Group VIII) 32 pp. 32 pp.

—FAX Communications—Teletex—X.225
CCITT RECMN T.62 BIS-89. Control Procedures for Teletex and G4 Facsimile Services Based on Recommendations X.215 and X.225—Terminal Equipment and Protocols for Telematic Services (Study Group VIII) 32 pp. 32 pp.

Control Rods
Use For: Flight Control Rods *See Also:* Control Cables (Mechanical)

—Aerospace
AECMA PREN2289-90. Rod Bodies, Flight Controls in Aluminium Alloys Technical Specification. 11 pp.
AECMA PREN2290-81. Rod Bodies Flight Control in Aluminium Alloys for Adjustable End Fittings—Dimensions. 7 pp.
CEN PREN 2544-91. Rod Assemblies for Flight Controls Technical Specifications. 10 pp.

—Aerospace Washers
AECMA PREN2328-80. Washers Tab for Flight Control Rods—Dimensions (C1/23-3). 5 pp.
BSI BS EN 2328-90. 1990 Washers, Tab for Flight Control Rods. Dimensions. 7 pp.
CEN EN 2328-88. Washers, Tab for Flight Control Rods Dimensions. 7 pp.

—Assemblies—Aerospace
CEN PREN 2644-92. Rod Assemblies for Flight Controls Technical Specification. 10 pp.

—Lock Washers—Aerospace
AECMA PREN2327-87. Washers, Lock with Radial Serrations in Alloy Steel Dimensions.
AECMA PREN2546-81. Washers, Lock, with Radial Serrations, in Corrosion Resisting Steel—Dimensions. 5 pp.
AECMA PREN2586-86. Washers, Lock for Flight Control Rods Dimensions. 3 pp.
AECMA PREN2596-84. Washers, Lock, with Radial Serrations, in Corrosion Resisting Steel, Cadmium Plated Dimensions. 5 pp.
BSI BS EN 2327-89. 1989 Washers, Lock with Radial Serrations in Alloy Steel. Dimensions. 8 pp.
BSI BS EN 2546-90. 1990 Washers, Lock with Radial Serrations in Corrosion Resisting Steel. Dimensions. 8 pp.
BSI BS EN 2586-91. 1991 Washers, Lock for Flight Control Rods—Dimensions. 9 pp.
BSI BS EN 2596-90. 1990 Washers, Lock, with Radial Serrations, in Corrosion Resisting Steel, Cadmium Plated. Dimensions. 8 pp.
CEN EN 2327-87. Washers, Lock with Radial Serrations in Alloy Steel Dimensions. 5 pp.
CEN PREN 2546-87. Washers, Lock with Radial Serrations in Corrosion Resisting Still Dimensions. 5 pp.
CEN EN 2546-88. Washers, Lock with Radial Serrations in Corrosions Resisting Steel Dimensions. 5 pp.
CEN EN 2586-91. Washers, Lock for Flight Control Rods—Dimensions. 5 pp.
CEN EN 2596-88. Washers, Lock, with Radial Serrations in Corrosion Resisting Steel, Cadmium Plated Dimensions. 7 pp.

—Rod Ends—Aerospace
AECMA PREN2067-89. Rod-Ends with Self-Aligning Ball Bearings Technical Specification. 18 pp.
AECMA PREN2068-89. Rod-Ends with Self-Lubricating Self-Aligning Bearings Technical Specification. 11 pp.
AECMA PREN2498-87. Rod-Ends Adjustable Self-Aligning Plain Bearing with Self-Lubricating Liner and Threaded Shank Dimendions and Loads.
AECMA PREN2790-86. Rod-Ends, Adjustable Self-Aligning Plain Bearing with Self-Lubricating Liner and Threaded Shank Dimensions and Loads. 5 pp.
AECMA PREN2791-86. Rod-Ends, Adjustable Single Fork and Threaded Shank Dimensions and Loads. 5 pp.
AECMA PREN2792-86. Rod Ends, Adjustable Double Fork and Threaded Shank Dimensions and Loads.
BSI BS EN 2498-90. 1990 Rod Ends, Adjustable Self-Aligning Plain Bearing, with Self-Lubricating Liner and Threaded Shank. Dimensions and Loads. 9 pp.

Control Rods *(Cont.)*
—Rod Ends—Aerospace *(Cont.)*
BSI BS EN 2515-90. 1990 Rod Ends, Adjustable Single Fork and Threaded Shank. Dimensions and Loads. 9 pp.
BSI BS EN 2587-90. 1990 Rod Ends, Adjustable Double Fork and Threaded Shank. Dimensions and Loads. 9 pp.
BSI BS EN 2790-91. 1991 Rod Ends, Adjustable, Self-Aligning Plain Bearing with Self-Lubricating Liner and Threaded Shank with Engagement: 1.5 x Thread Diameter—Dimensions and Loads. 11 pp.
BSI BS EN 2791-91. 1991 Rod Ends, Adjustable, Single Fork and Threaded Shank with Engagement: 1.5 x Thread Diameter—Dimensions and Loads. 11 pp.
BSI BS EN 2792-91. 1991 Rod Ends, Adjustable, Double Fork and Threaded Shank with Engagement: 1.5 x Thread Diameter—Dimensions and Loads. 11 pp.
BSI BS EN 3541-92. 1992 Rod Ends, Adjustable, Self-Aligning Ball Bearing with Threaded Shank—Dimensions, Torques, Clearances and Loads. 11 pp.
CEN EN 2498-90. Rod Ends, Adjustable Self-Aligning Plain Bearing with Self-Lubricating Lines and Threaded Shank Dimensions and Loads. 6 pp.
CEN PREN2515-89. Rod Ends, Adjustable Single Fork and Threaded Shank Dimensions and Loads. 6 pp.
CEN EN 2515-90. Rod Ends, Adjustable Single Fork and Threaded Shank Dimensions and Loads. 7 pp.
CEN PREN2587-89. Rod-Ends, Adjustable Double Fork and Threaded Shank Dimensions and Loads. 6 pp.
CEN EN 2587-90. Rod Ends Adjustable Double Fork and Threaded Shank Dimensions and Loads. 6 pp.
CEN EN 2790-91. Rod Ends, Adjustable Self-Aligning Plain Bearing with Self-Lubricating Liner and Threaded Shank with Engagement: 1,5 x Thread DO Dimensions and Loads. 7 pp.
CEN EN 2791-91. Rod Ends, Adjustable, Single Fork and Threaded Shank with Engagement: 1,5 x Thread DO Dimensions and Loads. 14 pp.
CEN EN 2792-91. Rod Ends, Adjustable, Double Fork and Threaded Shank with Engagement: 1,5 Thread DO Dimensions and Loads. 7 pp.
CEN EN 3541-92. Rod Ends, Adjustable, Self-Aligning Ball Bearing with Threaded Shank—Dimensions, Torques, Clearances and Loads. 7 pp.

—Rod Ends—Aircraft
AECMA PREN2515-84. Rod Ends, Adjustable Single Fork and Threaded Shank and Load. 6 pp.
AECMA PREN2587-84. Rod Ends, Adjustable Double Fork and Threaded Shank and Load. 6 pp.
AECMA PREN2601-90. Fork Ends, Adjustable Technical Specification. 7 pp.

—Silver Indium Cadmium Alloys
CNS J2050-82. Method for Determining of Silver, Indium, and Cadmium in Silver-Indium-Cadmium Alloys Control Rod (May)(8832).

Control Rooms
See Also: Control Systems Equipment
CNS C1044-81. Low-Voltage Control Center (Jun)(3989).

—Ergonomics
DIN ENGL 33414 Pt 1-85. Ergonomic Design of Control Rooms; Seated Work Stations; Terms and Definitions, Principles, Dimensions (Apr). 16 pp.

—Nuclear Power Plants—Ergonomics
IEC 964-89. Design for Control Rooms of Nuclear Power Plants First Edition. 133 pp.
IEC 1227-93. Nuclear Power Plants—Control Rooms—Operator Controls First Edition. 35 pp.

—Sound Broadcasting—Acoustics
CCIR QUESTION 81/10-90. Determination of the Acoustical Properties of Control Rooms and High-Quality Listening Rooms in Broadcasting—Questions Concerning Study Group 10—Broadcasting Service (Sound). 1 p.

Control Signals
Scope Note: For additional listings, use a more specific term *See Also:* Control and Indication Signals; Hand Signals; Signaling Systems

—CCITT No. 6 Signaling Systems—Glossaries
CCITT RECMN Q.255-89. 2.2 Signalling-System-Control Signals—Specifications of Signalling System No. 6 (Study Group XI) 2 pp. 2 pp.

Control Signals *(Cont.)*
—Signal Units—Formats/Codes—CCITT No. 6 Signaling Systems
CCITT RECMN Q.259-89. 3.3 Signalling-System-Control Signals—Specifications of Signalling Sytem No. 6 (Study Group XI) 4 pp. 4 pp.

—Telecommunication Circuits—Data Signals
CCITT RECMN X.71-89. Decentralized Terminal and Transit Control Signalling System on International Circuits Between Synchronous Data Networks—Data Communication Networks—Transmission, Signalling and Switching, Network Aspects, Maintenance and Administrative Arrangements. 35 pp.

Control Surfaces
See Also: Ailerons; Control Systems Equipment; Flaps (Control Surfaces); Flight Dynamics; Stability Testing

—Aerostats
CAA Chapter Q3-9 12.79. Structrual Deformation, Flutter and Vibration (Non-Rigid Airships). 1 p.

—Aircraft
CAA 820 (K). Section K Light Aeroplanes Flutter Fail Safe Criteria (Blue Papers). 11 pp.
CAA Chapter K3-9 10.92. Flutter Prevention and Structural Stiffness (Light Aeroplanes). 4 pp.
CAA Chapter K3-9 App 10.69. Flutter Prevention and Structural Stiffness (Light Aeroplanes). 6 pp.

—Aircraft—Design
CAA CAP 482 SUB-Part D 03.83. Design and Construction (Small Light Aeroplanes). 11 pp.

—Aircraft—Installation
CAA CAP 482 SUB-Part D 03.83. Design and Construction (Small Light Aeroplanes). 11 pp.

—Aircraft—Loads (Forces)
CAA CAP 482 SUB-Part C 03.83. Structure (Small Light Aeroplanes). 12 pp.

—Gliders
CAA JAR-22 SUBPART C. Structure (Joint Airworthiness Requirements). 17 pp.
CAA JAR-22 SUBPART D. Design and Construction (Joint Airworthiness Requirements). 15 pp.

Control Switches
See Also: Control Systems Equipment; Electric Switches; Limit Switches; Pilot Switches; Position Switches; Proximity Switches; Pushbutton Switches; Rotary Switches; Time Switches
SAA AS 1431.7-89. Low Voltages Switchgear and Controlgear—Control Circuit Devices and Switching Elements —Part 7: Additional Requirements for Control Switches with Positive-Opening Operation (This is a Joint Standard with SNZ NZS 1431.7). 3 pp.
SNZ NZS/AS 1431. 7-89. Low Voltage Switchgear and Controlgear—Control Circuit Devices and Switching Elements Part 7: Additional Requirements for Control Switches with Positive-Opening Operation (This is a Joint Standard with SAA AS 1431.7). 3 pp.

—Low Voltage
BSI BS 4794: Part 1-79. (WITHDRAWN) 1979 Amd 1 Control Switches (Switching Devices, Including Contactor Relays, for Control and Auxiliary Circuits for Voltages up to and Including 1000 V a.c. and 1200 V d.c.) Part 1: General Requirements. 28 pp.
BSI BS 4794: Sec 2.1-77. (WITHDRAWN) 1977 Control Switches (Switching Devices, Including Contactor Relays, for Control and Auxiliary Circuits for Voltages up to and Including 1000 V a.c. and 1200 V d.c.) Part 2: Special Requirements for Specific Types of. 12 pp.
BSI BS 6517-84. (WITHDRAWN) 1984 Low Voltage Switchgear and Controlgear for Industrial Use. Single Hole Mounted Control Switches and Indicator Lights. Mounting Dimensions (Superseded by BS EN 60947-5-1: 1992). 6 pp.
BSI BS EN 60947-5-1-01. 1992 Amd 1 Low-Voltage Switchgear and Controlgear Part 5: Control Circuit Devices and Switching Elements Section 1: Electromechanical Control Circuit Devices (AMD 7338) December 15, 1992 (E). 75 pp.
BSI BS EN 60947-5-1-02. 1992 Amd 2 Low-Voltage Switchgear and Controlgear Part 5: Control Circuit Devices and Switching Elements Section 1: Electromechanical Control Circuit Devices (AMD 7850) July 15, 1993 (E). 76 pp.
BSI BS EN 60947-5-1-92. 1992 Low-Voltage Switchgear and Controlgear Part 5: Control Circuit Devices and Switching Elements Section 1: Electromechanical Control Circuit Devices. 74 pp.
CENELEC HD 420-82. Control Switches (Low Voltage-Devices for Control and Auxiliary Circuits, Including Contactor Relays). 5 pp.

INDUSTRY STANDARDS

INTERNATIONAL AND NON-U.S. NATIONAL STANDARDS
SUBJECT INDEX

Control

Control Switches (Cont.)
—**Low Voltage** (Cont.)
CENELEC HD 420 S2-88. Control Switches (Low Voltage Switching—Devices for Control and Auxiliary Circuit, Including Contractor Relays). 6 pp.
CENELEC EN 50 007-81. Low Voltage Switchgear and Controlgear for Industrial Use: Single Hole Mounted Control Switches and Indicator Lights: Mounting Dimensions. 4 pp.
CENELEC EN 60947-5-1-91. Low-Voltage Switchgear and Controlgear Part 5: Control Circuit Devices and Switching Elements Section One—Electromechanical Control Circuit Devices. 6 pp.
CENELEC EN 60947-5-1-92. CORRIGENDUM Low-Voltage Switchgear and Controlgear Part 5: Control Circuit Devices and Switching Elements Section One—Electromechanical Control Circuit Devices. 1 p.
CNS C1079-81. General Rules for Control Switches (Jul)(7621).
CNS C3122-81. Method of Test for Control Switches (Jul)(7622).
DIN VDE 0660 Pt 200-82. Switchgear and Control Gear; Low Voltage Switchgear and Control Gear; Control Switches; General Requirements (Sept). 34 pp.
DIN VDE 0660 Pt 200 A1 (D)-87. Switchgear and Controlgear; Low-Voltage Switchgear and Controlgear; Control Circuit Devices and Switching Elements (Aug). 15 pp.
DIN VDE 0660 Pt 200 A2 (D)-89. Switchgear and Controlgear; Low-Voltage Switchgear and Controlgear; Control Circuit Devices and Switching Elements; Amendment 2 (Aug). 9 pp.
IEC 947 Pt 5-1-90. Low-Voltage Switchgear and Controlgear Part 5: Control Circuit Devices and Switching Elements Section One—Electromechanical Control Circuit Devices First Edition. 131 pp.
JIS C 4520-91. General Rules for Control Switches. 41 pp.

—**Low Voltage—Button**
CNS C3123-81. Method of Test for Button Control Switches (Jul)(7624).
CNS C4324-81. Button Control Switches (Jul)(7623).
DIN VDE 0660 Pt 201-82. Switchgear and Control Gear; Low Voltage Switchgear and Control Gear; Control Switches; Additional Specifications for Pushbuttons and Related Control Switches (Sept). 18 pp.
JIS C 4521-91. Button Control Switches. 21 pp.

—**Low Voltage—Cam**
CNS C3124-81. Method of Test for Cam Operated Control Switches (Jul)(7626).
CNS C4325-81. Cam Operated Control Switches (Jul)(7625).
JIS C 4522-91. Cam Operated Control Switches. 14 pp.

—**Low Voltage—Numbering**
BSI BS 5581-78. 1978 Specification for Low-Voltage Switchgear and Controlgear for Industrial Use. Terminal Marking and Distinctive Number for Particular Control Switches. 6 pp.

—**Low Voltage—Terminals—Identification Systems**
BSI BS 5581-78. 1978 Specification for Low-Voltage Switchgear and Controlgear for Industrial Use. Terminal Marking and Distinctive Number for Particular Control Switches. 6 pp.
CENELEC EN 50 013-77. Low Voltage Switchgear and Controlgear for Industrial use Terminal Marking and Distinctive Number for Particular Control Switches. 4 pp.

—**Ranges**
BSI BS 4177-01. 1992 Amd 1 Cooker Control Units (AMD 7750) July 15, 1993 (L). 45 pp.
BSI BS 4177-67. 1967 Amd 2 Cooker Control Units Rated at 30 Amperes and 45 Amperes, 250 Volts Single-Phase Only. 41 pp.

—**Rotary**
BSI BS 4794: Sec 2.1-77. (WITHDRAWN) 1977 Control Switches (Switching Devices, Including Contactor Relays, for Control and Auxiliary Circuits for Voltages up to and Including 1000 V a.c. and 1200 V d.c.) Part 2: Special Requirements for Specific Types of. 12 pp.
BSI BS 4794: Sec 2.2-77. (WITHDRAWN) 1977 Control Switches (Switching Devices, Including Contactor Relays, for Control and Auxiliary Circuits for Voltages up to and Including 1000 V a.c. and 1200 V d.c. Part 2: Special Requirements for Specific Types of. 9 pp.
DIN VDE 0660 Pt 201-82. Switchgear and Control Gear; Low Voltage Switchgear and Control Gear; Control Switches; Additional Specifications for Pushbuttons and Related Control Switches (Sept). 18 pp.

Control Switches (Cont.)
—**Rotary** (Cont.)
DIN VDE 0660 Pt 202-82. Switchgear and Control Gear; Low-Voltage Switchgear and Control Gear; Control Switches; Additional Specification for Rotary Switches (Sept). 14 pp.

—**Ships—Lighting—Explosive Atmospheres**
CNS F5033-86. Marine Control Switches for Explosion Proof (Aug)(8588).
JIS F 8846-90. Control Switches of Flameproof Light for Marine Use.

Control Systems and Equipment
Use: Control Systems Equipment

Control Systems Equipment
Scope Note: For additional listings, use a more specific term *Use For:* Computer Control; Control Devices *See Also:* Access Control Systems; Actuators; Automatic Control Equipment; Building Automation Systems; Circuit Breakers; Command and Control Systems; Computer Numerical Control; Contactors; Control Cables (Mechanical); Control Circuits; Control Consoles; Control Engineering; Control Panels; Control Rooms; Control Surfaces; Control Switches; Control Transformers; Control Units; Controlgear; Electric Switches; Electronic Control Equipment; Flame Detectors; Flow Regulators; Insect Control Equipment; Manual Controls; Network Management Systems; Numerical Control; Photoelectric Controls; Pneumatic Control Equipment; Pneumatic Controllers; Pneumatic Servomechanisms; Pressure Control; Pressure Regulators; Process Control Equipment; Process Control Systems; Protocols; Recording Instruments; Regulators; Relays; Remote Control Equipment; Rotary Switches; Safety Valves; Servovalves; Smoke Density Controls; Speed Controls; Steam Traps; Supervisory Control Systems; Surge Arresters; Switching Systems; Synchronizers; Temperature Controllers; Thermal Switches; Thermostatic Valves; Thermostats; Toggle Switches; Traffic Control Devices; Transfer Stations; Trim Controls; Voltage Regulators
BSI BS 6650-85. (WITHDRAWN) 1985 Control of the Operation of a Computer. 11 pp.
CSA CAN/CSA-C22. 2 NO 14-M91. Industrial Control Equipment; (Gen Instr 1 Thru 4). 142 pp.

—**Aerostats—Deflation**
CAA Chapter Q4-9 12.79. Envelope Design (Non-Rigid Airships). 2 pp.

—**Aerostats—Engines**
CAA Chapter Q5-7 12.79. Controls (Non-Rigid Airships). 2 pp.

—**Aircraft**
CAA LEAFLET 2-13 07.90. Control Systems. 13 pp.
NATO STANAG 3220 ED 4 AMD 6-74. Location, Actuation and Shape of Airframe Controls for Fixed Wing Aircraft. 11 pp.

—**Aircraft—Bowden—Assembly**
SBAC AS 2831 ISSUE 4. Assembly of Bowden Control. 1 p.

—**Aircraft—Design**
CAA Chapter K4-8 04.72. Control System Loads and Design (Light Aeroplanes). 7 pp.
CAA Chapter P4-2. Flight Crew Compartment Design (Provisional Airworthiness Requirements for Civil Powered Lift Aircraft). 16 pp.
CAA Chapter P4-8. Control System Design (Provisional Airworthiness Requirements for Civil Powered Lift Aircraft). 3 pp.
CAA CAP 482 SUB-Part D 03.83. Design and Construction (Small Light Aeroplanes). 11 pp.

—**Aircraft Engines**
CAA Chapter K5-7 04.74. Controls (Light Aeroplanes). 2 pp.
CAA Chapter P5-7. Control Systems (Provisional Airworthiness Requirements for Civil Powered Lift Aircraft). 6 pp.
CAA CAP 482 SUB-Part E 03.83. Power-Plant (Small Light Aeroplanes). 5 pp.

—**Aircraft Engines—Rotary Wing**
CAA CAP 524 SUB-Part E 12.86. Power-Plant (Rotorcraft). 27 pp.

—**Aircraft—Escape**
SBAC RS 433 ISSUE 1. Handle for Escape Hatch Control.

—**Aircraft—Identification Systems**
CAA CAP 482 SUB-Part G 03.83. Operating Limitations and Information (Small Light Aeroplanes). 6 pp.

Control Systems Equipment (Cont.)
—**Aircraft—Landing Gear**
CAA Chapter K4-8 04.72. Control System Loads and Design (Light Aeroplanes). 7 pp.

—**Aircraft—Loads (Forces)**
CAA Chapter P3-6. Control System Loads (Provisional Airworthiness Requirements for Civil Powered Lift Aircraft). 3 pp.
CAA CAP 482 SUB-Part C 03.83. Structure (Small Light Aeroplanes). 12 pp.

—**Aircraft—Lockwire**
NATO STANAG 3752 ED 2 AMD 3-89. Witness (Breaking) Wire for Aircraft Emergency Controls and Equipment. 8 pp.
NATO STANAG 3752 ED 2 AMD 4-89. Witness (Breaking) Wire for Aircraft Emergency Controls and Equipment. 9 pp.

—**Aircraft Propellers**
CAA Chapter K5-7 04.74. Controls (Light Aeroplanes). 2 pp.

—**Appliances**
BSI BS 3955-86. 1986 Amd 1 Electrical Controls for Household and Similar General Purposes. 84 pp.
CEN PREN 126-91. Multifunctional Controls for Gas Burning Appliances. 77 pp.
EC 84/530/EEC-84. Council Directive on the Approximation of the Laws of the Member States Relating to Common Provisions for Appliances Using Gaseous Fuels, Safety and Control Devices for These Appliances and Methods of Surveillance of Them. 11 pp.
EC 90/396/EEC-90. Council Directive on the Approximation of the Laws of the Member States Relating to Appliances Burning Gaseous Fuels (NEW APP). 16 pp.

—**Appliances—Gas—Lubricants**
DIN ENGL 3536-82. Lubricants for Gas Valves and Controls; Requirements, Testing (Nov). 5 pp.

—**Appliances—Portable**
SNZ NZS/AS 3197-93. Approval and Test Specifications—Portable Electric Control or Conditioning Devices (This is a Joint Standard with SAA AS 3197). 14 pp.

—**Dielectric Properties**
CNS C1043-81. Dielectric Properties of Control Device (Jun)(3988).

—**Electrical Codes**
CSA 1341A Bull. Electrical Bulletin 1341A April 5, 1984 to C22.2 NO 14. 2 pp.

—**Electrical Safety**
BSI BS EN 61010-1-93. 1993 Safety Requirements for Electrical Equipment for Measurement, Control, and Laboratory Use Part 1: General Requirements. 123 pp.
CENELEC EN 61010-1-93. Safety Requirements for Electrical Equipment for Measurement, Control and Laboratory Use Part 1: General Requirements (IEC 1010-1: 1990 + A1: 1992, Modified). 117 pp.
CSA CAN/CSA-C22. 2NO1010.1-92. Safety Requirements for Electrical Equipment for Measurement, Control, and Laboratory Use, Part 1: General Requirements (IEC 1010-1:1990); (Gen Instr 1). 160 pp.
DIN VDE 0411 Pt 100-86. Measurement and Control; Safety Requirements for Electrically Operated Measuring, Control and Laboratory Equipment; General Requirements; Identical with IEC 66E (Sec) 22 (Aug). 137 pp.
IEC 1010 Pt 1-90. Safety Requirements for Electrical Equipment for Measurement, Control and Laboratory Use Part 1: General Requirements First Edition (CENELEC EN 61010-1: 1993). 190 pp.

—**Electrical Spacings**
CSA CAN/CSA-C22. 2 NO 14-M91. Industrial Control Equipment; (Gen Instr 1 Thru 4). 142 pp.

—**Elevator Contactors—Endurance Testing**
CSA CAN/CSA-C22. 2 NO 14-M91. Industrial Control Equipment; (Gen Instr 1 Thru 4). 142 pp.

—**Elevators (Lifts)—Handicapped Persons**
CSA CAN/CSA-B355-M86. Elevating Devices for the Handicapped. 35 pp.
CSA CAN/CSA-B613-M87. Elevating Devices for the Handicapped in Private Residences; (Gen Instr 1). 30 pp.

—**Function Charts**
IEC 848-88. Preparation of Function Charts for Control Systems First Edition; (Corrigendum—Sept 1990). 104 pp.

INDEX and DIRECTORY of

INTERNATIONAL AND NON-U.S. NATIONAL STANDARDS
SUBJECT INDEX

Control

Control Systems Equipment *(Cont.)*
—Glossaries
BSI BS 1523: Part 1-67. 1967 Glossary of Terms Used in Automatic Controlling and Regulating Systems Part 1: Process and Kinetic Control. 39 pp.

—Ground Effect Machines
CAA 58.
CAA Chapter B6-3 01.91. Control Systems. 3 pp.

—Laboratory Thermometers—Liquid In Glass—Contact
DIN ENGL 12878-80. Electrical Laboratory Equipment; Adjustable Liquid-In-Glass Contact Thermometers and Control Devices; General and Safety Requirements, Testing (Dec). 6 pp.

—Laboratory Thermometers—Liquid In Glass—Contact—Safety
DIN ENGL 12878-80. Electrical Laboratory Equipment; Adjustable Liquid-In-Glass Contact Thermometers and Control Devices; General and Safety Requirements, Testing (Dec). 6 pp.

—Machinery—Design
CEN PREN 954-1-92. Safety of Machinery—Safety Related Parts of Control Systems—Part 1: General Principles for Design. 40 pp.

—Mining Equipment
BSI BS 534-90. 1990 Steel Pipes, Joints and Specials for Water and Sewage. 26 pp.
BSI BS 534-81. (WITHDRAWN) 1981 Amd 1 Steel Pipes and Specials for Water and Sewage. 17 pp.

—Naval Ships—Winterization
MOD UK NES 716-84. Requirements for Winterization on Surface Ships Issue 1 (03.84). 17 pp.
MOD UK NES 716-92. Requirements for Winterization on Surface Ships Issue 2 (07.92). 22 pp.
MOD UK NES 716-01. Requirements for Winterization on Surface Ships Issue 2 (07.92); Amendment 1. 23 pp.

—Nuclear Power Plants—Safety—Classification
IEC 1226-93. Nuclear Power Plants—Instrumentation and Control Systems Important for Safety—Classification First Edition. 47 pp.

—Nuclear Power Plants—Seismic Design
IEC 980-89. Recommended Practices for Seismic Qualification of Electrical Equipment of the Safety System for Nuclear Generating Stations First Edition. 83 pp.

—Nuclear Reactors
CSA CAN3-N290.4-M82. Requirements for the Reactor Regulating Systems of CANDU Nuclear Power Plants (R 1992); (Gen Instr 1). 22 pp.
IEC 639-79. Nuclear Reactors Use of the Protection System for Non-Safety Purposes First Edition. 18 pp.

—Power Supplies—Nuclear Power Plants
BSI BS 7674-93. 1993 Nuclear Power Plants—Instrumentation and Control Systems Important for Safety—Requirements for Electrical Supplies (IEC 1225: 1993) (Q). 24 pp.
IEC 1225-93. Nuclear Power Plants—Instrumentation and Control Systems Important for Safety—Requirements for Electrical Supplies First Edition. 49 pp.

—Reed Relays
JIS C 4523-85. Control Reed Relays. 31 pp.

—Resistance Welding Equipment
CNS C4159-84. Control Timer for Resistance Welding Machine of Lap Joint (Jul)(5027).
JIS C 9313-91. Control Equipments of Resistance Welding Machines for Lap Joint. 22 pp.

—Rolling Stock
JIS E 5004-91. Control Equipment for Electric Rolling Stock—Test Methods.

—Rotary Wing Aircraft
CAA 235 (G). Section G Rotorcraft Miscellaneous Amendments Derived from Section D, "Aeroplanes" (Blue Papers). 6 pp.
CAA Chapter G2-6 02.63. Handling—General (Rotorcraft). 3 pp.
CAA Chapter G2-8 App 08.82. Handling—Controllability (Rotorcraft). 2 pp.
CAA Chapter G3-6 01.75. Control System Loads (Rotorcraft). 2 pp.
CAA Chapter G4-2 12.80. Flight Crew Compartment Design (Rotorcraft). 3 pp.
CAA Chapter G5-7 06.76. Control Systems (Rotorcraft). 3 pp.

Control Systems Equipment *(Cont.)*
—Rotary Wing Aircraft *(Cont.)*
CAA CAP 524 SUB-Part D 12.86. Design and Construction (Rotorcraft). 23 pp.
NATO STANAG 3218 ED 6 AMD 2-86. Location, Actuation and Shape of Engine Controls and Switches. 21 pp.
NATO STANAG 3225 ED 4 AMD 7-75. Location, Actuation and Shape of Airframe Controls for Rotary Wing Aircraft. 10 pp.

—Rotary Wing Aircraft—Design
CAA 749 (G). Section G Rotocraft Chapter G-4 Control System Design. 2 pp.
CAA Chapter G4-8 01.75. Control System Division (Rotorcraft). 3 pp.
CAA Chapter G4-8 App 01.75. Control System Division (Rotorcraft). 5 pp.

—Rotary Wing Aircraft—Loads (Forces)
CAA CAP 524 SUB-Part C 12.86. Structure (Rotorcraft). 11 pp.

—Ships—Electrical Installations
CNS F5118-87. Device Function Numbers for Control Devices and Equipment of Electrical Installations in Ships (Jun)(11996).
JIS F 8076-86. Electrical Installations in Ships Part 504 Special Features—Control and Instrumentation (IEC 92-504-1974).
MOD UK NES 626-91. Control and Surveillance Equipment for Plant and System Installations Issue 2 (04.91). 63 pp.

—Ships—Environmental Testing
CNS F1009-84. General Rules for Environmental Tests of Control and Instrumentation Equipment for Marine Use (Apr)(10856).
JIS F 0807-89. General Rules for Environmental Tests of Control and Instrumentation Equipment for Marine Use.

—Slide Projectors
BSI BS 5817: Sec 5.2-84. 1984 Amd 1 Audio-Visual, Video and Television Equipment and Systems Part 5: Control, Synchronization and Address Codes Sec 5.2: Method of Operation for Pulsed Cue Tone Control Systems for Two Still Projectors (AMD 6492) January 31, 1991. 12 pp.
CENELEC HD 369.5.2-86. Audio-Visual, Video and Television Equipment and Systems Part 5: Control Synchronization and Address Codes Chapter II: Control Systems for Two Still Projectors Operating Practice. 2 pp.
IEC 574 Pt 5-2-83. Audio-Visual, Video and Television Equipment and Systems Part 5: Control, Synchronization and Address Codes Chapter II: Control Systems for Two Still Projectors Operating Practice First Edition. 21 pp.

—Stairlifts—Handicapped Persons
CSA CAN/CSA-B355-M86. Elevating Devices for the Handicapped. 35 pp.
CSA CAN/CSA-B613-M87. Elevating Devices for the Handicapped in Private Residences; (Gen Instr 1). 30 pp.

—Supersonic Transports
CAA STANDARD NO. 3-3&AP 03.76. Flying Controls. 4 pp.
CAA STANDARD NO. 7-3&AP 05.73. Flying Controls. 4 pp.

—Switchgear
CSA CAN/CSA-C22. 2 NO 31-M89. Switchgear Assemblies; (Gen Instr 1 Thru 3). 54 pp.
CSA 1169 Bull. Electrical Bulletin 1169 June 27, 1978 to C22.2 NO 31. 2 pp.

—Symbols
JIS Z 8204-83. Instrumentation Symbols. 64 pp.

Control Transformers
See Also: Control Systems Equipment; Transformers
DIN VDE 0550 Pt 3-69. Regulations for Small Transformers; Particular Regulations for Isolating and Control Transformers with Network Connection and Isolating Transformers Above 1000 V (Dec) (Partially Superseded by Draft 0550 Part 1/11.87). 10 pp.

Control Units
Use For: Computer Control Units; Computer Controllers; Computer Controls; Programming Control Units *See Also:* Computer Equipment; Control Systems Equipment; Controllers

—Defense Equipment—Hazard Analysis
MOD UK DSTAN 00-56-91. Hazard Analysis and Safety Classification of the Computer and Programmable Electronic System Elements of Defence Equipment Issue 1. 68 pp.

Control Units *(Cont.)*
—Nuclear Reactors
BSI BS 6078-81. 1981 Application of Digital Computers to Nuclear Reactor Instrumentation and Control. 18 pp.
IEC 643-79. Application of Digital Computers to Nuclear Reactor Instrumentation and Control First Edition. 36 pp.

Control Valves
Scope Note: For additional listings, use a more specific term *See Also:* Directional Control Valves; Float Valves; Hydraulic Valves; Pressure Control Valves; Process Control Valves; Proportional Valves; Sequence Valves; Valves

—Fire Fighting
CNS Z2055-87. Control Valve for Fire Fighting Use (Sep)(10763).

—Flow Rate
JIS B 8654-89. Test Methods for Electro-Hydraulic Proportional Series Flow Control Valves. 23 pp.

—Hydraulic—Controllers—Identification Systems
ISO 9461-92. Hydraulic Fluid Power—Identification of Valve Ports, Subplates, Control Devices and Solenoids First Edition. 7 pp.

—Hydraulic—Ports—Identification Systems
ISO 9461-92. Hydraulic Fluid Power—Identification of Valve Ports, Subplates, Control Devices and Solenoids First Edition. 7 pp.

—Hydraulic—Power Amplifiers
JIS B 8658-89. Rules of Electronic Amplifier for Electro-Hydraulic Control Valves. 7 pp.

—Hydraulic—Proportional Control
BSI BS 6697-86. 1986 Electrohydraulic Proportional Control Valves. 20 pp.

—Hydraulic—Subplates—Identification Systems
ISO 9461-92. Hydraulic Fluid Power—Identification of Valve Ports, Subplates, Control Devices and Solenoids First Edition. 7 pp.

—Mounting Surfaces
BSI BS 6494: Part 5-89. 1989 Hydraulic Fluid Power Valve Mounting Surfaces Part 5: Compensated Low-Control Valves. 21 pp.

—Plug—Gas
SAA AS A72-54. Control Plug Cocks for Low Pressure Gas Amdt 1 February 1959. 10 pp.

—Remote Control Equipment—Ships
JIS F 3024-78. Ships' Deck Stands for Controlling Valves.
JIS F 3024-71. Ships' Deck Stands for Controlling Valves. 16 pp.

—Water—Electrical Properties
IEC 730 Pt 2-8-92. Automatic Electrical Controls for Household and Similar Use Part 2: Particular Requirements for Electrically Operated Water Valves, Including Mechanical Requirements First Edition. 70 pp.

—Water Heaters
SAA AS 1357.2-92. Water Supply—Valves for Use with Unvented Water Heaters—Part 2: Control Valves (in Professional Packages 61A, 61B). 27 pp.
SNZ NZS 4608-92. Control Valves for Hot Water Systems. 24 pp.

—Water—Mechanical Properties
IEC 730 Pt 2-8-92. Automatic Electrical Controls for Household and Similar Use Part 2: Particular Requirements for Electrically Operated Water Valves, Including Mechanical Requirements First Edition. 70 pp.

—Water—Safety
IEC 730 Pt 2-8-92. Automatic Electrical Controls for Household and Similar Use Part 2: Particular Requirements for Electrically Operated Water Valves, Including Mechanical Requirements First Edition. 70 pp.

Control Wheels
See Also: Aircraft Equipment; Hand Wheels; Wheels
SBAC AS 3038-3047 ISSUE 1. Pilot's Control Wheel Assemblies. 1 p.
SBAC AS 3050 ISSUE 1. Pilots Control Wheel. 1 p.
SBAC AS 3153 ISSUE 2. Insertion Tube Switch Body Pilot's Control Wheel. 1 p.
SBAC AS 6579 ISSUE 1. Pilot's Control Wheel.

—Aircraft—Adjusters
SBAC AS 3122 ISSUE 1. Brake Lock Adjuster. 1 p.

Control Wheels (Cont.)

—Aircraft—Adjusters (Cont.)
SBAC AS 6582 ISSUE 1. Brake Lock Adjuster for Pilot's Control Wheel.

—Aircraft—Lever Switches—Pivots
SBAC AS 6594 ISSUE 1. Switch Lever Pivot Pilot's Control Wheel.

—Assembly
SBAC AS 6560-6569 ISSUE 1. Pilot's Control Wheel, Assemblies.

—Bearings—Bushings
SBAC AS 3121 ISSUE 2. Bearing Bush for Pilot's Control Wheel. 1 p.

—Brake Lock Shafts
SBAC AS 6583 ISSUE 1. Brake Lock, Shaft—Pilot's Control Wheel.

—Brakes
SBAC AS 3048 ISSUE 1. Pilot's Control Wheel Brake Assembly (Spade Type). 1 p.
SBAC AS 3049 ISSUE 1. Pilot's Control Wheel Brake Assembly (Horn Type). 1 p.
SBAC AS 6570 ISSUE 1. Pilot's Control Wheel, Brake Assembly—Spade Type.
SBAC AS 6571 ISSUE 1. Pilot's Control Wheel, Brake Assembly (Horn Type).
SBAC AS 6580 ISSUE 2. Brake off Stop Pilot's Control Wheel.
SBAC AS 6581 ISSUE 2. Brake on Stop Pilot's Control Wheel.

—Brakes—Locking Rings
SBAC AS 6586 ISSUE 1. Locking, Ring for Brake Stop for Pilot's Control Wheel.

—Fittings
SBAC AS 3055 ISSUE 2(I). End Piece for Pilot's Control Wheel. 1 p.

—Lever Switches
SBAC AS 3146 ISSUE 1. Switch Lever Auto Pilot's Switch Pilot's Control Wheel. 1 p.
SBAC AS 3147 ISSUE 1. Switch Lever Auto Pilot Switch Pilot's Control Wheel. 1 p.
SBAC AS 6597 ISSUE 1. Switch, Lever Auto Pilot Switch, Pilot's Control Wheel.

—Levers
SBAC AS 3052 ISSUE 2. Brake Lock Lever for Pilot's Control Wheel. 1 p.
SBAC AS 3123 ISSUE 3. Brake Lever for Pilot's Control Wheel. 1 p.
SBAC AS 3124 ISSUE 3. Brake Lever for Pilot's Control Wheel. 1 p.
SBAC AS 3125 ISSUE 2. Brake Lever for Pilot's Control Wheel. 1 p.

—Pushbuttons
SBAC AS 3058 ISSUE 1. Switch Assembly for Pilot's Control Wheel. 1 p.
SBAC AS 3059 ISSUE 1. Switch Assembly—Pilot's Control Wheel. 1 p.
SBAC AS 3060 ISSUE 1. Switch Assembly Pilot's Control Wheel. 1 p.
SBAC AS 3061 ISSUE 1. Switch Assembly Pilot's Control Wheel. 1 p.
SBAC AS 3069 ISSUE 3. Press Button Pilot's Control Wheel. 1 p.

—Pushbuttons—Inserts
SBAC AS 3068 ISSUE 1. Insert for Press Button Pilots Control Wheel. 1 p.

—Pushbuttons—Spring Covers
SBAC AS 3133 ISSUE 2. Spring Cover Pilot's Control Wheel Switch. 1 p.

—Rods—Fittings
SBAC AS 3126 ISSUE 4. Brake Push Rod End Pilot's Control Wheel. 1 p.

—Screws
SBAC AS 3070 ISSUE 1. Switch Contact Screw. 1 p.

—Shafts
SBAC AS 3053 ISSUE 1. Brake Lock Shaft—Pilot's Control Wheel. 1 p.

—Springs (Elastic)
SBAC AS 3054 ISSUE 2. Brake Lock Spring for Pilot's Control Wheel. 1 p.
SBAC AS 3071 ISSUE 3. Switch Spring for Pilot's Control Wheel. 1 p.
SBAC AS 3145 ISSUE 2. Switch Spring for Pilot's Control Wheel. 1 p.

—Stops
SBAC AS 3129 ISSUE 2. Brake off Stop Pilot's Control Wheel. 1 p.
SBAC AS 3130 ISSUE 2. Brake on Stop Pilot's Control Wheel. 1 p.

—Stops—Lockrings
SBAC AS 3131 ISSUE 1. Ring Locking for Brake Stop Pilot's Control Wheel. 1 p.

—Switch Contact Screws
SBAC AS 6593 ISSUE 1. Switch Contact, Screw for Pilot's Control Wheel.

—Switch Lever—Pivots
SBAC AS 3149 ISSUE 1. Switch Lever Pivot Pilots Control Wheel. 1 p.
SBAC AS 6594 ISSUE 1. Switch Lever Pivot Pilot's Control Wheel.

—Switch Lever—Spigots
SBAC AS 6592 ISSUE 2. Switch Lever, Spigot.

—Switches
SBAC AS 3158 ISSUE 1. Switch Body Assembly Pilots Control Wheel. 1 p.
SBAC AS 3305 ISSUE 1. Switch Body Assembly Pilots Control Wheel. 1 p.
SBAC AS 6572 ISSUE 2. Switch Assembly for Pilot's Control Wheel.
SBAC AS 6573 ISSUE 2. Switch Assembly: Pilot's Control Wheel.
SBAC AS 6574 ISSUE 1. Switch Assembly: Pilot's Control Wheel.
SBAC AS 6575 ISSUE 1. Switch Assembly: Pilot's Control Wheel.
SBAC AS 6576 ISSUE 1. Switch Assembly: Pilot's Control Wheel.
SBAC AS 6577 ISSUE 1. Switch Assembly: Pilot's Control Wheel.
SBAC AS 6589 ISSUE 1. Switch Body Assembly Pilot's Control Wheel.
SBAC AS 6590 ISSUE 2(I). Switch Body Assembly: Pilot's Control Wheel.
SBAC AS 6591 ISSUE 1. Switch, Lever: Auto-Pilot Switch, Pilot's Control Wheel.
SBAC AS 6595 ISSUE 1. Switch Body Assembly: Pilot's Control Wheel.

—Switches—Insertion Tubes
SBAC AS 3152 ISSUE 2. Insertion Tube Switch Body Pilot's Control Wheel. 1 p.
SBAC AS 3154 ISSUE 2. Insertion Tube—Switch Body Pilot's Control Wheel. 1 p.
SBAC AS 3155 ISSUE 2. Insertion Tube—Switch Body Pilot's Control Wheel. 1 p.
SBAC AS 3156 ISSUE 2. Insertion Tube Switch Body Pilot's Control Wheel. 1 p.
SBAC AS 3157 ISSUE 2. Insertion Tube Switch Body Pilot's Control Wheel. 1 p.

Controlgear
Use For: Electric Control Gear
See Also: Contactors; Control Systems Equipment; Switchgear

BSI BS 6423-83. 1983 Maintenance of Electrical Switchgear and Control-Gear for Voltages up to and Including 650V. 24 pp.
BSI BS 6423-01. 1983 Amd 1 Maintenance of Electrical Switchgear and Controlgear for Voltages up to and Including 1 kV (AMD 6812) February 28, 1992 (E). 26 pp.
DIN VDE 0660 Suppl. 1-82. Switchgear and Control Gear; Index of the Standards of the Series VDE 0660 (Sept). 6 pp.
DIN VDE 0660 Suppl. 2-85. Switchgear and Control Gear; Quoted and Further Standards in the Series of VDE 0660 (Dec). 18 pp.

— 1 kV
BSI BS 6423-01. 1983 Amd 1 Maintenance of Electrical Switchgear and Controlgear for Voltages up to and Including 1 kV (AMD 6812) February 28, 1992 (E). 26 pp.

—1-36 kV
BSI BS 6626-02. 1985 Amd 2 Code of Practice for Maintenance of Electrical Switchgear and Controlgear for Voltages Above 1 kV and up to and Including 36 kV (AMD 6813) December 24, 1991. 44 pp.

—1-72.5 kV
SAA AS 2086-84. High-Voltage a.c. Switchgear and Controlgear—Metal-Enclosed—Rated Voltages Above 1 kV up to and Including 72.5 kV. 25 pp.

—Assemblies
DIN VDE 0660 Pt 503-86. Switchgear and Control Gear; Low Voltage Switchgear and Control Gear Assemblies; Additional Specification for Cable Distribution Cabinets (July). 11 pp.
DIN VDE 0660 Pt 504 (D)-88. Switchgear and Controlgear; Low-Voltage Switchgear and Controlgear; Additional Specifications for Type-Tested Assemblies Intended to Be Installed in Places Where Unskilled Persons Have Access for Their Use (Nov). 26 pp.
IEC 439 Pt 4-90. Low-Voltage Switchgear and Controlgear Assemblies Part 4: Particular Requirements for Assemblies for Construction Sites (ACS) First Edition. 50 pp.
IEC 890-87. A Method of Temperature-Rise Assessment by Extrapolation for Partially Type-Tested Assemblies (PTTA) of Low-Voltage Switchgear and Controlgear First Edition. 54 pp.

—Assemblies—Low Voltage
BSI BS 5486: Part 1-90. 1990 Low-Voltage Switchgear and Controlgear Assemblies Part 1: Requirements for Type-Tested and Partially Type-Tested Assemblies. 75 pp.
BSI BS 5486: Part 1-01. 1990 Amd 1 Low-Voltage Switchgear and Controlgear Assemblies Part 1: Requirements for Type-Tested and Partially Type-Tested Assemblies (AMD 7389) January 15, 1993 (E). 84 pp.
BSI BS 5486: Part 1-02. 1990 Amd 2 Low-Voltage Switchgear and Controlgear Assemblies Part 1: Requirements for Type-Tested and Partially Type-Tested Assemblies (AMD 7774) June 15, 1993 (E). 90 pp.
BSI BS 5486: Part 1-86. (WITHDRAWN) 1986 Low-Voltage Switchgear and Controlgear Assemblies Part 1: Specification for Type-Tested Assemblies (General Requirements) (Superseded by BS 5486: Part 1: 1990 and BS 5486 Parts 11, 12, and 13: 1991). 68 pp.
BSI BS 5486: Part 12-89. 1989 Low-Voltage Switchgear and Controlgear Assemblies Part 12: Particular Requirements of Type-Tested Miniature Circuit-Breaker Boards. 12 pp.
BSI BS 5486: Part 12-79. (WITHDRAWN) 1979 Factory-Built Assemblies of Switchgear and Controlgear for Voltages up to and Including 1000 V a.c. and 1200 V d.c. Part 12: Particular Requirements for Miniature Circuit-Breaker Boards. 11 pp.
BSI BS 5486: Part 13-89. 1989 Amd 1 Low-Voltage Switchgear and Controlgear Assemblies Part 13: Specification for Particular Requirements of Consumer Units (AMD 6554) April 30, 1991. 19 pp.
BSI BS 5486: Part 13-79. (WITHDRAWN) 1979 Amd 1 Factory-Built Assemblies of Switchgear and Controlgear for Voltages up to and Including 1000 V a.c. and 1200 V d.c. Part 13: Particular Requirements for Consumer Units. 15 pp.
BSI BS EN 60439-4-91. 1991 Low-Voltage Switchgear and Controlgear Assemblies Part 4: Particular Requirements for Assemblies for Construction Sites (ACS). 32 pp.
CENELEC EN 60 439-1-90. Low-Voltage Switchgear and Controlgear Assemblies Part 1: Requirements for Type-Tested and Partially Type-Tested Assemblies. 20 pp.
CENELEC EN 60439-1/A1-93. AMD 1 Low-Voltage Switchgear and Controlgear Assemblies Part 1: Requirements for Type-Tested and Partially Type-Tested Assemblies (IEC 439-1:1985/A1:1991). 4 pp.
CENELEC PREN 60 439-4-90. Low-Voltage Switchgear and Controlgear Assemblies Part 4: Particular Requirements for Assemblies for Construction Sites (ACS). 29 pp.
CENELEC EN 60439-4-91. Low-Voltage Switchgear and Controlgear Assemblies Part 4: Particular Requirements for Assemblies for Construction Sites. 6 pp.
CENELEC EN 60439-4-91. Low-Voltage Switchgear and Controlgear Assemblies Part 4: Particular Requirements for Assemblies for Construction Sites (ACS). 8 pp.
CENELEC PREN 60439-5-92. Low-Voltage Switchgear and Controlgear Assemblies Part 5: Particular Requirements for Low Voltage Switchgear and Controlgear Assemblies Intended to Be Installed Outdoors on Public Places—Cable Distr. Cabinets for Power Distr. in Networks. 21 pp.
DIN VDE 0660 Pt 500-84. Deviations from IEC 439-1 1984; Low Voltage Switchgear and Controlgear Assemblies (Nov). 17 pp.
DIN VDE 0660 Pt 500 A2-86. Switchgear and Control Gear; Low Voltage Switchgear and Control Gear Assemblies: Amendment 2 (June). 4 pp.
IEC 439 Pt 1-92. Low-Voltage Switchgear and Controlgear Assemblies Part 1: Type-Tested and Partially Type-Tested Assemblies Third Edition; (CENELEC EN 60439/1/A1: 1993). 198 pp.

—Clearances
CNS C1046-82. Clearances and Creepage Distances for Control Gear (May)(4380).

—Connectors
CENELEC EN 50 043-85. Low-Voltage Switchgear and Controlgear for Industrial Use: Size Numbers and Gauges for Flat Connections. 6 pp.

INTERNATIONAL AND NON-U.S. NATIONAL STANDARDS
SUBJECT INDEX
Controlgear

Controlgear *(Cont.)*

—Electrical Enclosures

SAA AS 3132-91. Approval and Test Specification—Enclosures of Insulating Material for Switchgear and Controlgear Amdt 1 January/February 1993 (in Professional Package 28). 4 pp.

—Electrical Installations—Buildings

CENELEC PRHD 384.5. 53 S2-92. Electrical Installations of Buildings Part 5: Selection and Erection of Equipment Chapter 53: Switchgear and Controlgear Sections 532 and 536. 4 pp.

IEC 364 Pt 5-537-81. Electrical Installations of Buildings Part 5: Selection and Erection of Electrical Equipment Chapter 53: Switchgear and Controlgear Section 537—Devices for Isolation and Switching First Edition; (Amendment 1-1989). 20 pp.

—Electrical Installations—Construction—Fire Protection

IEC 364 Pt 5-53-86. Electrical Installations of Buildings Part 5: Selection and Erection of Electrical Equipment—Chapter 53: Switchgear and Controlgear First Edition; (Amendment 2-1992, Includes Amendment 1). 30 pp.

—Electrical Installations—Ships

IEC 92 Pt 302-80. Electrical Installations in Ships Part 302: Equipment—Switchgear and Controlgear Assemblies Third Edition; (Amendment 1-1989). 28 pp.

IEC 92 Pt 504-74. Electrical Installations in Ships Part 504: Special Features Control and Instrumentation First Edition; (Supplement A-1977). 72 pp.

JIS F 8065-86. Electrical Installations in Ships Part 302 Equipment—Switchgear and Controlgear Assemblies (IEC 92-302-1980).

—Generator Sets—Internal Combustion Piston Engines

ISO 8528 Pt 4-93. Reciprocating Internal Combustion Engine Driven Alternating Current Generating Sets—Part 4: Controlgear and Switchgear First Edition. 16 pp.

—Glossaries

BSI BS 4727:Pt2: Group 06-85. 1985 Electrotechnical, Power, Telecommunication, Electronics, Lighting and Colour Terms Part 2: Terms Particular to Power Engineering Group 06: Switchgear and Controlgear Terminology (Including Fuse Terminology). 27 pp.

IEC 50 Chap 441-84. International Electrotechnical Vocabulary Chapter 441: Switchgear, Controlgear and Fuses Second Edition. 102 pp.

SAA AS 1852.441-85. International Electrotechnical Vocabulary—Part 441: Switchgear, Controlgear and Fuses. 61 pp.

SNZ IEC 50: 50(441)-84. International Electrotechnical Vocabulary 50(441): Switchgear, Controlgear and Fuses. 90 pp.

—High Voltage

BSI BS 5227-92. 1992 A.C. Metal-Enclosed Switchgear and Control-Gear for Rated Voltages Above 1 kV and up to and Including 52 kV. 65 pp.

BSI BS 5227-84. 1984 A.C. Metal-Enclosed Switchgear and Control-Gear of Rated Voltage Above 1 Kv. 44 pp.

BSI BS 6581-85. 1985 Amd 1 Common Requirements for High-Voltage Switchgear and Controlgear Standards (AMD 6463) December 21, 1990. 43 pp.

BSI BS 6626-85. 1985 Amd 1 Code of Practice for Maintenance of Electrical Switchgear and Controlgear for Voltages Above 650V and up to and Including 36kV (AMD 6305) March 28, 1991. 42 pp.

BSI BS 6626-02. 1985 Amd 2 Code of Practice for Maintenance of Electrical Switchgear and Controlgear for Voltages Above 1 kV and up to and Including 36 kV (AMD 6813) December 24, 1991. 44 pp.

BSI BS 6878-88. 1988 Amd 1 High-Voltage Switchgear and Controlgear for Industrial Use. Cast Aluminium Alloy Enclosures for Gas-Filled High-Voltage Switchgear and Controlgear (EN 50 052: 1986) (AMD 6552) April 30, 1991). 21 pp.

BSI BS 7315-90. 1990 Wrought Aluminium and Aluminium Alloy Enclosures for Gas-Filled High Voltage Switchgear and Controlgear. 53 pp.

CENELEC HD 184 S4-89. A.C. Metal-Enclosed Switchgear and Controlgear for Rated Voltages Above 1kv and up to and Including 72.5kV. 15 pp.

CENELEC HD 187 S3-84. A.C. Metal-Enclosed Switchgear and Controlgear for Rated Voltages Above 1kV and up to and Including 72.5kV. 10 pp.

CENELEC HD 187 S4-89. A.C Metal-Enclosed Switchgear and Controlgear for Rated Voltages above 1KV and up to and Including 72.5 KV. 15 pp.

CENELEC HD 187 S5-92. A.C. Metal-Enclosed Switchgear and Control Gear for Rated Voltages Above 1 kV and up to and Including 52 kV. 6 pp.

Controlgear *(Cont.)*

—High Voltage *(Cont.)*

CENELEC HD 448 S2-89. Common clauses for high-voltage switchgear and controlgear standards. 4 pp.

DIN VDE 0670 Pt 3-81. AC Switchgear and Control Gear for Voltages Above 1 kV (Sept). 110 pp.

IEC 298-90. A.C. Metal-Enclosed Switchgear and Controlgear for Rated Voltages Above 1 kV and up to and Including 52 kV Third Edition; (Includes Amendment 1). 115 pp.

IEC 466-87. A.C. Insulation-Enclosed Switchgear and Controlgear for Rated Voltages Above 1 kV and up to and Including 38 kV Second Edition. 97 pp.

IEC 518-75. Dimensional Standardization of Terminals for High-Voltage Switchgear and Controlgear First Edition. 4 pp.

IEC 694-80. Common Clauses for High-Voltage Switchgear and Controlgear Standards First Edition; (Amendment 1-1985) (Amendment 2-1993). 97 pp.

IEC 932-88. Additional Requirements for Enclosed Switchgear and Controlgear from 1 kV to 72.5 kV to Be Used in Severe Climatic Conditions First Edition. 43 pp.

—High Voltage—Electrical Enclosures

BSI BS EN 50 068-91. 1991 Wrought Steel Enclosures for Gas-Filled High-Voltage Switchgear and Controlgear. 56 pp.

BSI BS EN 50069-91. 1991 Welded Composite Enclosures of Cast and Wrought Aluminium Alloys for Gas-Filled High-Voltage Switchgear and Controlgear. 18 pp.

CENELEC EN 50 052-86. Cast Aluminium Alloy Enclosures for Gas-Filled High-Voltage Switchgear and Control Gear. 16 pp.

CENELEC EN 50052/A1-90. AMD 1 Cast Aluminium Alloy Enclosures for Gas-Filled High-Voltage Switchgear and Control Gear. 6 pp.

CENELEC EN 50052 PRA2-93. AMD prA2 Cast Aluminium Alloy Enclosures for Gas-Filled High-Voltage Switchgear and Controlgear. 6 pp.

CENELEC EN 50 064-89. Wrought Aluminium and Aluminium Alloy Enclosures for Gas-Filled High-Voltage Switchgear and Controlgear. 82 pp.

CENELEC EN 50064 PRA1-93. AMD prA1 Wrought Aluminium and Aluminium Alloy Enclosures for Gas-Filled High-Voltage Switchgear and Controlgear. 3 pp.

CENELEC EN 50 068-91. Wrought Steel Enclosures for Gas-Filled High-Voltage Switchgear and Controlgear. 86 pp.

CENELEC EN 50068 PRA1-93. AMD prA1 Wrought Steel Enclosures for Gas-Filled High-Voltage Switchgear and Controlgear. 2 pp.

CENELEC EN 50 069-91. Welded Composite Enclosure of Cost and Wrought Aluminum for Gas-Filled High-Voltage Switchgear and Control Gear. 27 pp.

CENELEC EN 50069 PRA1-93. AMD prA1 Welded Composite Enclosures of Cast and Wrought Aluminium Alloys for Gas-Filled High-Voltage Switchgear and Controlgear. 3 pp.

—High Voltage—Pressurized Partitions

BSI BS EN 50089-92. 1992 Cast Resin Partitions for Metal-Enclosed Gas-Filled High-Voltage Switchgear and Controlgear. 11 pp.

CENELEC EN 50089-92. Cast Resin Partitions for Metal Enclosed Gas-Filled High Voltage Switchgear and Controlgear. 15 pp.

—Insulated Cables

BSI BS 6231-90. 1990 PVC—Insulated Cables for Switchgear and Controlgear Wiring. 13 pp.

BSI BS 6231-01. 1990 Amd 1 PVC-Insulated Cables for Switchgear and Controlgear Wiring (AMD 6936) May 1, 1992. 18 pp.

BSI BS 6231-81. (WITHDRAWN) 1981 PVC—Insulated Cables for Switchgear and Controlgear Wiring. 12 pp.

MOD UK DSTAN 61-12: Part 13-01. Wires, Cords, and Cables, Electrical-Metric Units Part 13: Wires, Electrical (PVC Insulated Non-Sheathed Cables for Switchgear and Controlgear Wiring) Issue 2; Amendment 1. 22 pp.

—Insulation Resistance

CNS C3168-83. Method of Test for Control Gear Insulation (Jan)(9808).

JIS C 0704-81. Insulation Test for Control Gear. 22 pp.

—Leakage Paths

CNS C1046-82. Clearances and Creepage Distances for Control Gear (May)(4380).

—Low Voltage

BSI BS 5420-77. (WITHDRAWN) 1977 Degrees of Protection of Enclosures of Switchgear and Controlgear for Voltages up to and Including 1000 V a.C. and 1200 V d.c. (Superseded by BS EN 60947-1: 1992). 15 pp.

Controlgear *(Cont.)*

—Low Voltage *(Cont.)*

BSI BS 5824-80. 1980 Low Voltage Switchgear and Controlgear for Industrial Use. Mounting Rails. C-Profile and Accessories for the Mounting of Equipment. 10 pp.

BSI BS 5994-80. (WITHDRAWN) 1980 Industrial Low Voltage Switchgear and Control-Gear. Terminal Aperture Sizes for Unprepared Round Copper Conductors (Superseded by BS EN 60947-1: 1992). 7 pp.

BSI BS 6272-82. 1982 Low Voltage Switchgear and Controlgear for Industrial Use. Terminal Marking. Terminals for External Associated Electronic Circuit, Components and Contacts. 9 pp.

BSI BS 6733-87. (WITHDRAWN) 1987 Low-Voltage Switchgear and Controlgear for Industrial Use. Requirements Applicable to Terminals Concerning Cross-Sections of Connectable Conductors (Superseded by BS EN 60947-1: 1992). 6 pp.

BSI BS EN 60947-1-92. 1992 Low-Voltage Switchgear and Controlgear Part 1: General Rules (Supersedes BS 5994: 1980 & BS 6733: 1987). 147 pp.

BSI BS EN 60947-1-01. 1992 Amd 1 Low-Voltage Switchgear and Controlgear Part 1: General Rules (AMD 7336) December 15, 1992 (Supersedes BS 5994: 1980 & BS 6733: 1987) (E). 148 pp.

BSI BS EN 60947-1-02. 1992 Amd 2 Low-Voltage Switchgear and Controlgear Part 1: General Rules (AMD 7852) July 15, 1993 (E). 149 pp.

CENELEC HD 528 S1-89. A Method of Temperature—Rise Assessment by Extrapolation for Partially Type-Tested Assemblies (PTTA) of Low Voltage Switchgear and Controlgear. 3 pp.

CENELEC EN 50 005-76. Low Voltage Switchgear and Controlgear for Industrial Use: Terminal Marking and Distinctive Number: General Rules. 7 pp.

CENELEC EN 50 022-77. Low Voltage Switchgear and Controlgear for Industrial use Mounting Rails Top Hat Rails 35 mm Wide for Snap-on Mounting of Equipment. 7 pp.

CENELEC EN 50 023-77. Low Voltage Switchgear and Controlgear for Industrial use Mounting Rails Top Hat Rails 75 mm Wide for Snap-on Mounting of Equipment. 6 pp.

CENELEC EN 50 024-78. Low Voltage Switchgear and Controlgear for Industrial Use: Mounting Profile and Accessories for the Mounting of Equipment. 8 pp.

CENELEC EN 50 027-79. Industrial Voltage Switchgear and Controlgear: Terminal Aperture Sizes for Unprepared Round Copper Conductors. 5 pp.

CENELEC EN 50 042-80. Low Voltage Switchgear and Controlgear for Industrial Use: Terminal Marking: Terminals for External Associated Electronic Circuit Components and Contacts. 7 pp.

CENELEC EN 50 043-85. Low-Voltage Switchgear and Controlgear for Industrial Use: Size Numbers and Gauges for Flat Connections. 6 pp.

CENELEC EN 50 051-85. Low-Voltage Switchgear and Controlgear for Industrial Use Requirements Applicable to Terminals Concerning Cross-Sections of Connectable Conductors. 4 pp.

CENELEC EN 60947-1-91. Low-Voltage Switchgear and Controlgear Part 1: General Rules. 9 pp.

CENELEC EN 60947-1-91. Low-Voltage Switchgear and Controlgear Part 1: General Rules (Supersedes EN 50027: 1979 and EN 50051: 1985) (IEC 947-1: 1988; IEC/SC 17B (Not Appended)). 9 pp.

CENELEC EN 60947-1-92. CORRIGENDUM Low Voltage Switchgear and Controlgear Part 1: General Rules. 1 p.

CENELEC EN 60947-1/PRAA-93. AMD prAA Low-Voltage Switchgear and Controlgear Part 1: General Rules. 3 pp.

DIN VDE 0660 Pt 12-82. Switchgear and Control Gear; Low Voltage Switchgear and Control Gear; Terminals for Protective Conductors (Sept). 7 pp.

DIN VDE 0660 Pt 99-82. Switchgear and Control Gear; Low-Voltage Switchgear and Control Gear; Connectable Conductor Cross-Sections (Sept). 5 pp.

DIN VDE 0660 Pt 100 (D)-85. Switchgear and Control Gear; Low-Voltage Switchgear and Controlgear; General Rules; Terminology and Requirements; Identical to IEC 17B(CO)123-I and II (June). 101 pp.

DIN VDE 0660 Pt 100 A1 (D)-85. Low Voltage Switchgear and Controlgear Tests; Amendment 1 to Draft DIN VDE 0660 Part 100/06.85 Identical to IEC 17B(CO) 124 (July). 46 pp.

DIN VDE 0660 Pt 100 A2 (D)-85. Low Voltage Switchgear and Controlgear Tests; Amendment 2 to Draft DIN VDE 0660 Part 100/06.85 Identical to IEC 17B(CO) 125 (July). 8 pp.

DIN VDE 0660 Pt 100 A3 (D)-86. Low Voltage Switchgear and Controlgear Tests; Amendment 3 to Draft DIN VDE 0660 Part 100/06.86 Identical to IEC 17B(CO) 129 (June). 19 pp.

DIN VDE 0660 Pt 200-82. Switchgear and Control Gear; Low Voltage Switchgear and Control Gear; Control Switches; General Requirements (Sept). 34 pp.

INDUSTRY STANDARDS
637

INTERNATIONAL AND NON-U.S. NATIONAL STANDARDS SUBJECT INDEX

Controlgear (Cont.)

—Low Voltage (Cont.)
DIN VDE 0660 Pt 200 A1 (D)-87. Switchgear and Controlgear; Low-Voltage Switchgear and Controlgear; Control Circuit Devices and Switching Elements (Aug). 15 pp.

DIN VDE 0660 Pt 200 A2 (D)-89. Switchgear and Controlgear; Low-Voltage Switchgear and Controlgear; Control Circuit Devices and Switching Elements; Amendment 2 (Aug). 9 pp.

DIN VDE 0660 Pt 500-84. Deviations from IEC 439-1 1984; Low Voltage Switchgear and Controlgear Assemblies (Nov). 17 pp.

DIN VDE 0660 Pt 500 A2-86. Switchgear and Control Gear; Low Voltage Switchgear and Control Gear Assemblies: Amendment 2 (June). 4 pp.

DIN VDE 0660 Pt 504 (D)-88. Switchgear and Controlgear; Low-Voltage Switchgear and Controlgear; Additional Specifications for Type-Tested Assemblies Intended to Be Installed in Places Where Unskilled Persons Have Access for Their Use (Nov). 26 pp.

IEC 439 Pt 4-90. Low-Voltage Switchgear and Controlgear Assemblies Part 4: Particular Requirements for Assemblies for Construction Sites (ACS) First Edition. 50 pp.

IEC 715-81. Dimensions of Low-Voltage Switchgear and Controlgear Standardized Mounting on Rails for Mechanical Support of Electrical Devices in Switchgear and Controlgear Installations First Edition. 45 pp.

IEC 890-87. A Method of Temperature-Rise Assessment by Extrapolation for Partially Type-Tested Assemblies (PTTA) of Low-Voltage Switchgear and Controlgear First Edition. 54 pp.

IEC 947 Pt 1-88. Low-Voltage Switchgear and Controlgear Part 1: General Rules First Edition; (Corrigendum—June 1989) (Corrigendum—Jan 1992) (CENELEC EN 60947-1: 1991). 247 pp.

IEC 947 Pt 5-1-90. Low-Voltage Switchgear and Controlgear Part 5: Control Circuit Devices and Switching Elements Section One—Electromechanical Control Circuit Devices First Edition. 131 pp.

IEC 947 Pt 7-1-89. Low-Voltage Switchgear and Controlgear Part 7: Ancillary Equipment Section One—Terminal Blocks for Copper Conductors First Edition. 43 pp.

SAA AS 1136.1-88. Low Voltage Switchgear and Controlgear Assemblies—Part 1: General Requirements Amdt 1 March 1989. 59 pp.

SAA AS 1431.1-89. Low Voltage Switchgear and Controlgear—Control Circuit Devices and Switching Elements—Part 1: General Requirements (This is a Joint Standard with SNZ NZS 1431.1). 27 pp.

SAA AS 3650-88. Low Voltage Switchgear and Controlgear—Common Requirements (This is a Joint Standard with SANZ NZS 3650). 61 pp.

SNZ NZS/AS 1136. 1-88. Low Voltage Switchgear and Controlgear Assembiles—Part 1: General Requirements Amend: 1 (This is a Joint Standard with SAA AS 1136.1). 59 pp.

SNZ NZS/AS 1431. 1-89. Low Voltage Switchgear and Controlgear—Control Circuit Devices and Switching Elements Part 1: General Requirements (This is a Joint Standard with SAA AS 1431.1). 27 pp.

SNZ NZS/AS 3650-88. Low Voltage Switchgear and Controlgear—Common Requirements (This is a Joint Standard with SAA AS 3650). 61 pp.

—Low Voltage—Busways
BSI BS EN 60439-2-93. 1993 Low-Voltage Switchgear and Controlgear Assemblies Part 2: Particular Requirements for Busbar Trunking Systems (Busways) (AMD 7775) July 15, 1993 (E). 27 pp.

CENELEC EN 60439-2-93. Low-Voltage Switchgear and Controlgear Assemblies Part 2: Particular Requirements for Busbar Trunking Systems (Busways) (IEC 439-2: 1987 + A1: 1991, Modified). 8 pp.

CENELEC EN 60439-2-93. Low-Voltage Switchgear and Controlgear Assemblies Part 2: Particular Requirements for Busbar Trunking Systems (Busways) (IEC 439-2: 1987 + A1: 1991, Modified). 15 pp.

IEC 439 Pt 2-87. Low-Voltage Switchgear and Controlgear Assemblies Part 2: Particular Requirements for Busbar Trunking Systems (Busways) Second Edition; (Amendment 1-1991) (CENELEC EN 60439-2: 1993). 42 pp.

—Low Voltage—Junction Boxes
DIN VDE 0660 Pt 505-90. Switchgear and Controlgear; Low-Voltage Switchgear and Controlgear Assemblies; Specification for Domestic Connection Boxes and Fuseboxes (July). 38 pp.

—Low Voltage—Multiposition Switches
DIN ENGL 43697-82. Low Voltage Switchgear and Controlgear; Multi-Position Switches (Apr). 4 pp.

Controlgear (Cont.)

—Low Voltage—Railroads
DIN VDE 0660 Pt 14-82. Switchgear and Control Gear; Low Voltage Switchgear and Control Gear; Additional Requirements for Born Vehicles (Sept). 11 pp.

—Low Voltage—Switching Devices—Control and Protective Equipment
IEC 947 Pt 6-2-92. Low-Voltage Switchgear and Controlgear Part 6: Multiple Function Equipment Section Two—Control and Protective Switching Devices (or Equipment) (CPS) First Edition; (CENELEC EN 60947-6-2: 1993). 115 pp.

—Low Voltage—Top Hat Rails
BSI BS 5584-78. 1978 Low-Voltage Switch Gear and Controlgear for Industrial Use. Mounting Rails. Top Hat Rails 35 mm Wide for Snap-On Mounting of Equipment. 9 pp.

BSI BS 5585-78. 1978 Low-Voltage Switch Gear and Controlgear for Industrial Use. Mounting Rails. Top Hat Rails 75 mm Wide for Snap-On Mounting of Equipment. 8 pp.

—Naval Ships
MOD UK NES 635-89. Requirements for Electric Motors and Control Gear Associated with Domestic Workshops and Similar Machinery Issue 3 (01.89). 21 pp.

MOD UK NES 635-93. Requirements for Electric Motors and Control Gear Associated with Domestic Workshops and Similar Machinery Issue 4 (01.93). 29 pp.

MOD UK NES 636-89. Requirements for AC and DC Starting and Control Gear Issue 2 (10.89). 40 pp.

—Noise
DIN ENGL 45635 Pt 12-78. Measurement of Noise Emitted by Machines; Airbourne Noise Measurement, Enveloping Surface Method; Electrical Switchgear and Control Gear (Mar). 5 pp.

—Rails—Mounting
BSI BS 5824-80. 1980 Low Voltage Switchgear and Controlgear for Industrial Use. Mounting Rails. C-Profile and Accessories for the Mounting of Equipment. 10 pp.

CENELEC EN 50 022-77. Low Voltage Switchgear and Controlgear for Industrial use Mounting Rails Top Hat Rails 35 mm Wide for Snap-on Mounting of Equipment. 7 pp.

CENELEC EN 50 023-77. Low Voltage Switchgear and Controlgear for Industrial use Mounting Rails Top Hat Rails 75 mm Wide for Snap-on Mounting of Equipment. 6 pp.

CENELEC EN 50 024-78. Low Voltage Switchgear and Controlgear for Industrial Use: Mounting Profile and Accessories for the Mounting of Equipment. 8 pp.

—Ships—Generator Sets—Internal Combustion Piston Engines
ISO 8528 Pt 4-93. Reciprocating Internal Combustion Engine Driven Alternating Current Generating Sets—Part 4: Controlgear and Switchgear First Edition. 16 pp.

—Submarines
MOD UK NES 635-89. Requirements for Electric Motors and Control Gear Associated with Domestic Workshops and Similar Machinery Issue 3 (01.89). 21 pp.

MOD UK NES 635-93. Requirements for Electric Motors and Control Gear Associated with Domestic Workshops and Similar Machinery Issue 4 (01.93). 29 pp.

MOD UK NES 636-89. Requirements for AC and DC Starting and Control Gear Issue 2 (10.89). 40 pp.

—Symbols
CNS C1020-57. Symbols of Switch Gears and Control Gears (Jan)(404) (R 1971).

—Symbols—Diagrams
BSI BS 3939: Part 7-85. 1985 Graphical Symbols for Electrical Power, Telecommunications and Electronics Diagrams Part 7: Switchgear, Controlgear and Protective Devices (G). 32 pp.

IEC 617 Pt 7-83. Graphical Symbols for Diagrams Part 7: Switchgear, Controlgear and Protective Devices First Edition. 58 pp.

SAA AS 1102.107-89. Graphical Symbols for Electrotechnology—Part 107: Switchgear, Controlgear and Protective Devices. 40 pp.

SNZ IEC 617: Part 7-83. Graphical Symbols for Diagrams Part 7: Switchgear, Controlgear and Protective Devices. 56 pp.

Controlgear (Cont.)

—Terminals
BSI BS 5994-80. (WITHDRAWN) 1980 Industrial Low Voltage Switchgear and Control-Gear. Terminal Aperture Sizes for Unprepared Round Copper Conductors (Superseded by BS EN 60947-1: 1992). 7 pp.

BSI BS 6733-87. (WITHDRAWN) 1987 Low-Voltage Switchgear and Controlgear for Industrial Use. Requirements Applicable to Terminals Concerning Cross-Sections of Connectable Conductors (Superseded by BS EN 60947-1: 1992). 6 pp.

CENELEC EN 50 027-79. Industrial Voltage Switchgear and Controlgear: Terminal Aperture Sizes for Unprepared Round Copper Conductors. 5 pp.

CENELEC EN 50 051-85. Low-Voltage Switchgear and Controlgear for Industrial Use Requirements Applicable to Terminals Concerning Cross-Sections of Connectable Conductors. 4 pp.

DIN VDE 0660 Pt 12-82. Switchgear and Control Gear; Low Voltage Switchgear and Control Gear; Terminals for Protective Conductors (Sept). 7 pp.

DIN VDE 0660 Pt 99-82. Switchgear and Control Gear; Low-Voltage Switchgear and Control Gear; Connectable Conductor Cross-Sections (Sept). 5 pp.

IEC 518-75. Dimensional Standardization of Terminals for High-Voltage Switchgear and Controlgear First Edition. 4 pp.

—Terminals—Identification Systems
BSI BS 5472-77. 1977 Low Voltage Switchgear and Controlgear for Industrial Use. Terminal Marking and Distinctive Number. General Rules. 9 pp.

BSI BS 6272-82. 1982 Low Voltage Switchgear and Controlgear for Industrial Use. Terminal Marking. Terminals for External Associated Electronic Circuit, Components and Contacts. 9 pp.

CENELEC EN 50 005-76. Low Voltage Switchgear and Controlgear for Industrial Use: Terminal Marking and Distinctive Number: General Rules. 7 pp.

CENELEC EN 50 042-80. Low Voltage Switchgear and Controlgear for Industrial Use: Terminal Marking: Terminals for External Associated Electronic Circuit Components and Contacts. 7 pp.

Controlled Atmospheres

See Also: Clean Rooms; Cold Storage; Environmental Conditions; Humidity; Humidity Test Chambers; Preservation; Test Chambers

BSI BS 4194-67. (WITHDRAWN) 1967 Design Requirements and Testing of Controlled Atmosphere Laboratories. 24 pp.

BSI BS 5295: Part 1-89. 1989 Environmental Cleanliness in Enclosed Spaces Part 1: Specification for Clean Rooms and Clean Air Devices (N). 15 pp.

BSI BS 5295: Part 1-01. 1989 Amd 1 Environmental Cleanliness in Enclosed Spaces Part 1: Clean Rooms and Clean Air Devices (AMD 6602) December 21, 1990. 17 pp.

BSI BS 5295: Part 2-89. 1989 Environmental Cleanliness in Enclosed Spaces Part 2: Method for Specifying the Design, Construction and Commissioning of Clean Rooms and Clean Air Devices (N). 14 pp.

BSI BS 5295: Part 2-01. 1989 Amd 1 Environmental Cleanliness in Enclosed Spaces Part 2: Method for Specifying the Design, Construction and Commissioning of Clean Rooms and Clean Air Devices (AMD 6603) December 21, 1990. 15 pp.

BSI BS 5295: Part 3-89. 1989 Environmental Cleanliness in Enclosed Spaces Part 3: Guide to Operational Procedures and Disciplines Applicable to Clean Rooms and Clean Air Devices. 8 pp.

SAA AS 1386.1-89. Cleanrooms and Clean Workstations—Part 1: Principles of Clean Space Control. 22 pp.

—Constant Test Atmospheres
DIN ENGL 50015-75. Atmospheres and Their Technical Application; Constant Test Atmospheres (Aug). 2 pp.

Controllers

Scope Note: For additional listings, use a more specific term *Use For:* Electric Controllers
See Also: Analyzers; Automatic Control Equipment; Building Automation Systems; Bus Controllers; Combustion Control Equipment; Computer Numerical Control; Control Systems Equipment; Control Units; Cryostats; Floppy Disk Controllers; Gateways; Indicating Instruments; Interrupt Controllers; Microcontrollers; Motor Controllers; Peripheral Controllers; Photoelectric Controls; Pneumatic Controllers; Pneumatic Servomechanisms; Power Controllers; Pressure Regulators; Priority Interrupt Controllers; Process Control Systems; Programmable Controllers; Recording Instruments; Regulators; Remote Control Equipment; Revolution Counters; Servomotors; Servovalves; Speed Control Testing; Speed Controls; Steam Traps; Temperature

Controllers (Cont.)

See Also: (Cont.)
Controllers; Thermostatic Equipment; Thermostatic Valves; Thermostats; Voltage Regulators

—Drilling Equipment
BSI BS 787: Part 4-72. 1972 Mining Type Flameproof Gate-End Boxes Part 4: Gate-End Boxes for Drilling Machines (For Use on 3-Phase A.C. Circuits up to X 650 V). 24 pp.
BSI BS 5126: Part 4-82. 1982 Mining Type Flameproof Supply and Control Units for Use on Systems up to 1100V Part 4: Units to Power Drilling Machines. 16 pp.

—Earth Stations
CEPT T/CCTS 2-85. Lignes Directrices Pour La Specification Des Equipements De Petites Stations Terriennes. 9 pp.

—Electric Fences
BSI BS 2632-80. (WITHDRAWN) 1980 Amd 1 Mains-Operated Electric Fence Controllers (Superseded by BS EN 61011: 1993). 36 pp.
BSI BS 6167-81. (WITHDRAWN) 1981 Amd 1 Battery-Operated Electric Fence Controllers Not Suitable for Connection to the Supply Mains (Superseded by BS EN 61011-2: 1993). 28 pp.
BSI BS 6369-83. (WITHDRAWN) 1983 Amd 1 Battery-Operated Electric Fence Contollers Suitable for Connection to the Supply Mains (Superseded by BS EN 61011-1: 1993). 40 pp.
CENELEC HD 92-74. Mains Operated Electric Fence Controllers. 5 pp.
SNZ NZS 2210-68. Battery Operated Electric Fence Controllers Amend: 1, 1972. 16 pp.
SNZ NZS 6203: Part 1-87. Specification for Agricultural Electric Fencing (Safety Requirements Part 1: Alternating Current Mains-Operated Electric Fence Energizers Amend: 1, 1989; 2, 1991. 24 pp.

—Electric Fences—Peak Discharge
CSA CAN/CSA-C22. 2 NO 103-M92. Electric Fence Controllers; (Gen Instr 1 Thru 2). 59 pp.

—Electric Fences—Safety
BSI BS EN 61011-93. 1993 Electric Fence Energizers Safety Requirements for Mains-Operated Electric Fence Energizers (Supersedes BS 2632: 1980) (E). 116 pp.
BSI BS EN 61011-1-93. 1993 Electric Fence Energizers Safety Requirements for Battery-Operated Electric Fence Energizers Suitable for Connection to the Supply Mains (Supersedes BS 6369: 1983) (E). 24 pp.
BSI BS EN 61011-2-93. 1993 Electric Fence Energizers Safety Requirements for Battery-Operated Electric Fence Energizers Not for Connection to the Supply Mains (Supersedes BS 6167: 1981) (E). 27 pp.
CENELEC PREN 61011-91. Electric Fence Energizers. Safety Requirements for Mains-Operated Electric Fence Energizers. 111 pp.
CENELEC EN 61011-92. Electric Fence Energizers Safety Requirements for Mains-Operated Electric Fence Energizers (IEC 1011: 1989, Modified). 107 pp.
CENELEC EN 61011/PRAA-92. AMD A Electric Fence Energizers—Safety Requirements for Mains-Operated Electric Fence Energizers. 2 pp.
CENELEC PREN 61011-1-90. Electric Fence Energizers Part 1: Safety Requirements for Battery-Operated Electric Fence Energizers Suitable for Connection to the Supply Mains.
CENELEC EN 61011-1-92. Electric Fence Energizers Part 1: Safety Requirements for Battery-Operated Electric Fence Energizers Suitable for Connection to the Supply Mains. 8 pp.
CENELEC EN 61011-1-92. Electric Fence Energizers Safety Requirements for Battery-Operated Electric Fence Energizers Suitable for Connection to the Supply Mains. 10 pp.
CENELEC EN 61011-1/PRAA-92. AMD A Electric Fence Energizers Safety Requirements for Battery-Operated Electric Fence Energizers Suitable for Connection to the Supply Mains. 2 pp.
CENELEC PREN 61011-2-91. Electric Fence Energizers Part 2: Safety Requirements for Battery-Operated Electric Fence Energizers Not for Connection to the Supply Mains. 27 pp.
CENELEC EN 61011-2-92. Electric Fence Energizers Part 2: Safety Requirements for Battery-Operated Electric Fence Energizers Not for Connection to the Supply Mains. 7 pp.
CENELEC EN 61011-2/PRAA-92. AMD A Electric Fence Energizers Safety Requirements for Battery-Operated Electric Fence Energizers Not for Connection to the Supply Mains. 2 pp.
IEC 1011-89. Electric Fence Energizers Safety Requirements for Mains-Operated Electric Fence Energizers First Edition; (Amendment 1-1991) (Corrigendum—June 1993) (Amendment 2-1993) (CENELEC EN 61011:1992). 268 pp.

Controllers (Cont.)

—Electric Fences—Safety (Cont.)
IEC 1011 Pt 1-89. Electric Fence Energizers Safety Requirements for Battery-Operated Electric Fence Energizers Suitable for Connection to the Supply Mains First Edition; (Amendment 2-1993, Incorporating Amendment 1). 45 pp.
IEC 1011 Pt 2-90. Electric Fence Energizers Safety Requirements for Battery-Operated Electric Fence Energizers Not for Connection to the Supply Mains First Edition; (Amendment 2-1993, Incorporating Amendment 1). 51 pp.

—Electric Fences—Timed Sinusoidal
CSA CAN/CSA-C22. 2 NO 103-M92. Electric Fence Controllers; (Gen Instr 1 Thru 2). 59 pp.
CSA 1169 Bull. Electrical Bulletin 1169 June 27, 1978 to C22.2 NO 103. 2 pp.

—Electric Vehicles
IEC 786-84. Controllers for Electric Road Vehicles First Edition. 30 pp.

—Hydraulic Valves—Identification Systems
ISO 9461-92. Hydraulic Fluid Power—Identification of Valve Ports, Subplates, Control Devices and Solenoids First Edition. 7 pp.

—Process Control Equipment
BSI BS 5558: Part 1-89. 1989 Controllers with Analogue Signals for Use in Industrial-Process Control Systems Part 1: Methods for Evaluating Performance (Renumbered as BS EN 60546-1: 1993). 36 pp.
BSI BS 5558: Part 1-01. (WITHDRAWN) 1989 Amd 1 Controllers with Analogue Signals for Use in Industrial-Process Control Systems Part 1: Methods for Evaluating Performance (IEC 546-1: 1987) (AMD 7546) February 15, 1993 (Renumbered as BS EN 60546-1: 1993) (Q). 37 pp.
BSI BS 5558: Part 2-89. 1989 Controllers with Analogue Signals for Use in Industrial-Process Control Systems Part 2: Guide to Inspection and Routine Testing (Renumberded as BS EN 60546-2: 1993). 12 pp.
BSI BS 5558: Part 2-01. (WITHDRAWN) 1989 Amd 1 Controllers with Analogue Signals for Use in Industrial-Process Control Systems Part 2: Guide to Inspection and Routine Testing (IEC 546-2:1987) (AMD 7547) February 15, 1993 (Renumbered as BS EN 60546-2: 1993) (Q). 13 pp.
BSI BS EN 60546-1-93. 1993 Amd 2 Controllers with Analogue Signals for Use in Industrial-Process Control Systems Part 1: Methods of Evaluating the Performance (IEC 546-1: 1987) (AMD 7861) July 15, 1993 (Q). 49 pp.
BSI BS EN 60546-2-93. 1993 Amd 2 Controllers with Analogue Signals for Use in Industrial-Process Control Systems Part 2: Guidance for Inspection and Routine Testing (IEC 546-2: 1987) (AMD 7862) July 15, 1993 (Q). 21 pp.
CENELEC HD 530.1 S1-89. Controllers with Analogue Signals for Use in Industrial—Process Control Systems Part 1: Methods of Evaluating the Performance. 4 pp.
CENELEC HD 530.2 S1-89. Controllers with Analogue Signals for Use in Industrial—Process Control Systems Part 2: Guidance for Inspection and Routine Testing. 3 pp.
CENELEC EN 60546-1-93. Controllers with Analogue Signals for Use in Industrial-Process Control Systems Part 1: Methods of Evaluating the Performance (IEC 546-1: 1987). 41 pp.
CENELEC EN 60546-2-93. Controllers with Analogue Signals for Use in Industrial-Process Control Systems Part 2: Guidance for Inspection and Routine Testing (IEC 546-2: 1987). 11 pp.
IEC 546 Pt 1-87. Controllers with Analogue Signals for Use in Industrial-Process Control Systems Part 1: Methods of Evaluating the Performance Second Edition; (CENELEC EN 60546-1: 1993). 65 pp.
IEC 546 Pt 2-87. Controllers with Analogue Signals for Use in Industrial-Process Control Systems Part 2: Guidance for Inspection and Routine Testing First Edition; (CENELEC EN 60546-2: 1993). 18 pp.
JIS C 1801-86. Method of Evaluating the Performance of Controllers with Analogue Signals for Use in Industrial Process Control. 22 pp.

—Robots—Symbols
JIS B 8434-91. Industrial Robots—Identification Symbols for Operator Controls. 26 pp.

Controls, Manual
Use: Manual Controls

Convair CV240 Aircraft
See Also: Aircraft

—Accidents
CAA. Convair CV240, 340 and 440. 1 p.

Convair CV340 Aircraft
See Also: Aircraft

—Accidents
CAA. Convair CV240, 340 and 440. 1 p.

Convair CV440 Aircraft
See Also: Aircraft

—Accidents
CAA. Convair CV240, 340 and 440. 1 p.

Convair CV580 Aircraft
See Also: Aircraft

—Accidents
CAA. Convair CV580 and 640 Series (World Airline Accident Summary). 1 p.
CAA. Convair CV580 and 640 Series. 1 p.

Convair CV640 Aircraft
See Also: Aircraft

—Accidents
CAA. Convair CV580 and 640 Series (World Airline Accident Summary). 1 p.
CAA. Convair CV580 and 640 Series. 1 p.

Convair CV880 Aircraft
See Also: Aircraft

—Accidents
CAA. Convair CV880 and CV990 Series (World Airline Accident Summary). 1 p.
CAA. Convair CV880 and CV990 Series. 1 p.

Convair CV990 Aircraft
See Also: Aircraft

—Accidents
CAA. Convair CV880 and CV990 Series (World Airline Accident Summary). 1 p.
CAA. Convair CV880 and CV990 Series. 1 p.

Convair PBY 5 Catalina Aircraft
See Also: Aircraft

—Accidents
CAA. Convair PBY5 Catalina. 1 p.
CAA. Convair CV240, 340 and 440 (World Airline Accident Summary). 3 pp.
CAA. Convair PBY5 Catalina. 1 p.

Convection Heaters
Use: Convectors

Convection Ovens
See Also: Appliances; Cooking Appliances; Ovens
CGSB CAN/CGSB-52.23-M88. Ovens, Convection, Electric. 11 pp.

—Safety
BSI BS 3456: Sec 102.42-89. (WITHDRAWN) 1989 Safety of Household Electrical Appliances Part 102: Particular Requirements Section 102.42: Commercial Electric Forced Convention Ovens (Superseded by BS 3456: Section 202.42: 1990). 19 pp.
BSI BS 3456: Sec 202.42-90. (WITHDRAWN) 1990 Safety of Household Electrical Appliances Part 202: Particular Requirements Section 202.42: Commercial Electric Forced Convection Ovens (Supersedes BS 3456: Section 102.42: 1989). 22 pp.
BSI BS 3456: Sec 202.42-01. 1990 Amd 1 Safety of Household and Similar Electrical Appliances Part 202: Particular Requirements Section 202.42: Commercial Electric Forced Convection Ovens (AMD 7831) July 15, 1993 (L). 38 pp.
BSI BS 5386: Part 6-91. 1991 Gas Burning Appliances Part 6: Specification for Domestic Gas Cooking Appliances with Forced-Convection Ovens. 9 pp.
BSI BS 5386: Part 6-01. 1991 Amd 1 Gas Burning Appliances Part 6: Specification for Domestic Gas Cooking Appliances with Forced-Convection Ovens (AMD 6926) February 28, 1992. 10 pp.
CENELEC HD 288 S1-87. Safety of Household and Similar Electric Appliances—Part 2: Particular Requirements for Commercial Electric Forced Convection Ovens. 9 pp.
CENELEC EN 60 335-2-42-89. Safety of Household and Similar Electrical Appliances Part 2 Particular Requirements for Commercial Electric Forced Convection Ovens. 9 pp.
CENELEC EN 60335-2-42/A1-92. AMD 1 Safety of Household and Similar Electrical Appliances Part 2 Particular Requirements for Commercial Electric Forced Convection Ovens. 6 pp.
IEC 335 Pt 2-42-87. Safety of Household and Similar Electrical Appliances Part 2: Particular Requirements for Commercial Electric Forced Convection Ovens Second Edition; (Amendment 1-1990). 48 pp.

INTERNATIONAL AND NON-U.S. NATIONAL STANDARDS
SUBJECT INDEX

Convectors
Use For: Convection Heaters *See Also:* Air Handling Equipment; Heaters; Heating Equipment; Space Heaters
- BSI BS 3561-62. 1962 Amd 2 Non-Domestic Space Heaters Burning Town Gas. 70 pp.
- BSI BS 4096-67. 1967 Amd 1 Non-Domestic Space Heaters Burning Liquefied Petroleum Gases. 76 pp.
- CEN PREN 613-91. Independent Gas-Fired Convection Heaters. 90 pp.
- DIN VDE 0720 Pt 2 FA-79. Particular Specification for Room Heaters and Similar Appliances; Partial Modification A (Feb). 3 pp.
- DIN VDE 0722-83. Electrical Equipment on Non-Electrically Heated Cooking and Heating Appliances (Apr). 31 pp.
- JIS A 4004-87. Convectors, Baseboard-Heaters and Panel-Radiators. 24 pp.
- JIS A 4007-87. Fan Convector. 30 pp.

—**Electric**
- CSA C22.2 NO 46-M1988. Electric Air-Heaters. 76 pp.
- CSA 1169 Bull. Electrical Bulletin 1169 June 27, 1978 to C22.2 NO 46. 2 pp.

—**Installation**
- BSI BS 5871-80. (WITHDRAWN) 1980 Amd 2 Installation of Gas Fires, Convectors and Fire/Back Boilers (2nd Family Gas) (AMD 4638) March 29, 1985 (Superseded by BS 5871: Parts 1&2: 1991). 22 pp.
- BSI BS 5871: Part 1-91. 1991 Installation of Gas Fires, Convector Heaters, Fire/Back Boilers and Decorative Fuel Effect Gas Appliances Part 1: Gas Fires, Convector Heaters and Fire/Back Boilers (1st, 2nd and 3rd Family Gases). 32 pp.
- BSI BS 8303-86. 1986 Amd 1 Code of Practice for Installation of Domestic Heating and Cooking Appliances Burning Solid Mineral Fuels (AMD 5723) October 30, 1987. 27 pp.

—**Noise**
- BSI BS 6686: Sec 2.2-90. 1990 Methods for Determination of Airborne Acoustical Noise Emitted by Household and Similar Electrical Appliances Part 2: Particular Requirements Section 2.2: Forced Draught Convection Heaters. 13 pp.
- CENELEC HD 423.2.2 S1-88. Test Code for the Determination of Airborne Acoustical Noise Emitted by Household and Similar Electrical Appliances-Part 2: Particular Requirements for Forced Draught Convection Heaters. 2 pp.
- IEC 704 Pt 2-2-85. Test Code for the Determination of Airborne Acoustical Noise Emitted by Household and Similar Electrical Appliances Part 2: Particular Requirements for Forced Draught Convection Heaters First Edition. 22 pp.

—**Portable**
- BSI CP 3008-70. 1970 Use of Transportable Industrial Space Heaters. 18 pp.

—**Radiation Capacity**
- JIS A 1403-88. Convectors, Baseboard-Heaters, Panel-Radiators, Cast-Iron Radiators and Similar Appliances-Testing Method of Steam Heating Capacity. 9 pp.

—**Safety**
- BSI BS 5258: Part 13-86. 1986 Safety of Domestic Gas Appliances Part 13: Convector Heaters. 33 pp.
- CEN PREN 525-91. Requirements for Non-Domestic Direct Gas-Fired Forced Convection Air Heaters for Space Heating. 74 pp.

—**Thermal Efficiency**
- BSI BS 6332: Part 4-83. 1983 Thermal Performance of Domestic Gas Appliances Part 4: Specification for Thermal Performance of Independent Convector Heaters. 12 pp.

—**Thermal Measurement**
- CEN PREN 442-91. Radiators, Convectors and Similar Appliances—Testing and Rating Standard. 83 pp.
- ISO 3148-75. Radiators, Convectors and Similar Appliances—Determination of Thermal Output—Test Method Using Air-Cooled Closed Booth First Edition. 5 pp.
- ISO 3149-75. Radiators, Convectors and Similar Appliances—Determination of Thermal Output—Test Method Using Liquid-Cooled Closed Booth First Edition. 5 pp.
- ISO 3150-75. Radiators, Convectors and Similar Appliances—Calculation of Thermal Output and Presentation of Results First Edition. 4 pp.

Conversion Coatings
Use For: Chemical Coatings; Chemical Conversion Coatings *See Also:* Anodic Coatings; Chromate Coatings; Coatings; Hard Surfacing; Iron Phosphate Coatings (Made From Iron Phosphate); Manganese Phosphate Coatings (Made From Manganese Phosphate); Metal Coatings (Made From Metal); Metal Coatings (On Metal); Nonmetallic Coatings; Organic Coatings; Paints; Phosphate Coatings; Zinc Phosphate Coatings (Made From Zinc Phosphate)

—**Aluminum**
- ISO 10546-93. Chemical Conversion Coatings—Rinsed and Non-Rinsed Chromate Conversion Coatings on Aluminium and Aluminium Alloys First Edition. 10 pp.

—**Aluminum—Aerospace**
- ISO 8081-85. Aerospace Process—Chemical Conversion Coating for Aluminium Alloys—General Purpose First Edition. 6 pp.

—**Aluminum—Aircraft**
- JIS W 1110-90. Chemical Conversion Coatings on Aluminum and Aluminum Alloys for Aircraft.

—**Aluminum—Brushed**
- CGSB 31-GP-101MA-89. Chemical Conversion Films for Aluminum and Aluminum Alloys. 9 pp.

—**Aluminum—Dipped**
- CGSB 31-GP-101MA-89. Chemical Conversion Films for Aluminum and Aluminum Alloys. 9 pp.

—**Aluminum—Salt Spray Testing**
- CGSB 31-GP-0A METH 14.4A-62. Methods of Testing Corrosion-Prevention Materials and Processes Salt Spray Resistance of Conversion Coatings (Supersedes 14.4 15 March 1957). 1 p.

—**Aluminum—Sprayed**
- CGSB 31-GP-101MA-89. Chemical Conversion Films for Aluminum and Aluminum Alloys. 9 pp.

—**Atmospheric Corrosion Testing**
- BSI BS 7561-92. 1992 Method for Corrosion Tests in Artificial Atmospheres at Very Low Concentrations of Polluting Gas(es) (ISO 10062: 1991) (V). 14 pp.
- ISO 10062-91. Corrosion Tests in Artificial Atmosphere at Very Low Concentrations of Polluting Gas(es) First Edition. 12 pp.

—**Corrosion Testing**
- BSI BS 5466: Part 7-82. 1982 Methods for Corrosion Testing of Metallic Coatings Part 7: Guidance on Stationary Outdoor Exposure Corrosion Tests. 12 pp.
- ISO 4543-81. Metallic and Other Non-Organic Coatings—General Rules for Corrosion Tests Applicable for Storage Conditions First Edition. 7 pp.

—**Engineering Drawings—Designations**
- DIN ENGL 50960 Pt 2-86. Electroplated and Chemical Coatings; Indications on Drawings (Feb) (Together with DIN 50960 Part 1, February 1986 Edition, Supersedes DIN 50960, June 1963 Edition Withdrawn in July 1976). 6 pp.

—**Iron**
- CGSB CAN/CGSB-31.114-M91. Black Oxide Conversion Coatings for Ferrous Metals. 8 pp.

—**Metals**
- BSI BS 4479: Part 3-90. 1990 Design of Articles That Are to Be Coated Part 3: Recommendations for Conversion Coatings. 10 pp.

—**Metals—Salt Spray Testing**
- BSI BS 5466: Part 9-86. 1986 Methods for Corrosion Testing of Metallic Coatings Part 9: Saline Droplets Corrosion Test (SD Test). 8 pp.
- ISO 4536-85. Metallic and Non-Organic Coatings on Metallic Substrates—Saline Droplets Corrosion Test (SD Test) First Edition. 6 pp.

—**Metals—Weight Measurement**
- CGSB 31-GP-0A METH 18.2-57. Methods of Testing Corrosion-Prevention Materials and Processes Coating Weight (Stripping by Chromic Acid); (Re-Edited April 1982). 1 p.

—**Paints (Aluminum)—Adhesion Testing**
- CGSB 31-GP-0A METH 7.2-57. Methods of Testing Corrosion-Prevention Materials and Processes Adhesion of Paint to Chemical Films; (Amended June 1974) (Re-Edited March 1982). 1 p.

—**Paints (Enamel)—Salt Spray Testing**
- CGSB 31-GP-0A METH 14.7A-62. Methods of Testing Corrosion-Prevention Materials and Processes Salt Spray Resistance of Painted Parts; (Supersedes 14.7 15 March 1957) (Amended September 1974) (Re-Edited March 1982). 1 p.

—**Technical Documents—Designations**
- DIN ENGL 50960 Pt 1-86. Electroplated and Chemical Coatings; Designation and Specification in Technical Documents (Feb) (Together with DIN 50960 Part 2, February 1986 Edition, Supersedes DIN 50960, June 1963 Edition, Withdrawn in July 1976). 6 pp.

Conversion Facilities, Message
Use: Message Conversion Systems

Conversion Formulae, Cutting Tools
Use: Cutting Tools—Conversion Formulae

Conversion Tables
Use: Tables (Data)

Converters
Scope Note: For additional listings, use a more specific term *Use For:* Electric Converters *See Also:* Analog to Digital Converters; Asynchronous to Synchronous Converters; Digital to Analog Converters; Down Converters; Level Translators; Power Converters; Television Standards Converters; Time to Amplitude Converters; Up Converters; Voltage Converters; Voltage to Frequency Converters

—**Arc Welding Equipment**
- DIN VDE 0540-65. Regulations for Direct Current Arc-Welding; Generators and Convertors (Feb). 19 pp.

—**Radio Equipment—Marine**
- CNS C4302-81. Marine Converters and Inverters for Wireless Sets (Apr)(7212).

Convex Cutters
Use: Milling Cutters—Convex

Conveyor Belts
Use For: Transport Belts *See Also:* Belt Conveyors; Belts (Machinery); Elevator Belts
- BSI BS 490: Part 1-90. 1990 Conveyor and Elevator Belting Part 1: Rubber and Plastics Conveyor Belting of Textile Construction for General Use. 12 pp.
- CEN PREN 808-92. Conveyor Belts—Method for the Determination of the Transverse Flexibility and Throughability. 7 pp.
- CEN PREN 873-92. Light Conveyor Belts—Principal Characteristics and Applications. 4 pp.
- CNS K4076-83. Rubber Belts for Conveyors (Feb)(10023). 3 pp.
- CNS K6751-83. Method of Test for Rubber Belts for Conveyors (Feb)(10024).
- ISO 251-87. Conveyor Belts—Widths and Lengths Second Edition. 4 pp.
- ISO 432-89. Ply Type Conveyor Belts—Characteristics of Construction Second Edition. 4 pp.
- ISO 3870-76. Conveyor Belts (Fabric Carcass), with Length Between Pulley Centres up to 300 m, for Loose Bulk Materials—Adjustment of Take-up Device First Edition. 5 pp.
- JIS K 6322-88. Ply Construction Conveyor Belts. 26 pp.
- JIS K 6369-79. Steel Cord Conveyor Belts. 21 pp.
- SAA AS 1332-91. Conveyor Belting—Textile Reinforced. 9 pp.
- SNZ NZS/BS 490: Part 1-85. Conveyor and Elevator Belting Part 1: Specification for Rubber and Plastics Conveyor Belting of Textile Construction for General Use. 8 pp.

—**Adhesion Testing**
- BSI BS 490: Sec 10.4-83. 1983 Conveyor and Elevator Belting Part 10: Testing for Physical Properties Section 10.4: Method for Determination of Adhesion Strength of Rubber and Plastics Belting of Textile Construction. 4 pp.
- ISO 252-88. Conveyor Belts—Ply Adhesion Between Constitutive Elements—Test Methods and Requirements Second Edition. 10 pp.
- ISO 8094-84. Steel Cord Conveyor Belts—Adhesion Strength Test of the Cover to the Core Layer First Edition. 4 pp.
- SNZ NZS/BS 490: Pt10:Sec10.4-83. Conveyor and Elevator Belting Part 10: Testing and Physical Properties Section 10.4: Method for Determination of Adhesion Strength of Rubber and Plastics Belting of Textile Construction. 2 pp.

—**Bonding Strength**
- ISO 7623-84. Steel Cord Conveyor Belts—Cord-to-Coating Bond Test First Edition. 4 pp.

—**Covers**
- BSI BS 490: Part 2-75. 1975 Conveyor and Elevator Belting Part 2: Rubber and Plastics Belting of Textile Construction for Use on Bucket Elevators. 18 pp.

INTERNATIONAL AND NON-U.S. NATIONAL STANDARDS
SUBJECT INDEX
Conveyors

Conveyor Belts (Cont.)

—Covers (Cont.)
ISO 583-90. Conveyor Belts with a Textile Carcass—Tolerances on Total Thickness and Thickness of Covers—Direct Measurement Method Second Edition. 7 pp.
ISO 10247-90. Conveyor Belts—Characteristics of Covers—Classification First Edition; (Corrected and Reprinted 1991). 7 pp.
SNZ NZS/BS 490: Part 2-75. Conveyor and Elevator Belting Part 2: Rubber and Plastics Belting of Textile Construction for Use on Bucket Elevators. 15 pp.

—Covers—Aging Testing
SAA AS 1334.6-85. Methods of Testing Conveyor and Elevator Belting—Part 6: Determination of Resistance of Covers to Ageing. 2 pp.

—Electrical Resistance
BSI BS 490: Sec 11.4-90. (WITHDRAWN) 1990 Conveyor and Elevator Belting Part 11: Methods of Test for Safety Section 11.4: Determination of Electrical Conductivity (Superseded by BS EN 20284: 1993). 7 pp.
BSI BS EN 20284-93. 1993 Conveyor Belts—Electrical Conductivity—Specification and Method of Test (ISO 284: 1982) (Q). 10 pp.
CEN EN 20284-93. Conveyor Belts—Electrical Conductivity—Specification and Method of Test (ISO 284: 1982). 5 pp.
CSA CAN/CSA-M422-M87. Fire-Performance and Antistatic Requirements for Conveyor Belting; (Gen Instr 1). 26 pp.
ISO 284-82. Conveyor Belts—Electrical Conductivity—Specification and Method of Test Second Edition; (CEN EN 20284: 1993). 5 pp.
SNZ NZS/BS 490: Pt11:Sec11.4-90. Conveyor and Elevator Belting Part 11: Methods of Test for Safety Section 11.4: Determination of Electrical Conductivity. 6 pp.

—Elongation
BSI BS 490: Part 1-90. 1990 Conveyor and Elevator Belting Part 1: Rubber and Plastics Conveyor Belting of Textile Construction for General Use. 12 pp.
BSI BS 490: Sec 10.2-83. 1983 Conveyor and Elevator Belting Part 10: Testing for Physical Properties Section 10.2: Method for Determination of Full Thickness Tensile Strength and Elongation of Rubber and Plastics Conveyor Belting of Textile Construction. 4 pp.
BSI BS 490: Sec 10.5-84. 1984 Conveyor and Elevator Belting Part 10: Testing for Physical Properties Section 10.5: Method for Determination of Tensile Strength and Elongation at Break of Rubber Covers. 2 pp.
ISO 283-90. Conveyor Belts—Full Thickness Tensile Strength and Elongation—Specifications and Method of Test Second Edition. 11 pp.
ISO 7622 Pt 1-84. Steel Cord Conveyor Belts—Longitudinal Traction Test—Part 1: Measurement of Elongation First Edition. 4 pp.
SNZ NZS/BS 490: Part 1-85. Conveyor and Elevator Belting Part 1: Specification for Rubber and Plastics Conveyor Belting of Textile Construction for General Use. 8 pp.
SNZ NZS/BS 490: Pt10:Sec10.2-83. Conveyor and Elevator Belting Part 10: Testing and Physical Properties Section 10.2: Method for Determination of Full Thickness Tensile Strength and Elongation of Rubber and Plastics Conveyor Belting of Textile Construction. 3 pp.
SNZ NZS/BS 490: Pt10:Sec10.5-84. Conveyor and Elevator Belting Part 10: Testing and Physical Properties Section 10.5: Method for Determination of Tensile Strength and Elongation at Break of Rubber Covers. 2 pp.

—Fire Resistance
CSA CAN/CSA-M422-M87. Fire-Performance and Antistatic Requirements for Conveyor Belting; (Gen Instr 1). 26 pp.

—Flammability Testing
BSI BS 490: Sec 11.1-90. (WITHDRAWN) 1990 Conveyor and Elevator Belting Part 11: Methods of Test for Safety Section 11.1: Laboratory Flame Tests (Superseded by BS EN 20340: 1973). 12 pp.
BSI BS EN 20340-93. 1993 Conveyor Belts—Flame Retardation—Specifications and Test Method (ISO 340: 1988) (Q). 15 pp.
CEN EN 20340-93. Conveyor Belts—Flame Retardation—Specifications and Test Method (ISO 340: 1988). 9 pp.
CNS K4078-83. Flame-Resistant Rubber Belts for Conveyors (Feb)(10026). 2 pp.
ISO 340-88. Conveyor Belts—Flame Retardation—Specifications and Test Method Second Edition; (CEN EN 20340: 1993). 10 pp.
JIS K 6324-77. Qualitative Standard for Flame Resistance of Conveyor Belts (R 1986). 6 pp.

Conveyor Belts (Cont.)

—Flammability Testing (Cont.)
SAA AS 1334.11-88. Methods of Testing Conveyor and Elevator Belting—Part 11: Determination of Ignitability and Maximum Surface Temperature of Belting Subjected to Friction Amdt 1 July 1989. 4 pp.
SNZ NZS/BS 490: Pt11:Sec11.1-90. Conveyor and Elevator Belting Part 11: Method of Test for Safety Section 11.1: Laboratory Flame Tests. 12 pp.

—Glossaries
ISO 5284-86. Conveyor Belts—List of Equivalent Terms Second Edition. 56 pp.
SAA AS 4035-92. Conveyor and Elebator Belting—Glossary of Terms (Supersedes AS B255—1969). 6 pp.

—Heat Testing
ISO 4195 Pt 1-87. Conveyor Belts—Heat Resistance—Part 1: Test Method First Edition. 4 pp.
ISO 4195 Pt 2-88. Conveyor Belts—Heat Resistance—Part 2: Specifications First Edition. 4 pp.

—Joint Sealants
BSI BS 6593-85. 1985 Code of Practice for on-Site Non-Mechanical Jointing of Plied Textile and Steel Reinforced Conveyor Belting. 26 pp.

—Modulus of Elasticity
BSI BS 490: Sec 10.8-90. 1990 Conveyor and Elevator Belting Part 10: Testing for Physical Properties Section 10.8: Method for Determination of Elastic Modulus (Including Permanent and Elastic Elongation). 8 pp.
ISO 9856-89. Conveyor Belts—Determination of Elastic Modulus First Edition. 7 pp.
SNZ NZS/BS 490: Pt10:Sec10.8-90. Conveyor and Elevator Belting Part 10: Testing and Physical Properties Section 10.8: Method for Determination of Elastic Modulus (Including Permanent and Elastic Elongation). 8 pp.

—Physical Properties
BSI BS 490: Sec 10.1-83. 1983 Conveyor and Elevator Belting Part 10: Testing for Physical Properties Section 10.1: Introduction. 4 pp.
BSI BS 490: Sec 10.7-84. 1984 Conveyor and Elevator Belting Part 10: Testing for Physical Properties Section 10.7: Method for Determination of Length of an Endless Belt. 2 pp.
SNZ NZS/BS 490: Pt10:Sec10.1-83. Conveyor and Elevator Belting Part 10: Testing and Physical Properties Section 10.1: Introduction. 2 pp.
SNZ NZS/BS 490: Pt10:Sec10.7-84. Conveyor and Elevator Belting Part 10: Testing and Physical Properties Section 10.7: Method for Determination of Length of an Endless Belt. 1 p.

—Sampling
ISO 282-92. Conveyor Belts—Sampling Second Edition. 4 pp.

—Storage and Handling
BSI BS 5767-79. 1979 Guide for Storage and Handling of Conveyor Belts. 4 pp.
ISO 5285-78. Conveyor Belts—Guide to Storage and Handling First Edition. 4 pp.

—Tear Strength
SAA AS 1334.8-82. Methods of Testing Conveyor and Elevator Belting—Part 8: Determination of Resistance to Tear Propagation and Resistance of Carcass to Tearing. 3 pp.

—Tear Testing
ISO 505-82. Conveyor Belts—Tear Propagation Resistance of the Carcass—Method of Test Second Edition. 5 pp.

—Tensile Strength
BSI BS 490: Sec 10.2-83. 1983 Conveyor and Elevator Belting Part 10: Testing for Physical Properties Section 10.2: Method for Determination of Full Thickness Tensile Strength and Elongation of Rubber and Plastics Conveyor Belting of Textile Construction. 4 pp.
BSI BS 490: Sec 10.5-84. 1984 Conveyor and Elevator Belting Part 10: Testing for Physical Properties Section 10.5: Method for Determination of Tensile Strength and Elongation at Break of Rubber Covers. 2 pp.
SNZ NZS/BS 490: Pt10:Sec10.2-83. Conveyor and Elevator Belting Part 10: Testing and Physical Properties Section 10.2: Method for Determination of Full Thickness Tensile Strength and Elongation of Rubber and Plastics Conveyor Belting of Textile Construction. 3 pp.
SNZ NZS/BS 490: Pt10:Sec10.5-84. Conveyor and Elevator Belting Part 10: Testing and Physical Properties Section 10.5: Method for Determination of Tensile Strength and Elongation at Break of Rubber Covers. 2 pp.

Conveyor Belts (Cont.)

—Tensile Testing
ISO 7622 Pt 2-84. Steel Cord Conveyor Belts—Longitudinal Traction Test—Part 2: Measurement of Tensile Strength First Edition. 5 pp.

—Thickness Measurement
BSI BS 490: Sec 10.2-83. 1983 Conveyor and Elevator Belting Part 10: Testing for Physical Properties Section 10.2: Method for Determination of Full Thickness Tensile Strength and Elongation of Rubber and Plastics Conveyor Belting of Textile Construction. 4 pp.
BSI BS 490: Sec 10.3-84. 1984 Conveyor and Elevator Belting Part 10: Testing for Physical Properties Section 10.3: Methods for Measurement of Overall Thickness and Cover Thickness. 4 pp.
ISO 7590-90. Steel Cord Conveyor Belts—Cover Thickness Measurement First Edition. 6 pp.
SNZ NZS/BS 490: Pt10:Sec10.2-83. Conveyor and Elevator Belting Part 10: Testing and Physical Properties Section 10.2: Method for Determination of Full Thickness Tensile Strength and Elongation of Rubber and Plastics Conveyor Belting of Textile Construction. 3 pp.
SNZ NZS/BS 490: Pt10:Sec10.3-84. Conveyor and Elevator Belting Part 10: Testing and Physical Properties Section 10.3: Methods for Measurement of Overall Thickness and Cover Thickness. 3 pp.

—Transition Measurement
ISO 5293-81. Conveyor Belts—Formula for Transition Distance on Three Equal Length Idler Rollers First Edition. 5 pp.
ISO TR10357-89. Conveyor Belts—Formula for Transition Distance on Three Equal Length Idler Rollers (New Method) First Edition. 10 pp.

—Troughed—Flexibility
BSI BS 490: Sec 10.6-83. 1983 Conveyor and Elevator Belting Part 10: Testing for Physical Properties Section 10.6: Determination of Tolerability of Conveyor Belts. 6 pp.
ISO 703-88. Conveyor Belts—Troughability—Characteristics of Transverse Flexibility and Test Method Second Edition. 6 pp.
SNZ NZS/BS 490: Pt10:Sec10.6-83. Conveyor and Elevator Belting Part 10: Testing and Physical Properties Section 10.6: Method for Determination of Troughability of Conveyor Belts. 4 pp.

Conveyor Chains
See Also: Chains

BSI BS 4116: Part 4-92. 1992 Conveyor Chains, Their Attachments and Associated Chain Wheels Part 4: Specification for Chains and Attachments (British Series) (Q). 14 pp.
DIN ENGL 8167 Pt 1-86. ISO-Type M Conveyor Chains with Solid Pins; Single Strand Chains and Double Strand Chains (Mar). 4 pp.
DIN ENGL 8167 Pt 3-86. ISO-Type MT Conveyor Chains with Solid Pins and Deep Plates (Mar). 2 pp.
DIN ENGL 8168 Pt 1-86. ISO-Type MC Conveyor Chains with Hollow Pins; Single Strand Chains (Mar). 3 pp.
DIN ENGL 8168 Pt 3-86. ISO-Type MCT Conveyor Chains with Hollow Pins and Deep Plates (Mar). 2 pp.

—Attachments
BSI BS 4116: Part 4-92. 1992 Conveyor Chains, Their Attachments and Associated Chain Wheels Part 4: Specification for Chains and Attachments (British Series) (Q). 14 pp.

—Link—Calibrated
DIN ENGL 762 Pt 1-82. Calibrated and Tested Round Steel Link Chains for Continuous Conveyors; Grade 2, Pitch 5d (Jan). 4 pp.
DIN ENGL 762 Pt 2-82. Calibrated and Tested Round Steel Link Chains for Continuous Conveyors; Grade 3, Pitch 5d (Jan). 4 pp.
DIN ENGL 764 Pt 1-82. Calibrated and Tested Round Steel Link Chains for Continuous Conveyors; Grade 2, Pitch 3,5d (Jan). 3 pp.
DIN ENGL 764 Pt 2-82. Calibrated and Tested Round Steel Link Chains for Continuous Conveyors; Grade 3, Pitch 3,5d (Jan). 3 pp.
DIN ENGL 22252-83. Calibrated and Tested Round Steel Link Chains for Conveyors and Machines Used in Mining (Sept) (Supersedes DIN 22252 Parts 1 and 2, December 1973 Editions). 12 pp.

—Roller—Double Pitch
JIS B 1803-83. Double-Pitch Conveyor Roller Chains, Attachments and Chain Wheels. 37 pp.

Conveyors
Use For: Continuous Handling Equipment
See Also: Airslide Conveyors; Belt Conveyors; Bucket Elevators; Chain Conveyors; Drum Feeders; Elevators (Conveyors); Hydraulic Conveyors; Materials Handling Equipment; Passenger Conveyors; Plate

INTERNATIONAL AND NON-U.S. NATIONAL STANDARDS
SUBJECT INDEX

Conveyors

Conveyors (Cont.)
See Also: (Cont.)
Conveyors; Pneumatic Conveyors; Push Bar Conveyors; Roller Conveyors; Rollers (Mechanical); Scoops (Shovels); Scraper Conveyors; Screw Conveyors; Skids; Slat Conveyors; Vertical Conveyors; Vibrating Conveyors; Vibrating Feeders

—Glossaries
BSI BS 3810: Part 2-65. 1965 Amd 1 Glossary of Terms Used in Materials Handling Part 2: Terms Used in Connection with Conveyors and Elevators (Excluding Pneumatic and Hydraulic Handling). 29 pp.

BSI BS 6100: SUBSEC 3.2.2-84. 1984 Glossary of Building and Civil Engineering Terms Part 3: Services Section 3.2: Internal Communication and Transport Subsection 3.2.2: Internal Transport. 7 pp.

BSI BS 6100: SUBSEC 3.2.2-01. 1984 Amd 1 Building and Civil Engineering Terms Part 3: Services Section 3.2: Internal Communication and Transport Subsection 3.2.2: Internal Transport (AMD 7258) August 15, 1992. 8 pp.

ISO 2148-74. Continuous Handling Equipment—Nomenclature First Edition. 70 pp.

JIS B 0140-86. Glossary of Terms Relating to Conveyors (Part 1 Conveyors). 21 pp.

JIS B 0141-86. Glossary of Terms Relating to Conveyors (Part 2 Conveyor Parts and Accessories). 37 pp.

SNZ NZS 2000: Part 2-66. Glossary of Terms Used in Materials Handling Part 2: Terms Used in Connection with Conveyors and Elevators (Excluding Pneumatic and Hydraulic Handling) (Reconfirmed 1978). 28 pp.

—Mobile
ISO 2406-74. Continuous Mechanical Handling Equipment—Mobile and Portable Conveyors—Constructional Specifications First Edition. 5 pp.

ISO 5049 Pt 1-80. Mobile Continuous Bulk Handling Equipment—Part 1: Rules for the Design of Structures First Edition. 48 pp.

—Noise
DIN ENGL 45635 Pt 45-88. Measurement of Airborne Noise Emitted by Machines; Enveloping Surface Method; Continuous Handling Equipment (June). 9 pp.

—Portable
ISO 2406-74. Continuous Mechanical Handling Equipment—Mobile and Portable Conveyors—Constructional Specifications First Edition. 5 pp.

—Safety
BSI BS 5667: Part 1-79. 1979 Continuous Mechanical Handling Equipment. Safety Requirements Part 1: General. 15 pp.

ISO 1819-77. Continuous Mechanical Handling Equipment—Safety Code—General Rules First Edition. 13 pp.

ISO 7149-82. Continuous Handling Equipment—Safety Code—Special Rules First Edition. 26 pp.

SAA AS 1755-86. Conveyors—Design, Construction, Installation and Operation—Safety Requirements. 60 pp.

Cooker Hoods
Use: Range Hoods

Cookers
Use: Ranges

Cooking Appliances
Scope Note: For additional listings, use a more specific term *See Also:* Appliances; Barbecues; Blenders; Camp Stoves; Coffee Makers; Convection Ovens; Cooking Pans; Cooking Utensils; Cookware; Deep Fat Fryers; Food Preparation Equipment; Food Processing Equipment; Frying Utensils; Griddles; Grills (Appliances); Hot Plates (Warmers); Kettles; Microwave Ovens; Ovens; Pots (Electric); Pressure Cookers; Range Hoods; Ranges; Rice Cookers; Roasters; Steam Tables; Steamers; Teakettles; Toaster Ovens; Toasters; Waffle Irons

BSI BS 3300-85. 1985 Amd 1 Paraffin Unflued Space Heating and Cooking Appliances for Domestic Use (AMD 6251) July 31, 1989. 27 pp.

CSA 852 Bull. Electrical Bulletin 852 November 22, 1971 to C22.2 NO 64. 1 p.

CSA 1412 Bull. Electrical Bulletin 1412 August 22, 1983 to C22.2 NO 64. 6 pp.

SAA AS 3162-90. Approval and Test Specification—Electric Kitchen Machines for Household Use (This is a Joint Standard with SANZ NZS 3162) Amdt 1 January 1991 Amdt 2 March 1992 Amdt 3 October 1992 Amdt 4 June 1993 (in Prof. Package 28). 1 p.

Cooking Appliances (Cont.)
SAA AS 3172-90. Approval and Test Specification—Electric Cooking Appliances for Household Use (This is a Joint Standard with SANZ NZS 3172:1990) Amdt 1 November 1990 Amdt 2 January 1991 Amdt 3 March 1992 Amdt 4 October 1992 Amdt 5 June 1993 (in Prof. Package 28). 1 p.

SNZ NZS/BS 3300-85. Paraffin Unflued Space Heating and Cooking Appliances for Domestic Use Amend: 1. 28 pp.

—Burners—Gas
BSI BS 5314: Part 2-76. (WITHDRAWN) 1976 Amd 2 Gas Heated Catering Equipment Part 2: Boiling Burners (AMD 5397) July 31, 1987 (Superseded by BS EN 203-1: 1993). 27 pp.

—Comprehensive Testing
SNZ NZS/AS 3172-90. Approval and Test Specification—Electric Cooking Appliances for Household Use Amend: 1; 2; 3; 4, 1992 (This is a Joint Standard with SAA AS 3172). 19 pp.

—Electric
BSI BS 3456: Sec 202.14-90. 1990 Safety of Household Electrical Appliances Part 202: Particular Requirements Section 202.14: Electric Kitchen Machines. 31 pp.

BSI BS 3456: Sec 202.14-02. 1990 Amd 2 Safety of Household and Similar Electrical Appliances Part 202: Particular Requirements Section 202.14: Electric Kitchen Machines (AMD 7532) January, 15, 1993 (L). 50 pp.

BSI BS 3456: Sec A1-66. (WITHDRAWN) 1966 Amd 4 Safety of Household Electrical Appliances Part A: Heating and Cooking Appliances Section A1: General Requirements. 57 pp.

BSI BS 3456: Sec A4-71. (WITHDRAWN) 1971 Amd 9 Safety of Household Electrical Appliances Part A: Heating and Cooking Appliances Section A4: Electrically Heated Blankets (Withdrawn, superseded by BS En 60967: 1991). 67 pp.

CSA C22.2 NO 1335.2.14-93. Portable Electrical Motor-Operated and Heating Appliances: Particular Requirements for Electrical Motor-Operated Kitchen Appliances; (Gen Instr 1). 63 pp.

DIN VDE 0720 Pt 2 ZE-82. Electric Cooking and Heating Appliances for Domestic and Similar Purposes; Electric Catering Equipment with Heating Elements for Commercial Use (May). 23 pp.

DIN VDE 0722-83. Electrical Equipment on Non-Electrically Heated Cooking and Heating Appliances (Apr). 31 pp.

SAA AS 1996-77. Special Requirements for Electrically Operated Cooking and Heating Appliances for Marine Use Amdt 1 June 1984 Reconfirmed 1984. 16 pp.

—Electric—Household—Safety
BSI BS EN 60335-2-6-91. 1991 Safety of Household and Similar Electrical Appliances Part 2: Particular Requirements Section 2.6: Cooking Ranges, Cooking Tables, Ovens and Similar Appliances for Household Use. 52 pp.

BSI BS EN 60335-2-6-03. 1991 Amd 3 Safety of Household and Similar Electrical Appliances Part 2: Particular Requirements Section 2.6: Cooking Ranges, Cooking Tables, Ovens and Similar Appliances for Household Use (AMD 7916) October 15, 1993 (L). 84 pp.

BSI BS EN 60335-2-9-91. 1991 Safety of Household and Similar Electrical Appliances Part 2: Particular Requirements Section 2.9: Toasters, Grills, Roasters and Similar Appliances. 26 pp.

BSI BS EN 60335-2-9-01. 1991 Amd 1 Safety of Household and Similar Electrical Appliances Part 2: Particular Requirements Section 2.9: Toasters, Grills, Roasters and Similar Appliances (AMD 6897) April 1, 1992. 32 pp.

BSI BS EN 60335-2-9-02. 1991 Amd 2 Safety of Household and Similar Electrical Appliances Part 2: Particular Requirements Section 2.9: Toasters, Grills, Roasters and Similar Appliances (AMD 7584) March 15, 1993 (L). 42 pp.

CENELEC EN 60 335-2-9-90. Safety of Household and Similar Electrical Appliances Part 2: Particular Requirements for Toasters, Grills, Roasters and Similar Appliances. 8 pp.

CENELEC EN 60335-2-9 /A2-92. AMD 2 Safety of Household and Similar Electrical Appliances Part 2: Particular Requirments for Toasters, Grills, Roasters and Similar Appliances. 5 pp.

CSA C22.2 NO 1335.1-93. Portable Electrical Motor-Operated and Heating Appliances: General Requirements; (Gen Instr 1). 208 pp.

CSA C22.2 NO 1335.2.9-93. Portable Electrical Motor-Operated and Heating Appliances: Particular Requirements for Portable Electric Cooking Appliances; (Gen Instr 1). 40 pp.

IEC 335 Pt 2-9-93. Safety of Household and Similar Electrical Appliances Part 2: Particular Requirements for Toasters, Grills, Roasters and Similar Appliances Fourth Edition. 46 pp.

Cooking Appliances (Cont.)
—Electric—Liquids
CSA C22.2 NO 1335.2.15-93. Portable Electrical Motor-Operated and Heating Appliances: Particular Requirements for Liquid-Heating Appliances; (Gen Instr 1). 36 pp.

—Electric—Safety
CENELEC HD 251-75. General Specification for Electric Cooking and Heating Appliances for Household and Similar Purposes. 9 pp.

CENELEC HD 251.2-78. General Specification for Electric Cooking and Heating Appliances for Household and Similar Purposes. 3 pp.

CENELEC HD 265 S2-84. Safety of Household and Similar Electrical Appliances—Particular Requirements for Appliances for Toasters, Grills, Roasters and Similar Appliances. 8 pp.

CENELEC HD 270 S1-83. Safety of Household and Similar Electrical Appliances—Particular Requirements for Microwave Cooking Appliances (Superseded by EN 41003). 17 pp.

CENELEC HD 275 S1-86. Safety of Household and Similar Electrical Appliances—Part 2: Particular Requirements for Cooking Ranges, Cooking Tables, Ovens and Similar Appliances for Household Use. 30 pp.

IEC 335 Pt 2-14-84. Safety of Household and Similar Electrical Appliances Part 2: Particular Requirements for Electric Kitchen Machines Second Edition; (Amendment 1-1989) (Amendment 2-1990). 69 pp.

—Gas—Installation
BSI BS 6172-90. 1990 Installation of Domestic Gas Cooking Appliances (1st, 2nd and 3rd Family Gases). 12 pp.

BSI BS 6172-82. 1982 Installation of Domestic Gas Cooking Appliances (2nd Family Gases). 8 pp.

—Gas—Liquids
BSI BS 5314: Part 6-76. (WITHDRAWN) 1976 Amd 2 Gas Heated Catering Equipment Part 6: Bulk Liquid Heaters (AMD 5402) July 31, 1987 (Superseded by BS EN 203-1: 1993). 18 pp.

CSA CAN/CSA-C22. 2 NO 64-M91. Household Cooking and Liquid-Heating Appliances; (Gen Instr 1 Thru 2). 91 pp.

CSA 852 Bull. Electrical Bulletin 852 November 22, 1971 to C22.2 NO 64. 1 p.

CSA 1412 Bull. Electrical Bulletin 1412 August 22, 1983 to C22.2 NO 64. 6 pp.

—Gas—Portable—Gas Cylinders
JIS S 2148-91. Gas Cylinders for Portable Gas Cooker. 19 pp.

—Gas—Safety
BSI BS 5258: Part 2-75. (OBSOLESCENT) 1975 Amd 2 Safety of Domestic Gas Appliances Part 2: Cooking Appliances. 34 pp.

CENELEC PRHD 1003-88. Heating in Contact with the Front of the Domestic Cooking Appliances Burning Gas. 10 pp.

CENELEC HD 1003-90. Heating in Contact with the Front of the Domestic Cooking Appliances Burning Gas. 10 pp.

—Household
CSA CAN/CSA-C22. 2 NO 64-M91. Household Cooking and Liquid-Heating Appliances; (Gen Instr 1 Thru 2). 91 pp.

—Installation
BSI BS 8303-86. 1986 Amd 1 Code of Practice for Installation of Domestic Heating and Cooking Appliances Burning Solid Mineral Fuels (AMD 5723) October 30, 1987. 27 pp.

—Liquids
SNZ NZS 6315-89. Approval and Test Specification—Particular Requirements for Appliances for Heating Liquids Amend: 1, 1990; 2, 1991; 3, 1992 (NZS 6315:1989/AS 3313-1989). 15 pp.

—Portable
BSI BS 3879-81. 1981 Amd 3 Portable Appliances Operating at Vapour Pressure from Liquefied Petroleum Gas Containers. 14 pp.

—Safety
BSI BS 3456: Sec 102.6-89. (WITHDRAWN) 1989 Amd 2 Safety of Household and Similar Electrical Appliances Part 102: Particular Requirements Section 102.6: Cooking Ranges, Cooking Tables, Ovens and Similar Appl. for Household Use (AMD 6567) April 30, 1991. 46 pp.

BSI BS 3456: Sec 102.9-91. (WITHDRAWN) 1991 Safety of Household and Similar Electrical Appliances Part 102: Particular Requirements Section 102.9: Toasters, Grills, Roasters and Similar Appliances (Superseded by BS EN 60335-2-9: 1991). 26 pp.

INDEX and DIRECTORY of

INTERNATIONAL AND NON-U.S. NATIONAL STANDARDS
SUBJECT INDEX
Cooking

Cooking Appliances *(Cont.)*
—**Safety** *(Cont.)*
BSI BS 3456: Sec 102.25-88. (WITHDRAWN) 1988 Safety of Household Electrical Appliances Part 102: Particular Requirements Section 102.25: Appliances for Heating Food by Means of Microwave Energy (Superseded by BS EN 60335-2-25: 1991). 39 pp.
BSI BS 3456: Sec 202.14-03. 1990 Amd 3 Safety of Household and Similar Electrical Appliances Part 202: Particular Requirements Section 202.14: Electric Kitchen Machines (AMD 7800) June 15, 1993 (SUPERSEDES BS 3456: SECTION 3. 12: 1979). 57 pp.
BSI BS 3456: Sec 3.17-80. (WITHDRAWN) 1980 Amd 2 1972-1981 Edition. Specification for Safety of Household Electrical Appliances: Part 3: Complete Particular Specif-ications: Section 3.17: Toasters. Grills, Waffle Irons, Roasters & Other Dry Cooking Appliances (Superseded by BS 3456: Section 102.9: 1991). 48 pp.
BSI BS EN 60335-2-6-01. 1991 Amd 1 Safety of Household and Similar Electrical Appliances Part 2: Particular Requirements Section 2.6: Cooking Ranges, Cooking Tables, Ovens and Similar Appliances for Household Use (AMD 7677) July 15, 1993 (L). 70 pp.
BSI BS EN 60335-2-6-02. 1991 Amd 2 Safety of Household and Similar Electrical Appliances Part 2: Particular Requirements Section 2.6: Cooking Ranges, Cooking Tables, Ovens and Similar Appliances for Household Use (AMD 7915) September 15, 1993 (L). 81 pp.
BSI BS EN 60335-2-15-91. 1991 Safety of Household and Similar Electrical Appliances Part 2: Particular Requirements Section 2.15: Appliances for Heating Liquids (L). 29 pp.
BSI BS EN 60335-2-15-02. 1991 Amd 2 Safety of Household and Similar Electrical Appliances Part 2: Particular Requirements Section 2.15: Appliances for Heating Liquids (AMD 7281) October 15, 1992 (L). 52 pp.
BSI BS EN 60335-2-15-03. 1991 Amd 3 Safety of Household and Similar Electrical Appliances Part 2: Particular Requirements Section 2.15: Appliances for Heating Liquids (AMD 7695) July 15, 1993 (L). 58 pp.
CENELEC HD 261-80. Particular Specification for Kitchen Machines. 7 pp.
CENELEC HD 261 S1-84. Particular Specification for Kitchen Machines (Replaced by EN 60 335-2-14-1988). 3 pp.
CENELEC HD 261.2-77. Particular Specification for Kitchen Machines. 5 pp.
CENELEC HD 261.3-78. Particular Specification for Kitchen Machines. 5 pp.
CENELEC HD 261.4 S1-88. Particular Specification for Kitchen Machines. 5 pp.
CENELEC HD 265 S2-84. Safety of Household and Similar Electrical Appliances—Particular Requirements for Appliances for Toasters, Grills, Roasters and Similar Appliances. 8 pp.
CENELEC HD 265 S2/A1-88. AMD 1 Safety of Household and Similar Electrical Appliances—Particular Requirements for Appliances for Toasters, Grills, Roasters and Similar Appliances. 4 pp.
CENELEC HD 275 S1-86. Safety of Household and Similar Electrical Appliances—Part 2: Particular Requirements for Cooking Ranges, Cooking Tables, Ovens and Similar Appliances for Household Use. 30 pp.
CENELEC HD 275 S1/A1-89. AMD 1 Safety of Household and Similar Electrical Appliances—Part 2: Particular Requirements for Cooking Ranges, Cooking Tables, Ovens and Similar Appliances for Household Use. 9 pp.
CENELEC EN 60335-2-6-90. Safety of Household and Similar Electrical Appliances Part 2: Particular Requirements for Cooking Ranges, Cooking Tables, Ovens and Similar Appliances for Household Use (Supercedes HD 275 S1: 1986 and Its Amendment 1: 1988). 38 pp.
CENELEC EN 60335-2-6-91. CORRIGENDUM Safety of Household and Similar Electrical Appliances Part 2: Particular Requirements for Cooking Ranges, Cooking Tables, Ovens and Similar Appliances for Household Use. 2 pp.
CENELEC EN 60335-2-6 /A2-92. AMD 2 Safety of Household and Similar Electrical Appliances Part 2: Particular Requirements for Cooking Ranges, Cooking Tables, Ovens and Similar Appliances for Household Use; (IEC 335-2-6:1986/A2: 1990). 6 pp.
CENELEC EN 60 335-2-14-88. Safety of Household and Similar Electrical Appliances Part 2: Particular Requirements for Electric Kitchen Machines (Replaces HD 261 S1-1984). 11 pp.
CENELEC EN 60335-2-14/A1-90. AMD 1 Safety of Household and Similar Electrical Appliances Part 2: Particular Requirements for Electric Kitchen Machines; (Replaces HD 261 S1:1984). 3 pp.
CENELEC EN 60335-2-14/A51-91. AMD 51 Safety of Household and Similar Electrical Appliances Part 2: Particular Requirements for Electric Kitchen Machines; (Replaces HD 261 S1:1984). 6 pp.
CENELEC EN 60335-2-14/A52-92. AMD 52 Safety of Household and Similar Electrical Appliances Part 2: Particular Requirements for Electric Kitchen Machines. 4 pp.
CENELEC EN 60335-2-15-90. Safety of Household and Similar Electrical Appliances Part 2: Particular Requirements for Heating Liquids. 10 pp.
CENELEC EN 60335-2-15/A1-91. AMD 1 Safety of Household and Similar Electrical Appliances Part 2: Particular Requirements for Heating Liquids. 10 pp.
CENELEC EN 60335-2-15/A2-92. AMD 2 Safety of Household and Similar Electrical Appliances Part 2: Particular Requirements for Heating Liquids. 7 pp.
CENELEC EN 60335-2-15/A52-92. AMD 52 Safety of Household and Similar Electrical Appliances Part 2: Particular Requirements for Appliances for Heating Liquids. 4 pp.
DIN VDE 0700 Pt 6 A8 (D)-89. Safety of Household and Similar Electrical Appliances; Cooking Ranges, Cooking Tables, Ovens and Similar Appliances for Household Use; Amendment 8 to Draft DIN 57700 Part 6/VDE 0700 Part 6/02.84 (Apr). 15 pp.
IEC 335 Pt 2-6-86. Safety of Household and Similar Electrical Appliances Part 2: Part-icular Requirements for Cooking Ranges, Cooking Tables, Ovens and Sim-ilar Appliances for Household Use Third Edition; (Amendment 2-1990, Incorporating Amendment 1) (Amendment 3-1992). 106 pp.
IEC 335 Pt 2-15-86. Safety of Household and Similar Electrical Appliances Part 2: Particular Requirements for Appliances for Heating Liquids Third Edition; (Corrigendum—Dec 1986) (Amendment 1-1988) (Amendment 2-1990) (Amendment 3-1992). 62 pp.
IEC 335 Pt 2-16-86. Safety of Household and Similar Electrical Appliances Part 2: Particular Requirements for Food Waste Disposers Third Edition. 37 pp.
IEC 335 Pt 2-64-91. Safety of Household and Similar Electrical Appliances Part 2: Particular Requirements for Commercial Electric Kitchen Machines First Edition. 44 pp.
JIS B 9651-88. Design Rules for Safety and Sanitation of Baking Machinery. 29 pp.
JIS B 9652-88. Design Rules for Safety and Sanitation of Cake Making Machinery. 35 pp.

—**Ships**
IEC 92 Pt 307-80. Electrical Installations in Ships Part 307: Equipment—Heating and Cooking Appliances Third Edition. 18 pp.
JIS F 8070-86. Electrical Installations in Ships Part 307 Equipment—Heating and Cooking Appliances (IEC 92-307-1980).

Cooking Pans
Use For: Cooking Pots; Pans (Cooking)
See Also: Boiling Pans; Cooking Appliances; Cooking Utensils

—**Boiling—Electric—Commercial—Safety**
BSI BS 3456: Sec 102.39-89. (WITHDRAWN) 1989 Safety of Household Electrical Appliances Part 102: Particular Requirements Section 102.39: Commercial Electric Multi-Purpose Cooking Pans (Superseded by BS 3456: Section 202.39: 1990). 16 pp.
BSI BS 3456: Sec 202.39-90. (WITHDRAWN) 1990 Safety of Household Electrical Appliances Part 202: Particular Requirements Section 202.39: Commercial Electric Multi-Purpose Cooking Pans (Supersedes BS 3456: Section 102.39: 1989). 20 pp.
BSI BS 3456: Sec 202.39-01. 1990 Amd 1 Safety of Household and Similar Electrical Appliances Part 202: Particular Requirements Section 202.39: Commercial Electric Multi-Purpose Cooking Pans (AMD 7616) March 15, 1993 (L). 28 pp.
BSI BS 3456: Sec 202.39-02. 1990 Amd 2 Safety of Household and Similar Electrical Appliances Part 202: Particular Requirements Section 202.39: Commercial Electric Multi-Purpose Cooking Pans (AMD 7830) July 15, 1993 (L). 44 pp.
BSI BS 5784: Part 4-84. 1984 Amd 1 Safety of Electrical Commerical Catering Equipment Part 4: Multi-Purpose Cooking Pans. 57 pp.
BSI BS EN 60335-2-47-03. 1992 Amd 3 Safety of Household and Similar Electrical Appliances Part 2: Particular Requirements Section 2.47: Commercial Electric Boiling Pans (AMD 7425) November 15, 1992. 46 pp.
BSI BS EN 60335-2-47-92. 1992 Safety of Household and Similar Electrical Appliances Part 2: Particular Requirements Section 2.47: Commercial Electric Boiling Pans (L). 23 pp.
CENELEC HD 287 S1-87. Safety of Household and Similar Electrical Appliances—Part 2: Particular Requirements for Commercial Electric Multi-Purpose Cooking Pans (Replaced by EN 335-2-39-1989). 9 pp.
CENELEC EN 60 335-2-39-89. Safety of Household and Similar Electrical Appliances Part 2: Particular Requirements for Commercial Electric Multi-Purpose Cooking Pans (Replaces HD 287 S1-1987). 10 pp.
CENELEC EN 60335-2-39/A1-92. AMD 1 Safety of Household and Similar Electrical Appliances Part 2: Particular Requirements for Commercial Electric Multi-Purpose Cooking Pans; (Replaces HD 287 S1:1987). 5 pp.
CENELEC EN 60 335-2-47-90. Safety of Household and Similar Electrical Appliances Part 2 Particular Requirements for Commercial Electric Boiling Pans. 10 pp.
CENELEC EN 60335-2-47/A1-92. AMD 1 Safety of Household and Similar Electrical Appliances Part 2 Particular Requirements for Commercial Electric Boiling Pans. 5 pp.
CENELEC EN 60335-2-47/A52-92. AMD 52 Safety of Household and Similar Electrical Appliances Part 2 Particular Requirements for Commercial Electric Boiling Pans. 7 pp.
CENELEC EN 60335-2-47/A52-92. AMD 52 Safety of Household and Similar Electrical Appliances Part 2: Particular Requirements for Commercial Electric Boiling Pans. 2 pp.
IEC 335 Pt 2-39-86. Safety of Household and Similar Electrical Appliances Part 2: Particular Requirements for Commercial Electric Multi-Purpose Cooking Pans Second Edition; (Amendment 1-1990). 48 pp.
IEC 335 Pt 2-47-87. Safety of Household and Similar Electrical Appliances Part 2: Particular Requirements for Commercial Electric Boiling Pans First Edition; (Amendment 1-1990). 46 pp.

—**Electric**
CNS C4464-90. Electric Pots (Oct)(12625).

—**Handles**
BSI BS 6743-87. 1987 Amd 1 Performance of Handles and Handle Assemblies Attached to Cookware. 20 pp.

Cooking Pots
Use: Cooking Pans; Cooking Utensils

Cooking Tables
Use: Ranges

Cooking Testing
See Also: Testing

—**Dry Noodles**
CNS N6112-79. Method of Test for Dry Noodles (Sep)(4992). 2 pp.

—**Rice**
ISO 6648-93. Rice—Determination of Viscoelastic Properties at Various Stages of Cooking—Method Using a Viscoelastograph. 7 pp.

Cooking Utensils
Use For: Culinary Utensils *See Also:* Appliances; Boiling Pans; Cooking Appliances; Cooking Pans; Cookware; Cutlery; Food Processing Equipment; Frying Utensils; Ladles (Utensils); Microwave Cookware; Steam Tables
BSI BS 7069-88. 1988 Cockware. 11 pp.

—**Aluminum**
CEN PREN 851-92. Aluminium and Aluminium Alloys—Wrought Products—Circle and Circle Stock for Culinary Utensil Applications. 11 pp.

—**Aluminum Alloys**
CEN PREN 851-92. Aluminium and Aluminium Alloys—Wrought Products—Circle and Circle Stock for Culinary Utensil Applications. 11 pp.

—**Ceramic—Cadmium Content**
ISO 8391 Pt 1-86. Ceramic Cookware in Contact with Food—Release of Lead and Cadmium—Part 1: Method of Test First Edition. 6 pp.
ISO 8391 Pt 2-86. Ceramic Cookware in Contact with Food—Release of Lead and Cadmium—Part 2: Permissible Limits First Edition. 4 pp.

—**Ceramic—Lead Content**
ISO 8391 Pt 1-86. Ceramic Cookware in Contact with Food—Release of Lead and Cadmium—Part 1: Method of Test First Edition. 6 pp.
ISO 8391 Pt 2-86. Ceramic Cookware in Contact with Food—Release of Lead and Cadmium—Part 2: Permissible Limits First Edition. 4 pp.

—**Enamels—Thermal Shock**
BSI BS 1344: Part 16-75. 1975 Methods of Testing Vitreous Enamel Finishes Part 16: Resistance to Thermal Shock of Coatings on Cooking Utensils. 7 pp.
ISO 2747-73. Vitreous and Porcelain Enamels—Enamelled Cooking Utensils—Determination of Resistance to Thermal Shock First Edition. 5 pp.

INDUSTRY STANDARDS

Cooking Utensils (Cont.)

—Glass—Hygiene
CNS Z6087-88. Method of Test for the Hygiene of Food Utensils, Containers and Packages Products (Jan)(12218).

—Handles
BSI BS 6743-87. 1987 Amd 1 Performance of Handles and Handle Assemblies Attached to Cookware. 20 pp.

—Metal—Hygiene
CNS Z6088-88. Method of Test for the Hygiene of Food Utensils, Containers and Packages Metallic (Jan)(12219).

—Plastic—Hygiene
CNS Z6089-88. Method of Test for the Hygiene of Food Utensils, Containers and Packages Plastic Products (General Regulation) (Jan)(12220).
CNS Z6090-88. Method of Test for the Hygiene of Food Utensils, Containers and Packages Plastic Products (Classified Regulation) (Jan)(12221).

—Plastic—Microwave Ovens
JIS S 2033-91. Plastics Food Receptacles for Microwave Oven Use. 10 pp.

—Safety
BSI BS 3456: Sec 102.15-87. 1987 Safety of Household Electrical Appliances Part 102: Particular Requirements Section 102.15: Appliances for Heating Liquids. 28 pp.
CENELEC HD 264 S2-84. Safety of Household and Similar Electrical Appliances—Particular Requirements for Appliances for Heating Liquids. 8 pp.
CENELEC HD 287 S1-87. Safety of Household and Similar Electrical Appliances—Part 2: Particular Requirements for Commercial Electric Multi-Purpose Cooking Pans (Replaced by EN 335-2-39-1989). 9 pp.
CENELEC HD 287 S1/A1-90. AMD 1 Safety of Household and Similar Electrical Appliances—Part 2: Particular Requirements for Commercial Electric Multi-Purpose Cooking Pans. 5 pp.
CENELEC EN 60335-2-47/A51-91. AMD 51 Safety of Household and Similar Electrical Appliances Part 2 Particular Requirements for Commercial Electric Boiling Pans. 7 pp.

—Sizes
SAA AS 1408-73. Size Coding Scheme for Cooking Utensils. 8 pp.

—Steel—Enameled
CNS R2097-79. Enamel Tank (Nov)(5041). 1 p.

—Tin Coatings
BSI BS 3788-84. 1984 Tin Coated Finish for Culinary Utensils. 4 pp.

Cookware

See Also: Cooking Appliances; Cooking Utensils; Food Service Equipment
SNZ NZS 5702-91. Specification for Cookware Amend: A, 1991. 12 pp.

—Handles
BSI BS 6743-87. 1987 Amd 1 Performance of Handles and Handle Assemblies Attached to Cookware. 20 pp.
SNZ NZS 5701-91. Specification for Performance of Handles and Handle Assemblies Attached to Cookware Amend: 1; A, 1991. 16 pp.

Cool Rooms

See Also: Refrigeration Systems

—Construction—Submarines
MOD UK NES 111-91. Requirements for the Insulation and Fittings for the Refrigeration Spaces in Surface Ships and Submarines Issue 2 (06.91). 35 pp.

—Fittings—Submarines
MOD UK NES 111-91. Requirements for the Insulation and Fittings for the Refrigeration Spaces in Surface Ships and Submarines Issue 2 (06.91). 35 pp.

—Insulation—Submarines
MOD UK NES 111-91. Requirements for the Insulation and Fittings for the Refrigeration Spaces in Surface Ships and Submarines Issue 2 (06.91). 35 pp.

Coolant Pumps

See Also: Pumps; Refrigeration Equipment

—Machine Tools
BSI BS 3766-79. 1979 Coolant Pumps for Machine Tools. 10 pp.

—Machine Tools—Nominal Capacity
DIN ENGL 5440-78. Pumps; Coolant Pumps for Machine Tools; Nominal Capacity, Dimensions (Aug). 3 pp.

Coolants

See Also: Air Cooling; Coolers; Cooling Systems; Cutting Fluids; Engine Coolants; Nuclear Reactors; Refrigerants

—Hoses—Automotive
BSI BS 7038-88. 1988 Rubber Coolant Hoses and Tubing for Use on Private Cars and Light Commercial Vehicles. 12 pp.
BSI BS AU 108-65. 1965 Amd 1 Plain and Reinforced Hoses of Rubber. 8 pp.
ISO 4081-87. Rubber—Coolant Hoses and Tubing for Use on Private Cars and Light Commercial Vehicles—Specification First Edition. 10 pp.

—Hydrogen
IEC 842-88. Guide for Application and Operation of Turbine-Type Synchronous Machines Using Hydrogen as a Coolant First Edition. 23 pp.

—Radio Equipment
MOD UK DSTAN 68-61-79. Coolant Fluid, Inhibited: Radio Equipment Joint Service Designation: AL-26 Issue 1. 14 pp.

—Test Papers—Aircraft
MOD UK DTD-629-46. Coolant Test Papers (Reprinted May 1961). 6 pp.

—Tubes—Automotive
BSI BS 7038-88. 1988 Rubber Coolant Hoses and Tubing for Use on Private Cars and Light Commercial Vehicles. 12 pp.
ISO 4081-87. Rubber—Coolant Hoses and Tubing for Use on Private Cars and Light Commercial Vehicles—Specification First Edition. 10 pp.

Coolers

Scope Note: For additional listings, use a more specific term *See Also:* Air Coolers; Beverage Coolers; Compressors (Pressure Equipment); Coolants; Cooling Equipment; Cooling Systems; Cooling Towers; Dehumidifiers; Evaporative Coolers; Fluid Coolers; Freezers; Milk Coolers; Oil Coolers; Refrigeration Equipment; Refrigerators; Unit Coolers; Water Coolers
JIS S 2048-92. Coolers. 11 pp.

—Copper Strips—Aircraft
MOD UK DTD-607-44. Copper Strip for Radiators and Coolers (Reprinted January 1954). 2 pp.

Cooling Coils

See Also: Cooling Systems
BSI BS 5141: Part 1-75. 1975 Air Heating and Cooling Coils Part 1: Method of Testing for Rating of Cooling Coils. 53 pp.

Cooling Equipment

See Also: Air Conditioners; Coolers; Cooling Systems; Refrigerants; Refrigeration Equipment; Refrigeration Systems

—Temperature Controllers
CSA C22.2 NO 24-93. Temperature-Indicating and-Regulating Equipment; (Gen Instr 1). 88 pp.

—Volt Ampere Limits
CSA 551 Bull. Electrical Bulletin 551 July 19, 1962 to C22.2 NO 117. 1 p.

Cooling Systems

Use For: Water Cooling Systems
See Also: Absorbers (Equipment); Air Conditioners; Air Cooling; Blowers (Ventilation); Compressors (Pressure Equipment); Condensing Units; Coolants; Coolers; Cooling Coils; Cooling Equipment; Desupereaters; Ducts; Environmental Engineering; Evaporative Coolers; Exhaust Systems; Fans; Heat Exchangers; Heat Pumps; Heat Sinks; Heating, Ventilating and Air Conditioning Equipment; Pipelines; Radiators (Heating); Refrigerants; Refrigerated Transportation; Refrigeration Transportation Equipment; Refrigeration Equipment; Refrigeration Systems; Registers (Air Circulation); Unit Coolers; Ventilation Equipment; Vents; Water Coolers

—Aerostats
CAA Chapter Q5-4 12.79. Cooling Systems (Non-Rigid Airships). 3 pp.

—Aircraft—Flight Testing
CAA 600 (K). Section K Light Aeroplanes Powerplant Installations—Cooling Systems Chapter K5-4 (Blue Papers). 7 pp.
CAA Chapter K5-4 10.92. Cooling Systems (Light Aeroplanes). 3 pp.
CAA Chapter K5-4 App 10.92. Cooling Systems. 1 p.

—Automotive—Hoses
CNS D2046-86. Coolant System Hose for Automobiles (Aug)(6162).
JIS D 2602-82. Water Hose for Automobiles.
JIS D 2602-77. Water Hoses for Automobiles. 12 pp.

—Automotive—Pressure Caps
ISO 9133-88. Passenger Cars—Engine Cooling Systems—Threaded Pressure Caps and Their Seats on Filler Necks First Edition. 4 pp.
ISO 9817-91. Passenger Cars—Engine Cooling Systems—Dimensions of Pressure Caps and Their Ramp Seats on Filler Necks First Edition. 6 pp.
ISO 9818-91. Passenger Cars—Engine Cooling Systems—Test Methods and Marking of Pressure Caps First Edition. 7 pp.

—Automotive—Pressure Caps—Identification Systems
ISO 9818-91. Passenger Cars—Engine Cooling Systems—Test Methods and Marking of Pressure Caps First Edition. 7 pp.

—Combat Vehicles—Hoses
NATO STANAG 4043 ED 1 AMD 4-58. Cooling System Hoses for Wheeled Transport Tactical Vehicles. 7 pp.

—Construction
CSA CAN/CSA-C22. 2 NO 236-M90. Heating and Cooling Equipment; (Gen Instr 1) (ANSI/UL 1995). 136 pp.

—Evaporative Coolers—Motor Operated
CSA C22.2 NO 104-M1983. Humidifiers and Evaporative Coolers; (Gen Instr 1 Thru 2). 37 pp.

—Expansion Vessels
BSI BS 7074: Part 3-89. 1989 Application, Selection and Installation of Expansion Vessels and Ancillary Equipment for Sealed Water Systems Part 3: Code of Practice for Chilled and Condenser Systems. 20 pp.

—Hoverplatforms
CAA Chapter C4 01.74. Systems and Machinery. 1 p.

—Humidifiers—Motor Operated
CSA C22.2 NO 104-M1983. Humidifiers and Evaporative Coolers; (Gen Instr 1 Thru 2). 37 pp.

—Internal Combustion Engines—Glossaries
BSI BS 7016: Part 5-92. 1992 Components and Systems of Reciprocating Internal Combustion Engines Part 5: Glossary of Terms for Cooling Systems (ISO 7967-5: 1992). 19 pp.
ISO 7967 Pt 5-92. Reciprocating Internal Combustion Engines—Vocabulary of Components and Systems—Part 5: Cooling Systems First Edition. 20 pp.

—Internal Combustion Engines—Hoses
BSI BS 2952-58. 1958 Amd 2 Rubber Hose for I.C. Engine Cooling Systems. 13 pp.

—Rotary Wing Aircraft
CAA Chapter G5-4 06.76. Cooling Systems (Rotorcraft). 1 p.
CAA CAP 524 SUB-Part E 12.86. Power-Plant (Rotorcraft). 27 pp.
CAA CAP524 SUB-Part E 12.86. (Rotorcraft). 19 pp.

—Rotary Wing Aircraft—Comprehensive Testing
CAA Chapter G5-4 App 06.76. Cooling System Tests (Rotorcraft). 4 pp.

—Ships—Automatic Control Equipment
JIS F 0805-79. Methods of Onboard Test on Automatic Control of Cooling Water System for Smaller Ships.

—Temperature Indicators
CSA C22.2 NO 24-93. Temperature-Indicating and-Regulating Equipment; (Gen Instr 1). 88 pp.

—Thermostats
CSA C22.2 NO 24-93. Temperature-Indicating and-Regulating Equipment; (Gen Instr 1). 88 pp.

—Water Formed Deposits—Sampling
BSI BS 2455: Part 1-73. 1973 Methods of Sampling and Examining Deposits from Boilers and Associated Industrial Plant Part 1: Water-Side Deposits. 31 pp.
BSI BS 2455: Part 1-01. 1973 Amd 1 Methods of Sampling and Examining Deposits from Boilers and Associated Industrial Plant Part 1: Water-Side Deposits (AMD 7753) October 15, 1993 (Q). 32 pp.

Cooling Towers
Use For: Water Cooling Towers See Also: Coolers; Evaporative Coolers; Refrigeration Equipment; Water Coolers
BSI BS 4485: Part 2-88. 1988 Water Cooling Towers Part 2: Methods for Performance Testing. 36 pp.
DIN ENGL 1947-89. Thermal Performance Acceptance Testing of Water Cooling Towers; (VDI Code of Practice) (May). 23 pp.
JIS B 8609-81. Performance Tests of Mechanical Draft Cooling Tower. 29 pp.

—Design
BSI BS 4485: Part 3-88. 1988 Water Cooling Towers Part 3: Code of Practice for Thermal and Functional Design. 51 pp.
BSI BS 4485: Part 4-75. 1975 Water Cooling Towers Part 4: Structural Design of Cooling Towers. 34 pp.

—Glossaries
BSI BS 4485: Part 1-69. 1969 Water Cooling Towers Part 1: Glossary of Terms. 7 pp.

—Noise
DIN ENGL 45635 Pt 46-85. Measurement of Noise Emitted by Machines; Airborne Noise Emission; Enveloping Surface Method; Cooling Towers (June). 4 pp.

Cooling Water
See Also: Industrial Water; Water

—Lead Acid Batteries—Submarines
NATO STANAG 4391 ED 1 AMD 0-91. Cooling Water for Submarine Main Lead Acid Batteries. 4 pp.

—Sampling
DIN ENGL 38402 Pt 22-91. German Standard Methods for the Examination of Water, Waste Water and Sludge; General Information (Group A); Sampling of Cooling Water for Industrial Use (A 22) (June). 5 pp.

Coombs' Test
Use: Antiglobulin Testing

Cooperation in Space
Use For: COSPAS See Also: Satellite Communications; Search and Rescue
ICAO Circular 185. Satellite-Aided Search and Rescue—the COSPAS-SARSAT System—1986; (Erratum 10/03/86). 48 pp.

Coordinate Measuring Instruments
Use For: Measuring Machines, Coordinate
See Also: Measuring Instruments; Plotters

—Accuracy Testing
BSI BS 6808: Part 2-87. 1987 Coordinate Measuring Machines Part 2: Methods for Verifying Performance. 10 pp.
JIS B 7440-87. Test Code for Accuracy of Coordinate Measuring Machines. 29 pp.

—Glossaries
BSI BS 6808: Part 1-87. 1987 Coordinate Measuring Machines Part 1: Glossary of Terms. 19 pp.

Coordinate Systems
—Drafting
DIN ENGL 406 Pt 3-75. Dimensioning in Drawings; Dimensioning by Co-Ordinates (July). 8 pp.
DIN ENGL 406 Pt 4-80. Dimensioning on Drawings; Dimensioning for Programming by Machine (Dec). 4 pp.
DIN ENGL 406 Pt 4 Suppl. 1-80. Dimensioning on Drawings; Dimensioning for Programming by Machine; Example of Application (Dec). 5 pp.

—Radiation Therapy Equipment
CENELEC PREN 61217-93. Guide to a Co-Ordinate System, Including Marking of Scales, for Movement and Location of Radiotherapy Equipment. 68 pp.
CENELEC PREN 61217-93. Guide to a Co-Ordinate System, Including Marking of Scales, for Movement and Location of Radiotherapy Equipment. 65 pp.

—Robots
BSI BS 7228: Part 2-91. 1991 Industrial Robots Part 2: Guide to Definitions of Coordinate Systems and Motions (ISO 9787: 1990). 10 pp.
BSI BS 7228: Part 2-01. 1991 Amd 1 Industrial Robots Part 2: Guide to Definitions of Coordinate Systems and Motions (ISO 9787: 1990) (AMD 7297) September 15, 1992. 15 pp.
CEN EN 29787-92. Manipulating Industrial Robots—Coordinate Systems and Motions. 11 pp.
DIN ENGL EN 29787-91. Manipulating Industrial Robots; Coordinate Systems and Motions; (ISO 9787: 1990) (Sept). 8 pp.

Coordinate Systems (Cont.)
—Robots (Cont.)
ISO 9787-90. Manipulating Industrial Robots—Coordinate Systems and Motions First Edition. 9 pp.
JIS B 8437-90. Coordinate System and Motion Nomenclature for Industrial Robots. 10 pp.
JTC1 9787-90. Manipulating Industrial Robots—Coordinate Systems and Motions First Edition. 9 pp.
SAA AS 3986-91. Manipulating Industrial Robots—Coordinate Systems and Motions (ISO 9787:1990). 5 pp.

Coordinated Universal Time
Use For: Universal Time Coordinated; UTC
See Also: Time; Time Scales
CCIR RECMN 460-4-86. Standard-Frequency and Time-Signal Emissions—Section 7B—Specifications for the Standard-Frequency and Time-Signal Services. 4 pp.
CCIR RECMN 535-1-82. Use of the Term UTC—Section 7B—Specifications for the Standard-Frequency and Time-Signal Services. 1 p.
CCIR RECMN 808-92. Broadcasting of Time and Date Information in Coded Form. 5 pp.

—Abbreviations
CCIR RECMN 536-78. Time-Scale Notations—Section 7B—Specifications for the Standard-Frequency and Time-Signal Services. 2 pp.

—Documentation—Telecommunication Services
CCITT RECMN B.11-89. Legal Time; Use of the Term UTC—Terms and Definitions Abbreviations on Means of Expression (Series B) General Telecommunications Statistics (Series C) 1 pp. 1 p.

—Reference Clocks
CCIR Report 898-2-90. Operational Experience with Reference Clocks in Time Systems—Section 7D—Characterization of Sources and Time Scales Formation. 10 pp.

—Telecommunication Services
CCITT RECMN B.11-89. Legal Time; Use of the Term UTC—Terms and Definitions Abbreviations on Means of Expression (Series B) General Telecommunications Statistics (Series C) 1 pp. 1 p.

Coordination Areas
See Also: Area; Coordination Contours; Coordination Distances; Radio Frequency Interference
CCIR Decision 87-89. Determination of the Coordination Area (Appendix 28 of the Radio Regulations) This Text May be Found in the Annex to Part 2 of Volumes IV and IX—Volume V—Propagation in Non-Ionized Media. 1 p.
CCIR Decision 87-89. Determination of the Coordination Area (Appendix 28 of the Radio Regulations) This Text May be Found in Volumes IV/IX, Part 2—Annex 3 to Volume VIII—Mobile, Radiodetermination, Amateur and Related Satellite Services. 1 p.
CCIR Decision 87-89. Determination of the Coordination Area (Appendix 28 of the Radio Regulations)—Annex to Volume IX-1—Fixed Service Using Radio-Relay Systems. 1 p.
CCIR Decision 87-89. Determination of the Coordination Area (Appendix 28 of the Radio Regulations)—Volume X-1—Broadcasting Service (Sound). 1 p.
CCIR Decision 87-89. Determination of the Coordination Area (Appendix 28 of the Radio Regulations) (See Annex to Vols. IV/IX-2)—Annex to Volume XI-1—Broadcasting Service (Television). 1 p.

—Bidirectional—Earth Stations—Radio Wave Propagation
CCIR Report 1010-86. Propagation Data for Bi-Directional Coordination of Earth Stations—Section 5G—Propagation Data Required for the Evaluation of Interference: Space and Terrestrial Systems. 5 pp.

—Earth Stations
CCIR RECMN 848-92. Determination of the Coordination Area of a Transmitting Earth Station Using the Same Frequency Band as Receiving Earth Stations in Bidirectionally Allocated Frequency Bands. 9 pp.
CCIR RECMN 850-92. Coordination Areas Using Predetermined Coordination Distances. 6 pp.

—Earth Stations—Fixed Satellite Services
CCIR Report 1163-90. Coordination Area of an Earth Station of the Fixed-Satellite Service Sharing the Same Frequency Band with the Radionavigation Service—Section 8D—Radiodetermination, Global Maritime Distress and Safety System and Related Subjects. 2 pp.

Coordination Areas (Cont.)
—Earth Stations—Fixed Satellite Services (Cont.)
CCIR QUESTION 5/12-90. Coordination Area of an Earth Station of the Fixed-Satellite Service Sharing the Same Frequency Band with the Radionavigation Service—Questions Concerning Study Group 12—Inter-Service Sharing and Compatibility. 1 p.

—Earth Stations—Geostationary Space Stations
CCIR RECMN 847-92. Determination of the Coordination Area of an Earth Station Operating with a Geostationary Space Station and Using the Same Frequency Band as a System in a Terrestrial Service. 30 pp.

—Earth Stations—Mobile Stations (Communications)
CCIR QUESTION 4/12-90. Coordination Between an Earth Station and Mobile Stations in the Mobile Services—Questions Concerning Study Group 12—Inter-Service Sharing and Compatibility. 1 p.

—Earth Stations—Nongeostationary Space Stations
CCIR RECMN 849-92. Determination of Coordination Area for Earth Stations Operating with Non-Geostationary Spacecraft in Bands Shared with Terrestrial Services. 10 pp.

—Earth Stations—Radio Wave Propagation
CCIR RECMN 620-86. Propagation Data Required for the Calculation of Coordination Distances—Section 5G—Propagation Data Required for the Evaluation of Interference: Space and Terrestrial Systems. 1 p.
CCIR Report 724-2-86. Propagation Data Required for the Evaluation of Coordination Distance in the Frequency Range 1-40 GHz—Section 5G—Propagation Data Required for the Evaluation of Interference: Space and Terrestrial Systems. 41 pp.
CCIR RECMN 620-1-92. Propagation Data Required for the Calculation of Coordination Distances in the Frequency Range 1-40 GHz. 9 pp.

—Mobile Satellite Communications—Geostationary Orbits
CCIR Report 1185-90. Technical Aspects of Coordination Among Mobile Satellite Systems Using the Geostationary Satellite Orbit—Section 8H—Efficient Use of the Radio Spectrum Characteristics and Sharing of Frequency Resources. 12 pp.

Coordination Commitee For Vocabulary
Use: CCV

Coordination Contours
See Also: Coordination Areas; Coordination Distances; Dimensions; Maps; Radio Frequency Interference

—Earth Stations
CCIR Report 773-78. Concept of Coordination and Protection Contours for Use in the Coordination of Mobile Earth Stations—Section 8H—Efficient Use of the Radio Spectrum Characteristics and Sharing of Frequency Resources. 6 pp.

—Earth Stations—Mobile Satellite Communications
CCIR Report 773-78. Concept of Coordination and Protection Contours for Use in the Coordination of Mobile Earth Stations—Section 8H—Efficient Use of the Radio Spectrum Characteristics and Sharing of Frequency Resources. 6 pp.

Coordination Distances
See Also: Coordination Areas; Coordination Contours; Dimensions; Radio Frequency Interference

—Earth Stations
CCIR RECMN 850-92. Coordination Areas Using Predetermined Coordination Distances. 6 pp.

—Mobile Satellite Communications
CCIR Report 1179-90. Methodology for the Derivation of Interference and Sharing Criteria for the Mobile-Satellite Services—Section 8H—Efficient Use of the Radio Spectrum Characteristics and Sharing of Frequency Resources. 13 pp.

Copal Ester Gum
Use: Ester Gum—Copal

Copal Varnishes
See Also: Varnishes

INTERNATIONAL AND NON-U.S. NATIONAL STANDARDS
SUBJECT INDEX

Copal Varnishes (Cont.)
—Shells (Ammunition)
MOD UK DSTAN 80-176-93. Varnish, Copal, Type QX Issue 1. 17 pp.
MOD UK CS 1844A. Varnish Copal, LF Quality (Superseded by Def Stan 80-176).

Copiers
Use: Copying Machines

Coping Saws
See Also: Fretsaws; Hand Tools; Saws
—Blades
MOD UK DSTAN 51-11: Part 10-75. Hand Tools, General Purpose Part 10: Saw, Coping and Blade, Coping Saw Issue 1. 9 pp.
—Frames
MOD UK DSTAN 51-11: Part 10-75. Hand Tools, General Purpose Part 10: Saw, Coping and Blade, Coping Saw Issue 1. 9 pp.

Copings
Use For: Wall Copings *See Also:* Walls
—Concrete/Clayware/Stone
BSI BS 5642: Part 2-83. 1983 Sills and Copings Part 2: Coping of Precast Concrete, Cast Stone, Clayware, Slate and Natural Stone. 9 pp.

Copolymers
Scope Note: For additional listings, use a more specific term *See Also:* Ethylene Copolymers; Ethylene Propylene Copolymers; Polymers; Vinyl Copolymers
—Chlorine Content
JIS K 7229-87. Determination of Chlorine in Chlorine-Containing Polymers, Copolymers and Their Compounds.
—Paints—Binders
ISO 7143-82. Binders for Paints and Varnishes—Aqueous Dispersions of Polymers and Copolymers—General Methods of Test First Edition. 3 pp.
—PH
ISO 1148-80. Plastics—Aqueous Dispersions of Polymers and Copolymers—Determination of pH Second Edition. 3 pp.
—Sieve Analysis
ISO 4576-78. Plastics—Aqueous Dispersions of Homopolymers and Copolymers—Determination of Gross Particle Content by Sieve Analysis First Edition. 5 pp.
—Temperature Change Testing
ISO 2115-76. Plastics—Aqueous Dispersions of Polymers and Copolymers—Determination of White Point Temperature and Minimum Film-Forming Temperature Second Edition. 6 pp.
—Thermal Shock
ISO 1147-88. Plastics—Polymer Dispersions—Freeze-Thaw Cycle Stability Test Second Edition. 4 pp.
—Varnishes—Binders
ISO 7143-82. Binders for Paints and Varnishes—Aqueous Dispersions of Polymers and Copolymers—General Methods of Test First Edition. 3 pp.

Copper
Scope Note: For additional listings, see also specific products made from copper *See Also:* Blister Copper; Cement Copper; Copper Coatings (Made From Copper); Copper Coatings (On Copper); Copper Content Analysis; Copper Number; Copper Wire; Electrolytic Copper; Magnesium Copper; Metals; Nonferrous Metals; Tough Pitch Copper
CNS K7158-66. Chemical Reagent (Copper) (Mar)(1658).
JIS K 8660-61. Copper.
—Aircraft
SBAC RS 116 (V). S.B.A.C. Protective Treatment Chart.
—Aluminum Content
JIS H 1057-87. Methods for Determination of Aluminium in Copper and Copper Alloys. 9 pp.
—Antimony Content—Spectrophotometry
ISO 5956-84. Copper and Copper Alloys—Determination of Antimony Content—Rhodamine B Spectrometric Method First Edition. 4 pp.
—Arsenic Content
JIS H 1059-87. Methods for Determination of Arsenic in Copper and Copper Alloys. 10 pp.

Copper (Cont.)
—Arsenic Content—Photometry
ISO 3220-75. Copper and Copper Alloys—Determination of Arsenic—Photometric Method First Edition. 4 pp.
—Bars
BSI BS 1433-70. 1970 Amd 3 Copper for Electrical Purposes. Rod and Bar. 34 pp.
BSI BS 1434-85. 1985 Copper for Electrical Purposes: Copper Sections in Bars, Blanks and Segments for Commutators. 14 pp.
CNS H3124-86. Copper and Copper Alloy-Rods and Bars (Sep)(10442). 15 pp.
DIN ENGL 17672 Pt 2-74. Bars of Copper and Wrought Copper Alloys; Technical Conditions of Delivery (June). 3 pp.
ISO 1637-87. Wrought Copper and Copper Alloy Rod and Bar—Technical Conditions of Delivery Second Edition. 10 pp.
JIS H 3250-92. Copper and Copper Alloy Rods and Bars. 23 pp.
JIS H 3510-92. Oxygen Free Copper Sheet, Plate, Strip, Seamless Pipe and Tube, Rod, Bar and Wire for Electron Devices. 19 pp.
—Bars—Drawn—Hexagonal
ISO 3490-84. Wrought Copper and Copper Alloys—Drawn Hexagonal Bars—All Minus Tolerances on Width Across Flats and Form Tolerances First Edition. 5 pp.
ISO 7757-84. Wrought Copper and Copper Alloys—Drawn Hexagonal Bars—Symmetric Plus and Minus Tolerances on Width Across Flats and Form Tolerances First Edition. 5 pp.
—Bars—Drawn—Rectangular
DIN ENGL 1759-74. Rectangular Bars of Copper and Wrought Copper Alloys; Drawn, with Sharp Edges; Dimensions, Permissible Variations, Static Values (June). 12 pp.
DIN ENGL 46433-59. Rectangular Wires and Rectangular Bars; Drawn, with Radiused Edges; Dimensions (Nov). 8 pp.
DIN ENGL 46433 Extr. 1-59. Rectangular Wires and Rectangular Bars; Drawn, with Radiused Edges; Dimensions; Selected Sizes for Electrical Machines and Switchgear (Nov). 8 pp.
DIN ENGL 46433 Extr. 3-59. Rectangular Bars; Drawn, with Radiused Edges; Dimensions; Selected Sizes for Switchgears (Nov). 2 pp.
ISO 6958-84. Wrought Copper and Copper Alloys—Drawn Rectangular Bars—Dimension and Form Tolerances First Edition. 6 pp.
—Bars—Drawn—Round
ISO 3489-84. Wrought Copper and Copper Alloys—Drawn Round Bars—All Minus Tolerances on Diameter and Form Tolerances First Edition. 5 pp.
ISO 7756-84. Wrought Copper and Copper Alloys—Drawn Round Bars—Symmetric Plus and Minus Tolerances on Diameter and Form Tolerances First Edition. 5 pp.
—Bars—Drawn—Square
ISO 3491-84. Wrought Copper and Copper Alloys—Drawn Square Bars—All Minus Tolerances on Width Across Flats and Form Tolerances First Edition. 5 pp.
ISO 7758-84. Wrought Copper and Copper Alloys—Drawn Square Bars—Symmetric Plus and Minus Tolerances on Width Across Flats and Form Tolerances First Edition. 5 pp.
—Bars—Electrical Properties
DIN ENGL 17672 Pt 1-83. Wrought Copper and Copper Alloy Rod and Bar; Properties (Dec). 11 pp.
—Bars—Extruded—Hexagonal
ISO 3488-82. Wrought Copper and Copper Alloys—Extruded Round, Square or Hexagonal Bars—Dimensions and Tolerances First Edition. 4 pp.
—Bars—Extruded—Round
ISO 3488-82. Wrought Copper and Copper Alloys—Extruded Round, Square or Hexagonal Bars—Dimensions and Tolerances First Edition. 4 pp.
—Bars—Extruded—Square
ISO 3488-82. Wrought Copper and Copper Alloys—Extruded Round, Square or Hexagonal Bars—Dimensions and Tolerances First Edition. 4 pp.
—Bars—Mechanical Properties
DIN ENGL 17672 Pt 1-83. Wrought Copper and Copper Alloy Rod and Bar; Properties (Dec). 11 pp.
—Bars—Rolled
SAA AS 1566-85. Copper and Copper Alloys—Rolled Flat Products. 12 pp.

Copper (Cont.)
—Bars—Wrought
SAA AS 1567-85. Copper and Copper Alloys—Wrought Rods, Bars and Sections. 14 pp.
—Billets—Chemical Analysis
ISO 1811 Pt 1-88. Copper and Copper Alloys—Selection and Preparation of Samples for Chemical Analysis—Part 1: Sampling of Cast Unwrought Products First Edition. 7 pp.
—Bismuth Content—Spectrometry
ISO 5959-84. Copper and Copper Alloys—Determination of Bismuth Content—Diethyldithiocarbamate Spectrometric Method First Edition. 5 pp.
—Charcoal Fabrics
MOD UK DSTAN 83-84-93. Charcoal Cloth Impregnated with Copper and Chromium, Type 1303 Issue 1. 28 pp.
—Chemical Analysis
CNS H2073-85. Method of Chemical Analysis for Copper Products (May)(11260).
CNS H2076-88. General Rules of Chemical Analysis for Copper Products and Copper Alloys (Dec)(11942).
CNS K6359-73. Chemical Analysis of Copper in Natural Rubber (May)(3568).
ISO 1336-80. Wrought Coppers (Having Minimum Copper Contents of 97,5 %)—Chemical Composition and Forms of Wrought Products First Edition. 4 pp.
ISO 1337-80. Wrought Coppers (Having Minimum Copper Contents of 99,85 %)—Chemical Composition and Forms of Wrought Products First Edition. 4 pp.
JIS H 1012-91. General Rules for Chemical Analysis of Copper and Copper Alloys.
JIS H 1201-77. Methods for Chemical Analysis of Copper Products.
—Chromium Content—Atomic Absorption Spectrometry
ISO 4744-84. Copper and Copper Alloys—Determination of Chromium Content—Flame Atomic Absorption Spectrometric Method First Edition. 5 pp.
—Chromium Content—Atomic Absorption Spectrophotometry
BSI BS 6721: Part 4-86. 1986 Sampling and Analysis of Copper and Copper Alloys Part 4: Method for Determination of Chromium in Copper and Copper Alloys by Flame Atomic Absorption Spectrophotometry. 7 pp.
—Cobalt Content—Absorptiometric Analysis
JIS H 1060-89. Methods for Determination of Cobalt in Copper and Copper Alloys. 12 pp.
—Cobalt Content—Atomic Absorption Spectrometry
JIS H 1060-89. Methods for Determination of Cobalt in Copper and Copper Alloys. 12 pp.
—Copper Content—Electrolytic Analysis
ISO 1553-76. Unalloyed Copper Containing Not Less Than 99,90 % of Copper—Determination of Copper Content—Electrolytic Method First Edition. 6 pp.
SAA AS K208.1-70. Methods for the Analysis of Unalloyed Copper—Part 1: Method for the Electrolytic Determination of Copper in Unalloyed Copper Containing Not Less Than 99.85 Percent Copper Amdt 1 April 1971. 7 pp.
—Copper Content—Spectrophotometry
ISO 1553-76. Unalloyed Copper Containing Not Less Than 99,90 % of Copper—Determination of Copper Content—Electrolytic Method First Edition. 6 pp.
—Electrical Insulation—Corrosion
CNS C3148-88. Method of Test for Copper Corrosion of Heat Shrinkable Tubing for Electrical Insulation (Sep)(8795).
—Electrical Resistance
CNS C1002-80. Resistance of Copper (Jul)(32).
CNS C1132-84. Resistance and Conductivity of Copper Materials for Electrical Purposes (Jun)(10913).
JIS C 3001-81. Resistance of Copper Materials for Electrical Purposes. 6 pp.
—Extruded—Copper Content
JIS H 1051-92. Methods for Determination of Copper in Copper and Copper Alloys. 22 pp.
—Fertilizers
CNS N3097-88. Chelated Copper Fertilizer (Jul)(12031).

INTERNATIONAL AND NON-U.S. NATIONAL STANDARDS
SUBJECT INDEX

Copper

Copper (Cont.)

—Forgings
BSI BS 2872-89. 1989 Copper and Copper Alloy Forging Stock and Forgings. 24 pp.
DIN ENGL 17673 Pt 2-73. Drop Forgings of Copper and Wrought Copper Alloys; Technical Conditions of Delivery (Dec). 3 pp.
DIN ENGL 17673 Pt 4-74. Drop Forgings of Copper and Wrought Copper Alloys; Permissible Variations (Jan). 7 pp.
DIN ENGL 17678 Pt 2-73. Hand Forgings of Copper and Wrought Copper Alloys; Technical Conditions of Delivery (Nov). 3 pp.
DIN ENGL 17678 Pt 4-73. Hand Forgings of Copper and Wrought Copper Alloys; Permissible Variations (Dec). 2 pp.
SAA AS 1568-85. Copper and Copper Alloys—Forging Stock and Forgings. 7 pp.

—Forgings—Design
DIN ENGL 17673 Pt 3-74. Drop Forgings of Copper and Wrought Copper Alloys; Design Principles (Jan). 7 pp.
DIN ENGL 17678 Pt 3-73. Hand Forgings of Copper and Wrought Copper Alloys; Basis of Design (Dec). 2 pp.

—Forgings—Electrical Properties
DIN ENGL 17673 Pt 1-83. Wrought Copper and Copper Alloy Drop Forgings; Properties (Dec). 4 pp.
DIN ENGL 17678 Pt 1-83. Wrought Copper and Copper Alloy Hand Forgings; Properties (Dec). 3 pp.

—Forgings—Mechanical Properties
DIN ENGL 17673 Pt 1-83. Wrought Copper and Copper Alloy Drop Forgings; Properties (Dec). 4 pp.
DIN ENGL 17678 Pt 1-83. Wrought Copper and Copper Alloy Hand Forgings; Properties (Dec). 3 pp.
ISO 1640-74. Wrought Copper Alloys—Forgings—Mechanical Properties First Edition. 4 pp.

—Glossaries
BSI BS 6931-88. 1988 Glossary of Terms for Copper and Copper Alloys. 18 pp.
ISO 197 Pt 1-83. Copper and Copper Alloys—Terms and Definitions—Part 1: Materials First Edition. 5 pp.
ISO 197 Pt 2-83. Copper and Copper Alloys—Terms and Definitions—Part 2: Unwrought Products (Refinery Shapes) First Edition. 3 pp.
ISO 197 Pt 3-83. Copper and Copper Alloys—Terms and Definitions—Part 3: Wrought Products First Edition. 8 pp.
ISO 197 Pt 5-80. Copper and Copper Alloys—Terms and Definitions—Part 5: Methods of Processing and Treatment First Edition. 3 pp.

—Grain Size Analysis
BSI BS 7428-91. 1991 Estimating the Average Grain Size of Copper and Copper Alloys (ISO 2624: 1990). 12 pp.
CNS H2069-82. Methods for Estimating the Average Grain Size Test of Wrought Copper and Copper Base Alloys (Oct)(9504).
ISO 2624-90. Copper and Copper Alloys—Estimation of Average Grain Size Second Edition. 11 pp.
JIS H 0501-86. Methods for Estimating Average Grain Size of Wrought Copper and Copper-Alloys. 8 pp.

—High Conductivity—Tensile Testing
BSI DD 79-81. 1981 Method for Performing the Spiral Elongation Test on High Conductivity Copper. 7 pp.
ISO TR4745-78. High Conductivity Copper—Spiral Elongation Test First Edition. 5 pp.

—Hollow Sections—Wrought
SAA AS 1567-85. Copper and Copper Alloys—Wrought Rods, Bars and Sections. 14 pp.

—Hydrogen Embrittlement
BSI BS 5899-80. 1980 Method for Hydrogen Embrittlement Test for Copper. 4 pp.
ISO 2626-73. Copper—Hydrogen Embrittlement Test First Edition. 4 pp.

—Ingots
SAA AS 1565-85. Copper and Copper Alloys—Ingots and Castings. 36 pp.

—Iron Content
CNS H2087-89. Methods of Determination for Iron in Copper and Copper Alloys (Jun)(12533).
JIS H 1054-84. Methods for Determination of Iron in Copper and Copper Alloys. 15 pp.

—Lead Content
CNS H2086-89. Methods of Determination for Lead in Copper and Copper Alloys (Jun)(12532).
JIS H 1053-84. Methods for Determination of Lead in Copper and Copper Alloys. 17 pp.

Copper (Cont.)

—Lead Content—Volumetric Analysis
ISO 3112-75. Copper and Copper Alloys—Determination of Lead—Extracting Titration Method First Edition. 6 pp.

—Manganese Content
JIS H 1055-87. Methods for Determination of Manganese in Copper and Copper Alloys. 9 pp.

—Manganese Content—Spectrophotometry
ISO 2543-73. Copper and Copper Alloys—Determination of Manganese—Spectrophotometric Method First Edition. 4 pp.

—Materials—Designation Codes
ISO 1190 Pt 1-82. Copper and Copper Alloys—Code of Designation—Part 1: Designation of Materials First Edition. 3 pp.

—Mineral Supplements—Animal Feed
CNS N4064-80. Method of Test for the Purity of Copper Salts in Mineral Supplements (For Feeding) (Dec)(6712).

—Nickel Content
JIS H 1056-87. Methods for Determination of Nickel in Copper and Copper Alloys. 17 pp.

—Oxygen Free—Adhesive Strength
BSI BS 5909-80. 1980 Scale Adhesion Test for Oxygen-Free Copper. 2 pp.
ISO 4746-77. Oxygen-Free Copper—Scale Adhesion Test First Edition. 3 pp.

—Peppermint Oil—Corrosion
CNS K6110-90. Method of Test for Copper Strip Corrosion of Peppermint Oil (Aug)(1219).

—Petroleum Products—Corrosion
BSI BS 2000: Part 154-93. 1993 Methods of Test for Petroleum and Its Products Part 154: Petroleum Products—Corrosiveness to Copper—Copper Strip Test (ISO 2160: 1985) (W). 8 pp.
BSI BS 2000: Part 154-82. 1982 Petroleum and Its Products Part 154: Detection of Copper Corrosion from Petroleum Products by the Copper Strip Tarnish Test. 10 pp.
BSI BS 6924-88. 1988 Method for Determination of Corrosiveness of Liquefied Petroleum Gases to Copper. 7 pp.
CNS K6249-74. Method of Test for Corrosion of Liquefied Petroleum Gases (Copper Strip) (Jun)(2750).
ISO 2160-85. Petroleum Products—Corrosiveness to Copper—Copper Strip Test Second Edition. 9 pp.
ISO 6251-82. Liquefied Petroleum Gases—Corrosiveness to Copper—Copper Strip Test First Edition. 6 pp.
JIS K 2513-91. Petroleum Products—Corrosiveness to Copper—Copper Strip Test. 15 pp.

—Phosphorus Content
JIS H 1058-87. Methods for Determination of Phosphorus in Copper and Copper Alloys. 10 pp.

—Phosphorus Content—Spectrometry
ISO 4741-84. Copper and Copper Alloys—Determination of Phosphorus Content—Molybdovanadate Spectrometric Method First Edition. 5 pp.

—Phosphorus Content—Spectrophotometry
SAA AS 1696.1-87. Copper—Part 1: Determination of Phosphorous—Spectrophotometric Method. 3 pp.

—Plates
CNS H3132-84. Copper and Copper Alloy Sheets, Plates, Strips and Coiled Sheets (Oct)(11073).
DIN ENGL 17670 Pt 2-69. Plate, Sheet and Strip of Copper and Wrought Copper Alloys; Technical Conditions of Delivery (June). 3 pp.
ISO 1634 Pt 1-87. Wrought Copper and Copper Alloy Plate, Sheet and Strip Part 1: Technical Conditions of Delivery for Plate, Sheet and Strip for General Purposes First Edition. 11 pp.
JIS H 3100-92. Copper and Copper Alloy Sheets, Plates and Strips. 60 pp.
JIS H 3510-92. Oxygen Free Copper Sheet, Plate, Strip, Seamless Pipe and Tube, Rod, Bar and Wire for Electron Devices. 19 pp.

—Plates—Electrical Properties
DIN ENGL 17670 Pt 1-83. Wrought Copper and Copper Alloy Plate, Sheet and Strip; Properties (Dec). 16 pp.

—Plates—Mechanical Properties
DIN ENGL 17670 Pt 1-83. Wrought Copper and Copper Alloy Plate, Sheet and Strip; Properties (Dec). 16 pp.

—Plates—Rolled
SAA AS 1566-85. Copper and Copper Alloys—Rolled Flat Products. 12 pp.

Copper (Cont.)

—Plates—Wrought
BSI BS 2875-69. 1969 Copper and Copper Alloys. Plate. 38 pp.

—Refinery Shapes
BSI BS 6017-81. 1981 Amd 2 Copper Refinery Shapes (AMD 5725) March 31, 1988. 12 pp.
DIN ENGL 1708-73. Copper; Cathodes and Refinery Shapes (Jan). 4 pp.
ISO 197 Pt 2-83. Copper and Copper Alloys—Terms and Definitions—Part 2: Unwrought Products (Refinery Shapes) First Edition. 3 pp.
ISO 431-81. Copper Refinery Shapes Second Edition. 9 pp.

—Residual Stress
ISO 196-78. Wrought Copper and Copper Alloys—Detection of Residual Stress—Mercury (I) Nitrate Test First Edition. 3 pp.

—Rods
BSI BS 1433-70. 1970 Amd 3 Copper for Electrical Purposes. Rod and Bar. 34 pp.
BSI BS 2874-86. 1986 Copper and Copper Alloy Rods and Sections (Other Than Forging Stock). 23 pp.
CNS H3069-88. Copper Rods for Electrical Purposes (Dec)(4745).
CNS H3124-86. Copper and Copper Alloy-Rods and Bars (Sep)(10442). 15 pp.
DIN ENGL 17672 Pt 2-74. Bars of Copper and Wrought Copper Alloys; Technical Conditions of Delivery (June). 3 pp.
ISO 1637-87. Wrought Copper and Copper Alloy Rod and Bar—Technical Conditions of Delivery Second Edition. 10 pp.
JIS H 3250-92. Copper and Copper Alloy Rods and Bars. 23 pp.
JIS H 3510-92. Oxygen Free Copper Sheet, Plate, Strip, Seamless Pipe and Tube, Rod, Bar and Wire for Electron Devices. 19 pp.

—Rods—Drawn
DIN ENGL 1756-69. Round Rod of Copper and Wrought Copper Alloys; Drawn; Dimensions (July). 4 pp.

—Rods—Drawn—Hexagonal
DIN ENGL 1763-69. Hexagon Rod of Copper and Wrought Copper Alloys; Drawn, with Sharp Edges; Dimensions (July). 4 pp.

—Rods—Drawn—Square
DIN ENGL 1761-69. Square Rod of Copper and Copper Wrought Alloys; Drawn, with Sharp Edges; Dimensions (July). 4 pp.

—Rods—Electrical Properties
DIN ENGL 17672 Pt 1-83. Wrought Copper and Copper Alloy Rod and Bar; Properties (Dec). 11 pp.

—Rods—Extruded—Round
DIN ENGL 1782-69. Round Rod of Copper and Wrought Copper Alloys; Extruded; Dimensions (July). 4 pp.

—Rods—Mechanical Properties
DIN ENGL 17672 Pt 1-83. Wrought Copper and Copper Alloy Rod and Bar; Properties (Dec). 11 pp.

—Rods—Wrought
SAA AS 1567-85. Copper and Copper Alloys—Wrought Rods, Bars and Sections. 14 pp.

—Rolled—Copper Content
JIS H 1051-92. Methods for Determination of Copper in Copper and Copper Alloys. 22 pp.

—Sampling
ISO 1811 Pt 1-88. Copper and Copper Alloys—Selection and Preparation of Samples for Chemical Analysis—Part 1: Sampling of Cast Unwrought Products First Edition. 7 pp.
ISO 1811 Pt 2-88. Copper and Copper Alloys—Selection and Preparation of Samples for Chemical Analysis—Part 2: Sampling of Wrought Products and Castings First Edition. 5 pp.

—Scrap—Classification
JIS H 2109-86. Classification Standard of Copper and Copper Alloy Scraps. 15 pp.

—Scrap—Copper Content
CNS M3185-85. Method for Sampling and Method for Determination of Copper Content and Moisture of Copper Scraps for Smelting (Dec)(11457).
JIS M 8082-76. Method of Sampling and Methods of Determination of Copper Content and Moisture of Copper Scraps for Smelting.

INDUSTRY STANDARDS

Copper (Cont.)

—Scrap—Moisture Content
CNS M3185-85. Method for Sampling and Method for Determination of Copper Content and Moisture of Copper Scraps for Smelting (Dec)(11457).
JIS M 8082-76. Method of Sampling and Methods of Determination of Copper Content and Moisture of Copper Scraps for Smelting.

—Sections
BSI BS 2874-86. 1986 Copper and Copper Alloy Rods and Sections (Other Than Forging Stock). 23 pp.
DIN ENGL 17674 Pt 2-63. Copper and Wrought Copper Alloy Extruded Sections; Technical Conditions of Delivery (June). 3 pp.

—Sections—Extruded
DIN ENGL 17674 Pt 4-63. Copper and Copper Wrought Alloy Extruded Sections; Extruded, Permissible Variations (June). 3 pp.

—Sections—Extruded—Design
DIN ENGL 17674 Pt 3-63. Copper and Wrought Copper Alloy Extruded Sections; Design (June). 3 pp.

—Sections—Extruded—Drawn
DIN ENGL 17674 Pt 5-63. Copper and Wrought Copper Alloy Extruded Sections; Drawn; Permissible Variations (June). 3 pp.

—Sections—Extruded—Electrical Properties
DIN ENGL 17674 Pt 1-83. Wrought Copper and Copper Alloy Extruded Sections; Properties (Dec). 3 pp.

—Sections—Extruded—Mechanical Properties
DIN ENGL 17674 Pt 1-83. Wrought Copper and Copper Alloy Extruded Sections; Properties (Dec). 3 pp.

—Sections—Wrought
SAA AS 1567-85. Copper and Copper Alloys—Wrought Rods, Bars and Sections. 14 pp.

—Semifinished Products
DIN ENGL 1787-73. Copper; Half-Finished Products (Jan). 4 pp.
SAA AS 1279-85. Copper Refinery Shapes. 11 pp.

—Sheets
CNS H3132-84. Copper and Copper Alloy Sheets, Plates, Strips and Coiled Sheets (Oct)(11073).
DIN ENGL 17650-88. Copper Sheet and Strip for Use in Building Construction; Technical Delivery Conditions (Dec). 4 pp.
DIN ENGL 17670 Pt 2-69. Plate, Sheet and Strip of Copper and Wrought Copper Alloys; Technical Conditions of Delivery (June). 3 pp.
ISO 1634 Pt 1-87. Wrought Copper and Copper Alloy Plate, Sheet and Strip Part 1: Technical Conditions of Delivery for Plate, Sheet and Strip for General Purposes First Edition. 11 pp.
JIS H 3100-92. Copper and Copper Alloy Sheets, Plates and Strips. 60 pp.
JIS H 3510-92. Oxygen Free Copper Sheet, Plate, Strip, Seamless Pipe and Tube, Rod, Bar and Wire for Electron Devices. 19 pp.

—Sheets—Cold Rolled
DIN ENGL 1751-73. Sheet and Sheet Cut to Length of Copper and Wrought Copper Alloys; Cold Rolled; Dimensions (June). 5 pp.
ISO 3486-80. Wrought Copper and Copper Alloys—Cold-Rolled Flat Products Delivered in Straight Lengths (Sheet)—Dimensions and Tolerances First Edition. 5 pp.
ISO 3487-80. Wrought Copper and Copper Alloys—Cold-Rolled Flat Products in Coils or on Reels (Strip)—Dimensions and Tolerances First Edition. 4 pp.

—Sheets—Electrical Properties
DIN ENGL 17670 Pt 1-83. Wrought Copper and Copper Alloy Plate, Sheet and Strip; Properties (Dec). 16 pp.

—Sheets—Mechanical Properties
DIN ENGL 17670 Pt 1-83. Wrought Copper and Copper Alloy Plate, Sheet and Strip; Properties (Dec). 16 pp.

—Sheets—Rolled
BSI BS 2870-80. 1980 Rolled Copper and Copper Alloys. Sheet, Strip and Foil. 32 pp.
BSI BS 4608-70. 1970 Copper for Electrical Purposes. Rolled Sheet, Strip and Foil. 20 pp.
SAA AS 1566-85. Copper and Copper Alloys—Rolled Flat Products. 12 pp.

—Silicon Content—Absorptiometric Analysis
JIS H 1061-89. Methods for Determination of Silicon in Copper and Copper Alloys. 10 pp.

Copper (Cont.)

—Silicon Content—Gravimetric Analysis
JIS H 1061-89. Methods for Determination of Silicon in Copper and Copper Alloys. 10 pp.

—Spectrochemical Analysis
JIS H 1291-77. Method for Atomic Absorption Spectrochemical Analysis of Copper and Copper Alloys (R 1986). 15 pp.

—Standard Solutions
JIS K 0010-92. Copper Standard Solution. 13 pp.

—Strips
BSI BS 1432-87. 1987 Copper for Electrical Purposes: High Conductivity Copper Rectangular Conductors with Drawn or Rolled Edges. 18 pp.
CNS H3132-84. Copper and Copper Alloy Sheets, Plates, Strips and Coiled Sheets (Oct)(11073).
DIN ENGL 17650-88. Copper Sheet and Strip for Use in Building Construction; Technical Delivery Conditions (Dec). 4 pp.
DIN ENGL 17670 Pt 2-69. Plate, Sheet and Strip of Copper and Wrought Copper Alloys; Technical Conditions of Delivery (June). 3 pp.
ISO 1634 Pt 1-87. Wrought Copper and Copper Alloy Plate, Sheet and Strip Part 1: Technical Conditions of Delivery for Plate, Sheet and Strip for General Purposes First Edition. 11 pp.
JIS H 3100-92. Copper and Copper Alloy Sheets, Plates and Strips. 60 pp.
JIS H 3510-92. Oxygen Free Copper Sheet, Plate, Strip, Seamless Pipe and Tube, Rod, Bar and Wire for Electron Devices. 19 pp.

—Strips—Cold Rolled
DIN ENGL 1791-73. Strip and Strip Cut to Length of Copper and Wrought Copper Alloys; Cold Rolled; Dimensions (June). 5 pp.
ISO 3487-80. Wrought Copper and Copper Alloys—Cold-Rolled Flat Products in Coils or on Reels (Strip)—Dimensions and Tolerances First Edition. 4 pp.

—Strips—Coolers—Aircraft
MOD UK DTD-607-44. Copper Strip for Radiators and Coolers (Reprinted January 1954). 2 pp.

—Strips—Electrical Properties
DIN ENGL 17670 Pt 1-83. Wrought Copper and Copper Alloy Plate, Sheet and Strip; Properties (Dec). 16 pp.

—Strips—Mechanical Properties
DIN ENGL 17670 Pt 1-83. Wrought Copper and Copper Alloy Plate, Sheet and Strip; Properties (Dec). 16 pp.

—Strips—Radiators—Aircraft
MOD UK DTD-607-44. Copper Strip for Radiators and Coolers (Reprinted January 1954). 2 pp.

—Strips—Rolled
BSI BS 2870-80. 1980 Rolled Copper and Copper Alloys. Sheet, Strip and Foil. 32 pp.
BSI BS 4608-70. 1970 Copper for Electrical Purposes. Rolled Sheet, Strip and Foil. 20 pp.
SAA AS 1566-85. Copper and Copper Alloys—Rolled Flat Products. 12 pp.

—Sulfur Content—Volumetric Analysis
ISO 7266-84. Copper and Copper Alloys—Determination of Sulfur Content—Combustion Titrimetric Method First Edition. 7 pp.

—Surface Finishing—Adhesive Bonding
ISO 4588-89. Adhesives—Preparation of Metal Surfaces for Adhesive Bonding First Edition. 8 pp.

—Tellurium Content
JIS H 1064-92. Method for Determination of Tellurium in Copper. 6 pp.

—Tempering—Designation Codes
ISO 1190 Pt 2-82. Copper and Copper Alloys—Code of Designation—Part 2: Designation of Tempers First Edition. 4 pp.

—Test Specimens
ISO R1811-71. Chemical Analysis of Copper and Copper Alloys Sampling of Copper Refinery Shapes First Edition. 4 pp.
ISO 4739-85. Wrought Copper and Copper Alloy Products—Selection and Preparation of Specimens and Test Pieces for Mechanical Testing First Edition. 6 pp.

—Tin Content
JIS H 1052-84. Methods for Determination of Tin in Copper and Copper Alloys. 20 pp.

—Tin Content—Spectrometry
ISO 4751-84. Copper and Copper Alloys—Determination of Tin Content—Spectrometric Method First Edition. 5 pp.

Copper (Cont.)

—Tubes
BSI BS 1977-76. 1976 High Conductivity Copper Tubes for Electrical Purposes. 10 pp.
BSI BS 2871: Part 2-72. 1972 Amd 1 Copper and Copper Alloys. Tubes Part 2: Tubes for General Purposes. 30 pp.
DIN ENGL 17671 Pt 2-69. Tubes of Copper and Wrought Copper Alloys; Technical Conditions of Delivery (June). 3 pp.
ISO 274-75. Copper Tubes of Circular Section—Dimensions First Edition. 5 pp.

—Tubes—Carbon Content
CEN PREN 723-92. Wrought Copper—Combustion Methods for Determination of Surface Carbon on Tubes. 9 pp.

—Tubes—Electrical Properties
DIN ENGL 17671 Pt 1-83. Wrought Copper and Copper Alloy Tubes; Properties (Dec). 8 pp.

—Tubes—Marine
MOD UK NES 837-88. Requirements for Copper Tubes Issue 1 (09.88). 28 pp.

—Tubes—Mechanical Properties
DIN ENGL 17671 Pt 1-83. Wrought Copper and Copper Alloy Tubes; Properties (Dec). 8 pp.
ISO 1635-74. Wrought Copper and Copper Alloys—Round Tubes for General Purposes—Mechanical Properties First Edition. 5 pp.

—Tubes—Seamless
CEN PREN 1057-93. Copper and Copper Alloys—Seamless, Round Copper Tubes for Water and Gas in Sanitary and Heating Applications. 19 pp.
CNS H2049-80. Method of Test for Copper and Copper Alloy Seamless Pipes and Tubes (Jan)(5128).
JIS H 3510-92. Oxygen Free Copper Sheet, Plate, Strip, Seamless Pipe and Tube, Rod, Bar and Wire for Electron Devices. 19 pp.
SAA AS 1569-85. Copper and Copper Alloys—Seamless Tubes for Heat Exchangers. 11 pp.
SAA AS 1571-85. Copper—Seamless Tubes for Airconditioning and Refrigeration. 6 pp.
SAA AS 1572-85. Copper and Copper Alloys—Seamless Tubes for Engineering Purposes. 11 pp.

—Tubes—Seamless—Drawn
DIN ENGL 1754 Pt 1-69. Copper Tubes; Seamless Drawn; Dimension Ranges and Coordination of Tolerances (Aug). 5 pp.
DIN ENGL 1754 Pt 2-69. Copper Tubes; Seamless Drawn; Preferred Dimensions for General Purposes (Aug). 6 pp.

—Tubes—Threads
BSI BS 61-69. 1969 Amd 1 Threads for Light Gauge Copper Tubes and Fittings. 19 pp.

—Wire Bars
CNS H3120-83. Copper Wire Bars (Jun)(10347). 3 pp.

—Wood Preservatives
BSI BS 4072: Part 1-87. 1987 Amd 1 Wood Preservation by Means of Copper/Chromium/Arsenic Compositions Part 1: Preservatives (AMD 6200) June 30, 1989. 5 pp.

—Zinc Content—Atomic Absorption Spectrometry
BSI BS 6721: Part 5-86. 1986 Sampling and Analysis of Copper and Copper Alloys Part 5: Method for Determination of Zinc in Copper and Copper Alloys by Flame Atomic Absorption Spectrophotometry. 7 pp.
ISO 4740-85. Copper and Copper Alloys—Determination of Zinc Content—Flame Atomic Absorption Spectrometric Method First Edition. 6 pp.

Copper Accelerated Acetic Acid Salt Spray Testing
Use: Salt Spray Testing

Copper Acetate
Use: Cupric Acetate

Copper Alloy Castings
See Also: Castings; Copper Alloys; Copper Castings; Metal Products

—Sand—Composition
DIN ENGL 17658-73. Copper-Nickel Casting Alloys; Castings (June). 3 pp.

Copper Alloy Conductor Wire
Use: Copper Conductors

Copper Alloys
Scope Note: For additional listings, see also specific products made from copper alloys

Copper Alloys (Cont.)

See Also: Alloys; Aluminum Bronzes; Beryllium Copper Alloys; Brasses; Brazing Alloys; Bronzes; Chromium Cupronickel; Copper Alloy Castings; Copper Aluminum Alloys; Copper Beryllium Alloys; Copper Coatings (Made From Copper); Copper Coatings (On Copper); Copper Nickel Alloys; Copper Nickel Manganese Alloys; Copper Nickel Zinc Alloys; Copper Silicon Alloys; Copper Tin Alloys; Copper Zinc Alloys; Cupronickel; Gilding Metals; Leaded Brasses; Leaded Bronzes; Nickel Copper Alloys; Nickel Silver Alloys; Nonferrous Alloys; Phosphor Copper

—Aircraft
SBAC RS 116 (V). S.B.A.C. Protective Treatment Chart.

—Aluminum Content
BSI BS 1748: Parts 1-5-61. 1961 Methods for the Analysis of Copper Alloys Parts 1-5: Determination of Copper, Lead, Iron, Aluminium and Nickel in Copper Alloys. 23 pp.
JIS H 1057-87. Methods for Determination of Aluminium in Copper and Copper Alloys. 9 pp.

—Aluminum Content—Volumetric Analysis
ISO 3110-75. Copper Alloys—Determination of Aluminium as Alloying Element—Volumetric Method First Edition. 5 pp.

—Antimony Content—Spectrophotometry
ISO 5956-84. Copper and Copper Alloys—Determination of Antimony Content—Rhodamine B Spectrometric Method First Edition. 4 pp.

—Arsenic Content
JIS H 1059-87. Methods for Determination of Arsenic in Copper and Copper Alloys. 10 pp.

—Arsenic Content—Photometry
ISO 3220-75. Copper and Copper Alloys—Determination of Arsenic—Photometric Method First Edition. 4 pp.

—Bars
CNS H3124-86. Copper and Copper Alloy-Rods and Bars (Sep)(10442). 15 pp.
DIN ENGL 17672 Pt 2-74. Bars of Copper and Wrought Copper Alloys; Technical Conditions of Delivery (June). 3 pp.
ISO 1637-87. Wrought Copper and Copper Alloy Rod and Bar—Technical Conditions of Delivery Second Edition. 10 pp.
JIS H 3250-92. Copper and Copper Alloy Rods and Bars. 23 pp.

—Bars—Aerospace
BSI 2TA 22-73. 1973 Bars and Sections for Machining of Titanium-Copper Alloy (Tensile Strength 540-770 N/Square mm). 2 pp.
BSI TA 53-73. 1973 Amd 1 Bars and Sections for Machining of Titanium-Copper Alloy (Tensile Strength 650-880 N/Square mm) (Limiting Ruling Section 75 mm). 3 pp.

—Bars—Drawn—Hexagonal
ISO 3490-84. Wrought Copper and Copper Alloys—Drawn Hexagonal Bars—All Minus Tolerances on Width Across Flats and Form Tolerances First Edition. 5 pp.
ISO 7757-84. Wrought Copper and Copper Alloys—Drawn Hexagonal Bars—Symmetric Plus and Minus Tolerances on Width Across Flats and Form Tolerances First Edition. 5 pp.

—Bars—Drawn—Rectangular
DIN ENGL 1759-74. Rectangular Bars of Copper and Wrought Copper Alloys; Drawn, with Sharp Edges; Dimensions, Permissible Variations, Static Values (June). 12 pp.
ISO 6958-81. Wrought Copper and Copper Alloys—Drawn Rectangular Bars—Dimension and Form Tolerances First Edition. 6 pp.

—Bars—Drawn—Round
ISO 3489-84. Wrought Copper and Copper Alloys—Drawn Round Bars—All Minus Tolerances on Diameter and Form Tolerances First Edition. 5 pp.
ISO 7756-84. Wrought Copper and Copper Alloys—Drawn Round Bars—Symmetric Plus and Minus Tolerances on Diameter and Form Tolerances First Edition. 5 pp.

—Bars—Drawn—Square
ISO 3491-84. Wrought Copper and Copper Alloys—Drawn Square Bars—All Minus Tolerances on Width Across Flats and Form Tolerances First Edition. 5 pp.
ISO 7758-84. Wrought Copper and Copper Alloys—Drawn Square Bars—Symmetric Plus and Minus Tolerances on Width Across Flats and Form Tolerances First Edition. 5 pp.

Copper Alloys (Cont.)

—Bars—Electrical Properties
DIN ENGL 17672 Pt 1-83. Wrought Copper and Copper Alloy Rod and Bar; Properties (Dec). 11 pp.

—Bars—Extruded—Hexagonal
ISO 3488-82. Wrought Copper and Copper Alloys—Extruded Round, Square or Hexagonal Bars—Dimensions and Tolerances First Edition. 4 pp.

—Bars—Extruded—Round
ISO 3488-82. Wrought Copper and Copper Alloys—Extruded Round, Square or Hexagonal Bars—Dimensions and Tolerances First Edition. 4 pp.

—Bars—Extruded—Square
ISO 3488-82. Wrought Copper and Copper Alloys—Extruded Round, Square or Hexagonal Bars—Dimensions and Tolerances First Edition. 4 pp.

—Bars—Mechanical Properties
DIN ENGL 17672 Pt 1-83. Wrought Copper and Copper Alloy Rod and Bar; Properties (Dec). 11 pp.

—Bars—Rolled
SAA AS 1566-85. Copper and Copper Alloys—Rolled Flat Products. 12 pp.

—Bars—Wrought
SAA AS 1567-85. Copper and Copper Alloys—Wrought Rods, Bars and Sections. 14 pp.

—Beryllium Content—Absorptiometric Analysis
JIS H 1063-89. Methods for Determination of Beryllium in Copper Alloys. 12 pp.

—Beryllium Content—Atomic Absorption Spectrometry
JIS H 1063-89. Methods for Determination of Beryllium in Copper Alloys. 12 pp.

—Beryllium Content—Emission Spectrometry
JIS H 1063-89. Methods for Determination of Beryllium in Copper Alloys. 12 pp.

—Bismuth Content—Spectrometry
ISO 5959-84. Copper and Copper Alloys—Determination of Bismuth Content—Diethyldithiocarbamate Spectrometric Method First Edition. 5 pp.

—Cadmium Content
BSI BS 1825-52. 1952 Determination of Cadmium in Copper-Cadmium Alloys (Electrolytic Method). 8 pp.

—Cadmium Content—Atomic Absorption Spectrometry
ISO 5960-84. Copper Alloys—Determination of Cadmium Content—Flame Atomic Absorption Spectrometric Method First Edition. 6 pp.
SAA AS 1515.5-87. Copper Alloys—Part 5: Determination of Cadmium—Flame Atomic Absorption Spectrometric Method. 3 pp.

—Chemical Analysis
CNS H2076-88. General Rules of Chemical Analysis for Copper Products and Copper Alloys (Dec)(11942).
ISO 1187-83. Special Wrought Copper Alloys—Chemical Composition and Forms of Wrought Products First Edition. 5 pp.
JIS H 1012-91. General Rules for Chemical Analysis of Copper and Copper Alloys.
JIS H 1414-76. Methods of Chemical Analysis for Manganin (R 1984). 16 pp.

—Chromium Content—Atomic Absorption Spectrometry
ISO 4744-84. Copper and Copper Alloys—Determination of Chromium Content—Flame Atomic Absorption Spectrometric Method First Edition. 5 pp.

—Chromium Content—Atomic Absorption Spectrophotometry
BSI BS 6721: Part 4-86. 1986 Sampling and Analysis of Copper and Copper Alloys Part 4: Method for Determination of Chromium in Copper and Copper Alloys by Flame Atomic Absorption Spectrophotometry. 7 pp.

—Chromium Content—Volumetric Analysis
ISO 6437-84. Copper Alloys—Determination of Chromium Content—Titrimetric Method First Edition. 5 pp.

—Cobalt Content—Absorptiometric Analysis
JIS H 1060-89. Methods for Determination of Cobalt in Copper and Copper Alloys. 12 pp.

Copper Alloys (Cont.)

—Cobalt Content—Atomic Absorption Spectrometry
JIS H 1060-89. Methods for Determination of Cobalt in Copper and Copper Alloys. 12 pp.

—Copper Content
BSI BS 1748: Parts 1-5-61. 1961 Methods for the Analysis of Copper Alloys Parts 1-5: Determination of Copper, Lead, Iron, Aluminium and Nickel in Copper Alloys. 23 pp.
CNS H2084-89. Methods of Determination for Copper in Copper Alloys (Jun)(12530).

—Copper Content—Electrolytic Analysis
ISO 1554-76. Wrought and Cast Copper Alloys—Determination of Copper Content—Electrolytic Method First Edition. 4 pp.
SAA AS 1515.4-78. Copper Alloys—Part 4: Method for the Electrolytic Determination of Copper in Wrought and Cast Copper Alloys Reconfirmed 1989. 10 pp.

—Corrosion Testing
ISO 6957-88. Copper Alloys—Ammonia Test for Stress Corrosion Resistance First Edition. 6 pp.

—Extruded—Copper Content
JIS H 1051-92. Methods for Determination of Copper in Copper and Copper Alloys. 22 pp.

—Forgings
BSI BS 2872-89. 1989 Copper and Copper Alloy Forging Stock and Forgings. 24 pp.
DIN ENGL 17673 Pt 2-73. Drop Forgings of Copper and Wrought Copper Alloys; Technical Conditions of Delivery (Dec). 3 pp.
DIN ENGL 17678 Pt 2-73. Hand Forgings of Copper and Wrought Copper Alloys; Technical Conditions of Delivery (Nov). 3 pp.
SAA AS 1568-85. Copper and Copper Alloys—Forging Stock and Forgings. 7 pp.

—Forgings—Aerospace
BSI B 23-91. 1991 Copper-Aluminium-Nickel-Iron Alloy Rods, Sections, Forging Stock and Forgings. 7 pp.
BSI 2TA 23-73. 1973 Forging Stock of Titanium-Copper Alloy (Tensile Strength 540-770 N/Square mm). 2 pp.
BSI 2TA 24-73. 1973 Forgings of Titanium-Copper Alloy (Tensile Strength 540-770 N/Square mm). 2 pp.
BSI TA 54-73. 1973 Amd 1 Forging Stock of Titanium-Copper Alloy (Tensile Strength 650-880 N/Square mm) (Limiting Ruling Section 75 mm). 3 pp.
BSI TA 55-73. 1973 Amd 1 Forgings of Titanium-Copper Alloy (Tensile Strength 650-880 N/Square mm) (Limiting Ruling Section 75 mm). 3 pp.

—Forgings—Design
DIN ENGL 17673 Pt 3-74. Drop Forgings of Copper and Wrought Copper Alloys; Design Principles (Jan). 7 pp.
DIN ENGL 17678 Pt 3-73. Hand Forgings of Copper and Wrought Copper Alloys; Basis of Design (Dec). 2 pp.

—Forgings—Electrical Properties
DIN ENGL 17673 Pt 1-83. Wrought Copper and Copper Alloy Drop Forgings; Properties (Dec). 4 pp.
DIN ENGL 17678 Pt 1-83. Wrought Copper and Copper Alloy Hand Forgings; Properties (Dec). 3 pp.

—Forgings—Heat Treated—Aerospace
BSI B 25-91. 1991 Copper-Nickel-Silicon Alloy Rods, Sections, Forging Stock and Forgings (Heat Treated). 7 pp.

—Forgings—Mechanical Properties
DIN ENGL 17673 Pt 1-83. Wrought Copper and Copper Alloy Drop Forgings; Properties (Dec). 4 pp.
DIN ENGL 17678 Pt 1-83. Wrought Copper and Copper Alloy Hand Forgings; Properties (Dec). 3 pp.
ISO 1640-74. Wrought Copper Alloys—Forgings—Mechanical Properties First Edition. 4 pp.

—Glossaries
BSI BS 6931-88. 1988 Glossary of Terms for Copper and Copper Alloys. 18 pp.
DIN ENGL 1718-59. Copper Alloys; Definitions (Nov). 2 pp.
ISO 197 Pt 1-83. Copper and Copper Alloys—Terms and Definitions—Part 1: Materials First Edition. 5 pp.
ISO 197 Pt 2-83. Copper and Copper Alloys—Terms and Definitions—Part 2: Unwrought Products (Refinery Shapes) First Edition. 3 pp.

Copper Alloys (Cont.)

—Glossaries (Cont.)
ISO 197 Pt 3-83. Copper and Copper Alloys—Terms and Definitions—Part 3: Wrought Products First Edition. 8 pp.
ISO 197 Pt 4-83. Copper and Copper Alloys—Terms and Definitions—Part 4: Castings First Edition. 3 pp.
ISO 197 Pt 5-80. Copper and Copper Alloys—Terms and Definitions—Part 5: Methods of Processing and Treatment First Edition. 3 pp.
JIS H 7001-89. Glossary of Terms Used in Shape Memory Alloys. 14 pp.

—Grain Size Analysis
BSI BS 7428-91. 1991 Estimating the Average Grain Size of Copper and Copper Alloys (ISO 2624: 1990). 12 pp.
CNS H2069-82. Methods for Estimating the Average Grain Size Test of Wrought Copper and Copper Base Alloys (Oct)(9504).
ISO 2624-90. Copper and Copper Alloys—Estimation of Average Grain Size Second Edition. 11 pp.
JIS H 0501-86. Methods for Estimating Average Grain Size of Wrought Copper and Copper-Alloys. 8 pp.

—Hollow Sections—Wrought
SAA AS 1567-85. Copper and Copper Alloys—Wrought Rods, Bars and Sections. 14 pp.

—Ingots
SAA AS 1565-85. Copper and Copper Alloys—Ingots and Castings. 36 pp.

—Ingots—Castings
DIN ENGL 17656-73. Copper Casting Alloys; Ingot Metals; Composition (June). 5 pp.

—Ingots—Chemical Analysis
DIN ENGL 17656-73. Copper Casting Alloys; Ingot Metals; Composition (June). 5 pp.

—Iron Content
BSI BS 1748: Parts 1-5-61. 1961 Methods for the Analysis of Copper Alloys Parts 1-5: Determination of Copper, Lead, Iron, Aluminium and Nickel in Copper Alloys. 23 pp.
CNS H2087-89. Methods of Determination for Iron in Copper and Copper Alloys (Jun)(12533).
JIS H 1054-84. Methods for Determination of Iron in Copper and Copper Alloys. 15 pp.

—Iron Content—Spectrophotometry
ISO 1812-76. Copper Alloys—Determination of Iron Content—1,10-Phenanthroline Spectrophotometric Method First Edition. 4 pp.

—Iron Content—Volumetric Analysis
ISO 4748-84. Copper Alloys—Determination of Iron Content—Na2EDTA Titrimetric Method First Edition. 4 pp.

—Lead Content
BSI BS 1748: Parts 1-5-61. 1961 Methods for the Analysis of Copper Alloys Parts 1-5: Determination of Copper, Lead, Iron, Aluminium and Nickel in Copper Alloys. 23 pp.
CNS H2086-89. Methods of Determination for Lead in Copper and Copper Alloys (Jun)(12532).
JIS H 1053-84. Methods for Determination of Lead in Copper and Copper Alloys. 17 pp.

—Lead Content—Atomic Absorption Spectrometry
ISO 4749-84. Copper Alloys—Determination of Lead Content—Flame Atomic Absorption Spectrometric Method First Edition. 6 pp.
SAA AS K209.1-70. Methods for the Analysis of Copper Alloys—Part 1: Lead in Copper Alloys (Atomic Absorption Spectrometric Method). 5 pp.

—Lead Content—Volumetric Analysis
ISO 3112-75. Copper and Copper Alloys—Determination of Lead—Extracting Titration Method First Edition. 6 pp.

—Manganese Content
JIS H 1055-87. Methods for Determination of Manganese in Copper and Copper Alloys. 9 pp.

—Manganese Content—Atomic Absorption Spectrometry
SAA AS K209 Part 2-71. Methods for the Analysis of Copper Alloys—Part 2: Manganese in Copper Alloys (Atomic Absorption Spectrometric Method). 5 pp.

—Manganese Content—Spectrophotometry
ISO 2543-73. Copper and Copper Alloys—Determination of Manganese—Spectrophotometric Method First Edition. 4 pp.

—Master Alloys—Chemical Analysis
DIN ENGL 17657-73. Copper Master Alloys; Composition (Mar). 4 pp.

Copper Alloys (Cont.)

—Materials—Designation Codes
ISO 1190 Pt 1-82. Copper and Copper Alloys—Code of Designation—Part 1: Designation of Materials First Edition. 3 pp.

—Nickel Content
BSI BS 1748: Parts 1-5-61. 1961 Methods for the Analysis of Copper Alloys Parts 1-5: Determination of Copper, Lead, Iron, Aluminium and Nickel in Copper Alloys. 23 pp.
JIS H 1056-87. Methods for Determination of Nickel in Copper and Copper Alloys. 17 pp.

—Nickel Content—Gravimetric Analysis
ISO 4742-84. Copper Alloys—Determination of Nickel Content—Gravimetric Method First Edition. 4 pp.

—Nickel Content—Spectrophotometry
ISO 1810-76. Copper Alloys—Determination of Nickel (Low Contents)—Dimethylglyoxime Spectrophotometric Method First Edition. 4 pp.

—Nickel Content—Volumetric Analysis
ISO 4743-84. Copper Alloys—Determination of Nickel Content—Titrimetric Method First Edition. 5 pp.

—Phosphorus Content
BSI BS 1748: Part 8-60. 1960 Methods for the Analysis of Copper Alloys Part 8: Determination of Phosphorus in Copper Alloys (Photometric Method). 8 pp.
JIS H 1058-87. Methods for Determination of Phosphorus in Copper and Copper Alloys. 10 pp.

—Phosphorus Content—Absorptiometric Analysis
MOD UK M 9201/59. Determination of Phosphorous in Copper Alloys: Absorptiometric Method (No Information).

—Phosphorus Content—Spectrometry
ISO 4741-84. Copper and Copper Alloys—Determination of Phosphorus Content—Molybdovanadate Spectrometric Method First Edition. 5 pp.

—Plates
CNS H3132-84. Copper and Copper Alloy Sheets, Plates, Strips and Coiled Sheets (Oct)(11073).
DIN ENGL 17670 Pt 2-69. Plate, Sheet and Strip of Copper and Wrought Copper Alloys; Technical Conditions of Delivery (June). 3 pp.
ISO 1634 Pt 1-87. Wrought Copper and Copper Alloy Plate, Sheet and Strip Part 1: Technical Conditions of Delivery for Plate, Sheet and Strip for General Purposes First Edition. 11 pp.
JIS H 3100-92. Copper and Copper Alloy Sheets, Plates and Strips. 60 pp.

—Plates—Electrical Properties
DIN ENGL 17670 Pt 1-83. Wrought Copper and Copper Alloy Plate, Sheet and Strip; Properties (Dec). 16 pp.

—Plates—Mechanical Properties
DIN ENGL 17670 Pt 1-83. Wrought Copper and Copper Alloy Plate, Sheet and Strip; Properties (Dec). 16 pp.

—Plates—Rolled
SAA AS 1566-85. Copper and Copper Alloys—Rolled Flat Products. 12 pp.

—Plates—Wrought
BSI BS 2875-69. 1969 Copper and Copper Alloys. Plate. 38 pp.

—Refinery Shapes
ISO 197 Pt 2-83. Copper and Copper Alloys—Terms and Definitions—Part 2: Unwrought Products (Refinery Shapes) First Edition. 3 pp.

—Residual Stress
ISO 196-78. Wrought Copper and Copper Alloys—Detection of Residual Stress—Mercury (I) Nitrate Test First Edition. 3 pp.

—Rods
BSI BS 2874-86. 1986 Copper and Copper Alloy Rods and Sections (Other Than Forging Stock). 23 pp.
CNS H3124-86. Copper and Copper Alloy-Rods and Bars (Sep)(10442). 15 pp.
DIN ENGL 17672 Pt 2-74. Bars of Copper and Wrought Copper Alloys; Technical Conditions of Delivery (June). 3 pp.
ISO 1637-87. Wrought Copper and Copper Alloy Rod and Bar—Technical Conditions of Delivery Second Edition. 10 pp.
JIS H 3250-92. Copper and Copper Alloy Rods and Bars. 23 pp.

Copper Alloys (Cont.)

—Rods—Aerospace
BSI B 23-91. 1991 Copper-Aluminium-Nickel-Iron Alloy Rods, Sections, Forging Stock and Forgings. 7 pp.
BSI B 24-91. 1991 Copper-Tin-Phosphorous Alloy Rods and Sections. 5 pp.

—Rods—Cold Worked—Heat Treated—Aerospace
BSI B 26-91. 1991 Copper-Nickel-Silicon Alloy Rods and Sections (Cold Worked and Heat Treated). 5 pp.

—Rods—Drawn
DIN ENGL 1756-69. Round Rod of Copper and Wrought Copper Alloys; Drawn; Dimensions (July). 4 pp.

—Rods—Drawn—Hexagonal
DIN ENGL 1763-69. Hexagon Rod of Copper and Wrought Copper Alloys; Drawn, with Sharp Edges; Dimensions (July). 4 pp.

—Rods—Drawn—Square
DIN ENGL 1761-69. Square Rod of Copper and Copper Wrought Alloys; Drawn, with Sharp Edges; Dimensions (July). 4 pp.

—Rods—Electrical Properties
DIN ENGL 17672 Pt 1-83. Wrought Copper and Copper Alloy Rod and Bar; Properties (Dec). 11 pp.

—Rods—Extruded
DIN ENGL 1782-69. Round Rod of Copper and Wrought Copper Alloys; Extruded; Dimensions (July). 4 pp.

—Rods—Heat Treated—Aerospace
BSI B 25-91. 1991 Copper-Nickel-Silicon Alloy Rods, Sections, Forging Stock and Forgings (Heat Treated). 7 pp.

—Rods—Mechanical Properties
DIN ENGL 17672 Pt 1-83. Wrought Copper and Copper Alloy Rod and Bar; Properties (Dec). 11 pp.

—Rods—Wrought
SAA AS 1567-85. Copper and Copper Alloys—Wrought Rods, Bars and Sections. 14 pp.

—Rolled—Copper Content
JIS H 1051-92. Methods for Determination of Copper in Copper and Copper Alloys. 22 pp.

—Sampling
ISO 1811 Pt 1-88. Copper and Copper Alloys—Selection and Preparation of Samples for Chemical Analysis—Part 1: Sampling of Cast Unwrought Products First Edition. 7 pp.
ISO 1811 Pt 2-88. Copper and Copper Alloys—Selection and Preparation of Samples for Chemical Analysis—Part 2: Sampling of Wrought Products and Castings First Edition. 5 pp.

—Scraps—Classification
JIS H 2109-86. Classification Standard of Copper and Copper Alloy Scraps. 15 pp.

—Sections
BSI BS 2874-86. 1986 Copper and Copper Alloy Rods and Sections (Other Than Forging Stock). 23 pp.

—Sections—Aerospace
BSI B 23-91. 1991 Copper-Aluminium-Nickel-Iron Alloy Rods, Sections, Forging Stock and Forgings. 7 pp.
BSI B 24-91. 1991 Copper-Tin-Phosphorous Alloy Rods and Sections. 5 pp.
BSI 2TA 22-73. 1973 Bars and Sections for Machining of Titanium-Copper Alloy (Tensile Strength 540-770 N/Square mm). 2 pp.
BSI TA 53-73. 1973 Amd 1 Bars and Sections for Machining of Titanium-Copper Alloy (Tensile Strength 650-880 N/Square mm) (Limiting Ruling Section 75 mm). 3 pp.

—Sections—Cold Worked—Heat Treated—Aerospace
BSI B 26-91. 1991 Copper-Nickel-Silicon Alloy Rods and Sections (Cold Worked and Heat Treated). 5 pp.

—Sections—Extruded
DIN ENGL 17674 Pt 2-63. Copper and Wrought Copper Alloy Extruded Sections; Technical Conditions of Delivery (June). 3 pp.
DIN ENGL 17674 Pt 4-63. Copper and Copper Wrought Alloy Extruded Sections; Extruded, Permissible Variations (June). 3 pp.

INTERNATIONAL AND NON-U.S. NATIONAL STANDARDS
SUBJECT INDEX
Copper

Copper Alloys (Cont.)

—Sections—Extruded—Design
DIN ENGL 17674 Pt 3-63. Copper and Wrought Copper Alloy Extruded Sections; Design (June). 3 pp.

—Sections—Extruded—Drawn
DIN ENGL 17674 Pt 5-63. Copper and Wrought Copper Alloy Extruded Sections; Drawn; Permissible Variations (June). 3 pp.

—Sections—Extruded—Electrical Properties
DIN ENGL 17674 Pt 1-83. Wrought Copper and Copper Alloy Extruded Sections; Properties (Dec). 3 pp.

—Sections—Extruded—Mechanical Properties
DIN ENGL 17674 Pt 1-83. Wrought Copper and Copper Alloy Extruded Sections; Properties (Dec). 3 pp.
ISO 1639-74. Wrought Copper Alloys—Extruded Sections—Mechanical Properties First Edition. 4 pp.

—Sections—Heat Treated—Aerospace
BSI B 25-91. 1991 Copper-Nickel-Silicon Alloy Rods, Sections, Forging Stock and Forgings (Heat Treated). 7 pp.

—Sections—Wrought
SAA AS 1567-85. Copper and Copper Alloys—Wrought Rods, Bars and Sections. 14 pp.

—Semifinished Products
SAA AS 1279-85. Copper Refinery Shapes. 11 pp.

—Semifinished Products—Chemical Analysis
DIN ENGL 17666-83. Low Alloy Wrought Copper Alloys; Composition (Dec). 7 pp.

—Sheets
CNS H3132-84. Copper and Copper Alloy Sheets, Plates, Strips and Coiled Sheets (Oct)(11073).
DIN ENGL 17670 Pt 2-69. Plate, Sheet and Strip of Copper and Wrought Copper Alloys; Technical Conditions of Delivery (June). 3 pp.
ISO 1634 Pt 1-87. Wrought Copper and Copper Alloy Plate, Sheet and Strip Part 1: Technical Conditions of Delivery for Plate, Sheet and Strip for General Purposes First Edition. 11 pp.
JIS H 3100-92. Copper and Copper Alloy Sheets, Plates and Strips. 60 pp.

—Sheets—Cold Rolled
DIN ENGL 1751-73. Sheet and Sheet Cut to Length of Copper and Wrought Copper Alloys; Cold Rolled; Dimensions (June). 5 pp.
ISO 3486-80. Wrought Copper and Copper Alloys—Cold-Rolled Flat Products Delivered in Straight Lengths (Sheet)—Dimensions and Tolerances First Edition. 5 pp.
ISO 3487-80. Wrought Copper and Copper Alloys—Cold-Rolled Flat Products in Coils or on Reels (Strip)—Dimensions and Tolerances First Edition. 4 pp.

—Sheets—Electrical Properties
DIN ENGL 17670 Pt 1-83. Wrought Copper and Copper Alloy Plate, Sheet and Strip; Properties (Dec). 16 pp.

—Sheets—Mechanical Properties
DIN ENGL 17670 Pt 1-83. Wrought Copper and Copper Alloy Plate, Sheet and Strip; Properties (Dec). 16 pp.

—Sheets—Rolled
BSI BS 2870-80. 1980 Rolled Copper and Copper Alloys. Sheet, Strip and Foil. 32 pp.
SAA AS 1566-85. Copper and Copper Alloys—Rolled Flat Products. 12 pp.

—Silicon Content
BSI BS 1748: Part 7-60. 1960 Methods for the Analysis of Copper Alloys Part 7: Determination of Silicon in Copper Alloys (Photometric Method). 8 pp.

—Silicon Content—Absorptiometric Analysis
JIS H 1061-89. Methods for Determination of Silicon in Copper and Copper Alloys. 10 pp.

—Silicon Content—Gravimetric Analysis
JIS H 1061-89. Methods for Determination of Silicon in Copper and Copper Alloys. 10 pp.

—Silver Content—Atomic Absorption Spectrometry
SAA AS 1515.3-89. Copper Alloys—Part 3: Determination of Silver Content—Flame Atomic Absorption Spectrometric Method. 3 pp.

Copper Alloys (Cont.)

—Solderability
DIN ENGL 32506 Pt 2-81. Testing of Solderability for Soft Soldering; Vertical Dipping Test for Specimens of Copper Alloys; Testing; Assessment (July). 5 pp.

—Spectrochemical Analysis
JIS H 1291-77. Method for Atomic Absorption Spectrochemical Analysis of Copper and Copper Alloys (R 1986). 15 pp.

—Stress Corrosion Cracking
DIN ENGL 50916 Pt 2-85. Testing of Copper Alloys; Stress Corrosion Cracking Test Using Ammonia; Testing of Components (Sept). 3 pp.

—Strips
CNS H3132-84. Copper and Copper Alloy Sheets, Plates, Strips and Coiled Sheets (Oct)(11073).
DIN ENGL 1777-86. Wrought Copper Alloy Strip for Springs; Technical Delivery Conditions (Jan). 8 pp.
DIN ENGL 17670 Pt 2-69. Plate, Sheet and Strip of Copper and Wrought Copper Alloys; Technical Conditions of Delivery (June). 3 pp.
ISO 1634 Pt 1-87. Wrought Copper and Copper Alloy Plate, Sheet and Strip Part 1: Technical Conditions of Delivery for Plate, Sheet and Strip for General Purposes First Edition. 11 pp.
JIS H 3100-92. Copper and Copper Alloy Sheets, Plates and Strips. 60 pp.

—Strips—Cold Rolled
DIN ENGL 1791-73. Strip and Strip Cut to Length of Copper and Wrought Copper Alloys; Cold Rolled; Dimensions (June). 5 pp.
ISO 3487-80. Wrought Copper and Copper Alloys—Cold-Rolled Flat Products in Coils or on Reels (Strip)—Dimensions and Tolerances First Edition. 4 pp.

—Strips—Detonators
MOD UK DSTAN 95-10-01. Cap Copper Alloy Strip for Detonator Cups and Percussion Caps Issue 1; Amendment 1. 14 pp.

—Strips—Electrical Properties
DIN ENGL 17670 Pt 1-83. Wrought Copper and Copper Alloy Plate, Sheet and Strip; Properties (Dec). 16 pp.

—Strips—Mechanical Properties
DIN ENGL 17670 Pt 1-83. Wrought Copper and Copper Alloy Plate, Sheet and Strip; Properties (Dec). 16 pp.

—Strips—Rolled
BSI BS 2870-80. 1980 Rolled Copper and Copper Alloys. Sheet, Strip and Foil. 32 pp.
SAA AS 1566-85. Copper and Copper Alloys—Rolled Flat Products. 12 pp.

—Sulfur Content—Volumetric Analysis
ISO 7266-84. Copper and Copper Alloys—Determination of Sulfur Content—Combustion Titrimetric Method First Edition. 7 pp.

—Surface Finishing—Adhesive Bonding
ISO 4588-89. Adhesives—Preparation of Metal Surfaces for Adhesive Bonding First Edition. 8 pp.

—Tempering—Designation Codes
ISO 1190 Pt 2-82. Copper and Copper Alloys—Code of Designation—Part 2: Designation of Tempers First Edition. 4 pp.

—Test Specimens
ISO R1811-71. Chemical Analysis of Copper and Copper Alloys Sampling of Copper Refinery Shapes First Edition. 4 pp.
ISO 4739-85. Wrought Copper and Copper Alloy Products—Selection and Preparation of Specimens and Test Pieces for Mechanical Testing First Edition. 6 pp.

—Tin Content
BSI BS 1748: Part 6-60. 1960 Methods for the Analysis of Copper Alloys Part 6: Determination of Tin in Copper Alloys (Nickel Coil Reduction Method). 8 pp.
CNS H2085-89. Methods of Determination for Tin in Copper Alloys (Jun)(12531).
JIS H 1052-84. Methods for Determination of Tin in Copper and Copper Alloys. 20 pp.

—Tin Content—Spectrometry
ISO 4751-84. Copper and Copper Alloys—Determination of Tin Content—Spectrometric Method First Edition. 5 pp.

—Tin Content—Volumetric Analysis
ISO 3111-75. Copper Alloys—Determination of Tin as Alloying Element—Volumetric Method First Edition. 7 pp.

Copper Alloys (Cont.)

—Transformation Temperature
JIS H 7101-89. Method for Determining the Transformation Temperatures of Shape Memory Alloys. 9 pp.

—Tubes
BSI BS 2871: Part 2-72. 1972 Amd 1 Copper and Copper Alloys. Tubes Part 2: Tubes for General Purposes. 30 pp.
DIN ENGL 17671 Pt 2-69. Tubes of Copper and Wrought Copper Alloys; Technical Conditions of Delivery (June). 3 pp.

—Tubes—Aerospace
BSI B 27-91. 1991 Copper-Zinc-Aluminium-Nickel-Silicon Alloy Tube. 7 pp.

—Tubes—Electrical Properties
DIN ENGL 17671 Pt 1-83. Wrought Copper and Copper Alloy Tubes; Properties (Dec). 8 pp.

—Tubes—Mechanical Properties
DIN ENGL 17671 Pt 1-83. Wrought Copper and Copper Alloy Tubes; Properties (Dec). 8 pp.

—Tubes—Round—Mechanical Properties
ISO 1635-74. Wrought Copper and Copper Alloys—Round Tubes for General Purposes—Mechanical Properties First Edition. 5 pp.

—Tubes—Seamless
CNS H2049-80. Method of Test for Copper and Copper Alloy Seamless Pipes and Tubes (Jan)(5128).
SAA AS 1569-85. Copper and Copper Alloys—Seamless Tubes for Heat Exchangers. 11 pp.
SAA AS 1572-85. Copper and Copper Alloys—Seamless Tubes for Engineering Purposes. 11 pp.

—Tubes—Seamless—Drawn
DIN ENGL 1755 Pt 1-69. Wrought Copper Alloy Tubes; Seamless Drawn; Dimension Ranges and Coordination of Tolerances (Aug). 7 pp.
DIN ENGL 1755 Pt 2-69. Copper Wrought Alloy Tubes; Seamless Drawn; Preferred Dimensions for General Purposes (Aug). 12 pp.

—Wrought—Aerospace
BSI B 100-92. 1992 Inspection, Testing and Acceptance of Wrought Copper Alloys. 25 pp.

—Wrought—Residual Stress
DIN ENGL 50911-80. Testing of Copper Alloys; Mercurous Nitrate Test (June). 3 pp.

—X-Ray Fluorescence Spectrometry
JIS H 1292-84. Method for Fluorescent X-Ray Analysis of Copper Alloys. 7 pp.

—Zinc Content
BSI BS 1748: Part 9-63. 1963 Methods for the Analysis of Copper Alloys Part 9: Determination of Zinc in Copper Alloys. 11 pp.

—Zinc Content—Atomic Absorption Spectrometry
BSI BS 6721: Part 5-86. 1986 Sampling and Analysis of Copper and Copper Alloys Part 5: Method for Determination of Zinc in Copper and Copper Alloys by Flame Atomic Absorption Spectrophotometry. 7 pp.
ISO 4740-85. Copper and Copper Alloys—Determination of Zinc Content—Flame Atomic Absorption Spectrometric Method First Edition. 6 pp.
JIS H 1062-89. Methods for Determination of Zinc in Copper Alloys. 16 pp.

—Zinc Content—Spectrochemical Analysis
JIS H 1062-89. Methods for Determination of Zinc in Copper Alloys. 16 pp.

—Zinc Content—Volumetric Analysis
JIS H 1062-89. Methods for Determination of Zinc in Copper Alloys. 16 pp.

Copper Aluminum Alloys
Scope Note: For additional listings, see also specific products made from copper aluminum alloys
Use For: Aluminum Copper Alloys
See Also: Aluminum Bronzes; Copper Alloys

—Wrought
ISO 428-83. Wrought Copper-Aluminium Alloys—Chemical Composition and Forms of Wrought Products Second Edition. 5 pp.

Copper Arsenate
Use For: Ammoniacal Copper Arsenate; Chromated Copper Arsenate

INDUSTRY STANDARDS

Copper

Copper Arsenate (Cont.)
—Wood Preservatives
CNS K1187-81. Wood Preservatives (Ammoniacal Copper Arsenate) (ACA, Chemonite) (May)(4135). 1 p.
CNS K1189-80. Wood Preservatives (Chromated Copper Arsenate) (CCA) (Apr)(4137). 2 pp.
CNS K6406-81. Method of Test for Ammoniacal Copper Arsenate (Wood Preservatives) (Jul)(4199).
CNS K6408-81. Method of Test for Chromated Copper Arsenate (Wood Preservatives) (Jul)(4201).

Copper Bars
Use: Copper—Bars

Copper Bearing Alloys
Use: Bearing Alloys—Copper

Copper Beryllium Alloys
Scope Note: For additional listings, see also specific products made from copper beryllium alloys
See Also: Copper Alloys

—Rods—Extruded—Chemical Analysis
BSI B 32-91. 1991 Copper-Beryllium Alloy Rods, Sections and Parts (Solution Treated (W) and Precipitated). 5 pp.
BSI B 32-01. 1991 Amd 1 Copper-Beryllium Alloy Rods, Sections and Parts (Solution Treated (W) and Precipitated) (AMD 7119) April 1, 1992. 6 pp.

—Rods—Extruded—Mechanical Properties
BSI B 32-91. 1991 Copper-Beryllium Alloy Rods, Sections and Parts (Solution Treated (W) and Precipitated). 5 pp.
BSI B 32-01. 1991 Amd 1 Copper-Beryllium Alloy Rods, Sections and Parts (Solution Treated (W) and Precipitated) (AMD 7119) April 1, 1992. 6 pp.

—Rods—Forged—Chemical Analysis
BSI B 32-91. 1991 Copper-Beryllium Alloy Rods, Sections and Parts (Solution Treated (W) and Precipitated). 5 pp.
BSI B 32-01. 1991 Amd 1 Copper-Beryllium Alloy Rods, Sections and Parts (Solution Treated (W) and Precipitated) (AMD 7119) April 1, 1992. 6 pp.

—Rods—Forged—Mechanical Properties
BSI B 32-91. 1991 Copper-Beryllium Alloy Rods, Sections and Parts (Solution Treated (W) and Precipitated). 5 pp.
BSI B 32-01. 1991 Amd 1 Copper-Beryllium Alloy Rods, Sections and Parts (Solution Treated (W) and Precipitated) (AMD 7119) April 1, 1992. 6 pp.

—Rods—Rolled—Chemical Analysis
BSI B 32-91. 1991 Copper-Beryllium Alloy Rods, Sections and Parts (Solution Treated (W) and Precipitated). 5 pp.
BSI B 32-01. 1991 Amd 1 Copper-Beryllium Alloy Rods, Sections and Parts (Solution Treated (W) and Precipitated) (AMD 7119) April 1, 1992. 6 pp.

—Rods—Rolled—Mechanical Properties
BSI B 32-91. 1991 Copper-Beryllium Alloy Rods, Sections and Parts (Solution Treated (W) and Precipitated). 5 pp.
BSI B 32-01. 1991 Amd 1 Copper-Beryllium Alloy Rods, Sections and Parts (Solution Treated (W) and Precipitated) (AMD 7119) April 1, 1992. 6 pp.

—Sections—Extruded—Chemical Analysis
BSI B 32-91. 1991 Copper-Beryllium Alloy Rods, Sections and Parts (Solution Treated (W) and Precipitated). 5 pp.
BSI B 32-01. 1991 Amd 1 Copper-Beryllium Alloy Rods, Sections and Parts (Solution Treated (W) and Precipitated) (AMD 7119) April 1, 1992. 6 pp.

—Sections—Extruded—Mechanical Properties
BSI B 32-91. 1991 Copper-Beryllium Alloy Rods, Sections and Parts (Solution Treated (W) and Precipitated). 5 pp.
BSI B 32-01. 1991 Amd 1 Copper-Beryllium Alloy Rods, Sections and Parts (Solution Treated (W) and Precipitated) (AMD 7119) April 1, 1992. 6 pp.

—Sections—Forged—Chemical Analysis
BSI B 32-91. 1991 Copper-Beryllium Alloy Rods, Sections and Parts (Solution Treated (W) and Precipitated). 5 pp.
BSI B 32-01. 1991 Amd 1 Copper-Beryllium Alloy Rods, Sections and Parts (Solution Treated (W) and Precipitated) (AMD 7119) April 1, 1992. 6 pp.

—Sections—Forged—Mechanical Properties
BSI B 32-91. 1991 Copper-Beryllium Alloy Rods, Sections and Parts (Solution Treated (W) and Precipitated). 5 pp.
BSI B 32-01. 1991 Amd 1 Copper-Beryllium Alloy Rods, Sections and Parts (Solution Treated (W) and Precipitated) (AMD 7119) April 1, 1992. 6 pp.

Copper Beryllium Alloys (Cont.)
—Sections—Rolled—Chemical Analysis
BSI B 32-91. 1991 Copper-Beryllium Alloy Rods, Sections and Parts (Solution Treated (W) and Precipitated). 5 pp.
BSI B 32-01. 1991 Amd 1 Copper-Beryllium Alloy Rods, Sections and Parts (Solution Treated (W) and Precipitated) (AMD 7119) April 1, 1992. 6 pp.

—Sections—Rolled—Mechanical Properties
BSI B 32-91. 1991 Copper-Beryllium Alloy Rods, Sections and Parts (Solution Treated (W) and Precipitated). 5 pp.
BSI B 32-01. 1991 Amd 1 Copper-Beryllium Alloy Rods, Sections and Parts (Solution Treated (W) and Precipitated) (AMD 7119) April 1, 1992. 6 pp.

—Springs—Drawn—Chemical Analysis
BSI B 33-91. 1991 Copper-Beryllium Alloy Wire and Springs (Solution Treated (W) and Precipitated). 5 pp.
BSI B 33-01. 1991 Amd 1 Copper-Beryllium Alloy Wire and Springs (Solution Treated (W) and Precipitated) (AMD 7120) April 1, 1992. 6 pp.

—Springs—Drawn—Mechanical Properties
BSI B 33-91. 1991 Copper-Beryllium Alloy Wire and Springs (Solution Treated (W) and Precipitated). 5 pp.
BSI B 33-01. 1991 Amd 1 Copper-Beryllium Alloy Wire and Springs (Solution Treated (W) and Precipitated) (AMD 7120) April 1, 1992. 6 pp.

—Strips
CNS H3127-85. Copper-Beryllium Alloy Plate, Sheet, Strip and Rolled Bar (Jan)(10879).
JIS H 3130-92. Copper Beryllium Alloy, Phosphor Bronze and Nickel Silver Sheets, Plates and Strips for Springs. 19 pp.

—Strips—Cold Rolled—Chemical Analysis
BSI B 29-91. 1991 Copper-Beryllium Alloy Strip, Foil and Parts (Solution Treated, Cold Rolled: Quarter Hard (W(1/4H)) and Precipitated) (S). 5 pp.
BSI B 29-01. 1991 Amd 1 Copper-Beryllium Alloy Strip, Foil and Parts (Solution Treated, Cold Rolled: Quarter Hard (W(1/4H)) and Precipitated) (AMD 7116) April 1, 1992. 6 pp.
BSI B 30-91. 1991 Copper-Beryllium Alloy Strip, Foil and Parts (Solution Treated, Cold Rolled: Half Hard (W(1/2H)) and Precipitated). 5 pp.
BSI B 30-01. 1991 Amd 1 Copper-Beryllium Alloy Strip, Foil and Parts (Solution Treated, Cold Rolled: Half Hard (W(1/2H)) and Precipitated) (AMD 7117) April 1, 1992. 6 pp.
BSI B 31-91. 1991 Copper-Beryllium Alloy Strip, Foil and Parts (Solution Treated, Cold Rolled: Full Hard (W(H)) and Precipitated). 5 pp.
BSI B 31-01. 1991 Amd 1 Copper-Beryllium Alloy Strip, Foil and Parts (Solution Treated, Cold Rolled: Full Hard (W(H)) and Precipitated) (AMD 7118) April 1, 1992. 6 pp.

—Strips—Cold Rolled—Mechanical Properties
BSI B 29-91. 1991 Copper-Beryllium Alloy Strip, Foil and Parts (Solution Treated, Cold Rolled: Quarter Hard (W(1/4H)) and Precipitated) (S). 5 pp.
BSI B 29-01. 1991 Amd 1 Copper-Beryllium Alloy Strip, Foil and Parts (Solution Treated, Cold Rolled: Quarter Hard (W(1/4H)) and Precipitated) (AMD 7116) April 1, 1992. 6 pp.
BSI B 30-91. 1991 Copper-Beryllium Alloy Strip, Foil and Parts (Solution Treated, Cold Rolled: Half Hard (W(1/2H)) and Precipitated). 5 pp.
BSI B 30-01. 1991 Amd 1 Copper-Beryllium Alloy Strip, Foil and Parts (Solution Treated, Cold Rolled: Half Hard (W(1/2H)) and Precipitated) (AMD 7117) April 1, 1992. 6 pp.
BSI B 31-91. 1991 Copper-Beryllium Alloy Strip, Foil and Parts (Solution Treated, Cold Rolled: Full Hard (W(H)) and Precipitated). 5 pp.
BSI B 31-01. 1991 Amd 1 Copper-Beryllium Alloy Strip, Foil and Parts (Solution Treated, Cold Rolled: Full Hard (W(H)) and Precipitated) (AMD 7118) April 1, 1992. 6 pp.

—Strips—Rolled—Chemical Analysis
BSI B 28-91. 1991 Copper-Beryllium Alloy Strip, Foil and Parts (Solution Treated (W) and Precipitated). 5 pp.
BSI B 28-01. 1991 Amd 1 Copper-Beryllium Alloy Strip, Foil and Parts (Solution Treated (W) and Precipitated) (AMD 7115) April 1, 1992. 6 pp.

—Strips—Rolled—Mechanical Properties
BSI B 28-91. 1991 Copper-Beryllium Alloy Strip, Foil and Parts (Solution Treated (W) and Precipitated). 5 pp.
BSI B 28-01. 1991 Amd 1 Copper-Beryllium Alloy Strip, Foil and Parts (Solution Treated (W) and Precipitated) (AMD 7115) April 1, 1992. 6 pp.

Copper Beryllium Alloys (Cont.)
—Wire—Drawn—Chemical Analysis
BSI B 33-91. 1991 Copper-Beryllium Alloy Wire and Springs (Solution Treated (W) and Precipitated). 5 pp.
BSI B 33-01. 1991 Amd 1 Copper-Beryllium Alloy Wire and Springs (Solution Treated (W) and Precipitated) (AMD 7120) April 1, 1992. 6 pp.

—Wire—Drawn—Mechanical Properties
BSI B 33-91. 1991 Copper-Beryllium Alloy Wire and Springs (Solution Treated (W) and Precipitated). 5 pp.
BSI B 33-01. 1991 Amd 1 Copper-Beryllium Alloy Wire and Springs (Solution Treated (W) and Precipitated) (AMD 7120) April 1, 1992. 6 pp.

Copper Borate
Use: Copper Metaborate

Copper Carbonate
Use For: Cupric Carbonate
CNS K7165-66. Chemical Reagent (Cupric Carbonate) (Mar)(1665).

Copper Castings
See Also: Copper Alloy Castings; Metal Products
BSI BS 1400-85. 1985 Amd 1 Copper Alloy Ingots and Copper Alloy and High Conductivity Copper Castings (AMD 6634) September 28, 1990. 39 pp.
BSI BS 3071-86. 1986 Nickel-Copper Alloy Castings. 8 pp.
JIS H 5100-90. Copper Castings. 9 pp.
SAA AS 1565-85. Copper and Copper Alloys—Ingots and Castings. 36 pp.
SNZ NZS/BS 1400-85. Specification for Copper Alloy Ingots and Copper Alloy and High Conductivity Copper Castings. 40 pp.

—Acceptance Testing—Marine
MOD UK NES 863-93. Requirements for the Classification, Dimensions, Tolerances and General Standards of Acceptance for Copper and Nickel Alloy Castings Issue 1 (05.93). 16 pp.
MOD UK NES 745: Part 1-90. Classification, Inspection Requirements and Acceptance Standards for Castings Part 1: Copper and Nickel Alloy Castings Issue 1 (04.90). 19 pp.

—Aerospace
BSI HC 502-81. 1981 Copper Base Aluminium-Nickel-Iron Alloy Castings (620 MPa) (Al 9.5, Ni 5, Fe 5). 2 pp.

—Bearings
CNS H3130-84. Copper-Lead Alloy Castings for Bearing (Aug)(10983).

—Centrifugal—Aerospace
BSI HC 100-72. 1972 Amd 3 Inspection and Testing Procedure for Iron, Nickel, Copper, Cobalt and Refractory Metal Base Alloy Castings (AMD 6574) May 28, 1991. 17 pp.

—Chill
DIN ENGL 17655-81. Unalloyed and Low Alloy Copper Materials for Casting; Castings (Nov). 4 pp.

—Classification—Marine
MOD UK NES 863-93. Requirements for the Classification, Dimensions, Tolerances and General Standards of Acceptance for Copper and Nickel Alloy Castings Issue 1 (05.93). 16 pp.
MOD UK NES 745: Part 1-90. Classification, Inspection Requirements and Acceptance Standards for Castings Part 1: Copper and Nickel Alloy Castings Issue 1 (04.90). 19 pp.

—Copper Content
JIS H 1051-92. Methods for Determination of Copper in Copper and Copper Alloys. 22 pp.

—Glossaries
ISO 197 Pt 4-83. Copper and Copper Alloys—Terms and Definitions—Part 4: Castings First Edition. 3 pp.

—Ingots
BSI BS 1400-85. 1985 Amd 1 Copper Alloy Ingots and Copper Alloy and High Conductivity Copper Castings (AMD 6634) September 28, 1990. 39 pp.
DIN ENGL 17656-73. Copper Casting Alloys; Ingot Metals; Composition (June). 5 pp.
SNZ NZS/BS 1400-85. Specification for Copper Alloy Ingots and Copper Alloy and High Conductivity Copper Castings. 40 pp.

—Inspection—Marine
MOD UK NES 745: Part 1-90. Classification, Inspection Requirements and Acceptance Standards for Castings Part 1: Copper and Nickel Alloy Castings Issue 1 (04.90). 19 pp.

INTERNATIONAL AND NON-U.S. NATIONAL STANDARDS
SUBJECT INDEX
Copper

Copper Castings *(Cont.)*

—Investment—Aerospace

BSI HC 100-72. 1972 Amd 3 Inspection and Testing Procedure for Iron, Nickel, Copper, Cobalt and Refractory Metal Base Alloy Castings (AMD 6574) May 28, 1991. 17 pp.

—Sampling

ISO 1811 Pt 2-88. Copper and Copper Alloys—Selection and Preparation of Samples for Chemical Analysis—Part 2: Sampling of Wrought Products and Castings First Edition. 5 pp.

—Sand

DIN ENGL 17655-81. Unalloyed and Low Alloy Copper Materials for Casting; Castings (Nov). 4 pp.

—Sand—Aerospace

BSI HC 100-72. 1972 Amd 3 Inspection and Testing Procedure for Iron, Nickel, Copper, Cobalt and Refractory Metal Base Alloy Castings (AMD 6574) May 28, 1991. 17 pp.

—Welding—Repair—Marine

MOD UK NES 771-91. Requirements Procedure and Inspection for Weld Repair of Copper Alloy and Nickel Alloy Castings Issue 1 (07.91). 35 pp.

Copper Chromate

Use: Cupric Chromate

Copper-Chrome-Arsenic

See Also: Hazardous Materials

—Highway Transportation—Emergency Procedures

SAA AS 1678.6.0. 011-88. Emergency Procedure Guides—Transport—Part 6.0.011: Copper Chrome Arsenic (C.C.A. Wood Preservative).

Copper Clad Laminates

See Also: Laminates

—Printed Circuit Base Materials

BSI BS 4584: Part 1-83. 1983 Amd 1 Metal-Clad Base Materials for Printed Wiring Boards Part 1: Methods of Test (IEC 249-1: 1982) (AMD 5988) July 31, 1989. 48 pp.
BSI BS 4584: Part 1-02. 1983 Amd 2 Metal-Clad Base Materials for Printed Wiring Boards Part 1: Methods of Test (IEC 249-1: 1982) (AMD 6791) February 28, 1992. 57 pp.
CNS C5173-85. General Rules of Copper-Clad Laminates for Printed Circuits (Dec)(10554).
CNS C6206-85. Method of Test for Copper-Clad Laminates for Printed Circuits (Dec)(10555).
CNS C7157-85. Copper-Clad Laminates for Printed Circuits (Paper Base, Epoxy Resin) (Dec)(10556).
CNS C7158-85. Copper-Clad Laminates for Printed Circuits (Synthetic Fiber Fabric Base, Epoxy Resin) (Dec)(10557).
CNS C7159-85. Copper-Clad Laminates for Printed Circuits (Glass Fabric Base, Epoxy Resin) (Dec)(10558).
CNS C7160-85. Copper-Clad Laminates for Printed Circuits (Paper Base, Phenolic Resin) (Dec)(10559).
CNS C7162-85. Epoxy Resin-Impregnated Glass Cloth (Pre-Preg) for Multilayer Printed Circuits (Dec)(10561).
JIS C 6471-90. Test Methods of Copper-Clad Laminates for Flexible Printed Wiring Boards. 25 pp.
JIS C 6472-90. Copper-Clad Laminates for Flexible Printed Wiring Boards (Polyester Film, Polyimide Film). 11 pp.
JIS C 6480-89. General Rules of Copper-Clad Laminates for Printed Wiring Boards. 9 pp.
JIS C 6481-90. Test Methods of Copper-Clad Laminates for Printed Wiring Boards. 33 pp.
JIS C 6482-91. Copper-Clad Laminates for Printed Wiring Boards—Paper Base, Epoxy Resin. 14 pp.
JIS C 6483-91. Copper-Clad Laminates for Printed Wiring Boards—Synthetic Fiber Fabric Base, Epoxy Resin. 13 pp.
JIS C 6484-91. Copper-Clad Laminates for Printed Wiring Boards—Glass Fabric Base, Epoxy Resin. 13 pp.
JIS C 6485-91. Copper-Clad Laminates for Printed Wiring Boards—Paper Base, Phenolic Resin. 14 pp.
JIS C 6486-90. Thin Copper-Clad Laminates for Multilayer Printed Wiring Boards—Glass Fabric Base, Epoxy Resin. 14 pp.
JIS C 6488-89. Copper-Clad Laminates for Printed Wiring Boards (Glass Cloth Surfaces, Cellulose Paper Core, Epoxy Resin). 13 pp.
JIS C 6489-89. Copper-Clad Laminates for Printed Wiring Boards (Glass Cloth Surfaces, Nonwoven Glass Core, Epoxy Resin). 13 pp.
JIS C 6511-92. Test Methods of Copper Foil for Printed Wiring Boards. 11 pp.
JIS C 6512-92. Electrodeposited Copper Foil for Printed Wiring Boards. 9 pp.
JIS H 8646-91. Electroless Copper Platings. 15 pp.

Copper Clad Laminates *(Cont.)*

—Printed Circuit Base Materials—Copper Foil

BSI BS 4584: Sec 103.2-90. 1990 Metal-Clad Base Materials for Printed Circuits Part 103: Materials Used in Connection with Printed Circuits Section 103.2: Copper Foil for Use in the Manufacture of Copper-Clad Base Materials. 12 pp.

IEC 249 Pt 3A-76. Metal-Clad Base Materials for Printed Circuits Part 3: Special Materials Used in Connection with Printed Circuits Specification No. 2 Specification for Copper Foil for Use in the Manufacture of Copper-Clad Base Materials (This is a Supplement). 26 pp.

JIS C 6512-92. Electrodeposited Copper Foil for Printed Wiring Boards. 9 pp.

—Printed Circuit Base Materials—Epoxide Cellulose Paper

BSI BS 4584: Sec 102.3-90. 1990 Metal-Clad Base Materials for Printed Circuits Part 102: Copper-Clad Base Materials Section 102.3: Epoxide Cellulose Paper Copper-Clad Laminated Sheet of Defined Flammability (Vertical Burning Test). 12 pp.

BSI BS 4584: Sec 102.9-90. 1990 Metal-Clad Base Materials for Printed Circuits Part 102: Copper-Clad Base Materials Section 102.9: Epoxide Cellulose Paper Core, Epoxide Glass Cloth Surf. Copper-Clad Lamin. Sheet of Defined Flamm. (Vertical Burning Test). 12 pp.

BSI BS 4584: Sec 102.9-01. 1990 Amd 1 Metal-Clad Base Materials for Printed Wiring Boards Part 102: Copper-Clad Base Materials Section 102.9: Specification for Epoxide Cellulose Paper Core, Epoxide Glass Cloth Surfaces Copper-Clad Laminated Sheet of. 15 pp.

CENELEC HD 313.2.3 S2-90. Base Materials for Printed Circuits Part 2: Specifications Specification No.3: Epoxide Cellulose Paper Copper-Clad Laminated Sheet of Defined Flammability (Vertical Burning Test). 3 pp.

CENELEC HD 313.2.4 S2-90. Base Materials for Printed Circuits Part 2: Specifications Specification No.4: Epoxide Woven Glass Fabric Copper-Clad Laminated Sheet, General Purpose Grade. 3 pp.

CENELEC HD 313.2.9 S2-90. Base Materials for Printed Circuits Part 2: Specifications Specification No.9: Epoxide Cellulose Paper Core, Epoxide Glass Cloth Surfaces Copper-Clad Laminated Sheet of Defined Flammability (Vertical Burning Test). 4 pp.

CENELEC HD 313.2.9 S3-91. Base Materials for Printed Circuits Part 2: Specifications Specification No. 9: Epoxide Cellulose Paper Core, Epoxide Glass Cloth Surfaces Copper-Clad Laminated Sheet of Defined Flammability (Vertical Burning Test). 6 pp.

IEC 249 Pt 2-3-87. Base Materials for Printed Circuits Part 2: Specifications Specification No. 3: Epoxide Cellulose Paper Copper-Clad Laminated Sheet of Defined Flammability (Vertical Burning Test) Second Edition; (Amendment 2-1993, Incorporating Amendment 1). 37 pp.

IEC 249 Pt 2-9-87. Base Materials for Printed Circuits Part 2: Specifications Specification No. 9: Epoxide Cellulose Paper Core, Epoxide Glass Cloth Surfaces Copper-Clad Laminated Sheet of Defined Flammability (Vertical Burning Test) First Ed.; (Amd 3-1993, Incorp. Amds 1 and 2). 38 pp.

—Printed Circuit Base Materials—Glass Fabric

BSI BS 4584: Part 3-72. (WITHDRAWN) 1972 Amd 1 Metal-Clad Base Materials for Printed Circuits Part 3: Epoxide Woven Glass Fabric Copper-Clad Laminated Sheet, Flame Retardant Grade: EP-GC-Cu-3 (S) (Superseded by BS 4584: Section 102.5: 1990). 18 pp.

BSI BS 4584: Part 11-77. (WITHDRAWN) 1977 Metal-Clad Base Materials For Printed Circuits Part 11: Bonding Sheet Material for Use in the Fabrication of Multi-layer Printed Boards. EP-GC-11 (Superseded by BS 4584: Section 103.1: 1990). 13 pp.

BSI BS 4584: Sec 102.4-90. 1990 Metal-Clad Base Materials for Printed Circuits Part 102: Copper-Clad Base Materials Section 102.4: Epoxide Woven Glass Fabric Copper-Clad Laminated Sheet, General Purpose Grade. 12 pp.

BSI BS 4584: Sec 102.5-90. 1990 Metal-Clad Base Materials for Printed Circuits Part 102: Copper-Clad Base Materials Section 102.5: Epoxide Woven Glass Fabric Copper-Clad Laminated Sheet of Defined Flammability (Vertical Burning Test). 14 pp.

BSI BS 4584: Sec 102.9-90. 1990 Metal-Clad Base Materials for Printed Circuits Part 102: Copper-Clad Base Materials Section 102.9: Epoxide Cellulose Paper Core, Epoxide Glass Cloth Surf. Copper-Clad Lamin. Sheet of Defined Flamm. (Vertical Burning Test). 12 pp.

Copper Clad Laminates *(Cont.)*

—Printed Circuit Base Materials—Glass Fabric *(Cont.)*

BSI BS 4584: Sec 102.9-01. 1990 Amd 1 Metal-Clad Base Materials for Printed Wiring Boards Part 102: Copper-Clad Base Materials Section 102.9: Specification for Epoxide Cellulose Paper Core, Epoxide Glass Cloth Surfaces Copper-Clad Laminated Sheet of. 15 pp.
BSI BS 4584: Sec 102.10-90. 1990 Metal-Clad Base Materials for Printed Circuits Part 102: Copper-Clad Base Materials Section 102.10: Epoxide Non-Woven/Woven Glass Reinforced Copper-Clad Laminated Sheet of Defined Flammability (Vertical Burning Test). 12 pp.
BSI BS 4584:Sec 102.10-01. 1990 Amd 1 Metal-Clad Base Materials for Printed Wiring Boards Part 102: Copper-Clad Base Materials Section 102.10: Specification for Epoxide Non-Woven/Woven Glass Reinforced Copper-Clad Laminated Sheet of Defined Flammability. 15 pp.
BSI BS 4584: Sec 102.11-90. 1990 Metal-Clad Base Materials for Printed Circuits Part 102: Copper-Clad Base Materials Section 102. 11: Thin Epoxide Woven Glass Fabric Copper-Clad Laminated Sheet, General Purpose Grade, for Use in the Fabrication of Multilayer Prtd Boards. 10 pp.
BSI BS 4584:Sec 102.11-01. 1990 Amd 1 Metal-Clad Base Materials for Printed Wiring Boards Part 102: Copper-Clad Base Materials Section 102.11: Specification for Thin Epoxide Woven Glass Fabric Copper-Clad Laminated Sheet, General Purpose Grade, for Use. 13 pp.
BSI BS 4584: Sec 102.12-90. 1990 Metal-Clad Base Materials for Printed Circuits Part 102: Copper-Clad Base Materials Section 102. 12: Thin Epoxide Woven Glass Fabric Copper-Clad Lamin. Sheet of Defined Flammability, for Use in the Fabrication of Multilayer Prtd Boards. 10 pp.
BSI BS 4584:Sec 102.12-01. 1990 Amd 1 Metal-Clad Base Materials for Printed Wiring Boards Part 102: Copper-Clad Base Materials Section 102.12: Specification for Thin Epoxide Woven Glass Fabric Copper-Clad Laminated Sheet of Defined Flammability,. 12 pp.
BSI BS 4584: Sec 102.16-92. 1992 Metal-Clad Base Materials for Printed Wiring Boards Part 102: Copper-Clad Base Materials Sec 102.16: Spec for Polyimide Woven Glass Fabric Copper-Clad Laminated Sheet of Defined Flammability (Vertical Burning Test) (IEC 249-2-16: 1992). 19 pp.
BSI BS 4584: Sec 102.17-92. 1992 Metal-Clad Base Materials for Printed Wiring Boards Part 102: Copper-Clad Base Materials Section 102.17: Specification for Thin Polyimide Woven Glass Fabric Copper-Clad Laminated Sheet of Defined Flammability for Use in. 16 pp.
BSI BS 4584: Sec 102.18-92. 1992 Metal-Clad Base Materials for Printed Wiring Boards Part 102: Copper-Clad Base Materials Section 102.18: Specification for Bismaleimide/Triazine Modified Epoxide Woven Glass Fabric Copper-Clad Laminated Sheet of. 19 pp.
BSI BS 4584: Sec 102.19-92. 1992 Metal-Clad Base Materials for Printed Wiring Boards Part 102: Copper-Clad Base Materials Section 102. 19: Spec. for Thin Bismaleimide/Triazine Modified Epoxide Glass Fabric Copper-Clad Laminated Sheets of Defined Flammability for. 16 pp.
CENELEC HD 313.2.3 S2-90. Base Materials for Printed Circuits Part 2: Specifications Specification No.3: Epoxide Cellulose Paper Copper-Clad Laminated Sheet of Defined Flammability (Vertical Burning Test). 3 pp.
CENELEC HD 313.2.4 S1-89. Base Materials for Printed Circuits-Part 2: Specifications. Specification No.4: Epoxide Woven Glass Fabric Copper-Clad Laminated Sheet, General Purpose Grade. 3 pp.
CENELEC HD 313.2.4 S2-90. Base Materials for Printed Circuits Part 2: Specifications Specification No.4: Epoxide Woven Glass Fabric Copper-Clad Laminated Sheet, General Purpose Grade. 3 pp.
CENELEC HD 313.2.5 S1-89. Base Materials for Printed Circuits-Part 2: Specifications Specification No.5: Epoxide Woven Glass Fabric Copper-Clad, Laminated Sheet of Defined Flammability (Vertical Burning Test). 3 pp.
CENELEC HD 313.2.5 S2-90. Base Materials for Printed Circuits Part 2: Specifications Specification No.5: Epoxide Woven Glass Fabric Copper-Clad Laminated Sheet of Defined Flammability (Vertical Burning Test). 3 pp.
CENELEC HD 313.2.9 S1-89. Base Materials for Printed Circuits-Part 2: Specifications Specification No 9: Epoxide Cellulose Paper Core, Epoxide Glass Cloth Surfaces Copper-Clad Laminated Sheet of Defined Flammability (Vehicle Burning Test). 3 pp.
CENELEC HD 313.2.9 S2-90. Base Materials for Printed Circuits Part 2: Specifications Specification No.9: Epoxide Cellulose Paper Core, Epoxide Glass Cloth Surfaces Copper-Clad Laminated Sheet of Defined Flammability (Vertical Burning Test). 4 pp.

Copper Clad Laminates *(Cont.)*

—Printed Circuit Base Materials—Glass Fabric *(Cont.)*

CENELEC HD 313.2.9 S3-91. Base Materials for Printed Circuits Part 2: Specifications Specification No. 9: Epoxide Cellulose Paper Core, Epoxide Glass Cloth Surfaces Copper-Clad Laminated Sheet of Defined Flammability (Vertical Burning Test). 6 pp.

CENELEC HD 313.2.10 S1-89. Base Materials for Printed Circuits-Part 2: Specifications Specification No 10: Epoxide Non-Woven/Woven Glass Reinforced Copper-Clad Laminated Sheet of Defined Flammability (Vertical Burning Test). 3 pp.

CENELEC HD 313.2.10 S2-90. Base Materials for Printed Circuits Part 2: Specifications Specification No. 10: Epoxide Non-Woven Glass Reinforced Copper-Clad Laminated Sheet of Defined Flammability (Vertical Burning Test). 3 pp.

CENELEC HD 313.2.10 S3-91. Base Materials for Printed Circuits Part 2: Specifications Specification No. 10: Epoxide Non-Woven/Woven Glass Reinforced Copper-Clad Laminated Sheet of Defined Flammability (Vertical Burning Test). 6 pp.

CENELEC HD 313.2.11 S1-89. Base Materials for Printed Circuits-Part 2: Specifications Specification No 11: Thin Epoxide Woven Glass Fabric Copper-Clad Laminated Sheet, General Purpose Grade for Use in the Fabrication of Multilayer Printed Boards. 3 pp.

CENELEC HD 313.2.11 S2-90. Base Materials for Printed Circuits Part 2: Specifications Specification No. 11: Thin Epoxide Woven Glass Fabric Copper-Clad Laminated Sheet, General Purpose Grade, for Use in the Fabrication of Multilayer Printed Boards. 3 pp.

CENELEC HD 313.2.12 S1-89. Base Materials for Printed Circuits-Part 2: Specifications Specification No 12: Thin Epoxide Woven Glass Fabric Copper-Clad Laminated Sheet of Defined Flammability for Use in the Fabrication of Multilayer Printed Boards. 3 pp.

CENELEC HD 313.2.12 S2-90. Base Materials for Printed Circuits Part 2: Specifications Specification No. 12: Thin Epoxide Woven Glass Fabric Copper-Clad Laminated Sheet of Defined Flammability, for Use in the Fabrication of Multilayer Printed Boards. 3 pp.

CNS C7161-85. Thin Copper-Clad Laminates for Multilayer Printed Circuits (Glass Fabric Base, Epoxy Resin) (Dec)(10560).

IEC 249 Pt 2-4-87. Base Materials for Printed Circuits Part 2: Specifications Specification No. 4: Epoxide Woven Glass Fabric Copper-Clad Laminated Sheet, General Purpose Grade Second Edition; (Amendment 3-1993, Incorporating Amendment 1 and 2). 39 pp.

IEC 249 Pt 2-5-87. Base Materials for Printed Circuits Part 2: Specifications Specification No. 5: Epoxide Woven Glass Fabric Copper-Clad Laminated Sheet of Defined Flammability (Vertical Burning Test) Second Edition; (Amd. 3-1993, Incorporating Amendment 1 and 2). 43 pp.

IEC 249 Pt 2-9-87. Base Materials for Printed Circuits Part 2: Specifications Specification No. 9: Epoxide Cellulose Paper Core, Epoxide Glass Cloth Surfaces Copper-Clad Laminated Sheet of Defined Flammability (Vertical Burning Test) First Ed.; (Amd 3-1993, Incorp. Amds 1 and 2). 38 pp.

IEC 249 Pt 2-10-87. Base Materials for Printed Circuits Part 2: Specifications Spec. No. 10: Epoxide Non-Woven/Woven Glass Reinforced Copper-Clad Laminated Sheet of Defined Flammability (Vertical Burning Test) First Ed.; (Amendment 3-1993, Incorporating Amendment 1 and 2). 39 pp.

IEC 249 Pt 2-11-87. Base Materials for Printed Circuits Part 2: Specifications Spec. No. 11: Thin Epoxide Woven Glass Fabric Copper-Clad Laminated Sheet, General Purpose Grade, for Use in the Fabrication of Multilayer Printed Boards First Edition (Amendment 2-1993, Incorporating Amd 1). 33 pp.

IEC 249 Pt 2-12-87. Base Materials for Printed Circuits Part 2: Specifications Spec. No. 12: Thin Epoxide Woven Glass Fabric Copper-Clad Laminated Sheet of Defined Flammability, for Use in the Fabrication of Multi-layer Printed Boards Second Ed.; (Amd 2-1993, Incorporating Amd 1). 35 pp.

IEC 249 Pt 2-16-92. Base Materials for Printed Circuits Part 2: Specifications Specification No. 16: Polyimide Woven Glass Fabric Copper-Clad Laminated Sheet of Defined Flammability (Vertical Burning Test) First Edition; (Amendment 1-1993). 41 pp.

IEC 249 Pt 2-17-92. Base Materials for Printed Circuits Part 2: Specifications Spec. No. 17: Thin Polyimide Woven Glass Fabric Copper-Clad Laminated Sheet of Defined Flammability for Use in the Fabrication of Multilayer Printed Board First Ed.; (Cor—June 1992) (Amd 1-1993). 40 pp.

Copper Clad Laminates *(Cont.)*

—Printed Circuit Base Materials—Glass Fabric *(Cont.)*

IEC 249 Pt 2-18-92. Base Materials for Printed Circuits Part 2: Specifications Specification No. 18: Bismaleimide/Triazine Modified Epoxide Woven Glass Fabric Copper-Clad Laminated Sheet of Defined Flammability (Vertical Burning Test) First Edition; (Amendment 1-1993). 41 pp.

IEC 249 Pt 2-19-92. Base Materials for Printed Circuits Part 2: Specifications Specification No. 19: Thin Bismaleimide/Triazine Modified Epoxide Woven Glass Fabric Copper-Clad Laminated Sheet of Defined Flammability for Use in the Fabrication of Multilayer Printed. 37 pp.

JIS R 3423-87. Finished Glass Fabrics Used for Copper Clad Laminates and Other Electric Purposes.

MOD UK DSTAN 59-50-01. Requirements for Plastics Sheet, Laminated Copper Clad, Epoxide Resin Bonded, Woven Glass Fabric Base-Fire-Retardant (Metal Clad Base Materials for Printed Circuits) Issue 1; Amendment 1. 18 pp.

—Printed Circuit Base Materials—Phenolic Cellulose Paper

BSI BS 4584: Part 5-72. 1972 Metal-Clad Base Materials for Printed Circuits Part 5: Phenolic Cellulose Paper Copper-Clad Laminated Sheet of Medium Electrical Quality: PF-CP-Cu-5. 11 pp.

BSI BS 4584: Part 6-72. 1972 Metal-Clad Base Materials for Printed Circuits Part 6: Phenolic Cellulose Paper Copper-Clad Laminated Sheet of Medium Electrical Quality, Flame Retardant Grade: PF-CP-Cu-6. 12 pp.

BSI BS 4584: Part 7-89. (WITHDRAWN) 1989 Metal-Clad Base Materials for Printed Circuits Part 7: Phenolic Cellulose Paper Copper-Clad Laminated Sheet of High Electrical Quality (Superseded by BS 4584: Section 102.1: 1990). 12 pp.

BSI BS 4584: Sec 102.1-90. 1990 Metal-Clad Base Materials for Printed Circuits Part 102: Copper-Clad Base Materials Section 102.1: Phenolic Cellulose Paper Copper-Clad Laminated Sheet, High Electrical Quality. 15 pp.

BSI BS 4584: Sec 102.2-90. 1990 Metal-Clad Base Materials for Printed Circuits Part 102: Copper-Clad Base Materials Section 102.2: Phenolic Cellulose Paper Copper-Clad Laminated Sheet, Economic Quality. 13 pp.

BSI BS 4584:Sec 102.2-01. 1990 Amd 1 Metal-Clad Base Materials for Printed Wiring Boards Part 102: Copper-Clad Base Materials Section 102.2: Specification for Phenolic Cellulose Paper Copper-Clad Laminated Sheet, Economic Quality. 17 pp.

BSI BS 4584: Sec 102.6-90. 1990 Metal-Clad Base Materials for Printed Circuits Part 102: Copper-Clad Base Materials Section 102.6: Phenolic Cellulose Paper Copper-Clad Laminated Sheet of Defined Flammability (Horizontal Burning Test). 16 pp.

BSI BS 4584: Sec 102.7-90. 1990 Metal-Clad Base Materials for Printed Circuits Part 102: Copper-Clad Base Materials Section 102.7: Phenolic Cellulose Copper-Clad Laminated Sheet of Defined Flammability (Vertical Burning Test). 12 pp.

BSI BS 4584: Sec 102.14-90. 1990 Metal-Clad Base Materials for Printed Circuits Part 102: Copper-Clad Base Materials Section 102. 14: Specification for Phenolic Cellulose Paper Copper-Clad Lamin. Sheet of Defined Flammability (Vert. Burning Test) and of Economic Quality (S). 15 pp.

BSI BS 4584:Sec 102.14-01. 1990 Amd 1 Metal-Clad Base Materials for Printed Wiring Boards Part 102: Copper-Clad Base Materials Section 102.14: Specification for Phenolic Cellulose Paper Copper-Clad Laminated Sheet of Defined Flammability (Vertical. 16 pp.

CENELEC HD 313.2.1-87. Base Materials for Printed Circuits Part 2: Specifications Specification No 1: Phenolic Cellulose Paper Copper—Clad Laminated Sheet, High Electrical Quality. 3 pp.

CENELEC HD 313.2.1 S2-90. Base Materials for Printed Circuits Part 2 Specifications Specification No 1: Phenolic Cellulose Paper Copper-Clad Laminated Sheet High Electrical Quality. 3 pp.

CENELEC HD 313.2.2-87. Base Materials for Printed Circuits Part 2: Specifications Specification No:2 Phenolic Cellulose Paper Copper-Clad Laminated Sheet, Economic Quality.

CENELEC HD 313.2.2 S2-90. Base Materials for Printed Circuits Part 2: Specifications Specification No.2: Phenolic Cellulose Paper Copper-Clad Laminated Sheet, Economic Quality. 3 pp.

CENELEC HD 313.2.2 S3-92. Base Materials for Printed Circuits Part 2: Specifications Specification No. 2: Phenolic Cellulose Paper Copper-Clad Laminated Sheet, Economic Quality. 5 pp.

CENELEC HD 313.2.6-87. Base Materials for Printed Circuits Part 2: Specifications Specification No 6: Phenolic Cellulose Paper Copper-Clad Laminated Sheet of Defined Flammability (Horizontal Burning Test). 3 pp.

Copper Clad Laminates *(Cont.)*

—Printed Circuit Base Materials—Phenolic Cellulose Paper *(Cont.)*

CENELEC HD 313.2.6 S2-90. Base Materials for Printed Circuits Part 2: Specifications Specification No. 6: Phenolic Cellulose Paper Copper-Clad Laminated Sheet of Defined Flammability (Horizontal Burning Test). 3 pp.

CENELEC HD 313.2.7 S2-90. Base Materials for Printed Circuits Part 2: Specifications Specification No.7: Phenolic Cellulose Paper Copper-Clad Laminated Sheet of Defined Flammability (Vertical Burning Test). 4 pp.

CENELEC HD 313.2.14 S1-90. Base Materials for Printed Circuits Part 2: Specifications Specification No 14: Phenolic Cellulose Paper Copper-Clad Laminated Sheet of Defined Flammability (Vertical Turning Test) Economic Quality. 3 pp.

CENELEC HD 313.2.14 S2-92. Base Materials for Printed Circuits Part 2: Specifications Specification No. 14: Phenolic Cellulose Paper Copper-Clad Laminated Sheet of Defined Flammability (Vertical Burning Test), Economic Quality. 5 pp.

CENELEC HD 323.2.32 S2-91. Basic Environmental Testing Procedures Part 2: Tests Test ed: Free Fall (Superseded by EN 60068-2-32: 1993). 5 pp.

IEC 249 Pt 2-1-85. Base Materials for Printed Circuits Part 2: Specifications Specification No. 1: Phenolic Cellulose Paper Copper-Clad Laminated Sheet, High Electrical Quality Second Edition; (Amendment 2-1993, Incorporating Amendment 1). 35 pp.

IEC 249 Pt 2-2-85. Base Materials for Printed Circuits Part 2: Specifications Specification No. 2: Phenolic Cellulose Paper Copper-Clad Laminated Sheet, Economic Quality Second Edition; (Amendment 3-1993, Incorporating Amendment 1 and 2). 35 pp.

IEC 249 Pt 2-6-85. Base Materials for Printed Circuits Part 2: Specifications Specification No. 6: Phenolic Cellulose Paper Copper-Clad Laminated Sheet of Defined Flammability (Horizontal Burning Test) Second Edition; (Amendment 2-1993, Incorporating Amendment 1). 37 pp.

IEC 249 Pt 2-7-87. Base Materials for Printed Circuits Part 2: Specifications Specification No. 7: Phenolic Cellulose Paper Copper-Clad Laminated Sheet of Defined Flammability (Vertical Burning Test) Second Edition; (Amendment 2-1993, Incorporating Amendment 1). 37 pp.

IEC 249 Pt 2-14-88. Base Materials for Printed Circuits Part 2: Specifications Spec. No. 14: Phenolic Cellulose Paper Copper-Clad Laminated Sheet of Defined Flammability (Vertical Burning Test), Economic Quality First Edition; (Amendment 3-1993, Incorporating Amd 1 & 2). 39 pp.

—Printed Circuit Base Materials—Polyester Film

BSI BS 4584: Sec 102.8-90. 1990 Metal-Clad Base Materials for Printed Circuits Part 102: Copper-Clad Base Materials Section 102.8: Flexible Copper-Clad Polyester (PETP) Film. 11 pp.

CENELEC HD 313.2.8 S1-89. Base Materials for Printed Circuits-Part 2: Specifications Specification No. 8: Flexible Copper-Clad Polyester (PETP) Film. 3 pp.

IEC 249 Pt 2-8-87. Base Materials for Printed Circuits Part 2: Specifications Specification No. 8: Flexible Copper-Clad Polyester (PETP) Film First Edition; (Amendment 1-1993). 31 pp.

—Printed Circuit Base Materials—Polyimide Film

BSI BS 4584: Part 10-77. (WITHDRAWN) 1977 Metal-Clad Base Materials for Printed Circuits Part 10: Flexible Copper-Clad Polyimide Film: P1-F-Cu-10 (Superseded by BS 4584: Section 102.15: 1991). 17 pp.

BSI BS 4584: Sec 102.13-90. 1990 Metal-Clad Base Materials for Printed Circuits Part 102: Copper-Clad Base Materials Section 102.13: Flexible Copper-Clad Polyimide Film, General Purpose Grade. 12 pp.

BSI BS 4584: Sec 102.15-91. 1991 Metal-Clad Base Materials for Printed Wiring Boards Part 102: Copper-Clad Base Materials Section 102. 15: Specification for Flexible Copper-Clad Polyimide Film of Defined Flammability (IEC 249-2-15: 1987). 14 pp.

CENELEC HD 313.2.13 S1-89. Base Materials for Printer Circuits-Part 2: Specifications Specification No. 13: Flexible Copper-Clad Polyimide Film, General Purpose Grade. 3 pp.

CENELEC HD 313.2.15 S1-91. Base Materials for Printed Circuits Part 2: Specifications Specification No. 15: Flexible Copper-Clad Polymide Film, of Defined Flammability. 5 pp.

IEC 249 Pt 2-13-87. Base Materials for Printed Circuits Part 2: Specifications Specification No. 13: Flexible Copper-Clad Polyimide Film, General Purpose Grade First Edition; (Amendment 1-1993). 35 pp.

INTERNATIONAL AND NON-U.S. NATIONAL STANDARDS
SUBJECT INDEX

Copper

Copper Clad Laminates *(Cont.)*
—Printed Circuit Base Materials—Polyimide Film *(Cont.)*
IEC 249 Pt 2-15-87. Base Materials for Printed Circuits Part 2: Specifications Specification No. 15: Flexible Copper-Clad Polyimide Film, of Defined Flammability First Edition; (Amendment 1-1993). 35 pp.

Copper Clad Nickel-Iron Alloy Wire
Use: Dumet Wire

Copper Coatings (Made From Copper)
Scope Note: Includes coatings made from copper alloys *Use For:* Copper Paints (Made From Copper) *See Also:* Coatings; Copper; Copper Alloys; Metal Coatings (Made From Metal)

—Electroless—Printed Circuits
JIS H 8646-91. Electroless Copper Platings. 15 pp.

—Electroplated
BSI BS 4493-69. 1969 Copper Salts for Electroplating. 25 pp.

CNS H3060-87. Electroplated Coatings of Nickel and Chromium (Sep)(4157).

CNS H3109-82. Recommended Practice for Copper-Nickel-Chromium Plating (Jun)(8977).

CNS H3137-87. Electroplated Coatings of Copper (Oct)(11392).

JIS H 8617-91. Electroplated Coatings of Nickel and Chromium. 17 pp.

—Oleoresinous—Wood—Marine—Exterior
CNS K2022-86. Oleoresinous Bottom Paint for Wooden Ship (Copper Paint) (May)(778). 1 p.

CNS K6856-86. Method of Test for Oleoresinous Bottom Paint for Wooden Ship (Copper Paint) (May)(11586).

—Plastic—Electroplated
CNS H3136-87. Electroplated Coatings of Copper Nickel Chromium on Plastics (Oct)(11391).

—Steel—Electroplated
DIN ENGL 50967-91. Electrodeposited Coatings of Nickel Plus Chromium and Copper Plus Nickel Plus Chromium (Jan). 8 pp.

—Zinc—Electroplated
CNS H3106-86. Operation Standard for Chromium Plating on Zinc Alloys (Feb)(8508).

DIN ENGL 50967-91. Electrodeposited Coatings of Nickel Plus Chromium and Copper Plus Nickel Plus Chromium (Jan). 8 pp.

JIS H 9125-65. Operation Standard for Chromium Plating on Zinc Alloys (R 1983). 9 pp.

Copper Coatings (On Copper)
Scope Note: Includes coatings on copper alloys *See Also:* Coatings; Copper; Copper Alloys; Metal Coatings (On Metal)

—Cadmium—Electrodeposited
MOD UK DTD-904C-63. Cadmium Plating (Reprinted December 1966, Incorporating Amendment No. 1) (Superseded by Def Stan 03-19). 5 pp.

—Cadmium—Electrodeposited—Aircraft
AECMA PREN2133-81. Cadmium Plating of Steels with Maximum Specified Tensile Strength Less Than or Equal to 1450 MPa and Copper and Copper Alloys (C7/SC4/D). 14 pp.

—Cadmium—Electroplated—Aerospace
CEN PREN 2133-92. Cadmium Plating of Steels with Maximum Specified Tensile Strength Equal to or Less Than 1450 MPa Copper, Copper Alloys and Nickel Alloys. 16 pp.

—Chromium—Electroplated
BSI BS 1224-70. 1970 Amd 1 Electroplated Coatings of Nickel and Chromium. 29 pp.

BSI BS 3382: Parts 3 & 4-65. 1965 Amd 1 Electroplated Coatings on Threaded Components Part 3: Nickel or Nickel Plus Chromium on Steel Components Part 4: Nickel or Nickel Plus Chromium on Copper and Copper Alloy (Including Brass) Components. 33 pp.

—Lacquers (Acrylic)
MOD UK DSTAN 80-24-73. Lacquer, Acrylic (Incralac) Issue 1. 7 pp.

—Nickel—Electroplated
BSI BS 1224-70. 1970 Amd 1 Electroplated Coatings of Nickel and Chromium. 29 pp.

Copper Coatings (On Copper) *(Cont.)*
—Nickel—Electroplated *(Cont.)*
BSI BS 3382: Parts 3 & 4-65. 1965 Amd 1 Electroplated Coatings on Threaded Components Part 3: Nickel or Nickel Plus Chromium on Steel Components Part 4: Nickel or Nickel Plus Chromium on Copper and Copper Alloy (Including Brass) Components. 33 pp.

DIN ENGL 50967-91. Electrodeposited Coatings of Nickel Plus Chromium and Copper Plus Nickel Plus Chromium (Jan). 8 pp.

DIN ENGL 50968-91. Electrodeposited Coatings of Nickel and Nickel Plus Copper (Jan). 5 pp.

ISO 1458-88. Metallic Coatings—Electrodeposited Coatings of Nickel Second Edition. 16 pp.

—Oxide
CGSB CAN/CGSB-31.112-M90. Black Finish for Copper Alloys. 6 pp.

—Silver—Electroplated
BSI BS 3382: Parts 5 & 6-67. 1967 Electroplated Coatings on Threaded Components Part 5: Tin on Copper and Copper Alloy (Including Brass) Components Part 6: Silveron Copper and Copper Alloy (Including Brass) Components. 33 pp.

BSI BS 4290-89. 1989 Electroplated Coatings of Silver and Silver Alloys for Decorative and Protective Purposes. 10 pp.

—Tin—Electroplated
BSI BS 3382: Parts 5 & 6-67. 1967 Electroplated Coatings on Threaded Components Part 5: Tin on Copper and Copper Alloy (Including Brass) Components Part 6: Silveron Copper and Copper Alloy (Including Brass) Components. 33 pp.

JIS H 8624-90. Electroplated Coatings of Tin-Lead Alloys. 8 pp.

—Tin—Lead Content—Quantitative Analysis
BSI BS 6534-84. 1984 Quantitative Determination of Lead in Tin Coatings. 8 pp.

Copper Conductor Wire
Use: Copper Conductors

Copper Conductors
Use For: Copper Alloy Conductor Wire; Copper Conductor Wire; Electrical Copper Wire
See Also: Copper Wire; Electric Conductors; Electric Wire; Power Line Hardware; Wire; Wiring Devices

BSI BS 3573-90. 1990 Polyolefin Copper-Conductor Telecommunication Cables. 22 pp.

BSI BS 4109-70. 1970 Amd 1 Copper for Electrical Purposes. Wire for General Electrical Purposes and for Insulated Cables and Flexible Cords. 24 pp.

BSI BS 6207-91. 1991 Mineral-Insulated Copper-Sheathed Cables with Copper Conductors. 18 pp.

BSI BS 6360-91. 1991 Amd 1 Conductors in Insulated Cables and Cords (AMD 6769) July 31, 1991. 20 pp.

BSI BS 6360-02. 1991 Amd 2 Conductors in Insulated Cables and Cords (AMD 7637) May 15, 1993 (F). 21 pp.

CNS C2002-87. Hard-Drawn Copper Wire, Single Solid (Non-Insulated) (Jul)(666). 3 pp.

CNS C2030-87. Annealed Electrical Copper Wire (Jul)(1364). 4 pp.

CNS C2034-87. Half-Hard-Drawn Copper Wires, Solid (Non-Insulated) (Jul)(1368). 3 pp.

CNS C2046-84. Soft-Copper Binding Wire (May)(2549).

CNS C2052-84. Polyvinly Chloride Insulated Copper Wire (Flat or Twist) (May)(2900).

CNS C2068-84. Rectangular Copper Wires for Electrical Purposes (Dec)(5749). 6 pp.

CNS C2069-90. Annealed Copper Wire for Magnet Wires (Aug)(5750).

CNS C3026-87. Testing Standard of Electrical Copper Wires (Jul)(1370). 3 pp.

CNS C3080-87. Method of Test for Electrical Copper and Aluminium Wires (Jul)(5745).

CNS H3120-83. Copper Wire Bars (Jun)(10347). 3 pp.

DIN ENGL 40500 Pt 4-73. Copper for Electrical Engineering; Wires of Copper and Copper-Silver Alloy; Technical Conditions of Delivery (Sept). 7 pp.

DIN ENGL 46431-70. Round Copper Wires for Electrical Purposes; Precision Drawn; Dimensions (June). 2 pp.

DIN ENGL 46433-59. Rectangular Wires and Rectangular Bars; Drawn, with Radiused Edges; Dimensions (Nov). 8 pp.

DIN ENGL 46433 Extr. 1-59. Rectangular Wires and Rectangular Bars; Drawn, with Radiused Edges; Dimensions; Selected Sizes for Electrical Machines and Switchgear (Nov). 8 pp.

DIN ENGL 46433 Extr. 3-59. Rectangular Bars; Drawn, with Radiused Edges; Dimensions; Selected Sizes for Switchgears (Nov). 2 pp.

DIN ENGL 47600 Pt 3-74. Cast Metal Joint Boxes for Power Cables up to 10 kV; Correlation of Joint Boxes with Paper-Insulated Cables; Correlation of Bare Stranded Copper Bonding Wire (Oct). 2 pp.

Copper Conductors *(Cont.)*
DIN ENGL 48203 Pt 1-84. Copper Wires and Copper Stranded Conductors; Technical Delivery Conditions (Mar). 4 pp.

DIN ENGL 48203 Pt 2-84. Wrought Copper Alloy (Bz) Wires and Conductors; Technical Delivery Conditions (Mar). 4 pp.

IEC 28-25. International Standard of Resistance for Copper Second Edition. 9 pp.

IEC 55 Pt 1-78. Paper-Insulated Metal-Sheathed Cables for Rated Voltages up to 18/30 kV (with Copper or Aluminium Conductors and Excluding Gas-Pressure and Oil-Filled Cables) Part 1: Tests Fourth Edition; (Amendment 1-1989). 32 pp.

IEC 55 Pt 2-81. Paper-Insulated Metal-Sheathed Cables for Rated Voltages up to 18/30 kV (with Copper or Aluminium Conductors and Excluding Gas-Pressure and Oil-Filled Cables) Part 2: General and Construction Requirements First Edition; (Amendment 1-1989). 85 pp.

IEC 708 Pt 2-81. Low-Frequency Cables with Polyolefin Insulation and Moisture Barrier Polyolefin Sheath Part 2: Unit Type, Filled, Moisture Barrier Polyethylene Sheathed Cables with Copper Conductors and Solid or Cellular Insulation First Edition; (Amendment 1-1983). 19 pp.

IEC 708 Pt 3-81. Low-Frequency Cables with Polyolefin Insulation and Moisture Barrier Polyolefin Sheath Part 3: Unit Type, Unfilled, Moisture Barrier Polyethylene Sheathed Cables with Copper Conductors and Solid or Cellular Insulation First Edition; (Amendment 1-1983). 19 pp.

IEC 708 Pt 4-81. Low-Frequency Cables with Polyolefin Insulation and Moisture Barrier Polyolefin Sheath Part 4: Unit Type, Unfilled, Moisture Barrier Poly-ethylene Sheathed Cables with Copper Conductors, Solid Insulation and In-tegral Suspension Strand First Edition; (Amendment 1-1983). 17 pp.

IEC 719-92. Calculation of the Lower and Upper Limits for the Average Outer Dimensions of Cables with Circular Copper Conductors and of Rated Voltages up to and Including 450/750 V Second Edition. 19 pp.

IEC 798-84. Copper Alloy Used for Equipment Wires First Edition. 9 pp.

JIS C 3002-92. Testing Methods of Electrical Copper and Aluminium Wires. 12 pp.

JIS C 3006-83. Methods of Test for Fiber or Paper Insulated Copper and Aluminum Winding Wires. 9 pp.

JIS C 3101-76. Hard-Drawn Copper Wires for Electrical Purposes. 7 pp.

JIS C 3102-84. Annealed Copper Wires for Electrical Purposes. 7 pp.

JIS C 3104-84. Rectangular Copper Wires for Electrical Purposes. 8 pp.

—Aerospace
AECMA PREN2083-77. Copper and Copper Alloy Conductors for Electrical Cables (C2/01). 4 pp.

BSI G 231-83. (WITHDRAWN) 1983 Conductors for General Purpose Aircraft Electrical Cables and Aerospace Applications (Superseded by 2G 231: 1990). 6 pp.

ISO 2635-79. Aircraft—Conductors for General Purpose Aircraft Electrical Cables and Aerospace Applications—Dimensions and Characteristics First Edition. 4 pp.

—Aircraft
AECMA PREN2084-80. Electric Cables for General Purpose with Conductors in Copper or Copper Alloy—Technical Specification (C2/02). 25 pp.

—Automotive
ISO 6722 Pt 1-84. Road Vehicles—Unscreened Low-Tension Cables—General Requirements and Test Methods Second Edition. 12 pp.

ISO 6722 Pt 2-85. Road Vehicles—Unscreened Low-Tension Cables—Part 2: Cable Classes, Applicable Tests and Special Requirements Second Edition. 5 pp.

JIS C 3406-87. Low-Voltage Cables for Automobile. 12 pp.

—Braided
MOD UK DSTAN 61-12: Part 20-80. Wires, Cords, and Cables, Electrical-Metric Units Part 20: Braids, Wire Issue 1. 7 pp.

—Compression Joints
BSI BS 4579: Part 1-70. 1970 Amd 3 Performance of Mechanical and Compression Joints in Electric Cable and Wire Connectors Part 1: Compression Joints in Copper Conductors (AMD 4687) March 29, 1985. 12 pp.

BSI BS 4579: Part 2-73. 1973 Amd 1 Performance of Mechanical and Compression Joints in Electric Cable and Wire Connectors Part 2: Compression Joints in Nickel, Iron and Platted Copper Conductors. 11 pp.

INDUSTRY STANDARDS

Copper Conductors *(Cont.)*

—Connectors

CSA C22.2 NO 65-93. Wire Connectors; (Gen Instr 1). 67 pp.

—Connectors—Compressed

BSI BS 4579: Part 1-70. 1970 Amd 3 Performance of Mechanical and Compression Joints in Electric Cable and Wire Connectors Part 1: Compression Joints in Copper Conductors (AMD 4687) March 29, 1985. 12 pp.

BSI BS 4579: Part 2-73. 1973 Amd 1 Performance of Mechanical and Compression Joints in Electric Cable and Wire Connectors Part 2: Compression Joints in Nickel, Iron and Platted Copper Conductors. 11 pp.

—Cotton Covered

BSI BS 4927: Part 1-74. 1974 Paper-, Cotton-, Silk-or Rayon-Covered Copper Conductors Part 1: Round Wire. 18 pp.

CNS C2093-83. Cotton Covered Copper Wires and Silk Covered Copper Wires (Feb)(7353).

—Cotton Covered—Windings

JIS C 3204-88. Fiber or Paper Covered Copper Winding Wires. 18 pp.

—Electric Terminals—Quick Connect—Flat Cables—Safety

IEC 1210-93. Connecting Devices—Flat Quick-Connect Terminations for Electrical Copper Conductors—Safety Requirements First Edition. 50 pp.

—Electric Terminals—Twist On

BSI BS EN 60998-2-4-93. 1993 Connecting Devices for Low-Voltage Circuits for Household and Similar Purposes Part 2-4: Particular Requirements for Twist-on Connecting Devices (IEC 998-2-4: 1993) (E). 27 pp.

CENELEC EN 60998-2-4-93. Connecting Devices for Low-Voltage Circuits for Household and Similar Purposes Part 2-4: Particular Requirements for Twist-on Connecting Devices (IEC 998-2-4: 1993). 22 pp.

IEC 998 Pt 2-4-93. Connecting Devices for Low-Voltage Circuits for Household and Similar Purposes Part 2-4: Particular Requirements for Twist-on Connecting Devices First Edition; (Supersedes 685 Pt 2-4) (CENELEC EN 60998-2-4: 1993). 44 pp.

—Electron Tubes—Chemical Analysis

BSI BS 3839-78. 1978 Oxygen-Free High-Conductivity Copper for Electronic Tubes and Semi-Conductor Devices. 8 pp.

JIS H 1202-82. Method for Chemical Analysis of Oxygen-Free Copper Products for Electron Tubes (R 1988). 54 pp.

JIS H 3510-92. Oxygen Free Copper Sheet, Plate, Strip, Seamless Pipe and Tube, Rod, Bar and Wire for Electron Devices. 19 pp.

—Elongation

DIN VDE 0472 Pt 623-83. Testing of Cables, Wires and Flexible Cords; Elongation at Break of the Copper Conductor (Jan). 4 pp.

—Enameled

BSI PD 6451-70. (WITHDRAWN) 1970 Metric Dimensions for Enamelled Round Copper Wires. 8 pp.

CNS C2084-85. General Rules for Enameled Copper Wires (Jul)(6554). 8 pp.

JIS C 3003-84. Methods of Test for Enamelled Copper and Enamelled Aluminium Wires; (Erratum). 28 pp.

JIS C 3053-88. General Rules for Enameled Copper Wires.

JIS C 3053-75. General Rules for Enameled Copper Wires (R 1983). 16 pp.

—Enameled—Fiberglass—Braided—Windings

BSI BS 6811: Sec 7.0-93. 1993 Winding Wires Part 7: Specifications for Particular Types of Glass-Fibre Braided Bare or Enamelled Rectangular Copper Winding Wires Section 7.0: General Requirements (IEC 317-0-5: 1992). 10 pp.

BSI BS 6811: Sec 7.1-93. 1993 Winding Wires Part 7: Specifications for Particular Types of Glass-Fibre Braided Bare or Enamelled Rectangular Copper Winding Wires Section 7.1: Glass-Fibre Braided, Polyester or Polyesterimide Varnish-Treated, Bare or Enamelled Rectangular. 10 pp.

BSI BS 6811: Sec 7.2-93. 1993 Winding Wires Part 7: Specifications for Particular Types of Glass-Fibre Braided Bare or Enamelled Rectangular Copper Winding Wires Section 7.2: Glass-Fibre Braided, Silicone Varhish-Treated, Bare or Enamelled Rectangular Copper Wire,. 10 pp.

IEC 317 Pt 0-5-92. Specifications for Particular Types of Winding Wires Part 0: General Requirements Section 5—Glass-Fibre Braided, Bare or Enamelled Rectangular Copper Wire First Edition. 46 pp.

Copper Conductors *(Cont.)*

—Enameled—Fiberglass Covered—Windings

BSI BS 6811: Sec 6.0-93. 1993 Winding Wires Part 6: Specifications for Particular Types of Glass-Fibre Wound Bare or Enamelled Rectangular Copper Winding Wires Section 6.0: General Requirements (IEC 317-0-4: 1990). 23 pp.

BSI BS 6811: Sec 6.0-01. 1993 Amd 1 Winding Wires Part 6: Specifications for Particular Types of Glass-Fibre Wound Bare or Enamelled Rectangular Copper Winding Wires Section 6.0: General Requirements (IEC 317-0-4: 1990) (AMD 7906) September 15, 1993 (E). 24 pp.

BSI BS 6811: Sec 6.3-93. 1993 Winding Wires Part 6: Specifications for Particular Types of Glass-Fibre Wound Bare or Enamelled Rectangular Copper Winding Wires Section 6.3: Glass-Fibre Wound, Silicone Varnish-Treated, Bare or Enamelled Rectangular Copper Wire, Temperature. 13 pp.

CENELEC HD 555.0.4 S1-92. Specifications for Particular Types of Winding Wires Part 0: General Requirements Section 4: Glass-Fibre Wound Bare or Enammelled Rectangular Copper Wire. 5 pp.

CENELEC HD 555.33 S1-92. Specifications for Particular Types of Winding Wires Part 33: Glass-Fibre Wound, Silicone Varnish-Treated, Bare or Enamelled Rectangular Copper Wire, Temperature Index 200. 5 pp.

IEC 317 Pt 0-4-90. Specifications for Particular Types of Winding Wires Part 0: General Requirements Section 4: Glass-Fibre Wound Bare or Enamelled Rectangular Copper Wire First Edition; (Amendment 1-1992) (Amendment 2-1993). 61 pp.

IEC 317 Pt 33-90. Specifications for Particular Types of Winding Wires Part 33: Glass-Fibre Wound, Silicone Varnish-Treated, Bare or Enamelled Rectangular Copper Wire, Temperature Index 200 First Edition. 19 pp.

—Enameled—Silk Covered—Windings

IEC 317 Pt 11-90. Specifications for Particular Types of Winding Wires Part 11: Bunched Solderable Polyurethane Enamelled Round Copper Wires, Class 130, with Silk Covering Second Edition; (Amendment 1-1993). 47 pp.

—Enameled—Windings

BSI BS 6811: Sec 3.0-93. 1993 Winding Wires Part 3: Specifications for Particular Types of Enamelled Round Copper Winding Wires Section 3.0: General Requirements (IEC 317-0-1: 1990) (E). 28 pp.

BSI BS 6811: Sec 3.0-01. 1993 Amd 1 Winding Wires Part 3: Specifications for Particular Types of Enamelled Round Copper Winding Wires Section 3.0: General Requirements (IEC 317-0-1: 1990) (AMD 7903) September 15, 1993 (E). 30 pp.

BSI BS 6811: Sec 3.5-93. 1993 Winding Wires Part 3: Specifications for Particular Types of Enamelled Round Copper Winding Wires Section 3.5: Polyvinyl Acetal Enamelled Round Copper Wire, Class 105 (IEC 317-1: 1990) (E). 13 pp.

BSI BS 6811: Sec 3.7-93. 1993 Winding Wires Part 3: Specifications for Particular Types of Enamelled Round Copper Winding Wires Section 3.7: Polyvinyl Acetal Enamelled Round Copper Wire, Class 120 (IEC 317-12: 1990). 13 pp.

BSI BS 6811: Sec 3.8-93. 1993 Winding Wires Part 3: Specifications for Particular Types of Enamelled Round Copper Winding Wires Section 3.8: Polyimide Enamelled Round Copper Wire, Class 220 (IEC 317-7: 1990) (E). 13 pp.

BSI BS 6811: Sec 3.11-93. 1993 Winding Wires Part 3: Specifications for Particular Types of Enamelled Round Copper Winding Wires Section 3.11: Solderable Polyesterimide Enamelled Round Copper Wire, Class 180 (IEC 317-23: 1990). 14 pp.

BSI BS 6811: Sec 4.1-93. 1993 Winding Wires Part 4: Specifications for Particular Types of Enamelled Rectangular Copper Winding Wires Section 4.1: General Requirements (IEC 317-0-2: 1990). 23 pp.

BSI BS 6811: Sec 4.1-91. 1991 Winding Wires Part 4: Specifications for Particular Types of Rectangular Winding Wires Section 4.1: General Requirements (IEC 317-0-2: 1990). 21 pp.

BSI BS 6811: Sec 4.5-93. 1993 Winding Wires Part 4: Specifications for Particular Types of Enamelled Rectangular Copper Winding Wires Section 4.5: Polyvinyl Acetal Enamelled Rectangular Copper Wire, Class 105 (IEC 317-17: 1990). 13 pp.

BSI BS 6811: Sec 4.6-93. 1993 Winding Wires Part 4: Specifications for Particular Types of Enamelled Rectangular Copper Winding Wires Section 4.6: Polyvinyl Acetal Enamelled Rectangular Copper Wire, Class 120 (IEC 317-18: 1990). 13 pp.

BSI BS 6811: Sec 4.7-93. 1993 Winding Wires Part 4: Specifications for Particular Types of Enamelled Rectangular Copper Winding Wires Section 4.7: Polyimide Enamelled Rectangular Copper Wire, Class 220 (IEC 317-30: 1990). 12 pp.

BSI BS 6811: Sec 6.0-93. 1993 Winding Wires Part 6: Specifications for Particular Types of Glass-Fibre Wound Bare or Enamelled Rectangular Copper Winding Wires Section 6.0: General Requirements (IEC 317-0-4: 1990). 23 pp.

Copper Conductors *(Cont.)*

—Enameled—Windings *(Cont.)*

BSI BS 6811: Sec 6.0-01. 1993 Amd 1 Winding Wires Part 6: Specifications for Particular Types of Glass-Fibre Wound Bare or Enamelled Rectangular Copper Winding Wires Section 6.0: General Requirements (IEC 317-0-4: 1990) (AMD 7906) September 15, 1993 (E). 24 pp.

BSI BS 6811: Sec 6.1-93. 1993 Winding Wires Part 6: Specifications for Particular Types of Glass-Fibre Wound Bare or Enamelled Rectangular Copper Winding Wires Section 6.1: Glass-Fibre Wound, Polyester or Polyesterimide Varnish-Treated, Bare or Enamelled Rectangular. 13 pp.

BSI BS 6811: Sec 6.2-93. 1993 Winding Wires Part 6: Specifications for Particular Types of Glass-Fibre Wound Bare or Enamelled Rectangular Copper Winding Wires Section 6.2: Glass-Fibre Wound, Polyester or Polyesterimide Varnish-Treated, Bare or Enamelled Rectangular. 13 pp.

BSI BS 6811: Sec 6.3-93. 1993 Winding Wires Part 6: Specifications for Particular Types of Glass-Fibre Wound Bare or Enamelled Rectangular Copper Winding Wires Section 6.3: Glass-Fibre Wound, Silicone Varnish-Treated, Bare or Enamelled Rectangular Copper Wire, Temperature. 13 pp.

DIN ENGL 46416 Pt 4-77. Winding Wires; Round Copper Wires, Insulated; Enamelled, Heat Resistant with a Temperature Index of 155, Type W 155, Technical Conditions of Delivery (Apr). 7 pp.

DIN ENGL 46435-77. Winding Wires; Round Copper Wires, Insulated; Enamelled; Dimensions and DC Resistances (Apr). 4 pp.

DIN ENGL 46453 Pt 1-77. Winding Wires; Round Copper Wires, Enamelled; Test and Measuring Methods (Apr). 15 pp.

DIN ENGL 46453 Pt 2-77. Winding Wires; Round Copper Wires, Enamelled; Determination of Limiting Temperature (Apr). 5 pp.

DIN ENGL 46453 Pt 5-77. Winding Wires; Round Copper Wires, Enamelled; Testing of Resistance to Transformer Oil in the Presence of Water (Apr). 2 pp.

IEC 317 Pt 0-1-90. Specifications for Particular Types of Winding Wires Part 0: General Requirements Section 1: Enamelled Round Copper Wire First Edition; (Corrigendum—March 1991) (Amendment 1-1992) (Amendment 2-1993). 78 pp.

IEC 317 Pt 0-2-90. Specifications for Particular Types of Winding Wires Part 0: General Requirements Section 2: Enamelled Rectangular Copper Wire First Edition; (Amendment 1-1992) (Amendment 2-1993). 59 pp.

IEC 317 Pt 0-4-90. Specifications for Particular Types of Winding Wires Part 0: General Requirements Section 4: Glass-Fibre Wound Bare or Enamelled Rectangular Copper Wire First Edition; (Amendment 1-1992) (Amendment 2-1993). 61 pp.

IEC 317 Pt 1-90. Specifications for Particular Types of Winding Wires Part 1: Polyvinyl Acetal Enamelled Round Copper Wire, Class 105 Third Edition. 19 pp.

IEC 317 Pt 3-90. Specifications for Particular Types of Winding Wires Part 3: Polyester Enamelled Round Copper Wire, Class 155 Second Edition. 19 pp.

IEC 317 Pt 7-90. Specifications for Particular Types of Winding Wires Part 7: Polyimide Enamelled Round Copper Wire, Class 220 Third Edition. 19 pp.

IEC 317 Pt 8-90. Specifications for Particular Types of Winding Wires Part 8: Polyesterimide Enamelled Round Copper Wire, Class 180 Third Edition. 21 pp.

IEC 317 Pt 12-90. Specifications for Particular Types of Winding Wires Part 12: Polyvinyl Acetal Enamelled Round Copper Wire, Class 120 Second Edition. 19 pp.

IEC 317 Pt 17-90. Specifications for Particular Types of Winding Wires Part 17: Polyvinyl Acetal Enamelled Rectangular Copper Wire, Class 105 Second Edition. 19 pp.

IEC 317 Pt 18-90. Specifications for Particular Types of Winding Wires Part 18: Polyvinyl Acetal Enamelled Rectangular Copper Wire, Class 120 Second Edition. 19 pp.

IEC 317 Pt 21-90. Specifications for Particular Types of Winding Wires Part 21: Solderable Polyurethane Enamelled Round Copper Wire Overcoated with Polyamide, Class 155 Second Edition. 23 pp.

IEC 317 Pt 22-90. Specifications for Particular Types of Winding Wires Part 22: Polyester or Polyesterimide Enamelled Round Copper Wire Overcoated with Polyamide, Class 180 Second Edition. 23 pp.

IEC 317 Pt 23-90. Specifications for Particular Types of Winding Wires Part 23: Solderable Polyesterimide Enamelled Round Copper Wire, Class 180 Second Edition. 21 pp.

IEC 317 Pt 28-90. Specifications for Particular Types of Winding Wires Part 28: Polyesterimide Enamelled Rectangular Copper Wire, Class 180 First Edition. 17 pp.

INTERNATIONAL AND NON-U.S. NATIONAL STANDARDS
SUBJECT INDEX
Copper

Copper Conductors *(Cont.)*
—Enameled—Windings *(Cont.)*
IEC 317 Pt 29-90. Specifications for Particular Types of Winding Wires Part 29: Polyester or Polyesterimide Overcoated with Polyamide-imide Enamelled Rectangular Copper Wire, Class 200 First Edition. 17 pp.

IEC 317 Pt 30-90. Specifications for Particular Types of Winding Wires Part 30: Polyimide Enamelled Rectangular Copper Wire, Class 220 First Edition. 17 pp.

IEC 317 Pt 31-90. Specifications for Particular Types of Winding Wires Part 31: Glass-Fibre Wound, Polyester or Polyesterimide Varnish-Treated, Bare or Enamelled Rectangular Copper Wire, Temperature Index 180 First Edition. 19 pp.

IEC 317 Pt 32-90. Specifications for Particular Types of Winding Wires Part 32: Glass-Fibre Wound, Polyester or Polyesterimide Varnish-Treated, Bare or Enamelled Rectangular Copper Wire, Temperature Index 155 First Edition. 19 pp.

IEC 317 Pt 33-90. Specifications for Particular Types of Winding Wires Part 33: Glass-Fibre Wound, Silicone Varnish-Treated, Bare or Enamelled Rectangular Copper Wire, Temperature Index 200 First Edition. 19 pp.

IEC 317 Pt 34-90. Specifications for Particular Types of Winding Wires Part 34: Polyester Enamelled Round Copper Wire, Class 130 First Edition. 19 pp.

MOD UK DSTAN 61-12: Part 14-76. Wires, Cords, and Cables, Electrical-Metric Units Part 14: Wires, Electrical (Enamel Insulated Round Winding Wires and Non-Insulated Tinned Copper Wires) Issue 1. 14 pp.

MOD UK DSTAN 61-12: Part 14-01. Wires, Cords, and Cables, Electrical-Metric Units Part 14: Wires, Electrical (Enamel Insulated Round Winding Wires and Non-Insulated Tinned Copper Wires) Issue 1; Amendment 1. 15 pp.

SAA AS 1194.1-84. Winding Wires—Part 1: Enamelled Round Copper Winding Wires. 48 pp.

SAA AS 1194.2-83. Winding Wires—Part 2: Enamelled Rectangular Copper Winding Wires. 28 pp.

—Enameled—Windings—Refrigeration Equipment
IEC 317 Pt 10-72. Specifications for Particular Types of Winding Wires Part 10: Enamelled Round Copper Wires with a Temperature Index of 180 for Use in Refrigerant Systems First Edition; (Supplement A-1978). 57 pp.

—Enameled—Windings—Self Fluxing
BSI BS 6811: Sec 3.1-93. 1993 Winding Wires Part 3: Specifications for Particular Types of Enamelled Round Copper Winding Wires Section 3.1: Solderable Polyurethane Enamelled Round Copper Wire, Class 130 (IEC 317-4: 1990). 14 pp.

BSI BS 6811: Sec 3.1-91. 1991 Winding Wires Part 3: Specifications for Particular Types of Round Copper Winding Wires Section 3.1: Polyurethane Base Enamel with Solderable Properties. Class 130. 15 pp.

BSI BS 6811: Sec 3.4-93. 1993 Winding Wires Part 3: Specifications for Particular Types of Enamelled Round Copper Winding Wires Section 3.4: Solderable Polyurethane Enamelled Round Copper Wire, Class 130, with a Bonding Layer (IEC 317-2: 1990) (E). 14 pp.

BSI BS 6811: Sec 3.9-93. 1993 Winding Wires Part 3: Specifications for Particular Types of Enamelled Round Copper Winding Wires Section 3.9: Solderable Polyurethane Enamelled Round Copper Wire, Class 155 (IEC 317-20: 1990). 14 pp.

CENELEC HD 555.2 S1-90. Specifications for Particular Types of Winding Wires Part 2: Solderable Polyurethane Enamelled Round Copper Winding Wire, Class 130, with a Bonding Layer. 4 pp.

CENELEC HD 555.2 S2-92. Specifications for Particular Types of Winding Wires Part 2: Solderable Polyurethane Enamelled Round Copper Wire, Class 130, with a Bonding Layer. 5 pp.

CENELEC HD 555.4 S1-90. Specifications for Particular Types of Winding Wires Part 4: Solderable Polyurethane Enamelled Round Copper Winding Wire, Class 130. 3 pp.

CENELEC HD 555.4 S2-92. Specifications for Particular Types of Winding Wires Part 4: Solderable Polyurethane Enamelled Round Copper Wire, Class 130. 5 pp.

CENELEC HD 555.19 S1-90. Specifications for Particular Types of Winding Wires Part 19: Solderable Polyurethane Enamelled Round Copper Winding Wire Overcoated with Polyamide, Class 130. 3 pp.

CENELEC HD 555.20 S1-90. Specifications for Particular Types of Winding Wires Part 20: Solderable Polyurethene Enamelled Round Copper Winding Wire, class 155. 3 pp.

Copper Conductors *(Cont.)*
—Enameled—Windings—Self Fluxing *(Cont.)*
IEC 317 Pt 2-90. Specifications for Particular Types of Winding Wires Part 2: Solderable Polyurethane Enamelled Round Copper Wire, Class 130, with a Bonding Layer Third Edition. 21 pp.

IEC 317 Pt 4-90. Specifications for Particular Types of Winding Wires Part 4: Solderable Polyurethane Enamelled Round Copper Wire, Class 130 Third Edition. 21 pp.

IEC 317 Pt 19-90. Specifications for Particular Types of Winding Wires Part 19: Solderable Polyurethane Enamelled Round Copper Wire Overcoated with Polyamide, Class 130 Second Edition. 23 pp.

IEC 317 Pt 20-90. Specifications for Particular Types of Winding Wires Part 20: Solderable Polyurethane Enamelled Round Copper Wire, Class 155 Second Edition. 21 pp.

—Fiber Insulated—Windings
JIS C 3006-83. Methods of Test for Fiber or Paper Insulated Copper and Aluminum Winding Wires. 9 pp.

JIS C 3204-88. Fiber or Paper Covered Copper Winding Wires. 18 pp.

—Fiberglass Coated
BSI BS 4799: Part 1-72. 1972 Varnish-Bonded Glass-Lapped Copper Conductors Part 1: Round Wire. 21 pp.

BSI BS 4799: Part 2-72. (WITHDRAWN) 1972 Varnish-Bonded Glass-Lapped Copper Conductors Part 2: Rectangular Conductors (Refer to IEC 317-31, 32 and 33). 18 pp.

—Fiberglass Covered
CNS C2055-86. Glass-Fiber-Covered Round Copper Wire (Jan)(3206).

CNS C2056-86. Double Glass-Fiber-Covered Rectangular Copper Wire (Jan)(3207).

—Fiberglass Covered—Braided
BSI BS 4801: Part 2-72. 1972 Varnish-Bonded Glass Braided Copper Conductors Part 2: Rectangular Conductors. 18 pp.

—Fiberglass Covered—Braided—Windings
BSI BS 6811: Sec 7.0-93. 1993 Winding Wires Part 7: Specifications for Particular Types of Glass-Fibre Braided Bare or Enamelled Rectangular Copper Winding Wires Section 7.0: General Requirements (IEC 317-0-5: 1992). 10 pp.

BSI BS 6811: Sec 7.1-93. 1993 Winding Wires Part 7: Specifications for Particular Types of Glass-Fibre Braided Bare or Enamelled Rectangular Copper Winding Wires Section 7.1: Glass-Fibre Braided, Polyester or Polyesterimide Varnish-Treated, Bare or Enamelled Rectangular. 10 pp.

BSI BS 6811: Sec 7.2-93. 1993 Winding Wires Part 7: Specifications for Particular Types of Glass-Fibre Braided Bare or Enamelled Rectangular Copper Winding Wires Section 7.2: Glass-Fibre Braided, Silicone Varhish-Treated, Bare or Enamelled Rectangular Copper Wire,. 10 pp.

IEC 317 Pt 0-5-92. Specifications for Particular Types of Winding Wires Part 0: General Requirements Section 5—Glass-Fibre Braided, Bare or Enamelled Rectangular Copper Wire First Edition. 46 pp.

IEC 317 Pt 39-92. Specifications for Particular Types of Winding Wires Part 39: Glass-Fibre Braided, Polyester or Polyesterimide Varnish-Treated, Bare or Enamelled Rectangular Copper Wire, Temperature Index 80 First Edition. 20 pp.

IEC 317 Pt 40-92. Specifications for Particular Types of Winding Wires Part 40: Glass-Fibre Braided, Silicone Varnish-Treated, Bare or Enamelled Rectangular Copper Wire, Temperature Index 200 First Edition. 20 pp.

—Fiberglass Covered—Windings
BSI BS 6811: Sec 6.0-93. 1993 Winding Wires Part 6: Specifications for Particular Types of Glass-Fibre Wound Bare or Enamelled Rectangular Copper Winding Wires Section 6.0: General Requirements (IEC 317-0-4: 1990). 23 pp.

BSI BS 6811: Sec 6.0-01. 1993 Amd 1 Winding Wires Part 6: Specifications for Particular Types of Glass-Fibre Wound Bare or Enamelled Rectangular Copper Winding Wires Section 6.0: General Requirements (IEC 317-0-4: 1990) (AMD 7906) September 15, 1993 (E). 24 pp.

BSI BS 6811: Sec 6.3-93. 1993 Winding Wires Part 6: Specifications for Particular Types of Glass-Fibre Wound Bare or Enamelled Rectangular Copper Winding Wires Section 6.3: Glass-Fibre Wound, Silicone Varnish-Treated, Bare or Enamelled Rectangular Copper Wire, Temperature. 13 pp.

CENELEC HD 555.33 S1-92. Specifications for Particular Types of Winding Wires Part 33: Glass-Fibre Wound, Silicone Varnish-Treated, Bare or Enamelled Rectangular Copper Wire, Temperature Index 200. 5 pp.

Copper Conductors *(Cont.)*
—Fiberglass Covered—Windings *(Cont.)*
IEC 317 Pt 0-4-90. Specifications for Particular Types of Winding Wires Part 0: General Requirements Section 4: Glass-Fibre Wound Bare or Enamelled Rectangular Copper Wire First Edition; (Amendment 1-1992) (Amendment 2-1993). 61 pp.

IEC 317 Pt 33-90. Specifications for Particular Types of Winding Wires Part 33: Glass-Fibre Wound, Silicone Varnish-Treated, Bare or Enamelled Rectangular Copper Wire, Temperature Index 200 First Edition. 19 pp.

JIS C 3204-88. Fiber or Paper Covered Copper Winding Wires. 18 pp.

—Fire Testing
BSI BS 4066: Part 2-89. 1989 Tests on Electric Cables Under Fire Conditions Part 2: Method of Test on a Single Small Vertical Insulated Wire or Cable. 8 pp.

IEC 332 Pt 2-89. Tests on Electric Cables Under Fire Conditions Part 2: Test on a Single Small Vertical Insulated Copper Wire or Cable First Edition. 13 pp.

—Fluorescent Lighting
CNS C2072-89. 1000 V Grade Insulated Wires for Fluorescent Lamps (Jul)(6063). 6 pp.

—Hard Drawn—Overhead Line
SAA AS 1736-91. Conductors—Bare Overhead—Hard-Drawn Copper.

SAA AS 1746-91. Conductors—Bare Overhead—Hard-Drawn Copper.

—Hard Drawn—Telegraphy
BSI BS 174-70. (WITHDRAWN) 1970 Hard-Drawn Copper and Copper-Cadmium Wire for Telegraph and Telephone Purposes. 18 pp.

—Hard Drawn—Telephone Systems
BSI BS 174-70. (WITHDRAWN) 1970 Hard-Drawn Copper and Copper-Cadmium Wire for Telegraph and Telephone Purposes. 18 pp.

—Heat Resistant—Aircraft
ISO 2032-73. Heat-Resisting Equipment Wires for Aircraft First Edition. 6 pp.

—Lead Coated
BSI BS 4393-91. 1991 Solderable Tin or Tin-Lead Coated Wires for Component Terminations. 18 pp.

—Local Cables
IEC 488-74. Dimensions of Copper Conductors in Local Cables First Edition. 4 pp.

—Oleoresinous Enameled
CNS C2038-84. Oleo-Resinous Enameled Copper Wires (Dec)(2182). 8 pp.

—Overhead Power Lines
BSI BS 125-70. 1970 Hard-Drawn Copper and Copper-Cadmium Conductors for Overhead Power Transmission Purposes. 20 pp.

BSI BS 2755-70. 1970 Copper and Copper-Cadmium Stranded Conductors for Overhead Electric Traction Systems. 18 pp.

—Paper Covered
BSI BS 4653: Part 2-70. 1970 Paper Covered Copper Conductors Part 2: Rectangular Conductors. 12 pp.

BSI BS 4927: Part 1-74. 1974 Paper-, Cotton-, Silk- or Rayon-Covered Copper Conductors Part 1: Round Wire. 18 pp.

—Paper Covered—Windings
BSI BS 6811: Sec 8.1-93. 1993 Winding Wires Part 8: Specifications for Particular Types of Rectangular Copper Winding Wires Section 8.1: Paper Covered Rectangular Copper Wire (IEC 317-27: 1990). 23 pp.

BSI BS 6811: Sec 8.1-01. 1993 Amd 1 Winding Wires Part 8: Specifications for Particular Types of Rectangular Copper Winding Wires Section 8.1: Paper Covered Rectangular Copper Wire (IEC 317-27: 1990) (AMD 7907) September 15, 1993 (E). 25 pp.

CENELEC HD 555.27 S1-90. Specification for Particular Types of Winding Wires Part 27: Paper Covered Rectangular Copper Winding Wire. 3 pp.

CENELEC HD 555.27 S2-92. Specifications for Particular Types of Winding Wires Part 27: Paper Covered Rectangular Copper Wire. 5 pp.

IEC 317 Pt 27-90. Specifications for Particular Types of Winding Wires Part 27: Paper Covered Rectangular Copper Wire Second Edition; (Amendment 1-1993). 45 pp.

—Paper Insulated—Windings
JIS C 3006-83. Methods of Test for Fiber or Paper Insulated Copper and Aluminum Winding Wires. 9 pp.

INDUSTRY STANDARDS

Copper
INTERNATIONAL AND NON-U.S. NATIONAL STANDARDS
SUBJECT INDEX

Copper Conductors *(Cont.)*

—Polyamide-imide Enameled—Windings

BSI BS 6811: Sec 3.3-93. 1993 Winding Wires Part 3: Spec. for Particular Types of Enamelled Round Copper Winding Wires Sec. 3.3: Polyester or Polyesterimide Overcoated with Polyamide-imide Enamelled Round Copper Wire, Class 200 (IEC 317-13: 1990). 14 pp.

BSI BS 6811: Sec 3.3-89. 1989 Winding Wires Part 3: Particular Types of Round Copper Winding Section 3.3: Polyester or Polyesterimide, Coated with Polyamide-Imide Enamel. Class 200. 15 pp.

BSI BS 6811: Sec 3.12-93. 1993 Winding Wires Part 3: Specifications for Particular Types of Enamelled Round Copper Winding Wires Section 3.12: Polyamide-Imide Enamelled Round Copper Wire, Class 200 (IEC 317-26: 1990). 13 pp.

BSI BS 6811: Sec 3.16-93. 1993 Winding Wires Part 3: Specifications for Particular Types of Enamelled Round Copper Winding Wires Section 3.16: Polyester or Polyesterimide Overcoated with Polyamide-Imide Enamelled Round Copper Wire, Class 200, with. 11 pp.

CENELEC HD 555.13 S1-90. Specifications for Particular Types of Winding Wires Part 13: Polyester (imide) Overcoated with Polyamide-Imide, Enamelled Round Copper Winding Wire, Class 200. 3 pp.

CENELEC HD 555.13 S2-92. Specifications for Particular Types of Winding Wires Part 13: Polyester or Polyesterimide Overcoated with Polyamide-Imide Enamelled Round Copper Wire, Class 200. 5 pp.

CENELEC HD 555.26 S1-90. Specifications for Particular Types of Winding Wires Part 26: Polyamide-Imide Enamelled Round Copper Winding Wire, Class 200. 3 pp.

CENELEC HD 555.26 S2-92. Specifications for Particular Types of Winding Wires Part 26: Polyamide-Imide Enamelled Round Copper Wire, Class 200. 5 pp.

IEC 317 Pt 26-90. Specifications for Particular Types of Winding Wires Part 26: Polyamide-Imide Enamelled Round Copper Wire, Class 200 Second Edition. 19 pp.

IEC 317 Pt 38-92. Specifications for Particular Types of Winding Wires Part 38: Polyester or Polyesterimide Overcoated with Polyamide-Imide Enamelled Round Copper Wire, Class 200, with a Bonding Layer First Edition. 22 pp.

—Polyester-amide-imide Enameled

CNS C2169-85. Polyester Amide Imide Enameled Wires (Oct)(11383).

—Polyester Enameled

BSI BS 4720: Part 2-71. (WITHDRAWN) 1971 Enamelled Copper Conductors Temperature Index 155 (Modified Polyester Base) Part 2: Rectangular Conductors (Refer to IEC 317-16). 26 pp.

CNS C2039-85. Polyester Enameled Copper Wires (Jul)(2183). 10 pp.

—Polyester Enameled—Fiberglass Covered—Windings

BSI BS 6811: Sec 6.1-93. 1993 Winding Wires Part 6: Specifications for Particular Types of Glass-Fibre Wound Bare or Enamelled Rectangular Copper Winding Wires Section 6.1: Glass-Fibre Wound, Polyester or Polyesterimide Varnish-Treated, Bare or Enamelled Rectangular. 13 pp.

BSI BS 6811: Sec 6.2-93. 1993 Winding Wires Part 6: Specifications for Particular Types of Glass-Fibre Wound Bare or Enamelled Rectangular Copper Winding Wires Section 6.2: Glass-Fibre Wound, Polyester or Polyesterimide Varnish-Treated, Bare or Enamelled Rectangular. 13 pp.

—Polyester Enameled—Windings

BSI BS 6811: Sec 3.10-93. 1993 Winding Wires Part 3: Specifications for Particular Types of Enamelled Round Copper Winding Wires Section 3.10: Polyester or Polyesterimide Enamelled Round Copper Wire Overcoated with Polyamide, Class 180 (IEC 317-22: 1990). 15 pp.

BSI BS 6811: Sec 3.16-93. 1993 Winding Wires Part 3: Specifications for Particular Types of Enamelled Round Copper Winding Wires Section 3.16: Polyester or Polyesterimide Overcoated with Polyamide-Imide Enamelled Round Copper Wire, Class 200, with. 11 pp.

BSI BS 6811: Sec 4.4-93. 1993 Winding Wires Part 4: Specifications for Particular Types of Enamelled Rectangular Copper Wire Winding Wires Section 4.4: Polyester Enamelled Rectangular Copper Wire, Class 155 (IEC 317-16: 1990). 12 pp.

CENELEC HD 555.3 S1-92. Specifications for Particular Types of Winding Wires Part 3: Polyester Enammelled Round Copper Wire, Class 155. 5 pp.

CENELEC HD 555.16 S1-90. Specifications for Particular Types of Winding Wires Part 16: Polyester Enamelled Rectangular Copper Winding Wire, Class 155. 3 pp.

Copper Conductors *(Cont.)*

—Polyester Enameled—Windings *(Cont.)*

CENELEC HD 555.16 S2-92. Specifications for Particular Types of Winding Wires Part 16: Polyester Enammelled Rectangular Copper Wire, Class 155. 5 pp.

CENELEC HD 555.22 S1-90. Specifications for Particular Types of Winding Wires Part 22: Polyester (Imide) Enamelled Round Copper Winding Wire Overcoated with Polyamide, Class 180. 3 pp.

CENELEC HD 555.22 S2-92. Specifications for Particular Types of Winding Wires Part 22: Polyester or Polyesterimide Enamelled Round Copper Wire Overcoated with Polyamide, Class 180. 5 pp.

CENELEC HD 555.34 S1-92. Specifications for Particular Types of Winding Wires Part 34: Polyester Enamelled Round Copper Wire, Class 130. 5 pp.

IEC 317 Pt 16-90. Specifications for Particular Types of Winding Wires Part 16: Polyester Enamelled Rectangular Copper Wire, Class 155 Second Edition. 17 pp.

IEC 317 Pt 34-90. Specifications for Particular Types of Winding Wires Part 34: Polyester Enamelled Round Copper Wire, Class 130 First Edition. 19 pp.

IEC 317 Pt 38-92. Specifications for Particular Types of Winding Wires Part 38: Polyester or Polyesterimide Overcoated with Polyamide-Imide Enamelled Round Copper Wire, Class 200, with a Bonding Layer First Edition. 22 pp.

—Polyester—Fiberglass Covered—Windings

BSI BS 6811: Sec 6.1-93. 1993 Winding Wires Part 6: Specifications for Particular Types of Glass-Fibre Wound Bare or Enamelled Rectangular Copper Winding Wires Section 6.1: Glass-Fibre Wound, Polyester or Polyesterimide Varnish-Treated, Bare or Enamelled Rectangular. 13 pp.

BSI BS 6811: Sec 6.2-93. 1993 Winding Wires Part 6: Specifications for Particular Types of Glass-Fibre Wound Bare or Enamelled Rectangular Copper Winding Wires Section 6.2: Glass-Fibre Wound, Polyester or Polyesterimide Varnish-Treated, Bare or Enamelled Rectangular. 13 pp.

—Polyester Insulated—Enameled

CNS C2171-85. Modified Polyester Insulated Polyamideimide Overcoated Enameled Copper Wires (Oct)(11385).

—Polyester Insulated—Polyamide-Imide Enameled

CNS C2171-85. Modified Polyester Insulated Polyamideimide Overcoated Enameled Copper Wires (Oct)(11385).

—Polyester-Nylon Enameled

CNS C2168-85. Polyester-Nylon Enameled Copper Wires (Oct)(11382).

—Polyesterimide Enameled

CNS C2100-86. Polyester-Imide Enameled Copper Wire (Jan)(7966).

—Polyesterimide Enameled—Fiberglass Covered—Windings

BSI BS 6811: Sec 6.1-93. 1993 Winding Wires Part 6: Specifications for Particular Types of Glass-Fibre Wound Bare or Enamelled Rectangular Copper Winding Wires Section 6.1: Glass-Fibre Wound, Polyester or Polyesterimide Varnish-Treated, Bare or Enamelled Rectangular. 13 pp.

BSI BS 6811: Sec 6.2-93. 1993 Winding Wires Part 6: Specifications for Particular Types of Glass-Fibre Wound Bare or Enamelled Rectangular Copper Winding Wires Section 6.2: Glass-Fibre Wound, Polyester or Polyesterimide Varnish-Treated, Bare or Enamelled Rectangular. 13 pp.

CENELEC HD 555.31 S1-92. Specifications for Particular Types of Winding Wires Part 31: Glass-Fibre Wound, Polyester or Polyesterimide Varnish-Treated, Bare or Enamelled Rectangular Copper Wire, Temperature Index 180. 5 pp.

CENELEC HD 555.32 S1-92. Specifications for Particular Types of Winding Wires Part 32: Glass-Fibre Wound, Polyester or Polyesterimide Varnish-Treated, Bare or Enamelled Rectangular Copper Wire, Temperature Index 155. 5 pp.

IEC 317 Pt 31-90. Specifications for Particular Types of Winding Wires Part 31: Glass-Fibre Wound, Polyester or Polyesterimide Varnish-Treated, Bare or Enamelled Rectangular Copper Wire, Temperature Index 180 First Edition. 19 pp.

IEC 317 Pt 32-90. Specifications for Particular Types of Winding Wires Part 32: Glass-Fibre Wound, Polyester or Polyesterimide Varnish-Treated, Bare or Enamelled Rectangular Copper Wire, Temperature Index 155 First Edition. 19 pp.

Copper Conductors *(Cont.)*

—Polyesterimide Enameled—Windings

BSI BS 6811: Sec 3.2-93. 1993 Winding Wires Part 3: Specifications for Particular Types of Enamelled Round Copper Winding Wires Section 3.2: Polyesterimide Enamelled Round Copper Wire, Class 180 (IEC 317-8: 1990). 14 pp.

BSI BS 6811: Sec 3.10-93. 1993 Winding Wires Part 3: Specifications for Particular Types of Enamelled Round Copper Winding Wires Section 3.10: Polyester or Polyesterimide Enamelled Round Copper Wire Overcoated with Polyamide, Class 180 (IEC 317-22: 1990). 15 pp.

BSI BS 6811: Sec 3.11-93. 1993 Winding Wires Part 3: Specifications for Particular Types of Enamelled Round Copper Winding Wires Section 3.11: Solderable Polyesterimide Enamelled Round Copper Wire, Class 180 (IEC 317-23: 1990). 14 pp.

BSI BS 6811: Sec 3.14-93. 1993 Winding Wires Part 3: Specifications for Particular Types of Enamelled Round Copper Winding Wires Section 3.14: Solderable Polyesterimide Enamelled Round Copper Wire, Class 180, with a Bonding Layer (IEC 317-36: 1992). 11 pp.

BSI BS 6811: Sec 3.15-93. 1993 Winding Wires Part 3: Specifications for Particular Types of Enamelled Round Copper Winding Wires Section 3.15: Polyesterimide Enamelled Round Copper Wire, Class 180, with a Bonding Layer (IEC 317-37: 1992). 11 pp.

BSI BS 6811: Sec 3.16-93. 1993 Winding Wires Part 3: Specifications for Particular Types of Enamelled Round Copper Winding Wires Section 3.16: Polyester or Polyesterimide Overcoated with Polyamide-Imide Enamelled Round Copper Wire, Class 200, with. 11 pp.

BSI BS 6811: Sec 4.2-91. 1991 Winding Wires Part 4: Specifications for Particular Types of Rect. Winding Wires Section 4.2: Polyester or Polyesterimide Overcoated with Polyamide-Imide Enamel-led Rectangular Copper Wire, Class 200 (IEC 317-29: 1990). 8 pp.

BSI BS 6811: Sec 4.3-91. 1991 Winding Wires Part 4: Specifications for Particular Types of Rectangular Winding Wires Section 4.3: Polyesterimide Base Enamel, Class 180 (IEC 317-28: 1990). 8 pp.

CENELEC HD 555.8 S1-90. Specifications for Particular Types of Winding Wires Part 8: Polyesterimide Enamelled Round Copper Winding Wire, Class 180. 4 pp.

CENELEC HD 555.8 S2-92. Specifications for Particular Types of Winding Wires Part 8: Polyesterimide Enamelled Round Copper Wire, Class 180. 5 pp.

CENELEC HD 555.22 S1-90. Specifications for Particular Types of Winding Wires Part 22: Polyester (Imide) Enamelled Round Copper Winding Wire Overcoated with Polyamide, Class 180. 3 pp.

CENELEC HD 555.22 S2-92. Specifications for Particular Types of Winding Wires Part 22: Polyester or Polyesterimide Enamelled Round Copper Wire Overcoated with Polyamide, Class 180. 5 pp.

CENELEC HD 555.23 S1-90. Specifications for Particular Types of Winding Wires Part 23: Solderable Polyester (Imide) Enamelled Round Copper Winding Wire, Class 180. 3 pp.

CENELEC HD 555.23 S2-92. Specifications for Particular Types of Winding Wires Part 23: Solderable Polyesterimide Enamelled Round Copper Wire, Class 180. 5 pp.

CENELEC HD 555.28 S1-92. Specifications for Particular Types of Winding Wires Part 28: Polyesterimide Enamelled Rectangular Copper Wire, Class 180. 5 pp.

CENELEC HD 555.29 S1-92. Specifications for Particular Types of Winding Wires Part 29: Polyester or Polyesterimide Overcoated with Polyamide-Imide Enamelled Rectangular Copper Wire, Class 200. 5 pp.

CENELEC HD 555.32 S1-92. Specifications for Particular Types of Winding Wires Part 32: Glass-Fibre Wound, Polyester or Polyesterimide Varnish-Treated, Bare or Enamelled Rectangular Copper Wire, Temperature Index 155. 5 pp.

IEC 317 Pt 8-90. Specifications for Particular Types of Winding Wires Part 8: Polyesterimide Enamelled Round Copper Wire, Class 180 Third Edition. 21 pp.

IEC 317 Pt 13-90. Specifications for Particular Types of Winding Wires Part 13: Polyester or Polyesterimide Overcoated with Polyamide-Imide Enamelled Round Copper Wire, Class 200 Second Edition. 21 pp.

IEC 317 Pt 22-90. Specifications for Particular Types of Winding Wires Part 22: Polyester or Polyesterimide Enamelled Round Copper Wire Overcoated with Polyamide, Class 180 Second Edition. 23 pp.

IEC 317 Pt 23-90. Specifications for Particular Types of Winding Wires Part 23: Solderable Polyesterimide Enamelled Round Copper Wire, Class 180 Second Edition. 21 pp.

Copper Conductors (Cont.)

—Polyesterimide Enameled—Windings (Cont.)

IEC 317 Pt 28-90. Specifications for Particular Types of Winding Wires Part 28: Polyesterimide Enamelled Rectangular Copper Wire, Class 180 First Edition. 17 pp.

IEC 317 Pt 29-90. Specifications for Particular Types of Winding Wires Part 29: Polyester or Polyesterimide Overcoated with Polyamide-imide Enamelled Rectangular Copper Wire, Class 200 First Edition. 17 pp.

IEC 317 Pt 32-90. Specifications for Particular Types of Winding Wires Part 32: Glass-Fibre Wound, Polyester or Polyesterimide Varnish-Treated, Bare or Enamelled Rectangular Copper Wire, Temperature Index 155 First Edition. 19 pp.

IEC 317 Pt 36-92. Specifications for Particular Types of Winding Wires Part 36: Solderable Polyesterimide Enamelled Round Copper Wire, Class 180, with a Bonding Layer First Editon. 22 pp.

IEC 317 Pt 37-92. Specifications for Particular Types of Winding Wires Part 37: Polyesterimide Enamelled Round Copper Wire, Class 180, with a Bonding Layer First Edition. 22 pp.

IEC 317 Pt 38-92. Specifications for Particular Types of Winding Wires Part 38: Polyester or Polyesterimide Overcoated with Polyamide-Imide Enamelled Round Copper Wire, Class 200, with a Bonding Layer First Edition. 22 pp.

—Polyesterimide—Fiberglass Covered—Windings

BSI BS 6811: Sec 6.1-93. 1993 Winding Wires Part 6: Specifications for Particular Types of Glass-Fibre Wound Bare or Enamelled Rectangular Copper Winding Wires Section 6.1: Glass-Fibre Wound, Polyester or Polyesterimide Varnish-Treated, Bare or Enamelled Rectangular. 13 pp.

BSI BS 6811: Sec 6.2-93. 1993 Winding Wires Part 6: Specifications for Particular Types of Glass-Fibre Wound Bare or Enamelled Rectangular Copper Winding Wires Section 6.2: Glass-Fibre Wound, Polyester or Polyesterimide Varnish-Treated, Bare or Enamelled Rectangular. 13 pp.

CENELEC HD 555.31 S1-92. Specifications for Particular Types of Winding Wires Part 31: Glass-Fibre Wound, Polyester or Polyesterimide Varnish-Treated, Bare or Enamelled Rectangular Copper Wire, Temperature Index 180. 5 pp.

CENELEC HD 555.32 S1-92. Specifications for Particular Types of Winding Wires Part 32: Glass-Fibre Wound, Polyester or Polyesterimide Varnish-Treated, Bare or Enamelled Rectangular Copper Wire, Temperature Index 155. 5 pp.

IEC 317 Pt 31-90. Specifications for Particular Types of Winding Wires Part 31: Glass-Fibre Wound, Polyester or Polyesterimide Varnish-Treated, Bare or Enamelled Rectangular Copper Wire, Temperature Index 180 First Edition. 19 pp.

IEC 317 Pt 32-90. Specifications for Particular Types of Winding Wires Part 32: Glass-Fibre Wound, Polyester or Polyesterimide Varnish-Treated, Bare or Enamelled Rectangular Copper Wire, Temperature Index 155 First Edition. 19 pp.

—Polyesterimide—Windings

CENELEC HD 555.32 S1-92. Specifications for Particular Types of Winding Wires Part 32: Glass-Fibre Wound, Polyester or Polyesterimide Varnish-Treated, Bare or Enamelled Rectangular Copper Wire, Temperature Index 155. 5 pp.

IEC 317 Pt 32-90. Specifications for Particular Types of Winding Wires Part 32: Glass-Fibre Wound, Polyester or Polyesterimide Varnish-Treated, Bare or Enamelled Rectangular Copper Wire, Temperature Index 155 First Edition. 19 pp.

—Polyimide Enameled—Windings

BSI BS 4663: Part 1-71. (WITHDRAWN) 1971 Amd 1 Enamelled Copper Conductors, Temperature Index 220 (Aromatic Polymide Base) Part 1: Round Wire (Refer to IEC 317-7). 57 pp.

BSI BS 6811: Sec 3.8-93. 1993 Winding Wires Part 3: Specifications for Particular Types of Enamelled Round Copper Winding Wires Section 3.8: Polyimide Enamelled Round Copper Wire, Class 220 (IEC 317-7: 1990) (E). 13 pp.

BSI BS 6811: Sec 4.7-93. 1993 Winding Wires Part 4: Specifications for Particular Types of Enamelled Rectangular Copper Winding Wires Section 4.7: Polyimide Enamelled Rectangular Copper Wire, Class 220 (IEC 317-30: 1990). 12 pp.

CENELEC HD 555.7 S1-90. Specifications for Particular Types of Winding Wires Part 7: Polyimide Enamelled Round Copper Winding Wire, Class 220. 3 pp.

CENELEC HD 555.7 S2-92. Specifications for Particular Types of Winding Wires Part 7: Polyimide Enamelled Round Copper Wire, Class 220. 5 pp.

IEC 317 Pt 7-90. Specifications for Particular Types of Winding Wires Part 7: Polyimide Enamelled Round Copper Wire, Class 220 Third Edition. 19 pp.

IEC 317 Pt 30-90. Specifications for Particular Types of Winding Wires Part 30: Polyimide Enamelled Rectangular Copper Wire, Class 220 First Edition. 17 pp.

—Polyolefin Insulated—Communication Cables

BSI BS 3573-90. 1990 Polyolefin Copper-Conductor Telecommunication Cables. 22 pp.

—Polytetrafluoroethylene Insulated—Aerospace

BSI G 210-70. (WITHDRAWN) 1970 Amd 7 PTFE Insulated Equipment Wires (with Silver Plated Copper Conductors) (Superseded by BS 2G 210: 1991). 20 pp.

BSI 2G 210-91. 1991 PTFE Insulated Equipment Wires and Cables, Single-and Multi-Core, with Silver Plated Copper Conductors (190 Degrees C) or Nickel Plated Copper Conductors (260 Degrees C). 28 pp.

BSI 2G 210-01. 1991 Amd 1 PTFE Insulated Equipment Wires and Cables, Single-and Multi-Core, with Silver Plated Copper Conductors (190 Degrees C) or Nickel Plated Copper Conductors (260 Degrees C) (AMD 7223) July 15, 1992. 29 pp.

—Polytetrafluoroethylene Insulated—Copper Braided

CNS C2161-83. Polytetrafluoroethylene (Teflon) Insulated Copper Braided Electrical Wire (Oct)(10613).

—Polytetrafluoroethylene Insulated—Copper Braided—Polyamide Jacket

CNS C2163-83. Polytetrafluoroethylene (Teflon) Insulated Copper Braided Polyamide Jacketed Electrical Wire (Oct)(10615).

—Polytetrafluoroethylene Insulated—Copper Braided—PVC Jacket

CNS C2164-83. Polytetrafluoroethylene (Teflon) Insulated Copper Braided Polyvinylchloride Jacketed Electrical Wire (Oct)(10616).

—Polyurethane Enameled

CNS C2041-85. Polyurethane Enameled Copper Wires (Jul)(2185).

—Polyurethane Enameled—Self-Bonding

CNS C2074-86. Self-Bonding Polyurethane Enameled Copper Wires (Jan)(6066). 9 pp.

—Polyurethane Enameled—Silk Covered—Windings

BSI BS 6811: Sec 3.6-93. 1993 Winding Wires Part 3: Specifications for Particular Types of Enamelled Round Copper Winding Wires Section 3.6: Bunched Solderable Polyurethane Enamelled Round Copper Wire, Class 130, with Silk Covering (IEC 317-11: 1990). 23 pp.

BSI BS 6811: Sec 3.6-01. 1993 Amd 1 Winding Wires Part 3: Specifications for Particular Types of Enamelled Round Copper Winding Wires Sec. 3.6: Bunched Solderable Polyurethane Enamelled Round Copper Wire, Class 130, with Silk Covering (IEC 317-11: 1990) (AMD 7904) Sept. 15, 1993 (E). 26 pp.

CENELEC HD 555.11 S1-92. Specifications for Particular Types of Winding Wires Part 11: Bunched Solderable Polyurethane Enamelled Round Copper Wires, Class 130, with Silk Covering. 5 pp.

—Polyurethane Enameled—Windings

BSI BS 6811: Sec 3.1-93. 1993 Winding Wires Part 3: Specifications for Particular Types of Enamelled Round Copper Winding Wires Section 3.1: Solderable Polyurethane Enamelled Round Copper Wire, Class 130 (IEC 317-4: 1990). 14 pp.

BSI BS 6811: Sec 3.1-91. 1991 Winding Wires Part 3: Specifications for Particular Types of Round Copper Winding Wires Section 3.1: Polyurethane Base Enamel with Solderable Properties. Class 130. 15 pp.

BSI BS 6811: Sec 3.4-93. 1993 Winding Wires Part 3: Specifications for Particular Types of Enamelled Round Copper Winding Wires Section 3.4: Solderable Polyurethane Enamelled Round Copper Wire, Class 130, with a Bonding Layer (IEC 317-2: 1990) (E). 14 pp.

BSI BS 6811: Sec 3.9-93. 1993 Winding Wires Part 3: Specifications for Particular Types of Enamelled Round Copper Winding Wires Section 3.9: Solderable Polyurethane Enamelled Round Copper Wire, Class 155 (IEC 317-20: 1990). 14 pp.

BSI BS 6811: Sec 3.13-93. 1993 Winding Wires Part 3: Specifications for Particular Types of Enamelled Round Copper Winding Wires Section 3.13: Solderable Polyurethane Enamelled Round Copper Wire, Class 155, with a Bonding Layer (IEC 317-35: 1992). 11 pp.

CENELEC HD 555.2 S1-90. Specifications for Particular Types of Winding Wires Part 2: Solderable Polyurethane Enamelled Round Copper Winding Wire, Class 130, with a Bonding Layer. 4 pp.

CENELEC HD 555.2 S2-92. Specifications for Particular Types of Winding Wires Part 2: Solderable Polyurethane Enamelled Round Copper Wire, Class 130, with a Bonding Layer. 5 pp.

CENELEC HD 555.4 S1-90. Specifications for Particular Types of Winding Wires Part 4: Solderable Polyurethane Enamelled Round Copper Winding Wire, Class 130. 3 pp.

CENELEC HD 555.4 S2-92. Specifications for Particular Types of Winding Wires Part 4: Solderable Polyurethane Enamelled Round Copper Wire, Class 130. 5 pp.

CENELEC HD 555.19 S1-90. Specifications for Particular Types of Winding Wires Part 19: Solderable Polyurethane Enamelled Round Copper Winding Wire Overcoated with Polyamide, Class 130. 3 pp.

CENELEC HD 555.19 S2-92. Specifications for Particular Types of Winding Wires Part 19: Solderable Polyurethane Enamelled Round Copper Wire Overcoated with Polyamide, Class 130. 5 pp.

CENELEC HD 555.20 S1-90. Specifications for Particular Types of Winding Wires Part 20: Solderable Polyurethane Enamelled Round Copper Winding Wire, class 155. 3 pp.

CENELEC HD 555.20 S2-92. Specifications for Particular Types of Winding Wires Part 20: Solderable Polyurethane Enamelled Round Copper Wire, Class 155. 5 pp.

CENELEC HD 555.21 S1-90. Specifications for Particular Types of Winding Wires Part 21: Solderable Polyrethene Enamelled Round Copper Winding Wire Overcoated with Polyamide, Class 155. 3 pp.

CENELEC HD 555.21 S2-92. Specifications for Particular Types of Winding Wires Part 21: Solderable Polyurethane Enamelled Round Copper Wire Overcoated with Polyamide, Class 155. 5 pp.

IEC 317 Pt 2-90. Specifications for Particular Types of Winding Wires Part 2: Solderable Polyurethane Enamelled Round Copper Wire, Class 130, with a Bonding Layer Third Edition. 21 pp.

IEC 317 Pt 4-90. Specifications for Particular Types of Winding Wires Part 4: Solderable Polyurethane Enamelled Round Copper Wire, Class 130 Third Edition. 21 pp.

IEC 317 Pt 19-90. Specifications for Particular Types of Winding Wires Part 19: Solderable Polyurethane Enamelled Round Copper Wire Overcoated with Polyamide, Class 130 Second Edition. 23 pp.

IEC 317 Pt 20-90. Specifications for Particular Types of Winding Wires Part 20: Solderable Polyurethane Enamelled Round Copper Wire, Class 155 Second Edition. 21 pp.

IEC 317 Pt 21-90. Specifications for Particular Types of Winding Wires Part 21: Solderable Polyurethane Enamelled Round Copper Wire Overcoated with Polyamide, Class 155 Second Edition. 23 pp.

IEC 317 Pt 35-92. Specifications for Particular Types of Winding Wires Part 35: Solderable Polyurethane Enamelled Round Copper Wire, Class 155, with a Bonding Layer First Edition. 22 pp.

—Polyurethane-Nylon Enameled

CNS C2170-85. Polyurethane-Nylon Enameled Copper Wires (Oct)(11384).

—Polyvinyl Acetal Enameled

BSI BS 4516: Part 2-71. (WITHDRAWN) 1971 Enamelled Copper Conductors (Polyvinyl Acetal Base with High Mechanical Properties) Part 2: Rectangular Conductors. 19 pp.

—Polyvinyl Acetal Enameled—Windings

BSI BS 6811: Sec 3.5-93. 1993 Winding Wires Part 3: Specifications for Particular Types of Enamelled Round Copper Winding Wires Section 3.5: Polyvinyl Acetal Enamelled Round Copper Wire, Class 105 (IEC 317-1: 1990) (E). 13 pp.

BSI BS 6811: Sec 3.7-93. 1993 Winding Wires Part 3: Specifications for Particular Types of Enamelled Round Copper Winding Wires Section 3.7: Polyvinyl Acetal Enamelled Round Copper Wire, Class 120 (IEC 317-12: 1990). 13 pp.

BSI BS 6811: Sec 4.5-93. 1993 Winding Wires Part 4: Specifications for Particular Types of Enamelled Rectangular Copper Winding Wires Section 4.5: Polyvinyl Acetal Enamelled Rectangular Copper Wire, Class 105 (IEC 317-17: 1990). 13 pp.

Copper Conductors (Cont.)

—Polyvinyl Acetal Enameled—Windings (Cont.)
- BSI BS 6811: Sec 4.6-93. 1993 Winding Wires Part 4: Specifications for Particular Types of Enamelled Rectangular Copper Winding Wires Section 4.6: Polyvinyl Acetal Enamelled Rectangular Copper Wire, Class 120 (IEC 317-18: 1990). 13 pp.
- CENELEC HD 555.1 S1-90. Specifications for Particular Types of Winding Wires Part 1: Polyvinyl Acetal Enamelled Round Copper Winding Wire, Class 105. 4 pp.
- CENELEC HD 555.1 S2-92. Specifications for Particular Types of Winding Wires Part 1: Polyvinyl Acetal Enamelled Round Copper Wire, Class 105. 5 pp.
- CENELEC HD 555.12 S1-90. Specifications for Particular Types of Winding Wires Part 12: Polyvinyl Acetal Enamelled Round Copper Winding Wire, Class 120. 3 pp.
- CENELEC HD 555.12 S2-92. Specifications for Particular Types of Winding Wires Part 12: Polyvinyl Acetal Enamelled Round Copper Wire, Class 120. 5 pp.
- CENELEC HD 555.17 S1-90. Specifications for Particular Types of Winding Wires Part 17: Polyvinyl Acetal Enamelled Retangular Copper Winding Wire, Class 105. 3 pp.
- CENELEC HD 555.17 S2-92. Specifications for Particular Types of Winding Wires Part 17: Polyvinyl Acetal Enamelled Rectangular Copper Wire, Class 105. 5 pp.
- CENELEC HD 555.18 S1-90. Specifications for Particular Types of Winding Wires Part 18: Polyvinyl Acetal Enamelled Rectangular Copper Winding Wire, Class 120. 3 pp.
- CENELEC HD 555.18 S2-92. Specifications for Particular Types of Winding Wires Part 18: Polyvinyl Acetal Enamelled Rectangular Copper Wire, Class 120. 5 pp.
- IEC 317 Pt 1-90. Specifications for Particular Types of Winding Wires Part 1: Polyvinyl Acetal Enamelled Round Copper Wire, Class 105 Third Edition. 19 pp.
- IEC 317 Pt 12-90. Specifications for Particular Types of Winding Wires Part 12: Polyvinyl Acetal Enamelled Round Copper Wire, Class 120 Second Edition. 19 pp.
- IEC 317 Pt 17-90. Specifications for Particular Types of Winding Wires Part 17: Polyvinyl Acetal Enamelled Rectangular Copper Wire, Class 105 Second Edition. 19 pp.
- IEC 317 Pt 18-90. Specifications for Particular Types of Winding Wires Part 18: Polyvinyl Acetal Enamelled Rectangular Copper Wire, Class 120 Second Edition. 19 pp.

—Polyvinyl Enameled
- CNS C2040-85. Polyvinyl Formal Enameled Copper Wires (Jul)(2184). 11 pp.
- CNS C2085-85. Polyvinyl Formal Enameled Rectangular Copper Wire (Jul)(6555). 6 pp.

—PVC Insulated
- CNS C2052-84. Polyvinly Chloride Insulated Copper Wire (Flat or Twist) (May)(2900).

—Rayon Covered
- BSI BS 4927: Part 1-74. 1974 Paper-, Cotton-, Silk-or Rayon-Covered Copper Conductors Part 1: Round Wire. 18 pp.

—Rectangular
- CNS C2068-89. Rectangular Copper Wires for Electrical Purposes (Jul)(5749).

—Resistance—Communication Cables
- IEC 344-80. Guide to the Calculation of Resistance of Plain and Coated Copper Conductors of Low-Frequency Cables and Wires Second Edition; (Amendment 1-1985). 16 pp.

—Semiconductor Devices—Chemical Analysis
- BSI BS 3839-78. 1978 Oxygen-Free High-Conductivity Copper for Electronic Tubes and Semi-Conductor Devices. 8 pp.

—Silk Covered
- BSI BS 4927: Part 1-74. 1974 Paper-, Cotton-, Silk-or Rayon-Covered Copper Conductors Part 1: Round Wire. 18 pp.
- CNS C2093-83. Cotton Covered Copper Wires and Silk Covered Wires (Feb)(7353).

—Silk Covered—Windings
- JIS C 3204-88. Fiber or Paper Covered Copper Winding Wires. 18 pp.

—Stranded
- CENELEC HD 383 S2-86. Conductors of Insulated Cables Guide to the Dimensional Limits of Circular Conductors. 2 pp.

Copper Conductors (Cont.)

—Stranded (Cont.)
- CENELEC HD 383 S2/A2-93. Conductors of Insulated Cables First Supplement: Guide to the Dimensional Limits of Circular Conductors (IEC 228:1978 + IEC 228A: 1982, Modified). 6 pp.
- CNS C2031-87. Soft Copper Wires, Stranded (Non-Insulated) (Jul)(1365). 2 pp.
- IEC 228-78. Conductors of Insulated Cables Second Edition; (Errata—Sept 1979) (Supplement A-1982) (Amendment 1-1993) (CENELEC HD 383 S2/A2: 1993). 52 pp.

—Telephone
- CNS C1030-80. Dimensions of Copper-Silicon Alloy Wire (for Telephone Use) (Jul)(677).
- CNS C1031-80. Performance of Copper-Silicon Alloy Wire (for Telephone Use) (Jul)(678).

—Terminal Blocks
- DIN VDE 0611 Pt 1-77. VDE Specification for Terminal Blocks for Connecting Copper Conductors up to 1000V a.c. and 1200 V d.c.; Transit Terminal Blocks up to 240mm2 (Nov). 21 pp.
- DIN VDE 0611 Pt 4-91. Terminal Blocks for Connecting Copper Conductors Distribution Terminal Blocks up to 6 mm2 (Feb). 24 pp.

—Tin Coated
- BSI BS 4393-91. 1991 Solderable Tin or Tin-Lead Coated Wires for Component Terminations. 18 pp.
- CNS C2005-87. Tin Coated Annealed Copper Wire (Non-Insulated) (Jul)(670). 4 pp.
- CNS C2032-87. Tinned Hard-Drawn Copper Wires, Solid (Non-Insulated) (Jul)(1366). 3 pp.
- DIN ENGL 40500 Pt 5-83. Copper for Electrical Purposes; Tinned Wire; Technical Delivery Conditions (June). 4 pp.
- JIS C 3151-76. Tin Coated Hard-Drawn Copper Wires. 7 pp.
- JIS C 3152-84. Tin Coated Annealed Copper Wires. 7 pp.
- MOD UK DSTAN 61-12: Part 14-76. Wires, Cords, and Cables, Electrical-Metric Units Part 14: Wires, Electrical (Enamel Insulated Round Winding Wires and Non-Insulated Tinned Copper Wires) Issue 1. 14 pp.
- MOD UK DSTAN 61-12: Part 14-01. Wires, Cords, and Cables, Electrical-Metric Units Part 14: Wires, Electrical (Enamel Insulated Round Winding Wires and Non-Insulated Tinned Copper Wires) Issue 1; Amendment 1. 15 pp.

—Tin Coated—Dry Batteries
- CNS C2045-80. Connecting Wire for Dry Battery (Tentative) (Jul)(2548). 1 p.

—Tin Coated—Stranded
- CNS C2007-87. Tinned Soft Copper Wire Stranded (Non-Insulated) (Jul)(672). 4 pp.
- CNS C2033-87. Tinned Hard-Drawn Copper Wire, Stranded (Non-Insulated) (Jul)(1367). 2 pp.

—Trolley
- BSI BS 23-70. 1970 Copper and Copper-Cadmium Trolley and Contact Wire for Electric Traction. 18 pp.
- JIS E 2101-90. Hard-Drawn Grooved Trolley Wires.
- JIS E 2101-77. Hard Drawn Copper Grooved Trolley Wires (R 1980). 8 pp.
- JIS E 2102-90. Hard-Drawn Copper Round Trolley Wires.
- JIS E 2102-77. Hard Drawn Copper Round Trolley Wires (R 1980). 7 pp.

—Windings
- BSI BS 6811: Sec 4.1-93. 1993 Winding Wires Part 4: Specifications for Particular Types of Enamelled Rectangular Copper Winding Wires Section 4.1: General Requirements (IEC 317-0-2: 1990). 23 pp.
- BSI BS 6811: Sec 4.1-91. 1991 Winding Wires Part 4: Specifications for Particular Types of Rectangular Winding Wires Section 4.1: General Requirements (IEC 317-0-2: 1990). 21 pp.
- BSI BS 6811: Sec 6.0-93. 1993 Winding Wires Part 6: Specifications for Particular Types of Glass-Fibre Wound Bare or Enamelled Rectangular Copper Winding Wires Section 6.0: General Requirements (IEC 317-0-4: 1990). 23 pp.
- BSI BS 6811: Sec 6.0-01. 1993 Amd 1 Winding Wires Part 6: Specifications for Particular Types of Glass-Fibre Wound Bare or Enamelled Rectangular Copper Winding Wires Section 6.0: General Requirements (IEC 317-0-4: 1990) (AMD 7906) September 15, 1993 (E). 24 pp.
- BSI BS 6811: Sec 6.1-93. 1993 Winding Wires Part 6: Specifications for Particular Types of Glass-Fibre Wound Bare or Enamelled Rectangular Copper Winding Wires Section 6.1: Glass-Fibre Wound, Polyester or Polyesterimide Varnish-Treated, Bare or Enamelled Rectangular. 13 pp.

Copper Conductors (Cont.)

—Windings (Cont.)
- BSI BS 6811: Sec 6.2-93. 1993 Winding Wires Part 6: Specifications for Particular Types of Glass-Fibre Wound Bare or Enamelled Rectangular Copper Winding Wires Section 6.2: Glass-Fibre Wound, Polyester or Polyesterimide Varnish-Treated, Bare or Enamelled Rectangular. 13 pp.
- BSI BS 6811: Sec 6.3-93. 1993 Winding Wires Part 6: Specifications for Particular Types of Glass-Fibre Wound Bare or Enamelled Rectangular Copper Winding Wires Section 6.3: Glass-Fibre Wound, Silicone Varnish-Treated, Bare or Enamelled Rectangular Copper Wire, Temperature. 13 pp.
- CENELEC HD 42.3 S2-78. Basic Dimensions of Winding Wires—Part 3: Dimensions of Conductors of Rectangular Copper Winding Wires. 2 pp.
- CENELEC HD 555.31 S1-92. Specifications for Particular Types of Winding Wires Part 31: Glass-Fibre Wound, Polyester or Polyesterimide Varnish-Treated, Bare or Enamelled Rectangular Copper Wire, Temperature Index 180. 5 pp.
- CENELEC HD 555.32 S1-92. Specifications for Particular Types of Winding Wires Part 32: Glass-Fibre Wound, Polyester or Polyesterimide Varnish-Treated, Bare or Enamelled Rectangular Copper Wire, Temperature Index 155. 5 pp.
- CENELEC HD 555.33 S1-92. Specifications for Particular Types of Winding Wires Part 33: Glass-Fibre Wound, Silicone Varnish-Treated, Bare or Enamelled Rectangular Copper Wire, Temperature Index 200. 5 pp.
- IEC 317 Pt 0-2-90. Specifications for Particular Types of Winding Wires Part 0: General Requirements Section 2: Enamelled Rectangular Copper Wire First Edition; (Amendment 1-1992) (Amendment 2-1993). 59 pp.
- IEC 317 Pt 0-4-90. Specifications for Particular Types of Winding Wires Part 0: General Requirements Section 4: Glass-Fibre Wound Bare or Enamelled Rectangular Copper Wire First Edition; (Amendment 1-1992) (Amendment 2-1993). 61 pp.
- IEC 317 Pt 31-90. Specifications for Particular Types of Winding Wires Part 31: Glass-Fibre Wound, Polyester or Polyesterimide Varnish-Treated, Bare or Enamelled Rectangular Copper Wire, Temperature Index 180 First Edition. 19 pp.
- IEC 317 Pt 32-90. Specifications for Particular Types of Winding Wires Part 32: Glass-Fibre Wound, Polyester or Polyesterimide Varnish-Treated, Bare or Enamelled Rectangular Copper Wire, Temperature Index 155 First Edition. 19 pp.
- IEC 317 Pt 33-90. Specifications for Particular Types of Winding Wires Part 33: Glass-Fibre Wound, Silicone Varnish-Treated, Bare or Enamelled Rectangular Copper Wire, Temperature Index 200 First Edition. 19 pp.
- JIS C 3103-84. Annealed Copper Wires for Winding Wires. 7 pp.

Copper Content Analysis

See Also: Copper

—Aluminum
- BSI BS 1728: Part 1-51. 1951 Amd 1 Methods for the Analysis of Aluminium and Aluminium Alloys Part 1: Determination of Copper. 9 pp.
- JIS H 1354-72. Methods for Determination of Copper in Aluminium and Aluminium Alloy.

—Aluminum—Absorptiometric Analysis
- BSI BS 1728: Part 5-53. 1953 Amd 1 Methods for the Analysis of Aluminium and Aluminium Alloys Part 5: Determination of Copper (Absorptiometric Method). 8 pp.

—Aluminum Alloys
- BSI BS 1728: Part 1-51. 1951 Amd 1 Methods for the Analysis of Aluminium and Aluminium Alloys Part 1: Determination of Copper. 9 pp.
- JIS H 1354-72. Methods for Determination of Copper in Aluminium and Aluminium Alloy.

—Aluminum Alloys—Absorptiometric Analysis
- BSI BS 1728: Part 5-53. 1953 Amd 1 Methods for the Analysis of Aluminium and Aluminium Alloys Part 5: Determination of Copper (Absorptiometric Method). 8 pp.

—Aluminum Alloys—Atomic Absorption Spectrophotometry
- BSI BS 1728: Part 23-76. 1976 Methods for Analysis of Aluminium and Aluminium Alloys: Part 23: Copper (Atomic Absorption Method). 4 pp.
- ISO 3980-77. Aluminium and Aluminium Alloys—Determination of Copper—Atomic Absorption Spectrophotometric Method First Edition. 5 pp.

—Aluminum Alloys—Electrolytic Analysis
- ISO 796-73. Aluminium Alloys—Determination of Copper—Electrolytic Method First Edition. 7 pp.

INTERNATIONAL AND NON-U.S. NATIONAL STANDARDS
SUBJECT INDEX
Copper

Copper Content Analysis (Cont.)

—Aluminum Alloys—Photometry
ISO 795-76. Aluminium and Aluminium Alloys—Determination of Copper Content—Oxalyldihydrazide Photometric Method First Edition. 6 pp.

—Aluminum—Atomic Absorption Spectrophotometry
BSI BS 1728: Part 23-76. 1976 Methods for the Analysis of Aluminium and Aluminium Alloys: Part 23: Copper (Atomic Absorption Method). 4 pp.

ISO 3980-77. Aluminium and Aluminium Alloys—Determination of Copper—Atomic Absorption Spectrophotometric Method First Edition. 5 pp.

—Aluminum—Electrolytic Analysis
ISO 796-73. Aluminium Alloys—Determination of Copper—Electrolytic Method First Edition. 7 pp.

—Aluminum—Photometry
ISO 795-76. Aluminium and Aluminium Alloys—Determination of Copper Content—Oxalyldihydrazide Photometric Method First Edition. 6 pp.

—Ammonium Sulfate—Photometry
ISO 3333-75. Ammonium Sulphate for Industrial Use—Determination of Copper Content—Zinc Dibenzyldithiocarbamate Photometric Method First Edition. 5 pp.

—Animal Fats—Colorimetry
BSI BS 684: Sec 2.16-76. 1976 Methods of Analysis of Fats and Fatty Oils Part 2: Other Methods Section 2.16: Determination of Copper. Colorimetric Method. 2 pp.

—Blister Copper
CNS M3043-81. Methods for Determination of Copper in Blister Coppers (Apr)(7291).

JIS M 8125-76. Method for Determination of Copper in Blister Copper (R 1984). 6 pp.

—Boric Acids—Photometry
BSI BS 5688: Part 7-79. 1979 Orthoboric Acid (Boric Acid), Diboron Trioxide (Boric Oxide), Disodium Tetraborates, Sodium Perborates and Crude Sodium Borates for Industrial Use Part 7: Determination of Copper Content of Boric Acid, Boric Oxide and Disodium Tetraborates. 4 pp.

ISO 2215-72. Boric Acid, Boric Oxide and Disodium Tetraborates for Industrial Use—Determination of Copper Content—Zinc Dibenzyldithiocarbamate Photometric Method First Edition. 4 pp.

—Boron Oxides—Photometry
BSI BS 5688: Part 7-79. 1979 Orthoboric Acid (Boric Acid), Diboron Trioxide (Boric Oxide), Disodium Tetraborates, Sodium Perborates and Crude Sodium Borates for Industrial Use Part 7: Determination of Copper Content of Boric Acid, Boric Oxide and Disodium Tetraborates. 4 pp.

ISO 2215-72. Boric Acid, Boric Oxide and Disodium Tetraborates for Industrial Use—Determination of Copper Content—Zinc Dibenzyldithiocarbamate Photometric Method First Edition. 4 pp.

—Cast Iron—Atomic Absorption Spectrometry
BSI BS 6200: SUB SEC 3.12.3-86. 1986 Sampling and Analysis of Iron, Steel and Other Ferrous Metals Part 3: Methods of Analysis Section 3.12: Determination of Copper Subsection 3.12.3: Steel and Cast Iron: Flame Atomic Absorption Spectrometric Method. 8 pp.

BSI BS 6200: SUB SEC 3.12.3-01. 1986 Amd 1 Sampling and Analysis of Iron, Steel and Other Ferrous Metals Part 3: Methods of Analysis Sec 3.12: Determination of Copper Subsec 3.12.3: Steel and Cast Iron: Flame Atomic Absorption Spectrometric Method (ISO 4943: 1985) (AMD 7071). 13 pp.

CEN EN 24 943-90. Chemical Analysis of Ferrous Metal—Determination of Copper Content—Flame Atomic Absorption Spectrometric Method. 2 pp.

DIN ENGL EN 24943-92. Determination of Copper Content of Steel and Cast Iron by Flame Atomic Absorption Spectrometry; (ISO 4943: 1985) (Oct). 9 pp.

ISO 4943-85. Steel and Cast Iron—Determination of Copper Content—Flame Atomic Absorption Spectrometric Method First Edition. 8 pp.

—Cast Iron—Spectrophotometry
BSI BS 6200: SUB SEC 3.12.2-85. 1985 Sampling and Analysis of Iron, Steel and Other Ferrous Metals Part 3: Methods of Analysis Section 3.12: Determination of Copper Subsection 3.12.2: Steel and Cast Iron: Spectrophotometric Method. 8 pp.

BSI BS 6200: SUB SEC 3.12.2-01. 1985 Amd 1 Sampling and Analysis of Iron, Steel and Other Ferrous Metals Part 3: Methods of Analysis Sec 3.12: Determination of Copper Subsec 3.12.2: Steel and Cast Iron: Spectrophotometric Meth (ISO 4946: 1984) (AMD 7070) February 28, 1992. 13 pp.

Copper Content Analysis (Cont.)

—Cast Iron—Spectrophotometry (Cont.)
CEN EN 24 946-90. Determination of Copper Content in Steel and Cast Iron. 2 pp.

DIN ENGL EN 24946-92. Determination of Copper Content of Steel and Cast Iron by the 2,2'-Diquinolyl Spectrophotometric Method (ISO 4946: 1984) (Nov). 9 pp.

ISO 4946-84. Steel and Cast Iron—Determination of Copper Content—2,2'-Diquinolyl Spectrophotometric Method First Edition; (Corrected and Reprinted -1986). 8 pp.

SAA AS 1050.17-88. Methods for the Analysis of Iron and Steel—Part 17: Determination of Copper Content—Spectrophotometric Method. 4 pp.

—Cast Iron—Volumetric Analysis
BSI BS 6200: SUB SEC 3.12.1-86. 1986 Sampling and Analysis of Iron, Steel and Other Ferrous Metals Part 3: Methods of Analysis Section 3.12: Determination of Copper Subsection 3.12.1: Steel and Cast Iron: Volumetric Method. 4 pp.

BSI BS 6200: SUB SEC 3.12.4-86. 1986 Sampling and Analysis of Iron, Steel and Other Ferrous Metals Part 3: Methods of Analysis Section 3.12: Determination of Copper Subsection 3.12.4: Cast Iron: Volumetric Method. 4 pp.

—Copper Alloys
CNS H2084-89. Methods of Determination for Copper in Copper Alloys (Jun)(12530).

—Copper Alloys—Electrolytic Analysis
BSI BS 1748: Parts 1-5-61. 1961 Methods for the Analysis of Copper Alloys Parts 1-5: Determination of Copper, Lead, Iron, Aluminium and Nickel in Copper Alloys. 23 pp.

BSI BS 1748: Parts 11-12-64. 1964 Methods for the Analysis of Copper Alloys Parts 11-12: The Determination of Copper and Lead in Leaded Bronze Alloys. 12 pp.

ISO 1554-76. Wrought and Cast Copper Alloys—Determination of Copper Content—Electrolytic Method First Edition. 4 pp.

SAA AS 1515.4-78. Copper Alloys—Part 4: Method for the Electrolytic Determination of Copper in Wrought and Cast Copper Alloys Reconfirmed 1989. 10 pp.

—Copper Alloys—Products
JIS H 1051-92. Methods for Determination of Copper in Copper and Copper Alloys. 22 pp.

—Copper Castings
JIS H 1051-92. Methods for Determination of Copper in Copper and Copper Alloys. 22 pp.

—Copper—Electrolytic Analysis
SAA AS K208.1-70. Methods for the Analysis of Unalloyed Copper—Part 1: Method for the Electrolytic Determination of Copper in Unalloyed Copper Containing Not Less Than 99.85 Percent Copper Amdt 1 April 1971. 7 pp.

—Copper—Products
JIS H 1051-92. Methods for Determination of Copper in Copper and Copper Alloys. 22 pp.

—Cordage
CGSB 40-GP-1M METH 8-78. Methods of Sampling and Testing Cordage Determination of Copper Content. 1 p.

—Dairy Products—Photometry
BSI BS 6394: Part 1-83. 1983 Trace Elements in Milk and Milk Products Part 1: Method for the Determination of Copper Content (Reference Method). 9 pp.

ISO 5738-80. Milk and Milk Products—Determination of Copper Content—Photometric Reference Method First Edition; (Erratum—Nov 1981). 9 pp.

—Drinking Water—Voltametry
DIN ENGL 38406 Pt 16-90. German Standard Methods for the Examination of Water, Waste Water and Sludge; Cations (Group E); Determination of Zinc, Cadmium, Lead, Copper, Thallium, Nickel, Cobalt by Voltammetry (E16) (Mar). 8 pp.

—Dyes
MOD UK M 818/63. Examination of Dyestuffs for Use in Pyrotechnic Compositions and HE Substitutes.

—Elastomers—Atomic Absorption Spectroscopy
DIN ENGL 53569 Pt 2-72. Testing of Rubber and Elastomers; Determination of Copper Content; Determination by Atomic Absorption Spectroscopy (Nov). 7 pp.

Copper Content Analysis (Cont.)

—Elastomers—Photometry
DIN ENGL 53569 Pt 1-72. Testing of Rubber and Elastomers; Determination of Copper Content; Photometric Method (Nov). 4 pp.

—Fabrics—Atomic Absorption Spectrophotometry
CGSB CAN/CGSB-4.2 NO.42-M91. Textile Test Methods Copper Content of Textiles. 6 pp.

—Ferrotungsten
CNS G2154-83. Methods of Chemical Analysis for Copper in Ferrotungsten (Jan)(9865).

—Fertilizers
EC COM(88) 562-88. Proposal for a Council Directive on the Approximation of the Laws of the Member States in Respect of the Trace Elements Boron, Cobalt, Copper, Iron, Manganese, Molybdenum and Zinc Contained in Fertilizers. 20 pp.

EC 89/530/EEC-89. Council Directive Supplementing and Amending Directive 76/116/EEC in Respect of the Trace Elements Boron, Cobalt, Copper Iron, Manganese, Molybdenum and Zinc Contained in Fertilizers. 9 pp.

—Fruits—Photometry
ISO 3094-74. Fruit and Vegetable Products—Determination of Copper Content—Photometric Method First Edition. 4 pp.

—Gallium Arsenide
JIS H 1191-91. Methods for Chemical Analysis of Gallium Arsenide. 57 pp.

—Ground Water—Voltametry
DIN ENGL 38406 Pt 16-90. German Standard Methods for the Examination of Water, Waste Water and Sludge; Cations (Group E); Determination of Zinc, Cadmium, Lead, Copper, Thallium, Nickel, Cobalt by Voltammetry (E16) (Mar). 8 pp.

—Hydrofluoric Acid—Emission Spectroscopy
DIN ENGL 50451 Pt 2-90. Determination of Cobalt, Chromium, Copper, Iron and Nickel as Impurities in Hydrofluoric Acid for Use in Semiconductor Technology by Plasma-Induced Emission Spectrometry (Oct). 2 pp.

—Iron
CNS G2234-84. Method of Determination for Copper in Iron and Steel (Dec)(11165).

JIS G 1219-81. Methods for Determination of Copper in Iron and Steel.

—Iron Ores—Absorptiometric Analysis
CNS M3079-81. Method for Determination of Copper in Iron Ores (BCOD Absorptiometric Method) (Oct)(8047).

JIS M 8218-83. Methods for Determination of Copper in Iron Ores. 14 pp.

—Iron Ores—Atomic Absorption Spectrometry
BSI BS 7020: Sec 10.2-88. 1988 Analysis of Iron Ores Part 10: Methods for the Determination of Copper Content Section 10.2: Flame Atomic Absorption Spectrometric Method. 12 pp.

ISO 4693-86. Iron Ores—Determination of Copper Content—Flame Atomic Absorption Spectrometric Method First Edition. 10 pp.

JIS M 8218-83. Methods for Determination of Copper in Iron Ores. 14 pp.

—Iron Ores—Spectrophotometry
BSI BS 7020: Sec 10.1-88. 1988 Analysis of Iron Ores Part 10: Methods for the Determination of Copper Content Section 10.1: Spectrophotometric Method. 12 pp.

ISO 5418-84. Iron Ores—Determination of Copper Content—2,2'-Biquinolyl Spectrophotometric Method First Edition. 10 pp.

—Iron Ores—Volumetric Analysis
CNS M3078-81. Method for Determination of Copper in Iron Ores (Sodium Thiosulfate Titration Method) (Oct)(8046).

JIS M 8218-83. Methods for Determination of Copper in Iron Ores. 14 pp.

—Latex—Photometry
BSI BS 7164: Sec 28.2-90. 1990 Chemical Tests for Raw and Vulcanized Rubber Part 28: Methods for Determination of Copper Content Section 28.2: Photometric Method (ISO 8053: 1986). 10 pp.

ISO 8053-86. Rubber and Latex—Determination of Copper Content—Photometric Method First Edition. 6 pp.

Copper Content Analysis (Cont.)

—Lead Alloys—Photometry
BSI BS 3908: Part 4-67. 1967 Methods for the Sampling and Analysis of Lead and Lead Alloys Part 4: Copper in Lead and Lead Alloys (Photometric Method). 8 pp.

—Lead—Photometry
BSI BS 3908: Part 4-67. 1967 Methods for the Sampling and Analysis of Lead and Lead Alloys Part 4: Copper in Lead and Lead Alloys (Photometric Method). 8 pp.

—Lead—Polarographic Analysis
MOD UK M 9331/63. Determination of Bismuth, Copper, Cadmium, and Zinc in Pure Lead (Polarographic Method) (No Information) (Withdrawn).

—Liquid Fuels—Atomic Absorption Spectrometry
DIN ENGL 51404 Pt 2-85. Testing of Lubricants and Fuels; Determination of Copper Content of Lubricating Oils and Liquid Fuels; Direct Determination by Atomic Absorption Spectrometry (May). 4 pp.

—Lubricating Oils—Atomic Absorption Spectrometry
DIN ENGL 51404 Pt 2-85. Testing of Lubricants and Fuels; Determination of Copper Content of Lubricating Oils and Liquid Fuels; Direct Determination by Atomic Absorption Spectrometry (May). 4 pp.

—Magnesium Alloys
JIS H 1336-76. Methods for Determination of Copper in Magnesium Alloys.

—Magnesium Alloys—Photometry
BSI BS 3907: Part 3-66. 1966 Methods for the Analysis of Magnesium and Magnesium Alloys Part 3: Copper in Magnesium and Magnesium Alloys (Photometric Method). 8 pp.

ISO 794-76. Magnesium and Magnesium Alloys—Determination of Copper Content—Oxalyldihydrazide Photometric Method First Edition. 5 pp.

—Magnesium—Photometry
BSI BS 3907: Part 3-66. 1966 Methods for the Analysis of Magnesium and Magnesium Alloys Part 3: Copper in Magnesium and Magnesium Alloys (Photometric Method). 8 pp.

ISO 794-76. Magnesium and Magnesium Alloys—Determination of Copper Content—Oxalyldihydrazide Photometric Method First Edition. 5 pp.

—Magnets—Volumetric Analysis
BSI BS 6200: SUB SEC 3.12.5-86. 1986 Sampling and Analysis of Iron, Steel and Other Ferrous Metals Part 3: Methods of Analysis Section 3.2: Determination of Copper Subsection 3.12.5: Permanent Magnet Alloys: Volumetric Method. 4 pp.

—Manganese Ores—Absorptiometric Analysis
JIS M 8242-89. Methods for Determination of Copper in Manganese Ores. 11 pp.

—Manganese Ores—Atomic Absorption Spectrometry
ISO 5889-83. Manganese Ores and Concentrates—Determination of Aluminium, Copper, Lead and Zinc Contents—Flame Atomic Absorption Spectrometric Method First Edition. 6 pp.

JIS M 8242-89. Methods for Determination of Copper in Manganese Ores. 11 pp.

—Manganese Ores—Spectrometry
ISO 4294-84. Manganese Ores and Concentrates—Determination of Copper Content—Extraction-Spectrometric and Spectrometric Methods Second Edition. 7 pp.

—Milk—Photometry
BSI BS 6394: Part 1-83. 1983 Trace Elements in Milk and Milk Products Part 1: Method for the Determination of Copper Content (Reference Method). 9 pp.

ISO 5738-80. Milk and Milk Products—Determination of Copper Content—Photometric Reference Method First Edition; (Erratum—Nov 1981). 9 pp.

—Nickel
JIS H 1424-79. Methods for Determination of Copper in Nickel Materials for Electron Tubes.

—Nickel Alloys—Atomic Absorption Spectrometry
BSI BS 7455: Part 4-91. 1991 Analysis of Nickel Alloys by Flame Atomic Absorption Spectrometry Part 4: Method for the Determination of Copper (ISO 7530-4: 1990). 8 pp.

Copper Content Analysis (Cont.)

—Nickel Alloys—Atomic Absorption Spectrometry (Cont.)
ISO 7530 Pt 4-90. Nickel Alloys—Flame Atomic Absorption Spectrometric Analysis—Part 4: Determination of Copper Content First Edition. 6 pp.

—Nickel—Atomic Absorption Spectrometry
BSI BS 6783: Part 1-86. 1986 Sampling and Analysis of Nickel, Ferronickel and Nickel Alloys Part 1: Method for Determination of Silver, Bismuth, Cadmium, Cobalt, Copper, Iron, Manganese, Lead and Zinc in Nickel by Flame Atomic Absorption Spectrometry. 15 pp.

ISO 6351-85. Nickel—Determination of Silver, Bismuth, Cadmium Cobalt, Copper, Iron, Manganese, Lead and Zinc Contents—Flame Atomic Absorption Spectrometric Method First Edition. 13 pp.

—Nickel Castings—Electrogravimetric Analysis
JIS H 1272-88. Methods for Determination of Copper in Nickel and Nickel Alloy Castings. 8 pp.

—Nickel Castings—Spectrochemical Analysis
JIS H 1272-88. Methods for Determination of Copper in Nickel and Nickel Alloy Castings. 8 pp.

—Nickel—Photometry
BSI BS 3727: Part 6-64. 1964 Methods for the Analysis of Nickel for Use in Electronic Tubes and Valves Part 6: Determination of Copper (Photometric Method). 8 pp.

—Ores
CNS M3073-81. General Rules for Determination of Copper in Ores (Oct) (8041).

JIS M 8121-82. Methods for Determination of Copper in Ores.

—Ores—Absorptiometric Analysis
CNS M3076-81. Method for Determination of Copper in Ores (Cupric Ammonia Absorptiometric Method) (Oct)(8044).

—Ores—Atomic Absorption Analysis
CNS M3077-81. Method for Determination of Copper in Ores (Atomic Absorption Method) (Oct)(8045).

—Ores—Electrogravimetric Analysis
CNS M3074-81. Method for Determination of Copper in Ores (Electrogravimetric Method) (Oct)(8042).

—Ores—Volumetric Analysis
CNS M3075-81. Method for Determination of Copper in Ores (Sodium Thiosulfate Titration Method) (Oct)(8043).

—Organic Coatings—Quantitative Analysis
CNS K6804-27-89. Method of Test for Organic Coating (Chemical Analysis) — Quantative Test of Copper in Coating (Jan)(10880-27).

—Papers—Atomic Absorption Spectrophotometry
CPPA G.34P-92. Determination of Sodium, Magnesium, Calcium, Manganese, Iron, Copper and Cadmium in Wood, Pulp or Paper by Atomic Absorption Spectrophotometry. 3 pp.

—Precipitation—Voltametry
DIN ENGL 38406 Pt 16-90. German Standard Methods for the Examination of Water, Waste Water and Sludge; Cations (Group E); Determination of Zinc, Cadmium, Lead, Copper, Thallium, Nickel, Cobalt by Voltammetry (E16) (Mar). 8 pp.

—Pulps—Atomic Absorption Spectrometry
BSI BS 4897: Part 2-83. 1983 Trace Metal Contents of Pulps Part 2: Method for Determination of Copper Content by Extraction—Photometric and Flame Atomic Absorption Spectrometric Methods. 8 pp.

ISO 778-82. Pulps—Determination of Copper Content—Extraction-Photometric and Flame Atomic Absorption Spectrometric Methods First Edition. 6 pp.

—Pulps—Atomic Absorption Spectrophotometry
CPPA G.34P-92. Determination of Sodium, Magnesium, Calcium, Manganese, Iron, Copper and Cadmium in Wood, Pulp or Paper by Atomic Absorption Spectrophotometry. 3 pp.

Copper Content Analysis (Cont.)

—Pulps—Photometry
BSI BS 4897: Part 2-83. 1983 Trace Metal Contents of Pulps Part 2: Method for Determination of Copper Content by Extraction—Photometric and Flame Atomic Absorption Spectrometric Methods. 8 pp.

ISO 778-82. Pulps—Determination of Copper Content—Extraction-Photometric and Flame Atomic Absorption Spectrometric Methods First Edition. 6 pp.

—Rubber—Atomic Absorption Spectrometry
BSI BS 7164: Sec 28.1-90. 1990 Chemical Tests for Raw and Vulcanized Rubber Part 28: Methods for Determination of Copper Content Section 28.1: Atomic Absorption Spectrometry (ISO 6101-3: 1988). 11 pp.

ISO 6101 Pt 3-88. Rubber—Determination of Metal Content by Atomic Absorption Spectrometry—Part 3: Determination of Copper Content First Edition. 8 pp.

—Rubber—Atomic Absorption Spectroscopy
DIN ENGL 53569 Pt 2-72. Testing of Rubber and Elastomers; Determination of Copper Content; Determination by Atomic Absorption Spectroscopy (Nov). 7 pp.

—Rubber—Photometry
BSI BS 903: Part B17-59. (WITHDRAWN) 1959 Methods of Testing Vulcanized Rubber Part B17: Determination of Total Copper (Superseded by BS 7164: Section 28.2: 1990). 8 pp.

BSI BS 7164: Sec 28.2-90. 1990 Chemical Tests for Raw and Vulcanized Rubber Part 28: Methods for Determination of Copper Content Section 28.2: Photometric Method (ISO 8053: 1986). 10 pp.

DIN ENGL 53569 Pt 1-72. Testing of Rubber and Elastomers; Determination of Copper Content; Photometric Method (Nov). 4 pp.

ISO 8053-86. Rubber and Latex—Determination of Copper Content—Photometric Method First Edition. 6 pp.

—Sodium Borates—Photometry
BSI BS 5688: Part 7-79. 1979 Orthoboric Acid (Boric Acid), Diboron Trioxide (Boric Oxide), Disodium Tetraborates, Sodium Perborates and Crude Sodium Borates for Industrial Use Part 7: Determination of Copper Content of Boric Acid, Boric Oxide and Disodium Tetraborates. 4 pp.

ISO 2215-72. Boric Acid, Boric Oxide and Disodium Tetraborates for Industrial Use—Determination of Copper Content—Zinc Dibenzyldithiocarbamate Photometric Method First Edition. 4 pp.

ISO 5934-80. Crude Sodium Borates for Industrial Use—Determination of Alkali-Soluble Copper and Manganese Contents—Zinc Bis (Dibenzyldithiocarbamate) and Formaldehyde Oxime Photometric Methods First Edition. 7 pp.

—Sodium Carbonates—Spectrophotometry
BSI BS 6070: Part 7-81. 1981 Methods of Sampling and Test for Sodium Carbonate for Industrial Use Part 7: Determination of Copper Content. 3 pp.

—Sodium Chloride—Photometry
BSI BS 7319: Part 7-90. 1990 Analysis of Sodium Chloride for Industrial Use Part 7: Method for Determination of Copper Content. 7 pp.

—Sodium Hydroxide—Spectrophotometry
BSI BS 6075: Part 12-81. 1981 Sampling and Test for Sodium Hydroxide for Industrial Use Part 12: Determination of Copper Content. 4 pp.

—Solders—Photometry
BSI BS 3338: Part 4-61. 1961 Methods for Sampling and Analysis of Tin and Tin Alloys Part 4: Determination of Copper in Ingot Tin and Tin-Lead Solders (Photometric Method). 8 pp.

—Steels
CNS G2234-84. Method of Determination for Copper in Iron and Steel (Dec)(11165).

JIS G 1219-81. Methods for Determination of Copper in Iron and Steel.

—Steels—Atomic Absorption Spectrometry
BSI BS 6200: SUB SEC 3.12.3-86. 1986 Sampling and Analysis of Iron, Steel and Other Ferrous Metals Part 3: Methods of Analysis Section 3.12: Determination of Copper Subsection 3.12.3: Steel and Cast Iron: Flame Atomic Absorption Spectrometric Method. 8 pp.

BSI BS 6200: SUB SEC 3.12.3-01. 1986 Amd 1 Sampling and Analysis of Iron, Steel and Other Ferrous Metals Part 3: Methods of Analysis Sec 3.12: Determination of Copper Subsec 3.12.3: Steel and Cast Iron: Flame Atomic Absorption Spectrometric Method (ISO 4943: 1985) (AMD 7071). 13 pp.

INTERNATIONAL AND NON-U.S. NATIONAL STANDARDS
SUBJECT INDEX
Copper

Copper Content Analysis *(Cont.)*

—Steels—Atomic Absorption Spectrometry *(Cont.)*
CEN EN 24 943-90. Chemical Analysis of Ferrous Metal—Determination of Copper Content—Flame Atomic Absorption Spectrometric Method. 2 pp.

DIN ENGL EN 24943-92. Determination of Copper Content of Steel and Cast Iron by Flame Atomic Absorption Spectrometry; (ISO 4943: 1985) (Oct). 9 pp.

ISO 4943-85. Steel and Cast Iron—Determination of Copper Content—Flame Atomic Absorption Spectrometric Method First Edition. 8 pp.

—Steels—Spectrophotometry
BSI BS 6200: SUB SEC 3.12.2-85. 1985 Sampling and Analysis of Iron, Steel and Other Ferrous Metals Part 3: Methods of Analysis Section 3.12: Determination of Copper Subsection 3.12.2: Steel and Cast Iron: Spectrophotometric Method. 8 pp.

BSI BS 6200: SUB SEC 3.12.2-01. 1985 Amd 1 Sampling and Analysis of Iron, Steel and Other Ferrous Metals Part 3: Methods of Analysis Sec 3.12: Determination of Copper Subsec 3.12.2: Steel and Cast Iron: Spectrophotometric Meth (ISO 4946: 1984) (AMD 7070) February 28, 1992. 13 pp.

CEN EN 24 946-90. Determination of Copper Content in Steel and Cast Iron. 2 pp.

DIN ENGL EN 24946-92. Determination of Copper Content of Steel and Cast Iron by the 2,2'-Diquinolyl Spectrophotometric Method (ISO 4946: 1984) (Nov). 9 pp.

ISO 4946-84. Steel and Cast Iron—Determination of Copper Content—2,2'-Diquinolyl Spectrophotometric Method First Edition; (Corrected and Reprinted -1986). 7 pp.

SAA AS 1050.17-88. Methods for the Analysis of Iron and Steel—Part 17: Determination of Copper Content—Spectrophotometric Method. 4 pp.

—Steels—Volumetric Analysis
BSI BS 6200: SUB SEC 3.12.1-86. 1986 Sampling and Analysis of Iron, Steel and Other Ferrous Metals Part 3: Methods of Analysis Section 3.12: Determination of Copper Subsection 3.12.1: Steel and Cast Iron: Volumetric Method. 4 pp.

—Surface Waters—Voltametry
DIN ENGL 38406 Pt 16-90. German Standard Methods for the Examination of Water, Waste Water and Sludge; Cations (Group E); Determination of Zinc, Cadmium, Lead, Copper, Thallium, Nickel, Cobalt by Voltammetry (E16) (Mar). 8 pp.

—Tantalum
JIS H 1686-76. Methods for Determination of Copper in Tantalum (R 1986). 8 pp.

—Tin Alloys—Photometry
BSI BS 3338: Part 4-61. 1961 Methods for the Sampling and Analysis of Tin and Tin Alloys Part 4: Determination of Copper in Ingot Tin and Tin-Lead Solders (Photometric Method). 8 pp.

—Tin—Photometry
BSI BS 3338: Part 4-61. 1961 Methods for the Sampling and Analysis of Tin and Tin Alloys Part 4: Determination of Copper in Ingot Tin and Tin-Lead Solders (Photometric Method). 8 pp.

BSI BS 3338: Part 6-61. 1961 Methods for the Sampling and Analysis of Tin and Tin Alloys Part 6: Determination of Copper in High Purity Ingot Tin (Photometric Method). 8 pp.

—Unalloyed Copper—Electrolytic Analysis
ISO 1553-76. Unalloyed Copper Containing Not Less Than 99,90 % of Copper—Determination of Copper Content—Electrolytic Method First Edition. 6 pp.

—Unalloyed Copper—Spectrophotometry
ISO 1553-76. Unalloyed Copper Containing Not Less Than 99,90 % of Copper—Determination of Copper Content—Electrolytic Method First Edition. 6 pp.

—Vegetable Fats—Colorimetry
BSI BS 684: Sec 2.16-76. 1976 Methods of Analysis of Fats and Fatty Oils Part 2: Other Methods Section 2.16: Determination of Copper. Colorimetric Method. 2 pp.

—Vegetable Oils
CNS N6105-78. Method of Test for Edible Vegetable Oils (Determination of Lead and Copper) (Aug)(4529). 3 pp.

—Vegetables—Photometry
ISO 3094-74. Fruit and Vegetable Products—Determination of Copper Content—Photometric Method First Edition. 4 pp.

Copper Content Analysis *(Cont.)*

—Waste Water—Atomic Absorption Spectrophotometry
CNS K9089-82. Method of Test for Copper in Industrial Waste Water Absorption Spectrophotometric Method (Jun)(8998).

—Waste Water—Polarographic Analysis
CNS K9090-82. Method of Test for Copper in Industrial Waste Water Polargraphic Method (Jun)(8999).

—Waste Water—Spectrophotometry
CNS K9088-82. Method of Test for Copper in Industrial Waste Water Spectrophotometric Method (Jun)(8997).

—Water
BSI BS 2690: Part 1-64. 1964 Water Used in Industry Part 1: Copper and Iron. 16 pp.

CNS K9056-80. Method of Test for Concentrating and Measuring Trace Quantities of Copper in High Purity Water Used in the Electronic Industry (Dec)(6673).

—Water—Atomic Absorption Spectrometry
BSI BS 6068: Sec 2.29-87. 1987 Water Quality Part 2: Physical, Chemical and Bio-Chemical Methods Section 2.29: Determination of Cobalt, Nickel, Copper, Zinc, Cadmium and Lead: Flame Atomic Absorption Spectrometric Methods. 13 pp.

CNS K9030-80. Method of Test for Copper in Water (Atomic Absorption, Direct) (Apr)(5464).

DIN ENGL 38406 Pt 7-91. German Standard Methods for the Examination of Water, Waste Water and Sludge; Cations (Group E); Determination of Copper by Atomic Absorption Spectrometry (AAS) (E 7) (Sept). 5 pp.

ISO 8288-86. Water Quality—Determination of Cobalt, Nickel, Copper, Zinc, Cadmium and Lead—Flame Atomic Absorption Spectrometric Methods First Edition. 13 pp.

JIS K 0553-90. Testing Methods for Determination of Metallic Elements in Highly Purified Water. 27 pp.

—White Metals—Electrolytic Analysis
BSI BS 3338: Part 15-65. 1965 Methods for the Sampling and Analysis of Tin and Tin Alloys Part 15: Determination of Copper and Lead in White Metal Bearing Alloys (Electrodeposition Method). 8 pp.

—Wood—Atomic Absorption Spectrophotometry
CPPA G.34P-92. Determination of Sodium, Magnesium, Calcium, Manganese, Iron, Copper and Cadmium in Wood, Pulp or Paper by Atomic Absorption Spectrophotometry. 3 pp.

—Wood Preservatives—Atomic Absorption Spectrophotometry
BSI BS 5666: Part 3-91. 1991 Methods of Analysis of Wood Preservatives and Treated Timber Part 3: Quantitative Analysis of Preservatives and Treated Timber Containing Copper/Chromium/Arsenic Formulations. 14 pp.

BSI BS 5666: Part 4-79. 1979 Amd 1 Wood Preservatives and Treated Timber Part 4: Quantitative Analysis of Preservatives and Treated Timber Containing Copper Naphthenate. 5 pp.

—Wood Preservatives—Colorimetric Analysis
BSI BS 5666: Part 3-91. 1991 Methods of Analysis of Wood Preservatives and Treated Timber Part 3: Quantitative Analysis of Preservatives and Treated Timber Containing Copper/Chromium/Arsenic Formulations. 14 pp.

BSI BS 5666: Part 4-79. 1979 Amd 1 Wood Preservatives and Treated Timber Part 4: Quantitative Analysis of Preservatives and Treated Timber Containing Copper Naphthenate. 5 pp.

—Zinc Alloys—Atomic Absorption Spectrometry
DIN ENGL 50551-90. Determination of the Lead, Cadmium and Copper Content of Zinc and Zinc Alloys by Atomic Absorption Spectrometry (Oct). 5 pp.

SAA AS 1329.5-80. Methods for the Analysis of Zinc and Zinc Alloys—Part 5: Determination of Copper Content (0.0001 Percent to 0.0025 Percent)—Flame Atomic Absorption Spectrometric Method. 10 pp.

SAA AS 1329.6-81. Methods for the Analysis of Zinc and Zinc Alloys—Part 6: Determination of Copper Content (0.25 Percent to 1.25 Percent)—Flame Atomic Absorption Spectrometric Method. 5 pp.

—Zinc Alloys—Electrolytic Analysis
BSI BS 3630: Part 13-72. 1972 Methods for the Sampling and Analysis of Zinc and Zinc Alloys Part 13: Copper in Zinc Alloys (Alloy B) (Electrolytic Method). 8 pp.

Copper Content Analysis *(Cont.)*

—Zinc Alloys—Electrolytic Analysis *(Cont.)*
ISO 1976-75. Zinc Alloys—Determination of Copper Content—Electrolytic Method First Edition. 4 pp.

—Zinc Alloys—Photometry
BSI BS 3630: Part 9-69. 1969 Methods for the Sampling and Analysis of Zinc and Zinc Alloys: Copper in Ingot Zinc and Zinc Alloy A (Photometric Method). 11 pp.

—Zinc—Atomic Absorption Spectrometry
DIN ENGL 50551-90. Determination of the Lead, Cadmium and Copper Content of Zinc and Zinc Alloys by Atomic Absorption Spectrometry (Oct). 5 pp.

—Zinc Ingots—Photometry
BSI BS 3630: Part 9-69. 1969 Methods for the Sampling and Analysis of Zinc and Zinc Alloys: Copper in Ingot Zinc and Zinc Alloy A (Photometric Method). 11 pp.

—Zinc—Spectrophotometry
ISO 1053-75. Zinc—Determination of Copper Content—Spectrophotometric Method First Edition. 6 pp.

—Zirconium
JIS H 1657-85. Methods for Determination of Copper in Zirconium and Zirconium Alloys. 11 pp.

—Zirconium Alloys
JIS H 1657-85. Methods for Determination of Copper in Zirconium and Zirconium Alloys. 11 pp.

Copper Foil
See Also: Foils; Metal Products

—Cold Rolled
BSI B 29-91. 1991 Copper-Beryllium Alloy Strip, Foil and Parts (Solution Treated, Cold Rolled: Quarter Hard (W(1/4H)) and Precipitated) (S). 5 pp.

BSI B 29-01. 1991 Amd 1 Copper-Beryllium Alloy Strip, Foil and Parts (Solution Treated, Cold Rolled: Quarter Hard (W(1/4H)) and Precipitated) (AMD 7116) April 1, 1992. 6 pp.

BSI B 30-91. 1991 Copper-Beryllium Alloy Strip, Foil and Parts (Solution Treated, Cold Rolled: Half Hard (W(1/2H)) and Precipitated). 5 pp.

BSI B 30-01. 1991 Amd 1 Copper-Beryllium Alloy Strip, Foil and Parts (Solution Treated, Cold Rolled: Half Hard (W(1/2H)) and Precipitated) (AMD 7117) April 1, 1992. 6 pp.

BSI B 31-91. 1991 Copper-Beryllium Alloy Strip, Foil and Parts (Solution Treated, Cold Rolled: Full Hard (W(H)) and Precipitated). 5 pp.

BSI B 31-01. 1991 Amd 1 Copper-Beryllium Alloy Strip, Foil and Parts (Solution Treated, Cold Rolled: Full Hard (W(H)) and Precipitated) (AMD 7118) April 1, 1992. 6 pp.

—Printed Circuit Base Materials
BSI BS 4584: Sec 103.2-90. 1990 Metal-Clad Base Materials for Printed Circuits Part 103: Materials Used in Connection with Printed Circuits Section 103.2: Copper Foil for Use in the Manufacture of Copper-Clad Base Materials. 6 pp.

IEC 249 Pt 3A-76. Metal-Clad Base Materials for Printed Circuits Part 3: Special Materials Used in Connection with Printed Circuits Specification No. 2 Specification for Copper Foil for Use in the Manufacture of Copper-Clad Base Materials (This is a Supplement). 26 pp.

JIS C 6511-92. Test Methods of Copper Foil for Printed Wiring Boards. 11 pp.

JIS C 6512-92. Electrodeposited Copper Foil for Printed Wiring Boards. 9 pp.

—Rolled
BSI BS 2870-80. 1980 Rolled Copper and Copper Alloys. Sheet, Strip and Foil. 32 pp.

BSI BS 4608-70. 1970 Copper for Electrical Purposes. Rolled Sheet, Strip and Foil. 20 pp.

BSI B 28-91. 1991 Copper-Beryllium Alloy Strip, Foil and Parts (Solution Treated (W) and Precipitated). 5 pp.

BSI B 28-01. 1991 Amd 1 Copper-Beryllium Alloy Strip, Foil and Parts (Solution Treated (W) and Precipitated) (AMD 7115) April 1, 1992. 6 pp.

SAA AS 1566-85. Copper and Copper Alloys—Rolled Flat Products. 12 pp.

Copper Inorganic Compounds
Scope Note: Use a more specific term *See:* Copper Metaborate; Copper Sulfide; Cupric Nitrate; Cupric Oxide; Cupric Salicylate; Cupric Sulfate; Cuprous Oxide

Copper Metaborate
Use For: Copper Borate

INTERNATIONAL AND NON-U.S. NATIONAL STANDARDS
SUBJECT INDEX

Copper

Copper Metaborate *(Cont.)*
—Wood Preservatives
- CNS K1277-85. Wood Preservatives: Chromated Copper Borate (CCB) (May)(11261).
- CNS K6809-85. Method of Test for Chromated Copper Borate (Wood Preservatives) (May)(11262).

Copper Naphthenate
- BSI BS 3769-64. 1964 Copper Naphthenate and Copper Naphthenate Concentrates. 20 pp.
- CNS K6401-81. Method of Test for Copper Naphthenate (Jul)(4132).

—Wood Preservatives
- BSI BS 5056-74. 1974 Copper Naphthenate Wood Preservatives. 11 pp.
- CNS K1184-81. Wood Preservatives (Copper Naphthenate) (May)(4131). 1 p.
- SNZ NZS 2234-68. Specification for Copper Naphthenate and Copper Naphthenate Dilutions. 23 pp.

Copper Nickel Alloys
Scope Note: For additional listings, see also specific products made from copper nickel alloys
See Also: Chromium Cupronickel; Copper Alloys; Copper Nickel Manganese Alloys; Copper Nickel Zinc Alloys; Cupronickel

—Forgings
- BSI B 25-01. 1991 Amd 1 Copper-Nickel-Silicon Alloy Rods, Sections, Forging Stock and Forgings (Heat Treated) (AMD 7647) April 15, 1993 (S). 8 pp.

—Forgings—Marine
- MOD UK NES 779: Part 2-89. Requirements for 90/10 Copper Nickel Alloy Material Part 2: Forgings, Forging Stock, Rods and Sections Issue 1 (10.89). 24 pp.
- MOD UK NES 780: Part 2-01. Requirements for 70/30 Copper Nickel Alloy Material Part 2: Forgings, Forging Stock, Rods and Sections Issue 2 (10.88); Amendment 1. 25 pp.

—Plates—Marine
- MOD UK NES 779: Part 1-87. Requirements for 90/10 Copper Nickel Alloy Material Part 1: Sheet, Strip and Plate Issue 1 (11.87). 17 pp.
- MOD UK NES 780: Part 1-87. Requirements for 70/30 Copper Nickel Alloy Material Part 1: Sheet, Strip and Plate Issue 1 (06.87). 17 pp.

—Rods
- BSI B 25-01. 1991 Amd 1 Copper-Nickel-Silicon Alloy Rods, Sections, Forging Stock and Forgings (Heat Treated) (AMD 7647) April 15, 1993 (S). 8 pp.

—Rods—Marine
- MOD UK NES 779: Part 2-89. Requirements for 90/10 Copper Nickel Alloy Material Part 2: Forgings, Forging Stock, Rods and Sections Issue 1 (10.89). 24 pp.
- MOD UK NES 780: Part 2-01. Requirements for 70/30 Copper Nickel Alloy Material Part 2: Forgings, Forging Stock, Rods and Sections Issue 2 (10.88); Amendment 1. 25 pp.

—Sand Castings—Composition
- DIN ENGL 17658-73. Copper-Nickel Casting Alloys; Castings (June). 3 pp.

—Sections
- BSI B 25-01. 1991 Amd 1 Copper-Nickel-Silicon Alloy Rods, Sections, Forging Stock and Forgings (Heat Treated) (AMD 7647) April 15, 1993 (S). 8 pp.

—Sections—Marine
- MOD UK NES 779: Part 2-89. Requirements for 90/10 Copper Nickel Alloy Material Part 2: Forgings, Forging Stock, Rods and Sections Issue 1 (10.89). 24 pp.
- MOD UK NES 780: Part 2-01. Requirements for 70/30 Copper Nickel Alloy Material Part 2: Forgings, Forging Stock, Rods and Sections Issue 2 (10.88); Amendment 1. 25 pp.

—Semifinished Products—Chemical Analysis
- DIN ENGL 17664-83. Wrought Copper Alloys; Copper-Nickel Alloys; Composition (Dec). 6 pp.

—Sheets—Marine
- MOD UK NES 779: Part 1-87. Requirements for 90/10 Copper Nickel Alloy Material Part 1: Sheet, Strip and Plate Issue 1 (11.87). 17 pp.
- MOD UK NES 780: Part 1-87. Requirements for 70/30 Copper Nickel Alloy Material Part 1: Sheet, Strip and Plate Issue 1 (06.87). 17 pp.

—Strips—Marine
- MOD UK NES 779: Part 1-87. Requirements for 90/10 Copper Nickel Alloy Material Part 1: Sheet, Strip and Plate Issue 1 (11.87). 17 pp.

Copper Nickel Alloys *(Cont.)*
—Strips—Marine *(Cont.)*
- MOD UK NES 780: Part 1-87. Requirements for 70/30 Copper Nickel Alloy Material Part 1: Sheet, Strip and Plate Issue 1 (06.87). 17 pp.

—Tubes—Marine
- MOD UK NES 779: Part 3-01. Requirements for 90/10 Copper Nickel Alloy Material Part 3: Tubes Issue 1 (10.87); Amendment 1. 34 pp.
- MOD UK NES 780: Part 3-86. Requirements for 70/30 Copper Nickel Alloy Material Part 3: Tubes Issue 2 (09.86). 28 pp.

—Wrought
- ISO 429-83. Wrought Copper-Nickel Alloys—Chemical Composition and Forms of Wrought Products Second Edition. 4 pp.

Copper Nickel Manganese Alloys
See Also: Copper Alloys; Copper Nickel Alloys

—Forgings—Marine
- MOD UK NES 835-88. Requirements for High Strength Copper Nickel Manganese Alloy Forgings, Forging Stock Rods and Sections Issue 1 (12.88). 23 pp.

—Rods—Marine
- MOD UK NES 835-88. Requirements for High Strength Copper Nickel Manganese Alloy Forgings, Forging Stock Rods and Sections Issue 1 (12.88). 23 pp.

—Sections—Marine
- MOD UK NES 835-88. Requirements for High Strength Copper Nickel Manganese Alloy Forgings, Forging Stock Rods and Sections Issue 1 (12.88). 23 pp.

Copper Nickel Zinc Alloys
Scope Note: For additional listings, see also specific products made from copper nickel zinc alloys
See Also: Copper Alloys; Copper Nickel Alloys; Nickel Silver Alloys

—Bars
- CNS H3123-85. Copper-Nickel-Zinc (Nickel Silver) Alloy Rods, Bars and Wires (Dec)(10441).

—Plates
- CNS H3122-85. Copper-Nickel-Zinc (Nickel Silver) Alloy Sheets, Plates and Strips (Dec)(10440).

—Rods
- CNS H3123-85. Copper-Nickel-Zinc (Nickel Silver) Alloy Rods, Bars and Wires (Dec)(10441).

—Semifinished Products—Chemical Analysis
- DIN ENGL 17663-83. Wrought Copper Alloys; Copper-Nickel-Zinc Alloys; (Nickel Silver); Composition (Dec). 6 pp.

—Sheets
- CNS H3122-85. Copper-Nickel-Zinc (Nickel Silver) Alloy Sheets, Plates and Strips (Dec)(10440).

—Strips
- CNS H3122-85. Copper-Nickel-Zinc (Nickel Silver) Alloy Sheets, Plates and Strips (Dec)(10440).

—Wrought
- ISO 430-83. Wrought Copper-Nickel-Zinc Alloys—Chemical Composition and Forms of Wrought Products Second Edition. 4 pp.

Copper Number
See Also: Copper

—Paperboard
- CNS P3074-85. Method of Test for Copper Number of Pulp, Paper and Paperboard (Sep)(11374).

—Papers
- CNS P3074-85. Method of Test for Copper Number of Pulp, Paper and Paperboard (Sep)(11374).

—Pulps
- CNS P3074-85. Method of Test for Copper Number of Pulp, Paper and Paperboard (Sep)(11374).
- CPPA G.22-92. Copper Number of Bleached Pulp. 2 pp.

Copper Ores
See Also: Niccolite

—Dry Matter Content
- JIS M 8101-88. Methods for Sampling, Preparation and Determination of Moisture Content of Non-Ferrous Metal Bearing Ores. 25 pp.

Copper Ores *(Cont.)*
—Moisture Content
- JIS M 8101-88. Methods for Sampling, Preparation and Determination of Moisture Content of Non-Ferrous Metal Bearing Ores. 25 pp.

—Sampling
- CNS M3032-81. Methods for Sampling of Metal Bearing Ores of Copper, Lead, Zinc, Tin, Gold, Silver and Others (Apr)(7280).
- JIS M 8101-88. Methods for Sampling, Preparation and Determination of Moisture Content of Non-Ferrous Metal Bearing Ores. 25 pp.

Copper Oxide Black
Use: Cupric Oxide

Copper Paints (Made From Copper)
Use: Copper Coatings (Made From Copper)

Copper Pipes
Scope Note: Includes pipes made from copper alloys
See Also: Pipes
- SNZ NZS 3501-76. Specification for Copper Tubes for Water, Gas, and Sanitation Amend: 1, 1983; 2; 3. 13 pp.
- SNZ NZS 3502-76. Specification for Copper and Copper Alloy Tubes for General Engineering Purposes (Reconfirmed 1983). 20 pp.

—Gas
- BSI BS 2871: Part 1-71. 1971 Amd 2 Copper and Copper Alloys. Tubes Part 1: Copper Tubes for Water, Gas and Sanitation. 18 pp.
- BSI BS 6891-88. 1988 Installation of Low Pressure Gas Pipework of up to 28mm (R1) in Domestic Premises (2nd Family Gas). 18 pp.
- SAA AS 1432-90. Copper Tubes for Plumbing, Gasfitting and Drainage Applications. 9 pp.

—Seamless
- CNS H2049-80. Method of Test for Copper and Copper Alloy Seamless Pipes and Tubes (Jan)(5128).
- JIS H 3510-92. Oxygen Free Copper Sheet, Plate, Strip, Seamless Pipe and Tube, Rod, Bar and Wire for Electron Devices. 19 pp.

—Seamless—Drawn
- DIN ENGL 1754 Pt 3-74. Copper Tubes; Seamless Drawn; Preferred Dimensions for Pipelines (Apr). 4 pp.
- DIN ENGL 1755 Pt 3-69. Copper Wrought Alloy Tubes; Seamless Drawn; Preferred Dimensions for Pipelines (Aug). 8 pp.
- DIN ENGL 1786-80. Seamless Drawn Copper Tubes for Piping Systems (May). 7 pp.

—Seamless—Eddy Current Testing
- JIS H 0502-86. Method of Eddy Current Testing for Copper and Copper-Alloy Pipes and Tubes. 6 pp.

—Sewer
- BSI BS 2871: Part 1-71. 1971 Amd 2 Copper and Copper Alloys. Tubes Part 1: Copper Tubes for Water, Gas and Sanitation. 18 pp.

—Ships
- BSI BS MA 60-86. 1986 Summary and Application of Copper and Copper Alloy Tubes for Ships' Pipework Systems. 11 pp.
- JIS F 0506-89. Application of Copper Pipes for Marine Use.

—Springs (Elastic)
- BSI BS 5431-76. 1976 Bending Springs for Use with Copper Tubes for Water, Gas and Sanitation. 7 pp.

—Unions
- JIS F 7440-89. Copper Pipe 20 K Brazed Type Unions Marine Use.

—Water
- BSI BS 2871: Part 1-71. 1971 Amd 2 Copper and Copper Alloys. Tubes Part 1: Copper Tubes for Water, Gas and Sanitation. 18 pp.
- SAA AS 1432-90. Copper Tubes for Plumbing, Gasfitting and Drainage Applications. 9 pp.

Copper Powder
See Also: Metal Powders
- CNS H3152-89. Copper Powder (Jun)(12534).
- CNS K2128-82. Copper Powder (Jan)(8424). 1 p.
- JIS H 2114-83. Copper Powder. 8 pp.

—Insoluble Matter Content
- BSI BS 5600: Sec 2.9-80. (WITHDRAWN) 1980 Powder Metallurgical Materials and Products Part 2: Methods of Sampling and Testing Metallic Powders Section 2.9: Determination of Acid-Insoluble Content in Iron, Copper, Tin and Bronze Powders. 4 pp.
- BSI BS EN 24496-93. 1993 Metallic Powders—Determination of Acid Insoluble Content in Iron, Copper, Tin and Bronze Powders (ISO 4496: 1978) (V). 10 pp.

Copper Powder (Cont.)

—Insoluble Matter Content (Cont.)
CEN EN 24496-93. Metallic Powders—Determination of Acid Insoluble Content in Iron, Copper, Tin and Bronze Powders (ISO 4496: 1978). 5 pp.
ISO 4496-78. Metallic Powders—Determination of Acid-Insoluble Content in Iron, Copper, Tin and Bronze Powders First Edition (CEN EN 24496: 1993). 5 pp.

Copper Silicon Alloys
See Also: Copper Alloys

—Bars—Aircraft
MOD UK DTD-498-42. Silicon-Nickel-Copper Alloy Bars and Forgings (Reprinted October 1961, Incorporating Amendment No. 1). 6 pp.
MOD UK DTD-504-42. Copper-Nickel-Silicon Alloy Bars (Reprinted August 1965, Incorporating Amendment No. 1). 4 pp.

—Forgings—Aircraft
MOD UK DTD-498-42. Silicon-Nickel-Copper Alloy Bars and Forgings (Reprinted October 1961, Incorporating Amendment No. 1). 6 pp.

Copper Strip Corrosion
See Also: Corrosion; Corrosion Prevention; Corrosion Testing

—Greases
BSI BS 2000: Part 112-93. 1993 Methods of Test for Petroleum and Its Products Part 112: Determination of Corrosiveness to Copper of Lubricating Grease—Copper Strip Method (Supersedes BS 4455: 1969). 3 pp.
BSI BS 2000: Part 112-82. 1982 Petroleum and Its Products Part 112: Corrosive Substances in Grease (Copper Strip Test). 4 pp.

—Petroleum Products
BSI BS 2000: Part 154-93. 1993 Methods of Test for Petroleum and Its Products Part 154: Petroleum Products—Corrosiveness to Copper—Copper Strip Test (ISO 2160: 1985) (W). 8 pp.
BSI BS 2000: Part 154-82. 1982 Petroleum and Its Products Part 154: Detection of Copper Corrosion from Petroleum Products by the Copper Strip Tarnish Test. 10 pp.
ISO 2160-85. Petroleum Products—Corrosiveness to Copper—Copper Strip Test Second Edition. 9 pp.
JIS K 2513-91. Petroleum Products—Corrosiveness to Copper—Copper Strip Test. 15 pp.

Copper Sulfate
Use: Cupric Sulfate

Copper Sulfate (Anhydrous)
Use: Cupric Sulfate

Copper Sulfide

—Moisture Content—Gravimetric Analysis
ISO 9599-91. Copper, Lead and Zinc Sulfide Concentrates—Determination of Hygroscopic Moisture in the Analysis Sample—Gravimetric Method First Edition. 8 pp.

Copper Tin Alloys
Scope Note: For additional listings, see also specific products made from copper tin alloys
See Also: Copper Alloys; Gunmetal; Phosphor Bronzes

—Wrought
ISO 427-83. Wrought Copper-Tin Alloys—Chemical Composition and Forms of Wrought Products Second Edition. 4 pp.

Copper Wire
Scope Note: Includes wire made from copper alloys
See Also: Copper; Copper Conductors; Copper Wire Fittings; Electric Wire; Hookup Wire; Wire
BSI BS 2873-69. 1969 Copper and Copper Alloys. Wire. 32 pp.
CNS H3123-85. Copper-Nickel-Zinc (Nickel Silver) Alloy Rods, Bars and Wires (Dec)(10441).
CNS H3125-83. Copper and Copper Alloy Wires (Jul)(10443). 4 pp.
CNS H3134-85. Copper-Beryllium Alloy Wire (Jan)(11188).
CNS K7169-66. Chemical Reagent (Cupric Oxide, Wire) (Mar)(1669).
DIN ENGL 1757-74. Wire of Copper and Wrought Copper Alloys, Drawn; Dimensions (June). 7 pp.
DIN ENGL 17652-82. Copper Drawing Stock (June). 6 pp.
DIN ENGL 17677 Pt 2-74. Wires of Copper and Wrought Copper Alloys; Technical Conditions of Delivery (Aug). 2 pp.
DIN ENGL 17682-79. Round Spring Wires Made of Wrought Copper Alloys; Strength Properties; Technical Conditions of Delivery (Aug). 4 pp.

Copper Wire (Cont.)
ISO 1638-87. Wrought Copper and Copper Alloy Wire—Technical Conditions of Delivery Second Edition. 8 pp.
ISO 3492-82. Wrought Copper and Copper Alloys—Drawn Round Wire—Tolerances on Diameter First Edition. 5 pp.
JIS H 3260-92. Copper and Copper Alloy Wires. 15 pp.
JIS H 3270-92. Copper Beryllium Alloy, Phosphor Bronze and Nickel Silver Rods, Bars and Wires. 22 pp.
SAA AS 1573-85. Copper and Copper Alloys—Wire for Engineering Purposes. 10 pp.
SAA AS 1574-84. Copper and Copper Alloys—Wire for Electrical Purposes. 9 pp.

—Automotive
ISO 6722 Pt 3-93. Road Vehicles—Unscreened Low-Tension Cables—Part 3: Conductor Sizes and Dimensions for Thick-Wall Insulated Cables Second Edition. 6 pp.

—Electron Tubes
JIS H 3510-92. Oxygen Free Copper Sheet, Plate, Strip, Seamless Pipe and Tube, Rod, Bar and Wire for Electron Devices. 19 pp.

—Mechanical Properties
DIN ENGL 17677 Pt 1-83. Wrought Copper and Copper Alloy Wire; Properties (Dec). 7 pp.

—Spools
BSI BS 5561-79. 1979 Spools for Copper Wire. 8 pp.

—Springs
DIN ENGL 2076-84. Round Spring Wire; Dimensions, Masses, Permissible Deviations (Dec). 5 pp.

—Springs—Weight
DIN ENGL 2076-84. Round Spring Wire; Dimensions, Masses, Permissible Deviations (Dec). 5 pp.

—Tin Coatings
DIN ENGL 51213-70. Testing of Metallic Coatings on Wires; Coatings of Tin or Zinc (Dec). 5 pp.

—Winding—Polyesterimide Enameled
BSI BS 6811: Sec 3.2-91. 1991 Winding Wires Part 3: Specifications for Particular Types of Round Copper Winding Wires Section 3.2: Polyesterimide Base Enamel. Class 180. 15 pp.

Copper Wire Fittings
See Also: Copper Wire

—Aircraft
BSI G 204-67. 1967 Amd 1 Copper Terminal Ends for Crimping to Electric Cables with Copper Conductors. 13 pp.

—Crimp Sleeves
JIS C 2806-91. Non-Insulated Crimp-Type Sleeves for Copper Conductors. 18 pp.

—Telegraphy
BSI BS 177-70. (WITHDRAWN) 1970 Copper and Copper-Cadmium Tapes and Binders for Telegraph and Telephone Purposes. 12 pp.
BSI BS 181-70. 1970 Copper-Cadmium Jointing Sleeves for Telegraph and Telephone Purposes. 8 pp.

—Telephone Systems
BSI BS 177-70. (WITHDRAWN) 1970 Copper and Copper-Cadmium Tapes and Binders for Telegraph and Telephone Purposes. 12 pp.
BSI BS 179-70. (WITHDRAWN) 1970 Copper Jointing Sleeves for Telegraph and Telephone Purposes. 10 pp.
BSI BS 181-70. 1970 Copper-Cadmium Jointing Sleeves for Telegraph and Telephone Purposes. 8 pp.

Copper Zinc Alloys
Scope Note: For additional listings, see also specific products made from copper zinc alloys
See Also: Brasses; Copper Alloys

—Leaded—Wrought
ISO 426 Pt 2-83. Wrought Copper-Zinc Alloys—Chemical Composition and Forms of Wrought Products—Part 2: Leaded Copper-Zinc Alloys Second Edition. 5 pp.

—Nonleaded—Wrought
ISO 426 Pt 1-83. Wrought Copper-Zinc Alloys—Chemical Composition and Forms of Wrought Products—Part 1: Non-Leaded and Special Copper-Zinc Alloys Second Edition. 5 pp.

Coprocessors

—32-Bit—HCMOS
CECC CECC 90 110-020 ISSUE 2-90. BS CECC 90 110-020; 32 Bit Floating-Point Coprocessor (En). 76 pp.
CECC CECC 90 110-024 ISSUE 1-89. BS CECC 90 110-024; 32 Bit Floating-Point Coprocessor (En). 76 pp.

Copy Machines
Use: Copying Machines

Copy Papers
Use: Reproduction Papers

Copy Preparation
Use For: Manuscript Preparation
See Also: Proofreading
BSI BS 5261: Part 1-75. 1975 Amd 1 Copy Preparation and Proof Correction Part 1: Recommendations for Preparation of Typescript Copy for Printing. 26 pp.

—Mathematics
BSI BS 5261: Part 3-89. 1989 Amd 1 Copy Preparation and Proof Correction Part 3: Marks for Mathematical Copy Preparation and Mathematical Proof Correction and Their Use (AMD 6619) December 21, 1990. 22 pp.

—Symbols
BSI BS 5261: Part 2-76. 1976 Amd 2 Copy Preparation and Proof Correction Part 2: Specification for Typographic Requirements, Marks for Copy Preparation and Proof Correction, Proofing Procedure. 19 pp.
BSI BS 5261C-76. 1976 Copy Preparation and Proof Correction Part 2: Specification for Typographic Requirements, Marks for Copy Preparation and Proof Correction, Proofing Procedure 5261C: is an Extract from BS 5261 Part 2 (G). 8 pp.
ISO 5776-83. Graphic Technology—Symbols for Text Correction First Edition. 7 pp.

Copying Machines
Use For: Copiers; Duplicating Machines; Photocopying Machines *See Also:* Duplicators; Office Machines; Optical Equipment; Photocopies; Reproduction (Copying)
BSI BS 6050-81. 1981 Minimum Information to Be Included in Specification Sheets of Document Copying Machines. 7 pp.
BSI BS ISO/IEC 11159-92. 1992 Information Technology—Office Equipment—Minimum Information to Be Included in Specification Sheets—Copying Machines (S). 19 pp.
IEC 11159-92. Information Technology—Office Equipment—Minimum Information to Be Included in Specification Sheets—Copying Machines First Edition. 17 pp.
ISO 4232 Pt 2-80. Office Machines—Minimum Information to Be Included in Specifications Sheets—Part 2: Document Copying Machines First Edition. 6 pp.
ISO 11159-92. Information Technology—Office Equipment—Minimum Information to Be Included in Specification Sheets—Copying Machines First Edition. 17 pp.
JTC1 4232 Pt 2-80. Office Machines—Minimum Information to Be Included in Specifications Sheets—Part 2: Document Copying Machines First Edition. 6 pp.
JTC1 11159-92. Information Technology—Office Equipment—Minimum Information to Be Included in Specification Sheets—Copying Machines First Edition. 17 pp.
OSI ISO 4232-2-80. Office Machines—Minimum Information to Be Included in Specification Sheets—Part 2: Document Copying Machines. 6 pp.

—Diazo
JIS B 9504-77. Printing Width for Diazo Copying Machines.
JIS B 9505-77. Standard Form of Specification for Diazo Copying Machines.
JIS B 9506-77. Methods of Test for Diazo Copying Machines.

—Glossaries
BSI BS 5479: Part 4-85. 1985 Duplicators and Document Copying Machines Part 4: Glossary of Terms for Document Copying Machines. 18 pp.
JIS B 0137-85. Glossary of Terms for Copying Machines. 34 pp.
JTC1 5138 Sec 02-80. Office Machines—Vocabulary—Section 02: Duplicators First Edition. 31 pp.

—Spirit—Duplicating Liquids
CGSB CAN/CGSB-53.7-92. Duplicating Liquid, Direct-Process, Spirit Type. 8 pp.

Copying Machines (Cont.)
—Stencil
BSI BS 5479: Part 2-77. 1977 Duplicators and Document Copying Machines Part 2: Minimum Overprint and Attachment Features of Stencils for Duplicators. 7 pp.

—Symbols
JIS B 0139-87. Graphic Symbols for Copying Machines. 59 pp.

—Test Charts
JIS B 9523-87. Monochrome Test Chart for Document Copying Machines. 16 pp.
JIS B 9524-89. Test Chart for Full Colour Copying Machines.

Copying Tools
See Also: Tools

—Tool Holders—Indexable Inserts
BSI BS 4193: Part 7-90. 1990 Hardmetal Insert Tooling Part 7: Dimensions for Single Point Turning and Copying Tool Holders for Indexable Inserts. 12 pp.
ISO 5610-89. Single-Point Tool Holders for Turning and Copying, for Indexable Inserts—Dimensions AMENDMENT 1: Tool Holders Style H, J and V, with Rhombic V-Shape Indexable Inserts Third Edition; (Amendment 1-1993). 15 pp.

Copyrights
See Also: Patents
EURO 1989-17. Patents, Trade Marks and Copyright in the European Community. 9 pp.
EURO ECO/IPR/85/1 5460-84. Intellectual Property Rights in International Co-Operative Projects. 88 pp.

—Defense Contracts
MOD UK DEFCON 90-89. Copyright Clause 12/89. 1 p.
MOD UK DEFCON 91-92. Intellectual Property Rights in Software 10/92. 9 pp.

—Standards Preparation
EURO MEMORANDUM 8-92. Standardization and Intellectual Property Rights (IPR). 20 pp.

CORAL
Use For: Computer On-Line Real-Time Applications Language
BSI BS 5905-80. 1980 Amd 1 Computer Programming Language CORAL 66. 34 pp.

Cord Clamps
Use: Cable Clamps

Cord Reels
Use: Cable Reels

Cord Sets (Electric)
Use: Cords (Electric)

Cord Switches
See Also: Cords (Electric); Electric Switches; Switches

—Appliances
BSI BS EN 61058-2-1-93. 1993 Switches for Appliances Part 2: Particular Requirements Section 2.1: Cord Switches (L). 26 pp.
CENELEC EN 61058-2-1-93. Switches for Appliances Part 2-1: Particular Requirements for Cord Switches (IEC 1058-2-1:1992). 21 pp.
IEC 1058 Pt 2-1-92. Switches for Appliances Part 2-1: Particular Requirements for Cord Switches First Edition; (CENELEC EN 61058-2-1: 1993). 44 pp.

Cordage
Scope Note: For additional listings, use a more specific term *Use For:* Bonding Flexibles; Cords (Woven) *See Also:* Cables (Mechanical); Cords (Electric); Fibers; Fourrageres; Ropes; Tire Cords; Twines; Yarns
MOD UK DSTAN 40-6-75. Guide to the Use of Metric Units for Cordage Issue 1. 4 pp.

—Aircraft—Parachute Holders
SBAC AS 426 ISSUE 3(I). Retaining Cord—Parachute Stowage. 1 p.

—Aramid—Aircraft
SBAC AS 44445 (V). Retaining Cord Assembly (Nomex).

—Braided—Elastomeric
BSI BS 7141: Part 5-90. 1990 Narrow Fabrics Part 5: Elastic Flat Braids Containing Natural Rubber. 8 pp.

Cordage (Cont.)
—Braided—Nylon—Aerospace
BSI F 132-87. 1987 Amd 1 Braided Nylon Cord with Specified Dimensional Stability for Aerospace Purposes. 12 pp.
BSI F 135-87. (WITHDRAWN) 1987 Braided Nylon Cord with Unspecified Dimensional Stability for Aerospace Purposes (Superseded by 2F 135: 1992). 8 pp.
BSI 2F 135-92. 1992 Braided Nylon Cord with Unspecified Dimensional Stability for Aerospace Purposes. 12 pp.

—Breaking Load
CGSB 40-GP-1M METH 5-78. Methods of Sampling and Testing Cordage Breaking Strength. 1 p.

—Classification
BSI BS 1000: (679)-83. 1983 Universal Decimal Classification (UDC). English Full Edition (679): Industries Based on Various Processable Materials. Cable and Cordage Industries. Stone Industries. 15 pp.
SNZ NZS/BS 1000 (679)-83. Universal Decimal Classification Industries Based on Various Processable Materials. Cable and Cordage Industries. Stone Industry. 16 pp.

—Coir
CNS L3054-74. Method of Test for Coir Cordage (Mar)(2504).
CNS L4031-74. Coir Cordage (Mar)(2503).

—Color Coding
BSI BS 6033-81. 1981 Colour Code for Identification of the Constituent Material of Ropes and Cordage. 4 pp.

—Comprehensive Testing
BSI BS 5053-85. 1985 Cordage and Webbing Slings and for Fibre Cores for Wire Ropes. 22 pp.
SNZ NZS/BS 5053-85. Methods of Test for Cordage and Webbing Slings and for Fibre Cores for Wire Ropes. 24 pp.

—Copper—Aircraft
SBAC AGS 2097 ISSUE 2. Bonding, Flexibles—General.

—Copper Content
CGSB 40-GP-1M METH 8-78. Methods of Sampling and Testing Cordage Determination of Copper Content. 1 p.

—Cotton
CGSB 40-GP-9MA-89. Cord, Braided Cotton Cover with Cotton or Jute Core. 12 pp.

—Diameters
CGSB 40-GP-1M METH 3-78. Methods of Sampling and Testing Cordage Diameter. 1 p.

—Elongation
CGSB CAN/CGSB-4.2 NO.10-M87. Textile Test Methods Elongation. 9 pp.

—Flax
BSI BS 6125-81. 1981 Natural Fibre Cords, Lines and Twines. 10 pp.

—Flax—Aerospace
BSI 7F 35-85. 1985 Amd 1 Braided Linen (Flax) Lacing Cord for Aerospace Purposes (AMD 6068) October 31, 1989. 6 pp.
BSI 4F 59-85. 1985 Amd 1 Braided Linen (Flax) Cord for Aerospace Purposes. 6 pp.

—Fungus Resistance Testing
SAA AS 1157.3-78. Methods of Testing Materials for Resistance to Fungal Growth—Part 3: Resistance of Cordage and Yarns to Fungal Growth. 6 pp.

—Glossaries
BSI BS 3724-91. 1991 Terms Relating to Fibre Ropes and Cordage. 32 pp.
ISO 1968-73. Ropes and Cordage—Vocabulary First Edition. 7 pp.

—Hemp
BSI BS 6125-81. 1981 Natural Fibre Cords, Lines and Twines. 10 pp.
CGSB 40-GP-5MA-89. Hemp Cordage; (Corrigendum—Sept 1990). 13 pp.

—Jute
CGSB 40-GP-9MA-89. Cord, Braided Cotton Cover with Cotton or Jute Core. 12 pp.

—Linear Density
CGSB 40-GP-1M METH 4-78. Methods of Sampling and Testing Cordage Linear Density. 1 p.

—Manila Fiber Content—Swett Test
CGSB 40-GP-1M METH 9-78. Methods of Sampling and Testing Cordage Nonmanila Fiber. 1 p.

Cordage (Cont.)
—Nonmetallic
MOD UK DSTAN 40-7-75. Selection, Use, Care, Inspection and Maintenance of Non-Metallic Ropes and Cords Issue 1. 24 pp.

—Nonmetallic—Inspection
MOD UK DSTAN 40-7-75. Selection, Use, Care, Inspection and Maintenance of Non-Metallic Ropes and Cords Issue 1. 24 pp.

—Nonmetallic—Maintenance
MOD UK DSTAN 40-7-75. Selection, Use, Care, Inspection and Maintenance of Non-Metallic Ropes and Cords Issue 1. 24 pp.

—Nylon—Aircraft
MOD UK DTD-5620-01. Braided Nylon Cord; Amendment 1. 10 pp.

—Packaging
BSI BS 1133: Sec 13-54. (OBSOLESCENT) 1954 Amd 1 Packaging Code Section 13: Twines and Cords for Packaging. 18 pp.

—Polyethylene Terephthalate—Aircraft
SBAC AS 41091 ISSUE 4(I). Cord Assembly, Retaining (Terylene).

—Rayon/Nylon—Aircraft
MOD UK DTD-5524-58. Rayon/Nylon Braided Cord. 1 p.

—Rubber—Aerospace
BSI 2F 70-83. (WITHDRAWN) 1983 Amd 1 Heavy Duty Braided Rubber Cord (AMD 5356) September 30,1986 (Superseded by BS 3F 70: 1991). 7 pp.
BSI 3F 70-91. 1991 Heavy Duty Braided Rubber Cord. 9 pp.
BSI 2F 71-83. (WITHDRAWN) 1983 Amd 1 Light Duty Braided Rubber Cord (AMD 5355) October 31, 1986. 7 pp.
BSI 3F 71-91. 1991 Light Duty Braided Rubber Cord. 10 pp.

—Rubber—Aerospace—Assemblies
BSI 2SP 170-171-85. 1985 Amd 1 Braided Rubber Cord Assemblies for Aerospace Use. 5 pp.

—Sampling
BSI BS 5053-85. 1985 Cordage and Webbing Slings and for Fibre Cores for Wire Ropes. 22 pp.
CGSB 40-GP-1M METH 1-78. Methods of Sampling and Testing Cordage Sampling. 1 p.
CGSB 40-GP-1M METH 2-78. Methods of Sampling and Testing Cordage Conditioning of Samples. 1 p.
CGSB 40-GP-1M METH 3-78. Methods of Sampling and Testing Cordage Diameter. 1 p.
SNZ NZS/BS 5053-85. Methods of Test for Cordage and Webbing Slings and for Fibre Cores for Wire Ropes. 24 pp.

—Solvent Extraction
CGSB 40-GP-1M METH 7-78. Methods of Sampling and Testing Cordage Solvent Extractable Material. 1 p.

—Telecommunication Equipment—Naval Ships
MOD UK NES 563-81. Guide to Internal Communication Equipment Cord and Cordage Issue 2 (01.81). 102 pp.
MOD UK NES 563-93. Guide to Internal Communication Equipment Cords and Cordage Issue 3 (05.93). 69 pp.

—Telecommunication Equipment—Submarines
MOD UK NES 563-81. Guide to Internal Communication Equipment Cord and Cordage Issue 2 (01.81). 102 pp.
MOD UK NES 563-93. Guide to Internal Communication Equipment Cords and Cordage Issue 3 (05.93). 69 pp.

—Turn
CGSB 40-GP-1M METH 6-78. Methods of Sampling and Testing Cordage Turn. 1 p.

Cordless Telephone Sets
Use: Cordless Telephones

Cordless Telephones
Use For: Cordless Telephone Sets
See Also: Telephones
BSI BS 6833: Part 2-92. 1992 Apparatus Using Cordless Attachments (Excluding Cellular Radio Apparatus) for Connection to Analogue Interfaces of Public Switched Telephone Networks Part 2: Spec. for Cordless Telephone Apparatus Using Radio Links. 63 pp.

INTERNATIONAL AND NON-U.S. NATIONAL STANDARDS
SUBJECT INDEX
Cords

Cordless Telephones *(Cont.)*
BSI BS 6833: Part 2-87. 1987 Apparatus Using Cordless Attachments (Excluding Cellular Radio Apparatus) for Connection to Analogue Interfaces of Public Switched Telephone Networks Part 2: Cordless Telephone Apparatus Using Radio Links. 56 pp.
CCIR Report 1025-1-90. Technical and Operating Characteristics of Cordless Telephones—Section 8A—Land Mobile Service and Related Subjects. 18 pp.
CENELEC PRI-ETS 300 131-90. Radio Equipment and Systems Second Generation Cordless Telephones Common Air Interface Specification to be Used for the Interworking Between Cordless Telephone Apparatus Including Public Access Services. 145 pp.
CENELEC PRI-ETS 300 131-93. Radio Equipment and Systems (RES); Common Air Interface Specification to Be Used for the Interworking Between Cordless Telephone Apparatus in the Frequency Band 864,1 MHz to 868,1 MHz, Including Public Access Services. 266 pp.
CENELEC PRI-ETS 300 235-92. Radio Equipment and Systems (RES); Technical Characteristics, Test Conditions and Methods of Measurement for Radio Aspects of Cordless Telephones CT1. 36 pp.
CEPT T/CS 34-18-84. Aspects Techniques Du Telephone Sans Cordon. 18 pp.
CEPT T/CS 34-18 E-84. Cordless Telephone Technical Aspects. 18 pp.
CEPT T/R 24-03-87. Caracteristiques Radioelectriques Des Telephones Sans Cordon. 26 pp.
CEPT T/R 24-03 E-88. Radio Characteristics of Cordless Telephones. 27 pp.
CEPT T/SF 30-83. Aspects, Services Et Facilities Des Telephones Sans Cordon. 4 pp.
CEPT T/SF 42 E-86. Services and Facilities Aspects of 2nd Generation Cordless Telephones. 4 pp.
CSA CAN/CSA-C22. 2 NO 225-M90. Telecommunication Equipment; (Gen Instr 1 Thru 3). 115 pp.
ETSI PRETS 300 131-90. Radio Equipment and Systems Second Generation Cordless Telephones Common Air Interface Specification to be Used for the Interworking Between Cordless Telephone Apparatus Including Public Access Services. 144 pp.
ETSI PRI-ETS 300 131-93. Radio Equipment and Systems (RES); Common Air Interface Specification to Be Used for the Interworking Between Cordless Telephone Apparatus in the Frequency Band 864,1 MHz to 868,1 MHz, Including Public Access Services. 266 pp.
ETSI PRI-ETS 300 131-90. Radio Equipment and Systems Second Generation Cordless Telephones Common Air Interface Specification to be Used for the Interworking Between Cordless Telephone Apparatus Including Public Access Services. 145 pp.
ETSI PRI-ETS 300 235-92. Radio Equipment and Systems (RES); Technical Characteristics, Test Conditions and Methods of Measurement for Radio Aspects of Cordless Telephones CT1. 36 pp.

—**Public Switched Telephone Networks**
BSI PD 6570: Part 2-92. 1992 Apparatus Using Cordless Attachments (Excluding Cellular Radio Apparatus) for Connection to Analogue Interfaces of Public Switched Telephone Networks Run by Certain Public Telecommunications Operators Part 2:. 64 pp.
BSI PD 6570: Part 6-92. 1992 Apparatus Using Cordless Attachments (Excluding Cellular Radio Apparatus) for Connection to Analogue Interfaces of Public Switched Telephone Networks Run by Certain Public Telecommunications Operators Part 6:. 61 pp.
BSI PD 6572-92. 1992 Apparatus Using Cordless Attachments (Excluding Cellular Radio Apparatus) for Connection to Analogue Interfaces of Public Switched Telephone Networks Part 6: Specification for Cordless Portable Parts and Cordless Fixed Parts. 62 pp.
CCIR Report 1025-1-90. Technical and Operating Characteristics of Cordless Telephones—Section 8A—Land Mobile Service and Related Subjects. 18 pp.

Cords (Electric)
Use For: Fixture Wire; Portable Cords (Electric); Service Cords (Electric); **See Also:** Cables (Electric); Cord Switches; Cordage; Extension Cords; Heater Cords; Lighting Chains; Power Supply Cords; Power Systems; Telephone Cords; Tinsel Cords
BSI BS 6141-91. 1991 Insulated Cables and Flexible Cords for Use in High Temperature Zones. 26 pp.
BSI BS 6141-01. 1991 Amd 1 Insulated Cables and Flexible Cords for Use in High Temperature Zones (AMD 7567) May 15, 1993 (F). 32 pp.
BSI BS 6141-81. 1981 Amd 2 Insulated Cables and Flexible Cords for Use in High Temperature Zones (AMD 6228) April 30, 1990. 27 pp.
CENELEC EN 60 799-87. Cord Sets. 14 pp.
CNS C3191-89. Method of Test for Cord Sets (Nov)(10918).

Cords (Electric) *(Cont.)*
CNS C4412-87. General Rule for Cord Sets (Jun)(10917).
CNS C4412-1-85. General Use Cord Sets (Feb)(10917-1).
CSA CAN/CSA-C22. 2 NO 21-M90. Cord Sets and Power-Supply Cords; (Gen Instr 1 Thru 2). 74 pp.
CSA 895C Bull. Electrical Bulletin 895C July 28, 1978 to C22.2 NO 21. 2 pp.
CSA 983B Bull. Electrical Bulletin 983B April 25, 1977 to C22.2 NO 38. 39 pp.
DIN VDE 0472 Suppl. 1-87. Testing of Cables, Wires and Flexible Cords; List of Standards in the DIN VDE 0472 Series (June). 18 pp.
DIN VDE 0472 Pt 1-87. Testing of Cables, Wires and Flexible Cords; General Requirements (June). 13 pp.
DIN VDE 0472 Pt 401-84. Testing of Cables and Insulated Flexible Cords; Outer Dimensions (June). 8 pp.
IEC 541-76. Comparative Information on IEC and North-American Flexible Cord Types First Edition. 27 pp.
IEC 799-84. Cord Sets First Edition; (Amendment 1-1993). 21 pp.
MOD UK DSTAN 61-12: Part 1-01. Wires, Cords, and Cables, Electrical-Metric Units Part 1: Cables, Electrical (Insulated Flexible Cords and Flexible Cables) Issue 4; Amendment 5. 47 pp.
MOD UK DSTAN 61-12: Part 7-86. Wires, Cords, and Cables, Electrical-Metric Units Part 7: Cords, Electrical General Requirements Issue 2. 31 pp.
MOD UK DSTAN 61-12: Part 11-01. Wires, Cords, and Cables, Electrical-Metric Units Part 11: Imperial/Metric Cross Reference Index Issue 2; Amendment 1. 89 pp.
SAA AS 1660. Methods of Test for Electric Cables, Cords and Conductors Complete Set in Binder.
SAA AS 3191-91. Approval and Test Specification—Electric Flexible Cords Amdt 1 June 1992 (in Professional Package 28). 36 pp.
SNZ NZS/AS 3191-91. Approval and Test Specification for Electric Flexible Cords (This is a Joint Standard with SAA AS 3191). 43 pp.
SNZ NZS 6217-89. Specification for Cord Extension Sockets. 16 pp.
SNZ NZS 6218-89. Specification for Plug Socket Adaptors. 16 pp.
SNZ NZS 6219-89. Specification for Electrical Portable Outlet Devices. 16 pp.

— **300 V**
CSA 983B Bull. Electrical Bulletin 983B April 25, 1977 to C22.2 NO 38. 39 pp.
CSA CAN/CSA-C22. 2 NO 49-92. Flexible Cords and Cables; (Gen Instr 1). 121 pp.
CSA 983B Bull. Electrical Bulletin 983B April 25, 1977 to C22.2 NO 49. 39 pp.
CSA 1378 Bull. Electrical Bulletin 1378 August 23, 1982 to C22.2 NO 49. 4 pp.

— **600 V**
CSA 983B Bull. Electrical Bulletin 983B April 25, 1977 to C22.2 NO 38. 39 pp.
CSA CAN/CSA-C22. 2 NO 49-92. Flexible Cords and Cables; (Gen Instr 1). 121 pp.
CSA 983B Bull. Electrical Bulletin 983B April 25, 1977 to C22.2 NO 49. 39 pp.
CSA 1378 Bull. Electrical Bulletin 1378 August 23, 1982 to C22.2 NO 49. 4 pp.

—**Adapters**
CNS C4412-5-85. Adapter Cord Sets (Feb)(10917-5).

—**Adhesive Strength**
DIN VDE 0472 Pt 618-83. Testing of Cables, Wires and Flexible Cords; Adhesive Strength (Jan). 6 pp.

—**Aerospace—Rubber Insulated—Braided—Ferrules**
BSI 2SP 168-169-85. 1985 Ferrules and Assembly Wires for Braided Rubber Cord Assemblies for Aerospace Use. 4 pp.

—**Aerospace—Rubber Insulated—Braided—Steel Wire**
BSI 2SP 168-169-85. 1985 Ferrules and Assembly Wires for Braided Rubber Cord Assemblies for Aerospace Use. 4 pp.

—**Aging Testing**
DIN VDE 0472 Pt 303-90. Testing of Cables, Wires and Flexible Cords; Ageing Procedures (May). 15 pp.

—**Appliances—Three Conductor**
CSA CAN/CSA-C22. 2 NO 21-M90. Cord Sets and Power-Supply Cords; (Gen Instr 1 Thru 2). 74 pp.
CSA 895B Bull. Electrical Bulletin 895B May 12, 1976 to C22.2 NO 21. 2 pp.

—**Bend Testing**
DIN VDE 0472 Pt 610-85. Testing of Cables, Wires and Flexible Cords; Bending Test at Low Temperature (Jan). 6 pp.

Cords (Electric) *(Cont.)*
—**Breaking Length**
DIN VDE 0472 Pt 626-83. Testing of Cables, Wires and Flexible Cords; Breaking Length (Jan). 5 pp.

—**Cable Reels**
CSA CAN/CSA-C22. 2 NO 21-M90. Cord Sets and Power-Supply Cords; (Gen Instr 1 Thru 2). 74 pp.

—**Color Coding**
BSI PD 2379-90. 1990 Register of Colours of Manufacturers' Identification Threads for Electric Cables and Cords. 25 pp.
CENELEC HD 27-78. Colours of the Cores of Flexible Cables and Cords. 7 pp.
DIN VDE 0293-90. Identification of Cores and Cables and Flexible Cords Used in Power Installations with Voltage Ratings up to 1000 V (Jan). 14 pp.
IEC 173-64. Colours of the Cores of Flexible Cables and Cords First Edition. 5 pp.
SNZ NZS/BSPD 2379-90. Register of Colours of Manufacturers' Identification Threads for Electric Cables and Cords. 26 pp.

—**Cross Reference Index**
MOD UK DSTAN 61-12: Part 90-89. Wires, Cords and Cables of Assessed Quality Part 90: Detail Specifications Issue 1. 5 pp.
MOD UK DSTAN 61-12: Part 90-93. Wires, Cords and Cables Electrical Metric Units Part 90: Detail Specification Cross Reference Index Issue 2. 5 pp.
MOD UK DSTAN 61-12: Part 90-01. Wires, Cords and Cables Electrical Metric Units Part 90: Detail Specification Cross Reference Index Issue 2; Amendment 1. 6 pp.

—**Current Ratings**
CNS C1116-90. Security Current of Rubber Insulated Flexible Cords (Feb)(9827).
DIN VDE 0100 Pt 523-81. Installation of Power Plant with Rated Voltages up to 1000 V; Dimensioning of Cables and Cords; Mechanical Strength, Voltage Drop and Current Carrying Capacity (June). 10 pp.
DIN VDE 0298 Pt 2-79. Application of Cables and Flexible Cords in Power Installations; Recommended Values for the Current Carrying of Cables with Rated Voltages Uo/U up to 18/30 kV (Nov). 73 pp.
DIN VDE 0298 Pt 2 (D)-90. Application of Cables and Leads in Power Installations; Recommended Current-Carrying Capacities for Cables with Rated Voltages V/V up to 18/30 kV (July). 54 pp.

—**Data Processing Equipment**
DIN VDE 0814-81. Cords for Telecommunication Systems and Information Processing Systems (Oct). 28 pp.
DIN VDE 0891 Pt 1-90. Use of Cables and Insulated Cords for Telecommunications and Information Processing Systems; General Provisions (May). 27 pp.
DIN VDE 0891 Pt 4-81. Use of Cables and Insulated Cords for Telecommunications and Information Processing Installations; Special Guidance on Cords According to DIN 57 814/VDE 0814 (VDE Guide) (Dec). 8 pp.

—**Detonators—PVC Insulated**
MOD UK DSTAN 61-12: Part 17-01. Wires, Cords, and Cables, Electrical-Metric Units Part 17: Cables, Special Purpose, Electrical (for Detonator Firing Circuits) Issue 1; Amendment 1. 11 pp.

—**Dielectric Loss Factor**
DIN VDE 0472 Pt 505-83. Testing of Cables and Insulated Flexible Cords; Loss Factor, Dielectric Loss Coefficient and Leakage (Apr). 9 pp.

—**Dielectric Strength**
DIN VDE 0472 Pt 508-86. Testing of Insulated Cables and Flexible Cords; Dielectric Strength of Insulated Cables and Flexible Cords for Power Installations (May). 8 pp.
DIN VDE 0472 Pt 509-86. Testing of Cables, Wires and Flexible Cords; Dielectric Strength on Cables, Wires and Cords for Telecommunications and Information Processing Systems (Oct). 11 pp.

—**Door Openers**
CSA CAN/CSA-C22. 2 NO 49-92. Flexible Cords and Cables; (Gen Instr 1). 121 pp.
CSA 427 Bull. Electrical Bulletin 427 March 14, 1958 to C22.2 NO 49. 1 p.

—**Effective Capacitance**
DIN VDE 0472 Pt 504-83. Testing of Cables, Wires and Flexible Cords; Effective Capacitance (Apr). 5 pp.

—**Electric Outlets**
CNS C4412-19-85. General-Purpose Nonlocking Plugs and Receptacles for Cord Sets (Feb)(10917-19).
CNS C4412-20-85. Specific-Purpose Locking Plugs and Receptacles for Cord Sets (Feb)(10917-20).

INDUSTRY STANDARDS

INTERNATIONAL AND NON-U.S. NATIONAL STANDARDS
SUBJECT INDEX

Cords

Cords (Electric) (Cont.)

—Electric Outlets (Cont.)
CNS C4412-21-85. Other Plugs and Receptacles for Cord Sets (Feb)(10917-21).

—Electric Resistance
DIN VDE 0472 Pt 501-83. Testing of Cables and Insulated Cords; Conductor Resistance (Apr). 7 pp.
DIN VDE 0472 Pt 510-84. Testing of Cables and Insulated Flexible Cords; Resistance to Direct Current (Sept). 6 pp.

—Electrical Codes
CSA 1334 Bull. Electrical Bulletin 1334 August 28, 1981 to C22.2 NO 16. 1 p.
CSA 851 Bull. Electrical Bulletin 851 November 17, 1971 to C22.2 NO 38. 2 pp.
CSA 675A Bull. Electrical Bulletin 675A January 2, 1968 to C22.2 NO 49. 1 p.
CSA 675B Bull. Electrical Bulletin 675B June 20, 1968 to C22.2 NO 49. 1 p.
CSA 851 Bull. Electrical Bulletin 851 November 17, 1971 to C22.2 NO 49. 2 pp.
CSA 959X Bull. Electrical Bulletin 959X May 14, 1976 to C22.2 NO 49. 1 p.
CSA 1293 Bull. Electrical Bulletin 1293 November 7, 1980 to C22.2 NO 49. 3 pp.
CSA 1334 Bull. Electrical Bulletin 1334 August 28, 1981 to C22.2 NO 49. 1 p.

—Electrical Equipment
CSA C22.2 NO 0.1-M1985. General Requirements for Double-Insulated Equipment; (Gen Instr 1). 19 pp.

—Electrical Installations
DIN VDE 0100 Pt 730-86. Erection of Power Installations with Nominal Voltages up to 1000 V; Laying of Cables and Cords in Hollow Walls and in Buildings and Structures Made of Chiefly Combustible Building Materials According to DIN 4102 (Feb). 7 pp.
DIN VDE 0245 Pt 1 (D)-90. Cables & Cords for Electrical & Electronic Equipment in Power Installations; General Requirements (Oct). 15 pp.
DIN VDE 0298 Pt 1-82. Application of Cables and Flexible Cords in Power Installations; General Requirement for Cables with Rated Voltages Uo/U up to 18/30 kV (Nov). 29 pp.

—Electrical Insulating Bushings
CNS C4214-80. Drop Cord Bushing (Aug)(6047).

—Electrical Properties
IEC 885 Pt 1-87. Electrical Test Methods for Electric Cables Part 1: Electrical Tests for Cables, Cords and Wires for Voltages up to and Including 450/750 V First Edition. 13 pp.
SAA AS 1660.3-93. Electrical Tests. 22 pp.

—Electrical Protection Equipment
DIN VDE 0100 Pt 430-81. Installation of Power Plant with Rated Voltages up to 1000 V; Protection of Cables and Cords Against Undue Temperature Rise (June). 22 pp.
DIN VDE 0100 Pt 430 Suppl. 1-91. Erection of Power Installations with Nominal Voltages up to 1000V; Protection of Cables and Cords Against Overcurrent; Recommended Values for Current-Carrying Capacity Iz and the Allocation of Overcurrent. 17 pp.

—Fire Testing
DIN VDE 0472 Pt 804-89. Testing of Cables, Wires and Flexible Cords; Burning. 17 pp.

—Flammability Testing—Identification Systems
CSA CAN/CSA-C22. 2 NO 49-92. Flexible Cords and Cables; (Gen Instr 1). 121 pp.

—Flat—Neoprene Sheathed
DIN VDE 0250 Pt 809-85. Cables, Wires and Flexible Cords for Power Installations; Flat Ordinary Tough-Polychloroprene-Sheathed Flexible Cable (May). 11 pp.

—Flat—PVC Insulated
DIN VDE 0281 Pt 302-85. PVC Cables, Wires and Flexible Cords for Power Installation; Flat Twin Flexible Cord (Apr). 7 pp.

—Glass Insulated
BSI BS 6500-90. 1990 Amd 1 Insulated Flexible Cords and Cables (AMD 6759) September 30, 1991 (F). 51 pp.
BSI BS 6500-02. 1990 Amd 2 Insulated Flexible Cords and Cables (AMD 6866) April 1, 1992. 56 pp.

—Glossaries
DIN VDE 0289 Pt 1-88. Definition for Insulated Cables and Cords; General Definitions (Mar) (Together with DIN VDE 0289 P.2/03.88 and DIN VDE 0289 P.4/03.88 Supersedes DIN 57289 Part 100/VDE 0289 Part 100/11.79). 11 pp.

Cords (Electric) (Cont.)

—Glossaries (Cont.)
DIN VDE 0289 Pt 4-88. Definitions for Cables, Wires and Flexible Cords for Power Installations Testing (Mar) (Together with DIN VDE 0289 Part 1 and DIN VDE 0289 Part 2 Supersedes DIN 57 289 Part 100/VDE 0289 Part 100). 6 pp.
DIN VDE 0289 Pt 5-88. Definitions for Cables, Wires and Flexible Cords for Power Installations; Lengths (Mar). 3 pp.

—Halogen Free
DIN VDE 0472 Pt 815-89. Testing of Cables, Wires and Flexible Cords; Non-Halogen Verification (Mar). 14 pp.

—Identification Systems
CSA 1368 Bull. Electrical Bulletin 1368 April 7, 1982 to C22.2 NO 38. 2 pp.
CSA 1368 Bull. Electrical Bulletin 1368 April 7, 1982 to C22.2 NO 49. 2 pp.
DIN VDE 0293-90. Identification of Cores and Cables and Flexible Cords Used in Power Installations with Voltage Ratings up to 1000 V (Jan). 14 pp.

—Impact Testing
DIN VDE 0472 Pt 611-85. Testing of Cables and Insulated Flexible Cords; Cold Impact Test (Jan). 7 pp.

—Insulated
CSA 1334 Bull. Electrical Bulletin 1334 August 28, 1981 to C22.2 NO 16. 1 p.

—Insulation—Thickness Measurement
DIN VDE 0472 Pt 402-84. Testing of Insulated Cables, Wires and Flexible Cords; Thickness of Insulation and Armouring (May). 12 pp.

—Leakage
DIN VDE 0472 Pt 505-83. Testing of Cables and Insulated Flexible Cords; Loss Factor, Dielectric Loss Coefficient and Leakage (Apr). 9 pp.

—Lighting Chains—300 V
BSI BS 6726-91. 1991 Festoon and Temporary Lighting Cables and Cords. 27 pp.
BSI BS 6726-01. 1991 Amd 1 Festoon and Temporary Lighting Cables and Cords (AMD 6939) May 1, 1992. 31 pp.
BSI BS 6726-02. 1991 Amd 2 Festoon and Temporary Lighting Cables and Cords (AMD 7888) September 15, 1993 (F). 37 pp.

—Loss Factor
DIN VDE 0472 Pt 505-83. Testing of Cables and Insulated Flexible Cords; Loss Factor, Dielectric Loss Coefficient and Leakage (Apr). 9 pp.

—Medical Electrical Equipment
CSA CAN/CSA-C22. 2NO601.1-M90. Medical Electrical Equipment, Part 1: General Requirements for Safety; (Gen Instr 1). 240 pp.

—Medical Electrical Equipment—Three Conductor
CSA CAN/CSA-C22. 2 NO 21-M90. Cord Sets and Power-Supply Cords; (Gen Instr 1 Thru 2). 74 pp.
CSA 895B Bull. Electrical Bulletin 895B May 12, 1976 to C22.2 NO 21. 2 pp.

—Office Machines—Three Conductor
CSA CAN/CSA-C22. 2 NO 21-M90. Cord Sets and Power-Supply Cords; (Gen Instr 1 Thru 2). 74 pp.
CSA 895B Bull. Electrical Bulletin 895B May 12, 1976 to C22.2 NO 21. 2 pp.

—Outdoor
CNS C4412-3-85. Outdoor-Use Cord Sets (Feb)(10917-3).

—PVC Insulated
BSI BS 6500-90. 1990 Amd 1 Insulated Flexible Cords and Cables (AMD 6759) September 30, 1991 (F). 51 pp.
BSI BS 6500-02. 1990 Amd 2 Insulated Flexible Cords and Cables (AMD 6866) April 1, 1992. 56 pp.
CENELEC HD 21.5 S2-81. Polyvinyl Chloride Insulated Cables of Rated Voltages up to and Including 450/750 V—Part 5: Flexible Cables. 19 pp.
CENELEC HD 21.5 S2/A3-90. AMD 3 Polyvinyl Chloride Insulated Cables of Rated Voltages up to and Including 450/750 V—Part 5: Flexible Cables. 4 pp.
CNS C2053-89. Polyvinyl Chloride Insulated Flexible Cords (Nov)(3199).
DIN VDE 0281 Pt 1-85. PVC Cables, Wires and Flexible Cords for Power Installation; General Requirements (Apr). 13 pp.
IEC 227 Pt 5-79. Polyvinyl Chloride Insulated Cables of Rated Voltages up to and Including 450/750 V Part 5: Flexible Cables (Cords) First Edition; (Amendment 1-1987). 38 pp.

Cords (Electric) (Cont.)

—PVC Insulated (Cont.)
JIS C 3306-87. Polyvinyl Chloride Insulated Flexible Cords. 14 pp.
JIS C 3307-87. 600 V Grade Polyvinyl Chloride Insulated Wires. 11 pp.
MOD UK DSTAN 61-12: 90/001-88. Cables, Special Purpose, Electrical (Digital Data Transmission) Issue 2. 10 pp.
SNZ NZS 6402-73. Specification for PVC-Insulated Flexible Cords. 23 pp.

—PVC Insulated—PVC Sheathed
BSI BS 6500-90. 1990 Amd 1 Insulated Flexible Cords and Cables (AMD 6759) September 30, 1991 (F). 51 pp.
BSI BS 6500-02. 1990 Amd 2 Insulated Flexible Cords and Cables (AMD 6866) April 1, 1992. 56 pp.
DIN VDE 0281 Pt 401-85. PVC Cables, Wires and Flexible Cords for Power Installation; PVC-Sheathed Flexible Cord 03VV (Apr). 9 pp.
DIN VDE 0281 Pt 402-88. PVC Cables, Wires and Flexible Cords for Power Installation; PVC-Sheathed Flexible Cord 05VV (Mar). 9 pp.

—Quality Assurance
MOD UK DSTAN 61-12: Part 0-92. Wires Cords and Cables Electrical—Metric Units Part 0: General Requirements and Test Methods for Qualification Approval Generic Specification Issue 1. 62 pp.
MOD UK DSTAN 61-12: Part 0-01. Wires Cords and Cables Electrical—Metric Units Part 0: General Requirements and Test Methods for Qualification Approval Generic Specification Issue 1; Amendment 1. 65 pp.

—Recreational Vehicles
CNS C4412-6-85. Recreational-Vehicle Cord Sets (Feb)(10917-6).

—Rubber Insulated
CENELEC HD 22.4 S2-83. Rubber Insulated Cables of Rated Voltages up to and Including 450/750 V—Part 4: Cords and Flexible Cables. 22 pp.
CENELEC HD 22.4 S2-92. Rubber Insulated Cables of Rated Voltages up to and Including 450/750 V Part 4: Cords and Flexible Cables (Reprint Incorporating A1 to A5). 26 pp.
CENELEC HD 22.4 S2/A6-92. AMD 6 Rubber Insulated Cables of Rated Voltages up to and Including 450/750 V Part 4: Cords and Flexible Cables. 4 pp.
CNS C2001-87. Rubber Insulated Flexible Cords (Sep)(546).
CNS F5005-86. Protective Rubber-Like Sheaths of Flexible Cord for Marine Use (Aug)(7907).
DIN VDE 0282 Pt 1-85. Rubber Cables, Wires and Flexible Cords for Power Installation; General Requirements (Apr). 16 pp.
IEC 245 Pt 4-80. Rubber Insulated Cables of Rated Voltage up to and Including 450/750 V Part 4: Cords and Flexible Cables First Edition; (Amendment 2-1988, Incorporating Amendment 1). 56 pp.
JIS C 3301-87. Rubber Insulated Flexible Cords. 21 pp.
MOD UK DSTAN 61-12: 90/001-88. Cables, Special Purpose, Electrical (Digital Data Transmission) Issue 2. 10 pp.

—Rubber Insulated—Braided
BSI BS 6500-90. 1990 Amd 1 Insulated Flexible Cords and Cables (AMD 6759) September 30, 1991 (F). 51 pp.
BSI BS 6500-02. 1990 Amd 2 Insulated Flexible Cords and Cables (AMD 6866) April 1, 1992. 56 pp.
DIN VDE 0282 Pt 801-85. Rubber Cables, Wires and Flexible Cords for Power Installation; Braided Flexible Cord (Apr). 6 pp.

—Saponification Number
DIN VDE 0472 Pt 704-89. Testing of Insulated Cables and Cords; Saponification Value (Mar) (Supersedes DIN 57 472 Part 704). 6 pp.

—Ships
CNS C2110-88. General Rules for Cables and Flexible Cords for Electrical Equipment of Ships (Apr)(9124).
CNS C3161-88. Method of Test for Cables and Flexible Cords for Electrical Equipment of Ships (Apr)(9125).
CNS F5005-86. Protective Rubber-Like Sheaths of Flexible Cord for Marine Use (Aug)(7907).
JIS C 3410-87. Cables and Flexible Cords for Electrical Equipment of Ships. 86 pp.

—Solderability
DIN VDE 0472 Pt 808-84. Testing of Cables, Wires and Flexible Cords; Tinning, Solderability and Soldering Shrinkage (VDE Specification) (Feb). 7 pp.

INTERNATIONAL AND NON-U.S. NATIONAL STANDARDS
SUBJECT INDEX

Cores

Cords (Electric) *(Cont.)*
—**Splitting Testing**
DIN VDE 0472 Pt 627-83. Testing of Cables, Wires and Flexible Cords; Separability of Flat Twin Flexible Cords (VDE Specification) (Jan). 4 pp.

—**Tear Strength**
DIN VDE 0472 Pt 613-85. Testing of Cables, Wires and Flexible Cords; Tear Resistance (Nov) (Partial Replacement for VDE 0472/09.71 and VDE 0472d/12.77). 5 pp.

—**Telecommunication Equipment**
DIN VDE 0814-81. Cords for Telecommunication Systems and Information Processing Systems (Oct). 28 pp.
DIN VDE 0891 Pt 1-90. Use of Cables and Insulated Cords for Telecommunications and Information Processing Systems; General Provisions (May). 27 pp.
DIN VDE 0891 Pt 4-81. Use of Cables and Insulated Cords for Telecommunications and Information Processing Installations; Special Guidance on Cords According to DIN 57 814/VDE 0814 (VDE Guide) (Dec). 8 pp.
MOD UK DSTAN 61-12: Pt 7:Sec 1-01. Wires, Cords, and Cables, Electrical-Metric Units Part 7: Section 1: Cordage Issue 1; Amendment 1. 54 pp.
MOD UK DSTAN 61-12: Pt 7:Sec 2-75. Wires, Cords, and Cables, Electrical—Metric Units Part 7: Section 2: Cords, Telecommunication Issue 1. 27 pp.

—**Telecommunication Equipment—Bend Testing**
DIN VDE 0472 Pt 603-89. Testing of Cables, Wires and Flexible Cords; Bending Behaviour (VDE Specification) (July). 17 pp.

—**Telecommunication Equipment—Naval Ships**
MOD UK NES 563-81. Guide to Internal Communication Equipment Cord and Cordage Issue 2 (01.81). 102 pp.
MOD UK NES 563-93. Guide to Internal Communication Equipment Cords and Cordage Issue 3 (05.93). 69 pp.

—**Telecommunication Equipment—Submarines**
MOD UK NES 563-81. Guide to Internal Communication Equipment Cord and Cordage Issue 2 (01.81). 102 pp.
MOD UK NES 563-93. Guide to Internal Communication Equipment Cords and Cordage Issue 3 (05.93). 69 pp.

—**Telecommunication Systems**
CSA CAN/CSA-C22. 2 NO 233-M89. Cords and Cord Sets for Communication Systems; (Gen Instr 1 Thru 2). 31 pp.

—**Telecommunication Systems—Multiconductor**
CSA CAN/CSA-C22. 2 NO 233-M89. Cords and Cord Sets for Communication Systems; (Gen Instr 1 Thru 2). 31 pp.

—**Telecommunication Systems—Two Conductor**
CSA CAN/CSA-C22. 2 NO 233-M89. Cords and Cord Sets for Communication Systems; (Gen Instr 1 Thru 2). 31 pp.

—**Testing**
SAA AS 1660.4-86. Methods of Test for Electrical Cables, Cords and Conductors—Part 4: Complete Cable and Flexible Cord. 6 pp.

—**Tin Coatings**
DIN VDE 0472 Pt 808-84. Testing of Cables, Wires and Flexible Cords; Tinning, Solderability and Soldering Shrinkage (VDE Specification) (Feb). 7 pp.

—**Vacuum Cleaners**
CNS C4412-2-85. 2-Wire Flexible Cord Types General-Use Cord Sets for Vacuum Cleaner (SV) (Feb)(10917-2).

Cords (Fiber Optics)
Use: Fiber Optic Cables

Cords (Uniform)
Use: Fourrageres

Cords (Woven)
Use: Cordage

Corduroy
See Also: Fabrics
CNS L4033-66. Corduroy Cloth (Both Side) (Tentative) (Sep)(2562)(R 1973).

Corduroy *(Cont.)*
—**Cotton**
CNS L4020-80. Cotton Velvet and Cotton Corduroy Fabrics (Unfinished) (Oct)(2205).
CNS L4021-80. Cotton Velvet and Cotton Corduroy Fabrics (Finished) (Oct)(2206).

Core Bits
Use For: Core Drill Bits *See Also:* Bits (Tools); Core Drilling Equipment; Core Drills; Drill Bits
CNS M2028-80. Diamond Core Bits for Core Drills (Mar)(5355).
CNS M2030-80. Tungsten Carbide Bits and Bit Blanks for Core Drills (Mar)(5357).
CNS M2031-80. Shot Bits for Core Drills (Mar)(5358).
JIS M 1401-82. Diamond Core Bits.
JIS M 1403-82. Tungsten Carbide Bits and Bit Blanks for Core Drills.

—**Shot**
CNS M2031-80. Shot Bits for Core Drills (Mar)(5358).

—**Tungsten Carbide**
CNS M2030-80. Tungsten Carbide Bits and Bit Blanks for Core Drills (Mar)(5357).
JIS M 1403-82. Tungsten Carbide Bits and Bit Blanks for Core Drills.

Core Drill Bits
Use: Core Bits

Core Drilling Equipment
See Also: Core Bits; Core Drills; Drilling Equipment
BSI BS 4019: Part 1-74. (WITHDRAWN) 1974 Amd 1 Core Drilling Equipment Basic Equipment (Supeseded by BS 4019: Parts 3 & 4: 1993). 136 pp.

—**Diamond**
ISO 8866-91. Rotary Core Diamond Drilling Equipment—System C First Edition; (Corrigendum 1-1991) (Corrigendum 2-1992). 52 pp.

—**Diamond—Wireline**
ISO 10098-92. Wireline Diamond Core Drilling Equipment—System CSSK First Edition. 34 pp.

—**Glossaries**
CNS M4007-85. Glossary of Terms for Core Drilling Equipments and Tools (Sep) (11368).
JIS M 0103-68. Glossary of Terms for Core Drilling Equipments and Tools.

—**Seamless Pipes**
DIN ENGL 4940-65. Seamless Drill Pipes for Water and Rock Drillings According to the Percussive Drilling and Core Drilling Process (Feb). 2 pp.

Core Drills
See Also: Core Bits; Core Drilling Equipment; Drills
—**Bit Blanks**
CNS M2030-80. Tungsten Carbide Bits and Bit Blanks for Core Drills (Mar)(5357).
JIS M 1403-82. Tungsten Carbide Bits and Bit Blanks for Core Drills.

—**Casings**
CNS M2038-80. Casings for Core Drills (Mar)(5365).
JIS M 1411-82. Casing for Core Drills.

—**Core Barrel Heads**
CNS M2035-80. Core Barrel Heads for Core Drills (Mar) (5362).
JIS M 1408-82. Core Barrel Heads for Core Drills.

—**Core Barrels**
CNS M2034-80. Core Barrel for Core Drills (Mar)(5361).

—**Core Shells**
CNS M2032-80. Core Shells for Core Drills (Mar)(5359).
CNS M2033-80. Core Shell Couplings for Core Drills (Mar)(5360).
JIS M 1405-82. Core Shells for Core Drills.
JIS M 1406-82. Core Shell Couplings for Core Drills.

—**Core Tubes**
JIS M 1407-82. Core Tubes for Core Drills.

—**Diamond**
BSI BS 4019: Part 3-93. 1993 Rotary Core Drilling Equipment Part 3: Specification for System A. Metric Units (ISO 3551-1: 1992) (Q). 84 pp.
BSI BS 4019: Part 4-93. 1993 Rotary Core Drilling Equipment Part 4: Specification for System A. Inch Units (ISO 3551-2: 1992) (Q). 84 pp.
ISO 3551 Pt 1-92. Rotary Core Diamond Drilling Equipment—System A—Part 1: Metric Units First Edition. 84 pp.

Core Drills *(Cont.)*
—**Diamond** *(Cont.)*
ISO 3551 Pt 2-92. Rotary Core Diamond Drilling Equipment—System A—Part 2: Inch Units First Edition. 84 pp.
ISO 3552 Pt 1-92. Rotary Core Diamond Drilling Equipment—System B—Part 1: Metric Units First Edition. 48 pp.
ISO 3552 Pt 2-92. Rotary Core Diamond Drilling Equipment—System B—Part 2: Inch Units First Edition. 48 pp.

—**High Speed Steels—Morse Taper Shanks**
CNS B3395-80. Morse Taper Shank Core Drills of High Speed Steel (Dec)(6762).

—**High Speed Steels—Parallel Shanks**
CNS B3394-80. Parallel Shank Core Drills of High Speed Steel (Dec)(6761).

—**Morse Taper Shanks**
BSI BS 328: Part 3-83. 1983 Drills and Reamers Part 3: Dimensions of Core Drills. 7 pp.
ISO 7079-81. Core Drills with Parallel Shanks and with Morse Taper Shanks First Edition. 6 pp.
SNZ NZS/BS 328: Part 3-83. Drills and Reamers Part 3: Specification for the Dimensions of Core Drills. 4 pp.

—**Parallel Shanks**
BSI BS 328: Part 3-83. 1983 Drills and Reamers Part 3: Dimensions of Core Drills. 7 pp.
ISO 7079-81. Core Drills with Parallel Shanks and with Morse Taper Shanks First Edition. 6 pp.
SNZ NZS/BS 328: Part 3-83. Drills and Reamers Part 3: Specification for the Dimensions of Core Drills. 4 pp.

—**Reaming Shells**
CNS M2029-80. Diamond Reaming Shells for Core Drills (Mar) (5356).
JIS M 1402-82. Diamond Reaming Shells.

—**Rod Couplings**
CNS M2037-80. Drill Rod Couplings for Core Drills (Mar) (5364).
JIS M 1410-82. Drill Rod Couplings for Core Drills.

—**Rods**
CNS M2036-80. Drill Rods for Core Drills (Mar)(5363).
JIS M 1409-82. Drill Rods for Core Drills.

—**Tips—Tungsten Carbide**
CNS B3450-84. Tungsten-Carbide Tip Core Drills (Dec)(11159).

Core Loss
Use For: Iron Loss *See Also:* Magnetic Measurement
—**High Frequency**
JIS H 7153-91. Measuring Method for High Frequency Core Loss in Amorphous Magnetic Cores. 11 pp.

Core Plugs (Pipes)
Use: Pipe Plugs

Core Wire
See Also: Magnet Wire; Wire
—**Coated Electrodes**
CNS G3181-88. Core Wires for Covered Electrodes (May)(8968).
JIS G 3523-80. Core Wires for Covered Electrode. 7 pp.

—**Coated Electrodes—Wire Rods**
CNS G3033-88. Wire Rods for Core Wire of Covered Electrode (May)(2067). 3 pp.
JIS G 3503-80. Wire Rods for Core Wire of Covered Electrode. 6 pp.

—**Stainless Steel**
CNS G3190-88. Stainless Steel Wire Rods for Welding (May)(9269).
JIS G 4316-91. Stainless Steel Wire Rods for Welding. 8 pp.

Cores (Electrical)
Use: Magnetic Cores

Cores (Waveguides)
See Also: Cladding (Waveguides); Optical Fibers; Optical Waveguides; Waveguides
—**Optical Fibers**
DIN VDE 0472 Pt 212-86. Testing of Cables, Wires and Flexible Cords; Core and Cladding Dimensions or Optical Fibres (Aug). 5 pp.

INDUSTRY STANDARDS

Cores

Cores (Waveguides) (Cont.)
—Optical Fibers—Diameters
CCITT RECMN G.651-89. Characteristics of a 50/125 Micrometer Multimode Graded Index Optical Fibre Cable—Transmission Media Characteristics (Study Group XV) 30 pp. 30 pp.

Coriander
See Also: Coriander Oil; Herbs
BSI BS 7087: Part 6-90. 1990 Herbs and Spices Ready for Food Use Part 6: Coriander Seeds (Whole and Ground). 9 pp.
ISO 2255-80. Coriander, Whole or Ground (Powdered)—Specification First Edition. 6 pp.

Coriander Oil
See Also: Coriander
CNS K5118-81. Oil of Coriander (Nov)(8130).
ISO 3516-80. Oil of Coriander First Edition. 4 pp.

Cork
Use For: Manufactured Cork; Virgin Cork
See Also: Wood
ISO 8724-89. Cork Decorative Panels—Specification First Edition. 6 pp.

—Agglomerated—Bulk Density
ISO 2189-86. Expanded Pure Agglomerated Cork—Determination of Bulk Density Second Edition. 3 pp.

—Agglomerated—Deformation
ISO 2191-72. Cork—Expanded Pure Agglomerated—Deformation Under Constant Pressure First Edition. 4 pp.

—Agglomerated—Expansion Joints
ISO 3867-82. Agglomerated Cork Material of Expansion Joints for Construction and Building—Test Methods First Edition. 7 pp.

—Agglomerated—Expansion Joints—Packaging
ISO 3869-81. Agglomerated Cork—Filler Material of Expansion Joints for Construction and Buildings—Characteristics, Sampling and Packing First Edition. 3 pp.

—Agglomerated—Moisture Content
ISO 2066-86. Expanded Pure Agglomerated Cork—Determination of Moisture Content Second Edition. 4 pp.

—Agglomerated—Packaging
ISO 2219-89. Expanded Pure Agglomerated Cork for Thermal Insulation—Characteristics, Sampling and Packaging Second Edition. 4 pp.

—Agglomerated—Sampling
ISO 2219-89. Expanded Pure Agglomerated Cork for Thermal Insulation—Characteristics, Sampling and Packaging Second Edition. 4 pp.

—Agglomerated—Sealing
ISO 9392-89. Agglomerated Cork Discs—Sealing Behaviour First Edition. 6 pp.

—Agglomerated—Washers
ISO 4711-87. Agglomerated Cork Discs—Specifications First Edition. 4 pp.

—Carbonized
CNS A2035-86. Carbonized Cork Board (Dec)(2314). 3 pp.
CNS A3056-86. Method of Test for Carbonized Cork Boards (Dec)(2315).

—Commercially Dry—Packaging
ISO 1215-86. Commercially Dry Virgin Cork, Ramassage, Gleanings, Corkwood Refuse and Corkwaste—Definitions and Packaging Second Edition. 3 pp.
ISO 1216-90. Corkwood in Planks—Grading, Classification and Packing Second Edition. 4 pp.

—Composition
ISO 4714-86. Composition Cork—Specifications First Edition. 3 pp.
ISO 7322-86. Cork—Composition Cork—Test Methods First Edition. 5 pp.
ISO 9148-87. Composition Cork in Rolls for Decoration—Test Methods First Edition. 4 pp.
ISO 9149-87. Composition Cork in Rolls for Decoration—Specifications First Edition. 4 pp.

—Composition—Gaskets
BSI BS 4332-89. 1989 Phenol-Formaldehyde Resin-Bonded Cork Jointing. 12 pp.
ISO 4708-85. Cork—Composition Cork Gasket Material—Test Methods First Edition. 4 pp.
ISO 4709-85. Cork—Composition Cork Gasket Material—Specifications First Edition. 4 pp.

Cork (Cont.)
—Composition—Sheets—Marine
MOD UK NES 804: Part 1-89. Requirements for Cork Thermal Insulation Material Part 1: Cork Products Composition Cork Sheet Fire Retarded Issue 2 (04.89). 13 pp.

—Composition—Soles (Footwear)
ISO 9986-90. Composition Cork for Shoe Outsoles First Edition. 6 pp.

—Floor Coverings
BSI BS 8203-87. 1987 Code of Practice for Installation of Sheet and Tile Flooring. 31 pp.
ISO 3810-87. Floor Tiles of Agglomerated Cork—Methods of Test Second Edition; (Corrected and Reprinted -1992). 6 pp.

—Gaskets
CNS O1013-64. Cork Binder Discs (May)(2302). 2 pp.

—Glossaries
BSI BS 6100: Sec 4.5-84. 1984 Glossary of Building and Civil Engineering Terms Part 4: Forest Products Section 4.5: Cork. 7 pp.
ISO 633-86. Cork—Vocabulary First Edition. 15 pp.

—Granulated
BSI BS ISO 1997-92. 1992 Granulated Cork and Cork Powder—Classification, Properties and Packing. 9 pp.
ISO 1997-92. Granulated Cork and Cork Powder—Classification, Properties and Packing Second Edition. 6 pp.

—Granulated—Bulk Density
ISO 2031-91. Granulated Cork—Determination of Bulk Density Second Edition. 5 pp.

—Granulated—Moisture Content
ISO 2190-88. Granulated Cork—Determination of Moisture Content Second Edition. 4 pp.

—Granulated—Sampling
ISO 2067-89. Granulated Cork—Sampling Second Edition. 4 pp.

—Granulated—Sieve Analysis
ISO 2030-90. Granulated Cork—Size Analysis by Mechanical Sieving Second Edition. 4 pp.

—Modulus of Rupture
ISO 2077-79. Pure Expanded Corkboard—Determination of the Modulus of Rupture by Bending First Edition; (Amendment Slip-1980). 4 pp.

—Moisture Content
ISO 2386-88. Corkwood in Planks, Virgin Cork, Ramassage, Gleanings, Corkwood Refuse and Corkwaste—Determination of Moisture Content Second Edition. 4 pp.

—Powdered
BSI BS ISO 1997-92. 1992 Granulated Cork and Cork Powder—Classification, Properties and Packing. 9 pp.
ISO 1997-92. Granulated Cork and Cork Powder—Classification, Properties and Packing Second Edition. 6 pp.

—Rubber Bonded—Gaskets—Aircraft
BSI 2F 66-82. 1982 Rubber Bonded Cork Sheets. 7 pp.

—Sampling
ISO 2385-93. Corkwood in Planks, Virgin Cork, Cleanings, Cork Pieces, Corkwood Refuse and Corkwaste—Sampling to Determine Moisture Content Second Edition. 5 pp.

—Sections—Marine
MOD UK NES 804: Part 2-89. Requirements for Cork Thermal Insulation Material Part 2: Cork Products Cork Slabs Non-Fire Retarded Issue 2 (04.89). 13 pp.
MOD UK NES 804: Part 2-01. Requirements for Cork Thermal Insulation Material Part 2: Cork Products Cork Slabs Non-Fire Retarded Issue 2 (04.89); Amendment 1. 14 pp.

—Sheets—Floor Coverings
BSI BS 6263: Part 2-91. 1991 Care and Maintenance of Floor Surfaces Part 2: Code of Practice for Resilient Sheet and Tile Flooring. 14 pp.
BSI BS 6826-87. 1987 Linoleum and Cork Carpet Sheet and Tiles. 12 pp.

—Sheets—Gaskets
CNS K6099-59. Method of Test for Cork Sheet for Gasket (Sep)(1068).
CNS K8001-59. Cork Sheet for Gasket (Sep)(1067)(R 1971).

Cork (Cont.)
—Sheets—Oil Resistant
CNS C4109-85. Oil Resistant Cork Sheet (Oct)(3422).

—Slabs—Marine
MOD UK NES 804: Part 2-89. Requirements for Cork Thermal Insulation Material Part 2: Cork Products Cork Slabs Non-Fire Retarded Issue 2 (04.89). 13 pp.
MOD UK NES 804: Part 2-01. Requirements for Cork Thermal Insulation Material Part 2: Cork Products Cork Slabs Non-Fire Retarded Issue 2 (04.89); Amendment 1. 14 pp.

—Stoppers—Absorption
ISO 9727-91. Cylindrical Stoppers of Natural Cork—Physical Tests—Reference Methods First Edition. 12 pp.

—Stoppers—Bacteria Count Methods
ISO 10718-93. Cork Stoppers—Enumeration of Colony-Forming Units of Yeasts, Moulds and Bacteria Capable of Growth in an Alcoholic Medium First Edition. 5 pp.

—Stoppers—Bulk Density
ISO 9727-91. Cylindrical Stoppers of Natural Cork—Physical Tests—Reference Methods First Edition. 12 pp.

—Stoppers—Capillarity
ISO 9727-91. Cylindrical Stoppers of Natural Cork—Physical Tests—Reference Methods First Edition. 12 pp.

—Stoppers—Compression Testing
ISO 9727-91. Cylindrical Stoppers of Natural Cork—Physical Tests—Reference Methods First Edition. 12 pp.

—Stoppers—Extraction Analysis
ISO 9727-91. Cylindrical Stoppers of Natural Cork—Physical Tests—Reference Methods First Edition. 12 pp.

—Stoppers—Fungi—Count Methods
ISO 10718-93. Cork Stoppers—Enumeration of Colony-Forming Units of Yeasts, Moulds and Bacteria Capable of Growth in an Alcoholic Medium First Edition. 5 pp.

—Stoppers—Moisture Content
ISO 9727-91. Cylindrical Stoppers of Natural Cork—Physical Tests—Reference Methods First Edition. 12 pp.

—Stoppers—Penetration Resistance
ISO 9727-91. Cylindrical Stoppers of Natural Cork—Physical Tests—Reference Methods First Edition. 12 pp.

—Thermal Conductivity
ISO 2582-78. Cork and Cork Products—Determination of Thermal Conductivity—Hot Plate Method First Edition. 9 pp.

Corkboard
Use: Cork

Corks
Use: Stoppers

Corkwood
Use: Cork

Corn
Use For: Maize See Also: Cereals; Cornmeal

—Canned
CNS N5096-87. Canned Baby Corn or Young Corn (Sep)(3184). 4 pp.
CNS N5220-86. Canned Sweet Corn (Sep)(11712). 3 pp.

—Fatty Acids
CNS K1262-81. Distilled Corn Fatty Acids (May)(7430).

—Grading
CNS N1066-89. Corn (Nov)(2432). 2 pp.

—Insoluble Matter Content
ISO 8129 Pt 1-84. Fruits, Vegetables and Derived Products—Determination of Alcohol-Insoluble Solids Content—Part 1: Method for Fresh or Quick-Frozen Maize First Edition. 4 pp.

—Moisture Content
BSI BS 4317: Part 15-81. 1981 Methods of Test for Cereals and Pulses Part 15: Determination of Moisture Content of Maize (Milled and Whole). 14 pp.

INTERNATIONAL AND NON-U.S. NATIONAL STANDARDS
SUBJECT INDEX
Corrosion

Corn *(Cont.)*
—Moisture Content *(Cont.)*
ISO 6540-80. Maize—Determination of Moisture Content (on Milled Grains and on Whole Grains) First Edition. 14 pp.

—Zearalenone Content
BSI BS 5766: Part 13-86. (WITHDRAWN) 1986 Methods for Analysis of Animal Feeding Stuffs Part 13: Determination of Zearalenone. 8 pp.
ISO 6870-85. Animal Feeding Stuffs—Determination of Zearalenone Content First Edition. 7 pp.

Corn Grinders
See Also: Appliances; Food Processing Equipment; Grinders

—Safety
DIN VDE 0700 Pt 33 A2 (D)-85. Safety of Household and Similar Electrical Appliances; Coffee and Grain Mills: Amendment 2 (Sept). 4 pp.

Corn Mills
Use: Corn Grinders

Corn Oil
Use For: Maize Oil *See Also:* Oils; Vegetable Oils
CNS K5105-81. Crude Corn Oil (Sep)(7416).
CNS N5110-89. Edible Maize Oil (Edible Corn Oil) (Sep)(3527). 2 pp.
CNS N6074-83. Methods of Test for Edible Oils and Fats (General Rules) (Mar)(3639). 2 pp.

Corner Fittings
See Also: Lifting Equipment; Shipping Containers
BSI BS 3951: Sec 1.2-85. 1985 Amd 1 Freight Containers Part 1: General Section 1.2: Corner Fittings for Series 1 Freight Containers (AMD 6330) July 31, 1991. 28 pp.
ISO 1161-84. Series 1 Freight Containers—Corner Fittings—Specification Fourth Edition; (Corrigendum 1-1990). 26 pp.
JIS Z 1616-79. Corner Fittings of Freight Containers for International Trade (R 1984). 10 pp.
SAA AS E45-69. Corner Fittings for Freight Containers (Incorporating Amdt 1) (Withdrawn).
SNZ NZS/ISO 1161-84. Series 1 Freight Containers. Corner Fittings. Specifications. 11 pp.

Corner Joints
Scope Note: Use a more specific term
See: Construction Joints; Joints

Cornmeal
See Also: Animal Feed; Barley Hull Meal; Corn; Leucaena Meal; Oil Meal; Shell Meal

—Animal Feed
CNS N2010-86. Corn Meal (for Feeding) (Mar)(2290). 1 p.

Corona-Discharge Tubes
Use: Corona Stabilizer Tubes

Corona Stabilizer Tubes
See Also: Electron Tubes; Gas Filled Tubes
MOD UK DSTAN 59-60: Part 4-73. Valves, Electronic (Electronic Tubes) (Listed on EPIC Database) Part 4: Corona Stabilizer Tubes Issue 1. 16 pp.
MOD UK DSTAN 59-60: Pt 4:Sec E-74. Valves, Electronic (Electronic Tubes) (Listed on EPIC Database) Part 4: Corona Stabilizer Tubes Section E: Detail Specifications Issue 1. 4 pp.
MOD UK DSTAN 59-60: 04/001-71. Valve Electronic, Corona Stabilzer Tube Issue 1. 14 pp.
MOD UK DSTAN 59-60: 04/002-71. Valve Electronic, Corona Stabilizer Tube Issue 1. 15 pp.
MOD UK DSTAN 59-60: 04/003-71. Valve Electronic, Corona Stabilizer Tube Issue 1. 15 pp.
MOD UK DSTAN 59-60: 04/004-71. Valve Electronic, Corona Stabilizer Tube Issue 1. 13 pp.

—Electrical Measurement
IEC 151 Pt 19-69. Measurements of the Electrical Properties of Electronic Tubes and Valves Part 19: Methods of Measurement on Corona Stabilizers First Edition. 17 pp.

—Quality Assurance
BSI BS 9020-70. (WITHDRAWN) 1970 Amd 1 Corona Stabilizer Tubes of Assessed Quality: Generic Data and Methods of Test. 16 pp.
BSI BS 9021-70. (WITHDRAWN) 1970 Amd 1 Rules for the Preparation of Detail Specifications for Corona Stabilizer Tubes of Assessed Quality. 15 pp.

Correcting Slips
CNS K6410-78. Method of Test for Correcting Slip (Mar)(4278).
CNS K8014-78. Correcting Slip (Mar)(4277).

Correction Fluids, Typewriter
Use: Typewriters—Correction Fluids

Correction Papers, Typewriter
Use: Typewriters—Correction Tapes

Correction Tapes, Typewriter
Use: Typewriters—Correction Tapes

Correction Time
—Mortars
DIN ENGL 18555 Pt 8-87. Testing of Mortars Containing Mineral Binders; Freshly Mixed Mortar; Determination of Workability Time and Correction Time of Thin-Bed Mortar for Use with Masonry (Nov). 2 pp.

Corrodkote Testing
See Also: Accelerated Testing; Atmospheric Corrosion Testing; Corrosion Testing; Immersion Testing; Salt Spray Testing

—Metal Coatings—Corrosion Testing
BSI BS 5466: Part 5-79. 1979 Methods for Corrosion Testing of Metallic Coatings Part 5: Corrodkote Test (CORR Test). 7 pp.
ISO 4540-80. Metallic Coatings—Coatings Cathodic to the Substrate—Rating of Electroplated Test Specimens Subjected to Corrosion Tests First Edition. 18 pp.
ISO 4541-78. Metallic and Other Non-Organic Coatings—Corrodkote Corrosion Test (CORR Test) First Edition. 5 pp.

Corrosion
See Also: Anodic Coatings; Cathodic Protection Equipment; Cavitation Corrosion; Coatings; Copper Strip Corrosion; Corrosion Prevention; Corrosion Testing; Electrochemical Corrosion; Electrochemistry; Electrolysis; Electrolytic Corrosion; Etching; Failure (Quality Control); Finishes; Humidity; Metal Coatings (Made From Metal); Passivation; Scale (Corrosion); Stress Corrosion Cracking

—Aircraft by Fabric Impurities
CAA 786 (G). Section G Rotorcraft Compartment Fire Precautions Chapter G4-3 (Blue Papers). 15 pp.
CAA Chapter G4-3 App #3. Testing for Corrosive Impurities in Textiles (Rotocraft). 2 pp.

—Aircraft by Fluids
CAA Chapter G4-1 App #2 01.75. Protection Against Corrosion and Other Effects of the Presence of Fluids (Rotocraft). 2 pp.
CAA Chapter K4-1 App 04.74. Protection Against Corrosion and Other Effects of the Presence of Fluids (Light Aeroplanes). 3 pp.
CAA Chapter Q4-1 App 12.79. Protection Against Corrosion and Other Effects of the Presence of Fluids (Non-Rigid Airships). 4 pp.

—Alternating Current
CCITT RECMN L.8-89. Corrosion Caused by Alternating Current—Construction, Installation and Protection of Cable and Other Elements of Outside Plant (Study Group VI) 1 pp. 1 p.

—Brake Linings
CNS D3178-87. Method of Test of Seizure to Ferrous Mating Surface Due to Corrosion for Brake Linings and Pads of Automobiles (Dec)(12174).
JIS D 4414-86. Test Procedure of Seizure to Ferrous Mating Surface Due to Corrosion for Brake Linings and Pads of Automobiles. 8 pp.

—Brake Pads
CNS D3178-87. Method of Test of Seizure to Ferrous Mating Surface Due to Corrosion for Brake Linings and Pads of Automobiles (Dec)(12174).
JIS D 4414-86. Test Procedure of Seizure to Ferrous Mating Surface Due to Corrosion for Brake Linings and Pads of Automobiles. 8 pp.

—Copper by Aromatic Hydrocarbons
CNS K6259-74. Method of Test for Copper Corrosion of Aromatic Hydrocarbons (Jun)(2760).

—Copper by Electrical Insulation
CNS C3148-88. Method of Test for Copper Corrosion of Heat Shrinkable Tubing for Electrical Insulation (Sep)(8795).

—Copper by Liquefied Petroleum Gases
BSI BS 6924-88. 1988 Method for Determination of Corrosiveness of Liquefied Petroleum Gases to Copper. 7 pp.
CNS K6249-74. Method of Test for Corrosion of Liquefied Petroleum Gases (Copper Strip) (Jun)(2750).
ISO 6251-82. Liquefied Petroleum Gases—Corrosiveness to Copper—Copper Strip Test First Edition. 6 pp.

Corrosion *(Cont.)*
—Copper by Peppermint Oil
CNS K6110-90. Method of Test for Copper Strip Corrosion of Peppermint Oil (Aug)(1219).

—Fire Effluents
IEC 695 Pt 5-1-93. Fire Hazard Testing Part 5: Assessment of Potential Corrosion Damage by Fire Effluent—Section 1: General Guidance. 25 pp.

—Glossaries
BSI BS 6918-90. 1990 Terms for Corrosion of Metals and Alloys. 20 pp.
DIN ENGL 50900 Pt 1-82. Corrosion of Metals; Terminology; General Concepts (Apr). 6 pp.
ISO 8044-89. Corrosion of Metals and Alloys—Vocabulary Second Edition; (Amendment 1-1993). 28 pp.

—Ground Electrodes
DIN VDE 0151-86. Material & Minimum Dimensions of Earth Electrodes with Respect to Corrosion (June). 26 pp.

—Materiel
MOD UK DSTAN 00-50-82. Guide to Chemical Environmental Contaminants and Corrosion Affecting the Design of Military Materiel Issue 1. 11 pp.

—Metal by Vulcanized Rubber
BSI BS 903: Part A37-87. 1987 Methods of Testing Vulcanized Rubber Part A37: Determination of Adhesion to and Corrosion of Metals. 10 pp.
ISO 6505-84. Rubber, Vulcanized—Determination of Adhesion to, and Corrosion of, Metals First Edition. 6 pp.

—Pipelines
DIN ENGL 50929 Pt 3-85. Corrosion of Metals; Probability of Corrosion of Metallic Materials When Subject to Corrosion from the Outside; Buried and Underwater Pipelines and Structural Components (Sept). 12 pp.

—Refractory Materials by Slag
CNS R3156-87. Method of Test for Slag Corrosion of Refractories (Sep) (12109).

—Service Equipment—Buildings
DIN ENGL 50929 Pt 2-85. Corrosion of Metals; Probability of Corrosion of Metallic Materials When Subject to Corrosion from the Outside; Service Components Inside Buildings (Sept). 4 pp.

—Structural Members
DIN ENGL 50929 Pt 3-85. Corrosion of Metals; Probability of Corrosion of Metallic Materials When Subject to Corrosion from the Outside; Buried and Underwater Pipelines and Structural Components (Sept). 12 pp.

Corrosion Control
Use: Corrosion Prevention

Corrosion Inhibitor Content Analysis
See Also: Corrosion Inhibitors

—Metal Conditioners
CGSB 31-GP-0A METH 39.1-57. Methods of Testing Corrosion-Prevention Materials and Processes Inhibiting Agents in Non-Inhibited Metal Conditioners. 1 p.
CGSB 31-GP-0A METH 39.2-57. Methods of Testing Corrosion-Prevention Materials and Processes Inhibiting Agents in Inhibited Metal Conditioners. 1 p.

—Rust Removers
CGSB 31-GP-0A METH 39.1-57. Methods of Testing Corrosion-Prevention Materials and Processes Inhibiting Agents in Non-Inhibited Metal Conditioners. 1 p.
CGSB 31-GP-0A METH 39.2-57. Methods of Testing Corrosion-Prevention Materials and Processes Inhibiting Agents in Inhibited Metal Conditioners. 1 p.

Corrosion Inhibitors
Use For: Rust Inhibitors *See Also:* Additives; Antioxidants; Corrosion Inhibitor Content Analysis; Corrosion Prevention; Metals; Penetrating Oils; Retarders (Materials)
BSI BS 4959-74. 1974 Recommendations for Corrosion and Scale Prevention in Engine Cooling Water Systems. 19 pp.
CNS Z6044-82. Method of Test for Volatile Corrosion Inhibitor (Feb)(8554).
JIS Z 1519-86. Volatile Corrosion Inhibitor. 14 pp.

—Aircraft Fuels
MOD UK TS 10067E. Fluid AL-38 (Superseded by Def Stan 68-150).

INDUSTRY STANDARDS

Corrosion

Corrosion Inhibitors (Cont.)

—Aircraft Fuels (Cont.)
MOD UK DERD 2461: APL. Approved Products List of Aircraft Materials to Specification DERD 2461 Issue 1.
SBAC TS 98 ISSUE 2. Tables of Aerospace Fuels, Corrosion Preventatives, and Miscellaneous Products.

—Alkalinity
CNS K6481-80. Method of Test for Reserve Alkalinity of Engine Antifreezes Antirusts and Coolants (May)(5584).

—Ash Content
CNS K6484-80. Method of Test for Ash Content of Engine Antifreezes, Antirusts and Coolants (May)(5587).

—Chromate Coatings
BSI BS 6338-82. 1982 Chromate Conversion Coatings on Electroplated Zinc and Cadmium Coatings. 4 pp.
MOD UK DEF-130-61. Chromate Passivation of Cadmium and Zinc Surfaces (Reprinted March 1977, Incorporating Amendment No. 1). 4 pp.

—Corrosion Testing
CGSB 31-GP-0A METH 11.1-57. Methods of Testing Corrosion-Prevention Materials and Processes Corrosion (Cold Application Compounds). 2 pp.
CGSB 31-GP-0A METH 11.17-74. Methods of Testing Corrosion-Prevention Materials and Processes Corrosion. 1 p.

—Crude Oils—Hydraulic Equipment—Aerospace
MOD UK DTD-5540B-75. Corrosion Preventive Oil: Hydraulic System NATO Code Number: C-635 Joint Service Designation: PX-26. 8 pp.

—Drying
CGSB 31-GP-0A METH 8.1-57. Methods of Testing Corrosion-Prevention Materials and Processes Drying. 1 p.

—Electrical Equipment
MOD UK TS 10151. Corrosion Preventive and Water Displacing Fluid for Use on Electrical Equipment PX-29.

—Engine Coolants
BSI BS 4959-74. 1974 Recommendations for Corrosion and Scale Prevention in Engine Cooling Water Systems. 19 pp.
BSI BS 5117: Part 0-92. 1992 Testing Corrosion Inhibiting, Engine Coolant Concentrate ('Antifreeze') Part 0: General Introduction. 12 pp.
BSI BS 5117: Part 0-85. 1985 Testing Corrosion Inhibiting, Engine Coolant Concentrate ('Antifreeze') Part 0: General Introduction. 10 pp.
BSI BS 5117: Sec 1.1-85. 1985 Testing Corrosion Inhibiting, Engine Coolant Concentrate ('Antifreeze') Part 1: Methods of Test for Determination of Physical and Chemical Properties Section 1.1: Ancillary Procedures. 6 pp.
BSI BS 5117: Sec 2.1-85. 1985 Testing Corrosion Inhibiting, Engine Coolant Concentrate ('Antifreeze') Part 2: Methods of Test for Corrosion Inhibition Performance Section 2.1: General Procedures. 8 pp.
BSI BS 5117: Sec 2.2-85. 1985 Testing Corrosion Inhibiting, Engine Coolant Concentrate ('Antifreeze') Part 2: Methods of Test for Corrosion Inhibition Performance Section 2.2: Glassware Tests. 6 pp.
BSI BS 5117: Sec 2.3-85. 1985 Testing Corrosion Inhibiting, Engine Coolant Concentrate ('Antifreeze') Part 2: Methods of Test for Corrosion Inhibition Performance Section 2.3: Recirculating Rig Test. 23 pp.
BSI BS 5117: Sec 2.4-85. 1985 Testing Corrosion Inhibiting, Engine Coolant Concentrate ('Antifreeze') Part 2: Methods of Test for Corrosion Inhibition Performance Section 2.4: Static Engine Test. 8 pp.
BSI BS 5117: Sec 2.5-85. 1985 Testing Corrosion Inhibiting, Engine Coolant Concentrate ('Antifreeze') Part 2: Methods of Test for Corrosion Inhibition Performance Section 2.5: Field Test. 5 pp.
BSI BS 5117: Sec 2.6-92. 1992 Testing Corrosion Inhibiting, Engine Coolant Concentrate ('Antifreeze') Part 2: Methods of Test for Corrosion Inhibition Performance Section 2.6: Test for Corrosion of Cast Aluminium Alloys Under Heat-Transfer Conditions. 10 pp.
BSI BS 6580-92. 1992 Corrosion Inhibiting, Engine Coolant Concentrate ('Antifreeze'). 10 pp.
BSI BS 6580-85. 1985 Corrosion Inhibiting, Engine Coolant Concentrate. 8 pp.
JIS K 2408-90. Corrosion Inhibitors for Engine Coolant. 22 pp.

—Environmental Testing
CGSB 31-GP-0A METH 16.1-57. Methods of Testing Corrosion-Prevention Materials and Processes Accelerated Weathering; (Re-Edited April 1982). 2 pp.

Corrosion Inhibitors (Cont.)

—Environmental Testing (Cont.)
CGSB 31-GP-0A METH 17.1-57. Methods of Testing Corrosion-Prevention Materials and Processes Shed Storage. 2 pp.

—Evaporation Loss
CGSB 31-GP-0A METH 2.3-62. Methods of Testing Corrosion-Prevention Materials and Processes Loss by Evaporation. 1 p.

—Ferroalloys—Humidity
CGSB 31-GP-0A METH 13.1-57. Methods of Testing Corrosion-Prevention Materials and Processes Humidity Test. 9 pp.

—Film Forming Capacity
MOD UK DSTAN 05-50: Part 23-01. Methods for Testing Fuels, Lubricants and Associated Products Part 23: Film Forming Properties of Corrosion Preventive Fluids Issue 1; Amendment 1. 9 pp.

—Flow Measurement
CGSB 31-GP-0A METH 23.1-57. Methods of Testing Corrosion-Prevention Materials and Processes Flow Point. 1 p.

—Furniture—Coatings
CNS S2136-86. Method of Test for Antirust Property of Furniture Coating (Aug)(11685).

—Gas Turbine Engines
MOD UK DSTAN 68-10-89. Corrosion Preventive Compound: Water Displacing NATO Code: C-634 Joint Service Designation: PX-24 Issue 3. 28 pp.
MOD UK DSTAN 68-10-01. Corrosion Preventive Compound: Water Displacing NATO Code: C-634 Joint Service Designation: PX-24 Issue 3; Amendment 1. 29 pp.

—Greases
DIN ENGL 51802-90. Testing Lubricating Greases for Their Corrosion-Inhibiting Properties by the SKF Emcor Method (Apr). 5 pp.
MOD UK DSTAN 91-34-01. Grease, Sea Water Resisting NATO Code No. G-460 Joint Service Designation: XG-286 Issue 1; Amendment 2. 12 pp.
MOD UK CS 3120. Protective PX-19.

—Hydrocarbon Content
CGSB 31-GP-0A METH 49.1-57. Methods of Testing Corrosion-Prevention Materials and Processes Hydrocarbons Volatile with Steam; (Re-Edited April 1982). 1 p.
CGSB 31-GP-0A METH 50.1-57. Methods of Testing Corrosion-Prevention Materials and Processes Hydrocarbons Absorbed in Sulfuric Acid. 1 p.

—Impregnated Papers
CNS Z5110-82. Volatile Corrosion Inhibitor Treated Paper (Jul)(9196).
CNS Z6049-82. Method of Test for Volatile Corrosion Inhibitor Treated Paper (Jul)(9197).
JIS Z 1535-86. Volatile Corrosion Inhibitor Treated Paper. 15 pp.

—Interchangeability
NATO STANAG 1135 ED 3 AMD 6-83. Interchangeability of Fuels, Lubricants and Associated Products Used by the Armed Forces of the North Atlantic Treaty Nations. 82 pp.

—Lubricants
BSI BS 6413: Part 8-88. 1988 Lubricants, Industrial Oils and Related Products (Class L) Part 8: Classification for Family R (Temporary Protection Against Corrosion). 6 pp.
DIN ENGL 51360 Pt 1-85. Testing of Cooling Lubricants; Determination of Corrosion Preventing Characteristics of Cooling Lubricants Mixed with Water; Herbert Corrosion Test (Aug). 5 pp.
DIN ENGL 51360 Pt 2-81. Testing of Cooling Lubricants; Determination of Corrosion Preventing Characteristics of Cooling Lubricants Mixed with Water; Chip/Filter Paper Method (July). 4 pp.

—Lubricating Oils
DIN ENGL 51386 Pt 1-86. Testing of Corrosion Preventive Oils in a Condensation Water Alternating Atmosphere (Mar). 5 pp.
DIN ENGL 51394-84. Testing of Lubricants; Testing of Low-Viscosity Lubricating Oils for Oxidation and Corrosion Inhibiting Properties (Oct). 5 pp.
MOD UK DSTAN 80-34-01. Corrosion Preventive Compound, Oil Film Type Joint Service Designation: PX-4 Issue 2; Amendment 1. 38 pp.

—Lubricating Oils—Engine Cylinders—Aircraft
MOD UK DTD-791C-72. Corrosion Preventive Oil, Aircraft Piston Engine: Static Preservation, Upper Cylinder NATO Code Number: C-613 Joint Service Designation: PX-13. 6 pp.

Corrosion Inhibitors (Cont.)

—Metals
MOD UK DSTAN 03-30: Part 4-92. Treatments for the Protection of Metal Parts of Service Stores and Equipment Against Corrosion Part 4: Schedule of Corrosion Preventives Issue 1. 14 pp.
MOD UK DSTAN 03-30: Part 5-92. Treatments for the Protection of Metal Parts of Service Stores and Equipment Against Corrosion Part 5: Index, Related Documents and Sources Issue 1. 17 pp.
MOD UK DG-8: CONTENT LIST. Defence Guide: Treatments for the Protection of Metal Parts of Service Stores and Equipments Against Corrosion.
MOD UK DG-8: INTRODUCTION. Defence Guide: Treatments for the Protection of Metal Parts of Service Stores and Equipments Against Corrosion: Introduction Issue 2.
MOD UK DG-8: Part 3. Defence Guide: Treatments for the Protection of Metal Parts of Service Stores and Equipments Against Corrosion Part 3: List of Service Schedules Issue 2.

—Metals—Adhesion Testing
CGSB 31-GP-0A METH 7.1-57. Methods of Testing Corrosion-Prevention Materials and Processes Low Temperature Adhesion. 2 pp.

—Metals—Packaging Materials
BSI BS 1133: Sec 6-66. 1966 Packaging Code Section 6: Temporary Protection of Metal Surfaces Against Corrosion (During Transport and Storage). 119 pp.
BSI BS 1133: SUBSEC 6.1-91. 1991 Packaging Code Section 6: Protection of Metal Surfaces Against Corrosion During Transport and Storage Subsection 6.1: Cleaning and Drying of Metal Surfaces. 20 pp.
BSI BS 1133: SUBSEC 6.2-91. 1991 Packaging Code Section 6: Protection of Metal Surfaces Against Corrosion During Transport and Storage Subsection 6.2: Temporary Protectives and Their Application. 20 pp.
BSI BS 7541-92. 1992 Temporary Protectives for the Protection of Metal Surfaces Against Corrosion During Transport and Storage (H). 18 pp.

—Mineral Oils
MOD UK TS 10035A. PX 25 Corrosion Preventive Oil: Enclosed Ferrous Metal Systems, Contact and Volatile Corrosion Inhibited (Withdrawn).

—Mineral Oils—Ferroalloys
MOD UK TS 10035A. PX 25 Corrosion Preventive Oil: Enclosed Ferrous Metal Systems, Contact and Volatile Corrosion Inhibited (Withdrawn).

—Motor Oils
DIN ENGL 51585-71. Testing of Lubricants; Testing of Corrosion-Protection Properties of Steam-Turbine Oils and Hydraulic Oils Containing Additives (Dec). 4 pp.
MOD UK DSTAN 91-40-81. Corrosion Preventive Oil, Aircraft Engine: Piston, Metallic NATO Code No: C-615 Joint Service Designation: PX-27 Issue 2. 10 pp.

—Motor Oils—Aircraft Engines
MOD UK DTD-791C-72. Corrosion Preventive Oil, Aircraft Piston Engine: Static Preservation, Upper Cylinder NATO Code Number: C-613 Joint Service Designation: PX-13. 6 pp.
MOD UK DERD 2493-01. Lubricating Oil: Aircraft Turbine Engines: Synthetic Type (D.Eng.R.D.2487) Reclaimed for Bench Test and Flight Usage Issue 3; Amendment 2. 9 pp.
MOD UK DERD 2497. Lubricating Oil, Aircraft Turbine Engine, Synthetic Issue 3.

—Oils
CNS K5130-83. Finger Print Removing Rust Preventive Oil (Jul)(10445).
CNS K5131-83. Regular Rust Preventive Oil (Jul)(10446).
CNS K5132-83. Solvent Diluted Rust Preventive Oil (Jul)(10447).
CNS K5134-83. Mist Type Rust Preventive Oil (Jul)(10449).

—Paints—Adhesion Testing
CGSB 31-GP-0A METH 7.3-57. Methods of Testing Corrosion-Prevention Materials and Processes Paint Adherence (Tape Test); (Amended September 1974) (Re-Edited March 1982). 1 p.

—Petrolatum
CNS K5133-83. Petrolatum for Rust Prevention (Jul)(10448).

—Reinforcing Steels
CEN PREN 990-92. Test Methods for Verification of Corrosion Protection of Reinforcement in Autoclaved Aerated Concrete and Lightweight Aggregate Concrete with Open Structure. 5 pp.
CNS A2232-88. Corrosion Inhibitor for Reinforcing Steel in Concrete (Nov)(12456).

INTERNATIONAL AND NON-U.S. NATIONAL STANDARDS
SUBJECT INDEX
Corrosion

Corrosion Inhibitors (Cont.)

—**Reinforcing Steels** (Cont.)
- CNS A3299-88. Method of Test for Corrosion Inhibitor for Reinforcing Steel in Concrete (Nov)(12457).
- JIS A 6205-87. Corrosion Inhibitor for Reinforcing Steel in Concrete. 17 pp.

—**Removability**
- CGSB 31-GP-0A METH 19.1-57. Methods of Testing Corrosion-Prevention Materials and Processes Removability. 2 pp.

—**Salt Spray Testing**
- CGSB 31-GP-0A METH 14.1-57. Methods of Testing Corrosion-Prevention Materials and Processes Salt Spray (Fog). 1 p.

—**Small Arms**
- MOD UK TS 10164. PX-31 Corrosion Preventive Compound: Hard Film, Cold Application.

—**Sprayed**
- CGSB 31-GP-0A METH 10.1-57. Methods of Testing Corrosion-Prevention Materials and Processes Sprayability. 2 pp.

—**Stability Testing**
- CGSB 31-GP-0A METH 9.17-62. Methods of Testing Corrosion-Prevention Materials and Processes Stability. 1 p.

—**Thermal Stability**
- CGSB 31-GP-0A METH 9.1-57. Methods of Testing Corrosion-Prevention Materials and Processes Stability (Recovery from Low Temperature) (Cold Application Compounds). 1 p.
- CGSB 31-GP-0A METH 9.2-57. Methods of Testing Corrosion-Prevention Materials and Processes Storage Stability (Cold Application Compounds). 1 p.

—**Thickness Measurement**
- CGSB 31-GP-0A METH 18.1-57. Methods of Testing Corrosion-Prevention Materials and Processes Film Thickness. 1 p.

—**Tools**
- MOD UK TS 10164. PX-31 Corrosion Preventive Compound: Hard Film, Cold Application.

—**Water—Displacement**
- CGSB 31-GP-0A METH 12.1-57. Methods of Testing Corrosion-Prevention Materials and Processes Water Displacement and Water Stability. 2 pp.

Corrosion Prevention

Use For: Corrosion Control *See Also:* Anodic Coatings; Cathodic Protection Equipment; Cleaning; Coatings; Copper Strip Corrosion; Corrosion; Corrosion Inhibitors; Metals; Packaging; Passivation; Preservation; Sewage Treatment; Surface Finishing; Water Treatment

- CNS H2055-86. Method of Test for Steel Spray Deposits (Feb)(8289).
- JIS H 8302-90. Steel Spray Deposits. 10 pp.
- JIS H 8651-78. Processes for Corrosion Protection of Magnesium Alloys. 7 pp.
- JIS H 8664-90. Test Methods for Steel Spray Deposits. 15 pp.
- MOD UK DTD-911C-63. Protection of Magnesium-Rich Alloys Against Corrosion (Reprinted January 1967). 6 pp.

—**Adhesive Tapes**
- MOD UK DSTAN 80-119-88. Adhesive, Non Corrosive and Water Resistant Issue 1. 13 pp.

—**Aircraft Fuels**
- SBAC TS 98 ISSUE 2. Tables of Aerospace Fuels, Corrosion Preventatives, and Miscellaneous Products.

—**Bimetallic Contacts—Design**
- BSI PD 6484-79. 1979 Commentary on Corrosion at Bimetallic Contact and Its Alleviation. 29 pp.

—**Communication Cables**
- CCITT RECMN K.29-92. Coordinated Protection Schemes for Telecommunication Cables Below Ground (Study Group V) 7 pp. 7 pp.

—**Electrical Installations**
- DIN VDE 0150-83. Protection Against Corrosion Due to Stray Currents from DC Installations (Apr). 40 pp.

—**Engine Coolants**
- BSI BS 4959-74. 1974 Recommendations for Corrosion and Scale Prevention in Engine Cooling Water Systems. 19 pp.

—**Glossaries**
- JIS Z 0103-78. Glossary of Terms Used in Rust and Corrosion Preventive Technology.

Corrosion Prevention (Cont.)

—**Ground Electrodes**
- CENELEC PREN 50114-93. Materials and Size Requirements for Earth Electrodes from the Corrosion Point of View. 23 pp.

—**Iron**
- SAA AS 2312-85. Guide to the Protection of Iron and Steel Against Exterior Atmospheric Corrosion (This is a Joint Standard with SANZ NZS 2312). 59 pp.
- SNZ NZS/AS 2312-85. Guide to the Protection of Iron and Steel Against Exterior Atmospheric Corrosion (This is a Joint Standard with SAA AS 2312). 59 pp.

—**Iron Pipes**
- DIN ENGL 30675 Pt 2-85. External Corrosion Protection of Buried Pipes; Protection of Ductile Iron Pipelines (Apr). 4 pp.

—**Iron Structures**
- SNZ NZMP 2312-87. Commentary on AS 2312: 1984 Guide to the Protection of Iron and Steel Against Exterior Atmospheric Corrosion. 12 pp.

—**Magnesium Alloys**
- CNS H3117-83. Processes for Corrosion Protection of Magnesium Alloys (Feb)(10008).
- JIS H 8651-78. Processes for Corrosion Protection of Magnesium Alloys. 7 pp.
- MOD UK DTD-911C-63. Protection of Magnesium-Rich Alloys Against Corrosion (Reprinted January 1967). 6 pp.

—**Metals**
- BSI BS 7195-89. 1989 Guide for Prevention of Corrosion of Metals Caused by Vapours from Organic Materials. 14 pp.
- MOD UK DSTAN 03-30: Part 1-91. Treatments for the Protection of Metal Parts of Service Stores and Equipment Against Corrosion Part 1: Advice to Designers on Corrosion and Its Prevention Issue 1. 49 pp.

—**Metals—Marine**
- MOD UK NES 1005-87. Code of Practice for Protective Finishes Issue 1 (04.87). 100 pp.
- MOD UK NES 1005-93. Code of Practice for Protective Finishes Issue 2 (07.93). 101 pp.

—**Metals—Naval Ships**
- MOD UK NES 738-87. Metals and Corrosion Guide Issue 1 (02.87). 113 pp.
- MOD UK NES 738-92. Metals and Corrosion Guide Issue 2 (08.92). 123 pp.

—**Metals—Submarines**
- MOD UK NES 738-87. Metals and Corrosion Guide Issue 1 (02.87). 113 pp.
- MOD UK NES 738-92. Metals and Corrosion Guide Issue 2 (08.92). 123 pp.

—**Packaging**
- CNS Z5120-82. Corrosion Preventive Packaging Method (General Rule) (Dec)(9761).

—**Petrolatum Tapes**
- JIS Z 1902-87. Petrolatum Tapes for Corrosion Protection. 13 pp.

—**PVC Film—Adhesive Tapes**
- JIS Z 1901-88. Pressure Sensitive Adhesive Polyvinyl Chloride Tapes for Corrosion Protection. 9 pp.

—**Rust**
- CNS Z6028-80. Method of Test for Storage of Rust Preventing Oil (Dec)(6728).

—**Rust—Busways**
- CSA C22.2 NO 27-1973. Busways; (Amd 1 June 1987). 19 pp.
- CSA 1091 Bull. Electrical Bulletin 1091 February 8, 1977 to C22.2 NO 27. 1 p.
- CSA 1169 Bull. Electrical Bulletin 1169 June 27, 1978 to C22.2 NO 27. 2 pp.

—**Rust—Glossaries**
- JIS Z 0103-78. Glossary of Terms Used in Rust and Corrosion Preventive Technology.

—**Rust—Greases**
- BSI BS 2000: Part 220-93. 1993 Methods of Test for Petroleum and Its Products Part 220: Determination of Rust Prevention Characteristics of Lubricating Greases (W). 6 pp.
- BSI BS 2000: Part 220-82. 1982 Petroleum and Its Products Part 220: Dynamic Anti-Rust Test for Lubricating Greases. 7 pp.

—**Rust—Lubricating Oils**
- MOD UK DSTAN 05-50: Part 43-89. Methods for Testing Fuels, Lubricants and Associated Products Part 43: Rust Inhibiting Properties of Oils Issue 1. 9 pp.

Corrosion Prevention (Cont.)

—**Rust—Motor Oils**
- BSI BS 2000: Part 135-93. 1993 Methods of Test for Petroleum and Its Products Part 135: Determination of Rust-Preventing Characteristics of Steam Turbine Oil in the Presence of Water (W). 7 pp.
- BSI BS 2000: Part 135-83. 1983 Petroleum and Its Products Part 135: Rust Preventing Characteristics of Steam-Turbine Oil in the Presence of Water. 8 pp.
- CNS K6324-72. Method of Test for Rust Prevention Characteristics of Steam-Turbine Oil in the Presence of Water (Oct)(3385).

—**Sea Water Systems—Naval Ships**
- MOD UK NES 781-90. Process and Procedure Requirements for the Protection of Sea—Water Systems Using Sodium Dimethydithiocarbamate Issue 3 (09.90). 11 pp.

—**Solar Water Heaters**
- ISO TR10217-89. Solar Energy—Water Heating Systems—Guide to Material Selection with Regard to Internal Corrosion First Edition. 8 pp.

—**Steel Pipes**
- DIN ENGL 30675 Pt 1-85. External Corrosion Protection of Buried Pipes; Protection of Steel Pipelines (Apr). 5 pp.

—**Steel Structures**
- DIN ENGL 55928 Pt 1-91. Corrosion Protection of Steel Structures by the Application of Organic or Metallic Coatings; General, Concepts and Corrosion Loads (May). 8 pp.
- DIN ENGL 55928 Pt 2-91. Corrosion Protection of Steel Structures by the Application of Organic or Metallic Coatings; Designing for the Prevention of Corrosion (May). 7 pp.
- DIN ENGL 55928 Pt 3-78. Corrosion Protection of Steel Structures by Organic and Metallic Coatings; Planning of Corrosion Protection (Nov). 5 pp.
- DIN ENGL 55928 Pt 4-91. Corrosion Protection of Steel Structures by the Application of Organic or Metallic Coatings; Preparation and Testing of Surfaces (May). 16 pp.
- DIN ENGL 55928 Pt 4 Suppl 1 A1-91. Corrosion Protection of Steel Structures by the Application of Organic or Metallic Coatings; Preparation and Testing of Surfaces Representative Photographic Examples Amendment 1 to Supplement 1 to DIN 55928 Part 4 (May). 5 pp.
- DIN ENGL 55928 Pt 4 Suppl 2 A1-91. Corrosion Protection of Steel Structures by the Application of Organic or Metallic Coatings; Preparation and Testing of Surfaces Representative Photographic Examples of Surfaces After Localized Mechanical Grinding (Standard Preparation Grade PMa). 3 pp.
- DIN ENGL 55928 Pt 5-91. Corrosion Protection of Steel Structures by the Application of Organic or Metallic Coatings; Coating Materials and Protective Systems (May). 24 pp.
- DIN ENGL 55928 Pt 6-91. Corrosion Protection of Steel Structures by the Application of Organic or Metallic Coatings; Execution and Inspection of Corrosion Protection Work (May). 7 pp.
- DIN ENGL 55928 Pt 7-91. Corrosion Protection of Steel Structures by the Application of Organic or Metallic Coatings; Reference Areas (May). 7 pp.
- DIN ENGL 55928 Pt 8-80. Corrosion Protection of Steel Structures by Organic and Metallic Coatings; Corrosion Protection of Thin-Walled Structural Members (Light Gauge Steel Construction) (Mar) (Supersedes Parts of DIN 55928). 6 pp.
- SNZ NZMP 2312-87. Commentary on AS 2312: 1984 Guide to the Protection of Iron and Steel Against Exterior Atmospheric Corrosion. 12 pp.

—**Steels**
- SAA AS 2312-85. Guide to the Protection of Iron and Steel Against Exterior Atmospheric Corrosion (This is a Joint Standard with SANZ NZS 2312). 59 pp.
- SNZ NZS/AS 2312-85. Guide to the Protection of Iron and Steel Against Exterior Atmospheric Corrosion (This is a Joint Standard with SAA AS 2312). 59 pp.

—**Storage Tanks—Underground**
- DIN ENGL 6607-91. External Corrosion Protection of Underground Steel Storage Tanks; Requirements and Testing (Jan). 5 pp.

—**Structural Steels—Sections**
- BSI DD 24-73. (WITHDRAWN) 1973 Amd 1 Recommendations for Methods of Protection Against Corrosion on Light Section Steel Used in Building. 35 pp.

—**Valves**
- DIN ENGL 30677 Pt 2-88. External Corrosion Protection of Buried Valves; Heavy-Duty Thermoset Plastics Coatings (Sept). 6 pp.

INDUSTRY STANDARDS

Corrosion

INTERNATIONAL AND NON-U.S. NATIONAL STANDARDS
SUBJECT INDEX

Corrosion Prevention (Cont.)

—**Water Heaters**

BSI BS 7593-92. 1992 Treatment of Water in Domestic Hot Water Central Heating Systems. 10 pp.

DIN ENGL 4753 Pt 4-82. Water Heating Installations for Drinking Water and Service Water; Corrosion Protection on the Water Side by Thermosetting, Resin-Bonded Lining Materials; Requirements and Testing (July). 4 pp.

DIN ENGL 4753 Pt 5-82. Water Heating Installations for Drinking Water and Service Water; Corrosion Protection on the Water Side by Natural or Synthetic Rubber Films; Requirements and Testing (July). 7 pp.

—**Water Supply Installations**

DIN ENGL 1988 Pt 7-88. Drinking Water Supply Systems; Prevention of Corrosion and Scaling; (DVGW Code of Practice) (Dec). 7 pp.

—**Water Tanks**

MOD UK DTD-909-36. Protection of the Interior of Drinking Water Tanks Against Corrosion (Reprinted March 1963). 1 p.

Corrosion Protection
Use: Corrosion Prevention

Corrosion Resistance Testing
Use: Corrosion Testing

Corrosion Resistant Castings
See Also: Castings; Corrosion Resistant Steel Castings

BSI BS 1591-75. 1975 Corrosion Resisting High Silicon Iron Castings. 8 pp.

Corrosion Resistant Coatings
Scope Note: See the subheading Corrosion Inhibitive under specific types of coatings

Corrosion Resistant Paints
Scope Note: See the subheading Corrosion Inhibitive under specific types of paints

Corrosion Resistant Pumps
Scope Note: See the subheading Corrosion Resistant under specific types of pumps

Corrosion Resistant Steel Castings
See Also: Castings; Corrosion Resistant Castings; Corrosion Resistant Steels; Steel Castings

—**Aerospace**

BSI HC 103-74. 1974 Amd 1 23% Chromium-Nickel-Tungsten Corrosion-Resisting Steel Castings (Cr 23, Ni 11, W 3). 3 pp.

BSI HC 104-74. 1974 Amd 1 19% Chromium-10% Nickel Niobium-Stabilized Corrosion-Resisting Steel Castings (460 MPa). 3 pp.

BSI HC 105-74. 1974 Amd 1 18% Chromium-11% Nickel-2.5% Molybdenum Niobium-Stabilized Corrosion-Resisting Steel Castings. 3 pp.

—**Investment**

BSI BS 3146: Part 2-75. 1975 Amd 1 Investment Castings in Metal Part 2: Corrosion and Heat Resisting Steels, Nickel and Cobalt Base Alloys. 28 pp.

—**Investment—Aerospace**

MOD UK DTD-5259-67. Chromium-Nickel Corrosion-Resisting Steel Investment Castings (Not Stabilised) (Tensile Strength 47 kgf /mm2) (Not to Be Used for Applications at Temperatures Exceeding 350 Degrees C). 2 pp.

MOD UK DTD-5269-67. Chromium-Nickel Corrosion-Resisting Steel Investment Castings (Niobium Stabilized) (Tensile Strength 47 kgf /mm2). 2 pp.

MOD UK DTD-5279-68. Chromium-Nickel-2.5 Per Cent Molybdenum Heat-Resisting and Corrosion-Resisting Steel Investment Castings (50 Hbar) (High Temperature Properties Not Verified). 2 pp.

MOD UK DTD-5289-68. Chromium-Nickel-3.5 Per Cent Molybdenum Heat-Resisting and Corrosion-Resisting Steel Investment Castings (50 Hbar) (High Temperature Properties Not Verified). 2 pp.

—**Precipitation Hardening—Aerospace**

BSI HC 106-74. 1974 Amd 2 Precipitation Hardening Chromium-Nickel-Copper Corrosion-Resisting Steel Castings (1250-1500 MPa) (S) (AMD 5049) May 30, 1986. 5 pp.

Corrosion Resistant Steel Coatings
See Also: Coatings; Corrosion Resistant Steels; Metal Coatings (On Metal); Steel Coatings

—**Cadmium—Electroplated**

MOD UK DTD-904C-63. Cadmium Plating (Reprinted December 1966, Incorporating Amendment No. 1) (Superseded by Def Stan 03-19). 5 pp.

Corrosion Resistant Steel Coatings (Cont.)

—**Epoxy**

CGSB CAN/CGSB-1.207-M91. Low Temperature Curing Epoxy Coating. 11 pp.

Corrosion Resistant Steels
Scope Note: For additional listings, see also specific products made from corrosion resistant steel

See Also: Alloy Steels; Austenitic Stainless Steels; Chromium Molybdenum Steels; Chromium Molybdenum Vanadium Steels; Chromium Steels; Coatings; Corrosion Resistant Steel Castings; Corrosion Resistant Steel Coatings; Ferritic Stainless Steels; Ferroalloys; Heat Resistant Steels; Low Alloy Steels; Martensitic Stainless Steels; Metal Coatings (On Metal); Nickel Alloys; Nickel Chromium Steels; Nickel Steels; Stainless Steels; Steel Coatings; Steels

—**Bars**

CNS G3197-87. Corrosion-Resisting and Heat-Resisting Superalloy Bars (Apr) (9604).

JIS G 4901-91. Corrosion-Resisting and Heat-Resisting Superalloy Bars. 15 pp.

—**Bars—Aerospace**

BSI 5S 80-76. (WITHDRAWN) 1976 Amd 2 High Chromium-Nickel Corrosion-Resisting Steel Billets, Bars, Forgings and Parts (880-1080 MPa; Limiting Ruling Section 100 mm) (Superseded by BS 6S 80: 1990). 6 pp.

BSI 6S 80-90. 1990 Amd 1 High Chromium-Nickel Corrosion-Resisting Steel Forging Stock, Bars, Forgings and Parts (880-1080 MPa, Limiting Ruling Section 100 mm) (AMD 6396) September 30, 1991. 4 pp.

BSI 2S 130-76. 1976 Amd 1 18/9 Chromium-Nickel Corrosion-Resisting Steel (Niobium Stabilized) Billets, Bars, Forgings and Parts (540 MPa; Limiting Ruling Section 150 mm). 5 pp.

—**Bars—Bright—Aerospace**

BSI 2S 137-76. 1976 High Chromium-Nickel Corrosion-Resisting Steel Bright Bars (Free Machining) (880-1080 MPa; Limiting Ruling Section 70 mm). 4 pp.

BSI S 160-90. 1990 Chromium-Nickel Corrosion-Resisting Steel—Bright Bars and Parts (540 MPa, Limiting Ruling Section 160mm) (Cr18, Ni10.5, Controlled Nitrogen Content). 3 pp.

BSI S 161-90. 1990 Chromium-Nickel-Molybdenum Corrosion-Resisting Steel Bright Bars and Parts (540 MPa, Limiting Ruling Section 160mm) (Cr17.5, Ni12.5, Mo2.25, Controlled Nitrogen Content). 3 pp.

—**Bars—Precipitation Hardened—Aerospace**

BSI 2S 143-02. 1976 Amd 2 Chromium-Nickel-Copper-Molybdenum Corrosion-Resisting Steel (Precipitation Hardening) Billets, Bars, Forgings and Parts (930-1080 MPa) (AMD 6775) December 24, 1991. 6 pp.

BSI 2S 145-03. 1976 Amd 3 Chromium-Nickel-Copper-Molybdenum Corrosion-Resisting Steel (Precipitation Hardening) Billets, Bars, Forgings and Parts (1270-1470 MPa) (AMD 6777) December 24, 1991. 5 pp.

—**Billets—Aerospace**

BSI 5S 80-76. (WITHDRAWN) 1976 Amd 2 High Chromium-Nickel Corrosion-Resisting Steel Billets, Bars, Forgings and Parts (880-1080 MPa; Limiting Ruling Section 100 mm) (Superseded by BS 6S 80: 1990). 6 pp.

BSI 2S 130-76. 1976 Amd 1 18/9 Chromium-Nickel Corrosion-Resisting Steel (Niobium Stabilized) Billets, Bars, Forgings and Parts (540 MPa; Limiting Ruling Section 150 mm). 5 pp.

—**Billets—Precipitation Hardening—Aerospace**

BSI 2S 143-02. 1976 Amd 2 Chromium-Nickel-Copper-Molybdenum Corrosion-Resisting Steel (Precipitation Hardening) Billets, Bars, Forgings and Parts (930-1080 MPa) (AMD 6775) December 24, 1991. 6 pp.

BSI 2S 145-03. 1976 Amd 3 Chromium-Nickel-Copper-Molybdenum Corrosion-Resisting Steel (Precipitation Hardening) Billets, Bars, Forgings and Parts (1270-1470 MPa) (AMD 6777) December 24, 1991. 5 pp.

—**Bolts—Aerospace**

MOD UK DTD-5056-68. Chromium-Nickel Corrosion Resisting Steel for Cold Headed Bolts and Set Screws (Suitable for Cold Forming). 4 pp.

—**Cold Drawn—Tubes—Aerospace**

BSI 2T 68-80. 1980 Cold Drawn 18/10 Chromium-Nickel Corrosion-Resisting Steel Tube (Niobium Stabilized: 800 N/Square mm) (Weldable). 2 pp.

Corrosion Resistant Steels (Cont.)

—**Forgings—Aerospace**

BSI 5S 80-76. (WITHDRAWN) 1976 Amd 2 High Chromium-Nickel Corrosion-Resisting Steel Billets, Bars, Forgings and Parts (880-1080 MPa; Limiting Ruling Section 100 mm) (Superseded by BS 6S 80: 1990). 6 pp.

BSI 6S 80-90. 1990 Amd 1 High Chromium-Nickel Corrosion-Resisting Steel Forging Stock, Bars, Forgings and Parts (880-1080 MPa, Limiting Ruling Section 100 mm) (AMD 6396) September 30, 1991. 4 pp.

BSI 2S 130-76. 1976 Amd 1 18/9 Chromium-Nickel Corrosion-Resisting Steel (Niobium Stabilized) Billets, Bars, Forgings and Parts (540 MPa; Limiting Ruling Section 150 mm). 5 pp.

—**Forgings—Precipitation Hardening—Aerospace**

BSI 2S 143-02. 1976 Amd 2 Chromium-Nickel-Copper-Molybdenum Corrosion-Resisting Steel (Precipitation Hardening) Billets, Bars, Forgings and Parts (930-1080 MPa) (AMD 6775) December 24, 1991. 6 pp.

BSI 2S 145-03. 1976 Amd 3 Chromium-Nickel-Copper-Molybdenum Corrosion-Resisting Steel (Precipitation Hardening) Billets, Bars, Forgings and Parts (1270-1470 MPa) (AMD 6777) December 24, 1991. 5 pp.

—**Hydrogen Peroxide**

MOD UK DEF-61-58. Selection and Treatment of Corrosion Resisting Steels for Use with Concentrated Hydrogen Peroxide (H.T.P.) (Withdrawn). 5 pp.

—**Passivation—Aerospace**

JIS W 1105-87. Passivation Treatments for Corrosion-Resisting Steel of Aerospace Use.

—**Plates**

BSI BS 1501: Part 3-90. 1990 Amd 1 Steels for Pressure Purposes: Plates, Sheet and Strip Part 3: Specification for Corrosion-and Heat-Resisting Steels (AMD 6744) July 31, 1991. 24 pp.

BSI BS 1501: Part 3-02. 1990 Amd 2 Steels for Pressure Purposes: Plates, Sheet and Strip Part 3: Specification for Corrosion—and Heat-Resisting Steels (AMD 6868) October 31, 1991. 28 pp.

BSI BS 1501: Part 3-03. 1990 Amd 3 Steels for Pressure Purposes Part 3: Specification for Corrosion—and Heat-Resisting Steels: Plates, Sheet and Strip (AMD 7026) June 15, 1992 (Q). 29 pp.

CNS G3198-87. Corrosion-Resisting and Heat-Resisting Superalloy Sheets and Plates (Apr)(9606).

JIS G 4902-91. Corrosion-Resisting and Heat-Resisting Superalloy Plates and Sheets. 18 pp.

SNZ NZS/BS 1501: Part 3-90. Steels for Pressure Purposes: Plates Part 3: Specification for Corrosion- and Heat-Resisting Steels Amend: 1, 1991; 2, 1991. 20 pp.

—**Precipitation Hardened**

MOD UK DSTAN 95-16-82. Manufacture of Armament Components in Weldable Chromium—Nickel-Copper—Molybdenum Precipitation Hardening Corrosion Resisting Steel Issue 1. 6 pp.

—**Precipitation Hardened—Bars**

MOD UK DSTAN 95-14-82. Bars, Billets, Plates and Forgings, Weldable Chromium—Nickel-Copper—Molybdenum Precipitation Hardening Corrosion Resisting Steel Issue 1. 7 pp.

—**Precipitation Hardened—Bars—Aerospace**

BSI 2S 143-76. 1976 Amd 1 Chromium-Nickel-Copper-Molybdenum Corrosion-Resisting Steel (Precipitation Hardening) Billets, Bars, Forgings and Parts (930-1080 MPa). 5 pp.

BSI 2S 143-02. 1976 Amd 2 Chromium-Nickel-Copper-Molybdenum Corrosion-Resisting Steel (Precipitation Hardening) Billets, Bars, Forgings and Parts (930-1080 MPa) (AMD 6775) December 24, 1991. 6 pp.

BSI 2S 144-76. 1976 Amd 3 Chromium-Nickel-Copper-Molybdenum Corrosion-Resisting Steel (Precipitation Hardening) Billets, Bars, Forgings and Parts (1130-1330 MPa) (AMD 5608) November 30, 1987. 7 pp.

BSI 2S 144-04. 1976 Amd 4 Chromium-Nickel-Copper-Molybdenum Corrosion-Resisting Steel (Precipitation Hardening) Billets, Bars, Forgings and Parts (1130-1330 MPa) (AMD 6774) December 24, 1991. 8 pp.

BSI 2S 145-76. 1976 Amd 2 Chromium-Nickel-Copper-Molybdenum Corrosion-Resisting Steel (Precipitation Hardening) Billets, Bars, Forgings and Parts (1270-1470 MPa) (AMD 5607) November 30, 1987. 5 pp.

BSI 2S 145-03. 1976 Amd 3 Chromium-Nickel-Copper-Molybdenum Corrosion-Resisting Steel (Precipitation Hardening) Billets, Bars, Forgings and Parts (1270-1470 MPa) (AMD 6777) December 24, 1991. 5 pp.

INTERNATIONAL AND NON-U.S. NATIONAL STANDARDS
SUBJECT INDEX
Corrosion

Corrosion Resistant Steels (Cont.)
—Precipitation Hardened—Billets
MOD UK DSTAN 95-14-82. Bars, Billets, Plates and Forgings, Weldable Chromium—Nickel—Copper—Molybdenum Precipitation Hardening Corrosion Resisting Steel Issue 1. 7 pp.

—Precipitation Hardened—Billets—Aerospace
BSI 2S 143-76. 1976 Amd 1 Chromium-Nickel-Copper-Molybdenum Corrosion-Resisting Steel (Precipitation Hardening) Billets, Bars, Forgings and Parts (930-1080 MPa). 5 pp.
BSI 2S 143-02. 1976 Amd 2 Chromium-Nickel-Copper-Molybdenum Corrosion-Resisting Steel (Precipitation Hardening) Billets, Bars, Forgings and Parts (930-1080 MPa) (AMD 6775) December 24, 1991. 6 pp.
BSI 2S 144-76. 1976 Amd 3 Chromium-Nickel-Copper-Molybdenum Corrosion-Resisting Steel (Precipitation Hardening) Billets, Bars, Forgings and Parts (1130-1330 MPa) (AMD 5608) November 30, 1987. 7 pp.
BSI 2S 144-04. 1976 Amd 4 Chromium-Nickel-Copper-Molybdenum Corrosion-Resisting Steel (Precipitation Hardening) Billets, Bars, Forgings and Parts (1130-1330 MPa) (AMD 6774) December 24, 1991. 8 pp.
BSI 2S 145-76. 1976 Amd 2 Chromium-Nickel-Copper-Molybdenum Corrosion-Resisting Steel (Precipitation Hardening) Billets, Bars, Forgings and Parts (1270-1470 MPa) (AMD 5607) November 30, 1987. 5 pp.
BSI 2S 145-03. 1976 Amd 3 Chromium-Nickel-Copper-Molybdenum Corrosion-Resisting Steel (Precipitation Hardening) Billets, Bars, Forgings and Parts (1270-1470 MPa) (AMD 6777) December 24, 1991. 5 pp.

—Precipitation Hardened—Forgings
MOD UK DSTAN 95-14-82. Bars, Billets, Plates and Forgings, Weldable Chromium—Nickel—Copper—Molybdenum Precipitation Hardening Corrosion Resisting Steel Issue 1. 7 pp.

—Precipitation Hardened—Forgings—Aerospace
BSI 2S 143-76. 1976 Amd 1 Chromium-Nickel-Copper-Molybdenum Corrosion-Resisting Steel (Precipitation Hardening) Billets, Bars, Forgings and Parts (930-1080 MPa). 5 pp.
BSI 2S 143-02. 1976 Amd 2 Chromium-Nickel-Copper-Molybdenum Corrosion-Resisting Steel (Precipitation Hardening) Billets, Bars, Forgings and Parts (930-1080 MPa) (AMD 6775) December 24, 1991. 6 pp.
BSI 2S 144-76. 1976 Amd 3 Chromium-Nickel-Copper-Molybdenum Corrosion-Resisting Steel (Precipitation Hardening) Billets, Bars, Forgings and Parts (1130-1330 MPa) (AMD 5608) November 30, 1987. 7 pp.
BSI 2S 144-04. 1976 Amd 4 Chromium-Nickel-Copper-Molybdenum Corrosion-Resisting Steel (Precipitation Hardening) Billets, Bars, Forgings and Parts (1130-1330 MPa) (AMD 6774) December 24, 1991. 8 pp.
BSI 2S 145-76. 1976 Amd 2 Chromium-Nickel-Copper-Molybdenum Corrosion-Resisting Steel (Precipitation Hardening) Billets, Bars, Forgings and Parts (1270-1470 MPa) (AMD 5607) November 30, 1987. 5 pp.
BSI 2S 145-03. 1976 Amd 3 Chromium-Nickel-Copper-Molybdenum Corrosion-Resisting Steel (Precipitation Hardening) Billets, Bars, Forgings and Parts (1270-1470 MPa) (AMD 6777) December 24, 1991. 5 pp.

—Precipitation Hardened—Plates
MOD UK DSTAN 95-14-82. Bars, Billets, Plates and Forgings, Weldable Chromium—Nickel—Copper—Molybdenum Precipitation Hardening Corrosion Resisting Steel Issue 1. 7 pp.
MOD UK DSTAN 95-15-82. Sheet, Strip and Plate Weldable Chromium—Nickel—Copper—Molybdenum Precipitation Hardening Corrosion Resisting Steel Issue 1. 6 pp.

—Precipitation Hardened—Sheets
MOD UK DSTAN 95-15-82. Sheet, Strip and Plate Weldable Chromium—Nickel—Copper—Molybdenum Precipitation Hardening Corrosion Resisting Steel Issue 1. 6 pp.

—Precipitation Hardened—Strips
MOD UK DSTAN 95-15-82. Sheet, Strip and Plate Weldable Chromium—Nickel—Copper—Molybdenum Precipitation Hardening Corrosion Resisting Steel Issue 1. 6 pp.

—Rods—Aerospace
MOD UK DTD-5036-01. Low Carbon Chromium-Nickel Corrosion-Resisting Steel Wire, Rivets and Split Pins (Weldable); Amendment 1. 5 pp.

Corrosion Resistant Steels (Cont.)
—Rods—Aircraft
MOD UK DTD-161A-58. Corrosion-Resisting Steel Rod and Wire (Suitable for Locking Wire) (Reprinted September 1959). 2 pp.

—Screws—Aerospace
MOD UK DTD-5056-68. Chromium-Nickel Corrosion Resisting Steel for Cold Headed Bolts and Set Screws (Suitable for Cold Forming). 4 pp.

—Sheets
CNS G3198-87. Corrosion-Resisting and Heat-Resisting Superalloy Sheets and Plates (Apr)(9606).
JIS G 4902-91. Corrosion-Resisting and Heat-Resisting Superalloy Plates and Sheets. 18 pp.

—Sheets—Aerospace
BSI S 527-69. 1969 Amd 1 Softened 18/10 Chromium-Nickel Corrosion-Resisting Steel Sheet and Strip (Niobium Stabilized: 54 Hbar). 4 pp.
BSI S 536-70. 1970 Low Carbon 18/10 Chromium-Nickel Corrosion-Resisting Steel Sheet and Strip (50 Hbar). 3 pp.
BSI S 537-70. 1970 Low Carbon 17/12 Chromium-Nickel-Molybdenum Corrosion-Resisting Steel Sheet and Strip (50 Hbar). 4 pp.

—Sheets—Cold Rolled—Aerospace
BSI S 525-69. 1969 Amd 2 Cold-Rolled 18/10 Chromium-Nickel Corrosion-Resisting Steel Sheet and Strip (Niobium Stabilized: 80 Hbar). 5 pp.

—Sheets—Precipitation Hardened—Aerospace
BSI S 532-69. 1969 Amd 1 Chromium-Nickel-Copper-Molybdenum Corrosion-Resisting Steel Sheet and Strip (Precipitation Hardening: 98/118 Hbar). 3 pp.
BSI S 533-69. 1969 Amd 1 Chromium-Nickel-Copper-Molybdenum Corrosion-Resisting Steel Sheet and Strip (Precipitation Hardening: 118/137 H Bar). 3 pp.

—Strips—Aerospace
BSI S 527-69. 1969 Amd 1 Softened 18/10 Chromium-Nickel Corrosion-Resisting Steel Sheet and Strip (Niobium Stabilized: 54 Hbar). 4 pp.
BSI S 536-70. 1970 Low Carbon 18/10 Chromium-Nickel Corrosion-Resisting Steel Sheet and Strip (50 Hbar). 3 pp.
BSI S 537-70. 1970 Low Carbon 17/12 Chromium-Nickel-Molybdenum Corrosion-Resisting Steel Sheet and Strip (50 Hbar). 4 pp.

—Strips—Aircraft
MOD UK DTD-271-35. Non-Corrodible Steel Strips (Suitable for Magneto Contact Breaker Springs) (Reprinted March 1962). 2 pp.

—Strips—Cold Rolled—Aerospace
BSI S 525-69. 1969 Amd 2 Cold-Rolled 18/10 Chromium-Nickel Corrosion-Resisting Steel Sheet and Strip (Niobium Stabilized: 80 Hbar). 5 pp.

—Strips—Precipitation Hardened—Aerospace
BSI S 532-69. 1969 Amd 1 Chromium-Nickel-Copper-Molybdenum Corrosion-Resisting Steel Sheet and Strip (Precipitation Hardening: 98/118 Hbar). 3 pp.
BSI S 533-69. 1969 Amd 1 Chromium-Nickel-Copper-Molybdenum Corrosion-Resisting Steel Sheet and Strip (Precipitation Hardening: 118/137 H Bar). 3 pp.

—Structural
CEN PREN 10 155-91. Structural Steels with Improved Atmospheric Corrosion Resistance Technical Conditions. 32 pp.
CEN EN 10155-93. Structural Steels with Improved Atmospheric Corrosion Resistance—Technical Delivery Conditions. 21 pp.
CNS G2042-86. Method of Test for Hot-Rolled Atmospheric Corrosion Resisting Steels for Welded Structure (Jun)(4270).
CNS G2047-78. Method of Test for Superior Atmospheric Corrosion Resisting Rolled Steels (Oct)(4621). 3 pp.
CNS G3108-78. Superior Atmospheric Corrosion Resisting Rolled Steels (Oct)(4620). 2 pp.
ISO 4952-81. Structural Steels with Improved Atmospheric Corrosion Resistance First Edition. 10 pp.
ISO 4995-91. Hot-Rolled Steel Sheet of Structural Quality Second Edition. 12 pp.
ISO 4996-91. Hot-Rolled Steel Sheet of High Yield Stress Structural Quality Second Edition. 12 pp.
ISO 4997-91. Cold-Reduced Steel Sheet of Structural Quality Second Edition. 11 pp.
ISO 5952-83. Continuously Hot-Rolled Steel Sheet of Structural Quality with Improved Atmospheric Corrosion Resistance First Edition; DoD Adopted. 11 pp.
JIS G 3114-88. Hot-Rolled Atmospheric Corrosion Resisting Steels for Welded Structure. 18 pp.

Corrosion Resistant Steels (Cont.)
—Structural (Cont.)
JIS G 3125-87. Superior Atmospheric Corrosion Resisting Rolled Steels.
JIS G 3125-77. Superior Atmospheric Corrosion Resisting Rolled Steels. 10 pp.

—Structural—Hot Rolled
CNS G3099-86. Hot-Rolled Atmospheric Corrosion Resisting Steels for Welded Structure (Jun)(4269). 4 pp.

—Tubes—Aerospace
BSI 2T 66-80. 1980 18/10 Chromium-Nickel Corrosion-Resisting Steel Tube (Niobium Stabilized: 550 MPa) (Weldable). 2 pp.
BSI 2T 66-01. 1980 Amd 1 18/10 Chromium-Nickel Corrosion-Resisting Steel Tube (Niobium Stabilized: 550 MPa) (Weldable) (AMD 6853) November 29, 1991. 3 pp.
BSI T 72-73-77. 1977 18/10 Chromium-Nickel Corrosion-Resisting Steel Tube for Hydraulic Purposes (Niobium/Titanium Stabilized: 550 MPa). 7 pp.
BSI T 72-73-01. 1977 Amd 1 18/10 Chromium-Nickel Corrosion-Resisting Steel Tube for Hydraulic Purposes (Niobium/Titanium Stabilized: 550 MPa) (AMD 6852) December 24, 1991. 8 pp.

—Tubes—Aircraft
MOD UK DTD-97B-50. 28-Ton 12 Per Cent Chromium Corrosion-Resistant Steel Tubes. 1 p.
MOD UK DTD-203B-50. 50-Ton 12 Per Cent Chromium Corrosion-Resistant Steel Tubes (Reprinted August 1952). 2 pp.

—Tubes—Hydraulic—Aerospace
BSI T 72-73-01. 1977 Amd 1 18/10 Chromium-Nickel Corrosion-Resisting Steel Tube for Hydraulic Purposes (Niobium/Titanium Stabilized: 550 MPa) (AMD 6852) December 24, 1991. 8 pp.

Corrosion Resistant Valves
Scope Note: See the subheading Corrosion Resistant under specific types of valves

Corrosion Testing
Use For: Stress Corrosion Testing
See Also: Accelerated Testing; Acid Resistance Testing; Alkali Resistance Testing; Atmospheric Corrosion Testing; Copper Strip Corrosion; Corrodkote Testing; Corrosion; Destructive Testing; Environmental Testing; Filiform Corrosion Testing; Immersion Testing; Intergranular Corrosion Testing; Materials Testing; Metals; Pitting Testing; Salt Spray Testing; Staining

—Alloys
ISO 7539 Pt 1-87. Corrosion of Metals and Alloys—Stress Corrosion Testing—Part 1: General Guidance on Testing Procedures First Edition. 14 pp.
ISO 7539 Pt 2-89. Corrosion of Metals and Alloys—Stress Corrosion Testing—Part 2: Preparation and Use of Bent-Beam Specimens First Edition. 10 pp.
ISO 7539 Pt 3-89. Corrosion of Metals and Alloys—Stress Corrosion Testing—Part 3: Preparation and Use of U-Bend Specimens First Edition. 7 pp.
ISO 7539 Pt 4-89. Corrosion of Metals and Alloys—Stress Corrosion Testing—Part 4: Preparation and Use of Uniaxially Loaded Tension Specimens First Edition. 7 pp.
ISO 7539 Pt 5-89. Corrosion of Metals and Alloys—Stress Corrosion Testing—Part 5: Preparation and Use of C-Ring Specimens First Edition. 13 pp.
ISO 7539 Pt 6-89. Corrosion of Metals and Alloys—Stress Corrosion Testing—Part 6: Preparation and Use of Pre-Cracked Specimens First Edition. 34 pp.
ISO 7539 Pt 7-89. Corrosion of Metals and Alloys—Stress Corrosion Testing—Part 7: Slow Strain Rate Testing First Edition. 7 pp.

—Aluminum Alloys
AECMA PREN2720-90. Test Method for Metallic Materials Testing of Susceptibility to Exfoliation Corrosion in 2XXX and 7XXX Series Wrought Aluminium Alloy Products for Aerospace Constructions. 1 p.
CEN PREN 2004 (Part 4)-92. Test Methods for Aluminium and Aluminium Alloy Products Part 4—Stress Corrosion Test by Alternate Immersion for High Strength Aluminium Alloy Wrought Products. 14 pp.
CEN PREN 2716-90. Test Method for Susceptibility to Intergranular Corrosion of Wrought Products in 2XXX Series Aluminium Alloys. 4 pp.
CEN PREN 2717-90. Test Method for Susceptibility to Intergranular Corrosion of Wrought Products in 5XXX Series Aluminium Alloys with a Magnesium Content /3.5%. 7 pp.

INDUSTRY STANDARDS

INTERNATIONAL AND NON-U.S. NATIONAL STANDARDS
SUBJECT INDEX — Corrosion

Corrosion Testing (Cont.)

—**Anodic Coatings**
CNS H2063-82. Methods of Test for Corrosion Resistance of Anodic Oxidation Coatings on Aluminium and Aluminium Alloys (Jan)(8410). 4 pp.
JIS H 8681-88. Test Methods for Corrosion Resistance of Anodic Oxidation Coatings on Aluminium and Aluminium Alloys. 23 pp.

—**Anodic Coatings—Aluminum**
SAA AS 2039.3.1-78. Methods for Testing Anodic Oxidation Coatings on Aluminium and Aluminium Alloys —Part 3.1: Corrosion Tests—Neutral Salt Spray Test (NSS Test) of Anodic Oxidation Coatings Reconfirmed 1990. 6 pp.
SAA AS 2039.3.2-78. Methods for Testing Anodic Oxidation Coatings on Aluminium and Aluminium Alloys —Part 3.2: Corrosion Tests—Copper Accelerated Acetic Acid Salt Spray Test (CASS Test) of Anodic Oxidation Coatings Reconfirmed 1990. 6 pp.
SAA AS 2039.3.3-78. Methods for Testing Anodic Oxidation Coatings on Aluminium and Aluminium Alloys—Part 3.3: Corrosion Tests—Mortar Test for Clear Organic Coatings on Anodic Oxidation Coatings Reconfirmed 1990. 2 pp.

—**Austenitic Stainless Steels**
DIN ENGL 50921-84. Corrosion of Metals; Testing of Austenitic Stainless Steels for Resistance to Local Corrosion in Highly Oxidizing Acids; Corrosion Test in Nitric Acid Medium by Measurement of Loss in Mass (Huey Test) (Oct). 3 pp.

—**Bearing Lubricants**
ISO TR10129-93. Plain Bearings—Testing of Bearing Metals—Resistance to Corrosion by Lubricants Under Static Conditions First Edition. 7 pp.

—**Brazed Joints**
JIS Z 3195-71. Method of Wet Corrosion Test for Brazed Joint. 6 pp.
JIS Z 3196-72. Method of Gaseous Corrosion Test for Brazed Joint (R 1979). 9 pp.

—**Cable Insulation—Combustion**
DIN VDE 0472 Pt 813-83. Testing of Cables, Wires and Flexible Cords; Corrosivity of Combustion Gases (Aug). 6 pp.

—**Cleaning Agents**
CGSB 31-GP-0A METH 11.4A-62. Methods of Testing Corrosion-Prevention Materials and Processes Corrosion (Cleaning Compounds) (Supersedes 11.4 15 March 1957). 1 p.
CGSB 31-GP-0A METH 11.5-57. Methods of Testing Corrosion-Prevention Materials and Processes Corrosion (Emulsion Cleaners); (Amended September 1966) (Re-Edited March 1982). 1 p.

—**Connectors**
SBAC TS 333 ISSUE 1. Test for Electrical Connectors Corrosion.

—**Copper Alloys**
ISO 6957-88. Copper Alloys—Ammonia Test for Stress Corrosion Resistance First Edition. 6 pp.

—**Corrosion Inhibitors**
CGSB 31-GP-0A METH 11.1-57. Methods of Testing Corrosion-Prevention Materials and Processes Corrosion (Cold Application Compounds). 2 pp.
CGSB 31-GP-0A METH 11.17-74. Methods of Testing Corrosion-Prevention Materials and Processes Corrosion. 1 p.

—**Degreasers (Cleaning Agents)**
CGSB 31-GP-0A METH 11.11-57. Methods of Testing Corrosion-Prevention Materials and Processes Corrosion. 1 p.

—**Enamels**
DIN ENGL 51174-88. Corrosion Testing of Enamel in Closed Systems (May). 4 pp.

—**Fire Alarms**
CEN EN 54-5-76. Components of Automatic Fire Detection Systems Part 5 Heat Sensitive Detectors—Point Detectors Containing a Static Element. 20 pp.
CEN EN 54 (Part 5)-88. AMD 1 Components of Automatic Fire Detection Systems Part 5 Heat Sensitive Detectors—Point Detectors Containing a Static Element. 21 pp.
CEN EN 54-6-82. Components of Automatic Fire Detection Systems—Part 6: Heat-Sensitive Detectors; Rate-of-Rise Point Detectors Without a Static Element. 12 pp.
CEN EN 54-6-88. AMD 1 Components of Automatic Fire Detection Systems—Part 6: Heat-Sensitive Detectors; Rate-of-Rise Point Detectors Without a Static Element. 4 pp.

Corrosion Testing (Cont.)

—**Fire Alarms** (Cont.)
CEN EN 54-7-88. Components of Automatic Fire Detection Systems—Part 7: Point-Type Smoke Detectors; Detectors Using Scattered Light, Transmitted Light or Ionization. 30 pp.
CEN EN 54-8-88. Components of Automatic Fire Detection Systems—Part 8: Hig Temperature Heat Detectors. 23 pp.
CEN EN 54-9-82. Components of Automatic Fire Fetection Systems Part 9 Fire Sensitivity Test. 10 pp.

—**Fire Extinguishers**
CEN EN 3-5-84. Portable Fire Extinguishers—Part 5: Complementary Requirements and Tests. 17 pp.
CEN PREN 3-5-93. Fight Against Fire—Portable Fire Extinguishers—Part 5: Complementary Requirements and Tests. 23 pp.

—**Glass Containers**
CNS R3031-73. Method of Test for Chemical Corrosion Resistance of Glass Containers (Nov)(996).

—**Gold Coatings (Made From Gold)**
ISO 3160 Pt 2-92. Watch Cases and Accessories—Gold Alloy Coverings—Part 2: Determination of Fineness, Thickness, Corrosion Resistance and Adhesion Second Edition. 12 pp.

—**Hoses**
SAA AS 1180.7G-72. Methods of Test for Hose Made from Elastomeric Materials—Part 7G: Corrosion Resistance Reconfirmed 1988.

—**Insulating Oils—Sulfur Content**
DIN ENGL 51353-85. Testing of Insulating Oils; Detection of Corrosive Sulfur; Silver Strip Test (Dec). 3 pp.

—**Lacquers**
SAA AS 1580.452. 2-92. Paints and Related Materials—Methods of Test—Part 452.2: Resistance to Corrosion-Salt Droplet Test (in Professional Packages 30, 39). 5 pp.
SNZ NZS/AS 1580. 452.2-92. Methods of Test for Paints and Related Materials Part 452.2: Resistance to Corrosion—Salt Droplet Test (This is a Joint Standard with SAA AS 1580.452.2). 5 pp.

—**Metal Adhesives—Sampling**
SAA AS 1321.2-75. Methods for the Sampling and Testing of Adhesives—Part 2: Corrosive Effect of Set Adhesives on Metals Reconfirmed 1989. 7 pp.

—**Metal Coatings (Made From Metal)**
BSI BS 3745-70. 1970 Evaluation of Results of Accelerated Corrosion Tests on Metallic Coatings. 17 pp.
BSI BS 5466: Part 4-79. 1979 Methods for Corrosion Testing of Metallic Coatings Part 4: Thioacetamide Test (TAA Test). 7 pp.
BSI BS 5466: Part 6-82. 1982 Methods for Corrosion Testing of Metallic Coatings Part 6: Ratings of Results of Corrosion Tests on Electroplated Coatings Cathodic to the Substrate. 20 pp.
BSI BS 5466: Part 8-86. 1986 Methods for Corrosion Testing of Metallic Coatings Part 8: Sulphur Dioxide Test with General Condensation of Moisture. 8 pp.
BSI BS 5466: Part 10-92. 1992 Methods for Corrosion Testing of Metallic Coatings Part 10: Rating of Test Specimens with Coatings Anodic to the Substrate (ISO 8403: 1991). 14 pp.
ISO 1462-73. Metallic Coatings—Coatings Other Than Those Anodic to the Basis Metal—Accelerated Corrosion Tests—Method for the Evaluation of the Results First Edition. 6 pp.
ISO 4538-78. Metallic Coatings—Thioacetamide Corrosion Test (TAA Test) First Edition. 5 pp.
ISO 4540-80. Metallic Coatings—Coatings Cathodic to the Substrate—Rating of Electroplated Test Specimens Subjected to Corrosion Tests First Edition. 18 pp.
ISO 4543-81. Metallic and Other Non-Organic Coatings—General Rules for Corrosion Tests Applicable for Storage Conditions First Edition. 7 pp.
ISO 6988-85. Metallic and Other Non-Organic Coatings—Sulfur Dioxide Test with General Condensation of Moisture First Edition. 6 pp.
ISO 8403-91. Metallic Coatings—Coatings Anodic to the Substrate—Rating of Test Specimens Subjected to Corrosion Tests First Edition. 11 pp.
JIS H 8502-88. Methods of Corrosion Resistance Test for Metallic Coatings. 32 pp.
SAA AS 1247-91. Metallic Coatings—Coatings Other Than Those Anodic to the Basis Metal—Accelerated Corrosion Tests—Method for the Evaluation of the Results (ISO 1462:1973). 4 pp.

Corrosion Testing (Cont.)

—**Metal Coatings (On Metal)**
CGSB 1-GP-71 METH 95.1-82. Methods of Testing Paints and Pigments Examination of Metal Panels for Effects of Exposure to Corrosion Tests. 1 p.
DIN ENGL 50928-85. Corrosion of Metals; Testing and Assessment of the Corrosion Protection of Coated Metallic Materials in Contact with Aqueous Corrosive Agents (Sept). 8 pp.

—**Metals**
BSI BS 6980: Part 1-88. 1988 Stress Corrosion Testing Part 1: Guide to Testing Procedures. 15 pp.
BSI BS 6980: Part 2-90. 1990 Stress Corrosion Testing Part 2: Method for the Preparation and Use of Bent-Beam Specimens (ISO 7539-2: 1989). 12 pp.
BSI BS 6980: Part 3-90. 1990 Stress Corrosion Testing Part 3: Method for the Preparation and Use of U-Bend Specimens (ISO 7539-3: 1989). 8 pp.
BSI BS 6980: Part 4-90. 1990 Stress Corrosion Testing Part 4: Method for the Preparation and Use of Uniaxially Loaded Tension Specimens (ISO 7539-4: 1989). 9 pp.
BSI BS 6980: Part 5-90. 1990 Stress Corrosion Testing Part 5: Method for the Preparation and Use of C-Ring Specimens (ISO 7539-5: 1989). 15 pp.
BSI BS 6980: Part 6-90. 1990 Stress Corrosion Testing Part 6: Method for the Preparation and Use of Pre-Cracked Specimens (ISO 7539-6: 1989). 36 pp.
BSI BS 6980: Part 7-90. 1990 Stress Corrosion Testing Part 7: Method for Slow Strain Rate Testing (ISO 7539-7: 1989). 10 pp.
DIN ENGL 50905 Pt 1-87. Corrosion of Metals; Corrosion Testing; Principles (Jan). 5 pp.
DIN ENGL 50905 Pt 2-87. Corrosion of Metals; Corrosion Testing; Corrosion Characteristics Under Uniform Corrosion Attack (Jan). 4 pp.
DIN ENGL 50905 Pt 3-87. Corrosion of Metals; Corrosion Testing; Corrosion Characteristics Under Nonuniform and Localized Corrosion Attack Without Mechanical Stress (Jan). 4 pp.
DIN ENGL 50917 Pt 2-87. Corrosion of Metals; Testing Under Natural Conditions; Testing in Sea Water (Feb). 7 pp.
DIN ENGL 50920 Pt 1-85. Corrosion of Metals; Corrosion Testing in Flowing Liquids; General (Oct). 8 pp.
ISO 7539 Pt 1-87. Corrosion of Metals and Alloys—Stress Corrosion Testing—Part 1: General Guidance on Testing Procedures First Edition. 14 pp.
ISO 7539 Pt 2-89. Corrosion of Metals and Alloys—Stress Corrosion Testing—Part 2: Preparation and Use of Bent-Beam Specimens First Edition. 10 pp.
ISO 7539 Pt 3-89. Corrosion of Metals and Alloys—Stress Corrosion Testing—Part 3: Preparation and Use of U-Bend Specimens First Edition. 7 pp.
ISO 7539 Pt 4-89. Corrosion of Metals and Alloys—Stress Corrosion Testing—Part 4: Preparation and Use of Uniaxially Loaded Tension Specimens First Edition. 7 pp.
ISO 7539 Pt 5-89. Corrosion of Metals and Alloys—Stress Corrosion Testing—Part 5: Preparation and Use of C-Ring Specimens First Edition. 13 pp.
ISO 7539 Pt 6-89. Corrosion of Metals and Alloys—Stress Corrosion Testing—Part 6: Preparation and Use of Pre-Cracked Specimens First Edition. 34 pp.
ISO 7539 Pt 7-89. Corrosion of Metals and Alloys—Stress Corrosion Testing—Part 7: Slow Strain Rate Testing First Edition. 7 pp.
SAA AS 1580.481. 4-75. Paints and Related Materials—Methods of Test—Part 481.4: Assessment of Corrosion of an Underlying Iron or Steel Surface (Superseded by AS 1580.481.3—1992).
SNZ NZS/AS 1580. 481.4-75. Methods of Test for Paints and Related Materials Part 481.4: Assessment of Corrosion of an Underlying Iron or Steel Surface (This is a Joint Standard with SAA AS 1580.481.4). 2 pp.

—**Metals—Sea Water**
SAA AS 4036-92. Corrosion of Metal—Dissimilar Metals in Contact in Seawater. 21 pp.

—**Nonferrous Alloys**
MOD UK DSTAN 01-2: Part 2-91. Guide to Engineering Alloys Used in Navy Service: Data Sheets Part 2: Non Ferrous Alloys (NF) Issue 2. 102 pp.

—**Oils**
MOD UK DSTAN 05-50: Part 31-01. Methods for Testing Fuels, Lubricants and Associated Products Part 31: Corrosion Preventive Properties of Oils by the Steel/Brass Bimetallic Couple Method Issue 1; Amendment 1. 11 pp.

—**Paint Removers**
CGSB 31-GP-0A METH 11.18-74. Methods of Testing Corrosion-Prevention Materials and Processes Corrosiveness of Paint Remover Residues. 1 p.

—**Paints**
BSI BS AU 148: Part 10-69. 1969 Methods of Test for Motor Vehicle Paints Part 10: Resistance to Heat and Corrosion. 1 p.

INTERNATIONAL AND NON-U.S. NATIONAL STANDARDS
SUBJECT INDEX
Corrugated

Corrosion Testing (Cont.)

—Paints (Cont.)
CEN PREN3665-89. Paints and Varnishes Filiform Corrosion Resistance Test. 8 pp.
SAA AS 1580.452. 2-92. Paints and Related Materials—Methods of Test—Part 452.2: Resistance to Corrosion-Salt Droplet Test (in Professional Packages 30, 39). 5 pp.
SNZ NZS/AS 1580. 452.2-92. Methods of Test for Paints and Related Materials Part 452.2: Resistance to Corrosion—Salt Droplet Test (This is a Joint Standard with SAA AS 1580.452.2). 5 pp.

—Refractory Materials—Slags
JIS R 2214-75. Testing Method for Slag Corrosion of Refractories Using Crucibles (R 1988). 6 pp.

—Rust Preventive Oils
CNS K6790-83. Method of Test for Corrosive Properties of Rust Preventive Oil (Jul)(10451).

—Soil Analysis
BSI BS 1377: Part 9-90. 1990 Methods of Test for Soils for Civil Engineering Purposes Part 9: In-Situ Tests. 66 pp.

—Soldering Fluxes
DIN ENGL 8527 Pt 1-70. Fluxes for Soft Soldering Heavy Metals; Testing (June). 4 pp.

—Soldering Fluxes—Residues
ISO 9455 Pt 12-92. Soft Soldering Fluxes—Test Methods—Part 12: Steel Tube Corrosion Test First Edition. 7 pp.

—Stainless Steels
CNS G2171-83. Method of Forty Two Percent Magnesium Chloride Stress Corrosion Test for Stainless Steels (Apr)(10171).
CNS G2172-83. Method of Ferric Chloride Corrosion Test for Stainless Steels (Apr)(10172).
JIS G 0580-86. Method of Electrochemical Potentiokinetic Reactivation Ratio Measurement for Stainless Steels. 7 pp.

—Steels
BSI DD 207-92. 1992 Method of Test for Soluble Iron Corrosion Products to Assess Surface Cleanliness of Steel Substrates Before Application of Paints and Related Products (ISO/TR 8502-1: 1991) (V). 12 pp.
CEN PREN 3026-92. Aerospace Series Test Method for Dry Film Lubricants Corrosion Test on Steels Specimens. 2 pp.
ISO TR8502 Pt 1-91. Preparation of Steel Substrates Before Application of Paints and Related Products—Tests for the Assessment of Surface Cleanliness—Part 1: Field Test for Soluble Iron Corrosion Products First Edition. 10 pp.

—Steels—Swaged Pipe Couplings—Aircraft
SBAC RS 740 ISSUE 1. British Metric Swaged Pipe Coupling (AS 43020 and AS 43090 Series) Designers's Guide.

—Tents
CNS S2146-90. Method of Test for Tent: Testing for Resistance Against Corrosion (Mar)(12691). 1 p.

—Terminal Blocks—Cracking (Fracturing)
CSA C22.2 NO 158-1987. Terminal Blocks; (Gen Instr 1 Thru 4). 32 pp.

—Test Specimens—Corrosion Removal
BSI BS 7545-91. 1991 Method for Removal of Corrosion Products from Corrosion Test Specimens of Metals and Alloys (ISO 8407: 1991). 12 pp.
ISO 8407-91. Corrosion of Metals and Alloys—Removal of Corrosion Products from Corrosion Test Specimens First Edition. 10 pp.

—Toys
CNS Z8016-7-86. Method of Test for Toy Safety (Testing for Corrosive Substance Content) (Jun)(4798-7). 1 p.

—Varnishes
CEN PREN3665-89. Paints and Varnishes Filiform Corrosion Resistance Test. 8 pp.
SAA AS 1580.452. 2-92. Paints and Related Materials—Methods of Test—Part 452.2: Resistance to Corrosion-Salt Droplet Test (in Professional Packages 30, 39). 5 pp.
SNZ NZS/AS 1580. 452.2-92. Methods of Test for Paints and Related Materials Part 452.2: Resistance to Corrosion—Salt Droplet Test (This is a Joint Standard with SAA AS 1580.452.2). 5 pp.

Corrugated Board
See Also: Corrugated Fiberboard; Corrugated Paperboard; Fiberboard; Laminates; Liner Board; Packaging Materials; Paperboard; Papers
CNS P2045-87. Corrugating Medium (Sep)(2955). 2 pp.
JIS P 3904-85. Corrugating Media. 8 pp.

Corrugated Board (Cont.)
MOD UK DSTAN 81-46-01. Board, Corrugated, Double-Faced (Types A, B and C) Issue 1; Amendment 1. 15 pp.

—Adhesion Testing
CNS Z6013-88. Method of Test for Adhesion of Corrugated Board by Selective Separation (Sep)(3327).

—Boxes (Containers)
MOD UK TS 415C. Board, Corrugated (Two Flutes, Three Linears) Type 1, Type 2 and Type 3.
MOD UK TS 417D. Board, Corrugated (Three -Flute, Four-Liner) Heavy Duty Grade.
MOD UK TS 10015. Board Corrugated, Double Faced, 'E' Flute.
MOD UK TS 10244. Board, Corrugated, (Three Flute, Four-Linear), Medium Grade.

—Boxes (Containers)—Fire Resistant
MOD UK TS 10280. Board, Corrugated, (Three Flute, Four Liner) Fire Resistant.

—Boxes (Containers)—Glossaries
CNS Z5085-81. Terminology of Materials and Accessories for Corrugated Boxes (May)(7458).
CNS Z5096-81. Terminology of Various Types of Corrugated Boxes (Oct)(8071).

—Burst Testers
CPPA M-18-84. Perkins Model 'AH' Mullen Tester. 1 p.

—Bursting Strength
CPPA D.19P-90. Bursting Strength of Board. 2 pp.
SAA AS 1301.438S-89. Methods of Test for Pulp and Paper (Metric Units)—Part 438s: Bursting Strength of Paperboard and Corrugated Fibreboard (This is a Joint Standard with SANZ NZS 1301). 7 pp.
SNZ NZS/AS 1301. 438S-89. Methods of Test for Pulp and Paper Bursting Strength of Paperboard and Corrugated Fibreboard (This is a Joint Standard with SAA AS 1301.438S). 7 pp.

—Compression Testing
BSI BS 6844-87. 1987 Determination of the Flat Crush Resistance of Corrugating Medium After Laboratory Fluting (CMT Concora Medium Test). 8 pp.
CNS P3024-85. Method of Test for Ring Crush of Corrugating Medium and Paperboard (Aug)(2956). 4 pp.
CNS P3088-87. Method of Test for Flat Crush of Corrugating Medium (CMT Test) (Sep)(12107).
CNS Z6082-86. Method of Test for Edgewise Compressive Strength of Corrugated Fiberboard-Butterfly Form Specimen (Jan)(11486).
CPPA D.20P-80. Flat Crush Test of Corrugated Board. 2 pp.
CPPA D.24P-88. Flat Crush Test of Corrugating Medium. 3 pp.
CPPA D.30P-77. Edgewise Compression Resistance of Fluted Corrugating Medium. 2 pp.
ISO 7263-85. Corrugating Medium—Determination of the Flat Crush Resistance After Laboratory Fluting First Edition. 7 pp.
SAA AS 1301.429S-89. Methods of Test for Pulp and Paper (Metric Units)—Part 429s: Flat Crush Resistance of Corrugated Board (This is a Joint Standard with SANZ NZS 1301). 3 pp.
SAA AS 1301.434S-91. Methods of Test for Pulp and Paper—Part 434S: Crush Resistance of Corrugating Medium Amdt 1 August 1991 (This is a Joint Standard with SANZ NZS 1301.434S).
SAA AS 1301.439S-91. Methods of Test for Pulp and Paper —Part 439s: Bendsten Roughness of Paper and Paperboard (This is a Joint Standard with SANZ NZS 1301). 8 pp.
SAA AS 1301.444S-92. Methods of Test for Pulp and Paper—Part 444s: Edgewise Compression Resistance of Corrugated Fibreboard. 3 pp.
SNZ NZS/AS 1301. 429S-89. Methods of Test for Pulp and Paper Flat Crush Resistance of Corrugated Board (This is a Joint Standard with SAA AS 1301.429S). 3 pp.
SNZ NZS/AS 1301. 434S-91. Methods of Test for Pulp and Paper Crush Resistance of Corrugating Medium (This is a Joint Standard with SAA AS 1301.434S). 4 pp.
SNZ NZS/AS 1301. 439S-91. Methods of Test for Pulp and Paper Bendsten Roughness of Paper and Paperboard (This is a Joint Standard with SAA AS 1301.439S). 8 pp.
SNZ NZS/AS 1301. 444S-92. Methods of Test for Pulp and Paper Edgewise Compression Resistance of Corrugated Fibreboard (This is a Joint Standard with SAA AS 1301.444S). 3 pp.

—Containers—Canned Foods
CNS Z5011-88. Corrugated Paperboard Containers for Canned Food (Dec)(1161). 3 pp.

Corrugated Board (Cont.)

—Containers—Ply Separation
CPPA D.10U-77. Ply Separation of Combined Container Board. 1 p.

—Containers—Waterproof Adhesives—Adhesion Testing
CPPA D.10U-77. Ply Separation of Combined Container Board. 1 p.

—Hardness Testing
SAA AS 1301.445S-89. Methods of Test for Pulp and Paper (Metric Units)—Part 445s: Hardness of Corrugated Board Amdt 1 June 1990 (This is a Joint Standard with SANZ NZS 1301). 2 pp.
SNZ NZS/AS 1301. 445S-89. Methods of Test for Pulp and Paper Hardness of Corrugated Board (This is a Joint Standard with SAA AS 1301.445S). 2 pp.

—Liner Adhesion
SAA AS 1301.430S-89. Methods of Test for Pulp and Paper (Metric Units)—Part 430s: Linear Adhesion of Corrugated Board (This is a Joint Standard with SANZ NZS 1301). 3 pp.
SNZ NZS/AS 1301. 430S-89. Methods of Test for Pulp and Paper Liner Adhesion of Corrugated Board (This is a Joint Standard with SAA AS 1301.430S). 3 pp.

—Puncture Resistance
CNS Z6080-86. Method of Test for Puncture of Paperboard and Corrugated Fiberboard (Jan)(11484).
JIS P 8134-76. Testing Method for Puncture of Paperboard (R 1984). 8 pp.

—Shipping Containers
CNS P2009-88. Corrugated Paperboard for Outer Packaging (Dec)(1454). 4 pp.
JIS Z 1506-85. Corrugated Shipping Containers. 9 pp.

—Tear Strength
CPPA D.17U-86. Edge Tear Test for Corrugated Board Scope. 2 pp.

—Thickness Measurement
SAA AS 1301.P426 S-88. Methods of Test for Pulp and Paper (Metric Units)—Part P426s: Thickness of Single Sheets of Paper, Paperboard and Corrugated Fibreboard (This is a Joint Standard with SANZ NZS 1301). 3 pp.
SNZ NZS/AS 1301. P426S-88. Methods of Test for Pulp and Paper Thickness of Single Sheets of Paper, Paperboard and Corrugated Fibreboard (This is a Joint Standard with SAA AS 1301.P426S). 3 pp.

—Waterproof
CNS Z5119-82. Waterproof Corrugated Fiberboard (Dec)(9760). 5 pp.

Corrugated Fiberboard
See Also: Corrugated Board; Corrugated Paperboard; Fiberboard
MOD UK DSTAN 81-107-93. Board, Corrugated (Three-Flute, Four-Liner) Heavy Duty Grade Issue 1. 13 pp.
MOD UK DSTAN 81-108-93. Fibreboard Corrugated Double Faced 'E' Flute Issue 1. 11 pp.

—Adhesion Testing
JIS Z 0402-88. Test Method for Adhesion of Corrugated Fibreboard. 7 pp.

—Bonding Strength—Water Resistance
ISO 3038-75. Corrugated Fibreboard—Determination of the Water Resistance of the Glue Bond by Immersion First Edition. 6 pp.

—Boxes (Containers)
CGSB CAN/CGSB-43.22-92. Corrugated Fibreboard Products. 9 pp.
JIS Z 1507-89. Types of Corrugated Fibreboard Boxes. 16 pp.
MOD UK DSTAN 81-9-89. Cartons, Fibreboard, Fixed Joint, Double-Faced Corrugated Issue 3. 10 pp.
MOD UK DSTAN 81-28-89. Cartons, Fibreboard, Fixed Joint, Multi-Wall, Corrugated Issue 3. 11 pp.

—Boxes (Containers)—Compression Testing
CNS Z6016-72. Method of Test for Compressive Strength of Corrugated Fiberboard Container (Oct)(3511).

—Color Printing
JIS Z 0501-76. Standard for Colour Printed on Corrugated Fibreboard (R 1984). 6 pp.

—Compression Testing
BSI BS 6036-85. 1985 Method for Determination of the Edgewise Crush Resistance of Corrugated Fibre Board. 4 pp.
CNS Z6081-86. Method of Test for Edgewise Compressive Strength of Corrugated Fiberboard-Rectangle Form Specimen (Jan)(11485).

INDUSTRY STANDARDS

Corrugated Fiberboard (Cont.)
—Compression Testing (Cont.)
CNS Z6083-86. Method of Test for Edgewise Compressive Strength of Corrugated Fiberboard-Butterfly Form Specimen (Jan)(11487).
CNS Z6084-86. Method of Test for Flat Crush of Corrugated Fiberboard (Jan)(11488).
ISO 3035-82. Single-Faced and Single-Wall Corrugated Fibreboard—Determination of Flat Crush Resistance Second Edition. 4 pp.
ISO 3037-82. Corrugated Fibreboard—Determination of Edgewise Crush Resistance Second Edition. 4 pp.
JIS Z 0401-85. Method of Compression Test for Corrugated Fibreboard.

—Glossaries
CNS Z5035-77. Glossary of Terms Used in Corrugated Fiberboard (Aug)(3363).

—Papers—Weight Measurement
ISO 3039-75. Corrugated Fibreboard—Determination of the Grammage of the Component Papers After Separation First Edition. 4 pp.

—Puncture Resistance
CNS Z6080-86. Method of Test for Puncture of Paperboard and Corrugated Fiberboard (Jan)(11484).

—Shipping Containers
JIS Z 1516-85. Corrugated Fibreboards for Shipping Containers. 9 pp.

—Shipping Containers—Compression Testing
BSI BS 4686-71. 1971 Determination of the Flat Crush Resistance of Single Faced and Single Wall Corrugated Fibreboard. 8 pp.

—Thickness Measurement
BSI BS 4817-72. 1972 Amd 1 Determination of the Thickness of Corrugated Fibreboard. 9 pp.

—Waterproof
JIS Z 1537-89. Water Proof Corrugated Fibreboards. 8 pp.

Corrugated Paperboard
See Also: Corrugated Board; Corrugated Fiberboard; Paperboard

—Bursting Strength
JIS P 8131-77. Testing Method for Bursting Strength of Paper and Paperboard by Mullen High-Pressure Tester. 7 pp.

—Containers—Fruits
CNS Z5033-88. Corrugated Paperboard Containers for Fresh Fruits (Dec)(3247). 4 pp.

—Partitions
CNS P2021-85. Corrugated Paperboard (for Partitions) (Aug)(2053). 1 p.

—Shipping Containers
CNS Z5020-88. Corrugated Paperboard Containers for Outer Packaging (Dec)(2354). 11 pp.

Corrugated Papers
See Also: Packaging Papers; Papers
MOD UK DEF-1253-57. Paper, Corrugated, Single Faced, Coarse Flute (Reprinted January 1976, Incorporating Amendment No. 1). 4 pp.

Corundum
Use: Aluminum Oxide

Cosmetics
Use For: Toiletries See Also: Perfumes; Sunscreens
EC COM(93) 239-93. Re-Examined Proposal for a Council Directive Amending for the Sixth Time Directive 76/768/EEC on the Approximation of the Laws of the Member States Relating to Cosmetic Products. 5 pp.
EC 76/768/EEC-76. Council Directive on the Approximation of the Laws of the Member States Relating to Cosmetic Products. 32 pp.
EC 79/661/EEC-79. Council Directive Amending Directive 76/768/EEC on the Approximation of the Laws of the Member States Relating to Cosmetic Products. 1 p.
EC 82/147/EEC-82. Commission Directive Adapting to Technical Progress Annex II to Council Directive 76/768/EEC on the Approximation of the Laws of the Member States Relating to Cosmetic Products. 2 pp.
EC 82/368/EEC-82. Council Directive Amending for the Second Time Directive 76/768/EEC on the Approximation of the Laws of the Member States Relating to Cosmetic Products. 32 pp.
EC 83/191/EEC-83. Second Commission Directive Adapting to Technical Progress Annexes II, III, IV and V to Council Directive 76/768/EEC on the Approximation of the Laws of the Member States Relating to Cosmetic Products. 3 pp.

Cosmetics (Cont.)
EC 83/341/EEC-83. Third Commission Directive Adapting to Technical Progress Annexes II, III and V of Council Directive 76/768/EEC on the Approximation of the Laws of the Member States Relating to Cosmetic Products. 1 p.
EC 83/496/EEC-83. Fourth Commission Directive Adapting to Technical Progress Annex VI to Council Directive 76/768/EEC on the Approximation of the Laws of the Member States Relating to Cosmetic Products. 2 pp.
EC 84/415/EEC-84. Fifth Commission Directive Adapting to Technical Progress Annexes II, III, IV, V and VI to Council Directive 76/768/EEC on the Approximation of the Laws of the Member States Relating to Cosmetic Products. 3 pp.
EC 85/391/EEC-85. Sixth Commission Directive Adapting to Technical Progress Annexes II, III, IV, V and VI to Council Directive 76/768/EEC on the Approximation of the Laws of the Member States Relating to Cosmetic Products. 2 pp.
EC 85/490/EEC-85. Fourth Commission Directive on the Approximation of the Laws of the Member States Relating to Methods of Analysis Necessary for Checking the Composition of Cosmetic Products. 16 pp.
EC 86/179/EEC-86. Seventh Commission Directive Adapting to Technical Progress Annexes II, III, IV and V to Council Directive 76/768/EEC on the Approximation of the Laws of the Member States Relating to Cosmetic Products. 9 pp.
EC 86/199/EEC-86. Eighth Commission Directive Adapting to Technical Progress Annexes II, IV and VI to Council Directive 76/768/EEC on the Approximation of the Laws of the Member States Relating to Cosmetic Products. 8 pp.
EC 87/137/EEC-87. Ninth Commission Directive Adapting to Technical Progress Annexes II, III, IV, V and VI to Council Directive 76/768/EEC on the Approximation of the Laws of the Member States Relating to Cosmetic Products. 2 pp.
EC 88/233/EEC-88. Tenth Commission Directive Adapting to Technical Progress Annexxes II, III, IV and VI to Council Directive 76/768/EEC on the Approximation of the Laws of the Member States Relating to Cosmetic Products. 4 pp.
EC 89/174/EEC-89. Eleventh Commission Directive Adapting to Technical Progress Annexes II, III, IV, V, VI and VII to Council Directive 76/768/EEC on the Approximation of the Laws of the Member States Relating to Cosmetic Products. 4 pp.
EC 89/679/EEC-89. Council Directive Amending for the Fifth Time Directive 76/768/EEC on the Approximation of the Laws of the Member States Relating to Cosmetic Products. 1 p.
EC 90/121/EEC-90. Twelfth Commission Directive Adapting to Technical Progress Annexes II, III, IV and VI to Council Directive 76/768/EEC on the Approximation of the Laws of the Member States Relating to Cosmetic Products. 3 pp.
EC 91/184/EEC-91. Thirteenth Commission Directive Adapting to Technical Progress Annexes II, III, IV, V, VI and VII to Council Directive 76/768/EEC on the Approximation of the Laws of the Member States Relating to Cosmetic Products. 4 pp.
EC 93/35/EEC-93. Council Directive Amending for the Sixth Time Directive 76/768/EEC on the Approximation of the Laws of the Member States Relating to Cosmetic Products. 6 pp.
EC 93/47/EEC-93. Sixteenth Commission Directive Adapting to Technical Progress Annexes II, III, V, VI and VII to Council Directive 76/768/EEC on the Approximation of the Laws of the Member States Relating to Cosmetic Products. 3 pp.

—Acidity
CNS S2073-82. Method of Hygienic Test for Cosmetics Value, Acidity and Alkalinity (Jun)(9036).

—Alkali Content
EC 83/514/EEC-83. Third Commission Directive on the Approximation of the Laws of the Member States Relating to Methods of Analysis Necessary for Checking the Composition of Cosmetic Products. 38 pp.

—Alkalinity
CNS S2073-82. Method of Hygienic Test for Cosmetics Value, Acidity and Alkalinity (Jun)(9036).

—Ammonia Content
EC 83/514/EEC-83. Third Commission Directive on the Approximation of the Laws of the Member States Relating to Methods of Analysis Necessary for Checking the Composition of Cosmetic Products. 38 pp.

—Arsenic Content
CNS S2087-82. Methods of Hygienic Test for Cosmetics — Arsenic (Oct)(9541).

—Bergapten Content
CNS S2095-82. Method of Hygienic Test for Cosmetics Bergapten (Dec)(9753).

Cosmetics (Cont.)
—Bismuth Ethyl Camphorate Content
CNS S2103-83. Method of Hygienic Test for Cosmetics Bismuth Salts (Jun)(10381).

—Borates Content
CNS S2080-82. Method of Hygienic Test for Cosmetics Boric Acid and Borate (Sep)(9440).

—Boric Acid Content
CNS S2080-82. Method of Hygienic Test for Cosmetics Boric Acid and Borate (Sep)(9440).

—Cadmium Content
CNS S2089-83. Methods of Hygienic Test for Cosmetics — Cadmium (Jun)(9543).

—Chloramine-T Content
EC 83/514/EEC-83. Third Commission Directive on the Approximation of the Laws of the Member States Relating to Methods of Analysis Necessary for Checking the Composition of Cosmetic Products. 38 pp.

—Chlorate Content
EC 85/490/EEC-85. Fourth Commission Directive on the Approximation of the Laws of the Member States Relating to Methods of Analysis Necessary for Checking the Composition of Cosmetic Products. 16 pp.

—Chlorobutanol Content
EC 85/490/EEC-85. Fourth Commission Directive on the Approximation of the Laws of the Member States Relating to Methods of Analysis Necessary for Checking the Composition of Cosmetic Products. 16 pp.

—Chromium Content
CNS S2102-83. Method of Hygienic Test for Cosmetics Chromium (Jun)(10380).

—Classification
BSI BS 1000: (687)-84. 1984 Universal Decimal Classification (UDC). English Full Edition (687): Clothing Industry. Beauty Culture Industries. 30 pp.
SNZ NZS/BS 1000 (687)-84. Universal Decimal Classification Clothing Industry. Beauty Culture Industries. 32 pp.

—Estrogens
CNS S2086-82. Methods of Hygienic Test for Cosmetics — Estrogenic Hormones (Oct)(9540).

—Ethyl Aminobenzoates Content
CNS S2097-83. Methods of Hygienic Test for Cosmetics Ethyl p-Aminoibenzoate and Ethyl O-Aminobenzoate (Jan)(9914).

—Fluoride Content
CNS S2088-82. Methods of Hygienic Test for Cosmetics — Fluorides (Oct)(9542).

—Formaldehyde Content
CNS S2084-82. Methods of Hygienic Test for Cosmetics — Formaldehyde (Oct)(9538).
EC 82/434/EEC-82. Second Commission Directive on the Approximation of the Laws of the Member States Relating to Methods of Analysis Necessary for Checking the Composition of Cosmetic Products. 28 pp.
EC 90/207/EEC-90. Commission Directive Amending the Second Directive 82/434/EEC on the Approximation of the Laws of the Member States Relating to Methods of Analysis Necessary for Checking the Composition of Cosmetic Products. 10 pp.

—Glyceryl Aminobenzoate Content
EC 85/490/EEC-85. Fourth Commission Directive on the Approximation of the Laws of the Member States Relating to Methods of Analysis Necessary for Checking the Composition of Cosmetic Products. 16 pp.

—Halophenol Bactericides—Hygiene
CNS S2081-82. Methods of Hygienic Test for Cosmetics Halophenol Germicide (Sep)(9441).

—Hexachlorophene Content
EC 83/514/EEC-83. Third Commission Directive on the Approximation of the Laws of the Member States Relating to Methods of Analysis Necessary for Checking the Composition of Cosmetic Products. 38 pp.

—Hydroxyquinoline Content
EC 83/514/EEC-83. Third Commission Directive on the Approximation of the Laws of the Member States Relating to Methods of Analysis Necessary for Checking the Composition of Cosmetic Products. 38 pp.

INTERNATIONAL AND NON-U.S. NATIONAL STANDARDS
SUBJECT INDEX

Cosmetics (Cont.)

—Hydroxyquinoline Sulfate Content
EC 83/514/EEC-83. Third Commission Directive on the Approximation of the Laws of the Member States Relating to Methods of Analysis Necessary for Checking the Composition of Cosmetic Products. 38 pp.

—Identification Systems
EC 83/574/EEC-83. Council Directive Amending for the Third Time Directive 76/768/EEC on the Approximation of the Laws of the Member States Relating to Cosmetic Products. 5 pp.

EC 88/667/EEC-88. Council Directive Amending for the Fourth Time Directive 76/768/EEC on the Approximation of the Laws of the Member States Relating to Cosmetic Products. 3 pp.

—Lead Content
CNS S2104-84. Method of Hygienic Test for Cosmetics Lead (Jan)(10382).

—Lecithin Content
CNS S2096-83. Methods of Hygienic Test for Cosmetics Lecithin.

—Liquefied Gases—Hygiene
CNS S2074-82. Method of Hygienic Test for Cosmetics Liquefied Gases Used as Aerosol Propellants (Jun)(9037).

—Mafenide Content
CNS S2082-82. Methods of Hygienic Test for Cosmetics Homosulfamine (Sep)(9442).

—Mercury Content
CNS S2114-84. Method of Hygienic for Cosmetics (Jan)(10762).

—Methanol Content
CNS S2094-82. Methods of Hygienic Test for Cosmetics Methanol (Dec)(9752).

EC 82/434/EEC-82. Second Commission Directive on the Approximation of the Laws of the Member States Relating to Methods of Analysis Necessary for Checking the Composition of Cosmetic Products. 28 pp.

—Methylene Chloride Content
EC 83/514/EEC-83. Third Commission Directive on the Approximation of the Laws of the Member States Relating to Methods of Analysis Necessary for Checking the Composition of Cosmetic Products. 38 pp.

—Nitrite Content
EC 82/434/EEC-82. Second Commission Directive on the Approximation of the Laws of the Member States Relating to Methods of Analysis Necessary for Checking the Composition of Cosmetic Products. 28 pp.

—Nitromethane Content
EC 83/514/EEC-83. Third Commission Directive on the Approximation of the Laws of the Member States Relating to Methods of Analysis Necessary for Checking the Composition of Cosmetic Products. 38 pp.

—Organic Colors
CNS S2077-83. Method of Hygienic Test for Cosmetics Organic Coloring Materials with Multiple Coal-Tal Colors (Jun)(9189).

CNS S2078-82. Method of Hygienic Test for Cosmetics Organic Coloring Materials Containing Liquid Oils (Jul)(9190).

CNS S2079-82. Method of Hygienic Test for Cosmetics Organic Coloring Materials Semi-Solid Products Containing Waxes (Jul)(9191).

—PH Content
CNS S2073-82. Method of Hygienic Test for Cosmetics Value, Acidity and Alkalinity (Jun)(9036).

—Phenolsulfonic Acid Content
EC 80/1335/EEC-80. First Commission Directive on the Approximation of the Laws of the Member States Relating to Methods of Analysis Neccessary for Checking the Composition of Cosmetic Products. 20 pp.

—Phenylmercuric Compounds Content
EC 83/514/EEC-83. Third Commission Directive on the Approximation of the Laws of the Member States Relating to Methods of Analysis Necessary for Checking the Composition of Cosmetic Products. 38 pp.

—Potassium Hydroxide Content
EC 80/1335/EEC-80. First Commission Directive on the Approximation of the Laws of the Member States Relating to Methods of Analysis Neccessary for Checking the Composition of Cosmetic Products. 20 pp.

Cosmetics (Cont.)

—Quinine Content
EC 85/490/EEC-85. Fourth Commission Directive on the Approximation of the Laws of the Member States Relating to Methods of Analysis Necessary for Checking the Composition of Cosmetic Products. 16 pp.

—Sampling
EC 80/1335/EEC-80. First Commission Directive on the Approximation of the Laws of the Member States Relating to Methods of Analysis Neccessary for Checking the Composition of Cosmetic Products. 20 pp.

—Sodium Hydroxide Content
EC 80/1335/EEC-80. First Commission Directive on the Approximation of the Laws of the Member States Relating to Methods of Analysis Neccessary for Checking the Composition of Cosmetic Products. 20 pp.

—Sodium Iodate Content
EC 85/490/EEC-85. Fourth Commission Directive on the Approximation of the Laws of the Member States Relating to Methods of Analysis Necessary for Checking the Composition of Cosmetic Products. 16 pp.

—Sulfite Content
EC 85/490/EEC-85. Fourth Commission Directive on the Approximation of the Laws of the Member States Relating to Methods of Analysis Necessary for Checking the Composition of Cosmetic Products. 16 pp.

—Thioglycolic Acid Content
CNS S2085-82. Methods of Hygienic Test for Cosmetics — Thioglycolic Acid and Its Salts (Oct)(9539).

—Trichloroethane Content
EC 83/514/EEC-83. Third Commission Directive on the Approximation of the Laws of the Member States Relating to Methods of Analysis Necessary for Checking the Composition of Cosmetic Products. 38 pp.

—Zinc Content
EC 80/1335/EEC-80. First Commission Directive on the Approximation of the Laws of the Member States Relating to Methods of Analysis Neccessary for Checking the Composition of Cosmetic Products. 20 pp.

EC 87/143/EEC-87. Commission Directive Amending the First Directive 80/1335/EEC on the Approximation of the Laws of the Member States Relating to Methods of Analysis Necessary for Checking the Composition of Cosmetic Products. 1 p.

Cosmic Radiation
Use: Cosmic Rays

Cosmic Rays

—Supersonic Transports—Safety
CAA STANDARD NO. 7-1&AP 07.69. Air Conditioning Systems and Cosmic Radiation. 13 pp.

COSPAS
Use: Cooperation in Space

Cost Analysis
Use For: Cost Studies *See Also:* Economics; Expenses

—Life Cycle—Aircraft
EURO PSC/87/18535-87. Life Cycle Cost Perspectives—Military Aircraft Applications. 48 pp.

—Satellite Communications—Regional
CCITT FASCICLE II.1-88. General Tariff Principles—Charging and Accounting in International Telecommunications Services Series D Recommendations. 371 pp.

—Supergroups—Regional
CCITT FASCICLE II.1-88. General Tariff Principles—Charging and Accounting in International Telecommunications Services Series D Recommendations. 371 pp.

—Telecommunication Services—Tariffs
CCITT RECMN D.5-89. Cost and Value of Services Rendered as Factors in the Fixing of Rates—General Tariff Principles—Charging and Accounting in International Telecommunications Services (Study Group III) 2 pp. 2 pp.

—Telephone Circuits—Regional
CCITT FASCICLE II.1-88. General Tariff Principles—Charging and Accounting in International Telecommunications Services Series D Recommendations. 371 pp.

Cost Analysis (Cont.)

—Telephone Services—Europe
CCITT FASCICLE II.1-88. General Tariff Principles—Charging and Accounting in International Telecommunications Services Series D Recommendations. 371 pp.

—Telephone Services—Handbooks
CCITT FASCICLE II.1-88. General Tariff Principles—Charging and Accounting in International Telecommunications Services Series D Recommendations. 371 pp.

—Telephone Services—Mediterranean Basin
CCITT FASCICLE II.1-88. General Tariff Principles—Charging and Accounting in International Telecommunications Services Series D Recommendations. 371 pp.

—Telephone Services—Regional
CCITT FASCICLE II.1-88. General Tariff Principles—Charging and Accounting in International Telecommunications Services Series D Recommendations. 371 pp.

—Telex Communications—Europe
CCITT FASCICLE II.1-88. General Tariff Principles—Charging and Accounting in International Telecommunications Services Series D Recommendations. 371 pp.

—Telex Communications—Handbooks
CCITT FASCICLE II.1-88. General Tariff Principles—Charging and Accounting in International Telecommunications Services Series D Recommendations. 371 pp.

—Telex Communications—Mediterranean Basin
CCITT FASCICLE II.1-88. General Tariff Principles—Charging and Accounting in International Telecommunications Services Series D Recommendations. 371 pp.

—Telex Communications—Regional
CCITT FASCICLE II.1-88. General Tariff Principles—Charging and Accounting in International Telecommunications Services Series D Recommendations. 371 pp.

Cost Estimates

—Aircraft—Maintenance
EURO PSC/88/20184-89. Discussion Paper on the Subject "Prediction of Maintenance Costs". 25 pp.

—Aircraft—Manufacturing—Glossaries
EURO CTI/86/16946 /D,E,F-85. Definition of an AECMA-Cost-Weight and AECMA-Cost-Hours for Aircraft Manufacture. 25 pp.

—Aircraft—Passenger
ICAO Circular 236. Investment Requirements for Aircraft Fleets and for Airport and Route Facility Infrastructure to the Year 2010—1992. 39 pp.

—Aircraft—Tooling
EURO CTI/83/12301-82. Tooling Costs Investigation—Final Report. 112 pp.

—Airports
ICAO Circular 236. Investment Requirements for Aircraft Fleets and for Airport and Route Facility Infrastructure to the Year 2010—1992. 39 pp.

Cost Plus Contracts
Use: Cost Reimbursement Contracts

Cost Reimbursement Contracts
Use For: Cost Plus Contracts
See Also: Construction Contracts; Incentive Contracts

—Production—Cost Management
MOD UK DEFCON Guide no.4 (PROV)-82. Cost Management of Unpriced Production Contracts 9/82. 4 pp.

—Research and Development—Monitoring
MOD UK DEFCON Guide no. 2-82. Planning and Monitoring of Minor Development Contracts 9/82. 5 pp.

—Research and Development—Planning
MOD UK DEFCON Guide no. 2-82. Planning and Monitoring of Minor Development Contracts 9/82. 5 pp.

Cost Studies
Use: Cost Analysis

Costs
Use: Expenses

INTERNATIONAL AND NON-U.S. NATIONAL STANDARDS
SUBJECT INDEX

Cots
See Also: Beds; Furniture
—Camping
BSI BS 3298: Part 1-60. (WITHDRAWN) 1960 Amd 1 Camp Beds Part 1: Spring Steel and Woven Fabric Camp Beds. 21 pp.
—Canvas
CNS L4054-72. Canvas for Cot (Tentative) (Oct)(3487).

Cots, Children's
Use: Cribs (Furniture)

Cottage Rolls
Use: Pork—Cottage Rolls

Cotter Pins
See Also: Fasteners; Pins
BSI BS 1574-72. 1972 Amd 1 Split Cotter Pins; Metric and Inch Series. 16 pp.
CNS B2067-83. Fastenings of Taper Pins and Cotter Pins (Aug)(399).
SAA AS 1236-73. Split Cotter Pins (Metric Series). 12 pp.
—Aircraft—Diameters
SBAC RS 722 ISSUE 1. Selection of Cotter Pins —Metric.
—Aircraft—Hole Size
SBAC RS 722 ISSUE 1. Selection of Cotter Pins —Metric.
—Aircraft—Split
SBAC RS 722 ISSUE 1. Selection of Cotter Pins —Metric.
—Bicycles
BSI BS 6102: Part 7-82. 1982 Cycles Part 7: Cotter Pins and Assembly of the Axle/Cotter Pin/Crank. 6 pp.
ISO 6693-81. Cycles—Cotter Pin and Assembly of the Axle/Cotter Pin/Crank First Edition. 5 pp.
JIS D 9432-80. Chain Adjusters and Crank Cotter Pins for Bicycles.
—Collars (Mechanical Components)
CNS B2007-72. Collar (Using Cotter Pin or Taper Pin) (Jun)(122).
—Split
MOD UK DSTAN 53-10-01. Pins, Cotter, Split Issue 2; Amendment 2. 45 pp.
SAA AS B175-63. Split Cotter Pins. 8 pp.
—Split—Corrosion Resistant—Aircraft
BSI SP 90-61. 1961 Amd 3 Corrosion-Resisting Steel Split Cotter Pins for Aircraft. 4 pp.
—Split—Rolling Stock
CNS E1046-85. Split Cotters for Railway Rolling Stock (Jun)(11283).
JIS E 4113-73. Split Cotters for Railway Rolling Stock.

Cotters
See Also: Spring Cotters
—Keys
CNS B3032-83. Flat Fitting Keys and Sliding Cotter (May)(169).
—Taper Shanks
DIN ENGL 1806-64. Taper Shanks for Cotter Retention (June). 2 pp.

Cotton
Use For: Raw Cotton
—Facial
CNS S1210-88. Facial Cotton (Jul)(12216).
CNS S2140-88. Method of Test for Facial Cotton (Jul)(12217).
—Trash Content
BSI BS 2889-67. 1967 Determination of Trash Content of Cotton and Waste. 11 pp.

Cotton/Cellulose Blend Fabrics
See Also: Cotton Fabrics; Fabrics
—Cuprammonium Viscosity
SAA AS 2001.3.6-80. Methods of Test for Textiles—Part 3: Chemical Tests—Part 3.6: Determination of Cuprammonium Fluidity of Cotton and Cellulosic Man-Made Fibres Corrig.. 8 pp.

Cotton Cloth
Use: Cotton Fabrics

Cotton Fabrics
Scope Note: For additional listings, see also specific products made from cotton *Use For:* Cotton Cloth
See Also: Cheesecloth; Cotton/Cellulose Blend Fabrics; Cotton Fibers; Cotton/Polyester Blend Fabrics; Fabrics; Wool/Cotton Blend Fabrics
CNS L3022-82. Method of Test for Cotton Fabrics (Sep)(1479).
CNS N1002-65. Cotton (Apr)(83)(R 1973).
MOD UK DSTAN 83-48-01. Cloths, Plain Weave, Cotton Issue 3; Amendment 6. 43 pp.
MOD UK DSTAN 83-81-93. Primed Cotton Fabrics Type QX, No 2 or Type QX No 3 Issue 1. 12 pp.
—Aerospace
BSI 7F 8-92. 1992 140 g /m2 Mercerized Cotton Fabrics and Serrated Edge Strip for Aerospace Purposes. 9 pp.
BSI 2F 114-75. 1975 Amd 1 80 G Cotton Madapolam Fabric. 3 pp.
BSI F 141-92. 1992 Closely Woven Cotton Fabrics (L28 and L34) for Aerospace Purposes. 14 pp.
—Balloons
BSI 3F 57-75. (WITHDRAWN) 1975 Scoured Cotton Fabric for Inflatable Equipments (Superseded by BS 4F 57: 1990). 3 pp.
BSI 4F 57-90. 1990 Scoured Cotton Fabric for Inflatable Equipment for Aerospace Purposes. 4 pp.
—Bandoleers—Ammunition
MOD UK DSTAN 83-62-01. Cloth, Drill, Cotton, Khaki, for Bandoliers Issue 1; Amendment 1. 15 pp.
—Canvas—Aerospace
BSI 6F 8-75. (WITHDRAWN) 1975 Mercerized 140 G Cotton Fabric and Serrated Edge Strip (Superseded by Bf 7F 8: 1992). 3 pp.
BSI 5F 37-81. (WITHDRAWN) 1981 570 G/Square Meters and 650 G/Square Meters Cotton Fabrics (Superseded by BS 6F 37: 1990). 4 pp.
BSI 5F 55-82. (WITHDRAWN) 1982 345 G/Square Metres and 415 G/Square Metres Cotton Fabrics (Superseded by BS 6F 55: 1991). 5 pp.
BSI 6F 55-91. 1991 345g/m2 and 415g/m2 Cotton Canvas Fabrics Suitable for Aerospace Purposes. 6 pp.
MOD UK DTD-575A-58. Scoured Cotton Fabric. 1 p.
—Colorfastness Testing
MOD UK DSTAN 83-63-01. Cloth, Cellular, Cotton Issue 1; Amendment 2. 17 pp.
—Colors
JIS L 0600-58. Standard Colour of Cotton Fabrics for Export.
—Components—Ammunition
MOD UK DSTAN 83-36-79. Cloths, Plain Woven, Cotton, Type QX Issue 2. 10 pp.
—Corduroy
CNS L4020-80. Cotton Velvet and Cotton Corduroy Fabrics (Unfinished) (Oct)(2205).
CNS L4021-80. Cotton Velvet and Cotton Corduroy Fabrics (Finished) (Oct)(2206).
—Grading
CNS L4001-81. Cotton Fabrics (Unfinished) (May)(819). 2 pp.
CNS L4019-77. Cotton Fabrics (Finished) (Jun)(2204). 3 pp.
—Laundering Testing—Surfactants
BSI BS 5377: Part 1-88. 1988 Assessment of Laundering Effects by Means of Cotton Control Cloth Part 1: Preparation and Use of the Cotton Control Cloth. 10 pp.
BSI BS 5377: Part 2-90. 1990 Assessment of Laundering Effects by Means of a Cotton Control Cloth Part 2: Method of Analysis and Test for the Unsoiled Control Cloth. 36 pp.
ISO 2267-86. Surface Active Agents—Evaluation of Certain Effects of Laundering—Methods of Preparation and Use of Unsoiled Cotton Control Cloth Third Edition. 9 pp.
ISO 4312-89. Surface Active Agents—Evaluation of Certain Effects of Laundering—Methods of Analysis and Test for Unsoiled Cotton Control Cloth Second Edition. 37 pp.
—Life Rafts—Aircraft
BSI 3F 57-75. (WITHDRAWN) 1975 Scoured Cotton Fabric for Inflatable Equipments (Superseded by BS 4F 57: 1990). 3 pp.
BSI 4F 57-90. 1990 Scoured Cotton Fabric for Inflatable Equipment for Aerospace Purposes. 4 pp.
BSI 3F 122-89. 1989 Two Ply Cotton Fabric Interlayered with Polychloroprene for Liferaft Packs. 4 pp.
—Medical Supplies
CGSB 4-GP-119MA-90. Utility Cotton Cloth, 155 g/m2. 7 pp.

Cotton Fabrics (Cont.)
—Satin
MOD UK DSTAN 83-55-01. Cloths, Satin Drill, Cotton Issue 3; Amendment 4. 25 pp.
—Serge
CNS L4034-66. Cotton Serge (Tentative) (Sep)(2563)(R 1973).
—Sizing (Surface Treatment)
BSI BS 4032-78. 1978 Determination of Certain Water-Or Alkali-Soluble Additives in Cellulosic or Synthetic Fibres, Yarns and Fabrics or Yarns and Fabrics Made from Blends of Such Fibres. 4 pp.
—Terry—Oil Filters—Ships
MOD UK DSTAN 83-57-01. Cloth, Terry, Cotton, (for Filters) Issue 1; Amendment 1. 11 pp.
—Twill
MOD UK DSTAN 83-39-01. Cloths, 3 x 1 Twill, Cotton Issue 3; Amendment 2. 25 pp.
MOD UK DSTAN 83-39-01. Cloths, 2 x 2 and 2 x 1 Twill, Cotton Issue 3; Amendment 5. 25 pp.
—Uniforms
BSI BS 1771: Part 2-90. 1990 Fabrics for Uniforms and Workwear Part 2: Specification for Fabrics from Cellulosic Fibres, Synthetic Fibres and Blends. 11 pp.
MOD UK DSTAN 83-68-79. Cloths, Velvet, Cotton and Viscose, and Cloths, Velveteen, Cotton Issue 1. 12 pp.
—Velvet
CNS L4020-80. Cotton Velvet and Cotton Corduroy Fabrics (Unfinished) (Oct)(2205).
CNS L4021-80. Cotton Velvet and Cotton Corduroy Fabrics (Finished) (Oct)(2206).
—Velvet—Uniforms
MOD UK DSTAN 83-68-79. Cloths, Velvet, Cotton and Viscose, and Cloths, Velveteen, Cotton Issue 1. 12 pp.
—Velveteen—Uniforms
MOD UK DSTAN 83-68-79. Cloths, Velvet, Cotton and Viscose, and Cloths, Velveteen, Cotton Issue 1. 12 pp.
—Wool—Serge—Linings
CGSB 4-GP-84MA-90. Cotton/Wool Serge Cloth, 330 g/m2; (Corrigendum—Aug 1990). 10 pp.

Cotton Fibers
See Also: Cotton Fabrics; Dissolving Pulps; Fibers
CNS L3224-88. Method of Test for Cotton Fiber (Nov)(12471).
CNS N1002-65. Cotton (Apr)(83)(R 1973).
ISO 4913-81. Textiles—Cotton Fibres—Determination of Length (Span Length) and Uniformity Index First Edition. 5 pp.
JIS L 1019-77. Testing Methods for Cotton Fibre.
—Bales—Density
BSI BS 6874-87. 1987 Dimensions and Density of Cotton Bales. 4 pp.
ISO 8115-86. Cotton Bales—Dimensions and Density First Edition. 4 pp.
—Binary Mixtures—Quantitative Analysis
CGSB CAN/CGSB-4.2 NO.14.1-M88. Textile Test Methods Quantitative Analysis of Fibre Mixtures—Binary Mixtures Containing Wool —70% Sulfuric Acid Method. 12 pp.
CGSB CAN/CGSB-4.2 NO.14.7-M88. Textile Test Methods Quantitative Analysis of Fibre Mixtures—Binary Mixtures Containing Acetate Fibres—Acetone Method. 12 pp.
CGSB CAN/CGSB-4.2 NO.14.8-M88. Textile Test Methods Quantitative Analysis of Fibre Mixtures—Binary Mixtures Containing Triacetate Fibres—Dichloromethane Method. 12 pp.
CGSB CAN/CGSB-4.2 NO.14.12-M88. Textile Test Methods Quantitative Analysis of Fibre Mixtures—Binary Mixtures Containing Nylon 6 or Nylon 6,6—80% Formic Acid Method. 12 pp.
—Bobbins
CNS L2013-81. Warping Bobbin for Cotton (Mar)(589).
JIS L 6407-58. Warping Bobbin for Cotton. 5 pp.
—Breaking Load
BSI BS 5116-74. 1974 Method of Test for the Determination of Breaking Tenacity of Flat Bundles of Cotton Fibres. 11 pp.
ISO 3060-74. Cotton Fibres—Determination of Breaking Tenacity of Flat Bundles First Edition. 8 pp.
—Colorimetry—Lighting
ISO 4911-80. Textiles—Cotton Fibres—Equipment and Artificial Lighting for Cotton Classing Rooms First Edition. 9 pp.

Cotton Fibers (Cont.)
—Cuprammonium Viscosity
BSI BS 2610-78. 1978 Amd 1 Method of Test for the Determination of the Cuprammonium Fluidity of Cotton and Certain Cellulosic Man-Made Fibres (AMD 4965) November 29, 1985. 13 pp.
CGSB CAN/CGSB-4.2 NO.17-M90. Textile Test Methods Cuprammonium Fluidity of Textile Cellulose. 13 pp.

—Fat Content
BSI BS 3477-62. 1962 Amd 1 Determination of Oils, Fats and Waxes in Cotton. 8 pp.

—Gas Permeability
BSI BS 3181: Part 1-87. 1987 Method for the Determination of Cotton Fiber Properties the Airflow Method Part 1: Determination of Micronaire Value by the Single Compression Airflow Method. 8 pp.
ISO 2403-72. Textiles—Cotton Fibres—Determination of Micronaire Value First Edition. 8 pp.

—Maturity Testing
BSI BS 3085-81. 1981 Evaluation of the Maturity of Cotton Fibres (Microscopic Method). 12 pp.
DIN ENGL 53943 Pt 1-86. Testing of Textiles; Cotton Fibre Maturity; Concepts (Oct). 3 pp.
DIN ENGL 53943 Pt 3-83. Testing of Textiles; Cotton Fibre Maturity; Determination of the Percent Maturity; Microscopic Method Using a Polarisation Microscope (Mar). 2 pp.
ISO 4912-81. Textiles—Cotton Fibres—Evaluation of Maturity—Microscopic Method First Edition. 10 pp.
ISO 10306-93. Textiles—Cotton Fibres —Evaluation of Maturity by the Air Flow Method First Edition. 9 pp.

—Micronaire Value
SAA AS 2001.2.11-81. Methods of Test for Textiles—Part 2: Physical Tests—Part 2.11: Determination of Micronaire Value of Cotton Fibres. 4 pp.

—Oil Content
BSI BS 3477-62. 1962 Amd 1 Determination of Oils, Fats and Waxes in Cotton. 8 pp.

—Sampling
CNS L3126-81. Method of Sampling for Cotton Fibers (Jun) (7501).

—Thread (Textiles)
BSI BS 7318-90. 1990 Industrial Sewing Threads Made from Linen (Flax) or Cotton. 13 pp.
CGSB 4-GP-80MA-90. Cotton Thread. 11 pp.
CNS L3072-85. Method of Test for Cotton Sewing Threads (May)(3729).
CNS L4063-85. Cotton Sewing Thread (May)(3728).
JIS L 2101-78. Cotton Sewing Thread.

—Thread (Textiles)—Ammunition
MOD UK DSTAN 83-51-75. Threads, Cotton, Type QX Issue 1. 11 pp.

—Thread (Textiles)—Mercerizing—Shrinkage
ISO 6836-83. Surface Active Agents—Mercerizing Agents—Evaluation of the Activity of Wetting Products for Mercerization by Determination of the Shrinkage Rate of Cotton First Edition. 7 pp.

—Wax Content
BSI BS 3477-62. 1962 Amd 1 Determination of Oils, Fats and Waxes in Cotton. 8 pp.

—Yarns—Aerospace—Construction
BSI F 128-85. (WITHDRAWN) 1985 Cotton Webbing Suitable for Aerospace Purposes (Warp Yarn Nominally R 310 Tex) (Superseded by BS 2F 128: 1992). 8 pp.

—Yarns—Aerospace—Identification Systems
BSI F 128-85. (WITHDRAWN) 1985 Cotton Webbing Suitable for Aerospace Purposes (Warp Yarn Nominally R 310 Tex) (Superseded by BS 2F 128: 1992). 8 pp.

—Yarns—Linear Density
SAA AS 1506.1-78. Preferred Linear Density of Yarn in Tex Units—Part 1: Cotton. 8 pp.

Cotton Linters
Use: Linters

Cotton/Polyester Blend Fabrics
See Also: Cotton Fabrics; Fabrics; Polyester Fabrics
MOD UK DSTAN 83-65-83. Cloths, Polyester and Cotton Blends (Reprinted 14 December 1990, Incorporating Amendments 1-9) Issue 2. 25 pp.
MOD UK DSTAN 83-65-01. Cloths, Polyester and Cotton Blends (Reprinted 14 December 1990, Incorporating Amendments 1-9) Issue 2; Amendment 10. 26 pp.
MOD UK DSTAN 83-65-02. Cloths, Polyester and Cotton Blends (Reprinted December, 1990—Incorporating Amendments 1-9) Issue 2; Amendment 11. 27 pp.

—Canvas—Aerospace
MOD UK DSTAN 83-80-91. Polyester/Cotton Core-Spun Canvas Issue 1. 14 pp.

—Components—Ammunition
MOD UK DSTAN 83-43-01. Cloths, Plain Weave, Polyester and Cotton, Type QX Issue 2; Amendment 1. 15 pp.

—Pajamas
CGSB CAN/CGSB-4.180-M91. Polyester/Cotton Plain Cloth, 150 g/m2. 7 pp.

—Poplin
CGSB 4-GP-117MA-88. Cloth, Polyester/Cotton Poplin, 170 g/m2. 8 pp.

Cotton Spun Yarns
Use: Spun Yarns—Cotton

Cotton Tapes
Use: Tapes—Cotton Fabric

Cotton Wadding
Use: Wadding—Cotton

Cottonseed
See Also: Cottonseed Oil; Oilseeds

—Fatty Acids
CNS K1264-81. Fractionated and Distilled Cottonseed Fatty Acids (May)(7432).

—Gossypol Content
BSI BS 5766: Part 12-86. 1986 Methods for Analysis of Animal Feeding Stuffs Part 12: Determination of Gossypol. 6 pp.
CNS N4071-81. Determination of Free Gossypol in Cottonseed Meal of Cake (Feed Grade)(Feb)(7067).
ISO 6866-85. Animal Feeding Stuffs—Determination of Free and Total Gossypol First Edition. 5 pp.

Cottonseed Cake
Use: Oil Meal

Cottonseed Oil
See Also: Cottonseed; Oils; Vegetable Oils
CNS K5106-81. Crude Cottonseed Oil (Sep)(7417).
CNS N5144-89. Edible Cottonseed Oil (Sep)(4832). 2 pp.

Cottonseed Oil Content Analysis
—Vegetable Oils
BSI BS 684: Sec 2.29-78. 1978 Methods of Analysis of Fats and Fatty Oils Part 2: Other Methods Section 2.29: Cotton Seed Oil Test (Halphen). 2 pp.

Couches
See Also: Furniture; Seats; Settees; Sofas
CNS S1013-68. Couch (Jan)(906).

—Medical Equipment
BSI BS 2838: Part 1-88. 1988 Examination and General Treatment Couches Part 1: Specification for Fixed Height Couches. 14 pp.
BSI BS 2838: Part 2-91. 1991 Examination and General Treatment Couches Part 2: Specification for Variable Height Couches. 19 pp.

—Sofas
CNS S1053-68. Long Sofa Couch (Apr)(2844).

Coulombmeters
See Also: Electric Measuring Instruments; Measuring Instruments

—Call Duration
CCITT RECMN E.261-89. Devices for Measuring and Recording Call Durations—Telephone Network and ISDN—Operation, Numbering, Routing and Mobile Service (Study Group II) 3 pp. 3 pp.

Coulometric Analysis
See Also: Coulometric Titration; Electrometric Analysis

—Crude Oil
JIS K 2609-90. Crude Petroleum and Petroleum Products—Determination of Nitrogen Content. 50 pp.

—Manganese Ores
JIS M 8237-82. Methods for Determination of Sulfur in Manganese Ores. 20 pp.

Coulometric Analysis (Cont.)
—Metal Coatings—Thickness Measurement
BSI BS 5411: Part 4-86. 1986 Methods of Test for Metallic and Related Coatings Part 4: Coulometric Method for the Measurement of Coating Thickness. 12 pp.
CNS H2043-87. Measurement of Thickness of Metallic Coatings by the Coulometric Method (Sep)(4161).
DIN ENGL 50955-83. Measurement of Coating Thicknesses; Measurement of Thickness of Metallic Coatings by Local Anodic Dissolution; Coulometric Method (Dec) (Supersedes DIN 50932, June 1971 Edition). 6 pp.
ISO 2177-85. Metallic Coatings—Measurement of Coating Thickness—Coulometric Method by Anodic Dissolution Second Edition; (Corrected and Reprinted -1986). 11 pp.

—Nickel Castings
JIS H 1277-88. Methods for Determination of Sulfur in Nickel and Nickel Alloy Castings. 10 pp.

—Petroleum Products
JIS K 2609-90. Crude Petroleum and Petroleum Products—Determination of Nitrogen Content. 50 pp.

—Plutonium
CNS J2038-82. Plutonium by Controlled-Potential Coulometry (Jan)(8417).

—Sediments
DIN ENGL 38414 Pt 17-89. German Standard Methods for the Examination of Water, Waste Water and Sludge; Sludge and Sediments (Group S); Determination of Strippable and Extractable Organically Bound Halogens (S 17) (Nov). 5 pp.
DIN ENGL 38414 Pt 18-89. German Standard Methods for the Examination of Water, Waste Water and Sludge; Sludge and Sediments (Group S); Determination of Adsorbed Organically Bound Halogens (AOX) (S18) (Nov). 4 pp.

—Silicon Carbide
DIN ENGL 51075 Pt 2-84. Testing of Ceramic Materials; Chemical Analysis of Silicon Carbide; Direct Determination of the Free Carbon Content (Mar). 7 pp.
DIN ENGL 51075 Pt 3-82. Testing of Ceramic Materials; Chemical Analysis of Silicon Carbide; Determination of the Total Carbon Content (Oct). 3 pp.

—Sludge (Sewage)
DIN ENGL 38414 Pt 17-89. German Standard Methods for the Examination of Water, Waste Water and Sludge; Sludge and Sediments (Group S); Determination of Strippable and Extractable Organically Bound Halogens (S 17) (Nov). 5 pp.
DIN ENGL 38414 Pt 18-89. German Standard Methods for the Examination of Water, Waste Water and Sludge; Sludge and Sediments (Group S); Determination of Adsorbed Organically Bound Halogens (AOX) (S18) (Nov). 4 pp.

—Steels
ISO TR4830 Pt IV-78. Steel—Determination of Low Carbon Contents—Part IV: Coulometric Method After Combustion First Edition. 8 pp.

Coulometric Titration
See Also: Coulometric Analysis
CNS K0029-87. General Rules for Method of Potentiometric Amperometric and Coulometric Titrations (Oct)(12128).
JIS K 0113-90. General Rules for Methods of Potentiometric, Amperometric, Coulometric, and Karl-Fischer Titrations. 25 pp.

—Aromatic Hydrocarbons
CNS K6521-80. Method of Test for Bromine Index of Aromatic Hydrocarbons by Coulometric Titration (Jul)(5838).

Coumarone Indene Resin
CNS K1116-66. Coumarone-Indene Resin (Jun)(2614)(R 1971).
CNS K6233-72. Method of Test for Coumarone Indene Resin (Jan)(2615).

Count Methods, Bacteria
Use: Bacteria Count Methods

Counter/Dividers
See Also: Decade Counter/Dividers; Ripple Counter/Divider/Oscillators; Ripple Counter/Dividers
CECC CECC 90 101-030 ISSUE 1-87. UTE C 86-213/B 50; Digital Integrated Circuits in Accordance with FS 90 101; 54/64/74 92A; Divide-By-Twelve Counter (En). 16 pp.
CECC CECC 90 103-043 ISSUE 1-81. BS CECC 90 103-043; Divide-By-Twelve Counters (En) AMD 1 (En). 20 pp.

Counter/Dividers *(Cont.)*

CECC CECC 90 104-171 ISSUE 1-81. BS CECC 90 104-171; Presettable Divide-by-N Counter (En). 26 pp.

CECC CECC 90 104-232 ISSUE 1. NL CECC 90 104-232 Issue 1; Digital Integrated Circuits in Accordance with FS 90 104; HEC/HEF 4059B; Programmable Divide-By-N Counter (En). 9 pp.

—24 Stage

CECC CECC 90 104-176 ISSUE 1-81. BS CECC 90 104-176; 24 Stage Counter/Divider (En). 27 pp.

CECC CECC 90 104-176 ISSUE 1-86. CEI CECC 90 104-176; 24-Stage Counter/Divider (En). 2 pp.

—4-Bit

CECC CECC 90 104-177 ISSUE 1-81. BS CECC 90 104-177; Programmable Divide-by-N 4-Bit Counter (En). 27 pp.

CECC CECC 90 104-177 ISSUE 1-86. CEI CECC 90 104-177; Programmable Divide-by-N 4-Bit Counter (En). 2 pp.

—4 Stage

CECC CECC 90 104-020 ISSUE 2. NL CECC 90 104-020 Issue 2; Digital Integrated Circuits in Accordance with FS 90 104; HEC/HEF 4022B; 4-Stage Divide-by-8 Johnson Counter (En). 11 pp.

Counter/Timer Modules

See Also: Communication Interfaces

—Programmable

CECC CECC 90 104-238 ISSUE 1. NL CECC 90 104-238 Issue 1; Digital Integrated Circuits in Accordance with FS 90 104; HEC/HEC 4753B; Universal Timer Module (En). 14 pp.

Counter Tubes

See Also: Electron Tubes; Proportional Counter Tubes

MOD UK DSTAN 59-60: Part 3-77. Valves, Electronic (Electronic Tubes) (Listed on EPIC Database) Part 3: Counter, Selector, and Indicator Tubes Issue 2. 29 pp.

—Quality Assurance

BSI BS 9015-70. (WITHDRAWN) 1970 Amd 1 Counter and Indicator Tubes of Assessed Quality: Generic Data and Methods of Test. 12 pp.

Counterbore Nuts

See Also: Fasteners; Nuts (Fasteners)

—Aerospace—Anchor—Locking

CEN PREN 2862-90. Nuts Anchor, Self-Locking, Fixed, 90 Degrees Corner, with Counterbore, in Alloy Steel, Cadmium Plated, MO2 Lubricated Classification: 1100 MPa (at Ambient Temperature) /235 Degrees Celsius. 6 pp.

CEN PREN 2862-93. Aerospace Series Nuts, Anchor, Self-Locking, Fixed, 90 Degree Corner, with Counterbore, in Alloy Steel, Cadmium Plated, MoS2 Lubricated Classification: 1 100 MPa (at Ambient Temperature)/235 Degrees C. 5 pp.

CEN PREN 2862-93. CORRIGENDUM Aerospace Series Nuts, Anchor, Self-Locking, Fixed, 90 Degree Corner, with Counterbore, in Alloy Steel, Cadmium Plated, MoS2 Lubricated Classification: 1 100 MPa (at Ambient Temperature) /235 Degrees C. 1 p.

CEN PREN 2863-90. Nuts, Anchor Self-Locking, Fixed, 90 Degrees Corner, with Counterbore, in Heat Resisting Steel, Passivated, MOS2 Lubricated Classification: 1100 MPa (at Ambient Temperature) /315 Degrees Celsius. 6 pp.

CEN PREN 2863-93. Aerospace Series Nuts, Anchor, Self-Locking, Fixed, 90 Degree Corner, with Counterbore, in Heat Resisting Steel, MoS2 Lubricated Classification: 1 100 MPa (at Ambient Temperature)/315 Degrees C. 5 pp.

CEN PREN 2863-93. CORRIGENDUM Aerospace Series Nuts, Anchor, Self-Locking, Fixed, 90 Degree Corner, with Counterbore, in Heat Resisting Steel, MoS2 Lubricated Classification: 1 100 MPa (at Ambient Temperature) /315 Degrees C. 1 p.

CEN PREN 2865-93. Aerospace Series Nuts, Anchor, Self-Locking, Floating, Two Lug, with Counterbore, in Heat Resisting Steel, MoS2 Lubricated Classification: 1 100 MPa (at Ambient Temperature)/315 Degrees C. 5 pp.

CEN PREN 2865-93. CORRIGENDUM Aerospace Series Nuts, Anchor, Self-Locking, Floating, Two Lug, with Counterbore, in Heat Resisting Steel, MoS2 Lubricated Classification: 1 100 MPa (at Ambient Temperature) /315 Degrees C. 1 p.

CEN PREN 2866-92. Nuts, Anchor, Self-Locking, Floating, One Lug, with Counterbore, in Steel, Cadmium Plated, MoS2 Lubricated Classification: 1 110 MPa (at Ambient Temperature) /235 Degrees C. 5 pp.

Counterbore Nuts *(Cont.)*

—Aerospace—Anchor—Locking *(Cont.)*

CEN PREN 2867-92. Nuts, Anchor, Self-Locking, Floating, One Lug, with Counterbore, in Heat Resisting Steel, MoS2 Lubricated Classification: 1 110 MPa (at Ambient Temperature) /315 Degrees C. 5 pp.

CEN PREN 3435-93. Aerospace Series Nuts, Anchor, Self-Locking, Floating, Two Lug, Reduced Series, with Counterbore, in Heat Resisting Steel, MoS2 Lubricated Classification: 1 100 MPa (at Ambient Temperature)/315 Degrees C. 5 pp.

CEN PREN 3435-93. CORRIGENDUM Aerospace Series Nuts, Anchor, Self-Locking, Floating, Two Lug, Reduced Series, with Counterbore, in Heat Resisting Steel, MoS2 Lubricated Classification: 1 100 MPa (at Ambient Temperature) /315 Degrees C. 1 p.

CEN PREN 3537-90. Nuts, Anchor, Self-Locking, Fixed, Two Lug, with Counterbore, in Heat Resisting Steel, Passivated, MoS2 Lubricated Classification:1100 MPa (at Ambient Temperature) /315 Degrees Celsius. 6 pp.

CEN PREN 3537-93. Aerospace Series Nuts, Anchor, Self-Locking, Fixed, Two Lug, with Counterbore, in Heat Resisting Steel, MoS2 Lubricated Classification: 1 100 MPa (at Ambient Temperature)/315 Degrees C. 5 pp.

CEN PREN 3537-93. CORRIGENDUM Aerospace Series Nuts, Anchor, Self-Locking, Fixed, Two Lug, with Counterbore, in Heat Resisting Steel, MoS2 Lubricated Classification: 1 100 MPa (at Ambient Temperature) /315 Degrees C. 1 p.

CEN PREN 3538-90. Nuts, Anchor, Self-Locking, Fixed, Two Lug, Reduced Series with Counterbore, in Heat Resisting Steel Passivated, MoS 2 Lubricated Classification: 100 MPa (at Ambient Temperature) /315 Degrees Celsius. 6 pp.

CEN PREN 3538-93. Aerospace Series Nuts, Anchor, Self-Locking, Fixed, Two Lug, Reduced Series, with Counterbore, in Heat Resisting Steel, MoS2 Lubricated Classification: 1 100 MPa (at Ambient Temperature)/315 Degrees C. 5 pp.

CEN PREN 3538-93. CORRIGENDUM Aerospace Series Nuts, Anchor, Self-Locking, Fixed, Two Lug, Reduced Series, with Counterbore, in Heat Resisting Steel, MoS2 Lubricated Classification: 1 100 MPa (at Ambient Temperature) /315 Degrees C. 1 p.

CEN PREN 3539-90. Nuts Anchor, Self-Locking, One Lug, Fixed, with Counterbore, in Heat Resisting Steel, MoS2 Lubricated Classification: 1100 MPa (at Ambient Temperature) /315 Degrees Celsius. 5 pp.

CEN PREN 3539-93. Aerospace Series Nuts, Anchor, Self-Locking, One Lug, Fixed, with Counterbore, in Heat Resisting Steel, MoS2 Lubricated Classification: 1 100 MPa (at Ambient Temperature) /315 Degrees C. 5 pp.

CEN PREN 3712-91. Nuts, Anchor, Self-Locking, One Lug Fixed, Reduced Series, with Counterbore, in Steel, Cadmium Plated, MoS2 Lubricated Classification: 1100 MPa (at Ambient Temperature)/ 235 Degrees C. 6 pp.

CEN PREN 3714-93. Aerospace Series Nuts, Anchor, Self-Locking, Floating, Two Lug, with Counterbore, in Heat Resisting Steel, Silver Plated Classification: 1 100 MPa (at Ambient Temperature)/425 Degrees C. 5 pp.

CEN PREN 3714-93. CORRIGENDUM Aerospace Series Nuts, Anchor, Self-Locking, Floating, Two Lug, with Counterbore, in Heat Resisting Steel, Silver Plated Classification: 1 100 MPa (at Ambient Temperature) /425 Degrees C. 1 p.

CEN PREN 3750-90. Nuts, Anchor, Self-Locking, Fixed 90 Degrees Corner, Reduced Series with Counterbore, in Heat Resisting Steel, Passivated, MoS2 Lubricated/Classification:100 MPa (at Ambient Temperature) /315 Degrees Celsius. 6 pp.

CEN PREN 3750-93. Aerospace Series Nuts, Anchor, Self-Locking, Fixed, 90 Degree Corner, Reduced Series, with Counterbore, in Heat Resisting Steel, MoS2 Lubricated Classification: 1 100 MPa (at Ambient Temperature)/315 Degrees C. 5 pp.

CEN PREN 3750-93. CORRIGENDUM Aerospace Series Nuts, Anchor, Self-Locking, Fixed, 90 Degree Corner, Reduced Series, with Counterbore, in Heat Resisting Steel, MoS2 Lubricated Classification: 1 100 MPa (at Ambient Temperature) /315 Degrees C. 1 p.

CEN PREN 3751-90. Nuts, Anchor, Self-Locking, Fixed, Closed Corner, Reduced Series, with Counterbore, in Heat Resisting Steel, Passivated, MoS2 Lubricated Classification: 1100 MPa (at Ambient Temperature)/315 Degrees Celsius. 6 pp.

CEN PREN 3751-93. Aerospace Series Nuts, Anchor, Self-Locking, Fixed, Closed Corner, Reduced Series, with Counterbore, in Heat Resisting Steel, MoS2 Lubricated Classification: 1 100 MPa (at Ambient Temperature)/315 Degrees C. 5 pp.

CEN PREN 3751-93. CORRIGENDUM Aerospace Series Nuts, Anchor, Self-Locking, Fixed, Closed Corner, Reduced Series, with Counterbore, in Heat Resisting Steel, MoS2 Lubricated Classification: 1 100 MPa (at Ambient Temperature) /315 Degrees C. 1 p.

Counterbore Nuts *(Cont.)*

—Aerospace—Anchor—Locking *(Cont.)*

CEN PREN 3753-90. Nuts, Anchor, Self-Locking, Fixed, Closed Corner, with Counterbore, in Alloy Steel, Cadmium Plated, MoS2 Lubricated Classification:1100 MPa (at Ambient Temperature) /235 Degrees Celsius. 6 pp.

CEN PREN 3753-93. Aerospace Series Nuts, Anchor, Self-Locking, Fixed, 60 Degree Corner, with Counterbore, in Alloy Steel, Cadmium Plated, MoS2 Lubricated Classification: 1 100 MPa (at Ambient Temperature)/235 Degrees C. 5 pp.

CEN PREN 3753-93. CORRIGENDUM Aerospace Series Nuts, Anchor, Self-Locking, Fixed, 60 Degree Corner, with Counterbore, in Alloy Steel, Cadmium Plated, MoS2 Lubricated Classification: 1 100 MPa (at Ambient Temperature) /235 Degrees C. 1 p.

CEN PREN 3754-90. Nuts, Anchor, Self-Locking, Fixed, Closed Corner, with Counterbore, in Heat Resisting Steel, Passivated, MoS2 Lubricated Classification: 100 MPa (at Ambient Temperature) /315 Degrees Celsius. 6 pp.

CEN PREN 3754-93. Aerospace Series Nuts, Anchor, Self-Locking, Fixed, 60 Degree Corner, with Counterbore, in Heat Resisting Steel, MoS2 Lubricated Classification: 1 100 MPa (at Ambient Temperature)/315 Degrees C. 5 pp.

CEN PREN 3754-93. CORRIGENDUM Aerospace Series Nuts, Anchor, Self-Locking, Fixed, 60 Degree Corner, with Counterbore, in Heat Resisting Steel, MoS2 Lubricated Classification: 1 100 MPa (at Ambient Temperature) /315 Degrees C. 1 p.

CEN PREN 3757-93. Aerospace Series Nuts, Anchor, Self-Locking, Floating, Self-Aligning, Two Lug, in Heat Resisting Steel, MoS2 Lubricated Classification: 900 MPa (at Ambient Temperature) /315 Degrees C. 5 pp.

CEN PREN 3757-93. CORRIGENDUM Aerospace Series Nuts, Anchor, Self-Locking, Floating, Self-Aligning, Two Lug, in Heat Resisting Steel, MoS2 Lubricated Classification: 900 MPa (at Ambient Temperature) /315 Degrees C. 1 p.

CEN PREN 3768-90. Nuts, Anchor, Self-Locking, One Lug, Fixed, Reduced Series, with Counterbore, in Heat Resisting Steel, MoS2 Lubricated Classification: 1100 MPa (at Ambient Temperature) /315 Degrees Celsius. 5 pp.

CEN PREN 3768-93. Aerospace Series Nuts, Anchor, Self-Locking, One Lug, Fixed, Reduced Series, with Counterbore, in Heat Resisting Steel, MoS2 Lubricated Classification: 1 100 MPa (at Ambient Temperature) /315 Degrees C. 5 pp.

CEN PREN 3834-91. Nuts, Anchor, Self-Locking, Floating, Two Lug, Incremental Counterbore, in Heat Resisting Steel, MoS2 Lubricated Classification: 900 MPa (at Ambient Temperature)/315 Degrees C. 7 pp.

CEN PREN 3834-93. Aerospace Series Nuts, Anchor, Self-Locking, Floating, Two Lug, Incremental Counterbore, in Heat Resisting Steel, MoS2 Lubricated Classification: 900 MPa (at Ambient Temperature) /315 Degrees C. 5 pp.

CEN PREN 4084-93. Aerospace Series Nuts, Anchor, Self-Locking, Fixed, Two Lug, with Counterbore, in Alloy Steel, Cadmium Plated, MoS2 Lubricated Classification: 1 100 MPa (at Ambient Temperature) /235 Degrees C. 5 pp.

DIN ENGL LN 29982-78. Aerospace; Nuts Anchor; Self-Locking; Deep Counterbore; Double Lug for Temperatures up to 235 Degrees C (Mar). 3 pp.

DIN ENGL LN 29983-77. Nuts Anchor; Self-Locking; Deep Counterbore; Double Lug Reduced for Temperatures up to 235 Degrees C (July). 3 pp.

DIN ENGL LN 29984-77. Nuts Anchor; Self-Locking; Deep Counterbore; Single Lug for Temperatures up to 235 Degrees C (July). 3 pp.

DIN ENGL LN 29985-77. Nuts Anchor; Self-Locking; Floating; Deep Counterbore; Double Lug for Temperatures up to 235 Degrees C (Dec). 3 pp.

DIN ENGL LN 29987-77. Nuts Anchor; Self-Locking; Floating; Deep Counterbore; Single Lug for Temperatures up to 235 Degrees C (Dec). 3 pp.

DIN ENGL LN 29988-77. Nuts Anchor; Self-Locking; Deep Counterbore; Corner Lug for Temperatures up to 235 Degrees C (July). 3 pp.

DIN ENGL LN 29989-77. Nuts Anchor; Self-Locking; Deep Counterbore; Corner Lug Reduced for Temperatures up to 235 Degrees C (July). 3 pp.

DIN ENGL LN 29990-77. Anchor Nuts; Self-Locking, Deep Counterbore, Double Lug, for Temperatures up to 315 Degrees C and up to 425 Degrees C (July). 3 pp.

DIN ENGL LN 29992-77. Nuts, Anchor; Self-Locking, Deep Counterbore, Single Lug for Temperatures up to 315 Degrees C and up to 425 Degrees C (July). 3 pp.

DIN ENGL LN 29996-77. Nuts Anchor; Self-Locking; Deep Counterbore; Single Lug Reduced for Temperatures up to 315 Degrees C and up to 425 Degrees C (May). 3 pp.

ISO 3168-86. Aerospace—Self-Locking, Fixed, Single-Lug Anchor Nuts with Counterbore, Classification 1 100 MPa/235 Degrees Celsius First Edition. 5 pp.

INTERNATIONAL AND NON-U.S. NATIONAL STANDARDS
SUBJECT INDEX

Counters

Counterbore Nuts (Cont.)
—Aerospace—Anchor—Locking (Cont.)

ISO 3191-85. Aerospace—Self-Locking, Fixed, Single-Lug Anchor Nuts, Reduced Series, with Counterbore, Strength Classification 1 100 MPa and Maximum Operating Temperature 235 Degrees Celsius First Edition. 5 pp.

ISO 3209-89. Aerospace—Nuts, Anchor, Self-Locking, Floating, Two-Lug, with Counter-bore, with MJ Threads, Coated or Uncoated, Classification 1 100 MPa/235 Degrees Celsius, 1 100 MPa/315 Degrees Celsius or 1 100 MPa/425 Degrees Celsius—Dimensions First Edition. 6 pp.

ISO 3221-89. Aerospace—Nuts, Anchor, Self-Locking, Fixed, 90 Degree Corner, with Counterbore, with MJ Threads, Coated or Uncoated, Classification 1 100 MPa/235 Degrees Celsius, 1 100 MPa/315 Degrees Celsius or 1 100 MPa/425 Degrees Celsius—Dimensions First Edition. 5 pp.

ISO 3222-89. Aerospace—Nuts, Anchor, Self-Locking, Fixed, Closed Corner, Reduced Series, with Counterbore, with MJ Threads, Coated or Uncoated, Classification 1 100 MPa/235 Degrees C, 1 100 MPa/315 Degrees C or 1 100 MPa/425 Degrees C—Dimensions First Edition. 5 pp.

ISO 3223-89. Aerospace—Nuts, Anchor, Self-Locking, Fixed, Two-Lug, with Counter-bore, with MJ Threads, Coated or Uncoated, Classification 1 100 MPa/235 Degrees Celsius, 1 100 MPa/315 Degrees Celsius or 1 100 MPa/425 Degrees Celsius—Dimensions First Edition. 6 pp.

ISO 3224-85. Aerospace—Self-Locking, Floating, Single-Lug Anchor Nuts, with Counterbore, Classification 1 100 MPa/235 Degrees Celsius First Edition. 5 pp.

ISO 3225-85. Aerospace—Self-Locking, Fixed, Two Lug Anchor Nuts, Reduced Series, with Counterbore, Strength Classification 1 100 MPa and Maximum Operating Temperature 235 Degrees Celsius First Edition. 5 pp.

ISO 7332-83. Metric Fasteners for Aerospace Construction—Nuts, Anchor, Self-Locking, Floating, Two Lug, Reduced Series, with Counterbore—Strength Classification 1 100 MPa—Maximum Operating Temperature 235 Degrees Celsius First Edition. 6 pp.

ISO 8940-88. Aerospace—Nuts, Anchor, Self-Locking, Sealing, Floating, Two-Lug, with Counterbore, Classifications 900 MPa/ 120 Degrees Celsius, 900 MPa/175 Degrees Celsius and 900 MPa/235 Degrees Celsius Dimensions First Edition. 4 pp.

ISO 9156-89. Aerospace—Nuts, Anchor, Self-Locking, Fixed, 90 Degree Corner, Reduced Series, with Counter-bore, with MJ Threads, Coated or Uncoated, Classification 1 100 MPa/235 Degrees Celsius, 1 100 MPa/315 Degrees Celsius or 1 100 MPa/425 Degrees Celsius—Dimensions First Edition. 5 pp.

—Aerospace—Anchor—Locking—Miniature

DIN ENGL LN 29986-77. Anchor Nuts; Self-Locking, Floating, Deep Counterbore, Double Lug, Miniature, for Temperatures up to 235 Degrees C (May). 3 pp.

DIN ENGL LN 29991-77. Anchor Nuts; Self-Locking, Deep Counterbore, Double Lug, Miniature for Temperatures up to 315 Degrees C and up to 425 Degrees C (July). 3 pp.

—Aerospace—Channel—Locking

DIN ENGL LN 29993-74. Gang Channel Nuts; Deep Counterbore, Self-Locking, Floating for Temperatures up to 120 Degrees C (Oct). 2 pp.

—Aerospace—Double Hexagonal—Locking

AECMA PREN2908-86. Nuts, Self Locking, Bi Hexagonal, Deep Counterbored, in Heat Resisting Steel FE-PA92HT (A286) Unplated Classification 1100 MPa/650 Degrees Celsius. 4 pp.

AECMA PREN2909-86. Nuts, Self Locking, Bi Hexagonal, Deep Counterbored, in Heat Resisting Steel FE-PA92HT (A286) Silver Plated Classification 1100 MPa/650 Degrees Celsius. 4 pp.

CEN PREN 3721-91. Nuts, Bihexagonal, Self-Locking, Deep Counterbored, in Heat Resisting Steel FE-PA92HT (A286), MoS Coated Classification:1100 MpA (at Ambient Temperature)/Degrees C. 5 pp.

CEN PREN 3843-93. Aerospace Series Nuts, Bi-Hexagonal, Self-Locking, with Counterbore, in Heat Reisisting Steel, Passivated Classification: 1 100 MPa (at Ambient Temperature) /650 Degrees C. 5 pp.

—Aerospace—Hexagonal—Locking

ISO 8538-86. Aerospace—Self-Locking Hexagon Nuts with Counterbore and Captive Washer, Classification 1 100 MPa/235 Degrees Celsius First Edition. 5 pp.

SBAC TS 135 ISSUE 1. All Metal Self-Locking Nuts (MJ Threads) Reduced Hexagon—Deep Counterbored Tensile Strength:—1100 MPa at Ambient Temperature.

Counterbore Nuts (Cont.)
—Aerospace—Locking

CEN PREN 3824-90. Nuts, Blind, Self-Locking, with Counterbore, in Heat Resisting Steel, MoS2 Lubricated Classification: 600 MPa (at Ambient Temperature)/315 Degrees C. 9 pp.

—Aircraft—Anchor—Locking

AECMA PREN2264-82. Nuts, Anchor, Self Locking Two Lug, Floating Incremental Counterbore Classification 1100 MPa/235 Degrees Celsius. 6 pp.

AECMA PREN2752-86. Nuts, Anchor, Self Locking Fixed, Two Lug, Reduced Series, with Counterbore Classification 1100 MPa/235 Degrees Celsius. 4 pp.

AECMA PREN2753-87. Nuts, Anchor, Self Locking One Lug, with Counterbore Classification 1100 MPa/235 Degrees Celsius. 6 pp.

AECMA PREN2754-87. Nuts, Anchor, Self Locking Two Lug, Floating with Counterbore Classification 1100 MPa/235 Degrees Celsius. 6 pp.

BSI BS EN 2752-90. 1990 Nuts, Anchor, Self Locking Fixed, Two Lug, Reduced Series with Counterbore. Classification: 1100 MPa/235 Degrees C. 8 pp.

CEN EN 2752-89. Nuts, Anchor, Self Locking Fixed, Two Lug, Reduced Series, with Counterbore Classification: 1100 MPa/235 Degrees C. 5 pp.

—Aircraft—Double Hexagonal—Locking

SBAC AS 27820-829 (V). Nut, Self-Locking Extended Washer, Double Hexagon Deep Counterbored.

—Aircraft—Hexagonal—Locking

SBAC AS 52600-699 ISSUE 1. Nut—Reduced—Hexagon—Self Locking—Deep Counterbored—Metric Series.

SBAC AS 54361-54460 (V). Nut—Reduced Hexagon—Self Locking—Deep Counterbored—Inch Series.

Counterbored
See Also: Washers (Fasteners)

—Aerospace

CEN PREN 3821-91. Washers, with Cylindrical Counterbore, in Heat Resisting Steel, Passivated. 6 pp.

CEN PREN 3821-93. Aerospace Series Washers, Chamfered, with Counterbore, in Heat Resisting Steel, Passivated. 4 pp.

DIN ENGL LN 9028-80. Aerospace; Counterbored Washers (Apr). 3 pp.

Counterbores
See Also: Tools

—Bolts

JIS B 1001-85. Diameter of Clearance Holes and Counterbores for Bolts and Screws. 6 pp.

—Cap Screws

DIN ENGL 74 Pt 2-91. Counterbores (Holes) for Cheese Head, Pan Head and Hexagon Socket Head Cap Screws (May). 5 pp.

DIN ENGL 974 Pt 1-91. Diameters of Counterbores (Holes) for Cheese Head, Pan Head and Hexagon Socket Head Cap Screws (May). 4 pp.

—Carbide Tips

DIN ENGL 8011-63. Carbide Tips for Reamers, Countersinks, Counterbores and End Mills (Sept). 1 p.

—Cheese Head Screws

CNS B1130-83. Counterbores for Hexagon Socket Head Screws and Slotted Cheese Head Screws (Mar)(4807).

DIN ENGL 74 Pt 2-91. Counterbores (Holes) for Cheese Head, Pan Head and Hexagon Socket Head Cap Screws (May). 5 pp.

DIN ENGL 974 Pt 1-91. Diameters of Counterbores (Holes) for Cheese Head, Pan Head and Hexagon Socket Head Cap Screws (May). 4 pp.

—Drafting—Simplified Representations

DIN ENGL 30-70. Drawing Practice; Simplified Presentations (Dec). 2 pp.

—Hexagonal Head Bolts

DIN ENGL 74 Pt 3-91. Counterbores (Holes) for Hexagon Bolts and Nuts (May). 6 pp.

DIN ENGL 974 Pt 2-91. Diameters of Counterbores (Holes) for Hexagon Bolts and Nuts (May). 4 pp.

JIS B 4236-89. Counterbores for Hexagon Socket Head Bolts. 9 pp.

—Hexagonal Head Screws

CNS B1130-83. Counterbores for Hexagon Socket Head Screws and Slotted Cheese Head Screws (Mar)(4807).

CNS B1131-79. Counterbores for Hexagon Screws and Nuts (Apr)(4808).

DIN ENGL 74 Pt 2-91. Counterbores (Holes) for Cheese Head, Pan Head and Hexagon Socket Head Cap Screws (May). 5 pp.

Counterbores (Cont.)
—Hexagonal Head Screws (Cont.)

DIN ENGL 974 Pt 1-91. Diameters of Counterbores (Holes) for Cheese Head, Pan Head and Hexagon Socket Head Cap Screws (May). 4 pp.

—Hexagonal Nuts

CNS B1131-79. Counterbores for Hexagon Screws and Nuts (Apr)(4808).

DIN ENGL 74 Pt 3-91. Counterbores (Holes) for Hexagon Bolts and Nuts (May). 6 pp.

DIN ENGL 974 Pt 2-91. Diameters of Counterbores (Holes) for Hexagon Bolts and Nuts (May). 4 pp.

—Machine Screws

JIS B 4233-89. Machine Screw Counterbores and Countersinks. 10 pp.

—Morse Taper Shanks—Pilots

BSI BS 328: Part 5-83. 1983 Drills and Reamers Part 5: Dimensions of Countersinks and Counterbores. 13 pp.

BSI BS 328: Part 5-01. 1983 Amd 1 Drills and Reamers Part 5: Specification for the Dimensions of Counterbores with Morse Taper Shanks and Countersinks with Parallel Shanks Without Solid Pilots and Morse Taper Shanks (AMD 7211) July 15, 1992 (F). 15 pp.

CNS B3397-80. Morse Taper Shank Counterbores with Detachable Pilot (Dec)(6764).

DIN ENGL 375-75. Counterbores with Morse Taper Shank and Detachable Pilot (Aug). 6 pp.

ISO 4207-77. Counterbores with Morse Taper Shanks and Detachable Pilots First Edition. 3 pp.

SNZ NZS/BS 328: Part 5-83. Drills and Reamers Part 5: Specification for the Dimensions of Countersinks and Counterbores. 9 pp.

—Pan Head Screws

DIN ENGL 74 Pt 2-91. Counterbores (Holes) for Cheese Head, Pan Head and Hexagon Socket Head Cap Screws (May). 5 pp.

DIN ENGL 974 Pt 1-91. Diameters of Counterbores (Holes) for Cheese Head, Pan Head and Hexagon Socket Head Cap Screws (May). 4 pp.

—Parallel Shanks—Pilots

BSI BS 328: Part 5-83. 1983 Drills and Reamers Part 5: Dimensions of Countersinks and Counterbores. 13 pp.

BSI BS 328: Part 5-01. 1983 Amd 1 Drills and Reamers Part 5: Specification for the Dimensions of Counterbores with Morse Taper Shanks and Countersinks with Parallel Shanks Without Solid Pilots and Morse Taper Shanks (AMD 7211) July 15, 1992 (F). 15 pp.

BSI BS 328: Part 7-92. 1992 Drills and Reamers Part 7: Counterbores with Parallel Shanks and Solid Pilots (ISO 4206: 1991). 6 pp.

CNS B3396-80. Parallel Shank Counterbores with Soild Pilot (Dec)(6763).

DIN ENGL 373-75. Counterbores with Parallel Shank and Solid Pilot (Aug). 4 pp.

ISO 4206-91. Counterbores with Parallel Shanks and Solid Pilots Second Edition. 6 pp.

SNZ NZS/BS 328: Part 5-83. Drills and Reamers Part 5: Specification for the Dimensions of Countersinks and Counterbores. 9 pp.

—Pilots

DIN ENGL 1868-75. Detachable Pilots for Counterbores and Countersinks (Aug). 2 pp.

ISO 4208-77. Detachable Pilots for Use with Counterbores and 90 Degrees Countersinks—Dimensions First Edition. 3 pp.

—Screws

JIS B 1001-85. Diameter of Clearance Holes and Counterbores for Bolts and Screws. 6 pp.

—Slotted Head Screws

CNS B1130-83. Counterbores for Hexagon Socket Head Screws and Slotted Cheese Head Screws (Mar)(4807).

—Socket Head Screws

CNS B1130-83. Counterbores for Hexagon Socket Head Screws and Slotted Cheese Head Screws (Mar)(4807).

DIN ENGL 74 Pt 2-91. Counterbores (Holes) for Cheese Head, Pan Head and Hexagon Socket Head Cap Screws (May). 5 pp.

DIN ENGL 974 Pt 1-91. Diameters of Counterbores (Holes) for Cheese Head, Pan Head and Hexagon Socket Head Cap Screws (May). 4 pp.

Counterpanes
Use: Bedspreads

Counters (Instruments)
Scope Note: Use a more specific term *See:* Binary Counter/Dividers; Binary Counters; Counter/Dividers; Counter/Timer Modules; Decade Counters; Digital Counters; Down Counters; Electromagnetic Counters;

Counters

Counters (Instruments) (Cont.)
See: (Cont.)
Frequency Divider/Digital Timers; Frequency Dividers; Geiger Mueller Tubes; Impulse Noise Counters; Interruption Counters; Light Scattering Automatic Particle Counters; Phase Detector/ Counters; Pulse Counters; Scintillation Counters; Sodium Iodide Scintillation Counters; Up/Down Counters

Countersink Cutters
Use: Countersinks

Countersinks
Use For: Countersink Cutters *See Also:* Cutting Tools; Milling Cutters
CNS B3056-80. Countersink 60 Degrees (Sep)(229).
CNS B3057-80. Countersink 90 Degrees (Sep)(230).
CNS B3058-80. Countersink 120 Degrees (Dec)(231).
JIS B 4231-88. Countersinks. 10 pp.

—**Bolts—Agricultural Equipment**
CNS B2355-82. Countersunk Double-Rib Bolts and Counter Sinkings for Agricultural Equipments (Oct)(4698).

—**Bone Screws—Identification Systems**
BSI BS 3531: Sec 5.7-91. 1991 Implants for Osteosynthesis Part 5: Bone Screws and Auxiliary Equipment Section 5.7: Specification for Drill Bits, Taps and Countersink Cutters (ISO 9714-1: 1991). 10 pp.
ISO 9714 Pt 1-91. Orthopaedic Drilling Instruments—Part 1: Drill Bits, Taps and Countersink Cutters First Edition. 8 pp.

—**Bone Screws—Materials**
BSI BS 3531: Sec 5.7-91. 1991 Implants for Osteosynthesis Part 5: Bone Screws and Auxiliary Equipment Section 5.7: Specification for Drill Bits, Taps and Countersink Cutters (ISO 9714-1: 1991). 10 pp.
ISO 9714 Pt 1-91. Orthopaedic Drilling Instruments—Part 1: Drill Bits, Taps and Countersink Cutters First Edition. 8 pp.

—**Bone Screws—Mechanical Properties**
BSI BS 3531: Sec 5.7-91. 1991 Implants for Osteosynthesis Part 5: Bone Screws and Auxiliary Equipment Section 5.7: Specification for Drill Bits, Taps and Countersink Cutters (ISO 9714-1: 1991). 10 pp.
ISO 9714 Pt 1-91. Orthopaedic Drilling Instruments—Part 1: Drill Bits, Taps and Countersink Cutters First Edition. 8 pp.

—**Carbide Tips**
DIN ENGL 8011-63. Carbide Tips for Reamers, Countersinks, Counterbores and End Mills (Sept). 1 p.

—**Countersunk Head Rivets**
DIN ENGL 1863-62. Countersinks for Countersunk Head Rivets (Jan). 2 pp.
SBAC AS 4695 ISSUE 3(I). Rivet Countersunk 90 Degree. 1 p.
SBAC AS 4696 ISSUE 3(I). Rivet Countersunk 120 Degree. 1 p.

—**Countersunk Head Screws**
CNS B1128-83. Countersinks for Countersunk Head Screws (Feb)(4805).
CNS B1129-79. Countersinks for Countersunk Head Screws for Fine Mechanics (Apr)(4806).
DIN ENGL 66-90. Countersinks for Countersunk Head Screws with ISO 7721 Common Head Style (Apr). 3 pp.

—**Countersunk Head Screws—Designations**
DIN ENGL 74 Pt 1-80. Countersinks for Countersunk Head Screws (Dec). 4 pp.
DIN ENGL 75 Pt 1-72. Countersinks for Countersunk Head Screws; Old Type (Apr). 3 pp.

—**Morse Taper Shanks**
BSI BS 328: Part 5-83. 1983 Drills and Reamers Part 5: Dimensions of Countersinks and Counterbores. 13 pp.
BSI BS 328: Part 5-01. 1983 Amd 1 Drills and Reamers Part 5: Specification for the Dimensions of Counterbores with Morse Taper Shanks and Countersinks with Parallel Shanks Without Solid Pilots and Morse Taper Shanks (AMD 7211) July 15, 1992 (F). 15 pp.
DIN ENGL 347-62. Countersinks with 120 Degree Point Angle (Mar). 2 pp.
ISO 3293-75. Morse Taper Shank Countersinks for Angles 60 Degrees, 90 Degrees and 120 Degrees Inclusive First Edition. 3 pp.
SNZ NZS/BS 328: Part 5-83. Drills and Reamers Part 5: Specification for the Dimensions of Countersinks and Counterbores. 9 pp.

Countersinks (Cont.)

—**Morse Taper Shanks—Pilots**
DIN ENGL 1867-75. 90 Degree Countersinks with Morse Taper Shank and Detachable Pilot (June). 3 pp.
ISO 4204-77. Countersinks, 90 Degrees, with Morse Taper Shanks and Detachable Pilots First Edition. 4 pp.

—**Morse Taper Shanks—Rivets**
DIN ENGL 1863-62. Countersinks for Countersunk Head Rivets (Jan). 2 pp.

—**Parallel Shanks**
BSI BS 328: Part 5-83. 1983 Drills and Reamers Part 5: Dimensions of Countersinks and Counterbores. 13 pp.
BSI BS 328: Part 5-01. 1983 Amd 1 Drills and Reamers Part 5: Specification for the Dimensions of Counterbores with Morse Taper Shanks and Countersinks with Parallel Shanks Without Solid Pilots and Morse Taper Shanks (AMD 7211) July 15, 1992 (F). 15 pp.
DIN ENGL 347-62. Countersinks with 120 Degree Point Angle (Mar). 2 pp.
ISO 3294-75. Parallel Shank Countersinks for Angles 60 Degrees, 90 Degrees and 120 Degrees Inclusive First Edition. 3 pp.
SNZ NZS/BS 328: Part 5-83. Drills and Reamers Part 5: Specification for the Dimensions of Countersinks and Counterbores. 9 pp.

—**Parallel Shanks—Pilots**
BSI BS 328: Part 6-92. 1992 Drills and Reamers Part 6: Countersinks, 90 Degrees, with Parallel Shanks and Solid Pilots (ISO 4205: 1991). 7 pp.
DIN ENGL 1866-75. 90 Degree Countersinks with Parallel Shank and Solid Pilot (June). 3 pp.
ISO 4205-91. Countersinks, 90 Degrees, with Parallel Shanks and Solid Pilots Second Edition. 6 pp.

—**Pilots**
DIN ENGL 1868-75. Detachable Pilots for Counterbores and Countersinks (Aug). 2 pp.
ISO 4208-77. Detachable Pilots for Use with Counterbores and 90 Degrees Countersinks—Dimensions First Edition. 3 pp.

Countersunk Head Bolts
See Also: Bolts; Countersunk Head Screws
BSI BS 4933-73. (OBSOLESCENT) 1973 ISO Metric Black Cup and Countersunk Head Bolts and Screws with Hexagon Nuts. 28 pp.
CEN PREN 3543-90. Bolts, 100 Degrees Countersunk Normal Head, Offset Cruciform-Ribbed Recess, Close Tolerance Shank, Short Thread in Heat and Corrosion Resisting Steel, Passivated Classification: 100 MPa (at Ambient Temperature) /425 Degrees Celsius. 8 pp.
CNS B2355-82. Countersunk Double-Rib Bolts and Counter Sinkings for Agricultural Equipments (Oct)(4698).
SAA AS B108-52. Black Cup and Countersunk Bolts, Nuts and Washers Being BS 325:1947 (Including Amendment of May 1954), Endorsed Subject to Australian Amendment Amdt 1 July 1952 Amdt 2 May 1961 Amdt 3 June 1962 Corrig. November 1967. 16 pp.

—**Aerospace**
CEN PREN 3304-92. Bolts, 100 Degrees Countersunk Reduced Head, Offset Cruciform Recess, Close Tolerance Shank, Short Thread, in Titanium Alloy, MoS2 Lubricated Classification: 1100 MPa (at Ambient Temperature)/315 Degrees C. 7 pp.
CEN PREN 3305-92. Bolts, 100 Degree Countersunk Reduced Head, Offset Cruciform Recess, Close Tolerance Shank, Short Thread, in Alloy Steel, Cadmium Plated Classification: 1100 MPa (at Ambient Temperature)/235 Degrees C. 8 pp.
CEN PREN 3381-90. Bolts, 100 Degrees Countersunk Normal Head, Offset Cruciform—Ribbed Recess, Close Tolerance Shank Short Thread in Titanium Anodized, MoS Lubricated Classification: 100 MPa (at Ambient Temperature) /315 Degrees Celsius. 8 pp.
CEN PREN 6024-91. Bolts, 100 Degree Countersunk Reduced Head, Offset Cruciform Recess, Close Tolerance Shank, Short Thread, Titanium Alloy, Lubricated Classification: 1100 MPa (at Ambient Temperature) /315 Degrees C Inch Series. 9 pp.

—**Aerospace—Flushness Control—Gaging**
BSI A 273-78. 1978 Gauging Practice for 100 Degree Countersunk Head Fasteners for Flushness Control. 7 pp.

—**Aerospace—Locking Collars**
DIN ENGL LN 65047 Pt 1-78. Aerospace; Bolts, Close Tolerance, in Titanium Alloy, with Self-Locking Collars in Aluminium Alloy; Procurement Specification; Bolts (May). 17 pp.

Countersunk Head Bolts (Cont.)
—**Aerospace—Locking Collars** (Cont.)
DIN ENGL LN 65047 Pt 2-78. Aerospace; Bolts, Close Tolerance, in Titanium Alloy, with Self-Locking Collars in Aluminium Alloy; Procurement Specification; Collars (May). 10 pp.

—**Aerospace—Phillips—Nonmagnetic**
DIN ENGL LN 9136-81. Aerospace; Bolts, Countersunk Head, Cross-Recessed, Non Magnetic (Dec). 4 pp.

—**Aerospace—Recessed**
AECMA PREN3381-89. Bolts, 100 Degrees Countersunk Normal Head, Offset Enciform-Ribbed Recess, Close Tolerance Shank, Short Thread, in Titanium Anodized Classification: 1100 MPa/315 Degrees Celsius. 7 pp.
AECMA PREN3543-89. Bolts, 100 Degrees Countersunk Normal Head, Offset Cruciform-Ribbed Recess, Close Tolerance Shank, Short Thread, in Corrosion Resisting Steel, Passivated Classification: 1100 MPa/425 Degrees Celsius. 8 pp.
BSI A 266-A 271-77. 1977 Amd 1 100 Degree Countersunk Head Titanium Alloy Bolts 160 000 Lbf/Square Inches (1100 MPa) with Hi-Torque Speed Drive Recesses. 11 pp.
CEN PREN 3760-90. Bolts, 100 Degress C Countersunk Normal Head, Offset Cruciform-Ribbed Recess, Threaded to Head, in Heat and Corrosion Resisting Steel, Passivated Classification: 1100 MPa (at Ambient Temperature)/425 Degrees C. 8 pp.

—**Aerospace—Slotted**
AECMA PREN2653-89. Bolts, 100 Degrees Countersunk Normal Head, Slotted, Threaded to Head, in Corrosion Resisting Steel, Passivated Classification: 600 MPa/425 Degrees Celsius. 7 pp.
AECMA PREN2654-89. Bolts, 100 Degrees Countersunk Normal Head, Slotted, Threaded to Head, in Brass, Tin Plated Classification: 280 MPa/ 80 Degrees Celsius. 7 pp.
CEN PREN 2653-91. Bolts, 100 Degrees Countersunk Head, Slotted, Threaded to Head, in Corrosion Resisting Steel, Passivated Classification: 600 MPa (at Ambient Temperature) /425 Degrees C. 7 pp.
CEN PREN 2654-91. Bolts, 100 Degrees Countersunk Normal Head, Slotted Threaded to Head, in Brass, Tin Plated Classification: 380 MPa (at Ambient Temperature)/80 Degrees C. 7 pp.

—**Aerospace—TORQ-SET**
DIN ENGL LN 29956-80. Aerospace; Bolts Countersunk Head TORQ-SET Recess; Short Thread Length; Titanium Alloy (Aug). 4 pp.

—**Aerospace—Torque Recessed**
BSI A 272-77. 1977 Hi-Torque Speed Drive Recess: Dimensions and Gauging for Countersunk Head Fasteners. 3 pp.

—**Agricultural Equipment—Double Nibbed**
ISO 5713-90. Equipment for Working the Soil—Fixing Bolts for Soil Working Elements First Edition. 8 pp.

—**Agricultural Equipment—Oval Head**
ISO 5713-90. Equipment for Working the Soil—Fixing Bolts for Soil Working Elements First Edition. 8 pp.

—**Agricultural Equipment—Round Head**
ISO 5713-90. Equipment for Working the Soil—Fixing Bolts for Soil Working Elements First Edition. 8 pp.

—**Aircraft**
BSI 2A 173-62. 1962 Amd 6 100 Degrees Countersunk Head Steel Bolts (Unified Treads) for Aircraft. 19 pp.
BSI 2A 175-62. 1962 Amd 2 100 Degree Countersunk Head Aluminium Alloy Bolts (Unified Threads) for Aircraft. 15 pp.
SBAC AS 4563 ISSUE 7. Bolt, 90 Degree Countersunk Head Aluminium Alloy. 1 p.
SBAC AS 4599-4600 ISSUE 3(I). 120 Degree Countersunk Head Bolt. 2 pp.
SBAC AS 6208 ISSUE 2(I). 120 Degree Countersunk Head, Bolt (6 UNC). 1 p.
SBAC AGS 3818 (V). 'Jo-Bolt' Fastners 100 Degree Countersunk Head.

—**Aircraft—Corrosion Resistant**
BSI 2A 174-62. 1962 Amd 2 100 Degree Countersunk Head Corrosion-Resisting Steel Bolts (Unified Threads) for Aircraft. 15 pp.

—**Aircraft—Oval Head**
SBAC AS 4564 ISSUE 6. Bolt Raised 90 Degree Countersunk Head Aluminium Alloy. 1 p.

—**Aircraft—Phillips**
SBAC AS 2929 ISSUE 2(I). Bolt M.S.-"Phillip's" 120 Degree Countersunk Head. 1 p.
SBAC AS 3294 ISSUE 7. Bolt—Phillip's 90 Degree Countersunk Head. 1 p.
SBAC AS 3297 ISSUE 7. Bolt—Phillip's 120 Degree Countersunk Head. 1 p.

INTERNATIONAL AND NON-U.S. NATIONAL STANDARDS
SUBJECT INDEX
Countersunk

Countersunk Head Bolts (Cont.)

—Aircraft—Phillips—Flat Head
SBAC AS 2926 ISSUE 2(I). Bolt M.S.—"Phillip's" 90 Degree Countersunk Head. 1 p.

—Aircraft—Phillips—Oval Head
SBAC AS 2927 ISSUE 2(I). Bolt M.S.—"Phillip's" Raised 90 Degree Countersunk Head. 1 p.
SBAC AS 2928 ISSUE 2(I). Bolt M.S.—"Phillip's" Raised 120 Degree Countersunk Head. 1 p.
SBAC AS 3295 ISSUE 7. Bolt—"Phillip's" Raised 90 Degree Countersunk Head. 1 p.
SBAC AS 3296 ISSUE 7. Bolt—"Phillip's" Raised 120 Degree Countersunk Head. 1 p.

—Aircraft—Recessed
AECMA PREN3035-87. Bolts, 100 Degrees Countersunk Head, Torq-Setrm Recess, Close Tolerance Shank, Short Thread, in Steel Cadmium Plated Classification 1100 MPa/235 Degrees Celsius. 7 pp.
BSI A 211-61. 1961 Amd 4 100 Degree Countersunk Head Steel Bolts (Unified Threads and Cruciform Recesses) for Aircraft. 19 pp.
BSI 2A 230-74. 1974 Amd 2 100 Degree Countersunk Head Steel Bolts (160 000 Lbf/Square Inches (1100 MPa) with Hi-Torque Speed Drive Recesses) (AMD 6151) May 31, 1990. 22 pp.
BSI A 232-68. 1968 Amd 4 100 Degree Countersunk Head Steel Bolts (160 000 Lbf/Square Inches (110 hbar) with Torq-Set Recesses). 21 pp.

—Aircraft—Slotted—Flat Head—High Strength
SBAC AS 1242 ISSUE 18. Bolt, 90 Degree Countersunk Head H.T.S.. 1 p.

—Aircraft—Slotted—Oval Head
SBAC AS 2921 ISSUE 6. Bolt, Raised 90 Degree Countersunk Head Stainless Steel. 1 p.

—Aircraft—Tanks (Containers)
SBAC AGS 159 (V). Repair Bolts for DE-BERGUE Riveted Tanks—C'S'K, Head.

—Double Nip
CNS B2354-82. Countersunk Bolts with Double Nips (Oct)(4697).

—Flat Head
DIN ENGL 604-81. Flat Countersunk Nib Bolts (Oct). 4 pp.

—Flat Head—Square Neck
DIN ENGL 605-81. Flat Countersunk Square Neck Bolts with Long Square (Oct). 4 pp.
DIN ENGL 608-81. Flat Countersunk Square Neck Bolts with Short Square (Oct). 4 pp.

—Nip
CNS B2237-81. Countersunk Head Nip Bolts (Apr)(4417).

—Square Head
CNS B2238-81. Countersunk Square Bolts with Long Square (Apr)(4418).
CNS B2239-81. Countersunk Square Bolts with Short Square (Apr)(4419).

—Square Neck
BSI BS 4933-73. (OBSOLESCENT) 1973 ISO Metric Black Cup and Countersunk Head Bolts and Screws with Hexagon Nuts. 28 pp.

—Structural—Slotted
CNS B2240-81. Countersunk Head Bolts with Forged Slot for Steel Structures (Apr)(4420).

Countersunk Head Rivets
See Also: Rivets; Round Head Rivets
BSI BS 641-51. 1951 Amd 3 Dimensions of Small Rivets for General Purposes. 20 pp.
BSI BS 4894-93. 1993 Bifurcated Rivets for General Purpose Use (Metric Series). 10 pp.
BSI BS 4894-73. 1973 Bifurcated Rivets for General Purpose Use. 8 pp.
CNS B2560-81. Countersunk Head Rivets, Nominal Diameter 10 to 36mm (May)(7338).
CNS B2562-81. Countersunk Head Rivets Nominal Diameter 1 to 8mm (May)(7340).
DIN ENGL 302-77. Countersunk Head Rivets; Nominal Diameters 10 to 36 mm (July). 2 pp.
DIN ENGL 661-77. Countersunk Head Rivets; Nominal Diameters 1 to 8 mm (July). 4 pp.
SAA AS B118-53. Dimensions of Small Rivets for General Purposes Being BS 641:1951 (Including Amendment of July 1955), Endorsed Without Australian Amendment Amdt 1 September 1961. 18 pp.
SBAC AS 3362 ISSUE 5. Close Tolerance, Rivet 90 Degree Countersunk Head. 1 p.
SBAC AS 3363 ISSUE 5. Close Tolerance, Rivet 120 Degree Countersunk Head. 1 p.

Countersunk Head Rivets (Cont.)
SBAC AS 4695 ISSUE 3(I). Rivet Countersunk 90 Degree. 1 p.
SBAC AS 4696 ISSUE 3(I). Rivet Countersunk 120 Degree. 1 p.

—Aerospace
AECMA PREN2550-85. Rivets, Solid, Countersunk Head in Aluminium EN2114. 6 pp.
AECMA PREN2551-87. Rivets, Solid, Countersunk Head with Dome in Aluminium Alloy EN 2627. 6 pp.
AECMA PREN2552-87. Rivets, Solid, Countersunk Head with Dome in Aluminium Alloy EN 2627 Anodised. 6 pp.
AECMA PREN2553-85. Rivets, Solid, Countersunk Head with Dome in Aluminium Alloy EN2116. 6 pp.
AECMA PREN2555-85. Rivets, Solid, Countersunk Head with Dome in Aluminium Alloy EN2117. 6 pp.
AECMA PREN2556-85. Rivets, Solid, Countersunk Head with Dome in Aluminium Alloy EN2117—Anodised. 6 pp.
AECMA PREN3137-89. Rivets, Solid, 100 Degrees Normal Countersunk Head with Dome, in Corrosion Resisting Steel EN 2470, Passivated. 5 pp.
AECMA PREN3393-90. Rivets, Solid, 100 Degrees Normal Countersunk Head with Dome, in Aluminium Alloy 2117, Metric Series. 6 pp.
AECMA PREN3394-90. Rivets, Solid, 100 Degrees Normal Countersunk Head, in Aluminium Alloy 2117, Metric Series. 6 pp.
AECMA PREN3642-90. Rivets, Solid, 100 Degrees Normal Countersunk Head with Dome, in Titanium T1-P02, Anodized. 6 pp.
AECMA PREN3643-90. Rivets, Solid, 100 Degrees Normal Countersunk Head, in Titanium T1-P02, Anodized. 6 pp.
BSI BS EN 2550-93. 1993 Rivets, Solid, 100 Degree Normal Countersunk Head, in Aluminium 1050A, Inch Based Series. 11 pp.
BSI BS EN 2551-93. 1993 Rivets, Solid, 100 Degree Normal Countersunk Head with Dome, in Aluminium Alloy 2117, Inch Based Series. 11 pp.
BSI BS EN 2552-93. 1993 Rivets, Solid, 100 Degree Normal Countersunk Head with Dome, in Aluminium Alloy 2117, Anodized or Chromated, Inch Based Series. 11 pp.
BSI BS EN 2553-93. 1993 Rivets, Solid, 100 Degree Normal Countersunk Head with Dome, in Aluminium Alloy 2017A, Inch Based Series. 12 pp.
BSI BS EN 2555-93. 1993 Rivets, Solid, 100 Degree Normal Countersunk Head with Dome, in Aluminium Alloy 5056A, Inch Based Series. 11 pp.
BSI BS EN 2556-93. 1993 Rivets, Solid, 100 Degree Normal Countersunk Head with Dome, in Aluminium Alloy 5056A, Anodized or Chromated, Inch Based Series. 11 pp.
BSI 2SP 68-71-73. 1973 Amd 2 100 Degree Countersunk Precision Head Aluminium and Aluminium Alloy Rivets for Aircraft. 15 pp.
BSI 2SP 142-143-73. 1973 Solid Rivets with 100 Degree Countersunk Truncated Radiused Head Made from L 86 (Sp 142) and L 37 (Sp 143) Materials. 8 pp.
CEN PREN2550-89. Rivets, Solid, 100 Degrees Normal Countersunk Head, in Aluminium EN 2114. 6 pp.
CEN EN 2550-92. Aerospace Series—Rivets, Solid, 100 Degrees Normal Countersunk Head, in Aluminium 1050A, Inch Based Series. 6 pp.
CEN EN 2551-92. Aerospace Series—Rivets, Solid, 100 Degrees Normal Countersunk Head with Dome, in Aluminium Alloy 2117, Inch Based Series. 6 pp.
CEN EN 2552-92. Areospace Series—Rivets, Solid, 100 Degrees Normal Countersunk Head with Dome, in Aluminium Alloy 2117, Anodized or Chromated, Inch Based Series. 6 pp.
CEN PREN2553-89. Rivets, Solid, 100 Degrees Normal Countersunk Head with Dome, in Aluminium Alloy EN 2116. 6 pp.
CEN EN 2553-92. Aerospace Series—Rivets, Solid, 100 Degrees Normal Countersunk Head with Dome, in Aluminium Alloy 2017A, Inch Based Series. 7 pp.
CEN PREN2555-89. Rivets, Solid, 100 Degrees Normal Countersunk Head with Dome, in Aluminum Alloy EN 2116. 6 pp.
CEN EN 2555-92. Aerospace Series—Rivets, Solid, 100 Degrees Normal Countersunk Head with Dome, in Aluminium Alloy 5056A, Inch Based Series. 6 pp.
CEN PREN2556-89. Rivets, Solid, 100 Degrees Normal Countersunk Head with Dome, in Aluminum Alloy EN 2117. 6 pp.
CEN EN 2556-92. Aerospace Series—Rivets, Solid, 100 Degrees Normal Countersunk Head with Dome, inAluminium Alloy 5056A, Anodized or Chromated, Inch Based Series. 6 pp.
CEN PREN 3137-91. Rivets, Solid, 100 Degree Normal Countersunk Head with Dome, in Corrosion Resisting Steel FE-PA11, Passivated. 6 pp.

Countersunk Head Rivets (Cont.)

—Aerospace (Cont.)
CEN PREN 3138-91. Rivets, Solid, 100 Degree Normal Countersunk Head with Dome, in Corrosion Resisting Steel FE-PA92HT, Passivated. 6 pp.
CEN PREN 3139-91. Rivets, Solid, 100 Degree Normal Countersunk Head, in Corrosion Resisting Steel FE-PA11, Passivated. 6 pp.
CEN PREN 3140-91. Rivets, Solid, 100 Degree Normal Countersunk Head, in Corrosion Resisting Steel FE-PA92HT, Passivated. 6 pp.
CEN PREN 3370-92. Rivets, Solid, 100 Degree Normal Countersunk Head with Dome, in Aluminium Alloy 7050, Anodized or Chromated, Inch Based Series. 7 pp.
CEN PREN 3395-90. Rivets, Solid, 100 Degree Normal Countersunk Head with Dome, in Aluminum Alloy 2017A, Metric Series. 6 pp.
CEN PREN 3395-91. Rivets, Solid, 100 Degree Normal Countersunk Head with Dome, in Aluminium Alloy 2017A, Metric Series. 6 pp.
CEN PREN 3396-90. Rivets, Solid, 100 Degree Normal Countersunk Head with Dome, in Aluminium Alloy 5056A, Metric Series. 6 pp.
CEN PREN 3396-91. Rivets, Solid, 100 Degree Normal Countersunk Head with Dome, in Aluminium Alloy 5056A, Metric Series. 6 pp.
CEN PREN 3399-90. Rivets Solid, 100 Degrees Normal Countersunk Head, in Aluminum Alloy 5056A, Metric Series. 6 pp.
CEN PREN 3399-91. Rivets, Solid, 100 Degree Normal Countersunk Head, in Aluminium Alloy 5056A, Metric Series. 6 pp.
CEN PREN 3400-91. Rivets, Solid, 100 Degree Normal Countersunk Head, in Aluminium Alloy 5056A, Anodized or Chromated, Metric Series. 6 pp.
CEN PREN 3401-92. Rivets, Solid, 100 Degree Normal Countersunk Head, in Aluminium Alloy 7050, Metric Series. 6 pp.
CEN PREN 3402-92. Rivets, Solid, 100 Degree Normal Countersunk Head, in Aluminium Alloy 7050, Anodized or Chromated, Metric Series. 6 pp.
CEN PREN 3405-92. Rivets, Solid, 100 Degree Normal Countersunk Head with Dome, in Aluminium Alloy 7050, Anodized or Chromated, Metric Series. 6 pp.
CEN PREN 3406-91. Rivets, Solid, 100 Degree Normal Countersunk Head with Dome, in Aluminium Alloy 2117, Anodized or Chromated, Metric Series. 6 pp.
CEN PREN 3407-91. Rivets, Solid, 100 Degree Normal Countersunk Head, in Aluminium alloy 2117, Anodized or Chromated, Metric Series. 6 pp.
CEN PREN 3408-90. Rivets, Solid, 100 Degree Normal Countersunk Head, in Aluminum Alloy 2017A, Metric Series. 6 pp.
CEN PREN 3408-91. Rivets, Solid, 100 Degree Normal Countersunk Head, in Aluminium Alloy 2017A, Metric Series. 6 pp.
CEN PREN 3410-93. Aerospace Series Rivets, Solid, 100 Degrees Normal Countersunk Head, in Aluminium 1050A, Metric Series. 5 pp.
CEN PREN 3412-90. Rivets, Solid, 100 Degrees Countersunk Head with Dome, in Aluminum Alloy 5056A, Anodized or Chromated, Metric Series. 6 pp.
CEN PREN 3412-91. Rivets, Solid, 100 Degrees Normal Countersunk Head with Dome, in Aluminium Alloy 5056A, Anodized or Chromated, Metric Series. 6 pp.
CEN PREN 3413-92. Rivets, Solid, 100 Degrees Normal Countersunk Head with Dome, in Aluminium Alloy 7050, Metric Series. 6 pp.
CEN PREN 3418-90. Rivets, Solid 100 Degrees Countersunk Head in Nickel Base Alloy NI-P11, Cadmium Plated, Metric Series. 6 pp.
CEN PREN 3418-91. Rivets, Solid, 100 Degree Normal Countersunk Head in Nickel Base Alloy NI-P11, Cadmium Plated, Metric Series. 6 pp.
CEN PREN 3419-90. Rivets, Solid, 100 Degrees Normal Countersunk Head in Nickel Base Alloy NI-P11, Metric Series. 6 pp.
CEN PREN 3419-91. Rivets, Solid, 100 Degrees Normal Countersunk Head, in Nickel Base Alloy NI-P11, Metric Series. 6 pp.
CEN PREN 3738-91. Rivets, Solid, 100 Degree Normal Countersunk Head with Dome, in Titanium Alloy Ti 45,5 Cb, Metric Series. 6 pp.
CEN PREN 3739-91. Rivets, Solid, 100 Degree Normal Countersunk Head, in Titanium Alloy Ti 45,5 Cb, Metric Series. 6 pp.

—Aerospace—Flushness Control—Gaging
BSI A 273-78. 1978 Gauging Practice for 100 Degree Countersunk Head Fasteners for Flushness Control. 7 pp.

—Aerospace—Torque Recesses
BSI A 272-77. 1977 Hi-Torque Speed Drive Recess: Dimensions and Gauging for Countersunk Head Fasteners. 3 pp.

INDUSTRY STANDARDS

INTERNATIONAL AND NON-U.S. NATIONAL STANDARDS
SUBJECT INDEX

Countersunk

Countersunk Head Rivets (Cont.)
—Aircraft
- BSI SP 86-59. 1959 100 Degree Countersunk Head Steel Rivets for Aircraft. 5 pp.
- BSI 2SP 87-88-73. 1973 100 Degree Countersunk Head High Nickel-Copper Alloy Rivets for Aircraft. 10 pp.
- SBAC AS 462 ISSUE 5(I). Rivet, Countersunk 90 Degree. 1 p.
- SBAC AS 463 ISSUE 1. Rivet, Countersunk 120 Degree. 1 p.
- SBAC AS 465 ISSUE 5(I). Rivet, Countersunk 120 Degree. 1 p.
- SBAC AS 467 ISSUE 5. Rivet, Csk, 90 Degree. 1 p.
- SBAC AS 2918 ISSUE 3. Close Tolerance Rivet (90 Degree Countersunk Head). 1 p.
- SBAC AS 2919 ISSUE 4. Close Tolerance Rivet (120 Degree Countersunk Head). 1 p.
- SBAC AS 16400-599 ISSUE 2. Rivet, Solid, 100 Degree Countersunk Head, DTD 5036, (Corrosion and Heat Resistant Steel).
- SBAC AS 16600-799 ISSUE 2. Rivet, Solid, 100 Degree Countersunk Head, BS.HR 504, (Heat Resistant Alloy).
- SBAC AS 46786-787 (V). Rivet, Blind, Self-Plugging, Flush Break, Mechanically Locked Spindle, Aluminium Alloy, 100 Degree CSK Head.
- SBAC AS 46790 ISSUE 2. Rivet, Blind, Self-Plugging, Flush Break, Mechanically Locked Stem, Corrosion-Resisting Steel, 100 Degree CSK Head, Cadmium Plated.
- SBAC AS 46791 ISSUE 1. Rivet, Blind, Self-Plugging, Flush Break, Mechanically Locked Stem, Corrosion-Resisting Steel, 100 Degree CSK Head, Unplated.
- SBAC AS 63250-255 ISSUE 1. Rivet Blind, Self-Plugging, Flush Break, Visible Mechanically Locked Stem, Aluminium Alloys 100 Degree Countersunk Head.
- SBAC AS 63270-271 ISSUE 1. Rivet Blind, Self-Plugging, Flush Break, Visible Mechanically Locked Stem, Corrosion Resisting Steel 100 Degree Countersunk Head.
- SBAC AGS 2041 ISSUE 5. Rivets, Chobert, Countersunk Head, 120 Degree (Steel) D.T.D. 720 Countersunk Head.
- SBAC AGS 2044 ISSUE 4. Rivets, Chobert, Countersunk Head, 120 Degree, Aluminium Alloy (L.37) Countersunk Head.
- SBAC AGS 2046 ISSUE 6. Rivets, Chobert, 120 Degree, Aluminium Alloy Countersunk Head.
- SBAC AGS 2049 ISSUE 6(I). Rivets, Tucker, Aluminium Alloy 'Pop' C'sk, Head.
- SBAC AGS 2051 ISSUE 7. Rivets, Tucker, Monel 'Pop', Countersunk Head.
- SBAC AGS 2054 ISSUE 4(I). Rivets, Tucker Aluminium Alloy Cup Countersunk Head.
- SBAC AGS 2056 ISSUE 3. Rivet Tucker Monel Cup, Countersunk Head.
- SBAC AGS 2060 ISSUE 5. Rivets, Tucker, Monel 'Pop', Countersunk Head Unplated.
- SBAC AGS 2066 ISSUE 7. Avdel Self Sealing 100 Degree Countersunk Rivets, (Aluminium Alloy).
- SBAC AGS 2067 ISSUE 6. Chobert, 100 Degree Countersunk, Rivet Steel.
- SBAC AGS 2068 ISSUE 5. Chobert, 100 Degree Countersunk Rivets, (Aluminium Alloy).
- SBAC AGS 2069 ISSUE 4(I). Chobert 100 Degree Countersunk Rivets.
- SBAC AGS 2070 ISSUE 4. Rivets, Tucker 100 Degree Monel 'Pop', Countersunk Head, (Plated).
- SBAC AGS 2071 ISSUE 4. Rivets, Tucker, 100 Degree Monel 'Pop', Countersunk Head, (Unplated).
- SBAC AGS 2073 ISSUE 2(I). Rivets, Tucker Aluminium Alloy 'Pop' Countersunk Head.
- SBAC AGS 3921 ISSUE 2. Rivet Blind, 100 Degree Countersunk Head, Self Plugging, Interference Lock (Avdel), C.R. Steel.
- SBAC AGS 3923 (V). Rivet Blind, 100 Degree Countersunk Head, Self Plugging, Interference Lock(Avdel), C.R. Steel, Cadmium Coated.
- SBAC AGS 4716 (V)(I). Rivet Self Sealing 120 Degree Countersunk Avdel, Aluminium Alloy.
- SBAC AGS 4718 (V). Rivet, Tucker 'Pop', Countersunk Head, Aluminium Alloy.

—Countersinks
- DIN ENGL 1863-62. Countersinks for Countersunk Head Rivets (Jan). 2 pp.

—Flat Head
- CNS B2576-81. Flat Countersunk Head Rivets Nominal Diameters 3 to 5mm (Jun)(7564).

—Tubular
- BSI BS 4895-81. 1981 Semi-Tubular Rivets for General Purpose Use. 8 pp.
- CNS B2580-81. Semi-Tubular Countersunk Head Rivets Nominal Diameters 1.6 to 10mm (Jun)(7568).

Countersunk Head Screws
See Also: Countersunk Head Bolts; Screws
- BSI BS 4933-73. (OBSOLESCENT) 1973 ISO Metric Black Cup and Countersunk Head Bolts and Screws with Hexagon Nuts. 28 pp.

Countersunk Head Screws (Cont.)
- BSI BS EN 27721-92. 1992 Countersunk Head Screws—Head Configuration and Gauging (ISO 7721: 1983). 10 pp.
- CEN EN 27721-91. Countersunk Head Screws—Head Configuration and Gauging. 3 pp.

—Aeronautical—Machine
- BSI 2A 204-63. 1963 Amd 2 2A 206, 2A 208, A 217-225 Machine Screws and Nuts (Unified Threads) Below 1/4 in for Aeronautical Purposes. 12 pp.

—Aerospace
- BSI A 260-A 265-76. 1976 Amd 2 Specification for Metric Screws. 8 pp.
- CEN PREN 3782-91. Holes for 100 Degrees Countersunk Head Screws Design Standard. 7 pp.
- DIN ENGL LN 29940-80. Aerospace; Countersunk Head Screws, with TORQ-SET Recess; Non-Magnetizable (Mar). 3 pp.
- DIN ENGL LN 65019-77. Bolts, Countersunk Head with TORQ-SET Recess (Dec). 3 pp.

—Aerospace—Hole Size
- CEN PREN 3782-92. Holes for 100 Degrees Countersunk Head Screws Design Standard. 5 pp.

—Aerospace—Phillips
- DIN ENGL LN 9136-81. Aerospace; Bolts, Countersunk Head, Cross-Recessed, Non Magnetic (Dec). 4 pp.
- DIN ENGL LN 9438-83. Aerospace; Screws, Countersunk Head with Cross Recess (Nov). 4 pp.
- ISO 5856-91. Aerospace—Screws, 100 Degree Normal Counter-sunk Head, Internal off-set Cruciform Ribbed Drive, Normal Shank, Short or Medium Length MJ Threads, Metallic Material, Coated or Uncoated, Strength Classes Less Than or Equal to 1 100 MPa—Dimensions First Edition. 6 pp.

—Aerospace—Phillips—Nonmagnetic
- DIN ENGL LN 9136-81. Aerospace; Bolts, Countersunk Head, Cross-Recessed, Non Magnetic (Dec). 4 pp.

—Aerospace—Recessed
- AECMA PREN2939-86. Screws, 100 Degrees Countersunk Head, Torq-Set Recess, Thread to Head, in Heat Resisting Steel FE-PA92HT (A286) Unplated Classification 900 MPa/650 Degrees Celsius. 5 pp.
- AECMA PREN2940-86. Screws, 100 Degrees Countersunk Head, Torq-Set Recess, Thread to Head, in Heat Resisting Steel FE-PA92HT (A286) Silver Plated Classification 900 MPa/650 Degrees Celsius. 5 pp.
- BSI A 272-77. 1977 Hi-Torque Speed Drive Recess: Dimensions and Gauging for Countersunk Head Fasteners. 3 pp.
- CEN PREN 3306-91. Screws, 100 Degree Countersunk Normal Head, Offset Cruciform-Ribbed Recess, Threaded to Head, in Titanium Alloy, MoS2 Lubricated Classification: 1100 MPa (at Ambient Temperature)/315 Degrees C. 6 pp.
- CEN PREN 3760-92. Screws, 100 Degree Countersunk Normal Head, Offset Cruciform-Ribbed Recess, Threaded to Head, in Heat and Corrosion Resisting Steel, Passivated Classification: 1 100 MPa (at Ambient Temperature) /425 Degrees C. 5 pp.
- CEN PREN 6024-92. Screws, 100 Degree Countersunk Reduced Head, Offset Cruciform Recess, Close Tolerance Shank, Short Thread, in Titanium Alloy, Anodized, MoS2 Lubricated, Classification: 1100 MPa (at Ambient Temperature) /315 Degrees C Inch Series. 8 pp.

—Aerospace—Slotted
- AECMA PREN2652-87. Screws, 100 Degrees Countersunk Head, Slotted, Fully Threaded, in Steel Cadmium Plated Classification 900 MPa/235 Degrees C.

—Aerospace—TORQ-SET
- DIN ENGL LN 29787-74. Screws Countersunk Head TORQ-SET Recess; Fully Threaded to Head; Titanium Alloy (Dec). 3 pp.
- DIN ENGL LN 29794-80. Aerospace; Screws, Countersunk Head with TORQ-SET Recess, Nearly Threaded to the Head (Mar). 4 pp.
- DIN ENGL LN 65019-77. Bolts, Countersunk Head with TORQ-SET Recess (Dec). 3 pp.

—Aircraft—Phillips
- SBAC AS 2993 ISSUE 4. Screw-"Phillip's" 90 Degree Countersunk Head. 1 p.
- SBAC AS 2996 ISSUE 3. Screw—"Phillip's" 120 Degree Countersunk Head. 1 p.

—Aircraft—Phillips—Flat
- SBAC AS 50600-699 (V). Screw, 100 Degree Countersunk Head, Torq-Set ACR Recess BS.HR650. 112-40 UNJC-3A. BS4084.
- SBAC AS 50700-799 (V). Screw, 100 Degree Countersunk Head, Torq-Set ACR Recess BS.HR650. 138-32 UNJC-3A. BS4084.

Countersunk Head Screws (Cont.)
—Aircraft—Phillips—Flat (Cont.)
- SBAC AS 50800-899 (V). Screw, 100 Degree Countersunk Head, Torq-Set ACR Recess BS.HR650. 164-32 UNJC-3A. BS4084.
- SBAC AS 50900-999 (V). Screw, 100 Degree Countersunk Head, Torq-Set ACR Recess BS.HR650. 190-32 UNJF-3A. BS4084.
- SBAC AS 51000-099 (V). Screw, 100 Degree Countersunk Head, Torq-Set ACR Recess BS.HR650. 250-28 UNJF-3A. BS4084.

—Aircraft—Phillips—Oval
- SBAC AS 2994 ISSUE 3. Screw-"Phillip's" Raised 90 Countersunk Head. 1 p.
- SBAC AS 2995 ISSUE 3. Screw-"Phillip's" Raised 120 Degree Countersunk Head. 1 p.

—Aircraft—Recessed—Wood
- BSI SP 128-133-65. 1965 Wood Screws for Aircraft. 5 pp.

—Cheese Head—Electric Contacts
- CNS B2233-81. Cheese Head Countersunk Screws for Electric Contact Set (Apr)(4413). 3 pp.

—Countersinks
- CNS B1128-83. Countersinks for Countersunk Head Screws (Feb)(4805).
- CNS B1129-79. Countersinks for Countersunk Head Screws for Fine Mechanics (Apr)(4806).
- DIN ENGL 66-90. Countersinks for Countersunk Head Screws with ISO 7721 Common Head Style (Apr). 3 pp.

—Countersinks—Designations
- DIN ENGL 74 Pt 1-80. Countersinks for Countersunk Head Screws (Dec). 4 pp.
- DIN ENGL 75 Pt 1-72. Countersinks for Countersunk Head Screws; Old Type (Apr). 3 pp.

—Hexagonal—Socket
- BSI BS 4168: Part 8-82. 1982 Amd 1 Hexagon Socket Screws and Wrench Keys; Metric Series Part 8: Specification for Hexagon Socket Countersunk Head Screws (AMD 5569) August 28, 1987. 9 pp.
- CNS B2292-81. Hexagon Socket Countersunk Head Screws (Apr)(4558). 4 pp.
- DIN ENGL 7991-86. Hexagon Socket Countersunk Head Cap Screws (Jan). 6 pp.
- SNZ NZS/BS 4168: Part 8-82. Hexagon Socket Screws and Wrench Keys; Metric Series Part 8: Specification for Hexagon Socket Countersunk Head Screws. 8 pp.

—Phillips
- ISO 7047-83. Cross Recessed Raised Countersunk Head Screws (Common Head Style)—Product Grade A First Edition. 6 pp.

—Phillips—Flat
- CNS B2294-81. Cross Recessed Countersunk (Flat) Head Screws (Apr)(4560). 4 pp.
- DIN ENGL 965-90. Cross Recessed Countersunk Flat Head Screws (Aug). 5 pp.
- DIN ENGL 7987-72. Cross Recessed Countersunk (Flat) Head Screws; (Countersunk Heads of Type Hitherto Used) (May). 2 pp.
- ISO 7046-83. Cross Recessed Countersunk Flat Head Screws (Common Head Style)—Product Grade A and Property Class 4.8 Only First Edition. 6 pp.
- ISO 7046 Pt 2-90. Cross Recessed Countersunk Flat Head Screws (Common Head Style)—Grade A—Part 2: Steel of Property Class 8.8, Stainless Steel and Non-Ferrous Metals First Edition. 7 pp.
- ISO 7721 Pt 2-90. Countersunk Flat Head Screws—Part 2: Penetration Depth of Cross Recesses First Edition. 7 pp.

—Phillips—Flat—Machine
- BSI BS 7433: Part 2-91. 1991 Cross Recessed Machine Screws Part 2: Specification for Countersunk Flat Head Screws of Product Grade A and Property Class 4.8 Only. 6 pp.

—Phillips—Flat—Tapping
- CNS B2185-87. Cross Recessed Countersunk (Flat) Head Tapping Screws (Jul)(4306). 5 pp.
- DIN ENGL 7982-90. Cross Recessed Countersunk Head Tapping Screws (Aug). 4 pp.
- ISO 7050-83. Cross Recessed Countersunk (Flat) Head Tapping Screws (Common Head Style) First Edition. 6 pp.
- MOD UK DSTAN 53-76-01. Screws, Tapping, Thread Forming (Self Tapping Screws) Issue 2; Amendment 1. 22 pp.

—Phillips—Flat—Wood
- CNS B2297-81. Cross Recessed Countersunk (Flat) Head Wood Screws (Apr)(4563). 3 pp.

—Phillips—Oval
- CNS B2295-81. Cross Recessed Countersunk (Oval) Head Screws (Apr)(4561). 4 pp.

INTERNATIONAL AND NON-U.S. NATIONAL STANDARDS
SUBJECT INDEX

Couplers

Countersunk Head Screws (Cont.)
—Phillips—Oval (Cont.)
 DIN ENGL 966-90. Cross Recessed Raised Countersunk Head Screws (Aug). 5 pp.
 DIN ENGL 7988-72. Cross Recessed Raised Countersunk (Oval) Head Screws; (Raised Countersunk Heads of Type Hitherto Used) (May). 2 pp.

—Phillips—Oval—Machine
 BSI BS 7433: Part 3-91. 1991 Cross Recessed Machine Screws Part 3: Specification for Raised Countersunk Head Screws Product Grade A. 8 pp.

—Phillips—Oval—Tapping
 CNS B2186-87. Cross Recesses Raised Countersink (Oval) Head Tapping Screws (Oct)(4307). 5 pp.
 DIN ENGL 7983-90. Cross Recessed Raised Countersunk Head Tapping Screws (Aug). 4 pp.
 ISO 7051-83. Cross Recessed Raised Countersunk (Oval) Head Tapping Screws First Edition. 6 pp.

—Phillips—Oval—Wood
 CNS B2298-81. Cross Recessed Raised Countersunk (Oval) Head Wood Screws (Apr)(4564). 3 pp.
 DIN ENGL 7995-84. Cross Recessed Raised Countersunk Head Wood Screws (Dec). 3 pp.

—Phillips—Wood
 DIN ENGL 7997-84. Cross Recessed Countersunk Head Wood Screws (Dec). 3 pp.

—Recessed
 DIN ENGL EN 27721-92. Countersunk Head Screws; Head Configuration and Gauging; (ISO 7721: 1983) (Feb) (Supersedes DIN ISO 7721, May 1985 Edition). 6 pp.
 ISO 7721-83. Countersunk Head Screws—Head Configuration and Gauging First Edition. 5 pp.

—Slotted
 DIN ENGL 63-72. Slotted Countersunk Head Screws; Small Head; (Countersunk Heads of Type Hitherto Used) (May). 3 pp.
 DIN ENGL 925-86. Slotted Countersunk Head Screws with Full Dog Point (Sept). 3 pp.
 DIN ENGL 7969-89. Slotted Countersunk Head Screws for Structural Steel Bolting for Supply with or Without Nut (Oct). 5 pp.
 DIN ENGL EN 27721-92. Countersunk Head Screws; Head Configuration and Gauging; (ISO 7721: 1983) (Feb) (Supersedes DIN ISO 7721, May 1985 Edition). 6 pp.
 ISO 7721-83. Countersunk Head Screws—Head Configuration and Gauging First Edition. 5 pp.

—Slotted—Flat
 CNS B2208-80. Slotted Countersunk (Flat) Head Screws for Fine Mechanics (MO.4 to M1.4) (Jul)(4363).
 CNS B2231-81. Slotted Countersunk (Flat) Head Screws (Apr)(4411).
 CNS B2232-81. Slotted Countersunk (Flat) Head Screws with Full Dog Point (Apr)(4412).
 DIN ENGL 963-90. Slotted Countersunk Head Screws (Aug). 6 pp.
 DIN ENGL 8245-72. Slotted Countersunk (Flat) Head Screws for Fine Mechanics M 0.4 to M 1.4 (Oct). 1 p.
 ISO 2009-83. Slotted Countersunk Head Screws (Common Head Style)—Product Grade A Second Edition. 6 pp.

—Slotted—Flat—Tapping
 CNS B2182-87. Slotted Countersunk (Flat) Head Tapping Screws (Jul)(4303). 5 pp.
 DIN ENGL 7972-90. Slotted Countersunk Head Tapping Screws (Aug). 3 pp.
 ISO 1482-83. Slotted Countersunk (Flat) Head Tapping Screws (Common Head Style) First Edition. 5 pp.
 MOD UK DSTAN 53-76-01. Screws, Tapping, Thread Forming (Self Tapping Screws) Issue 2; Amendment 1. 22 pp.

—Slotted—Flat—Wood
 DIN ENGL 97-86. Slotted Countersunk (Flat) Head Wood Screws (Dec). 2 pp.

—Slotted—Machine
 CNS B2733-88. Slotted Countersunk Head Machine Screws for Precision Instruments (Apr)(9678).
 CNS B2734-82. Slotted Raised Countersunk Head Machine Screws for Precision Instruments (Dec)(9679).

—Slotted—Oval
 CNS B2234-81. Slotted Raised Countersunk (Oval) Head Screws (Apr)(4414).
 CNS B2235-81. Slotted Countersunk (Oval) Head Screws with Full Dog Point (Apr)(4415).
 DIN ENGL 924-86. Slotted Raised Countersunk Head Screws with Full Dog Point (Sept). 3 pp.
 DIN ENGL 964-90. Slotted Raised Countersunk Head Screws (Aug). 6 pp.

Countersunk Head Screws (Cont.)
—Slotted—Oval (Cont.)
 ISO 2010-83. Slotted Raised Countersunk Head Screws (Common Head Style)—Product Grade A Second Edition. 6 pp.

—Slotted—Oval—Tapping
 CNS B2183-87. Slotted Raised Countersunk (Oval) Head Tapping Screws (Jul)(4304). 5 pp.
 DIN ENGL 7973-90. Slotted Raised Countersunk Head Tapping Screws (Aug). 3 pp.
 ISO 1483-83. Slotted Raised Countersunk (Oval) Head Tapping Screws (Common Head Style) First Edition. 5 pp.

—Slotted—Oval—Wood
 DIN ENGL 95-86. Slotted Raised Countersunk (Oval) Head Wood Screws (Dec). 2 pp.

—Tapping—Flat—Recessed
 BSI BS 7496: Part 2-91. 1991 Cross-Recessed Tapping Screws Part 2: Specification for Countersunk Flat Head Screws. 8 pp.

Countersunk Nuts
See Also: Nuts (Fasteners)
 SAA AS B108-52. Black Cup and Countersunk Bolts, Nuts and Washers Being BS 325:1947 (Including Amendment of May 1954), Endorsed Subject to Australian Amendment Amdt 1 July 1952 Amdt 2 May 1961 Amdt 3 June 1962 Corrig. November 1967. 16 pp.

—Aircraft—Anchor—Locking
 SBAC AGS 2009 ISSUE 14(I). Stiffnuts—Double Anchor—Countersunk.
 SBAC AGS 2014 ISSUE 14(I). Stiffnuts—Floating Anchor—Countersunk.
 SBAC AGS 2020 ISSUE 13(I). Stiffnuts—Single Anchor—Countersunk.

—Aircraft—Hexagonal—Locking
 SBAC AGS 2003 ISSUE 14(I). Stiffnuts—Hexagon—Countersunk.

—Aircraft—Locking
 SBAC AGS 2017 ISSUE 14(I). Stiffnuts—Stripnut—Countersunk.

Countersunk Washers
See Also: Washers (Fasteners)

—Aerospace
 CEN PREN 3835-91. Washers 100 Degrees Countersunk, in Heat Resisting Steel, Passivated. 6 pp.

Countersurveillance
—Tanks (Combat Vehicles)
 NATO STANAG 4319 ED 1 AMD 0-91. Countersurveillance Requirements for Future Main Battle Tanks Infrared/Thermal Aspects. 19 pp.

Countertops
Use For: Bench Tops; CounterTops; Worktops
Use: Countertops *See Also:* Cabinets (Furniture)

—Kitchen—Safety
 CEN PREN 1153-93. Kitchen Furniture—Safety Requirements and Test Methods for Built-in and Free Standing Kitchen Cabinets and Work Tops. 16 pp.

—Kitchen—Sizes
 CEN PREN 1116-93. Kitchen Furniture—Co-Ordinating Sizes for Kitchen Furniture and Kitchen Appliances. 8 pp.

Counting Dishes
 JIS Z 4401-81. Counting Dishes.

Counting Rate Meters
Use For: Rate Meters; Ratemeters *See Also:* Kerma Rate Meters; Radiation Measuring Instruments; Radiation Meters
 IEC 650-79. Analogue Counting Ratemeters Characteristics and Test Methods First Edition. 80 pp.
 IEC 739-83. Digital Counting Ratemeters Characteristics and Test Methods First Edition. 77 pp.
 IEC 808-85. Complementary Instrumentation for Counting Ratemeters Characteristics and Test Methods First Edition. 107 pp.

Counting Tubes
Use: Counter Tubes

Countries
Scope Note: Use a more specific term *See:* Border Control; China

Country Codes
See Also: Codes; Digits (Telecommunications); Mobile Country Codes; Provider Codes; Provider Oriented Codes
 BSI BS 5374-89. 1989 Codes for the Representation of Names of Countries. 56 pp.
 BSI BS 6879-87. 1987 Codes for the Representation of Names of Counties and Similar Areas. 30 pp.
 CGSB CAN/CGSB-200.4-89. Codes for the Representation of Names of Countries (ISO 3166-1988); (Amendment 1 Thru 2 April 1990) (Amendment 3 Thru 14 July 1991) (Amendment 15 Thru 17 March 1992) (Amendment 18 June 1993) (Amendment 19 August 1993). 103 pp.

—Allocation—Nonstandard Facilities
 CCITT RECMN T.35-91. Procedure for the Allocation of CCITT Defined Codes for Non-Standard Facilities (Study Group VIII) 8 pp. 8 pp.

—Message Handling Systems
 CCITT RECMN F.401-92. Message Handling Services: Naming and Addressing for Public Message Handling Services (Study Group I) 21 pp. 21 pp.
 CCITT RECMN F.401-89. Message Handling Services: Naming and Addressing for Public Message Handling Services—Message Handling and Directory Services—Operations and Definition of Service (Study Group I) 11 pp. 11 pp.

—Mobile Satellite Communications—INMARSAT
 CCITT RECMN E.215-89. Telephone/ISDN Numbering Plan for the Mobile-Satellite Services of INMARSAT—Telephone Network and ISDN—Operation, Numbering, Routing and Mobile Service (Study Group II)14 pp. 14 pp.

—Numbering Plans
 CCITT RECMN Q.11-89. Numbering Plan for the International Telephone Service—General Recommendations on Telephone Switching and Signalling—Functions and Information Flows for Services in the ISDN —Supplements (Study Group XI) 8 pp. 8 pp.
 CCITT RECMN Q.11 BIS-89. Numbering Plan for the ISDN Era—General Recommendations on Telephone Switching and Signalling—Functions and Information Flows for Services in the ISDN —Supplements (Study Group XI) 6 pp. 6 pp.
 CCITT RECMN Q.11 TER-89. Timetable for Coordinated Implementation of the Full Capability of the Numbering Plan for the ISDN Era (Recommendation E.164)—General Recommendations on Telephone Switching and Signalling—Functions and Information Flows for Services in the ISDN —Supplements. 2 pp.
 CCITT RECMN Q.103-89. Numbering Used—General Recommendations on Telephone Switching and Signalling—Functions and Information Flows for Services in the ISDN—Supplements (Study Group XI) 1 pp. 1 p.
 JIS X 0304-88. Code for the Representation of Numbers of Countries.

—Numbering Plans—Integrated Services Digital Networks
 CCITT RECMN E.164-91. Numbering Plan for the ISDN Era (Study Group II) 19 pp (Same as Recmn I.331). 19 pp.
 CCITT RECMN E.164-89. Numbering Plan for the ISDN Era—Telephone Network and ISDN—Operation, Numbering, Routing and Mobile Service (Study Group II) 6 pp (Same as Recmn I.331). 6 pp.

—Numbering Plans—Telephone Services
 CCITT RECMN E.163-89. Numbering Plan for the International Telephone Service—Telephone Network and ISDN—Operation, Numbering, Routing and Mobile Service (Study Group II) 8 pp. 8 pp.

Couplers
See Also: Acoustic Couplers; Antenna Couplers; Cable Couplers; Couplings; Hybrid Junctions; Microwave Couplers; Reference Couplers; Star Couplers
 IEC 320 Pt 2-2-90. Appliance Couplers for Household and Similar General Purposes Part 2: Interconnection Couplers for Household and Similar Equipment First Edition; (Corrigendum—Aug 1990). 69 pp.
 JIS C 7709-89. Lamp Caps and Holders. 126 pp.

—Appliances
 BSI BS 196-61. 1961 Amd 4 Protected-Type Non-Reversible Plugs, Socket-Outlets, Cable-Couplers and Appliance Couplers, with Earthing Contacts for Single Phase a.c. Circuits up to 250 Volts. 64 pp.
 BSI BS 6991-90. 1990 6/10A, Two-Pole Weather-Resistant Couplers for Household, Commercial and Light Industrial Equipment. 32 pp.

INDUSTRY STANDARDS

Couplers (Cont.)
—Appliances (Cont.)
BSI BS EN 60320-2-2-92. 1992 Appliance Couplers for Household and Similar General Purposes Part 2: Specification for Individual Types of Coupler Section 2.2: Interconnection Couplers for Household and Similar Equipment. 37 pp.

CENELEC EN 60320-2-2-91. Appliance Couplers for Household and Similar General Purposes Part 2: Interconnection Couplers for Household and Similar Equipment. 7 pp.

CNS C4293-90. Appliance Couplers for Domestic and Similar Use (Nov)(6797).

IEC 320 Pt 2-2-90. Appliance Couplers for Household and Similar General Purposes Part 2: Interconnection Couplers for Household and Similar Equipment First Edition; (Corrigendum—Aug 1990). 69 pp.

JIS C 8358-88. Appliance Couplers for Domestic and Similar Use. 19 pp.

—Cables (Electric)—Fillers—Mining Equipment
BSI BS 7383-90. 1990 Cold-Pour Resin-Based Compound for Use as a Filling Medium in Terminating Cables in Enclosures for Voltages Not Exceeding 11 kV for Use in Coal Mines. 14 pp.

—Electric Conduits
DIN ENGL 49020-59. Conduit for Electrical Wiring; Screwed Steel Conduit; Plain Conduit; Couplers (Jan). 2 pp.

—Industrial
BSI BS 4343-92. 1992 Plugs, Socket-Outlets and Couplers for Industrial Purposes Part 2: Dimensional Interchangeability Requirements for Pin and Contact-Tube Accessories of Harmonized Configurations (E). 74 pp.

BSI BS 4343-01. 1992 Amd 1 Plugs, Socket-Outlets and Couplers for Industrial Purposes Part 2: Dimensional Interchangeability Requirements for Pin and Contact-Tube Accessories of Harmonized Configurations (AMD 7890) August 15, 1993 (E). 75 pp.

BSI BS 4343-68. 1968 Amd 1 Industrial Plugs, Socket-Outlets and Couplers for a.c and d.c. Supplies. 139 pp.

BSI BS EN 60309-1-92. 1992 Plugs, Socket-Outlets and Couplers for Industrial Purposes Part 1: General Requirements. 67 pp.

BSI BS EN 60309-1-01. 1992 Amd 1 Plugs, Socket-Outlets and Couplers for Industrial Purposes Part 1: General Requirements (AMD 7889) August 15, 1993 (E). 68 pp.

BSI BS EN 60309-2-92. 1992 Plugs, Socket-Outlets and Couplers for Industrial Purposes Part 2: Dimensional Interchangeability Requirements for Pin and Contact-Tube Accessories of Harmonized Configurations (E). 74 pp.

BSI BS EN 60309-2-01. 1992 Amd 1 Plugs, Socket-Outlets and Couplers for Industrial Purposes Part 2: Dimensional Interchangeability Requirements for Pin and Contact-Tube Accessories of Harmonized Configurations (AMD 7890) August 15, 1993 (E). 75 pp.

CENELEC HD 196-75. Plugs, Sockets—Outlets and Coupless for Industrial Purposes. 7 pp.

CENELEC EN 60309-1-92. Plugs, Socket-Outlets and Couplers for Industrial Purposes Part 1: General Requirements. 8 pp.

CENELEC EN 60309-2-92. Plugs, Socket-Outlets and Couplers for Industrial Purposes Part 2: Dimensional Interchangeability Requirements for Pin and Contact-Tube Accessories. 7 pp.

IEC 309 Pt 1-88. Plugs, Socket-Outlets and Couplers for Industrial Purposes Part 1: General Requirements Second Edition; (Corrigendur.—March 1992). 110 pp.

IEC 309 Pt 2-89. Plugs, Socket-Outlets and Couplers for Industrial Purposes Part 2: Dimensional Interchangeability Requirements for Pin and Contact-Tube Accessories Second Edition; (Corrigendum—April 1992). 95 pp.

—Jumper Cables
CNS E1017-82. Jumper Couplers (Dec)(9698).
JIS E 4202-91. Jumper Couplers.

—Lighting—Interchangeability
BSI BS 7001-88. 1988 Interchangeability and Safety of a Standardized Luminaire Supporting Coupler. 14 pp.

BSI BS 7001-01. 1988 Amd 1 Interchangeability and Safety of a Standardized Luminaire Supporting Coupler (AMD 7625) June 15, 1993 (E). 23 pp.

—Lighting—Safety
BSI BS 6972-88. 1988 General Requirements for Luminaire Supporting Couplers for Domestic, Light Industrial and Commercial Use. 22 pp.

BSI BS 7001-88. 1988 Interchangeability and Safety of a Standardized Luminaire Supporting Coupler. 14 pp.

Couplers (Cont.)
—Lighting—Safety (Cont.)
BSI BS 7001-01. 1988 Amd 1 Interchangeability and Safety of a Standardized Luminaire Supporting Coupler (AMD 7625) June 15, 1993 (E). 23 pp.

—Scaffolds
BSI BS 1139: Part 2-82. (WITHDRAWN) 1982 Amd 1 Metal Scaffolding Part 2: Couplers and Fittings for Use in Tubular Scaffolding (AMD 5533) August 28, 1987 (Superseded by BS 1139: Sections 2.1 & 2.2: 1991). 10 pp.

BSI BS 1139: Sec 2.1-91. 1991 Metal Scaffolding Part 1: Tubes for Use in Scaffolding Section 2.1: Steel Couplers, Loose Spigots and Base-Plates for Use in Working Scaffolds and Falsework Made of Steel Tubes (Supersedes BS 1139: Part 2: 1982). 22 pp.

BSI BS 1139: Sec 2.2-91. 1991 Metal Scaffolding Part 1: Tubes for Use In Scaffolding Section 2.2: Steel and Aluminium Couplers, Fittings and Accessories for Use in Tubular Scaffolding (Supersedes BS 1139: Part 2: 1982). 11 pp.

CEN EN 74-88. Couplers, Loose Spigots and Base-Plates for Use in Working Scaffolds and Falsework Made of Steel Tubes; Requirements and Test Procedures. 17 pp.

SAA AS 1576.2-91. Scaffolding—Part 2: Couplers and Accessories Amdt 1 November 1992 (Supersedes AS 1575—1974 (in Part)). 31 pp.

SAA AS 1576.3-91. Scaffolding—Part 3: Prefabricated and Tube-and-Coupler Scaffolding Amdt 1 November 1992 (Supersedes AS 1575—1974 (in Part) and AS 1576—1974 (in Part)). 18 pp.

SAA AS 1576.3 Supp 1-91. Scaffolding—Part 3: Prefabricated and Tube-and-Coupler Scaffolding—Supplement 1: Metal Tube-and-Coupler Scaffolding—Deemed to Comply Amdt 1 November 1992 (Supplement to AS 1576.3 —1991) (Supersedes AS 1576—1974 (in Part)). 10 pp.

SNZ NZS/AS 1576. 2-91. Scaffolding Part 2: Couplers and Accessories (This is a Joint Standard with SAA AS 1576.2). 34 pp.

SNZ NZS/AS 1576. 3-91. Scaffolding Part 3: Prefabricated and Tube-and-Coupler Scaffolding (This is a Joint Standard with SAA AS 1576.3). 16 pp.

SNZ NZS/AS 1576. 3 Supp 1-91. Scaffolding Part 3: Prefabricated and Tube-and-Coupler Scaffolding—Metal Tube-and-Coupler Scaffolding—Deemed to Comply (Supplement to AS 1576.3-1991) (This is a Joint Standard with SAA AS 1576.3). 9 pp.

—Welding Cables—Safety
CENELEC HD 433-83. Safety Requirements for Arc Welding Equipment Coupling Devices for Welding Cables. 10 pp.

—Welding Equipment—Safety
IEC 501-75. Safety Requirements for Arc Welding Equipment—Plugs, Socket-Outlets and Couplers for Welding Cables First Edition. 13 pp.

—Welding Transformers—Resistance Welding
DIN ENGL 44767-76. Appliance Coupling for the Primary Connection of Resistance Welding Transformers; 200 A; 550 V (Feb). 2 pp.

Couplers (Mechanical)
Use: Couplings

Coupling Balls
Use: Ball Couplings

Coupling Capacitors
See Also: Capacitors

BSI BS 7578-92. 1992 Coupling Capacitors and Capacitor Dividers (IEC 358: 1990). 48 pp.

CENELEC HD 597 S1-92. Coupling Capacitors and Capacitor Dividers. 6 pp.

CENELEC HD 597 S1-92. CORRIGENDUM Coupling Capacitors and Capacitor Dividers. 2 pp.

DIN VDE 0560 Pt 2-70. Regulations for Capacitors; Coupling Capacitors for Voltages up to 1000 V and Power up to 0.5 kvar (May). 25 pp.

IEC 358-90. Coupling Capacitors and Capacitor Dividers Second Edition. 84 pp.

Coupling Nuts
Use: Union Nuts

Coupling Pins
See Also: Pins

—Fifth Wheels
BSI BS AU 1B-77. (WITHDRAWN) 1977 Dimensions of 50 mm Fifth Wheel King Pin of Semi-Trailers (Superseded by BS AU 1C: 1989). 4 pp.

BSI BS AU 1C-89. 1989 Dimensions of '50' Fifth Wheel King Pin for Semi-Trailers. 4 pp.

Coupling Pins (Cont.)
—Fifth Wheels (Cont.)
BSI BS AU 2A-70. 1970 Dimensions of 3 1/2-4 1/2 Inch Diameter Fifth Wheel King Pin for Use with Extra Heavy Duty Semi-Trailers. 2 pp.

CNS D2081-86. Shape and Dimensions of Fifth Wheel Coupling Pins for Semi-Trailers (Aug)(7030).

ISO 337-81. Road Vehicles—50 Semi-Trailer Fifth Wheel Coupling Pin—Basic and Mounting/Interchangeability Dimensions Second Edition; (Corrigendum 1-1990). 5 pp.

ISO 4086-82. Road Vehicles—90 Semi-Trailer Fifth Wheel Coupling Pin—Basic and Mounting/Interchangeability Dimensions Second Edition. 4 pp.

JIS D 6602-85. Shapes and Dimensions of Fifth Wheel Coupling Pins for Semi-Trailers. 9 pp.

—Fifth Wheels—Strength Testing
BSI BS AU 235: Part 2-91. 1991 Fifth Wheel Couplings for Commercial Vehicles Part 2: Method of Testing Strength of Coupling Pins (ISO 8716: 1988). 6 pp.

ISO 8716-88. Commercial Road Vehicles—Fifth Wheel Coupling Pins—Strength Tests First Edition. 4 pp.

Coupling Screws
See Also: Couplings; Screws

—Assemblies
CNS B2383-82. Coupling Screws with Collared Nut and Washers for Double-Ties (Oct)(4948).

Couplings
Scope Note: For additional listings, use a more specific term *Use For:* Couplers (Mechanical); Drawgear *See Also:* Anchors (Fasteners); Automatic Couplings; Ball Couplings; Bolts; Clevis; Clips; Clutches; Compression Couplings; Connectors; Couplers; Coupling Screws; Drawbars; Duct Joints; Electromagnetic Couplings; Fasteners; Filling Connectors; Flexible Couplings; Fuel Couplings; Hose Couplings; Hose Fittings; Hydraulic Couplings; Joints; Linchpins; Pins; Pipe Couplings; Pipe Fittings; Quick Disconnect Couplings; Quick Release Fasteners; Retaining Rings; Rivets; Rod Couplings; Screws; Shaft Couplings; Sleeves (Fittings); Splines; Stud Couplings; Tube Couplings; Yokes

—Aerospace
BSI 2L 87-71. 1971 Hexagonal Bars for Nuts, Couplings and Hollow Machined Parts of Al-Cu-Mg-Si-Mn Alloy (Solution and Precipita-tion Treated) (Free from Peripheral and Asymmetric Coarse Grain) (MW 14 mm, MW 36mm Across Flats) (Cu 4.4, Mg 0.5, Si 0.7, Mn 0.8). 12 pp.

—Aerospace—V Band—Duct Joints
NATO STANAG 3312 ED 4 AMD 5-75. Profile Dimensions of Flanges for V-Band Couplings for Piping and Ducting. 18 pp.

—Aerospace—V Band—Flanges
NATO STANAG 3312 ED 4 AMD 5-75. Profile Dimensions of Flanges for V-Band Couplings for Piping and Ducting. 18 pp.

—Aircraft Engines—Motor Oils
BSI C 19-86. 1986 Aircraft Pressure Replenishment Connection for Engine Oil (Inch Dimensions). 4 pp.

—Aircraft—Fuel Dispensing Equipment—Blanking Caps
SBAC AS 2941 ISSUE 2. Blanking Cap for 1 1/2 Inch Fuel Coupling. 1 p.

SBAC AS 2946 ISSUE 2. Blanking Cap for 2 1/2 Inch Fuel Coupling. 1 p.

—Aircraft—Fuel Dispensing Equipment—Flanged—Half
SBAC AS 2952 ISSUE 8. G.A. Flanged Half for 1 1/2 Inch Fuelling Coupling (Non-Self Sealing Type). 1 p.

SBAC AS 2953 ISSUE 8. G.A. Flanged Half for 1 1/2 Inch Re-Fuelling Coupling (Self Sealing Type). 1 p.

SBAC AS 2969 ISSUE 5. G.A. Flanged Half for 2 1/2 Inch Re-Fuelling Coupling (Open Ended Type). 1 p.

SBAC AS 2970 (V). G.A. Flanged Half for 2 1/2 Inch Re-Fuelling Coupling (Self Sealing Type). 2 pp.

—Aircraft—Fuel Dispensing Equipment—Springs (Elastic)
SBAC AS 2959 ISSUE 5. Spring. 1 p.

—Aircraft—Fuel Dispensing Equipment—Valves
SBAC AS 2958 ISSUE 4. Valve. 1 p.
SBAC AS 2979 ISSUE 3. Valve. 1 p.
SBAC AS 2980 ISSUE 3. Spring. 1 p.

INTERNATIONAL AND NON-U.S. NATIONAL STANDARDS
SUBJECT INDEX

Couplings (Cont.)

—Aircraft—Oxygen Supply Equipment
BSI C 20-86. 1986 Aircraft Gaseous Oxygen Replenishment Connection (Inch Dimensions). 4 pp.
NATO STANAG 3296 ED 3 AMD 4-85. Aircraft Gaseous Oxygen Replenishment Couplings. 16 pp.

—Aircraft—Oxygen Supply Equipment—Fluid Power
ISO 8775-88. Aerospace—Gaseous Oxygen Replenishment Connection for Use in Fluid Systems (New Type)—Dimensions (Inch Series) First Edition. 6 pp.

—Aircraft—Refueling Equipment
NATO STANAG 3447 ED 3 AMD 2-90. Aerial Refuelling Equipment Dimensional and Functional Characteristics. 10 pp.

—Aircraft—Tire Valves
NATO STANAG 3209 ED 5 AMD 2-86. Tyre Valve Couplings. 20 pp.
NATO STANAG 3209 ED 5 AMD 3-86. Tyre Valve Couplings. 20 pp.

—Arc Welding Cables
IEC 974 Pt 12-92. Arc-Welding Equipment Part 12: Coupling Devices for Welding Cables First Edition. 26 pp.

—Automotive—Air Brakes
MOD UK DSTAN 25-1-93. Vehicle Braking Systems, Hose Assemblies and Air Brake Coupling Adaptors, Brake, Intervehicular Issue 2. 10 pp.

—Automotive—Commercial—Drawbars
BSI BS AU 219-88. 1988 Dimensions of 40mm Drawbar Couplings Between Commercial Road Vehicles and Trailers. 7 pp.
BSI BS AU 220-88. 1988 Dimension of 50mm Drawbar Couplings Between Commercial Road Vehicles and Trails. 7 pp.
ISO 1102-86. Commercial Road Vehicles—Mechanical Connections Between Towing Vehicles and Trailers—50 mm Drawbar Coupling Second Edition. 7 pp.
ISO 8755-86. Commercial Road Vehicles—Mechanical Connections Between Towing Vehicles and Trailers—40 mm Drawbar Coupling First Edition. 7 pp.

—Automotive—Commercial—Drawbars—Strength Testing
BSI BS AU 239-91. 1991 Method of Testing Strength of Drawbar Couplings and Eyes for Hinged Drawbars (ISO 8718: 1988). 5 pp.
ISO 8718-88. Commercial Road Vehicles—Drawbar Couplings and Eyes for Hinged Drawbars—Strength Test First Edition. 4 pp.

—Automotive—Commercial—Power Takeoffs
BSI BS AU 203A-88. 1988 Dimensions of Couplings Between Power Take offs (PTOs) and Ancillary Driven Units on Commercial Road Vehicles. 5 pp.
ISO 7653-85. Road Vehicles—Commercial Vehicles—Couplings Between Power Take-offs (PTO) and Ancillary Driven Units First Edition; (Corrected and Reprinted -1986). 5 pp.

—Automotive—Drawbars
ISO 11406-93. Commercial Road Vehicles —Mechanical Coupling Between Towing Vehicles with Rear-Mounted Coupling and Drawbar Trailers—Interchangeability First Edition. 8 pp.

—Automotive—Towing Attachments
BSI BS AU 241-91. 1991 Mechanical Coupling Between Tractors and Semi-Trailers: Interchangeability (ISO 1726: 1989). 11 pp.
ISO 1726-89. Road Vehicles—Mechanical Coupling Between Tractors and Semi-Trailers—Interchangeability Second Edition. 8 pp.

—Cultivators—Rollers
BSI BS 7006-89. 1989 Trailed Agricultural Rollers. 7 pp.
ISO 8912-86. Equipment for Working the Soil—Roller Sections—Coupling and Section Width First Edition. 5 pp.

—Dental Instruments
BSI BS 7077: Sec 2.2-89. 1989 Dental Handpieces and Accessories Part 2: Accessories Section 2.2: Specification for Coupling Dimensions (ISO 3964: 1982). 4 pp.
BSI BS 7077: Sec 2.2-01. 1989 Amd 1 Dental Handpieces and Accessories Part 2: Accessories Section 2.2: Specification for Coupling Dimensions (ISO 3964: 1982) (AMD 6497) April 30, 1991. 7 pp.
CEN EN 23 964-89. Dentistry: Dental Handpieces; Coupling Dimensions. 6 pp.

Couplings (Cont.)

—Dental Instruments (Cont.)
CSA CAN/CSA-Z349. 29-35.2-M92. Dental Instruments and Equipment—Volume 1 (CAN/CSA-Z349.29-M92 to CAN/CSA-Z349.35.2 M92); (Gen Instr 1). 183 pp.
ISO 3964-82. Dental Handpieces—Coupling Dimensions Second Edition. 4 pp.

—Electromagnetic Compatibility
DIN VDE 0846 Pt 12 (D)-90. Measuring Apparatus for Assessment of Electromagnetic Compatibility; Coupling Equipment (Jan). 18 pp.

—Fifth Wheels—Strength Testing
BSI BS AU 235: Part 1-89. 1989 Fifth Wheel Couplings for Commercial Vehicles Part 1: Test Conditions and Strength Requirements. 6 pp.
ISO 8717-88. Commercial Road Vehicles—Fifth Wheel Couplings—Strength Tests First Edition. 6 pp.

—Irrigation Equipment
ISO 8224 Pt 2-91. Traveller Irrigation Machines—Part 2: Softwall Hose and Couplings—Test Methods First Edition. 7 pp.

—Nuts (Fasteners)
CNS B2748-83. Nut, Coupling (Jun)(10303).

—Rolling Stock
BSI BS 3710-64. (OBSOLESCENT) 1964 Brake Hose Couplings for Locomotives and Rolling Stock. 29 pp.

—Safety
DIN ENGL 31005-85. Safety Design of Technical Products; Locking Devices; Couplings (Apr). 6 pp.

—Semitrailers
BSI BS AU 30B-70. 1970 Amd 1 Standard Automatic Roller-Type Coupling for Semi-Trailers of 7 Tons to 14 3/4 Tons Laden Weight. 3 pp.
BSI BS AU 146B-70. 1970 Amd 1 Standard Automatic Roller-Type Coupling for Semi-Trailers Not Exceeding 7 1/4 Tons Laden Weight. 3 pp.

—Semitrailers—Brake Lines
BSI BS AU 138A-80. 1980 Dimensions of Contact Type Couplings for Air Pressure Braking Systems on Trailers and Semi-Trailers and Their Towing Vehicles, and the Arrangement of These Couplings on Articulated and Drawbar Combinations. 10 pp.

—Shaft Ends
CNS B2008-83. Cylindrical Shaft End (for Assembling Coupling, Pulley and Gear) (May)(147).
CNS B2009-83. Taper Shaft End (for Assembling Coupling and Gear) (May)(148).

—Ships—Fuel Dispensing Equipment
ISO 3926-80. Shipbuilding—Inland Navigation—Couplings for Oil and Fuel Reception—Mating Dimensions Second Edition. 5 pp.
NATO STANAG 1223 ED 2 AMD 2-83. Standardization of the Breakable Spool Coupling for Astern and Abeam Fuelling at Sea. 11 pp.

—Tractor Trucks—Brake Lines
CNS D2099-86. Air Brake Line Couplings Between Truck Tractors and Trailers (Jan)(7882).
JIS D 6605-84. Air Brake Line Couplings Between Truck Tractors and Trailers. 9 pp.

—Trailers—Brake Lines
BSI BS AU 5-63. 1963 Dimensions of a Contact Type Coupling for Vacuum and Pressure Braking Systems on Trailers and Semi-Trailers. 3 pp.
BSI BS AU 138A-80. 1980 Dimensions of Contact Type Couplings for Air Pressure Braking Systems on Trailers and Semi-Trailers and Their Towing Vehicles, and the Arrangement of These Couplings on Articulated and Drawbar Combinations. 10 pp.
CNS D2099-86. Air Brake Line Couplings Between Truck Tractors and Trailers (Jan)(7882).
JIS D 6605-84. Air Brake Line Couplings Between Truck Tractors and Trailers. 9 pp.

—Trucks—Towing Attachments
BSI BS AU 166-77. 1977 Mounting of Mechanical Coupling Devices on Rear Cross Members of Trucks. 4 pp.
ISO 3584-75. Road Vehicles—Mounting of Mechanical Coupling Devices on Rear Cross Members of Trucks First Edition. 4 pp.

—Tugboats
ISO 7545-83. Shipbuilding and Marine Structures—Inland Navigation—Single-Lock Automatic Couplings for Push Tows First Edition. 8 pp.

—Vacuum Equipment
JIS B 8365-88. Dimensions of Clamped-Type Vacuum Couplings. 7 pp.

Cover Papers
See Also: Papers

—Antique Finish
CGSB 9-GP-38M-79. Paper, Cover, Antique Finish, Standard for; (Amendment 1 Feb 1983). 11 pp.

—Kraft
CGSB 9-GP-31MA-87. Paper, Golden Kraft, for Envelopes. 8 pp.
CGSB 9-GP-32MA-87. Paper, Kraft, for File Folders. 8 pp.

—Weight Measurement
CGSB CAN/CGSB-9.63-M78. Grammage of Fine Papers. 6 pp.

Cover Plates (Electrical)
Use: Wall Plates

Coveralls
Use For: Jump Suits See Also: Clothing; Overalls

—Boys'
CGSB CAN/CGSB-49.49-M83. Jump Suits and Coveralls, Children's Sizes 2 to 6X, Regular Range—Dimensions. 13 pp.

—Fire Resistant—Mining
DIN ENGL 23320 Pt 2-88. Flameproof Clothing for the Mining Industry; Safety Requirements and Testing; One-Piece Coveralls (May). 2 pp.

—Girls'
CGSB CAN/CGSB-49.49-M83. Jump Suits and Coveralls, Children's Sizes 2 to 6X, Regular Range—Dimensions. 13 pp.
CGSB CAN/CGSB-49.78-M85. Jump Suits and Coveralls, Girls' Sizes 7 to 16, Regular Range—Dimensions. 12 pp.

—Infants'
CGSB CAN/CGSB-49.301-M84. Jump Suits, Coveralls and Rompers, Infants'—Dimensions; (Amendment 1 May 1985). 11 pp.

—Juniors'
CGSB CAN/CGSB-49.217-92. Juniors, Misses and Women's Canada Standard Sizes Jumpsuits—Dimensions. 17 pp.

—Men's
CGSB CAN/CGSB-38.15-M91. Men's Industrial Coveralls. 15 pp.
CGSB 38-GP-108M-79. Coveralls, Men's, Industrial—Dimensions, Standard for. 10 pp.

—Men's—Polyester/Cotton
CGSB 38-GP-45M-79. Coveralls, Men's Industrial, Polyester/Cotton, Standard for. 15 pp.

—Polyester/Cotton
CGSB CAN/CGSB-4.141-M90. Polyester/Cotton Twill Cloth, 225 g/m2; (Corrigendum—Aug 1990). 10 pp.

—Women's
CGSB CAN/CGSB-49.217-92. Juniors, Misses and Women's Canada Standard Sizes Jumpsuits—Dimensions. 17 pp.

Covered Electrodes
Use: Coated Electrodes

Covering Power
See Also: Contrast Ratio

—Paints
BSI BS 3900: Part D7-83. 1983 Methods of Test for Paints Group D: Optical Tests on Paint Films Part D7: Determination of Hiding Power of White and Light-Coloured Paints by the Kubelka-Munk Method. 33 pp.
ISO 2814-73. Paints and Varnishes—Comparison of Contrast Ratio (Hiding Power) of Paints of the Same Type and Colour First Edition. 4 pp.
ISO 6504 Pt 1-83. Paints and Varnishes—Determination of Hiding Power—Part 1: Kubelka-Munk Method for White and Light Coloured Paints First Edition. 33 pp.

—Pigments—Colorimetry
DIN ENGL 55987-81. Testing of Pigments; Determination of Hiding Power Value of Pigmented Media; Colorimetric Method (Feb). 4 pp.

Coverplates (Electrical)
Use: Wall Plates

Covers
Scope Note: See the subheading Covers under the specific product See Also: Electric Outlet Covers; Manhole Covers; Seal Covers; Valve Covers

Cows

Cows Milk
Use: Milk

CPPA
Use For: Canadian Pulp and Paper Association
See Also: Standards

—**Standards Preparation—Testing**
CPPA Z.1-92. Preparation and Adoption of Standard Testing Methods. 5 pp.

—**Testing—Bibliographies**
CPPA. Technical Section Canadian Pulp and Paper Association Standard Testing Methods. 3 pp.
CPPA. Technical Section Canadian Pulp and Paper Association Useful Methods. 2 pp.

CPU
Use: Central Processing Units

CR
Use: Chloroprene Rubber

Crab
See Also: Crustacea; Shellfish

—**Canned**
CNS N5109-85. Canned Crab Meat (Jan)(3526). 3 pp.

Crab Shell Meal
Use: Shell Meal

Crack Growth
Use: Cracking (Fracturing)

Crack Opening Displacement Testing
Use: Cracking (Fracturing)

Crack Propagation
Use: Cracking (Fracturing)

Crack Resistance
Use: Cracking (Fracturing)

Crack Testing
Use: Cracking (Fracturing)

Cracking (Fracturing)
Use For: Crack Testing; Craze Resistance; Crazing
See Also: Acoustic Emissions; Checking Resistance; Environmental Stress Cracking; Failure (Quality Control); Fracture Strength; Stress Corrosion Cracking

BSI BS 5762-79. (WITHDRAWN) 1979 Methods for Crack Opening Displacement (COD) Testing (Withdrawn, Superseded by BS 7448: Part 1: 1991). 18 pp.

—**Adhesives**
BSI BS 5350: Part H2-82. 1982 Methods of Test for Adhesives Group H: Physical Tests on Hot Melt Adhesives Part H2: Determination of Low Temperature Flexibility or Cold Crack Temperature. 4 pp.

—**Anodic Coatings**
CNS H2062-89. Methods of Test for Resistance to Cracking by Deforming of Anodic Oxidation Coatings on Aluminium and Aluminium Alloys (Apr)(8409).
ISO 3211-77. Anodizing of Aluminium and Its Alloys—Assessment of Resistance of Anodic Oxide Coatings to Cracking by Deformation Second Edition. 4 pp.
JIS H 8684-77. Test Method for Resistance to Cracking by Deforming of Anodic Oxidation Coatings on Aluminium and Aluminium Alloys. 8 pp.

—**Cable Insulation**
BSI BS 6469: Sec 3.1-92. 1992 Insulation and Sheathing Materials of Electric Cables Part 3: Methods of Test Specific to PVC Compounds Section 3.1: Pressure Test at High Temperature —Tests for Resistance to Cracking. 21 pp.
CENELEC HD 505.1.2 S1-88. Common Test Methods for Insulating and Sheathing Materials of Electric Cables Part 1: Methods for General Application Section Two—Thermal Ageing Methods. 3 pp.
CENELEC HD 505.1.2 S2-91. Common Test Methods for Insulating and Sheathing Materials of Electric Cables Part 1: Methods for General Application Section Two—Thermal Ageing Methods. 5 pp.
CENELEC HD 505.3.1 S1-88. Common Test Methods for Insulating and Sheathing Materials of Electric Cables Part 2: Methods Specific to PVC Compounds Section one—Pressure Test at High Temperature—Tests for Resistance to Cracking. 3 pp.

Cracking (Fracturing) *(Cont.)*
—**Cable Insulation** *(Cont.)*
IEC 811 Pt 3-1-85. Common Test Methods for Insulating and Sheathing Materials of Electric Cables Part 3: Methods Specific to PVC Compounds Section One—Pressure Test At High Temperature—Tests for Resistance to Cracking First Edition; (Corrigendum—May 1986). 31 pp.

—**Cables (Electric)**
BSI BS 6469: Sec 99.1-92. 1992 Insulating and Sheathing Materials of Electric Cables Part 99: Test Methods Used in the United Kingdom but Not Specified in Parts 1 to 5 Section 99: Non-Electrical Tests. 14 pp.

—**Ceramic Tiles**
BSI BS 6431: Part 17-83. 1983 Ceramic Floor and Wall Tiles Part 17: Method for Determination of Glazing Resistance. Glazed Tiles. 8 pp.
BSI BS 6431: Part 17-01. 1983 Amd 1 Ceramic Floor and Wall Tiles Part 17: Method for Determination of Glazing Resistance. Glazed Tiles (AMD 7106) July 15, 1992. 11 pp.
CEN EN 105-81. Ceramic Tiles: Determination of Crazing Resistance; Glazed Tiles. 6 pp.
CEN EN 105-91. Ceramic Tiles—Determination of Crazing Resistance—Glazed Tiles. 5 pp.

—**Ceramics**
BSI BS 7134: Sec 1.1-89. 1989 Testing of Engineering Ceramics Part 1: General and Textural Properties Section 1.1: Method for Determination of the Presence of Cracks and Other Defects by Dye Penetration Tests. 4 pp.

—**Coated Fabrics**
SAA AS 1441.14-73. Methods of Test for Coated Fabrics—Part 14: Method for Determination of Resistance to Cold Cracking.

—**Corrosion Testing—Terminal Blocks**
CSA C22.2 NO 158-1987. Terminal Blocks; (Gen Instr 1 Thru 4). 32 pp.

—**Elastomers—Flexure**
DIN ENGL 53522 Pt 2-79. Testing of Rubber and Elastomers; Flexing Endurance Test; Determination of Resistance to Flex Cracking (Jan). 2 pp.

—**Enamels—Thermal Shock**
DIN ENGL 51167-85. Testing of Vitreous and Porcelain Enamels; Determination of Crack Formation Temperature in the Thermal Shock Testing of Enamels for the Chemical Industry (Aug). 6 pp.

—**Filler Metal**
DIN ENGL 50129-73. Testing of Metallic Materials; Testing of Welding Filler Metals for Liability to Cracking (Oct). 4 pp.

—**Glazes—Aircraft**
AECMA PREN2155-19-81. Test Methods for Transparent Materials for Aircraft Glazing—Part 19—Determination of Craze Resistance. 5 pp.
AECMA PREN2155-21-81. Test Methods for Transparent Materials for Aircraft Glazing—Part 21—Determination of Resistance to Crack Propagation Less Than Factor Greater Than. 7 pp.
AECMA PREN2342-82. Heat and Crazing Resistant Acrylic Sheets for Aircraft Glazing—Technical Specification. 12 pp.
BSI BS EN 2155-21-89. 1989 Test Methods for Transparent Materials for Aircraft Glazing Part 21: Determination of Resistance to Crack Propagation (K Factor). 11 pp.
CEN EN 2155 (Part 21)-89. Test Methods for Transparent Materials for Aircraft Glazing Part 21: Determination of Resistance to Crack Propagation (K Factor). 8 pp.

—**Glazing Compounds**
CGSB CAN2-19.0-M77METH 9.3-78. Methods of Testing Putty, Caulking and Sealing Compounds Oil Bleeding and Spotting of Face Glazing Compounds. 1 p.
CGSB CAN2-19.0-M77METH19.1-78. Methods of Testing Putty, Caulking and Sealing Compounds Susceptibility of Glazing Compounds to Cracking. 1 p.

—**Lacquers—Temperature Change Testing**
CGSB 1-GP-71 METH 121.2-79. Methods of Testing Paints and Pigments Resistance to Temperature Changes Cold-Check Resistance. 1 p.

—**Leather**
CNS K6675-81. Method of Test for Grain Crack of Leather (Jul)(7701).
ISO 3378-75. Leather—Determination of Resistance to Grain Cracking, and of Crack Index First Edition. 6 pp.

Cracking (Fracturing) *(Cont.)*
—**Metals**
BSI DD 186-91. 1991 Determination of Threshold Stress Intensity Factors and Fatigue Crack Growth Rates in Metallic Materials. 44 pp.

—**Nuts (Fasteners)**
BSI BS EN 493-92. 1992 Fasteners—Surface Discontinuities—Nuts. 22 pp.
CEN PREN 493-91. Fasteners—Surface Discontinuities—Nuts. 19 pp.
CEN EN 493-92. Fasteners—Surface Discontinuities—Nuts. 19 pp.

—**Oxide Coatings**
CNS H2062-89. Methods of Test for Resistance to Cracking by Deforming of Anodic Oxidation Coatings on Aluminium and Aluminium Alloys (Apr)(8409).
JIS H 8684-77. Test Method for Resistance to Cracking by Deforming of Anodic Oxidation Coatings on Aluminium and Aluminium Alloys. 8 pp.

—**Paints**
SAA AS 1580.409. 1-92. Paints and Related Materials—Methods of Test—Part 409.1: Resistance to Mudcracking (in Professional Packages 30, 39). 2 pp.
SAA AS 1580.481. 1.8-91. Paints and Related Materials—Methods of Test—Part 481: Coatings—Part 481.1: Assessment of Individual Defects of Exposed Films —Part 481.1.8: Exposed to Weathering—Degree of Cracking (Supersedes AS 1580.481.1—1975 (in Part)) (in Professional Packages 30, 39). 4 pp.
SNZ NZS/AS 1580. 409.1-92. Methods of Test for Paints and Related Materials Part 409.1: Resistance to Mudcracking (This is a Joint Standard with SAA AS 1580.409.1). 2 pp.
SNZ NZS/AS 1580. 481.1.8-91. Methods of Test for Paints and Related Materials Part 481.1.8: Coatings—Exposed to Weathering—Degree of Cracking (This is a Joint Standard with SAA AS 1580.481.1.8). 4 pp.

—**Paperboard**
SAA AS 1301.P428 RP-83. Methods of Test for Pulp and Paper (Metric Units)—Part P428rp: Cracking Resistance of Paperboard (This is a Joint Standard with SANZ NZS 1301). 3 pp.
SNZ NZS/AS 1301. P428RP-83. Methods of Test for Pulp and Paper Cracking Resistance of Paperboard (This is a Joint Standard with SAA AS 1301.P428RP). 3 pp.

—**Plastic Sheets**
BSI BS 2782:Pt1: METH 150D-76. 1976 Methods of Testing Plastics Part 1: Thermal Properties Method 150D: Cold Crack Temperature of Film and Thin Sheeting. 4 pp.
DIN ENGL 53488-63. Testing of Plastic Sheets; Hole Test (Sept). 2 pp.

—**Rubber—Flexure**
DIN ENGL 53522 Pt 2-79. Testing of Rubber and Elastomers; Flexing Endurance Test; Determination of Resistance to Flex Cracking (Jan). 2 pp.

—**Sealing Materials**
CGSB CAN2-19.0-M77METH19.2-78. Methods of Testing Putty, Caulking and Sealing Compounds Crack and Bubbling Resistance of Sealing Compounds. 1 p.

—**Steels**
SAA AS 1544.5-81. Methods for Impact Tests on Metals—Part 5: Assessment of Fracture Surface Appearance on Steel (R 1993). 10 pp.

—**Tableware—Vitreous China**
BSI BS 4034-90. 1990 Vitrified Hotelware. 7 pp.

—**Toys—Fabrics**
CNS Z8016-8-86. Method of Test for Toy Safety (Testing for Wet/Dry Crocking of Textile Materials) (Jun)(4798-8). 2 pp.

—**Varnishes**
BSI BS 3900: Part E1-70. 1970 Methods of Test for Paints Group E: Mechanical Tests on Paint Films Part E1: Bend Test (Cylindrical Mandrel). 4 pp.
BSI BS 3900: Part E4-76. 1976 Methods of Test for Paints Group E: Mechanical Tests on Paint Films Part E4: Cupping Test. 3 pp.
ISO 1519-73. Paints and Varnishes—Bend Test (Cylindrical Mandrel) First Edition. 7 pp.
ISO 1520-73. Paints and Varnishes—Cupping Test First Edition. 5 pp.

—**Vulcanized Rubber**
BSI BS 903: Part A10-84. 1984 Methods of Testing Vulcanized Rubber Part A10: Determination of Flex Cracking (De Mattia). 7 pp.

Cracking (Fracturing) (Cont.)
—Vulcanized Rubber (Cont.)
BSI BS 903: Part A11-85. 1985 Methods of Testing Vulcanized Rubber Part A11: Determination of Resistance to Cut Growth (De Mattia-Type Machine). 8 pp.
BSI BS 903: Part A43-90. 1990 Physical Testing of Rubber Part A43: Method for Determination of Resistance to Ozone Cracking (Static Strain Test) (ISO 1431-1: 1989). 14 pp.
BSI BS 903: Part A44-83. 1983 Methods of Testing Vulcanized Rubber Part A44: Determination of Resistance to Ozone Cracking (Dynamic Strain Test). 10 pp.
CNS K6746-83. Method of Test for Ozone Cracking of Vulcanized Rubber (Feb)(10018).
DIN ENGL 53509 Pt 1-90. Determination of Resistance of Rubber to Ozone Cracking Under Static Strain (May) (ISO 1431-1: 1989). 7 pp.
ISO 132-83. Rubber, Vulcanized—Determination of Flex Cracking (De Mattia) Second Edition. 5 pp.
ISO 133-83. Rubber, Vulcanized—Determination of Crack Growth (De Mattia) Third Edition. 6 pp.
ISO 1431 Pt 1-89. Rubber, Vulcanized or Thermoplastic—Resistance to Ozone Cracking—Part 1: Static Strain Test Third Edition; (DIN 53509 Part 1). 10 pp.
ISO 1431 Pt 2-82. Rubber, Vulcanized—Resistance to Ozone Cracking—Part 2: Dynamic Strain Test First Edition. 8 pp.
SAA AS 1683.18-81. Methods of Test for Elastomers—Part 18: Vulcanized Rubbers—Determination of Resistance to Flex Cracking (De Mattia Type Machine). 5 pp.
SAA AS 1683.24-84. Methods of Test for Elastomers—Part 24: Rubber—Vulcanized—Determination of Resistance to Ozone Cracking—Static Strain Test. 5 pp.
SAA AS 1683.25-84. Methods of Test for Elastomers—Part 25: Rubber—Vulcanized—Determination of Resistance to Ozone Cracking—Dynamic Strain Test. 5 pp.

—Welded Joints
JIS Z 3153-93. Method of T-Joint Weld Cracking Test. 6 pp.
JIS Z 3154-93. Method of Controlled Thermal Severity Weld Cracking Test. 8 pp.
JIS Z 3155-93. Method of FISCO Test. 8 pp.
JIS Z 3157-93. Method of U-Groove Weld Cracking Test. 8 pp.
JIS Z 3158-93. Method of Y-Groove Weld Cracking Test. 9 pp.
JIS Z 3159-93. Method of H-Type Restrained Weld Cracking Test. 7 pp.

Cracking Testing
Use: Cracking (Fracturing)

Cradles
See Also: Holders

—Aircraft—Bolts
SBAC AS 1895 ISSUE 3. Bolt, Wing—Oxygen Cradle. 1 p.
SBAC AS 6246 ISSUE 2. Bolt Wing—Oxygen Cradle. 1 p.

—Aircraft—Nuts (Fasteners)
SBAC AS 6240 ISSUE 2(I). Wing Nut—Oxygen Cradle. 1 p.

—Aircraft—Oxygen Bottles
SBAC AS 1890 ISSUE 3. Cradle for Oxygen Bottle. 1 p.

—Aircraft—Oxygen Bottles—Assembly
SBAC AS 1889 ISSUE 3. Assembly of Cradle for Oxygen Bottle. 1 p.
SBAC AS 6245 ISSUE 1. Assembly of Cradle for Oxygen Bottle. 1 p.

—Aircraft—Pins
SBAC AS 1897 ISSUE 3. Draw Pin, Inner-Oxygen Cradle. 1 p.
SBAC AS 1898 ISSUE 4. Draw Pin, Outer-Oxygen Cradle. 1 p.

—Aircraft—Springs
SBAC AS 1899 ISSUE 3. Spring for Oxygen Cradle. 1 p.

—Aircraft—Straps
SBAC AS 1891 ISSUE 2. Strap for Oxygen Cradle. 1 p.
SBAC AS 1893 ISSUE 3. Strap for Oxygen Cradle. 1 p.
SBAC AS 6262 ISSUE 1. Strap for Oxygen Cradle. 1 p.

Cradles (Furniture)
See Also: Furniture
CEN PREN 1130-2-93. Furniture—Cribs and Cradles for Domestic Use—Part 2: Test Methods. 12 pp.

Cradles (Furniture) (Cont.)
CNS S2032-80. Method of Test for Baby Cradle (Aug)(6262). 2 pp.

—Safety
CEN PREN 1130-1-93. Furniture—Cribs and Cradles for Domestic Use—Part 1: Safety Requirements. 6 pp.

Cradles (Hoists)
See Also: Construction Equipment; Suspended Scaffolds
BSI BS 2830-73. 1973 Suspended Safety Chairs and Cradles for Use in the Construction Industry. 12 pp.

Crafts
Use: Handicrafts

Crampons
See Also: Mountaineering Equipment

—Safety
CEN PREN 893-92. Mountaineering Equipment—Crampons—Safety Requirements, Testing, Marking. 10 pp.

Cranberries
See Also: Food; Fruits

—Sauce
CGSB 32.236M-89. Jams, Marmalades, Jellies and Cranberry Sauce. 9 pp.

Cranes
Use For: Liftcranes *See Also:* Booms (Equipment); Bridge Cranes; Cable Cranes; Construction Equipment; Craneways; Crawler Cranes; Davits; Derricks; Gantry Cranes; Hoists; Industrial Equipment; Jib Cranes; Lifting Equipment; Portal Cranes; Railroad Cranes; Tower Cranes; Truck Cranes; Wheeled Cranes
ISO 4301 Pt 1-86. Cranes and Lifting Appliances—Classification—Part 1: General Second Edition. 6 pp.
ISO 4310-81. Cranes—Test Code and Procedures First Edition. 6 pp.
ISO 7363-86. Cranes and Lifting Appliances—Technical Characteristics and Acceptance Documents First Edition. 14 pp.
ISO 9374 Pt 1-89. Cranes—Information to Be Provided—Part 1: General First Edition. 4 pp.
SAA AS 1418.1-86. Cranes (Including Hoists and Winches) (Known as the SAA Crane Code)—Part 1: General Requirements Amdt 1 November 1987. 147 pp.
SAA AS 1418.5-90. Cranes (Including Hoists and Winches) (Known as the SAA Crane Code)—Part 5: Mobile, Vehicle-Loading and Vehicle-Tow Cranes. 27 pp.
SAA AS 1418.12-91. Cranes (Including Hoists and Winches) (Known as the SAA Crane Code)—Part 12: Crane Collector Systems. 27 pp.

—Acceptance Testing
DIN ENGL 15030-77. Lifting Equipment; Acceptance Testing of Crane Installations; Principles (Nov). 5 pp.

—Axles—Seal Covers
DIN ENGL 15092-82. Cranes; Driving Wheel Units and Idler Wheel Units; Sealing Covers (July). 6 pp.

—Bearings—Covers
CNS B1288-82. Cranes; Covers for Balland Roller Bearings; Connecting Dimensions (Aug)(9223).

—Cabins
ISO 8566 Pt 1-92. Cranes—Cabins—Part 1: General First Edition. 7 pp.

—Color Coding—Safety
DIN ENGL 15026-78. Lifting Appliances; Marking of Points of Hazard (Jan). 2 pp.

—Controls
DIN ENGL 15025-78. Cranes; Direction of Actuation and Arrangement of Controls in Crane Cabins (Jan). 7 pp.

—Design
BSI BS 2573: Part 1-83. 1983 Rules for Design Cranes Part 1: Classification, Stress Calculations and Design Criteria for Structures. 72 pp.
BSI BS 2573: Part 2-80. 1980 Amd 2 Rules for the Design of Cranes Part 2: Classification, Stress Calculations and Design of Mechanisms (AMD 5013) July 31, 1986. 24 pp.
CNS B1216-80. Specification for the Design of Crane Structures (Sep)(6426).
ISO 8686 Pt 1-89. Cranes—Design Principles for Loads and Load Combinations—Part 1: General First Edition. 31 pp.

Cranes (Cont.)
—Design (Cont.)
JIS B 8821-76. Specification for the Design of Crane Structures (R 1983). 51 pp.
SNZ NZS/BS 2573: Part 1-83. Rules for the Design of Cranes Part 1: Specification for Classification, Stress Calculations and Design Criteria for Structures. 74 pp.
SNZ NZS/BS 2573: Part 2-80. Rules for the Design of Cranes Part 2: Specification for Classification, Stress Calculations and Design of Mechanisms. 24 pp.

—Drive Axles—Seal Covers
DIN ENGL 15092-82. Cranes; Driving Wheel Units and Idler Wheel Units; Sealing Covers (July). 6 pp.

—Driving Wheels
DIN ENGL 15090-82. Cranes; Driving Wheel Units and Idler Wheel Units; Assembly (July). 6 pp.

—Driving Wheels—Bearing Rings
DIN ENGL 15094-82. Cranes; Driving Wheel Units and Idler Wheel Units; Bearing Cage Rings (July). 3 pp.

—Driving Wheels—Traveling Wheels
DIN ENGL 15093-82. Cranes; Driving Wheel Units and Idler Wheel Units; Travelling Wheels (July). 5 pp.

—Driving Wheels—Traveling Wheels—Shafts
DIN ENGL 15091-82. Cranes; Driving Wheel Units and Idler Wheel Units; Travelling Wheel Shafts (July). 5 pp.

—Electric
CSA C22.2 NO 33-M1984. Construction and Test of Electric Cranes and Hoists (R 1992); (Gen Instr 1 Thru 2). 30 pp.
CSA 1169 Bull. Electrical Bulletin 1169 June 27, 1978 to C22.2 NO 33. 2 pp.

—Electric—Construction
CSA C22.2 NO 33-M1984. Construction and Test of Electric Cranes and Hoists (R 1992); (Gen Instr 1 Thru 2). 30 pp.
CSA 1169 Bull. Electrical Bulletin 1169 June 27, 1978 to C22.2 NO 33. 2 pp.

—Electric—Slip Friction Clutches
CSA C22.2 NO 33-M1984. Construction and Test of Electric Cranes and Hoists (R 1992); (Gen Instr 1 Thru 2). 30 pp.
CSA 891 Bull. Electrical Bulletin 891 August 24, 1972 to C22.2 NO 33. 1 p.
CSA 1169 Bull. Electrical Bulletin 1169 June 27, 1978 to C22.2 NO 33. 2 pp.

—Gear Reducers
CNS B1243-81. Crans; Reduction Gears; Connecting Dimensions Contour Gauges Output Torques (Apr)(7191).

—Glossaries
BSI BS 3810: Part 4-68. 1968 Glossary of Terms Used in Materials Handling Part 4: Terms Used in Connection with Cranes. 29 pp.
CNS B1162-80. Glossary of Terms Relating to Cranes (Part 1 Kinds of Cranes) (May)(5510).
CNS B1163-80. Glossary of Terms Relating to Cranes (Part 2 Ability of Cranes) (Jul)(5675).
CNS B1164-80. Glossary of Terms Relating to Cranes (Part 3 Motion of Cranes) (Jul)(5676).
CNS B1165-80. Glossary of Terms Relating to Cranes (Part 4 Mechanical Part of Cranes) (Jul)(5677).
CNS B1166-80. Glossary of Terms Relating to Cranes (Part 5 Structure of Cranes) (Jul)(5678).
DIN ENGL 15001 Pt 1-73. Cranes; Vocabulary, Classification According to Type (Nov). 12 pp.
DIN ENGL 15001 Pt 2-75. Cranes; Vocabulary, Classification According to Application (July). 4 pp.
DIN ENGL 15003-70. Lifting Appliances; Load Suspending Devices, Loads and Forces; Definitions (Feb). 3 pp.
JIS B 0135-72. Glossary of Terms Relating to Cranes (Part 1 Kinds of Cranes).
JIS B 0136-74. Glossary of Terms Relating to Cranes (Part 2 Ability and Structure of Cranes).

—Guided
SAA AS 1418.6-88. Cranes (Including Hoists and Winches) (Known as the SAA Crane Code)—Part 6: Guided Storing and Retrieving Appliances. 50 pp.

—Hand Signals
SNZ NZS 5818-82. Specification for Hand Signals for the Direction of Cranes and Similar Lifting Devices Including Helicopters (Other Than for Cargo Handling in Wharf Areas) (Reconfirmed 1989). 12 pp.

—Handles
CNS B3436-83. Handles for Cranes (May)(10237).

Cranes (Cont.)

—Idlers
DIN ENGL 15090-82. Cranes; Driving Wheel Units and Idler Wheel Units; Assembly (July). 6 pp.

—Idlers—Bearing Rings
DIN ENGL 15094-82. Cranes; Driving Wheel Units and Idler Wheel Units; Bearing Cage Rings (July). 3 pp.

—Idlers—Traveling Wheels
DIN ENGL 15093-82. Cranes; Driving Wheel Units and Idler Wheel Units; Travelling Wheels (July). 5 pp.

—Idlers—Traveling Wheels—Shafts
DIN ENGL 15091-82. Cranes; Driving Wheel Units and Idler Wheel Units; Travelling Wheel Shafts (July). 5 pp.

—Indicating Instruments
BSI BS 7262-90. 1990 Automatic Safe Load Indicators. 24 pp.

—Inspection
BSI BS 7121: Part 2-91. 1991 Code of Practice for Safe Use of Cranes Part 2: Inspection, Testing and Examination. 23 pp.

—Insulated Conductors
CNS C4311-81. Contact Conductor with Insulating Cover for Crane (Jun)(7598).

—Loads (Forces)
ISO 8686 Pt 1-89. Cranes—Design Principles for Loads and Load Combinations—Part 1: General First Edition. 31 pp.

—Loads (Forces)—Glossaries
DIN ENGL 1305-88. Mass, as Weighed Value, Force, Weight Force, Weight, Load; Concepts (Jan). 2 pp.

—Manual Controls—Color Coding
ISO 7296 Pt 1-91. Cranes—Graphic Symbols —Part 1: General First Edition. 18 pp.

—Manual Controls—Symbols
ISO 7296 Pt 1-91. Cranes—Graphic Symbols —Part 1: General First Edition. 18 pp.

—Measurement Systems—Accuracy Testing
ISO 9373-89. Cranes and Related Equipment—Accuracy Requirements for Measuring Parameters During Testing First Edition. 4 pp.

—Mobile
BSI BS 1757-86. 1986 Power-Driven Mobile Cranes (Q). 51 pp.
ISO 4301 Pt 2-85. Lifting Appliances—Classification—Part 2: Mobile Cranes First Edition. 4 pp.
ISO 4310-81. Cranes—Test Code and Procedures First Edition. 6 pp.
MOD UK DEF-1137-66. Cranes, Road Mobile, Fully Slewing (over 6 and up to 15 Tons Inclusive Maximum Free Load). 12 pp.
MOD UK DEF-1422-A-66. Cranes, Road Mobile, Fully Slewing (2 to 6 Tons Inclusive Maximum Free Load). 11 pp.
MOD UK DEF-1448-65. Cranes, Truck Mounted, Fully Slewing. 12 pp.

—Mobile—Control Systems and Equipment
ISO 7752 Pt 2-85. Lifting Appliances—Controls—Layout and Characteristics—Part 2: Basic Arrangement and Requirements for Mobile Cranes First Edition; (Addendum 1-1986). 12 pp.

—Mobile—Glossaries
ISO 4306 Pt 2-85. Lifting Appliances—Vocabulary—Part 2: Mobile Cranes First Edition. 9 pp.
JIS D 6304-73. Glossary of Terms Relating to Mobile Cranes (R 1984). 34 pp.

—Mobile—Safety
BSI CP 3010-72. 1972 Amd 1 Safe Use of Cranes (Mobile Cranes, Tower Cranes, and Derrick Cranes). 89 pp.
CSA Z150-1974. Safety Code for Mobile Cranes; (Supplement 1 1977). 63 pp.

—Mobile—Sheaves
ISO 8087-85. Mobile Cranes—Drum and Sheave Sizes First Edition. 3 pp.

—Mobile—Stability Testing
DIN ENGL 15019 Pt 2-79. Cranes; Stability for Non-Rail Mounted Mobile Cranes; Test Loading and Calculation (June). 7 pp.
ISO 4305-91. Mobile Cranes—Determination of Stability Second Edition. 8 pp.

—Mobile—Steel—Construction
DIN ENGL 15018 Pt 2-84. Cranes; Steel Structures; Principles of Design and Construction (Nov). 9 pp.

—Mobile—Steel—Design
DIN ENGL 15018 Pt 3-84. Cranes; Principles Relating to Steel Structures; Design of Cranes on Vehicles (Nov). 12 pp.

—Mobile—Steel—Loads (Forces)
DIN ENGL 15018 Pt 1-84. Cranes; Steel Structures; Verification and Analyses (Nov). 38 pp.

—Mobile—Tires
CGSB 20-GP-5D-80. Tires, Pneumatic, Low Speed, Off Highway, Standard for. 23 pp.

—Mobile—Wire Ropes
ISO 4308 Pt 2-88. Cranes and Lifting Appliances—Selection of Wire Ropes—Part 2: Mobile Cranes—Coefficient of Utilization First Edition. 4 pp.

—Operators—Training
ISO 9926 Pt 1-90. Cranes—Training of Drivers—Part 1: General First Edition. 7 pp.

—Pulleys
DIN ENGL 15062 Pt 1-82. Cranes; Rope Pulleys; Selected Ranges and Correlation of Diameters and Overall Widths (July). 4 pp.
DIN ENGL 15417-82. Cranes; Bottom Blocks; Type D Rope Pulleys with Plain Bearings (July). 4 pp.
DIN ENGL 15418 Pt 1-82. Cranes; Bottom Blocks; Type C Rope Pulleys with Deep Groove Ball Bearing, Without Internal Sleeve (July). 4 pp.
DIN ENGL 15421 Pt 1-82. Cranes; Bottom Blocks; Type B Rope Pulleys with Deep Groove Ball Bearing and Internal Sleeve (July). 4 pp.
DIN ENGL 15422 Pt 1-82. Cranes; Bottom Blocks; Type A Rope Pulleys with Cylindrical Rolling Bearing and Internal Sleeve (July). 4 pp.

—Pulleys—Bearings
DIN ENGL 15062 Pt 2-82. Cranes; Rope Pulleys; Dimensions of Hubs and Bearings (July). 4 pp.

—Pulleys—Hubs
DIN ENGL 15062 Pt 2-82. Cranes; Rope Pulleys; Dimensions of Hubs and Bearings (July). 4 pp.

—Pulleys—Sealing Covers
DIN ENGL 15418 Pt 3-82. Cranes; Bottom Blocks; Sealing Covers for Type C Rope Pulleys with Deep Groove Ball Bearing, Without Internal Sleeve (July). 2 pp.
DIN ENGL 15421 Pt 3-82. Cranes; Bottom Blocks; Sealing Covers for Type B Rope Pulleys with Deep Groove Ball Bearing and Internal Sleeve (July). 2 pp.
DIN ENGL 15422 Pt 3-82. Cranes; Bottom Blocks; Sealing Covers for Type A Rope Pulleys with Cylindrical Rolling Bearing and Internal Sleeve (July). 2 pp.

—Pulleys—Sleeves
DIN ENGL 15418 Pt 2-82. Cranes; Bottom Blocks; Spacer Sleeves for Type C Rope Pulleys with Deep Groove Ball Bearing, Without Internal Sleeve (July). 2 pp.
DIN ENGL 15421 Pt 2-82. Cranes; Bottom Blocks; Internal Sleeves and Spacer Sleeves for Type B Rope Pulleys with Deep Groove Ball Bearing and Internal Sleeve (July). 4 pp.
DIN ENGL 15422 Pt 2-82. Cranes; Bottom Blocks; Internal Sleeves and Spacer Sleeves for Type A Rope Pulleys with Cylindrical Rolling Bearing and Internal Sleeve (July). 3 pp.

—Rail Wheels—Tires
BSI BS 3037: Part 1-58. 1958 Amd 1 Tyres for Crane Rail Wheels Part 1: Double-Flanged Parallel-Tread Tyres (AMD 6061) July 31, 1989. 13 pp.
BSI BS 3037: Part 2-75. 1975 Amd 1 Tyres for Crane Rail Wheels Part 2: Forged or Rolled Steel Double-Flanged Rail Wheels and Tyres (Metric Units) (AMD 6060) May 31, 1989. 11 pp.

—Rails—Steel—Hot Rolled
DIN ENGL 536 Pt 1-74. Crane Rails; Type A (With Foot Flange); Dimensions, Static Values, Steel Grades (Dec). 2 pp.
DIN ENGL 536 Pt 2-74. Crane Rails; Type F (Flat); Dimensions, Static Values, Steel Grades (Dec). 2 pp.

—Safety
BSI BS 5744-79. 1979 Code of Practice for Safe Use of Cranes (Partially Replaced by BS 7121: Part 1: 1989). 48 pp.
BSI BS 7121: Part 1-89. 1989 Code of Practice for Safe Use of Cranes Part 1: General. 38 pp.
BSI BS 7121: Part 2-91. 1991 Code of Practice for Safe Use of Cranes Part 2: Inspection, Testing and Examination. 23 pp.
CNS B4044-84. Safe Working Load, Rated Speed and Slewing Radii of Cranes (Sep)(6643).
JIS B 8820-73. Safe Working Loads, Rated Speeds and Slewing Radii of Cranes (R 1983). 9 pp.

—Sheave Blocks
DIN ENGL 15408-82. Cranes Twin Sheave Bottom Blocks; Assembly (July). 4 pp.
DIN ENGL 15409-82. Cranes; Quadruple Sheave Bottom Blocks; Assembly (July). 4 pp.

—Ships
JIS F 2104-87. Ships' Cranes for General Use.
JIS F 2104-74. Ships' Cranes for General Use. 6 pp.
MOD UK NES 113: Part 2-90. Requirements for Mechanical Handling Part 2: Fixed Cranes and Gantries Design Parameters Issue 2 (10.90). 17 pp.

—Stability Testing
DIN ENGL 15019 Pt 1-79. Cranes; Stability for All Cranes Except Non-Rail Mounted Mobile Cranes and Except Floating Cranes (Sept). 6 pp.
ISO 4304-87. Cranes Other Than Mobile and Floating Cranes—General Requirements for Stability First Edition. 6 pp.

—Steel—Design
DIN ENGL 15018 Pt 2-84. Cranes; Steel Structures; Principles of Design and Construction (Nov). 9 pp.

—Steel—Loads (Forces)
DIN ENGL 15018 Pt 1-84. Cranes; Steel Structures; Verification and Analyses (Nov). 38 pp.

—Stresses
BSI BS 2573: Part 2-80. 1980 Amd 2 Rules for the Design of Cranes Part 2: Classification, Stress Calculations and Design of Mechanisms (AMD 5013) July 31, 1986. 24 pp.
SNZ NZS/BS 2573: Part 2-80. Rules for the Design of Cranes Part 2: Specification for Classification, Stress Calculations and Design of Mechanisms. 24 pp.

—Technical Manuals—Driving
ISO 9928 Pt 1-90. Cranes—Crane Driving Manual—Part 1: General First Edition. 6 pp.

—Wheels
BSI BS 3037: Part 2-75. 1975 Amd 1 Tyres for Crane Rail Wheels Part 2: Forged or Rolled Steel Double-Flanged Rail Wheels and Tyres (Metric Units) (AMD 6060) May 31, 1989. 11 pp.

—Wind Loads
ISO 4302-81. Cranes—Wind Load Assessment First Edition. 8 pp.

—Wire Rope
ISO 4308 Pt 1-86. Cranes and Lifting Appliances—Selection of Wire Ropes—Part 1: General Second Edition. 7 pp.

Craneways
See Also: Cranes

—Prestressed Concrete—Design
DIN ENGL 4212-86. Reinforced Concrete and Prestressed Concrete Craneways; Design and Construction (Jan). 11 pp.

—Reinforced Concrete—Design
DIN ENGL 4212-86. Reinforced Concrete and Prestressed Concrete Craneways; Design and Construction (Jan). 11 pp.

—Steel—Design
DIN ENGL 4132-81. Craneways; Steel Structures; Principles for Calculation, Design and Construction (Feb). 32 pp.
DIN ENGL 4132 Suppl. 1-81. Craneways; Steel Structures; Principles for Calculation, Design and Construction; Explanations (Feb). 12 pp.

Crank Presses
See Also: Presses; Punches (Tools)

—Accuracy Testing
CNS B7045-81. Accuracy Test Standard for Crank Press (Tentative) (Nov)(3157).

Crankcase Oil
See Also: Crankcases; Gear Oils; Lubricants; Oils

—Caterpillars (Vehicles)
CNS K6959-89. Method of Test for Evaluating the Performance of Crankcase Lubricants with Single Cylinder Engine-Caterpillar 1H2 Test Method (Jan)(12482).
CNS K6960-89. Method of Test for Evaluating the Performance of Crankcase Lubricants with Single Cylinder Engine-Caterpillar 1G2 Test Method (Jan)(12483).

Crankcases
See Also: Crankcase Oil; Internal Combustion Engines; Piston Engines

INTERNATIONAL AND NON-U.S. NATIONAL STANDARDS
SUBJECT INDEX

Crankcases (Cont.)
—Paints (Cellulose Nitrate)—Oil Resistant
MOD UK DSTAN 80-37-73. Paint, Finishing, Cellulose Nitrate, Oil-Resisting Types: Brushing Spraying Issue 1. 7 pp.

Cranked Link Chains
See Also: Chains; Link Chains

—Drag
ISO 6971-82. Welded Steel Type Cranked Link Drag Chains and Chain Wheels First Edition. 14 pp.

—Mill
ISO 6972-82. Welded Steel Type Cranked Link Mill Chains and Chain Wheels First Edition. 21 pp.

—Roller—Transmissions (Power Sources)—Breaking Load
DIN ENGL 8182-90. Cranked Link Roller Chains (Rotary Chains) (July). 2 pp.

—Transmissions (Power Sources)
ISO 3512-92. Heavy-Duty Cranked-Link Transmission Chains Second Edition. 14 pp.

Cranks
—Bicycles—Pedals
BSI BS 6102: Part 14-91. 1991 Cycles Part 14: Specification for Dimensions of Bottom Bracket Axle and Crank Assembly with Square End Fittings (ISO 6695: 1991). 7 pp.
ISO 6695-91. Cycles—Pedal Axle and Crank Assembly with Square End Fitting—Assembly Dimensions First Edition. 4 pp.

—Hand
CNS B3104-54. Hand Crank, Curved (Jul)(322)(R 1973).
CNS B3105-54. Hand Crank, Straight (Jul)(323)(R 1973).

—Motorcycles—Pedals
BSI BS 6102: Part 14-91. 1991 Cycles Part 14: Specification for Dimensions of Bottom Bracket Axle and Crank Assembly with Square End Fittings (ISO 6695: 1991). 7 pp.
ISO 6695-91. Cycles—Pedal Axle and Crank Assembly with Square End Fitting—Assembly Dimensions First Edition. 4 pp.

Crankshafts
See Also: Internal Combustion Engines; Shafts (Machine Elements)

—Magnetic Particle Testing
BSI BS 5138-74. 1974 Amd 1 Magnetic Particle Flaw Inspection of Finished Machined Solid Forged and Drop Stamped Crank-Shafts. 24 pp.

Crash Beacons
Use: Crash Locator Beacons

Crash Landing
Use: Aircraft Landings—Crash

Crash Locator Beacons
Use For: ELT See Also: Aircraft; Aircraft Accidents; Aircraft Safety; Emergency Position Indicating Radio Beacons; Marker Beacons; Radio Transmitters

—Helicopters
CAA Spec. NO. 16 ISSUE 1.

—Sonar—Aircraft
CAA Spec. NO. 12 ISSUE 1.

Crates (Shipping Containers)
See Also: Boxes (Containers); Cartons; Containers; Livestock Crates

—Wood
BSI BS 1133: Sec 8-91. 1991 Packaging Code Section 8: Wooden Boxes, Cases and Crates (H). 89 pp.
BSI BS 1133: Sec 8-81. 1981 Packaging Code Section 8: Wooden Containers. 40 pp.
MOD UK DSTAN 81-71-01. Cases, Wood, Packing, Reusable Issue 1; Amendment 1. 47 pp.

Crawler Cranes
See Also: Cranes; Crawler Tractors
CNS D3150-83. Test Code of Truck Cranes, Wheel Cranes and Crawler Cranes (Jan)(9853).
JIS D 6301-76. Standard for Construction and Performance of Truck Cranes, Wheel Cranes and Crawler Cranes. 25 pp.
JIS D 6302-76. Standard Form of Specifications for Truck Cranes, Wheel Cranes and Crawler Cranes (R 1983). 26 pp.
JIS D 6303-76. Test Code of Truck Cranes, Wheel Cranes and Crawler Cranes. 12 pp.

Crawler Loaders
See Also: Loaders
JIS A 8421-90. Terminology and Commercial Specifications on Loaders. 29 pp.

—Glossaries
JIS A 8421-90. Terminology and Commercial Specifications on Loaders. 29 pp.

—Loads (Forces)
BSI BS 6819-87. (WITHDRAWN) 1987 Measurement of Tool Forces and Tipping Loads of Loaders Used for Earth-Moving (Superseded by BS 6911: Part 2: 1990). 11 pp.
BSI BS 6911: Part 2-90. 1990 Testing Earth-Moving Machinery Part 2: Measurement of Tool Forces and Tipping Loads of Loaders. 12 pp.
BSI BS 6912: Part 2-89. 1989 Safety of Earth Moving Machinery Part 2: Rated Operating Load for Crawler and Wheel Loaders. 4 pp.
ISO 5998-86. Earth-Moving Machinery—Rated Operating Load for Crawler and Wheel Loaders Second Edition. 3 pp.
ISO 8313-89. Earth-Moving Machinery—Loaders—Methods of Measuring Tool Forces and Tipping Loads Second Edition. 10 pp.
JIS A 8917-92. Earth-Moving Machinery—Rated Operating Load for Crawler and Wheel Loaders.

—Manual Controls
BSI BS 6211-82. 1982 Amd 1 Operator's Controls for Earth-Moving Machinery: Crawler Tractors and Crawler Loaders. 5 pp.
ISO 7095-82. Earth-Moving Machinery—Crawler Tractors and Crawler Loaders—Operator's Controls Second Edition. 4 pp.
JIS A 8918-92. Earth-Moving Machinery—Crawler Tractors and Crawler Loaders—Operators Controls.

—Roll Over Protective Structures
BSI BS 5527: Part 1-87. 1987 Roll-Over Protective Structures on Earth-Moving Machinery: Laboratory Tests and Preformance Requirements Part 1: Crawler, Wheel Loaders and Tractors, Backhoe Loaders, Graders, Tractor Scrapers and Articulated Steel Dumpers. 19 pp.
ISO 3471 Pt 1-86. Earth-Moving Machinery—Roll-over Protective Structures—Laboratory Tests and Performance Requirements—Part 1: Crawler, Wheel Loaders and Tractors, Backhoe Loaders, Graders, Tractor Scrapers, Articulated Steer Dumpers First Edition. 18 pp.

—Tool Forces
BSI BS 6819-87. (WITHDRAWN) 1987 Measurement of Tool Forces and Tipping Loads of Loaders Used for Earth-Moving (Superseded by BS 6911: Part 2: 1990). 11 pp.
BSI BS 6911: Part 2-90. 1990 Testing Earth-Moving Machinery Part 2: Measurement of Tool Forces and Tipping Loads of Loaders. 12 pp.
ISO 8313-89. Earth-Moving Machinery—Loaders—Methods of Measuring Tool Forces and Tipping Loads Second Edition. 10 pp.

Crawler Tractors
See Also: Bulldozers; Caterpillars (Vehicles); Crawler Cranes; Tractors
BSI BS 2800-57. (WITHDRAWN) 1957 Amd 2 Tests for Industrial Crawler and Wheeled Tractors (Excluding Units Designed for Materials Handling in Factories). 28 pp.
CNS D1022-82. Standard Form of Specification of Crawler Tractors (Mar)(8570).
JIS A 8420-88. Terminology and Commercial Specifications on Tractor. 37 pp.
JIS D 0003-77. Standard Form of Specifications of Crawler Tractors.
JIS D 6503-82. Testing Methods of Crawler Tractors. 24 pp.

—Bulldozer Blades—Capacity Measurement
BSI BS 6911: Part 1-88. 1988 Testing Earth-Moving Machinery Part 1: Method for Determination of Volumetric Rating of Crawler and Wheel Tractor Dozer Blades. 7 pp.
ISO 9246-88. Earth-Moving Machinery—Crawler and Wheel Tractor Dozer Blades—Volumetric Ratings First Edition. 7 pp.

—Bulldozers
JIS D 6507-76. Test Method for Bulldozer on Crawler Tractors (R 1979). 17 pp.

—Drawbars
JIS D 6106-92. Dimensions of Drawbars for Crawler Tractors. 5 pp.

—Glossaries
BSI BS 6914: Part 6-89. 1989 Amd 1 Terminology (Including Definitions of Dimensions and Symbols) for Earth-Moving Machinery Part 6: Terminology for Self-Propelled Crawler and Wheel Tractors and Their Equipment. 23 pp.

Crawler Tractors (Cont.)
—Glossaries (Cont.)
ISO 6747-88. Earth-Moving Machinery—Tractors—Terminology and Commercial Specifications Second Edition; (Amendment 1-1989). 26 pp.
JIS A 8420-88. Terminology and Commercial Specifications on Tractor. 37 pp.

—Manual Controls
BSI BS 6211-82. 1982 Amd 1 Operator's Controls for Earth-Moving Machinery: Crawler Tractors and Crawler Loaders. 5 pp.
ISO 7095-82. Earth-Moving Machinery—Crawler Tractors and Crawler Loaders—Operator's Controls Second Edition. 4 pp.

—Roll Over Protective Structures
BSI BS 5527: Part 1-87. 1987 Roll-Over Protective Structures on Earth-Moving Machinery: Laboratory Tests and Preformance Requirements Part 1: Crawler, Wheel Loaders and Tractors, Backhoe Loaders, Graders, Tractor Scrapers and Articulated Steel Dumpers. 19 pp.
CNS A3126-81. Earth-Moving Machinery-Roll-Over Protective Structures for Driver Laboratory Tests and Performance Requirements (Sep)(7849).
ISO 3471 Pt 1-86. Earth-Moving Machinery—Roll-over Protective Structures—Laboratory Tests and Performance Requirements—Part 1: Crawler, Wheel Loaders and Tractors, Backhoe Loaders, Graders, Tractor Scrapers, Articulated Steer Dumpers First Edition. 18 pp.
JIS A 8910-89. Earth-Moving Machinery—Roll-Over Protective Structures—Laboratory Tests and Performance Requirements. 25 pp.

—Shovels
JIS D 6505-82. Testing Method of Loaders. 40 pp.

Crayons
See Also: Pastels
CNS K6376-75. Method of Test for Crayons (Feb)(3762).
CNS S1066-75. Crayon (Feb)(2984). 1 p.
CNS S1076-71. Crayons, Paper Bonded (Apr)(3283).
JIS S 6026-92. Crayons and Oil Pastels. 10 pp.

—Lumber
CGSB CAN/CGSB-53.29-92. Lumber Marking Crayon. 7 pp.

Craze Resistance
Use: Cracking (Fracturing)

Crazing
Use: Cracking (Fracturing)

Cream (Dairy Products)
See Also: Beverages; Dairy Products; Milk
CNS N5086-85. Edible Cream (Dec)(2878). 2 pp.

—Acidity—Volumetric Analysis
BSI BS 1741: Sec 10.1-89. 1989 Methods for Chemical Analysis of Liquid Milk and Cream Part 10: Determination of Titratable Acidity Section 10.1: Method for Liquid Milk. 4 pp.

—Chemical Analysis
BSI BS 1741-63. (WITHDRAWN) 1963 Amd 4 The Chemical Analysis of Liquid Milk and Cream (W) (Superseded by Various Parts and Sections of the Revision of BS 1741). 19 pp.
SNZ NZS 2246-69. Methods for the Chemical Analysis of Liquid Milk and Cream. 24 pp.

—Fat Content
SNZ NZS 501: Part 2-71. Gerber Method for the Determination of Fat in Milk and Milk Products Part 2: Methods Amend: A, 1982. 28 pp.

—Fat Content—Gravimetric Analysis
BSI BS 1741: Part 4-87. 1987 Methods for Chemical Analysis of Liquid Milk and Cream Part 4: Determination of Fat Content of Cream. 12 pp.
ISO 2450-85. Cream—Determination of Fat Content—Gravimetric Method (Reference Method) Second Edition. 10 pp.

—Lactose Content
BSI BS 1741: Sec 7.1-89. 1989 Methods for Chemical Analysis of Liquid Milk and Cream Part 7: Determination of Lactose Content Section 7.1: Reference Method. 4 pp.

—Microbiological Analysis
BSI BS 4285: Sec 5.2-89. 1989 Microbiological Examination for Dairy Purposes Part 5: Ancillary Methods Section 5.2: Methylene Blue Reduction Test for Cream and Ice Cream (W). 4 pp.
SAA AS 1095.2.2-81. Microbiological Methods for the Dairy Industry Part 2: Methods for the Examination of Specific Dairy Products Part 2.2: Cream. 2 pp.

INDUSTRY STANDARDS

Cream (Dairy Products) (Cont.)

—Sampling

BSI BS 1741: Part 1-87. 1987 Methods for Chemical Analysis of Liquid Milk and Cream Part 1: General Introduction Including Preparation of Samples. 4 pp.

—Solids Content

BSI BS 1741: Part 2-90. 1990 Methods for Chemical Analysis of Liquid Milk and Cream Part 2: Determination of Total Solids Content of Liquid Milk, Cream and Unsweetened Condensed Milk. 7 pp.

ISO 6731-89. Milk, Cream and Evaporated Milk—Determination of Total Solids Content (Reference Method) First Edition. 6 pp.

—Test Specimens

CNS N6056-80. Method of Test for Milk and Milk Products (General Rules) (Aug)(3440). 2 pp.

Cream of Tartar

Use: Potassium Bitartrate

Credenzas

See Also: Furniture; Office Furniture; Shelves

—Modular—With Bookcase

CGSB 44.153-85. Credenza, Modular, Single Case with Bookcase. 36 pp.

Credit Card Calling

Use: Credit Card Calling Services

Credit Card Calling Services

Use For: Charge Card Calling Services; International Telecommunication Charge Card Services
See Also: Credit Cards; Operator Assisted Calls; Reverse Charging; Telephone Services; Universal Personal Telecommunication Services

CCITT RECMN E.116-92. International Telecommunication Charge Card Service (Study Group I) 6 pp. 6 pp.

—Integrated Services Digital Networks

CCITT RECMN I.256-89. Charging Supplementary Services—Integrated Services Digital Network (ISDN)—General Structure and Service Capabilities (Study Group XVIII) 16 pp. 16 pp.

CCITT RECMN I.256. 1-88. Credit Card Calling—Integrated Services Digital Network (ISDN)—General Structure and Service Capabilities (Study Group XVIII) 1 pp. 1 p.

—Tariffs

CCITT RECMN D.174-89. Conventional Transmission of Information Necessary for Billing and Accounting Regarding Collect and Credit Card Calls—General Tariff Principles—Charging and Accounting in International Telecommunications Services (Study Group III) 1 pp. 1 p.

CCITT RECMN E.277-89. Conventional Transmission of Information Necessary for the Collection of Charges and the Accounting Regarding Collect and Credit Card Calls—Telephone Network and ISDN—Operation, Numbering, Routing and Mobile Service (Study Group II) 1pp(Same as Recmn D.174). 1 p.

—Validation (Security)—Automated

CEPT T/SF 68 E-92. International Validation of Automatic Telecommunications Charge Cards. 5 pp.

Credit Cards

See Also: Credit Card Calling Services; Identification Cards; Magnetic Cards

—Conference Calling

CCITT RECMN E.116-89. International Telephone Credit Cards for Use in a Non-Automated Environment—Telephone Network and ISDN—Operation, Numbering, Routing and Mobile Service (Study Group II) 3 pp. 3 pp.

—Formsets

CSA Z243.32-M1983. Formsets for Use with Embossed Cards and Imprinting Equipment; (Gen Instr 1). 16 pp.

—Magnetic

JIS X 6301-79. Credit Cards with Magnetic Stripe.
JIS X 6302-79. Credit Cards—Magnetic Stripe Encoded.

—Point of Sale Systems—Printing

JIS X 9102-81. Printing Specifications for Optical Character Recognition of POS Systems—Credit Cards.

—Telephone

CCITT RECMN E.116-92. International Telecommunication Charge Card Service (Study Group I) 6 pp. 6 pp.

Credit Cards (Cont.)

—Telephone (Cont.)

CCITT RECMN E.116-89. International Telephone Credit Cards for Use in a Non-Automated Environment—Telephone Network and ISDN—Operation, Numbering, Routing and Mobile Service (Study Group II) 3 pp. 3 pp.

CCITT RECMN E.118-92. International Telecommunication Charge Card (Study Group I) 9 pp. 9 pp.

CEPT T/SF 29-83. Services Et Facilities Des Cartes De Telephone Debitrices. 2 pp.

CEPT T/SF 39 E-85. Credit Telephone Cards (Not Referring to General-Purpose Credit Cards). 2 pp.

CEPT T/SF 51 E-87. Use of Cards Which Require Identification in Telecommunication Services. 2 pp.

—Telephone—Automated

CCITT RECMN E.116-92. International Telecommunication Charge Card Service (Study Group I) 6 pp. 6 pp.

CCITT RECMN E.118-89. Automated International Telephone Credit Card System—Telephone Network and ISDN—Operation, Numbering, Routing and Mobile Service (Study Group II) 7 pp. 7 pp.

—Telephone—Automated—Accounting

CCITT RECMN E.118-89. Automated International Telephone Credit Card System—Telephone Network and ISDN—Operation, Numbering, Routing and Mobile Service (Study Group II) 7 pp. 7 pp.

—Telephone—Automated—Numbering Systems

CCITT RECMN E.118-89. Automated International Telephone Credit Card System—Telephone Network and ISDN—Operation, Numbering, Routing and Mobile Service (Study Group II) 7 pp. 7 pp.

—Telephone—Automated—Tariffs

CCITT RECMN D.120-89. Collection Charges Applied to Automated Telephone Credit Cards—General Tariff Principles—Charging and Accounting in International Telecommunications Services (Study Group III) 1 pp. 1 p.

CCITT RECMN E.118-89. Automated International Telephone Credit Card System—Telephone Network and ISDN—Operation, Numbering, Routing and Mobile Service (Study Group II) 7 pp. 7 pp.

—Telephone—Automated—Validation (Security)

CCITT RECMN E.113-89. Validation Procedures for an Automated International Telephone Credit Card System—Telephone Network and ISDN—Operation, Numbering, Routing and Mobile Service (Study Group II) 6 pp. 6 pp.

—Telephone—Document Formats—Registration

CCITT RECMN E.118-89. Automated International Telephone Credit Card System—Telephone Network and ISDN—Operation, Numbering, Routing and Mobile Service (Study Group II) 7 pp. 7 pp.

—Telephone—Numbering Systems

CCITT RECMN E.118-92. International Telecommunication Charge Card (Study Group I) 9 pp. 9 pp.

—Telephone—Numbering Systems—Accounting

CCITT RECMN E.116-89. International Telephone Credit Cards for Use in a Non-Automated Environment—Telephone Network and ISDN—Operation, Numbering, Routing and Mobile Service (Study Group II) 3 pp. 3 pp.

—Telephone—Person to Person Calls

CCITT RECMN E.116-89. International Telephone Credit Cards for Use in a Non-Automated Environment—Telephone Network and ISDN—Operation, Numbering, Routing and Mobile Service (Study Group II) 3 pp. 3 pp.

—Telephone—Station to Station Calls

CCITT RECMN E.116-89. International Telephone Credit Cards for Use in a Non-Automated Environment—Telephone Network and ISDN—Operation, Numbering, Routing and Mobile Service (Study Group II) 3 pp. 3 pp.

—Telephone—Tariffs

CCITT RECMN E.118-92. International Telecommunication Charge Card (Study Group I) 9 pp. 9 pp.

Credit Institutions

See Also: Balance of Payments; Banking; Financial Institutions

Credit Institutions (Cont.)

EC COM(87) 715-88. Proposal for a Second Council Directive on the Coordination of Laws, Regulations and Administrative Provisions Relating to the Taking-up and Pursuit of the Business of Credit Institutions and Amending Directive 77/780/EEC. 60 pp.

EC 89/646/EEC-89. Second Council Directive on the Coordination of Laws, Regulations and Administrative Provisions Relating to the Taking up and Pursuit of the Business of Credit Institutions and Amending Directive 77/780/EEC. 13 pp.

EC 92/30/EEC-92. Council Directive on the Supervision of Credit Institutions on a Consolidated Basis. 8 pp.

—Capital Funds

EC COM(86) 169/2-86. Proposal for a Council Directive on the Own Funds of Credit Institutions. 17 pp.

EC COM(88) 15-88. Amended Proposal for a Council Directive on the Own Funds of Credit Institutions. 18 pp.

EC COM(88) 194-88. Proposal for a Council Directive on a Solvency Ratio for Credit Institutions. 39 pp.

EC 89/299/EEC-89. Council Directive on the Own Funds of Credit Institutions. 5 pp.

EC 89/647/EEC-89. Council Directive on a Solvency Ratio for Credit Institutions. 9 pp.

—Deposit-Guarantee Schemes

EC COM(88) 4-88. Amended Proposal for a Council Directive Concerning the Reorganization and the Winding-up of Credit Institutions and Deposit-Guarantee Schemes. 53 pp.

—Liquidation

EC COM(85) 788-85. Proposal for a Council Directive on the Coordination of Laws, Regulations and Administrative Provisions Relating to the Reorganization and the Winding-up of Credit Institutions. 53 pp.

EC COM(88) 4-88. Amended Proposal for a Council Directive Concerning the Reorganization and the Winding-up of Credit Institutions and Deposit-Guarantee Schemes. 53 pp.

—Mortgages

EC COM(84) 730-85. Proposal for a Council Directive on the Freedom of Establishment and the Free Supply of Services in the Field of Mortgage Credit. 27 pp.

EC COM(87) 255-87. Amended Proposal for a Council Directive on the Freedom of Establishment and the Free Supply of Services in the Field of Mortgage Credit. 19 pp.

—Reorganization

EC COM(85) 788-85. Proposal for a Council Directive on the Coordination of Laws, Regulations and Administrative Provisions Relating to the Reorganization and the Winding-up of Credit Institutions. 53 pp.

EC COM(88) 4-88. Amended Proposal for a Council Directive Concerning the Reorganization and the Winding-up of Credit Institutions and Deposit-Guarantee Schemes. 53 pp.

—Solvency Ratio

EC COM(88) 194-88. Proposal for a Council Directive on a Solvency Ratio for Credit Institutions. 39 pp.

EC 89/647/EEC-89. Council Directive on a Solvency Ratio for Credit Institutions. 9 pp.

Credit Insurance

See Also: Insurance

EC 87/343/EEC-87. Council Directive Amending, as Regards Credit Insurance and Suretyship Insurance, First Directive 73/239/EEC on the Coordination of Laws, Regulations and Administrative Provisions Relating to the Taking-up and Pursuit of the Business of Direct Insurance. 5 pp.

Creep (Slipping)

Use: Slippage

Creep (Strain)

Use: Creep Properties

Creep Properties

Scope Note: For additional listings, use a more specific term *Use For:* Creep (Strain); Creep Testing
See Also: Cold Flow; Compression Testing; Creep Rupture Strength; Dimensional Stability; Ductility; Fatigue Testing; High Temperature Testing; Plastic Deformation; Plastic Properties; Residual Stress; Rheological Properties; Stresses; Tensile Testing

—Adhesives

BSI BS 5350: Part C7-90. 1990 Methods of Test for Adhesives Group C: Adhesively Bonded Joints: Mechanical Tests Part C7: Determination of Creep and Resistance to Sustained Application of Force. 4 pp.

INTERNATIONAL AND NON-U.S. NATIONAL STANDARDS
SUBJECT INDEX

Creep Properties (Cont.)

—Cellular Materials
DIN ENGL 53425-65. Testing of Rigid Foams; Time-Dependent Creep Compression Test Under Heat (Sept). 2 pp.

—Cellular Plastics
BSI BS 4443: Part 3-88. 1988 Methods of Test for Flexible Cellular Materials Part 3: Method 8, Determination of Creep Method 9, Determination of Dynamic Cushioning Performance. 12 pp.
ISO 10066-91. Flexible Cellular Polymeric Materials—Determination of Creep in Compression First Edition. 6 pp.

—Concretes
BSI BS 6319: Part 11-93. 1993 Testing of Resin and Polymer/Cement Compositions for Use in Construction Part 11: Methods for Determination of Creep in Compression and in Tension. 12 pp.
SAA AS 1012.16-74. Methods of Testing Concrete—Part 16: Determination of Creep of Concrete Cylinders in Compression. 9 pp.

—Dampproof Courses
BSI DD 86: Part 2-84. 1984 Damp-Proof Courses Part 2: Method of Test for Creep Deformation. 6 pp.

—Fiberboard
CEN PREN 1156-93. Wood-Based Panels—Determination of Duration of Load and Creep Factors. 11 pp.

—Footwear—Adhesives
BSI BS 5131: Sec 1.1-91. 1991 Methods of Test for Footwear and Footwear Materials Part 1: Adhesives Section 1.1: Resistance of Adhesive Joints to Heat (Creep Test). 9 pp.
BSI BS 5131: SUB SEC 1.1.1-76. (WITHDRAWN) 1976 Amd 1 Methods of Test for Footwear and Footwear Materials Part 1: Adhesives Section 1.1: Resistance of Adhesive Joints to Heat and to Peeling Subsec 1.1.1: Resist. to Heat (Creep Test) (Superseded by BS 5131: Section 1.1: 1991). 5 pp.
BSI BS 5131: Sec 1.3-91. 1991 Methods of Test for Footwear and Footwear Materials Part 1: Adhesives Section 1.3: Preparation of Test Assemblies Using Adhesives (Other Than Hot Melt Adhesives) for Heat Resistance (Creep) and Peel Tests. 15 pp.

—Geotextiles
BSI BS 6906: Part 5-91. 1991 Geotextiles Part 5: Determination of Creep. 10 pp.

—Metals
BSI BS 3500: Part 3-69. 1969 Methods for Creep and Rupture Testing of Metals Part 3: Tensile Creep Testing. 19 pp.
JIS Z 2271-78. Method of Tensile Creep Test for Metallic Materials. 8 pp.

—Mortars
BSI BS 6319: Part 11-93. 1993 Testing of Resin and Polymer/Cement Compositions for Use in Construction Part 11: Methods for Determination of Creep in Compression and in Tension. 12 pp.

—Particle Board
CEN PREN 1156-93. Wood-Based Panels—Determination of Duration of Load and Creep Factors. 11 pp.

—Plastics
BSI BS 4618: Sec 1.1-70. 1970 Recommendations for the Presentation of Plastics Design Data Part 1: Mechanical Properties Section 1.1: Creep. 7 pp.
BSI BS 4618: SUBSEC 1.1.1-70. 1970 Recommendations for the Presentation of Plastics Design Data Part 1: Mechanical Properties Section 1.1: Creep Subsection 1.1.1: Creep in Uniaxial Tension or Compression (with Particular Reference to Solid Plastics). 3 pp.
BSI BS 4618: SUBSEC 1.1.2-76. 1976 Amd 1 Recommendations for the Presentation of Plastics Design Data Part 1: Mechanical Properties Section 1.1: Creep Subsection 1.1.2: Creep in Flexureat Low Strains. 5 pp.
BSI BS 4618: SUBSEC 1.1.3-74. 1974 Recommendations for the Presentation of Plastics Design Data Part 1: Mechanical Properties Section 1.1: Creep Subsection 1.1.3: Creep Lateral Contraction Ratio (Poisson's Ratio). 3 pp.
DIN ENGL 53444-90. Testing the Tensile Creep of Plastics (Jan). 4 pp.
DIN ENGL 54852-86. Testing of Plastics; Determination of Flexural Creep of Plastics by Three-Point Loading and Four-Point Loading (Sept). 8 pp.
ISO 6602-85. Plastics—Determination of Flexural Creep by Three-Point Loading First Edition. 8 pp.
JIS K 7115-86. Testing Method for Tensile Creep of Plastics. 20 pp.

Creep Properties (Cont.)

—Plywood
CEN PREN 1156-93. Wood-Based Panels—Determination of Duration of Load and Creep Factors. 11 pp.

—Refractory Materials
BSI BS 1902: Sec 4.10-90. 1990 Methods for Testing Refractory Materials Part 4: Properties Measured Under an Applied Stress Section 4.10: Determination of Creep in Compression (Method 1902-410). 16 pp.
ISO 3187-89. Refractory Products—Determination of Creep in Compression First Edition. 12 pp.

—Steels
ISO R204-61. Non-Interrupted Creep Testing of Steel at Elevated Temperatures First Edition. 8 pp.

—Storage Tanks
CEN PREN 978-92. Underground Tanks of Glass-Reinforced Plastics (GRP)—Determination of Creep Factor and Ageing Factor. 10 pp.

—Textile Floor Coverings
CEN PREN 995-93. Textile Floorcoverings—Assessment of the Creep of the Backings—Test Simulating the Leg of a Heavy Piece of Furniture. 5 pp.

—Vulcanized Rubber
BSI BS 903: Part A15-90. 1990 Physical Testing of Rubber Part A15: Method for Determination of Creep in Compression or Shear (ISO 8013: 1988). 18 pp.
ISO 8013-88. Rubber, Vulcanized—Determination of Creep in Compression or Shear First Edition. 15 pp.

—Wood
CNS O2030-81. Method of Test for Creep of Wood (Mar) (7173). 4 pp.
JIS Z 2118-63. Method of Creep Test for Wood (R 1969). 6 pp.

Creep Rate
Use: Creep Properties

Creep Rupture Strength
Use For: Rupture Strength; Stress Rupture Strength
See Also: Creep Properties; Modulus of Rupture

—Adhesives
JIS K 6859-80. Testing Method for Creep Rupture of Adhesive Bonds. 11 pp.

—Hard Metals
BSI BS 5600: Sec 4.8-87. (WITHDRAWN) 1987 Powder Metallurgical Materials and Products Part 4: Methods of Testing and Chemical Analysis of Hardmetals Section 4.8: Determination of Transverse Rupture Strength (Superseded by BS EN 23326: 1993). 4 pp.
BSI BS EN 23327-93. 1993 Hardmetals—Determination of Transverse Rupture Strength (ISO 3327: 1982) (V). 9 pp.
CEN EN 23327-93. Hardmetals—Determination of Transverse Rupture Strength (ISO 3327: 1982). 4 pp.
ISO 3327-82. Hardmetals—Determination of Transverse Rupture Strength Second Edition (CEN EN 23327: 1993). 4 pp.

—Metals
BSI BS 3500: Part 1-69. 1969 Methods for Creep and Rupture Testing of Metals Part 1: Tensile Rupture Testing. 20 pp.
BSI 4A 4:Pt1: Sec 3-67. 1967 Amd 1 Test Pieces and Test Methods for Metallic Materials for Aircraft Part 1: Tensile Tests Section 3: Uninterrupted Creep and Rupture Tests. 17 pp.
CEN PREN 2002-5-91. Test Methods for Metallic Materials—Part 5—Uninterrupted Creep and Rupture Testing. 21 pp.
JIS Z 2272-78. Method of Creep Rupture Test for Metallic Materials. 8 pp.

—Plastic Fittings
CNS K6640-81. Method of Test for Short-Time Rupture Strength of Plastic Rope, Tubing and Fittings (Mar)(7051).

—Plastic Rope
CNS K6640-81. Method of Test for Short-Time Rupture Strength of Plastic Rope, Tubing and Fittings (Mar)(7051).

—Plastic Tubing
CNS K6640-81. Method of Test for Short-Time Rupture Strength of Plastic Rope, Tubing and Fittings (Mar)(7051).

Creep Rupture Strength (Cont.)

—Sintered Materials
BSI BS 5600: Sec 3.8-79. 1979 Powder Metallurgical Materials and Products Part 3: Methods of Testing Metal Materials and Products, Excluding Hardmetals Section 3.8: Determination of Transverse Rupture Strength. 4 pp.
ISO 3325-75. Sintered Metal Materials, Excluding Hardmetals—Determination of Transverse Rupture Strength First Edition. 4 pp.

—Steels
BSI BS 3228: Part 2-60. (WITHDRAWN) 1960 Procedures for Obtaining Properties of Steel at Elevated Temperatures Part 2: Rupture Strength (Superseded by ISO 6303). 8 pp.
BSI BS 3228: Part 3-60. (WITHDRAWN) 1960 Procedures for Obtaining Properties of Steel at Elevated Temperatures Part 3: Creep Strength (Superseded by ISO 6303). 8 pp.
BSI PD 6525: Part 1-90. 1990 Elevated Temperature Properties for Steels for Pressure Purposes Part 1: Stress Rupture Properties. 96 pp.
ISO R206-61. Creep Stress Rupture Testing of Steel at Elevated Temperatures First Edition. 8 pp.
ISO 6303-81. Pressure Vessel Steels Not Included in ISO 2604, Parts 1 to 6—Derivation of Long-Time Stress Rupture Properties First Edition. 7 pp.
ISO TR7468-81. Summary of Average Stress Rupture Properties of Wrought Steels for Boilers and Pressure Vessels First Edition. 50 pp.

Creep Rupture Testing
Use: Creep Rupture Strength

Creep Testers
See Also: Testing Equipment

—Force
BSI BS 1610: Part 3-90. 1990 Materials Testing Machines and Force Verification Equipment: Part 3: Grading of the Forces Applied by Deadweight and Lever Creep Testing Machines. 13 pp.

—Metals
DIN ENGL 51226-77. Material Testing Machines; Creep Testing Machines for Tensile Stress of Metallic Materials (Dec). 3 pp.

Creep Testing
Use: Creep Properties

Creepage Distances
Use: Leakage Paths

Crematoria
See Also: Buildings; Cemeteries

—Model Bylaws
SNZ NZS 9201: Chapter 14-72. Model General Bylaws Chapter 14: Cemeteries and Crematoria (Reconfirmed 1980). 27 pp.

Creosote
See Also: Coal Tar; Tars

—Coal Tar
CNS K5062-77. Coal-Tar for Blending with Creosote (Jun)(4128).

—Crude Oils
CNS K5063-77. Petroleum Oil for Blending with Creosote (Jun)(4129).

—Solvents
CNS K5034-90. Petroleum Solvent for Creosote Blending (Aug)(2265). 1 p.

—Wood Preservatives
BSI BS 144: Part 1-90. 1990 Wood Preservation Using Coal Tar Creosotes Part 1: Preservative. 23 pp.
BSI BS 144: Part 2-90. 1990 Wood Preservation Using Coal Tar Creosotes Part 2: Methods of Timber Treatment. 12 pp.
CEN PREN 1014-1-93. Wood Preservation—Creosote and Creosoted Timber—Methods of Sampling and Analysis—Part 1: Procedure for Sampling Creosote. 11 pp.
SAA AS 1143-73. High Temperature Creosote for the Preservation of Timber Reconfirmed 1988. 31 pp.

Creosote Oil
Use For: Coal Tar Creosote
JIS K 2425-83. Methods for Testing Creosote Oil, Prepared Tar and Tar Pitch.
JIS K 2439-83. Creosote Oil—Prepared Tar—Tar Pitch. 6 pp.

—Acidity
CNS K6074-57. Method of Test for Acidity of Creosote Oil (Jul)(917)(R 1971).

INDUSTRY STANDARDS

Creosote Oil (Cont.)

—Benzene Content
CNS K6071-57. Method of Test for Benzene Insoluble Substances of Creosote Oil (Jul)(914)(R 1971).

—Carbon Residue Testing
CNS K6072-57. Method of Test for Carbon Residue of Creosote Oil (Jul)(915)(R 1971).

—Crossties—Pressure Methods
JIS A 9104-92. Crossties Treated with Creosote Oil by Pressure Processes. 7 pp.

—Crystallization
CNS K6073-57. Method of Test for Crystallization of Creosote Oil (Jul)(916)(R 1971).

—Density
CNS K6067-57. Method of Sampling for Specific Gravity of Creosote Oil (Jul)(910)(R 1971).

—Distillation Methods
CNS K6070-57. Method of Test for Fractional Distillation of Creosote Oil (Jul)(913)(R 1971).

—Insoluble Matter Content
CNS K6071-57. Method of Test for Benzene Insoluble Substances of Creosote Oil (Jul)(914)(R 1971).

—Moisture Content
CNS K6068-57. Method of Test for Moisture of Creosote Oil (Jul)(911)(R 1971).

—Poles (Supports)
SNZ NZS/ASTM D390-86. Standard Specification for Coal-Tar Creosote for the Preservative Treatment of Piles, Poles, and Timbers for Marine, Land and Fresh Water Use. 2 pp.

—Poles (Supports)—Pressure Methods
JIS A 9101-92. Wood Poles Treated with Creosote Oil by Pressure Processes. 6 pp.

—Sampling
CNS K6066-57. Method of Sampling for Creosote Oil (Jul)(909)(R 1971).

—Structural Timber
SNZ NZS/ASTM D390-86. Standard Specification for Coal-Tar Creosote for the Preservative Treatment of Piles, Poles, and Timbers for Marine, Land and Fresh Water Use. 2 pp.

—Timber—Pressure Method
BSI BS 913-73. (WITHDRAWN) 1973 Wood Preservation by Means of Pressure Creosoting (W) (Superseded by BS 144: Parts 1 & 2: 1990). 16 pp.

—Viscosity
CNS K6069-57. Method of Test for Viscosity of Creosote Oil (Jul)(912)(R 1971).

—Wood Preservatives
BSI BS 144-73. (WITHDRAWN) 1973 Amd 1 Coal Tar Creosote for the Preservation of Timber (Superseded by BS 144: Parts 1 & 2: 1990). 15 pp.
BSI BS 3051-72. (WITHDRAWN) 1972 Coal Tar Creosotes for Wood Preservation (Other Than Creosotes to BS 144) (Superseded by BS 144: Parts 1&2: 1990). 26 pp.
CNS K5012-81. Wood Preservatives (Creosote) (May)(764). 1 p.

—Wooden Piles
SNZ NZS/ASTM D390-86. Standard Specification for Coal-Tar Creosote for the Preservative Treatment of Piles, Poles, and Timbers for Marine, Land and Fresh Water Use. 2 pp.

Crepe Papers
See Also: Electrical Insulating Papers; Papers; Tissue Papers

—Breaking Load
CPPA D.32P-90. Dry and Wet Tensile Breaking Strength and Stretch of Creped Papers. 2 pp.

—Bulk Density
BSI BS 7387-91. 1991 Method for Determination of the Bulking Thickness, Apparent Bulk Density, Compressibility and Compressibility Index of Soft Creped Tissue Paper. 12 pp.

—Compression Testing
BSI BS 7387-91. 1991 Method for Determination of the Bulking Thickness, Apparent Bulk Density, Compressibility and Compressibility Index of Soft Creped Tissue Paper. 12 pp.

—Electrical—Cellulosic
BSI BS 5626: Sec 3.3-82. 1982 Cellulosic Papers for Electrical Purposes Part 3: Specifications for Individual Materials Section 3.3: Crepe Paper. 8 pp.
IEC 554 Pt 3-3-80. Specification for Cellulosic Papers for Electrical Purposes Part 3: Specification for Individual Materials Sheet 3: Crepe Paper First Edition. 15 pp.

—Kraft
MOD UK DEF-1252-A-64. Paper, Kraft, Creped (Anti-Bleed) (Reprinted May, 1970 Incorporating Amendments Nos. 1 and 2). 5 pp.

—Kraft—Waterproof
MOD UK DEF-1239-A-66. Paper, Creped, Kraft Union, Reinforced. 6 pp.

—Kraft—Waxed
MOD UK DEF-1238-57. Paper, Kraft, Creped, Waxed (Reprinted June 1971, Incorporating Amendment No. 1). 3 pp.

—Kraft—Wet Strong
MOD UK DEF-1247-01. Paper, Kraft, Creped, High Wet-Strength; Amendment 1. 5 pp.

—Sterilization
BSI BS 6254-89. 1989 Creped Sterilization Paper for Medical Use. 10 pp.

—Stretching
CPPA D.32P-90. Dry and Wet Tensile Breaking Strength and Stretch of Creped Papers. 2 pp.

—Thickness Measurement
BSI BS 7387-91. 1991 Method for Determination of the Bulking Thickness, Apparent Bulk Density, Compressibility and Compressibility Index of Soft Creped Tissue Paper. 12 pp.

—Water Absorption
CPPA F.7P-90. Absorption of Water by Lightweight Creped Papers. 2 pp.

Cresol
Use For: Methyl Phenol See Also: Cresol Content Analysis; Hazardous Materials; Phenol
CNS K1085-64. M-Cresol for Industrial Use (May)(2327)(R 1971).
CNS K1087-64. Cresols (May)(2329)(R 1971).
JIS K 8304-88. o-Cresol.
JIS K 8305-88. m-Cresol.
JIS K 8306-88. p-Cresol.

—Crystallization
ISO 1897 Pt 11-83. Phenol, o-Cresol, m-Cresol, p-Cresol, Cresylic Acid and Xylenols for Industrial Use—Methods of Test—Part 11: Determination of Crystallizing Point (Excluding Cresylic Acid and Xylenols) First Edition. 6 pp.
ISO 2208-73. Phenol, o-Cresol, m-Cresol and p-Cresol for Industrial Use—Determination of Crystallizing Point After Drying with a Molecular Sieve First Edition. 4 pp.

—Distillation Methods
CNS K6214-65. Method of Fractional Distillation Test for Crude Methyl Phenol, M-Methyl Phenol and Coal Tar (High Boiling Point) (Sep)(2496)(R 1971).

—Hydrogen Sulfide Content
CNS K6219-65. Method of Test for Hydrogen Sulfide in Crude Methyl Phenol and M-Methyl Phenol (Sep)(2501)(R 1971).

—Impurities Content—Visual Testing
ISO 1897 Pt IV-77. Phenol, o-Cresol, m-Cresol, p-Cresol, Cresylic Acid and Xylenols for Industrial Use—Methods of Test—Part IV: Visual Test for Impurities Insoluble in Sodium Hydroxide Solution (Excluding Cresylic Acid and Xylenols) First Edition. 6 pp.

—Karl Fischer Method
ISO 1897 Pt I-77. Phenol, o-Cresol, m-Cresol, p-Cresol, Cresylic Acid and Xylenols for Industrial Use—Methods of Test—Part I: General First Edition. 4 pp.

—Neutral Oil Content—Volumetric Analysis
ISO 1897 Pt III-77. Phenol, o-Cresol, m-Cresol, p-Cresol, Cresylic Acid and Xylenols for Industrial Use—Methods of Test—Part III: Determination of Neutral Oils and Pyridine Bases First Edition. 7 pp.

—Pyridine Base Content—Volumetric Analysis
ISO 1897 Pt III-77. Phenol, o-Cresol, m-Cresol, p-Cresol, Cresylic Acid and Xylenols for Industrial Use—Methods of Test—Part III: Determination of Neutral Oils and Pyridine Bases First Edition. 7 pp.

—Quantitative Analysis
CNS K6218-65. Method of Quantitative Analysis of M-Methyl Phenol (Sep)(2500)(R 1971).

—Residue Content—Evaporation
ISO 1897 Pt 10-82. Phenol, o-Cresol, m-Cresol, p-Cresol, Cresylic Acid and Xylenols for Industrial Use—Methods of Test—Part 10: Determination of Dry Residue After Evaporation on a Water Bath (Excluding Cresylic Acid and Xylenols) First Edition. 4 pp.

—Water Content—Dean and Stark Method
ISO 1897 Pt II-77. Phenol, o-Cresol, m-Cresol, p-Cresol, Cresylic Acid and Xylenols for Industrial Use—Methods of Test—Part II: Determination of Water-Dean and Stark Method First Edition. 7 pp.

Cresol Content Analysis
See Also: Cresol

—Cresylic Acids
ISO 1897 Pt IX-77. Phenol, o-Cresol, m-Cresol, p-Cresol, Cresylic Acid and Xylenols for Industrial Use—Methods of Test—Part IX: Determination of m-Cresol Content (Cresylic Acid Only) First Edition. 6 pp.

—Phenol Formaldehyde Resins
DIN ENGL 53748-70. Chemical Analysis of Phenol-Formaldehyde Resins, Phenoplastic Moulding Materials and Moulded Materials (July). 3 pp.

—Phenolic Resins
DIN ENGL 53748-70. Chemical Analysis of Phenol-Formaldehyde Resins, Phenoplastic Moulding Materials and Moulded Materials (July). 3 pp.

—Waste Water
CNS K9095-82. Method of Test for P-Cresols in Industrial Waste Water (Jul)(9182).

—Xylenols
ISO 1897 Pt VIII-77. Phenol, o-Cresol, m-Cresol, p-Cresol, Cresylic Acid and Xylenols for Industrial Use—Methods of Test—Part VIII: Determination of o-Cresol Content (Cresylic Acid and Xylenols Only) First Edition. 7 pp.

Cresol Red
CNS K7159-66. Chemical Reagent (Cresol Red) (Mar)(1659).
JIS K 8308-72. Cresol Red.

Cresol Red Index

—Soybeans
BSI BS 4325: Part 7-80. 1980 Analysis of Oilseed Residues Part 7: Determination of Cresol Red Index in Soya Bean Products. 4 pp.
ISO 5514-79. Soya Bean Products—Determination of Cresol Red Index First Edition. 4 pp.

o-Cresolphthalein
JIS K 8307-72. o-Cresolphthalein.

o-Cresotic Acid
Use For: o-Cresotinic Acid
CNS K2177-87. O-Cresotinic Acid (2-Hydroxy-3-Methylbenzoic Acid) (Dec)(12187). 3 pp.

o-Cresotinic Acid
Use: o-Cresotic Acid

Cresylic Acids
See Also: Phenol
JIS K 2422-87. Methods for Testing Phenols. 54 pp.
JIS K 2437-87. Phenols (Phenol-Cresol and Cresylic Acid-Xylenols).

—Color Testing
ISO 1897 Pt VII-77. Phenol, o-Cresol, m-Cresol, p-Cresol, Cresylic Acid and Xylenols for Industrial Use—Methods of Test—Part VII: Measurement of Colour (Cresylic Acid and Xylenols Only) First Edition. 4 pp.

—m-Cresol Content
ISO 1897 Pt IX-77. Phenol, o-Cresol, m-Cresol, p-Cresol, Cresylic Acid and Xylenols for Industrial Use—Methods of Test—Part IX: Determination of m-Cresol Content (Cresylic Acid Only) First Edition. 6 pp.

—o-Cresol Content
ISO 1897 Pt VIII-77. Phenol, o-Cresol, m-Cresol, p-Cresol, Cresylic Acid and Xylenols for Industrial Use—Methods of Test—Part VIII: Determination of o-Cresol Content (Cresylic Acid and Xylenols Only) First Edition. 7 pp.

—Density
ISO 1897 Pt I-77. Phenol, o-Cresol, m-Cresol, p-Cresol, Cresylic Acid and Xylenols for Industrial Use—Methods of Test—Part I: General First Edition. 4 pp.

INTERNATIONAL AND NON-U.S. NATIONAL STANDARDS
SUBJECT INDEX

Cresylic Acids (Cont.)
—Distillation Methods
ISO 1897 Pt 12-83. Phenol, o-Cresol, m-Cresol, p-Cresol, Cresylic Acid and Xylenols for Industrial Use—Methods of Test—Part 12: Determination of Distillation Characteristics (Cresylic Acid and Xylenols Only) First Edition. 4 pp.

—Hydrogen Sulfide Content
ISO 1897 Pt VI-77. Phenol, o-Cresol, m-Cresol, p-Cresol, Cresylic Acid and Xylenols for Industrial Use—Methods of Test—Part VI: Test for Absence of Hydrogen Sulphide (Cresylic Acid and Xylenols Only) First Edition. 4 pp.

—Karl Fischer Method
ISO 1897 Pt I-77. Phenol, o-Cresol, m-Cresol, p-Cresol, Cresylic Acid and Xylenols for Industrial Use—Methods of Test—Part I: General First Edition. 4 pp.

—Neutral Oil Content—Volumetric Analysis
ISO 1897 Pt III-77. Phenol, o-Cresol, m-Cresol, p-Cresol, Cresylic Acid and Xylenols for Industrial Use—Methods of Test—Part III: Determination of Neutral Oils and Pyridine Bases First Edition. 7 pp.

—Pyridine Base Content—Volumetric Analysis
ISO 1897 Pt III-77. Phenol, o-Cresol, m-Cresol, p-Cresol, Cresylic Acid and Xylenols for Industrial Use—Methods of Test—Part III: Determination of Neutral Oils and Pyridine Bases First Edition. 7 pp.

—Residue Content—Distillation Methods
ISO 1897 Pt 13-83. Phenol, o-Cresol, m-Cresol, p-Cresol, Cresylic Acid and Xylenols for Industrial Use—Methods of Test—Part 13: Determination of Residue on Distillation (Cresylic Acid and Xylenols Only) First Edition. 4 pp.

—Water Content—Dean and Stark Method
ISO 1897 Pt II-77. Phenol, o-Cresol, m-Cresol, p-Cresol, Cresylic Acid and Xylenols for Industrial Use—Methods of Test—Part II: Determination of Water-Dean and Stark Method First Edition. 7 pp.

Cribs (Furniture)
Use For: Baby Beds; Cots, Children's
See Also: Bassinets; Beds; Furniture
CEN PREN 1130-2-93. Furniture—Cribs and Cradles for Domestic Use—Part 2: Test Methods. 12 pp.
CNS S1201-86. Baby Bed (Aug)(11676).
CNS S2137-87. Method of Test for Baby Bed (Jan)(11821).
JIS S 1103-84. Baby Bed.

—Comprehensive Testing
ISO 7175 Pt 2-88. Furniture—Children's Cots—Safety Requirements and Testing—Part 2: Test Methods First Edition. 11 pp.

—Hospitals
BSI BS 1694-90. 1990 Amd 5 Hospital Ward Cots for Children. 16 pp.

—Mattresses
BSI BS 1877: Part 10-82. 1982 Amd 2 Domestic Bedding Part 10: Specification for Mattresses and Bumpers for Children's Cots, Perambulators and Similar Domestic Articles (AMD 6593) April 30, 1991. 12 pp.
BSI BS 1877: Part 10-03. 1982 Amd 3 Domestic Bedding Part 10: Specification for Mattresses and Bumpers for Children's Cots, Perambulators and Similar Domestic Articles (AMD 6949) February 28, 1992. 13 pp.

—Portable
SNZ NZS 5844-89. Carry Cots and Stands (Safety Requirements) Amend: A, 1989. 12 pp.

—Quality Assurance
CEN PREN 716-2-92. Furniture—Children's and Nursery Furniture—Children's Cots and Folding Cots for Domestic Use—Part 2: Test Methods. 18 pp.

—Safety
BSI BS 1753-87. 1987 Amd 1 Safety Requirements for Children's Cots for Domestic Use (AMD 6265) January 31, 1990. 18 pp.
BSI BS 1753-02. 1987 Amd 2 Safety Requirements for Children's Cots for Domestic Use (AMD 7657) June 15, 1993. 21 pp.
BSI BS 7423-91. 1991 Safety Requirements for Children's Travel Cots of Internal Base Length Not Less Than 900 mm. 16 pp.
BSI BS 7423-01. 1991 Amd 1 Safety Requirements for Children's Travel Cots of Internal Base Length Not Less Than 900 mm (AMD 7656) June 15, 1993. 17 pp.

Cribs (Furniture) (Cont.)
—Safety (Cont.)
CEN PREN 716-1-92. Furniture—Children's and Nursery Furniture—Children's Cots and Folding Cots for Domestic Use—Part 1: Safety Requirements. 9 pp.
CEN PREN 1130-1-93. Furniture—Cribs and Cradles for Domestic Use—Part 1: Safety Requirements. 6 pp.
SAA AS 2172-91. Cots for Household Use (Safety Requirements). 15 pp.
SNZ NZS 5810-92. Cots for Household Use (Safety Requirements) Amend: 1, 1992. 15 pp.

Cricket Balls
See Also: Cricket Equipment
—Leather
BSI BS 5993-87. 1987 Leather-Covered Cricket Balls. 15 pp.

Cricket Equipment
See Also: Cricket Balls; Sports Equipment
—Protective Clothing
BSI BS 6183: Part 1-81. 1981 Protective Equipment for Cricketers Part 1: Specification for Batting Gloves, Leg Guards and Boxes. 11 pp.

Crimp Contacts
Use For: Solderless Contacts *See Also:* Electric Contacts
BSI BS 5310: Part 2A-76. 1976 Amd 1 Hand Crimping Tools for the Termination of Electrical Cables and Wires for Low Frequency and Radio Frequency Applications Part 2: Hand Crimping Tools (Fixed Die, Sizes A to E), for Radio Frequency Connectors and Concentric Contacts. 9 pp.
BSI BS 5310: Part 2B-81. 1981 Hand Crimping Tools for the Termination of Electrical Cables and Wires for Low Frequency and Radio Frequency Applications Part 2B: Hand Crimping Tools (Removable and Inter-Changeable Dies, Sizes A to G & Q to S) for Radio Freq Conn & Conc Contact. 8 pp.
BSI BS 5310: Part 3A-88. 1988 Hand Crimping Tools for the Termination of Electrical Cables and Wires for Low Frequency and Radio Frequency Applications Part 3A: Hand Crimping Tools for Contacts of Electrical Connectors. 24 pp.
IEC 203-66. Dimensions of the Crimp Area of Machined Crimp Type Contacts First Edition. 5 pp.

—Aerospace
CEN PREN 3155-004-91. Electrical-Contacts, Used in Elements of Connection Part 004—Contact, Electrical, Male 004 Type A, Crimp, Class T Product Standard. 11 pp.
CEN PREN 3155-004-93. Aerospace Series Electrical Contacts Used in Elements of Connection Part 004—Contacts, Electrical, Male 004, Type A, Crimp, Class T Product Standard. 7 pp.
CEN PREN 3155-005-91. Electrical Contacts Used in Elements of Connection Part 005—Contact, Electrical, Female 005, Type A, Crimp, Class T Product Standard. 11 pp.
CEN PREN 3155-005-93. Aerospace Series Electrical Contacts Used in Elements of Connection Part 005—Contacts, Electrical, Female 005, Type A, Crimp, Class T Product Standard. 7 pp.
CEN PREN 3155-014-91. Electrical Contacts Used in Elements of Connection Part 014-Contacts, Electrical, Male 014 Type A, Crimp, Class S Product Standard. 8 pp.
CEN PREN 3155-014-93. Aerospace Series Electrical Contacts Used in Elements of Connection Part 014—Contacts, Electrical, Male 014, Type A, Crimp, Class S Product Standard. 6 pp.
CEN PREN 3155-015-91. Electrical Contacts Used in Elements of Connection Part 015—Contacts, Electrical, Female 015, Type A, Crimp, Class S Product Standard. 8 pp.
CEN PREN 3155-015-93. Aerospace Series Electrical Contacts Used in Elements of Connection Part 015—Contacts, Electrical, Female 015, Type A, Crimp, Class S Product Standard. 6 pp.
CEN PREN 3155-018-91. Electrical Contacts Used in Elements of Connection Part 018—Contacts, Electrical, Male 018, Type A, Crimp, Class S Product Standard. 11 pp.
CEN PREN 3155-018-93. Aerospace Series Electrical Contacts Used in Elements of Connection Part 018—Contacts, Electrical, Male 018, Type A, Crimp, Class S Product Standard. 8 pp.
CEN PREN 3155-019-91. Electrical Contacts Used in Elements of Connection Part 019-Contacts, Electrical, Female 019, Type A, Crimp, Class S Product Standard. 11 pp.
CEN PREN 3155-019-93. Aerospace Series Electrical Contacts Used in Elements of Connection Part 019—Contacts, Electrical, Female 019, Type A, Crimp, Class S Product Standard. 8 pp.

Crimp Contacts (Cont.)
—Aerospace (Cont.)
CEN PREN 3155-022-91. Electrical Contacts Used in Elements of Connection Part 022—Contacts, Electrical, Male 022, Type A, Crimp, Class R Product Standard. 7 pp.
CEN PREN 3155-022-93. Aerospace Series Electrical Contacts Used in Elements of Connection Part 022—Contacts, Electrical, Male 022, Type A, Crimp, Class R Product Standard. 5 pp.
CEN PREN 3155-023-91. Electrical Contacts Used in Elements of Connection Part 023—Contacts, Electrical, Female 023, Type A, Crimp, Class R Product Standard. 8 pp.
CEN PREN 3155-023-93. Aerospace Series Electrical Contacts Used in Elements of Connection Part 023—Contacts, Electrical, Female 023, Type A, Crimp, Class R Product Standard. 8 pp.
CEN PREN 3155-026-91. Electrical Contacts Used in Elements of Connection Part 026—Contacts, Electrical, Male 026, Type A, Crimp, Class R Product Standard. 9 pp.
CEN PREN 3155-026-93. Aerospace Series Electrical Contacts Used in Elements of Connection Part 026—Contacts, Electrical, Male 026, Type A, Crimp, Class R Product Standard. 7 pp.
CEN PREN 3155-027-91. Electrical Contacts Used in Elements of Connection Part 027—Contacts, Electrical, Female 027, Type A, Crimp, Class R Product Standard. 9 pp.
CEN PREN 3155-027-93. Aerospace Series Electrical Contacts Used in Elements of Connection Part 027—Contacts, Electrical, Female 027, Type A, Crimp, Class R Product Standard. 9 pp.
CEN PREN 3155-042-93. Aerospace Series Electrical Contacts Used in Elements of Connection Part 042—Contacts, Electrical, Triaxial, Size 08, Male 042, Type D, Solder, Class P Product Standard. 11 pp.

—Aerospace—Connectors
CEN PREN 2591 (Part D17)-92. Elements of Electrical and Optical Connection Test Methods Part D17—Tensile Strength (Crimped Connection). 4 pp.

—Aerospace—Connectors—Deformation
CEN PREN 2591 (Part E3)-92. Elements of Electrical and Optical Connection Test Methods Part E3—Contact Deformation After Crimping. 4 pp.

—Aircraft—Cables (Electric)
BSI 5G 178: Part 1-93. 1993 Crimped Joints for Aircraft Electrical Cables and Wires Part 1: Specification for Design Requirements (Including Tests) for Components and Tools. 15 pp.
BSI 4G 178: Part 1-84. (WITHDRAWN) 1984 Amd 1 Crimped Joints for Aircraft Electrical Cables and Wires Part 1: Design Requirements (Including Tests) for Components and Tools (Superseded by 5G 178: Part 1: 1993). 11 pp.
BSI 4G 178: Part 2-86. 1986 Crimped Joints for Aircraft Electrical Cables and Wires Part 2: Control of Crimping (Including User Control Tests). 10 pp.
BSI G 184-G188-64. (OBSOLESCENT) 1964 Amd 1 Aluminium Terminal Ends and Inline Connectors for Hexagonal Crimping to Aircraft Aluminium Electric Cables. 17 pp.
BSI G 204-67. 1967 Amd 1 Copper Terminal Ends for Crimping to Electric Cables with Copper Conductors. 13 pp.
ISO 1965-73. Aluminium Terminal Ends for Crimping to Aircraft Aluminium Electrical Cables First Edition. 4 pp.
ISO 1966-73. Crimped Joints for Aircraft Electrical Cables First Edition. 9 pp.

—Aircraft—Cables (Electric)—Thermocouples
BSI 2G 215: Part 2-89. 1989 Extension Cables for Aircraft Nickel-Chromium and Nickel-Aluminium Thermocouple Part 2: Terminations. 6 pp.
BSI G 220-76. (WITHDRAWN) 1976 Amd 1 Crimped Terminations to Stranded Flexible Thermocouple Cables of the Nickel-Chromium and Nickel-Aluminium Types (Superseded by BS 2G 215: Part 2: 1989). 5 pp.
ISO 8056 Pt 2-88. Aircraft—Nickel-Chromium and Nickel-Aluminium Thermocouple Extension Cables—Part 2: Terminations—General Requirements and Tests First Edition. 7 pp.
ISO 8056 Pt 3-87. Aircraft—Nickel-Chromium and Nickel-Aluminium Thermocouple Extension Cables—Part 3: Crimp-Type Ring Terminal Ends—Dimensions First Edition. 4 pp.

—Aircraft—Connectors
SBAC TS 355 ISSUE 1. Test for Electrical Connectors Contact Deformation After Crimping.

Crimp

Crimp Contacts (Cont.)
—Aircraft—Electric Wire
BSI 5G 178: Part 1-93. 1993 Crimped Joints for Aircraft Electrical Cables and Wires Part 1: Specification for Design Requirements (Including Tests) for Components and Tools (S). 15 pp.

BSI 4G 178: Part 1-84. (WITHDRAWN) 1984 Amd 1 Crimped Joints for Aircraft Electrical Cables and Wires Part 1: Design Requirements (Including Tests) for Components and Tools (Superseded by 5G 178: Part 1: 1993). 11 pp.

BSI 4G 178: Part 2-86. 1986 Crimped Joints for Aircraft Electrical Cables and Wires Part 2: Control of Crimping (Including User Control Tests). 10 pp.

—Aircraft—Identification Systems
ISO 8843-91. Aircraft—Crimp-Removable Contacts for Electrical Connectors—Identification System First Edition. 5 pp.

—Circular Connectors—Quality Assurance
BSI BS 9522 N0001-82. 1982 Amd 1 Detail Specification for MultiContact Circular Electrical Connectors. Bayonet Coupling with Front Release Rear Removable Crimp Contacts. Full Assessment (AMD 6821) September 30, 1991. 48 pp.

BSI BS 9522 N0003-90. 1990 Multi-Contact Circular Electrical Connectors Bayonet Coupling Non-Barrier Sealed, Environment Resistant with Rear Insertable, Rear Release, Rear Removable Crimp Contacts Also Barrier Sealed with Non-Removable Solder Contact Styles. 142 pp.

—Connectors
IEC 807 Pt 3-90. Rectangular Connectors for Frequencies Below 3 MHz Part 3: Detail Specification for a Range of Connectors with Trapezoidal Shaped Metal Shells and Round Contacts—Removable Crimp Contact Types with Closed Crimp Barrels, Rear Insertion/Rear Extraction First Edition. 61 pp.

—Copper Conductors
CNS C4118-90. Non-Insulated Crimp-Style Terminals for Copper Conductors (Dec)(3434).

CNS C4185-90. Non-Insulated Crimp-Style Connecting Sleeves for Copper Conductors (Dec)(5518).

Crimping
See Also: Bulking; Crimping Tools

—Connectors
MOD UK DSTAN 59-71: Part 0-85. Crimped Electrical Connections Part 0: Guide to the Crimping of Electrical Connections for Low Frequency and Radio Frequency Applications Issue 1. 15 pp.

—Rigidity—Yarns
BSI BS 6663-86. 1986 Determination of Crimp Rigidity of Textured Nylon Yarns. 7 pp.

—Yarns
DIN ENGL 53840 Pt 1-83. Testing of Textiles; Determination of Parameters for the Crimp of Textured Filament Yarns; Filament Yarns with a Linear Density of up to 500 dtex (Nov). 6 pp.

DIN ENGL 53840 Pt 2-83. Testing of Textiles; Determination of Parameters for the Crimp of Textured Filament Yarns; Filament Yarns with a Linear Density Exceeding 500 dtex (Nov). 5 pp.

DIN ENGL 53852-78. Testing of Textiles; Determination of the Length Ratios (Crimp) of Yarns in Fabrics (Aug). 4 pp.

—Yarns—Fabrics
BSI BS 2863-84. 1984 Determination Of Crimp of Yarn in Fabric. 4 pp.

ISO 7211 Pt 3-84. Textiles—Woven Fabrics—Construction—Methods of Analysis—Part 3: Determination of Crimp of Yarn in Fabric First Edition. 5 pp.

Crimping Compounds
—Aluminum Wire
MOD UK DTD-5503-54. Cable Crimping Compound (for Aluminium Cables) (Reprinted December 1965). 1 p.

Crimping Tools
See Also: Crimping; Hand Tools; Tools

BSI 5G 178: Part 1-93. 1993 Crimped Joints for Aircraft Electrical Cables and Wires Part 1: Specification for Design Requirements (Including Tests) for Components and Tools (S). 15 pp.

BSI 4G 178: Part 1-84. (WITHDRAWN) 1984 Amd 1 Crimped Joints for Aircraft Electrical Cables and Wires Part 1: Design Requirements (Including Tests) for Components and Tools (Superseded by 5G 178: Part 1: 1993). 11 pp.

CNS B3463-88. Crimming Tools (Wire Strippers) (Jun)(12334). 7 pp.

Crimping Tools (Cont.)
CNS C3076-80. Method of Test for Compression Tools for Wire Connectors of Interior Wiring (Apr)(5418).

CNS C4174-80. Compression Tools for Wire Connectors of Interior Wiring (Apr)(5417).

JIS C 9711-90. Compression Tools for Wire Connectors of Interior Wiring. 22 pp.

—Cables (Electric)
CEN PREN 2242-91. Control of Tools Used for Crimping of Electric Cables with Conductors Defined by EN 2083 and EN 2346. 10 pp.

—Coaxial Connectors
BSI BS 5310: Part 2A-76. 1976 Amd 1 Hand Crimping Tools for the Termination of Electrical Cables and Wires for Low Frequency and Radio Frequency Applications Part 2: Hand Crimping Tools (Fixed Die, Sizes A to E), for Radio Frequency Connectors and Concentric Contacts. 9 pp.

BSI BS 5310: Part 2B-81. 1981 Hand Crimping Tools for the Termination of Electrical Cables and Wires for Low Frequency and Radio Frequency Applications Part 2B: Hand Crimping Tools (Removable and Inter-Changeable Dies, Sizes A to G & Q to S) for Radio Freq Conn & Conc Contact. 8 pp.

—Connectors
BSI BS 5310: Part 1-76. 1976 Hand Crimping Tools for the Termination of Electrical Cables and Wires for Low Frequency and Radio Frequency Applications Part 1: General Requirements and Tests. 7 pp.

—Crimp Contacts
BSI BS 5310: Part 2A-76. 1976 Amd 1 Hand Crimping Tools for the Termination of Electrical Cables and Wires for Low Frequency and Radio Frequency Applications Part 2: Hand Crimping Tools (Fixed Die, Sizes A to E), for Radio Frequency Connectors and Concentric Contacts. 9 pp.

BSI BS 5310: Part 2B-81. 1981 Hand Crimping Tools for the Termination of Electrical Cables and Wires for Low Frequency and Radio Frequency Applications Part 2B: Hand Crimping Tools (Removable and Inter-Changeable Dies, Sizes A to G & Q to S) for Radio Freq Conn & Conc Contact. 8 pp.

BSI BS 5310: Part 3A-88. 1988 Hand Crimping Tools for the Termination of Electrical Cables and Wires for Low Frequency and Radio Frequency Applications Part 3A: Hand Crimping Tools for Contacts of Electrical Connectors. 24 pp.

—Hand
DIN ENGL 41641 Pt 1-83. Tools for Solderless Electrical Connections; Hand Crimping Tools; Concepts, Requirements, Testing (Jan). 6 pp.

—Marine
MOD UK NES 640-88. Guide to the Selection of Hand Held Crimping Tools Issue 1 (11.88). 31 pp.

—User Certification
CNS C7109-80. Tools, Crimping, Solderless Wiring Devices Procedures for User Certification (May)(5551).

Crinkle Washers
See Also: Spring Lock Washers; Washers (Fasteners)

BSI BS 3401-61. 1961 Amd 1 Beryllium Copper Crinkle Washers. 7 pp.

BSI BS 4463-69. 1969 Crinkle Washers for General Engineering Purposes (Metric Series). 11 pp.

CNS B2178-78. Spring Washers Curved or Crinkle Type (Mar)(4252).

CNS B2227-78. Spring Washers (Jun)(4406). 3 pp.

DIN ENGL 137-87. Curved and Wave Spring Washers (Oct). 3 pp.

SNZ NZS/BS 4463-69. Specification for Crinkle Washers for General Engineering Purposes. Metric Series. 12 pp.

—Aerospace
BSI SP 134-138-68. 1968 Copper Beryllium Crinkle Washers. 4 pp.

BSI SP 139-140-69. 1969 Amd 1 Corrosion-Resisting Steel Crinkle Washers. 4 pp.

—Screw Assemblies
DIN ENGL 6900 Pt 2-90. Screw and Washer Assemblies; Coarse Threaded Screws with Captive Wave Spring Washer (Dec). 4 pp.

Critical Micelle Concentration Analysis
—Cationic Surfactants
ISO 6840-82. Cationic Surface Active Agents (Hydrochlorides and Hydrobromides)—Determination of Critical Micellization Concentration—Method by Measurement of Counter Ion Activity First Edition. 5 pp.

—Surfactants—Surface Tension
ISO 4311-79. Anionic and Non-Ionic Surface Active Agents—Determination of the Critical Micellization Concentration—Method by Measuring Surface Tension with a Plate, Stirrup or Ring First Edition. 5 pp.

Critical Micellization Concentration
Use: Critical Micelle Concentration Analysis

Criticality
See Also: Nuclear Reactor Safety

—Fissionable Materials
BSI BS 3598-70. 1970 Recommendations for Criticality Safety in Handling and Processing Fissile Materials. 12 pp.

ISO 1709-75. Nuclear Energy—Fissile Materials—Principles of Criticality Safety in Handling and Processing First Edition. 5 pp.

—Warning Systems
IEC 860-87. Warning Equipment for Criticality Accidents First Edition. 41 pp.

ISO 7753-87. Nuclear Energy—Performance and Testing Requirements for Criticality Detection and Alarm Systems First Edition. 12 pp.

CRM
Use: Collision Avoidance—Aircraft

CRO
Use: Oscilloscopes

Crochet Hooks
JIS B 9093-78. Crochet Needles for Hand Knitting Machines (R 1983). 5 pp.

Crochet Needles
Use: Crochet Hooks

Crop Dryers
Use For: Grain Dryers *See Also:* Agricultural Equipment; Dryers; Separators (Mechanical)

BSI BS 3986-66. 1966 Methods of Test for Agricultural Grain Driers. 45 pp.

JIS B 9219-87. Standard Form of Specifications for Circulation Type Grain Dryers. 10 pp.

—Comprehensive Testing
BSI BS 3986-91. 1991 Methods of Test for Drying Performance of Agricultural Grain Dryers. 29 pp.

BSI BS 3986-66. 1966 Methods of Test for Agricultural Grain Driers. 45 pp.

—Oil—Portable
CSA B140.8-1967. Portable Industrial Oil-Fired Heaters (R 1991). 22 pp.

Cross Bitts
Use: Bitts

Cross Connect Equipment
See Also: Telecommunication Equipment

—Synchronous—64 kbits
CCITT RECMN G.796-92. Characteristics of a 64 kBit/s Cross-Connect Equipment with 2048 kBit/s Access Ports (Study Group XV) 13 pp. 13 pp.

Cross Cores
Use: Magnetic Cores

Cross Country Skis
Use: Skis—Cross Country

Cross Cut Testing
See Also: Physical Testing

—Paints
BSI BS 3900: Part E6-92. 1992 Methods of Test for Paints Part E6: Cross-Cut Test (ISO 2409: 1992). 13 pp.

BSI BS 3900: Part E6-74. 1974 Amd 1 Methods of Test for Paints Group E: Mechanical Tests on Paint Films Part E6: Cross-Cut Test. 6 pp.

ISO 2409-92. Paints and Varnishes—Cross-Cut Test Second Edition. 12 pp.

—Varnishes
BSI BS 3900: Part E6-92. 1992 Methods of Test for Paints Part E6: Cross-Cut Test (ISO 2409: 1992). 13 pp.

INTERNATIONAL AND NON-U.S. NATIONAL STANDARDS
SUBJECT INDEX

Cross Cut Testing *(Cont.)*
—Varnishes *(Cont.)*
BSI BS 3900: Part E6-74. 1974 Amd 1 Methods of Test for Paints Group E: Mechanical Tests on Paint Films Part E6: Cross-Cut Test. 6 pp.
ISO 2409-92. Paints and Varnishes—Cross-Cut Test Second Edition. 12 pp.

Cross Field Amplifier Tubes
Use: Crossed-Field Amplifier Tubes

Cross Modulation
See Also: Crosstalk; Intermodulation; Modulation; Radio Frequency Interference
—**Ionospheric Propagation**
CCIR Report 574-2-86. Ionospheric Cross-Modulation—Section 6A—Ionospheric Properties. 9 pp.
—**Ionospheric—Sound Broadcasting**
CCIR RECMN 498-2-90. Ionospheric Cross-Modulation in the LF and MF Broadcasting Bands—Section 10A-1—Amplitude-Modulation Sound Broadcasting in Bands 5 (LF), 6 (MF) and 7 (HF). 9 pp.
CCIR QUESTION 95/10-90. Ionospheric Cross-Modulation—Limitation of Radiated Power in LF and MF Broadcasting—Questions Concerning Study Group 10—Broadcasting Service (Sound). 1 p.

Cross Polarization
See Also: Polarization (Waves); Polarization Discrimination
—**Atmospheric—Radio Wave Propagation**
CCIR Report 722-3-90. Cross-Polarization Due to the Atmosphere—Section 5C—Effects of the Atmosphere (Radiometeorology). 11 pp.
—**Earth Stations—Antennas**
CCIR Report 1141-90. Polarization Discrimination in Interference Calculation—Section 4D1—Permissible Levels of Interference. 20 pp.
—**Very Small Aperture Terminals—Earth Stations**
CCIR RECMN 727-92. Cross-Polarization Isolation from Very Small ApertureTerminals (VSATs)—Section 4B1—Systems Aspects. 1 p.

Cross Recessed Bolts
Use: Phillips Bolts

Cross Recessed Screws
Use: Phillips Screws

Cross Servicing (Military)
Use: Interdepartmental Procurement

Crossarms
See Also: Poles (Supports); Power Poles
SAA AS O20-48. Grading Rules for Cross-Arms.
—**Hardwoods**
SAA AS O61-55. Cross-Arms from Eastern and South-Eastern Australian Hardwoods. 7 pp.
—**Wood**
CSA O116-1969. Power and Communication Sawn Wood Crossarms. 43 pp.
SNZ NZS 485-69. Specification for New South Wales Hardwoods. 38 pp.

Crosscut Saws
See Also: Hand Tools; Saws
MOD UK DSTAN 51-11: Part 24-76. Hand Tools, General Purpose Part 24: Saws, Hand, Crosscutting Saw, Hand, Ripping and Saws, Tenon Issue 1. 5 pp.
—**Saw Blades**
CNS B1339-83. Woodworking Crosscut Saws (Tooth Shapes) (Dec)(10682).
CNS B3445-83. Saw Blades for Woodworking Crosscut Saws (Dec)(10681).

Crossdyeing
—**Wool Fibers—Colorfastness Testing**
CNS L3182-82. Method of Test for Colour Fastness to Cross-Dyeing of Wool (Aug)(9305).
JIS L 0872-75. Testing Method for Colour Fastness to Cross-Dyeing of Wool.

Crossed-Field Amplifier Tubes
Use For: Cross Field Amplifier Tubes
See Also: Microwave Tubes
—**Electrical Measurement**
IEC 235 Pt 9-75. Measurement of the Electrical Properties of Microwave Tubes Part 9: Crossed-Field Amplifier Tubes First Edition. 23 pp.

Crossed-Field Amplifier Tubes *(Cont.)*
—**Preferred Products List**
CECC CECC MUAHAG Vol 12 IS 1-90. Preferred Products List; Microwave Components (En, Fr, Ge). 76 pp.

Crosses (Pipe Fittings)
Use: Pipe Crosses

Crosses (Tube Fittings)
Use: Tube Fittings

Crosspoint Switches
See Also: Electronic Switches; Linear Circuits; Switches
CECC CECC 90 104-209 ISSUE 1-83. BS CECC 90 104-209; 4 x 4 Crosspoint Switch with Control Memory (En). 30 pp.
CECC CECC 90 104-209 ISSUE 1-86. CEI-CECC 90 104-209; 4 x 4 Crosspoint Switch with Control Memory (En). 2 pp.

Crosstalk
Use For: Linear Crosstalk *See Also:* Cross Modulation; Electromagnetic Interference; Electromagnetic Noise; Interference Cancellation Systems; Radio Frequency Interference
—**Carrier Systems—Telephone Transmission**
CCITT RECMN G.221-89. Overall Recommendations Relating to Carrier-Transmission Systems—International Analogue Carrier Systems (Study Group XV) 2 pp. 2 pp.
—**Communication Terminal Equipment—Groups**
CCITT RECMN G.232-89. 12-Channel Terminal Equipments—International Analogue Carrier Systems (Study Group XV) 13 pp. 13 pp.
CCITT RECMN G.235-89. 16-Channel Terminal Equipments—International Analogue Carrier Systems (Study Group XV) 4 pp. 4 pp.
—**Echo Suppressors**
CCITT RECMN G.164-89. Echo Suppressors—General Characteristics of International Telephone Connections and Circuits (Study Groups XII and XV) 36 pp. 36 pp.
—**Fiber Optics**
CNS C6286-87. Method of Test for Fiber Optic Devices (FOTP-42 Optical Crosstalk in Fiber Optic Components) (Jun)(11992).
—**Frequency Translation Equipment**
CCITT RECMN G.233-89. Recommendations Concerning Translating Equipments—International Analogue Carrier Systems (Study Group XV) 10 pp. 10 pp.
—**Groups—Carrier Systems**
CCITT RECMN G.313-89. Open-Wire Lines for Use with 12-Channel Carrier Systems—International Analogue Carrier Systems (Study Group XV) 3 pp. 3 pp.
—**Maritime Mobile Satellite Communications—Automatic Services**
CCITT RECMN G.473-89. Interconnection of a Maritime Mobile Satellite System with the International Automatic Switched Telephone Service; Transmission Aspects—International Analogue Carrier Systems (Study Group XV) 10 pp. 10 pp.
—**Sound-Program Circuits—Carrier Systems**
CCITT RECMN J.18-89. Crosstalk in Sound-Programme Circuits Set up on Carrier Systems—Line Transmission of Non-Telephone Signals—Transmission of Sound-Programme and Television Signals (Study Group XV) 3 pp. 3 pp.
—**Telephone Cables**
CCITT RECMN G.611-89. Characteristics of Symmetric Cable Pairs for Analogue Transmission—Transmission Media Characteristics (Study Group XV) 5 pp. 5 pp.
CCITT RECMN G.612-89. Characteristics of Symmetric Cable Pairs Designed for the Transmission of Systems with Bit Rates of the Order of 6 to 34 Mbit/s—Transmission Media Characteristics (Study Group XV) 4 pp. 4 pp.
CCITT RECMN G.613-89. Characteristics of Symmetric Cable Pairs Usable Wholly for the Transmission of Digital Systems with a Bit Rate of up to 2 Mbits—Transmission Media Characteristics (Study Group XV) 4 pp. 4 pp.
CCITT RECMN G.614-89. Characteristics of Symmetric Pair Star-Quad Cables Designed Earlier for Analogue Transmission Systems and Being Used Now for Digital System Transmission at Bit Rates of 6 to 34 Mbit/s—Transmission Media Characteristics (Study Group XV) 5 pp. 5 pp.

Crosstalk *(Cont.)*
—**Telephone Cables** *(Cont.)*
CCITT RECMN G.621-89. Characteristics of 0.7/2.9 mm Coaxial Cable Pairs—Transmission Media Characteristics (Study Group XV) 4 pp. 4 pp.
CCITT RECMN G.622-89. Characteristics of 1.2/4.4 mm Coaxial Cable Pairs—Transmission Media Characteristics (Study Group XV) 8 pp. 8 pp.
CCITT RECMN G.623-89. Characteristics of 2.6/9.5 mm Coaxial Cable Pairs—Transmission Media Characteristics (Study Group XV) 8 pp. 8 pp.
CCITT FASCICLE III.5-89. Digital Networks, Digital Sections and Digital Line Systems—Recommendations G.801—G.961. 292 pp.
—**Telephone Circuits**
CCITT FASCICLE III.1. General Characteristics of International Telephone Connections and Circuits Recommendations G.101-G.181. 332 pp.
CCITT RECMN G.134-89. Linear Crosstalk—General Characteristics of International Telephone Connections and Circuits (Study Groups XII and XV) 2 pp. 2 pp.
CCITT RECMN G.151-89. General Performance Objectives Applicable to All Modern International Circuits and National Extension Circuits—General Characteristics of International Telephone Connections and Circuits (Study Groups XII and XV) 6 pp. 6 pp.
—**Telephone Circuits—Measurement**
CCITT RECMN G.134-89. Linear Crosstalk—General Characteristics of International Telephone Connections and Circuits (Study Groups XII and XV) 2 pp. 2 pp.
—**Telephone Circuits—Sound-Program Circuits**
CCITT FASCICLE III.6-89. Line Transmission of Non-Telephone Signals Transmission of Sound-Programme and Television Signals—Series H and J Recommendations. 237 pp.
—**Telephone Exchanges—Measurement**
CCITT RECMN G.134-89. Linear Crosstalk—General Characteristics of International Telephone Connections and Circuits (Study Groups XII and XV) 2 pp. 2 pp.
—**Telephone Networks—Hypothetical Reference Model**
CCITT RECMN G.105-89. Hypothetical Reference Connection for Crosstalk Studies—General Characteristics of International Telephone Connections and Circuits (Study Groups XII and XV) 6 pp. 6 pp.
—**Telephone Repeaters**
CCITT RECMN G.611-89. Characteristics of Symmetric Cable Pairs for Analogue Transmission—Transmission Media Characteristics (Study Group XV) 5 pp. 5 pp.
—**Telephone Repeaters—Carrier Systems**
CCITT RECMN G.312-89. Intermediate Repeaters for Open-Wire Carrier Systems Conforming to Recommendation G.311—International Analogue Carrier Systems (Study Group XV) 2 pp. 2 pp.
CCITT RECMN G.313-89. Open-Wire Lines for Use with 12-Channel Carrier Systems—International Analogue Carrier Systems (Study Group XV) 3 pp. 3 pp.
CCITT RECMN G.322-89. General Characteristics Recommended for Systems on Symmetric Pair Cables —International Analogue Carrier Systems (Study Group XV) 8 pp. 8 pp.
—**Telephone Signals**
CCITT RECMN G.227-89. Conventional Telephone Signal—International Analogue Carrier Systems (Study Group XV) 3 pp. 3 pp.
—**Telephone Transmission—Quality of Service**
CCITT RECMN P.16-89. Subjective Effects of Direct Crosstalk; Tresholds of Audibility and Intelligibility—Telephone Transmission Quality (Study Group XII) 8 pp. 8 pp.
—**Through Connection Filters**
CCITT RECMN G.242-89. Through-Connection of Groups, Supergroups, Etc.—Analogue Carrier Systems (Study Group XV) 7 pp. 7 pp.
—**Through Group Filters**
CCITT RECMN G.242-89. Through-Connection of Groups, Supergroups, Etc.—Analogue Carrier Systems (Study Group XV) 7 pp. 7 pp.
—**Through Mastergroup Filters**
CCITT RECMN G.242-89. Through-Connection of Groups, Supergroups, Etc.—Analogue Carrier Systems (Study Group XV) 7 pp. 7 pp.

Crosstalk (Cont.)
—Through Supergroup Filters
CCITT RECMN G.242-89. Through-Connection of Groups, Supergroups, Etc.—Analogue Carrier Systems (Study Group XV) 7 pp. 7 pp.
—Through Supermastergroup Filters
CCITT RECMN G.242-89. Through-Connection of Groups, Supergroups, Etc.—Analogue Carrier Systems (Study Group XV) 7 pp. 7 pp.
—Through 15 Supergroup Assembly Filters
CCITT RECMN G.242-89. Through-Connection of Groups, Supergroups, Etc.—Analogue Carrier Systems (Study Group XV) 7 pp. 7 pp.

Crossties
See Also: Railroad Equipment
—Wood Preservatives—Creosote Oil—Pressure Method
JIS A 9104-92. Crossties Treated with Creosote Oil by Pressure Processes. 7 pp.

Crowbars (Electrical)
See Also: Electrical Protection Equipment; Overvoltage Protection Equipment
—Triggered
MOD UK DSTAN 59-60: 90/119-88. Valve Electronic, Triggered Crowbar Device Issue 1. 12 pp.

Crowbars (Tools)
Use For: Claw Bars; Pry Bars; Wrecking Bars
See Also: Tools
CGSB 39-GP-45A-74. Bars, Steel, Standard for; (Amendment 1 Mar 1975). 18 pp.
CNS B3202-84. Crowbars (Dec)(3150).
CNS B3467-88. Crowbars (Jun)(12338).
—Fire Fighting
CNS B3413-84. Crowbars (for Fire Fighting Use) (Dec)(7344).
—Tires
CNS B3437-88. Tire Iron (Jun)(10420).

Crown Bottle Openers
Use: Bottle Openers

Crown Gears
See Also: Gears
CNS B2557-81. Remote Control Gear with Transmission Tubing for Manual Operation; Crown Gear Units Slewable (Apr)(7188).
—Deck Machinery
CNS B2592-81. Remote Control Gear with Transmission Tubing for Manual Operation; Deck Operation Devices with Slewable Crown Gear Unit Welding-In Type (Jun)(7589).
—Glossaries
CNS B1191-84. Definition and Denominations for Glossary of Gear Terms — the Basic Crown Gears (May)(5733).
—Hoods
CNS B2589-81. Remote Control Gear with Transmission Tubing for Manual Operation; Protecting Hoods; for Crown Gear Units Slewable (Jun)(7586).

Crown Glass
See Also: Glass; Glassware
—Borosilicate—Optical
CNS R2147-86. Borosilicate Crown Optical Glass (Feb) (8541).

Crown Head Bolts
See Also: Bolts
CNS A2100-81. Crowning Bolts (Mar)(6996).

Crown Metal Bearings
See Also: Bearings
—Marine
MOD UK NES 836-88. Requirements for Crown Metal Cast Bearings Issue 1 (08.88). 16 pp.

Crown Nuts
Use: Cap Nuts

Crowns (Dental)
Use: Dental Materials

CRT
Use: Cathode Ray Tubes

CRT Displays
Use: Cathode Ray Tube Displays

Crucible Induction Furnaces
Use: Induction Furnaces—Crucible

Crucibles
See Also: Containers; Laboratory Equipment
—Glass
CNS R2055-69. Glass Melting Crucibles (Jul)(3060).
JIS R 2801-60. Crucible for Glass.
—Graphite
JIS R 2701-77. Graphite Crucible and Its Accessories (R 1985). 10 pp.
—High Frequency
CNS R2105-80. High Frequency Combustion Crucibles for Chemical Analysis (Sep)(6519).
JIS R 1308-80. High Frequency Combustion Crucibles for Chemical Analysis.
—Platinum
CNS K0026-82. Platinum Crucibles (Nov)(9624).
JIS H 6201-86. Platinum Crucibles for Chemical Analysis.
JIS H 6201-76. Platinum Crucibles. 8 pp.
—Porcelain
CNS R2123-81. Porcelain Crucibles for Chemical Analysis (Jan) (6960).
ISO 1772-75. Laboratory Crucibles in Porcelain and Silica First Edition. 6 pp.
JIS R 1301-87. Porcelain Crucibles for Chemical Analysis. 7 pp.
—Refractories—Slag Corrosion
JIS R 2214-75. Testing Method for Slag Corrosion of Refractories Using Crucibles (R 1988). 6 pp.
—Silica
ISO 1772-75. Laboratory Crucibles in Porcelain and Silica First Edition. 6 pp.

Cruciform Recessed Bolts
Use: Phillips Bolts

Cruciform Recessed Screws
Use: Phillips Screws

Crude Fats
Use: Fats

Crude Oil
Use For: Petroleum; Petroleum Liquid; Petroleum Oil
See Also: Bunker Fuel Oil; Fossil Fuels; Fuel Oils; Natural Gas; Oil Pipelines; Oil Wells; Petroleum Gases; Petroleum Industry; Waxes
JIS K 2601-90. Testing Methods for Crude Petroleum. 70 pp.
—Ash Content
CEN EN 7-74. Determination of Ash from Petroleum Products. 5 pp.
JIS K 2272-85. Testing Methods for Ash and Sulfated Ash of Crude Oil and Petroleum Products.
—Asphaltene Content
BSI BS 2000: Part 143-93. 1993 Methods of Test for Petroleum and Its Products Part 143: Determination of Asphaltenes (Heptane Insolubles) (W). 6 pp.
BSI BS 2000: Part 143-85. 1985 Petroleum and Its Products Part 143: Asphaltenes in Petroleum Products (Precipitation with Normal Heptane). 8 pp.
—Carbon Residue Testing
JIS K 2270-90. Crude Petroleum and Petroleum Products—Determination of Carbon Residue. 21 pp.
—Chemical Analysis—Distillation Methods
DIN ENGL 51567-73. Testing of Mineral Oils; Determination of the Composition of Petroleum; Fractional Distillation According to Grosse-Oetringhaus (Sept). 9 pp.
—Cloud Point
CNS K6914-87. Method of Test for Cloud Point of Petroleum Oils (Jul)(12015).
JIS K 2269-87. Testing Methods for Pour Point and Cloud Point of Crude Oil and Petroleum Products. 19 pp.
—Corrosion Inhibitors—Hydraulic Equipment—Aerospace
MOD UK DTD-5540B-75. Corrosion Preventive Oil: Hydraulic System NATO Code Number: C-635 Joint Service Designation: PX-26. 8 pp.
—Creosote
CNS K5063-77. Petroleum Oil for Blending with Creosote (Jun)(4129).

Crude Oil (Cont.)
—Demulsibility
ISO 6614-83. Petroleum Oils and Synthetic Fluids—Determination of Demulsibility Characteristics First Edition. 6 pp.
—Density
BSI BS 4699-85. 1985 Methods for Determination of Density or Relative Density of Petroleum Products (Pyknometer Methods). 16 pp.
BSI BS 4714-80. 1980 Laboratory Determination of Density or Relative Density of Crude Petroleum and Liquid Petroleum Products (Hydrometer Method). 9 pp.
CNS K6916-90. Method of Test for Density, Relative Density, or API Gravity of Crude Petroleum and Liquid Petroleum Products by Hydrometer Method (Aug)(12017).
ISO 3675-93. Crude Petroleum and Liquid Petroleum Products—Laboratory Determination of Density or Relative Density—Hydrometer Method Second Edition. 11 pp.
ISO 3838-83. Crude Petroleum and Liquid or Solid Petroleum Products—Determination of Density or Relative Density—Capillary-Stoppered Pyknometer and Graduated Bicapillary Pyknometer Methods First Edition. 13 pp.
JIS K 2249-87. Testing Methods for Density of Crude Oil and Petroleum Products, and Petroleum Measurement Tables Based on Reference Temperature 15 Degrees C.
—Emulsification
CNS K6340-73. Method of Test for Emulsion Characteristics of Petroleum Oils and Synthetic Fluids (Feb)(3518).
—Flash Point
JIS K 2265-89. Testing Methods for Flash Point of Crude Oil and Petroleum Products. 35 pp.
—Hydrocarbon Content—Volumetric Analysis
BSI BS 7340-91. 1991 Schedule for Compressibility Factors for Petroleum Hydrocarbons in the Density Range 638kg/m3 to 1074 kg/m3. 14 pp.
ISO 9770-89. Crude Petroleum and Petroleum Products—Compressibility Factors for Hydrocarbons in the Range 638 kg/m3 to 1 074 kg/m3 First Edition; (Corrected and Reprinted -1989). 11 pp.
—Nickel Content—Atomic Absorption Spectroscopy
BSI BS 2000: Part 288-93. 1993 Methods of Test for Petroleum and Its Products Part 288: Determination of Nickel, Sodium and Vanadium—Atomic Absorption Spectroscopy Method (W). 5 pp.
BSI BS 2000: Part 288-83. 1983 Petroleum and Its Products Part 288: Sodium, Nickel and Vanadium in Fuel Oils and Crude Oils by Atomic Absorption Spectroscopy. 6 pp.
—Nitrogen Content
JIS K 2609-90. Crude Petroleum and Petroleum Products—Determination of Nitrogen Content. 50 pp.
—Pour Point (Viscosity)
BSI BS 2000: Part 15-93. 1993 Methods of Test for Petroleum and Its Products Part 15: Determination of Pour Point. 5 pp.
BSI BS 2000: Part 15-82. 1982 Petroleum and Its Products Part 15: Pour Point of Petroleum Oils. 8 pp.
JIS K 2269-87. Testing Methods for Pour Point and Cloud Point of Crude Oil and Petroleum Products. 19 pp.
—Reserves
JIS M 1006-75. Calculation of Reserves for Crude Oil and Natural Gas.
—Sampling
CNS K6108-60. Method of Sampling for Petroleum Oil and Petroleum Products (May)(1217)(R 1971).
DIN ENGL 51750 Pt 2-90. Sampling of Liquid Petroleum Products (Dec). 5 pp.
JIS K 2251-91. Crude Petroleum and Petroleum Products—Sampling. 52 pp.
—Sediment Content
BSI BS 2882-80. (WITHDRAWN) 1980 Determination of Water and Sediment in Crude Petroleum and Fuel Oils (Centrifuge Method). 8 pp.
BSI BS 4382-80. 1980 Determination of Sediment in Crude Petroleum and Fuel Oils (Extraction Method). 6 pp.
CNS K6576-80. Method of Test for Sediment in Crude and Fuel Oils by Extraction (Aug)(6357).
CNS K6577-80. Method of Test for Water and Sediment in Crude Oils and Fuel Oils by Centrifuge (Aug)(6358).

INTERNATIONAL AND NON-U.S. NATIONAL STANDARDS
SUBJECT INDEX

Crude Oil (Cont.)
—Sediment Content (Cont.)
ISO 3734-76. Crude Petroleum and Fuel Oils—Determination of Water and Sediment—Centrifuge Method First Edition. 6 pp.

ISO 3735-75. Crude Petroleum and Fuel Oils—Determination of Sediment—Extraction Method First Edition. 5 pp.

—Sediment Content—Centrifuging
DIN ENGL 51793-71. Testing of Liquid Fuels; Determination of Water Content and Sediments in Fuel Oils and Crude Oils; Centrifuge Method (July). 3 pp.

ISO 9030-90. Crude Petroleum—Determination of Water and Sediment—Centrifuge Method First Edition. 12 pp.

—Sodium Content—Atomic Absorption Spectroscopy
BSI BS 2000: Part 288-93. 1993 Methods of Test for Petroleum and Its Products Part 288: Determination of Nickel, Sodium and Vanadium—Atomic Absorption Spectroscopy Method (W). 5 pp.

BSI BS 2000: Part 288-83. 1983 Petroleum and Its Products Part 288: Sodium, Nickel and Vanadium in Fuel Oils and Crude Oils by Atomic Absorption Spectroscopy. 6 pp.

—Sulfur Content
DIN ENGL 51400 Pt 1-78. Testing of Mineral Oils and Fuels; Determination of the Sulfur Content (Total Sulfur); General Working Principles (Feb). 6 pp.

DIN ENGL 51400 Pt 3-78. Testing of Mineral Oils and Fuels; Determination of the Sulfur Content (Total Sulfur); Combustion According to Schoniger; Thorin-Sulfonazo-III Titration (Feb). 3 pp.

DIN ENGL 51400 Pt 4-90. Determination of Total Sulfur Content of Gaseous Petroleum Products by the Lingener Combustion Method (Oct). 8 pp.

DIN ENGL 51400 Pt 8-78. Testing of Mineral Oils and Fuels; Determination of Sulfur Content (Total Sulfur); Nickel Reduction Method; Dithizone Titration (Feb). 4 pp.

JIS K 2541-92. Crude Oil and Petroleum Products—Determination of Sulfur Content. 77 pp.

—Sulfur Content—Analyzers
JIS B 7995-84. Automatic Analyzers for Sulphur in Crude Oil and Petroleum Products. 24 pp.

—Vanadium Content—Atomic Absorption Spectroscopy
BSI BS 2000: Part 288-93. 1993 Methods of Test for Petroleum and Its Products Part 288: Determination of Nickel, Sodium and Vanadium—Atomic Absorption Spectroscopy Method (W). 5 pp.

BSI BS 2000: Part 288-83. 1983 Petroleum and Its Products Part 288: Sodium, Nickel and Vanadium in Fuel Oils and Crude Oils by Atomic Absorption Spectroscopy. 6 pp.

—Vapor Pressure
CEN PREN12-88. Pretroleum Products; Determination of Reid Vapour Pressure; Wet Method. 26 pp.

ISO 3007-86. Petroleum Products—Determination of Vapour Pressure—Reid Method Second Edition. 15 pp.

JIS K 2258-87. Testing Method for Vapor Pressure of Crude Oil and Petroleum Products (Reid Method). 21 pp.

—Viscosity
JIS K 2283-83. Testing Methods for Kinematic Viscosity and Calculating Method for Viscosity Index of Crude Oil and Petroleum Products. 75 pp.

—Water Content
BSI BS 2882-80. (WITHDRAWN) 1980 Determination of Water and Sediment in Crude Petroleum and Fuel Oils (Centrifuge Method). 8 pp.

BSI BS 4385: Part 1-91. 1991 Methods for Determination of Water in Crude Petroleum, Petroleum Products and Bituminous Materials by Distillation Part 1: Method for Crude Petroleum (ISO 9020: 1990). 17 pp.

CNS K6577-80. Method of Test for Water and Sediment in Crude Oils and Fuel Oils by Centrifuge (Aug)(6358).

ISO 3734-76. Crude Petroleum and Fuel Oils—Determination of Water and Sediment—Centrifuge Method First Edition. 6 pp.

ISO 9029-90. Crude Petroleum—Determination of Water—Distillation Method First Edition. 12 pp.

JIS K 2275-89. Testing Methods for Water Content of Crude Oil and Petroleum Products. 36 pp.

—Water Content—Centrifuging
DIN ENGL 51793-71. Testing of Liquid Fuels; Determination of Water Content and Sediments in Fuel Oils and Crude Oils; Centrifuge Method (July). 3 pp.

Crude Oil (Cont.)
—Water Content—Centrifuging (Cont.)
ISO 9030-90. Crude Petroleum—Determination of Water and Sediment—Centrifuge Method First Edition. 12 pp.

Crude Protein Content Analysis
Use: Protein Content Analysis

Crushed Stone
See Also: Aggregates; Rocks

—Concretes
CNS A1027-80. Crushed Stone for Concrete (Aug)(6299). 6 pp.

JIS A 5005-87. Crushed Stone for Concrete. 9 pp.

—Impact Testing
DIN ENGL 52115 Pt 2-88. Determination of Impact Resistance of Mineral Aggregates; Testing of Crushed Stone (Aug) (This Standard, Together with DIN 52115 Part 1, August 1988 Edition, Supersedes Parts of DIN 52109, October 1939 Edition, Withdrawn in 1985). 2 pp.

—Paving Materials
CNS A1026-86. Crushed Stone for Road Construction (Dec)(6298).

JIS A 5001-88. Crushed Stone for Road Construction. 9 pp.

—Sampling
CNS A3004-59. Method of Sampling for Block Stone, Crushed Stone, Gravel Sand and Furnace Slag (Oct)(485) (R 1970).

Crusher Gages
See Also: Gages; Manometers

—Ballistics
MOD UK DSTAN 13-36: Part 8-91. Ballistic Standardization of Gun Ammunition Part 8: Crushers and Crusher Gauges Issue 1. 32 pp.

—Pressure Measurement
NATO STANAG 4113 ED 2 AMD 0-76. NATO Crusher Gauge. 19 pp.

NATO STANAG 4113 ED 3 AMD 0-93. Pressure Measurement by Crusher Gauges. 4 pp.

NATO STANAG 4113 ED 4 AMD 0-00. Pressure Measurement by Crusher Gauges. 4 pp.

Crushers
See Also: Materials Handling Equipment

—Plastic
CNS B7194-82. Method of Test for Plastic Crusher (Jul)(9070). 2 pp.

Crushing Strength Testing
Use: Compression Testing

Crustacea
See Also: Aquatic Organisms; Crab; Daphnia; Fish; Mollusks; Shellfish; Shrimps

—Health Inspection
EC 91/67/EEC-91. Council Directive Concerning the Animal Health Conditions Governing the Placing on the Market of Aquaculture Animals and Products. 21 pp.

—Microbiological Analysis
SAA AS 1766.3.5-83. Methods for the Microbiological Examination of Food—Part 3: Examination of Specific Products—Part 3.5: Molluscs, Crustaceans and Fish, and Products Thereof. 4 pp.

Crutches
See Also: Axilla Crutches; Disability Equipment; Medical Equipment

—Elbow
BSI BS 4988-90. 1990 Elbow Crutches. 12 pp.

SNZ NZS 5831: Part 2-89. Walking Aids for the Disabled Part 2: Specification for Elbow Crutches. 12 pp.

—Wooden
BSI BS 4997-91. 1991 Wooden Axilla Crutches. 14 pp.

JIS T 9204-82. Wooden Axilla Crutches.

Cryogenic Containers
See Also: Containers; Cylinders (Containers)

—Airborne—Identification Systems—Interdepartmental Procurement
NATO STANAG 3056 ED 5 AMD 3-83. Marking of Airborne and Ground Gas and Cryogenic Fluid Containers. 7 pp.

Cryogenic Containers (Cont.)
—Airborne—Identification Systems—Interdepartmental Procurement (Cont.)
NATO STANAG 3056 ED 5 AMD 4-83. Marking of Airborne and Ground Gas and Cryogenic Fluid Containers. 9 pp.

Cryogenic Coolers
Use: Cryogenic Refrigerators

Cryogenic Fluid Storage Equipment
See Also: Cryogenic Fluids; Cryogenics

—Storage Tanks
BSI BS 5387-76. (WITHDRAWN) 1976 Vertical Cylindrical Welded Storage Tanks for Low-Temperature Services: Double-Wall Tanks for Temperature Down to-196 Degrees Celsius (Superseded by BS 7777: Parts 1, 2, 3 and 4: 1993). 57 pp.

BSI BS 5429-76. 1976 Code of Practice for Safe Operation of Small-Scale Storage Facilities for Cryogenic Liquids. 9 pp.

BSI BS 7777: Part 4-93. 1993 Flat-Bottomed, Vertical, Cylindrical Storage Tanks for Low Temperature Service Part 4: Specification for the Design and Construction of Single Containment Tanks for the Storage of Liquid Oxygen, Liguid Nitrogen or Liquid Argon (Q). 16 pp.

Cryogenic Fluids
Use For: Cryogenic Liquids *See Also:* Cryogenic Fluid Storage Equipment; Fluids

—Safety
SAA AS 1894-76. Code of Practice for the Safe Handling of Cryogenic Fluids Amdt 1 October 1979. 36 pp.

—Tank Trucks
SAA AS 2809.6-91. Road Tank Vehicles for Dangerous Goods—Part 6: Tankers for Cryogenic Liquids. 16 pp.

Cryogenic Liquids
Use: Cryogenic Fluids

Cryogenic Refrigerators
Use For: Cryogenic Coolers *See Also:* Refrigerators

—Integral Stirling—Military
NATO STANAG 4345 ED 1 AMD 0-89. (DRAFT) 1/3 Watt Integral Stirling Cryogenic Cooler. 9 pp.

NATO STANAG 4346 ED 1 AMD 0-89. (DRAFT) Dispositif De Refroidissement Cryogenique Type Stirling Integral 1 W. 9 pp.

—Split Stirling—Military
NATO STANAG 4343 ED 1 AMD 0-89. (DRAFT) 1/3 Watt Split Stirling Cryogenic Cooler. 9 pp.

NATO STANAG 4344 ED 1 AMD 0-89. 1 Watt Split Stirling Cryogenic Cooler. 9 pp.

Cryogenic Valves
See Also: Valves

BSI BS 6364-84. 1984 Amd 1 Valves for Cryogenic Service (AMD 6591) April 30, 1991. 13 pp.

Cryogenics
See Also: Cryogenic Fluid Storage Equipment

—Metals—Tensile Testing
JIS Z 2277-90. Tensile Testing Method for Metallic Materials in Liquid Helium. 15 pp.

Cryolite
See Also: Aluminum; Aluminum Ores; Minerals; Ores; Potassium Cryolite; Sodium Fluoride

MOD UK DSTAN 68-59-90. Cryolite Issue 2. 8 pp.

—Aluminum Content
MOD UK M 874/68. Examination of Cryolite.

—Aluminum Content—Atomic Absorption Analysis
CNS M3118-83. Method for Determining Aluminium Content in Natural and Artificial Cryolite (Atomic Absorption Method) (Mar)(10117).

—Aluminum Content—Atomic Absorption Spectrophotometry
BSI BS 5050-74. 1974 Methods of Test for Cryolite. 20 pp.

ISO 2830-73. Cryolite, Natural and Artificial—Determination of Aluminium Content—Atomic Absorption Method First Edition. 5 pp.

—Aluminum Content—Gravimetric Analysis
BSI BS 5050-74. 1974 Methods of Test for Cryolite. 20 pp.

INDUSTRY STANDARDS

INTERNATIONAL AND NON-U.S. NATIONAL STANDARDS
SUBJECT INDEX

Cryolite

Cryolite (Cont.)
—**Aluminum Content—Gravimetric Analysis** *(Cont.)*
- CNS M3117-83. Method for Determining Aluminium Content in Natural and Artificial Cryolite (Oxine (8-Hydroxyquinoline) Gravimetric Method) (Mar)(10116).
- CNS M3118-83. Method for Determining Aluminium Content in Natural and Artificial Cryolite (Atomic Absorption Method) (Mar)(10117).
- ISO 2367-72. Cryolite (Natural and Artificial)—Determination of Aluminium Content—8-Hydroxyquinoline Gravimetric Method First Edition. 5 pp.

—**Calcium Content—Atomic Absorption Spectrophotometry**
- BSI BS 5050: Part 8-80. 1980 Methods of Test for Cryolite Part 8: Determination of Calcium Content. 7 pp.
- CNS M3119-83. Method for Determining Calcium Content in Natural and Artificial Cryolite (Flame Atomic Absorption Method) (Mar)(10118).
- ISO 3391-76. Cryolite, Natural and Artificial—Determination of Calcium Content—Flame Atomic Absorption Method First Edition. 6 pp.

—**Fluoride Content**
- CNS M3120-83. Method for Determining Free Fluorides Content in Natural and Artificial Cryolite (Conventional Test) (Mar)(10119).
- ISO 4277-77. Cryolite, Natural and Artificial—Conventional Test for Evaluation of Free Fluorides Content First Edition. 5 pp.

—**Fluorine Content**
- BSI BS 5050-74. 1974 Methods of Test for Cryolite. 20 pp.
- MOD UK M 874/68. Examination of Cryolite.

—**Fluorine Content—Willard Winter Method**
- CNS M3114-83. Method for Determining Fluorine Content in Natural and Artificial Cryolite (Modified Willard Method) (Mar)(10113).
- ISO 1693-76. Cryolite, Natural and Artificial—Determination of Fluorine Content—Modified Willard-Winter Method First Edition. 8 pp.

—**Insoluble Matter Content**
- MOD UK M 874/68. Examination of Cryolite.

—**Iron Content**
- MOD UK M 874/68. Examination of Cryolite.

—**Iron Content—Photometry**
- BSI BS 5050-74. 1974 Methods of Test for Cryolite. 20 pp.
- CNS M3115-83. Method for Determining Iron Content in Natural and Artificial Cryolite (1.10-Phenanthroline Photometric Method) (Mar)(10114)).
- ISO 1694-76. Cryolite, Natural and Artificial—Determination of Iron Content—1,10-Phenanthroline Photometric Method First Edition. 7 pp.

—**Moisture Content—Gravimetric Analysis**
- BSI BS 5050: Part 10-80. 1980 Methods of Test for Cryolite Part 10: Determination of Moisture Content (Gravimetric Method). 7 pp.
- CNS M3122-83. Method for Determining Moisture Content in Natural and Artificial Cryolite and Aluminium Fluoride for Industrial Use (Gravimetric Method) (Mar)(10121).
- ISO 3393-76. Cryolite, Natural and Artificial, and Aluminium Fluoride for Industrial Use—Determination of Moisture Content—Gravimetric Method First Edition. 4 pp.

—**Moisture Content—Karl Fischer Method**
- BSI BS 5050: Part 9-80. 1980 Methods of Test for Cryolite Part 9: Determination of Moisture Content (Karl Fischer Method). 10 pp.

—**PH**
- MOD UK M 874/68. Examination of Cryolite.

—**Phosphorus Content—Atomic Absorption Spectrometry**
- BSI BS 4993: Part 10-83. 1983 Methods of Test for Aluminium Fluoride for Industrial Use Parts 10 & 13: Determination for Phosphorus Content. 7 pp.
- BSI BS 5050: Part 13-83. 1983 Methods of Test for Cryolite Part 13: Determination of Phosphorus Content. 7 pp.
- ISO 6374-81. Cryolite, Natural and Artificial, and Aluminium Fluoride for Industrial Use—Determination of Phosphorus Content—Atomic Absorption Spectrometric Method After Extraction First Edition. 6 pp.

Cryolite (Cont.)
—**Phosphorus Content—Photometry**
- CNS M3124-83. Method for Determining Phosphorus Content in Natural and Artificial Cryolite and Aluminium Fluoride for Industrial Use (Reduced Molybdophosphate Photometric Method) (Mar)(10123).
- ISO 5930-79. Cryolite, Natural and Artificial, and Aluminium Fluoride for Industrial Use—Determination of Phosphorus Content—Reduced Molybdophosphate Photometric Method First Edition. 6 pp.

—**Sampling**
- CNS M3112-83. Preparation and Storage of Test samples of Natural and Artificial Cryolite (Mar)(10111).
- ISO 1619-76. Cryolite, Natural and Artificial—Preparation and Storage of Test Samples First Edition. 4 pp.

—**Sieve Analysis**
- MOD UK M 874/68. Examination of Cryolite.

—**Silica Content—Spectrophotometry**
- BSI BS 5050-74. 1974 Methods of Test for Cryolite. 20 pp.
- CNS M3113-83. Method for Determining Silica Content in Natural and Artificial Cryolite (Reduced Molybdosilicate Spectrophotometric Method) (Mar)(10112).
- ISO 1620-76. Cryolite, Natural and Artificial—Determination of Silica Content—Molybdosilicate Spectrophotometric Method First Edition. 7 pp.

—**Sodium Content**
- MOD UK M 874/68. Examination of Cryolite.

—**Sodium Content—Atomic Absorption Spectrophotometry**
- BSI BS 5050-74. 1974 Methods of Test for Cryolite. 20 pp.
- CNS M3116-83. Method for Determining Sodium Content in Natural and Artificial Cryolite (Flame Emission and Atomic Absorption Spectrophotometric Method) (Mar)(10115).
- ISO 2366-74. Cryolite, Natural and Artificial—Determination of Sodium Content—Flame Emission and Atomic Absorption Spectrophotometric Methods First Edition. 6 pp.

—**Sodium Content—Atomic Emission Spectrophotometry**
- CNS M3116-83. Method for Determining Sodium Content in Natural and Artificial Cryolite (Flame Emission and Atomic Absorption Spectrophotometric Method) (Mar)(10115).
- ISO 2366-74. Cryolite, Natural and Artificial—Determination of Sodium Content—Flame Emission and Atomic Absorption Spectrophotometric Methods First Edition. 6 pp.

—**Sulfate Content—Gravimetric Analysis**
- BSI BS 5050: Part 11-80. 1980 Methods of Test for Cryolite Part 11: Determination of Sulphate Content. 8 pp.
- CNS M3123-83. Method for Determining Sulphate Content in Natural and Artificial Cryolite and Aluminium Fluoride for Industrial Use (Barium Sulphate Gravimetric Method) (Mar)(10122).
- ISO 4280-77. Cryolite, Natural and Artificial, and Aluminium Fluoride for Industrial Use—Determination of Sulphate Content—Barium Sulphate Gravimetric Method First Edition; (Erratum—Sept 1978). 6 pp.

—**Sulfur Content**
- MOD UK M 874/68. Examination of Cryolite.

—**Sulfur Content—X-Ray Fluorescence Spectrometry**
- BSI BS 5050: Part 12-80. 1980 Methods of Test for Cryolite Part 12: Determination of Sulphur Content (X-Ray Fluorescence Spectrometric Method). 8 pp.
- CNS M3125-83. Method for Determining Sulphur Content in Natural and Artificial Cryolite and Aluminium Fluoride for Industrial Use (X-Ray Fluorescence Spectrometric Method) (Mar)(10124).
- ISO 5938-79. Cryolite, Natural and Artificial, and Aluminium Fluoride for Industrial Use—Determination of Sulphur Content—X-Ray Fluorescence Spectrometric Method First Edition; (Amendment Slip-1980). 7 pp.

—**Test Specimens—Storage**
- BSI BS 5050-74. 1974 Methods of Test for Cryolite. 20 pp.

—**Visual Inspection**
- MOD UK M 874/68. Examination of Cryolite.

—**Volatile Matter Content**
- MOD UK M 874/68. Examination of Cryolite.

Cryolite (Cont.)
—**Water Content—Electrolytic Analysis**
- CNS M3121-83. Method for Determining Water Content in Natural and Artificial Cryolite and Aluminium Fluoride for Industrial Use (Electrometric Method) (Mar)(10120).
- ISO 3392-76. Cryolite, Natural and Artificial, and Aluminium Fluoride for Industrial Use—Determination of Water Content—Electrometric Method First Edition. 7 pp.

Cryostats
See Also: Electric Switches; Low Temperature Testing; Regulators; Thermostats

—**End Caps—Germanium Detectors**
- IEC 937-88. Cryostat End-Cap Dimensions for Germanium Semiconductor Detectors for Gamma-Ray Specatrometers First Edition. 11 pp.

Cryosurgery
- CNS T1006-81. Cryosurgery (Jul)(5936).

Cryptographic Check Values
Use: Message Authentication Codes

Crystal Controlled Oscillators
Use: Crystal Oscillators

Crystal Defects
Use For: Thermal Spikes *See Also:* Defects; Surface Defects

—**Silicon—Etching**
- DIN ENGL 50434-86. Testing of Materials for Semiconductor Technology; Detection of Crystal Defects in Monocrystalline Silicon Using Etching Techniques on (111) and (100) Surfaces (Feb). 9 pp.

—**Silicon—X-Ray Topography**
- DIN ENGL 50443 Pt 1-88. Testing of Materials for Use in Semiconductor Technology; Detection of Crystal Defects and Inhomogeneities in Silicon Single Crystals by X-Ray Topography (July). 10 pp.

Crystal Filters
Use For: Piezoelectric Filters *See Also:* Bandpass Filters; Filter Crystal Units; Piezoelectric Devices; Quartz Crystal Filters; Radio Frequency Filters
- BSI BS 5069: Part 2-83. 1983 Dimensions of Piezoelectric Devices Part 2: Standard Outlines for Piezoelectric Filters. 23 pp.
- CECC EN 167 000-93. Generic Specification: Piezoelectric Filters. 32 pp.
- CNS C7202-88. Crystal Filters (Apr)(12257).
- IEC 368 Pt 1-92. Piezoelectric Filters Part 1: General Information, Standard Values and Test Conditions Third Edition. 65 pp.
- IEC 368 Pt 3-91. Piezoelectric Filters Part 3: Standard Outlines Second Edition. 30 pp.
- JIS C 6703-90. General Rules of Crystal Filters. 56 pp.

—**Military—Preferred Products List**
- CECC CECC MUAHAG Vol 11 IS 2-92. Preferred Products List; Filters (En, Fr, Ge). 88 pp.

—**Quality Assurance**
- BSI BS 9600-83. 1983 Piezoelectric Crystal Filters of Assessed Quality: Generic Data and Methods of Test. 23 pp.
- BSI BS EN 167000-93. 1993 Generic Specification: Piezoelectric Filters (T). 38 pp.
- CENELEC EN 167000-93. Generic Specification: Piezoelectric Filters. 32 pp.
- CENELEC PREN 167 101-92. Blank Detail Specification: Piezoelectric Filters (Capability Approval). 9 pp.

Crystal Glass
Use: Lead Glass

Crystal Holders
Use For: Quartz Crystal Holders
See Also: Piezoelectric Devices

—**Outlets**
- IEC 149 Pt 3-75. Sockets and Accessories for Electronic Plug-in Devices Part 3: Sockets for Crystal Holders First Edition; (Supplement A-1976). 43 pp.

Crystal Oscillators
See Also: Microwave Oscillators; Oven Controlled Crystal Oscillators; Quartz Crystal Oscillators; Temperature Compensated Crystal Oscillators; Voltage Controlled Crystal Oscillators
- CNS C7198-88. Quartz Crystal Units for Oscillators (for 1—125 MHz) (Apr)(12253).
- CNS C7199-88. Quartz Crystal Units for Oscillators (for 200—1000 kHz) (Apr)(12254).

INTERNATIONAL AND NON-U.S. NATIONAL STANDARDS
SUBJECT INDEX

Crystal Oscillators (Cont.)
IEC 679 Pt 3-89. Quartz Crystal Controlled Oscillators Part 3: Standard Outlines and Lead Connections First Edition; (Supplement A-1991). 54 pp.

—HF—CMOS Compatible—Preferred Products List
CECC CECC MUAHAG Vol 6 IS 3-91. Preferred Products List; Piezoelectric Devices (En, Fr, Gr). 29 pp.

—HF—LS/TTL Compatible—Preferred Products List
CECC CECC MUAHAG Vol 6 IS 3-91. Preferred Products List; Piezoelectric Devices (En, Fr, Gr). 29 pp.

—HF—Sine Wave—Preferred Products List
CECC CECC MUAHAG Vol 6 IS 3-91. Preferred Products List; Piezoelectric Devices (En, Fr, Gr). 29 pp.

—HF—TTL/CMOS Compatible—Preferred Products List
CECC CECC MUAHAG Vol 6 IS 3-91. Preferred Products List; Piezoelectric Devices (En, Fr, Gr). 29 pp.

—HF—TTL Compatible—Preferred Products List
CECC CECC MUAHAG Vol 6 IS 3-91. Preferred Products List; Piezoelectric Devices (En, Fr, Gr). 29 pp.

—Quality Assurance
BSI BS 9611 N004-77. 1977 Solder-Sealed Quartz Crystal Units of Assessed Quality for Oscillator Applications. General Application Category. 12 pp.

BSI BS 9611 N005-77. 1977 Detail Specification for Solder-Sealed Quartz Crystal Units of Assessed Quality for Oscillator Applications. Full Assessment Level. 10 pp.

BSI BS 9611 N006-77. 1977 Detail Specification for Solder-Sealed Quartz Crystal Units of Assessed Quality for Oscillator Applications. Full Assessment Level. 9 pp.

BSI BS 9611 N007-77. 1977 Detail Specification for Solder-Sealed Quartz Crystal Units of Assessed Quality for Oscillator Applications. Full Assessment Level. 10 pp.

BSI BS 9611 N008-77. 1977 Detail Specification for Solder-Sealed Quartz Crystal Units of Assessed Quality for Oscillator Applications. Full Assessment Level. 9 pp.

BSI BS 9611 N009-77. 1977 Detail Specification for Solder-Sealed Quartz Crystal Units of Assessed Quality for Oscillator Applications. Full Assessment Level. 8 pp.

BSI BS 9612 N004-77. 1977 Detail Specification for Cold-Welded Quartz Crystal Units of Assessed Quality for Oscillator Applications. Full Assessment Level. 12 pp.

BSI BS 9612 N005-77. 1977 Detail Specification for Cold-Welded Quartz Crystal Units of Assessed Quality for Oscillator Applications. Frequency Ranges 0.8 to 20 MHz, and 3 to 30 MHz, Narrow Temperature Range. 11 pp.

BSI BS 9612 N006-77. 1977 Detail Specification for Cold-Welded Quartz Crystal Units of Assessed Quality for Oscillator Applications. Frequency Range 17 to 75 MHz, Wide Temperature Range. 9 pp.

BSI BS 9612 N007-77. 1977 Detail Specification for Cold-Welded Quartz Crystal Units of Assessed Quality for Oscillator Applications. Frequency Range 17 to 75 MHz, Narrow Temperature Range. 9 pp.

BSI BS 9612 N008-77. 1977 Detail Specification for Cold-Welded Quartz Crystal Units of Assessed Quality for Oscillator Applications. Frequency Range 50 to 125 MHz, Wide Temperature Range. 9 pp.

BSI BS 9612 N009-77. 1977 Detail Specification for Cold-Welded Quartz Crystal Units of Assessed Quality for Oscillator Applications. Frequency Range 50 to 125 MHz, Narrow Temperature Range. 8 pp.

BSI BS 9612 N010-79. 1979 Detail Specification for Cold-Welded Quartz Crystal Units for Oscillator Applications. Frequency Range 6.0 to 25 MHz, Wide Temperature Range. 7 pp.

BSI BS 9612 N016-79. 1979 Detail Specification for Resistance Welded Quartz Crystal Units for Oscillator Applications. Frequency Ranges 0.8 to 20 MHz and 3 to 30 MHz, Wide Temperature Range Level. 10 pp.

BSI BS 9612 N017-79. 1979 Detail Specification for Resistance Welded Quartz Crystal Units for Oscillator Applications. Frequency Ranges 0.8 to 20 MHz and 3 to 30 MHz, Narrow Temperature Range. 8 pp.

BSI BS 9612 N018-79. 1979 Detail Specification for Resistance Welded Quartz Crystal Units for Oscillator Applications. Frequecy Range 17 to 75 MHz, Wide Temperature Range. 2 pp.

Crystal Oscillators (Cont.)
—Quality Assurance (Cont.)
BSI BS 9612 N019-79. 1979 Detail Specification for Resistance Welded Quartz Crystal Units for Oscillator Applications. Frequency Range 17 to 75 MHz, Narrow Temperature Range. 7 pp.

BSI BS 9612 N020-79. 1979 Detail Specification for Resistance Welded Quartz Crystal Units for Oscillator Applications. Frequency Range 50 to 125 MHz, Wide Temperature Range. 7 pp.

BSI BS 9612 N021-79. 1979 Detail Specification for Resistance Welded Quartz Crystal Units for Oscillator Applications. Frequency Range 50 to 125 MHz, Narrow Temperature Range. 7 pp.

—Radio
CNS C7179-86. Crystal Unit Quartz CR-84/u for Radio Oscillation (Jul)(11624).

—VHF—TTL Compatible—Preferred Products List
CECC CECC MUAHAG Vol 6 IS 3-91. Preferred Products List; Piezoelectric Devices (En, Fr, Gr). 29 pp.

Crystal Ovens
See Also: Quartz Crystal Units

CNS C7201-88. Ovens for Quartz Crystal Units (Apr)(12256).

JIS C 6702-92. Ovens for Quartz Crystal Units. 26 pp.

Crystal Resonators
Use For: Piezoelectric Resonators
See Also: Piezoelectric Devices

—Ceramic
IEC 483-76. Guide to Dynamic Measurements of Piezoelectric Ceramics with High Electromechanical Coupling First Edition. 37 pp.

IEC 642-79. Piezoelectric Ceramic Resonators and Resonator Units for Frequency Control and Selection Chapter I: Standard Values and Conditions Chapter II: Measuring and Test Conditions First Edition; (Amendment 1-1992). 59 pp.

IEC 642 Pt 3-92. Piezoelectric Ceramic Resonators Part 3: Standard Outlines First Edition. 12 pp.

IECQ PQC 65-85. Generic Specification for: Piezoelectric Ceramic Resonators for Use in Electronic Equipment. 23 pp.

IECQ PQC 66-85. Sectional Specification for: Piezoelectric Ceramic Resonators for Low-Frequency (LF). 23 pp.

IECQ PQC 67-85. Blank Detail Specification for: Piezoelectric Ceramic Resonators for Low-Frequency (LF) Assessment Level E. 12 pp.

IECQ PQC 67/JP 0001-87. Detail Specification for: Piezoelectric Ceramic Resonators for Low-Frequency (LF) Assessment Level E. 10 pp.

IECQ PQC 67/SG 0001-90. Detail Specification for: Piezoelectric Ceramic Resonators for Low-Frequency (LF) Assessment Level E. 10 pp.

IECQ PQC 68-85. Sectional Specification for: Piezoelectric Ceramic Resonators for High-Frequency (HF). 23 pp.

IECQ PQC 69-85. Blank Detail Specification for: Piezoelectric Ceramic Resonators for High-Frequency (HF) Assessment Level E. 12 pp.

IECQ PQC 69/JP 0001-87. Detail Specification for: Piezoelectric Ceramic Resonators for High-Frequency (HF) Assessment Level E. 10 pp.

IECQ PQC 69/JP 0002-90. Detail Specification for: Piezoelectric Ceramic Resonators for High-Frequency (HF) Assessment Level E. 12 pp.

Crystal Structure
Use For: Crystallinity *See Also:* Microstructures

—Steels
BSI BS 131: Part 5-65. 1965 Methods for Notched Bar Tests Part 5: Determination of Crystallinity. 12 pp.

Crystal Violet
Use: Methyl Violet

Crystallinity
Use: Crystal Structure

Crystallization
See Also: Heat of Crystallization; Inoculation; Sedimentation; Solidification Points; Solubility

ISO 1392-77. Determination of Crystallizing Point—General Method First Edition. 6 pp.

—Benzene
ISO 5278-80. Benzene—Determination of Crystallizing Point First Edition; (Erratum—Dec 1981). 6 pp.

—Caprolactam
ISO 7060-82. Caprolactam for Industrial Use—Determination of Crystallizing Point First Edition. 3 pp.

Crystallization (Cont.)
—Chemical Products
BSI BS 4633 & 4634-70. 1970 Determination of Crystallizing Point. Determination of Melting Point and/or Melting Range. 12 pp.

BSI BS 4633 & 4634-01. 1970 Amd 1 Method for the Determination of Crystallizing Point Method for the Determination of Melting Point and/or Melting Range (AMD 7715) July 15, 1993 (L). 14 pp.

—Classification
BSI BS 1000: (54)-72. 1972 Amd 2 Universal Decimal Classification (UDC). English Full Edition (54): Chemistry. Crystallography. Mineralogy. 117 pp.

SNZ NZS/BS 1000 (54)-72. Universal Decimal Classification Chemistry. Crystallography. Mineralogy. 120 pp.

—Creosote Oil
CNS K6073-57. Method of Test for Crystallization of Creosote Oil (Jul)(916)(R 1971).

—Dextrose
CNS N5153-79. Crystalline Dextrose (May)(4867). 2 pp.

—Epoxy Resins
ISO 4895-87. Plastics—Liquid Epoxide Resins—Determination of Tendency to Crystallize First Edition. 4 pp.

—Fats
BSI BS 684: Sec 1.13-76. 1976 Methods of Analysis of Fats and Fatty Oils Part 1: Physical Methods Section 1.13: Determination of Cooling Curve. 4 pp.

—Glassy Alloys
JIS H 7151-91. Method of Determining the Crystallization Temperatures of Amorphous Metals. 7 pp.

—Naphthalene
BSI BS 5962-80. 1980 Naphthalene (Including Test Methods). 14 pp.

—Phenols
ISO 1897 Pt 11-83. Phenol, o-Cresol, m-Cresol, p-Cresol, Cresylic Acid and Xylenols for Industrial Use—Methods of Test—Part 11: Determination of Crystallizing Point (Excluding Cresylic Acid and Xylenols) First Edition. 6 pp.

ISO 2208-73. Phenol, o-Cresol, m-Cresol and p-Cresol for Industrial Use—Determination of Crystallizing Point After Drying with a Molecular Sieve First Edition. 4 pp.

—Rubber
BSI BS 5294-76. 1976 Determination of Crystallization Effects in Rubbers by Hardness Measurements. 7 pp.

ISO 3387-78. Rubbers—Determination of Crystallization Effects by Hardness Measurements First Edition. 7 pp.

—Vulcanized Rubber
BSI BS 5294-76. 1976 Determination of Crystallization Effects in Rubbers by Hardness Measurements. 7 pp.

ISO 6471-83. Rubber, Vulcanized—Determination of Crystallization Effects Under Compression First Edition. 8 pp.

Crystallizing Point
Use: Crystallization

CSA
Use For: Canadian Standards Association
See Also: Standards

—Certification—Directories
CSA Volume I. List of Certified Electrical Equipment Volume I Electrical Service Equipment, Wiring Hardware and Controls. 1127 pp.

CSA Volume II. List of Certified Electrical Equipment Volume II Electrically-Equipped Consumer, Commercial and Industrial Products. 1527 pp.

CSA Volume IV. List of Certified Electrical Equipment Volume IV Electrical Components and Accessories. 1087 pp.

—Trade Directories—Data Elements
CGSB CAN/CGSB-200.9-91. Trade Data Elements Directory (UNTDED 1990). 326 pp.

CSMA/CD
Use: Carrier Sense Multiple Access/CD

CSMA/CD (Ethernet)
Use: Ethernet (BTN)

CTDAS
Use: Canadian Trade Document Alignment System

INDUSTRY STANDARDS

Cubeb Oil
See Also: Essential Oils
CNS K5095-80. Oil of Cubeb (Sep)(6466).
ISO 3214-74. Oil of Litsea Cubeba First Edition. 4 pp.
ISO 3756-76. Oil of Cubeb First Edition. 3 pp.

Cucumbers
See Also: Chayotes; Food; Gourds; Pumpkins; Vegetables

—Grading
CNS N1030-82. Grades and Packaging of Cucumber (Jan)(2097). 3 pp.
CNS N1108-83. Grades of Baby Cucumber (Apr)(10192). 2 pp.

—Packaging
CNS N1030-82. Grades and Packaging of Cucumber (Jan)(2097). 3 pp.

—Preserved
CNS N5150-84. Brine-Cured Cucumbers (Jul)(4838). 1 p.

—Refrigerated Transportation
ISO 7560-83. Cucumbers—Guide to Storage and Refrigerated Transport First Edition. 4 pp.

—Storage
ISO 7560-83. Cucumbers—Guide to Storage and Refrigerated Transport First Edition. 4 pp.

CUG
Use: Closed User Groups

Culinary Utensils
Use: Cooking Utensils

Cultivators
Use For: Tillers (Agricultural Equipment)
See Also: Agricultural Equipment; Hoes; Plows; Soils; Tines
CNS B4003-80. Power Tiller (15KW (20HP) or Less) (Tentative) (Jul)(2057).
CNS B7031-80. Method of Test for Power Tiller (Tentative) (Jul)(2181).
CNS B7047-80. Method of Test for Power Cultivator (Jul)(3470).

—Blades
CNS B4038-80. Blades for Tillers (Aug)(6013).
ISO 4197-89. Equipment for Working the Soil—Hoe Blades—Fixing Dimensions Second Edition. 4 pp.
ISO 8945-89. Equipment for Working the Soil—Rotary Cultivator Blades—Fixing Dimensions First Edition. 4 pp.
JIS B 9210-88. Blades for Tillers. 7 pp.

—Disks
BSI BS 5822-79. 1979 Agricultural Discs: Classification, Main Fixing Dimensions and General Requirements. 14 pp.
ISO 4002 Pt I-79. Equipment for Sowing and Planting—Disks—Part I: Concave Disks Type D1—Dimensions First Edition. 4 pp.
ISO 4002 Pt II-77. Equipment for Sowing and Planting—Disks—Part II: Flat Disks Type D2 with Single Bevel—Dimensions First Edition. 4 pp.
ISO TR4122-77. Equipment for Working the Soil—Dimensions of Flat Disks—Type A First Edition. 3 pp.
ISO 5679-79. Equipment for Working the Soil—Disks—Classification, Main Fixing Dimensions and Specifications First Edition. 9 pp.

—Electric
CSA CAN/CSA-C22. 2 NO 147-M90. Motor-Operated Gardening Appliances; (Gen Instr 1 Thru 3). 85 pp.

—Hitches
CNS B4031-80. Dimensions of Hitch for Power Tillers (Aug)(6006).
JIS B 9209-76. Dimensions of Hitch for Power Tillers. 4 pp.

—Rollers—Couplers
BSI BS 7006-89. 1989 Trailed Agricultural Rollers. 7 pp.
ISO 8912-86. Equipment for Working the Soil—Roller Sections—Coupling and Section Width First Edition. 5 pp.

—Safety
CEN PREN 708-92. Safety Requirements for Agricultural and Forestry Machinery—Soil Working Machines with Powered Tools. 11 pp.
DIN ENGL 11001 Pt 6-80. Agricultural Machines and Tractors; Implements for Soil Cultivation, Sowing, Plant Treatment and Fertilizing; Special Technical Safety Requirements and Testing (Aug). 4 pp.

Cultivators (Cont.)
—Safety (Cont.)
DIN ENGL 11001 Pt 9-81. Agricultural Machines and Tractors; Motorized Drag Hoes and Walking Tractors with Mounted Rotary Cultivators; Special Technical Safety Requirements and Testing (Feb). 4 pp.
ISO 4254 Pt 5-92. Tractors and Machinery for Agriculture and Forestry—Technical Means for Ensuring Safety—Part 5: Power-Driven Soil-Working Equipment First Edition. 5 pp.

—Shovels
ISO 5680-79. Equipment for Working Soil—Tines and Shovels for Cultivators—Main Fixing Dimensions First Edition. 5 pp.

—Trailers
CNS B4014-80. Trailers for Power Tiller (Jul)(3196).
JIS B 9207-77. Trailers for Power Tiller. 11 pp.

—Trailers—Axles and Brakes
CNS B4037-80. Axles with Brakes of Trailer for Power Tiller (Aug)(6012).
JIS B 9208-76. Axles with Brakes of Trailer for Power Tiller (R 1979). 9 pp.

—Yokes
ISO 6880-83. Machinery for Agriculture—Trailed Units of Shallow Tillage Equipment—Main Dimensions and Attachment Points First Edition. 4 pp.

Culture Dishes
Use: Petri Dishes

Culture Media
See Also: Agar; In Vitro Analysis; Microbiological Analysis
SAA AS 1095.3A-76. Microbiological Methods for the Dairy Industry—Part 3: Methods of Examination for Specific Groups of Microorganisms—Part 3A: Appendix A: Preparation of Media and Diluents. 22 pp.
SAA AS 1766.5-82. Methods for the Microbiological Examination of Food—Part 5: Preparation of Media, Diluents and Reagents. 20 pp.

—Cells
JIS K 3604-90. Medium for Tissue Culture (Minimum Essential Medium). 16 pp.

—Cells—Medical Equipment—Biological Hazards
BSI BS 5736: Part 10-88. 1988 Amd 1 Evaluation of Medical Devices for Biological Hazards Part 10: Method of Test for Toxicity to Cells in Culture of Extracts from Medical Devices (AMD 6215) July 31, 1989. 10 pp.

—Glossaries
DIN ENGL 58942 Pt 1-84. Medical Microbiology; Culture Media Used in Bacteriology; Terminology (May). 2 pp.

—Transport Systems
DIN ENGL 58942 Pt 4-86. Medical Microbiology; Culture Media Used in Bacteriology; Transport Systems for Specimens Containing Bacteria; Physical, Chemical and Biological Requirements (Oct). 3 pp.

—Urinalysis—Dip Slides
DIN ENGL 58942 Pt 3-85. Medical Microbiology; Culture Media Used in Bacteriology; Dip Slides for Bacteriological Urine Testing (Sept). 3 pp.

Culture Media, Blood
Use: Blood Culture Media

Culture Tubes
Use: Test Tubes

Cultured Dairy Products
Use: Dairy Products—Cultured

Culvert Pipe Joints
See Also: Pipe Joints

—Ogee—Concrete
BSI BS 5911: Part 3-82. (WITHDRAWN) 1982 Precast Concrete Pipes and Fittings for Drainage and Sewerage Part 3: Ogee Jointed Concrete Pipes, Bends and Junctions, Unreinforced or Reinforced with Steel Cages or Hoops (Superseded by BS 5911: Part 110: 1992). 19 pp.

Culverts (Pipes)
See Also: Drainpipes; Highways; Roads

Culverts (Pipes) (Cont.)
—Concrete
BSI BS 5911: Part 3-82. (WITHDRAWN) 1982 Precast Concrete Pipes and Fittings for Drainage and Sewerage Part 3: Ogee Jointed Concrete Pipes, Bends and Junctions, Unreinforced or Reinforced with Steel Cages or Hoops (Superseded by BS 5911: Part 110: 1992). 19 pp.
BSI BS 5911: Part 100-88. 1988 Amd 1 Precast Concrete Pipes and Fittings for Drainage and Sewerage Part 100: Unreinforced and Reinforced Pipes and Fittings with Flexible Joints (AMD 6269) December 22, 1989. 39 pp.
BSI BS 5911: Part 100-02. 1988 Amd 2 Precast Concrete Pipes, Fittings and Ancillary Products Part 100: Specification for Unreinforced and Reinforced Pipes and Fittings with Flexible Joints (AMD 7588) April 15, 1993 (R). 40 pp.
BSI BS 5911: Part 101-88. 1988 Precast Concrete Pipes and Fittings for Drainage and Sewerage Part 101: Glass Composite Concrete (GCC) Pipes and Fittings with Flexible Joints. 32 pp.
BSI BS 5911: Part 110-92. 1992 Precast Concrete Pipes, Fittings and Ancillary Products Part 110: Specification for Ogee Pipes and Fittings (Including Perforated). 34 pp.
BSI BS 5911: Part 120-89. 1989 Precast Concrete Pipes and Fittings for Drainage and Sewerage Part 120: Reinforced Jacking Pipes with Flexible Joints. 25 pp.
BSI DD 76: Part 2-83. 1983 Precast Concrete Pipes of Composite Construction Part 2: Precast Concrete Pipes Strengthened by Chopped Zinc-Coated Steel Fibres. 20 pp.
SAA AS 1597.1-74. Precast Reinforced Concrete Box Culverts—Part 1: Small Culverts (Not Exceeding 1 200mm Width and 900mm Depth). 19 pp.

—Concrete Blocks
JIS A 5328-90. Reinforced Concrete Built-up Culvert Blocks. 10 pp.

Cumin
See Also: Herbs
BSI BS 7087: Part 5-90. 1990 Herbs and Spices Ready for Food Use Part 5: Dried Cumin (Whole and Ground). 9 pp.
ISO 6465-84. Whole Cumin (Cuminum Cyminum Linnaeus)—Specification First Edition. 5 pp.

Cumulative Sum Chart
Use: Cusum Charts

Cup Head Bolts
See Also: Bolts
DIN ENGL 607-81. Cup Head Nib Bolts (Oct). 4 pp.

—Square Neck
BSI BS 4933-73. (OBSOLESCENT) 1973 ISO Metric Black Cup and Countersunk Head Bolts and Screws with Hexagon Nuts. 28 pp.
SAA AS 1390-74. Metric Cup Head Bolts (This is a Joint Standard with SNZ NZS 1390). 22 pp.
SNZ NZS/AS 1390-74. Metric Cup Head Bolts (This is a Joint Standard with SAA AS 1390). 22 pp.

Cup Wheels
Use: Grinding Wheels

Cupboards
See Also: Cabinets (Furniture); Fume Cupboards

—Door Bolts
CNS A2099-81. Cupboard Bolts (Mar)(6995). 3 pp.

—Hardware
BSI BS 3827: Part 4-67. (WITHDRAWN) 1967 Glossary of Terms Relating to Builders' Hardware Part 4: Door Drawer, Cupboard and Gate Furniture (Superseded by BS 6100: Subsection 1.3.6:1991). 24 pp.

Cupellation
See Also: Assaying

—Gold Alloys
ISO 11426-93. Determination of Gold in Gold Jewellery Alloys—Cupellation Method (Fire Assay) First Edition. 7 pp.

Cupferron
Use For: Nitrosophenylhydroxylamine Ammonium Salt
CNS K7161-66. Chemical Reagent (Cupferron)(Ammonium Nitrosophenylhydroxylamine) (Mar)(1661).
JIS K 8289-78. Cupferron (Nitrosophenylhydroxylamine Ammonium Salt).

Cupping Testers
See Also: Testing Equipment

INTERNATIONAL AND NON-U.S. NATIONAL STANDARDS
SUBJECT INDEX

Cupping Testers *(Cont.)*
—Erichsen
CNS B6066-83. Erichsen Cupping Testers (Jan)(8764).
CNS B7190-82. Method of Test for Erichsen Cupping Testers (May)(8765).
CNS Z8007-87. Method of Erichsen Cupping Test (Jun)(3464).
JIS B 7729-76. Erichsen Cupping Testers (R 1979). 10 pp.

Cupping Tests
Use: Ductility

Cuprammonium Fluidity
Use: Cuprammonium Viscosity

Cuprammonium Viscosity
Use For: Cuprammonium Fluidity
See Also: Rheological Properties; Viscosity
—Cellulose Fibers
BSI BS 2610-78. 1978 Amd 1 Method of Test for the Determination of the Cuprammonium Fluidity of Cotton and Certain Cellulosic Man-Made Fibres (AMD 4965) November 29, 1985. 13 pp.
CGSB CAN/CGSB-4.2 NO.17-M90. Textile Test Methods Cuprammonium Fluidity of Textile Cellulose. 13 pp.
—Cotton Fibers
BSI BS 2610-78. 1978 Amd 1 Method of Test for the Determination of the Cuprammonium Fluidity of Cotton and Certain Cellulosic Man-Made Fibres (AMD 4965) November 29, 1985. 13 pp.
CGSB CAN/CGSB-4.2 NO.17-M90. Textile Test Methods Cuprammonium Fluidity of Textile Cellulose. 13 pp.
—Linen
BSI BS 3090-78. 1978 Amd 1 Method of Test for the Determination of the Cuprammonium Fluidity of Linen Materials. 9 pp.
—Rayon
CGSB CAN/CGSB-4.2 NO.17-M90. Textile Test Methods Cuprammonium Fluidity of Textile Cellulose. 13 pp.

Cupric Acetate
Use For: Copper Acetate *See Also:* Acetates
CNS K7162-66. Chemical Reagent (Cupric Acetate) (Mar)(1662).
JIS K 8370-79. Cupric Acetate, Monohydrate.

Cupric Ammonium Chloride
Use: Ammonium Cupric Chloride

Cupric Bromide
CNS K7164-66. Chemical Reagent (Cupric Bromide) (Mar)(1664).

Cupric Carbonate
Use: Copper Carbonate

Cupric Chloride
CNS K7166-66. Chemical Reagent (Cupric Chloride) (Mar)(1666).
JIS K 8145-78. Copper Chloride, Dihydrate.

Cupric Chromate
Use For: Acid Copper Chromate; Copper Chromate
—Propellants
MOD UK DSTAN 68-99-88. Copper Chromate, Basic Issue 1. 8 pp.
—Wood Preservatives
CNS K1186-81. Wood Preservatives (Acid Copper Chromate) (ACC, Celcure) (May)(4134). 1 p.
CNS K6405-81. Method of Test for Acid Copper Chromate (Wood Preservatives) (Jul)(4198).

Cupric Cyanide
CNS K1104-66. Cupric Cyanide (Industrial Grade) (Sep)(2525)(R 1972).
CNS K6225-66. Method of Test for Cupric Cyanide of Industrial Grade (Sep)(2551).

Cupric Diammonium Chloride Dihydrate
JIS K 8146-84. Copper (II) Diammonium Chloride Dihydrate.

Cupric Nitrate
CNS K7167-66. Chemical Reagent (Cupric Nitrate) (Mar)(1667).
JIS K 8560-88. Copper (2) Nitrate Trihydrate.

Cupric Oxide
Use For: Copper Oxide Black
CNS K7168-66. Chemical Reagent (Cupric Oxide, Powder) (Mar)(1668).

Cupric Oxide *(Cont.)*
JIS K 8422-61. Cupric Oxide.
MOD UK DSTAN 68-114-90. Cupric Oxide, Types A and B Issue 1. 10 pp.
MOD UK TS 565A. Cupric Oxide Types A and B (Superseded by Def Stan 68-141).
—Wire
CNS K7169-66. Chemical Reagent (Cupric Oxide, Wire) (Mar)(1669).

Cupric Potassium Chloride
See Also: Potassium Chloride
—Reagents
CNS K7170-65. Chemical Reagent (Cupric Potassium Chloride) (Jan)(1670).

Cupric Salicylate
Use For: Basic Cupric Salicylate
MOD UK TS 10118. Basic Cupric Salicylate (Withdrawn).

Cupric Sulfate
Use For: Cupric Sulphate; Fehling's Solution A
CNS K1076-63. Cupric Sulfate (Industrial Grade) (Jul)(2199)(R 1968).
CNS K7171-65. Chemical Reagent (Cupric Sulfate) (Jan)(1671).
CNS K7172-65. Chemical Reagent (Cupric Sulfate, Anhydrous) (Jan)(1672).
JIS K 1433-86. Copper Sulfate for Industrial Use. 13 pp.
JIS K 8983-80. Copper (II) Sulfate Pentahydrate.
JIS K 8984-80. Copper (II) Sulfate (Anhydrous).
—Fertilizer
CNS N3096-88. Copper Sulfate, Fertilizer Grade (Jul)(12030).
—Standard Solutions
JIS K 0010-92. Copper Standard Solution. 13 pp.
—Water Treatment
MOD UK DSTAN 68-20-71. Cupric Sulphate, Pentahydrate, Technical Issue 1. 9 pp.
—Wood Preservatives—Poles—Bucherie Methods
JIS A 9105-89. Wood Poles Treated with Copper Sulfate by Butt Pressure Process. 12 pp.

Cupric Sulphate
Use: Cupric Sulfate

Cupronickel
See Also: Copper Alloys; Copper Nickel Alloys
—Chemical Analysis
CNS H2083-88. Method of Chemical Analysis for Cupro-Nickel and Nickel Silver (Apr)(12259).
JIS H 1231-77. Methods for Chemical Analysis of Cupro-Nickel and Nickel Silver.

Cuprous Chloride
CNS K7173-65. Chemical Reagent (Cuprous Chloride) (Jan)(1673).
JIS K 8138-78. Cuprous Chloride.

Cuprous Oxide
CNS K2057-76. Cuprous Oxide (for Paint Use) (Jun)(3952). 1 p.
CNS K6392-82. Method of Test for Cuprous Oxide (Oct)(3953).
—Antifouling Coatings
MOD UK TS 10241. Cuprous Oxide for Antifouling Paints.
—Paints
CNS K2057-76. Cuprous Oxide (for Paint Use) (Jun)(3952). 1 p.

Cups (Tableware)
See Also: Tableware
—Aluminum
CNS H3015-86. Aluminium Cups (Jul)(757).
—Stainless Steel
CNS S1211-88. Stainless Steel Thermal Cup (May)(12323).

Curatives
Use: Curing Agents

Curb Boxes
Use For: Surface Boxes *See Also:* Gas Pipelines; Water Pipelines
BSI BS 5834: Part 2-83. 1983 Surface Boxes, and Guards for Underground Stopvalves for Gas and Waterworks Purposes Part 2: Small Surface Boxes. 10 pp.

Curb Boxes *(Cont.)*
BSI BS 5834: Part 3-85. 1985 Surface Boxes and Guards for Underground Stopvalves for Gas and Waterworks Purposes Part 3: Large Surface Boxes. 14 pp.
—Covers
SNZ NZS/BS 750-84. Specification for Underground Fire Hydrants and Surface Box Frames and Covers. 16 pp.
—Foundation Boxes
BSI BS 5834: Part 1-85. 1985 Surface Boxes and Guards for Underground Stopvalves for Gas and Waterworks Purposes Part 1: Guards, Including Foundation Units. 11 pp.
—Frames
SNZ NZS/BS 750-84. Specification for Underground Fire Hydrants and Surface Box Frames and Covers. 16 pp.
—Guards
BSI BS 5834: Part 1-85. 1985 Surface Boxes and Guards for Underground Stopvalves for Gas and Waterworks Purposes Part 1: Guards, Including Foundation Units. 11 pp.
—Underground Chambers
BSI BS 5834: Part 4-89. 1989 Surface Boxes and Guards for Underground Stopvalves for Gas and Waterworks Purposes Part 4: Preformed Chambers. 22 pp.

Curb-Stop Boxes
Use: Curb Boxes

Curbs
Use For: Kerbs *See Also:* Roads
—Asphalt
BSI BS 5931-80. 1980 Machine Laid In-Situ Edge Details for Paved Areas. 12 pp.
—Concrete
BSI BS 5931-80. 1980 Machine Laid In-Situ Edge Details for Paved Areas. 12 pp.
CNS A2053-85. Concrete Curbs (Feb) (3930). 3 pp.
CNS A2057-84. L-Type Concrete and Reinforced Concrete Curb-Gutters (Sep)(4065). 4 pp.
CNS A3066-85. Method of Test for Concrete Curbs (Feb)(3931). 2 pp.
CNS A3070-84. Method of Test for Concrete and Reinforced Concrete Curb-Gutters (L Type) (Sep)(4066).
DIN ENGL 483-81. Concrete Kerbs (Aug). 5 pp.
JIS A 5306-88. Concrete Curb-Gutters and Reinforced Concrete Curb-Gutters. 10 pp.
JIS A 5307-87. Concrete Curbs. 11 pp.
SAA AS A175-70. Concrete Kerbs and Channels (Gutters) Corrig.. 15 pp.
—Concrete—Bend Testing
CNS A3066-85. Method of Test for Concrete Curbs (Feb)(3931). 2 pp.
—Concrete—Water Absorption
CNS A3066-85. Method of Test for Concrete Curbs (Feb)(3931). 2 pp.
—Precast Concrete
BSI BS 340-79. (WITHDRAWN) 1979 Precast Concrete Kerbs, Channels, Edgings and Quadrants (Superseded by BS 7263: Part 1: 1990). 12 pp.
BSI BS 7263: Part 1-90. 1990 Precast Concrete Flags, Kerbs, Channels Edgings and Quadrants Part 1: Specification. 19 pp.
—Precast Concrete—Installation
BSI BS 7263: Part 2-90. 1990 Precast Concrete Flags, Kerbs, Channels Edgings and Quadrants Part 2: Code of Practice for Laying. 12 pp.
—Reinforced Concrete
CNS A2057-84. L-Type Concrete and Reinforced Concrete Curb-Gutters (Sep)(4065). 4 pp.
CNS A3070-84. Method of Test for Concrete and Reinforced Concrete Curb-Gutters (L Type) (Sep)(4066).
JIS A 5306-88. Concrete Curb-Gutters and Reinforced Concrete Curb-Gutters. 10 pp.
—Stone
BSI BS 435-75. 1975 Amd 1 Dressed Natural Stone Kerbs, Channels, Quadrants and Setts. 11 pp.
DIN ENGL 482-88. Natural Stone Kerbs (Sept). 4 pp.

Curcumin
JIS K 8297-78. Curcumin.

Curd Content Analysis
—Butter
SAA AS 2300.7.2-91. Methods of Chemical and Physical Testing for the Dairying Industry—Part 7: Butter—Part 7.2: Determination of Fat and Curd (Supersedes AS 1739 —1975 (in Part)). 2 pp.

Curemeters
See Also: Measuring Instruments

—Vulcanized Rubber
BSI BS 903: Sec 60.2-92. 1992 Physical Testing of Rubber Part A60: Curemetering Section 60.2: Method for the Determination of Vulcanization Characteristics Using an Oscillating Disc Curemeter (ISO 3417: 1991) (V). 18 pp.
DIN ENGL 53529 Pt 1-83. Testing of Rubber and Elastomers; Measurement of Vulcanization Characteristics (Curometry); General Working Principles (Mar). 6 pp.
ISO 3417-91. Rubber—Measurement of Vulcanization Characteristics with the Oscillating Disc Curemeter Second Edition. 15 pp.
ISO 6502-91. Rubber—Measurement of Vulcanization Characteristics with Rotorless Curemeters Second Edition. 10 pp.

Curettes
See Also: Dental Equipment; Medical Equipment; Medical Instruments; Surgical Instruments
CNS T1028-80. Medical Sharp Curettes (Aug)(6287).
JIS T 2604-85. Medical Sharp Curettes.

—Dental
CNS T3041-81. Dental Curettes (Oct)(8063).
JIS T 5413-88. Dental Curettes. 5 pp.

—Urethral
CNS T1039-81. Urethral Curettes (Jan)(6970).
JIS T 2614-88. Urethral Curettes. 7 pp.

—Uterine
CNS T1035-81. Uterine Curettes Sharp (Jan)(6966).
CNS T1036-81. Uterine Curettes (Jan)(6967).
JIS T 2609-80. Uterine Curettes Shar
JIS T 2610-80. Uterine Curettes.

Curing
Scope Note: Use a more specific term
See: Preservation; Steam Curing

Curing Agents
Use For: Curing Compounds *See Also:* Concretes; Lead 2-Ethylhexanoate
CNS A2032-86. Liquid Membrane-Forming Compounds for Curing Concrete (Dec)(2178). 4 pp.

—Water Retention
CNS A3138-82. Method of Test for Water Retention by Concrete Curing Material (Jan)(8188).

—Water Retention Efficiency
BSI BS 7542-92. 1992 Method of Test for Curing Compounds for Concrete. 12 pp.
BSI DD 147-87. (WITHDRAWN) 1987 Method of Test for Curing Compounds for Concrete (Superseded by BS 7542: 1992). 10 pp.

Curing Compounds
Use: Curing Agents

Curl (Materials)
See Also: Dimensional Stability

—Fine Papers
CNS P3026-89. Method of Test for Degree of Curl and Sizing of Paper (Curl Test) (Aug)(3318). 2 pp.

—Papers
CPPA D.18U-77. Curl of Paper at Various Relative Humidities. 2 pp.

—Photographic Films
BSI BS 5846-80. 1980 Methods for Determining the Curl of Photographic Film. 10 pp.
ISO 4330-87. Photography—Determination of the Curl of Photographic Film Second Edition. 8 pp.
JIS K 7619-88. Determination of the Curl of Photographic Film.

Curling Irons
Use For: Hair Curling Appliances
See Also: Appliances; Hair Care Appliances; Hair Dryers
CNS C4399-87. Hair Curling Appliances (Aug)(10607).
CSA C22.2 NO 36-M1989. Hairdressing Equipment; (Gen Instr 1 Thru 4). 79 pp.
CSA 1169 Bull. Electrical Bulletin 1169 June 27, 1978 to C22.2 NO 36. 2 pp.
JIS C 9616-77. Hair Curling Appliances.

Curometers
Use: Curemeters

Currencies
Use For: Money *See Also:* Accounting; Banking; Capital Movements; Cash Registers; European Monetary System; Financial Institutions; Securities

—Codes
BSI BS 5716-81. (WITHDRAWN) 1981 Codes for the Representation of Currencies and Funds (Superseded by ISO 4217: 1990 (Ed 4) Which Will Be Dual-Numbered as BS 7095). 47 pp.
CGSB CAN/CGSB-200.6-91. Codes for the Representation of Currencies and Funds (ISO 4217:1990); (Amendment 1 September 1992) (Amendment 2 October 1992) (Amendment 3 December 1992) (Amendment 4 December 1992) (Amendment 5 December 1992) (Amendment 6 April 1993). 54 pp.
ISO 4217-90. Codes for the Representation of Currencies and Funds Fourth Edition. 36 pp.

—European Communities
EURO 1991 Mar. Community and the Third World. 9 pp.

Current Amplifiers
Scope Note: Use a more specific term
See: Amplifiers; DC Amplifiers

Current Carrying Capacity
Use: Current Ratings

Current Interrupters
Use: Current Limiters

Current Limiters
Use For: Current Interrupters *See Also:* Circuit Breakers
CNS C4363-82. Current Limiter (Jul)(9098).
JIS C 8368-91. Current Limiters. 25 pp.

Current Limiting Circuit Breakers
See Also: Circuit Breakers

—Molded Case—0-115 V AC
CSA CAN/CSA-C22. 2 NO 5.1-M91. Moulded Case Circuit Breakers; (Gen Instr 1 Thru 3). 126 pp.

—Molded Case—0-115 V DC
CSA CAN/CSA-C22. 2 NO 5.1-M91. Moulded Case Circuit Breakers; (Gen Instr 1 Thru 3). 126 pp.

—Molded Case—120-240 V AC
CSA CAN/CSA-C22. 2 NO 5.1-M91. Moulded Case Circuit Breakers; (Gen Instr 1 Thru 3). 126 pp.

—Molded Case—120-240 V DC
CSA CAN/CSA-C22. 2 NO 5.1-M91. Moulded Case Circuit Breakers; (Gen Instr 1 Thru 3). 126 pp.

—Molded Case—250-500 V AC
CSA CAN/CSA-C22. 2 NO 5.1-M91. Moulded Case Circuit Breakers; (Gen Instr 1 Thru 3). 126 pp.

—Molded Case—250-500 V DC
CSA CAN/CSA-C22. 2 NO 5.1-M91. Moulded Case Circuit Breakers; (Gen Instr 1 Thru 3). 126 pp.

—Molded Case—600 V AC
CSA CAN/CSA-C22. 2 NO 5.1-M91. Moulded Case Circuit Breakers; (Gen Instr 1 Thru 3). 126 pp.

—Molded Case—600 V DC
CSA CAN/CSA-C22. 2 NO 5.1-M91. Moulded Case Circuit Breakers; (Gen Instr 1 Thru 3). 126 pp.

Current Limiting Fuses
Use For: Cable Protectors *See Also:* Fuses (Electric); HRC Fuses

—High Voltage
BSI BS 2692: Part 1-86. 1986 Amd 2 Fuses for Voltages Exceeding 1000 V a.c. Part 1: Current Limiting Fuses (AMD 6072) January 31, 1990. 78 pp.
BSI BS 2692: Part 1-03. 1986 Amd 3 Fuses for Voltages Exceeding 1000 V a.c. Part 1: Specification for Current-Limiting Fuses (IEC 282-1: 1985) (AMD 6923) April 1, 1992. 81 pp.
CENELEC HD 492.1 S1-88. High-Voltage Fuses—Part 1: Current-Limiting Fuses.
CENELEC HD 492.1 S2-89. High-Voltage Fuses Part 1: Current Limiting Fuses. 3 pp.
CENELEC HD 492.1 S2/A1-90. AMD 1 High-Voltage Fuses Part 1: Current Limiting Fuses. 4 pp.
IEC 282 Pt 1-85. High-Voltage Fuses Part 1: Current-Limiting Fuses Third Edition; (Amendment 1-1988) (Amendment 2-1992). 161 pp.
JIS C 4604-88. High-Voltage Current-Limiting Fuses. 63 pp.
SAA AS 1033.2-88. High Voltage Fuses (For Rated Voltages Exceeding 1000 V)—Part 2: Current-Limiting (Powder-Filled) Type. 36 pp.

Current Limiting Fuses *(Cont.)*
—High Voltage—Short Circuit Power Factor
IEC 282 Pt 3-76. High-Voltage Fuses Part 3: Determination of Short-Circuit Power Factor for Testing Current-Limiting Fuses and Expulsion and Similar Fuses First Edition. 10 pp.

Current Limiting Transformers
Use: Current Transformers

Current Meters (Fluids)
Use: Flowmeters

Current Noise
See Also: Electrical Measurement; Noise (Spurious Signals)

—Fixed Resistors
BSI BS 4119-67. 1967 Method of Measurement of Current Noise Generated in Fixed Resistors. 14 pp.
IEC 195-65. Method of Measurement of Current Noise Generated in Fixed Resistors First Edition. 27 pp.

Current Operated Earth Leakage Devices
Use: Residual Current Devices

Current Ratings
Use For: Ampacity
CNS C1015-68. Standard Electric Current Rating (Apr)(317).
IEC 59-38. IEC Standard Current Ratings First Edition. 3 pp.
IEC 287-82. Calculation of the Continuous Current Rating of Cables (100% Load Factor) Second Edition; (Amendment 3-1993, Including Amendments 1 and 2). 156 pp.

—Cables (Electric)
DIN VDE 0100 Pt 523-81. Installation of Power Plant with Rated Voltages up to 1000 V; Dimensioning of Cables and Cords; Mechanical Strength, Voltage Drop and Current Carrying Capacity (June). 10 pp.
DIN VDE 0298 Pt 2-79. Application of Cables and Flexible Cords in Power Installations; Recommended Values for the Current Carrying of Cables with Rated Voltages Uo/U up to 18/30 kV (Nov). 73 pp.
DIN VDE 0298 Pt 2 (D)-90. Application of Cables and Leads in Power Installations; Recommended Current-Carrying Capacities for Cables with Rated Voltages V/V up to 18/30 kV (July). 54 pp.
DIN VDE 0298 Pt 4-88. Application of Cables and Insulated Conductors in Power Plant (Feb) (Partially Supersedes: DIN 57100 Part 523/VDE 0100 Part 523/06.81). 43 pp.
IEC 1042-91. Method for Calculating Reduction Factors for Groups of Cables in Free Air, Protected from Solar Radiation First Edition. 26 pp.

—Cables (Electric)—Automotive
BSI BS AU 88A-85. 1985 Recommendations for Ratings for Light Duty Cables for Automobile Use. 4 pp.
CNS D1051-83. Current Capacity of Low Tension Cables for Automobiles (Apr)(10158).

—Cords (Electric)
CNS C1116-90. Security Current of Rubber Insulated Flexible Cords (Feb)(9827).
DIN VDE 0100 Pt 523-81. Installation of Power Plant with Rated Voltages up to 1000 V; Dimensioning of Cables and Cords; Mechanical Strength, Voltage Drop and Current Carrying Capacity (June). 10 pp.
DIN VDE 0298 Pt 2-79. Application of Cables and Flexible Cords in Power Installations; Recommended Values for the Current Carrying of Cables with Rated Voltages Uo/U up to 18/30 kV (Nov). 73 pp.
DIN VDE 0298 Pt 2 (D)-90. Application of Cables and Leads in Power Installations; Recommended Current-Carrying Capacities for Cables with Rated Voltages V/V up to 18/30 kV (July). 54 pp.

—Dry Reed Switches
CNS C6165-87. Method of Test for Dry Reed Switches (Carry Current) (Jul)(8672).

—Power Lines—Overhead
IEC 826-91. Loading and Strength of Overhead Transmission Lines Second Edition; (Corrigendum—Sept 1991). 236 pp.

Current Reference Diodes
See Also: Current Regulator Diodes; Diodes; Semiconductor Devices

INTERNATIONAL AND NON-U.S. NATIONAL STANDARDS
SUBJECT INDEX

Current Reference Diodes (Cont.)
CECC EN 150 013-91. Blank Detail Specification: Current Regulator and Current Reference Diodes (Supersedes CECC 50 013 Issue 1: 1984). 11 pp.

—**Quality Assurance**
BSI BS CECC 50013-84. (WITHDRAWN) 1984 Current Regulator and Current Reference Diodes (Renumbered as BS EN 150013: 1993). 12 pp.
BSI BS EN 150013-93. 1993 Amd 1 Blank Detail Specification: Current Regulator and Current Reference Diodes (AMD 7594) February 1993 (T). 16 pp.
CENELEC EN 150 013-91. Blank Detail Specification: Current Regulator and Current Reference Diodes. 11 pp.

Current Regulator Diodes
See Also: Current Reference Diodes; Diodes; Semiconductor Devices
BSI BS 6493: Sec 1.3-86. 1986 Semiconductor Devices Part 1: Discrete Devices Section 1.3: Recommendations for Signal (Including Switching) and Regulator Diodes. 52 pp.
BSI BS 6493: Sec 1.3-01. 1986 Amd 1 Semiconductor Devices Part 1: Discrete Devices Section 1.3: Recommendations for Signal (Including Switching) and Regulator Diodes (IEC 747-3: 1985) (AMD 7205) September 15, 1992. 58 pp.
CECC EN 150 013-91. Blank Detail Specification: Current Regulator and Current Reference Diodes (Supersedes CECC 50 013 Issue 1: 1984). 11 pp.
IEC 747 Pt 3-85. Semiconductor Devices Discrete Devices Part 3: Signal (Including Switching) and Regulator Diodes First Edition; (Amendment 1-1991). 118 pp.
MOD UK DSTAN 59-61: 90/219-83. Current Regulator Diode Issue 1. 25 pp.
MOD UK DSTAN 59-61: 90/220-83. Current Regulator Diode Issue 1. 13 pp.
MOD UK DSTAN 59-61: 90/221-83. Current Regulator Diode Issue 1. 14 pp.

—**Quality Assurance**
BSI BS CECC 50013-84. (WITHDRAWN) 1984 Current Regulator and Current Reference Diodes (Renumbered as BS EN 150013: 1993). 12 pp.
BSI BS EN 150013-93. 1993 Amd 1 Blank Detail Specification: Current Regulator and Current Reference Diodes (AMD 7594) February 1993 (T). 16 pp.
CENELEC EN 150 013-91. Blank Detail Specification: Current Regulator and Current Reference Diodes. 11 pp.

Current Relays
See Also: Overcurrent Relays; Protective Relays; Relays; Sensitive Relays
CNS C4397-83. Residual Current Sensing and Relaying Equipment (Oct)(10603).
JIS C 8374-91. Residual Current Sensing and Relaying Equipment. 23 pp.

Current Sensing Relays
Use: Current Relays

Current Sensitive Relays
Use: Current Relays

Current Transformers
See Also: Instrument Transformers; Transformers; Voltage Transformers
BSI BS 3938-73. (OBSOLESCENT) 1973 Current Transformers (Superseded by BS 7626: 1993). 44 pp.
BSI BS 7626-93. 1993 Current Transformers (Supersedes BS 3938: 1973) (E). 49 pp.
CENELEC HD 553 S1-92. Current Transformers. 9 pp.
CNS C4037-86. Current Transformers (Jul)(1329).
CNS C4434-85. Summating Current Transformer for Multilines (Dec)(11435).
DIN VDE 0414 Pt 1-70. Specifications for Instrument Transformers; Part 1: General Specifications (Dec). 30 pp.
DIN VDE 0414 Pt 1A-78. Specifications for Instrument Transformers; Amendment to; Part 1: General Specifications (Feb). 3 pp.
IEC 185-87. Current Transformers Second Edition; (Amendment 1-1990). 80 pp.
SAA AS 1675-86. Current Transformers—Measurement and Protection. 48 pp.
SNZ BS 3938-73. Specification for Current Transformers. 44 pp.

—**Aircraft**
BSI 2G 127-67. 1967 Power and Current Transformers for Use in Aircraft Electrical Power Supply Systems. 12 pp.

Current Transformers (Cont.)
—**Inductive—Electrical Protection Equipment**
IEC 44 Pt 6-92. Instrument Transformers Part 6: Requirements for Protective Current Transformers for Transient Performance First Edition. 90 pp.

—**Mounting Hardware—Indoor/Outdoor**
CSA C22.2 NO 115-M1989. Meter-Mounting Devices; (Gen Instr 1 Thru 3). 55 pp.
CSA 1169 Bull. Electrical Bulletin 1169 June 27, 1978 to C22.2 NO 115. 2 pp.

—**Symbols—Architectural Drawings**
CNS C1095-87. Symbols of Potential Transformers and Current Transformers of Interior Wiring Diagrams for Architectural Plans (Dec)(9106).

Curry Powder
See Also: Spices
CNS N5188-81. Spices and Condiments—Curry Powder—Specification (Oct)(8049). 2 pp.
ISO 2253-86. Curry Powder—Specification Second Edition. 5 pp.

Curtain Coating Machines
See Also: Papermaking Equipment
—**Glossaries**
BSI BS 4361: Part 28-90. 1990 Woodworking Machines Part 28: Nomenclature for Curtain Coating Machines. 11 pp.
ISO 9414-89. Woodworking Machines—Curtain Coating Machines—Nomenclature First Edition. 9 pp.

Curtains
Use: Draperies

Curtiss C 46 Commando Aircraft
See Also: Aircraft
—**Accidents**
CAA. Curtiss C46 Commando (World Airline Accident Summary). 6 pp.
CAA. Curtis C46 Commando. 1 p.

Curved Spring Washers
Use: Spring Lock Washers

Cushioning Materials
Use: Packaging Materials

Cushions
See Also: Furniture; Headrests; Pillows
—**Carpets**
CGSB 4-GP-36M-78. Carpet Underlay, Fibre Type, Standard for. 9 pp.
CGSB 20-GP-23M-78. Cushion, Carpet, Flexible Polymeric Material, Standard for. 10 pp.

—**Cellular Plastic**
CNS Z7036-83. Plastic Foam Cushion for Sick Bed (Sep)(3463).

—**Coconut Fiber—Beds**
CNS Z7191-84. Coconut Fibre Cushion for Sick Bed (Dec)(11179).

—**Foam Rubber**
CNS K4073-83. Latex Foam Rubber for Cushion (Jan)(9890).
CNS K6734-83. Method of Test for Latex Foam Rubber for Cushion (Jan)(9891).
JIS K 6382-78. Latex Foam Rubber for Cushion (R 1984). 11 pp.

—**Personal Flotation Devices**
CGSB CAN/CGSB-65.3-M88. Life-Saving Cushions. 14 pp.

—**Polyurethane Foams**
CNS K3035-78. Flexible Urethane Foam for Cushion (Jul)(4451). 2 pp.
CNS K6428-78. Method of Test for Flexible Urethane Foam (for Cushion) (Jul)(4452).
JIS K 6401-80. Flexible Urethane Foam for Cushion. 13 pp.

—**Rubber—Clamps—Aerospace**
AECMA PREN2901-89. Clamps, Loop (P Type) with Rubber Cushioning in Corrosion Resisting Steel Dimensions-Masses. 6 pp.
AECMA PREN2902-89. Clamps, Loop (P type) with Rubber Cushioning in Aluminium Alloy Dimensions-Masses. 6 pp.
AECMA PREN2905-89. Rubber Cushionings for Clamps Dimensions-Masses. 5 pp.

—**Rubber—Clamps—Aircraft**
SBAC AS 46760 (V). Clamp, Loop Corrosion Resisting Steel Fluorosilicone Rubber Cushion.

Cushions (Cont.)
—**Rubber—Clamps—Aircraft** (Cont.)
SBAC AS 46761 (V). Clamp, Loop, Aluminium Alloy Fluorosilicone Rubber Cushion.
SBAC AS 46782 (V). Clamp Loop, Corrosion Resistant Steel Fluorosilicone Rubber Cushion (Imperial Dimensions).
SBAC AS 46783 (V). Clamp Loop, Aluminium Alloy Fluorosilicone Rubber Cushion (Imperial Dimensions).
SBAC AS 61900-934 (V) (I). Clamp, Loop Style, Cushion Reinforced (Electrical Harness) (for Use with 5mm or 0.190 Bolts).
SBAC AS 61935-969 (V) (I). Clamp, Loop Style, Cushion Reinforced (Electrical Harness) (for Use with 6 mm or 0.250 Bolts).
SBAC AS 62200-234 (V). Clamp, Loop Style (Snap Lock) Cushion Reinforced (Electrical Harness) (for 5.00mm & 0.1900 Inch Fasteners).
SBAC AS 62250-284 ISSUE 2. Clamp, Loop Style (Snap Lock) Cushion Reinforced (Electrical Harness) (for 6.00mm & 0.2500 Inch Fasteners).
SBAC AS 62403-432 (V). Clamp, Loop Style, Box Cushion, Assembly of.
SBAC AS 62500-534 (V). Clamp, Loop Style, Cushion Reinforced (Electrical Harness) (for Use with 5mm or 0.190 Bolts).
SBAC AS 62600-634 (V). Clamp, Loop Style, Cushion Reinforced (Electrical Harness) (for Use with 6 mm or 0.250 Bolts).
SBAC AS 62700-742 (V). Clamp Loop Style (Grounding Strip) Cushion Reinforced (Electrical Harness) (for Use with 5mm or 0.190 Bolts).
SBAC AS 62800-842 ISSUE 2. Clamp Loop Style (Grounding Strip) Cushion Reinforced (Electrical Harness) (for Use with 6mm or 0.250 Bolts).

—**Rubber—Medical**
CNS T2005-80. Rubber Cushions for Patient (Jul)(5946).
JIS T 9105-75. Rubber Patient Cushions.

—**Safety Belts**
BSI BS AU 185-83. 1983 Seat Belt Booster Cushions. 18 pp.

—**Silicone Rubber—Aircraft**
SBAC TS 147 ISSUE 2. Silicone, Elastomer, Glassfibre Fabric Reinforced Cushion.

—**Urethane Foams—Naval Ships**
MOD UK NES 137-91. Requirements for Urethane Foam Components for Mattresses, Cushions and Upholstery Issue 2 (09.91). 15 pp.
MOD UK NES 137-01. Requirements for Urethane Foam Components for Mattresses, Cushions and Upholstery Issue 2 (09.91); Amendment 1. 20 pp.

Cuspidors
Use: Spittoons

Customer Premises Equipment
See Also: Communication Terminal Equipment; Telecommunication Equipment
—**Integrated Services Digital Networks**
CENELEC ETR 034-91. Business Telecommunications (BT); Approval Requirements for Complex Customer Premises Apparatus and Installations Connected to the Public ISDN (Including Principles for the Application of the Essential Requirements to Any Apparatus). 39 pp.
CSA CAN/CSA-T540-M92. Integrated Services Digital Network (ISDN) Primary Rate—Customer Installation Metallic Interfaces—Layer 1 Specification (ANSI T1.408-1990); (Gen Instr 1). 69 pp.
ETSI ETR 034-91. Business Telecommunications (BT); Approval Requirements for Complex Customer Premises Apparatus and Installations Connected to the Public ISDN (Including Principles for the Application of the Essential Requirements to Any Apparatus). 39 pp.

—**Power Supplies**
CEPT T/L 02-10-87. Alimentation D'Equipement De Telecommunication En Service Dans Les Installations D'Abonnes. 9 pp.
CEPT T/L 02-10 E-87. Power Supplies for Telecommunications Equipment Installed in Customers Premises. 9 pp.

Customs Duties
Use: Import Tariffs

Customs Regulations
—**Luggage—Rail Transportation**
EC 89/339/EEC-89. Council Decisions Accepting on Behalf of the Community the Recommendation of 5 June 1962 of the Customs Cooperation Council Concerning the Customs Treatment of Registered Baggage Carried by Rail as Amended of 21 June. 1 p.

INTERNATIONAL AND NON-U.S. NATIONAL STANDARDS
SUBJECT INDEX

Customs

Customs Regulations (Cont.)
—Shipping—Food
MOD UK DEFCON 20-82. HM Customs and Excise Transhipment Procedure 2/82. 1 p.

Customs Seals
See Also: Seals
—Ships
ISO 6205-88. Inland Navigation Vessels—Customs Sealing Systems—Basic Technical Requirements First Edition. 8 pp.

Cusum Charts
Use For: Cumulative Sum Chart *See Also:* Control Charts
—Data Analysis
BSI BS 5703: Part 1-80. 1980 Amd 1 Guide to Data Analysis and Quality Control Using Cusum Techniques Part 1: Introduction to Cusum Charting. 35 pp.
BSI BS 5703: Part 2-80. 1980 Amd 1 Guide to Data Analysis and Quality Control Using Cusum Techniques Part 2: Decision Rules and Statistical Tests for Cusum Charts and Tabulations. 44 pp.
BSI BS 5703: Part 3-81. 1981 Guide to Data Analysis and Quality Control Using Cusum Techniques Part 3: Cusum Methods for Process/Quality Control by Measurement. 48 pp.
BSI BS 5703: Part 4-82. 1982 Guide to Data Analysis and Quality Control Using Cusum Techniques Part 4: Cusums for Counted/Attributes Data. 39 pp.

—Process Control
BSI BS 5700-84. 1984 Amd 1 Process Control Using Quality Control Chart Methods and Cusum Techniques. 53 pp.

—Quality Control
BSI BS 5703: Part 1-80. 1980 Amd 1 Guide to Data Analysis and Quality Control Using Cusum Techniques Part 1: Introduction to Cusum Charting. 35 pp.
BSI BS 5703: Part 2-80. 1980 Amd 1 Guide to Data Analysis and Quality Control Using Cusum Techniques Part 2: Decision Rules and Statistical Tests for Cusum Charts and Tabulations. 44 pp.
BSI BS 5703: Part 3-81. 1981 Guide to Data Analysis and Quality Control Using Cusum Techniques Part 3: Cusum Methods for Process/Quality Control by Measurement. 48 pp.
BSI BS 5703: Part 4-82. 1982 Guide to Data Analysis and Quality Control Using Cusum Techniques Part 4: Cusums for Counted/Attributes Data. 39 pp.

Cutback Asphalt Cements
Use: Asphalt Cements

Cutback Asphalt Coatings
Use: Asphalt Coatings (Made From Asphalt)

Cutback Bitumens
Use: Bitumens

Cutlery
See Also: Cooking Utensils; Flatware; Spoons
BSI BS 5577-84. 1984 Amd 2 Table Cutlery. 17 pp.
BSI BS 6813-87. 1987 Amd 1 Trade Cutlery for Use with Food. 12 pp.

—Glossaries
ISO 4481-77. Cutlery and Flatware—Nomenclature First Edition. 42 pp.

—Silver Coatings—Electroplated
CNS S1185-87. Electroplated Coatings of Silver for Cutlery, Flat-Ware and Hollow-Ware (Oct)(11316).

—Silver Plated
ISO 8442-88. Stainless Steel and Silver-Plated Table Cutlery—Requirements First Edition. 15 pp.

—Stainless Steel
ISO 8442-88. Stainless Steel and Silver-Plated Table Cutlery—Requirements First Edition. 15 pp.

Cutoff Call Probability
See Also: Cutoff Call Ratio; Telephone Services
—Telephone Services
CCITT RECMN E.850-92. Connection Retainability Objective for the International Telephone Service—Telephone Network and ISDN—Quality of Service, Network Management and Traffic Engineering (Study Group II) 7 pp. 7 pp.

—Telephone Services—Estimation
CCITT RECMN E.850-92. Connection Retainability Objective for the International Telephone Service—Telephone Network and ISDN—Quality of Service, Network Management and Traffic Engineering (Study Group II) 7 pp. 7 pp.

Cutoff Call Ratio
See Also: Cutoff Call Probability; Telephone Services
—Telephone Services
CCITT RECMN E.428-92. Connection Retention—Telephone Network and ISDN of Service, Network Management and Traffic Engineering (Study Group II) 4 pp. 4 pp.

Cutoff Valves
Use: Stop Valves

Cutoff Wavelength
—Optical Fibers
CCITT RECMN G.652-89. Characteristics of a Single-Mode Optical Fibre Cable—Transmission Media Characteristics (Study Group XV) 34 pp. 34 pp.
CCITT RECMN G.654-89. Characteristics of a 1550 mm Wavelength Loss-Minimized Single-Mode Optical Fibre Cable—Transmission Media Characteristics (Study Group XV) 3 pp. 3 pp.

—Optical Fibers—Graphs (Charts)
CCITT FASCICLE III.3. Transmission Media Characteristics Recommendations G.601—G.654. 134 pp.

Cutout Boxes
See Also: Electrical Enclosures; Fuse Boxes; Fuses (Electric)
—Hinged Covers
CSA C22.2 NO 40-M1989. Cutout, Junction, and Pull Boxes; (Gen Instr 1 Thru 2). 46 pp.
CSA 896 Bull. Electrical Bulletin 896 September 18, 1972 to C22.2 NO 40. 1 p.

Cutters
Use: Cutting Tools

Cutters (Tools)
Use: Cutting Tools

Cutting
See Also: Chipping; Cutting Tools; Gas Cutting; Laser Cutting; Thermal Cutting
—Glossaries
SNZ NZS/AS 2812-85. Welding, Brazing and Cutting of Metals—Glossary of Terms (This is a Joint Standard with SAA AS 2812). 98 pp.

—Safety
SNZ NZS 4781-73. Code of Practice for Safety in Welding and Cutting Amend: 1, 1976. 44 pp.

Cutting Equipment
Use: Cutting Tools

Cutting Fluids
See Also: Coolants; Cutting Tools; Lubricants
CNS K5060-74. Water Soluble Cutting Fluid (Tentative) (Jan)(3698).
JIS K 2241-86. Cutting Fluid. 28 pp.
MOD UK DSTAN 91-26-76. Cutting Fluid: Compounded NATO Code No: 0-216 Joint Service Designation: ZX-6 Cutting Fluid: Mineral Joint Service Designation: ZX—27 Issue 1 (Withdrawn). 6 pp.
MOD UK DSTAN 91-70-90. Cutting Fluid, Soluble, Biostable Joint Service Designation ZX-9 Issue 1. 25 pp.
MOD UK DSTAN 91-70-01. Cutting Fluid, Soluble, Biostable Joint Service Designation ZX-9 Issue 1; Amendment 1. 26 pp.

Cutting Machines
Use For: Clipper Machines; Clipping Machines; Tipping Machines; Trimming Machines
See Also: Machine Tools; Nibblers; Tools
—Bolt Heads
CNS B7107-78. Testing Standard for (Bolt Head) Trimming Machine (Nov)(4668).

—Safety
CEN PREN 706-92. Safety Requirements for Agricultural and Forestry Machinery—Vine Shoot Tipping Machines. 12 pp.

Cutting Pliers
See Also: Nippers; Pliers
BSI BS 3087: Part 7-86. 1986 Amd 1 Pliers and Nippers Part 7: Specification for Dimensions of Lever Assisted Side Cutting Pliers, End and Diagonal Cutting Nippers (AMD 6539) July 31, 1991 (Supersedes BS 3087: 1959). 12 pp.
CNS B3146-86. Cutting Pliers (for General Use) (Feb)(1131). 2 pp.
CNS B3149-87. Cutting Pliers (Apr)(1134). 4 pp.

Cutting Pliers (Cont.)
ISO 5747-84. Pliers and Nippers—Lever Assisted Side Cutting Pliers, End and Diagonal Cutting Nippers—Dimensions First Edition. 6 pp.
JIS B 4614-75. Slip Joint Combination Pliers with Cutters. 7 pp.
JIS B 4623-87. Cutting Pliers. 9 pp.
SAA AS/NZS 4056. 6-92. Pliers and Nippers—Part 6: Lever Assisted Side Cutting Pliers, End and Diagonal Cutting Nippers—Dimensions and Test Values (ISO 5747:1984). 4 pp.
SNZ NZS/AS 4056. 6-92. Pliers and Nippers Part 6: Lever Assisted Side-Cutting Pliers, End and Diagonal Cutting Nippers—Dimensions and Test Values (This is a Joint Standard with SAA AS 4056.6). 4 pp.

—Nonsparking
CNS M2066-86. Non-Sparking Beryllium Copper Alloy Tools Cutting Pliers (Nov) (5914).

—Wire
CNS B3149-87. Cutting Pliers (Apr)(1134). 4 pp.

Cutting Tools
Scope Note: For additional listings, use a more specific term *Use For:* Cutters; Cutters (Tools); Cutting Equipment *See Also:* Adzes; Axes; Band Saws; Bill Hooks; Bits (Tools); Boring Tools; Broaches; Broaching Machines; Burrs (Tools); Cable Cutters; Carbide Tools; Center Drills; Circular Saws; Countersinks; Cutting; Cutting Fluids; Dental Instruments; Diamond Tools; Die Cutters; Drilling Equipment; Drills; Electric Discharge Machine Tools; Food Cutters; Gas Cutting Equipment; Grinders; Grinding Wheels; Groove Cutters; Guillotine Shears; Hand Tools; High Speed Steel Tools; Hobbing Machines; Knives; Lathes; Machine Tools; Meat Cutters; Milling Cutters; Milling Machines; Mortising Machines; Paper Cutters; Pipe Cutters; Planers (Tools); Pneumatic Tools; Reamers; Saws; Scissors; Shears (Hand Tools); Sheeters; Shredders; Single Point Cutting Tools; Slicers; Taps (Threading Tools); Tool Life; Tools; Turning Tools; Turret Lathes; Veneer Clippers; Wire Cutters; Woodworking Equipment
CNS B3201-85. Cutter for Steel Bars (Dec)(3100).
CNS B7213-82. Method of Test for Cutting Device (Oct)(9474).

—Blowpipes
DIN ENGL 8543 Pt 5-86. Blowpipes for Oxy-Fuel Gas Processes; Machine Cutting Blowpipes for Fuel Gas/Oxygen; Types, Concepts, Requirements, Marking, Testing (Sept). 11 pp.
ISO 5172-77. Manual Blowpipes for Welding and Cutting First Edition. 13 pp.

—Bourdon Tubes
BSI BS 6752-86. 1986 Bourdon Tube Pressure Gauges Used in Welding, Cutting and Related Processes. 12 pp.
DIN ENGL 8549-86. Bourdon-Tube Pressure Gauges Used in Welding, Cutting and Related Processes with 50 mm or 63 mm Diameter Casing (Dec). 8 pp.
ISO 5171-80. Pressure Gauges Used in Welding, Cutting and Related Processes First Edition; (Erratum—Dec 1981). 10 pp.

—Coatings
MOD UK CS 2486B. Protective PX 15.

—Construction Equipment—Road—Joint Cutters—Safety
CEN PREN 500-5-91. Mobile Road Construction Machines—Safety—Part 5: Specific Requirements for Joint Cutters. 6 pp.

—Conversion Formulae
ISO 3002 Pt 2-82. Basic Quantities in Cutting and Grinding—Part 2: Geometry of the Active Part of Cutting Tools—General Conversion Formulae to Relate Tool and Working Angles First Edition. 38 pp.

—Energy
ISO 3002 Pt 4-84. Basic Quantities in Cutting and Grinding—Part 4: Forces, Energy, Power First Edition. 14 pp.

—Forces
ISO 3002 Pt 4-84. Basic Quantities in Cutting and Grinding—Part 4: Forces, Energy, Power First Edition. 14 pp.

—Geometric and Kinematic Quantities
ISO 3002 Pt 3-84. Basic Quantities in Cutting and Grinding—Part 3: Geometric and Kinematic Quantities in Cutting First Edition. 12 pp.

Cutting Tools (Cont.)

—Glossaries

DIN ENGL 9869 Pt 2-69. Terms for Pressworking Tools; Cutting Tools (Nov). 8 pp.

DIN ENGL 9870 Pt 2-72. Terms for Stamping Practice; Production Processes and Tools for Severing (Oct). 10 pp.

ISO 3002 Pt 1-82. Basic Quantities in Cutting and Grinding—Part 1: Geometry of the Active Part of Cutting Tools—General Terms, Reference Systems, Tool and Working Angles, Chip Breakers Second Edition; (Amendment 1-1992). 62 pp.

ISO 3002 Pt 2-82. Basic Quantities in Cutting and Grinding—Part 2: Geometry of the Active Part of Cutting Tools—General Conversion Formulae to Relate Tool and Working Angles First Edition. 38 pp.

JIS B 0170-73. Glossary of Common Terms of Cutting Tools.

—Hose Fittings

DIN ENGL 8544-81. Probe Couplings for Equipment and Gas Hoses for Welding, Cutting and Related Processes; Couplings for Hoses from 4 to 10 mm Internal Diameter (Nov). 7 pp.

ISO 3253-75. Hose Connections for Equipment for Welding, Cutting and Related Processes First Edition. 6 pp.

—Indexable Inserts

ISO 6987 Pt 2-90. Indexable Inserts for Cutting Tools—Hardmetal (Carbide) Inserts with Rounded Corners, with Partly Cylindrical Fixing Hole—Part 2: Dimensions of Inserts with 11 Degree Normal Clearance First Edition. 12 pp.

ISO TR6987 Pt 3-90. Indexable Inserts for Cutting Tools—Hardmetal (Carbide) Inserts with Rounded Corners, with Partly Cylindrical Fixing Hole—Part 3: V-Shape Inserts First Edition. 9 pp.

—Indexable Inserts—Designations

BSI BS 4193: Part 1-93. 1993 Hardmetal Insert Tooling Part 1: Specification for Designation of Indexable Inserts for Cutting Tools (ISO 1832: 1991). 20 pp.

BSI BS 4193: Part 1-86. 1986 Hardmetal Insert Tooling Part 1: Designation of Indexable Inserts for Cutting Tools. 16 pp.

ISO 1832-91. Indexable Inserts for Cutting Tools—Designation Third Edition. 18 pp.

JIS B 4120-85. Indexable Inserts for Cutting Tools-Designation. 21 pp.

—Inserts

CNS B3191-76. Inserts in Cutting Material (Dec)(3006).

—Packaging

MOD UK DEF-1244-57. Packaging of Engineers' Cutting Tools (Superseded by Def Stan 81-41: Part 2). 4 pp.

—Power

ISO 3002 Pt 4-84. Basic Quantities in Cutting and Grinding—Part 4: Forces, Energy, Power First Edition. 14 pp.

—Quick Disconnect Couplings—Stop Valves

ISO 7289-90. Quick-Action Couplings with Shut-off Valve for Welding, Cutting and Allied Processes First Edition. 10 pp.

—Safety

DIN VDE 0700 Pt 30-83. Test Certificate, Safety Test of Household & Similar Electrical Appliances. Portable Electric Heating Tools (Jan). 26 pp.

DIN VDE 0700 Pt 30 A1 (D)-84. Safety of Household and Similar Electrical Appliances; Particular Specifications for Portable Electric Heating Tools (VDE Specification): Amendment 1 (Nov). 19 pp.

—Shanks

JIS B 4002-90. Square Portion of Shanks for Small Tools. 8 pp.

Cuttlefish

See Also: Mollusks; Snails; Squid

—Dried

CNS N5050-63. Dried Cuttlefish (for Export) (Tentative) (R1973) (Nov)(2246). 2 pp.

—Dried—Sampling

CNS N6042-81. Method of Test for Dried Squids and Cuttle Fish (Aug)(2299). 2 pp.

—Dried—Visual Inspection

CNS N6042-81. Method of Test for Dried Squids and Cuttle Fish (Aug)(2299). 2 pp.

Cyanates

Scope Note: Use a more specific term *See:* Mercuric Thiocyanate; Toluene Diisocyanates

Cyanidation

Use: Cyanide Process

Cyanidation Precipitates

Use: Cyanide Process

Cyanide Content Analysis

—Animal Feed

CNS N4063-80. Method of Test for Cyanide in Feedstuff (Dec)(6711).

—Drinking Water—Photometry

DIN ENGL 38405 Pt 14-88. German Standard Methods for the Examination of Water, Waste Water and Sludge; Anions (Group D); Determination of Cyanides in Drinking Water, and in Groundwater and Surface Water with Low Pollution Levels (D 14) (Dec). 7 pp.

—Ground Water—Photometry

DIN ENGL 38405 Pt 14-88. German Standard Methods for the Examination of Water, Waste Water and Sludge; Anions (Group D); Determination of Cyanides in Drinking Water, and in Groundwater and Surface Water with Low Pollution Levels (D 14) (Dec). 7 pp.

—Polishes

CGSB 25-GP-1M METH 19.1-84. Methods of Sampling and Testing Waxes and Polishes Cyanides. 1 p.

—Surface Waters—Photometry

DIN ENGL 38405 Pt 14-88. German Standard Methods for the Examination of Water, Waste Water and Sludge; Anions (Group D); Determination of Cyanides in Drinking Water, and in Groundwater and Surface Water with Low Pollution Levels (D 14) (Dec). 7 pp.

—Waste Water

CNS K9077-82. Method of Test for Cyanide in Industrial Waste Water Pyridine Pyrazolone Method (Jan)(8418).

CNS K9078-82. Method of Test for Cyanide in Industrial Waste Water Mercury Thiocyanate Method (Jan)(8419).

CNS K9079-84. Method of Test for Cyano-Compounds in Industrial Waste Water (Silver Nitrate Method) (Dec)(8420).

—Waste Water—Photometry

DIN ENGL 38405 Pt 13-81. German Standard Methods for the Analysis of Water, Waste Water and Sludge; Anions (Group D); Determination of Cyanides (D 13) (Feb). 13 pp.

—Waste Water—Volumetric Analysis

DIN ENGL 38405 Pt 13-81. German Standard Methods for the Analysis of Water, Waste Water and Sludge; Anions (Group D); Determination of Cyanides (D 13) (Feb). 13 pp.

—Water—Diffusion

BSI BS 6068: Sec 2.20-86. 1986 Water Quality Part 2: Physical, Chemical and Bio-Chemical Methods Section 2.20: Methods for Determination of Cyanide by Diffusion at pH6. 9 pp.

ISO 6703 Pt 4-85. Water Quality—Determination of Cyanide—Part 4: Determination of Cyanide by Diffusion at pH 6 First Edition. 9 pp.

—Water—Photometry

BSI BS 6068: Sec 2.17-86. 1986 Water Quality Part 2: Physical, Chemical and Bio-Chemical Methods Section 2.17: Methods for Determination of Total Cyanide. 14 pp.

BSI BS 6068: Sec 2.18-86. 1986 Water Quality Part 2: Physical, Chemical and Bio-Chemical Methods Section 2.18: Methods for Determination of Easily Liberatable Cyanide. 14 pp.

DIN ENGL 38405 Pt 13-81. German Standard Methods for the Analysis of Water, Waste Water and Sludge; Anions (Group D); Determination of Cyanides (D 13) (Feb). 13 pp.

ISO 6703 Pt 1-84. Water Quality—Determination of Cyanide—Part 1: Determination of Total Cyanide First Edition. 15 pp.

ISO 6703 Pt 2-84. Water Quality—Determination of Cyanide—Part 2: Determination of Easily Liberatable Cyanide First Edition. 15 pp.

—Water—Volumetric Analysis

BSI BS 6068: Sec 2.17-86. 1986 Water Quality Part 2: Physical, Chemical and Bio-Chemical Methods Section 2.17: Methods for Determination of Total Cyanide. 14 pp.

BSI BS 6068: Sec 2.18-86. 1986 Water Quality Part 2: Physical, Chemical and Bio-Chemical Methods Section 2.18: Methods for Determination of Easily Liberatable Cyanide. 14 pp.

DIN ENGL 38405 Pt 13-81. German Standard Methods for the Analysis of Water, Waste Water and Sludge; Anions (Group D); Determination of Cyanides (D 13) (Feb). 13 pp.

ISO 6703 Pt 1-84. Water Quality—Determination of Cyanide—Part 1: Determination of Total Cyanide First Edition. 15 pp.

ISO 6703 Pt 2-84. Water Quality—Determination of Cyanide—Part 2: Determination of Easily Liberatable Cyanide First Edition. 15 pp.

—Waxes

CGSB 25-GP-1M METH 19.1-84. Methods of Sampling and Testing Waxes and Polishes Cyanides. 1 p.

Cyanide Process

—Gold Content

CNS M3039-81. Methods for Determination of Gold and Silver in Cyanidation Precipitates (Apr)(7287).

—Moisture Content

CNS M3045-81. Methods for Sampling and Determination of Moisture Content of Cyanidation Precipitates (May)(7442).

—Sampling

CNS M3045-81. Methods for Sampling and Determination of Moisture Content of Cyanidation Precipitates (May)(7442).

—Silver Content

CNS M3039-81. Methods for Determination of Gold and Silver in Cyanidation Precipitates (Apr)(7287).

Cyanides

Scope Note: For additional listings, use a more specific term *See Also:* Cadmium Cyanide; Cupric Cyanide; Hazardous Materials; Mercuric Cyanide; Potassium Cyanide; Silver Cyanide; Sodium Cyanide; Sodium Nitroprusside; Zinc Cyanide

—Highway Transportation—Emergency Procedures

SAA AS 1678.6.0. 009-84. Emergency Procedure Guides—Transport—Part 6.0.009: Cyanides, Inorganic.

Cyanmethaemoglobin

Use: Cyanmethemoglobin

Cyanmethemoglobin

Use For: Cyanmethaemoglobin; Haemoglobin Cyanide

—Blood Analysis—Photometry

BSI BS 3985-78. 1978 Specification for Cyanmethemoglobin (Haemiglobincyanide) Solution for Photometric Haemoglobinometry. 7 pp.

Cyanoacrylate Adhesives

See Also: Adhesives

JIS K 6861-77. Testing Methods for Alpha Particle-Cyanoacrylate Adhesives.

MOD UK DSTAN 80-146-91. Adhesive, Ethyl and Methyl Cyanoacrylate, Types 1, 2 and 3 Issue 1. 10 pp.

Cyanogen Bromide

CNS K7174-65. Chemical Reagent (Cyanogen Bromide) (Jan)(1674).

JIS K 8508-61. Cyanogen Bromide.

Cyanogen Chloride Content Analysis

—Water

BSI BS 6068: Sec 2.19-86. 1986 Water Quality Part 2: Physical, Chemical and Bio-Chemical Methods Section 2.19: Methods for Determination of Cyanogen Chloride. 9 pp.

ISO 6703 Pt 3-84. Water Quality—Determination of Cyanide—Part 3: Determination of Cyanogen Chloride First Edition. 8 pp.

Cycle Holders

Use: Bicycle Racks

Cycle Life Testing

Use: Cyclic Testing

Cycle Racks

Use: Bicycle Racks

Cycles

See Also: Bicycles; Exercise Bicycles; Motorcycles

—Chains

ISO 9633-92. Cycle Chains—Characteristics and Test Methods First Edition. 13 pp.

Cycles (Cont.)

—Classification
JIS D 9111-86. Classification and Essential Characteristics of Cycles.

—Glossaries
JIS D 9101-91. Cycles—Terminology.

—Lamps—Photometry
BSI BS 6102: Part 3-86. 1986 Amd 1 Cycles Part 3: Photometric and Physical Requirements of Lighting Equipment. 23 pp.

Cyclic Redundancy Check
See Also: Error Control; Telecommunication Services

—Frame Alignment—Frame Structure
CCITT RECMN G.706-91. Frame Alignment and Cyclic Redundancy Check (CRC) Procedures Relating to Basic Frame Structures Defined in Recommendation G.704 (Study Group XVIII) 19 pp. 19 pp.
CCITT RECMN G.706-89. Frame Alignment and Cyclic Redundancy Check (CRC) Procedures Relating to Basic Frame Structures Defined in Recommendation G.704—General Aspects of Digital Transmission Systems; Terminal Equipments (Study Groups XV and XVIII) 7 pp. 7 pp.

Cyclic Testing
See Also: Life (Durability)

—Damp Heat
CNS C6334-89. Basic Environmental Testing Procedures, Part 2: Tests, Test Db: Damp Heat, Cycle (12+12 —Hour Cycle) (Jul)(12565).

—Fiberboard
BSI BS EN 321-93. 1993 Fibreboards—Cyclic Tests in Humid Conditions (R). 10 pp.
CEN PREN 321-90. Fibreboards—Wet Cyclic Test. 6 pp.
CEN EN 321-93. Fibreboards—Cyclic Tests in Humid Conditions. 5 pp.

—Humidity
CNS C6335-89. Basic Environmental Testing Procedures, Part 2: Tests, Test Z/AD: Composite Temperature /Humidity Cyclic Test (Jul)(12566).

—Temperature
CNS C6335-89. Basic Environmental Testing Procedures, Part 2: Tests, Test Z/AD: Composite Temperature /Humidity Cyclic Test (Jul)(12566).

Cyclohexane
See Also: Benzene Hexachloride; Hazardous Materials; n-Hexane; Hydrocarbons
CNS K7175-65. Chemical Reagent (Cyclohexane) (Jan)(1675).
JIS K 8464-84. Cyclohexane.
MOD UK DSTAN 68-16-70. Cyclohexane, Technical Issue 1. 6 pp.

—Benzene Content—Gas Chromatography
CNS K6651-81. Method of Test for Purity and Benzene Content of Cyclohexane by Gas Chromatography (Mar)(7170).

—Purity—Gas Chromatography
CNS K6651-81. Method of Test for Purity and Benzene Content of Cyclohexane by Gas Chromatography (Mar)(7170).

Cyclohexane-995
See Also: Hydrocarbons
CNS K1249-80. Cyclohexane-995 (Jul)(5828).

Cyclohexanols
See Also: Alcohols
CNS K7176-65. Chemical Reagent (Cyclohexanol) (Jan)(1676).
JIS K 8462-78. Cyclohexanol.

Cyclohexanone
Use For: Anone; Ketohexamethylene; Nadone; Pimelic Ketone *See Also:* Hazardous Materials; Ketones
CNS K2206-88. Cyclohexanone (Apr)(12274). 2 pp.
CNS K7177-65. Chemical Reagent (Cyclohexanone) (Jan)(1677).
JIS K 4171-63. Cyclohexanone.
JIS K 8463-81. Cyclohexanone.

N-Cyclohexyl-2-Benzothiazyl Sulfenamide
CNS K4049-79. Rubber Vulcanization Accelertor CBS (N-Cyclo-Hexyl-2-Benzothiazyl Sulfenamide) (Oct)(5005).

Cyclohexylamine Content Analysis

—Water
CNS K9053-80. Method of Test for Cyclohexylamine in Water (Sep)(6476).

—Water—Spectrophotometry
BSI BS 2690: Part 115-83. 1983 Water Used in Industry Part 115: Cyclohexylamine; Spectrophotometric Method. 4 pp.

Cyclonite
Use For: Trimethylene Trinitroamine
See Also: Explosives
JIS K 4816-69. Trimethylene Trinitroamine.

Cyclotetramethylenetetranitramine
Use: Octogene

Cylinder Blocks
Use For: Engine Blocks *See Also:* Cylinder Heads; Engine Cylinders; Engines; Pistons

—Diesel Engines—Preservation
MOD UK DSTAN 81-72-90. Preservation of Cylinder Blocks (with or Without Pistons) and Cylinder Heads Issue 1. 9 pp.

—Gasoline Engines—Preservation
MOD UK DSTAN 81-72-90. Preservation of Cylinder Blocks (with or Without Pistons) and Cylinder Heads Issue 1. 9 pp.

Cylinder Bores

—Fluid Power Cylinders
ISO 3320-87. Fluid Power Systems and Components—Cylinder Bores and Piston Rod Diameters—Metric Series Second Edition. 4 pp.
ISO 3321-75. Fluid Power Systems and Components—Cylinder Bores and Piston Rod Diameters—Inch Series First Edition. 4 pp.

—Fluid Power Cylinders—Port Thread Sizes
ISO 7180-86. Pneumatic Fluid Power—Cylinders—Bore and Port Thread Sizes First Edition. 4 pp.

Cylinder Gages
See Also: Cylinders (Containers)
CNS B6031-82. Bore Gauge (Cylinder Gauges) (Jan)(4754).
JIS B 7515-82. Cylinder Gauges. 17 pp.

—Gas Cylinders—Valves
CEN PREN 628-92. Gas Cylinders—Requirements for Gauge Inspection of Tapered Threads for Connection of Valves to Gas Cylinders of 25.8 mm Nominal Diameter. 17 pp.

Cylinder Heads
See Also: Cylinder Blocks; Engine Cylinders; Engine Valves

—Diesel Engines—Preservation
MOD UK DSTAN 81-72-90. Preservation of Cylinder Blocks (with or Without Pistons) and Cylinder Heads Issue 1. 9 pp.

—Gaskets
CNS D2090-81. Cylinder Head Gaskets for Automobile Engines (Jul)(7667).
JIS D 3105-92. Cylinder Head Gaskets for Automobile Engines. 26 pp.

—Gasoline Engines—Preservation
MOD UK DSTAN 81-72-90. Preservation of Cylinder Blocks (with or Without Pistons) and Cylinder Heads Issue 1. 9 pp.

—Spark Plugs—Housings
ISO 1919-88. Road Vehicles—M14 X 1,25 Spark-Plugs with Flat Seating and Their Cylinder Head Housing Fourth Edition. 7 pp.
ISO 2344-92. Road Vehicles—M14 X 1,25 Spark-Plugs with Conical Seating and Their Cylinder Head Housing Sixth Edition. 8 pp.
ISO 2345-87. Road Vehicles—M18 X 1,5 Spark-Plugs with Conical Seating and Their Cylinder Head Housing Fourth Edition. 7 pp.
ISO 2346-91. Road Vehicles—M14 x 1, 25 Compact Spark-Plugs with Flat Seating and Their Cylinder Head Housing Fourth Edition. 7 pp.
ISO 2347-87. Road Vehicles—M14 X 1,25 Compact Spark-Plugs with Conical Seating and Their Cylinder Head Housing Fourth Edition. 7 pp.
ISO 2704-82. Road Vehicles—Spark Plugs M 10 X 1 with Flat Seating and Their Cylinder Head Housing Third Edition. 6 pp.
ISO 2705-91. Road Vehicles—M12 X 1,25 Spark-Plugs with Flat Seating and Their Cylinder Head Housing Fourth Edition. 7 pp.

Cylinder Heads (Cont.)

Spark Plugs—Housings (Cont.)
ISO 3412-92. Road Vehicles—Screened and Waterproof Spark-Plugs and Their Connections—Types 1A and 1B Third Edition. 6 pp.
ISO 3895-86. Road Vehicles—Screened and Waterproof Spark-Plug and Its Connection—Type 2 Third Edition. 5 pp.
ISO 3896-86. Road Vehicles—Screened and Waterproof Spark-Plug and Its Connection—Type 3 Third Edition. 5 pp.

Cylinder Locks
See Also: Locks (Security)

—Doors
CNS A2052-91. Cylindrical Locks and Tubular Locks for Doors (Feb)(3928). 5 pp.
CNS A3065-88. Method of Test for Cylindrical Locks and Tubular Locks for Doors (Apr)(3929). 2 pp.
JIS A 5535-76. Cylindrical Locks and Tubular Locks for Doors.

Cylinder Valves
See Also: Cylinders (Containers); Gas Cylinders; Valves

—Acetylene
CNS B5077-90. Valve for Dissolved Acetylene Cylinder (May)(4152).

Cylinders
See: Engine Cylinders; Fluid Power Cylinders; Gas Cylinders; Hydraulic Cylinders; Measuring Cylinders; Pneumatic Cylinders; Shipping Containers

Cylinders (Containers)
Scope Note: For additional listings, use a more specific term *Use For:* Canisters (Containers)
See Also: Cryogenic Containers; Cylinder Gages; Cylinder Valves; Engine Cylinders; Fluid Power Cylinders; Gas Cylinders; Hydraulic Cylinders; Measuring Cylinders; Pneumatic Cylinders; Shipping Containers

—Hazardous Materials
CSA CAN/CSA-B339-88. Cylinders, Spheres, and Tubes for the Transportation of Dangerous Goods; (Gen Instr 1 Thru 3). 112 pp.
CSA CAN/CSA-B340-M88. Selection and Use of Cylinders, Spheres, Tubes, and Other Containers for the Transportation of Dangerous Goods, Class 2; (Gen Instr 1 Thru 3). 50 pp.

—Lubricating Oils
CNS K5054-88. Cylinder Oil (Jan)(2979). 2 pp.

—Protective Masks
NATO STANAG 4155 ED 1 AMD 0-81. NBC Protective Mask and Filter Canister Screw Threads. 11 pp.

Cylindrical Connectors
Use: Circular Connectors

Cylindrical Cutters
Use: Milling Cutters—Cylindrical

Cylindrical Gears
See Also: Gears; Spur Gears
CNS B1240-84. Addendum Modification of Cylindrical Gears (Aug)(7186).
CNS B2420-84. Cylindrical Gears for Fine Mechanics; Radial Setting of Tool (Aug)(5277).
ISO 1340-76. Cylindrical Gears—Information to Be Given to the Manufacturer by the Purchaser in Order to Obtain the Gear Required First Edition. 4 pp.

—Gear Cutters—Pinion Type
CNS B1212-80. Pinion-Type Cutters for Cylindrical Gears; Tolerances (Sep)(6419).

—Gear Hobs
CNS B3410-81. Metal-Cutting Tools; Hobs for Cylindrical Gears with Clutch Drive Slot or Keyway (Apr)(7194).

—Gear Teeth
CNS B2415-84. Accuracy of Cylindrical Gears Teeth; Tolerance for Errors of Cumulative Circular Pitch (Aug)(5267).
CNS B2416-84. Accuracy of Cylindrical Gears Teeth; Tolerance for Errors of Composite (Aug)(5268).
CNS B2426-84. Accuracy of Cylindrical Gears Teeth; Tolerance for Errors of Tooth (Aug)(5283).
CNS B2428-84. Accuracy of Cylindrical Gear Teeth; Tolerance for Errors of Individual Parameters (Aug)(5285).
DIN ENGL 3961-78. Tolerances for Cylindrical Gear Teeth; Bases (Aug). 12 pp.

INTERNATIONAL AND NON-U.S. NATIONAL STANDARDS
SUBJECT INDEX

Cylindrical Gears (Cont.)
—Gear Teeth (Cont.)
DIN ENGL 3962 Pt 1-78. Tolerances for Cylindrical Gear Teeth; Tolerances for Deviations of Individual Parameters (Aug). 18 pp.
DIN ENGL 3962 Pt 2-78. Tolerances for Cylindrical Gear Teeth; Tolerances for Tooth Trace Deviations (Aug). 2 pp.
DIN ENGL 3962 Pt 3-78. Tolerances for Cylindrical Gear Teeth; Tolerances for Pitch-Span Deviations (Aug). 2 pp.
DIN ENGL 3963-78. Tolerances for Cylindrical Gear Teeth; Tolerances for Working Deviations (Aug). 18 pp.
ISO TR4467-82. Addendum Modification of the Teeth of Cylindrical Gears for Speed-Reducing and Speed-Increasing Gear Pairs First Edition. 12 pp.
JIS B 1741-77. Tooth Contact Marking of Gears. 10 pp.

—Gear Teeth—Backlash—Glossaries
CNS B1189-84. Definitions and Denominations for Glossary of Gear Terms — Errors for the Axial Positions and Backlash for a Cylindrical Gear Teeth (May)(5731).

—Gear Teeth—Glossaries
CNS B1185-84. Definitions and Denominations for Glossary of Gear Terms — Cylindrical Gear Teeth (May)(5727).
CNS B1187-84. Definitions and Denominations for Glossary of Gear Terms — Deviation, Tolerances and Change Factor for Cylindrical Gear Teeth (May)(5729).
CNS B1188-84. Definitions and Denominations for Glossary of Gear Terms — Errors for Cylindrical Gear Teeth (May)(5730).

—Gear Teeth—Loads (Forces)
CNS B1202-84. Calculation of Load Capacity on Tooth Root for Cylindrical Gears (Aug)(5982).
CNS B1203-84. Calculation of Load Capacity on Tooth Flank for Cylindrical Gears (Aug)(5983).
CNS B1206-84. Calculation of Load Capacity for Cylindrical and Bevel Gears-Tooth Form Factor YF (Aug)(5986).

—Glossaries
CNS B1176-84. Glossary of Gears Geometrical Definitions (Cylindrical Gears and Cylindrical Gear Pairs) (Feb)(5718).
CNS B1186-84. Definitions and Denominations for Glossary of Gear Terms — Cylindrical Gear Pair (May)(5728).
DIN ENGL 3998 Pt 2-76. Denominations on Gears and Gear Pairs; Cylindrical Gears and Gear Pairs (Sept). 17 pp.

—Involute Gear Teeth
CNS B2421-84. Basic Rack of Cylindrical Gears with Involute Teeth for Fine Mechanics (Aug)(5278).

—Involute Gear Teeth—Engineering Drawings
DIN ENGL 3966 Pt 1-78. Information on Gear Teeth in Drawings; Information on Involute Teeth for Cylindrical Gears (Aug). 8 pp.

—Involute Gear Teeth—Inspection
ISO TR10064 Pt 1-92. Cylindrical Gears—Code of Inspection Practice—Part 1: Inspection of Corresponding Flanks of Gear Teeth First Edition. 63 pp.

—Involute Gear Teeth—Profiles
CNS B1249-84. Basic Rack Tooth Profile of Gear Tools for Cylindrical Gears with Involute Teeth for Fine Mechanics (Aug)(7572).
CNS B2023-84. Basic Rack of Cylindrical Gears with Involute Teeth for General and Heavy Engineering (Aug)(184).
DIN ENGL 867-86. Basic Rack Tooth Profiles for Involute Teeth of Cylindrical Gears for General Engineering and Heavy Engineering (Feb). 3 pp.

—Involute Gear Teeth—Thickness Measurement
DIN ENGL 3977-81. Measuring Element Diameters for the Radial or Diametral Dimension for Testing Tooth Thickness of Cylindrical Gears (Feb). 7 pp.

—Loads (Forces)—Capacity Measurement
CNS B2519-84. Calculation of Load Capacity for Cylindrical and Bevel Gears-Load Sharing Factor Y Sub Epsilon and Overlap Ration Sub Epsilon Sub Beta (Aug)(6407).
CNS B2522-84. Calculation of Load Capacity for Cylindrical and Bevel Gears-Pinion Single Tooth Contact Factor ZB Wheel Single Tooth Contact Factor ZD and Traverse Contact Ratio Sub Epsilon Sub Beta (Aug)(6410).
CNS B2523-84. Calculation of Load Capacity for Cylindrical and Bevel Gears-Contact Ration Factor Z Sub Epsilon (Aug)(6411).

Cylindrical Gears (Cont.)
—Loads (Forces)—Capacity Measurement (Cont.)
CNS B2524-84. Calculation of Load Capacity for Cylindrical and Bevel Gears-Helix Angle Factor Y Beta Ray (Aug)(6412).

—Master
JIS B 1751-76. Master Cylindrical Gears.
JIS B 1751-69. Master Cylindrical Gears. 13 pp.

—Modules
CNS B2022-84. Series of Modules for Cylindrical Gears (Aug)(183).

—Shafts (Machine Elements)
CNS B2424-84. Deviation of Center Distance and Parallelism Tolerances for Axes of Cases for Cylindrical Gears (Aug)(5281).
DIN ENGL 3964-80. Deviations of Shaft Centre Distances and Shaft Position; Tolerances of Casings for Cylindrical Gears (Nov). 4 pp.

Cylindrical Grinders
See Also: Grinders; Machine Tools
CNS B1120-79. Specified Items of Machine Tools Cylindrical Grinder and Universal Grinder (Apr)(4670-23). 2 pp.
CNS B1124-79. Specified Items of Machine Tools (Centerless Grinder) (Apr)(4670-27). 1 p.

—Acceptance Testing—External
BSI BS 4656: Part 9-74. 1974 Amd 1 Accuracy of Machine Tools and Methods of Test Part 9: External Cylindrical Grinding Machines with Reciprocating Table. 16 pp.
BSI BS 4656: Part 24-93. 1993 Accuracy of Machine Tools and Methods of Test Part 24: Specification for Cylindrical External Centreless Grinding Machines (ISO 3875: 1990). 20 pp.
BSI BS 4656: Part 24-80. 1980 Accuracy of Machine Tools and Methods of Test Part 24: Cylindrical External Centreless Grinding Machines. 12 pp.
DIN ENGL 8630-88. Machine Tools; External Cylindrical Grinding Machines with a Movable Workpiece Table; Acceptance Conditions (May). 10 pp.
DIN ENGL 8634-79. Machine Tools; External Cylindrical Centreless Grinding Machines; Acceptance Conditions (Sept). 10 pp.
ISO 2433-84. Acceptance Conditions for External Cylindrical Grinding Machines with a Movable Table—Testing of Accuracy Second Edition. 13 pp.
ISO 3875-90. Acceptance Conditions for External Cylindrical Centreless Grinding Machines—Testing of the Accuracy Second Edition. 18 pp.

—Acceptance Testing—Internal
BSI BS 4656: Part 8-74. 1974 Amd 1 Accuracy of Machine Tools and Methods of Test Part 8: Internal Cylindrical Grinding Machines with Horizontal Spindle. 15 pp.
DIN ENGL 8631 Pt 1-79. Machine Tools; Internal Cylindrical Grinding Machines with Horizontal Spindle; Acceptance Conditions (July). 10 pp.
DIN ENGL 8631 Pt 2-79. Machine Tools; Internal Cylindrical Grinding Machines with Horizontal Spindle; Machines Having a Surfacing Wheel Slide; Acceptance Conditions (July). 5 pp.
DIN ENGL 8637-90. Machine Tools; Grinding Spindles with Cartridge Diameters up to 200 mm; Acceptance Conditions (May). 6 pp.
ISO 2407-84. Acceptance Conditions for Internal Cylindrical Grinding Machines with Horizontal Spindle—Testing of Accuracy Second Edition. 15 pp.

—Accuracy Testing
CNS B7100-78. Accuracy Inspection of Centerless Grinding Machines (Jul)(4441). 7 pp.
JIS B 6220-90. External Cylindrical Centreless Grinding Machines—Test Code for Performance and Accuracy. 14 pp.

—Accuracy Testing—External
JIS B 6212-86. Test Code for Performance and Accuracy of External Cylindrical and Universal Grinding Machines. 28 pp.

—Accuracy Testing—Internal
CNS B7088-81. Accuracy Inspection of Internal Cylindrical Grinding Machines (Oct)(4261). 4 pp.
JIS B 6211-86. Test Code for Performance and Accuracy of Internal Cylindrical Grinding Machines. 24 pp.

—External
CNS B7094-81. Accuracy Inspection of External Cylindrical and Universal Grinding Machines (Oct)(4299).
CNS B7157-80. Test Code for Performance of Internal Cylindrical Grinding Machines (Oct)(6553).

Cylindrical Grinders (Cont.)
—Glossaries
DIN ENGL 69718 Pt 1-76. Machine Tools; External Cylindrical Grinding Machines; Definitions and Terms (May). 6 pp.
DIN ENGL 69718 Pt 3-77. Machine Tools; Centreless External Cylindrical Grinding Machines; Definitions and Terms (Feb). 8 pp.

—Internal
CNS B7186-82. Test Code for Performance of External Cylindrical Centerless Grinding Machine (Apr)(8648).

—Numerical Control—External
JIS B 6335-86. Test Code for Performance and Accuracy of Numerically Controlled External Cylindrical and Universal Grinding Machines. 33 pp.

Cylindrical Roller Bearings
See Also: Needle Roller Bearings; Roller Bearings
BSI BS 292: Part 1-82. 1982 Amd 1 Rolling Bearings: Ball Bearings, Cylindrical and Spherical Roller Bearings Part 1: Dimensions of Ball Bearings, Cylindrical and Spherical Roller Bearings (Metric Series). 36 pp.
BSI BS 292: Part 2-82. 1982 Amd 1 Rolling Bearings: Ball Bearings, Cylindrical and Spherical Roller Bearings Part 2: Dimensions of Ball Bearings and Cylindrical Roller Bearings (Inch Series). 17 pp.
CNS B2542-81. Cylindrical Roller Bearings; with Tapper Bore 1:12 with Snap Ring Groove (Mar)(7105).
CNS B2547-81. Cylindrical Roller Bearings; with Parallel Bore Snap Ring Groove (Mar)(7110).
CNS B2548-81. Cylindrical Roller Bearings; Outer Race with Cage and Rollers (Mar)(7111).
CNS B2549-81. Cylindrical Roller Bearings; Inner Race with Cage and Rollers (Mar)(7112).
CNS B2612-81. Double Row Cylindrical Roller Bearings; Full Type Roller Bearing; Dimensions Series 48 and 49 (Oct)(7949).
CNS B3253-81. Cylindrical Roller Bearings Angular Rings (Oct)(4034).
DIN ENGL 5402 Pt 1-73. Rolling Bearing Components; Cylindrical Rollers (June). 2 pp.
JIS B 1533-86. Cylindrical Roller Bearings. 29 pp.
SNZ NZS/BS 292: Part 1-82. Rolling Bearings: Ball Bearings, Cylindrical and Spherical Roller Bearings Part 1: Specification for Dimensions of Ball Bearings, Cylindrical and Spherical Roller Bearings (Metric Series) Amend: 1. 33 pp.
SNZ NZS/BS 292: Part 2-82. Rolling Bearings: Ball Bearings, Cylindrical and Spherical Roller Bearings Part 2: Specification for Dimensions of Ball Bearings and cylindrical Roller Bearings (Inch Series Amend: 1. 12 pp.

—Double Row
CNS B2543-81. Cylindrical Roller Bearings; with Lipped Inner Race Double-Row Type (Mar)(7106).

—Internal Clearance
BSI BS 6107: Part 3-92. 1992 Rolling Bearings: Tolerances Part 3: Specification for Radial Internal Clearance (ISO 5753: 1991). 12 pp.
BSI BS 6107: Part 3-81. 1981 Rolling Bearings: Tolerances Part 3: Radial Internal Clearance. 9 pp.
ISO 5753-91. Rolling Bearings—Radial Internal Clearance Second Edition. 9 pp.

—Radial
CNS B2611-81. Radial Cylindrical Roller Bearings; Roller and Outer Ring Assembly, Type E (Oct)(7948).

—Single Row
CNS B2473-85. Cylindrical Roller Bearings; with Ribs on Outer Ring, Single-Row (Mar)(5502).
CNS B2541-81. Cylindrical Roller Bearings; with Lipped Inner Ring Single-Row Type (Mar)(7104).

—Thrust
CNS B2629-81. Thrust Cylindrical Roller Bearings; Single Direction (Nov)(8101).
DIN ENGL 722-87. Rolling Bearings; Single Direction Thrust Cylindrical Roller Bearings (Aug). 3 pp.

—Thrust Collars
BSI BS 5646: Part 6-80. 1980 Rolling Bearings—Accessories Part 6: Dimensions of Separate Thrust Collars for Cylindrical Roller Bearings. 3 pp.
CNS B2613-81. Cylindrical Roller Bearings; Separate Thrust Collar; E Types (Oct)(7950).
ISO 246-78. Cylindrical Roller Bearings—Separate Thrust Collars—Boundary Dimensions First Edition. 3 pp.

Cypress Wood
Scope Note: For additional listings, see the subheading Cypress Wood under specific products made from Cypress Wood See Also: Softwoods; Wood

INDUSTRY STANDARDS

Cypress Wood *(Cont.)*
—**Grading**
CNS O1028-83. Scaling and Grading Rules of Taiwan Cypress Lumber for Export (for Taiwan Area) (Apr)(10199). 5 pp.

—**Scaling**
CNS O1028-83. Scaling and Grading Rules of Taiwan Cypress Lumber for Export (for Taiwan Area) (Apr)(10199). 5 pp.

Cyrillic Alphabet
Use: Cyrillic Characters

Cyrillic Characters
See Also: Graphic Characters

—**Coded Character Sets**
CEN ENV 41508-90. Information Systems Interconnection—East European Graphic Character Repertoires and Their Coding. 11 pp.
ECMA ECMA 113-88. 8-Bit Single-Byte Coded Graphic Character Sets, Latin/Cyrillic Alphabet. 15 pp.
IEC 8859 Pt 5-88. Information Processing—8-Bit Single-Byte Coded Graphic Character Sets—Part 5: Latin/Cyrillic Alphabet First Edition. 10 pp.
ISO 8859 Pt 5-88. Information Processing—8-Bit Single-Byte Coded Graphic Character Sets—Part 5: Latin/Cyrillic Alphabet First Edition. 10 pp.
JTC1 8859-88. Information Processing—8-Bit Single-Byte Coded Graphic Character Sets—Part 5: Latin/Cyrillic Alphabet First Edition. 10 pp.
OSI ISO DIS 6937-8-87. Coded Character Sets for Text Communication—Part 8: Cyrillic Graphic Characters. 13 pp.
OSI ISO IEC 8859-5-88. Information Processing—8-Bit Single-Byte Coded Graphic Character Sets—Part 5: Latin/Cyrillic Alphabet. 10 pp.
OSI ISO DIS 8859-5-86. Information Processing—8-Bit Single-Byte Coded Graphic Character Sets—Part 5: Latin/Cyrillic Alphabet. 18 pp.
OSI ISO 8859 2-87. Information Processing-8-Bit Single-Byte Coded Graphic Character Sets-Part 2: Latin Alphabet No. 2.
OSI ISO DP 8859 3-86. Information Processing-8-Bit Single Byte Coded Graphic Character Sets-Part 3: Latin Alphabet No. 3.
OSI ISO DP 8859 4-86. Information Processing-8-Bit Single Byte Coded Graphic Character Sets-Part 4: Latin Alphabet No. 4.
OSI ISO DIS 8859 5-86. Information Processing-8-Bit Single-Byte Coded Graphic Characters Sets-Part 5: Latin/Cyrillic Alphabet.
OSI ISO DIS 8859 8-87. Information Processing-8-Bit Single-Byte Coded Graphic Character Sets-Part 8: Latin/Hebrew Alphabet.

—**Coded Character Sets—Bibliographic References**
BSI BS 6474: Part 4-86. 1986 Coded Character Sets for Bibliographic Information Interchange Part 4: Extension of the Cyrillic Alphabet Coded Character Set. 7 pp.
ISO 5427-84. Extension of the Cyrillic Alphabet Coded Character Set for Bibliographic Information Interchange First Edition. 6 pp.

—**Engineering Drawings**
ISO 3098 Pt 4-84. Technical Drawings—Lettering—Part 4: Cyrillic Characters First Edition. 7 pp.

—**Romanization**
ISO 9-86. Documentation—Transliteration of Slavic Cyrillic Characters into Latin Characters First Edition. 10 pp.

Cysteine Content Analysis
—**Wool Fibers—Colorimetric Analysis**
ISO 2913-75. Wool—Colorimetric Determination of Cystine Plus Cysteine in Hydrolysates First Edition. 6 pp.
ISO 2915-75. Wool—Determination of Cysteic Acid Content of Wool Hydrolysates by Paper Electrophoresis and Colorimetry First Edition. 7 pp.

—**Wool Fibers—Electrophoresis**
ISO 2915-75. Wool—Determination of Cysteic Acid Content of Wool Hydrolysates by Paper Electrophoresis and Colorimetry First Edition. 7 pp.

L-Cysteine Monohydrochloride Monohydrate
JIS K 8470-91. L-Cysteine Hydrochloride, Monohydrate.

L-Cystine
JIS K 9048-91. L-Cystine.

Cystoscopes
See Also: Medical Electrical Equipment; Medical Equipment

Cystoscopes *(Cont.)*
JIS T 4407-88. Cystoscopes. 8 pp.

Cytocompatibility
Use: Biocompatibility

Cytotoxicity
Use: Toxicity

C02 Analyzers
Use: Carbon Dioxide Analyzers

D/A Converters
Use: Digital to Analog Converters

D Rings
Use For: Dee Rings; Dees *See Also:* Rings
CNS Z7062-79. 25mm D Ring (Apr)(4793).

—**Aircraft—Fuel Filters**
SBAC AS 3273 ISSUE 1. 'D', Ring Filter Fuel Type 'B'. 1 p.

D Subminiature Connectors
See Also: Connectors; Printed Circuit Connectors; Rack and Panel Connectors; Rectangular Connectors
MOD UK DSTAN 59-35: Pt 3:Sec 8-01. Lead Assembly, Electrical (Connectors, Electrical) Part 3: Connectors for d.c. and Low Frequency Applications (See Also EPIC Database) Section 8: Pattern 9523 F0002 List of Items Conforming to BS 9523 F0002 Issue 1; Amendment 2. 21 pp.

D-Type Bistables
Use: D-Type Flip Flops

D-Type Flip Flops
Use For: D-Type Bistables *See Also:* D-Type Registers; Digital Circuits; Flip Flops
CECC CECC 90 102-021 ISSUE 1-81. BS 9420 F466 to 478 Issue A; Dual 'D' Type Positive Edge Triggered Bistables (En). 9 pp.
CECC CECC 90 103-125 ISSUE 1-81. BS CECC 90 103-125; D-Type Bistable (En) AMD 1 (En). 20 pp.
CECC CECC 90 106-002 ISSUE 1-85. UTE C 86-217; Digital Integrated Circuits in Accordance with FS 90 106; 54/74 ALS 74, 54/74 ALS 74A; D Type Positive-Edge-Triggered Bistables (EN, Fr) ADD 3 (En, Fr). 18 pp.
CECC CECC 90 106-009 ISSUE 1-85. UTE C 86-217; Digital Integrated Circuits in Accordance with FS 90 106; 54 ALS 174, 74 ALS 174; D Type Bistable (En, Fr) ADD 3 (En, Fr). 17 pp.
CECC CECC 90 106-055 ISSUE 1-85. UTE C 86-217; Digital Integrated Circuits in Accordance with FS 90 106; 54 ALS 175, 74 ALS 175; D Type Bistable (En, Fr) ADD 3 (En, Fr). 17 pp.
CECC CECC 90 107-010 ISSUE 2-89. UTE C 86-218 ADD 2/FA 10; Digital Integrated Circuits in Accordance with FS 90 107; 54/74 F 74; D-Type Positive Edge-Triggered Bistable (En, Fr) ADD 3 (En, Fr). 11 pp.
CECC CECC 90 107-024 ISSUE 2-89. UTE C 86-218 ADD 3/FA 24; Digital Integrated Circuits in Accordance with FS 90 107; 54/74 F 174; D-Type Bistable (En, Fr) ADD 3 (En, Fr). 11 pp.
CECC CECC 90 109-606 ISSUE 1-86. Digital Integrated Circuits in Accordance with FS 90 109; 54 HC 74, 74 HC 74; Dual D-Type Bistable with Preset and Clear (En, Fr). 6 pp.
CECC CECC 90 109-624 ISSUE 1-86. Digital Integrated Circuits in Accordance with FS 90 109; 54 HC 175, 74 HC 175; D-Type Bistable with Clear (En, Fr). 7 pp.
CECC CECC 90 109-625 ISSUE 1-86. Digital Integrated Circuits in Accordance with FS 90 109; 54 HC 273, 74 HC 273; D-Type Bistable with Clear (En, Fr). 6 pp.
MOD UK DSTAN 59-62: 90/020-69. Integrated Circuit Issue 1. 16 pp.
MOD UK DSTAN 59-62: 90/021-68. Integrated Circuit Issue 1. 2 pp.

—**Dual**
CECC CECC 90 101-025 ISSUE 1-87. UTE C 86-213/B 45; Digital Integrated Circuits in Accordance with FS 90 101; 54/64/74 74; Dual D-Type Bistable with Preset and Clear (En, Fr). 16 pp.
CECC CECC 90 103-034 ISSUE 1-81. BS CECC 90 103-034; Dual D-Type Positive-Edge Triggered Bistable with Preset and Clear (En) AMD 1 (En). 18 pp.
CECC CECC 90 104-112 ISSUE 1-81. BS CECC 90 104-112; Dual Type D Bistable (En). 24 pp.
CECC CECC 90 104-112 ISSUE 1-86. CEI CECC 90 104-112; Dual Type D Bistable (En). 2 pp.
CECC CECC 90 109-742 ISSUE 1-87. Digital Integrated Circuits in Accordance with FS 90 109; 54 HCT 74, 74 HCT 74; Dual D-Type Bistable with Preset and Clear (En, Fr). 6 pp.

D-Type Flip Flops *(Cont.)*
—**Dual—Preferred Products List**
CECC CECC MUAHAG Vol 7 IS 8-92. Preferred Products List; Active Microcircuits (En, Fe, Ge). 89 pp.

—**Hex**
CECC CECC 90 101-052 ISSUE 1-87. UTE C 86-213/B 72; Digital Integrated Circuits in Accordance with FS 90 101 54/64/74 174; Hex D-Type Bistable (En). 14 pp.
CECC CECC 90 102-031 ISSUE 1-81. BS CECC 90 102-031; Hex 'D' Type Bistables with Clear (En) AMD 1 (En). 20 pp.
CECC CECC 90 103-077 ISSUE 1-81. BS CECC 90 103-077; Hex D-Type Bistables with Common Clock and Clear (En) AMD 1 (En). 19 pp.
CECC CECC 90 104-097 ISSUE 2. NL CECC 90 104-097 Issuc 2; Digital Integrated Circuits in Accordance with FS 90 104; HEC/HEF 40174B; Hex D-Type Flip-Flop (En). 8 pp.
CECC CECC 90 104-193 ISSUE 1-81. BS CECC 90 104-193 Issue 1; Hex Type D Bistable (En). 25 pp.
CECC CECC 90 104-193 ISSUE 1-86. CEI CECC 90 104-193 Issue 1; Hex Type "D" Bistable (En). 2 pp.
CECC CECC 90 109-609 ISSUE 1-86. Digital Integrated Circuits in Accordance with FS 90 109; 54 HC 174, 74 HC 174; Hex D-Type Bistable with Clear (En, Fr). 7 pp.
CECC CECC 90 109-748 ISSUE 1-87. Digital Integrated Circuits in Accordance with FS 90 109; 54 HC 378, 74 HC 378; Hex D-Type Bistable with Clock Enable (En, Fr). 6 pp.
CECC CECC 90 109-791 ISSUE 1-87. Detail Specification: Digital Integrated Circuits Silicon Monolithic C MOC, Cavity or Non-Cavity Packages; Type(s) 54/74 HCT 174 Bascule D-Type Bistable with Clear Assessment Levels P, Y, L (En, Fr, Ge). 10 pp.

—**Hex—Preferred Products List**
CECC CECC MUAHAG Vol 7 IS 8-92. Preferred Products List; Active Microcircuits (En, Fe, Ge). 89 pp.

—**Octal**
CECC CECC 90 102-068 ISSUE 1-89. UTE C 86-216/F04; Octal D-Type Bistable Edge-Triggered Bistable Circuits and 3-State Outputs (En, Fr). 11 pp.
CECC CECC 90 103-102 ISSUE 1-81. BS CECC 90 103-102; Octal D-Type Bistable with Clear (En) AMD 1 (En). 18 pp.
CECC CECC 90 103-123 ISSUE 1-81. BS CECC 90 103-122; Octal D-Type Latches with 3 Level Outputs (En) AMD 1 (En). 22 pp.
CECC CECC 90 103-124 ISSUE 1-81. BS CECC 90 103-123; Octal D-Type Edge-Triggered Bistable Circuits with Three Level Outputs (En) AMD 1 (En). 19 pp.
CECC CECC 90 104-231 ISSUE 1. NL CECC 90 104-231 Issue 1; Digital Integrated Circuits in Accordance with FS 90 104; HEC/HEF 40374B; Octal D-Type Flip-Flop with 3-State Outputs (En). 10 pp.
CECC CECC 90 106-020 ISSUE 2-86. UTE C 86-217 ADD 2/GA 20; Digital Integrated Circuits in Accordance with FS 90 106; 54 ALS 374, 74 ALS 374; Octal D-Type Edge-Triggered Bistable with 3 State Non Inverting Outputs (En, Fr) ADD 3 (En, Fr). 16 pp.
CECC CECC 90 106-022 ISSUE 2-86. UTE C 86-217 ADD 2/GA 22; Digital Integrated Circuits in Accordance with FS 90 106; 54 ALS 574 74, ALS 574; Octal D-Type Edge-Triggered Bistable with 3 State Non Inverting Outputs (En, Fr) ADD 3 (En, Fr). 16 pp.
CECC CECC 90 106-041 ISSUE 2-86. UTE C 86-217 ADD 2/GA 42; Digital Integrated Circuits in Accordance with FS 90 106; 54 ALS 576, 74 ALS 576; Octal D-Type Edge-Triggered Bistable with 3 State Inverting Outputs (En, Fr) ADD 3 (En, Fr). 16 pp.
CECC CECC 90 106-057 ISSUE 2-86. UTE C 86-217 ADD 2/GA 46; Digital Integrated Circuits in Accordance with FS 90 106; 54 ALS 273, 74 ALS 273; Octal D Type Edge Triggered Bistable (En, Fr) ADD 3 (En, Fr). 16 pp.
CECC CECC 90 107-041 ISSUE 2-89. UTE C 86-218 ADD 3/FA 41; Digital Integrated Circuits in Accordance with FS 90 107; 54/74 F 374; Octal D-Type Bistable with 3-State Non Inverting Outputs (En, Fr) ADD 3 (En, Fr). 9 pp.
CECC CECC 90 109-613 ISSUE 1-86. Digital Integrated Circuits in Accordance with FS 90 109; 54 HC 374, 74 HC 374; 3-State Octal D-Type Bistable (En, Fr). 6 pp.
CECC CECC 90 109-614 ISSUE 1-86. Digital Integrated Circuits in Accordance with FS 90 109; 54 HC 534, 74 HC 534; Octal 3-State Inverting D-Type Bistable (En, Fr). 6 pp.

INTERNATIONAL AND NON-U.S. NATIONAL STANDARDS
SUBJECT INDEX

Dairy

D-Type Flip Flops *(Cont.)*
—Octal *(Cont.)*
CECC CECC 90 109-685 ISSUE 1-87. Digital Integrated Circuits in Accordance with FS 90 109; 54/74 HC 377 S; Octal D-Type Bistable (En, Fr). 6 pp.

CECC CECC 90 109-704 ISSUE 1-87. Digital Integrated Circuits in Accordance with FS 90 109; 54 HCT 374, 74 HCT 374; 3-State Octal D-Type Bistable (En, Fr). 6 pp.

CECC CECC 90 109-725 ISSUE 1-87. Digital Integrated Circuits in Accordance with FS 90 109; 54 HC 564, 74 HC 564; Octal 3-State Inverting D-Type Bistable (En, Fr). 7 pp.

CECC CECC 90 109-726 ISSUE 1-87. Digital Integrated Circuits in Accordance with FS 90 109; 54 HC 574, 74 HC 574; 3-State Octal D-Type Bistable (En, Fr). 7 pp.

CECC CECC 90 109-750 ISSUE 1-87. Digital Integrated Circuits in Accordance with FS 90 109; 54 HC 377T, 74 HC 377T; Octal D-Type Bistable with Clock Enable (En, Fr). 6 pp.

CECC CECC 90 109-751 ISSUE 1-87. Digital Integrated Circuits in Accordance with FS 90 109; 54 HC 377P, 74 HC 377P; Octal D-Type Bistable with Data Enable (En, Fr). 6 pp.

CECC CECC 90 109-827 ISSUE 1-89. Digital Integrated Circuits; Silicon Monolithic C MOS, Cavity or Non-Cavity Packages; Type(s); 54/74 HCT 273 D-Type Bistable with Clear (En, Fr). 9 pp.

CECC CECC 90 109-859 ISSUE 1-90. Digital Integrated Circuits; Silicon Monolithic C MOS, Cavity or Non-Cavity Packages; Type(s) 54/74 HCT 377P Octal D-Type Bistable with Data Enable Assessment Levels P, Y, L (En, Fr, Ge). 9 pp.

CECC CECC 90 109-862 ISSUE 1-90. Digital Integrated Circuits; Silicon Monolithic C MOS, Cavity or Non-Cavity Packages; Type(s); 54/74 HCT 534 Octal 3-State Inverting D-Type Bistable; Assessment Levels P, Y, L (En, Fr). 10 pp.

CECC CECC 90 109-864 ISSUE 1-90. Digital Integrated Circuits; Silicon Monolithic C MOS, Cavity or Non-Cavity Packages; Type(s); 54/74 HCT 564 Octal 3-State Inverting D-Type Bistable; Assessment Levels P, Y, L (En, Fr). 10 pp.

CECC CECC 90 109-866 ISSUE 1-90. Digital Integrated Circuits; Silicon Monolithic C MOS, Cavity or Non-Cavity Packages; Type(s); 54/74 HCT 574 3-State Octal D-Type Bistable; Assessment Levels P, Y, L (En, Fr). 10 pp.

—Octal—Preferred Products List
CECC CECC MUAHAG Vol 7 IS 8-92. Preferred Products List; Active Microcircuits (En, Fe, Ge). 89 pp.

—Quad
CECC CECC 90 101-053 ISSUE 1-87. UTE C 86-213/B 73; Digital Integrated Circuits in Accordance with FS 90 101 54/64/74 175; Quad D-Type Bistable (En). 14 pp.

CECC CECC 90 102-055 ISSUE 1-82. BS CECC 90 102-055; Quad D-Type Bistables (En). 19 pp.

CECC CECC 90 103-126 ISSUE 1-81. BS CECC 90 103-126; Quad D-Type Bistable (En) AMD 1 (En). 20 pp.

CECC CECC 90 103-138 ISSUE 1-81. BS CECC 90 103-138; Quad D-Type Bistables (En). 18 pp.

CECC CECC 90 104-051 ISSUE 2. NL CECC 90 104-051 Issue 2; Digital Integrated Circuits in Accordance with FS 90 104; HEC/HEF 4076B; Quadruple D-Type Register with 3-State Outputs (En). 10 pp.

CECC CECC 90 104-098 ISSUE 2. NL CECC 90 104-098 Issue 2; Digital Integrated Circuits in Accordance with FS 90 104; HEC/HEF 40175B; Quadruple D-Type Flip-Flop (En). 8 pp.

CECC CECC 90 104-173 ISSUE 1. BS CECC 90 104-173; Quad Type 'D' Bistable (En). 25 pp.

CECC CECC 90 107-025 ISSUE 2-89. UTE C 86-218 ADD 3/FA 25; Digital Integrated Circuits in Accordance with FS 90 107; 54/74 F 175; D-Type Bistable (En, Fr) ADD 3 (En, Fr). 11 pp.

CECC CECC 90 109-711 ISSUE 1-87. Digital Integrated Circuits in Accordance with FS 90 109; 54/74 HC 173; Quad D-Type Bistable with 3-State Outputs (En, Fr). 7 pp.

CECC CECC 90 109-749 ISSUE 1-87. Digital Integrated Circuits in Accordance with FS 90 109; 54 HC 379, 74 HC 379; Quad D-Type Bistable with Clock Enable (En, Fr). 6 pp.

CECC CECC 90 109-790 ISSUE 1-87. Detail Specification: Digital Integrated Circuits Silicon Monolithic C MOS, Cavity Or Non-Cavity Packages; Type(s) 54/74 HCT 173 Quad D-Type Bistable with 3-State Outputs Assessment Levels P, Y, L (En, Fr, Ge). 10 pp.

CECC CECC 90 109-792 ISSUE 1-87. Digital Integrated Circuits; Silicon Monolithic C MOS, Cavity or Non-Cavity Packages; Type(s); 54/74 HCT 175; D-Type Bistable with Clear (En, Fr). 10 pp.

D-Type Flip Flops *(Cont.)*
—Quad—Preferred Products List
CECC CECC MUAHAG Vol 7 IS 8-92. Preferred Products List; Active Microcircuits (En, Fe, Ge). 89 pp.

—4-Bit
CECC CECC 90 106-148 ISSUE 1-86. UTE C 86-217 ADD 3/GA 95 Digital Integrated Circuits in Accordance with FS 90 106; 54 ALS 874, 74 ALS 874; 4 Bit D-Type Edge-Triggered Bistable with 3 State Non Inverting Outputs (En, Fr) ADD 3 (En, Fr). 16 pp.

D-Type Latches
See Also: Digital Circuits; Latches (Circuits)
—Octal
CECC CECC 90 102-067 ISSUE 1-89. UTE C 86-216/F03 BS CECC 90 102-038 Issue 1; Octal D-Type Transparent Latches with 3-State Outputs (En, Fr). 11 pp.

CECC CECC 90 103-122 ISSUE 1-81. Octal D-Type Latches with 3 Level Outputs (En) AMD 1 (En). 23 pp.

CECC CECC 90 106-019 ISSUE 2-86. UTE C 86-217 ADD 2/GA 19; Digital Integrated Circuits in Accordance with FS 90 106; 54 ALS 373, 74 ALS 373; Octal D-Type Transparent Latch with 3 State Non Inverting Outputs (En, Fr) ADD 3 (En, Fr). 16 pp.

CECC CECC 90 106-021 ISSUE 2-86. UTE C 86-217 ADD 2/GA 21; Digital Integrated Circuits in Accordance with FS 90 106; 54 ALS 573, 74 ALS 573; Octalo D-Type Transparent Latch with 3 State Non Inverting Outputs (En, Fr) ADD 3 (En, Fr). 16 pp.

CECC CECC 90 106-040 ISSUE 2-86. UTE C 86-217 ADD 2/GA 40; Digital Integrated Circuits in Accordance with FS 90 106; 54/74 ALS 563; Octal D-Type Transparent Latches with 3 State Inverting Outputs (En, Fr) ADD 3 (En, Fr). 16 pp.

CECC CECC 90 106-042 ISSUE 2-86. UTE C 86-217 ADD 2/GA 55; Digital Integrated Circuits in Accordance with FS 90 106; 54/74 ALS 580; Octal D-Type Transparent Latches with 3 State Inverting Outputs (En, Fr) ADD 3 (En, Fr). 16 pp.

CECC CECC 90 109-626 ISSUE 1-86. Digital Integrated Circuits in Accordance with FS 90 109; 54/74 HC 373; 3-State Octal D-Type Latch (En, Fr). 6 pp.

CECC CECC 90 109-674 ISSUE 1-87. Digital Integrated Circuits in Accordance with FS 90 109; 54/74 HC 573; 3 State Octal D Type Transparent Latch (En, Fr). 7 pp.

CECC CECC 90 109-695 ISSUE 1-87. Digital Integrated Circuits in Accordance with FS 90 109; 54/74 HCT 373; 3-State Octal D-Type Latch (En, Fr). 6 pp.

CECC CECC 90 109-861 ISSUE 1-90. Digital Integrated Circuits; Silicon Monolithic C MOS, Cavity or Non-Cavity Packages; Type(s); 54/74 HCT 533 3-State Octal Inverting D-Type Transparent Latch; Assessment Levels P, Y, L (En, Fr). 10 pp.

CECC CECC 90 109-863 ISSUE 1-90. Digital Integrated Circuits; Silicon Monolithic C MOS, Cavity or Non-Cavity Packages; Type(s); 54/74 HCT 563 3-State Octal Inverting D-Type Transparent Latch; Assessment Levels P, Y, L (En, Fr). 10 pp.

CECC CECC 90 109-865 ISSUE 1-90. Digital Integrated Circuits; Silicon Monolithic C MOS, Cavity or Non-Cavity Packages; Type(s); 54/74 HCT 573 3-State Octal D-Type Transparent Latch; Assessment Levels P, Y, L (En, Fr). 10 pp.

—Octal—Inverting
CECC CECC 90 109-673 ISSUE 1-87. Digital Integrated Circuits in Accordance with FS 90 109; 54/74 HC 563; 3 State Octal Inverting D Type Transparent Latch (En, Fr). 6 pp.

CECC CECC 90 109-675 ISSUE 1-87. Digital Integrated Circuits in Accordance with FS 90 109; 54/74 HC 533; 3 State Octal Inverting D Type Transparent Latch (En, Fr). 7 pp.

—Octal—Preferred Products List
CECC CECC MUAHAG Vol 7 IS 8-92. Preferred Products List; Active Microcircuits (En, Fe, Ge). 89 pp.

—Quad
CECC CECC 90 104-032 ISSUE 2. NL CECC 90 104-032 Issue 2; Digital Integrated Circuits in Accordance with FS 90 104; HEC/HEF 4042B; Quadruple D-Latch (En). 9 pp.

—4-Bit
CECC CECC 90 106-147 ISSUE 1-86. UTE C 86-217 ADD 3/GA 94; Digital Integrated Circuits in Accordance with FS 90 106; 54 ALS 873, 74 ALS 873; 4 Bit D-Type Latch with 3 State Outputs (En, Fr) ADD 3 (En, Fr). 16 pp.

D-Type Registers
See Also: D-Type Flip Flops
—Quad
CECC CECC 90 103-076 ISSUE 1-82. BS CECC 90 103-076; Three-Level Quad D-Registers (En) AMD 1 (En). 21 pp.

—4-Bit
CECC CECC 90 101-051 ISSUE 1-87. UTE C 86-213/B 71; Digital Integrated Circuits in Accordance with FS 90 101 54/64/74 173; 4-Bit D-Type Register with 3-State Outputs (En). 19 pp.

D-Weighting
Use: Aircraft Noise—Frequency Weighting

DAC
Use: Digital to Analog Converters

DACS
Use: Digital Access and Cross Connect Systems

Dadoes
See Also: Wall Coverings
—Ships—Paint
MOD UK DSTAN 80-151-91. Paint, Finishing, for Interior Decks and Diodes Types: Brushing Spraying Issue 1. 15 pp.

Dairies
See Also: Dairy Equipment; Dairy Products; Farms
—Hygiene—Inspection—Sampling
ISO 8086-86. Dairy Plant—Hygiene Conditions—General Guidance on Inspection and Sampling Procedures First Edition. 8 pp.

—Hygiene—Visual Inspection
ISO 8086-86. Dairy Plant—Hygiene Conditions—General Guidance on Inspection and Sampling Procedures First Edition. 8 pp.

—Microbiological Analysis
BSI BS 4285: Part 0-90. 1990 Microbiological Examination for Dairy Purposes Part 0: General Introduction. 8 pp.

BSI BS 4285: Sec 1.2-84. 1984 Microbiological Examination for Dairy Purposes Part 1: Guide to General Procedures Section 1.2: Diluents, Media and Apparatus and Their Preparation and Sterilization. 8 pp.

BSI BS 4285: Part 4-91. 1991 Microbiological Examination for Dairy Purposes Part 4: Methods for Assessment of Hygienic Conditions. 16 pp.

—Milking Areas—Design
BSI BS 5502: Part 49-90. 1990 Design of Buildings and Structures for Agriculture Part 49: Code of Practice for Design and Construction of Milking Premises. 11 pp.

—Milking Areas—Safety
SNZ NZS/BS 5539-78. Specification for Safety Requirements in Rotary Milking Parlours. 16 pp.

—Planning and Management
DIN ENGL 11487-86. Dairy Plant; Planning, Project Management and Implementation (Jan). 3 pp.

Dairy Cattle
See Also: Beef Cattle
—Health Inspection
EC 85/397/EEC-85. Council Directive on Health and Animal-Health Problems Affecting Ultra-Community Trade in Heat-Treated Milk. 17 pp.

—Milk Recording
ISO 1546-81. Procedure for Milk Recording for Cows First Edition. 9 pp.

Dairy Equipment
See Also: Cheese Molds; Dairies; Dairy Products; Milk Coolers; Milking Machines; Pasteurizers
—Cleaning
BSI BS 5226-91. 1991 Code of Practice for Equipment and Procedures for the Cleaning and Disinfecting of Milking Machine Installations. 33 pp.

SAA AS 1162-91. Cleaning and Sanitizing Dairy Factory Equipment. 21 pp.

—Cleaning Agents
BSI BS 5305-84. 1984 Cleaning and Disinfecting of Plant and Equipment Used in the Dairying Industry. 62 pp.

SAA AS 1389-73. Acidic Detergents for Use in the Dairying Industry Amdt 1 March 1976. 8 pp.

SAA AS 1398-82. Iodophors for Use in the Dairying Industry. 14 pp.

SAA AS 1400-81. Heavy-Duty Alkaline Detergents for 'In-Place' Cleaning in Dairy Factories. 6 pp.

INDUSTRY STANDARDS

Dairy Equipment *(Cont.)*

—Cleaning Agents—Elastomer Sealants
DIN ENGL 11483 Pt 2-84. Dairy Installations; Cleaning and Disinfection; Consideration of the Action on Sealing Materials (Feb). 2 pp.

—Cleaning Agents—Stainless Steel
DIN ENGL 11483 Pt 1-83. Dairy Installations; Cleaning and Disinfection; Consideration of the Action on Stainless Steel (Jan). 3 pp.

—Detergents
SAA AS 1803-82. General Purpose Detergents for Use in the Dairying Industry. 16 pp.

—Iodophors
SNZ NZS 8341-72. Specification for Dairy Iodophors. 12 pp.

—Pipe Fittings—Stainless Steel
BSI BS 1864-66. (WITHDRAWN) 1966 Stainless Steel Milk Pipes and Fittings (Recessed Ring Joint Type) (Superseded by BS 4825: Part 5: 1991). 25 pp.

—Pipelines—Welding
SNZ NZS 4703-85. Welder Qualification Tests for Stainless Steel Pipe for the Dairy Industry. 18 pp.

—Pipes—Stainless Steel
BSI BS 1864-66. (WITHDRAWN) 1966 Stainless Steel Milk Pipes and Fittings (Recessed Ring Joint Type) (Superseded by BS 4825: Part 5: 1991). 25 pp.

—Sanitation
SAA AS 1162-91. Cleaning and Sanitizing Dairy Factory Equipment. 21 pp.

—Tanks (Containers)
BSI BS 3441-89. 1989 Tanks for the Transport of Milk and Liquid Milk Products. 12 pp.
DIN ENGL 11482-87. Dairy Machinery; Storage Tanks for Milk and Liquid Dairy Products (May). 5 pp.
SAA AS 1187-88. Refrigerated Bulk Milk Tanks. 13 pp.

—Thermometers
DIN ENGL 11857-57. Dairying Machines; Dial Thermometer; Principal Dimensions and Design Details (July). 1 p.
SAA AS 1030-73. Dial-Type General Purpose Thermometers for Use in the Dairying Industry (R 1993). 9 pp.
SAA AS 1031-73. Metal-Cased Mercury-In-Glass Thermometers for Use in the Dairying Industry (R 1993). 12 pp.

—Water Heaters
SNZ NZS 4604-90. Dairy-Type Thermal Storage Electric Water Heaters. 8 pp.

—Water Heaters—Installation
SNZ NZS 4605-78. Code of Practice for the Installation of Dairy-Type Thermal Storage Electric Water Heaters. 11 pp.

Dairy Products

Use For: Milk Products *See Also:* Agricultural Products; Amido Black Method; Animal Products; Butter; Buttermilk; Caseins; Cheeses; Condensed Milk; Cream (Dairy Products); Dairies; Eggs; Evaporated Milk; Food; Ice Cream; Milk; Powdered Milk; Skim Milk; Whey; Yogurt
EC EEC/231/87-87. Council Regulation Amending Regulation (EEC) No 804/68 on the Common Organization of the Market in Milk and Milk Products. 1 p.
EC EEC/374/92-92. Commission Regulation Amending the Annex to Council Regulation (EEC) No 804/68 on the Common Organization of the Market in Milk and Milk Products. 1 p.

—Acidimetry
CNS N6057-72. Method of Test for Milk and Milk Products (Titration of Acidity) (Oct)(3441). 1 p.

—Acidity—Volumetric Analysis
CNS N6195-88. Method of Test for Milk and Milk Products: Determination of Titratable Acidity (Sep)(11026). 1 p.

—Bacillus Cereus—Count Methods
BSI BS 4285: Sec 3.12-89. 1989 Microbiological Examination for Dairy Purposes Part 3: Methods for Detection and/or Enumeration of Specific Groups of Microorganisms Section 3.12: Enumeration of Bacillus Cereus. 6 pp.

—Bacteria
SAA AS 1095.3.7-79. Microbiological Methods for the Dairy Industry—Part 3: Methods of Examination for Specific Groups of Microorganisms —Part 3.7: Bacterial Spores. 2 pp.

Dairy Products *(Cont.)*

—Bacteria Count Methods
BSI BS 4285: Sec 3.2-91. 1991 Microbiological Examination for Dairy Purposes Part 3: Methods for Detection and/or Enumeration of Specific Groups of Microorganisms Sec 3.2: Enumeration of Thermoduric Bacteria (W). 5 pp.
BSI BS 4285: Sec 3.2-01. 1991 Amd 1 Microbiological Examination for Dairy Purposes Part 3: Methods for Detection and/or Enumeration of Specific Groups of Microorganisms Sec 3.2: Enumeration of Thermoduric Bacteria (AMD 6985) February 28, 1992 (W). 6 pp.
BSI BS 4285: Sec 3.3-86. 1986 Microbiological Examination for Dairy Purposes Part 3: Methods for Detection and/or Enumeration of Specific Groups of Microorganisms Section 3.3: Enumeration of Aerobic Bacterial Spores. 4 pp.
CNS N6068-85. Method of Test for Milk and Milk Products (Test of Bacteria) (Oct)(3452). 5 pp.
DIN ENGL 10192 Pt 1-84. Microbiological Analysis of Milk; Determination of Bacterial Count; Reference Method (Apr). 4 pp.
DIN ENGL 10192 Pt 2-83. Microbiological Analysis of Milk; Determination of Bacterial Count; Simplified Plate Count Method According to Koch (Feb). 2 pp.
DIN ENGL 10192 Pt 3-83. Microbiological Analysis of Milk; Determination of Bacterial Count; Plate-Loop Method (Feb). 3 pp.
DIN ENGL 10195 Pt 1-85. Microbiological Analysis of Milk; Microcolony Count; Enumeration of Microcolonies by Stereomicroscopic Counts (Dec). 3 pp.

—Bacteria—Microscopic Analysis
BSI BS 4285: Sec 2.5-89. 1989 Microbiological Examination for Dairy Purposes Part 2: Methods of General Application for Enumeration of Microorganisms Section 2.5: Enumeration of Bacteria by Direct Microscopic Counts. 6 pp.

—Chemical Analysis
BSI BS 1743-68. 1968 Amd 4 Methods for the Analysis of Dried Milk and Dried Milk Products (AMD 6154) December 21, 1990. 23 pp.

—Clostridium Count Methods
BSI BS 4285: Sec 3.13-91. 1991 Microbiological Examination for Dairy Purposes Part 3: Methods for Detection and/or Enumeration of Specific Groups of Microorganisms Section 3.13: Enumeration of Clostridium Perfringens. 6 pp.

—Coliform Bacteria—Count Methods
BSI BS 4285: Sec 3.7-87. 1987 Microbiological Examination for Dairy Purposes Part 3: Methods for Detection and/or Enumeration of Specific Groups of Microorganisms Section 3.7: Enumeration of Coliform Bacteria. 4 pp.
ISO 5541 Pt 1-86. Milk and Milk Products—Enumeration of Coliforms—Part 1: Colony Count Technique at 30 Degrees Celsius First Edition. 9 pp.
ISO 5541 Pt 2-86. Milk and Milk Products—Enumeration of Coliforms—Part 2: Most Probable Number Technique at 30 Degrees Celsius First Edition. 10 pp.

—Containers—Microbiological Analysis
SAA AS 1095.5.2-79. Microbiological Methods for the Dairy Industry-Part 5: Methods for the Assessment of Dairy Plant Hygiene—Part 5.2: Assessment of the Microbial Condition of Retail Containers for Dairy Products. 5 pp.

—Copper Content—Photometry
BSI BS 6394: Part 1-83. 1983 Trace Elements in Milk and Milk Products Part 1: Method for the Determination of Copper Content (Reference Method). 9 pp.
ISO 5738-80. Milk and Milk Products—Determination of Copper Content—Photometric Reference Method First Edition; (Erratum—Nov 1981). 9 pp.

—Cultured—Microbiological Analysis
SAA AS 1766.3.12-92. Food Microbiology—Part 3: Examination of Specific Products—Part 3.12: Cultured Dairy Products (Supersedes AS 1095.2.11 —1978). 3 pp.

—Density
CNS N6058-72. Method of Test for Milk and Milk Products (Determination of Specific Gravity) (Oct)(3442). 3 pp.

—Escherichia Coli
SAA AS 1095.3.10-84. Microbiological Methods for the Dairy Industry—Part 3: Methods of Examination for Specific Groups of Microorganisms —Part 3.10: Escherichia Coli—Direct Plate Method. 3 pp.

Dairy Products *(Cont.)*

—Escherichia Coli—Count Methods
BSI BS 4285: Sec 3.8-88. 1988 Microbiological Examination for Dairy Purposes Part 3: Methods for Detection and/or Enumeration of Specific Groups of Microorganisms Section 3.8: Enumeration of Presumptive Escherichia Coli. 6 pp.

—Fat Content
SNZ NZS 501: Part 2-71. Gerber Method for the Determination of Fat in Milk and Milk Products Part 2: Methods Amend: A, 1982. 28 pp.

—Fat Content—Baby Foods—Gravimetric Analysis
BSI BS 7142: Part 1-89. 1989 Analysis of Milk-Based Products Part 1: Determination of Fat Content of Milk-Based Infant Foods by the Rose-Gottlieb Gravimetric Method. 12 pp.
BSI BS 7142: Part 2-89. 1989 Analysis of Milk-Based Products Part 2: Determination of Fat Content of Milk-Based Infant Foods by the Weibull-Berntrop Gravimetric Method. 8 pp.
ISO 8262 Pt 1-87. Milk Products and Milk-Based Foods—Determination of Fat Content by the Weibull-Berntrop Gravimetric Method (Reference Method)—Part 1: Infant Foods First Edition. 8 pp.
ISO 8381-87. Milk-Based Infant Foods—Determination of Fat Content—Rose-Gottlieb Gravimetric Method (Reference Method) First Edition. 12 pp.

—Fat Content—Gas Chromatography
CNS N6203-86. Method of Detection for Vegetable Fat in Milk Fat (Oct)(11737). 3 pp.

—Fat Content—Gravimetric Analysis
BSI BS 1743: Part 11-86. 1986 Analysis of Dried Milk and Dried Milk Products Part 11: Determination of Fat Content (Gravimetric Reference Method). 10 pp.
BSI BS 5522-77. 1977 Milk and Milk Products. Determination of Fat Content. Mojonnier-Type Fat Extraction Flasks. 6 pp.
ISO 1736-85. Dried Milk, Dried Whey, Dried Buttermilk and Dried Butter Serum—Determination of Fat Content—Gravimetric Method (Reference Method) Second Edition. 10 pp.
ISO 3889-77. Milk and Milk Products—Determination of Fat Content—Mojonnier-Type Fat Extraction Flasks First Edition. 5 pp.
ISO 8262 Pt 3-87. Milk Products and Milk-Based Foods—Determination of Fat Content by the Weibull-Berntrop Gravimetric Method (Reference Method)—Part 3: Special Cases First Edition. 8 pp.

—Fat Content—Sampling
CNS N6203-86. Method of Detection for Vegetable Fat in Milk Fat (Oct)(11737). 3 pp.

—Fat Content—Volumetric Analysis
BSI BS 696: Part 1-89. 1989 Gerber Method for the Determination of Fat in Milk and Milk Products Part 1: Apparatus. 23 pp.
BSI BS 696: Part 2-89. 1989 Gerber Method for the Determination of Fat in Milk and Milk Products Part 2: Methods. 12 pp.
CNS N6060-86. Method of Test for Milk and Milk Products (Determination of Fat Content) (Oct)(3444). 2 pp.

—Fungi
SAA AS 1095.3.3-82. Microbiological Methods for the Dairy Industry—Part 3: Methods of Examination for Specific Groups of Microorganisms —Part 3.3: Yeasts and Moulds. 2 pp.

—Fungi—Count Methods
BSI BS 4285: Sec 3.6-86. 1986 Microbiological Examination for Dairy Purposes Part 3: Methods Part 3: Methods for Detection and/or Enumeration of Specific Groups of Microorganisms Section 3.6: Enumeration of Yeasts and Moulds. 4 pp.

—Glucose Content—Enzymatic Activity
DIN ENGL 10326-86. Determination of Sucrose and Glucose Content in Milk Products and Ice Cream; Enzymatic Method (Feb). 5 pp.

—Handling
EC EEC/252/92-92. Commission Regulation Amending Council Regulation (EEC) No 206/91 Concerning the Exclusion of Milk Products from Inward Processing Arrangements and of Recourse to Certain Usual Forms of Handling. 1 p.

—Health Requirements
EC 92/46/EEC-92. Council Directive Laying down the Health Rules for the Production and Placing on the Market of Raw Milk, Heat-Treated Milk and Milk-Based Products. 32 pp.
EC 92/47/EEC-92. Council Directive on the Conditions for Granting Temporary and Limited Derogations from Specific Community Health Rules on the Production and Placing on the Market of Milk and Milk-Based Products. 2 pp.

INTERNATIONAL AND NON-U.S. NATIONAL STANDARDS
SUBJECT INDEX

Dairy

Dairy Products (Cont.)

—Import Tariffs

EC EEC/735/79-79. Commission Regulation Fixing the Import Levies on Milk and Milk Products. 3 pp.

—Incubation Conditions—Microbiological Analysis

BSI BS 4285: Sec 1.3-84. 1984 Microbiological Examination for Dairy Purposes Part 1: Guide to General Procedures Section 1.3: Procedures for Obtaining Incubation Conditions. 4 pp.

—Insoluble Matter Content

BSI BS 1743: Part 3-88. 1988 Analysis of Dried Milk and Dried Milk Products Part 3: Determination of Insolubility Index. 10 pp.

ISO 8156-87. Dried Milk and Dried Milk Products—Determination of Insolubility Index First Edition. 10 pp.

—Inspection by Attributes

BSI BS 7037: Part 1-88. 1988 Sampling Plans for Milk Products Part 1: Inspection by Attributes. 18 pp.

ISO 5538-87. Milk and Milk Products—Sampling—Inspection by Attributes First Edition. 16 pp.

—Inspection by Variables

BSI BS 7037: Part 2-89. 1989 Sampling Plans for Milk and Milk Products Part 2: Inspection by Variables. 8 pp.

ISO 8197-88. Milk and Milk Products—Sampling—Inspection by Variables First Edition. 9 pp.

—Iron Content—Spectrometry

BSI BS 6394: Part 2-85. 1985 Trace Elements in Milk and Milk Products Part 2: Method for the Determination of Iron Content (Reference Method). 9 pp.

ISO 6732-85. Milk and Milk Products—Determination of Iron Content—Spectrometric Method (Reference Method) First Edition. 8 pp.

—Lactose Content

CNS N6061-88. Method of Test for Milk and Milk Products (Determination of Lactose) (Sep)(3445). 4 pp.

—Licensing

EC EEC/507/92-92. Commission Regulation Adopting Definitive Measures on the Issuing of STM Licences for Milk and Milk Products in Regard to Spain. 1 p.

—Medicinal Products—Residues

EC COM(93) 287-93. Proposal for a Council Regulation (EEC). 12 pp.

EC EEC/2377/90-90. Council Regulation Laying Down a Community Procedure for the Establishment of Maximum Residue Limits of Veterinary Medicinal Products in Foodstuffs of Animal Origin. 5 pp.

—Microbiological Analysis

BSI BS 4285: Part 0-90. 1990 Microbiological Examination for Dairy Purposes Part 0: General Introduction. 8 pp.

BSI BS 4285: Sec 1.2-84. 1984 Microbiological Examination for Dairy Purposes Part 1: Guide to General Procedures Section 1.2: Diluents, Media and Apparatus and Their Preparation and Sterilization. 8 pp.

BSI BS 4285: Part 4-91. 1991 Microbiological Examination for Dairy Purposes Part 4: Methods for Assessment of Hygienic Conditions. 16 pp.

DIN ENGL 10178 Pt 1-88. Microbiological Analysis of Milk; Determination of Coagulase-Positive Staphylococci by a Method Involving Selective Enrichment (June). 6 pp.

SAA AS 1095. Microbiological Methods for the Dairy Industry (Complete Set in Binder) Note: See Also AS 2300 Series.

SAA AS 1095.2.7-86. Microbiological Methods for the Dairy Industry—Part 2: Methods for the Examination of Specific Dairy Products—Part 2. 7: Microbiological Examination of Long Shelf Life Dairy Products in Hermetically Sealed Containers. 5 pp.

SAA AS 1095.3A-76. Microbiological Methods for the Dairy Industry—Part 3: Methods of Examination for Specific Groups of Microorganisms —Part 3A: Appendix A: Preparation of Media and Diluents. 22 pp.

—Microorganism Content

SAA AS 1766.2.14-91. Method for the Microbiological Examination of Food—Part 2: Examination for Specific Organisms—Part 2.14: Psychrotropic Organisms in Dairy Products. 2 pp.

Dairy Products (Cont.)

—Microorganisms—Count Methods

BSI BS 4285: Sec 2.1-84. 1984 Microbiological Examination for Dairy Purposes Part 2: Methods of General Application for Enumeration of Microorganisms Sec 2.1: Enumeration of Microorganisms by Poured Plate Technique for Colony Count. 4 pp.

BSI BS 4285: Sec 2.3-84. 1984 Microbiological Examination for Dairy Purposes Part 2: Methods of General Application for Enumeration of Microorganisms Sec 2.3: Enumeration of Microorganisms by Surface Plate Technique for Colony Count. 8 pp.

BSI BS 4285: Sec 2.6-84. 1984 Microbiological Examination for Dairy Purposes Part 2: Methods of General Application for Enumeration of Microorganisms Sec 2.6: Enumeration of Microorganisms by Most Probable Number Technique. 6 pp.

BSI BS 4285: Sec 3.1-86. 1986 Microbiological Examination for Dairy Purposes Part 3: Methods for Detection and/or Enumerations of Specific Groups of Microorganisms Sec 3.1: Enumeration of Psychrotrophic Microorganisms. 4 pp.

BSI BS 4285: SUBSEC 3.4.1-87. 1987 Microbiological Examination for Dairy Purposes Part 3: Methods for Detection and/or Enumeration of Specific Groups of Microorganisms Section 3.4: Enumeration of Lipolytic Micro-organisms Subsec 3.4.1: Victoria Blue Method. 4 pp.

BSI BS 4285: SUBSEC 3.4.2-89. 1989 Microbiological Examination for Dairy Purposes Part 3: Methods for Detection and or Enumeration of Specific Groups of Microorganisms Part 3.4: Enumeration of Lipolytic Microorganisms Subsec 3.4.2: Tributyrin Agar Method. 4 pp.

BSI BS 4285: Sec 3.5-88. 1988 Microbiological Examination for Dairy Purposes Part 3: Methods for Detection and/or Enumeration of Specific Groups of Microorganisms Section 3.5: Enumeration of Proteolytic Microorganisms. 4 pp.

BSI BS 4285: Sec 3.14-88. 1988 Microbiological Examination for Dairy Purposes Part 3: Methods for Detection and/or Enumeration of Specific Groups of Microorganisms Section 3.14: Enumeration of Mesophilic Microorganisms. 4 pp.

ISO 6610-92. Milk and Milk Products—Enumeration of Colony-Forming Units of Micro-Organisms—Colony-Count Technique at 30 Degrees Celsius First Edition. 8 pp.

—Moisture Content

CNS N6059-72. Methods of Test for Milk and Milk Products (Determination of Moisture Content) (Oct)(3443). 1 p.

—Nitrogen Content—Kjeldahl Method

SAA AS 2300.1.2. 1-91. Methods of Chemical and Physical Testing for the Dairying Industry—Part 1: General Methods and Principles—Part 1.2.1: Determination of Nitrogen—Reference Kjeldahl Method. 7 pp.

—Packaging—Aluminum Foil

SAA AS 1813-75. Aluminium Foil/Vegetable Parchment Laminates for Wrapping Dairy Products. 8 pp.

—Packaging—Cans

ISO TR8610-84. Light Gauge Metal Containers—Round Vent-Hole Cans with Soldered Ends for Milk and Milk Products—Capacities and Related Diameters First Edition. 2 pp.

—Packaging—Greaseproof Papers

SAA AS 1764-75. Vegetable Parchment for Wrapping Dairy Products. 24 pp.

SAA AS 1813-75. Aluminium Foil/Vegetable Parchment Laminates for Wrapping Dairy Products. 8 pp.

SAA AS 1814-75. High Wet-Strength Greaseproof Paper for Wrapping Dairy Products. 24 pp.

—Packaging—Polyethylene

SAA AS 1326 Supp 1-72. Polyethylene (Polythene) Film for Packaging and Allied Purposes (Metric Units). 7 pp.

—Pesticide Residues

EC 86/363/EEC-86. Council Directive on the Fixing of Maximum Levels for Pesticide Residues in and on Foodstuffs of Animal Origin. 5 pp.

EC 93/57/EEC-93. Council Directive Amending the Annexes to Directives 86/362/EEC and 86/363/EEC on the Fixing of Maximum Levels for Pesticide Residues in and on Cereals and Foodstuffs of Animal Origin Respectively. 5 pp.

—Phosphatase Activity

CNS N6063-80. Method of Test for Milk and Milk Products (Test of Phosphatase) (Aug)(3447). 2 pp.

Dairy Products (Cont.)

—Phosphatase Activity (Cont.)

ISO 3356-75. Milk and Dried Milk, Buttermilk and Buttermilk Powder, Whey and Whey Powder—Determination of Phosphatase Activity (Reference Method) First Edition. 5 pp.

—Plastic Containers

SAA AS 1404-74. Single-Use Rigid and Semi-Rigid Plastics Containers for Dairy Products. 8 pp.

—Protein Content

CNS N6065-88. Methods of Test for Milk and Milk Products: the Quantative Analysis of Protein (Sep)(3449). 2 pp.

—Salmonella Content

BSI BS 4285: Sec 3.9-87. 1987 Microbiological Examination for Dairy Purposes Part 3: Methods for Detection and/or Enumeration of Specific Groups of Microorganisms Section 3.9: Detection of Salmonella. 15 pp.

BSI BS 4285: SUB SEC 3.9.1-01. 1987 Amd 1 Microbiological Examination for Dairy Purposes Part 3: Methods for Detection and/or Enumeration of Specific Groups of Microorganisms Section 3.9: Detection of Salmonella Subsection 3.9.1: Reference Method. 17 pp.

BSI BS 4285: SUB SEC 3.9.2-92. 1992 Microbiological Examination for Dairy Purposes Part 3: Methods for Detection and/or Enumeration of Specific Groups of Microorganisms Section 3.9: Dectection of Salmonella Subsec 3.9.2: Screening Method Using Electrical Conductance. 14 pp.

ISO 6785-85. Milk and Milk Products—Detection of Salmonella First Edition; (Corrected and Reprinted -1985). 15 pp.

—Sampling

BSI BS 809-85. 1985 Methods for Sampling Milk and Milk Products. 32 pp.

BSI BS 1743: Part 1-86. 1986 Analysis of Dried Milk and Dried Milk Products Part 1: General Introduction, Including Preparation of Laboratory Samples. 4 pp.

CGSB 32.175M-89. Code of Practice for Sampling Dairy Products. 9 pp.

ISO 707-85. Milk and Milk Products—Methods of Sampling First Edition. 32 pp.

SAA AS 1166-92. Milk and Milk Products—Methods of Sampling (ISO 707:1985). 28 pp.

SAA AS 2300.7.1-91. Methods of Chemical and Physical Testing for the Dairying Industry—Part 7: Butter—Part 7.1: General Information and Preparation of Samples (Supersedes AS 1739—1975 (in Part)). 2 pp.

—Sampling—Microbiological Analysis

BSI BS 4285: Sec 1.1-91. 1991 Microbiological Examination for Dairy Purposes Part 1: Guide to General Procedures Section 1.1: Sampling and Preparation of Samples. 9 pp.

—Scorched Particles Content

BSI BS 1743: Part 19-85. 1985 Analysis of Dried Milk and Dried Milk Products Part 19: Determination of Scorched Particles Content. 7 pp.

CNS N6196-88. Method of Test for Milk and Milk Products: Determination of Scorched Particles (Sep)(11028). 2 pp.

—Sediment Content

CNS N6067-85. Method of Test for Milk and Milk Products (Determination of Sediment) (Oct)(3451). 1 p.

—Sodium Chloride

SAA AS 2093-77. Salt for Use in the Manufacture of Dairy Products (R 1993). 28 pp.

—Sodium Hypochlorite

SAA AS 1087-91. Sodium Hypochlorite Solutions for Use in the Dairying Industry. 11 pp.

—Solids Content

CNS N6064-72. Methods of Test for Milk and Milk Products (Determination of Total Solids) (Oct)(3448). 1 p.

—Solubility

BSI BS 1743: Part 2-80. 1980 Analysis of Dried Milk and Dried Milk Products Part 2: Determination of the Solubility of Dried Milk, Dried Whey and Dried Buttermilk (Reference Method). 4 pp.

—Staphylococcus

SAA AS 1095.3.2-86. Microbiological Methods for the Dairy Industry—Part 3: Methods of Examination for Specific Groups of Microorganisms—Part 3.2: Coagulase-Positive Staphylococci. 4 pp.

—Staphylococcus Content—Dilution

DIN ENGL 10178 Pt 2-88. Microbiological Analysis of Milk; Determination of Coagulase-Positive Staphylococci by the Dilution Method (Apr). 4 pp.

INDUSTRY STANDARDS

Dairy Products (Cont.)

—Staphylococcus Content—Selective Enrichment
DIN ENGL 10178 Pt 1-88. Microbiological Analysis of Milk; Determination of Coagulase-Positive Staphylococci by a Method Involving Selective Enrichment (June). 6 pp.

—Staphylococcus—Count Methods
BSI BS 4285: SUB SEC 3.10.1-89. 1989 Microbiological Examination for Dairy Purposes Part 3: Methods for Detection and/or Enumeration of Specific Groups of Microorganisms Sec 3.10: Staphylococcus Aureus Subsection 3.10.1: Enumeration Using the Colony Count Technique. 6 pp.
BSI BS 4285: SUB SEC 3.10.2-89. 1989 Microbiological Examination for Dairy Purposes Part 3: Methods for Detection and/or Enumeration of Specific Groups of Microorganisms Sec 3.10: Staphylococcus Aureus Subsection 3.10.2: Detection. 6 pp.
BSI BS 4285: SUB SEC 3.10.3-89. 1989 Microbiological Examina-tion for Dairy Purposes Part 3: Methods for Detection and/or Enumeration of Specific Groups of Microorganisms Section 3.10: Staphylo-Coccus Aureus Subsection 3.10.3: Enumeration Using the Most Probable Number Technique. 6 pp.
DIN ENGL 10178 Pt 3-88. Microbiological Analysis of Milk; Determination of Coagulase-Positive Staphylococci; Colony Count Technique (Apr). 3 pp.

—Streptococcus—Count Methods
BSI BS 4285: Sec 3.11-85. 1985 Microbiological Examination for Dairy Purposes Part 3: Methods for Detection and/or Enumeration of Specific Groups of Microorganisms Section 3.11: Detection and Enumeration of Faecal Streptococci. 4 pp.

—Sucrose Content
CNS N6062-80. Method of Test for Milk and Milk Products (Determination of Surcrose) (Aug)(3446). 4 pp.

—Sucrose Content—Enzymatic Activity
DIN ENGL 10326-86. Determination of Sucrose and Glucose Content in Milk Products and Ice Cream; Enzymatic Method (Feb). 5 pp.

—Tanks (Containers)
BSI BS 3441-89. 1989 Tanks for the Transport of Milk and Liquid Milk Products. 12 pp.
DIN ENGL 11482-87. Dairy Machinery; Storage Tanks for Milk and Liquid Dairy Products (May). 5 pp.

—Test Specimens
CNS N6056-80. Method of Test for Milk and Milk Products (General Rules) (Aug)(3440). 2 pp.
ISO 8261-89. Milk and Milk Products—Preparation of Test Samples and Dilutions for Microbiological Examination First Edition. 9 pp.

—Thermophilic Organisms
SAA AS 1095.3.9-78. Microbiological Methods for the Dairy Industry—Part 3: Methods of Examination for Specific Groups of Microorganisms —Part 3.9: Thermophilic Organisms (Withdrawn).

—Yeasts
SAA AS 1095.3.3-82. Microbiological Methods for the Dairy Industry—Part 3: Methods of Examination for Specific Groups of Microorganisms —Part 3.3: Yeasts and Moulds. 2 pp.

—Yeasts—Count Methods
BSI BS 4285: Sec 3.6-86. 1986 Microbiological Examination for Dairy Purposes Part 3: Methods Part 3: Methods for Detection and/or Enumeration of Specific Groups of Microorganisms Section 3.6: Enumeration of Yeasts and Moulds. 4 pp.
ISO 6611-92. Milk and Milk Products—Enumeration of Colony-Forming Units of Yeasts and/or Moulds—Colony-Count Technique at 25 Degrees C First Edition. 9 pp.

Dakota Aircraft
See Also: Aircraft

—Antenna Positions
CAA. Dakota (Approved Aerial Positions). 1 p.

Dammar
Use For: Gum Dammar

—Content Analysis
MOD UK M 9535/68. Examination of Gum, Dammar, and Gum, Dammar, Lead Free.

—Residue-On-Ignition Determination
MOD UK M 9535/68. Examination of Gum, Dammar, and Gum, Dammar, Lead Free.

Dammar (Cont.)

—Sampling
MOD UK M 9535/68. Examination of Gum, Dammar, and Gum, Dammar, Lead Free.

—Saponification Number
MOD UK M 9535/68. Examination of Gum, Dammar, and Gum, Dammar, Lead Free.

—Sieve Analysis
MOD UK M 9535/68. Examination of Gum, Dammar, and Gum, Dammar, Lead Free.

Damp Heat Testing
See Also: Environmental Testing; High Temperature Testing; Humidity; Temperature Testing
CNS C6334-89. Basic Environmental Testing Procedures, Part 2: Tests, Test Db: Damp Heat, Cycle (12+12 —Hour Cycle) (Jul)(12565).
JIS C 0027-88. Basic Environmental Testing Procedures Part 2: Tests, Test Db: Damp Heat, Cyclic (12 + 12-Hour Cycle). 12 pp.

—Connectors
CEN PREN 2591 (Part C3)-92. Elements of Electrical and Optical Connection Test Methods Part C3—Cold/Low Pressure and Damp Heat. 6 pp.
CEN PREN 2591 (Part C21)-92. Elements of Electrical and Optical Connection Test Methods Part C21—Damp Heat, Cyclic Test. 4 pp.
CEN PREN 2591-FC3-93. Aerospace Series Elements of Electrical and Optical Connection Test Methods Part FC3—Optical Elements Cold/Low Pressure and Damp Heat. 3 pp.
SBAC TS 346 ISSUE 1. Test for Electrical Connectors Damp Heat—Cyclic Test.

—Electrical Components
BSI BS 2011: Part 2.1CA-77. 1977 Amd 1 Basic Environmental Testing Procedures Part 2.1: Tests Part2.1Ca: Test Ca. Damp Heat, Steady State (AMD 5362) March 31, 1987. 4 pp.
BSI BS 2011: Part 2.1CB-90. 1990 Basic Environmental Testing Procedures Part 2.1: Tests Part 2.1Cb: Damp Heat, Steady State, Primarily for Equipment. 8 pp.
BSI BS 2011: Part 2.1DB-81. 1981 Amd 1 Basic Environmental Testing Procedures Part 2.1: Tests Part 2.1Db: Test Db and Guidance: Damp Heat, Cyclic (12 + 12-Hour Cycle). 13 pp.
BSI BS 2011: Part 2.1Z/AD-77. 1977 Basic Environmental Testing Procedures Part 2.1: Tests Part 2.1Z/AD: Test Z/AD. Composite Temperature/Humidity Cyclic Test. 12 pp.
BSI BS 2011: Pt 2.1Z/AMD-77. 1977 Basic Environmental Testing Procedures Part 2.1: Tests Part 2.1Z/AMD. Combined Sequential Cold, Low Air Pressure and Damp Heat Test. 4 pp.
BSI BS 2011: Pt 2.2C & D-90. 1990 Basic Environmental Testing Procedures Part 2.2: Guidance Part 2.2C and D: Guidance for Damp Heat Tests Guidance. 20 pp.
CENELEC HD 323.2.3 S2-87. Basic Environmental Testing Procedures—Part 2: Tests. Test Ca: Damp Heat, Steady State. 3 pp.
CENELEC HD 323.2.28 S1-88. Basic Environmental Testing Procedures—Part 2: Tests. Guidance for Damp Heat Tests. 3 pp.
CENELEC HD 323.2.30 S3-88. Basic Environmental Testing Procedures—Part 2: Tests. Test Db and Guidance: Damp Heat, Cyclic (12 + 12-Hour Cycle). 3 pp.
CENELEC HD 323.2.38 S1-76. Basic Environmental Testing Procedures—Part 2: Tests. Test Z/AD: Composite Temperature/Humidity Cyclic Test. 2 pp.
CENELEC HD 323.2.39 S1-76. Basic Environmental Testing Procedures-Part 2: Tests. Test Z/Amd: Combined Sequential Cold, Low Air Pressure, and Damp Heat Test. 2 pp.
CENELEC HD 323.2.56 S1-90. Basic Environmental Testing Procedures Part 2: Tests Test Cb; Damp Heat, Steady State, Primarly for Equipment. 3 pp.
IEC 68 Pt 2-3-69. Basic Environmental Testing Procedures Part 2: Tests—Test Ca: Damp Heat, Steady State Third Edition; (Amendment 1-1984). 13 pp.
IEC 68 Pt 2-28-90. Environmental Testing Part 2: Tests—Guidance for Damp Heat Tests Third Edition. 37 pp.
IEC 68 Pt 2-30-80. Basic Environmental Testing Procedures Part 2: Tests—Test Db and Guidance: Damp Heat, Cyclic (12 + 12-Hour Cycle) Second Edition; (Amendment 1-1985). 25 pp.
IEC 68 Pt 2-38-74. Basic Environmental Testing Procedures Part 2: Tests—Test Z/AD: Composite Temperature/Humidity Cyclic Test First Edition. 23 pp.
IEC 68 Pt 2-39-76. Basic Environmental Testing Procedures Part 2: Tests—Test Z/AMD: Combined Sequential Cold, Low Air Pressure, and Damp Heat Test First Edition. 12 pp.

Damp Heat Testing (Cont.)

—Electrical Components (Cont.)
IEC 68 Pt 2-56-88. Environmental Testing Part 2: Tests—Test Cb: Damp Heat, Steady State, Primarily for Equipment First Edition. 19 pp.
JIS C 0022-87. Basic Environmental Testing Procedures Part 2: Tests, Test Ca: Damp Heat, Steady State. 7 pp.
JIS C 0032-90. Environmental Testing Test Cb: Damp Heat, Steady State, Primarily for Equipment. 9 pp.
SAA AS 1099.2DA-71. Basic Environmental Testing Procedures for Electrotechnology—Part 2: Tests—Part 2Da: Damp Heat, 24-Hour Cycle Reconfirmed 1985. 2 pp.
SAA AS 1099.2DB-82. Basic Environmental Testing Procedures for Electrotechnology—Part 2: Tests—Part 2Db: Damp Heat, Cyclic (12 + 12 Hours). 7 pp.
SAA AS 1099.2Z/A D-80. Basic Environmental Testing Procedures for Electrotechnology—Part 2: Tests—Part 2Z/AD: Composite Temperature/Humidity Cyclic Test Reconfirmed 1985. 8 pp.
SAA AS 1099.2Z/A MD-80. Basic Environmental Testing Procedures for Electrotechnology—Part 2: Tests—Part 2Z/AMD: Combined Sequential Cold, Low Air Pressure and Damp Heat Test Reconfirmed 1985. 2 pp.
SAA AS 1099.2.3-90. Basic Environmental Testing Procedures for Electrotechnology—Part 2: Tests—Part 2.3: Test Ca—Damp Heat, Steady State (IEC 68-2-3:1969). 3 pp.
SAA AS 1099.3.3-82. Basic Environmental Testing Procedures for Electrotechnology—Part 3: Background Information—Part 3.3: Section 3-Guidance for Damp Heat Tests. 11 pp.
SNZ IEC 68: Part 2-3-69. Basic Environmental Testing Procedures Part 2-3: Test Ca: Damp Heat, Steady State. 9 pp.
SNZ IEC 68: Part 2-28-90. Basic Environmental Testing Procedures Part 2-28: Tests—Guidance for Damp Heat Tests. 33 pp.
SNZ IEC 68: Part 2-30-80. Basic Environmental Testing Procedures Part 2-30: Test Db and Guidance: Damp Heat, Cyclic (12 + 12-Hour Cycle) Amend: 1. 19 pp.
SNZ IEC 68: Part 2-38-74. Basic Environmental Testing Procedures Part 2-38: Test 2/AD: Composite Temperature/Humidity Cyclic Test. 20 pp.
SNZ IEC 68: Part 2-39-76. Basic Environmental Testing Procedures Part 2-39: Test 2/AMD: Combined Sequential Cold, Low Air Pressure, and Damp Heat Test. 8 pp.

—Electronic Components
CNS C6010-89. Damp Heat (Steady State) Testing Procedures for Electronic Components (Jun)(3623).
CNS C6012-85. Damp Heat (Cyclic) Testing Procedures for Electronic Components (Apr)(3625). 5 pp.

—Paperboard
BSI BS 6388: Part 3-87. 1987 Accelerated Ageing of Paper and Board Part 3: Method for Moist Heat Treatment at 80 Degrees Celsius and 65 Percent Relative Humidity. 8 pp.
ISO 5630 Pt 3-86. Paper and Board—Accelerated Ageing—Part 3: Moist Heat Treatment at 80 Degrees Celsius and 65% Relative Humidity First Edition. 7 pp.

—Papers
BSI BS 6388: Part 3-87. 1987 Accelerated Ageing of Paper and Board Part 3: Method for Moist Heat Treatment at 80 Degrees Celsius and 65 Percent Relative Humidity. 8 pp.
ISO 5630 Pt 3-86. Paper and Board—Accelerated Ageing—Part 3: Moist Heat Treatment at 80 Degrees Celsius and 65% Relative Humidity First Edition. 7 pp.

—Plywood
CNS O2035-82. Method of Test for Wet Heating of Special Plywood (Jan)(8323). 1 p.

—Semiconductor Devices
CNS C6072-88. Environmental Testing Methods and Endurance Testing Methods for Discrete Semiconductor Devices (Damp Heat Test) (Nov)(6117).

Damp-Heat Tests
Use: Damp Heat Testing

Damper Valves
See Also: Dampers; Hydraulic Valves; Valves

—Aerodynamic Testing
BSI BS 6821-88. 1988 Methods for Aerodynamic Testing of Dampers and Valves. 11 pp.
ISO 7244-84. Air Distribution and Air Diffusion—Aerodynamic Testing of Dampers and Valves First Edition. 8 pp.

INTERNATIONAL AND NON-U.S. NATIONAL STANDARDS
SUBJECT INDEX

Damper Valves (Cont.)
—Rolling Stock—Vibration Isolators
CNS E1028-83. Rubber Vibration Isolators for Oil Dampers of Railway Rolling Stock (Mar)(10066).
JIS E 4715-85. Rubber Vibration Isolators for Oil Dampers of Railway Rolling Stock.

Dampers
See Also: Air Handling Equipment; Fire Dampers; Heating Equipment; Mufflers

—Aerodynamic Testing
BSI BS 6821-88. 1988 Methods for Aerodynamic Testing of Dampers and Valves. 11 pp.
ISO 7244-84. Air Distribution and Air Diffusion—Aerodynamic Testing of Dampers and Valves First Edition. 8 pp.

—Oil Burners
CSA B140.14-M1979. Automatic Flue-Pipe Dampers for Use with Oil-Fired Appliances (R 1991). 20 pp.

—Oil Burners—Rolling Stock
JIS E 4205-91. Oil Dampers for Railway Rolling Stock.

—Ships
JIS F 7113-75. Marine Ventilation Dampers.

—Sound Power
CEN EN 25135-91. Acoustics—Determination of Sound Power Levels of Noise from Air Terminal Devices, High/Low Velocity/Pressure Assemblies, Dampers and Valves by Measurement in a Reverberation Room. 3 pp.
DIN ENGL EN 25135-91. Determination of Sound Power Levels of Noise from Air Terminal Devices, High/Low Velocity/Pressure Assemblies, Dampers and Valves by Measurement in a Reverberation Room (ISO 5135: 1984) (Nov). 15 pp.
ISO 5135-84. Acoustics—Determination of Sound Power Levels of Noise from Air Terminal Devices, High/Low Velocity/Pressure Assemblies, Dampers and Valves by Measurement in a Reverberation Room First Edition. 13 pp.

Damping, Mechanical
Use: Mechanical Damping

Damping, Vibration
Use: Vibration Damping

Damping Compounds
See Also: Damping Fluids

—Vibration—Ships
MOD UK NES 338: Part 2-82. Shock & Vibration Mounts Part 2: Damping Compound for Type 'X' Mounts Issue 1 (04.82). 13 pp.

Damping Fluids
See Also: Damping Compounds; Mechanical Damping; Vibration Damping

—Complex Modulus—Graphic Methods
BSI BS 7544-91. 1991 Method for Graphical Presentation of the Complex Modulus of Damping Materials (ISO 10112: 1991) (V). 15 pp.
ISO 10112-91. Damping Materials—Graphical Presentation of the Complex Modulus First Edition. 13 pp.

—Dimethyl Silicone
MOD UK DSTAN 91-46-79. Damping Fluids: Dimethyl Silicone Issue 1. 11 pp.

Damping Tests
Use: Mechanical Damping

Dampproof Courses
Use: Waterproof Courses

Dampproofing
Use: Waterproofing

Dampproofing Coatings
Use: Waterproof Coatings

Dams
Scope Note: For additional listings, use a more specific term See Also: Barriers; Cofferdams; Dikes; Hydraulic Structures; Hydroelectric Power Plants; Reservoirs (Lakes); Weirs

—Plants
DIN ENGL 19700 Pt 10-86. Dam Plants; General Specifications (Jan) (This Standard, Together with DIN 19700 Part 11, Supersedes DIN 19700 Part 1, December 1965 Edition, Together with DIN 19700 Part 12, Supersedes DIN 19700 Part 99, October 1980 Edition, and Together with DIN 19700 Part 13,. 15 pp.

Dams (Cont.)
—Plants (Cont.)
DIN ENGL 19700 Pt 11-86. Dam Plants; Dams (Jan) (This Standard, Together with DIN 19700 Part 10, January 1986 Edition, Supersedes DIN 19700 Part 1, December 1965 Edition). 14 pp.

Danger Lamps
Use: Warning Lights

Dangerous Goods
Use: Hazardous Materials

Daphnia
See Also: Crustacea; Shellfish

—Waste Water—Toxicity
DIN ENGL 38412 Pt 11-82. German Standard Methods for the Examination of Water, Waste Water and Sludge; Test Methods Using Water Organisms (Group L); Determination of the Effect on Microcrustacea of Substances Contained in Water (Daphina Short-Time Test) (L 11) (Oct). 4 pp.
DIN ENGL 38412 Pt 30-89. German Standard Methods for the Examination of Water, Waste Water and Sludge; Bio-Assays (Group L); Determining the Tolerance of Daphnia to the Toxicity of Waste Water by Way of a Dilution (Mar). 4 pp.

—Water—Toxicity
BSI BS 6068: Sec 5.1-90. 1990 Water Quality Part 5: Methods for Biological Testing Section 5.1: Determination of the Inhibition of the Mobility of Daphnia Magna Staus (Cladocera, Crustacea). 12 pp.
DIN ENGL 38412 Pt 11-82. German Standard Methods for the Examination of Water, Waste Water and Sludge; Test Methods Using Water Organisms (Group L); Determination of the Effect on Microcrustacea of Substances Contained in Water (Daphina Short-Time Test) (L 11) (Oct). 4 pp.
ISO 6341-89. Water Quality—Determination of the Inhibition of the Mobility of Daphnia Magna Straus (Cladocera, Crustacea) Second Edition. 11 pp.

Darkrooms
See Also: Photographic Processing Equipment

—Lighting
CSA C22.2 NO 118-1959. Construction and Test of Picture Machines and Appliances (R 1992); (Erratum April 1959) (Rev 1-5 March 1969). 21 pp.
CSA 649J Bull. Electrical Bulletin 649J November 17, 1975 to C22.2 NO 118. 1 p.
CSA 1169 Bull. Electrical Bulletin 1169 June 27, 1978 to C22.2 NO 118. 2 pp.

—Meters
CSA C22.2 NO 118-1959. Construction and Test of Picture Machines and Appliances (R 1992); (Erratum April 1959) (Rev 1-5 March 1969). 21 pp.
CSA 649J Bull. Electrical Bulletin 649J November 17, 1975 to C22.2 NO 118. 1 p.
CSA 1169 Bull. Electrical Bulletin 1169 June 27, 1978 to C22.2 NO 118. 2 pp.

—Radiology—Safelight Conditions
IEC 1223 Pt 2-3-93. Evaluation and Routine Testing in Medical Imaging Departments—Part 2-3: Constancy Tests—Darkroom Safelight Conditions First Edition. 36 pp.

—Safelight Conditions
ISO 8374-86. Photography—Determination of ISO Safelight Conditions First Edition. 8 pp.

—Timers
CSA C22.2 NO 118-1959. Construction and Test of Picture Machines and Appliances (R 1992); (Erratum April 1959) (Rev 1-5 March 1969). 21 pp.
CSA 649J Bull. Electrical Bulletin 649J November 17, 1975 to C22.2 NO 118. 1 p.
CSA 1169 Bull. Electrical Bulletin 1169 June 27, 1978 to C22.2 NO 118. 2 pp.

Darlington Phototransistors
Use: Photodarlingtons

Darlington Transistor Arrays
Use: Transistor Arrays—Darlington

Darlington Transistors
See Also: Photodarlingtons

—NPN—Epitaxial—LF
CECC CECC 50 003-025. BS CECC 50 003-025 Issue 1; Case Rated Low Frequency Amplification Darlington Bipolar Transistors (En). 10 pp.
CECC CECC 50 003-026. BS CECC 50 003-026 Issue 1; Case Rated Low Frequency Amplification Darlington Bipolar Transistors (En). 10 pp.

Darlington Transistors (Cont.)
—NPN—Epitaxial—LF (Cont.)
CECC CECC 50 003-036 ISSUE 1-80. BS CECC 50 003-036; Case Rated Low Frequency Amplification Bipolar Darlington Transistor (En). 8 pp.

—NPN—LF
CECC CECC 50 002-124. BS CECC 50 002-124 Issue 1; NPN Silicon Ambient Rated Low Frequency Amplification Transistors in Plastic Encapsulation (En). 7 pp.
CECC CECC 50 002-241 ISSUE 1-88. BS CECC 50 002-241 Issue 1; NPN Silicon Darlington Connected Ambient Rated Low Frequency Amplification Transistors in Plastic Encapsulation (En). 7 pp.
CECC CECC 50 003-005-006 IS 2-78. BS CECC 50 003-005-006; NPN Silicon Case Rated Low Frequency Amplification Darlington Connected Transistors in Hermetically Sealed Metal Case (En) AMD 1 (En). 8 pp.
CECC CECC 50 003-007-008 IS 2-77. BS CECC 50 003-007-008; NPN Silicon Case Rated Low Frequency Amplification Darlington Connected Transistors in Hermetically Sealed Metal Case (En) AMD 1 (En). 8 pp.
CECC CECC 50 003-053 ISSUE 1-88. BS CECC 50 003-053; NPN Silicon Case Rated Low Ferquency Amplification Darlington Connected Transistors in Hermetically Sealed Metal Case (En). 9 pp.

—NPN/PNP—Epitaxial—LF
CECC CECC 50 003-030 ISSUE 1-79. BS CECC 50 003-030; Case Rated Low Frequencey Amplification Darlington Bipolar Transistors (En). 13 pp.

—NPN—Power—Preferred Products List
CECC CECC MUAHAG Vol 9 IS 3-90. Preferred Products List; Semiconductors (En, Fr, Ge) AMD 1 (En, Fr, Ge). 51 pp.

—NPN—Power—Switching
CECC CECC 50 004-120 ISSUE 1-83. BS CECC 50 004-120; Case Rated Darlington Power Transistors for Switching Applications (En). 25 pp.
CECC CECC 50 004-138 ISSUE 1-83. BS CECC 50 004-138; Case Rated Darlington Connected Power Transistors for High Voltage Switching Applications (En). 26 pp.
CECC CECC 50 004-152 ISSUE 1-86. BS CECC 50 004-152; Case Rated Darlington Connected Power Transistors for High Voltage Switching Applications (En). 28 pp.

—NPN—Switching
CECC CECC 50 004-073 ISSUE 1-80. BS CECC 004-073; N-P-N Ambient Rated Silicon Planar Darlington Transistor for Switching Application, in Hermetically Sealed Metal Case (En) AMD (En). 10 pp.
CECC CECC 50 004-155 ISSUE 2-89. NL CECC 50 004-155; NPN Darlington, Silicon, Ambient-Rated Switching Transistors in Plastic Encapsulation (En). 9 pp.

—PNP—Epitaxial—LF
CECC CECC 50 003-027. BS CECC 50 003-027 Issue 1; Case Rated Low Frequency Amplification Darlington Bipolar Transistor (En). 10 pp.
CECC CECC 50 003-028. BS CECC 50 003-028 Issue 1; Case Rated Low Frequency Amplification Darlington Bipolar Transistor (En). 10 pp.
CECC CECC 50 003-029. UTE C 86 613-029 Edition 2; Case Rated Bipolar Transistors for Low Frequency Amplification (Fr). 6 pp.
CECC CECC 50 003-037 ISSUE 1-80. BS CECC 50 003-037; Case Rated Low Frequency Amplification Bipolar Darlington Transistor (En). 8 pp.

—PNP—Power—Preferred Products List
CECC CECC MUAHAG Vol 9 IS 3-90. Preferred Products List; Semiconductors (En, Fr, Ge) AMD 1 (En, Fr, Ge). 51 pp.

Dassault Mystere (Fan Jet Falcon) Aircraft
See Also: Aircraft

—Accidents
CAA. Dassault Mystere/Fan Jet Falcon (World Airline Accident Summary). 1 p.
CAA. Dassault Mystere/Fan Jet Falcon 10, 100, 20, 200, 50 and 900. 1 p.

—Foreign Airworthiness Directives
CAA. Dassult Aviation Fan Jet Falcon and Mystere-Falcon 200 Series Aircraft. 5 pp.

Dassault Mystere-Falcon 900 Aircraft
Use For: Dassault Falcon 900 Aircraft; Mystere-Falcon 900 Aircraft See Also: Aircraft

—Certification
CAA. Dassault Mystere Falcon 900. 11 pp.

Dassault Mystere-Falcon 900 Aircraft (Cont.)
—**Foreign Airworthiness Directives**
CAA. Dassult Aviation Mystere-Falcon 900 Series. 1 p.

Dassualt Falcon 900 Aircraft
Use: Dassault Mystere-Falcon 900 Aircraft

Data Acquisition Equipment
Use: Data Acquisition Systems

Data Acquisition Subsystems
Use: Data Acquisition Systems

Data Acquisition Systems
Use For: Data Acquisition Equipment; Data Acquisition Subsystems *See Also:* Data Analysis; Data Collection; Data Logging; Data Management; Data Processing; Data Processing Equipment; Data Transmission; Document Acquisition; FASTBUS Systems

—**Codes**
NATO STANAG 4177 ED 1 AMD 1-86. Codification of Items of Supply—Uniform System of Data Acquisition. 10 pp.
NATO STANAG 4177 ED 2 AMD 0-90. Codification of Items of Supply—Uniform System of Data Acquisition. 8 pp.

—**Fires**
ISO TR7248-85. Fire Data—Collection and Presentation System First Edition. 15 pp.

—**Glossaries**
JIS X 0705-89. Documentation and Information-Vocabulary—Acquisition Identification, and Analysis of Documents and Data.

Data Analysis
See Also: Data Acquisition Systems; Data Collection; Data Logging; Data Organization; Data Processing; Statistical Analysis

—**Glossaries**
JIS X 0705-89. Documentation and Information-Vocabulary—Acquisition Identification, and Analysis of Documents and Data.

—**Network Performance—Telecommunication Networks**
CCITT RECMN E.880-89. Field Data Collection and Evaluation on the Performance of Equipment, Networks and Services—Telephone Network and ISDN—Quality of Service, Network Management and Traffic Engineering (Study Group II) 7 pp. 7 pp.

—**Quality of Service—Telex Communications**
CCITT RECMN F.70-89. Evaluating the Quality of the International Telex Service—Telegraph and Mobile Services Operations and Quality of Service (Study Group I) 3 pp. 3 pp.

—**Telecommunication Equipment**
CCITT RECMN E.880-89. Field Data Collection and Evaluation on the Performance of Equipment, Networks and Services—Telephone Network and ISDN—Quality of Service, Network Management and Traffic Engineering (Study Group II) 7 pp. 7 pp.

—**Telecommunication Services**
CCITT RECMN E.880-89. Field Data Collection and Evaluation on the Performance of Equipment, Networks and Services—Telephone Network and ISDN—Quality of Service, Network Management and Traffic Engineering (Study Group II) 7 pp. 7 pp.

—**Telecommunication Traffic**
CCITT RECMN E.503 (REV 1)-92. Traffic Measurement Data Analysis (Study Group II) 9 pp. 9 pp.
CCITT RECMN E.503-89. Traffic Measurement Data Analysis—Telephone Network and ISDN—Quality of Service, Network Management and Traffic Engineering (Study Group II) 4 pp. 4 pp.

—**Telecommunication Traffic—Man-Machine Language**
CCITT RECMN E.503 (REV 1)-92. Traffic Measurement Data Analysis (Study Group II) 9 pp. 9 pp.
CCITT RECMN E.503-89. Traffic Measurement Data Analysis—Telephone Network and ISDN—Quality of Service, Network Management and Traffic Engineering (Study Group II) 4 pp. 4 pp.

Data Banks
—**Ionospheric Propagation**
CCIR Decision 83-89. HF Measurements and Data Banks—Annex to Volume VI—Propagation in Ionized Media. 2 pp.
CCIR QUESTION 42/6-90. Measurements and Data Banks—Questions Concerning Study Group 6—Radio Wave Propagation in Ionized Media. 1 p.

—**Radio Wave Propagation**
CCIR Report 1144-90. Data Banks to Support Evaluation of Prediction Methods—Section 5A—Texts of General Interest. 25 pp.

—**Signal Intensity—Ionospheric Propagation**
CCIR QUESTION 42/6-90. Measurements and Data Banks—Questions Concerning Study Group 6—Radio Wave Propagation in Ionized Media. 1 p.

—**Sky Waves—Signal Intensity—Ionospheric Propagation**
CCIR OPINION 68-1-90. Data Bank of HF Sky-Wave Signal Intensity Measurements—Volume VI—Propagation in Ionized Media. 1 p.

—**Tropospheric Propagation**
CCIR RECMN 311-5-90. Acquisition, Presentation and Analysis of Data in Studies of Tropospheric-Wave Propagation—Section 5A—Texts of General Interest. 1 p.
CCIR RECMN 311-6-92. Acquisition, Presentation and Analysis of Data in Studies of Tropospheric Propagation. 43 pp.

Data Base Management Systems
Use For: DBMS *See Also:* Data Bases; Data Processing; Data Processing Equipment; Tactical Data Systems
ECMA ECMA TR/58-92. Databases and Networking. 26 pp.
ECMA ECMA TR/59-92. Object-Oriented Databases. 12 pp.

—**Conceptual Schema—Glossaries**
ISO TR9007-87. Information Processing Systems—Concepts and Terminology for the Conceptual Schema and the Information Base First Edition. 181 pp.
JTC1 TR9007-87. Information Processing Systems—Concepts and Terminology for the Conceptual Schema and the Information Base First Edition. 181 pp.
OSI ISO TR 9007-87. Concepts and Terminology for the Conceptual Schema and the Information Base.

Data Bases
See Also: Data Base Management Systems

—**Frequency Management—Computer-Aided**
CCIR Report 1110-90. Frequency Management Data Base Systems Using Small Computers—Section 1A—Spectrum Engineering and Computer-Aided Principles and Techniques. 11 pp.

—**Terrain—Radio Wave Propagation**
CCIR Decision 102-89. Propagation Predictions Based on Digital Terrain Data Bases Including Terrain Roughness and Man-Made Structures—Volume V—Propagation in Non-Ionized Media. 2 pp.

Data Blocks
See Also: Data Processing; Information Interchange; Record Layouts; Variable Block Formats

—**Shiplines**
ISO 7838-84. Shipbuilding—Shiplines—Formats and Data Organization First Edition. 4 pp.

Data Broadcasting
See Also: Broadcasting; Data Communication; Data Transmission; Program Delivery Control; Telesoftware; Television Broadcasting
CCIR Decision 72-1-89. Data Broadcasting Services—Volume X-1—Broadcasting Service (Sound). 2 pp.
CCIR OPINION 83-1-90. Data Broadcasting Services—Volume XI-1—Broadcasting Service (Television). 1 p.
CCIR Report 802-3-90. Additional Services Using Broadcasting Channels—Section 11B—Ancillary Television Services. 13 pp.
CCIR Report 1207-90. Reference Model for Data Broadcasting—Section 11B—Ancillary Television Services. 6 pp.
CCIR Decision 72-1-89. Data Broadcasting Services—Annex to Volume XI-1—Broadcasting Service (Television). 2 pp.

—**Broadcasting Satellite Services**
CCIR RECMN 712-90. High-Quality Sound/Data Standards for the Broadcasting-Satellite Service in the 12 GHz Band—Section 10/11B—Systems. 1 p.

Data Broadcasting (Cont.)
—**Broadcasting Satellite Services (Cont.)**
CCIR RECMN 712-1-92. High-Quality Sound/Data Standards for the Broadcasting-Satellite Service in the 12 GHz Band—Section 10/11B—Systems. 48 pp.
CCIR Report 1228-90. High Quality Sound/Data Standards for the Broadcasting Satellite Service in the 12 GHz Band—Section 10/11B—Systems. 34 pp.
CCIR Decision 51-4-89. Satellite Broadcasting of High Definition Television (HDTV) Signals and Accomodation of Several Audio and/or Data Signals Either Associated with Television Signals or for Sound/Data Broadcasting in Terrestrial and Satellite Broadcasting. 6 pp.

—**Error Control**
CCIR Report 1210-90. Error-Protection Strategies for Data Broadcasting Services—Section 11B—Ancillary Television Services. 28 pp.

—**High Definition Television Systems**
CCIR Report 1225-90. Data Broadcasting Systems and Services in and HDTV Environment—Section 11B—Ancillary Television Services. 8 pp.

—**Narrowband Channels—Ancillary Services**
CCIR QUESTION 74/11-90. Additional Services Provided by Data Broadcasting in a Television or Narrow-Band Channel—Questions Concerning Study Group 11—Broadcasting Service (Television). 1 p.

—**Open Systems Interconnection—Reference Model**
CCIR RECMN 807-92. Reference Model for Data Broadcasting. 4 pp.

—**Quality Assurance**
CCIR Report 956-2-90. Data Broadcasting Systems: Signal and Service Quality, Field Trials and Theoretical Studies—Section 11B—Ancillary Television Services. 10 pp.

—**Signals—Distortion—Vestigial Sideband Transmission**
CCIR QUESTION 56/11-90. Distortion of Television and Data-Broadcasting Signals in Emission and Reception, Caused by the Use of Vestigial-Sideband Techniques—Questions Concerning Study Group 11—Broadcasting Service (Television). 1 p.

—**Signals—Reception—Receiving Antennas**
CCIR QUESTION 68/11-90. Recommended Characteristics for Individual, Collective and Cable Distribution Antenna Systems for Domestic Reception of Television and Data Broadcasting Signals from Terrestrial Transmitters—Questions Concerning Study Group 11-Broadcasting Service. 1 p.

—**Television Channels**
CCIR QUESTION 29-3/11-86. Broadcasting of Still Pictures and Other Information Intended for the Public Using a Television or Narrow-Band Channel—Questions Concerning Study Group 11—Broadcasting Service (Television). 1 p.

—**Television Channels—Ancillary Services**
CCIR QUESTION 74/11-90. Additional Services Provided by Data Broadcasting in a Television or Narrow-Band Channel—Questions Concerning Study Group 11—Broadcasting Service (Television). 1 p.

—**Television Channels—Multiplexing**
CCIR QUESTION 72/11-90. Specification for Multiplex Broadcasting of Information in the Television Channel—Questions Concerning Study Group 11—Broadcasting Service (Television). 1 p.

Data Broadcasting Services
Use: Data Broadcasting

Data Buffers
Use: Buffers (Data)

Data Bus Couplers
Use For: DBC *See Also:* Connectors; Data Buses

—**Aerospace**
CEN PREN 3567-001-91. Couplers for Use in Multiplex Data Bus Systems in Accordance with Stanag 3838 Part 001—Technical Specification. 16 pp.
CEN PREN 3567-001-92. Aerospace Series Couplers for Use in Multiplex Data Bus Systems in Accordance with STANAG 3838 Part 001—Technical Specification. 12 pp.
CEN PREN 3567-003-91. Couplers for Use in Multiplex Data Bus Systems in Accordance with Stanag 3838 Part 003—Single In-Line Couplers Product Standard. 6 pp.

INTERNATIONAL AND NON-U.S. NATIONAL STANDARDS
SUBJECT INDEX

Data

Data Bus Couplers *(Cont.)*
—Aerospace *(Cont.)*

CEN PREN 3567-003-92. Aerospace Series Couplers for Use in Multiplex Data Bus Systems in Accordance with STANAG 3838 In-Line Couplers Product Standard. 5 pp.

CEN PREN 3567-004-91. Couplers for Use in Multiplex Data Bus Systems in Accordance with Stanag 3838 Part 004—Double In-Line Couplers Product Standard. 6 pp.

CEN PREN 3567-004-92. Aerospace Series Couplers for Use in Multiplex Data Bus Systems in Accordance with STANAG 3838 Part 004—Double In-Line Couplers Product Standard. 5 pp.

Data Bus Systems
Use: Data Buses

Data Buses
Use For: Field Buses *See Also:* Backplanes; Bus Drivers; Busbars; Communication Interfaces; Data Bus Couplers; Data Processing Equipment; Data Transmission; Domestic Digital Bus Systems; FASTBUS Systems; Interfaces

CEN PREN 3910-90. High Speed Data Transmission Under Stanag 3838 or Fiber Optic Equivalent Control. 100 pp.

CEN PREN 3910-92. High Speed Data Transmission Under STANAG 3838 or Fiber Optic Equivalent Control. 93 pp.

— 8-Bit—Interfaces—Microprocessors

BSI BS IEC 796: Part 1-90. 1990 Microprocessor System Bus—8-Bit and 16-Bit Data (MULTIBUS I) Part 1: Functional Description with Electrical and Timing Specifications. 59 pp.

BSI BS IEC 796: Part 2-90. 1990 Microprocessor System Bus—8-Bit and 16-Bit Data (MULTIBUS I) Part 2: Mechanical and Pin Descriptions for the System Bus Configuration, with Edge Connectors (Direct). 12 pp.

CENELEC HD 593.1 S1-92. Microprocessor System Bus—8-Bit and 16-Bit Data (MULTIBUS I) Part 1: Functional Description with Electrical and Timing Specifications. 5 pp.

CENELEC HD 593.2 S1-92. Microprocessor System Bus—8-Bit and 16-Bit Data (MULTIBUS I) Part 2: Mechanical and Pin Descriptions for the System Bus Configuration, with Edge Connectors (Direct). 4 pp.

CENELEC HD 593.3 S1-91. Microprocessor System BUS I, 8-Bit and 16-Bit Data Part 3: Mechanical and Pin Descriptions for the Eurocard Configuration with Pin and Socket (Indirect) Connectors. 6 pp.

CENELEC HD 593.3 S1-92. CORRIGENDUM Microprocessor System BUS I, 8-Bit and 16-Bit Data Part 3: Mechanical and Pin Descriptions for the Eurocard Configuration with Pin and Socket (Indirect) Connectors. 3 pp.

IEC 796 Pt 1-90. Microprocessor System BUS—8-Bit and 16-Bit Data (MULTIBUS 1) Part 1: Functional Description with Electrical and Timing Specifications First Edition. 116 pp.

IEC 796 Pt 2-90. Microprocessor System BUS—8-Bit and 16-Bit Data (MULTIBUS 1) Part 2: Mechanical and Pin Descriptions for the System Bus Configuration, with Edge Connectors (Direct) First Edition. 22 pp.

IEC 796 Pt 3-90. Microprocessor System BUS I, 8-Bit and 16-Bit Data Part 3: Mechanical and Pin Descriptions for the Eurocard Configuration with Pin and Socket (Indirect) Connectors First Edition. 23 pp.

JTC1 796 Pt 1-90. Microprocessor System BUS—8-Bit and 16-Bit Data (MULTIBUS 1) Part 1: Functional Description with Electrical and Timing Specifications First Edition. 116 pp.

JTC1 796 Pt 2-90. Microprocessor System BUS—8-Bit and 16-Bit Data (MULTIBUS 1) Part 2: Mechanical and Pin Descriptions for the System Bus Configuration, with Edge Connectors (Direct) First Edition. 22 pp.

JTC1 796 Pt 3-90. Microprocessor System BUS I, 8-Bit and 16-Bit Data Part 3: Mechanical and Pin Descriptions for the Eurocard Configuration with Pin and Socket (Indirect) Connectors First Edition. 23 pp.

— 16-Bit—Interfaces—Microprocessors

BSI BS IEC 796: Part 1-90. 1990 Microprocessor System Bus—8-Bit and 16-Bit Data (MULTIBUS I) Part 1: Functional Description with Electrical and Timing Specifications. 59 pp.

BSI BS IEC 796: Part 2-90. 1990 Microprocessor System Bus—8-Bit and 16-Bit Data (MULTIBUS I) Part 2: Mechanical and Pin Descriptions for the System Bus Configuration, with Edge Connectors (Direct). 12 pp.

CENELEC HD 593.1 S1-92. Microprocessor System Bus—8-Bit and 16-Bit Data (MULTIBUS I) Part 1: Functional Description with Electrical and Timing Specifications. 5 pp.

Data Buses *(Cont.)*
— 16-Bit—Interfaces—Microprocessors *(Cont.)*

CENELEC HD 593.2 S1-92. Microprocessor System Bus—8-Bit and 16-Bit Data (MULTIBUS I) Part 2: Mechanical and Pin Descriptions for the System Bus Configuration, with Edge Connectors (Direct). 4 pp.

CENELEC HD 593.3 S1-91. Microprocessor System BUS I, 8-Bit and 16-Bit Data Part 3: Mechanical and Pin Descriptions for the Eurocard Configuration with Pin and Socket (Indirect) Connectors. 6 pp.

CENELEC HD 593.3 S1-92. CORRIGENDUM Microprocessor System BUS I, 8-Bit and 16-Bit Data Part 3: Mechanical and Pin Descriptions for the Eurocard Configuration with Pin and Socket (Indirect) Connectors. 3 pp.

IEC 796 Pt 1-90. Microprocessor System BUS—8-Bit and 16-Bit Data (MULTIBUS 1) Part 1: Functional Description with Electrical and Timing Specifications First Edition. 116 pp.

IEC 796 Pt 2-90. Microprocessor System BUS—8-Bit and 16-Bit Data (MULTIBUS 1) Part 2: Mechanical and Pin Descriptions for the System Bus Configuration, with Edge Connectors (Direct) First Edition. 22 pp.

IEC 796 Pt 3-90. Microprocessor System BUS I, 8-Bit and 16-Bit Data Part 3: Mechanical and Pin Descriptions for the Eurocard Configuration with Pin and Socket (Indirect) Connectors First Edition. 23 pp.

JTC1 796 Pt 1-90. Microprocessor System BUS—8-Bit and 16-Bit Data (MULTIBUS 1) Part 1: Functional Description with Electrical and Timing Specifications First Edition. 116 pp.

JTC1 796 Pt 2-90. Microprocessor System BUS—8-Bit and 16-Bit Data (MULTIBUS 1) Part 2: Mechanical and Pin Descriptions for the System Bus Configuration, with Edge Connectors (Direct) First Edition. 22 pp.

JTC1 796 Pt 3-90. Microprocessor System BUS I, 8-Bit and 16-Bit Data Part 3: Mechanical and Pin Descriptions for the Eurocard Configuration with Pin and Socket (Indirect) Connectors First Edition. 23 pp.

— 32-Bit—Microprocessors—NuBus Compatible

IEC DIS 10860.2-93. Information Technology—Simple 32-Bit Backplane Bus: NuBus ***CD-ROM ONLY***. 100 pp.

ISO DIS 10860.2-93. Information Technology—Simple 32-Bit Backplane Bus: NuBus ***CD-ROM ONLY***. 100 pp.

JTC1 DIS10860.2-93. Information Technology—Simple 32-Bit Backplane Bus: NuBus ***CD-ROM ONLY***. 100 pp.

—Aerospace

BSI G 242-89. 1989 Data Bus Interconnecting Systems. 16 pp.

—Buffers—Octal—Preferred Products List

CECC CECC MUAHAG Vol 7 IS 8-92. Preferred Products List; Active Microcircuits (En, Fe, Ge). 89 pp.

—Buffers—Quad—Preferred Products List

CECC CECC MUAHAG Vol 7 IS 8-92. Preferred Products List; Active Microcircuits (En, Fe, Ge). 89 pp.

—Control and Status Registers

IEC DIS 13213-92. Information Technology—Standard Control and Status Register (CSR) Architecture for Microcomputer Buses ***CD-ROM ONLY***. 134 pp.

ISO DIS 13213-92. Information Technology—Standard Control and Status Register (CSR) Architecture for Microcomputer Buses ***CD-ROM ONLY***. 134 pp.

JTC1 DIS13213-92. Information Technology—Standard Control and Status Register (CSR) Architecture for Microcomputer Buses ***CD-ROM ONLY***. 134 pp.

—Futurebus—Logical Layer—Microprocessors

IEC DIS 10857.2-92. Information Technology—Microprocessor Systems—Futurebus+—Logical Protocol Specification ***CD-ROM ONLY***. 205 pp.

ISO DIS 10857.2-92. Information Technology—Microprocessor Systems—Futurebus+—Logical Protocol Specification ***CD-ROM ONLY***. 205 pp.

JTC1 DIS10857.2-92. Information Technology—Microprocessor Systems—Futurebus+—Logical Protocol Specification ***CD-ROM ONLY***. 205 pp.

Data Buses *(Cont.)*
—Interfaces—Data Circuit Terminating Equipment

CEPT T/CD 02-05-86. Specifications Techniques D'une Interface De Bus En Vue D'une Utilisation Sur Equipement De Communication De Donnees. 15 pp.

CEPT T/CD 02-05 E-86. Engineering Requirements for a Bus Interface to Be Used in Data Communication Equipment. 15 pp.

—Interfaces—Microprocessors—VMEbus Compatible

BSI BS IEC 823-90. 1990 Microprocessor System Bus (VMSbus)—Serial Sub-System Bus of the IEC 821 Bus (VMEbus). 163 pp.

IEC 823-90. Microprocessor System Bus (VMSbus)—Serial Sub-System Bus of the IEC 821 Bus (VMEbus) First Edition. 324 pp.

JTC1 823-90. Microprocessor System Bus (VMSbus)—Serial Sub-System Bus of the IEC 821 Bus (VMEbus) First Edition. 324 pp.

OSI ISO/IEC DIS 10857-90. Information Technology—Microprocessor Systems—Futurebus+-Logical Layer Specifications. 239 pp.

—Microcomputers—Control and Status Registers

IEC DIS 13213-92. Information Technology—Standard Control and Status Register (CSR) Architecture for Microcomputer Buses ***CD-ROM ONLY***. 134 pp.

ISO DIS 13213-92. Information Technology—Standard Control and Status Register (CSR) Architecture for Microcomputer Buses ***CD-ROM ONLY***. 134 pp.

JTC1 DIS13213-92. Information Technology—Standard Control and Status Register (CSR) Architecture for Microcomputer Buses ***CD-ROM ONLY***. 134 pp.

—Microprocessors—VICbus Compatible

IEC DIS 11458-92. Information Technology—VICbus: an Inter-Crate Bus for the IEC 821 Bus (VMEbus) ***CD-ROM ONLY***. 103 pp.

ISO DIS 11458-92. Information Technology—VICbus: an Inter-Crate Bus for the IEC 821 Bus (VMEbus) ***CD-ROM ONLY***. 103 pp.

JTC1 DIS11458-92. Information Technology—VICbus: an Inter-Crate Bus for the IEC 821 Bus (VMEbus) ***CD-ROM ONLY***. 103 pp.

—Microprocessors—VMEbus Compatible

BSI BS IEC 821-91. 1991 IEC 821 VMEbus—Microprocessor System Bus for 1 Byte to 4 Byte Data. 564 pp.

IEC 821-91. IEC 821 VMEbus—Microprocessor System Bus for 1 Byte to 4 Byte Data Second Edition. 563 pp.

ISO 821-91. IEC 821 VMEbus—Microprocessor System Bus for 1 Byte to 4 Byte Data Second Edition. 563 pp.

JTC1 821-91. IEC 821 VMEbus—Microprocessor System Bus for 1 Byte to 4 Byte Data Second Edition. 563 pp.

OSI ISO/IEC DIS 821-89. IEC 821 VME Bus—Microprocessor System Bus for 1 to 4 Byte Data. 282 pp.

—Multiplex—Aerospace

CEN PREN 2591 (Part G6)-92. Elements of Electrical and Optical Connection Test Methods Part G6—Transmission Test. 3 pp.

—Multiplex—Aircraft

MOD UK DSTAN 00-18: Pt 1:Sec 2-01. Avionic Data Transmission Interface Systems Part 1: Guide to Interface Systems Section 2: Guide to the Serial, Time Division, Command/Response Multiplex Data Bus Standard Issue 1; Amendment 3. 213 pp.

MOD UK DSTAN 00-18: Pt 1:Sec 2-91. Avionic Data Transmission Interface Systems Part 1: Guide to Interface Systems Section 2: Guide to the Serial, Time Division, Command/Response Multiplex Data Bus Standard Issue 2. 294 pp.

—Multiplex—Weapons Systems

NATO STANAG 3838 ED 1 AMD 3-81. Digital Time Division Command/Response Multiplex Data Bus. 54 pp.

NATO STANAG 3838 ED 1 AMD 4-81. Digital Time Division Command/Response Multiplex Data Bus. 50 pp.

NATO STANAG 3838 ED 1 AMD 5-81. Digital Time Division Command/Response Multiplex Data Bus. 51 pp.

NATO STANAG 3838 ED 1 AMD 6-81. Digital Time Division Command/Response Multiplex Data Bus. 52 pp.

Data Cabinet Safes
Use: Data Safes

INDUSTRY STANDARDS

Data
INTERNATIONAL AND NON-U.S. NATIONAL STANDARDS
SUBJECT INDEX

Data Carriers

Use: Data Storage Devices

Data Cartridges

See Also: Cartridges; Computer Equipment; Computer Tapes; Magnetic Materials; Magnetic Tapes; Tape Cartridges

BSI BS 4783: Part 5-91. 1991 Storage, Transportation and Maintenance of Media for Use in Data Processing and Information Storage Part 5: Recommendations for 12.7 mm Magnetic Tape Cartridges for Data Interchange, Recording at 1491 Data Bytes Per Millimetre on 18 Tracks. 13 pp.

BSI BS 7062-88. 1988 Amd 1 Construction and Use of 12.7 mm (0.5 in) Wide Magnetic Tape Cartridges for Data Interchange, Recording at 1491 Data Bytes per Millimetre (37 871 Data Bytes per Inch), on 18 Tracks (ISO 9661: 1988) (AMD 6362) June 28, 1991. 48 pp.

BSI BS 7062-02. 1988 Amd 2 Construction and Use of 12.7 mm (0.5 in) Wide Magnetic Tape Cartridges for Data Interchange, Recording at 1491 Data Bytes per Millimetre (37 871 Data Bytes per Inch), on 18 Tracks (ISO 9661: 1988) (AMD 6861) January 31, 1992. 53 pp.

BSI BS ISO 6098-84. 1984 Information Processing—Self-Loading Cartridges for 12,7 mm (0.5 in) Wide Magnetic Tape. 17 pp.

IEC 9661 COR 1. Corrigendum 1 Information Processing—Data Interchange on 12,7 mm (0.5 in) Wide Magnetic Tape Cartridges —18 Tracks, 1 491 Data Bytes per Millimetre (37 871 Data Bytes per Inch); (1990). 1 p.

IEC 11559-93. Information Technology—Data Interchange on 12,7 mm Wide 18-Track Magnetic Tape Cartridges —Extended Format First Edition. 71 pp.

ISO 6098-84. Information Processing—Self-Loading Cartridges for 12,7 mm (0.5 in) Wide Magnetic Tape Second Edition. 14 pp.

ISO 11559-93. Information Technology—Data Interchange on 12,7 mm Wide 18-Track Magnetic Tape Cartridges —Extended Format First Edition. 71 pp.

OSI ISO 6098-84. Information Processing-Self Loading Cartridges for 12,7mm (0.5in) Wide Magnetic Tape. 14 pp.

—Format

ECMA ECMA 182-92. Data Interchange on 12,7 mm 48-Track Magnetic Tape Cartridges DLT 1 Format. 65 pp.

IEC 11559-93. Information Technology—Data Interchange on 12,7 mm Wide 18-Track Magnetic Tape Cartridges —Extended Format First Edition. 71 pp.

IEC DIS 13421-93. Information Technology—Data Interchange on 12,7 mm 48-Track Magnetic Tape Cartridges—DLT 1 Format ***CD-ROM ONLY***. 76 pp.

ISO 11559-93. Information Technology—Data Interchange on 12,7 mm Wide 18-Track Magnetic Tape Cartridges —Extended Format First Edition. 71 pp.

ISO DIS 13421-93. Information Technology—Data Interchange on 12,7 mm 48-Track Magnetic Tape Cartridges—DLT 1 Format ***CD-ROM ONLY***. 76 pp.

JTC1 DIS13421-93. Information Technology—Data Interchange on 12,7 mm 48-Track Magnetic Tape Cartridges—DLT 1 Format ***CD-ROM ONLY***. 76 pp.

—Handling

BSI BS 4783: Part 4-88. 1988 Storage, Transportation and Maintenance of Magnetic Media for Use in Data Processing and Information Storage Part 4: Recommendations for Magnetic Tape Cartridges and Cassettes. 10 pp.

BSI BS 4783: Part 4-01. 1988 Amd 1 Storage, Transportation and Maintenance of Media for Use in Data Processing and Information Storage Part 4: Recommendations for Magnetic Tape Cartridges and Cassettes (AMD 6890) February 28, 1992. 11 pp.

—Helical Scan—Format

BSI BS ISO/IEC 10777-91. 1991 Information Technology—3,81 mm Wide Magnetic Tape Cartridge for Information Interchange—Helical Scan Recording—DDS Format. 112 pp.

BSI BS ISO/IEC 11321-92. 1992 Information Technology—3,81 mm Wide Magnetic Tape Cartridge for Information Interchange—Helical Scan Recording —DATA/DAT Format (S). 139 pp.

BSI BS ISO/IEC 11557-92. 1992 Information Technology—3,81 mm Wide Magnetic Tape Cartridge for Information Interchange—Helical Scan Recording —DDS-DC Format Using 60 m and 90 m Length Tapes (S). 114 pp.

ECMA ECMA 150-92. 3.81 mm Wide Magnetic Tape Cartridge for Information Interchange Helical Scan Recording DDS-DC Format Using 60 m and 90 m Length Tapes. 115 pp.

Data Cartridges *(Cont.)*

—Helical Scan—Format *(Cont.)*

ECMA ECMA 150-91. 3.81 mm Wide Magnetic Tape Cartridge for Information Interchange Helical Scan Recording DDS-DC Format. 115 pp.

ECMA ECMA 169-92. 8 mm Wide Magnetic Tape Cartridge Dual Azimuth Format for Information Interchange Helical Scan Recording. 78 pp.

ECMA ECMA 170-92. 3,81 mm Wide Magnetic Tape Cartridge for Information Interchange Helical Scan Recording DDS Format Using 60 m and 90 m Length Tapes. 109 pp.

IEC 11319-93. Information Technology—8 mm Wide Magnetic Tape Cartridge for Information Interchange—Helical Scan Recording First Edition. 71 pp.

IEC 10777-91. Information Technology—3,81 mm Wide Magnetic Tape Cartridge for Information Interchange—Helical Scan Recording —DDS Format First Edition. 110 pp.

IEC 11321-92. Information Technology—3,81 mm Wide Magnetic Tape Cartridge for Information Interchange—Helical Scan Recording —DATA/DAT Format First Edition. 136 pp.

IEC 11557-92. Information Technology—3,81 mm Wide Magnetic Tape Cartridge for Information Interchange—Helical Scan Recording —DDS-DC Format Using 60 m and 90 m Length Tapes First Edition. 112 pp.

IEC DIS 12246-92. Information Technology—8 mm Wide Magnetic Tape Cartridge Dual Azimuth Format for Information Interchange—Helical Scan Recording ***CD-ROM ONLY***. 84 pp.

IEC DIS 12247-92. Information Technology—3,81 mm Wide Magnetic Tape Cartridge for Information Interchange—Helical Scan Recording —DDS Format Using 60 m and 90 m Length Tapes ***CD-ROM ONLY***. 110 pp.

IEC DIS 12248-92. Information Technology—3,81 mm Wide Magnetic Tape Cartridge for Information Interchange—Helical Scan Recording —DATA/DAT-DC Format Using 60 m and 90 m Length Tapes ***CD-ROM ONLY***. 144 pp.

ISO 10777-91. Information Technology—3,81 mm Wide Magnetic Tape Cartridge for Information Interchange—Helical Scan Recording —DDS Format First Edition. 110 pp.

ISO 11319-93. Information Technology—8 mm Wide Magnetic Tape Cartridge for Information Interchange —Helical Scan Recording First Edition. 71 pp.

ISO 11321-92. Information Technology—3,81 mm Wide Magnetic Tape Cartridge for Information Interchange—Helical Scan Recording —DATA/DAT Format First Edition. 136 pp.

ISO 11557-92. Information Technology—3,81 mm Wide Magnetic Tape Cartridge for Information Interchange—Helical Scan Recording —DDS-DC Format Using 60 m and 90 m Length Tapes First Edition. 112 pp.

ISO DIS 12246-92. Information Technology—8 mm Wide Magnetic Tape Cartridge Dual Azimuth Format for Information Interchange—Helical Scan Recording ***CD-ROM ONLY***. 84 pp.

ISO DIS 12247-92. Information Technology—3,81 mm Wide Magnetic Tape Cartridge for Information Interchange—Helical Scan Recording —DDS Format Using 60 m and 90 m Length Tapes ***CD-ROM ONLY***. 110 pp.

ISO DIS 12248-92. Information Technology—3,81 mm Wide Magnetic Tape Cartridge for Information Interchange—Helical Scan Recording —DATA/DAT-DC Format Using 60 m and 90 m Length Tapes ***CD-ROM ONLY***. 144 pp.

JTC1 10777-91. Information Technology—3,81 mm Wide Magnetic Tape Cartridge for Information Interchange—Helical Scan Recording —DDS Format First Edition. 110 pp.

JTC1 11321-92. Information Technology—3,81 mm Wide Magnetic Tape Cartridge for Information Interchange—Helical Scan Recording —DATA/DAT Format First Edition. 136 pp.

JTC1 11557-92. Information Technology—3,81 mm Wide Magnetic Tape Cartridge for Information Interchange—Helical Scan Recording —DDS-DC Format Using 60 m and 90 m Length Tapes First Edition. 112 pp.

JTC1 DIS12246-92. Information Technology—8 mm Wide Magnetic Tape Cartridge Dual Azimuth Format for Information Interchange—Helical Scan Recording ***CD-ROM ONLY***. 84 pp.

JTC1 DIS12247-92. Information Technology—3,81 mm Wide Magnetic Tape Cartridge for Information Interchange—Helical Scan Recording —DDS Format Using 60 m and 90 m Length Tapes ***CD-ROM ONLY***. 110 pp.

JTC1 DIS12248-92. Information Technology—3,81 mm Wide Magnetic Tape Cartridge for Information Interchange—Helical Scan Recording —DATA/DAT-DC Format Using 60 m and 90 m Length Tapes ***CD-ROM ONLY***. 144 pp.

Data Cartridges *(Cont.)*

—Helical Scan—Format *(Cont.)*

SAA AS 4062-92. Information Technology—3.81mm Wide Magnetic Tape Cartridge for Information Interchange—Helical Scan Recording—DDS Format (ISO/IEC 10777:1991) (in Professional Package 26A). 100 pp.

—Helical Scan—Magnetic Measurement

BSI BS ISO/IEC 11321-92. 1992 Information Technology—3,81 mm Wide Magnetic Tape Cartridge for Information Interchange—Helical Scan Recording —DATA/DAT Format (S). 139 pp.

BSI BS ISO/IEC 11557-92. 1992 Information Technology—3,81 mm Wide Magnetic Tape Cartridge for Information Interchange—Helical Scan Recording —DDS-DC Format Using 60 m and 90 m Length Tapes (S). 114 pp.

ECMA ECMA 150-92. 3.81 mm Wide Magnetic Tape Cartridge for Information Interchange Helical Scan Recording DDS-DC Format Using 60 m and 90 m Length Tapes. 115 pp.

ECMA ECMA 150-91. 3.81 mm Wide Magnetic Tape Cartridge for Information Interchange Helical Scan Recording DDS-DC Format. 115 pp.

ECMA ECMA 169-92. 8 mm Wide Magnetic Tape Cartridge Dual Azimuth Format for Information Interchange Helical Scan Recording. 78 pp.

ECMA ECMA 170-92. 3,81 mm Wide Magnetic Tape Cartridge for Information Interchange Helical Scan Recording DDS Format Using 60 m and 90 m Length Tapes. 109 pp.

IEC 11319-93. Information Technology—8 mm Wide Magnetic Tape Cartridge for Information Interchange—Helical Scan Recording First Edition. 71 pp.

IEC 11321-92. Information Technology—3,81 mm Wide Magnetic Tape Cartridge for Information Interchange—Helical Scan Recording —DATA/DAT Format First Edition. 136 pp.

IEC 11557-92. Information Technology—3,81 mm Wide Magnetic Tape Cartridge for Information Interchange—Helical Scan Recording —DDS-DC Format Using 60 m and 90 m Length Tapes First Edition. 112 pp.

IEC DIS 12246-92. Information Technology—8 mm Wide Magnetic Tape Cartridge Dual Azimuth Format for Information Interchange—Helical Scan Recording ***CD-ROM ONLY***. 84 pp.

IEC DIS 12247-92. Information Technology—3,81 mm Wide Magnetic Tape Cartridge for Information Interchange—Helical Scan Recording —DDS Format Using 60 m and 90 m Length Tapes ***CD-ROM ONLY***. 110 pp.

IEC DIS 12248-92. Information Technology—3,81 mm Wide Magnetic Tape Cartridge for Information Interchange—Helical Scan Recording —DATA/DAT-DC Format Using 60 m and 90 m Length Tapes ***CD-ROM ONLY***. 144 pp.

ISO 11319-93. Information Technology—8 mm Wide Magnetic Tape Cartridge for Information Interchange —Helical Scan Recording First Edition. 71 pp.

ISO 11321-92. Information Technology—3,81 mm Wide Magnetic Tape Cartridge for Information Interchange—Helical Scan Recording —DATA/DAT Format First Edition. 136 pp.

ISO 11557-92. Information Technology—3,81 mm Wide Magnetic Tape Cartridge for Information Interchange—Helical Scan Recording —DDS-DC Format Using 60 m and 90 m Length Tapes First Edition. 112 pp.

ISO DIS 12246-92. Information Technology—8 mm Wide Magnetic Tape Cartridge Dual Azimuth Format for Information Interchange—Helical Scan Recording ***CD-ROM ONLY***. 84 pp.

ISO DIS 12247-92. Information Technology—3,81 mm Wide Magnetic Tape Cartridge for Information Interchange—Helical Scan Recording —DDS Format Using 60 m and 90 m Length Tapes ***CD-ROM ONLY***. 110 pp.

ISO DIS 12248-92. Information Technology—3,81 mm Wide Magnetic Tape Cartridge for Information Interchange—Helical Scan Recording —DATA/DAT-DC Format Using 60 m and 90 m Length Tapes ***CD-ROM ONLY***. 144 pp.

JTC1 11321-92. Information Technology—3,81 mm Wide Magnetic Tape Cartridge for Information Interchange—Helical Scan Recording —DATA/DAT Format First Edition. 136 pp.

JTC1 11557-92. Information Technology—3,81 mm Wide Magnetic Tape Cartridge for Information Interchange—Helical Scan Recording —DDS-DC Format Using 60 m and 90 m Length Tapes First Edition. 112 pp.

JTC1 DIS12246-92. Information Technology—8 mm Wide Magnetic Tape Cartridge Dual Azimuth Format for Information Interchange—Helical Scan Recording ***CD-ROM ONLY***. 84 pp.

JTC1 DIS12247-92. Information Technology—3,81 mm Wide Magnetic Tape Cartridge for Information Interchange—Helical Scan Recording —DDS Format Using 60 m and 90 m Length Tapes ***CD-ROM ONLY***. 110 pp.

INTERNATIONAL AND NON-U.S. NATIONAL STANDARDS
SUBJECT INDEX
Data

Data Cartridges *(Cont.)*

—Helical Scan—Magnetic Measurement *(Cont.)*

JTC1 DIS12248-92. Information Technology—3,81 mm Wide Magnetic Tape Cartridge for Information Interchange—Helical Scan Recording —DATA/DAT-DC Format Using 60 m and 90 m Length Tapes ***CD-ROM ONLY***. 144 pp.

—Helical Scan—Physical Properties

BSI BS ISO/IEC 10777-91. 1991 Information Technology—3,81 mm Wide Magnetic Tape Cartridge for Information Interchange—Helical Scan Recording—DDS Format. 112 pp.
BSI BS ISO/IEC 11321-92. 1992 Information Technology—3,81 mm Wide Magnetic Tape Cartridge for Information Interchange—Helical Scan Recording —DATA/DAT Format (S). 139 pp.
BSI BS ISO/IEC 11557-92. 1992 Information Technology—3,81 mm Wide Magnetic Tape Cartridge for Information Interchange—Helical Scan Recording —DDS-DC Format Using 60 m and 90 m Length Tapes (S). 114 pp.
ECMA ECMA 150-92. 3.81 mm Wide Magnetic Tape Cartridge for Information Interchange Helical Scan Recording DDS-DC Format Using 60 m and 90 m Length Tapes. 115 pp.
ECMA ECMA 150-91. 3.81 mm Wide Magnetic Tape Cartridge for Information Interchange Helical Scan Recording DDS-DC Format. 115 pp.
ECMA ECMA 169-92. 8 mm Wide Magnetic Tape Cartridge Dual Azimuth Format for Information Interchange Helical Scan Recording. 78 pp.
ECMA ECMA 170-92. 3,81 mm Wide Magnetic Tape Cartridge for Information Interchange Helical Scan Recording DDS Format Using 60 m and 90 m Length Tapes. 109 pp.
IEC 11319-93. Information Technology—8 mm Wide Magnetic Tape Cartridge for Information Interchange—Helical Scan Recording First Edition. 71 pp.
IEC 10777-91. Information Technology—3,81 mm Wide Magnetic Tape Cartridge for Information Interchange—Helical Scan Recording —DDS Format First Edition. 110 pp.
IEC 11321-92. Information Technology—3,81 mm Wide Magnetic Tape Cartridge for Information Interchange—Helical Scan Recording —DATA/DAT Format First Edition. 136 pp.
IEC 11557-92. Information Technology—3,81 mm Wide Magnetic Tape Cartridge for Information Interchange—Helical Scan Recording —DDS-DC Format Using 60 m and 90 m Length Tapes First Edition. 112 pp.
IEC DIS 12246-92. Information Technology—8 mm Wide Magnetic Tape Cartridge Dual Azimuth Format for Information Interchange—Helical Scan Recording ***CD-ROM ONLY***. 84 pp.
IEC DIS 12247-92. Information Technology—3,81 mm Wide Magnetic Tape Cartridge for Information Interchange—Helical Scan Recording —DDS Format Using 60 m and 90 m Length Tapes ***CD-ROM ONLY***. 110 pp.
IEC DIS 12248-92. Information Technology—3,81 mm Wide Magnetic Tape Cartridge for Information Interchange—Helical Scan Recording —DATA/DAT-DC Format Using 60 m and 90 m Length Tapes ***CD-ROM ONLY***. 144 pp.
ISO 10777-91. Information Technology—3,81 mm Wide Magnetic Tape Cartridge for Information Interchange—Helical Scan Recording —DDS Format First Edition. 110 pp.
ISO 11319-93. Information Technology—8 mm Wide Magnetic Tape Cartridge for Information Interchange —Helical Scan Recording First Edition. 71 pp.
ISO 11321-92. Information Technology—3,81 mm Wide Magnetic Tape Cartridge for Information Interchange—Helical Scan Recording —DATA/DAT Format First Edition. 136 pp.
ISO 11557-92. Information Technology—3,81 mm Wide Magnetic Tape Cartridge for Information Interchange—Helical Scan Recording —DDS-DC Format Using 60 m and 90 m Length Tapes First Edition. 112 pp.
ISO DIS 12246-92. Information Technology—8 mm Wide Magnetic Tape Cartridge Dual Azimuth Format for Information Interchange—Helical Scan Recording ***CD-ROM ONLY***. 84 pp.
ISO DIS 12247-92. Information Technology—3,81 mm Wide Magnetic Tape Cartridge for Information Interchange—Helical Scan Recording —DDS Format Using 60 m and 90 m Length Tapes ***CD-ROM ONLY***. 110 pp.
ISO DIS 12248-92. Information Technology—3,81 mm Wide Magnetic Tape Cartridge for Information Interchange—Helical Scan Recording —DATA/DAT-DC Format Using 60 m and 90 m Length Tapes ***CD-ROM ONLY***. 144 pp.
JTC1 10777-91. Information Technology—3,81 mm Wide Magnetic Tape Cartridge for Information Interchange—Helical Scan Recording —DDS Format First Edition. 110 pp.

Data Cartridges *(Cont.)*

—Helical Scan—Physical Properties *(Cont.)*

JTC1 11321-92. Information Technology—3,81 mm Wide Magnetic Tape Cartridge for Information Interchange—Helical Scan Recording —DATA/DAT Format First Edition. 136 pp.
JTC1 11557-92. Information Technology—3,81 mm Wide Magnetic Tape Cartridge for Information Interchange—Helical Scan Recording —DDS-DC Format Using 60 m and 90 m Length Tapes First Edition. 112 pp.
JTC1 DIS12246-92. Information Technology—8 mm Wide Magnetic Tape Cartridge Dual Azimuth Format for Information Interchange—Helical Scan Recording ***CD-ROM ONLY***. 84 pp.
JTC1 DIS12247-92. Information Technology—3,81 mm Wide Magnetic Tape Cartridge for Information Interchange—Helical Scan Recording —DDS Format Using 60 m and 90 m Length Tapes ***CD-ROM ONLY***. 110 pp.
JTC1 DIS12248-92. Information Technology—3,81 mm Wide Magnetic Tape Cartridge for Information Interchange—Helical Scan Recording —DATA/DAT-DC Format Using 60 m and 90 m Length Tapes ***CD-ROM ONLY***. 144 pp.
SAA AS 4062-92. Information Technology—3.81mm Wide Magnetic Tape Cartridge for Information Interchange—Helical Scan Recording—DDS Format (ISO/IEC 10777:1991) (in Professional Package 26A). 100 pp.

—Helical Scan—Recording Methods

BSI BS ISO/IEC 10777-91. 1991 Information Technology—3,81 mm Wide Magnetic Tape Cartridge for Information Interchange—Helical Scan Recording—DDS Format. 112 pp.
BSI BS ISO/IEC 11321-92. 1992 Information Technology—3,81 mm Wide Magnetic Tape Cartridge for Information Interchange—Helical Scan Recording —DATA/DAT Format (S). 139 pp.
BSI BS ISO/IEC 11557-92. 1992 Information Technology—3,81 mm Wide Magnetic Tape Cartridge for Information Interchange—Helical Scan Recording —DDS-DC Format Using 60 m and 90 m Length Tapes (S). 114 pp.
ECMA ECMA 150-92. 3.81 mm Wide Magnetic Tape Cartridge for Information Interchange Helical Scan Recording DDS-DC Format Using 60 m and 90 m Length Tapes. 115 pp.
ECMA ECMA 150-91. 3.81 mm Wide Magnetic Tape Cartridge for Information Interchange Helical Scan Recording DDS-DC Format. 115 pp.
ECMA ECMA 169-92. 8 mm Wide Magnetic Tape Cartridge Dual Azimuth Format for Information Interchange Helical Scan Recording. 78 pp.
ECMA ECMA 170-92. 3,81 mm Wide Magnetic Tape Cartridge for Information Interchange Helical Scan Recording DDS Format Using 60 m and 90 m Length Tapes. 109 pp.
IEC 11319-93. Information Technology—8 mm Wide Magnetic Tape Cartridge for Information Interchange—Helical Scan Recording First Edition. 71 pp.
IEC 10777-91. Information Technology—3,81 mm Wide Magnetic Tape Cartridge for Information Interchange—Helical Scan Recording —DDS Format First Edition. 110 pp.
IEC 11321-92. Information Technology—3,81 mm Wide Magnetic Tape Cartridge for Information Interchange—Helical Scan Recording —DATA/DAT Format First Edition. 136 pp.
IEC 11557-92. Information Technology—3,81 mm Wide Magnetic Tape Cartridge for Information Interchange—Helical Scan Recording —DDS-DC Format Using 60 m and 90 m Length Tapes First Edition. 112 pp.
IEC DIS 12246-92. Information Technology—8 mm Wide Magnetic Tape Cartridge Dual Azimuth Format for Information Interchange—Helical Scan Recording ***CD-ROM ONLY***. 84 pp.
IEC DIS 12247-92. Information Technology—3,81 mm Wide Magnetic Tape Cartridge for Information Interchange—Helical Scan Recording —DDS Format Using 60 m and 90 m Length Tapes ***CD-ROM ONLY***. 110 pp.
IEC DIS 12248-92. Information Technology—3,81 mm Wide Magnetic Tape Cartridge for Information Interchange—Helical Scan Recording —DATA/DAT-DC Format Using 60 m and 90 m Length Tapes ***CD-ROM ONLY***. 144 pp.
ISO 10777-91. Information Technology—3,81 mm Wide Magnetic Tape Cartridge for Information Interchange—Helical Scan Recording —DDS Format First Edition. 110 pp.
ISO 11319-93. Information Technology—8 mm Wide Magnetic Tape Cartridge for Information Interchange —Helical Scan Recording First Edition. 71 pp.
ISO 11321-92. Information Technology—3,81 mm Wide Magnetic Tape Cartridge for Information Interchange—Helical Scan Recording —DATA/DAT Format First Edition. 136 pp.

Data Cartridges *(Cont.)*

—Helical Scan—Recording Methods *(Cont.)*

ISO 11557-92. Information Technology—3,81 mm Wide Magnetic Tape Cartridge for Information Interchange—Helical Scan Recording —DDS-DC Format Using 60 m and 90 m Length Tapes First Edition. 112 pp.
ISO DIS 12246-92. Information Technology—8 mm Wide Magnetic Tape Cartridge Dual Azimuth Format for Information Interchange—Helical Scan Recording ***CD-ROM ONLY***. 84 pp.
ISO DIS 12247-92. Information Technology—3,81 mm Wide Magnetic Tape Cartridge for Information Interchange—Helical Scan Recording —DDS Format Using 60 m and 90 m Length Tapes ***CD-ROM ONLY***. 110 pp.
ISO DIS 12248-92. Information Technology—3,81 mm Wide Magnetic Tape Cartridge for Information Interchange—Helical Scan Recording —DATA/DAT-DC Format Using 60 m and 90 m Length Tapes ***CD-ROM ONLY***. 144 pp.
JTC1 10777-91. Information Technology—3,81 mm Wide Magnetic Tape Cartridge for Information Interchange—Helical Scan Recording —DDS Format First Edition. 110 pp.
JTC1 11321-92. Information Technology—3,81 mm Wide Magnetic Tape Cartridge for Information Interchange—Helical Scan Recording —DATA/DAT Format First Edition. 136 pp.
JTC1 11557-92. Information Technology—3,81 mm Wide Magnetic Tape Cartridge for Information Interchange—Helical Scan Recording —DDS-DC Format Using 60 m and 90 m Length Tapes First Edition. 112 pp.
JTC1 DIS12246-92. Information Technology—8 mm Wide Magnetic Tape Cartridge Dual Azimuth Format for Information Interchange—Helical Scan Recording ***CD-ROM ONLY***. 84 pp.
JTC1 DIS12247-92. Information Technology—3,81 mm Wide Magnetic Tape Cartridge for Information Interchange—Helical Scan Recording —DDS Format Using 60 m and 90 m Length Tapes ***CD-ROM ONLY***. 110 pp.
JTC1 DIS12248-92. Information Technology—3,81 mm Wide Magnetic Tape Cartridge for Information Interchange—Helical Scan Recording —DATA/DAT-DC Format Using 60 m and 90 m Length Tapes ***CD-ROM ONLY***. 144 pp.
SAA AS 4062-92. Information Technology—3.81mm Wide Magnetic Tape Cartridge for Information Interchange—Helical Scan Recording—DDS Format (ISO/IEC 10777:1991) (in Professional Package 26A). 100 pp.

—Helical Scan—Signal Quality

BSI BS ISO/IEC 10777-91. 1991 Information Technology—3,81 mm Wide Magnetic Tape Cartridge for Information Interchange—Helical Scan Recording—DDS Format. 112 pp.
BSI BS ISO/IEC 11321-92. 1992 Information Technology—3,81 mm Wide Magnetic Tape Cartridge for Information Interchange—Helical Scan Recording —DATA/DAT Format (S). 139 pp.
BSI BS ISO/IEC 11557-92. 1992 Information Technology—3,81 mm Wide Magnetic Tape Cartridge for Information Interchange—Helical Scan Recording —DDS-DC Format Using 60 m and 90 m Length Tapes (S). 114 pp.
ECMA ECMA 150-92. 3.81 mm Wide Magnetic Tape Cartridge for Information Interchange Helical Scan Recording DDS-DC Format Using 60 m and 90 m Length Tapes. 115 pp.
ECMA ECMA 150-91. 3.81 mm Wide Magnetic Tape Cartridge for Information Interchange Helical Scan Recording DDS-DC Format. 115 pp.
ECMA ECMA 169-92. 8 mm Wide Magnetic Tape Cartridge Dual Azimuth Format for Information Interchange Helical Scan Recording. 78 pp.
ECMA ECMA 170-92. 3,81 mm Wide Magnetic Tape Cartridge for Information Interchange Helical Scan Recording DDS Format Using 60 m and 90 m Length Tapes. 109 pp.
IEC 10777-91. Information Technology—3,81 mm Wide Magnetic Tape Cartridge for Information Interchange—Helical Scan Recording —DDS Format First Edition. 110 pp.
IEC 11321-92. Information Technology—3,81 mm Wide Magnetic Tape Cartridge for Information Interchange—Helical Scan Recording —DATA/DAT Format First Edition. 136 pp.
IEC 11557-92. Information Technology—3,81 mm Wide Magnetic Tape Cartridge for Information Interchange—Helical Scan Recording —DDS-DC Format Using 60 m and 90 m Length Tapes First Edition. 112 pp.
IEC DIS 12246-92. Information Technology—8 mm Wide Magnetic Tape Cartridge Dual Azimuth Format for Information Interchange—Helical Scan Recording ***CD-ROM ONLY***. 84 pp.
IEC DIS 12247-92. Information Technology—3,81 mm Wide Magnetic Tape Cartridge for Information Interchange—Helical Scan Recording —DDS Format Using 60 m and 90 m Length Tapes ***CD-ROM ONLY***. 110 pp.

INDUSTRY STANDARDS
721

INTERNATIONAL AND NON-U.S. NATIONAL STANDARDS
SUBJECT INDEX

Data Cartridges (Cont.)
—Helical Scan—Signal Quality (Cont.)

IEC DIS 12248-92. Information Technology—3,81 mm Wide Magnetic Tape Cartridge for Information Interchange—Helical Scan Recording —DATA/DAT-DC Format Using 60 m and 90 m Length Tapes ***CD-ROM ONLY***. 144 pp.

ISO 10777-91. Information Technology—3,81 mm Wide Magnetic Tape Cartridge for Information Interchange—Helical Scan Recording —DDS Format First Edition. 110 pp.

ISO 11321-92. Information Technology—3,81 mm Wide Magnetic Tape Cartridge for Information Interchange—Helical Scan Recording —DATA/DAT Format First Edition. 136 pp.

ISO 11557-92. Information Technology—3,81 mm Wide Magnetic Tape Cartridge for Information Interchange—Helical Scan Recording —DDS-DC Format Using 60 m and 90 m Length Tapes First Edition. 112 pp.

ISO DIS 12246-92. Information Technology—8 mm Wide Magnetic Tape Cartridge Dual Azimuth Format for Information Interchange—Helical Scan Recording ***CD-ROM ONLY***. 84 pp.

ISO DIS 12247-92. Information Technology—3,81 mm Wide Magnetic Tape Cartridge for Information Interchange—Helical Scan Recording —DDS Format Using 60 m and 90 m Length Tapes ***CD-ROM ONLY***. 110 pp.

ISO DIS 12248-92. Information Technology—3,81 mm Wide Magnetic Tape Cartridge for Information Interchange—Helical Scan Recording —DATA/DAT-DC Format Using 60 m and 90 m Length Tapes ***CD-ROM ONLY***. 144 pp.

JTC1 10777-91. Information Technology—3,81 mm Wide Magnetic Tape Cartridge for Information Interchange—Helical Scan Recording —DDS Format First Edition. 110 pp.

JTC1 11321-92. Information Technology—3,81 mm Wide Magnetic Tape Cartridge for Information Interchange—Helical Scan Recording —DATA/DAT Format First Edition. 136 pp.

JTC1 11557-92. Information Technology—3,81 mm Wide Magnetic Tape Cartridge for Information Interchange—Helical Scan Recording —DDS-DC Format Using 60 m and 90 m Length Tapes First Edition. 112 pp.

JTC1 DIS12246-92. Information Technology—8 mm Wide Magnetic Tape Cartridge Dual Azimuth Format for Information Interchange—Helical Scan Recording ***CD-ROM ONLY***. 84 pp.

JTC1 DIS12247-92. Information Technology—3,81 mm Wide Magnetic Tape Cartridge for Information Interchange—Helical Scan Recording —DDS Format Using 60 m and 90 m Length Tapes ***CD-ROM ONLY***. 110 pp.

JTC1 DIS12248-92. Information Technology—3,81 mm Wide Magnetic Tape Cartridge for Information Interchange—Helical Scan Recording —DATA/DAT-DC Format Using 60 m and 90 m Length Tapes ***CD-ROM ONLY***. 144 pp.

SAA AS 4062-92. Information Technology—3.81mm Wide Magnetic Tape Cartridge for Information Interchange—Helical Scan Recording—DDS Format (ISO/IEC 10777:1991) (in Professional Package 26A). 100 pp.

—Identification Systems

ECMA ECMA 182-92. Data Interchange on 12,7 mm 48-Track Magnetic Tape Cartridges DLT 1 Format. 65 pp.

IEC DIS 13421-93. Information Technology—Data Interchange on 12,7 mm 48-Track Magnetic Tape Cartridges—DLT 1 Format ***CD-ROM ONLY***. 76 pp.

ISO DIS 13421-93. Information Technology—Data Interchange on 12,7 mm 48-Track Magnetic Tape Cartridges—DLT 1 Format ***CD-ROM ONLY***. 76 pp.

JTC1 DIS13421-93. Information Technology—Data Interchange on 12,7 mm 48-Track Magnetic Tape Cartridges—DLT 1 Format ***CD-ROM ONLY***. 76 pp.

—Magnetic Measurement

ECMA ECMA 182-92. Data Interchange on 12,7 mm 48-Track Magnetic Tape Cartridges DLT 1 Format. 65 pp.

IEC DIS 13421-93. Information Technology—Data Interchange on 12,7 mm 48-Track Magnetic Tape Cartridges—DLT 1 Format ***CD-ROM ONLY***. 76 pp.

ISO DIS 13421-93. Information Technology—Data Interchange on 12,7 mm 48-Track Magnetic Tape Cartridges—DLT 1 Format ***CD-ROM ONLY***. 76 pp.

JTC1 DIS13421-93. Information Technology—Data Interchange on 12,7 mm 48-Track Magnetic Tape Cartridges—DLT 1 Format ***CD-ROM ONLY***. 76 pp.

Data Cartridges (Cont.)
—Physical Properties

ECMA ECMA 182-92. Data Interchange on 12,7 mm 48-Track Magnetic Tape Cartridges DLT 1 Format. 65 pp.

IEC 11559-93. Information Technology—Data Interchange on 12,7 mm Wide 18-Track Magnetic Tape Cartridges —Extended Format First Edition. 71 pp.

IEC DIS 13421-93. Information Technology—Data Interchange on 12,7 mm 48-Track Magnetic Tape Cartridges—DLT 1 Format ***CD-ROM ONLY***. 76 pp.

ISO 11559-93. Information Technology—Data Interchange on 12,7 mm Wide 18-Track Magnetic Tape Cartridges —Extended Format First Edition. 71 pp.

ISO DIS 13421-93. Information Technology—Data Interchange on 12,7 mm 48-Track Magnetic Tape Cartridges—DLT 1 Format ***CD-ROM ONLY***. 76 pp.

JTC1 DIS13421-93. Information Technology—Data Interchange on 12,7 mm 48-Track Magnetic Tape Cartridges—DLT 1 Format ***CD-ROM ONLY***. 76 pp.

—Recording Methods

ECMA ECMA 182-92. Data Interchange on 12,7 mm 48-Track Magnetic Tape Cartridges DLT 1 Format. 65 pp.

IEC DIS 13421-93. Information Technology—Data Interchange on 12,7 mm 48-Track Magnetic Tape Cartridges—DLT 1 Format ***CD-ROM ONLY***. 76 pp.

ISO DIS 13421-93. Information Technology—Data Interchange on 12,7 mm 48-Track Magnetic Tape Cartridges—DLT 1 Format ***CD-ROM ONLY***. 76 pp.

JTC1 DIS13421-93. Information Technology—Data Interchange on 12,7 mm 48-Track Magnetic Tape Cartridges—DLT 1 Format ***CD-ROM ONLY***. 76 pp.

—Signal Quality

ECMA ECMA 182-92. Data Interchange on 12,7 mm 48-Track Magnetic Tape Cartridges DLT 1 Format. 65 pp.

IEC DIS 13421-93. Information Technology—Data Interchange on 12,7 mm 48-Track Magnetic Tape Cartridges—DLT 1 Format ***CD-ROM ONLY***. 76 pp.

ISO DIS 13421-93. Information Technology—Data Interchange on 12,7 mm 48-Track Magnetic Tape Cartridges—DLT 1 Format ***CD-ROM ONLY***. 76 pp.

JTC1 DIS13421-93. Information Technology—Data Interchange on 12,7 mm 48-Track Magnetic Tape Cartridges—DLT 1 Format ***CD-ROM ONLY***. 76 pp.

—Storage

BSI BS 4783: Part 4-88. 1988 Storage, Transportation and Maintenance of Magnetic Media for Use in Data Processing and Information Storage Part 4: Recommendations for Magnetic Tape Cartridges and Cassettes. 10 pp.

BSI BS 4783: Part 4-01. 1988 Amd 1 Storage, Transportation and Maintenance of Media for Use in Data Processing and Information Storage Part 4: Recommendations for Magnetic Tape Cartridges and Cassettes (AMD 6890) February 28, 1992. 11 pp.

Data Channel Equipment
Use: Data Communication Equipment

Data Channel Failure Detectors
—Signaling Links—CCITT No. 6 Signaling Systems

CCITT RECMN Q.275-89. 6.5 Data Channel Failure Detection—Specifications of Signalling System No. 6 (Study Group XI) 1 pp. 1 p.

Data Channels
See Also: Communication Channels; Data Transmission

—Digital

SAA AS 4064-92. Information Processing—Digital Channel Aggregation (N x 64) (in Professional Package 26A). 17 pp.

—Formats—Digital Audio Interfaces

CCIR RECMN 776-92. Format for User Data Channel of the Digital Audio Interface—Section 10C—Audio-Frequency Characteristics of Sound-Broadcasting Signals. 18 pp.

Data Circuit Multiplication Equipment
See Also: Data Circuits

Data Circuit Multiplication Equipment (Cont.)
—International Switching Centers—Signaling Systems

CCITT RECMN Q.50-89. Signalling Between Circuit Multiplication Equipments (CME) and International Switching Centres (ISC)—General Recommendations on Telephone Switching and Signalling—Functions and Information Flows for Services in the ISDN—Supplements. 17 pp.

Data Circuit Terminating Equipment
Use For: DCE *See Also:* Communication Interfaces; Data Circuits; Data Communication Equipment; Data Encryption; Data Terminal Equipment; Interfaces; Open Systems Interconnection; Tandem Data Circuits

CCITT RECMN X.21 BIS-89. Use on Public Data Networks of Data Terminal Equipment (DTE) Which is Designed for Interfacing to Synchronous V-Series Modems—Data Communication Networks: Services and Facilities, Interfaces (Study Group VII) 13 pp. 13 pp.

—Analog Networks

CEPT T/CD 01-01-81. Specifications Techniques Generales Pour Un Equipment De Terminaison Du Circuit De Donnees Susceptible D'Etre Utilise Sur Les Reseaux Analogiques et Numeriques. 24 pp.

CEPT T/CD 01-01 E-87. General Engineering Requirements for Data Circuit Terminating Equipment for Analogue and Digital Networks. 35 pp.

—Back-Up Facilities

BSI BS 6639-90. 1990 Guide to DTE/DCE Interface Control Operation Using the 25 Pole Connector. 12 pp.

ISO 8480-87. Information Processing—Data Communication—DTE/DCE Interface Back-up Control Operation Using the 25-Pole Connector First Edition. 9 pp.

JTC1 8480-87. Information Processing—Data Communication—DTE/DCE Interface Back-up Control Operation Using the 25-Pole Connector First Edition. 9 pp.

OSI ISO 8480-87. Information Processing—Data Communication—DTE/DCE Interface Back-Up Control Operation Using the 25-Pole Connector.

—Connectors—Glossaries

CNS C5183-88. Interface Between Data Terminal Equipment and Data Circuit-Terminating Equipment Employing Serial Binary Data Interchange (Jul)(10691).

—Connectors—Interfaces

BSI BS 6514-84. (WITHDRAWN) 1984 Implementation of V24 or RS232 as an Asynchronous Local Interface. 4 pp.

BSI BS 6623: Part 1-85. (WITHDRAWN) 1985 DTE/DCE Interface Connectors and Pin Assignments Part 1: Specification for 25-Pin Connector (Superseded by BS ISO 2110: 1989). 17 pp.

BSI BS 6623: Part 2-86. 1986 DTE/DCE Interface Connectors and Pin Assignments Part 2: Specification for 37-Pin Connector. 25 pp.

BSI BS 6623: Part 3-86. 1986 DTE/DCE Interface Connectors and Pin Assignments Part 3: 15-Pin Connector. 20 pp.

BSI BS 6623: Part 4-86. (WITHDRAWN) 1986 DTE/DCE Interface Connectors and Pin Assignments Part 4: 34-Pin Connector (Superseded by BS ISO/IEC 2593: 1993). 8 pp.

BSI BS ISO 2110-89. 1989 Information Technology—Data Communication—25-Pole DTE/DCE Interface Connector and Contact Number Assignments. 18 pp.

BSI BS ISO 2110-01. 1989 Amd 1 Information Technology—Data Communication—25-Pole DTE/DCE Interface Connector and Contact Number Assignments (AMD 7165) July 15, 1992. 26 pp.

BSI BS ISO 4902-89. 1989 Information Technology—Data Communication—37 Pole DTE/DCE Interface Connector and Contact Number Assignments. 24 pp.

BSI BS ISO 4903-89. 1989 Information Technology—Data Communication—15-Pole DTE/DCE Interface Connector and Contact Number Assignments. 22 pp.

BSI BS ISO/IEC 2593-93. 1993 Information Technology—Telecommunications and Information Exchange Between Systems—34-Pole DTE/DCE Interface Connector Mateability Dimensions and Contact Number Assignments (S). 23 pp.

CEPT T/CD 01-14-86. Specification Fonctionnelle D'Equipement De Terminaison De Circuit De Donnees. 36 pp.

CEPT T/CD 01-14 E-86. Specification of Equipment Practice for Data Transmission Equipment. 36 pp.

CNS C5177-88. Interface Between Data Terminal Equipment and Data Circuit-Terminating Equipment Employing Serial Binary Data Interchange Rules (Jul)(10685).

INTERNATIONAL AND NON-U.S. NATIONAL STANDARDS
SUBJECT INDEX Data

Data Circuit Terminating Equipment *(Cont.)*

—Connectors—Interfaces *(Cont.)*

CNS C5180-88. Interface Between Data Terminal Equipment and Data Circuit-Terminating Equipment Employing Serial Binary Data Interchange Functional Description of Interchange Circuits (Jul)(10688).

CNS C5181-88. Interface Between Data Terminal Equipment and Data Circuit-Terminating Equipment Employing Serial Binary Data Interchange Interface for Selected Communication System Configurations (Jul)(10689).

CNS C5182-88. Interface Between Data Terminal Equipment and Data Circuit-Terminating Equipment Employing Serial Binary Data Interchange Recommendations and Explanatory Notes (Jul)(10690).

CSA Z243.12.2-1981. Data Communication—15-Pin DTE/DCE Interface Connector and Pin Assignments (ISO 4903-1980 has Been Adopted as a CSA Standard) (R 1988). 25 pp.

IEC 2593-93. Information Technology—Telecommunications and Information Exchange Between Systems—34-Pole DTE/DCE Interface Connector Mateability Dimensions and Contact Number Assignments Third Edition. 21 pp.

IEC 11569-93. Information Technology—Telecommunications and Information Exchange Between Systems—26-Pole Interface Connector Mateability Dimensions and Contact Number Assignments First Edition. 11 pp.

ISO 2110-89. Information Technology—Data Communication—25-Pole DTE/DCE Inter-face Connector and Con-tact Number Assignments Amendment 1: Interface Connector and Contact Assignments for a DTE/DCE Interface for Data Signalling Rates Above 20 000 Bit/s Third Ed.; (Amendment 1-1991). 22 pp.

ISO 2593-93. Information Technology—Telecommunications and Information Exchange Between Systems—34-Pole DTE/DCE Interface Connector Mateability Dimensions and Contact Number Assignments Third Edition. 21 pp.

ISO 4902-89. Information Technology—Data Communication—37-Pole DTE/DCE Interface Connector and Contact Number Assignments Second Edition. 20 pp.

ISO 4903-89. Information Technology—Data Communication—15-Pole DTE/DCE Interface Connector and Contact Number Assignments Second Edition. 20 pp.

ISO TR7477-85. Data Communication—Arrangements for DTE to DTE Physical Connection Using V.24 and X.24 Interchange Circuits First Edition. 8 pp.

ISO 8481-86. Data Communication—DTE to DTE Physical Connection Using X.24 Interchange Circuits with DTE Provided Timing First Edition. 5 pp.

ISO 11569-93. Information Technology—Telecommunications and Information Exchange Between Systems—26-Pole Interface Connector Mateability Dimensions and Contact Number Assignments First Edition. 11 pp.

JIS X 5101-82. Interface Between Date Circuit Terminating Equipment (DCE) and Data Terminal Equipment (DTE) (25-Pin Interface).

JIS X 5102-82. Interface Between Date Circuit Terminating Equipment (DCE) and Data Terminal Equipment (DTE) (15-Pin Interface).

JIS X 5103-82. Interface Between Date Circuit Terminating Equipment (DCE) and Data Terminal Equipment (DTE) (37/9-Pin Interface).

JTC1 2110-89. Information Technology—Data Communication—25-Pole DTE/DCE Inter-face Connector and Con-tact Number Assignments Amendment 1: Interface Connector and Contact Assignments for a DTE/DCE Interface for Data Signalling Rates Above 20 000 Bit/s Third Ed.; (Amendment 1-1991). 22 pp.

JTC1 2593-93. Information Technology—Telecommunications and Information Exchange Between Systems—34-Pole DTE/DCE Interface Connector Mateability Dimensions and Contact Number Assignments Third Edition. 21 pp.

JTC1 4902-89. Information Technology—Data Communication—37-Pole DTE/DCE Interface Connector and Contact Number Assignments Second Edition. 20 pp.

JTC1 4903-89. Information Technology—Data Communication—15-Pole DTE/DCE Interface Connector and Contact Number Assignments Second Edition. 20 pp.

JTC1 TR7477-85. Data Communication—Arrangements for DTE to DTE Physical Connection Using V.24 and X.24 Interchange Circuits First Edition. 8 pp.

JTC1 8481-86. Data Communication—DTE to DTE Physical Connection Using X.24 Interchange Circuits with DTE Provided Timing First Edition. 5 pp.

OSI ISO 2110-89. Information Technology—Date Communication—25-Pole DTE/DCE Interface Connector and Contact Number Assignment—Amd 1: Interface Connector and Contact Assignments for a DTE/DCE Interface for Data Signalling Rates Above 20 000 bits/s. 23 pp.

Data Circuit Terminating Equipment *(Cont.)*

—Connectors—Interfaces *(Cont.)*

OSI ISO 2110 DAM 1-89. Interface Connector and Contact Number Assignments for a DTE/DCE Interface for Data Signalling Rates Above 20 000 bits per Second. 5 pp.

OSI ISO/IEC DIS 2593-89. Information Technology—Data Communication—34-Pole DTE/DCE Interface Connector and Contact Number Assignments. 14 pp.

OSI ISO 2593-84. Data Communication—34 Pin DTE/DCE Interface Connector and Pin Assignments. 7 pp.

OSI ISO 4902-89. Information Technology—Date Communication—37 Pole DTE/DCE Interface Connector and Contact Number Assignments. 20 pp.

OSI ISO 4903-89. Information Technology—Data Communication—15-Pole DTE/DCE Interface Connector and Contact Number Assignments. 20 pp.

OSI ISO TR 7477-85. Data Communication—Arrangements for DTE to DTE Physical Connection Using V.24 and X.24 Interchange Circuits.

SAA AS 2748-91. Information Technology—Data Communication—25-Pole DTE/DCE Interface Connector and Contact Number Assignments (ISO/IEC 2110:1989) (in Professional Packages 26A, 26C). 13 pp.

SAA AS 3613-91. Information Technology—Data Communications—15-Pole DTE/DCE Interface Connector and Contact Number Assignments (ISO 4903:1989) (in Professional Packages 26A, 26C). 17 pp.

SNZ NZS/ISO 4902-80. Data Communication—37-Pin and 9-Pin DTE/DCE Interface Connectors and Pin Assignments. 24 pp.

SNZ NZS/ISO 4903-89. Information Technology—Data Communications—15-Pole DTE/DCE Interface Connector and Contact Number Assignments. 17 pp.

SNZ NZS/ISO TR 7477-85. Data Communication—Arrangements for DTE to DTE Physical Connection V. 8 pp.

SNZ NZS/ISO 8481-86. Data Communication—DTE to DTE Physical Connection Using X.24 Interchange Circuits with DTE Provided Timing. 3 pp.

—Connectors—Mechanical Properties

CNS C5179-88. Interface Between Data Terminal Equipment and Data Circuit-Terminating Equipment Employing Serial Binary Data Interchange Mechanical Characteristics (Jul)(10687).

—Connectors—Safety

CEPT T/CD 04-03-85. Exigences De Securite a L'Interface ETTD-ETCD. 3 pp.

CEPT T/CD 04-03 E-86. Safety Requirements for the DTE-DCE Interface. 3 pp.

—Connectors—Signal Quality

CNS C5178-88. Interface Between Data Terminal Equipment and Data Circuit-Terminating Equipment Employing Binary Data Interchange Characteristics (Jul)(10686).

CNS C5224-88. Signal Quality at Interface Between Data Terminal Equipment and Synchronous Data Circuit -Terminating Equipment for Serial Data Transmission (Aug)(12396).

CNS C5225-88. Signal Quality Using Serial Data Transmission at the Interface with Non-Synchronous Data Circuit-Terminating Equipment (Aug)(12397).

—Data Buses

CEPT T/CD 02-05-86. Specifications Techniques D'une Interface De Bus En Vue D'une Utilisation Sur Equipement De Communication De Donnees. 15 pp.

CEPT T/CD 02-05 E-86. Engineering Requirements for a Bus Interface to Be Used in Data Communication Equipment. 15 pp.

—Digital Networks

CEPT T/CD 01-01-81. Specifications Techniques Generales Pour Un Equipment De Terminaison Du Circuit De Donnees Susceptible D'Etre Utilise Sur Les Reseaux Analogiques et Numeriques. 24 pp.

CEPT T/CD 01-01 E-87. General Engineering Requirements for Data Circuit Terminating Equipment for Analogue and Digital Networks. 35 pp.

—Error Control

CCITT RECMN V.41-89. Code-Independent Error-Control System—Data Communication over the Telephone Network (Study Group XVII) 10 pp. 10 pp.

—Error Correction—Asynchronous to Synchronous Converters

CCITT RECMN V.42-89. Error-Correcting Procedures for DCEs Using Asynchronous-to-Synchronous Conversion—Data Communication over the Telephone Network (Study Group XVII) 75 pp. 75 pp.

Data Circuit Terminating Equipment *(Cont.)*

—Error Correction—Data Compression

CCITT RECMN V.42 BIS-90. Data Compression Procedures for Data Circuit Terminating Equipment (DCE) Using Error Correction Procedures (Study Group XVII) 30 pp. 30 pp.

—Interchange Circuits—Glossaries

CCITT RECMN X.24-89. List of Definitions for Interchange Circuits Between Data Terminal Equipment (DTE) and Data Circuit-Terminating Equipment (DCE) on Public Data Networks—Data Communication Networks: Services and Facilities, Interfaces (Study Group VII) 6 pp. 6 pp.

—Interfaces

CCITT RECMN V.230-89. General Data Communications Interface Layer 1 Specification—Data Communication over the Telephone Network (Study Group XVII) 43 pp. 43 pp.

—Interfaces—Data Terminal Equipment

CCITT RECMN X.20-89. Interface Between Data Terminal Equipment (DTE) and Data Circuit-Terminating Equipment (DCE) for Start-Stop Transmission Services on Public Data Networks—Data Communication Networks: Services and Facilities, Interfaces (Study Group VII) 27 pp. 27 pp.

CCITT RECMN X.20 BIS-89. Use on Public Data Networks of Data Terminal Equipment (DTE) Which is Designed for Interfacing to Asynchronous Duplex V-Series Modems—Data Communication Networks: Services and Facilities, Interfaces (Study Group VII) 7 pp. 7 pp.

CENELEC ETR 017-92. Terminal Equipment (TE); Requirements for the Attachment of Data Terminal Equipment Comprising CCITT V. Series Type Interfaces to Data Circuit-Terminating Equipment for Low and Medium Data Signalling Rates. 37 pp.

ETSI ETR 017-92. Terminal Equipment (TE); Requirements for the Attachment of Data Terminal Equipment Comprising CCITT V. Series Type Interfaces to Data Circuit-Terminating Equipment for Low and Medium Data Signalling Rates. 37 pp.

—Interfaces—Data Terminal Equipment—Modems

CCITT RECMN X.20-89. Interface Between Data Terminal Equipment (DTE) and Data Circuit-Terminating Equipment (DCE) for Start-Stop Transmission Services on Public Data Networks—Data Communication Networks: Services and Facilities, Interfaces (Study Group VII) 27 pp. 27 pp.

CCITT RECMN X.20 BIS-89. Use on Public Data Networks of Data Terminal Equipment (DTE) Which is Designed for Interfacing to Asynchronous Duplex V-Series Modems—Data Communication Networks: Services and Facilities, Interfaces (Study Group VII) 7 pp. 7 pp.

CCITT RECMN X.21-89. Interface Between Data Terminal Equipment (DTE) and Data Circuit-Terminating Equipment (DCE) for Synchronous Operation on Public Data Networks—Data Communication Networks: Services and Facilities, Interfaces (Study Group VII) 56 pp. 56 pp.

CCITT RECMN X.21 BIS-89. Use on Public Data Networks of Data Terminal Equipment (DTE) Which is Designed for Interfacing to Synchronous V-Series Modems—Data Communication Networks: Services and Facilities, Interfaces (Study Group VII) 13 pp. 13 pp.

—Interfaces—Data Terminal Equipment—Packet Switched Networks

BSI BS ISO/IEC 10588-93. 1993 Information Technology—Use of X.25 Packet Layer Protocol in Conjunction with X.21/X.21 bis to Provide the OSI Connection-Mode Network Service (S). 18 pp.

CCITT RECMN X.25-89. Interface Between Data Terminal Equipment (DTE) and Data Circuit-Terminating Equip. (DCE) for Terminals Operating in the Packet Mode and Connected to Public Data Networks by Dedicated CI—DataCommunication Networks: Services and Facilities, Interfaces (Study Group VII) 160 pp. 160 pp.

CCITT RECMN X.613-92. Information Technology—Use of X.25 Packet Layer Protocol in Conjunction with X.21/X.21 bis to Provide the OSI Connection-Mode Network Service 15 pp (ISO/CCIT Common Text). 15 pp.

IEC 10588-93. Information Technology—Use of X.25 Packet Layer Protocol in Conjunction with X.21/X.21 bis to Provide the OSI Connection-Mode Network Service First Edition; (CCITT RECMN X.613). 15 pp.

ISO 10588-93. Information Technology—Use of X.25 Packet Layer Protocol in Conjunction with X.21/X.21 bis to Provide the OSI Connection-Mode Network Service First Edition; (CCITT RECMN X.613). 15 pp.

INDUSTRY STANDARDS 723

Data Circuit Terminating Equipment (Cont.)

—Interfaces—Data Terminal Equipment—Packet Switched Networks (Cont.)

JTC1 10588-93. Information Technology—Use of X.25 Packet Layer Protocol in Conjunction with X.21/X.21 bis to Provide the OSI Connection-Mode Network Service First Edition; (CCITT RECMN X.613). 15 pp.

—Interfaces—Data Terminal Equipment—Public Data Networks

BSI BS ISO/IEC 10588-93. 1993 Information Technology—Use of X.25 Packet Layer Protocol in Conjunction with X.21/X.21 bis to Provide the OSI Connection-Mode Network Service (S). 18 pp.

CCITT RECMN X.21-92. Interface Between Data Terminal Equipment and Data Circuit-Terminating Equipment for Synchronous Operation on Public Data Networks (Study Group VII) 58 pp. 58 pp.

CCITT RECMN X.21-89. Interface Between Data Terminal Equipment (DTE) and Data Circuit-Terminating Equipment (DCE) for Synchronous Operation on Public Data Networks—Data Communication Networks: Services and Facilities, Interfaces (Study Group VII) 56 pp. 56 pp.

CCITT RECMN X.21 BIS-89. Use on Public Data Networks of Data Terminal Equipment (DTE) Which is Designed for Interfacing to Synchronous V-Series Modems—Data Communication Networks: Services and Facilities, Interfaces (Study Group VII) 13 pp. 13 pp.

CCITT RECMN X.22-89. Multiplex DTE/DCE Interface for User Classes 3-6—Data Communication Networks: Services and Facilities, Interfaces (Study Group VII) 6 pp. 6 pp.

CCITT RECMN X.613-92. Information Technology—Use of X.25 Packet Layer Protocol in Conjunction with X.21/X.21 bis to Provide the OSI Connection-Mode Network Service 15 pp (ISO/CCIT Common Text). 15 pp.

IEC 10588-93. Information Technology—Use of X.25 Packet Layer Protocol in Conjunction with X.21/X.21 bis to Provide the OSI Connection-Mode Network Service First Edition; (CCITT RECMN X.613). 15 pp.

ISO 10588-93. Information Technology—Use of X.25 Packet Layer Protocol in Conjunction with X.21/X.21 bis to Provide the OSI Connection-Mode Network Service First Edition; (CCITT RECMN X.613). 15 pp.

JTC1 10588-93. Information Technology—Use of X.25 Packet Layer Protocol in Conjunction with X.21/X.21 bis to Provide the OSI Connection-Mode Network Service First Edition; (CCITT RECMN X.613). 15 pp.

—Interfaces—Data Terminal Equipment—Start-Stop Transmission

CCITT RECMN X.20-89. Interface Between Data Terminal Equipment (DTE) and Data Circuit-Terminating Equipment (DCE) for Start-Stop Transmission Services on Public Data Networks—Data Communication Networks: Services and Facilities, Interfaces (Study Group VII) 27 pp. 27 pp.

CCITT RECMN X.20 BIS-89. Use on Public Data Networks of Data Terminal Equipment (DTE) Which is Designed for Interfacing to Asynchronous Duplex V-Series Modems—Data Communication Networks: Services and Facilities, Interfaces (Study Group VII) 7 pp. 7 pp.

—Interfaces—FAX Machines—Public Data Networks

CCITT RECMN X.38-92. G3 Facsimile Equipment/DCE Interface for G3 Facsimile Equipment Accessing the Facsimile Packet Assembly/Disassembly Facility (FPAD) in a Public Data Network Situated in the Same Country (Study Group VII) 49 pp. 49 pp.

—Interfaces—Packet Switched Networks

CCITT RECMN X.32-89. Interface Between Data Terminal Equipment (DTE) and Data Circuit-Terminating Equipment (DCE) for Terminals Operating in the Packet Mode and Accessing a Packet Switched Public Data Network Through a Public Switched Telephone Network or an Integrated Services. 57 pp.

CENELEC ETR 059-93. Terminal Equipment (TE); Technical Requirements for Packet Switched Public Data Network (PSPDN) Indirect Access (Based on CCITT Recommendation X.32). 33 pp.

ETSI ETR 059-93. Terminal Equipment (TE); Technical Requirements for Packet Switched Public Data Network (PSPDN) Indirect Access (Based on CCITT Recommendation X.32). 33 pp.

Data Circuit Terminating Equipment (Cont.)

—Interfaces—Public Data Networks

CCITT RECMN X.32-89. Interface Between Data Terminal Equipment (DTE) and Data Circuit-Terminating Equipment (DCE) for Terminals Operating in the Packet Mode and Accessing a Packet Switched Public Data Network Through a Public Switched Telephone Network or an Integrated Services. 57 pp.

CENELEC ETR 059-93. Terminal Equipment (TE); Technical Requirements for Packet Switched Public Data Network (PSPDN) Indirect Access (Based on CCITT Recommendation X.32). 33 pp.

ETSI ETR 059-93. Terminal Equipment (TE); Technical Requirements for Packet Switched Public Data Network (PSPDN) Indirect Access (Based on CCITT Recommendation X.32). 33 pp.

—Interfaces—Serial Data Transmission

CNS C5180-88. Interface Between Data Terminal Equipment and Data Circuit-Terminating Equipment Employing Serial Binary Data Interchange Functional Description of Interchange Circuits (Jul)(10688).

CNS C5181-88. Interface Between Data Terminal Equipment and Data Circuit-Terminating Equipment Employing Serial Binary Data Interchange Interface for Selected Communication System Configurations (Jul)(10689).

CNS C5182-88. Interface Between Data Terminal Equipment and Data Circuit-Terminating Equipment Employing Serial Binary Data Interchange Recommendations and Explanatory Notes (Jul)(10690).

—Public Data Networks—Maintenance

CCITT RECMN X.150-89. Principles of Maintenance Testing for Public Data Networks Using Data Terminal Equipment (DTE) and Data Circuit-Terminating Equipment (DCE) Test Loops—Data Communication Networks—Transmission, Signalling and Switching, Network Aspects, Maintenance and Administrative. 7 pp.

—Signal Quality

BSI BS 6638: Part 1-87. (WITHDRAWN) 1987 Guide to Transmission Signal Quality at DTE/DCE Interfaces Part 1: Start-Stop Signal Quality (Superseded by BS ISO/IEC 7480: 1991). 13 pp.

BSI BS 6638: Part 2-89. 1989 Guide to Transmission Signal Quality at DTE/DCE Interfaces Part 2: Synchronous Transmission. 15 pp.

BSI BS ISO/IEC 7480-91. 1991 Information Technology—Telecommunications and Information Exchange Between Systems—Start-Stop Transmission Signal Quality at DTE/DCE Interfaces. 17 pp.

IEC 7480-91. Information Technology—Telecommunications and Information Exchange Between Systems—Start-Stop Transmission Signal Quality at DTE/DCE Interfaces Second Edition. 14 pp.

ISO 7480-91. Information Technology—Telecommunications and Information Exchange Between Systems—Start-Stop Transmission Signal Quality at DTE/DCE Interfaces Second Edition. 14 pp.

ISO 9543-89. Information Processing Systems—Information Exchange Between Systems—Synchronous Transmission Signal Quality at DTE/DCE Interfaces First Edition. 14 pp.

JTC1 7480-91. Information Technology—Telecommunications and Information Exchange Between Systems—Start-Stop Transmission Signal Quality at DTE/DCE Interfaces Second Edition. 14 pp.

JTC1 9543-89. Information Processing Systems—Information Exchange Between Systems—Synchronous Transmission Signal Quality at DTE/DCE Interfaces First Edition. 14 pp.

OSI ISO/IEC DIS 7480-91. Information Technology—Telecommunications and Information Exchange Between Systems—Start/ Stop Transmission Quality at DTE/DCE Interfaces. 16 pp.

OSI ISO 7480-84. Information Processing-Start-Stop Transmission Signal Quality at DTE/DCE Interfaces. 12 pp.

OSI ISO DIS 9543-87. Information Processing Systems-Information Exchange Between Systems—Synchronous Transmission Signal Quality at DTE/DCE Interfaces. 20 pp.

SNZ NZS/ISO 7480-84. Information Processing—Start-Stop Transmission Signal Quality at DTE/DCE Interfaces. 10 pp.

—Telephone Networks

CEPT T/CD 01-12 E-83. Concerning the Specification of Engineering Requirements for Three Types of Plug-In DCE's Operating with a User Data Signalling Rate of 2,400 Bits/s. 90 pp.

Data Circuit Terminating Equipment (Cont.)

—Telex Communications—Data Terminal Equipment

CCITT RECMN S.16-89. Connection to the Telex Network of an Automatic Terminal Using a V.24 (1) DCE/DTE Interface—Telegraph Services Terminal Equipment (Study Group IX) 10 pp. 10 pp.

—Telex Communications—Data Terminal Equipment—Answering/Calling

CCITT RECMN S.19-89. Calling and Answering in the Telex Network with Automatic Terminal Equipment—Telegraph Services Terminal Equipment (Study Group IX) 3 pp. 3 pp.

Data Circuits

See Also: Data Circuit Multiplication Equipment; Data Circuit Terminating Equipment; Tandem Data Circuits; Telecommunication Circuits

—Bit Error

CCITT RECMN O.153-89. Basic Parameters for the Measurement of Error Performance at Bit Rates Below the Primary Rate—Specifications for Measuring Equipment (Study Group IV) 5 pp. 5 pp.

—Circuit Noise

CCITT RECMN G.143-89. Circuit Noise and the Use of Compandors—General Characteristics of International Telephone Connections and Circuits (Study Groups XII and XV) 5 pp. 5 pp.

—Group Delay

CCITT RECMN O.82-89. Group-Delay Measuring Equipment for the Range 5 to 600 kHz—Specifications for Measuring Equipment (Study Group IV) 6 pp. 6 pp.

—High Frequency

CCIR Report 1132-90. Use of Coding Diversity on HF Data Circuits—Section 3Cb—Data Transmission. 4 pp.

—Impulse Noise—Pulse Code Modulation Systems

CCITT FASCICLE III.1. General Characteristics of International Telephone Connections and Circuits Recommendations G.101-G.181. 332 pp.

Data Codes

See Also: Codes; Data Transmission; Encoded Data Transmission; Time and Address Codes; Time Codes

—Accounting—Maritime Mobile Services

CCITT RECMN D.91-91. Transmission in Encoded Form of Maritime Telecommunications Accounting Information (Study Group III) 13 pp. 13 pp.

CCITT RECMN D.91-89. Transmission in Encoded Form of Maritime Telecommunications Accounting Information—General Tariff Principles—Charging and Accounting in International Telecommunications Services (Study Group III) 8 pp. 8 pp.

—Aeronautical Information Services

ICAO 8400. Procedures for Air Navigation Services ICAO Abbreviations and Codes Fourth Edition—1989; (Incorporating Amendments 1 to 20) (Amendment 21 7/1/93). 133 pp.

—Binary Number Systems

CCITT RECMN V.1-89. Equivalence Between Binary Notation Symbols and the Significant Conditions of a Two-Condition Code—Data Communication Over the Telephone Network (Study Group XVII) 2 pp. 2 pp.

—CCITT Group 3—FAX Machines—Procedures

CCITT RECMN T.35-89. Procedure for the Allocation of CCITT Members' Codes—Terminal Equipment and Protocols for Telematic Services (Study Group VIII) 5 pp. 5 pp.

—CCITT Group 4—FAX Communications

CCITT RECMN F.184-89. Operational Provisions for the International Public Facsimile Service Between Subscriber Stations with Group 4 Facsimile Machines (Telefax 4)—Telematic, Data Transmission and Teleconference Services—Operations and Quality of Service (Study Group I) 9 pp. 9 pp.

—CCITT Group 4—FAX Machines

CCITT RECMN T.6-89. Facsimile Coding Schemes and Coding Control Functions for Group 4 Facsimile Apparatus—Terminal Equipment and Protocols for Telematic Services (Study Group VIII) 10 pp. 10 pp.

INTERNATIONAL AND NON-U.S. NATIONAL STANDARDS
SUBJECT INDEX

Data

Data Codes (Cont.)
—CCITT Group 4—FAX Machines—Procedures
CCITT RECMN T.35-89. Procedure for the Allocation of CCITT Members' Codes—Terminal Equipment and Protocols for Telematic Services (Study Group VIII) 5 pp. 5 pp.

—Container Equipment Data Exchange (CEDEX)
ISO 9897 Pt 1-90. Freight Containers—Container Equipment Data Exchange (CEDEX)—Part 1: General Communication Codes First Edition. 63 pp.

—Digital Recording—Audio and Video
IEC DIS 11172-92. Information Technology—Coding of Moving Pictures and Associated Audio for Digital Storage Media up to About 1,5 Mbit/s ***CD-ROM ONLY***. 344 pp.
IEC DIS 11544-92. Information Technology—Coded Representation of Picture and Audio Information—Progressive Bi-Level Image Compression ***CD-ROM ONLY***. 85 pp.
ISO DIS 11172-92. Information Technology—Coding of Moving Pictures and Associated Audio for Digital Storage Media up to About 1,5 Mbit/s ***CD-ROM ONLY***. 344 pp.
ISO DIS 11544-92. Information Technology—Coded Representation of Picture and Audio Information—Progressive Bi-Level Image Compression ***CD-ROM ONLY***. 85 pp.
JTC1 DIS11172-92. Information Technology—Coding of Moving Pictures and Associated Audio for Digital Storage Media up to About 1,5 Mbit/s ***CD-ROM ONLY***. 344 pp.

—Identification—Banking
BSI BS 7104-90. 1990 Procedure for Allocating a Bank Identifier Code (BIC). 11 pp.
ISO 9362-87. Banking—Banking Telecommunication Messages—Bank Indentifier Codes First Edition. 7 pp.
ISO 10383-92. Codes for Exchanges and Regulated Markets—Market Identifier Codes (MIC) First Edition. 6 pp.
JTC1 9362-87. Banking—Banking Telecommunication Messages—Bank Indentifier Codes First Edition. 7 pp.
OSI ISO 9362-87. Banking—Banking Telecommunication Messages—Banker Indentifier Codes. 5 pp.

—Identification—Data Elements
IEC DIS 7826 Pt 1-93. Information Technology—General Structure for the Interchange of Data Item Code Values—Part 1: Identification of Coding Schemes for Data Elements (Types) ***CD-ROM ONLY***. 11 pp.
ISO DIS 7826 Pt 1-93. Information Technology—General Structure for the Interchange of Data Item Code Values—Part 1: Identification of Coding Schemes for Data Elements (Types) ***CD-ROM ONLY***. 11 pp.
JTC1 DIS 7826 Pt 1-93. Information Technology—General Structure for the Interchange of Data Item Code Values—Part 1: Identification of Coding Schemes for Data Elements (Types) ***CD-ROM ONLY***. 11 pp.

—Identification—Human Sexual Representation
BSI BS 5249: Part 2-78. 1978 Representation of Elements of Data in Interchanges Using Data Processing Systems Part 2: Representation of Human Sexes. 4 pp.
CNS C5127-82. Information Interchange — Representation of Human Sexes (Jan)(8381).
ISO 5218-77. Information Interchange—Representation of Human Sexes First Edition. 3 pp.
JIS X 0303-71. Male-Female Code.
JTC1 5218-77. Information Interchange—Representation of Human Sexes First Edition. 3 pp.
OSI ISO 5218-77. Information Interchange—Representation of Human Sexes. 3 pp.

—Identification—Languages
OSI ISO 639-88. Code for the Representation of Names of Languages. 20 pp.

—Identification—Organizations
BSI BS ISO 6523-84. 1984 Data Interchange—Structure for the Identification of Organizations (S). 14 pp.
IEC 6523 Draft AMD 1. Data Interchange—Structure for the Identification of Organizations Amendment 1: Guidance on the Use of ISO 6523; (1993) ***CD-ROM ONLY***. 3 pp.
ISO 6523-84. Data Interchange—Structure for the Identification of Organizations First Edition. 11 pp.
ISO 6523 Draft AMD 1. Data Interchange—Structure for the Identification of Organizations Amendment 1: Guidance on the Use of ISO 6523; (1993) ***CD-ROM ONLY***. 3 pp.

Data Codes (Cont.)
—Identification—Organizations (Cont.)
JTC1 6523-84. Data Interchange—Structure for the Identification of Organizations First Edition. 11 pp.
JTC1 6523 Draft AMD 1. Data Interchange—Structure for the Identification of Organizations Amendment 1: Guidance on the Use of ISO 6523; (1993) ***CD-ROM ONLY***. 3 pp.
OSI ISO 6523-84. Data Interchange-Structure for the Identification of Organizations. 11 pp.

—Mobile Satellite Communications
CCIR Report 509-5-90. Modulation and Coding Technique for Mobile Satellite Service—Section 8I—Technical and Operating Characteristics of Mobile Satellite Services. 23 pp.

—Mobile Stations (Communications)
CCIR Report 509-5-90. Modulation and Coding Technique for Mobile Satellite Service—Section 8I—Technical and Operating Characteristics of Mobile Satellite Services. 23 pp.

—Registration—Data Elements
IEC DIS 7826 Pt 2-93. Information Technology—General Structure for the Interchange of Data Item Code Values—Part 2: Registration of Coding Schemes for Data Elements (Types) ***CD-ROM ONLY***. 13 pp.
ISO DIS 7826 Pt 2-93. Information Technology—General Structure for the Interchange of Data Item Code Values—Part 2: Registration of Coding Schemes for Data Elements (Types) ***CD-ROM ONLY***. 13 pp.
JTC1 DIS 7826 Pt 2-93. Information Technology—General Structure for the Interchange of Data Item Code Values—Part 2: Registration of Coding Schemes for Data Elements (Types) ***CD-ROM ONLY***. 13 pp.

—Structure
OSI ISO DP 7826-87. Data Interchange—General Structure for the Interchange of Coded Representations. 13 pp.

Data Collection
Use For: Field Data Collection *See Also:* Data Acquisition Systems; Data Analysis; Data Logging; Data Processing; Data Transmission

—Earth Exploration Satellites
CCIR Report 538-4-90. Earth Exploration Satellites Satellites for Location of Platforms and for Data Collection—Section 2F—Earth Exploration Satellites. 11 pp.
CCIR QUESTION 142/7-90. Radiocommunications for Earth Exploration Satellites Data Collection and Position Location Systems—Questions Concerning Study Group 7—Science Services. 1 p.

—Network Performance—Telecommunication Networks
CCITT RECMN E.880-89. Field Data Collection and Evaluation on the Performance of Equipment, Networks and Services—Telephone Network and ISDN—Quality of Service, Network Management and Traffic Engineering (Study Group II) 7 pp. 7 pp.

—Telecommunication Equipment
CCITT RECMN E.880-89. Field Data Collection and Evaluation on the Performance of Equipment, Networks and Services—Telephone Network and ISDN—Quality of Service, Network Management and Traffic Engineering (Study Group II) 7 pp. 7 pp.

—Telecommunication Services
CCITT RECMN E.880-89. Field Data Collection and Evaluation on the Performance of Equipment, Networks and Services—Telephone Network and ISDN—Quality of Service, Network Management and Traffic Engineering (Study Group II) 7 pp. 7 pp.

Data Communication
Use For: Data Exchange *See Also:* Bit Error; Bit Rates; Circuit Switched Data Transmission Services; Connectionless Broadband Data Service; Data Broadcasting; Data Transmission; End of Message Codes; International Reference Alphabet; NODEF; Packet Switched Data Transmission Services; Public Data Communication; Public Data Transmission; SELCAL Services; Service Messages; Start of Message Codes; Store and Forward Mode; Store and Forward Units

BSI BS EN 61107-92. 1992 Data Exchange for Meter Reading, Tariff and Load Control—Direct Local Data Exchange (IEC 1107: 1992). 37 pp.
CENELEC EN 61107-92. Data Exchange for Meter Reading, Tariff and Load Control—Direct Local Data Exchange. 36 pp.

Data Communication (Cont.)
—Basic Mode Control Procedures
CNS C5240-90. Information Processing—Basic Mode Control Procedures for Data Communication Systems (Oct)(12796).

—Basic Mode Control Procedures—Code Independent Information
CNS C5241-90. Data Communication—Basic Mode Control Procedures—Code Independent Information (Nov)(12797).

—Global Maritime Distress and Safety Systems
CCIR Report 1043-86. Characteristics of a Data Exchange System for Use with Maritime Navigation and Radiocommunication Equipment—Section 8D—Radiodetermination, Global Maritime Distress and Safety System and Related Subjects. 7 pp.

—Glossaries
BSI BS 4727:Pt3: Group 03-92. 1992 Glossary of Electrotechnical, Power, Telecommunication, Electronics, Lighting and Colour Terms Part 3: Terms Particular to Telecommunications and Electronics Group 03: Telegraphy, Facsimile and Data Communication (G). 77 pp.
IEC DIS 2382 Pt 9-93. Data Processing—Vocabulary—Part 09: Data Communication ***CD-ROM ONLY***. 51 pp.
ISO DIS 2382 Pt 9-93. Data Processing—Vocabulary—Part 09: Data Communication ***CD-ROM ONLY***. 51 pp.
JIS X 0009-87. Glossary of Terms Used in Information Processing (Data Communication).
JTC1 DIS2382 Pt 9-93. Data Processing—Vocabulary—Part 09: Data Communication ***CD-ROM ONLY***. 51 pp.

—Interchange Circuits—Electrical Measurement
CCITT RECMN X.26-89. Electrical Characteristics for Unbalanced Double-Current Interchange Circuits for General Use with Integrated Circuit Equipment in the Field of Data Communications —Data Communication Networks: Services and Facilities, Interfaces (Study Group VII) 1 pp. 1 p.
CCITT RECMN X.27-89. Electrical Characteristics for Balanced Double-Current Interchange Circuits for General Use with Integrated Cir. Equipment in the Field of Data Communications—Data Comm. Networks:Services and Facilities, Inter. (Study Group VII) 1 pp. 1 p.

—Maritime Mobile Services
CCIR QUESTION 55-2/8-90. Development and Future Implementation of Data Exchange Systems and Ship Movement Telemetry and Telecommand Systems—Questions Concerning Study Group 8—Mobile, Radiodetermination, Amateur and Related Satellite Services. 1 p.
CCIR QUESTION 76-1/8-90. Data Communication in the Maritime Mobile Service—Questions Concerning Study Group 8—Mobile, Radiodetermination, Amateur and Related Satellite Services. 1 p.

—Maritime Mobile Services—HF
CCIR Report 1158-90. Data Communication in the Maritime Mobile Services Using MF, HF and VHF Frequencies—Section 8B —Maritime Mobile Service: Telegraphy and Related Subjects. 8 pp.

—Maritime Mobile Services—MF
CCIR Report 1158-90. Data Communication in the Maritime Mobile Services Using MF, HF and VHF Frequencies—Section 8B —Maritime Mobile Service: Telegraphy and Related Subjects. 8 pp.

—Maritime Mobile Services—VHF
CCIR Report 1158-90. Data Communication in the Maritime Mobile Services Using MF, HF and VHF Frequencies—Section 8B —Maritime Mobile Service: Telegraphy and Related Subjects. 8 pp.

—Radio Navigation Equipment
CCIR Report 1043-86. Characteristics of a Data Exchange System for Use with Maritime Navigation and Radiocommunication Equipment—Section 8D—Radiodetermination, Global Maritime Distress and Safety System and Related Subjects. 7 pp.

—Ships
CCIR QUESTION 55-2/8-90. Development and Future Implementation of Data Exchange Systems and Ship Movement Telemetry and Telecommand Systems—Questions Concerning Study Group 8—Mobile, Radiodetermination, Amateur and Related Satellite Services. 1 p.

—Telephone Networks—Glossaries
CCITT RECMN V.7-89. Definitions of Terms Concerning Data Communication over the Telephone Network—Data Communication over the Telephone Network (Study Group XVII) 3 pp. 3 pp.

INDUSTRY STANDARDS

INTERNATIONAL AND NON-U.S. NATIONAL STANDARDS
SUBJECT INDEX

Data

Data Communication *(Cont.)*
—Units of Measurement—Quantity—Abbreviations
   CCITT RECMN B.14-89. Terms and Abbreviations for Information Quantities in Telecommunications—Terms and Definitions Abbreviations and Acronyms Recommendations on Means of Expression (Series B) General Telecommunications Statistics (Series C) 2 pp. 2 pp.

—Units of Measurement—Quantity—Glossaries
   CCITT RECMN B.14-89. Terms and Abbreviations for Information Quantities in Telecommunications—Terms and Definitions Abbreviations and Acronyms Recommendations on Means of Expression (Series B) General Telecommunications Statistics (Series C) 2 pp. 2 pp.

—Units of Measurement—Quantity—Symbols
   CCITT RECMN B.14-89. Terms and Abbreviations for Information Quantities in Telecommunications—Terms and Definitions Abbreviations and Acronyms Recommendations on Means of Expression (Series B) General Telecommunications Statistics (Series C) 2 pp. 2 pp.

Data Communication Equipment
Use For: Data Channel Equipment
See Also: Broadband Amplifiers; Communication Interfaces; Communication Networks; Data Circuit Terminating Equipment; Data Links; Data Transmission; Data Transmission Equipment; Encoders; Modems; Multiplexers; Store and Forward Units; Voice/Data Systems

—Connectors
   ISO TR7477-85. Data Communication—Arrangements for DTE to DTE Physical Connection Using V.24 and X.24 Interchange Circuits First Edition. 8 pp.
   ISO 8481-86. Data Communication—DTE to DTE Physical Connection Using X.24 Interchange Circuits with DTE Provided Timing First Edition. 5 pp.
   JTC1 TR7477-85. Data Communication—Arrangements for DTE to DTE Physical Connection Using V.24 and X.24 Interchange Circuits First Edition. 8 pp.
   JTC1 8481-86. Data Communication—DTE to DTE Physical Connection Using X.24 Interchange Circuits with DTE Provided Timing First Edition. 5 pp.
   OSI ISO TR 7477-85. Data Communication—Arrangements for DTE to DTE Physical Connection Using V.24 and X.24 Interchange Circuits.
   SNZ NZS/ISO TR 7477-85. Data Communication—Arrangements for DTE to DTE Physical Connection V. 8 pp.
   SNZ NZS/ISO 8481-86. Data Communication—DTE to DTE Physical Connection Using X.24 Interchange Circuits with DTE Provided Timing. 3 pp.

—Data Safes—Fire Resistant
   CEN PREN 1047-1-93. Secure Storage Units—Classification and Methods of Test for Resistance to Fire—Part 1: Data Cabinets. 16 pp.

—Glossaries
   IEC 50 Chap 721-91. International Electrotechnical Vocabulary Chapter 721: Telegraphy, Facsimile and Data Communication First Edition. 320 pp.

—Meteorological Satellites
   CCIR QUESTION 141/7-90. Command and Data Communication Systems for Meteorological Satellites—Questions Concerning Study Group 7—Science Services. 1 p.

Data Communication Networks
Use: Communication Networks

Data Communication Services
Use: Data Communication

Data Communication Systems
Use: Data Transmission

Data Compact Disks
Use: CD ROMs

Data Compression
See Also: Information Interchange

—Algorithms
   BSI BS ISO/IEC 11558-92. 1992 Information Technology—Data Compression for Information Interchange—Adaptive Coding with Embedded Dictionary—DCLZ Algorithm. 18 pp.
   ECMA ECMA 151-91. Data Compression for Information Interchange Adaptive Coding with Embedded Dictionary DCLZ Algorithm. 14 pp.

Data Compression *(Cont.)*
—Algorithms *(Cont.)*
   ECMA ECMA 159-91. Data Compression for Information Interchange Binary Arithmetic Coding Algorithm. 21 pp.
   IEC 11558-92. Information Technology—Data Compression for Information Interchange—Adaptive Coding with Embedded Dictionary—DCLZ Algorithm First Edition. 15 pp.
   IEC DIS 11576-92. Information Technology—Procedure for the Registration of Algorithms for the Lossless Compression of Data ***CD-ROM ONLY***. 7 pp.
   IEC DIS 12042-92. Information Technology—Data Compression for Information Interchange—Binary Arithmetic Coding Algorithm ***CD-ROM ONLY***. 25 pp.
   ISO 11558-92. Information Technology—Data Compression for Information Interchange—Adaptive Coding with Embedded Dictionary—DCLZ Algorithm First Edition. 15 pp.
   ISO DIS 11576-92. Information Technology—Procedure for the Registration of Algorithms for the Lossless Compression of Data ***CD-ROM ONLY***. 7 pp.
   ISO DIS 12042-92. Information Technology—Data Compression for Information Interchange—Binary Arithmetic Coding Algorithm ***CD-ROM ONLY***. 25 pp.
   JTC1 DIS11576-92. Information Technology—Procedure for the Registration of Algorithms for the Lossless Compression of Data ***CD-ROM ONLY***. 7 pp.

—Data Circuit Terminating Equipment—Error Correction
   CCITT RECMN V.42 BIS-90. Data Compression Procedures for Data Circuit Terminating Equipment (DCE) Using Error Correction Procedures (Study Group XVII) 30 pp. 30 pp.

—Digital Recording—Audio and Video
   IEC DIS 11544-92. Information Technology—Coded Representation of Picture and Audio Information—Progressive Bi-Level Image Compression ***CD-ROM ONLY***. 85 pp.
   ISO DIS 11544-92. Information Technology—Coded Representation of Picture and Audio Information—Progressive Bi-Level Image Compression ***CD-ROM ONLY***. 85 pp.

—Image Processing
   IEC DIS 10918 Pt 1-92. Information Technology—Digital Compression and Coding of Continuous-Tone Still Images—Part 1: Requirements and Guidelines ***CD-ROM ONLY***. 203 pp.
   IEC DIS 10918 Pt 2-93. Information Technology—Digital Compression and Coding of Continuous-Tone Still Images—Part 2: Compliance Testing ***CD-ROM ONLY***. 77 pp.
   ISO DIS 10918 Pt 1-92. Information Technology—Digital Compression and Coding of Continuous-Tone Still Images—Part 1: Requirements and Guidelines ***CD-ROM ONLY***. 203 pp.
   ISO DIS 10918 Pt 2-93. Information Technology—Digital Compression and Coding of Continuous-Tone Still Images—Part 2: Compliance Testing ***CD-ROM ONLY***. 77 pp.
   JTC1 DIS10918 Pt 1-92. Information Technology—Digital Compression and Coding of Continuous-Tone Still Images—Part 1: Requirements and Guidelines ***CD-ROM ONLY***. 203 pp.
   JTC1 DIS10918 Pt 2-93. Information Technology—Digital Compression and Coding of Continuous-Tone Still Images—Part 2: Compliance Testing ***CD-ROM ONLY***. 77 pp.

Data Descriptive Files (Information Interchange)
Use: Information Interchange—Data Descriptive Files

Data Elements
See Also: Data Processing
   IEC DIS 11179 Pt 3-93. Information Technology—Coordination of Data Element Standardization—Part 3: Basic Attributes of Data Elements (Types) ***CD-ROM ONLY***. 40 pp.
   ISO DIS 11179 Pt 3-93. Information Technology—Coordination of Data Element Standardization—Part 3: Basic Attributes of Data Elements (Types) ***CD-ROM ONLY***. 40 pp.
   JTC1 DIS 11179 Pt 3-93. Information Technology—Coordination of Data Element Standardization—Part 3: Basic Attributes of Data Elements (Types) ***CD-ROM ONLY***. 40 pp.

—Data Codes—Identification
   IEC DIS 7826 Pt 1-93. Information Technology—General Structure for the Interchange of Data Item Code Values—Part 1: Identification of Coding Schemes for Data Elements (Types) ***CD-ROM ONLY***. 11 pp.

Data Elements *(Cont.)*
—Data Codes—Identification *(Cont.)*
   ISO DIS 7826 Pt 1-93. Information Technology—General Structure for the Interchange of Data Item Code Values—Part 1: Identification of Coding Schemes for Data Elements (Types) ***CD-ROM ONLY***. 11 pp.
   JTC1 DIS 7826 Pt 1-93. Information Technology—General Structure for the Interchange of Data Item Code Values—Part 1: Identification of Coding Schemes for Data Elements (Types) ***CD-ROM ONLY***. 11 pp.

—Data Codes—Registration
   IEC DIS 7826 Pt 2-93. Information Technology—General Structure for the Interchange of Data Item Code Values—Part 2: Registration of Coding Schemes for Data Elements (Types) ***CD-ROM ONLY***. 13 pp.
   ISO DIS 7826 Pt 2-93. Information Technology—General Structure for the Interchange of Data Item Code Values—Part 2: Registration of Coding Schemes for Data Elements (Types) ***CD-ROM ONLY***. 13 pp.
   JTC1 DIS 7826 Pt 2-93. Information Technology—General Structure for the Interchange of Data Item Code Values—Part 2: Registration of Coding Schemes for Data Elements (Types) ***CD-ROM ONLY***. 13 pp.

Data Encoding
Use: Encoding

Data Encryption
See Also: Data Circuit Terminating Equipment; Data Terminal Equipment; Encoding; Message Authentication Codes; Open Systems Interconnection

—Algorithms
   CNS C5164-83. Data Encryption Algorithm (Sep)(10545).
   IEC 9979-91. Data Cryptographic Techniques—Procedures for the Registration of Cryptographic Algorithms First Edition. 8 pp.
   ISO 9979-91. Data Cryptographic Techniques—Procedures for the Registration of Cryptographic Algorithms First Edition. 8 pp.
   JTC1 9979-91. Data Cryptographic Techniques—Procedures for the Registration of Cryptographic Algorithms First Edition. 8 pp.

—Algorithms—Banking
   BSI BS 7102: Part 1-89. 1989 Recommendations for Algorithms for Use in Banking Message Authentication Part 1: Data Encryption Algorithm (DEA). 6 pp.
   BSI BS ISO 10126-2-91. 1991 Banking—Procedures for Message Encipherment (Wholesale)—Part 2: DEA Algorithm (G). 10 pp.
   ISO 8731 Pt 1-87. Banking—Approved Algorithms for Message Authentication—Part 1: DEA First Edition. 4 pp.
   ISO 10126 Pt 2-91. Banking—Procedures for Message Encipherment (Wholesale)—Part 2: DEA Algorithm First Edition. 7 pp.
   JTC1 8731 Pt 1-87. Banking—Approved Algorithms for Message Authentication—Part 1: DEA First Edition. 4 pp.
   JTC1 10126 Pt 2-91. Banking—Procedures for Message Encipherment (Wholesale)—Part 2: DEA Algorithm First Edition. 7 pp.

—Banking
   BSI BS ISO 10126-1-91. 1991 Banking—Procedures for Message Encipherment (Wholesale) Part 1: General Principles. 26 pp.
   ISO 10126 Pt 1-91. Banking—Procedures for Message Encipherment (Wholesale)—Part 1: General Principles First Edition. 23 pp.
   JTC1 10126 Pt 1-91. Banking—Procedures for Message Encipherment (Wholesale)—Part 1: General Principles First Edition. 23 pp.

—Digital Signatures
   OSI ISO DP 9796-87. Data Cryptographic Techniques—Digital Signature Scheme with Shadow Using a Public-Key System. 9 pp.

—Physical Layer
   BSI BS 7112-91. 1991 Achievement of Interoperability and Security by Use of Encipherment at the Physical Layer of OSI in Telecommunication Systems Conveying Automatic Data Processing Information (ISO 9160: 1988). 20 pp.
   ISO 9160-88. Information Processing—Data Encipherment—Physical Layer Interoperability Requirements First Edition. 17 pp.
   JIS X 5051-90. Data Encipherment—Physical Layer Interoperability Requirements (ISO 9160:1988).
   JTC1 9160-88. Information Processing—Data Encipherment—Physical Layer Interoperability Requirements First Edition. 17 pp.

INTERNATIONAL AND NON-U.S. NATIONAL STANDARDS
SUBJECT INDEX
Data

Data Encryption (Cont.)
—Physical Layer (Cont.)
OSI ISO 9160-88. Information Processing—Data Encipherment—Physical Layer Interoperability Requirements. 17 pp.

Data Exchange
Use: Data Communication

Data Handling
Use: Data Processing

Data Highways
See Also: Communication Networks; Data Links; Local Area Networks

—Distributed Processing Systems
IEC 954-90. Process Data Highway, Types A and B (PROWAY A and B), for Distributed Process Control Systems First Edition. 327 pp.
IEC 955-89. Process Data Highway, Type C (PROWAY C), for Distributed Process Control Systems First Edition; (Amendment 1-1992). 526 pp.

Data Interchange
Use: Information Interchange

Data Link Layer (OSI)
See Also: Data Links; Network Protocol (OSI); Open Systems Interconnection

BSI BS ISO/IEC 8886-92. 1992 Information Technology—Telecommunications and Information Exchange Between Systems—Data Link Service Definition for Open Systems Interconnection. 43 pp.
CCITT RECMN X.212-88. Data Link Service Definition for Open Systems Interconnection for CCITT Applications—Data Communication Networks—Open Systems Interconnection (OSI) Model and Notation, Service Definition (Study Group VII) 45 pp (Corrigenda-Oct 1992). 45 pp.
CCITT RECMN X.212-88. Data Link Service Definition for Open Systems Interconnection for CCITT Applications—Data Communication Networks—Open Systems Interconnection (OSI) Model and Notation, Service Definition (Study Group VII) 42 pp. 42 pp.
IEC 8886-92. Information Technology—Telecommunications and Information Exchange Between Systems—Data Link Service Definition for Open Systems Interconnection First Edition. 41 pp.
IEC DIS 11575-92. Information Technology—Telecommunications and Information Exchange Between Systems—Protocol Mappings for the OSI Data Link Service ***CD-ROM ONLY***. 18 pp.
ISO 8886-92. Information Technology—Telecommunications and Information Exchange Between Systems—Data Link Service Definition for Open Systems Interconnection First Edition. 41 pp.
ISO DIS 11575-92. Information Technology—Telecommunications and Information Exchange Between Systems—Protocol Mappings for the OSI Data Link Service ***CD-ROM ONLY***. 18 pp.
JTC1 8886-92. Information Technology—Telecommunications and Information Exchange Between Systems—Data Link Service Definition for Open Systems Interconnection First Edition. 41 pp.
JTC1 DIS11575-92. Information Technology—Telecommunications and Information Exchange Between Systems—Protocol Mappings for the OSI Data Link Service ***CD-ROM ONLY***. 18 pp.
OSI ISO/IEC DIS 8882-2-90. Information Technology Telecommunications and Information Exchange Between Systems—X.25-DTE Conformance Testing—Part 2: Data Link Layer Test Suite. 225 pp.
OSI ISO DP 8882-2-87. OSI-X.25-DTE Conformance Testing-Part 2: Data Link Layer Conformance Tests. 112 pp.

—Base Station Systems—Interfaces
CENELEC GSM 04.05-92. MS—BSS Data Link Layer General Aspects. 27 pp.
CENELEC GSM 08.56-92. BSC-BTS Specification. 14 pp.
ETSI GSM 04.05-92. MS—BSS Data Link Layer General Aspects. 27 pp.
ETSI GSM 08.56-92. BSC-BTS Specification. 14 pp.

—Broadcast Procedures—ISDN—DSS1
CCITT RECMN Q.921-89. ISDN User-Network Interface—Data Link Layer Specification—Digital Subscriber Signalling System No. 1 (DSS 1), Data Link Layer (Study Group XI) 123 pp (Same as Recmn I.441). 123 pp.

—D Channel—Private Switched Networks—Protocols
CENELEC PRETS 300 169-91. Data Link Layer Protocol for the D-Channel of the Interfaces at the Reference Point Between Terminal Equipment and Private Telecommunications Networks. 28 pp.

Data Link Layer (OSI) (Cont.)
—D Channel—Private Switched Networks—Protocols (Cont.)
CENELEC PRI-ETS 300 169-92. Private Telecommunications Network (PTN); Signalling at the S-Reference Point Data Link Layer Protocol. 25 pp.
CENELEC I-ETS 300 169-92. Private Telecommunication Network (PTN); Signalling at the S-Reference Point Data Link Layer Protocol. 24 pp.
ETSI I-ETS 300 169-92. Private Telecommunication Network (PTN); Signalling at the S-Reference Point Data Link Layer Protocol. 24 pp.
ETSI PRI-ETS 300 169-92. Private Telecommunications Network (PTN); Signalling at the S-Reference Point Data Link Layer Protocol. 25 pp.
ETSI PRETS 300 169-91. Data Link Layer Protocol for the D-Channel of the Interfaces at the Reference Point Between Terminal Equipment and Private Telecommunications Networks. 28 pp.

—DSS1
CCITT FASCICLE VI.10-88. Digital Subscriber Signalling System No. 1 (DSS 1), Data Link Layer. Q.920-Q.921 (Study Group XI). 148 pp.

—Error Control—Public Data Networks
CCITT RECMN X.141-89. General Principles for the Detection and Correction of Errors in Public Data Networks—Data Communication Networks—Transmission, Signalling and Switching, Network Aspects, Maintenance and Administrative Arrangements (Study Group VII) 8 pp. 8 pp.

—Fiber Distributed Data Interface
OSI ISO DIS 9314-2-88. Information Processing Systems-Fibre Distribution Data Interface (FDDI) Part 2: Media Access Control (MAC). 48 pp.

—Frame Mode Bearer Services—ISDN—DSS1
CCITT RECMN Q.922-92. ISDN Data Link Layer Specification for Frame Mode Bearer Services (Study Group XI) 112 pp. 112 pp.

—Glossaries
OSI ISO DIS 8886-2-87. Information Processing Systems—Data Communication—Data Link Service Definition for Open Systems Interconnection.

—High Level Data Link Control—Conformance
BSI BS ISO/IEC 8882-2-92. 1992 Information Technology—Telecommunications and Information Exchange Between Systems—X.25 DTE Conformance Testing—Part 2: Data Link Layer Conformance Test Suite. 241 pp.
IEC 8882 Pt 2-92. Information Technology—Telecommunications and Information Exchange Between Systems—X.25 DTE Conformance Testing—Part 2: Data Link Layer Conformance Test Suite First Edition. 241 pp.
ISO 8882 Pt 2-92. Information Technology—Telecommunications and Information Exchange Between Systems—X.25 DTE Conformance Testing—Part 2: Data Link Layer Conformance Test Suite First Edition. 241 pp.
JTC1 8882-92. Information Technology—Telecommunications and Information Exchange Between Systems—X.25 DTE Conformance Testing—Part 2: Data Link Layer Conformance Test Suite First Edition. 241 pp.

—Interfaces—European Digital Cordless Telecommunications
CENELEC PRETS 300 175-4-91. Radio Equipment and Systems Digital European Cordless Telecommunications Common Interface Part 4: Data Link Control Layer (DE/RES 3001-4). 137 pp.
CENELEC PRETS 300 175-4-92. Radio Equipment and Systems (RES); Digital European Cordless Telecommunications (DECT) Common Interface Part 4: Data Link Control Layer. 43 pp.
CENELEC ETS 300 175-4-92. Radio Equipment and Systems (RES); Digital European Cordless Telecommunications (DECT) Common Interface Part 4: Data Link Control Layer. 126 pp.
ETSI ETS 300 175-4-92. Radio Equipment and Systems (RES); Digital European Cordless Telecommunications (DECT) Common Interface Part 4: Data Link Control Layer. 126 pp.
ETSI PRETS 300 175-4-92. Radio Equipment and Systems (RES); Digital European Cordless Telecommunications (DECT) Common Interface Part 4: Data Link Control Layer. 43 pp.
ETSI PRETS 300 175-4-91. Radio Equipment and Systems Digital European Cordless Telecommunications Common Interface Part 4: Data Link Control Layer (DE/RES 3001-4). 137 pp.

Data Link Layer (OSI) (Cont.)
—Interfaces—ISDN
CCITT RECMN I.440-89. ISDN User-Network Interface Data Link Layer—General Aspects—Integrated Services Digital Network (ISDN)—Overall Network Aspects and Functions, ISDN User-Network Interfaces (Study Group XVIII) 1 pp (Same as Recmn Q.920). 1 p.
CCITT RECMN I.441-89. ISDN User-Network Interface, Data Link Layer Specification—Integrated Services Digital Network (ISDN)—Overall Network Aspects and Functions, ISDN User-Network Interfaces (Study Group XVIII) 1 pp (Same as Recmn Q.921). 1 p.
CENELEC PRETS 300 125-90. Integrated Services Digital Network (ISDN); User-Network Interface Data Link Layer Specifications Application of CCITT Recommendations Q.920/I.440 and Q.921/I.441. 162 pp.
CENELEC ETS 300 125-91. Integrated Services Digital Network (ISDN); User-Network Interface Data Link Layer Specification Application of CCITT Recommendations Q.920/I.440 and Q.921/I.441. 161 pp.
ETSI PRETS 300 125-90. Integrated Services Digital Network (ISDN); User-Network Interface Data Link Layer Specifications—Application of CCITT Recommendations Q/920/I.440 and Q.921/I.441—((Formerly ETS(CC))). 162 pp.
ETSI ETS 300 125-91. Integrated Services Digital Network (ISDN); User-Network Interface Data Link Layer Specification Application of CCITT Recommendations Q.920/I.440 and Q.921/I.441. 161 pp.

—Interfaces—ISDN—NATO Reference Model
NATO STANAG 4462 ED 1 AMD 0-00. (Draft) NATO Reference Model for Open Systems Interconnection/Integrated Services Digital Network (ISDN) Layer 2 (Data Link Layer) Specification for ISDN Basic and Primary Rate Access at the S/T Reference Point. 17 pp.

—Interfaces—Mobile Stations (Communications)—Base Station Systems
CENELEC PRI-ETS 300 021-90. European Digital Cellular Telecommunications System (Phase 1); Mobile Station-Base Station System (MS-BSS) Interface Data Link Layer Specification. 83 pp.
CENELEC PRI-ETS 300 021-91. European Digital Cellular Telecommunications System (Phase 1); Mobile Station—Base Station System (MS-BSS) Interface Data Link Layer Specification. 86 pp.
CENELEC PRI-ETS 300 021-92. European Digital Cellular Telecommunications System (Phase 1); Mobile Station—Base Station System (MS-BSS) Interface Data Link Layer Specification. 86 pp.
CENELEC I-ETS 300 021-92. European Digital Cellular Telecommunications System (Phase 1); Mobile Station—Base Station System (MS-BSS) Interface Data Link Layer Specification (GSM 04.06). 86 pp.
CENELEC GSM 04.06-92. See PRI-ETS 300 021. 87 pp.
ETSI PRI-ETS 300 021-90. European Digital Cellular Telecommunications System (Phase 1); Mobile Station—Base Station System (MS-BSS) Interface Data Link Specification (GSM 04. 06). 83 pp.
ETSI I-ETS 300 021-92. European Digital Cellular Telecommunications System (Phase 1); Mobile Station—Base Station System (MS-BSS) Interface Data Link Layer Specification (GSM 04.06). 86 pp.
ETSI PRI-ETS 300 021-92. European Digital Cellular Telecommunications System (Phase 1); Mobile Station—Base Station System (MS-BSS) Interface Data Link Layer Specification. 86 pp.
ETSI GSM 04.06-92. See PRI-ETS 300 021. 87 pp.

—Interfaces—PBX—ISDN
CEPT T/S 49-20 E-87. ISDN ISPBX-ISPBX Interface Layer 2 Specification Application of CEPT Recommendation T/S 46-20. 9 pp.

—Link Access Procedures—D Channel—ISDN—DSS1
CCITT RECMN Q.920-89. ISDN User-Network Interface Data Link Layer—General Aspects—Digital Subscriber Signalling System No.1 (DSS 1), Data Link Layer (Study Group XI) 16 pp (Same as Recmn I.440). 16 pp.
CCITT RECMN Q.921-89. ISDN User-Network Interface—Data Link Layer Specification—Digital Subscriber Signalling System No.1 (DSS 1), Data Link Layer (Study Group XI) 123 pp (Same as Recmn I.441). 123 pp.

—Link Control Procedures
BSI BS 5397: Part 6-90. 1990 High-Level Data Link Control Procedures for Data Communication Part 6: Multilink Procedures. 15 pp.
CSA CAN/CSA-Z243.52-88. Information Processing Systems—Data Communication—Multilink Procedures (ISO 7478-1987). 23 pp.

INDUSTRY STANDARDS

INTERNATIONAL AND NON-U.S. NATIONAL STANDARDS
SUBJECT INDEX

Data

Data Link Layer (OSI) *(Cont.)*

—Link Control Procedures *(Cont.)*

ECMA ECMA 141-90. Data Link Layer Protocol at the Q Reference Point for the Signalling Channel Between Two Private Telecommunication Network Exchanges. 24 pp.

IEC 7478-89. Corrigendum 1 Information Processing Systems—Data Communication—Multilink Procedures. 2 pp.

ISO 7478-87. Information Processing Systems—Data Communication—Multilink Procedures First Edition; (Corrigendum 1-1989). 16 pp.

JTC1 7478-87. Information Processing Systems—Data Communication—Multilink Procedures First Edition; (Corrigendum 1-1989). 16 pp.

OSI ISO 7478. Information Processing Systems—Data Communication—Multilink Procedures.

SNZ NZS/ISO 7478-87. Information Processing Systems—Data Communication—Multilink Procedures. 11 pp.

—Local Area Networks

BSI DD ENV 41114-93. 1993 Information Technology—Functional Standard for Profile T/A52—Local Area Network—Token Bus (COTS + CLNS) (S). 66 pp.

ECMA ECMA 82-84. Local Area Networks (CSMA/CD Baseband) Link Layer. 40 pp.

—Management Information

IEC DIS 10742-92. Information Technology—Telecommunications and Information Exchange Between Systems—Elements of Management Information Related to OSI Data Link Layer Standards ***CD-ROM ONLY***. 81 pp.

ISO DIS 10742-92. Information Technology—Telecommunications and Information Exchange Between Systems—Elements of Management Information Related to OSI Data Link Layer Standards ***CD-ROM ONLY***. 81 pp.

JTC1 DIS10742-92. Information Technology—Telecommunications and Information Exchange Between Systems—Elements of Management Information Related to OSI Data Link Layer Standards ***CD-ROM ONLY***. 81 pp.

—Media Access Control Layer (OSI)

OSI ISO/IEC DIS 10039-89. Information Technology—Telecommunications and Information Exchange Between Systems—Medium Access Control Service Definition. 20 pp.

—NATO Reference Model

NATO STANAG 4252 ED 1 AMD 0-00. (DRAFT) NATO Reference Model for Open System Interconnection Layer 2 (Data Link Layer) Service Definition. 12 pp.

NATO STANAG 4262 ED 1 AMD 0-00. (DRAFT) NATO Reference Model for Open System Interconnection Layer 2 (Data Link Layer) Protocol Specification. 17 pp.

—Packet Switched Data Transmission Services

OSI ISO/IEC DISP 10609-9-90. International Standardized Profiles TB, TC TD and TE—Connection-Mode Trans. Serv. over Conn.-Mode Network Serv.—Pt.9: Subnetwork-Type Dependent Reqmts. for Network, Data Link and Phys. Layer Concerning Permanent Access to a PSDN Using Virtual Call. 34 pp.

—Packet Switched Networks—Conformance

BSI BS ISO/IEC 8882-2-92. 1992 Information Technology—Telecommunications and Information Exchange Between Systems—X.25 DTE Conformance Testing—Part 2: Data Link Layer Conformance Test Suite. 241 pp.

CENELEC ETR 002-90. Network Aspects (NA); Guidelines for the Provision of X.75 Links at Data Rates Higher Than 64 kbit/s. 4 pp.

ETSI ETR 002-90. Network Aspects (NA); Guidelines for the Provision of X.75 Links at Data Rates Higher Than 64 kbit/s. 4 pp.

IEC 8882 Pt 2-92. Information Technology—Telecommunications and Information Exchange Between Systems—X.25 DTE Conformance Testing—Part 2: Data Link Layer Conformance Test Suite First Edition. 241 pp.

IEC 8882 Pt 3 Draft AMD 1. Information Technology—Telecommunications and Information Exchange Between Systems—X.25-DTE Conformance Testing—Part 3: Packet Layer Conformance Test Suite Amendment 1: Use of Data Link Service Primitives; (1992) ***CD-ROM ONLY***. 9 pp.

ISO 8882 Pt 2-92. Information Technology—Telecommunications and Information Exchange Between Systems—X.25 DTE Conformance Testing—Part 2: Data Link Layer Conformance Test Suite First Edition. 241 pp.

Data Link Layer (OSI) *(Cont.)*

—Packet Switched Networks—Conformance *(Cont.)*

ISO 8882 Pt 3 Draft AMD 1. Information Technology—Telecommunications and Information Exchange Between Systems—X.25-DTE Conformance Testing—Part 3: Packet Layer Conformance Test Suite Amendment 1: Use of Data Link Service Primitives; (1992) ***CD-ROM ONLY***. 9 pp.

JTC1 8882-92. Information Technology—Telecommunications and Information Exchange Between Systems—X.25 DTE Conformance Testing—Part 2: Data Link Layer Conformance Test Suite First Edition. 241 pp.

—Point to Point Procedures—ISDN—DSS1

CCITT RECMN Q.921-89. ISDN User-Network Interface—Data Link Layer Specification—Digital Subscriber Signalling System No.1 (DSS 1), Data Link Layer (Study Group XI) 123 pp (Same as Recmn I.441). 123 pp.

—Private Telecommunication Network Exchanges—Signaling Systems

CENELEC PRETS 300 170-91. Data Link Layer Protocol at the Q Reference Point for the Signalling Channel Between Two Private Telecommunication Network Exchanges. 27 pp.

CENELEC PRI-ETS 300 170-92. Private Telecommunications Network (PTN); Inter-Exchange Signalling Data Link Layer Protocol. 23 pp.

CENELEC I-ETS 300 170-92. Private Telecommunication Network (PTN); Inter-Exchange Signalling Data Link Layer Protocol. 22 pp.

ETSI PRETS 300 170-91. Data Link Layer Protocol at the Q Reference Point for the Signalling Channel Between Two Private Telecommunication Network Exchanges (Standard ECMA-141). 26 pp.

ETSI I-ETS 300 170-92. Private Telecommunication Network (PTN); Inter-Exchange Signalling Data Link Layer Protocol. 22 pp.

ETSI PRI-ETS 300 170-92. Private Telecommunications Network (PTN); Inter-Exchange Signalling Data Link Layer Protocol. 23 pp.

ETSI PRETS 300 170-91. Data Link Layer Protocol at the Q Reference Point for the Signalling Channel Between Two Private Telecommunication Network Exchanges. 27 pp.

—Protocol Implementation Conformance Statement—ISDN—DSS1

CENELEC PRI-ETS 300 305-93. Integrated Services Digital Network (ISDN); Digital Subscriber Signalling System No. One (DSS1) Protocol Implementation Conformance Statement (PICS) Proforma for Basic-Access User for Data-Link-Layer Protocol for General Application. 19 pp.

CENELEC PRI-ETS 300 306-93. Integrated Services Digital Network (ISDN); Digital Subscriber Signalling System No. One (DSS1) Protocol Implementation Conformance Statement (PICS) Proforma for Primary-Rate-Access User for Data-Link-Layer Protocol for General Application. 20 pp.

CENELEC PRI-ETS 300 307-93. Integrated Services Digital Network (ISDN); Digital Subscriber Signalling System No. One (DSS1) Protocol Implementation Conformance Statement (PICS) Proforma for Basic-Access Network for Data-Link-Layer Protocol for General Application. 20 pp.

CENELEC PRI-ETS 300 308-93. Integrated Services Digital Network (ISDN); Digital Subscriber Signalling System No. One (DSS1) Protocol Implementation Conformance Statement (PICS) Proforma for Primary-Rate-Access Network for Data-Link-Layer Protocol for General Application. 19 pp.

ETSI PRI-ETS 300 305-93. Integrated Services Digital Network (ISDN); Digital Subscriber Signalling System No. One (DSS1) Protocol Implementation Conformance Statement (PICS) Proforma for Basic-Access User for Data-Link-Layer Protocol for General Application. 19 pp.

ETSI PRI-ETS 300 306-93. Integrated Services Digital Network (ISDN); Digital Subscriber Signalling System No. One (DSS1) Protocol Implementation Conformance Statement (PICS) Proforma for Primary-Rate-Access User for Data-Link-Layer Protocol for General Application. 20 pp.

ETSI PRI-ETS 300 307-93. Integrated Services Digital Network (ISDN); Digital Subscriber Signalling System No. One (DSS1) Protocol Implementation Conformance Statement (PICS) Proforma for Basic-Access Network for Data-Link-Layer Protocol for General Application. 20 pp.

ETSI PRI-ETS 300 308-93. Integrated Services Digital Network (ISDN); Digital Subscriber Signalling System No. One (DSS1) Protocol Implementation Conformance Statement (PICS) Proforma for Primary-Rate-Access Network for Data-Link-Layer Protocol for General Application. 19 pp.

Data Link Layer (OSI) *(Cont.)*

—Protocol Implementation Extra Information for Testing—ISDN—DSS1

CENELEC PRI-ETS 300 309-93. Integrated Services Digital Network (ISDN); Digital Subscriber Signalling System No. One (DSS1) Protocol Implementation eXtra Information for Testing (PIXIT) Proforma for Basic-Access User for Data-Link-Layer Protocol for General Application. 11 pp.

CENELEC PRI-ETS 300 310-93. Integrated Services Digital Network (ISDN); Digital Subscriber Signalling System No. One (DSS1) Protocol Implementation eXtra Information for Testing (PIXIT) Proforma for Primary-Rate-Access User for Data-Link-Layer Protocol for General Application. 11 pp.

CENELEC PRI-ETS 300 311-93. Integrated Services Digital Network (ISDN); Digital Subscriber Signalling System No. One (DSS1) Protocol Implementation eXtra Information for Testing (PIXIT) Proforma for Basic-Access Network for Data-Link-Layer Protocol for General Application. 11 pp.

CENELEC PRI-ETS 300 312-93. Integrated Services Digital Network (ISDN); Digital Subscriber Signalling System No. One (DSS1) Protocol Implementation eXtra Information for Testing (PIXIT) Proforma for Primary-Rate-Access Network for Data-Link-Layer Protocol for General Application. 11 pp.

ETSI PRI-ETS 300 309-93. Integrated Services Digital Network (ISDN); Digital Subscriber Signalling System No. One (DSS1) Protocol Implementation eXtra Information for Testing (PIXIT) Proforma for Basic-Access User for Data-Link-Layer Protocol for General Application. 11 pp.

ETSI PRI-ETS 300 310-93. Integrated Services Digital Network (ISDN); Digital Subscriber Signalling System No. One (DSS1) Protocol Implementation eXtra Information for Testing (PIXIT) Proforma for Primary-Rate-Access User for Data-Link-Layer Protocol for General Application. 11 pp.

ETSI PRI-ETS 300 311-93. Integrated Services Digital Network (ISDN); Digital Subscriber Signalling System No. One (DSS1) Protocol Implementation eXtra Information for Testing (PIXIT) Proforma for Basic-Access Network for Data-Link-Layer Protocol for General Application. 11 pp.

ETSI PRI-ETS 300 312-93. Integrated Services Digital Network (ISDN); Digital Subscriber Signalling System No. One (DSS1) Protocol Implementation eXtra Information for Testing (PIXIT) Proforma for Primary-Rate-Access Network for Data-Link-Layer Protocol for General Application. 11 pp.

—Subnetworks—Packet Switched Networks—Virtual Call Services

BSI BS ISO/IEC ISP 10609-9-92. 1992 Information Technology—International Standardized Profiles TB, TC, TD and TE—Connection-Mode Transport Service over Connection-Mode Network Service—Part 9: Subnetwork-Type Dependent Requirements for Network Layer, Data. 34 pp.

IEC ISP10609 Pt 9-92. Information Technology—International Standardized Profiles TB, TC, TD and TE—Connection-Mode Transport Service over Connection-Mode Network Service—Part 9: Subnetwork-Type Dependent Requirements for Network Layer, Data Link Layer and Physical. 32 pp.

ISO ISP10609 Pt 9-92. Information Technology—International Standardized Profiles TB, TC, TD and TE—Connection-Mode Transport Service over Connection-Mode Network Service—Part 9: Subnetwork-Type Dependent Requirements for Network Layer, Data Link Layer and Physical. 32 pp.

JTC1 ISP 10609 Pt 9-92. Information Technology—International Standardized Profiles TB, TC, TD and TE—Connection-Mode Transport Service over Connection-Mode Network Service—Part 9: Subnetwork-Type Dependent Requirements for Network Layer, Data Link Layer and Physical. 32 pp.

Data Link Systems
Use: Data Links

Data Links
Use For: Analog Links; Data Link Systems; Data Transmitting Sets; Digital Links; Digital Transmission Links; Multilinks *See Also:* Bit Rates; Communication Interfaces; Data Communication Equipment; Data Highways; Data Link Layer (OSI); Data Processing Equipment; Data Transmission; Digital Transmission Systems; High Level Data Link Control; Link Access Procedures; Link Control Procedures; Open Systems Interconnection; Radio Relay Systems; Telecommunication Equipment; Telemetry

INTERNATIONAL AND NON-U.S. NATIONAL STANDARDS
SUBJECT INDEX
Data

Data Links *(Cont.)*

—Access Points—Maintenance

CCITT RECMN M.120-89. Access Points for Maintenance—General Maintenance Principles—Maintenance of International Transmission Systems and Telephone Circuits (Study Group IV) 2 pp. 2 pp.

—Connectors

CEN PREN 2591 (Part G)-92. Elements of Electrical and Optical Connection Test Methods Part G-. 4 pp.

—Data Relay Satellites

CCIR Report 982-86. Data Relay Satellites for the Earth Exploration Satellite Service—Section 2D—Data Relay Satellites. 9 pp.

—Digital Exchanges—Terminating

CCITT RECMN G.705-89. Characteristics Required to Terminate Digital Links on a Digital Exchange—General Aspects of Digital Transmission Systems; Terminal Equipments (Study Group XV and XVIII) 5 pp. 5 pp.

—Interfaces—Integrated Services Digital Networks

CEPT T/S 46-20-87. Specification De La Couche De Liaisonde Donnees a L'Interface Usager-Reseau Dans Un RNIS; Application Des Recommendations CCITT Q.920/I.440 Et Q.921/I.441. 188 pp.

CEPT T/S 46-20 E-87. ISDN User-Network Interface Data Link Layer Specification Application of CCITT Recommendations Q.920/I.440 and Q.921/I.441. 168 pp.

CSA CAN/CSA-T542-90. Integrated Services Digital Network (ISDN)—Data-Link Layer Signalling Specification for Application at the User-Network Interface (ANSI T1.602-1989). 199 pp.

—Interfaces—Protocols—Integrated Services Digital Networks

ECMA ECMA 105-90. Data Link Layer Protocol for the D-Channel of the Interfaces at the Reference Point Between Terminal Equipment and Private Telecommunication Networks. 68 pp.

—Interfaces—Protocols—Private Networks

ECMA ECMA 105-90. Data Link Layer Protocol for the D-Channel of the Interfaces at the Reference Point Between Terminal Equipment and Private Telecommunication Networks. 68 pp.

—Noise

CCITT RECMN G.226-89. Noise on a Real Link—International Analogue Carrier Systems (Study Group XV) 2 pp. 2 pp.

—Reference Clocks—Plesiochronous

CCITT RECMN G.811-89. Timing Requirements at the Outputs of Primary Reference Clocks Suitable for Plesichronous Operation of International Digital Links—Dig. Networks, Dig. Sections and Dig. Line Systems (Study Groups XV and XVIII) 6 pp. 6 pp.

—Research Satellites—Protection

CCIR RECMN 609-86. Protection Criteria for Telecommunication Links for Manned and Unmanned Near-Earth Research Satellites—Section 2E—Space Research. 1 p.

CCIR RECMN 609-1-92. Protection Criteria for Telecommunication Links for Manned and Unmanned Near-Earth Research Satellites—Section 2E—Space Research. 4 pp.

—Slave Clocks—Plesiochronous

CCITT RECMN G.812-89. Timing Requirements at the Outputs of Slave Clocks Suitable for Plesichronous Operation of International Digital Links Networks, Digital Sections and Digital Line Systems (Study Groups XV and XVIII) 6 pp. 6 pp.

—Transmission Restoration Systems

CCITT RECMN G.180-89. Characteristics of N + M Type Direct Transmission Restoration Systems for Use on Digital Sections, Links or Equipment—General Characteristics of International Telephone Connections and Circuits (Study Groups XII and XV) 11 pp. 11 pp.

CCITT RECMN G.181-89. Characteristics of 1 + 1 Type Restoration Systems for Use on Digital Transmission Links—General Characteristics of International Telephone Connections and Circuits (Study Groups XII and XV) 3 pp. 3 pp.

—Voice Frequency Telegraphy

CCITT RECMN H.22-89. Transmission Requirements of International Voice-Frequency Telegraph Links (at 50, 100 and 200 Baud)—Line Transmission of Non-Telephone Signals—Transmission of Sound-Programme and Television Signals (Study Group XV) 1 pp. 1 p.

Data Links *(Cont.)*

—Voice Frequency Telegraphy *(Cont.)*

CCITT RECMN R.30-89. Transmission Characteristic for International VFT Links—Telegraph Transmission (Study Group IX) 2 pp. 2 pp.

—Voice Frequency Telegraphy—Electrical Measurement

CCITT RECMN M.830-89. Routine Measurements to Be Made on International Voice-Frequency Telegraph Links—Maintenance of International Telegraph, Phototelegraph and Leased Circuits—Maintenance of the International Public Telephone Network—Maintenance of Maritime. 1 p.

—Voice Frequency Telegraphy—Maintenance

CCITT RECMN M.830-89. Routine Measurements to Be Made on International Voice-Frequency Telegraph Links—Maintenance of International Telegraph, Phototelegraph and Leased Circuits—Maintenance of the International Public Telephone Network—Maintenance of Maritime. 1 p.

—Voice Frequency Telegraphy—Measurement—Maintenance

CCITT RECMN M.820-89. Periodicity of Routine Tests on International Voice-Frequency Telegraph Links—Maintenance of International Telegraph, Phototelegraph and Leased Circuits—Maintenance of the International Public Telephone Network—Maintenance of Maritime. 2 pp.

—Voice Frequency Telegraphy—Pulse Code Modulation

CCITT RECMN R.30-89. Transmission Characteristic for International VFT Links—Telegraph Transmission (Study Group IX) 2 pp. 2 pp.

—Voice Frequency Telegraphy—Telegraph Circuits—Line Up

CCITT RECMN M.810-89. Setting up and Lining up an International Voice-Frequency Telegraph Link for Public Telegraph Circuits (for 50, 100 and 200 Baud Modulation Rates)—Maintenance of International Telegraph, Phototelegraph and Leased Circuits—Maintenance of Maritime Satellite and Data. 11 pp.

Data Logging

Use For: Recording of Data *See Also:* Data Acquisition Systems; Data Analysis; Data Collection; Logging Stations (Data Transmission); Measurement; Measuring Instruments; Recording Instruments

—Call Duration

CCITT RECMN E.261-89. Devices for Measuring and Recording Call Durations—Telephone Network and ISDN—Operation, Numbering, Routing and Mobile Service (Study Group II) 3 pp. 3 pp.

—Call Duration—Automatic Services

CCITT RECMN E.260-89. Basic Technical Problems Concerning the Measurement and Recording of Call Durations—Telephone Network and ISDN—Operation, Numbering, Routing and Mobile Service (Study Group II) 4 pp. 4 pp.

—Call Duration—Telephone Services

CCITT RECMN E.260-89. Basic Technical Problems Concerning the Measurement and Recording of Call Durations—Telephone Network and ISDN—Operation, Numbering, Routing and Mobile Service (Study Group II) 4 pp. 4 pp.

—Format—Weapons Systems

MOD UK NES 1044-92. Requirements for the Logical Format for Weapons Equipment Data Recording Issue 1 (09.92). 38 pp.

—Network Performance—Telecommunication Networks

CCITT RECMN E.880-89. Field Data Collection and Evaluation on the Performance of Equipment, Networks and Services—Telephone Network and ISDN—Quality of Service, Network Management and Traffic Engineering (Study Group II) 7 pp. 7 pp.

—Telecommunication Equipment

CCITT RECMN E.880-89. Field Data Collection and Evaluation on the Performance of Equipment, Networks and Services—Telephone Network and ISDN—Quality of Service, Network Management and Traffic Engineering (Study Group II) 7 pp. 7 pp.

Data Logging *(Cont.)*

—Telecommunication Services

CCITT RECMN E.880-89. Field Data Collection and Evaluation on the Performance of Equipment, Networks and Services—Telephone Network and ISDN—Quality of Service, Network Management and Traffic Engineering (Study Group II) 7 pp. 7 pp.

—Telex Communications

CCITT RECMN F.68-89. Establishment of the Automatic Intercontinental Telex Network—Telegraph and Mobile Services Operations and Quality of Service (Study Group I) 6 pp. 6 pp.

Data Management

See Also: Data Acquisition Systems

—Aircraft—L Band Avionics

EUROCAE ED-25 07.76. MPS for Experimental AEROSAT L-Band Avionics. 75 pp.

—Reference Models

IEC DIS 10032-91. Information Technology—Reference Model of Data Management ***CD-ROM ONLY***. 69 pp.

ISO DIS 10032-91. Information Technology—Reference Model of Data Management ***CD-ROM ONLY***. 69 pp.

—Software

SAA AS 4042-92. Software Configuration Management Plans (IEEE 828—1990). 9 pp.

SAA AS 4043-92. Software Configuration Management (IEEE 1042—1987). 84 pp.

—Statistical Analysis

JIS Z 9041-68. Presentation and Reduction of Data. 50 pp.

Data Manipulation Languages

Scope Note: Use a more specific term *See:* NDL; Programming Languages; SQL

Data Media

Use: Data Storage Devices

Data Messages

Use: Protocol Data Units

Data Modems

Use: Modems

Data Network Identification Codes

Use For: DNIC *See Also:* Codes; Identification Systems; Telecommunication Systems

—Integrated Services Digital Networks

CCITT RECMN E.167-89. ISDN Network Identification Codes—Telephone Network and ISDN—Operation, Numbering, Routing and Mobile Service (Study Group II) 2 pp. 2 pp.

—Integrated Services Digital Networks—Closed User Groups

CCITT RECMN E.167-89. ISDN Network Identification Codes—Telephone Network and ISDN—Operation, Numbering, Routing and Mobile Service (Study Group II) 2 pp. 2 pp.

Data Networks

See Also: Circuits; Data Terminal Equipment; Mesh Networks; Packet Switched Data Transmission Services; Packet Switched Networks

—Anisochronous—Data Terminal Equipment

CCITT RECMN X.52-89. Method of Encoding Anisochronous Signals Into a Synchronous User Bearer—Data Communication Networks—Transmission, Signalling and Switching, Network Aspects, Maintenance and Administrative Arrangements (Study Group VII) 3 pp. 3 pp.

—Anisochronous—Quality Assurance

CCITT RECMN R.121-89. Standard Limits of Transmission Quality for Start-Stop User Classes of Service 1 and 2 on Anisochronous Data Networks—Telegraph Transmission (Study Group IX) 1 pp. 1 p.

—Anisochronous—Telecommunication Circuits—Start-Stop Transmission

CCITT RECMN X.70-89. Terminal and Transit Control Signalling System for Start-Stop Services on International Circuits Between Anisochronous Data Networks—Data Communication Networks—Transmission, Signalling and Switching, Network Aspects, Maintenance and Administrative. 33 pp.

INDUSTRY STANDARDS

INTERNATIONAL AND NON-U.S. NATIONAL STANDARDS
SUBJECT INDEX

Data

Data Networks *(Cont.)*

—Interfaces—Data Signaling Rates—10 Bit Envelopes
- CCITT RECMN X.51-89. Fundamental Parameters of a Multiplexing Scheme for the International Interface Between Synchronous Data Networks Using 10-Bit Envelope Structure—Data Communication Networks—Transmission, Signalling and Switching, Network Aspects, Maintenance and. 5 pp.
- CCITT RECMN X.51 BIS-89. Fundamental Parameters of a 48-kBit/s User Data Signalling Rate Transmission Scheme for the International Interface Between Synchronous Data Networks Using 10-Bit Envelope Structure—Data Communication Networks—Transmission, Signalling and. 2 pp.

—Interfaces—Multiplexing—10 Bit Envelopes
- CCITT RECMN X.51-89. Fundamental Parameters of a Multiplexing Scheme for the International Interface Between Synchronous Data Networks Using 10-Bit Envelope Structure—Data Communication Networks—Transmission, Signalling and Switching, Network Aspects, Maintenance and. 5 pp.
- CCITT RECMN X.51 BIS-89. Fundamental Parameters of a 48-kBit/s User Data Signalling Rate Transmission Scheme for the International Interface Between Synchronous Data Networks Using 10-Bit Envelope Structure—Data Communication Networks—Transmission, Signalling and. 2 pp.

—Interworking—Mobile Satellite Communications
- CCIR Report 1176-90. Interworking Between the Mobile Satellite Systems and the Terrestrial Networks for Data Transmission Services—Section 8G—Availability, Performance Objectives and Interworking with Terrestrial Networks. 8 pp.
- CCIR QUESTION 89-1/8-90. Compatibility for Interworking Between the Mobile-Satellite Systems and Terrestrial Networks Including ISDN—Questions Concerning Study Group 8—Mobile, Radiodetermination, Amateur and Related Satellite Services. 1 p.

—Synchronous—Communication Channels—8 Bit Envelope
- CCITT RECMN X.57-89. Method of Transmitting a Single Lower Speed Data Channel on a 64 kBit/s Data Stream—Data Communication Networks—Transmission, Signalling and Switching, Network Aspects, Maintenance and Administrative Arrangements (Study Group VII) 2 pp. 2 pp.

—Synchronous—Digital Multiplexers—1544 kbit/s
- CCITT RECMN G.734-89. Characteristics of Synchronous Digital Multiplex Equipment Operating at 1544 kbit/s—General Aspects of Digital Transmission Systems; Terminal Equipments (Study Groups XV and XVIII) 4 pp. 4 pp.

—Synchronous—Digital Multiplexers—2048 kbit/s
- CCITT RECMN G.736-89. Characteristics of a Synchronous Digital Multiplex Equipment Operating at 2048 kbit/s—General Aspects of Digital Transmission Systems; Terminal Equipments (Study Groups XV and XVIII) 5 pp. 5 pp.

—Synchronous—Interfaces—Data Signaling Rates
- CCITT RECMN X.50-89. Fundamental Parameters of a Multiplexing Scheme for the International Interface Between Synchronous Data Networks—Data Communication Networks—Transmission, Signalling and Switching, Network Aspects, Maintenance and Administrative Arrangements. 7 pp.
- CCITT RECMN X.50 BIS-89. Fundamental Parameters of a 48-kBit/s User Data Signalling Rate Transmission Scheme for the International Interface Between Synchronous Data Networks—Data Communication Networks—Transmission, Signalling and Switching, Network Aspects, Maintenance. 1 p.

—Synchronous—Interfaces—Multiplexing
- CCITT RECMN X.50-89. Fundamental Parameters of a Multiplexing Scheme for the International Interface Between Synchronous Data Networks—Data Communication Networks—Transmission, Signalling and Switching, Network Aspects, Maintenance and Administrative Arrangements. 7 pp.
- CCITT RECMN X.50 BIS-89. Fundamental Parameters of a 48-kBit/s User Data Signalling Rate Transmission Scheme for the International Interface Between Synchronous Data Networks—Data Communication Networks—Transmission, Signalling and Switching, Network Aspects, Maintenance. 1 p.

Data Networks *(Cont.)*

—Synchronous—Non-Envelope—Multiplexing—Interworking
- CCITT RECMN X.58-89. Fundamental Parameters of a Multiplexing Scheme for the International Interface Between Synchronous Non-Switched Data Networks Using No Envelope Structure—Data Communication Networks—Transmission, Signalling and Switching, Network Aspects, Maintenance and. 3 pp.

—Synchronous—Satellite Communications—10 Bit Envelope
- CCITT RECMN X.56-89. Interface Between Synchronous Data Networks Using an 8 + 2 Envelope Structure and Single Channel Per Carrier (SCPC) Satellite Channels—Data Communication Networks—Transmission, Signalling and Switching, Network Aspects, Maintenance and Administrative. 5 pp.

—Synchronous—Satellite Communications—8 Bit Envelope
- CCITT RECMN X.55-89. Interface Between Synchronous Data Networks Using a 6 + 2 Envelope Structure and Single Channel Per Carrier (SCPC) Satellite Channels—Data Communication Networks—Transmission, Signalling and Switching, Network Aspects, Maintenance and Administrative. 3 pp.

—Synchronous—Telecommunication Circuits—Control Signals
- CCITT RECMN X.71-89. Decentralized Terminal and Transit Control Signalling System on International Circuits Between Synchronous Data Networks—Data Communication Networks—Transmission, Signalling and Switching, Network Aspects, Maintenance and Administrative Arrangements. 35 pp.

Data Organization
Use For: Organization of Data *See Also:* Data Analysis; Data Processing

—Glossaries
- JIS X 0004-89. Glossary of Terms Used in Information Processing (Organization of Data).

—Sequential Files—Record Groups
- CNS C5227-89. Information Processing Program Flow for Processing Sequential Files in Terms of Record Groups (Dec)(12644).

Data Processing
Use For: Data Handling; Information Processing
See Also: Automation; Computer Programming; Computers; Control Functions; Data Acquisition Systems; Data Analysis; Data Base Management Systems; Data Blocks; Data Collection; Data Elements; Data Organization; Data Processing Equipment; Data Transmission; Decision Tables; Editing Systems; Encoding; Information Interchange; Information Theory; Open Systems Interconnection; Programming Languages; Record Layouts; Software; Standard Page Description Language; Symbols; Systems Engineering; Systems Manuals; Text Processing; Video Data Digital Processing

- ECMA MEMENTO-92. Ecma Memento 1992. 96 pp.
- ECMA MEMENTO-93. Ecma Memento 1993. 116 pp.
- ISO 1155-78. Information Processing—Use of Longitudinal Parity to Detect Errors in Information Messages Second Edition. 3 pp.
- ISO 1177-85. Information Processing—Character Structure for Start/Stop and Synchronous Character Oriented Transmission Second Edition. 4 pp.
- JIS X 4211-92. Information Processing.
- JTC1 1155-78. Information Processing—Use of Longitudinal Parity to Detect Errors in Information Messages Second Edition. 3 pp.
- JTC1 1177-85. Information Processing—Character Structure for Start/Stop and Synchronous Character Oriented Transmission Second Edition. 4 pp.
- OSI ISO 1155-78. Information Processing—Use of Longitudinal Parity to Detect Errors in Information Messages. 3 pp.
- OSI ISO 1177-85. Information Processing—Character Structure for Start/Stop and Synchronous Character Oriented Transmission. 4 pp.
- SNZ NZS/ISO 1155-78. Information Processing—Use of Longitudinal Parity to Detect Errors in Information Messages. 1 p.

—Aeronautical Navigation Charts
- NATO STANAG 3721 ED 3 AMD 1-88. Automatic Data Processing (ADP) Master File for Land Maps and Aeronautical Charts. 23 pp.

—Arithmetic Operations—Glossaries
- JIS X 0002-87. Glossary of Terms Used in Information Processing (Arithmetic and Logic Operations).

Data Processing *(Cont.)*

—Arithmetic Operations—Glossaries *(Cont.)*
- OSI ISO 2382-2-76. Data Processing—Vocabulary—Section 02: Arithmetic and Logic Operations. 37 pp.
- SAA AS 1189.2-82. Data Processing—Vocabulary—Part 2: Arithmetic and Logic Operations. 21 pp.

—Bank Cards
- BSI BS 7096-90. 1990 Guide to Design and Use of Bank Cards with a Magnetic Stripe That Employs Track 3. 17 pp.
- CEN EN 24 909-90. Bank Cards Magnetic Stripe Data Content for Track 3. 3 pp.
- ISO 4909-87. Bank Cards—Magnetic Stripe Data Content for Track 3 Second Edition. 13 pp.
- JTC1 4909-87. Bank Cards—Magnetic Stripe Data Content for Track 3 Second Edition. 13 pp.
- OSI ISO 4909-87. Bankcards—Magnetic Stripe Data Content for Track 3. 15 pp.

—Bibliographic Data Elements
- BSI BS 7124: Part 1-89. 1989 Bibliographic Data Element Directory Part 1: Interloan Applications. 26 pp.
- ISO 8459 Pt 1-88. Documentation—Bibliographic Data Element Directory—Part 1: Interloan Applications First Edition. 24 pp.

—Computer Assisted—Glossaries
- BSI DD 179-89. 1989 Glossary of Terms Used in Computer-Assisted Publishing. 48 pp.
- ISO TR9544-88. Information Processing—Computer-Assisted Publishing—Vocabulary First Edition. 48 pp.
- JTC1 TR9544-88. Information Processing—Computer-Assisted Publishing—Vocabulary First Edition. 48 pp.
- OSI ISO/TR 9544-88. Information Processing—Computer Assisted Publishing—Vocabulary. 48 pp.
- SAA AS 3877-91. Manipulating Industrial Robots—Vocabulary (ISO/TR 8373:1988). 9 pp.

—Documentation
- ISO 5127 Pt 1-83. Documentation and Information—Vocabulary—Part 1: Basic Concepts First Edition. 25 pp.

—Flow Charts—Symbols
- BSI BS 4058-87. 1987 Data Processing Flow Chart Symbols, Rules and Conventions. 31 pp.
- CNS C5035-83. Flowchart Symbols for Information Processing (Feb)(5204).
- ISO 5807-85. Information Processing—Documentation Symbols and Conventions for Data, Program and System Flowcharts, Program Network Charts and System Resources Charts First Edition. 29 pp.
- JTC1 5807-85. Information Processing—Documentation Symbols and Conventions for Data, Program and System Flowcharts, Program Network Charts and System Resources Charts First Edition. 29 pp.
- OSI ISO 5807-85. Information Processing—Documentation Symbols and Conventions for Data, Program and System Flowcharts, Programnet-work Charts and System Resources Charts. 29 pp.

—Glossaries
- BSI BS ISO 2382/1-84. 1984 Information Technology—Vocabulary 2382/1: Fundamental Terms. 23 pp.
- BSI BS ISO 2382/II-76. 1976 Information Technology—Vocabulary 2382/II: Processing Units. 26 pp.
- BSI BS ISO 2382/4-87. 1987 Information Technology—Vocabulary 2382/4: Organization of Data. 26 pp.
- BSI BS ISO 2382/5-89. 1989 Information Technology—Vocabulary 2382/5: Representation of Data. 29 pp.
- BSI BS ISO 2382/6-87. 1987 Information Technology—Vocabulary 2382/6: Preparation and Handling of Data. 22 pp.
- BSI BS ISO 2382/8-86. 1986 Information Technology—Vocabulary 2382/8: Control, Integrity and Security. 25 pp.
- BSI BS ISO 2382/9-84. 1984 Information Technology—Vocabulary 2382/9: Data Communication. 32 pp.
- BSI BS ISO 2382/X-79. 1979 Information Technology—Vocabulary 2382/X: Operating Techniques and Facilities. 27 pp.
- BSI BS ISO 2382/11-87. 1987 Information Technology—Vocabulary 2382/11: Processing Units. 26 pp.
- BSI BS ISO 2382/XIV-78. 1978 Information Technology—Vocabulary Part 14: Reliability, Maintenance and Availability. 20 pp.
- BSI BS ISO 2382/XVI-78. 1978 Information Technology—Vocabulary 2382/XVI: Information Theory. 26 pp.
- BSI BS ISO 2382/18-87. 1987 Information Technology—Vocabulary Part 18: Distributed Data Processing. 20 pp.
- BSI BS ISO/IEC 2382-20-90. 1990 System Development. 27 pp.

INTERNATIONAL AND NON-U.S. NATIONAL STANDARDS
SUBJECT INDEX
Data

Data Processing *(Cont.)*
—**Glossaries** *(Cont.)*

BSI BS ISO/IEC TR 12382-92. 1992 Permuted Index of the Vocabulary of Information Processing (S). 251 pp.
BSI BS ISO/IEC TR 12382-89. 1989 Permuted Index of the Vocabulary of Information Processing (S). 285 pp.
CNS C5139-85. Data Processing Vocabulary (Part 1: Fundamental Terms) (Jul)(9359).
CNS C5140-85. Data Processing Vocabulary (Part 2: Arithmetic and Logic Operations) (Jul)(9360).
CNS C5141-89. Data Processing Vocabulary (Part 3: Equipment Technology) (Oct)(9361).
CNS C5142-90. Data Processing Vocabulary (Part 4: Organization of Data) (Jan)(9362).
CNS C5143-90. Data Processing Vocabulary (Part 5: Representation of Data) (Jan)(9691).
CNS C5144-89. Data Processing Vocabulary (Part 6: Preparation and Handling of Data) (Oct)(9692).
CNS C5152-85. Data Processing Vocabulary (Part 10: Operating Techniques and Facilities) (Aug)(10239).
CNS C5153-90. Data Processing Vocabulary (Part 11: Control, Input-Output and Arithmetic Equipment) (Nov)(10240).
CNS C5155-85. Data Processing Vocabulary (Part 14: Reliability, Maintenance and Availability) (Aug)(10242).
CNS C5202-85. Data Processing Vocabulary (Part 9: Data Communication) (Nov)(11404).
CSA CAN3-Z243.27. 1-79. Data Processing—Vocabulary—Fundamental Terms. 35 pp.
CSA CAN3-Z243.27. 4-81. Data Processing—Vocabulary—Organization of Data. 38 pp.
CSA CAN/CSA-Z243.58-92. Information Technology Vocabulary. 612 pp.
IEC DIS 2382 Pt 1.2-92. Information Technology—Vocabulary—Part 1: Fundamental Terms (Revision of ISO 2382-1:1984) ***CD-ROM ONLY***. 36 pp.
IEC DIS 2382 Pt 9-93. Data Processing—Vocabulary—Part 09: Data Communication ***CD-ROM ONLY***. 51 pp.
IEC 2382 Pt 20-90. Information Technology—Vocabulary—Part 20: System Development First Edition. 23 pp.
IEC TR12382-92. Permuted Index of the Vocabulary of Information Technology Second Edition. 249 pp.
ISO 2382 Pt 1-84. Data Processing—Vocabulary—Part 01: Fundamental Terms Second Edition. 20 pp.
ISO DIS 2382 Pt 1.2-92. Information Technology—Vocabulary—Part 1: Fundamental Terms (Revision of ISO 2382-1:1984) ***CD-ROM ONLY***. 36 pp.
ISO 2382 Sec II-76. Data Processing—Vocabulary—Section 02: Arithmetic and Logic Operations First Edition. 38 pp.
ISO 2382 Pt 4-87. Information Processing Systems—Vocabulary—Part 04: Organization of Data Second Edition. 23 pp.
ISO 2382 Pt 5-89. Information Processing Systems—Vocabulary—Part 05: Representation of Data Second Edition. 26 pp.
ISO 2382 Pt 6-87. Information Processing Systems—Vocabulary—Part 06: Preparation and Handling of Data Second Edition. 19 pp.
ISO 2382 Pt 8-86. Information Processing Systems—Vocabulary—Part 08: Control, Integrity and Security First Edition. 22 pp.
ISO 2382 Pt 9-84. Data Processing—Vocabulary—Part 09: Data Communication First Edition. 29 pp.
ISO DIS 2382 Pt 9-93. Data Processing—Vocabulary—Part 09: Data Communication ***CD-ROM ONLY***. 51 pp.
ISO 2382 Sec X-79. Data Processing—Vocabulary—Section 10: Operating Techniques and Facilities First Edition. 23 pp.
ISO 2382 Pt 11-87. Information Processing Systems—Vocabulary—Part 11: Processing Units Second Edition. 23 pp.
ISO 2382 Sec XIV-78. Data Processing—Vocabulary—Section 14: Reliability, Maintenance and Availability First Edition. 17 pp.
ISO 2382 Sec XVI-78. Data Processing—Vocabulary—Section 16: Information Theory First Edition. 22 pp.
ISO 2382 Pt 18-87. Information Processing Systems—Vocabulary—Part 18: Distributed Data Processing First Edition. 18 pp.
ISO 2382 Pt 20-90. Information Technology—Vocabulary—Part 20: System Development First Edition. 23 pp.
ISO DIS 2382 Pt 23-92. Information Processing Systems—Vocabulary Part 23: Text Processing ***CD-ROM ONLY***. 33 pp.
ISO 5127 Pt 1-83. Documentation and Information—Vocabulary—Part 1: Basic Concepts First Edition. 25 pp.
ISO TR12382-92. Permuted Index of the Vocabulary of Information Technology Second Edition. 249 pp.
JIS X 0001-87. Glossary of Terms Used in Information Processing.
JIS X 0004-89. Glossary of Terms Used in Information Processing (Organization of Data).
JIS X 0005-90. Glossary of Terms Used in Information Processing (Representation of Data).
JIS X 0006-89. Glossary of Terms Used in Information Processing (Preparation and Handling of Data).
JIS X 0008-87. Glossary of Terms Used in Information Processing (Control, Integrity and Security).
JIS X 0010-87. Glossary of Terms Used in Information Processing (Operating Techniques and Facilities).
JIS X 0014-87. Glossary of Terms Used in Information Processing (Reliability, Maintenance and Availability).
JTC1 2382 Pt 1-84. Data Processing—Vocabulary—Part 01: Fundamental Terms Second Edition. 20 pp.
JTC1 DIS2382 Pt 1.2-92. Information Technology—Vocabulary—Part 1: Fundamental Terms (Revision of ISO 2382-1:1984) ***CD-ROM ONLY***. 36 pp.
JTC1 2382 Sec II-76. Data Processing—Vocabulary—Section 02: Arithmetic and Logic Operations First Edition. 38 pp.
JTC1 2382 Pt 4-87. Information Processing Systems—Vocabulary—Part 04: Organization of Data Second Edition. 23 pp.
JTC1 2382 Pt 5-89. Information Processing Systems—Vocabulary—Part 05: Representation of Data Second Edition. 26 pp.
JTC1 2382 Pt 6-87. Information Processing Systems—Vocabulary—Part 06: Preparation and Handling of Data Second Edition. 19 pp.
JTC1 2382 Pt 8-86. Information Processing Systems—Vocabulary—Part 08: Control, Integrity and Security First Edition. 22 pp.
JTC1 2382 Pt 9-84. Data Processing—Vocabulary—Part 09: Data Communication First Edition. 29 pp.
JTC1 DIS2382 Pt 9-93. Data Processing—Vocabulary—Part 09: Data Communication ***CD-ROM ONLY***. 51 pp.
JTC1 2382 Sec X-79. Data Processing—Vocabulary—Section 10: Operating Techniques and Facilities First Edition. 23 pp.
JTC1 2382 Pt 11-87. Information Processing Systems—Vocabulary—Part 11: Processing Units Second Edition. 23 pp.
JTC1 2382 Sec XIV-78. Data Processing—Vocabulary—Section 14: Reliability, Maintenance and Availability First Edition. 17 pp.
JTC1 2382 Sec XVI-78. Data Processing—Vocabulary—Section 16: Information Theory First Edition. 22 pp.
JTC1 2382 Pt 18-87. Information Processing Systems—Vocabulary—Part 18: Distributed Data Processing First Edition. 18 pp.
JTC1 2382-90. Information Technology—Vocabulary—Part 20: System Development First Edition. 23 pp.
JTC1 TR12382-92. Permuted Index of the Vocabulary of Information Technology Second Edition. 249 pp.
OSI ISO/DIS 2382-1-90. Information Technology—Vocabulary—Part 1: Fundamental Terms. 66 pp.
OSI ISO 2382-1-84. Data Processing—Vocabulary—Part 1: Fundamental Terms. 20 pp.
OSI ISO 2382-4-87. Information Processing Systems—Vocabulary—Part 4: Organization of Data. 23 pp.
OSI ISO 2382-5-89. Information Processing Systems—Vocabulary—Part 5: Representation of Data. 26 pp.
OSI ISO 2382-6-87. Information Processing Systems—Vocabulary—Part 6: Preparation and Handling of Data. 20 pp.
OSI ISO 2382-8-86. Information Processing Systems—Vocabulary—Part 8: Control, Integrity and Security. 10 pp.
OSI ISO 2382-9-84. Data Processing—Vocabulary—Part 09: Data Communication. 29 pp.
OSI ISO 2382-10-79. Data Processing—Vocabulary—Section 10: Operating Techniques and Facilities. 23 pp.
OSI ISO 2382-16-78. Data Processing—Vocabulary—Part 16: Information Theory. 22 pp.
OSI ISO 2382-18-87. Information Processing Systems—Vocabulary—Part 18: Distributed Data Processing. 11 pp.
OSI ISO/IEC 2382-20-90. Information Technology Vocabulary—Part 20: System Development. 23 pp.
OSI ISO/IEC TR 12382-89. Permuted Index of the Vocabulary of Information Processing. 251 pp.
SAA AS 1189. Data Processing (Complete Set in Binder). 245 pp.
SAA AS 1189.0-91. Data Processing—Vocabulary—Part 0: Consolidated Index (ISO/IEC/TR 12382:1989) (in Professional Package 26A).
SAA AS 1189.1-91. Data Processing—Vocabulary—Part 1: Fundamental Terms (ISO 2382-1:1984) (In Professional Package 26). 7 pp.
SAA AS 1189.4-91. Data Processing—Vocabulary—Part 4: Organization of Data (ISO 2382-4:1987) (In Professional Package 26). 8 pp.
SAA AS 1189.5-91. Data Processing—Vocabulary—Part 5: Representation of Data (ISO 2382-5:1989) (In Professional Package 26). 9 pp.
SAA AS 1189.6-91. Data Processing—Vocabulary—Part 6: Preparation and Handling of Data (ISO 2383-6:1987) (In Professional Package 26). 7 pp.
SAA AS 1189.8-91. Data Processing—Vocabulary—Part 8: Control, Integrity and Security (ISO 2382-8:1986) (In Professional Package 26). 9 pp.
SAA AS 1189.10-82. Data Processing—Vocabulary—Part 10: Operating Techniques and Facilities. 12 pp.
SAA AS 1189.11-91. Data Processing—Vocabulary—Part 11: Processing Units (ISO 2382-11:1987) (In Professional Package 26). 10 pp.
SAA AS 1189.14-82. Data Processing—Vocabulary—Part 14: Reliability, Maintenance and Availability. 9 pp.
SAA AS 1189.18-91. Data Processing—Vocabulary—Part 18: Distributed Data Processing (ISO 2382-18:1987) (In Professional Package 26). 7 pp.
SAA AS 1189.20-91. Data Processing—Vocabulary—. 7 pp.

—**Graphics—Glossaries**

SAA AS 1189.13-87. Data Processing—Vocabulary—Part 13: Computer Graphics. 11 pp.

—**International Private Leased Circuits**

CCITT RECMN D.1-91. General Principles for the Lease of International (Continental and Intercontinental) Private Telecommunication Circuits and Networks (Study Group III) 11 pp. 11 pp.
CCITT RECMN D.1-89. General Principles for the Lease of International (Continental and Intercontinental) Private Telecommunication Circuits—General Tariff Principles—Charging and Accounting in International Telecommunications. 8 pp.

—**International Trade**

BSI BS EN 27372-92. 1992 Trade Data Interchange. Trade Data Elements Directory (ISO 7372: 1986). 8 pp.
ISO 7372-86. Trade Data Interchange—Trade Data Elements Directory First Edition. 3 pp.
OSI ISO 7372-86. Trade Data Interchange—Trade Data Elements Directory. 5 pp.

—**Logic Operations—Glossaries**

JIS X 0002-87. Glossary of Terms Used in Information Processing (Arithmetic and Logic Operations).
OSI ISO 2382-2-76. Data Processing—Vocabulary—Section 02: Arithmetic and Logic Operations. 37 pp.
SAA AS 1189.2-82. Data Processing—Vocabulary—Part 2: Arithmetic and Logic Operations. 21 pp.

—**Maps**

NATO STANAG 3721 ED 3 AMD 1-88. Automatic Data Processing (ADP) Master File for Land Maps and Aeronautical Charts. 23 pp.

—**Stock Control**

BSI BS 5729: Part 4-81. (WITHDRAWN) 1981 Guide to Stock Control Part 4: Data Processing. 20 pp.

—**Videotex Communications**

CCITT RECMN F.300-89. Videotex Service—Telematic, Data Transmission and Teleconference Services—Operations and Quality of Service (Study Group I) 25 pp. 25 pp.

Data Processing Equipment

Scope Note: For additional listings, use a more specific term *Use For:* Data Processing Systems; Information Processing Equipment *See Also:* Central Processing Units; Character Imaging Devices; Computer Tapes; Computers; Consoles; Continuous Forms; Data Acquisition Systems; Data Base Management Systems; Data Buses; Data Links; Data Processing; Data Terminal Equipment; Disk Packs; Image Processing Systems; Information Systems; Keyboards; Magnetic Cards; Magnetic Cores; Office Machines; Open Systems Interconnection; Peripheral Equipment; Plotters; Printers; Word Processors

CENELEC HD 40 001-85. Section 2: Requirements for Information Technology Equipment. 10 pp.
ECMA MEMENTO-92. Ecma Memento 1992. 96 pp.
ECMA MEMENTO-93. Ecma Memento 1993. 116 pp.

—**Cable Television Equipment**

CSA CAN/CSA-C22. 2 NO 220-M91. Information Processing and Business Equipment; (Gen Instr 1). 119 pp.

—**Cables (Electric)**

DIN VDE 0813-88. Distribution Wires for Telecommunication and Data Processing Systems (Nov). 25 pp.
DIN VDE 0815-85. Wiring Cables for Telecommunication and Information Processing Systems (Sept). 51 pp.
DIN VDE 0815 A1-88. Wiring Cables for Telecommunication and Data Processing Systems: Amendment 1 (May). 6 pp.

Data Processing Equipment (Cont.)
—Cables (Electric) (Cont.)
DIN VDE 0816 Pt 1-88. Outdoor Cables for Telecommunications and Information Processing Systems; Cables with Insulation and Sheath of PE of Unit Construction (Feb) (Together with DIN VDE 0816 Part 2/02.88 and DIN VDE 0816 Part 3/02.88 Replaces DIN 57816/VDE 0816/02.79). 39 pp.

DIN VDE 0816 Pt 2-88. Outdoor Cables for Telecommunication and Information Processing Systems; Signalling and Measuring Cables, Mine Cables (Feb) (Together with DIN VDE 0816 Part 1/02.88 and DIN VDE 0816 Part 3/02.88, Replaces DIN 57816/VDE 0816/02.79). 37 pp.

DIN VDE 0816 Pt 3-88. Outdoor Cables for Telecommunications and Information Processing Systems; Cables with Paper Insulation (Feb) (Together with DIN VDE 0816 Part 1/02.88 and DIN VDE 0816 Part 2/02.88 Replaces DIN 57816/02.79). 58 pp.

DIN VDE 0817-84. Cables with Stranded Conductors for Increased Mechanical Stress for Telecommunications Systems and Data Processing Systems (Apr). 22 pp.

DIN VDE 0891 Pt 1-90. Use of Cables and Insulated Cords for Telecommunications and Information Processing Systems; General Provisions (May). 27 pp.

DIN VDE 0891 Pt 6-90. Use of Cables and Insulated Cords for Telecommunications and Information Processing Installations; Special Requirements for Outside Cables According to DIN VDE 0816 Part 1 to Part 3 (May). 31 pp.

—Cables (Electric)—Stranded Conductors
DIN VDE 0817-84. Cables with Stranded Conductors for Increased Mechanical Stress for Telecommunications Systems and Data Processing Systems (Apr). 22 pp.

—Cables (Electric)—Wiring
DIN VDE 0891 Pt 5-85. Use of Cables and Insulated Cords for Telecommunications and Information Processing Installations. Special Guidance on Wiring Cables and Cords According to DIN/VDE 0815 (Nov). 11 pp.

—Capacitors—Electrical Insulation
CSA CAN/CSA-C22. 2 NO 220-M91. Information Processing and Business Equipment; (Gen Instr 1). 119 pp.

—Certification
EURO 1987 Dec. IT Certification in Europe. 213 pp.

—Circuit Switched Networks—Interfaces
ECMA ECMA 103-87. Physical Layer at the Basic Access Interface Between Data Processing Equipment and Private Circuit Switching Networks. 61 pp.

ECMA ECMA 104-85. Physical Layer at the Primary Rate Access Interface Between Data Processing Equipment and Private Circuit Switching Networks. 20 pp.

ECMA ECMA 105-90. Data Link Layer Protocol for the D-Channel of the Interfaces at the Reference Point Between Terminal Equipment and Private Telecommunication Networks. 68 pp.

ECMA ECMA 106-91. Layer 3 Protocol for Signalling over the D-Channel of Interfaces at the S Reference Point Between Terminal Equipment and Private Telecommunication Networks for the Control of Circuit-Switched Calls. 39 pp.

ECMA ECMA-TR 34-86. Maintenance at the Interface Between Data Processing Equipment and Private Switching Network. 61 pp.

ECMA ECMA-TR 45-87. Information Interchange for the Remote Maintenance at the Interface Between Data Processing Equipment and Private Switching Networks.. 95 pp.

—Classification
BSI BS 1000: (68/681.3)-82. 1982 Universal Decimal Classification (UDC). English Full Edition (68/681.3): Finished Articles in General. Precision Mechanisms. Horology. Instrumentation. Data Processing Equipment. 31 pp.

SNZ NZS/BS 1000 (68/681.3)-82. Universal Decimal Classification Finished Articles in General. Precision Mechanisms. Horology. Instrumentation. Data Processing Equipment. 32 pp.

—Communication Cables
DIN VDE 0891 Pt 1-90. Use of Cables and Insulated Cords for Telecommunications and Information Processing Systems; General Provisions (May). 27 pp.

—Connectors
DIN VDE 0627-86. Connectors and Plug-And-Socket Devices for Rated Voltages up to 1000 V a.c. and 1200 V d.c. and Rated Currents up to 500 A for Each Pole (June). 36 pp.

—Cords (Electric)
DIN VDE 0814-81. Cords for Telecommunication Systems and Information Processing Systems (Oct). 28 pp.

DIN VDE 0891 Pt 1-90. Use of Cables and Insulated Cords for Telecommunications and Information Processing Systems; General Provisions (May). 27 pp.

DIN VDE 0891 Pt 4-81. Use of Cables and Insulated Cords for Telecommunications and Information Processing Installations; Special Guidance on Cords According to DIN 57 814/VDE 0814 (VDE Guide) (Dec). 8 pp.

—Double Insulated
CSA CAN/CSA-C22. 2 NO 220-M91. Information Processing and Business Equipment; (Gen Instr 1). 119 pp.

—Electric Wire
DIN VDE 0812-88. Equipment Wires with Solid and Stranded Conductors for Telecommunication Systems and Data Processing Systems (Nov) (Replaces VDE 0812a/12.77). 25 pp.

DIN VDE 0891 Pt 9-90. Use of Cables and Insulated Cords for Telecommunications and Information Processing Systems; Special Guidelines for Equipment Wires with Solid and Stranded Conductors with Extended Temperature Range According to DIN VDE 0881 (May). 15 pp.

—Electrical Grounding
IEC 364 Pt 7-707-84. Electrical Installations of Buildings Part 7: Requirements for Special Installations or Locations Section 707—Earthing Requirements for the Installation of Data Processing Equipment First Edition. 23 pp.

—Electrical Insulation
CSA CAN/CSA-C22. 2 NO 220-M91. Information Processing and Business Equipment; (Gen Instr 1). 119 pp.

—Electromagnetic Interference—Immunity
CENELEC PREN 55 101-4-90. Immunity Requirements for Information Technology Equipment—Part 4: Conducted Immunity (See PREN 55024-4). 14 pp.

—Electromagnetic Radiation
CSA C108.8-M1983. Electromagnetic Emissions from Data Processing Equipment and Electronic Office Machines (R 1989); (Amd 1-5 September 1983). 20 pp.

—Electrostatic Discharge—Immunity
CENELEC PREN 55024-2-89. Immunity Requirements for Information Technology Equipment Part 2: Electrostatic Discharge Requirements (See EN 55101-2). 44 pp.

CENELEC PREN 55024-3-89. Immunity Characteristics of Information Technology Equipment Part 3: Immunity to Radiated Radio Frequency Disturbances (See EN 55101-3). 20 pp.

CENELEC EN 55101-2-89. (Draft) Immunity Requirements for Information Technology Equipment Part 2: Electrostatic Discharge Requirements (See PREN 55024-2). 44 pp.

CENELEC EN 55101-3-89. (Draft) Immunity Characteristics of Information Technology Equipment. Part 3: Immunity to Radiated Radio Frequency Disturbances (See PREN 55024-3). 20 pp.

—Fire Protection
BSI BS 6266-92. 1992 Fire Protection for Electronic Data Processing Installations. 35 pp.

BSI BS 6266-82. 1982 Fire Protection for Electronic Data Processing Installations (Formerly CP 95). 16 pp.

—Glossaries
BSI BS ISO 2382/3-87. 1987 Information Technology—Vocabulary 2382/3: Equipment Technology. 21 pp.

CNS C5153-90. Data Processing Vocabulary (Part 11: Control, Input-Output and Arithmetic Equipment) (Nov)(10240).

CNS C5154-90. Data Processing Vocabulary (Part 12: Data Media, Storage and Related Equipment) (Nov)(10241).

CNS C5235-90. Information Processing Systems—Vocabulary (Part 8: Control, Integrity and Security) (May)(12717).

ISO 2382 Pt 3-87. Information Processing Systems—Vocabulary—Part 03: Equipment Technology Second Edition. 18 pp.

JIS X 0003-89. Glossary of Terms Used in Information Processing (Equipment Technology).

JIS X 0010-87. Glossary of Terms Used in Information Processing (Operating Techniques and Facilities).

JTC1 2382 Pt 3-87. Information Processing Systems—Vocabulary—Part 03: Equipment Technology Second Edition. 18 pp.

OSI ISO 2382-3-87. Information Processing Systems—Vocabulary—Part 03: Equipment Technology. 18 pp.

SAA AS 1189.3-91. Data Processing—Vocabulary—Part 3: Equipment Technology (ISO 2382-3:1987) (In Professional Package 26). 7 pp.

SAA AS 1189.10-82. Data Processing—Vocabulary—Part 10: Operating Techniques and Facilities. 12 pp.

—Hookup Wire
DIN VDE 0881-86. Jumper Wires and Stranded Hook-Up Wires with Extended Temperature Range for Telecommunications Systems and Data Processing Systems (Mar). 34 pp.

DIN VDE 0891 Pt 2-90. Use of Cables and Insulated Cords for Telecommunications and Information Processing Systems; Special Provisions for Hook-Up Wires with Solid and Stranded Conductors in Accordance with DIN VDE 0812 (May). 6 pp.

—Information Interchange—Radio Links
NATO STANAG 4202 ED 2 AMD 0-88. Transmission Envelope Characteristics for High Reliability Data Exchange Between Land Tactical Data Processing Equipment over Single Channel Radio Links. 30 pp.

—Interfaces
BSI DD 153-90. 1990 S5/8 Serial Interface for the Interconnection of Data Processing Equipment. 17 pp.

—Interfaces—Naval
NATO STANAG 4146 ED 1 AMD 2-76. Interim Standard Specifications for Input /Output Interfaces in NATO Naval Data Handling Equipments (Withdrawn). 157 pp.

—Interfaces—Private Automatic Branch Exchanges
ECMA ECMA-TR 24-85. Interfaces Between Data Processing Equipment and Private Automatic Bench Exchange. Circuit Switching Application. 63 pp.

—Interfaces—Private Networks
ECMA ECMA 103-87. Physical Layer at the Basic Access Interface Between Data Processing Equipment and Private Circuit Switching Networks. 61 pp.

ECMA ECMA 104-85. Physical Layer at the Primary Rate Access Interface Between Data Processing Equipment and Private Circuit Switching Networks. 20 pp.

ECMA ECMA-TR 34-86. Maintenance at the Interface Between Data Processing Equipment and Private Switching Network. 61 pp.

ECMA ECMA-TR 45-87. Information Interchange for the Remote Maintenance at the Interface Between Data Processing Equipment and Private Switching Networks. 95 pp.

—Interfaces—Protocols—Private Networks
ECMA ECMA 106-91. Layer 3 Protocol for Signalling over the D-Channel of Interfaces at the S Reference Point Between Terminal Equipment and Private Telecommunication Networks for the Control of Circuit-Switched Calls. 39 pp.

—Jumpers (Electric)
DIN VDE 0881-86. Jumper Wires and Stranded Hook-Up Wires with Extended Temperature Range for Telecommunications Systems and Data Processing Systems (Mar). 34 pp.

—Keyboards
BSI BS 4822-80. 1980 Keyboard Arrangements of the Graphic Characters of the United Kingdom 7-Bit Data Code, for Data Processing. 8 pp.

CNS C5150-83. Keyboard for International Information Processing Interchange Using 7-Bit Coded Character Set (Alphanumeric Area) (Apr)(10153).

CSA CAN/CSA-Z243.200-92. Canadian Keyboard Standard for the English and French Languages; (Gen Instr 1). 43 pp.

ISO 2530-75. Keyboard for International Information Processing Interchange Using the ISO 7-Bit Coded Character Set—Alphanumeric Area First Edition. 9 pp.

JIS X 6002-80. Keyboard Layout for Information Processing Using the JIS 7 Bit Coded Character Set.

JTC1 2530-75. Keyboard for International Information Processing Interchange Using the ISO 7-Bit Coded Character Set—Alphanumeric Area First Edition. 9 pp.

OSI ISO 2530-75. Keyboard for International Information Processing Interchange Using the ISO 7-Bit Coded Character Set-Alphanumeric Area. 8 pp.

INTERNATIONAL AND NON-U.S. NATIONAL STANDARDS
SUBJECT INDEX
Data

Data Processing Equipment (Cont.)
—Local Area Networks
CSA CAN/CSA-C22. 2 NO 220-M91. Information Processing and Business Equipment; (Gen Instr 1). 119 pp.

—Measurement—Ozone
ECMA ECMA-TR 56-91. Information Technology Equipment Recommended Measuring Method for Ozone Emission. 17 pp.

—Noise
BSI BS 7135: Part 3-89. 1989 Noise Emitted by Computer and Business Equipment Part 3: Method for Determining and Verifying Declared Noise Emission Values. 11 pp.
ISO 9296-88. Acoustics—Declared Noise Emission Values of Computer and Business Equipment First Edition. 11 pp.
JTC1 9296-88. Acoustics—Declared Noise Emission Values of Computer and Business Equipment First Edition. 11 pp.

—Power Supplies
CSA 1402C Bull. Electrical Bulletin 1402C April 11, 1989 to C22.2 No 220. 29 pp.

—Radio Frequency Interference
BSI BS 6527-88. 1988 Limits and Methods of Measurement of Radio Interference Characteristics of Information Technology Equipment. 22 pp.
CEN PREN 55 101-4-90. Immunity Requirements for Information Technology Equipment—Conducted Immunity. 13 pp.
CENELEC EN 55 022-88. Limits and Methods of Measurement of Radio Interference Characteristics of Information Technology Equipment. 11 pp.
DIN VDE 0871 Pt 2A2 (D)-88. Radio Interference Suppression of Radio-Frequency-Equipment; Limits and Methods of Measurement of Radio Interference Characteristics of Information Technology Equipment: Amendment 2 German Version prAM 1 to EN 55022: 1987 (June). 11 pp.
DIN VDE 0878 Pt 30 (D)-89. Electromagnetic Compatibility of Information Technology and Telecommunications Equipment; Limits and Methods of Measurement of Radio Interference Characteristics of Information Technology Equipment (Nov) (Supersedes Draft DIN 0871 Part 20/03.87). 25 pp.
IEC CISPR 22-85. Limits and Methods of Measurement of Radio Interference Characteristics of Information Technology Equipment First Edition. 29 pp.
SAA AS/NZS 3548-92. Limits and Methods of Measurement of Radio Interference Characteristics of Information Technology Equipment (IEC/CISPR 22:1985) (Supersedes AS 3548—1988). 16 pp.
SNZ NZS/AS 3548-92. Limits and Methods of Measurement of Radio Interference Characteristics of Information Technology Equipment (This is a Joint Standard with SAA AS 3548). 16 pp.

—Radio Frequency Interference—Suppression
DIN VDE 0871 Pt 100 (D)-84. Radio-Interference Suppression of Data Processing Machines and Electronic Office Machines (Apr). 38 pp.

—Rectangular Connectors
IEC 807 Pt 6-88. Rectangular Connectors for Frequencies Below 3 MHz Part 6: Detail Specification for a Range of Rectangular Connectors with a Size 20 (7.5 A) Round Contacts Having Polarized Guides—Fixed Solder Contact Types First Edition. 63 pp.

—Ribbon Cables
DIN VDE 0811-85. Ribbon Cables with Round Conductors with a Pitch of 1.27 mm for Telecommunications Systems and Data Processing Systems (Nov). 23 pp.

—Safety
BSI BS 6204-82. (WITHDRAWN) 1982 Amd 1 Safety of Data Processing Equipment (AMD 6062) June 30, 1989 (Superseded by BS 7002: 1989). 68 pp.
BSI BS 7002-92. 1992 Safety of Information Technology Equipment, Including Electrical Business Equipment (S). 195 pp.
BSI BS 7002-02. (WITHDRAWN) 1989 Amd 2 Safety of Information Technology Equipment Including Electrical Business Equipment (AMD 6974) January 31, 1992 (Superseded by BS EN 60950: 1992). 180 pp.
BSI BS EN 60950-92. 1992 Safety of Information Technology Equipment, Including Electrical Business Equipment (Supersedes BS 7002: 1989). 195 pp.
BSI BS EN 60950-01. 1992 Amd 1 Safety of Information Technology Equipment, Including Electrical Business Equipment (S). 212 pp.
CENELEC PREN 50116-93. Information Technology Equipment—Routine Electrical Safety Testing in Production (Possible New British Standard). 4 pp.

Data Processing Equipment (Cont.)
—Safety (Cont.)
CENELEC EN 60 950-88. Safety of Information Technology Equipment Including Electrical Business Equipment. 141 pp.
CENELEC EN 60950/A1-90. AMD 1 Safety of Information Technology Equipment Including Electrical Business Equipment. 5 pp.
CENELEC EN 60950/A2-91. AMD 2 Safety of Information Technology Equipment Including Electrical Business Equipment. 5 pp.
CENELEC EN 60950-92. Safety of Information Technology Equipment, Including Electrical Business Equipment. 15 pp.
CENELEC EN 60950-92. Safety of Information Technology Equipment, Including Electrical Business Equipment; (IEC 950: 1991, Modified). 184 pp.
CENELEC EN 60950/PRAA-92. AMD A Safety of Information Technology Equipment Including Electrical Business Equipment. 4 pp.
CENELEC EN 60950/A1-93. AMD 1 Safety of Information Technology Equipment, Including Electrical Business Equipment (IEC 950: 1991/A1: 1992). 186 pp.
CSA CAN/CSA-C22. 2 NO 950-M89. Safety of Information Technology Equipment, Including Electrical Business Equipment; (Gen Instr 1) (Supersedes C22.2 No 220). 183 pp.
DIN VDE 0701 Pt 240-86. Repair, Modification and Inspection of Electrical Appliances; Safety Requirements for Data Processing Equipment and Office Machines (Apr). 10 pp.
DIN VDE 0701 Pt 240 (D)-90. Repair, Modification & Inspection of Electrical Appliances; Information Technology Equipment (Feb). 12 pp.
DIN VDE 0805 (D)-84. Deviations from IEC 435/2nd Ed (1983); DIN IEC 435 A1/VDE 0805 A1 Draft; Safety of Data Processing Equipment (Nov). 36 pp.
DIN VDE 0805 A1 (D)-84. Safety of Data Processing Equipment: Amendment 1 Draft (Nov). 5 pp.
ECMA ECMA 83-85. Safety Requirements for DTE-DCE Interface. 5 pp.
ECMA ECMA 129-88. Safety of Information Technology Equipment (ITE). 204 pp.
ECMA ECMA 166-92. Information Technology Equipment—Routine Electrical Safety Testing in Production. 7 pp.
ECMA ECMA-TR 28-85. Safety Verification (SAVE) Report—ECMA 57/IEC 435. 52 pp.
ECMA ECMA TR 39-92. Compliance Verification Report (Cover) ECMA 129—IEC 950—EN60950. 77 pp.
ECMA ECMA-TR 39-88. Compliance Verification (Cover) Report IEC 950. 63 pp.
IEC 950-91. Safety of Information Technology Equipment, Including Electrical Business Equipment Second Edition; (Amendment 1-1992) (Amendment 2-1993) (CENELEC EN 60950:1992) (CENELEC EN 60950/A1: 1993). 372 pp.
SNZ NZS/AS 3260-93. Approval and Test Specification—Safety of Information Technology Equipment Including Electrical Business Equipment (This is a Joint Standard with SAA AS 3260). 175 pp.
SNZ NZS 6661-89. Safety of Information Technology Equipment, Including Electrical Business Equipment Amend: A. 245 pp.

—Serial Interfaces—Naval
NATO STANAG 4153 ED 1 AMD 3-83. Standard Specification for an Asynchronous Serial Data Interface for Connection to Data Networks in NATO Naval Systems. 44 pp.
NATO STANAG 4153 ED 1 AMD 4-83. Standard Specification for an Asynchronous Serial Data Interface for Point to Point Connections and for Connection to Data Networks in NATO Naval Systems. 44 pp.
NATO STANAG 4153 ED 2 AMD 0-00. Standard for NATO Naval Intra-Ship Asynchronous Serial Data Interface for Point-to-Point and Data Network Connections. 41 pp.
NATO STANAG 4156 ED 1 AMD 1-85. Standard Specifications for a Serial Data Interface for Synchronous Connections to a Data Network (Withdrawn). 40 pp.

—Ships—Design
JIS F 9001-87. Basic Design for Integrated Information Systems on Shi

—Static Electricity
ECMA ECMA-TR 23-84. Electrostatic Discharge Susceptibility. 12 pp.
ECMA ECMA-TR 40-87. Electrostatic Discharge Immunity Testing of Information Technology Equipment. 21 pp.

—Surges—Immunity
CENELEC EN 55024-4-90. Immunity Requirements for Information Technology Equipment—Part 4: Conducted Immunity (See PREN 55101-4). 14 pp.

Data Processing Equipment (Cont.)
—Surges—Immunity (Cont.)
CENELEC PREN 55024-4-93. Immunity of Information Technology Equipment (ITE) Part 4: Immunity to Electrical Fast Transients/Bursts. 6 pp.

—Switchboards—Communication Cables
DIN VDE 0891 Pt 3-90. Use of Cables and Insulated Cords for Telecommunications and Information Processing Installations; Particular Requirements for Switchboard Cables According to DIN VDE 0813 (May). 5 pp.

—Telecommunication Equipment
CSA CAN/CSA-C22. 2 NO 220-M91. Information Processing and Business Equipment; (Gen Instr 1). 119 pp.
DIN VDE 0815-85. Wiring Cables for Telecommunication and Information Processing Systems (Sept). 51 pp.
DIN VDE 0815 A1-88. Wiring Cables for Telecommunication and Data Processing Systems: Amendment 1 (May). 6 pp.

—Twisted Pair Cables—Network Interconnection
BSI BS 7248-90. 1990 Multipoint Interconnection of Data Communications Equipment by Twisted Pair Cable. 20 pp.
IEC DIS 8482-92. Information Technology—Telecommunications and Information Exchange Between Systems—Twisted Pair Multipoint Interconnections (Revision of First Edition (ISO 8482:1987)) ***CD-ROM ONLY***. 18 pp.
ISO 8482-87. Information Processing Systems—Data Communication—Twisted Pair Multipoint Interconnections First Edition. 19 pp.
ISO DIS 8482-92. Information Technology—Telecommunications and Information Exchange Between Systems—Twisted Pair Multipoint Interconnections (Revision of First Edition (ISO 8482:1987)) ***CD-ROM ONLY***. 18 pp.
JTC1 8482-87. Information Processing Systems—Data Communication—Twisted Pair Multipoint Interconnections First Edition. 19 pp.
JTC1 DIS8482-92. Information Technology—Telecommunications and Information Exchange Between Systems—Twisted Pair Multipoint Interconnections (Revision of First Edition (ISO 8482:1987)) ***CD-ROM ONLY***. 18 pp.
OSI ISO 8482-87. Information Processing Systems—Data Communication—Twisted Pair Multipoint InterConnections. 19 pp.
OSI ISO DIS 8482-86. Data Communications—Twisted Pair Multipoint Interconnections. 23 pp.
SNZ NZS/ISO 8482-87. Information Processing Systems—Data Communication—Twisted Pair Multipoint Interconnections. 16 pp.

Data Processing Services
Use: Data Processing

Data Processing Systems
Use: Data Processing Equipment

Data Relay Satellites
See Also: Communications Satellites; Satellites
CCIR Report 848-2-90. Characteristics of Data Relay Satellite Systems—Section 2D—Data Relay Satellites. 15 pp.
CCIR Report 982-86. Data Relay Satellites for the Earth Exploration Satellite Service—Section 2D—Data Relay Satellites. 9 pp.

—Data Links
CCIR Report 982-86. Data Relay Satellites for the Earth Exploration Satellite Service—Section 2D—Data Relay Satellites. 9 pp.

—Frequency Band Sharing—Longitude Separation Angle
CCIR Report 983-86. Minimum Longitude Separation Angle Necessary to Share Frequencies Between Two Data Relay Satellites—Section 2D—Data Relay Satellites. 6 pp.

—Frequency Bands
CCIR Report 848-2-90. Characteristics of Data Relay Satellite Systems—Section 2D—Data Relay Satellites. 15 pp.

—Geostationary—Space Communications—Frequency Band Sharing
CCIR Report 847-1-86. Data Relay Satellites Sharing with Other Services in Bands 9 and 10—Section 2D—Data Relay Satellites. 7 pp.

—Near Earth
CCIR Report 848-2-90. Characteristics of Data Relay Satellite Systems—Section 2D—Data Relay Satellites. 15 pp.

INDUSTRY STANDARDS

INTERNATIONAL AND NON-U.S. NATIONAL STANDARDS SUBJECT INDEX

Data

Data Relay Satellites (Cont.)

—Radio Communications—Frequency Band Sharing

CCIR QUESTION 118/7-90. Data Relay Satellite Systems and Factors Which Affect Frequency Sharing with Other Services—Questions Concerning Study Group 7—Science Services. 2 pp.

—Radio Links—Earth Stations—Spacecraft

CCIR QUESTION 117/7-90. Radio Links Between Earth Stations and Spacecraft by Means of Data Relay Satellites—Questions Concerning Study Group 7—Science Services. 1 p.

—Space Research Systems—Frequency Band Sharing

CCIR Report 846-82. Data Relay Satellites Sharing with Other Space Research Systems Near 2 GHz—Section 2D—Data Relay Satellites. 5 pp.

Data Safes

Use For: Computer Data Safes; Data Cabinet Safes
See Also: Safes

—Fire Resistant

CEN PREN 1047-1-93. Secure Storage Units—Classification and Methods of Test for Resistance to Fire—Part 1: Data Cabinets. 16 pp.

Data Selector/Multiplexers, Analog

Use: Analog Multiplexers

Data Selector/Multiplexers, Digital

Use: Digital Multiplexers

Data Selectors

Use: Matrix Switches

Data Sheets

See Also: Documents; Pilatus Britten-Norman Islander BN 2T-2R Aircraft

—Certification—Aircraft

CAA NOTICE #15 ISSUE 3. U.K. Certification of Foreign Aircraft of MIWA Not Exceeding 5700 kg (Airworthiness Notices). 4 pp.
CAA NOTICE #43 ISSUE 2. Aircraft Type Certificates Data Sheets (Airworthiness Notices). 2 pp.
CAA. 53.
CAA Chapter A3-3 01.91. Type Certificates for Hovercraft. 4 pp.
CAA Chapter A3-3 App 01.91. Type Certificates for Hovercraft. 3 pp.
CAA Chapter A3-4 01.74. Type Certificates for Hovercraft Items. 5 pp.
CAA Chapter A3-4 App 01.91. Type Certificates for Hovercraft Items. 1 p.
CAA. Contents Issue 104 (Type Certificate Data Sheets). 3 pp.
CAA. Foreword (Type Certificate Data Sheets). 3 pp.
CAA BA1 ISSUE 7. Beagle 121 Series 1, Series 2 and Series 3. 4 pp.
CAA BA2 ISSUE 27. Glos-Airtourer T3, 115, 150, Super 150. 4 pp.
CAA BA3 ISSUE 10. BAC One Eleven 475 and 500 Series, 475EZ, 476FM, 479FU, 481FW, 485GD, 487GK, 488GH, 492GM, 500EN, 501EX, 509EW, 501EW, 510ED, 515FB, 516FP, 517FE, 518FG, 520FN, 521FH, 523FJ, 524FF, 525FT, 525/1FT, 527FK, 528FL, 529FR, 530FX, 531FS, 537GF, 539GL, 560RB. 5 pp.
CAA BA4 ISSUE 4. HP 137 Jetstream, Mark 1 and SAL Jetstream Series 200. 5 pp.
CAA BA5 ISSUE 5. Saunders ST27 and ST27-100. 4 pp.
CAA BA6 ISSUE 15. Britten-Norman Trislander BN2A Mark III, BN2A Mark III-1 and BN2A Mark III-2. 4 pp.
CAA BA7 ISSUE 8. British Aerospace Scottish Division Bulldog Series 100, Models 101, 102, 103, 104 Series 120, Models 122, 123, 125, 126, 127, 128, 129, 1210. 5 pp.
CAA BA8 ISSUE 10. Pilatus Britten-Norman Ltd Islander BN2A-8,-9,-20,-21,-26,-27, Islander BN2B-20,-21,-26,-27, Islander BN2T,-2,-2R. 6 pp.
CAA BA9 ISSUE 11. Corporate Jets Ltd, HS 125 Series 600B Model F600B; HS Series 400B Models 401B, 403B, F400B, HS 125 Series 700B, BAe 125 Series 800B, 1000B. 8 pp.
CAA BA10 ISSUE 10. BAC/SNIA Concorde Type 1 Variant 102. 4 pp.
CAA BA11 ISSUE 1. Short Brothers SD3 Variant 100, 200, 300, 400, 500. 7 pp.
CAA BA12 ISSUE 1. NDN Aircraft NDN-1. 3 pp.
CAA BA13 ISSUE 6. Short Brothers & Harland SC-5. 3 pp.
CAA BA15 ISSUE 10. British Aerospace PLC Jetstream Series 3100 Jetstream Series 3200. 11 pp.
CAA BA16 ISSUE 11. British Aerospace BAe 146 Series 100, Series 200, Series 300. 5 pp.
CAA BA17 ISSUE 1. Slingsby Aviation Ltd T67. 4 pp.
CAA BA18 ISSUE 1. NDN Aircraft NDN-IT. 3 pp.
CAA BA19 ISSUE 2. Trago Mills Ltd SAH1. 3 pp.
CAA BA21 ISSUE 3. Optica Industries Ltd. 3 pp.
CAA BA22 ISSUE 2. ARV Aviation Ltd, Super 2. 2 pp.
CAA BA23 ISSUE 2. British Aerospace ATP. 4 pp.
CAA BA24. NAC-6 Fieldmaster. 3 pp.
CAA BAS1 ISSUE 1. Airship Industries. 2 pp.
CAA BAS2 ISSUE 1. Airship Industries. 3 pp.
CAA BAS3 ISSUE 2. Cameron Balloons Ltd Skystar. 2 pp.
CAA BAS4 ISSUE 1. Thunder and Colt, GA-42. 2 pp.
CAA BAS5. Cameron Balloons Ltd DG-14. 2 pp.
CAA BG1 ISSUE 1. Slingsby T.53B. 3 pp.
CAA BG2 ISSUE 1. Slingsby T.59D. 3 pp.
CAA BG3. Slingsby T.65A Vega. 2 pp.
CAA BR1 ISSUE 4. Westland 30 Series 100, Series 100-60. 3 pp.
CAA FA1 ISSUE 11. Piper PA31, PA31-350, PA31P, PA31T1. 5 pp.
CAA FA2 ISSUE 1. Boeing 737-200 Series, Model 204, 204C, 222, 204 (ADV), 219, 236,2S3, 2S3, 2K9, 2L9, 2M8, 2Q8, 2T5, 2T7, 2U4 (ADV), 219DA, 2T4, Boeing 737-300 Series Model 3T5. 8 pp.
CAA FA3 ISSUE 1. Canadair CL44 Series, Model CL44 D4. 2 pp.
CAA FA4 ISSUE 3. Beech 60. 2 pp.
CAA FA5 ISSUE 1. Boeing 747-100 & 200 Series, Models 136 and 236B. 2 pp.
CAA FA6 ISSUE 2. Beech 100. 3 pp.
CAA FA7 ISSUE 1. Grumman G-1159 Gulfstream II & III. 4 pp.
CAA FA8 ISSUE 11. LET Model L-410AG Turbolet. 3 pp.
CAA FA9 ISSUE 4. Lockheed Model L1011-385-1 Configurations 193A, 193K, 193N, Model L1011-385-1-15 Configurations 193N, 193T, 193U, 293C and Model L1011-385-3 Configuration 193V. 6 pp.
CAA FA10 ISSUE 1. McDonnell Douglas DC-10, Models 10 and 30. 3 pp.
CAA FA11 ISSUE 2. Boeing 727-100 Series, Model 727-46. 2 pp.
CAA FA12 ISSUE 6. Fan Jet Falcon Series E and F. 3 pp.
CAA FA13 ISSUE 2. Cessna Models 500 and 550. 4 pp.
CAA FA15 ISSUE 2. Beech Super King Air Model 200. 5 pp.
CAA FA16. McDonnell Douglas DC-8 Model 8F-54. 3 pp.
CAA FA19 ISSUE 4. McDonnell Douglas DC-9 Series 10 Model DC-9-15. 4 pp.
CAA FA20 ISSUE 2. Embraer EMB-110. 4 pp.
CAA FA21 ISSUE 1. Fokker F28 Mk 4000. 4 pp.
CAA FA22 ISSUE 1. Embraer EMB-121. 3 pp.
CAA FA23 ISSUE 1. Gates Learjet 35A and 36A. 3 pp.
CAA FA24 ISSUE 1. Airbus A300 B4-203. 3 pp.
CAA FA25 ISSUE 1. Fairchild Swearingen Merlin IIIB. 2 pp.
CAA FA26 ISSUE 1. De Havilland Canada DHC-7. 2 pp.
CAA FA27 ISSUE 2. Piper PA42. 3 pp.
CAA FA28 ISSUE 1. Boeing 757 Model 236, 2T7. 3 pp.
CAA FA29 ISSUE 1. Gulfstream. 2 pp.
CAA FA31 ISSUE 1. Airbus Industrie A310-203. 3 pp.
CAA FA32 ISSUE 1. Beech Model 99 Airliner. 2 pp.
CAA FA33 ISSUE 1. Boeing 767-200 Model 204. 3 pp.
CAA FA34 ISSUE 1. Dornier 228. 3 pp.
CAA FA35 ISSUE 2. SAAB-Scania AB.SAAB-Fairchild 340A. 2 pp.
CAA FA36 ISSUE 2. De Havilland Canada, DHC-8, Series—100. 3 pp.
CAA FA37 ISSUE 1. Embraer, EMB-120. 2 pp.
CAA FA38 ISSUE 1. Mitsubishi MU-300. 2 pp.
CAA FA43 ISSUE 1. Fairchild Aircraft, Model SA227-AC, Metro 111. 2 pp.
CAA FA47 ISSUE 1. ATR 42, Model 300. 4 pp.
CAA FA48 ISSUE 1. Canadair Challenger, Model CL-600-2B16, Variant 601-3A. 2 pp.
CAA FA49. Harbin Y12 (11). 2 pp.
CAA FR1 ISSUE 2. Bell Model 212. 2 pp.
CAA FR2 ISSUE 3. Sikorsky S-58T Series. 3 pp.
CAA FR3 ISSUE 4. Messerschmitt-Bolkow-Blohm GmbH BO 105 D, DB, DS, DBS-4. 2 pp.
CAA FR4 ISSUE 1. Aerospatiale,Gazelle SA341G and Gazelle SA341G Series 1. 3 pp.
CAA FR5 ISSUE 3. Aerospatiale SE3160 Alouette III, SA315B Lama, SA316B Alouette III, SA316C Alouette III, SA391B Alouette III. 3 pp.
CAA FR6 ISSUE 2. Aerospatiale Puma SA330G and SA330J. 3 pp.
CAA FR7 ISSUE 3. Bell 206L. 3 pp.
CAA FR8 ISSUE 7. Aerospatiale AS350B Ecureuil. 2 pp.
CAA FR9 ISSUE 6. Aerospatiale SA365C, SA365C1, SA365C2 and SA365N Dauphin. 4 pp.
CAA FR10 ISSUE 4. Sikorsky S-76A, S-76B. 3 pp.
CAA FR11 ISSUE 1. Agusta A109A. 2 pp.
CAA FR12 ISSUE 1. Bell Model 222. 2 pp.
CAA FR13 ISSUE 2. Boeing Vertol 234LR. 2 pp.
CAA FR14 ISSUE 3. Model 412. 2 pp.
CAA FR15 ISSUE 2. Aerospatiale AS 355F Ecureuil II. 2 pp.
CAA FR16 ISSUE 1. Aerospatiale AS 332L Supa Puma. 3 pp.
CAA FR17 ISSUE 1. Bell Model 214ST. 2 pp.
CAA FR19. Messerschmitt-Bolkow-Blohm GmbH BK 117. 2 pp.
CAA FR20 ISSUE 1. Eurocopter France AS332L2 Super Puma Mk.2. 2 pp.

—Certification—Ground Effect Machines

CAA 53.
CAA Chapter A3-3 01.91. Type Certificates for Hovercraft. 4 pp.
CAA Chapter A3-3 App 01.91. Type Certificates for Hovercraft. 3 pp.
CAA Chapter A3-4 01.74. Type Certificates for Hovercraft Items. 5 pp.
CAA Chapter A3-4 App 01.91. Type Certificates for Hovercraft Items. 1 p.

Data Signaling Rates

See Also: Data Signals; Signaling Systems; Units of Measurement

—Interchange Circuits—Data Terminal Equipment

CCITT RECMN V.31-89. Electrical Characteristics for Single-Current Interchange Circuits Controlled by Contact Closure—Data Communication over the Telephone Network (Study Group XVII) 3 pp. 3 pp.
CCITT RECMN V.31 BIS-89. Electrical Characteristics for Single-Current Interchange Circuits Using Optocouplers—Data Communication over the Telephone Network (Study Group XVII) 3 pp. 3 pp.

—Interchange Circuits—Data Terminal Equipment—Optoisolators

CCITT RECMN V.31-89. Electrical Characteristics for Single-Current Interchange Circuits Controlled by Contact Closure—Data Communication over the Telephone Network (Study Group XVII) 3 pp. 3 pp.
CCITT RECMN V.31 BIS-89. Electrical Characteristics for Single-Current Interchange Circuits Using Optocouplers—Data Communication over the Telephone Network (Study Group XVII) 3 pp. 3 pp.

—Interfaces—Data Networks

CCITT RECMN X.50-89. Fundamental Parameters of a Multiplexing Scheme for the International Interface Between Synchronous Data Networks—Data Communication Networks—Transmission, Signalling and Switching, Network Aspects, Maintenance and Administrative Arrangements. 7 pp.
CCITT RECMN X.50 BIS-89. Fundamental Parameters of a 48-kBit/s User Data Signalling Rate Transmission Scheme for the International Interface Between Synchronous Data Networks—Data Communication Networks—Transmission, Signalling and Switching, Network Aspects, Maintenance. 1 p.
CCITT RECMN X.51-89. Fundamental Parameters of a Multiplexing Scheme for the International Interface Between Synchronous Data Networks Using 10-Bit Envelope Structure—Data Communication Networks—Transmission, Signalling and Switching, Network Aspects, Maintenance and. 5 pp.
CCITT RECMN X.51 BIS-89. Fundamental Parameters of a 48-kBit/s User Data Signalling Rate Transmission Scheme for the International Interface Between Synchronous Data Networks Using 10-Bit Envelope Structure—Data Communication Networks—Transmission, Signalling and. 2 pp.

—Leased Circuits—Modems—Public Data Networks

CCITT RECMN V.36-89. Modems for Synchronous Data Transmission Using 60-108 kHz Group Band Circuits—Data Communication over the Telephone Network (Study Group XVII) 10 pp. 10 pp.

—Modems—Data Transmission

CCITT RECMN V.37-89. Synchronous Data Transmission at a Data Signalling Rate Higher Than 72 kBit/s Using 60-108 kHz Group Band Circuits—Data Communication over the Telephone Network (Study Group XVII) 11 pp. 11 pp.

INTERNATIONAL AND NON-U.S. NATIONAL STANDARDS
SUBJECT INDEX
Data

Data Signaling Rates *(Cont.)*

—Telephone Circuits—Data Transmission

CCITT RECMN V.6-89. Standardization of Data Signalling Rates for Synchronous Data Transmission on Leased Telephone-Type Circuits—Data Communication over the Telephone Network (Study Group XVII) 2 pp. 2 pp.

—Telephone Circuits—Modems—Telephone Networks

CCITT RECMN V.32-89. Family of 2-Wire, Duplex Modems Operating at Data Signalling Rates of up to 9600 Bit/s for Use on the General Switched Telephone Network and on Leased Telephone-Type Circuits—Data Communication over the Telephone Network (Study Group XVII) 17 pp. 17 pp.

CCITT RECMN V.32 BIS-91. Duplex Modem Operating at Data Signalling Rates of up to 14 400 Bit/s for Use on the General Switched Telephone Network and on Leased Point-to-Point 2-Wire Telephone-Type Circuits (Study Group XVII) 25 pp. 25 pp.

CCITT RECMN V.33-89. 14 400 Bits per Second Modem Standardized for Use on Point-to-Point 4-Wire Leased Telephone-Type Circuits—Data Communication over the Telephone Network (Study Group XVII) 16 pp. 16 pp.

—Telephone Networks—Data Transmission

CCITT RECMN V.5-89. Standardization of Data Signalling Rates for Synchronous Data Transmission in the General Switched Telephone Network—Data Communication over the Telephone Network (Study Group XVII) 1 pp. 1 p.

—Test Sets

CCITT RECMN V.57-89. Comprehensive Data Test Set for High Data Signalling Rates—Data Communication over the Telephone Network (Study Group XVII) 1 pp. 1 p.

Data Signalling Rates
Use: Data Signaling Rates

Data Signals
See Also: Data Signaling Rates; Signaling Systems

—Broadcasting—Distortion—Vestigial Sideband Transmission

CCIR QUESTION 56/11-90. Distortion of Television and Data-Broadcasting Signals in Emission and Reception, Caused by the Use of Vestigial-Sideband Techniques—Questions Concerning Study Group 11-Broadcasting Service (Television). 1 p.

—Broadcasting—Reception—Receiving Antenna

CCIR QUESTION 68/11-90. Recommended Characteristics for Individual, Collective and Cable Distribution Antenna Systems for Domestic Reception of Television and Data Broadcasting Signals from Terrestrial Transmitters—Questions Concerning Study Group 11-Broadcasting Service. 1 p.

—Sound-Program Signals—Time Division Multiplexing

CCIR QUESTION 53/CMTT-90. Standards for the Use of an Auxiliary Data Channel Time-Multiplexed with Digital Sound-Programme Signals—Questions Concerning the CMTT CCIR/CCITT Joint Study Group for Television and Sound Transmission. 1 p.

—Video Signals—Multiplexing

CCIR QUESTION 21-2/CMTT-86. Standards for Transmission of Signals with Multiplexing of Video, Sound and Other Types of Signal—Questions Concerning the CMTT CCIR/CCITT Joint Study Group for Television and Sound Transmission. 1 p.

Data Stations
Scope Note: Use a more specific term *See:* Logging Stations (Data Transmission)

Data Storage Devices
Use For: Data Carriers; Data Media; Store and Retrieve Units *See Also:* CD ROMs; CD WOMs; Computer Tapes; Data Processing Equipment; Disk Packs; Magnetic Cards; Magnetic Cores; Magnetic Disks; Magnetic Tapes; Optical Disks; Punched Cards

—EPROMs—Information Interchange

ECMA ECMA 167-92. Volume and File Structure of Write-Once and Rewritable Media Using Non-Sequential Recording for Information Interchange. 120 pp.

IEC DIS 13346-93. Information Technology—Volume and File Structure of Write-Once and Rewritable Media Using Non-Sequential Recording for Information Interchange ***CD-ROM ONLY***. 123 pp.

Data Storage Devices *(Cont.)*

—EPROMs—Information Interchange *(Cont.)*

ISO DIS 13346-93. Information Technology—Volume and File Structure of Write-Once and Rewritable Media Using Non-Sequential Recording for Information Interchange ***CD-ROM ONLY***. 123 pp.

JTC1 DIS13346-93. Information Technology—Volume and File Structure of Write-Once and Rewritable Media Using Non-Sequential Recording for Information Interchange ***CD-ROM ONLY***. 123 pp.

—Glossaries

JIS X 0012-90. Glossary of Terms Used in Information Processing (Data Media, Storage and Related Equipment).

—ROMs—Information Interchange

ECMA ECMA 167-92. Volume and File Structure of Write-Once and Rewritable Media Using Non-Sequential Recording for Information Interchange. 120 pp.

IEC DIS 13346-93. Information Technology—Volume and File Structure of Write-Once and Rewritable Media Using Non-Sequential Recording for Information Interchange ***CD-ROM ONLY***. 123 pp.

ISO DIS 13346-93. Information Technology—Volume and File Structure of Write-Once and Rewritable Media Using Non-Sequential Recording for Information Interchange ***CD-ROM ONLY***. 123 pp.

JTC1 DIS13346-93. Information Technology—Volume and File Structure of Write-Once and Rewritable Media Using Non-Sequential Recording for Information Interchange ***CD-ROM ONLY***. 123 pp.

—Teletex Communications—Telecommunication Networks

CCITT RECMN F.203-89. Network Based Storage for the Teletex Service—Telematic, Data Transmission and Teleconference Services—Operations and Quality of Service (Study Group I) 5 pp. 5 pp.

—Telex Communications—Status Enquiry Function

CCITT RECMN F.89-92. Status Enquiry Function in the International Telex Service (Study Group I) 6 pp. 6 pp.

Data Terminal Equipment
Use For: Automatic Terminal Equipment; DTE; Terminal Equipment *See Also:* Communication Terminal Equipment; Data Encryption; Data Networks; Data Processing Equipment; Display Devices; Interfaces; Open Systems Interconnection; Terminal Identification; Video Display Terminals; Voice/Data Systems

CEPT T/GSI 05-01-85. Support Des Equipements Terminaux De Traitement De Donnees Des Series X Et V Pour Le Reseau Numerique Avec Integration De Services. 1 p.

CEPT T/GSI 05-01 E-85. Support of X and V Series DTE's. 1 p.

EC 91/263/EEC-91. Council Directive on the Approximation of the Laws of the Member States Concerning Telecommunications Terminal Equipment, Including the Mutual Recognition of Their Conformity (NEW APP). 19 pp.

OSI ISO DIS 8882-1-87. Information Processing Systems—X.25 DTE Conformance Testing—Part 1: General Principles. 17 pp.

OSI ISO/IEC DIS 8882-1.2-91. Information Technology—X.25 DTE Conformance Testing—Part 1: General Principles. 10 pp.

—Adapters

CEPT T/GSI 04-01 E-87. Interface Structures and Access Capabilities. 6 pp.

CEPT T/GSI 04-04-87. Adaptateurs De Terminaux. 1 p.

CEPT T/GSI 04-04 E-87. Terminal Adaptors. 1 p.

—Audiovisual Equipment—Digital Channels

CCITT RECMN H.242-90. System for Establishing Communication Between Audiovisual Terminals Using Digital Channels up to 2 Mbit/s (Study Group XV) 35 pp. 35 pp.

—Backup Facilities

BSI BS 6639-90. 1990 Guide to DTE/DCE Interface Control Operation Using the 25 Pole Connector. 12 pp.

ISO 8480-87. Information Processing—Data Communication—DTE/DCE Interface Back-up Control Operation Using the 25-Pole Connector First Edition. 9 pp.

Data Terminal Equipment *(Cont.)*

—Backup Facilities *(Cont.)*

JTC1 8480-87. Information Processing—Data Communication—DTE/DCE Interface Back-up Control Operation Using the 25-Pole Connector First Edition. 9 pp.

OSI ISO 8480-87. Information Processing—Data Communication—DTE/DCE Interface Back-Up Control Operation Using the 25-Pole Connector.

—Connectors—Safety

CEPT T/CD 04-03-85. Exigences De Securite a L'Interface ETTD-ETCD. 3 pp.

CEPT T/CD 04-03 E-86. Safety Requirements for the DTE-DCE Interface. 3 pp.

—Control Functions—Integrated Services Digital Networks

CEPT T/SF 31-08 E-89. User Control Functions of ISDN Terminal Equipment. 6 pp.

—Data Link Layer

OSI ISO/IEC DIS 8882-2-90. Information Technology Telecommunications and Information Exchange Between Systems—X.25-DTE Conformance Testing—Part 2: Data Link Layer Test Suite. 225 pp.

OSI ISO DP 8882-2-87. OSI-X.25-DTE Conformance Testing-Part 2: Data Link Layer Conformance Tests. 112 pp.

—Data Link Layer—High Level Data Link Control—Conformance

BSI BS ISO/IEC 8882-2-92. 1992 Information Technology—Telecommunications and Information Exchange Between Systems—X.25 DTE Conformance Testing—Part 2: Data Link Layer Conformance Test Suite. 241 pp.

IEC 8882 Pt 2-92. Information Technology—Telecommunications and Information Exchange Between Systems—X.25 DTE Conformance Testing—Part 2: Data Link Layer Conformance Test Suite First Edition. 241 pp.

ISO 8882 Pt 2-92. Information Technology—Telecommunications and Information Exchange Between Systems—X.25 DTE Conformance Testing—Part 2: Data Link Layer Conformance Test Suite First Edition. 241 pp.

JTC1 8882-92. Information Technology—Telecommunications and Information Exchange Between Systems—X.25 DTE Conformance Testing—Part 2: Data Link Layer Conformance Test Suite First Edition. 241 pp.

—Data Link Layer—Packet Switched Networks—Conformance

BSI BS ISO/IEC 8882-2-92. 1992 Information Technology—Telecommunications and Information Exchange Between Systems—X.25 DTE Conformance Testing—Part 2: Data Link Layer Conformance Test Suite. 241 pp.

IEC 8882 Pt 2-92. Information Technology—Telecommunications and Information Exchange Between Systems—X.25 DTE Conformance Testing—Part 2: Data Link Layer Conformance Test Suite First Edition. 241 pp.

ISO 8882 Pt 2-92. Information Technology—Telecommunications and Information Exchange Between Systems—X.25 DTE Conformance Testing—Part 2: Data Link Layer Conformance Test Suite First Edition. 241 pp.

JTC1 8882-92. Information Technology—Telecommunications and Information Exchange Between Systems—X.25 DTE Conformance Testing—Part 2: Data Link Layer Conformance Test Suite First Edition. 241 pp.

—Data Networks

CCITT RECMN X.52-89. Method of Encoding Anisochronous Signals Into a Synchronous User Bearer—Data Communication Networks—Transmission, Signalling and Switching, Network Aspects, Maintenance and Administrative Arrangements (Study Group VII) 3 pp. 3 pp.

—Design

CENELEC ETR 005-90. Terminal Equipment (TE); Technical Requirements for Data Terminal Equipment for Connection to High Speed Digital Fixed-Connection Services. 44 pp.

ETSI ETR 005-90. Terminal Equipment (TE); Technical Requirements for Data Terminal Equipment for Connection to High Speed Digital Fixed-Connection Services. 44 pp.

—Digital Transmission Systems

CCITT FASCICLE III.4-89. General Aspects of Digital Transmission Systems Terminal Equipments—Recommendations G.700-G.795. 621 pp.

CCITT RECMN G.700-89. Framework of the Series G.700, G.800 and G.900 Recommendations—General Aspects of Digital Transmission Systems; Terminal Equipments (Study Groups XV and XVIII) 4 pp. 4 pp.

INDUSTRY STANDARDS

INTERNATIONAL AND NON-U.S. NATIONAL STANDARDS SUBJECT INDEX

Data Terminal Equipment *(Cont.)*

—Disabled Persons
CEPT T/SF 3-77. Services Et Facilities Pour Les Personnes Handicapees. 7 pp.

—Document Transfer and Manipulation
CCITT RECMN T.431-89. Document Transfer and Manipulation (DTAM)—Services and Protocols—Introduction and General Principles—Terminal Equipment and Protocols for Telematic Services (Study Group VIII) 15 pp. 15 pp.

—Environmental Conditions
CEPT T/CS 34-06-81. Conditions D'Environnement De L'Equipement Terminal Installe Chez L'Abonne. 2 pp.

—Error Control
CCITT RECMN V.41-89. Code-Independent Error-Control System—Data Communication over the Telephone Network (Study Group XVII) 10 pp. 10 pp.

—Facility Identifier—Videotex Communications
CENELEC PRETS 300 076-90. Terminal Equipment (TE): Videotex Terminal Facility Identifier (TFI). 8 pp.
CENELEC ETS 300 076-90. Terminal Equipment (TE); Videotex Terminal Facility Identifier (TFI). 10 pp.
CENELEC PRETS 300 076 RV 1-91. Terminal Equipment (TE); Videotex Terminal Facility Identifier (TFI). 15 pp.
CENELEC PRETS 300 076 RV 1-92. Terminal Equipment (TE); Videotex Terminal Facility Identifier (TFI). 19 pp.
CENELEC ETS 300 076-92. Terminal Equipment (TE); Videotex Terminal Facility Identifier (TFI). 19 pp.
CEPT T/TE 06-05 E-88. Videotex Presentation Layer Data Syntax Terminal Facility Identifier. 4 pp.
ETSI PRETS 300 076-88. Videotex Presentation Layer Data Syntax Terminal Facility Identifier (Formerly ETS(H), T/GT10, TE(88) 66 & T/TE06-05). 10 pp.
ETSI ETS 300 076-92. Terminal Equipment (TE); Videotex Terminal Facility Identifier (TFI). 19 pp.
ETSI PRETS 300 076 RV 1-92. Terminal Equipment (TE); Videotex Terminal Facility Identifier (TFI). 19 pp.
ETSI PRETS 300 076 RV 1-91. Terminal Equipment (TE); Videotex Terminal Facility Identifier (TFI). 15 pp.
ETSI ETS 300 076-90. Terminal Equipment (TE); Videotex Terminal Facility Identifier (TFI). 10 pp.

—Fault Isolation—Test Loops
BSI BS 7249-90. 1990 Procedures for Automatic Fault Isolation Using Test Loops. 8 pp.
ISO 9067-87. Information Processing Systems—Data Communication—Automatic Fault Isolation Procedures Using Test Loops First Edition. 7 pp.
JTC1 9067-87. Information Processing Systems—Data Communication—Automatic Fault Isolation Procedures Using Test Loops First Edition. 7 pp.
OSI ISO 9067-87. Information Processing Systems—Data Communication—Automatic Fault Isolation Procedures Using. 7 pp.
OSI ISO DIS 9067-86. Information Processing Systems—Data Communication—Automatic Fault Isolation Procedures Using. 10 pp.
SNZ NZS/ISO 9067-87. Information Processing Systems—Data Communication—Automatic Fault Isolation Procedures Using Test Loops. 4 pp.

—FAX Machines
CCITT FASCICLE VII.3. Terminal Equipment and Protocols for Telematic Services—Recommendations T.0-T.63. 494 pp.
CCITT FASCICLE VII.6. Terminal Equipment and Protocols for Telematic Services—Recommendations T.400–T.418. 482 pp.
CCITT FASCICLE VII.7. Terminal Equipment and Protocols for Telematic Services—Recommendations T.431–T.564. 329 pp.
CENELEC PRETS 300 242-92. Terminal Equipment (TE); Group 3 Facsimile Equipment. 43 pp.
CENELEC ETS 300 242-92. Terminal Equipment (TE); Group 3 Facsimile Equipment. 43 pp.
ETSI ETS 300 242-92. Terminal Equipment (TE); Group 3 Facsimile Equipment. 43 pp.
ETSI PRETS 300 242-92. Terminal Equipment (TE); Group 3 Facsimile Equipment. 43 pp.

—Fiber Optics
JIS C 6110-90. General Rules of Transmitting and/or Receiving Modules for Low Speed Fiber Optic Transmission. 16 pp.
JIS C 6111-90. Test Methods of Transmitting and/or Receiving Modules for Low Speed Fiber Optic Transmission. 27 pp.

Data Terminal Equipment *(Cont.)*

—Integrated Services Digital Networks
CCITT RECMN I.470-89. Relationship of Terminal Functions to ISDN—Integrated Services Digital Network (ISDN)—Overall Network Aspects and Functions, ISDN User-Network Interfaces (Study Group XVIII) 4 pp. 4 pp.
CENELEC ETS 300 103-90. Integrated Services Digital Network (ISDN); Support of CCITT Recommendation X.21, X.21 bis and X.20 bis Based Data Terminal Equipments (DTEs) by an ISDN Synchronous and Asynchronous Terminal Adaption Functions. 105 pp.
CEPT NET 3 Part 1-88. Approval Requirements for Terminal Equipment to Connect to Integrated Services Digital Network (ISDN) Using ISDN Basic Access Part 1: Layers 1 and 2 Aspects. 93 pp.
CEPT NET 3 Part 2-89. Approval Requirements for Terminal Equipment to Connect to ISDN Using ISDN Basic Access, Layer 3 Aspect (ETS P (Rev.1)). 225 pp.
CEPT T/CAC S 10.8 E-92. User Control Functions of ISDN Terminal Equipment. 8 pp.
CEPT T/S 46-40-87. Support Des Equipments Terminaux De Traitement De Donnees (ETTD) Dutype X.21, X.21 Bis X.20 Par Le Reseau Numerique a Integration De Services (RNIS) Fonction D'Adaptation Des Terminaux Synchrones Et Asynchrones. 59 pp.
CEPT T/S 46-40 E-87. Support of X.21, X.21 Bis and X.20 Bis Based Data Terminal Equipments (DTEs) by an Integrated Services Digital Network (ISDN) Synchronous and Asynchronous Terminal Adaptation Function. 59 pp.
CEPT T/S 54-08-87. Maintenance Des Acces Et Installations Terminales D'Abonnes Du RNIS. 22 pp.
CSA CAN/CSA-T552-M91. Integrated Services Digital Network (ISDN) Terminal Adaptation Using Statistical Multiplexing (ANSI T1.612-1990); (Gen Instr 1). 49 pp.
ECMA ECMA-TR 51-90. Requirements for Access to Integrated Voice and Data Local and Metropolitan Area Networks. 62 pp.
ETSI ETS 300 103-90. Integrated Services Digital Network (ISDN); Support of CCITT Recommendation X.21, X.21 bis and X.20 bis Based Data Terminal Equipments (DTEs) by an ISDN Synchronous and Asynchronous Terminal Adaption Functions. 105 pp.

—Integrated Services Digital Networks—Interfaces
CCITT RECMN E.331-91. Minimum User-Terminal Interface for a Human User Entering Address Information into an ISDN Terminal (Study Group I) 9 pp. 9 pp.

—Integrated Services Digital Networks—Interworking
CCITT RECMN I.320-89. ISDN Protocol Reference Model—Integrated Services Digital Network (ISDN)—Overall Network Aspects and Functions, ISDN User-Network Interfaces (Study Group XVIII) 10 pp. 10 pp.

—Interchange Circuits—Data Signaling Rates
CCITT RECMN V.31-89. Electrical Characteristics for Single-Current Interchange Circuits Controlled by Contact Closure—Data Communication over the Telephone Network (Study Group XVII) 3 pp. 3 pp.
CCITT RECMN V.31 BIS-89. Electrical Characteristics for Single-Current Interchange Circuits Using Optocouplers—Data Communication over the Telephone Network (Study Group XVII) 3 pp. 3 pp.

—Interchange Circuits—Data Signaling Rates—Optoisolators
CCITT RECMN V.31-89. Electrical Characteristics for Single-Current Interchange Circuits Controlled by Contact Closure—Data Communication over the Telephone Network (Study Group XVII) 3 pp. 3 pp.
CCITT RECMN V.31 BIS-89. Electrical Characteristics for Single-Current Interchange Circuits Using Optocouplers—Data Communication over the Telephone Network (Study Group XVII) 3 pp. 3 pp.

—Interchange Circuits—Glossaries
CCITT RECMN V.24-89. List of Definitions for Interchange Circuits Between Data Terminal Equipment (DTE) and Data Circuit-Terminating Equipment (DCE)—Data Communication over the Telephone Network (Study Group XVII) 18 pp. 18 pp.

Data Terminal Equipment *(Cont.)*

—Interchange Circuits—Glossaries *(Cont.)*
CCITT RECMN X.24-89. List of Definitions for Interchange Circuits Between Data Terminal Equipment (DTE) and Data Circuit-Terminating Equipment (DCE) on Public Data Networks—Data Communication Networks: Services and Facilities, Interfaces (Study Group VII) 6 pp. 6 pp.

—Interchange Circuits—Telephone Answering Equipment
CCITT RECMN V.25-89. Automatic Answering Equipment and/or Parallel Automatic Calling Equipment on the General Switched Telephone Network Including Procedures for Disabling of Echo Control Devices for Both Manually and Automatically Established Calls—Data. 11 pp.
CCITT RECMN V.25 BIS-89. Automatic Calling and/or Answering Equipment on the General Switched Telephone Network (GSTN) Using the 100-Series Interchange Circuits—Data Communication over the Telephone Network (Study Group XVII) 23 pp. 23 pp.

—Interfaces
CCITT RECMN V.230-89. General Data Communications Interface Layer 1 Specification—Data Communication over the Telephone Network (Study Group XVII) 43 pp. 43 pp.

—Interfaces—Connectors
BSI BS 6514-84. (WITHDRAWN) 1984 Implementation of V24 or RS232 as an Asynchronous Local Interface. 4 pp.
BSI BS 6623: Part 1-85. (WITHDRAWN) 1985 DTE/DCE Interface Connectors and Pin Assignments Part 1: Specification for 25-Pin Connector (Superseded by BS ISO 2110: 1989). 17 pp.
BSI BS 6623: Part 2-86. 1986 DTE/DCE Interface Connectors and Pin Assignments Part 2: Specification for 37-Pin Connector. 25 pp.
BSI BS 6623: Part 3-86. 1986 DTE/DCE Interface Connectors and Pin Assignments Part 3: 15-Pin Connector. 20 pp.
BSI BS 6623: Part 4-86. (WITHDRAWN) 1986 DTE/DCE Interface Connectors and Pin Assignments Part 4: 34-Pin Connector (Superseded by BS ISO/IEC 2593: 1993). 8 pp.
BSI BS 6640: Part 1-87. 1987 Arrangements for DTE to DTE Physical Connection Part 1: General Arrangements for DTE to DTE Physical Connection Using V.24 and X.24 Interchange Circuits. 11 pp.
BSI BS 6640: Part 2-87. 1987 Arrangements for DTE to DTE Physical Connection Part 2: DTE to DTE Physical Connection Using X.24 Interchange Circuits with DTE Providing Timing. 7 pp.
BSI BS ISO 2110-89. 1989 Information Technology—Data Communication—25-Pole DTE/DCE Interface Connector and Contact Number Assignments. 18 pp.
BSI BS ISO 2110-01. 1989 Amd 1 Information Technology—Data Communication—25-Pole DTE/DCE Interface Connector and Contact Number Assignments (AMD 7165) July 15, 1992. 26 pp.
BSI BS ISO 4902-89. 1989 Information Technology—Data Communication—37 Pole DTE/DCE Interface Connector and Contact Number Assignments. 24 pp.
BSI BS ISO 4903-89. 1989 Information Technology—Data Communication—15-Pole DTE/DCE Interface Connector and Contact Number Assignments. 22 pp.
BSI BS ISO/IEC 2593-93. 1993 Information Technology—Telecommunications and Information Exchange Between Systems—34-Pole DTE/DCE Interface Connector Mateability Dimensions and Contact Number Assignments. 23 pp.
CSA Z243.12.2-1981. Data Communication—15-Pin DTE/DCE Interface Connector and Pin Assignments (ISO 4903-1980 has Been Adopted as a CSA Standard) (R 1988). 25 pp.
IEC 2593-93. Information Technology—Telecommunications and Information Exchange Between Systems—34-Pole DTE/DCE Interface Connector Mateability Dimensions and Contact Number Assignments Third Edition. 21 pp.
IEC 11569-93. Information Technology—Telecommunications and Information Exchange Between Systems—26-Pole DTE/DCE Interface Connector Mateability Dimensions and Contact Number Assignments First Edition. 11 pp.
ISO 2110-89. Information Technology—Data Communication—25-Pole DTE/DCE Inter-face Connector and Con-tact Number Assignments Amendment 1: Interface Connector and Contact Assignments for a DTE/DCE Interface for Data Signalling Rates Above 20 000 Bit/s Third Ed.; (Amendment 1-1991). 22 pp.
ISO 2593-93. Information Technology—Telecommunications and Information Exchange Between Systems—34-Pole DTE/DCE Interface Connector Mateability Dimensions and Contact Number Assignments Third Edition. 21 pp.

INTERNATIONAL AND NON-U.S. NATIONAL STANDARDS
SUBJECT INDEX
Data

Data Terminal Equipment *(Cont.)*

—Interfaces—Connectors *(Cont.)*

ISO 4902-89. Information Technology—Data Communication—37-Pole DTE/DCE Interface Connector and Contact Number Assignments Second Edition. 20 pp.
ISO 4903-89. Information Technology—Data Communication—15-Pole DTE/DCE Interface Connector and Contact Number Assignments Second Edition. 20 pp.
ISO TR7477-85. Data Communication—Arrangements for DTE to DTE Physical Connection Using V.24 and X.24 Interchange Circuits First Edition. 8 pp.
ISO 8481-86. Data Communication—DTE to DTE Physical Connection Using X.24 Interchange Circuits with DTE Provided Timing First Edition. 5 pp.
ISO 11569-93. Information Technology—Telecommunications and Information Exchange Between Systems—26-Pole Interface Connector Mateability Dimensions and Contact Number Assignments First Edition. 11 pp.
JIS X 5101-82. Interface Between Date Circuit Terminating Equipment (DCE) and Data Terminal Equipment (DTE) (25-Pin Interface).
JIS X 5102-82. Interface Between Date Circuit Terminating Equipment (DCE) and Data Terminal Equipment (DTE) (15-Pin Interface).
JIS X 5103-82. Interface Between Date Circuit Terminating Equipment (DCE) and Data Terminal Equipment (DTE) (37/9-Pin Interface).
JTC1 2110-89. Information Technology—Data Communication—25-Pole DTE/DCE Inter-face Connector and Con-tact Number Assignments Amendment 1: Interface Connector and Contact Assignments for a DTE/DCE Interface for Data Signalling Rates Above 20 000 Bit/s Third Ed.; (Amendment 1-1991). 22 pp.
JTC1 2593-93. Information Technology—Telecommunications and Information Exchange Between Systems—34-Pole DTE/DCE Interface Connector Mateability Dimensions and Contact Number Assignments Third Edition. 21 pp.
JTC1 4902-89. Information Technology—Data Communication—37-Pole DTE/DCE Interface Connector and Contact Number Assignments Second Edition. 20 pp.
JTC1 4903-89. Information Technology—Data Communication—15-Pole DTE/DCE Interface Connector and Contact Number Assignments Second Edition. 20 pp.
JTC1 TR7477-85. Data Communication—Arrangements for DTE to DTE Physical Connection Using V.24 and X.24 Interchange Circuits First Edition. 8 pp.
JTC1 8481-86. Data Communication—DTE to DTE Physical Connection Using X.24 Interchange Circuits with DTE Provided Timing First Edition. 5 pp.
OSI ISO 2110-89. Information Technology—Date Communication—25-Pole DTE/DCE Interface Connector and Contact Number Assignment—Amd 1: Interface Connector and Contact Assignments for a DTE/DCE Interface for Data Signalling Rates Above 20 000 bits/s. 23 pp.
OSI ISO 2110 DAM 1-89. Interface Connector and Contact Number Assignments for a DTE/DCE Interface for Data Signalling Rates Above 20 000 bits per Second. 5 pp.
OSI ISO/IEC DIS 2593-89. Information Technology—Data Communication—34-Pole DTE/DCE Interface Connector and Contact Number Assignments. 14 pp.
OSI ISO 2593-84. Data Communication—34 Pin DTE/DCE Interface Connector and Pin Assignments. 7 pp.
OSI ISO 4902-89. Information Technology—Date Communication—37 Pole DTE/DCE Interface Connector and Contact Number Assignments. 20 pp.
OSI ISO 4903-89. Information Technology—Data Communication—15-Pole DTE/DCE Interface Connector and Contact Number Assignments. 20 pp.
OSI ISO TR 7477-85. Data Communication—Arrangements for DTE to DTE Physical Connection Using V.24 and X.24 Interchange Circuits.
OSI ISO 8481-86. Data Communication—DTE to DTE Physical Connection Using X.24 Interchange Circuits with DTE Provided Timing. 5 pp.
SAA AS 2748-91. Information Technology—Data Communication—25-Pole DTE/DCE Interface Connector and Contact Number Assignments (ISO/IEC 2110:1989) (in Professional Packages 26A, 26C). 13 pp.
SAA AS 3613-91. Information Technology—Data Communications—15-Pole DTE/DCE Interface Connector and Contact Number Assignments (ISO 4903:1989) (in Professional Packages 26A, 26C). 17 pp.
SNZ NZS/ISO 4902-80. Data Communication—37-Pin and 9-Pin DTE/DCE Interface Connectors and Pin Assignments. 24 pp.
SNZ NZS/ISO 4903-89. Information Technology—Data Communications—15-Pole DTE/DCE Interface Connector and Contact Number Assignments. 17 pp.
SNZ NZS/ISO TR 7477-85. Data Communication—Arrangements for DTE to DTE Physical Connection V. 8 pp.

Data Terminal Equipment *(Cont.)*

—Interfaces—Connectors *(Cont.)*

SNZ NZS/ISO 8481-86. Data Communication—DTE to DTE Physical Connection Using X.24 Interchange Circuits with DTE Provided Timing. 3 pp.

—Interfaces—Data Circuit Terminating Equipment

CCITT RECMN X.20-89. Interface Between Data Terminal Equipment (DTE) and Data Circuit-Terminating Equipment (DCE) for Start-Stop Transmission Services on Public Data Networks—Data Communication Networks: Services and Facilities, Interfaces (Study Group VII) 27 pp. 27 pp.
CCITT RECMN X.20 BIS-89. Use on Public Data Networks of Data Terminal Equipment (DTE) Which is Designed for Interfacing to Asynchronous Duplex V-Series Modems—Data Communication Networks: Services and Facilities, Interfaces (Study Group VII) 7 pp. 7 pp.
CENELEC ETR 017-92. Terminal Equipment (TE); Requirements for the Attachment of Data Terminal Equipment Comprising CCITT V. Series Type Interfaces to Data Circuit-Terminating Equipment for Low and Medium Data Signalling Rates. 37 pp.
ETSI ETR 017-92. Terminal Equipment (TE); Requirements for the Attachment of Data Terminal Equipment Comprising CCITT V. Series Type Interfaces to Data Circuit-Terminating Equipment for Low and Medium Data Signalling Rates. 37 pp.

—Interfaces—Data Circuit Terminating Equipment—Modems

CCITT RECMN X.20-89. Interface Between Data Terminal Equipment (DTE) and Data Circuit-Terminating Equipment (DCE) for Start-Stop Transmission Services on Public Data Networks—Data Communication Networks: Services and Facilities, Interfaces (Study Group VII) 27 pp. 27 pp.
CCITT RECMN X.20 BIS-89. Use on Public Data Networks of Data Terminal Equipment (DTE) Which is Designed for Interfacing to Asynchronous Duplex V-Series Modems—Data Communication Networks: Services and Facilities, Interfaces (Study Group VII) 7 pp. 7 pp.
CCITT RECMN X.21-89. Interface Between Data Terminal Equipment (DTE) and Data Circuit-Terminating Equipment (DCE) for Synchronous Operation on Public Data Networks—Data Communication Networks: Services and Facilities, Interfaces (Study Group VII) 56 pp. 56 pp.
CCITT RECMN X.21 BIS-89. Use on Public Data Networks of Data Terminal Equipment (DTE) Which is Designed for Interfacing to Synchronous V-Series Modems—Data Communication Networks: Services and Facilities, Interfaces (Study Group VII) 13 pp. 13 pp.

—Interfaces—Data Circuit Terminating Equipment—PSDN

BSI BS ISO/IEC 10588-93. 1993 Information Technology—Use of X.25 Packet Layer Protocol in Conjunction with X.21/X.21 bis to Provide the OSI Connection-Mode Network Service (S). 18 pp.
CCITT RECMN X.25-89. Interface Between Data Terminal Equipment (DTE) and Data Circuit-Terminating Equip. (DCE) for Terminals Operating in the Packet Mode and Connected to Public Data Networks by Dedicated CI—DataCommunication Networks: Services and Facilities, Interfaces (Study Group VII) 160 pp. 160 pp.
CCITT RECMN X.613-92. Information Technology—Use of X.25 Packet Layer Protocol in Conjunction with X.21/X.21 bis to Provide the OSI Connection-Mode Network Service 15 pp (ISO/CCIT Common Text). 15 pp.
CENELEC PRETS 300 123-90. Attachment Requirements for Data Terminal Equipment (DTE) to Connect to Packet Switched Public Data Networks (PSPDN) Using CCITT Recommendation X.25 (1984) Interface Requirements Applicable to DTEs Subscribing to Modulo 128 Operation. 7 pp.
ETSI PRETS 300 123-90. Attachment Requirements for Data Terminal Equipment (DTE) to Connect to Packet Switched Public Data Networks (PSPDN) Using CCITT Recommendation X.25 (1984) Interface Requirements Applicable to DTE's Subscribing to Modulo 128 Operation. 7 pp.
ETSI ETS 300 123-91. Attachment Reqmts for Data Terminal Equip. (DTE) to Connect to Packet Switched Publ. Data Networks (PSPDN) Using CCITT Recommendation X.25 (1984) Interface Reqmts Appl. to DTEs Sub. to Link Access Procedure Balanced (LAPB) Extended (Modulo 128) Operation. 8 pp.
IEC 10588-93. Information Technology—Use of X.25 Packet Layer Protocol in Conjunction with X.21/X.21 bis to Provide the OSI Connection-Mode Network Service First Edition; (CCITT RECMN X.613). 15 pp.

Data Terminal Equipment *(Cont.)*

—Interfaces—Data Circuit Terminating Equipment—PSDN *(Cont.)*

ISO 10588-93. Information Technology—Use of X.25 Packet Layer Protocol in Conjunction with X.21/X.21 bis to Provide the OSI Connection-Mode Network Service First Edition; (CCITT RECMN X.613). 15 pp.
JTC1 10588-93. Information Technology—Use of X.25 Packet Layer Protocol in Conjunction with X.21/X.21 bis to Provide the OSI Connection-Mode Network Service First Edition; (CCITT RECMN X.613). 15 pp.

—Interfaces—Data Circuit Terminating Equipment—Public Networks

BSI BS ISO/IEC 10588-93. 1993 Information Technology—Use of X.25 Packet Layer Protocol in Conjunction with X.21/X.21 bis to Provide the OSI Connection-Mode Network Service (S). 18 pp.
CCITT RECMN X.21-92. Interface Between Data Terminal Equipment and Data Circuit-Terminating Equipment for Synchronous Operation on Public Data Networks (Study Group VII) 58 pp. 58 pp.
CCITT RECMN X.21-89. Interface Between Data Terminal Equipment (DTE) and Data Circuit-Terminating Equipment (DCE) for Synchronous Operation on Public Data Networks—Data Communication Networks: Services and Facilities, Interfaces (Study Group VII) 56 pp. 56 pp.
CCITT RECMN X.21 BIS-89. Use on Public Data Networks of Data Terminal Equipment (DTE) Which is Designed for Interfacing to Synchronous V-Series Modems—Data Communication Networks: Services and Facilities, Interfaces (Study Group VII) 13 pp. 13 pp.
CCITT RECMN X.22-89. Multiplex DTE/DCE Interface for User Classes 3-6—Data Communication Networks: Services and Facilities, Interfaces (Study Group VII) 6 pp. 6 pp.
CCITT RECMN X.613-92. Information Technology—Use of X.25 Packet Layer Protocol in Conjunction with X.21/X.21 bis to Provide the OSI Connection-Mode Network Service 15 pp (ISO/CCIT Common Text). 15 pp.
IEC 10588-93. Information Technology—Use of X.25 Packet Layer Protocol in Conjunction with X.21/X.21 bis to Provide the OSI Connection-Mode Network Service First Edition; (CCITT RECMN X.613). 15 pp.
ISO 10588-93. Information Technology—Use of X.25 Packet Layer Protocol in Conjunction with X.21/X.21 bis to Provide the OSI Connection-Mode Network Service First Edition; (CCITT RECMN X.613). 15 pp.
JTC1 10588-93. Information Technology—Use of X.25 Packet Layer Protocol in Conjunction with X.21/X.21 bis to Provide the OSI Connection-Mode Network Service First Edition; (CCITT RECMN X.613). 15 pp.

—Interfaces—Data Circuit Terminating Equipment—Start-Stop

CCITT RECMN X.20-89. Interface Between Data Terminal Equipment (DTE) and Data Circuit-Terminating Equipment (DCE) for Start-Stop Transmission Services on Public Data Networks—Data Communication Networks: Services and Facilities, Interfaces (Study Group VII) 27 pp. 27 pp.
CCITT RECMN X.20 BIS-89. Use on Public Data Networks of Data Terminal Equipment (DTE) Which is Designed for Interfacing to Asynchronous Duplex V-Series Modems—Data Communication Networks: Services and Facilities, Interfaces (Study Group VII) 7 pp. 7 pp.

—Interfaces—Integrated Services Digital Networks

CCITT RECMN I.463-89. Support of Data Terminal Equipments (DTEs) with V-Series Type Interfaces by an Integrated Services Digital Network (ISDN)—Integrated Services Network (ISDN)—Overall Network Aspects and Functions, ISDN User-Network Interfaces (Study Group XVIII) 1 pp. 1 p.
CCITT RECMN I.465-89. Support by an ISDN of Data Terminal Equipment with V-Series Type Interfaces with Provision for Statisical Multiplexing—Integrated Services Digital Network Aspects and Functions, ISDN User-Network Interfaces (Study Group XVIII) 1 pp (Same as Recmn V.120). 1 p.
CCITT RECMN V.110-92. Support of Data Terminal Equipments with V-Series Type Interfaces by an Integrated Services Digital Network (Study Group XVII) 60 pp. 60 pp.
CCITT RECMN V.110-89. Support of Data Terminal Equipments (DTEs) with V-Series Type Interfaces by an Integrated Services Digital Network (ISDN)—Data Communication Over the Telephone Network (Study Group XVII) 47 pp (Same as Recmn I.463). 47 pp.

INDUSTRY STANDARDS
737

Data Terminal Equipment (Cont.)

—Interfaces—Integrated Services Digital Networks (Cont.)

CCITT RECMN V.120-92. Support by an ISDN of Data Terminal Equipment with V-Series Type Interfaces with Provision for Statistical Multiplexing (Study Group XVII) 38 pp. 38 pp.

CCITT RECMN V.120-89. Support by an ISDN of Data Terminal Equipment with V-Series Type Interfaces with Provision for Statistical Multiplexing—Data Communication over the Telephone Network (Study Group XVII) 26 pp (Same as Recmn I.465). 26 pp.

—Interfaces—Interchange Circuits—Electrical Measurement

CCITT RECMN V.10-89. Electrical Characteristics for Unbalanced Double-Current Interchange Circuits for General Use with Integrated Circuit Equipment in the Field of Data Communications—Data Communication Over the Telephone Network (Study Group XVII) 17 pp. 17 pp.

CCITT RECMN V.11-89. Electrical Characteristics for Balanced Double-Current Interchange Circuits for General Use with Integrated Circuit Equipment in the Field of Data Communications—Data Communication Over the Telephone Network (Study Group XVII) 12 pp. 12 pp.

—Interfaces—Open Network Provision

CENELEC PRETS 300 290-92. Business Telecommunications (BT); Open Network Provision (ONP) Technical Requirements; 64 kbit/s Digital Unstructured Leased Line (D64U) Terminal Equipment Interface. 28 pp.

ETSI PRETS 300 290-92. Business Telecommunications (BT); Open Network Provision (ONP) Technical Requirements; 64 kbit/s Digital Unstructured Leased Line (D64U) Terminal Equipment Interface. 28 pp.

—Interfaces—Packet Switched Networks

CCITT RECMN X.32-89. Interface Between Data Terminal Equipment (DTE) and Data Circuit-Terminating Equipment (DCE) for Terminals Operating in the Packet Mode and Accessing a Packet Switched Public Data Network Through a Public Switched Telephone Network or an Integrated Services. 57 pp.

CENELEC ETR 059-93. Terminal Equipment (TE); Technical Requirements for Packet Switched Public Data Network (PSPDN) Indirect Access (Based on CCITT Recommendation X.32). 33 pp.

ETSI ETR 059-93. Terminal Equipment (TE); Technical Requirements for Packet Switched Public Data Network (PSPDN) Indirect Access (Based on CCITT Recommendation X.32). 33 pp.

—Interfaces—Public Data Networks

CCITT RECMN X.20-89. Interface Between Data Terminal Equipment (DTE) and Data Circuit-Terminating Equipment (DCE) for Start-Stop Transmission Services on Public Data Networks—Data Communication Networks: Services and Facilities, Interfaces (Study Group VII) 27 pp. 27 pp.

CCITT RECMN X.20 BIS-89. Use on Public Data Networks of Data Terminal Equipment (DTE) Which is Designed for Interfacing to Asynchronous Duplex V-Series Modems—Data Communication Networks: Services and Facilities, Interfaces (Study Group VII) 7 pp. 7 pp.

CCITT RECMN X.32-89. Interface Between Data Terminal Equipment (DTE) and Data Circuit-Terminating Equipment (DCE) for Terminals Operating in the Packet Mode and Accessing a Packet Switched Public Data Network Through a Public Switched Telephone Network or an Integrated Services. 57 pp.

CENELEC PRETS 300 124-90. Attachment Requirements for Data Terminal Equipment (DTE) to Connect to Packet Switched Public Data Networks (PSPDN) Using CCITT Recommendation X.25 (1984) Interface Requirements Applicable to DTEs Subscribing to Multilink Operation. 20 pp.

CENELEC ETR 059-93. Terminal Equipment (TE); Technical Requirements for Packet Switched Public Data Network (PSPDN) Indirect Access (Based on CCITT Recommendation X.32). 33 pp.

CEPT NET 1-88. Approval Requirements for Data Terminal Equipment to Connect to Circuit Switched Public Data Networks and Leased Circuits Using CCITT Recommendation X.21 Interface (T/TE 04-07E). 67 pp.

CEPT NET 2-88. Approval Requirements for Data Terminal Equipment to Connect to Packet Switched Public Data Networks Using CCITT Recommendation X.25 (1984) Interface (T/TE 04-06E). 99 pp.

ETSI ETR 059-93. Terminal Equipment (TE); Technical Requirements for Packet Switched Public Data Network (PSPDN) Indirect Access (Based on CCITT Recommendation X.32). 33 pp.

Data Terminal Equipment (Cont.)

—Interfaces—Voice/Data Systems—Local Area Networks

ECMA ECMA-TR 51-90. Requirements for Access to Integrated Voice and Data Local and Metropolitan Area Networks. 62 pp.

—Interfaces—Voice/Data Systems—Metropolitan Area Networks

ECMA ECMA-TR 51-90. Requirements for Access to Integrated Voice and Data Local and Metropolitan Area Networks. 62 pp.

—Local Area Networks

BSI BS 6531: Part 4-84. (WITHDRAWN) 1984 10 Mbps Slotted Ring Local Area Network Part 4: Specification for Basic and Enhanced Class Nodes with Type 1 Node/DTE Interface (Superseded by BS ISO 8802-7: 1991). 31 pp.

BSI BS 6532: Part 1-84. (WITHDRAWN) 1984 Data Terminal Equipment for Attachment to 10 Mbps Slotted Ring Local Area Network Part 1: Specification for Media Access Control Procedures for Data Terminal Equipment (Superseded by BS ISO 8802-7: 1991). 30 pp.

BSI BS 6532: Part 2-84. (WITHDRAWN) 1984 Data Terminal Equipment for Attachment to 10 Mbps Slotted Ring Local Area Network Part 2: Specification for Implementation Requirements for Media Access Control in General Purpose Data Terminal Equipment. 16 pp.

—Mobile Radio Services

CENELEC PRI-ETS 300 041-90. European Digital Cellular Telecommunications System (phase 1): General on Terminal Adaption Functions for Mobile Stations. 54 pp.

CENELEC PRI-ETS 300 041-91. European Digital Cellular Telecommunications System (Phase 1); General on Terminal Adaptation Functions for Mobile Stations. 57 pp.

CENELEC PRI-ETS 300 041-92. European Digital Cellular Telecommunications System (Phase 1); General on Terminal Adaptation Functions for Mobile Stations. 54 pp.

CENELEC I-ETS 300 041-92. European Digital Cellular Telecommunications System (Phase 1); General on Terminal Adaptation Functions for Mobile Stations (GSM 07.01). 53 pp.

CENELEC PRI-ETS 300 041-93. European Digital Cellular Telecommunications System (Phase 1); General on Terminal Adaptation Functions for Mobile Stations (GSM 07.01). 53 pp.

CENELEC PRI-ETS 300 042-90. European Digital Cellular Telecommunications System (Phase 1): Terminal Adaptation Functions for Services Using Asynchronous Bearer Capabilities. 19 pp.

CENELEC PRI-ETS 300 042-91. European Digital Cellular Telecommunications System (Phase 1); Terminal Adaption Functions for Services Using Asynchronous Bearer Capabilities. 22 pp.

CENELEC PRI-ETS 300 042-92. European Digital Cellular Telecommunications System (Phase 1); Terminal Adaptation Functions for Services Using Asynchronous Bearer Capabilities. 22 pp.

CENELEC I-ETS 300 042-92. European Digital Cellular Telecommunications System (Phase 1); Terminal Adaptation Functions for Services Using Asynchronous Bearer Capabilities (GSM 07.02). 21 pp.

CENELEC PRI-ETS 300 042-93. European Digital Cellular Telecommunications System (Phase 1); Terminal Adaptation Functions for Services Using Asynchronous Bearer Capabilities (GSM 07.02). 22 pp.

CENELEC PRI-ETS 300 043-90. European Digital Cellular Telecommunications System (Phase 1); Terminal Adaptation Functions for Services Using Synchronous Bearer Capabilities. 38 pp.

CENELEC PRI-ETS 300 043-91. European Digital Cellular Telecommunications System (Phase 1); Terminal Adaptation Functions for Services Using Asynchronous Bearer Capabilities. 41 pp.

CENELEC PRI-ETS 300 043-92. European Digital Cellular Telecommunications System (Phase 1); Terminal Adaptation Functions for Services Using Synchronous Bearer Capabilities. 41 pp.

CENELEC I-ETS 300 043-92. European Digital Cellular Telecommunications System (Phase 1); Terminal Adaptation Functions for Services Using Synchronous Bearer Capabilities (GSM 07.03). 40 pp.

CENELEC PRI-ETS 300 043-93. European Digital Cellular Telecommunications System (Phase 1); Terminal Adaption Functions for Services Using Synchronous Bearer Capabilities (GSM 07.03). 40 pp.

CENELEC GSM 07.01-92. See PRI-ETS 300 041. 69 pp.

CENELEC GSM 07.02-92. See PRI-ETS 300 042. 28 pp.

CENELEC GSM 07.03-93. See PRI-ETS 300 043. 48 pp.

Data Terminal Equipment (Cont.)

—Mobile Radio Services (Cont.)

ETSI PRI-ETS 300 041-90. European Digital Cellular Telecommunications System (Phase 1); General on Terminal Adaptation Functions for Mobile Stations (GSM 07. 01). 54 pp.

ETSI PRI-ETS 300 042-90. European Digital Cellular Telecommunications System (Phase 1); Terminal Adaptation Functions for Services Using Asynchronous Bearer Capabilities (GSM 07.02). 19 pp.

ETSI PRI-ETS 300 043-90. European Digital Cellular Telecommunications System (Phase 1); Terminal Adaptation Functions for Services Using Synchronous Bearer Capabilities (GSM 07.03). 38 pp.

ETSI I-ETS 300 041-92. European Digital Cellular Telecommunications System (Phase 1); General on Terminal Adaptation Functions for Mobile Stations (GSM 07.01). 53 pp.

ETSI PRI-ETS 300 041-92. European Digital Cellular Telecommunications System (Phase 1); General on Terminal Adaptation Functions for Mobile Stations. 54 pp.

ETSI PRI-ETS 300 041-93. European Digital Cellular Telecommunications System (Phase 1); General on Terminal Adaptation Functions for Mobile Stations (GSM 07.01). 53 pp.

ETSI PRI-ETS 300 041-91. European Digital Cellular Telecommunications System (Phase 1); General on Terminal Adaptation Functions for Mobile Stations. 57 pp.

ETSI I-ETS 300 042-92. European Digital Cellular Telecommunications System (Phase 1); Terminal Adaptation Functions for Services Using Asynchronous Bearer Capabilities (GSM 07.02). 21 pp.

ETSI PRI-ETS 300 042-92. European Digital Cellular Telecommunications System (Phase 1); Terminal Adaptation Functions for Services Using Asynchronous Bearer Capabilities. 22 pp.

ETSI PRI-ETS 300 042-93. European Digital Cellular Telecommunications System (Phase 1); Terminal Adaption Functions for Services Using Asynchronous Bearer Capabilities (GSM 07.02). 22 pp.

ETSI PRI-ETS 300 042-91. European Digital Cellular Telecommunications System (Phase 1); Terminal Adaption Functions for Services Using Asynchronous Bearer Capabilities. 22 pp.

ETSI I-ETS 300 043-92. European Digital Cellular Telecommunications System (Phase 1); Terminal Adaptation Functions for Services Using Synchronous Bearer Capabilities (GSM 07.03). 40 pp.

ETSI PRI-ETS 300 043-92. European Digital Cellular Telecommunications System (Phase 1); Terminal Adaptation Functions for Services Using Synchronous Bearer Capabilities. 41 pp.

ETSI PRI-ETS 300 043-93. European Digital Cellular Telecommunications System (Phase 1); Terminal Adaptation Functions for Services Using Asynchronous Bearer Capabilities (GSM 07.03). 40 pp.

ETSI PRI-ETS 300 043-91. European Digital Cellular Telecommunications System (Phase 1); Terminal Adaptation Functions for Services Using Asynchronous Bearer Capabilities. 41 pp.

ETSI GSM 07.01-92. See PRI-ETS 300 041. 69 pp.

ETSI GSM 07.02-92. See PRI-ETS 300 042. 28 pp.

ETSI GSM 07.03-93. See PRI-ETS 300 043. 48 pp.

—Modems—Remote Control Circuits

CCITT RECMN V.13-89. Simulated Carrier Control—Data Communication over the Telephone Network (Study Group XVII) 2 pp. 2 pp.

—Numerical Control—Interfaces

CNS C5159-83. Interface Between Numerical Control Equipment and Data Terminal Equipment Employing Parallel Binary Data Interchange (Jun)(10330).

—Overvoltage Protection—Integrated Services Digital Networks

CCITT RECMN K.22-89. Overvoltage Resistibility of Equipment Connected to an ISDN T/S Bus—Protection Against Interference (Study Group V) 3 pp. 3 pp.

—Packet Assembly/Disassembly Devices—Information Interchange

CCITT RECMN X.29-89. Procedures for the Exchange of Control Information and User Data Between a Packet Assembly/Disassembly (PAD) Facility and a Packet Mode DTE or Another Pad—Data Communication Networks: Services and Facilities, Interfaces (Study Group VII) 15 pp. 15 pp.

INTERNATIONAL AND NON-U.S. NATIONAL STANDARDS
SUBJECT INDEX Data

Data Terminal Equipment *(Cont.)*

—Packet Assembly/Disassembly Devices—Information Interchange *(Cont.)*

CCITT RECMN X.39-92. Procedures for the Exchange of Control Information and User Data Between a Facsimile Packet Assembly/Disassembly (FPAD) Facility and a Packet Mode Data Terminal Equipment (DTE) or Another FPAD (Study Group VII) 20 pp. 20 pp.

—Packet Level Protocol

BSI BS 7219-89. (WITHDRAWN) 1989 X.25 Packet Level Protocol for Data Terminal Equipment for Information Processing Systems (Superseded by ISO/IEC 8208: 1990). 155 pp.

BSI BS ISO/IEC 8208-90. 1990 Information Technology—Data Communications—X.25 Packet Layer Protocol for Data Terminal Equipment. 144 pp.

CSA CAN/CSA-Z243.38-91. Information Technology—Data Communications—X.25 Packet Layer Protocol for Data Terminal Equipment (ISO/IEC 8208:1990); (Gen Instr 1). 141 pp.

IEC 8208-90. Information Technology—Data Communications—X.25 Packet Layer Protocol for Data Terminal Equipment AMENDMENT 1: Alternative Logical Channel Identifier Assignment Second Edition; (Amendment 1-1900). 174 pp.

IEC DIS 8208-93. Information Technology—Data Communications—X.25 Packet Later Protocol for Data Terminal Equipment; (Revision of Second Edition (ISO 8208:1990)) ***CD-ROM ONLY***. 179 pp.

ISO 8208-90. Information Technology—Data Communications—X.25 Packet Layer Protocol for Data Terminal Equipment AMENDMENT 1: Alternative Logical Channel Identifier Assignment Second Edition; (Amendment 1-1990). 174 pp.

ISO DIS 8208-93. Information Technology—Data Communications—X.25 Packet Later Protocol for Data Terminal Equipment; (Revision of Second Edition (ISO 8208:1990)) ***CD-ROM ONLY***. 179 pp.

JTC1 8208-90. Information Technology—Data Communications—X.25 Packet Layer Protocol for Data Terminal Equipment AMENDMENT 1: Alternative Logical Channel Identifier Assignment Second Edition; (Amendment 1-1990). 177 pp.

JTC1 DIS8208-93. Information Technology—Data Communications—X.25 Packet Later Protocol for Data Terminal Equipment; (Revision of Second Edition (ISO 8208:1990)) ***CD-ROM ONLY***. 179 pp.

OSI ISO/IEC 8208/DAD 3-88. Information Processing Systems—Data Communications—X.25 Packet Level Protocol for Data Terminal Equipment Addendum 3: Conformance Requirements. 69 pp.

OSI ISO IEC 8208-90. Information Technology—Data Communications—X. 25 Packet Layer Protocol for Data Terminal Equipment. 130 pp.

OSI ISO 8208-87. Information Processing Systems-Data Communications-X.25 Packet Level Protocol for Data Terminal Equipment.

OSI ISO IEC DIS 8882-3-89. Information Technology—X.25—DTE Conformance Testing—Part 3: Packet Level Conformance Test Suite. 764 pp.

OSI ISO DP 8882-3-86. X.25-DTE Conformance Testing: Part 3 Packet Level Conformance Suite (See ISO IEC Dis 8882-3). 176 pp.

OSI ISO DP 9068-86. Open Systems Interconnection-Provision of the Connectionless Network Service Using ISO 8208.

SNZ NZS/ISO 8208-87. Information Processing Systems—Data Communications—X.25 Packet Level Protocol for Data Terminal Equipment. 147 pp.

—Packet Level Protocol—Identification

CCITT RECMN X.244-89. Procedure for the Exchange of Protocol Identification During Virtual Call Establishment on Packet Switched Public Data Networks—Data Communication Networks—Open Systems Interconnection (OSI) Protocol Specifications, Conformance Testing. 2 pp.

—Packet Level Protocol—Integrated Services Digital Networks

BSI BS ISO/IEC 9574-89. 1989 Information Technology—Telecommunications and Information Exchange Between Systems—Provision of the OSI Connection-Mode Network Service by Packet Mode Terminal Equipment Connected to an Inter-grated Services Digital Network (ISDN). 18 pp.

CCITT RECMN X.612-92. Information Technology—Provision of the OSI Connection-Mode Network Service by Packet-Mode Terminal Equipment Connected to an Integrated Services Digital Network (ISDN) 21 pp (ISO/CCIT Common Text). 21 pp.

CCITT RECMN X.612-90. Provision of the OSI Connection-Mode Network Service by Packet Mode Terminal Equipment Connected to an Integrated Services Digital Network (ISDN) for CCITT Applications (Study Group VII) 19 pp. 19 pp.

Data Terminal Equipment *(Cont.)*

—Packet Level Protocol—Integrated Services Digital Networks *(Cont.)*

CENELEC PRETS 300 007-90. Integrated Services Digital Network (ISDN): Support of Packet Mode Terminal Equipment by an ISDN (T/S 46-50). 116 pp.

CENELEC PRETS 300 007-91. Integrated Services Digital Network (ISDN); Support of Packet-Mode Terminal Equipment by an ISDN. 104 pp.

CENELEC ETS 300 007-91. Integrated Services Digital Network (ISDN); Support of Packet-Mode Terminal Equipment by an ISDN. 103 pp.

ETSI PRETS 300 007-90. Integrated Services Digital Network (ISDN); Support Packet Mode Terminal Equipment by an ISDN Also Known as T/S 46-50. 116 pp.

ETSI ETS 300 007-91. Integrated Services Digital Network (ISDN); Support of Packet-Mode Terminal Equipment by an ISDN. 103 pp.

ETSI PRETS 300 007-91. Integrated Services Digital Network (ISDN); Support of Packet-Mode Terminal Equipment by an ISDN. 104 pp.

ETSI PRETS 300 007-90. Integrated Services Digital Network (ISDN): Support of Packet Mode Terminal Equipment by an ISDN (T/S 46-50). 116 pp.

IEC 9797-89. Data Cryptographic Techniques—Data Integrity Mechanism Using a Cryptographic Check Function Employing a Block Cipher Algorithm First Edition. 7 pp.

ISO 9574-92. Information Technology—Provision of the OSI Connection-Mode Network Service by Packet Mode Terminal Equipment to an Integrated Services Digital Network (ISDN) Second Edition. 23 pp.

JTC1 9574-92. Information Technology—Provision of the OSI Connection-Mode Network Service by Packet Mode Terminal Equipment to an Integrated Services Digital Network (ISDN) Second Edition. 23 pp.

OSI ISO 8880-2 DAM 1-91. Telecommunications and Information Exchange Between Systems—Protocol Combinations to Provide and Support and Support the OSI Network Service—Part 2: Provision and Support of the Connection-Mode Network Service-Amendment 1: Addition of the ISDN Environment.

OSI ISO IEC 9574-89. Information Technology—Telecommunications and Information Exchange Between Systems—Provision of the OSI Connection-Mode Network Service by Packet Mode Terminal Equipment Connected to an Integrated Services Digital Network (ISDN). 15 pp.

OSI ISO IEC DIS 9574-88. Information Processing Systems—Data Communications—Provision of the OSI Connection—Mode Network Service by Packet Mode Terminal Equipment Connected to an Integrated Services Digital Network. 16 pp.

—Packet Level Protocol—Network Services

BSI BS ISO/IEC 10588-93. 1993 Information Technology—Use of X.25 Packet Layer Protocol in Conjunction with X.21/X.21 bis to Provide the OSI Connection-Mode Network Service (S). 18 pp.

CCITT RECMN X.613-92. Information Technology—Use of X.25 Packet Layer Protocol in Conjunction with X.21/X.21 bis to Provide the OSI Connection-Mode Network Service 15 pp (ISO/CCIT Common Text). 15 pp.

IEC 10177-93. Information Technology—Telecommunications and Information Exchange Between Systems—Provision of the Connection-Mode Network Internal Layer Service by Intermediate Systems Using ISO/IEC 8208, the X.25 Packet Layer Protocol First Edition. 34 pp.

IEC 10588-93. Information Technology—Use of X.25 Packet Layer Protocol in Conjunction with X.21/X.21 bis to Provide the OSI Connection-Mode Network Service First Edition; (CCITT RECMN X.613). 15 pp.

ISO 10177-93. Information Technology—Telecommunications and Information Exchange Between Systems—Provision of the Connection-Mode Network Internal Layer Service by Intermediate Systems Using ISO/IEC 8208, the X.25 Packet Layer Protocol First Edition. 34 pp.

ISO 10588-93. Information Technology—Use of X.25 Packet Layer Protocol in Conjunction with X.21/X.21 bis to Provide the OSI Connection-Mode Network Service First Edition; (CCITT RECMN X.613). 15 pp.

JTC1 10177-93. Information Technology—Telecommunications and Information Exchange Between Systems—Provision of the Connection-Mode Network Internal Layer Service by Intermediate Systems Using ISO/IEC 8208, the X.25 Packet Layer Protocol First Edition. 34 pp.

JTC1 10588-93. Information Technology—Use of X.25 Packet Layer Protocol in Conjunction with X.21/X.21 bis to Provide the OSI Connection-Mode Network Service First Edition; (CCITT RECMN X.613). 15 pp.

Data Terminal Equipment *(Cont.)*

—Packet Level Protocol—Network Services—Telephone Networks

BSI BS ISO/IEC 10732-93. 1993 Information Technology—Use of X.25 Packet Layer Protocol to Provide the OSI Connection-Mode Network Service over the Telephone Network (S). 14 pp.

CCITT RECMN X.614-92. Information Technology—Use of X.25 Packet Layer Protocol to Provide the OSI Connection-Mode Network Service over the Telephone Network 11 pp (ISO/CCIT Common Text). 11 pp.

IEC 10732-93. Information Technology—Use of X.25 Packet Layer Protocol to Provide the OSI Connection-Mode Network Service over the Telephone Network First Edition; (CCITT RECMN X.614). 12 pp.

ISO 10732-93. Information Technology—Use of X.25 Packet Layer Protocol to Provide the OSI Connection-Mode Network Service over the Telephone Network First Edition; (CCITT RECMN X.614). 12 pp.

JTC1 10732-93. Information Technology—Use of X.25 Packet Layer Protocol to Provide the OSI Connection-Mode Network Service over the Telephone Network First Edition; (CCITT RECMN X.614). 12 pp.

—Packet Switched Networks

IEC 10177-93. Information Technology—Telecommunications and Information Exchange Between Systems—Provision of the Connection-Mode Network Internal Layer Service by Intermediate Systems Using ISO/IEC 8208, the X.25 Packet Layer Protocol First Edition. 34 pp.

ISO 10177-93. Information Technology—Telecommunications and Information Exchange Between Systems—Provision of the Connection-Mode Network Internal Layer Service by Intermediate Systems Using ISO/IEC 8208, the X.25 Packet Layer Protocol First Edition. 34 pp.

JTC1 10177-93. Information Technology—Telecommunications and Information Exchange Between Systems—Provision of the Connection-Mode Network Internal Layer Service by Intermediate Systems Using ISO/IEC 8208, the X.25 Packet Layer Protocol First Edition. 34 pp.

—Packet Switched Networks—Conformance

BSI BS ISO/IEC 8882-3-91. 1991 Information Technology—Telecommunications and Information Exchange Between Systems—X.25-DTE Conformance Testing—Part 3: Packet Layer Conformance Suite (S). 701 pp.

IEC 8882 Pt 3-91. Information Technology—Telecommunications and Information Exchange Between Systems—X.25-DTE Conformance Testing—Part 3: Packet Layer Conformance Test Suite First Edition. 698 pp.

ISO 8882 Pt 3-91. Information Technology—Telecommunications and Information Exchange Between Systems—X.25-DTE Conformance Testing—Part 3: Packet Layer Conformance Test Suite First Edition. 698 pp.

JTC1 8882-91. Information Technology—Telecommunications and Information Exchange Between Systems—X.25-DTE Conformance Testing—Part 3: Packet Layer Conformance Test Suite First Edition. 698 pp.

—Packet Switched Networks—High Level Data Link Control—LAPB

IEC DIS 7776-93. Information Technology—Telecommunications and Information Exchange Between Systems—High-Level Data Link Control Procedures—Description of the X.25 LAPB-Compatible DTE Data Link Procedures ***CD-ROM ONLY***. 44 pp.

ISO 7776-86. Information Processing Systems—Data Communi-cation—High-Level Data Link Control Procedures—Description of the X.25 LAPB-Compatible DTE Data Link Procedures Amendment 1: Conformance Requirements First Ed.; (Corrigendum 1-2:1989) (Corrigendum 3-1991) (Amendment 1-1992). 47 pp.

ISO DIS 7776-93. Information Technology—Telecommunications and Information Exchange Between Systems—High-Level Data Link Control Procedures—Description of the X.25 LAPB-Compatible DTE Data Link Procedures ***CD-ROM ONLY***. 44 pp.

JTC1 7776-86. Information Processing Systems—Data Communi-cation—High-Level Data Link Control Procedures—Description of the X.25 LAPB-Compatible DTE Data Link Procedures Amendment 1: Conformance Requirements First Ed.; (Corrigendum 1-2:1989) (Corrigendum 3-1991) (Amendment 1-1992). 31 pp.

JTC1 DIS7776-93. Information Technology—Telecommunications and Information Exchange Between Systems—High-Level Data Link Control Procedures—Description of the X.25 LAPB-Compatible DTE Data Link Procedures ***CD-ROM ONLY***. 44 pp.

INDUSTRY STANDARDS 739

Data Terminal Equipment (Cont.)

—Packet Switched Networks—High Level Data Link Control—LAPB (Cont.)

OSI ISO 7776-86. Information Processing Systems—Data Communication—High-Level Data Link Control Procedures—Description of the X.25 LAPB—Compatible DTE Data Link Procedures (with Technical Corrigendums 1-3). 33 pp.

—Performance Parameters—Packet Switched Public Data Networks

CCITT RECMN X.139-92. Echo, Drop, Generator and Test DTEs for Measurement of Performance Values in Public Data Networks when Providing International Packet-Switched Services (Study Group VII) 12 pp. 16 pp.

—Private Automatic Branch Exchanges—Automatic Redial

CEPT T/SF 26-83. Exigences D'Exploitation Et Les Procedures Pour La Commande Du Dernier Numerocompose (Enregistrement Automatique)-Exigences Relatives a La Solution Adoptee Pour Le Terminal. 6 pp.

—Public Data Networks

CEN ENV 41 901-87. Information Systems Interconnection: X.28-Mode Procedures; X.3 Character-Mode Access; X.28 Character-Mode Access. 43 pp.

CENELEC ETS 300 123-91. Attachment Reqmts for Data Terminal Equip. (DTE) to Connect to Packet Switched Publ. Data Networks (PSPDN) Using CCITT Recommendation X.25 (1984) Interface Reqmts Appl. to DTEs Sub. to Link Access Procedure Balanced (LAPB) Extended (Modulo 128) Operation. 8 pp.

CENELEC PRETS 300 124-90. Attachment Requirements for Data Terminal Equipment (DTE) to Connect to Packet Switched Public Data Networks (PSPDN) Using CCITT Recommendation X.25 (1984) Interface Requirements Applicable to DTEs Subscribing to Multilink Operation. 20 pp.

CENELEC ETS 300 124-91. Attachment Requirements for Data Terminal Equipment (DTE) to Connect to Packet Switched Public Data Networks (PSPDN) Using CCITT Recommendation X.25 (1984) Interface Requirements Applicable to DTEs Subscribing to Multilink Operation. 21 pp.

CEPT NET 1-88. Approval Requirements for Data Terminal Equipment to Connect to Circuit Switched Public Data Networks and Leased Circuits Using CCITT Recommendation X.21 Interface (T/TE 04-07E). 67 pp.

CEPT NET 2-88. Approval Requirements for Data Terminal Equipment to Connect to Packet Switched Public Data Networks Using CCITT Recommendation X.25 (1984) Interface (T/TE 04-06E). 99 pp.

ECMA ECMA-TR 16-83. Interface Characteristics for a DTE to Operate with European REC.X.25 Networks. 15 pp.

ETSI PRETS 300 124-90. Attachment Requirements for Data Terminal Equipment (DTE) to Connect to Packet Switched Public Data Networks (PSPDN) Using CCITT Recommendation X.25 (1984) Interface Requirements Applicable to DTE's Subscribing to Multilink Operation. 20 pp.

ETSI ETS 300 124-91. Attachment Requirements for Data Terminal Equipment (DTE) to Connect to Packet Switched Public Data Networks (PSPDN) Using CCITT Recommendation X.25 (1984) Interface Requirements Applicable to DTEs Subscribing to Multilink Operation. 21 pp.

—Public Data Networks—Access—Accounting

CEPT T/PGT 35 E-90. Accounting Principles for the Use of X.32 Access Method. 1 p.

—Public Data Networks—Maintenance

CCITT RECMN X.150-89. Principles of Maintenance Testing for Public Data Networks Using Data Terminal Equipment (DTE) and Data Circuit-Terminating Equipment (DCE) Test Loops—Data Communication Networks—Transmission, Signalling and Switching, Network Aspects, Maintenance and Administrative. 7 pp.

—Public Data Transmission

CCITT RECMN X.10-89. Categories of Access for Data Terminal Equipment (DTE) to Public Data Transmission Services—Data Communication Networks: Services and Facilities, Interfaces (Study Group VII) 8 pp. 8 pp.

—Public Land Mobile Networks

CCITT RECMN Q.1000-89. Structure of the Q.1000-Series Recommendations for Public Land Mobile Networks—Public Land Mobile Network Interworking with ISDN and PSTN (Study Group XI) 3 pp. 3 pp.

Data Terminal Equipment (Cont.)

—Serial Data Transmission—Interfaces

CNS C5177-88. Interface Between Data Terminal Equipment and Data Circuit-Terminating Equipment Employing Serial Binary Data Interchange Rules (Jul)(10685).

CNS C5180-88. Interface Between Data Terminal Equipment and Data Circuit-Terminating Equipment Employing Serial Binary Data Interchange Functional Description of Interchange Circuits (Jul)(10688).

CNS C5181-88. Interface Between Data Terminal Equipment and Data Circuit-Terminating Equipment Employing Serial Binary Data Interchange Interface for Selected Communication System Configurations (Jul)(10689).

CNS C5182-88. Interface Between Data Terminal Equipment and Data Circuit-Terminating Equipment Employing Serial Binary Data Interchange Recommendations and Explanatory Notes (Jul)(10690).

—Serial Data Transmission—Interfaces—Glossaries

CNS C5183-88. Interface Between Data Terminal Equipment and Data Circuit-Terminating Equipment Employing Serial Binary Data Interchange (Jul)(10691).

—Serial Data Transmission—Interfaces—Mechanical Properties

CNS C5179-88. Interface Between Data Terminal Equipment and Data Circuit-Terminating Equipment Employing Serial Binary Data Interchange Mechanical Characteristics (Jul)(10687).

—Serial Data Transmission—Interfaces—Signal Quality

CNS C5178-88. Interface Between Data Terminal Equipment and Data Circuit-Terminating Equipment Employing Binary Data Interchange Characteristics (Jul)(10686).

CNS C5224-88. Signal Quality at Interface Between Data Terminal Equipment and Synchronous Data Circuit -Terminating Equipment for Serial Data Transmission (Aug)(12396).

CNS C5225-88. Signal Quality Using Serial Data Transmission at the Interface with Non-Synchronous Data Circuit-Terminating Equipment (Aug)(12397).

—Signal Quality

BSI BS 6638: Part 1-87. (WITHDRAWN) 1987 Guide to Transmission Signal Quality at DTE/DCE Interfaces Part 1: Start-Stop Signal Quality (Superseded by BS ISO/IEC 7480: 1991). 13 pp.

BSI BS 6638: Part 2-89. 1989 Guide to Transmission Signal Quality at DTE/DCE Interfaces Part 2: Synchronous Transmission. 15 pp.

BSI BS ISO/IEC 7480-91. 1991 Information Technology—Telecommunications and Information Exchange Between Systems—Start-Stop Transmission Signal Quality at DTE/DCE Interfaces. 17 pp.

IEC 7480-91. Information Technology—Telecommunications and Information Exchange Between Systems—Start-Stop Transmission Signal Quality at DTE/DCE Interfaces Second Edition. 14 pp.

ISO 7480-91. Information Technology—Telecommunications and Information Exchange Between Systems—Start-Stop Transmission Signal Quality at DTE/DCE Interfaces Second Edition. 14 pp.

ISO 9543-89. Information Processing Systems—Information Exchange Between Systems—Synchronous Transmission Signal Quality at DTE/DCE Interfaces First Edition. 14 pp.

JTC1 7480-91. Information Technology—Telecommunications and Information Exchange Between Systems—Start-Stop Transmission Signal Quality at DTE/DCE Interfaces Second Edition. 14 pp.

JTC1 9543-89. Information Processing Systems—Information Exchange Between Systems—Synchronous Transmission Signal Quality at DTE/DCE Interfaces First Edition. 14 pp.

OSI ISO/IEC DIS 7480-91. Information Technology—Telecommunications and Information Exchange Between Systems—Start/ Stop Transmission Quality at DTE/DCE Interfaces. 16 pp.

OSI ISO 7480-84. Information Processing-Start-Stop Transmission Signal Quality at DTE/DCE Interfaces. 12 pp.

OSI ISO DIS 9543-87. Information Processing Systems-Information Exchange Between Systems—Synchronous Transmission Signal Quality at DTE/DCE Interfaces. 20 pp.

SNZ NZS/ISO 7480-84. Information Processing—Start-Stop Transmission Signal Quality at DTE/DCE Interfaces. 10 pp.

—Signaling Systems

CEPT T/SF 34 E-84. Customer Equipment Aspects of Terminal to Terminal Signalling. 2 pp.

Data Terminal Equipment (Cont.)

—Single Channel Per Carrier, Phase-Shift-Keying—Earth Stations

IEC 835 Pt 3-9-93. Methods of Measurement for Equipment Used in Digital Microwave Radio Transmission Systems Part 3: Measurement on Satellite Earth Stations Section 9—Terminal Equipment SCP-PSK First Edition. 58 pp.

—Start-Stop Transmission—International Telegraph Alphabet No. 5

CCITT RECMN S.31-89. Transmission Characteristics for Start-Stop Data Terminal Equipment Using International Alphabet No. 5—Telegraph Services Terminal Equipment (Study Group IX) 2 pp. 2 pp.

—Telegraph Equipment

CCITT FASCICLE VII.1. Telegraph Transmission—Series R Recommendations Telegraph Services Terminal Equipment Series S Recommendations. 280 pp.

—Teletex Communications

CCITT FASCICLE VII.3. Terminal Equipment and Protocols for Telematic Services—Recommendations T.0-T.63. 494 pp.

CCITT RECMN T.60-89. Terminal Equipment for Use in the Teletex Service—Terminal Equipment and Protocols for Telematic Services (Study Group VIII) 20 pp. 20 pp.

CCITT FASCICLE VII.6. Terminal Equipment and Protocols for Telematic Services—Recommendations T.400–T.418. 482 pp.

CCITT FASCICLE VII.7. Terminal Equipment and Protocols for Telematic Services—Recommendations T.431–T.564. 329 pp.

OSI ISO DP 9071-2-87. Text and Office Systems—Basic and Optional Requirements—Part 2: Text Communication Terminals. 11 pp.

—Teletex Communications—Interworking

CCITT RECMN U.70-89. Telex Service Signals for Telex to Teletex Interworking—Telegraph Switching (Study Group IX) 3 pp. 3 pp.

—Teletex Communications—Mixed Mode

CCITT RECMN T.561-89. Terminal Characteristics for Mixed Mode of Operation MM—Terminal Equipment and Protocols for Telematic Services (Study Group VIII) 12 pp. 12 pp.

—Teletex Communications—Processable Mode Number One

CCITT RECMN T.562-89. Terminal Characteristics for Teletex Processable Mode PM.1—Terminal Equipment and Protocols for Telematic Services (Study Group VIII) 7 pp. 7 pp.

—Teletex Communications—Verification (CCITT)

CCITT RECMN T.63-89. Provisions for Verification of Teletex Terminal Compliance—Terminal Equipment and Protocols for Telematic Services (Study Group VIII) 20 pp. 20 pp.

—Telex Communications

BSI BS 7362-90. 1990 Telex Terminal Apparatus for Connection to Public Telecommunications Operators Telex Networks. 51 pp.

—Telex Communications—Answerback

CCITT RECMN F.73-90. Operational Principles for Communication Between Terminals of the International Telex Service and Data Terminal Equipment on Packet Switched Public Data Networks (Study Group I) 9 pp. 9 pp.

CCITT RECMN F.73-89. Operational Principles for Communication Between Terminals on Telex Networks and Data Terminal Equipment on Packet Switched Public Data Networks—Telegraph and Mobile Services Operations and Quality of Service (Study Group I) 5 pp. 5 pp.

CCITT RECMN U.75-89. Automatic Called Telex Answerback Check—Telegraph Switching (Study Group IX) 3 pp. 3 pp.

—Telex Communications—Data Circuit Terminating—Answering/Calling

CCITT RECMN S.19-89. Calling and Answering in the Telex Network with Automatic Terminal Equipment—Telegraph Services Terminal Equipment (Study Group IX) 3 pp. 3 pp.

—Telex Communications—Data Circuit Terminating Equipment

CCITT RECMN S.16-89. Connection to the Telex Network of an Automatic Terminal Using a V.24 (1) DCE/DTE Interface—Telegraph Services Terminal Equipment (Study Group IX) 10 pp. 10 pp.

Data Terminal Equipment (Cont.)

—Telex Communications—Inquiry Characters
CCITT RECMN F.73-90. Operational Principles for Communication Between Terminals of the International Telex Service and Data Terminal Equipment on Packet Switched Public Data Networks (Study Group I) 9 pp. 9 pp.

CCITT RECMN F.73-89. Operational Principles for Communication Between Terminals on Telex Networks and Data Terminal Equipment on Packet Switched Public Data Networks—Telegraph and Mobile Services Operations and Quality of Service (Study Group I) 5 pp. 5 pp.

—Telex Communications—Interworking
CCITT RECMN U.70-89. Telex Service Signals for Telex to Teletex Interworking—Telegraph Switching (Study Group IX) 3 pp. 3 pp.

—Telex Communications—Not Ready Condition
CCITT RECMN U.45-89. Response to the Not-Ready Condition of the Telex Terminal—Telegraph Switching (Study Group IX) 3 pp. 3 pp.

—Telex Communications—Packet Switched Public Data Networks
CCITT RECMN F.73-90. Operational Principles for Communication Between Terminals of the International Telex Service and Data Terminal Equipment on Packet Switched Public Data Networks (Study Group I) 9 pp. 9 pp.

CCITT RECMN F.73-89. Operational Principles for Communication Between Terminals on Telex Networks and Data Terminal Equipment on Packet Switched Public Data Networks—Telegraph and Mobile Services Operations and Quality of Service (Study Group I) 5 pp. 5 pp.

—Telex Communications—Signaling Systems
CCITT RECMN U.40-89. Reactions by Automatic Terminals Connected to the Telex Network in the Event of Ineffective Call Attempts or Signalling Incidents—Telegraph Switching (Study Group IX) 4 pp. 4 pp.

—Telex Equipment—Answerback
CCITT RECMN S.23-89. Automatic Request of the Answerback of the Terminal of the Calling Party, by the Telex Terminal of the Called Party or by the International Network—Telegraph Services Terminal Equipment (Study Group IX) 2 pp. 2 pp.

—Telex Equipment—Automatic Clearing
CCITT RECMN S.20-89. Automatic Clearing Procedure for a Telex Terminal—Telegraph Services Terminal Equipment (Study Group IX) 2 pp. 2 pp.

—Telex Equipment—Prerecorded Messages
CCITT RECMN S.22-89. "Conversation Impossible" and or Pre-Recorded Message in Response to J/Bell Signals from a Telex Terminal—Telegraph Services Terminal Equipment (Study Group IX) 2 pp. 2 pp.

—Terminal Adapters—Integrated Services Digital Networks
CCITT RECMN I.461-89. Support of X.21, X.21 bis and X.20 bis Based Data Terminal Equipments (DTEs) by an Integrated Services Digital Network (ISDN)—Integrated Services Digital Network (ISDN)—Overall Network Aspects and Functions, ISDN User-Network Interfaces (Study Group XVIII) 1 pp. 1 p.

CCITT RECMN X.30-89. Support of X.21,X.21 bis and X.20 bis Based Data Terminal Equipments (DTEs) by an Integrated Services Digital Network (ISDN)—Data Communication Networks: Services and Facilities, Interfaces (Study Group VII) 43 pp (Same as Recmn I.461). 43 pp.

CENELEC PRETS 300 077-90. Integrated Services Digital Network (ISDN): Attachment Requirements for Terminal Adaptors to Connect to an ISDN at the S/T reference Point (T/TE 04-10) (Candidate NET 7). 187 pp.

CENELEC PRETS 300 104-90. Integrated Services Digital Network (ISDN); Attachment Requirements for Terminal Equipment to Connect to an ISDN Using ISDN Basic Access Layer 3 Aspects. 212 pp.

CENELEC ETS 300 104-91. Integrated Services Digital Network (ISDN); Attachment Requirements for Terminal Equipment to Connect to an ISDN Using ISDN Basic Access Layer 3 Aspects. 177 pp.

CENELEC PRETS 300 153-91. Integrated Services Digital Network (ISDN) Attachment Requirements for Terminal Equipment to Connect to an ISDN Using ISDN Basic Access (T/TE 04-08). 62 pp.

CENELEC PRETS 300 153-92. Integrated Services Digital Network (ISDN); Attachment Requirements for Terminal Equipment to Connect to an ISDN Using ISDN Basic Access. 69 pp.

CENELEC ETS 300 153-92. Integrated Services Digital Network (ISDN); Attachment Requirements for Terminal Equipment to Connect to an ISDN Using ISDN Basic Access (Candidate NET 3 Part 1). 70 pp.

CENELEC PRETS 300 156-91. Integrated Services Digital Network (ISDN); Attachment Requirements for Terminal Equipment to Connect to an ISDN Using ISDN Primary Rate Access (T/TE 04-24). 38 pp.

CENELEC PRETS 300 156-92. Integrated Services Digital Network (ISDN); Attachment Requirements for Terminal Equipment to Connect to an ISDN Using ISDN Primary Rate Access. 41 pp.

CENELEC ETS 300 156-92. Integrated Services Digital Network (ISDN); Attachment Requirements for Terminal Equipment to Connect to an ISDN Using ISDN Primary Rate Access (Candidate NET 5). 41 pp.

ETSI PRETS 300 077-90. Integrated Services Digital Network (ISDN); Attachment Requirements for Terminal Adaptors to Connect to an ISDN at the S/T Reference Point (T/TE 04-10) (Candidate NET 7). 186 pp.

ETSI PRETS 300 104-90. Integrated Services Digital Network (ISDN); Attachment Requirements for Terminal Equipment to Connect to an ISDN Using ISDN Basic Access Layer 3 Aspects (The Text of This Draft ETS May Be Utilized, Wholly or in Part, for the Establishment of NET 3 Part 2). 212 pp.

ETSI PRETS 300 153-91. Integrated Services Digital Network (ISDN); Attachment Requirements for Terminal Equipment to Connect to an ISDN Using ISDN Basic Access (T/TE 04-08).

ETSI PRETS 300 156-91. Integrated Services Digital Network (ISDN); Attachment Requirements for Terminal Equipment to Connect to an ISDN Using ISDN Primary Rate Access (T/TE 04-24).

ETSI PRETS 300 077-90. Integrated Services Digital Network (ISDN); Attachment Requirements for Terminal Adaptors to Connect to an ISDN at the S/T reference Point (T/TE 04-10) (Candidate NET 7). 187 pp.

ETSI ETS 300 104-91. Integrated Services Digital Network (ISDN); Attachment Requirements for Terminal Equipment to Connect to an ISDN Using ISDN Basic Access Layer 3 Aspects. 177 pp.

ETSI PRETS 300 104-90. Integrated Services Digital Network (ISDN); Attachment Requirements for Terminal Equipment to Connect to an ISDN Using ISDN Basic Access Layer 3 Aspects. 212 pp.

ETSI ETS 300 153-92. Integrated Services Digital Network (ISDN); Attachment Requirements for Terminal Equipment to Connect to an ISDN Using ISDN Basic Access (Candidate NET 3 Part 1). 70 pp.

ETSI PRETS 300 153-92. Integrated Services Digital Network (ISDN); Attachment Requirements for Terminal Equipment to Connect to an ISDN Using ISDN Basic Access. 69 pp.

ETSI PRETS 300 153-91. Integrated Services Digital Network (ISDN) Attachment Requirements for Terminal Equipment to Connect to an ISDN Using ISDN Basic Access (T/TE 04-08). 62 pp.

ETSI ETS 300 156-92. Integrated Services Digital Network (ISDN); Attachment Requirements for Terminal Equipment to Connect to an ISDN Using ISDN Primary Rate Access (Candidate NET 5). 41 pp.

ETSI PRETS 300 156-92. Integrated Services Digital Network (ISDN); Attachment Requirements for Terminal Equipment to Connect to an ISDN Using ISDN Primary Rate Access. 41 pp.

ETSI PRETS 300 156-91. Integrated Services Digital Network (ISDN); Attachment Requirements for Terminal Equipment to Connect to an ISDN Using ISDN Primary Rate Access (T/TE 04-24). 38 pp.

—Universal Personal Telecommunication Services
CENELEC ETR 055-5-92. Universal Personal Telecommunication (UPT); the Service Concept Part 5: UPT Terminals and UPT Access Devices. 11 pp.

ETSI ETR 055-5-92. Universal Personal Telecommunication (UPT); the Service Concept Part 5: UPT Terminals and UPT Access Devices. 11 pp.

—Video Telephone Services
CCITT RECMN F.720-92. Videotelephony Services—General (Study Group I) 11 pp. 11 pp.

—Videoconferencing Services
CCITT RECMN F.730-92. Videoconference Service—General (Study Group I) 16 pp. 16 pp.

—Videotex Communications
CENELEC PRI-ETS 300 236-92. Terminal Equipment (TE); Syntax-Based Videotex Protocol Terminal Conformance Testing. 457 pp.

CENELEC PRI-ETS 300 236-93. Terminal Equipment (TE); Syntax-Based Videotex Protocol Terminal Conformance Testing. 475 pp.

CENELEC I-ETS 300 236-93. Terminal Equipment (TE); Syntax-Based Videotex Protocol Terminal Conformance Testing. 475 pp.

ETSI I-ETS 300 236-93. Terminal Equipment (TE); Syntax-Based Videotex Protocol Terminal Conformance Testing. 475 pp.

ETSI PRI-ETS 300 236-93. Terminal Equipment (TE); Syntax-Based Videotex Protocol Terminal Conformance Testing. 475 pp.

ETSI PRI-ETS 300 236-92. Terminal Equipment (TE); Syntax-Based Videotex Protocol Terminal Conformance Testing. 457 pp.

—Videotex Communications—Audio Syntax
CENELEC PRETS 300 149-91. Terminal Equipment (TE) Videotex Audio Syntax (T/TE 06-07). 19 pp.

CENELEC PRETS 300 149-92. Terminal Equipment (TE); Videotex Audio Syntax. 18 pp.

CENELEC ETS 300 149-92. Terminal Equipment (TE); Videotex Audio Syntax. 18 pp.

ETSI PRETS 300 149-91. Terminal Equipment (TE); Videotex—Audio Syntax (T/TE 06-07).

ETSI ETS 300 149-92. Terminal Equipment (TE); Videotex Audio Syntax. 18 pp.

ETSI PRETS 300 149-92. Terminal Equipment (TE); Videotex Audio Syntax. 18 pp.

—Videotex Communications—Interworking
CENELEC PRETS 300 105-90. Terminal Equipment (TE): Videotex Interworking (T/TE 06-20). 14 pp.

CENELEC PRETS 300 105-91. Terminal Equipment (TE); International Videotex Interworking (T/TE 06-20). 15 pp.

CENELEC ETS 300 105-91. Terminal Equipment (TE); International Videotex Interworking. 14 pp.

CENELEC PRETS 300 106-90. Terminal Equipment (TE); International Interworking Between a Terminal and a Host (T/TE 06-21). 9 pp.

CENELEC PRETS 300 106-91. Terminal Equipment (TE); International Videotex Interworking Between a Terminal and a Host (T/TE 06-21). 21 pp.

CENELEC ETS 300 106-91. Terminal Equipment (TE); International Videotex Interworking Between a Terminal and a Host. 20 pp.

CENELEC PRETS 300 107-90. Terminal Equipment (TE); International Interworking Between Gateways (T/TE 06-22). 16 pp.

CENELEC PRETS 300 107-91. Terminal Equipment (TE); International Interworking Between Gateways (T/TE 06-22). 14 pp.

CENELEC ETS 300 107-91. Terminal Equipment (TE); International Videotex Interworking Between Gateways. 13 pp.

ETSI PRETS 300 105-90. Terminal Equipment (TE); Videotex Interworking (T/TE 06-20). 31 pp.

ETSI PRETS 300 106-90. Terminal Equipment (TE)—International Interworking Between a Terminal and a Host (T/TE 06-21). 11 pp.

ETSI PRETS 300 107-90. Terminal Equipment (TE)—International Interworking Between Gateways (T/TE 06-22). 18 pp.

ETSI ETS 300 105-91. Terminal Equipment (TE); International Videotex Interworking. 14 pp.

ETSI PRETS 300 105-90. Terminal Equipment (TE): Videotex Interworking (T/TE 06-20). 14 pp.

ETSI ETS 300 106-91. Terminal Equipment (TE); International Videotex Interworking Between a Terminal and a Host. 20 pp.

ETSI PRETS 300 106-91. Terminal Equipment (TE); International Videotex Interworking Between a Terminal and a Host (T/TE 06-21). 21 pp.

ETSI PRETS 300 106-90. Terminal Equipment (TE); International Interworking Between a Terminal and a Host (T/TE 06-21). 9 pp.

ETSI ETS 300 107-91. Terminal Equipment (TE); International Videotex Interworking Between Gateways. 13 pp.

ETSI ETS 300 108-92. Integrated Services Digital Network (ISDN); Circuit-Mode 64 kbit/s Unrestricted 8 kHz Structured Bearer Service Category Service Description. 14 pp.

—Videotex Equipment
CCITT RECMN F.300-89. Videotex Service—Telematic, Data Transmission and Teleconference Services—Operations and Quality of Service (Study Group I) 25 pp. 25 pp.

—Videotex Equipment—Telesoftware
CCITT RECMN F.300-89. Videotex Service—Telematic, Data Transmission and Teleconference Services—Operations and Quality of Service (Study Group I) 25 pp. 25 pp.

—Voice/Data Systems—Integrated Services Digital Networks
CSA CAN/CSA-T531-M91. Acoustic-to-Digital and Digital-to-Acoustic Transmission Requirements for ISDN Terminals; (Gen Instr 1). 47 pp.

INTERNATIONAL AND NON-U.S. NATIONAL STANDARDS
SUBJECT INDEX

Data

Data Transfer Rates
Use For: Data Transmission Rates *See Also:* Bit Rates; Data Transmission

—Signaling Links—CCITT No. 6 Signaling Systems
CCITT RECMN Q.273-89. 6.2 Data Transmission Rate—Specifications of Signalling System No. 6 (Study Group XI) 1 pp. 1 p.

Data Transmission
Use For: Data Communication Systems; Data Transmission Systems; Digital Data Transmission; International Data Transmission Services; Synchronous Data Transmission *See Also:* Automatic Request-Repeat Equipment; Automatic Request-Repeat Mode; Bit Rates; Circuit Switched Data Transmission Services; Data Acquisition Systems; Data Broadcasting; Data Buses; Data Channels; Data Codes; Data Collection; Data Communication; Data Communication Equipment; Data Links; Data Processing; Data Transfer Rates; Dropout; Echo Path Loss; EDIFACT; Encoded Data Transmission; End of Message Codes; End of Transmission Codes; Fast Select Services; Forward Error Correction; Half Duplex Transmission; Information Interchange; Information Theory; Interchange Circuits; International Reference Alphabet; Message Formats; Message Handling Systems; Message Routing; Message Stores; Message Switching; Mixed Mode; Multiplexing; Open Systems Interconnection; Packet Assembly/Disassembly Devices; Packet Switched Data Transmission Services; Packet Switched Networks; Phase Jitter; Processable Mode Number One; Protocols; Public Data Communication; Public Data Networks; Public Data Transmission; Return Loss (Data Transmission); SELCAL Services; Stability Loss (Transmission); Start of Message Codes; Store and Forward Mode; Store and Forward Units; Synchronous Digital Hierarchy; Telecommunication Connections; Telecommunication Equipment; Telecommunication Systems; Telemetry; Teleservices; Time Division Multiplexers; Timing Jitter; Transceivers; Transmission Loss; Transmission Medium Requirements; Transmission Restoration; Transmission Restoration Systems; Unbalance (Data Transmission); Voice/Data Systems

BSI BS 4505: Part 2-90. 1990 Amd 1 Digital Data Transmission Part 2: Character Structure for Start/Stop and Synchronous Transmission. 7 pp.
BSI BS 4505: Part 3-81. (OBSOLESCENT) 1981 Digital Data Transmission Part 3: The Use of Longitudinal Parity to Detect Errors in Information Messages. 3 pp.
CCIR Volume 3 TOC-90. Table of Contents. 2 pp.
CCIR Volume 3 Index-90. Numerical Index of Texts. 1 p.
CCIR Volume 3 TERMS-90. Terms of Reference of Study Group 3. 7 pp.
CEN PREN 3758-92. Simplex High Speed Data Transmission System. 58 pp.
CEN PREN 3910-90. High Speed Data Transmission Under Stanag 3838 or Fiber Optic Equivalent Control. 100 pp.
CEN PREN 3910-92. High Speed Data Transmission Under STANAG 3838 or Fiber Optic Equivalent Control. 93 pp.
CEPT T/L 03-09 E-88. Digital Hierarchy Bit Rates. 1 p.
CEPT T/L 03-16 E-88. Transmission in a 64 kBit/s Channel on Audio Signal Other Than Using A-Law and U-Law. 1 p.
CNS C5105-83. Information Processing — Use of Longitudinal Parity to Detect Error in Information Messages (Feb)(7221).
CNS C5106-89. Information Processing — Character Structure for Start/Stop and Synchronous Transmission (Mar)(7222).
CNS C5146-83. Synchronous High-Speed Data Signaling Rates Between Data Terminal Equipment and Data Communication Equipment (Feb)(9979).
CNS C5160-89. Information Interchange — the Bit Sequencing for Serial-by-Bit Data Transmission (Jul)(10331).
CNS C5162-89. Information Interchange — Character Structure and Character Parity Sense for Parallel-by-Bit Data Communication (Jul)(10427).
JIS X 5001-82. Character Structure on the Transmission Circuits and Horizontal Parity Method.

—Accounting—Telecommunication Administrations
CCITT RECMN E.252-89. Mode of Application of the Flat-Rate Price Procedure Set Forthe in Recommendations D.67 and D.150 for Remuneration of Facilities Made Available to the Administrations of Other Countries—Telephone Network and ISDN—Operation, Numbering, Routing and Mobile. 1 p.

Data Transmission (Cont.)
—Acoustic Couplers
CCITT RECMN V.15-89. Use of Acoustic Coupling for Data Transmission—Data Communication over the Telephone Network (Study Group XVII) 2 pp. 2 pp.

—Aerial Reconnaissance—Imagery
NATO STANAG 7023 ED 1 AMD 0-93. Air Reconnaissance Imagery Data Architecture. 190 pp.

—Audiovisual Equipment—Message Channel
CEPT T/N 34-02 E-88. Message Channel Specification, Layers 1-6 for Use in Audiovisual Communication. 7 pp.

—Bank Cards
OSI ISO 9992-1-90. Financial Transaction Cards—Messages Between the Integrated Circuit Card and the Card Accepting Device—Part 1: Concepts and Structures. 10 pp.

—Banking
BSI BS 6601-85. 1985 Mail Payment Orders. 8 pp.
BSI BS 6695-86. 1986 Nostro Accounts Reconciliation in Banking. 14 pp.
BSI BS 7099: Part 1-89. (WITHDRAWN) 1989 Procedures for Telecommunication of Funds Transfer Messages Between Banks Part 1: Vocabulary and Data Elements (Superseded by BS EN 27982-1: 1991). 23 pp.
BSI BS 7101-89. (WITHDRAWN) 1989 Procedures for Protecting Authentic Wholesale Messages Between Financial Institutions (Superseded by BS ISO 8730: 1990). 11 pp.
BSI BS 7102: Part 1-89. 1989 Recommendations for Algorithms for Use in Banking Message Authentication Part 1: Data Encryption Algorithm (DEA). 6 pp.
BSI BS 7102: Part 2-89. (WITHDRAWN) 1989 Recommendations for Algorithms for Use in Banking Message Authentication Part 2: Message Authenticator Algorithm (MAA) (Superseded by BS ISO 8731-2: 1992). 12 pp.
BSI BS 7104-90. 1990 Procedure for Allocating a Bank Identifier Code (BIC). 11 pp.
BSI BS EN 27982: Part 1-91. 1991 Bank Telecommunication—Funds Transfer Messages Part 1: Vocabulary and Data Elements (ISO 7982-1: 1987). 28 pp.
BSI BS ISO 8730-90. 1990 Banking—Requirements for Message Authentication (Wholesale). 37 pp.
BSI BS ISO 9564-1-91. 1991 Banking—Personal Identification Number Management and Security—Part 1: PIN Protection Principles and Techniques (G). 38 pp.
BSI BS ISO 9564-2-91. 1991 Banking—Personal Identification Number Management and Security—Part 2: Approved Algorithm(s) for PIN Encipherment (G). 7 pp.
BSI BS ISO 9807-91. 1991 Banking and Related Financial Services—Requirements for Message Authentication (Retail) (G). 19 pp.
BSI BS ISO 10126-1-91. 1991 Banking—Procedures for Message Encipherment (Wholesale) Part 1: General Principles. 26 pp.
CEN EN 27 982 (Part 1)-91. Bank Telecommunication Funds Transfer Messages — Part 1 Vocabulary and Data Elements. 2 pp.
CEPT T/SF 57 E-87. Electronic Funds Transfer Telecommunications Services. 17 pp.
ISO 6260-84. Mail Payment Orders First Edition. 7 pp.
ISO 6680-87. International Cheque Remittance First Edition. 8 pp.
ISO 7341-85. Banking—Nostro Accounts Reconciliation First Edition. 13 pp.
ISO 7982 Pt 1-87. Bank Telecommunication—Funds Transfer Messages—Part 1: Vocabulary and Data Elements First Edition. 23 pp.
ISO 8730-90. Banking—Requirements for Message Authentication (Wholesale) Second Edition. 31 pp.
ISO 8731 Pt 1-87. Banking—Approved Algorithms for Message Authentication—Part 1: DEA First Edition. 4 pp.
ISO 9362-87. Banking—Banking Telecommunication Messages—Bank Indentifier Codes First Edition. 7 pp.
ISO 9564 Pt 1-91. Banking—Personal Identification Number Management and Security—Part 1: PIN Protection Principles and Techniques First Edition. 35 pp.
ISO 9564 Pt 2-91. Banking—Personal Identification Number Management and Security—Part 2: Approved Algorithm(s) for PIN Encipherment First Edition. 5 pp.
ISO 9807-91. Banking and Related Financial Services—Requirements for Message Authentication (Retail) First Edition. 16 pp.
ISO 10126 Pt 1-91. Banking—Procedures for Message Encipherment (Wholesale)—Part 1: General Principles First Edition. 23 pp.

Data Transmission (Cont.)
—Banking (Cont.)
ISO 10383-92. Codes for Exchanges and Regulated Markets—Market Identifier Codes (MIC) First Edition. 6 pp.
ISO 11131-92. Banking and Related Financial Services—Sign-on Authentication First Edition. 15 pp.
JTC1 7982 Pt 1-87. Bank Telecommunication—Funds Transfer Messages—Part 1: Vocabulary and Data Elements First Edition. 23 pp.
JTC1 8730-90. Banking—Requirements for Message Authentication (Wholesale) Second Edition. 31 pp.
JTC1 8731 Pt 1-87. Banking—Approved Algorithms for Message Authentication—Part 1: DEA First Edition. 4 pp.
JTC1 9362-87. Banking—Banking Telecommunication Messages—Bank Indentifier Codes First Edition. 7 pp.
JTC1 9564 Pt 1-91. Banking—Personal Identification Number Management and Security—Part 1: PIN Protection Principles and Techniques First Edition. 35 pp.
JTC1 9564 Pt 2-91. Banking—Personal Identification Number Management and Security—Part 2: Approved Algorithm(s) for PIN Encipherment First Edition. 5 pp.
JTC1 9807-91. Banking and Related Financial Services—Requirements for Message Authentication (Retail) First Edition. 16 pp.
JTC1 10126 Pt 1-91. Banking—Procedures for Message Encipherment (Wholesale)—Part 1: General Principles First Edition. 23 pp.
OSI ISO 6260-84. Mail Payment Orders. 7 pp.
OSI ISO 6680-87. International Cheque Remittance. 8 pp.
OSI ISO 7341-85. Banking—Nostro Accounts Reconciliation. 13 pp.
OSI ISO 8730-90. Banking—Requirements for Message Authentication (Wholesale). 31 pp.
OSI ISO 8730-86. Banking—Requirements for Message Authentication (Wholesale). 9 pp.
OSI ISO 8730-86. Banking—Requirements for Message Authentication (Wholesale).
OSI ISO/DIS 8908. Banking and Related Financial Services—Vocabulary and Data Elements. 135 pp.
OSI ISO 9362-87. Banking—Banking Telecommunication Messages—Banker Indentifier Codes. 5 pp.
OSI ISO DP 9777-87. Forms for Confirming Foreign Exchange Deals. 14 pp.
OSI ISO DP 9778-87. Forms for Confirming Loan/Deposit Contracts. 21 pp.

—Banking—Telex Communications
BSI BS 7098-89. 1989 Method of Formatting Inter-Bank Telexes for Messages Relating to the Transfer of Funds or Other Financial Messages. 55 pp.
ISO 7746-88. Banking—Telex Formats for Inter-Bank Messages First Edition. 52 pp.
JTC1 7746-88. Banking—Telex Formats for Inter-Bank Messages First Edition. 52 pp.
OSI ISO 7746-88. Banking—Telex Formats for Interbank Messages. 52 pp.

—Binary Number Systems—Codes
CCITT RECMN V.1-89. Equivalence Between Binary Notation Symbols and the Significant Conditions of a Two-Condition Code—Data Communication Over the Telephone Network (Study Group XVII) 2 pp. 2 pp.

—Cables (Electric)
MOD UK DSTAN 61-12: Part 23-01. Wires, Cords, and Cables, Electrical—Metric Units Part 23: General Requirements for Cables, Special Purpose, Electrical (Digital Data Transmission) Issue 1; Amendment 1. 25 pp.
MOD UK DSTAN 61-12: 90/002-01. Cables, Special Purpose, Electrical (Digital Data Transmission) Issue 2; Amendment 1. 14 pp.

—Call Control Procedures—Public Networks—Interworking
CCITT RECMN X.301-89. Description of the General Arrangements for Call Control Within a Subnetwork and Between Subnetworks for the Provision of Data Transmission Services—Data Communication Networks—Interworking Between Networks, Mobile Data Transmission Systems, Internetwork. 72 pp.

—Cooperation—CCITT
CCITT RECMN A.20-89. Collaboration with Other International Organizations over Data Transmission—Data Communication over the Telephone Network (Study Group XVII) 2 pp. 2 pp.

Data Transmission (Cont.)

—Designations
CCITT RECMN M.140-89. Designations of International Circuits, Groups, Group and Line Links, Digital Blocks, Digital Paths, Data Transmission Systems and Related Information—General Maintenance Principles—Maintenance of International Transmission Systems and Telephone Circuits. 53 pp.

—Distortion Analyzers—Maintenance
CCITT RECMN V.52-89. Characteristics of Distortion and Error-Rate Measuring Apparatus for Data Transmission—Data Communication over the Telephone Network (Study Group XVII) 1 pp. 1 p.

—Electronic Chart Display Systems—Updates
CCIR Report 1165-90. Transmission of Digital Data for the Updating of Electronic Chart Display Systems (ECDIS)—Section 8D—Radiodetermination, Global Maritime Distress and Safety System and Related Subjects. 15 pp.
CCIR QUESTION 98/8-90. Transmission of Digital Data for the Updating of Electronic Chart Display Systems (ECDIS)—Questions Concerning Study Group 8—Mobile, Radiodetermination, Amateur and Related Satellite Services. 1 p.
CCIR RECMN 826-92. Transmission of Information for Updating Electronic Chart Display and Information Systems (ECDIS). 2 pp.

—Frame Structure
CEPT T/L 03-10 E-88. Synchronous Frame Structures, Frame Alignment and CRC Procedures Used at Primary and Secondary Hierarchical Levels. 1 p.
CEPT T/TR 01-03-86. Structure De Trame a 64 Kbit/s Pour Application Multimedias. 8 pp.
CEPT T/TR 01-03 E-86. Frame Structure at 64 kBit/s for Multimedia Applications. 8 pp.

—Frequency Bands
CEPT T/R 10-01 E-93. Wide Band Data Transmission Systems Using Spread-Spectrum Technology in the 2.5 GHz Band. 2 pp.
CEPT T/R 10-01 E-92. Relating to the Harmonized Radio Frequency Bands for Wide Band Data Transmission Systems Using Spread Spectrum Technology. 2 pp.

—Gateways—Military Communications
NATO STANAG 4213 ED 1 AMD 0-88. The NATO Multi-Channel Tactical Digital Gateway —Data Transmission Standards. 24 pp.
NATO STANAG 4213 ED 2 AMD 0-89. (DRAFT) The NATO Multi-Channel Tactical Digital Gateway—Data Transmission Standards. 27 pp.

—Glossaries
CEPT T/L 03-08 E-88. Vocabulary of Digital Transmission and Multiplexing and Pulse Code Modulating (PCM) Terms. 1 p.
IEC DIS 2382 Pt 27-92. Information Technology—Vocabulary—Part 27: Office Automation ***CD-ROM ONLY***. 22 pp.
ISO DIS 2382 Pt 27-92. Information Technology—Vocabulary—Part 27: Office Automation ***CD-ROM ONLY***. 22 pp.
JTC1 DIS2382 Pt 27-92. Information Technology—Vocabulary—Part 27: Office Automation ***CD-ROM ONLY***. 22 pp.
SAA AS 1189.9-85. Data Processing—Vocabulary—Part 9: Data Communication. 12 pp.

—Group Links
CCITT RECMN H.14-89. Characteristics of Group Links for the Transmission of Wide-Spectrum Signals—Line Transmission of Non-Telephone Signals—Transmission of Sound-Programme and Television Signals (Study Group XV) 1 pp. 1 p.
CCITT RECMN H.52-89. Transmission of Wide-Spectrum Signals (Data, Facsimile, Etc.) on Wideband Group Links—Line Transmission of Non-Telephone Signals—Transmission of Sound-Programme and Television Signals (Study Group XV) 1 pp. 1 p.
CCITT RECMN M.900-89. Use of Leased Group and Supergroup Links for Wide-Spectrum Signal Transmission (Data, Facsimile, Etc.)—Maintenance of International Telegraph, Phototelegraph and Leased Circuits—Maintenance of the International Public Telephone Network-. 4 pp.
CCITT RECMN M.910-89. Setting up and Lining up an International Leased Group Link for Wide-Spectrum Signal Transmission—Maintenance of International Telegraph, Phototelegraph and Leased Circuits—Maintenance of the International Public Telephone Network-. 4 pp.

—High Frequency—Network Architecture
CCIR Report 995-86. Network Architecture for HF Data Transmission—Section 3Cb—Data Transmission. 4 pp.

Data Transmission (Cont.)

—Identification Cards
BSI BS 7109: Part 3-90. 1990 Guide to Design and Use of Identification Cards Having Integrated Circuits with Contacts Part 3: Electronic Signals and Transmission Protocols. 21 pp.
BSI BS 7109: Part 3-01. 1990 Amd 1 Guide to Design and Use of Identification Cards Having Integrated Circuits with Contacts Part 3: Electronic Signals and Transmission Protocols (ISO/IEC 7816-3: 1989) (AMD 7325) November 15, 1992. 25 pp.
CEN EN 27816-3-92. Identification Cards—Integrated Circuit(s) Cards with Contacts—Part 3: Electronic Signals and Transmission Protocols. 21 pp.
IEC 7816 Pt 3-89. Identification Cards—Integrated Circuit(s) Cards with Contacts—Part 3: Electronic Signals and Transmission Protocols Amendment 1: Protocol Type T = 1, Asynchronous Half Duplex Block Transmission Protocol First Edition; (Amendment 1-1992). 35 pp.
ISO 7816 Pt 3-89. Identification Cards—Integrated Circuit(s) Cards with Contacts—Part 3: Electronic Signals and Transmission Protocols Amendment 1: Protocol Type T = 1, Asynchronous Half Duplex Block Transmission Protocol First Edition; (Amendment 1-1992). 35 pp.
JTC1 7816-89. Identification Cards—Integrated Circuit(s) Cards with Contacts—Part 3: Electronic Signals and Transmission Protocols Amendment 1: Protocol Type T = 1, Asynchronous Half Duplex Block Transmission Protocol First Edition; (Amendment 1-1992). 19 pp.
OSI ISO/IEC 7816-3 DAM 1-90. Identification Cards—Integrated Circuit(s) Cards with Contacts—Part 3: Electronic Signals and Transmission Protocols AMENDMENT 1. 27 pp.
OSI ISO IEC 7816-3-89. Identification Cards—Integrated Circuit(s) Cards with Contacts—Part 3: Electronic Signals and Transmission Protocols. 19 pp.
OSI ISO DIS 7816-3-87. Identification Cards—Integrated Circuit Cards with Contacts—Part 3: Electronic Signals and Exchange Protocols. 28 pp.
OSI ISO DIS 7816-3-87. Identification Cards—Integrated Circuit Cards with Contacts—Part 3: Electronic Signals and Exchange Protocols. 28 pp.
OSI ISO DIS 7816-3-89. Identification Cards—Integrated Circuit(s) Cards with Contacts—Part 3: Electronic Signals and Transmission Protocols. 19 pp.
OSI ISO IEC 7816-3-89. Identification Cards—Integrated Circuit(s) Cards with Contacts—Part 3: Electronic Signals and Transmission Protocols. 19 pp.

—Impulse Noise
CCITT RECMN H.16-89. Characteristics of an Impulsive-Noise Measuring Instrument for Wideband Data Transmission—Line Transmission of Non-Telephone Signals—Transmission of Sound-Programme and Television Signals (Study Group XV) 1 pp (Same as Recmn O.72). 1 p.
CCITT RECMN O.72-89. Characteristics of an Impulsive Noise Measuring Instrument for Wideband Data Transmissions—Specifications for Measuring Equipment (Study Group IV) 1 pp (Same as Recmn H.16). 1 p.

—Interfaces
CENELEC PRETS 300 166-91. Transmission and Multiplexing Physical/Electrical Characteristics of Hierarchical Digital Interfaces for Equipment Using the 2048 Kbit/s-Based Plesiochronous or Synchronous Digital Hierarchies (DE/TM-3002). 8 pp.
CENELEC PRETS 300 166-93. Transmission and Multiplexing (TM); Physical and Electrical Characteristics of Hierarchial Digital Interfaces for Equipment Using the 2 048 kbit/s—Based Plesiochronous or Synchronous Digital Hierarchies. 10 pp.
ETSI PRETS 300 166-91. Transmission and Multiplexing Physical/Electrical Characteristics of Hierarchical Digital Interfaces for Equipment Using the 2048 Kbit/s-Based Plesiochronous or Synchronous Digital Hierarchies (DE/TM-3002). 8 pp.
ETSI PRETS 300 166-93. Transmission and Multiplexing (TM); Physical and Electrical Characteristics of Hierarchical Digital Interfaces for Equipment Using the 2 048 kbit/s—Based Plesiochronous or Synchronous Digital Hierarchies. 10 pp.
ETSI PRETS 300 166-91. Transmission and Multiplexing Physical/Electrical Characteristics of Hierarchical Digital Interfaces for Equipment Using the 2048 Kbit/s-Based Plesiochronous or Synchronous Digital Hierarchies (DE/TM-3002). 8 pp.

—Interfaces—Protocols
CENELEC PRETS 300 150-91. Transmission and Multiplexing Protocol Suites for Q Interfaces for Management of Transmission Systems (DE/TM-2001). 47 pp.

Data Transmission (Cont.)

—Interfaces—Protocols (Cont.)
CENELEC PRETS 300 150-92. Transmission and Multiplexing (TM); Protocol Suites for Q Interfaces for Management of Transmission Systems. 44 pp.
CENELEC ETS 300 150-92. Transmission and Multiplexing (TM); Protocol Suites for Q Interfaces for Management of Transmission Systems. 47 pp.
ETSI PRETS 300 150-91. Transmission and Multiplexing—Protocol Suites for Q Interfaces for Management of Transmission Systems (DE/TM-2001).
ETSI ETS 300 150-92. Transmission and Multiplexing (TM); Protocol Suites for Q Interfaces for Management of Transmission Systems. 47 pp.
ETSI PRETS 300 150-92. Transmission and Multiplexing (TM); Protocol Suites for Q Interfaces for Management of Transmission Systems. 44 pp.
ETSI PRETS 300 150-91. Transmission and Multiplexing Protocol Suites for Q Interfaces for Management of Transmission Systems (DE/TM-2001). 47 pp.
MOD UK NES 1028: Part 2-85. Standard for Inter-System Communications Protocols Part 2: Rules and Protocols for Inter—Communication Across Digital Highways Issue 1 (03.85). 47 pp.

—International Alphabet No. 5—Public Data Networks
CCITT RECMN X.4-89. General Structure of Signals of International Alphabet No. 5 Code for Character Oriented Data Transmission over Public Data Networks—Data Communication Networks: Services and Facilities, Interfaces (Study Group VII) 2 pp. 2 pp.

—International Leased Circuits
CCITT RECMN M.1015-89. Types of Transmission on Leased Circuits—Maintenance of International Telegraph, Phototelegraph and Leased Circuits—Maintenance of the International Public Telephone Network—Maintenance of Maritime Satellite and Data Transmission Systems. 3 pp.

—ISDN
CCITT RECMN F.353-89. Provision of Telematic and Data Transmission Services on Integrated Services Digital Network (ISDN)—Telematic, Data Transmission and Teleconference Services—Operations and Quality of Service (Study Group I) 5 pp. 5 pp.
CCITT RECMN I.326-89. Reference Configuration for Relative Network Resource Requirements—Integrated Services Digital Network (ISDN)—Overall Network Aspects and Functions, ISDN User-Network Interfaces (Study Group XVIII) 3 pp. 3 pp.
CEPT T/L 05-03 E-88. Digital Transmission System on Metallic Local Lines for ISDN Basic Rate Access. 25 pp.
CEPT T/SF 58 E-87. Telematic and Data Transmission Services in the ISDN. 24 pp.

—ISDN—Circuit Switched Public Data Networks—Interworking
CCITT RECMN I.540-89. General Arrangements For Interworking Between Circuit Switched Public Data Networks (CSPDNs) and Integrated Services Digital Networks (ISDNs) for the Provision of Data Transmission—Integrated Services Digital Network (ISDN)—Internetwork Interfaces and Maintenance. 1 p.
CCITT RECMN X.321-89. General Arrangements for Interworking Between Circuit Switched Public Data Networks (CSPDNs) and Integrated Service Digital Networks (ISDNs) for the Provision of Data Transmission Services—Data Communication Networks—Interworking Between Networks, Mobile Data. 7 pp.

—ISDN—Interworking
CCITT RECMN X.320-89. General Arrangements for Interworking Between Integrated Services Digital Networks (ISDNs) for the Provision of Data Transmission Services—Data Communication Networks—Interworking Between Networks, Mobile Data Transmission Systems, Internetwork Management. 8 pp.

—ISDN—Public Data Networks
CCITT RECMN X.2-89. International Data Transmission Services and Optional User Facilities in Public Data Networks and (ISDNs)—Data Communication Networks: Services and Facilities, Interfaces (Study Group VII) 12 pp. 12 pp.

—Land Mobile Services
CCIR Report 903-2-90. Digital Transmission in the Land Mobile Service—Section 8A—Land Mobile Service and Related Subjects. 34 pp.

INTERNATIONAL AND NON-U.S. NATIONAL STANDARDS
SUBJECT INDEX

Data Transmission (Cont.)

—Leased Circuits—Circuit Noise
CCITT RECMN G.143-89. Circuit Noise and the Use of Compandors—General Characteristics of International Telephone Connections and Circuits (Study Groups XII and XV) 5 pp. 5 pp.

—Maritime Mobile Satellite Communications—Public Data Networks
CCITT RECMN F.122-89. Operational Procedures for the Maritime Satellite Data Transmission Service—Telegraph and Mobile Services Operations and Quality of Service (Study Group I) 8 pp. 8 pp.

—Military Geographic Information—Forms
NATO STANAG 3986 ED 1 AMD 1-87. Digital Data File Transmittal Form for Geographic Information. 14 pp.
NATO STANAG 3986 ED 1 AMD 2-87. Digital Data File Transmittal Form for Geographic Information. 15 pp.
NATO STANAG 3986 ED 1 AMD 3-87. Digital Data File Transmittal Form for Geographic Information. 16 pp.

—Mobile Satellite Communications—INMARSAT
CCIR Report 1173-90. Technical and Operational Considerations for Aeronautical Mobile-Satellite Communications—Section 8F—Frequencies, Orbits and Systems. 40 pp.

—Mobile Satellite Communications—INMARSAT—Public Data Networks
CCITT RECMN F.122-89. Operational Procedures for the Maritime Satellite Data Transmission Service—Telegraph and Mobile Services Operations and Quality of Service (Study Group I) 8 pp. 8 pp.

—Mobile Satellite Communications—Interworking
CCIR Report 1176-90. Interworking Between the Mobile Satellite Systems and the Terrestrial Networks for Data Transmission Services—Section 8G—Availability, Performance Objectives and Interworking with Terrestrial Networks. 8 pp.
CCIR QUESTION 89-1/8-90. Compatibility for Interworking Between the Mobile-Satellite Systems and Terrestrial Networks Including ISDN—Questions Concerning Study Group 8—Mobile, Radiodetermination, Amateur and Related Satellite Services. 1 p.
CCITT RECMN X.350-89. General Interworking Requirements to Be Met for Data Transmission in International Public Mobile Satellite Systems—Data Communication Networks—Interworking Between Networks, Mobile Data Transmission Systems, Internetwork Management (Study Group VII) 9 pp. 9 pp.

—Mobile Satellite Communications—UHF
CCIR Report 1183-90. Technical and Operational Considerations for the Land Mobile-Satellite Service (LMSS)—Section 8F—Frequencies, Orbits and Systems. 32 pp.

—Modems
CCITT RECMN V.35-89. Data Transmission at 48 Kilobits per Second Using 60-108 kHz Group Band Circuits—Data Communication over the Telephone Network (Study Group XVII) 1 pp. 1 p.

—Modems—Data Signaling Rates
CCITT RECMN V.37-89. Synchronous Data Transmission at a Data Signalling Rate Higher Than 72 kBit/s Using 60-108 kHz Group Band Circuits—Data Communication over the Telephone Network (Study Group XVII) 11 pp. 11 pp.

—Modems—Electrocardiograph Recorders
CCITT RECMN V.16-89. Medical Analogue Data Transmission Modems—Data Communication over the Telephone Network (Study Group XVII) 7 pp. 7 pp.

—Modems—Telephone Networks
CCITT RECMN V.19-89. Modems for Parallel Data Transmission Using Telephone Signalling Frequencies—Data Communication over the Telephone Network (Study Group XVII) 4 pp. 4 pp.
CCITT RECMN V.20-89. Parallel Data Transmission Modems Standardized for Universal Use in the General Switched Telephone Network—Data Communication over the Telephone Network (Study Group XVII) 6 pp. 6 pp.

—Monitors—Remote—Thermal Containers
ISO 10368-92. Freight Thermal Containers—Remote Condition Monitoring First Edition. 95 pp.

Data Transmission (Cont.)

—Multichannel—Television Broadcasting
CCIR QUESTION 77/10-90. Transmission of Two or More Sound Programmes and/or Information Channels in Television—Questions Concerning Study Group 10—Broadcasting Service (Sound). 2 pp.

—Multiplexing—Broadcasting Satellite Services
CCIR Report 954-2-90. Multiplexing Methods for the Emission of Several Digital Audio Signals and Also Data Signals in Broadcasting—Section 10/11B—Systems. 8 pp.

—Naval Communications
NATO STANAG 4222 ED 1 AMD 0-90. Standard Specification for Digital Representation of Shipboard Data Parameters. 95 pp.

—Open Systems Interconnection
BSI BS 6568: Part 1-88. 1988 Amd 1 Reference Model of Open Systems Interconnection Part 1: Basic Reference Model (Incorporating Connectionless-Mode Transmission) (ISO 7498-1984) (AMD 6112) April 30, 1991. 62 pp.

—Packet Switched Networks—Interworking
CCITT RECMN X.327-89. General Arrangements for Interworking Between Packet Switched Public Data Networks (PSPDNs) and Private Data Networks for the Provision of Data Transmission Services—Data Communication Networks—Interworking Between Networks, Mobile Data Transmission. 9 pp.

—Private Switched Networks—Interworking
CCITT RECMN X.327-89. General Arrangements for Interworking Between Packet Switched Public Data Networks (PSPDNs) and Private Data Networks for the Provision of Data Transmission Services—Data Communication Networks—Interworking Between Networks, Mobile Data Transmission. 9 pp.

—Public Data Networks
CCITT FASCICLE VIII.3-89. Data Communications Networks Transmission, Signalling and Switching, Network Aspects, Maintenance and Administrative Arrangements Recommendations X.40—X.181 (Study Group VII). 481 pp.
CEPT T/SF 13-81. Procedure D'Agrement Pour Les Materiels De Teleinformatique Ou Les Equipements De Transmission De Donnees Appartenant a Des Personnes Privees Et Susceptibles D'Etre Directement Connectes Aux Reseaux Publics Ou Aux Installations Des Administrations. 3 pp.
CEPT T/T 1-74. Services Publics De Transmissions De Donnees. 1 p.

—Public Data Networks—Interworking
CCITT RECMN X.300-89. General Principles for Interworking Between Public Networks, and Between Public Networks and Other Networks for the Provision of Data Transmission Services—Data Communication Networks—Interworking Between Networks, Mobile Data Transmission Systems, Internetwork. 50 pp.

—Public Packet Switched Networks
CCITT RECMN F.601-89. Service and Operational Principles for Packet-Switched Public Data Networks—Telematic, Data Transmission and Teleconference Services—Operations and Quality of Service (Study Group I) 5 pp. 5 pp.

—Public Packet Switched Networks—Accounting—Europe
CCITT RECMN D.306R-91. Remuneration of Public Packet-Switched Data Transmission Networks Between the Countries of Europe and the Mediterranean Basin (Study Group III) 5 pp. 5 pp.
CCITT RECMN D.306R-89. Remuneration of Public Packet-Switched Data Transmission Networks Between the Countries of Europe and the Mediterranean Basin—General Tariff Principles—Charging and Accounting in International Telecom. Serv. (Study Group III) 1 pp. 1 p.
CEPT T/CAC E13 E-92. General Accounting Principles for the International Public Packet-Switched Data Transmission Service. 1 p.

—Public Packet Switched Networks—Accounting—Mediterranean Basin
CCITT RECMN D.306R-91. Remuneration of Public Packet-Switched Data Transmission Networks Between the Countries of Europe and the Mediterranean Basin (Study Group III) 5 pp. 5 pp.
CCITT RECMN D.306R-89. Remuneration of Public Packet-Switched Data Transmission Networks Between the Countries of Europe and the Mediterranean Basin—General Tariff Principles—Charging and Accounting in International Telecom. Serv. (Study Group III) 1 pp. 1 p.

Data Transmission (Cont.)

—Public Packet Switched Networks—Circuit Switched Data Networks
CCITT RECMN X.322-89. General Arrangements for Interworking Between Packet Switched Public Data Networks (PSPDNs) and Circuit Switched Public Data Networks (CSPDNs) for the Provision of Data Transmission Services—Data Communication Networks—Interworking Between Networks, Mobile. 7 pp.

—Public Packet Switched Networks—Interworking
CCITT RECMN X.323-89. General Arrangements for Interworking Between Packet Switched Public Data Networks (PSPDNs)—Data Communication Networks—Interworking Between Networks, Mobile Data Transmission Systems, Internetwork Management (Study Group VII) 3 pp. 3 pp.

—Public Packet Switched Networks—Mobile Satellite—Interworking
CCITT RECMN X.324-89. General Arrangements for Interworking Between Packet Switched Public Data Networks (PSPDNs) and Public Mobile Systems for the Provision of Data Transmission Services—Data Communication Networks—Interworking Between Networks, Mobile Data Transmission. 5 pp.

—Public Packet Switched Networks—Routing
CCITT RECMN F.601-89. Service and Operational Principles for Packet-Switched Public Data Networks—Telematic, Data Transmission and Teleconference Services—Operations and Quality of Service (Study Group I) 5 pp. 5 pp.

—Public Packet Switched Networks—Tariffs—Europe
CCITT RECMN D.306R-91. Remuneration of Public Packet-Switched Data Transmission Networks Between the Countries of Europe and the Mediterranean Basin (Study Group III) 5 pp. 5 pp.
CCITT RECMN D.306R-89. Remuneration of Public Packet-Switched Data Transmission Networks Between the Countries of Europe and the Mediterranean Basin—General Tariff Principles—Charging and Accounting in International Telecom. Serv. (Study Group III) 1 pp. 1 p.

—Public Packet Switched Networks—Tariffs—Mediterranean Basin
CCITT RECMN D.306R-91. Remuneration of Public Packet-Switched Data Transmission Networks Between the Countries of Europe and the Mediterranean Basin (Study Group III) 5 pp. 5 pp.
CCITT RECMN D.306R-89. Remuneration of Public Packet-Switched Data Transmission Networks Between the Countries of Europe and the Mediterranean Basin—General Tariff Principles—Charging and Accounting in International Telecom. Serv. (Study Group III) 1 pp. 1 p.

—Public Switched Telephone Networks—Maintenance
CCITT RECMN M.729-89. Organization of the Maintenance of International Public Switched Telephone Circuits Used For Data Transmission—General Maintenance Principles—Maintenance of International Transmission Systems and Telephone Circuits (Study Group IV) 2 pp. 2 pp.

—Public Telephone Networks—International Alphabet No. 5
CCITT RECMN V.4-89. General Structure of Signals of International Alphabet No. 5 Code for Character Oriented Data Transmission over Public Telephone Networks—Data Communication over the Telephone Network (Study Group XVII) 2 pp. 2 pp.

—Quality of Service
CCITT FASCICLE II.5-88. Telematic, Data Transmission and Teleconference Service—Operations and Quality of Service Recommendations F.160—F.353, F.600, F.601, F.710—F.730. 154 pp.

—Radio Beacons—Frequency Shift Keying
CCIR Report 1037-86. Choice Between the FSK and MSK Techniques for Data Transmission from Maritime Radiobeacons—Section 8D—Radiodetermination, Global Maritime Distress and Safety System and Related Subjects. 2 pp.

—Radio Beacons—Minimum Shift Keying
CCIR Report 1037-86. Choice Between the FSK and MSK Techniques for Data Transmission from Maritime Radiobeacons—Section 8D—Radiodetermination, Global Maritime Distress and Safety System and Related Subjects. 2 pp.

INTERNATIONAL AND NON-U.S. NATIONAL STANDARDS
SUBJECT INDEX

Data

Data Transmission (Cont.)
—Radio Circuits—Independant Sideband
- CCIR QUESTION 145/9-90. Characteristics Required for Single-Sideband and Independent-Sideband Systems Used for High-Speed Data Transmission over HF Radio Circuits—Questions of Study Group 9 Fixed Service. 1 p.

—Radio Circuits—Single Sideband
- CCIR QUESTION 145/9-90. Characteristics Required for Single-Sideband and Independent-Sideband Systems Used for High-Speed Data Transmission over HF Radio Circuits—Questions of Study Group 9 Fixed Service. 1 p.

—Radio Circuits—Voice Frequency Telegraphy—Frequency Shift Keying
- CCIR RECMN 456-70. Data Transmission at 1200/600 Bit/s Over HF Circuits When Using Multi-Channel Voice-Frequency Telegraph Systems and Frequency-Shift Keying—Section 3Cb—Data Transmission. 3 pp.

—Radio Circuits—Voice Frequency Telegraphy—Phase Shift Keying
- CCIR RECMN 763-92. Data Transmission at 2400/1200/600/300/150/75 Bit/s over HF Circuits Using Multi-Channel Voice-Frequency Telegraphy and Phase-Shift Keying—Section 3Cb—Data Transmission. 5 pp.
- CCIR Report 864-1-86. Data Transmission at 2400/1200/600/300/150/75 Bit/s Over HF Circuits Using Multi-Channel Voice-Frequency Telegraphy and Phase-Shift Keying—Section 3Cb—Data Transmission. 5 pp.

—Radio Relay Systems
- CENELEC PRETS 300 197-92. Transmission and Multiplexing (TM); Parameters for Radio Relay Systems for the Transmission of Digital Signals and Analogue Video Signals Operating at 38 GHz (DE/TM-4001). 33 pp.
- CENELEC PRETS 300 198-92. Transmission and Multiplexing (TM); Parameters of Radio Relay Systems for the Transmission of Digital Signals and Analogue Video Signals Operating at 23 GHz (DE/TM-4003). 34 pp.
- ETSI PRETS 300 197-92. Transmission and Multiplexing (TM); Parameters for Radio Relay Systems for the Transmission of Digital Signals and Analogue Video Signals Operating at 38 GHz (DE/TM-4001). 33 pp.
- ETSI PRETS 300 198-92. Transmission and Multiplexing (TM); Parameters of Radio Relay Systems for the Transmission of Digital Signals and Analogue Video Signals Operating at 23 GHz (DE/TM-4003). 34 pp.

—Remote Control—Protocols
- BSI BS 7404: Sec 5.1-91. 1991 Telecontrol Equipment and Systems Part 5: Transmission Protocols Section 5.1: Specification for Transmission Frame Formats (IEC 870-5-1: 1990). 42 pp.
- BSI BS 7404: Sec 5.2-92. 1992 Telecontrol Equipment and Systems Part 5: Transmission Protocols Section 5.2: Specification for Link Transmission Procedures (IEC 870-5-2: 1992). 51 pp.
- BSI BS EN 60870-5-3-92. 1992 Telecontrol Equipment and Systems Part 5: Transmission Protocols Section 5.3: Specification for General Structure of Application Data. 27 pp.
- CENELEC EN 60870-5-3-92. Telecontrol Equipment and Systems Part 5: Transmission Protocols Section 3: General Structure of Application Data. 5 pp.
- IEC 870 Pt 5-1-90. Telecontrol Equipment and Systems Part 5: Transmission Protocols Section One—Transmission Frame Formats First Edition. 82 pp.
- IEC 870 Pt 5-2-92. Telecontrol Equipment and Systems Part 5: Transmission Protocols Section 2: Link Transmission Procedures First Edition. 102 pp.
- IEC 870 Pt 5-3-92. Telecontrol Equipment and Systems Part 5: Transmission Protocols Section 3: General Structure of Application Data First Edition. 46 pp.
- IEC 870 Pt 5-4-93. Telecontrol Equipment and Systems Part 5: Transmission Protocols Section 4: Definition and Coding of Application Information Elements First Edition. 49 pp.

—Securities
- BSI BS 6636: Part 1-85. (WITHDRAWN) 1985 Message Types for Securities Part 1: Messages for Receipt/Delivery (Superseded by BS ISO 7775: 1991). 23 pp.
- BSI BS 6636: Part 2-87. (WITHDRAWN) 1987 Message Types for Securities Part 2: Messages for Orders to Buy/Sell (Superseded by BS ISO 7775: 1991). 19 pp.
- BSI BS 6822-87. 1987 Format for Transmission of Certificate Numbers of Securities. 7 pp.
- ISO 8532-86. Securities—Format for Transmission of Certificate Numbers First Edition. 6 pp.

Data Transmission (Cont.)
—Securities (Cont.)
- JTC1 8532-86. Securities—Format for Transmission of Certificate Numbers First Edition. 6 pp.
- OSI ISO 7775-1-84. Securities—Standard Scheme for Message Types—Part: Receipt—Delivery. 23 pp.
- OSI ISO 7775-2-86. Securities—Scheme for Message Types—Part 2: Order to Buy-Sell. 17 pp.
- OSI ISO 8532-86. Securities—Format for Transmission of Certificate Numbers. 6 pp.

—Serial Highways
- MOD UK DSTAN 00-19-87. ASWE Serial Highway Issue 2. 74 pp.

—Signaling Links—CCITT No. 6 Signaling Systems
- CCITT RECMN Q.274-89. 6.3 Transmission Methods—Specifications of Signalling System No. 6 (Study Group XI) 8 pp. 8 pp.

—Sound Broadcasting—AM
- CCIR RECMN 706-90. Data System in Monophonic AM Sound Broadcasting (AMDS)—Section 10A-1—Amplitude-Modulation Sound-Broadcasting in Bands 5 (LF), 6 (MF) and 7 (HF). 3 pp.
- CCIR RECMN 706-1-92. Data System in Monophonic AM Sound Broadcasting (AMDS)—Section 10A-1—Amplitude-Modulation Sound-Broadcasting in Bands 5 (LF), 6 (MF) and 7 (HF). 4 pp.

—Sound Broadcasting—FM
- CCIR QUESTION 102/10-90. Transmission of Data Information as an Alternative to the Main Programme in Frequency-Modulation Sound Broadcasting—Questions Concerning Study Group 10—Broadcasting Service (Sound). 1 p.

—Space Communications
- ISO 11103-91. Space Data and Information Transfer Systems—Radio Metric and Orbit Data First Edition. 4 pp.
- ISO 11104-91. Space Data and Information Transfer Systems—Time Code Formats First Edition. 4 pp.
- JTC1 11103-91. Space Data and Information Transfer Systems—Radio Metric and Orbit Data First Edition. 4 pp.
- JTC1 11104-91. Space Data and Information Transfer Systems—Time Code Formats First Edition. 4 pp.

—Standards
- CCITT RECMN A.20-89. Collaboration with Other International Organizations over Data Transmission—Data Communication over the Telephone Network (Study Group XVII) 2 pp. 2 pp.

—Subnetworks—Internal Utilities
- CCITT RECMN X.302-89. Description of the General Arrangements for Internal Network Utilities Within a Subnetwork and Intermediate Utilities Between Subnetworks for the Provision of Data Transmission Services—Data Communication Networks—Interworking Between Networks, Mobile. 7 pp.

—Supergroup Links
- CCITT RECMN H.15-89. Characteristics of Supergroup Links for the Transmission of Wide-Spectrum Signals—Line Transmission of Non-Telephone Signals—Transmission of Sound-Programme and Television Signals (Study Group XV) 1 pp. 1 p.
- CCITT RECMN H.53-89. Transmission of Wide-Spectrum Signals (Data, Etc.) over Wideband Supergroup Links—Line Transmission of Non-Telephone Signals—Transmission of Sound-Programme and Television Signals (Study Group XV) 1 pp. 1 p.
- CCITT RECMN M.900-89. Use of Leased Group and Supergroup Links for Wide-Spectrum Signal Transmission (Data, Facsimile, Etc.)—Maintenance of International Telegraph, Phototelegraph and Leased Circuits—Maintenance of the International Public Telephone Network-. 4 pp.

—Tariffs (Telecommunications)
- BSI BS EN 61142-93. 1993 Data Exchange for Meter Reading, Tariff and Load Control—Local Bus Data Exchange (IEC 1142: 1993) (S). 173 pp.
- CENELEC EN 61142-93. Data Exchange for Meter Reading, Tariff and Load Control—Local Bus Data Exchange (IEC 1142:1993). 4 pp.
- IEC 1107-92. Data Exchange for Meter Reading, Tariff and Load Control—Direct Local Data Exchange First Edition. 66 pp.
- IEC 1142-93. Data Exchange for Meter Reading, Tariff and Load Control—Local Bus Data Exchange First Edition; (CENELEC EN 61142:1993). 341 pp.

Data Transmission (Cont.)
—Tariffs—Telecommunication Administrations
- CCITT RECMN D.160-89. Mode of Application of the Flat-Rate Price Procedure Set Forth in Recommendation D.67 and Recommendation D.150 for Remuneration of Facilities Made Available to the Administrations of Other Countries—General Tariff Principles—Charging and Accounting. 5 pp.
- CCITT RECMN E.252-89. Mode of Application of the Flat-Rate Price Procedure Set Forthe in Recommendations D.67 and D.150 for Remuneration of Facilities Made Available to the Administrations of Other Countries—Telephone Network and ISDN—Operation, Numbering, Routing and Mobile. 1 p.

—Telecommunication Equipment
- CENELEC PRETS 300 147-91. Transmission and Multiplexing (TM); Synchronous Digital Hierarchy (SDH) Multiplexing Structure. 11 pp.
- CENELEC ETS 300 147-92. Transmission and Multiplexing (TM); Synchronous Digital Hierarchy (SDH) Multiplexing Structure. 10 pp.
- CENELEC PRETS 300 167-91. Transmission and Multiplexing Functional Characteristics of 2 Mbit/s Interfaces (DE/TM-3006). 10 pp.
- CENELEC PRETS 300 167-93. Transmission and Multiplexing (TM); Functional Characteristics of 2 048 kbit/s Interfaces. 9 pp.
- ETSI PRETS 300 147-91. Transmission and Multiplexing—Synchronous Digital Hierarchy—Multiplexing Structure (DE/TM-3001).
- ETSI PRETS 300 167-91. Transmission and Multiplexing Functional Characteristics of 2 Mbit/s Interfaces (DE/TM-3006).
- ETSI ETS 300 147-92. Transmission and Multiplexing (TM); Synchronous Digital Hierarchy (SDH) Multiplexing Structure. 10 pp.
- ETSI PRETS 300 147-91. Transmission and Multiplexing (TM); Synchronous Digital Hierarchy (SDH) Multiplexing Structure. 11 pp.
- ETSI PRETS 300 167-93. Transmission and Multiplexing (TM); Functional Characteristics of 2 048 kbit/s Interfaces. 9 pp.
- ETSI PRETS 300 167-91. Transmission and Multiplexing Functional Characteristics of 2 Mbit/s Interfaces (DE/TM-3006). 10 pp.

—Telegraph Distortion—Maintenance
- CCITT RECMN V.50-89. Standard Limits for Transmission Quality of Data Transmission—Data Communication over the Telephone Network (Study Group XVII) 2 pp. 2 pp.

—Telephone Circuits—Data Signaling Rates
- CCITT RECMN V.6-89. Standardization of Data Signalling Rates for Synchronous Data Transmission on Leased Telephone-Type Circuits—Data Communication over the Telephone Network (Study Group XVII) 2 pp. 2 pp.

—Telephone Circuits—Maintenance
- CCITT RECMN V.51-89. Organization of the Maintenance of International Telephone-Type Circuits Used for Data Transmission—Data Communication over the Telephone Network (Study Group XVII) 1 pp. 1 p.
- CCITT RECMN V.53-89. Limits for the Maintenance of Telephone-Type Circuits Used for Data Transmission—Data Communication over the Telephone Network (Study Group XVII) 4 pp. 4 pp.

—Telephone Lines—Power Levels
- CCITT RECMN H.51-89. Power Levels for Data Transmission over Telephone Lines—Line Transmission of Non-Telephone Signals—Transmission of Sound-Programme and Television Signals (Study Group XV) 1pp. 1 p.
- CCITT RECMN V.2-89. Power Levels for Data Transmission over Telephone Lines—Data Communication over the Telephone Network (Study Group XVII) 2 pp. 2 pp.

—Telephone Networks
- CCITT FASCICLE VIII.1-89. Data Communication over the Telephone Network Series V Recommendations (Study Group XVII). 530 pp.

—Telephone Networks—Data Signaling Rates
- CCITT RECMN V.5-89. Standardization of Data Signalling Rates for Synchronous Data Transmission in the General Switched Telephone Network—Data Communication over the Telephone Network (Study Group XVII) 1 pp. 1 p.

—Telephone Services
- CCITT RECMN E.140-89. Principles for the Operation of International Telephone Services—Telephone Network and ISDN—Operation, Numbering, Routing and Mobile Service (Study Group II) 3 pp. 3 pp.

INDUSTRY STANDARDS

Data

Data Transmission (Cont.)

—Television Channels—Sound Transmission
CCIR QUESTION 47-1/10-86. Transmission of Sound and Information Signals Accompanying the Video Signal in a Television Channel—Questions Concerning Study Group 10—Broadcasting Service (Sound). 1 p.

—Telex Communications
CCITT RECMN S.15-89. Use of the Telex Network for Data Transmission at 50 Bauds—Telegraph Services Terminal Equipment (Study Group IX) 4 pp. 4 pp.

—Time Division Multiplexers—Telegraphy
CCITT RECMN R.111-89. Code and Speed Independent TDM System for Anisochronous Telegraph and Data Transmission—Telegraph Transmission (Study Group IX) 9 pp. 9 pp.

—Time Division Multiplexers—Telegraphy—Bit Interleaving
CCITT RECMN R.101-89. Code and Speed Dependent TDM System for Anisochronous Telegraph and Data Transmission Using Bit Interleaving—Telegraph Transmission (Study Group IX) 15 pp. 15 pp.
CCITT RECMN R.102-89. 4800 Bit/s Code and Speed Dependent and Hybrid TDM Systems for Anisochronous Telegraph and Data Transmission Using Bit Interleaving—Telegraph Transmission (Study Group IX) 11 pp. 11 pp.
CCITT RECMN R.112-89. TDM Hybrid System for Anisochronous Telegraph and Data Transmission Using Bit Interleaving—Telegraph Transmission (Study Group IX) 3 pp. 3 pp.

—Tractors—Agricultural
BSI BS 7436: Part 1-91. 1991 Data Transfer Between Agricultural Tractors and Attached Machinery Part 1: Specification for Data Bus Protocol. 15 pp.

—Units of Measurement—Quantity—Abbreviations
CCITT RECMN B.14-89. Terms and Abbreviations for Information Quantities in Telecommunications—Terms and Definitions Abbreviations and Acronyms Recommendations on Means of Expression (Series B) General Telecommunications Statistics (Series C) 2 pp. 2 pp.

—Units of Measurement—Quantity—Glossaries
CCITT RECMN B.14-89. Terms and Abbreviations for Information Quantities in Telecommunications—Terms and Definitions Abbreviations and Acronyms Recommendations on Means of Expression (Series B) General Telecommunications Statistics (Series C) 2 pp. 2 pp.

—Units of Measurement—Quantity—Symbols
CCITT RECMN B.14-89. Terms and Abbreviations for Information Quantities in Telecommunications—Terms and Definitions Abbreviations and Acronyms Recommendations on Means of Expression (Series B) General Telecommunications Statistics (Series C) 2 pp. 2 pp.

—Videoconferencing Services—Encoder/Decoders
CCITT RECMN H.120-89. Codecs for Videoconferencing Using Primary Digital Group Transmission—Line Transmission of Non-Telephone Signals—Transmission of Sound-Programme and Television Signals (Study Group XV) 59 pp. 59 pp.

—Videoconferencing Services—Hypothetical Reference Model
CCITT RECMN H.110-89. Hypothetical Reference Connections for Videoconferencing Using Primary Digital Group Transmission—Line Transmission of Non-Telephone Signals—Transmission of Sound-Programme and Television Signals (Study Group XV) 6 pp. 6 pp.

—Voice Band—Adaptive Delta Pulse Code Modulation
CCITT RECMN G.113-89. Transmission Impairments—General Characteristics of International Telephone Connections and Circuits (Study Groups XII and XV) 22 pp. 22 pp.

—Voice Band—Adaptive Delta Pulse Code Modulation—Bit Rate Encoding
CCITT RECMN G.724-89. Characteristics of a 48-Channel Low Bit Rate Encoding Primary Multiplex Operating at 1544 kbit/s—General Aspects of Digital Transmission Systems; Terminal Equipments (Study Groups XV and XVIII) 6 pp. 6 pp.

Data Transmission (Cont.)

—Voice Band—Attenuation Distortion
CCITT RECMN G.113-89. Transmission Impairments—General Characteristics of International Telephone Connections and Circuits (Study Groups XII and XV) 22 pp. 22 pp.

—Voice Band—Circuit Noise
CCITT RECMN G.113-89. Transmission Impairments—General Characteristics of International Telephone Connections and Circuits (Study Groups XII and XV) 22 pp. 22 pp.

—Voice Band—Dropout
CCITT RECMN G.113-89. Transmission Impairments—General Characteristics of International Telephone Connections and Circuits (Study Groups XII and XV) 22 pp. 22 pp.

—Voice Band—Echo
CCITT RECMN G.113-89. Transmission Impairments—General Characteristics of International Telephone Connections and Circuits (Study Groups XII and XV) 22 pp. 22 pp.

—Voice Band—Envelope Delay Distortion
CCITT RECMN G.113-89. Transmission Impairments—General Characteristics of International Telephone Connections and Circuits (Study Groups XII and XV) 22 pp. 22 pp.

—Voice Band—Error Correction
CCITT RECMN G.113-89. Transmission Impairments—General Characteristics of International Telephone Connections and Circuits (Study Groups XII and XV) 22 pp. 22 pp.

—Voice Band—Gain Hits
CCITT RECMN G.113-89. Transmission Impairments—General Characteristics of International Telephone Connections and Circuits (Study Groups XII and XV) 22 pp. 22 pp.

—Voice Band—Harmonic Distortion
CCITT RECMN G.113-89. Transmission Impairments—General Characteristics of International Telephone Connections and Circuits (Study Groups XII and XV) 22 pp. 22 pp.

—Voice Band—Impulse Noise
CCITT RECMN G.113-89. Transmission Impairments—General Characteristics of International Telephone Connections and Circuits (Study Groups XII and XV) 22 pp. 22 pp.

—Voice Band—Intermodulation Distortion
CCITT RECMN G.113-89. Transmission Impairments—General Characteristics of International Telephone Connections and Circuits (Study Groups XII and XV) 22 pp. 22 pp.

—Voice Band—Nonlinear Distortion
CCITT RECMN G.113-89. Transmission Impairments—General Characteristics of International Telephone Connections and Circuits (Study Groups XII and XV) 22 pp. 22 pp.

—Voice Band—Phase Hits
CCITT RECMN G.113-89. Transmission Impairments—General Characteristics of International Telephone Connections and Circuits (Study Groups XII and XV) 22 pp. 22 pp.

—Voice Band—Phase Jitter
CCITT RECMN G.113-89. Transmission Impairments—General Characteristics of International Telephone Connections and Circuits (Study Groups XII and XV) 22 pp. 22 pp.

—Voice Band—Pulse Code Modulation
CCITT RECMN G.711-89. Pulse Code Modulation (PCM) of Voice Frequencies—General Aspects of Digital Transmission Systems; Terminal Equipments (Study Groups XV and XVIII) 10 pp. 10 pp.

—Voice Band—Pulse Code Modulation Multiplexers
CCITT RECMN G.731-89. Primary PCM Multiplex Equipment for Voice Frequencies—General Aspects of Digital Transmission Systems; Terminal Equipments (Study Groups XV and XVIII) 1 pp. 1 p.

—Voice Band—Pulse Code Modulation—2 Wire Interfaces
CCITT RECMN G.712-92. Transmission Performance Characteristics of Pulse Code Modulation (Study Group XV) 33 pp (Replaces G.713, G.714, and G.715). 33 pp.
CCITT RECMN G.713-89. Performance Characteristics of PCM Channels Between 2-Wire Interfaces at Voice Frequencies—General Aspects of Digital Transmission Systems; Terminal Equipments (Study Groups XV and XVIII) 12 pp (Replaced by Recmn G.712). 12 pp.

Data Transmission (Cont.)

—Voice Band—Pulse Code Modulation—4 Wire Interfaces
CCITT RECMN G.712-92. Transmission Performance Characteristics of Pulse Code Modulation (Study Group XV) 33 pp (Replaces G.713, G.714, and G.715). 33 pp.
CCITT RECMN G.712-89. Performance Characteristics of PCM Channels Between 4-Wire Interfaces at Voice Frequencies—General Aspects of Digital Transmission Systems; Terminal Equipments (Study Groups XV and XVIII) 10 pp. 10 pp.

—Voice Band—Quantization Distortion
CCITT RECMN G.113-89. Transmission Impairments—General Characteristics of International Telephone Connections and Circuits (Study Groups XII and XV) 22 pp. 22 pp.

—Voice Band—Switching Discontinuities—Fixed Satellite Services
CCIR RECMN 730-92. Compensation of the Effects of Switching Discontinuities for Voice Band Data and of Doppler Frequency-Shifts in the Fixed-Satellite Service—Section 4B2—Performance and Availability. 12 pp.

—Voice Band—Transmission Loss
CCITT RECMN G.113-89. Transmission Impairments—General Characteristics of International Telephone Connections and Circuits (Study Groups XII and XV) 22 pp. 22 pp.

Data Transmission Equipment

See Also: Data Communication Equipment; Open Systems Interconnection; Radio Data Systems; Store and Forward Units; Telecommunication Equipment

CCITT RECMN M.1300-89. International Data Transmission Systems Operating at 2400 Bit/s and Above—Maintenance of International Telegraph, Phototelegraph and Leased Circuits—Maintenance of the International Public Telephone Network—Maintenance of Maritime. 3 pp.
CEPT T/CD 01-07-83. Specifications Techniques D'un Modem Pour Transmission Parallele De Donnees Utilisant Les Frequences De Signalisation Des Postes Telephoniques Pour Utilisation Sur Le Reseau Telephonique General Avec Commutation. 6 pp.
CEPT T/CD 01-07 E-85. Engineering Requirements for a Parallel Data Transmission Modem Using Telephone Signalling Frequencies for Use in the GSTN. 6 pp.
CEPT T/CD 01-08-83. Specifications Techniques Applicables a Un Modem Pour Transmission Parallele De Donnees Utilisable Dans Le Reseau Telephonique General Avec Commutation. 9 pp.
CEPT T/CD 01-08 E-85. Engineering Requirements for a Parallel Data Transmission Modem for Use in the GSTN. 9 pp.

—Aircraft
BSI G 183-63. 1963 Amd 1 Interconnections in Analogue Data Transmission Systems in Aircraft. 15 pp.

—Avionics—Interfaces
MOD UK DSTAN 00-18: Pt 1:Sec 3. Avionic Data Transmission Interface Systems Part 1: Guide to Interface Systems Section 3: Guide to the Simplex and Half Duplex Serial Digital Transmission Interface Systems Standard Issue 2. 69 pp.
MOD UK DSTAN 00-18: Pt 1:Sec 5-91. Avionic Data Transmission Interface Systems Part 1: Guide to Interface Systems Section 5: Guide to the Fibre Optic Interface Standardization Issue 2. 156 pp.
MOD UK DSTAN 00-18: Pt 2:Supp A-90. Avionic Data Transmission Interface Systems Part 2: Serial, Time Division, Command/Response Multiplex Data Bus Supplement A: Fibre Optic Supplement for a Point to Point Link Issue 1. 15 pp.
MOD UK DSTAN 00-18: Pt 2:Supp B-90. Avionic Data Transmission Interface Systems Part 2: Serial, Time Division, Command/Response Multiplex Data Bus Supplement B: Fibre Optic Supplement for a Single Transmissive Star Issue 1. 15 pp.
MOD UK DSTAN 00-18: Pt 2:Supp C-90. Avionic Data Transmission Interface Systems Part 2: Serial, Time Division, Command/Response Multiplex Data Bus Supplement C: Fibre Optic Supplement for a Single Reflective Star Issue 1. 15 pp.
MOD UK DSTAN 00-18: Pt 2:Supp D-90. Avionic Data Transmission Interface Systems Part 2: Serial, Time Division, Command/Response Multiplex Data Bus Supplement D: Fibre Optic Supplement for a Multi-Local Transmissive Star Issue 1. 20 pp.

—Communication Channels—Numbering
CCITT RECMN M.1320-89. Numbering of Channels in Data Transmission Systems—Maintenance of International Telegraph, Phototelegraph and Leased Circuits—Maintenance of the International Public Telephone Network—Maintenance of Maritime Satellite and Data. 2 pp.

INTERNATIONAL AND NON-U.S. NATIONAL STANDARDS
SUBJECT INDEX

DC

Data Transmission Equipment *(Cont.)*
—Communication Networks—Interfaces—Interworking
CCITT FASCICLE VIII.6. Data Communication Networks Interworking Between Networks, Mobile Data Transmission Systems, Internetwork Management Recommendations X.300-X.370 (Study Group VII). 240 pp.

—Designations
CCITT RECMN M.1370-89. Setting up and Lining up of International Data Transmission Systems Operating at 48 kBit/s and Above—Maintenance of International Telegraph, Phototelegraph and Leased Circuits—Maintenance of the International Public Telephone Network-. 6 pp.

—Faults
CCITT RECMN M.1355-89. Maintenance of International Data Transmission Systems Operating in the Range 2.4 to 14.4 kBit/s—Maintenance of International Telegraph, Phototelegraph and Leased Circuits—Maintenance of the International Public Telephone Network-. 2 pp.
CCITT RECMN M.1375-89. Maintenance of International Data Transmission Systems Operating at 48 kBit/s and Above—Maintenance of International Telegraph, Phototelegraph and Leased Circuits—Maintenance of the International Public Telephone Network-. 4 pp.

—Hydrometry—Open Channel Flow
BSI BS 3680: Part 8F-86. 1986 Measurement of Liquid Flow in Open Channels: Measuring Instruments and Equipment: Hydrometric Data Transmission Systems: General. 21 pp.
ISO 6419 Pt 1-84. Hydrometric Data Transmission Systems—Part 1: General First Edition. 22 pp.

—Line Up
CCITT RECMN M.1350-89. Setting up, Lining up and Characteristics of International Data Transmission Systems Operating in the Range 2.4 kBit/s to 14.4 kBit/s—Maintenance of International Telegraph, Phototelegraph and Leased Circuits—Maintenance of the International Public. 3 pp.
CCITT RECMN M.1370-89. Setting up and Lining up of International Data Transmission Systems Operating at 48 kBit/s and Above—Maintenance of International Telegraph, Phototelegraph and Leased Circuits—Maintenance of the International Public Telephone Network-. 6 pp.

—Maintenance
CCITT FASCICLE IV.2. Maintenance of International Telegraph, Phototelegraph and Leased Circuits Maintenance of Maritime Satellite and Data Transmission Systems Recommendations M.800—M.1375 (Study Group IV). 130 pp.
CCITT RECMN M.1300-89. International Data Transmission Systems Operating at 2400 Bit/s and Above—Maintenance of International Telegraph, Phototelegraph and Leased Circuits—Maintenance of the International Public Telephone Network—Maintenance of Maritime. 3 pp.
CCITT RECMN M.1355-89. Maintenance of International Data Transmission Systems Operating in the Range 2.4 to 14.4 kBit/s—Maintenance of International Telegraph, Phototelegraph and Leased Circuits—Maintenance of the International Public Telephone Network-. 2 pp.
CCITT RECMN M.1375-89. Maintenance of International Data Transmission Systems Operating at 48 kBit/s and Above—Maintenance of International Telegraph, Phototelegraph and Leased Circuits—Maintenance of the International Public Telephone Network-. 4 pp.

—Mobile Radio Services
CENELEC HD 466.6 S1-89. Methods of Measurement for Radio Equipment Used in the Mobile Services Part 6: Selective—Calling and Data Equipment. 4 pp.
CENELEC HD 466.6 S2-92. Methods of Measurement for Radio Equipment Used in the Mobile Services Part 6: Selective-Calling Equipment (IEC 489-6:1987 + A1: 1989). 5 pp.
IEC 489 Pt 6-87. Methods of Measurement for Radio Equipment Used in the Mobile Services Part 6: Selective-Calling and Data Equipment Second Edition; (Amendment 1-1989) (Amendment 2-1991) (CENELEC HD 466.6 S2: 1992). 350 pp.

—Set Up
CCITT RECMN M.1350-89. Setting up, Lining up and Characteristics of International Data Transmission Systems Operating in the Range 2.4 kBit/s to 14.4 kBit/s—Maintenance of International Telegraph, Phototelegraph and Leased Circuits—Maintenance of the International Public. 3 pp.

Data Transmission Equipment *(Cont.)*
—Set Up *(Cont.)*
CCITT RECMN M.1370-89. Setting up and Lining up of International Data Transmission Systems Operating at 48 kBit/s and Above—Maintenance of International Telegraph, Phototelegraph and Leased Circuits—Maintenance of the International Public Telephone Network-. 6 pp.

—Wideband
CENELEC PRETS 300 328-93. Radio Equipment and Systems (RES); Wideband Data Transmission Systems Technical Characteristics and Test Conditions for Data Transmission Equipment Operating in the 2,4 GHz ISM Band and Using Spread Spectrum Modulation Techniques. 29 pp.
ETSI PRETS 300 328-93. Radio Equipment and Systems (RES); Wideband Data Transmission Systems Technical Characteristics and Test Conditions for Data Transmission Equipment Operating in the 2,4 GHz ISM Band and Using Spread Spectrum Modulation Techniques. 29 pp.

Data Transmission Prefixes
Use: Prefixes (Communications)—Data Transmission

Data Transmission Rates
Use: Data Transfer Rates

Data Transmission Services
Use: Data Transmission

Data Transmission Systems
Use: Data Transmission; Data Transmission Equipment

Data Transmitters
Use: Data Transmission Equipment

Data Transmitting Sets
Use: Data Links

Database Management Systems
Use: Data Base Management Systems

Datafax Communications
See Also: FAX Communications; Telecommunication Services; Teleinformatic Services
CCITT RECMN A.21-89. Collaboration with Other International Organizations on CCITT-Defined Telematic Services—Terminal Equipment and Protocols for Telematic Services (Study Group VIII) 2 pp. 2 pp.

Date and Time Codes
Use: Time Codes

Datebooks
Use: Appointment Books

Datum (Elevation)
See Also: Datum (Geodetic)
—Engineering Drawings
ISO 5459-81. Technical Drawings—Geometrical Tolerancing—Datums and Datum-Systems for Geometrical Tolerances First Edition. 19 pp.

—Military Operations
NATO STANAG 2211 ED 5 AMD 0-91. Geodetic Datums, Ellipsoids, Grids and Grid References. 97 pp.
NATO STANAG 2211 ED 5 AMD 1-91. Geodetic Datums, Ellipsoids, Grids and Grid References. 75 pp.
NATO STANAG 2211 ED 5 AMD 2-91. Geodetic Datums, Ellipsoids, Grids and Grid References. 89 pp.

Datum (Geodetic)
See Also: Datum (Elevation); Latitude; Longitude; Mapping; Spherical Earth Models
—Military Operations
NATO STANAG 2211 ED 5 AMD 0-91. Geodetic Datums, Ellipsoids, Grids and Grid References. 97 pp.
NATO STANAG 2211 ED 5 AMD 1-91. Geodetic Datums, Ellipsoids, Grids and Grid References. 75 pp.
NATO STANAG 2211 ED 5 AMD 2-91. Geodetic Datums, Ellipsoids, Grids and Grid References. 89 pp.

—Trig Lists
NATO STANAG 2210 ED 3 AMD 6-69. Trig Lists (Lists of Geodetic Data). 10 pp.

Datum Plane
Use: Datum (Elevation)

Davits
See Also: Cranes; Ships
BSI BS MA 41-75. 1975 Amd 1 Two Tonnes General Purpose Davits. 11 pp.
CNS F3129-81. Ships' Jib Davits for General Use (Aug)(7791).
CNS F3139-81. Ships' Davits for General Use (Oct)(7984).
JIS F 2103-87. Ships' Davits for General Use.
JIS F 2103-74. Ships' Davits for General Use. 8 pp.
MOD UK NES 113: Part 8-89. Requirements for Mechanical Handling Part 8: Davits and Derricks Issue 2 (11.89). 14 pp.

—Manhole Covers
DIN ENGL 28124 Pt 4-89. Davits for Manhole Closures (Mar). 4 pp.

Daylight Testing
See Also: Daylighting; Environmental Testing; Light Testing; Solar Radiation
—Fabrics—Colorfastness
CGSB CAN/CGSB-4.2 NO.18.2-M90. Textile Test Methods Textiles—Tests for Colourfastness—Part B01: Colourfastness to Light: Daylight (ISO 105-B01:1988). 20 pp.
CNS L3026-82. Method of Test for Colour Fastness to Sunlight and Daylight (Jun)(1493).
JIS L 0841-92. Testing Methods for Colour Fastness to Daylight. 12 pp.

—Glass—Laminated
JIS R 3106-85. Testing Method on Transmittance and Reflectance for Daylight and Solar Radiation and Solar Heat Gain Coefficient of Flat Glass. 19 pp.

—Glass—Plate
CNS R3161-88. Method of Test on Transmittance and Reflectance for Daylight and Solar Radiation and Solar Heat Gain Coefficient of Flat Glass (Jul)(12381).
JIS R 3106-85. Testing Method on Transmittance and Reflectance for Daylight and Solar Radiation and Solar Heat Gain Coefficient of Flat Glass. 19 pp.

—Photographic Film—Light Source
JIS K 7602-84. Sensitometric Illuminants for Photography.

—Plastics
BSI BS 2782:Pt5: METH 540A-77. 1977 Methods of Testing Plastics Part 5: Optical and Colour Properties, Weathering Method 540A: Determination of Resistance to Change upon Exposure Under Glass to Daylight. 12 pp.
BSI BS 2782:Pt5: METH 540C-88. 1988 Methods of Testing Plastics Part 5: Optical and Colour Properties Method 540C: Determination of Ultraviolet Radiation Intensity Using Polysulphane Film. 8 pp.
ISO 877-76. Plastics—Determination of Resistance to Change upon Exposure Under Glass to Daylight First Edition. 10 pp.

Daylighting
See Also: Daylight Testing; Lighting; Sky Brightness
DIN ENGL 5034 Pt 1-83. Daylight in Interiors; General Requirements (Feb). 5 pp.
DIN ENGL 5034 Pt 2-85. Daylight in Interiors; Principles (Feb). 13 pp.

—Buildings—Design
BSI BS 8206: Part 2-92. 1992 Lighting for Buildings Part 2: Code of Practice for Daylighting (R). 43 pp.
BSI BS 8206: Part 2-01. 1992 Amd 1 Lighting for Buildings Part 2: Code of Practice for Daylighting (AMD 7391) October 15, 1992. 45 pp.
BSI DD 73-82. (WITHDRAWN) 1982 Basic Data for the Design of Buildings: Daylight (Superseded by BS 8206: Part 2: 1992). 103 pp.

—Sensitometry
CNS Z9005-80. Light Sources for Use in Sensitometric Exposure of the Spectral Distribution of Daylight (Jan)(5146).

DBC
Use: Data Bus Couplers

DBMS
Use: Data Base Management Systems

DC Aircraft
Scope Note: See listings under Douglas and/or McDonnell Douglas

DC Amplifiers
Use For: Direct Current Amplifiers
See Also: Amplifiers
IEC 527-75. Direct Current Amplifiers; Characteristics and Test Methods First Edition. 69 pp.

INTERNATIONAL AND NON-U.S. NATIONAL STANDARDS SUBJECT INDEX

DC Equipment
Use For: Direct Current Equipment
See Also: Electrical Equipment
CNS C4446-87. Direct Current Machines (Apr)(11894).

—Symbols
CNS C1021-57. Symbols of D.C. Machines (Jan)(405) (R 1971).

DC Generators
Use For: Direct Current Generators *See Also:* DC Power Supplies; Generators; Turbine Generators
CNS C1005-88. Scheduled Information to Be Given on Enquiring and Ordering of Electric Machines (Small and Medium Size D. C. Generators) (Jan)(102).
CNS C1006-88. Scheduled Information to Be Given on Enquiring and Ordering of Electric Machines (Small and Medium Size D. C. Motors) (Jan)(103).

—Aircraft
BSI 2G 134-70. 1970 D.C. Generators. 9 pp.

—Arc Welding Equipment
DIN VDE 0540-65. Regulations for Direct Current Arc-Welding; Generators and Convertors (Feb). 19 pp.

—Automobiles
CNS D3045-87. Mounting Dimensions of D.C. Generators for Automobiles (Oct)(5322).
CNS D3063-87. Method of Test for DC Generators for Automobiles (Oct)(6170).
JIS D 5201-73. Mounting Dimensions of DC Generators for Automobiles (R 1984). 10 pp.

—Design—Interoperability
NATO STANAG 4134 ED 1 AMD 3-77. Electrical Characteristics of Rotating 28 Volt D.C. Generating Sets. 11 pp.
NATO STANAG 4134 ED 2 AMD 0-91. Electrical Characteristics of Rotating 28-Volt DC Generating Sets. 10 pp.
NATO STANAG 4134 ED 2 AMD 1-91. Electrical Characteristics of Rotating 28-Volt DC Generating Sets. 13 pp.

—Safety
CSA CAN/CSA-C22. 2 NO 100-92. Motors and Generators; (Gen Instr 1). 74 pp.
CSA 1169 Bull. Electrical Bulletin 1169 June 27, 1978 to C22.2 NO 100. 2 pp.

—Ships
MOD UK NES 631-91. Requirements for Main DC Generators Issue 3 (09.91). 15 pp.

—Submarines
MOD UK NES 631-91. Requirements for Main DC Generators Issue 3 (09.91). 15 pp.

DC Motors
Use For: Direct Current Motors *See Also:* Motors; Universal Motors

—Aircraft
BSI 2G 146-66. 1966 D.C. Motors for Aircraft. 5 pp.

—Controllers—Semiconductor Power Converters—Ratings
IEC 1136 Pt 1-92. Semiconductor Power Convertors—Adjustable Speed Electric Drive Systems—General Requirements Part 1: Rating Specifications, Particularly for d.c. Motor Drives First Edition. 71 pp.

—Generators—Preferred Products List
CECC CECC MUAHAG Vol 13 IS 1-91. Preferred Products List; Servo Components (En, Fr, Ge). 27 pp.

—Motor Protectors
CSA C22.2 NO 77-1988. Motors with Inherent Overheating Protection; (Gen Instr 1). 23 pp.
CSA 1169 Bull. Electrical Bulletin 1169 June 27, 1978 to C22.2 NO 77. 2 pp.

—Rotating Machines
BSI BS 5000: Part 25-81. 1981 Amd 2 Rotating Electrical Machines of Particular Types or for Particular Applications Part 25: D.C. Mill Auxiliary Motors (AMD 5631) July 29, 1988. 17 pp.

—Safety
CSA CAN/CSA-C22. 2 NO 100-92. Motors and Generators; (Gen Instr 1). 74 pp.
CSA 1169 Bull. Electrical Bulletin 1169 June 27, 1978 to C22.2 NO 100. 2 pp.

—Servo—Torque—Preferred Products List
CECC CECC MUAHAG Vol 13 IS 1-91. Preferred Products List; Servo Components (En, Fr, Ge). 27 pp.

DC Motors (Cont.)

—Tachometer Generators
CECC CECC MUAHAG Vol 13 IS 1-91. Preferred Products List; Servo Components (En, Fr, Ge). 27 pp.

—Tachometer Generators—Brushless—Limited Rotation—PPL
CECC CECC MUAHAG Vol 13 IS 1-91. Preferred Products List; Servo Components (En, Fr, Ge). 27 pp.

DC Power Supplies
See Also: DC Generators; Power Supplies; Switching Power Supplies
CNS C4440-86. D.C. Power Supply for Shore Use (May)(11569).

—Low Voltage
IEC 1204-93. Low-Voltage Power Supply Devices, d.c. Output—Performance Characteristics and Safety Requirements First Edition. 59 pp.

—Low Voltage—Safety
IEC 1204-93. Low-Voltage Power Supply Devices, d.c. Output—Performance Characteristics and Safety Requirements First Edition. 59 pp.

—Stabilized
BSI BS 5654: Part 2-79. 1979 Stabilized Power Supplies, d.c. Output Part 2: Method of Specifying Rating and Performance. 12 pp.
BSI BS 5654: Part 4-79. 1979 Stabilized Power Supplies d.c. Output Part 4: Test Other Than Radio-Frequency Interference. 44 pp.
IEC 478 Pt 2-86. Stabilized Power Supplies, D.C. Output Part 2: Rating and Performance Second Edition. 28 pp.
IEC 478 Pt 3-89. Stabilized Power Supplies, D.C. Output Part 3: Reference Levels and Measurement of Conducted Electromagnetic Interference (EMI) Second Edition. 20 pp.
IEC 478 Pt 4-76. Stabilized Power Supplies, D.C. Output Part 4: Tests Other Than Radio-Frequency Interference First Edition. 78 pp.

—Stabilized—Electromagnetic Interference
IEC 478 Pt 3-89. Stabilized Power Supplies, D.C. Output Part 3: Reference Levels and Measurement of Conducted Electromagnetic Interference (EMI) Second Edition. 20 pp.
IEC 478 Pt 5-93. Stabilized Power Supplies, d.c. Output Part 5: Measurement of the Magnetic Component of the Reactive Near Field First Ediiton. 23 pp.

—Stabilized—Glossaries
BSI BS 5654: Part 1-79. 1979 Stabilized Power Supplies, d.c. Output Part 1: Terms and Definitions. 21 pp.
IEC 478 Pt 1-74. Stabilized Power Supplies, D.C. Output Part 1: Terms and Definitions First Edition. 38 pp.

dc Power Transmission
Use: Power Transmission—DC

DC Signaling Systems
Use For: Direct Current Signaling Systems
See Also: Signaling Systems

—Telephone Sets—Pushbutton
CEPT T/CS 46-03-85. Signalisation Pour Poste a Clavier Combinant Les Signaux Multifrequences De Base Et Une Signalisation Par Courant Continu. 11 pp.
CEPT T/CS 46-03 E-86. Signalling System for Push-Button Telephones Combining Basic MF Signalling with Direct Current Signalling. 9 pp.

DC Signals
Use For: Analog Direct Current Signals; Direct Current Signals

—Mining Equipment
BSI BS 5754-87. 1987 Amd 1 Electrical Analogue and State Signals for Use in Coal Mines (AMD 6481) November 30, 1990. 11 pp.

—Process Control Systems
BSI BS 5863: Part 1-84. 1984 Analogue Signals for Process Control Systems Part 1: Specification for Direct Current Signals. 6 pp.
CENELEC HD 452.1-84. Analogue Signals for Process Control Systems Part 1: Direct Current Signals. 2 pp.
CNS C5234-90. Binary Direct Voltage Signals for Process Measurement and Control Systems (May)(12714).
IEC 381 Pt 1-82. Analogue Signals for Process Control Systems Part 1: Direct Current Signals Second Edition. 12 pp.

dc Voltage Signals
Use: Direct Voltage Signals

DCE
Use: Data Circuit Terminating Equipment

DCME
Use: Digital Circuit Multiplication Equipment

DCMS
Use: Digital Circuit Multiplication Equipment

DDI
Use: Direct Dialing In Services

DDT
Use For: Dichlorodiphenyltrichloroethane
See Also: Insecticides
CNS K1026-57. DDT for Industrial Use (Dec)(374)(R 1968).
CNS K6036-62. Method of Test for DDTT of Industrial Grade and Its Products (Sep)(635)(R 1968).
SAA AS N27-58. DDT for Insecticidal Preparations Corrig. 14 pp.
SAA AS 1870.1D-76. Standard for Development—Pesticides for Agricultural Use—Part 1D: DDT-Dichlorodiphenyl-trichloroethane. 18 pp.
SAA AS 1870.2D-76. Standard for Development—Pesticides for Agricultural Use—Part 2D: BHC and BHC-DDT. 30 pp.

—Cockroaches
MOD UK CS 3099A. DDT Cockroach Powder (Withdrawn).

—Solvents
CNS K5031-64. Petroleum Solvent for DDT (May)(2261). 1 p.

De Havilland Canada DHC-8-102 Aircraft
See Also: Aircraft

—Certification
CAA. De Havilland Canada DHC-8-102. 14 pp.

De Havilland Canada 2 Beaver Aircraft
See Also: Aircraft

—Foreign Airworthiness Directives
CAA. De Havilland DHC-2 (Foreign Airworthiness Directives). 5 pp.

De Havilland Canada 2 Turbo Beaver Aircraft
See Also: Aircraft

—Foreign Airworthiness Directives
CAA. De Havilland DHC-2 (Foreign Airworthiness Directives). 5 pp.

De Havilland Canada 6 Twin Otter Aircraft
See Also: Aircraft

—Accidents
CAA. DHC-6 Twin Otter (World Airline Accident Summary). 1 p.
CAA. DHC-6 Twin Otter. 3 pp.

—Antenna Positions
CAA. DHC-6 Twin Otter (Approved Aerial Positions). 1 p.

—Certification
CAA. DHC-6 Twin Otter Series 110, 210 and 310. 2 pp.

—Foreign Airworthiness Directives
CAA. De Havilland DHC-6 Twin Otter (Foreign Airworthiness Directives). 11 pp.

De Havilland Canada 7 Aircraft
See Also: Aircraft

—Accidents
CAA. DHC-7 (World Airline Accident Summary). 1 p.

—Antenna Positions
CAA. DHC-7 (Approved Aerial Positions). 1 p.

—Certification
CAA. De Havilland Canada DHC-7. 12 pp.

—Data Sheets
CAA FA26 ISSUE 1. De Havilland Canada DHC-7. 2 pp.

INTERNATIONAL AND NON-U.S. NATIONAL STANDARDS
SUBJECT INDEX

De Havilland Canada 7 Aircraft (Cont.)
—**Foreign Airworthiness Directives**
CAA. De Havilland DHC-7 Series Aircraft (Foreign Airworthiness Directives). 7 pp.

De Havilland Canada 8 Aircraft
See Also: Aircraft
—**Certification**
CAA. De Havilland Canada DHC-8. 14 pp.
—**Data Sheets**
CAA FA36 ISSUE 2. De Havilland Canada, DHC-8, Series—100. 2 pp.
—**Foreign Airworthiness Directives**
CAA. De Havilland DHC-8 Series Aircraft. 7 pp.

Dean and Stark Method
See Also: Distillation Methods
BSI BS 756-52. 1952 Amd 2 Dean and Stark Apparatus. 30 pp.
SAA AS R25-62. Dean and Stark Apparatus Being BS 756:1952, (Including Amendment of December 1953), Endorsed Subject to Australian Amendment Amdt 1 February 1962. 25 pp.
—**Bituminous Coatings**
MOD UK M 3004/67. Protective PX-2 and Composition Rust Preventative Type B.
—**Coal Tar Pitch**
BSI BS 6043: Sec 1.2-81. 1981 Methods of Sampling and Test for Carbonaceous Materials Used in Aluminium Manufacture Part 1: Electrode Pitch Section 1.2: Determination of Water Content (Dean and Stark Method). 8 pp.
ISO 5939-80. Carbonaceous Materials for the Production of Aluminium—Pitch for Electrodes—Determination of Water Content—Azeotropic Distillation (Dean and Stark) Method First Edition. 7 pp.
—**Paints**
BSI BS 3900: Part B1-65. (OBSOLESCENT) 1965 Methods of Test for Paints Group B: Tests Involving Chemical Examination of Liquid Paints and Dried Paint Films Part B1: Determination of Water by the Dean and Stark Method. 2 pp.
—**Phenols**
ISO 1897 Pt II-77. Phenol, o-Cresol, m-Cresol, p-Cresol, Cresylic Acid and Xylenols for Industrial Use—Methods of Test—Part II: Determination of Water-Dean and Stark Method First Edition. 7 pp.
—**Soaps**
MOD UK M 9532/67. Examination of Soap, Saddle, Glycerine (No Information) (Withdrawn).

Death
—**Causes—Statistical Classifications**
NATO STANAG 2050 ED 5 AMD 2-89. Statistical Classification of Diseases, Injuries and Causes of Death. 28 pp.

Decade/Biquinary Counter/Latches
See Also: Decade Counters; Latches (Circuits)
CECC CECC 90 102-057 ISSUE 1-82. BS CECC90 102-057; Presettable Decade/Biquinary Counters/Latches (En). 20 pp.
—**BCD**
CECC CECC 90 103-085 ISSUE 1-82. BS CECC 90 103-085; Presettable Decade/Bi-Quinary Counters/Latches (En) AMD 1 (En). 21 pp.

Decade Counter/Dividers
See Also: Counter/Dividers; Decade Counters; Linear Circuits
CECC CECC 90 104-170 ISSUE 1-81. BS CECC 90 104-170; Decade Counter/Divider (En). 26 pp.
CECC CECC 90 104-170 ISSUE 1-86. CEI CECC 90 104-170; Decade Counter/Divider (En). 2 pp.
CECC CECC 90 109-634 ISSUE 1-86. Digital Integrated Circuits in Accordance with FS 90 109; 54/74 HC 4017; Decade Counter/Divider (En, Fr). 6 pp.
CECC CECC 90 109-869 ISSUE 1-90. Digital Integrated Circuits; Silicon Monolithic C MOS, Cavity or Non-Cavity Packages; Type(s); 54/74 HCT 4017 Decade Counter/Divider; Assessment Levels P, Y, L (En, Fr). 9 pp.
—**Preferred Products List**
CECC CECC MUAHAG Vol 7 IS 8-92. Preferred Products List; Active Microcircuits (En, Fe, Ge). 89 pp.

Decade Counter/Registers
See Also: Decade Counters

Decade Counter/Registers (Cont.)
—**Asynchronous**
CECC CECC 90 109-836 ISSUE 1-89. Digital Integrated Circuits; Silicon Monolithic C MOS, Cavity or Non-Cavity Packages; Type(s) 54/74 HC 690 Asynchronous Clear Decade Counter /Register with 3-State Outputs Assessment Levels P, Y, L (En, Fr, Ge). 11 pp.
—**Synchronous**
CECC CECC 90 109-808 ISSUE 1-88. Digital Integrated Circuits in Accordance with FS 90 109; 54/74 HC 692; Synchronous Clear Decade Counter/Register with 3-State Outputs (En, Fr). 8 pp.
—**Synchronous/Asynchronous—Count Up/Down**
CECC CECC 90 109-848 ISSUE 1-90. Digital Integrated Circuits; Silicon Monolithic C MOS, Cavity or Non-Cavity Packages; Type(s) 54/74 HC 696 Synchronous Up/Down Decade Counter/Register with Asynchronous Clear and 3-State Outputs Assessment Levels P, Y, L (En, Fr, Ge). 11 pp.
—**Synchronous—Count Up/Down**
CECC CECC 90 109-850 ISSUE 1-90. Digital Integrated Circuits; Silicon Monolithic C MOS, Cavity or Non-Cavity Packages; Type(s) 54/74 HC 968 Synchronous Up/Down Decade Counter/Register with Synchronous Clear and 3-State Outputs Assessment Levels P, Y, L (En, Fr, Ge). 11 pp.

Decade Counters
See Also: Binary/Decade Counters; Decade/Biquinary Counter/Latches; Decade Counter/Dividers; Decade Counter/Registers; Digital Circuits
CECC CECC 90 103-041 ISSUE 1-81. BS CECC 90 103-041; Decade Divide-By-Two and Divide-By-Five Counters, Supply to Pins 5 and 10 (En) AMD 1 (En). 21 pp.
CECC CECC 90 103-107 ISSUE 1-81. BS CECC 90 103-107; Decade Divide-by-2 and Divide-by-5 Counters with Supply to Pins 7 and 14 (En) AMD 1 (En). 21 pp.
CECC CECC 90 104-080 ISSUE 2. NL CECC 90 104-080 Issue 2; Digital Integrated Circuits in Accordance with FS 90 104; HEC/HEF 4534B; Real Time 5-Decade Counter (En). 13 pp.
—**Asynchronous**
CECC CECC 90 109-815 ISSUE 1-88. Digital Integrated Circuits in Accordance with FS 90 109; 54/74 HC 4518; Dual Decade Counter with Asynchronous Clear (En, Fr). 6 pp.
—**Asynchronous—BCD Output**
CECC CECC 90 104-237 ISSUE 1. NL CECC 90 104-237 Issue 1; Digital Integrated Circuits in Accordance with FS 90 104; HEC/HEF 4737B; HEC/HEF 4737V; Quadruple Static Decade Counters (En). 12 pp.
—**BCD**
CECC CECC 90 101-029 ISSUE 1-87. UTE C 86-213/B 49; Digital Integrated Circuits in Accordance with FS 90 101; 54/64/74 90A; BCD Decade Counters (En). 16 pp.
—**BCD Output—Count Up/Down**
CECC CECC 90 104-124 ISSUE 1-81. BS CECC 90 104-124; Silicon Complementary MOS with (B) Buffered Outputs and Cavity Packaging (En). 14 pp.
CECC CECC 90 104-124 ISSUE 1-86. CEI CECC 90 104-124; Silicon Complementary MOS with (B) Buffered Outputs Cavity and Non Cavity Packaging (En). 2 pp.
—**Dual**
CECC CECC 90 103-128 ISSUE 1-81. BS CECC 90 103-128; Dual Decade Counter (En) AMD 1 (En). 20 pp.
CECC CECC 90 109-730 ISSUE 1-87. Digital Integrated Circuits in Accordance with FS 90 109; 54 HC 390, 74 HC 390; Dual Decade Counter (En, Fr). 7 pp.
CECC CECC 90 109-847 ISSUE 1-90. Digital Integrated Circuits; Silicon Monolithic C MOS, Cavity or Non-Cavity Packages; Type(s) 54/74 HCT 390 Dual Decade Counter Assessment Levels P, Y, L (En, Fr, Ge). 10 pp.
—**Dual—Preferred Products List**
CECC CECC MUAHAG Vol 7 IS 8-92. Preferred Products List; Active Microcircuits (En, Fe, Ge). 89 pp.
—**Preferred Products List**
CECC CECC MUAHAG Vol 7 IS 8-92. Preferred Products List; Active Microcircuits (En, Fe, Ge). 89 pp.

Decade Counters (Cont.)
—**Real Time**
CECC CECC 90 104-210 ISSUE 1-82. BS CECC 90 104-210; Real Time 5 Decade Counter with 3-State Outputs (En). 32 pp.
CECC CECC 90 104-210 ISSUE 1-86. CEI-CECC 90 104-210; Real Time 5 Decade Counter with 3-State Outputs (En). 2 pp.
—**Synchronous**
CECC CECC 90 104-191 ISSUE 1-81. BS CECC 90 104-191 Issue 1; Synchronous Decade Counter with Synchronous Clear (En). 27 pp.
CECC CECC 90 104-191 ISSUE 1-86. CEI CECC 90 104-191 Issue 1; Synchronous Decade Counter with Synchronous Clear (En). 2 pp.
CECC CECC 90 109-645 ISSUE 1-86. Digital Integrated Circuits in Accordance with FS 90 109; 54/74 HC 162; Synchronous Decade Counter with Synchronous Clear (En, Fr). 7 pp.
CECC CECC 90 109-786 ISSUE 1-87. Detail Specification: Digital Integrated Circuits; Silicon Monolithic C MOS, Cavity or Non-Cavity Packages; Type(s) 54/74 HC 162 Synchronous Decade Counter with Synchronous Clear; Assessment Levels P, Y, L (En, Fr). 9 pp.
—**Synchronous/Asynchronous**
CECC CECC 90 104-189 ISSUE 1-81. BS CECC 90 104-189 Issue 1; Synchronous Decade Counter with Asynchronous Clear (En). 26 pp.
CECC CECC 90 104-189 ISSUE 1-86. CEI CECC 90 104-189 Issue 1; Synchronous Decade Counter with Asynchronous Clear (En). 2 pp.
CECC CECC 90 109-630 ISSUE 1-86. Digital Integrated Circuits in Accordance with FS 90 109; 54/74 HC 160; Synchronous Decade Counter with Asynchronous Clear (En, Fr). 6 pp.
—**Synchronous/Asynchronous—4-Bit**
CECC CECC 90 104-093 ISSUE 2. NL CECC 90 104-093 Issue 2; Digital Integrated Circuits in Accordance with FS 90 104; HEC/HEF 40160B; 4-Bit Synchronous Decade Counter with Asynchronous Reset (En). 14 pp.
CECC CECC 90 104-095 ISSUE 2. NL CECC 90 104-095 Issue 2; Digital Integrated Circuits in Accordance with FS 90 104; HEC/HEF 40162B; 4-Bit Synchronous Decade Counter with Synchronous Reset (En). 14 pp.
CECC CECC 90 106-068 ISSUE 1-85. UTE C 86-217; Digital Integrated Circuits in Accordance with FS 90 106; 54/74 ALS 160, 54/74 ALS 160A; Synchronous 4-Bit Decade Counters with Asynchronous Clear (En, Fr) ADD 3 (En, Fr). 17 pp.
—**Synchronous—Count Up/Down**
CECC CECC 90 109-728 ISSUE 1-87. Digital Integrated Circuits in Accordance with FS 90 109; 54 HC 192, 74 HC 192; Synchronous Up/Down Decade Counter (En, Fr). 6 pp.
CECC CECC 90 109-794 ISSUE 1-87. Digital Integrated Circuits; Silicon Monolithic C MOS, Cavity or Non-Cavity Packages; Type(s) 54/74 HCT 192; Synchronous Up/Down (En, Fr). 9 pp.
—**Synchronous—4-Bit**
CECC CECC 90 102-054 ISSUE 1-82. BS CECC 90 102-054; 4-Bit Decade Counters with Synchronous Clear (En). 20 pp.
CECC CECC 90 103-067 ISSUE 1-82. BS CECC 90 103-067; Decade Direct Clear, Synchronous 4-Bit Counters (En) AMD 1 (En). 22 pp.
CECC CECC 90 103-068 ISSUE 1-82. BS CECC 90 103-068; Binary Direct Clear, Synchronous 4-Bit Counters (En). 22 pp.
CECC CECC 90 103-069 ISSUE 1-82. BS CECC 90 103-069; 4-Bit Synchronous Decade Counter with Synchronous Clear (En). 23 pp.
CECC CECC 90 106-036 ISSUE 2-86. UTE C 86-217 ADD 2/GA 36; Digital Integrated Circuits in Accordance with FS 90 106; 54/74 ALS 560, 54/74 ALS 560A; Synchronous 4-Bit Decade Counter with 3 State Outputs (En, Fr) ADD 3 (En, Fr). 18 pp.
CECC CECC 90 106-069 ISSUE 1-85. UTE C 86-217; Digital Integrated Circuits in Accordance with FS 90 106; 54/74 ALS 162, 54/74 ALS 162A; Synchronous 4-Bit Decade Counters with Synchronous Clear (En, Fr) ADD 3 (En, Fr). 17 pp.
CECC CECC 90 106-074 ISSUE 1-85. UTE C 86-217; Digital Integrated Circuits in Accordance with FS 90 106; 54/74 ALS 168/168B, 54/74 ALS 168U; Synchronous 4-Bit Up/Down Decade Counter (En, Fr) ADD 3 (En, Fr). 17 pp.
CECC CECC 90 107-019 ISSUE 2-89. UTE C 86-218 ADD 2/FA 19; Digital Integrated Circuits in Accordance with FS 90 107; 54/74 F 160A; Synchronous 4-Bit Decade Counters with Asynchronous Clear (En, Fr) ADD 3 (En, Fr). 11 pp.

INDUSTRY STANDARDS

Decade Counters (Cont.)

—Synchronous—4-Bit—Count Up/Down
CECC CECC 90 102-029 ISSUE 1-81. BS CECC 90 102-029; 4-Bit Up/Down Synchronous Decade Counter (En) AMD 1 (En). 21 pp.
CECC CECC 90 103-169 ISSUE 1-81. BS CECC 90 103-169; Synchronous 4-Bit Up/Down Decade Counter (En) AMD 1 (En). 23 pp.
CECC CECC 90 106-029 ISSUE 1-85. UTE C 86-217; Digital Integrated Circuits in Accordance with FS 90 106; 54 ALS 190, 74 ALS 190; Synchronous 4-Bit Up/Down Decade Counter (En, Fr) ADD 3 (En, Fr). 17 pp.
CECC CECC 90 106-031 ISSUE 1-85. UTE C 86-217; Digital Integrated Circuits in Accordance with FS 90 106; 54/74 ALS 192, 54/74 ALS 192U; Synchronous 4-Bit Up/Down Decade Counter (En, Fr) ADD 3 (En, Fr). 18 pp.
CECC CECC 90 106-038 ISSUE 2-86. UTE C 86-217 ADD 2/GA 38; Digital Integrated Circuits in Accordance with FS 90 106; 54/74 ALS 568, 54/74 ALS 568A; Synchronous 4-Bit Up/Down Decade Counter with 3 State Outputs (En, Fr) ADD 3 (En, Fr). 18 pp.
CECC CECC 90 107-026 ISSUE 2-89. UTE C 86-218 ADD 3/FA 26; Digital Integrated Circuits in Accordance with FS 90 107; 54/74 F 190; Synchronous 4-Bit Up/Down Decade Counter (En, Fr) ADD 3 (En, Fr). 11 pp.
CECC CECC 90 107-027 ISSUE 2-89. UTE C 86-218 ADD 3/FA 27; Digital Integrated Circuits in Accordance with FS 90 107; 54/74 F 192; Synchronous 4-Bit Up/Down Decade Counter (En, Fr) ADD 3 (En, Fr). 11 pp.
CECC CECC 90 109-752 ISSUE 1-87. Digital Integrated Circuits in Accordance with FS 90 109; 54 HC 190, 74 HC 190; Synchronous 4-Bit Up/Down Decade Counter (En, Fr). 7 pp.
CECC CECC 90 109-896 ISSUE 1-90. Digital Integrated Circuits; Silicon Monolithic C MOS, Cavity or Non-Cavity Packages; Type(s); 54/74 HCT 190 Synchronous 4-Bit Up/Down Decade Counter; Assessment Levels P, Y, L (En, Fr). 10 pp.

—Synchronous—4-Bit—Preferred Products List
CECC CECC MUAHAG Vol 7 IS 8-92. Preferred Products List; Active Microcircuits (En, Fe, Ge). 89 pp.

—Synchronous—8-Bit—Count Down
CECC CECC 90 109-885 ISSUE 1-90. Digital Integrated Circuits; Silicon Monolithic C MOS, Cavity or Non-Cavity Packages; Type(s); 54/74 HC 40102 8-Bit Synchronous Decade Down Counter; Assessment Levels P, Y, L (En, Fr). 10 pp.
CECC CECC 90 109-887 ISSUE 1-90. Digital Integrated Circuits; Silicon Monolithic C MOS, Cavity or Non-Cavity Packages; Type(s); 54/74 HCT 40102 8-Bit Synchronous Decade Down Counter; Assessment Levels P, Y, L (En, Fr). 10 pp.

—4-Bit—Count Up/Down
CECC CECC 90 104-099 ISSUE 2. NL CECC 90 104-099 Issue 2; Digital Integrated Circuits in Accordance with FS 90 104; HEC/HEF 40192B; 4-Bit Up/Down Decade Counter (En). 13 pp.

—4-Bit—Dual
CECC CECC 90 103-180 ISSUE 1-82. BS 90 103-180; Dual 4-Bit Decade Counters with Set-to-9 and Clear Inputs (En). 19 pp.

Decade Resistors
See Also: Resistors; Variable Resistors

—DC—Voltage Dividers
CNS C5037-80. Decade Resistive Voltage Dividers (Direct Current Type) (Feb)(5207).

Decade Transformers
See Also: Transformers

—Voltage Dividers
CNS C5038-80. Requirements for Decade Transformers Dividers (Voltage Type) (Feb)(5208).

Decals
See Also: Labels
CGSB 62-GP-2B-65. Decalcomanias: General Purpose, for Use on Exterior and Interior Surfaces, Specification for; (Amendment 2 Oct 1970). 10 pp.
CGSB 62-GP-9M-80. Prefabricated Markings, Positionable, Exterior, for Aircraft, Ground Equipment and Facilities, Standard for. 17 pp.

—Aircraft
CGSB 62-GP-3M-80. Prefabricated Markings, Exterior, for Aircraft, Standard for. 18 pp.
CGSB 62-GP-9M-80. Prefabricated Markings, Positionable, Exterior, for Aircraft, Ground Equipment and Facilities, Standard for. 17 pp.

Decals (Cont.)

—Ground Vehicles
CGSB 62-GP-6M-82. Prefabricated Transparent-Base Markings, General-Purpose, for Use on Exterior and Interior Surfaces, Standard for. 12 pp.
CGSB 62-GP-7M-82. Prefabricated "Mirrorized" Markings, General Purpose, for Use on Exterior and Interior Surfaces, Standard for. 12 pp.
CGSB 62-GP-8M-80. Prefabricated Opaque Markings, General Purpose, for Use on Exterior and Interior Surfaces, Standard for. 12 pp.
CGSB 62-GP-9M-80. Prefabricated Markings, Positionable, Exterior, for Aircraft, Ground Equipment and Facilities, Standard for. 17 pp.

—Signal Devices
CGSB 62-GP-8M-80. Prefabricated Opaque Markings, General Purpose, for Use on Exterior and Interior Surfaces, Standard for. 12 pp.

n-Decane
CNS K0005-88. High Purity Decane (Sep)(6220).

Decanes
See Also: Hydrocarbons
JIS K 0508-83. High Purity Decane.

Decarburizing
Scope Note: See the subheading Decarburizing under specific types of metals

Decating
See Also: Fabrics

—Fabrics—Colorfastness Testing
CNS L3184-82. Method of Test for Colour Fastness to Decatizing (Aug)(9307).
ISO 105 Pt E10-87. Textiles—Tests for Colour Fastness—Part E10: Colour Fastness to Decatizing Second Edition. 4 pp.
JIS L 0874-75. Testing Method for Colour Fastness to Decatizing (R 1983). 8 pp.

Decatizing
Use: Decating

Decay Heat

—Nuclear Fuels—Light Water Reactors
ISO 10645-92. Nuclear Energy—Light Water Reactors—Calculation of the Decay Heat Power in Nuclear Fuels First Edition. 16 pp.

Decca Navigator Systems
See Also: Navigational Aids; Radio Navigation; Radio Navigation Equipment

—Radio Frequency Interference
CCIR Report 915-2-90. Interference Between Fixed, Maritime Mobile and Radionavigation Services in the Bands Between 70 kHz and 130 kHz—Section 8D—Radiodetermination, Global Maritime Distress and Safety System and Related Subjects. 17 pp.

—Receivers—Ships
IEC 1135-92. Decca Navigator System: Receivers for Ships Minimum Performance Standards—Methods of Testing and Required Test Results First Edition. 30 pp.

Deceleration Testing
See Also: Acceleration Testing; Braking Testing; Impact Testing

—Carburetors
CNS D3094-89. Deceleration Test Code of Carburetors for Automobiles (Jan)(8251).

Deceleration Valves
Use For: Speed Control Valves See Also: Flow Control Valves; Valves

—Pneumatic
JIS B 8376-82. Speed Control Valves for Pneumatic Use. 14 pp.

Decibels
See Also: Nepers; Power Levels; Signal to Distortion Ratio; Signal to Interference Ratio; Signal to Noise Ratio; Sound Pressure; Voltage

—Telecommunication Equipment—Glossaries
CCIR RECMN 574-3-90. Use of the Decibel and the Neper in Telecommunications—Section C—Other Means of Expression. 11 pp.

—Telecommunication Services—Glossaries
CCIR RECMN 574-3-90. Use of the Decibel and the Neper in Telecommunications—Section C—Other Means of Expression. 11 pp.

Decibels (Cont.)

—Telecommunication Systems—Glossaries
CCIR RECMN 574-3-90. Use of the Decibel and the Neper in Telecommunications—Section C—Other Means of Expression. 11 pp.

Deciduous Woods
Use: Hardwoods

Decimal Notation
Use: Numbers—Decimal Notation

Decision Tables
See Also: Data Processing; Tables (Data)
BSI BS 5487-86. 1986 Single-Hit Decision Tables. 17 pp.
ISO 5806-84. Information Processing—Specification of Single-Hit Decision Tables First Edition. 16 pp.
JIS X 0125-86. Decision Tables.
JTC1 5806-84. Information Processing—Specification of Single-Hit Decision Tables First Edition. 16 pp.
OSI ISO 5806-84. Information Processing—Specification of Single-Hit Decision Tables. 16 pp.

Deck Machinery
See Also: Decks; Ships
ISO 7825-85. Shipbuilding—Deck Machinery—General Requirements First Edition. 4 pp.

—Crown Gears—Remote Control
CNS B2592-81. Remote Control Gear with Transmission Tubing for Manual Operation; Deck Operation Devices with Slewable Crown Gear Unit Welding-In Type (Jun)(7589).

—Gears—Remote Control
CNS B2591-81. Remote Control Gear with Transmission Tubing for Manual Operation; Deck Operation Devices Welding-In Type (Jun)(7588).

—Glossaries
BSI BS MA 30-73. 1973 Glossary of Terms and Graphical Symbols for Ships' Deck Machinery. 16 pp.
ISO 3828-84. Shipbuilding and Marine Structures—Deck Machinery—Vocabulary Second Edition. 28 pp.

—Safety
BSI BS MA 93-81. 1981 Code of Practice for Structural Safety of Deck Machinery Installation. 7 pp.

—Symbols
BSI BS MA 30-73. 1973 Glossary of Terms and Graphical Symbols for Ships' Deck Machinery. 16 pp.

—Warping Ends—Profiles
ISO 6482-80. Shipbuilding—Deck Machinery—Warping End Profiles First Edition. 5 pp.

Deck Pipes
See Also: Pipes

—Ships
MOD UK NES 512: Part 11-01. Guide to Cables, Electrical and Associated Items Part 11: Glands, Grommets and Deck Tubes Issue 2 (03.91); Amendment 1. 81 pp.

—Ships—Covers
CNS F3094-81. Ships' Rope Hole Covers (Apr)(7237).
JIS F 2010-82. Ships' Rope Hole Covers.
JIS F 2010-70. Ships' Rope Hole Covers. 6 pp.

—Ships—Mooring Rope
CNS F3025-80. Mooring Pipes (Mar)(5333).
JIS F 2007-76. Mooring Pipes. 7 pp.
JIS F 2030-78. Single Point Mooring Pipes.

—Ships—Mountings
CNS F3062-80. Ships' Deck and Bulkhead Pieces for Small Size Copper Tubes (Sep)(6442).

—Ships—Scupper Grilles
JIS F 3015-77. Gratings of Ships' Scupper Pipes (R 1984). 11 pp.

—Ships—Sounding Pipes
CNS F3051-80. Deck Pieces for Sounding Pipes (Aug)(6325).
JIS F 3002-82. Deck Pieces for Sounding Pipes.
JIS F 3002-68. Deck Pieces for Sounding Pipes (R 1977). 5 pp.

—Submarines
MOD UK NES 512: Part 11-01. Guide to Cables, Electrical and Associated Items Part 11: Glands, Grommets and Deck Tubes Issue 2 (03.91); Amendment 1. 81 pp.

Decks
See Also: Deck Machinery; Floors; Roofing; Ships

INTERNATIONAL AND NON-U.S. NATIONAL STANDARDS
SUBJECT INDEX

Decks (Cont.)

—Boats (Marine)—Guardrails
CEN PREN 711-92. Inland Navigation Vessels—Railings for Decks—Requirements, Types. 9 pp.

—Bricks
DIN ENGL 4159-78. Structurally Cooperating Bricks for Floors and Wall Panels (Apr). 9 pp.
DIN ENGL 4160-78. Structurally Non-Cooperating Bricks for Floors (Aug). 6 pp.

—Metal—Epoxy Coatings—Nonskid
CGSB 1-GP-192MA-85. Deck Coating, Nonslip, Epoxy. 14 pp.

—Metal—Paints—Nonskid
MOD UK DSTAN 80-73-77. Paint, Non-Slip, Deck Type: Brushing Issue 1. 13 pp.

—Metal—Polyurethane Coatings—Nonskid
CGSB 1-GP-200MA-85. Deck Coating, Nonslip, Polyurethane. 14 pp.

—Roof
DIN ENGL 18530-87. Solid Structural Decks for Roofs; Design and Construction (Mar). 6 pp.

—Roof—Aluminum Alloy
CNS A2229-88. Components for Aluminium Alloy Roof-Decks (Sep)(12414).
CNS A3296-88. Method of Test for Components for Aluminium Alloy Roof-Decks (Sep)(12415).

—Roof—Steel
CNS A2127-87. Steel Roof-Decks (Dec)(8339).
CNS A3139-87. Method of Test for Steel Roof-Decks (Dec)(8340).
JIS A 6514-90. Components for Metal Roof-Decks. 25 pp.

—Ships
MOD UK NES 115-01. Details and List of Weatherdeck and Side Arrangements for Surface Ships Issue 1 (04.84); Amendment 1. 54 pp.

—Ships—Awnings
MOD UK NES 727-88. Guide to the Design and Manufacture of Awnings and Associated Screens and Covers Issue 1 (12.88). 28 pp.

—Ships—Coverings
MOD UK NES 727-88. Guide to the Design and Manufacture of Awnings and Associated Screens and Covers Issue 1 (12.88). 28 pp.

—Ships—Coverings—Flammability Testing
MOD UK NES 750-81. Fire Tests for Deck Coverings with Under Deck Heat Source Issue 2 (04.81) (Withdrawn). 12 pp.

—Ships—Coverings—Rubber
MOD UK NES 2073-91. Specification for Low Fire Hazard Solid Rubber Deck—Covering Issue 1 (04.91). 11 pp.

—Ships—Draperies
MOD UK NES 727-88. Guide to the Design and Manufacture of Awnings and Associated Screens and Covers Issue 1 (12.88). 28 pp.

—Ships—Guardrails
ISO 3674-76. Shipbuilding—Inland Vessels—Deck Rail First Edition. 8 pp.
ISO 5480-79. Shipbuilding—Guardrails for Cargo Ships First Edition. 8 pp.

—Ships—Lighting
CNS F3218-83. Deck Lights (May)(10250).
CNS F5067-86. Boat Deck Light (May)(10337).
JIS F 2406-76. Deck Lights.
JIS F 2406-68. Deck Lights. 5 pp.
JIS F 8413-83. Boat Deck Lights.

—Ships—Paint
MOD UK DSTAN 80-151-91. Paint, Finishing, for Interior Decks and Diodes Types: Brushing Spraying Issue 1. 15 pp.

—Ships—Paint—Slip Resistant
MOD UK DSTAN 80-134-89. Paint System, Anti-Slip for Ships' Decking, Two Pack Type 1: Medium Texture Type 2: Rough Texture Issue 1. 28 pp.
MOD UK DSTAN 80-134-01. Paint System, Anti-Slip for Ships' Decking, Two Pack Type 1: Medium Texture Type 2: Rough Texture Issue 1; Amendment 1. 25 pp.

—Ships—Painting
MOD UK NES 753-01. Requirements for Preservation and Coating of Flight Decks and Weather Decks Issue 3 (02.89); Amendment 2. 30 pp.
MOD UK NES 758-91. Requirements for Weatherwork Preparation and Painting of Surface Ships issue 4 (11.91). 31 pp.

Decks (Cont.)

—Ships—Rope Guides—End Rollers
CNS F3091-81. Cast Iron Deck End Rollers (Apr)(7234).
CNS F3092-81. Steel Place Deck End Rollers (Apr)(7235).
CNS F3107-81. Small Size Cast Iron Deck-End Rollers (May)(7372).
CNS F3108-81. Small Size Plate Deck-End Rollers (May)(7373).
CNS F3118-81. Ships' Horizontal Rollers (Jul)(7690).
CNS F3120-81. Ships' Small Size Stand Rollers (Jul)(7692).
JIS F 2003-90. Cast Iron Deck End Rollers.
JIS F 2003-68. Cast Iron Deck End Rollers. 6 pp.
JIS F 2004-90. Steel Plate Deck End Rollers.
JIS F 2004-76. Steel Plate Deck-End Rollers. 8 pp.
JIS F 2019-79. Cast Iron Deck End Rollers (Small Size).
JIS F 2019-76. Small Size Cast Iron Deck-End Rollers. 7 pp.
JIS F 2020-79. Steel Plate Deck End Rollers (Small Size).
JIS F 2020-76. Small Size Steel Plate Deck-End Rollers. 8 pp.
JIS F 2022-76. Ships' Horizontal Rollers. 8 pp.
JIS F 2024-75. Ships' Small Size Stand Rollers. 6 pp.

—Ships—Screens (Protectors)
MOD UK NES 727-88. Guide to the Design and Manufacture of Awnings and Associated Screens and Covers Issue 1 (12.88). 28 pp.

—Ships—Scupper Grilles
JIS F 3015-77. Gratings of Ships' Scupper Pipes (R 1984). 11 pp.

—Ships—Shafts—Mountings
JIS F 3008-77. Deck and Bulkhead Pieces for Transmission Shaft (R 1984). 8 pp.

—Ships—Stuffing Boxes
CNS F5045-83. Marine Cable Glands for Bulkhead and Deck (Jan)(9854).
JIS F 8802-87. Marine Cable Glands for Bulkhead and Deck.

—Ships—Tracks
BSI BS MA 28-73. 1973 Dimensions of Tracks for Mast and Deck Fittings. 3 pp.

—Structural Timber
SAA AS O80-63. Decking Timbers from Eastern and South-Eastern Australian Hardwoods. 8 pp.

—Submarines—Coverings
MOD UK NES 157: Part 1-89. Requirements for Deck Coverings Part 1: General Requirements Issue 2 (11.89). 22 pp.
MOD UK NES 157: Part 2-01. Requirements for Deck Coverings Part 2: Deck Preparation and Laying Procedures Issue 2 (11.89); Amendment 1. 36 pp.

—Submarines—Coverings—Rubber
MOD UK NES 2073-91. Specification for Low Fire Hazard Solid Rubber Deck—Covering Issue 1 (04.91). 11 pp.

—Thermal Insulation
CGSB CAN/CGSB-51.31-M84. Thermal Insulation, Mineral Fibre Board for Above Roof Decks. 9 pp.

—Wood—Epoxy Coatings—Nonskid
CGSB 1-GP-192MA-85. Deck Coating, Nonslip, Epoxy. 14 pp.

—Wood—Pitch
MOD UK TS 10198A. Marine Glue Pitch.

—Wood—Polyurethane Coatings—Nonskid
CGSB 1-GP-200MA-85. Deck Coating, Nonslip, Polyurethane. 14 pp.

Decoder/Demultiplexers
Use: Digital Demultiplexers

Decoder/Drivers
Use: Driver/Decoders

Decoders
Use For: Descramblers *See Also:* Driver/Decoders; Encoder/Decoders; Encoders; Latch/Decoder/Drivers; Latch/Decoders

—Audio Systems
CNS C5061-89. Decoders (Type 1) for Reproducing Matrix Quadraphonic Disc Records (Oct)(6139).
CNS C5062-89. Decoders (Type II) for Reproducing Matrix Quadraphonic Disc Records (Oct)(6140).
IEC 11172 Pt 3-93. Information Technology—Coding of Moving Pictures and Associated Audio for Digital Storage Media at up to About 1,5 Mbit/s—Part 3: Audio First Edition. 157 pp.

Decoders (Cont.)

—Audio Systems (Cont.)
ISO 11172 Pt 3-93. Information Technology—Coding of Moving Pictures and Associated Audio for Digital Storage Media at up to About 1,5 Mbit/s—Part 3: Audio First Edition. 157 pp.
JTC1 11172 Pt 3-93. Information Technology—Coding of Moving Pictures and Associated Audio for Digital Storage Media at up to About 1,5 Mbit/s—Part 3: Audio First Edition. 157 pp.

—Audio/Video Systems
IEC 11172 Pt 1-93. Information Technology—Coding of Moving Pictures and Associated Audio for Digital Storage Media at up to About 1,5 Mbit/s—Part 1: Systems First Edition. 60 pp.
ISO 11172 Pt 1-93. Information Technology—Coding of Moving Pictures and Associated Audio for Digital Storage Media at up to About 1,5 Mbit/s—Part 1: Systems First Edition. 60 pp.
JTC1 11172 Pt 1-93. Information Technology—Coding of Moving Pictures and Associated Audio for Digital Storage Media at up to About 1,5 Mbit/s—Part 1: Systems First Edition. 60 pp.

—Integrated Circuit
CECC CECC 90 104-025 ISSUE 2. NL CECC 90 104-025 Issue 2; Digital Integrated Circuits in Accordance with FS 90 104; HEC/HEF 4028B; 1-of-10 Decoder (En). 7 pp.

—Integrated Circuit—BCD to Decimal
CECC CECC 90 104-123 ISSUE 1-81. BS CECC 90 104-123; Silicon Complementray MOS with (B) Buffered Outputs and Cavity Packaging (En). 24 pp.
CECC CECC 90 104-123 ISSUE 1-86. BS CECC 90 104-123; Silicon Complementary MOS with (B) Buffered Outputs Cavity and Non Cavity Packaging (En). 2 pp.
CECC CECC 90 109-672 ISSUE 1-87. Digital Integrated Circuits in Accordance with FS 90 109; 54/74 HC 42; BCD to Decimal Decoder (En, Fr). 5 pp.
CECC CECC 90 109-768 ISSUE 1-87. Digital Integrated Circuits in Accordance with FS 90 109; 54/74 HCT 42; BCD to Decimal Decoder (En, Fr). 5 pp.
CECC CECC 90 109-834 ISSUE 1-89. Digital Integrated Circuits; Silicon Monolithic C MOS, Cavity or Non-Cavity Packages; Type(s) 54/74 HC 4028 BDC to Decimal Decoder Assessment Levels P, Y, L (En, Fr, Ge). 8 pp.

—Integrated Circuit—BCD to Decimal—Preferred Products List
CECC CECC MUAHAG Vol 7 IS 8-92. Preferred Products List; Active Microcircuits (En, Fe, Ge). 89 pp.

—Integrated Circuit—BCD to 7-Segment
CECC CECC 90 104-223 ISSUE 1-85. BS CECC 90 104-223; BCD to Seven Segment Decoder (En). 29 pp.
CECC CECC 90 104-223 ISSUE 1-86. CEI-CECC 90 104-223; BCD-to-Seven Segment Decoder (En). 2 pp.

—Integrated Circuit—Binary
CECC CECC 90 104-123 ISSUE 1-81. BS CECC 90 104-123; Silicon Complementray MOS with (B) Buffered Outputs and Cavity Packaging (En). 24 pp.
CECC CECC 90 104-123 ISSUE 1-86. BS CECC 90 104-123; Silicon Complementary MOS with (B) Buffered Outputs Cavity and Non Cavity Packaging (En). 2 pp.

—Integrated Circuit—1 of 4
CECC CECC 90 107-014 ISSUE 2-89. UTE C 86-218 ADD 2/FA 14 Digital Integrated Circuits in Accordance with FS 90 107;54/74 F 139; Decoder (En, Fr) ADD 3 (En, Fr). 9 pp.

—Integrated Circuit—3 to 8 Lines
CECC CECC 90 109-608 ISSUE 1-86. Digital Integrated Circuits in Accordance with FS 90 109; 54 HC 138, 74 HC 138; 3 to 8 Line Decoder (En, Fr). 5 pp.
CECC CECC 90 109-684 ISSUE 1-87. Digital Integrated Circuits in Accordance with FS 90 109; 54/74 HC 259; 8-Bit Addressable Latch and 3 to 8 Line Decoder (En, Fr). 7 pp.
CECC CECC 90 109-735 ISSUE 1-87. Digital Integrated Circuits in Accordance with FS 90 109; 54 HC 238, 74 HC 238; 3 to 8 Line Decoder (En, Fr). 5 pp.
CECC CECC 90 109-777 ISSUE 1-87. Digital Integrated Circuits in Accordance with FS 90 109; 54/74 HCT 138; 3 to 8 Line Decoder (En, Fr). 5 pp.

—Integrated Circuit—4 to 16 Lines
CECC CECC 90 101-044 ISSUE 1-87. UTE C 86-213/B 64; Digital Integrated Circuits in Accordance with FS 90 101; 54/64/74 154; 4-to-16 Line Decoder (En). 12 pp.

INDUSTRY STANDARDS

INTERNATIONAL AND NON-U.S. NATIONAL STANDARDS
SUBJECT INDEX

Decoders

Decoders *(Cont.)*
—**Integrated Circuit—4 to 16 Lines** *(Cont.)*
CECC CECC 90 109-667 ISSUE 1-87. Digital Integrated Circuits in Accordance with FS 90 109; 54/74 HC 154; 4 to 16 Line Decoder (En, Fr). 5 pp.
CECC CECC 90 109-781 ISSUE 1-87. Digital Integrated Circuits in Accordance with FS 90 109; 54/74 HCT 154; 4 to 16 Lines Decoder (En, Fr). 5 pp.

—**Integrated Circuit—4 to 16 Lines—4-Bit**
CECC CECC 90 104-159 ISSUE 1-81. BS CECC 90 104-159; 4-Bit Latch/4-to 16 Line Decoder (En). 26 pp.
CECC CECC 90 104-159 ISSUE 1-86. CEI CECC 90 104-159; 4-Bit Latch /4 to 16 Line Decoder (En). 2 pp.
CECC CECC 90 104-160 ISSUE 1-81. BS CECC 90 104-160; 4-Bit Latch/4-to-16 Line Decoder (En). 26 pp.
CECC CECC 90 104-160 ISSUE 1-86. CEI CECC 90 104-160; 4-Bit Latch /4 to 16 Line Decoder (En). 2 pp.

—**MAC/Packet—Electrical Measurement**
IEC 1079 Pt 5-93. Methods of Measurement on Receivers for Satellite Broadcast Transmissions in the 12 GHz Band Part 5: Electrical Measurements on Decoder Units for MAC/Packet Systems First Edition. 156 pp.

—**MAC/Packet—Electromagnetic Compatibility**
CCIR Report 1101-90. Preliminary Studies of Radio Interference Limits for MAC/Packet Decoders—Section 1D—Spectrum Utilization and Applications. 27 pp.

—**Pulse Code Modulation Measuring Instruments**
CCITT RECMN O.133-89. Equipment for Measuring the Performance of PCM Encoders and Decoders—Specifications for Measuring Equipment (Study Group IV) 26 pp. 26 pp.

—**Video Systems**
IEC 11172 Pt 2-93. Information Technology—Coding of Moving Pictures and Associated Audio for Digital Storage Media at up to About 1,5 Mbit/s—Part 2: Video First Edition. 122 pp.
ISO 11172 Pt 2-93. Information Technology—Coding of Moving Pictures and Associated Audio for Digital Storage Media at up to About 1,5 Mbit/s—Part 2: Video First Edition. 122 pp.
JTC1 11172 Pt 2-93. Information Technology—Coding of Moving Pictures and Associated Audio for Digital Storage Media at up to About 1,5 Mbit/s—Part 2: Video First Edition. 122 pp.

Decontamination
See Also: Contamination; Radiation Decontamination
—**Calcium Hypochlorite**
MOD UK DSTAN 68-132-89. Calcium Hypochlorite Issue 1. 11 pp.
NATO STANAG 4323 ED 1 AMD 0-00. Specification of Calcium Hypochlorite Used for Biological and Chemical Decontamination. 14 pp.
NATO STANAG 4323 ED 1 AMD 0-92. Specification of Calcium Hypochlorite Used for Biological and Chemical Decontamination. 13 pp.

Decontamination Index, Radioactive
Use: Radioactive Decontamination Index

Decontamination Kits
—**Personal**
MOD UK DSTAN 42-12-71. Decontamination Kits, Personal Issue 1. 5 pp.

Decoquinate Content Analysis
—**Animal Feed**
CNS N4082-82. Method of Test for Feed Additives: Determination of Decoquinate (Jan) (8318). 2 pp.

Decorative Coatings
Use: Coatings

Decorative Lamps
Use: Decorative Lighting

Decorative Lighting
See Also: Landscape Lighting; Lighting; Lighting Chains
CNS C4106-87. Decorative-Lighting Outfits (Dec)(3329).
CNS C4346-82. Decorative Lamp (May)(8800).
CSA C22.2 NO 37-M1989. Christmas Tree and Other Decorative Lighting Outfits; (Gen Instr 1 Thru 2). 37 pp.

Decorative Lighting *(Cont.)*
CSA 1439 Bull. Electrical Bulletin 1439 May 30, 1986 to C22.2 NO 37. 2 pp.

—**Safety**
SNZ NZS/AS 3152-92. Approval and Test Specification—Decorative Lighting Outfits AMENDMENT No. 1, 1993. 6 pp.

Decorative Panels
See Also: Panels; Wall Coverings
—**Cork**
ISO 8724-89. Cork Decorative Panels—Specification First Edition. 6 pp.

—**Kitchens**
CEN PREN 1116-93. Kitchen Furniture—Co-Ordinating Sizes for Kitchen Furniture and Kitchen Appliances. 8 pp.

—**Naval Ships**
MOD UK NES 130: Part 1-91. Requirements for Decorative Linings and Composite Bulkheads Part 1: Decorative Linings Issue 3 (07.91). 20 pp.
MOD UK NES 130: Part 1-01. Requirements for Decorative Linings and Composite Bulkheads Part 1: Decorative Linings Issue 3 (07.91); Amendment 1. 28 pp.

—**Submarines**
MOD UK NES 130: Part 1-91. Requirements for Decorative Linings and Composite Bulkheads Part 1: Decorative Linings Issue 3 (07.91). 20 pp.
MOD UK NES 130: Part 1-01. Requirements for Decorative Linings and Composite Bulkheads Part 1: Decorative Linings Issue 3 (07.91); Amendment 1. 28 pp.

Decorative Sheet Laminates
Use: Sheet Laminates

DECT
Use: European Digital Cordless Telecommunications

Dedicated Networks
See Also: Dedicated Services; Digital Satellite Dedicated Networks; Telecommunication Networks

—**Digital—Sound-Program Signals—Multiplexing**
CCIR QUESTION 54/CMTT-90. Standards for Multiplexing and Routing of Digital Sound-Programme Signals in Dedicated Digital Networks—Questions Concerning the CMTT CCIR/CCITT Joint Study Group for Television and Sound Transmission. 1 p.

—**Digital—Sound-Program Signals—Routing**
CCIR QUESTION 54/CMTT-90. Standards for Multiplexing and Routing of Digital Sound-Programme Signals in Dedicated Digital Networks—Questions Concerning the CMTT CCIR/CCITT Joint Study Group for Television and Sound Transmission. 1 p.

—**Integrated Services Digital Networks—Interworking**
CCITT RECMN I.500-89. General Structure of the ISDN Interworking Recommendations—Integrated Services Digital Network (ISDN)—Internetwork Interfaces and Maintenance Principles (Study Group XVIII) 4 pp. 4 pp.
CCITT RECMN I.510-89. Definitions and General Principles for ISDN Interworking—Integrated Services Digital Network (ISDN)—Internetwork Interfaces and Maintenance Principles (Study Group XVIII) 12 pp. 12 pp.

—**Sound Studios—Interfaces**
CCIR QUESTION 61/CMTT-90. Standards for Digital Interfaces Between Studio and Dedicated or Integrated Services Digital Networks—Questions Concerning the CMTT CCIR/CCITT Joint Study Group for Television and Sound Transmission. 1 p.

Dedicated Services
See Also: Dedicated Networks; Digital Satellite Dedicated Networks; Telecommunication Equipment

—**Alarm Transmission Systems**
IEC 839 Pt 5-4-91. Alarm Systems Part 5: Requirements for Alarm Transmission Systems Section 4: Alarm Transmission Systems Using Dedicated Alarm Transmission Paths First Edition. 16 pp.

Dedicated Services *(Cont.)*
—**Integrated Services Digital Networks—Interfaces**
CCITT RECMN I.410-89. General Aspects and Principles Relating to Recommendations on ISDN User-Network Interfaces—Integrated Services Digital Network (ISDN)—Overall Network Aspects and Functions, ISDN User -Network Interfaces (Study Group XVIII) 3 pp. 3 pp.

Dee Head Bolts
See Also: Bolts
—**Aircraft**
SBAC AS 26100-199 ISSUE 4. Bolt, Close Tolerance—Dee Head, BSS.159, 190-32 UNJF-3A MOD BS4084.
SBAC AS 26200-299 ISSUE 4. Bolt, Close Tolerance—Dee Head, BSS.159, 250-28 UNJF-3A.MOD. B.S.4084.
SBAC AS 26300-399 ISSUE 4. Bolt, Close Tolerance—Dee Head, BSS.159, 3125-24 UNJF-3A.MOD. B.S.4084.
SBAC AS 26400-499 ISSUE 4. Bolt, Close Tolerance—Dee Head, BS.S159, 375-24 UNJF-3A.MOD. B.S.4084.
SBAC AS 26500-599 ISSUE 4. Bolt, Close Tolerance—Dee Head, BS.S159, 4375-20 UNJF-3A.MOD. B.S.4084.
SBAC AS 26600-699 ISSUE 3. Bolt, Close Tolerance—Dee Head, BS.S159, 500-20 UNJF-3A.MOD. B.S.4084.
SBAC AS 26700-799 ISSUE 3. Bolt, Close Tolerance—Dee Head, BSHR601. 190-32 UNJF-3A. MOD. B.S.4084.
SBAC AS 26800-899 ISSUE 3. Bolt, Close Tolerance—Dee Head, BSHR601, 250-28 UNJF-3A. MOD. B.S.4084.
SBAC AS 26900-999 ISSUE 3. Bolt, Close Tolerance—Dee Head, BS HR601, 3125-24 3A.MOD. BS.4084.
SBAC AS 27000-099 ISSUE 3. Bolt, Close Tolerance—Dee Head, BSHR601, 375-24 UNJF-3A.MOD. B.S.4084.
SBAC AS 27100-199 ISSUE 3. Bolt, Close Tolerance—Dee Head, BSHR601, 4375-20 UNJF-3A.MOD. B.S.4084.
SBAC AS 27200-299 ISSUE 3. Bolt, Close Tolerance—Dee Head, BSHR601, 500-20 UNJF-3A.MOD. B.S.4084.
SBAC AS 28100-199 ISSUE 3. Bolt, Close Tolerance—Large Dee Head, BS. HR650, 190-32 UNJF-3A. MOD. B.S.4084.
SBAC AS 28200-299 ISSUE 3. Bolt, Close Tolerance—Large Dee Head, BS. HR650, 250-28 UNJF-3A. MOD. B.S.4084.
SBAC AS 28300-399 ISSUE 3. Bolt, Close Tolerance—Large Dee Head, BS. HR650, 3125-24 UNJF-3A. MOD. B.S.4084.
SBAC AS 28400-499 ISSUE 3. Bolt, Close Tolerance—Large Dee Head, BS. HR650, 375-24 UNJF-3A. MOD. B.S.4084.
SBAC AS 28500-599 ISSUE 3. Bolt, Close Tolerance—Large Dee Head, BS. HR650, 4375-20 UNJF-3A. MOD. B.S.4084.
SBAC AS 28600-699 ISSUE 3. Bolt, Close Tolerance—Large Dee Head, BS. HR650, 500-20 UNJF-3A. MOD. B.S.4084.
SBAC AS 29000-099 ISSUE 3. Bolt, Externally Relieved Body Dee Head, DTD5066, 10(.190)-32 UNJF-3A BS4084.
SBAC AS 29100-199 ISSUE 3. Bolt, Externally Relieved Body—Dee Head, DTD 5066. 250-28 UNJF-3A BS4084.
SBAC AS 29200-299 ISSUE 3. Bolt, Externally Relieved Body—Dee Head, BS 5159, 3125-24 UNJF-3A BS4084.
SBAC AS 29300-399 ISSUE 3. Bolt, Externally Relieved Body—Dee Head, BS 5159, 375-24 UNJF-3A BS4084.
SBAC AS 29400-499 ISSUE 3. Bolt, Externally Relieved Body—Dee Head, BS5159, 4375-20 UNJF-3A BS4084.
SBAC AS 29500-599 ISSUE 3. Bolt, Externally Relieved Body—Dee Head, BS5159, 500-20 UNJF-3A BS4084.
SBAC AS 29600-699 ISSUE 2(I). Bolt, Externally Relieved Body—Dee Head, DTD5077, 10(.190)-32 UNJF-3A BS4084.
SBAC AS 29700-799 ISSUE 2(I). Bolt, Externally Relieved Body—Dee Head, DTD5077. 250-28 UNJF-3A BS4084.
SBAC AS 29800-899 ISSUE 2(I). Bolt, Externally Relieved Body—Dee Head, DTD5077, 3125-24 UNJF-3A BS4084.
SBAC AS 29900-999 ISSUE 2(I). Bolt, Externally Relieved Body—Dee Head, DTD5077, 375-24 UNJF-3A BS4084.
SBAC AS 30000-099 ISSUE 2(I). Bolt, Externally Relieved Body—Dee Head, DTD5077, 4375-20 UNJF-3A BS4084.
SBAC AS 30100-199 ISSUE 2(I). Bolt, Externally Relieved Body—Dee Head, DTD5077, 500-20 UNJF-3A BS4084.

Dee Head Bolts (Cont.)
—Aircraft (Cont.)
SBAC AS 32100-199 ISSUE 3. Bolt, Close Tolerance—Dee Head, BS.HR650. 190-32 UNJF-3A MOD. BS4084.
SBAC AS 32200-299 ISSUE 3. Bolt, Close Tolerance—Dee Head, BS.HR650, 250-28 UNJF-3A MOD. BS4084.
SBAC AS 32300-399 ISSUE 3. Bolt, Close Tolerance—Dee Head, BS.HR650, 3125-24 UNJF-3A MOD. BS4084.
SBAC AS 32400-499 ISSUE 3. Bolt, Close Tolerance—Dee Head, BS.HR650, 375-24 UNJF-3A MOD. BS4084.
SBAC AS 32500-599 ISSUE 3. Bolt, Close Tolerance—Dee Head, BS.HR650, 4375-20 UNJF-3A MOD. BS4084.
SBAC AS 32600-699 ISSUE 3. Bolt, Close Tolerance—Dee Head, BS.HR650, 5000-20 UNJF-3A MOD. BS4084.

Dee Rings
Use: D Rings

Deemphasis Networks
See Also: Communication Networks; Equalizers (Electrical)
—Television Systems—Signal to Noise Ratio
CCIR Report 637-3-86. Signal-to-Noise Ratio in Television Effect of the De-Emphasis Network, Alone or in Combination with a Weighting Network —Section CMTT A—Television Transmission Standards and Performance Objectives. 6 pp.

Deep Fat Fryers
Use For: Brat Pans; Bratt Pans See Also: Cooking Appliances; Food Processing Equipment; Food Service Equipment; Frying Utensils
BSI BS 5314: Part 13-82. (WITHDRAWN) 1982 Amd 1 Gas Heated Catering Equipment Part 13: Brat Pans (AMD 5409) July 31, 1987 (Superseded by BS EN 203-1: 1993). 22 pp.
CSA CAN/CSA-C22. 2 NO 64-M91. Household Cooking and Liquid-Heating Appliances; (Gen Instr 1 Thru 2). 91 pp.
—Commercial—Electric
CGSB CAN/CGSB-52.15-M86. Fryer, Electric, Commercial (Replaces 52-GP-16M). 11 pp.
CSA C22.2 NO 109-M1981. Commercial Cooking Appliances; (Gen Instr 1 Thru 3). 36 pp.
CSA 1169 Bull. Electrical Bulletin 1169 June 27, 1978 to C22.2 NO 109. 2 pp.
—Commercial—Electric—Safety
BSI BS 3456: Sec 102.37-89. (WITHDRAWN) 1989 Safety of Household Electrical Appliances Part 102: Particular Requirements Section 102.37: Commercial Electric Deep Fat Fryers (Superseded by BS 3456: Section 202.37: 1990). 18 pp.
BSI BS 3456: Sec 202.37-90. (WITHDRAWN) 1990 Safety of Household Electrical Appliances Part 202: Particular Requirements Section 202.37: Commercial Electric Deep Fat Fryers (Supersedes BS 3456: Section 102.37: 1989). 22 pp.
BSI BS 3456: Sec 202.37-01. 1990 Amd 1 Safety of Household and Similar Electrical Appliances Part 202: Particular Requirements Section 202.37: Commercial Electric Deep Fat Fryers (AMD 7811) June 15, 1993 (L). 41 pp.
BSI BS 5784: Part 2-84. 1984 Amd 1 Safety of Electrical Commercial Catering Equipment Part 2: Specification for Deep Fat Fryers. 58 pp.
CENELEC HD 285 S1-87. Safety of Household and Similar Electrical Appliances—Part 2: Particular Requirements for Commercial Electric Deep Fat Fryers. 10 pp.
CENELEC EN 60 335-2-37-89. Safety of Household and Similar Electrical Appliances Part 2: Particular Requirements for Commercial Electric Deep Fat Fryers. 10 pp.
CENELEC EN 60335-2-37/A1-92. AMD 1 Safety of Household and Similar Electrical Appliances Part 2: Particular Requirements for Commercial Electric Deep Fat Fryers. 5 pp.
IEC 335 Pt 2-37-86. Safety of Household and Similar Electrical Appliances Part 2: Particular Requirements for Commercial Electric Deep Fat Fryers Second Edition; (Amendment 1-1990). 48 pp.
—Electric—Safety
BSI BS 3456: Sec 3.16-80. (WITHDRAWN) 1980 1972-1981 Edition. Specification for Safety of Household Electrical Appliances: Part 3: Complete Particular Sepcifications: Section 3.16: Frying Pans, Deep Fat Fryers and Similar Appliances (Superseded by BS 3456: Section 102.13: 1991). 46 pp.
BSI BS EN 60335-2-13-91. 1991 Safety of Household and Similar Electrical Appliances Part 2: Particular Requirements Section 2.13: Frying Pans, Deep Fat Fryers and Similar Appliances (L). 24 pp.
BSI BS EN 60335-2-13-01. 1991 Amd 1 Safety of Household and Similar Electrical Appliances Part 2: Particular Requirements Section 2.13: Frying Pans, Deep Fat Fryers and Similar Appliances (AMD 7585) March 15, 1993 (L). 30 pp.
CENELEC HD 279 S1-86. Safety of Household and Similar Electrical Appliances-Part 2: Particular Requirements for Frying Pans, Deep Fat Fryers and Similar Appliances. 10 pp.
CENELEC EN 60 335-2-13-90. Safety of Household and Similar Electrical Appliances Part 2: Particular Requirements For Frying Pans, Deep Fat Fryers and Similar Appliances. 7 pp.
CENELEC EN 60335-2-13/A1-92. AMD 1 Safety of Household and Similar Electrical Appliances Part 2: Particular Requirements for Frying Pans, Deep Fat Fryers and Similar Appliances. 4 pp.
—Gas
BSI BS 6350-83. 1983 Gas Heated Fish and Chip Frying Ranges. 17 pp.

Deep Well Pumps
See Also: Centrifugal Pumps; Pumps; Submersible Pumps; Well Pumps; Wells
JIS B 8318-89. Electric Deep Well Pumps. 23 pp.
—Submersible
CNS B4064-85. Submersible Motor-Pumps for Deep Well (Aug)(11327). 19 pp.
JIS B 8324-85. Submersible Motor-Pumps for Deep Well.
JIS B 8324-66. Submersible Motor-Pumps for Deep Well (R 1969). 30 pp.
—Submersible—Three-Phase Motors
CNS C4428-88. Low-Voltage Three-Phase Squirrel-Cage Induction Motors (for Submersible Motor-Pumps for Deep Well)(Jun)(11330).

Dees
Use: D Rings

DEET
Use: Diethyltoluamide

Defects
Use For: Flaws See Also: Crystal Defects; Linting; Porosity; Surface Defects; Voids
—Brazed Joints—Classification
DIN ENGL 8515 Pt 1-79. Defects in Metallic Solder Joints; Brazing Joints and High-Temperature Solder Joints; Classification; Designations; Explanatory Notes (June). 8 pp.
—Castings—Glossaries
BSI BS 2737-56. 1956 Terminology of Internal Defects in Castings as Revealed by Radiography. 35 pp.
—Coffee
BSI BS 5752: Part 4-80. 1980 Coffee and Coffee Products Part 4: Green Coffee: Olfactory and Visual Examination and Determination of Foreign Matter and Defects. 5 pp.
BSI BS 7683-93. 1993 Guide to Defects of Green Coffee (ISO 10470: 1993) (W). 29 pp.
ISO 4149-80. Green Coffee—Olfactory and Visual Examination and Determination of Foreign Matter and Defects First Edition. 4 pp.
ISO 10470-93. Green Coffee—Defect Reference Chart First Edition. 29 pp.
—Contraceptive Diaphragms
ISO 8009 Pt 4-85. Reusable Rubber Contraceptive Diaphragms—Part 4: Freedom from Visible Defects First Edition. 5 pp.
—Doors
BSI BS 5277-76. 1976 Doors. Measurement of Defects of General Flatness of Door Leaves. 6 pp.
BSI BS 5278-76. 1976 Doors. Measurement of Dimensons and of Defects of Squareness of Doors Leaves. 6 pp.
—Dried Foods
CNS N6183-84. Method of Test for Salted Prunes (May)(10887). 2 pp.
—Enamels—Low Voltage Testing
BSI BS 1344: Part 20-87. 1987 Methods of Testing Vitreous Enamel Finishes Part 20: Low Voltage Test for Detecting and Locating Defects. 5 pp.
ISO 8289-86. Vitreous and Porcelain Enamels—Low Voltage Test for Detecting and Locating Defects First Edition. 4 pp.
—Fabrics
CNS L3076-77. Testing Method for the Woven Fabric Defects (Jun)(3854). 4 pp.
CNS L3086-80. Method of Test for Nonwoven Fabrics Defects (May)(5618).

Defects (Cont.)
—Fabrics (Cont.)
CNS L3209-84. Method of Test for the Weft Knitted Fabric Defects (Jul)(10968).
CNS L4076-83. Nonwoven Interlining Fabric (Jun)(5617). 2 pp.
CNS L4152-84. Weft Knitted Fabrics (Finished) (Jul)(10970). 2 pp.
—Fabrics—Glossaries
BSI BS 7343-90. 1990 Glossary of Terms for Defects in Knitted Fabrics. 22 pp.
ISO 8498-90. Woven Fabrics—Description of Defects—Vocabulary First Edition. 27 pp.
ISO 8499-90. Knitted Fabrics—Description of Defects—Vocabulary First Edition. 22 pp.
SAA AS 1083-71. Glossary of Terms for Defects in Woven and Knitted Textile Piece Goods Corrig. Reconfirmed 1986. 23 pp.
—Firebricks
BSI BS 1902: Sec 3.13-85. 1985 Amd 1 Methods for Testing Refractory Materials Part 3: General and Textural Prop. Sec 3.13: Mst. of Corner and Edge Defects and Other Sur. Imperf. of Refractory Bricks and Blocks (Att-ributive Prop.) (Method 1902-313) (AMD 6343) June 28, 1991. 10 pp.
—Flame Cutting—Classification
DIN ENGL 8518-74. Defects of Flame Cuts and Plasma Cuts; Classification, Designations, Definitions (Nov). 8 pp.
—Fruits
CNS N4031-72. Method of Test for Ponkan, Tankan and Satsuma Oranges (Jun)(3352). 2 pp.
CNS N4032-72. Method of Test for Pomelo (Shaddocks) (Jun)(3353). 2 pp.
CNS N4033-84. Method of Test for Pineapple (Sep)(3354). 2 pp.
CNS N4034-72. Method of Test for Papaya (Jun)(3355). 2 pp.
CNS N4035-72. Method of Test for Watermelons (Jun)(3356). 2 pp.
CNS N4036-72. Method of Test for Tomatoes (Jun)(3357). 2 pp.
—Fusion Welding—Radiography—Glossaries
SAA AS Z5.2-68. Glossary of Metal Welding Terms and Definitions—Part 2: Terminology of and Abbreviations for Fusion Weld Imperfections as Revealed by Radiography Being BS 499:Part 3:1965 Endorsed Subject to Australian Amendment. 130 pp.
—Fusion Welds—Metals—Classification
BSI BS EN 26520-92. 1992 Imperfections in Metallic Fusion Welds, with Explanations (ISO 6520: 1982). 19 pp.
CEN EN 26520-91. Classification of Imperfections in Metallic Fusion Welds, with Explanations. 3 pp.
DIN ENGL EN 26520-91. Imperfections in Metallic Fusion Welds; Classification and Terminlology (ISO 6520: 1982) (English Version of DIN EN 26 520) (Dec) (Supersedes DIN 8524 Part 1, July 1986 Edition). 16 pp.
ISO 6520-82. Classification of Imperfections in Metallic Fusion Welds, with Explanations First Edition. 15 pp.
—Logs (Wood)—Classification
ISO 4473-88. Coniferous and Broadleaved Tree Sawlogs—Visible Defects—Classification First Edition. 4 pp.
—Logs (Wood)—Glossaries
ISO 4474-89. Coniferous and Broadleaved Tree Sawlogs—Visible Defects—Terms and Definitions First Edition. 19 pp.
—Logs (Wood)—Measurement
ISO 4475-89. Coniferous and Broadleaved Tree Sawlogs—Visible Defects—Measurement First Edition. 16 pp.
—Metal Coatings—Classification
DIN ENGL 50903-67. Metallic Coatings; Pores, Inclusions, Blisters and Cracks; Definitions (Jan). 2 pp.
—Plasma Arc Cutting—Classification
DIN ENGL 8518-74. Defects of Flame Cuts and Plasma Cuts; Classification, Designations, Definitions (Nov). 8 pp.
—Plywood—Classification
BSI BS 6566: Part 6-85. 1985 Plywood Part 6: Limits of Defects for the Classification of Plywood by Appearance. 17 pp.
—Polyester Fabrics
CNS L4081-80. Polyester Fabrics (Unfinished) (Jul) (5893). 2 pp.
CNS L4083-80. Polyester Filament Fabrics (Unfinished) (Jul)(5895). 2 pp.

INTERNATIONAL AND NON-U.S. NATIONAL STANDARDS
SUBJECT INDEX

Defects

Defects (Cont.)
—**Polyester Fabrics** (Cont.)
CNS L4084-80. Polyester Filament Fabrics (Finished) (Jul) (5896). 3 pp.

—**Post Design Services—Electronic Equipment**
MOD UK DSTAN 05-125: Chapter 7-01. Modification and Change Procedure Issue 2; Amendment 1. 72 pp.

—**Radiography**
CEN PREN 444-90. Non-Destructive Testing —General Principles for Radiographic Examination of Metallic Materials by X-and Gamma-Rays. 15 pp.

—**Soldered Joints—Classification**
DIN ENGL 8515 Pt 1-79. Defects in Metallic Solder Joints; Brazing Joints and High-Temperature Solder Joints; Classification; Designations; Explanatory Notes (June). 8 pp.

—**Spices**
CNS N4039-72. Method of Test for Gingers (Jun)(3360). 2 pp.

—**Vegetables**
CNS N4037-72. Method of Test for Garlic (Jun)(3358). 2 pp.
CNS N4038-72. Method of Test for Shallot (or Scallion) (Jun)(3359). 2 pp.

—**Welded Joints—Classification**
DIN ENGL 8524 Pt 2-79. Defects in Metallic Welded Joints; Pressure Welded Joints; Classification; Designations; Explanatory Notes (Mar). 10 pp.
DIN ENGL 8524 Pt 3-75. Defects in Metallic Welded Joints; Cracks; Classification; Denominations; Explanatory Notes (Aug). 4 pp.

—**Welded Joints—Plastic—Classification**
DIN ENGL 32502-85. Imperfections in Plastic Welded Joints; Classification, Terminology (July). 8 pp.

—**Welded Joints—Steels—Classification**
BSI BS EN 25817-92. 1992 Acr-Welded Joints in Steel—Guidance on Quality Levels for Imperfections (ISO 5817: 1992). 18 pp.
CEN EN 25817-92. Arc-Welded Joints in Steel—Guidance on Quality Levels for Imperfections. 13 pp.
DIN ENGL EN 25817-92. Arc-Welded Joints in Steel; Guidance on Quality Levels for Imperfections (ISO 5817: 1992) (Sep) (Supersedes DIN 8563 Part 3, October 1985 Edition). 15 pp.
ISO 5817-92. Arc-Welded Joints in Steel—Guidance on Quality Levels for Imperfections First Edition; (Corrected and Reprinted -1992). 14 pp.

Defense Batteries
Use: Batteries—Defense Contracts

Defense Contract Forms
Use: Forms (Paper)—Defense Contracts

Defense Contracts
Scope Note: For defense contracts pertaining to specific products or materials, see the material or product *Use For:* Ministry of Defence Contracts
See Also: Contractors; Demand Order Contracts; Fixed Price Contracts; Government Supplied Property; Incentive Contracts; Invitation for Bids; Production Contracts; Research and Development Contracts; Running Contracts; Service Contracts; Sole Source Contracts

MOD UK FORM GC/STORES/1-79. Standard Conditions of Government Contracts for Stores Purchases. 48 pp.
MOD UK DEFCON MEMORANDUM-86. Note to Users of Defcons 313A and 313B 2/86. 1 p.
MOD UK DEFCON 112A-89. Basic Set of Conditions of Contract 8/89. 1 p.
MOD UK DEFCON 112A-92. Basic Set of Conditions of Contract—Aircraft Equipment 9/92. 1 p.
MOD UK DEFCON 112AN-90. Basic Set of Conditions for Navy Production Contracts 7/90. 5 pp.
MOD UK DEFCON 112CB-89. Conditions of Contract for RNSD Eaglescliffe 6/89. 1 p.
MOD UK DEFCON 112G-88. Basic Set of Conditions of Contract—General Stores 11/88. 1 p.
MOD UK DEFCON 112 NSRP-88. NSRP Equipment Procurement Conditions of Contract 8/88. 17 pp.
MOD UK DEFCON 112R-91. Basic Set of Conditions of Contract—General Stores 8/91. 2 pp.
MOD UK DEFCON 123-90. General Conditions of Contract 9/90. 5 pp.
MOD UK DEFCON 123-91. General Conditions of Contract 11/91. 5 pp.

Defense Contracts (Cont.)
MOD UK DEFCON 142-92. Overhead Rates and CP/CE Ratios Required for Calendar Year 19 1/92. 2 pp.
MOD UK DEFCON 159 (GLA)-92. Acceptance of Tender Letter 3/92. 1 p.
MOD UK. Selling to the MOD.

—**Acceptance of Offer of Contract—Forms**
MOD UK DEFCON 10-89. Acknowledgement of Receipt and Terms of Contract Form 9/89. 1 p.
MOD UK DEFCON 10-91. Acceptance of Offer of Contract 7/91. 1 p.

—**Acknowledgement of Receipt and Terms of Contract—Forms**
MOD UK DEFCON 10-89. Acknowledgement of Receipt and Terms of Contract Form 9/89. 1 p.
MOD UK DEFCON 10A-90. Acknowledgement of Receipt and Terms of Contract Form 10/90. 1 p.

—**Advice and Inspection Note Signature—Forms**
MOD UK DEFCON 5-89. MOD Form 640—Advice and Inspection Note Signature by MOD QAR 12/89. 3 pp.
MOD UK DEFCON 5-92. MOD Forms 640—Advice and Inspection Notes 11/92. 1 p.

—**Bank or Insurance Company Guarantee—Forms**
MOD UK DEFCON 24A-90. Specimen—for Drafting Purposes Only Guarantee Given by a Bank or Insurance Company 1/90. 2 pp.

—**Commercial Exploitation**
MOD UK DEFCON 70-89. Commercial Exploitation—Annual Reminder 8/89. 1 p.

—**Competitive Business Practices**
MOD UK DEFCON 176B-90. MOD Requirements for Competition in Sub-Contracting (Competitive Main Contract) 9/90. 1 p.
MOD UK DEFCON Guide no. 8-86. Code of Practice for Competitive Sub-Contracting 5/86. 8 pp.

—**Contract Modifications**
MOD UK DEFCON 56-78. Authorisation of Alterations Navy Contracts D of C (SE) 3/78. 2 pp.

—**Contractor—Bankruptcy**
MOD UK DEFCON 515-93. Bankruptcy and Insolvency 3/93. 2 pp.

—**Contractor—Failure of Performance**
MOD UK DEFCON 92-90. Failure of Performance 8/90. 1 p.

—**Contractor—Insolvency**
MOD UK DEFCON 515-93. Bankruptcy and Insolvency 3/93. 2 pp.

—**Contractor Liability**
MOD UK DEFCON 6-76. Liabilities During Running Trials 2/76. 2 pp.

—**Copyrights**
MOD UK DEFCON 90-89. Copyright Clause 12/89. 1 p.

—**Copyrights—Software**
MOD UK DEFCON 91-92. Intellectual Property Rights in Software 10/92. 9 pp.

—**Design Rights**
MOD UK DEFCON 15-74. Design Rights 8/74. 2 pp.
MOD UK DEFCON 15A-74. Design Rights A 8/74. 2 pp.
MOD UK DEFCON 177-80. Design Rights and Patents (Sub-Contractors) Agreement 3/80. 3 pp.

—**Diversion Orders**
MOD UK DEFCON 113-89. Diversion Orders 3/89. 4 pp.

—**Drawings, Specifications, and Manufacturing Data**
MOD UK DEFCON 19-76. Free User, Maintenance and Supply of Drawings 1/76. 1 p.

—**Financial Disclosure—Quarterly Report Forms**
MOD UK DEFCON 136-75. Quarterly Financial Report 11/75. 1 p.

—**Financing**
MOD UK DEFCON Guide no. 9-86. Financing Terms for Defence Contracts 10/86. 5 pp.

—**International Agreements**
MOD UK DEFCON 126-73. International Collaboration Clause 6/73. 2 pp.

Defense Contracts (Cont.)
—**Parent Company Guarantee of a Subsidiary—Forms**
MOD UK DEFCON 24-90. Specimen—for Drafting Purposes Only Guarantee Given by a Parent Company in Respect of a Subsidiary 1/90. 2 pp.

—**Patents**
MOD UK DEFCON 14-82. Inventions and Designs—Crown Rights and Ownership of Patents and Registered Designs 10/82. 3 pp.
MOD UK DEFCON 14A-77. Collaboration—Ownership of Patents and Registered Designs 4/77. 2 pp.
MOD UK DEFCON 177-80. Design Rights and Patents (Sub-Contractors) Agreement 3/80. 3 pp.

—**Procurement—Coding**
MOD UK DEFCON 90-89. Copyright Clause 12/89. 1 p.
MOD UK DEFCON 91-92. Intellectual Property Rights in Software 10/92. 9 pp.

—**Progress Payments**
MOD UK DEFCON 35-73. Progress Payments 4/73. 2 pp.

—**Quality Assurance**
MOD UK DEFCON 123BA-81. General Conditions of Contract (Local Purchase) 8/81. 2 pp.
MOD UK DEFCON 123BA-92. General Conditions of Contract (Local Purchase) 6/92. 4 pp.
MOD UK DSTAN 05-61: Part 1-87. Quality Assurance Procedural Requirements Part 1: Concessions and Production Permits Issue 2. 9 pp.

—**Quality Assurance—Materiel**
NATO STANAG 4107 ED 5 AMD 2-86. Mutual Acceptance of Government Quality Assurance. 30 pp.

—**Repair**
MOD UK DEFCON 112CC-88. Conditions of Contract (Repair) for RNSD Eaglescliffe 7/88. 2 pp.
MOD UK DEFCON 112 REPAIR-89. Basic Set of Conditions of Contract—Repairs (Other Than for DOS Contracts) 1/89. 2 pp.
MOD UK DEFCON 112 REPAIR-92. Conditions of Contract for Repair 11/92. 3 pp.

—**Security Measures**
MOD UK DEFCON 39-76. Security Measures (Sub-Contracts) 5/76. 1 p.

—**Small Value Orders**
MOD UK DEFCON 112AJ-89. Conditions of Contract for Small Value Orders 4/89. 1 p.
MOD UK DEFCON 112AJ-92. Conditions of Contract for Small Value Orders 2/92. 1 p.

Defense Materials
Use: Materiel

Defense Quality Assurance Board Technical Bulletins
Use: Quality Assurance—Bulletins

Defibrillators
See Also: Medical Electrical Equipment; Medical Equipment
JIS T 1355-87. Defibrillators. 25 pp.

—**Cardiac—Monitors—Safety**
CSA CAN/CSA-C22. 2N601.2.4M90. Medical Electrical Equipment, Part 2: Particular Requirements for the Safety of Cardiac Defibrillators and Cardiac Defibrillator-Monitors; (Gen Instr 1). 78 pp.

—**Cardiac—Safety**
BSI BS 5724: Sec 2.4-85. 1985 Medical Electrical Equipment Part 2: Particular Requirements for Safety Section 2.4: Cardiac Defibrillators and Cardiac Defibrillator-Monitors. 43 pp.
CSA CAN/CSA-C22. 2N601.2.4M90. Medical Electrical Equipment, Part 2: Particular Requirements for the Safety of Cardiac Defibrillators and Cardiac Defibrillator-Monitors; (Gen Instr 1). 78 pp.
IEC 601 Pt 2-4-83. Medical Electrical Equipment Part 2: Particular Requirements for the Safety of Cardiac Defibrillators and Cardiac Defibrillator-Monitors First Edition. 80 pp.

—**Connectors**
CSA CAN/CSA-C22. 2NO601.1-M90. Medical Electrical Equipment, Part 1: General Requirements for Safety; (Gen Instr 1). 240 pp.

—**Identification Systems**
CSA CAN/CSA-C22. 2NO601.1-M90. Medical Electrical Equipment, Part 1: General Requirements for Safety; (Gen Instr 1). 240 pp.

INTERNATIONAL AND NON-U.S. NATIONAL STANDARDS
SUBJECT INDEX
Deformation

Defibrillators (Cont.)
—Testing
CSA CAN/CSA-C22. 2NO601.1-M90. Medical Electrical Equipment, Part 1: General Requirements for Safety; (Gen Instr 1). 240 pp.

Deflecting Electrodes
—Cathode Ray Tubes
IEC 236-74. Methods for the Designation of Electrostatic Deflecting Electrodes of Cathode-Ray Tubes Second Edition. 10 pp.

Deflection
See Also: Bending; Deformation; Diffraction; Extensibility; Reflection; Torsion

—Agricultural Equipment
BSI BS 4220: Part 1-74. 1974 Methods of Test for Seats on Agricultural Wheeled Tractors Part 1: Tests on Artificial Track. 23 pp.

—Earthmoving Equipment—Protective Structures
BSI BS 5495-80. 1980 Amd 1 Laboratory Evaluations of Roll-Over and Falling-Object Protective Structures: The Deflection-Limiting Volume for Earth-Moving Machinery (ISO 3164: 1979) (AMD 3711) October 30, 1981. 9 pp.
BSI BS 5495-02. (WITHDRAWN) 1980 Amd 2 Laboratory Evaluations of Roll-Over and Falling-Object Protective Structures: the Deflection-Limiting Volume for Earth-Moving Machinery (ISO 3164: 1979) (AMD 6878) February 28, 1992 (Superseded by BS 6912: Part 7: 1992). 14 pp.
BSI BS 6912: Part 7-92. 1992 Safety of Earth-Moving Machinery Part 7: Specification for Laboratory Evaluations of Roll-Over and Falling-Object Protective Structures: the Deflection-Limiting Volume for Earth-Moving Machinery (ISO 3164: 1992). 7 pp.
CNS A4006-81. Deflection Limiting Volume of Earth-Moving Machinery-Roll-Over Protective Structure for Driver (Sep)(7850).
CNS A4024-88. Earth-Moving Machinery Laboratory Evaluations of Roll-Over and Falling-Object Protective Structures Specifications for the Deflection-Limiting Volume (Dec)(9955).
ISO 3164-92. Earth-Moving Machinery—Laboratory Evaluations of Roll-over and Falling-Object Protective Structures—Specifications for Deflection-Limiting Volume Fourth Edition; (Incorporating Amendment 1). 6 pp.

—Plastics—Molding Materials
BSI BS 2782:Pt1: METH 115A-78. 1978 Methods of Testing Plastics Part 1: Thermal Properties Method 115A: Plastics Yield. 2 pp.
SAA AS 1369-73. Method for the Determination of the Deflection Temperature of Plastics Subjected to an Applied Stress Corrig. Reconfirmed 1986. 7 pp.

—Polyurethane Foams
SAA AS 2282.8-91. Methods for Testing Flexible Cellular Polyurethane—Part 8: Deflection of Force Deflection. 3 pp.

—Presses
BSI BS 6979-89. 1989 Amd 1 Method for Measurement of Static Deflection of of Power Presses (AMD 6296) May 31, 1990. 21 pp.

—Ships
MOD UK NES 145-86. Code of Practice for Taking Breakage Issue 2 (01.86). 15 pp.

—Submarines
MOD UK NES 145-86. Code of Practice for Taking Breakage Issue 2 (01.86). 15 pp.

—Thermometals—Thermal Curvature
DIN ENGL 1715 Pt 2-83. Thermostat Metals; Testing the Specific Thermal Curvature (Nov). 4 pp.

Deflection Temperature
See Also: Temperature

—Ebonite—Bending Stress
BSI BS 2782:Pt1: METH 121A-B-91. 1991 Methods of Testing Plastics Part 1: Thermal Properties: Methods 121A and 121B: Determination of Temperature of Deflection of Plastics Under a Bending Stress (ISO 75: 1987) (V). 8 pp.
BSI BS 2782:Pt1: METH 121A-C-76. (WITHDRAWN) 1976 Methods of Testing Plastics Part 1: Thermal Properties: Method 121A-C: (A) Determ of Temp of Deflection Under a Bending Stress of 1.8 MPa of Plastics and Ebonite.(B) Determ of Temp of Deflec Under a Bending Stress of 0.45 MPa of Plastic & Ebonite. 4 pp.
CNS K6617-80. Plastics and Ebonite Determination of Temperature of Deflection Under Load (Dec)(6683).

Deflection Temperature (Cont.)
—Ebonite—Bending Stress (Cont.)
ISO 75 Pt 2-93. Plastics—Determination of Temperature of Deflection Under Load—Part 2: Plastics and Ebonite First Edition; (Cancels and Replaces ISO 75:1987). 6 pp.

—Glazes—Aircraft
AECMA PREN2155-13-89. Test Methods for Transparent Materials for Aircraft Glazing Part 13—Determination of Temperature at Deflection Under Load. 8 pp.
BSI BS EN 2155-13-93. 1993 Test Methods for Transparent Materials for Aircraft Glazing Part 13: Determination of Temperature at Deflection Under Load (S). 13 pp.
CEN EN 2155-13-93. Aerospace Series Test Methods for Transparent Materials for Aircraft Glazing Part 13: Determination of Temperature at Deflection Under Load. 8 pp.

—Mortars—Bending Stress
BSI BS 6319: Part 10-87. 1987 Testing for Resin Composition for Use in Construction Part 10: Method for Measurement of Temperature of Deflection Under a Bending Stress. 4 pp.

—Plastics
DIN ENGL 53462-87. Testing of Plastics; Martens Method of Determining the Temperature of Deflection Under a Bending Stress (Jan) (Supersedes March 1965 Edition and July 1968 Edition of DIN 53458). 5 pp.

—Plastics—Bending Stress
BSI BS 2782:Pt1: METH 121A-B-91. 1991 Methods of Testing Plastics Part 1: Thermal Properties: Methods 121A and 121B: Determination of Temperature of Deflection of Plastics Under a Bending Stress (ISO 75: 1987) (V). 8 pp.
BSI BS 2782:Pt1: METH 121C-92. 1992 Methods of Testing Plastics Part 1: Thermal Properties: Method 121C: Determination of Temperature of Deflection Under a Bending Stress for Rigid Thermosetting Resin Bonded Laminated Plastics Sheet (V). 7 pp.
BSI BS 2782:Pt1: METH 121A-C-76. (WITHDRAWN) 1976 Methods of Testing Plastics Part 1: Thermal Properties: Method 121A-C: (A) Determ of Temp of Deflection Under a Bending Stress of 1.8 MPa of Plastics and Ebonite.(B) Determ of Temp of Deflec Under a Bending Stress of 0.45 MPa of Plastic & Ebonite. 4 pp.
CNS K6617-80. Plastics and Ebonite Determination of Temperature of Deflection Under Load (Dec)(6683).
ISO 75 Pt 1-93. Plastics—Determination of Temperature of Deflection Under Load—Part 1: General Test Method First Edition; (Cancels and Replaces ISO 75:1987). 8 pp.
ISO 75 Pt 2-93. Plastics—Determination of Temperature of Deflection Under Load—Part 2: Plastics and Ebonite First Edition; (Cancels and Replaces ISO 75:1987). 6 pp.
JIS K 7207-83. Testing Method for Deflection Temperature of Rigid Plastics Under Load. 16 pp.

—Polymer Concretes—Bending Stress
BSI BS 6319: Part 10-87. 1987 Testing for Resin Composition for Use in Construction Part 10: Method for Measurement of Temperature of Deflection Under a Bending Stress. 4 pp.

—Reinforced Plastics—Bending Stress
ISO 75 Pt 3-93. Plastics—Determination of Temperature of Deflection Under Load—Part 3: High Strength Thermosetting Laminates and Long-Fibre-Reinforced Plastics First Edition; (Cancels and Replaces ISO 75:1987). 6 pp.

—Thermosetting Resins—Bending Stress
ISO 75 Pt 3-93. Plastics—Determination of Temperature of Deflection Under Load—Part 3: High Strength Thermosetting Laminates and Long-Fibre-Reinforced Plastics First Edition; (Cancels and Replaces ISO 75:1987). 6 pp.

Deflocculating
—Surfactants—Dyes
DIN ENGL 53908 Pt 1-82. Testing of Surface Active Agents and Textile Auxiliaries; Determination of Dispersing Power or Flocculation Preventive Power of Surface Active Agents When Added to Dyestuffs in Acidic Medium (Aug). 5 pp.
DIN ENGL 53908 Pt 3-82. Testing of Surface Active Agents and Textile Auxiliaries; Determination of Dispersing Power of Flocculation Preventive Power of Surface Active Agents When Added to Dyestuffs in Neutral Medium (Aug). 5 pp.

Deformation
Use For: Distortion See Also: Bending; Dilatation; Elongation; Extensibility; Plastic Deformation; Torsion; Twisting; Warpage

Deformation (Cont.)
—Bitumens
BSI DD 185-90. 1990 Method for Determination of Creep Stiffness of Bitumen Aggregate Mixtures Subject to Confined Uniaxial Loading. 12 pp.
DIN ENGL 1996 Pt 11-81. Testing of Bituminous Materials for Road Building and Related Purposes; Determination of Marshall Stability and of Marshall Flow Value (July). 4 pp.

—Buildings
ISO 4356-77. Bases for the Design of Structures—Deformations of Buildings at the Serviceability Limit States First Edition. 21 pp.

—Ceramics
BSI BS 7134: Sec 3.5-89. 1989 Testing of Engineering Ceramics Part 3: Thermo-Mechanical Properties Section 3.5: Methods of Determination of Pyroplastic Deformation (Sagging). 10 pp.
CEN ENV 820-2-92. Advanced Technical Ceramics—Methods of Test for Monolithic Ceramics—Thermo-Mechanical Properties—Part 2: Determination of Selfloaded Deformation. 10 pp.

—Cork
ISO 2191-72. Cork—Expanded Pure Agglomerated—Deformation Under Constant Pressure First Edition. 4 pp.

—Crimp Contacts
CEN PREN 2591 (Part E3)-92. Elements of Electrical and Optical Connection Test Methods Part E3—Contact Deformation After Crimping. 4 pp.

—Doors
CEN EN 108-80. Methods of Testing Doors: Test for Deformation of the Leaf in Its Plane. 4 pp.
CEN EN 129-84. Methods of Testing Doors: Test for Deformation in Torsion of the Door Leaves. 4 pp.
CEN EN 130-84. Methods of Testing Doors: Test for the Change in Stiffness of the Door Leaves by Repeated Torsion. 7 pp.
DIN ENGL EN 129-90. Methods of Testing Doors; Test for Deformation in Torsion of Door Leaves (Nov). 6 pp.
DIN ENGL EN 130-90. Methods of Testing Doors; Test for Change in Stiffness of the Door Leaves by Repeated Torsion (Nov). 7 pp.
JIS A 1522-89. Doors and Windows—Test Method for Mechanical Deformation of Edge Rail. 11 pp.

—Eye Protectors
CEN PREN 168-89. Personal Eye Protection: Non-Optical Test Methods. 24 pp.

—Mortars
DIN ENGL 18555 Pt 2-82. Testing of Mortars Containing Mineral Binders; Freshly Mixed Mortars Containing Aggregates of Dense Structure (Heavy Aggregates); Determination of Consistence, Bulk Density and Air Content (Sept). 4 pp.

—Paving Materials—Creep Testing
BSI DD 185-90. 1990 Method for Determination of Creep Stiffness of Bitumen Aggregate Mixtures Subject to Confined Uniaxial Loading. 12 pp.

—Plastic Pipes—PVC
CNS K6641-81. Method of Test for Rigid Polyvinyl Chloride Corrugated Pipes (Mar)(7057). 6 pp.

—Plastics
CNS K6686-81. Method of Test for Deformation of Plastics Under Load (Jul)(7734).

—Polyurethane Athletic Fields
CNS K6591-86. Method of Test for Polyurethane (PU) Athletic Installation Material (May)(6483). 6 pp.

—PVC
BSI BS 2782:Pt1: METH 122A-76. 1976 Methods of Testing Plastics Part 1: Thermal Properties Method 122A: Determination of Deformation Under Heat of Flexible Polyvinyl Chloride Compound. 2 pp.

—Refractory Materials
SAA AS 1774.26-85. Refractories and Refractory Materials—Physical Test Methods—Part 26: Creep in Compression—Deformation Under Constant Load and Constant Temperature. 7 pp.

—Roofing Tiles
SAA AS 4046.1-92. Methods of Testing Roof Tiles—Part 1: Determination of Distortion (in Professional Packages 20, 21, 30, 41, 58, 62, 63, 64, 65, 66, 67, 68, 69) (Supersedes AS 1757—1989 (in Part) and AS 2049—1989 (in Part)). 2 pp.

—Sandwich Structures—Tensile Testing
DIN ENGL 53292-82. Testing of Sandwiches; Tensile Test Perpendicular to the Faces (Feb). 4 pp.

INDUSTRY STANDARDS

Deformation (Cont.)

—Shell Molding—Refractory
BSI BS 1902: Sec 10.1-93. 1993 Methods of Testing Refractory Materials Part 10: Investment Casting Shell Mould Systems Section 10.1: Determination of Resistance to Deformation at Elevated Temperatures (F). 13 pp.

—Skis—Alpine
ISO 6265-92. Alpine Skis—Determination of Deformation Load and Breaking Load Second Edition. 7 pp.

—Supersonic Transports
CAA STANDARD NO. 4-7 08.69. Flutter and Structural Distortion. 4 pp.

—Welded Joints
DIN ENGL 50120 Pt 1-75. Testing of Steel; Tensile Test on Welded Joints; Fusion Welded Butt Joints (Sept). 6 pp.

—Windows
JIS A 1522-89. Doors and Windows—Test Method for Mechanical Deformation of Edge Rail. 11 pp.

Defrosters
See Also: Automotive Equipment

—Windshields—Automotive
CNS D3124-82. Method of Performance Test for Windshield Defrosting and Defogging Systems for Passenger Cars (Jun)(8945).
CNS D3125-82. Method of Performance Test for Windshield Defrosting and Defogging Systems for Passenger Trucks (Jun)(8946).
EC 78/317/EEC-77. Council Directive on the Approximation of the Laws of the Member States Relating to the Defrosting and Demisting Systems of Glazed Surfaces of Motor Vehicles. 22 pp.
ISO 3468-89. Passenger Cars—Windscreen Defrosting Systems—Test Method Second Edition. 9 pp.
ISO 5898-87. Road Vehicles—Rear-Window Defrosting System for Passenger Cars—Test Method First Edition. 7 pp.
JIS D 4501-84. Windscreen Defrosting Systems for Passenger Cars. 9 pp.
JIS D 4503-85. Windscreen Defrosting Systems for Trucks and Buses. 9 pp.

—Windshields—Trucks
CNS D3125-82. Method of Performance Test for Windshield Defrosting and Defogging Systems for Passenger Trucks (Jun)(8946).
JIS D 4503-85. Windscreen Defrosting Systems for Trucks and Buses. 9 pp.

Defrosters, Windshield
Use: Windshields—Defrosters

Degaussing Systems
See Also: Mine Countermeasures

—Design—Ships
MOD UK NES 612-88. Guide to the Design of Degaussing Systems Issue 3 (06.88). 20 pp.
MOD UK NES 612-91. Guide to the Design of Degaussing Systems Issue 4 (12.91). 20 pp.

—Design—Submarines
MOD UK NES 612-88. Guide to the Design of Degaussing Systems Issue 3 (06.88). 20 pp.
MOD UK NES 612-91. Guide to the Design of Degaussing Systems Issue 4 (12.91). 20 pp.

—Installation—Ships
MOD UK NES 613-88. Requirements for the Installation of Degaussing Systems in HM Ships Issue 3 (06.88). 30 pp.
MOD UK NES 613-91. Requirements for the Installation of Degaussing Systems in H M Ships Issue 4 (12.91). 31 pp.
MOD UK NES 615-88. Requirements for the Installation of Degaussing Systems in Merchant Ships Issue 3 (06.88). 33 pp.
MOD UK NES 615-91. Requirements for the Installation of Degaussing Systems in Merchant Ships Issue 4 (12.91). 34 pp.

—Installation—Submarines
MOD UK NES 613-88. Requirements for the Installation of Degaussing Systems in HM Ships Issue 3 (06.88). 30 pp.
MOD UK NES 613-91. Requirements for the Installation of Degaussing Systems in H M Ships Issue 4 (12.91). 31 pp.

—Minesweepers (Ships)—Acoustic Ranges
NATO STANAG 1203 ED 3 AMD 1-85. Degaussing and Acoustic Ranging Information Concerning North Atlantic Treaty Minesweepers and Minehunters—AMP-4A(A) (Navy). 4 pp.

Degaussing Systems (Cont.)

—Minesweepers (Ships)—Acoustic Ranges (Cont.)
NATO STANAG 1203 ED 3 AMD 2-85. Degaussing and Acoustic Ranging Information Concerning North Atlantic Treaty Minesweepers and Minehunters—AMP-4(A) (Navy). 5 pp.

—Ships
MOD UK NES 614-88. Requirements for Degaussing Equipment for HM Surface Ships Issue 3 (06.88). 32 pp.
MOD UK NES 614-91. Requirements for Degaussing Equipment for H M Surface Ships Issue 4 (12.91). 31 pp.
MOD UK NES 616-88. Requirements for Degaussing Equipment for Merchant Ships Issue 3 (06.88). 14 pp.
MOD UK NES 616-91. Requirements for Degaussing Equipment for Merchant Ships Issue 4 (12.91). 16 pp.
NATO STANAG 1333 ED 1 AMD 0-88. Protection of Vessels from Electromagnetic Mines—AMP-14. 5 pp.
NATO STANAG 1333 ED 1 AMD 1-88. Protection of Vessels from Electromagnetic Mines—AMP-14. 5 pp.

Degradation of Service
Use For: Service Degradation
See Also: Telecommunication Services; Telephone Services
CCITT FASCICLE I.2-88. Opinions and Resolutions Recommendations on the Organisation and Working Procedures of CCITT (Series A). 64 pp.

—Fixed Satellite Services
CCIR RECMN 496-2-86. Limits of Power Flux-Density of Radionavigation Transmitters to Protect Space Station Receivers in the Fixed-Satellite Service in the 14 GHz Band—Section 8D—Radiodetermination, Global Maritime Distress and Safety System and Related Subjects. 1 p.
CCIR RECMN 496-3-92. Limits of Power Flux-Density of Radionavigation Transmitters to Protect Space Station Receivers in the Fixed-Satellite Service in the 14 GHz Band—Section 8D—Radiodetermination, Global Maritime Distress and Safety System and Related Subjects. 2 pp.

Degreasers (Cleaning Agents)
See Also: Cleaning Agents; Solvents

—Alkalinity
MOD UK M 9522/66. Examination of Degreasing Compound, Alkaline, Type A (Superseded by Def Stan 68-160).

—Classification
EC 83/265/EEC-83. Council Directive Amending Directive 77/728/EEC on the Approximation of the Laws, Regulations and Administrative Provisions of the Member States Relating to the Classification, Packaging and Labelling of Paints, Varnishes, Printing Inks, Adhesives. 7 pp.

—Corrosion Testing
CGSB 31-GP-0A METH 11.11-57. Methods of Testing Corrosion-Prevention Materials and Processes Corrosion. 1 p.

—Identification Systems
EC 83/265/EEC-83. Council Directive Amending Directive 77/728/EEC on the Approximation of the Laws, Regulations and Administrative Provisions of the Member States Relating to the Classification, Packaging and Labelling of Paints, Varnishes, Printing Inks, Adhesives. 7 pp.

—Insoluble Matter Content
MOD UK M 9522/66. Examination of Degreasing Compound, Alkaline, Type A (Superseded by Def Stan 68-160).

—Metals
CGSB 31-GP-202M-89. Grease Removing Cleaning Compound Solvent; (Corrigendum—November 1992). 9 pp.
MOD UK DSTAN 68-160-93. Degreasing Compound, Types A and C Issue 1. 17 pp.
MOD UK DSTAN 68-189-93. Solvent for Grease, Emulsifiable Issue 1. 12 pp.
MOD UK TS 414. Solvent for Grease, Emulsifiable.
MOD UK CS 3081A. Degreasing Compound, Type C (Superseded by Def Stan 68-160).
MOD UK CS 3110. Degreasing Compound, Alkaline, Type A (Superseded by Def Stan 68-160).
MOD UK TS 10320. Hydrocarbon Solvent for Cleaning and Degreasing Applications.

Degreasers (Cleaning Agents) (Cont.)

—Packaging
EC 83/265/EEC-83. Council Directive Amending Directive 77/728/EEC on the Approximation of the Laws, Regulations and Administrative Provisions of the Member States Relating to the Classification, Packaging and Labelling of Paints, Varnishes, Printing Inks, Adhesives. 7 pp.

—Stability Testing
CGSB 31-GP-0A METH 9.15-57. Methods of Testing Corrosion-Prevention Materials and Processes Stability (Accelerated Oxidation). 2 pp.

Degreasing
See Also: Vapor Degreasing

—Trichloroethane
CGSB 31-GP-213MA-91. Vapour Degreasing and Cold Cleaning Inhibited 1,1,1-Trichloroethane; (Corrigendum—Oct 1991). 10 pp.

Degree of Delignification
Use: Delignification

Degree of Polymerization
Use: Polymerization

Degree of Protection (Electrical)
See Also: Electrical Enclosures; Electrical Protection Equipment

—Ships
CNS F5090-84. General Requirement for Degree of Protection and Inspection of Enclosures for Marine Electrical Apparatus (Apr)(10854).
JIS F 8007-89. General Requirements for Degree of Protection and Inspection of Enclosures for Marine Electrical Apparatus.

Degumming

—Fabrics—Colorfastness Testing
ISO 105 Pt X08-87. Textiles—Tests for Colour Fastness—Part X08: Colour Fastness to Degumming Third Edition. 4 pp.

—Silk—Colorfastness Testing
CNS L3160-82. Method of Test for Colour Fastness to Degumming (Mar)(8618).
JIS L 0865-75. Testing Method for Colour Fastness to Degumming (R 1983). 6 pp.

DeHavilland DHC-2 Beaver MK1 Aircraft
See Also: Aircraft

—Antenna Positions
CAA. DHC-2 Beaver MK1. 1 p.

DeHavilland DHC-4 Aircraft
See Also: Aircraft

—Accidents
CAA. DHC-4 (World Airline Accident Summary). 1 p.

DeHavilland DHC-5 Aircraft
See Also: Aircraft

—Accidents
CAA. DHC-5 (World Airline Accident Summary). 1 p.

DeHavilland DHC-8 Aircraft
See Also: Aircraft

—Accidents
CAA. DHC-8 (World Airline Accident Summary). 1 p.

DeHavilland DHC-8 Dash 8 Series 300
See Also: Aircraft

—Certification
CAA. De Havilland DHC-8-311. 16 pp.

Dehumidifiers
See Also: Absorbers (Equipment); Condensing Units; Coolers; Humidifiers; Humidity; Humidity Indicators; Separators (Mechanical)
BSI BS 4788-89. 1989 Rating and Performance of Dehumidifiers Incorporating Electrically Operated Refrigeration Systems. 8 pp.
CEN PREN 810-92. Dehumidifiers—Dehumidifiers with Electrically Driven Compressors—Rating Tests, Marking, Operational Requirements and Technical Data Sheets. 19 pp.
CNS C4461-89. Dehumidifiers (Feb)(12492).
CSA C22.2 NO 92-1971. Dehumidifiers (R 1992); (Rev 1-3 April 1973). 32 pp.

INTERNATIONAL AND NON-U.S. NATIONAL STANDARDS
SUBJECT INDEX

Dehumidifiers (Cont.)
CSA 1169 Bull. Electrical Bulletin 1169 June 27, 1978 to C22.2 NO 92. 2 pp.
JIS C 9617-92. Dehumidifiers. 28 pp.
MOD UK DSTAN 44-2: Part 1-83. Desiccant Containers, Dehumidifier and Indicators, Humidity, Plug Part 1: General Requirements Issue 3. 37 pp.
MOD UK DSTAN 44-2: Part 1-93. Desiccant Containers, Dehumidifier Part 1: Guide to the Use of Desiccants Issue 4. 17 pp.
MOD UK DSTAN 44-2: Part 2-01. Desiccant Containers, Dehumidifier and Indicators, Humidity, Plug Part 2: Desiccant Containers, Dehumidifier (Tubular Type) Issue 3; Amendment 1. 27 pp.
MOD UK DSTAN 44-2: Part 2-93. Desiccant Containers, Dehumidifier Part 2: General and Test Requirements Issue 4. 24 pp.
MOD UK DSTAN 44-2: Part 3-01. Desiccant Containers, Dehumidifier and Indicators, Humidity, Plug Part 3: Desiccant Containers, Dehumidifier (Breather Tube Type) Issue 3; Amendment 1. 43 pp.
MOD UK DSTAN 44-2: Part 3-93. Desiccant Containers, Dehumidifier Part 3: Desiccator and Indicator Types Issue 4. 20 pp.
MOD UK DSTAN 44-2: Part 4-83. Desiccant Containers, Dehumidifier and Indicators, Humidity, Plug Part 4: Desiccant Containers, Dehumidifier (Sachet Type, with Sight Window) (Sachet Type, Without Sight Window) Issue 2. 15 pp.

—Data Sheets
CEN PREN 810-92. Dehumidifiers—Dehumidifiers with Electrically Driven Compressors—Rating Tests, Marking, Operational Requirements and Technical Data Sheets. 19 pp.

—Household—Electrical Grounding
CSA 649A Bull. Electrical Bulletin 649A January 10, 1967 to C22.2 NO 92. 1 p.

—Loads (Forces)
CSA 1072 Bull. Electrical Bulletin 1072 September 14, 1976 to C22.2 NO 92. 1 p.

—Safety
CENELEC PREN 60 335-2-402-89. Safety of Household and Similar Electrical Appliances Part 2: Particular Requirements for Electric Heat Pumps, Air Conditioners and Dehumidifiers for Commercial and Light Industrial Applications. 37 pp.
DIN VDE 0700 Pt 258-88. Safety of Household and Similar Electrical Appliances; Dehumidifiers (May). 24 pp.

Dehydrated Foods
Use: Dried Foods

Dehydrated Milk
Use: Powdered Milk

Dehydration
See Also: Drying

—Dairy Products
BSI BS 770: Part 10-86. 1986 Chemical Analysis of Cheese Part 10: Determination of Total Solids Content (Reference Method). 5 pp.
BSI BS 1741: Part 2-90. 1990 Methods for Chemical Analysis of Liquid Milk and Cream Part 2: Determination of Total Solids Content of Liquid Milk, Cream and Unsweetened Condensed Milk. 7 pp.
BSI BS 5086: Part 2-84. 1984 Analysis of Butter Part 2: Methods for Determination of Water, Solids-Non-Fat and Fat Contents (Reference Method). 4 pp.
CNS N6064-72. Methods of Test for Milk and Milk Products (Determination of Total Solids) (Oct)(3448). 1 p.
ISO 2920-74. Whey Cheese—Determination of Dry Matter Content (Reference Method) First Edition. 4 pp.
ISO 3727-77. Butter—Determination of Water, Solids-Not-Fat and Fat Contents on the Same Test Portion (Reference Method) First Edition. 5 pp.
ISO 3728-77. Ice-Cream and Milk Ice—Determination of Total Solids Content (Reference Method) First Edition. 5 pp.
ISO 5534-85. Cheese and Processed Cheese—Determination of Total Solids Content (Reference Method) First Edition. 4 pp.
ISO 6731-89. Milk, Cream and Evaporated Milk—Determination of Total Solids Content (Reference Method) First Edition. 6 pp.
ISO 6734-89. Sweetened Condensed Milk—Determination of Total Solids Content (Reference Method) First Edition. 6 pp.

—Fats
BSI BS 684: Sec 1.10-82. 1982 Methods of Analysis of Fats and Fatty Oils Part 1: Physical Methods Section 1.10: Determination of Moisture and Other Volatile Matter. 4 pp.

Dehydration (Cont.)
—Fats (Cont.)
CNS N6077-82. Methods of Test for Edible Oils and Fats (Determination of Moisture and Volatile Matter) (Jan)(3642). 1 p.
ISO 662-80. Animal and Vegetable Fats and Oils—Determination of Moisture and Volatile Matter Content First Edition; (Erratum—Nov 1981). 5 pp.

—Fruits
ISO 1026-82. Fruit and Vegetable Products—Determination of Dry Matter Content by Drying Under Reduced Pressure and of Water Content by Azeotropic Distillation First Edition. 6 pp.

—Latex
DIN ENGL 53563-79. Testing of Latex; Determination of Total Solids Content (May). 2 pp.

—Pulps
BSI BS 4502-80. 1980 Determination of Dry-Matter Content of Pulps. 4 pp.
BSI BS 5878-80. 1980 Determination of Stock Concentration of Pulps (Rapid Method). 4 pp.
ISO 638-78. Pulps—Determination of Dry Matter Content First Edition. 4 pp.
ISO 4119-78. Pulps—Determination of Stock Concentration (Rapid Method) First Edition. 4 pp.

—Vegetable Oils
CNS N6077-82. Methods of Test for Edible Oils and Fats (Determination of Moisture and Volatile Matter) (Jan)(3642). 1 p.

—Vegetables
ISO 1026-82. Fruit and Vegetable Products—Determination of Dry Matter Content by Drying Under Reduced Pressure and of Water Content by Azeotropic Distillation First Edition. 6 pp.

Dehydroacetic Acid Content Analysis
—Food—Gas Chromatography
CNS N6190-84. Method of Test for Preservatives in Food (Jun)(10949). 8 pp.

—Food—Thin Layer Chromatography
CNS N6190-84. Method of Test for Preservatives in Food (Jun)(10949). 8 pp.

Deicer/Antiicing Fluids
Use For: Antiicing/Deicing Fluids; Deicing/Antiicing Fluids See Also: Antiicing Additives; Antiicing Equipment; Deicers; Deicing Equipment

—Aircraft
ISO 11075-93. Aerospace—Aircraft De-Icing/Anit-Icing Newtonian Fluids, ISO Type 1 First Edition. 12 pp.
ISO 11076-93. Aerospace—Aircraft De-Icing/Anti-Icing Methods with Fluids First Edition. 14 pp.

Deicers
Use For: Deicing Fluids See Also: Antiicing Additives; Antiicing Equipment; Deicer/Antiicing Fluids; Deicing; Deicing Equipment; Ice Prevention
CGSB 3-GP-525MA-88. Isopropanol. 8 pp.

—Aircraft
CAA CAP 512 Part 1. Maintenance Aspects (Civil Air Publications: Ground De-Icing of Aircraft). 1 p.
CGSB 3.856M-90. Fluid, Deicing-Defrosting Concentrate for Aircraft Surfaces; (Corrigendum—June 1991). 9 pp.
MOD UK DTD-406B-61. De-Icing Fluid AL-5. 1 p.

—Aircraft—Carburetors
CAA Chapter K5-5 App 04.74. Ice Protection Systems (Light Aeroplanes). 1 p.
CAA Chapter Q5-5 App 12.79. Ice Protection Systems (Non-Rigid Airships). 1 p.

—Aircraft—Tanks (Containers)
SBAC AS 2469-2470 ISSUE 2. Tanks—1 Gallon (for De-Icing Fluid and Drinking Water). 1 p.
SBAC AS 2471-2472 ISSUE 3. Tanks—2 1/2 Gallons (for De-Icing Fluid and Drinking Water). 1 p.
SBAC AS 2473-2474 ISSUE 3. Tanks—4 Gallons (for De-Icing Fluid and Drinking Water). 1 p.
SBAC AS 2475-2476 ISSUE 3. Tanks—6 Gallons (for De-Icing Fluid and Drinking Water). 1 p.
SBAC AS 2477-2478 ISSUE 3. Tanks—7 Gallons (for De-Icing Fluid and Drinking Water). 1 p.
SBAC AS 2479-2480 ISSUE 3. Tanks—8 Gallons (for De-Icing Fluid and Drinking Water). 1 p.

—Aircraft—Tanks (Containers)—Labels
SBAC AS 2484-2489 ISSUE 2. Label for De-Icing Fluid Tanks. 1 p.

—Airport Runways
MOD UK TS 10228A. Ice Control Agents for Aircraft Runways.

Deicers (Cont.)
—Automotive
JIS K 2397-91. Deicing and Defrosting Fluids for Automobiles. 19 pp.

—Lacquers—Solvent Resistance Testing
CGSB 1-GP-71 METH 108.3-79. Methods of Testing Paints and Pigments Solvent Resistance Resistance of Lacquers to De-Icing Fluid; (Amended September 1980) (Re-Edited September 1982). 1 p.

—Sodium Chloride—Pavement
CGSB CAN/CGSB-15.9-92. Sodium Chloride, Pavement De-Icer. 7 pp.

Deicing
See Also: Deicers; Ice Prevention

—Aircraft
CAA. Contents (Civil Air Publications: Ground De-Icing of Aircraft). 1 p.
CAA. Foreword (Civil Air Publications: Ground De-Icing of Aircraft). 1 p.
CAA CAP 512 Part 1. Maintenance Aspects (Civil Air Publications: Ground De-Icing of Aircraft). 1 p.
CAA CAP 512 Part 2. Operational Aspects (Civil Air Publications: Ground De-Icing of Aircraft). 7 pp.
CAA CAP 512 Part 3. Common Practices or Suggested Practices for Safe Cold Weather Operations (Civil Air Publications: Ground De-Icing of Aircraft). 3 pp.
CAA CAP 512 Appendix. General Information Relating to Ground and Flight Operations in Conditions Conductive to Aircraft Icing (Civil Air Publications: Ground De-Icing of Aircraft). 6 pp.

Deicing/Antiicing Fluids
Use: Deicer/Antiicing Fluids

Deicing Equipment
Use For: Deicing Systems and Equipment
See Also: Deicer/Antiicing Fluids; Deicers

—Aircraft—Self Propelled Machinery
ISO 11077-93. Aerospace—Self-Propelled De-Icing/Anti-Icing Vehicles—Functional Requirements First Edition. 9 pp.

Deicing Fluids
Use: Deicers

Deicing Systems and Equipment
Use: Deicing Equipment

Delamination
—Adhesives—Resistance Testing
BSI BS EN 302-2-92. 1992 Adhesives for Load-Bearing Timber Structures: Test Methods Part 2: Determination of Resistance to Delamination (Laboratory Method) (Supersedes BS 1204: Parts 1 & 2: 1979). 10 pp.
BSI BS EN 302-2-01. 1992 Amd 1 Adhesives for Load-Bearing Timber Structures: Test Methods Part 2: Determination of Resistance to Delamination (Laboratory Method) (AMD 7719) May 15, 1993 (W). 11 pp.
CEN PREN 302 (Part 2)-89. Adhesives for Load-Bearing Timber Structures: Polycondensation Adhesives of the Phenolic and Aminoplastic Types—Test Methods—Part 2: Determination of Resistance to Delamination (Laboratory Method). 10 pp.
CEN EN 302-2-92. Adhesives for Load-Bearing Timber Structures—Test Methods—Part 2: Determination of Resistance to Delamination (Laboratory Method). 9 pp.
DIN ENGL EN 302 Pt 2-92. Test Methods for Adhesives for Loadbearing Timber Structures; Determination of Resistance to Delamination (Laboratory Method) (Aug). 5 pp.

—Cables (Electric)—Aircraft
CEN PREN 3475 (Part 402)-92. Cables, Electrical, Aircraft Use Test Methods Part 402—Shrinkage and Delamination. 2 pp.
CEN PREN 3475 (Part 403)-92. Cables, Electrical, Aircraft Use Test Methods Part 403—Delamination and Blocking. 3 pp.

—Laminated Wood
CEN PREN 391-90. Glued Laminated Timber-Delamination Test of Glue Lines. 8 pp.

—Plywood
CNS O2031-81. Method of Test for Soak Delamination of Plywood (Oct)(8059). 2 pp.

—Textile Floor Coverings
BSI BS 7399-91. 1991 Determination of Delamination Force of Textile Floor Coverings. 10 pp.

INDUSTRY STANDARDS

Delay

See Also: Call Release Time; Call Set Up Time; Delay Distortion; End Delay; Exchange Call Set Up Delay; Group Delay; Incoming Response Delay; Postselection Delay; Preselection Delay; Propagation Time; Signaling Systems; Through Connection Delay; Time Delay; Time Shift; Transmission Delay

—Antennas
CCIR Report 1017-86. Characterization of Signal Delays in Antennas—Section 7C—Systems for Dissemination and Comparison. 2 pp.

—Communication Channels—Radio—Error Correction
CCITT RECMN U.22-89. Signals Indicating Delay in Transmission on Calls Set up by Means of Synchronous Systems with Automatic Error Correction by Repetition—Telegraph Switching (Study Group IX) 2 pp. 2 pp.

—Communication Channels—Telex—Error Correction
CCITT RECMN U.22-89. Signals Indicating Delay in Transmission on Calls Set up by Means of Synchronous Systems with Automatic Error Correction by Repetition—Telegraph Switching (Study Group IX) 2 pp. 2 pp.

Delay Distortion
See Also: Delay; Distortion (Electrical); Envelope Delay Distortion

—Echo Cancellers
CCITT RECMN G.165-89. Echo Cancellers—General Characteristics of International Telephone Connections and Circuits (Study Groups XII and XV) 23 pp. 23 pp.

—Echo Suppressors
CCITT RECMN G.164-89. Echo Suppressors—General Characteristics of International Telephone Connections and Circuits (Study Groups XII and XV) 36 pp. 36 pp.

Delay Lines
See Also: Integrated Circuits

—Active—Preferred Products List
CECC CECC MUAHAG Vol 4 IS 3-90. Preferred Products List; Magnetic Components and Ferrite Materials (En, Fr, Ge). 46 pp.

—Electromagnetic—Glossaries
CNS C5071-80. Definition for Electromagnetic Delay Lines (Aug)(6320).

—Passive—Preferred Products List
CECC CECC MUAHAG Vol 4 IS 3-90. Preferred Products List; Magnetic Components and Ferrite Materials (En, Fr, Ge). 46 pp.

—Video—Charge Coupled Devices
MOD UK DSTAN 59-62: 90/174-82. Charge Coupled Device Issue 1. 35 pp.

Delay Relays
Use: Time Delay Relays

Delay Switches
Use: Time Delay Switches

Delay Timers

—Control Circuits
JIS C 4551-81. Motor Driven Timer. 27 pp.

Delignification
Use For: Degree of Delignification

—Pulps
BSI BS 4498-82. 1982 Determination of the Kappa Number of Pulp (Degree of Delignification). 8 pp.
CPPA G.10U-77. Chlorite Delignification of Cellulosic Materials. 1 p.
ISO 3260-82. Pulps—Determination of Chlorine Consumption (Degree of Delignification) Second Edition. 5 pp.

Delta Rays
Use: Ionizing Radiation

Delta Starters
Use For: Star Delta Starters *See Also:* Starters
BSI BS 587-57. 1957 Amd 6 Motor Starters and Controllers. 62 pp.
BSI BS 4941: Part 2-77. (WITHDRAWN) 1977 Motor Starters for Voltages up to and Including 1000 V a.c. and 1200 V d.c. Part 2: Reduced Voltage a.c. Starters, Star Delta Starters (Superseded by BS EN 60947-4-1: 1992). 21 pp.

Delta Starters *(Cont.)*
DIN VDE 0660 Pt 106-82. Switchgear and Controlgear; Low-Voltage Switchgear and Controlgear; Low-Voltage Motor Starters, Reduced Voltage: a.c. Starters, Star Delta Starters (Sept). 4 pp.
SAA AS 1202.2-76. A.C. Motor Starters (up to and Including 1000 V)—Part 2: Star Delta Starter. 24 pp.

—AC
DIN VDE 0660 Pt 106-82. Switchgear and Controlgear; Low-Voltage Switchgear and Controlgear; Low-Voltage Motor Starters, Reduced Voltage: a.c. Starters, Star Delta Starters (Sept). 4 pp.

Demagnetization Curve
DIN ENGL 50470-80. Testing of Permanent Magnet Materials; Determination of the Demagnetization Curve and Permanent Permeability in a Yoke; Inductive Method (Sept). 8 pp.

Demand Assignment Circuits
See Also: Satellite Communication Equipment; Telephone Circuits

—Line Up
CCITT RECMN M.675-89. Lining up and Maintaining International Demand Assignment Circuits (Spade)—General Maintenance Principles—Maintenance of International Transmission Systems and Telephone Circuits (Study Group IV) 6 pp. 6 pp.

—Maintenance
CCITT RECMN M.675-89. Lining up and Maintaining International Demand Assignment Circuits (Spade)—General Maintenance Principles—Maintenance of International Transmission Systems and Telephone Circuits (Study Group IV) 6 pp. 6 pp.

Demand Assignment Signaling Systems
See Also: Satellite Communication Equipment; Signaling Systems
CCITT RECMN Q.48-89. Demand Assignment Signalling Systems—General Recommendations on Telephone Switching and Signalling—Functions and Information Flows for Services in the ISDN—Supplements (Study Group XI) 4 pp. 4 pp.

Demand Meters
See Also: Electric Power Meters; Meters; Watt Hour Meters

—Mounting Hardware—Indoor/Outdoor
CSA C22.2 NO 115-M1989. Meter-Mounting Devices; (Gen Instr 1 Thru 3). 55 pp.
CSA 1169 Bull. Electrical Bulletin 1169 June 27, 1978 to C22.2 NO 115. 2 pp.

Demand Operating
See Also: International Exchanges; Teleinformatic Services; Telephone Services

—Manual—Busy Hour
CCITT RECMN E.510-89. Determination of the Number of Circuits in Manual Operation—Telephone Network and ISDN—Quality of Service, Network Management and Traffic Engineering (Study Group II) 5 pp. 5 pp.

—Manual—Circuit Groups
CCITT RECMN E.510-89. Determination of the Number of Circuits in Manual Operation—Telephone Network and ISDN—Quality of Service, Network Management and Traffic Engineering (Study Group II) 5 pp. 5 pp.

—Manual—Maritime Mobile Services—Radiotelephony
CCITT RECMN F.110-89. Operational Provisions for the Maritime Mobile Service—Telegraph and Mobile Services Operations and Quality of Service (Study Group I) 21 pp (Same as Recmn E.200). 21 pp.

—Manual—Maritime Mobile Services—Radioteletype
CCITT RECMN F.110-89. Operational Provisions for the Maritime Mobile Service—Telegraph and Mobile Services Operations and Quality of Service (Study Group I) 21 pp (Same as Recmn E.200). 21 pp.

—Manual—Telecommunication Circuits
CCITT RECMN E.143-89. Demand Operating of International Circuits—Telephone Network and ISDN—Operation, Numbering, Routing and Mobile Service (Study Group II) 1 pp. 1 p.

Demand Operating *(Cont.)*

—Manual—Telephone Circuits
CCITT RECMN E.510-89. Determination of the Number of Circuits in Manual Operation—Telephone Network and ISDN—Quality of Service, Network Management and Traffic Engineering (Study Group II) 5 pp. 5 pp.

—Manual—Telephone Networks
CCITT RECMN E.110-89. Organization of the International Telephone Network—Telephone Network and ISDN—Operation, Numbering, Routing and Mobile Service (Study Group II) 1 pp. 1 p.

—Manual—Telephone Services—Public Data Transmission
CCITT RECMN F.600-89. Service and Operational Principles for Public Data Transmission Services—Telematic, Data Transmission and Teleconference Services—Operations and Quality of Service (Study Group I) 4 pp. 4 pp.

—Manual—Telephone Services—Tariffs
CCITT RECMN D.100-89. Charging for International Calls in Manual or Semi-Automatic Operating—General Tariff Principles—Charging and Accounting in International Telecommunications Services (Study Group III) 2 pp. 2 pp.

—Manual—Telex Communications
CCITT RECMN F.60-89. Operational Provisions for the International Telex Service—Telegraph and Mobile Services Operations and Quality of Service (Study Group I) 17 pp. 17 pp.

—Manual—Telex Communications—Call Duration—Tariffs
CCITT RECMN F.61-89. Operational Provisions Relating to the Chargeable Duration of a Telex Call—Telegraph and Mobile Services Operations and Quality of Service (Study Group I) 1 pp. 1 p.

—Manual—Transit Exchanges
CCITT RECMN E.147-89. Manually Operated International Transit Traffic—Telephone Network and ISDN—Operation, Numbering, Routing and Mobile Service (Study Group II) 2 pp. 2 pp.

—Semiautomatic—Answering Time
CCITT RECMN E.142-89. Time-to-Answer by Operators—Telephone Network and ISDN—Operation, Numbering, Routing and Mobile Service (Study Group II) 1 pp. 1 p.

—Semiautomatic—Circuit Groups
CCITT RECMN E.520-89. Number of Circuits to be Provided in Automatic and/or Semiautomatic Operation, Without Overflow Facilities—Telephone Network and ISDN—Quality of Service, Network Management and Traffic Engineering (Study Group II) 3 pp. 3 pp.

—Semiautomatic—Maritime Mobile Services—Radiotelephony
CCITT RECMN F.110-89. Operational Provisions for the Maritime Mobile Service—Telegraph and Mobile Services Operations and Quality of Service (Study Group I) 21 pp (Same as Recmn E.200). 21 pp.

—Semiautomatic—Maritime Mobile Services—Radioteletype
CCITT RECMN F.110-89. Operational Provisions for the Maritime Mobile Service—Telegraph and Mobile Services Operations and Quality of Service (Study Group I) 21 pp (Same as Recmn E.200). 21 pp.

—Semiautomatic—Quality of Service
CCITT RECMN E.144-89. Advantages of Semiautomatic International Service—Telephone Network and ISDN—Operation, Numbering, Routing and Mobile Service (Study Group II) 2 pp. 2 pp.

—Semiautomatic—Routing—Documentation
CCITT RECMN E.149-89. Presentation of Routing Data—Telephone Network and ISDN—Operation, Numbering, Routing and Mobile Service (Study Group II) 4 pp. 4 pp.

—Semiautomatic—Telephone Circuits
CCITT RECMN E.520-89. Number of Circuits to be Provided in Automatic and/or Semiautomatic Operation, Without Overflow Facilities—Telephone Network and ISDN—Quality of Service, Network Management and Traffic Engineering (Study Group II) 3 pp. 3 pp.

—Semiautomatic—Telephone Networks
CCITT RECMN E.110-89. Organization of the International Telephone Network—Telephone Network and ISDN—Operation, Numbering, Routing and Mobile Service (Study Group II) 1 pp. 1 p.

Demand Operating (Cont.)
—Semiautomatic—Telephone Services—Accounting
CCITT RECMN D.178-89. Monthly Accounts for Semi-Automatic Telephone Calls (Ordinary and Urgent Calls, with or Without Special Facilities)—General Tariff Principles—Charging and Accounting in International Telecommunications Services (Study Group III) 1 pp. 1 p.

—Semiautomatic—Telephone Services—Public Data Transmission
CCITT RECMN F.600-89. Service and Operational Principles for Public Data Transmission Services—Telematic, Data Transmission and Teleconference Services—Operations and Quality of Service (Study Group I) 4 pp. 4 pp.

—Semiautomatic—Telephone Services—Tariffs
CCITT RECMN D.100-89. Charging for International Calls in Manual or Semi-Automatic Operating—General Tariff Principles—Charging and Accounting in International Telecommunications Services (Study Group III) 2 pp. 2 pp.
CCITT RECMN D.178-89. Monthly Accounts for Semi-Automatic Telephone Calls (Ordinary and Urgent Calls, with or Without Special Facilities)—General Tariff Principles—Charging and Accounting in International Telecommunications Services (Study Group III) 1 pp. 1 p.

—Semiautomatic—Telex Communications
CCITT RECMN F.60-89. Operational Provisions for the International Telex Service—Telegraph and Mobile Services Operations and Quality of Service (Study Group I) 17 pp. 17 pp.

—Semiautomatic—Telex Communications—Call Duration—Tariffs
CCITT RECMN F.61-89. Operational Provisions Relating to the Chargeable Duration of a Telex Call—Telegraph and Mobile Services Operations and Quality of Service (Study Group I) 1 pp. 1 p.

—Semiautomatic—Time Consistent Busy Hour
CCITT RECMN E.520-89. Number of Circuits to be Provided in Automatic and/or Semiautomatic Operation, Without Overflow Facilities—Telephone Network and ISDN—Quality of Service, Network Management and Traffic Engineering (Study Group II) 3 pp. 3 pp.

Demand Order Contracts
Scope Note: For demand order contracts pertaining to specific products or materials, see the material or product *See Also:* Defense Contracts
MOD UK DEFCON 112CA-83. Conditions of Contract for Demand Order Contracts Placed by Navy Branches 9/83. 6 pp.
MOD UK DEFCON 112CF-83. Conditions of Contract for Demand or Running Contracts—RNSPDC Eaglescliffe 9/83. 2 pp.
MOD UK DEFCON 112 DOS-85. Basic Set of Conditions of Contract for Demand Order Contracts 7/85. 4 pp.

Demeton
See Also: Demeton Methyl; Pesticides; Thiometon
SAA AS 1870.8D-77. Standard for Development—Pesticides for Agricultural Use—Part 8D: Demeton, Demeton-S-Methyl and Thiometon. 26 pp.

Demeton Methyl
Use For: Demeton-S-Methyl *See Also:* Demeton; Pesticides
SAA AS 1870.8D-77. Standard for Development—Pesticides for Agricultural Use—Part 8D: Demeton, Demeton-S-Methyl and Thiometon. 26 pp.

Demeton-S-Methyl
Use: Demeton Methyl

Demineralizers
Use For: Water Softeners *See Also:* Separators (Mechanical)
—Water Supply Installations
DIN ENGL 19636-89. Water Softeners for Use in Drinking Water Supply Installations; Requirements and Testing (DVGW Code of Practice) (July). 8 pp.

Demisters
See Also: Automotive Equipment; Condensing Units
—Windshields—Automotive
EC 78/317/EEC-77. Council Directive on the Approximation of the Laws of the Member States Relating to the Defrosting and Demisting Systems of Glazed Surfaces of Motor Vehicles. 22 pp.

Demisters (Cont.)
—Windshields—Automotive (Cont.)
ISO 3470-89. Passenger Cars—Windscreen Demisting Systems—Test Method Second Edition. 7 pp.
ISO 5897-87. Road Vehicles—Rear-Window Demisting System for Passenger Cars—Test Method First Edition. 7 pp.
JIS D 4502-84. Windscreen Demisting Equipment for Passenger Cars. 10 pp.
JIS D 4504-85. Windscreen Demisting Systems for Trucks and Buses. 10 pp.

Demisters, Windshield
Use: Windshields—Demisters

Demodulators
See Also: FAX Modules; Frequency Demodulators; Frequency Modulators
—Glossaries
BSI BS 4727:Pt3: Group 14-92. 1992 Electrotechnical, Power, Telecommunication, Electronics, Lighting and Colour Terms Part 3: Terms Particular to Telecommunications and Electronics Group 14: Oscillations, Signals and Related Devices (IEC 50(702): 1992) (G). 235 pp.
IEC 50 Chap 702-92. International Electrotechnical Vocabulary Chapter 702: Oscillations, Signals and Related Devices First Edition. 241 pp.

—Sound Transmission—Noise
CCITT RECMN J.74-89. Methods for Measuring the Transmission Characteristics of Translating Equipments—Line Transmission of Non-Telephone Signals—Transmission of Sound-Programme and Television Signals (Study Group XV) 2 pp. 2 pp.

—Vestigial Sideband—Television Transmitters
CENELEC HD 236.8 S2-87. Methods of Measurement for Radio Transmitters—Part 8: Vestigial-Sideband Demodulators for use in Conjunction with Transmitters or Transposers for Monochrome or Colour Television. 2 pp.
IEC 244 Pt 8-93. Methods of Measurement for Radio Transmitters Part 8: Performance Characteristics of Vestigial-Sideband Demodulators Used for Testing Television Transmitters and Transposers Second Edition. 56 pp.

—Vestigial Sideband—Transposers
CENELEC HD 236.8 S2-87. Methods of Measurement for Radio Transmitters—Part 8: Vestigial-Sideband Demodulators for use in Conjunction with Transmitters or Transposers for Monochrome or Colour Television. 2 pp.
IEC 244 Pt 8-93. Methods of Measurement for Radio Transmitters Part 8: Performance Characteristics of Vestigial-Sideband Demodulators Used for Testing Television Transmitters and Transposers Second Edition. 56 pp.

Demolition
See Also: Construction; Explosions; Explosives
BSI BS 6187-82. 1982 Code of Practice for Demolition. 30 pp.
SAA AS 2601-91. Demolition of Structures (in Professional Packages 21, 30, 66). 31 pp.

—Engineering Drawings
ISO 7518-83. Technical Drawings—Construction Drawings—Simplified Representation of Demolition and Rebuilding First Edition. 4 pp.

—Interchangeable—Accessories
NATO STANAG 2818 ED 1 AMD 8-71. Characteristics of Demolition Accessories to Determine Their Operational Interchangeability. 168 pp.

—Interchangeable—Symbols
NATO STANAG 2315 ED 3 AMD 4-80. Symbol to Denote the Operational Interchangeability of Ammunition and Demolition Accessories (Short Title: NATO Symbol of Inter-changeability). 9 pp.

—Noise
SNZ NZS 6803P-84. Measurement and Assessment of Noise from Construction, Maintenance and Demolition Work. 32 pp.

—Safety
CSA S350-M1980. Code of Practice for Safety in Demolition of Structures. 42 pp.

—Sites—Electrical Installations
IEC 364 Pt 7-704-89. Electrical Installations of Buildings Part 7: Requirements for Special Installations or Locations Section 704—Construction and Demolition Site Installations First Edition. 15 pp.

Demolition (Cont.)
—Sites—Noise Reduction—Legislation
BSI BS 5228: Part 2-84. 1984 Code of Practice for Noise Control on Construction and Demolition Sites Part 2: Guide to Noise Control Legislation for Construction and Demolition, Including Road Construction and Maintenance. 8 pp.

Demulsification
See Also: Emulsification; Emulsifying Agents; Emulsions
—Emulsified Asphalts
CNS K6780-83. Method of Test for Demulsibility of Emulsified Asphalts (Jun)(10363). 2 pp.

—Hydraulic Fluids
DIN ENGL 51599-75. Testing of Lubricating Oils; Determination of Demulsification Capacity According to the Stirring Method (Oct). 3 pp.

—Lubricating Oils
BSI BS 2000: Part 19-93. 1993 Methods of Test for Petroleum and Its Products Part 19: Determination of Demulsibility Characteristics of Lubricating Oil. 4 pp.
BSI BS 2000: Part 19-90. 1990 Petroleum and Its Products Part 19: Demulsification Number of Lubricating Oil. 6 pp.
DIN ENGL 51599-75. Testing of Lubricating Oils; Determination of Demulsification Capacity According to the Stirring Method (Oct). 3 pp.
JIS K 2520-91. Petroleum Products—Lubricating Oils—Determination of Demulsibility Characteristics. 17 pp.

—Petroleum Products
ISO 6614-83. Petroleum Oils and Synthetic Fluids—Determination of Demulsibility Characteristics First Edition. 6 pp.

—Synthetic Oils
ISO 6614-83. Petroleum Oils and Synthetic Fluids—Determination of Demulsibility Characteristics First Edition. 6 pp.

Demultiplexer/Multiplexers
Use: Multiplexer/Demultiplexers

Demultiplexers
See Also: Digital Demultiplexers; Muldexes; Multiplexer/Demultiplexers; Pulse Code Modulation Measuring Instruments
—Fiber Optics
IEC 875 Pt 3-92. Fibre Optic Branching Devices Part 3: Sectional Specification—Wavelength Selective Branching Devices Second Edition (IECQ QC 810200). 82 pp.
IECQ QC 810200-92. Fibre Optic Branching Devices Part 3: Sectional Specification—Wavelength Selective Branching Devices (IEC 875-3 ED 2). 83 pp.

Denaturant Content Analysis
—Ethanol
CNS K6713-82. Method of Test for Volatile Denaturants in Alcohol (Feb)(8525).

Denial Measure
See Also: Military Operations; Tactical Warfare; Warfare
—Military Supplies
NATO STANAG 2113 ED 5 AMD 1-91. Denial of a Unit's Military Equipment and Supplies to an Enemy. 7 pp.

—Military Vehicles
NATO STANAG 2113 ED 5 AMD 1-91. Denial of a Unit's Military Equipment and Supplies to an Enemy. 7 pp.

—Weapons Systems
NATO STANAG 2113 ED 5 AMD 1-91. Denial of a Unit's Military Equipment and Supplies to an Enemy. 7 pp.

Densimeters
Use: Densitometers

Densitometers
Use For: Density Gradient Columns
See Also: Densitometry; Measuring Instruments; Photometers; Photometry
—Ionizing Radiation
CENELEC HD 435-83. Density Meters Utilizing Ionizing Radiation Definitions and Test Methods. 2 pp.
IEC 692-80. Density Meters Utilizing Ionizing Radiation Definitions and Test Methods First Edition. 29 pp.

INTERNATIONAL AND NON-U.S. NATIONAL STANDARDS
SUBJECT INDEX
Densitometry

Densitometry
See Also: Densitometers; Photometry

—Aerosols—Volatile Matter Content
CNS Z6053-82. Method of Test for Volatile Content of Aerosol Products (Densimetric Method) (Sep)(9448).

—Photographic Film
BSI BS 6872-87. 1987 Visual Densities of Diazo and Vesicular Second-Generation Microforms. 4 pp.

CGSB CAN/CGSB-72.21-M89. Micrographics—Diazo and Vesicular Films—Visual Density—Specifications (ISO 8126-1986). 12 pp.

ISO 8126-86. Micrographics—Diazo and Vesicular Films—Visual Density—Specifications First Edition. 5 pp.

JIS K 7616-85. Determination of Residual Chemicals in Processed Photographic Films and Papers—Methylene Blue Photometric Method and Silver Sulphide Densitometric Method.

—Photographic Processing
BSI BS 5706-79. 1979 Determination of Thiosulphate and Other Residual Chemicals in Processed Photographic Films, Plates and Papers: Methylene Blue Photometric Method and Silver Sulphide Densitometric Method. 8 pp.

ISO 417-93. Photography—Determination of Residual Thiosulfate and Other Related Chemicals in Processed Photographic Materials—Methods Using Iodine-Amylose, Methylene Blue and Silver Sulfide Second Edition. 22 pp.

Density
Use For: Absolute Density; API Gravity; Proctor Density; Proctor Soil Test; Relative Density; Specific Gravity *See Also:* Absolute Humidity; Absorptance; Absorptivity; Bulk Density; Density Columns; Electrical Properties; Hydrometers; Linear Density; Mass; Opacity; Optical Density; Particle Density; Porosity; Pour Density; Pycnometers; Pycnometric Analysis; Surface Density; Tap Density; Translucence; Transmissivity; Transmittance; Turbidity

—Abrasives
CNS R3099-81. Method of Test for Specific Gravity of Artificial Abrasives (Jun)(7531). 2 pp.

JIS R 6125-76. Testing Method for Specific Gravity of Artificial Abrasives (R 1979). 6 pp.

—Acetaldehyde
ISO 2513-74. Acetaldehyde for Industrial Use—Determination of Density at 15 Degrees Celsius First Edition. 4 pp.

—Adhesives
BSI BS 5350: Part B1-78. 1978 Amd 1 Methods of Test for Adhesives Group B: Adhesives Part B1: Determination of Density (AMD 6015) July 31, 1989 (W). 8 pp.

CEN PREN 542-91. Adhesives—Determination of Density. 7 pp.

CEN PREN 543-91. Adhesives—Determination of Aparent Density of Powder and Granule Adhesives. 6 pp.

CNS K6441-80. Method of Measurement for Specific Gravity of Liquid Adhesives (Jan)(5131).

—Admixtures
CEN PREN 480 (Part 7)-91. Admixtures for Concrete, Mortar and Grout—Test Methods—Part 7: Determination of the Density of Liquid Admixtures. 4 pp.

—Aggregates
BSI BS 812: Part 2-75. 1975 Amd 1 Testing Aggregates Part 2: Physical Properties. 24 pp.

BSI BS 3681: Part 2-73. (WITHDRAWN) 1973 Methods for the Sampling and Testing of Lightweight Aggregates for Concrete Part 2: Metric Units (Superseded by BS 3797: 1990). 19 pp.

CEN PREN 992-92. Determination of the Dry Density of Lightweight Aggregate Concrete with Open Structure. 3 pp.

CNS A3006-86. Method of Test for Specific Gravity and Absorption of Fine Aggregate (Aug)(487). 6 pp.

CNS A3007-72. Method of Test for Specific Gravity and Absorption of Coarse Aggregate (Jun)(488).

JIS A 1109-76. Method of Test for Specific Gravity and Absorption of Fine Aggregate. 7 pp.

JIS A 1110-89. Method of Test for Specific Gravity and Absorption of Coarse Aggregates. 6 pp.

SNZ NZS 4407: Part 3.7.1-91. Methods of Sampling and Testing Road Aggregates Part 3: Methods of Testing Road Aggregates—Laboratory Tests Test 3: The Solid Density of Aggregate Particles. Test 3.7.1: Pycnometer Method for Particles Passing the 19 mm Test Sieve. 3 pp.

SNZ NZS 4407: Part 3.7.2-91. Methods of Sampling and Testing Road Aggregates Part 3: Methods of Testing Road Aggregates—Laboratory Tests Test 3.7: The Solid Density of Aggregate Particles. Test 3.7.2: Immersion in Water Method for Coarse Aggregate. 4 pp.

SNZ NZS 4407: Part 4.1.1-91. Methods of Sampling and Testing Road Aggregates Part 4: Methods of Testing Road Aggregates—Field Tests Test 4.1: The Density of Compacted Aggregate Test 4.1.1: Sand Replacement Method. 7 pp.

SNZ NZS 4407: Pt 4.2.1-91. Methods of Sampling and Testing Road Aggregates Part 4: Methods of Testing Road Aggregates—Field Tests Test 4.2: Theield Water Content and Field Dry Density of Compacted Aggregate Test 4.2.1: Method Using a Nuclear Surface Moisture-Density Gauge-. 3 pp.

SNZ NZS 4407: Part 4.2.2-91. Methods of Sampling and Testing Road Aggregates Part 4: Methods of Testing Road Aggregates—Field Tests Test 4.2: The Field Water Content and Field Dry Density of Compacted Aggregate Test 4.2.2: Method Using a Nuclear Surface Moisture-Density Gauge—Backscatter Mode. 3 pp.

SNZ NZS 4407: Part 4.2.3-91. Methods of Sampling and Testing Road Aggregates Part 4: Methods of Testing Road Aggregates—Field Tests Test 4.2: The Field Water Content and Field Dry Density of Compacted Aggregate Test 4.2.3: Calibration of Nuclear Surface Moisture-Density Gauge for Field Use. 3 pp.

SNZ NZS 4407: Part 4.2.4-91. Methods of Sampling and Testing Road Aggregates Part 4: Methods of Testing Road Aggregates—Field Tests Test 4.2: The Field Water Content and Field Dry Density of Compacted Aggregate Test 4.2.4: Calibration of Nuclear Surface Moisture-Density Gauge Using Standard Blocks. 6 pp.

—Aluminum Oxide
BSI BS 4140: Part 8-86. 1986 Methods of Test for Aluminium Oxide Part 8: Determination of Absolute Density Using a Pyknometer. 6 pp.

BSI BS 4140: Part 10-86. 1986 Methods of Test for Aluminium Oxide Part 10: Determination of Untamped Density. 4 pp.

ISO 901-76. Aluminium Oxide Primarily Used for the Production of Aluminium—Determination of Absolute Density—Pyknometer Method First Edition. 6 pp.

ISO 903-76. Aluminium Oxide Primarily Used for the Production of Aluminium—Determination of Untamped Density First Edition. 4 pp.

—Aluminum Stearate
CNS K6320-72. Method of Test for Aluminum Stearate of Industrial Grade (Temporary Standard) (Jun)(3381). 1 p.

—Ammonium Hydroxide—Gravimetric Analysis
BSI BS 4651: Part 1-88. 1988 Ammonia Solution Part 1: Methods for Determination of Density at 20 Degrees Celsius. 4 pp.

—Antifreezes
CNS K6483-80. Method of Test for Specific Gravity of Engine Antifreezes (May)(5586).

—Beeswax
MOD UK M 2012/71. Examination of Beeswax GS and Beeswax, Lead Free.

MOD UK M 9517/65. Examination of Beeswax and Beeswax, Lead Free (LAG 931A, No Information).

—Benzene
ISO 5281-80. Aromatic Hydrocarbons—Benzene, Xylene and Toluene—Determination of Density at 20 Degrees Celsius First Edition. 11 pp.

—Bitumens
CNS A3149-87. Method of Test for Bulk Specific Gravity and Density of Compacted Bituminous Mixtures Using Paraffin-Coated Specimens (Aug)(8757).

CNS A3150-87. Method of Test for Theoretical Maximum Specific Gravity of Bituminous Paving Mixtures (Aug)(8758).

CNS A3151-87. Method of Test for Bulk Specific Gravity and Density of Compacted Bituminous Mixtures Using Saturated Surface-Dry Specimens (Aug)(8759).

—Brake Linings—Automotive
CNS D3182-87. Method of Test of Specific Gravity for Brake Linings and Pads of Automobiles (Dec)(12178).

JIS D 4417-86. Test Procedure of Specific Gravity for Brake Linings and Pads of Automobiles. 7 pp.

—Brake Pads—Automotive
CNS D3182-87. Method of Test of Specific Gravity for Brake Linings and Pads of Automobiles (Dec)(12178).

JIS D 4417-86. Test Procedure of Specific Gravity for Brake Linings and Pads of Automobiles. 7 pp.

—Buildings
ISO 9194-87. Bases for Design of Structures—Actions Due to the Self-Weight of Structures, Non-Structural Elements and Stored Materials—Density First Edition. 12 pp.

—Bunker Fuel Oils
CPPA J.24(D)P-77. API Gravity of Bunker "C" Oil. 1 p.

—Cables (Electric)
BSI BS 6469: Sec 1.3-92. 1992 Insulation and Sheathing Materials of Electric Cables Part 1: Methods of Test for General Application Section 1.3: Methods for Determining the Density—Water Absorption Test—Shrinkage Test. 20 pp.

CENELEC HD 505.1.3 S2-91. Common Test Methods for Insulating and Sheathing Materials of Electric Cables Part 1: Methods for General Application Section Three—Methods for Determining the Density Water Absorption Tests-. 5 pp.

DIN VDE 0472 Pt 601-83. Testing of Cables, Wires and Flexible Cords; Density (Aug). 8 pp.

—Carbon Fibers
AECMA PREN2557-88. Carbon Fibre Preimpregnates Test Method for the Determination of Mass Per Unit Area. 7 pp.

AECMA PREN2559-88. Carbon Fibre Preimpregnates Test Method for the Determination of the Resin and Fibre Content and the Mass of Fibre Per Unit Area. 13 pp.

BSI BS 7658: Part 1-93. 1993 Carbon Fibre Part 1: Method for Determination of Density (ISO 10119: 1992) (V). 14 pp.

ISO 10119-92. Carbon Fibre—Determination of Density First Edition. 11 pp.

—Cements
BSI BS 4550: Sec 3.2-78. (WITHDRAWN) 1978 Methods of Testing Cement Part 3: Physical Tests Section 3.2: Density Test (Superseded by BS EN 196-6: 1992). 4 pp.

—Ceramic Powders
CEN PREN 725-7-92. Advanced Technical Ceramics—Methods of Test for Ceramic Powders—Part 7: Determination of the Absolute Density. 10 pp.

—Ceramics
BSI BS 7134: Sec 1.2-89. 1989 Testing of Engineering Ceramics Part 1: General and Textural Properties Section 1.2: Methods of Determination of Density and Porosity. 10 pp.

CEN PREN 623-2-91. Methods of Testing Advanced Technical Ceramics—General and Textural Properties—Part 2: Determination of Density and Porosity. 15 pp.

CNS R3057-69. Method of Test for Specific Gravity of Fired Fine Ceramic and Porcelain Material (Jan)(2893)(R 1975).

—Chemical Products
BSI BS 4522-88. 1988 Method for Determination of Absolute Density at 20 Degrees of Liquid Chemical Products for Industrial Use. 4 pp.

CNS K0015-82. Method for Determining Specific Gravity of Chemical Products (May)(8834).

CNS K6560-80. Determination of Density of Liquid Chemical Products (at 20 Degrees Celsius) (Aug)(6205).

ISO 758-76. Liquid Chemical Products for Industrial Use—Determination of Density at 20 Degrees Celsius First Edition. 4 pp.

JIS K 0061-92. Test Methods for Density and Relative Density of Chemical Products. 60 pp.

—Chloroform
BSI BS 4774-72. 1972 Methods of Test for Chloroform for Industrial Use. 12 pp.

—Citronella Oil
CNS K6063-65. Method of Test for Citronella Oil (Aug)(817)(R 1971). 3 pp.

CNS K6598-80. Determination of the Density and Relative Density of Essential Oils (Oct)(6571).

—Coal
SAA AS 1038.21-83. Methods for the Analysis and Testing of Coal and Coke—Part 21: Determination of the Relative Density and Apparent Relative Density of Hard Coal (Superseded (in Part) by AS 1038.21.2—1992). 10 pp.

INTERNATIONAL AND NON-U.S. NATIONAL STANDARDS
SUBJECT INDEX

Density (Cont.)

—Coal Tar
CNS K6211-65. Method for Determination of Specific Gravity of Coal Tar (Sep)(2493)(R 1971).

—Coal Tar Pitch
ISO 6999-83. Carbonaceous Materials for the Production of Aluminium—Pitch for Electrodes—Determination of Density—Pyknometric Method First Edition. 5 pp.

—Coatings
DIN ENGL 53217 Pt 1-91. Determination of Density of Paints, Varnishes and Similar Coating Materials; Survey of Test Methods (Mar). 2 pp.
DIN ENGL 53217 Pt 2-91. Determination of Density of Paints, Varnishes and Similar Coating Materials by the Pyknometer Method (Mar). 5 pp.
DIN ENGL 53217 Pt 3-91. Determination of Density of Paints, Varnishes and Similar Coating Materials by the Displacement Float Method (Mar). 4 pp.
DIN ENGL 53217 Pt 5-91. Determination of Density of Paints, Varnishes and Similar Coating Materials by the Vibration Method (Mar). 4 pp.

—Coatings—Hydrometry
DIN ENGL 53217 Pt 4-91. Determination of Density of Paints, Varnishes and Similar Coating Materials by the Hydrometer Method (Mar). 3 pp.

—Coconut Oil
CNS K6180-74. Method of Test for Coconut Oil (Technical Grade) (Oct)(2241). 5 pp.

—Coke
BSI BS 1016: Sec 108.5-92. 1992 Methods for Analysis and Testing of Coal and Coke Part 108: Tests Special to Coke Section 108.5: Determination of Density and Porosity. 10 pp.
BSI BS 6043: Sec 2.6-86. 1986 Methods of Sampling and Test for Carbonaceous Materials Used in Aluminium Manufacture Part 2: Electrode Coke Section 2.6: Determination of the Density in Xylene of Calcined Coke. 8 pp.
CNS M3179-84. Method for Determination of True Relative Density Apparent Relative Density and Porosity of Coke (Nov)(11121).
ISO 1014-85. Coke—Determination of True Relative Density, Apparent Relative Density and Porosity Second Edition. 5 pp.
ISO 8004-85. Carbonaceous Materials for the Production of Aluminium—Calcined Coke and Calcined Carbon Products—Determination of the Density in Xylene—Pyknometric Method First Edition. 6 pp.

—Concretes
BSI BS 1881: Part 107-83. 1983 Amd 2 Testing Concrete Part 107: Method for Determination of Density of Compacted Fresh Concrete (AMD 6722) July 31, 1991. 8 pp.
BSI BS 1881: Part 114-83. 1983 Amd 2 Testing Concrete Part 114: Methods for Determination of Density of Hardened Concrete (AMD 6721) July 31, 1991. 11 pp.
BSI BS 1881: Part 129-92. 1992 Testing Concrete Part 129: Method for Determination of Density of Partially Compacted Semi-Dry Fresh Concrete. 10 pp.
BSI BS 6319: Part 5-84. 1984 Testing of Resin Compositions for Use in Construction Part 5: Methods for Determination of Sensitivity of Hardened Resin Compositions. 4 pp.
CEN PREN 678-92. Determination of the Dry Density of Autoclaved Aerated Concrete. 5 pp.
ISO 6275-82. Concrete, Hardened—Determination of Density First Edition. 4 pp.
ISO 6276-82. Concrete, Compacted Fresh—Determination of Density First Edition. 3 pp.
SNZ NZS 3112: Part 3-86. Methods of Test for Concrete Part 3: Tests on Hardened Concrete Other Than for Strength. 11 pp.

—Creosote Oil
CNS K6067-57. Method of Sampling for Specific Gravity of Creosote Oil (Jul)(910)(R 1971).

—Crude Oils
BSI BS 4699-85. 1985 Methods for Determination of Density or Relative Density of Petroleum Products (Pyknometer Methods). 16 pp.
BSI BS 4714-80. 1980 Laboratory Determination of Density or Relative Density of Crude Petroleum and Liquid Petroleum Products (Hydrometer Method). 9 pp.
CNS K6916-90. Method of Test for Density, Relative Density, or API Gravity of Crude Petroleum and Liquid Petroleum Products by Hydrometer Method (Aug)(12017).
ISO 3675-93. Crude Petroleum and Liquid Petroleum Products—Laboratory Determination of Density or Relative Density—Hydrometer Method Second Edition. 11 pp.

Density (Cont.)

—Crude Oils (Cont.)
ISO 3838-83. Crude Petroleum and Liquid or Solid Petroleum Products—Determination of Density or Relative Density—Capillary-Stoppered Pyknometer and Graduated Bicapillary Pyknometer Methods First Edition. 13 pp.
JIS K 2249-87. Testing Methods for Density of Crude Oil and Petroleum Products, and Petroleum Measurement Tables Based on Reference Temperature 15 Degrees C.

—Dairy Products
BSI BS 734: Part 1-73. 1973 Amd 1 Density Hydrometers for Use in Milk Part 1: Measurement of the Density of Milk Using a Hydrometer. 18 pp.
BSI BS 734: Part 2-59. 1959 Amd 1 Density Hydrometers for Use in Milk Part 2: Methods. 734C is an Extract From BS 734 Part 2. 32 pp.
CNS N6058-72. Methods of Test for Milk and Milk Products (Determination of Specific Gravity) (Oct)(3442). 3 pp.

—Dyes
MOD UK M 818/63. Examination of Dyestuffs for Use in Pyrotechnic Compositions and HE Substitutes.

—Enamels
CGSB 1-GP-71 METH 2.1-78. Methods of Testing Paints and Pigments Density (Weight Per Gallon). 2 pp.
CGSB 1-GP-71 METH 2.2-74. Methods of Testing Paints and Pigments Specific Gravity. 2 pp.
CNS K6030-72. Method of Test for Enamel (Jun)(627). 7 pp.

—Epoxy Resins
JIS K 7232-86. Testing Methods for Specific Gravity of Epoxide Resins and Hardeners. 16 pp.

—Epoxy Resins—Hardeners
JIS K 7232-86. Testing Methods for Specific Gravity of Epoxide Resins and Hardeners. 16 pp.

—Essential Oils
ISO 279-81. Essential Oils—Determination of Relative Density at 20 Degree C (Reference Method) First Edition. 4 pp.

—Extenders
BSI BS 3483: Part B8-93. 1993 Methods for Testing Pigments for Paints Part B8: Determination of Density (Pyknometer Method) (ISO 787-10: 1993) (W). 9 pp.
BSI BS 3483: Part B8-82. 1982 Methods for Testing Pigments in Paints Part B8: Determination of Density (Pyknometer Method) 4 Degrees Centigrate. 6 pp.
BSI BS 3483: Part B9-80. 1980 Methods for Testing Pigments in Paints Part B9: Determination of Density Relative to Water (Using a Centrifuge). 4 pp.
ISO 787 Pt 10-93. General Methods of Test for Pigments and Extenders—Part 10: Determination of Density—Pyknometer Method Second Edition. 9 pp.

—Fats
CNS N6079-82. Methods of Test for Edible Oils and Fats (Determination of Specific Gravity) (Jan)(3644). 1 p.

—Fiberboard
BSI BS EN 323-93. 1993 Wood-Based Panels—Determination of Density (R). 10 pp.
CEN PREN 323-89. Wood-Based Panels—Determination of Density of Test Pieces. 7 pp.
CEN EN 323-93. Wood-Based Panels—Determination of Density. 5 pp.
ISO 9427-89. Wood-Based Panels—Determination of Density First Edition. 4 pp.

—Fiberglass Reinforced Plastics
CNS K6662-81. Method of Test for Fiber Glass Products (May)(7397). 9 pp.

—Fire Wire and Ribbon—Electronic Equipment
CNS C6096-80. Method of Test for Density of Fire Wire and Ribbon for Electronic Devices (Aug)(6316).

—Firebricks
CNS R3013-86. Method of Test for Apparant Porosity Water Absorption and Specific Gravity of Refractory Bricks (Oct)(619).
JIS R 2205-92. Testing Method for Apparent Porosity, Water Absorption and Specific Gravity of Refractory Bricks. 8 pp.

—Gaseous Fuels
CNS K6245-71. Method for Determination of Specific Gravity of Gas Fuels (Jul)(2705).

Density (Cont.)

—Gases
BSI BS 3156: Part 3-68. (WITHDRAWN) 1968 Methods for the Analysis of Fuel Gases Part 3: Combustion Characteristics. 24 pp.
DIN ENGL 51858-82. Gaseous Fuels and Other Gases; Calculation of Gross and Nett Calorific Values and of Relative Density of Gas Mixtures (Nov). 7 pp.
JIS K 2301-92. Fuel Gases and Natural Gas—Methods for Chemical Analysis and Testing. 116 pp.

—Gases—Glossaries
CNS Z7183-84. Definitions of Terms Relating to Density and Specific Gravity of Solids Liquids and Gases (Oct)(11027).

—Glass Fabrics
ISO 4602-78. Textile Glass—Woven Fabrics—Determination of Number of Yarns per Unit Length of Warp and Weft. First Edition. 4 pp.

—Glazes—Aircraft
AECMA PREN2155-01-84. Test Methods for Transparent Materials for Aircraft Glazing Part 1—Determination of the Density and Relative Density. 6 pp.

—Glycerol
BSI BS 5711: Part 4-79. 1979 Methods of Sampling and Test for Glycerol Part 4: Determination of Density and Relative Density. 6 pp.
ISO 2099-72. Purified Glycerol for Industrial Use—Determination of Density at 20 Degrees Celsius First Edition. 6 pp.

—Hard Metals
BSI BS 5600: Sec 4.4-79. (WITHDRAWN) 1979 Powder Metallurgical Materials and Products Part 4: Methods of Testing and Chemical Analysis of Hardmetals Section 4.4: Determination of Density (Superseded by BS EN 23369: 1993). 6 pp.
BSI BS EN 23369-93. 1993 Impermeable Sintered Metal Materials and Hardmetals—Determination of Density (ISO 3369: 1975) (V). 10 pp.
CEN EN 23369-93. Impermeable Sintered Metal Materials and Hardmetals—Determination of Density (ISO 3369: 1975). 5 pp.
ISO 3369-75. Impermeable Sintered Metal Materials and Hardmetals—Determination of Density First Edition (CEN EN 23369: 1993). 5 pp.

—Herbicides
MOD UK M 9529/66. Examination of Solution Weedkiller Chlorate.

—High Explosives
MOD UK M 216/91. Determination of Density as Applied to High Explosives in Biscuit and Pellet Form.

—Hydraulic Cements
CNS R3122-85. Method of Test for Density of Hydraulic Cement (May)(11272).

—Hydrocarbons
BSI BS 6665-85. 1985 Method for Determination of Density or Relative Density of Liquefied Petroleum Gases and Other Light Hydrocarbons by Pressure Hydrometer Method. 10 pp.
ISO 3993-84. Liquefied Petroleum Gas and Light Hydrocarbons—Determination of Density or Relative Density—Pressure Hydrometer Method First Edition. 8 pp.
JIS K 0519-78. Testing Method for Density of High-Purity Hydrocarbons by Bingham Pycnometer.

—Hydrochloric Acid
BSI BS 976-87. 1987 Density-Composition Tables for Aqueous Solutions of Hydrochloric Acid. 34 pp.

—Ink Removers
MOD UK M 2003/59. Examination of Solvent, Cleaning.

—Insulating Firebricks
CNS R3141-86. Method of Test for Specific Gravity and True Porosity of Insulating Fire Bricks (Oct)(11738).
JIS R 2614-85. Testing Method for Specific Gravity and True Porosity of Insulating Fire Bricks. 6 pp.

—Ion Exchangers
DIN ENGL 54410-85. Testing of Ion Exchangers; Determination of Density (Oct). 2 pp.

—Iron Ores
JIS M 8717-88. Test Methods for Determination of Density of Iron Ores. 9 pp.

—Lacquers
CNS K6386-76. Method of Test for Lacquer in Surface of Leather Shoes (Mar) (3922). 2 pp.
CNS K6665-81. Method of Test for Density of Paint Varnish, Lacquer and Related Products (May)(7405).

INDUSTRY STANDARDS

Density (Cont.)

—Lacquers (Cont.)
SAA AS 1580.202. 1-80. Paints and Related Materials—Methods of Test—Part 202.1: Density (R 1991). 2 pp.
SNZ NZS/AS 1580. 202.1-80. Methods of Test for Paints and Related Materials Part 202.1: Density (This is a Joint Standard with SAA AS 1580.202.1). 2 pp.

—Latex
BSI BS 6057: Sec 3.7-82. 1982 Rubber Latices Part 3: Methods of Test Section 3.7: Determination of Density of Natural Rubber Latices. 4 pp.
DIN ENGL 53597-72. Latex; Determination of Density (Sept). 2 pp.
ISO 705-74. Natural Rubber Latex—Determination of Density First Edition. 5 pp.

—Linseed Oil
CNS K6055-75. Method of Test for Boiled Linseed Oil (Feb)(769). 4 pp.

—Liquefied Petroleum Gas—Hydrometry
BSI BS 6665-85. 1985 Method for Determination of Density or Relative Density of Liquefied Petroleum Gases and Other Light Hydrocarbons by Pressure Hydrometer Method. 10 pp.
ISO 3993-84. Liquefied Petroleum Gas and Light Hydrocarbons—Determination of Density or Relative Density—Pressure Hydrometer Method First Edition. 8 pp.

—Liquids
CNS Z8070-88. Method for Measuring Specific Gravity of Liquid (Oct)(12450).
JIS Z 8804-76. Methods of Measuring Specific Gravity of Liquid (R 1989). 12 pp.

—Liquids—Glossaries
CNS Z7183-84. Definitions of Terms Relating to Density and Specific Gravity of Solids Liquids and Gases (Oct)(11027).

—Masonry
CEN PREN 772-13-92. Methods of Test for Masonry Units—Part 13: Determination of Net and Gross Dry Density of Masonry Units. 9 pp.

—Masonry—Stone
CEN PREN 772-4-92. Methods of Test for Masonry Units—Part 4: Determination of Real Density and Bulk Density for Natural Stone Masonry Units. 7 pp.

—Metals—Aerospace
AECMA PREN6018-90. Test Methods for Metallic Materials Determination of Density Accordintg to Displacement Method. 2 pp.

—Microfilm
BSI BS ISO 6200-91. 1991 Micrographics—First Generation Silver-Gelatin Microforms of Source Documents—Density Specifications (H). 10 pp.
ISO 6200-90. Micrographics—First Generation Silver-Gelatin Microforms of Source Documents—Density Specifications Second Edition. 7 pp.

—Milk
CNS N6058-72. Methods of Test for Milk and Milk Products (Determination of Specific Gravity) (Oct)(3442). 3 pp.

—Mineral Aggregates
DIN ENGL 52102-88. Determination of Absolute Density, Dry Density, Compactness and Porosity of Natural Stone and Mineral Aggregates (Aug). 10 pp.

—Molding Materials
CNS K6395-76. Method of Test for Polystyrene Moulding Materials (Sep)(4013). 7 pp.

—Mortars
BSI BS 6319: Part 5-84. 1984 Testing of Resin Compositions for Use in Construction Part 5: Methods for Determination of Sensitivity of Hardened Resin Compositions. 4 pp.

—Natural Gas
DIN ENGL 51858-82. Gaseous Fuels and Other Gases; Calculation of Gross and Nett Calorific Values and of Relative Density of Gas Mixtures (Nov). 7 pp.
ISO 6976-83. Natural Gas—Calculation of Calorific Value, Density and Relative Density First Edition. 16 pp.

—Nitric Acid
BSI BS 975-87. 1987 Density-Composition Tables for Aqueous Solutions of Nitric Acid. 58 pp.
ISO 2990-74. Nitric Acid for Industrial Use—Evaluation of the Nitric Acid Concentration by Measurement of Density First Edition. 5 pp.

Density (Cont.)

—Oleic Acid
MOD UK M 9545/68. Examination of Oleic Acid, Special, Lead Free.

—Paints
BSI BS 3900: Part A12-75. 1975 Methods of Test for Paints Group A: Tests on Liquid Paints (Excluding Chemical Tests) Part A12: Determination of Density. 3 pp.
CGSB 1-GP-71 METH 2.1-78. Methods of Testing Paints and Pigments Density (Weight Per Gallon). 2 pp.
CGSB 1-GP-71 METH 2.3-74. Methods of Testing Paints and Pigments Specific Gravity Viscous Liquids. 1 p.
CNS K6143-87. Method of Test for Traffic Paints (May)(1334). 19 pp.
CNS K6665-81. Method of Test for Density of Paint Varnish, Lacquer and Related Products (May)(7405).
DIN ENGL 53217 Pt 1-91. Determination of Density of Paints, Varnishes and Similar Coating Materials; Survey of Test Methods (Mar). 2 pp.
DIN ENGL 53217 Pt 2-91. Determination of Density of Paints, Varnishes and Similar Coating Materials by the Pyknometer Method (Mar). 5 pp.
DIN ENGL 53217 Pt 3-91. Determination of Density of Paints, Varnishes and Similar Coating Materials by the Displacement Float Method (Mar). 4 pp.
DIN ENGL 53217 Pt 5-91. Determination of Density of Paints, Varnishes and Similar Coating Materials by the Vibration Method (Mar). 4 pp.
ISO 2811-74. Paints and Varnishes—Determination of Density First Edition. 6 pp.
SAA AS 1580.202. 1-80. Paints and Related Materials—Methods of Test—Part 202.1: Density (R 1991). 2 pp.
SAA AS 1580.202. 2-80. Paints and Related Materials—Methods of Test—Part 202.2: Density of Water-Dispersed Paints Subject to Foaming (R 1991). 2 pp.
SNZ NZS/AS 1580. 202.1-80. Methods of Test for Paints and Related Materials Part 202.1: Density (This is a Joint Standard with SAA AS 1580.202.1). 2 pp.
SNZ NZS/AS 1580. 202.2-80. Methods of Test for Paints and Related Materials Part 202.2: Density of Water-Dispersed Paints Subject to Foaming (This is a Joint Standard with SAA AS 1580.202.2). 2 pp.

—Paperboard
BSI BS 3983-89. 1989 Method for Determination of Thickness and Apparent Bulk Density or Apparent Sheet Density of Paper and Board. 10 pp.
CPPA D.4-92. Thickness and Apparent Density of Paper and Paperboard. 2 pp.
ISO 534-88. Paper and Board—Determination of Thickness and Apparent Bulk Density or Apparent Sheet Density Second Edition; (Supersedes 438). 9 pp.

—Papers
BSI BS 3983-89. 1989 Method for Determination of Thickness and Apparent Bulk Density or Apparent Sheet Density of Paper and Board. 10 pp.
CPPA D.4-92. Thickness and Apparent Density of Paper and Paperboard. 2 pp.
ISO 534-88. Paper and Board—Determination of Thickness and Apparent Bulk Density or Apparent Sheet Density Second Edition; (Supersedes 438). 9 pp.

—Particle Board
BSI BS EN 323-93. 1993 Wood-Based Panels—Determination of Density (R). 10 pp.
CEN PREN 323-89. Wood-Based Panels—Determination of Density of Test Pieces. 7 pp.
CEN EN 323-93. Wood-Based Panels—Determination of Density. 5 pp.
ISO 9427-89. Wood-Based Panels—Determination of Density First Edition. 4 pp.

—Peppermint Oil
CNS K6102-72. Method of Test for Crude and De-Mentholized Peppermint Oil (Jun)(1094). 2 pp.

—Petroleum Products
BSI BS 4699-85. 1985 Methods for Determination of Density or Relative Density of Petroleum Products (Pyknometer Methods). 16 pp.
BSI BS 4714-80. 1980 Laboratory Determination of Density or Relative Density of Crude Petroleum and Liquid Petroleum Products (Hydrometer Method). 9 pp.
BSI BS 6665-85. 1985 Method for Determination of Density or Relative Density of Liquefied Petroleum Gases and Other Light Hydrocarbons by Pressure Hydrometer Method. 10 pp.
CNS K6916-90. Method of Test for Density, Relative Density, or API Gravity of Crude Petroleum and Liquid Petroleum Products by Hydrometer Method (Aug)(12017).
DIN ENGL 51757-84. Testing of Petroleum and Related Materials; Determination of Density (Jan). 22 pp.

Density (Cont.)

—Petroleum Products (Cont.)
ISO 3675-93. Crude Petroleum and Liquid Petroleum Products—Laboratory Determination of Density or Relative Density—Hydrometer Method Second Edition. 11 pp.
ISO 3838-83. Crude Petroleum and Liquid or Solid Petroleum Products—Determination of Density or Relative Density—Capillary-Stoppered Pyknometer and Graduated Bicapillary Pyknometer Methods First Edition. 13 pp.
ISO 3993-84. Liquefied Petroleum Gas and Light Hydrocarbons—Determination of Density or Relative Density—Pressure Hydrometer Method First Edition. 8 pp.
JIS K 2249-87. Testing Methods for Density of Crude Oil and Petroleum Products, and Petroleum Measurement Tables Based on Reference Temperature 15 Degrees C.

—Phenolic Resins
CNS K6273-74. Method of Test for Phenol Formaldehyde Resin Molding Compounds (Oct)(2988). 11 pp.

—Phenols
CNS K6217-65. Method for Determination of Specific Gravity of Phenols (Sep)(2499)(R 1971).
ISO 1897 Pt I-77. Phenol, o-Cresol, m-Cresol, p-Cresol, Cresylic Acid and Xylenols for Industrial Use—Methods of Test—Part I: General First Edition. 4 pp.

—Phosphoric Acids
CNS K6235-72. Method of Test for Phosphoric Acid (Industrial Use) (Jun)(2620). 2 pp.

—Photographic Chemicals
ISO 10349 Pt 11-92. Photography—Photographic-Grade Chemicals—Test Methods—Part 11: Determination of Specific Gravity First Edition. 6 pp.
ISO 10349 Pt 12-92. Photography—Photographic-Grade Chemicals—Test Methods—Part 12: Determination of Density First Edition. 6 pp.

—Pigments
BSI BS 3483: Part B8-93. 1993 Methods for Testing Pigments for Paints Part B8: Determination of Density (Pyknometer Method) (ISO 787-10: 1993) (W). 9 pp.
BSI BS 3483: Part B8-82. 1982 Methods for Testing Pigments in Paints Part B8: Determination of Density (Pyknometer Method) 4 Degrees Centigrate. 6 pp.
BSI BS 3483: Part B9-80. 1980 Methods for Testing Pigments in Paints Part B9: Determination of Density Relative to Water (Using a Centrifuge). 4 pp.
CGSB 1-GP-71 METH 2.1-78. Methods of Testing Paints and Pigments Density (Weight Per Gallon). 2 pp.
CGSB 1-GP-71 METH 2.3-74. Methods of Testing Paints and Pigments Specific Gravity Viscous Liquids. 1 p.
ISO 787 Pt 10-93. General Methods of Test for Pigments and Extenders—Part 10: Determination of Density—Pyknometer Method Second Edition. 9 pp.
ISO 787 Pt 23-79. General Methods of Test for Pigments and Extenders—Part 23: Determination of Density (Using a Centrifuge to Remove Entrained Air) First Edition. 5 pp.

—Pipe Fittings
ISO 3477-81. Polypropylene (PP) Pipes and Fittings—Density—Determination and Specification Second Edition. 3 pp.
ISO 3514-76. Chlorinated Polyvinyl Chloride (CPVC) Pipes and Fittings—Specification and Determination of Density First Edition. 3 pp.
ISO 4439-79. Unplasticized Polyvinyl Chloride (PVC) Pipes and Fittings—Determination and Specification of Density First Edition. 3 pp.
ISO 4451-80. Polyethylene (PE) Pipes and Fittings—Determination of Reference Density of Uncoloured and Black Polyethylenes First Edition; (Erratum—Nov 1980). 7 pp.

—Pitch (Material)
BSI BS 6043: Sec 1.7-88. 1988 Methods of Sampling and Test for Carbonaceous Materials Used in Aluminium Manufacture Part 1: Electrode Pitch Section 1.7: Determination of Apparent Density (Bouyancy Method). 7 pp.

—Plastic Pipes
BSI M 56-85. 1985 Methods for Determination of Density and Relative Density of Polytetrafluoroethylene (PTFE) Tubing for Aerospace Applications. 8 pp.
ISO 3477-81. Polypropylene (PP) Pipes and Fittings—Density—Determination and Specification Second Edition. 3 pp.

INTERNATIONAL AND NON-U.S. NATIONAL STANDARDS
SUBJECT INDEX

Density (Cont.)

Plastic Pipes (Cont.)
ISO 3514-76. Chlorinated Polyvinyl Chloride (CPVC) Pipes and Fittings—Specification and Determination of Density First Edition. 3 pp.
ISO 4439-79. Unplasticized Polyvinyl Chloride (PVC) Pipes and Fittings—Determination and Specification of Density First Edition. 3 pp.
ISO 4451-80. Polyethylene (PE) Pipes and Fittings—Determination of Reference Density of Uncoloured and Black Polyethylenes First Edition; (Erratum—Nov 1980). 7 pp.
ISO 7258-84. Polytetrafluoroethylene (PTFE) Tubing for Aerospace Applications—Methods for the Determination of the Density and Relative Density First Edition. 8 pp.

Plastics
BSI BS 2782:Pt6: METH 620A-D-91. 1991 Methods of Testing Plastics Part 6: Dimensional Properties Method 620A to 620D: Determination of Density and Relative Density of Non-Cellular Plastics (ISO 1183: 1987. 13 pp.
BSI BS 4370: Part 1-88. 1988 Methods of Test for Rigid Cellular Materials Part 1: Methods 1-5. 12 pp.
BSI BS 4443: Part 1-88. 1988 Amd 1 Methods of Test for Flexible Cellular Materials Part 1: Methods 1 to 6. 16 pp.
BSI BS 4618: Sec 5.1-70. 1970 Recommendations for the Presentation of Plastics Design Data Part 5: Other Properties Section 5.1: Density. 2 pp.
CNS K6687-81. Method of Test for Density of Plastics by the Density-Gradient Technique (Jul)(7735).
DIN ENGL 53479-76. Testing of Plastics and Elastomers; Determination of Density (July). 5 pp.
ISO 1183-87. Plastics—Methods for Determining the Density and Relative Density of Non-Cellular Plastics First Edition. 11 pp.
JIS K 7112-80. Methods for Determining the Density the Specific Gravity of Plastics. 15 pp.
SAA AS 1193-89. Plastics—Methods for Determining the Density and Relative Density of Non-Cellular Plastics. 8 pp.

Plywood
BSI BS EN 323-93. 1993 Wood-Based Panels—Determination of Density (R). 10 pp.
CEN PREN 323-89. Wood-Based Panels—Determination of Density of Test Pieces. 7 pp.
CEN EN 323-93. Wood-Based Panels—Determination of Density. 5 pp.
CNS O2037-87. Method of Test for Specific Gravity of Plywood (May)(8325). 1 p.
ISO 9427-89. Wood-Based Panels—Determination of Density First Edition. 4 pp.

Polyester Resins
CNS K6731-82. Method of Test for Liquid Unsaturated Polyester Resin (Dec)(9716). 11 pp.

Polymers
ISO 8962-87. Plastics—Polymer Dispersions—Determination of Density First Edition. 8 pp.

Polyurethane Athletic Fields
CNS K6591-86. Method of Test for Polyurethane (PU) Athletic Installation Material (May)(6483). 6 pp.

Porcelain
CNS R3057-69. Method of Test for Specific Gravity of Fired Fine Ceramic and Porcelain Material (Jan)(2893)(R 1975).

Potassium Silicates—Hydrometry
BSI BS 6092: Part 1-81. 1981 Sampling and Test for Sodium and Potassium Silicates for Industrial Use Part 1: Determination of Solution Density at 20 Degrees Celsius. 6 pp.
ISO 1687-76. Sodium and Potassium Silicates for Industrial Use—Determination of Density at 20 Degrees Celsius of Products in Solution—Method Using Density Hydrometer and Method Using Pyknometer First Edition. 5 pp.

Pulpwood
JIS P 8014-76. Testing Method for Density of Pulpwood and Woodchips (R 1985). 8 pp.

Refractory Materials
BSI BS 1902: Sec 3.4-81. 1981 Methods for Testing Refractory Materials Part 3: General and Textural Properties Section 3.4: Determination of True Density (Method 1902-304). 4 pp.
BSI BS 1902: Sec 3.5-81. 1981 Methods for Testing Refractory Materials Part 3: General and Textural Properties Section 3.5: Determination of Powder Density (Method 1902-305). 4 pp.
CEN PREN 993-2-93. Method of Test for Dense Shaped Refractory Products—Part 2: Determination of True Density. 9 pp.

Density (Cont.)

Refractory Materials (Cont.)
CNS R3014-83. Method of Test for True Specific Gravity of Refractory Material by Water Immersion (Apr) (620).
ISO 5018-83. Refractory Materials—Determination of True Density First Edition. 6 pp.
SAA AS 1774.5-89. Refractories and Refractory Materials—Physical Test Methods—Part 5: Determination of Density, Porosity and Water Absorption. 3 pp.
SAA AS 1774.6-92. Refractories and Refractory Materials—Physical Test Methods—Part 6: Determination of True Density. 3 pp.

Resins
BSI BS 2782:Pt6: METH 620E-80. 1980 Methods of Testing Plastics Part 6: Dimensional Properties Method 620E: Determination of Density of Liquid Resins by the Pycnometer Method. 4 pp.
ISO 1675-85. Plastics—Liquid Resins—Determination of Density by the Pyknometer Method Second Edition. 4 pp.

Rocks
DIN ENGL 52102-88. Determination of Absolute Density, Dry Density, Compactness and Porosity of Natural Stone and Mineral Aggregates (Aug). 10 pp.

Roofing
SAA AS 3991.3-92. Methods of Testing Flat Cellulose-Cement Sheets—Part 3: Determination of Density (Oven-Dried). 1 p.

Rubber
CNS K6458-80. Determination of Density for Vulcanized Rubbers (Mar)(5341).
DIN ENGL 53479-76. Testing of Plastics and Elastomers; Determination of Density (July). 5 pp.
ISO 2781-88. Rubber, Vulcanized—Determination of Density Third Edition. 6 pp.

Screeds
DIN ENGL 272-86. Testing of Magnesium Oxychloride Screeds (Feb). 8 pp.

Sintered Materials
BSI BS 5600: Sec 3.1-79. (WITHDRAWN) 1979 Powder Metallurgical Materials and Products Part 3: Methods of Testing Sintered Metal Materials and Products, Excluding Hardmetals Section 3.1: Determination of Density for Impermeable Sintered Metal Materials. 6 pp.
BSI BS 5600: Sec 3.2-88. 1988 Amd 1 Powder Metallurgical Materials and Products Part 3: Methods of Testing Sintered Metal Materials and Products, Excluding Hardmetals Sec 3.2: Determin. of Density Oil Content and Porosity (AMD 6496) December 21, 1990. 9 pp.
BSI BS EN 23369-93. 1993 Impermeable Sintered Metal Materials and Hardmetals—Determination of Density (ISO 3369: 1975) (V). 10 pp.
CEN EN 23369-93. Impermeable Sintered Metal Materials and Hardmetals—Determination of Density (ISO 3369: 1975). 5 pp.
CNS Z8041-82. Method for Determining Density of Metal Powder Sintered Materials (Jul)(9205).
ISO 2738-87. Permeable Sintered Metal Materials—Determination of Density, Oil Content, and Open Porosity Second Edition; (Supersedes 2737). 8 pp.
ISO 3369-75. Impermeable Sintered Metal Materials and Hardmetals—Determination of Density First Edition (CEN EN 23369: 1993). 5 pp.
JIS Z 2505-89. Method for Determination of Density of Metal Powder Sintered Materials. 6 pp.

Sodium Silicates
BSI BS 6092: Part 1-81. 1981 Sampling and Test for Sodium and Potassium Silicates for Industrial Use Part 1: Determination of Solution Density at 20 Degrees Celsius. 6 pp.
ISO 1687-76. Sodium and Potassium Silicates for Industrial Use—Determination of Density at 20 Degrees Celsius of Products in Solution—Method Using Density Hydrometer and Method Using Pyknometer First Edition. 5 pp.

Soil Analysis
BSI BS 1377: Part 9-90. 1990 Methods of Test for Soils for Civil Engineering Purposes Part 9: In-Situ Tests. 66 pp.
CNS A3089-88. Method of Test for Specific Gravity of Soils (Aug)(5090).
DIN ENGL 18125 Pt 1-86. Soil Analysis; Determination of Soil Density; Laboratory Tests (May). 4 pp.
DIN ENGL 18125 Pt 2-86. Soil Analysis; Determination of Soil Density; Field Tests (May). 9 pp.
DIN ENGL 18126-89. Soil Analysis; Determination of Maximum and Minimum Dry Densities of Non-Cohesive Soils (Sept). 10 pp.

Density (Cont.)

Soil Analysis (Cont.)
DIN ENGL 19683 Pt 11-73. Methods of Soil Analysis for Water Management for Agricultural Purposes; Physical Laboratory Tests; Determination of Density (Apr). 2 pp.
JIS A 1202-90. Test Method for Density of Soil Particles. 7 pp.
JIS A 1214-90. Test Method for Soil Density by the Sand Replacement Method. 12 pp.
SAA AS 1289.E1.1-77. Methods of Testing Soil for Engineering Purposes—Part E1.1: Soil Compaction and Density Tests—Determination of the Dry Density/Moisture Content Relation of a Soil Using Standard Compaction—Standard Method (Superseded by AS 1289.5.1.1—1992).
SAA AS 1289.E1.2-77. Methods of Testing Soil for Engineering Purposes—Part E1.2: Soil Compaction and Density Tests—Determination of the Dry Density/Moisture Content Relation of a Soil Using Standard Compaction—Subsidiary Method. 1 p.
SAA AS 1289.E2.1-77. Methods of Testing Soil for Engineering Purposes—Part E2.1: Soil Compaction and Density Tests—Determination of the Dry Density/Moisture Content Relation of a Soil Using Modified Compaction—Standard Method (Superseded by AS 1289.5.2.1—1993).
SAA AS 1289.E2.2-77. Methods of Testing Soil for Engineering Purposes—Part E2.2: Soil Compaction and Density Tests—Determination of the Dry Density/Moisture Content Relation of a Soil Using Modified Compaction—Subsidiary Method (Superseded by AS 1289.5.2.1—1993).
SAA AS 1289.E3.1-77. Methods of Testing Soil for Engineering Purposes—Part E3.1: Soil Compaction and Density Tests—Determination of the Field Dry Density of a Soil—Sand Replacement Method Using a Sand-Cone Pouring Apparatus Corrig. (Superseded by AS 1289.5.3.1—1993).
SAA AS 1289.E3.2-77. Methods of Testing Soil for Engineering Purposes—Part E3.2: Soil Compaction and Density Tests—Determination of the Field Dry Density of a Soil—Sand Replacement Method Using a Sand Pouring Apparatus (Superseded by AS 1289.5.3.2—1993).
SAA AS 1289.E3.3-77. Methods of Testing Soil for Engineering Purposes—Part E3.3: Soil Compaction and Density Tests—Determination of the Field Dry Density of a Soil—Core Cutter Method for Fine-Grained Soils (Withdrawn).
SAA AS 1289.E3.4-77. Methods of Testing Soil for Engineering Purposes—Part E3.4: Soil Compaction and Density Tests—Determination of the Field Dry Density of a Soil—Balloon Densometer Method (Withdrawn).
SAA AS 1289.E3.5-82. Methods of Testing Soil for Engineering Purposes—Part E3.5: Soil Compaction and Density Tests—Determination of the Field Dry Density of a Soil—Water Replacement Method. 3 pp.
SAA AS 1289.E4.1-82. Methods of Testing Soil for Engineering Purposes—Part E4.1: Soil Compaction and Density Tests—Dry Density Ratio—Normal Method (Superseded by AS 1289.5.4.1—1993).
SAA AS 1289.E5.1-77. Methods of Testing Soil for Engineering Purposes—Part E5.1: Soil Compaction and Density Tests—Determination of Minimum and Maximum Dry Density of a Cohesionless Material. 3 pp.
SAA AS 1289.E6.1-81. Methods of Testing Soil for Engineering Purposes—Part E6.1: Soil Compaction and Density Tests—Compaction Control Test—Density Index Method for a Cohesionless Material. 1 p.
SAA AS 1289.E7.1-83. Methods of Testing Soil for Engineering Purposes—Part E7.1: Soil Compaction and Density Tests—Compaction Control Test—Hilf Density Ratio—Rapid Method (Superseded by AS 1289.5.7.1—1993).
SAA AS 1289.E8.1-84. Methods of Testing Soil for Engineering Purposes—Part E8.1: Soil Compaction and Density Tests—Determination of Field Moisture Content and Field Dry Density of a Soil—Method Using a Nuclear Surface Moisture-Density Guage—Direct Transmission Mode. 2 pp.
SAA AS 1289.E8.2-84. Methods of Testing Soil for Engineering Purposes—Part E8.2: Soil Compaction and Density Tests—Determination of Field Moisture Content and Field Dry Density of a Soil—Method Using a Nuclear Surface Moisture-Density Guage—Backscatter Mode.
SNZ NZS 4402: Pt 2:TEST 2.7.1-86. Methods of Testing Soils for Civil Engineering Purposes Part 2: Soil Classification Tests. Section 2.7: Determination of the Solid Density of Soil Particles. Test 2.7.1: Method for Coarse, Medium and Fine Soils. 4 pp.
SNZ NZS 4402: Pt 2:TEST 2.7.2-86. Methods of Testing Soils for Civil Engineering Purposes Part 2: Soil Classification Tests. Section 2.7: Determination of the Solid Density of Soil Particles. Test 2.7.2: Method for Medium and Fine Soils. 4 pp.

INDUSTRY STANDARDS

Density (Cont.)
—Soil Analysis (Cont.)
SNZ NZS 4402: Pt 4:TEST 4.2.1-88. Methods of Testing Soils for Civil Engineering Purposes Part 4: Soil Compaction Tests. Section 4.2: Determination of the Minimum and Maximum Dry Densities and Relative Density of a Cohesionless Soil. Test 4.2.1: Minimum Dry Density. 3 pp.

SNZ NZS 4402: Pt 4:TEST 4.2.2-88. Methods of Testing Soils for Civil Engineering Purposes Part 4: Soil Compaction Tests. Section 4.2: Determination of the Minimum and Maximum Dry Densities and Relative Density of a Cohesion-less Soil. Test 4.2.2: Maximum Dry Density. 6 pp.

SNZ NZS 4402: Pt 4:TEST 4.2.3-88. Methods of Testing Soils for Civil Engineering Purposes Part 4: Soil Compaction Tests. Section 4.2: Determination of the Minimum and Maximum Dry Densities and Relative Density of a Cohesion-less Soil. Test 4.2.3: Relative Density. 2 pp.

—Soil Analysis—Balloon Densometer
SNZ NZS 4402: Pt 5:TEST 5.1.2-86. Methods of Testing Soils for Civil Engineering Purposes Part 5: Soil Density Tests. Section 5.1: Determination of the Density of Soil. Test 5.1.2: Balloon Densometer Method for the Determination of the In Situ Density. 8 pp.

—Soil Analysis—Calibration
SAA AS 1289.E8.3-84. Methods of Testing Soil for Engineering Purposes—Part E8.3: Soil Compaction and Density Tests—Calibration of Nuclear Surface Moisture-Density Guage for Field Use (Withdrawn). 2 pp.

SAA AS 1289.E8.4-86. Methods of Testing Soil for Engineering Purposes—Part E8.4: Soil Compaction and Density Tests—Calibration of Nuclear Surface Moisture-Density Guage Using Standard Blocks. 5 pp.

—Soil Analysis—Sampling Tube
SNZ NZS 4402: Pt 5:TEST 5.1.3-86. Methods of Testing Soils for Civil Engineering Purposes Part 5: Soil Density Tests. Section 5.1: Determination of the Density of Soil. Test 5.1.3: Sampling Tube Method for the Determination of the In Situ Density. 4 pp.

—Soil Analysis—Sand Replacement Method
SNZ NZS 4402: Pt 5:TEST 5.1.1-86. Methods of Testing Soils for Civil Engineering Purposes Part 5: Soil Density Tests. Section 5.1: Determination of the Density of Soil. Test 5.1.1: Sand Replacement Method for the Determination of In Situ Density. 9 pp.

—Soil Analysis—Water Displacement
SNZ NZS 4402: Pt 5:TEST 5.1.5-86. Methods of Testing Soils for Civil Engineering Purposes Part 5: Soil Density Tests. Section 5.1: Determination of the Density of Soil. Test 5.1.5: Water Displacement Method. 5 pp.

—Soil Analysis—Water Immersion
SNZ NZS 4402: Pt 5:TEST 5.1.4-86. Methods of Testing Soils for Civil Engineering Purposes Part 5: Soil Density Tests. Section 5.1: Determination of the Density of Soil. Test 5.1.4: Immersion in Water Method. 6 pp.

—Soldering Fluxes
MOD UK M 9542/70. Examination of Flux, Soldering (Zinc Chloride Solution).

—Solids
CNS Z8071-88. Method for Measuring Specific Gravity of Solid (Oct)(12451).

JIS Z 8807-76. Measuring Methods for Specific Gravity of Solid.

—Solids—Glossaries
CNS Z7183-84. Definitions of Terms Relating to Density and Specific Gravity of Solids Liquids and Gases (Oct)(11027).

—Soy Sauce
CNS N6008-88. Methods of Test for Soy Sauce (Nov)(955). 7 pp.

—Stabilized Soil Analysis
BSI BS 1924-75. (WITHDRAWN) 1975 Methods of Test for Stabilized Soils (R) (Superseded by BS 1924: Parts 1 & 2: 1990). 98 pp.

BSI BS 1924: Part 2-90. 1990 Stabilized Materials for Civil Engineering Purposes Part 2: Methods of Test for Cement-Stabilized and Lime-Stabilized Materials. 104 pp.

—Steels
DIN ENGL 50463-85. Testing of Steel; Determination of the Density of Iron/Silicon Alloy Magnetic Sheet (July). 4 pp.

Density (Cont.)
—Structural Timber
CEN PREN 384-90. Structural Timber—Determination of Characteristic Valves of Mechanical Properties and Density. 13 pp.

—Sulfuric Acid
BSI BS 753-87. 1987 Density-Composition Tables for Aqueous Solutions of Sulphuric Acid. 67 pp.

—Suspended Sediments
BSI BS 3680: Part 10D-86. 1986 Measurement of Liquid Flow in Open Channels Part 10D: Sediment Transport: Methods for Determination of Concentration, Particle Size Distribution and Relative Density of Sediment in Streams and Canals. 35 pp.

ISO 4365-85. Liquid Flow in Open Channels—Sediment in Streams and Canals—Determination of Concentration, Particle Size Distribution and Relative Density First Edition. 32 pp.

—Tables (Data)
CNS Z7144-83. Pound-Kilogram Conversion Tables (Jan)(9920).

CNS Z7162-83. Pounds per Square Inch-Kilograms per Square Centimetre Conversion Tables (Jun)(10399).

SAA AS 1377.4-73. Conversion Tables—Part 4: Mass, Density, Concentration and Flow Rate (Incorporating Amdt 1). 28 pp.

—Thermal Insulation
CNS K6224-65. Method of Test for Foam Polystyrene Heat Insulating Material (Sep)(2536). (R 1971). 3 pp.

—Titanium Tetrachloride
MOD UK M 888/72. Examination of Titanium Tetrachloride (Withdrawn).

—Toluene
ISO 5281-80. Aromatic Hydrocarbons—Benzene, Xylene and Toluene—Determination of Density at 20 Degrees Celsius First Edition. 11 pp.

—Uranium Dioxide Pellets
ISO 9279-92. Uranium Dioxide Pellets—Determination of Density and Total Porosity—Mercury Displacement Method First Edition. 7 pp.

—Urea Resins
CNS K6272-74. Method of Test for Urea Resin Molding Compounds (Oct)(2986). 9 pp.

—Varnishes
CNS K6056-70. Method of Test for Baking Varnish (Jan)(771). 6 pp.

CNS K6057-70. Method of Test for Clear Varnish for Baking Varnish (Jan)(773). 3 pp.

CNS K6665-81. Method of Test for Density of Paint Varnish, Lacquer and Related Products (May)(7405).

DIN ENGL 53217 Pt 1-91. Determination of Density of Paints, Varnishes and Similar Coating Materials; Survey of Test Methods (Mar). 2 pp.

DIN ENGL 53217 Pt 2-91. Determination of Density of Paints, Varnishes and Similar Coating Materials by the Pyknometer Method (Mar). 5 pp.

DIN ENGL 53217 Pt 3-91. Determination of Density of Paints, Varnishes and Similar Coating Materials by the Displacement Float Method (Mar). 4 pp.

DIN ENGL 53217 Pt 5-91. Determination of Density of Paints, Varnishes and Similar Coating Materials by the Vibration Method (Mar). 4 pp.

ISO 2811-74. Paints and Varnishes—Determination of Density First Edition. 5 pp.

SAA AS 1580.202. 1-80. Paints and Related Materials—Methods of Test—Part 202.1: Density (R 1991). 2 pp.

SNZ NZS/AS 1580. 202.1-80. Methods of Test for Paints and Related Materials Part 202.1: Density (This is a Joint Standard with SAA AS 1580.202.1). 2 pp.

—Vulcanized Rubber
BSI BS 903: Part A1-80. 1980 Amd 1 Methods of Testing Vulcanized Rubber Part A1: Determination of Density. 6 pp.

SAA AS 1683.4-92. Methods of Test for Elastomers—Part 4: Rubber, Vulcanized—Determination of Density (ISO 2781:1988). 3 pp.

—Water—Pycnometric Analysis
CNS K9051-80. Method of Test for Specific Gravity of Water (Pycnometer Method) (Aug)(6233).

—Wood
CNS O2002-81. Method of Test for Specific Gravity of Wood (Mar)(451). 1 pp.

CPPA A.8P-90. Basic Density of Wood. 2 pp.

ISO 3131-75. Wood—Determination of Density for Physical and Mechanical Tests First Edition. 4 pp.

JIS Z 2102-57. Method of Measuring Average Width of Annual Rings, Moisture Content and Specific Gravity of Wood (R 1966). 4 pp.

SAA AS 1080.3-81. Methods of Test for Timber—Part 3: Determination of Density. 4 pp.

Density (Cont.)
—Wood—Chips
CPPA A.1H-64. Basic Density of Wood. 2 pp.

JIS P 8014-76. Testing Method for Density of Pulpwood and Woodchips (R 1985). 8 pp.

SAA AS 1301.P1S-79. Methods of Test for Pulp and Paper (Metric Units)—Part P1s: Basic Density of Wood Chips (This is a Joint Standard with SANZ NZS 1301).

SNZ NZS/AS 1301. P1S-79. Methods of Test for Pulp and Paper Basic Density of Wood Chips (This is a Joint Standard with SAA AS 1301.P1S). 2 pp.

—Wood—Disks
CPPA A.1H-64. Basic Density of Wood. 2 pp.

—Wood Panels
CEN PREN 1058-93. Wood Based Panels—Determination of Characteristic Values of Mechanical Properties and Density. 8 pp.

—Xylenes
ISO 5281-80. Aromatic Hydrocarbons—Benzene, Xylene and Toluene—Determination of Density at 20 Degrees Celsius First Edition. 11 pp.

—Zinc Stearate
CNS K6296-71. Method of Test for Zinc Stearate Powder (Jun)(3215). 2 pp.

Density Bottles
Use: Pycnometers

Density Columns
Use For: Concentration Gradient Density Columns
See Also: Density
—Solids
BSI BS 3715-64. (WITHDRAWN) 1964 Concentration Gradient Density Columns (Superseded by BS 2782: Methods 620A to 620D: 1991). 18 pp.

Density Flasks
Use: Flasks—Density

Density Gages
Use: Densitometers

Density Gradient Columns
Use: Densitometers

Density Hydrometric Analysis
Use: Hydrometry

Density Indicators
Use: Densitometers

Density Modulated Tubes
See Also: Electron Tubes; Microwave Tubes; Power Tetrodes; Power Tubes; Transmitting Tubes; Triodes
CECC CECC 45 000 ISSUE 1-75. Generic Specification: Space-Charge Controlled Tubes (En, Fr, Ge). 34 pp.

Density/Specific Gravity Meters
Use: Hydrometers

Densometers (Gurley)
Use: Gurley Testers

Dental Amalgamators
Use: Amalgamators—Dental

Dental Burrs
Use: Dental Instruments—Burrs

Dental Chairs
See Also: Chairs (Seats); Dental Equipment
BSI BS EN 27493-91. 1991 Dental Operator's Stool (ISO 7493-1985). 12 pp.

CEN EN 27 493-90. Dentistry—Dental Operator's Stool. 7 pp.

CSA CAN/CSA-Z349. 29-35.2-M92. Dental Instruments and Equipment—Volume 1 (CAN/CSA-Z349.29-M92 to CAN/CSA-Z349.35.2 M92); (Gen Instr 1). 183 pp.

ISO 6875-88. Dental Patient Chair First Edition. 13 pp.

ISO 7493-85. Dental Operator's Stool First Edition. 7 pp.

Dental Chisels
Use: Dental Instruments—Chisels

Dental Elevators
See Also: Dental Equipment
CNS T3036-81. Dental Elevators (Jul)(7756).

JIS T 5407-79. Dental Elevators (R 1985). 6 pp.

Dental Engines
See Also: Dental Equipment; Engines

INTERNATIONAL AND NON-U.S. NATIONAL STANDARDS
SUBJECT INDEX
Dental

Dental Engines (Cont.)
—Electric
CNS T3020-80. Electric Dental Engines (Dec)(6724).
JIS T 5109-79. Electric Dental Engines (R 1985). 13 pp.

—Electric—Belts
CNS T3017-80. Belts for Electric Dental Engine (Dec)(6721).
JIS T 5106-76. Belts for Electric Dental Engines (R 1986). 5 pp.

—Electric—Mountings
CNS T3018-80. Stand for Electric Dental Engine (Dec)(6722).
CNS T3019-80. Bracket Arm for Electric Dental Engine (Dec)(6723).
CNS T3020-80. Electric Dental Engines (Dec)(6724).
JIS T 5107-88. Stands for Electric Dental Engine. 5 pp.
JIS T 5108-88. Bracket Arms for Electric Dental Engine. 5 pp.
JIS T 5109-79. Electric Dental Engines (R 1985). 13 pp.

—Electric—Slip Joints
CNS T3016-80. Wrist and Slip-Joint K4 for Electric Dental Engine (Dec)(6720).
JIS T 5105-89. Wrist and Slip-Joint K4 for Electric Dental Engine. 5 pp.

—Electric—Wrists
CNS T3016-80. Wrist and Slip-Joint K4 for Electric Dental Engine (Dec)(6720).
JIS T 5105-89. Wrist and Slip-Joint K4 for Electric Dental Engine. 5 pp.

Dental Equipment
See Also: Amalgamators; Curettes; Dental Chairs; Dental Elevators; Dental Engines; Dental Instruments; Dental Materials; Dental Treatment Units; Dentistry; Dentures; Electrosurgical Equipment; Medical Electrical Equipment; Medical Equipment; Oral Irrigators; Orthodontic Equipment
MOD UK DSTAN 65-5-77. Medical, Surgical and Dental Instruments, Equipment and Supplies Issue 2. 8 pp.
MOD UK DSTAN 65-6-77. Medical and Dental Supply Procedures Issue 3. 5 pp.

—Glossaries
BSI BS 4492-83. (WITHDRAWN) 1983 Dental Terms (Superseded by BS EN 21942-1: 1992 and Subsequent Parts). 118 pp.
CEN EN 21942-1-91. Dental Vocabulary—Part 1: General and Clinical Terms. 17 pp.
CEN EN 21942-2-92. Dental Vocabulary—Part 2: Dental Materials (ISO 1942-2: 1989). 18 pp.
ISO 1942 Pt 1-89. Dental Vocabulary—Part 1: General and Clinical Terms Second Edition; (Amendment 1-1992) (Amendment 2-1992) (Amendment 5-1993). 40 pp.
ISO 1942 Pt 4-89. Dental Vocabulary—Part 4: Dental Equipment Second Edition; (Amendment 1-1992). 20 pp.
ISO 1942 Pt 5-89. Dental Vocabulary—Part 5: Terms Associated with Testing Second Edition. 17 pp.

—Gloves
SAA AS 4011-92. Examination Gloves for General Medical and Dental Use (in Professional Packages 17B, 17G). 10 pp.

—Hypodermic Needles
SAA AS 1264-85. Dental Equipment—Cartridge Hypodermic Needles (Sterile), Single-Use (This is a Joint Standard with SANZ NZS 1264). 17 pp.
SNZ NZS/AS 1264-85. Dental Equipment—Cartridge Hypodermic Needles (Sterile), Single Use (This is a Joint Standard with SAA AS 1264). 17 pp.

—Identification Systems
CSA CAN/CSA-Z349. 29-35.2-M92. Dental Instruments and Equipment—Volume 1 (CAN/CSA-Z349.29-M92 to CAN/CSA-Z349.35.2 M92); (Gen Instr 1). 183 pp.
ISO 4073-80. Dental Equipment—Items of Dental Equipment at the Working Place—Identification System First Edition; (Corrected and Reprinted -1981). 7 pp.

—Interdepartmental Procurement
NATO STANAG 2128 ED 3 AMD 3-82. Medical and Dental Supply Procedures. 6 pp.
NATO STANAG 2128 ED 4 AMD 0-91. Medical and Dental Supply Procedures. 5 pp.

—Radiographic Films
BSI BS 2585-75. 1975 Sizes of Film for Dental Radiography. 7 pp.
BSI BS 6358-83. 1983 Determination of ISO Speed and Average Gradient of Direct-Exposure Medical and Dental Radiographic Film/Process Combinations. 11 pp.

Dental Equipment (Cont.)
—Radiographic Films (Cont.)
ISO 3665-76. Photography—Intra-Oral Dental Radiographic Film—Specification First Edition. 9 pp.
ISO 5799-91. Photography—Direct-Exposing Medical and Dental Radiographic Film/Process Systems—Determination of ISO Speed and ISO Average Gradient Second Edition. 14 pp.
JIS K 7615-85. Determination of ISO Speed and ISO Average Gradient of Direct-Exposure Radiographic Film for Medical and Dental Use.
SAA AS 1139-85. Dental Radiographic Film. 9 pp.

—Spittoons
CNS T3044-81. Dental Spitton (Oct)(8066).
JIS T 5902-79. Dental Spittoon (R 1985). 7 pp.

—Symbols
BSI BS 7665-93. 1993 Graphical Symbols for Dental Equipment (ISO 9687: 1993) (N). 26 pp.
ISO 9687-93. Dental Equipment—Graphical Symbols First Ediiton. 24 pp.

Dental Implants
See Also: Dental Materials; Surgical Implants
ISO TR 11175-93. Dental Implants—Guidelines for Developing Dental Implants First Edition. 6 pp.

Dental Instruments
See Also: Chisels; Cutting Tools; Dental Equipment; Dental Treatment Units; Drills; Medical Electrical Equipment; Medical Equipment; Medical Instruments
BSI BS 6828: Part 2-87. 1987 Amd 1 Dental Rotary Instruments Part 2: Nominal Sizes and Designations (AMD 6499) March 28, 1991. 5 pp.
BSI BS 6828: Part 9-87. 1987 Amd 1 Dental Rotary Instruments Part 9: Methods of Test (AMD 6677) May 31, 1991. 11 pp.
CEN EN 22 157-90. Dentistry Dental Rotary Instruments Nominal Sizes and Designation. 6 pp.
NATO STANAG 2127 ED 3 AMD 2-89. Medical, Surgical and Dental Instruments, Epuipment and Supplies. 14 pp.

—Applicators—Gingival
JIS T 5415-88. Gingival Applicators. 5 pp.

—Broaches
BSI BS 6815: Part 1-93. 1993 Dental Root Canal Instruments Part 1: Specification for Files, Reamers, Barbed Broaches, Rasps, Paste Carriers, Explorers and Cotton Broaches. 24 pp.
CNS T3024-81. Dental Barbed Broaches (Jan)(6964).
CNS T3025-81. Dental Smooth Broaches (Jan)(6965).
ISO 3630 Pt 1-92. Dental Root-Canal Instruments—Part 1: Files, Reamers, Barbed Broaches, Rasps, Paste Carriers, Explorers and Cotton Broaches First Edition. 25 pp.
JIS T 5206-88. Dental Broaches. 7 pp.

—Broaches—Holders
CNS T3038-81. Dental Broach Holders (Jul)(7758).
JIS T 5409-78. Dental Broach Holders (R 1987). 6 pp.

—Burrs
CNS T3021-81. Dental Burs (Jan)(6961).
CSA CAN3-Z349.2-M79. Dental Burs and Cutters—Fitting Dimensions (Adopted ISO 1797-1976). 12 pp.
JIS T 5201-76. Dental Burs (R 1986). 12 pp.

—Burrs—Carbide
BSI BS 6828: Part 3-87. 1987 Dental Rotary Instruments Part 3: Steel and Carbide Burs (ISO 3823/1: 1986). 16 pp.
BSI BS 6828: Part 3-01. 1987 Amd 1 Dental Rotary Instruments Part 3: Specification for Steel and Carbide Burs (ISO 3823/1: 1986) (AMD 7022) May 1, 1992. 21 pp.
BSI BS 6828: Part 4-87. 1987 Dental Rotary Instruments Part 4: Steel and Carbide Finishing Burs (ISO 3823/2: 1986). 17 pp.
BSI BS 6828: Part 4-01. 1987 Amd 1 Dental Rotary Instruments Part 4: Specification for Steel and Carbide Finishing Burs (ISO 3823/2: 1986) (AMD 7020) May 1, 1992. 23 pp.
CEN PREN 23 823-1-90. Dental Rotary Instruments—Part 1: Steel and Carbide Burs. 3 pp.
CEN EN 23823-1-91. Dental Rotary Instruments—Part 1: Steel and Carbide Burs. 2 pp.
CEN PREN 23 823-2-90. Dental Rotary Instruments—Part 2: Steel and Carbide Finishing Burs. 3 pp.
CEN EN 23823-2-91. Dental Rotary Instruments—Part 2: Steel and Carbide Finishing Burs. 3 pp.
ISO 3823 Pt 1-86. Dental Rotary Instruments—Part 1: Steel and Carbide Burs First Edition. 15 pp.
ISO 3823 Pt 2-86. Dental Rotary Instruments—Part 2: Steel and Carbide Finishing Burs First Edition. 16 pp.

Dental Instruments (Cont.)
—Burrs—Steel
BSI BS 6828: Part 3-87. 1987 Dental Rotary Instruments Part 3: Steel and Carbide Burs (ISO 3823/1: 1986). 16 pp.
BSI BS 6828: Part 3-01. 1987 Amd 1 Dental Rotary Instruments Part 3: Specification for Steel and Carbide Burs (ISO 3823/1: 1986) (AMD 7022) May 1, 1992. 21 pp.
BSI BS 6828: Part 4-87. 1987 Dental Rotary Instruments Part 4: Steel and Carbide Finishing Burs (ISO 3823/2: 1986). 17 pp.
BSI BS 6828: Part 4-01. 1987 Amd 1 Dental Rotary Instruments Part 4: Specification for Steel and Carbide Finishing Burs (ISO 3823/2: 1986) (AMD 7020) May 1, 1992. 23 pp.
CEN PREN 23 823-1-90. Dental Rotary Instruments—Part 1: Steel and Carbide Burs. 3 pp.
CEN EN 23823-1-91. Dental Rotary Instruments—Part 1: Steel and Carbide Burs. 2 pp.
CEN PREN 23 823-2-90. Dental Rotary Instruments—Part 2: Steel and Carbide Finishing Burs. 3 pp.
CEN EN 23823-2-91. Dental Rotary Instruments—Part 2: Steel and Carbide Finishing Burs. 3 pp.
ISO 3823 Pt 1-86. Dental Rotary Instruments—Part 1: Steel and Carbide Burs First Edition. 15 pp.
ISO 3823 Pt 2-86. Dental Rotary Instruments—Part 2: Steel and Carbide Finishing Burs First Edition. 16 pp.

—Chisels
BSI BS 2965-70. 1970 Amd 1 Dental Chisels, Excavators, Probes and Scalers. 45 pp.
CNS T3032-81. Dental Chisel (May)(7453).
JIS T 5403-79. Dental Chisel (R 1985). 5 pp.
SAA AS 1086-71. Dental Chisels, Excavators, Probes and Scalers Being BS 2965:1970 (Including Amd 799) Endorsed, Subject to Australian Amendment 1 Amdt 1 May 1971 Reconfirmed 1986. 40 pp.

—Cutting Tools
CEN EN 28 325-90. Dentistry—Dental Rotary Instruments—Test Methods. 4 pp.
CSA CAN3-Z349.2-M79. Dental Burs and Cutters—Fitting Dimensions (Adopted ISO 1797-1976). 12 pp.
ISO 2157-92. Dental Rotary Instruments—Nominal Diameters and Designation Code Number Third Edition. 7 pp.
ISO 8325-85. Dental Rotary Instruments—Test Methods First Edition. 7 pp.

—Cutting Tools—Carbide
BSI BS 6828: Sec 8.2-87. 1987 Dental Rotary Instruments Part 8: Cutters Section 8.2: Carbide Laboratory Cutters. 9 pp.
BSI BS 6828: Sec 8.2-01. 1987 Amd 1 Dental Rotary Instruments Part 8: Cutters Section 8.2: Specification for Carbide Laboratory Cutters (AMD 6346) September 30, 1991. 14 pp.
BSI BS 6828: Sec 8.3-92. 1992 Dental Rotary Instruments Part 8: Cutters Section 8.3: Specification for Carbide Laboratory Cutters for Milling Machines (ISO 7787-3: 1991). 10 pp.
CEN EN 27 787 (Part 2)-90. Dentistry—Dental Rotary Instruments—Cutters Part 2: Carbide Laboratory Cutters. 4 pp.
CSA CAN/CSA-Z349. 29-35.2-M92. Dental Instruments and Equipment—Volume 1 (CAN/CSA-Z349.29-M92 to CAN/CSA-Z349.35.2 M92); (Gen Instr 1). 183 pp.
ISO 7787 Pt 2-92. Dental Rotary Instruments—Cutters—Part 2: Carbide Laboratory Cutters Second Edition. 14 pp.
ISO 7787 Pt 3-91. Dental Rotary Instruments—Cutters—Part 3: Carbide Laboratory Cutters for Milling Machines First Edition. 7 pp.

—Cutting Tools—Diamond
BSI BS 6828: Part 6-87. 1987 Amd 1 Dental Rotary Instruments Part 6: Specification for Diamond Inmstruments (AMD 6675) May 31, 1991. 26 pp.
CEN EN 27 711-90. Dentistry—Dental Rotary Instruments—Diamond Instruments. 4 pp.
CSA CAN/CSA-Z349. 29-35.2-M92. Dental Instruments and Equipment—Volume 1 (CAN/CSA-Z349.29-M92 to CAN/CSA-Z349.35.2 M92); (Gen Instr 1). 183 pp.
ISO 7711-84. Dental Rotary Instruments—Diamond Instruments First Edition. 22 pp.

—Cutting Tools—Diamond—Color Coding
BSI BS 6828: Sec 6.3-93. 1993 Dental Rotary Instruments Part 6: Diamond Instruments Section 6.3: Specification for Grit Sizes, Designation and Colour Code (ISO 7711-3: 1992). 7 pp.
ISO 7711 Pt 3-92. Dental Rotary Instruments—Diamond Instruments—Part 3: Grit sizes, Designation and Colour Code First Edition. 5 pp.

—Cutting Tools—Diamond—Designations
BSI BS 6828: Sec 6.3-93. 1993 Dental Rotary Instruments Part 6: Diamond Instruments Section 6.3: Specification for Grit Sizes, Designation and Colour Code (ISO 7711-3: 1992). 7 pp.

INDUSTRY STANDARDS

Dental Instruments (Cont.)
—Cutting Tools—Diamond—Designations (Cont.)
ISO 7711 Pt 3-92. Dental Rotary Instruments—Diamond Instruments—Part 3: Grit sizes, Designation and Colour Code First Edition. 5 pp.

—Cutting Tools—Diamond—Sizes
BSI BS 6828: Sec 6.3-93. 1993 Dental Rotary Instruments Part 6: Diamond Instruments Section 6.3: Specification for Grit Sizes, Designation and Colour Code (ISO 7711-3: 1992). 7 pp.
ISO 7711 Pt 3-92. Dental Rotary Instruments—Diamond Instruments—Part 3: Grit sizes, Designation and Colour Code First Edition. 5 pp.

—Cutting Tools—Disks—Bore Diameters
ISO 10323-91. Dental Rotary Instruments—Bore Diameters for Discs and Wheels First Edition. 4 pp.

—Cutting Tools—Disks—Diamond
BSI BS 6828: Sec 6.2-93. 1993 Dental Rotary Instruments Part 6: Diamond Instruments Section 6.2: Specification for Discs (ISO 7711-2: 1992). 12 pp.
ISO 7711 Pt 2-92. Dental Rotary Instruments—Diamond Instruments—Part 2: Discs First Edition. 10 pp.

—Cutting Tools—Identification Systems
BSI BS 6828: Sec 5.1-87. (WITHDRAWN) 1987 Dental Rotary Instruments Part 5: Number Coding System Section 5.1: Specification for General Characteristics (ISO 6360/1-1985) (N). 24 pp.
BSI BS 6828: Sec 5.1-01. 1987 Amd 1 Dental Rotary Instruments Part 5: Number Coding System Section 5.1: Specification for General Characteristics (ISO 6360/1-1985) (AMD 6672) May 31, 1991. 29 pp.
BSI BS 6828: Sec 5.2-87. 1987 Dental Rotary Instruments Part 5: Number Coding System Section 5.2: Shape and Specific Characteristics. 81 pp.
CEN EN 26 360 (Part 1)-90. Dentistry—Dental Rotary Instruments—Number Coding System—Part 1: General Characteristics. 4 pp.
CEN EN 26 360 (Part 2)-91. Dental Rotary Instruments, Number Coding System—Part 2: Shape and Specific Characteristics—Addendum 1. 3 pp.
CSA CAN/CSA-Z349. 29-35.2-M92. Dental Instruments and Equipment—Volume 1 (CAN/CSA-Z349.29-M92 to CAN/CSA-Z349.35.2 M92); (Gen Instr 1). 183 pp.
ISO 6360 Pt 1-85. Dental Rotary Instruments—Number Coding System—Part 1: General Characteristics First Edition; (Addendum 1-1988). 26 pp.
ISO 6360 Pt 2-86. Dental Rotary Instruments—Number Coding System Shape and Specific Characteristics First Edition; (Amendment 1-1991). 106 pp.
ISO 8170-85. Dental Hand Instruments—Hand-Held Cutting Instruments for Restorative Dentistry—Designation and Marking of Dimensional Characteristics First Edition. 10 pp.

—Cutting Tools—Shanks
BSI BS 6828: Part 1-87. 1987 Amd 1 Dental Rotary Instruments Part 1: Shanks (AMD 6498) November 30, 1990. 13 pp.
CEN EN 21 797-89. Dentistry, Dental Rotary Instruments; Shanks. 6 pp.

—Cutting Tools—Silicone Carbide Points
CNS T3029-81. Dental Carborundum Points (Mar)(7086).
JIS T 5210-85. Dental Carborundum Points. 6 pp.

—Cutting Tools—Steel
BSI BS 6828: Sec 8.1-87. 1987 Amd 1 Dental Rotary Instruments Part 8: Cutters Section 8.1: Specification for Steel Laboratory Cutters (AMD 6676) May 31, 1991. 12 pp.
CEN EN 27 787 (Part 1)-90. Dentistry—Dental Rotary Instruments—Cutters—Part 1: Steel Laboratory Cutters. 4 pp.
CSA CAN/CSA-Z349. 29-35.2-M92. Dental Instruments and Equipment—Volume 1 (CAN/CSA-Z349.29-M92 to CAN/CSA-Z349.35.2 M92); (Gen Instr 1). 183 pp.
ISO 7787 Pt 1-84. Dental Rotary Instruments—Cutters Part 1: Steel Laboratory Cutters First Edition. 8 pp.

—Cutting Tools—Wheels
CNS T3028-81. Dental Carborundum Wheels (Mar)(7085).
JIS T 5209-85. Dental Carborundum Wheels. 7 pp.

—Cutting Tools—Wheels—Bore Diameters
ISO 10323-91. Dental Rotary Instruments—Bore Diameters for Discs and Wheels First Edition. 4 pp.

—Enlargers
BSI BS EN 23630-2-92. 1992 Dental Root Canal Instruments Part 2: Specification for Enlargers (ISO 3630-2: 1986). 12 pp.

Dental Instruments (Cont.)
—Enlargers (Cont.)
CEN EN 23630-2-91. Dental Root Canal Instruments—Part 2: Enlargers. 3 pp.
ISO 3630 Pt 2-86. Dental Root Canal Instruments—Part 2: Enlargers First Edition. 7 pp.

—Excavators
BSI BS 2965-70. 1970 Amd 1 Dental Chisels, Excavators, Probes and Scalers. 45 pp.
CNS T3033-81. Dental Spoon Excavators (May)(7454).
JIS T 5404-79. Dental Spoon Excavators (R 1985). 5 pp.
SAA AS 1086-71. Dental Chisels, Excavators, Probes and Scalers Being BS 2965:1970 (Including Amd 799) Endorsed, Subject to Australian Amendment 1 Amdt 1 May 1971 Reconfirmed 1986. 40 pp.

—Explorers
BSI BS 6815: Part 1-93. 1993 Dental Root Canal Instruments Part 1: Specification for Files, Reamers, Barbed Broaches, Rasps, Paste Carriers, Explorers and Cotton Broaches. 24 pp.
ISO 3630 Pt 1-92. Dental Root-Canal Instruments—Part 1: Files, Reamers, Barbed Broaches, Rasps, Paste Carriers, Explorers and Cotton Broaches First Edition. 25 pp.

—Files
BSI BS 6815: Part 1-93. 1993 Dental Root Canal Instruments Part 1: Specification for Files, Reamers, Barbed Broaches, Rasps, Paste Carriers, Explorers and Cotton Broaches. 24 pp.
BSI BS 6828: Part 7-87. (WITHDRAWN) 1987 Dental Rotary Instruments Part 7: Laboratory Abrasive Instruments (Superseded by BS EN 27786: 1992). 7 pp.
BSI BS EN 27786-92. 1992 Dental Rotary Instruments—Laboratory Abrasive Instruments (ISO 7786: 1990). 14 pp.
CEN EN 27786-91. Dental Rotary Instruments—Laboratory Abrasive Instruments. 3 pp.
ISO 3630 Pt 1-92. Dental Root-Canal Instruments—Part 1: Files, Reamers, Barbed Broaches, Rasps, Paste Carriers, Explorers and Cotton Broaches First Edition. 25 pp.
ISO 7786-90. Dental Rotary Instruments—Laboratory Abrasive Instruments Second Edition; (Incorporating Draft Addendum 1-1988). 11 pp.

—Files—Bone
CNS T3037-81. Dental Bone Files (Jul)(7757).
JIS T 4903-85. Bone Files.
JIS T 5408-89. Dental Bone Files. 6 pp.

—Filling—Plastic
CNS T3034-81. Dental Plastic Filling Instruments (May)(7455).
JIS T 5405-79. Dental Plastic Filling Instruments (R 1985). 5 pp.

—Forceps
BSI BS 4750-71. (WITHDRAWN) 1971 Dental Extracting Forceps (Performance Requirements) (Superseded by BS 7549: Part 1: 1992). 8 pp.
BSI BS 5211-75. (WITHDRAWN) 1973 Popular Patterns of Dental Extracting Forceps (Superseded by BS 7549: Part 1: 1992). 29 pp.
BSI BS 7549: Part 1-92. 1992 Dental Extraction Forceps Part 1: Specification for Screw and Pin Joint Types (ISO 9173-1: 1991). 24 pp.
CNS T3030-81. Dental Pincettes (Mar)(7087).
CNS T3039-81. Dental Forceps (Nov)(8061).
ISO 9173 Pt 1-91. Dental Extraction Forceps Part 1: Screw and Pin Joint Types First Edition. 23 pp.
JIS T 5401-84. Dental Pincettes. 7 pp.
JIS T 5410-88. Dental Forceps. 5 pp.

—Glossaries
ISO 1942 Pt 3-89. Dental Vocabulary—Part 3: Dental Instruments Second Edition; (Amendment 1-2:1992). 42 pp.
ISO 1942 Pt 5-89. Dental Vocabulary—Part 5: Terms Associated with Testing Second Edition. 17 pp.

—Hand Pieces
BSI BS 7077: Sec 1.2-92. 1992 Dental Handpieces and Accessories Part 1: Dental Handpieces Section 1.2: Specification for Straight and Geared Angle Handpieces (ISO 7785-2: 1991). 12 pp.
CNS T3043-81. Dimensions for Dental Hand-Pieces (Oct)(8065). 3 pp.
ISO 7785 Pt 2-91. Dental Handpieces—Part 2: Straight and Geared Angle Handpieces First Edition. 9 pp.
JIS T 5901-89. Dimension for Dental Hand-Piece. 6 pp.

—Hand Pieces—Couplings
BSI BS 4222-67. 1967 Dimensions of the Transmission Coupling (Slip-Joint) to Dental Handpieces. 8 pp.
BSI BS 7077: Sec 2.2-89. 1989 Dental Handpieces and Accessories Part 2: Accessories Section 2.2: Specification for Coupling Dimensions (ISO 3964: 1982). 4 pp.

Dental Instruments (Cont.)
—Hand Pieces—Couplings (Cont.)
BSI BS 7077: Sec 2.2-01. 1989 Amd 1 Dental Handpieces and Accessories Part 2: Accessories Section 2.2: Specification for Coupling Dimensions (ISO 3964: 1982) (AMD 6497) April 30, 1991. 7 pp.
CEN EN 23 964-89. Dentistry: Dental Handpieces; Coupling Dimensions. 6 pp.
CSA CAN/CSA-Z349. 29-35.2-M92. Dental Instruments and Equipment—Volume 1 (CAN/CSA-Z349.29-M92 to CAN/CSA-Z349.35.2 M92); (Gen Instr 1). 183 pp.
ISO 3964-82. Dental Handpieces—Coupling Dimensions Second Edition. 4 pp.

—Hand Pieces—Fittings
BSI BS 7077: Sec 2.1-89. 1989 Dental Handpieces and Accessories Part 2: Accessories Section 2.1: Specification for Hose Connectors. 7 pp.
BSI BS 7077: Sec 2.1-01. 1989 Amd 1 Dental Handpieces and Accessories Part 2: Accessories Section 2.1: Specification for Hose Connectors (AMD 6679) May 31, 1991. 10 pp.
CEN EN 29 168-90. Dentistry—Dental Handpieces—Hose Connectors. 4 pp.
ISO 9168-91. Dental Handpieces—Hose Connectors Second Edition; (Incorporating Draft Addendum 1-1989). 9 pp.

—Hand Pieces—High Speed
BSI BS 7077: Sec 1.1-93. 1993 Dental Handpieces and Accessories Part 1: Handpieces Section 1.1: Specification for High-Speed Air-Turbine Handpieces (ISO 7785-1: 1992). 12 pp.
ISO 7785 Pt 1-92. Dental Handpieces—Part 1: High-Speed Air-Turbine Handpieces First Edition. 9 pp.

—Hand Pieces—Symbols
BSI BS 7665-93. 1993 Graphical Symbols for Dental Equipment (ISO 9687: 1993) (N). 26 pp.
ISO 9687-93. Dental Equipment—Graphical Symbols First Ediiton. 24 pp.

—Identification Systems
CSA CAN/CSA-Z349. 29-35.2-M92. Dental Instruments and Equipment—Volume 1 (CAN/CSA-Z349.29-M92 to CAN/CSA-Z349.35.2 M92); (Gen Instr 1). 183 pp.
ISO 4073-80. Dental Equipment—Items of Dental Equipment at the Working Place—Identification System First Edition; (Corrected and Reprinted -1981). 7 pp.
ISO 7494-90. Dental Unit First Edition. 13 pp.

—Knives—Surgical
JIS T 5412-88. Dental Operating Knives. 6 pp.

—Mandrels
CNS T3023-81. Dental Mandrel (Jan)(6963).
JIS T 5204-84. Dental Mandrels. 6 pp.

—Mirrors
BSI BS 4681-91. 1991 Reusable Metal Dental Mirrors and Handles (ISO 9873: 1990). 12 pp.
CNS T3045-81. Mouth Mirror and Mouth Mirror Holder (Oct)(8067).
ISO 9873-90. Reusable Metal Dental Mirrors and Handles First Edition. 10 pp.
JIS T 5903-84. Dental Mirrors and Dental Mirror Holders. 6 pp.

—Paste Carriers
BSI BS 6815: Part 1-93. 1993 Dental Root Canal Instruments Part 1: Specification for Files, Reamers, Barbed Broaches, Rasps, Paste Carriers, Explorers and Cotton Broaches. 24 pp.
ISO 3630 Pt 1-92. Dental Root-Canal Instruments—Part 1: Files, Reamers, Barbed Broaches, Rasps, Paste Carriers, Explorers and Cotton Broaches First Edition. 25 pp.

—Probes
BSI BS 2965-70. 1970 Amd 1 Dental Chisels, Excavators, Probes and Scalers. 45 pp.
CNS T3031-81. Dental Explorers (May)(7452).
CSA CAN/CSA-Z349. 29-35.2-M92. Dental Instruments and Equipment—Volume 1 (CAN/CSA-Z349.29-M92 to CAN/CSA-Z349.35.2 M92); (Gen Instr 1). 183 pp.
ISO 7492-83. Dental Explorers First Edition. 5 pp.
JIS T 5402-79. Dental Explorers (R 1985). 5 pp.
SAA AS 1086-71. Dental Chisels, Excavators, Probes and Scalers Being BS 2965:1970 (Including Amd 799) Endorsed, Subject to Australian Amendment 1 Amdt 1 May 1971 Reconfirmed 1986. 40 pp.

—Punches (Tools)
SAA AS 1283-74. Dental Rubber Dam Punch Reconfirmed 1988. 6 pp.

INTERNATIONAL AND NON-U.S. NATIONAL STANDARDS
SUBJECT INDEX

Dental

Dental Instruments *(Cont.)*

—Rasps
BSI BS 6815: Part 1-93. 1993 Dental Root Canal Instruments Part 1: Specification for Files, Reamers, Barbed Broaches, Rasps, Paste Carriers, Explorers and Cotton Broaches. 24 pp.
ISO 3630 Pt 1-92. Dental Root-Canal Instruments—Part 1: Files, Reamers, Barbed Broaches, Rasps, Paste Carriers, Explorers and Cotton Broaches First Edition. 25 pp.

—Reamers
BSI BS 6815: Part 1-93. 1993 Dental Root Canal Instruments Part 1: Specification for Files, Reamers, Barbed Broaches, Rasps, Paste Carriers, Explorers and Cotton Broaches. 24 pp.
CNS T3022-81. Pulp Canal Reamers (Jan)(6962).
CNS T3026-81. Pulp Canal Reamers for Contra-Angle Hand Piece (Mar)(7083).
CNS T3027-81. Peeso's Root Reamers for Dental Use (Mar)(7084).
ISO 3630 Pt 1-92. Dental Root-Canal Instruments—Part 1: Files, Reamers, Barbed Broaches, Rasps, Paste Carriers, Explorers and Cotton Broaches First Edition. 25 pp.
JIS T 5202-79. Pulp Canal Reamers (R 1985). 7 pp.
JIS T 5207-79. Pulp Canal Reamers for Angle Hand-Piece (R 1985). 9 pp.
JIS T 5208-79. Peeso's Root Reamers for Dental Use (R 1985). 7 pp.

—Root Canal
BSI BS 6815-87. 1987 Dental Root Canal Instruments. 19 pp.
BSI BS EN 23630-2-92. 1992 Dental Root Canal Instruments Part 2: Specification for Enlargers (ISO 3630-2: 1986). 12 pp.
CEN EN 23630-2-91. Dental Root Canal Instruments—Part 2: Enlargers. 3 pp.
CNS T3022-81. Pulp Canal Reamers (Jan)(6962).
CNS T3026-81. Pulp Canal Reamers for Contra-Angle Hand Piece (Mar)(7083).
CNS T3027-81. Peeso's Root Reamers for Dental Use (Mar)(7084).
ISO 3630 Pt 2-86. Dental Root Canal Instruments—Part 2: Enlargers First Edition. 7 pp.
JIS T 5202-79. Pulp Canal Reamers (R 1985). 7 pp.
JIS T 5207-79. Pulp Canal Reamers for Angle Hand-Piece (R 1985). 9 pp.
JIS T 5208-79. Peeso's Root Reamers for Dental Use (R 1985). 7 pp.

—Root Canal—Designations
BSI BS 6815: Part 1-93. 1993 Dental Root Canal Instruments Part 1: Specification for Files, Reamers, Barbed Broaches, Rasps, Paste Carriers, Explorers and Cotton Broaches. 24 pp.
ISO 3630 Pt 1-92. Dental Root-Canal Instruments—Part 1: Files, Reamers, Barbed Broaches, Rasps, Paste Carriers, Explorers and Cotton Broaches First Edition. 25 pp.

—Scalers
BSI BS 2965-70. 1970 Amd 1 Dental Chisels, Excavators, Probes and Scalers. 45 pp.
CNS T3035-81. Dental Scalers (May)(7456).
JIS T 5206-88. Dental Broaches. 7 pp.
JIS T 5406-79. Dental Scalers (R 1985). 5 pp.
SAA AS 1086-71. Dental Chisels, Excavators, Probes and Scalers Being BS 2965:1970 (Including Amd 799) Endorsed, Subject to Australian Amendment 1 Amdt 1 May 1971 Reconfirmed 1986. 40 pp.

—Scissors—Crown
CNS T3040-81. Dental Crown Scissors (Oct)(8062).
JIS T 5411-88. Dental Crown Scissors. 5 pp.

—Shanks
ISO 1797 Pt 1-92. Dental Rotary Instruments—Shanks—Part 1: Shanks Made of Metals First Edition. 10 pp.
ISO 1797 Pt 2-92. Dental Rotary Instruments—Shanks—Part 2: Shanks Made of Plastics First Edition. 8 pp.

—Shanks—Quality Assurance
ISO 1797 Pt 1-92. Dental Rotary Instruments—Shanks—Part 1: Shanks Made of Metals First Edition. 10 pp.
ISO 1797 Pt 2-92. Dental Rotary Instruments—Shanks—Part 2: Shanks Made of Plastics First Edition. 8 pp.

—Spatulas—Wax
CNS T3042-81. Dental Wax Spatula (Oct)(8064).
JIS T 5414-88. Dental Wax Spatulas. 4 pp.

—Symbols
BSI BS 7665-93. 1993 Graphical Symbols for Dental Equipment (ISO 9687: 1993) (N). 26 pp.
ISO 9687-93. Dental Equipment—Graphical Symbols First Ediiton. 24 pp.

—Tweezers
BSI BS 6136-81. 1981 Dental Tweezers. 11 pp.

Dental Materials
See Also: Dental Equipment; Dental Implants; Dentistry; Dentures; Orthodontic Equipment; Toothpastes
BSI PD 6502-82. (WITHDRAWN) 1982 Guide to the Use of Dental Materials. 42 pp.

—Agar
BSI BS 5794-79. (WITHDRAWN) 1979 Dental Agar Impression Material (Superseded by BS EN 21 564: 1991). 9 pp.
BSI BS EN 21 564-91. 1991 Dental Agar Impression Material (N). 12 pp.
CEN EN 21 564-90. Dentistry—Agar Impression Material. 9 pp.
CNS T3059-86. Agar Impression Material (Aug)(11687).
ISO 1564-76. Agar Impression Material First Edition. 8 pp.
JIS T 6512-90. Dental Agar Impression Material. 13 pp.

—Alginates
BSI BS 4269: Part 2-91. 1991 Dental Elastic Impression Material Part 2: Specification for Alginate Impression Material (ISO 1563: 1990). 16 pp.
BSI BS 4269: Part 2-01. 1991 Amd 1 Dental Elastic Impression Material Part 2: Specification for Alginate Impression Material (ISO 1563: 1990) (AMD 7037) May 1, 1992. 21 pp.
CEN EN 21563-91. Alginate Dental Impression Material. 3 pp.
CNS T3058-86. Alginate Dental Impression Material (Aug)(11686).
CSA CAN3-Z349.6-M82. Alginate Dental Impression Material; (Gen Instr 1). 23 pp.
CSA CAN/CSA-Z349.38-93. Dental Alginate Impression Material (ISO 1563-1990); (Gen Instr 1). 39 pp.
ISO 1563-90. Dental Alginate Impression Material Second Edition. 15 pp.
JIS T 6505-89. Dental Alginate Impression Material. 11 pp.

—Alloys—Corrosion
ISO TR 10271-93. Dentistry—Determination of Tarnish and Corrosion of Metals and Alloys First Edition. 13 pp.

—Alloys—Tarnish
ISO TR 10271-93. Dentistry—Determination of Tarnish and Corrosion of Metals and Alloys First Edition. 13 pp.

—Amalgams
BSI BS 2938-85. (WITHDRAWN) 1985 Dental Amalgam Alloy (Superseded by BS EN 21559: 1992). 10 pp.
BSI BS 4227-86. 1986 Amd 1 Specification for Dental Mercury (AMD 6327) June 28, 1991. 5 pp.
BSI BS 4227-02. 1986 Amd 2 Dental Mercury (ISO 1560: 1985) (AMD 7023) May 1, 1992. 10 pp.
BSI BS EN 21559-92. 1992 Alloys for Dental Amalgam (ISO 1559: 1986) (Supersedes BS 2938: 1985). 11 pp.
CEN EN 21559-91. Dentistry—Alloys for Dental Amalgam. 7 pp.
CEN EN 21560-91. Dentistry—Dental Mercury. 3 pp.
CNS T3009-80. Dental Silver Amalgam Alloy (Oct)(6619).
CNS T3012-80. Dental Mercuries (Oct)(6622). 1 p.
ISO 1559-86. Dentistry—Alloys for Dental Amalgam Second Edition. 6 pp.
ISO 1560-85. Dental Mercury Second Edition. 4 pp.
JIS T 6109-91. Alloys for Dental Amalgam. 9 pp.
JIS T 6112-89. Dental Mercury. 5 pp.

—Brazing Alloys
BSI BS EN 29333-92. 1992 Dental Brazing Materials (ISO 9333: 1990). 11 pp.
CEN PREN 29 333-90. Dental Brazing Materials. 3 pp.
CEN EN 29333-91. Dental Brazing Materials. 6 pp.
CSA CAN/CSA-Z349.45-93. Dental Brazing Materials (ISO 9333-1990); (Gen Instr 1). 31 pp.
ISO 9333-90. Dental Brazing Materials First Edition. 7 pp.

—Cements—Glass Ionomer
BSI BS 6039-81. 1981 Dental Glass Ionomer Cements. 12 pp.

—Cements—Glass Polyalkenoate
CSA CAN/CSA-Z349.47-93. Dental Water-Based Cements (ISO 9917-1991); (Gen Instr 1). 40 pp.
ISO 9917-91. Dental Water-Based Cements First Edition; (Corrigendum 1-1993). 18 pp.
JIS T 6607-92. Dental Glass Polyalkenoate Cement.

—Cements—Silicate
BSI BS 3365: Part 1-80. 1980 Dental Silicate Cement and Dental Silico-Phosphate Cement Part 1: Dental Silicate Cement (Hand-Mixed). 13 pp.

Dental Materials *(Cont.)*

—Cements—Silicate *(Cont.)*
BSI BS 3365: Part 3-81. 1981 Dental Silicate Cement and Dental Silico-Phosphate Cement Part 3: Capsulated Cements. 16 pp.
CNS T3053-82. Dental Silicate Cement (May)(8870).
CSA CAN/CSA-Z349.47-93. Dental Water-Based Cements (ISO 9917-1991); (Gen Instr 1). 40 pp.
ISO 9917-91. Dental Water-Based Cements First Edition; (Corrigendum 1-1993). 18 pp.
JIS T 6603-87. Dental Silicate Cement. 7 pp.

—Cements—Silicophosphate
BSI BS 3365: Part 2-87. 1987 Dental Silicate Cement and Dental Silico-Phosphate Cement Part 2: Dental Silico-Phosphate Cements (Hand Mixed). 16 pp.
BSI BS 3365: Part 3-81. 1981 Dental Silicate Cement and Dental Silico-Phosphate Cement Part 3: Capsulated Cements. 16 pp.
CSA CAN/CSA-Z349.47-93. Dental Water-Based Cements (ISO 9917-1991); (Gen Instr 1). 40 pp.
ISO 9917-91. Dental Water-Based Cements First Edition; (Corrigendum 1-1993). 18 pp.

—Cements—Zinc Oxide/Eugenol
BSI BS 7214-89. 1989 Dental Zinc Oxide/Eugenol Cements and Zinc Oxide Non-Eugenol Cements (ISO 3107: 1988). 14 pp.
BSI BS 7214-01. 1989 Amd 1 Dental Zinc Oxide/Eugenol Cements and Zinc Oxide Non-Eugenol Cements (ISO 3107: 1988) (AMD 7039) May 1, 1992. 19 pp.
CEN EN 23107-91. Dental Zinc Oxide/Eugenol Cements and Zinc Oxide Non-Eugenol Cements. 3 pp.
ISO 3107-88. Dental Zinc Oxide/Eugenol Cements and Zinc Oxide Non-Eugenol Cements Second Edition; (Supersedes 3106). 13 pp.

—Cements—Zinc Phosphate
BSI BS 3364-80. 1980 Dental Zinc Phosphate Cement. 14 pp.
CNS T3052-82. Dental Zinc Phosphate Cement (May)(8869).
CSA CAN/CSA-Z349.47-93. Dental Water-Based Cements (ISO 9917-1991); (Gen Instr 1). 40 pp.
ISO 9917-91. Dental Water-Based Cements First Edition; (Corrigendum 1-1993). 18 pp.
JIS T 6602-85. Dental Zinc Phosphate Cement. 9 pp.

—Cements—Zinc Polycarboxylate
BSI BS 6814-87. 1987 Dental Zinc Polycarboxylate Cements. 12 pp.
BSI BS 6814-01. 1987 Amd 1 Dental Zinc Polycarboxylate Cements (ISO 4104: 1984) (AMD 6344) February 28, 1992. 17 pp.
CSA CAN/CSA-Z349.47-93. Dental Water-Based Cements (ISO 9917-1991); (Gen Instr 1). 40 pp.
ISO 9917-91. Dental Water-Based Cements First Edition; (Corrigendum 1-1993). 18 pp.
JIS T 6606-90. Dental Zinc Polycarboxylate Cements. 9 pp.

—Ceramic
ISO 6872-84. Dental Ceramic First Edition. 16 pp.

—Ceramic—Restorations—Casting
CSA CAN/CSA-Z349.46-93. Dental Ceramic Fused to Metal Restorative Materials (ISO 9693-1991); (Gen Instr 1). 33 pp.
ISO 9693-91. Dental Ceramic Fused to Metal Restorative Materials First Edition. 10 pp.

—Cleaning Agents—Pulp Canal
JIS T 5205-88. Dental Pulp Canal Cleansers. 7 pp.

—Cobalt Alloy—Wire
CNS T3004-80. Cobalt-Chromium Alloy Wires for Dental Use (Aug)(6398).
JIS T 6104-87. Cobalt-Chromium Alloy Wires for Dental Use. 6 pp.

—Cobalt Chromium Alloy—Casting
CNS T3015-80. Dental Casting Cobalt Chromium Alloy (Oct)(6625).
JIS T 6115-85. Dental Casting Cobalt Chromium Alloy.

—Copper Alloy
BSI BS EN 21559-92. 1992 Alloys for Dental Amalgam (ISO 1559: 1986) (Supersedes BS 2938: 1985). 11 pp.
CEN EN 21559-91. Dentistry—Alloys for Dental Amalgam. 7 pp.
ISO 1559-86. Dentistry—Alloys for Dental Amalgam Second Edition. 6 pp.

—Denture Linings
BSI BS 7589: Part 1-92. 1992 Resilient Lining Materials for Removable Dentures Part 1: Specification for Short-Term Materials (ISO 10139-1: 1991). 11 pp.
ISO 10139 Pt 1-91. Dentistry—Resilient Lining Materials for Removable Dentures—Part 1: Short-Term Materials First Edition. 9 pp.

INDUSTRY STANDARDS

Dental Materials (Cont.)

—Duplicating
CNS T3061-86. Dental Duplicating Materials (Aug)(11689).
SAA AS 1097-72. Dental Duplicating Material Reconfirmed 1986. 16 pp.

—Elastomeric—Impression Compounds
BSI BS 4269: Part 1-92. 1992 Dental Elastic Impression Material Part 1: Specification for Elastomeric Impression Materials (ISO 4823: 1992). 19 pp.
BSI BS 4269: Part 1-87. 1987 Dental Elastic Impression Material Part 1: Elastomeric Impression Material. 16 pp.
CSA CAN/CSA-Z349.39-93. Dental Elastomeric Impression Materials (ISO 4823-1992); (Gen Instr 1). 40 pp.
ISO 4823-92. Dental Elastomeric Impression Materials Second Edition. 17 pp.
JIS T 6513-91. Dental Elastomeric Impression Materials. 16 pp.

—Fusible Alloy
CNS T3010-80. Dental Fusible Alloy (Oct)(6620).
JIS T 6110-84. Dental Fusible Alloy. 4 pp.

—Fusible Alloy—Restorations—Casting
CSA CAN/CSA-Z349.46-93. Dental Ceramic Fused to Metal Restorative Materials (ISO 9693-1991); (Gen Instr 1). 33 pp.
ISO 9693-91. Dental Ceramic Fused to Metal Restorative Materials First Edition. 10 pp.

—Glossaries
BSI BS 4492-83. (WITHDRAWN) 1983 Dental Terms (Superseded by BS EN 21942-1: 1992 and Subsequent Parts). 118 pp.
BSI BS EN 21942-1-92. 1992 Dental Vocabulary Part 1: General and Clinical Terms (ISO 1942-1: 1989). 21 pp.
BSI BS EN 21942-2-93. 1993 Dental Vocabulary Part 2: Dental Materials (ISO 1942-2: 1989) (N). 23 pp.
CEN EN 21942-1-91. Dental Vocabulary—Part 1: General and Clinical Terms. 17 pp.
CEN EN 21942-2-92. Dental Vocabulary—Part 2: Dental Materials (ISO 1942-2: 1989). 18 pp.
ISO 1942 Pt 1-89. Dental Vocabulary—Part 1: General and Clinical Terms Second Edition; (Amendment 1-1992) (Amendment 2-1992) (Amendment 5-1993). 40 pp.
ISO 1942 Pt 2-89. Dental Vocabulary—Part 2: Materials Second Edition; (Amendment 1-2:1992) (CEN EN 21942-2:1992). 34 pp.
ISO 1942 Pt 5-89. Dental Vocabulary—Part 5: Terms Associated with Testing Second Edition. 17 pp.

—Gold Alloy
CNS T3005-80. Dental Wrought Gold-Silver-Palladium Alloy (Aug)(6399).
JIS T 6105-91. Dental Wrought Gold-Silver-Palladium Alloys. 9 pp.
SAA AS 1625-76. Dental Wrought Gold Alloys Reconfirmed 1989 (This is a Joint Standard with SANZ NZS 1625). 6 pp.
SNZ NZS/AS 1625-76. Dental Wrought Gold Alloys (Reconfirmed 1989) (This is a Joint Standard with SAA AS 1625). 6 pp.

—Gold Alloy—Casting
BSI BS 4425-88. 1988 Amd 1 Dental Casting Gold Alloys (N) (ISO 1562-1984) (AMD 6469) April 30, 1991. 9 pp.
BSI BS 6042-01. 1990 Amd 1 Dental Semi-Precious Metal Casting Alloys (ISO 8891: 1990) (AMD 7019) May 1, 1992. 16 pp.
BSI BS EN 27490-92. 1992 Dental Gypsum-Bonded Casting Investments for Gold Alloys (ISO 7490: 1990). 24 pp.
CEN EN 21 562-89. Dentistry Dental Casting Gold Alloys. 8 pp.
CEN PREN 28 891-90. Dental Casting Alloys with Noble Metal Content of 25 % up to but Not Including 75 %. 3 pp.
CEN EN 28891-91. Dental Casting Alloys with Noble Metal Content of 25 % up to but Not Including 75 %. 3 pp.
CNS T3006-80. Dental Casting Gold-Silver-Palladium Alloy (Oct)(6616).
CNS T3013-80. Dental Casting 14K Gold Alloy (Oct)(6623).
CNS T3014-80. Plus Metals for Dental Casting 14K Gold Alloy (Oct)(6624).
CSA CAN/CSA-Z349.44-93. Dental Casting Alloys with Noble Metal Content of 25% up to but Not Including 75% (ISO 8891-1990); (Gen Instr 1). 32 pp.
ISO 1562-84. Dental Casting Gold Alloys Second Edition. 6 pp.
ISO 8891-90. Dental Casting Alloys with Noble Metal Content of 25 % up to but Not Including 75 % First Edition. 9 pp.
JIS T 6106-91. Dental Casting Gold-Silver-Palladium Alloys. 9 pp.
JIS T 6113-87. Dental Casting 14K Gold Alloy. 6 pp.
JIS T 6114-87. Plus Metals for Dental Casting 14K Gold Alloys. 5 pp.
JIS T 6116-90. Dental Casting Gold Alloys. 6 pp.

—Gold Alloy—Casting Investments
CSA CAN/CSA-Z349.43-93. Dental Gypsum-Bonded Casting Investments for Gold Alloys (ISO 7490-1990); (Gen Instr 1). 43 pp.
ISO 7490-90. Dental Gypsum-Bonded Casting Investments for Gold Alloys First Edition. 20 pp.

—Gold Alloy—Solders
BSI BS 3384-84. 1984 Dental Gold Solders. 7 pp.
CNS T3007-80. Gold-Silver-Palladium Alloy Solders for Dental Use (Oct)(6617).
JIS T 6107-91. Dental Gold-Silver-Palladium Alloy Solders. 10 pp.
JIS T 6117-91. Dental Gold Alloy Solders. 6 pp.

—Gypsum
BSI BS 7013-89. 1989 Dental Gypsum Products (ISO 6873: 1983). 14 pp.
BSI BS 7013-01. 1989 Amd 1 Dental Gypsum Products (ISO 6873: 1983) (AMD 7038) May 1, 1992. 17 pp.
CEN EN 26873-91. Dentistry—Dental Gypsum Products. 3 pp.
ISO 6873-83. Dental Gypsum Products First Edition. 10 pp.

—Implants
ISO TR10451-91. Dental Implants—State of the Art—Survey of Materials First Edition. 22 pp.

—Impression Compounds
BSI BS 3886-65. 1965 Dental Impression Compounds. 12 pp.
BSI BS 4284-81. 1981 Dental Impression Pastes (Zinc Oxide Type). 8 pp.
JIS T 6504-85. Dental Impression Compound. 7 pp.
SAA AS 1241-73. Dental Shellac Baseplates Reconfirmed 1986. 9 pp.
SAA AS 1917-76. Dental Modelling Compounds Reconfirmed 1989 (This is a Joint Standard with SANZ NZS 1917). 10 pp.
SAA AS 1963-77. Dental Impression Paste Reconfirmed 1989 (This is a Joint Standard with SANZ NZS 1963). 11 pp.
SNZ NZS/AS 1917-76. Dental Modelling Compounds (Reconfirmed 1989) (This is a Joint Standard with SAA AS 1917). 10 pp.
SNZ NZS/AS 1963-77. Dental Impression Paste (Reconfirmed 1989) (This is a Joint Standard with SAA AS 1963). 11 pp.

—Investment Compounds—Gypsum Bonded
CSA CAN/CSA-Z349.43-93. Dental Gypsum-Bonded Casting Investments for Gold Alloys (ISO 7490-1990); (Gen Instr 1). 43 pp.
ISO 7490-90. Dental Gypsum-Bonded Casting Investments for Gold Alloys First Edition. 20 pp.

—Mercury
BSI BS 4227-86. 1986 Amd 1 Specification for Dental Mercury (AMD 6327) June 28, 1991. 5 pp.
BSI BS 4227-02. 1986 Amd 2 Dental Mercury (ISO 1560: 1985) (AMD 7023) May 1, 1992. 10 pp.
CEN EN 21560-91. Dentistry—Dental Mercury. 3 pp.
CNS T3012-80. Dental Mercuries (Oct)(6622). 1 p.
ISO 1560-85. Dental Mercury Second Edition. 4 pp.
JIS T 6112-89. Dental Mercury. 5 pp.

—Mercury Amalgams
CEN PREN 21 560-90. Dentistry—Dental Mercury. 3 pp.
CSA CAN/CSA-Z349. 16TO28-M92. Filling, Restorative, and Prosthodontic Materials for Dentistry—Volume 1; (Gen Instr 1). 137 pp.

—Metals—Base—Casting
BSI BS 3366: Part 1-88. 1988 Amd 1 Dental Base Metal Casting Alloys Part 1: Specification for Denture Base Alloys (ISO 6871-1987) (EN 26 871: 1989) (AMD 6673) May 31, 1991. 13 pp.
BSI BS 3366: Part 2-76. 1976 Dental Base Metal Casting Alloys Part 2: Inlay, Crown and Bridge Alloys. 8 pp.
CEN EN 26 871-90. Dentistry—Dental Base Metal Casting Alloys. 4 pp.
CSA CAN/CSA-Z349.40-93. Dental Base Metal Casting Alloys (ISO 6871-1987); (Gen Instr 1). 32 pp.
ISO 6871-87. Dental Base Metal Casting Alloys First Edition. 8 pp.

—Metals—Corrosion
ISO TR 10271-93. Dentistry—Determination of Tarnish and Corrosion of Metals and Alloys First Edition. 13 pp.

—Metals—Precious—Wire
BSI BS 3520-62. 1962 Dental Wrought Precious Metal Alloy Wire. 8 pp.

—Metals—Restorations—Casting
CSA CAN/CSA-Z349.46-93. Dental Ceramic Fused to Metal Restorative Materials (ISO 9693-1991); (Gen Instr 1). 33 pp.
ISO 9693-91. Dental Ceramic Fused to Metal Restorative Materials First Edition. 10 pp.

—Metals—Semiprecious—Casting
BSI BS 6042-90. 1990 Dental Semi-Precious Metal Casting Alloys. 11 pp.

—Metals—Tarnish
ISO TR 10271-93. Dentistry—Determination of Tarnish and Corrosion of Metals and Alloys First Edition. 13 pp.

—Nickel Chromium Alloy—Plates
CNS T3002-80. Nickel-Chromium Alloy Plates for Dental Use (Aug)(6396).
JIS T 6102-87. Nickel-Chromium Alloy Plates for Dental Use.

—Nickel Chromium Alloy—Wire
CNS T3001-80. Nickel-Chromium Alloy Wires for Dental Use (Aug)(6395).
JIS T 6101-87. Nickel-Chromium Alloy Wires for Dental Use.

—Plaster
CNS T3054-87. Dental Plaster (Mar)(8871).
JIS T 6604-87. Dental Plaster. 8 pp.
SNZ NZS/AS 1651-74. Dental Impression Plaster (Reconfirmed 1988) (This is a Joint Standard with SAA AS 1651). 10 pp.
SNZ NZS/AS 1652-74. Dental Laboratory Plaster (Reconfirmed 1988) (This is a Joint Standard with SAA AS 1652). 10 pp.

—Platinum Alloy—Casting
BSI BS 6042-01. 1990 Amd 1 Dental Semi-Precious Metal Casting Alloys (ISO 8891: 1990) (AMD 7019) May 1, 1992. 16 pp.
CEN PREN 28 891-90. Dental Casting Alloys with Noble Metal Content of 25 % up to but Not Including 75 %. 3 pp.
CEN EN 28891-91. Dental Casting Alloys with Noble Metal Content of 25 % up to but Not Including 75 %. 3 pp.
CSA CAN/CSA-Z349.44-93. Dental Casting Alloys with Noble Metal Content of 25% up to but Not Including 75% (ISO 8891-1990); (Gen Instr 1). 32 pp.
ISO 8891-90. Dental Casting Alloys with Noble Metal Content of 25 % up to but Not Including 75 % First Edition. 9 pp.

—Polymers—Bridges
BSI BS 7651-93. 1993 Dental Polymer-Based Crown and Bridge Materials (ISO 10477: 1992). 14 pp.
ISO 10477-92. Dentistry—Polymer-Based Crown and Bridge Materials First Edition. 12 pp.

—Polymers—Color Stability
BSI BS 6790-87. 1987 Determination of Colour Stability of Dental Polymeric Materials (ISO 7491: 1985). 4 pp.
BSI BS 6790-01. 1987 Amd 1 Determination of Colour Stability of Dental Polymeric Materials (ISO 7491: 1985) (AMD 7021) May 1, 1992. 9 pp.
CEN PREN 27 491-90. Dentistry—Dental Materials—Determination of Colour Stability of Dental Polymeric Materials. 3 pp.
CEN EN 27491-91. Dentistry—Dental Materials—Determination of Colour Stability of Dental Polymeric Materials. 3 pp.
ISO 7491-85. Dental Materials—Determination of Colour Stability of Dental Polymeric Materials First Edition. 4 pp.

—Polymers—Crowns
BSI BS 7651-93. 1993 Dental Polymer-Based Crown and Bridge Materials (ISO 10477: 1992) (N). 14 pp.
ISO 10477-92. Dentistry—Polymer-Based Crown and Bridge Materials First Edition. 12 pp.

—Porcelain—Jacket Crowns
BSI BS 5612-78. 1978 Dental Porcelains for Jacket Crowns. 14 pp.

—Resins
BSI BS 5199-89. 1989 Resin-Based Dental Filling Materials (Class B) (ISO 4049: 1988). 15 pp.
BSI BS 5199-01. 1989 Amd 1 Resin-Based Dental Filling Materials (Class B) (ISO 4049: 1988) (AMD 7513) January 15, 1993. 20 pp.
ISO 4049-88. Dentistry—Resin-Based Filling Materials Second Edition; (Corrigendum 1-1992). 16 pp.

—Resins—Acrylic
CNS T3049-82. Dental Self-Curing Acrylic Resin for Filling (Mar)(8636).

INTERNATIONAL AND NON-U.S. NATIONAL STANDARDS
SUBJECT INDEX

Dental Materials (Cont.)

—Resins—Acrylic—Crowns
  CNS T3050-82. Dental Heat Curing Acrylic Resin for Crown (Mar)(8637).
  JIS T 6508-85. Dental Heat Curing Acrylic Resin for Crown. 8 pp.
  JIS T 6509-85. Dental Self-Curing Acrylic Resin for Crown. 7 pp.

—Resins—Composite—Filling
  CNS T3060-86. Direct Filling Composite Resins (Aug)(11688).

—Resins—Sealants
  BSI BS 7180-89. (WITHDRAWN) 1989 Dental Resin-Based Pit and Fissure Sealants (ISO 6874: 1988) (Renumbered as BS EN 26874: 1992). 10 pp.
  BSI BS EN 26874-92. 1992 Amd 1 Dental Resin-Based Pit and Fissure Sealants (ISO 6874: 1988) (AMD 7512) February 15, 1993 (N). 16 pp.
  CEN EN 26874-92. Dental Resin-Based Pit and Fissure Sealants (ISO 6874: 1988). 11 pp.
  CSA CAN/CSA-Z349.41-93. Dental Resin-Based Pit and Fissure Sealants (ISO 6874-1988); (Gen Instr 1). 33 pp.
  ISO 6874-88. Dental Resin-Based Pit and Fissure Sealants First Edition; (CEN EN 26874:1992). 10 pp.

—Rubber Dam
  CNS T3064-86. Dental Rubber Dam (Oct)(11746).
  SAA AS 1022-71. Dental Rubber Dam Amdt 1 March 1973 Reconfirmed 1986. 6 pp.

—Safety
  BSI BS 5828-89. 1989 Biological Assessment of Dental Materials. 24 pp.
  ISO TR7405-84. Biological Evaluation of Dental Materials First Edition. 54 pp.

—Sealing—Root Canals
  BSI BS 6934-88. 1988 Amd 1 Dental Root Canal Sealing Materials (AMD 6674) May 31, 1991. 10 pp.
  CEN EN 26 876-90. Dentistry—Dental Root Canal Sealing Materials. 2 pp.
  CSA CAN/CSA-Z349.42-93. Dental Root Canal Sealing Materials (ISO 6876-1986); (Gen Instr 1). 30 pp.
  ISO 6876-86. Dental Root Canal Sealing Materials First Edition. 6 pp.

—Silicate
  ISO 3851-77. Capsulated Dental Silicate and Silico-Phosphate Filling Materials First Edition. 13 pp.

—Silicophosphate
  ISO 3851-77. Capsulated Dental Silicate and Silico-Phosphate Filling Materials First Edition. 13 pp.

—Silver Alloy
  BSI BS EN 21559-92. 1992 Alloys for Dental Amalgam (ISO 1559: 1986) (Supersedes BS 2938: 1985). 11 pp.
  CEN EN 21559-91. Dentistry—Alloys for Dental Amalgam. 7 pp.
  CNS T3009-80. Dental Silver Amalgam Alloy (Oct)(6619).
  ISO 1559-86. Dentistry—Alloys for Dental Amalgam Second Edition. 6 pp.

—Silver Alloy—Casting
  CNS T3008-80. Dental Casting Silver Alloy (Oct)(6618).
  JIS T 6108-87. Dental Casting Silver Alloys. 6 pp.

—Silver—Solder
  CNS T3011-80. Dental Silver Solders (Oct)(6621).
  JIS T 6111-89. Dental Silver Solders. 6 pp.

—Stainless Steel—Wire
  CNS T3003-80. Stainless Steel Wires for Dental Use (Aug)(6397).
  JIS T 6103-87. Stainless Steel Wires for Dental Use. 6 pp.

—Stone
  CNS T3055-87. Dental Stone (Mar)(8872).
  JIS T 6605-87. Dental Stone. 7 pp.
  SNZ NZS/AS 1616-74. Dental Artificial Stone (Reconfirmed 1989) (This is a Joint Standard with SAA AS 1616). 10 pp.

—Tapes
  BSI BS 3507-76. 1976 Orthodontic Wire and Tape and Dental Ligature Wire. 8 pp.

—Temporary
  CNS T3048-82. Dental Temporary Stopping (Jan)(8332).
  JIS T 6507-87. Dental Temporary Stopping. 5 pp.

Dental Materials (Cont.)

—Tin Alloy
  BSI BS EN 21559-92. 1992 Alloys for Dental Amalgam (ISO 1559: 1986) (Supersedes BS 2938: 1985). 11 pp.
  CEN EN 21559-91. Dentistry—Alloys for Dental Amalgam. 7 pp.
  ISO 1559-86. Dentistry—Alloys for Dental Amalgam Second Edition. 6 pp.

—Waxes
  SAA AS 1583-73. Dental Sticky Wax Reconfirmed 1986. 7 pp.

—Waxes—Casting
  BSI BS 3508-62. (WITHDRAWN) 1962 Dental Inlay Casting Wax (Superseded by BS EN 21561: 1992). 15 pp.
  BSI BS EN 21561-92. 1992 Dental Inlay Casting Wax (ISO 1561: 1975). 12 pp.
  CEN PREN 21 561-90. Dental Inlay Casting Wax. 3 pp.
  CNS T3062-86. Dental Inlay Casting Wax (Oct)(11744).
  ISO 1561-75. Dental Inlay Casting Wax First Edition. 7 pp.
  JIS T 6503-85. Dental Inlay Casting Wax. 6 pp.
  SAA AS 1453-73. Dental Modelling Waxes Reconfirmed 1986 (This is a Joint Standard with SNZ NZS 1453). 12 pp.
  SAA AS 1582-73. Dental Inlay Casting Waxes Reconfirmed 1986. 7 pp.
  SNZ NZS/AS 1453-73. Dental Modelling Waxes (Reconfirmed 1986) (This is a Joint Standard with SAA AS 1453). 12 pp.

—Wire
  BSI BS 3507-76. 1976 Orthodontic Wire and Tape and Dental Ligature Wire. 8 pp.

—Zinc Oxide/Eugenol
  BSI BS 4284-81. 1981 Dental Impression Pastes (Zinc Oxide Type). 8 pp.

Dental Rotary Instruments
Use: Dental Instruments—Cutting Tools

Dental Syringes
Use: Syringes—Dental

Dental Treatment Units
Use For: Dental Units *See Also:* Dental Equipment; Dental Instruments
  ISO 7494-90. Dental Unit First Edition. 13 pp.

—Instruments—Identification Systems
  ISO 7494-90. Dental Unit First Edition. 13 pp.

Dental Units
Use: Dental Treatment Units

Dentistry
See Also: Dental Equipment; Dental Materials; Medical Equipment

—Glossaries
  BSI BS 4492-83. (WITHDRAWN) 1983 Dental Terms (Superseded by BS EN 21942-1: 1992 and Subsequent Parts). 118 pp.
  BSI BS EN 21942-1-92. 1992 Dental Vocabulary Part 1: General and Clinical Terms (ISO 1942-1: 1989). 21 pp.
  CEN EN 21942-1-91. Dental Vocabulary—Part 1: General and Clinical Terms. 17 pp.
  CEN EN 21942-2-92. Dental Vocabulary—Part 2: Dental Materials (ISO 1942-2: 1989). 18 pp.
  CSA CAN/CSA-Z349.1-87. Dental Terminology and Definitions. 99 pp.
  ISO 1942 Pt 1-89. Dental Vocabulary—Part 1: General and Clinical Terms Second Edition; (Amendment 1-1992) (Amendment 2-1992) (Amendment 5-1993). 40 pp.
  ISO 1942 Pt 5-89. Dental Vocabulary—Part 5: Terms Associated with Testing Second Edition. 17 pp.

Dentures
Use For: Artificial Teeth; False Teeth; Partial Dentures *See Also:* Dental Equipment; Dental Materials; Prosthetic Devices

—Acrylic Resins
  BSI BS 3990-80. 1980 Amd 2 Acrylic Resin Teeth. 12 pp.
  CNS T3046-82. Acrylic Resin Teeth for Denture (Jan)(8329).
  CNS T3057-86. Acrylic Denture Base Resin (Apr)(11557).
  CNS T4001-82. Method of Test for Acrylic Resin Teeth for Denture (Jan)(8330).
  JIS T 6501-87. Acrylic Denture Base Resin. 8 pp.
  SAA AS 1043-71. Acrylic Denture Base Resin Reconfirmed 1986. 14 pp.

Dentures (Cont.)

—Acrylic Resins (Cont.)
  SNZ NZS/AS 1626-74. Acrylic Teeth (Reconfirmed 1989) (This is a Joint Standard with SAA AS 1626). 14 pp.

—Base Plates
  BSI BS 2487-89. 1989 Denture Base Polymers. 12 pp.
  CNS T3056-86. Dental Base Plates (Apr)(11556).
  CNS T3063-86. Base Plate Wax (Oct)(11745).
  JIS T 6502-87. Dental Base Plate Wax. 5 pp.
  JIS T 6510-85. Dental Base Plates. 6 pp.
  SAA AS 1241-73. Dental Shellac Baseplates Reconfirmed 1986. 9 pp.

—Casting Investment
  BSI BS 5189: Part 1-75. (WITHDRAWN) 1975 Dental Casting Investments Part 1: Gypsum-Bonded Investment Materials (Superseded by BS EN 27490: 1992). 13 pp.
  BSI BS 5189: Part 2-75. 1975 Dental Casting Investments Part 2: Phosphate-Bonded Dental Casting Investments. 13 pp.
  BSI BS 5189: Part 3-75. 1975 Dental Casting Investments Part 3: Ethyl Silicate-Bonded Dental Casting Investments. 12 pp.
  CNS T3051-82. Dental Casting Investment (May)(8868).
  JIS T 6601-89. Gypsum-Bonded Dental Investments for Casting. 8 pp.

—Porcelain
  BSI BS 6817-91. 1991 Porcelain Denture Teeth. 10 pp.
  CEN PREN 261-87. Dentistry; Porcelain Denture Teeth. 5 pp.
  CEN HD 341-91. Dentistry—Porcelain Denture Teeth; (ISO 4824:1981). 2 pp.
  CNS T3047-82. Porcelain Teeth for Plate (Jan)(8331).
  ISO 4824-93. Dentistry—Ceramic Denture Teeth Second Edition. 8 pp.
  JIS T 6511-89. Porcelain Denture Teeth. 10 pp.

—Removable—Linings
  BSI BS 7589: Part 1-92. 1992 Resilient Lining Materials for Removable Dentures Part 1: Specification for Short-Term Materials (ISO 10139-1: 1991). 11 pp.
  ISO 10139 Pt 1-91. Dentistry—Resilient Lining Materials for Removable Dentures—Part 1: Short-Term Materials First Edition. 9 pp.

—Resins
  BSI BS 2487-89. 1989 Denture Base Polymers. 12 pp.
  BSI BS 6747-87. 1987 Amd 1 Orthodontic Resins (AMD 6663) April 30 1991. 15 pp.
  ISO 1567-88. Dentistry—Denture Base Polymers Second Edition. 13 pp.
  ISO 3336-93. Dentistry—Synthetic Polymer Teeth Second Edition. 12 pp.

—Resins—Base Plates
  BSI BS 2487-89. 1989 Denture Base Polymers. 12 pp.

—Resins—Polymethyl Methacrylate
  JIS T 6506-89. Acrylic Resin Teeth. 12 pp.

—Waxes—Base Plates
  CNS T3063-86. Base Plate Wax (Oct)(11745).
  JIS T 6502-87. Dental Base Plate Wax. 5 pp.

Dependability
Use: Reliability

Deposit Gages
See Also: Deposition Rates; Gages; Particulate Monitors
  BSI BS 1747: Part 1-69. 1969 Methods for the Measurement of Air Pollution Part 1: Deposit Gauges. 17 pp.

Deposition Rates
See Also: Deposit Gages
  ISO 9225-92. Corrosion of Metals and Alloys—Corrosivity of Atmospheres—Measurement of Pollution First Edition. 14 pp.

—Sulfur Dioxide—Atmospheric Corrosion Testing
  ISO 9225-92. Corrosion of Metals and Alloys—Corrosivity of Atmospheres—Measurement of Pollution First Edition. 14 pp.

Depth Gages
See Also: Depth Micrometers

—Dial
  CGSB 39-GP-19M-79. Calipers and Gages, Vernier and Dial, Standard for. 22 pp.

—TORQ-SET Recesses
  DIN ENGL LN 29999 Pt 2-74. TORQ-SET Recesses; Penetration Gauges (Aug). 3 pp.

INDUSTRY STANDARDS

Depth Gages *(Cont.)*
—Vernier
CGSB 39-GP-19M-79. Calipers and Gages, Vernier and Dial, Standard for. 22 pp.

Depth Micrometers
See Also: Depth Gages; Micrometers
BSI BS 6468-84. 1984 Depth Micrometers. 8 pp.
CGSB 39-GP-18M-79. Calipers and Gages, Micrometer, Standard for. 24 pp.
DIN ENGL 863 Pt 2-81. Micrometers; Built-In Micrometers, Micrometer Depth Gauges; Concepts, Requirements, Testing (June). 4 pp.
JIS B 7544-81. Depth Micrometers.

Depth Sounders
Use: Sonic Depth Finders

Derailleurs
See Also: Gearshifts
JIS D 9428-87. Derailleur for Bicycles.

Derrick Booms
Use: Derricks

Derricks
Use For: Drilling Derricks *See Also:* Booms (Equipment); Cranes; Hoists
BSI BS 327-64. 1964 Amd 1 Power-Driven Derrick Cranes. 65 pp.

—Bearings—Ships
ISO 6045-87. Shipbuilding and Marine Structures—Bearings for Derrick Goosenecks—Assemblies and Components First Edition. 19 pp.

—Booms—Head Pieces—Ships
CNS F3151-81. Boom Rest Head Pieces (Nov)(8115).
JIS F 2205-87. Boom Rest Head Pieces.
JIS F 2205-76. Boom Rest Head Pieces. 8 pp.

—Booms—Ships
CNS F3142-81. Ships' Steel Plate Derrick Booms (Oct)(7987).
CNS F3154-81. Ships' Light Load Derrick Booms (Nov)(8118).
JIS F 2201-90. Ships' Derrick Booms.
JIS F 2201-75. Ships' Steel Plate Derrick Booms. 17 pp.
JIS F 2251-90. Ships' Light Load Derrick Booms.
JIS F 2251-76. Ships' Light Load Derrick Booms. 8 pp.

—Brackets—Ships
CNS F3149-81. Ships' Derrick Topping Brackets (Nov)(8113).
CNS F3150-81. Ships' Derrick Gooseneck Brackets (Nov)(8114).
CNS F3152-81. Ships' Light Load Derrick Topping Brackets (Nov)(8116).
CNS F3153-81. Ships' Light Load Derrick Gooseneck Brackets (Nov)(8117).
JIS F 2202-76. Ships' Derrick Topping Brackets. 8 pp.
JIS F 2203-73. Ships' Derrick Gooseneck Brackets. 6 pp.
JIS F 2206-76. Ships' Light Load Derrick Topping Brackets.
JIS F 2206-70. Ships' Light Load Derrick Topping Brackets. 4 pp.
JIS F 2207-76. Ships' Light Load Derrick Gooseneck Brackets (R 1984). 7 pp.

—Clamps—Ships
BSI BS MA 17-72. 1972 Derrick Boom Clamp. 7 pp.

—Design—Loads (Forces)
DIN ENGL 4111-84. Drilling Derricks; Design Principles (Mar). 8 pp.

—Design—Ships
BSI BS MA 48-76. 1976 Code of Practice for Design and Operation of Ships' Derrick Rigs. 35 pp.

—Fittings
DIN ENGL 82003 Pt 1-72. Cargo Lifting Gear; Accessories and Fittings; Summary (Feb). 12 pp.
DIN ENGL 82003 Pt 2-72. Cargo Lifting Gear; Accessories and Fittings; Technical Conditions of Delivery (Feb). 4 pp.

—Fittings—Ships
BSI BS MA 81-80. 1980 Ships' Derrick Fittings. 16 pp.

—Head Fittings—Ships
ISO 8148-85. Shipbuilding and Marine Structures—Derrick Boom Headfittings—Fixed Type First Edition. 7 pp.

—Heel Fittings—Ships
ISO 6044-85. Shipbuilding and Marine Structures—Derrick Boom Heel Fittings—Main Dimensions First Edition. 4 pp.

Derricks *(Cont.)*
—Rope Cleats—Ships
BSI BS MA 11-71. 1971 Horncleats. 3 pp.
CNS F3075-80. Horn Cleats (Dec)(6827).
CNS F3077-80. Ships' Derrick Guy Cleats (Dec)(6829).
JIS F 3414-74. Horn Cleats. 5 pp.
JIS F 3416-74. Ships' Derrick Guy Cleats. 5 pp.

—Safety—Codes of Practice
BSI CP 3010-72. 1972 Amd 1 Safe Use of Cranes (Mobile Cranes, Tower Cranes, and Derrick Cranes). 89 pp.

—Ships
MOD UK NES 113: Part 8-89. Requirements for Mechanical Handling Part 8: Davits and Derricks Issue 2 (11.89). 14 pp.

—Trunnions—Ships
ISO 8314-87. Shipbuilding and Marine Structures—Trunnion Pieces for Span Bearings and Lead Block Bearings First Edition. 7 pp.

Derricks (Swing)
Use: Gangways

Derris Roots
CNS K6077-65. Method of Test for Derris Roots (Aug)(921) (R 1971). 3 pp.

Derusters
Use: Rust Removers

Desalination Plants
Use For: Water Desalting Plants *See Also:* Water Supply

—Ships
MOD UK NES 328-85. Desalination Plants Issue 3 (09.85). 52 pp.
MOD UK NES 366-79. Trials and Inspections of Desalination Machinery Issue 1 (09.79). 17 pp.

—Submarines
MOD UK NES 328-85. Desalination Plants Issue 3 (09.85). 52 pp.

Descaling Compounds
See Also: Cleaning Agents

—Food Preparation Equipment
CGSB CAN/CGSB-2.43-93. Scale-Removing Compound for Lime Deposits. 9 pp.

—Hydrogen Chloride—Hot Water Installations
MOD UK DSTAN 68-27-77. Scale Removing Compound (for Hot Water Installations) Issue 2. 6 pp.
MOD UK DSTAN 68-27-92. Scale Removing Compound (for Hot Water Installations) Issue 3. 10 pp.

—Ships
MOD UK DSTAN 68-18-71. Evaporator Feed Treatment and Descaling Chemicals for Shipboard Use Issue 1. 8 pp.

Descramblers
Use: Decoders

Desford Trainer Aircraft
See Also: Aircraft

—Antenna Positions
CAA. Desford Trainer (Approved Aerial Positions). 1 p.

Desiccants
Use: Drying Agents

Desiccators
See Also: Separators (Mechanical)

—Glass—Chemical Analysis
CNS R2132-86. Glass Desiccator for Chemical Analysis (Mar) (7308).

—Vacuum—Laboratory
BSI BS 3423-86. 1986 Recommendations for Design of Glass Vacuum Vessels (Including Desiccators) for Laboratory Use. 18 pp.

Design
Scope Note: For design of specific products or systems, see the subheadings Design or Design Loads under the product or system. For additional references, consult the following list *See:* Barrier Free Design; CAD; Ergonomics; Experimental Design; Loads (Forces); Seismic Design; Structural Design

Design Rights
EURO ECO/IPR/85/1 5460-84. Intellectual Property Rights in International Co-Operative Projects. 88 pp.

—Defense Contracts
MOD UK DEFCON 15-74. Design Rights 8/74. 2 pp.
MOD UK DEFCON 15A-74. Design Rights A 8/74. 2 pp.
MOD UK DEFCON 177-80. Design Rights and Patents (Sub-Contractors) Agreement 3/80. 3 pp.

Design Speed
See Also: Speed

—Aerostats
CAA Chapter Q3-2 12.79. Design Airspeeds and Manoeuvres (Non-Rigid Airships). 1 p.

—Aircraft
CAA Chapter P3-2. Flight Manoeuvering Loads and Design Airspeeds (Provisional Airworthiness Requirements for Civil Powered Lift Aircraft). 13 pp.
CAA STANDARD NO. 4-3 08.69. Design Airspeeds, Gust Cases and Flight Manoeuvring Loads. 13 pp.

—Tractors
EC 74/152/EEC-74. Council Directive on the Approximation of the Laws of the Member States Relating to the Maximum Design Speed of and Load Platforms for Wheeled Agricultural or Forestry Tractors. 3 pp.
EC 88/412/EEC-88. Commission Directive Adapting to Technical Progress Council Directive 74/152/EEC on the Approximation of the Laws of the Member States Relating to the Maximum Design Speed and Load Platforms of Wheeled Agricultural and Forestry Tractors. 1 p.

Designations
Scope Note: See specific products

Desk Blotters
CGSB CAN/CGSB-53.57-M88. Pad, Desk, Blotter. 11 pp.

Desk Calendars
Use: Calendars—Desk

Desk Fans
Use: Fans—Desk

Desks
See Also: Furniture; Office Furniture; Tables (Stands)

—Classrooms
JIS S 1015-74. Sizes and Dimensions of Fixed Desk and Chair for Lecture Room.
JIS S 1016-76. Fixed Desk and Chair for Lecture Room.
JIS S 1021-91. School Furniture (Desks and Chairs for Class Room). 23 pp.
JIS S 1086-91. School Furniture (Double Desks for Class Room). 20 pp.

—Household—Student
CNS S1199-86. Student Desk for Domestic (Aug)(11674).
JIS S 1061-84. Student Desks for Domestic Use.

—Lighting
MOD UK DSTAN 62-8-73. Light, Desk Issue 1. 13 pp.
MOD UK DSTAN 62-8: Addendum. Light, Desk. 3 pp.

—Modular—Shelves
CGSB 44.168-85. Shelf, Assembly: Modular Desk 6030. 10 pp.

—Modular—Typewriter—Steel
CGSB 44.175-85. Desk, Typewriter, Modular, 48 x 24. 27 pp.

—Modular—Typewriter—Wood
CGSB 44.175-85. Desk, Typewriter, Modular, 48 x 24. 27 pp.

—Office
BSI BS 5940: Part 1-80. 1980 Office Furniture Part 1: Design and Dimensions of Office Workstations, Desks, Tables and Chairs. 14 pp.
CEN PREN 527-1-91. Office Furniture—Tables and Desks—Part 1: Dimensions. 7 pp.
CNS S1011-66. Office Desk (Mar)(904).
DIN ENGL 4549-82. Office Furniture; Desks, Machine Operator's Desks and Visual Display Unit (VDU) Desks; Dimensions (Nov). 3 pp.

—Office—Modular—Steel
CGSB 44.151-85. Runoff, Desk, Flush, Modular. 42 pp.

INTERNATIONAL AND NON-U.S. NATIONAL STANDARDS
SUBJECT INDEX

Desks *(Cont.)*
—Office—Modular—Steel *(Cont.)*
CGSB 44.152-85. Runoff, Secretarial, Desk, Drop, Modular. 27 pp.

—Office—Modular—Wood
CGSB 44.151-85. Runoff, Desk, Flush, Modular. 42 pp.
CGSB 44.152-85. Runoff, Secretarial, Desk, Drop, Modular. 27 pp.

—Office—Steel
CNS S1071-84. Steel Desk (Office Furniture) (Feb)(3088).
JIS S 1031-91. Office Furniture—Steel Desks. 15 pp.

—Office—Wood
CNS S1042-81. Office Wooden Desk (May)(2833).
JIS S 1023-89. Office Wooden Furniture (Desk, Table and Chair).

—Office—Writing
CNS S1151-84. Standard Size of Writing Desks for Office (Feb)(9639).
JIS S 1010-78. Standard Size of Writing Desks for Office.

Desktop Publishing
Use For: Computer Assisted Publishing
See Also: Publishing

—Glossaries
BSI DD 179-89. 1989 Glossary of Terms Used in Computer-Assisted Publishing. 48 pp.
ISO TR9544-88. Information Processing—Computer-Assisted Publishing—Vocabulary First Edition. 48 pp.
JTC1 TR9544-88. Information Processing—Computer-Assisted Publishing—Vocabulary First Edition. 48 pp.
OSI ISO/TR 9544-88. Information Processing—Computer Assisted Publishing—Vocabulary. 48 pp.
SAA AS 3877-91. Manipulating Industrial Robots—Vocabulary (ISO/TR 8373:1988). 9 pp.

—NATO Allied Publications
NATO STANAG 1421 ED 1 AMD 0-92. (Draft) Desktop Publishing Standards for Allied Publications—APP-32 Volume I. 3 pp.

Desoldering Equipment
See Also: Tools

—Portable
CSA C22.2 NO 122-M1989. Hand-Held Electrically Heated Tools; (Gen Instr 1 Thru 2). 50 pp.
CSA 1169 Bull. Electrical Bulletin 1169 June 27, 1978 to C22.2 NO 122. 2 pp.

Desorbable Gas Content Analysis
—Coal Seams
SAA AS 3980-91. Guide to the Determination of Desorbable Gas Content of Coal Seams—Direct Method (in Professional Package 32). 11 pp.

Destination Indicators (Telegraphy)
See Also: Address Codes; Identification Systems; Message Headers; Origin Indicators (Telegraphy); Telegraphy

—Telegraph Repeaters
CCITT RECMN F.31-89. Telegram Retransmission System—Telegraph and Mobile Services Operations and Quality of Service (Study Group I) 12 pp. 12 pp.
CCITT RECMN F.96-89. List of Destination Indicators—Telegraph and Mobile Services Operations and Quality of Service (Study Group I) 2 pp. 2 pp.

—Telegraph Repeaters—Documentation
CCITT RECMN F.96-89. List of Destination Indicators—Telegraph and Mobile Services Operations and Quality of Service (Study Group I) 2 pp. 2 pp.

Destination Network Codes
—Numbering Plans—Integrated Services Digital Networks
CCITT RECMN E.164-91. Numbering Plan for the ISDN Era (Study Group II) 19 pp (Same as Recmn I.331). 19 pp.
CCITT RECMN E.164-89. Numbering Plan for the ISDN Era—Telephone Network and ISDN—Operation, Numbering, Routing and Mobile Service (Study Group II) 6 pp (Same as Recmn I.331). 6 pp.

Destroyers
See Also: Naval Ships; Warships

Destroyers *(Cont.)*
—Seakeeping
NATO STANAG 4154 ED 2 AMD 2-86. General Criteria and Common Procedures for Seakeeping Performance Assessment. 49 pp.

Destructive Testing
See Also: Corrosion Testing

—Fusion Welded Joints
BSI BS 709-83. 1983 Methods of Destructive Testing Fusion Welded Joints and Weld Metal in Steel. 24 pp.

—Plain Bearings—Bonding Strength
BSI BS 7585: Part 2-92. 1992 Metallic Multilayer Plain Bearings Part 2: Method for Destructive Testing of Bond for Bearing Metal Layer Thicknesses Greater Than or Equal to 2 mm (ISO 4386-2: 1992). 10 pp.
ISO 4386 Pt 2-82. Plain Bearings—Metallic Multilayer Plain Bearings—Part 2: Destructive Testing of Bond for Bearing Metal Layer Thicknesses Greater Than or Equal to 2 mm First Edition. 6 pp.

—Shunt Power Capacitors
IEC 831 Pt 2-88. Shunt Power Capacitors of the Self-Healing Type for a.c. Systems Having a Rated Voltage up to and Including 660 V Part 2: Ageing Test, Self-Healing Test and Destruction Test First Edition; (Amendment 1-1991) (Amendment 2-1993). 37 pp.
IEC 931 Pt 2-92. Shunt Power Capacitors of the Non-Self-Healing Type for a.c. Systems Having a Rated Voltage up to and Including 1 000 V Part 2: Ageing Test and Destruction Test First Edition. 20 pp.

—Weld Metal
BSI BS 709-83. 1983 Methods of Destructive Testing Fusion Welded Joints and Weld Metal in Steel. 24 pp.

Desuperheaters
Use For: Steam Desuperheaters *See Also:* Cooling Systems; Superheater Tubes

—Refrigerants
CSA CAN/CSA-C22. 2 NO 236-M90. Heating and Cooling Equipment; (Gen Instr 1) (ANSI/UL 1995). 136 pp.

Detaching
—Paints—Mechanical Testing
BSI BS 3900: Part E1-70. 1970 Methods of Test for Paints Group E: Mechanical Tests on Paint Films Part E1: Bend Test (Cylindrical Mandrel). 4 pp.
BSI BS 3900: Part E4-76. 1976 Methods of Test for Paints Group E: Mechanical Tests on Paint Films Part E4: Cupping Test. 3 pp.
ISO 1519-73. Paints and Varnishes—Bend Test (Cylindrical Mandrel) First Edition. 7 pp.
ISO 1520-73. Paints and Varnishes—Cupping Test First Edition. 5 pp.

—Varnishes—Mechanical Testing
BSI BS 3900: Part E1-70. 1970 Methods of Test for Paints Group E: Mechanical Tests on Paint Films Part E1: Bend Test (Cylindrical Mandrel). 4 pp.
BSI BS 3900: Part E4-76. 1976 Methods of Test for Paints Group E: Mechanical Tests on Paint Films Part E4: Cupping Test. 3 pp.
ISO 1519-73. Paints and Varnishes—Bend Test (Cylindrical Mandrel) First Edition. 7 pp.
ISO 1520-73. Paints and Varnishes—Cupping Test First Edition. 5 pp.

Detector Diodes
See Also: Microwave Diodes; Semiconductor Devices
BSI BS 6493: Sec 1.4-92. 1992 Semiconductor Devices Part 1: Discrete Devices Section 1.4: Recommendations for Microwave Diodes and Transistors (IEC 747-4: 1991). 83 pp.
IEC 747 Pt 4-91. Semiconductor Devices Discrete Devices Part 4: Microwave Diodes and Transistors First Edition. 163 pp.
MOD UK DSTAN 59-61: 90/118-72. Semiconductor Device, Diode Issue 1. 13 pp.
MOD UK DSTAN 59-61: 90/119-72. Semiconductor Device, Microwave Diode Issue 1. 13 pp.
MOD UK DSTAN 59-61: 90/129-72. Semiconductor Device, Diode Issue 1. 15 pp.

—Quality Assurance
BSI BS 9322-71. 1971 Amd 2 Rules for the Preparation of Detail Specifications for Semiconductor Devices of Assessed Quality: Microwave Detector Diodes (AMD 6169) December 31, 1989. 9 pp.

Detector Diodes *(Cont.)*
—Reliability Assured
BSI BS 9322-71. 1971 Amd 2 Rules for the Preparation of Detail Specifications for Semiconductor Devices of Assessed Quality: Microwave Detector Diodes (AMD 6169) December 31, 1989. 9 pp.

—Schottky Barrier—Preferred Products List
CECC CECC MUAHAG Vol 9 IS 3-90. Preferred Products List; Semiconductors (En, Fr, Ge) AMD 1 (En, Fr, Ge). 51 pp.

Detectors
Scope Note: Use a more specific term *See:* Active Detectors; Alarm Systems; Analyzers; Burglar Detectors; Combustible Gas Detectors; Data Channel Failure Detectors; Flame Detectors; Floor Mat Detectors; Gas Detectors; Germanium Detectors; Heat Detectors; Hydrophones; Indicating Instruments; Infrared Detectors; Leak Detectors; Lie Detectors; Measuring Instruments; Microwave Motion Detectors; Monitors; Neutron Counters; Passive Detectors; Passive Infrared Detectors; Photodetectors; Pressure Transducers; Radiation Measuring Instruments; Radio Frequency Motion Detectors; Resistance Temperature Detectors; Revolution Counters; Semiconductor Devices; Smoke Detectors; Telecommunication Equipment; Telecommunication Systems; Ultrasonic Motion Detectors; Vibration Detection Systems; Warning Systems

Detergent Alkylate
Use: Dodecylbenzene

Detergent Content Analysis
—Water
BSI BS 2690: Part 11-71. 1971 Water Used in Industry Part 11: Anionic, Cationic and Non-Ionic Detergents and Oil. 23 pp.

Detergent Resistance Testing
See Also: Laundering Testing

—Enamels
BSI BS 1344: Part 5-84. 1984 Methods of Testing Vitreous Enamel Finishes Part 5: Determination of Resistance to Hot Detergent Solutions Used for Washing Textiles. 7 pp.
BSI BS 1344: Part 19-84. 1984 Methods of Testing Vitreous Enamel Finishes Part 19: Apparatus for Determination of Resistance to Hot Detergent Solutions Used for Washing Textiles. 8 pp.
ISO 4533-83. Vitreous and Porcelain Enamels—Determination of Resistance to Hot Detergent Solutions Used for Washing Textiles First Edition. 5 pp.
ISO 4535-83. Vitreous and Porcelain Enamels—Apparatus for Determination of Resistance to Hot Detergent Solutions Used for Washing Textiles First Edition. 7 pp.

—Hoses
SAA AS 1180.7D-72. Methods of Test for Hose Made from Elastomeric Materials—Part 7D: Resistance to Detergent Reconfirmed 1988.

—Prints
ISO 2840-74. Prints and Printing Inks—Determination of the Resistance of Prints to Detergents First Edition. 4 pp.

Detergents
Use For: Synthetic Detergents; Washing Powders
See Also: Cleaning Agents; Dishwashing Compounds; Laundry Detergents; Lubricating Oils; Nonionic Surfactants; Soaps; Surfactants
BSI BS 3762: Part 0-91. 1991 Analysis of Formulated Detergents Part 0: General Introduction. 6 pp.
CGSB CAN/CGSB-2.34-M90. Guide to the Selection and Use of Soaps and Detergents. 24 pp.
CGSB 2-GP-11M METH 20.3-88. Methods of Testing and Analysis of Soaps and Detergents Cleaning Efficiency. 8 pp.
EC 86/94/EEC-86. Council Directive Amending for the Second Time Directive 73/404/EEC on the Approximation of the Laws of the Member States Relating to Detergents. 1 pp.
JIS K 3362-90. Testing Methods for Synthetic Detergent. 76 pp.

—Abrasion Testing
CGSB 2-GP-11M METH 15.1-83. Methods of Testing and Analysis of Soaps and Detergents Abrasion Test. 1 p.

INDUSTRY STANDARDS

Detergents (Cont.)

—Alcohol Content—Gas Chromatography
BSI BS 3762: Sec 3.24-88. 1988 Analysis of Formulated Detergents Part 3: Quantitative Test Methods Section 3.24: Methods for Determination of Low Molecular Mass Alcohols Content. 6 pp.

—Alkanolamides Content
BSI BS 3762: Sec 3.8-86. 1986 Analysis of Formulated Detergents Part 3: Quantitative Test Methods Section 3.8: Method for Determination of Alkanalamides Content. 4 pp.

—Alkylbenzene Sulfonates Content
BSI BS 3762: Sec 3.10-89. 1989 Analysis of Formulated Detergents Part 3: Quantitative Test Methods Section 3.10: Methods for Determination of Short-Chain Alkylbenzenesulphonates Content. 6 pp.

—Ammonia Content—Volumetric Analysis
BSI BS 3762: Sec 3.12-85. 1985 Analysis of Formulated Detergents Part 3: Quantitative Test Methods Section 3.12: Method for Determination of Ammonia Content. 4 pp.

—Angle of Repose
ISO 4324-77. Surface Active Agents—Powders and Granules—Measurement of the Angle of Repose First Edition. 5 pp.

—Anionic Active Matter Content
BSI BS 3762: Sec 3.3-87. 1987 Analysis of Formulated Detergents Part 3: Quantitative Test Methods Section 3.3: Method for Determination of Hydrolizable and Non-Hydrolizable Anionic-Active Matter Content After Hydrolysis Under Acid Conditions. 5 pp.

ISO 2868-73. Surface Active Agents—Detergents—Anionic-Active Matter Stable to Acid Hydrolysis—Determination of Trace Amounts First Edition. 4 pp.

ISO 2869-73. Surface Active Agents—Detergents—Anionic-Active Matter Hydrolyzable Under Alkaline Conditions—Determination of Hydrolyzable and Non-Hydrolyzable Anionic-Active Matter First Edition. 4 pp.

ISO 2870-86. Surface Active Agents—Detergents—Determination of Anionic-Active Matter Hydrolyzable and Non-Hydrolyzable Under Acid Conditions Second Edition. 5 pp.

—Anionic Active Matter Content—Volumetric Analysis
BSI BS 3762: Sec 3.1-90. 1990 Analysis of Formulated Detergents Part 3: Quantitative Test Methods Section 3.1: Method for Determination for Anionic-Active Matter Content. 8 pp.

BSI BS 3762: Sec 3.2-83. 1983 Analysis of Formulated Detergents Part 3: Quantitative Test Methods Section 3.2: Method for Determination of Hydrolizable and Non-Hydrolizable Anionic-Active Matter Content After Hydrolysis Under Alkaline. 5 pp.

CGSB 2-GP-11M METH 17.3-89. Methods of Testing and Analysis of Soaps and Detergents Anionic Active Ingredient (Cationic Titration Method); (Amendment 2 May 1989). 4 pp.

ISO 2271-89. Surface Active Agents—Detergents—Determination of Anionic-Active Matter by Manual or Mechanical Direct Two-Phase Titration Procedure Second Edition. 8 pp.

—Automotive
MOD UK TS 10302A. Detergent Solution, for the Cleaning of Vehicles and Power Packs in Pressure Washing Machines.

—Bar—Sampling
CGSB 2-GP-11M METH 1.1-83. Methods of Testing and Analysis of Soaps and Detergents Sampling-Bars and Cakes. 1 p.

—Biodegradability
CNS K6436-79. Method of Test for Biodegradability of Synthetic Detergents (Sep)(4984).

JIS K 3363-90. Testing Method for Biodegradability of Synthetic Detergent. 14 pp.

—Boron Content—Volumetric Analysis
BSI BS 3762: Sec 3.13-83. 1983 Analysis of Formulated Detergents Part 3: Quantitative Test Methods Section 3.13: Method of Determination of Total Boron Content. 6 pp.

ISO 6835-81. Surface Active Agents—Washing Powders—Determination of Total Boron Content—Titrimetric Method First Edition. 5 pp.

—Bulk Density
BSI BS 3762: Sec 4.2-86. 1986 Analysis of Formulated Detergents Part 4: Physical Test Methods Section 4.2: Method for Determination of Apparent Bulk Density. 5 pp.

Detergents (Cont.)

—Bulk Density (Cont.)
ISO 697-81. Surface Active Agents—Washing Powders—Determination of Apparent Density—Method by Measuring the Mass of a Given Volume Second Edition. 5 pp.

—Carbonate Content
BSI BS 3762: Sec 3.14-85. 1985 Analysis of Formulated Detergents Part 3: Quantitative Test Methods Section 3.14: Method for Determination of Carbonate Content. 4 pp.

CGSB 2-GP-11M METH 26.1-83. Methods of Testing and Analysis of Soaps and Detergents Carbonates. 1 p.

—Carboxylmethyl Cellulose Content
BSI BS 3762: Sec 3.15-85. 1985 Analysis of Formulated Detergents Part 3: Quantitative Test Methods Section 3.15: Method for Determination of Carboxymethyl Cellulose Content. 4 pp.

—Cationic Active Matter Content
BSI BS 3762: Sec 3.4-01. 1991 Amd 1 Analysis of Formulated Detergents Part 3: Quantitative Test Methods Section 3.4: Method for Determination of Lower Molecular Mass Cationic-Active Matter Content (ISO 2871-2: 1990) (AMD 7161) July 15, 1992. 9 pp.

BSI BS 3762: Sec 3.26-89. 1989 Analysis of Formulated Detergents Part 3: Quantitative Test Methods Section 3.26: Method for Determination of High Molecular Mass Cationic-Active Matter Content. 4 pp.

ISO 2871 Pt 1-88. Surface Active Agents—Detergents—Determination of Cationic-Active Matter Content—Part 1: High-Molecular-Mass Cationic-Active Matter First Edition. 5 pp.

ISO 2871 Pt 2-90. Surface Active Agents—Detergents—Determination of Cationic-Active Matter Content—Part 2: Cationic-Active Matter of Low Molecular Mass (Between 200 and 500) First Edition. 7 pp.

—Cationic Active Matter Content—Volumetric Analysis
BSI BS 3762: Sec 3.4-91. 1991 Analysis of Formulated Detergents Part 3: Quantitative Test Methods Section 3.4: Method for Determination of Lower Molecular Mass Cationic-Active Matter Content (ISO 2871-2: 1990). 8 pp.

—Chemical Analysis
CNS K6438-79. Method of Chemical Analysis for Synthetic Detergents (Sep)(4986).

SAA AS 4010-92. Analysis of Detergents. 14 pp.

—Chloride Content—Volumetric Analysis
BSI BS 3762: Sec 3.16-85. 1985 Analysis of Formulated Detergents Part 3: Quantitative Test Methods Section 3.16: Method for Determination of Chloride Content. 4 pp.

—Chlorine Content—Volumetric Analysis
BSI BS 3762: Sec 3.18-86. 1986 Analysis of Formulated Detergents Part 3: Quantitative Test Methods Section 3.18: Method for Determination of Chlorine Oxidizing Agents Content. 4 pp.

—Clear Point
BSI BS 3762: Sec 4.1-89. 1989 Analysis of Formulated Detergents Part 4: Physical Test Methods Section 4.1: Method for Determination of Clear Point of Liquid Detergents. 3 pp.

—Containers—Fiberboard
BSI BS 5167-78. 1978 Packages for Washing and Cleaning Powders. Dimensions and Volumes of Cartons and Drums from Fibreboard. 9 pp.

CEN EN 23-1-78. Packages for Washing and Cleaning Powders—Part 1: Dimensions and Volumes of Cartons and Drums from Fibreboard. 7 pp.

—Containers—Laminates
MOD UK TS 10263. Metal-Foil/Plastics Laminate for Detergent Sachets.

—Dairy Equipment
SAA AS 1389-73. Acidic Detergents for Use in the Dairying Industry Amdt 1 March 1976. 8 pp.

SAA AS 1400-81. Heavy-Duty Alkaline Detergents for 'In-Place' Cleaning in Dairy Factories. 6 pp.

SAA AS 1803-82. General Purpose Detergents for Use in the Dairying Industry. 16 pp.

—Dry Matter Content
CGSB 2-GP-11M METH 17.2-93. Methods of Testing and Analysis of Soaps and Detergents Determination of Surfactant Content; (Amendment 4 March 1993). 3 pp.

—Dyeing
CNS K8009-71. Liquid Detergent (for Printing and Dyeing Processes) (Tentative) (Jan)(3163).

Detergents (Cont.)

—EDTA Content—Volumetric Analysis
BSI BS 3762: Sec 3.23-90. 1990 Analysis of Formulated Detergents Part 3: Quantitative Test Methods Section 3.23: Method for Determination of Chelating Agents Content. 6 pp.

ISO 4325-90. Soaps and Detergents—Determination of Chelating Agent Content—Titrimetric Method Second Edition. 4 pp.

—Flakes—Sampling
CGSB 2-GP-11M METH 1.2-83. Methods of Testing and Analysis of Soaps and Detergents Sampling-Flakes, Chips and Powders. 1 p.

—Foaming Power
CGSB 2-GP-11M METH 27.1-89. Methods of Testing and Analysis of Soaps and Detergents Foaming Ability; (Amendment 3 August 1989). 2 pp.

—Food Preparation Equipment
CNS S1085-75. Detergents for Vegetables and Kitchen Utensils (Jul)(3800).

—Formaldehyde Content
BSI BS 3762: Sec 3.11-89. 1989 Analysis of Formulated Detergents Part 3: Quantitative Test Methods Section 3.11: Method for Determination of Free Formaldehyde Content. 4 pp.

—Granular—Sampling
CGSB 2-GP-11M METH 1.2-83. Methods of Testing and Analysis of Soaps and Detergents Sampling-Flakes, Chips and Powders. 1 p.

CGSB 2-GP-11M METH 1.6-83. Methods of Testing and Analysis of Soaps and Detergents Sampling-Granular Materials. 1 p.

—Hard Water—Stability Testing
CGSB 2-GP-11M METH 19.1-83. Methods of Testing and Analysis of Soaps and Detergents Stability to Hard Water. 1 p.

—Insoluble Matter Content
CGSB 2-GP-11M METH 9.2-83. Methods of Testing and Analysis of Soaps and Detergents Alcohol Insoluble Matter in Synthetic Detergents. 1 p.

CGSB 2-GP-11M METH 10.2-83. Methods of Testing and Analysis of Soaps and Detergents Water Insoluble Matter in Synthetic Detergents. 1 p.

CGSB 2-GP-11M METH 10.3-83. Methods of Testing and Analysis of Soaps and Detergents Water Insoluble Matter in Synthetic Detergents. 1 p.

—Kitchen
JIS K 3370-79. Synthetic Detergents for Kitchen.

—Liquid
CGSB CAN/CGSB-2.107-92. General Purpose Built Liquid Detergent. 7 pp.

—Liquid—Germicidal
CGSB CAN/CGSB-2.160-M87. Detergent, Germicidal, General Purpose, Liquid. 9 pp.

—Liquid—Sampling
CGSB 2-GP-11M METH 1.3-83. Methods of Testing and Analysis of Soaps and Detergents Sampling-Liquids. 1 p.

—Metal Cleaners
CNS K6226-66. Method of Test for Detergent After Metal Pickling (Jan)(2553)(R 1971).

CNS K8008-66. Detergent (for Metal Pickling) (R 1971)(Sep)(2552).

—Nonionic Active Matter Content—Gravimetric Analysis
BSI BS 3762: Sec 3.7-86. 1986 Analysis of Formulated Detergents Part 3: Quantitative Test Methods Section 3.7: Method for Determination of Total Non-Ionic Matter Content. 4 pp.

—Nonionic—Biodegradability
CNS K6437-79. Method of Test for Biodegradability of Nonionic Surface Active Agent (Sep)(4985).

—Nonionic—Liquid Concentrate
CGSB CAN/CGSB-2.175-93. Concentrated Nonionic, Unbuilt Liquid Detergent. 8 pp.

—Organic Matter Content—Gravimetric Analysis
BSI BS 3762: Sec 3.5-86. 1986 Analysis of Formulated Detergents Part 3: Quantitative Test Methods Section 3.5: Methods for Determination of Total Organic Matter Content. 4 pp.

—Oxygen Content—Volumetric Analysis
BSI BS 3762: Sec 3.17-83. 1983 Analysis of Formulated Detergents Part 3: Quantitative Test Methods Section 3.17: Method for Determination of Active Oxygen Content. 4 pp.

INTERNATIONAL AND NON-U.S. NATIONAL STANDARDS
SUBJECT INDEX

Detergents (Cont.)
—Oxygen Content—Volumetric Analysis (Cont.)
ISO 4321-77. Washing Powders—Determination of Active Oxygen Content—Titrimetric Method First Edition. 4 pp.

—Paste—Sampling
CGSB 2-GP-11M METH 1.4-83. Methods of Testing and Analysis of Soaps and Detergents Sampling-Pastes. 1 p.

—PH
CGSB 2-GP-11M METH 18.1-83. Methods of Testing and Analysis of Soaps and Detergents pH Value. 1 p.
MOD UK M 9521/65. Examination of Detergent Solution for Plastic Food Utensils.

—Phosphate Content
BSI BS 3762: Sec 3.20-87. (WITHDRAWN) 1987 Analysis of Formulated Detergents Part 3: Quantitative Test Methods Section 3.20: Method for Determination of Phosphate Types. 6 pp.
CGSB 2-GP-11M METH 25.1-83. Methods of Testing and Analysis of Soaps and Detergents Phosphates. 2 pp.

—Phosphorus Oxide Content—Gravimetric Analysis
BSI BS 3762: Sec 3.19-83. 1983 Analysis of Formulated Detergents Part 3: Quantitative Test Methods Section 3.19: Method for Determination of Total Phosphorus Oxide Content. 6 pp.
ISO 4313-76. Washing Powders—Determination of Total Phosphorus (V) Oxide Content—Quinoline Phosphomolybdate Gravimetric Method First Edition. 5 pp.

—Physical Properties
CNS K6439-79. Method of Test for Physical Properties of Synthetic Detergents (Sep)(4987).

—Powder
CGSB CAN/CGSB-2.141-92. General-Purpose Powder Cleaning Compound. 7 pp.
MOD UK NES 2008-01. Specification for the Production of a Powdered Detergent (NSN 7930-99-768-1311) for General Purpose Cleaning Issue 1 (09.84); Amendment 2. 19 pp.

—Powder—Sampling
CGSB 2-GP-11M METH 1.2-83. Methods of Testing and Analysis of Soaps and Detergents Sampling-Flakes, Chips and Powders. 1 p.

—Power Packs
MOD UK TS 10302A. Detergent Solution, for the Cleaning of Vehicles and Power Packs in Pressure Washing Machines.

—Printing
CNS K8009-71. Liquid Detergent (for Printing and Dyeing Processes) (Tentative) (Jan)(3163).

—Qualitative Analysis
BSI BS 3762: Part 2-89. 1989 Analysis of Formulated Detergents Part 2: Qualitative Test Methods. 4 pp.
CGSB 2-GP-11M METH 30.1-83. Methods of Testing and Analysis of Soaps and Detergents Qualitative Test for Type of Synthetic Detergent (Supersedes 30.1a of May 1968). 1 p.

—Quaternary Compound Content
MOD UK M 9521/65. Examination of Detergent Solution for Plastic Food Utensils.

—Residue Content
MOD UK M 9521/65. Examination of Detergent Solution for Plastic Food Utensils.

—Rinsing Properties
CGSB 2-GP-11M METH 31.1-83. Methods of Testing and Analysis of Soaps and Detergents Rinsing Properties. 1 p.

—Sampling
BSI BS 3762: Part 1-83. 1983 Analysis of Formulated Detergents Part 1: Method of Sample Division. 10 pp.
CGSB 2-GP-11M METH 2-83. Methods of Testing and Analysis of Soaps and Detergents Laboratory Preparation of Samples (Supersedes 2.1, 2.2, 2.3, 2.4 of October 1958). 1 p.
CNS K6720-82. Method of Sampling for Synthetic Detergents (Mar)(8603).
ISO 607-80. Surface Active Agents and Detergents—Methods of Sample Division First Edition. 9 pp.
ISO 8212-86. Soaps and Detergents—Techniques of Sampling During Manufacture First Edition. 15 pp.

—Sea Water—Stability Testing
CGSB 2-GP-11M METH 19.2-83. Methods of Testing and Analysis of Soaps and Detergents Stability to Sea Water. 1 p.

Detergents (Cont.)
—Silica Content—Gravimetric Analysis
BSI BS 3762: Sec 3.21-86. 1986 Analysis of Formulated Detergents Part 3: Quantitative Test Methods Section 3.21: Method for Determination of Total Silica Content. 4 pp.
ISO 8215-85. Surface Active Agents—Washing Powders—Determination of Total Silica Content—Gravimetric Method First Edition. 5 pp.

—Silicate Content
CGSB 2-GP-11M METH 24.1-83. Methods of Testing and Analysis of Soaps and Detergents Silicates. 1 p.

—Soap Content
BSI BS 3762: Sec 3.9-86. 1986 Analysis of Formulated Detergents Part 3: Quantitative Test Methods Section 3.9: Method for Determination of Soap Content. 4 pp.

—Soap Content—Qualitative Analysis
CGSB 2-GP-11M METH 29.1-83. Methods of Testing and Analysis of Soaps and Detergents Soap (Qualitative Test) (Supersedes 29.1a of May 1968). 1 p.

—Solubility
CGSB 2-GP-11M METH 32.1-83. Methods of Testing and Analysis of Soaps and Detergents Solubility. 1 p.

—Soluble Matter Content
BSI BS 3762: Sec 3.6-86. 1986 Analysis of Formulated Detergents Part 3: Quantitative Test Methods Section 3.6: Method for Determination of Matter Soluble in Light Petroleum. 4 pp.

—Sulfate Content—Gravimetric Analysis
BSI BS 3762: Sec 3.22-86. 1986 Analysis of Formulated Detergents Part 3: Quantitative Test Methods Section 3.22: Method for Determination of Inorganic Sulphate Content. 6 pp.
ISO 8214-85. Surface Active Agents—Washing Powders—Determination of Inorganic Sulfates—Gravimetric Method First Edition. 5 pp.

—Surfactants
SAA AS 3867-91. The Qualitative Identification of Surfactant Components in Detergents. 6 pp.

—Volatile Matter Content
BSI BS 3762: Sec 3.25-89. 1989 Analysis of Formulated Detergents Part 3: Quantitative Test Methods Section 3.25: Methods for Determination of Water and Volatile Matter Content. 6 pp.

—Water Content
BSI BS 3762: Sec 3.25-89. 1989 Analysis of Formulated Detergents Part 3: Quantitative Test Methods Section 3.25: Methods for Determination of Water and Volatile Matter Content. 6 pp.

Detonating Cords
See Also: Detonating Fuses; Detonators; Explosives
CNS Z2036-88. Detonating Cord (Sep)(7532).
CNS Z3019-88. Method of Test for Detonating Cord (Sep)(7533).

Detonating Fuses
See Also: Detonating Cords; Fuses (Ordnance)
JIS K 4820-78. Detonating Fuses.

Detonation Flame Spraying Services
See Also: Flame Spraying; Services
MOD UK DTD-941-69. Surface Coating of Parts by Use of Detonation, Flame and Plasma Spraying Processes. 2 pp.

—Aerospace
MOD UK DTD-941-69. Surface Coating of Parts by Use of Detonation, Flame and Plasma Spraying Processes. 2 pp.

Detonators
Use For: Percussion Caps See Also: Detonating Cords; Explosives; Initiators (Explosives)
CNS Z2005-88. General Detonator for Engineering Use (Sep)(1427).
MOD UK DSTAN 95-10-01. Cap Copper Alloy Strip for Detonator Cups and Percussion Caps Issue 1; Amendment 1. 14 pp.

—Cables (Electric)
MOD UK DSTAN 61-12: Part 17-01. Wires, Cords, and Cables, Electrical-Metric Units Part 17: Cables, Special Purpose, Electrical (for Detonator Firing Circuits) Issue 1; Amendment 1. 11 pp.

—Cords (Electric)
MOD UK DSTAN 61-12: Part 17-01. Wires, Cords, and Cables, Electrical-Metric Units Part 17: Cables, Special Purpose, Electrical (for Detonator Firing Circuits) Issue 1; Amendment 1. 11 pp.

Detonators (Cont.)
—Electric Wire
MOD UK DSTAN 61-12: Part 17-01. Wires, Cords, and Cables, Electrical-Metric Units Part 17: Cables, Special Purpose, Electrical (for Detonator Firing Circuits) Issue 1; Amendment 1. 11 pp.

—Electrical
CNS Z2006-88. Electrical Detonator (Sep)(1428).

—Electrical—Delay
CNS Z2069-88. Delay Electric Detonator for Engineering Use (Sep)(12429).

—Mining
CNS M2013-73. Generator Type Explosion Initiators (Dec)(3680).

—Permissible
JIS K 4821-71. Permissible Detonator.

Deuterium Oxide
Use For: Heavy Water
CNS J2018-81. Method of Test for Deuterium Oxide (May)(7395).
CNS J3001-81. Deuterium Oxide (Mar)(7155). 1 p.

—Impurities—Potassium Permanganate
CNS J2009-81. Method of Test for Consumption of Potassium Permanganate by Impurities in Deuterium Oxide (Mar)(7154).

Deutsches Institut fuer Normung
Use: DIN

Devarda's Alloy
Use: Devarda's Metal

Devarda's Metal
CNS K7178-65. Chemical Reagent (Devarda Metal) (Jan)(1678).
JIS K 8653-85. Devarda's Alloy (Devarda's Metal).

Developing Machines (Photography)
Use: Photographic Processors

Development
Use: Research and Development

Development Contracts
Use: Research and Development Contracts

Devitrified Glass
Use: Glass Ceramics

Dew
—Doors
CNS A3234-87. Method of Test for Dew Condensation of Windows and Doors (Oct)(11525).
JIS A 1514-82. Test Method of Dew Condensation for Windows and Doors (R 1987). 15 pp.

—Windows
CNS A3234-87. Method of Test for Dew Condensation of Windows and Doors (Oct)(11525).
JIS A 1514-82. Test Method of Dew Condensation for Windows and Doors (R 1987). 15 pp.

Dew Point Hygrometers
See Also: Hygrometers
DIN ENGL 50012 Pt 4-86. Climates and Their Technical Application; Methods of Measuring Humidity; Dew Point Hygrometer (Jan). 2 pp.

—Natural Gas
BSI BS 3156: SUB SEC 10.3.1-87. 1987 Methods for the Analysis of Fuel Gases Part 10: General Methods Section 10.3: Determina-tion of Water Dew Point and Relative Humidity Subsection 10.3.1: Prin-ciples of Cooled Surface Condensation Hygrometers and General Description of Their Use. 8 pp.
ISO 6327-81. Gas Analysis—Determination of the Water Dew Point of Natural Gas—Cooled Surface Condensation Hygrometers First Edition. 7 pp.

Dextrin Content Analysis
—Glucose
CNS N6054-72. Method of Test for Solid Glucose (Jun)(3349). 5 pp.
CNS N6055-72. Method of Test for Powdered Glucose (Jun)(3351). 3 pp.

—Maltose Syrups
CNS N6073-82. Method of Test for Malt Sugar Syrup (May)(3584). 5 pp.

Dextrins
See Also: Starches
CNS K6169-65. Method of Test for Dextrine of Industrial Grade (Sep)(2189)(R 1971).

INDUSTRY STANDARDS

INTERNATIONAL AND NON-U.S. NATIONAL STANDARDS
SUBJECT INDEX

Dextrins

Dextrins *(Cont.)*
CNS K7179-65. Chemical Reagent (Dextrin) (Jan)(1679).
CNS K8005-66. Dextrin (Industrial Grade) (Sep)(2188)(R 1973).
JIS K 8646-61. Dextrin.

Dextrose
Use: Glucose

Dezincification
Scope Note: See the subheading Dezincification under specific types of metals

DH 104 Riley Dove Aircraft
See Also: Aircraft
—**Antenna Positions**
  CAA. DH104/Dove. 1 p.
—**Mandatory Aircraft Modifications and Inspections Summaries**
  CAA. DH 104 Riley Dove Series. 4 pp.

DH 112/Venom MK1 Aircraft
See Also: Aircraft
—**Antenna Positions**
  CAA. DH 112/Venom MK1. 1 p.

DH 80A Puss Moth Aircraft
See Also: Aircraft
—**Antenna Positions**
  CAA. DH80A Puss Moth (Approved Aerial Positions). 1 p.

DH 82A Tiger Moth Aircraft
See Also: Aircraft
—**Antenna Positions**
  CAA. DH82A Tiger Moth (Approved Aerial Positions). 1 p.

DH 85 Leopard Moth Aircraft
See Also: Aircraft
—**Antenna Positions**
  CAA. DH85 Leopard Moth (Approved Aerial Positions). 1 p.

DH 87B Aircraft
See Also: Aircraft
—**Antenna Positions**
  CAA. DH87B (Approved Aerial Positions). 1 p.

DH 89A Rapide Aircraft
See Also: Aircraft
—**Antenna Positions**
  CAA. DH89A Rapide (Approved Aerial Positions). 1 p.

DH 90 Dragonfly Aircraft
See Also: Aircraft
—**Antenna Positions**
  CAA. DH90/DH90A Dragonfly (Approved Aerial Positions). 1 p.

DH 90A Dragonfly Aircraft
See Also: Aircraft
—**Antenna Positions**
  CAA. DH90/DH90A Dragonfly (Approved Aerial Positions). 1 p.

di-Boron Trioxide
Use: Boric Anhydrides

Di-2-Ethyl-Hexyl Phthalate
Use: Di(2-Ethylhexyl) Phthalate

Di(2-Ethylhexyl) Phthalate
Use For: Bis(2-Ethylhexyl) Phthalate; Di-2-Ethyl-Hexyl Phthalate *See Also:* Di-N-Octyl Phthalate
CNS K1179-83. Di-2-Ethyl-Hexyl Phthalate (Nov)(4010).
JIS K 6753-77. Di-2-Ethylhexyl Phthalate (R 1980). 5 pp.

Di (2-Ethylhexyl) Sebacate
Use: Bis (2-Ethylhexyl) Sebacate

Di(Methylcyclohexyl) Phthalate
MOD UK DSTAN 68-15-70. Di (Methylcylcohexyl) Phthalate, Technical and Di (Methylcyclohexyl) Phthalate, Technical (Lead—Free) Issue 1. 7 pp.

N,N'-di-beta-Naphthyl-p-Phenylenediamine
Use For: DNPD
CNS K4043-87. Rubber Antioxidant DNPD (N, N-Di—B—Naphthyl P—Phenylenediamine) (Mar)(4861).
CNS K6902-87. Method of Test for Rubber Antioxidant DNPD Phenylenediamine) (Mar)(11883).

Di-N-Octyl Phthalate
Use For: Dinormal Octyl Phthalate *See Also:* Di(2-Ethylhexyl) Phthalate
CNS K1274-83. Dinormal Octyl Phthalate (Nov)(10657).
JIS K 6754-77. Dinormal Octyl Phthalate (R 1980). 5 pp.

Di-ortho-toylguanidine
Use For: DOTG
CNS K4042-78. Rubber Vulcanization Accelerator DOTG (Dio-Toyl Guanidine) (Oct)(4636).

Di-Sodium Hydrogen Phosphate
Use: Sodium Phosphate, Dibasic

Diacetin
Use For: Glyceryl Diacetate
BSI BS 1594-50. (WITHDRAWN) 1950 Amd 1 Diacetin (Glyceryl Diacetate). 11 pp.

Diacetone Alcohol
See Also: Acetone; Alcohols; Hazardous Materials
BSI BS 549-70. 1970 Diacetone Alcohol. 12 pp.
CNS K1226-80. Diacetone Alcohol (Feb)(5232).
ISO 2517-74. Diacetone Alcohol for Industrial Use—List of Methods of Test First Edition. 4 pp.
—**Acidity—Volumetric Analysis**
  ISO 2887-73. secButyl Alcohol, Methyl Ethyl Ketone, isoButyl Methyl Ketone, isoAmyl Ethyl Ketone, Diacetone Alcohol, and Hexylene Glycol for Industrial Use—Determination of Acidity to Phenolphthalein—Volumetric Method First Edition. 4 pp.
—**Miscibility**
  ISO 2518-74. Diacetone Alcohol and Hexylene Glycol for Industrial Use—Test for Miscibility with Water First Edition. 4 pp.

Diacetyl Monoxime
Use: 2,3-Butanedione Monoxime

Diacetylmethane
Use: Acetylacetone

Diagnostic Agents
Scope Note: Use a more specific term *See:* Barium Sulfate; Drugs; Dyes

Diagnostic Equipment
Use For: Diagnostic Systems *See Also:* Laboratory Equipment; Self Diagnostic Routine; Testing Equipment
—**Automatic—Communication Equipment**
  CCIR Report 993-86. Automatic Diagnosis of HF Equipment—Section 3Ad—Operational Questions. 4 pp.
—**Automotive—Glossaries**
  BSI BS AU 206: Part 1-86. 1986 Diagnostic Systems for Road Vehicles Part 1: Glossary. 3 pp.
  ISO 4092-88. Road Vehicles—Diagnostic Systems for Motor Vehicles—Vocabulary Third Edition; (Corrigendum 1-1991). 5 pp.
—**Automotive—Symbols**
  BSI BS AU 206: Part 2-86. 1986 Diagnostic Systems for Road Vehicles Part 2: Graphical Symbols for Diagnostic Testers. 11 pp.
  ISO 7639-85. Road Vehicles—Diagnostic Systems—Graphical Symbols First Edition. 12 pp.

Diagnostic Imaging Equipment
Use For: Diagnostic Imaging Installations
See Also: Medical Electrical Equipment; Medical Equipment; X-Ray Computed Tomographs; X-Ray Equipment; X-Ray Film Changers
CSA CAN/CSA-C22. 2 NO 114-M90. Diagnostic Imaging and Radiation Therapy Equipment; (Gen Instr 1). 122 pp.
CSA 1169 Bull. Electrical Bulletin 1169 June 27, 1978 to C22.2 NO 114. 2 pp.
IEC 1223 Pt 1-93. Evaluation and Routine Testing in Medical Imaging Departments—Part 1: General Aspects First Edition. 46 pp.

Diagnostic Imaging Installations
Use: Diagnostic Imaging Equipment

Diagnostic Instruments (Medical)
Use For: Clinical Investigation Instruments
See Also: Electrocardiographs; Electroencephalographs; Medical Electrical Equipment; Medical Equipment; Medical Instruments; Probes (Medical); Speculum; Tongue Depressors; Ultrasonic Medical Equipment; Vectorcardiographs
—**Lamps**
  BSI BS 1929-53. (OBSOLESCENT) 1953 Screw Threads and External Dimensions of Lamps for Endoscopic and Diagnostic Instruments. 8 pp.
—**Medical Electrical Equipment**
  CSA C22.2 NO 125-M1984. Electromedical Equipment (R 1992); (Gen Instr 1 Thru 2). 68 pp.
  CSA 1169 Bull. Electrical Bulletin 1169 June 27, 1978 to C22.2 NO 125. 2 pp.

Diagnostic Systems
Use: Diagnostic Equipment

Diagnostic Testing
—**Automotive—Electronic Systems**
  BSI BS AU 206: Part 4-86. 1986 Diagnostic Systems for Road Vehicles Part 4: General Requirements for Testing Electronic Systems. 5 pp.
  ISO 8093-85. Road Vehicles—Diagnostic Testing of Electronic Systems First Edition. 5 pp.
  ISO 9141-89. Road Vehicles—Diagnostic Systems—Requirements for Interchange of Digital Information First Edition. 16 pp.
—**Automotive—Ingition Systems**
  BSI BS AU 206: Part 3-86. 1986 Diagnostic Systems for Road Vehicles Part 3: General Requirements for Testing Ignition Systems. 8 pp.
  ISO 7342-82. Road Vehicles—Diagnostic Systems—Equipment for Ignition Systems Testing First Edition. 8 pp.
—**Earthmoving Equipment**
  BSI BS 5635-82. (WITHDRAWN) 1982 Service Instrumentation of Earth-Moving Machinery (Superseded by BS 6913: Part 4: 1990). 11 pp.
  BSI BS 6913: Part 4-90. 1990 Operation and Maintenance of Earth-Moving Machinery Part 4: Recommendations for Service Instrumentation. 12 pp.
  ISO 6012-89. Earth-Moving Machinery—Service Instrumentation Third Edition. 10 pp.

Diagrams
Scope Note: For additional listings, see the subheading Diagrams under specific products or applications *Use For:* Figures *See Also:* Charts; Circuit Diagrams; Drawings; Flow Charts; Graphic Arts; Kinematic Diagrams; Logic Diagrams; Polar Diagrams; Schematic Diagrams; Symbols; Visual Aids
BSI BS 5070: Part 1-88. 1988 Engineering Diagram Drawing Practice Part 1: Recommendations for General Principles. 15 pp.
CNS B1001-11-89. Engineering Drawing Representation of Diagram (Mar)(3-11).
SAA AS 1103.5-78. Diagrams, Charts and Tables for Electrotechnology—Part 5: Preparation of Interconnection Diagrams and Tables. 4 pp.
SNZ NZS/AS 1103. 5-78. Diagrams, Charts and Tables for Electrotechnology Part 5: Preparation of Interconnection Diagrams and Tables (This is a Joint Standard with SAA AS 1103.5). 8 pp.
SNZ BS 5070-74. Drawing Practice for Engineering Diagrams. 65 pp.
SNZ BS 5070: Part 1-88. Engineering Diagram Drawing Practice Part 1: Recommendations for General Principles. 16 pp.
—**Designations**
  CENELEC HD 246.2-75. Diagrams, Charts, Tables Part 2: Item Designation. 2 pp.
—**Glossaries**
  CENELEC HD 246.1-75. Diagrams, charts, tables Part 1: Definitions and classification. 2 pp.

Dial Depth Gages
Use: Depth Gages—Dial

Dial Gages
Use For: Dial Gauges; Lever Gauges
See Also: Valves
BSI BS 907-65. 1965 Dial Gauges for Linear Measurement. 20 pp.
CNS B6027-89. Dial Gauge Reading in 0.01mm (Dec)(4176).
CNS B6028-89. Dial Gauge Reading in 0.001mm (Dec)(4177).
DIN ENGL 878-83. Dial Gauges (Oct). 5 pp.
DIN ENGL 879 Pt 1-83. Dial Indicator with Mechanical Indication (Oct). 4 pp.
DIN ENGL 879 Pt 3-75. Dial Indicators with Electric Limit Contacts (Nov). 2 pp.
DIN ENGL 2270-85. Lever Gauges (Apr). 5 pp.

Dial Gages *(Cont.)*
ISO R463-65. Dial Gauges Reading in 0.01 mm, 0.001 in and 0.0001 in First Edition. 7 pp.
JIS B 7503-92. Dial Gauges. 16 pp.
JIS Z 8306-64. General Rules for Scale of Industrial Instruments.

—**Depth**
CGSB 39-GP-19M-79. Calipers and Gages, Vernier and Dial, Standard for. 22 pp.

Dial Gauges
Use: Dial Gages

Dial Indicators
Use: Dials

Dial Scales
See Also: Scales (Indicators)

—**Graduated**
BSI BS 6903-87. 1987 Graduated Dial Scales for Ophthalmic Instruments. 4 pp.
ISO 8429-86. Optics and Optical Instruments—Ophthalmology—Graduated Dial Scale First Edition. 3 pp.

Dial Tones
See Also: PABX Internal Dial Tones; Second Dial Tones; Special Dial Tones; Tones (Telephone Services)
CCITT RECMN Q.35-89. Technical Characteristics of Tones for the Telephone Service—General Recommendations on Telephone Switching and Signalling—Functions and Information Flows for Services in the ISDN —Supplements (Study Group XI) 9 pp. 9 pp.
CENELEC PRETS 300 295-92. Human Factors (HF); Specification of Characteristics of Telephone Services Tones When Locally Generated in Terminals. 9 pp.
ETSI PRETS 300 295-92. Human Factors (HF); Specification of Characteristics of Telephone Services Tones When Locally Generated in Terminals. 9 pp.

—**Distortion (Electrical)**
CCITT RECMN E.180-89. Technical Characteristics of Tones for the Telephone Service—Telephone Network and ISDN—Operation, Numbering, Routing and Mobile Service (Study Group II) 9 pp. 9 pp.

—**Frequency Bands**
CCITT RECMN E.180-89. Technical Characteristics of Tones for the Telephone Service—Telephone Network and ISDN—Operation, Numbering, Routing and Mobile Service (Study Group II) 9 pp. 9 pp.

—**Private Automatic Branch Exchanges**
CEPT T/SF 50 E-87. PABX Internal Dial Call. 1 p.

Diallyl Phthalate
Use For: Polydiallyl Phthalate

—**Molding Materials**
JIS K 6918-77. Diallyl Phthalate Molding Compounds (R 1980). 16 pp.

Dials
Use For: Dial Indicators *See Also:* Display Devices; Indicating Instruments
BSI BS 2795-81. 1981 Amd 1 Dial Test Indicators (Lever Type) for Linear Measurement. 9 pp.
BSI BS 3693A-64. (WITHDRAWN) 1964 Recommendations for the Design of Scales and Indexes: Part A: Recommended Form of Digits for Use on Dials and Scales. 4 pp.
BSI BS 3693B-64. (WITHDRAWN) 1964 Recommendations for the Design of Scales and Indexes: Part B: Geometric Construction of the Recommended Form of Digits for Use on Dials and Scales. 7 pp.
CGSB 39-GP-39-59. Indicators; Dial, Mechanical, and Accessories, Specification for. 29 pp.
CNS B6030-82. Lever Type Dial Test Indicators (Jan)(4753).
JIS B 7533-90. Dial Test Indicators (Lever Type). 15 pp.

—**Aluminum—Photosensitive**
CGSB 62-GP-10-69. Photosensitive Anodized Aluminum Markings: Plates and Foil, Photographic, Standard for. 17 pp.

—**Telex Communications**
CCITT RECMN U.2-89. Standardization of Dials and Dial Pulse Generators for the International Telex Service—Telegraph Switching (Study Group IX) 1 pp. 1 p.

—**Trim Controls—Aircraft**
SBAC AS 877 ISSUE 4. Zero Marker—Trim Control. 1 p.
SBAC AS 891 ISSUE 6. Dial—Trim Control. 1 p.

Dials *(Cont.)*
—**Trim Controls—Aircraft** *(Cont.)*
SBAC AS 1234 ISSUE 3. Dial Assembly—Trim Control. 1 p.
SBAC AS 1235 ISSUE 2. Dial Assembly—Trim Control. 1 p.

Dialysis Equipment
Use For: Peritoneal Dialysis Equipment
See Also: Medical Electrical Equipment; Medical Equipment

—**Safety**
BSI BS 5724: Sec 2.29:-92. 1992 Medical Electrical Equipment Part 2: Particular Requirements for Safety Section 2.29: Specification for Peritoneal Dialysis Equipment (ALSO KNOWN AS BS EN 50072: 1992). 16 pp.
BSI BS EN 50072-92. 1992 Medical Electrical Equipment Part 2: Particular Requirements for Safety Section 2.29: Specification for Peritoneal Dialysis Equipment. 16 pp.
CENELEC PREN 50 072-89. Medical Electrical Equipment Particular Requirements for the Safety of Peritoneal Dialysis Equipment. 11 pp.
CENELEC EN 50072-92. Medical Electrical Equipment—Particular Requirements for the Safety of Peritoneal Dialysis Equipment. 12 pp.

Diameters
Scope Note: For diameters of specific products, see the subheading Diameters under the specific product. For additional references, consult the following list
See Also: Hole Size; Index of Cooperation
CNS Z7002-44. Standard Diameter (Jun)(2)(R 1972).

Diamines
See Also: N-Isopropyl-N'-Phenyl-p-Phenylenediamine

—**Amine Content—Potentiometric Analysis**
CNS K6689-81. Total, Primary, Secondary and Tertiary Amine Values of Fatty Amines, Amidoamines and Diamines by Referee Potentiometric Method (Sep)(7913).

—**Iodine Number**
CNS K6682-81. Method of Test for Iodine Value of Fatty Amines, Amidoamines, and Diamines (Jul)(7726).

4,4'-Diamino-2,2-stilbenedisulfonic Acid
Use: Amsonic Acid

Diaminostilbene Disulfonic Acid
Use: Amsonic Acid

4,4-Diaminostilbenedisulfonic Acid
Use: Amsonic Acid

Diammonium Hydrogen Citrate
JIS K 8284-80. Diammonium Hydrogen Citrate.

Diammonium Hydrogenphosphate
Use: Ammonium Phosphate, Dibasic

Diamond Chisels
See Also: Chisels; Diamond Tools
CGSB 39-GP-43A-74. Chisels, Hand, Cutting, and Swaging (For Metals), Standard for; (Amendment 1 Dec 1978). 15 pp.

Diamond Cup Wheels
Use: Diamond Wheels

Diamond Dressers
See Also: Diamond Tools
JIS B 4134-83. Single Point Diamond Dressing Tools. 10 pp.
JIS B 4135-84. Impregnated Diamond Dressing Tools. 9 pp.
JIS B 4136-91. Chisel Edge Diamond Dressing Tools. 14 pp.

Diamond Drills
See Also: Diamond Tools; Diamond Wheels; Rotary Drilling Equipment

—**Coring**
BSI BS 4019: Part 3-93. 1993 Rotary Core Drilling Equipment Part 3: Specification for System A. Metric Units (ISO 3551-1: 1992) (Q). 84 pp.
BSI BS 4019: Part 4-93. 1993 Rotary Core Drilling Equipment Part 4: Specification for System A. Inch Units (ISO 3551-2: 1992) (Q). 84 pp.
ISO 3551 Pt 1-92. Rotary Core Diamond Drilling Equipment—System A—Part 1: Metric Units First Edition. 84 pp.
ISO 3551 Pt 2-92. Rotary Core Diamond Drilling Equipment—System A—Part 2: Inch Units First Edition. 84 pp.

Diamond Drills *(Cont.)*
—**Coring** *(Cont.)*
ISO 3552 Pt 1-92. Rotary Core Diamond Drilling Equipment—System B—Part 1: Metric Units First Edition. 48 pp.
ISO 3552 Pt 2-92. Rotary Core Diamond Drilling Equipment—System B—Part 2: Inch Units First Edition. 48 pp.
ISO 8866-91. Rotary Core Diamond Drilling Equipment—System C First Edition; (Corrigendum 1-1991) (Corrigendum 2-1992). 52 pp.

—**Electric—Portable**
CSA CAN/CSA-C22. 2 NO71.1-M89. Portable Electric Tools; (Gen Instr 1 Thru 4). 89 pp.

—**Glossaries**
BSI BS 4019: Part 3-93. 1993 Rotary Core Drilling Equipment Part 3: Specification for System A. Metric Units (ISO 3551-1: 1992) (Q). 84 pp.
BSI BS 4019: Part 4-93. 1993 Rotary Core Drilling Equipment Part 4: Specification for System A. Inch Units (ISO 3551-2: 1992) (Q). 84 pp.
ISO 3551 Pt 1-92. Rotary Core Diamond Drilling Equipment—System A—Part 1: Metric Units First Edition. 84 pp.
ISO 3551 Pt 2-92. Rotary Core Diamond Drilling Equipment—System A—Part 2: Inch Units First Edition. 84 pp.
ISO 3552 Pt 1-92. Rotary Core Diamond Drilling Equipment—System B—Part 1: Metric Units First Edition. 48 pp.
ISO 3552 Pt 2-92. Rotary Core Diamond Drilling Equipment—System B—Part 2: Inch Units First Edition. 48 pp.

—**Safety**
IEC 1029 Pt 2-6-93. Safety of Transportable Motor-Operated Electric Tools Part 2: Particular Requirements for Diamond Drills with Water Supply First Edition. 24 pp.

Diamond Saws
See Also: Circular Saws; Diamond Tools; Saws
BSI BS 2064: Part 2-76. 1976 Diamond Abrasive Products Part 2: Diamond Abrasive Circular Saws and Frame Saws. 12 pp.

—**Blades**
BSI BS 2064: Part 3-90. 1990 Diamond Abrasive Products Part 3: Dimensions of Steel Blades for Segmented Saws for Cutting of Stone and Masonry. 8 pp.
ISO 6105-88. Abrasive Products—Segmented Saws for Machining of Stone and Masonry Cutting—Dimensions of Steel Blades First Edition. 8 pp.

—**Glossaries**
BSI BS 5831-79. 1979 Designation and Multilingual Nomenclature for Diamond or Cubic Boron Nitride Grinding Wheels and Saws. 20 pp.
ISO 6104-79. Abrasive Products—Diamond or Cubic Baron Nitride Grinding Wheels and Saws—General Survey, Designation and Multilingual Nomenclature First Edition. 16 pp.

—**Safety**
IEC 1029 Pt 2-7-93. Safety of Transportable Motor-Operated Electric Tools Part 2: Particular Requirements for Diamond Saws with Water Supply First Edition. 24 pp.

Diamond Tools
See Also: Cutting Tools; Diamond Chisels; Diamond Dressers; Diamond Drills; Diamond Saws; Diamond Wheels; Diamonds (Industrial); Tools

—**Dental Instruments**
BSI BS 6828: Part 6-87. 1987 Amd 1 Dental Rotary Instruments Part 6: Specification for Diamond Inmstruments (AMD 6675) May 31, 1991. 26 pp.
BSI BS 6828: Sec 6.2-93. 1993 Dental Rotary Instruments Part 6: Diamond Instruments Section 6.2: Specification for Discs (ISO 7711-2: 1992). 12 pp.
CEN EN 27 711-90. Dentistry—Dental Rotary Instruments—Diamond Instruments. 4 pp.
CSA CAN/CSA-Z349. 29-35.2-M92. Dental Instruments and Equipment—Volume 1 (CAN/CSA-Z349.29-M92 to CAN/CSA-Z349.35.2 M92); (Gen Instr 1). 183 pp.
ISO 7711-84. Dental Rotary Instruments—Diamond Instruments First Edition. 22 pp.
ISO 7711 Pt 2-92. Dental Rotary Instruments—Diamond Instruments—Part 2: Discs First Edition. 10 pp.

—**Dies—Wire Drawing**
JIS B 4132-83. Diamond Dies for Wire Drawing. 10 pp.
JIS B 4133-83. Polycrystalline Diamond Dies for Wire Drawing. 10 pp.

Diamond Wheels
Use For: Diamond Cup Wheels *See Also:* Diamond Drills; Diamond Tools; Grinding Wheels; Wheels
BSI BS 2064: Part 1-76. 1976 Amd 1 Diamond Abrasive Products Part 1: Diamond Abrasive Grinding Wheels (AMD 5935) May 31, 1989. 21 pp.
ISO 6168-80. Abrasive Products—Diamond or Cubic Boron Nitride Grinding Wheels—Dimensions First Edition. 17 pp.
JIS B 4131-82. Diamond or Cubic Boron Nitride Grinding Wheels.
SAA AS B209-69. Diamond Abrasive Wheels and Hones. 39 pp.

—Glossaries
BSI BS 5831-79. 1979 Designation and Multilingual Nomenclature for Diamond or Cubic Boron Nitride Grinding Wheels and Saws. 20 pp.
ISO 6104-79. Abrasive Products—Diamond or Cubic Boron Nitride Grinding Wheels and Saws—General Survey, Designation and Multilingual Nomenclature First Edition. 16 pp.

Diamonds (Industrial)
See Also: Carbon; Diamond Tools; Grinding Wheels; Minerals; Ores

—Grain Size Analysis
BSI BS 5851-80. 1980 Grain Sizes of Diamond or Cubic Boron Nitride. 18 pp.
ISO 6106-79. Abrasive Products—Grain Sizes of Diamond or Cubic Boron Nitride First Edition. 8 pp.
JIS B 4130-82. Grain Sizes of Diamond or Cubic Boron Nitride.

Dianisidine
CNS K2174-87. Dianisidine (4, 4-Diamino-3, 3-Dimethoxybiphenyl) (Dec)(12184). 2 pp.
JIS K 4106-77. Dianisidine (4, 4'-Diamino-3, 3'-Dimethoxybiphenyl).

Diantipyrylmethane
JIS K 9565-86. Diantipyrylmethane Monohydrate (4,4'-methylenediantipyrine Monohydrate).

Diapers
Use For: Babies' Napkins; Nappies
See Also: Incontinent Care Products
BSI BS 5815: Part 2-88. 1988 Sheets, Sheeting, Pillowslips, Towels, Napkins, Counterpanes and Continental Quilt Secondary Covers Suitable for Use in the Public Sector Part 2: Towels and Napkins. 8 pp.

—Cotton
BSI BS 5815-79. (WITHDRAWN) 1979 Cotton and Man-Made Fibre Blend Sheeting, Sheets, Pillowslips, Towels and Napkins for Use in the Public Sector (Superseded by BS 5815: Part 1: 1989, Part 2: 1988, and Part 3: 1991). 11 pp.

—Covers
JIS L 4211-81. Diaper Cover.

—Disposable
CNS P2071-90. Disposable Diaper for Baby (Jun)(12639). 2 pp.
CNS P3102-80. Method of Test for Disposable Diaper for Baby (Jun)(12640). 4 pp.

—Polyester/Cotton
BSI BS 5815-79. (WITHDRAWN) 1979 Cotton and Man-Made Fibre Blend Sheeting, Sheets, Pillowslips, Towels and Napkins for Use in the Public Sector (Superseded by BS 5815: Part 1: 1989, Part 2: 1988, and Part 3: 1991). 11 pp.

Diaphragm Flowmeters
Use For: Dry Meters *See Also:* Flowmeters

—Gas
BSI BS 4161: Part 5-90. 1990 Amd 1 Gas Meters Part 5: Specification for Diaphragm Meters for Working Pressures up to 7 Bar (AMD 6751) September 30, 1991. 38 pp.
BSI BS 4161: Part 5-02. 1990 Amd 2 Gas Meters Part 5: Specification for Diaphragm Meters for Working Pressures up to 7 Bar (AMD 7450) January 15, 1993. 40 pp.

Diaphragm Gages
See Also: Differential Pressure Gages; Vacuum Gages
DIN ENGL 16005-87. General Purpose Pressure Gauges with Elastic Pressure-Responsive Elements; Requirements and Testing (Feb). 7 pp.
JIS B 7546-83. Diaphragm Seal Pressure Gauges. 21 pp.

Diaphragm Meters
Use: Diaphragm Flowmeters

Diaphragm Pumps
See Also: Reciprocating Pumps

—Fuel—Automotive
CNS D2119-90. Mechanical Fuel Pumps for Automobiles (Jan)(8386).
CNS D2120-82. Fuel Pump Diaphragms for Automobiles (Jan)(8387).
CNS D3187-90. Method of Test for Fuel Pumps Diaphragms for Automobiles (Jan)(12656).
JIS D 3601-88. Mechanical Fuel Pumps for Automobiles. 15 pp.

Diaphragm Valves
See Also: Valves
BSI BS 1212: Part 3-90. 1990 Float Operated Valves Part 3: Specification for Diaphragm Type Float Operated Valves (Plastics Bodied) for Cold Water Services Only (Excluding Floats). 23 pp.
BSI BS 5156-85. 1985 Amd 1 Diaphragm Valves for General Purposes (AMD 6513) June 29, 1990. 24 pp.
DIN ENGL 3441 Pt 3-84. Unplasticized Polyvinyl Chloride (PVC-U) Valves; Diaphragm Valves; Dimensions (Aug). 3 pp.
DIN ENGL 3442 Pt 3-87. Polypropylene (PP) Valves; Diaphragm Valves; Dimensions (July). 3 pp.

—Fire Hydrants
BSI BS 5041. Part 1-87. 1987 Amd 1 Fire Hydrant System Equipment Part 1: Landing Valves for Wet Risers (AMD 5912) September 30, 1988. 15 pp.

—Float
BSI BS 1212: Part 2-90. 1990 Float Operated Valves (Excluding Floats) Part 2: Diaphragm Type Float Operated Valves (Copper Alloy Body) (Excluding Floats). 24 pp.
BSI BS 1212: Part 2-70. (WITHDRAWN) 1970 Amd 4 Ballvalves (Excluding Floats) Part 2: Diaphragm Type (Brass Body). 35 pp.

—Pipelines—Flanged
CEN PREN 558-1-91. Metal Valves for Use in Flanged Pipe Systems—Face-to-Face and Centre-to-Face Dimensions—Part 1: General. 5 pp.
CEN PREN 558-2-91. Metal Valves for Use in Flanged Pipe Systems—Face-to-Face and Centre-to-Face Dimensions—Part 2: PN Designated Valves. 13 pp.
CEN PREN 558-3-91. Metal Valves for Use in Flanged Pipe Systems—Face-to-Face and Centre-to-Face Dimensions—Part 3: Class-Designation Valves. 17 pp.
ISO 5752-82. Metal Valves for Use in Flanged Pipe Systems—Face-to-Face and Centre-to-Face Dimensions Second Edition. 13 pp.

Diaphragm Walls
Use: Retaining Walls

Diaphragms (Contraceptive)
Use: Contraceptive Diaphragms

Diaphragms (Mechanics)
See Also: Flow Measurement Equipment; Orifice Plates

—Asbestos Fabric—Electrolysis
CNS R2180-85. Asbestos Cloth Diaphragms for Electrolysis of Water (May)(11269).
JIS R 3456-79. Asbestos Cloth Diaphragms for Electrolysis of Water. 8 pp.

—Expansion Vessels—Water Heaters
BSI BS 4814-90. 1990 Expansion Vessels Using an Internal Diaphragm, for Sealed Hot Water Heating Systems. 12 pp.
BSI BS 6144-90. 1990 Expansion Vessels Using an Internal Diaphragm, for Unvented Hot Water Supply Systems. 13 pp.

—Leather—Gas Meters
BSI BS 2797-76. 1976 Amd 1 Leathers for Gas Meter Diaphrams. 9 pp.

—Rubber—Appliances
BSI BS EN 278-91. 1991 Rubber Materials for Diaphragms in Domestic Appliances Using Combustible Gases up to 200 mbar. 11 pp.
CEN PREN 278-88. Rubber Materials for Diaphrams in Domestic Appliances Using Combustible Gases. 10 pp.
CEN EN 278-91. Rubber Materials for Diaphrams in Domestic Appliances Using Combustible Gases up to 200 mbar. 12 pp.

Diapositives
Use: Projection Slides

Diathermy Machines
Use For: Short Wave Therapy Equipment
See Also: Electrotherapy Equipment; Medical Electrical Equipment; Medical Equipment

Diathermy Machines *(Cont.)*
SAA AS 3130-88. Approval and Test Specification—Beauty Therapy Equipment (This is a Joint Standard with SANZ NZS 3130). 16 pp.
SAA AS 3209-86. Approval and Test Specification—Shortwave Therapy Equipment (This is a Joint Standard with SANZ NZS 3209) (Withdrwan). 12 pp.
SNZ NZS/AS 3130-88. Approval and Test Specification—Beauty Therapy Equipment (This is a Joint Standard with SAA AS 3130). 16 pp.
SNZ NZS/AS 3209-86. Approval and Test Specification—Short-Wave Therapy Equipment (This is a Joint Standard with SAA AS 3209). 12 pp.

—Microwave
JIS T 1353-84. Microwave Therapy Equipment.
SAA AS 3210-86. Approval and Test Specification—Microwave Therapy Equipment (This is a Joint Standard with SANZ NZS 3210) (Withdrawn). 9 pp.
SNZ NZS/AS 3210-86. Approval and Test Specification—Microwave Therapy Equipment (This is a Joint Standard with SAA AS 3210). 9 pp.

—Safety
BSI BS 5724: Sec 2.3-92. 1992 Medical Electrical Equipment Part 2: Particular Requirements for Safety Section 2.3: Specification for Short-Wave Therapy Equipment (IEC 601-2-3: 1991) (N). 24 pp.
BSI BS 5724: Sec 2.3-01. 1992 Amd 1 Medical Electrical Equipment Part 2: Particular Requirements for Safety Section 2.3: Specification for Short-Wave Therapy Equipment (BS EN 60601-2-3: 1993) (AMD 7879) August 15, 1993 (N). 32 pp.
BSI BS 5724: Sec 2.3-83. 1983 Medical Electrical Equipment Part 2: Particular Requirements for Safety Section 2.3: Short Wave Therapy Equipment. 20 pp.
BSI BS EN 60601-2-3-93. 1993 Amd 1 Medical Electrical Equipment Part 2: Particular Requirements for Safety Section 2.3: Specification for Short-Wave Therapy Equipment (IEC 60-2-3: 1991) (AMD 7879) August 15, 1993 (N). 32 pp.
CENELEC HD 395.2.3-85. Medical Electrical Equipment Part 2: Particular Requirements for the Safety of Short-Wave Therapy Equipment. 3 pp.
CENELEC EN 60601-2-3-93. Medical Electrical Equipment Part 2: Particular Requirements for the Safety of Short-Wave Therapy Equipment (IEC 601-2-3: 1991). 20 pp.
CSA CAN/CSA-C22. 2N601.2.3-92. Medical Electrical Equipment Part 2: Particular Requirements for the Safety of Short-Wave Therapy Equipment (IEC 601-2-3:1991); (Gen Instr 1). 36 pp.
IEC 601 Pt 2-3-91. Medical Electrical Equipment Part 2: Particular Requirements for the Safety of Short-Wave Therapy Equipment Second Edition; (CAN/CSA-C22.2 No. 601.2.3-92) (CENELEC EN 60601-2-3: 1993). 43 pp.
SAA AS 3200.2.3-92. Approval and Test Specification—Medical Electrical Equipment—Part 2: Particular Requirements for Safety—Part 2.3: Short-Wave Therapy Equipment (IEC 601-2-3:1991) (Supersedes AS 3209—1986 (Which Will Remain Current)) (in Professional. 17 pp.

—Ultrasonic
SAA AS 3211-86. Approval and Test Specification—Ultrasonic Therapy Equipment (This is a Joint Standard with SANZ NZS 3211) (Withdrawn). 12 pp.
SNZ NZS/AS 3211-86. Approval and Test Specification—Ultrasonic Therapy Equipment (This is a Joint Standard with SAA AS 3211). 12 pp.

Diatomaceous Earth
Use For: Diatomaceous Silica
JIS K 8330-61. Diatomaceous Earth.

—Pigments
CNS K2117-86. Diatomaceous Silica Pigment (Feb)(5882). 1 p.
CNS K6583-86. Method of Test for Diatomaceous Silica Pigment (Feb)(6364).

—Water Filters
MOD UK TS 10074B. Diatomaceous Earth for Water Filtration Equipment.

Diatomaceous Silica
Use: Diatomaceous Earth

Diazinon (BTN)
See Also: Pesticides
SAA AS 1870.7D-77. Standard for Development—Pesticides for Agricultural Use—Part 7D: Diazinon. 20 pp.

Diazo Copying Machines
Use: Copying Machines—Diazo

Diazo Films
See Also: Photographic Films

INTERNATIONAL AND NON-U.S. NATIONAL STANDARDS
SUBJECT INDEX
Die

Diazo Films (Cont.)
—Stability Testing
ISO 8225-87. Photography—Ammonia Processed Diazo Photographic Film—Specification for Stability First Edition. 14 pp.

—Visual Density
BSI BS 6872-87. 1987 Visual Densities of Diazo and Vesicular Second-Generation Microforms. 4 pp.
CGSB CAN/CGSB-72.21-M89. Micrographics—Diazo and Vesicular Films—Visual Density—Specifications (ISO 8126-1986). 12 pp.
ISO 8126-86. Micrographics—Diazo and Vesicular Films—Visual Density—Specifications First Edition. 5 pp.

Diazo Papers
See Also: Papers
CNS P2035-84. Diazo Sensitized Paper (Feb)(2440). 3 pp.
CNS P2067-84. Base Paper for Diazo-Type Sensitized Paper (Oct)(10777). 2 pp.
JIS P 4504-78. Diazotype Base Paper.
JIS P 4505-79. Diazo Sensitized Papers.

Dibasic Acid Esters
Scope Note: Use a more specific term See: Sodium Oxalate

Dibenzbyl Peroxides
Use: Benzoyl Peroxide

2,6-Dibromo-N-chloro-p-benzoquinoneimine
Use: 2,6-Dibromoquinone-4-Chlorimide

2,6-Dibromoquinone-4-Chlorimide
Use For: 2,6-Dibromo-N-chloro-p-benzoquinoneimine
CNS K7182-65. Chemical Reagent (2,6-Dibromoquinone-Chloroimide) (Jan)(1682).
JIS K 8491-81. 2,6-Dibromo-N-Chloro-p-Benzoquinone Monoimine.

Dibutyl Phthalate
CNS K1115-83. Dibutyl Phthalate (99 Percent Grade) (Nov)(2612).
CNS K1275-83. Dibutyl Phthalate (Nov)(10658).
JIS K 6752-77. Dibutyl Phthalate (D.B.P.) (R 1980). 5 pp.

Dibutyltin Dilaurate
MOD UK DSTAN 68-120-90. Dibutyltin Dilaurate Issue 1. 10 pp.
MOD UK TS 406A. Dibutyltin Dilaurate (Superseded by Def Stan 68-120).

Dibutyltin Dilaurate Content Analysis
Use For: Butynorate Content Analysis
—Animal Feed
CNS N4081-82. Method of Test for Butynorate in Feed (Jan)(8317). 2 pp.

Dichlofluanid
See Also: Pesticides
SAA AS 1870.13D-79. Standard for Development—Pesticides for Agricultural Use—Part 13D: Dichlofluanid. 13 pp.

1,4-Dichloro-2-nitrobenzene
JIS K 4103-82. Chloronitrobenzenes (o-Chloronitrobenzene-p-Chloronitrobenzene). 1,4-Dichloro-2-Nitrobenzene. 1-Chloro-1,4-Dinitobenzene).

o-Dichlorobenzene
See Also: Chlorobenzene
ISO 1698-77. o-Dichlorobenzene for Industrial Use—List of Methods of Test First Edition. 4 pp.
JIS K 4102-83. Chlorobenzenes (Chlorobenzene.o-Dichlorobene.p-Dichlorobenzene).

p-Dichlorobenzene
ISO 1699-77. d-Dichlorobenzene for Industrial Use—List of Methods of Test First Edition. 3 pp.
JIS K 4102-83. Chlorobenzenes (Chlorobenzene.o-Dichlorobene.p-Dichlorobenzene).

Dichlorodifluoromethane
Use: Halon

Dichlorodiphenyltrichloroethane
Use: DDT

sym-Dichloroethane
Use: Ethylene Dichloride

1,2-Dichloroethane
Use: Ethylene Dichloride

Dichlorofluorescein
CNS K7183-65. Chemical Reagent (Dichlorofluorescein) (Jan)(1683).
JIS K 8466-72. Dichlorofluorescein.

2,6-Dichloroindophenol Sodium
Use For: 2,6-Dichlorophenol-indophenol Sodium
CNS K7184-65. Chemical Reagent (2,6-Dichlorophenol-Indophenol Sodium) (Jan)(1684).
JIS K 8469-78. 2,6-Dichloroindophenol Sodium, Hydrate 2,6-Dichloro-4-((4-Hydroxypheny) Imino)-2, 5-Cyclohexadien-1-One Sodium Salt.

Dichloromethane
Use: Methylene Chloride

Dichloromethane Content Analysis
Use: Methylene Chloride Content Analysis

Dichlorophen
BSI BS 4926-73. 1973 Dichlorophen. 11 pp.

2,6-Dichlorophenol-indophenol Sodium
Use: 2,6-Dichloroindophenol Sodium

2,6-Dichloroquinone Chlorimide
See Also: Reagents
CNS K7185-65. Chemical Reagent (2,6-Dichloroquinone-Chlorimidie) (Jan)(1685).

Dichlorotetrafluoroethane
Use For: FLON 114; Freon-114 (R)
See Also: Chlorofluorocarbons
JIS K 1528-82. Fluoroethanes Trichlorotrifluoroethane (FLON 113) Dichlorotetrafluoro-ethane (FLON 114).

Dichromate Number
—Acetic Acid
ISO 753 Pt 7-81. Acetic Acid for Industrial Use—Methods of Test—Part 7: Determination of Dichromate Index First Edition. 4 pp.

Dicofol
—Plants (Botany)—Prohibited Use
EC 90/533/EEC-90. Council Directive Amending the Annex to Directive 79/117/EEC Prohibiting the Placing on the Market and Use of Plant Protection on the Market and Use of Plant Protection Product Containing Certain Active Substances. 1 p.

Dictating Machines
Use For: Dictation Equipment See Also: Office Machines
BSI BS 3738: Part 3-80. 1980 Dictation Equipment Part 3: Basic Operating Requirements. 8 pp.
ISO 1730-80. Dictation Equipment—Basic Operating Requirements Second Edition. 7 pp.
JTC1 1730-80. Dictation Equipment—Basic Operating Requirements Second Edition. 7 pp.
OSI ISO 1730-80. Dictation Equipment—Basic Operating Requirements. 7 pp.

—Glossaries
BSI BS 3738: Part 1-80. 1980 Dictation Equipment Part 1: Glossary of Terms. 13 pp.
ISO 5138 Sec 01-78. Office Machines—Vocabulary—Section 01: Dictation Equipment First Edition. 27 pp.
JTC1 5138 Sec 01-78. Office Machines—Vocabulary—Section 01: Dictation Equipment First Edition. 27 pp.
OSI ISO 5138-1-78. Office Machines—Vocabulary—Section 01: Dictation Equipment. 27 pp.

—Symbols
BSI BS 3738: Part 2-80. 1980 Dictation Equipment Part 2: Specification for Symbols. 8 pp.
CNS C5198-85. Dictation Equipment—Symbols (Mar)(11222).
ISO 4062-77. Dictation Equipment—Symbols First Edition. 7 pp.
JTC1 4062-77. Dictation Equipment—Symbols First Edition. 7 pp.
OSI ISO 4062-77. Dictation Equipment—Symbols. 7 pp.

Dictation Equipment
Use: Dictating Machines

Dictionaries
Scope Note: For dictionaries about specific subjects, products, or materials, see the subheading Glossaries under specific subjects, products or materials

Dicyandiamide
See Also: Fertilizers
—Aqueous Solutions
MOD UK M 812/77. Examination of Dicyandiamide.

Dicyandiamide (Cont.)
—Content Analysis
MOD UK M 812/77. Examination of Dicyandiamide.
—Melting Points
MOD UK M 812/77. Examination of Dicyandiamide.
—Sieve Analysis
MOD UK M 812/77. Examination of Dicyandiamide.
—Visual Inspection
MOD UK M 812/77. Examination of Dicyandiamide.

Die Blocks
See Also: Die Holders; Dies
—Hammer Forging
JIS B 6470-83. Die Blocks for Closed-Die Hammer Forging (R 1987). 9 pp.
—Press Forging
JIS B 6471-87. Die Blocks for Closed-Die Press Forging. 8 pp.

Die Bushings
Use For: Piercing Die Bushes See Also: Bushings
DIN ENGL 9845-79. Piercing Die Bushes and Punch Guide Bushes (Feb). 5 pp.
JIS B 5063-89. Die Bushings for Press Dies. 7 pp.
—Press Dies
JIS B 5063-89. Die Bushings for Press Dies. 7 pp.

Die Casting Machinery
CNS B3369-80. Master Dies for Die Casting Metal Mold (Aug)(5995).
JIS B 5101-89. Holding Blocks and Cavity Inserts for Die Casting. 9 pp.
—Safety
CEN PREN 869-92. Safety Requirements for Metal Diecasting Units. 48 pp.

Die Castings
See Also: Castings; Metal Products; Steel Castings
CNS B1041-87. Permissible Deviations in Dimensions Without Tolerance Indication (Die Castings) (Jun)(4022).
DIN ENGL 1687 Pt 4-86. Heavy Metal Alloy Raw Castings; Pressure Die Castings; General Tolerances, Machining Allowances (Aug). 5 pp.
DIN ENGL 1688 Pt 4-86. Light Metal Alloy Raw Castings; Pressure Die Castings; General Tolerances, Machining Allowances (Aug). 4 pp.
JIS B 0409-80. Permissible Deviations in Dimensions Without Tolerance Indication for Die Castings. 7 pp.
—Aircraft—Screws
SBAC AS 2989 (V). Locking Screw—Die Casting. 2 pp.
—Aluminum Alloy
DIN ENGL 1725 Pt 2-86. Aluminium Alloys; Casting Alloys; Sand Casting; Gravity Die Casting; Pressure Die Casting; Investment Casting (Feb). 15 pp.
JIS H 5302-90. Aluminium Alloys Die Castings. 10 pp.
MOD UK DEF-30-A-69. Aluminium Alloy Pressure Die Castings (Control of Manufacture and Inspection). 7 pp.
MOD UK DEF-166-69. Aluminium Alloy Gravity Die Casting (Control of Manufacture and Inspection). 7 pp.
—Aluminum Alloy—Ingots
JIS H 2118-90. Aluminium-Base Alloys in Ingot for Die Castings. 7 pp.
—Aluminum Bronze—Aircraft
MOD UK DTD-412-40. Aluminium-Bronze Sand or Die Castings (Reprinted April 1963, Incorporating Amendment No. 1). 2 pp.
—Brass—Ingots
BSI DD 187-90. 1990 Dezincification Resistant Brass Die Castings and Ingots for Die Castings. 11 pp.
—Insert Nuts
CNS B2343-84. Insert Nuts Heavy Type for Plastics Moulding and Pressure Die Castings with Dead Hole (Jul)(4665).
—Light Metal Alloy
DIN ENGL 1688 Pt 3-80. Light Metal Alloy Raw Castings; Gravity Die Castings; General Tolerances, Machining Allowances (Oct). 4 pp.
—Magnesium Alloy
DIN ENGL 1729 Pt 2-73. Magnesium Alloys; Casting Alloys; Sand Castings, Gravity Die Castings, Pressure Die Castings (July). 8 pp.
JIS H 5303-91. Magnesium Alloy Die Castings. 7 pp.

INDUSTRY STANDARDS

Die Castings (Cont.)

—Magnesium Alloy—Ingots
JIS H 2222-91. Magnesium Alloy Ingots for Die Castings. 7 pp.

—Tin Alloy
DIN ENGL 1742-71. Tin Alloys for Pressure Die Castings (July). 1 p.

—Zinc Alloy
BSI BS 1004-72. 1972 Amd 2 Zinc Alloys for Die Casting and Zinc Alloy Die Castings (AMD 4733) March 29 1985. 20 pp.
BSI BS 5338-76. 1976 Code of Practice for Zinc Alloy Pressure Die Casting for Engineering. 15 pp.
CNS H3044-90. Zinc Alloy Die Castings (Dec)(3334).
DIN ENGL 1743 Pt 2-78. High Purity Zinc Casting Alloys; Pressure-, Sand-and Gravity Die Castings (Apr). 3 pp.
JIS H 5301-90. Zinc Alloys Die Castings. 8 pp.
MOD UK DEF-17-B-69. Zinc Alloy Pressure Die Castings (Control of Manufacture and Inspection). 6 pp.

—Zinc Alloy—Chemical Analysis
JIS H 1551-64. Methods for Chemical Analysis of Die Casting Zinc Alloy.

—Zinc Alloy—Ingots
BSI BS 1004-72. 1972 Amd 2 Zinc Alloys for Die Casting and Zinc Alloy Die Castings (AMD 4733) March 29 1985. 20 pp.
CNS H3043-90. Zinc Alloy Ingot for Die Castings (Dec)(3333).
JIS H 2201-57. Zinc Alloy Ingot for Die Casting (R 1971). 4 pp.

—Zinc Alloy—Spectrochemical Analysis
JIS H 1560-76. Methods for Emission Spectrochemical Analysis of Die Casting Zinc Alloy.

—Zinc Alloy—Spectrography
BSI BS 1225-70. 1970 Recommended Methods for the Spectrographic Analysis of High Purity Zinc and Zinc Alloys for Die Casting. 20 pp.
BSI BS 3630: Part 4-63. 1963 Methods for the Sampling and Analysis of Zinc and Zinc Alloys Part 4: Sampling of Ingot Zinc, Zinc Alloy Ingots and Zinc Alloy Die Castings for Spectrographic Analysis. 8 pp.
MOD UK M 9301/62. Spectrographic Analysis of Zinc Die-Casting Alloy A (No Information).

—Zinc—Spectrography
BSI BS 1225-70. 1970 Recommended Methods for the Spectrographic Analysis of High Purity Zinc and Zinc Alloys for Die Casting. 20 pp.

Die Cutters
See Also: Cutting Tools; Dies
JIS B 9604-88. General Rules for Testing of Die Cutters. 23 pp.

Die Forgings
Scope Note: For additional listing, see the subheading Forgings under the specific metal
Use For: Closed Die Forging; Open Die Forgings; Split Die Forgings *See Also:* Forgings; Metal Products
CNS B1316-83. Steel Die Forgings Minimum Wall Thickness for Different Cross-Sectional Forms (Jun)(10307).
CNS B1317-83. Steel Die Forgings Machining Allowances, Radius and Drafts (Jun)(10308).
CNS B1318-83. Steel Die Forgings Tolerances and Permissible Variations (Jun)(10309).
CNS B1319-83. Steel Die Forgings Forging in Horizontal Upsetter (Jun)(10310).
CNS B1321-83. Steel Die Forgings Technical Conditions of Delivery (Aug)(10488).
CNS B1354-87. Machining Allowance for Open Die Forgings (Sep)(12078).
CNS B2788-87. Minimum Wall Thickness and Height for Different Cross Sections of Hot Forging Closed Die (Sep)(12075).
CNS B2789-87. Die War Tolerance for Hot Forging Closed Die (Sep)(12076).
CNS B2790-87. Minimum Rounded Radio and Drafts of Hot Forging Closed Die (Sep)(12077).
JIS B 0418-85. Machining Allowance for Open Die Forgings. 15 pp.
JIS B 0706-89. Corner and Fillet Radii of Forged Parts (Hot and Warm Forgings). 11 pp.

—Aerospace
AECMA PREN2082-01-86. Aluminium Alloy Forging Stock and Forgings—Technical Specification—Part 1—General Requirements. 10 pp.
AECMA PREN2094-82. Aluminium Alloy 7009 T736—Die Forgings A Less Than or Equal to 150 mm (C5/40). 3 pp.

Die Forgings (Cont.)

—Aerospace (Cont.)
AECMA PREN2207-77. Steel FE-PL43S—900 MPa Less Than or Equal to Rm Less Than or Equal to 1100 MPa—Hand Anddie Forgings De Less Than or Equal to 40 mm (C5/47). 3 pp.
AECMA PREN2208-77. Steel FE-PL43S—650 MPa Less Than or Equal to Rm Less Than or Equal to 850 MPa—Hand Anddie Forgings De Less Than or Equal to 150 mm (C5/47). 3 pp.
AECMA PREN2222-77. Steel FE-PL31—Hardened and Tempered—Hand and Die Forgings (C5/47). 3 pp.
AECMA PREN2225-77. Steel FE-PL32—Hardened and Tempered—Hand and Die Forgings (C5/47). 3 pp.
AECMA PREN2252-77. Steel FE-PL52S—1080 MPa Less Than or Equal to Rm Less Than or Equal to 1250 MPa—Hand and Die Gorgings De Less Than or Equal to 100 mm (C5/47). 3 pp.
AECMA PREN2681-84. Aluminium Alloy 7010-T736—Die Forgings a Less Than or Equal to 150 mm. 4 pp.
AECMA PREN2688-84. Aluminium Alloy 7050-T736—Die Forgings a Less Than or Equal to 150 mm (C5/40). 4 pp.
AECMA PREN3339-88. Aluminium Alloy (7010) Solution Treated and Artificially Aged (T76) Die Forgings a Less Than or Equal to 200 mm. 5 pp.
AECMA PREN3340-88. Aluminium Alloy (7050) Solution Treated and Artificially Aged (T76) Die Forgings a Less Than or Equal to 200 mm. 5 pp.
BSI BS EN 2082: Part 1-91. 1991 Aluminium Alloy Forging Stock and Forgings. Technical Specification Part 1: General Requirements. 13 pp.
BSI BS EN 2082: Part 1-01. 1991 Amd 1 Aluminium Alloy Forging Stock and Forgings Technical Specification Part 1: General Requirements (AMD 7707) October 15, 1993 (S). 19 pp.
BSI L 161-76. 1976 Hand and Die Forgings of Aluminium-Zinc-Magnesium-Copper-Chromium Alloy (Solution Treated and Artificially Aged to an Overaged Condition) (Zn 5.6, Mg 2.5, Cu 1.6, Cr 0.22). 4 pp.
CEN EN 2082-1-89. Aluminium Alloy Forging Stock and Forgings Technical Specification Part 1: General Requirements. 10 pp.
CEN EN 2082 Part 1-89. Aerospace Series Aluminium Alloy Forging Stock and Forgings Technical Specification Part 1: General Requirements. 12 pp.
CEN PREN 2858 Part 1-93. Aerospace Series Titanium and Titanium Alloys Forging Stock and Forgings Technical Specification Part 1—General Requirements. 9 pp.
CEN PREN 2858 Part 2-93. Aerospace Series Titanium and Titanium Alloys Forging Stock and Forgings Technical Specification Part 2—Forging Stock. 11 pp.
CEN PREN 2858 Part 3-93. Aerospace Series Titanium and Titanium Alloys Forging Stock and Forgings Technical Specification Part 3—Pre-Production amd Production Forgings. 10 pp.

—Engineering Drawings
CNS B1315-83. Steel Die Forgings Directive Items for Forging Drawings (Jun)(10306).

—Flash Cavities
CNS B2787-87. Flash Cavity of Hot Forging Closed Die (Sep)(12074).

—Materials
CNS B2786-87. Die Material of Hot Forging Closed Die (Sep)(12073).

Die Grinders
See Also: Grinders

—Safety
CEN PREN 792-9-92. Handheld Non-Electric Power Tools—Safety Requirements—Part 9: Die Grinders. 9 pp.

Die Holders
See Also: Die Blocks; Tool Holders

—Press Dies
CNS B3363-85. Punch Holder and Die Holder for Press Dies (Jul)(5989).
JIS B 5004-75. Punch Holder and Die Holder for Press Dies.

Die Sets
See Also: Dies; Tool Holders; Tools

—Press Dies
CNS B3364-87. Die Sets for Press Dies (Apr)(5990).
CNS B3368-87. Ball Side Die Sets for Press Dies (Apr)(5994).
CNS B7155-87. Accuracy Test for Press Die Sets (Apr)(6406).
JIS B 5013-89. Die Sets for Press Dies. 12 pp.
JIS B 5060-89. Steel Die Sets for Press Dies. 23 pp.

Die Sets (Cont.)

—Press Dies—Accuracy Testing
JIS B 5031-75. Accuracy Test for Press Die Sets (R 1987). 12 pp.

—Press Dies—Safety
CNS Z1017-87. Safety Code for Power Press (Design Construction and Setting of Dies) (Oct)(3420).

Die Sinking Cutters
Use: Milling Cutters—Die Sinking

Die Stocks
See Also: Threading Dies; Tool Holders
BSI BS 1127: Part 1-90. 1990 Screwing Dies and Dienuts Part 1: Hand—and Machine-Operated Circular Screwing Dies and Hand-Operated Die Stocks. 17 pp.
CEN EN 22 568-89. Hand-and Machine-Operated Circular Screwing Dies and Hand-Operated Die Stocks. 2 pp.
CGSB 46-GP-6-60. Dies; Threadcutting and Rethreading, Collets and Diestocks. 13 pp.
DIN ENGL EN 22568-90. Hand and Machine-Operated Circular Screwing Dies and Hand-Operated Die Stocks (ISO 2568:1988) (June) (Supersedes DIN 223 Parts 1, 3, and 10, September 1979 Editions, and DIN 225, December 1987 Edition). 14 pp.
ISO 2568-88. Hand-and Machine-Operated Circular Screwing Dies and Hand-Operated Die Stocks Second Edition. 14 pp.
MOD UK DSTAN 51-16-77. Diestocks for ISO Metric Coarse Pitch Thread Cutting Dies Issue 1. 5 pp.

Dieldrin Content Analysis

—Wood Preservatives—Colorimetric Analysis
BSI BS 5666: Part 6-83. 1983 Amd 1 Wood Preservatives and Treated Timber: Part 6: Quantitative Analysis or Preservative Solution and Treated Timber Containing Pentachloro-phenol, Phenachloro-phenyl Laurate, GS Hexachlorocyclohexane and Dieldrin (AMD 6224). 11 pp.

—Wood Preservatives—Gas Chromatography
BSI BS 5666: Part 6-83. 1983 Amd 1 Wood Preservatives and Treated Timber: Part 6: Quantitative Analysis or Preservative Solution and Treated Timber Containing Pentachloro-phenol, Phenachloro-phenyl Laurate, GS Hexachlorocyclohexane and Dieldrin (AMD 6224). 11 pp.

Dielectric Breakdown
Use For: Electric Breakdown *See Also:* Breakdown Voltage; Dielectric Strength; Electrical Faults; Electrical Properties; Treeing

—Adhesive Tapes
SAA AS 1635.15.1-74. Methods of Testing Pressure Sensitive Adhesive Tape—Part 15.1: Dielectric Breakdown and Proof Voltage. 4 pp.

—Anodic Coatings
BSI BS 6161: Part 15-87. 1987 Methods of Test for Anodic Oxidation Coatings on Aluminium and Its Alloys Part 15: Determination of Electrical Breakdown Potential. 4 pp.
ISO 2376-72. Anodization (Anodic Oxidation) of Aluminium and Its Alloys—Insulation Check by Measurement of Breakdown Potential First Edition. 3 pp.

—Phenolic Resins
CNS K6273-74. Method of Test for Phenol Formaldehyde Resin Molding Compounds (Oct)(2988). 11 pp.

—Urea Resins
CNS K6272-74. Method of Test for Urea Resin Molding Compounds (Oct)(2986). 9 pp.

Dielectric Constant
See Also: Permittivity

—Electrical Insulation
DIN ENGL 53483 Suppl. 1-55. Testing of Insulating Materials; Determination of the Dielectric Constant (Relative Permittivity) and of the Loss Factor; Measuring Equipment (Oct). 4 pp.

—Electrical Insulation Tubing—Heat Shrinkable
CNS C3140-88. Method for Determining Dielectric Constant and Dielectric Dissipation Factor of Heat Shrinkable Tubing for Electrical Insulation (Sep)(8787).

INTERNATIONAL AND NON-U.S. NATIONAL STANDARDS
SUBJECT INDEX

Dielectric

Dielectric Constant *(Cont.)*

—Magnetic Materials
CNS C6113-81. Method of Test for Complex Dielectric Constant of Nonmetallic Magnetic Materials at Microwave Frequencies (Jul)(7634).

—Polyethylene
CNS K6661-81. Method of Test for Dielectric Constant and Dissipation Factor of Polyethylene by Liquid Displacement Procedure (May)(7396).

Dielectric Dissipation Factor
Use For: Dissipation Factor; Loss Tangent
See Also: Dielectric Heating; Dielectric Loss; Dielectric Strength; Electrical Properties; Electrical Resistance

—Electrical Insulating Liquids
BSI BS 5737-79. 1979 Measurement of Relative Premittivity, Dielectric Dissipation Factor and d.c. Resistivity of Insulating Liquids. 19 pp.
IEC 247-78. Measurement of Relative Permittivity, Dielectric Dissipation Factor and D.C. Resistivity of Insulating Liquids Second Edition. 35 pp.

—Electrical Insulation
BSI BS 4542-70. 1970 Determination of Loss Tangent and Permittivity of Electrical Insulating Materials in Sheet Form (Lynch Method). 12 pp.
BSI BS 7663-93. 1993 Methods of Test for Determination of Permittivity and Dissipation Factor of Electrical Insulating Material in Sheet or Tubular Form (F). 18 pp.
DIN VDE 0303 Pt 4-69. Regulations for Electrical Tests on Insulating Materials; Part 4: Determination of Dielectric Properties (Dec). 38 pp.
IEC 250-69. Recommended Methods for the Determination of the Permittivity and Dielectric Dissipation Factor of Electrical Insulating Materials at Power, Audio and Radio Frequencies Including Metre Wavelengths First Edition. 56 pp.
IEC 377 Pt 1-73. Recommended Methods for the Determination of the Dielectric Properties of Insulating Materials at Frequencies Above 300 MHz Part 1: General First Edition. 26 pp.
IEC 377 Pt 2-77. Methods for the Determination of the Dielectric Properties of Insulating Materials at Frequencies Above 300 MHz Part 2: Resonance Methods First Edition. 40 pp.

—Electrical Insulation—Tubing—Heat Shrinkable
CNS C3140-88. Method for Determining Dielectric Constant and Dielectric Dissipation Factor of Heat Shrinkable Tubing for Electrical Insulation (Sep)(8787).

—Plastics
BSI BS 903: Part C3-82. 1982 Methods of Testing Vulcanized Rubber Part C3: Determinations of Loss Tangent and Permittivity at Power and Audio Frequencies Incorporating BS 2782: Part 2: Method 240A and 240B. 11 pp.
BSI BS 2782:Pt2: METH 241A-84. 1984 Methods of Testing Plastics Part 2: Electrical Properties Method 241A: Determination of Effect of Polyvinyl Chloride Compound on the Loss Tangent of Polyethylene. 2 pp.
BSI BS 4618: Sec 2.2-70. 1970 Recommendations for the Presentation of Plastics Design Data Part 2: Electrical Properties Section 2.2: Loss Tangent. 6 pp.

—Polyethylene
CNS K6661-81. Method of Test for Dielectric Constant and Dissipation Factor of Polyethylene by Liquid Displacement Procedure (May)(7396).

—Rubber
BSI BS 903: Part C3-82. 1982 Methods of Testing Vulcanized Rubber Part C3: Determinations of Loss Tangent and Permittivity at Power and Audio Frequencies Incorporating BS 2782: Part 2: Method 240A and 240B. 11 pp.

Dielectric Filters
See Also: Bandpass Filters; Ceramic Filters; Electric Filters; Filters

—Passive—Electromagnetic Interference—Military—PPL
CECC CECC MUAHAG Vol 11 IS 2-92. Preferred Products List; Filters (En, Fr, Ge). 88 pp.

Dielectric Heating
See Also: Dielectric Dissipation Factor; Heating Equipment; Induction Heating Equipment

—Safety
CENELEC HD 491.9 S1-91. Safety in Electroheat Installations Part 9: Particular Requirements for High-Frequency Dielectric Heating Installations. 6 pp.

Dielectric Heating *(Cont.)*

—Safety *(Cont.)*
IEC 519 Pt 9-87. Safety in Electroheat Installations Part 9: Particular Requirements for High-Frequency Dielectric Heating Installations First Edition. 24 pp.

Dielectric Loss
See Also: Dielectric Dissipation Factor; Dielectric Strength; Dielectric Testing; Electrical Properties

—Electrical Insulation
DIN VDE 0303 Pt 4-69. Regulations for Electrical Tests on Insulating Materials; Part 4: Determination of Dielectric Properties (Dec). 38 pp.

—Windings
IEC 894-87. Guide for a Test Procedure for the Measurement of Loss Tangent of Coils and Bars for Machine Windings First Edition; (Corrigendum—July 1987). 24 pp.

Dielectric Loss Factor

—Cables (Electric)
DIN VDE 0472 Pt 505-83. Testing of Cables and Insulated Flexible Cords; Loss Factor, Dielectric Loss Coefficient and Leakage (Apr). 9 pp.

—Cords (Electric)
DIN VDE 0472 Pt 505-83. Testing of Cables and Insulated Flexible Cords; Loss Factor, Dielectric Loss Coefficient and Leakage (Apr). 9 pp.

—Electrical Insulation
DIN ENGL 53483 Pt 1-69. Testing of Insulating Materials; Determination of Dielectric Properties; Definitions; General Information (July). 4 pp.
DIN ENGL 53483 Pt 2-70. Testing of Insulating Materials; Determination of Dielectric Properties; Testing at Standard Frequencies of 50 Hz, 1 kHz, 1 MHz (Mar). 8 pp.
DIN ENGL 53483 Pt 3-69. Testing of Insulating Materials; Determination of Dielectric Properties; Measuring Cells for Liquids for Frequencies up to 100 MHz (July). 3 pp.

—Electrical Insulation—Glossaries
DIN ENGL 53483 Pt 1-69. Testing of Insulating Materials; Determination of Dielectric Properties; Definitions; General Information (July). 4 pp.

Dielectric Properties
Scope Note: Use a more specific term
See: Dielectric Constant; Dielectric Dissipation Factor; Dielectric Loss; Dielectric Loss Factor; Dielectric Strength; Insulation Resistance

Dielectric Strength
Use For: Electric Strength *See Also:* Dielectric Breakdown; Dielectric Dissipation Factor; Dielectric Loss; Dielectric Testing; Electrical Properties; Insulation Resistance

—Brooders
CSA C22.2 NO 102-1958. Construction and Test of Brooders and Incubators (R 1992). 41 pp.

—Communication Cables
DIN VDE 0472 Pt 509-86. Testing of Cables, Wires and Flexible Cords; Dielectric Strength on Cables, Wires and Cords for Telecommunications and Information Processing Systems (Oct). 11 pp.

—Connectors
SBAC TS 323 ISSUE 1. Test for Electrical Connectors Dielectric Withstand.

—Control Cables (Electric)
CSA CAN/CSA-C22. 2 NO 239-M91. Control and Instrumentation Cables; (Gen Instr 1 Thru 3). 50 pp.

—Cords (Electric)
DIN VDE 0472 Pt 508-86. Testing of Insulated Cables and Flexible Cords; Dielectric Strength of Insulated Cables and Flexible Cords for Power Installations (May). 8 pp.

—Electric Wire
CSA 1277 Bull. Electrical Bulletin 1277 July 17, 1980 to C22.2 NO 116. 2 pp.
CSA 1277 Bull. Electrical Bulletin 1277 July 17, 1980 to C22.2 NO 127. 2 pp.

—Electrical Insulation
BSI BS 5874-80. 1980 Amd 1 Determination of the Electric Strength of Insulating Oils. 10 pp.
CENELEC HD 559.1 S1-91. Methods of Test for Electric Strength of Solid Insulating Materials Part 1: Tests at Power Frequencies. 6 pp.
CENELEC HD 559.2 S1-91. Methods of Test for Electric Strength of Solid Insulating Materials—Part 2: Additional Requirements for Tests Using Direct Voltage. 6 pp.

Dielectric Strength *(Cont.)*

—Electrical Insulation *(Cont.)*
CNS C3091-80. Method of Test for Electric Strength of Solid Insulating Materials (Aug)(6040).
DIN VDE 0303 Pt 2-74. Electrical Tests of Insulating Materials; Breakdown Voltage; Electric Strength (Nov). 23 pp.
IEC 243 Pt 1-88. Methods of Test for Electric Strength of Solid Insulating Materials Part 1: Tests at Power Frequencies First Edition; (Corrigendum—March 1989). 41 pp.
IEC 243 Pt 2-90. Methods of Test for Electric Strength of Solid Insulating Materials Part 2: Additional Requirements for Tests Using Direct Voltage First Edition. 17 pp.
JIS C 2110-75. Testing Methods for Electric Strength of Solid Insulating Materials. 13 pp.

—Electrical Insulation—Plastics
SAA AS 1255.3-73. Methods of Test for Electrical Characteristics of Solid Plastics Insulating Materials—Part 3: Determination of Electric Strength at Power Frequencies. 19 pp.

—Electrical Insulation—Tubing—Heat Shrinkable
CNS C3138-88. Method of Test for Electric Strength of Heat Shrinkable Tubing for Electrical Insulation (Sep)(8785).

—Incubators
CSA C22.2 NO 102-1958. Construction and Test of Brooders and Incubators (R 1992). 41 pp.

—Instrumentation Cables
CSA CAN/CSA-C22. 2 NO 239-M91. Control and Instrumentation Cables; (Gen Instr 1 Thru 3). 50 pp.

—Insulating Oils
IEC 156-63. Method for the Determination of the Electric Strength of Insulating Oils First Edition. 15 pp.

—Lead Wire
CSA 1277 Bull. Electrical Bulletin 1277 July 17, 1980 to C22.2 NO 116. 2 pp.

—Medical Electrical Equipment
CSA CAN/CSA-C22. 2NO601.1-M90. Medical Electrical Equipment, Part 1: General Requirements for Safety; (Gen Instr 1). 240 pp.

—Storage Water Heaters
CSA CAN/CSA-C22. 2 NO 110-M90. Construction and Test of Electric Storage-Tank Water Heaters; (Gen Instr 1). 39 pp.
CSA 1169 Bull. Electrical Bulletin 1169 June 27, 1978 to C22.2 NO 110. 2 pp.

—Telephone Cables
CCITT RECMN G.611-89. Characteristics of Symmetric Cable Pairs for Analogue Transmission—Transmission Media Characteristics (Study Group XV) 5 pp. 5 pp.
CCITT RECMN G.621-89. Characteristics of 0.7/2.9 mm Coaxial Cable Pairs—Transmission Media Characteristics (Study Group XV) 4 pp. 4 pp.
CCITT RECMN G.622-89. Characteristics of 1.2/4.4 mm Coaxial Cable Pairs—Transmission Media Characteristics (Study Group XV) 8 pp. 8 pp.
CCITT RECMN G.623-89. Characteristics of 2.6/9.5 mm Coaxial Cable Pairs—Transmission Media Characteristics (Study Group XV) 8 pp. 8 pp.

—Telephone Cables—Lightning Protection
CCITT RECMN K.13-89. Induced Voltages in Cables with Plastic-Insulated Conductors—Protection Against Interference (Study Group V) 2 pp. 2 pp.

—Telephone Repeaters
CCITT RECMN G.611-89. Characteristics of Symmetric Cable Pairs for Analogue Transmission—Transmission Media Characteristics (Study Group XV) 5 pp. 5 pp.

—Vulcanized Rubber
BSI BS 903: Part C4-83. 1983 Methods of Testing Vulcanized Rubber Part C4: Determination of Electric Strength (Incorporating BS 2782: Methods 220 & 221). 11 pp.

Dielectric Testing
See Also: Dielectric Loss; Dielectric Strength

—Dry Reed Switches
CNS C6150-87. Method of Test for Dry Reed Switches (Dielectric) (Jul)(8228).

—Electrical Insulation
DIN VDE 0303 Pt 4-69. Regulations for Electrical Tests on Insulating Materials; Part 4: Determination of Dielectric Properties (Dec). 38 pp.

INDUSTRY STANDARDS

Dielectric Testing (Cont.)

—**Fire Alarms**
CEN EN 54-5-76. Components of Automatic Fire Detection Systems Part 5 Heat Sensitive Detectors—Point Detectors Containing a Static Element. 20 pp.
CEN EN 54 (Part 5)-88. AMD 1 Components of Automatic Fire Detection Systems Part 5 Heat Sensitive Detectors—Point Detectors Containing a Static Element. 21 pp.
CEN EN 54-7-88. Components of Automatic Fire Detection Systems—Part 7: Point-Type Smoke Detectors; Detectors Using Scattered Light, Transmitted Light or Ionization. 30 pp.
CEN EN 54-8-88. Components of Automatic Fire Detection Systems—Part 8: Hig Temperature Heat Detectors. 23 pp.
CEN EN 54-9-82. Components of Automatic Fire Fetection Systems Part 9 Fire Sensitivity Test. 10 pp.

—**Fire Extinguishers**
CEN EN 3-2-78. Portable Fire Extinguishers—Part 2. 7 pp.
CEN PREN 3-2-93. Fight Against Fire—Portable Fire Extinguishers—Part 2: Retention of Charge, Dielectric Test, Compaction, Special Provision. 9 pp.

—**High Voltage**
DIN VDE 0432 Pt 1-78. High-Voltage Test Techniques; General Definitions and Test Requirements (Oct). 23 pp.
DIN VDE 0432 Pt 2-78. High-Voltage Test Techniques; Test Procedures (Oct). 48 pp.

—**Low Voltage Installations**
BSI BS 7640: Part 1-93. 1993 High-Voltage Test Techniques for Low-Voltage Equipment Part 1: Definitions, Test and Procedure Requirements (IEC 1180-1: 1992) (E). 31 pp.
IEC 1180 Pt 1-92. High-Voltage Test Techniques for Low-Voltage Equipment Part 1: Definitions, Test and Procedure Requirements First Edition. 60 pp.

—**Metal Working Equipment—Industrial**
CSA C22.2 NO 105-1953. Construction and Test of Electrical Equipment for Woodworking Machinery (R 1992). 18 pp.
CSA 1169 Bull. Electrical Bulletin 1169 June 27, 1978 to C22.2 NO 105. 2 pp.

—**Metering Centers (Electric)**
CSA C22.2 NO 229-M1988. Switching and Metering Centres; (Gen Instr 1). 41 pp.

—**Power Transformers**
BSI BS 171: Part 3-87. 1987 Power Transformers Part 3: Insulation Levels and Dielectric Tests. 35 pp.
CENELEC HD 398.3-86. Power Transformers Insulation Levels and Dielectric Tests. 10 pp.
IEC 76 Pt 3-80. Power Transformers Part 3: Insulation Levels and Dielectric Tests First Edition; (Amendment 1-1981). 81 pp.
IEC 76 Pt 3-1-87. Power Transformers Part 3: Insulation Levels and Dielectric Tests External Clearances in Air First Edition. 26 pp.
SAA AS 2374.3.0-82. Power Transformers—Part 3.0: Insulation Levels and Dielectric Tests (Redesignated by Amdt 1 September 1992 from AS 2374.3—1982). 1 p.
SAA AS 2374.3.1-92. Power Transformers—Part 3: Insulation Levels and Dielectric Tests—Part 3.1: External Clearances in Air. 13 pp.

—**Terminal Blocks**
CSA C22.2 NO 158-1987. Terminal Blocks; (Gen Instr 1 Thru 4). 32 pp.

—**Transformers**
CSA C22.2 NO 66-1988. Speciality Transformers; (Gen Instr 1). 53 pp.
CSA 1169 Bull. Electrical Bulletin 1169 June 27, 1978 to C22.2 NO 66. 2 pp.

—**Woodworking Equipment—Industrial**
CSA C22.2 NO 105-1953. Construction and Test of Electrical Equipment for Woodworking Machinery (R 1992). 18 pp.
CSA 1169 Bull. Electrical Bulletin 1169 June 27, 1978 to C22.2 NO 105. 2 pp.

Dielectrics
Use For: Liquid Dielectrics *See Also:* Capacitor Papers; Capacitors; Cladding (Waveguides); Electrical Insulating Liquids; Electrical Insulating Oils; Electrical Insulation

—**Dimethyl Silicone**
MOD UK DSTAN 91-46-79. Damping Fluids: Dimethyl Silicone Issue 1. 11 pp.

—**Sampling**
BSI BS 5263-75. 1975 Method for Sampling Liquid Dielectrics. 11 pp.

Dielectrics (Cont.)
—**Sampling (Cont.)**
IEC 475-74. Method of Sampling Liquid Dielectrics First Edition. 25 pp.

—**Water Content—Karl Fischer Method**
BSI BS 6725-86. 1986 Determination of Water in Liquid Dielectrics by Automatic Coulometric Karl Fischer Titration. 16 pp.

Diemakers' Squares
Use: Squares (Instruments)

Diene Resins
Scope Note: Use a more specific term *See:* Butyl Rubber; Styrene Butadiene Resins; Styrene Butadiene Rubber; Thermosetting Resins

Dies
See Also: Die Blocks; Die Cutters; Die Sets; Drawing Dies; Hand Tools; Heading Dies; Metal Working; Molds (Casting); Press Dies; Punches (Tools); Thread Rolling Dies; Threading Dies
BSI BS 1127: Part 5-90. 1990 Screwing Dies and Dienuts Part 5: Dimensions of Hexagonal Dienuts. 7 pp.
BSI BS 3821: Part 4-82. 1982 Hardmetal Dies and Associated Hardmetal Tools Part 4: Dimensions and Tolerances of As-Sintered Hard-Metal Pellets Used in Heading Dies. 7 pp.
ISO 5407-81. As-Sintered Hardmetal Pellets Used in Heading Dies—Dimensions and Tolerances First Edition. 6 pp.
ISO 7226-88. Hexagonal Dienuts Second Edition. 6 pp.

—**Angle Pins**
BSI BS 7564: Part 8-92. 1992 Tools for Moulding Part 8: Angle Pins—Basic Dimensions (ISO 8404: 1986). 7 pp.
CNS B2784-87. Angle Pin of Die Casting Molds (Sep)(12071).
ISO 8404-86. Angle Pins—Basic Dimensions First Edition. 4 pp.

—**Defense Contracts**
MOD UK DEFCON 22-76. Moulds and Dies (Certificate and List) 6/76. 2 pp.
MOD UK DEFCON 23-76. Special Jigs, Tools, Etc 6/76. 1 p.

—**Dienuts**
BSI BS 1127-74. (WITHDRAWN) 1974 Amd 1 Circular Machine Screwing Dies, Hand Screwing Dies and Hexagon Dienuts (Superseded by BS 1127: Parts 1, 2, 3, 4, and 5: 1990). 22 pp.
BSI BS 1127: Part 2-90. 1990 Screwing Dies and Dienuts Part 2: Hand—and Machine-Operated Circular Screwing Dies for Taper Pipe Threads R Series. 8 pp.
BSI BS 1127: Part 5-90. 1990 Screwing Dies and Dienuts Part 5: Dimensions of Hexagonal Dienuts. 7 pp.
ISO 7226-88. Hexagonal Dienuts Second Edition. 6 pp.

—**Ejector Pins**
BSI BS 7564: Part 5-92. 1992 Tools for Moulding Part 5: Ejector Pins with Cylindrical Head—Dimensions (ISO 6751: 1986). 6 pp.
BSI BS 7564: Part 6-92. 1992 Tools for Moulding Part 6: Flat Ejector Pins (ISO 8693: 1987). 8 pp.
BSI BS 7564: Part 7-92. 1992 Tools for Moulding Part 7: Shouldered Ejector Pins (ISO 8694: 1987). 7 pp.
CNS B2782-87. Ejector Rod of Die Casting Molds (Sep)(12069).
CNS B3371-80. Ejector Pins of Dies for Die Casting (Aug)(5997).
ISO 6751-86. Ejector Pins with Cylindrical Head—Dimensions Second Edition. 4 pp.
ISO 8693-87. Tools for Moulding—Flat Ejector Pins First Edition. 5 pp.
ISO 8694-87. Tools for Moulding—Shouldered Ejector Pins First Edition. 4 pp.
JIS B 5103-89. Ejector Pins of Dies for Die Casting. 8 pp.

—**Ejector Sleeves**
BSI BS 7564: Part 9-92. 1992 Tools for Moulding Part 9: Ejector Sleeves with Cylindrical Head—Basic Series for General Purposes (ISO 8405: 1986). 7 pp.
ISO 8405-86. Ejector Sleeves with Cylindrical Head—Basic Series for General Purposes First Edition. 4 pp.

—**Electric Conduits—Steel**
BSI BS 1813-74. 1974 Dimensions of Conduit Dies, Diestocks and Guides. 13 pp.

—**Guide Bushings**
CNS B3373-80. Guide Pin Bushings of Dies for Die Casting (Aug)(5999).

Dies (Cont.)
—**Guide Bushings (Cont.)**
JIS B 5105-89. Guide Pin Bushes of Dies for Die Casting. 8 pp.

—**Guide Pins**
CNS B2783-87. Ejector Plate Guide Pin of Die Casting Molds (Sep)(12070).
CNS B3370-80. Guide Pins of Dies for Die Casting (Aug)(5996).
JIS B 5102-89. Guide Pins of Dies for Die Casting. 9 pp.

—**Identification Systems**
BSI BS 3821: Part 1-74. 1974 Hardmetal Dies and Associated Hardmetal Tools Part 1: Designation Marking of As-Sintered Pellets and Finished Dies. 12 pp.

—**Invitation for Bids**
MOD UK DEFCON 47V-86. Invitation to Tender Supplementary Conditions—Jigs, Tools, Etc 10/86. 1 p.

—**Locating Pins**
CNS B2785-87. Locating Pin of Die Casting Molds (Sep)(12072).

—**Mandrels**
ISO 1651-74. Tube Drawing Mandrels First Edition; (Amendment 1-1982). 7 pp.

—**Master**
CNS B3369-80. Master Dies for Die Casting Metal Mold (Aug)(5995).
JIS B 5101-89. Holding Blocks and Cavity Inserts for Die Casting. 9 pp.

—**Return Pins**
CNS B3372-80. Return Pins of Dies for Die Casting (Aug)(5998).
JIS B 5104-89. Return Pins of Dies for Die Casting. 7 pp.

—**Sintered—Pellets**
BSI BS 3821: Part 4-82. 1982 Hardmetal Dies and Associated Hardmetal Tools Part 4: Dimensions and Tolerances of As-Sintered Hard-Metal Pellets Used in Heading Dies. 7 pp.
ISO 2804-73. Wire, Bar or Tube Drawing Dies—as Sintered Pellets of Hard Metal (Carbide)—Dimensions First Edition. 8 pp.
ISO 5407-81. As-Sintered Hardmetal Pellets Used in Heading Dies—Dimensions and Tolerances First Edition. 6 pp.

—**Supports**
CNS B2781-87. Support of Die Casting Molds (Sep)(12068).

Diesel Engine Oils
Use: Diesel Motor Oil

Diesel Engines
Use For: Compression Ignition Engines
See Also: Fuel Injectors; Internal Combustion Engines; Marine Engines; Piston Pins
CNS D3026-68. Method of Test for Diesel Engines of Automobiles (Aug)(2745).
MOD UK DSTAN 28-2-01. Engines, Diesel for General Purpose Applications Issue 5; Amendment 1. 16 pp.
MOD UK DSTAN 28-2-91. Engines, Diesel for General Purpose Applications Issue 6. 13 pp.

—**Acceptance Testing**
NATO STANAG 4195 ED 1 AMD 1-85. NATO Standard Engine Laboratory Test for Diesel and Gasoline Engines and Gas Turbine Engines. 6 pp.

—**Agricultural Equipment**
JIS B 8012-77. Small Size Water Cooled Diesel Engines for Land Use (R 1982). 9 pp.
JIS B 8016-77. Small Size Forced Air-Cooled Diesel Engines for Land Use (R 1982). 10 pp.
JIS B 8018-89. Test Method of Performance of Small Size Diesel Engines for Land Use. 25 pp.

—**Air Cooled**
CNS B4011-69. Air-Cooled Diesel Engines for Land Use (Small Size) (Jul)(3002). 6 pp.
JIS B 8016-77. Small Size Forced Air-Cooled Diesel Engines for Land Use (R 1982). 10 pp.

—**Automotive**
BSI BS AU 141A-71. 1971 Performance of Diesel Engines for Road Vehicles. 20 pp.

—**Automotive—Air Pollution**
EC 88/436/EEC-88. Council Directive Amending Directive 70/220/EEC on the Approximation of the Laws of the Member States Relating to Measures to be Taken Against Air Pollution by Gases from Engines of Motor Vehicles (Restriction of Particulate Pollutant). 18 pp.

INTERNATIONAL AND NON-U.S. NATIONAL STANDARDS
SUBJECT INDEX

Diesel

Diesel Engines (Cont.)

—Automotive—Fuel Filters
ISO 7310-93. Diesel Engines—Heads for Spin-on Fuel Filters with Horizontal Flange—Mounting and Connecting Dimensions Second Edition. 7 pp.
ISO 7311-93. Diesel Engines—Heads for Fuel Filters with Vertical Flange—Mounting and Connecting Dimensions Second Edition. 9 pp.

—Constant Speed
CNS B7043-69. Method of Test for Diesel Engine of Constant Speed (Jul)(3005).
JIS B 8014-76. Methods of Test for Constant Speed Diesel Engines.
JIS B 8014-64. Test Code for Constant Speed Diesel Engines. 17 pp.

—Construction Equipment
CNS D1044-83. Standard Form of Specification of Diesel Engines for Construction Machineries (Jan)(9851).
JIS D 0006-86. Standard Form of Specifications of Diesel Engines for Construction Machinery. 16 pp.
JIS D 1005-86. Testing Methods of Diesel Engines for Construction Machinery. 22 pp.

—Cylinder Blocks—Preservation
MOD UK DSTAN 81-72-90. Preservation of Cylinder Blocks (with or Without Pistons) and Cylinder Heads Issue 1. 9 pp.

—Cylinder Heads—Preservation
MOD UK DSTAN 81-72-90. Preservation of Cylinder Blocks (with or Without Pistons) and Cylinder Heads Issue 1. 9 pp.

—Exhaust Gas Analyzers
CNS D4009-83. Reflection Type Smokemeters for Measuring Smoke Concentration of Exhaust Gas for Diesel Automobiles (Jan)(9845).
CNS D4010-83. Light Extinction Type Smokemeters for Measuring Smoke Concentration of Exhaust Gas for Diesel Automobiles (Jan)(9846).
JIS D 8004-86. Reflection Type Smokemeters for Automobile Diesel Engine.
JIS D 8004-71. Reflection Type Smokemeters for Measuring Carbon Concentration of Diesel Smoke for Diesel Automobiles. 12 pp.
JIS D 8005-73. Light Extinction Type Smokemeters for Measuring Automotive Diesel Exhaust.

—Exhaust Gases
CNS D3174-88. Method of Test for Exhaust Smoke Under Noload and Rapid Acceleration for Diesel Engine Automobiles (Sep)(11644). 10 pp.
CNS D3175-88. Method of Test for Exhaust Smoke Under Full Load and Steady Speed for Diesel Engine Automobiles (Sep)(11645). 9 pp.
EC 88/77/EEC-87. Council Directive on the Approximation of the Laws of the Member States Relating to the Measures to be Taken against the Emission of Gaseous Pollutants from Diesel Engines for Use in Vehicles. 29 pp.
EC 91/542/EEC-91. Council Directive Amending Directive 88/77/EEC on the Approximation of the Laws of the Member States Relating to the Measures to Be Taken against the Emission of Gaseous Pollutants from Diesel Engines for Use in Vehicles. 19 pp.

—Exhaust Gases—Carbon Content
JIS D 1101-85. Diesel Engine Smoke Measurement. 5 pp.

—Exhaust Gases—Opacity
ISO 3173-74. Road Vehicles—Apparatus for Measurement of the Opacity of Exhaust Gas from Diesel Engines Operating Under Steady State Conditions First Edition. 29 pp.
ISO TR4011-76. Road Vehicles—Apparatus for Measurement of the Opacity of Exhaust Gas from Diesel Engines First Edition. 17 pp.
ISO 7644-88. Road Vehicles—Measurement of Opacity of Exhaust Gas from Compression-Ignition (Diesel) Engines—Lug-Down Test First Edition. 6 pp.
ISO 7645-88. Road Vehicles—Measurement of Opacity of Exhaust Gas from Compression-Ignition (Diesel) Engines—Steady Single-Speed Test First Edition. 4 pp.

—Exhaust Gases—Smoke Content
ISO TR9310-87. Road Vehicles—Smoke Measurement of Compression-Ignition (Diesel) Engines—Survey of Short in-Service Tests First Edition. 18 pp.

—Fuel Filters
BSI BS 4552: Part 1-79. 1979 Amd 1 Fuel Filters, Strainers and Sedimentors for Compression-Ignition Engines Part 1: Methods of Test. 43 pp.
BSI BS 4552: Part 2-80. 1980 Amd 1 Fuel Filters, Strainers and Sedimentors for Compression-Ignition Engines Part 2: Method of Classification. 5 pp.

Diesel Engines (Cont.)

—Fuel Filters (Cont.)
BSI BS 4552: Part 3-92. 1992 Fuel Filters, Strainers and Sedimentors for Compression-Ignition Engines Part 3: Specification for Mounting and Connecting Dimensions of Spin-on Fuel Filters (ISO 7654: 1991). 12 pp.
CNS D3139-82. Method of Test of Fuel Filters for Automobile Diesel Engines (Sep)(9373).
CNS D3164-84. Fuel Filter Test Method for Automotive Gasoline Engines (Mar)(10800).
ISO 4020 Pt 1-79. Road Vehicles—Fuel Filters for Automotive Compression Ignition Engines—Part 1: Test Methods First Edition; (Corrected and Reprinted -1980). 33 pp.
ISO 4020 Pt 2-79. Road Vehicles—Fuel Filters for Automotive Compression Ignition Engines—Part 2: Test Values and Classification First Edition. 6 pp.
ISO 7654-91. Road Vehicles—Spin-on Fuel Filters for Compression-Ignition Engines—Mounting and Connecting Dimensions Second Edition. 9 pp.
ISO 7774-84. Road Vehicles—Compression Ignition Engines—Single Fuel Filters with Horizontal Flange and Centre Bolt Fixing—Mounting and Connecting Dimensions First Edition. 6 pp.
JIS D 1608-82. Fuel Filter Test Methods for Automotive Gasoline Engines. 11 pp.
JIS D 1617-79. Test Method of Fuel Filters for Automobile Diesel Engines.

—Glow Plugs
ISO 6550-89. Road Vehicles—M12 X 1,25 and M14 X 1,25 Sheath-Type Glow-Plugs with Conical Seating and Their Cylinder Head Housing Second Edition. 6 pp.
ISO 7578-86. Road Vehicles—Sheath-Type Glow Plugs—General Requirements and Test Methods Second Edition. 5 pp.

—Industrial Equipment
JIS B 8018-89. Test Method of Performance of Small Size Diesel Engines for Land Use. 25 pp.

—Lubricating Oils—Marine
MOD UK DSTAN 91-22-01. Lubricating Oil, Naval Diesel: Severe Service, Grade 40 OMD-113 NATO 0-278 Lubricating Oil, Naval Diesel: Severe Service, Grade 10 Issue 2; Amendment 3. 24 pp.

—Military Vehicles
MOD UK DSTAN 28-10-92. Engines, Diesel and Multi-Fuel, for Military Vehicle Applications Issue 1. 9 pp.

—Military Vehicles—Type Testing
MOD UK DSTAN 28-9-01. Type Testing of Diesel and Multi-Fuel Engines for Military Vehicle Applications Issue 1; Amendment 2. 29 pp.

—Mining Equipment
BSI BS 6680-85. 1985 Flameproof Equipment for Diesel Engines for the Use in Coal Mines and Other Mines Susceptible to Firedamp. 16 pp.
SAA AS 3584-91. Diesel Engine Systems for Underground Coal Mines (in Professional Package 32). 28 pp.

—Paints (Enamel)—Interior/Exterior—Heat Resistant
CGSB CAN/CGSB-1.76-M91. Interior and Exterior Heat-Resistant Enamel. 8 pp.

—Preservation
MOD UK DSTAN 81-70-01. Preservation of Engines (Gasoline and Diesel) Issue 1; Amendment 1. 22 pp.

—Rolling Stock
CNS E2006-82. General Rules for the Diesel Railcar on Completion of Construction (Feb)(8486).
JIS E 5303-89. Test Methods for Diesel Engines of Railway Rolling Stock.

—Sedimentors
BSI BS 4552: Part 1-79. 1979 Amd 1 Fuel Filters, Strainers and Sedimentors for Compression-Ignition Engines Part 1: Methods of Test. 43 pp.
BSI BS 4552: Part 2-80. 1980 Amd 1 Fuel Filters, Strainers and Sedimentors for Compression-Ignition Engines Part 2: Method of Classification. 5 pp.

—Shaft Horsepower
JIS D 1001-82. Engine Power Test Code for Road Vehicles. 20 pp.

—Ships—Fuel Systems
MOD UK NES 320-91. Requirements for Design and Installation of Fuel Systems for Gas Turbines and Diesel Engines in Surface Ships Issue 2 (07.91). 112 pp.

—Ships—Output Computation
JIS F 0404-85. Calculation Methods for Main Diesel Engine Output at Sea Trial.

Diesel Engines (Cont.)

—Ships—Propulsion
JIS F 4301-89. Water Cooled Four Cycle Marine Diesel Engines for Propelling Use.
MOD UK NES 313-89. Diesel Engines for Marine Propulsion and Auxiliary Machinery Issue 2 (10.89). 78 pp.

—Starting Fluids
MOD UK DSTAN 29-6-72. Cold Starting Aids (Fluid Type) for Gasoline and Diesel Engines Issue 1 (Withdrawn). 11 pp.

—Strainers
BSI BS 4552: Part 1-79. 1979 Amd 1 Fuel Filters, Strainers and Sedimentors for Compression-Ignition Engines Part 1: Methods of Test. 43 pp.
BSI BS 4552: Part 2-80. 1980 Amd 1 Fuel Filters, Strainers and Sedimentors for Compression-Ignition Engines Part 2: Method of Classification. 5 pp.

—Submarines—Propulsion
MOD UK NES 313-89. Diesel Engines for Marine Propulsion and Auxiliary Machinery Issue 2 (10.89). 78 pp.

—Tractors—Exhaust Gases
EC 77/537/EEC-77. Council Directive on the Approximation of the Laws of the Member States Relating to the Measures to Be Taken Against the Emission of Pollutants from Diesel Engines for Use in Wheeled Agricultural or Forestry Tractors. 22 pp.

—Type Testing
MOD UK DSTAN 28-4-01. Type Testing of Diesel and Gasoline Engines for General Purpose Applications Issue 3; Corrigendum. 23 pp.
MOD UK DSTAN 28-5-89. Type Testing of Engines for General Purpose Applications Using Diesel and AVTUR Fuels Issue 2. 26 pp.

—Water Cooled
CNS B4007-66. Water-Cooled Diesel Engines for Land Use (Small Size) (Dec)(2695). 5 pp.
JIS B 8012-77. Small Size Water Cooled Diesel Engines for Land Use in (R 1982). 9 pp.

—Water Cooled—Ships—Generators
JIS F 4306-87. Water-Cooled Four Cycle Marine Diesel Engines for Electric Generator.

Diesel Fuels

See Also: Alcohol Fuels; Automotive Fuels; Fuel Oils; Gasohol; Liquid Fuels; Petroleum Products
BSI BS 2869: Part 1-88. 1988 Amd 1 Fuel Oils for Non-Marine Use Part 1: Automotive Diesel Fuel (Class A7) (AMD 6507) December 21, 1990 (Supersedes BS 2869: 1983). 10 pp.
CGSB CAN/CGSB-3.6-M90. Automotive Diesel Fuel. 17 pp.
CNS K5024-89. Diesel Oil (Oct)(1471).
DIN ENGL 51601-86. Liquid Fuels; Diesel Fuel; Minimum Requirements (Feb). 3 pp.

—Asphaltene Content
BSI BS 2000: Part 143-93. 1993 Methods of Test for Petroleum and Its Products Part 143: Determination of Asphaltenes (Heptane Insolubles) (W). 6 pp.
BSI BS 2000: Part 143-85. 1985 Petroleum and Its Products Part 143: Asphaltenes in Petroleum Products (Precipitation with Normal Heptane). 8 pp.

—Automotive
BSI BS EN 590-93. 1993 Automotive Diesel Fuel (W). 15 pp.
CEN PREN 590-91. Automotive Fuels—Diesel—Requirements and Methods of Test. 17 pp.
MOD UK DSTAN 91-7-01. Reference Gasoline, High Lead Reference Gasoline, Low Lead Reference Diesel Fuel, High Sulphur Issue 3 Amendment 2. 12 pp.
MOD UK DSTAN 91-7-93. Reference Diesel, High Sulphur Issue 4. 12 pp.

—Carbon Residue Testing
DIN ENGL 51551-86. Testing of Lubricants and Liquid Fuels; Determination of Conradson Carbon Residue (Mar). 4 pp.

—Cetane Number
BSI BS 2000: Part 218-82. (OBSOLESCENT) 1982 Amd 1 Petroleum and Its Products Part 218: Calculation of Cetane Index of Diesel Fuels (Range 55 and Above). 5 pp.
BSI BS 2000: Part 364-85. 1985 Petroleum and Its Products Part 364: Calculated Cetane Index of Diesel Fuels. 6 pp.

—Cold Filter Plugging Point
BSI BS 6188-82. 1982 Determination of Cold Filter Plugging Point of Diesel and Domestic Heating Fuels. 14 pp.

INDUSTRY STANDARDS

Diesel Fuels (Cont.)

—Cold Filter Plugging Point (Cont.)
CEN EN 116-81. Determination of Cold Filter Plugging Point of Diesel and Domestic Heating Fuels. 13 pp.

—Combat Vehicles—Low Temperature
NATO STANAG 2415 ED 1 AMD 2-90. Procedures for Operation of Mechanical Ground Equipment to Minimize Diesel Fuel Problems at Low Ambient Temperature. 9 pp.

—Ignition Quality
BSI BS 5580-78. 1978 Diesel Fuels. Determination of Ignition Quality. Cetane Method. 4 pp.
DIN ENGL 51773-71. Testing of Liquid Fuels; Determination of Ignition Quality (Cetane Number) of Diesel Fuels (July). 4 pp.
ISO 5165-92. Diesel Fuels—Determination of Ignition Quality—Cetane Method Second Edition. 11 pp.

—Low Temperature Testing
BSI BS 6380-83. (WITHDRAWN) 1983 Amd 1 Low Temperature Properties and Cold Weather Use of Diesel Fuels and Gas Oils (Classes A1, A2 and D of BS 2869) (AMD 5452) September 30, 1986. 16 pp.

—Military Vehicles—Low Temperature
MOD UK DSTAN 91-9-80. Diesel Fuel: Military NATO Code No: F-54 Joint Service Designation: DIESO MILITARY Diesel Fuel: Sub-Zero Joint Service Designation: DIESO SUB-ZERP Issue 3. 8 pp.
MOD UK DSTAN 91-9-93. Diesel Fuel, Military NATO Code: F-54 Joint Service Designation: DIESO MILITARY Issue 4. 12 pp.
NATO STANAG 2415 ED 1 AMD 2-90. Procedures for Operation of Mechanical Ground Equipment to Minimize Diesel Fuel Problems at Low Ambient Temperature. 9 pp.

—Mining Equipment
CGSB CAN/CGSB-3.16-M88. Mining Diesel Fuel. 14 pp.

—Rolling Stock
CGSB CAN/CGSB-3.18-92. Diesel Fuel for Locomotive Type Medium Speed Diesel Engines. 12 pp.

—Storage Stability
MOD UK DSTAN 05-50: Part 40-87. Methods for Testing Fuels, Lubricants and Associated Products Part 40: Storage Stability of Diesel Fuels Issue 1. 10 pp.

Diesel Locomotives
Use: Locomotives—Diesel

Diesel Motor Oil
Use For: Diesel Engine Oils *See Also:* Lubricating Oils; Motor Oils; Oils
MOD UK DSTAN 91-43-01. Lubricating Oil, Engine: Moderate Duty, Diesel Engine Service: NATO Code No: O-176 (Grade 10) Joint Service Designation: OMD-40 NATO Code No: O-178 (Grade 20) Joint Service Designation: OMD-60 NATO Code No: O-180 (Grade 30) Joint Service Designation: OMD-110. 12 pp.
MOD UK DSTAN 91-43-93. Lubricating Oil, Engine: Grade 50 NATO Code: O-182 Joint Service Designation: OMD-330 Issue 2. 14 pp.
MOD UK TS 10033E. Lubricating Oil, Engine: OMD-30, OMD-80 and OMD-85.

—Fuel Dilution—Gas Chromatography
DIN ENGL 51380-90. Determination of Readily Volatile Components in Used Automotive Engine Oils by Gas Chromatography (Nov). 2 pp.

Diesel Rolling Stock
Use: Rolling Stock

Dietary Foods
See Also: Food
EC COM(86) 91-86. Proposal for a Council Directive on the Approximation of the Laws of the Member States Relating to Foodstuffs Intended for Particular Nutritional Uses. 13 pp.
EC COM(87) 241-87. Amended Proposal for a Council Directive on the Approximation of the Laws of the Member States Relating to Foodstuffs Intended for Particular Nutritional Uses (COM(86)91 Final of 22 April 1986). 6 pp.
EC 89/398/EEC-89. Council Directive on the Approximation of the Laws of the Member States Relating to Foodstuffs Intended for Particular Nutritional Uses (For Rel Stds See TEST 89/398). 6 pp.

Dietary Foods (Cont.)

—Low Sodium
CNS N5196-82. Special Dietary Foods with Low Sodium Contents (Including Salt Substitutes) (Aug)(9318). 2 pp.

Diethanolamine
Use For: Diethylolamine; 2,2'-Iminodiethanol
CNS K7186-65. Chemical Reagent (Diethanolamine) (Jan)(1686).
JIS K 8453-88. 2,2'-Iminodiethauol (Diethanolamine).

Diethyl Ether
Use: Ethyl Ether

Diethyl Phthalate
CNS K1237-80. Diethylphthalate (Apr)(5452).
CNS K6474-80. Method of Test for Diethylphthalate (Apr)(5455).

Diethylamine
See Also: Hazardous Materials

—Storage and Handling—Information Cards
SAA AS 2508.3.01 0-82. Safe Storage and Handling Information Cards for Hazardous Materials—Part 3.010: Amines (Isopropylamine, Diethylamine, Triethylamine) Double Sided Card.

Diethylaniline
Use For: N,N-Diethylaniline
CNS K2181-87. N,N-Diethylaniline (Dec)(12197). 5 pp.
JIS K 4112-81. N-Substituted Anilines (N, N-Dimethylaniline.N, N-Diethylaniline. N-Ethylaniline. N-Methylaniline. N-Benzyl-N-Ethylaniline).

N,N-Diethylaniline
Use: Diethylaniline

Diethylene Dioxide
Use: Dioxane

Diethylene Glycol
CNS K1180-77. Diethylene Glycol (Apr)(4086).

Diethylolamine
Use: Diethanolamine

Diethyltoluamide
Use For: DEET

—Insect Repellents
CGSB CAN/CGSB-15.19-92. Insect Repellent, Diethyltoluamide. 8 pp.
MOD UK TS 475A. Insect Repellant.

Differential Amplifiers
See Also: Amplifiers; Common Mode Rejection; Differential Comparators; Differential Operational Amplifiers; Integrated Circuits; Operational Amplifiers
MOD UK DSTAN 59-62: 90/155-71. Integrated Circuit Issue 1. 11 pp.

Differential Comparator Amplifiers
Use: Differential Comparators

Differential Comparators
See Also: Comparators; Differential Amplifiers; Voltage Comparators

—Quality Assurance
BSI BS 9461-74. (WITHDRAWN) 1974 Amd 1 Rules for the Preparation of Detail Specifications for Integrated Circuits of Assessed Quality: Differential Comparator Amplifiers. General Application Category. 11 pp.

Differential Modulation
Use For: Differential Transmission
See Also: Modulation; Radio Transmission

—Radio Beacons—Global Navigation Satellite Systems
CCIR RECMN 823-92. Technical Characteristics of Differential Transmissions for Global Navigation Satellite Systems (GNSS) from Maritime Radio Beacons in the Frequency Band 285-325 kHz (283.5-315 kHz in Region 1). 7 pp.

—Radio Beacons—Global Positioning Systems
CCIR Report 1166-90. Technical Characteristics of GPS Differential Transmissions from Maritime Radiobeacons—Section 8D—Radiodetermination, Global Maritime Distress and Safety System and Related Subjects. 5 pp.

Differential Operational Amplifiers
See Also: Amplifiers; Differential Amplifiers; Operational Amplifiers

Differential Operational Amplifiers (Cont.)
CECC CECC 90 202-001 ISSUE 1-88. BS CECC 90 202-001; Silicon Monolithic Integrated Circuit Ultra-Low Noise Precision Operational Amplifier (En). 16 pp.
CECC CECC 90 202-003 ISSUE 1-87. BS CECC 90 202-003; Silicon Monolithic Integrated Circuit Low Noise Precision DIFET Operational Amplifier (En). 16 pp.
CECC CECC 90 202-005 ISSUE 1-89. UTE C 86-111/A1; Analogue Integrated Circuits in Accordance with FS 90 202; 155 (En, Fr). 23 pp.
CECC CECC 90 202-006 ISSUE 1-89. UTE C 86-111/A2; Analogue Integrated Circuits in Accordance with FS 90 202; 155A (En, Fr). 23 pp.
CECC CECC 90 202-007 ISSUE 1-89. UTE C 86-111/A3; Analogue Integrated Circuits in Accordance with FS 90 202; 156 (En, Fr). 23 pp.
CECC CECC 90 202-008 ISSUE 1-89. UTE C 86-111/A4; Analogue Integrated Circuits in Accordance with FS 90 202; 156A (En, Fr). 23 pp.
CECC CECC 90 202-009 ISSUE 1-89. UTE C 86-111/A5; Analogue Integrated Circuits in Accordance with FS 90 202; 157 (En, Fr). 23 pp.
CECC CECC 90 202-010 ISSUE 1-89. UTE C 86-111/A6; Analogue Integrated Circuits in Accordance with FS 90 202; 157A (En, Fr). 23 pp.
CECC CECC 90 202-011 ISSUE 1-89. UTE C 86-111/A7; Analogue Integrated Circuits in Accordance with FS 90 202; 255 (En, Fr). 23 pp.
CECC CECC 90 202-012 ISSUE 1-89. UTE C 86-111/A8; Analogue Integrated Circuits in Accordance with FS 90 202; 256 (En, Fr). 23 pp.
CECC CECC 90 202-014 ISSUE 1-89. UTE C 86-111/A10; Analogue Integrated Circuits in Accordance with FS 90 202 (En, Fr). 21 pp.
CECC CECC 90 202-015 ISSUE 1-89. UTE C 86-111/A11; Analogue Integrated Circuits in Accordance with FS 90 202 (En, Fr). 21 pp.
CECC CECC 90 202-016 ISSUE 1-89. UTE C 86-111/A12; Analogue Integrated Circuits in Accordance with FS 90 202 (En, Fr). 21 pp.
CECC CECC 90 202-017 ISSUE 1-89. UTE C 86-111 ADD 1/A13; Analogue Integrated Circuits in Accordance with FS 90 202 (En, Fr). 21 pp.
CECC CECC 90 202-018 ISSUE 1-89. UTE C 86-111 ADD 1/A14; Analogue Integrated Circuits in Accordance with FS 90 202 (En, Fr). 21 pp.
CECC CECC 90 202-019 ISSUE 1-89. UTE C 86-111 ADD 1/A15; Analogue Integrated Circuits in Accordance with FS 90 202 (En, Fr). 21 pp.
CECC CECC 90 202-020 ISSUE 1-89. UTE C 86-111 ADD 1/A16; Analogue Integrated Circuits in Accordance with FS 90 202 (En, Fr). 21 pp.
CECC CECC 90 202-021 ISSUE 1-89. UTE C 86-111 ADD 1/A17; Analogue Integrated Circuits in Accordance with FS 90 202 (En, Fr). 21 pp.
CECC CECC 90 202-022 ISSUE 1-89. UTE C 86-111 ADD 1/A18; Analogue Integrated Circuits in Accordance with FS 90 202 (En, Fr). 21 pp.
CECC CECC 90 202-023 ISSUE 1-89. UTE C 86-111 ADD 1/A19; Analogue Integrated Circuits in Accordance with FS 90 202 (En, Fr). 21 pp.
CECC CECC 90 202-024 ISSUE 1-89. UTE C 86-111 ADD 1/A20; Analogue Integrated Circuits in Accordance with FS 90 202 (En, Fr). 23 pp.
CECC CECC 90 202-025 ISSUE 1-89. UTE C 86-111 ADD 1/A21; Analogue Integrated Circuits in Accordance with FS 90 202 (En, Fr). 23 pp.
CECC CECC 90 202-026 ISSUE 1-89. UTE C 86-111 ADD 1/A22; Analogue Integrated Circuits in Accordance with FS 90 202 (En, Fr). 21 pp.
CECC CECC 90 202-027 ISSUE 1-89. UTE C 86-111 ADD 1/A23; Analogue Integrated Circuits in Accordance with FS 90 202 (En, Fr). 21 pp.
CECC CECC 90 202-028 ISSUE 1-89. UTE C 86-111 ADD 1/A24; Analogue Integrated Circuits in Accordance with FS 90 202 (En, Fr). 21 pp.
CECC CECC 90 202-029 ISSUE 1-89. UTE C 86-111 ADD 2/A25; Analogue Integrated Circuits in Accordance with FS 90 202 (En, Fr). 22 pp.
CECC CECC 90 202-030 ISSUE 1-89. UTE C 86-111 ADD 2/A26; Analogue Integrated Circuits in Accordance with FS 90 202 (En, Fr). 21 pp.
CECC CECC 90 202-031 ISSUE 1-89. UTE C 86-111 ADD 2/A27; Analogue Integrated Circuits in Accordance with FS 90 202 (En, Fr). 21 pp.
CECC CECC 90 202-032 ISSUE 1-89. UTE C 86-111 ADD 2/A28; Analogue Integrated Circuits in Accordance with FS 90 202 (En, Fr). 21 pp.
CECC CECC 90 202-033 ISSUE 1-89. UTE C 86-111 ADD 2/A29; Analogue Integrated Circuits in Accordance with FS 90 202 (En, Fr). 22 pp.
CECC CECC 90 202-034 ISSUE 1-89. UTE C 86-111 ADD 2/A30; Analogue Integrated Circuits in Accordance with FS 90 202 (En, Fr). 22 pp.
CECC CECC 90 202-035 ISSUE 1-89. UTE C 86-111 ADD 2/A31; Analogue Integrated Circuits in Accordance with FS 90 202 (En, Fr). 25 pp.
CECC CECC 90 202-036 ISSUE 1-89. UTE C 86-111 ADD 2/A32; Analogue Integrated Circuits in Accordance with FS 90 202 (En, Fr). 22 pp.

Differential Operational Amplifiers (Cont.)

CECC CECC 90 202-037 ISSUE 1-89. UTE C 86-111 ADD 2/A33; Analogue Integrated Circuits in Accordance with FS 90 202 (En, Fr). 22 pp.

CECC CECC 90 202-038 ISSUE 1-89. UTE C 86-111 ADD 2/A34; Analogue Integrated Circuits in Accordance with FS 90 202 (En, Fr). 25 pp.

CECC CECC 90 202-039 ISSUE 1-89. UTE C 86-111 ADD 2/A35; Analogue Integrated Circuits in Accordance with FS 90 202 (En, Fr). 22 pp.

CECC CECC 90 202-040 ISSUE 1-89. UTE C 86-111 ADD 2/A36; Analogue Integrated Circuits in Accordance with FS 90 202 (En, Fr). 22 pp.

CECC CECC 90 202-042 ISSUE 1-90. UTE C 86-111 ADD 3/A37; Operational Amplifiers with J FET Inputs (En, Fr). 14 pp.

CECC CECC 90 202-043 ISSUE 1-90. UTE C 86-111 ADD 3/A38; Analogue Integrated Circuits; Selon/In Accordance with: FS 90202 (En, Fr). 12 pp.

MOD UK DSTAN 59-62: 90/158-71. Integrated Circuit Issue 1. 12 pp.

Differential Pressure

See Also: Differential Pressure Gages; Differential Pressure Transmitters; Pitot Tubes; Pressure

—Aircraft

BSI 3G 100:Pt 2: SUBSEC 3.4-72. 1972 General Requirements for Equipment for Use in Aircraft Part 2: All Equipment Section 3: Environmental Conditions Subsection 3.4: Differential Pressure Requirements. 6 pp.

—Airspeed Indicators

JIS W 0202-79. Table of Differential Pressure for Airspeed Indicators.

—Valves

DIN ENGL 19208-88. Flow Measurement; Mating Dimensions and Application of Shut-Off Valves for Differential Pressure Transducers and Differential Pressure Piping (Aug). 3 pp.

Differential Pressure Flowmeters

Use: Flowmeters—Differential Pressure

Differential Pressure Gages

See Also: Absolute Pressure Gages; Diaphragm Gages; Differential Pressure; Differential Pressure Transmitters; Flowmeters; Gages; Pressure Gages

—Aircraft—Fuel Filters

NATO STANAG 3583 ED 2 AMD 3-74. Standards of Accuracy for Differential Pressure Gauges for Aviation Fuel Filters and Filter Separators. 8 pp.

NATO STANAG 3583 ED 3 AMD 0-93. Standards of Accuracy for Differential Pressure Gauges for Aviation Fuel Filters and Filter Separator Vessels. 7 pp.

NATO STANAG 3583 ED 3 AMD 1-93. Standards of Accuracy for Differential Pressure Gauges for Aviation Fuel Filters and Filter Separator Vessels. 8 pp.

—Aircraft—Fuel Separators/Filters

NATO STANAG 3583 ED 2 AMD 3-74. Standards of Accuracy for Differential Pressure Gauges for Aviation Fuel Filters and Filter Separators. 8 pp.

NATO STANAG 3583 ED 3 AMD 0-93. Standards of Accuracy for Differential Pressure Gauges for Aviation Fuel Filters and Filter Separator Vessels. 7 pp.

NATO STANAG 3583 ED 3 AMD 1-93. Standards of Accuracy for Differential Pressure Gauges for Aviation Fuel Filters and Filter Separator Vessels. 8 pp.

—Stop Valves

DIN ENGL 19212-80. Methods for Measurement of Fluid Flow; Threaded Connections for Shut-Off Units and for Differential Pressure Measuring Instruments (Aug). 4 pp.

Differential Pressure Gauges

Use: Differential Pressure Gages

Differential Pressure Indicators

Use: Differential Pressure Gages

Differential Pressure Transmitters

See Also: Differential Pressure; Differential Pressure Gages; Gage Pressure Transmitters; Pressure Transmitters; Transmitters

BSI BS 6174-82. 1982 Amd 1 Differential Pressure Transmitters with Electrical Outputs. 15 pp.

—Process Control Equipment

JIS C 1031-90. Methods of Evaluating the Performance of Pressure and Differential Pressure Transmitters for Use in Industrial-Process Control Systems. 21 pp.

Differential Relays

See Also: Relays

—Percentage

BSI BS 142: Sec 4.1-84. 1984 Electrical Protection Relays Part 4: Requirements for Multi-Input Energizing Quantity Relays Section 4.1: Biased (Percentage) Differential Relays. 12 pp.

IEC 255 Pt 13-80. Electrical Relays Part 13: Biased (Percentage) Differential Relays First Edition. 27 pp.

Differential Sensitivity

See Also: Echo Suppressors; Sensitivity (Electrical); Signal Levels

CCITT RECMN G.164-89. Echo Suppressors—General Characteristics of International Telephone Connections and Circuits (Study Groups XII and XV) 36 pp. 36 pp.

Differential Thermometers

Use: Bimetallic Thermometers

Differential Transformer Transducers

See Also: Transducers

—Angular—Synchros—Preferred Products List

CECC CECC MUAHAG Vol 13 IS 1-91. Preferred Products List; Servo Components (En, Fr, Ge). 27 pp.

Differential Transmission

Use: Differential Modulation

Differential Voltage Comparators

Use: Voltage Comparators

Diffraction

See Also: Attenuation; Deflection; Electromagnetic Radiation; Optical Properties; Refraction; Sound Transmission

—Field Strength—Ground Wave Propagation

CCIR RECMN 526-1-82. Propagation by Diffraction—Section 5B—Effects of the Ground (Including Ground-Wave Propagation). 1 p.

CCIR RECMN 526-2-92. Propagation by Diffraction. 16 pp.

—Radio Wave Propagation

CCIR Report 715-3-90. Propagation by Diffraction—Section 5B—Effects of the Ground (Including Ground-Wave Propagation). 16 pp.

Diffuse Transmission Density

Use: Optical Density

Diffusivity

Use For: Thermal Diffusivity *See Also:* Thermal Transmittance; Thermodynamic Properties

—Ceramics

CEN PREN 821-2-92. Advanced Technical Ceramics—Thermo-Physical Properties of Monolithic Ceramics—Part 2: Determination of Thermal Diffusivity by the Laser Flash (or Heat Pulse) Method. 20 pp.

JIS R 1611-91. Testing Methods of Thermal Diffusivity, Specific Heat Capacity, and Thermal Conductivity for High Performance Ceramics by Laser Flash Method. 16 pp.

Difluorodichloromethane

Use: Halon

Digestion (Decomposition)

—Water

ISO 10048-91. Water Quality—Determination of Nitrogen—Catalytic Digestion After Reduction with Devarda's Alloy First Edition. 8 pp.

Diggers

Use For: Digging Bars *See Also:* Tools

CGSB 39-GP-45A-74. Bars, Steel, Standard for; (Amendment 1 Mar 1975). 18 pp.

Digging Bars

Use: Diggers

Digital Access and Cross Connect Equipment

Use: Digital Access and Cross Connect Systems

Digital Access and Cross Connect Systems

Use For: DACS; Digital Access and Cross Connect Equipment *See Also:* Carrier Systems; Telephone Systems

Digital Access and Cross Connect Systems (Cont.)

CENELEC PRETS 300 010-1-90. Synchronous Cross Connect Equipment 64 and NX 64 Kbit/s Cross Connection Rate 2048 Kbit/s Access Ports Part 1: Core Functions and Characteristics (L 03-17). 19 pp.

CENELEC PRETS 300 010-1-92. Transmission and Multiplexing (TM); Synchronous Cross Connect Equipment 64 and n x 64 kbit/s Cross Connection Rate 2 048 kbit/s Access Ports Part 1: Core Functions and Characteristics. 24 pp.

CENELEC ETS 300 010-1-92. Transmission and Multiplexing (TM); Synchronous Cross Connect Equipment 64 and n x 64 kbit/s Cross Connection Rate 2 048 kbit/s Access Ports Part 1: Core Functions and Characteristics. 24 pp.

ETSI PRETS 300 010-1-90. Synchronous Cross Connect Equipment 64 and x 64 Kbit/s Crossconnection Rate 2048 Kbit/s Access Parts Part 1: Core Functions and Characteristics. 19 pp.

ETSI ETS 300 010-1-92. Transmission and Multiplexing (TM); Synchronous Cross Connect Equipment 64 and n x 64 kbit/s Cross Connection Rate 2 048 kbit/s Access Ports Part 1: Core Functions and Characteristics. 24 pp.

ETSI PRETS 300 010-1-92. Transmission and Multiplexing (TM); Synchronous Cross Connect Equipment 64 and n x 64 kbit/s Cross Connection Rate 2 048 kbit/s Access Ports Part 1: Core Functions and Characteristics. 24 pp.

ETSI PRETS 300 010-1-90. Synchronous Cross Connect Equipment 64 and NX 64 Kbit/s Cross Connection Rate 2048 Kbit/s Access Ports Part 1: Core Functions and Characteristics (L 03-17). 19 pp.

Digital Audio Coding

See Also: Acoustic Signals; Encoding

CCIR Report 1068-86. Digital Audio Coding Standards—Section 10C—Audio-Frequency Characteristics of Sound-Broadcasting Signals. 3 pp.

CCIR Report 1199-90. Low Bit-Rate Digital Audio Coding Systems—Section 10C—Audio-Frequency Characteristics of Sound-Broadcasting Signals. 8 pp.

CCIR Decision 98-89. Low Bit-Rate Digital Audio Coding Systems—Volume X-1—Broadcasting Service (Sound). 3 pp.

CCIR Decision 98-89. Low Bit-Rate Digital Audio Coding Systems—Annex to Volume XII—Television and Sound Transmission (CMTT). 3 pp.

IEC 11172 Pt 3-93. Information Technology—Coding of Moving Pictures and Associated Audio for Digital Storage Media at up to About 1,5 Mbit/s—Part 3: Audio First Edition. 157 pp.

ISO 11172 Pt 3-93. Information Technology—Coding of Moving Pictures and Associated Audio for Digital Storage Media at up to About 1,5 Mbit/s—Part 3: Audio First Edition. 157 pp.

JTC1 11172 Pt 3-93. Information Technology—Coding of Moving Pictures and Associated Audio for Digital Storage Media at up to About 1,5 Mbit/s—Part 3: Audio First Edition. 157 pp.

—Broadcasting Satellite Services

CCIR RECMN 651-86. Digital PCM Coding for the Emission of High-Quality Sound Signals in Satellite Broadcasting (15 kHz Nominal Bandwidth)—Section 10/11B—Systems. 2 pp.

CCIR Report 953-2-90. Digital Coding for the Emission of High-Quality Sound Signals in Satellite Broadcasting (15 kHz Nominal Bandwidth)—Section 10/11B—Systems. 18 pp.

—Sound Broadcasting

CCIR QUESTION 86/10-90. Digital Audio Coding Standards—Questions Concerning Study Group 10—Broadcasting Service (Sound). 2 pp.

CCIR QUESTION 87/10-90. Digital Coding in Broadcasting for Emission of Sound Signals—Questions Concerning Study Group 10—Broadcasting Service (Sound). 1 p.

Digital Audio Tapes

Use: Audio Tapes—Digital

Digital Blocks

See Also: Communication Channels

—Designations

CCITT RECMN M.140-89. Designations of International Circuits, Groups, Group and Line Links, Digital Blocks, Digital Paths, Data Transmission Systems and Related Information—General Maintenance Principles—Maintenance of International Transmission Systems and Telephone Circuits. 53 pp.

—Numbering—Transmission Systems

CCITT RECMN M.410-89. Numbering of Digital Blocks in Transmission Systems—General Maintenance Principles—Maintenance of International Transmission Systems and Telephone Circuits (Study Group IV) 3 pp. 3 pp.

INTERNATIONAL AND NON-U.S. NATIONAL STANDARDS
SUBJECT INDEX

Digital

Digital Blocks *(Cont.)*
—Set Up
CCITT RECMN M.555-89. Bringing International Digital Blocks, Paths and Sections into Service—General Maintenance Principles—Maintenance of International Transmission Systems and Telephone Circuits (Study Group IV) 5 pp. 5 pp.

Digital Channels
Use: Communication Channels—Digital

Digital Circuit Multiplication Equipment
Use For: DCME; DCMS *See Also:* Circuit Multiplication Equipment; Synchronous Digital Hierarchy; Telecommunication Circuits; Telecommunication Equipment; Telephone Circuits

—Adaptive Delta Pulse Code Modulation
CCITT RECMN G.723-89. Extensions of Recommendation G.721 Adaptive Differential Pulse Code Modulation to 24 and 40 kbit/s for Digital Circuit Multiplication Equipment Application—General Aspects of Digital Transmission Systems; Terminal Equipments (Stdy Grp XV/XVIII)18 pp. 18 pp.

—Adaptive Delta Pulse Code Modulation—Digital Speech Interpolation
CCITT RECMN G.763-91. Digital Circuit Multiplication Equipment Using 32 kBit/s ADPCM and Digital Speech Interpolation (Study Group XV) 368 pp. 368 pp.
CCITT RECMN G.763-89. Digital Circuit Multiplication Equipment Using 32 kBit/s ADPCM and Digital Speech Interpolation—General Aspects of Digital Transmission Systems; Terminal Equipments (Study Groups XV and XVIII) 37 pp. 37 pp.

—Documents
CCITT FASCICLE III.4-89. General Aspects of Digital Transmission Systems Terminal Equipments—Recommendations G.700-G.795. 621 pp.

—FAX Modules
CCITT RECMN G.766-92. Facsimile Demodulation/Remodulation for DCME (Study Group XV) 115 pp. 115 pp.

—Telephone Transmission—Acoustic Testing
CCITT RECMN P.84-89. Subjective Listening Test Method for Evaluating Digital Circuit Multiplication and Packetized Voice Systems—Telephone Transmission Quality (Study Group XII) 27 pp. 27 pp.

Digital Circuit Multiplication Systems
Use: Digital Circuit Multiplication Equipment

Digital Circuits
Use For: Digital Integrated Circuits *See Also:* AND Buffer Gates; AND Gates; AND OR Gates; AND OR INVERT Gates; AND OR SELECT Gates; Arithmetic Logic Unit/Function Generators; Binary Counters; Bus Registers; Bus Transceiver/Registers; Bus Transceivers; Carry Save Adders; Charge Transfer Devices; D-Type Flip Flops; D-Type Latches; Decade Counters; Digital Demultiplexers; Exclusive NOR Gates; Exclusive OR Gates; Flip Flops; Frequency Divider/Digital Timers; Frequency Dividers; Full Adders; Gate Circuits; Integrated Circuits; J-K Flip Flops; J-K Master/Slave Flip Flops; Latches (Circuits); Logic Circuits; Look Ahead Carry Generators; Magnitude Comparators; Microprocessors; Monostable Multivibrators; Multivibrators; NAND Gates; NOR Gates; NOR INVERT Gates; Octal Counters; OR Gates; Parity Generator/Checkers; Parity Generators; Parity Trees; R-S Latches; RAMs; Rate Multipliers; S-R Latches; Schmitt Triggers; Shift Registers; Shift/Storage Registers; Synchronous Digital Hierarchy

BSI BS 6493: Sec 2.2-86. 1986 Semiconductor Devices Part 2: Integrated Circuits Section 2.2: Recommendations for Digital Integrated Circuits. 143 pp.
BSI BS 6493: Sec 2.2-01. 1986 Amd 1 Semiconductor Devices Part 2: Integrated Circuits Section 2.2: Recommendations for Digital Integrated Circuits (IEC 748-2: 1985) (AMD 7003) July 15, 1992. 166 pp.
IEC 748 Pt 2-85. Semiconductor Devices Integrated Circuits Part 2: Digital Integrated Circuits First Edition; (Amendment 1-1991). 312 pp.
IECQ QC 790132/BY 0001-BY0010-90. Detail Specification for Integrated Circuits from Family K155. 2 pp.

Digital Circuits *(Cont.)*
IECQ QC 790132/RF 0001-RF0011-90. Detail Specification for Integrated Circuits from Family K1102AP SK1102AP4—SK1102AP14. 2 pp.
IECQ QC 790132/SU 0001-90. Detail Specification for Electronic Components; Digital Integrated Circuits Series KI55 (IEC 748-2-1 ED 1). 174 pp.

—AS TTL Schottky
CECC CECC 90 106 ISSUE 2-87. Family Specification: TTL Advanced Low Power Schottky Digital Integrated Circuits (En, Fr). 36 pp.

—Backdriving
MOD UK DSTAN 00-53-91. Safe Operating Limits for Backdriving Issue 1. 22 pp.

—Bipolar
JIS C 7041-82. Measuring Methods for Bipolar Digital Integrated Circuits. 19 pp.
MOD UK DSTAN 59-62: Part 2-01. Microcircuits Electronic (Integrated Circuits) (Listed on EPIC Database) Part 2: Bipolar Logic Families Issue 2; Amendment 1. 27 pp.

—Bipolar—Quality Assurance
BSI BS 9440-73. (WITHDRAWN) 1973 Amd 1 Rules for the Preparation of Detail Specifications for Integrated Circuits of Assessed Quality: Complex Digital Circuits Bipolar. General Application Category. 16 pp.
BSI BS QC 790132-92. 1992 Bipolar Monolithic Digital Integrated Circuit Gates (Excluding Uncommitted Logic Arrays) (T). 16 pp.

—Cells—Quality Assurance
CENELEC PREN 190 117-92. Blank Detail Specification: Digital Standard Cells Under Capability Approval. 10 pp.

—CMOS
MOD UK DSTAN 59-62: Part 3-83. Microcircuits Electronic (Integrated Circuits) (Listed on EPIC Database) Part 3: CMOS Logic Families Issue 1. 17 pp.

—CMOS—Preferred Products List
CECC CECC MUAHAG Vol 7 IS 8-92. Preferred Products List; Active Microcircuits (En, Fe, Ge). 89 pp.

—CMOS—Quality Assurance
BSI BS CECC 90104-90. 1990 C.Mos Digital Integrated Circuits. Series 4000B and 4000UB: Family Specification. 24 pp.

—DTL—Quality Assurance
BSI BS 9404-72. (WITHDRAWN) 1972 Amd 1 Rules for the Preparation of Detail Specifications for Integrated Circuits of Assessed Quality: DTL Digital Inverter Circuits. General Application Category. 17 pp.

—Electric Endurance Testing—Quality Assurance
BSI BS CECC 90100-86. (WITHDRAWN) 1986 Amd 2 Harmonized System of Quality Assessment for Electronic Components: Sectional Specification: Digital Monolithic Integrated Circuits (AMD 5954) March 28, 1991 (Renumbered as BS EN 190100: 1993). 69 pp.

—Electrical Measurement—Quality Assurance
BSI BS CECC 90100-86. (WITHDRAWN) 1986 Amd 2 Harmonized System of Quality Assessment for Electronic Components: Sectional Specification: Digital Monolithic Integrated Circuits (AMD 5954) March 28, 1991 (Renumbered as BS EN 190100: 1993). 69 pp.

—HCMOS
CECC CECC 90 109 ISSUE 2-89. Family Specification: Digital HC MOS Circuits; Series HC/HCT HCU (En, Fr). 74 pp.
IECQ QC 790109-92. Semiconductor Devices Integrated Circuits Part 2: Digital Integrated Circuits Section Two—Family Specification for HCMOS Digital Integrated Circuits, Series 54/74 HC, 54/74 HCT, 54/74 HCU (IEC 748-2-2 ED 1). 48 pp.
IECQ QC 790130-92. Semiconductor Devices Integrated Circuits Part 2: Digital Integrated Circuits Section Three—Blank Detail Specification for HCMOS Digital Integrated Circuits (Series 54/74 HC, 54/74 HCT, 54/74 HCU) (IEC 748-2-3 ED 1). 30 pp.

—HCMOS—Quality Assurance
BSI BS QC 790109-92. 1992 Family Specification HCMOS Digital Integrated Circuits, Series 54/74 IIC, 54/74 HCT, 54/74 HCU (T). 23 pp.

Digital Circuits *(Cont.)*
—HCMOS—Quality Assurance *(Cont.)*
BSI BS QC 790130-92. 1992 Blank Detail Specification HCMOS Digital Integrated Circuits (Series 54/74 HC, 54/74 HCT, 54/74 HCU) (IEC 748-2-3: 1992) (T). 13 pp.
IEC 748 Pt 2-2-92. Semiconductor Devices Integrated Circuits Part 2: Digital Integrated Circuits Section Two—Family Specification for HCMOS Digital Integrated Circuits, Series 54/74 HC, 54/74 HCT, 54/74 HCU First Edition (IECQ QC 790109). 48 pp.
IEC 748 Pt 2-3-92. Semiconductor Devices Integrated Circuits Part 2: Digital Integrated Circuits Section Three—Blank Detail Specification for HCMOS Digital Integrated Circuits (Series 54/74 HC, 54/74 HCT, 54/74 HCU) First Edition (IECQ QC 790130). 30 pp.

—MOS
CENELEC PREN 190 116-92. Family Specification: AC MOS Digital Integrated Circuits. 34 pp.
IECQ QC 790104-92. Semiconductor Devices Integrated Circuits Part 2: Digital Integrated Circuits Section Four—Family Specification for Complementary MOS Digital Integrated Circuits, Series 4 000 B and 4 000 UB (IEC 748-2-4 ED 1). 32 pp.
IECQ QC 790131-92. Semiconductor Devices Integrated Circuits Part 2: Digital Integrated Circuits Section Five—Blank Detail Specification for Complementary MOS Digital Integrated Circuits (Series 4 000 B and 4 000 UB) (IEC 748-2-5 ED 1). 32 pp.
JIS C 7042-82. Measuring Methods for MOS Digital Integrated Circuits. 24 pp.

—MOS—Quality Assurance
BSI BS QC 790104-92. 1992 Family Specification Complementary MOS Digital Integrated Circuits, Series 4000 B and 4000 UB (IEC 748-2-4: 1992). 16 pp.
BSI BS QC 790131-92. 1992 Blank Detail Specification Complementary MOS Digital Integrated Circuits (Series 4000 B and 4000 UB). 14 pp.
BSI BS CECC 90109-90. 1990 Digital Integrated HC MOS Circuits Series HC/HCT/HCU: Family Specification. 36 pp.
IEC 748 Pt 2-4-92. Semiconductor Devices Integrated Circuits Part 2: Digital Integrated Circuits Section Four—Family Specification for Complementary MOS Digital Integrated Circuits, Series 4 000 B and 4 000 UB First Edition (IECQ QC 790104). 34 pp.
IEC 748 Pt 2-5-92. Semiconductor Devices Integrated Circuits Part 2: Digital Integrated Circuits Section Five—Blank Detail Specification for Complementary MOS Digital Integrated Circuits (Series 4 000 B and 4 000 UB) First Edition (IECQ QC 790131). 32 pp.

—Quality Assurance
BSI BS 9490-75. 1975 Amd 3 Rules for the Preparation of Detail Specifications for Digital Integrated Circuits of Assessed Quality: Full Assessment Level. 71 pp.
BSI BS 9492-78. (OBSOLESCENT) 1978 Amd 1 Rules for the Preparation of Detail Specifications for Capacitively Coupled Digital Integrated Circuits of Assessed Quality. Full Assessment Level. 10 pp.

—Reliability Assured
JIS C 7310-79. General Rules for Reliability Assured Digital Semiconductor Integrated Circuits (R 1984). 14 pp.

—S TTL
CECC CECC 90 102 ISSUE 2-89. Family Specification: TTL Schottky Digital Integrated Circuits; Series 54S, 64S, 74S, 84S (En, Fr). 36 pp.
CECC CECC 90 103 ISSUE 3-88. Family Specification: Digital Integrated TTL Low Power Schottky Circuits; Series 54 LS, 64 LS, 74 LS, 84 LS (En, Fr, Ge). 44 pp.

—TTL
CECC CECC 90 101 ISSUE 2-88. Family Specification: Digital Integrated TTL Circuits; Series 54, 64, 74, 84 (En, Fr, Ge). 48 pp.
CECC CECC 90 107 ISSUE 2-87. Family Specification: TTL Fast Digital Integrated Circuits (En, Fr, Ge). 54 pp.
CECC CECC 90 108 ISSUE 2-87. Family Specification: TTL Advanced Schottky Digital Integrated Circuits (En, Fr). 34 pp.

—TTL—Preferred Products List
CECC CECC MUAHAG Vol 7 IS 8-92. Preferred Products List; Active Microcircuits (En, Fe, Ge). 89 pp.

INDEX and DIRECTORY of

INTERNATIONAL AND NON-U.S. NATIONAL STANDARDS
SUBJECT INDEX

Digital Circuits *(Cont.)*
—TTL—Quality Assurance
BSI BS 9403-72. (WITHDRAWN) 1972 Amd 1 Rules for the Preparation of Detail Specifications for Integrated Circuits of Assessed Quality: TTL Digital Inverter Circuits. General Application Category. 15 pp.

BSI BS 9405-72. (WITHDRAWN) 1972 Amd 1 Rules for the Preparation of Detail Specifications for Integrated Circuits of Assessed Quality: TTL Digital Interconnected Gate Circuits. General Application Category. 19 pp.

BSI BS 9420-72. (WITHDRAWN) 1972 Amd 1 Rules for the Preparation of Detail Specifications for Integrated Circuits of Assessed Quality: TTL Digital Bistable Circuits. General Application Category. 14 pp.

BSI BS CECC 90101-90. 1990 Digital Integrated TTL Circuits. Series 54, 64, 74, 84: Family Specification. 19 pp.

BSI BS CECC 90102-90. 1990 Digital Integrated TTL Schottky Circuits. Series 54S, 64S, 74S, 84S: Family Specification. 19 pp.

BSI BS CECC 90103-90. 1990 Amd 1 Harmonized System of Quality Assessment for Electronic Components Family Specification Digital Integrated TTL Low Power SCHOTTKY Circuits Series 54LS, 64LS, 74LS, 84LS (AMD 6764) June 28, 1991. 26 pp.

BSI BS CECC 90106-87. 1987 TTL Advanced Low Power SCHOTTKY Digital Intergrated Circuits. 19 pp.

BSI BS CECC 90107-87. 1987 TTL Fast Digital Intergrated Circuits. 19 pp.

Digital Communication Systems
Use: Digital Communications

Digital Communications
Use For: Digital Communication Systems
See Also: Message Handling Systems; Pulse Communications; Telecommunication Systems

—Cables (Electric)
BSI BS 7630-93. 1993 Symmetric Pair/Quad and Multicore Cables for Digital Communication (S). 21 pp.

CENELEC HD 608 S1-92. Generic Specification for Symmetric Pair/Quad and Multicore Cables for Digital Communication. 35 pp.

—Reference Clocks
CCIR Report 898-2-90. Operational Experience with Reference Clocks in Time Systems—Section 7D—Characterization of Sources and Time Scales Formation. 10 pp.

Digital Computers
Use: Computers

Digital Counters
See Also: Binary Counter/Dividers

—Audiovisual Equipment
BSI BS 5817: Part 13-90. 1990 Audio-Visual, Video and Television Equipment and Systems Part 13: Digital Counters for Audio Cassette Systems. 8 pp.

CENELEC HD 369.13-84. Audio-Visual Video and Television Equipment and Systems Part 13: Digital Counter for Audio Cassette Systems. 2 pp.

IEC 574 Pt 13-82. Audio-Visual, Video and Television Equipment and Systems Part 13: Digital Counter for Audio Cassette Systems First Edition. 10 pp.

—BCD
CECC CECC 90 104-222 ISSUE 1-85. BS CECC 90 104-222; Three-Digit BCD Counter (En). 35 pp.

CECC CECC 90 104-222 ISSUE 1-86. CEI-CECC 90 104-222; Silicon with (B) Buffered Outputs Three-Digit BCD Counter (En). 2 pp.

—Preferred Products List
CECC CECC MUAHAG Vol 7 IS 8-92. Preferred Products List; Active Microcircuits (En, Fe, Ge). 89 pp.

—12 Stage
CECC CECC 90 104-027 ISSUE 2. NL CECC 90 104-027 Issue 2; Digital Integrated Circuits in Accordance with FS 90 104; HEC/HEF 4030B; 12-Stage Counter (En). 6 pp.

Digital Data Compact Disks
Use: CD ROMs

Digital Data Selector/Multiplexers
Use: Digital Multiplexers

Digital Data Signals
Use: Digital Signals

Digital Data Transmission
Use: Data Transmission

Digital Demultiplexers
Use For: Decoder/Demultiplexers
See Also: Demultiplexers; Digital Circuits

CECC CECC 90 106-007 ISSUE 1-85. UTE C 86-217; Digital Integrated Circuits in Accorcance with FS 90 106; 54 ALS 138, 74 ALS 138; Decoder/Demultiplexer (En, Fr) ADD 3 (En, Fr). 15 pp.

—1 of 4
CECC CECC 90 104-181 ISSUE 1-81. BS CECC 90 104-181; Dual Binary to 1-of-4 Decoder/Demultiplexer (En). 23 pp.

CECC CECC 90 104-181 ISSUE 1-86. CEI CECC 90 104-181; Dual Binary to 1 of 4 Decoder/Demultiplexer (En). 2 pp.

CECC CECC 90 104-199 ISSUE 1-86. CEI CECC 90 104-199; Silicon Complementary MOS. with (B) Buffered Outputs Cavity and Non Cavity Packaging (En). 2 pp.

—1 of 4—Preferred Products List
CECC CECC MUAHAG Vol 7 IS 8-92. Preferred Products List; Active Microcircuits (En, Fe, Ge). 89 pp.

—1 of 8
CECC CECC 90 107-013 ISSUE 2-89. UTE C 86-218 ADD 2/FA 13; Digital Integrated Circuits in Accordance with FS 90 107; 54/74 F 138; Decoder/Demultiplexer (En, Fr) ADD 3 (En, Fr). 9 pp.

—1 of 8—Preferred Products List
CECC CECC MUAHAG Vol 7 IS 8-92. Preferred Products List; Active Microcircuits (En, Fe, Ge). 89 pp.

—1 to 4 Lines
CECC CECC 90 101-045 ISSUE 1-87. UTE C 86-213/B 65; Digital Integrated Circuits in Accordance with FS 90 101; 54/64/74 155; Dual 2-to-4 Line Decoder or 1-to-4 Line Demultiplexer (En). 16 pp.

CECC CECC 90 104-084 ISSUE 2. NL CECC 90 104-084; Digital Integrated Circuits in Accordance with FS 90 104; HEC/HEF 4555B; Dual 1-of-4 Decoder/Demultiplexer (En). 7 pp.

CECC CECC 90 104-085 ISSUE 2. NL CECC 90 104-085 Issue 2; Digital Integrated Circuits in Accordance with FS 90 104; HEC/HEF 4556B; Dual 1-of-4 Decoder/Demultiplexer (En). 7 pp.

CECC CECC 90 104-199 ISSUE 1-81. BS CECC 90 104-199; Silicon Complementary MOS. with (B) Buffered Outputs and Cavity Packaging (En). 24 pp.

—2-Line
CECC CECC 90 109-831 ISSUE 1-89. Digital Integrated Circuits; Silicon Monolithic C MOS, Cavity or Non-Cavity Packages; Type(s) 54/74 HC 4053 Triple 2-Channel Analog Multiplexer /Demultiplexer Assessment Levels P, Y, L. 17 pp.

—2 to 4 Lines
CECC CECC 90 102-051 ISSUE 1-82. BS CECC 90 102-051; Dual 2-Line to 4-Line Decoders/Demultiplexers (En). 19 pp.

CECC CECC 90 103-136 ISSUE 1-81. BS CECC 90 103-136; Dual 2-Line to 4-Line Decoders/Demultiplexers with Totem Pole Outputs (En) AMD 1 (En). 20 pp.

CECC CECC 90 106-053 ISSUE 1-85. UTE C 86-217; Digital Integrated Circuits in Accordance with FS 90 106; 54 ALS 139,74 ALS 139; Decoder/Demultiplexer (En, Fr) ADD 3 (En, Fr). 15 pp.

CECC CECC 90 109-681 ISSUE 1-87. Digital Integrated Circuits in Accordance with FS 90 109; 54/74 HC 139; 2 to 4 Line Decoder/Demultiplexer (En, Fr). 6 pp.

CECC CECC 90 109-804 ISSUE 1-88. Digital Integrated Circuits in Accordance with FS 90 109; 54/74 HC 155; 2 to 4 Line Decoder (En, Fr). 5 pp.

CECC CECC 90 109-900 ISSUE 1-90. Digital Integrated Circuits; Silicon Monolithic C MOS, Cavity or Non-Cavity Packages; Type(s); 54/74 HCT 139 2 to 4 Line Decoder/Demultiplexer; Assessment Levels P, Y, L (En, Fr). 9 pp.

—2 to 4 Lines—Preferred Products List
CECC CECC MUAHAG Vol 7 IS 8-92. Preferred Products List; Active Microcircuits (En, Fe, Ge). 89 pp.

—3 to 8 Lines
CECC CECC 90 102-050 ISSUE 1-82. BS CECC 90 102-050; 3-to-8 Line Decoders/Demultiplexers (En). 19 pp.

CECC CECC 90 109-680 ISSUE 1-87. Digital Integrated Circuits in Accordance with FS 90 109; 54/74 HC 137; 3 to 8 Line Decoders/Demultiplexers (En, Fr). 6 pp.

CECC CECC 90 109-738 ISSUE 1-87. Digital Integrated Circuits in Accordance with FS 90 109; 54 HC 237, 74 HC 237; 3 to 8 Line Decoder/Demultiplexer (En, Fr). 6 pp.

Digital Demultiplexers *(Cont.)*
—3 to 8 Lines *(Cont.)*
CECC CECC 90 109-818 ISSUE 1-88. Digital Integrated Circuits in Accordance with FS 90 109; 54/74 HCT 238; 3 to 8 Line Decoder (En, Fr). 6 pp.

CECC CECC 90 109-899 ISSUE 1-90. Digital Integrated Circuits; Silicon Monolithic C MOS, Cavity or Non-Cavity Packages; Type(s); 54/74 HCT 137 3 to 8 Line Decoder/Demultiplexer; Assessment Levels P, Y, L (En, Fr). 9 pp.

CECC CECC 90 109-901 ISSUE 1-90. Digital Integrated Circuits; Silicon Monolithic C MOS, Cavity or Non-Cavity Packages; Type(s); 54/74 HCT 237 3 to 8 Line Decoder/Demultiplexer; Assessment Levels P, Y, L (En, Fr). 9 pp.

—3 to 8 Lines—Preferred Products List
CECC CECC MUAHAG Vol 7 IS 8-92. Preferred Products List; Active Microcircuits (En, Fe, Ge). 89 pp.

—4-Line
CECC CECC 90 109-830 ISSUE 1-89. DIgital Integrated Circuits; Silicon Monolithic C MOS, Cavity or Non-Cavity Packages; 54/74 HC 4052 Dual 4-Channel Analog Multiplexer /Demultiplexer Assessment Levels P, Y, L (En, Fr, Ge). 17 pp.

—4 to 16 Lines
CECC CECC 90 109-721 ISSUE 1-87. Digital Integrated Circuits in Accordance with FS 90 109; 54/74 HC 4514; 4 to 16 Line Decoder/Demultiplexer with Input Latches (En, Fr). 6 pp.

CECC CECC 90 109-722 ISSUE 1-87. Digital Integrated Circuits in Accordance with FS 90 109; 54/74 HC 4515; 4 to 16 Line Decoder/Demultiplexer with Input Latches (En, Fr). 6 pp.

CECC CECC 90 109-891 ISSUE 1-90. Digital Integrated Circuits; Silicon Monolithic C MOS, Cavity or Non-Cavity Packages; Type(s); 54/74 HCT 4514 4 to 16 Line Decoder/Demultiplexer with Input Latches; Assessment Levels P, Y, L (En, Fr). 9 pp.

CECC CECC 90 109-892 ISSUE 1-90. Digital Integrated Circuits; Silicon Monolithic C MOS, Cavity or Non-Cavity Packages; Type(s); 54/74 HCT 4515 4 to 16 Line Decoder/Demultiplexer with Input Latches; Assessment Levels P, Y, L (En, Fr). 10 pp.

Digital European Cordless Telecommunications
Use: European Digital Cordless Telecommunications

Digital Exchanges
See Also: Telephone Exchanges; Transmission Delay

CCITT RECMN G.142-89. Transmission Characteristics of Exchanges—General Characteristics of International Telephone Connections and Circuits (Study Groups XII and XV) 6 pp. 6 pp.

CEPT T/CS 62-01-86. Introduction, Domaine D'Application Et Fonctions De Base Dans Les Commutateurs Numeriques Principaux D'Abonne Ou Mixtes. 4 pp.

CEPT T/CS 62-01 E-86. Introduction, Field of Application and Basic Functions for Digital Local and Combined Exchanges. 4 pp.

CEPT T/CS 68-01-86. Introduction Domaine D'Application Et Fonctions De Base Pour Commutateurs De Transit Numeriques. 3 pp.

CEPT T/CS 68-01 E-86. Introduction, Field of Application and Basic Functions for Digital Transit Exchanges. 3 pp.

CEPT T/S 60-00 E-88. Introduction and Field of Application for Digital Local, Transit and Combined Exchanges. 1 p.

—Automatic Transmission Measuring
CEPT T/CS 54-04-81. Methode De Mesure Automatique De Transmission Entre Commutateurs Numeriques. 12 pp.

CEPT T/CS 54-04 E-86. Automatic Transmission Supervision and Measuring Between Digital Exchanges. 9 pp.

—Automatic Transmission Supervision
CEPT T/CS 54-04-81. Methode De Mesure Automatique De Transmission Entre Commutateurs Numeriques. 12 pp.

CEPT T/CS 54-04 E-86. Automatic Transmission Supervision and Measuring Between Digital Exchanges. 9 pp.

—Call Handling
CEPT T/CS 62-03-86. Connexions, Signalisation, Commande, Traitement Des Appels Et Fonctions Auxiliaires Pour Commutateurs Numeriques Principaux D'Abonne Ou Mixtes. 2 pp.

CEPT T/CS 62-03 E-86. Connections, Signalling, Control, Call Handling and Ancillary Functions for Digital Local and Combined Exchanges. 2 pp.

INDUSTRY STANDARDS

Digital Exchanges (Cont.)
—Call Handling (Cont.)
CEPT T/CS 68-03-86. Connexions, Signalisation, Commande, Traitement Des Appels Et Fonctions Auxiliaires Pour Commutateurs De Transit Numeriques. 1 p.
CEPT T/CS 68-03 E-86. Connections, Signalling, Control, Call Handling and Ancillary Functions for Digital Transit Exchanges. 1 p.

—Connectors
CEPT T/CS 62-03-86. Connexions, Signalisation, Commande, Traitement Des Appels Et Fonctions Auxiliaires Pour Commutateurs Numeriques Principaux D'Abonne Ou Mixtes. 2 pp.
CEPT T/CS 62-03 E-86. Connections, Signalling, Control, Call Handling and Ancillary Functions for Digital Local and Combined Exchanges. 2 pp.
CEPT T/CS 68-03-86. Connexions, Signalisation, Commande, Traitement Des Appels Et Fonctions Auxiliaires Pour Commutateurs De Transit Numeriques. 1 p.
CEPT T/CS 68-03 E-86. Connections, Signalling, Control, Call Handling and Ancillary Functions for Digital Transit Exchanges. 1 p.
CEPT T/S 62-20 E-88. Digital Exchange Connections, Signalling and Ancillary Functions. 5 pp.

—Data Links—Terminating
CCITT RECMN G.705-89. Characteristics Required to Terminate Digital Links on a Digital Exchange—General Aspects of Digital Transmission Systems; Terminal Equipments (Study Group XV and XVIII) 5 pp. 5 pp.

—Design
CCITT RECMN Q.541-89. Digital Exchange Design Objectives—General—Digital Local, Transit, Combined and International Exchanges in Integrated Digital Networks and Mixed Analogue-Digital Networks—Supplements (Study Group XI) 6 pp. 6 pp.
CCITT RECMN Q.542-89. Digital Exchange Design Objectives—Operations and Maintenance—Digital Local, Transit, Combined and International Exchanges in Integrated Digital Networks and Mixed Analogue-Digital Networks—Supplements (Study Group XI) 21 pp. 21 pp.
CCITT RECMN Q.543-89. Digital Exchange Performance Design Objectives—Digital Local, Transit, Combined and International Exchanges in Integrated Digital Networks and Mixed Analogue-Digital Networks—Supplements (Study Group XI) 40 pp. 40 pp.
CEPT T/CS 62-04-86. Objectifs Nominaux De Qualite Et De Disponibilite Pour Commutateurs Numeriques Principaux D'Abonne Ou Mixtes. 5 pp.
CEPT T/CS 62-04 E-86. Performance and Availability Design Ojectives for Digital Local and Combined Exchanges. 5 pp.
CEPT T/CS 68-04-86. Objectifs Nominaux De Qualite Et De Disponibilite Pour Commutateurs De Transit Numeriques. 5 pp.
CEPT T/CS 68-04 E-86. Performance and Availability Design Objectives for Digital Transit Exchanges. 5 pp.
CEPT T/S 64-30 E-88. Digital Exchange Performance Design Objectives. 1 p.

—Dialing Services
CEPT T/SF 28-83. Exigences D'Exploitation Relatives Aux Appareils Emetteurs Automatiques De Numeros. 2 pp.

—Documentation
CEPT T/CS 01-10-84. Exigence De Documentation Des Systemes De Commutation Numeriques. 12 pp.
CEPT T/CS 01-10 E-84. Documentation Requirements for Digital Switching Systems. 12 pp.

—Echo Path Loss
CCITT RECMN G.142-89. Transmission Characteristics of Exchanges—General Characteristics of International Telephone Connections and Circuits (Study Groups XII and XV) 6 pp. 6 pp.

—Grade of Service
CCITT RECMN E.543-89. Grades of Service in Digital International Telephone Exchanges—Telephone Network and ISDN—Quality of Service, Network Management and Traffic Engineering (Study Group II) 4 pp. 4 pp.

—Grade of Service—Exchange Call Set Up Delay
CCITT RECMN E.543-89. Grades of Service in Digital International Telephone Exchanges—Telephone Network and ISDN—Quality of Service, Network Management and Traffic Engineering (Study Group II) 4 pp. 4 pp.

Digital Exchanges (Cont.)
—Grade of Service—Incoming Response Delay
CCITT RECMN E.543-89. Grades of Service in Digital International Telephone Exchanges—Telephone Network and ISDN—Quality of Service, Network Management and Traffic Engineering (Study Group II) 4 pp. 4 pp.

—Grade of Service—Internal Loss Probability
CCITT RECMN E.543-89. Grades of Service in Digital International Telephone Exchanges—Telephone Network and ISDN—Quality of Service, Network Management and Traffic Engineering (Study Group II) 4 pp. 4 pp.

—Grade of Service—Measurement
CCITT RECMN E.543-89. Grades of Service in Digital International Telephone Exchanges—Telephone Network and ISDN—Quality of Service, Network Management and Traffic Engineering (Study Group II) 4 pp. 4 pp.

—Grade of Service—Through Connection Delay
CCITT RECMN E.543-89. Grades of Service in Digital International Telephone Exchanges—Telephone Network and ISDN—Quality of Service, Network Management and Traffic Engineering (Study Group II) 4 pp. 4 pp.

—Integrated Digital Networks
CCITT RECMN Q.500-89. Digital Local, Combined, Transit and International Exchanges Introduction and Field of Application—Digital Local, Transit, Combined and International Exchanges in Integrated Digital Networks and Mixed Analogue-Digital Networks—Supplements (Study Group XI) 4 pp. 4 pp.
CCITT RECMN Q.521-89. Exchange Functions—Digital Local, Transit, Combined and International Exchanges in Integrated Digital Networks and Mixed Analogue-Digital Networks—Supplements (Study Group XI) 3 pp. 3 pp.

—Integrated Services Digital Networks
CCITT RECMN Q.500-89. Digital Local, Combined, Transit and International Exchanges Introduction and Field of Application—Digital Local, Transit, Combined and International Exchanges in Integrated Digital Networks and Mixed Analogue-Digital Networks—Supplements (Study Group XI) 4 pp. 4 pp.
CCITT RECMN Q.522-89. Digital Exchange Connections, Signalling and Ancillary Functions—Digital Local, Transit, Combined and International Exchanges in Integrated Digital Networks and Mixed Analogue-Digital Networks—Supplements (Study Group XI) 16 pp. 16 pp.
CEPT T/GSI 03-09-87. Commutation. 1 p.
CEPT T/GSI 03-09 E-87. Switching. 1 p.

—Interfaces
CCITT RECMN Q.511-89. Exchange Interfaces Towards Other Exchanges—Digital Local, Transit, Combined and International Exchanges in Integrated Digital Networks and Mixed Analogue-Digital Networks—Supplements (Study Group XI) 4 pp. 4 pp.
CEPT T/CS 62-02-86. Interfaces Pour Commutateurs Numeriques Principaux D'Abonne Ou Mixtes. 2 pp.
CEPT T/CS 62-02 E-86. Interfaces for Digital Local and Combined Exchanges. 2 pp.
CEPT T/CS 68-02-86. Interfaces Pour Commutateurs De Transit Numeriques. 1 p.
CEPT T/CS 68-02 E-86. Interfaces for Digital Transit Exchanges. 1 p.
CEPT T/S 61-10 E-88. Exchange Interfaces Towards Other Exchanges. 1 p.
CEPT T/S 61-20 E-88. Exchange Interfaces to Subscribers. 9 pp.

—Interfaces—Operations, Administration and Maintenance
CCITT RECMN Q.513-89. Exchange Interfaces for Operations, Administration and Maintenance—Digital Local, Transit, Combined and International Exchanges in Integrated Digital Networks and Mixed Analogue-Digital Networks—Supplements (Study Group XI) 4 pp. 4 pp.

—Interfaces—Subscriber Access
CCITT RECMN Q.512-89. Exchange Interfaces for Subscriber Access—Digital Local, Transit, Combined and International Exchanges in Integrated Digital Networks and Mixed Analogue-Digital Networks—Supplements (Study Group XI) 9 pp. 9 pp.

Digital Exchanges (Cont.)
—Interfaces—Transmission
CCITT RECMN Q.552-89. Transmission Characteristics at 2-Wire Analogue Interfaces of Digital Exchange—Digital Local, Transit, Combined and International Exchanges in Integrated Digital Networks and Mixed Analogue-Digital Networks—Supplements (Study Group XI) 25 pp. 25 pp.
CCITT RECMN Q.553-89. Transmission Characteristics at 4-Wire Analogue Interfaces of a Digital Exchange—Digital Local, Transit, Combined and International Exchanges in Integrated Digital Networks and Mixed Analogue-Digital Networks—Supplements (Study Group XI) 12 pp. 12 pp.
CCITT RECMN Q.554-89. Transmission Characteristics at Digital Interfaces of a Digital Exchange—Digital Local, Transit, Combined and International Exchanges in Integrated Digital Networks and Mixed Analogue-Digital Networks—Supplements (Study Group XI) 3 pp. 3 pp.

—Maintenance
CCITT RECMN Q.542-89. Digital Exchange Design Objectives—Operations and Maintenance—Digital Local, Transit, Combined and International Exchanges in Integrated Digital Networks and Mixed Analogue-Digital Networks—Supplements (Study Group XI) 21 pp. 21 pp.
CEPT T/CS 62-06-86. Fonctions D'Exploitation Et De Maintenance Pour Commutateurs Numeriques Principaux D'Abonne Ou Mixtes. 2 pp.
CEPT T/CS 62-06 E-86. Digital Local and Combined Exchange Operation and Maintenance Functions. 2 pp.
CEPT T/CS 68-06-86. Fonctions D'Exploitation Et De Maintenance Pour Commutateurs De Transit Numeriques. 1 p.
CEPT T/CS 68-06 E-86. Digital Transit Exchange Operation and Maintenance Functions. 1 p.

—Measurement
CCITT RECMN Q.544-89. Digital Exchange Measurements—Digital Local, Transit, Combined and International Exchanges in Integrated Digital Networks and Mixed Analogue-Digital Networks—Supplements (Study Group XI) 15 pp. 15 pp.

—Mixed Analog/Digital Networks
CCITT RECMN Q.500-89. Digital Local, Combined, Transit and International Exchanges Introduction and Field of Application—Digital Local, Transit, Combined and International Exchanges in Integrated Digital Networks and Mixed Analogue-Digital Networks—Supplements (Study Group XI) 4 pp. 4 pp.
CCITT RECMN Q.521-89. Exchange Functions—Digital Local, Transit, Combined and International Exchanges in Integrated Digital Networks and Mixed Analogue-Digital Networks—Supplements (Study Group XI) 3 pp. 3 pp.

—Operation
CCITT RECMN Q.542-89. Digital Exchange Design Objectives—Operations and Maintenance—Digital Local, Transit, Combined and International Exchanges in Integrated Digital Networks and Mixed Analogue-Digital Networks—Supplements (Study Group XI) 21 pp. 21 pp.
CEPT T/CS 62-06-86. Fonctions D'Exploitation Et De Maintenance Pour Commutateurs Numeriques Principaux D'Abonne Ou Mixtes. 2 pp.
CEPT T/CS 62-06 E-86. Digital Local and Combined Exchange Operation and Maintenance Functions. 2 pp.
CEPT T/CS 68-06-86. Fonctions D'Exploitation Et De Maintenance Pour Commutateurs De Transit Numeriques. 1 p.
CEPT T/CS 68-06 E-86. Digital Transit Exchange Operation and Maintenance Functions. 1 p.

—Pulse Code Modulation
CCITT RECMN G.142-89. Transmission Characteristics of Exchanges—General Characteristics of International Telephone Connections and Circuits (Study Groups XII and XV) 6 pp. 6 pp.

—Return Loss
CCITT RECMN G.142-89. Transmission Characteristics of Exchanges—General Characteristics of International Telephone Connections and Circuits (Study Groups XII and XV) 6 pp. 6 pp.

—Sidetone
CCITT RECMN G.121-89. Loudness Ratings (LRs) of National Systems—General Characteristics of International Telephone Connections and Circuits (Study Groups XII and XV) 17 pp. 17 pp.
CCITT RECMN G.142-89. Transmission Characteristics of Exchanges—General Characteristics of International Telephone Connections and Circuits (Study Groups XII and XV) 6 pp. 6 pp.

Digital Exchanges *(Cont.)*

—Signal Levels
CCITT RECMN G.142-89. Transmission Characteristics of Exchanges—General Characteristics of International Telephone Connections and Circuits (Study Groups XII and XV) 6 pp. 6 pp.

—Signaling Systems
CEPT T/CS 62-03-86. Connexions, Signalisation, Commande, Traitement Des Appels Et Fonctions Auxiliaires Pour Commutateurs Numeriques Principaux D'Abonne Ou Mixtes. 2 pp.
CEPT T/CS 62-03 E-86. Connections, Signalling, Control, Call Handling and Ancillary Functions for Digital Local and Combined Exchanges. 2 pp.
CEPT T/CS 68-03-86. Connexions, Signalisation, Commande, Traitement Des Appels Et Fonctions Auxiliaires Pour Commutateurs De Transit Numeriques. 1 p.
CEPT T/CS 68-03 E-86. Connections, Signalling, Control, Call Handling and Ancillary Functions for Digital Transit Exchanges. 1 p.
CEPT T/S 62-20 E-88. Digital Exchange Connections, Signalling and Ancillary Functions. 5 pp.

—Signaling Systems—Interworking
CCITT RECMN Q.522-89. Digital Exchange Connections, Signalling and Ancillary Functions—Digital Local, Transit, Combined and International Exchanges in Integrated Digital Networks and Mixed Analogue-Digital Networks—Supplements (Study Group XI) 16 pp. 16 pp.

—Stability Loss
CCITT RECMN G.142-89. Transmission Characteristics of Exchanges—General Characteristics of International Telephone Connections and Circuits (Study Groups XII and XV) 6 pp. 6 pp.

—Synchronous Digital Hierarchy—Frame Structure
CCITT RECMN G.704-91. Synchronous Frame Structures Used at Primary and Secondary Hierarchical Levels (Study Group XVIII) 35 pp. 35 pp.
CCITT RECMN G.704-89. Synchronous Frame Structures Used at Primary and Secondary Hierarchical Levels—General Aspects of Digital Transmission Systems; Terminal Equipments (Study Groups XV and XVIII) 21 pp. 21 pp.

—Traffic Units
CCITT RECMN Q.544-89. Digital Exchange Measurements—Digital Local, Transit, Combined and International Exchanges in Integrated Digital Networks and Mixed Analogue-Digital Networks—Supplements (Study Group XI) 15 pp. 15 pp.

—Transmission
CCITT RECMN Q.551-89. Transmission Characteristics of Digital Exchanges—Digital Local, Transit, Combined and International Exchanges in Integrated Digital Networks and Mixed Analogue-Digital Networks—Supplements (Study Group XI) 14 pp. 14 pp.

—Transmission Delay
CCITT RECMN G.142-89. Transmission Characteristics of Exchanges—General Characteristics of International Telephone Connections and Circuits (Study Groups XII and XV) 6 pp. 6 pp.

—Transmission Loss
CCITT RECMN G.142-89. Transmission Characteristics of Exchanges—General Characteristics of International Telephone Connections and Circuits (Study Groups XII and XV) 6 pp. 6 pp.

Digital Hierarchy

See Also: Synchronous Digital Hierarchy

—Bit Rates
CCITT RECMN G.702-89. Digital Hierarchy Bit Rates—General Aspects of Digital Transmission Systems; Terminal Equipments (Study Groups XV and XVIII) 4 pp. 4 pp.

—Bit Rates—Interfaces—Electrical Characteristics
CCITT RECMN G.703-91. Physical/Electrical Characteristics of Hierarchical Digital Interfaces (Study Group XVIII) 42 pp. 42 pp.
CCITT RECMN G.703-89. Physical/Electrical Characteristics of Hierarchical Digital Interfaces—General Aspects of Digital Transmission Systems; Terminal Equipments (Study Groups XV and XVIII) 31 pp. 31 pp.

—Bit Rates—Interfaces—Physical Characteristics
CCITT RECMN G.703-91. Physical/Electrical Characteristics of Hierarchical Digital Interfaces (Study Group XVIII) 42 pp. 42 pp.

Digital Hierarchy *(Cont.)*

—Bit Rates—Interfaces—Physical Characteristics *(Cont.)*
CCITT RECMN G.703-89. Physical/Electrical Characteristics of Hierarchical Digital Interfaces—General Aspects of Digital Transmission Systems; Terminal Equipments (Study Groups XV and XVIII) 31 pp. 31 pp.

—Telecommunication Networks—Interworking—Speech Encoding Laws
CCITT RECMN G.802-89. Interworking Between Networks Based on Different Digital Hierarchies and Speech Encoding Laws—Digital Networks, Digital Sections and Digital Line Systems (Study Groups XV and XVIII) 5 pp. 5 pp.

Digital Integrated Circuits
Use: Digital Circuits

Digital Line Sections
See Also: Digital Sections

—Bit Rates—3152 kbit/s
CCITT RECMN G.931-89. Digital Line Sections at 3152 kBit/s—Digital Networks, Digital Sections and Digital Line Systems (Study Groups XV and XVIII) 2 pp. 2 pp.

—Digital Transmission Systems
CCITT FASCICLE III.5-89. Digital Networks, Digital Sections and Digital Line Systems—Recommendations G.801—G.961. 292 pp.

—Maintenance
CCITT RECMN M.550-89. Performance Limits for Bringing into Service and Maintenance of Digital Paths, Sections, and Line Sections—General Maintenance Principles—Maintenace of International Transmission Systems and Telephone Circuits—(Study Group IV) 16 pp. 16 pp.

—Optical Fibers
CEPT T/TR 05-01-83. Relative a Des Systemes De Ligne Numeriques Sur Fibres Optiques a 8 Mbit/s Fontionnant Entre 820 Et 880 Nm. 4 pp.
CEPT T/TR 05-02-83. Relative a Des Systemes De Ligne Numerique Sur Fibres Optiques a 34 Mbit/s Fonctionnant Entre 820 Et 880 Nm. 4 pp.

—Telecommunication Networks
CEPT T/L 03-05-81. Sections Et Systemes De Ligne Numeriques Sur Abies a 2048, 8448, 34368, 139264 Et 564992 KBits/s. 1 p.
CEPT T/L 03-05 E-81. Digital Sections and Line Systems at 2,048, 8,448, 34,368, 139,264 and 564, 992 KBit/s. 1 p.

Digital Line Systems
See Also: Transmission Systems

—Customer Administration
CENELEC PRI-ETS 300 291-93. Network Aspects (NA); Functional Specification of Customer Administration (CA) on the Operations System/ Network Element (OS/NE) Interface. 131 pp.
ETSI PRI-ETS 300 291-93. Network Aspects (NA); Functional Specification of Customer Administration (CA) on the Operations System/Network Element (OS/NE) Interface. 131 pp.

—Fiber Optic Cables
CCITT RECMN G.950-89. General Considerations on Digital Line Systems—Digital Networks, Digital Sections and Digital Line Systems (Study Groups XV and XVIII) 2 pp. 2 pp.

—Fiber Optic Cables—Synchronous Digital Hierarchy
CCITT RECMN G.958-90. Digital Line Systems Based on the Synchronous Digital Hierarchy for Use on Optical Fibre Cables (Study Group XV) 28 pp. 28 pp.

—Fiber Optic Cables—1544 kbit/s
CCITT RECMN G.955-89. Digital Line Systems Based on the 1544 kBit/s Hierarchy on Optical Fibre Cables—Digital Networks, Digital Sections and Digital Line Systems (Study Groups XV and XVIII) 8 pp. 8 pp.

—Fiber Optic Cables—2048 kbit/s
CCITT RECMN G.956-89. Digital Line Systems Based on the 2048 kBit/s Hierarchy on Optical Fibre Cables—Digital Networks, Digital Sections and Digital Line Systems (Study Groups XV and XVIII) 8 pp. 8 pp.

Digital Line Systems *(Cont.)*

—Frequency Division Multiplexers—Bearer Services
CCITT RECMN G.941-89. Digital Line Systems Provided by FDM Transmission Bearers—Digital Networks, Digital Sections and Digital Line Systems (Study Groups XV and XVIII) 4 pp. 4 pp.

—Integrated Digital Networks
CCITT FASCICLE III.5-89. Digital Networks, Digital Sections and Digital Line Systems—Recommendations G.801—G.961. 292 pp.

—Telecommunication Networks
CCITT RECMN G.901-89. General Considerations on Digital Networks and Digital Line Systems—Digital Networks, Digital Sections and Digital Line Systems (Study Groups XV and XVIII) 3 pp. 3 pp.

—Telephone Cables
CCITT RECMN G.950-89. General Considerations on Digital Line Systems—Digital Networks, Digital Sections and Digital Line Systems (Study Groups XV and XVIII) 2 pp. 2 pp.

—Telephone Cables—1544 kbit/s
CCITT RECMN G.951-89. Digital Line Systems Based on the 1544 kBit/s Hierarchy on Symmetric Pair Cables—Digital Networks, Digital Sections and Digital Line Systems (Study Groups XV and XVIII) 8 pp. 8 pp.
CCITT RECMN G.953-89. Digital Line Systems Based on the 1544 kBit/s Hierarchy on Coaxial Pair Cables—Digital Networks, Digital Sections and Digital Line Systems (Study Groups XV and XVIII) 6 pp. 6 pp.

—Telephone Cables—2048 kbit/s
CCITT RECMN G.952-89. Digital Line Systems Based on the 2048 kBit/s Hierarchy on Symmetric Pair Cables—Digital Networks, Digital Sections and Digital Line Systems (Study Groups XV and XVIII) 8 pp. 8 pp.
CCITT RECMN G.954-89. Digital Line Systems Based on the 2048 kBit/s Hierarchy on Coaxial Pair Cables—Digital Networks, Digital Sections and Digital Line Systems (Study Groups XV and XVIII) 13 pp. 13 pp.

Digital Links
Use: Data Links

Digital Loopbacks
CCITT RECMN M.125-89. Digital Loopback Mechanisms—General Maintenance Principles—Maintenance of International Transmission Systems and Telephone Circuits (Study Group IV) 5 pp. 5 pp.

Digital Measuring Instruments
Use: Electric Measuring Instruments

Digital Microwave Radios
See Also: Microwave Equipment; Radio Equipment

—Earth Stations
BSI BS 7573: Sec 3.12-93. 1993 Methods of Measurement for Equipment Used in Digital Microwave Radio Transmission Systems Part 3: Measurements on Satellite Earth Stations Section 3.12: Overall System Performance (IEC 835-3-12: 1993) (S). 7 pp.
CENELEC EN 60835-3-1-92. Methods of Measurement for Equipment Used in Digital Microwave Radio Transmission Systems Part 3: Measurements on Satellite Earth Stations Section One: General; (IEC 835-3-1: 1990). 5 pp.
CENELEC EN 60835-3-1-92. Methods of Measurement for Equipment Used in Digital Microwave Radio Transmission Systems Part 3: Measurements on Satellite Earth Stations Section One: General (IEC 835-3-1:1990). 5 pp.
IEC 835 Pt 3-1-90. Methods of Measurement for Equipment Used in Digital Microwave Radio Transmission Systems Part 3: Measurements on Satellite Earth Stations Section One—General First Edition; (CENELEC EN 60835-3-1: 1992). 9 pp.
IEC 835 Pt 3-12-93. Methods of Measurement for Equipment Used in Digital Microwave Radio Transmission Systems Part 3: Measurements on Satellite Earth Stations Section 12: Overall System Performance First Edition. 14 pp.

—Earth Stations—Adaptive Equalizers
IEC 835 Pt 2-8-93. Methods of Measurement for Equipment Used in Digital Microwave Radio Transmissions Systems Part 2: Measurements on Terrestrial Radio-Relay Systems Section 8: Adaptive Equalizer First Edition. 44 pp.

Digital Microwave Radios *(Cont.)*

—Earth Stations—Data Terminal Equipment

IEC 835 Pt 3-9-93. Methods of Measurement for Equipment Used in Digital Microwave Radio Transmission Systems Part 3: Measurement on Satellite Earth Stations Section 9—Terminal Equipment SCP-PSK First Edition. 58 pp.

—Earth Stations—Digital Signal Processing

IEC 835 Pt 2-5-93. Methods of Measurement for Equipment Used in Digital Microwave Radio Transmission Systems Part 2: Measurements on Terrestrial Radio-Relay Systems Section 5: Digital Signal Processing Sub-System First Edition. 48 pp.

—Earth Stations—Low Noise Amplifiers

BSI BS 7573: Sec 3.4-93. 1993 Methods of Measurement for Equipment Used in Digital Microwave Radio Transmission Systems Part 3: Measurements on Satellite Earth Stations Section 3.4: Low Noise Amplifier (IEC 835-3-4: 1993) (S). 8 pp.

IEC 835 Pt 3-4-93. Methods of Measurement for Equipment Used in Digital Microwave Radio Transmission Systems Part 3: Measurements on Satellite Earth Stations Section 4: Low Noise Amplifier First Edition. 20 pp.

—Earth Stations—Relay Systems

BSI BS 7573: Sec 1.2-93. 1993 Methods of Measurement for Equipment Used in Digital Microwave Radio Transmission Systems Part 1: Measurements Common to Terrestrial Radio-Relay Systems and Satellite Earth Stations Section 1.2: Bacic Characteristics (IEC 835-1-2: 1992) (S). 21 pp.

BSI BS 7573: Sec 2.10-93. 1993 Methods of Measurement for Equipment Usedin Digital Microwave Radio Transmission Systems Part 2: Measurements on Terrestrial Radio-Relay Systems Section 2.10: Overall System Performance (IEC 835-2-10: 1992) (S). 15 pp.

BSI BS EN 60835-2-10-93. 1993 Methods of Measurement for Equipment Used in Digital Microwave Radio Transmission Systems Part 2: Measurements on Terrestrial Radio-Relay Systems Section 2.10: Overall System Performance (IEC 835-2-10: 1992) (S). 15 pp.

CENELEC EN 60835-1-1-92. Methods of Measurement for Equipment Used in Digital Microwave Radio Transmission Systems Part 1: Measurements Common to Terrestrial Radio-Relay Systems and Satellite Earth Stations Section One: General; (IEC 835-1-1: 1990). 5 pp.

CENELEC EN 60835-1-1-92. Methods of Measurement for Equipment Used in Digital Microwave Radio Transmission Systems Part 1: Measurements Common to Terrestrial Radio-Relay Systems and Satellite Earth Stations Section One: General (IEC 835-1-1:1990). 5 pp.

CENELEC EN 60835-2-1-92. Methods of Measurement for Equipment Used in Digital Microwave Radio Transmission Systems Part 2: Measurements on Terrestrial Radio-Relay Systems Section One: General (IEC 835-2-1:1990). 5 pp.

CENELEC EN 60835-2-10-93. Methods of Measurement for Equipment Used in Digital Microwave Radio Transmission Systems Part 2: Measurements on Terrestrial Radio-Relay Systems Section Ten—Overall System Performance (IEC 835-2-10:1992). 5 pp.

CENELEC EN 60835-2-10-93. Methods of Measurement for Equipment Used in Digital Microwave Radio Transmission Systems Part 2: Measurements on Terrestrial Radio-Relay Systems Section Ten—Overall System Performance (IEC 835-2-10: 1992). 9 pp.

IEC 835 Pt 1-1-90. Methods of Measurement for Equipment Used in Digital Microwave Radio Transmission Systems Part 1: Measurements Common to Terrestrial Radio-Relay Systems and Satellite Earth Stations Section One—General First Edition; (CENELEC EN 60835-1-1: 1992). 37 pp.

IEC 835 Pt 1-2-92. Methods of Measurement for Equipment Used in Digital Microwave Radio Transmission Systems Part 1: Measurements Common to Terrestrial Radio-Relay Systems and Satellite Earth Stations Section 2: Basic Characteristics First Edition. 43 pp.

IEC 835 Pt 2-1-90. Methods of Measurement for Equipment Used in Digital Microwave Radio Transmission Systems Part 2: Measurements on Terrestrial Radio-Relay Systems Section One—General First Edition; (CENELEC EN 60835-2-1: 1992). 9 pp.

IEC 835 Pt 2-10-92. Methods of Measurement for Equipment Used in Digital Microwave Radio Transmission Systems Part 2: Measurements on Terrestrial Radio-Relay Systems Section Ten—Overall System Performance First Edition; (CENELEC EN 60835-2-10: 1993). 26 pp.

Digital Microwave Radios *(Cont.)*

—Earth Stations—Transceivers

BSI BS 7573: Sec 2.4-93. 1993 Methods of Measurement for Equipment Used in Digital Microwave Radio Transmission Systems Part 2: Measurements on Terrestrial Radio-Relay Systems Section 2.4: Transmitter/Receiver Including Modulator/Demodulator (IEC 835-2-4: 1993) (S). 23 pp.

IEC 835 Pt 2-4-93. Methods of Measurement for Equipment Used in Digital Microwave Radio Transmission Systems Part 2: Measurements on Terrestrial Radio-Relay Systems Section 4: Transmitter/Receiver Including Modulator/Demodulator First Edition. 48 pp.

—Earth Stations—Transmission

BSI BS 7573: Sec 1.3-93. 1993 Methods of Measurement for Equipment Used in Digital Microwave Radio Transmission Systems Part 1: Measurements Common to Terrestrial Radio-Relay Systems and Satellite Earth Stations Sect. 1.3: Transmission Characteristics (IEC 835-1-3: 1992) (S). 22 pp.

BSI BS 7573: Sec 1.4-93. 1993 Methods of Measurement for Equipment Used in Digital Microwave Radio Transmission Systems Part 1: Measurements Common to Terrestrial Radio-Relay Systems and Satellite Earth Stations Sec. 1.4: Transmission Performance (IEC 835-1-4: 1992) (S). 17 pp.

IEC 835 Pt 1-3-92. Methods of Measurement for Equipment Used in Digital Microwave Radio Transmission Systems Part 1: Measurements Common to Terrestrial Radio-Relay Systems and Satellite Earth Stations Section 3: Transmission Characteristics First Edition. 37 pp.

IEC 835 Pt 1-4-92. Methods of Measurement for Equipment Used in Digital Microwave Radio Transmission Systems Part 1: Measurements Common to Terrestrial Radio-Relay Systems and Satellite Earth Stations Section 4: Transmission Performance First Edition. 31 pp.

—Relay Systems

BSI BS 7573: Sec 1.2-93. 1993 Methods of Measurement for Equipment Used in Digital Microwave Radio Transmission Systems Part 1: Measurements Common to Terrestrial Radio-Relay Systems and Satellite Earth Stations Section 1.2: Bacic Characteristics (IEC 835-1-2: 1992) (S). 21 pp.

BSI BS 7573: Sec 2.10-93. 1993 Methods of Measurement for Equipment Usedin Digital Microwave Radio Transmission Systems Part 2: Measurements on Terrestrial Radio-Relay Systems Section 2.10: Overall System Performance (IEC 835-2-10: 1992) (S). 15 pp.

BSI BS EN 60835-2-10-93. 1993 Methods of Measurement for Equipment Used in Digital Microwave Radio Transmission Systems Part 2: Measurements on Terrestrial Radio-Relay Systems Section 2.10: Overall System Performance (IEC 835-2-10: 1992) (S). 15 pp.

CENELEC EN 60835-1-1-92. Methods of Measurement for Equipment Used in Digital Microwave Radio Transmission Systems Part 1: Measurements Common to Terrestrial Radio-Relay Systems and Satellite Earth Stations Section One: General; (IEC 835-1-1: 1990). 5 pp.

CENELEC EN 60835-1-1-92. Methods of Measurement for Equipment Used in Digital Microwave Radio Transmission Systems Part 1: Measurements Common to Terrestrial Radio-Relay Systems and Satellite Earth Stations Section One: General (IEC 835-1-1:1990). 5 pp.

CENELEC EN 60835-2-1-92. Methods of Measurement for Equipment Used in Digital Microwave Radio Transmission Systems Part 2: Measurements on Terrestrial Radio-Relay Systems Section One: General; (IEC 835-2-1: 1990). 5 pp.

CENELEC EN 60835-2-1-92. Methods of Measurement for Equipment Used in Digital Microwave Radio Transmission Systems Part 2: Measurements on Terrestrial Radio-Relay Systems Section One: General (IEC 835-2-1:1990). 5 pp.

CENELEC EN 60835-2-10-93. Methods of Measurement for Equipment Used in Digital Microwave Radio Transmission Systems Part 2: Measurements on Terrestrial Radio-Relay Systems Section Ten—Overall System Performance (IEC 835-2-10:1992). 5 pp.

CENELEC EN 60835-2-10-93. Methods of Measurement for Equipment Used in Digital Microwave Radio Transmission Systems Part 2: Measurements on Terrestrial Radio-Relay Systems Section Ten—Overall System Performance (IEC 835-2-10: 1992). 9 pp.

IEC 835 Pt 1-1-90. Methods of Measurement for Equipment Used in Digital Microwave Radio Transmission Systems Part 1: Measurements Common to Terrestrial Radio-Relay Systems and Satellite Earth Stations Section One—General First Edition; (CENELEC EN 60835-1-1: 1992). 37 pp.

Digital Microwave Radios *(Cont.)*

—Relay Systems *(Cont.)*

IEC 835 Pt 1-2-92. Methods of Measurement for Equipment Used in Digital Microwave Radio Transmission Systems Part 1: Measurements Common to Terrestrial Radio-Relay Systems and Satellite Earth Stations Section 2: Basic Characteristics First Edition. 43 pp.

IEC 835 Pt 2-1-90. Methods of Measurement for Equipment Used in Digital Microwave Radio Transmission Systems Part 2: Measurements on Terrestrial Radio-Relay Systems Section One—General First Edition; (CENELEC EN 60835-2-1: 1992). 9 pp.

IEC 835 Pt 2-10-92. Methods of Measurement for Equipment Used in Digital Microwave Radio Transmission Systems Part 2: Measurements on Terrestrial Radio-Relay Systems Section Ten—Overall System Performance First Edition; (CENELEC EN 60835-2-10: 1993). 26 pp.

—Relay Systems—Branching Networks

BSI BS 7573: Sec 2.3-93. 1993 Methods of Measurement for Equipment Used in Digital Microwave Radio Transmission Systems Part 2: Measurements on Terrestrial Radio-Relay Systems Section 2.3: RF Branching Networks (IEC 835-2-3: 1992) (S). 11 pp.

IEC 835 Pt 2-3-92. Methods of Measurement for Equipment Used in Digital Microwave Radio Transmission Systems Part 2: Measurements on Terrestrial Radio-Relay Systems Section 3: RF Branching Networks First Edition. 20 pp.

—Relay Systems—Transmission

BSI BS 7573: Sec 1.3-93. 1993 Methods of Measurement for Equipment Used in Digital Microwave Radio Transmission Systems Part 1: Measurements Common to Terrestrial Radio-Relay Systems and Satellite Earth Stations Sect. 1.3: Transmission Characteristics (IEC 835-1-3: 1992) (S). 22 pp.

BSI BS 7573: Sec 1.4-93. 1993 Methods of Measurement for Equipment Used in Digital Microwave Radio Transmission Systems Part 1: Measurements Common to Terrestrial Radio-Relay Systems and Satellite Earth Stations Sec. 1.4: Transmission Performance (IEC 835-1-4: 1992) (S). 17 pp.

IEC 835 Pt 1-3-92. Methods of Measurement for Equipment Used in Digital Microwave Radio Transmission Systems Part 1: Measurements Common to Terrestrial Radio-Relay Systems and Satellite Earth Stations Section 3: Transmission Characteristics First Edition. 37 pp.

IEC 835 Pt 1-4-92. Methods of Measurement for Equipment Used in Digital Microwave Radio Transmission Systems Part 1: Measurements Common to Terrestrial Radio-Relay Systems and Satellite Earth Stations Section 4: Transmission Performance First Edition. 31 pp.

Digital Modulation

See Also: Modulation

—Broadcasting Satellite Services

CCIR QUESTION 92/11-90. Digital Techniques in the Broadcasting-Satellite Service (Sound and Television)—Questions Concerning Study Group 11—Broadcasting Service (Television). 1 p.

—Carrier Systems—Fixed Satellite Services

CCIR RECMN 446-2-78. Carrier Energy Dispersal for Systems Employing Angle Modulation by Analogue Signals or Digital Modulation in the Fixed-Satellite Service—Section 4C—Earth Station and Baseband Characteristics —Earth Station Antennas —Maintenance of Earth Stations. 1 p.

CCIR RECMN 446-3-92. Carrier Energy Dispersal for Systems Employing Angle Modulation by Analogue Signals or Digital Modulation in the Fixed-Satellite Service—Section 4C—Earth Station and Baseband Characteristics —Earth Station Antennas —Maintenance of Earth Stations. 20 pp.

—Fixed Satellite Services—Energy Dispersal

CCIR Report 384-6-90. Energy Dispersal in the Fixed-Satellite Service—Section 4C—Earth Station and Baseband Characteristics—Earth Station Antennas—Maintenance of Earth Stations. 28 pp.

—Fixed Satellite Services—Radio Frequency Interference

CCIR QUESTION 50/4-90. Interference Criteria and Calculation Methods for Networks in the Fixed-Satellite Service Using Digital Modulation—Questions of Study Group 4 Fixed-Satellite Service. 1 p.

—Sound Recording

CCIR QUESTION 91/10-90. Sound Recording Using Digital Modulation—Questions Concerning Study Group 10—Broadcasting Service (Sound). 1 p.

INTERNATIONAL AND NON-U.S. NATIONAL STANDARDS
SUBJECT INDEX
Digital

Digital Modulation *(Cont.)*

—Television Systems

CCIR QUESTION 25-3/11-90. Standards for Television Systems Using Digital Modulation—Questions Concerning Study Group 11—Broadcasting Service (Television). 1 p.

Digital Monolithic Integrated Circuits

Use: Integrated Circuits

Digital Multiplex Equipment

Use: Digital Multiplexers

Digital Multiplexer/Registers

See Also: Digital Multiplexers

—8-Line

CECC CECC 90 109-842 ISSUE 1-89. Digital Integrated Circuits; Silicon Monolithic C MOS, Cavity or Non-Cavity Packages; Type(s) 54/74 HCT 354 8-Channel Multiplexer /Register with 3-State Outputs Assessment Levels P, Y, L (En, Fr, Ge). 10 pp.

CECC CECC 90 109-843 ISSUE 1-90. Digital Integrated Circuits; Silicon Monolithic C MOS, Cavity or Non-Cavity Packages; Type(s) 54/74 HCT 356 8-Channel Multiplexer/Register with 3-State Outputs Assessment Levels P, Y, L (En, Fr, Ge). 10 pp.

Digital Multiplexers

Use For: Data Selector/Multiplexers, Digital; Digital Data Selector/Multiplexers; Digital Multiplex Equipment *See Also:* Digital Multiplexer/Registers; Multiplexers

CEPT T/L 03-03-81. Multiplex Avec Acces De Donnees (64 Kbit/s Oun X 64 Kbit/s). 1 p.

CEPT T/L 03-03 E-81. Multiplex Equipment Providing Digital Access (at 64 KBit/s or N X 64 KBit/s). 1 p.

CEPT T/L 03-15 E-88. Digital Multiplex Equipments for International Interworking Between Networks Based on Different Digital Hierarchies. 1 p.

—Bit Rates—Fourth Order—139,264 kbit/s

CCITT RECMN G.751-89. Digt. Multiplex Eqpts. Operating at the Third Order Bit Rate of 34 368 kbit/s and the Fourth Order Bit Rate of 139 264 kbit/s and Using Positive Justification—General Aspects of Digital Transmission Systems; Terminal Equipments (Study Groups XV and XVIII) 13 pp. 13 pp.

CCITT RECMN G.754-89. Fourth Order Digital Multiplex Equipment Operating at 139 264 kbit/s and Using Positive/Zero/Negative Justification—General Aspects of Digital Transmission Systems; Terminal Equipments (Study Groups XV and XVIII) 3 pp. 3 pp.

—Bit Rates—Second Order—2048 kbit/s

CCITT RECMN G.747-89. Second Order Digital Multiplex Equipment Operating at 6312 kbit/s and Multiplexing Three Tributaries at 2048 kbit/s—General Aspects of Digital Transmission Systems; Terminal Equipments (Study Groups XV and XVIII) 7 pp. 7 pp.

—Bit Rates—Second Order—6312 kbit/s

CCITT RECMN G.747-89. Second Order Digital Multiplex Equipment Operating at 6312 kbit/s and Multiplexing Three Tributaries at 2048 kbit/s—General Aspects of Digital Transmission Systems; Terminal Equipments (Study Groups XV and XVIII) 7 pp. 7 pp.

CCITT RECMN G.752-89. Characteristics of Digital Multiplex Equipments Based on a Second Order Bit Rate of 6312 kbit/s and Using Positive Justification—General Aspects of Digital Transmission Systems; Terminal Equipments (Study Groups XV and XVIII) 10 pp. 10 pp.

—Bit Rates—Third Order—34,368 kbit/s

CCITT RECMN G.751-89. Digt. Multiplex Eqpts. Operating at the Third Order Bit Rate of 34 368 kbit/s and the Fourth Order Bit Rate of 139 264 kbit/s and Using Positive Justification—General Aspects of Digital Transmission Systems; Terminal Equipments (Study Groups XV and XVIII) 13 pp. 13 pp.

CCITT RECMN G.753-89. Third Order Digital Multiplex Equipment Operating at 34 368 kbit/s and Using Positive/Zero/Negative Justification—General Aspects of Digital Transmission Systems; Terminal Equipments (Study Groups XV and XVIII) 4 pp. 4 pp.

—Bit Rates—Three Tributaries—139,264 kbit/s

CCITT RECMN G.755-89. Digital Multiplex Eqpt. Operating at 139 264 kBit/s and Multiplexing Three Tributaries at 44 736 kBit/s—General Aspects of Digital Transmission Systems; Terminal Equipments (Study Groups XV and XVIII) 7 pp. 7 pp.

Digital Multiplexers *(Cont.)*

—Bit Rates—Three Tributaries—44,736 kbit/s

CCITT RECMN G.755-89. Digital Multiplex Eqpt. Operating at 139 264 kBit/s and Multiplexing Three Tributaries at 44 736 kBit/s—General Aspects of Digital Transmission Systems; Terminal Equipments (Study Groups XV and XVIII) 7 pp. 7 pp.

—Data Networks—1544 kbit/s

CCITT RECMN G.734-89. Characteristics of Synchronous Digital Multiplex Equipment Operating at 1544 kbit/s—General Aspects of Digital Transmission Systems; Terminal Equipments (Study Groups XV and XVIII) 4 pp. 4 pp.

—Data Networks—2048 kbit/s

CCITT RECMN G.736-89. Characteristics of a Synchronous Digital Multiplex Equipment Operating at 2048 kbit/s—General Aspects of Digital Transmission Systems; Terminal Equipments (Study Groups XV and XVIII) 5 pp. 5 pp.

—Integrated Circuit

CECC CECC 90 101-041 ISSUE 1-87. UTE C 86-213/B 61; Digital Integrated Circuits in Accordance with FS 90 101; 54/64/74 150; 16-Bit Data Selector (En). 13 pp.

CECC CECC 90 101-042 ISSUE 1-87. UTE C 86-213/B 62; Digital Integrated Circuits in Accordance with FS 90 101; 54/64/74 151A; 8-Bit Data Selector (En). 13 pp.

CECC CECC 90 102-052 ISSUE 1-82. BS CECC 90 102-052; Data Selector/Multiplexers (En). 18 pp.

CECC CECC 90 103-109 ISSUE 1-81. BS CECC 90 103-109; Quadruple 2-Input Multiplexer with Storage (En) AMD 1 (En). 18 pp.

CECC CECC 90 103-131 ISSUE 1-81. BS CECC 90 103-131; Quadruple 2-Input Multiplexers with Storage (En) AMD 1 (En). 20 pp.

CECC CECC 90 104-066 ISSUE 2. NL-CECC 90 104-066 Issue 2; Digital Integrated Circuits in Accordance with FS 90 104; HEC/HEF 4512B; 8-Input Multiplexer with 3-State Output (En). 9 pp.

CECC CECC 90 104-072 ISSUE 2. NL CECC 90 104-072 Issue 2; Digital Integrated Circuits in Accordance with FS 90 104; HEC/HEF 4519B; Quadruple 2-Input Multiplexer (En). 7 pp.

CECC CECC 90 104-081 ISSUE 2. NL CECC 90 104-081 Issue 2; Digital Integrated Circuits in Accordance with FS 90 104; HEC/HEF 4539B; Dual 4-Input Multiplexer (En). 7 pp.

CECC CECC 90 106-033 ISSUE 2-86. UTE C 86-217 ADD 2/GA 33; Digital Integrated Circuits in Accordance with FS 90 106; 54/74 ALS 257, 54/74 ALS 257U; Quad Data Selector/Multiplexer with 3 State Non Inverting Outputs (En, Fr) ADD 3 (En, Fr). 16 pp.

CECC CECC 90 106-054 ISSUE 1-85. UTE C 86-217; Digital Integrated Circuits in Accordance with FS 90 106; 54 ALS 153, 74 ALS 153; Dual 4 to 1 Multiplexer (En, Fr) ADD 3 (En, Fr). 15 pp.

CECC CECC 90 106-056 ISSUE 1-85. UTE C 86-217; Digital Integrated Circuits in Accordance with FS 90 106; 54/74 ALS 258, 54/74 ALS 258U; Quad Data Selector/Multiplexer with 3 State Inverting Outputs (En, Fr) ADD 3 (En, Fr). 16 pp.

CECC CECC 90 106-077 ISSUE 1-85. UTE C 86-217; Digital Integrated Circuits in Accordance with FS 90 106; 54/74 ALS 251, 54/74 ALS 251U; 8 to 1 Data Selector/Multiplexer with 3 State Outputs (En, Fr) ADD 3 (En, Fr). 17 pp.

CECC CECC 90 106-078 ISSUE 1-85. UTE C 86-217; Digital Integrated Circuits in Accordance with FS 90 106; 54/74 ALS 253, 54/74 ALS 253U; 4 to 1 Data Selector/Multiplexer with 3 State Outputs (En, Fr) ADD 3 (En, Fr). 15 pp.

CECC CECC 90 106-081 ISSUE 1-85. UTE C 86-217; Digital Integrated Circuits in Accordance with FS 90 106; 54/74 ALS 352; 4 to 1 Data Selector/Multiplexer with Inverting Outputs (En, Fr) ADD 3 (En, Fr). 15 pp.

CECC CECC 90 106-082 ISSUE 1-85. UTE C 86-217; Digital Integrated Circuits in Accordance with FS 90 106; 54/74 ALS 353, 54/74 ALS 353U; 4 to 1 Data Selector/Multiplexer with 3 State Inverting Outputs (En, Fr) ADD 3 (En, Fr). 15 pp.

CECC CECC 90 107-015 ISSUE 2-89. UTE C 86-218 ADD 2/FA 15; Digital Integrated Circuits in Accordance with FS 90 107; 54/74 F 151, 54/74 F 151A; Multiplexer (En, Fr) ADD 3 (En, Fr). 10 pp.

CECC CECC 90 107-016 ISSUE 2-89. UTE C 86-218 ADD 2/FA 16 Digital Integrated Circuits in Accordance with FS 90 107; 54/74 F 153; Multiplexer (En, Fr) ADD 3 (En, Fr). 9 pp.

CECC CECC 90 107-017 ISSUE 2-89. UTE C 86-218 ADD 2/FA 17; Digital Integrated Circuits in Accordance with FS 90 107; 54/74 F 157, 54/74 157A; Multiplexer (En, Fr) ADD 3 (En, Fr). 9 pp.

CECC CECC 90 107-018 ISSUE 2-89. UTE C 86-218 ADD 2/FA 18; Digital Integrated Circuits in Accordance with FS 90 107; 54/74 F 158, 54/74 F 158A; Multiplexer (En, Fr) ADD 3 (En, Fr). 9 pp.

Digital Multiplexers *(Cont.)*

—Integrated Circuit *(Cont.)*

CECC CECC 90 107-035 ISSUE 2-89. UTE C 86-218 ADD 3/FA 35; Digital Integrated Circuits in Accordance with FS 90 107; 54/74 F 251—54/74 F 251A; 8-Input Multiplexer with 3-State Outputs (En, Fr) ADD 3 (En, Fr). 9 pp.

CECC CECC 90 107-036 ISSUE 2-89. UTE C 86-218 ADD 3/FA 36; Digital Integrated Circuits in Accordance with FS 90 107; 54/74 F 253; Multiplexer with 3-State Outputs (En, Fr) ADD 3 (En, Fr). 9 pp.

CECC CECC 90 107-037 ISSUE 2-89. UTE C 86-218 ADD 3/FA 37; Digital Integrated Circuits in Accordance with FS 90 107; 54/74 F 257—54/74 F 257A; Multiplexer with 3-State Outputs (En, Fr) ADD 3 (En, Fr). 9 pp.

CECC CECC 90 107-042 ISSUE 2-89. UTE C 86-218 ADD 3/FA 42; Digital Integrated Circuits in Accordance with FS 90 107; 54/74 F 352; Multiplexer (En, Fr) ADD 3 (En, Fr). 9 pp.

CECC CECC 90 107-043 ISSUE 2-89. UTE C 86-218; Digital Integrated Circuits in Accordance with FS 90 107; 54/74 F 353; Multiplexer with 3-State Outputs (En, Fr) ADD 3 (En, Fr). 9 pp.

CECC CECC 90 109-628 ISSUE 1-86. Digital Integrated Circuits in Accordance with FS 90 109; 54/74 HC 257; Data Selector Multiplexer with 3 State Non Inverting Outputs (En, Fr). 6 pp.

CECC CECC 90 109-641 ISSUE 1-86. Digital Integrated Circuits in Accordance with FS 90 109; 54/74 HC 151; Multiplexer (En, Fr). 6 pp.

CECC CECC 90 109-642 ISSUE 1-86. Digital Integrated Circuits in Accordance with FS 90 109; 54/74 HC 153; Multiplexer (En, Fr). 6 pp.

CECC CECC 90 109-643 ISSUE 1-86. Digital Integrated Circuits in Accordance with FS 90 109; 54/74 HC 157; Multiplexer (En, Fr). 6 pp.

CECC CECC 90 109-644 ISSUE 1-86. Digital Integrated Circuits in Accordance with FS 90 109; 54/74 HC 158; Multiplexer (En, Fr). 6 pp.

CECC CECC 90 109-683 ISSUE 1-87. Digital Integrated Circuits in Accordance with FS 90 109; 54/74 HC 251; 8-Channel Multiplexer with 3 State Outputs (En, Fr). 6 pp.

CECC CECC 90 109-714 ISSUE 1-87. Digital Integrated Circuits in Accordance with FS 90 109; 54 HC 298, 74 HC 298; Multiplexer with Storage (En, Fr). 5 pp.

CECC CECC 90 109-715 ISSUE 1-87. Digital Integrated Circuits in Accordance with FS 90 109; 54 HC 352, 74 HC 352; Data Selector/Multiplexer (En, Fr). 6 pp.

CECC CECC 90 109-716 ISSUE 1-87. Digital Integrated Circuits in Accordance with FS 90 109; 54 HC 353, 74 HC 353; Data Selector/Multiplexer with 3-State Outputs (En, Fr). 7 pp.

CECC CECC 90 109-724 ISSUE 1-87. Digital Integrated Circuits in Accordance with FS 90 109; 54 HC 253, 74 HC 253; Dual 4 Channel 3-State Multiplexer (En, Fr). 6 pp.

CECC CECC 90 109-732 ISSUE 1-87. Digital Integrated Circuits in Accordance with FS 90 109; 54 HC 152, 74 HC 152; Selector/Multiplexer (En, Fr). 6 pp.

CECC CECC 90 109-754 ISSUE 1-87. Digital Integrated Circuits in Accordance with FS 90 109; 54 HC 258, 74 HC 258; Selector/Multiplexer with 3-State Outputs (En, Fr). 6 pp.

CECC CECC 90 109-779 ISSUE 1-87. Digital Integrated Circuits in Accordance with FS 90 109; 54/74 HCT 151; Multiplexer (En, Fr). 5 pp.

CECC CECC 90 109-780 ISSUE 1-87. Digital Integrated Circuits in Accordance with FS 90 109; 54/74 HCT 153; Multiplexer (En, Fr). 6 pp.

CECC CECC 90 109-782 ISSUE 1-87. Digital Integrated Circuits in Accordance with FS 90 109; 54/74 HCT 157; Multiplexer (En, Fr). 5 pp.

CECC CECC 90 109-783 ISSUE 1-87. Detail Specification: Digital Integrated Circuits; Silicon Monolithic C MOS, Cavity or Non-Cavity Packages; Type(s) 54/74 HCT 158 Multiplexer; Assessment Levels P, Y, L (En, Fr). 8 pp.

CECC CECC 90 109-824 ISSUE 1-89. Digital Integrated Circuits; Silicon Monolithic C MOS, Cavity or Non-Cavity Packages; Type(s); 54/74 HCT 257 Data Selector Multiplexer with 3-State Non-Inverting Outputs (En, Fr). 9 pp.

CECC CECC 90 109-825 ISSUE 1-89. Digital Integrated Circuits; Silicon Monolithic C MOS, Cavity or Non-Cavity Packages; Type(s); 54/74 HCT 258 Selector/Multiplexer with 3 State Outputs (En, Fr). 9 pp.

—Integrated Circuit—1 of 16

CECC CECC 90 104-067 ISSUE 2. NL CECC 90 104-067 Issue 2; Digital Integrated Circuits in Accordance with FS 90 104; HEC/HEF 4514B; 1-of-16 Decoder/Demultiplexer with Input Latches (En). 8 pp.

INDUSTRY STANDARDS

789

Digital Multiplexers *(Cont.)*

—Integrated Circuit—1 of 16 *(Cont.)*
CECC CECC 90 104-068 ISSUE 2. NL CECC 90 104-068 Issue 2; Digital Integrated Circuits in Accordance with FS 90 104; HEC/HEF 4515B; 1-of-16 Decoder/Demultiplexer with Input Latches (En). 8 pp.

—Integrated Circuit—2-Input—Preferred Products List
CECC CECC MUAHAG Vol 7 IS 8-92. Preferred Products List; Active Microcircuits (En, Fe, Ge). 89 pp.

—Integrated Circuit—2-Line
CECC CECC 90 102-026 ISSUE 1-81. BS CECC 90 102-026; 1-Quadruple 2-Line-to-1-Line Data Selectors/Multiplexers (En) AMD 1 (En). 19 pp.
CECC CECC 90 103-065 ISSUE 1-81. BS CECC 90 103-065; Quad 2-Line to 1-Line Data Selector/Multiplexer (Non-Inverting) (En) AMD 1 (En). 19 pp.
CECC CECC 90 103-098 ISSUE 1-81. BS CECC 90 103-098; Quad 2-Line to 1-Line Data Selector/Multiplexers with Three Level Outputs (En) AMD 1 (En). 21 pp.
CECC CECC 90 103-099 ISSUE 1-82. BS CECC 90 103-099; Quadruple 2-Line to 1-Line Data Selector/Multiplexers with Three Level Outputs (En) AMD 1 (En). 20 pp.

—Integrated Circuit—2 to 1 Lines
CECC CECC 90 101-046 ISSUE 1-87. UTE C 86-213/B 66; Digital Integrated Circuits in Accordance with FS 90 101; 54/64/74 157; Quad 2-to-1 Line Data Selector/Multiplexer (En). 13 pp.
CECC CECC 90 102-027 ISSUE 1-82. BS CECC 90 102-027; Quadruple 2-Line-to-1-Line Data Selectors/Multiplexers (En) AMD 1 (En). 18 pp.
CECC CECC 90 102-060 ISSUE 1-82. BS CECC 90 102-060; Quad 2-Line to 1-Line Data Selector/Multiplexer with Non-Inverting Three Level Outputs (En). 20 pp.
CECC CECC 90 102-061 ISSUE 1-82. BS CECC 90 102-061; Quad 2-Line to 1-Line Data Selector/Multiplexer with Inverting Three Level Output (En). 18 pp.
CECC CECC 90 103-027 ISSUE 1-81. BS CECC 90 103-027; Quadruple 2-Line-to-1-Line Data Select/Multiplexers (En) AMD 1 (En). 18 pp.
CECC CECC 90 103-066 ISSUE 1-81. BS CECC 90 103-066; Quad 2-Line to 1-Line Data Selector/Multiplexer with Inverted Outputs (En) AMD 1 (En). 19 pp.

—Integrated Circuit—2 to 1 Lines—Preferred Products List
CECC CECC MUAHAG Vol 7 IS 8-92. Preferred Products List; Active Microcircuits (En, Fe, Ge). 89 pp.

—Integrated Circuit—2 to 4 Lines—Preferred Products List
CECC CECC MUAHAG Vol 7 IS 8-92. Preferred Products List; Active Microcircuits (En, Fe, Ge). 89 pp.

—Integrated Circuit—4 Channel
CECC CECC 90 109-822 ISSUE 1-88. Digital Integrated Circuits in Accordance with FS 90 109; 54/74 HCT 253; Dual 4-Channel 3-State Multiplexer (En, Fr). 6 pp.
CECC CECC 90 109-823 ISSUE 1-89. Digital Integrated Circuits; Silicon Monolithic C MOS, Cavity or Non-Cavity Packages; Type(s); 54/74 HCT 253B Dual 4-Channel 3-State Multiplexer with Bus Driver Outputs (En, Fr). 9 pp.

—Integrated Circuit—4-Line
CECC CECC 90 103-097 ISSUE 1-81. BS CECC 90 103-097; Dual 4-Line to 1-Line Data Selectors/Multiplexers with 3-Level Outputs (En) AMD 1 (En). 22 pp.
CECC CECC 90 104-219 ISSUE 1-85. BS CECC 90 104-219; Dual 4 Channel Data Selector/Multiplexer (En). 29 pp.

—Integrated Circuit—4 to 1 Lines
CECC CECC 90 101-043 ISSUE 1-87. UTE C 86-213/B 63; Digital Integrated Circuits in Accordance with FS 90 101; 54/64/74 153; Dual 4-To-1 Line Data Selector/Multiplexer (En). 12 pp.
CECC CECC 90 102-053 ISSUE 1-82. BS CECC 90 102-053; Dual 4-Line-to-1-Line Data Selectors/Multiplexers (En). 17 pp.
CECC CECC 90 103-118 ISSUE 1-81. BS CECC 90 103-118; Dual 4-Line-to-1-Line Data Selector/Multiplexer (En) AMD 1 (En). 20 pp.
CECC CECC 90 103-119 ISSUE 1-82. BS CECC 90 103-119; Dual 4-Line-to-1-Line Data Selector/Multiplexer with Inverted 3-Level Outputs (En). 21 pp.

Digital Multiplexers *(Cont.)*

—Integrated Circuit—4 to 1 Lines—Preferred Products List
CECC CECC MUAHAG Vol 7 IS 8-92. Preferred Products List; Active Microcircuits (En, Fe, Ge). 89 pp.

—Integrated Circuit—8 Channel
CECC CECC 90 109-821 ISSUE 1-88. Digital Integrated Circuits in Accordance with FS 90 109; 54/74 HCT 251; 8-Channel Multiplexer with 3-State Outputs (En, Fr). 6 pp.

—Integrated Circuit—8-Line
CECC CECC 90 104-158 ISSUE 1-81. BS CECC 90 104-158; 8 Channel Data Selector with Three State Outputs (En). 26 pp.
CECC CECC 90 104-158 ISSUE 1-86. CEI CECC 90 104-158; 8 Channel Data Selector with Three State Outputs (En). 2 pp.

—Integrated Circuit—8 to 1 Lines
CECC CECC 90 102-059 ISSUE 1-82. BS CECC 90 102-059; 8-Line to 1-Line Data Selector/Multiplexers with Three Level Outputs (En). 20 pp.
CECC CECC 90 103-096 ISSUE 1-81. BS CECC 90 103-096; 8-Line to 1-Line Data Selector/Multiplexers with Three Level Outputs (En) AMD 1 (En). 22 pp.

—Integrated Circuit—8 to 1 Lines—Preferred Products List
CECC CECC MUAHAG Vol 7 IS 8-92. Preferred Products List; Active Microcircuits (En, Fe, Ge). 89 pp.

—Military Communications—Gateways
NATO STANAG 4207 ED 1 AMD 2-83. The NATO Multi-Channel Tactical Digital Gateway—Multiplex Group Framing Standards. 28 pp.
NATO STANAG 4207 ED 2 AMD 0-89. (DRAFT) NATO Multi-Channel Tactical Digital Gateway—Multiplex Group Framing Standards. 26 pp.

—Military Communications—Gateways—Signaling Format
NATO STANAG 4208 ED 1 AMD 2-83. The NATO Multi-Channel Tactical Digital Gateway —Signalling Standards. 51 pp.
NATO STANAG 4208 ED 2 AMD 0-89. (DRAFT) The NATO Multi-Channel Tactical Digital Gateway—Signalling Standards. 51 pp.

—Synchronous Digital Hierarchy
CCITT RECMN G.781-90. Structure of Recommendations on Multiplexing Equipment for the Synchronous Digital Hierarchy (SDH) (Study Group XV) 5 pp. 5 pp.
CCITT RECMN G.782-90. Types and General Characteristics of Synchronous Digital Hierarchy (SDH) Multiplexing Equipment (Study Group XV) 27 pp. 27 pp.
CEPT T/CD 02-01-83. Caracteristiques Techniques Pour un Multiplexeur Numerique Synchrone Fonctiionnant a 64 KBit/s Et Utilisant Une Structure D'Enveloppe a 8 Bits (6+2). 14 pp.
CEPT T/CD 02-01 E-86. Specification of Engineering Requirements for a Synchronous Digital Multiplexer Operating at 64 KBit/s Using an 8 Bit (6+2) Envelope Structure. 14 pp.
CEPT T/CD 02-02-83. Caracteristiques Techniques Pour Un Multiplexeur Numerique Synchrone Fonctionnant a 64 KBit/s, Utilisant Une Structure D'Enveloppe a 10 Bits (8+2). 17 pp.
CEPT T/CD 02-02 E-86. Specification of Engineering Requirements for a Synchronous Digital Multiplexer Operating at 64 KBit/s Using a 10 Bit (8+2) Envelope Structure. 17 pp.
CEPT T/CD 02-04 E-88. Engineering Requirements for a Synchronous Digital Multiplexer for Use with Non Envelope Structured Data (Revised in Edinburgh 1988). 12 pp.
CEPT T/CD 02-04-86. Specifications Techniques D'un Multiplexeur Numerique Synchrone Pour Utilisation Avec Donnees Sans Structure D'Enveloppe. 12 pp.
CEPT T/CD 02-04 E-86. Engineering Requirements for a Synchronous Digital Multiplexer for Use with Non Envelope Structured Data. 12 pp.

—Synchronous Digital Hierarchy—Frame Structure
CCITT RECMN G.704-91. Synchronous Frame Structures Used at Primary and Secondary Hierarchical Levels (Study Group XVIII) 35 pp. 35 pp.
CCITT RECMN G.704-89. Synchronous Frame Structures Used at Primary and Secondary Hierarchical Levels—General Aspects of Digital Transmission Systems; Terminal Equipments (Study Groups XV and XVIII) 21 pp. 21 pp.

Digital Multiplexers *(Cont.)*

—Synchronous Digital Hierarchy—Functional Blocks
CCITT RECMN G.783-90. Characteristics of Synchronous Digital Hierarchy (SDH) Multiplexing Equipment Functional Blocks (Study Group XV) 65 pp. 65 pp.

—Synchronous Digital Hierarchy—Interfaces
CCITT RECMN G.783-90. Characteristics of Synchronous Digital Hierarchy (SDH) Multiplexing Equipment Functional Blocks (Study Group XV) 65 pp. 65 pp.

Digital Paths
See Also: Digital Sections; Hypothetical Reference Digital Paths

—Access Points—Maintenance
CCITT RECMN M.120-89. Access Points for Maintenance—General Maintenance Principles—Maintenance of International Transmission Systems and Telephone Circuits (Study Group IV) 2 pp. 2 pp.

—Bit Error
CCITT RECMN O.152-89. Error Performance Measuring Equipment for 64 kBit/s Paths—Specifications for Measuring Equipment (Study Group IV) 3 pp. 3 pp.

—Designations
CCITT RECMN M.140-89. Designations of International Circuits, Groups, Group and Line Links, Digital Blocks, Digital Paths, Data Transmission Systems and Related Information—General Maintenance Principles—Maintenance of International Transmission Systems and Telephone Circuits. 53 pp.

—Maintenance
CCITT RECMN M.550-89. Performance Limits for Bringing into Service and Maintenance of Digital Paths, Sections, and Line Sections—General Maintenance Principles—Maintenace of International Transmission Systems and Telephone Circuits—(Study Group IV) 16 pp. 16 pp.

—Set Up
CCITT RECMN M.555-89. Bringing International Digital Blocks, Paths and Sections into Service—General Maintenance Principles—Maintenance of International Transmission Systems and Telephone Circuits (Study Group IV) 5 pp. 5 pp.
CCITT RECMN M.2110-92. Bringing into Service International Digital Paths, Sections and Transmission Systems (Study Group IV) 9 pp. 9 pp.

Digital Radio Systems
See Also: Digital Satellite Systems; Pulse Communications; Radio Systems

CENELEC ETR 019-91. Transmission and Multiplexing Specification of New Generation High-Capacity Digital Radio Systems. 42 pp.
ETSI ETR 019-91. Transmission and Multiplexing Specification of New Generation High-Capacity Digital Radio Systems. 42 pp.

—Frequency Bands
CEPT T/R 22-05 E-93. Frequencies for Mobile Digital Trunked Radio Systems. 1 p.

—Interference—Frequency Band Sharing—Fixed Satellites
CCIR Report 877-1-86. Interference Criteria for Digital Radio-Relay Systems Sharing Frequency Bands with the Fixed-Satellite Service—Section 4/9A—Sharing Conditions. 6 pp.

—Radio Relay Systems—ISDN—Availability
CCIR RECMN 697-1-91. Error Performance and Availability Objectives for the Local-Grade Portion at Each End of an ISDN Connection Utilizing Digital Radio-Relay Systems—Section 9A—Performance Objectives, Propagation and Interference Effects. 5 pp.

—Radio Relay Systems—ISDN—Bit Error Rate
CCIR RECMN 697-90. Error Performance Objectives for the Local-Grade Portion at Each End of an ISDN Connection Utilizing Digital Radio-Relay Systems—Section 9A—Performance Objectives, Propagation and Interference Effects. 2 pp.
CCIR RECMN 697-1-91. Error Performance and Availability Objectives for the Local-Grade Portion at Each End of an ISDN Connection Utilizing Digital Radio-Relay Systems—Section 9A—Performance Objectives, Propagation and Interference Effects. 5 pp.

INTERNATIONAL AND NON-U.S. NATIONAL STANDARDS
SUBJECT INDEX
Digital

Digital Readouts
See Also: Display Devices; Indicating Instruments
CNS C7211-90. 1200 bps Numerical-Display (Dec)(12814).

—Color Coding
BSI BS 4099: Part 2-77. (WITHDRAWN) 1977 Amd 1 Colours of Indicator Lights, Push Buttons, Annunciators and Digital Readouts: Flashing Lights, Annunciators and Digital Readouts (Superseded by BS EN 60073: 1993). 8 pp.

Digital Recorders
See Also: Recording Instruments
BSI BS 7528: Part 1-91. 1991 Digital Recorders for Measurements in High-Voltage Impulse Tests Part 1: Requirements for Digital Recorders (Renumbered as BS EN 61083-1: 1993). 32 pp.

—Impulse Voltage Testing
BSI BS 7528: Part 1-91. 1991 Digital Recorders for Measurements in High-Voltage Impulse Tests Part 1: Requirements for Digital Recorders (Renumbered as BS EN 61083-1: 1993). 32 pp.
BSI BS EN 61083-1-93. 1993 Amd 1 Digital Recorders for Measurements in High-Voltage Impulse Tests Part 1: Requirements for Digital Recorders (AMD 7925) September 15, 1993 (S). 41 pp.
CENELEC EN 61083-1-93. Digital Recorders for Measurements in High-Voltage Impulse Tests Part 1: Requirements for Digital Recorders (IEC 1083-1: 1991, Modified). 32 pp.
IEC 1083 Pt 1-91. Digital Recorders for Measurements in High-Voltage Impulse Tests Part 1: Requirements for Digital Recorders First Edition (CENELEC EN 61083-1: 1993). 58 pp.

Digital Recording
See Also: High Density Digital Recording; Recording Instruments

—Audio Cassettes
IEC 1119 Pt 1-92. Digital Audio Tape Cassette System (DAT) Part 1: Dimensions and Characteristics First Edition. 226 pp.
IEC 1119 Pt 3-92. Digital Audio Tape Cassette System (DAT) Part 3: DAT Tape Properties First Edition. 22 pp.

—Audio Cassettes—Calibration
IEC 1119 Pt 2-91. Digital Audio Tape Cassette System Part 2: DAT Calibration Tape First Edition. 13 pp.

—Audio Cassettes—Professional
IEC 1119 Pt 5-93. Digital Audio Tape Cassette System (DAT) Part 5: DAT for Professional Use First Edition. 40 pp.

—Audio Cassettes—Serial Copy Management Systems
IEC 1119 Pt 6-92. Digital Audio Tape Cassette System (DAT) Part 6: Serial Copy Management System First Edition. 18 pp.

—Audio Cassettes—Sound-Program Exchange
CCIR RECMN 777-92. International Exchange of Digital Audio Recordings. 2 pp.

—Audio Tapes
CCIR Report 950-2-90. Digital Recording of Audio Signals—Section 10/11F—Exchange of Recorded Sound Programmes. 2 pp.

—Audio Tapes—Encoding
CNS C5232-90. Sampling Rate and Source Encoding for Professional Digital Audio Recording (May)(12712).

—Audio Tapes—Sampling
CNS C5232-90. Sampling Rate and Source Encoding for Professional Digital Audio Recording (May)(12712).

—Audio Tapes—Sound-Program Exchange
CCIR RECMN 648-86. Digital Recording of Audio Signals—Section 10/11F—Exchange of Recorded Sound Programmes. 1 p.

—Audio/Video Systems
IEC 11172 Pt 1-93. Information Technology—Coding of Moving Pictures and Associated Audio for Digital Storage Media at up to About 1,5 Mbit/s—Part 1: Systems First Edition. 60 pp.
IEC 11172 Pt 2-93. Information Technology—Coding of Moving Pictures and Associated Audio for Digital Storage Media at up to About 1,5 Mbit/s—Part 2: Video First Edition. 122 pp.
IEC 11172 Pt 3-93. Information Technology—Coding of Moving Pictures and Associated Audio for Digital Storage Media at up to About 1,5 Mbit/s—Part 3: Audio First Edition. 157 pp.

Digital Recording (Cont.)
—Audio/Video Systems (Cont.)
ISO 11172 Pt 1-93. Information Technology—Coding of Moving Pictures and Associated Audio for Digital Storage Media at up to About 1,5 Mbit/s—Part 1: Systems First Edition. 60 pp.
ISO 11172 Pt 2-93. Information Technology—Coding of Moving Pictures and Associated Audio for Digital Storage Media at up to About 1,5 Mbit/s—Part 2: Video First Edition. 122 pp.
ISO 11172 Pt 3-93. Information Technology—Coding of Moving Pictures and Associated Audio for Digital Storage Media at up to About 1,5 Mbit/s—Part 3: Audio First Edition. 157 pp.
JTC1 11172 Pt 1-93. Information Technology—Coding of Moving Pictures and Associated Audio for Digital Storage Media at up to About 1,5 Mbit/s—Part 1: Systems First Edition. 60 pp.
JTC1 11172 Pt 2-93. Information Technology—Coding of Moving Pictures and Associated Audio for Digital Storage Media at up to About 1,5 Mbit/s—Part 2: Video First Edition. 122 pp.
JTC1 11172 Pt 3-93. Information Technology—Coding of Moving Pictures and Associated Audio for Digital Storage Media at up to About 1,5 Mbit/s—Part 3: Audio First Edition. 157 pp.

—Data Codes—Audio and Video
IEC DIS 11172-92. Information Technology—Coding of Moving Pictures and Associated Audio for Digital Storage Media up to About 1,5 Mbit/s ***CD-ROM ONLY***. 344 pp.
ISO DIS 11172-92. Information Technology—Coding of Moving Pictures and Associated Audio for Digital Storage Media up to About 1,5 Mbit/s ***CD-ROM ONLY***. 344 pp.
JTC1 DIS11172-92. Information Technology—Coding of Moving Pictures and Associated Audio for Digital Storage Media up to About 1,5 Mbit/s ***CD-ROM ONLY***. 344 pp.

—Data Codes—Audio and Video—Data Compression
IEC DIS 11544-92. Information Technology—Coded Representation of Picture and Audio Information—Progressive Bi-Level Image Compression ***CD-ROM ONLY***. 85 pp.
ISO DIS 11544-92. Information Technology—Coded Representation of Picture and Audio Information—Progressive Bi-Level Image Compression ***CD-ROM ONLY***. 85 pp.

—Magnetic Tape—Television
CCIR RECMN 657-1-90. Digital Television Tape Recording Standards for the International Exchange of Television Programmes on Magnetic Tape—Section 10/11G—Exchange of Recorded Television Programmes. 65 pp.
CCIR RECMN 657-2-92. Digital Television Tape Recording Standards for the International Exchange of Television Programmes on Magnetic Tape—Section 10/11G—Exchange of Recorded Television Programmes. 16 pp.

—Magnetic Tapes
BSI BS ISO/IEC 1863-90. 1990 Information Processing—9-Track, 12,7 mm (0,5 in) Wide Magnetic Tape for Information Interchange Using NRZI at 32 ftpmm (800 ftpi)—32 cpmm (800 cpi). 14 pp.
ISO 1861-75. Information Processing—7-Track, 12,7 mm (0.5 in) Wide Magnetic Tape for Information Interchange Recorded at 8 rpmm (200 rpi) First Edition. 9 pp.
ISO 1862-75. Information Processing—9-Track, 12,7 mm (0.5 in) Wide Magnetic Tape for Information Interchange Recorded at 8 rpmm (200 rpi) First Edition. 9 pp.
ISO 1863-90. Information Processing—9-Track, 12,7 mm (0,5 in) Wide Magnetic Tape for Information Interchange Using NRZ1 at 32 ftpmm (800 ftpi)—32 cpmm (800 cpi) Second Edition. 11 pp.
JTC1 1861-75. Information Processing—7-Track, 12,7 mm (0.5 in) Wide Magnetic Tape for Information Interchange Recorded at 8 rpmm (200 rpi) First Edition. 9 pp.
JTC1 1862-75. Information Processing—9-Track, 12,7 mm (0.5 in) Wide Magnetic Tape for Information Interchange Recorded at 8 rpmm (200 rpi) First Edition. 9 pp.
JTC1 1863-90. Information Processing—9-Track, 12,7 mm (0,5 in) Wide Magnetic Tape for Information Interchange Using NRZ1 at 32 ftpmm (800 ftpi)—32 cpmm (800 cpi) Second Edition. 11 pp.
OSI ISO 1861-75. Information Processing—7-Track, 12,7mm (0.5in) Wide Magnetic Tape for Information Interchange Recorded at 8 rpmm (200rpi). 9 pp.
OSI ISO 1862-75. Information Processing—9-Track, 12,7mm (0.5in) Wide Magnetic Tape for Information Interchange Recorded at 8 rpmm (200rpi). 9 pp.
OSI ISO/IEC 1863-90. Information Processing—9-Track, 12,7mm (0.5in) Wide Magnetic Tape for Information Interchange Using NRZI at 32ftpmm (800 ftpi)-32cpmm (800cpi). 11 pp.

Digital Recording (Cont.)
—Magnetic Tapes (Cont.)
SAA AS 1009-83. Information Processing—9-Track, 12.7 mm Wide Magnetic Tape for Information Interchange Recorded at 32 rpmm. 12 pp.
SAA AS 1011-86. Information Processing—Unrecorded 12.7 mm (0.5 in) Wide Magnetic Tape for Information Interchange—32 ftpmm (800 ftpi) NRZ-1,126 ftpmm (3 200 ftpi) Phase Encoded and 356 ftpmm (9 042 ftpi) NRZ-1. 11 pp.

—Magnetic Tapes—Cassette
BSI BS 5079-81. 1981 Amd 1 Information Interchange on 3.81 mm (0.150 in) Magnetic Tape Cassette at 4 cpmm (100 cpi), Phase Encoded at 63 ftpmm (1600 ftpi). 28 pp.
ISO 3407-83. Information Processing—Information Interchange on 3,81 mm (0.150 in) Magnetic Tape Cassette at 4 cpmm (100 cpi), Phase Encoded at 63 ftpmm (1 600 ftpi) Second Edition. 24 pp.
JTC1 3407-83. Information Processing—Information Interchange on 3,81 mm (0.150 in) Magnetic Tape Cassette at 4 cpmm (100 cpi), Phase Encoded at 63 ftpmm (1 600 ftpi) Second Edition. 24 pp.
OSI ISO 3407-83. Information Processing—Information Interchange on 3,81 mm (0.150 in) Magnetic Tape Cassette at 4 cpmm (100 cpi), Phase Encoded at 63 ft pmm (1 600 ft pi). 24 pp.

—Magnetic Tapes—High Definition Television Systems
CCIR Decision 59-3-89. Television Programmes on Digital Tape and on Film—Annex to Volumes X and XI—Part 3—Sound and Television Recording. 3 pp.

—Video Cassettes—Television Program Exchange
CCIR RECMN 779-92. Operating Practices for 4:2:2 Digital Television Recording. 5 pp.

Digital Satellite Dedicated Networks
See Also: Dedicated Networks; Dedicated Services; Digital Satellite Systems
CCIR Report 1134-90. Digital Satellite Dedicated Networks—Section 4B1—Systems Aspects. 20 pp.
CCIR QUESTION 53/4-90. Digital Satellite Communication Systems for Dedicated/User Oriented Networks—Questions of Study Group 4 Fixed-Satellite Service. 1 p.

—Error Control
CCIR Report 1134-90. Digital Satellite Dedicated Networks—Section 4B1—Systems Aspects. 20 pp.

—Grade of Service
CCIR Report 1134-90. Digital Satellite Dedicated Networks—Section 4B1—Systems Aspects. 20 pp.

Digital Satellite Systems
See Also: Digital Radio Systems; Digital Satellite Dedicated Networks; Satellite Communications; Satellites; Telecommunication Systems
CCIR Decision 70-1-89. Implementation of Digital Satellite Systems—Annex to Volume IV-1—Fixed-Satellite Service. 1 p.

—Interference—Radio Communications—Frequency Band Sharing
CCIR Report 793-1-86. Derivation of Interference Criteria for Digital Systems in the Fixed-Satellite Service Sharing Bands with Terrestrial Systems—Section 4/9A—Sharing Conditions. 3 pp.

—Radio Relay Systems—EIRR—Frequency Band Sharing
CCIR Report 790-1-82. E.i.r.p. and Power Limits for Terrestrial Radio-Relay Transmitters Sharing with Digital Satellite Systems in Bands Between 11 to 14 GHz and Around 30 GHz—Section 4/9A—Sharing Conditions. 9 pp.

Digital Sections
See Also: Digital Line Sections; Digital Paths

—Bit Rates—2048 kbit/s
CCITT RECMN G.921-89. Digital Sections Based on the 2048 kBit/s Hierarchy—Digital Networks, Digital Sections and Digital Line Systems (Study Groups XV and XVIII) 5 pp. 5 pp.

—Integrated Digital Networks
CCITT FASCICLE III.5-89. Digital Networks, Digital Sections and Digital Line Systems—Recommendations G.801—G.961. 292 pp.

—Integrated Services Digital Networks
CCITT RECMN G.960-89. Digital Section for ISDN Basic Rate Access—Digital Networks, Digital Sections and Digital Line Systems (Study Groups XV and XVIII) 38 pp. 38 pp.

INDUSTRY STANDARDS

Digital Sections (Cont.)

—Integrated Services Digital Networks (Cont.)

CENELEC PRETS 300 297-93. Integrated Services Digital Network (ISDN); Access Digital Section for ISDN Basic Rate. 45 pp.

ETSI PRETS 300 297-93. Integrated Services Digital Network (ISDN); Access Digital Section for ISDN Basic Rate. 45 pp.

—Maintenance

CCITT RECMN M.550-89. Performance Limits for Bringing into Service and Maintenance of Digital Paths, Sections, and Line Sections—General Maintenance Principles—Maintenace of International Transmission Systems and Telephone Circuits—(Study Group IV) 16 pp. 16 pp.

—Set Up

CCITT RECMN M.555-89. Bringing International Digital Blocks, Paths and Sections into Service—General Maintenance Principles—Maintenance of International Transmission Systems and Telephone Circuits (Study Group IV) 5 pp. 5 pp.

CCITT RECMN M.2110-92. Bringing into Service International Digital Paths, Sections and Transmission Systems (Study Group IV) 9 pp. 9 pp.

—Telecommunication Networks

CCITT RECMN G.901-89. General Considerations on Digital Sections and Digital Line Systems—Digital Networks, Digital Sections and Digital Line Systems (Study Groups XV and XVIII) 3 pp. 3 pp.

—Transmission Restoration Systems

CCITT RECMN G.180-89. Characteristics of N + M Type Direct Transmission Restoration Systems for Use on Digital Sections, Links or Equipment—General Characteristics of International Telephone Connections and Circuits (Study Groups XII and XV) 11 pp. 11 pp.

Digital Selective Calling

Use: SELCAL Systems—Digital

Digital Signal Processing

Use For: DSP *See Also:* Signal Processing

—Earth Stations—Digital Microwave Radios

IEC 835 Pt 2-5-93. Methods of Measurement for Equipment Used in Digital Microwave Radio Transmission Systems Part 2: Measurements on Terrestrial Radio-Relay Systems Section 5: Digital Signal Processing Sub-System First Edition. 48 pp.

—Glossaries

BSI BS 4727:Pt3: Group 14-92. 1992 Electrotechnical, Power, Telecommunication, Electronics, Lighting and Colour Terms Part 3: Terms Particular to Telecommunications and Electronics Group 14: Oscillations, Signals and Related Devices (IEC 50(702): 1992) (G). 235 pp.

IEC 50 Chap 702-92. International Electrotechnical Vocabulary Chapter 702: Oscillations, Signals and Related Devices First Edition. 241 pp.

—Monitoring—Radio Spectra

CCIR Report 1107-90. Digital Signal Processing Methods in Spectrum Monitoring—Section 1C—Spectrum Monitoring Techniques. 23 pp.

CCIR QUESTION 73/1-90. Digital Signal Processing Methods in Radio Monitoring—Questions Concerning Study Group 1 —Spectrum Management Techniques (Spectrum Engineering, Planning, Sharing, Monitoring and Utilization). 1 p.

Digital Signal Processors

See Also: Microcomputers; Microprocessors

—Microprocessors—Preferred Products List

CECC CECC MUAHAG Vol 7 IS 8-92. Preferred Products List; Active Microcircuits (En, Fe, Ge). 89 pp.

Digital Signals

Use For: Digital Data Signals

—Television—Multiplexing—Protection Ratio

CCIR Report 1084-86. Protection Ratios for Digital Data Signals Multiplexed with Television Signals—Section 11E—Planning of Television Networks, Protection Ratios, Television Receivers and Antennas. 3 pp.

—Television—Test Signals

CCIR RECMN 801-92. Test Signals for Digitally Encoded Colour Television Signals Conforming with Recommendations 601 and 656. 12 pp.

Digital Sound Recording

Use: Sound Recording and Reproduction Equipment—Digital

Digital Speech Interpolation

Use For: DSI *See Also:* Speech Interpolation; Speech Interpolation Equipment

CCITT FASCICLE VI.1. General Recommendations on Telephone Switching and Signalling—Functions and Information Flows for Services in the ISDN—Supplements Recommendations Q1—Q118 Bis. 502 pp.

—DCME—Adaptive Delta Pulse Code Modulation

CCITT RECMN G.763-91. Digital Circuit Multiplication Equipment Using 32 kBit/s ADPCM and Digital Speech Interpolation (Study Group XV) 368 pp. 368 pp.

CCITT RECMN G.763-89. Digital Circuit Multiplication Equipment Using 32 kBit/s ADPCM and Digital Speech Interpolation—General Aspects of Digital Transmission Systems; Terminal Equipments (Study Groups XV and XVIII) 37 pp. 37 pp.

Digital Subscriber Signaling System No. 1

Use: Digital Subscriber Signaling Systems

Digital Subscriber Signaling Systems

Use For: DSS1 *See Also:* Signaling Systems

—Call Holding

CENELEC PRETS 300 141-90. Integrated Services Digital Network (ISDN): Call Hold (HOLD) Supplementary Service Digital Subscriber Signalling One (DSS1) Protocol (T/S 46-33S). 11 pp.

CENELEC PRETS 300 141-91. Integrated Services Digital Network (ISDN); Call Hold (HOLD) Supplementary Service Digital Subscriber Signalling System No. One (DSS1) Protocol. 30 pp.

CENELEC ETS 300 141-92. Integrated Services Digital Network (ISDN); Call Hold (HOLD) Supplementary Service Digital Subscriber Signalling System No. One (DSS1) Protocol. 19 pp.

ETSI PRETS 300 141-90. Integrated Services Digital Network (ISDN); Call Hold (HOLD) Supplementary Service Digital Subscriber Signalling One (DSS1) Protocol (T/S)46-33S). 11 pp.

ETSI ETS 300 141-92. Integrated Services Digital Network (ISDN); Call Hold (HOLD) Supplementary Service Digital Subscriber Signalling System No. One (DSS1) Protocol. 19 pp.

ETSI PRETS 300 141-91. Integrated Services Digital Network (ISDN); Call Hold (HOLD) Supplementary Service Digital Subscriber Signalling System No. One (DSS1) Protocol. 30 pp.

—Call Waiting

CCITT RECMN Q.953-92. Stage 3 Description for Call Completion Supplementary Services Using DSS 1 Section 1—Call Waiting (Study Group XI) 14 pp. 14 pp.

—Calling Line Identification—Malicious Calls

CENELEC PRETS 300 130-90. Integrated Services Digital Network (ISDN); Malicious Call Identification (MCID) Supplementary Service Digital Subscriber Signalling One (DSSI) Protocol (T/S 46-33N). 14 pp.

CENELEC PRETS 300 130-91. Integrated Services Digital Network (ISDN); Malicious Call Identification (MCID) Supplementary Service Digital Subscriber Signalling System No. One (DSS1) Protocol. 16 pp.

CENELEC ETS 300 130-92. Integrated Services Digital Network (ISDN); Malicious Call Identification (MCID) Supplementary Service Digital Subscriber Signalling System No. One (DSS1) Protocol. 15 pp.

ETSI PRETS 300 130-90. Intergrated Services Digital Network (ISDN); Malicious Call Identification (MCID) Supplementary Service Digital Subscriber Signalling One (DSS1) Protocol (T/S 46-33N). 14 pp.

ETSI ETS 300 130-92. Integrated Services Digital Network (ISDN); Malicious Call Identification (MCID) Supplementary Service Digital Subscriber Signalling System No. One (DSS1) Protocol. 15 pp.

ETSI PRETS 300 130-91. Integrated Services Digital Network (ISDN); Malicious Call Identification (MCID) Supplementary Service Digital Subscriber Signalling System No. One (DSS1) Protocol. 16 pp.

—Calling Line Identification Restriction

CENELEC PRETS 300 093-90. Integrated Services Digital Network (ISDN) Calling Line Identification Restriction (CLIR) Supplementary Service Digital Subscriber Signalling One (DSSI) Protocol (T/S 46-33D). 9 pp.

Digital Subscriber Signaling Systems (Cont.)

—Calling Line Identification Restriction (Cont.)

CENELEC PRETS 300 093-91. Integrated Services Digital Network (ISDN); Calling Line Identification Restriction (CLIR) Supplementary Service Digital Subscriber Signalling System No. One (DSS1) Protocol. 14 pp.

CENELEC ETS 300 093-92. Integrated Services Digital Network (ISDN); Calling Line Identification Restriction (CLIR) Supplementary Service Digital Subscriber Signalling System No. One (DSS1) Protocol. 12 pp.

ETSI PRETS 300 093-90. Integrated Services Digital Network (ISDN) Calling Line Identification Restriction (CLIR) Supplementary Service—Digital Subscriber Signalling One (DSS1) Protocol (T/S 46-33D). 9 pp.

ETSI ETS 300 093-92. Integrated Services Digital Network (ISDN); Calling Line Identification Restriction (CLIR) Supplementary Service Digital Subscriber Signalling System No. One (DSS1) Protocol. 12 pp.

ETSI PRETS 300 093-91. Integrated Services Digital Network (ISDN); Calling Line Identification Restriction (CLIR) Supplementary Service Digital Subscriber Signalling System No. One (DSS1) Protocol. 14 pp.

ETSI ETS 300 094-92. Integrated Services Digital Network (ISDN); Connected Line Identification Presentation (COLP) Supplementary Service Description. 16 pp.

—Closed User Groups

CCITT RECMN Q.955-92. Stage 3 Description for Community of Interest Supplementary Services Using DSS 1 Section 1—Closed User Group (Study Group XI) 24 pp. 24 pp.

CENELEC PRETS 300 138-90. Integrated Services Digital Network (ISDN): Closed User Group (CUG) Supplementary Service Digital Subscriber Signalling One (DSS1) Protocol (T/S 46-33H). 29 pp.

CENELEC PRETS 300 138-91. Integrated Services Digital Network (ISDN); Closed User Group (CUG) Supplementary Service Digital Subscriber Signalling No. One (DSS1) Protocol. 30 pp.

CENELEC ETS 300 138-92. Integrated Services Digital Network (ISDN); Closed User Group (CUG) Supplementary Service Digital Subscriber Signalling System No. One (DSS1) Protocol. 29 pp.

ETSI PRETS 300 138-90. Intergrated Services Digital Network (ISDN); Closed User Group (CUG) Supplementary Service Digital Subscriber Signalling One (DSS1) Protocol (T/S 46-33H). 29 pp.

ETSI ETS 300 138-92. Integrated Services Digital Network (ISDN); Closed User Group (CUG) Supplementary Service Digital Subscriber Signalling System No. One (DSS1) Protocol. 29 pp.

ETSI PRETS 300 138-91. Integrated Services Digital Network (ISDN); Closed User Group (CUG) Supplementary Service Digital Subscriber Signalling No. One (DSS1) Protocol. 30 pp.

—Connected Line Identification Restriction

CENELEC PRETS 300 098-90. Integrated Services Digital Network (ISDN) Connected Line Identification Restriction (COLR) Supplementary Service Digital Subscriber Signalling One (DSS1) Protocol (T/S 46-33M). 9 pp.

CENELEC PRETS 300 098-91. Integrated Services Digital Network (ISDN); Connected Line Identification Restriction (COLR) Supplementary Service Digital Subscriber Signalling System No. One (DSS1) Protocol. 15 pp.

CENELEC ETS 300 098-92. Integrated Services Digital Network (ISDN); Connected Line Identification Restriction (COLR) Supplementary Service Digital Subscriber Signalling System No. One (DSS1) Protocol. 12 pp.

ETSI PRETS 300 098-90. Integrated Services Digital Network (ISDN)—Connected Line Identification Restriction (COLR) Supplementary Service Digital Subscriber Signalling One (DSS1) (T/S 46-33M). 9 pp.

ETSI PRETS 300 098-91. Integrated Services Digital Network (ISDN); Connected Line Identification Restriction (COLR) Supplementary Service Digital Subscriber Signalling System No. One (DSS1) Protocol. 15 pp.

—Data Link Layer

CCITT FASCICLE VI.10-88. Digital Subscriber Signalling System No. 1 (DSS 1), Data Link Layer. Q.920-Q.921 (Study Group XI). 148 pp.

—ISDN—Add On Conference

CENELEC PRETS 300 185-91. Integrated Services Digital Network (ISDN); Conference Call Add-on (CONF) Supplementary Service Digital Subscriber Signalling System No. One (DSS1) Protocol (T/S 46-33J1). 38 pp.

Digital Subscriber Signaling Systems *(Cont.)*

—ISDN—Add On Conference *(Cont.)*

CENELEC PRETS 300 185-92. Integrated Services Digital Network (ISDN); Conference Call, Add-on (CONF) Supplementary Service Digital Subscriber Signalling System No. One (DSS1) Protocol. 44 pp.

CENELEC ETS 300 185-93. Integrated Services Digital Network (ISDN); Conference Call, Add-on (CONF) Supplementary Service Digital Subscriber Signalling System No. One (DSS1) Protocol. 44 pp.

ETSI ETS 300 185-93. Integrated Services Digital Network (ISDN); Conference Call, Add-on (CONF) Supplementary Service Digital Subscriber Signalling System No. One (DSS1) Protocol. 44 pp.

ETSI PRETS 300 185-92. Integrated Services Digital Network (ISDN); Conference Call, Add-on (CONF) Supplementary Service Digital Subscriber Signalling System No. One (DSS1) Protocol. 44 pp.

ETSI PRETS 300 185-91. Integrated Services Digital Network (ISDN); Conference Call Add-on (CONF) Supplementary Service Digital Subscriber Signalling System No. One (DSS1) Protocol (T/S 46-33J1). 38 pp.

—ISDN—Advice of Charge Services

CENELEC PRETS 300 182-91. Integrated Services Digital Network (ISDN); Advice of Charge (AOC) Supplementary Service Digital Subscriber Signalling System No. One (DSS1) Protocol (T/S 46-33K). 37 pp.

CENELEC PRETS 300 182-92. Integrated Services Digital Network (ISDN); Advice of Charge (AOC) Supplementary Service Digital Subscriber Signalling System No. One (DSS1) Protocol. 41 pp.

CENELEC PRETS 300 182-92. Integrated Services Digital Network (ISDN); Advice of Charge (AOC) Supplementary Service Digital Subscriber Signalling System No. One (DSS1) Protocol. 41 pp.

ETSI ETS 300 182-93. Integrated Services Digital Network (ISDN); Advice of Charge (AOC) Supplementary Service Digital Subscriber Signalling System No. One (DSS1) Protocol. 41 pp.

ETSI PRETS 300 182-92. Integrated Services Digital Network (ISDN); Advice of Charge (AOC) Supplementary Service Digital Subscriber Signalling System No. One (DSS1) Protocol. 41 pp.

ETSI PRETS 300 182-91. Integrated Services Digital Network (ISDN); Advice of Charge (AOC) Supplementary Service Digital Subscriber Signalling System No. One (DSS1) Protocol (T/S 46-33K). 37 pp.

—ISDN—Call Waiting

CENELEC PRETS 300 058-90. Integrated Services Digital Network (ISDN); Call Waiting (CW) Supplementary Service Digital Subscriber Signalling One (DSS1) Protocol (T/S 46-33f). 14 pp.

CENELEC PRETS 300 058-91. Integrated Services Digital Network (ISDN); Call Waiting (CW) Supplementary Service Digital Subscriber Signalling System No. One (DSS1) Protocol. 18 pp.

CENELEC ETS 300 058-91. Integrated Services Digital Network (ISDN); Call Waiting (CW) Supplementary Service Digital Subscriber Signalling System No. One (DSS1) Protocol. 17 pp.

ETSI PRETS 300 058-90. Integrated Services Digital Network (ISDN); Call Waiting (CW) Supplementary Service Digital Subscriber Signalling One (DSS1) Protocol (T/S 46-33F). 14 pp.

ETSI ETS 300 058-91. Integrated Services Digital Network (ISDN); Call Waiting (CW) Supplementary Service Digital Subscriber Signalling System No. One (DSS1) Protocol. 17 pp.

ETSI PRETS 300 058-90. Integrated Services Digital Network (ISDN); Call Waiting (CW) Supplementary Service Digital Subscriber Signalling One (DSS1) Protocol (T/S 46-33f). 14 pp.

—ISDN—Calling Line Identification Presentation

CENELEC PRETS 300 092-90. Integrated Services Digital Network (ISDN) Calling Line Identification, Presentation (CLIP) Supplementary Service Digital Subscriber Signalling One (DSS1) Protocol (T/S 46-33C). 18 pp.

CENELEC PRETS 300 092-91. Integrated Services Digital Network (ISDN); Calling Line Identification Presentation (CLIP) Supplementary Service Digital Subscriber Signalling System No. One (DSS1) Protocol. 4 pp.

CENELEC ETS 300 092-92. Integrated Services Digital Network (ISDN); Calling Line Identification Presentation (CLIP) Supplementary Service Digital Subscriber Signalling No. One (DSS1) Protocol. 22 pp.

CENELEC ETS 300 092 PRA1-92. AMD 1 Integrated Services Digital Network (ISDN); Calling Line Identification Presentation (CLIP) Supplementary Service Digital Subscriber Signalling System No. One (DSS1) Protocol. 4 pp.

Digital Subscriber Signaling Systems *(Cont.)*

—ISDN—Calling Line Identification Presentation *(Cont.)*

CENELEC ETS 300 092/A1-93. AMD 1 Integrated Services Digital Network (ISDN); Calling Line Identification Presentation (CLIP) Supplementary Service Digital Subscriber Signalling System No. One (DSS1) Protocol. 4 pp.

ETSI PRETS 300 092-90. Integrated Services Digital Network (ISDN) Calling Line Identification Presentation (CLIP) Supplementary Service—Digital Subscriber Signalling One (DSS1) Protocol (T/S 46-33C). 18 pp.

ETSI ETS 300 092/A1-93. AMD 1 Integrated Services Digital Network (ISDN); Calling Line Identification Presentation (CLIP) Supplementary Service Digital Subscriber Signalling System No. One (DSS1) Protocol. 4 pp.

ETSI ETS 300 092 PRA1-92. AMD 1 Integrated Services Digital Network (ISDN); Calling Line Identification Presentation (CLIP) Supplementary Service Digital Subscriber Signalling System No. One (DSS1) Protocol. 4 pp.

ETSI ETS 300 092-92. Integrated Services Digital Network (ISDN); Calling Line Identification Presentation (CLIP) Supplementary Service Digital Subscriber Signalling No. One (DSS1) Protocol. 22 pp.

ETSI PRETS 300 092-91. Integrated Services Digital Network (ISDN); Calling Line Identification Presentation (CLIP) Supplementary Service Digital Subscriber Signalling System No. One (DSS1) Protocol. 4 pp.

—ISDN—Connected Line Identification Presentation

CENELEC PRETS 300 097-90. Integrated Services Digital Network (ISDN) Connected Line Identification Presentation (COLP) Supplementary Service Digital Subscriber One (DSS1) Protocol (T/S 46-33L). 17 pp.

CENELEC PRETS 300 097-91. Integrated Services Digital Network (ISDN); Connected Line Identification Presentation (COLP) Supplementary Service Digital Subscriber Signalling System No. One (DSS1) Protocol. 24 pp.

CENELEC ETS 300 097-92. Integrated Services Digital Network (ISDN); Connected Line Identification Presentation (COLP) Supplementary Service Digital Subscriber Signalling System No. One (DSS1) Protocol. 21 pp.

ETSI PRETS 300 097-90. Integrated Services Digital Network (ISDN) Connected Line Identification, Presentation (COLP) Supplementary Service—Digital Subscriber Signalling One (DSSI) Protocol (T/S 46-33L). 17 pp.

ETSI ETS 300 097-92. Integrated Services Digital Network (ISDN); Connected Line Identification Presentation (COLP) Supplementary Service Digital Subscriber Signalling System No. One (DSS1) Protocol. 21 pp.

ETSI PRETS 300 097-91. Integrated Services Digital Network (ISDN); Connected Line Identification Presentation (COLP) Supplementary Service Digital Subscriber Signalling System No. One (DSS1) Protocol. 24 pp.

ETSI ETS 300 098-92. Integrated Services Digital Network (ISDN); Connected Line Identification Restriction (COLR) Supplementary Service Digital Subscriber Signalling System No. One (DSS1) Protocol. 12 pp.

—ISDN—Data Link Layer (OSI)—Frame Mode Bearer Services

CCITT RECMN Q.922-92. ISDN Data Link Layer Specification for Frame Mode Bearer Services (Study Group XI) 112 pp. 112 pp.

—ISDN—Data Link Layer (OSI)—PICS

CENELEC PRI-ETS 300 305-93. Integrated Services Digital Network (ISDN); Digital Subscriber Signalling System No. One (DSS1) Protocol Implementation Conformance Statement (PICS) Proforma for Basic-Access User for Data-Link-Layer Protocol for General Application. 19 pp.

CENELEC PRI-ETS 300 306-93. Integrated Services Digital Network (ISDN); Digital Subscriber Signalling System No. One (DSS1) Protocol Implementation Conformance Statement (PICS) Proforma for Primary-Rate-Access User for Data-Link-Layer Protocol for General Application. 20 pp.

CENELEC PRI-ETS 300 307-93. Integrated Services Digital Network (ISDN); Digital Subscriber Signalling System No. One (DSS1) Protocol Implementation Conformance Statement (PICS) Proforma for Basic-Access Network for Data-Link-Layer Protocol for General Application. 20 pp.

Digital Subscriber Signaling Systems *(Cont.)*

—ISDN—Data Link Layer (OSI)—PICS *(Cont.)*

CENELEC PRI-ETS 300 308-93. Integrated Services Digital Network (ISDN); Digital Subscriber Signalling System No. One (DSS1) Protocol Implementation Conformance Statement (PICS) Proforma for Primary-Rate-Access Network for Data-Link-Layer Protocol for General Application. 19 pp.

ETSI PRI-ETS 300 305-93. Integrated Services Digital Network (ISDN); Digital Subscriber Signalling System No. One (DSS1) Protocol Implementation Conformance Statement (PICS) Proforma for Basic-Access User for Data-Link-Layer Protocol for General Application. 19 pp.

ETSI PRI-ETS 300 306-93. Integrated Services Digital Network (ISDN); Digital Subscriber Signalling System No. One (DSS1) Protocol Implementation Conformance Statement (PICS) Proforma for Primary-Rate-Access User for Data-Link-Layer Protocol for General Application. 20 pp.

ETSI PRI-ETS 300 307-93. Integrated Services Digital Network (ISDN); Digital Subscriber Signalling System No. One (DSS1) Protocol Implementation Conformance Statement (PICS) Proforma for Basic-Access Network for Data-Link-Layer Protocol for General Application. 20 pp.

ETSI PRI-ETS 300 308-93. Integrated Services Digital Network (ISDN); Digital Subscriber Signalling System No. One (DSS1) Protocol Implementation Conformance Statement (PICS) Proforma for Primary-Rate-Access Network for Data-Link-Layer Protocol for General Application. 19 pp.

—ISDN—Data Link Layer (OSI)—PIXIT

CENELEC PRI-ETS 300 309-93. Integrated Services Digital Network (ISDN); Digital Subscriber Signalling System No. One (DSS1) Protocol Implementation eXtra Information for Testing (PIXIT) Proforma for Basic-Access User for Data-Link-Layer Protocol for General Application. 11 pp.

CENELEC PRI-ETS 300 310-93. Integrated Services Digital Network (ISDN); Digital Subscriber Signalling System No. One (DSS1) Protocol Implementation eXtra Information for Testing (PIXIT) Proforma for Primary-Rate-Access User for Data-Link-Layer Protocol for General Application. 11 pp.

CENELEC PRI-ETS 300 311-93. Integrated Services Digital Network (ISDN); Digital Subscriber Signalling System No. One (DSS1) Protocol Implementation eXtra Information for Testing (PIXIT) Proforma for Basic-Access Network for Data-Link-Layer Protocol for General Application. 11 pp.

CENELEC PRI-ETS 300 312-93. Integrated Services Digital Network (ISDN); Digital Subscriber Signalling System No. One (DSS1) Protocol Implementation eXtra Information for Testing (PIXIT) Proforma for Primary-Rate-Access Network for Data-Link-Layer Protocol for General Application. 11 pp.

ETSI PRI-ETS 300 309-93. Integrated Services Digital Network (ISDN); Digital Subscriber Signalling System No. One (DSS1) Protocol Implementation eXtra Information for Testing (PIXIT) Proforma for Basic-Access User for Data-Link-Layer Protocol for General Application. 11 pp.

ETSI PRI-ETS 300 310-93. Integrated Services Digital Network (ISDN); Digital Subscriber Signalling System No. One (DSS1) Protocol Implementation eXtra Information for Testing (PIXIT) Proforma for Primary-Rate-Access User for Data-Link-Layer Protocol for General Application. 11 pp.

ETSI PRI-ETS 300 311-93. Integrated Services Digital Network (ISDN); Digital Subscriber Signalling System No. One (DSS1) Protocol Implementation eXtra Information for Testing (PIXIT) Proforma for Basic-Access Network for Data-Link-Layer Protocol for General Application. 11 pp.

ETSI PRI-ETS 300 312-93. Integrated Services Digital Network (ISDN); Digital Subscriber Signalling System No. One (DSS1) Protocol Implementation eXtra Information for Testing (PIXIT) Proforma for Primary-Rate-Access Network for Data-Link-Layer Protocol for General Application. 11 pp.

—ISDN—Data Link Layer (OSI)—User Network Interfaces

CCITT RECMN Q.920-89. ISDN User-Network Interface Data Link Layer—General Aspects—Digital Subscriber Signalling System No.1 (DSS 1), Data Link Layer (Study Group XI) 16 pp (Same as Recmn I.440). 16 pp.

CCITT RECMN Q.921-89. ISDN User-Network Interface—Data Link Layer Specification—Digital Subscriber Signalling System No.1 (DSS 1), Data Link Layer (Study Group XI) 123 pp (Same as Recmn I.441). 123 pp.

Digital
INTERNATIONAL AND NON-U.S. NATIONAL STANDARDS
SUBJECT INDEX

Digital Subscriber Signaling Systems *(Cont.)*

—ISDN—Direct Dialing In Services

CCITT RECMN Q.951-92. Stage 3 Description for Number Identification Supplementary Services Using DSS 1 Section 1—Direct-Dialling-in (DDI) Section 2—Multiple Subscriber Number (MSN) Section 8—Sub-Addressing (SUB) (Study Group XI) 17 pp. 17 pp.

CENELEC PRETS 300 064-90. Integrated Services Digital Network (ISDN); Direct Dialling in (DDI) Supplementary Service Digital Subscriber Signalling One (DSS1) Protocol (T/S 46-33A). 8 pp.

CENELEC PRETS 300 064-91. Integrated Services Digital Network (ISDN); Direct Dialling In (DDI) Supplementary Service Digital Subscriber Signalling System No. One (DSS1) Protocol. 10 pp.

CENELEC ETS 300 064-91. Integrated Services Digital Network (ISDN); Direct Dialling In (DDI) Supplementary Service Digital Subscriber Signalling System No. One (DSS1) Protocol. 10 pp.

ETSI PRETS 300 064-90. Integrated Services Digital Network (ISDN) Direct Dialling in (DDI) Supplementary Service Digital Subscriber Signalling One (DSS1) Protocol (T/S 46-33A). 8 pp.

ETSI ETS 300 064-91. Integrated Services Digital Network (ISDN); Direct Dialling In (DDI) Supplementary Service Digital Subscriber Signalling System No. One (DSS1) Protocol. 10 pp.

ETSI PRETS 300 064-91. Integrated Services Digital Network (ISDN); Direct Dialling In (DDI) Supplementary Service Digital Subscriber Signalling System No. One (DSS1) Protocol. 10 pp.

—ISDN—European Automatic Freephone Service

CENELEC PRETS 300 210-92. Integrated Services Digital Network (ISDN); Freephone (FPH) Supplementary Service Digital Subscriber Signalling System No. One (DSS1) Protocol. 22 pp.

ETSI PRETS 300 210-92. Integrated Services Digital Network (ISDN); Freephone (FPH) Supplementary Service Digital Subscriber Signalling System No. One (DSS1) Protocol. 22 pp.

—ISDN—Interworking

CCITT RECMN Q.699-89. Interworking Between the Digital Subscriber Signalling System Layer 3 Protocol and the Signalling System No. 7 ISDN User Part—Interworking of Signalling Systems (Study Group XI) 66 pp. 66 pp.

—ISDN—Multiple Subscriber Numbers

CCITT RECMN Q.951-92. Stage 3 Description for Number Identification Supplementary Services Using DSS 1 Section 1—Direct-Dialling-in (DDI) Section 2—Multiple Subscriber Number (MSN) Section 8—Sub-Addressing (SUB) (Study Group XI) 17 pp. 17 pp.

—ISDN—Network Layer—User Network Interfaces

CCITT RECMN Q.930-89. ISDN User-Network Interface Layer 3—General Aspects—Digital Subscriber Signalling System No.1 (DSS 1), Network Layer, User-Network Management (Study Group XI) 4 pp (Same as Recmn I.450). 4 pp.

CCITT RECMN Q.931-89. ISDN User-Network Interface Layer 3 Specification for Basic Call Control—Digital Subscriber Signalling System No. 1 (DSS 1), Network Layer, User-Network Management (Study Group XI) 356 pp (Same as Recmn I.451). 356 pp.

CCITT RECMN Q.932-89. Generic Procedures for the Control of ISDN Supplementary Services—Digital Subscriber Signalling System No. 1 (DSS 1), Network Layer, User-Network Management (Study Group XI) 61 pp (Same as Recmn I.452). 61 pp.

CCITT RECMN Q.940-89. ISDN User-Network Interface Protocol for Management—General Aspects Subscriber Signalling System No. 1 (DSS 1), Network Layer, User-Network Management (Study Group XI) 11 pp. 11 pp.

—ISDN—Protocols—Call Control Procedures

CENELEC PRI-ETS 300 314-93. Integrated Services Digital Network (ISDN); Digital Subscriber Signalling System No. One (DSS1) Protocol Implementation Conformance Statement (PICS) Proforma for Basic-Access User for Signalling-Network-Layer Protocol for Circuit-Mode Basic Call Control. 51 pp.

CENELEC PRI-ETS 300 315-93. Integrated Services Digital Network (ISDN); Digital Sub. Signalling Sys. No. One (DSS1) Protocol Implementation Conformance Statement (PICS) Proforma for Primary-Rate-Access User for Signalling-Network-Layer Protocol for Circuit-Mode Basic Call Control. 52 pp.

Digital Subscriber Signaling Systems *(Cont.)*

—ISDN—Protocols—Call Control Procedures *(Cont.)*

CENELEC PRI-ETS 300 316-93. Integrated Services Digital Network (ISDN); Digital Subscriber Signalling System No. One (DSS1) Protocol Implementation Conformance Statement (PICS) Proforma for Basic-Access Network for Signalling-Network-Layer Protocol for Circuit-Mode Basic Call Control. 47 pp.

CENELEC PRI-ETS 300 317-93. Integrated Services Digital Network (ISDN); Digital Subscriber Signalling System No. One (DSS1) Protocol Impl. Conformance Statement (PICS) Proforma for Primary-Rate-Access Network for Signalling-Network-Layer Protocol for Circuit-Mode Basic Call Control. 46 pp.

CENELEC PRI-ETS 300 318-93. Integrated Services Digital Network (ISDN); Digital Subscriber Signalling System No. One (DSS1) Protocol Implementation eXtra Information for Testing (PIXIT) Proforma for Basic-Access User for Signalling-Network-Layer Protocol for Circuit-Mode Basic Call Control. 13 pp.

CENELEC PRI-ETS 300 319-93. Integrated Services Digital Network (ISDN); Digital Subscriber Signalling System No. One (DSS1) Protocol Impl. eXtra Information for Testing (PIXIT) Proforma for Primary-Rate-Access User for Signalling-Network-Layer Protocol for Circuit-Mode Basic Call Control. 13 pp.

CENELEC PRI-ETS 300 320-93. Integrated Services Digital Network (ISDN); Digital Subscriber Signalling System No. One (DSS1) Protocol Implementation eXtra Information for Testing (PIXIT) Proforma for Basic-Access Network for Signalling-Network-Layer Protocol for Circuit-Mode Basic Call Control. 13 pp.

CENELEC PRI-ETS 300 321-93. Integrated Services Digital Network (ISDN); Digital Sub. Signalling Sys. No. One (DSS1) Protocol Implementation eXtra Information for Testing (PIXIT) Proforma for Primary-Rate-Access Network for Signalling-Network-Layer Protocol for Circuit-Mode Basic Call Control. 13 pp.

ETSI PRI-ETS 300 314-93. Integrated Services Digital Network (ISDN); Digital Subscriber Signalling System No. One (DSS1) Protocol Implementation Conformance Statement (PICS) Proforma for Basic-Access User for Signalling-Network-Layer Protocol for Circuit-Mode Basic Call Control. 51 pp.

ETSI PRI-ETS 300 315-93. Integrated Services Digital Network (ISDN); Digital Sub. Signalling Sys. No. One (DSS1) Protocol Implementation Conformance Statement (PICS) Proforma for Primary-Rate-Access User for Signalling-Network-Layer Protocol for Circuit-Mode Basic Call Control. 52 pp.

ETSI PRI-ETS 300 316-93. Integrated Services Digital Network (ISDN); Digital Subscriber Signalling System No. One (DSS1) Protocol Implementation Conformance Statement (PICS) Proforma for Basic-Access Network for Signalling-Network-Layer Protocol for Circuit-Mode Basic Call Control. 47 pp.

ETSI PRI-ETS 300 317-93. Integrated Services Digital Network (ISDN); Digital Subscriber Signalling System No. One (DSS1) Protocol Impl. Conformance Statement (PICS) Proforma for Primary-Rate-Access Network for Signalling-Network-Layer Protocol for Circuit-Mode Basic Call Control. 46 pp.

ETSI PRI-ETS 300 318-93. Integrated Services Digital Network (ISDN); Digital Subscriber Signalling System No. One (DSS1) Protocol Implementation eXtra Information for Testing (PIXIT) Proforma for Basic-Access User for Signalling-Network-Layer Protocol for Circuit-Mode Basic Call Control. 13 pp.

ETSI PRI-ETS 300 319-93. Integrated Services Digital Network (ISDN); Digital Subscriber Signalling System No. One (DSS1) Protocol Impl. eXtra Information for Testing (PIXIT) Proforma for Primary-Rate-Access User for Signalling-Network-Layer Protocol for Circuit-Mode Basic Call Control. 13 pp.

ETSI PRI-ETS 300 320-93. Integrated Services Digital Network (ISDN); Digital Subscriber Signalling System No. One (DSS1) Protocol Implementation eXtra Information for Testing (PIXIT) Proforma for Basic-Access Network for Signalling-Network-Layer Protocol for Circuit-Mode Basic Call Control. 13 pp.

ETSI PRI-ETS 300 321-93. Integrated Services Digital Network (ISDN); Digital Sub. Signalling Sys. No. One (DSS1) Protocol Implementation eXtra Information for Testing (PIXIT) Proforma for Primary-Rate-Access Network for Signalling-Network-Layer Protocol for Circuit-Mode Basic Call Control. 13 pp.

Digital Subscriber Signaling Systems *(Cont.)*

—ISDN—Protocols—Keyboards

CENELEC PRETS 300 122-90. Integrated Services Digital Network (ISDN); Generic Keyboard Protocol for the Support of Supplementary Services Digital Subscriber Signaling One (DSSI) Protocol (T/S 46-32A). 8 pp.

CENELEC PRETS 300 122-91. Integrated Services Digital Network (ISDN); Generic Keypad Protocol for the Support of Supplementary Services Digital Subscriber Signalling No. One (DSS1) Protocol. 10 pp.

CENELEC ETS 300 122-92. Integrated Services Digital Network (ISDN); Generic Keypad Protocol for the Support of Supplementary Services Digital Subscriber Signalling System No. One (DSS1) Protocol. 10 pp.

ETSI ETS 300 122-92. Integrated Services Digital Network (ISDN); Generic Keypad Protocol for the Support of Supplementary Services Digital Subscriber Signalling System No. One (DSS1) Protocol. 10 pp.

—ISDN—Protocols—Multiple Subscriber Numbers

CENELEC PRETS 300 052-90. Integrated Services Digital Network (ISDN): Multiple Subscriber Number (MSN) Supplementary Service Digital Subscriber Signalling One (DSS1) Protocol (T/S 46-33B). 9 pp.

CENELEC PRETS 300 052-91. Integrated Services Digital Network (ISDN); Multiple Subscriber Number (MSN) Supplementary Service Digital Subscriber Signalling System No. One (DSS1) Protocol. 13 pp.

CENELEC ETS 300 052-91. Integrated Services Digital Network (ISDN); Multiple Subscriber Number (MSN) Supplementary Service Digital Subscriber Signalling System No. One (DSS1) Protocol. 14 pp.

ETSI PRETS 300 052-90. Integrated Services Digital Network (ISDN); Multiple Subscriber Number (MSN) Supplementary Service Digital Subscriber Signalling One (DSS1) Protocol (T/S 46-33B). 9 pp.

ETSI ETS 300 052-91. Integrated Services Digital Network (ISDN); Multiple Subscriber Number (MSN) Supplementary Service Digital Subscriber Signalling System No. One (DSS1) Protocol. 14 pp.

ETSI PRETS 300 052-91. Integrated Services Digital Network (ISDN); Multiple Subscriber Number (MSN) Supplementary Service Digital Subscriber Signalling System No. One (DSS1) Protocol. 13 pp.

ETSI PRETS 300 052-90. Integrated Services Digital Network (ISDN): Multiple Subscriber Number (MSN) Supplementary Service Digital Subscriber Signalling One (DSS1) Protocol (T/S 46-33B). 9 pp.

—ISDN—Protocols—Supplementary Services

CENELEC PRETS 300 195-92. Integrated Services Digital Network (ISDN); Supplementary Service Interactions Digital Subscriber Signalling System No. One (DSS1) Protocol. 41 pp.

CENELEC PRETS 300 196-92. Integrated Services Digital Network (ISDN); Generic Functional Protocol for the Support of Supplementary Services Digital Subscriber Signalling System No. One (DSS1) Protocol. 88 pp.

CENELEC PRETS 300 196-93. Integrated Services Digital Network (ISDN); Generic Functional Protocol for the Support of Supplementary Services Digital Subscriber Signalling System No. One (DSS1) Protocol. 108 pp.

CENELEC PRETS 300 207-92. Integrated Services Digital Network (ISDN); Diversion Supplementary Services Digital Subscriber Signalling No. One (DSS1) Protocol. 72 pp.

ETSI PRETS 300 195-92. Integrated Services Digital Network (ISDN); Supplementary Service Interactions Digital Subscriber Signalling System No. One (DSS1) Protocol. 41 pp.

ETSI PRETS 300 196-93. Integrated Services Digital Network (ISDN); Generic Functional Protocol for the Support of Supplementary Services Digital Subscriber Signalling System No. One (DSS1) Protocol. 108 pp.

ETSI PRETS 300 196-92. Integrated Services Digital Network (ISDN); Generic Functional Protocol for the Support of Supplementary Services Digital Subscriber Signalling System No. One (DSS1) Protocol. 88 pp.

ETSI PRETS 300 207-92. Integrated Services Digital Network (ISDN); Diversion Supplementary Services Digital Subscriber Signalling No. One (DSS1) Protocol. 72 pp.

—ISDN—Subaddressing

CCITT RECMN Q.951-92. Stage 3 Description for Number Identification Supplementary Services Using DSS 1 Section 1—Direct-Dialling-in (DDI) Section 2—Multiple Subscriber Number (MSN) Section 8—Sub-Addressing (SUB) (Study Group XI) 17 pp. 17 pp.

INTERNATIONAL AND NON-U.S. NATIONAL STANDARDS
SUBJECT INDEX

Digital

Digital Subscriber Signaling Systems (Cont.)

—ISDN—Subaddressing (Cont.)
CENELEC PRETS 300 061-90. Integrated Services Digital Network (ISDN); Subaddressing (SUB) Supplementary Service Digital Subscriber Signalling One (DSS1) Protocol (T/S 46-331). 9 pp.

CENELEC PRETS 300 061-91. Integrated Services Digital Network (ISDN); Subaddressing (SUB) Supplementary Service Digital Subscriber Signalling System No. One (DSS1) Protocol. 11 pp.

CENELEC ETS 300 061-91. Integrated Services Digital Network (ISDN); Subaddressing (SUB) Supplementary Service Digital Subscriber Signalling System No. One (DSS1) Protocol. 11 pp.

ETSI PRETS 300 061-90. Integrated Services Digital Network (ISDN); Subaddressing (SUB) Supplementary Service Digital Subscriber Signalling One (DSS1) Protocol (T/S 46-331). 9 pp.

ETSI ETS 300 061-91. Integrated Services Digital Network (ISDN); Subaddressing (SUB) Supplementary Service Digital Subscriber Signalling System No. One (DSS1) Protocol. 11 pp.

ETSI PRETS 300 061-91. Integrated Services Digital Network (ISDN); Subaddressing (SUB) Supplementary Service Digital Subscriber Signalling System No. One (DSS1) Protocol. 11 pp.

ETSI PRETS 300 061-90. Integrated Services Digital Network (ISDN); Subaddressing (SUB) Supplementary Service Digital Subscriber Signalling One (DSS1) Protocol (T/S 46-331). 9 pp.

—ISDN—Telephony
CENELEC PRETS 300 267-92. Integrated Services Digital Network (ISDN); Telephony 1 kHz and Videotelephony Teleservices Digital Subscriber Signalling System No. One (DSS1). 36 pp.

ETSI PRETS 300 267-92. Integrated Services Digital Network (ISDN); Telephony 1 kHz and Videotelephony Teleservices Digital Subscriber Signalling System No. One (DSS1). 36 pp.

—ISDN—Terminal Portability
CENELEC PRETS 300 055-90. Integrated Services Digital Network (ISDN); Terminal Portability (TP) Supplementary Service Digital Subscriber Signalling One (DSS1) Protocol (T/S 46-33E). 8 pp.

CENELEC PRETS 300 055-91. Integrated Services Digital Network (ISDN); Terminal Portability (TP) Supplementary Service Digital Subscriber Signalling System No. One (DSS1) Protocol. 11 pp.

CENELEC PRETS 300 055-91. Integrated Services Digital Network (ISDN); Terminal Portability (TP) Supplementary Service Digital Subscriber Signalling System No. One (DSS1) Protocol. 10 pp.

ETSI PRETS 300 055-90. Integrated Services Digital Network (ISDN); Terminal Portability (TP) Supplementary Service Digital Subscriber Signalling One (DSS1) Protocol (T/S 46-33E). 8 pp.

ETSI PRETS 300 055-91. Integrated Services Digital Network (ISDN); Terminal Portability (TP) Supplementary Service Digital Subscriber Signalling System No. One (DSS1) Protocol. 11 pp.

ETSI PRETS 300 055-91. Integrated Services Digital Network (ISDN); Terminal Portability (TP) Supplementary Service Digital Subscriber Signalling System No. One (DSS1) Protocol. 10 pp.

ETSI PRETS 300 055-90. Integrated Services Digital Network (ISDN); Terminal Portability (TP) Supplementary Service Digital Subscriber Signalling One (DSS1) Protocol (T/S 46-33E). 8 pp.

—ISDN—Three Party Services
CENELEC PRETS 300 188-92. Integrated Services Digital Network (ISDN); Three Party (3PTY) Supplementary Service Digital Subscriber Signalling System No. One (DSS1) Protocol. 24 pp.

CENELEC PRETS 300 188-93. Integrated Services Digital Network (ISDN); Three-Party (3PTY) Supplementary Service Digital Subscriber Signalling System No. One (DSS1) Protocol. 39 pp.

ETSI PRETS 300 188-93. Integrated Services Digital Network (ISDN); Three-Party (3PTY) Supplementary Service Digital Subscriber Signalling System No. One (DSS1) Protocol. 39 pp.

—ISDN—User to User Signaling
CENELEC PRETS 300 286-93. Integrated Services Digital Network (ISDN); User-to-User Signalling (UUS) Supplementary Service Digital Subscriber Signalling System No. One (DSS1) Protocol. 79 pp.

ETSI PRETS 300 286-93. Integrated Services Digital Network (ISDN); User-to-User Signalling (UUS) Supplementary Service Digital Subscriber Signalling System No. One (DSS1) Protocol. 79 pp.

—ISDN—Video Telephone Services
CENELEC PRETS 300 267-92. Integrated Services Digital Network (ISDN); Telephony 1 kHz and Videotelephony Teleservices Digital Subscriber Signalling System No. One (DSS1). 36 pp.

Digital Subscriber Signaling Systems (Cont.)

—ISDN—Video Telephone Services (Cont.)
ETSI PRETS 300 267-92. Integrated Services Digital Network (ISDN); Telephony 1 kHz and Videotelephony Teleservices Digital Subscriber Signalling System No. One (DSS1). 36 pp.

—Network Layer—User Network Interfaces
CCITT FASCICLE VI.11-88. Digital Subscriber Signalling System No. 1 (DSS 1), Network Layer, User Network Management Recommendations Q.930—Q.940 (Study Group XI). 441 pp.

Digital Switching Systems
See Also: Circuit Switching Systems; Switching Systems

—Documentation
CEPT T/CS 01-10-84. Exigence De Documentation Des Systemes De Commutation Numeriques. 12 pp.

CEPT T/CS 01-10 E-84. Documentation Requirements for Digital Switching Systems. 12 pp.

—Telephone Exchanges—Documentation
CEPT T/CS 01-10-84. Exigence De Documentation Des Systemes De Commutation Numeriques. 12 pp.

CEPT T/CS 01-10 E-84. Documentation Requirements for Digital Switching Systems. 12 pp.

Digital Television
See Also: Television Systems

CCIR Decision 60-3-89. Digital Television Standards—Annex to Volume XI-1—Broadcasting Service (Television). 4 pp.

— 525 Lines—Television Studios—Encoding
CCIR RECMN 601-2-90. Encoding Parameters of Digital Television for Studios—Section 11F—Digital Methods of Transmitting Television Information. 10 pp.

CCIR RECMN 601-3-92. Encoding Parameters of Digital Television for Studios—Section 11F—Digital Methods of Transmitting Television Information. 11 pp.

— 525 Lines—Video Signals—Interfaces
CCIR RECMN 656-86. Interfaces for Digital Component Video Signals in 525-Line and 625-Line Television Systems—Section 11F—Digital Methods of Transmitting Television Information. 13 pp.

CCIR RECMN 656-1-92. Interfaces for Digital Component Video Signals in 525-Line and 625-Line Television Systems Operating at the 4:2:2 Level of Recommendation 601—Section 11F—Digital Methods of Transmitting Television Information. 16 pp.

CCIR Report 1088-1-90. Interfaces for Digital Video Signals in 525-Line and 625-Line Television Systems—Section 11F—Digital Methods of Transmitting Television Information. 7 pp.

CCIR RECMN 799-92. Interfaces for Digital Component Video Signals in 525-Line and 625-Line Television Systems Operating at the 4:4:4 Level of Recommendation 601. 17 pp.

— 625 Lines—Television Studios—Encoding
CCIR RECMN 601-2-90. Encoding Parameters of Digital Television for Studios—Section 11F—Digital Methods of Transmitting Television Information. 10 pp.

CCIR RECMN 601-3-92. Encoding Parameters of Digital Television for Studios—Section 11F—Digital Methods of Transmitting Television Information. 11 pp.

— 625 Lines—Video Signals—Interfaces
CCIR RECMN 656-86. Interfaces for Digital Component Video Signals in 525-Line and 625-Line Television Systems—Section 11F—Digital Methods of Transmitting Television Information. 13 pp.

CCIR RECMN 656-1-92. Interfaces for Digital Component Video Signals in 525-Line and 625-Line Television Systems Operating at the 4:2:2 Level of Recommendation 601—Section 11F—Digital Methods of Transmitting Television Information. 16 pp.

CCIR Report 1088-1-90. Interfaces for Digital Video Signals in 525-Line and 625-Line Television Systems—Section 11F—Digital Methods of Transmitting Television Information. 7 pp.

CCIR RECMN 799-92. Interfaces for Digital Component Video Signals in 525-Line and 625-Line Television Systems Operating at the 4:4:4 Level of Recommendation 601. 17 pp.

—High Definition
CCIR QUESTION 47/11-90. Standards for Digital High Definition Television—Questions Concerning Study Group 11—Broadcasting Service (Television). 1 p.

Digital Television (Cont.)

—Impairments—Measurement
CCIR Report 1212-90. Measurements and Test Signals for Digitally Encoded Colour Television Signals—Section 11F—Digital Methods of Transmitting Television Information. 15 pp.

—Layered Architectures—Open Systems Interconnection
CCIR Report 1223-90. Layered Model Approach for Digital Television—Section 11F—Digital Methods of Transmitting Television Information. 6 pp.

CCIR QUESTION 46/11-90. Application of a Layered Model to Digital Television Chains—Questions Concerning Study Group 11—Broadcasting Service (Television). 1 p.

—Studios—Sync Signals
CCIR Report 1219-90. Synchronizing Signals for the Component Digital Studio—Section 11F—Digital Methods of Transmitting Television Information. 3 pp.

—Television Studios—Synchronizing Reference Signals
CCIR RECMN 711-90. Synchronizing Reference Signals for the Component Digital Studio —Section 11F—Digital Methods of Transmitting Television Information. 2 pp.

CCIR RECMN 711-1-92. Synchronizing Reference Signals for the Component Digital Studio —Section 11F—Digital Methods of Transmitting Television Information. 2 pp.

Digital Timers
See Also: Timers

—Programmable
CECC CECC 90 104-218 ISSUE 1-86. CEI-CECC 90 104-218; Silicon with (B) Buffered Outputs Programmable Timer (En). 2 pp.

Digital Timers/Frequency Dividers
Use: Frequency Divider/Digital Timers

Digital-To-Acoustic Transmission Systems
Use: Voice/Data Systems

Digital to Analog Converters
Use For: D/A Converters; DAC *See Also:* Analog to Digital Converters; Converters; Integrated Circuits; Plotters; Voltage Converters

—Integrated Circuit—CMOS
MOD UK DSTAN 59-62: 90/175-01. Integrated Circuit Issue 1; Amendment 1. 25 pp.

—Integrated Circuit—Preferred Products List
CECC CECC MUAHAG Vol 7 IS 8-92. Preferred Products List; Active Microcircuits (En, Fe, Ge). 89 pp.

—Integrated Circuit—12-Bit—Microprocessor Compatible—PPL
CECC CECC MUAHAG Vol 7 IS 8-92. Preferred Products List; Active Microcircuits (En, Fe, Ge). 89 pp.

—Integrated Circuit—12-Bit—Preferred Products List
CECC CECC MUAHAG Vol 7 IS 8-92. Preferred Products List; Active Microcircuits (En, Fe, Ge). 89 pp.

—Integrated Circuit—16-Bit—Preferred Products List
CECC CECC MUAHAG Vol 7 IS 8-92. Preferred Products List; Active Microcircuits (En, Fe, Ge). 89 pp.

—Integrated Circuit—8-Bit—Microprocessor Compatible—PPL
CECC CECC MUAHAG Vol 7 IS 8-92. Preferred Products List; Active Microcircuits (En, Fe, Ge). 89 pp.

—Integrated Circuit—8-Bit—Preferred Products List
CECC CECC MUAHAG Vol 7 IS 8-92. Preferred Products List; Active Microcircuits (En, Fe, Ge). 89 pp.

Digital Transmission
See Also: Digital Transmission Systems

CCIR Report 996-86. Mode of Operation and Modulation Rate in Relation to Circuit Length and Multipath Characteristics—Section 3Cd—Performance of Digital Transmission Systems. 5 pp.

INDUSTRY STANDARDS

Digital Transmission (Cont.)

—Acoustic Signals
CCIR Report 1236-90. Digital Transmission of Medium-and High-Quality Sound Signals at Low Bit Rates—Section CMTT C—Transmission Standards and Performance Objectives for Sound-Programme Channels. 8 pp.
CCIR QUESTION 51-2/10-90. Standards for Digital Techniques for Sound in Broadcasting—Questions Concerning Study Group 10—Broadcasting Service (Sound). 1 p.

—Acoustic Signals—Communication Channels
CCIR RECMN 724-90. Transmission of Digital Studio Quality Sound Signals over H1 Channels—Section CMTT C—Transmission Standards and Performance Objectives for Sound-Programme Channels. 9 pp.

—Fixed Satellite Services
CCIR Report 1139-90. General System and Performance Aspects of Digital Transmission in the Fixed-Satellite Service—Section 4B1—Systems Aspects. 19 pp.
CCIR QUESTION 52/4-90. Characteristics for International Digital Transmission Links in the Fixed-Satellite Service—Questions of Study Group 4 Fixed-Satellite Service. 1 p.

—Fixed Satellite Services—Error Performance
CCIR Report 1139-90. General System and Performance Aspects of Digital Transmission in the Fixed-Satellite Service—Section 4B1—Systems Aspects. 19 pp.

—Fixed Satellite Services—Hypothetical Reference Digital Path
CCIR RECMN 521-2-86. Hypothetical Reference Digital Path for Systems Using Digital Transmission in the Fixed-Satellite Service—Section 4B2—Performance and Availability. 2 pp.

—Fixed Satellite Services—Radio Frequency Interference
CCIR Report 867-2-90. Maximum Permissible Interference in Single-Channel-per-Carrier and Intermediate Rate Digital Transmissions in Networks of the Fixed-Satellite Service—Section 4D1—Permissible Levels of Interference. 23 pp.

—Mobile Satellite Communications—Multipath Propagation
CCIR Report 762-2-82. Effects of Multipath on Digital Transmission over Links in the Maritime Mobile-Satellite Service—Section 8I—Technical and Operating Characteristics of Mobile Satellite Services. 16 pp.

—Modulation Rate
CCIR Report 996-86. Mode of Operation and Modulation Rate in Relation to Circuit Length and Multipath Characteristics—Section 3Cd—Performance of Digital Transmission Systems. 5 pp.

—Pulse Code Modulation Channels
CCITT RECMN G.712-92. Transmission Performance Characteristics of Pulse Code Modulation (Study Group XV) 33 pp (Replaces G.713, G.714, and G.715). 33 pp.

—Radio Circuits—Error Control
CCIR Report 435-70. Error Statistics and Error Control in Digital Transmission Over Operating Radio Circuits—Section 3Cd—Performance of Digital Transmission Systems. 4 pp.

—Radio Circuits—Error Statistics
CCIR Report 435-70. Error Statistics and Error Control in Digital Transmission Over Operating Radio Circuits—Section 3Cd—Performance of Digital Transmission Systems. 4 pp.

—Simulcast
CCIR Report 1022-1-90. Multi-Transmitter Radio Systems Using Quasi-Synchronous (Simulcast) Transmission in the Land Mobile Service—Section 8A—Land Mobile Service and Related Subjects. 18 pp.

—Sound
CCIR Decision 18-6-89. Digital Systems for the Transmission of Sound-Programme and Television Signals—Volume X-1—Broadcasting Service (Sound). 5 pp.

—Sound Broadcasting
CCIR QUESTION 51-2/10-90. Standards for Digital Techniques for Sound in Broadcasting—Questions Concerning Study Group 10—Broadcasting Service (Sound). 1 p.

—Sound Broadcasting—Broadcast Receivers
CCIR RECMN 774-92. Digital Sound Broadcasting to Vehicular, Portable and Fixed Receivers Using Terrestrial Transmitters in the VHF/UHF Bands—Section 10B—Frequency-Modulation Sound Broadcasting in Bands 8 (VHF) and 9 (UHF). 7 pp.
CCIR RECMN 789-92. Digital Sound Broadcasting to Vehicular, Portable and Fixed Receivers for BSS (Sound) in the Frequency Range 500-3 000 MHz. 7 pp.

—Sound—Hypothetical Reference Circuits
CCIR QUESTION 60/CMTT-90. Definition of a Digital Hypothetical Reference Connection for Digital Sound-Programme Transmission—Questions Concerning the CMTT CCIR/CCITT Joint Study Group for Television and Sound Transmission. 1 p.

—Sound-Program Signals
CCIR RECMN 659-86. Digital Transmission of Sound Programmes—General Principles—Section CMTT C—Transmission Standards and Performance Objectives for Sound-Programme Channels. 1 p.
CCIR Report 647-4-90. Digital Transmission of Sound-Programme Signals—Section CMTT C—Transmission Standards and Performance Objectives for Sound-Programme Channels. 21 pp.
CCIR Report 1095-1-90. Guidelines Regarding Transmission of Digital Sound-Programme Signals of Digital Studio Quality—Section CMTT C—Transmission Standards and Performance Objectives for Sound-Programme Channels. 3 pp.
CCIR Decision 18-6-89. Digital Systems for the Transmission of Sound-Programme and Television Signals—Annex to Volume XII—Television and Sound Transmission (CMTT). 5 pp.
CCIR QUESTION 52/CMTT-90. Standards for the Digital Transmission of Sound-Programme Signals—Questions Concerning the CMTT CCIR/CCITT Joint Study Group for Television and Sound Transmission. 1 p.
CCIR QUESTION 57/CMTT-90. Transmission of Digital Sound-Programme Signals with Digital Studio Quality (Contribution Circuits)—Questions Concerning the CMTT CCIR/CCITT Joint Study Group for Television and Sound Transmission. 1 p.
CCIR QUESTION 65/CMTT-90. Digital Transmission of Sound-Programme Signals Operational Characteristics—Questions Concerning the CMTT CCIR/CCITT Joint Study Group for Television and Sound Transmission. 1 p.

—Sound-Program Signals—Broadcasting Satellite Services—Feeder Link
CCIR QUESTION 59/CMTT-90. Transmission of Digital Sound-Programme Signals over Broadcasting Satellite Feeder Links—Questions Concerning the CMTT CCIR/CCITT Joint Study Group for Television and Sound Transmission. 1 p.

—Sound-Program Signals—Communication Channels
CCIR RECMN 718-90. Digital Transmission of High-Quality Sound-Programme Signals on Distribution Circuits Using 480 kbit/s (496 kbit/s) per Audio Channel—Section CMTT C—Transmission Standards and Performance Objectives for Sound-Programme Channels. 9 pp.
CCIR Decision 77-1-89. Transmission of Digital Sound Programme Signals of Digital Studio Quality on Circuits Using the H1 Channel—Annex to Volume XII—Television and Sound Transmission (CMTT). 3 pp.
CCIR QUESTION 58/CMTT-90. Standards for the Digital Transmission of Sound-Programme Signals on One, Two or Three 64 kbit/s Channels—Questions Concerning the CMTT CCIR/CCITT Joint Study Group for Television and Sound Transmission. 2 pp.

—Sound-Program Signals—Integrated Services Digital Networks
CCIR QUESTION 56/CMTT-90. Specific Requirements for the Integrated Services Digital Network (ISDN) Carrying Sound-Programme Services—Questions Concerning the CMTT CCIR/CCITT Joint Study Group for Television and Sound Transmission. 1 p.

—Sound Program Signals—Long Distances
CCIR QUESTION 18-2/CMTT-86. Standards for Digital Transmission of Sound-Programme Signals over Long Distances—Questions Concerning the CMTT CCIR/CCITT Joint Study Group for Television and Sound Transmission. 1 p.

—Sound-Program Signals—Monitoring
CCIR QUESTION 64/CMTT-90. Digital Transmission of Sound-Programme Signals Methods of Monitoring and Measurement—Questions Concerning the CMTT CCIR/CCITT Joint Study Group for Television and Sound Transmission. 1 p.

—Sound-Program Signals—Sampling Frequencies
CCIR RECMN 606-1-90. Sampling Frequency to be Used for the Digital Transmission of High-Quality Sound-Programme Signals—Section CMTT C—Transmission Standards and Performance Objectives for Sound-Programme Channels. 1 p.

—Television
CCIR Decision 18-6-89. Digital Systems for the Transmission of Sound-Programme and Television Signals—Volume X-1—Broadcasting Service (Sound). 5 pp.

—Television—B-ISDN
CCIR Report 1241-90. Liaison Between the CMTT and Other Study Groups of the CCITT on Matters Related to Digital Transmission—Section CMTT A—Television Transmission Standards and Performance Objectives. 17 pp.

—Television Signals
CCIR Report 646-4-90. Digital or Mixed Analogue-and-Digital Transmission of Television Signals—Section CMTT A—Television Transmission Standards and Performance Objectives. 21 pp.
CCIR Decision 18-6-89. Digital Systems for the Transmission of Sound-Programme and Television Signals—Annex to Volume XII—Television and Sound Transmission (CMTT). 5 pp.
CCIR Decision 67-2-89. Digital Transmission of Component-Coded Television and High Definition Television Signals—Annex to Volume XII—Television and Sound Transmission (CMTT). 3 pp.
CCIR QUESTION 34/CMTT-90. Digital Transmission of Television Signals—Questions Concerning the CMTT CCIR/CCITT Joint Study Group for Television and Sound Transmission. 1 p.
CCIR QUESTION 37/CMTT-90. Performance of Long-Distance Digital Circuits Carrying Broadcasting Television Signals—Questions Concerning the CMTT CCIR/CCITT Joint Study Group for Television and Sound Transmission. 1 p.
CCIR QUESTION 47/CMTT-90. Measurement Methods for Television Transmission Systems Using Digital or Analogue and Digital Modulation—Questions Concerning the CMTT CCIR/CCITT Joint Study Group for Television and Sound Transmission. 1 p.
CCIR RECMN 798-92. Digital Television Terrestrial Broadcasting in the VHF/UHF Bands. 24 pp.

—Television Signals—Availability
CCIR QUESTION 49/CMTT-90. Availability Objectives Applicable to Television Transmission Systems Using Digital Modulation—Questions Concerning the CMTT CCIR/CCITT Joint Study Group for Television and Sound Transmission. 1 p.

—Television Signals—Bit Rates—140 Mbit/s
CCIR RECMN 721-90. Transmission of Component-Coded Digital Television Signals for Contribution-Quality Applications at Bit Rates Near 140 Mbit/s—Section CMTT A—Television Transmission Standards and Performance Objectives. 12 pp.
CCIR RECMN 721-1-92. Transmission of Component-Coded Digital Television Signals for Contribution-Quality Applications at Bit Rates Near 140 Mbit/s—Section CMTT A—Television Transmission Standards and Performance Objectives. 14 pp.
CCIR Report 1234-90. Digital Transmission of Component-Coded Television Signals at Bit Rates near 68 Mbit/s and 140 Mbit/s—Section CMTT A—Television Transmission Standards and Performance Objectives. 12 pp.

—Television Signals—Bit Rates—30-34 Mbit/s
CCIR Report 1235-90. Digital Transmission of Component-Coded Television Signals at 30-34 Mbit/s and 45 Mbit/s—Section CMTT A—Television Transmission Standards and Performance Objectives. 43 pp.

—Television Signals—Bit Rates—30-45 Mbit/s
CCIR RECMN 723-90. Transmission of Component-Coded Digital Television Signals for Contribution-Quality Applications at the Third Hierarchical Level of CCITT Recommendation G.702—Section CMTT A—Television Transmission Standards and Performance Objectives. 14 pp.
CENELEC PRETS 300 174-91. Network Aspects Digital Coding of Component Television Signals for Contribution Quality Applications in the Range 34-45 Mbit/s. 84 pp.
CENELEC ETS 300 174-92. Network Aspects (NA); Digital Coding of Component Television Signals for Contribution Quality Applications in the Range 34-45 Mbit/s. 83 pp.

Digital Transmission (Cont.)

—Television Signals—Bit Rates—30-45 Mbit/s (Cont.)

ETSI PRETS 300 174-91. Network Aspects Digital Coding of Component Television Signals for Contribution Quality Applications in the Range 34-45 Mbit/s.

ETSI ETS 300 174-92. Network Aspects (NA); Digital Coding of Component Television Signals for Contribution Quality Applications in the Range 34-45 Mbit/s. 83 pp.

ETSI PRETS 300 174-91. Network Aspects Digital Coding of Component Television Signals for Contribution Quality Applications in the Range 34-45 Mbit/s. 84 pp.

—Television Signals—Bit Rates—45 Mbit/s

CCIR Report 1235-90. Digital Transmission of Component-Coded Television Signals at 30-34 Mbit/s and 45 Mbit/s—Section CMTT A—Television Transmission Standards and Performance Objectives. 43 pp.

—Television Signals—Bit Rates—68 Mbit/s

CCIR Report 1234-90. Digital Transmission of Component-Coded Television Signals at Bit Rates near 68 Mbit/s and 140 Mbit/s—Section CMTT A—Television Transmission Standards and Performance Objectives. 12 pp.

—Television Signals—Contribution Networks

CCIR RECMN 800-92. User Requirements for the Transmission Through Contribution and Primary Distribution Networks of Digital Television Signals Defined According to the 4:2:2 Standard of Recommendation 601. 5 pp.

—Television Signals—High Definition Television Systems

CCIR Decision 67-2-89. Digital Transmission of Component-Coded Television and High Definition Television Signals—Annex to Volume XII—Television and Sound Transmission (CMTT). 3 pp.

CCIR QUESTION 72/CMTT-90. Digital Transmission of High Definition Television (HDTV) Signals—Questions Concerning the CMTT CCIR/CCITT Joint Study Group for Television and Sound Transmission. 1 p.

—Television Signals—Impairments

CCIR QUESTION 36/CMTT-90. Laws of Addition for Impairments Associated with All-Digital and Mixed Analogue-and-Digital Transmission of Television Signals—Questions Concerning the CMTT CCIR/CCITT Joint Study Group for Television and Sound Transmission. 1 p.

—Television Signals—Impairments—Protection

CCIR Report 967-2-90. Digital Television: Transmission Impairments and Methods of Protection—Section CMTT A—Television Transmission Standards and Performance Objectives. 8 pp.

—Television Signals—Long Distances

CCIR RECMN 604-2-90. Digital Television Transmission over Long Distances—General Principles—Section CMTT A—Television Transmission Standards and Performance Objectives. 1 p.

CCIR RECMN 658-1-90. Mixed Analogue-and-Digital Transmission of Analogue Composite Television Signals over Long Distances—Section CMTT A—Television Transmission Standards and Performance Objectives. 3 pp.

—Television Signals—Primary Distribution Networks

CCIR RECMN 800-92. User Requirements for the Transmission Through Contribution and Primary Distribution Networks of Digital Television Signals Defined According to the 4:2:2 Standard of Recommendation 601. 5 pp.

—Television Signals—Quality Assurance

CCIR QUESTION 35/CMTT-90. Quality Parameters for Television Transmission Systems Using Digital or Analogue-and-Digital Modulation—Questions Concerning the CMTT CCIR/CCITT Joint Study Group for Television and Sound Transmission. 1 p.

—Television Signals—Secondary Distribution

CCIR Report 1239-90. Standards for Digital Secondary Distribution Systems—Section CMTT A—Television Transmission Standards and Performance Objectives. 4 pp.

Digital Transmission (Cont.)

—Television Signals—Time Division Multiplexing

CCIR QUESTION 71/CMTT-90. Digital Transmission of Signals Comprising a Time-Division Multiplex Television Programme—Questions Concerning the CMTT CCIR/CCITT Joint Study Group for Television and Sound Transmission. 1 p.

Digital Transmission Links

Use: Data Links

Digital Transmission Networks

Use: Digital Transmission Systems

Digital Transmission Systems

Use For: Digital Transmission Networks
See Also: Communication Networks; Data Links; Digital Transmission; Phase Shift Keying; Pulse Code Modulation Systems; Telecommunication Equipment; Transmission Systems

CCIR Report 996-86. Mode of Operation and Modulation Rate in Relation to Circuit Length and Multipath Characteristics—Section 3Cd—Performance of Digital Transmission Systems. 5 pp.

—Attenuation Distortion

CCITT RECMN G.622-89. Characteristics of 1.2/4.4 mm Coaxial Cable Pairs—Transmission Media Characteristics (Study Group XV) 8 pp. 8 pp.

CCITT RECMN G.623-89. Characteristics of 2.6/9.5 mm Coaxial Cable Pairs—Transmission Media Characteristics (Study Group XV) 8 pp. 8 pp.

—Bit Error

CCITT RECMN O.151-89. Error Performance Measuring Equipment for Digital Systems at the Primary Bit Rate and Above—Specifications for Measuring Equipment (Study Group IV) 5 pp. 5 pp.

—Broadcasting Studios—Interfaces

CCIR Report 1069-86. Digital Audio Interface for the Interconnection of Digital Broadcasting Studios with Digital Transmission Networks—Section 10C—Audio-Frequency Characteristics of Sound-Broadcasting Signals. 2 pp.

—Data Terminal Equipment

CCITT FASCICLE III.4-89. General Aspects of Digital Transmission Systems Terminal Equipments—Recommendations G.700-G.795. 621 pp.

CCITT RECMN G.700-89. Framework of the Series G.700, G.800 and G.900 Recommendations—General Aspects of Digital Transmission Systems; Terminal Equipments (Study Groups XV and XVIII) 4 pp. 4 pp.

—Digital Line Sections

CCITT FASCICLE III.5-89. Digital Networks, Digital Sections and Digital Line Systems—Recommendations G.801-G.961. 292 pp.

—Hypothetical Reference Model

CCITT RECMN G.801-89. Digital Transmission Models—Digital Networks, Digital Sections and Digital Line Systems (Study Groups XV and XVIII) 6 pp. 6 pp.

—In Service Code Violation Monitors

CCITT RECMN O.161-89. In-Service Code Violation Monitors for Digital Systems—Specifications for Measuring Equipment (Study Group IV) 4 pp. 4 pp.

—Land Mobile Services

CCIR Report 903-2-90. Digital Transmission in the Land Mobile Service—Section 8A—Land Mobile Service and Related Subjects. 34 pp.

CCIR Report 1021-86. Equipment Characteristics for Digital Transmission in the Land Mobile Services—Section 8A—Land Mobile Service and Related Subjects. 4 pp.

CCIR QUESTION 40-2/8-86. Digital Transmission in the Land Mobile Service—Questions Concerning Study Group 8—Mobile, Radiodetermination, Amateur and Related Satellite Services. 1 p.

—Monitoring

CCITT RECMN O.162-89. Equipment to Perform in Service Monitoring on 2048 kBit/s Signals—Specifications for Measuring Equipment (Study Group IV) 6 pp. 6 pp.

CCITT RECMN O.163-89. Equipment to Perform In-Service Monitoring on 1544 kBit/s Signals—Specifications for Measuring Equipment (Study Group IV) 5 pp. 5 pp.

Digital Transmission Systems (Cont.)

—Pulse Code Modulation—Multiplexing—Glossaries

CCITT RECMN G.701-89. Vocabulary of Digital Transmission and Multiplexing, and Pulse Code Modulation (PCM) Terms—General Aspects of Digital Transmission Systems; Terminal Equipments (Study Groups XV and XVIII) 34 pp. 34 pp.

—Radiotelephone Circuits

CCIR Report 1127-90. Digital Transmission Systems for HF Radiotelephone Circuits—Section 3B—Radiotelephony. 14 pp.

—Set Up

CCITT RECMN M.2110-92. Bringing into Service International Digital Paths, Sections and Transmission Systems (Study Group IV) 9 pp. 9 pp.

—Sound Broadcasting

CCIR OPINION 51-74. Study of Digital Techniques by CCIR Study Groups and the CMTT—Volume X-1—Broadcasting Service (Sound). 1 p.

—Sound Broadcasting—Quality Assurance—Subjective Assessment

CCIR Report 799-2-86. Subjective Assessment of Quality of Sound in Broadcasting Using Digital Techniques—Section 10C—Audio-Frequency Characteristics of Sound-Broadcasting Signals. 3 pp.

CCIR QUESTION 85/10-90. Subjective Assessment of Sound Quality in Broadcasting Using Digital Techniques—Questions Concerning Study Group 10—Broadcasting Service (Sound). 1 p.

—Symbols

CNS C1143-5-89. Graphical Symbols for Use on Electrical and Electronic Equipment (Digital Transmission) (Feb)(12491-5).

—Telephone Cables

CCITT RECMN G.612-89. Characteristics of Symmetric Cable Pairs Designed for the Transmission of Systems with Bit Rates of the Order of 6 to 34 Mbit/s—Transmission Media Characteristics (Study Group XV) 4 pp. 4 pp.

CCITT RECMN G.613-89. Characteristics of Symmetric Cable Pairs Usable Wholly for the Transmission of Digital Systems with a Bit Rate of up to 2 Mbits—Transmission Media Characteristics (Study Group XV) 4 pp. 4 pp.

CCITT RECMN G.614-89. Characteristics of Symmetric Pair Star-Quad Cables Designed Earlier for Analogue Transmission Systems and Being Used Now for Digital System Transmission at Bit Rates of 6 to 34 Mbit/s—Transmission Media Characteristics (Study Group XV) 5 pp. 5 pp.

CCITT RECMN G.621-89. Characteristics of 0.7/2.9 mm Coaxial Cable Pairs—Transmission Media Characteristics (Study Group XV) 4 pp. 4 pp.

CCITT RECMN G.622-89. Characteristics of 1.2/4.4 mm Coaxial Cable Pairs—Transmission Media Characteristics (Study Group XV) 8 pp. 8 pp.

—Telephone Cables—Integrated Services Digital Networks

CCITT RECMN G.961-89. Digital Transmission System on Metallic Local Lines for ISDN Basic Rate Access—Digital Networks, Digital Sections and Digital Line Systems (Study Groups XV and XVIII) 119 pp. 119 pp.

—Timing Jitter

CCITT RECMN O.171-89. Timing Jitter Measuring Equipment for Digital Systems—Specifications for Measuring Equipment (Study Group IV) 10 pp. 10 pp.

Digitonin

See Also: Reagents

CNS K7187-65. Chemical Reagent (Digitonin) (Jan)(1687).

JIS K 8452-78. Digitonin.

Digits (Telecommunications)

See Also: Country Codes; Discriminating Digits; International Prefixes; Language Digits

—Limitation—International Numbers (Telephone)

CCITT RECMN E.163-89. Numbering Plan for the International Telephone Service—Telephone Network and ISDN—Operation, Numbering, Routing and Mobile Service (Study Group II) 8 pp. 8 pp.

—Limitation—National Significant Numbers

CCITT RECMN E.163-89. Numbering Plan for the International Telephone Service—Telephone Network and ISDN—Operation, Numbering, Routing and Mobile Service (Study Group II) 8 pp. 8 pp.

INTERNATIONAL AND NON-U.S. NATIONAL STANDARDS
SUBJECT INDEX

Digits (Telecommunications) (Cont.)
—Registers (Switching)—Capacity
CCITT RECMN E.163-89. Numbering Plan for the International Telephone Service—Telephone Network and ISDN—Operation, Numbering, Routing and Mobile Service (Study Group II) 8 pp. 8 pp.

o,p-Dihydroxyazo-p-Nitrobenzene
See Also: Benzene
CNS K7295-68. Chemical Reagent (O,p-Dihydroxyazo-p-Nitrobenzene) (Nov)(1794).

1,2-Dihydroxybenzene-3,5-disulfonic Acid Disodium Salt
See Also: Alcohols
JIS K 9566-91. Tiron (1,2-Dihydroxybenzene-3,5-disulfonic Acid Disodium Salt.

p-Dihydroxybenzene
Use: Hydroquinone

Diiodofluorescein
CNS K7188-65. Chemical Reagent (Diiodofluorescein) (Jan)(1688).

Diisocyanate Content Analysis
—Polyurethane Resins
DIN ENGL 55956-82. Binders for Paints and Varnishes; Determination of Monomeric Diisocyanates in Isocyanate Resins (Nov). 3 pp.

4,4'-Diisocyanatodiphenylmethane
Use: Diphenylmethane-4,4'-Diisocyanate

2,4-Diisocyanototoluene
Use: Toluene-2,4-Diisocyanate

Dikes
See Also: Dams; Earthwork
—Protection
DIN ENGL 19657-73. Protection of Water Courses, Dikes and Coastal Dunes; Guidelines (Sept). 28 pp.
—Structural Engineering—Construction Contracts
DIN ENGL 18310-88. Tendering and Performance Stipulations in Contracts for Construction Works (VOB); Part C: General Technical Specifications in Contracts for Construction Works (ATV); Revetment Works (Sept) (This Standard, Together with DIN 18299,. 4 pp.

Dilatation
See Also: Deformation; Volume
—Animal Fats
ISO 8293-90. Animal and Vegetable Fats and Oils—Determination of Dilatation First Edition. 10 pp.
—Animal Oils
ISO 8293-90. Animal and Vegetable Fats and Oils—Determination of Dilatation First Edition. 10 pp.
—Vegetable Fats
ISO 8293-90. Animal and Vegetable Fats and Oils—Determination of Dilatation First Edition. 10 pp.
—Vegetable Oils
ISO 8293-90. Animal and Vegetable Fats and Oils—Determination of Dilatation First Edition. 10 pp.

Dilatometry
—Coal
BSI BS 1016: Part 12-80. (WITHDRAWN) 1980 Methods for the Analysis and Testing of Coal and Coke Part 12: Caking and Swelling Properties of Coal (Superseded by BS 1016: Sections 107.1, 107.2: 1991 and 107.3: 1990). 23 pp.
BSI BS 1016: Sec 107.3-90. 1990 Methods for the Analysis and Testing of Coal and Coke Part 107: Caking and Swelling Properties of Coal Section 107.3: Determination of Swelling Properties Using a Dilatometer. 14 pp.
ISO 349-75. Hard Coal—Audibert-Arnu Dilatometer Test First Edition. 13 pp.
ISO 8264-89. Hard Coal—Determination of the Swelling Properties Using a Dilatometer First Edition. 13 pp.
SAA AS 1038.12.3-84. Methods for the Analysis and Testing of Coal and Coke—Part 12.3: Determination of the Dilatometer Characteristics of Higher Rank Coal. 9 pp.
—Fats
BSI BS 684: Sec 1.12-90. 1990 Methods of Analysis of Fats and Fatty Oils Part 1: Physical Methods Section 1.12: Determination of Dilation. 13 pp.

Dilatometry (Cont.)
—Refractory Materials
BSI BS 1902: Sec 4.8-85. (WITHDRAWN) 1985 Methods for Testing Refractory Materials Part 4: Properties Measured Under an Applied Stress Section 4.8: Determination of Refractorines-Under-Load (Rising Temperature Test:Dilatometer Method) (Method 1902-408). 11 pp.

Dilators (Surgical)
See Also: Surgical Instruments
—Uterine
CNS T1059-82. Uterine Cervical Dilator (Jan)(8336).
JIS T 2608-80. Uterine Dilaters.

Diluent Gases
See Also: Gases
—Oxygen Content
JIS K 0225-90. Testing Methods for Determination of Trace Components in Diluent Gas and Zero Gas. 29 pp.

Diluents
See Also: Additives; Dispersants; Exhaust Gases; Paint Thinners; Solvents
SAA AS 1766.5-82. Methods for the Microbiological Examination of Food—Part 5: Preparation of Media, Diluents and Reagents. 20 pp.
—Odors
CNS K6519-80. Method of Test for Odor of Volatile Solvents and Dilutents (Jul)(5836).

Dimefox
See Also: Pesticides
SAA AS 1870.16D-79. Standard for Development—Pesticides for Agricultural Use—Part 16D: Dimefox and Schradan. 16 pp.

Dimensional Changes
Use: Dimensional Stability

Dimensional Codes
See Also: Codes; Dimensions
—Motor Vehicles
ISO 7656-93. Commercial Road Vehicles—Dimensional Codes First Edition. 35 pp.

Dimensional Measurement
Use: Dimensions

Dimensional Measuring Instruments
Scope Note: For additional listings, use a more specific term See Also: Compasses (Drawing); Distance Measuring Equipment; Extensometers; Gage Blocks; Length Measuring Instruments; Measuring Instruments; Measuring Tapes; Micrometers; Plug Gages; Protractors; Ring Gages; Rulers; Scales (Indicators); Straightedges; Toolmakers Flats
—Graduations
BSI BS 4484: Part 1-69. 1969 Measuring Instruments for Constructional Works Part 1: Metric Graduation and Figuring of Instruments for Linear Measurement. 12 pp.
SNZ NZS 6504: Part 1-75. Specification for Measuring Instruments for Constructional Works Part 1: Metric Graduation and Figuring of Instruments for Linear Measurement Amend: A, 1975; B, 1979. 12 pp.
—Inspection Equipment
DIN ENGL 2272 Pt 1-85. Inspection Equipment Used in Dimensional Metrology; Requirements; Auxiliary Measuring Devices (Aug). 5 pp.
—Roundness
BSI BS 3730: Part 2-82. 1982 Assessment of Departures from Roundness Part 2: Characteristics of Stylus Instruments for Measuring Variations in Radius (Including Guidance on Use and Calibration). 19 pp.
JIS B 7451-91. Roundness Measuring Machines. 41 pp.

Dimensional Stability
Use For: Longitudinal Reversion See Also: Creep Properties; Curl (Materials); Shrinkage; Stability Testing
—Bitumens
CEN PREN 1107-93. Bitumen Sheets for Waterproofing—Determination of Dimensional Stability at Elevated Temperature. 11 pp.
DIN ENGL 1996 Pt 17-90. Testing of Asphalt; Determination of Dimensional Stability When Heated (Nussel Deformation Index) (Nov). 2 pp.

Dimensional Stability (Cont.)
—Bonded Fabrics
CGSB CAN/CGSB-4.2 NO.66-M91. Textile Test Methods Dimensional Change and Appearance After Dry Cleaning of Coated, Bonded, Laminated and Fused Fabrics. 9 pp.
—Brake Pads
BSI BS AU 180: Part 4-82. 1982 Brake Linings Part 4: Method for Determining Effects of Heat on Dimensions and Form of Disc Brake Pads. 6 pp.
ISO 6313-80. Road Vehicles—Brake Linings—Effects of Heat on Dimensions and Form of Disc Brake Pads—Test Procedure First Edition. 6 pp.
—Cellular Materials
DIN ENGL 53424-78. Testing of Rigid Cellular Materials; Determination of Dimensional Stability at Elevated Temperatures with Flexural Load and with Compressive Load (Dec). 3 pp.
—Cellular Plastics
BSI BS 4370: Part 1-88. 1988 Methods of Test for Rigid Cellular Materials Part 1: Methods 1-5. 12 pp.
ISO 2796-86. Cellular Plastics, Rigid—Test for Dimensional Stability Third Edition. 5 pp.
—Clothing
BSI BS 4931-86. 1986 Preparation, Marking and Measuring of Textile Fabrics, Garments and Fabric Assemblies in Tests for Assessing Dimensional Change. 8 pp.
—Coated Fabrics
BSI BS 3424: Part 17-87. 1987 Testing Coated Fabrics Part 17: Methods 20. Method for Determination of Dimensional Stability to Water Immersion. 4 pp.
BSI BS 3424: Part 20-87. 1987 Testing Coated Fabrics Part 20: Methods for Determination of Dimensional Changes on Mechanical Relaxation at Zero Tension. 4 pp.
CGSB CAN/CGSB-4.2 NO.66-M91. Textile Test Methods Dimensional Change and Appearance After Dry Cleaning of Coated, Bonded, Laminated and Fused Fabrics. 9 pp.
—Coated Fabrics—Laundering Testing
BSI BS 3424: Part 36-93. 1993 Testing Coated Fabrics Part 36: Method 39. Method for Determination of the Dimensional Stability of Coated Fabrics to Domestic Washing. 7 pp.
—Color Filters
BSI BS 3944: Part 1-92. 1992 Colour and Diffusion Filter Material for Theatre, Television and Similar Entertainment Purposes Part 1: Specification for Flammability and Dimensional Stability. 8 pp.
—Concrete Blocks
CEN PREN 772-14-92. Methods of Test for Masonry Units—Part 14: Determination of Moisture Movement of Aggregate Concrete Masonry Units. 9 pp.
CEN PREN 772-19-93. Methods of Test for Masonry Units—Part 19: Determination of Dimensional Stability of Large Clay Masonry Units. 7 pp.
—Elastic Fabrics
CGSB CAN/CGSB-4.2 NO.56.1-M87. Textile Test Methods Unidirectional Extension and Recovery Properties of Elastic Fabrics. 12 pp.
—Fabrics
BSI BS 2819-90. 1990 Determination of Bow, Skew and Lengthway Distortion in Woven and Knitted Fabrics. 8 pp.
BSI BS 4736-85. 1985 Method for Determination of Dimensional Changes of Fabrics Induced by Cold-Water Immersion. 7 pp.
BSI BS 4931-86. 1986 Preparation, Marking and Measuring of Textile Fabrics, Garments and Fabric Assemblies in Tests for Assessing Dimensional Change. 8 pp.
BSI BS 4961: Part 1-80. 1980 Determination of Dimensional Stability of Textiles to Dry Cleaning in Tetrachloroethylene Part 1: Machine Method. 7 pp.
BSI BS 5807-87. 1987 Determination of Dimensional Change of Textiles in Domestic Washing and Drying. 4 pp.
BSI BS 7552: Part 2-92. 1992 Effect of Dry Heat on Fabrics Under Low Pressure Part 2: Method for Determination of Dimensional Change in Fabrics Exposed to Dry Heat (ISO 9866-2: 1991). 8 pp.
CGSB CAN/CGSB-4.2 NO.24-M91. Textile Test Methods Colourfastness and Dimensional Change in Commercial Laundering. 7 pp.
CGSB CAN/CGSB-4.2 NO.24.1-M88. Textile Test Methods Dimensional Change in Washing of Woven Fabrics—Accelerated Method. 9 pp.
CGSB CAN/CGSB-4.2 NO.24.2-M91. Textile Test Methods Dimensional Change in Commercial Type Laundering of Textiles (Washwheel). 13 pp.
CGSB CAN/CGSB-4.2 NO.25.1-M90. Textile Test Methods Dimensional Change in Wetting. 9 pp.

INTERNATIONAL AND NON-U.S. NATIONAL STANDARDS
SUBJECT INDEX

Dimensional

Dimensional Stability (Cont.)
—**Fabrics** (Cont.)
CGSB CAN/CGSB-4.2 NO.25.2-M89. Textile Test Methods Dimensional Change of Textile Fabrics to Open-Head Steaming. 8 pp.
CGSB CAN/CGSB-4.2 NO.30-M90. Textile Test Methods Dimensional Change in Dry Cleaning. 9 pp.
CGSB CAN/CGSB-4.2 NO.67-M90. Textile Test Methods Dimensional Change and Appearance After Laundering of Coated, Bonded, Laminated and Fused Fabrics. 11 pp.
CNS L3083-80. Method of Test for Dimensional Change of Fabrics Induced by Free-Stream (May)(5614).
ISO 675-79. Textiles—Woven Fabrics—Determination of Dimensional Change on Commercial Laundering Near the Boiling Point First Edition; (Corrigendum 1-1990). 7 pp.
ISO 3175-79. Textiles—Determination of Dimensional Change on Dry Cleaning in Perchlorethylene—Machine Method Second Edition. 6 pp.
ISO 5077-84. Textiles—Determination of Dimensional Change in Washing and Drying First Edition. 5 pp.
ISO 7771-85. Textiles—Determination of Dimensional Changes of Fabrics Induced by Cold-Water Immersion First Edition. 6 pp.
ISO 9866 Pt 1-91. Textiles—Effect of Dry Heat on Fabrics Under Low Pressure—Part 1: Procedure for Dry-Heat Treatment of Fabrics First Edition. 5 pp.
ISO 9866 Pt 2-91. Textiles—Effect of Dry Heat on Fabrics Under Low Pressure—Part 2: Determination of Dimensional Change in Fabrics Exposed to Dry Heat First Edition. 6 pp.
SAA AS 1287.2-72. Methods for the Determination of Dimensional Change in Fabrics—Part 2: Relaxation in Steam Reconfirmed 1989. 7 pp.
SAA AS 1287.4-74. Methods for the Determination of Dimensional Change in Fabrics—Part 4: Dimensional Stability to Heat. 6 pp.
SAA AS 2001.5.1-87. Methods of Test for Textiles—Part 5: Dimensional Change—Part 5.1: General Requirements. 7 pp.
SAA AS 2001.5.3-85. Methods of Test for Textiles—Part 5: Dimensional Change—Part 5.3: Determination of Relaxation in Aqueous Solution. 2 pp.
SAA AS 2001.5.4-87. Methods of Test for Textiles—Part 5: Dimensional Change—Part 5.4: Determination of Dimensional Change in Laundering of Textile Fabrics and Garments—Automatic Machine Method (in Professional Packages 62, 69). 5 pp.
SAA AS 2001.5.5-87. Method of Test for Textiles—Part 5: Dimensional Change—Part 5.5: Determination of Dimensional Change in Laundering of Textile Fabrics and Garments—Cubex Machine Method. 3 pp.
SAA AS 2001.5.7-86. Method of Test for Textiles—Part 5: Dimensional Change—Part 5.7: Determination of Dimensional Change on Dry Cleaning in Perchloroethylene Excluding Finishing—Machine Method. 3 pp.

—**Fishing Nets**
BSI BS 5171-74. 1974 Netting Yarns: Determination of Change in Length After Immersion in Water. 8 pp.
ISO 3090-74. Netting Yarns—Determination of Change in Length After Immersion in Water First Edition. 4 pp.

—**Floor Coverings**
CEN PREN 434-90. Resilient Floor Coverings—Determination of Dimensional Stability and Curling After Exposure to Heat. 6 pp.
CEN PREN 662-92. Resilient Floorcoverings—Determination of Curling on Exposure to Moisture. 5 pp.
DIN ENGL 51962-77. Testing of Organic Floor Coverings (Except Textile Floor Coverings); Determination of the Change in Dimensions After Heat Treatment (May). 2 pp.

—**Floor Coverings—Linoleum**
CEN PREN 669-92. Linoleum Floorcoverings—Determination of the Dimensional Changes of Tiles Caused by Atmospheric Humidity Changes. 5 pp.

—**Floor Coverings—Textile**
BSI BS 4052-87. 1987 Determination of Thickness Loss of Textile Floor Coverings Under Dynamic Loading. 6 pp.
BSI BS 4098-75. 1975 Amd 1 Determination of Thickness, Compression and Recovery Characteristics of Textile Floor Coverings. 11 pp.
BSI BS 4682: Part 2-88. 1988 Amd 1 Methods of Test for the Dimensional Stability of Textile Floor Coverings Part 2: Determination of Dimensional Changes Due to Changes in Ambient Humidity. 4 pp.

Dimensional Stability (Cont.)
—**Floor Coverings—Textile** (Cont.)
BSI BS 4682: Part 3-81. 1981 Methods of Test for the Dimensional Stability of Textile Floor Coverings Part 3: Determination of Dimensional Changes After Exposure to Heat. 4 pp.
BSI BS 4682: Part 4-81. 1981 Methods of Test for the Dimensional Stability of Textile Floor Coverings Part 4: Dimensional Changes After Immersion in Water. 4 pp.
BSI BS 4939-87. 1987 Determination of Thickness Loss of Textile Floor Coverings After Prolonged Heavy Static Loading. 6 pp.
CNS L3094-80. Textile Floor Coverings — Determination of Thickness Loss After Prolonged, Heavy Static Loading (Oct)(6584).
CNS L3095-80. Textile Floor Coverings — Determination of Thickness Loss After Brief, Moderate Static Loading (Oct)(6585).
CNS L3096-80. Textile Floor Coverings — Determination of Thickness Loss Under Dynamic Loading (Oct) (6586).
CNS L3108-81. Dimensional Stability of Textile Floor Coverings — Determination of Extension Under Mechanical Action (Mar) (7059).
ISO 2094-86. Textile Floor Coverings—Determination of Thickness Loss Under Dynamic Loading Second Edition. 5 pp.
ISO 3415-86. Textile Floor Coverings—Determination of Thickness Loss After Brief, Moderate Static Loading Second Edition. 4 pp.
ISO 3416-86. Textile Floor Coverings—Determination of Thickness Loss After Prolonged, Heavy Static Loading Second Edition. 5 pp.

—**Fused Fabrics**
CGSB CAN/CGSB-4.2 NO.66-M91. Textile Test Methods Dimensional Change and Appearance After Dry Cleaning of Coated, Bonded, Laminated and Fused Fabrics. 9 pp.

—**Hardboards**
ISO TR7469-81. Dimensional Stability of Hardboard First Edition. 5 pp.

—**Laminated Fabrics**
CGSB CAN/CGSB-4.2 NO.66-M91. Textile Test Methods Dimensional Change and Appearance After Dry Cleaning of Coated, Bonded, Laminated and Fused Fabrics. 9 pp.

—**Masonry—Stone**
CEN PREN 772-12-92. Methods of Test for Masonry Units—Part 12: Determination of Length Change During Moisture Movement in Autoclaved Aerated Concrete Masonry Units. 8 pp.

—**Mortars**
CEN PREN 1015-13-93. Methods of Test for Mortar for Masonry—Part 13: Determination of Dimensional Stability of Hardened Mortars. 9 pp.

—**Nylon Braid—Aerospace**
BSI F 132-87. 1987 Amd 1 Braided Nylon Cord with Specified Dimensional Stability for Aerospace Purposes. 12 pp.

—**Paperboard**
BSI BS 5880-90. 1990 Method for Determination of Water Absorption of Paper and Board After Immersion in Water. 8 pp.
BSI BS 6712: Part 1-86. 1986 Measurement of Hygroexpansivity of Paper and Board Part 1: Method for Measurement of Hygroexpansivity up to a Maximum Relative Humidity 68%. 7 pp.
BSI BS 6712: Part 2-91. 1991 Measurement of Hygroexpansivity of Paper and Board Part 2: Method for Measurement of Hygroexpansivity up to a Maximum Relative Humidity of 86 Percent (ISO 8226-2: 1990). 9 pp.
ISO 5637-89. Paper and Board—Determination of Water Absorption After Immersion in Water Second Edition. 6 pp.
ISO 8226 Pt 1-85. Paper and Board—Measurement of Hygroexpansivity—Part 1: Hygroexpansivity up to a Maximum Relative Humidity of 68 % First Edition. 5 pp.
ISO 8226 Pt 2-90. Paper and Board—Measurement of Hygroexpansivity—Part 2: Hygroexpansivity up to a Maximum Relative Humidity of 86 % First Edition. 7 pp.

—**Papers**
BSI BS 5879-80. 1980 Measurement of Dimensional Change of Paper After Immersion in Water. 4 pp.
BSI BS 5880-90. 1990 Method for Determination of Water Absorption of Paper and Board After Immersion in Water. 8 pp.
BSI BS 6712: Part 1-86. 1986 Measurement of Hygroexpansivity of Paper and Board Part 1: Method for Measurement of Hygroexpansivity up to a Maximum Relative Humidity 68%. 7 pp.

Dimensional Stability (Cont.)
—**Papers** (Cont.)
BSI BS 6712: Part 2-91. 1991 Measurement of Hygroexpansivity of Paper and Board Part 2: Method for Measurement of Hygroexpansivity up to a Maximum Relative Humidity of 86 Percent (ISO 8226-2: 1990). 9 pp.
ISO 5635-78. Paper—Measurement of Dimensional Change After Immersion in Water First Edition; (Erratum—Aug 1982). 5 pp.
ISO 5637-89. Paper and Board—Determination of Water Absorption After Immersion in Water Second Edition. 6 pp.
ISO 8226 Pt 1-85. Paper and Board—Measurement of Hygroexpansivity—Part 1: Hygroexpansivity up to a Maximum Relative Humidity of 68 % First Edition. 5 pp.
ISO 8226 Pt 2-90. Paper and Board—Measurement of Hygroexpansivity—Part 2: Hygroexpansivity up to a Maximum Relative Humidity of 86 % First Edition. 7 pp.

—**Photographic Film**
ISO 6221-91. Photography—Films and Papers—Determination of Dimensional Change Second Edition. 16 pp.
JIS K 7620-88. Photography-Photographic Films and Papers—Determination of the Dimensional Change Characteristics (ISO 6221-1980).

—**Photographic Papers**
ISO 6221-91. Photography—Films and Papers—Determination of Dimensional Change Second Edition. 16 pp.
JIS K 7620-88. Photography-Photographic Films and Papers—Determination of the Dimensional Change Characteristics (ISO 6221-1980).

—**Plastic Pipes**
BSI BS 2782:Pt11: METH 1102A-81. 1981 Methods of Testing Plastics Part 11: Thermoplastic Pipes, Fittings and Valves Method 1102A: Longitudinal Reversion of Pipes: Immersion Bath Method. 2 pp.
BSI BS 2782:Pt11: METH 1102B-81. 1981 Methods of Testing Plastics Part 11: Thermoplastic Pipes, Fittings and Valves Method 1102B: Longitudinal Reversion of Pipes: Oven Method. 2 pp.
CNS K6593-80. Polyethylene (PE) Pipes — Determination of Longitudinal Reversion — Liquid Bath Immersion Method (Sep)(6486).
CNS K6594-80. Polyethylene (PP) Pipes Determination of Longitudinal Reversion (Sep)(6487).
CNS K6595-80. Unplasticized Polyvinyl Chloride (PVC) Pipes Determination of Longitudinal Reversion — Liquid Bath Immersion Method (Sep)(6488).
ISO 2505-81. Unplasticized Polyvinyl Chloride (PVC) Pipes—Longitudinal Reversion—Test Methods and Specification Second Edition. 5 pp.
ISO 2506-81. Polyethylene Pipes (PE)—Longitudinal Reversion—Test Methods and Specification Second Edition. 5 pp.
ISO 3478-75. Polypropylene (PP) Pipes—Determination of Longitudinal Reversion First Edition. 4 pp.
ISO 3480-76. Polypropylene (PP) Pipes —Maximum Permissible Longitudinal Reversion First Edition. 3 pp.

—**Plastic Sheets**
BSI BS 2782:Pt6: METH 641A-83. 1983 Amd 1 Methods of Testing Plastics Part 6: Dimensional Properties Method 641A: Determination of Dimensional Stability at 100 Degrees Celsius of Flexible Polyvinyl Chloride Sheet (AMD 6282) February 28, 1991. 5 pp.
DIN ENGL 53377-69. Testing of Plastic Films; Determination of Dimensional Stability (May). 2 pp.

—**Plastics**
BSI BS 4618: Sec 5.2-70. 1970 Recommendations for the Presentation of Plastics Design Data Part 5: Other Properties Section 5.2: Change in Linear Dimensions with Moisture Absorption. 2 pp.

—**Refractory Materials**
BSI BS 1902: Sec 5.9-85. 1985 Methods for Testing Refractory Materials Part 5: Refractory and Thermal Properties Section 5.9: Determination of Permanent Change in Dimensions on Heating of Shaped Insulating Products (Method 1902-509). 8 pp.
BSI BS 1902: Sec 5.10-86. 1986 Methods for Testing Refractory Materials Part 5: Refractory and Thermal Properties Section 5.10: Determination of Permanent Change in Dimensions on Heating of Dense Shaped Products. 10 pp.
CNS R3149-87. Method of Test for the Rate of Linear Change of Castable Refractories (May) (11975).
CNS R3152-87. Method of Test for Permanent Linear Change on Light Weight Castable Refractories (May)(11978).
ISO 2477-87. Shaped Insulating Refractory Products—Determination of Permanent Change in Dimensions on Heating Second Edition. 6 pp.

INDUSTRY STANDARDS

Dimensional Stability (Cont.)
—Refractory Materials (Cont.)
ISO 2478-87. Dense Shaped Refractory Products—Determination of Permanent Change in Dimensions on Heating Second Edition. 7 pp.
JIS R 2554-76. Method of Test for Permanent Linear Change on Castable Refractories.
JIS R 2555-81. Testing Method for the Rate of Linear Change of Castable Refractories.
JIS R 2654-75. Method of Test for Permanent Linear Change on Light Weight Castable Refractories.
SAA AS 1774.13-89. Refractories and Refractory Materials—Physical Test Methods—Part 13: Permanent Dimensional Change. 6 pp.

—Thermal Insulation
DIN ENGL 52271-81. Testing of Mineral Fibre Insulating Materials; Behaviour at Elevated Temperatures (June). 3 pp.

—Thread (Textiles)
SAA AS 2001.5.6-87. Method of Test for Textiles—Part 5: Dimensional Change—Part 5.6: Determination of Dimensional Change of Yarns and Sewing Threads. 2 pp.

—Wood—Shrinkage
ISO 4469-81. Wood—Determination of Radial and Tangential Shrinkage First Edition. 4 pp.
ISO 4858-82. Wood—Determination of Volumetric Shrinkage First Edition. 5 pp.

—Wood—Swelling
ISO 4859-82. Wood—Determination of Radial and Tangential Swelling First Edition. 4 pp.
ISO 4860-82. Wood—Determination of Volumetric Swelling First Edition. 5 pp.

—Wool Fabrics
BSI BS 1955-81. 1981 Determination of Dimensional Changes of Wool-Containing Fabrics During Washing. 10 pp.

—Yarns
SAA AS 2001.5.6-87. Method of Test for Textiles—Part 5: Dimensional Change—Part 5.6: Determination of Dimensional Change of Yarns and Sewing Threads. 2 pp.

Dimensional Stability Testing
Use: Dimensional Stability

Dimensional Tolerances
Use: Tolerances

Dimensionless Numbers
Use For: Dimensionless Parameters; Nondimensional Parameters *See Also:* Heat Transfer Equipment; Mach Number

—Transport Properties—Symbols
BSI BS 5775: Part 12-93. 1993 Quantities, Units and Symbols Part 12: Characteristic Numbers (ISO 31-12: 1992) (G). 14 pp.
BSI BS 5775: Part 12-82. 1982 Amd 1 Specifications for Quantities, Units and Symbols Part 12: Dimensionless Parameters (AMD 5857) July 31, 1989. 11 pp.
CNS Z7196-12-85. Dimensionless Parameters (Sep)(11296-12).
ISO 31 Pt 12-92. Quantities and Units—Part 12: Characteristic Numbers Third Edition. 12 pp.

Dimensionless Parameters
Use: Dimensionless Numbers

Dimensions
Use For: Dimensional Measurement; Linear Dimensions *See Also:* Angular Distance; Coordination Contours; Coordination Distances; Dimensional Codes; Dimensional Stability; Height; Hole Size; Length Measurement; Protection Contours; Units of Measurement; Width Measurement
BSI BS 308: Part 2-85. 1985 Amd 1 Engineering Drawing Practice Part 2: Recommendations for Dimensioning and Tolerancing of Size (AMD 6711) April 30, 1991. 55 pp.
BSI BS 308: Part 2-02. 1985 Amd 2 Engineering Drawing Practice Part 2: Recommendations for Dimensioning and Tolerancing of Size (AMD 7157) September 15, 1992. 64 pp.
BSI BS 5775: Part 1-93. 1993 Quantities, Units and Symbols Part 1: Space and Time (ISO 31-1: 1992) (G). 19 pp.
BSI BS 5775: Part 1-79. 1979 Amd 1 Specification for Quantities, Units and Symbols Part 1: Space and Time (AMD 5846) July 31, 1989. 17 pp.
CNS B1001-1-88. Engineering Drawing Dimensioning (Dec)(3-1).
CNS B1046-87. General Rules for Permissible Deviations in Dimensions Without Tolerance Indication (Sep)(4027).
CNS Z7196-1-85. Quantities and Units of Space and Time (Sep)(11296-1).
CSA CAN/CSA-B78.2-M91. Dimensioning and Tolerancing of Technical Drawings. 147 pp.
CSA B97.1-1970. Standard Tolerances for Linear Dimensions, Inch and Metric (R 1992); (Amd 1-3 January 1982). 18 pp.
DIN ENGL 406 Pt 1-77. Dimensioning in Drawings; Kinds (Apr). 3 pp.
DIN ENGL 406 Pt 2-81. Dimensioning on Drawings; Rules (Aug). 20 pp.
DIN ENGL 406 Pt 3-75. Dimensioning in Drawings; Dimensioning by Co-Ordinates (July). 8 pp.
IEC Guide 103-80. Guide on Dimensional Co-Ordination First Edition. 44 pp.
ISO 31 Pt 1-92. Quantities and Units—Part 1: Space and Time Second Edition. 17 pp.
ISO 129-85. Technical Drawings—Dimensioning—General Principles, Definitions, Methods of Execution and Special Indications First Edition; (Supersedes 2595). 15 pp.
ISO 370-75. Toleranced Dimensions—Conversion from Inches into Millimetres and Vice Versa First Edition. 8 pp.
ISO 10579-93. Technical Drawings—Dimensioning and Tolerancing—Non-Rigid Parts First Edition. 7 pp.
JIS B 0405-91. General Tolerances—Part 1: Tolerances for Linear and Angular Dimensions Without Individual Tolerance Indications (ISO 2768-1:1989). 9 pp.
JIS B 0621-84. Definitions and Designations of Geometrical Deviations. 23 pp.
JIS Z 8317-84. Technical Drawing—Dimensioning. 36 pp.
SNZ NZS/BS 308: Part 2-85. Engineering Drawing Practice Part 2: Recommendations for Dimensioning and Tolerancing of Size (Withdrawn—Superseded by NZS/AS 1100 Parts 101-501).

—Glossaries
BSI BS 6100: SUBSEC 1.5.1-84. 1984 Amd 1 Glossary of Building and Civil Engineering Terms Part 1: General and Miscellaneous Section 1.5: Operations; Associated Plant and Equipment; Sub Section 1.5.1 Coordinat-ion of Dimensions; Tolerances and Accuracy. 9 pp.
BSI BS 6100: SUBSEC 1.5.1-02. 1984 Amd 2 Glossary of Building and Civil Engineering Terms Part 1: General and Miscellaneous Section 1.5: Operations; Associated Plant and Equipment Subsec 1.5.1: Coordination of Dimensions; Tolerances and Accuracy (AMD 7240) August 15, 1992. 10 pp.
DIN ENGL 2257 Pt 1-82. Terminology Used in Dimensional Metrology; Units, Activities, Checking Instruments; Metrological Concepts (Nov). 6 pp.
SAA AS 1233-72. Glossary of Terms for Dimensional Coordination (Bound with AS 1234—1972).

5,5-Dimethyl-1,3-Cyclohexanedione
See Also: Ketones
JIS K 8497-80. 5,5-Dimethyl-1, 3-Cyclohexanedione.

N,N-Dimethyl-p-Phenylenediamine Dihydrochloride
CNS K7591-81. Chemical Reagent (Dimethyl-P-Phenylenediamine Dihydrochloride) (Oct)(8029).
JIS K 8193-87. N,n-Dimethyl-p-Phenylene-Diammonium Dichloride (Dimethyl-p-Phenylenediamine Hydrochloride).

Dimethyl Silicone
—Damping Fluids
MOD UK DSTAN 91-46-79. Damping Fluids: Dimethyl Silicone Issue 1. 11 pp.

—Dielectrics
MOD UK DSTAN 91-46-79. Damping Fluids: Dimethyl Silicone Issue 1. 11 pp.

—Heat Transfer Fluids
MOD UK DSTAN 91-46-79. Damping Fluids: Dimethyl Silicone Issue 1. 11 pp.

—Transducer Fluids
MOD UK DSTAN 91-46-79. Damping Fluids: Dimethyl Silicone Issue 1. 11 pp.

Dimethyl Sulfate
Use For: Methyl Sulfate *See Also:* Esters; Sulfates
CNS K7195-65. Chemical Reagent (Dimethyl Sulfate) (Jan)(1695).

Dimethyl Sulfoxide
JIS K 9702-83. Dimethyl Sulfoxide.

2',4'-Dimethylacetoacetanilide
JIS K 4173-81. Acetoacetanilides (Acetoacetanilide 2-Methylaceto-acetanilide 2',4'-Dimethylaceto-acetanilide.2'-Chloroacetoacetanilide).

p-Dimethylaminoazobenzene
Use For: Methyl Yellow

p-Dimethylaminoazobenzene (Cont.)
CNS K7189-65. Chemical Reagent (P-Dimethylaminoazoben-zene) (Jan)(1689).
JIS K 8494-72. Methyl Yellow (p-Dimethylaminoazobenzene).

p-Dimethylaminobenzaldehyde
See Also: Aldehydes
CNS K7190-65. Chemical Reagent (P-Dimethylaminobenzal-dehyde) (Jan)(1690).
JIS K 8496-85. p-Dimethylamino-benzaldehyde.

p-Dimethylaminobenzalrhodanine
CNS K7191-65. Chemical Reagent (P-Dimethylaminobenzal-rhodanine) (Jan)(1691).

5-(4-Dimethylaminobenzilidene)-2-Thioxo-4-Thiazolidinone
See Also: Ketones
JIS K 8495-81. 5-(4-Dimethylaminoben-zilidene)-2-Thioxo-4-Thiazolidinone.

Dimethylaniline
Use: Xylidine

N,N-Dimethylaniline
CNS K2180-87. N,N-Dimethylaniline (Dec)(12196). 5 pp.
JIS K 4112-81. N-Substituted Anilines (N, N-Dimethylaniline.N, N-Diethylaniline. N-Ethylaniline. N-Methylaniline. N-Benzyl-N-Ethylaniline).
JIS K 8493-80. N,N-Dimethylaniline.

3,3'-Dimethylbenzidine Dihydrochloride
Use For: o-Tolidine Dihydrochloride
JIS K 8669-62. o-Tolidine Dihydrochloride (3,3'-Dimethylbenzidine Dihydrochloride).

3,3'Dimethylbenzidine
Use: o-Tolidine

Dimethylformamide
Use: N,N-Dimethylformamide

N,N-Dimethylformamide
Use For: Dimethylformamide
CNS K7193-65. Chemical Reagent (Dimethylformamide) (Jan)(1693).
JIS K 8500-76. N,N-Dimethylformamide.

Dimethylglyoxime
Use For: 2,3-Butenediondioxime
CNS K7194-65. Chemical Reagent (Dimethylglyoxime) (Jan)(1694).
JIS K 8498-78. Dimethylglyoxime, 2,3-Butenediondioxime.

Dimethylketone
Use: Acetone

Dimetric Projections
Use: Axonometric Projections

Dimetridazole Content Analysis
—Animal Feed
CNS N4122-84. Method of Test for Feed Additives: Determination of Dimetridazole (Nov)(11128).

Dimmers
Use For: Light Dimmers; Variable Control Switches
See Also: Control Circuits; Lamps; Lighting

—Electrical Codes
CSA 1207 Bull. Electrical Bulletin 1207 January 3, 1979 to C22.2 NO 55. 2 pp.
CSA 1207 Bull. Electrical Bulletin 1207 January 3, 1979 to C22.2 NO 184. 2 pp.

—Electronic
DIN VDE 0632 AQ-79. Switches up to 750 V 63 A; Particular Requirements for Electronic Actuating Switches up to 250 V a.c. and 16 A: Amendment Q (Apr). 32 pp.
DIN VDE 0632 AR-79. Switches up to 750 V 63 A; Particular Requirements for Electronic Switches up to 250 V a.c. and 16 A: Amendment R (Apr). 31 pp.

—Headlamps—Automotive
CNS D2033-87. Dimmer Switches for Automobiles (Aug)(5789).

—Instrument Lighting—Ships
CNS F5049-83. Dimmers for Marine Instrument Illumination (Jan)(9858).
JIS F 8852-78. Dimmers for Marine Instrument Illumination.

INTERNATIONAL AND NON-U.S. NATIONAL STANDARDS
SUBJECT INDEX

DIP

Dimmers *(Cont.)*
—Ships
CNS F5048-83. Dimmers for Marine Lamps (Jan)(9857).
JIS F 8851-77. Dimmers for Marine Lamps.

—Tungsten Lamps
BSI BS 5518-77. 1977 Amd 2 Electronic Variable Control Switches (Dimmer Switches) for Tungsten Filament Lighting. 22 pp.

DIN
Use For: Deutsches Institut fuer Normung; German Standards Institute *See Also:* Standards

—Standards Preparation
DIN ENGL 820 Pt 1-86. Standards Work; Principles (Jan). 5 pp.
DIN ENGL 820 Pt 4-86. Standards Work; Working Procedure (Jan). 8 pp.

—Standards Preparation—Glossaries
DIN ENGL 820 Pt 3-75. Standardization Procedure; Definitions (Mar). 16 pp.

DIN Connectors
Use For: DIN 41612 Connectors; Eurocard Connectors *See Also:* Connectors; Printed Circuit Connectors

—Blade Contact
DIN ENGL 41622 Pt 1-65. Multipole Connectors with Blade Contacts 3 x 1 mm; Dimensions (Jan). 6 pp.

DIN 41612 Connectors
Use: DIN Connectors

Dinghies
See Also: Boats (Marine)
BSI BS 3501-62. 1962 Amd 1 Dinghy Buoyancy Equipment. 23 pp.

Dinitolmide Content Analysis
—Animal Feed
CNS N4094-82. Method of Test for Feed Additives: Determination of Zoalene (Apr)(8731).

Dinitrobenzene
See Also: Benzene
CNS K2237-88. Dinitrobenzenes (Sep)(12419).
JIS K 4148-82. Dinitrobenzenes (m-Dinitrobenzene. 2,4-Dinitrotoluene).
JIS K 8482-78. 1,3-Dinitrobenzene.

m-Dinitrobenzene
See Also: Benzene
CNS K7680-83. Chemical Reagent (M-Dinitrobenzene) (May)(10284).

1,3-Dinitrobenzene
Use: Dinitrobenzene

3,5-Dinitrobenzoyl Chloride
CNS K7196-65. Chemical Reagent (3,5-Dinitrobenzoyl Chloride) (Jan)(1696).

2,4-Dinitrochlorobenzene
Use: 1-Chloro-2,4-dinitrobenzene

2,4-Dinitrofluorobenzene
Use: 1-Fluoro-2,4-Dinitrobenzene

2,4-Dinitrophenol
CNS K7198-65. Chemical Reagent (2,4-Dinitrophenol) (Jan)(1698).

2,5-Dinitrophenol
See Also: Phenol
CNS K7675-83. Chemical Reagent (2,5-Dinitrophenol) (May)(10279).

2,4-Dinitrophenylhydrazine
CNS K7199-65. Chemical Reagent (2,4-Dinitrophenyl-hydrazine) (Jan)(1699).
JIS K 8480-83. 2,4-Dinitrophenyl-hydrazine.

N,N-Dinitrosopentamethylenetetramine
CNS K1191-77. N, N-Dinitrosopentamethylenetetramine Blowing Agent (Jun)(4140).
CNS K6403-77. Method of Test for N,N'-Dinitrosopentameth-ylenete-Tramine Blowing Agent (Jun)(4141).

Dinitrotoluene
Use For: DNT *See Also:* Dynamite; Explosives
CNS K1233-80. Dinitrotoluene (Solid) (Apr)(5447).
CNS K6471-80. Method of Test for (Solid) Dinitrotoluene (Apr)(5448).
JIS K 4148-82. Dinitrobenzenes (m-Dintrobenzene. 2,4-Dinitrotoluene).

Dinitrotoluene *(Cont.)*
—Propellants—Delivery Systems
NATO STANAG 4041 ED 1 AMD 4-60. Specification for DNT (Dinitrotoluene) for Propellants. 12 pp.

2,4-Dinitrotoluene
Use: Dinitrotoluene

Dinormal Octyl Phthalate
Use: Di-N-Octyl Phthalate

Dinoseb
—Plants (Botany)—Prohibited Use
EC 90/533/EEC-90. Council Directive Amending the Annex to Directive 79/117/EEC Prohibiting the Placing on the Market and Use of Plant Protection on the Market and Use of Plant Protection Product Containing Certain Active Substances. 1 p.

Diode Array Detectors
Use: Photodetectors

Diode Lasers
See Also: Laser Diodes; Laser Equipment; Lasers
IECQ QC 720100-91. Semiconductor Devices Part 12: Sectional Specification for Optoelectronic Devices (IEC 747-12 ED 1). 44 pp.

Diode Switches
Use For: Diode Switching Arrays *See Also:* Diodes; Electric Switches; Switches
CECC CECC 50 001-014-017. BS CECC 50 001-014-017; Silicon Ambient Rated Switching Diode Array in Dual-in-Line or Flat-Pack Ceramic/Glass Hermetically Sealed Encapsulation (En) AMD 1 (En). 9 pp.
CECC CECC 50 001-018. BS CECC 50 001-018; Silicon Ambient Rated Switching Diode Array in Dual-in-Line Ceramic/ Glass Hermetically Sealed Encapsulation (En) AMD 1 (En). 7 pp.

Diode Switching Arrays
Use: Diode Switches

Diode Transitor Logic Circuits
Use: Digital Circuits—DTL

Diodes
Scope Note: For additional listings, use a more specific term *Use For:* Semiconductor Diodes
See Also: Avalanche Diodes; Avalanche Photodiodes; Avalanche Rectifier Diodes; Coaxial Diodes; Current Reference Diodes; Current Regulator Diodes; Detector Diodes; Diode Switches; Discrete Components; Electron Tubes; Electronic Components; Fast Recovery Rectifiers; Germanium Diodes; Gunn Diodes; Gunn Oscillators; IMPATT Diodes; Infrared Emitting Diodes; Laser Diodes; Light Emitting Diode Arrays; Light Emitting Diodes; Microwave Diodes; Mixer Diodes; NIP Diodes; Photodiodes; PIN Diodes; Power Diodes; Rectifier Diodes; Rectifiers; Reverse Recovery Time; Schottky Barrier Diodes; Schottky Rectifier Diodes; Selenium Rectifiers; Semiconductor Devices; Semiconductors; Signal Diodes; Silicon Diodes; Silicon Rectifiers; Snap Off Diodes; Step Recovery Diodes; Switching Diodes; Transient Suppressor Diodes; Transmit Receive Limiters; Tuning Diodes; Varactor Diodes; Voltage Reference Diodes; Voltage Regulator Diodes; Zener Diodes
CECC CECC 50 001-046. DIN 45 930 Teil 102; Diode BAW 76 (Ge). 6 pp.
MOD UK DSTAN 59-61: Part 2-01. Semiconductor Devices (Listed on EPIC Database) Part 2: Diodes Issue 3; Amendment 3. 130 pp.
MOD UK DSTAN 59-61: 80/043-86. Semiconductor Device Diode Issue 1. 17 pp.
MOD UK DSTAN 59-61: 90/017-71. Semiconductor Device Set Issue 1. 3 pp.
MOD UK DSTAN 59-61: 90/180-77. Semiconductor Device, Diode Issue 1. 18 pp.

—Parameter Measurement—Thermal Environments
CNS C5189-84. Thermal Equilibrium Conditions for Measurement of Diode Static Parameters (Nov)(11106).

—Quality Assurance
BSI BS CECC 50000:Supp1-83. 1983 Discrete Semiconductor Devices: Generic Specification Supplement 1: CECC Assessed Process Average. 7 pp.
BSI BS CECC 50001-81. (WITHDRAWN) 1981 General Purpose Semiconductor Diodes: Blank Detail Specification (Renumbered as BS EN 150001: 1993). 10 pp.

Diodes *(Cont.)*
—Quality Assurance *(Cont.)*
BSI BS EN 150001-93. 1993 Amd 1 Blank Detail Specification: General Purpose Semiconductor Diodes (AMD 7590) February 1993 (T). 14 pp.
CENELEC EN 150 001-91. Blank Detail Specification: General Purpose Semiconductor Diodes. 8 pp.

—Reverse Recovery Time
CNS C5074-80. Measurement of Reverse Recovery Time for Semiconductor Diodes (Oct)(6560).
CNS C5075-80. Characterization of a Reverse Recovery Test Fixture for Semiconductor Diodes (Oct)(6561).

—Transient Response
CNS C6153-82. Forward Transient Measurement on Semiconductor Diodes (Jan)(8383).

Dioxane
Use For: Diethylene Dioxide
CNS K7200-65. Chemical Reagent (Dioxane) (Jan)(1700).
JIS K 8461-76. 1,4-Dioxane (Diethylene Dioxide).

Dioxides
Scope Note: Use a more specific term *See:* Carbon Dioxide; Chlorine Dioxide; Manganese Dioxide; Nitrogen Dioxide; Plutonium Dioxide; Silica; Sulfur Dioxide; Titanium Dioxide

DIP Carrier Sockets
Use: DIP Sockets

DIP Packages
See Also: Electronic Component Packaging; Integrated Circuits; Printed Circuit Boards; Semiconductor/Integrated Circuit Packaging

—Stick Magazines
BSI BS 6062: Part 4-91. 1991 Packaging of Electronic Components for Automatic Handling Part 4: Specification for Stick Magazines for Dual-in-Line Packages (IEC 286-4: 1991). 8 pp.
IEC 286 Pt 4-91. Packaging of Components for Automatic Handling Part 4: Stick Magazines for Dual-in-Line Packages First Editon. 17 pp.

DIP Sockets
Use For: DIP Carrier Sockets; Dual In-Line Package Sockets *See Also:* Sockets (Electric)

—Quality Assurance
BSI BS 9501-90. (OBSOLESCENT) 1990 In-Line and Pin-Grid Array Sockets. 23 pp.

DIP Switches
Use For: Dual In-Line Package Switches; Dual In-Line Packaging Switches *See Also:* Switches
IECQ QC 960500-91. Electromechanical Switches for Use in Electronic Equipment Part 3: Sectional Specification for In-Line Package Switches (IEC 1020-3 ED 1). 34 pp.
IECQ QC 960501-91. Electromechanical Switches for Use in Electronic Equipment Part 3: Sectional Specification for In-Line Package Switches Section 1—Blank Detail Specification (IEC 1020-3-1 ED 1). 26 pp.

—Preferred Products List
CECC CECC MUAHAG Vol 8 IS 4-91. Preferred Products List; Switches (En, Ge, Fr). 29 pp.

—Quality Assurance
BSI BS QC 960500-92. 1992 Electromechanical Switches for Use in Electronic Equipment Sectional Specification for in-Line Package Switches (IEC 1020-3: 1991). 16 pp.
BSI BS QC 960501-92. 1992 Electromechanical Switches for Use in Electronic Equipment Blank Detail Specification for in-Line Package Switches (IEC 1020-3-1: 1991). 10 pp.
IEC 1020 Pt 3-91. Electromechanical Switches for Use in Electronic Equipment Part 3: Sectional Specification for In-Line Package Switches First Edition; (IECQ QC 960500). 34 pp.
IEC 1020 Pt 3-1-91. Electromechanical Switches for Use in Electronic Equipment Part 3: Sectional Specification for In-Line Package Switches Section 1—Blank Detail Specification; First Edition; (IECQ QC 960501). 26 pp.

—Sealed—Preferred Products List
CECC CECC MUAHAG Vol 8 IS 4-91. Preferred Products List; Switches (En, Ge, Fr). 29 pp.

—Soldering Lugs—Sealed—Printed Circuit Mount—PPL
CECC CECC MUAHAG Vol 8 IS 4-91. Preferred Products List; Switches (En, Ge, Fr). 29 pp.

INDUSTRY STANDARDS

Dipcoats
Use: Immersion Coatings

Diphenyl Ketone
Use: Benzophenones

Diphenylamine
Use For: N-Phenylbenzeneamine
CNS K1219-80. Industrial Diphenylamine (N-Phenylbenzeneamine) (Feb)(5220).
CNS K6453-80. Testing Method for Technical Diphenylamine (N-Phenlbenzeneamine) (Feb)(5221).
CNS K7201-66. Chemical Reagent (Diphenylamine) (Jun)(1701).
JIS K 4153-86. Diphenylamine.
JIS K 8487-78. Diphenylamine, N-Phenylbenzeneamine.

—**Propellants—Stability Testing**
MOD UK DSTAN 05-51-01. Stability Test Procedures and Requirements for Propellants Stabilized with Diphenylamine, Ethyl Centralite or Mixtures of Both Issue 1; Amendment 1. 19 pp.
NATO STANAG 4117 ED 2 AMD 4-73. Stability Test Procedures and Requirements for Propellants Stabilized with Diphenylamine, Ethyl Centralite or Mixtures of Both. 44 pp.

Diphenylbenzidine
See Also: Ketones
CNS K7202-66. Chemical Reagent (Diphenylbenzidine) (Jun)(1702).

sym-Diphenylcarbazide
Use For: 1,5-Diphenylcarbohydrazide
CNS K7203-66. Chemical Reagent (S-Diphenylcarbazide) (Jun)(1703).
JIS K 8488-76. 1,5-Diphenylcarbo-hydrazide.

Diphenylcarbazone
CNS K7204-66. Chemical Reagent (Diphenylcarbazone) (Jun)(1704).
JIS K 8489-76. s-Diphenylcarbazone.

1,5-Diphenylcarbohydrazide
Use: sym-Diphenylcarbazide

1,3-Diphenylguanidine
—**Accelerators (Materials)**
CNS K4036-78. Rubber Vulcanization Accelerator DPG (1, 3-Diphenyl Guanidine) (Oct)(4630).

Diphenylmethane Diisocyanate
Use For: Methylene bis-(phenylene)diisocyanate

—**Highway Transportation—Emergency Procedures**
SAA AS 1678.6.1. 016-84. Emergency Procedure Guides—Transport—Part 6.1.016: Diphenylmethane DI-ISO-Cyanate (Methylene Bis-(Phenylene) Diisocyanate) (MDI).

Diphenylmethane-4,4'-Diisocyanate
Use For: 4,4'-Diisocyanatodiphenylmethane; MDI; Methylenebis(Phenylene) Diisocyanate; Methylenebis(4-Phenylene Isocyanate); Methylenebisphenylene Diisocyanate; Methylenediphenyl Isocyanate
See Also: Hazardous Materials
MOD UK TS 10216. 4-4' Di-Isocyanato Diphenyl Methane, Synonyms: Methyl-Bis-4 Phenyl Di-Isocyanate (MDI) Diphenyl Methane Di-Isocyanate.
MOD UK TS 10247A. 4-4' Di-Isocyanato Diphenyl Methane (MDI), Technical Grade.

Diphosphorus Pentoxide
Use: Phosphorus Pentoxide

Dipotassium Hydrogen Phosphate
Use: Potassium Phosphate Dibasic

Dipotassium Phosphate
Use: Potassium Phosphate Dibasic

Dipped Coatings
Use: Immersion Coatings

Dipropylene Glycols
CNS K1230-80. Dipropylene Glycol (Feb)(5236).

Dipsticks
See Also: Level Indicators; Measuring Instruments

—**Aircraft**
SBAC AS 6263 ISSUE 1. Dip Stick. 1 p.

—**Cap Assemblies—Aircraft**
SBAC AS 3104-3105 ISSUE 1. Assembly of Cap and Dip Stick. 1 p.
SBAC AS 6268-6269 ISSUE 1(I). Assembly of Cap and Dipstick. 1 p.

alpha,alpha'-Dipyridyl
Use For: 2,2'-Bipyridine
CNS K7205-66. Chemical Reagent (Alpha, Alpha—Dipyridyl) (Jun)(1705).
JIS K 8486-78. 2,2'-Bipyridyl, 2,2'-Bipyridine.

Diquat
Use For: Diquat Dibromide *See Also:* Pesticides
SAA AS 1870.9D-78. Standard for Development—Pesticides for Agricultural Use—Part 9D: Diquat and Paraquat. 14 pp.

Diquat Dibromide
Use: Diquat

Direct Arc Furnaces
See Also: Electric Arc Furnaces; Furnaces
CENELEC HD 598 S1-92. Test Methods for Direct Arc Furnaces. 6 pp.
IEC 676-80. Test Methods for Direct Arc Furnaces First Edition. 35 pp.

Direct Current Amplifiers
Use: DC Amplifiers

Direct Current Equipment
Use: DC Equipment

Direct Current Generators
Use: DC Generators

Direct Current Motors
Use: DC Motors

Direct Current Power Transmission
Use: Power Transmission—DC

Direct Current Relays
Scope Note: See the subheading DC under specific types of relays

Direct Current Signaling Systems
Use: DC Signaling Systems

Direct Current Signals
Use: DC Signals

Direct Dialing In Services
Use For: DDI *See Also:* Private Automatic Branch Exchanges; Telephone Dialing Services; Telephone Services

—**Integrated Services Digital Networks**
CCITT RECMN E.164-91. Numbering Plan for the ISDN Era (Study Group II) 19 pp (Same as Recmn I.331). 19 pp.
CCITT RECMN E.164-89. Numbering Plan for the ISDN Era—Telephone Network and ISDN—Operation, Numbering, Routing and Mobile Service (Study Group II) 6 pp (Same as Recmn I.331). 6 pp.
CCITT RECMN I.251-89. Number Identification Supplementary Services—Integrated Services Digital Network (ISDN)—General Structure and Service Capabilities (Study Group XVIII) 28 pp. 28 pp.
CCITT RECMN I.251. 1 (REV 1)-92. Direct-Dialling-in (Study Group I) 7 pp. 7 pp.
CENELEC PRETS 300 062-90. Integrated Services Digital Network (ISDN); Direct Dialling in (DDI) Supplementary Service Description (T/NAI(89) 19). 10 pp.
CENELEC PRETS 300 062-91. Integrated Services Digital Network (ISDN); Direct Dialling In (DDI) Supplementary Service Service Description. 14 pp.
CENELEC ETS 300 062-91. Integrated Services Digital Network (ISDN); Direct Dialling in (DDI) Supplementary Service Service Description. 14 pp.
CENELEC PRETS 300 063-90. Integrated Services Digital Network (ISDN); Direct Dialling in (DDI) Supplementary Service Functional Capabilities and Information Flows (T/S 22-27). 11 pp.
CENELEC PRETS 300 063-91. Integrated Services Digital Network (ISDN); Direct Dialling In (DDI) Supplementary Service Functional Capabilities and Information Flows. 14 pp.
CENELEC ETS 300 063-91. Integrated Services Digital Network (ISDN); Direct Dialing in (DDI) Supplementary Service Functional Capabilities and Information Flows. 14 pp.
CENELEC PRETS 300 064-90. Integrated Services Digital Network (ISDN); Direct Dialling in (DDI) Supplementary Service Digital Subscriber Signalling One (DSS1) Protocol (T/S 46-33A). 8 pp.
CENELEC PRETS 300 064-91. Integrated Services Digital Network (ISDN); Direct Dialling in (DDI) Supplementary Service Digital Subscriber Signalling System No. One (DSS1) Protocol. 10 pp.
CENELEC ETS 300 064-91. Integrated Services Digital Network (ISDN); Direct Dialling In (DDI) Supplementary Service Digital Subscriber Signalling System No. One (DSS1) Protocol. 10 pp.

Direct Dialing In Services (Cont.)
—**Integrated Services Digital Networks (Cont.)**
ETSI PRETS 300 062-90. Integrated Services Digital Network (ISDN); Direct Dialling in (DDT) Supplementary Service—Service Description (T/NA 1 (89) 19). 10 pp.
ETSI PRETS 300 063-90. Integrated Services Digital Network (ISDN); Direct Dialling in (DDI) Supplementary Service Functional Capabilities and Information Flows (T/S 22-27). 11 pp.
ETSI PRETS 300 064-90. Integrated Services Digital Network (ISDN) Direct Dialling in (DDI) Supplementary Service Digital Subscriber Signalling One (DSS1) Protocol (T/S 46-33A). 8 pp.
ETSI ETS 300 062-91. Integrated Services Digital Network (ISDN); Direct Dialling in (DDI) Supplementary Service Service Description. 14 pp.
ETSI PRETS 300 062-90. Integrated Services Digital Network (ISDN); Direct Dialling in (DDI) Supplementary Service Description (T/NAI(89) 19). 10 pp.
ETSI ETS 300 063-91. Integrated Services Digital Network (ISDN); Direct Dialing in (DDI) Supplementary Service Functional Capabilities and Information Flows. 14 pp.
ETSI ETS 300 064-91. Integrated Services Digital Network (ISDN); Direct Dialling in (DDI) Supplementary Service Digital Subscriber Signalling System No. One (DSS1) Protocol. 10 pp.
ETSI PRETS 300 064-91. Integrated Services Digital Network (ISDN); Direct Dialling In (DDI) Supplementary Service Digital Subscriber Signalling System No. One (DSS1) Protocol. 10 pp.

—**Integrated Services Digital Networks—CCITT No. 7 Signaling**
CCITT RECMN Q.730-89. ISDN Supplementary Services—Specifications of Signalling System No. 7 (Study Group XI) 59 pp. 59 pp.
CCITT RECMN Q.731-91. Stage 3 Description for Number Identification Supplementary Services Using Signalling System No. 7 Section 1—Direct Dialling in (DDI) Section 8—Sub-Addressing (SUB)—Specifications of Signalling System No. 7 (Study Group XI) 12 pp. 12 pp.

—**Integrated Services Digital Networks—DSS1**
CCITT RECMN Q.951-92. Stage 3 Description for Number Identification Supplementary Services Using DSS 1 Section 1—Direct-Dialling-in (DDI) Section 2—Multiple Subscriber Number (MSN) Section 8—Sub-Addressing (SUB) (Study Group XI) 17 pp. 17 pp.

—**Integrated Services Digital Networks—End of Address**
CCITT RECMN E.164-91. Numbering Plan for the ISDN Era (Study Group II) 19 pp (Same as Recmn I.331). 19 pp.
CCITT RECMN E.164-89. Numbering Plan for the ISDN Era—Telephone Network and ISDN—Operation, Numbering, Routing and Mobile Service (Study Group II) 6 pp (Same as Recmn I.331). 6 pp.

—**Integrated Services Digital Networks—Functions/Information Flows**
CCITT RECMN Q.81-91. Stage 2 Description for Number Identification Supplementary Services (Study Group XI) 46 pp. 46 pp.
CCITT RECMN Q.81-89. Number Identification Supplementary Services—General Recommendations on Telephone Switching and Signalling—Functions and Information Flows for Services in the ISDN—Supplements (Study Group XI) 30 pp. 30 pp.

Direct Inward Dialing Services
See Also: Telephone Dialing Services; Telephone Services

—**Private Automatic Branch Exchanges**
CEPT T/CS 20-22-83. Selection Directe D'un Poste Supplementaire. 2 pp.
CEPT T/CS 20-22 E-83. Direct Dialling-in (DDI). 1 p.

Direct Printers
Use: Telegraph Printers

Direct Relations
Use: Telephone Relations—Direct

Direct Stress Fatigue Testing
Use: Fatigue Testing

Direct Tension Indicators
Use: Tension Indicators (Fasteners)

INTERNATIONAL AND NON-U.S. NATIONAL STANDARDS
SUBJECT INDEX

Direct View Storage Tubes
Use For: Display Storage Tubes; Image Storage Tubes *See Also:* Cathode Ray Storage Tubes; Cathode Ray Tubes
CECC EN 111 100-91. Sectional Specification: Display Storage Tubes (Supersedes CECC 11 100 Issue 1: 1984). 16 pp.

—Glossaries
CNS C5065-80. Display Storage Tube Nomenclature (Aug))(6313).

—Quality Assurance
BSI BS CECC 11101-84. 1984 Display Storage Tubes: Blank Detail Specification. 14 pp.
BSI BS EN 111100-92. 1992 Amd 1 Sectional Specification: Display Storage Tubes (AMD 7601) February 15, 1993 (T). 21 pp.
CENELEC EN 111100-91. Sectional Specification: Display Storage Tubes. 16 pp.

Direct Voltage Signals
Use For: Analog Direct Voltage Signals; Binary Direct Voltage Signals

—Process Control Systems
BSI BS 5863: Part 2-80. 1980 Analogue Signals for Process Control Systems Part 2: Direct Voltage Signals. 7 pp.
CENELEC HD 557 S1-90. Binary Direct Voltage Signals for Process Measurement and Control Systems. 3 pp.
IEC 381 Pt 2-78. Analogue Signals for Process Control Systems Part 2: Direct Voltage Signals First Edition. 12 pp.
IEC 946-88. Binary Direct Voltage Signals for Process Measurement and Control Systems First Edition. 17 pp.

Direction Finders
Use For: Radio Compasses; Radio Direction Finders *See Also:* Automatic Direction Finders; Homing Systems; Navigational Aids; Radio Navigation Equipment; Radio Receivers
MOD UK DSTAN 58-5-78. Military Characteristics for Ground and Surface UHF Direction Finding Equipment Issue 1. 4 pp.

—Distress Frequency—Ships
CCIR RECMN 428-3-90. Direction-Finding and/or Homing in the 2 MHz Band on Board Ships—Section 8C—Maritime Mobile Service; Telephony and Related Subjects. 3 pp.

—Global Maritime Distress and Safety Systems
CCIR Report 744-2-86. Use of Class J3E Emissions for Distress and Safety Purposes—Section 8C—Maritime Mobile Service; Telephony and Related Subjects. 7 pp.

—Maritime—Radio Frequency Interference
CCIR Report 913-2-90. Technical Characteristics of Maritime Radiobeacons—Section 8D—Radiodetermination, Global Maritime Distress and Safety System and Related Subjects. 16 pp.

—Monitoring—Radio Spectra
CCIR Report 372-6-90. Direction Finding at Monitoring Stations—Section 1C—Spectrum Monitoring Techniques. 18 pp.
CCIR RECMN 854-92. Direction Finding at Monitoring Stations of Signals Below 30 MHz. 1 p.

—Symbols
CNS C1143-7-89. Graphical Symbols for Use on Electrical and Electronic Equipment (Radar Control Desk, Radio Direction Finder) (Feb)(12491-7).

Direction Indicator Lamps
Use: Signal Lights

Direction of Movement
Scope Note: For additional listings, see specific items or equipment *See Also:* Orientation
JIS Z 8907-87. Geometrical Orientation and Directions of Movements.

—Controls—Machine Tools
CNS B7124-79. Direction for Operation of Machine Tool Controls (Dec)(5062).
ISO 447-84. Machine Tools—Direction of Operation of Controls Second Edition. 4 pp.
JIS B 6011-67. Direction of Operation of Machine Tool Controls.

—Ergonomics
DIN ENGL 33417-87. Description of Position, Orientation and Direction of Movement of Objects (Aug). 4 pp.

—Glossaries
ISO 1503-77. Geometrical Orientation and Directions of Movements First Edition. 37 pp.

Direction of Movement *(Cont.)*
—Machine Tools—Symbols
CNS B1135-80. Symbols for Indication Appearing on Machine Tools—Symbols of Moving Direction of Moving Parts (Jan)(5092). 6 pp.

Directional Antennas
See Also: Antennas
CCIR RECMN 162-2-70. Use of Directional Antennas in the Bands 4 to 28 MHz—Section 3Ab—Antennas Characteristics. 3 pp.

—Broadcasting
CCIR Report 356-3-90. Use of Directional Antennas in the Bands 4 to 28 MHz—Section 3Ab—Antennas Characteristics. 11 pp.

—Cellular Mobile Radio Equipment
CCIR Report 1155-90. Adaptation of Mobile Radiocommunication Technology to the Needs of Developing Countries—Section 8A—Land Mobile Service and Related Subjects. 15 pp.

—Fixed Services (Radio Communications)
CCIR Report 356-3-90. Use of Directional Antennas in the Bands 4 to 28 MHz—Section 3Ab—Antennas Characteristics. 11 pp.

—Glossaries
CCIR RECMN 162-2-70. Use of Directional Antennas in the Bands 4 to 28 MHz—Section 3Ab—Antennas Characteristics. 3 pp.

—HF
CCIR RECMN 162-2-70. Use of Directional Antennas in the Bands 4 to 28 MHz—Section 3Ab—Antennas Characteristics. 3 pp.
CCIR Report 356-3-90. Use of Directional Antennas in the Bands 4 to 28 MHz—Section 3Ab—Antennas Characteristics. 11 pp.

—HF—Fixed Services (Radio Communications)
CCIR RECMN 162-3-92. Use of Directional Transmitting Antennas in the Fixed Service Operating in Bands Below About 30 MHz—Section 3Ab—Antenna Characteristics. 12 pp.
CCIR QUESTION 150/9-90. Use of Directional Antennas in the Bands 4 to 27.5 MHz Limitation of Radiation Outside the Direction Necessary for the Service—Questions of Study Group 9 Fixed Service. 1 p.

—MF/HF—Fixed Services (Radio Communications)
CCIR QUESTION 151/9-90. Use of Directional Antennas in the Bands Below 30 MHz in the Fixed Service—Questions of Study Group 9 Fixed Service. 1 p.

—MF—Mobile Radio Equipment
CCIR Report 1030-86. Use of Directional Antennas in the MF Band Allocated to the Maritime Mobile Service to Improve Spectrum Efficiency—Section 8B—Maritime Mobile Service: Telegraphy and Related Subjects. 3 pp.

—Mobile Radio Equipment
CCIR Report 1155-90. Adaptation of Mobile Radiocommunication Technology to the Needs of Developing Countries—Section 8A—Land Mobile Service and Related Subjects. 15 pp.

Directional Control Valves
Use For: Five Way Valves; Four Way Valves; Three Way Valves; Two Way Valves *See Also:* Bypass Valves; Control Valves; Hydraulic Valves; Measuring Instruments; Pneumatic Valves; Proportional Valves; Valve Subplates; Valves

—Aircraft—Check—Safety
SBAC AGS 2099 ISSUE 1. Method of Assembling Safety Guard and Securing Ring on Non Return Valve.

—Aircraft—Rings
SBAC AGS 2087 ISSUE 1. Non Return Valve Safety Guard Securing Ring.

—Electrohydraulic
JIS B 8657-89. Test Methods for Electro-Hydraulic Proportional Directional Bypass Flow Control Valves. 36 pp.

—Hydraulic—Clamping
BSI BS 6494: Part 4-89. 1989 Hydraulic Fluid Power Valve Mounting Surfaces Part 4: Clamping Dimensions of Four-Port Sizes 03 and 05, Modular Stack Valves and Directional Control Valves. 6 pp.
ISO 7790-86. Hydraulic Fluid Power—Four-Port Modular Stack Valves and Four-Port Directional Control Valves, Sizes 03 and 05—Clamping Dimensions First Edition. 4 pp.

Directional Control Valves *(Cont.)*
—Hydraulic—Mounting Surfaces
BSI BS 6494: Part 1-84. 1984 Hydraulic Fluid Power Valve Mounting surfaces Part 1: Four-Port Directional Control Valves. 22 pp.
ISO 4401-80. Hydraulic Fluid Power—Four-Port Directional Control Valves—Mounting Surfaces First Edition. 10 pp.

—Pneumatic—Mounting Surfaces
BSI BS 7389: Part 1-90. 1990 Pneumatic Fluid Power Valve Mounting Surfaces Part 1: Five-Port Directional Control Valves (Without Electrical Connector). 14 pp.
ISO 5599 Pt 1-89. Pneumatic Fluid Power—Five-Port Directional Control Valves—Part 1: Mounting Interface Surfaces Without Electrical Connector Second Edition. 11 pp.
ISO 5599 Pt 2-90. Pneumatic Fluid Power—Five-Port Directional Control Valves—Part 2: Mounting Interface Surfaces with Optional Electrical Connector First Edition. 15 pp.

—Pneumatic—Mounting Surfaces—Codes
ISO 5599 Pt 3-90. Pneumatic Fluid Power—Five-Port Directional Control Valves—Part 3: Code System for Communication of Valve Functions First Edition. 13 pp.

—Rolling Stock
JIS E 4105-91. Three-Way Cocks for Railway Rolling Stock.

Directional Gyros
Use: Gyrocompasses

Directional Gyroscopes
Use: Gyrocompasses

Directional Relays
See Also: Measuring Relays; Relays
BSI BS 142: Sec 4.2-84. 1984 Electrical Protection Relays Part 4: Requirements for Multi-Input Energizing Quantity Relays Section 4.2: Directional Relays and Power Relays. 12 pp.
IEC 255 Pt 12-80. Electrical Relays Part 12: Directional Relays and Power Relays with Two Input Energizing Quantities First Edition. 29 pp.

—Ground—Sets
JIS C 4609-90. Directional Ground Relay Set for 6.6 kV Consumer. 24 pp.

Directivity (Electromagnetic)
—Antennas—Television Broadcasting
CCIR RECMN 419-2-90. Directivity of Antennas in the Reception of Television Broadcasting—Section 11E—Planning of Television Networks, Protection Ratios Television Receivers and Antennas. 1 p.
CCIR RECMN 419-3-92. Directivity and Polarization Discrimination of Antennas in the Reception of Television Broadcasting—Section 11E—Planning of Television Networks, Protection Ratios Television Receivers and Antennas. 3 pp.

Directories
See Also: Indexes (Documentation); Telephone Directories
ISO 2146-88. Documentation—Directories of Libraries, Archives, Information and Documentation Centres, and Their Data Bases Second Edition. 28 pp.

—CSA—Data Elements
CGSB CAN/CGSB-200.9-91. Trade Data Elements Directory (UNTDED 1990). 326 pp.

—CSA—Electrical Equipment—Certification
CSA Volume I. List of Certified Electrical Equipment Volume I Electrical Service Equipment, Wiring Hardware and Controls. 1127 pp.
CSA Volume II. List of Certified Electrical Equipment Volume II Electrically-Equipped Consumer, Commercial and Industrial Products. 1527 pp.
CSA Volume IV. List of Certified Electrical Equipment Volume IV Electrical Components and Accessories. 1087 pp.

—Document Formats—Telex Call Numbers
CCITT RECMN F.60-89. Operational Provisions for the International Telex Service—Telegraph and Mobile Services Operations and Quality of Service (Study Group I) 17 pp. 17 pp.

—European Communities
EURO 1988 11TH EDITION. Directory of Community Legislation in Force and Other Acts of the Community Institutions—Volume 11. 165 pp.
EURO 1989 13TH EDITION. Directory of Community Legislation in Force and Other Acts of the Community Institutions-Volume 1. 813 pp.
EURO 1989 14TH ED VOLUME 1. Directory of Community Legislation in Force and Other Acts of the Community Institutions-Volume 1. 752 pp.

INDUSTRY STANDARDS

INTERNATIONAL AND NON-U.S. NATIONAL STANDARDS
SUBJECT INDEX

Directories

Directories (Cont.)
—European Communities (Cont.)
EURO 1989 13TH ED. Directory of Community Legislation in Force and Other Acts of the Community Institutions-Volume II. 192 pp.
EURO 1989 14TH ED VOLUME II. Directory of Community Legislation in Force and Other Acts of the Community Institutions—Volume II. 178 pp.
EURO 1990-1 Volume I. Directory of Community Legislation in Force and Other Acts of the Community Institutions-Volume I. 761 pp.
EURO 1991-1 Volume I. Directory of Community Legislation in Force and Other Acts of the Community Institutions Volume I. 853 pp.
EURO 1992-1 Volume I. Directory of Community Legislation in Force and Other Acts of the Community Institutions Volume I. 847 pp.
EURO 1992-1 Volume I. Directory of Community Legislation in Force and Other Acts of the Community Institutions Volume I. 847 pp.
EURO 1993-1 Volume I. Directory of Community Legislation in Force and Other Acts of the Community Institutions Volume I Analytical Register. 859 pp.
EURO 1990-2 Volume II. Directory of Community Legislation in Force and Other Acts of the Community Institutions-Volume II. 181 pp.
EURO 1991-2 Volume II. Directory of Community Legislation in Force and Other Acts of the Community Institutions Volume II. 198 pp.
EURO 1992-2 Volume II. Directory of Community Legislation in Force and Other Acts of the Community Institutions Volume II. 195 pp.
EURO 1992-2 Volume II. Directory of Community Legislation in Force and Other Acts of the Community Institutions Volume II. 197 pp.
EURO 1993-2 Volume II. Directory of Community Legislation in Force and Other Acts of the Community Institutions Volume II Chronological Index Alphabetical Index. 198 pp.

—FAX Communications
CCITT RECMN F.180-89. General Operational Provisions for the International Public Facsimile Service Between Subscribers' Stations (Telefax)—Telematic, Data Transmission and Teleconference Services —Operations and Quality of Service (Study Group I) 5 pp. 5 pp.

—FAX Communications—Accounting
CCITT RECMN F.180-89. General Operational Provisions for the International Public Facsimile Service Between Subscribers' Stations (Telefax)—Telematic, Data Transmission and Teleconference Services —Operations and Quality of Service (Study Group I) 5 pp. 5 pp.

—IEC
IEC. IEC Directory 1993. 320 pp.

—INMARSAT Mobile Numbers
CCITT RECMN E.215-89. Telephone/ISDN Numbering Plan for the Mobile-Satellite Services of INMARSAT—Telephone Network and ISDN—Operation, Numbering, Routing and Mobile Service (Study Group II)14 pp. 14 pp.
CCITT RECMN F.125-89. Telex Numbering Plan for the Mobile-Satellite Services of INMARSAT—Telegraph and Mobile Services Operations and Quality of Service (Study Group I) 9 pp. 9 pp.

—International Sound-Program Centers
CCITT RECMN D.180-89. Occasional Provision of Circuits for International Sound-and Television-Programme Transmissions—General Tariff Principles—Charging and Accounting in International Telecommunications Services (Study Group III) 13 pp. 13 pp.

—International Television Program Centers
CCITT RECMN D.180-89. Occasional Provision of Circuits for International Sound-and Television-Programme Transmissions—General Tariff Principles—Charging and Accounting in International Telecommunications Services (Study Group OIII) 13 pp. 13 pp.

—Program Booking Centers
CCITT RECMN D.180-89. Occasional Provision of Circuits for International Sound-and Television-Programme Transmissions—General Tariff Principles—Charging and Accounting in International Telecommunications Services (Study Group III) 13 pp. 13 pp.

—Public Data Networks
CEPT T/SF 37 E-85. Contents Od Directories for Public Data Networks. 3 pp.

—Suppliers—Rubber—Aircraft
SBAC TS 101 ISSUE 2. Register of Rubber Suppliers.

Directories (Cont.)
—Teletex Communications
CCITT RECMN F.220-89. Service Requirements Unique to the Processable Mode Number One (PM1) Used Within the Teletex Service—Telematic, Data Transmission and Teleconference Services—Operations and Quality of Service (Study Group I) 6 pp. 6 pp.
CCITT RECMN F.230-89. Service Requirements Unique to the Mixed Mode (MM) Used Within the Teletex Service—Telematic, Data Transmission and Teleconference Services—Operations and Quality of Service (Study Group I) 5 pp. 5 pp.

—Telex Call Numbers
CCITT RECMN F.60-92. Operational Provisions for the International Telex Service (Study Group I) 34 pp. 34 pp.
CCITT RECMN F.60-89. Operational Provisions for the International Telex Service—Telegraph and Mobile Services Operations and Quality of Service (Study Group I) 17 pp. 17 pp.

Directors (Aircraft)
Use: Automatic Control Equipment—Aircraft

Directory Assistance
See Also: Directory Services; Information Services; Telephone Directories; Telephone Services

—Freephone Services
CCITT RECMN E.152-89. International Freephone Service (IFS)—Telephone Network and ISDN—Operation, Numbering, Routing and Mobile Service (Study Group II) 12 pp. 12 pp.

—Operators'—Alphanumeric Character Sets
CCITT RECMN E.115-91. Computerized Information Service for Telephone Subscriber Numbers in Foreign Countries (Directory Assistance), Reserved for Operators (Study Group I) 19 pp. 19 pp.
CCITT RECMN E.115-89. Computerized Information Service for Telephone Subscriber Numbers in Foreign Countries (Directory Assistance), Reserved for Operators—Telephone Network and ISDN—Operation, Numbering, Routing and Mobile Service (Study Group II) 11 pp. 11 pp.

—Operators'—Computerized
CCITT RECMN E.115-91. Computerized Information Service for Telephone Subscriber Numbers in Foreign Countries (Directory Assistance), Reserved for Operators (Study Group I) 19 pp. 19 pp.
CCITT RECMN E.115-89. Computerized Information Service for Telephone Subscriber Numbers in Foreign Countries (Directory Assistance), Reserved for Operators—Telephone Network and ISDN—Operation, Numbering, Routing and Mobile Service (Study Group II) 11 pp. 11 pp.

—Operators'—Message Formats
CCITT RECMN E.115-91. Computerized Information Service for Telephone Subscriber Numbers in Foreign Countries (Directory Assistance), Reserved for Operators (Study Group I) 19 pp. 19 pp.
CCITT RECMN E.115-89. Computerized Information Service for Telephone Subscriber Numbers in Foreign Countries (Directory Assistance), Reserved for Operators—Telephone Network and ISDN—Operation, Numbering, Routing and Mobile Service (Study Group II) 11 pp. 11 pp.

—Operators'—Tariffs
CCITT RECMN E.115-91. Computerized Information Service for Telephone Subscriber Numbers in Foreign Countries (Directory Assistance), Reserved for Operators (Study Group I) 19 pp. 19 pp.
CCITT RECMN E.115-89. Computerized Information Service for Telephone Subscriber Numbers in Foreign Countries (Directory Assistance), Reserved for Operators—Telephone Network and ISDN—Operation, Numbering, Routing and Mobile Service (Study Group II) 11 pp. 11 pp.

Directory Databases
Use For: International Directory Databases
See Also: Directory Entries; Directory Systems
CCITT RECMN F.500-92. International Public Directory Services (Study Group I) 39 pp. 39 pp.
CCITT RECMN F.500-89. International Public Directory Services—Message Handling Services—Operations and Definition of Service (Study Group I) 32 pp. 32 pp.

—Bearer Services—Circuit Switched Public Data Networks
CCITT RECMN E.115-91. Computerized Information Service for Telephone Subscriber Numbers in Foreign Countries (Directory Assistance), Reserved for Operators (Study Group I) 19 pp. 19 pp.

Directory Databases (Cont.)
—Bearer Services—Circuit Switched Public Data Networks (Cont.)
CCITT RECMN E.115-89. Computerized Information Service for Telephone Subscriber Numbers in Foreign Countries (Directory Assistance), Reserved for Operators—Telephone Network and ISDN—Operation, Numbering, Routing and Mobile Service (Study Group II) 11 pp. 11 pp.

—Bearer Services—Packet Switched Public Data Networks
CCITT RECMN E.115-91. Computerized Information Service for Telephone Subscriber Numbers in Foreign Countries (Directory Assistance), Reserved for Operators (Study Group I) 19 pp. 19 pp.
CCITT RECMN E.115-89. Computerized Information Service for Telephone Subscriber Numbers in Foreign Countries (Directory Assistance), Reserved for Operators—Telephone Network and ISDN—Operation, Numbering, Routing and Mobile Service (Study Group II) 11 pp. 11 pp.

—Bearer Services—Public Switched Telephone Networks
CCITT RECMN E.115-91. Computerized Information Service for Telephone Subscriber Numbers in Foreign Countries (Directory Assistance), Reserved for Operators (Study Group I) 19 pp. 19 pp.
CCITT RECMN E.115-89. Computerized Information Service for Telephone Subscriber Numbers in Foreign Countries (Directory Assistance), Reserved for Operators—Telephone Network and ISDN—Operation, Numbering, Routing and Mobile Service (Study Group II) 11 pp. 11 pp.

—Distributed Processing Systems
CCITT RECMN X.518-89. The Directory—Procedures for Distributed Operation—Data Communication Networks Directory (Study Group VII) 59 pp. 59 pp.

—Maintenance
CCITT RECMN F.500-92. International Public Directory Services (Study Group I) 39 pp. 39 pp.
CCITT RECMN F.500-89. International Public Directory Services—Message Handling Services—Operations and Definition of Service (Study Group I) 32 pp. 32 pp.

Directory Entries
See Also: Directory Databases; Directory Services; Directory Systems; Record Layouts
CCITT RECMN F.500-92. International Public Directory Services (Study Group I) 39 pp. 39 pp.
CCITT RECMN F.500-89. International Public Directory Services—Message Handling Services—Operations and Definition of Service (Study Group I) 32 pp. 32 pp.

—Attributes
CCITT RECMN F.500-92. International Public Directory Services (Study Group I) 39 pp. 39 pp.
CCITT RECMN F.500-89. International Public Directory Services—Message Handling Services—Operations and Definition of Service (Study Group I) 32 pp. 32 pp.

Directory Information Base
Use: Directory Databases

Directory Management Domain
See Also: Directory System Agents; Directory Systems; Directory User Agents
CCITT RECMN F.500-92. International Public Directory Services (Study Group I) 39 pp. 39 pp.
CCITT RECMN F.500-89. International Public Directory Services—Message Handling Services—Operations and Definition of Service (Study Group I) 32 pp. 32 pp.

Directory Services
Use For: Public Directory Services
See Also: Application Services (OSI); Directory Assistance; Directory Entries; Directory System Agents; Directory Systems; Directory User Agents; Open Systems Interconnection; Telecommunication Services; Telephone Services
BSI BS ISO/IEC 9594-1-90. 1990 Information Technology—Open Systems Interconnection—the Directory—Part 1: Overview of Concepts, Models, and Services. 18 pp.
BSI BS ISO/IEC 9594-2-90. 1990 Information Technology—Open Systems Interconnection—the Directory—Part 2: Models (S). 30 pp.
BSI BS ISO/IEC 9594-2-01. 1990 Amd 1 Information Technology—Open Systems Interconnection—the Directory—Part 2: Models (AMD 7146) June 15, 1992. 33 pp.

Directory Services *(Cont.)*

BSI BS ISO/IEC 9594-2-02. 1990 Amd 2 Information Technology—Open Systems Interconnection—the Directory—Part 2: Models (AMD 7471) November 15, 1992 (Technical Corr 2). 36 pp.
BSI BS ISO/IEC 9594-3-90. 1990 Information Technology—Open Systems Interconnection—the Directory—Part 3: Abstract Service Definition. 33 pp.
BSI BS ISO/IEC 9594-3-01. 1990 Amd 1 Information Technology—Open Systems Interconnection—the Directory—Part 3: Abstract Service Definition (AMD 7147) July 15, 1992 (S). 36 pp.
BSI BS ISO/IEC 9594-3-02. 1990 Amd 2 Information Technology—Open Systems Interconnection—the Directory—Part 3: Abstract Service Definition (AMD 7472) November 15, 1992 (Technical Corr 2) (S). 39 pp.
BSI BS ISO/IEC 9594-3-03. 1990 Amd 3 Information Technology—Open Systems Interconnection—the Directory—Part 3: Abstract Service Definition (AMD 7622) February 15, 1993 (Technical Corr 3). 42 pp.
BSI BS ISO/IEC 9594-3-04. 1990 Amd 4 Information Technology—Open Systems Interconnection—the Directory—Part 3: Abstract Service Definition (AMD 7803) August 15, 1993 (Technical Corr 4) (S). 46 pp.
BSI BS ISO/IEC 9594-4-90. 1990 Information Technology—Open Systems Interconnection—The Directory—Part 4: Procedures for Distributed Operation. 55 pp.
BSI BS ISO/IEC 9594-4-01. 1990 Amd 1 Information Technology—Open Systems Interconnection—the Directory—Part 4: Procedures for Distributed Operation (AMD 7148) June 15, 1992 (S). 60 pp.
BSI BS ISO/IEC 9594-4-02. 1990 Amd 2 Information Technology—Open Systems Interconnection—the Directory—Part 4: Procedures for Distributed Operation (AMD 7473) November 15, 1992 (Technical Corr 2). 62 pp.
BSI BS ISO/IEC 9594-4-03. 1990 Amd 3 Information Technology—Open Systems Interconnection—the Directory—Part 4: Procedures for Distributed Operation (AMD 7804) August 15, 1993 (Technical Corr 3) (S). 67 pp.
BSI BS ISO/IEC 9594-5-90. 1990 Information Technology—Open Systems Interconnection—the Directory—Part 5: Protocol Specifications. 19 pp.
BSI BS ISO/IEC 9594-5-01. 1990 Amd 1 Information Technology—Open Systems Interconnection—the Directory—Part 5: Protocol Specifications (AMD 7623) February 15, 1993 (Technical Corr 1). 23 pp.
BSI BS ISO/IEC 9594-6-90. 1990 Information Technology—Open Systems Interconnection—the Directory—Part 6: Selected Attribute Types. 25 pp.
BSI BS ISO/IEC 9594-7-90. 1990 Information Technology—Open Systems Interconnection—the Directory—Part 7: Selected Object Classes. 19 pp.
BSI BS ISO/IEC 9594-7-01. 1990 Amd 1 Information Technology—Open Systems Interconnection—the Directory—Part 7: Selected Object Classes (AMD 7149) June 15, 1992 (S). 21 pp.
BSI BS ISO/IEC 9594-7-02. 1990 Amd 2 Information Technology—Open Systems Interconnection—the Directory—Part 7: Selected Object Classes (AMD 7474) November 15, 1992 (Technical Corr 2). 24 pp.
BSI BS ISO/IEC 9594-8-90. 1990 Information Technology—Open Systems Interconnection—the Directory—Part 8: Authentication Framework. 33 pp.
BSI BS ISO/IEC 9594-8-01. 1990 Amd 1 Information Technology—Open Systems Interconnection—the Directory—Part 8: Authentication Framework (AMD 7151) July 15, 1992. 36 pp.
BSI DD ENV 41210-91. 1991 Information Systems Interconnection—Directory Access. 25 pp.
BSI DD ENV 41212-92. 1992 Information Systems Interconnection—Directory System Protocol (S). 28 pp.
BSI DD ENV 41215-93. 1993 Information Technology—Functional Standard for Profile A/DI32—Directory—Behaviour of DSAs for Distributed Operations (S). 34 pp.
BSI DD ENV 41512-91. 1991 Information Systems Interconnection—Common Directory Use. 25 pp.
CCITT FASCICLE II.6-88. Message Handling and Directory Services—Operations and Definition of Service Recommendations F.400—F.422, F.500. 184 pp.
CCITT RECMN F.500-92. International Public Directory Services (Study Group I) 39 pp. 39 pp.
CCITT RECMN F.500-89. International Public Directory Services—Message Handling Services—Operations and Definition of Service (Study Group I) 32 pp. 32 pp.
CEN ENV 41210-91. Information Systems Interconnection Directory Access. 23 pp.
CEN ENV 41212-92. Information Systems Interconnection—Directory System Protocol. 26 pp.
CEN ENV 41215-92. Information Technology—Functional Standard for Profile A/DI32—Directory—Behaviour of DSAs for Distributed Operations. 31 pp.

Directory Services *(Cont.)*

CEN ENV 41512-91. Information Systems Interconnection Common Directory Use. 23 pp.
CSA CAN/CSA-Z243. 177.1-91. Information Technology—Open Systems Interconnection—the Directory—Part 1: Overview of Concepts, Models and Services (ISO/IEC 9594-1:1990); (Gen Instr 1). 26 pp.
CSA CAN/CSA-Z243. 177.2-91. Information Technology—Open Systems Interconnection—the Directory—Part 2: Models (ISO/IEC 9594-2:1990); (Gen Instr 1). 38 pp.
CSA CAN/CSA-Z243. 177.3-91. Information Technology—Open Systems Interconnection—the Directory—Part 3: Abstract Service Definition (ISO/IEC 9594-3:1990); (Gen Instr 1). 42 pp.
CSA CAN/CSA-Z243. 177.4-91. Information Technology—Open Systems Interconnection—the Directory—Part 4: Procedures for Distributed Operation (ISO/IEC 9594-4:1990); (Gen Instr 1). 62 pp.
CSA CAN/CSA-Z243. 177.5-91. Information Technology—Open Systems Interconnection—the Directory—Part 5: Protocol Specifications (ISO/IEC 9594-5:1990); (Gen Instr 1). 27 pp.
CSA CAN/CSA-Z243. 177.6-91. Information Technology—Open Systems Interconnection—the Directory—Part 6: Selected Attribute Types (ISO/IEC 9594-6:1990); (Gen Instr 1). 34 pp.
CSA CAN/CSA-Z243. 177.7-91. Information Technology—Open Systems Interconnection—the Directory—Part 7: Selected Object Classes (ISO/IEC 9594-7:1990); (Gen Instr 1). 26 pp.
CSA CAN/CSA-Z243. 177.8-91. Information Technology—Open Systems Interconnection—the Directory—Part 8: Authentication Framework (ISO/IEC 9594-8:1990); (Gen Instr 1). 41 pp.
IEC DIS 7811 Pt 3-92. Identification Cards—Recording Technique—Part 3: Location of Embossed Characters on ID-1 Cards (Revision of First Edition (ISO 7811-3:1985)) ***CD-ROM ONLY***. 8 pp.
IEC 9594 Pt 1-90. Information Technology—Open Systems Interconnection—The Directory—Part 1: Overview of Concepts, Models and Services First Edition. 16 pp.
IEC 9594 Pt 1 Draft AMD 1. Information Technology—Open Systems Interconnection—The Directory—Part 1: Overview of Concepts, Models and Services Amendment 1: Replication, Schema and Access Control; (1992) ***CD-ROM ONLY***. 7 pp.
IEC 9594 Pt 2-90. Information Technology—Open Systems Interconnection—The Directory—Part 2: Models First Edition; (Corrigendum 1-1991) (Corrigendum 2-1992). 31 pp.
IEC 9594 Pt 2 Draft AMD 1. Information Technology—Open Systems Interconnection—The Directory—Part 2: Models Amendment 1: Access Control; (1992) ***CD-ROM ONLY***. 42 pp.
IEC 9594 Pt 2 Draft AMD 2. Information Technology—Open Systems Interconnection—The Directory—Part 2: Models Amendment 2: Schema; (1992) ***CD-ROM ONLY***. 77 pp.
IEC 9594 Pt 2 Draft AMD 3. Information Technology—Open Systems Interconnection—The Directory—Part 2: Models Amendment 3: Replication; (1992) ***CD-ROM ONLY***. 58 pp.
IEC 9594 Pt 3-90. Information Technology—Open Systems Interconnection—The Directory—Part 3: Abstract Service Definition First Edition; (Corrigendum 1-1991) (Corrigendum 2-3:1992) (Corrigendum 4-1993). 39 pp.
IEC 9594 Pt 3 Draft AMD 1. Information Technology—Open Systems Interconnection—The Directory—Part 3: Abstract Service Definition Amendment 1: Access Control; (1992) ***CD-ROM ONLY***. 24 pp.
IEC 9594 Pt 4-90. Information Technology—Open Systems Interconnection—The Directory—Part 4: Procedures of Distributed Operation First Edition; (Corrigendum 1-1991) (Corrigendum 2-1992) (Corrigendum 4-1993). 61 pp.
IEC 9594 Pt 4 Draft AMD 1. Information Technology—Open Systems Interconnection—The Directory—Part 4: Procedures for Distributed Operations Amendment 1: Access Control; (1992) ***CD-ROM ONLY***. 5 pp.
IEC 9594 Pt 4 Draft AMD 2. Information Technology—Open Systems Interconnection—The Directory—Part 4: Procedures for Distributed Operations Amendment 2: Replication, Schema and Enhanced Search; (1992) ***CD-ROM ONLY***. 90 pp.
IEC 9594 Pt 5-90. Information Technology—Open Systems Interconnection—The Directory—Part 5: Protocol Specifications First Edition; (Corrigendum 1-1992). 19 pp.
IEC 9594 Pt 5 Draft AMD 1. Information Technology—Open Systems Interconnection—The Directory—Part 5: Protocol Specifications Amendment 1: Replication; (1992) ***CD-ROM ONLY***. 23 pp.
IEC 9594 Pt 6-90. Information Technology—Open Systems Interconnection—The Directory—Part 6: Selected Attribute Types First Edition. 23 pp.

Directory Services *(Cont.)*

IEC 9594 Pt 6 Draft AMD 1. Information Technology—Open Systems Interconnection—The Directory—Part 6: Selected Attribute Types Amendment 1: Schema; (1992) ***CD-ROM ONLY***. 22 pp.
IEC 9594 Pt 7-90. Information Technology—Open Systems Interconnection—The Directory—Part 7: Selected Object Classes First Edition; (Corrigendum 1-1991) (Corrigendum 2-1992). 20 pp.
IEC 9594 Pt 7 Draft AMD 1. Information Technology—Open Systems Interconnection—The Directory—Part 7: Selected Object Classes Amendment 1: Schema; (1992) ***CD-ROM ONLY***. 14 pp.
IEC 9594 Pt 8-90. Information Technology—Open Systems Interconnection—The Directory—Part 8: Authentication Framework First Edition; (Corrigendum 1-1991). 33 pp.
IEC 9594 Pt 8 Draft AMD 1. Information Technology—Open Systems Interconnection—The Directory—Part 8: Authentication Framework Amendment 1: Access Control; (1992) ***CD-ROM ONLY***. 5 pp.
IEC DIS 9594 Pt 9-92. Information Technology—Open Systems Interconnection—The Directory—Part 9: Replication ***CD-ROM ONLY***. 37 pp.
ISO 9594 Pt 1-90. Information Technology—Open Systems Interconnection—The Directory—Part 1: Overview of Concepts, Models and Services First Edition. 16 pp.
ISO 9594 Pt 1 Draft AMD 1. Information Technology—Open Systems Interconnection—The Directory—Part 1: Overview of Concepts, Models and Services Amendment 1: Replication, Schema and Access Control; (1992) ***CD-ROM ONLY***. 7 pp.
ISO 9594 Pt 2-90. Information Technology—Open Systems Interconnection—The Directory—Part 2: Models First Edition; (Corrigendum 1-1991) (Corrigendum 2-1992). 31 pp.
ISO 9594 Pt 2 Draft AMD 1. Information Technology—Open Systems Interconnection—The Directory—Part 2: Models Amendment 1: Access Control; (1992) ***CD-ROM ONLY***. 42 pp.
ISO 9594 Pt 2 Draft AMD 2. Information Technology—Open Systems Interconnection—The Directory—Part 2: Models Amendment 2: Schema; (1992) ***CD-ROM ONLY***. 77 pp.
ISO 9594 Pt 2 Draft AMD 3. Information Technology—Open Systems Interconnection—The Directory—Part 2: Models Amendment 3: Replication; (1992) ***CD-ROM ONLY***. 58 pp.
ISO 9594 Pt 3-90. Information Technology—Open Systems Interconnection—The Directory—Part 3: Abstract Service Definition First Edition; (Corrigendum 1-1991) (Corrigendum 2-3:1992) (Corrigendum 4-1993). 39 pp.
ISO 9594 Pt 3 Draft AMD 1. Information Technology—Open Systems Interconnection—The Directory—Part 3: Abstract Service Definition Amendment 1: Access Control; (1992) ***CD-ROM ONLY***. 24 pp.
ISO 9594 Pt 4-90. Information Technology—Open Systems Interconnection—The Directory—Part 4: Procedures of Distributed Operation First Edition; (Corrigendum 1-1991) (Corrigendum 2-1992) (Corrigendum 3-1993). 61 pp.
ISO 9594 Pt 4 Draft AMD 1. Information Technology—Open Systems Interconnection—The Directory—Part 4: Procedures for Distributed Operations Amendment 1: Access Control; (1992) ***CD-ROM ONLY***. 5 pp.
ISO 9594 Pt 4 Draft AMD 2. Information Technology—Open Systems Interconnection—The Directory—Part 4: Procedures for Distributed Operations Amendment 2: Replication, Schema and Enhanced Search; (1992) ***CD-ROM ONLY***. 90 pp.
ISO 9594 Pt 5-90. Information Technology—Open Systems Interconnection—The Directory—Part 5: Protocol Specifications First Edition; (Corrigendum 1-1992). 19 pp.
ISO 9594 Pt 5 Draft AMD 1. Information Technology—Open Systems Interconnection—The Directory—Part 5: Protocol Specifications Amendment 1: Replication; (1992) ***CD-ROM OLNY***. 23 pp.
ISO 9594 Pt 6-90. Information Technology—Open Systems Interconnection—The Directory—Part 6: Selected Attribute Types First Edition. 23 pp.
ISO 9594 Pt 6 Draft AMD 1. Information Technology—Open Systems Interconnection—The Directory—Part 6: Selected Attribute Types Amendment 1: Schema; (1992) ***CD-ROM ONLY***. 22 pp.
ISO 9594 Pt 7-90. Information Technology—Open Systems Interconnection—The Directory—Part 7: Selected Object Classes First Edition; (Corrigendum 1-1991) (Corrigendum 2-1992). 20 pp.
ISO 9594 Pt 7 Draft AMD 1. Information Technology—Open Systems Interconnection—The Directory—Part 7: Selected Object Classes Amendment 1: Schema; (1992) ***CD-ROM ONLY***. 14 pp.

INDUSTRY STANDARDS
805

Directory
INTERNATIONAL AND NON-U.S. NATIONAL STANDARDS
SUBJECT INDEX

Directory Services *(Cont.)*

ISO 9594 Pt 8-90. Information Technology—Open Systems Interconnection—The Directory—Part 8: Authentication Framework First Edition; (Corrigendum 1-1991). 33 pp.
ISO 9594 Pt 8 Draft AMD 1. Information Technology—Open Systems Interconnection—The Directory—Part 8: Authentication Framework Amendment 1: Access Control; (1992) ***CD-ROM ONLY***. 5 pp.
ISO DIS 9594 Pt 9-92. Information Technology—Open Systems Interconnection—The Directory—Part 9: Replication ***CD-ROM ONLY***. 37 pp.
JIS X 5731-90. Information Processing Systems—Open Systems Interconnection—the Directory—Part 1: Overview of Concepts, Models and Service (ISO/IEC/DIS 9594-1).
JTC1 9594-90. Information Technology—Open Systems Interconnection—The Directory—Part 1: Overview of Concepts, Models and Services First Edition. 16 pp.
JTC1 9594 Pt1 Draft AMD 1. Information Technology—Open Systems Interconnection—The Directory—Part 1: Overview of Concepts, Models and Services Amendment 1: Replication, Schema and Access Control; (1992) ***CD-ROM ONLY***. 7 pp.
JTC1 9594-90. Information Technology—Open Systems Interconnection—The Directory—Part 2: Models First Edition; (Corrigendum 1-1991) (Corrigendum 2-1992). 2 pp.
JTC1 9594 Pt2 Draft AMD 1. Information Technology—Open Systems Interconnection—The Directory—Part 2: Models Amendment 1: Access Control; (1992) ***CD-ROM ONLY***. 42 pp.
JTC1 9594 Pt2 Draft AMD 2. Information Technology—Open Systems Interconnection—The Directory—Part 2: Models Amendment 2: Schema; (1992) ***CD-ROM ONLY***. 77 pp.
JTC1 9594 Pt2 Draft AMD 3. Information Technology—Open Systems Interconnection—The Directory—Part 2: Models Amendment 3: Replication; (1992) ***CD-ROM ONLY***. 58 pp.
JTC1 9594-90. Information Technology—Open Systems Interconnection—The Directory—Part 3: Abstract Service Definition First Edition; (Corrigendum 1-1991) (Corrigendum 2-3:1992) (Corrigendum 4-1993). 2 pp.
JTC1 9594 Pt3 Draft AMD 1. Information Technology—Open Systems Interconnection—The Directory—Part 3: Abstract Service Definition Amendment 1: Access Control; (1992) ***CD-ROM ONLY***. 24 pp.
JTC1 9594-90. Information Technology—Open Systems Interconnection—The Directory—Part 4: Procedures of Distributed Operation First Edition; (Corrigendum 1-1991) (Corrigendum 2-1992) (Corrigendum 4-1993). 3 pp.
JTC1 9594 Pt4 Draft AMD 1. Information Technology—Open Systems Interconnection—The Directory—Part 4: Procedures for Distributed Operations Amendment 1: Access Control; (1992) ***CD-ROM ONLY***. 5 pp.
JTC1 9594 Pt4 Draft AMD 2. Information Technology—Open Systems Interconnection—The Directory—Part 4: Procedures for Distributed Operations Amendment 2: Replication, Schema and Enhanced Search; (1992) ***CD-ROM ONLY***. 90 pp.
JTC1 9594-90. Information Technology—Open Systems Interconnection—The Directory—Part 5: Protocol Specifications First Edition; (Corrigendum 1-1992). 3 pp.
JTC1 9594 Pt5 Draft AMD 1. Information Technology—Open Systems Interconnection—The Directory—Part 5: Protocol Specifications Amendment 1: Replication; (1992) ***CD-ROM OLNY***. 23 pp.
JTC1 9594-90. Information Technology—Open Systems Interconnection—The Directory—Part 6: Selected Attribute Types First Edition. 23 pp.
JTC1 9594 Pt6 Draft AMD 1. Information Technology—Open Systems Interconnection—The Directory—Part 6: Selected Attribute Types Amendment 1: Schema; (1992) ***CD-ROM ONLY***. 22 pp.
JTC1 9594-90. Information Technology—Open Systems Interconnection—The Directory—Part 7: Selected Object Classes First Edition; (Corrigendum 1-1991) (Corrigendum 2-1992). 20 pp.
JTC1 9594 Pt7 Draft AMD 1. Information Technology—Open Systems Interconnection—The Directory—Part 7: Selected Object Classes Amendment 1: Schema; (1992) ***CD-ROM ONLY***. 14 pp.
JTC1 9594-90. Information Technology—Open Systems Interconnection—The Directory—Part 8: Authentication Framework First Edition; (Corrigendum 1-1991). 33 pp.
JTC1 9594 Pt8 Draft AMD 1. Information Technology—Open Systems Interconnection—The Directory—Part 8: Authentication Framework Amendment 1: Access Control; (1992) ***CD-ROM ONLY***. 5 pp.

Directory Services *(Cont.)*

OSI ISO/IEC 9594-1-90. Information Technology—Open Systems Interconnection—the Directory—Part 1: Overview of Concepts, Models and Services. 16 pp.
OSI ISO DIS 9594-1-88. Information Processing Systems—Open System Interconnection—The Directory—Part 1: Overview of Concepts, Models and Service. 17 pp.
OSI ISO/IEC 9594-2-90. Information Technology—Open Systems Interconnection—the Directory—Part 2: Models. 27 pp.
OSI ISO IEC DIS 9594-2-88. Information Processing Systems—Open System Interconnection—The Directory—Part 2: Models. 29 pp.
OSI ISO/IEC 9594-3-90. Information Technology—Open Systems Interconnection—the Directory—Part 3: Abstract Service Definition. 31 pp.
OSI ISO IEC DIS 9594-3-88. Information Processing Systems—Open System Interconnection—The Directory—Part 3: Abstract Service Definition. 34 pp.
OSI ISO/IEC 9594-4-90. Information Technology—Open Systems Interconnection—the Directory—Part 4: Procedures for Distributed Operation. 52 pp.
OSI ISO IEC DIS 9594-4-88. Information Processing Systems-Open Systems Interconnection—Part 4: Procedures for Distributed Operations. 74 pp.
OSI ISO/IEC 9594-5-90. Information Technology—Open Systems Interconnection—the Directory—Part 5: Protocol Specifications. 16 pp.
OSI ISO IEC DIS 9594-5-88. Information Processing Systems—Open Systems Interconnection—The Directory—Part 5: Protocol Specifications. 19 pp.
OSI ISO/IEC 9594-6-90. Information Technology—Open Systems Interconnection—the Directory—Part 6: Selected Attribute Types. 22 pp.
OSI ISO IEC DIS 9594-6-88. Information Processing Systems—Open System Interconnection—The Directory—Part 6: Selected Attribute Types. 26 pp.
OSI ISO/IEC 9594-7-90. Information Technology—Open Systems Interconnection—the Directory—Part 7: Selected Object Classes. 17 pp.
OSI ISO IEC DIS 9594-7-88. Information Processing Systems—Open Systems Interconnection—The Directory—Part 7: Selected Object Classes. 17 pp.
OSI ISO/IEC 9594-8-90. Information Technology—Open Systems Interconnection—the Directory—Part 8: Authentication Framework. 31 pp.
OSI ISO DIS 9594-8. Information Processing Systems—Open Systems Interconnection—The Directory—Part 8: Authentication Framework. 31 pp.
SAA AS 4019.1-92. Information Technology—Open Systems Interconnection—The Directory—Part 1: Overview of Concepts, Models and Services (ISO/IEC 9594-1:1990) (in Professional Package 26A). 11 pp.
SAA AS 4019.2-92. Information Technology—Open Systems Interconnection—The Directory—Part 2: Models (ISO/IEC 9594-2:1990) (in Professional Package 26A). 22 pp.
SAA AS 4019.3-92. Information Technology—Open Systems Interconnection—The Directory—Part 3: Abstract Service Definition (ISO/IEC 9594:1990) (in Professional Package 26A). 26 pp.
SAA AS 4019.4-92. Information Technology—Open Systems Interconnection—The Directory—Part 4: Procedures for Distributed Operation (ISO/IEC 9594-4:1990) (in Professional Package 26A). 46 pp.
SAA AS 4019.5-92. Information Technology—Open Systems Interconnection—The Directory—Part 5: Protocol Specifications (ISO/IEC 9594-5:1990) (in Professional Package 26A). 12 pp.
SAA AS 4019.6-92. Information Technology—Open Systems Interconnection—The Directory—Part 6: Selected Attribute Types (ISO/IEC 9594-6:1990) (in Professional Package 26A). 18 pp.
SAA AS 4019.7-92. Information Technology—Open Systems Interconnection—The Directory—Part 7: Selected Object Classes (ISO/IEC 9594-7:1990) (in Professional Package 26A). 12 pp.
SAA AS 4019.8-92. Information Technology—Open Systems Interconnection—The Directory—Part 8: Authentication Framework (ISO/IEC 9594-8:1990) (in Professional Package 26A). 26 pp.
SNZ NZS/ISO-IEC 9594.1-90. Information Technology—Open Systems Interconnection—The Directory Part 1: Overview of Concepts, Models and Services. 11 pp.
SNZ NZS/ISO-IEC 9594.2-90. Information Technology—Open Systems Interconnection—The Directory Part 2: 1990 Models. 22 pp.
SNZ NZS/ISO-IEC 9594.3-90. Information Technology—Open Systems Interconnection—The Directory Part 3: Abstract Service Definition. 26 pp.
SNZ NZS/ISO-IEC 9594.4-90. Information Technology—Open Systems Interconnection—The Directory Part 4: Procedures for Distributed Operation. 46 pp.
SNZ NZS/ISO-IEC 9594.5-90. Information Technology—Open Systems Interconnection—The Directory Part 5: Protocol Specifications. 12 pp.

Directory Services *(Cont.)*

SNZ NZS/ISO-IEC 9594.6-90. Information Technology—Open Systems Interconnection—The Directory Part 6: Selected Attribute Types. 18 pp.
SNZ NZS/ISO-IEC 9594.7-90. Information Technology—Open Systems Interconnection—The Directory Part 7: Selected Object Classes. 12 pp.
SNZ NZS/ISO-IEC 9594.8-90. Information Technology—Open Systems Interconnection—The Directory Part 8: Authentication Framework. 26 pp.

—Abbreviations

CCITT RECMN F.500-92. International Public Directory Services (Study Group I) 39 pp. 39 pp.
CCITT RECMN F.500-89. International Public Directory Services—Message Handling Services—Operations and Definition of Service (Study Group I) 32 pp. 32 pp.

—Communication Networks

CCITT FASCICLE VIII.8. Data Communication Networks Directory Recommendation X.500—X.521 (Study Group VII). 231 pp.

—Distribution Lists (Message Handling)

CCITT RECMN F.400-92. Message Handling Services: Message Handling System and Service Overview (Study Group I) 82 pp (Same as Recmn X.400). 82 pp.
CCITT RECMN F.400-89. Message Handling System and Service Overview—Message Handling and Directory Services—Operations and Definition of Service (Study Group I) 73 pp. 73 pp.
CCITT RECMN X.400-93. Message Handling Services: Message Handling System and Service Overview (Study Group VII) 82 pp (Same as Recmn F.400). 82 pp.

—Error Control

CCITT RECMN F.500-92. International Public Directory Services (Study Group I) 39 pp. 39 pp.
CCITT RECMN F.500-89. International Public Directory Services—Message Handling Services—Operations and Definition of Service (Study Group I) 32 pp. 32 pp.

—Glossaries

CCITT RECMN F.500-92. International Public Directory Services (Study Group I) 39 pp. 39 pp.
CCITT RECMN F.500-89. International Public Directory Services—Message Handling Services—Operations and Definition of Service (Study Group I) 32 pp. 32 pp.

—Interpersonal Messaging Services

CCITT RECMN F.420-92. Message Handling Services: Public Interpersonal Messaging Service (Study Group I) 16 pp. 16 pp.
CCITT RECMN F.420-89. Message Handling Services: the Public Interpersonal Messaging Service—Message Handling and Directory Services—Operations and Definition of Service (Study Group I) 15 pp. 15 pp.

—Message Handling Systems

CCITT FASCICLE II.6-88. Message Handling and Directory Services—Operations and Definition of Service Recommendations F.400—F.422, F.500. 184 pp.
CCITT RECMN F.400-92. Message Handling Services: Message Handling System and Service Overview (Study Group I) 82 pp (Same as Recmn X.400). 82 pp.
CCITT RECMN F.400-89. Message Handling System and Service Overview—Message Handling and Directory Services—Operations and Definition of Service (Study Group I) 73 pp. 73 pp.
CCITT RECMN X.400-93. Message Handling Services: Message Handling System and Service Overview (Study Group VII) 82 pp (Same as Recmn F.400). 82 pp.

—Public Land Mobile Networks

CENELEC GSM 02.14-92. European Digital Cellular Telecommunication System (Phase 1); Service Directory. 4 pp.
ETSI GSM 02.14-92. European Digital Cellular Telecommunication System (Phase 1); Service Directory. 4 pp.

—Quality of Service

CCITT RECMN F.500-92. International Public Directory Services (Study Group I) 39 pp. 39 pp.
CCITT RECMN F.500-89. International Public Directory Services—Message Handling Services—Operations and Definition of Service (Study Group I) 32 pp. 32 pp.

Directory System Agents

See Also: Application Services (OSI); Directory Management Domain; Directory Services; Directory Systems; Directory User Agents; Open Systems Interconnection
CCITT RECMN F.500-92. International Public Directory Services (Study Group I) 39 pp. 39 pp.

806
INDEX and DIRECTORY of

Directory System Agents *(Cont.)*
CCITT RECMN F.500-89. International Public Directory Services—Message Handling Services—Operations and Definition of Service (Study Group I) 32 pp. 32 pp.

—Message Handling Systems
CCITT RECMN F.400-92. Message Handling Services: Message Handling System and Service Overview (Study Group I) 82 pp (Same as Recmn X.400). 82 pp.

CCITT RECMN F.400-89. Message Handling System and Service Overview—Message Handling and Directory Services—Operations and Definition of Service (Study Group I) 73 pp. 73 pp.

CCITT RECMN X.400-93. Message Handling Services: Message Handling System and Service Overview (Study Group VII) 82 pp (Same as Recmn F.400). 82 pp.

Directory Systems
See Also: Directory Databases; Directory Entries; Directory Management Domain; Directory Services; Directory System Agents; Directory User Agents; Message Handling Systems; Telecommunication Systems

—Message Handling Services
CCITT RECMN F.500-92. International Public Directory Services (Study Group I) 39 pp. 39 pp.

CCITT RECMN F.500-89. International Public Directory Services—Message Handling Services—Operations and Definition of Service (Study Group I) 32 pp. 32 pp.

—Military Communications
NATO STANAG 5046 ED 2 AMD 0-85. The NATO Military Communications Directory System. 24 pp.

Directory User Agents
Use For: DUA *See Also:* Application Services (OSI); Directory Management Domain; Directory Services; Directory System Agents; Directory Systems; Open Systems Interconnection; User Agents

CCITT RECMN F.500-92. International Public Directory Services (Study Group I) 39 pp. 39 pp.

CCITT RECMN F.500-89. International Public Directory Services—Message Handling Services—Operations and Definition of Service (Study Group I) 32 pp. 32 pp.

—Message Handling Systems
CCITT RECMN F.400-92. Message Handling Services: Message Handling System and Service Overview (Study Group I) 82 pp (Same as Recmn X.400). 82 pp.

CCITT RECMN F.400-89. Message Handling System and Service Overview—Message Handling and Directory Services—Operations and Definition of Service (Study Group I) 73 pp. 73 pp.

CCITT RECMN X.400-93. Message Handling Services: Message Handling System and Service Overview (Study Group VII) 82 pp (Same as Recmn F.400). 82 pp.

Dirt Content Analysis
Use: Impurities Content Analysis

Disability Equipment
Scope Note: Includes assistive equipment manufactured for use by disabled persons. For equipment which has been modified to provide access to disabled persons, see the listings under Handicapped Accessible Equipment *See Also:* Canes; Crutches; Handicapped Accessible Equipment; Handicapped Persons; Hearing Aids; Medical Electrical Equipment; Medical Equipment; Prosthetic Devices; Walkers; Wheelchair Lifts; Wheelchairs

—Automotive
CSA CAN/CSA-D409-92. Motor Vehicles for the Transportation of Persons with Physical Disabilities; (Gen Instr 1 Thru 2). 52 pp.

—Classification
ISO 9999-92. Technical Aids for Disabled Persons—Classification First Edition. 63 pp.

—Glossaries
JIS T 0102-91. Glossary of Terms Used in Technical Systems and Aids for Disabled or Handicapped Persons.

Disabled Persons
Use: Handicapped Persons

Disaccharides
Scope Note: Use a more specific term *See:* Maltose; Sucrose

Disaster Relief Operations
Use: Emergency Relief Operations

Disc Brakes
Use: Disk Brakes

Disc Clutches
Use: Disk Clutches

Disc Grinders
Use: Disk Grinders

Disc Seal Tubes
Use: Disk Seal Tubes

Disc Spacers
Use: Spacers (Mechanical)

Disc Springs
Use: Disk Springs

Disc Wheels
Use: Disk Wheels

Discharge Coefficient
Use For: Coefficient of Discharge

—Safety Valves
JIS B 8225-86. Measuring Methods for Coefficient of Discharge of Safety Valves. 27 pp.

Discharge Gages
See Also: Vacuum Gages
JIS Z 8752-89. Measuring Methods of Low Pressures by Hot Cathode and Cold Cathode Ionization Gauges. 8 pp.

Discharge Lamps
See Also: Flashtubes; Fluorescent Lighting; Glow Lamps; High Intensity Discharge Lamps; Incandescent Lighting; Lamps; Lighting; Luminous Tube Signs; Mercury Vapor Lamps; Metal Halide Lamps; Neon Tubes; Sodium Vapor Lamps; Ultraviolet Lamps; Xenon Lamps

DIN VDE 0712 Pt 1-71. Regulations for Discharge Lamp Accessories for Rated Voltage up to 1000 V; Part 1: General Regulations (June). 50 pp.

DIN VDE 0712 Pt 1A1 (D)-78. Regulations for Discharge Lamps for a Supply Voltage up to 1000 V Part 1: General Regulations (VDE Specification) (Oct). 15 pp.

DIN VDE 0712 Pt 2 A2 (D)-83. Equipment for Discharge Lamps for a Supply Voltage up to 1000V; Ballasts: Amendment 2 (VDE Specifications) (Oct). 9 pp.

MOD UK DSTAN 62-6-01. Lamps, Electric Issue 2; Amendment 1. 109 pp.

—Ballasts (Electric)
BSI BS 4782-71. (WITHDRAWN) 1971 Amd 3 Ballasts for Discahrge Lamps (Excluding Ballasts for Tubular Flourescent Lamps) (Superseded by BS EN 60922: 1991 and BS EN 60923: 1991). 67 pp.

BSI BS EN 60922-91. 1991 General and Safety Requirements for Ballasts for Discharge Lamps (Excluding Tubular Fluorescent Lamps) (E). 38 pp.

BSI BS EN 60922-01. 1991 Amd 1 General and Safety Requirements for Ballasts for Discharge Lamps (Excluding Tubular Fluorescent Lamps) (AMD 7173) September 15, 1992. 50 pp.

CENELEC EN 60922-91. Ballasts for Discharge Lamps (Excluding Tubular Fluorescent Lamps) General and Safety Requirements. 6 pp.

CENELEC EN 60922/A1-92. AMD 1 Ballasts for Discharge Lamps (Excluding Tubular Fluorescent Lamps) General and Safety Requirements. 6 pp.

CENELEC EN 60923/A1-92. AMD 1 Ballasts for Discharge Lamps (Excluding Tubular Fluorescent Lamps)—Performance Requirements. 5 pp.

CSA CAN/CSA-C22. 2 NO 74-92. Equipment for Use with Electric Discharge Lamps (Incorporating Electrical Bulletin Nos. 523F, 753, 846, 1124A, 1125A, 1325, and 1326); (Gen Instr 1). 82 pp.

DIN VDE 0712 Pt 2-71. Accessories for Discharge Lamps with Nominal Voltage up to 1000 V; Part 2: Special Specifications for Ballasts (June) (Valid Only in Conjunction with 0712 Part 1/6.71). 50 pp.

DIN VDE 0712 Pt 2 A2 (D)-83. Equipment for Discharge Lamps for a Supply Voltage up to 1000V; Ballasts: Amendment 2 (VDE Specifications) (Oct). 9 pp.

DIN VDE 0712 Pt 2A (D)-78. Equipment for Discharge Lamps for a Supply Voltage up to 1000V: Particular Specification for Ballasts (VDE Specification) (Oct). 16 pp.

IEC 922-89. Ballasts for Discharge Lamps (Excluding Tubular Fluorescent Lamps) General and Safety Requirements First Edition; (Amendment 2-1992, Includes Amendment 1). 89 pp.

—Capacitors
BSI BS 4017-79. (WITHDRAWN) 1979 Capacitors for Use in Tubular Fluorescent, High Pressure Mercury, and Low Pressure Sodium Vapour Discharge Lamp Circuits (SUPERSEDED BY BS EN 61048 & BS EN 61049). 22 pp.

Discharge Lamps *(Cont.)*
—Capacitors *(Cont.)*
BSI BS EN 61048-93. 1993 Capacitors for Use in Tubular Fluorescent and Other Discharge Lamp Circuits—General and Safety Requirements (E). 31 pp.

BSI BS EN 61049-93. 1993 Capacitors for Use in Tubular Fluorescent and Other Discharge Lamp Circuits—Performance Requirements (E). 21 pp.

CENELEC EN 61048-93. Capacitors for Use in Tubular Fluorescent and Other Discharge Lamp Circuits—General and Safety Requirements (IEC 1048:1991 + Corrigendum January 1992). 6 pp.

CENELEC EN 61048-93. Capacitors for Use in Tubular Fluorescent and Other Discharge Lamp Circuits—General and Safety Requirements (IEC 1048: 1991, Modified + Corrigendum: 1992). 25 pp.

CENELEC EN 61049-93. Capacitors for Use in Tubular Fluorescent and Other Discharge Lamp Circuits Performance Requirements (IEC 1049: 1991 + Corrigendum 1992, Modified). 14 pp.

DIN VDE 0560 Pt 6-86. Capacitors; Capacitors with Ratings up to 1.5 kvar for Installations with Discharge Lamps, and in Particular Fluorescent Lamps (Jan). 32 pp.

IEC 1048-91. Capacitors for Use in Tubular Fluorescent and Other Discharge Lamp Circuits General and Safety Requirements First Edition; (Replaces 566) (CENELEC EN 61048:1993). 54 pp.

IEC 1049-91. Capacitors for Use in Tubular Fluorescent and Other Discharge Lamp Circuits Performance Requirements First Edition; (Replaces 566) (CENELEC EN 61049: 1993). 30 pp.

—Chromaticity
CNS Z1042-84. Chromaticity Domain of Discharge Lamps (Mar)(10840).

JIS Z 9113-73. Chromaticity Domain of Discharge Lamps.

—Cold Cathode—Transformers
SAA AS 3143-82. Approval and Test Specification—for Transformers for Cold-Cathode Electric Discharge Lamps and Lighting Systems (This is a Joint Standard with SANZ NZS 3143). 4 pp.

SNZ NZS/AS 3143-82. Approval and Test Specification for Transformers for Cold-Cathode Electric Discharge Lamps and Lighting Systems (This is a Joint Standard with SAA AS 3143). 4 pp.

—Electric Switches
CSA C22.2 NO 111-M1986. General Use Switches (R 1992); (Gen Instr 1 Thru 2). 40 pp.

—Installation
DIN VDE 0128-81. Installations of Electric Discharge Lamps with Voltages Exceeding 1000 V (June). 42 pp.

—Lampholders
CSA CAN/CSA-C22. 2 NO 74-92. Equipment for Use with Electric Discharge Lamps (Incorporating Electrical Bulletin Nos. 523F, 753, 846, 1124A, 1125A, 1325, and 1326); (Gen Instr 1). 82 pp.

—Procurement—Ships
MOD UK NES 2004-88. Specification for Mercury Fluorescent Discharge Lamp (MBF/U) NSN 6240-99-996-7878 Issue 2 (12.88). 8 pp.

MOD UK NES 2004-92. Specification for Mercury Fluorescent Discharge Lamp (MBF/U) NSN 6240-99-996-7878 Issue 3 (01.92). 8 pp.

—Procurement—Submarines
MOD UK NES 2004-88. Specification for Mercury Fluorescent Discharge Lamp (MBF/U) NSN 6240-99-996-7878 Issue 2 (12.88). 8 pp.

MOD UK NES 2004-92. Specification for Mercury Fluorescent Discharge Lamp (MBF/U) NSN 6240-99-996-7878 Issue 3 (01.92). 8 pp.

—Self Ballasted
BSI BS 7199-89. 1989 Self-Ballasted Lamps for General Lighting Services. Performance Requirements (Renumbered as BS EN 60969: 1993). 11 pp.

BSI BS 7199-01. 1989 Amd 1 Self-Ballasted Lamps for General Lighting Services Performance Requirements (IEC 969: 1988) (AMD 7013) July 15, 1992 (Renumbered as BS EN 60969: 1993) (E). 19 pp.

BSI BS EN 60969-93. 1993 Self-Ballasted Lamps for General Lighting Services. Performance Requirements (AMD 7717) May 15, 1993. 29 pp.

CENELEC HD 594 S1-91. Self-Ballasted Lamps for General Lighting Services—Performance Requirements. 6 pp.

CENELEC EN 60969-93. Self-Ballasted Lamps for General Lighting Services Performance Requirements (IEC 969:1988) (Supersedes HD 594 S1:1991). 5 pp.

CENELEC EN 60969-93. Self-Ballasted Lamps for General Lighting Services Performance Requirements (IEC 969:1988). 12 pp.

Discharge Lamps (Cont.)

—Self Ballasted (Cont.)
CENELEC EN 60969/A1-93. AMD 1 Self-Ballasted Lamps for General Lighting Services Performance Requirements (IEC 969:1988/A1: 1991). 4 pp.
IEC 969-88. Self-Ballasted Lamps for General Lighting Services Performance Requirements First Edition; (Amendment 1-1991) (CENELEC EN 60969/A1: 1993). 24 pp.

—Self Ballasted—Safety
BSI BS 7173-89. 1989 Amd 1 Self-Ballasted Lamps for General Lighting Services Safety Requirements (AMD 6707) September 30, 1991. 24 pp.
BSI BS 7173-02. 1989 Amd 2 Self-Ballasted Lamps for General Lighting Services Safety Requirements (AMD 7716) May 15, 1993. 30 pp.
CENELEC EN 60968-90. Self-Ballasted Lamps for General Lighting Services Safety Requirements. 5 pp.
CENELEC EN 60968/A1-93. AMD 1 Self-Ballasted Lamps for General Lighting Services Safety Requirements (IEC 968:1988/A1: 1991). 4 pp.
IEC 968-88. Self-Ballasted Lamps for General Lighting Services Safety Requirements First Edition; (Amendment 1-1991) (CENELEC EN 60968/A1: 1993). 34 pp.

—Starters
BSI BS EN 60926-91. 1991 General and Safety Requirements for Starting Devices (Other Than Glow Starters). 36 pp.
BSI BS EN 60927-91. 1991 Performance Requirements for Starting Devices (Other Than Glow Starters). 22 pp.
BSI BS EN 60927-01. 1991 Amd 1 Performance Requirements for Starting Devices (Other Than Glow Starters) (AMD 6882) May 1, 1992. 52 pp.
CENELEC EN 60 926-90. Starting Devices (Other Than Glow Starters) General and Safety Requirements. 6 pp.
CENELEC EN 60927-90. Starting Devices (Other Than Glow Starters) Performance Requirements. 16 pp.
CENELEC EN 60927/A1-91. AMD 1 Starting Devices (Other Than Glow Starters) Performance Requirements. 5 pp.
IEC 926-90. Starting Devices (Other Than Glow Starters) General and Safety Requirements First Edition; (Amendment 1-1992) (Amendment 2-1993). 85 pp.
IEC 927-88. Starting Devices (Other Than Glow Starters) Performance Requirements First Edition; (Corrigendum—Dec 1989) (Amendment 1-1990). 60 pp.

—Starters—Holders
CSA CAN/CSA-C22. 2 NO 74-92. Equipment for Use with Electric Discharge Lamps (Incorporating Electrical Bulletin Nos. 523F, 753, 846, 1124A, 1125A, 1325, and 1326); (Gen Instr 1). 82 pp.

—Tubular
DIN VDE 0713 Pt 1-85. Equipment for Tubular Discharge Lamp Installations over 1000 V; General Specification (Sept). 34 pp.
DIN VDE 0713 Pt 3-85. Equipment for Tubular Discharge Lamp Installations over 1000 V; Discharge Lamp Appliances (Sept). 7 pp.
DIN VDE 0713 Pt 5-85. Equipment for Tubular Discharge Lamp Installations over 1000 V; Discharge Lamps (Sept). 5 pp.

—Tubular—Fittings
DIN VDE 0713 Pt 4-85. Equipment for Tubular Discharge Lamp Installations over 1000 V; Small Fittings (Sept). 8 pp.

—Tubular—Ground Fault Circuit Interrupters
DIN VDE 0713 Pt 2-85. Equipment for Tubular Discharge Installations over 1000 V; Earth Leakage Protection (Sept). 8 pp.

—Tubular—Transformers
BSI BS EN 61050-92. 1992 Transformers for Tubular Discharge Lamps Having a No-Load Output Voltage Exceeding 1000 V (Generally Called Neon-Transformers) General and Safety Requirements. 42 pp.
CENELEC HD 388-80. Transformers for Tubular Discharge Lamps Having a No-Load Output Voltage Exceeding 1000 V (Generally Called Neon-Transformers). 46 pp.
CENELEC EN 61050-92. Transformers for Tubular Discharge Lamps Having a No-Load Output Voltage Exceeding 1000 V (Generally Called Neon-Transformers)—General and Safety Requirements. 6 pp.
IEC 1050-91. Transformers for Tubular Discharge Lamps Having a No-Load Output Voltage Exceeding 1 000 V (Generally Called Neon-Transformers) General and Safety Requirements First Edition; (Corrigendum—March 1992). 77 pp.

Discharge Lamps (Cont.)

—Tubular—Transformers—Safety
BSI BS EN 61050-92. 1992 Transformers for Tubular Discharge Lamps Having a No-Load Output Voltage Exceeding 1000 V (Generally Called Neon-Transformers) General and Safety Requirements. 42 pp.
CENELEC EN 61050-92. Transformers for Tubular Discharge Lamps Having a No-Load Output Voltage Exceeding 1000 V (Generally Called Neon-Transformers)—General and Safety Requirements. 6 pp.
IEC 1050-91. Transformers for Tubular Discharge Lamps Having a No-Load Output Voltage Exceeding 1 000 V (Generally Called Neon-Transformers) General and Safety Requirements First Edition; (Corrigendum—March 1992). 77 pp.

Discharge Stacks

—Fluid Inlets—Design
BSI BS 6367-83. 1983 Amd 1 Drainage of Roofs and Paved Areas. 58 pp.

Discharge Tubes
Use: Gas Discharge Tubes

Discharging Agents
See Also: Dyes; Stripping

—Hydrosulfites
JIS K 1476-80. Hydrosulfites for Discharging and Bleaching Agents. 9 pp.

Discoloration
See Also: Color

—Coated Fabrics—Cigarette Smoke
ISO 6449-82. Rubber or Plastics Coated Fabrics—Determination of Discolouration by Cigarette Smoke First Edition. 4 pp.

—Paints
SAA AS 1580.481. 1.2-91. Paints and Related Materials—Methods of Test—Pt. 481: Coatings —Pt. 481.1: Assessment of IndividualDefects of Exposed Films-Pt. 481. 1.2: Exposed to Weather-ing—Discolouration (Including Bronzing) (Supersedes AS1580.481. .1—1975 (in Part)) (in Prof. Packages 30, 39). 2 pp.
SNZ NZS/AS 1580. 481.1.2-91. Methods of Test for Paints and Related Materials Part 481.1.2: Coatings—Exposed to Weathering—Discolouration (Including Bronzing) (This is a Joint Standard with SAA AS 1580.481.1.2). 2 pp.

Discolouration
Use: Discoloration

Disconnect Switches
Use: Disconnecting Switches

Disconnecting Switches
Use For: Alternating Current Disconnectors; Disconnectors; Hookstick Switches; Isolator Switches; Switch Disconnectors; Switch Stick Switches
See Also: Air Switches; Bypass Switches; Circuit Breakers; Electric Switches; Electrical Protection Equipment; Fuses (Electric); Isolating Switches; Load Break Switches; Power Systems; Switchgear
BSI BS 5253-90. 1990 Alternation Current Disconnectors and Earthing Switches. 47 pp.
BSI BS EN 60947-3-92. 1992 Low-Voltage Switchgear and Controlgear Part 3: Switches, Disconnectors, Switch-Disconnectors and Fuse-Combination Units. 62 pp.
BSI BS EN 60947-3-01. 1992 Amd 1 Low-Voltage Switchgear and Controlgear Part 3: Switches, Disconnectors, Switch-Disconnectors and Fuse-Combination Units (AMD 7854) July 15, 1993 (E). 63 pp.
CENELEC HD 408-79. Alternating Current Disconnectors (Isolators) and Earthing Switches. 3 pp.
CENELEC HD 408 S2-90. Alternating Current Disconnectors and Earthing Switches. 4 pp.
CENELEC EN 60947-3-92. Low-Voltage Switchgear and Controlgear Part 3: Switches, Disconnectors, Switch-Disconnectors and Fuse-Combination Units. 8 pp.
CENELEC EN 60947-3-92. CORRIGENDUM Low-Voltage Switchgear and Controlgear Part 3: Switches, Disconnectors, Switch Diconnectors and Fuse-Combination Units. 1 p.
DIN VDE 0660 Pt 107 A1 (D)-91. Switchgear and Controlgear; Low Voltage Switchgear and Controlgear; Part 3: Switches, Disconnectors, Switch-Disconnectors and Fuse-Combination Units (Nov). 5 pp.
DIN VDE 0660 Pt 107 A2 (D)-92. Switchgear and Controlgear; Low-Voltage Switchgear and Controlgear; Part 3: Switches, Disconnectors and Fuse Combination Units (Feb). 5 pp.

Disconnecting Switches (Cont.)

IEC 129-84. Alternating Current Disconnectors and Earthing Switches Third Edition; (Amendment 1-1993). 92 pp.
IEC 947 Pt 3-90. Low-Voltage Switchgear and Controlgear Part 3: Switches, Disconnectors, Switch-Disconnectors and Fuse-Combination Units First Edition; (Corrigendum—Dec 1991). 115 pp.
SAA AS 1306-85. High Voltage a.c. Switchgear and Controlgear—Disconnectors (Isolators) and Earthing Switches. 28 pp.
SAA AS 1775-84. Low Voltage Switchgear and Controlgear—Air-Break Switches, Isolators and Fuse-Combination Units (up to and Including 1 000 V a.c. and 1 200 V d.c.) Amdt 1 January 1985. 38 pp.

— 52 kV—Bus Transfer Current Switching
IEC 1128-92. Alternating Current Disconnectors Bus-Transfer Current Switching by Disconnectors First Edition. 26 pp.

—Air
BSI BS 5419-77. (WITHDRAWN) 1977 Amd 1 Air-Break Switches, Air-Break Disconnectors, Air-Break Switch Disconnectors and Fuse Combination Units for Voltages up to and Including 1000 V a.c. and 1200 V d.c. (Superseded by BS EN 60947-3: 1992). 51 pp.

—Capacitors
BSI BS 7631-93. 1993 Internal Fuses and Internal Overpressure Disconnectors for Shunt Capacitors (IEC 593: 1977) (E). 16 pp.
BSI BS 7632-93. 1993 Internal Fuses and Internal Overpressure Disconnectors for Capacitors for Inductive Heat Generating Plants (IEC 594: 1977) (E). 16 pp.
IEC 593-77. Internal Fuses and Internal Overpressure Disconnectors for Shunt Capacitors First Edition; (Amendment 1-1980) (Amendment 2-1986). 36 pp.
IEC 594-77. Internal Fuses and Internal Overpressure Disconnectors for Capacitors for Inductive Heat Generating Plants First Edition; (Amendment 2-1987). 36 pp.

—Fusible
BSI BS EN 60947-3-92. 1992 Low-Voltage Switchgear and Controlgear Part 3: Switches, Disconnectors, Switch-Disconnectors and Fuse-Combination Units. 62 pp.
BSI BS EN 60947-3-01. 1992 Amd 1 Low-Voltage Switchgear and Controlgear Part 3: Switches, Disconnectors, Switch-Disconnectors and Fuse-Combination Units (AMD 7854) July 15, 1993 (E). 63 pp.
CENELEC EN 60947-3-92. Low-Voltage Switchgear and Controlgear Part 3: Switches, Disconnectors, Switch-Disconnectors and Fuse-Combination Units. 8 pp.
CENELEC EN 60947-3-92. CORRIGENDUM Low-Voltage Switchgear and Controlgear Part 3: Switches, Disconnectors, Switch Diconnectors and Fuse-Combination Units. 1 p.
DIN VDE 0660 Pt 107 A1 (D)-91. Switchgear and Controlgear; Low Voltage Switchgear and Controlgear; Part 3: Switches, Disconnectors, Switch-Disconnectors and Fuse-Combination Units (Nov). 5 pp.
DIN VDE 0660 Pt 107 A2 (D)-92. Switchgear and Controlgear; Low-Voltage Switchgear and Controlgear; Part 3: Switches, Disconnectors and Fuse Combination Units (Feb). 5 pp.
IEC 947 Pt 3-90. Low-Voltage Switchgear and Controlgear Part 3: Switches, Disconnectors, Switch-Disconnectors and Fuse-Combination Units First Edition; (Corrigendum—Dec 1991). 115 pp.

—High Voltage
SAA AS 1025.1-84. High-Voltage a.c. Switchgear and Controlgear—Switches and Switch-Disconnectors—Part 1: For Rated Voltages Above 1 kV and Less Than 52 kV. 16 pp.
SAA AS 1025.2-89. High-Voltage a.c. Switchgear and Controlgear—Switches and Switch-Disconnectors—Part 2: For Rated Voltages of 52 kV and Above. 43 pp.

—Hook Bars
CNS C3106-80. Method of Test for Hook Bars for Disconnecting Switch Operation (Dec)(6654).
CNS C4266-80. Hook Bars for Disconnecting Switch Operation (Dec)(6653).
JIS C 4510-91. Hook Bars for Disconnecting Switch Operation. 7 pp.

—Indoor
CNS C3113-81. Method of Test for 6.6KV Disconnecting Switches for Indoor Use (Mar)(7122).
CNS C4298-81. 6.6KV Disconnecting Switches for Indoor Use (Mar)(7121).
JIS C 4606-71. 6.6 kV Disconnecting Switches for Indoor Use (R 1982). 16 pp.

INTERNATIONAL AND NON-U.S. NATIONAL STANDARDS
SUBJECT INDEX

Dishwashers

Disconnecting Switches (Cont.)

—Low Voltage
CENELEC HD 422-82. Low-Voltage Air-Break Switches, Air-Break Disconnectors, Air-Break Switch-Disconnectors and Fuse-Combination Units. 7 pp.

—Motors
BSI BS EN 60947-3-92. 1992 Low-Voltage Switchgear and Controlgear Part 3: Switches, Disconnectors, Switch-Disconnectors and Fuse-Combination Units. 62 pp.

BSI BS EN 60947-3-01. 1992 Amd 1 Low-Voltage Switchgear and Controlgear Part 3: Switches, Disconnectors, Switch-Disconnectors and Fuse-Combination Units (AMD 7854) July 15, 1993 (E). 63 pp.

CENELEC EN 60947-3-92. Low-Voltage Switchgear and Controlgear Part 3: Switches, Disconnectors, Switch-Disconnectors and Fuse-Combination Units. 8 pp.

CENELEC EN 60947-3-92. CORRIGENDUM Low-Voltage Switchgear and Controlgear Part 3: Switches, Disconnectors, Switch Diconnectors and Fuse-Combination Units. 1 p.

DIN VDE 0660 Pt 107 A1 (D)-91. Switchgear and Controlgear; Low Voltage Switchgear and Controlgear; Part 3: Switches, Disconnectors, Switch-Disconnectors and Fuse-Combination Units (Nov). 5 pp.

DIN VDE 0660 Pt 107 A2 (D)-92. Switchgear and Controlgear; Low-Voltage Switchgear and Controlgear; Part 3: Switches, Disconnectors and Fuse Combination Units (Feb). 5 pp.

IEC 947 Pt 3-90. Low-Voltage Switchgear and Controlgear Part 3: Switches, Disconnectors, Switch-Disconnectors and Fuse-Combination Units First Edition; (Corrigendum—Dec 1991). 115 pp.

Disconnectors
Use: Disconnecting Switches

Discrete Components
Scope Note: For additional listings, use a more specific term *See Also:* Capacitors; Diodes; Electronic Components; Resistors; Transistors

—Environmental Conditions—Industrial
ISO TR10450-91. Industrial Automation Systems and Integration—Operating Conditions for Discrete Part Manufacturing—Equipment in Industrial Environments First Edition. 13 pp.

Discrete Semiconductor Devices
Use: Semiconductor Devices

Discriminating Digits
See Also: Language Digits; Telephone Systems
CCITT RECMN Q.104-89. Language Digit or Discriminating Digit—General Recommendations on Telephone Switching and Signalling—Functions and Information Flows for Services in the ISDN—Supplements (Study Group XI) 2 pp. 2 pp.

Diseases
See Also: Medicine

—Statistical Classifications
NATO STANAG 2050 ED 5 AMD 2-89. Statistical Classification of Diseases, Injuries and Causes of Death. 28 pp.

Dish Warmers
Use: Food Warmers

Dish Washers
Use: Dishwashers

Dishes
Use: Tableware

Dishes (Laboratory)
Use: Laboratory Dishes

Dishwashers
See Also: Appliances
IEC 436-81. Methods for Measuring the Performance of Electric Dishwashers Second Edition; (Corrigendum—Sept 1982) (Amendment 1-1984) (Amendment 2-1992). 64 pp.

—Commercial
CGSB CAN/CGSB-52.35-M88. Dishwashing Machines, Commercial, Undercounter Type, Heat Sanitizing, Stationary Rack, Electric. 12 pp.

CGSB CAN/CGSB-52.36-M88. Dishwashing Machines, Commercial, Single Tank, Heat Sanitizing, Stationary Rack, Electrically, Steam or Gas Heated. 14 pp.

CGSB CAN/CGSB-52.37-M88. Dishwashing Machines, Commercial, Single Tank, Heat Sanitizing, Conveyor Rack Type, Electrically, Steam or Gas Heated. 16 pp.

Dishwashers (Cont.)

—Commercial (Cont.)
CGSB CAN/CGSB-52.38-M88. Dishwashing Machines, Commercial, Multiple Tank, Heat Sanitizing, Conveyor Rack Type, Electrically, Steam or Gas Heated. 17 pp.

CGSB CAN/CGSB-52.39-M88. Dishwashing Machines, Commercial, Multiple Tank, Heat Sanitizing, Rackless (Flight) Type, Electrically, Steam or Gas Heated. 17 pp.

CSA C22.2 NO 53-1968. Electric Washing Machines (Superceded in Part by C22.2 NO 169); (Rev 1-2 May 1969). 31 pp.

CSA 776 Bull. Electrical Bulletin 776 November 19, 1969 to C22.2 NO 53. 3 pp.

CSA 1169 Bull. Electrical Bulletin 1169 June 27, 1978 to C22.2 NO 53. 2 pp.

CSA C22.2 NO 168-M1981. Commercial Dishwashing Machines (R 1992); (Gen Instr 1). 31 pp.

—Commercial—Safety
BSI BS EN 60335-2-58-93. 1993 Safety of Household and Similar Electrical Appliances Part 2: Particular Requirements Section 2.58: Commercial Electric Dishwashing Machines (L). 33 pp.

DIN VDE 0700 Pt 231-84. Safety of Household & Similar Electrical Appliances; Commercial Dishwashers (VDE Specification) (Aug) (Supplements or Replaces Corresponding Sections in DIN 57700 Part 1/VDE 0700 Part 1). 34 pp.

—Commercial—Water Heaters—Safety
BSI BS 5809-80. 1980 Amd 1 Safety and Efficiency of the Gas Heating Equipment of Commercial Dishwashing Machines. 17 pp.

—Drainage Systems
CEN PREN 128-89. Performance Criteria and Requirements for Sanitary Appliances. 4 pp.

—Hose Assemblies
ISO 6804-91. Rubber Hoses and Hose Assemblies for Washing-Machines and Dishwashers —Specification for Inlet Hoses Second Edition. 8 pp.

—Hoses
ISO 6804-91. Rubber Hoses and Hose Assemblies for Washing-Machines and Dishwashers —Specification for Inlet Hoses Second Edition. 8 pp.

—Household
BSI BS 3999: Part 11-72. (WITHDRAWN) 1972 Methods of Measuring the Performance of Household Electrical Appliances Part 11: Dishwashing Machines. 21 pp.

CSA C22.2 NO 53-1968. Electric Washing Machines (Superceded in Part by C22.2 NO 169); (Rev 1-2 May 1969). 31 pp.

CSA 776 Bull. Electrical Bulletin 776 November 19, 1969 to C22.2 NO 53. 3 pp.

CSA 1169 Bull. Electrical Bulletin 1169 June 27, 1978 to C22.2 NO 53. 2 pp.

SAA AS 2007-88. Performance of Household Electrical Appliances—Dishwashers. 21 pp.

SAA AS 3184-90. Approval and Test Specifications—Electric Dishwashing Machines (This is a Joint Standard with SANZ NZS 3184). 7 pp.

SNZ NZS/AS 3184-90. Approval and Test Specification—Electrical Dishwashers for Household Use (This is a Joint Standard with SAA AS 3184). 7 pp.

—Household—Energy Consumption
CENELEC HD 378-78. Methods to be Used for Measuring Energy Consumption of Electric Automatic Dishwashers for Cold Water Supply Only, for Household Use and for the Purpose of Informing Consumers of It. 5 pp.

CSA CAN/CSA-C373-92. Energy Consumption Test Methods for Household Dishwashers; (Gen Instr 1). 21 pp.

—Household—Noise
BSI BS 6686: Sec 2.3-91. 1991 Methods for Determination of Airborne Acoustical Noise Emitted by Household and Similar Electrical Appliances Part 2: Particular Requirements Section 2.3: Dishwashers. 17 pp.

CENELEC HD 423.2.3 S1-89. Test Code for the Determination of Airborne Acoustical Noise Emitted by Household and Similar Electrical Appliances Part 2: Particular Requirements for Dishwashers. 3 pp.

CENELEC HD 423.2.3 S1-90. Test Code for the Determination by Airborne Acoustical Noise Emitted by Household and Similar Electrical Appliances Part 2: Particular Requirements for Dishwashers. 4 pp.

CENELEC HD 423.2.3 S1/A1-90. AMD 1 Test Code for the Determination by Airborne Acoustical Noise Emitted by Household and Similar Electrical Appliances Part 2: Particular Requirements for Dishwashers. 4 pp.

Dishwashers (Cont.)

—Household—Noise (Cont.)
IEC 704 Pt 2-3-87. Test Code for the Determination of Airborne Acoustical Noise Emitted by Household and Similar Electrical Appliances Part 2: Particular Requirements for Dishwashers First Edition; (Corrigendum—Dec 1989). 30 pp.

—Household—Safety
BSI BS 3456: Sec 102.5-88. (WITHDRAWN) 1988 Safety of Household Electrical Appliances Part 102: Particular Requirements Section 102.5: Dishwashers (Superseded by BS 3456: Section 202.5: 1990). 24 pp.

BSI BS 3456: Sec 202.5-90. 1990 Safety of Household Electrical Appliances Part 202: Particular Requirements Section 202.5: Dishwashers. 28 pp.

BSI BS 3456: Sec 202.5-01. 1990 Amd 1 Safety of Household and Similar Electrical Appliances Part 202: Particular Requirements Section 202.5: Dishwashers (AMD 6899) April 1, 1992 (L). 36 pp.

BSI BS 3456: Sec 202.5-03. 1990 Amd 3 Safety of Household and Similar Electrical Appliances Part 202: Particular Requirements Section 202.5: Dishwashers (AMD 7663) June 15, 1993 (SUPERSEDES BS 3456: SECTION 102. 5: 1988). 56 pp.

BSI BS 3456: Sec 202.5-02. 1990 Amd 2 Safety of Household and Similar Electrical Appliances Part 202: Particular Requirements Section 202.5: Dishwashers (AMD 7348) November 15, 1992 (Supersedes BS 3456: Section 102.5: 1988). 50 pp.

CENELEC HD 257 S2-84. Safety of Household and Similar Electrical Apliances Particular Requirements for Dishwashers. 17 pp.

CENELEC EN 60 335-2-5-89. Safety of Household and Similar Electrical Appliances—Part 2: Particular Requirements for Dishwashers. 12 pp.

CENELEC EN 60 335-2-5/A1-90. AMD 1 Safety of Household and Similar Electrical Appliances—Part 2: Particular Requirements for Dishwashers. 7 pp.

CENELEC EN 60 335-2-5/A2-90. AMD 2 Safety of Household and Similar Electrical Appliances—Part 2: Particular Requirements for Dishwashers. 5 pp.

CENELEC EN 60335-2-5/A3-92. AMD 3 Safety of Household and Similar Electrical Appliances Part 2: Particular Requirements for Dishwashers (IEC 335-2-5:1984/A3:1990). 5 pp.

CENELEC EN 60335-2-58-93. Safety of Household and Similar Electrical Appliances Part 2: Particular Requirements for Commercial Electric Dishwashing Machines (IEC 335-2-58: 1990, Modified). 25 pp.

DIN VDE 0700 Pt 600 (D)-88. Safety of Household and Similar Electrical Appliances; Connection of Washing Machines, Dishwashers and Tumble Dryers to the Supply Mains; German Version prHD 274 S2: 1987 (Oct). 37 pp.

IEC 335 Pt 2-5-92. Safety of Household and Similar Electrical Appliances Part 2: Particular Requirements for Dishwashers Fourth Edition; (CENELEC EN 60335-2-5/A3:1992). 44 pp.

IEC 335 Pt 2-58-90. Safety of Household and Similar Electrical Appliances Part 2: Particular Requirements for Commercial Electric Dishwashing Machines First Edition (CENELEC EN 60335-2-58: 1993). 53 pp.

—Household—0-250 V
CSA CAN/CSA-C22. 2 NO 167-M90. Household Dishwashing Machines; (Gen Instr 1). 56 pp.

CSA 1303 Bull. Electrical Bulletin 1303 January 26, 1981 to C22.2 NO 167. 2 pp.

—Noise
SAA AS 2991.2-91. Acoustics—Method for the Determination of Airborne Noise Emitted by Household and Similar Electrical Appliances—Part 2: Particular Requirements for Dishwashers. 9 pp.

—Symbols
CNS C1143-11-89. Graphical Symbols for Use on Electrical and Electronic Equipment (Laundry Machine, Dish Washer) (Feb)(12491-11).

—Water Pipelines—Connection
BSI BS 6614-91. (WITHDRAWN) 1991 Safety Devices and Water Supply Connections for Washing Machines, Dishwashers and Tumbler Dryers Connected to the Water Supply Mains (Superseded by BS EN 50084: 1992). 18 pp.

BSI BS 6614-85. 1985 Safety Devices and Water Supply Connections for Washing Machines and Dishwashers Connected to the Water Supply Mains. 25 pp.

BSI BS EN 50084-92. 1992 Safety of Household and Similar Electrical Appliances. Requirements for the Connection of Washing Machines, Dishwashers and Tumbler Dryers to the Water Supply Mains (L). 24 pp.

CENELEC HD 274 S1-86. Requirements for the Connection of Washing Machines and Dishwashers to the Water Supply Mains. 41 pp.

INDUSTRY STANDARDS

INTERNATIONAL AND NON-U.S. NATIONAL STANDARDS
SUBJECT INDEX

Dishwashers (Cont.)
—Water Pipelines—Connection (Cont.)
CENELEC HD 274 S2-90. Safety of Household and Similiar Electrical Appliances Requirements for the Connection of Washing Machines, Dishwashers and Tumble Dryers to the Water Supply Mains. 24 pp.
CENELEC EN 50084-92. Safety of Household and Similar Electrical Appliances—Requirements for the Connection of Washing Machines, Dishwashers and Tumbler Dryers to the Water Mains. 21 pp.

—Water Pipelines—Safety
BSI BS 6614-85. 1985 Safety Devices and Water Supply Connections for Washing Machines and Dishwashers Connected to the Water Supply Mains. 25 pp.
BSI BS EN 50084-92. 1992 Safety of Household and Similar Electrical Appliances. Requirements for the Connection of Washing Machines, Dishwashers and Tumbler Dryers to the Water Mains (L). 24 pp.
CENELEC HD 274 S2-90. Safety of Household and Similiar Electrical Appliances Requirements for the Connection of Washing Machines, Dishwashers and Tumble Dryers to the Water Supply Mains. 24 pp.
CENELEC EN 50084-92. Safety of Household and Similar Electrical Appliances—Requirements for the Connection of Washing Machines, Dishwashers and Tumbler Dryers to the Water Mains. 21 pp.

Dishwashing Compounds
See Also: Cleaning Agents; Detergents; Soaps
BSI BS 6585-85. 1985 Guide for Comparative Testing of Performance of Detergents for Domestic Machine Dishwashing. 11 pp.
ISO 7535-84. Surface Active Agents—Detergents for Domestic Machine Dishwashing—Guide for Comparative Testing of Performance First Edition. 11 pp.

—Granular—Machine Washing
CGSB CAN/CGSB-2.20-M87. Machine Dishwashing Compound. 13 pp.
CGSB CAN/CGSB-2.21-92. Chlorinated Machine Dishwashing Compound. 10 pp.

—Liquid—Hand Washing
BSI BS 6584-85. 1985 Guide for Comparative Testing of Performance of Detergents for Hand Dishwashing. 10 pp.
CGSB CAN/CGSB-2.19-92. Hand Dishwashing Compound. 8 pp.
ISO 4198-84. Surface Active Agents—Detergents for Hand Dishwashing—Guide for Comparative Testing of Performance First Edition. 10 pp.
SAA AS 1999-77. Liquid Detergents for Household Hand Dishwashing Amdt 1 July 1978. 8 pp.

—Rinse Additives—Machine Washing
CGSB CAN/CGSB-2.22-92. Concentrated Dishwashing Rinse Additive. 8 pp.

Dishwashing Machines
Use: Dishwashers

Disinfectants
Use For: Antiseptic; Antiseptics
See Also: Antimicrobial Agents; Drugs
BSI BS 2462-86. 1986 Amd 1 Black and White Disinfectant Fluids. 10 pp.
BSI BS 5197-76. 1976 Amd 2 Aromatic Disinfectant Fluids (AMD 5382) October 31, 1986. 10 pp.
SNZ NZS 8302-80. Reference Method for the Evaluation of Disinfectants. 8 pp.

—Agricultural
BSI BS 6734-86. 1986 Method for Determination of the Antimicrobial Efficacy of Disinfectants for Veterinary and Agricultural Use. 8 pp.

—Ammonium Compounds
BSI BS 6424-90. 1990 QAC Based Aromatic Disinfectant Fluids. 4 pp.
BSI BS 6471-84. 1984 Amd 1 Determination of Antimicrobial Value of QAC Disinfectant Formulations. 9 pp.

—Ammonium Compounds—Suspension Testing
BSI BS 3286-60. 1960 Method for Laboratory Evaluation of Disinfectant Activity of Quaternary Ammonium Compounds by Suspension Test Procedure. 20 pp.

—Bactericidal Activity—Suspension Testing
CEN PREN 1040-93. Chemical Disinfectants and Antiseptics—Basic Bactericidal Activity—Test Method and Requirement. 22 pp.

—Calcium Hypochlorite
CGSB CAN/CGSB-15.32-93. Calcium Hypochlorite. 8 pp.
MOD UK DSTAN 68-132-89. Calcium Hypochlorite Issue 1. 11 pp.

Disinfectants (Cont.)
—Calcium Hypochlorite (Cont.)
MOD UK TS 10202A. Calcium Hypochlorite (Superseded by Def Stan 68-132).

—Chemical
BSI BS 7152-91. 1991 Guide to Choice of Chemical Disinfectants. 13 pp.

—Chick-Martin Test
BSI BS 808-86. 1986 Assessing the Efficacy of Disinfectants by the Modified Chick-Martin Test. 10 pp.

—Dairy Equipment
BSI BS 5305-84. 1984 Cleaning and Disinfecting of Plant and Equipment Used in the Dairying Industry. 62 pp.

—Dairy Equipment—Elastomeric Sealant
DIN ENGL 11483 Pt 2-84. Dairy Installations; Cleaning and Disinfection; Consideration of the Action on Sealing Materials (Feb). 2 pp.

—Dairy Equipment—Stainless Steel
DIN ENGL 11483 Pt 1-83. Dairy Installations; Cleaning and Disinfection; Consideration of the Action on Stainless Steel (Jan). 3 pp.

—Food Hygiene
BSI DD 177-88. 1988 Method of Test for the Antimicrobial Activity of Disinfectants in Food Hygiene. 9 pp.

—Glossaries
BSI BS 5283-86. 1986 Glossary of Terms Relating to Disinfectants. 4 pp.

—Medical Equipment—Concentration
BSI BS 6905-87. 1987 Method for Estimation of Concentration of Disinfectants Used in 'Dirty' Conditions in Hospitals by the Modified Kelsey-Sykes Test. 11 pp.

—Rideal-Walker Coefficient
BSI BS 541-85. 1985 Method for Determination of the Rideal-Walker Coefficient of Disinfectants. 11 pp.

—Sodium Hypochlorites
CGSB CAN/CGSB-15.31-93. Sodium Hypochlorite. 7 pp.
MOD UK CS 3137. Sodium Hypochlorite Solution (Withdrawn).

—Tar Acid Content
BSI BS 3265-60. 1960 Amd 1 Determination of Tar Acids in Black and White Disinfectant Fluids. 17 pp.

—Veterinary
BSI BS 6734-86. 1986 Method for Determination of the Antimicrobial Efficacy of Disinfectants for Veterinary and Agricultural Use. 8 pp.

Disinfection Equipment
Use: Sterilizers

Disinfection Systems
Use: Sterilizers

Disinfection
See Also: Disintegrators

—Mechanical Pulps
CPPA C.8P-84. Latency Removal by Hot Disintegration (Domtar Method). 2 pp.

—Pulps
CPPA C.6H-90. Pulp Evaluation—Disintegrator Method. 2 pp.
CPPA C.9P-90. Pulp Evaluation—25 Minute Disintegrator Method. 2 pp.
CPPA C.10P-90. Pulp Disintegration for Unbeaten Properties. 2 pp.
ISO 5263-79. Pulps—Laboratory Wet Disintegration First Edition. 7 pp.

—Refractory Materials
CNS R3113-83. Method of Test for Disintegration of Refractories in an Atmosphere of Carbon Monoxide (Apr)(10200).
SAA AS 1774.9-92. Refractories and Refractory Materials—Physical Test Methods—Part 9: Determination of Resistance to the Disintegrating Effect of Carbon Monoxide. 4 pp.

Disintegration Testing
Use: Disintegration

Disintegrators
See Also: Comminution; Disintegration; Grinders; Paper Machines; Shredders

Disintegrators (Cont.)
—Pulps
CPPA M-14-75. British Standard Disintegrator. 1 p.

Disk Brakes
See Also: Air Brakes; Automotive Equipment; Brake Pads; Brakes; Drum Brakes; Hydraulic Brakes

—Cylinders—Boots—Automotive
JIS D 2611-84. Rubber Boots of Hydraulic Disc Brakes for Automobiles.

—Cylinders—Seals—Automotive
CNS D2176-83. Elastomeric Seals of Hydraulic Disc Brake Cylinders for Automobiles (Dec)(10711).
ISO 4930-78. Road Vehicles—Elastomeric Seals for Hydraulic Disc Brake Cylinders Using a Non-Petroleum Base Hydraulic Brake Fluid (Service Temperature 150 Degrees Celsius Maximum) First Edition; (Erratum—June 1979). 9 pp.
ISO 6119-80. Road Vehicles—Elastomeric Seals for Hydraulic Disc Brake Cylinders Using a Non-Petroleum Base Hydraulic Brake Fluid (Service Temperature 120 Degrees Celcius Maximum) First Edition. 8 pp.
ISO 7632-85. Road Vehicles—Elastomeric Seals for Hydraulic Disc Brake Cylinders Using a Petroleum Base Hydraulic Brake Fluid (Service Temperature 120 Degrees Celsius Max.) First Edition. 8 pp.
JIS D 2609-82. Elastomeric Seals of Hydraulic Disc Brake Cylinders for Automobiles.

Disk Cartridges
Use For: Flexible Disk Cartridges
See Also: Computer Tapes; Disk Drives (Computer); Information Interchange; Magnetic Disks; Magnetic Materials; Video Disks
BSI BS 6958: Part 1-88. 1988 Amd 1 Guide to Construction and Use of 90mm (3.5 in) Flexible Disk Cartridges for Data Interchange, Using Modified Frequency Modulation Recording at 7958 ftprad, 5.3 tpmm (135 tpi) on 80 Tracks on Each Side Part 1: Dimensional, Physical and Magnetic. 30 pp.
BSI BS 6958: Part 1-02. 1988 Amd 2 Construction and Use of 90mm (3.5 in) Flexible Disk Cartridges for Data Interchange, Using Modified Frequency Modulation Recording at 7958 ftprad, 5.3 tpmm (135 tpi) on 80 Tracks on Each Side Part 1: Dimensional, Physical and Magnetic. 36 pp.
BSI BS ISO/IEC 10090-92. 1992 Information Technology—90 mm Optical Disk Cartridges, Rewritable and Read Only, for Data Interchange. 107 pp.
BSI BS ISO/IEC 11560-92. 1992 Information Technology—Information Interchange on 130 mm Optical Disk Cartridges Using the Magneto-Optical Effect, for Write Once, Read Multiple Functionality (S). 92 pp.
ECMA ECMA 54-82. Data Interchange on 200 mm Flexible Disk Cartridges Using Two-Frequency Recording at 13262 ftprad on One Side. 46 pp.
ECMA ECMA 59-79. Data Interchange on 200 mm Flexible Disk Cartridges Using Two-Frequency Recording at 13262 ftprad on Both Sides. 44 pp.
ECMA ECMA 69-81. Data Interchange on 200 mm Flexible Disk Cartridges Using MFM Recording at 7958 ftprad on Both Sides. 48 pp.
ECMA ECMA 70-86. Data Interchange on 130 mm Flexible Disk Cartridges Using MFM Recording at 7958 ftprad on 40 Tracks on Each Side. 60 pp.
ECMA ECMA 78-86. Data Interchange on 130 mm Flexible Disk Cartridges Using MFM Recording at 7958 ftprad on 80 Tracks on Each Side. 59 pp.
ECMA ECMA 99-85. Data Interchange on 130 mm Flexible Disk Cartridges Using MFM Recording at 13262 ftprad on Both Sides 3.8 Tracks Per mm. 57 pp.
ECMA ECMA 100-88. Data Interchange on 90 mm Flexible Disk Cartridges Using MFM Recording at 7958 Ftprad on Both Sides 5.3 Tracks per mm. 57 pp.
ECMA ECMA 125-87. Data Interchange on 90mm Flexible Disk Cartridges Using MFM Recording at 15916 ftprad on 80 Tracks on Each Side. 57 pp.
ECMA ECMA 153-91. Information Interchange on 130mm Optical Disk Cartridges of the Write Once, Read Multiple (WORM) Type, Using the Magneto-Optical Effect. 94 pp.
ECMA ECMA 154-91. Data Interchange on 90 mm Optical Disk Cartridges Read Only and Rewritable, M.O.. 105 pp.
IEC 10090-92. Information Technology—90 mm Optical Disk Cartridges, Rewritable and Read Only, for Data Interchange First Edition. 105 pp.
IEC 11560-92. Information Technology—Information Interchange on 130 mm Optical Disk Cartridges Using the Magneto-Optical Effect, for Write Once, Read Multiple Functionality First Edition. 90 pp.
ISO 10090-92. Information Technology—90 mm Optical Disk Cartridges, Rewritable and Read Only, for Data Interchange First Edition. 105 pp.

INTERNATIONAL AND NON-U.S. NATIONAL STANDARDS
SUBJECT INDEX
Disk

Disk Cartridges *(Cont.)*

ISO 11560-92. Information Technology—Information Interchange on 130 mm Optical Disk Cartridges Using the Magneto-Optical Effect, for Write Once, Read Multiple Functionality First Edition. 90 pp.

JIS X 6201-91. 200 mm Flexible Disk Cartridges. 20 pp.

JIS X 6211-86. 130 mm Flexible Disk Cartridges.

JIS X 6221-87. 90 mm Flexible Disk Cartridges (7958 ftprad). 33 pp.

JIS X 6223-87. 90 mm Flexible Disk Cartridges (13262/15916 ftprad). 32 pp.

JTC1 7487 Pt 3-86. Information Processing—Data Interchange on 130 mm (5.25 in) Flexible Disk Cartridges Using Modified Frequency Modulation Recording at 7 958 ftprad, 1,9 tpmm (48 tpi), on Both Sides—Part 3: Track Format B Second Edition; (Corrigendum 1-1991). 13 pp.

JTC1 10090-92. Information Technology—90 mm Optical Disk Cartridges, Rewritable and Read Only, for Data Interchange First Edition. 105 pp.

JTC1 11560-92. Information Technology—Information Interchange on 130 mm Optical Disk Cartridges Using the Magneto-Optical Effect, for Write Once, Read Multiple Functionality First Edition. 90 pp.

OSI ISO/IEC DIS 10994-91. Information Technology—Data Interchange on 90mm Flexible Disk Cartridges Using MFM Recording at 31 831 ftprod on 80 Tracks on Each Side—ISO Type 303. 55 pp.

—Designations

BSI BS ISO/IEC 9983-89. 1989 Information Processing Systems—Designation of Unrecorded Flexible Disk Cartridges (S). 10 pp.

BSI BS ISO/IEC 9983-01. 1989 Amd 1 Information Processing Systems—Designation of Unrecorded Flexible Disk Cartridges (AMD 6862) October 31, 1991. 17 pp.

CEN EN 29 983-91. Information Processing Systems of Unrecorded Flexible Disk Cartridges. 3 pp.

IEC 9983-89. Information Processing Systems—Designation of Unrecorded Flexible Disk Cartridges First Edition. 7 pp.

ISO 9983-89. Information Processing Systems—Designation of Unrecorded Flexible Disk Cartridges First Edition. 7 pp.

JTC1 9983-89. Information Processing Systems—Designation of Unrecorded Flexible Disk Cartridges First Edition. 7 pp.

OSI ISO IEC 9983-89. Information Processing Systems—Designation of Unrecorded Flexible Disk Cartridges. 7 pp.

—File Structure

BSI BS 6542: Part 1-84. 1984 Amd 1 File Structure and Labelling of Flexible Disk Cartridges for Information Interchange Part 1: Sequential File Allocation Structure Throughout the Standard (AMD 5967) September 30, 1988. 27 pp.

BSI BS 6542: Part 2-88. 1988 File Structure and Labelling of Flexible Disk Cartridges for Information Interchange Part 2: Variable File Allocation Structure. 28 pp.

BSI BS 6542: Part 2-01. 1988 Amd 1 File Structure and Labelling of Flexible Disk Cartridges for Information Interchange Part 2: Specification for Variable File Allocation Structure (ISO 9293: 1987) (AMD 7046) February 28, 1992. 34 pp.

CEN EN 29 293-89. Information Processing Volume and File Structure of Flexible Disk Cartridges for Information Interchange. 2 pp.

DIN ENGL EN 29293-91. Information Processing; Volume and File Structure of Flexible Disk Cartridges for Information Interchange (ISO 9293: 1987) (Dec). 30 pp.

ECMA ECMA 91-84. File Structure and Labelling for Information Interchange. 49 pp.

ECMA ECMA 107-85. Volume and File Structure of Flexible Disk Cartridges for Information Interchange. 50 pp.

ISO 7665-83. Information Processing—File Structure and Labelling of Flexible Disk Cartridges for Information Interchange First Edition. 25 pp.

ISO 9293-87. Information Processing—Volume and File Structure of Flexible Disk Cartridges for Information Interchange First Edition. 30 pp.

JIS X 0603-88. File Structure and Labelling of Flexible Disk Cartridges for Information Interchange. 38 pp.

JIS X 0605-90. Volume and File Structure of Flexible Disk Cartridges for Information Interchange (ISO 9293:1987).

JTC1 7665-83. Information Processing—File Structure and Labelling of Flexible Disk Cartridges for Information Interchange First Edition. 25 pp.

JTC1 9293-87. Information Processing—Volume and File Structure of Flexible Disk Cartridges for Information Interchange First Edition. 30 pp.

Disk Cartridges *(Cont.)*

—Handling

BSI BS 4783: Part 1-88. 1988 Storage, Transportation and Maintenance of Magnetic Media for Use in Data Processing and Information Storage Part 1: Recommendations for Disk Packs, Storage Modules and Disk Cartridges. 25 pp.

BSI BS 4783: Part 1-01. 1988 Amd 1 Storage, Transportation and Maintenance of Media for Use in Data Processing and Information Storage Part 1: Recommendations for Disk Packs, Storage Modules and Disk Cartridges (AMD 6887) February 28, 1992. 26 pp.

BSI BS 4783: Part 3-88. 1988 Storage, Transportation and Maintenance of Magnetic Media for Use in Data Processing and Information Storage Part 3: Recommendations for Flexible Disk Cartridges. 17 pp.

BSI BS 4783: Part 3-01. 1988 Amd 1 Storage, Transportation and Maintenance of Media for Use in Data Processing and Information Storage Part 3: Recommendations for Flexible Disk Cartridges (AMD 6889) February 28, 1992. 18 pp.

—Identification Systems

BSI BS 6542: Part 1-84. 1984 Amd 1 File Structure and Labelling of Flexible Disk Cartridges for Information Interchange Part 1: Sequential File Allocation Structure Throughout the Standard (AMD 5967) September 30, 1988. 27 pp.

ECMA ECMA 91-84. File Structure and Labelling for Information Interchange. 49 pp.

ISO 7665-83. Information Processing—File Structure and Labelling of Flexible Disk Cartridges for Information Interchange First Edition. 25 pp.

JIS X 0603-88. File Structure and Labelling of Flexible Disk Cartridges for Information Interchange. 38 pp.

JTC1 7665-83. Information Processing—File Structure and Labelling of Flexible Disk Cartridges for Information Interchange First Edition. 25 pp.

—Magnetic Measurement

BSI BS 5356: Part 1-76. 1976 Magnetic Single-Disk Cartridges (Top Loading) for Data Processing Part 1: Mechanical and Magnetic Properties. 20 pp.

BSI BS 6957: Part 1-88. (WITHDRAWN) 1988 Guide to Construction & Use of 130mm (5.25 in) Flexible Disk Cartridges for Data Interchange, Using Modified Frequency Modulation Recording at 13 262 ftprad, on 80 Tracks on Each Side Part 1: Dimensional Phys. and Magnetic Char. (Renumbed as BS EN 28630-1: 1992). 20 pp.

BSI BS 6958: Part 1-88. 1988 Amd 1 Guide to Construction and Use of 90mm (3.5 in) Flexible Disk Cartridges for Data Interchange, Using Modified Frequency Modulation Recording at 7958 ftprad, 5.3 tpmm (135 tpi) on 80 Tracks on Each Side Part 1: Dimensional, Physical and Magnetic. 30 pp.

BSI BS 6958: Part 1-02. 1988 Amd 2 Construction and Use of 90mm (3.5 in) Flexible Disk Cartridges for Data Interchange, Using Modified Frequency Modulation Recording at 7958 ftprad, 5.3 tpmm (135 tpi) on 80 Tracks on Each Side Part 1: Dimensional, Physical and Magnetic. 36 pp.

BSI BS 7023: Part 1-88. 1988 Guide to Construction & Use of 130mm (5.25 in) Flexible Disk Cartridges for Data Interchange Using Modified Frequency Modulation Recording at 7958 ftprad, 3.8 tpmm (96 tpi) on Both Sides Part 1: Dimensional, Physical and Magnetic Characteristics. 20 pp.

BSI BS 7023: Part 1-01. 1988 Amd 1 Construction and Use of 130mm (5.25 in) Flexible Disk Cartridges for Data Interchange Using Modified Frequency Modulation Recording at 7958 ftprad, 3.8 tpmm (96 tpi) on Both Sides Part 1: Dimensional, Physical and Magnetic Characteristics. 25 pp.

BSI BS 7024: Part 1-88. 1988 Construction and Use of 130mm (5.25 in) Flexible Disk Cartridges for Data Interchange Using Modified Frequency Modulation Recording at 7958 ftprad, 1.9 tpmm (48 tpi) on Both Sides Part 1: Dimensional, Physical and Magnetic Characteristics. 20 pp.

BSI BS 7024: Part 1-01. 1988 Amd 1 Construction and Use of 130mm (5.25 in) Flexible Disk Cartridges for Data Interchange Using Modified Frequency Modulation Recording at 7958 ftprad, 1.9 tpmm (48 tpi) on Both Sides Part 1: Dimensional, Physical and Magnetic Characteristics. 25 pp.

BSI BS EN 28630-1-92. 1992 Construction and Use of 130 mm (5.25 in) Flexible Disk Cartridges for Data Interchange, Using Modified Frequency Modulation Recording at 13 262 ftprad, on 80 Tracks Each Side Part 1: Dimensional, Physical and Magnetic. 26 pp.

BSI BS ISO/IEC 9529-1-89. 1989 Information Processing Systems—Data Interchange on 90 mm (3, 5 in) Flexible Disk Cartridges Using Modified Frequency Modulation Recording at 15 916 ftprad, on 80 Tracks on Each Side Part 1: Dimensional, Physical and Magnetic. 47 pp.

Disk Cartridges *(Cont.)*

—Magnetic Measurement *(Cont.)*

BSI BS ISO/IEC 9529-1-01. 1989 Amd 1 Information Processing Systems—Data Interchange on 90 mm (3, 5 in) Flexible Disk Cartridges Using Modified Frequency Modulation Recording at 15 916 ftprad, on 80 Tracks on Each Side Part 1: Dimensional, Physical and Magnetic. 52 pp.

BSI BS ISO/IEC 10994-92. 1992 Information Technology—Data Interchange on 90 mm Flexible Disk Cartridges Using Modified Frequency Modulation Recording at 31 831 ftprad on 80 Tracks on Each Side—ISO Type 303. 54 pp.

CEN EN 27 487 (Part 1)-89. Information Processing Data Interchange on 130 mm (5.25) Flexible Disk Cartridges Using Modified Frequency Modulation Recording at 7 958 ftprad. 1.9 tpmm (48 tpi) on Both Sides—Part 1: Dimensional, Physical and Magnetic Characteristics. 21 pp.

CEN EN 28 378 (Part 1)-89. Information Processing Data Interchange on 130mm (5.25in) Flexible Disk Cartridges Using Modified Frequency Modulation Recording at 7 958 ftprad, 3.8tpmm (96tpi) on Both Sides Part 1: Dimensional, Physical and Magnetic Characteristics. 21 pp.

CEN EN 28630-1-92. Information Processing—Data Interchange on 130 mm (5.25 in) Flexible Disk Cartridges Using Modified Frequency Modulation Recording at 13 262 ftprad, on 80 Tracks Each Side Part 1 Dimensional, Physical and Magnetic Characteristics; (ISO 8630-1: 1987, Ed. 1). 19 pp.

CEN EN 28 860 (Part 1)-89. Information Processing Data Interchange on 90mm (3.5in) Flexible Disk Cartridges Using Modified Frequency Modulation of 7 958 ftprad 5.3tpmm (135tpi) on Both Sides Part 1: Dimensional Physical and Magnetic Characteristics. 2 pp.

CEN EN 29529-1-91. Information Processing Systems—Data Interchange on 90 mm (3,5 in) Flexible Disk Cartridges Using Modified Frequency Modulation Recordings at 15 916 ftprad, on 80 Tracks on Each Side—Part 1: Dimensional, Physical and Magnetic Characteristics. 3 pp.

DIN ENGL EN 27487 Pt 1-91. Information Processing; Data Interchange on 130 mm (5,25 in) Flexible Disk Cartridges Using Modified Frequency Modulation Recording at 7 958 ftprad, 1,9 tpmm (48 tpi), on Both Sides; Dimensional, Physical and Magnetic Characteristics (ISO 7487-1: 1985) (Dec). 20 pp.

DIN ENGL EN 28378 Pt 1-91. Information Processing; Data Interchange on 130 mm (5,25 in) Flexible Disk Cartridges Using Modified Frequency Modulation Recording at 7 958 ftprad, 3,8 tpmm (96 tpi), on Both Sides; Dimensional, Physical and Magnetic Characteristics (ISO 8378-1: 1986) (Dec). 19 pp.

DIN ENGL EN 28860 Pt 1-91. Information Processing; Data Interchange on 90 mm (3,5 in) Flexible Disk Cartridges Using Modified Frequency Modulation Recording at 7 958 ftprad on 80 Tracks on Each Side; Dimensional, Physical and Magnetic Characteristics (ISO 8860-1: 1987) (Dec). 28 pp.

DIN ENGL EN 29529 Pt 1-92. Information Processing Systems; Data Interchange on 90 mm (3, 5 in) Flexible Disk Cartridges Using Modified Frequency Modulation Recording at 15 916 ftprad, on 80 Tracks on Each Side; Dimensional, Physical and Magnetic Characteristics. 44 pp.

ECMA ECMA 38-73. Mechanical, Physical and Magnetic Characteristics of Interchangeable Single Disk Cartridges (Top Loaded). 44 pp.

ECMA ECMA 147-90. Data Interchange on 90mm Flexible Disk Cartridges Using MFM Recording at 31831 ftprad on 80 Tracks on Each Side—ISO Type 303. 53 pp.

IEC DIS 7487 Pt 1-92. Information Technology—Data Interchange on 130 mm (5,25 in) Flexible Disk Cartridges Using Modified Frequency Modulation Recording at 7 958 ftprad, 1,9 tpmm (48 tpi), on Both Sides—ISO Type 202—Part 1: Dimensional, Physical and Magnetic Character-istics ***CD-ROM ONLY***. 22 pp.

IEC 8860 Pt 1-87. Information Processing—Data Interchange on 90 mm (3.5 in) Flexible Disk Cartridges Using Modified Frequency Modu-lation Recording at 7 958 ftprad on 80 Tracks on Each Side—Part 1: Dimensional, Physical and Magnetic Character-istics (Corrigendum 1-1990). 29 pp.

IEC 9529 Pt 1-89. Information Processing Systems—Data Inter-change on 90 mm (3,5 in) Flexible Disk Cartridges Using Modified Frequency Modulation Recording at 15 916 ftprad, on 80 Tracks on Each Side—Part 1: Dimensional, Physical and Magnetic Characteristics First Edition. 45 pp.

IEC 10994-92. Information Technology—Data Interchange on 90 mm Flexible Disk Cartridges Using Modified Frequency Modulation Recording at 31 831 ftprad on 80 Tracks on Each Side—ISO Type 303 First Edition. 51 pp.

IEC DIS 13422-93. Information Technology—90 mm Flexible Disk Cartridges—10MB, 33 157 ftprad, for Sector Servo Tracking ***CD-ROM ONLY***. 68 pp.

INDUSTRY STANDARDS
811

Disk
INTERNATIONAL AND NON-U.S. NATIONAL STANDARDS
SUBJECT INDEX

Disk Cartridges *(Cont.)*

—Magnetic Measurement *(Cont.)*

ISO 3562-76. Information Processing—Interchangeable Magnetic Single Disk Cartridge (Top Loaded)—Physical and Magnetic Characteristics First Edition. 27 pp.

ISO 5654 Pt 1-84. Information Processing—Data Interchange on 200 mm (8 in) Flexible Disk Cartridges Using Two-Frequency Recording at 13 262 ftprad on One Side—Part 1: Dimensional, Physical and Magnetic Characteristics Second Edition. 19 pp.

ISO 6596 Pt 1-85. Information Processing—Data Interchange on 130 mm (5.25 in) Flexible Disk Cartridges Using Two-Frequency Recording at 7 958 ftprad, 1,9 tpmm (48 tpi), on One Side—Part 1: Dimensional, Physical and Magnetic Characteristics Second Edition. 19 pp.

ISO 7065 Pt 1-85. Information Processing—Data Interchange on 200 mm (8 in) Flexible Disk Cartridges Modified Frequency Modulation Recording at 13 262 ftprad, 1,9 tpmm (48 tpi), on Both Sides—Part 1: Dimensional, Physical and Magnetic Characteristics Second Edition. 19 pp.

ISO 7487 Pt 1-85. Information Processing—Data Interchange on 130 mm (5.25 in) Flexible Disk Cartridges Using Modified Frequency Modulation Recording at 7 958 ftprad, 1,9 tpmm (48 tpi), on Both Sides—Part 1: Dimensional, Physical and Magnetic Characteristics First Edition. 19 pp.

ISO DIS 7487 Pt 1-92. Information Technology—Data Interchange on 130 mm (5,25 in) Flexible Disk Cartridges Using Modified Frequency Modulation Recording at 7 958 ftprad, 1,9 tpmm (48 tpi), on Both Sides—ISO Type 202—Part 1: Dimensional, Physical and Magnetic Character-istics ***CD-ROM ONLY***. 22 pp.

ISO 7901-84. Information Processing—Unrecorded, Hard-Sectored, 130 mm (5.25 in) Flexible Disk Cartridges, One or Two-Sided Use—Dimensional, Physical, and Magnetic Characteristics First Edition. 4 pp.

ISO 8378 Pt 1-86. Information Processing—Data Interchange on 130 mm (5.25 in) Flexible Disk Cartridges Using Modified Frequency Modulation Recording at 7 958 ftprad, 3,8 tpmm (96 tpi), on Both Sides—Part 1: Dimensional, Physical and Characteristics First Edition. 19 pp.

ISO 8630 Pt 1-87. Information Processing—Data Interchange on 130 mm (5.25 in) Flexible Disk Cartridges Using Modified Frequency Modu-lation Recording at 13 262 ftprad, on 80 Tracks on Each Side—Part 1: Dimensional, Physical and Magnetic Character-istics First Edition; (CEN EN 28630-1:1992). 20 pp.

ISO 8860 Pt 1-87. Information Processing—Data Interchange on 90 mm (3.5 in) Flexible Disk Cartridges Using Modified Frequency Modu-lation Recording at 7 958 ftprad on 80 Tracks on Each Side—Part 1: Dimensional, Physical and Magnetic Character-istics (Corrigendum 1-1990). 29 pp.

ISO 9529 Pt 1-89. Information Processing Systems—Data Inter-change on 90 mm (3,5 in) Flexible Disk Cartridges Using Modified Frequency Modulation Recording at 15 916 ftprad, on 80 Tracks on Each Side—Part 1: Dimensional, Physical and Magnetic Characteristics First Edition. 45 pp.

ISO 10994-92. Information Technology—Data Interchange on 90 mm Flexible Disk Cartridges Using Modified Frequency Modulation Recording at 31 831 ftprad on 80 Tracks on Each Side—ISO Type 303 First Edition. 51 pp.

ISO DIS 13422-93. Information Technology—90 mm Flexible Disk Cartridges—10MB, 33 157 ftprad, for Sector Servo Tracking ***CD-ROM ONLY***. 68 pp.

JIS X 6157-80. Physical Characteristics of Interchangeable Magnetic Single Disk Cartridge (Top Loaded).

JIS X 6226-92. 90 mm Flexible Disk Cartridges—31831 ftprad. 46 pp.

JTC1 3562-76. Information Processing—Interchangeable Magnetic Single Disk Cartridge (Top Loaded)—Physical and Magnetic Characteristics First Edition. 27 pp.

JTC1 5654 Pt 1-84. Information Processing—Data Interchange on 200 mm (8 in) Flexible Disk Cartridges Using Two-Frequency Recording at 13 262 ftprad on One Side—Part 1: Dimensional, Physical and Magnetic Characteristics Second Edition. 19 pp.

JTC1 6596 Pt 1-85. Information Processing—Data Interchange on 130 mm (5.25 in) Flexible Disk Cartridges Using Two-Frequency Recording at 7 958 ftprad, 1,9 tpmm (48 tpi), on One Side—Part 1: Dimensional, Physical and Magnetic Characteristics Second Edition. 19 pp.

JTC1 7065 Pt 1-85. Information Processing—Data Interchange on 200 mm (8 in) Flexible Disk Cartridges Modified Frequency Modulation Recording at 13 262 ftprad, 1,9 tpmm (48 tpi), on Both Sides—Part 1: Dimensional, Physical and Magnetic Characteristics Second Edition. 19 pp.

JTC1 7487 Pt 1-85. Information Processing—Data Interchange on 130 mm (5.25 in) Flexible Disk Cartridges Using Modified Frequency Modulation Recording at 7 958 ftprad, 1,9 tpmm (48 tpi), on Both Sides—Part 1: Dimensional, Physical and Magnetic Characteristics First Edition. 19 pp.

JTC1 DIS7487 Pt 1-92. Information Technology—Data Interchange on 130 mm (5,25 in) Flexible Disk Cartridges Using Modified Frequency Modulation Recording at 7 958 ftprad, 1,9 tpmm (48 tpi), on Both Sides—ISO Type 202—Part 1: Dimensional, Physical and Magnetic Character-istics ***CD-ROM ONLY***. 22 pp.

JTC1 7901-84. Information Processing—Unrecorded, Hard-Sectored, 130 mm (5.25 in) Flexible Disk Cartridges, One or Two-Sided Use—Dimensional, Physical, and Magnetic Characteristics First Edition. 4 pp.

JTC1 8378 Pt 1-86. Information Processing—Data Interchange on 130 mm (5.25 in) Flexible Disk Cartridges Using Modified Frequency Modulation Recording at 7 958 ftprad, 3,8 tpmm (96 tpi), on Both Sides—Part 1: Dimensional, Physical and Characteristics First Edition. 19 pp.

JTC1 8630 Pt 1-87. Information Processing—Data Interchange on 130 mm (5.25 in) Flexible Disk Cartridges Using Modified Frequency Modulation Recording at 13 262 ftprad, on 80 Tracks on Each Side—Part 1: Dimensional, Physical and Magnetic Characteristics First Edition. 20 pp.

JTC1 8860-87. Information Processing—Data Interchange on 90 mm (3.5 in) Flexible Disk Cartridges Using Modified Frequency Modu-lation Recording at 7 958 ftprad on 80 Tracks on Each Side—Part 1: Dimensional, Physical and Magnetic Character-istics (Corrigendum 1-1990). 29 pp.

JTC1 9529-89. Information Processing Systems—Data Inter-change on 90 mm (3,5 in) Flexible Disk Cartridges Using Modified Frequency Modulation Recording at 15 916 ftprad, on 80 Tracks on Each Side—Part 1: Dimensional, Physical and Magnetic Characteristics First Edition. 45 pp.

JTC1 10994-92. Information Technology—Data Interchange on 90 mm Flexible Disk Cartridges Using Modified Frequency Modulation Recording at 31 831 ftprad on 80 Tracks on Each Side—ISO Type 303 First Edition. 51 pp.

JTC1 DIS13422-93. Information Technology—90 mm Flexible Disk Cartridges—10MB, 33 157 ftprad, for Sector Servo Tracking ***CD-ROM ONLY***. 68 pp.

OSI ISO 3562-76. Information Processing—Interchangeable Magnetic Single Disk Cartridge (Top Loaded)—Physical and Magnetic Characteristics. 27 pp.

OSI ISO 5654-1-84. Information Processing—Data Interchange on 200mm (8in) Flexible Disk Cartridges Using Two-Frequency Recording at 13 262 ftprad on One Side—Part 1: Dimensional, Physical and Magnetic Characteristics. 19 pp.

OSI ISO 6596-1-85. Information Processing—Data Interchange on 130mm (5.25in) Flexible Disk Cartridges Using Two-Frequency Recording Recording at 7 958 ftprad, 1,9tpmm on One Side—Characteristics. 19 pp.

OSI ISO 7065-1-85. Information Processing—Data Interchange on 200mm(8in) Flexible Disk Cartridges Using Modified Frequency Modulation Recording at 13 262 ftprad, 1,9tpmm (48tpi) on Both Sides—Part 1: Dimensional, Physical and Magnetic Characteristics. 19 pp.

OSI ISO 7901-84. Information Processing—Unrecorded, Hard-Sectored 130mm (5.25in) Flexible Disk Cartridges, One or Two-Sided Use—Dimensional, Physical and Magnetic Characteristics. 4 pp.

OSI ISO IEC 9529-1-89. Information Processing Systems—Data Interchange on 90 mm (3,5 in) Flexible Disk Cartridges Using Modified Frequency Modulation Recording at 15 916 ftprad, on 80 Tracks on Each Side—Part 1: Dimensional, Physical and Magnetic Characteristics. 45 pp.

SNZ NZS/AS 4116-93. Information Processing Systems—Data Interchange on 90 mm Flexible Disk Cartridges Using Modified Frequency Modulation Recording, at 31 831 ftprad on 80 Tracks on Each Side—ISO Type 303 (ISO/IEC 10994:1992) (This is a Joint Standard with SAA AS 4116). 45 pp.

—Mechanical Properties

BSI BS 5356: Part 1-76. 1976 Magnetic Single-Disk Cartridges (Top Loading) for Data Processing Part 1: Mechanical and Magnetic Properties. 20 pp.

BSI BS ISO/IEC 10994-92. 1992 Information Technology—Data Interchange on 90 mm Flexible Disk Cartridges Using Modified Frequency Modulation Recording at 31 831 ftprad on 80 Tracks on Each Side—ISO Type 303. 54 pp.

ECMA ECMA 38-73. Mechanical, Physical and Magnetic Characteristics of Interchangeable Single Disk Cartridges (Top Loaded). 44 pp.

Disk Cartridges *(Cont.)*

—Mechanical Properties *(Cont.)*

ECMA ECMA 147-90. Data Interchange on 90mm Flexible Disk Cartridges Using MFM Recording at 31831 ftprad on 80 Tracks on Each Side—ISO Type 303. 53 pp.

IEC 10994-92. Information Technology—Data Interchange on 90 mm Flexible Disk Cartridges Using Modified Frequency Modulation Recording at 31 831 ftprad on 80 Tracks on Each Side—ISO Type 303 First Edition. 51 pp.

IEC DIS 13422-93. Information Technology—90 mm Flexible Disk Cartridges—10MB, 33 157 ftprad, for Sector Servo Tracking ***CD-ROM ONLY***. 68 pp.

ISO 10994-92. Information Technology—Data Interchange on 90 mm Flexible Disk Cartridges Using Modified Frequency Modulation Recording at 31 831 ftprad on 80 Tracks on Each Side—ISO Type 303 First Edition. 51 pp.

ISO DIS 13422-93. Information Technology—90 mm Flexible Disk Cartridges—10MB, 33 157 ftprad, for Sector Servo Tracking ***CD-ROM ONLY***. 68 pp.

JIS X 6226-92. 90 mm Flexible Disk Cartridges—31831 ftprad. 46 pp.

JTC1 10994-92. Information Technology—Data Interchange on 90 mm Flexible Disk Cartridges Using Modified Frequency Modulation Recording at 31 831 ftprad on 80 Tracks on Each Side—ISO Type 303 First Edition. 51 pp.

JTC1 DIS13422-93. Information Technology—90 mm Flexible Disk Cartridges—10MB, 33 157 ftprad, for Sector Servo Tracking ***CD-ROM ONLY***. 68 pp.

SNZ NZS/AS 4116-93. Information Processing Systems—Data Interchange on 90 mm Flexible Disk Cartridges Using Modified Frequency Modulation Recording, at 31 831 ftprad on 80 Tracks on Each Side—ISO Type 303 (ISO/IEC 10994:1992) (This is a Joint Standard with SAA AS 4116). 45 pp.

—Optical

BSI BS EN 29171-1-93. 1993 Amd 1 Information Technology—130 mm Optical Disk Cartridge, Write Once, for Information Interchange—Part 1: Unrecorded Optical Disk Cartridge (AMD 7796) August 15, 1993 (S). 56 pp.

BSI BS EN 29171-2-93. 1993 Amd 1 Information Technology—130 mm Optical Disk Cartridge, Write Once, for Information Interchange—Part 2: Recording Format (AMD 7797) August 15, 1993 (S). 72 pp.

BSI BS ISO/IEC 9171-1-90. 1990 Information Technology—130 mm Optical Disk Cartridge, Write Once, for Information Interchange—Part 1: Unrecorded Optical Disk Cartridge (Renumbered by BS EN 29171-1: 1993). 50 pp.

BSI BS ISO/IEC 9171-2-90. 1990 Information Technology—130 mm Optical Disk Cartridge, Write Once, for Information Interchange—Part 2: Recording Format (Renumbered as BS EN 29171-2: 1993). 66 pp.

BSI BS ISO/IEC 10089-91. 1991 Information Technology—130 mm Rewritable Optical Disk Cartridge for Information Interchange. 107 pp.

CEN EN 29171-1-93. Information Technology—130 mm Optical Disk Cartridge, Write Once, for Information Interchange—Part 1: Unrecorded Optical Disk Cartridge (ISO/IEC 9171-1: 1990). 49 pp.

CEN EN 29171-2-93. Information Technology—130 mm Optical Disk Cartridge, Write Once, for Information Interchange—Part 2: Recording Format (ISO/IEC 9171-2: 1990). 65 pp.

ECMA ECMA 154-91. Data Interchange on 90 mm Optical Disk Cartridges Read Only and Rewritable, M.O.. 105 pp.

IEC 9171 Pt 1-90. Information Technology—130 mm Optical Disk Cartridge, Write Once, for Information Interchange—Part 1: Unrecorded Optical Disk Cartridge First Edition (CEN EN 29171-1: 1993). 47 pp.

IEC 9171 Pt 2-90. Information Technology—130 mm Optical Disk Cartridge, Write Once, for Information Interchange—Part 2: Recording Format First Edition (CEN 29171-2: 1993). 63 pp.

IEC 10089-91. Information Technology—130 mm Rewritable Optical Disk Cartridge for Information Interchange First Edition. 105 pp.

ISO 9171 Pt 1-90. Information Technology—130 mm Optical Disk Cartridge, Write Once, for Information Interchange—Part 1: Unrecorded Optical Disk Cartridge First Edition (CEN EN 29171-1: 1993). 47 pp.

ISO 9171 Pt 2-90. Information Technology—130 mm Optical Disk Cartridge, Write Once, for Information Interchange—Part 2: Recording Format First Edition (CEN EN 29171-2: 1993). 63 pp.

ISO 10089-91. Information Technology—130 mm Rewritable Optical Disk Cartridge for Information Interchange First Edition. 105 pp.

812
INDEX and DIRECTORY of

INTERNATIONAL AND NON-U.S. NATIONAL STANDARDS
SUBJECT INDEX
Disk

Disk Cartridges *(Cont.)*

—Optical *(Cont.)*

JTC1 9171-90. Information Technology—130 mm Optical Disk Cartridge, Write Once, for Information Interchange—Part 1: Unrecorded Optical Disk Cartridge First Edition. 47 pp.

JTC1 9171-90. Information Technology—130 mm Optical Disk Cartridge, Write Once, for Information Interchange—Part 2: Recording Format First Edition. 63 pp.

JTC1 10089-91. Information Technology—130 mm Rewritable Optical Disk Cartridge for Information Interchange First Edition. 105 pp.

OSI ISO/IEC 9171-1-90. Information Technology—130 mm Optical Disk Cartridge, Write Once, for Information Interchange—Part 1: Unrecorded Optical Disk Cartridge. 47 pp.

OSI ISO/IEC DIS 9171-1.2-89. Information Technology—130 mm Optical-Disk Cartridge, Write-Once, for Information Interchange—Part 1: Unrecorded Optical Disk Cartridge. 46 pp.

OSI ISO/IEC 9171-2-90. Information Technology—130 mm Optical Disk Cartridge, Write Once, for Information Interchange—Part 2: Recording Format. 63 pp.

OSI ISO/IEC DIS 9171-2.2-89. Information Technology—130 mm Optical-Disk Cartridge, Write-Once, for Information Interchange—Part 2: Recording Format. 54 pp.

OSI ISO IEC DIS 10089-90. Information Technology—130 mm Reqritable Optical Disk Cartridges for Information Interchange. 114 pp.

SAA AS 3956.1-91. Information Processing Systems—130 mm Optical Disk Cartridge, Write Once, for Information Interchange—Part 1: Unrecorded Optical Disk Cartridge (ISO/IEC 9171-1:1980) (in Professional Package 26A). 40 pp.

SAA AS 3956.2-91. Information Processing Systems—130 mm Optical Disk Cartridge, Write Once, for Information Interchange—Part 2: Recording Format (ISO/IEC 9171-2:1990) (in Professional Package 26A). 57 pp.

—Packaging

MOD UK DSTAN 81-52-89. Packaging of Pre-Recorded Magnetic Tapes and Flexible Discs Issue 1. 11 pp.

—Physical Properties

BSI BS 6957: Part 1-88. (WITHDRAWN) 1988 Guide to Construction & Use of 130mm (5.25 in) Flexible Disk Cartridges for Data Interchange, Using Modified Frequency Modulation Recording at 13 262 ftprad, on 80 Tracks on Each Side Part 1: Dimensional Phys. and Magnetic Char. (Renumbed as BS EN 28630-1: 1992). 20 pp.

BSI BS 6958: Part 1-88. 1988 Amd 1 Guide to Construction and Use of 90mm (3.5 in) Flexible Disk Cartridges for Data Interchange, Using Modified Frequency Modulation Recording at 7958 ftprad, 5.3 tpmm (135 tpi) on 80 Tracks on Each Side Part 1: Dimensional, Physical and Magnetic. 30 pp.

BSI BS 6958: Part 1-02. 1988 Amd 2 Construction and Use of 90mm (3.5 in) Flexible Disk Cartridges for Data Interchange, Using Modified Frequency Modulation Recording at 7958 ftprad, 5.3 tpmm (135 tpi) on 80 Tracks on Each Side Part 1: Dimensional, Physical and Magnetic. 36 pp.

BSI BS 7023: Part 1-88. 1988 Guide to Construction & Use of 130mm (5.25 in) Flexible Disk Cartridges for Data Interchange Using Modified Frequency Modulation Recording at 7958 ftprad, 3.8 tpmm (96 tpi) on Both Sides Part 1: Dimensional, Physical and Magnetic Characteristics. 20 pp.

BSI BS 7023: Part 1-01. 1988 Amd 1 Construction and Use of 130mm (5.25 in) Flexible Disk Cartridges for Data Interchange Using Modified Frequency Modulation Recording at 7958 ftprad, 3.8 tpmm (96 tpi) on Both Sides Part 1: Dimensional, Physical and Magnetic Characteristics. 25 pp.

BSI BS 7024: Part 1-88. 1988 Construction and Use of 130mm (5.25 in) Flexible Disk Cartridges for Data Interchange Using Modified Frequency Modulation Recording at 7958 ftprad, 1.9 tpmm (48 tpi) on Both Sides Part 1: Dimensional, Physical and Magnetic Characteristics. 20 pp.

BSI BS 7024: Part 1-01. 1988 Amd 1 Construction and Use of 130mm (5.25 in) Flexible Disk Cartridges for Data Interchange Using Modified Frequency Modulation Recording at 7958 ftprad, 1.9 tpmm (48 tpi) on Both Sides Part 1: Dimensional, Physical and Magnetic Characteristics. 25 pp.

BSI BS EN 28630-1-92. 1992 Construction and Use of 130 mm (5.25 in) Flexible Disk Cartridges for Data Interchange, Using Modified Frequency Modulation Recording at 13 262 ftprad, on 80 Tracks on Each Side Part 1: Dimensional, Physical and Magnetic. 26 pp.

Disk Cartridges *(Cont.)*

—Physical Properties *(Cont.)*

BSI BS ISO/IEC 9529-1-89. 1989 Information Processing Systems—Data Interchange on 90 mm (3, 5 in) Flexible Disk Cartridges Using Modified Frequency Modulation Recording at 15 916 ftprad, on 80 Tracks on Each Side Part 1: Dimensional, Physical and Magnetic. 47 pp.

BSI BS ISO/IEC 9529-1-01. 1989 Amd 1 Information Processing Systems—Data Interchange on 90 mm (3, 5 in) Flexible Disk Cartridges Using Modified Frequency Modulation Recording at 15 916 ftprad, on 80 Tracks on Each Side Part 1: Dimensional, Physical and Magnetic. 52 pp.

BSI BS ISO/IEC 10994-92. 1992 Information Technology—Data Interchange on 90 mm Flexible Disk Cartridges Using Modified Frequency Modulation Recording at 31 831 ftprad on 80 Tracks on Each Side—ISO Type 303. 54 pp.

CEN EN 27 487 (Part 1)-89. Information Processing Data Interchange on 130 mm (5.25) Flexible Disk Cartridges Using Modified Frequency Modulation Recording at 7 958 ftprad. 1.9 tpmm (48 tpi) on Both Sides—Part 1: Dimensional, Physical and Magnetic Characteristics. 21 pp.

CEN EN 28 378 (Part 1)-89. Information Processing Data Interchange on 130mm (5.25in) Flexible Disk Cartridges Using Modified Frequency Modulation Recording at 7 958 ftprad, 3.8tpmm (96tpi) on Both Sides Part 1: Dimensional, Physical and Magnetic Characteristics. 21 pp.

CEN EN 28630-1-92. Information Processing—Data Interchange on 130 mm (5.25 in) Flexible Disk Cartridges Using Modified Frequency Modulation Recording at 13 262 ftprad, on 80 Tracks Each Side Part 1 Dimensional, Physical and Magnetic Characteristics; (ISO 8630-1: 1987, Ed. 1). 19 pp.

CEN EN 28 860 (Part 1)-89. Information Processing Data Interchange on 90mm (3.5in) Flexible Disk Cartridges Using Modified Frequency Modulation of 7 958 ftprad 5.3tpmm (135tpi) on Both Sides Part 1: Dimensional Physical and Magnetic Characteristics. 2 pp.

CEN EN 29529-1-91. Information Processing Systems—Data Interchange on 90 mm (3,5 in) Flexible Disk Cartridges Using Modified Frequency Modulation Recordings at 15 916 ftprad, on 80 Tracks on Each Side—Part 1: Dimensional, Physical and Magnetic Characteristics. 3 pp.

DIN ENGL EN 27487 Pt 1-91. Information Processing; Data Interchange on 130 mm (5,25 in) Flexible Disk Cartridges Using Modified Frequency Modulation Recording at 7 958 ftprad, 1,9 tpmm (48 tpi), on Both Sides; Dimensional, Physical and Magnetic Characteristics (ISO 7487-1: 1985) (Dec). 20 pp.

DIN ENGL EN 28378 Pt 1-91. Information Processing; Data Interchange on 130 mm (5,25 in) Flexible Disk Cartridges Using Modified Frequency Modulation Recording at 7 958 ftprad, 3,8 tpmm (96 tpi), on Both Sides; Dimensional, Physical and Magnetic Characteristics (ISO 8378-1: 1986) (Dec). 19 pp.

DIN ENGL EN 28378 Pt 3-91. Information Processing; Data Interchange on 130 mm (5,25 in) Flexible Disk Cartridges Using Modified Frequency Modulation Recording at 7 958 ftprad, 3,8 tpmm (96 tpi), on Both Sides; Track Format B (ISO 8378-3: 1986) (Dec). 11 pp.

DIN ENGL EN 28860 Pt 1-91. Information Processing; Data Interchange on 90 mm (3,5 in) Flexible Disk Cartridges Using Modified Frequency Modulation Recording at 7 958 ftprad on 80 Tracks on Each Side; Dimensional, Physical and Magnetic Characteristics (ISO 8860-1: 1987) (Dec). 28 pp.

DIN ENGL EN 29529 Pt 1-92. Information Processing Systems; Data Interchange on 90 mm (3, 5 in) Flexible Disk Cartridges Using Modified Frequency Modulation Recording at 15 916 ftprad, on 80 Tracks on Each Side; Dimensional, Physical and Magnetic Characteristics. 44 pp.

ECMA ECMA 38-73. Mechanical, Physical and Magnetic Characteristics of Interchangeable Single Disk Cartridges (Top Loaded). 44 pp.

ECMA ECMA 147-90. Data Interchange on 90mm Flexible Disk Cartridges Using MFM Recording at 31831 ftprad on 80 Tracks on Each Side—ISO Type 303. 53 pp.

IEC DIS 7487 Pt 1-92. Information Technology—Data Interchange on 130 mm (5,25 in) Flexible Disk Cartridges Using Modified Frequency Modulation Recording at 7 958 ftprad, 1,9 tpmm (48 tpi), on Both Sides—ISO Type 202—Part 1: Dimensional, Physical and Magnetic Character-istics ***CD-ROM ONLY***. 22 pp.

IEC 8860 Pt 1-87. Information Processing—Data Interchange on 90 mm (3.5 in) Flexible Disk Cartridges Using Modified Frequency Modu-lation Recording at 7 958 ftprad on 80 Tracks on Each Side—Part 1: Dimensional, Physical and Magnetic Character-istics (Corrigendum 1-1990). 29 pp.

Disk Cartridges *(Cont.)*

—Physical Properties *(Cont.)*

IEC 9529 Pt 1-89. Information Processing Systems—Data Inter-change on 90 mm (3,5 in) Flexible Disk Cartridges Using Modified Frequency Modulation Recording at 15 916 ftprad, on 80 Tracks on Each Side—Part 1: Dimensional, Physical and Magnetic Characteristics First Edition. 45 pp.

IEC 10994-92. Information Technology—Data Interchange on 90 mm Flexible Disk Cartridges Using Modified Frequency Modulation Recording at 31 831 ftprad on 80 Tracks on Each Side—ISO Type 303 First Edition. 51 pp.

IEC DIS 13422-93. Information Technology—90 mm Flexible Disk Cartridges—10MB, 33 157 ftprad, for Sector Servo Tracking ***CD-ROM ONLY***. 68 pp.

ISO 3562-76. Information Processing—Interchangeable Magnetic Single Disk Cartridge (Top Loaded)—Physical and Magnetic Characteristics First Edition. 27 pp.

ISO 5654 Pt 1-84. Information Processing—Data Interchange on 200 mm (8 in) Flexible Disk Cartridges Using Two-Frequency Recording at 13 262 ftprad on One Side—Part 1: Dimensional, Physical and Magnetic Characteristics Second Edition. 19 pp.

ISO 6596 Pt 1-85. Information Processing—Data Interchange on 130 mm (5.25 in) Flexible Disk Cartridges Using Two-Frequency Recording at 7 958 ftprad, 1,9 tpmm (48 tpi), on One Side—Part 1: Dimensional, Physical and Magnetic Characteristics Second Edition. 19 pp.

ISO 7065 Pt 1-85. Information Processing—Data Interchange on 200 mm (8 in) Flexible Disk Cartridges Modified Frequency Modulation Recording at 13 262 ftprad, 1,9 tpmm (48 tpi), on Both Sides—Part 1: Dimensional, Physical and Magnetic Characteristics Second Edition. 19 pp.

ISO 7487 Pt 1-85. Information Processing—Data Interchange on 130 mm (5.25 in) Flexible Disk Cartridges Using Modified Frequency Modulation Recording at 7 958 ftprad, 1,9 tpmm (48 tpi), on Both Sides—Part 1: Dimensional, Physical and Magnetic Characteristics First Edition. 19 pp.

ISO DIS 7487 Pt 1-92. Information Technology—Data Interchange on 130 mm (5,25 in) Flexible Disk Cartridges Using Modified Frequency Modulation Recording at 7 958 ftprad, 1,9 tpmm (48 tpi), on Both Sides—ISO Type 202—Part 1: Dimensional, Physical and Magnetic Character-istics ***CD-ROM ONLY***. 22 pp.

ISO 7901-84. Information Processing—Unrecorded, Hard-Sectored, 130 mm (5.25 in) Flexible Disk Cartridges, One or Two-Sided Use—Dimensional, Physical, and Magnetic Characteristics First Edition. 4 pp.

ISO 8378 Pt 1-86. Information Processing—Data Interchange on 130 mm (5.25 in) Flexible Disk Cartridges Using Modified Frequency Modulation Recording at 7 958 ftprad, 3,8 tpmm (96 tpi), on Both Sides—Part 1: Dimensional, Physical and Characteristics First Edition. 19 pp.

ISO 8630 Pt 1-87. Information Processing—Data Interchange on 130 mm (5.25 in) Flexible Disk Cartridges Using Modified Frequency Modu-lation Recording at 13 262 ftprad, on 80 Tracks on Each Side—Part 1: Dimensional, Physical and Magnetic Character-istics First Edition; (CEN EN 28630-1:1992). 20 pp.

ISO 8860 Pt 1-87. Information Processing—Data Interchange on 90 mm (3.5 in) Flexible Disk Cartridges Using Modified Frequency Modu-lation Recording at 7 958 ftprad on 80 Tracks on Each Side—Part 1: Dimensional, Physical and Magnetic Character-istics (Corrigendum 1-1990). 29 pp.

ISO 9529 Pt 1-89. Information Processing Systems—Data Inter-change on 90 mm (3,5 in) Flexible Disk Cartridges Using Modified Frequency Modulation Recording at 15 916 ftprad, on 80 Tracks on Each Side—Part 1: Dimensional, Physical and Magnetic Characteristics First Edition. 45 pp.

ISO 10994-92. Information Technology—Data Interchange on 90 mm Flexible Disk Cartridges Using Modified Frequency Modulation Recording at 31 831 ftprad on 80 Tracks on Each Side—ISO Type 303 First Edition. 51 pp.

ISO DIS 13422-93. Information Technology—90 mm Flexible Disk Cartridges—10MB, 33 157 ftprad, for Sector Servo Tracking ***CD-ROM ONLY***. 68 pp.

JIS X 6157-80. Physical Characteristics of Interchangeable Magnetic Single Disk Cartridge (Top Loaded).

JIS X 6226-92. 90 mm Flexible Disk Cartridges—31831 ftprad. 46 pp.

JTC1 3562-76. Information Processing—Interchangeable Magnetic Single Disk Cartridge (Top Loaded)—Physical and Magnetic Characteristics First Edition. 27 pp.

INDUSTRY STANDARDS
813

Disk
INTERNATIONAL AND NON-U.S. NATIONAL STANDARDS
SUBJECT INDEX

Disk Cartridges *(Cont.)*

—Physical Properties *(Cont.)*

JTC1 5654 Pt 1-84. Information Processing—Data Interchange on 200 mm (8 in) Flexible Disk Cartridges Using Two-Frequency Recording at 13 262 ftprad on One Side—Part 1: Dimensional, Physical and Magnetic Characteristics Second Edition. 19 pp.

JTC1 6596 Pt 1-85. Information Processing—Data Interchange on 130 mm (5.25 in) Flexible Disk Cartridges Using Two-Frequency Recording at 7 958 ftprad, 1,9 tpmm (48 tpi), on One Side—Part 1: Dimensional, Physical and Magnetic Characteristics Second Edition. 19 pp.

JTC1 7065 Pt 1-85. Information Processing—Data Interchange on 200 mm (8 in) Flexible Disk Cartridges Modified Frequency Modulation Recording at 13 262 ftprad, 1,9 tpmm (48 tpi), on Both Sides—Part 1: Dimensional, Physical and Magnetic Characteristics Second Edition. 19 pp.

JTC1 7487 Pt 1-85. Information Processing—Data Interchange on 130 mm (5.25 in) Flexible Disk Cartridges Using Modified Frequency Modulation Recording at 7 958 ftprad, 1,9 tpmm (48 tpi), on Both Sides—Part 1: Dimensional, Physical and Magnetic Characteristics First Edition. 19 pp.

JTC1 DIS7487 Pt 1-92. Information Technology—Data Interchange on 130 mm (5,25 in) Flexible Disk Cartridges Using Modified Frequency Modulation Recording at 7 958 ftprad, 1,9 tpmm (48 tpi), on Both Sides—ISO Type 202—Part 1: Dimensional, Physical and Magnetic Character-istics ***CD-ROM ONLY***. 22 pp.

JTC1 7901-84. Information Processing—Unrecorded, Hard-Sectored, 130 mm (5.25 in) Flexible Disk Cartridges, One or Two-Sided Use—Dimensional, Physical, and Magnetic Characteristics First Edition. 4 pp.

JTC1 8378 Pt 1-86. Information Processing—Data Interchange on 130 mm (5.25 in) Flexible Disk Cartridges Using Modified Frequency Modulation Recording at 7 958 ftprad, 3,8 tpmm (96 tpi), on Both Sides—Part 1: Dimensional, Physical and Characteristics First Edition. 19 pp.

JTC1 8630 Pt 1-87. Information Processing—Data Interchange on 130 mm (5.25 in) Flexible Disk Cartridges Using Modified Frequency Modulation Recording at 13 262 ftprad, on 80 Tracks on Each Side—Part 1: Dimensional, Physical and Magnetic Characteristics First Edition. 20 pp.

JTC1 8860-87. Information Processing—Data Interchange on 90 mm (3.5 in) Flexible Disk Cartridges Using Modified Frequency Modu-lation Recording at 7 958 ftprad on 80 Tracks on Each Side—Part 1: Dimensional, Physical and Magnetic Character-istics (Corrigendum 1-1990). 29 pp.

JTC1 9529-89. Information Processing Systems—Data Inter-change on 90 mm (3,5 in) Flexible Disk Cartridges Using Modified Frequency Modulation Recording at 15 916 ftprad, on 80 Tracks on Each Side—Part 1: Dimensional, Physical and Magnetic Characteristics First Edition. 45 pp.

JTC1 10994-92. Information Technology—Data Interchange on 90 mm Flexible Disk Cartridges Using Modified Frequency Modulation Recording at 31 831 ftprad on 80 Tracks on Each Side—ISO Type 303 First Edition. 51 pp.

JTC1 DIS13422-93. Information Technology—90 mm Flexible Disk Cartridges—10MB, 33 157 ftprad, for Sector Servo Tracking ***CD-ROM ONLY***. 68 pp.

OSI ISO 3562-76. Information Processing—Interchangeable Magnetic Single Disk Cartridge (Top Loaded)—Physical and Magnetic Characteristics. 27 pp.

OSI ISO 5654-1-84. Information Processing—Data Interchange on 200mm (8in) Flexible Disk Cartridges Using Two-Frequency Recording at 13 262 ftprad on One Side—Part 1: Dimensional, Physical and Magnetic Characteristics. 19 pp.

OSI ISO 6596-1-85. Information Processing—Data Interchange on 130mm (5.25in) Flexible Disk Cartridges Using Two-Frequency Recording Recording at 7 958 ftprad, 1,9tpmm on One Side—Characteristics. 19 pp.

OSI ISO 7065-1-85. Information Processing—Data Interchange on 200mm(8in) Flexible Disk Cartridges Using Modified Frequency Modulation Recording at 13 262 ftprad, 1,9tpmm (48tpi) on Both Sides—Part 1: Dimensional, Physical and Magnetic Characteristics. 19 pp.

OSI ISO 7901-84. Information Processing—Unrecorded, Hard-Sectored 130mm (5.25in) Flexible Disk Cartridges, One or Two-Sided Use—Dimensional, Physical and Magnetic Characteristics. 4 pp.

OSI ISO IEC 9529-1-89. Information Processing Systems—Data Interchange on 90 mm (3,5 in) Flexible Disk Cartridges Using Modified Frequency Modulation Recording at 15 916 ftprad, on 80 Tracks on Each Side—Part 1: Dimensional, Physical and Magnetic Characteristics. 45 pp.

SNZ NZS/AS 4116-93. Information Processing Systems—Data Interchange on 90 mm Flexible Disk Cartridges Using Modified Frequency Modulation Recording, at 31 831 ftprad on 80 Tracks on Each Side—ISO Type 303 (ISO/IEC 10994:1992) (This is a Joint Standard with SAA AS 4116). 45 pp.

—Recording Format

BSI BS EN 29171-2-93. 1993 Amd 1 Information Technology—130 mm Optical Disk Cartridge, Write Once, for Information Interchange—Part 2: Recording Format (AMD 7797) August 15, 1993 (S). 72 pp.

BSI BS ISO/IEC 9171-2-90. 1990 Information Technology—130 mm Optical Disk Cartridge, Write Once, for Information Interchange—Part 2: Recording Format (Renumbered as BS EN 29171-2: 1993). 66 pp.

CEN EN 29171-2-93. Information Technology—130 mm Optical Disk Cartridge, Write Once, for Information Interchange—Part 2: Recording Format (ISO/IEC 9171-2: 1990). 65 pp.

IEC 9171 Pt 2-90. Information Technology—130 mm Optical Disk Cartridge, Write Once, for Information Interchange—Part 2: Recording Format First Edition (CEN 29171-2: 1993). 63 pp.

ISO 9171 Pt 2-90. Information Technology—130 mm Optical Disk Cartridge, Write Once, for Information Interchange—Part 2: Recording Format First Edition (CEN EN 29171-2: 1993). 63 pp.

JTC1 9171-90. Information Technology—130 mm Optical Disk Cartridge, Write Once, for Information Interchange—Part 2: Recording Format First Edition. 63 pp.

OSI ISO/IEC 9171-2-90. Information Technology—130 mm Optical Disk Cartridge, Write Once, for Information Interchange—Part 2: Recording Format. 63 pp.

OSI ISO/IEC DIS 9171-2.2-89. Information Technology—130 mm Optical-Disk Cartridge, Write-Once, for Information Interchange—Part 2: Recording Format. 54 pp.

OSI ISO/IEC DIS 10033-89. Information Technology—Text and Office Systems—Recording of Documents Conforming to ISO 8613 on Flexible Disk Cartridges Conforming to ISO 9293. 6 pp.

SAA AS 3956.2-91. Information Processing Systems—130 mm Optical Disk Cartridge, Write Once, for Information Interchange—Part 2: Recording Format (ISO/IEC 9171-2:1990) (in Professional Package 26A). 57 pp.

—Recording Methods

IEC DIS 13422-93. Information Technology—90 mm Flexible Disk Cartridges—10MB, 33 157 ftprad, for Sector Servo Tracking ***CD-ROM ONLY***. 68 pp.

ISO DIS 13422-93. Information Technology—90 mm Flexible Disk Cartridges—10MB, 33 157 ftprad, for Sector Servo Tracking ***CD-ROM ONLY***. 68 pp.

JTC1 DIS13422-93. Information Technology—90 mm Flexible Disk Cartridges—10MB, 33 157 ftprad, for Sector Servo Tracking ***CD-ROM ONLY***. 68 pp.

—Signal Quality

IEC DIS 13422-93. Information Technology—90 mm Flexible Disk Cartridges—10MB, 33 157 ftprad, for Sector Servo Tracking ***CD-ROM ONLY***. 68 pp.

ISO DIS 13422-93. Information Technology—90 mm Flexible Disk Cartridges—10MB, 33 157 ftprad, for Sector Servo Tracking ***CD-ROM ONLY***. 68 pp.

JTC1 DIS13422-93. Information Technology—90 mm Flexible Disk Cartridges—10MB, 33 157 ftprad, for Sector Servo Tracking ***CD-ROM ONLY***. 68 pp.

—Storage

BSI BS 4783: Part 1-88. 1988 Storage, Transportation and Maintenance of Magnetic Media for Use in Data Processing and Information Storage Part 1: Recommendations for Disk Packs, Storage Modules and Disk Cartridges. 25 pp.

BSI BS 4783: Part 1-01. 1988 Amd 1 Storage, Transportation and Maintenance of Media for Use in Data Processing and Information Storage Part 1: Recommendations for Disk Packs, Storage Modules and Disk Cartridges (AMD 6887) February 28, 1992. 26 pp.

BSI BS 4783: Part 3-88. 1988 Storage, Transportation and Maintenance of Magnetic Media for Use in Data Processing and Information Storage Part 3: Recommendations for Flexible Disk Cartridges. 17 pp.

BSI BS 4783: Part 3-01. 1988 Amd 1 Storage, Transportation and Maintenance of Media for Use in Data Processing and Information Storage Part 3: Recommendations for Flexible Disk Cartridges (AMD 6889) February 28, 1992. 18 pp.

Disk Cartridges *(Cont.)*

—Track Format

BSI BS 5356: Part 2-76. 1976 Magnetic Single-Disk Cartridges (Top Loading) for Data Processing Part 2: Track Format. 9 pp.

BSI BS 6957: Part 2-88. (WITHDRAWN) 1988 Guide to Construction & Use of 130mm (5.25 in) Flexible Disk Cartridges for Data Interchange, Using Modified Frequency Modulation Recording at 13 262 ftprad, on 80 Tracks on Each Side Part 2: Track Format A for 77 Tracks (Renumbered as BS EN 28630-2: 1992). 15 pp.

BSI BS 6957: Part 3-88. (WITHDRAWN) 1988 Guide to Construction & Use of 130mm (5.25 in) Flexible Disk Cartridges for Data Interchange, Using Modified Frequency Modulation Recording at 13 262 ftprad, on 80 Tracks on Each Side Part 3: Track Format B for 80 Tracks (Renumbered as BS EN 28630-3: 1992). 11 pp.

BSI BS 6958: Part 2-88. 1988 Amd 1 Construction and Use of 90mm (3.5 in) Flexible Disk Cartridges for Data Interchange, Using Mod. Freq. Modulation Record. at ftprad, 5.3 tpmm (135 tpi) on 80 Tracks on Each Side Part 2: Track Format (S) (AMD 6753) August 30, 1991. 15 pp.

BSI BS 6958: Part 2-02. 1988 Amd 2 Construction and Use of 90mm (3.5 in) Flexible Disk Cartridges for Data Interchange, Using Modified Frequency Modulation Recording at 7958 ftprad, 5.3 tpmm (135 tpi) on 80 Tracks on Each Side Part 2: Track Format (ISO 8860-2: 1987). 20 pp.

BSI BS 7023: Part 2-88. 1988 Guide to Construction & Use of 130mm (5.25 in) Flexible Disk Cartridges for Data Interchange Using Modified Frequency Modulation Recording at 7958 ftprad, 3.8 tpmm (96 tpi) on Both Sides Part 2: Track Format A. 16 pp.

BSI BS 7023: Part 2-01. 1988 Amd 1 Construction and Use of 130mm (5.25 in) Flexible Disk Cartridges for Data Interchange Using Modified Frequency Modulation Recording at 7958 ftprad, 3.8 tpmm (96 tpi) on Both Sides Part 2: Track Format A (ISO 8378/2: 1986) (AMD 7043). 21 pp.

BSI BS 7023: Part 3-88. 1988 Guide to Construction & Use of 130mm (5.25 in) Flexible Disk Cartridges for Data Interchange Using Modified Frequency Modulation Recording at 7958 ftprad, 3.8 tpmm (96 tpi) on Both Sides Part 3: Track Format B. 12 pp.

BSI BS 7023: Part 3-01. 1988 Amd 1 Construction & Use of 130mm (5.25 in) Flexible Disk Cartridges for Data Interchange Using Modified Frequency Modulation Recording at 7958 ftprad, 3.8 tpmm (96 tpi) on Both Sides Part 3: Track Format B (AMD 6467) October 31, 1991. 17 pp.

BSI BS 7024: Part 2-88. 1988 Guide to Construction & Use of 130mm (5.25 in) Flexible Disk Cartridges for Data Interchange Using Modified Frequency Modulation Recording at 7958 ftprad, 1.9 tpmm (48 tpi) on Both Sides Part 2: Track Format A. 16 pp.

BSI BS 7024: Part 2-01. 1988 Amd 1 Construction and Use of 130mm (5.25 in) Flexible Disk Cartridges for Data Interchange Using Mod Freq Modul Recording at 7958 ftprad, 1.9 tpmm (48 tpi) on Both Sides Part 2: Track Format A (ISO 7487/2: 1985) (AMD 7045) February 28, 1992. 21 pp.

BSI BS 7024: Part 3-88. 1988 Guide to Construction & Use of 130mm (5.25 in) Flexible Disk Cartridges for Data Interchange Using Modified Frequency Modulation Recording at 7958 ftprad, 1.9 tpmm (48 tpi) on Both Sides Part 3: Track Format (B). 12 pp.

BSI BS 7024: Part 3-01. 1988 Amd 1 Construction & Use of 130mm (5.25 in) Flexible Disk Cartridges for Data Interchange Using Modified Frequency Modulation Recording at 7958 ftprad, 1.9 tpmm (48 tpi) on Both Sides Part 3: Track Format B (AMD 6464) October 31, 1991. 17 pp.

BSI BS EN 28630-2-92. 1992 Construction and Use of 130mm (5.25 in) Flexible Disk Cartridges for Data Interchange, Using Modified Frequency Modulation Recording at 13 262 ftprad, on 80 Tracks Each Side Part 2: Track Format A for 7 Tracks (ISO 8630-2: 1987). 22 pp.

BSI BS EN 28630-3-92. 1992 Construction and Use of 130mm (5.25 in) Flexible Disk Cartridges for Data Interchange, Using Modified Frequency Modulation Recording at 13 262 ftprad, on 80 Tracks Each Side Part 3: Track Format B for 80 Tracks. 18 pp.

BSI BS ISO/IEC 9529-2-89. 1989 Track Format. 21 pp.

BSI BS ISO/IEC 9529-2-01. 1989 Amd 1 Information Processing Systems—Data Interchange on 90 mm (3, 5 in) Flexible Disk Cartridges Using Modified Frequency Modulation Recording at 15 916 ftprad, on 80 Tracks on Each Side Part 2: Track Format (AMD 7041) June 15, 1992. 26 pp.

BSI BS ISO/IEC 10994-92. 1992 Information Technology—Data Interchange on 90 mm Flexible Disk Cartridges Using Modified Frequency Modulation Recording at 31 831 ftprad on 80 Tracks on Each Side—ISO Type 303. 54 pp.

814

INDEX and DIRECTORY of

INTERNATIONAL AND NON-U.S. NATIONAL STANDARDS
SUBJECT INDEX

Disk

Disk Cartridges (Cont.)
—Track Format (Cont.)

CEN EN 27 487 (Part 2)-89. Information Processing Data Interchange on 130 mm (5.25in) Flexible Disk Cartridges Using Modified Frequency Modulation Recording at 7 958 ftprad, 19 tpmm (48tpi) on Both Sides Part 2: Track Format A. 18 pp.

CEN EN 27 487 (Part 3)-89. Information Processing Data Interchange on 130 mm (5.25in) Flexible Disk Cartridges Using Modified Frequency Recording at 7958 ftprad, 1.9tpmm (48tpi), on Both Sides—Part 3: Track Format B. 13 pp.

CEN EN 28 378 (Part 2)-89. Information Processing Data Interchange on 130mm (5.25in) Flexible Disk Cartridges Using Modified Frequency Modulation Recording at 7 958 ftprad 3.8tpmm (96 tpi), on Both Sides Part 2: Track Format A. 19 pp.

CEN EN 28 378 (Part 3)-89. Information Processing Data Interchange on 130mm (5.25in) Flexible Disk Cartridges Using Modified Frequency Modulation Recording at 7 958 ftprad 3.8tpmm (96 tpi) on Both sides Part 3: Track Format B. 13 pp.

CEN EN 28630-2-92. Information Processing—Data Interchange on 130 mm (5.25 in) Flexible Disk Cartridges Using Modified Frequency Modulation Recording at 13 262 ftprad, on 80 Tracks Each Side Part 2 Track Format A for 77 Tracks; (ISO 8630-2: 1987). 14 pp.

CEN EN 28630-3-92. Information Processing—Data Interchange on 130 mm (5.25 in) Flexible Disk Cartridges Using Modified Frequency Modulation Recording at 13 262 ftprad, on 80 Tracks Each Side Part 3 Track Format B for 80 Tracks; (ISO 8630-3: 1987). 10 pp.

CEN EN 28 860 (Part 2)-89. Information Processing Data Interchange on 90mm (3.5in) High Density Flexible Disk Cartridges Using Modified Frequency Modulation Recording at 7 958 ftprad 5.3tpmm (135itpi) on Both Sides Part 2: Track Format. 2 pp.

CEN EN 29529-2-91. Information Processing Systems—Data Interchange on 90 mm (3,5 in) Flexible Disk Cartridges Using Modified Frequency Modulation Recording at 15 916 ftprad, on 80 Tracks on Each Side—Part 2: Track Format. 3 pp.

DIN ENGL EN 27487 Pt 2-91. Information Processing; Data Interchange on 130 mm (5,25 in) Flexible Disk Cartridges Using Modified Frequency Modulation Recording at 7 958 ftprad, 1,9 tpmm (48 tpi), on Both Sides Track Format A (ISO 7487-2:1985) (Dec). 16 pp.

DIN ENGL EN 27487 Pt 3-91. Information Processing; Data Interchange on 130 mm (5,25 in) Flexible Disk Cartridges Using Modified Frequency Modulation Recording at 7 958 ftprad, 1,9 tpmm (48 tpi), on Both Sides; Track Format B (ISO 7487-3: 1986) (Dec). 11 pp.

DIN ENGL EN 28378 Pt 2-91. Information Processing; Data Interchange on 130 mm (5,25 in) Flexible Disk Cartridges Using Modified Frequency Modulation Recording at 7 958 ftprad, 3,8 tpmm (96 tpi), on Both Sides; Track Format A (ISO 8378-2: 1986) (Dec). 16 pp.

DIN ENGL EN 28860 Pt 2-91. Information Processing; Data Interchange on 90 mm (3,5 in) Flexible Disk Cartridges Using Modified Frequency Modulation Recording at 7 958 ftprad on 80 Tracks on Each Side; Track Format (ISO 8860-2: 1987) (Dec). 16 pp.

DIN ENGL EN 29529 Pt 2-92. Information Processing Systems; Data Interchange on 90 mm (3,5 in) Flexible Disk Cartridges Using Modified Frequency Modulation Recording at 15 916 ftprad, on 80 Tracks on Each Side; Track Format (ISO/IEC 9529-2: 1989) (Sept). 18 pp.

ECMA ECMA 39-73. Track Format Characteristics of Interchangeable Single Disk Cartridges (Top Loaded). 17 pp.

IEC 8630 Pt 2 COR 1. Corrigendum 1-Information Processing—Data Interchange on 130 mm (5.25 in) Flexible Disk Cartridges Using Modified Frequency Modulation Recording at 13 262 ftprad, on 80 Tracks on Each Side—Part 2: Track Format A for 77 Tracks; (1992). 1 p.

IEC 8630 Pt 3 COR 1. Corrigendum 1-Information Processing—Data Interchange on 130 mm (5.25 in) Flexible Disk Cartridges Using Modified Frequency Modulation Recording at 13 262 ftprad, on 80 Tracks on Each Side—Part 3: Track Format B for 80 Tracks; (1992). 1 p.

IEC 8860 Pt 2-87. Information Processing—Data Interchange on 90 mm (3.5 in) Flexible Disk Cartridges Using Modified Frequency Modulation Recording at 7 958 ftprad on 80 Tracks on Each Side—Part 2: Track Format First Edition; (Corrigendum 1-1990). 16 pp.

IEC 9529 Pt 2-89. Information Processing Systems—Data Inter-change on 90 mm (3,5 in) Flexible Disk Cartridges Using Modified Frequency Modulation Recording at 15 916 ftprad, on 80 Tracks on Each Side—Part 2: Track Format First Edition. 19 pp.

Disk Cartridges (Cont.)
—Track Format (Cont.)

IEC 10994-92. Information Technology—Data Interchange on 90 mm Flexible Disk Cartridges Using Modified Frequency Modulation Recording at 31 831 ftprad on 80 Tracks on Each Side—ISO Type 303 First Edition. 51 pp.

IEC DIS 13422-93. Information Technology—90 mm Flexible Disk Cartridges—10MB, 33 157 ftprad, for Sector Servo Tracking ***CD-ROM ONLY***. 68 pp.

ISO 3563-76. Information Processing—Interchangeable Magnetic Single-Disk Cartridge (Top Loaded)—Track Format First Edition. 11 pp.

ISO 5654 Pt 2-85. Information Processing—Data Interchange on 200 mm (8 in) Flexible Disk Cartridges Using Two-Frequency Recording at 13 262 ftprad, 1,9 tpmm (48 tpi), on One Side—Part 2: Track Format Second Edition. 12 pp.

ISO 6596 Pt 2-85. Information Processing—Data Interchange on 130 mm (5.25 in) Flexible Disk Cartridges Using Two-Frequency Recording at 7 958 ftprad, 1,9 tpmm (48 tpi), on One Side—Part 2: Track Format First Edition. 14 pp.

ISO 7065 Pt 2-85. Information Processing—Data Interchange on 200 mm (8 in) Flexible Disk Cartridges Using Modified Frequency Modulation Recording at 13 262 ftprad, 1,9 tpmm (48 tpi), on Both Sides—Part 2: Track Format First Edition. 15 pp.

ISO 7487 Pt 2-85. Information Processing—Data Interchange on 130 mm (5.25 in) Flexible Disk Cartridges Using Modified Frequency Modulation Recording at 7 958 ftprad, 1,9 tpmm (48 tpi), on Both Sides—Part 2: Track Format A First Edition. 16 pp.

ISO 7487 Pt 3-86. Information Processing—Data Interchange on 130 mm (5.25 in) Flexible Disk Cartridges Using Modified Frequency Modulation Recording at 7 958 ftprad, 1,9 tpmm (48 tpi), on Both Sides—Part 3: Track Format B Second Edition; (Corrigendum 1-1991). 13 pp.

ISO 8378 Pt 2-86. Information Processing—Data Interchange on 130 mm (5.25 in) Flexible Disk Cartridges Using Modified Frequency Modulation Recording at 7 958 ftprad, 3,8 tpmm (96 tpi), on Both Sides—Part 2: Track Format A First Edition. 16 pp.

ISO 8378 Pt 3-86. Information Processing—Data Interchange on 130 mm (5.25 in) Flexible Disk Cartridges Using Modified Frequency Modulation Recording at 7 958 ftprad, 3,8 tpmm (96 tpi), on Both Sides—Part 3: Track Format B Second Edition. 11 pp.

ISO 8630 Pt 2-87. Information Processing—Data Interchange on 130 mm (5.25 in) Flexible Disk Cartridges Using Modified Frequency Modu-lation Recording at 13 262 ftprad, on 80 Tracks on Each Side—Part 2: Track Format A for 77 Tracks First Edition; (Corrigendum 1-1992) (CEN EN 28530-2:1992). 17 pp.

ISO 8630 Pt 3-87. Information Processing—Data Interchange on 130 mm (5.25 in) Flexible Disk Cartridges Using Modified Frequency Modu-lation Recording at 13 262 ftprad, on 80 Tracks on Each Side—Part 3: Track Format B for 80 Tracks First Edition; (Corrigendum 1-1992) (CEN EN 28630-3:1992). 13 pp.

ISO 8860 Pt 2-87. Information Processing—Data Interchange on 90 mm (3.5 in) Flexible Disk Cartridges Using Modified Frequency Modulation Recording at 7 958 ftprad on 80 Tracks on Each Side—Part 2: Track Format First Edition; (Corrigendum 1-1990). 16 pp.

ISO 9529 Pt 2-89. Information Processing Systems—Data Interchange on 90 mm (3,5 in) Flexible Disk Cartridges Using Modified Frequency Modulation Recording at 15 916 ftprad, on 80 Tracks on Each Side—Part 2: Track Format First Edition. 19 pp.

ISO 10994-92. Information Technology—Data Interchange on 90 mm Flexible Disk Cartridges Using Modified Frequency Modulation Recording at 31 831 ftprad on 80 Tracks on Each Side—ISO Type 303 First Edition. 51 pp.

ISO DIS 13422-93. Information Technology—90 mm Flexible Disk Cartridges—10MB, 33 157 ftprad, for Sector Servo Tracking ***CD-ROM ONLY***. 68 pp.

JIS X 6202-91. Track Format for 200 mm Flexible Disk Cartridges. 23 pp.

JIS X 6212-90. Track Format of 130 mm Flexible Disk Cartridges —7958 ftprad. 15 pp.

JIS X 6213-90. Track Format for 130 mm High Density Flexible Disk Cartridges (13262 ft prad).

JIS X 6222-90. Track Format for 90 mm Flexible Disk Cartridges —7958 ftprad. 13 pp.

JIS X 6224-88. Track Format for 90 mm Flexible Disk Cartridges (13262 Ftprad).

JIS X 6225-90. Track Format for 90 mm Flexible Disk Cartridges (15916 ftprad). 17 pp.

JIS X 6226-92. 90 mm Flexible Disk Cartridges—31831 ftprad. 46 pp.

Disk Cartridges (Cont.)
—Track Format (Cont.)

JTC1 3563-76. Information Processing—Interchangeable Magnetic Single-Disk Cartridge (Top Loaded)—Track Format First Edition. 11 pp.

JTC1 5654 Pt 2-85. Information Processing—Data Interchange on 200 mm (8 in) Flexible Disk Cartridges Using Two-Frequency Recording at 13 262 ftprad, 1,9 tpmm (48 tpi), on One Side—Part 2: Track Format Second Edition. 12 pp.

JTC1 6596 Pt 2-85. Information Processing—Data Interchange on 130 mm (5.25 in) Flexible Disk Cartridges Using Two-Frequency Recording at 7 958 ftprad, 1,9 tpmm (48 tpi), on One Side—Part 2: Track Format First Edition. 14 pp.

JTC1 7065 Pt 2-85. Information Processing—Data Interchange on 200 mm (8 in) Flexible Disk Cartridges Using Modified Frequency Modulation Recording at 13 262 ftprad, 1,9 tpmm (48 tpi), on Both Sides—Part 2: Track Format First Edition. 15 pp.

JTC1 7487 Pt 2-85. Information Processing—Data Interchange on 130 mm (5.25 in) Flexible Disk Cartridges Using Modified Frequency Modulation Recording at 7 958 ftprad, 1,9 tpmm (48 tpi), on Both Sides—Part 2: Track Format A First Edition. 16 pp.

JTC1 8378 Pt 2-86. Information Processing—Data Interchange on 130 mm (5.25 in) Flexible Disk Cartridges Using Modified Frequency Modulation Recording at 7 958 ftprad, 3,8 tpmm (96 tpi), on Both Sides—Part 2: Track Format A First Edition. 16 pp.

JTC1 8378 Pt 3-86. Information Processing—Data Interchange on 130 mm (5.25 in) Flexible Disk Cartridges Using Modified Frequency Modulation Recording at 7 958 ftprad, 3,8 tpmm (96 tpi), on Both Sides—Part 3: Track Format B Second Edition. 11 pp.

JTC1 8630-87. Information Processing—Data Interchange on 130 mm (5.25 in) Flexible Disk Cartridges Using Modified Frequency Modulation Recording at 13 262 ftprad, on 80 Tracks on Each Side—Part 2: Track Format A for 77 Tracks First Edition; (Corrigendum 1-1992). 16 pp.

JTC1 8630-87. Information Processing—Data Interchange on 130 mm (5.25 in) Flexible Disk Cartridges Using Modified Frequency Modulation Recording at 13 262 ftprad, on 80 Tracks on Each Side—Part 3: Track Format B for 80 Tracks First Edition; (Corrigendum 1-1992). 12 pp.

JTC1 8860-87. Information Processing—Data Interchange on 90 mm (3.5 in) Flexible Disk Cartridges Using Modified Frequency Modulation Recording at 7 958 ftprad on 80 Tracks on Each Side—Part 2: Track Format First Edition; (Corrigendum 1-1990). 16 pp.

JTC1 9529-89. Information Processing Systems—Data Interchange on 90 mm (3,5 in) Flexible Disk Cartridges Using Modified Frequency Modulation Recording at 15 916 ftprad, on 80 Tracks on Each Side—Part 2: Track Format First Edition. 19 pp.

JTC1 10994-92. Information Technology—Data Interchange on 90 mm Flexible Disk Cartridges Using Modified Frequency Modulation Recording at 31 831 ftprad on 80 Tracks on Each Side—ISO Type 303 First Edition. 51 pp.

JTC1 DIS13422-93. Information Technology—90 mm Flexible Disk Cartridges—10MB, 33 157 ftprad, for Sector Servo Tracking ***CD-ROM ONLY***. 68 pp.

OSI ISO 3563-76. Information Processing—Interchangeable Magnetic Single-Disk Cartridge (Top Loaded)—Track Format. 11 pp.

OSI ISO 5654-2-85. Information Processing—Data Interchange on 200mm (8in) Flexible Disk Cartridges Using Two-Frequency Recording at 13 262 ftprad 1,9tpmm (48 tpi) on One Side—Part 2: Track Format. 12 pp.

OSI ISO 6596-2-85. Information Processing—Data Interchange on 130mm (5.25in) Flexible Disk Cartridges Using Two-Frequency Recording at 7 958 ftprad, 1,9tpmm (48tpi) on One Side—Part 2: Track Format. 14 pp.

OSI ISO 7065-2-85. Information Processing—Data Interchange on 200mm (8in) Flexible Disk Cartridges Using Modified Frequency Modulation Recording at 13 262 ftprad, 1,9tpmm (48 tpi) on Both Sides—Part 2: Track Format. 15 pp.

OSI ISO 8860-2-87. Information Processing—Data Interchange on 90mm (3.5 in) Flexible Disk Cartridges Using Modified Frequency Modulation Recording at 7 958 ft Prad on 80 Tracks on Each Side—Part 2: Track Format (with Technical Corrigendum 1). 19 pp.

OSI ISO IEC 9529-2-89. Information Processing Systems—Data Interchange on 90 mm (3,5 in) Flexible Disk Cartridges Using Modified Frequency Modulation Recording at 15 916 ftprad, on 80 Tracks on Each Side—Part 2: Track Format. 19 pp.

INDUSTRY STANDARDS

Disk — INTERNATIONAL AND NON-U.S. NATIONAL STANDARDS SUBJECT INDEX

Disk Cartridges (Cont.)
—Track Format (Cont.)
SNZ NZS/AS 4116-93. Information Processing Systems—Data Interchange on 90 mm Flexible Disk Cartridges Using Modified Frequency Modulation Recording, at 31 831 ftprad on 80 Tracks on Each Side—ISO Type 303 (ISO/IEC 10994:1992) (This is a Joint Standard with SAA AS 4116). 45 pp.

Disk Clutches
See Also: Clutches; Friction Clutches

JIS B 1401-76. Mechanical Multiple Disc Clutches (Wet Type). 12 pp.

—Automotive
CNS D2127-87. Clutch Disk Assemblies for Automobiles (Jun)(8675).

CNS D3112-87. Method of Test for Clutch Disk Assemblies for Automobiles (Jun)(8676).

—Hydraulic Equipment
JIS B 1402-76. Oil Hydraulic Multiple Disc Clutches (Wet Type) (R 1985). 10 pp.

—Internal Combustion Engines
BSI BS 3092-73. 1973 Main Friction Clutches, Main Power-Take-Off Assemblies and Associated Attachment for Internal Combustion Engines. 37 pp.

Disk Drives (Computer)
See Also: Compact Disk Players; Disk Cartridges; Floppy Disk Controllers; Magnetic Disks

—Interfaces
BSI BS EN 29315-92. 1992 Amd 0 Information Processing Systems—Interface Between Flexible Disk Cartridge Drives and Their Host Controllers (ISO 9315: 1989) (AMD 7668) April 15, 1993 (S). 29 pp.

BSI BS ISO 9315-89. (WITHDRAWN) 1989 Information Processing Systems—Interface Between Flexible Disk Cartridge Drives and Their Host Controllers (Renumbered as BS EN 29315: 1992). 23 pp.

BSI BS ISO/IEC 9318-2-90. 1990 Information Technology—Intelligent Peripheral Interface Part 2: Device Specific Command Set for Magnetic Disk Drives. 78 pp.

BSI BS ISO/IEC 9318-3-90. 1990 Information Technology—Intelligent Peripheral Interface Part 3: Device Generic Command Set for Magnetic and Optical Disk Drives. 285 pp.

CEN EN 29315-92. Information Processing Systems—Interface Between Flexible Disk Cartridge Drives and Their Host Controllers (ISO 9315: 1989). 22 pp.

IEC 9318 Pt 2-90. Information Technology—Intelligent Peripheral Interface Part 2: Device Specific Command Set for Magnetic Disk Drives First Edition. 76 pp.

IEC 9318 Pt 3-90. Information Technology—Intelligent Peripheral Interface Part 3: Device Generic Command Set for Magnetic and Optical Disk Drives First Edition. 279 pp.

ISO 9315-89. Information Processing Systems—Interface Between Flexible Disk Cartridge Drives and Their Host Controllers First Edition (CEN EN 29315: 1992). 20 pp.

ISO 9318 Pt 2-90. Information Technology—Intelligent Peripheral Interface Part 2: Device Specific Command Set for Magnetic Disk Drives First Edition. 76 pp.

ISO 9318 Pt 3-90. Information Technology—Intelligent Peripheral Interface Part 3: Device Generic Command Set for Magnetic and Optical Disk Drives First Edition. 279 pp.

JTC1 9315-89. Information Processing Systems—Interface Between Flexible Disk Cartridge Drives and Their Host Controllers First Edition. 20 pp.

JTC1 9318-90. Information Technology—Intelligent Peripheral Interface Part 2: Device Specific Command Set for Magnetic Disk Drives First Edition. 76 pp.

JTC1 9318-90. Information Technology—Intelligent Peripheral Interface Part 3: Device Generic Command Set for Magnetic and Optical Disk Drives First Edition. 279 pp.

JTC1 9318-90. Information Technology—Intelligent Peripheral Interface Part 4: Device Generic Command Set for Magnetic Tape Drives First Edition. 85 pp.

OSI ISO 9315-89. Information Processing Systems—Interface Between Flexible Disk Cartridge Drives and Their Host Controllers. 20 pp.

OSI ISO/IEC 9318-2-90. Information Technology—Intelligent Peripheral Interface Part 2: Device Specific Command Set for Magnetic Disk Drives. 76 pp.

OSI ISO/IEC 9318-3-90. Information Technology—Intelligent Peripheral Interface Part 3: Device Generic Command Set for Magnetic and Optical Disk Drives. 278 pp.

Disk Drives (Computer) (Cont.)
—Interfaces (Cont.)
OSI ISO/IEC 9318-4-90. Information Technology—Intelligent Peripheral Interface Part 4: Device Generic Command Set for Magnetic Tape Drives. 85 pp.

—Interfaces—Disk Controllers
JIS X 6052-90. Interface Between Flexible Disk Cartridge Drives and Their Host Controllers. 27 pp.

—Optical—Interfaces
IEC 9318 Pt 3-90. Information Technology—Intelligent Peripheral Interface Part 3: Device Generic Command Set for Magnetic and Optical Disk Drives First Edition. 279 pp.

ISO 9318 Pt 3-90. Information Technology—Intelligent Peripheral Interface Part 3: Device Generic Command Set for Magnetic and Optical Disk Drives First Edition. 279 pp.

JTC1 9318-90. Information Technology—Intelligent Peripheral Interface Part 3: Device Generic Command Set for Magnetic and Optical Disk Drives First Edition. 279 pp.

OSI ISO/IEC 9318-3-90. Information Technology—Intelligent Peripheral Interface Part 3: Device Generic Command Set for Magnetic and Optical Disk Drives. 278 pp.

Disk Grinders
See Also: Disk Sanders; Grinders

—Electric
CNS C4105-83. Portable Electric Disc Grinders (Oct)(3266). 8 pp.

JIS C 9611-90. Electric Disc Grinders. 28 pp.

—Electric—Portable
CNS C3183-85. Method of Test for Portable Electric Disc Grinders (Jan)(10610).

—Pneumatic
CNS B3259-82. Pneumatic Disc Grinders (Dec)(4071). 3 pp.

Disk Packs
See Also: Data Processing Equipment; Data Storage Devices; Magnetic Disks; Magnetic Materials; Memory (Data Storage)

BSI BS 5595: Part 1-78. 1978 Interchangeable Magnetic Twelve Disk Pack for Information Processing Part 1: 100 Mbytes Magnetic Twelve-Disk Pack. 44 pp.

BSI BS 5595: Part 2-81. 1981 Interchangeable Magnetic Twelve Disk Pack for Information Processing Part 2: 200 Mbytes Magnetic Twelve-Disk Pack. 46 pp.

ECMA ECMA 45-75. Data Interchange on Magnetic 12-Disk Packs (100 Mbytes). 71 pp.

ISO 4337-77. Information Processing—Interchangeable Magnetic Twelve-Disk Pack (100 Mbytes) First Edition. 43 pp.

ISO 5653-80. Information Processing—Interchangeable Magnetic Twelve-Disk Pack (200 Mbytes) First Edition. 44 pp.

JIS X 6156-82. Interchangeable Magnetic Twelve-Disk Pack (200M Bytes).

JTC1 4337-77. Information Processing—Interchangeable Magnetic Twelve-Disk Pack (100 Mbytes) First Edition. 43 pp.

JTC1 5653-80. Information Processing—Interchangeable Magnetic Twelve-Disk Pack (200 Mbytes) First Edition. 44 pp.

OSI ISO 4337-77. Information Processing—Interchangeable Magnetic Twelve-Disk Pack (100 M Bytes). 43 pp.

OSI ISO 5653-80. Information Processing—Interchangeable Magnetic Twelve-Disk Pack. 44 pp.

—Handling
BSI BS 4783: Part 1-88. 1988 Storage, Transportation and Maintenance of Magnetic Media for Use in Data Processing and Information Storage Part 1: Recommendations for Disk Packs, Storage Modules and Disk Cartridges. 25 pp.

BSI BS 4783: Part 1-01. 1988 Amd 1 Storage, Transportation and Maintenance of Media for Use in Data Processing and Information Storage Part 1: Recommendations for Disk Packs, Storage Modules and Disk Cartridges (AMD 6887) February 28, 1992. 26 pp.

—Magnetic Properties
BSI BS 4850: Part 1-72. 1972 Magnetic Six-Disk Pack for Data Processing Part 1: Mechanical and Magnetic Properties. 33 pp.

BSI BS 5359-76. 1976 Magnetic Eleven-Disk Packs for Data Processing. Mechanical and Magnetic Properties. 26 pp.

JIS X 6152-75. Magnetic Characteristics of Interchangeable Magnetic Six-Disk Packs.

JIS X 6154-80. Magnetic Characteristics of Interchangeable Magnetic Eleven-Disk Packs.

Disk Packs (Cont.)
—Mechanical Properties
BSI BS 5359-76. 1976 Magnetic Eleven-Disk Packs for Data Processing. Mechanical and Magnetic Properties. 26 pp.

—Physical Properties
BSI BS 4850: Part 1-72. 1972 Magnetic Six-Disk Pack for Data Processing Part 1: Mechanical and Magnetic Properties. 33 pp.

ISO 2864-74. Interchangeable Magnetic Six-Disk Pack—Physical and Magnetic Characteristics First Edition. 30 pp.

ISO 3564-76. Information Processing—Interchangeable Magnetic Eleven-Disk Pack—Physical and Magnetic Characteristics First Edition. 43 pp.

JIS X 6151-74. Physical Characteristics of Interchangeable Magnetic Six-Disk Packs.

JTC1 2864-74. Interchangeable Magnetic Six-Disk Pack—Physical and Magnetic Characteristics First Edition. 30 pp.

JTC1 3564-76. Information Processing—Interchangeable Magnetic Eleven-Disk Pack—Physical and Magnetic Characteristics First Edition. 43 pp.

OSI ISO 2864-74. Interchangeable Magnetic Six-Disk Pack—Physical and Magnetic Characteristics. 30 pp.

OSI ISO 3564-76. Information Processing—Interchangeable Magnetic Eleven-Disk Pack—Physical and Magnetic Characteristics. 43 pp.

—Recording Tracks
BSI BS 4850: Part 2-72. 1972 Amd 1 Magnetic Six-Disk Pack for Data Processing Part 2: Track Format. 15 pp.

—Storage
BSI BS 4783: Part 1-88. 1988 Storage, Transportation and Maintenance of Magnetic Media for Use in Data Processing and Information Storage Part 1: Recommendations for Disk Packs, Storage Modules and Disk Cartridges. 25 pp.

BSI BS 4783: Part 1-01. 1988 Amd 1 Storage, Transportation and Maintenance of Media for Use in Data Processing and Information Storage Part 1: Recommendations for Disk Packs, Storage Modules and Disk Cartridges (AMD 6887) February 28, 1992. 26 pp.

—Track Format
ISO 3561-76. Information Processing—Interchangeable Magnetic Six-Disk Pack—Track Format First Edition. 11 pp.

JTC1 3561-76. Information Processing—Interchangeable Magnetic Six-Disk Pack—Track Format First Edition. 11 pp.

OSI ISO 3561-76. Information Interchange—Interchangeable Magnetic Six-Disc Pack—Track Format. 11 pp.

Disk Sanders
See Also: Disk Grinders; Sanders (Tools); Woodworking Equipment

—Electric—Portable
MOD UK DSTAN 51-19: Part 2-92. Hand Tools, Powered Part 2: Electric Issue 1. 31 pp.

—Electric—Portable—Safety
BSI BS 2769: Sec 2.3-84. 1984 Hand-Held Electric-Motor-Operated Tools Part 2: Particular Requirements Section 2.3: Grinders, Polishers and Disc-Type Sanders. 16 pp.

BSI BS 2769: Sec 2.3-01. 1984 Amd 1 Hand-Held Electric Motor-Operated Tools Part 2: Particular Requirements Section 2.3: Specification for Grinders, Polishers and Disc-Type Sanders (AMD 6924) May 1, 1992. 17 pp.

—Glossaries
BSI BS 4361: Part 27-90. 1990 Woodworking Machines Part 27: Nomenclature for Disc Sanding Machines with Spindle in Fixed Position. 11 pp.

ISO 9375-89. Woodworking Machines—Disc Sanding Machines with Spindle in Fixed Position—Nomenclature First Edition. 9 pp.

—Pneumatic—Portable
CGSB 45-GP-3C-81. Grinders and Disk Sanders, Pneumatic, Portable, Standard for. 21 pp.

—Portable
CGSB 45-GP-4C-72. Grinders and Disk Sanders, Electric, Portable, Heavy Duty, Standard for. 9 pp.

—Safety
IEC 745 Pt 2-3-84. Safety of Hand-Held Motor-Operated Electric Tools Part 2: Particular Requirements for Grinders, Polishers and Disk-Type Sanders First Edition. 21 pp.

Disk Seal Tubes
Use For: Disc Seal Tubes *See Also:* Electron Tubes; Microwave Tubes

INTERNATIONAL AND NON-U.S. NATIONAL STANDARDS
SUBJECT INDEX

Display

Disk Seal Tubes *(Cont.)*
—**Electrical Properties**
IEC 235 Pt 3-72. Measurement of the Electrical Properties of Microwave Tubes Part 3: Disk Seal Tubes First Edition. 23 pp.

—**Power—Quality Assurance**
BSI BS CECC 45004-77. 1977 Disc Seal Power Tubes: Blank Detail Specification. 10 pp.

—**Triodes**
MOD UK DSTAN 59-60: 90/097-01. Valve, Electronic, Disc Seal Triode Issue 1; Amendment 1. 16 pp.

Disk Springs
See Also: Springs (Elastic)
CNS B1267-81. Disc Spring; Calculation (Sep)(7854).
CNS B2597-81. Disc Springs Dimensions and Quality Properties (Sep)(7853).
CNS B2684-82. Disc Springs (Jul)(9073).
JIS B 2706-78. Coned Disc Springs. 8 pp.

Disk Wafer Valves
Use: Butterfly Valves

Disk Wheels
See Also: Wheels
—**Automotive**
CNS D2087-81. Light Alloy Disc Wheels for Automobiles (Mar)(7135).
CNS D2130-82. Fastening Type and Dimension of Disc Wheel for Automobiles (May)(8815).
CNS D2156-82. Steel Disc Wheels for Automobiles (Oct)(9479). 34 pp.
JIS D 4103-89. Disc Wheels for Automobiles. 18 pp.
JIS D 4220-84. Fastening Methods and Dimensions of Disc Wheels for Automobiles.

—**Automotive—Fasteners**
DIN ENGL 74361 Pt 2-82. Disc Wheels for Motor Vehicles and Trailers; Fasteners for Stud Centring (Nov). 5 pp.
DIN ENGL 74361 Pt 3-79. Disc Wheels for Motor Vehicles and Trailers; Dimensions and Fasteners for Attachment with Centring on Wheel Bore (Nov). 4 pp.

—**Automotive—Joining**
CNS D2130-82. Fastening Type and Dimension of Disc Wheel for Automobiles (May)(8815).
JIS D 4220-84. Fastening Methods and Dimensions of Disc Wheels for Automobiles.

—**Trailers—Fasteners**
DIN ENGL 74361 Pt 2-82. Disc Wheels for Motor Vehicles and Trailers; Fasteners for Stud Centring (Nov). 5 pp.
DIN ENGL 74361 Pt 3-79. Disc Wheels for Motor Vehicles and Trailers; Dimensions and Fasteners for Attachment with Centring on Wheel Bore (Nov). 4 pp.

Disks
Scope Note: Use a more specific term *See:* Abrasive Disks; Compact Disks; Magnetic Disks; Video Disks

Disks (Storage)
Use: Magnetic Disks

Disodium Ethylenediaminetetraacetate
Use: Edetate Disodium

Disodium Hydrogen Phosphate
Use: Sodium Phosphate, Dibasic

Disodium Phosphate
Use: Sodium Phosphate, Dibasic

Disodium Tetraborates
Use: Sodium Borates

Dispatch Cases
Use: Briefcases

Dispatch Traffic
Use For: Radio Dispatch *See Also:* MOBITEX; Radio Communications; Telecommunication Traffic

—**Mobile Radio Equipment**
CCIR Report 741-3-90. Multi-Channel Land Mobile Systems for Dispatch Traffic (with or Without PSTN Interconnection)—Section 8A—Land Mobile Service and Related Subjects. 19 pp.

—**Mobile Radio Equipment—Public Switched Telephone Networks**
CCIR Report 741-3-90. Multi-Channel Land Mobile Systems for Dispatch Traffic (with or Without PSTN Interconnection)—Section 8A—Land Mobile Service and Related Subjects. 19 pp.

Dispatch Traffic *(Cont.)*
—**Radio Relay Systems**
CCIR Report 741-3-90. Multi-Channel Land Mobile Systems for Dispatch Traffic (with or Without PSTN Interconnection)—Section 8A—Land Mobile Service and Related Subjects. 19 pp.

Dispenser Weapons
See Also: Bombs (Ordnance); Weapons; Weapons Systems

—**Design—Safety**
NATO STANAG 3786 ED 2 AMD 3-85. Design Safety Principles and General Design Criteria for Airborne Dispenser Weapons. 20 pp.
NATO STANAG 3786 ED 3 AMD 0-92. Safety Design Requirements and General Design Guidance for Airborne Dispenser Weapons. 14 pp.
NATO STANAG 3786 ED 3 AMD 1-92. Safety Design Requirements and General Design Guidance for Airborne Dispenser Weapons. 15 pp.

Dispensing Equipment
Scope Note: Use a more specific term *See:* Drinking Fountains; Fuel Dispensing Equipment; Paper Towel Dispensers; Sprayers; Vending Machines

Dispensing Pipettes
Use: Pipettes

Dispersants
See Also: Admixtures; Diluents; Emulsification; Emulsifying Agents; Plasticizers; Plastisols; Surfactants

—**Latex—Cement Additives**
CNS A2168-83. Polymer Dispersions for Cement Modifier (Nov)(10639).
JIS A 6203-80. Polymer Dispersions for Cement Modifier. 10 pp.

—**Resins—Cement Additives**
CNS A2168-83. Polymer Dispersions for Cement Modifier (Nov)(10639).
JIS A 6203-80. Polymer Dispersions for Cement Modifier. 10 pp.

—**Silica**
MOD UK TS 10093A. Silica Powder, Amorphous.

Dispersion
Scope Note: Use a more specific term. For additional listings, see the subheading Dispersion under specific products or material *See Also:* Chromatic Dispersion

Displacement Compressors
See Also: Compressors (Pressure Equipment)
CNS B7237-83. Method of Test for Displacement Compressors (May)(10213).
CNS B7238-83. Method of Test for Air Vessel for Filling Use (Total Pressure Ratio Below 30) and for Small Size Displacement Compressor (Shaft Horsepower Below 5.5KW) (May)(10214).
JIS B 8341-83. Testing Methods for Displacement Compressors. 37 pp.

—**Acceptance Testing**
BSI BS 1571: Part 1-87. 1987 Testing of Positive Displacement Compressors and Exhausters Part 1: Methods for Acceptance Testing. 64 pp.
ISO 1217-86. Displacement Compressors—Acceptance Tests Second Edition; (Amendment 1-1988). 63 pp.
SNZ NZS/BS 1571: Part 1-87. Testing of Positive Displacement Compressors and Exhausters Part 1: Methods for Acceptance Testing. 64 pp.

Displacement Engines
Use: Piston Engines

Displacement Meters
—**Gas**
BSI BS 4161: Part 6-87. 1987 Amd 1 Gas Meters Part 6: Rotary Displacement and Turbine Meters for Gas Pressures up to 100 Bar. 14 pp.

—**Hydrocarbons**
BSI BS 6169: Part 1-81. 1981 Volumetric Measurement of Liquid Hydrocarbons Part 1: Displacement Meter Systems (Other Than Dispensing Pumps). 12 pp.
ISO 2714-80. Liquid Hydrocarbons—Volumetric Measurement by Displacement Meter Systems Other Than Dispensing Pumps First Edition; (Erratum—July 1982). 12 pp.

Display Cabinets, Refrigerated
Use: Refrigerated Display Cases

Display Cases, Refrigerated
Use: Refrigerated Display Cases

Display Devices
Use For: Displays; Electronic Displays; Readouts
See Also: Annunciators; Cathode Ray Tube Displays; Cathode Ray Tubes; Data Terminal Equipment; Dials; Digital Readouts; Display Drivers; Electroluminescent Displays; Electronic Chart Display Systems; Head Up Displays; Indicating Instruments; Indicator Lights; Instrument Panels; Light Emitting Diode Displays; Liquid Crystal Displays; Luminous Tube Signs; Multifunction Displays; Navigational Aids; Plotters; Raster Displays; Safety Parameter Display System; Three Dimensional Display Systems; Video Display Terminals

—**Aircraft—Avionics**
EUROCAE ED-64 12.88. Transport Category Airplane Electronic Display Systems Changes to Be Applied to FAA Advisory Circular No. 25-11 for Adoption as Jar AC. 17 pp.

—**Aircraft—Avionics—Color**
NATO STANAG 3940 ED 1 AMD 0-91. Aircraft Electronic Colour Display Systems. 7 pp.

—**Aircraft Cabins—Ergonomics**
NATO STANAG 3705 ED 3 AMD 0-92. Human Engineering Design Criteria for Controls and Displays in Aircrew Stations. 21 pp.
NATO STANAG 3705 ED 3 AMD 1-92. Human Engineering Design Criteria for Controls and Displays in Aircrew Stations. 22 pp.

—**Aircraft Cabins—Flight Crew**
CAA Chapter P4-2. Flight Crew Compartment Design (Provisional Airworthiness Requirements for Civil Powered Lift Aircraft). 16 pp.
NATO STANAG 3705 ED 2 AMD 1-81. Principles of Presentation of Information in Aircrew Stations. 11 pp.
NATO STANAG 3705 ED 2 AMD 3-81. Principles of Presentation of Information in Aircrew Stations. 10 pp.
NATO STANAG 3869 ED 1 AMD 4-80. Aircrew Station Control Panels. 26 pp.

—**Aircraft Cabins—Flight Crew—Antireflection Coatings**
NATO STANAG 3643 ED 2 AMD 7-73. Coating, Reflection Reducing for Glass Elements Used in Aircrew Station Displays. 6 pp.
NATO STANAG 3643 ED 2 AMD 9-73. Coating, Reflection Reducing for Glass Elements Used in Aircrew Station Displays. 6 pp.

—**Aircraft Cabins—Flight Crew—Glossaries**
NATO STANAG 3647 ED 3 AMD 1-86. Nomenclature in Aircrew Stations. 26 pp.
NATO STANAG 3647 ED 3 AMD 2-86. Nomenclature in Aircrew Stations. 22 pp.
NATO STANAG 3871 ED 1 AMD 2-85. NATO Glossary of Displays and Aircrew Stations (AI) Specialist Terminology and Abbreviations. 26 pp.
NATO STANAG 3871 ED 1 AMD 3-85. NATO Glossary of Displays and Aircrew Stations (AI) Specialist Terminology and Abbreviations. 38 pp.

—**Aircraft Cabins—Flight Crew—Symbols**
NATO STANAG 3329 ED 6 AMD 3-84. Numerials and Letters in Aircrew Stations. 14 pp.
NATO STANAG 3329 ED 6 AMD 4-84. Numerals and Letters in Aircrew Stations. 13 pp.

—**Aircraft Instruments—Engine Location**
NATO STANAG 3359 ED 5 AMD 3-85. Location and Arrangement of Engine Displays in Aircraft. 6 pp.

—**Aircraft Instruments—Flight Data**
NATO STANAG 3216 ED 5 AMD 3-84. Layout of Flight Data in Pilots' Displays. 7 pp.
NATO STANAG 3648 ED 2 AMD 5-75. Electronically and/or Optically Generated Aircraft Displays for Fixed Wing Aircraft. 11 pp.
NATO STANAG 3648 ED 2 AMD 6-75. Electronically and/or Optically Generated Aircraft Displays for Fixed Wing Aircraft. 11 pp.

—**Buses (Vehicles)—Destination**
CNS D2094-87. Destination Indicators for Buses (Jul)(7677).
JIS D 4705-88. Destination Indicators for Buses. 9 pp.

—**Dot Matrix Characters**
JIS X 9051-84. 16-Dots Matrix Character Patterns for Display Devices.
JIS X 9052-83. 24-Dots Matrix Character Patterns for Display Devices.

—**Ergonomics**
CEPT T/SF 40 E-85. Human Factor Aspects of Visual Display Terminals for Telecommunication Services. 11 pp.
ECMA ECMA 110-85. Ergonomics-Requirements for Monochromatic Visual Display Devices. 57 pp.

INDUSTRY STANDARDS

Display Devices (Cont.)

—Ergonomics (Cont.)
ECMA ECMA 126-87. Ergonomics-Requirements for Colour Visual Display Devices. 21 pp.
ECMA ECMA-TR 22-84. Ergonomics Recommendations for VDU Work Places. 32 pp.
MOD UK DSTAN 00-25: Part 7-86. Human Factors for Designers of Equipment Part 7: Visual Displays Issue 1. 37 pp.

—Health Aspects
ECMA ECMA-TR 33-85. Visual Displays-Health Aspects. 15 pp.

—High Definition Television Systems
CCIR QUESTION 70/11-90. Effect of Display Technology on the HDTV Standard—Questions Concerning Study Group 11—Broadcasting Service (Television). 1 p.

—Machinery—Ergonomics
CEN PREN 894-1-92. Safety of Machinery—Ergonomics Requirements for the Design of Displays and Control Actuators—Part 1: Human Interactions with Displays and Control Actuators. 19 pp.
CEN PREN 894-2-92. Safety of Machinery—Ergonomics Requirements for the Design of Displays and Control Actuators—Part 2: Displays. 24 pp.

—Nuclear Reactor Accidents
CSA CAN3-N290.6-M82. Requirements for Monitoring and Display of CANDU Nuclear Power Plant Status in the Event of an Accident (R 1992); (Gen Instr 1). 21 pp.

—Numerical
CNS C7211-90. 1200 bps Numerical-Display (Dec)(12814).

—Telex Equipment
CCITT RECMN S.21-89. Use of Display Screens in Telex Machines—Telegraph Services Terminal Equipment (Study Group IX) 2 pp. 2 pp.

Display Drivers
See Also: Computer Equipment; Display Devices; Driver/Decoders; Interface Circuits; Latch/Decoder/Drivers; Linear Circuits; Microcomputers; Microprocessors

—BCD to 7-Segment Decoder
CECC CECC 90 103-116 ISSUE 1-81. BS CECC 90 103-116; BCD-to-Seven Segment Decoders/Drivers with Open-Collector Outputs (En) AMD 1 (En). 23 pp.

—BCD to 7-Segment Decoder—Preferred Products List
CECC CECC MUAHAG Vol 7 IS 8-92. Preferred Products List; Active Microcircuits (En, Fe, Ge). 89 pp.

—LCD—BCD to 7-Segment Decoder
CECC CECC 90 109-720 ISSUE 1-87. Digital Integrated Circuits in Accordance with FS 90 109; 54 HC 4543, 74 HC 4543; BCD to 7 Segment LCD Driver (En, Fr). 7 pp.

Display Equipment
Scope Note: Use a more specific term
See: Audiovisual Equipment; Display Lighting; Screens (Projection); Signs

Display Lighting
Use For: Accent Lighting; Merchandising Lighting
See Also: Lighting
SNZ NZS 6703-84. Code of Practice for Interior Lighting Design. 82 pp.

Display Storage Tubes
Use: Direct View Storage Tubes

Displays
Use: Display Devices

Disposal
Scope Note: Use a more specific term
See: Explosive Ordnance Disposal; Solid Waste Disposal; Waste Disposal

Disruption of Service
Use: Interruption of Service

Dissertations
Use: Theses

Dissipation Factor
Use: Dielectric Dissipation Factor

Dissipators
Use: Heat Sinks

Dissolved Acetylene
Use: Acetylene

Dissolved Oxygen Meters
See Also: Oxygen
JIS K 0803-88. Recording Dissolved Oxygen Meter. 10 pp.

Dissolving Pulps
See Also: Cellulose; Chemical Pulps; Cotton Fibers; Kraft Pulps; Pulps; Sulfite Pulps
JIS P 2701-64. Dissolving Pul

—Comprehensive Testing
JIS P 8101-76. Testing Method for Dissolving Pulp (R 1984). 49 pp.

Distance Measuring Equipment
Use For: DME; Radio Distance Measuring Equipment *See Also:* Altimeters; Dimensional Measuring Instruments; Geodetic Instruments; Laser Instruments (Distance Measurement); Measuring Instruments; Surveying Instruments

—Accuracy Testing
BSI BS 7334: Part 8-92. 1992 Measuring Instruments for Building Construction Part 8: Methods for Determining Accuracy in Use of Electronic Distance-Measuring Instruments up to 150 m (ISO 8322-8: 1992). 26 pp.
ISO 8322 Pt 8-92. Building Construction—Measuring Instruments—Procedures for Determining Accuracy in Use—Part 8: Electronic Distance-Measuring Instruments up to 150 m First Edition. 24 pp.

—Aircraft
CAA Part 8 CAP 208. Distance Measuring Equipment (Civil Air Publications: Airborne Radio Apparatus). 8 pp.

—Aircraft—Accuracy Testing
EUROCAE ED-57 12.86. MPS for Distance Measuring Equipment (DME/N and DME/P). 96 pp.

—Aircraft—Electrical Properties
EUROCAE ED 54 01.87. MOPR for DME/N and DME/P Operating Within the Radio Frequency Range 960 to 1215 MHz. 144 pp.

—Aircraft—Environmental Testing
EUROCAE ED-54 01.87. MOPR for DME/N and DME/P Operating Within the Radio Frequency Range 960 to 1215 MHz. 144 pp.
EUROCAE ED-57 12.86. MPS for Distance Measuring Equipment (DME/N and DME/P). 96 pp.

—Aircraft—Maintainability
EUROCAE ED-57 12.86. MPS for Distance Measuring Equipment (DME/N and DME/P). 96 pp.

—Aircraft—Reliability
EUROCAE ED-57 12.86. MPS for Distance Measuring Equipment (DME/N and DME/P). 96 pp.

—Design
EUROCAE ED-54 01.87. MOPR for DME/N and DME/P Operating Within the Radio Frequency Range 960 to 1215 MHz. 144 pp.
EUROCAE ED-57 12.86. MPS for Distance Measuring Equipment (DME/N and DME/P). 96 pp.

—Marine
BSI BS 7408-91. 1991 Marine Speed and Distance Measuring Equipment (SDME). 13 pp.
CENELEC HD 600 S1-92. Marine Speed and Distance Measuring Equipment (SDME)—Operational and Performance Requirements Methods of Testing and Required Test Results. 6 pp.
IEC 1023-90. Marine Speed and Distance Measuring Equipment (SDME) Operational and Performance Requirements —Methods of Testing and Required Test Results First Edition. 22 pp.

Distance Pieces
See Also: Spacers (Mechanical)

—Aircraft—Control Columns
SBAC AS 862 ISSUE 7. Distance Piece—Control Column. 1 p.

—Aircraft—Parachutes
SBAC AS 480 ISSUE 3(I). Distance Piece Anti-Spin Parachute Jettison Slip. 1 p.

—Aircraft—Rudder Bars
SBAC AS 2210 ISSUE 5. Distance Piece (Rudder Bar). 1 p.

Distance Pieces (Cont.)

—Aircraft—Stirrups—Fuel Filters
SBAC AS 3245 ISSUE 1. Stirrup Distance, Piece—Filter Fuel Type 'A'. 1 p.
SBAC AS 3271 ISSUE 1. Stirrup Distance, Piece Filter Fuel Type 'B'. 1 p.

—Aircraft—Stop Valves—Floats
SBAC AS 4884 ISSUE 1. Distance Bush for Float Assembly. 1 p.

—Ship Hulls
JIS F 7131-87. Distance Pieces for Ships' Hull.

Distemper
See Also: Paints

—Aircraft—Camouflage
MOD UK DSTAN 80-110-85. Paint, Distemper, for Aircraft Issue 1. 10 pp.

Distillate Fuels
Use For: Petroleum Distillates *See Also:* Petroleum Products

—Cetane Number
CNS K6915-87. Method of Calculated Cetane Index of Distillate Fuels (Jul)(12016).
CNS K6972-90. Method for Calculated Cetane Index of Distillate Fuels (Four Variable Equation Method) (Aug)(12761).

—Distillation Methods
BSI BS 2000: Part 123-85. (OBSOLESCENT) 1985 Petroleum and Its Products: Part 123: Distillation of Petroleum Products. 29 pp.
BSI BS 7392-90. 1990 Method for Determination of Distillation Characteristics of Petroleum Products. 26 pp.
CNS K6109-74. Method of Test for Distillation of Petroleum Products (Mar)(1218).
DIN ENGL 51356-78. Testing of Lubricating Oils and Liquid Fuels; Determination of the Distillation Range Under Reduced Pressure According to Grosse-Oetringhaus (Feb). 7 pp.
ISO 3405-88. Petroleum Products—Determination of Distillation Characteristics Second Edition. 24 pp.
JIS K 2254-90. Petroleum Products—Determination of Distillation Characteristics. 56 pp.

—Lead Content—Complexometric Titrations
DIN ENGL 51769 Pt 5-84. Testing of Petroleum Products; Determination of Lead Content (Total Lead) of Gasolines; Complexometric Method (June). 7 pp.

—Marine
CGSB 3-GP-11MB-89. Naval Distillate Fuel (-6 Degrees Celsius Pour). 10 pp.
CGSB 3-GP-15MB-89. Naval Distillate Fuel (-18 Degrees Celsius Pour). 10 pp.
MOD UK DSTAN 91-4-01. Fuel, Naval Distillate, Low Pour Point NATO Code Number F-75 Joint Service Designation DIESO F-75 Fuel, Naval Distillate NATO Code Number F-76 Joint Service Designation DIESO F-76 Issue 5; Amendment 1. 20 pp.

—Thiol Content—Potentiometric Analysis
CNS K6547-80. Method of Test for Mercaptan Sulfur in Gasoline, Kerosine, Aviation Turbine and Distillate Fuels (Potentiometric Method) (Aug)(6192).

Distillate Oil
See Also: Oils

—Emulsified Asphalts—Distillation Methods
CNS K6794-83. Method of Test for Identification of Oil Distillate Obtained from Emulsified Asphalts by Micro-Distillation (Jul)(10455).

Distillation
Use: Distillation Methods

Distillation Apparatus
Use: Distillation Equipment

Distillation Equipment
See Also: Distillation Heads; Distillation Methods
BSI BS 658-89. 1989 Apparatus for the Determination of Distillation Range (Including Flasks and Receivers). 27 pp.

—Industrial
CSA C22.2 NO 88-1958. Construction and Test of Industrial Heating Equipment (R 1992). 22 pp.
CSA 1169 Bull. Electrical Bulletin 1169 June 27, 1978 to C22.2 NO 88. 2 pp.

—Industrial—Construction
CSA C22.2 NO 88-1958. Construction and Test of Industrial Heating Equipment (R 1992). 22 pp.

INTERNATIONAL AND NON-U.S. NATIONAL STANDARDS
SUBJECT INDEX
Distillation

Distillation Equipment (Cont.)
—Industrial—Construction (Cont.)
CSA 1169 Bull. Electrical Bulletin 1169 June 27, 1978 to C22.2 NO 88. 2 pp.

—Markham
BSI BS 1428: Part B2-60. 1960 Microchemical Apparatus: Group B: Apparatus for the Determination of Elements by Other Than Combustion Methods Part B2: Ammonia Distillation Apparatus (Markham). 8 pp.

Distillation Heads
See Also: Distillation Equipment

—Splash—Glass
BSI BS 6855-87. (WITHDRAWN) 1987 Design and Construction of Glass Splash Heads for Laboratory Use. 10 pp.

Distillation Methods
Use For: Fractional Distillation Methods
See Also: Azeotropic Distillation; Dean and Stark Method; Distillation Equipment; Feedwater Treatment; Resource Recovery; Vaporizing

—n-Butyl Acetate
BSI BS 551-90. 1990 Butyl Acetate for Industrial Use. 8 pp.

—Alcohols
ISO 1843 Pt 7-82. Higher Alcohols for Industrial Use—Methods of Test—Part 7: Determination of Distillation Yield First Edition. 4 pp.

—Amyl Acetate
BSI BS 552-70. 1970 Amyl Acetate. 12 pp.
BSI BS 552-01. 1970 Amd 1 Amyl Acetate (AMD 7218) September 15, 1992. 14 pp.

—Aromatic Hydrocarbons
CNS K6255-74. Method of Test for Distillation of Aromatic Hydrocarbons (Jun)(2756).

—Bitumens
BSI BS 2000: Part 27-93. 1993 Methods of Test for Petroleum and Its Products Part 27: Determination of Distillation Characteristics of Cutback Bitumen (Supersedes BS 4453: 1976). 6 pp.
BSI BS 2000: Part 27-82. 1982 Petroleum and Its Products Part 27: Distillation of Cut-Back Asphaltic (Bituminous) Products. 8 pp.
BSI BS 4385-80. (WITHDRAWN) 1980 Determination of Water in Petroleum Products and Bituminous Materials (Distillation Method) (Superseded by BS 4385: Part 1: 1991). 8 pp.
CNS K6339-73. Method of Test for Moisture Content in Petroleum Products and Bituminous Materials by Distillation (Feb)(3517). 4 pp.
DIN ENGL 1996 Pt 5-83. Testing of Bituminous Materials for Road Building and Related Purposes; Determination of Water Content (Apr). 4 pp.

—Chemical Products
CNS K0023-82. Method for Determining Distillation of Chemical Products (Jul)(9180).
JIS K 0066-92. Test Methods for Distillation of Chemical Products. 22 pp.

—Chloroform
BSI BS 4774-72. 1972 Methods of Test for Chloroform for Industrial Use. 12 pp.

—Cleaning Agents—Water Content
CGSB 2-GP-11M METH 13.2-83. Methods of Testing and Analysis of Soaps and Detergents Water Content (Distillation Method). 3 pp.

—Coal—Moisture Content
BSI BS 1016: Part 1-73. 1973 Methods for the Analysis and Testing of Coal and Coke Part 1: Total Moisture of Coal. 11 pp.

—Coal Tar
CNS K6213-65. Method of Fractional Distillation Test for Coal Tar (Sep)(2495)(R 1971).
CNS K6214-65. Method of Fractional Distillation Test for Crude Methyl Phenol, M-Methyl Phenol and Coal Tar (High Boiling Point) (Sep)(2496)(R 1971).

—Creosote Oil
CNS K6070-57. Method of Test for Fractional Distillation of Creosote Oil (Jul)(913)(R 1971).

—Cresylic Acids
ISO 1897 Pt 12-83. Phenol, o-Cresol, m-Cresol, p-Cresol, Cresylic Acid and Xylenols for Industrial Use—Methods of Test—Part 12: Determination of Distillation Characteristics (Cresylic Acid and Xylenols Only) First Edition. 4 pp.

Distillation Methods (Cont.)
—Cresylic Acids—Residue Content
ISO 1897 Pt 13-83. Phenol, o-Cresol, m-Cresol, p-Cresol, Cresylic Acid and Xylenols for Industrial Use—Methods of Test—Part 13: Determination of Residue on Distillation (Cresylic Acid and Xylenols Only) First Edition. 4 pp.

—Crude Oils
BSI BS 4385: Part 1-91. 1991 Methods for Determination of Water in Crude Petroleum, Petroleum Products and Bituminous Materials by Distillation Part 1: Method for Crude Petroleum (ISO 9020: 1990). 17 pp.
DIN ENGL 51567-73. Testing of Mineral Oils; Determination of the Composition of Petroleum; Fractional Distillation According to Grosse-Oetringhaus (Sept). 9 pp.
ISO 9029-90. Crude Petroleum—Determination of Water—Distillation Method First Edition. 12 pp.
JIS K 2275-89. Testing Methods for Water Content of Crude Oil and Petroleum Products. 36 pp.

—Emulsified Asphalts
CNS K6794-83. Method of Test for Identification of Oil Distillate Obtained from Emulsified Asphalts by Micro-Distillation (Jul)(10455).
CNS K6796-83. Method of Test for Examination of Residue Obtained from Distillation or Evaporation of Emulsified Asphalts (Jul)(10457).

—Emulsified Asphalts—Residue Content
CNS K6793-83. Method of Test for Residue by Distillation of Emulsified Asphalts (Jul)(10454).

—Essential Oils
CNS K6063-65. Method of Test for Citronella Oil (Aug)(817)(R 1971). 3 pp.
CNS K6703-81. Essential Oils Determination of Residue from Distillation Under Reduced Pressure (Nov)(8137).
ISO 5991-79. Essential Oils—Determination of Residue from Distillation Under Reduced Pressure First Edition. 5 pp.

—Ethanols—Hydrobon Content
ISO 1388 Pt 10-81. Ethanol for Industrial Use—Methods of Test—Part 10: Estimation of Hydrocarbons Content—Distillation Method First Edition. 7 pp.

—Ethyl Acetate
BSI BS 553-90. 1990 Ethyl Acetate for Industrial Use. 8 pp.

—Ethyl Ether
BSI BS 579-57. 1957 Diethyl Ether (Technical). 12 pp.

—Fluoride Content
CNS J2042-82. Fluoride by Distillation-Spectrophotometric Method (Feb)(8512).

—Gasoline
BSI BS 2000: Part 191-83. 1983 Petroleum and Its Products Part 191: Distillation of Natural Gasoline. 7 pp.

—Insecticides
MOD UK M 5003/66. Examination of Insecticide Concentrate Space Spray and Insecticide Space Spray.

—Isopropyl Acetate
BSI BS 1834-68. 1968 Isopropyl Acetate. 11 pp.
BSI BS 1834-01. 1968 Amd 1 Isopropyl Acetate (AMD 7219) September 15, 1992. 13 pp.

—Lignite
ISO 647-74. Brown Coals and Lignites—Determination of the Yields of Tar, Water, Gas and Coke Residue by Low Temperature Distillation First Edition. 8 pp.

—Lubricating Oils
DIN ENGL 51356-78. Testing of Lubricating Oils and Liquid Fuels; Determination of the Distillation Range Under Reduced Pressure According to Grosse-Oetringhaus (Feb). 7 pp.

—Mineral Oils
DIN ENGL 51356-78. Testing of Lubricating Oils and Liquid Fuels; Determination of the Distillation Range Under Reduced Pressure According to Grosse-Oetringhaus (Feb). 7 pp.
DIN ENGL 51432 Pt 1-82. Testing of Used Oils; Determination of the Percentage of Water, Solvent and Distillation Residue in Used Oil; Distillation Method (Apr). 4 pp.

—Molding Materials
DIN ENGL 53713-71. Testing of Plastics; Determination of the Water Content of Moulding Materials; Distillation Method (June). 2 pp.

Distillation Methods (Cont.)
—Organic Coatings—Chemical Analysis
CNS K6804-2-84. Method of Test for Organic Coating (Chemical Analysis) Distillation Test (Jun)(10880-2).

—Organic Liquids
BSI BS 4591-90. 1990 Method for Determination of Distillation Characteristics of Organic Liquids (Other Than Petroleum Products). 12 pp.

—Paints
CGSB 1-GP-71 METH 24.1-80. Methods of Testing Paints and Pigments Water Content Distillation Method. 1 p.

—Paperboard
CNS P3025-88. Method of Determining Moisture Content in Pulp and Paper (Sep)(3086). 4 pp.

—Papers—Moisture Content
CNS P3025-88. Method of Determining Moisture Content in Pulp and Paper (Sep)(3086). 4 pp.

—Perchloroethylene
BSI BS 1593-63. (WITHDRAWN) 1963 Amd 1 Perchloroethylene (Tetrachloroethylene). 13 pp.
SAA AS K105-66. Perchloroethylene (Tetrachloroethylene) Being BS 1593:1963, Endorsed Without Amendment. 12 pp.

—Petroleum Products
BSI BS 2000: Part 123-85. (OBSOLESCENT) 1985 Petroleum and Its Products: Part 123: Distillation of Petroleum Products. 29 pp.
BSI BS 4385-80. (WITHDRAWN) 1980 Determination of Water in Petroleum Products and Bituminous Materials (Distillation Method) (Superseded by BS 4385: Part 1: 1991). 8 pp.
BSI BS 7392-90. 1990 Method for Determination of Distillation Characteristics of Petroleum Products. 26 pp.
CNS K6109-74. Method of Test for Distillation of Petroleum Products (Mar)(1218).
CNS K6339-73. Method of Test for Moisture Content in Petroleum Products and Bituminous Materials by Distillation (Feb)(3517). 4 pp.
DIN ENGL 51356-78. Testing of Lubricating Oils and Liquid Fuels; Determination of the Distillation Range Under Reduced Pressure According to Grosse-Oetringhaus (Feb). 7 pp.
DIN ENGL 51567-73. Testing of Mineral Oils; Determination of the Composition of Petroleum; Fractional Distillation According to Grosse-Oetringhaus (Sept). 9 pp.
ISO 3405-88. Petroleum Products—Determination of Distillation Characteristics Second Edition. 24 pp.
ISO 3733-76. Petroleum Products and Bituminous Materials—Determination of Water—Distillation Method First Edition. 7 pp.
JIS K 2254-90. Petroleum Products—Determination of Distillation Characteristics. 56 pp.

—Petroleum Products—Water Content
JIS K 2275-89. Testing Methods for Water Content of Crude Oil and Petroleum Products. 36 pp.

—Reagents
CNS K7006-82. Chemical Reagent Boiling Point and Distillate Determination (Feb)(1506).

—Soldering Fluxes
ISO 9455 Pt 9-93. Soft Soldering Fluxes—Test Methods—Part 9: Determination of Ammonia Content First Edition. 8 pp.
MOD UK M 9542/70. Examination of Flux, Soldering (Zinc Chloride Solution).

—Tobacco—Water Content
CNS N4133-85. Method of Test for Tobacco-Determination of Water Content (Dec)(11459).
ISO 6488-81. Tobacco—Determination of Water Content (Reference Method) First Edition. 6 pp.

—Uranium Hexafluoride
CNS J2071-82. Method for Determining Metallic Impurities in UF6 Method by Carrier Distillation (Nov)(9614).

—Volatile Fluids
BSI BS 2000: Part 195-93. 1993 Methods of Test for Petroleum and Its Products Part 195: Determination of Distillation Characteristics of Volatile Organic Liquids (W). 8 pp.
BSI BS 2000: Part 195-90. 1990 Petroleum and Its Products Part 195: Distillation Range of Volatile Organic Liquids. 12 pp.
ISO 918-83. Volatile Organic Liquids for Industrial Use—Determination of Distillation Characteristics First Edition. 11 pp.

INDUSTRY STANDARDS

Distillation Methods (Cont.)
—Water
BSI BS 6068: Sec 2.7-84. 1984 Water Quality Part 2: Physical, Chemical and Bio-Chemical Methods Section 2.7: Determination of Ammonium: Distillation and Titration Method (SUPERSEDES BS 2690; PART 7: 1968). 6 pp.

CGSB 31-GP-0A METH 48.1-57. Methods of Testing Corrosion-Prevention Materials and Processes Water Content (Distillation Method). 3 pp.

ISO 5664-84. Water Quality—Determination of Ammonium—Distillation and Titration Method First Edition. 5 pp.

—Xylenols
ISO 1897 Pt 12-83. Phenol, o-Cresol, m-Cresol, p-Cresol, Cresylic Acid and Xylenols for Industrial Use—Methods of Test—Part 12: Determination of Distillation Characteristics (Cresylic Acid and Xylenols Only) First Edition. 4 pp.

—Xylenols—Residue Content
ISO 1897 Pt 13-83. Phenol, o-Cresol, m-Cresol, p-Cresol, Cresylic Acid and Xylenols for Industrial Use—Methods of Test—Part 13: Determination of Residue on Distillation (Cresylic Acid and Xylenols Only) First Edition. 4 pp.

Distortion
Use: Deformation

Distortion (Electrical)
Use For: Pulse Distortion *See Also:* Attenuation Distortion; Delay Distortion; Envelope Delay Distortion; Harmonic Distortion; Intermodulation Distortion; Isochronous Distortion; Jitter; Multimode Distortion; Multipath Distortion; Nonlinear Distortion; Phase Jitter; Quantizing Distortion; Signal to Distortion Ratio

—Cable Couplers
CEN PREN 2591 (Part G2)-92. Elements of Electrical and Optical Connection Test Methods Part G2—Measurement of Signal Distortion. 3 pp.

—Dial Tones
CCITT RECMN E.180-89. Technical Characteristics of Tones for the Telephone Service—Telephone Network and ISDN—Operation, Numbering, Routing and Mobile Service (Study Group II) 9 pp. 9 pp.

—Glossaries
CNS C5169-83. Glossary of Pulse Distortion and Jitter (Sep)(10550).

—Optical Fibers
CNS C6263-86. Method of Test for Fiber Optic Devices (FOTP-51 Pulse Distortion Measurement of Multimode Glass Optical Fiber Information Transmission Capacity) (Sep)(11707).

—Radiotelegraph Circuits
CCIR QUESTION 144/9-90. Efficency Factor and Telegraph Distortion on ARQ Circuits—Questions of Study Group 9 Fixed Service. 1 p.

—Telegraph Circuits
CCITT RECMN R.51-89. Standardized Text for Distortion Testing of the Code-Independent Elements of a Complete Circuit—Telegraph Transmission (Study Group IX) 2 pp. 2 pp.

CCITT RECMN R.51 BIS-89. Standardized Text for Testing the Elements of a Complete Circuit—Telegraph Transmission (Study Group IX) 1 pp. 1 p.

—Television Signals—Subjective Assessment
CCIR Report 1083-86. Advanced Statistical Processing of Subjective Test Results—Section 11D—Picture Quality and the Parameters Affecting It. 3 pp.

—Television Signals—Vestigial Sideband Transmission
CCIR Report 404-2-74. Distortion of Television Signals Due to the Use of Vestigial Sideband Emissions—Section 11D—Picture Quality and the Parameters Affecting It. 5 pp.

CCIR QUESTION 56/11-90. Distortion of Television and Data-Broadcasting Signals in Emission and Reception, Caused by the Use of Vestigial-Sideband Techniques—Questions Concerning Study Group 11—Broadcasting Service (Television). 1 p.

Distortion Analyzers
See Also: Analyzers; Harmonic Analyzers; Isochronous Distortion

—Data Transmission—Maintenance
CCITT RECMN V.52-89. Characteristics of Distortion and Error-Rate Measuring Apparatus for Data Transmission—Data Communication over the Telephone Network (Study Group XVII) 1 pp. 1 p.

Distortion Analyzers (Cont.)
—International Telegraph Circuits—Units of Measurement
CCITT RECMN R.5-89. Observation Conditions Recommended for Routine Distortion Measurements on International Telegraph Circuits—Telegraph Transmission (Study Group IX) 2 pp. 2 pp.

—Legislation—Units of Measurement
CCITT RECMN R.9-89. How the Laws Governing Distribution of Distortion Should Be Arrived at—Telegraph Transmission (Study Group IX) 2 pp. 2 pp.

—Telegraph Circuits—Calculation
CCITT RECMN R.11-89. Calculation of the Degree of Distortion of a Telegraph Circuit in Terms of the Degrees of Distortion of the Component Links—Telegraph Transmission (Study Group IX) 2 pp. 2 pp.

—Voice Frequency Telegraphy
CCITT RECMN R.53-89. Permissible Limits for the Degree of Distortion on an International 50-Baud/120-Hz VFT Channel (Frequency and Amplitude Modulation)—Telegraph Transmission (Study Group IX) 2 pp. 2 pp.

Distress and Safety Systems (Land)
Use: Global Land Distress and Safety Systems

Distress and Safety Systems (Maritime)
Use: Global Maritime Distress and Safety Systems

Distress Signal Devices
See Also: Signal Devices; Signal Lights; Smoke Signal Devices

—Pleasure Boats
SAA AS 2092-88. Pyrotechnic, Marine Distress Flares and Signals for Pleasure Craft Amdt 1 July 1989. 6 pp.

Distributed Process Control Systems
Use: Process Control Equipment

Distributed Processing Systems
See Also: Computer Systems; Process Control Systems

—Abstract Service—Message Handling Systems
CCITT RECMN X.407-89. Message Handling Systems: Abstract Service Definition Conventions—Data Communication Networks—Message Handling Systems (Study Group VII) 28 pp. 28 pp.

—Data Highways
IEC 954-90. Process Data Highway, Types A and B (PROWAY A and B), for Distributed Process Control Systems First Edition. 327 pp.

IEC 955-89. Process Data Highway, Type C (PROWAY C), for Distributed Process Control Systems First Edition; (Amendment 1-1992). 526 pp.

—Directory Databases
CCITT RECMN X.518-89. The Directory—Procedures for Distributed Operation—Data Communication Networks Directory (Study Group VII) 59 pp. 59 pp.

—Glossaries
CNS C5237-90. Information Processing Systems—Vocabulary (Part 18: Distributed Data Processing) (Aug)(12758).

JIS X 0018-89. Glossary of Terms Used in Information Processing (Distributed Data Processing).

—Local Area Networks
ECMA ECMA-TR 21-84. Local Area Networks Interworking Units for Distributed Systems. 13 pp.

Distributing Frames
Use For: Distribution Frames *See Also:* Panel Boards (Electrical); Telephone Equipment

—Group Links—Power Levels
CCITT RECMN G.233-89. Recommendations Concerning Translating Equipments—International Analogue Carrier Systems (Study Group XV) 10 pp. 10 pp.

—Supergroup Links—Power Levels
CCITT RECMN G.233-89. Recommendations Concerning Translating Equipments—International Analogue Carrier Systems (Study Group XV) 10 pp. 10 pp.

Distributing Frames (Cont.)
—Supermastergroup Links—Power Levels
CCITT RECMN G.233-89. Recommendations Concerning Translating Equipments—International Analogue Carrier Systems (Study Group XV) 10 pp. 10 pp.

—15 Supergroup Assembly Links—Power Levels
CCITT RECMN G.233-89. Recommendations Concerning Translating Equipments—International Analogue Carrier Systems (Study Group XV) 10 pp. 10 pp.

Distribution Boards
Use: Switchboards

Distribution Frames
Use: Distributing Frames

Distribution Lists (Message Handling)
See Also: Message Handling Systems; Originator/Recipient Names (Message Handling)

CCITT RECMN F.400-92. Message Handling Services: Message Handling System and Service Overview (Study Group I) 82 pp (Same as Recmn X.400). 82 pp.

CCITT RECMN F.400-89. Message Handling System and Service Overview—Message Handling and Directory Services—Operations and Definition of Service (Study Group I) 73 pp. 73 pp.

CCITT RECMN F.401-92. Message Handling Services: Naming and Addressing for Public Message Handling Services (Study Group I) 21 pp. 21 pp.

CCITT RECMN F.401-89. Message Handling Services: Naming and Addressing for Public Message Handling Services—Message Handling and Directory Services—Operations and Definition of Service (Study Group I) 11 pp. 11 pp.

CCITT RECMN X.400-93. Message Handling Services: Message Handling System and Service Overview (Study Group VII) 82 pp (Same as Recmn F.400). 82 pp.

—Directory Services
CCITT RECMN F.400-92. Message Handling Services: Message Handling System and Service Overview (Study Group I) 82 pp (Same as Recmn X.400). 82 pp.

CCITT RECMN F.400-89. Message Handling System and Service Overview—Message Handling and Directory Services—Operations and Definition of Service (Study Group I) 73 pp. 73 pp.

CCITT RECMN X.400-93. Message Handling Services: Message Handling System and Service Overview (Study Group VII) 82 pp (Same as Recmn F.400). 82 pp.

—Interpersonal Messaging Services
CCITT RECMN F.420-92. Message Handling Services: Public Interpersonal Messaging Service (Study Group I) 16 pp. 16 pp.

CCITT RECMN F.420-89. Message Handling Services: the Public Interpersonal Messaging Service—Message Handling and Directory Services—Operations and Definition of Service (Study Group I) 15 pp. 15 pp.

Distribution Panel Boards
Use: Panel Boards (Electrical)

Distribution Panelboards
Use: Panel Boards (Electrical)

Distribution Poles
Use: Power Poles

Distribution Systems
Use For: Supply Systems *See Also:* Gas Distribution; Materials Handling Equipment; Power Distribution Systems; Shipping; Transportation

—Electrical Components—Naval Ships
MOD UK NES 538-89. Guide to the Selection of Small Electrical Fittings and Components Issue 3 (08.89). 306 pp.

MOD UK NES 538-92. Guide to the Selection of Small Electrical Fittings and Components Issue 4 (10.92). 285 pp.

—Electrical Components—Submarines
MOD UK NES 538-89. Guide to the Selection of Small Electrical Fittings and Components Issue 3 (08.89). 306 pp.

MOD UK NES 538-92. Guide to the Selection of Small Electrical Fittings and Components Issue 4 (10.92). 285 pp.

—Electrical Equipment—Naval Ships
MOD UK NES 539-81. Guide to the Design of Supply Systems for Portable Electrical Equipment Issue 1 (06.81). 29 pp.

INTERNATIONAL AND NON-U.S. NATIONAL STANDARDS
SUBJECT INDEX

Distribution Systems (Cont.)
—**Electrical Equipment—Naval Ships (Cont.)**
MOD UK NES 539-93. Guide to the Design of Supply Systems for Portable Electrical Equipment Issue 2 (01.93). 26 pp.

—**Electrical Equipment—Submarines**
MOD UK NES 539-81. Guide to the Design of Supply Systems for Portable Electrical Equipment Issue 1 (06.81). 29 pp.
MOD UK NES 539-93. Guide to the Design of Supply Systems for Portable Electrical Equipment Issue 2 (01.93). 26 pp.

—**Fittings—Naval Ships**
MOD UK NES 538-89. Guide to the Selection of Small Electrical Fittings and Components Issue 3 (08.89). 306 pp.
MOD UK NES 538-92. Guide to the Selection of Small Electrical Fittings and Components Issue 4 (10.92). 285 pp.

—**Fittings—Submarines**
MOD UK NES 538-89. Guide to the Selection of Small Electrical Fittings and Components Issue 3 (08.89). 306 pp.
MOD UK NES 538-92. Guide to the Selection of Small Electrical Fittings and Components Issue 4 (10.92). 285 pp.

—**Glossaries**
JIS Z 0111-85. Glossary of Terms for Physical Distribution.

—**Military—Emergency Medical Services**
NATO STANAG 2105 ED 5 AMD 1-86. NATO Table of Military Equivalents (AMedP-1(D)). 8 pp.

—**Military Operations—Land Forces**
NATO STANAG 2034 ED 4 AMD 6-82. Land Forces Procedures for Allied Supply Transactions. 14 pp.

Distribution Transformers
See Also: Power Transformers; Transformers
CNS C3003-84. Testing Standard for Distribution Transformer (May)(599).
CNS C4010-79. Distribution Transformer (Nov)(598).

—**Dry Type—Loads (Electronic)**
CSA C9.1-M1981. Guide for Loading Dry-Type Distribution and Power Transformers; (Gen Instr 1). 18 pp.

—**Dry Type—Single Phase—Encapsulated**
JIS C 4306-91. 6 kV Encapsulated—Winding Distribution Transformers. 29 pp.

—**Dry Type—Three Phase—Encapsulated**
JIS C 4306-91. 6 kV Encapsulated—Winding Distribution Transformers. 29 pp.

—**Dry Type—Three Phase—High Voltage**
CENELEC HD 538.1 S1-92. Three-Phase Dry-Type Distribution Transformers 50 Hz, from 100 to 2 500 kVA, with Highest Voltage for Equipment not Exceeding 36 kV Part 1: General Requirements and Requirements for Transformers with High. Voltage for Equipment Not Exceeding 24 kV. 14 pp.

—**Junction Boxes**
BSI BS 6436-84. 1984 Ground Mounted Distribution Transformers for Cable Box or Unit Substation Connection. 11 pp.

—**Oil Filled— 5-15 kV**
JIS C 4304-77. 6 kV Oil-Immersed Distribution Transformers. 26 pp.

—**Oil Filled—Single Phase**
CSA CAN/CSA-C2-M91. Single-Phase and Three-Phase Distribution Transformers, Types ONAN and LNAN; (Gen Instr 1 Thru 2). 52 pp.
CSA C301.1-1976. Single-Phase Submersible Distribution Transformers Type ONAN (R 1987). 17 pp.

—**Oil Filled—Three Phase**
CSA CAN/CSA-C2-M91. Single-Phase and Three-Phase Distribution Transformers, Types ONAN and LNAN; (Gen Instr 1 Thru 2). 52 pp.
CSA C301.2-M1979. Three-Phase Submersible Distribution Transformers Type ONAN (R 1987). 26 pp.

—**Oil Filled—Three Phase— 15.5-34.5 kV**
CENELEC HD 428-83. Three-Phase Oil-Immersed Public Distribution Transformers, 50 Hz, from 50 to 2500 kVA, with Highest Voltage for Equipment Not Exceeding 24 kV. 15 pp.
CENELEC HD 428/A1-90. AMD 1 Three-Phase Oil-Immersed Public Distribution Transformers, 50 Hz, from 50 to 2500 kVA, with Highest Voltage for Equipment Not Exceeding 24 kV. 6 pp.

Distribution Transformers (Cont.)
—**Oil Filled—Three Phase— 50-2500 kV**
CENELEC HD 428.1 S1-92. Three Phase Oil-Immersed Dist. Transformers 50 Hz, from 50 to 2500 kVA with Highest Voltage for Equip. Not Exceeding 36 kV Part 1: General Requirements and Requirements for Transformers with Highest Voltage for Equip. Not Exceeding 24 kV. 4 pp.
CENELEC HD 428.1 S1-92. Three Phase Oil-Immersed Dist. Transformers 50 Hz, from 50 to 2500 kVA with Highest Voltage for Equip. Not Exceeding 36 kV Part 1: General Requirements and Requirements for Transformers with Highest Voltage for Equip. Not Exceeding 24 kV. 14 pp.

—**Single Phase—Pad Mounted**
CSA C227.1-1975. Single-Phase Pad Mounted Distribution Transformers (R 1987). 16 pp.

—**Single Phase—Pad Mounted—Low Profile**
CSA CAN/CSA-C227.3-M91. Low-Profile, Single-Phase, Dead-Front, Pad-Mounted Distribution Transformers; (Gen Instr 1 Thru 2). 29 pp.

—**Three Phase—Pad Mounted**
CSA CAN/CSA-C227.2-88. Three-Phase, Live-Front, Pad-Mounted Distribution Transformers; (Gen Instr 1). 32 pp.
CSA C227.4-M1978. Three-Phase Dead-Front Pad-Mounted Distribution Transformers (R 1987); (Erratum April 1979). 27 pp.

Distributors (Electrical)
See Also: Breaker Points; Commutators; Ignition Systems; Internal Combustion Engines; Internal Combustion Piston Engines; Piston Engines; Sewage Treatment; Sprayers; Spreaders

—**Automotive**
CNS D3043-87. Mounting Dimensions of Distributors for Automobiles (Oct)(5320).
JIS D 5202-80. Mounting Dimensions of Distributors for Automobiles. 9 pp.

—**Automotive—Breaker Points**
CNS D1004-87. Configuration and Mounting Dimension of Distributor Contact Breakers for Automobiles (Oct)(5559). 1 p.
JIS D 5206-71. Shape and Mounting Dimensions of Distributor Contact Breakers for Automobiles. 3 pp.

—**Automotive—Connectors**
CNS D1066-87. Dimension of High Tension Connection for Ignition Coils and Distributors for Automobiles (Oct)(12123).
CNS D2022-2-86. High Tension Cable Connection for Ignition Coils and Distributors for Automobiles (Nov)(5561-2).
CNS D2022-3-86. Low Tension Cable Connections for Ignition Coils and Distributors for Automobiles (Nov)(5561-3).
ISO 3553 Pt 1-87. Road Vehicles—High-Tension Connections for Ignition Coils and Distributors—Part 1: Socket-Type First Edition. 6 pp.
ISO 3553 Pt 2-87. Road Vehicles—High-Tension Connections for Ignition Coils and Distributors—Part 2: Plug-Type First Edition. 4 pp.

—**Automotive—Electric Terminals**
CNS D1066-87. Dimension of High Tension Connection for Ignition Coils and Distributors for Automobiles (Oct)(12123).
CNS D2022-2-86. High Tension Cable Connection for Ignition Coils and Distributors for Automobiles (Nov)(5561-2).
CNS D2022-3-86. Low Tension Cable Connections for Ignition Coils and Distributors for Automobiles (Nov)(5561-3).
CNS D2073-90. High Tension Cable Terminals of Distributor for Automobiles (LA Type) (Aug)(6911).
ISO 3553 Pt 1-87. Road Vehicles—High-Tension Connections for Ignition Coils and Distributors—Part 1: Socket-Type First Edition. 6 pp.
ISO 3553 Pt 2-87. Road Vehicles—High-Tension Connections for Ignition Coils and Distributors—Part 2: Plug-Type First Edition. 4 pp.

—**Automotive—Inspection**
CNS D3078-86. Method of Test for Distributors for Automobiles (Jun)(7896).
JIS D 1603-79. Inspection of Distributors for Automobiles (R 1987). 8 pp.

Distributors (Materials Handling)
See Also: Materials Handling Equipment

—**Binders (Materials)—Roads**
BSI BS 1707-89. 1989 Hot Binder Distributors for Road Surface Dressing. 14 pp.
BSI BS 1707-70. (WITHDRAWN) 1970 Hot Binder Distributors for Road Surface Dressing. 20 pp.

Disturbance of Service
Use: Interruption of Service

Ditching
Use: Aircraft Landings—Ditching

2,2'-Dithiobis(benzothiazole)
Use For: Benzothiazyl Disulfide; MBTS
CNS K4035-78. Rubber Vulcanization Accelerator MBTS (Benzothiozolyl Disulfide or Dibenzothiazyl Disulfide) (Oct)(4629).

Dithiooxamide
Use: Rubeanic Acid

Dithizone
Use For: Phenyldiazenecarbothioic Acid 2-Phenylhydrazide
CNS K7207-66. Chemical Reagent (Dithizone) (Jun)(1707).
JIS K 8490-78. Dithizone, Phenyldiazenecarbothioic Acid 2-Phenylhydrazide.

Divans
See Also: Beds
BSI BS 1877: Part 9-75. (WITHDRAWN) 1975 Amd 1 Domestic Bedding Part 9: Upholstered Wood-Framed Divans. 15 pp.

—**Flammability Testing**
BSI BS 6807-90. 1990 Methods of Test for Assessment of the Ignitability of Mattresses, Divans and Bed Bases with Primary and Secondary Sources of Ignition. 22 pp.
BSI BS 6807-01. 1990 Amd 1 Methods of Test for Assessment of the Ignitability of Mattresses, Divans and Bed Bases with Primary and Secondary Sources of Ignition (AMD 7551) January 15, 1993. 23 pp.
SNZ NZS 8720-92. Methods of Test for Assessment of the Ignitability of Mattresses, Divans and Bed Bases with Primary and Secondary Sources of Ignition Amend: A, 1992. 24 pp.

—**Ticking**
BSI BS 7337-90. 1990 Tickings (Supersedes BS 2036: 1962 & BS 2732: 1961 & BS 4820: 1972). 11 pp.

—**Webbing**
BSI BS 5127-74. 1974 Polyolefin Webbing for Divans and Bed Bases. 11 pp.

Diversion If Number Busy
Use: Call Diversion Services

Diversity (Communications)
See Also: Radio Communications; Radio Transmission

—**Radio Relay Systems**
CCIR RECMN 752-92. Diversity Techniques for Radio-Relay Systems—Section 9B2—System General Characteristics. 9 pp.
CCIR Report 376-6-90. Diversity Techniques for Radio-Relay Systems—Section 9B2—System General Characteristics. 11 pp.
CCIR QUESTION 106/9-90. Diversity Techniques for Radio-Relay Systems—Questions of Study Group 9 Fixed Service. 1 p.

Diversity Reception
See Also: Radio Communications; Radio Reception
CCIR Report 327-3-78. Diversity Reception—Section 3Ab—Antennas Characteristics. 5 pp.

Divided Wheels
Use: Wheels

Dividers (Drawing)
See Also: Drawing Equipment
CNS Z7108-82. Dividers for Precision Drawing Instruments (P) (May)(8889).
CNS Z7130-82. Divider for School Drawing Instruments (Dec)(9772).

—**Points**
CNS Z7111-82. Fixed Centre Spring Bow with Interchangeable Points for Precision Drawing Instruments (P) (May)(8892).
CNS Z7114-82. Divider Points for Precision Drawing Instruments (P) (Jun)(9044).
CNS Z7151-83. Divider Points for School Drawing Instruments (Jan)(9927).

—**Proportional**
CNS Z7127-82. Drawing Instruments (Proportional Divider) (Dec)(9769).

—**Reversible**
CNS Z7131-82. Reversible Spring Bow for School Drawing Instruments (Dec)(9773).

—**Spring**
BSI BS 3123-59. 1959 Spring Calipers and Spring Dividers. 11 pp.
CGSB CAN/CGSB-39.25-M88. Spring Dividers. 11 pp.

INDUSTRY STANDARDS

Dividers (Drawing) (Cont.)
—Spring (Cont.)
CNS Z7110-82. Spring-Bow Divider for Precision Drawing Instruments (May)(8891).
CNS Z7132-82. Spring Bow Divider for School Drawing Instruments (Dec)(9774).

Dividing Apparatus
Use For: Dividing Heads
BSI BS 4656: Part 12-79. 1979 Amd 1 Accuracy of Machine Tools and Methods of Test Part 12: Dividing Heads (AMD 5603) July 29, 1988. 9 pp.
CNS B2759-84. Universal Dividing Heads and Simple Dividing Heads (Jan)(10737).
DIN ENGL 8658 Pt 1-79. Machine Tools; Mechanical Dividing Heads; Acceptance Conditions (Nov). 5 pp.
ISO 5734-86. Acceptance Conditions of Mechanical Dividing Heads for Machine Tools—Testing of Accuracy Second Edition. 9 pp.
JIS B 6159-85. Universal Dividing Heads and Simple Dividing Heads. 16 pp.

Dividing Heads
Use: Dividing Apparatus

Diving
See Also: Diving Equipment; Scuba Diving
—Certification
SAA AS 4005.1-92. Training and Certification of Recreational Divers—Part 1: Minimum Entry-Level SCUBA Diving. 20 pp.
—Naval Operations
NATO STANAG 1372 ED 1 AMD 0-90. Allied Guide to Diving Operations—ADivP-1. 6 pp.
NATO STANAG 1372 ED 2 AMD 0-93. Allied Guide to Diving Operations—ADivP-1. 5 pp.
NATO STANAG 1372 ED 2 AMD 1-93. Allied Guide to Diving Operations—ADivP-1. 6 pp.
—Radiology—Dysbaric Osteonecrosis
NATO STANAG 1249 ED 1 AMD 3-81. The Method of Detection and Control of Dysbaric Osteonecrosis in Divers. 12 pp.
—Rescue Operations
NATO STANAG 1229 ED 1 AMD 4-81. Evacuation of Diving Accident Patients. 10 pp.
NATO STANAG 1229 ED 2 AMD 1-89. Evacuation of Diving Accident Patients. 13 pp.
—Training
SAA AS 4005.1-92. Training and Certification of Recreational Divers—Part 1: Minimum Entry-Level SCUBA Diving. 20 pp.

Diving Equipment
Use For: Scuba Diving Equipment
See Also: Aquatic Sports Equipment; Breathing Apparatus; Buoyancy Compensators; Diving; Fire Protection; Oxygen Supply Equipment; Sports Equipment; Wet Suits
SNZ NZS 5813-82. Underwater Swimming Equipment (Reconfirmed 1988). 24 pp.
—Breathing Apparatus
BSI BS 4001: Part 1-81. 1981 Recommendations for the Care and Maintenance of Underwater Part 1: Breathing Apparatus Compressed Air Open Circuit Type. 8 pp.
BSI BS 4001: Part 2-67. 1967 Amd 1 Recommendations for the Care and Maintenance of Underwater Breathing Apparatus Part 2: Standard Diving Equipment. 38 pp.
—Compressed Air
JIS S 7306-89. Compressed Air for SCUBA Diving. 9 pp.
—Face Shields
CNS S1214-89. Diving Mask (Feb)(12497). 2 pp.
CNS S2141-89. Method of Test for Diving Mask (Feb)(12498). 3 pp.
—Gas Cylinders
JIS S 7302-88. SCUBA Diving Goods—Cylinders. 10 pp.
—Gas Cylinders—Gas Purity
MOD UK DSTAN 68-75-01. Breathing Gas Purity for Diving Issue 1; Amendment 1. 9 pp.
MOD UK DSTAN 68-75-93. Breathing Gas Purity for Diving Issue 2. 11 pp.
—Gas Cylinders—Regulators
JIS S 7303-88. SCUBA Diving Goods—Cylinders Valves. 9 pp.
—Hoses
CNS K4069-82. Diving Hoses (Nov)(9618).
JIS K 6345-82. Rubber Hoses for Diving.
—Hoses—Regulators
JIS S 7304-88. SCUBA Diving Goods—Regulators. 13 pp.
—Lead Belts
CNS S1163-83. Nylon Black Waistband with Lead for Dive (May)(10293).
CNS S2101-83. Method of Test for Nylon Black Waistband with Lead for Dive (May)(10294).
—Medicine
NATO STANAG 1227 ED 1 AMD 4-80. Minimal Qualifications in Diving Medicine for Non Specialist Medical Officers in Support of Diving Operations. 6 pp.
—Safety
CSA CAN/CSA-Z275.2-92. Occupational Safety Code for Diving Operations; (Gen Instr 1). 64 pp.
SAA AS 2299-92. Occupational Diving (in Professional Package 47). 76 pp.
—Watches
ISO 6425-84. Divers' Watches Second Edition. 6 pp.

DMC
Use: Bulk Molding Compounds

DME
Use: Distance Measuring Equipment

DMM
Use: Multimeters

DNIC
Use: Data Network Identification Codes

DNPD
Use: N,N'-di-beta-Naphthyl-p-Phenylenediamine

DNT
Use: Dinitrotoluene

Do Not Disturb Services
See Also: Telephone Services
—Private Switched Networks
ECMA ECMA 193-93. Private Telecommunication Networks (PTN)—Specification, Functional Model and Information Flows—Do Not Disturb and Do Not Disturb Override Supplementary Services (DND(O)SD). 79 pp.
ECMA ECMA 194-93. Private Telecommunication Networks (PTN)—Inter-Exchange Signalling Protocol—Do Not Disturb and Do Not Disturb Override Supplementary Services (QSIG-DND(O)). 59 pp.

Dobbies
See Also: Textile Machinery
—Lags
ISO 573-76. Textile Machinery and Accessories—Dobby Lags and Pegs in Wood, Metal or Other Suitable Material—Dimensions First Edition. 4 pp.
JIS L 6515-81. Pattern Lags and Pegs for Dobbies.
JIS L 6515-62. Dobby Pegs and Lags for Silk Loom. 5 pp.
—Pattern Cards
JIS L 6518-82. Pattern Cards for Dobbies.
—Patterns—Paper
BSI BS 4266-77. 1977 Textile Machinery and Accessories. Paper Patterns for Dobbies. Dimensions. 3 pp.
ISO 576-76. Textile Machinery and Accessories—Paper Patterns for Dobbies—Dimensions First Edition. 4 pp.
—Pegs
ISO 573-76. Textile Machinery and Accessories—Dobby Lags and Pegs in Wood, Metal or Other Suitable Material—Dimensions First Edition. 4 pp.
JIS L 6515-81. Pattern Lags and Pegs for Dobbies.
JIS L 6515-62. Dobby Pegs and Lags for Silk Loom. 5 pp.

Dock Gates
Use: Docks—Gates

Docks
See Also: Drydocks; Quays; Ramps (Loading)
—Gates
BSI BS 6349: Part 3-88. 1988 Code of Practice for Maritime Structures; Part 3: Design of Dry Docks, Locks, Slipways and Shipbuilding Berths, Shiplifts and Dock and Lock Gates. 75 pp.

Doctor Testing
—Aromatic Hydrocarbons
ISO 5275-79. Aromatic Hydrocarbons—Test for Presence of Mercaptans (Thiols)—Doctor Test First Edition. 4 pp.
—Petroleum Products
CNS K6113-90. Method of Doctor Test for Petroleum Products (Aug)(1222).

Document Acquisition
See Also: Data Acquisition Systems; Documentation; Documents; Procurement
—Data Elements
BSI BS ISO 8459-2-92. 1992 Information and Documentation—Bibliographic Data Element Directory Part 2: Acquisitions Applications (S). 39 pp.
ISO 8459 Pt 2-92. Information and Documentation—Bibliographic Data Element Directory—Part 2: Acquisitions Applications First Edition. 37 pp.
—Glossaries
ISO 5127 Sec 3A-81. Information and Documentation—Vocabulary—Section 3A): Acquisition, Identification, and Analysis of Documents and Data First Edition. 30 pp.

Document Architecture
Use: Document Formats

Document Facsimile Terminals
Use: FAX Machines

Document Filing and Retrieval
Use: Application Layer (OSI)—Document Filing and Retrieval

Document Formats
See Also: Documents; Mixed Mode; Office Document Architecture; Processable Mode Number One
—Bureaufax Communications
CCITT RECMN F.170-92. Operational Provisions for the International Public Facsimile Service Between Public Bureaux (Bureaufax) (Study Group I) 9 pp. 9 pp.
CCITT RECMN F.170-89. Operational Provisions for the International Public Facsimile Service Between Public Bureaux (Bureaufax)—Telematic, Data Transmission and Teleconference Services—Operations and Quality of Service—(Study Group I) 8 pp. 8 pp.
—CCITT
CCITT FASCICLE I.2-88. Opinions and Resolutions Recommendations on the Organisation and Working Procedures of CCITT (Series A). 64 pp.
—CCITT—Machine Readable
CCITT FASCICLE I.2-88. Opinions and Resolutions Recommendations on the Organisation and Working Procedures of CCITT (Series A). 64 pp.
—Directories—Telex Call Numbers
CCITT RECMN F.60-89. Operational Provisions for the International Telex Service—Telegraph and Mobile Services Operations and Quality of Service (Study Group I) 17 pp. 17 pp.
—FAX Communications—CCITT Group 4
CCITT RECMN F.184-89. Operational Provisions for the International Public Facsimile Service Between Subscriber Stations with Group 4 Facsimile Machines (Telefax 4)—Telematic, Data Transmission and Teleconference Services—Operations and Quality of Service (Study Group I) 9 pp. 9 pp.
—Glossaries
CCITT RECMN A.22-89. Collaboration with Other International Organizations on Information Technology—Terminal Equipment and Protocols for Telematic Services (Study Group VIII) 2 pp. 2 pp.
—Invoices—Maritime Mobile Services
CCITT RECMN D.90-92. Charging, Accounting and Refunds in the Maritime Mobile Service (Study Group III) 23 pp. 23 pp.
CCITT RECMN D.90-89. Charging, Accounting and Refunds in the Maritime Mobile Service—General Tariff Principles—Charging and Accounting in International Telecommunications Services (Study Group III) 19 pp. 19 pp.
—Invoices—Telephone Services
CCITT RECMN D.170-89. Monthly Telephone and Telex Accounts—General Tariff Principles—Charging and Accounting in International Telecommunications Services (Study Group III) 9 pp (Same as Recmn E.270). 9 pp.

INTERNATIONAL AND NON-U.S. NATIONAL STANDARDS
SUBJECT INDEX

Documentation

Document Formats *(Cont.)*

—**Invoices—Telex Communications**
CCITT RECMN D.170-89. Monthly Telephone and Telex Accounts—General Tariff Principles—Charging and Accounting in International Telecommunications Services (Study Group III) 9 pp (Same as Recmn E.270). 9 pp.

—**Message Handling Systems—Message Text**
CCITT RECMN F.415-89. Message Handling Services: Intercommunication with Public Physical Delivery Services—Message Handling and Directory Services—Operations and Definition of Service (Study Group I) 15 pp (Erratum in Recmn F.410). 15 pp.

—**Orders (Sales Documents)—Freephone Services**
CCITT RECMN E.152-89. International Freephone Service (IFS)—Telephone Network and ISDN—Operation, Numbering, Routing and Mobile Service (Study Group II) 12 pp. 12 pp.

—**Registration—Telephone Credit Cards**
CCITT RECMN E.118-89. Automated International Telephone Credit Card System—Telephone Network and ISDN—Operation, Numbering, Routing and Mobile Service (Study Group II) 7 pp. 7 pp.

—**Technical Manuals—Foreign Visitors**
CCITT RECMN E.128-89. Leaflet to be Distributed to Foreign Visitors—Telephone Network and ISDN—Operation, Numbering, Routing and Mobile Service (Study Group II) 6 pp. 6 pp.

—**Teleinformatic Services—Processable Mode Number One**
CCITT RECMN T.502-91. Document Application Profile PM-11 for the Interchange of Character Content Documents in Processable and Formatted Forms (Study Group VIII) 50 pp. 50 pp.
CCITT RECMN T.502-89. Document Application Profile PM1 for the Interchange of Processable Form Documents—Terminal Equipment and Protocols for Telematic Services (Study Group VIII) 35 pp. 35 pp.

—**Telephone Directories—Foreign Visitors**
CCITT RECMN E.127-89. Pages in the Telephone Directory Intended for Foreign Visitors—Telephone Network and ISDN—Operation, Numbering, Routing and Mobile Service (Study Group II) 2 pp. 2 pp.

—**Telephone Directories—General Information**
CCITT RECMN E.126-89. Harmonization of the General Information Pages of the Telephone Directories Published by Administrations—Telephone Network and ISDN—Operation, Numbering, Routing and Mobile Service (Study Group II) 10 pp. 10 pp.

—**Teletex Communications**
CCITT RECMN F.200-92. Teletex Service (Study Group I) 28 pp. 28 pp.
CCITT RECMN F.200-89. Teletex Service—Telematic, Data Transmission and Teleconference Services—Operations and Quality of Service (Study Group I) 20 pp. 20 pp.

—**Teletex Communications—Service Observation**
CCITT RECMN F.200-92. Teletex Service (Study Group I) 28 pp. 28 pp.
CCITT RECMN F.200-89. Teletex Service—Telematic, Data Transmission and Teleconference Services—Operations and Quality of Service (Study Group I) 20 pp. 20 pp.

—**Transfer and Manipulation—Teleinformatic Services**
CCITT RECMN T.400-89. Introduction to Document Architecture, Transfer and Manipulation—Terminal Equipment and Protocols for Telematic Services (Study Group VIII) 5 pp. 5 pp.

—**Transit Routing—Telecommunication Administrations**
CCITT RECMN E.171-89. International Telephone Routing Plan—Telephone Network and ISDN—Operation, Numbering, Routing and Mobile Service (Study Group II) 10 pp. 10 pp.

—**Videotex Communications—Interworking**
CCITT RECMN T.504-89. Document Application Profile for Videotex Interworking—Terminal Equipment and Protocols for Telematic Services (Study Group VIII) 8 pp. 8 pp.

Document Formats *(Cont.)*

—**Yearbooks—Routing (Telecommunications)**
CCITT RECMN E.150-89. Publication of a "List of International Telephone Routes"—Telephone Network and ISDN—Operation, Numbering, Routing and Mobile Service (Study Group II) 3 pp. 3 pp.

Document Inserting Machines
Use: Mail Processing Machines

Document Layouts
Use: Document Formats

Document Printing Application
Use: Application Layer (OSI)—Document Printing Application

Document Processing Machines
Use: Mail Processing Machines

Document Reader Sorters

—**Papers**
DIN ENGL 6723 Pt 1-81. Papers for Data Processing; Uncoated and Untreated 90 g/m2 Paper for Document Reader Sorters; Requirements, Testing (Nov). 3 pp.

Document Storage and Retrieval Systems
See Also: Computer Assisted Retrieval Systems; Documentation; Libraries
ISO 838-74. Paper—Holes for General Filing Purposes—Specifications First Edition. 3 pp.

—**Documentation**
ISO 11442 Pt 4-93. Technical Product Documentation—Handling of Computer-Based Technical Information—Part 4: Document Management and Retrieval Systems First Edition. 8 pp.

—**Glossaries**
JIS X 0705-89. Documentation and Information-Vocabulary—Acquisition Identification, and Analysis of Documents and Data.

Document Style Semantics and Specification Language
Use: DSSSL

Document Transfer and Manipulation

—**Bulk Handling**
CCITT RECMN T.521-89. Communication Application Profile BT0 for Document Bulk Transfer Based on the Session Service—Terminal Equipment and Protocols for Telematic Services (Study Group VIII) 13 pp. 13 pp.
CCITT RECMN T.522-92. Communication Application Profile BT1 for Document Bulk Transfer (Study Group VIII) 7 pp. 7 pp.

—**Data Terminal Equipment**
CCITT RECMN T.431-89. Document Transfer and Manipulation (DTAM)—Services and Protocols—Introduction and General Principles—Terminal Equipment and Protocols for Telematic Services (Study Group VIII) 15 pp. 15 pp.

—**Protocols**
CCITT RECMN T.431-89. Document Transfer and Manipulation (DTAM)—Services and Protocols—Introduction and General Principles—Terminal Equipment and Protocols for Telematic Services (Study Group VIII) 15 pp. 15 pp.
CCITT RECMN T.432-89. Document Transfer and Manipulation (DTAM)—Services and Protocols—Service Definition—Terminal Equipment and Protocols for Telematic Services (Study Group VIII) 28 pp. 28 pp.
CCITT RECMN T.433-89. Document Transfer and Manipulation (DTAM)—Services and Protocols—Protocol Specification—Terminal Equipment and Protocols for Telematic Services (Study Group VIII) 87 pp. 87 pp.

—**Protocols—Operational Concept**
CCITT RECMN T.441-89. Document Transfer and Manipulation (DTAM)—Operational Structure—Terminal Equipment and Protocols for Telematic Services (Study Group VIII) 4 pp. 4 pp.

—**Protocols—Videotex Communications**
CCITT RECMN T.523-89. Communication Application Profile DM-1 for Videotex Interworking—Terminal Equipment and Protocols for Telematic Services (Study Group VIII) 19 pp. 19 pp.

Document Transfer and Manipulation *(Cont.)*

—**Teleinformatic Services**
CCITT RECMN T.431-89. Document Transfer and Manipulation (DTAM)—Services and Protocols—Introduction and General Principles—Terminal Equipment and Protocols for Telematic Services (Study Group VIII) 15 pp. 15 pp.
CCITT RECMN T.432-89. Document Transfer and Manipulation (DTAM)—Services and Protocols—Service Definition—Terminal Equipment and Protocols for Telematic Services (Study Group VIII) 28 pp. 28 pp.
CCITT RECMN T.433-89. Document Transfer and Manipulation (DTAM)—Services and Protocols—Protocol Specification—Terminal Equipment and Protocols for Telematic Services (Study Group VIII) 87 pp. 87 pp.

—**Teleinformatic Services—Operational Concept**
CCITT RECMN T.441-89. Document Transfer and Manipulation (DTAM)—Operational Structure—Terminal Equipment and Protocols for Telematic Services (Study Group VIII) 4 pp. 4 pp.

Documentation
See Also: Books; Document Acquisition; Document Storage and Retrieval Systems; Information Systems; Subject Indexing; Technical Writing
BSI BS 5848-80. 1980 Numbering of Divisions and Subdivisions in Written Documents (Point-Numbering). 4 pp.
ISO 2145-78. Documentation—Numbering of Divisions and Subdivisions in Written Documents Second Edition. 4 pp.

—**Agricultural Spraying Equipment**
BSI BS 6356: Part 7-93. 1993 Spraying Equipment for Crop Protection Part 7: Guide for Typical Data Sheet Layout (E). 20 pp.
ISO 10627 Pt 1-92. Agricultural Sprayers—Data Sheet—Part 1: Typical Layout First Edition. 18 pp.

—**Aircraft**
CAA LEAFLET 11-6 12.90. Use of 'B' Conditions Overseas. 4 pp.

—**Aircraft Equipment**
CAA LEAFLET 11-1 07.90. Materials, Parts or Appliances Approved to USA Technical Standards Orders (TSO's) for Part 21 Sub-Part 0. 3 pp.

—**Appliances—Aircraft**
CAA LEAFLET 11-1 07.90. Materials, Parts or Appliances Approved to USA Technical Standards Orders (TSO's) for Part 21 Sub-Part 0. 3 pp.

—**Bureaufax Communications—Tables (Data)**
CCITT RECMN F.170-89. Operational Provisions for the International Public Facsimile Service Between Public Bureaux (Bureaufax)—Telematic, Data Transmission and Teleconference Services—Operations and Quality of Service—(Study Group I) 8 pp. 8 pp.

—**CAD—Information Security**
ISO 11442 Pt 1-93. Technical Product Documentation—Handling of Computer-Based Technical Information—Part 1: Security Requirements First Edition. 7 pp.

—**CCIR**
CCIR RESOLUTION 24-7-90. Organization of CCIR Work—Volume XIV—Administrative Texts of the CCIR. 9 pp.

—**Coding Sheets—Defense Contracts**
MOD UK DEFCON 96-91. Coding Sheet for Procurement Documentation 2/91. 8 pp.
MOD UK DEFCON 96-92. Coding Sheet for Procurement Documentation 9/92. 8 pp.
MOD UK DEFCON 117-75. Conditions Relating to the Supply of Documentation for NATO Codification Purposes 1/75. 1 p.
MOD UK DEFCON 117A-75. Supply of Documentation for NATO Codification Purposes (Purchases Abroad) 3/75. 1 p.

—**Computer Systems**
JIS X 0126-87. Guidelines for the Documentation of Computer-Based Application Systems. 27 pp.

—**Construction**
BSI BS 4940-73. 1973 Recommendations for the Presentation of Technical Information About Products and Services in the Construction Industry. 29 pp.
SAA AS 1388-74. Recommendations for Trade and Technical Literature for the Building Industry. 10 pp.

INDUSTRY STANDARDS

INTERNATIONAL AND NON-U.S. NATIONAL STANDARDS
SUBJECT INDEX

Documentation

Documentation *(Cont.)*

—**Construction** *(Cont.)*
SNZ BS 4940-73. Recommendations for the Presentation of Technical Information About Products and Services in the Construction Industry. 32 pp.

—**Cooperation—CCIR—IEC**
CCIR RESOLUTION 23-3-90. Collaboration with the International Electrotechnical Commission on Graphical Symbols and Documentation Used in Telecommunications—Volume XIII—Vocabulary and Related Subjects. 1 p.

—**Defense Contracts**
MOD UK DEFCON 21-58. Drawings, Specifications and Manufacturing Data 1/58. 1 p.
MOD UK DEFCON 63-88. Supply of Technical Data 11/88. 1 p.

—**Destination Indicators—Telegraph Repeaters**
CCITT RECMN F.96-89. List of Destination Indicators—Telegraph and Mobile Services Operations and Quality of Service (Study Group I) 2 pp. 2 pp.

—**Document Storage and Retrieval Systems**
ISO 11442 Pt 4-93. Technical Product Documentation—Handling of Computer-Based Technical Information—Part 4: Document Management and Retrieval Systems First Edition. 8 pp.

—**Flight Operations**
CAA Chapter P2-8. Handling Controlability and Manoeuverability (Provisional Airworthiness Requirements for Civil Powered Lift Aircraft). 34 pp.

—**Gentex Networks—Routing—ITU**
CCITT RECMN F.20-89. International Gentex Service—Telegraph and Mobile Services Operations and Quality of Service (Study Group I) 5 pp. 5 pp.
CCITT RECMN F.93-89. Routing Table for Offices Connected to the Gentex Service—Telegraph and Mobile Services Operations and Quality of Service (Study Group I) 1 pp. 1 p.

—**Glossaries**
BSI BS 5408-76. 1976 Amd 1 Glossary of Documentation Terms. 86 pp.
BSI BS 6100: SUBSEC 1.5.6-88. 1988 Glossary of Building and Civil Engineering Terms Part 1: General and Miscellaneous Section 1.5: Operations: Associated Plant and Equipment Subsection 1.5.6: Documentation Excluding Drawings. 13 pp.
BSI BS 6100: SUBSEC 1.5.6-01. 1988 Amd 1 Glossary of Building and Civil Engineering Terms Part 1: General Section 1.5: Operations; Associated Plant and Equipment Subsection 1.5.6: Documentation Excluding Drawings (AMD 7244) August 15, 1992. 14 pp.
ISO 5127 Pt 1-83. Documentation and Information—Vocabulary—Part 1: Basic Concepts First Edition. 25 pp.
ISO 5127 Pt 2-83. Documentation and Information—Vocabulary—Part 2: Traditional Documents First Edition. 23 pp.
ISO 5127 Pt 3-88. Documentation and Information—Vocabulary—Part 3: Iconic Documents First Edition. 22 pp.
ISO 5127 Sec 3A-81. Information and Documentation—Vocabulary—Section 3A): Acquisition, Identification, and Analysis of Documents and Data First Edition. 30 pp.
ISO 5127 Pt 6-83. Documentation and Information—Vocabulary—Part 6: Documentary Languages First Edition. 30 pp.
ISO 5127 Pt 11-87. Documentation and Information—Vocabulary—Part 11: Audio-Visual Documents Second Edition. 15 pp.
JIS X 0701-89. Documentation and Information-Vocabulary—Basic Concepts.
JIS X 0702-89. Documentation and Information-Vocabulary—Traditional Documents.

—**Languages**
JIS X 0706-89. Documentation and Information-Vocabulary—Documentary Languages.

—**Laser Equipment**
ISO 11252-93. Lasers and Laser-Related Equipment—Laser Device —Minimum Requirements for Documentation First Edition. 6 pp.

—**Maintenance—Aircraft**
CAA CAP 482 SUB-Part G 03.83. Operating Limitations and Information (Small Light Aeroplanes). 6 pp.

—**Materials—Aircraft**
CAA LEAFLET 11-1 07.90. Materials, Parts or Appliances Approved to USA Technical Standards Orders (TSO's) for Part 21 Sub-Part 0. 3 pp.

Documentation *(Cont.)*

—**Merchant Ships—Naval**
NATO STANAG 1040 ED 14 AMD 0-89. Allied Naval Control of Shipping Manual—ATP-2 Volume 1. 5 pp.
NATO STANAG 1040 ED 15 AMD 0-92. Allied Naval Control of Shipping Manual—ATP-2 Volume 1. 5 pp.
NATO STANAG 1040 ED 16 AMD 0-92. Allied Naval Control of Shipping Manual—ATP-2 Vol I. 5 pp.
NATO STANAG 1040 ED 17 AMD 0-93. Allied Naval Control of Shipping Manual—ATP-2 Vol I. 5 pp.

—**Military—Cause of Death**
NATO STANAG 2132 ED 2 AMD 4-74. Documentation Relative to Medical Evacuation Treatment and Cause of Death of Patients. 17 pp.

—**Military—Medical**
NATO STANAG 2132 ED 2 AMD 4-74. Documentation Relative to Medical Evacuation Treatment and Cause of Death of Patients. 17 pp.

—**Nonferrous Metals**
BSI BS 6938-88. 1988 Recommendations for Documenation of the Preparation of Certified Reference Materials for Metallurgical Analysis of Non-Ferrous Metals. 11 pp.

—**Packaging**
MOD UK DSTAN 81-41: Part 4-84. Packaging of Defence Materiel Part 4: Documentation Issue 2; Amendment 3. 21 pp.

—**Post Design Services—Electronic Equipment**
MOD UK DSTAN 05-125: Chapter 3-01. Drawings and Documentation Issue 2; Amendment 1. 17 pp.

—**Restricted Use**
DIN ENGL 34-77. Protection Mark for Restricting the Use of Documents (Oct). 2 pp.

—**Routing—Automatic Services (Telephone)**
CCITT RECMN E.149-89. Presentation of Routing Data—Telephone Network and ISDN—Operation, Numbering, Routing and Mobile Service (Study Group II) 4 pp. 4 pp.

—**Routing—Semiautomatic Demand Operating**
CCITT RECMN E.149-89. Presentation of Routing Data—Telephone Network and ISDN—Operation, Numbering, Routing and Mobile Service (Study Group II) 4 pp. 4 pp.

—**Routing—Tables (Data)**
CCITT RECMN E.149-89. Presentation of Routing Data—Telephone Network and ISDN—Operation, Numbering, Routing and Mobile Service (Study Group II) 4 pp. 4 pp.

—**Shipping—Naval**
NATO STANAG 1040 ED 14 AMD 0-89. Allied Naval Control of Shipping Manual—ATP-2 Volume 1. 5 pp.
NATO STANAG 1040 ED 15 AMD 0-92. Allied Naval Control of Shipping Manual—ATP-2 Volume 1. 5 pp.
NATO STANAG 1040 ED 16 AMD 0-92. Allied Naval Control of Shipping Manual—ATP-2 Vol I. 5 pp.
NATO STANAG 1040 ED 17 AMD 0-93. Allied Naval Control of Shipping Manual—ATP-2 Vol I. 5 pp.

—**Software**
BSI BS 7137-89. 1989 User Documentation and Cover Information for Consumer Software Packages. 11 pp.
BSI BS 7649-93. 1993 Design and Preparation of Documentation for Users of Application Software (S). 108 pp.
BSI BS ISO/IEC TR 9294-90. 1990 Information Technology—Guidelines for the Management of Software Documentation. 15 pp.
IEC TR9294-90. Information Technology—Guidelines for the Management of Software Documentation First Edition. 12 pp.
ISO 6592-85. Information Processing—Guidelines for the Documentation of Computer-Based Application Systems First Edition. 20 pp.
ISO 9127-88. Information Processing Systems—User Documentation and Cover Information for Consumer Software Packages First Edition. 11 pp.
ISO TR9294-90. Information Technology—Guidelines for the Management of Software Documentation First Edition. 12 pp.
JIS X 0151-89. Information Processing Systems—User Documentation and Cover Information for Consumer Software Packages.

Documentation *(Cont.)*

—**Software** *(Cont.)*
JTC1 6592-85. Information Processing—Guidelines for the Documentation of Computer-Based Application Systems First Edition. 20 pp.
JTC1 9127-88. Information Processing Systems—User Documentation and Cover Information for Consumer Software Packages First Edition. 11 pp.
JTC1 TR9294-90. Information Technology—Guidelines for the Management of Software Documentation First Edition. 12 pp.
OSI ISO 6592-85. Information Processing—Guidelines for the Documentation of Computer-Based Application Systems. 19 pp.
OSI ISO 9127-88. Information Processing Systems—User Documentation and Cover Information for Consumer Software Packages. 11 pp.
OSI ISO/IEC TR 9294-90. Information Technology—Guidelines for the Management of Software Documentation. 12 pp.
SAA AS 3897-91. Information Processing—Guidelines for the Management of Software Documentation (ISO/IEC TR 9294:1990) (in Professional Package 26A). 7 pp.
SAA AS 3898-91. Information Processing—User Documentation and Cover Information for Consumer Software Packages (ISO 9127:1988) (in Professional Package 26A). 7 pp.
SNZ NZS/ISO 6592-85. Information Processing—Guidelines for the Documentation of Computer-Based Application Systems. 17 pp.
SNZ NZS/ISO 9127-88. Information Processing—User Documentation and Cover Information for Consumer Software Packages. 7 pp.

—**Steels—Inspection**
ISO 10474-91. Steel and Steel Products —Inspection Documents First Edition; (Corrected and Reprinted -1992). 9 pp.

—**Technical Information—Computer Based**
ISO 11442 Pt 2-93. Technical Product Documentation—Handling of Computer-Based Technical Information—Part 2: Original Documentation First Edition. 5 pp.

—**Technical Information—Computer Based—Product Design**
ISO 11442 Pt 3-93. Technical Product Documentation—Handling of Computer-Based Technical Information—Part 3: Phases in the Product Design Process First Edition. 5 pp.

—**Telecommunication Services—Coordinated Universal Time**
CCITT RECMN B.11-89. Legal Time; Use of the Term UTC—Terms and Definitions Abbreviations on Means of Expression (Series B) General Telecommunications Statistics (Series C) 1 pp. 1 p.

—**Telecommunication Systems—Symbols**
CCIR RECMN 461-4-90. Graphical Symbols and Rules for the Preparation of Documentation in Telecommunications—Section B—Graphical Symbols. 2 pp.
CCITT RECMN B.10-89. Graphical Symbols and Rules for the Preparation of Documentation in Telecommunications—Terms and Definitions Abbreviations and Acronyms Recommendations on Means of Expression (Series B) General Telecommunications Statistics (Ser.) 2 pp. 2 pp.

—**Telephone Exchanges**
CEPT T/CS 01-10-84. Exigence De Documentation Des Systemes De Commutation Numeriques. 12 pp.
CEPT T/CS 01-10 E-84. Documentation Requirements for Digital Switching Systems. 12 pp.

—**Telex Communications—Routing—Tables (Data)**
CCITT RECMN F.95-89. Table of International Telex Relations and Traffic—Telegraph and Mobile Services Operations and Quality of Service (Study Group I) 2 pp. 2 pp.

—**Telex Communications—Telecommunication Circuits—Tables (Data)**
CCITT RECMN F.95-89. Table of International Telex Relations and Traffic—Telegraph and Mobile Services Operations and Quality of Service (Study Group I) 2 pp. 2 pp.

—**Time Signals**
CCIR Report 896-1-86. Documentation of Changes in Transmitted Time Signals—Section 7C—Systems for Dissemination and Comparison. 1 p.

Documents

Scope Note: For additional listings, use a more specific term **See Also:** Abstracts; Archives; Audiovisual Documents; Banking Documents; Bibliographies; Books; Bulletins; Buyers Guides;

INTERNATIONAL AND NON-U.S. NATIONAL STANDARDS
SUBJECT INDEX

Documents (Cont.)
See Also: (Cont.)
Canadian Trade Document Alignment System; Catalogs; Construction Contracts; Data Sheets; Defense Contracts; Directories; Document Acquisition; Document Formats; Document Storage and Retrieval Systems; Documentation; Electronic Publications; Engineering Drawings; Engineering Reports; Export Documents; Flight Information Publication; Forms (Paper); Handbooks; Import Documents; Indexes (Documentation); Libraries; Logs (Reports); Military Geographic Documentation; Newsletters; Newspapers; Passports; Periodicals; Printed Forms; Processable Mode Number One; Publications; Purchase Orders; Reproduction Ratio (Copying); Reviews; Scientific Reports; Specification Circulars; Technical Reports; Theses; Trade Documents; Trade Literature; Translations; Trig Lists; Visas

—**Abbreviations**
NATO STANAG 1059 ED 5 AMD 0-88. National Distinguishing Letters for Use by NATO Forces. 5 pp.

—**Aircraft—Maintenance**
CAA Appendix 1 03.92. Example Documents. 18 pp.

—**Captured Enemy Equipment—Management**
NATO STANAG 2084 ED 5 AMD 3-80. Handling and Reporting of Captured Enemy Equipment and Documents. 22 pp.

—**Chemical Agent Casualties**
NATO STANAG 2917 ED 1 AMD 0-85. Chemical Casualty Assessment Exercise Publication (AXP-7). 5 pp.

—**Digital Circuit Multiplication Equipment**
CCITT FASCICLE III.4-89. General Aspects of Digital Transmission Systems Terminal Equipments—Recommendations G.700-G.795. 621 pp.

—**Electrical Components**
IEC 1082 Pt 1-91. Preparation of Documents Used in Electrotechnology Part 1: General Requirements First Edition; (Supersedes 113 Pt 1 and 3). 160 pp.

—**Flight Paths**
NATO STANAG 3961 ED 1 AMD 3-85. Application Form for Overflight and Landing Clearance. 10 pp.

—**Maintenance**
BSI BS 4971: Part 1-88. 1988 Repair and Allied Processes for the Conservation of Documents Part 1: Treatment of Sheets, Membranes and Seals. 19 pp.

—**Maps—Gazetters**
NATO STANAG 2213 ED 3 AMD 5-80. Gazetteers. 12 pp.

—**Marine Transportation**
NATO STANAG 2166 ED 3 AMD 1-88. Movements and Transport Documents Used for Movements by Ship. 21 pp.

—**Medical—Nuclear, Biological and Chemical Warfare**
NATO STANAG 2873 ED 2 AMD 0-82. Concept of Operations of Medical Support in Nuclear, Biological and Chemical Environments—AMedP-7(A). 7 pp.

—**Microfilming**
BSI BS ISO 6199-91. 1991 Micrographics—Microfilming of Documents on 16 mm and 35 mm Silver-Gelatin Type Microfilm—Operating Procedures (H). 26 pp.
ISO 6199-91. Micrographics—Microfilming of Documents on 16 mm and 35 mm Silver-Gelatin Type Microfilm—Operating Procedures First Edition. 21 pp.
ISO 10196-90. Micrographics—Recommendations for the Creation of Original Documents First Edition. 10 pp.

—**Military—Abbreviations**
NATO STANAG 2066 ED 4 AMD 6-79. Layout for Military Correspondence. 9 pp.
NATO STANAG 2066 ED 5 AMD 0-90. Layout for Military Correspondence. 9 pp.

—**Military—Distribution**
NATO STANAG 1033 ED 4 AMD 1-88. Minimum Distribution List of Allied Publications for Maritime Use—AAP-7(C). 5 pp.

—**Military—Layout**
NATO STANAG 2066 ED 4 AMD 6-79. Layout for Military Correspondence. 9 pp.
NATO STANAG 2066 ED 5 AMD 0-90. Layout for Military Correspondence. 9 pp.

n-Dodecane
CNS K0007-88. High Purity Dodecane (Sep)(6222).
JIS K 0510-83. High Purity Dodecane.

tert-Dodecyl Mercaptan
MOD UK TS 10292. Tertiary, Dodecyl Mercaptan.

Dodecylbenzene
Use For: Detergent Alkylate *See Also:* Benzene; Hydrocarbons
CNS K1107-65. Dodecyl Benzene for Industrial Use (Sep)(2529)(R 1971).

—**Hydrocarbon Content**
CNS K6266-68. Method of Test for Unsulfonatable Hydrocarbons in Detergent Alkylate (Oct)(2913)(R 1973).

Dodine
See Also: Pesticides
SAA AS 1870.14D-79. Standard for Development—Pesticides for Agricultural Use—Part 14D: Dodine. 9 pp.

Dog Step Ladders
See Also: Ladders; Step Irons; Stepladders

—**Ships**
CNS F3035-80. Ships' Foot Steps (May)(5568).
JIS F 2601-83. Ships' Dog Steps.
JIS F 2601-75. Ships' Footsteps. 4 pp.

—**Steel—Ships**
BSI BS 7468-91. 1991 Rungs for Dog-Step Ladders (ISO 9519: 1990). 8 pp.
BSI BS MA 39: Part 3-73. 1973 Amd 1 Ships' Ladders Part 3: Ladders, Steel Dog-Step. 4 pp.
ISO 5487-81. Shipbuilding—Steel Dog-Step Ladders First Edition. 4 pp.

Dogs (Mammals)
See Also: Muzzles (Pet)

—**Control—Model Bylaws**
SNZ NZS 9201: Chapter 12-72. Model General Bylaws Chapter 12: The Control of Dogs. 8 pp.

—**Hair**
CNS N1074-85. Dog Hair (May)(2819).

—**Muzzles**
BSI BS 7659-93. 1993 Dog Muzzles (N). 12 pp.

Dollies
Use For: Dolly Trucks *See Also:* Ground Vehicles; Hand Trucks; Industrial Equipment; Industrial Trucks; Lift Trucks; Materials Handling Equipment; Semitrailers

—**Axles—Springs**
CNS B2606-81. Industrial Trucks up to 20 km/h; Rubberboned-Springs for Axels of Handcarts and Trailers (Sep)(7867).

—**Gas Cylinders**
BSI BS 2718-79. 1979 Gas Cylinder Trolleys. 7 pp.

—**Glossaries**
BSI BS 3810: Part 1-64. 1964 Glossary of Terms Used in Materials Handling Part 1: Terms Used in Connection with Pallets, Stillages, Hand and Powered Trucks. 37 pp.
SNZ NZS 2000: Part 1-65. Glossary of Terms Used in Materials Handling Part 1: Terms Used in Connection with Pallets, Stillages, Hand and Powered Trucks Amend: A, 1970 (Reconfirmed 1978). 36 pp.

—**Wheels**
JIS B 8922-84. Industrial Wheels; (Erratum). 18 pp.

Dolly Trucks
Use: Dollies

Dolomite
See Also: Aggregates; Limestone
CNS M1007-63. Dolomite (Mar)(2080) (R 1973). 1 p.

—**Aluminum Oxide Content**
ISO 10058-92. Magnesites and Dolomites —Chemical Analysis First Edition. 23 pp.

—**Aluminum Oxide Content—Gravimetric Analysis**
CNS M3104-82. Method for Determining Aluminum Oxide in Dolomite (Gravimetric Method) (Sep)(9422).

—**Aluminum Oxide Content—Photometry**
DIN ENGL 52241 Pt 4-86. Analysis of Raw Materials Used in Glass Production; Chemical Analysis of Dolomite Containing Not Less Than 95% of Calcium Magnesium Carbonate; Determination of Aluminium Oxide (Nov). 2 pp.

Dolomite (Cont.)

—**Aluminum Oxide Content—Volumetric Analysis**
CNS M3103-82. Method for Determining Aluminum Oxide in Dolomite (EDTA Titration Method) (Sep)(9421).

—**Calcium Oxide Content**
ISO 10058-92. Magnesites and Dolomites —Chemical Analysis First Edition. 23 pp.

—**Calcium Oxide Content—Volumetric Analysis**
CNS M3106-82. Method for Determining Calcium Oxide in Dolomite (EDTA Titration Method) (Sep)(9424).
DIN ENGL 52241 Pt 8-86. Analysis of Raw Materials Used in Glass Production; Chemical Analysis of Dolomite Containing Not Less Than 95% of Calcium Magnesium Carbonate; Determination of Calcium Oxide and Magnesium Oxide (Nov). 2 pp.

—**Chemical Analysis**
DIN ENGL 52241 Pt 1-86. Analysis of Raw Materials Used in Glass Production; Chemical Analysis of Dolomite Containing Not Less Than 95% of Calcium Magnesium Carbonate; General Information and Test Report (Nov). 2 pp.
DIN ENGL 52241 Pt 2-86. Analysis of Raw Materials Used in Glass Production; Chemical Analysis of Dolomite Containing Not Less Than 95% of Calcium Magnesium Carbonate; Digestion Method (Nov). 2 pp.
JIS M 8851-83. Methods for Chemical Analysis of Dolomite (R 1990). 32 pp.

—**Chromium Oxide Content**
ISO 10058-92. Magnesites and Dolomites —Chemical Analysis First Edition. 23 pp.

—**Ignition Loss**
CNS M3101-82. Method of Determining Ignition Loss in Dolomite (Sep)(9419).
ISO 10058-92. Magnesites and Dolomites —Chemical Analysis First Edition. 23 pp.

—**Iron Oxide Content**
ISO 10058-92. Magnesites and Dolomites —Chemical Analysis First Edition. 23 pp.

—**Iron Oxide Content—Absorptiometric Analysis**
CNS M3105-82. Method for Determining Iron Oxide in Dolomite (O-Phenanthroline Absorptiometric Method) (Sep)(9423).

—**Iron Oxide Content—Photometry**
DIN ENGL 52241 Pt 5-86. Analysis of Raw Materials Used in Glass Production; Chemical Analysis of Dolomite Containing Not Less Than 95% of Calcium Magnesium Carbonate; Determination of Total Iron Content Calculated as Iron (III) Oxide (Nov). 2 pp.

—**Lithium Oxide Content**
ISO 10058-92. Magnesites and Dolomites —Chemical Analysis First Edition. 23 pp.

—**Magnesium Oxide Content**
ISO 10058-92. Magnesites and Dolomites —Chemical Analysis First Edition. 23 pp.

—**Magnesium Oxide Content—Volumetric Analysis**
CNS M3107-82. Method for Determining Magnesium Oxide in Dolomite (EDTA Indirect Titration Method) (Sep) (9425).
CNS M3108-82. Method for Determining Magnesium Oxide in Dolomite (EDTA Direct Titration Method) (Sep) (9426).
DIN ENGL 52241 Pt 8-86. Analysis of Raw Materials Used in Glass Production; Chemical Analysis of Dolomite Containing Not Less Than 95% of Calcium Magnesium Carbonate; Determination of Calcium Oxide and Magnesium Oxide (Nov). 2 pp.

—**Manganese Oxide Content**
ISO 10058-92. Magnesites and Dolomites —Chemical Analysis First Edition. 23 pp.

—**Manganese Oxide Content—Atomic Absorption Spectrometry**
DIN ENGL 52241 Pt 7-86. Analysis of Raw Materials Used in Glass Production; Chemical Analysis of Dolomite Containing Not Less Than 95% of Calcium Magnesium Carbonate; Determination of Manganese Content Calculated as MnO (Nov). 2 pp.

INDUSTRY STANDARDS

Dolomite (Cont.)

—Manganese Oxide Content—Photometry
DIN ENGL 52241 Pt 7-86. Analysis of Raw Materials Used in Glass Production; Chemical Analysis of Dolomite Containing Not Less Than 95% of Calcium Magnesium Carbonate; Determination of Manganese Content Calculated as MnO (Nov). 2 pp.

—Phosphorus Pentoxide Content—Absorptiometric Analysis
CNS M3109-82. Method for Determining Phosphorus Pentoxide in Dolomite (MIBK Absorptiometric Method) (Sep)(9427).

—Potassium Oxide Content
ISO 10058-92. Magnesites and Dolomites —Chemical Analysis First Edition. 23 pp.

—Silica Content
ISO 10058-92. Magnesites and Dolomites —Chemical Analysis First Edition. 23 pp.

—Silica Content—Gravimetric Analysis
CNS M3102-82. Method for Determining Silicon Dioxide in Dolomite (Gravimetric Method) (Sep)(9420).

—Silicon Oxide Content—Photometry
DIN ENGL 52241 Pt 3-86. Analysis of Raw Materials Used in Glass Production; Chemical Analysis of Dolomite Containing Not Less Than 95% of Calcium Magnesium Carbonate; Determination Silicon (IV) Oxide (Nov). 2 pp.

—Sodium Oxide Content
ISO 10058-92. Magnesites and Dolomites —Chemical Analysis First Edition. 23 pp.

—Titanium Dioxide Content
ISO 10058-92. Magnesites and Dolomites —Chemical Analysis First Edition. 23 pp.

—Titanium Oxide Content—Photometry
DIN ENGL 52241 Pt 6-86. Analysis of Raw Materials Used in Glass Production; Chemical Analysis of Dolomite Containing Not Less Than 95% of Calcium Magnesium Carbonate; Determination of Titanium (IV) Oxide (Nov). 2 pp.

—Water Filters
DIN ENGL 19621-73. Dolomite Filter Material for Water Treatment; Technical Conditions of Delivery (Oct). 5 pp.

—Water Treatment
CEN PREN 1017-93. Half-Burnt Dolomite Used for Treatment of Water Intended for Human Consumption. 12 pp.

Dolomite Plaster
See Also: Plaster
JIS A 6903-76. Dolomite Plaster.

Dolomite Refractories
See Also: Refractory Materials

—Chemical Analysis
BSI BS 1902: Sec 2.3-70. 1970 Amd 1 Methods for Testing Refractory Materials Part 2: Chemical Analysis (Wet Methods) Section 2.3: Chemical Analysis of Magnesites and Dolomites. 26 pp.

—Classification
CNS R1014-76. Standard Classification of Refractory Granular Dolomite (Jun)(3970).

—Ignition Loss
ISO 10058-92. Magnesites and Dolomites —Chemical Analysis First Edition. 23 pp.

Dolphins (Structures)
—Design
BSI BS 6349: Part 2-88. 1988 Code of Practice for Maritime Structures Part 2: Design of Quay Walls, Jetties and Dolphins. 110 pp.

Dome Nuts
Use: Cap Nuts

Domestic Digital Bus Systems
See Also: Data Buses
BSI BS 7536-91. 1991 Domestic Digital Bus (D2B) for Audio, Video and Audiovisual Systems (IEC 1030: 1991). 70 pp.
CENELEC EN 61030-93. Audio, Video and Audiovisual Systems Domestic Digital Bus (D2B) (IEC 1030:1991 + A1:1993). 5 pp.
IEC 1030-91. Audio, Video and Audiovisual Systems Domestic Digital Bus (D2B) First Edition; (Amendment 1-1993) (CENELEC EN 61030:1993). 164 pp.

Donkey Jackets
See Also: Protective Clothing
BSI BS 4171-81. 1981 Donkey Jackets. 8 pp.

Door Assemblies
Use: Doors

Door Bolts
See Also: Bolts; Buffer Bolts; Door Hardware; Latches (Fasteners)
CNS A2019-89. Sliding Bolts for Doors and Windows (Jun)(869). 2 pp.
CNS A2097-83. Pin Bolts, Coarse Type, Made of Steel (Jul)(6993). 2 pp.
CNS A2098-81. Pin Bolts: Made of Copper-Zinc Alloy (Mar)(6994). 4 pp.

—Cupboards
CNS A2099-81. Cupboard Bolts (Mar)(6995). 3 pp.

—Flush
CNS A2102-81. Flag-Form Hinges, Buffer and Flush Bolts for Steel Doors (Apr)(7185). 8 pp.
JIS A 5518-92. Fittings for Doors and Doorsets. 16 pp.

Door Chains
See Also: Chains; Door Hardware; Door Latches; Doors; Locks (Security)
CNS A2240-90. Chain Door Fasteners (Apr)(12694). 3 pp.
CNS A3314-90. Method of Test for Chain Door Fasteners (Apr)(12695). 1 p.
JIS A 5546-79. Chain Door Fasteners.

Door Closer/Holders
See Also: Door Closers; Door Hardware
CGSB CAN/CGSB-69.31-M89. Closer/Holder Release Devices (ANSI/BHMA A156.15-1981); (Amendment 1 March 1993). 20 pp.

Door Closers
See Also: Door Closer/Holders; Door Hardware; Door Openers; Doors
CGSB CAN/CGSB-69.20-M90. Door Controls (Closers) (ANSI/BHMA A156.4-1986). 39 pp.
CNS A2065-85. Door Closers (Feb)(4723). 3 pp.
CNS A3077-84. Method of Test for Checking Floor-Hinges and Door-Closers (Aug)(4725).
JIS A 1512-75. Testing Methods for Checking Floor-Hinges and Door-Closers.
JIS A 5544-75. Door Closers.

—Overhead
BSI BS 6459: Part 1-84. 1984 Door Closers Part 1: Mechanical Performance of Crank and Rack and Pinion Overhead Closers. 9 pp.

Door Frames
Use For: Door Surrounds *See Also:* Door Hardware; Doors

—Hardwoods
BSI BS 1285-80. 1980 Wood Surrounds for Steel Windows and Doors. 12 pp.

—Softwoods
BSI BS 1285-80. 1980 Wood Surrounds for Steel Windows and Doors. 12 pp.

—Wood
BSI BS 4787: Part 1-80. 1980 Amd 1 Internal and External Wood Doorsets, Door Leaves and Frames Part 1: Dimensional Requirements (AMD 4737) December 31, 1984. 11 pp.

Door Handles
Use: Doorknobs

Door Hardware
See Also: Door Bolts; Door Chains; Door Closer/Holders; Door Closers; Door Frames; Door Latches; Door Openers; Door Stops; Doorknobs; Doors
CGSB CAN/CGSB-69.32-M90. Auxiliary Hardware (ANSI/BHMA A156.16-1981) (Supersedes 69-GP-7M); (Amendment 1 May 1993). 59 pp.

Door Knobs
Use: Doorknobs

Door Latches
See Also: Door Chains; Door Hardware; Doors; Latches (Fasteners)
BSI BS 5872-80. 1980 Locks and Latches for Doors in Buildings. 22 pp.
CNS A2060-91. Locks and Latches for Doors in Buildings (Feb)(4349). 20 pp.
SNZ NZS/BS 5872-80. Specification for Locks and Latches for Doors in Buildings. 24 pp.

—Automotive
CNS D3162-83. Method of Test for Side Door Lock System of Automobiles (Dec)(10708).
JIS D 1620-88. Test Method of Side Door Latch Systems for Automobiles. 9 pp.

—Furniture
BSI BS 4951-73. 1973 Builders' Hardware: Lock and Latch Furniture (Doors). 17 pp.

Door Leaves
Use: Doors

Door Openers
Use For: Electric Door Operators *See Also:* Door Closers; Door Hardware; Doors; Gate Operators
CSA CAN/CSA-C22. 2 NO 68-92. Motor-Operated Appliances (Household and Commercial); (Gen Instr 1 Thru 2). 115 pp.
CSA 1169 Bull. Electrical Bulletin 1169 June 27, 1978 to C22.2 NO 68. 2 pp.
CSA CAN/CSA-C22. 2 NO 247-92. Operators and Systems of Doors, Gates, Draperies, and Louvres; (Gen Instr 1). 98 pp.

—Automatic—Cords (Electric)
CSA CAN/CSA-C22. 2 NO 49-92. Flexible Cords and Cables; (Gen Instr 1). 121 pp.
CSA 427 Bull. Electrical Bulletin 427 March 14, 1958 to C22.2 NO 49. 1 p.

Door Opening Devices
Use: Door Openers

Door Operators
Use: Door Openers

Door Profiles
Use: Door Frames

Door Stops
See Also: Door Hardware; Doors

—Adjusters
JIS A 5518-92. Fittings for Doors and Doorsets. 16 pp.

—Buses (Vehicles)
CNS D2092-87. Stoppers of Entrance Door for Buses (Jul)(7675).
JIS D 4707-76. Stoppers of Entrance Door for Buses (R 1984). 6 pp.

Door Surrounds
Use: Door Frames

Doorframes
Use: Door Frames

Doorknobs
Use For: Door Handles; Door Knobs
See Also: Door Hardware; Doors; Knobs
CNS A2014-89. Door Knobs (Jul)(864). 2 pp.
CNS A2017-89. Fixed Handles for Doors and Windows (with Mounted Plate) (Jul)(867). 2 pp.

Doors
Use For: Door Assemblies; Door Leaves
See Also: Access Doors; Aircraft Doors; Combination Doors; Door Chains; Door Closers; Door Frames; Door Hardware; Door Latches; Door Openers; Door Stops; Doorknobs; Emergency Exits; Fire Doors; Gates (Barriers); Hardware; Manholes; Wall Openings; Watertight Doors; Windows
BSI BS 5278-76. 1976 Doors. Measurement of Dimensons and of Defects of Squareness of Doors Leaves. 6 pp.
CEN PREN 951-1-92. Height, Width, Thickness and Squareness—Measurement Method—Part 1: Door Leaves. 6 pp.
CEN PREN 952-92. Door Leaves—General and Local Flatness—Measurement Method. 5 pp.
ISO 6443-80. Door Leaves—Measurement of Dimensions and of Defects of Squareness First Edition. 4 pp.
ISO 8248-85. Windows and Door Height Windows—Mechanical Tests First Edition. 22 pp.

—Acoustic Insulation
JIS A 1520-88. Method for Field Measurements of Sound Insulation of Windows and Doors. 15 pp.

—Air Permeability
ISO 8272-85. Doorsets—Air Permeability Test First Edition. 5 pp.

—Airtightness
CNS A3236-87. Method of Test for Air Tightness of Windows and Doors (Oct)(11527).
JIS A 1516-84. Test Method of Air Tightness for Windows and Doors.

INTERNATIONAL AND NON-U.S. NATIONAL STANDARDS
SUBJECT INDEX
Doors

Doors *(Cont.)*

—**Aluminum Alloy**
JIS A 4702-92. Door Sets. 33 pp.
SNZ NZS 3504-79. Specification for Aluminium Windows Amend: 1, 1988. 14 pp.

—**Aluminum Alloy—Doorstop Adjusters**
JIS A 5518-92. Fittings for Doors and Doorsets. 16 pp.

—**Aluminum Alloy—Flush Bolts**
JIS A 5518-92. Fittings for Doors and Doorsets. 16 pp.

—**Aluminum Alloy—Hinges**
JIS A 5518-92. Fittings for Doors and Doorsets. 16 pp.

—**Automatic**
CGSB CAN/CGSB-69.26-M90. Power-Operated Pedestrian Doors (ANSI/BHMA A156.10-1985). 45 pp.
CGSB CAN/CGSB-69.35-M89. Power Assist and Low Energy Power Operated Doors (ANSI/BHMA A156.19-1984); (Amendment 1 April 1993). 20 pp.
CSA CAN/CSA-C22. 2 NO 68-92. Motor-Operated Appliances (Household and Commercial); (Gen Instr 1 Thru 2). 115 pp.
CSA 1169 Bull. Electrical Bulletin 1169 June 27, 1978 to C22.2 NO 68. 2 pp.

—**Automatic—Safety**
BSI BS 7036-88. 1988 Code of Practice for Provision and Installation of Safety Devices for Automatic Power Operated Pedestrian Door Systems. 20 pp.

—**Bumpers—Buses (Vehicles)**
CNS D2093-87. Rubber Bumpers of Entrance Door for Buses (Jul)(7676).
JIS D 4706-88. Rubber Bumpers of Entrance Doors for Buses. 9 pp.

—**Burglar Resistant**
CNS S1204-86. Burglary Resistant Vault Doors and Modular Panels (Nov)(11770).
DIN ENGL 18103-83. Doors; Burglar Resistant Doors; Terminology, Requirements and Testing (Nov). 7 pp.

—**Closing Force**
JIS A 1519-88. Determination of Closing and Opening Forces for Windows and Doors. 9 pp.

—**Comprehensive Testing**
BSI DD 171-87. 1987 Guide to Specifying Performance Requirements for Hinged or Pivoted Doors (Including Test Methods). 26 pp.
CNS A3233-87. Method of Test for Windows and Doors—General Rule (Oct)(11524).
JIS A 1513-82. General Rule for Test Method of Windows and Doors. 7 pp.

—**Defects**
BSI BS 5277-76. 1976 Doors. Measurement of Defects of General Flatness of Door Leaves. 6 pp.
BSI BS 5278-76. 1976 Doors. Measurement of Dimensons and of Defects of Squareness of Doors Leaves. 6 pp.

—**Deformation**
CEN EN 108-80. Methods of Testing Doors: Test for Deformation of the Leaf in Its Plane. 4 pp.
CEN EN 129-84. Methods of Testing Doors: Test for Deformation in Torsion of the Door Leaves. 6 pp.
CEN EN 130-84. Methods of Testing Doors: Test for the Change in Stiffness of the Door Leaves by Repeated Torsion. 7 pp.
DIN ENGL EN 129-90. Methods of Testing Doors; Test for Deformation in Torsion of Door Leaves (Nov). 6 pp.
DIN ENGL EN 130-90. Methods of Testing Doors; Test for Change in Stiffness of the Door Leaves by Repeated Torsion (Nov). 7 pp.
JIS A 1522-89. Doors and Windows—Test Method for Mechanical Deformation of Edge Rail. 11 pp.

—**Dew**
CNS A3234-87. Method of Test for Dew Condensation of Windows and Doors (Oct)(11525).
JIS A 1514-82. Test Method of Dew Condensation for Windows and Doors (R 1987). 15 pp.

—**Environmental Testing**
CEN EN 79-77. Methods of Testing Doors: Behaviour of Door Leaves Placed Between Two Different Climates. 4 pp.
CEN PREN 1121-2-93. Behaviour Between Two Different Climates—Test Method—Part 2: Doors. 14 pp.
ISO 6445-89. Doors and Doorsets—Test of Behaviour Between Two Different Climates First Edition. 6 pp.
ISO 8273-85. Doors and Doorsets—Standard Atmospheres for Testing the Performance of Doors and Doorsets Placed Between Different Climates First Edition. 4 pp.

Doors *(Cont.)*

—**Extrusions**
CGSB 41-GP-19MA-78. Rigid Vinyl Extrusions for Windows and Doors, Standard for (R 1984). 13 pp.
CGSB 41-GP-20M-76. Extrusions, Vinyl, Nonrigid, for Windows and Doors, Standard for (R 1983). 9 pp.

—**Fire Testing**
CNS Z3007-72. Fire Tests of Door Assemblies (Oct)(3407). 6 pp.
ISO 3008-76. Fire-Resistance Tests—Door and Shutter Assemblies First Edition; (Amendment Slip-1976) (Amendment Slip-1977) (Erratum—March 1982) (Amendment 1-1984). 18 pp.

—**Folding—Closets—Mirrored**
CGSB CAN/CGSB-82.6-M86. Doors, Mirrored Glass, Sliding or Folding, Wardrobe. 11 pp.

—**Folding—Hardware**
CGSB CAN/CGSB-69.30-M90. Sliding and Folding Door Hardware (ANSI/BHMA A156.14-1985) (Supersedes 69-GP-6M). 36 pp.

—**Gas Permeability**
CEN PREN 1026-93. Windows and Doors—Air Permeability—Test Method. 8 pp.

—**Gaskets**
CNS A2154-88. Gaskets for Windows, Doors and Joints of Panel in Buildings (Dec)(10209).
CNS A2226-88. Sponge Gaskets for Windows, Doors and Joints of Panel in Buildings (Jul)(12351).
CNS A3186-88. Method of Test for Gaskets for Windows, Doors and Joints of Panel in Buildings (Dec)(10210).
CNS A3279-88. Method of Test for Sponge Gaskets for Windows, Doors and Joints of Panel in Building (Jul)(12352).
JIS A 5750-87. Sponge Gaskets for Windows, Doors and Joints of Panel in Buildings. 21 pp.
JIS A 5756-89. Gaskets for Windows, Doors and Joints of Panel in Buildings. 26 pp.

—**Glossaries**
BSI BS 6100: SUBSEC 1.3.5-88. 1988 Glossary of Building and Civil Engineering Terms Part 1: General and Miscellaneous Section 1.3: Parts of Construction Works Subsection 1.3.5: Doors, Windows and Openings. 8 pp.
BSI BS 6100: SUBSEC 1.3.5-01. 1988 Amd 1 Glossary of Building and Civil Engineering Terms Part 1: General and Miscellaneous Section 1.3: Parts of Construction Works Subsection 1.3.5: Doors, Windows and Openings (AMD 7235) August 15, 1992. 10 pp.
ISO 1804-72. Doors—Terminology First Edition. 10 pp.

—**Handles**
CNS A2017-89. Fixed Handles for Doors and Windows (with Mounted Plate) (Jul)(867). 2 pp.

—**Hardware—Glossaries**
BSI BS 3827: Part 4-67. (WITHDRAWN) 1967 Glossary of Terms Relating to Builders' Hardware Part 4: Door Drawer, Cupboard and Gate Furniture (Superseded by BS 6100: Subsection 1.3.6:1991). 24 pp.

—**Hinges**
CGSB CAN/CGSB-69.18-M90. Butts and Hinges (ANSI/BHMA A156.1-1981) (Supersedes 69-GP-1M); (Amendment 1 July 1993). 45 pp.
CGSB CAN/CGSB-69.23-M90. Template Hinge Dimensions (ANSI/BHMA A156.7-1981). 17 pp.
CGSB CAN/CGSB-69.36-M90. Strap and Tee Hinges and Hasps (ANSI/BHMA A156.20-1984); (Amendment 1 August 1993). 26 pp.
CNS A2007-89. Steel and Stainless Steel Butt Hinges (Jun)(857).
CNS A2085-86. Loose Pin Butt Hinges (Dec)(6536). 4 pp.
CNS A2087-86. Door Hinges (with Bushings or Washers) (Dec)(6538). 5 pp.
CNS A2102-81. Flag-Form Hinges, Buffer and Flush Bolts for Steel Doors (Apr)(7185). 8 pp.
CNS A2117-81. Spring Hinges for Doors, Single Action (Oct)(7937). 3 pp.
CNS A2118-81. Spring Hinges for Doors, Double Action (Oct)(7938).
CNS A3084-86. Method of Test for Steel and Stainless Steel Butt Hinges (Dec)(5084).
JIS A 5501-75. Wrought Steel and Wrought Stainless Steel Butt Hinges (R 1983). 11 pp.
JIS A 5510-71. Wrought Steel and Wrought Stainless Steel Loose Pin Butt Hinges (R 1987). 9 pp.
JIS A 5511-72. Door Hinges (with Bushing or Washers).
JIS A 5516-72. Door Hinges (with Ball Bearings).
JIS A 5518-92. Fittings for Doors and Doorsets. 16 pp.
SNZ NZS 844-51. Specification for Butt Hinges Amend: 1, 1965. 10 pp.

Doors *(Cont.)*

—**Hinges—Automotive**
CNS D2098-89. Door Hinges for Automobiles (Jul)(7681).
JIS D 1621-88. Test Method of Side Door Hinge Systems for Automobiles. 7 pp.

—**Hinges—Ball Bearings**
JIS A 5516-72. Door Hinges (with Ball Bearings).

—**Hinges—Bushings**
CNS A2087-86. Door Hinges (with Bushings or Washers) (Dec)(6538). 5 pp.
JIS A 5511-72. Door Hinges (with Bushing or Washers).

—**Hinges—Washers**
CNS A2087-86. Door Hinges (with Bushings or Washers) (Dec)(6538). 5 pp.
JIS A 5511-72. Door Hinges (with Bushing or Washers).

—**Humidity**
BSI BS 5369-87. 1987 Methods of Testing Doors; Behaviour Under Humidity Variations of Door Leaves Placed in Successive Uniform Climates. 7 pp.
CEN EN 43-85. Methods of Testing Doors: Behaviour Under Humidity Variations of Door Leaves Placed in Successive Uniform Climates. 5 pp.
DIN ENGL EN 43-90. Methods of Testing Doors; Behaviour Under Humidity Variations of Door Leaves Placed in Successive Uniform Climates (Nov). 3 pp.
ISO 6444-80. Door Leaves—Test of Behaviour Under Humidity Variations (Successive Uniform Climates) First Edition. 3 pp.

—**Impact Testing**
CEN EN 85-80. Methods of Testing Doors: Hard Body Impact Test on Door Leaves. 4 pp.
CEN EN 162-85. Methods of Testing Doors: Soft and Heavy Body Impact Test on Door Leaves. 6 pp.
CEN PREN 949-2-92. Resistance to Soft and Heavy Body Impact—Test Method—Part 2: Hinged, Pivoted or Sliding Doors. 6 pp.
CEN PREN 950-2-92. Resistance to Hard Body Impact—Test Method—Part 2: Door Leaves. 6 pp.
ISO 8270-85. Doorsets—Soft Heavy Body Impact Test First Edition. 4 pp.
ISO 8271-85. Door Leaves—Hard Body Impact Test First Edition. 4 pp.
JIS A 1518-86. Soft Heavy Body Impact Test for Doors and Doorsets. 9 pp.

—**Insulating Glass**
CNS A2217-87. Thermal Insulating Windows and Doors with Sealed Insulating Glasses (Sliding Windows and Doors) (Oct)(12115).

—**Interlock Switches—Automotive**
CNS D2055-87. Switches of Door Lamps for Automobiles (Jan)(6323).
JIS D 5813-76. Door Switches for Automobiles.

—**Lever Handle/Push Pads**
CEN PREN 1126-2-93. Building Hardware—Exit Devices—Part 2: Emergency Devices Operated by a Lever Handle or Push Pad—Specifications and Test Methods. 26 pp.

—**Loads (Forces)**
CEN PREN 947-2-92. Resistance to Vertical Load—Test Method—Part 2: Hinged or Pivoted Doors. 6 pp.
ISO 8275-85. Doorsets—Vertical Load Test First Edition. 4 pp.

—**Lock Cases**
CNS A2012-89. Lock Cases for Doors (Jul)(862). 2 pp.

—**Lock Covers**
CNS A2011-89. Coverplate for Door Locks (Jun)(861). 2 pp.
CNS A2013-89. Lock Mouth Covers for Door Locks (Jun)(863). 2 pp.

—**Locks (Security)**
BSI BS 5872-80. 1980 Locks and Latches for Doors in Buildings. 22 pp.
CNS A2052-91. Cylindrical Locks and Tubular Locks for Doors (Feb)(3928). 5 pp.
CNS A2060-91. Locks and Latches for Doors in Buildings (Feb)(4349). 20 pp.
CNS A3065-88. Method of Test for Cylindrical Locks and Tubular Locks for Doors (Apr)(3929). 2 pp.
IEC 730 Pt 2-12-93. Automatic Electrical Controls for Household and Similar Use Part 2: Particular Requirements for Electrically Operated Door Locks First Edition. 31 pp.
JIS A 5535-76. Cylindrical Locks and Tubular Locks for Doors.
JIS A 5546-79. Chain Door Fasteners.
SNZ NZS/BS 5872-80. Specification for Locks and Latches for Doors in Buildings. 24 pp.

INDUSTRY STANDARDS

INTERNATIONAL AND NON-U.S. NATIONAL STANDARDS
SUBJECT INDEX
Doors

Doors (Cont.)

—Locks (Security)—Automotive
BSI BS AU 209: Part 1-86. (WITHDRAWN) 1986 Vehicle Security Part 1: Mechanical Locking System for Passenger Cars and Car Derived Vehicles (Superseded by BS AU 209: Part 1a: 1992). 7 pp.
BSI BS AU 209: Part 1A-92. 1992 Vehicle Security Part 1a: Specification for Locking Systems for Passenger Cars and Car Derived Vehicles (E). 9 pp.

—Locks (Security)—Buses (Vehicles)
CNS D2107-87. Door Locks for Buses (Jul)(7969).

—Locks (Security)—Exit Devices
CGSB CAN/CGSB-69.19-M89. Exit Devices (ANSI/BHMA A156.3-1984). 27 pp.

—Locks (Security)—Furniture
BSI BS 4951-73. 1973 Builders' Hardware: Lock and Latch Furniture (Doors). 17 pp.

—Mechanical Properties
CEN EN 24-75. Doors: Measurement of Defects of General Flatness of Door Leaves. 4 pp.
CEN EN 25-75. Doors: Measurement of Dimensions and of Defects of Squareness of Door Leaves. 4 pp.
ISO 6442-81. Door Leaves—Measurement of Defects of General Flatness First Edition. 4 pp.
ISO 6443-80. Door Leaves—Measurement of Dimensions and of Defects of Squareness First Edition. 4 pp.

—Modular Construction
ISO 2776-74. Modular Co-Ordination—Co-Ordinating Sizes for Doorsets—External and Internal First Edition. 3 pp.

—Mounting Hardware—Construction Contracts
DIN ENGL 18357-88. Tendering and Performance Stipulations in Contracts for Construction Works (VOB); Part C: General Technical Specifications in Contracts for Construction Works (ATV); Mounting of Door and Window Hardware (Sept) (This Standard, Together with DIN 18299,. 6 pp.

—Nacelles—Aircraft Engine—Fire Extinguishers
BSI C 6-66. 1966 Aircraft Engine Nacelle Fire Extinguisher Doors. 1 p.
ISO 1021-80. Aircraft—Engine Nacelle Fire Extinguisher Apertures and Doors First Edition. 3 pp.

—Nominal Sizes
CNS A1019-78. Standard Normal Size of Opening Components for Buildings (May)(4347).
JIS A 0005-66. Standard Nominal Size of Opening Components for Buildings.

—Opening Force
JIS A 1519-88. Determination of Closing and Opening Forces for Windows and Doors. 9 pp.

—Overhead Holders
CGSB CAN/CGSB-69.24-M90. Door Controls—Overhead Holders (ANSI/BHMA A156.8-1982); (Amendment 1 March 1993). 22 pp.

—Power Drives—Safety
DIN VDE 0700 Pt 238-83. Safety of Household and Similar Electrical Appliances; Particular Requirements for Power Drive for Gates, Doors, Windows and Similar Equipment (Oct). 15 pp.

—Power Operated
CGSB CAN/CGSB-69.26-M90. Power-Operated Pedestrian Doors (ANSI/BHMA A156.10-1985). 45 pp.
CGSB CAN/CGSB-69.35-M89. Power Assist and Low Energy Power Operated Doors (ANSI/BHMA A156.19-1984); (Amendment 1 April 1993). 20 pp.
CSA CAN/CSA-C22. 2 NO 247-92. Operators and Systems of Doors, Gates, Draperies, and Louvres; (Gen Instr 1). 98 pp.

—Prehung
ISO 2776-74. Modular Co-Ordination—Co-Ordinating Sizes for Doorsets—External and Internal First Edition. 3 pp.

—Prehung—Aluminum Alloy
CNS A2227-88. Steel and Aluminium Alloy Door Sets for Entrance of Dwellings (Sep)(12410).
CNS A3294-88. Method of Test for Steel and Aluminium Alloy Door Sets for Entrance of Dwelling (Sep)(12411).

—Prehung—Aluminum Alloy—Doorstop Adjusters
JIS A 5518-92. Fittings for Doors and Doorsets. 16 pp.

Doors (Cont.)

—Prehung—Aluminum Alloy—Flush Bolts
JIS A 5518-92. Fittings for Doors and Doorsets. 16 pp.

—Prehung—Aluminum Alloy—Hinges
JIS A 5518-92. Fittings for Doors and Doorsets. 16 pp.

—Prehung—Aluminum—Buildings
JIS A 4712-86. Steel and Aluminium Door Sets for Entrance of Dwellings.

—Prehung—Burglar Resistant
ISO TR10476-90. Doorsets—Assessment of Burglar-Proofness First Edition. 7 pp.

—Prehung—Closing Force
ISO 8274-85. Doorsets—Determination of Closing Force First Edition. 4 pp.

—Prehung—Environmental Testing
ISO 8273-85. Doors and Doorsets—Standard Atmospheres for Testing the Performance of Doors and Doorsets Placed Between Different Climates First Edition. 4 pp.

—Prehung—Opening/Closing Testing
ISO 9379-89. Doorsets—Repeated Opening and Closing Test First Edition. 6 pp.

—Prehung—Residential Buildings—Installation
BSI BS 8213: Part 4-90. 1990 Windows, Doors and Rooflights Part 4: Code of Practice for the Installation of Replacement Windows and Doorsets in Dwellings. 26 pp.

—Prehung—Static Loads
ISO 8269-85. Doorsets—Static Loading Test First Edition. 8 pp.
JIS A 1521-88. Test Method of Diagonal Deformation by Static Load for Doorset. 11 pp.

—Prehung—Static Loads—Torsion
CEN PREN 948-2-92. Resistance to Static Torsion—Test Method—Part 2: Hinged or Pivoted Doors. 6 pp.
ISO 9381-89. Doorsets—Static Torsion Test First Edition. 4 pp.

—Prehung—Wood
BSI BS 4787: Part 1-80. 1980 Amd 1 Internal and External Wood Doorsets, Door Leaves and Frames Part 1: Dimensional Requirements (AMD 4737) December 31, 1984. 11 pp.

—Prehung—Wood—Hinges
JIS A 5518-92. Fittings for Doors and Doorsets. 16 pp.

—Push Bars—Alarmed
BSI BS 5725: Part 1-81. 1981 Emergency Exit Devices Part 1: Panic Bolts and Panic Latches Mechanically Operated by a Horizontal Push-Bar. 16 pp.

—Push/Touch Bars
CEN PREN 1125-1-93. Building Hardware—Exit Devices—Part 1: Panic Devices Operated by a Horizontal Bar—Specifications and Test Methods. 27 pp.

—Residential Buildings—Installation
BSI BS 8213: Part 1-91. 1991 Windows, Doors and Rooflights Part 1: Code of Practice for Safety in Use and During Cleaning of Windows and Doors (Including Guidance on Cleaning Materials and Methods). 18 pp.

—Rolling Stock
JIS E 7202-80. Shapes and Dimensions of Sash-Windows and Side Double Sliding Doors for Electric Commuter Cars.

—Rolling Stock—Engines
JIS E 6501-91. Door Engines for Railway Rolling Stock.

—Ships
BSI BS MA 6-70. 1970 Internal Doors and Frames. 6 pp.

—Ships—Cabins
JIS F 2334-87. Ships' Cabin Hollow Doors.
JIS F 2334-73. Ships' Cabin Hollow Door Units. 10 pp.

—Ships—Fiberglass Reinforced Plastic—Refrigeration Equipment
CNS F3221-83. Fiberglass Reinforced Plastic Doors of Provisions Refrigerating Chamber for Marine Use (May)(10253).

Doors (Cont.)

—Ships—Openings
ISO 3796-76. Shipbuilding—Clear Openings Through Frames for External Single-Leaf Doors First Edition. 3 pp.

—Ships—Steel
BSI BS MA 38-73. 1973 Amd 1 Weathertight and Spraytight Doors (Steel). 12 pp.
CNS F3045-80. Ships' Non-Watertight Steel Doors (Aug)(6177). 3 pp.
CNS F3213-83. Non-Watertight Steel Doors for Small Ships (Mar)(10074).
JIS F 2305-83. Ships' Non-Watertight Steel Doors.
JIS F 2305-75. Ships' Non-Watertight Steel Doors. 7 pp.
JIS F 2333-91. Non-Weathertight Steel Doors for Small Ships.
JIS F 2333-76. Small Ships' Non-Watertight Steel Doors (R 1984). 7 pp.

—Ships—Wood
BSI BS MA 37-73. 1973 External Doors (Wood) for Ships. 6 pp.

—Sliding
JIS A 4713-84. AMADO for Windows and Doors. 25 pp.

—Sliding—Glass
CGSB CAN/CGSB-82.1-M89. Sliding Doors. 50 pp.

—Sliding—Glass—Aluminum
BSI BS 5286-78. 1978 Amd 1 Aluminium Framed Sliding Glass Doors. 9 pp.

—Sliding—Glass—Residential Buildings—Energy Performance
CSA A440.2-93. Energy Performance Evaluation of Windows and Sliding Glass Doors; (Gen Instr 1). 53 pp.
CSA A440.3-93SP. User Guide to CSA Standard A440.2-93, Energy Performance Evaluation of Windows and Sliding Glass Doors; (Gen Instr 1). 33 pp.

—Sliding—Glass—Residential Buildings—Thermodynamic Properties
CSA A440.2-93. Energy Performance Evaluation of Windows and Sliding Glass Doors; (Gen Instr 1). 53 pp.
CSA A440.3-93SP. User Guide to CSA Standard A440.2-93, Energy Performance Evaluation of Windows and Sliding Glass Doors; (Gen Instr 1). 33 pp.

—Sliding—Hardware
CGSB CAN/CGSB-69.30-M90. Sliding and Folding Door Hardware (ANSI/BHMA A156.14-1985) (Supersedes 69-GP-6M). 36 pp.

—Sliding—Metal—Fireproof
JIS A 4902-83. Fireproof Metal Amado for Dwellings.

—Sliding—Metal—Waterproof
CNS A2228-88. Metal Sliding Doors and Windows of Dwellings for Weatherproof (Sep)(12412).
CNS A3295-88. Method of Test for Metal Sliding Doors and Windows of Dwellings for Weatherproof (Sep)(12413).

—Sliding—Mirrored—Closet
CGSB CAN/CGSB-82.6-M86. Doors, Mirrored Glass, Sliding or Folding, Wardrobe. 11 pp.

—Sliding—Power Assisted
CGSB CAN/CGSB-69.26-M90. Power-Operated Pedestrian Doors (ANSI/BHMA A156.10-1985). 45 pp.

—Sliding—Rails
CNS A2086-83. Sliding Door Rails (Jul)(6537). 3 pp.
JIS A 5509-71. Sliding Door Rails.

—Sliding—Sheaves
CNS A2088-86. Sheaves for Sliding Doors and Windows (Dec)(6539). 3 pp.
CNS A3114-86. Method of Test for Sheaves for Sliding Doors and Windows (Dec)(6540).
JIS A 5512-79. Sheaves for Sliding Doors and Windows.

—Sliding—Ships—Waterproof
CNS F3046-80. Watertight Sliding Doors (Aug)(6178).
JIS F 2314-90. Watertight Sliding Doors.
JIS F 2314-68. Watertight Sliding Doors (R 1971). 8 pp.

—Sliding—Ships—Waterproof—Indicating Instruments
CNS F3047-80. Indicators for Watertight Sliding Doors (Aug)(6179).
JIS F 2315-68. Ships' Watertight Sliding Door Indicators. 4 pp.

Doors (Cont.)
—Steel
BSI BS 1245-75. 1975 Metal Door Frames (Steel). 20 pp.
BSI BS 6510-84. 1984 Amd 1 Steel Windows, Sills, Window Boards and Doors. 35 pp.
CGSB CAN/CGSB-82.5-M88. Insulated Steel Doors. 19 pp.
CNS A2101-88. Steel Doors (Oct)(7184). 19 pp.
CNS A2227-88. Steel and Aluminium Alloy Door Sets for Entrance of Dwellings (Sep)(12410).
CNS A3294-88. Method of Test for Steel and Aluminium Alloy Door Sets for Entrance of Dwelling (Sep)(12411).
JIS A 4702-92. Door Sets. 33 pp.

—Steel—Buffer Bolts
CNS A2102-81. Flag-Form Hinges, Buffer and Flush Bolts for Steel Doors (Apr)(7185). 8 pp.

—Steel—Buildings
JIS A 4712-86. Steel and Aluminium Door Sets for Entrance of Dwellings.

—Steel—Flush Bolts
CNS A2102-81. Flag-Form Hinges, Buffer and Flush Bolts for Steel Doors (Apr)(7185). 8 pp.
JIS A 5518-92. Fittings for Doors and Doorsets. 16 pp.

—Steel—Hinges
CNS A2102-81. Flag-Form Hinges, Buffer and Flush Bolts for Steel Doors (Apr)(7185). 8 pp.
JIS A 5518-92. Fittings for Doors and Doorsets. 16 pp.

—Steel—Trim—Wood
BSI BS 1285-80. 1980 Wood Surrounds for Steel Windows and Doors. 12 pp.

—Swinging—Aluminum—Coatings
CNS A2105-88. Aluminum Doors (Oct)(7477). 22 pp.
CNS A3123-83. Method of Test for Transparent Synthetic Resin Coating Film of Aluminum Swinging Doors (Sep)(7478).

—Symbols
ISO R1226 Pt I-70. Symbolic Designation of Direction of Closing and Faces of Doors, Windows and Shutters Part I First Edition. 10 pp.
JIS A 0151-61. Symbols for Windows and Doors.

—Thermal Insulation—Heat Transfer Coefficient
DIN ENGL 52619 Pt 1-82. Testing of Thermal Insulation; Determination of Thermal Resistance and Overall Heat Transfer Coefficient of Windows; Measurement of the Whole Construction (Nov). 6 pp.

—Thermal Insulation—Thermal Resistance
DIN ENGL 52619 Pt 1-82. Testing of Thermal Insulation; Determination of Thermal Resistance and Overall Heat Transfer Coefficient of Windows; Measurement of the Whole Construction (Nov). 6 pp.

—Thermal Resistance
CNS A3197-87. Method of Test for Total Thermal Resistance for Windows and Doors (Oct)(10523).
JIS A 4710-89. Test Method of Thermal Resistance for Windows and Doors. 14 pp.

—Torsion
ISO 9380-90. Doorsets—Repeated Torsion Test First Edition. 6 pp.

—Tractors
EC 80/720/EEC-80. Council Directive on the Approximation of the Laws of the Member States Relating to the Operating Space, Access to the Driving Position and the Doors and Windows of Wheeled Agricultural or Forestry Tractors. 11 pp.
EC 88/414/EEC-88. Commission Directive Adapting to Technical Progress Council Directive 80/720/EEC on the Approximation of the Laws of the Member States Relating to the Operating Space, Access to the Driving Position and the Doors and Windows of Wheeled Agrl. and Forestry Tractors. 3 pp.

—Trailers
SNZ NZS 5412-82. Specification for Loading Door Aperture Dimensions of Road Vehicles Used for the Bulk Transportation of Meat. 4 pp.

—Trim
CGSB CAN/CGSB-69.22-M90. Architectural Door Trim (ANSI/BHMA A156.6-1986). 18 pp.

—Trucks
SNZ NZS 5412-82. Specification for Loading Door Aperture Dimensions of Road Vehicles Used for the Bulk Transportation of Meat. 4 pp.

Doors (Cont.)
—Wall Openings
DIN ENGL 18100-83. Doors; Wall Openings for Doors with Dimensions in Accordance with DIN 4172 (Oct). 4 pp.

—Watertightness
CEN PREN 1027-93. Windows and Doors—Watertightness—Test Method; (Will Supersede EN 86:1980). 12 pp.
CNS A3237-87. Method of Test for Water Tightness of Windows and Doors (Oct)(11528).
JIS A 1517-84. Test Method of Water Tightness for Windows and Doors. 9 pp.

—Weather Stripping
BSI BS 7386-90. 1990 Draughtstrips for the Draught Control of Existing Doors and Windows in Housing (Including Test Methods). 22 pp.

—Wind Resistance
CNS A3235-87. Method of Test for Wind Resistance of Windows and Doors (Oct)(11526).
ISO 6612-80. Windows and Door Height Windows—Wind Resistance Tests First Edition. 6 pp.
JIS A 1515-83. Test Method for Wind Resistance of Windows and Doors. 9 pp.

—Wood
BSI BS 459-88. 1988 Matchboarded Wooden Door Leaves for External Use. 10 pp.
BSI BS 1567-53. 1953 Amd 6 Wood Door Frames and Linings. 28 pp.
BSI CP 151: Part 1-57. 1957 Amd 1 Doors and Windows Including Frames and Linings Part 1: Wooden Doors. 53 pp.
CNS A2111-88. Wooden Panel Doors (Nov)(7931). 5 pp.
CNS A2112-88. Wooden Door Frames (Nov)(7932). 4 pp.
CSA CAN/CSA-O132. 2 Series-90. Wood Flush Doors; (Gen Instr 1). 68 pp.
CSA O132.4-M1980. Hinged Exterior Wood Door Frames; (Gen Instr 1). 14 pp.
CSA O132.5-M1992. Stile and Rail Wood Doors; (Gen Instr 1). 22 pp.
SNZ NZS 1158-65. Doors. 12 pp.
SNZ NZS 3619-79. Specification for Timber Windows. 15 pp.

—Wood—Engineering Drawings
CNS O1037-5-88. Technical Drawings for Woodworking—Woodworking for Windows and Doors (Dec)(11666-5).

—Wood—Hinges
JIS A 5518-92. Fittings for Doors and Doorsets. 16 pp.

—Wood—Installation
SAA AS 1909-84. Installation of Timber Doorsets. 4 pp.

—Wood—Linings
BSI BS 1567-53. 1953 Amd 6 Wood Door Frames and Linings. 28 pp.

—Wood—Metal Components
CNS A1005-89. General Rules for Metal Parts of Wooden Doors and Windows (May)(856).
CNS A2024-89. Method of Test for Metal Parts of Wooden Doors and Windows (Jun)(874). 3 pp.

Doorsets
Scope Note: See the subheading Prehung under specific types of Doors

Dopes
See Also: Coatings; Lacquers; Plasticizers; Sealants
—Thinners
CNS K2051-80. Dope Thinner (Feb)(3233). 2 pp.
MOD UK DSTAN 80-38-85. Thinners for: Paint Epoxy Two-Pack, Cellulose Nitrate Paints, Dopes and Lacquers Issue 2. 8 pp.

Doping (Fabrics)
CAA LEAFLET 2-9 07.90. Doping. 10 pp.
—Aircraft
BSI X 26-66. 1966 Doping and Finishing Schemes for Fabric Covered Aircraft. 7 pp.

Doppler Frequency
Use: Doppler Shift

Doppler Frequency Shift
Use: Doppler Shift

Doppler Radar
See Also: Doppler Shift; Radar

Doppler Radar (Cont.)
—Airborne—Velocity Transducers
NATO STANAG 4148 ED 1 AMD 2-76. Standard Method to Be Used in Specifying and Evaluating Performance Accuracies of Airborne Doppler Ground Velocity Sensors. 9 pp.

—Aircraft
CAA Part 12 CAP 208. Doppler Radar Navigation Apparatus (Civil Air Publications: Airborne Radio Apparatus). 5 pp.

—Alarm Systems
BSI BS 4737: Sec 3.4-78. 1978 Amd 2 Intruder Alarm Systems in Buildings Part 3: Components Section 3.4: Radiowave Doppler Detectors. 5 pp.

Doppler Shift
Use For: Doppler Frequency; Doppler Frequency Shift *See Also:* Doppler Radar; Frequencies

—Fixed Satellite Services
CCIR Report 214-4-86. Effects of Doppler Frequency-Shifts and Switching Discontinuities in the Fixed-Satellite Service—Section 4B2—Performance and Availability. 8 pp.

—Fixed Satellite Services—Compensation
CCIR RECMN 730-92. Compensation of the Effects of Switching Discontinuities for Voice Band Data and of Doppler Frequency-Shifts in the Fixed-Satellite Service—Section 4B2—Performance and Availability. 12 pp.

Doppler Ultrasound Systems
Use: Ultrasonic Equipment—Doppler

Dornier DO 27 Series Aircraft
See Also: Aircraft
—Foreign Airworthiness Directives
CAA. Dornier DO 27 Series. 2 pp.

Dornier DO 28 Aircraft
See Also: Aircraft
—Antenna Positions
CAA. Dornier DO 28. 1 p.
—Foreign Airworthiness Directives
CAA. Dornier DO 28 Series (Foreign Airworthiness Directives). 6 pp.

Dornier 228 Aircraft
See Also: Aircraft
—Accidents
CAA. Dornier 228 (World Airline Accident Summary). 1 p.
—Data Sheets
CAA FA34 ISSUE 1. Dornier 228. 3 pp.
—Foreign Airworthiness Directives
CAA. Dornier 228 Series (Foreign Airworthiness Directives). 3 pp.

Dornier 228-100 Aircraft
See Also: Aircraft
—Antenna Positions
CAA. Dornier DO 228-100 and-200. 1 p.
—Certification
CAA. Dornier 228-100 and 228-200. 8 pp.

Dornier 228-200 Aircraft
See Also: Aircraft
—Antenna Positions
CAA. Dornier DO 228-100 and-200. 1 p.
—Certification
CAA. Dornier 228-100 and 228-200. 8 pp.

Dosemeters
Use: Dosimeters

Dosimeters
Use For: Dosemeters *See Also:* Dosimetry; Ionization Chambers; Photographic Dosimeters; Radiation Measuring Instruments; Radiation Meters; Thermoluminescent Dosimeters
ISO 4071-78. Exposure Meters and Dosimeters—General Methods for Testing First Edition. 32 pp.

—Beta Radiation
IEC 846-89. Beta, X and Gamma Radiation Dose Equivalent and Dose Equivalent Rate Meters for Use in Radiation Protection First Edition. 81 pp.
IEC 1018-91. High Range Beta and Photon Dose and Dose Rate Portable Instruments for Emergency Radiation Protection Purposes First Edition. 45 pp.

Dosimeters (Cont.)

—Beta Radiation (Cont.)
JIS Z 4333-90. Portable Photon Ambient Dose Equivalent Ratemeters for Radiation Protection. 17 pp.

—Beta Radiation—Calibration
BSI BS 6689-86. 1986 Reference Beta Radiations for Calibrating Dosemeters and Doseratemeters and for Determining Their Response as a Function of Beta Radiation Energy. 12 pp.

ISO 6980-84. Reference Beta Radiations for Calibrating Dosemeters and Doseratemeters and for Determining Their Response as a Function of Beta Radiation Energy First Edition. 11 pp.

—Beta Radiation—Ceric Sulfate
CNS J2025-81. Method of Test for Absorbed Gamma and Electron Radiation Dose with the Ceric Sulfate Dosimeter (Jul)(7698).

—Beta Radiation—Cupric Sulfate
CNS J2024-81. Method of Test for Absorbed Gamma and Electron Radiation Dose with the Ferrous Sulfate-Cupric Sulfate Dosimeter (Jul)(7697).

—Beta Radiation—Ferrous Sulfate
CNS J2024-81. Method of Test for Absorbed Gamma and Electron Radiation Dose with the Ferrous Sulfate-Cupric Sulfate Dosimeter (Jul)(7697).

—Calibration
JIS Z 4511-91. Methods of Calibration for Exposure Meters and Dose-Equivalent Meters. 27 pp.

—Fricke—Gamma Radiation Dose Absorption
CNS J2026-81. Method of Test for Absorbed Gamma Radiation Dose in the Fricke Dosimeter (Jul)(7699).

—Gamma Radiation
IEC 846-89. Beta, X and Gamma Radiation Dose Equivalent and Dose Equivalent Rate Meters for Use in Radiation Protection First Edition. 81 pp.

IEC 951 Pt 3-89. Radiation Monitoring Equipment for Accident and Post-Accident Conditions in Nuclear Power Plants Part 3: High Range Area Gamma Radiation Dose Rate Monitoring Equipment First Edition. 29 pp.

IEC 1018-91. High Range Beta and Photon Dose and Dose Rate Portable Instruments for Emergency Radiation Protection Purposes First Edition. 45 pp.

IEC 1031-90. Design, Location and Application Criteria for Installed Area Gamma Radiation Dose Rate Monitoring Equipment for Use in Nuclear Power Plants During Normal Operation and Anticipated Operational Occurrences First Edition. 26 pp.

JIS Z 4332-92. General Requirements for Personal X-ray and Gamma-Ray Dosemeters. 12 pp.

—Gamma Radiation—Calibration
BSI BS 5869-80. 1980 Specification for X and Y Reference Radiations for Calibrating Dosemeters and Dose Ratemeters and for Determining Their Response as a Function of Photon Energy. 36 pp.

ISO 4037-79. X and Gamma Reference Radiations for Calibrating Dosemeters and Dose Ratemeters and for Determining Their Response as a Function of Photon Energy First Edition; (Addendum 1-1983) (Amendment 1-1983) (Addendum 2-1989). 61 pp.

—Gamma Radiation—Ceric Sulfate
CNS J2025-81. Method of Test for Absorbed Gamma and Electron Radiation Dose with the Ceric Sulfate Dosimeter (Jul)(7698).

—Gamma Radiation—Cupric Sulfate
CNS J2024-81. Method of Test for Absorbed Gamma and Electron Radiation Dose with the Ferrous Sulfate-Cupric Sulfate Dosimeter (Jul)(7697).

—Gamma Radiation—Ferrous Sulfate
CNS J2024-81. Method of Test for Absorbed Gamma and Electron Radiation Dose with the Ferrous Sulfate-Cupric Sulfate Dosimeter (Jul)(7697).

—Gamma Radiation—Fluoro Glass
JIS Z 4314-91. Radiophotoluminescent Glass Dosemeter Systems for X and Gamma Radiation Personal Monitoring.

—Gamma Radiation—Pocket
JIS Z 4308-91. Direct Reading Personal Dosemeters for X and Gamma Radiation. 15 pp.

—Ionization Chambers
CENELEC HD 534 S1-89. Medical Electrical Equipment Dosimeters with Ionization Chambers as Used in Radiotherapy. 4 pp.

IEC 731-82. Medical Electrical Equipment Dosimeters with Ionization Chambers as Used in Radiotherapy First Edition; (Amendment 1-1987). 157 pp.

—Ionization Chambers—Safety
CSA CAN/CSA-C22. 2N601.2.9-92. Medical Electrical Equipment Part 2: Particular Requirements for the Safety of Dosimeters Used in Radiotherapy with Electrically-Connected Radiation Detectors (IEC 601-2-9:1987); (Gen Instr 1). 29 pp.

IEC 601 Pt 2-9-87. Medical Electrical Equipment Part 2: Particular Requirements for the Safety of Dosimeters Used in Radiotherapy with Electrically-Connected Radiation Detectors First Edition; (CAN/CSA-C22.2 No. 601.2.9-92). 27 pp.

—Phantoms (Radiation)—Calibration
JIS Z 4331-89. Calibration Phantom for X-and Gamma-Ray Personal Dosemeters. 5 pp.

—Safety
BSI BS 5724: Sec 2.9-88. 1988 Medical Electrical Equipment Part 2: Particular Requirements for Safety Section 2.9: Dosimeters Used in Radiotherapy with Electrically-Connected Radiation Detectors. 13 pp.

BSI BS 5724: Sec 2.9-01. 1988 Amd 1 Medical Electrical Equipment Part 2: Particular Requirements for Safety Section 2.9: Specification for Dosimeters Used in Radiotherapy with Electrically-Connected Radiation Detectors (IEC 601-2-9: 1987). 14 pp.

CENELEC HD 395.2.9 S1-89. Medical Electrical Equipment-Part 2: Particular Requirements for the Safety of Dosimeters Used in Radiotherapy with Electrically-Connected Radiation Detectors. 3 pp.

CSA CAN/CSA-C22. 2N601.2.9-92. Medical Electrical Equipment Part 2: Particular Requirements for the Safety of Dosimeters Used in Radiotherapy with Electrically-Connected Radiation Detectors (IEC 601-2-9:1987); (Gen Instr 1). 29 pp.

IEC 601 Pt 2-9-87. Medical Electrical Equipment Part 2: Particular Requirements for the Safety of Dosimeters Used in Radiotherapy with Electrically-Connected Radiation Detectors First Edition; (CAN/CSA-C22.2 No. 601.2.9-92). 27 pp.

—X-Ray
IEC 846-89. Beta, X and Gamma Radiation Dose Equivalent and Dose Equivalent Rate Meters for Use in Radiation Protection First Edition. 81 pp.

JIS Z 4332-92. General Requirements for Personal X-ray and Gamma-Ray Dosemeters. 12 pp.

—X-Ray—Calibration
ISO 4037-79. X and Gamma Reference Radiations for Calibrating Dosemeters and Dose Ratemeters and for Determining Their Response as a Function of Photon Energy First Edition; (Addendum 1-1983) (Amendment 1-1983) (Addendum 2-1989). 61 pp.

—X-Ray—Fluoro Glass
JIS Z 4314-91. Radiophotoluminescent Glass Dosemeter Systems for X and Gamma Radiation Personal Monitoring.

—X-Ray—Pocket
JIS Z 4308-91. Direct Reading Personal Dosemeters for X and Gamma Radiation. 15 pp.

Dosimetry
Use For: Radiation Dosimetry *See Also:* Dosimeters

—International System of Units—Military Operations
NATO STANAG 2957 ED 2 AMD 0-88. International System (SI) Units Used by the Armed Forces in the Nuclear Field. 7 pp.

NATO STANAG 2957 ED 2 AMD 1-88. International System (SI) Units Used by the Armed Forces in the Nuclear Field. 8 pp.

—Radiation Protection and Safety
ISO 8963-88. Dosimetry of X and Gamma Reference Radiations for Radiation Protection over the Energy Range from 8 keV to 1,3 MeV First Edition. 11 pp.

Dosing Apparatus
Use: Metering Systems

Dot Matrix Characters
See Also: Symbols

—Display Devices
JIS X 9051-84. 16-Dots Matrix Character Patterns for Display Devices.

JIS X 9052-83. 24-Dots Matrix Character Patterns for Display Devices.

Dot Matrix Printers
Use For: Matrix Printers *See Also:* Computers; Information Interchange; Printers; Printing Presses

CNS C6290-87. Method of Test for Serical Impact Dot Matrix Printer (Jul)(12010).

Dot Matrix Printers (Cont.)

—Alphanumeric Character Sets
ECMA ECMA 42-73. Alphanumeric Character Set for 7 X 9 Matrix Printers. 14 pp.

—Alphanumeric Character Sets—OCR-A
ECMA ECMA 51-77. Implementation of the Numeric OCR-A Font with 9 X 9 Matrix Printers. 7 pp.

—Serial
CNS C6290-87. Method of Test for Serical Impact Dot Matrix Printer (Jul)(12010).

DOTG
Use: Di-ortho-toylguanidine

Double Hexagonal Bolts
Use For: Bihexagonal Head Bolts; Double Hexagonal Head Bolts; Twelve Point Head Bolts
See Also: Bolts; Hexagonal Head Bolts

—Aerospace
AECMA PREN3063-87. Bolts, Double Hexagon Head, Close Tolerance, Medium Thread Length in Heat Resisting Nickel Base Alloy NI-P101HT (Waspaloy) Classification 1210 MPa/735 Degrees Celsius, Uncoated. 5 pp.

AECMA PREN3327-89. Bolts, Double Hexagon Head, Close Tolerance Medium Thread Length in Heat Resisting Nickel Base Alloy NI-P100HT (Inconel 718), Uncoated Classification: 1275 MPa/650 Degrees Celsius. 6 pp.

AECMA PREN3328-89. Bolts, Double Hexagon Head Close Tolerance Medium Thread Length in Heat Resisting Steel FE-P1738 (FV535), Uncoated Classification: 1000 MPa/550 Degrees Celsius. 6 pp.

AECMA PREN3379-89. Bolts, Double Hexagon Head, Close Tolerance in Heat Resisting Nickel Base Alloy NI-P101HT (Waspaloy), Uncoated for Increased Height Nuts Classification: 1210 MPa/730 Degrees Celsius. 6 pp.

BSI A 241-73. 1973 Amd 3 General Requirements for Steel Protruding-Head Bolts of Tensile Strength 1250 MPa (180 000 Lbf/Square Inches) or Greater. 44 pp.

CEN PREN 2874-92. Bolts, Large Bihexagonal Head, Close Tolerance Normal Shank, Medium Length Thread, in Nickel Alloy, Passivated Classification: 1 550 MPa (at Ambient Temperature) /315 Degrees C. 7 pp.

CEN PREN 3907-92. Bolts, Double Hexagon Head, Normal Shank, Long Thread, in Titanium Alloy Tl-P63, MoS2 Coated Classification: 1 100 MPa (at Ambient Temperature)/350 Degrees C. 7 pp.

DIN ENGL LN 29551-81. Aerospace; Bolts, Bi-Hexagon for Temperatures up to 650 Degrees C (Dec). 4 pp.

DIN ENGL LN 29769-80. Aerospace; Bolts Bi-Hexagon; Nominal Tensile Strength of 1250 N/mm2; Medium Thread Length (Mar). 5 pp.

DIN ENGL LN 29934-83. Aerospace; Bolts Bi-Hexagon Close Tolerance; Nominal Tensile Strength of 1550 N/mm2; Medium Thread (Oct). 5 pp.

—Aerospace—Relieved Shank
AECMA PREN2925-86. Bolts with Double Hexagon Head, Relieved Shank, Long Thread, in Heat Resisting Steel FE-PA92HT (A286) Classification 900 MPa/650 Degrees Celsius. 6 pp.

AECMA PREN2926-86. Bolts with Double Hexagon Head, Relieved Shank, Long Thread, in Heat Resisting Steel FE-PA92HT (A286) Silver Plated Classification 900 MPa/650 Degrees Celsius. 6 pp.

AECMA PREN2927-86. Bolts with Double Hexagon Head, Relieved Shank, Long Thread, in Heat Resisting Nickel Base Alloy NI-P100HT (Inconel 718) Classification 1275 MPa/650 Degrees Celsius. 6 pp.

AECMA PREN2928-86. Bolts with Double Hexagon Head, Relieved Shank, Long Thread, in Heat Resisting Nickel Base Alloy NI-P100HT (Inconel 718) Silver Plated Classification 1275 MPa/650 Degrees Celsius. 6 pp.

AECMA PREN2929-86. Bolts with Double Hexagon Head, Relieved Shank, Long Thread, in Heat Resisting Nickel Base Alloy NI-P101HT (Waspaloy) Classification 1210 MPa/735 Degrees Celsius. 6 pp.

AECMA PREN2930-86. Bolts with Double Hexagon Head, Relieved Shank, Long Thread, in Heat Resisting Nickel Base Alloy NI-P101HT (Waspaloy) Silver Plated Classification 1210 MPa/735 Degrees Celsius. 6 pp.

AECMA PREN3323-90. Bolts with Double Hexagon Head, Relieved Shank, Long Thread, in Heat Resisting Steel FE-PM38 (FV535) Classification: 100 MPa/550 Degrees. 7 pp.

CEN PREN 2925-92. Bolts with Double Hexagon Head, Relieved Shank, Long Thread, in Heat Resisting Steel FE-PA92HT (A286) Classification: 900 MPa /650 Degrees C. 6 pp.

CEN PREN 2926-92. Bolts with Double Hexagon Head, Relieved Shank, Long Thread, in Heat Resisting Steel FE-PA92HT (A286), Silver Plated Classification: 900 MPa /650 Degrees C. 6 pp.

INTERNATIONAL AND NON-U.S. NATIONAL STANDARDS
SUBJECT INDEX
Double

Double Hexagonal Bolts *(Cont.)*

—Aerospace—Relieved Shank *(Cont.)*

CEN PREN 2927-92. Bolts with Double Hexagon Head, Relieved Shank, Long Thread, in Heat Resisting Nickel Base Alloy NI-P100HT (Inconel 718) Classification: 1 275 MPa /650 Degrees C. 6 pp.
CEN PREN 2928-92. Bolts with Double Hexagon Head, Relieved Shank, Long Thread, in Heat Resisting Nickel Base Alloy NI-P100HT (Inconel 718), Silver Plated Classification: 1 275 MPa /650 Degrees C. 6 pp.
CEN PREN 2929-92. Bolts with Double Hexagon Head, Relieved Shank, Long Thread, in Heat Resisting Nickel Base Alloy NI-P101HT (Waspaloy) Classification: 1 210 MPa /735 Degrees C. 6 pp.
CEN PREN 2930-92. Bolts with Double Hexagon Head, Relieved Shank, Long Thread, in Heat Resisting Nickel Base Alloy NI-P101HT (Waspaloy), Silver Plated Classification: 1 210 MPa /735 Degrees C. 6 pp.
CEN PREN 3686-91. Bolts with Double Hexagon Head, Relieved Shank, Long Thread, in Heat Resisting Steel, FE-PA92HT (A286), Silver Plated Classification: 1100 MPa/650 Degrees C. 7 pp.
CEN PREN 3724-92. Bolts, Double Hexagon Head, Relieved Shank, Long Thread, in Titanium Alloy TI-P63, MoS2 Coated Classification: 1 100 MPa (at Ambient Temperature). 6 pp.
CEN PREN 3832-91. Bolts, Double Hexagon Head, Relieved Shank, Long Thread, in Heat Resisting Nickel Base Alloy NI-P100HT (INCO 718), Uncoated Classification: 1550 MPa/650 Degrees C. 8 pp.
CEN PREN 3832-92. Aerospace Series Bolts, Double Hexagon Head, Relieved Shank, Long Thread, in Heat Resisting Nickel Base Alloy NI-P100HT (Inconel 718) Classification: 1 550 MPa (at Ambient Temperature)/650 Degrees C. 6 pp.
CENELEC PREN 3686-89. Bolts with Double Hexagon Lead, Relieved Shank, Long Thread, in Heat Resisting Steel FE-PA92HT (A286) Silver Plated Classification 1100 MPa/650 Degrees C. 6 pp.

—Aircraft

BSI 3A 228-79. 1979 Steel Double Hexagon Bolts 1250 MPa (180 000 Lbf/Square Inches) for Aircraft. 9 pp.
MOD UK DTD-5182-65. Bolt, Double Hexagon, External Wrenching—180,000 lbf/in2. 2 pp.
SBAC AS 54500-599 ISSUE 2. Bolt, Close Tolerance—Double Hexagon Extended Washer Head, INCO718, 0.1900-32 UNJF-3A MOD BS4084.
SBAC AS 54600-699 ISSUE 2. Bolt, Close Tolerance—Double Hexagon Extended Washer Head, INCO 718, 0.2500—28 UNJF—3A MOD BS4084.
SBAC AS 54800-899 ISSUE 2. Bolt, Close Tolerance—Double Hexagon Extended Washer Head, INCO 718, 03750—24 UNJF—3A MOD BS4084.
SBAC AS 54900-999 ISSUE 2. Bolt, Close Tolerance—Double Hexagon Extended Washer Head, INCO 718, 04375—20 UNJF—3A MOD BS4084.
SBAC AS 55000-099 ISSUE 2. Bolt, Close Tolerance—Doble Hexagon Extended Washer Head, INCO 718, 0.5000—20 UNJF—3A MOD BS4084.
SBAC AS 57100-199 ISSUE 1. Bolt, Close Tolerance—Double Hexagon Extended Washer head, Waspaloy 0.1900-32 UNJF—3A MOD BS 4084.
SBAC AS 57200-299 ISSUE 1. Bolt, Close Tolerance—Double Hexagon Extended Washer Head, Waspaloy, 0.2500—28 UNJF—3A MOD BS4084.
SBAC AS 57300-399 ISSUE 1. Bolt, Close Tolerance—Double Hexagon Extended Washer Head, Waspaloy, 0.3750—24 UNJF—3A MOD BS4084.
SBAC AS 57400-499 ISSUE 1. Bolt, Close Tolerance—Double Hexagon Extended Washer Head, Waspaloy, 0.4375—20 UNJF—3A MOD BS4084.
SBAC AS 57500-599 ISSUE 1. Bolt, Close Tolerance—Double Hexagon Extended Washer Head, Waspaloy, 0.4375-20 UNJF-3A MOD BS4084.
SBAC AS 57600-699 ISSUE 1. Bolt, Close Tolerance—Double Hexagon Extended Washer Head, Waspaloy, 0.5000-20 UNJF-3A MOD BS4084.
SBAC AS 61300-399 ISSUE 1. Bolt, Close Tolerance—Double Hexagon Extended Washer Head, FV535, 0.1900-32 UNJF-3A MOD BS4084.
SBAC AS 61400-499 ISSUE 1. Bolt, Close Tolerance—Double Hexagon Extended Washer Head, FV535, 0.2500-28 UNJF-3A MOD BS4084.
SBAC AS 61500-599 ISSUE 1. Bolt, Close Tolerance—Double Hexagon Extended Washer Head, FV535, 0.3125-24 UNJF-3A MOD BS4084.
SBAC AS 61600-699 ISSUE 1. Bolt, Close Tolerance—Double Hexagon Extended Washer Head, FV535, 0.3750-24 UNJF-3A MOD BS4084.
SBAC AS 61700-799 ISSUE 1. Bolt, Close Tolerance—Double Hexagon Extended Washer Head, FV535, 0.4375-20 UNJF-3A MOD BS4084.
SBAC AS 61800-899 ISSUE 1. Bolt, Close Tolerance—Double Hexagon Extended Washer Head, FV535, 0.5000-20 UNJF-3A MOD BS4084.

Double Hexagonal Bolts *(Cont.)*

—Aircraft—Externally Relieved Body—Lock Washer—Cup

SBAC AS 18100-199 ISSUE 3(I). Bolt, Machine-Double Hexagon Extended Washer Head, Cupwasher Locked, HR601, 372-24 UNS-3A.
SBAC AS 18200-299 ISSUE 3(I). Bolt, Externally Relieved Body Double Hexagon Extended Washer Head, Cupwasher Locked, DTD5066, 187-32 UNS-3A.
SBAC AS 18300-399 ISSUE 3(I). Bolt, Externally Relieved Body Double Hexagon Extended Washer Head, Cupwasher Locked, DTD5066, 247-28 UNS-3A.
SBAC AS 18400-499 ISSUE 3(I). Bolt, Externally Relieved Body—Extended Washer Head, Cupwasher Locked, DTD5066, 3095-24 UNS-3A.
SBAC AS 18500-599 ISSUE 3(I). Bolt, Externally Relieved Body—Double Hexagon Extended Washer Head, Cupwasher Locked, DTD5066, 372-24 UNS-3A.
SBAC AS 18600-699 ISSUE 3(I). Bolt, Externally Relieved Body—Double Hexagon Extended Washer Head, Cupwasher Locked,HR650, 187-32 UNS-3A.
SBAC AS 18700-799 ISSUE 3(I). Bolt, Externally Relieved Body—Double Hexagon Extended Washer Head, Cupwasher Locked, HR650, 247-28 UNS-3A.
SBAC AS 18800-899 ISSUE 3(I). Bolt, Externally Relieved Body—Double Hexagon Extended Washer Head, Cupwasher Locked, HR650, 3095-24 UNS-3A.
SBAC AS 18900-999 ISSUE 3(I). Bolt, Externally Relieved Body—Double Hexagon Extended Washer Head, Cupwasher Locked, HR650, 372-24 UNS-3A.
SBAC AS 19000-099 ISSUE 3(I). Bolt, Externally Relieved Body—Double Hexagon Extended Washer Head, Cupwasher Locked, HR601, 187-32 UNS-3A.
SBAC AS 19100-199 ISSUE 3(I). Bolt, Externally Relieved Body—Double Hexagon Extended Washer Head, Cupwasher Locked, HR601, 247-28 UNS-3A.
SBAC AS 19200-299 ISSUE 3(I). Bolt, Externally Relieved Body—Double Hexagon Extended Washer Head, Cupwasher Locked, HR601, 3095-24 UNS-3A.
SBAC AS 19300-399 ISSUE 3(I). Bolt, Externally Relieved Body—Double Hexagon Extended Washer Head, Cupwasher Locked, HR601, 372-24 UNS-3A.

—Aircraft—Externally Relieved Body—Washer

SBAC AS 14500-599 ISSUE 4(I). Bolt, Externally Relieved Body—Double Hexagon Extended Washer Head, D.T.D.5066, 187-32 UNS-3A.
SBAC AS 14600-699 ISSUE 4(I). Bolt, Externally Relieved Body—Double Hexagon Extended Washer Head, DTD5066, 247-28 UNS-3A.
SBAC AS 14700-799 ISSUE 4(I). Bolt, Externally Relieved Body—Double Hexagon Extended Washer Head, DTD5066, 3095-24 UNS. 3A.
SBAC AS 14800-899 ISSUE 4(I). Bolt, Externally Relieved Body—Double Hexagon Extended Washer Head, D.T.D.5066, 372-24 UNS-3A.
SBAC AS 14900-999 ISSUE 4(I). Bolt, Externally Relieved Body—Double Hexagon Extended Washer Head, HR650, 187-32 UNS-3A.
SBAC AS 15000-099 ISSUE 4(I). Bolt, Externally Relieved Body—Double Hexagon Extended Washer Head, HR650, 247-28 UNS-3A.
SBAC AS 15100-199 ISSUE 4(I). Bolt, Externally Relieved Body—Double Hexagon Extended Washer Head, HR650, 3095-24 UNS-3A.
SBAC AS 15200-299 ISSUE 4(I). Bolt, Externally Relieved Body—Double Hexagon Extended Washer Head, HR650, 372-24 UNS-3A.
SBAC AS 15300-399 ISSUE 4(I). Bolt, Externally Relieved Body—Double Hexagon Extended Washer Head, HR601, 187-32 UNS-3A.
SBAC AS 15400-499 ISSUE 4(I). Bolt, Externally Relieved Body—Double Hexagon Extended Washer Head, HR601, 247-28 UNS-3A.
SBAC AS 15500-599 ISSUE 4(I). Bolt, Externally Relieved Body—Double Hexagon Extended Washer Head, HR601, 3095-24 UNS-3A.
SBAC AS 15600-699 ISSUE 4(I). Bolt, Externally Relieved Body—Double Hexagon Extended Washer Head, HR601, 372-24 UNS-3A.
SBAC AS 23900-999 ISSUE 3. Bolt, Externally Relieved Body—Double Hexagon Extended Washer Head, BS.S.159, 190-32 UNJF-3A.B.S.4084. 1 p.
SBAC AS 24000-099 ISSUE 4. Bolt, Externally Relieved Body—Double Hexagon Extended Washer Head, BS.S.159, 250-28 UNJF-3A.B.S.4084. 1 p.
SBAC AS 24100-199 ISSUE 4. Bolt, Externally Relieved Body—Double Hexagon Extended Washer Head, BS.S.159, 3125-24 UNJF-3A. B.S.4084. 1 p.
SBAC AS 24200-299 ISSUE 4. Bolt, Externally Relieved Body—Double Hexagon Extended Washer Head, BS S.159, 375-24 UNJF-3A. BS. 4084. 1 p.
SBAC AS 24400-499 ISSUE 3. Bolt, Externally Relieved Body—Double Hexagon Extended Washer Head, D.T.D.5026, 10(.190)-32 UNJF-3A. B.S.4084. 1 p.
SBAC AS 24500-599 ISSUE 3. Bolt, Externally Relieved Body—Double Hexagon Extended Washer Head, D.T.D.5026, 250-28 UNJF-3A. B.S.4084. 1 p.

Double Hexagonal Bolts *(Cont.)*

—Aircraft—Externally Relieved Body—Washer *(Cont.)*

SBAC AS 24600-699 ISSUE 4. Bolt, Externally Relieved Body—Double Hexagon Extended Washer Head, BS.S159, 3125-24 UNJF-3A.B.S.4084. 1 p.
SBAC AS 24700-799 ISSUE 3. Bolt, Externally Relieved Body—Double Hexagon Extended Washer Head, D.T.D.5026, 375-24 UNJF-3A. B.S.4084. 1 p.
SBAC AS 24900-999 ISSUE 3. Bolt, Externally Relieved Body—Double Hexagon Extended Washer Head, BS S. 159, 375-24 UNJF-3A. BS 4084. 1 p.
SBAC AS 25000-099 ISSUE 3. Bolt, Externally Relieved Body—Double Hexagon Extended Washer Head, BSHR601,.250-28 UNJF-3A.B.S.4084. 1 p.
SBAC AS 25100-199 ISSUE 3. Bolt, Externally Relieved Body—Double Hexagon Extended Washer Head, BSHR601, 3125-24 UNJF-3A. B.S.4084. 1 p.
SBAC AS 25200-299 ISSUE 3. Bolt, Externally Relieved Body—Double Hexagon Extended Washer Head, BSHR601, 375-24 UNJF-3A. B.S.4084. 1 p.
SBAC AS 31400-499 ISSUE 3. Bolt, Externally Relieved Body—DoubleHexagon Extended Washer Head, BS.S159, 4375-20 UNJF-3A. BS.4084.
SBAC AS 31500-599 ISSUE 3. Bolt, Externally Relieved Body—Double Hexagon Extended Washer Head, BS.S159. 500-20 UNJF. 3A BS.4084.
SBAC AS 31600-699 ISSUE 2. Bolt, Externally Relieved Body—Double Hexagon Extended Washer Head, D.T.D.5026, 4375-20 UNJF-3A. BS.4084.
SBAC AS 31700-799 ISSUE 2. Bolt, Externally Relieved Body—Double Hexagon Extended Washer Head, D.T.D.5026, 500-20 UNJF-3A. B.S.4084.
SBAC AS 31800-899 ISSUE 2. Bolt, Externally Relieved Body—Double Hexagon Extended Washer Head D.T.D. 5077, 4375-20 UNJF.3A.BS.4084.
SBAC AS 31900-999 ISSUE 2. Bolt, Externally Relieved Body—Double Hexagon Extended Washer Head, BSHR601. 500-20 UNJF. 3A. BS.4084.

—Aircraft—Machine

SBAC AS 43200-299 (V). Bolt, Machine, Double Hexagon Extended Washer Head, Drilled, BS S 159, Silver Coated, 10(.190)-32 UNJF-3A. BS4084.
SBAC AS 43300-399 (V). Bolt, Machine, Double Hexagon Extended Washer Head, Drilled, BS S 159, Silver Coated, 250-28 UNJF-3A. BS4084.
SBAC AS 43400-499 (V). Bolt, Machine, Double Hexagon Extended Washer Head, Drilled, BS S159, Silver Coated, 3125-24 UNJF-3A. BS 4084.
SBAC AS 43500-599 (V). Bolt, Machine, Double Hexagon Extended Washer Head, Drilled, BS S159, Silver Coated, 375-24 UNJF-3A.BS4084.
SBAC AS 43600-699 (V). Bolt, Machine, Double Hexagon Extended Washer Head, Drilled, BS S 159, Silver Coated, 10(.190)-32 UNJF-3A. BS4084.
SBAC AS 43700-799 (V). Bolt, Machine, Double Hexagon Extended Washer Head, Drilled, BS S159, Silver Coated, 250-28 UNJF-3A. BS4084.
SBAC AS 43800-899 (V). Bolt, Machine, Double Hexagon Extended Washer Head, Drilled, BS.HR650, Silver Coated, 3125-24 UNJF-3A. BS4084.
SBAC AS 43900-999 (V). Bolt, Machine, Double Hexagon Extended Washer Head, Drilled, BS.HR650, Silver Coated, 375-24 UNJF-3A. BS4084.
SBAC AS 44000-099 (V). Bolt, Machine, Double Hexagon Extended Washer Head, Drilled, BS.HR601, Silver Coated, 10(.190)-32 UNJF-3A. BS4084.
SBAC AS 44100-199 (V). Bolt, Machine, Double Hexagon Extended Washer Head, Drilled, BS.HR601, Silver Coated, 250-28 UNJF-3A.BS4084.
SBAC AS 44200-299 (V). Bolt, Machine, Double Hexagon Extended Washer Head, Drilled, BS.HR601, Silver Coated, 3125-24 UNJF-3A.BS4084.

—Aircraft—Machine—Lock Washer—Cup

SBAC AS 17000-099 ISSUE 3(I). Bolt, Machine-Double Hexagon Extended Washer Head, Cupwasher Locked, DTD5066, 187-32 UNS-3A.
SBAC AS 17100-199 ISSUE 3(I). Bolt, Machine-Double Hexagon Extended Washer Head, Cupwasher Locked, DTD5066, 247-28 UNS-3A.
SBAC AS 17200-299 ISSUE 3(I). Bolt, Machine-Double Hexagon Extended Washer Head, Cupwasher Locked, DTD5066, 3095-24 UNS-3A.
SBAC AS 17300-399 ISSUE 3(I). Bolt, Machine-Double Hexagon Extended Washer Head, Cupwasher Locked, DTD5066, 372-24 UNS-3A.
SBAC AS 17400-499 ISSUE 3(I). Bolt, Machine-Double Hexagon Extended Washer Head, Cupwasher Locked, HR650, 187-32 UNS-3A.
SBAC AS 17500-599 ISSUE 3(I). Bolt, Machine-Double Hexagon Extended Washer Head, Cupwasher Locked, HR650, 247-28 UNS-3A.
SBAC AS 17600-699 ISSUE 3(I). Bolt, Machine-Double Hexagon Extended Washer Head, Cupwasher Locked, HR650, 3095-24 UNS-3A.
SBAC AS 17700-799 ISSUE 3(I). Bolt, Machine-Double Hexagon Extended Washer Head, Cupwasher Locked, HR650, 372-24 UNS-3A.

INDUSTRY STANDARDS
831

Double Hexagonal Bolts *(Cont.)*

—Aircraft—Machine—Lock Washer—Cup *(Cont.)*

SBAC AS 17800-899 ISSUE 3(I). Bolt, Machine-Double Hexagon Extended Washer Head, Cupwasher Locked, HR601, 187-32 UNS-3A.

SBAC AS 17900-999 ISSUE 3(I). Bolt, Machine-Double Hexagon Extended Washer Head, Cupwasher Locked, HR601, 247-28 UNS-3A.

SBAC AS 18000-099 ISSUE 3(I). Bolt, Machine-Double Hexagon Extended Washer Head, Cupwasher Locked, HR601, 3095-24 UNS-3A.

—Aircraft—Machine—Shank

SBAC AS 48014-099 ISSUE 1. Bolt, Machine, Double Hexagon Extended Washer Head, P.D. Shank, BS. HR650. 190-32 UNJF-3A. BS4084.

SBAC AS 48116-199 ISSUE 1. Bolt, Machine, Double Hexagon Extended Washer Head, P.D. Shank, BS. HR650. 250-28 UNJF-3A. BS4084.

SBAC AS 48218-299 ISSUE 1. Bolt, Machine, Double Hexagon Extended Washer Head, P.D. Shank, BS.HR 650. 3125-24 UNJF-3A. BS4084.

SBAC AS 48320-399 ISSUE 1. Bolt, Machine, Double Hexagon Extended Washer Head, P.D. Shank, BS. HR650. 375-24 UNJF-3A. BS4084.

SBAC AS 48400-499 ISSUE 2. Bolt, Machine, Double Hexagon Extended Washer Head, P.D. Shank, INCO 718 BS 4084.

SBAC AS 48500-599 ISSUE 2. Bolt, Machine, Double Hexagon Extended Washer Head, P.D. Shank, INCO 718 BS 4084.

SBAC AS 48600-699 ISSUE 2. Bolt, Machine, Double Hexagon Extended Washer Head, P.D. Shank, INCO 718 BS 4084.

SBAC AS 48700-799 ISSUE 2. Bolt, Machine, Double Hexagon Extended Washer Head, P.D. Shank, INCO 718 BS 4084.

SBAC AS 48800-899 ISSUE 2. Bolt, Machine, Double Hexagon Extended Washer Head, P.D. Shank, Waspaloy UNJF-3A. BS 4084.

SBAC AS 48900-999 ISSUE 2. Bolt, Machine, Double Hexagon Extended Washer Head, P.D. Shank, Waspaloy 0.3125-24 UNJF-3A. BS4084.

SBAC AS 49000-099 ISSUE 2. Bolt, Machine, Double Hexagon Extended Washer Head, P.D. Shank, Waspaloy 0.3750-24 UNJF-3A. BS4084.

SBAC AS 55500-599 ISSUE 1. Bolt, Machine, Double Hexagon Extended Washer Head, P.D. Shank, WASPALOY 0.4375-20 UNJF-3A.BS 4084.

SBAC AS 55600-699 ISSUE 1. Bolt, Machine, Double Hexagon Extended Washer Head, P.D. Shank, Waspaloy 0.500-20 UNJF-3A. BS4084.

SBAC AS 55900-999 ISSUE 1. Bolt, Machine, Double Hexagon Extended Washer Head, PD Shank, INCO 718 0.4375—20 UNJF—3A. BS4084.

SBAC AS 56000-099 ISSUE 1. Bolt, Machine, Double Hexagon Extended Washer Head, PD Shank, INCO 718 0.500—20 UNJF—3A. BS4084.

SBAC AS 60100-199 ISSUE 1. Bolt, Machine, Double Hexagon Extended Washer Head, P.D. Shank, FV535 0.1900-32UNJF-3A. BS4084.

SBAC AS 60200-299 ISSUE 1. Bolt, Machine, Double Hexagon Extended Washer Head, P.D. Shank, FV535 0.2500—28 UNJF—3A. BS4084.

SBAC AS 60300-399 ISSUE 1. Bolt, Machine, Double Hexagon Extended Washer Head, P.D. Shank, FV535 0.3125—24 UNJF—3A. BS4084.

SBAC AS 60400-499 ISSUE 1. Bolt, Machine, Double Hexagon Extended Washer Head, P.D. Shank, FV535 0.3750—24 UNJF—3A BS4084.

SBAC AS 60500-599 ISSUE 1. Bolt, Machine, Double Hexagon Extended Washer Head, P.D. Shank, FV535 0.4375—20 UNJF—3A. BS4084.

SBAC AS 60600-699 ISSUE 1. Bolt, Machine, Double Hexagon Extended Washer Head, P.D. Shank, FV535 0.5000—20 UNJF—3A. BS4084.

SBAC AS 62900-999 ISSUE 1. Bolt, Machine, Double Hexagon Extended Washer Head, P.D. Shank, Waspaloy 0.1900—32UNJF—3A. BS 4084.

—Aircraft—Machine—Washer

SBAC AS 13000-099 ISSUE 4(I). Bolt, Machine-Double Hexagon Extended Washer Head, DTD5066, 187-32 UNS. 3A.

SBAC AS 13100-199 ISSUE 5(I). Bolt, Machine-Double Hexagon Extended Washer Head, DTD5066, 28 UNS. 3A.

SBAC AS 13200-299 ISSUE 4(I). Bolt, Machine-Double Hexagon Extended Washer Head, DTD5066, 3095-24 UNS. 3A..

SBAC AS 13300-399 ISSUE 4(I). Bolt, Machine-Double Hexagon Extended Washer Head, DTD5066, 372-24 UNS. 3A.

SBAC AS 13400-499 ISSUE 4(I). Bolt, Machine-Double Hexagon Extended Washer Head, HR650, 187-32 UNS. 3A.

SBAC AS 13500-599 ISSUE 4(I). Bolt, Machine-Double Hexagon Extended Washer Head, HR650, 247-28 UNS. 3A.

SBAC AS 13600-699 ISSUE 4(I). Bolt, Machine-Double Hexagon Extended Washer Head, HR650, 3095-24 UNS. 3A.

Double Hexagonal Bolts *(Cont.)*

—Aircraft—Machine—Washer *(Cont.)*

SBAC AS 13700-799 ISSUE 4(I). Bolt, Machine-Double Hexagon Extended Washer Head, HR650, 372-24 UNS. 3A.

SBAC AS 13800-899 ISSUE 4(I). Bolt, Machine-Double Hexagon Extended Washer Head, HR601, 187-32 UNS. 3A.

SBAC AS 13900-999 ISSUE 4(I). Bolt, Machine-Double Hexagon Extended Washer Head, HR601, 247-28 UNS. 3A.

SBAC AS 14000-099 ISSUE 4(I). Bolt, Machine-Double Hexagon Extended Washer Head, HR601, 3095-24 UNS 3A.

SBAC AS 14100-199 ISSUE 4(I). Bolt, Machine-Double Hexagon Extended Washer Head, HR601, 372-24 UNS. 3A.

SBAC AS 20800-899 ISSUE 4. Bolt, Machine-Double Hexagon Extended Washer Head, BS.5159. 164-32-36 UNJF-3A. B.S.4084.

SBAC AS 20900-999 ISSUE 4. Bolt Machine Double Hexagon Extended Washer Head, BS.5159, 190-32 UNJF-3A-BS.4084.

SBAC AS 21000-099 ISSUE 4. Bolt, Machine-Double Hexagon Extended Washer Head, BS.5159. 250-28 UNJF-3A-B.S.4084.

SBAC AS 21100-199 ISSUE 4. Bolt, Machine-Double Hexagon Extended Washer Head, BS.5159, 3125-24 UNJF-3A.B.S.4084.

SBAC AS 21200-299 ISSUE 4. Bolt, Machine-Double Hexagon Extended Washer Head, BS.5159, 375-24 UNJF-3A.B.S.4084.

SBAC AS 21300-399 ISSUE 4. Bolt, Machine-Double Hexagon Extended Washer Head, D.T.D.5026 8(.164) -36 UNJF-3A. B.S.4084.

SBAC AS 21400-499 ISSUE 3. Bolt, Machine-Double Hexagon Extended Washer Head, BS HR 650. 190-32 UNJF-3A. B.S.4084.

SBAC AS 21500-599 ISSUE 3. Bolt, Machine-Double Hexagon Extended Washer Head, BS HR650. 250-28 UNJF-3A. B.S.4084.

SBAC AS 21600-699 ISSUE 3. Bolt, Machine-Double Hexagon Extended Washer Head, BS HR650, 3125-24 UNJF-3A. B.S.4084.

SBAC AS 21700-799 ISSUE 3. Bolt, Machine-Double Hexagon Extended Washer Head, BS HR650, 375-24 UNJF-3A. B.S.4084.

SBAC AS 21800-899 ISSUE 3(I). Bolt, Machine-Double Hexagon Extended Washer Head, D.T.D.5077, 8 (.164)-36 UNJF-3A.B.S. 4084.

SBAC AS 21900-999 ISSUE 3. Bolt, Machine-Double Hexagon Extended Washer Head, BSHR601. 190-32 UNJF-3A. B.S.4084.

SBAC AS 22000-099 ISSUE 3. Bolt, Machine Double Hexagon External Washer Head,BSHR601, 250-28 UNJF-3A. B.S.4084. 1 p.

SBAC AS 22100-199 ISSUE 3. Bolt, Machine-Double Hexagon Extended Washer Head, BSHR601, 3125-24 UNJF-3A. B.S.4084. 1 p.

SBAC AS 22200-299 ISSUE 3. Bolt, Machine-Double Hexagon Extended Washer Head, BSHR601, 375-24 UNJF-3A. B.S.4084. 1 p.

SBAC AS 30200-299 ISSUE 3. Bolt, Machine-Double Hexagon Extended Washer Head, BS.S.159, .4375-20 UNJF-3A.B.S.4084.

SBAC AS 30300-399 ISSUE 3. Bolt, Machine-Double Heaxgon Extended Washer Head, BS.S.159, 500-20 UNJF-3A.B.S.4084.

SBAC AS 30400-499 ISSUE 2. Bolt, Machine-Double Hexagon Extended Washer Head, BSHR650, 4375-20 UNJF-3A.B.S.4084.

SBAC AS 30500-599 ISSUE 2. Bolt, Machine-Double Hexagon Extended Washer Head, BSHR650, 500-20 UNJF-3A.B.S.4084.

SBAC AS 30600-699 ISSUE 2. Bolt, Machine-Double Hexagon Extended Washer Head, BSHR601, 4375-20 UNJF-3A.B.S.4084.

SBAC AS 30700-799 ISSUE 2. Bolt, Machine-Double Hexagon Extended Washer Head, BSHR601, .500-20 UNJF-3A.B.S.4084.

SBAC AS 44300-399 (V). Bolt, Machine, Double Hexagon Extended Washer Head, Drilled, BS.HR601, Silver Coated, 375-24 UNJF-3A. BS4084.

—Aircraft—Washers

SBAC AS 19400-499 ISSUE 3(I). Bolt, Close Tolerance—Double Hexagon Extended Washer Head, DTD5066, 187-32 UNS-3A.

SBAC AS 19500-599 ISSUE 3(I). Bolt, Close Tolerance—Double Hexagon Extended Washer Head, DTD5066, 247-28 UNS-3A.

SBAC AS 19600-699 ISSUE 3(I). Bolt, Close Tolerance—Double Hexagon Extended Washer Head, DTD5066, 3095-24 UNS-3A.

SBAC AS 19700-799 ISSUE 3(I). Bolt, Close Tolerance—Double Hexagon Extended Washer Head, DTD5066, 372-24 UNS-3A.

SBAC AS 19800-899 ISSUE 3(I). Bolt, Close Tolerance—Double Hexagon Extended Washer Head, HR650, 187-32 UNS-3A.

SBAC AS 19900-999 ISSUE 3(I). Bolt, Close Tolerance—Double Hexagon Extended Washer Head, HR650, 247-28 UNS-3A.

Double Hexagonal Bolts *(Cont.)*

—Aircraft—Washers *(Cont.)*

SBAC AS 20000-099 ISSUE 3(I). Bolt, Close Tolerance—Double Hexagon Extended Washer Head, HR650, 3095-24 UNS-3A.

SBAC AS 20100-199 ISSUE 3(I). Bolt, Close Tolerance—Double Hexagon Extended Washer Head, HR650, 372-24 UNS-3A.

SBAC AS 20200-299 ISSUE 3(I). Bolt, Close Tolerance—Double Hexagon Extended Washer Head, HR601, 187-32 UNS-3A.

SBAC AS 20300-399 ISSUE 3(I). Bolt, Close Tolerance—Double Hexagon Extended Washer Head, HR601, 247-28 UNS-3A.

SBAC AS 20400-499 ISSUE 3(I). Bolt, Close Tolerance—Double Hexagon Extended Washer Head, HR601, 3095-24 UNS-3A.

SBAC AS 20500-599 ISSUE 3(I). Bolt, Close Tolerance—Double Hexagon Extended Washer Head, BS.5159.

SBAC AS 22400-499 ISSUE 4. Bolt, Close Tolerance—Double Hexagon Extended Washer Head, BS.5159. 190-32 UNJF-3A.MOD.B.S. 4084. 1 p.

SBAC AS 22500-599 ISSUE 4. Bolt, Close Tolerance—Double Hexagon Extended Washer Head, BS.5159. 250-28 UNJF-3A.MOD.B.S. 4084. 1 p.

SBAC AS 22600-699 ISSUE 4. Bolt, Close Tolerance—Double Hexagon Extended Washer Head, BS.5159, 3125-24 UNJF-3A.MOD. B.S.4084. 1 p.

SBAC AS 22700-799 ISSUE 4. Bolt, Close Tolerance—Double Hexagon Extended Washer Head, BS.5159, 375-24 UNJF-3A.MOD.B.S. 4084. 1 p.

SBAC AS 22900-999 ISSUE 2. Bolt, Close Tolerance—Double Hexagon Extended Washer Head, BSS.159. 190-32 UNJF-3A.MOD.B.S. 4084. 1 p.

SBAC AS 23000-099 ISSUE 3. Bolt, Close Tolerance—Double Hexagon Extended Washer Head, BS.HR650, 250-28 UNJF-3A. MOD. B.S. 4084. 1 p.

SBAC AS 23100-199 ISSUE 2. Bolt, Close Tolerance—Double Hexagon Extended Washer Head, BS.HR650, 3125-24 UNJF-3A. MOD. B.S. 4084. 1 p.

SBAC AS 23200-299 ISSUE 3. Bolt, Close Tolerance—Double Hexagon Extended Washer Head, BS.HR650, 375-24 UNJF-3A.MOD.B.S. 4084. 1 p.

SBAC AS 23400-499 ISSUE 3. Bolt, Close Tolerance—Double Hexagon Extended Washer Head, BS.HR601, 10(.190)-32 UNJF-3A.MOD. B.S.4084. 1 p.

SBAC AS 23500-599 ISSUE 3. Bolt, Close Tolerance—Double Hexagon Extended Washer Head, BS.HR601, 250-28 UNJF-3A.MOD.B.S. 4084. 1 p.

SBAC AS 23600-699 ISSUE 3. Bolt, Close Tolerance—Double Hexagon Extended Washer Head, BS.HR601, 3125-24 UNJF-3A.MOD. B.S. 4084. 1 p.

SBAC AS 23700-799 ISSUE 3. Bolt, Close Tolerance—Double Hexagon Extended Washer Head, BS.HR601, 375-24 UNJF-3A.MOD.B.S. 4084. 1 p.

SBAC AS 30800-899 ISSUE 3. Bolt, Close Tolerance—Double Hexagon Extended Washer Head, BSS.159, 4375-20 UNJF-3A.MOD. BS.4084.

SBAC AS 30900-999 ISSUE 3. Bolt, Close Tolerance—Double Hexagon Extended Washer Head, BS.S159. 500-20 UNJF-3A.MOD. BS. 4084.

SBAC AS 31000-099 ISSUE 2. Bolt, Close Tolerance—Double Hexagon Extended Washer Head, BS.HR650. 4375-20 UNJF-3A.MOD. BS4084.

SBAC AS 31100-199 ISSUE 2. Bolt, Close Tolerance—Double Heaxgon Extended Washer Head, BS.HR650. 500-20 UNJF-3A.MOD. BS.4084.

SBAC AS 31200-299 ISSUE 2. Bolt, Close Tolerance—Double Hexagon Extended Washer Head, BS.HR601. 4375-20 UNJF-3A.MOD. BS.4084.

SBAC AS 31300-399 ISSUE 3. Bolt, Close Tolerance—Double Hexagon Extended Washer Head, BSHR601, 500-20 UNJF-3A.MOD. BS.4084.

Double Hexagonal Head Bolts

Use: Double Hexagonal Bolts

Double Hexagonal Head Screws

See Also: Screws

—Aerospace

ISO 4095-78. Fasteners for Aerospace Construction—Bi-Hexagonal Wrenching Configuration First Edition. 4 pp.

Double Hexagonal Nuts

Use For: Bihexagonal Nuts; Twelve Point Nuts
See Also: Hexagonal Nuts; Nuts (Fasteners)

—Aerospace

CEN PREN 3771-90. Nuts, Bi-Hexagonal, Plain with Wire Locking Holes, in Heat Resisting Steel, Passivated, MoS2 Lubricated Classification: 1550 MPa (at Ambient Temperature) /315 Degrees Celsius. 5 pp.

CEN PREN 3771-93. Aerospace Series Nuts, Bi-Hexagonal, Plain, in Heat Resisting Steel, MoS2 Lubricated Classification: 1 100 MPa (at Ambient Temperature) /315 Degrees C. 5 pp.

ISO 4095-78. Fasteners for Aerospace Construction—Bi-Hexagonal Wrenching Configuration First Edition. 4 pp.

Double Hexagonal Nuts (Cont.)
—Aerospace—Locking
AECMA PREN2906-86. Nuts, Self Locking, Bi Hexagonal, in Heat Resisting Steel FE-PA92HT (A286) Unplated Classification 1100 MPa/650 Degrees. 4 pp.

AECMA PREN2907-86. Nuts, Self Locking, Bi Hexagonal, in Heat Resisting Steel FE-PA92HT (A286) Silver Plated Classification 1100 MPa/650 Degrees Celsius. 4 pp.

AECMA PREN2908-86. Nuts, Self Locking, Bi Hexagonal, Deep Counterbored, in Heat Resisting Steel FE-PA92HT (A286) Unplated Classification 1100 MPa/650 Degrees Celsius. 4 pp.

AECMA PREN2909-86. Nuts, Self Locking, Bi Hexagonal, Deep Counterbored, in Heat Resisting Steel FE-PA92HT (A286) Silver Plated Classification 1100 MPa/650 Degrees Celsius. 4 pp.

AECMA PREN3012-88. Nuts, Self-Locking, Bihexagonal in Heat Resisting Nickel Base Alloy NI-P101HT (Waspaloy) Unplated Classification: 1210 MPa/730 Degrees Celsius. 5 pp.

AECMA PREN3013-88. Nuts, Self-Locking, Bihexagonal in Heat Resisting Nickel Base Alloy NI-P101HT (Waspaloy) Completely Silver Plated: Classification: 1210 MPa/730 Degrees Celsius. 5 pp.

AECMA PREN3239-88. Nuts, Self-Locking, Bihexagonal in Heat Resisting Nickel Base Alloy NI-P101HT (Waspaloy) Silver Plated Thread Classification: 1210 MPa/730 Degrees Celsius. 4 pp.

AECMA PREN3637-90. Nuts, Self-Locking, Bi-Hexagonal (Double Reduced), in Heat Resisting Nickel Base Alloy N1-P101HT (Waspaloy), Silver Plated Classification: 1210 MPa/730 Degrees Celsius. 5 pp.

CEN PREN 3713-91. Nuts, Anchor, Self-Locking, Bihexagonal, in Steel, Cadmium Plated, MoS2 Lubricated Classification: 1550 MPa (at Ambient Temperature)/235 Degrees C. 6 pp.

CEN PREN 3720-91. Nuts, Bihexagonal, Self-Locking, in Heat Resisting Steel FE-PA92HT (A286), MoS2 Coated Classification:1100 (at Ambient Temperature) /425 Degrees C. 5 pp.

CEN PREN 4011-93. Aerospace Series Nuts, Bihexagonal, Self-Locking, in Heat Resisting Nickel Base Alloy NI-P100HT (Inconel 718), Silver Plated Classification: 1 550 MPa (at Ambient Temperature) /600 Degrees C. 4 pp.

CEN PREN 4012-93. Aerospace Series Nuts, Bihexagonal, Self-Locking, in Heat Resisting Nickel Base Alloy NI-P100HT (Inconel 718), MoS2 Coated Classification: 1 550 MPa (at Ambient Temperature) /425 Degrees C. 4 pp.

DIN ENGL LN 29528-80. Aerospace; Nut Bihexagonal with Flange; Selflocking for Temperatures up to 235 Degrees C for Bolts and Screws with Nominal Tensile Strength up to 1800 N/mm2 (Nov). 2 pp.

DIN ENGL LN 29942-85. Aerospace; Nuts, Bi-Hexagon Flange, Deep Counterbore, Self-Locking for Temperatures up to 235 Degrees C for Bolts and Screws with Nominal Tensile Strength of 1550 and 1800 N/mm2 (Feb). 3 pp.

ISO 9199-87. Aerospace—Self-Locking Bihexagonal Nuts, Classificatons 1 100 MPa/650 Degrees Celsius, 1 250 MPa/760 Degrees Celsius, 1 550 MPa/235 Degrees Celsius and 1 550 MPa/650 Degrees Celsius—Dimensions First Edition. 4 pp.

—Aerospace—Locking—Counterbore
CEN PREN 3721-91. Nuts, Bihexagonal, Self-Locking, Deep Counterbored, in Heat Resisting Steel FE-PA92HT (A286), MoS Coated Classification:1100 MpA (at Ambient Temperature)/Degrees C. 5 pp.

CEN PREN 3772-90. Nuts, Bi-Hexagonal, with Flange, Deep Counterbore, Self Locking, in Heat Resisting Steel, Passivated Classification: 1100 MPa (at Ambient Temperature) /425 Degrees Celsius. 5 pp.

CEN PREN 3843-93. Aerospace Series Nuts, Bi-Hexagonal, Self-Locking, with Counterbore, in Heat Resisting Steel, Passivated Classification: 1 100 MPa (at Ambient Temperature) /650 Degrees C. 5 pp.

—Aircraft—Locking
SBAC AS 20620-639 ISSUE 8. Nut, Self-Locking, Extended Washer, Double Hexagon.

SBAC AS 42910-915 ISSUE 2. Nut, Self-Locking—Bi-Hexagon Waspaloy.

Doubling Machinery
Use: Twisting Machinery

Dough Molding Compounds
Use: Bulk Molding Compounds

Douglas C 47 Dakota Aircraft
See Also: Aircraft
—Accidents
CAA. Douglas DC-3/C47 Dakota (World Airline Accident Summary). 12 pp.

CAA. Douglas DC-3/C47 Dakota. 3 pp.

Douglas C 54 Aircraft
See Also: Aircraft
—Accidents
CAA. Douglas C54/DC4 (World Airline Accident Summary). 4 pp.

CAA. Douglas DC-4/C54. 1 p.

Douglas Dakota Aircraft
See Also: Aircraft
—Mandatory Aircraft Modifications and Inspections Summaries
CAA. Douglas DC-3 and Dakota. 7 pp.

Douglas DC 2 Aircraft
See Also: Aircraft
—Accidents
CAA. Douglas DC-2. 1 p.

Douglas DC 3 Aircraft
See Also: Aircraft
—Accidents
CAA. Douglas DC-3/C47 Dakota (World Airline Accident Summary). 12 pp.

CAA. Douglas DC-3/C47 Dakota. 3 pp.

—Mandatory Aircraft Modifications and Inspections Summaries
CAA. Douglas DC-3 and Dakota. 7 pp.

Douglas DC 4 Aircraft
See Also: Aircraft
—Accidents
CAA. Douglas C54/DC4 (World Airline Accident Summary). 4 pp.

CAA. Douglas DC-4/C54. 1 p.

Douglas DC 6 Aircraft
See Also: Aircraft
—Accidents
CAA. Douglas DC6 Series (World Airline Accident Summary). 3 pp.

CAA. Douglas DC-6 Series. 1 p.

Douglas DC 7 Aircraft
See Also: Aircraft
—Accidents
CAA. Douglas DC-7 Series (World Airline Accident Summary). 1 p.

CAA. Douglas DC-7 Series. 1 p.

Douglas DC 9 Series 10 Aircraft
See Also: Aircraft
—Certification
CAA. Douglas DC-9 Series 10 and 30. 13 pp.

Douglas DC 9 Series 30 Aircraft
See Also: Aircraft
—Certification
CAA. Douglas DC-9 Series 10 and 30. 13 pp.

Dovetail Saws
See Also: Saws; Woodworking Equipment
—Woodworking
BSI BS 3159: Part 2-93. 1993 Woodworking Saws for Hand Use Part 2: Specification for Tenon and Dovetail Saws. 16 pp.

BSI BS 3159: Part 2-62. 1962 Woodworking Saws for Hand Use Part 2: Tenon and Dovetail Saws. 16 pp.

Dowel Pins
Use: Dowels

Dowels
See Also: Fasteners; Locating Pins; Pins
—Aerospace
AECMA PREN3151-88. Plain Dowels in N1-P100HT (Inconel 718). 5 pp.

—Aircraft—Stop Valves
SBAC AS 4902 ISSUE 3. Dowel—Tubular. 1 p.

—Parallel—Hardened
BSI BS 7055-89. (WITHDRAWN) 1989 Hardened Parallel Pins (Dowel Pins) (Renumbered as BS EN 28734: 1992). 7 pp.

BSI BS EN 28734-92. 1992 Amd 1 Parallel Pins, Hardened (Dowel Pins) (ISO 8734: 1987) (AMD 7496) November 15, 1992. 14 pp.

CEN EN 28734-92. Parallel Pins, Hardened (Dowel Pins). 5 pp.

DIN ENGL EN 28734-92. Hardened Parallel Pins; (ISO 8734: 1987) (Oct) (Supersedes DIN 6325, October 1971 Edition). 7 pp.

ISO 8734-87. Parallel Pins, Hardened (Dowel Pins) First Edition. 6 pp.

JIS B 1355-90. Dowel Pins (Parallel Pins, Hardened). 10 pp.

—Parallel—Steel
BSI BS 1804: Part 1-66. 1966 Amd 2 Parallel Steel Dowel Pins Part 1: Inch Series. 13 pp.

BSI BS 1804: Part 2-68. 1968 Parallel Steel Dowel Pins Part 2: Metric Series. 12 pp.

BSI BS 1804: Part 3-73. 1973 Amd 1 Parallel Steel Dowel Pins Part 3: Parallel and Taper Types, Having Screw Threads for Extraction Purposes, Metric Series. 16 pp.

MOD UK DSTAN 53-81-01. Pins, Straight, Headless, Steel, Metric (Parallel Metric Steel Dowel Pins) Issue 1; Amendment 1. 9 pp.

—Press Dies
JIS B 5062-89. Dowel Pins for Press Dies. 6 pp.

—Taper—Steel
BSI BS 1804: Part 3-73. 1973 Amd 1 Parallel Steel Dowel Pins Part 3: Parallel and Taper Types, Having Screw Threads for Extraction Purposes, Metric Series. 16 pp.

—Wooden Structures
BSI BS EN 383-93. 1993 Timber Structures—Test Methods—Determination of Embedding Strength and Foundation Values for Dowel Type Fasteners (V). 13 pp.

CEN PREN 383-90. Timber Structures—Determination of Embedding Strength. 12 pp.

Down
See Also: Feathers
CGSB CAN/CGSB-139.2-M86. Determination of Composition of Mixtures of Feathers and Down, by Manual Sorting. 9 pp.

CGSB 139-GP-4M-84. Determination of Composition of Mixtures of Feathers and Down, by Mechanical Sorting, Standard for. 11 pp.

CNS N1009-85. Semi-Refined Waterfowl Feathers and Down (May)(820). 4 pp.

CNS N1034-85. Refined Waterfowl Feather and Down (May)(2119). 4 pp.

CNS N4011-85. Method of Test for Waterfowl Feathers and Down (May)(1483). 3 pp.

—Filling Power Index
BSI BS 5334-76. 1976 Amd 1 Method of Test for Filling Power Index of Down and Feathers. 5 pp.

—Glossaries
CGSB CAN/CGSB-139.1-M86. Feathers and Down-Terminology and Characteristics. 19 pp.

—Paddings
SAA AS 2479-87. Down and/or Feather Filling Materials and Filled Products (This is a Joint Standard with SANZ NZS 2479). 48 pp.

SNZ NZS/AS 2479-87. Down and/or Feather Filling Materials and Filled Products (This is a Joint Standard with SAA AS 2479). 48 pp.

Down Converters
Use For: Logic Level Down Converters
—Electrical Measurement—Satellite Communication Equipment
IEC 510 Pt 2-4-88. Methods of Measurement for Radio Equipment Used in Satellite Earth Stations Part 2: Measurements for Sub-Systems Section Four—up-and-Down-Converters First Edition. 38 pp.

—Incandescent Lighting
BSI BS EN 61046-92. 1992 D.C. or a.c. Supplied Electronic Step-Down Convertors for Filament Lamps. General and Safety Requirements. 31 pp.

BSI BS EN 61046-01. 1992 Amd 1 D.C. or a.c. Supplied Electronic Step-Down Convertors for Filament Lamps. General and Safety Requirements (IEC 1046: 1991) (AMD 7608) March 15, 1993. 40 pp.

BSI BS EN 61047-93. 1993 D.C. or a.c. Supplied Electronic Step-Down Convertors for Filament Lamps Performance Requirements (IEC 1047: 1991) (E). 22 pp.

CENELEC EN 61046-92. D.C. or a.c. Supplied Electronic Step-Down Convertors for Filament Lamps General and Safety Requirements. 2 pp.

CENELEC EN 61046/A1-92. AMD 1 D.C. or a.c Supplied Electronic Step-Down Convertors for Filament Lamps—General and Safety Requirements; (IEC 1046; 1991/A1: 1991). 5 pp.

Down Converters (Cont.)

—Incandescent Lighting (Cont.)

CENELEC EN 61047-92. D.C. or A.C. Supplied Electronic Step-Down Convertors for Filament Lamps—Performance Requirements; (IEC 1047: 1991). 5 pp.

IEC 1046-91. D.C. or a.c. Supplied Electronic Step-down Convertors for Filament Lamps General and Safety Requirements First Edition; (Amendment 1-1991) (CENELEC EN 61046/A1: 1992). 53 pp.

IEC 1047-91. D.C. or a.c. Supplied Electronic Step-Down Convertors for Filament Lamps Performance Requirements First Edition; (CENELEC EN 61047:1992). 36 pp.

—Incandescent Lighting—Safety

BSI BS EN 61046-92. 1992 D.C. or a.c. Supplied Electronic Step-Down Convertors for Filament Lamps. General and Safety Requirements. 31 pp.

BSI BS EN 61046-01. 1992 Amd 1 D.C. or a.c. Supplied Electronic Step-Down Convertors for Filament Lamps. General and Safety Requirements (IEC 1046: 1991) (AMD 7608) March 15, 1993. 40 pp.

CENELEC EN 61046-92. D.C. or a.c. Supplied Electronic Step-Down Convertors for Filament Lamps General and Safety Requirements. 2 pp.

CENELEC EN 61046/A1-92. AMD 1 D.C. or a.c Supplied Electronic Step-Down Convertors for Filament Lamps—General and Safety Requirements; (IEC 1046; 1991/A1: 1991). 5 pp.

IEC 1046-91. D.C. or a.c. Supplied Electronic Step-down Convertors for Filament Lamps General and Safety Requirements First Edition; (Amendment 1-1991) (CENELEC EN 61046/A1: 1992). 53 pp.

—Integrated Circuit

CECC CECC 90 109-719 ISSUE 1-87. Digital Integrated Circuits in Accordance with FS 90 109; 54/74 HC 4050; Non Inverting Bus Buffer/Logic Level Down Converter (En, Fr). 6 pp.

Down Counters

See Also: Up/Down Counters

—Binary—Synchronous—4-Bit

CECC CECC 90 104-075 ISSUE 2. NL CECC 90 104-075 Issue 2; Digital Integrated Circuits in Accordance with FS 90 104; HEC/HEF 4522B; Programmable 4-Bit BCD Down Counter (En). 14 pp.

CECC CECC 90 104-076 ISSUE 2. NL CECC 90 104-076 Issue 2; Digital Integrated Circuits in Accordance with FS 90 104; HEC/HEF 4526B; Programmable 4-Bit Binary Down Counter (En). 14 pp.

—Binary—Synchronous—8-Bit

CECC CECC 90 109-886 ISSUE 1-90. Digital Integrated Circuits; Silicon Monolithic C MOS, Cavity or Non-Cavity Packages; Type(s); 54/74 HC 40103 8-Bit Synchronous Binary Down Counter; Assessment Levels P, Y, L (En, Fr). 10 pp.

CECC CECC 90 109-888 ISSUE 1-90. Digital Integrated Circuits Silicon Monolithic C MOS, Cavity or Non-Cavity Packages; Type(s); 54/74 HCT 40103 8-Bit Synchronous Binary down Counter; Assessment Levels P, Y, L (En, Fr). 10 pp.

—Decade—Synchronous—8-Bit

CECC CECC 90 109-885 ISSUE 1-90. Digital Integrated Circuits; Silicon Monolithic C MOS, Cavity or Non-Cavity Packages; Type(s); 54/74 HC 40102 8-Bit Synchronous Decade Down Counter; Assessment Levels P, Y, L (En, Fr). 10 pp.

CECC CECC 90 109-887 ISSUE 1-90. Digital Integrated Circuits; Silicon Monolithic C MOS, Cavity or Non-Cavity Packages; Type(s); 54/74 HCT 40102 8-Bit Synchronous Decade Down Counter; Assessment Levels P, Y, L (En, Fr). 10 pp.

Down Filled Vests

Use: Vests—Down Filled

Down Links

Use: Downlinks

Downlinks

See Also: Feeder Links; Satellite Links

—Feeder Links—Noise—Broadcasting Satellite Services

CCIR RECMN 793-92. Partitioning of Noise Between Feeder Links for the Broadcasting-Satellite Service (BSS) and BSS Down Links. 5 pp.

Downlinks (Cont.)

—Geostationary Satellites—Frequency Bands

CCIR Report 557-2-86. Use of Frequency Bands Allocated to the Fixed-Satellite Service for Both the up Link and down Link Geostationary-Satellite Systems—Section 4D2—Coordination Methods. 9 pp.

Downspouts

See Also: Gutters

JIS A 5706-90. Unplasticized Polyvinyl Chloride Eaves Gutters and Downspouts. 11 pp.

—Fittings

SAA AS 1273.2-74. Unplasticized PVC (UPVC) Downpipe and Fittings for Rainwater—Part 2: Fittings Amdt 1 October 1979 (Superseded by AS 1273—1991).

—Metal

CEN PREN 612-91. Eaves Gutters and Rainwater Down-Pipes from Metal Sheet—Definitions, Classifications, Requirements and Testing. 13 pp.

—PVC—Unplasticized

SAA AS 1273.1-74. Unplasticized PVC (UPVC) Downpipe and Fittings for Rainwater—Part 1: Downpipe (Superseded by AS 1273—1991).

SAA AS 1273-91. Unplasticized PVC (UPVC) Downpipe and Fittings for Rainwater (Supersedes AS 1273.1—1974 and AS 1273.2—1974). 7 pp.

Dowty Rotol Propellers

Use: Aircraft Propellers—Dowty Rotol

Dozers

Use: Bulldozers

Dozers (Earthmoving Equipment)

Use: Bulldozers

DPA (Document Printing Application)

Use: Application Layer (OSI)—Document Printing Application

DQAB Bulletins

Use: Quality Assurance—Bulletins

Draft Gages

See Also: Manometers

—Ships

ISO 7606-88. Shipbuilding—Inland Navigation Vessels—Draught Scales First Edition. 8 pp.

MOD UK NES 164-90. Requirements for Draught Marks Issue 2 (02.90). 20 pp.

Draft Tubes

See Also: Hydraulic Turbines

—Appliances

JIS S 3025-92. Supply and Exhaust Pipes for Oil Burning Appliances. 17 pp.

Drafting

See Also: Drafting Films; Drafting Machines; Drawings; Engineering Drawings

JIS B 0021-84. Indications of Geometrical Tolerances on Drawings. 38 pp.

JIS B 0022-84. Datums and Datum-Systems for Geometrical Tolerances. 20 pp.

JIS B 0023-84. Maximum Material Principle. 19 pp.

JIS B 0031-82. Method of Indicating Surface Texture on Drawings. 22 pp.

SNZ NZMP 5905-81. Standard Technical Drawing Practice for School Certificate Courses in New Zealand. 31 pp.

SNZ NZMP 5906-84. Standard Technical Drawing Practice for Senior Courses in Schools and Technical Institutes. 74 pp.

—Antifriction Bearings

JIS B 0005-73. Drawing Office Practice for Rolling Bearings. 16 pp.

—Coordinate Systems

DIN ENGL 406 Pt 3-75. Dimensioning in Drawings; Dimensioning by Co-Ordinates (July). 8 pp.

DIN ENGL 406 Pt 4-80. Dimensioning on Drawings; Dimensioning for Programming by Machine (Dec). 4 pp.

DIN ENGL 406 Pt 4 Suppl. 1-80. Dimensioning on Drawings; Dimensioning for Programming by Machine; Example of Application (Dec). 5 pp.

—Counterbores

DIN ENGL 30-70. Drawing Practice; Simplified Presentations (Dec). 2 pp.

Drafting (Cont.)

—Fasteners

DIN ENGL 30-70. Drawing Practice; Simplified Presentations (Dec). 2 pp.

—Gears

DIN ENGL 37-61. Conventional and Simplified Representation of Gears and Gear Pairs (Dec). 4 pp.

JIS B 0003-89. Drawing Office Practice for Gears. 20 pp.

—Glossaries

ISO TR10623-91. Technical Product Documentation—Requirements for Computer-Aided Design and Draughting—Vocabulary First Edition. 30 pp.

JTC1 TR10623-91. Technical Product Documentation—Requirements for Computer-Aided Design and Draughting—Vocabulary First Edition. 30 pp.

—Integrated Circuits—Package Outlines

BSI BS 3934: Part 3-92. 1992 Mechanical Standardization of Semiconductor Devices Part 3: Recommendations for the Preparation of Outline Drawings of Integrated Circuits. 49 pp.

IEC 191 Pt 3-74. Mechanical Standardiza-tion of Semiconductor Devices Part 3: General Rules for the Preparation of Outline Drawings of Integrated Circuits Pin Grid Arrays First Edition; (Supplement A-1976) (Supplement B-1978) (Amendment 1-1983) (Supplement C-1987). 111 pp.

SAA AS C379 Part 3-78. Mechanical Standardization of Semiconductor Devices—Part 3: General Rules for the Preparation of Outline Drawings of Integrated Circuits. 24 pp.

—Microfilmed Drawings

CGSB CAN/CGSB-72.7-M88. Manual Drafting Requirements for Drawings to Be Microfilmed. 14 pp.

—Screw Threads

JIS B 0002-82. Drawing Office Practice for Screw Threads. 9 pp.

—Semiconductor Devices

BSI BS 3934: Part 1-92. 1992 Mechanical Standardization of Semiconductor Devices Part 1: Recommendations for the Preparation of Drawings of Semiconductor Devices. 45 pp.

BSI BS 3934: Part 2-92. 1992 Mechanical Standardization of Semiconductor Devices Part 2: Schedule of International Drawings Giving Dimensions (IEC 191-2: 1966). 315 pp.

BSI BS 3934: Part 10-92. 1992 Mechanical Standardization of Semiconductor Devices Part 10: Schedule of UK National Drawings Giving Dimensions. 93 pp.

IEC 191 Pt 1-66. Mechanical Standardization of Semiconductor Devices Part 1: Preparation of Drawings of Semiconductor Devices First Edition; (Supplement A-1969) (Supplement B-1970) (Supplement C-1974). 87 pp.

IEC 191 Pt 2-66. Mechanical Standardization of Semiconductor Devices Part 2: Dimensions First Edition; (Supplement A-1967) (Supplement B-1969) (Supplement C-1970) (Supplement D-1971) (Supplement E-1974) (Supplement F-1976) (Supplement G-H:1978). 374 pp.

SAA AS C379 Part 1-70. Mechanical Standardization of Semiconductor Devices—Part 1: Preparation of Drawings (IEC). 70 pp.

—Semiconductor Devices—Package Outlines

BSI BS 3934: Part 4-92. 1992 Mechanical Standardization of Semiconductor Devices Part 4: Recommendations for a Coding System and Classification into Forms of Package Outlines for Semiconductor Devices (IEC 191-4: 1987). 10 pp.

BSI BS 3934: Part 6-92. 1992 Mechanical Standardization of Semiconductor Devices Part 6: Specification for the Preparation of Outline Drawings of Surface Mounted Semiconductor Device Packages (IEC 191-6: 1990). 28 pp.

IEC 191 Pt 4-87. Mechanical Standardization of Semiconductor Devices Part 4: Coding System and Classification into Forms of Package Outlines for Semiconductor Devices First Edition. 13 pp.

IEC 191 Pt 6-90. Mechanical Standardization of Semiconductor Devices Part 6: General Rules for the Preparation of Outline Drawings of Surface Mounted Semiconductor Device Packages First Edition. 42 pp.

—Springs (Elastic)

JIS B 0004-76. Drawing Practice for Springs. 19 pp.

—Threads

DIN ENGL 30-70. Drawing Practice; Simplified Presentations (Dec). 2 pp.

Drafting Equipment

Scope Note: Use a more specific term

Drafting Equipment (Cont.)
See: Drafting; Drafting Films; Drafting Machines; Drafting Papers; Rulers

Drafting Films
See Also: Drafting; Drafting Papers
- ISO 9958 Pt 1-92. Draughting Media for Technical Drawings—Draughting Film with Polyester Base—Part 1: Requirements and Marking First Edition. 12 pp.
- ISO 9958 Pt 2-92. Draughting Media for Technical Drawings—Draughting Film with Polyester Base—Part 2: Determination of Properties First Edition. 10 pp.

—Identification Systems
- ISO 9958 Pt 1-92. Draughting Media for Technical Drawings—Draughting Film with Polyester Base—Part 1: Requirements and Marking First Edition. 12 pp.

Drafting Machines
Use For: Draughting Machines See Also: Drafting; Vector Plotters
- ISO 9962 Pt 2-92. Manually Operated Draughting Machines—Part 2: Characteristics, Performance, Inspection and Marking First Edition. 11 pp.
- JIS B 9513-83. Drafting Machines. 14 pp.

—Classification
- ISO 9962 Pt 1-92. Manually Operated Draughting Machines—Part 1: Definitions, Classification and Designation First Edition. 13 pp.

—Construction
- ISO 9962 Pt 2-92. Manually Operated Draughting Machines—Part 2: Characteristics, Performance, Inspection and Marking First Edition. 11 pp.

—Designations
- ISO 9962 Pt 1-92. Manually Operated Draughting Machines—Part 1: Definitions, Classification and Designation First Edition. 13 pp.

—Glossaries
- ISO 9179 Pt 1-88. Technical Drawings—Numerically Controlled Draughting Machines—Part 1: Vocabulary First Edition. 13 pp.
- ISO 9962 Pt 1-92. Manually Operated Draughting Machines—Part 1: Definitions, Classification and Designation First Edition. 13 pp.
- JIS B 3410-90. Glossary of Terms Used in Plotters Technical Drawing—Numerically Controlled Drafting Machines.

—Identification Systems
- ISO 9962 Pt 2-92. Manually Operated Draughting Machines—Part 2: Characteristics, Performance, Inspection and Marking First Edition. 11 pp.

—Inspection
- ISO 9962 Pt 2-92. Manually Operated Draughting Machines—Part 2: Characteristics, Performance, Inspection and Marking First Edition. 11 pp.

—Scales (Measuring Instruments)
- JIS B 9512-83. Scales for Drafting Machines. 18 pp.

Drafting Papers
See Also: Drafting Films; Drawing Papers; Papers

—Tracing
- BSI BS 7614-92. 1992 Natural Tracing Paper (H). 16 pp.
- ISO 9961-92. Draughting Media for Technical Drawings—Natural Tracing Paper First Edition. 14 pp.

Drafting Templates
Use: Templates—Drafting

Drag Chains
See Also: Chains
- ISO 6971-82. Welded Steel Type Cranked Link Drag Chains and Chain Wheels First Edition. 14 pp.

Drain Cocks
See Also: Cocks; Drain Valves

—Aircraft
- SBAC AGS 676 (V)(I). G.A. of a 1/4 Inch Bore Drain Cock (No 1 Size).
- SBAC AGS 677 (V)(I). G.A. of 1/2 Inch Bore Drain Cock (No 2 Size).
- SBAC AGS 3026 ISSUE 1. G.A. of 1/4 Inch Bore Drain Cock (No. 1 Size).
- SBAC AGS 3027 ISSUE 4. Body for Drain Cock.
- SBAC AGS 3030 ISSUE 1. G.A. of 1/2 Inch Bore Drain Cock (No. 2 Size).
- SBAC AGS 3031 ISSUE 1. Body for Drain Cock.

—Aircraft—Nuts (Fasteners)
- SBAC AGS 3029 ISSUE 1. Nut for Drain Cock.
- SBAC AGS 3033 ISSUE 1. Nut for Drain Cock.

—Aircraft—Pitot Tubes
- SBAC AS 40665 ISSUE 1. Cock, Poppet, Drain Assembly (Pitot).
- SBAC AS 40669 ISSUE 1. Cock, Poppet, Drain Assembly (Pitot).
- SBAC AS 40673 ISSUE 1. Cock, Poppet, Drain Assembly Pitot.

—Aircraft—Plugs
- SBAC AGS 3028 ISSUE 1. Plug for Drain Cock.
- SBAC AGS 3032 ISSUE 1. Plug for Drain Cock.

—Aircraft—Static Tubes
- SBAC AS 40667 ISSUE 1. Cock, Poppet, Drain Assembly (Static).
- SBAC AS 40671 ISSUE 1. Cock, Poppet, Drain Assembly (Static).
- SBAC AS 40675 ISSUE 1. Cock, Poppet, Drain Assembly (Static).

—Marine
- DIN ENGL 87001-67. Drain Cocks in Red Brass; Nominal Size R 1/4" to R 1"; ND 10: up to 80 Deg. C; ND 16: up to 40 Deg. C (Feb). 2 pp.
- DIN ENGL 87003 Pt 1-67. Lockable Drain Cocks in Red Brass; Nominal Size R 1/2" to R 1"; ND 6 up to 40 Deg. C; List of Sizes and Parts List (Feb). 2 pp.

—Rolling Stock
- CNS E1053-85. Drain Cocks for Railway Rolling Stock (Nov)(11411).
- JIS E 4101-91. Drain Cocks for Railway Rolling Stock.

Drain Plugs
See Also: Stoppers
- CNS B2377-82. Screwed Sealing Plugs for Drain Fittings of Copper-Zinc Alloy (Oct)(4850).
- CNS B2378-82. Screwed Sealing Plugs for Drain Fittings of High Density PE (Oct)(4851).

—Aircraft
- SBAC AS 2420 ISSUE 4. Plug—Drain Plug. 1 p.
- SBAC AS 6673 ISSUE 2. Plug—Drain Plug.

—Assembly—Aircraft
- SBAC AS 2419 ISSUE 5. Assembly—Drain Plug. 1 p.
- SBAC AS 6672 ISSUE 1. Assembly—Drain Plug.

—Cocks—Aircraft
- SBAC AGS 3032 ISSUE 1. Plug for Drain Cock.

—Earthmoving Equipment
- BSI BS 5796-87. 1987 Drain, Fill and Level Plugs for Earth-Moving Machinery. 8 pp.
- ISO 6302-86. Earth-Moving Machinery—Drain, Fill and Level Plugs Second Edition. 6 pp.

—Motorcycles
- CNS D2141-82. Thread Plug of Oil Drain Hole for Motorcycle (Jul)(9169).

—Seatings—Aircraft
- SBAC AS 2421 ISSUE 3. Seating—Drain Plug. 1 p.
- SBAC AS 6674 ISSUE 1. Seating—Drain Plug.

—Ships
- BSI BS MA 82-78. 1978 Shipbuilding: Oil and Water Drain Screws. 4 pp.
- ISO 5483-77. Shipbuilding—Oil and Water Drain Screws First Edition. 4 pp.
- JIS F 3006-77. Ships' Drain Plugs (R 1984). 11 pp.

—Ships' Bottom
- JIS F 3005-82. Ships' Bottom Plugs and Spanners.
- JIS F 3005-68. Ships' Bottom Plugs. 9 pp.

—Spanners—Ships' Bottom
- JIS F 3005-82. Ships' Bottom Plugs and Spanners.
- JIS F 3005-68. Ships' Bottom Plugs. 9 pp.

—Threaded
- SAA AS D29-73. Filler, Drain and Pipe Plugs with Dryseal American Standard Taper Pipe and Unified Threads for Automotive and Industrial Use Reconfirmed 1988. 20 pp.

—Threaded—Automotive
- SAA AS D29-73. Filler, Drain and Pipe Plugs with Dryseal American Standard Taper Pipe and Unified Threads for Automotive and Industrial Use Reconfirmed 1988. 20 pp.

—Washers—Aircraft
- SBAC AS 2422 ISSUE 3. Washer—Drain Plug. 1 p.

Drain Screws
Use: Drain Plugs

Drain Traps
Use: Traps (Drains)

Drain Valves
Use For: Draw-off Taps See Also: Drain Cocks; Taps (Valves); Valves

—Aircraft
- SBAC AS 2564 (V). Body. 1 p.
- SBAC AS 2574 (V). Body. 1 p.
- SBAC AS 2579 ISSUE 5. Valve. 1 p.
- SBAC AS 3300 ISSUE 2. Drain, Valve Assembly. 1 p.
- SBAC AS 3301 ISSUE 1. Valve. 1 p.
- SBAC AS 3334 ISSUE 2. Valve 1/4 Inch Bore Drain Valve. 1 p.
- SBAC AS 4585 ISSUE 1. Flush Water/Sediment Drain Valve. 1 p.
- SBAC AS 4586 ISSUE 1. Body Assembly. 1 p.
- SBAC AS 4587 ISSUE 1. Valve. 1 p.
- SBAC AS 4760 ISSUE 4. 1/4 Inch Bore Sediment Drain Valve. 1 p.
- SBAC AS 4761 ISSUE 1. Body (1/4 Inch Drain Valve). 1 p.
- SBAC AS 6296 (V). Valve Body. 2 pp.
- SBAC AS 6297 ISSUE 3. Valve. 1 p.
- SBAC AS 6361 ISSUE 4(I). Body—Drain Valve.
- SBAC AS 6370 ISSUE 3. Drainage Connection Drain Valve.
- SBAC AS 6617 ISSUE 1. Flush Water/Sediment Drain Valve.
- SBAC AS 6619 ISSUE 1. Valve.
- SBAC AS 6700 ISSUE 1. Assembly—Drain Valve.
- SBAC AS 6701 ISSUE 2. Body—Drain Valve.
- SBAC AS 8433 ISSUE 1. 1/4 Inch Bore Sediment Drain Valve, Unified.
- SBAC AS 8434 ISSUE 1. Body 1/4 Bore Drain Valve.

—Aircraft—Assembly
- SBAC AS 2560 ISSUE 5. Drain Valve Assembly 2 Inch. 1 p.
- SBAC AS 2573 ISSUE 6. Drain Valve Assembly 1 1/4 Inch. 1 p.
- SBAC AS 6270 ISSUE 2. Assembly of Flush Drain Valve—1 1/4 Inch Bore. 1 p.
- SBAC AS 6295 ISSUE 1. Assembly of Flush Drain Valve 2 Inch Bore. 1 p.
- SBAC AS 6360 ISSUE 3(I). Assembly—Drain Valve.
- SBAC AS 6618 ISSUE 1. Body Assembly.

—Aircraft—Cap Seals
- SBAC AS 6277 ISSUE 3. Cap, Seal. 1 p.
- SBAC AS 6301 ISSUE 3. Cap, Seal. 1 p.

—Aircraft—Caps (Lids)
- SBAC AS 2565 ISSUE 5. Screw Cap. 1 p.
- SBAC AS 2572 ISSUE 1. Cap Assembly. 1 p.
- SBAC AS 2575 ISSUE 4. Screw Cap. 1 p.
- SBAC AS 4762 ISSUE 1. Cap 1/4 Inch Bore Drain Valve. 1 p.
- SBAC AS 6274 ISSUE 3. Valve Cap. 1 p.
- SBAC AS 6299 ISSUE 3. Valve Cap. 2 pp.
- SBAC AS 8435 ISSUE 1. Cap 1/4 Bore Drain Valve.

—Aircraft—Chain Clips
- SBAC AS 3339 ISSUE 2. Chain Clip 1/4 Bore Drain Hole. 1 p.
- SBAC AS 8437 ISSUE 1. Chain, Clip 1/4 Inch Bore Drain Valve.

—Aircraft—Connections
- SBAC AS 6289 ISSUE 1. Tank Drainage Connection Drain Valve AS 6270, 1 1/4 Inch. 1 p.
- SBAC AS 6313 ISSUE 1. Tank Drainage Connection Drain Valve AS 6295, 2 Inches. 1 p.

—Aircraft—Guides
- SBAC AS 403 ISSUE 5(I). Valve Guide. 1 p.
- SBAC AS 2562 ISSUE 4. Valve Guide. 1 p.
- SBAC AS 6273 ISSUE 2. Valve Guide. 1 p.
- SBAC AS 6298 ISSUE 2. Valve Guide. 1 p.

—Aircraft—Guides—Bushings
- SBAC AS 2563 ISSUE 2. Bush for Valve Guide. 1 p.

—Aircraft—Helical Springs
- SBAC AS 6276 ISSUE 1. Cap, Spring. 1 p.
- SBAC AS 6300 ISSUE 1. Valve Spring. 1 p.
- SBAC AS 6366 ISSUE 3. Spring—Cap—Oil Drain Valve.

—Aircraft—Nuts (Fasteners)
- SBAC AS 3302 ISSUE 1. Union, Nut. 1 p.
- SBAC AS 3303 ISSUE 1. Release Nut. 1 p.

—Aircraft—Oil
- SBAC AS 6362 ISSUE 4. Valve—Oil Drain Valve.

—Aircraft—Rings
- SBAC AS 2800 (V). Tapped Ring for Drain Valve AS 2560. 1 p.
- SBAC AS 2801 (V). Tapped Ring for Drain Valve AS 2573. 2 pp.
- SBAC AS 6279-6286 (V). Tapped, Ring for Drain Valve. 2 pp.
- SBAC AS 6287 (V). Tapped Ring for Drain Valve. 2 pp.

Drain Valves (Cont.)

—Aircraft—Rings (Cont.)
SBAC AS 6303-6310 (V). Tapped, Ring for Drain Valve AS 6295. 2 pp.
SBAC AS 6311 (V). Tapped Ring for Drain Valve A.S.6295. 2 pp.

—Aircraft—Rings—Stampings
SBAC AS 6288 ISSUE 2. Stamping Tapped Ring for Drain Valve AS6270. 1 p.
SBAC AS 6312 ISSUE 2. Stamping Tapped Ring for Drain Valve AS 6295. 1 p.

—Aircraft—Seals
SBAC AS 4588 ISSUE 2. Seal. 1 p.
SBAC AS 6278 ISSUE 3. Valve, Seal. 1 p.
SBAC AS 6302 ISSUE 3. Valve Seal. 1 p.
SBAC AS 8436 ISSUE 1. Seal, 1/4 Inch Bore Drain Valve.

—Aircraft—Spindles
SBAC AS 3332 ISSUE 3. Valve, Spindle 1/4 Inch Bore Drain Valve. 1 p.
SBAC AS 3333 ISSUE 2. Insert 1/4 Inch Bore Drain Valve. 1 p.

—Aircraft—Springs
SBAC AS 4524 ISSUE 2. Spring—1/4 Inch Bore Drain Valve. 1 p.
SBAC AS 6275 ISSUE 1. Valve Spring. 1 p.

—Aircraft—Stops
SBAC AS 3304 ISSUE 1. Stop for Drain Valve. 1 p.

—Aircraft—Tanks (Containers)
SBAC AS 2580 ISSUE 5. Tank Drainage Connection—Drain Valve 2 Inch. 1 p.
SBAC AS 2584 ISSUE 5. Tank Drainage Connection—Drain Valve 1 1/4 Inch. 1 p.

—Aircraft—Washers (Fasteners)
SBAC AS 2567 ISSUE 3. Joint, Washer. 1 p.
SBAC AS 3337 ISSUE 2. Washer 1/4 Inch Bore Drain Valve. 1 p.

—Aircraft—Washers—Chain Tags
SBAC AS 3338 ISSUE 2. Chain Tag, Washer—1/4 Inch Bore Drain Valve Drain Valve. 1 p.

—Copper Alloy
SAA AS 1718-92. Water Supply—Copper Alloy Screw-Down Pattern Taps—Specified by Dimensions (in Professional Packages 50E, 61A, 61B). 34 pp.

—Marine—Fuel Tanks
CNS F3200-82. Marine Fuel Oil Tank Self-Closing Drain Valves (Oct)(9483).
JIS F 7398-89. Marine Fuel Oil Tank Self-Closing Drain Valves.

Drainage

Use For: Drainage Testing *See Also:* Irrigation; Soakaways; Subsurface Drainage; Surface Water; Tunneling (Excavations)
SAA MP52-93. Manual of Authorization Procedures for Plumbing and Drainage Products (in Professional Package 61A, 61B). 86 pp.

—Headboxes
CPPA D.1U-77. Drainage Test for Paper-Machine Headbox Stock. 1 p.

—Mining Equipment—Glossaries
BSI BS 3618: Sec 4-71. 1971 Glossary of Mining Terms Section 4: Drainage. 12 pp.

—Regulations—Models
SNZ NZS 1711-62. Model Land Drainage Bylaw Amend: 1, 1972. 11 pp.

Drainage Equipment

—Residential Buildings
CNS A3106-80. Method of Test for Drainage of Equipment Units for Dwellings (Aug)(5957).
JIS A 1705-76. Method of Test for Drainage of Equipment Units for Dwellings.

Drainage Pipes
Use: Drainpipes

Drainage Systems

See Also: Drainpipes; Drains; Flap Valves; Manholes
CEN PREN 752-2-92. Drain and Sewer Systems Outside Buildings—Part 2: Performance Requirements. 14 pp.

—Bathtubs
CEN PREN 128-89. Performance Criteria and Requirements for Sanitary Appliances. 4 pp.

—Bidets
CEN PREN 128-89. Performance Criteria and Requirements for Sanitary Appliances. 4 pp.

Drainage Systems (Cont.)

—Construction
DIN ENGL 1986 Pt 1-88. Site Drainage Systems; Principles, Design and Installation (June). 27 pp.

—Construction Contracts
DIN ENGL 18306-92. Construction Contract Procedures (VOB); Part C: General Technical Specifications in Construction Contracts (ATV); Underground Drainage (Dec). 4 pp.

—Design
CEN PREN 752-3-92. Drain and Sewer Systems Outside Buildings—Part 3: Planning. 19 pp.
CEN PREN 752-4-93. Drain and Sewer Systems Outside Buildings—Part 4: Hydraulic Design and Environmental Considerations. 27 pp.

—Dishwashers
CEN PREN 128-89. Performance Criteria and Requirements for Sanitary Appliances. 4 pp.

—Floor Drains
CEN PREN 128-89. Performance Criteria and Requirements for Sanitary Appliances. 4 pp.

—Glossaries
CEN PREN 752-1-92. Drain and Sewer Systems Outside Buildings—Part 1: General. 8 pp.

—Installation
DIN ENGL 1986 Pt 1-88. Site Drainage Systems; Principles, Design and Installation (June). 27 pp.

—Maintenance
DIN ENGL 1986 Pt 1-88. Site Drainage Systems; Principles, Design and Installation (June). 27 pp.
DIN ENGL 1986 Pt 3-82. Drainage and Sewerage Systems for Buildings and Plots of Land; Rules for Service and Maintenance (July). 4 pp.
DIN ENGL 1986 Pt 30-87. Site Drainage Systems; Maintenance (June). 2 pp.

—Ships
MOD UK NES 344-87. Feed Steam and Drain Systems in Ships Fitted with Auxiliary Boilers Issue 2 (11.87). 59 pp.
MOD UK NES 717-84. Requirements for Bilge, Sullage and Drain Tank Systems for Surface Ships Issue 2 (05.84). 40 pp.

—Shower Bases
CEN PREN 128-89. Performance Criteria and Requirements for Sanitary Appliances. 4 pp.

—Sinks
CEN PREN 128-89. Performance Criteria and Requirements for Sanitary Appliances. 4 pp.
CEN PREN 411-90. Sanitary Tapware—Waste Fittings for Sinks—General Technical Specifications. 16 pp.

—Sinks—Kitchen
CEN PREN 128-89. Performance Criteria and Requirements for Sanitary Appliances. 4 pp.

—Terminals—Engineering Drawings
ISO 6412 Pt 3-93. Technical Drawings—Simplified Representation of Pipelines—Part 3: Terminal Features of Ventilation and Drainage Systems First Edition. 8 pp.

—Urinals
CEN PREN 128-89. Performance Criteria and Requirements for Sanitary Appliances. 4 pp.

—Wash Basins
CEN PREN 128-89. Performance Criteria and Requirements for Sanitary Appliances. 4 pp.

—Washing Machines
CEN PREN 128-89. Performance Criteria and Requirements for Sanitary Appliances. 4 pp.

Drainage Testing
Use: Drainage

Drainboards
Use: Drains

Drainers
Use: Drains

Drainingboards
Use: Drains

Drainpipe Fittings

Use For: Soil Pipe Fittings *See Also:* Pipe Fittings; Traps (Drains)
CEN PREN 476-91. General Requirements for Components Used in Discharge Pipes, Drains and Sewers for Gravity Systems. 14 pp.
CEN PREN 773-92. General Requirements for Components Used in Hydraulically Pressurized Discharge Pipes, Drains and Sewers. 9 pp.

Drainpipe Fittings (Cont.)

DIN ENGL 19550-87. General Requirements for Pipes and Fittings for Buried Drains and Sewers (Oct). 6 pp.
DIN ENGL 19550 Pt 2-90. General Requirements for Drainage Pipework Inside Buildings (May). 6 pp.

—ABS
CSA CAN/CSA-B181.1-M90. ABS Drain, Waste, and Vent Pipe and Pipe Fittings; (Gen Instr 1). 50 pp.
DIN ENGL 19561-80. Pipes and Fittings of Acrylnitrile-Butadiene-Styrene (ABS) or Acrylester-Syrene-Acrylnitrile (ASA) with Sliding Socket Joint for Hot Water Resistant Drainage System (HT) Inside Buildings; Dimensions, Technical Conditions of Delivery (Mar). 21 pp.

—ABS—Flanges
CSA CAN/CSA-B181.1-M90. ABS Drain, Waste, and Vent Pipe and Pipe Fittings; (Gen Instr 1). 50 pp.

—Aluminum
BSI BS 2997-58. 1958 Amd 1 Aluminium Rainwater Goods. 44 pp.
CSA CAN/CSA-B281-M90. Aluminum Drain, Waste, and Vent Pipe and Components; (Gen Instr 1). 27 pp.

—ASA
DIN ENGL 19561-80. Pipes and Fittings of Acrylnitrile-Butadiene-Styrene (ABS) or Acrylester-Syrene-Acrylnitrile (ASA) with Sliding Socket Joint for Hot Water Resistant Drainage System (HT) Inside Buildings; Dimensions, Technical Conditions of Delivery (Mar). 21 pp.

—Asbestos Cement
BSI BS 569-73. 1973 Amd 1 Asbestos-Cement Rainwater Goods (AMD 5755) October 31, 1988. 35 pp.
BSI BS 3656-81. 1981 Amd 1 Asbestos-Cement Pipes, Joints and Fittings for Sewerage and Drainage. 13 pp.
CSA B127.1-M1977. Components for Use in Asbestos Cement Drain, Waste and Vent Systems; (Amd 1-3 March 1988) (Amd 4-11 December 1990). 23 pp.
CSA B127.11-M1982. Recommended Practice for the Installation of Asbestos Cement Drain, Waste, and Vent Pipe and Pipe Fittings; (Gen Instr 1). 15 pp.
ISO 881-80. Asbestos-Cement Pipes, Joints and Fittings for Sewerage and Drainage First Edition. 13 pp.

—Bathtubs
BSI BS 3380-82. 1982 Wastes (Excluding Skeleton Sink Wastes) and Bath Overflows. 12 pp.
BSI BS EN 274-93. 1993 Sanitary Tapware—Waste Fittings for Basins, Bidets and Baths—General Technical Specifications. 22 pp.
CEN PREN 274-88. Sanitary Tapware: Waste Fittings for Basins Bidets and Baths; General Technical Specifications. 29 pp.
CEN EN 274-92. Sanitary Tapware—Waste Fittings for Basins, Bidets and Baths—General Technical Specifications. 28 pp.
CEN EN 274-92. Sanitary Tapware—Waste Fittings for Basins, Bidets and Baths—General Technical Specifications. 17 pp.
JIS A 5711-78. Waste Fittings for Bathtubs (R 1983). 10 pp.

—Bell and Spigot—Vitreous Clay
CSA A60.1-M1976. Vitrified Clay Pipe (R 1992). 13 pp.

—Bends—Concrete
BSI BS 5911: Part 3-82. (WITHDRAWN) 1982 Precast Concrete Pipes and Fittings for Drainage and Sewerage Part 3: Ogee Jointed Concrete Pipes, Bends and Junctions, Unreinforced or Reinforced with Steel Cages or Hoops (Superseded by BS 5911: Part 110: 1992). 19 pp.

—Bends—Lead
DIN ENGL 1263-66. Lead Waste Pipes and Bends for Drainage Systems (Sept). 1 p.

—Bidets
BSI BS 3380-82. 1982 Wastes (Excluding Skeleton Sink Wastes) and Bath Overflows. 12 pp.
BSI BS EN 274-93. 1993 Sanitary Tapware—Waste Fittings for Basins, Bidets and Baths—General Technical Specifications. 22 pp.
CEN PREN 274-88. Sanitary Tapware: Waste Fittings for Basins Bidets and Baths; General Technical Specifications. 29 pp.
CEN EN 274-92. Sanitary Tapware—Waste Fittings for Basins, Bidets and Baths—General Technical Specifications. 28 pp.
CEN EN 274-92. Sanitary Tapware—Waste Fittings for Basins, Bidets and Baths—General Technical Specifications. 17 pp.

—Bituminized Fiber
CNS A2150-89. Fittings for Bituminized Fiber Drain and Sewer Pipe (Jul)(10045).

INTERNATIONAL AND NON-U.S. NATIONAL STANDARDS
SUBJECT INDEX
Drainpipe

Drainpipe Fittings *(Cont.)*

—Cast Iron
BSI BS 416: Part 2-90. 1990 Discharge and Ventilating Pipes and Fittings, Sand-Cast or Spun in Cast Iron Part 2: Specification for Socketless Systems. 30 pp.
BSI BS 437-78. 1978 Amd 1 Cast Iron Spigot and Socket Drain Pipes and Fittings. 42 pp.
BSI BS 460-64. 1964 Cast Iron Rainwater Goods. 29 pp.
BSI BS 6087-90. 1990 Amd 1 Flexible Joints for Grey or Ductile Cast Iron Drainpipes and Fittings (BS 437) and for Discharge and Ventilating Pipes and Fittings (BS 416) (AMD 6357) June 28, 1991 (R). 9 pp.
BSI BS 6087-81. (WITHDRAWN) 1981 Amd 1 Flexible Joints for Cast Iron Drainpipes and Fittings (BS 437) and for Cast Iron Soil, Waste and Ventilationg Pipes and Fittings (BS 416) (R). 8 pp.
CSA CAN/CSA-B70-M91. Cast Iron Soil Pipe, Fittings, and Means of Joining; (Gen Instr 1). 27 pp.
JIS G 5525-75. Cast-Iron Soil Pipes and Fittings. 22 pp.

—Cast Iron—Compression Testing
CSA CAN/CSA-B70-M91. Cast Iron Soil Pipe, Fittings, and Means of Joining; (Gen Instr 1). 27 pp.

—Cast Iron—Identification Systems
CSA CAN/CSA-B70-M91. Cast Iron Soil Pipe, Fittings, and Means of Joining; (Gen Instr 1). 27 pp.

—Cast Iron—Pressure Measurement
CSA CAN/CSA-B70-M91. Cast Iron Soil Pipe, Fittings, and Means of Joining; (Gen Instr 1). 27 pp.

—Ceramic
SNZ NZS 3302-83. Specification for Ceramic Pipes, Fittings and Joints (Reconfirmed 1991). 27 pp.

—Concrete
BSI BS 5911: Part 1-81. 1981 Amd 1 Precast Concrete Pipes and Fittings for Drainage and Sewerage Part 1: Specification for Pipes and Fittings with Flexible Joints and Manholes (AMD 4035) September 30, 1982. 42 pp.
BSI BS 5911: Part 100-88. 1988 Amd 1 Precast Concrete Pipes and Fittings for Drainage and Sewerage Part 100: Unreinforced and Reinforced Pipes and Fittings with Flexible Joints (AMD 6269) December 22, 1989. 39 pp.
BSI BS 5911: Part 100-02. 1988 Amd 2 Precast Concrete Pipes, Fittings and Ancillary Products Part 100: Specification for Unreinforced and Reinforced Pipes and Fittings with Flexible Joints (AMD 7588) April 15, 1993 (R). 40 pp.
BSI BS 5911: Part 101-88. 1988 Precast Concrete Pipes and Fittings for Drainage and Sewerage Part 101: Glass Composite Concrete (GCC) Pipes and Fittings with Flexible Joints. 32 pp.
BSI BS 5911: Part 110-92. 1992 Precast Concrete Pipes, Fittings and Ancillary Products Part 110: Specification for Ogee Pipes and Fittings (Including Perforated). 34 pp.

—Copper
BSI BS 1431-60. 1960 Wrought Copper and Wrought Zinc Rainwater Goods. 25 pp.

—Ductile Iron
ISO 7186-83. Ductile Iron Pipes and Accessories for Non-Pressure Pipe-Lines First Edition. 11 pp.

—Fiber Cement
CEN PREN 588-1-91. Fibre-Cement Pipes for Sewer and Drains—Part 1: Pipes, Joints and Fittings for Gravity Systems. 39 pp.

—Fused—Plastic
CSA CAN/CSA-B181.3-M86. Polyolefin Laboratory Drainage Systems (R 1992); (Gen Instr 1). 19 pp.

—Pitch Fiber
BSI BS 2760-73. 1973 Amd 1 Pitch-Impregnated Fibre Pipes and Fittings for Below and Above Ground Drainage (AMD 2675) June 30, 1978. 27 pp.

—Plain—Vitreous Clay
CSA A60.1-M1976. Vitrified Clay Pipe (R 1992). 13 pp.

—Plastic
BSI BS 4576: Part 1-89. 1989 Amd 1 Specification for Unplasticized Polyvinyl Chloride (PVC-U) Rainwater Goods and Accessories Part 1: Half-Round Gutters and Pipes of Circular Cross Section (AMD 6350) June 28, 1991. 17 pp.
BSI BS 4576: Part 1-70. (WITHDRAWN) 1970 Amd 3 Unplasticized PVC Rainwater Goods Part 1: Half-Round Gutters and Circular Pipe. 17 pp.

Drainpipe Fittings *(Cont.)*

—Plastic *(Cont.)*
BSI BS 4660-89. 1989 Unplasticized Polyvinyl Chloride (PVC-V) Pipes and Plastics Fittings of Nominal Sizes 110 and 160 for Below Ground Gravity Drainage and Sewerage. 22 pp.
BSI BS 4660-73. (WITHDRAWN) 1973 Amd 5 Unplasticized PVC Underground Drain Pipe and Fittings. 25 pp.
CGSB 41-GP-29MA-83. Tubing, Plastic, Corrugated, Drainage, Standard for. 17 pp.
CSA B182.11-1967SP. Recommended Practice for the Installation of Plastic Drain and Sewer Pipe and Pipe Fittings; (Rev 1 November 1976). 13 pp.
ISO TR7074-86. Performance Requirements for Plastics Pipes and Fittings for Use in Underground Drainage and Sewage First Edition. 31 pp.
SNZ NZS 7604-81. Specification for High Density Polyethylene Drain and Sewer Pipe and Fittings. 26 pp.

—Polyester—Fiberglass Reinforced
DIN ENGL 19565 Pt 1-89. Centrifugally Cast and Filled Polyester Resin Glass Fibre Reinforced (UP-GF) Pipes and Fittings for Buried Drains and Sewers; Dimensions and Technical Delivery Conditions (Mar). 24 pp.

—Polyethylene
DIN ENGL 19535 Pt 1-88. High-Density Polyethylene (HDPE) Pipes and Fittings for Hot-Water Resistant Drains and Sewers Inside Buildings; Dimensions (Mar) (This Standard, Together with DIN 19535 Part 2, April 1987 Edition, Supersedes DIN 19535, March 1977 Edition, Withdrawn in Jan. 1986). 12 pp.
DIN ENGL 19535 Pt 2-87. High-Density Polyethylene (HDPE) Pipes and Fittings for Hot Water Resistant (HT) Drains and Sewers Inside Buildings; Technical Delivery Conditions (Apr) (Supersedes Parts of the March 1977 Edition of DIN 19535, Withdrawn in January 1986). 6 pp.
DIN ENGL 19537 Pt 1-83. High Density Polyethylene (HDPE) Pipes and Fittings for Drains and Sewers; Dimensions (Oct). 19 pp.
DIN ENGL 19537 Pt 2-88. High-Density Polyethylene (HDPE) Pipes and Fittings for Drains and Sewers; Technical Delivery Conditions (Jan). 8 pp.
ISO 8770-91. High-Density Polyethylene (PE-HD) Pipes and Fittings for Soil and Waste Discharge (Low and High Temperature) Systems Inside Buildings—Specifications First Edition. 21 pp.

—Polyethylene—Corrugated
CSA CAN/CSA-B182.6-M92. Profile Polyethylene Sewer Pipe and Fittings; (Gen Instr 1). 37 pp.

—Polypropylene
DIN ENGL 19560-80. Pipes and Fittings of Polypropylene (PP) with Sliding Socket Joint for Hot Water Resistant Drainage Systems (HT) Inside Buildings; Dimensions, Technical Conditions of Delivery (Mar). 20 pp.
ISO 7671-91. Polypropylene (PP) Pipes and Fittings (Jointed by Means of Elastomeric Sealing Rings) for Soil and Waste Discharge (Low and High Temperature) Systems Inside Buildings—Specifications First Edition. 18 pp.

—PVC
BSI BS 4514-83. 1983 Amd 1 Unplasticized PVC Soil and Ventilating Pipes, Fittings and Accessories. 17 pp.
BSI BS 5481-77. 1977 Amd 2 Unplasticized PVC Pipe and Fittings for Gravity Sewers. 15 pp.
CSA CAN/CSA-B181.2-M90. PVC Drain, Waste, and Vent Pipe and Pipe Fittings; (Gen Instr 1). 56 pp.
CSA CAN/CSA-B182.2-M90. PVC Sewer Pipe and Fittings (PSM Type); (Gen Instr 1 Thru 4). 57 pp.
CSA CAN/CSA-B182.4-92. Profile PVC Sewer Pipe and Fittings; (Gen Instr 1). 41 pp.
ISO 3633-91. Unplasticized Poly(Vinyl Chloride) (PVC-U) Pipes and Fittings for Soil and Waste Discharge (Low and High Temperature) Systems Inside Buildings—Specifications First Edition. 20 pp.
ISO 4435-91. Unplasticized Poly(Vinyl Chloride) (PVC-U) Pipes and Fittings for Buried Drainage and Sewerage Systems—Specifications First Edition. 17 pp.
ISO TR7073-88. Recommended Techniques for the Installation of Unplasticized Poly(Vinyl Chloride) (PVC-U) Buried Drains and Sewers First Edition. 21 pp.
JIS K 6739-77. Unplasticized Polyvinyl Chloride Pipe Fittings for Drain (R 1980). 17 pp.
SNZ NZS 7642-71. Specification for Unplasticized PVC Soil and Ventilating Pipe, Fittings and Accessories Amend: A, 1971; 1 & 1A, 1974; 2 & 2A, 1975. 28 pp.
SNZ NZS 7649-88. Unplasticized PVC Sewer and Drain Pipe and Fittings. 24 pp.

Drainpipe Fittings *(Cont.)*

—PVC—Flanges
CSA CAN/CSA-B181.2-M90. PVC Drain, Waste, and Vent Pipe and Pipe Fittings; (Gen Instr 1). 56 pp.

—Screwed—Cast Iron
JIS B 2303-77. Screwed Drainage Fittings. 21 pp.

—Screwed—Copper Zinc Alloy
CNS B2377-82. Screwed Sealing Plugs for Drain Fittings of Copper-Zinc Alloy (Oct)(4850).

—Screwed—Polyethylene
CNS B2378-82. Screwed Sealing Plugs for Drain Fittings of High Density PE (Oct)(4851).

—Showers
CEN PREN 329-90. Sanitary Tapware—Waste Fittings for Shower Trays—General Technical Specifications. 21 pp.

—Sinks
BSI BS 3380-82. 1982 Wastes (Excluding Skeleton Sink Wastes) and Bath Overflows. 12 pp.

—Socket—Cast Iron
BSI BS 416-73. (WITHDRAWN) 1973 Amd 1 Cast Iron Spigot and Socket Soil, Waste and Ventilating Pipes (Sand Cast and Spun) and Fittings (Superseded by BS 416: Parts 1&2: 1990). 38 pp.
BSI BS 416: Part 1-90. 1990 Discharge and Ventilating Pipes and Fittings, Sand-Cast or Spun in Cast Iron Part 1: Specification for Spigot and Socket Systems. 28 pp.

—Soldered—Brass
CSA B158.1-1976. Cast Brass Solder Joint Drainage, Waste and Vent Fittings (R 1992). 13 pp.

—Spigot—Cast Iron
BSI BS 416-73. (WITHDRAWN) 1973 Amd 1 Cast Iron Spigot and Socket Soil, Waste and Ventilating Pipes (Sand Cast and Spun) and Fittings (Superseded by BS 416: Parts 1&2: 1990). 38 pp.
BSI BS 416: Part 1-90. 1990 Discharge and Ventilating Pipes and Fittings, Sand-Cast or Spun in Cast Iron Part 1: Specification for Spigot and Socket Systems. 28 pp.
DIN ENGL 19522 Pt 1-83. Cast Iron Spigot (SML) Drainpipes and Fittings; Dimensions (Feb). 14 pp.
DIN ENGL 19522 Pt 2-83. Cast Iron Spigot (SML) Drainpipes and Fittings; Technical Delivery Conditions (Feb). 6 pp.
ISO 6594-83. Cast Iron Drainage Pipes and Fittings—Spigot Series First Edition; (Corrected and Reprinted -1985). 14 pp.

—Steel
BSI BS 1091-63. 1963 Pressed Steel Gutters, Rainwater Pipes, Fittings and Accessories. 20 pp.

—Toilet Facilities—Aircraft
MOD UK DSTAN 16-18-74. Aircraft Toilet Flushing, Chemical Recharging, and Draining Connections Issue 1. 7 pp.

—Vitreous Clay
BSI BS 65-91. 1991 Vitrified Clay Pipes, Fittings and Ducts, Also Flexible Mechanical Joints for Use Solely with Surface Water Pipes and Fittings. 22 pp.
BSI BS 65-88. (WITHDRAWN) 1988 Vitrified Clay Pipes, Fittings, Joints and Ducts (Superseded by BS 65: 1991, BS EN 295-1, 295-2 and 295-3: 1991). 20 pp.
BSI BS 65-81. (WITHDRAWN) 1981 Amd 2 Vitrified Clay Pipes, Fittings and Joints (AMD 4394) November 30, 1983. 27 pp.
BSI BS EN 295: Part 1-91. 1991 Vitrified Clay Pipes and Fittings and Pipe Joints for Drains and Sewers Part 1: Requirements. 17 pp.
BSI BS EN 295: Part 2-91. 1991 Vitrified Clay Pipes and Fittings and Pipe Joints for Drains and Serwers: Part 2: Quality Control and Sampling. 20 pp.
BSI BS EN 295: Part 3-91. 1991 Vitrified Clay Pipes and Fittings and Pipe Joints for Drains and Serwers: Part 3: Test Methods. 30 pp.
CEN PREN 295 (Part 1)-89. Vitrified Clay Pipes and Fittings and Pipe Joints for Drains and Sewers; Part 1: Specifications. 16 pp.
CEN EN 295-1-91. Vitrified Clay Pipes and Fittings and Pipe Joints for Drains and Sewers—Part 1: Requirements. 13 pp.
CEN PREN 295 (Part 2)-90. Vitrified Clay Pipes and Fittings and Pipe Joints for Drains and Sewers—Part 2: Quality Control and Sampling. 21 pp.
CEN EN 295-2-91. Vitrified Clay Pipes and Fittings and Pipe Joints for Drains and Sewers—Part 2: Quality Control and Sampling. 16 pp.
CEN PREN 295 (Part 3)-90. Vitrified Clay Pipes and Fittings and Pipe Joints for Drains and Sewers—Part 3: Test Methods. 38 pp.
CEN EN 295-3-91. Vitrified Clay Pipes and Fittings and Pipe Joints for Drains and Sewers—Part 3: Test Methods. 26 pp.

INDUSTRY STANDARDS

Drainpipe Fittings (Cont.)

—Vitreous Clay (Cont.)

CEN PREN 295-4-92. Vitrified Clay Pipes and Fittings and Pipe Joints for Drains and Sewers—Part 4: Requirements for Special Fittings, Adaptors and Compatible Accessories. 15 pp.

CEN PREN 295-5-92. Vitrified Clay Pipes and Fittings and Pipe Joints for Drains and Sewers—Part 5: Requirements for Perforated Vitrified Clay Pipes and Fittings. 9 pp.

CSA A60.1-M1976. Vitrified Clay Pipe (R 1992). 13 pp.

—Wash Basins

BSI BS 3380-82. 1982 Wastes (Excluding Skeleton Sink Wastes) and Bath Overflows. 12 pp.

BSI BS EN 274-93. 1993 Sanitary Tapware—Waste Fittings for Basins, Bidets and Baths —General Technical Specifications. 22 pp.

CEN PREN 274-88. Sanitary Tapware: Waste Fittings for Basins Bidets and Baths; General Technical Specifications. 29 pp.

CEN EN 274-92. Sanitary Tapware—Waste Fittings for Basins, Bidets and Baths—General Technical Specifications. 28 pp.

CEN EN 274-92. Sanitary Tapware—Waste Fittings for Basins, Bidets and Baths—General Technical Specifications. 17 pp.

—Zinc

BSI BS 1431-60. 1960 Wrought Copper and Wrought Zinc Rainwater Goods. 25 pp.

Drainpipe Joints

See Also: Pipe Joints

CEN PREN 476-91. General Requirements for Components Used in Discharge Pipes, Drains and Sewers for Gravity Systems. 14 pp.

CSA CAN/CSA-B602-M90. Mechanical Couplings for Drain, Waste, and Vent Pipe and Sewer Pipe; (Gen Instr 1). 21 pp.

DIN ENGL 19543-82. General Requirements for Pipe Joints in Sewers and Drain Pipes (Aug). 4 pp.

—Asbestos Cement

BSI BS 3656-81. 1981 Amd 1 Asbestos-Cement Pipes, Joints and Fittings for Sewerage and Drainage. 13 pp.

ISO 881-80. Asbestos-Cement Pipes, Joints and Fittings for Sewerage and Drainage First Edition. 13 pp.

ISO 4488-79. Asbestos-Cement Pipes and Joints for Thrust-Boring and Pipe Jacking First Edition. 13 pp.

—Cast Iron

BSI BS 6087-90. 1990 Amd 1 Flexible Joints for Grey or Ductile Cast Iron Drainpipes and Fittings (BS 437) and for Discharge and Ventilating Pipes and Fittings (BS 416) (AMD 6357) June 28, 1991 (R). 9 pp.

BSI BS 6087-81. (WITHDRAWN) 1981 Amd 1 Flexible Joints for Cast Iron Drainpipes and Fittings (BS 437) and for Cast Iron Soil, Waste and Ventilationg Pipes and Fittings (BS 416) (R). 8 pp.

—Ceramic

SNZ NZS 3302-83. Specification for Ceramic Pipes, Fittings and Joints (Reconfirmed 1991). 27 pp.

—Clay

BSI BS 1196-89. 1989 Clayware Field Drain Pipes and Juntions. 8 pp.

BSI BS 1196-71. 1971 Clayware Field Drain Pipes. 8 pp.

SNZ BS 1196-89. Specification for Clayware Field Drain Pipes and Junctions. 4 pp.

—Concrete

BSI BS 5911: Part 100-88. 1988 Amd 1 Precast Concrete Pipes and Fittings for Drainage and Sewerage Part 100: Unreinforced and Reinforced Pipes and Fittings with Flexible Joints (AMD 6269) December 22, 1989. 39 pp.

BSI BS 5911: Part 100-02. 1988 Amd 2 Precast Concrete Pipes, Fittings and Ancillary Products Part 100: Specification for Unreinforced and Reinforced Pipes and Fittings with Flexible Joints (AMD 7588) April 15, 1993 (R). 40 pp.

BSI BS 5911: Part 101-88. 1988 Precast Concrete Pipes and Fittings for Drainage and Sewerage Part 101: Glass Composite Concrete (GCC) Pipes and Fittings with Flexible Joints. 32 pp.

BSI BS 5911: Part 120-89. 1989 Precast Concrete Pipes and Fittings for Drainage and Sewerage Part 120: Reinforced Jacking Pipes with Flexible Joints. 25 pp.

—Fiber Cement

CEN PREN 588-1-91. Fibre-Cement Pipes for Sewer and Drains—Part 1: Pipes, Joints and Fittings for Gravity Systems. 39 pp.

Drainpipe Joints (Cont.)

—Ogee—Concrete

BSI BS 5911: Part 3-82. (WITHDRAWN) 1982 Precast Concrete Pipes and Fittings for Drainage and Sewerage Part 3: Ogee Jointed Concrete Pipes, Bends and Junctions, Unreinforced or Reinforced with Steel Cages or Hoops (Superseded by BS 5911: Part 110: 1992). 19 pp.

—Polyethylene

ISO 8772-91. High-Density Polyethylene (PE-HD) Pipes and Fittings for Buried Drainage and Sewerage Systems—Specifications First Edition. 18 pp.

—Polypropylene

ISO 8773-91. Polypropylene (PP) Pipes and Fittings for Buried Drainage and Sewerage Systems—Specifications First Edition. 18 pp.

—Seals

SAA AS 1646-92. Elastomeric Seals for Waterworks Purposes and Drainage Purposes Amdt 1 March 1993 Amdt 2 October 1993. 21 pp.

—Vitreous Clay

BSI BS 65-91. 1991 Vitrified Clay Pipes, Fittings and Ducts, Also Flexible Mechanical Joints for Use Solely with Surface Water Pipes and Fittings. 22 pp.

BSI BS 65-88. (WITHDRAWN) 1988 Vitrified Clay Pipes, Fittings, Joints and Ducts (Superseded by BS 65: 1991, BS EN 295-1, 295-2 and 295-3: 1991). 20 pp.

BSI BS 65-81. (WITHDRAWN) 1981 Amd 2 Vitrified Clay Pipes, Fittings and Joints (AMD 4394) November 30, 1983. 27 pp.

BSI BS EN 295: Part 1-91. 1991 Vitrified Clay Pipes and Fittings and Pipe Joints for Drains and Sewers Part 1: Requirements. 17 pp.

BSI BS EN 295: Part 2-91. 1991 Vitrified Clay Pipes and Fittings and Pipe Joints for Drains and Serwers: Part 2: Quality Control and Sampling. 20 pp.

BSI BS EN 295: Part 3-91. 1991 Vitrified Clay Pipes and Fittings and Pipe Joints for Drains and Serwers: Part 3: Test Methods. 30 pp.

CEN PREN 295 (Part 1)-89. Vitrified Clay Pipes and Fittings and Pipe Joints for Drains and Sewers; Part 1: Specifications. 16 pp.

CEN EN 295-1-91. Vitrified Clay Pipes and Fittings and Pipe Joints for Drains and Sewers—Part 1: Requirements. 13 pp.

CEN PREN 295 (Part 2)-90. Vitrified Clay Pipes and Fittings and Pipe Joints for Drains and Sewers—Part 2: Quality Control and Sampling. 21 pp.

CEN EN 295-2-91. Vitrified Clay Pipes and Fittings and Pipe Joints for Drains and Sewers—Part 2: Quality Control and Sampling. 16 pp.

CEN PREN 295 (Part 3)-90. Vitrified Clay Pipes and Fittings and Pipe Joints for Drains and Sewers—Part 3: Test Methods. 38 pp.

CEN EN 295-3-91. Vitrified Clay Pipes and Fittings and Pipe Joints for Drains and Sewers—Part 3: Test Methods. 26 pp.

CEN PREN 295-5-92. Vitrified Clay Pipes and Fittings and Pipe Joints for Drains and Sewers—Part 5: Requirements for Perforated Vitrified Clay Pipes and Fittings. 9 pp.

Drainpipes

Use For: Rainwater Pipes; Soil Pipes
See Also: Culverts (Pipes); Drainage Systems; Drains; Floor Drains; Pipes; Traps (Drains)

CEN PREN 476-91. General Requirements for Components Used in Discharge Pipes, Drains and Sewers for Gravity Systems. 14 pp.

CEN PREN 773-92. General Requirements for Components Used in Hydraulically Pressurized Discharge Pipes, Drains and Sewers. 9 pp.

DIN ENGL 19550-87. General Requirements for Pipes and Fittings for Buried Drains and Sewers (Oct). 6 pp.

DIN ENGL 19550 Pt 2-90. General Requirements for Drainage Pipework Inside Buildings (May). 6 pp.

—ABS

CSA CAN/CSA-B181.1-M90. ABS Drain, Waste, and Vent Pipe and Pipe Fittings; (Gen Instr 1). 50 pp.

DIN ENGL 19561-80. Pipes and Fittings of Acrylnitrile-Butadiene-Styrene (ABS) or Acrylester-Syrene-Acrylnitrile (ASA) with Sliding Socket Joint for Hot Water Resistant Drainage System (HT) Inside Buildings; Dimensions, Technical Conditions of Delivery (Mar). 21 pp.

—Agricultural—Construction Contracts

DIN ENGL 18308-92. Construction Contract Procedures (VOB), Part C: General Technical Specifications in Construction Contracts (ATV); Land Drainage (Dec). 4 pp.

—Aluminum

BSI BS 2997-58. 1958 Amd 1 Aluminium Rainwater Goods. 44 pp.

Drainpipes (Cont.)

—Aluminum (Cont.)

CSA CAN/CSA-B281-M90. Aluminum Drain, Waste, and Vent Pipe and Components; (Gen Instr 1). 27 pp.

—Asbestos Cement

BSI BS 569-73. 1973 Amd 1 Asbestos-Cement Rainwater Goods (AMD 5755) October 31, 1988. 35 pp.

BSI BS 3656-81. 1981 Amd 1 Asbestos-Cement Pipes, Joints and Fittings for Sewerage and Drainage. 13 pp.

BSI BS 4624-81. 1981 Amd 1 Asbestos-Cement Building Products. 13 pp.

BSI BS 4624-02. 1981 Amd 2 Methods of Test for Asbestos-Cement Building Products (AMD 7307) October 15, 1992. 14 pp.

CGSB CAN/CGSB-34.22-M87. Pipe Asbestos-Cement, Drain; (Corrigendum). 12 pp.

CGSB CAN/CGSB-34.23-M87. Pipe, Asbestos-Cement, Sewer, House Connection. 11 pp.

CSA B127.1-M1977. Components for Use in Asbestos Cement Drain, Waste and Vent Systems; (Amd 1-3 March 1988) (Amd 4-11 December 1990). 23 pp.

CSA B127.11-M1982. Recommended Practice for the Installation of Asbestos Cement Drain, Waste, and Vent Pipe and Pipe Fittings; (Gen Instr 1). 15 pp.

ISO 881-80. Asbestos-Cement Pipes, Joints and Fittings for Sewerage and Drainage First Edition. 13 pp.

ISO 4488-79. Asbestos-Cement Pipes and Joints for Thrust-Boring and Pipe Jacking First Edition. 13 pp.

—Bituminized Fiber

CNS A2146-89. Homogeneous Bituminized Fiber Drain and Sewer Pipe (Jul)(10039).

CNS A2148-89. Laminated-Wall Bituminized Fiber Drain and Sewer Pipe (Jul)(10042).

—Bituminized Fiber—Perforated

CNS A2147-89. Perforated Homogeneous Bituminized Fiber Pipe for General Drainage (Jul)(10040).

CNS A2149-89. Perforated, Laminated-Wall Bituminized Fiber Pipe for General Drainage (Jul)(10043).

—Buildings

BSI BS 8000: Part 13-89. 1989 Workmanship on Building Sites Part 13: Code of Practice for Above Ground Drainage and Sanitary Appliances. 15 pp.

—Buildings—Structural Design

BSI BS 8301-85. 1985 Amd 2 Code of Practice for Building Drainage (AMD 6580) March 28, 1991. 70 pp.

BSI BS 8301: COMMENTARY-90. 1990 Commentary on BS 8301: 1985—British Standard Code of Practice for "Building Drainage" (Formerly CP301: 1971) for the Users of Vitrified Clay Pipes (R). 16 pp.

—Cast Iron

BSI BS 416-73. (WITHDRAWN) 1973 Amd 1 Cast Iron Spigot and Socket Soil, Waste and Ventilating Pipes (Sand Cast and Spun) and Fittings (Superseded by BS 416: Parts 1&2: 1990). 38 pp.

BSI BS 416: Part 1-90. 1990 Discharge and Ventilating Pipes and Fittings, Sand-Cast or Spun in Cast Iron Part 1: Specification for Spigot and Socket Systems. 28 pp.

BSI BS 416: Part 2-90. 1990 Discharge and Ventilating Pipes and Fittings, Sand-Cast or Spun in Cast Iron Part 2: Specification for Socketless Systems. 30 pp.

BSI BS 437-78. 1978 Amd 1 Cast Iron Spigot and Socket Drain Pipes and Fittings. 42 pp.

BSI BS 460-64. 1964 Cast Iron Rainwater Goods. 29 pp.

BSI BS 6087-90. 1990 Amd 1 Flexible Joints for Grey or Ductile Cast Iron Drainpipes and Fittings (BS 437) and for Discharge and Ventilating Pipes and Fittings (BS 416) (AMD 6357) June 28, 1991 (R). 9 pp.

BSI BS 6087-81. (WITHDRAWN) 1981 Amd 1 Flexible Joints for Cast Iron Drainpipes and Fittings (BS 437) and for Cast Iron Soil, Waste and Ventilationg Pipes and Fittings (BS 416) (R). 8 pp.

CSA CAN/CSA-B70-M91. Cast Iron Soil Pipe, Fittings, and Means of Joining; (Gen Instr 1). 27 pp.

DIN ENGL 19522 Pt 1-83. Cast Iron Spigot (SML) Drainpipes and Fittings; Dimensions (Feb). 14 pp.

DIN ENGL 19522 Pt 2-83. Cast Iron Spigot (SML) Drainpipes and Fittings; Technical Delivery Conditions (Feb). 6 pp.

ISO 6594-83. Cast Iron Drainage Pipes and Fittings—Spigot Series First Edition; (Corrected and Reprinted -1985). 14 pp.

JIS G 5525-75. Cast-Iron Soil Pipes and Fittings. 22 pp.

INTERNATIONAL AND NON-U.S. NATIONAL STANDARDS
SUBJECT INDEX
Drainpipes

Drainpipes *(Cont.)*

—**Cast Iron—Compression Testing**
CSA CAN/CSA-B70-M91. Cast Iron Soil Pipe, Fittings, and Means of Joining; (Gen Instr 1). 27 pp.

—**Cast Iron—Identification Systems**
CSA CAN/CSA-B70-M91. Cast Iron Soil Pipe, Fittings, and Means of Joining; (Gen Instr 1). 27 pp.

—**Cast Iron—Pressure Measurement**
CSA CAN/CSA-B70-M91. Cast Iron Soil Pipe, Fittings, and Means of Joining; (Gen Instr 1). 27 pp.

—**Ceramic**
SNZ NZS 3302-83. Specification for Ceramic Pipes, Fittings and Joints (Reconfirmed 1991). 27 pp.

—**Clay**
BSI BS 1196-89. 1989 Clayware Field Drain Pipes and Juntions. 8 pp.
BSI BS 1196-71. 1971 Clayware Field Drain Pipes. 8 pp.
SNZ BS 1196-89. Specification for Clayware Field Drain Pipes and Junctions. 4 pp.

—**Cleaning Agents**
CGSB CAN/CGSB-2.49-M87. Alkali Drain Cleaner. 10 pp.

—**Concrete**
BSI BS 1194-69. (WITHDRAWN) 1969 Concrete Porous Pipes for Under-Drainage (Superseded by BS 5911: Part 114: 1992). 24 pp.
BSI BS 5178-75. 1975 Prestressed Concrete Pipes for Drainage and Sewage. 17 pp.
BSI BS 5911: Part 1-81. 1981 Amd 1 Precast Concrete Pipes and Fittings for Drainage and Sewerage Part 1: Specification for Pipes and Fittings with Flexible Joints and Manholes (AMD 4035) September 30, 1982. 42 pp.
BSI BS 5911: Part 3-82. (WITHDRAWN) 1982 Precast Concrete Pipes and Fittings for Drainage and Sewerage Part 3: Ogee Jointed Concrete Pipes, Bends and Junctions, Unreinforced or Reinforced with Steel Cages or Hoops (Superseded by BSI 5911: Part 110: 1992). 19 pp.
BSI BS 5911: Part 100-88. 1988 Amd 1 Precast Concrete Pipes and Fittings for Drainage and Sewerage Part 100: Unreinforced and Reinforced Pipes and Fittings with Flexible Joints (AMD 6269) December 22, 1989. 39 pp.
BSI BS 5911: Part 100-02. 1988 Amd 2 Precast Concrete Pipes, Fittings and Ancillary Products Part 100: Specification for Unreinforced and Reinforced Pipes and Fittings with Flexible Joints (AMD 7588) April 15, 1993 (R). 40 pp.
BSI BS 5911: Part 101-88. 1988 Precast Concrete Pipes and Fittings for Drainage and Sewerage Part 101: Glass Composite Concrete (GCC) Pipes and Fittings with Flexible Joints. 32 pp.
BSI BS 5911: Part 110-92. 1992 Precast Concrete Pipes, Fittings and Ancillary Products Part 110: Specification for Ogee Pipes and Fittings (Including Perforated). 34 pp.
BSI BS 5911: Part 120-89. 1989 Precast Concrete Pipes and Fittings for Drainage and Sewerage Part 120: Reinforced Jacking Pipes with Flexible Joints. 25 pp.
SNZ NZS 3107-78. Specification for Precast Concrete Drainage and Pressure Pipes. 35 pp.

—**Construction Contracts**
DIN ENGL 18308-92. Construction Contract Procedures (VOB); Part C: General Technical Specifications in Construction Contracts (ATV); Land Drainage (Dec). 4 pp.

—**Copper**
BSI BS 1431-60. 1960 Wrought Copper and Wrought Zinc Rainwater Goods. 25 pp.
SAA AS 1432-90. Copper Tubes for Plumbing, Gasfitting and Drainage Applications. 9 pp.

—**Design**
BSI BS 6367-83. 1983 Amd 1 Drainage of Roofs and Paved Areas. 58 pp.
DIN ENGL 1986 Pt 1-88. Site Drainage Systems; Principles, Design and Installation (June). 27 pp.
DIN ENGL 18460-89. External Rainwater Pipes and Eaves Gutters; Concepts and Design Principles (May). 3 pp.

—**Ductile Iron**
ISO 7186-83. Ductile Iron Pipes and Accessories for Non-Pressure Pipe-Lines First Edition. 11 pp.

—**Elastomer Sealants**
DIN ENGL 4060-88. Elastomer Seals for Pipe Joints in Drains and Sewers; Requirements and Testing (Dec). 10 pp.

Drainpipes *(Cont.)*

—**Engineering Drawings—Symbols**
ISO 4067 Pt 1-84. Technical Drawings—Installations—Part 1: Graphical Symbols for Plumbing, Heating, Ventilation and Ducting First Edition. 12 pp.

—**Fiber Cement**
CEN PREN 588-1-91. Fibre-Cement Pipes for Sewer and Drains—Part 1: Pipes, Joints and Fittings for Gravity Systems. 39 pp.

—**Galvanized Steel**
BSI BS 3868-73. 1973 Prefabricated Drainage Stack Units: Galvanized Steel. 19 pp.

—**Glossaries**
BSI BS 4118-81. (WITHDRAWN) 1981 Amd 2 Glossary of Sanitation Terms (Superseded by BS 6100: Section 3.3: 1991). 41 pp.

—**Installation**
SNZ NZS 4452-86. Code of Practice for the Construction of Underground Pipe Sewers and Drains Amend: 1, 1992. 16 pp.

—**Installation—Plumbing—Construction Contracts**
DIN ENGL 18381-90. Tendering and Performance Stipulations in Contracts for Construction Works (VOB); Part C: General Technical Specifications in Contracts for Construction Works (ATV); Installation of Pipework for Gas, Water and Drainage Services (July). 7 pp.

—**Internal Diameters**
DIN ENGL 1986 Pt 2-78. Drainage and Sewerage Systems for Buildings and Plots of Land; Specifications for the Determination of the Internal Diameters and Nominal Widths of Pipelines (Sept). 28 pp.

—**Lead**
CSA B67-1972. Lead Service Pipe, Waste Pipe, Traps, Bends and Accessories; (Rev 1-2 October 1977). 14 pp.
DIN ENGL 1263-66. Lead Waste Pipes and Bends for Drainage Systems (Sept). 1 p.

—**Nominal Widths**
DIN ENGL 1986 Pt 2-78. Drainage and Sewerage Systems for Buildings and Plots of Land; Specifications for the Determination of the Internal Diameters and Nominal Widths of Pipelines (Sept). 28 pp.

—**Pitch Fiber**
BSI BS 2760-73. 1973 Amd 1 Pitch-Impregnated Fibre Pipes and Fittings for Below and Above Ground Drainage (AMD 2675) June 30, 1978. 27 pp.

—**Plastic**
BSI BS 4576: Part 1-89. 1989 Amd 1 Specification for Unplasticized Polyvinyl Chloride (PVC-U) Rainwater Goods and Accessories Part 1: Half-Round Gutters and Pipes of Circular Cross Section (AMD 6350) June 28, 1991. 17 pp.
BSI BS 4576: Part 1-70. (WITHDRAWN) 1970 Amd 3 Unplasticized PVC Rainwater Goods Part 1: Half-Round Gutters and Circular Pipe. 17 pp.
BSI BS 4962-89. 1989 Plastic Pipes and Fittings for Use as Subsoil Field Drains. 28 pp.
BSI BS 4962-82. (WITHDRAWN) 1982 Plastic Pipes for Use as Light Duty Sub-Soil Drains. 20 pp.
CGSB 41-GP-29MA-83. Tubing, Plastic, Corrugated, Drainage, Standard for. 17 pp.
CSA CAN/CSA-B182.2-M90. PVC Sewer Pipe and Fittings (PSM Type); (Gen Instr 1 Thru 4). 57 pp.
CSA CAN/CSA-B182.4-92. Profile PVC Sewer Pipe and Fittings; (Gen Instr 1). 41 pp.
CSA B182.11-1967SP. Recommended Practice for the Installation of Plastic Drain and Sewer Pipe and Pipe Fittings; (Rev 1 November 1976). 13 pp.
ISO TR7074-86. Performance Requirements for Plastics Pipes and Fittings for Use in Underground Drainage and Sewage First Edition. 31 pp.
SNZ NZS 7604-81. Specification for High Density Polyethylene Drain and Sewer Pipe and Fittings. 26 pp.

—**Polyester—Fiberglass Reinforced**
DIN ENGL 19565 Pt 1-89. Centrifugally Cast and Filled Polyester Resin Glass Fibre Reinforced (UP-GF) Pipes and Fittings for Buried Drains and Sewers; Dimensions and Technical Delivery Conditions (Mar). 24 pp.

—**Polyethylene**
DIN ENGL 4262 Pt 1-89. PVC-U and PE-HD Subsoil and Multi-Purpose Drain Pipes for Use in Road Construction and Civil Engineering; Requirements and Testing (Mar). 4 pp.

Drainpipes *(Cont.)*

—**Polyethylene** *(Cont.)*
DIN ENGL 19535 Pt 1-88. High-Density Polyethylene (HDPE) Pipes and Fittings for Hot-Water Resistant Drains and Sewers Inside Buildings; Dimensions (Mar) (This Standard, Together with DIN 19535 Part 2, April 1987 Edition, Supersedes DIN 19535, March 1977 Edition, Withdrawn in Jan. 1986). 12 pp.
DIN ENGL 19535 Pt 2-87. High-Density Polyethylene (HDPE) Pipes and Fittings for Hot Water Resistant (HT) Drains and Sewers Inside Buildings; Technical Delivery Conditions (Apr) (Supersedes Parts of the March 1977 Edition of DIN 19535, Withdrawn in January 1986). 6 pp.
DIN ENGL 19537 Pt 1-83. High Density Polyethylene (HDPE) Pipes and Fittings for Drains and Sewers; Dimensions (Oct). 19 pp.
DIN ENGL 19537 Pt 2-88. High-Density Polyethylene (HDPE) Pipes and Fittings for Drains and Sewers; Technical Delivery Conditions (Jan). 8 pp.
ISO 8770-91. High-Density Polyethylene (PE-HD) Pipes and Fittings for Soil and Waste Discharge (Low and High Temperature) Systems Inside Buildings —Specifications First Edition. 21 pp.
ISO 8772-91. High-Density Polyethylene (PE-HD) Pipes and Fittings for Buried Drainage and Sewerage Systems—Specifications First Edition. 18 pp.

—**Polyethylene—Corrugated**
CSA CAN/CSA-B182.6-M92. Profile Polyethylene Sewer Pipe and Fittings; (Gen Instr 1). 37 pp.

—**Polyolefin—Laboratory**
CSA CAN/CSA-B181.3-M86. Polyolefin Laboratory Drainage Systems (R 1992); (Gen Instr 1). 19 pp.

—**Polypropylene**
DIN ENGL 19560-80. Pipes and Fittings of Polypropylene (PP) with Sliding Socket Joint for Hot Water Resistant Drainage Systems (HT) Inside Buildings; Dimensions, Technical Conditions of Delivery (Mar). 20 pp.
ISO 7671-91. Polypropylene (PP) Pipes and Fittings (Jointed by Means of Elastomeric Sealing Rings) for Soil and Waste Discharge (Low and High Temperature) Systems Inside Buildings—Specifications First Edition. 18 pp.
ISO 8773-91. Polypropylene (PP) Pipes and Fittings for Buried Drainage and Sewerage Systems—Specifications First Edition. 18 pp.

—**PVC**
BSI BS 4514-83. 1983 Amd 1 Unplasticized PVC Soil and Ventilating Pipes, Fittings and Accessories. 17 pp.
BSI BS 4576: Part 1-89. 1989 Amd 1 Specification for Unplasticized Polyvinyl Chloride (PVC-U) Rainwater Goods and Accessories Part 1: Half-Round Gutters and Pipes of Circular Cross Section (AMD 6350) June 28, 1991. 17 pp.
BSI BS 4576: Part 1-70. (WITHDRAWN) 1970 Amd 3 Unplasticized PVC Rainwater Goods Part 1: Half-Round Gutters and Circular Pipe. 17 pp.
BSI BS 4660-89. 1989 Unplasticized Polyvinyl Chloride (PVC-V) Pipes and Plastics Fittings of Nominal Sizes 110 and 160 for Below Ground Gravity Drainage and Sewerage. 22 pp.
BSI BS 4660-73. (WITHDRAWN) 1973 Amd 5 Unplasticized PVC Underground Drain Pipe and Fittings. 25 pp.
BSI BS 5481-77. 1977 Amd 2 Unplasticized PVC Pipe and Fittings for Gravity Sewers. 15 pp.
BSI BS 5955: Part 6-80. 1980 Amd 2 Code of Practice for Plastics Pipework (Thermoplastics Materials) Part 6: Installation of Unplasticized PVC Pipework for Gravity Drains and Sewers. 16 pp.
CSA CAN/CSA-B181.2-M90. PVC Drain, Waste, and Vent Pipe and Pipe Fittings; (Gen Instr 1). 56 pp.
DIN ENGL 1187-82. Unplasticized Polyvinyl Chloride (PVC-U); Drainpipes; Dimensions, Requirements, Testing (Nov). 6 pp.
DIN ENGL 4262 Pt 1-89. PVC-U and PE-HD Subsoil and Multi-Purpose Drain Pipes for Use in Road Construction and Civil Engineering; Requirements and Testing (Mar). 4 pp.
ISO 3633-91. Unplasticized Poly(Vinyl Chloride) (PVC-U) Pipes and Fittings for Soil and Waste Discharge (Low and High Temperature) Systems Inside Buildings —Specifications First Edition. 20 pp.
ISO 4435-91. Unplasticized Poly(Vinyl Chloride) (PVC-U) Pipes and Fittings for Buried Drainage and Sewerage Systems—Specifications First Edition. 17 pp.
SNZ NZS 7642-71. Specification for Unplasticized PVC Soil and Ventilating Pipe, Fittings and Accessories Amend: A, 1971; 1 & 1A, 1974; 2 & 2A, 1975. 28 pp.
SNZ NZS 7649-88. Unplasticized PVC Sewer and Drain Pipe and Fittings. 24 pp.

INDUSTRY STANDARDS

Drainpipes (Cont.)

—PVC—Installation
ISO TR7024-85. Above-Ground Drainage—Recommended Practice and Techniques for the Installation of Unplasticized Polyvinyl Chloride (PVC-U) Sanitary Pipework for Above-Ground Systems Inside Buildings First Edition. 39 pp.
ISO TR7073-88. Recommended Techniques for the Installation of Unplasticized Poly(Vinyl Chloride) (PVC-U) Buried Drains and Sewers First Edition. 21 pp.

—Sealing Rings
CEN PREN 681-1-92. Elastomeric Seals—Materials Requirements for Pipe Joint Seals Used in Water and Drainage Applications—Part 1: Vulcanized Rubber. 21 pp.
CEN PREN 681-2-92. Elastomeric Seals—Material Requirements for Pipe Joint Seals Used in Water and Drainage Applications—Part 2: Thermoplastic Elastomers. 16 pp.
CEN PREN 681-3-92. Elastomeric Seals—Materials Requirements for Pipe Joints Seals Used in Drainage and Sewerage Applications—Part 3: Cellular Materials of Vulcanized Rubber. 15 pp.
CEN PREN 681-4-92. Elastomeric Seals—Material Requirements for Pipe Joint Seals Used in Drainage and Sewerage Applications—Part 4: Cast Polyurethane Sealing Elements. 11 pp.
ISO 4633-83. Rubber Seals—Joint Rings for Water Supply, Drainage and Sewerage Pipelines—Specification for Materials First Edition. 6 pp.

—Silencers—Ships
JIS F 7228-76. Marine Tube Type Drain Silencers.

—Steel
BSI BS 1091-63. 1963 Pressed Steel Gutters, Rainwater Pipes, Fittings and Accessories. 20 pp.

—Subsurface Drainage
BSI BS 8000: Part 14-89. 1989 Workmanship on Building Sites Part 14: Code of Practice for Below Ground Drainage. 24 pp.
DIN ENGL 1185 Pt 1-73. Drainage; Control of Sub-Surface Water Management by Draining with Pipes, Open-Ditch Drainage and Amelioration of the Subsoil; General Instructions and Special Cases (Dec). 17 pp.
DIN ENGL 1185 Pt 2-73. Drainage; Control of Sub-Surface Water Management by Drainage with Pipes, Open-Ditch Drainage and Amelioration of the Subsoil; Important Data for Planning and Dimensioning (Dec). 17 pp.
DIN ENGL 1185 Pt 3-73. Drainage; Control of Sub-Surface Water Management by Drainage with Pipes, Open-Ditch Drainage and Amelioration of the Subsoil; Construction (Dec). 6 pp.
DIN ENGL 1185 Pt 4-73. Drainage; Control of Sub-Surface Water Management by Drainage with Pipes, Open-Ditch Drainage and Amelioration of the Subsoil; Design and "As Completed" Drawings (Dec). 4 pp.
DIN ENGL 1185 Pt 5-73. Drainage; Control of Sub-Surface Water Management by Drainage with Pipes, Open-Ditch Drainage and Amelioration of the Subsoil; Maintenance (Dec). 4 pp.
DIN ENGL 4095-90. Planning, Design and Installation of Drainage Systems Protecting Structures Against Water in the Ground (June). 8 pp.

—Underground
DIN ENGL 19550-87. General Requirements for Pipes and Fittings for Buried Drains and Sewers (Oct). 6 pp.

—Units of Measurement
SAA AS 1686-74. Metric Units for Use in Water Supply, Sewerage and Drainage (Including Pumping). 10 pp.

—Vitreous Clay
BSI BS 65-91. 1991 Vitrified Clay Pipes, Fittings and Ducts, Also Flexible Mechanical Joints for Use Solely with Surface Water Pipes and Fittings. 22 pp.
BSI BS 65-88. (WITHDRAWN) 1988 Vitrified Clay Pipes, Fittings, Joints and Ducts (Superseded by BS 65: 1991, BS EN 295-1, 295-2 and 295-3: 1991). 20 pp.
BSI BS 65-81. (WITHDRAWN) 1981 Amd 2 Vitrified Clay Pipes and Joints (AMD 4394) November 30, 1983. 27 pp.
BSI BS EN 295: Part 1-91. 1991 Vitrified Clay Pipes and Fittings and Pipe Joints for Drains and Sewers Part 1: Requirements. 17 pp.
BSI BS EN 295: Part 2-91. 1991 Vitrified Clay Pipes and Fittings and Pipe Joints for Drains and Serwers: Part 2: Quality Control and Sampling. 20 pp.
BSI BS EN 295: Part 3-91. 1991 Vitrified Clay Pipes and Fittings and Pipe Joints for Drains and Serwers: Part 3: Test Methods. 30 pp.
CEN PREN 295 (Part 1)-89. Vitrified Clay Pipes and Fittings and Pipe Joints for Drains and Sewers; Part 1: Specifications. 16 pp.
CEN EN 295-1-91. Vitrified Clay Pipes and Fittings and Pipe Joints for Drains and Sewers—Part 1: Requirements. 13 pp.
CEN PREN 295 (Part 2)-90. Vitrified Clay Pipes and Fittings and Pipe Joints for Drains and Sewers—Part 2: Quality Control and Sampling. 21 pp.
CEN EN 295-2-91. Vitrified Clay Pipes and Fittings and Pipe Joints for Drains and Sewers—Part 2: Quality Control and Sampling. 16 pp.
CEN PREN 295 (Part 3)-90. Vitrified Clay Pipes and Fittings and Pipe Joints for Drains and Sewers—Part 3: Test Methods. 38 pp.
CEN EN 295-3-91. Vitrified Clay Pipes and Fittings and Pipe Joints for Drains and Sewers—Part 3: Test Methods. 26 pp.
CEN PREN 295-5-92. Vitrified Clay Pipes and Fittings and Pipe Joints for Drains and Sewers—Part 5: Requirements for Perforated Vitrified Clay Pipes and Fittings. 9 pp.
CSA A60.1-M1976. Vitrified Clay Pipe (R 1992). 13 pp.

—Zinc
BSI BS 1431-60. 1960 Wrought Copper and Wrought Zinc Rainwater Goods. 25 pp.

Drains
Use For: Drainboards; Drainers; Drainingboards; Wastes (Sanitary Appliances) *See Also:* Concrete Pipes; Drainage Systems; Drainpipes; Floor Drains; Gutters; Irrigation; Manholes; Plumbing Equipment; Sewers; Soakaways
JIS A 4421-91. Drain with Traps for Equipment Units. 9 pp.

—Inspection Chambers
BSI BS 7158-89. 1989 Plastics Inspection Chambers for Drains. 17 pp.

—Reinforced Concrete
JIS A 1140-78. Method of Test for Compressive Strength of Reinforced Concrete Street and House Inlet. 5 pp.

—Roofing
JIS A 5522-75. Roof Drains (R 1983). 11 pp.

—Ships
MOD UK NES 712-91. Requirements for Sewage and Grey-Water Systems for Surface Ships Issue 5 (09.91). 48 pp.

DRAMs
Use: Dynamic RAMs

Draperies
Use For: Curtains *See Also:* Fabrics
BSI BS 5867: Part 1-80. 1980 Fabrics for Curtains and Drapes Part 1: General Requirements. 4 pp.
BSI BS 6426-83. 1983 Made-up Curtains. 8 pp.
CGSB 4-GP-152MA-88. Drapes. 10 pp.

—Flammability Testing
BSI BS 5867: Part 2-80. 1980 Amd 1 Fabrics for Curtains and Drapes Part 2: Flammability Requirements. 8 pp.
CEN PREN 1101-93. Textiles and Textile Products—Burning Behaviour—Curtains and Drapes—Detailed Procedure to Determine the Ignitability of Vertically Oriented Specimens (Small Flame). 4 pp.
CEN PREN 1102-93. Textiles and Textile Products—Burning Behaviour—Curtains and Drapes—Detailed Procedure to Determine the Flame Spread of Vertically Oriented Specimens. 6 pp.

—Glass Fabric
JIS R 3418-91. Textile Glass Fabric Curtains for Casement and Drapery. 7 pp.

—Hooks
CNS S1113-81. Curtain Hooks (Mar)(7074). 1 p.

—Light Testing
JIS L 1055-87. Testing Methods for Light Blocking Effect of Curtain Materials. 8 pp.

—Power Operated
CSA CAN/CSA-C22. 2 NO 247-92. Operators and Systems of Doors, Gates, Draperies, and Louvres; (Gen Instr 1). 98 pp.

—Rails—Metal
JIS A 4802-76. Curtain Rails (Metal).

—Ships
MOD UK NES 727-88. Guide to the Design and Manufacture of Awnings and Associated Screens and Covers Issue 1 (12.88). 28 pp.

Draughting Machines
Use: Drafting Machines

Draughtstrips
Use: Weather Stripping

Draw Back Collets
Use: Collets

Draw Hooks
Use: Towing Hooks

Draw-off Taps
Use: Drain Valves

Draw Off Taps (Faucets)
Use: Faucets

Drawbar Pull
See Also: Drawbars
BSI BS 6419-83. 1983 Measurement of Drawbar Pull of Earth-Moving Machinery. 11 pp.
ISO 7464-83. Earth-Moving Machinery—Method of Test for the Measurement of Drawbar Pull First Edition. 10 pp.

Drawbars
Use For: Toolbars *See Also:* Couplings; Drawbar Pull; Linchpins; Towbars; Towing Attachments

—Anchors (Fasteners)—Automotive
BSI BS AU 188-83. 1983 Anchorages for Towing Ropes, Cables or Bars for Road Vehicles. 4 pp.
ISO 5422-82. Road Vehicles—Anchorages for Towing Ropes, Cables or Bars First Edition. 4 pp.

—Couplings—Automotive
BSI BS AU 138A-80. 1980 Dimensions of Contact Type Couplings for Air Pressure Braking Systems on Trailers and Semi-Trailers and Their Towing Vehicles, and the Arrangement of These Couplings on Articulated and Drawbar Combinations. 10 pp.
BSI BS AU 219-88. 1988 Dimensions of 40mm Drawbar Couplings Between Commercial Road Vehicles and Trailers. 7 pp.
BSI BS AU 220-88. 1988 Dimension of 50mm Drawbar Couplings Between Commercial Road Vehicles and Trails. 7 pp.
ISO 1102-86. Commercial Road Vehicles—Mechanical Connections Between Towing Vehicles and Trailers—50 mm Drawbar Coupling Second Edition. 7 pp.
ISO 8755-86. Commercial Road Vehicles—Mechanical Connections Between Towing Vehicles and Trailers—40 mm Drawbar Coupling First Edition. 7 pp.

—Couplings—Mechanical Properties
BSI BS AU 239-91. 1991 Method of Testing Strength of Drawbar Couplings and Eyes for Hinged Drawbars (ISO 8718: 1988). 5 pp.
ISO 8718-88. Commercial Road Vehicles—Drawbar Couplings and Eyes for Hinged Drawbars—Strength Test First Edition. 4 pp.

—Couplings—Tractors
ISO 11406-93. Commercial Road Vehicles—Mechanical Coupling Between Towing Vehicles with Rear-Mounted Coupling and Drawbar Trailers—Interchangeability First Edition. 8 pp.

—Couplings—Tractors—Aircraft
BSI C 15-60. 1960 Coupling Dimensions for Aircraft-To-Tractor Tow Bar Connections. 3 pp.

—Crawler Tractors
JIS D 6106-92. Dimensions of Drawbars for Crawler Tractors. 5 pp.

—Eyes—Aircraft
MOD UK DSTAN 25-6-73. Towing Vehicles and Towed Vehicular Equipment Issue 3. 10 pp.

—Eyes—Automotive
BSI BS AU 24A-89. 1989 Towing Connections for Trailers of up to 5000 kg Gross Mass. 8 pp.
BSI BS AU 29-64. 1964 Drawbar Eyes and Forecarriage Pins for Connections Between Trailers of 5 and 35 Tons Gross Weight and Towing Vehicle. 3 pp.

—Eyes—Mechanical Properties
BSI BS AU 239-91. 1991 Method of Testing Strength of Drawbar Couplings and Eyes for Hinged Drawbars (ISO 8718: 1988). 5 pp.
ISO 8718-88. Commercial Road Vehicles—Drawbar Couplings and Eyes for Hinged Drawbars—Strength Test First Edition. 4 pp.

—Eyes—Trailers
JIS D 6601-86. Shapes, Dimensions and Travel Angles of Drawbar Eyes for Full Trailers. 8 pp.

Drawbars *(Cont.)*

—Jacks (Lifts)—Agricultural Equipment
BSI BS 6792-86. 1986 Agricultural Trailer and Trailed Machinery Drawbar Jacks. 4 pp.

—Jacks (Lifts)—Trailers
BSI BS 6792-86. 1986 Agricultural Trailer and Trailed Machinery Drawbar Jacks. 4 pp.

—Lugs (Fasteners)—Automotive
BSI BS AU 24A-89. 1989 Towing Connections for Trailers of up to 5000 kg Gross Mass. 8 pp.

—Mechanical Properties
ISO 7641 Pt 1-83. Road Vehicles—Caravans and Light Trailers—Calculation of the Mechanical Strength of the Drawbar—Part 1: Steel Drawbars First Edition. 5 pp.

—Overrun Devices—Bolts—Trailers
BSI BS AU 190-83. 1983 Interface Dimensions and Fixing Bolts for the Mounting of Overrun Devices on Delta Shaped Drawbars of Trailers up to 2000 Kg. 4 pp.

—Pins—Automotive
BSI BS AU 24A-89. 1989 Towing Connections for Trailers of up to 5000 kg Gross Mass. 8 pp.
BSI BS AU 29-64. 1964 Drawbar Eyes and Forecarriage Pins for Connections Between Trailers of 5 and 35 Tons Gross Weight and Towing Vehicle. 3 pp.

—Tractors
BSI BS 5861-92. 1992 Rear-Mounted Power Take-Offs on Agricultural Tractors (ISO 500: 1991). 19 pp.
BSI BS 5861-80. 1980 Amd 1 Rear Mounted Power Take-Offs on Agricultural Tractors. 15 pp.
CNS D2083-86. Dimensions of Link Type Drawbars for Agricultural Wheeled Tractors (Jul)(7032).
ISO 500-91. Agricultural Tractors—Rear-Mounted Power Take-off—Types 1, 2 and 3 Second Edition. 16 pp.
ISO 6489 Pt 3-92. Agricultural Vehicles—Mechanical Connections on Towing Vehicles—Part 3: Tractor Drawbar First Edition. 6 pp.
ISO 9190-90. Lawn and Garden Ride-on (Riding) Tractors—Drawbar First Edition. 6 pp.
JIS D 6704-84. Dimensions of Hitches for Agricultural Tractors. 4 pp.
JIS D 6705-84. Dimensions of Link Type Drawbars for Agricultural Wheeled Tractors. 4 pp.
JIS D 6707-76. Testing Method for Drawbar Performance of Agricultural Tractors (R 1984). 8 pp.
SAA AS 1121-83. Guards for Agricultural Tractor PTO Drives. 12 pp.

—Trailers
SNZ NZS 5446-87. Code of Practice for Heavy Motor Vehicle Towing Connections: Drawbar Trailers Amend: 1, 1991. 20 pp.

—Trailers—Mechanical Properties
BSI BS AU 210: Part 1-87. 1987 Drawbars for Caravans and Light Trailers Part 1: Method for Calculation of the Mechanical Strength of Steel Drawbars. 6 pp.

Drawers

See Also: Furniture
CNS S2129-86. Method of Test for Performance of Drawer (Aug)(11678).

—Hardware
BSI BS 3827: Part 4-67. (WITHDRAWN) 1967 Glossary of Terms Relating to Builders' Hardware Part 4: Door Drawer, Cupboard and Gate Furniture (Superseded by BS 6100: Subsection 1.3.6:1991). 24 pp.

—Pin Tumbler Locks
CNS A3187-83. Method of Test for Drawer Lock (Pin Tumbler Type) (May)(10212).

Drawgear

Use: Couplings

Drawing Boards

See Also: Drawing Equipment
BSI BS 6381-83. 1983 Drawing Boards. 4 pp.
SAA AS 1974-76. Drawing Board Dimensions. 8 pp.

—Glossaries
DIN ENGL 32860-83. Mobile Drawing Boards; Terminology, Requirements, Testing (Feb). 4 pp.

—Stands (Supports)
BSI BS 3459-85. 1985 Stands for Drawing Boards. 8 pp.

Drawing Compounds

See Also: Drawing Dies; Metal Working Equipment

—Nickel Wire
ISO 9724-92. Nickel and Nickel Alloy Wire and Drawing Stock First Edition. 13 pp.

Drawing Dies

See Also: Dies; Drawing Compounds
BSI BS 3821: Part 2-74. 1974 Hardmetal Dies and Associated Hardmetal Tools Part 2: As-Sintered Pellets and Finished Dies for Drawing Round Wire. 15 pp.
BSI BS 3821: Part 3-74. 1974 Hardmetal Dies and Associated Hardmetal Tools Part 3: As-Sintered Pellets and Finished Dies for Drawing Round Bar. 15 pp.
ISO 1684-75. Wire, Bar and Tube Drawing Dies—Specifications First Edition; (Amendment 2-1982). 20 pp.
ISO 2804-73. Wire, Bar or Tube Drawing Dies—as Sintered Pellets of Hard Metal (Carbide)—Dimensions First Edition. 8 pp.

—Cemented Carbide
CNS B3194-89. Cemented Carbide Dies for Drawing Wires and Rods (Jun)(3093).
JIS B 4111-91. Cemented Carbide Dies for Drawing Wires and Rods. 9 pp.

—Diamond
JIS B 4132-83. Diamond Dies for Wire Drawing. 10 pp.
JIS B 4133-83. Polycrystalline Diamond Dies for Wire Drawing. 10 pp.

—Hard Metals
BSI BS 4276-68. 1968 Hard Metal for Wire, Bar and Tube Drawing Dies. 12 pp.
BSI BS 4276-01. 1968 Amd 1 Hard Metal for Wire, Bar and Tube Drawing Dies (AMD 7746) August 15, 1993 (F). 13 pp.

—Sintered—Pellets
BSI BS 3821: Part 2-74. 1974 Hardmetal Dies and Associated Hardmetal Tools Part 2: As-Sintered Pellets and Finished Dies for Drawing Round Wire. 15 pp.
BSI BS 3821: Part 3-74. 1974 Hardmetal Dies and Associated Hardmetal Tools Part 3: As-Sintered Pellets and Finished Dies for Drawing Round Bar. 15 pp.

Drawing Equipment

Use For: Drawing Office Equipment
See Also: Compasses (Drawing); Dividers (Drawing); Drawing Boards; Drawing Inks; Drawing Papers; Pencils; Plotters; Protractors; Rulers; Squares (Instruments); Telewriters
CNS Z7106-82. Drawing Instruments: Technical Requirements (May)(8887).

—Adjusting Screws
CNS B2380-83. Adjusting Screws for Drawing Instruments (Feb)(4945).

—Bushings
CNS Z7155-83. Threaded Bushing for Drawing Instruments (Jan)(9931).

—Center Tacks
CNS Z7126-82. Center Tacks, Horn Centres for Drawing Instruments (Dec)(9768).

—Center Wheels—Spring Bow
CNS Z7154-83. Spring Bow Center-Wheel for Drawing Instruments (Jan)(9930).

—Clamping Screws
CNS B2381-83. Clamping Screws for Drawing Instruments (Feb)(4946).

—Glossaries
CNS Z7105-82. Nomenclature of Drawing Instruments (Apr)(8752).

—Holders—Needles
CNS Z7156-83. Needle Holders for Drawing Instruments (Jan)(9932).

—Horn Centers
CNS Z7126-82. Center Tacks, Horn Centres for Drawing Instruments (Dec)(9768).

—Knurled Nuts
CNS B2364-84. Knurled Nuts for Drawing Instruments (Jul)(4731).

—Lengthening Bars
CNS Z7115-82. Lengthening Bar for Precision Drawing Instruments (P) (Jun)(9045).
CNS Z7152-83. Lengthening Bar for School Drawing Instruments (Jan)(9928).

—Needles
CNS Z7116-82. Needles for Drawing Instruments (Jun)(9046).

—Slotted Head Bolts
CNS B2306-81. Slotted Raised Countersunk Bolts (for Drawing Instruments (Apr)(4573).

—Tee Head Bolts
CNS B2305-81. T-Head Bolts (for Drawing Instruments (Apr)(4572).

Drawing Inks

See Also: Drawing Equipment; India Ink; Inks
CNS S1034-72. Drawing Ink (Tentative) (Jun)(2559).
CNS S2022-66. Method of Test for Drawing Inks (Sep)(2560)(R 1971).

Drawing Office Equipment

Use: Drawing Equipment

Drawing Papers

See Also: Art Papers; Drafting Papers; Drawing Equipment; Fine Papers; Papers; Tracing Papers; Writing Papers
BSI BS 3429-84. 1984 Size of Drawing Sheets. 4 pp.
CGSB CAN/CGSB-9.64-M79. Drawing Sheet Sizes. 11 pp.
CNS P2052-79. Drawing Paper for Student Use in General (Jun)(3748). 1 p.
ISO 5457-80. Technical Drawings—Sizes and Layout of Drawing Sheets First Edition. 8 pp.
ISO 9431-90. Construction Drawings—Spaces for Drawing and for Text, and Title Blocks on Drawing Sheets First Edition. 8 pp.
JIS P 3301-89. Drawing Papers. 11 pp.
JIS S 5508-76. Scratchpads and Drafting Papers.

—School
BSI BS 4448-88. 1988 School Exercise Books and Papers. 10 pp.

Drawing Stock

Use: Drawing Compounds

Drawings

Scope Note: For additional listings, use a more specific term *See Also:* Architectural Drawings; Charts; Diagrams; Drafting; Engineering Drawings; Flow Charts; Graphic Arts; Graphic Methods; Line Drawings; Visual Aids
DIN ENGL 7167-87. Relationship Between Tolerances of Size, Form, and Parallelism; Envelope Requirement Without Individual Indication on the Drawing (Jan) (This Standard, Together with DIN ISO 1101, March 1985 Edition, Supersedes DIN 7184 Part 1, May 1972 Edition). 2 pp.
MOD UK DSTAN 05-10-90. Drawing Procedure Issue 4. 44 pp.

—Annotations—Aerospace
CEN PREN 3074-91. Drawing Annotation for Composite Laminate Structure. 16 pp.

—Construction
BSI BS 1192: Part 1-84. 1984 Construction Drawing Practice Part 1: Recommendations for General Principles. 34 pp.
BSI BS 1192: Part 1-01. 1984 Amd 1 Construction Drawing Practice Part 1: Recommendations for General Principles (AMD 7031) June 15, 1992. 35 pp.

—Construction—CAD
BSI BS 1192: Part 5-90. 1990 Construction Drawing Practice Part 5: Guide to Structuring of Computer Graphic Information. 22 pp.
SAA AS 3883-91. Computer Graphics—Computer Aided Design (CAD)—Guide for Structuring of Computer Graphic Information (BS 1192:5:1990). 16 pp.

—Defense Contracts
MOD UK DEFCON 19-76. Free User, Maintenance and Supply of Drawings 1/76. 1 p.
MOD UK DEFCON 21-58. Drawings, Specifications and Manufacturing Data 1/58. 1 p.
MOD UK DEFCON 63-88. Supply of Technical Data 11/88. 1 p.

—Glossaries
JIS Z 8114-84. Glossary of Drawing Terms.

—Identification Systems
MOD UK DSTAN 05-53-01. Identification of Drawings for Their Retrieval Issue 1; Amendment 1. 24 pp.

—Landscaping
BSI BS 1192: Part 4-84. 1984 Construction Drawing Practice Part 4: Recommendations for Landscape Drawings. 40 pp.

Drawings (Cont.)
—Signaling Systems—Interworking
CCITT RECMN Q.605-89. Drawing Conventions—Interworking of Signalling Systems (Study Group XI) 6 pp. 6 pp.

Dredgers
Use: Dredging Equipment

Dredges
Use: Dredging Equipment

Dredging
See Also: Dredging Equipment
BSI BS 6349: Part 5-91. 1991 Maritime Structures Part 5: Code of Practice for Dredging and Land Reclamation. 133 pp.

Dredging Equipment
Use For: Dredges *See Also:* Construction Equipment; Dredging; Excavating Equipment; Scoops (Shovels)

—Anchors
JIS F 3991-75. Dredger's Anchors.

—Capacity Measurement
BSI BS MA 98: Part 1-85. 1985 Multi-Bucket Dredgers Part 1: Scale of Bucket Capacities. 2 pp.
ISO 7607-84. Shipbuilding—Inland Navigation—Multi-Bucket Dredgers—Scale of Bucket Capacities First Edition. 3 pp.

—Classification
BSI BS 7466-91. 1991 Dredgers (ISO 8385: 1990). 10 pp.
ISO 8358-91. Solid Fertilizers—Preparation of Samples for Chemical and Physical Analysis First Edition. 8 pp.
ISO 8385-90. Shipbuilding and Marine Structures—Dredgers—Classification First Edition. 8 pp.

—Construction Contracts
DIN ENGL 18311-92. Construction Contract Procedures (VOB); Part C; General Technical Specifications in Construction Contracts (ATV); Dredging Work (Dec). 5 pp.

—Discharge Pipe Sleeves—Rubber
JIS F 3995-91. Rubber Sleeves for Dredger Discharge Pipes.

—Discharge Pipes
JIS F 3993-90. Discharge Pipes for Dredger.
JIS F 3993-77. Dredger's Discharge Pipes (R 1985). 8 pp.

—Floats
JIS F 3994-90. Floaters of Discharge Pipes for Dredger.

—Glossaries
BSI BS 7473-91. 1991 Terms for Dredgers (ISO 8384: 1991). 24 pp.
ISO 8384-91. Shipbuilding and Marine Structures—Dredgers—Vocabulary First Edition. 23 pp.

—Sheaves
JIS F 3992-81. Dredger's Sheaves for General Use.
JIS F 3992-66. Dredger's Sheaves for General Use. 7 pp.

Dress Jackets
Use: Blazers

Dress Uniforms
See Also: Clothing; Combat Uniforms; Fourrageres; Uniforms

—Fabrics—Cotton
MOD UK DSTAN 83-68-79. Cloths, Velvet, Cotton and Viscose, and Cloths, Velveteen, Cotton Issue 1. 12 pp.

—Fabrics—Cotton—Velvet
MOD UK DSTAN 83-68-79. Cloths, Velvet, Cotton and Viscose, and Cloths, Velveteen, Cotton Issue 1. 12 pp.

—Fabrics—Cotton—Velveteen
MOD UK DSTAN 83-68-79. Cloths, Velvet, Cotton and Viscose, and Cloths, Velveteen, Cotton Issue 1. 12 pp.

—Fabrics—Viscose
MOD UK DSTAN 83-68-79. Cloths, Velvet, Cotton and Viscose, and Cloths, Velveteen, Cotton Issue 1. 12 pp.

—Firefighters'
CGSB 38-GP-27MA-90. Firefighter Dress Uniform (Federal Public Service). 26 pp.

Dress Uniforms (Cont.)
—Jackets
CGSB 38-GP-43M-78. Construction Details for Uniform Coats, Jackets, and Tunics (Fusible Interlining), Standard for. 29 pp.

—Jackets—Men's
CGSB 38-GP-110MA-83. Tunics and Jackets, Men's Uniform Dress Single-Breasted—Dimensions, Standard for. 10 pp.

—Pants
CGSB 38-GP-24M-79. Construction Details for Uniform Dress Pants, Standard for. 15 pp.

—Pants—Men's
CGSB CAN/CGSB-38.109-92. Men's Uniform Dress Trousers—Dimensions. 9 pp.

—Shirts—Men's
CGSB CAN/CGSB-38.17-92. Men's Uniform Dress Shirts. 24 pp.
CGSB 38-GP-102MB-83. Shirts, Men's, Dress and Uniform Dress—Dimensions, Standard for. 11 pp.

—Shirts—Women's
CGSB CAN/CGSB-38.48-93. Women's Dress and Uniform Dress Shirts. 24 pp.
CGSB 38-GP-111M-82. Shirts, Women's Dress and Uniform Dress—Dimensions, Standard for. 11 pp.

Dresses
Use For: Jumpers; Tunics *See Also:* Clothing

—Girls'
CGSB CAN/CGSB-49.46-M79. Dresses, Little Girls' and Girls', Regular and Chubby Ranges—Dimensions; (Amendment 2 Oct 1988). 14 pp.
CGSB CAN/CGSB-49.57-M91. Girls' Canada Standard Sizes 2 to 16 Regular Range, and Sizes 6-1/2 to 14-1/2 Chubby Range, Jumpers and Tunics—Dimensions. 10 pp.

—Hospital Workers'
CGSB CAN/CGSB-38.31-74. Dress, Uniform, Women's, Hospital and Institutional, Princess Style. 17 pp.

—Polyester/Cotton
CGSB CAN/CGSB-4.154-M91. Polyester/Cotton Poplin Cloth, 160 g/m2. 9 pp.

—Women's
CGSB CAN/CGSB-49.213-M84. Dresses, Junior, Misses and Women's Sizes—Dimensions. 18 pp.

Dressing Trays
See Also: Medical Equipment Trays; Trays
CNS T1017-80. Dressing Trays (Aug)(6276).
JIS T 3505-79. Dressing Trays.

Dressmakers' Models
Use: Mannequins

Dried Foods
Use For: Dehydrated Foods *See Also:* Food; Prunes; Raisins
BSI BS 1743: Part 3-88. 1988 Analysis of Dried Milk and Dried Milk Products Part 3: Determination of Insolubility Index. 10 pp.

—Apples
ISO 7701-86. Dried Apples—Specification First Edition. 8 pp.

—Beverage Powders
CGSB 32.283M-87. Beverage Powders, Fruit-Flavoured. 7 pp.

—Cherries
ISO 6755-84. Dried Sour Cherries—Specification First Edition. 6 pp.

—Condiments
CNS N5190-81. Flavored Condiments (Oct)(8055). 3 pp.

—Dairy Products—Chemical Analysis
BSI BS 1743-68. 1968 Amd 4 Methods for the Analysis of Dried Milk and Dried Milk Products (AMD 6154) December 21, 1990. 23 pp.

—Dairy Products—Fat Content
BSI BS 1743: Part 11-86. 1986 Analysis of Dried Milk and Dried Milk Products Part 11: Determination of Fat Content (Gravimetric Reference Method). 10 pp.
ISO 1736-85. Dried Milk, Dried Whey, Dried Buttermilk and Dried Butter Serum—Determination of Fat Content—Gravimetric Method (Reference Method) Second Edition. 10 pp.

Dried Foods (Cont.)
—Dairy Products—Insoluble Matter Content
BSI BS 1743: Part 3-88. 1988 Analysis of Dried Milk and Dried Milk Products Part 3: Determination of Insolubility Index. 10 pp.
ISO 8156-87. Dried Milk and Dried Milk Products—Determination of Insolubility Index First Edition. 10 pp.

—Dairy Products—Phosphatase Activity
ISO 3356-75. Milk and Dried Milk, Buttermilk and Buttermilk Powder, Whey and Whey Powder—Determination of Phosphatase Activity (Reference Method) First Edition. 5 pp.

—Dairy Products—Sampling
BSI BS 1743: Part 1-86. 1986 Analysis of Dried Milk and Dried Milk Products Part 1: General Introduction, Including Preparation of Laboratory Samples. 4 pp.

—Dairy Products—Test Specimens—Microbiological Analysis
DIN ENGL 10191 Pt 2-86. Microbiological Examination of Milk; Preparation of Samples of Dried Milk, Dried Whey, and Lactose (Oct). 2 pp.

—Fish
CNS N5050-63. Dried Cuttlefish (for Export) (Tentative) (R1973) (Nov)(2246). 2 pp.
CNS N5054-81. Dried Squids (Aug)(2298). 2 pp.
CNS N5066-81. Dried Bonito and Mackerel (Aug)(2379). 3 pp.
CNS N5073-81. Dried Herring and Horse Mackerel (Aug)(2538). 3 pp.

—Fish—Inspection
CNS N6045-81. Method of Test for Dried Shipjacks and Mackerels (Aug)(2417). 1 p.

—Fish—Larval
CNS N5175-81. Dehydrated Larval Fish, Chilled (Packaged) (Jan)(6945). 2 pp.

—Fish—Larval—Chemical Analysis
CNS N6152-81. Method of Test for Dehydrated Larval Fish, Chilled (Aug)(7839). 2 pp.

—Fish—Larval—Fluorescent Agents
CNS N6152-81. Method of Test for Dehydrated Larval Fish, Chilled (Aug)(7839). 2 pp.

—Fish—Larval—Microbiological Analysis
CNS N6152-81. Method of Test for Dehydrated Larval Fish, Chilled (Aug)(7839). 2 pp.

—Fish—Larval—Sampling
CNS N6152-81. Method of Test for Dehydrated Larval Fish, Chilled (Aug)(7839). 2 pp.

—Fish—Larval—Temperature Measurement
CNS N6152-81. Method of Test for Dehydrated Larval Fish, Chilled (Aug)(7839). 2 pp.

—Fish—Larval—Visual Inspection
CNS N6152-81. Method of Test for Dehydrated Larval Fish, Chilled (Aug)(7839). 2 pp.

—Fish—Sampling
CNS N6045-81. Method of Test for Dried Shipjacks and Mackerels (Aug)(2417). 1 p.

—Fruits
CGSB 32.276M-89. Dried Fruit. 9 pp.
CNS N5028-84. Preserved Fruits (May)(1346). 2 pp.
CNS N5202-83. Coconut Dried Prepared (Edible) (Feb)(10027). 2 pp.

—Fruits—Chemical Analysis
CNS N6049-85. Method of Test for Dehydrated Fruits and Vegetables (Mar)(2896). 2 pp.

—Fruits—Fumigant Residues
CNS N6049-85. Method of Test for Dehydrated Fruits and Vegetables (Mar)(2896). 2 pp.

—Fruits—Glossaries
ISO 4125-91. Dry Fruits and Dried Fruits—Definitions and Nomenclature Second Edition. 9 pp.

—Fruits—Impurities Content
CNS N6049-85. Method of Test for Dehydrated Fruits and Vegetables (Mar)(2896). 2 pp.

—Fruits—Microbiological Analysis
CNS N6049-85. Method of Test for Dehydrated Fruits and Vegetables (Mar)(2896). 2 pp.

—Fruits—Moisture Content
CNS N5027-85. Dehydrated Fruits and Vegetables (Mar)(1345). 3 pp.

INTERNATIONAL AND NON-U.S. NATIONAL STANDARDS
SUBJECT INDEX

Dried Foods (Cont.)

—Fruits—Packaging
CNS N5027-85. Dehydrated Fruits and Vegetables (Mar)(1345). 3 pp.

—Fruits—Sampling
CNS N6049-85. Method of Test for Dehydrated Fruits and Vegetables (Mar)(2896). 2 pp.

—Fruits—Sensory Analysis
CNS N6049-85. Method of Test for Dehydrated Fruits and Vegetables (Mar)(2896). 2 pp.

—Garlic
BSI BS 4585: Part 11-83. 1983 Methods of Test for Spices and Condiments Part 11: Determination of Volatile Organic Sulphur Compounds in Dehydrated Garlic. 10 pp.
BSI BS 6194-81. 1981 Dehydrated Garlic. 10 pp.
ISO 5560-83. Dehydrated Garlic—Specification Second Edition. 8 pp.
ISO 5567-82. Dehydrated Garlic—Determination of Volatile Organic Sulphur Compounds First Edition. 5 pp.

—Inspection
CNS N6036-87. Method of Test for Fried and Dried Shredded Meat (Jul)(2168). 2 pp.

—Meat
CNS N5042-88. Fried and Dried Season Pork Fiber (Nov)(2167). 3 pp.
CNS N5048-87. Cured Meat and Dried Smoked Liver (Jul)(2213). 3 pp.
CNS N5068-87. Dried, Sliced Meat (Jul)(2419). 2 pp.
CNS N6041-87. Method of Test for Cured Meat and Dried Smoked Liver (Jul)(2214). 6 pp.

—Meat—Chemical Analysis
CNS N6046-87. Method of Test for Dried Sliced Meat (Jul)(2420). 2 pp.

—Meat—Inspection
CNS N6046-87. Method of Test for Dried Sliced Meat (Jul)(2420). 2 pp.

—Meat—Sampling
CNS N6036-87. Method of Test for Fried and Dried Shredded Meat (Jul)(2168). 2 pp.
CNS N6046-87. Method of Test for Dried Sliced Meat (Jul)(2420). 2 pp.

—Microbiological Analysis
SAA AS 1766.3.3-91. Methods for the Microbiological Examination of Food—Part 3: Examination of Specific Products—Part 3.3: Dehydrated Foods Amdt 1 August 1993. 4 pp.

—Mint
ISO 2256-84. Dried Mint (Spearmint) (Mentha spicata Linnaeus Syn. Mentha viridis Linnaeus)—Specification First Edition. 5 pp.
ISO 5563-84. Dried Peppermint (Mentha Piperita Linnaeus)—Specification First Edition. 7 pp.

—Mollusks—Sampling
CNS N6042-81. Method of Test for Dried Squids and Cuttle Fish (Aug)(2299). 2 pp.

—Mollusks—Visual Inspection
CNS N6042-81. Method of Test for Dried Squids and Cuttle Fish (Aug)(2299). 2 pp.

—Mushrooms
CNS N5221-86. Forest Mushroom (Shiitake) Hot Air Dehydrated (Dec)(11809). 5 pp.

—Onions
BSI BS 6205-81. 1981 Dehydrated Onion. 8 pp.
ISO 5559-83. Dehydrated Onion—Specification Second Edition. 7 pp.

—Oregano
ISO 7925-85. Dried Oregano (Origanum Vulgare Linnaeus)—Whole or Ground Leaves—Specification First Edition. 5 pp.

—Peaches
ISO 7703-86. Dried Peaches—Specification First Edition. 8 pp.

—Pears
ISO 7702-86. Dried Pears—Specification First Edition. 8 pp.

—Poultry Meat
CNS N5047-79. Dry Salted Ducks (Jan)(2208). 4 pp.

—Poultry Meat—Inspection
CNS N6040-83. Method of Test for Dry Salted Ducks (Dec)(2209). 3 pp.

—Poultry Meat—Sampling
CNS N6040-83. Method of Test for Dry Salted Ducks (Dec)(2209). 3 pp.

Dried Foods (Cont.)

—Roe
CNS N5043-81. Dried Mullet Roe (Aug)(2169). 2 pp.

—Roe—Sampling
CNS N6037-81. Method of Test for Dried Mullet Roe (Aug)(2170). 2 pp.

—Roe—Sensory Analysis
CNS N6037-81. Method of Test for Dried Mullet Roe (Aug)(2170). 2 pp.

—Soups
CGSB 32.281M-91. Dehydrated Soup Mixes, Instant and Simmer. 8 pp.

—Vegetables
CNS N5071-72. Dehydrated Sweet Potato Strips (Jun)(2435). 2 pp.
CNS N6204-86. Method of Test for Forest Mushroom (Shiitake) Hot Air Dehydrated (Dec)(11810).

—Vegetables—Chemical Analysis
CNS N6049-85. Method of Test for Dehydrated Fruits and Vegetables (Mar)(2896). 2 pp.

—Vegetables—Fumigant Residues
CNS N6049-85. Method of Test for Dehydrated Fruits and Vegetables (Mar)(2896). 2 pp.

—Vegetables—Impurities Content
CNS N6049-85. Method of Test for Dehydrated Fruits and Vegetables (Mar)(2896). 2 pp.

—Vegetables—Microbiological Analysis
CNS N6049-85. Method of Test for Dehydrated Fruits and Vegetables (Mar)(2896). 2 pp.

—Vegetables—Moisture Content
CNS N5027-85. Dehydrated Fruits and Vegetables (Mar)(1345). 3 pp.

—Vegetables—Packaging
CNS N5027-85. Dehydrated Fruits and Vegetables (Mar)(1345). 3 pp.

—Vegetables—Sampling
CNS N6049-85. Method of Test for Dehydrated Fruits and Vegetables (Mar)(2896). 2 pp.

—Vegetables—Sensory Analysis
CNS N6049-85. Method of Test for Dehydrated Fruits and Vegetables (Mar)(2896). 2 pp.

—Wakame
CNS N6207-87. Method of Test for Wakame Dried (Apr)(11930). 2 pp.

Dried Paint Films
Use: Paints

Driers
Use For: Liquid Driers *See Also:* Drying Agents

—Paints
BSI BS 332-56. 1956 Amd 1 Liquid Driers for Oil Paints. 13 pp.
ISO 4619-80. Driers for Paints and Varnishes First Edition. 15 pp.

—Varnishes
ISO 4619-80. Driers for Paints and Varnishes First Edition. 15 pp.

Driers (Heating Equipment)
Use: Dryers

Drift Compensation
See Also: Common Channel Signaling Systems

—Signaling Links—CCITT No. 6 Signaling Systems
CCITT RECMN Q.279-89. 6.9 Drift Compensation—Specifications of Signalling System No. 6 (Study Group XI) 1 pp. 1 p.

Drift Expanding Testing
Use For: Ring Expanding Tests

—Metal Pipes
AECMA PREN2002-09-87. Test Methods for Metallic Materials—Part 9—Tube Drift Expanding Test. 5 pp.
DIN ENGL 50135-65. Testing of Metallic Materials; Drift Expanding Test for Tubes (Aug). 2 pp.
DIN ENGL 50138-79. Testing of Metallic Materials; Ring Tensile Test on Tubes (May). 2 pp.
ISO 8493-86. Metallic Materials—Tube—Drift Expanding Test First Edition. 4 pp.
ISO 8495-86. Metallic Materials—Tube—Ring Expanding Test First Edition. 4 pp.
ISO 8496-86. Metallic Materials—Tube—Ring Tensile Test First Edition. 4 pp.

Drift Expanding Testing (Cont.)

—Metal Tubes—Aircraft
BSI 4A 4-66. 1966 Amd 1 Test Pieces and Test Methods for Metallic Materials for Aircraft. 13 pp.

—Steel Pipes
DIN ENGL 50137-79. Testing of Steel; Ring Expanding Test on Tubes (June). 3 pp.

Drifts
Scope Note: Use a more specific term *See:* Ejector Drifts

Drill Bits
See Also: Auger Bits; Bits (Tools); Core Bits; Masonry Drill Bits; Rock Drills; Spot Drill Bits; Twist Drill Bits
BSI BS 2593-74. 1974 Rotary Drill Rods and Tungsten Carbide Tipped Rotary Drill Bits. 15 pp.

—Medical—Identification Systems
BSI BS 3531: Sec 5.7-91. 1991 Implants for Osteosynthesis Part 5: Bone Screws and Auxiliary Equipment Section 5.7: Specification for Drill Bits, Taps and Countersink Cutters (ISO 9714-1: 1991). 10 pp.
ISO 9714 Pt 1-91. Orthopaedic Drilling Instruments—Part 1: Drill Bits, Taps and Countersink Cutters First Edition. 8 pp.

—Medical—Materials
BSI BS 3531: Sec 5.7-91. 1991 Implants for Osteosynthesis Part 5: Bone Screws and Auxiliary Equipment Section 5.7: Specification for Drill Bits, Taps and Countersink Cutters (ISO 9714-1: 1991). 10 pp.
ISO 9714 Pt 1-91. Orthopaedic Drilling Instruments—Part 1: Drill Bits, Taps and Countersink Cutters First Edition. 8 pp.

—Medical—Mechanical Properties
BSI BS 3531: Sec 5.7-91. 1991 Implants for Osteosynthesis Part 5: Bone Screws and Auxiliary Equipment Section 5.7: Specification for Drill Bits, Taps and Countersink Cutters (ISO 9714-1: 1991). 10 pp.
ISO 9714 Pt 1-91. Orthopaedic Drilling Instruments—Part 1: Drill Bits, Taps and Countersink Cutters First Edition. 8 pp.

—Percussion
CNS B3449-84. Tungsten-Carbide Bits and Rod for Rock Drill Sets (Dec)(11158).
DIN ENGL 20379-90. Drill Bits with Tapered Connection for Percussive Wet Rock Drilling (Mar). 2 pp.
ISO 1717-74. Rock Drilling—Rotary Drill-Rods and Rotary Drill-Bits for Dry Drilling—Connecting Dimensions First Edition. 7 pp.
ISO 1718-91. Rock Drilling Equipment—Drill Rods with Tapered Connection for Percussive Drilling Second Edition. 6 pp.

—Rotary—Rock
CNS B3449-84. Tungsten-Carbide Bits and Rod for Rock Drill Sets (Dec)(11158).
CNS M2017-80. Screw Threads for Detachable Rock Bit (Feb)(5245).
CNS M2019-80. Rod Taper for Detachable Rock Bit (Feb)(5247).
DIN ENGL 20379-90. Drill Bits with Tapered Connection for Percussive Wet Rock Drilling (Mar). 2 pp.
ISO 1717-74. Rock Drilling—Rotary Drill-Rods and Rotary Drill-Bits for Dry Drilling—Connecting Dimensions First Edition. 7 pp.
ISO 1718-91. Rock Drilling Equipment—Drill Rods with Tapered Connection for Percussive Drilling Second Edition. 6 pp.
JIS M 3901-57. Screw Threads for Detachable Rock-Bit (R 1984). 7 pp.
JIS M 3903-79. Rod Taper for Detachable Rock Bit.

Drill Bushes
Use: Jig Bushings

Drill Bushings
Use: Jig Bushings

Drill Chucks
See Also: Chucks; Jacobs Tapers; Morse Tapers; Self Centering Chucks
CNS B3265-77. Chucks for Hand Drills and Breast Drills (May)(4100). 2 pp.

—Arbors
DIN ENGL 238 Pt 1-62. Drill Chuck Mounting Arrangements; Taper Arbors (June). 1 p.

—Key Type
DIN ENGL 6349 Pt 1-78. Key-Type Three-Jaw Drill Chucks; Form A with Fastening by Internal Cone (Mar). 3 pp.

INDUSTRY STANDARDS

Drill Chucks (Cont.)

—Key Type (Cont.)
DIN ENGL 6349 Pt 2-78. Key-Type Three-Jaw Drill Chucks; Form B, with Fastening by Threaded Hole (Mar). 3 pp.
DIN ENGL 6349 Pt 3-78. Key-Type Three-Jaw Drill Chucks; Form C with Fastening by Threaded Pin (Mar). 2 pp.
DIN ENGL 6349 Pt 5-78. Key-Type Three-Jaw Drill Chucks; Technical Conditions of Delivery (Mar). 5 pp.

—Machine Tools
CNS B3434-83. Drill Chuck for Machine Tools (Feb)(9967).
CNS B7230-83. Method of Test for Drill Chuck of Machine Tools (Feb)(9968).
JIS B 6001-92. Drill Chuck for Machine Tools. 13 pp.

—Portable Drills
CNS B3262-89. Chucks for Portable Electric Drills (Oct)(4096). 7 pp.
CNS B7081-89. Method of Test for Chucks of Portable Electric Drills (Oct)(4097).
CNS B7081-77. Method of Test for Chucks of Portable Electric Drills (May)(4097). 2 pp.
JIS B 4634-84. Chucks for Portable Electric Drills. 14 pp.

—Tapers
DIN ENGL 238 Pt 2-67. Drill Chuck Mounting Arrangements; Drill Chuck Tapers (Mar). 2 pp.

Drill Hammers
Use: Drilling Hammers

Drill Jigs
See Also: Jigs (Positioners)
DIN ENGL 6347-69. Drill Jigs with Hinged Cover (May). 2 pp.
DIN ENGL 6348-71. Drilling Jigs; Quick-Clamping (Jan). 4 pp.

Drill Pipes
Use For: Drillpipes See Also: Bits (Tools); Drilling Equipment; Drills; Pipes; Rotary Drilling Equipment

—Screw Threads
DIN ENGL 20314-62. Screw Threads for Drill Pipes with Diameters Exceeding 100 mm for Large Hole Borings in the Mining Industry (July). 1 p.

—Screw Threads—Rock Drilling
DIN ENGL 4941-65. Screw Threads for Drill Pipes for Seamless Drill Pipes for Water and Rock Drillings According to the Percussion Drilling and Core Drilling Method (Jan). 1 p.

—Screw Threads—Water Drive Drilling
DIN ENGL 4941-65. Screw Threads for Drill Pipes for Seamless Drill Pipes for Water and Rock Drillings According to the Percussion Drilling and Core Drilling Method (Jan). 1 p.

—Seamless—Core Drilling
DIN ENGL 4940-65. Seamless Drill Pipes for Water and Rock Drillings According to the Percussive Drilling and Core Drilling Process (Feb). 2 pp.

—Seamless—Percussion Drilling
DIN ENGL 4940-65. Seamless Drill Pipes for Water and Rock Drillings According to the Percussive Drilling and Core Drilling Process (Feb). 2 pp.

—Seamless—Steel
JIS G 3439-88. Seamless Steel Oil Well Casing, Tubing and Drill Pipe. 38 pp.
JIS G 3465-88. Seamless Steel Tubes for Drilling. 9 pp.

Drill Rigs
Use: Drilling Rigs

Drill Rods
See Also: Rods
BSI BS 2593-74. 1974 Rotary Drill Rods and Tungsten Carbide Tipped Rotary Drill Bits. 15 pp.

—Core Drills
CNS M2036-80. Drill Rods for Core Drills (Mar)(5363).
JIS M 1409-82. Drill Rods for Core Drills.

—Joints—Rock Drills
CNS M2052-80. Shape and Dimension of Joint in Extension Rod for Rock Drill (May) (5625).
JIS M 3914-76. Shape and Dimension of Joint in Extension Rod for Rock Drill.

—Rotary
ISO 1717-74. Rock Drilling—Rotary Drill-Rods and Rotary Drill-Bits for Dry Drilling—Connecting Dimensions First Edition. 7 pp.

Drill Rods (Cont.)

—Shanks
CNS M2047-80. Shape and Dimension of Shank for Drill Rod (May)(5620).
CNS M2048-80. Shape and Dimension of Shank for Drill Shank Rod (May)(5621).
JIS M 3907-74. Shape and Dimension of Shank for Drill Rod.
JIS M 3917-74. Shape and Dimension of Shank for Drill Shank Rod.

—Shanks—Rock Drills
CNS M2047-80. Shape and Dimension of Shank for Drill Rod (May)(5620).
JIS M 3907-74. Shape and Dimension of Shank for Drill Rod.

Drill Steels
See Also: Bits (Tools); Drills; Ferroalloys; Steels

—Extension Equipment—Rock Drilling
ISO 1721-74. Rock Drilling—Extension Drill-Steel Equipment for Percussive Long-Hole Drilling—Reverse-Buttress-Threaded Equipments 1 1/16 and 1 1/4 in (27 and 32 mm) First Edition. 8 pp.
ISO 1722-74. Rock Drilling—Extension Drill-Steel Equipment for Percussive Long-Hole Drilling—Reverse-Buttress-Threaded Equipments 1 1/2 to 2 1/2 in (38 to 64 mm) First Edition. 8 pp.
ISO 10207-91. Rock Drilling Equipment—Rope Threaded Drill Steel Equipment for Percussive Drilling, Nominal Sizes 22 mm to 38 mm First Edition; (Corrigendum 1-1991). 11 pp.
ISO 10208-91. Rock Drilling Equipment—Left-Hand Rope Threads First Edition. 5 pp.

—Hollow
CNS G2239-85. Method of Test for Hollow Drill Steels (Feb)(11208).
CNS G3225-85. Hollow Drill Steels (Feb) (11207).
JIS G 4410-84. Hollow Drill Steels. 10 pp.

—Hollow—Hexagonal—Chuck Bushings—Rock Drilling
ISO 723-91. Rock Drilling Equipment—Forged Collared Shanks and Corresponding Chuck Bushings for Hollow Hexagonal Drill Steels Second Edition. 6 pp.

—Hollow—Hexagonal—Rock Drilling
ISO 722-91. Rock Drilling Equipment—Hollow Drill Steels in Bar Form, Hexagonal and Round Third Edition. 4 pp.
ISO 723-91. Rock Drilling Equipment—Forged Collared Shanks and Corresponding Chuck Bushings for Hollow Hexagonal Drill Steels Second Edition. 6 pp.

—Hollow—Hexagonal—Shanks—Rock Drilling
ISO 723-91. Rock Drilling Equipment—Forged Collared Shanks and Corresponding Chuck Bushings for Hollow Hexagonal Drill Steels Second Edition. 6 pp.

—Hollow—Round—Rock Drilling
ISO 722-91. Rock Drilling Equipment—Hollow Drill Steels in Bar Form, Hexagonal and Round Third Edition. 4 pp.

—Insert
CNS M2051-80. Insert Drill Steel (May)(5624).
JIS M 3913-75. Insert Drill Steels.

—Pipes—Seamless
JIS G 3465-88. Seamless Steel Tubes for Drilling. 9 pp.

Drilling
See Also: Drill Pipes; Hole Size; Offshore Drilling; Rock Drilling; Screws

—Construction Contracts
DIN ENGL 18301-92. Construction Contract Procedures (VOB); Part C: General Technical Specifications in Construction Contracts (ATV); Drilling Work (Dec). 4 pp.

—Subsoil
DIN ENGL 4022 Pt 1-87. Subsoil and Groundwater; Classification and Description of Soil and Rock; Borehole Logging of Soil and Rock Not Involving Continuous Core Sample Recovery (Sept). 21 pp.

Drilling Derricks
Use: Derricks

Drilling Equipment
See Also: Bits (Tools); Core Drilling Equipment; Cutting Tools; Drill Pipes; Drilling Rigs; Drills; Dumpers; Dunes; Electric Discharge Machine Tools; Explosives; Fishing Tools; Machine Tools; Offshore Drilling Equipment; Percussion Drills; Reamers; Rock

Drilling Equipment (Cont.)
See Also: (Cont.)
Drilling Equipment; Rock Drills; Rotary Drilling Equipment; Surgical Drilling Equipment; Taps (Threading Tools); Upright Drilling Equipment

—Accuracy Testing
CNS B7121-82. Accuracy Inspection of Drilling and Milling Combined Machines (Dec)(4964).

—Bench
CNS B4059-83. Drilling Machines, Sensitive, Bench Type (Feb)(9962).
CNS B7228-83. Test Code for Performance of Drilling Machines, Sensitive, Bench Type (Feb)(9963).

—Bench—Accuracy Testing
CNS B7120-85. Test Code for Accuracy of Bench Drilling Machines (Jan)(4963).
CNS B7229-83. Test Code for Accuracy of Drilling Machines, Sensitive, Bench Type (Feb)(9964).

—Electric Controllers
BSI BS 787: Part 4-72. 1972 Mining Type Flameproof Gate-End Boxes Part 4: Gate-End Boxes for Drilling Machines (For Use on 3-Phase A.C. Circuits up to X 650 V). 24 pp.
BSI BS 5126: Part 4-82. 1982 Mining Type Flameproof Supply and Control Units for Use on Systems up to 1100V Part 4: Units to Power Drilling Machines. 16 pp.

—Glossaries
BSI BS 3618: Sec 7-73. 1973 Glossary of Mining Terms Section 7: Electrical Engineering and Lighting. 11 pp.
CNS B1221-80. Glossary of Terms Relating to Parts of Drilling Machine (Dec)(6745).

—Mining—Glossaries
BSI BS 3618: Sec 6-72. 1972 Glossary of Mining Terms Section 6: Drilling and Blasting. 25 pp.

—Radial
CNS B1106-79. Specified Items of Machine Tools Radial Drilling Machine (Jan)(4670-9). 1 p.
JIS B 6208-87. Test Code for Performance and Accuracy of Radial Drilling Machines. 18 pp.

—Radial—Accuracy Testing
CNS B7003-86. Test Code for Performance and Accuracy of Radial Drilling Machines (Feb)(96). 8 pp.

—Radial—Adjustable Arm—Acceptance Testing
BSI BS 4656: Part 10-75. 1975 Amd 1 Accuracy of Machine Tools and Methods of Test Part 10: Drilling Machines, Radial Type. 13 pp.
DIN ENGL 8625 Pt 1-76. Machine Tools; Radial Drilling Machines with Adjustable Arm; Acceptance Conditions (Jan). 5 pp.
ISO 2423-82. Acceptance Conditions for Radial Drilling Machines with the Arm Adjustable in Height—Testing of Accuracy Second Edition. 10 pp.

—Single Spindle—Accuracy Testing
BSI BS 4656: Part 15-76. 1976 Accuracy of Machine Tools and Methods of Test Part 15: Drilling Machines, Turret and Single Spindle Co-Ordinate Types. 14 pp.

—Spindle Noses
CNS B1218-80. Spindle Noses for Drilling Machines (Dec)(6740).
JIS B 6110-73. Spindle Noses for Drilling Machines (R 1984). 6 pp.

—Turret—Accuracy Testing
BSI BS 4656: Part 15-76. 1976 Accuracy of Machine Tools and Methods of Test Part 15: Drilling Machines, Turret and Single Spindle Co-Ordinate Types. 14 pp.

—Vertical—Box Type—Acceptance Testing
DIN ENGL 8626 Pt 1-76. Machine Tools; Vertical Drilling Machines; Box Type Drilling Machines; Acceptance Conditions (Jan). 5 pp.

—Vertical—Box Type—Accuracy Testing
ISO 2772 Pt I-73. Test Conditions for Box Type Vertical Drilling Machines—Testing of the Accuracy—Part I: Geometrical Tests First Edition. 5 pp.
ISO 2772 Pt II-74. Test Conditions for Box Type Vertical Drilling Machines—Testing of the Accuracy—Part II: Practical Test First Edition. 5 pp.

INTERNATIONAL AND NON-U.S. NATIONAL STANDARDS
SUBJECT INDEX

Drinking

Drilling Equipment *(Cont.)*

—Vertical—Double Column—Acceptance Testing
DIN ENGL 8626 Pt 5-80. Machine Tools; Vertical Drilling Machines; Double Column Coordinate Drilling Machines with Vertical Spindle and Table of Fixed Height; Acceptance Conditions (Jan). 10 pp.

—Vertical—Multispindle—Accuracy Testing
CNS B7233-83. Method of Test for Performance of Vertical Multi-Spindle Drilling Machines (Apr)(10149).
CNS B7234-83. Test Code for Accuracy of Vertical Multi-Spindle Drilling Machines (Apr)(10150).
CNS B7235-83. Method of Test for Performance of Jig Boring Machines (Apr)(10151).
JIS B 6221-87. Test Code for Performance and Accuracy of Vertical Multi-Spindle Drilling Machines. 17 pp.

—Vertical—Pillar Type—Acceptance Testing
DIN ENGL 8626 Pt 2-76. Machine Tools, Vertical Drilling Machines; Pillar Type Drilling Machines; Acceptance Conditions (Jan). 6 pp.

—Vertical—Pillar Type—Accuracy Testing
BSI BS 4656: Part 11-74. 1974 Accuracy of Machine Tools and Methods of Test Part 11: Drilling Machines, Vertical Floor Mounted Column and Pillar Types. 13 pp.
ISO 2773 Pt I-73. Test Conditions for Pillar Type Vertical Drilling Machines—Testing of the Accuracy—Part I: Geometrical Test First Edition. 5 pp.
ISO 2773 Pt II-73. Test Conditions for Pillar Type Vertical Drilling Machines—Testing of the Accuracy—Part II: Practical Test First Edition. 6 pp.

—Vertical—Single Column—Acceptance Testing
DIN ENGL 8626 Pt 3-80. Machine Tools; Vertical Drilling Machines; Single Column Coordinate Drilling Machines with Vertical Spindle and Knee of Variable Height; Acceptance Conditions (Jan). 10 pp.
DIN ENGL 8626 Pt 4-80. Machine Tools; Vertical Drilling Machines; Single Column Coordinate Drilling Machines with Vertical Spindle and Table of Fixed Height; Acceptance Conditions (Jan). 10 pp.

—Vertical—Single Column—Accuracy Testing
BSI BS 4656: Part 11-74. 1974 Accuracy of Machine Tools and Methods of Test Part 11: Drilling Machines, Vertical Floor Mounted Column and Pillar Types. 13 pp.

—Vertical—Single Spindle—Accuracy Testing
CNS B7214-83. Test Code for Performance of Numerically Controlled Turret and Single Spindle Drilling Machines with Vertical Spindle (Dec)(9475).
CNS B7215-86. Test Code for Accuracy of Numerically Controlled Turret and Single Spindle Drilling Machines with Vertical Spindle (Aug)(9476).
ISO 3190-75. Test Conditions for Turret and Single Spindle Co-Ordinate Drilling Machines with Vertical Spindle—Testing of the Accuracy First Edition. 14 pp.
ISO 3686-76. Test Conditions for Turret and Single Spindle Co-Ordinate Drilling and Boring Machines with Table of Fixed Height with Vertical Spindle—High Accuracy Machines—Testing of the Accuracy First Edition. 16 pp.
JIS B 6332-86. Test Code for Performance and Accuracy of Numerically Controlled Turret and Single Spindle Drilling Machines with Vertical Spindle. 22 pp.

—Vertical—Turret—Accuracy Testing
CNS B7214-83. Test Code for Performance of Numerically Controlled Turret and Single Spindle Drilling Machines with Vertical Spindle (Dec)(9475).
CNS B7215-86. Test Code for Accuracy of Numerically Controlled Turret and Single Spindle Drilling Machines with Vertical Spindle (Aug)(9476).
ISO 3190-75. Test Conditions for Turret and Single Spindle Co-Ordinate Drilling Machines with Vertical Spindle—Testing of the Accuracy First Edition. 14 pp.
ISO 3686-76. Test Conditions for Turret and Single Spindle Co-Ordinate Drilling and Boring Machines with Table of Fixed Height with Vertical Spindle—High Accuracy Machines—Testing of the Accuracy First Edition. 16 pp.
JIS B 6332-86. Test Code for Performance and Accuracy of Numerically Controlled Turret and Single Spindle Drilling Machines with Vertical Spindle. 22 pp.

Drilling Hammers
Use For: Drill Hammers *See Also:* Hand Tools; Mining Equipment

—Vibration
DIN ENGL 45675 Pt 4-87. Exposure to Mechanical Vibration Transmitted to the Hand-Arm System; Measurement of Vibration Induced by Drilling Hammers (Sept). 4 pp.

Drilling Rigs
Use For: Borer Rigs; Drill Rigs *See Also:* Drilling Equipment; Drills; Hoists

—Safety
CEN PREN 791-92. Drill Rigs—Safety. 63 pp.

Drillpipes
Use: Drill Pipes

Drills
Scope Note: For additional listings, use a more specific term *Use For:* Borers (Drills); Hand Drills *See Also:* Augers; Bits (Tools); Boring Tools; Breast Drills; Center Drills; Core Drills; Cutting Tools; Dental Instruments; Drill Pipes; Drill Steels; Drilling Equipment; Drilling Rigs; Hammer Drills; Hand Braces; Hand Tools; High Speed Steel Tools; Impact Drills; Masonry Drills; Pivot Drills; Pneumatic Drills; Pointing Drills; Ratchet Braces; Reamers; Rock Drills; Rotary Drills; Rotary Hammer Drills; Shell Drills; Tap Drills; Taps (Threading Tools); Tools; Twist Drills; Woodworking Equipment
BSI BS 2556-73. 1973 Amd 2 Hand and Breast Drills. 19 pp.
BSI BS 2556-03. 1973 Amd 3 Hand and Breast Drills (AMD 7729) July 15, 1993 (F). 20 pp.
CNS B3263-77. Hand Drills (May)(4098). 2 pp.

—Angles
CNS B1012-53. Drill Angles (Feb)(237) (R 1970).

—Carbide Tips
DIN ENGL 8010-63. Carbide Tips for Drills; Point Angle 115 Degree for Heavy Loading (Sept). 1 p.
DIN ENGL 8013-63. Carbide Tips for Drills; Point Angle 85 Degree for Light Loading (Sept). 1 p.

—Countersink—Morse Taper Shanks
CNS B3059-53. 30 Degree Countersink Drill (Feb)(232)(R 1970). 1 p.
CNS B3063-53. 75 Degree Countersink Drill (Feb)(236)(R 1970). 1 p.
CNS B3077-52. Adapter, Countersink Drill and Reamer, with Morse Taper Shank (Sep)(251)(R 1970).

—Countersink—Screws
CNS B3061-53. Countersink Drill (Apply to CNS ——— Whitworth Screws) (Feb)(234) (R 1970). 2 pp.

—Countersink—Square Head
CNS B3078-52. Adapter, Countersink Drill and Reamer, with Square Head (Sep)(252)(R 1970).

—Countersink—Square Head Shanks
CNS B3059-53. 30 Degree Countersink Drill (Feb)(232)(R 1970). 1 p.

—Countersink—Tapping Holes
CNS B3060-53. Countersink Drill, with Whitworth Tapping Hole (Feb)(233)(R 1970). 2 pp.
CNS B3062-53. Countersink Drill with Metric Tapping Hole (Feb)(235)(R 1970). 2 pp.

—Cylindrical Shanks
BSI BS 4193: Part 21-93. 1993 Hardmetal Insert Tooling Part 21: Specification for Cylindrical Shanks with a Parallel Flat for Drills with Indexable Inserts (ISO 9766: 1990). 6 pp.
ISO 9766-90. Drills with Indexable Inserts—Cylindrical Shanks with a Parallel Flat First Edition. 4 pp.

—Electric—Portable
BSI BS 2769: Sec 2.1-84. 1984 Hand-Held Electric-Motor-Operated Tools Part 2: Particular Requirements Section 2.1: Drills. 7 pp.
CGSB 45-GP-2C-72. Drills, Electric, Portable, Standard for. 9 pp.
CNS C4103-87. Portable Electric Drills (Jul)(3264). 4 pp.
CSA CAN/CSA-C22. 2 NO71.1-M89. Portable Electric Tools; (Gen Instr 1 Thru 4). 89 pp.
JIS C 9605-88. Portable Electric Drills. 26 pp.
MOD UK DSTAN 51-19: Part 2-92. Hand Tools, Powered Part 2: Electric Issue 1. 31 pp.

—Electric—Safety
IEC 745 Pt 2-1-89. Safety of Hand-Held Motor-Operated Electric Tools Part 2: Particular Requirements for Drills First Edition; (Amendment 1-1992). 37 pp.

Drills *(Cont.)*

—Glossaries
CNS B1357-89. Glossary of Terms for Drills (Oct)(12613).
JIS B 0171-89. Glossary of Terms for Drills.

—Mining Equipment
BSI BS 1090-70. 1970 Flameproof Hand-Held Electric Drilling Machines Primarily for Use in Mines. 12 pp.

—Safety
CENELEC HD 400.2A S1 PRA1-93. AMD 1 Hand-Held Motor Operated Tools Part II: Particular Specifications Section A: Drills. 4 pp.

Drinking Fountains
Use For: Water Drinking Fountains *See Also:* Water Coolers
CNS C3048-85. Method of Test for Drinking Water Fountain (Sep)(3911). 4 pp.
CNS C4129-85. Drinking Water Supply (Sep)(3910). 4 pp.

Drinking Troughs
Use: Watering Troughs

Drinking Water
Use For: Potable Water *See Also:* Mineral Water; Salt Water; Water Supply; Water Treatment Chemicals

—Aluminum Content—Spectrometry
DIN ENGL 38406 Pt 9-89. German Standard Methods for the Examination of Water, Waste Water and Sludge; Cations (Group E) Determination of Aluminium by Spectrometry (E9) (Feb). 6 pp.

—Aromatic Polycyclic Hydrocarbon Content—Thin Layer Chromatography
DIN ENGL 38409 Pt 13-81. German Standard Methods for the Analysis of Water, Waste Water and Sludge; Summary Action and Material; Characteristic Parameters (Group H); Determination of Polycyclic Aromatic Hydrocarbons (PAH) in Drinking Water (H 13—1 to 3) (June). 10 pp.

—Cadmium Content—Voltametry
DIN ENGL 38406 Pt 16-90. German Standard Methods for the Examination of Water, Waste Water and Sludge; Cations (Group E); Determination of Zinc, Cadmium, Lead, Copper, Thallium, Nickel, Cobalt by Voltammetry (E16) (Mar). 8 pp.

—Calcium Content—Atomic Absorption Spectrometry
DIN ENGL 38406 Pt 3-82. German Standard Methods for the Examination of Water, Waste Water and Sludge; Cations (Group E); Determination of Calcium and Magnesium (E3) (Sept). 7 pp.

—Cationic Surfactants Content
DIN ENGL 38409 Pt 20-89. German Standard Methods for the Examination of Water, Waste Water and Sludge; Parameters Characterizing Effects and Substances (Group H); Determination of Substances That React with Disulfine Blue (H 20) (July). 6 pp.

—Chlorine Content
CNS N6188-84. Method of Test for Bactericides in Water Test of Residual Chlorine (May)(10892). 3 pp.

—Cobalt Content—Voltametry
DIN ENGL 38406 Pt 16-90. German Standard Methods for the Examination of Water, Waste Water and Sludge; Cations (Group E); Determination of Zinc, Cadmium, Lead, Copper, Thallium, Nickel, Cobalt by Voltammetry (E16) (Mar). 8 pp.

—Color Testing
BSI BS 6920: Sec 2.3-90. 1990 Suitability of Non-Metallic Products for Use in Contact with Water Intended for Human Consumption with Regard to Their Effect on the Quality of the Water Part 2: Methods of Test Section 2.3: Appearance of Water. 7 pp.
SNZ NZS/BS 6920: Pt 2:Sec 2.3-90. Suitability of Non-Metallic Products for Use in Contact with Water Intended for Human Consumption with Regard to Their Effect on the Quality of Water Part 2: Methods of Test Section 2: Taste of Water Section 2.3: Appearance of Water. 8 pp.

INDUSTRY STANDARDS

Drinking Water (Cont.)

—Contamination

SAA AS 3855(INT)-91. Products for Use in Contact with Water Intended for Human Consumption with Regard to Their Effect on the Quality of Water Amdt 1 May 1992 (Redesignated AS 4020(Int)—1992 by This Amendment) (in Professional Packages 61A, 61B). 1 p.

SAA AS 4020(INT)-92. Products for Use in Contact with Water Intended for Human Consumption with Regard to Their Effect on the Quality of Water (Replaces AS 3855(Int)-1991).

—Copper Content—Voltametry

DIN ENGL 38406 Pt 16-90. German Standard Methods for the Examination of Water, Waste Water and Sludge; Cations (Group E); Determination of Zinc, Cadmium, Lead, Copper, Thallium, Nickel, Cobalt by Voltammetry (E16) (Mar). 8 pp.

—Cyanide Content—Photometry

DIN ENGL 38405 Pt 14-88. German Standard Methods for the Examination of Water, Waste Water and Sludge; Anions (Group D); Determination of Cyanides in Drinking Water, and in Groundwater and Surface Water with Low Pollution Levels (D 14) (Dec). 7 pp.

—Extraction Analysis

BSI BS 6920: Sec 2.5-90. 1990 Suitability of Non-Metallic Products for Use in Contact with Water Intended for Human Consumption with Regard to Their Effect on the Quality of the Water Part 2: Methods of Test Sec. 2.5: The Extraction of Subst. That May Be of Concern to Public Health. 10 pp.

SNZ NZS/BS 6920: Pt 2:Sec 2.5-90. Suitability of Non-Metallic Products for Use in Contact with Water Intended for Human Consumption with Regard to Their Effect on the Quality of Water Part 2: Methods of Test Section 2: Taste of Water Section 2.5: The Extraction of Substances That May Be of Concern. 12 pp.

—Filters

DIN ENGL 19632-87. Mechanical Filters for Drinking Water Installations; Requirements, Testing (DVGW Code of Practice) (Apr). 7 pp.

—Lead Content—Voltametry

DIN ENGL 38406 Pt 16-90. German Standard Methods for the Examination of Water, Waste Water and Sludge; Cations (Group E); Determination of Zinc, Cadmium, Lead, Copper, Thallium, Nickel, Cobalt by Voltammetry (E16) (Mar). 8 pp.

—Magnesium Content—Atomic Absorption Spectrometry

DIN ENGL 38406 Pt 3-82. German Standard Methods for the Examination of Water, Waste Water and Sludge; Cations (Group E); Determination of Calcium and Magnesium (E3) (Sept). 7 pp.

—Metals—Leach Testing

BSI BS 6920: Sec 2.6-90. 1990 Suitability of Non-Metallic Products for Use in Contact with Water Intended for Human Consumption with Regard to Their Effect on the Quality of the Water Part 2: Methods of Test Section 2.6: The Extraction of Metals. 8 pp.

BSI BS 6920: Part 3-90. 1990 Suitability of Non-Metallic Products for Use in Contact with Water Intended for Human Consumption with Regard to Their Effect on the Quality of the Water Part 3: High Temperature Tests. 6 pp.

SNZ NZS/BS 6920: Pt 2:Sec 2.6-90. Suitability of Non-Metallic Products for Use in Contact with Water Intended for Human Consumption with Regard to Their Effect on the Quality of Water Part 2: Methods of Test Section 2: Taste of Water Section 2.6: The Extraction of Metals. 8 pp.

SNZ NZS/BS 6920: Part 3-90. Suitability of Non-Metallic Products for Use in Contact with Water Intended for Human Consumption with Regard to Their Effect on the Quality of Water Part 3: High Temperature Tests. 8 pp.

—Metering Systems—DVGW Codes

DIN ENGL 19635-86. Dosing Apparatus for Drinking Water Treatment; Requirements, Testing, Operation; DVGW Code of Practice (June). 7 pp.

—Microbiological Analysis

BSI BS 6920: Sec 2.4-88. 1988 Suitability of Non-Metallic Products for Use in Contact with Water Intended for Human Consumption with Regard to Their Effect on the Quality of the Water Part 2: Methods of Test Section 2.4: Growth of Aquatic Microorganisms. 11 pp.

Drinking Water (Cont.)

—Microbiological Analysis (Cont.)

DIN ENGL 38411 Pt 1-83. German Standard Methods for the Examination of Water, Waste Water and Sludge; Microbiological Methods (Group K); Preparation for the Microbiological Examination of Water Samples (K 1) (Feb). 4 pp.

SNZ NZS/BS 6920: Pt 2:Sec 2.4-88. Suitability of Non-Metallic Products for Use in Contact with Water Intended for Human Consumption with Regard to Their Effect on the Quality of Water Part 2: Methods of Test Section 2: Taste of Water Section 2.4: Growth of Aquatic Microorganisms. 12 pp.

—Military Operations

NATO STANAG 2136 ED 2 AMD 7-74. Minimum Standards of Water Potability. 30 pp.

NATO STANAG 2885 ED 1 AMD 3-79. Procedures for the Treatment, Acceptability and Provision of Potable Water in the Field. 22 pp.

NATO STANAG 2885 ED 2 AMD 0-90. Emergency Supply of Water in War. 24 pp.

—Nickel Content—Voltametry

DIN ENGL 38406 Pt 16-90. German Standard Methods for the Examination of Water, Waste Water and Sludge; Cations (Group E); Determination of Zinc, Cadmium, Lead, Copper, Thallium, Nickel, Cobalt by Voltammetry (E16) (Mar). 8 pp.

—Nonmetallic Materials—High Temperature Testing

BSI BS 6920: Part 3-90. 1990 Suitability of Non-Metallic Products for Use in Contact with Water Intended for Human Consumption with Regard to Their Effect on the Quality of the Water Part 3: High Temperature Tests. 6 pp.

SNZ NZS/BS 6920: Part 3-90. Suitability of Non-Metallic Products for Use in Contact with Water Intended for Human Consumption with Regard to Their Effect on the Quality of Water Part 3: High Temperature Tests. 8 pp.

—Nuclear, Biological Chemical Warfare

NATO STANAG 2885 ED 1 AMD 3-79. Procedures for the Treatment, Acceptability and Provision of Potable Water in the Field. 22 pp.

—Plumbing Equipment

SAA AS 3855(INT)-92. Suitability of Plumbing Products for Contact with Potable Water (Expires 1 January 1997) (in Professional Packages 61A, 61B).

—Radioisotopes

DIN ENGL 38404 Pt 15-87. German Standard Methods for the Examination of Water, Waste Water and Sludge; Physical and Physico-Chemical Para-meters (Group C); Determination of Beta Activity per Unit Volume in Drinking Water, Ground Water, Surface Water and Waste Water (C 15) (Sept). 4 pp.

—Radioisotopes—Gamma Ray Spectrometry

DIN ENGL 38404 Pt 16-89. German Standard Methods for the Examination of Water, Waste Water and Sludge; Physical and Physicochemical Para-meters (Group C); Deter-mination of Radio-nuclides in Drinking Water, Ground Water, Surface Water and Waste Water by Y-Ray Spectro-metry (C 16) (Apr). 7 pp.

—Sampling

BSI BS 6068: Sec 6.5-91. 1991 Water Quality Part 6: Sampling Section 6.5: Guidance on Sampling of Drinking Water and Water Used for Food and Beverage Processing (ISO 5667-5: 1991). 13 pp.

BSI BS 6920: Sec 2.1-90. 1990 Suitability of Non-Metallic Products for Use in Contact with Water Intended for Human Consumption with Regard to their Effect on the Quality of the Water Part 2: Methods of Test Section 2.1: Samples for Testing. 8 pp.

DIN ENGL 38402 Pt 14-86. German Standard Methods for the Examination of Water, Waste Water and Sludge; General Information (Group A); Sampling of Untreated Water and Drinking Water (A14) (Mar). 6 pp.

ISO 5667 Pt 5-91. Water Quality—Sampling —Part 5: Guidance on Sampling of Drinking Water and Water Used for Food and Beverage Processing First Edition. 13 pp.

SNZ NZS/BS 6920: Pt 2:Sec 2.1-90. Suitability of Non-Metallic Products for Use in Contact with Water Intended for Human Consumption with Regard to Their Effect on the Quality of Water Part 2: Methods of Test Section 2: Taste of Water Section 2.1: Samples for Testing. 8 pp.

Drinking Water (Cont.)

—Selenium Content—Atomic Absorption Spectrometry

ISO 9965-93. Water Quality—Determination of Selenium—Atomic Absorption Spectrometric Method (Hydride Technique) First Edition. 8 pp.

—Silver Content—Atomic Absorption Spectrometry

DIN ENGL 38406 Pt 18-90. German Standard Methods for the Examination of Water, Waste Water and Sludge; Cations (Group E); Determination of Dissolved Silver by Atomic Absorption Spectrometry Using Electrothermal Atomization (E 18) (May). 5 pp.

—Taste Testing

BSI BS 6920: Sec 2.2-88. (WITHDRAWN) 1988 Suitability of Non-Metallic Products for Use in Contact with Water Intended for Human Consumption with Regard to Their Effect on the Quality of the Water Part 2: Methods of Test Section 2.2: Taste of Water (Superseded by. 10 pp.

BSI BS 6920: SUBSEC 2.2.1-90. 1990 Suitability of Non-Metallic Products for Use in Contact with Water Intended for Human Consumption with Regard to Their Effect on the Quality of the Water Part 2: Methods of Test Section 2.2: Taste of Water Subsec. 2.2.1 General Method of Test. 10 pp.

BSI BS 6920: SUBSEC 2.2.2-90. 1990 Suitability of Non-Met. Prod. for Use in Contact with Water Intd. for Human Cons. with Regard to Their Effect on the Qlty of the Water Pt 2: Meth. of Test Sec 2.2: Taste of Water Subsec. 2.2.2: Meth. of Testing Tastes Imparted to Water by Hoses. 7 pp.

BSI BS 6920:SUB Sec.2.2.3-90. 1990 Suitability of Non-Metallic Products for Use in Contact with Water Intended for Human Consumption with Regard to Their Effect on the Quality of Water Part 2: Methods of Test Section 2.2: Taste of Water Subsection 2.2.3: Method of Testing Tastes. 7 pp.

SNZ NZS/BS 6920: Pt2:Sec2.2.1-90. Suitability of Non-Metallic Products for Use in Contact with Water Intended for Human Consumption with Regard to Their Effect on the Quality of Water Part 2: Methods of Test Section 2: Taste of Water Subsection: 2.2.1: General Method of Test. 12 pp.

SNZ NZS/BS 6920: Pt2:Sec2.2.2-90. Suitability of Non-Metallic Products for Use in Contact with Water Intended for Human Consumption with Regard to Their Effect on the Quality of Water Part 2: Methods of Test Section 2: Taste of Water Subsection 2.2.2: Method of Testing Tastes Imparted to Water. 8 pp.

SNZ NZS/BS 6920: Pt2:Sec2.2.3-90. Suitability of Non-Metallic Products for Use in Contact with Water Intended for Human Consumption with Regard to Their Effect on the Quality of Water Part 2: Methods of Test Section 2: Taste of Water Subsection 2.2.3: Method of Testing Tastes Imparted to Water. 8 pp.

—Thallium Content—Voltametry

DIN ENGL 38406 Pt 16-90. German Standard Methods for the Examination of Water, Waste Water and Sludge; Cations (Group E); Determination of Zinc, Cadmium, Lead, Copper, Thallium, Nickel, Cobalt by Voltammetry (E16) (Mar). 8 pp.

—Tritium Content

DIN ENGL 38404 Pt 13-88. German Standard Methods for the Examination of Water, Waste Water and Sludge; Physical and Physicochemical Parameters (Group C); Determination of Tritium (C13) (May). 6 pp.

—Turbidity

BSI BS 6920: Sec 2.3-90. 1990 Suitability of Non-Metallic Products for Use in Contact with Water Intended for Human Consumption with Regard to Their Effect on the Quality of the Water Part 2: Methods of Test Section 2.3: Appearance of Water. 7 pp.

SNZ NZS/BS 6920: Pt 2:Sec 2.3-90. Suitability of Non-Metallic Products for Use in Contact with Water Intended for Human Consumption with Regard to Their Effect on the Quality of Water Part 2: Methods of Test Section 2: Taste of Water Section 2.3: Appearance of Water. 8 pp.

—Water Hardness

DIN ENGL 38409 Pt 6-86. German Standard Methods for the Examination of Water, Waste Water and Sludge; Summary Indices of Actions and Substances (Group H); Water Hardness (H 6) (Jan). 3 pp.

Drinking Water *(Cont.)*
—Zinc Content—Voltametry
DIN ENGL 38406 Pt 16-90. German Standard Methods for the Examination of Water, Waste Water and Sludge; Cations (Group E); Determination of Zinc, Cadmium, Lead, Copper, Thallium, Nickel, Cobalt by Voltammetry (E16) (Mar). 8 pp.

Drinking Water Coolers
Use: Water Coolers

Drinking Water Pipes
Use: Water Pipes

Drinking Water Supply Installations
Use: Water Supply Installations

Dripping Water Testing
Use: Water Resistance Testing

Drive Axles
Use For: Driving Axles *See Also:* Axles
—Cranes—Seal Covers
DIN ENGL 15092-82. Cranes; Driving Wheel Units and Idler Wheel Units; Sealing Covers (July). 6 pp.

Drive Pins
See Also: Pins
CNS Z1032-84. Drive Pin (Nov)(9644).
CNS Z3022-84. Method of Test for Drive Pin (Nov)(9645).

Drive Screws
Use For: Metallic Drive Screws *See Also:* Screws; Tapping Screws
BSI BS 4174-72. 1972 Self-Tapping Screws and Metallic Drive Screws. 52 pp.
SAA AS B194-70. Tapping and Metallic Drive Screws. 37 pp.
SNZ NZS/BS 4174-72. Specification for Self-Tapping Screws and Metallic Drive Screws. 52 pp.

Drive Shaft Assemblies
Use: Drive Shafts

Drive Shafts
Use For: Drive Shaft Assemblies; Transmission Shafts
See Also: Shafts (Machine Elements); Transmissions (Power Sources); Universal Joints

—Agricultural Equipment
BSI BS 3417: Part 1-84. 1984 1974 Power Take-Off Shafts and Guards for Tractors and Machinery for Agriculture and Forestry Part 1: Power Take-Off Drive Shafts. 8 pp.
BSI BS 3417: Part 2-84. 1984 1974 Power Take-Off Shafts and Guards for Tractors and Machinery for Agriculture and Forestry Part 2: Methods for Testing Guards. 8 pp.
BSI BS 3417: Part 3-86. 1986 Power Take-Off Shaft and Guards for Tractors and Machinery for Agriculture and Forestry Part 3: Supplementary Requirements to Part 1 'Specification for Power Take-Off Drive Shafts' and to Part 2 'Methods for Testing Guards' (Sup's BS 3417: 1974). 4 pp.
CEN PREN 1152-93. Tractors and Machinery for Agriculture and Forestry—Guards for Power Take-Off (PTO) Drive Shafts—Wear and Strength Tests. 12 pp.
ISO 5673-80. Agricultural Tractors—Power Take-off Drive Shafts for Machines and Implements First Edition. 7 pp.
ISO 5674 Pt 1-92. Tractors and Machinery for Agriculture and Forestry—Guards for Power Take-off (PTO) Drive-Shafts—Part 1: Strength Test First Edition. 9 pp.

—Agricultural Equipment—Wear Testing
CEN PREN 1152-93. Tractors and Machinery for Agriculture and Forestry—Guards for Power Take-Off (PTO) Drive Shafts—Wear and Strength Tests. 12 pp.
ISO 5674 Pt 2-92. Tractors and Machinery for Agriculture and Forestry—Guards for Power Take-off (PTO) Drive-Shafts—Part 2: Wear Test First Edition. 8 pp.

—Aircraft
SBAC RS 344 (V). Circular Flange and Drive 5.197 P.C.D. X. 500.
SBAC RS 345 (V). Circular Flange and Drive 5.197 P.C.D. X. 625.
SBAC RS 346 (V). Square Flange and Drive 3.812 Inch P.C.DIA. X. 375 Inch.
SBAC RS 501-506, 514-521 (V). Circular Flanges and Drives.

—Automotive
DIN ENGL 75532 Pt 2-79. Transmission of Rotary Motions; Flexible Drive Shafts (Apr). 2 pp.

Drive Shafts *(Cont.)*
—Cocks—Aircraft
SBAC RS 340 (V). 2 Bolt Flange and Drive (Primarily for Valves and Cocks).

—Electrical Equipment—Heights
DIN ENGL 747 Extr. 1-67. Shaft Heights for Driving and Driven Machines; Selection for Electrical Machines (June). 1 p.

—Electrical Machines—Aircraft
SBAC RS 333 (V). Circular Flange and Drive for Electrical Machines 5.000 Inch P.C.D. X. 500 Inch.
SBAC RS 334 (V). Circular Flange and Drive for Electrical Machines 5.000 Inch P.C.D. X. 5625 Inch.
SBAC RS 335 (V). Circular Flange and Drive for Electrical Machines 7.000 Inch P.C.D. X. 6875 Inch.
SBAC RS 336 (V). Circular Flange and Drive for Electrical Machines 8.000 Inch P.C.D. X. 8125 Inch.
SBAC RS 337 (V). Circular Flange and Drive for Electrical Machines 9.000 Inch P.C.D. X. 9375 Inch.

—Engine Starters—Aircraft
SBAC RS 308 (V). 4 Inch P.C.DIA. Starter Mounting Flange.
SBAC RS 309 (V). 5 Inch P.C.DIA. Starter Mounting Flange.
SBAC RS 310 (V). 5.750 Inch P.C.DIA. Starter Mounting Flange.

—Flanges—Aircraft
SBAC RS 300 (V). Circular Accessory Flange and Drive 4.290 Inch P.C.D. X. 500 Inch.
SBAC RS 301 (V). Circular Flange and Drive 4.290 Inch P.C.D. X. 626 Inches.
SBAC RS 302 (V). Circular Flange and Drive 5.000 Inch P.C.D. X. 750 Inches.
SBAC RS 303 (V). Circular Flange and Drive 6.000 Inch P.C.D. X. 875 Inches.
SBAC RS 304 (V). Circular Flange and Drive 7.000 Inch P.C.D X. 1.000 Inches.
SBAC RS 305 (V). Circular Flange and Drive 8.0 Inch P.C.D. X. 1.125 Inches.
SBAC RS 306 (V). Circular Flange and Drive 9.000 Inch P.C.D. X. 1.250 Inches.
SBAC RS 307 (V). Triangular Flange and Drive 3.450 Inch P.C.D. X. 375 Inches.
SBAC RS 314 (V). Circular Flange and Drive 4.290 Inch P.C.D. X. 500 Inch.
SBAC RS 316 (V). Circular Flange and Drive 6.000 Inch P.C.D. X. 625 Inch.
SBAC RS 317 (V). Circular Flange and Drive 7.000 Inch P.C.D. X. 750 Inch.
SBAC RS 318 (V). Circular Accessory Flange and Drive 8.000 Inch P.C.D. X. 875 Inch.
SBAC RS 319 (V). Circular Accessory Flange and Drive 9.000 Inch P.C.D X 1.000 Inch.
SBAC RS 320 (V). Circular Accessory Flange and Drive 10.5 Inch P.C.D. X. 1.125 Inch.
SBAC RS 321 (V). Circular Accessory Flange and Drive 12 Inch P.C.D. 1.250 Inch.
SBAC RS 327 (V). Square Flange and Drive 1.437 P.C.DIA..
SBAC RS 328 (V). Square Flange and Drive 2.000 P.C.DIA..
SBAC RS 329 (V). Square Flange and Drive 2.625 P.C.DIA..
SBAC RS 330 (V). Square Flange and Drive 3.125 P.C.DIA..
SBAC RS 331 (V). Square Flange and Drive 3.812 P.C.DIA..
SBAC RS 332 (V). Square Flange and Drive 4.312 P.C.DIA..

—Forestry Equipment
BSI BS 3417: Part 2-84. 1984 1974 Power Take-Off Shafts and Guards for Tractors and Machinery for Agriculture and Forestry Part 2: Methods for Testing Guards. 8 pp.
BSI BS 3417: Part 3-86. 1986 Power Take-Off Shaft and Guards for Tractors and Machinery for Agriculture and Forestry Part 3: Supplementary Requirements to Part 1 'Specification for Power Take-Off Drive Shafts' and to Part 2 'Methods for Testing Guards' (Sup's BS 3417: 1974). 4 pp.
CEN PREN 1152-93. Tractors and Machinery for Agriculture and Forestry—Guards for Power Take-Off (PTO) Drive Shafts—Wear and Strength Tests. 12 pp.
ISO 5674 Pt 1-92. Tractors and Machinery for Agriculture and Forestry—Guards for Power Take-off (PTO) Drive-Shafts—Part 1: Strength Test First Edition. 9 pp.

—Forestry Equipment—Wear Testing
CEN PREN 1152-93. Tractors and Machinery for Agriculture and Forestry—Guards for Power Take-Off (PTO) Drive Shafts—Wear and Strength Tests. 12 pp.

Drive Shafts *(Cont.)*
—Forestry Equipment—Wear Testing *(Cont.)*
ISO 5674 Pt 2-92. Tractors and Machinery for Agriculture and Forestry—Guards for Power Take-off (PTO) Drive-Shafts—Part 2: Wear Test First Edition. 8 pp.

—Ground Effect Machines
CAA Chapter B7-4 04.79. Transmission Items. 1 p.

—Heights
BSI BS 5186-75. 1975 Shaft Centre Heights. 9 pp.
CNS B4001-47. Height of Shaft, for Driving and Driven Lines (Mar)(146)(R 1973).
ISO 496-73. Driving and Driven Machines—Shaft Heights First Edition. 5 pp.
JIS B 0902-76. Shaft Heights for Driving and Driven Machines. 7 pp.

—Internal Combustion Engines
BSI BS 3092-73. 1973 Main Friction Clutches, Main Power-Take-Off Assemblies and Associated Attachment for Internal Combustion Engines. 37 pp.
BSI BS 5208-75. 1975 Dimensions of Power Take-Off Driving Shafts and Mounting Faces for Small Internal Combustion Engines. 13 pp.

—Involute Splines—Aircraft
SBAC RS 623 (V). Accessory Drives and Mounting Pads (Taper Flanges—Involute Splines).

—Involute Splines—Automotive
JIS D 2001-59. Involute Spline for Automobiles (R 1974). 41 pp.

—Joints—Automotive
DIN ENGL 75532 Pt 1-76. Transmission of Rotary Motions; Types of Connection to Gears, Intermediate Gears, Flexible Drive Shafts and Equipment (June). 8 pp.

—Mounting Pads—Aircraft
SBAC RS 623 (V). Accessory Drives and Mounting Pads (Taper Flanges—Involute Splines).

—Multispindle Heads
DIN ENGL 69001 Pt 11-81. Machine Tools; Multi-Spindle Heads; Assemblies; Drives; Types A to K (Oct). 22 pp.
DIN ENGL 69001 Pt 30-81. Machine Tools; Multi-Spindle Heads; Drive Shafts; Types A to F (Oct). 7 pp.
ISO 2912-73. Modular Units for Machine Tool Construction—Multi-Spindle Heads—Casing and Input Drive Shaft Dimensions First Edition; (Amendment 1-1982). 6 pp.

—Pads—Aircraft
SBAC RS 542 (V). 5 Inch P.C.U. Pad and Drive 4.400 Inch P.C.D. X. 650 Inch Three Oilway Unified Threads.
SBAC RS 543 (V). 6 Inch P.C.U. Pad and Drive 5.400 Inch P.C.D. X. 875 Inch Three Oilway Unified Threads.
SBAC RS 571-588 (V)(I). Accessory, Drive and Pad (Taper Flanges).
SBAC RS 589-594 (V)(I). Accessory Drives and Pads (Taper Flanges).

—Pinion Located—Aircraft
SBAC RS 341 ISSUE 3. Circular Pad and Drive (Pinion Located) 8.000 Inch P.C.DIA. X. 1.125 Inch.

—Propeller Control—Pads—Aircraft
SBAC RS 342 (V). 5 Inch Propeller Control Unit Pad and Drive 4.400 Inch P.C.D. X. 650 Inch Three Oilway.
SBAC RS 343 (V). 6 Inch Propeller Control Unit Pad and Drive 5.400 Inch P.C.D. X. 875 Inch Three Oilway.

—Speed Unit—Aircraft
SBAC RS 312 (V). 5 Inch Constant Speed Unit Pad and Drive 4.400 Inch PCD X. 650 Inch Twin Oilway.
SBAC RS 313 (V). 6 Inch Constant Speed Unit Flange and Shaft 5.400 Inch P.C.D. Inch X. 875 Inch Twin Oilway.

—Superchargers—Aircraft
SBAC RS 323 (V). Supercharger Mounting Flange 6.000 Inch CTS.
SBAC RS 324 (V). Supercharger Mounting Flange 2.4404 Inch (62 m/m) CTS..
SBAC RS 325 (V). Supercharger Mounting Flange 3.307 Inch (84 m/m) CTS..
SBAC RS 326 (V). Supercharger Mounting Flange 4.500 Inch CTS..

—Taper Flanges—Aircraft
SBAC RS 571-588 (V)(I). Accessory, Drive and Pad (Taper Flanges).

Drive

Drive Shafts (Cont.)
—Torsional Vibration
BSI BS 5514: Part 5-79. 1979 Reciprocating Internal Combustion Engines: Performance Part 5: Torsional Vibrations. 4 pp.
ISO 3046 Pt V-78. Reciprocating Internal Combustion Engines: Performance—Part V: Torsional Vibrations First Edition; (Erratum—April 1979). 4 pp.

—Tractors
BSI BS 3417: Part 1-84. 1984 1974 Power Take-Off Shafts and Guards for Tractors and Machinery for Agriculture and Forestry Part 1: Power Take-Off Drive Shafts. 8 pp.
ISO 5673-80. Agricultural Tractors—Power Take-off Drive Shafts for Machines and Implements First Edition. 7 pp.

—Trim Controls—Aircraft
SBAC AS 1241 ISSUE 2. Shaft, Trim Control—Direct Drive. 1 p.
SBAC AS 6226 ISSUE 1. Shaft Direct Drive Trim Control. 1 p.

Driven Piles
See Also: Piles

—Construction Contracts
DIN ENGL 18304-92. Construction Contract Procedures (VOB); Part C: General Technical Specifications in Construction Contracts (ATV); Pile Driving (Dec). 4 pp.

—Loads (Forces)
DIN ENGL 4026-75. Driven Piles; Manufacture, Dimensioning and Permissible Loading (Aug). 11 pp.
DIN ENGL 4026 Suppl. 1-75. Driven Piles; Manufacture, Dimensioning and Permissible Loading; Explanations (Aug). 6 pp.

Driver/Decoders
Use For: Decoder/Drivers *See Also:* Decoders; Display Drivers

—BCD to Decimal
CECC CECC 90 103-132 ISSUE 1-81. BS CECC 90 103-132; BCD-to-Decimal Decoder/Drivers (En) AMD 1 (En). 20 pp.

—BCD to 7-Segment
CECC CECC 90 101-022 ISSUE 1-87. UTE C 86-213/B 42; Digital Integrated Circuits in Accordance with FS 90 101; 54/64/74 47A; BCD to Seven Segment Decoders/Drivers with Open Collector Outputs (En). 20 pp.
CECC CECC 90 103-028 ISSUE 1-81. BS CECC 90 103-028; B-C-D To Seven Segment Decoder/Drivers with 15, OV Open Collector Outputs (En) AMD 1 (En). 19 pp.
CECC CECC 90 103-029 ISSUE 1-81. BS CECC 90 103-029; BCD/7 Segment Decoder/Driver with Open Collector Outputs (En) AMD 1 (En). 18 pp.
CECC CECC 90 103-093 ISSUE 1-82. BS CECC 90 103-093; BCD-Seven-Segment Decoders/Drivers (Active-Low Open Collector 15,OV Outputs) (En). 20 pp.
CECC CECC 90 103-094 ISSUE 1-82. BS CECC 90 103-094; BCD to-Seven-Segment Decoders/Drivers (Internal Pull-Up Outputs) (En). 20 pp.
CECC CECC 90 103-095 ISSUE 1-81. BS CECC 90 103-095; BCD-to-Seven-Segment Decoders/Drivers (Open Collector Outputs) (En) AMD 1 (En). 22 pp.
CECC CECC 90 103-116 ISSUE 1-81. BS CECC 90 103-116; BCD-to-Seven Segment Decoders/Drivers with Open-Collector Outputs (En) AMD 1 (En). 23 pp.
CECC CECC 90 103-133 ISSUE 1-82. BS CECC 90 103-133; BCD— to-Seven Segment Decoders/Drivers with Open Collector Outputs (En). 21 pp.
CECC CECC 90 103-134 ISSUE 1-81. BS CECC 90 103-134; BCD-to-Seven Segment Decoder Drivers (En) AMD 1 (En). 20 pp.

Driver Speakers
Use: Speakers

Drivers (Vehicles)
Use: Motor Vehicle Operators

Drives (Mechanical)
Use: Mechanical Drives

Driving Axles
Use: Drive Axles

Driving Pinions
Use: Pinions—Driving

Driving Squares
Use: Socket Wrenches

Driving Testing (Motor Vehicles)
Use: Road Testing

Driving Wheels
See Also: Wheels

—Cranes—Bearing Rings
DIN ENGL 15094-82. Cranes; Driving Wheel Units and Idler Wheel Units; Bearing Cage Rings (July). 3 pp.

—Cranes—Industrial Plants
DIN ENGL 15090-82. Cranes; Driving Wheel Units and Idler Wheel Units; Assembly (July). 6 pp.

—Cranes—Traveling Wheels
DIN ENGL 15093-82. Cranes; Driving Wheel Units and Idler Wheel Units; Travelling Wheels (July). 5 pp.

—Cranes—Traveling Wheels—Shafts
DIN ENGL 15091-82. Cranes; Driving Wheel Units and Idler Wheel Units; Travelling Wheel Shafts (July). 5 pp.

Drop Lights
Use: Trouble Lights

Drop-Point Apparatus
Use For: Ubbelohde Apparatus
BSI BS 894-56. 1956 Amd 1 Ubbelohde Apparatus for Flow and Drop Points. 17 pp.

Drop Testing
Use: Impact Testing

Drop Wire
See Also: Electric Conductors; Electric Wire; Wire

—Aluminum Conductor—PVC Insulated—PVC Sheathed
CNS C2104-82. Aluminum Conductor Polyvinyl Chloride Insulated Service Drop Wires (Mar)(8558).
JIS C 3371-87. Aluminium Conductor Polyvinyl Chloride Insulated Service Drop Wires. 15 pp.

—Polyethylene Insulated
CNS C2166-84. Polyethylene Insulated Self-Supporting Drop Wires (Dec)(11164).

—PVC Insulated
CNS C2097-87. Polyvinyl Chloride Insulated Drop Service Wires (Oct)(7781).
JIS C 3341-92. Polyvinyl Chloride Insulated Drop Service Wires. 11 pp.

—Transformers
CNS C2080-83. High-Voltage Drop Wire for Pole Transformer (Jul)(6073).
JIS C 3609-87. High-Voltage Drop Wire for Pole Transformer. 10 pp.

Drop Wires (Textile Machinery)
See Also: Textile Machinery
BSI BS 2609: Part 1-79. 1979 Warp Stop Materials Part 1: Dimensions of Drop Wires for Mechanical and Electrical Warp Stop Motions. 8 pp.
BSI BS 2609: Part 2-79. 1979 Warp Stop Materials Part 2: Dimensions of Closed-End Drop Wires for Mechanical and Electrical Warp Stop Motions and Automatic Drawing-In Machines. 8 pp.
ISO 441-78. Textile Machinery and Accessories—Drop Wires for Mechanical and Electrical Warp Stop Motions First Edition. 7 pp.
ISO 1150-78. Textile Machinery and Accessories—Closed-End Drop Wires for Mechanical and Electrical Warp Stop Motions and Automatic Drawing-in Machines First Edition. 7 pp.

—Looms
CNS L2030-82. Drop Wires for Looms (Mar)(8606).
JIS L 6212-86. Drop Wires for Looms.

Drop Zones
See Also: Air Drop Operations; Extraction Zones

—Tactical Air Transport Operations
NATO STANAG 3570 ED 3 AMD 4-86. Drop Zones and Extraction Zones—Criteria and Markings. 26 pp.
NATO STANAG 3570 ED 3 AMD 5-86. Drop Zones and Extraction Zones—Criteria and Markings. 27 pp.

—Tactical Air Transport Operations—Identification Systems
NATO STANAG 3570 ED 3 AMD 4-86. Drop Zones and Extraction Zones—Criteria and Markings. 26 pp.
NATO STANAG 3570 ED 3 AMD 5-86. Drop Zones and Extraction Zones—Criteria and Markings. 27 pp.

Dropout
See Also: Data Transmission; Signal Levels

—Voice Band Data Transmission
CCITT RECMN G.113-89. Transmission Impairments—General Characteristics of International Telephone Connections and Circuits (Study Groups XII and XV) 22 pp. 22 pp.

Dropped Ceilings
Use: Ceilings

Dropping Funnels
Use: Funnels—Glass—Dropping

Dropping Point
See Also: Greases; Thermodynamic Properties
BSI BS 2000: Part 132-90. 1990 Petroleum and Its Products Part 132: Dropping Point of Lubricating Grease (Supersedes BS 2877: 1976). 7 pp.
ISO 2176-72. Petroleum Products—Lubricating Grease—Determination of Dropping Point First Edition. 4 pp.

—Fats
BSI BS 684: Sec 1.4-76. 1976 Methods of Analysis of Fats and Fatty Oils Part 1: Physical Methods Section 1.4: Determination of Flow and Drop Points. 8 pp.

—Greases
BSI BS 2000: Part 132-93. 1993 Methods of Test for Petroleum and Its Products Part 132: Determination of Dropping Point of Lubricating Grease. 4 pp.

Drug Packaging
Use For: Pharmaceutical Packaging
See Also: Packaging
BSI BS 1679-65. (OBSOLESCENT) 1965 Amd 5 Containers for Pharmaceutical Dispensing. 31 pp.

—Child Resistant
BSI DD 30: Part 2-73. 1973 Resistance of Pharamaceutical Packages to Opening by Children Part 2: Non-Reclosable Unit Packages. 9 pp.

—Dispensing
BSI BS 1679: Part 1-76. 1976 Containers for Pharmaceutical Dispensing Part 1: Paperboard Containers for Strip and Blister Packs. 7 pp.

—Export
CNS Z5088-81. Packaging and Packing of Medicines for Export (Jun)(7540).

Drugs
Use For: Pharmaceuticals *See Also:* Anesthetics; Disinfectants; Histamine Content Analysis; Medicine Measures; Morphine; Ointments; Therapy; Vitamin Supplements

—Bottles
BSI BS 1679: Part 5-73. 1973 Amd 1 Containers for Pharmaceutical Dispensing Part 5: Eye Dropper Bottles. 10 pp.
BSI BS 1679: Part 6-84. 1984 Containers for Pharmaceutical Dispensing Part 6: Specification for Glass Medicine Bottles. 8 pp.
BSI BS 1679: Part 7-68. 1968 Containers for Pharmaceutical Dispensing Part 7: Ribbed Oval Glass Bottles. 13 pp.
BSI BS 1679: Part 8-92. 1992 Containers for Pharmaceutical Dispensing Part 8: Specification for Glass and Plastics Containers for Solid Dosage Forms, Semi-Solids and Powders (Supersedes BS 1679: 1965). 14 pp.
BSI BS 1679: Part 7-01. 1968 Amd 1 Containers for Pharmaceutical Dispensing Part 7: Ribbed Oval Glass Bottles (AMD 7203) August 15, 1992. 13 pp.
CNS R2074-72. Glass Bottles for Chemicals and Agricultural Drugs (Oct) (3502).
JIS R 3522-77. Glass Bottles for Drug.

—Classification
BSI BS 1000: (615)-79. 1979 Amd 1 Universal Decimal Classification (UDC). English Full Edition (615): Pharmaceutics. Therapeutics. Toxicology. 26 pp.
SNZ NZS/BS 1000 (615)-79. Universal Decimal Classification Pharmaceutics. Therapeutics. Toxicology. 28 pp.

—Containers
BSI BS 1679-65. (OBSOLESCENT) 1965 Amd 5 Containers for Pharmaceutical Dispensing. 31 pp.
BSI BS 1679: Part 1-76. 1976 Containers for Pharmaceutical Dispensing Part 1: Paperboard Containers for Strip and Blister Packs. 7 pp.

—Hospital Carts
CNS T2024-83. Cart, Medicine (Sep)(10582).

INTERNATIONAL AND NON-U.S. NATIONAL STANDARDS
SUBJECT INDEX

Drugs *(Cont.)*

—Medical Material—Recall Procedures
NATO STANAG 2907 ED 2 AMD 0-89. Procedures for Reporting and for Initial Disposition of Unsatisfactory Medical Materiel and Drugs. 11 pp.

—Packaging
BSI BS 1679-65. (OBSOLESCENT) 1965 Amd 5 Containers for Pharmaceutical Dispensing. 31 pp.

—Packaging—Child Resistant
BSI BS 7236-89. 1989 Code of Practice for Non-Reclosable Packaging for Solid Dose Units of Medicinal Products. 8 pp.
BSI DD 30: Part 2-73. 1973 Resistance of Pharamaceutical Packages to Opening by Children Part 2: Non-Reclosable Unit Packages. 9 pp.

—Packaging—Dispensing
BSI BS 1679: Part 1-76. 1976 Containers for Pharmaceutical Dispensing Part 1: Paperboard Containers for Strip and Blister Packs. 7 pp.

—Packaging—Export
CNS Z5088-81. Packaging and Packing of Medicines for Export (Jun)(7540).

—Plastic Containers
BSI BS 1679: Part 4-69. 1969 Containers for Pharmaceutical Dispensing Part 4: Plastic Containers for Tablets and Ointments. 17 pp.

—Syringes—Labels
CSA CAN/CSA-Z327-M91. Standard for User-Applied Drug Labels in Anaesthesia and Critical Care; (Gen Instr 1). 16 pp.

Drum Brakes

See Also: Air Brakes; Automotive Equipment; Brake Shoes; Brakes; Disk Brakes; Hydraulic Brakes; Wheel Cylinders

—Cylinders—Boots—Rubber—Automotive
ISO 4927-78. Road Vehicles—Elastomeric Boots for Drum Type Hydraulic Brake Wheel Cylinders Using a Non-Petroleum Base Hydraulic Brake Fluid (Service Temperature 120 Degrees Celsius Maximum) First Edition. 6 pp.
ISO 6117-80. Road Vehicles—Elastomeric Boots for Drum Type Hydraulic Brake Wheel Cylinders Using A Non-Petroleum Base Hydraulic Brake Fluid (Service Temperature 100 Degrees Celsius Maximum) First Edition. 6 pp.
ISO 7633-85. Road Vehicles—Elastomeric Boots for Drum Type Hydraulic Brake Wheel Cylinders Using a Petroleum Base Hydraulic Brake Fluid (Service Temperature 120 Degrees Celsius Max.) First Edition. 6 pp.

—Cylinders—O Rings—Rubber—Automotive
ISO 7630-85. Road Vehicles—Elastomeric O-Rings for Hydraulic Drum Brake Wheel Cylinders Using a Petroleum Base Hydraulic Brake Fluid (Service Temperature 120 Degrees Celsius Max.) First Edition. 7 pp.

Drum Chart Recorders
Use: Chart Recorders

Drum Conveyors
Use: Belt Conveyors

Drum Feeders
Use For: Tumbler Feeders *See Also:* Conveyors; Materials Handling Equipment

—Safety
BSI BS 5667: Part 6-79. 1979 Continuous Mechanical Handling Equipment. Safety Requirements Part 6: Loose Bulk Materials: Rotary Feeders Used in Pneumatic Handling. 3 pp.
BSI BS 5667: Part 7-79. 1979 Continuous Mechanical Handling Equipment. Safety Requirements Part 7: Loose Bulk Materials: Rotary Drum Feeders and Rotary Vane Feeders. 3 pp.
ISO 5033-77. Continuous Mechanical Handling Equipment for Loose Bulk Materials: Rotary Drum Feeders and Rotary Vane Feeders—Safety Code First Edition. 3 pp.

Drum Sanders

See Also: Sanders (Tools)
CNS B1303-84. Nominal Dimension of Drum Sanders (Jun)(10055).
CNS B7245-84. Test Code for Accuracy of Drum Sanders (Jun)(10225).
CNS B7263-84. Test Code for Performance of Drum Sanders (Mar)(10786).
JIS B 6545-91. Drum Sanders—Test and Inspection Methods. 13 pp.

Drums (Containers)

Scope Note: For additional listings, use a more specific term *See Also:* Barrels (Containers); Cans; Containers; Shipping Containers; Tanks (Containers)

—Cables (Ropes)—Aircraft
SBAC AS 2824 ISSUE 6. Drum Assembly. 1 p.
SBAC AS 2825 ISSUE 6. Drum. 1 p.

—Cleaning Agents—Fiberboard
BSI BS 5167-78. 1978 Packages for Washing and Cleaning Powders. Dimensions and Volumes of Cartons and Drums from Fibreboard. 9 pp.
CEN EN 23-1-78. Packages for Washing and Cleaning Powders—Part 1: Dimensions and Volumes of Cartons and Drums from Fibreboard. 7 pp.

—Detergents—Fiberboard
BSI BS 5167-78. 1978 Packages for Washing and Cleaning Powders. Dimensions and Volumes of Cartons and Drums from Fibreboard. 9 pp.
CEN EN 23-1-78. Packages for Washing and Cleaning Powders—Part 1: Dimensions and Volumes of Cartons and Drums from Fibreboard. 7 pp.

—Fiberboard
BSI BS 1133: SUBSEC 7.4-89. 1989 Packaging Code Section 7: Paper and Board Wrappers, Bags and Containers Subsection 7.4: Fibreboard Drums. 16 pp.
BSI BS 1596-74. 1974 Fibreboard Drums for Shipment of Goods Overseas. 9 pp.

—Fiberboard—Construction
BSI BS 1596-92. 1992 Fibreboard Drums (H). 11 pp.

—Flanges
JIS Z 1604-84. Plugs and Flanges for Steel Drums. 12 pp.

—Hazardous Materials—Fiberboard
CGSB CAN/CGSB-43.86-M90. Fibre Drums (TC-21C). 12 pp.
CGSB CAN/CGSB-43.87-M90. Fibre Drum for an Inside Plastic Container (TC-21P). 11 pp.

—Hazardous Materials—Polyethylene
CGSB CAN/CGSB-43.55-M87. Drums, Polyethylene, Reusable, Non-Removable Head, Without Overpack (TC-34). 11 pp.
CGSB CAN/CGSB-43.60-92. Non-Reusable Polyethylene Drum, Removable Head (TC-35). 9 pp.

—Hazardous Materials—Steel
CGSB CAN/CGSB-43.69-M90. Cylindrical Steel Overpack for Inside Plastic Container (TC-6D). 10 pp.
CGSB CAN/CGSB-43.96-M89. Steel Drums, Removable, or Non-Removable Head (TC-5B). 12 pp.
CGSB CAN/CGSB-43.97-M90. Stainless Steel Drums, Non-Removable Head (TC-5C). 12 pp.
CGSB CAN/CGSB-43.98-M90. Steel Drums, Removable, or Non-Removable Head (TC-6J). 11 pp.
CGSB CAN/CGSB-43.99-M88. Barrels or Drums, Steel (TC-6B). 11 pp.
CGSB CAN/CGSB-43.114-M90. Non-Reusable, Steel Drums, Removable Head (TC-37C). 10 pp.
CGSB CAN/CGSB-43.115-M90. Non-Reusable, Cylindrical, Steel Overpack for Inside Plastic Containers (TC-37M). 10 pp.
CGSB CAN/CGSB-43.116-M90. Non-Reusable Steel Drums with Polyethylene Liners (TC-37P). 11 pp.
CGSB CAN/CGSB-43.117-M89. Single-Trip Steel Drums, Removable or Non-Removable Head (TC-17C). 12 pp.
CGSB CAN/CGSB-43.118-M89. Single-Trip Steel Drums, Non-Removable Head (TC-17E). 12 pp.
CGSB CAN/CGSB-43.119-M89. Single-Trip Steel Drums, Non-Removable Head (TC-17F). 12 pp.
CGSB CAN/CGSB-43.121-M90. Single-Trip Steel Drums, Removable Head (TC-37A). 11 pp.
CGSB CAN/CGSB-43.122-M89. Single-Trip Steel Drums, Non-Removable Head (TC-37B). 11 pp.

—Liquids—Steel
JIS Z 1601-86. Steel Drums (for Liquid).
JIS Z 1601-76. Steel Drums (for Liquid). 10 pp.

—Metal
BSI BS 1133: SUBSEC 10.2-91. 1991 Packaging Code Section 10: Metal Containers Subsection 10.2: Metal Drums. 20 pp.

—Oil—Flanges—Steel
CNS Z5016-86. Plugs and Flanges for Steel Drums for Oil (Apr)(1322).

—Oil—Plugs—Steel
CNS Z5016-86. Plugs and Flanges for Steel Drums for Oil (Apr)(1322).

—Oil—Steel
CNS Z5016-86. Plugs and Flanges for Steel Drums for Oil (Apr)(1322).

Drums (Containers) *(Cont.)*

—Oil—Steel *(Cont.)*
CNS Z5112-86. Steel Drums for Oil (Apr)(9446).
CNS Z6002-86. Method of Test for Steel Drums for Oil (Apr)(1318).

—Plugs—Openers
CNS M2060-86. Non-Sparking Beryllium Copper Alloy Tools Drums Plug Openers (Nov)(5908).

—Plugs—Steel
JIS Z 1604-84. Plugs and Flanges for Steel Drums. 12 pp.

—Steel
BSI BS 814-74. 1974 Amd 1 Mild Steel and Tinplate Drums (Light Duty; Fixed Ends). 11 pp.
BSI BS 814: Part 1-87. 1987 Mild Steel and Tinplate Drums (Light Duty; Fixed Ends) Part 1: Steel Drums with a Total Capacity of 216.5 Litres. 8 pp.
BSI BS 1702-50. 1950 Amd 1 Mild Steel Drums (Heavy Duty-Fixed Ends). 18 pp.
BSI BS 2003-74. 1974 Amd 1 Mild Steel Drums (Light Duty; Removable Heads). 13 pp.
BSI BS 2003: Part 1-87. 1987 Mild Steel Drums (Light Duty: Removeable Heads) Part 1: Drums with a Total Capacity of 213 Liters. 8 pp.
CEN EN 209-86. Steel Drums: Open (Removable) Head Drums with a Total Capacity of 213 Litres. 5 pp.
CEN EN 210-86. Steel Drums: Tight Head Drums with a Total Capacity of 216,5 Litres. 5 pp.
DIN ENGL 6634-76. Means of Packaging; Rolled I-Steel for Drums; Dimensions (Jan). 1 p.
JIS Z 1600-88. Full Removable Head Steel Drums. 13 pp.
JIS Z 1604-84. Plugs and Flanges for Steel Drums. 12 pp.

—Tinplate
BSI BS 814-74. 1974 Amd 1 Mild Steel and Tinplate Drums (Light Duty; Fixed Ends). 11 pp.

Drums (Hoisting)
Use: Hoisting Drums

Drums (Materials Handling Equipment)
Use: Reels

Dry Cell Batteries
Use: Leclanche Batteries

Dry Cells
Use: Leclanche Batteries

Dry Cleaning

See Also: Dry Cleaning Fluids; Solvents

—Fabrics
BSI BS 5742-89. 1989 Textile Labels Requiring to Be Washed and/or Dry Cleaned. 8 pp.

—Fabrics—Colorfastness Testing
CGSB CAN/CGSB-4.2 NO.29.1-M89. Textile Test Methods Colourfastness to Dry Cleaning Solvent. 9 pp.
CNS L3155-82. Method of Test for Colour Fastness to Dry Cleaning (Jan)(8434).
ISO 105 Pt D01-87. Textiles—Tests for Colour Fastness—Part D01: Colour Fastness to Dry Cleaning Third Edition. 4 pp.
JIS L 0860-74. Testing Method for Colour Fastness to Dry Cleaning. 4 pp.

—Fabrics—Dimensional Stability
BSI BS 4961: Part 1-80. 1980 Determination of Dimensional Stability of Textiles to Dry Cleaning in Tetrachloroethylene Part 1: Machine Method. 7 pp.
CGSB CAN/CGSB-4.2 NO.30-M90. Textile Test Methods Dimensional Change in Dry Cleaning. 9 pp.
CGSB CAN/CGSB-4.2 NO.66-M91. Textile Test Methods Dimensional Change and Appearance After Dry Cleaning of Coated, Bonded, Laminated and Fused Fabrics. 9 pp.
ISO 3175-79. Textiles—Determination of Dimensional Change on Dry Cleaning in Perchlorethylene—Machine Method Second Edition. 6 pp.

—Leather
BSI BS 7269: Part 1-91. 1991 Drycleanability of Leather Garments Part 1: Specification for Drycleanability and for Appropriate Care Labels. 11 pp.
BSI BS 7269: Part 2-91. 1991 Drycleanability of Leather Garments Part 2: Method for Assessing Drycleanability. 12 pp.

INDUSTRY STANDARDS

Dry Cleaning (Cont.)

—Leather—Labels
BSI BS 7269: Part 1-91. 1991 Drycleanability of Leather Garments Part 1: Specification for Drycleanability and for Appropriate Care Labels. 11 pp.

—Yarns—Colorfastness Testing
CGSB CAN/CGSB-4.2 NO.29.1-M89. Textile Test Methods Colourfastness to Dry Cleaning Solvent. 9 pp.

Dry Cleaning Equipment
See Also: Cleaning Equipment and Supplies

DIN ENGL 11902-79. Laundry and Dry-Cleaning Machines; Testing of Output, Effect of Treatment and Consumption of Working Materials; Test Conditions and Methods (Sept). 4 pp.

ISO 8232-88. Closed-Circuit Dry-Cleaning Machines—Defining and Checking of Machine Characteristics First Edition. 8 pp.

—Glossaries
BSI BS ISO 8229-91. 1991 Operations and Baths Relating to Dry-Cleaning Machines—Vocabulary (L). 10 pp.

ISO 8229-91. Operations and Baths Relating to Dry-Cleaning Machines—Vocabulary First Edition. 8 pp.

Dry Cleaning Fluids
See Also: Dry Cleaning; Perchloroethylene; Solvents

—Solubility
ISO 6837-82. Surface Active Agents—Water Dispersing Power in Dry Cleaning Solvents First Edition. 7 pp.

Dry Heat Testing
See Also: Environmental Testing; High Temperature Testing

—Connectors
CEN PREN 2591 (Part C9)-92. Elements of Electrical and Optical Connection Test Methods Part C9—Dry Heat. 4 pp.

—Electrical Components
BSI BS 2011: Part 2.1B-77. (WITHDRAWN) 1977 Amd 1 Basic Environmental Testing Procedures Part 2.1: Tests Part 2.1B: Test B. Dry Heat (Superseded by BS EN 60068-2-2: 1993). 32 pp.

BSI BS 2011: Pt 2.1Z/BFC-84. 1984 Basic Environmental Testing Procedures Part 2.1: Tests Part 2.1Z/BFc: Test Z/BF Combined Dry Heat/Vibration (Sinusoidal) Tests for Both Heat-Dissipating and Non-Heat-Dissipating Specimens. 13 pp.

BSI BS 2011: Part 2.1Z/BM-77. 1977 Amd 1 Basic Environmental Testing Procedures Part 2.1: Tests Part 2.1Z/BM: Test Z/BM. Combined Dry Heat/Low Air Pressure Tests. 12 pp.

BSI BS 2011:Pt2. 2Z/AFC/BFC-86. 1986 Basic Environmental Testing Procedures Part 2.2: Guidance Part 2.2Z/AFc and Z/BFc: Test Z/AFc and Z/BFc. Guidance on Combined Temperature (Cold and Dry Heat) and Vibration (Sinusoidal) Tests. 8 pp.

BSI BS 2011: Part 3A & B-77. 1977 Amd 1 Basic Environmental Testing Procedures Part 3: Background Information Part 3A and B: Tests A (Cold) and Tests B (Dry Heat). 37 pp.

BSI BS 2011:Pt 3 A & B:Supp 1-80. 1980 Basic Environmental Testing Procedures Part 3: Background Information Part 3A and B: Tests A (Cold) and Tests B (Dry Heat): Supplement No. 1:. 5 pp.

BSI BS EN 60068-2-2-93. 1993 Environmental Testing Part 2: Tests Tests B. Dry Heat (IEC 68-2-2: 1974) (G). 39 pp.

CENELEC HD 323.2.41 S1-88. Basic Environmental Testing Procedures-Part 2: Tests. Test Z/BM: Combined Dry Heat/Low Air Pressure Tests. 3 pp.

CENELEC HD 323.3.1 S1-88. Basic Environmental Testing Procedures Part J: Background Information Section One—Cold and Dry Heat Tests. 2 pp.

CENELEC EN 60068-2-2-93. Basic Environmental Testing Procedures Part 2: Tests Tests B: Dry Heat (Includes Amendment A1: 1993) (Supersedes HD 323.2.2 S1: 1988) (IEC 68-2-2: 1974 + IEC 68-2-2A: 1976 + A1: 1993). 34 pp.

CNS C6021-89. Dry Heat Testing Procedures for Electronic Components (Jul)(3634).

IEC 68 Pt 2-2-74. Basic Environmental Testing Procedures Part 2: Tests—Tests B: Dry Heat Fourth Edition; (Supplement A-1976) (Amendment 1-1993) (CENE EN 60068-2-2: 1993 Includes Amendment A1: 1993). 74 pp.

IEC 68 Pt 2-41-76. Basic Environmental Testing Procedures Part 2: Tests—Test Z/BM: Combined Dry Heat/ Low Air Pressure Tests First Edition; (Amendment 1-1983). 24 pp.

Dry Heat Testing (Cont.)

—Electrical Components (Cont.)
IEC 68 Pt 2-51-83. Basic Environmental Testing Procedures Part 2: Tests Tests Z/BFc: Combined Dry Heat/Vibration (Sinusoidal) Tests for Both Heat-Dissipating and Non-Heat-Dissipating Specimens First Edition. 25 pp.

IEC 68 Pt 2-53-84. Basic Environmental Testing Procedures Part 2: Tests Guidance to Tests Z/AFc and Z/BFc: Combined Temperature (Cold and Dry Heat) and Vibration (Sinusoidal) Tests First Edition. 14 pp.

IEC 68 Pt 3-1-74. Basic Environmental Testing Procedures Part 3: Background Information Section One—Cold and Dry Heat Tests First Edition; (Supplement A-1978) (Corrigenda—March 1980). 73 pp.

JIS C 0021-87. Basic Environmental Testing Procedures Part 2: Tests, Tests B: Dry Heat. 35 pp.

JIS C 0031-90. Basic Environmental Testing Procedures Part 2: Tests-Test Z/BM Combined Dry Heat/Low Air Pressure Tests. 10 pp.

SAA AS 1099.2BA-80. Basic Environmental Testing Procedures for Electrotechnology—Part 2: Tests—Part 2Ba: Dry Heat Test for Non-Heat-Dissipating Specimens with Sudden Change of Temperature.

SAA AS 1099.2BB-80. Basic Environmental Testing Procedures for Electrotechnology—Part 2: Tests—Part 2Bb: Dry Heat Test for Non-Heat-Dissipating Specimens with Gradual Change of Temperature.

SAA AS 1099.2BC-80. Basic Environmental Testing Procedures for Electrotechnology—Part 2: Tests—Part 2Bc: Dry Heat Test for Heat-Dissipating Specimens with Sudden Change of Temperature.

SAA AS 1099.2BD-80. Basic Environmental Testing Procedures for Electrotechnology—Part 2: Tests—Part 2Bd: Dry Heat Test for Heat-Dissipating Specimens with Gradual Change of Temperature (Bound Together). 24 pp.

SAA AS 1099.2Z/B M-80. Basic Environmental Testing Procedures for Electrotechnology—Part 2: Tests—Part 2Z/BM: Combined Dry Heat/Low Air Pressure Tests Reconfirmed 1985. 6 pp.

SAA AS 1099.3.1-80. Basic Environmental Testing Procedures for Electrotechnology—Part 3: Background Information—Part 3.1: Section 1 Tests A and B-Cold and Dry Heat Tests. 32 pp.

SNZ IEC 68: Part 2-2-74. Basic Environmental Testing Procedures Part 2-2: Tests B: Dry Heat. 63 pp.

SNZ IEC 68: Part 2-2A-76. Basic Environmental Testing Procedures Part 2-2A: Tests B: Dry Heat. 2 pp.

SNZ IEC 68: Part 2-41-76. Basic Environmental Testing Procedures Part 2-41: Test Z/Bm: Combined Dry Heat/Low Air Pressure Tests Amend: 1. 19 pp.

SNZ IEC 68: Part 2-51-83. Basic Environmental Testing Procedures Part 2-51: Tests z/BFc: Combined Dry Heat/Vibration (Sinusoidal) Tests for Both Heat-Dissipating and Non-Heat -Dissipating Specimens. 21 pp.

SNZ IEC 68: Part 2-53-84. Basic Environmental Testing Procedures Part 2-53: Guidance to Tests Z/AFc and Z/BFc: Combined Temperature (Cold and Dry Heat) and Vibration (Sinusoidal) Tests. 10 pp.

SNZ IEC 68: Part 3-1-74. Basic Environmental Testing Procedures Part 3-1: Background Information. Section One —Cold and Dry Heat Tests. 59 pp.

SNZ IEC 68: Part 3-1A-78. Basic Environmental Testing Procedures Part 3-1A: Background Information. Section One —Cold and Dry Heat Tests. 7 pp.

—Fabrics
BSI BS 7552: Part 1-92. 1992 Effect of Dry Heat on Fabrics Under Low Pressure Part 1: Method for Dry Heat Treatment of Fabrics (ISO 9866-1: 1991). 8 pp.

BSI BS 7552: Part 2-92. 1992 Effect of Dry Heat on Fabrics Under Low Pressure Part 2: Method for Determination of Dimensional Change in Fabrics Exposed to Dry Heat (ISO 9866-2: 1991). 8 pp.

CNS L3178-82. Method of Test for Colour Fastness to Dry Heating (Jun)(9021).

ISO 9866 Pt 1-91. Textiles—Effect of Dry Heat on Fabrics Under Low Pressure—Part 1: Procedure for Dry-Heat Treatment of Fabrics First Edition. 5 pp.

ISO 9866 Pt 2-91. Textiles—Effect of Dry Heat on Fabrics Under Low Pressure—Part 2: Determination of Dimensional Change in Fabrics Exposed to Dry Heat First Edition. 6 pp.

JIS L 0879-75. Testing Method for Colour Fastness to Dry Heating (R 1983). 6 pp.

—Paints—Automotive
BSI BS AU 148: Part 9-69. 1969 Methods of Test for Motor Vehicle Paints Part 9: Resistance to Dry Heat. 2 pp.

Dry Heat Testing (Cont.)

—Paperboard
BSI BS 6388: Part 1-91. 1991 Accelerated Ageing of Paper and Board Part 1: Method for Dry Heat Treatment at 105 Degrees Celsius (ISO 5630-1: 1991). 8 pp.

BSI BS 6388: Part 2-87. 1987 Accelerated Ageing of Paper and Board Part 2: Method for Dry Heat Treatment at 120 Degrees C or 150 Degrees C. 7 pp.

ISO 5630 Pt 1-91. Paper and Board—Accelerated Ageing—Part 1: Dry Heat Treatment at 105 Degrees Celsius Second Edition. 7 pp.

ISO 5630 Pt 4-86. Paper and Board—Accelerated Ageing—Part 4: Dry Heat Treatment at 120 or 150 Degrees Celsius First Edition. 5 pp.

—Papers
BSI BS 6388: Part 1-91. 1991 Accelerated Ageing of Paper and Board Part 1: Method for Dry Heat Treatment at 105 Degrees Celsius (ISO 5630-1: 1991). 8 pp.

BSI BS 6388: Part 2-87. 1987 Accelerated Ageing of Paper and Board Part 2: Method for Dry Heat Treatment at 120 Degrees C or 150 Degrees C. 7 pp.

ISO 5630 Pt 1-91. Paper and Board—Accelerated Ageing—Part 1: Dry Heat Treatment at 105 Degrees Celsius Second Edition. 7 pp.

ISO 5630 Pt 4-86. Paper and Board—Accelerated Ageing—Part 4: Dry Heat Treatment at 120 or 150 Degrees Celsius First Edition. 5 pp.

Dry Ice
See Also: Refrigerants

—Blocks
CNS K1022-52. Dry Ice, Solid Blocks (Mar)(196)(R 1973).

Dry Matter Content Analysis

—Admixtures
CEN PREN 480 (Part 8)-91. Admixtures for Concrete, Mortar and Grout—Test Methods—Part 8: Determination of the Conventional Dry Material Content. 5 pp.

—Bauxite
JIS M 8110-75. Method for Sampling of Bauxite Ores. 39 pp.

—Butter—Solvent Extraction
BSI BS 5086: Part 2-84. 1984 Analysis of Butter Part 2: Methods for Determination of Water, Solids-Non-Fat and Fat Contents (Reference Method). 6 pp.

ISO 3727-77. Butter—Determination of Water, Solids-Not-Fat and Fat Contents on the Same Test Portion (Reference Method) First Edition. 5 pp.

—Cheeses—Dehydration
ISO 2920-74. Whey Cheese—Determination of Dry Matter Content (Reference Method) First Edition. 4 pp.

—Chicory Extracts
EC 79/1066/EEC-79. First Commission Directive Laying Down Community Methods of Analysis for Testing Coffee Extracts and Chicory Extracts. 12 pp.

—Coffee Extracts
EC 79/1066/EEC-79. First Commission Directive Laying Down Community Methods of Analysis for Testing Coffee Extracts and Chicory Extracts. 12 pp.

—Condensed Milk
EC 79/1067/EEC-79. First Commission Directive Laying Down Community Methods of Analysis for Testing Certain Partly or Wholly Dehydrated Preserved Milk for Human Consumption. 24 pp.

—Copper Ores
JIS M 8101-88. Methods for Sampling, Preparation and Determination of Moisture Content of Non-Ferrous Metal Bearing Ores. 25 pp.

—Detergents
CGSB 2-GP-11M METH 17.2-93. Methods of Testing and Analysis of Soaps and Detergents Determination of Surfactant Content; (Amendment 4 March 1993). 3 pp.

—Fruits—Dehydration
ISO 1026-82. Fruit and Vegetable Products—Determination of Dry Matter Content by Drying Under Reduced Pressure and of Water Content by Azeotropic Distillation First Edition. 6 pp.

—Glucose
EC 79/786/EEC-79. First Commission Directive Laying down Community Methods of Analysis for Testing Certain Sugars Intended for Human Consumption. 29 pp.

INTERNATIONAL AND NON-U.S. NATIONAL STANDARDS
SUBJECT INDEX

Dry

Dry Matter Content Analysis (Cont.)
—**Glucose Syrups**
EC 79/786/EEC-79. First Commission Directive Laying down Community Methods of Analysis for Testing Certain Sugars Intended for Human Consumption. 29 pp.

—**Gold Ores**
JIS M 8101-88. Methods for Sampling, Preparation and Determination of Moisture Content of Non-Ferrous Metal Bearing Ores. 25 pp.

—**Ice Cream—Solvent Extraction**
ISO 3728-77. Ice-Cream and Milk Ice—Determination of Total Solids Content (Reference Method) First Edition. 5 pp.

—**Ice Milk—Solvent Extraction**
ISO 3728-77. Ice-Cream and Milk Ice—Determination of Total Solids Content (Reference Method) First Edition. 5 pp.

—**Lead Ores**
JIS M 8101-88. Methods for Sampling, Preparation and Determination of Moisture Content of Non-Ferrous Metal Bearing Ores. 25 pp.

—**Nonferrous Metals—Bulk Flotation**
JIS M 8083-84. Methods for Sampling of Non-Ferrous Flotation Concentrates in Bulk. 21 pp.

—**Pulps**
DIN ENGL 54351-77. Testing of Pulp; Determination of the Dry Weight of Baled Pulp; Determination by Means of Testing Single Bales (Nov). 7 pp.
DIN ENGL 54352-77. Testing of Pulp; Determination of the Dry Content of Pulp Samples (Oct). 4 pp.

—**Pulps—Dehydration**
BSI BS 4502-80. 1980 Determination of Dry-Matter Content of Pulps. 4 pp.
BSI BS 5878-80. 1980 Determination of Stock Concentration of Pulps (Rapid Method). 4 pp.
ISO 638-78. Pulps—Determination of Dry Matter Content First Edition. 4 pp.
ISO 4119-78. Pulps—Determination of Stock Concentration (Rapid Method) First Edition. 4 pp.

—**Silver Ores**
JIS M 8101-88. Methods for Sampling, Preparation and Determination of Moisture Content of Non-Ferrous Metal Bearing Ores. 25 pp.

—**Sugars**
EC 79/786/EEC-79. First Commission Directive Laying down Community Methods of Analysis for Testing Certain Sugars Intended for Human Consumption. 29 pp.

—**Tin Ores**
JIS M 8101-88. Methods for Sampling, Preparation and Determination of Moisture Content of Non-Ferrous Metal Bearing Ores. 25 pp.

—**Vegetables—Dehydration**
ISO 1026-82. Fruit and Vegetable Products—Determination of Dry Matter Content by Drying Under Reduced Pressure and of Water Content by Azeotropic Distillation First Edition. 6 pp.

—**Zinc Ores**
JIS M 8101-88. Methods for Sampling, Preparation and Determination of Moisture Content of Non-Ferrous Metal Bearing Ores. 25 pp.

Dry Meters
Use: Diaphragm Flowmeters

Dry Milk
Use: Powdered Milk

Dry Noodles
Use For: Chinese Noodles; Chow Mein Noodles
See Also: Instant Noodles
CNS N5158-79. Dry Noodles (Sep)(4991). 2 pp.

—**Boric Acid Content**
CNS N6112-79. Method of Test for Dry Noodles (Sep)(4992). 2 pp.

—**Cooking Test**
CNS N6112-79. Method of Test for Dry Noodles (Sep)(4992). 2 pp.

—**Impurities Content**
CNS N6112-79. Method of Test for Dry Noodles (Sep)(4992). 2 pp.

—**Microbiological Analysis**
CNS N6112-79. Method of Test for Dry Noodles (Sep)(4992). 2 pp.

—**Moisture Content**
CNS N6112-79. Method of Test for Dry Noodles (Sep)(4992). 2 pp.

Dry Noodles (Cont.)
—**Protein Content**
CNS N6112-79. Method of Test for Dry Noodles (Sep)(4992). 2 pp.

—**Sampling**
CNS N6112-79. Method of Test for Dry Noodles (Sep)(4992). 2 pp.

Dry Pipe Systems
Scope Note: Use a more specific term *See:* Dry Pipe Valves; Fire Sprinkler Equipment

Dry Pipe Valves
See Also: Fire Alarm and Extinguishing Equipment; Fire Sprinkler Equipment; Valves

—**Accelerators (Mechanical)—Fire Alarm and Extinguishing Equipment**
ISO 6182 Pt 4-93. Fire Protection—Automatic Sprinkler Systems—Part 4: Requirements and Test Methods for Quick-Opening Devices First Edition. 11 pp.

—**Quick Opening—Fire Alarm and Extinguishing Equipment**
ISO 6182 Pt 3-93. Fire Protection—Automatic Sprinkler Systems—Part 3: Requirements and Test Methods for Dry Pipe Valves First Edition. 18 pp.
ISO 6182 Pt 4-93. Fire Protection—Automatic Sprinkler Systems—Part 4: Requirements and Test Methods for Quick-Opening Devices First Edition. 11 pp.

Dry Reed Contacts
See Also: Electric Contacts

—**Quality Assurance**
BSI BS CECC 18000-84. 1984 Dry Reed Change-over Contact Units Mechanically Biased: Generic Specification. 28 pp.

Dry Reed Relays
See Also: Mercury Wetted Contact Relays; Relays
CECC CECC 18 000 ISSUE 1-83. Generic Specification: Dry Reed Change-Over Contact Units Mechanically Biased (En, Fr, Ge). 78 pp.
CECC CECC 18 001 ISSUE 1-83. Blank Detail Specification: Dry Reed Change-Over Contact Units Mechanically Biased (General Application) (En, Fr, Ge). 51 pp.
CECC CECC 19 000 ISSUE 1-78. Generic Specification: Dry Reed Make Contact Units (En, Fr, Ge) AMD 1 (En, Fr, Ge) AMD 2 (En, Fr, Ge). 176 pp.
CECC CECC 19 002 ISSUE 1-82. Blank Detail Specification: Dry Reed Make Contact Units for General Applications (En, Fr, Ge). 48 pp.
IEC 255 Pt 9-79. Electrical Relays Part 9: Dry Reed Make Contact Units First Edition. 68 pp.

—**Mechanically Biased—Quality Assurance**
BSI BS CECC 18001-84. 1984 Dry Reed Change-Over Contact Units, Mechanically Biased, for General Application: Blank Detail Specification. 19 pp.

—**Quality Assurance**
BSI BS 9152-78. 1978 Amd 1 Rules for the Preparation of Detail Specifications for Reed Relays of Assessed Quality. 17 pp.
BSI BS CECC 19000-79. 1979 Amd 2 Dry Reed Make Contact Units: Generic Specification. 61 pp.
BSI BS CECC 19001-82. 1982 Dry Reed Make Contact Units for Telecommunication Applications: Blank Detail Specification. 17 pp.

—**Telecommunication Equipment**
BSI BS CECC 19001-82. 1982 Dry Reed Make Contact Units for Telecommunication Applications: Blank Detail Specification. 17 pp.
CECC CECC 19 001 ISSUE 1-82. Blank Detail Specification: Dry Reed Make Contact Units for Telecommunication Applications (En, Fr, Ge). 46 pp.

Dry Reed Switches
See Also: Electric Switches; Reed Switches
CNS C6146-87. Method of Test for Dry Reed Switches (General Rules) (Jul)(8224).
CNS C6149-87. Method of Test for Dry Reed Switches (Operation Parameters) (Jul)(8227).
CNS C6171-87. Method of Test for Dry Reed Switches (Physical Dimensions) (Jul)(9231).
CNS C6173-87. Method of Test for Dry Reed Switches (Detail Specification) (Jul)(9233).
CNS C6175-87. Method of Test for Dry Reed Switches (Application Notes) (Jul)(9235).

—**Capacitance Measurement**
CNS C6168-87. Method of Test for Dry Reed Switches (Capacitance) (Jul)(9228).

Dry Reed Switches (Cont.)
—**Contact Bounce**
CNS C6151-87. Method of Test for Dry Reed Switches (Operate, Bounce, Release and Transfer (SPDT) Time) (Jul)(8229).

—**Contact Stickiness**
CNS C6164-87. Method of Test for Dry Reed Switches (Contact Stickiness) (Jul)(8671).

—**Current Carrying Capacity**
CNS C6165-87. Method of Test for Dry Reed Switches (Carry Current) (Jul)(8672).

—**Dielectric Testing**
CNS C6150-87. Method of Test for Dry Reed Switches (Dielectric) (Jul)(8228).

—**Insulation Resistance**
CNS C6152-87. Method of Test for Dry Reed Switches (Insulation Resistance) (Jul)(8230).

—**Lead Finish Testing**
CNS C6169-87. Method of Test for Dry Reed Switches (Lead Finish and Terminal Strength) (Jul)(9229).

—**Life (Durability)**
CNS C6170-87. Method of Test for Dry Reed Switches (Contact Life Testing) (Jul)(9230).

—**Mechanical Shock**
CNS C6162-87. Method of Test for Dry Reed Switches (Shock) (Jul)(8669).

—**Mechanical Testing**
CNS C6148-87. Method of Test for Dry Reed Switches (Visual and Mechanical Inspection) (Jul)(8226).

—**Moisture**
CNS C6160-87. Method of Test for Dry Reed Switches (Internal Moisture) (Jul)(8667).

—**Operate Time**
CNS C6151-87. Method of Test for Dry Reed Switches (Operate, Bounce, Release and Transfer (SPDT) Time) (Jul)(8229).

—**Packaging**
CNS C6172-87. Method of Test for Dry Reed Switches (Preparation for Delivery) (Jul)(9232).

—**Quality Assurance**
CNS C6147-87. Method of Test for Dry Reed Switches (Quality Control and Quality Assurance Provisions) (Jul)(8225).

—**Release Time**
CNS C6151-87. Method of Test for Dry Reed Switches (Operate, Bounce, Release and Transfer (SPDT) Time) (Jul)(8229).

—**Seals**
CNS C6163-87. Method of Test for Dry Reed Switches (Seals) (Jul)(8670).

—**Terminal Strength**
CNS C6169-87. Method of Test for Dry Reed Switches (Lead Finish and Terminal Strength) (Jul)(9229).

—**Test Coils**
CNS C6174-87. Method of Test for Dry Reed Switches (Standard Test Coils, Single Insulation 130 Degrees C Wire) (Jul)(9234).

—**Transfer Time**
CNS C6151-87. Method of Test for Dry Reed Switches (Operate, Bounce, Release and Transfer (SPDT) Time) (Jul)(8229).

—**Vibration**
CNS C6161-87. Method of Test for Dry Reed Switches (Vibration) (Jul)(8668).

—**Visual Inspection**
CNS C6148-87. Method of Test for Dry Reed Switches (Visual and Mechanical Inspection) (Jul)(8226).

Dry Residue Content Analysis
Use: Residue Content Analysis

Dry Risers
Use: Fire Hydrants

Dry Rubber
Use: Rubber

Dry Sanding
Use: Sanding

Dry Type Transformers
Use For: Air Cooled Transformers

INDUSTRY STANDARDS

Dry Type Transformers (Cont.)

See Also: Transformers

CSA C9-M1981. Dry-Type Transformers; (Gen Instr 1). 38 pp.
DIN VDE 0550 Pt 1-69. Regulations for Small Transformers; General Regulations (Dec). 68 pp.
DIN VDE 0551-72. Regulations for Safety Transformers (May). 79 pp.
DIN VDE 0551E-75. Regulations for Safety Transformers: Amendment E (Sept). 9 pp.

— 0-600 V
CSA C22.2 NO 66-1988. Speciality Transformers; (Gen Instr 1). 53 pp.
CSA 1169 Bull. Electrical Bulletin 1169 June 27, 1978 to C22.2 NO 66. 2 pp.
CSA 1440 Bull. Electrical Bulletin 1440 June 23, 1986 to C22.2 NO 66. 2 pp.

—Distribution—Loads (Electric)
CSA C9.1-M1981. Guide for Loading Dry-Type Distribution and Power Transformers; (Gen Instr 1). 18 pp.

—Heating Circuits
CSA CAN/CSA-C22. 2 NO 47-M90. Air-Cooled Transformers (Dry Type); (Gen Instr 1 Thru 2). 32 pp.
CSA 1169 Bull. Electrical Bulletin 1169 June 27, 1978 to C22.2 NO 47. 2 pp.

—Identification Systems
CSA CAN/CSA-C22. 2 NO 47-M90. Air-Cooled Transformers (Dry Type); (Gen Instr 1 Thru 2). 32 pp.

—Lighting Circuits
CSA CAN/CSA-C22. 2 NO 47-M90. Air-Cooled Transformers (Dry Type); (Gen Instr 1 Thru 2). 32 pp.
CSA 1169 Bull. Electrical Bulletin 1169 June 27, 1978 to C22.2 NO 47. 2 pp.

—Live Parts
CSA CAN/CSA-C22. 2 NO 47-M90. Air-Cooled Transformers (Dry Type); (Gen Instr 1 Thru 2). 32 pp.

—Power Circuits
CSA CAN/CSA-C22. 2 NO 47-M90. Air-Cooled Transformers (Dry Type); (Gen Instr 1 Thru 2). 32 pp.
CSA 1169 Bull. Electrical Bulletin 1169 June 27, 1978 to C22.2 NO 47. 2 pp.

—Power—Loads (Electric)
CSA C9.1-M1981. Guide for Loading Dry-Type Distribution and Power Transformers; (Gen Instr 1). 18 pp.
IEC 905-87. Loading Guide for Dry-Type Power Transformers First Edition; (Corrigendum—April 1991). 48 pp.
SAA AS 3953-91. Loading Guide for Dry-Type Power Transformers (IEC 905:1987). 26 pp.

—Wiring Space
CSA 1166 Bull. Electrical Bulletin 1166 April 19, 1978 to C22.2 NO 47. 2 pp.

—601-999 V
CSA C22.2 NO 66-1988. Speciality Transformers; (Gen Instr 1). 53 pp.
CSA 1169 Bull. Electrical Bulletin 1169 June 27, 1978 to C22.2 NO 66. 2 pp.
CSA 1440 Bull. Electrical Bulletin 1440 June 23, 1986 to C22.2 NO 66. 2 pp.

Drydocks

See Also: Docks

NATO STANAG 1148 ED 2 AMD 3-78. Docking in Dry Dock or Floating Dock—ALP-5(A) (Navy). 16 pp.

—Design
BSI BS 6349: Part 3-88. 1988 Code of Practice for Maritime Structures; Part 3: Design of Dry Docks, Locks, Slipways and Shipbuilding Berths, Shiplifts and Dock and Lock Gates. 75 pp.

—Ships
MOD UK NES 850-01. Requirements for the Dry Docking, Slipping or Lifting of Mod Vessels Issue 1 (10.89); Amendment 1. 41 pp.

—Submarines
MOD UK NES 850-01. Requirements for the Dry Docking, Slipping or Lifting of Mod Vessels Issue 1 (10.89); Amendment 1. 41 pp.

Dryers

Scope Note: For additional listings, use a more specific term *Use For:* Driers (Heating Equipment)
See Also: Aggregate Dryers; Clothes Dryers; Compressed Air Dryers; Crop Dryers; Drying Ovens; Hair Dryers

Dryers (Cont.)

CNS B1305-83. Nominal Dimension of Roller Dryers (Mar)(10057).
CNS B4045-81. Roller Dryers (Jun)(7550).
JIS B 6547-91. Roller Dryer—Test and Inspection Methods. 9 pp.
JIS B 6550-91. Band Dryer—Test and Inspection Methods. 9 pp.

—Heat Balance
CNS R3121-85. Method for Calculating Heat Balance of Continuous Dryer for Mineral Materials and Their Related Products (Apr)(11250).
JIS R 0304-91. Heat Balancing of Continuous Dryer for Ores and Other Materials. 52 pp.

—Power Supply Cords
CNS C4412-9-85. Power-Supply Cords for Ranges and Dryers (Feb)(10917-9).

—Refrigeration Equipment
CSA C22.2 NO 140.3-M1987. Refrigerant-Containing Components for Use in Electrical Equipment (R 1993). 19 pp.

Drying

Use For: Drying Testing *See Also:* Air Drying; Caking; Dehydration; Spreaders; Tumbling

CGSB 1-GP-71 METH 44.1-81. Methods of Testing Paints and Pigments Baking Properties. 1 p.

—Adhesives
SAA AS 1937.3-77. Methods of Test for Sealers and Adhesives for Automotive Purposes—Part 3: Determination of Optimum Open Drying Time of Adhesives Reconfirmed 1989. 2 pp.

—Bituminous Coatings
MOD UK M 3004/67. Protective PX-2 and Composition Rust Preventative Type B.

—Coal
BSI BS 1016: Part 1-73. 1973 Methods for the Analysis and Testing of Coal and Coke Part 1: Total Moisture of Coal. 11 pp.

—Coke
BSI BS 1016: Part 2-73. 1973 Methods for the Analysis and Testing of Coal and Coke Part 2: Total Moisture of Coke. 8 pp.

—Corrosion Inhibitors
CGSB 31-GP-0A METH 8.1-57. Methods of Testing Corrosion-Prevention Materials and Processes Drying. 1 p.

—Drying Oils
CNS K6658-81. Method of Test for Drying Properties of Drying Oils (Apr)(7259).

—Elastomer Coatings
CGSB 1-GP-71 METH 5.12-74. Methods of Testing Paints and Pigments Drying Time Drying Time Over Elastomers. 1 p.

—Electrical Insulating Varnishes
MOD UK DEF-1053: METH NO. 38. Standard Methods of Testing Paint, Varnish, Lacquer and Related Products Method 38: Insulating Varnishes (Clear and Pigmented). Preparation of Specimens for Electrical Tests and for Assessment of Drying Properties (Withdrawn).

—Enamels
CGSB 1-GP-71 METH 5.1-78. Methods of Testing Paints and Pigments Drying Time General Method. 2 pp.
CNS K6030-72. Method of Test for Enamel (Jun)(627). 7 pp.
CNS K6031-87. Method of Test for Lacquer Enamel (May) (628). 6 pp.

—Fats
BSI BS 684: Sec 1.7-76. 1976 Methods of Analysis of Fats and Fatty Oils Part 1: Physical Methods Section 1.7: Determination of Surface-Drying Time. 2 pp.

—Lacquers
CGSB 1-GP-71 METH 5.1-78. Methods of Testing Paints and Pigments Drying Time General Method. 2 pp.
CNS K6386-76. Method of Test for Lacquer in Surface of Leather Shoes (Mar) (3922). 2 pp.
SAA AS 1580.101. 1-86. Paints and Related Materials—Methods of Test—Part 101.1: Air Drying Conditions (R 1992). 4 pp.
SAA AS 1580.401. 1-81. Paints and Related Materials—Methods of Test—Part 401.1: Surface Dry Condition (R 1992). 2 pp.
SAA AS 1580.401. 3-90. Paints and Related Materials—Methods of Test—Part 401.3: Drying Times Using a BK-Type Recorder. 4 pp.

Drying (Cont.)

—Lacquers (Cont.)
SNZ NZS/AS 1580. 101.1-86. Methods of Test for Paints and Related Materials Part 101.1: Air Drying Conditions (This is a Joint Standard with SAA AS 1580.101.1). 4 pp.
SNZ NZS/AS 1580. 401.1-81. Methods of Test for Paints and Related Materials Part 401.1: Surface Dry Condition (This is a Joint Standard with SAA AS 1580.401.1). 2 pp.
SNZ NZS/AS 1580. 401.3-90. Methods of Test for Paints and Related Materials Part 401.3: Drying Times Using a BK-Type Recorder (This ia a Joint Standard with SAA AS 1580.401.3). 4 pp.

—Linseed Oil
CNS K6055-75. Method of Test for Boiled Linseed Oil (Feb)(769). 4 pp.

—Mustard Seeds—Weight Loss
CNS N4059-80. Determination of Weight Loss on Mustard Seed (Oct)(6603). 1 p.

—Oleoresinous Paints
CNS K6032-86. Method of Test for Oleoresinous Deck Paint (May)(629). 4 pp.

—Oleoresinous Primers
CNS K6115-86. Method of Test for Oleoresinous Primer (Apr)(1249). 3 pp.

—Paints
BSI BS 3900: Part C2-71. 1971 Methods of Test for Paints Group C: Tests Associated with Paint Film Formation Part C2: Surface-Drying Time (Ballotini Method). 2 pp.
BSI BS 3900: Part C3-90. 1990 Methods of Test for Paints Part C3: Through Dry Test (W). 8 pp.
BSI BS 3900: Part C3-01. (WITHDRAWN) 1990 Amd 1 Methods of Test for Paints Part C3: Through Dry Test (ISO 9117: 1990) (AMD 6808) November 29, 1991 (Renumbered as BS EN 29117: 1992). 9 pp.
BSI BS EN 29117-92. 1992 Amd 2 Paints and Varnishes—Determination of Through-Dry Time—Method of Test (ISO 9117: 1990) (AMD 7501) December 15, 1992. 15 pp.
CEN EN 29117-92. Paints and Varnishes—Determination of Through-Dry State and Through-Dry Time—Method of Test; (ISO 9117: 1990). 6 pp.
CGSB 1-GP-71 METH 5.1-78. Methods of Testing Paints and Pigments Drying Time General Method. 2 pp.
CGSB 1-GP-71 METH 5.14-74. Methods of Testing Paints and Pigments Drying Time Complete Cure. 1 p.
DIN ENGL EN 29117-92. Paints and Varnishes; Determination of Through-Dry State and Through-Dry Time (ISO 9117:1990) (Sept). 9 pp.
ISO 1517-73. Paints and Varnishes—Surface-Drying Test—Ballotini Method First Edition. 4 pp.
ISO 9117-90. Paints and Varnishes—Determination of Through -Dry State and Through-Dry Time—Method of Test First Edition; (CEN EN 29117:1992). 8 pp.
SAA AS 1580.101. 1-86. Paints and Related Materials—Methods of Test—Part 101.1: Air Drying Conditions (R 1992). 4 pp.
SAA AS 1580.401. 1-81. Paints and Related Materials—Methods of Test—Part 401.1: Surface Dry Condition (R 1992). 2 pp.
SAA AS 1580.401. 3-90. Paints and Related Materials—Methods of Test—Part 401.3: Drying Times Using a BK-Type Recorder. 4 pp.
SNZ NZS/AS 1580. 101.1-86. Methods of Test for Paints and Related Materials Part 101.1: Air Drying Conditions (This is a Joint Standard with SAA AS 1580.101.1). 4 pp.
SNZ NZS/AS 1580. 401.1-81. Methods of Test for Paints and Related Materials Part 401.1: Surface Dry Condition (This is a Joint Standard with SAA AS 1580.401.1). 2 pp.
SNZ NZS/AS 1580. 401.3-90. Methods of Test for Paints and Related Materials Part 401.3: Drying Times Using a BK-Type Recorder (This ia a Joint Standard with SAA AS 1580.401.3). 4 pp.

—Paperboard
CNS P3025-88. Method of Determining Moisture Content in Pulp and Paper (Sep)(3086). 4 pp.

—Papers
CNS P3025-88. Method of Determining Moisture Content in Pulp and Paper (Sep)(3086). 4 pp.
CPPA G.3-73. Moisture in Paper. 1 p.

—Putty
CGSB CAN2-19.0-M77METH 2.2-78. Methods of Testing Putty, Caulking and Sealing Compounds Drying Time of Putty. 1 p.

Drying (Cont.)

—Traffic Paints
SAA AS 1580.401. 8-86. Paints and Related Materials—Methods of Test—Part 401.8: No-Pick-Up Time of Road Marking Paints (R 1992). 4 pp.
SNZ NZS/AS 1580. 401.8-86. Methods of Test for Paints and Related Materials Part 401.8: No-Pick-Up Time of Road Marking Paints (This is a Joint Standard with SAA AS 1580.401.8). 4 pp.

—Varnishes
BSI BS 3900: Part C2-71. 1971 Methods of Test for Paints Group C: Tests Associated with Paint Film Formation Part C2: Surface-Drying Time (Ballotini Method). 2 pp.
BSI BS 3900: Part C3-90. 1990 Methods of Test for Paints Part C3: Through Dry Test (W). 8 pp.
BSI BS 3900: Part C3-01. (WITHDRAWN) 1990 Amd 1 Methods of Test for Paints Part C3: Through Dry Test (ISO 9117: 1990) (AMD 6808) November 29, 1991 (Renumbered as BS EN 29117: 1992). 9 pp.
BSI BS EN 29117-92. 1992 Amd 2 Paints and Varnishes—Determination of Through-Dry Time—Method of Test (ISO 9117: 1990) (AMD 7501) December 15, 1992. 15 pp.
CEN EN 29117-92. Paints and Varnishes—Determination of Through-Dry State and Through-Dry Time—Method of Test; (ISO 9117: 1990). 6 pp.
CGSB 1-GP-71 METH 5.1-78. Methods of Testing Paints and Pigments Drying Time General Method. 2 pp.
DIN ENGL EN 29117-92. Paints and Varnishes; Determination of Through-Dry State and Through-Dry Time (ISO 9117:1990) (Sept). 9 pp.
ISO 1517-73. Paints and Varnishes—Surface-Drying Test—Ballotini Method First Edition. 4 pp.
ISO 9117-90. Paints and Varnishes—Determination of Through -Dry State and Through-Dry Time—Method of Test First Edition; (CEN EN 29117:1992). 8 pp.
SAA AS 1580.101. 1-86. Paints and Related Materials—Methods of Test—Part 101.1: Air Drying Conditions (R 1992). 4 pp.
SAA AS 1580.401. 1-81. Paints and Related Materials—Methods of Test—Part 401.1: Surface Dry Condition (R 1992). 2 pp.
SAA AS 1580.401. 3-90. Paints and Related Materials—Methods of Test—Part 401.3: Drying Times Using a BK-Type Recorder. 4 pp.
SNZ NZS/AS 1580. 101.1-86. Methods of Test for Paints and Related Materials Part 101.1: Air Drying Conditions (This is a Joint Standard with SAA AS 1580.101.1). 4 pp.
SNZ NZS/AS 1580. 401.1-81. Methods of Test for Paints and Related Materials Part 401.1: Surface Dry Condition (This is a Joint Standard with SAA AS 1580.401.1). 2 pp.
SNZ NZS/AS 1580. 401.3-90. Methods of Test for Paints and Related Materials Part 401.3: Drying Times Using a BK-Type Recorder (This ia a Joint Standard with SAA AS 1580.401.3). 4 pp.

—Waxes
CGSB 25-GP-1M METH 10.1-84. Methods of Sampling and Testing Waxes and Polishes Gloss and Drying Time (Supersedes 10.1e of September 1974). 2 pp.

Drying Agents

Use For: Desiccants *See Also:* Driers; Molecular Sieves; Silica Gel
BSI BS 3482-62. (WITHDRAWN) 1962 Amd 1 Methods of Test for Desiccants Used in Dynamic Dehumidification Equipment (Superseded by BS 3482: Parts 1 to 11: 1991). 20 pp.
BSI BS 3482: Part 1-91. 1991 Methods of Test for Desiccants Part 1: Sampling, and Preparation and Storage of Test Samples. 6 pp.
JIS K 1464-62. Desiccants, Activated for Industrial Dehumidification. 9 pp.
MOD UK DSTAN 44-2: Part 1-93. Desiccant Containers, Dehumidifier Part 1: Guide to the Use of Desiccants Issue 4. 17 pp.

—Absorptivity
BSI BS 3482: Part 6-91. 1991 Methods of Test for Desiccants Part 6: Determination of Adsorptive Capacity. 7 pp.

—Activated Alumina
MOD UK TS 10311. Desiccant, Activated Alumina, Pellets (Withdrawn).

—Activated Clay
BSI BS 7529-91. 1991 Desiccant Activated Clay (V). 7 pp.
MOD UK TS 487B. Desiccant, Activated Clay (Withdrawn).

Drying Agents (Cont.)

—Aluminum Oxide
BSI BS 2541-60. (WITHDRAWN) 1960 Amd 1 Activated Alumina for Use as Desiccant (Superseded by Various Parts of the 1991 Edition of BS 3482). 13 pp.

—Ammonia
BSI BS 3482: Part 8-91. 1991 Methods of Test for Desiccants Part 8: Estimation of Ammonia and Ammonium Compounds Content. 6 pp.

—Ammonium Acetate
BSI BS 3482: Part 8-91. 1991 Methods of Test for Desiccants Part 8: Estimation of Ammonia and Ammonium Compounds Content. 6 pp.

—Bulk Density
BSI BS 3482: Part 10-91. 1991 Methods of Test for Desiccants Part 10: Determination of Bulk Density (Dry Basis). 6 pp.

—Calcium Chlorides
CNS K7123-64. Chemical Reagent (Calcium Chloride, Anhydrous)(for Drying) (Jul)(1623).
JIS K 8124-75. Calcium Chloride (for Drying).

—Chloride Content
BSI BS 3482: Part 4-91. 1991 Methods of Test for Desiccants Part 4: Determination of Water-Soluble Chlorides Content. 5 pp.

—Cobaltous Chloride
BSI BS 3482: Part 9-91. 1991 Methods of Test for Desiccants Part 9: Determination of Cobalt Chloride Content. 5 pp.

—Dust Content
BSI BS 3482: Part 11-91. 1991 Methods of Test for Desiccants Part 11: Determination of Friability and Dust Content. 5 pp.

—Friability
BSI BS 3482: Part 11-91. 1991 Methods of Test for Desiccants Part 11: Determination of Friability and Dust Content. 5 pp.

—Moisture Content
BSI BS 3482: Part 2-91. 1991 Methods of Test for Desiccants Part 2: Determination of Moisture Content. 5 pp.

—Packaging
BSI BS 1133: Sec 19-86. 1986 Packaging Code Section 19: Use of Desiccants in Packaging. 12 pp.

—Packaging—Waterproofing
CNS Z5080-81. Method of Moisture-Proof Packaging (Using Desiccating Agent) (Mar)(7090).
JIS Z 0301-89. Method of Moisture-Proof Packaging. 10 pp.

—Particle Size Distribution
BSI BS 3482: Part 7-91. 1991 Methods of Test for Desiccants Part 7: Determination of Particle Size. 5 pp.

—pH Content
BSI BS 3482: Part 3-91. 1991 Methods of Test for Desiccants Part 3: Determination of pH. 5 pp.

—Silica Gel
BSI BS 7554-92. 1992 Beaded Desiccant Silica Gel (W). 7 pp.
MOD UK DSTAN 68-179-93. Silica Gel Indicator Type, Large Grain for Dynamic Air Drying Issue 1. 10 pp.
MOD UK DSTAN 80-22-72. Desiccant, Activated (Silica Gel for Dynamic Air Drying) Issue 1. 4 pp.
MOD UK CS 3069C. Silica Gel, Indicator Type, Large Grain for Dynamic Air-Drying (Superseded by Def Stan 68-179).
MOD UK TS 10208. Desiccant, Silica Gel, Beaded Form (Withdrawn).
MOD UK TS 10212. Desiccant, Silica Gel, Beaded Form, Indicator Type.

—Sodium Aluminum Silicate
MOD UK TS 10273. Desiccant, Beaded, Molecular-Sieve.

—Sulfate Content
BSI BS 3482: Part 5-91. 1991 Methods of Test for Desiccants Part 5: Determination of Water-Soluble Sulphates Content. 5 pp.

Drying Cabinets

Use: Clothes Dryers

Drying Oils

See Also: Boiled Oil; Driers; Linseed Oil; Oils; Oiticica Oil; Paints; Solvents; Soybean Oil; Tung Oil; Varnishes

Drying Oils (Cont.)

—Acetone Tolerance
CNS K6653-81. Method of Test for Acetone Tolerance of Heat-Bodied Drying Oils (Apr)(7254).

—Ash Content
CNS K6631-81. Method of Test for Ash in Drying Oils and Fatty Acid (Mar)(7040).

—Break—Quantitative Analysis
CNS K6654-81. Method of Test for Quantitative Determination of Break in Drying Oils (Apr)(7255).

—Colorimetry
CNS K6655-81. Method of Test for Color After Heating of Drying Oils (Apr)(7256).

—Drying
CNS K6658-81. Method of Test for Drying Properties of Drying Oils (Apr)(7259).

—Gel Time
CNS K6659-81. Method of Test for Gel Time of Drying Oils (Apr)(7260).

—Iodine Number
CNS K6646-81. Method of Test for Iodine Value of Drying Oils and Fatty Acid (Mar)(7165).
CNS K6671-81. Method of Test for Total Iodine Value of Drying Oils and Their Derivatives (May)(7435).

—Loss on Heating
CNS K6652-81. Method of Test for Loss on Heating of Drying Oils (Apr)(7253).

—Saponification Number
CNS K6633-81. Method of Test for Saponification Value of Drying Oils, Fatty Acids, and Polymerized Fatty Acids (Mar)(7042).

—Unsaponifiable Matter Content
CNS K6634-81. Method of Test for Unsaponifiable Matter in Drying Oils, Fatty Acids, and Polymerized Fatty Acids (Mar)(7043).

Drying Ovens

See Also: Dryers; Ovens

—Industrial
CSA C22.2 NO 88-1958. Construction and Test of Industrial Heating Equipment (R 1992). 22 pp.
CSA 1169 Bull. Electrical Bulletin 1169 June 27, 1978 to C22.2 NO 88. 2 pp.

—Industrial—Construction
CSA C22.2 NO 88-1958. Construction and Test of Industrial Heating Equipment (R 1992). 22 pp.
CSA 1169 Bull. Electrical Bulletin 1169 June 27, 1978 to C22.2 NO 88. 2 pp.

—Laboratory
BSI BS 2648-55. 1955 Amd 1 Performance Requirements for Electrically-Heated Laboratory Drying Ovens. 11 pp.

Drying Testing

Use: Drying

Drying Time

Use: Drying

Drywall (Gypsum)

Use: Gypsum Wallboard

Drywall Adhesives

See Also: Adhesives

—Studs (Fasteners)
CGSB CAN/CGSB-71.25-M88. Adhesives, for Bonding Drywall to Wood Framing and Metal Studs. 15 pp.

—Wood Framing
CGSB CAN/CGSB-71.25-M88. Adhesives, for Bonding Drywall to Wood Framing and Metal Studs. 15 pp.

DSC

Use: SELCAL Systems—Digital

DSI

Use: Digital Speech Interpolation

DSP

Use: Digital Signal Processing

DSSSL

Use For: Document Style Semantics and Specification Language
OSI ISO.IEC DIS 10179-91. Information Technology—Text and Office Systems—Document Style Semantics and Specification Language (DSSSL). 138 pp.

DSS1
Use: Digital Subscriber Signaling Systems

DTE
Use: Data Terminal Equipment

DTL Circuits
Use: Digital Circuits—DTL

DUA
Use: Directory User Agents

Dual Fuel Burners
Scope Note: See the subheading Dual Fuel under specific types of burners

Dual In-Line Package Sockets
Use: DIP Sockets

Dual In-Line Package Switches
Use: DIP Switches

Dual In-Line Packaging Switches
Use: DIP Switches

Dual Tone Multifrequency (Signaling)
Use: Multifrequency Signaling Systems—Dual Tone

Duck (Meat)
Use: Poultry Meat—Duck

Duct Fittings
See Also: Ducts

CNS C4223-82. General Rules for Fittings of Rigid Metal Conduits and Underfloor Ducts (Nov)(6079).
CNS C4242-87. Couplings for Underfloor Duct (Jan)(6098).
CNS C4244-87. Insert Fittings for Underfloor Duct (Jan)(6100).
ISO 4486-85. Asbestos-Cement Ventilation Ducts and Fittings—Dimensions and Characteristics First Edition. 9 pp.
JIS C 8355-91. Fittings for Underfloor Ducts. 9 pp.
JIS C 8359-91. General Rules for Fittings of Metal Conduits and Underfloor Ducts. 19 pp.

—**Asbestos Cement**

ISO 4486-85. Asbestos-Cement Ventilation Ducts and Fittings—Dimensions and Characteristics First Edition. 9 pp.

—**Flanged—Aircraft**

ISO 2563-74. Aircraft Ducting and Piping—Profile Dimensions for Flanges of V-Band Couplings First Edition. 3 pp.
MOD UK DSTAN 53-68-01. V-Flange Couplings for Aircraft Piping and Ducting (Metric) Issue 2; Amendment 2. 8 pp.

Duct Furnaces
Use: Duct Heaters

Duct Heaters
Use For: Duct Furnaces *See Also:* Heaters

—**Electric**

CSA C22.2 NO 155-M1986. Electric Duct Heaters (R 1992); (Gen Instr 1 Thru 2). 34 pp.

—**Electrical Codes**

CSA 1084 Bull. Electrical Bulletin 1084 December 3, 1976 to C22.2 NO 155. 2 pp.

Duct Joints
See Also: Clamps; Couplings; Joints

—**Couplings—V Band—Aerospace**

NATO STANAG 3312 ED 4 AMD 5-75. Profile Dimensions of Flanges for V-Band Couplings for Piping and Ducting. 18 pp.

Ductile Iron
Scope Note: For additional listings, see specific products made from ductile iron *Use For:* Nodular Iron; Spheroidal Graphite Cast Iron *See Also:* Cast Iron; Ferroalloys; Gray Iron; Malleable Cast Iron

—**Classification**

ISO 1083-87. Spheroidal Graphite Cast Iron—Classification Second Edition. 16 pp.

—**Mechanical Properties**

BSI BS 2789-85. 1985 Spheroidal Graphite or Nodular Graphite Cast Iron. 14 pp.

Ductile Iron Castings
Use For: Nodular Iron Castings *See Also:* Cast Iron; Castings; Ferroalloy Castings; Iron Castings; Metal Products; Steels

CNS B2118-83. Nodular Graphite Iron Casting (Aug)(2869).

Ductile Iron Castings (Cont.)

JIS G 5502-89. Spheroidal Graphite Iron Castings. 13 pp.
JIS G 5503-89. Austempered Spheroidal Graphite Iron Castings. 10 pp.
SAA AS 1831-85. Iron Castings—Spheroidal or Nodular Graphite Cast Iron. 13 pp.
SNZ NZS/AS 1831-85. Iron Castings—Spheroidal or Nodular Graphite Cast Iron (This is a Joint Standard with SAA AS 1831). 13 pp.

—**Raw**

DIN ENGL 1685 Pt 1-80. Raw Castings Made from Nodular Graphite Cast Iron; General Tolerances, Machining Allowances (Oct). 3 pp.

Ductile Iron Pipes
See Also: Iron Pipes; Pipes

BSI BS 4772-88. 1988 Amd 1 Ductile Iron Pipes and and Fittings (AMD 6270) November 30, 1989. 73 pp.
BSI BS 4772-80. 1980 Amd 4 Ductile Iron Pipes and and Fittings (AMD 5009) March 31, 1986 (Superseded by BS 4772: 1988). 42 pp.
CNS B7039-71. Testing Standard for Nodular Graphite Cast-Iron Pipes and Fittings (Jul)(2780).
CNS G2211-84. Method of Test for Ductile Cast Iron Pipes (Mar)(10809).
CNS G3219-84. Ductile Cast Iron Pipes (Mar)(10808).
DIN ENGL 2410 Pt 2-77. Pipes; Survey of Standards for Tubes of Ductile Cast Iron (Feb). 1 p.
JIS G 5526-89. Ductile Iron Pipes. 35 pp.

—**Bituminous Coatings**

DIN ENGL 30674 Pt 4-83. Coating of Ductile Cast Iron Pipes; Bitumen Coating (May). 3 pp.

—**Cement Mortar Linings**

ISO 4179-85. Ductile Iron Pipes for Pressure and Non-Pressure Pipelines—Centrifugal Cement Mortar Lining—General Requirements Second Edition. 6 pp.
ISO 6600-80. Ductile Iron Pipes—Centrifugal Cement Mortar Lining—Composition Controls of Freshly Applied Mortar First Edition. 4 pp.
JIS A 5314-84. Mortar Lining for Ductile Iron Pipes. 10 pp.

—**Epoxy Coatings—Powder**

JIS G 5528-84. Epoxy-Powder Coating for Interior of Ductile Iron Pipes and Fittings. 26 pp.

—**Nominal Size**

ISO 7186-83. Ductile Iron Pipes and Accessories for Non-Pressure Pipe-Lines First Edition. 11 pp.

—**Polyethylene Coatings**

CEN PREN 254-86. Preinsulated Boned Pipe Systems for Underground Hot Water Networks Pipe Assembly of Ductile Iron Pipes, Polyurethane, Thermal Insulation and Outer Casing of High Density Polythylene. 38 pp.
CEN EN 254-90. Preinsulated Bonded Pipe Systems for Underground Hot Water Networks—Pipe Assembly of Ductile Iron Pipes, Polyurethane Thermal Insuction and Outer Casing of High Density Polyethylene. 44 pp.
DIN ENGL 30674 Pt 1-82. Coating of Ductile Cast Iron Pipes; Polyethylene Coating (Sept). 8 pp.

—**Pressure**

ISO 2531-91. Ductile Iron Pipes, Fittings and Accessories for Pressure Pipelines Fourth Edition. 54 pp.
ISO 4179-85. Ductile Iron Pipes for Pressure and Non-Pressure Pipelines—Centrifugal Cement Mortar Lining—General Requirements Second Edition. 6 pp.
SAA AS 2280-91. Ductile Iron Pressure Pipes and Fittings. 52 pp.

—**Pressure—Flanged**

CNS B5023-82. Ductile Cast-Iron Flanged Piper for Pressure Main Lines (Jan)(833).

—**Pressure—Gas**

DIN ENGL 28600-83. Ductile Iron Pressure Pipes and Fittings for Gas and Water Pipelines; Technical Delivery Conditions (Jan). 12 pp.
DIN ENGL 28610 Pt 1-83. Ductile Iron Pressure Pipes with Socket with Cement-Mortar Lining for Gas and Water Pipelines; Dimensions and Masses (Jan). 8 pp.
DIN ENGL 28610 Pt 2-83. Ductile Iron Pressure Pipes with Socket for Gas and Water Pipelines Rated for Pressures over 4 Bar up to and Including 16 Bar; Dimensions and Masses (Jan). 6 pp.
DIN ENGL 28614-90. Ductile Iron Pipes with Cast-On Flanges for Use with Gas and Water Pipelines; Dimensions and Mass (Jan). 3 pp.
DIN ENGL 28615 Pt 1-90. Ductile Iron Pipes with Welded-On Flanges for Use with Gas and Water Pipelines; Dimensions and Mass (Jan) (Together with the January 1990 Edition of DIN 28615 Part 2, Supersedes the March 1976 Edition of DIN 28615). 3 pp.

Ductile Iron Pipes (Cont.)
—**Pressure—Gas** (Cont.)

DIN ENGL 28615 Pt 2-90. Ductile Iron Pipes with Screwed-On Flanges for Use with Gas and Water Pipelines; Dimensions and Mass (Jan) (Together with the January 1990 Edition of DIN 28615 Part 1, Supersedes the March 1976 Edition of DIN 28615). 3 pp.

—**Pressure—Water**

DIN ENGL 28600-83. Ductile Iron Pressure Pipes and Fittings for Gas and Water Pipelines; Technical Delivery Conditions (Jan). 12 pp.
DIN ENGL 28610 Pt 1-83. Ductile Iron Pressure Pipes with Socket with Cement-Mortar Lining for Gas and Water Pipelines; Dimensions and Masses (Jan). 8 pp.
DIN ENGL 28610 Pt 2-83. Ductile Iron Pressure Pipes with Socket for Gas and Water Pipelines Rated for Pressures over 4 Bar up to and Including 16 Bar; Dimensions and Masses (Jan). 6 pp.
DIN ENGL 28614-90. Ductile Iron Pipes with Cast-On Flanges for Use with Gas and Water Pipelines; Dimensions and Mass (Jan). 3 pp.
DIN ENGL 28615 Pt 1-90. Ductile Iron Pipes with Welded-On Flanges for Use with Gas and Water Pipelines; Dimensions and Mass (Jan) (Together with the January 1990 Edition of DIN 28615 Part 2, Supersedes the March 1976 Edition of DIN 28615). 3 pp.
DIN ENGL 28615 Pt 2-90. Ductile Iron Pipes with Screwed-On Flanges for Use with Gas and Water Pipelines; Dimensions and Mass (Jan) (Together with the January 1990 Edition of DIN 28615 Part 1, Supersedes the March 1976 Edition of DIN 28615). 3 pp.

—**Pressure—Water—Pressure Measurement**

DIN ENGL 4279 Pt 2-75. Testing of Pressure Pipelines for Water by Internal Pressure; Pressure Pipes Made of Ductile Cast Iron (Nov). 2 pp.
DIN ENGL 4279 Pt 3-90. Internal Hydrostatic Pressure Testing of Water Pressure Pipes; Ductile Cast Iron Pressure Pipes and Steel Pipes with Cement Mortar Lining (June). 4 pp.

—**Sewer**

DIN ENGL 19690-78. Technical Conditions of Delivery for Ductile Cast Iron Pipes and Special Castings for Sewers and Sewage Pipelines (July). 5 pp.
DIN ENGL 19691-78. Ductile Cast Iron Pipes with Plug Socket for Sewers and Sewage Pipelines; Dimensions (July). 2 pp.
DIN ENGL 19692 Pt 1-78. Special Castings of Ductile Cast Iron with Plug Sockets for Sewers and Sewage Pipelines; Double Socket Straight Access Pieces with Rectangular Aperture (RS Pieces); Summary (July). 1 p.
DIN ENGL 19692 Pt 2-78. Special Castings of Ductile Cast Iron with Plug Sockets for Sewers and Sewage Pipelines; Double Socket Straight Access Pieces with Rectangular Aperture (RS Pieces); Components (July). 3 pp.

—**Sleeves**

BSI BS 6076-81. 1981 Tubular Polyethylene Film for Use as Protective Sleeving for Buried Iron Pipes and Fittings. 4 pp.
DIN ENGL 30674 Pt 5-85. External Protection of Ductile Cast Iron Pipes; Polyethylene Sleeving (Mar). 4 pp.
ISO 8180-85. Ductile Iron Pipes—Polyethylene Sleeving First Edition. 4 pp.

—**Zinc Coatings (Made From Zinc)**

CSA CAN/CSA-B131.1-M88. External Zinc-Coating of Ductile Iron Pipe (ISO 8179-1985); (Gen Instr 1). 13 pp.
DIN ENGL 30674 Pt 3-82. Coating of Ductile Cast Iron Pipes; Zinc Coating with Protective Covering (Sept). 3 pp.
ISO 8179-85. Ductile Iron Pipes—External Zinc Coating First Edition. 4 pp.

Ductility
Use For: Cupping Tests *See Also:* Ball Cup Testing; Brittleness Testing; Compression Testing; Creep Properties; Dynamic Testing; Fatigue Testing; Hardness Testing; High Temperature Testing; Impact Testing; Plastic Properties; Reduction of Area; Shear Testing; Static Testing; Tensile Testing; Toughness

—**Bitumens**

CNS K6756-83. Method of Test for Ductility of Bituminous Materials (Mar)(10091).
DIN ENGL 52013-85. Testing of Bitumen; Determination of Ductility (July). 5 pp.

—**Metal Coatings (Made From Metal)**

BSI BS 5411: Part 15-88. 1988 Method of Test for Metallic and Related Coatings Part 15: Review of Methods of Measurement of Ductility. 35 pp.

Ductility (Cont.)

—Metal Coatings (Made From Metal) (Cont.)
ISO 8401-86. Metallic Coatings—Review of Methods of Measurement of Ductility First Edition. 34 pp.

—Paints
BSI BS 3900: Part E4-76. 1976 Methods of Test for Paints Group E: Mechanical Tests on Paint Films Part E4: Cupping Test. 3 pp.

ISO 1520-73. Paints and Varnishes—Cupping Test First Edition. 5 pp.

—Steel Wire
DIN ENGL 51214-77. Testing of Steel; Tensile Test for Round Wires Using Knotted Specimens (Feb). 1 p.

—Steels
CNS Z8047-84. Method of Conical Cup Test (Sep)(11040).

JIS Z 2249-63. Method of Conical Cup Test (R 1972). 5 pp.

—Varnishes
BSI BS 3900: Part E4-76. 1976 Methods of Test for Paints Group E: Mechanical Tests on Paint Films Part E4: Cupping Test. 3 pp.

ISO 1520-73. Paints and Varnishes—Cupping Test First Edition. 5 pp.

—Welded Joints—Bend Testing
DIN ENGL 50121 Pt 1-78. Testing of Metallic Materials; Technological Bending Test on Welded Joints and Weld Platings; Fusion Welded Joints (Jan). 9 pp.

DIN ENGL 50121 Pt 2-78. Testing of Metallic Materials; Technological Bending Test on Welded Joints and Weld Platings; Pressure Welded Joints (Jan). 6 pp.

—Welded Plates—Bend Testing
DIN ENGL 50121 Pt 3-78. Testing of Metallic Materials; Technological Bending Test on Welded Joints and Weld Platings; Fusion Welded Platings (Jan). 5 pp.

Ductility Testing
Use: Ductility

Ducting Components
Use: Ducts

Ducts
Use For: Ductwork; Vent Grilles *See Also:* Cooling Systems; Duct Fittings; Exhaust Systems; Fire Dampers; Flues; Materials Handling Equipment; Pipes; Ventilation Equipment; Vents

ISO 4486-85. Asbestos-Cement Ventilation Ducts and Fittings—Dimensions and Characteristics First Edition. 9 pp.

ISO 7807-83. Air Distribution—Straight Circular Sheet Metal Ducts with a Lock Type Spiral Seam and Straight Rectangular Sheet Metal Ducts—Dimensions First Edition. 4 pp.

JIS A 4009-77. Air Duct.

—Air Conditioners
CSA B228.1-1968. Pipes, Ducts, and Fittings for Residential Type Air Conditioning Systems. 21 pp.

—Air Supply—Aerostats
CAA Chapter Q6-2 12.79. Gas and Air Supply Systems (Non-Rigid Airships). 2 pp.

—Aircraft—Bead Profiles
MOD UK DSTAN 47-23-80. Bead Profile for Aircraft Piping and Ducting (Flexible Connections) Issue 1. 5 pp.

—Asbestos Cement
ISO 4486-85. Asbestos-Cement Ventilation Ducts and Fittings—Dimensions and Characteristics First Edition. 9 pp.

—Asbestos Rope
MOD UK DSTAN 53-99-85. Ropes, Asbestos and Packing (Braids), Asbestos: Dust Suppressed and Fibres, Asbestos Issue 1. 21 pp.

—Asbestos Yarns
MOD UK DSTAN 53-99-85. Ropes, Asbestos and Packing (Braids), Asbestos: Dust Suppressed and Fibres, Asbestos Issue 1. 21 pp.

—Barges
BSI BS MA 105: Part 3-87. 1987 Shipborne Barges Part 3: Principal Mating Dimensions for the Ventilating Systems. 4 pp.

ISO 8304-84. Shipbuilding—Shipborne Barges, Series 3—Ventilation System—Principal Mating Dimensions First Edition. 4 pp.

Ducts (Cont.)

—Building Services
BSI BS 8313-89. 1989 Code of Practice for Accommodation of Building Services in Ducts. 52 pp.

—Buildings—Fire Protection
BSI BS 5588: Part 9-89. 1989 Fire Precautions in the Design and Construction of Buildings Part 9: Code of Practice for Ventilation and Air Conditioning Ductwork. 25 pp.

—Engineering Drawings—Symbols
SAA AS 1101.5-84. Graphical Symbols for General Engineering—Part 5: Piping, Ducting and Mechanical Services for Buildings. 18 pp.

—Fire Testing
DIN ENGL 4102 Pt 6-77. Fire Behaviour of Building Materials and Building Components; Ventilation Ducts; Definitions, Requirements and Tests (Sept). 5 pp.

DIN ENGL 4102 Pt 11-85. Fire Behaviour of Building Materials and Building Components; Pipe Encasements, Pipe Bushings, Service Shafts and Ducts, and Barriers Across Inspection Openings; Terminology, Requirements and Testing (Dec). 12 pp.

ISO 6944-85. Fire Resistance Tests—Ventilation Ducts First Edition. 24 pp.

—Flow Rate
ISO 5221-84. Air Distribution and Air Diffusion—Rules to Methods of Measuring Air Flow Rate in an Air Handling Duct First Edition. 36 pp.

—Gas Filled—Double Bend—Design
DIN ENGL 25427 Pt 1-77. Design of Gas-Filled Double-Bend Ducts in Concrete Shields Against Gamma Radiation; Terms and Conditions (Oct). 3 pp.

DIN ENGL 25427 Pt 2-77. Design of Gas-Filled Double-Bend Ducts in Concrete Shields Against Gamma Radiation; Proportionment of the Duct and the Embedded Iron Layers for Point Source Radiation and Collimated Radiation (Oct). 6 pp.

—Glossaries
BSI BS 6100: Sec 2.8-90. 1990 Glossary of Building and Civil Engineering Terms Part 2: Civil Engineering Section 2.8: Pipelines and Ducts. 12 pp.

BSI BS 6100: Sec 2.8-01. 1990 Amd 1 Building and Civil Engineering Terms Part 2: Civil Engineering Section 2.8: Pipelines and Ducts (AMD 7255) August 15, 1992. 15 pp.

—Heating, Ventilating and Air Conditioning Equipment—Construction
SNZ NZS/SMACNA 15D-85. HVAC Duct Construction Standards-Metal and Flexible First Edition. 216 pp.

—Identification Systems
SAA AS 1345-82. Identification of the Contents of Piping, Conduits and Ducts (Incorporating Amdt 1) Amdt 2 March 1986. 18 pp.

—Jackets—PVC
CGSB 51-GP-53M-77. Jacketing, Polyvinyl Chloride Sheet, for Insulated Pipes, Vessels and Round Ducts, Standard for. 9 pp.

—Metal Sheets
DIN ENGL 24190-85. Ducting for Ventilation Equipment; Lock-Seamed and Welded Sheet Metal Ducts (Nov). 3 pp.

DIN ENGL 24191-85. Ducting for Ventilation Equipment; Lock-Seamed and Welded Sheet Metal Ducting Components (Nov). 7 pp.

—Metal Sheets—Tightness Testing
DIN ENGL 24194 Pt 1-85. Ducting Components for Ventilation Equipment; Tightness Testing of Sheet Metal Ducts and Sheet Metal Ducting Components (Nov). 2 pp.

—Particulates
BSI BS 893-78. 1978 Amd 1 Measurement of the Concentration of Particulate Material in Ducts Carrying Gases. 16 pp.

—Particulates—Gravimetric Analysis
BSI BS 6069: Sec 4.3-92. 1992 Characterization of Air Quality Part 4: Stationary Source Emissions Section 4.3: Method for the Manual Gravimetric Detn. of Concentration and Mass Flow Rate of Particulate Material in Gas-Carrying Ducts (ISO 9096: 1992). 37 pp.

ISO 9096-92. Stationary Source Emissions—Determination of Concentration and Mass Flow Rate of Particulate Material in Gas-Carrying Ducts—Manual Gravimetric Method First Edition. 36 pp.

Ducts (Cont.)

—Plastic—Chemical Resistant
CGSB 41-GP-22-69. Process Equipment: Reinforced Polyester, Chemical Resistant, Custom-Contact Molded, Standard for; (Amendment 3 Mar 1978). 32 pp.

—Plastic—Installation
CEN PREN 1046-93. Plastics Piping and Ducting Systems—Plastics Systems Outside Building Structures—Recommended Practice for Installation Above and Below Ground. 71 pp.

—PVC Sheets
CNS K3044-87. Rigid PVC Extrusion Sheet for Duct (Jul)(6479). 2 pp.

—Ships—Thermal Insulation
JIS F 0302-77. Standard Practice for Thermal Insulation Work for Small Ships' Airconditioning Ducts.

—Silencers
SAA AS 1277-83. Acoustics—Measurement Procedures for Ducted Silencers. 23 pp.

—Steel—Conveyors—Agricultural Equipment
BSI BS 4286: Part 2-70. 1970 Steel Ducting for Grain and Fodder Conveying Part 2: Metric Units. 11 pp.

—Thermal Insulation
CGSB CAN/CGSB-51.9-92. Mineral Fibre Thermal Insulation for Piping and Round Ducting. 9 pp.

CGSB CAN/CGSB-51.10-92. Mineral Fibre Board Thermal Insulation. 8 pp.

CGSB CAN/CGSB-51.11-92. Mineral Fibre Thermal Insulation Blanket. 8 pp.

CGSB 51-GP-29MA-88. Phenolic Thermal Insulation for Pipes and Ducts. 9 pp.

CGSB 51-GP-52MA-89. Vapor Barrier, Jacket and Facing Material for Pipe, Duct and Equipment Thermal Insulation. 10 pp.

—Underfloor
CNS C4241-87. Underfloor Ducts (Steel) (Jan)(6097).

JIS C 8351-85. Underfloor Ducts (Steel). 7 pp.

—Underfloor—Junction Boxes
CNS C4243-87. Junction Boxes for Underfloor Duct (Jan)(6099).

Ductwork
Use: Ducts

Dumbwaiters
Use: Service Lifts

Dumet Wire
Use For: Copper Clad Nickel-Iron Alloy Wire
See Also: Electric Wire; Incandescent Lighting; Sealing Wire; Seals; Wire

CNS H3155-90. Dumet Wires (Dec)(12823).

JIS H 4541-87. Dumet Wires. 11 pp.

Dummies (Testing)
Use: Test Dummies

Dump Trucks
Use For: Bottom Dump Haulers
See Also: Earthmoving Equipment; Ground Vehicles; Trucks

JIS D 6501-78. Testing Method of Dump Trucks.

—Capacity Measurement
ISO 6483-80. Earth-Moving Machinery—Dumper Bodies—Volumetric Rating First Edition. 7 pp.

JIS A 8804-92. Earth-Moving Machinery—Dumper Bodies—Volumetric Rating.

—Glossaries
BSI BS 6914: Part 4-90. 1990 Terminology (Including Definitions of Dimensions and Symbols) for Earth-Moving Machinery Part 4: Glossary for Dumpers. 15 pp.

ISO 7132-90. Earth-Moving Machinery—Dumpers—Terminology and Commercial Specifications Second Edition. 17 pp.

—Off Road
JIS A 8422-91. Terminology and Commercial Specification on off-Highway Dump Trucks. 24 pp.

JIS A 8803-84. Testing Methods of Off-Highway Dump Trucks. 67 pp.

—Off Road—Glossaries
JIS A 8422-91. Terminology and Commercial Specification on off-Highway Dump Trucks. 24 pp.

—Retarders (Devices)
ISO 10268-93. Earth-Moving Machinery—Retarders for Dumpers and Tractor-Scrapers—Performance Tests First Edition. 7 pp.

Dump

Dump Trucks (Cont.)
—Roll Over Protective Structures
ISO 3471 Pt 1-86. Earth-Moving Machinery—Rollover Protective Structures—Laboratory Tests and Performance Requirements—Part 1: Crawler, Wheel Loaders and Tractors, Backhoe Loaders, Graders, Tractor Scrapers, Articulated Steer Dumpers First Edition. 18 pp.

Dumper Trucks
Use: Dump Trucks

Dumpers
See Also: Drilling Equipment; Materials Handling Equipment
—Roll Over Protective Structures
BSI BS 5527: Part 1-87. 1987 Roll-Over Protective Structures on Earth-Moving Machinery: Laboratory Tests and Preformance Requirements Part 1: Crawler, Wheel Loaders and Tractors, Backhoe Loaders, Graders, Tractor Scrapers and Articulated Steel Dumpers. 19 pp.
ISO 3471 Pt 1-86. Earth-Moving Machinery—Rollover Protective Structures—Laboratory Tests and Performance Requirements—Part 1: Crawler, Wheel Loaders and Tractors, Backhoe Loaders, Graders, Tractor Scrapers, Articulated Steer Dumpers First Edition. 18 pp.
—Safety
CEN PREN 474-6-93. Earth-Moving Machinery—Safety—Part 6: Requirements for Dumpers and Site Carriers. 10 pp.

Dunes
Use For: Sand Dunes See Also: Drilling Equipment; Materials Handling Equipment; Sands; Shore Protection
—Coastal—Protection
DIN ENGL 19657-73. Protection of Water Courses, Dikes and Coastal Dunes; Guidelines (Sept). 28 pp.
—Coastal—Structural Engineering—Construction Contracts
DIN ENGL 18310-88. Tendering and Performance Stipulations in Contracts for Construction Works (VOB); Part C: General Technical Specifications in Contracts for Construction Works (ATV); Revetment Works (Sept) (This Standard, Together with DIN 18299,. 4 pp.

Duplicating Machines
Use: Copying Machines; Duplicators

Duplicating Masters
See Also: Reprography
—Carbon Paper—Sets
CGSB CAN/CGSB-53.36-92. Master Duplicating Paper Set. 8 pp.
—Engineering Drawings
DIN ENGL 6774 Pt 4-82. Technical Drawings; Rules for Preparation; Drawn Masters for Printing Purposes (Apr). 5 pp.

Duplicating Papers
Use: Reproduction Papers

Duplicators
Use For: Duplicating Machines See Also: Copying Machines; Photocopies; Reproduction (Copying)
BSI BS 5960-80. 1980 Minimum Information to Be Included in Specification Sheets of Duplicators. 6 pp.
ISO 4232 Pt 1-79. Office Machines—Minimum Information to Be Included in Specification Sheets—Part 1: Duplicators First Edition; (Erratum—Oct 1980). 6 pp.
JTC1 4232 Pt 1-79. Office Machines—Minimum Information to Be Included in Specification Sheets—Part 1: Duplicators First Edition; (Erratum—Oct 1980). 6 pp.
OSI ISO 4232-1-79. Office Machines—Minimum Information to Be Included in Specification Sheets—Part 1: Duplicators. 6 pp.
—Glossaries
BSI BS 5479: Part 3-81. 1981 Duplicators and Document Copying Machines Part 3: Glossary of Terms for Duplicators. 16 pp.
ISO 5138 Sec 02-80. Office Machines—Vocabulary—Section 02: Duplicators First Edition. 31 pp.
OSI ISO 5138-2-80. Office Machines—Vocabulary—Section 02: Duplicators. 31 pp.
—Registration
BSI BS 4588-70. 1970 Methods of Test for Determining the Registration Obtained on Duplicating Machines. 12 pp.

Duplicators (Cont.)
—Registration (Cont.)
ISO 3066-86. Duplicating Machines—Registration Second Edition. 5 pp.
JTC1 3066-86. Duplicating Machines—Registration Second Edition. 5 pp.
OSI ISO 3066-86. Duplicating Machines—Registration. 5 pp.
—Stencil
BSI BS 5479: Part 2-77. 1977 Duplicators and Document Copying Machines Part 2: Minimum Overprint and Attachment Features of Stencils for Duplicators. 7 pp.
CGSB CAN/CGSB-53.33-92. Stencil, Duplicating-Machine, Mimeograph. 7 pp.
ISO 2133-76. Stencils for Duplicators—Minimum Overprint and Attachment Features Second Edition. 6 pp.
ISO 3066-86. Duplicating Machines—Registration Second Edition. 5 pp.
JTC1 2133-76. Stencils for Duplicators—Minimum Overprint and Attachment Features Second Edition. 6 pp.
JTC1 3066-86. Duplicating Machines—Registration Second Edition. 5 pp.
OSI ISO 2133-76. Stencils for Duplicators—Minimum Overprint and Attachment Features. 6 pp.
OSI ISO 3066-86. Duplicating Machines—Registration. 5 pp.
—Stencil—Reproduction Papers
DIN ENGL 6734-65. Duplicating Paper for Stencil Duplicators; Material Types and Properties (Oct). 1 p.
—Symbols
IEC 6329-89. Duplicators and Document Copying Machines—Symbols First Edition. 19 pp.
ISO 6329-89. Duplicators and Document Copying Machines—Symbols First Edition. 19 pp.
JTC1 6329-89. Duplicators and Document Copying Machines—Symbols First Edition. 19 pp.
OSI ISO/IEC 6329-89. Duplicators and Document Copying Machines—Symbols. 18 pp.

Durable Press Fabrics
Use For: Permanent Press Fabrics See Also: Fabrics
—Laundering Testing
CGSB CAN/CGSB-4.2 NO.59.1-M88. Textile Test Methods Appearance After Repeated Domestic Laundering—Smoothness of Fabric. 11 pp.
ISO 7768-85. Textiles—Method for Assessing the Appearance of Durable Press Fabrics After Domestic Washing and Drying First Edition. 6 pp.
ISO 7769-92. Textiles—Method for Assessing the Appearance of Creases in Durable-Press Products After Domestic Washing and Drying Second Edition. 9 pp.
ISO 7770-85. Textiles—Method for Assessing the Appearance of Seams in Durable Press Products After Domestic Washing and Drying First Edition. 6 pp.

Durometer Hardness Testing
See Also: Dynamic Testing; Hardness Testing; Indentation Hardness Testing
—Ebonite
BSI BS 2782:Pt3: METH 365B-92. 1992 Methods of Testing Plastics Part 3: Mechanical Properties Method 365B: Determination of Indentation Hardness by Means of a Durometer (Shore Hardness) (ISO 868: 1985) (V). 6 pp.
BSI BS 2782:Pt3: METH 365B-81. 1981 Methods of Testing Plastics Part 3: Mechanical Properties Method 365B: Determination of Indentation Hardness by Means of a Durometer (Shore Hardness). 4 pp.
ISO 868-85. Plastics and Ebonite—Determination of Indentation Hardness by Means of a Durometer (Shore Hardness) Second Edition. 5 pp.
—Plastics
BSI BS 2782:Pt3: METH 365B-92. 1992 Methods of Testing Plastics Part 3: Mechanical Properties Method 365B: Determination of Indentation Hardness by Means of a Durometer (Shore Hardness) (ISO 868: 1985) (V). 6 pp.
BSI BS 2782:Pt3: METH 365B-81. 1981 Methods of Testing Plastics Part 3: Mechanical Properties Method 365B: Determination of Indentation Hardness by Means of a Durometer (Shore Hardness). 4 pp.
CNS K6968-89. Method of Test for Durometer Hardness of Plastics (Nov)(12628).
ISO 868-85. Plastics and Ebonite—Determination of Indentation Hardness by Means of a Durometer (Shore Hardness) Second Edition. 5 pp.
JIS K 7215-86. Testing Methods for Durometer Hardness of Plastics. 16 pp.

Durometer Hardness Testing (Cont.)
—Rollers—Plastic Covered
BSI BS 7442: Sec 3.2-91. 1991 Rubber or Plastics Covered Rollers Part 3: Methods of Test Section 3.2: Determination of Shore Hardness (ISO 7267-2: 1986) (V). 8 pp.
—Rollers—Rubber Covered
BSI BS 7442: Sec 3.2-91. 1991 Rubber or Plastics Covered Rollers Part 3: Methods of Test Section 3.2: Determination of Shore Hardness (ISO 7267-2: 1986) (V). 8 pp.
ISO 7267 Pt 2-86. Rubber-Covered Rollers—Determination of Apparent Hardness—Part 2: Shore-Type Durometer Method First Edition. 4 pp.
—Sealing Materials
CGSB CAN2-19.0-M77METH 8.1-78. Methods of Testing Putty, Caulking and Sealing Compounds Shore "A" Hardness. 1 p.
CGSB CAN2-19.0-M77METH 8.2-78. Methods of Testing Putty, Caulking and Sealing Compounds Shore "A" Hardness After Heat Aging. 1 p.

Durum Wheat
See Also: Cereals
—Pasta
ISO 7304-85. Durum Wheat Semolinas and Alimentary Pasta—Estimation of Cooking Quality of Spaghetti by Sensory Analysis First Edition. 12 pp.
—Sieve Analysis
ISO 5532-87. Durum Wheat—Determination of Proportion of Non-Wholly Vitreous Grains (Reference Method) Second Edition. 4 pp.

Dust
Use For: Dust Gages; Dust Particles; Dust Testing; Electrostatic Dust Samplers See Also: Aerosols; Flue Dust; Particulates; Sand
—Air Cleaners
JIS Z 8901-84. Dusts and Aerosols for Industrial Testing. 21 pp.
—Airborne—Gages
BSI BS 1747: Part 5-72. 1972 Methods for the Measurement of Air Pollution Part 5: Directional Dust Gauges. 21 pp.
—Atmosphere
CNS K9013-76. Method of Determination for Dust-Fall in Atmosphere (Mar)(3916).
CNS K9066-81. General Rule of Measuring Methods for Airbone Dust Concentration in Environmental Atmosphere (Apr)(7252).
JIS Z 8813-79. General Rules of Measuring Methods for Airborne Dust Concentration in Environmental Atmosphere.
JIS Z 8813-69. General Rules of Measuring Method for Airborne Dust Concentration in Environmental Atmosphere. 7 pp.
—Automotive
CNS D3069-81. General Rules of Dust Test for Automobile Parts (Mar)(7139).
JIS D 0207-77. General Rules of Dust Test for Automobile Parts.
—Clean Rooms
JIS B 9926-91. Clean Room—Test Methods for Dust Generation from Moving Mechanisms. 11 pp.
—Cleaning Agents
CGSB 31-GP-0A METH 45.1-57. Methods of Testing Corrosion-Prevention Materials and Processes Dusting Characteristics. 1 p.
—Connectors
CEN PREN 2591 (Part C8)-92. Elements of Electrical and Optical Connection Test Methods Part C8—Sand and Dust. 5 pp.
CNS C6256-86. Method of Test for Low Frequency (Below 3MHz) Electrical Connectors (TP-50 Sand and Dust Test) (May)(11573).
SBAC TS 330 ISSUE 1. Test for Electrical Connectors Dust and Sand Contamination.
—Electric Switches
CNS C6095-88. Method of Test for Electromechanical Switches (Sand and Dust) (Dec)(6158).
—Electrical Components
BSI BS 7527: Sec 2.5-91. 1991 Classification of Environmental Conditions Part 2: Environmental Conditions Appearing in Nature Section 2.5: Dust, Sand, Salt Mist (IEC 721 2-5: 1991). 21 pp.
IEC 721 Pt 2-5-91. Classification of Environmental Conditions Part 2: Environmental Conditions Appearing in Nature Section 5: Dust, Sand, Salt Mist First Edition. 38 pp.

INTERNATIONAL AND NON-U.S. NATIONAL STANDARDS SUBJECT INDEX

Dust *(Cont.)*

—Electrical Enclosures
IEC 1241 Pt 1-1-93. Electrical Apparatus for Use in the Presence of Combustible Dust—Part 1: Electrical Apparatus Protected by Enclosures—Section 1: Specification for Apparatus First Edition. 47 pp.
IEC 1241 Pt 1-2-93. Electrical Apparatus for Use in the Presence of Combustible Dust—Part 1: Electrical Apparatus Protected by Enclosures—Section 2: Selection, Installation, and Maintenance First Edition. 39 pp.

—Electrical Resistance
JIS B 9915-89. Measuring Methods for Dust Resistivity (with Parallel Electrodes). 15 pp.

—Electrical Resistivity
IEC 1241 Pt 2-2-93. Electrical Apparatus for Use in the Presence of Combustible Dust—Part 2: Test Methods—Section 2: Method for Determining the Electrical Resistivity of Dust in Layers First Edition. 21 pp.

—Explosive Atmospheres
SNZ NZS 6101: Part 2-90. Classification of Hazardous Areas Part 2: Combustible Dusts. 24 pp.

—Fiber Optics
CNS C6230-84. Method of Test for Fiber Optic Devices (Dust (Fine Sand) Test) (Jun)(10930).

—Filling Materials
BSI BS 3400-67. (WITHDRAWN) 1967 Methods of Test for Dust in Filling Materials. 20 pp.

—Newsprint
CPPA L.2U-77. Lint and Dust Test for Newsprint. 1 p.

—Radiation Monitors
JIS Z 4316-78. Radioactive Dust Monitors.

—Radioactive
JIS Z 4601-78. Radioactive Dust Samplers.

—Steels—Surface Analysis
BSI BS 7079: Part B3-93. 1993 Preparation of Steel Substrates Before Application of Paints and Related Products Group B: Methods for the Assessment of Surface Cleanliness Part B3: Assessment of Dust on Steel Surfaces Prepared for Painting (Pressure-Sensitive). 15 pp.
ISO 8502 Pt 3-92. Preparation of Steel Substrates Before Application of Paint and Related Products—Tests for the Assessment of Surface Cleanliness—Part 3: Assessment of Dust on Steel Surfaces Prepared for Painting (Pressure-Sensitive Tape Method) First Edition; (Corr. and Repr.—1993). 14 pp.

Dust Blouses
Use: Blouses—Dust

Dust Caps
Use: Dust Covers

Dust Chambers
Use: Dust Collectors

Dust Collectors
Use For: Dust Chambers; Dust Guards; Dust Separators *See Also:* Air Cleaners; Air Filters; Chimneys; Dust Filters; Pneumatic Conveyors; Separators (Mechanical)
CNS B1274-81. Expression of Specification for Dust Collectors (Oct)(7943).
CNS B7170-81. Method of Measuring Performance for Dust Collectors (Oct)(7942).
ISO 6584-81. Cleaning Equipment for Air and Other Gases—Classification of Dust Separators First Edition. 7 pp.
JIS B 9909-77. Expression of the Specification for Dust Collectors (R 1988). 17 pp.
JIS B 9910-77. Method of Measuring Performance for Dust Collectors. 15 pp.

—Axle Boxes—Rolling Stock
CNS E1038-85. Dust Guards for Plain Bearing Axle Boxes of Railway Rolling Stock (Feb)(11203).

—Inspection
CNS B7046-81. Inspection Standard of Mist Blowers and Desters (Jan)(3374).

Dust Content Analysis

—Drying Agents
BSI BS 3482: Part 11-91. 1991 Methods of Test for Desiccants Part 11: Determination of Friability and Dust Content. 5 pp.

—Flue Gases
CNS K9010-75. Determination for Location of Dust Content in Stack Gas (May)(3798).

Dust Cores

—Toroidal
CNS C4341-82. Toroidal Dust Cores of Fe-Si-Al Alloy (Jan)(8370).

Dust Covers
Use For: Dust Caps; Dust Lids; Dust Plugs; Dust Sheets; Dust Shields *See Also:* Guards (Protective)

—N Coaxial Connectors—Preferred Products List
CECC CECC MUAHAG Vol 3B IS 4-91. Preferred Products List; Connectors; R.F. and Fibre Optics (En, Fr, Ge). 65 pp.

Dust Explosions
See Also: Explosions

—Grain Elevators
CNS Z1020-88. Safety Code of Preventing Dust Explosion for Grain Elevator (Oct)(3364).

—Plastics—Safety
CNS Z1009-87. Code for Preventing Dust Explosion in Plastic Industry (Oct)(2502).

—Protection
BSI BS 6713: Part 1-86. 1986 Explosion Protection Systems Part 1: Method for Determination of Explosion Indices of Combustible Dusts in Air. 8 pp.
BSI BS 6713: Part 1-01. 1986 Amd 1 Explosion Protection Systems Part 1: Method for Determination of Explosion Indices of Combustible Dusts in Air (ISO 6184/1: 1985) (AMD 7083) February 28, 1992. 13 pp.
DIN ENGL EN 26184 Pt 1-91. Explosion Protection Systems; Determination of Explosion Indices of Combustible Dusts in Air (ISO 6184-1: 1985) (June). 8 pp.
ISO 6184 Pt 1-85. Explosion Protection Systems—Part 1: Determination of Explosion Indices of Combustible Dusts in Air First Edition. 7 pp.

Dust Filters
See Also: Dust Collectors; Protective Masks; Separators (Mechanical)

—Respirators—Portable
JIS T 8203-80. Portable Filter Type Dust and Fume Collectors.

Dust Free Rooms
Use: Clean Rooms

Dust Gages
Use: Dust

Dust Guards
Use: Dust Collectors

Dust Hoods
Use: Hoods (Headgear)—Dust

Dust Lids
Use: Dust Covers

Dust Mops
Use: Mops—Dust

Dust Particles
Use: Dust

Dust Plugs
Use: Dust Covers

Dust Pressed Tiles
Use: Ceramic Tiles

Dust Samplers
See Also: Particulate Monitors; Sampling

—Airborne—Electrostatic Testing
JIS Z 8810-79. Electrostatic Dust Sampler and Measuring Method for Mass of Airborne Dust by the Electrostatic Dust Sampler. 9 pp.

—Airborne—Workplace
JIS Z 8814-92. Low Volume Air Samplers and Methods for Measuring Mass Concentration of Airborne Dust by the Low Volume Air Samplers. 14 pp.

Dust Sampling
Use: Dust

Dust Separators
Use: Dust Collectors

Dust Sheets
Use: Dust Covers

Dust Shields
Use: Dust Covers

Dust Testing
Use: Dust

Dustbins
Use: Waste Containers

Dusters
CNS B4036-80. Hand Dusters (Aug)(6011).

Dusting (Paper)
Use: Linting

Dusting Brushes
See Also: Brushes (Cleaning/Polishing)
CGSB 22.24M-88. Bench Dusting Brush. 6 pp.

Dusting Powders

—Rubber—Adhesion Testing
MOD UK DSTAN 68-119-92. Powder, Dusting for Rubber, Types: 1 and 2 Issue 1. 15 pp.

Duties, Customs
Use: Import Tariffs

Duty Cycles
Use For: Duty Factors

—Emergency Position Indicating Radio Beacons—INMARSAT
CCIR RECMN 632-1-90. Transmission Characteristics of a Satellite Emergency Position-Indicating Radiobeacon (Satellite EPIRB) System Operating Through Geostationary Satellites in the 1.6 GHz Band—Section 8J—Technical and Operating Characteristics of Radiocommunications. 7 pp.

Duty Factors
Use: Duty Cycles

Duty Free Entry
See Also: Foreign Acquisition
MOD UK DEFCON 75-86. Waver of Import Duty 6/86. 1 p.

Dwellings
Use: Residential Buildings

Dye Content Analysis

—Food
CNS N6185-87. Method of Test for Dyes in Foods (Jun)(10889). 8 pp.

Dyeing
See Also: Color Matching; Dyes; Printing

—Detergents
CNS K8009-71. Liquid Detergent (for Printing and Dyeing Processes) (Tentative) (Jan)(3163).

Dyeing Equipment
See Also: Padders; Textile Finishing Machinery; Textile Machinery

—Beams (Textile Machinery)
BSI BS ISO 8116-9-91. 1991 Textile Machinery and Accessories—Beams for Winding—Part 9: Dyeing Beams for Textile Fabrics (L). 7 pp.
ISO 8116 Pt 9-91. Textile Machinery and Accessories—Beams for Winding—Part 9: Dyeing Beams for Textile Fabrics First Edition. 5 pp.

—Beams—Glossaries
ISO 1037-82. Textile Machinery and Accessories—Beams for Dyeing Slivers and Yarn—Terminology and Main Dimensions First Edition. 5 pp.

—Classification
BSI BS 4753: Part 3-74. (WITHDRAWN) 1974 Amd 1 Dyeing and Finishing Machines Part 3: Classification and Nomenclature. 25 pp.

—Cloth Guiders
JIS L 7507-82. Electric Cloth Guiders for Dyeing and Finishing.
JIS L 7508-92. Pneumatic Cloth Guiders for Dyeing and Finishing.
JIS L 7508-86. Pneumatic Cloth Guiders for Dyeing and Finishing. 15 pp.

—Designations
BSI BS ISO 10459-92. 1992 Textile Machinery—Dyeing and Finishing Machines—Designation of Operating Ranges of Component Parts (L). 7 pp.
ISO 10459-92. Textile Machinery—Dyeing and Finishing Machines—Designation of Operating Ranges of Component Parts First Edition. 5 pp.

INDUSTRY STANDARDS

Dyeing

Dyeing Equipment (Cont.)

—Drying Cylinders
- JIS L 7505-92. Drying Cylinders for Dyeing and Finishing.

—Glossaries
- BSI BS 4392: Part 5-78. 1978 Textile Machinery and Accessories. Definition of Left and Right Side Part 5: Right Side Dyeing and Finishing Machines. 4 pp.
- BSI BS 4753: Part 3-74. (WITHDRAWN) 1974 Amd 1 Dyeing and Finishing Machines Part 3: Classification and Nomenclature. 25 pp.
- BSI BS 4753: Part 4-85. (WITHDRAWN) 1985 Dyeing and Finishing Machines Part 4: Glossary of Terms for Ancillary Devices. 23 pp.
- ISO 1036-84. Textile Machinery—Dyeing and Finishing Machines—Definition of Left and Right Sides Second Edition. 4 pp.
- ISO 5248-82. Textile Machinery and Accessories—Dyeing and Finishing Machines—Vocabulary for Ancillary Devices First Edition. 23 pp.
- JIS L 0207-76. Glossary of Terms Used in Textile Industry (Dyeing and Finishing).
- JIS L 0308-85. Glossary of Terms Used in Dyeing and Finishing Machinery.

—Guide Rollers
- ISO 5249-88. Textile Machinery and Accessories—Guide Rollers for Dyeing and Finishing Machinery—Main Dimensions First Edition. 7 pp.
- JIS L 7506-81. Guide Rollers for Dyeing and Finishing.

—Nominal Speeds
- BSI BS ISO 10457-91. 1991 Textile Machinery—Dyeing and Finishing Machines—Nominal Speeds (L). 7 pp.
- ISO 10457-91. Textile Machinery—Dyeing and Finishing Machines—Nominal Speeds First Edition. 4 pp.

—Tubes—Cheese Dyeing
- BSI BS 2547: Part 7-79. 1979 Basic Dimensions of Cones and Tubes for Winding Textile Yarns Part 7: Perforated Cylindrical Tubes for Cheese Dyeing. 3 pp.
- ISO 574-79. Textile Machinery and Accessories—Perforated Cylindrical Tubes for Cheese Dyeing First Edition. 3 pp.

—Winders—Bars
- BSI BS ISO 10458-93. 1993 Textile Machinery—Square Bars for Winding Devices Relating to Dyeing and Finishing Machines—Dimensions (L). 7 pp.
- ISO 10458-93. Textile Machinery—Square Bars for Winding Devices Relating to Dyeing and Finishing Machines—Dimensions First Edition. 5 pp.

Dyeing Testing
- JIS K 4001-50. Sampling Method for Dyestuff.

—Anodic Coatings
- BSI BS 6161: Part 5-82. 1982 Methods of Test for Anodic Oxidation Coatings on Aluminium and Its Alloys Part 5: Estimation of Loss of Absorptive Power of Sealed Coatings: Dye Spot Test with Prior Acid Treatment. 7 pp.
- ISO 2143-81. Anodizing of Aluminium and Its Alloys—Estimation of Loss of Absorptive Power of Anodic Oxide Coatings After Sealing—Dye Spot Test with Prior Acid Treatment First Edition. 5 pp.

—Migration—Woven/Knitted Fabrics
- JIS L 1063-85. Testing Methods for Migration of Dyestuffs and Finishing Agents on Woven Fabrics and Knitted Fabrics.

—Oxide Coatings
- BSI BS 6161: Part 5-82. 1982 Methods of Test for Anodic Oxidation Coatings on Aluminium and Its Alloys Part 5: Estimation of Loss of Absorptive Power of Sealed Coatings: Dye Spot Test with Prior Acid Treatment. 7 pp.
- ISO 2143-81. Anodizing of Aluminium and Its Alloys—Estimation of Loss of Absorptive Power of Anodic Oxide Coatings After Sealing—Dye Spot Test with Prior Acid Treatment First Edition. 5 pp.

Dyes
Use For: Dyestuffs *See Also:* Colorants; Discharging Agents; Dyeing; Eosin Yellowish-(YS); Fabric Finishes; Fabrics; Fluorescent Brighteners; Furnish; Hair Dyes; Metallized Dyes; Methylene Blue; Nutgall; Oxidation Dye Content Analysis; Pigments; Sulfur Dyes; Toners

- CNS K6918-87. Method of Test for Organic Dyes Intermediates (Oct)(12127).

Dyes (Cont.)

—Chloride Content
- MOD UK DEF-1053: METH NO. 79. Standard Methods of Testing Paint, Varnish, Lacquer and Related Products. Indices Method 79: Determination of Water-Soluble Chloride Content in Dyestuffs (Withdrawn).

—Classification
- BSI BS 1000: (667)-79. 1979 Universal Decimal Classification (UDC). English Full Edition (667): Colour Industries (Dyes, Inks, Paints Etc.). 20 pp.
- SNZ NZS/BS 1000 (667)-79. Universal Decimal Classification Colour Industries (Dyes, Inks, Paints, Etc.). 20 pp.

—Explosives
- MOD UK DSTAN 68-58-01. Dyestuffs for Explosives Applications Issue 1; Amendment 3. 34 pp.

—Fabrics—Identification
- CNS L3215-84. Identification of Dyestuff Classes on Dyed Textiles (Oct) (11080).
- JIS L 1065-83. Identification of Dyestuff Classes on Dyed Textiles. 36 pp.

—Inventories
- MOD UK TS 50224. Dye, Green S.

—Lacquers
- MOD UK DEF-1430-64. Dyestuffs for R.D. Varnishes and Lacquers. 4 pp.

—Pyrotechnics
- MOD UK DSTAN 68-58-01. Dyestuffs for Explosives Applications Issue 1; Amendment 3. 34 pp.

—Pyrotechnics—Content Analysis
- MOD UK M 818/63. Examination of Dyestuffs for Use in Pyrotechnic Compositions and HE Substitutes.

—Pyrotechnics—Density
- MOD UK M 818/63. Examination of Dyestuffs for Use in Pyrotechnic Compositions and HE Substitutes.

—Pyrotechnics—Melting Points
- MOD UK M 818/63. Examination of Dyestuffs for Use in Pyrotechnic Compositions and HE Substitutes.

—Pyrotechnics—Sieve Analysis
- MOD UK M 818/63. Examination of Dyestuffs for Use in Pyrotechnic Compositions and HE Substitutes.

—Pyrotechnics—Visual Inspection
- MOD UK M 818/63. Examination of Dyestuffs for Use in Pyrotechnic Compositions and HE Substitutes.

—Solubility—Organic Solvents
- BSI BS 3483: Part B11-91. 1991 Methods for Testing Pigments for Paints Part B11: Determination of Solubility of Dyestuffs in Organic Solvents. 13 pp.

—Solubility—Organic Solvents—Gravimetric Analysis
- ISO 7579-90. Dyestuffs—Determination of Solubility in Organic Solvents—Gravimetric Method First Edition. 11 pp.

—Surfactants—Dispersion
- DIN ENGL 53908 Pt 1-82. Testing of Surface Active Agents and Textile Auxiliaries; Determination of Dispersing Power or Flocculation Preventive Power of Surface Active Agents When Added to Dyestuffs in Acidic Medium (Aug). 5 pp.
- DIN ENGL 53908 Pt 3-82. Testing of Surface Active Agents and Textile Auxiliaries; Determination of Dispersing Power of Flocculation Preventive Power of Surface Active Agents When Added to Dyestuffs in Neutral Medium (Aug). 5 pp.

—Unsulfonated Primary Aromatic Amines Content
- DIN ENGL 55610-86. Testing of Pigments and Solvent-Soluble Dyestuffs; Determination of Unsulfonated Primary Aromatic Amines (Sept). 5 pp.

—Varnishes
- MOD UK DEF-1430-64. Dyestuffs for R.D. Varnishes and Lacquers. 4 pp.

Dyestuffs
Use: Dyes

Dynamas
Use: Generators

Dynamic Load Testing
Use: Dynamic Loads

Dynamic Loads
See Also: Ice Loads; Snow Loads; Wind Loads

—Antifriction Bearings—Ratings
- CNS B1168-80. Load Carrying Capacity of Rolling Bearings, Definitions, Load Rating, Methods of Load and Rating Life (Jul)(5683).

—Automotive—Towing Attachments—Ball Couplings
- ISO 3853-77. Road Vehicles—Caravans and Light Trailers—Towing Brackets and Coupling Balls—Strength Test First Edition. 6 pp.

—Ball Screws—Ratings
- BSI BS 6101: Part 1-81. 1981 Machine Tool Ball Screws Part 1: Methods of Calculating Dynamic Load and Life Ratings. 6 pp.

—Bearings—Ratings
- BSI BS 5512: Part 1-77. (WITHDRAWN) 1977 Rolling Bearings. Dynamic Load Ratings and Rating Life Part 1: Calculation Methods (Superseded by BS 5512: 1991). 13 pp.
- BSI BS 5512-91. 1991 Dynamic Load Ratings and Rating Life of Rolling Bearings (ISO 281: 1990). 17 pp.
- ISO 281-90. Rolling Bearings—Dynamic Load Ratings and Rating Life First Edition. 16 pp.
- ISO TR8646-85. Explanatory Notes on ISO 281/1-1977 First Edition. 66 pp.
- JIS B 1518-92. Dynamic Load Ratings and Rating Life for Rolling Bearings. 25 pp.

—Buildings
- DIN ENGL 1055 Pt 3-71. Design Loads for Buildings; Live Loads (June). 7 pp.

—Parachutes—Gliders
- CEN PREN 926-92. Paraglider—Qualification of Paragliders—Structural Strength Tests. 8 pp.

—Seamless Pipes
- DIN ENGL 2445 Pt 1-74. Seamless Steel Tubes for Dynamic Loads; Hot Finished Tubes; Nominal Pressures 100 to 400 (Nov). 4 pp.
- DIN ENGL 2445 Suppl. 1-74. Seamless Steel Tubes for Dynamic Loads; Basis for Calculation of Straight Tubes (Nov). 4 pp.

Dynamic RAMs
Use For: DRAMs; Dynamic Random Access Memories *See Also:* RAMs

- BSI BS 6493: Sec 2.2-86. 1986 Semiconductor Devices Part 2: Integrated Circuits Section 2.2: Recommendations for Digital Integrated Circuits. 143 pp.
- BSI BS 6493: Sec 2.2-01. 1986 Amd 1 Semiconductor Devices Part 2: Integrated Circuits Section 2.2: Recommendations for Digital Integrated Circuits (IEC 748-2: 1985) (AMD 7003) July 15, 1992. 166 pp.
- CECC CECC 90 112-002 ISSUE 1-87. UTE C 86 253-002; MOS Read/Write Dynamic Memories Silicon Monolithic Circuits; Tupe 4164; Electrical Variants; 4164-12/4164-15/4164-20; Temperature/Package Variants (Fr). 29 pp.
- CECC CECC 90 112-003 ISSUE 1-87. UTE C 86 253-003; MOS Read/Write Dynamic Memories Silicon Monolithic Circuits; Type 4256; Electrical Variants; 4256-12/4256-15/4256-20; Temperature/Package Variants (Fr). 27 pp.
- CECC CECC 90 112-006 ISSUE 1-89. 1 Megabit (1 048 576 x 1 Bit) Dynamic RAM; HYB 511000-85, HYB 511000-10, HYB 511000-12, HYF 511000-12, HYC 511000-12 (En). 36 pp.
- IEC 748 Pt 2-85. Semiconductor Devices Integrated Circuits Part 2: Digital Integrated Circuits First Edition; (Amendment 1-1991). 312 pp.

—CMOS
- CECC CECC 90 112-004 ISSUE 1-88. HYB 511000-85/-10/-12; 1 048 576 X 1 Bit; Dynamic RAM (En). 36 pp.

—MOS
- CECC CECC 90 112 ISSUE 1-86. Blank Detail Specification: MOS Read/Write Dynamic Memories Silicon Monolithic Circuits (En, Fr, Ge). 77 pp.

—NMOS
- CECC CECC 90 112-001 ISSUE 1-87. UTE C 86 253-001; MOS Read/Write Dynamic Memories Silicon Monolithic Circuits; Type 4416; Electrical Variants; 4416-12/4416-15/4416-20; Temperature/Package Variants (Fr). 27 pp.

—Preferred Products List
- CECC CECC MUAHAG Vol 7 IS 8-92. Preferred Products List; Active Microcircuits (En, Fe, Ge). 89 pp.

INTERNATIONAL AND NON-U.S. NATIONAL STANDARDS
SUBJECT INDEX

Dynamic RAMs *(Cont.)*
—**Quality Assurance**
BSI BS CECC 90112-87. 1987 MOS Read/Write Dynamic Memories Silicon Monolithic Circuits. 28 pp.

Dynamic Random Access Memories
Use: Dynamic RAMs

Dynamic Testing
See Also: Compression Testing; Ductility; Durometer Hardness Testing; Dynamic Loads; Fatigue Testing; Hardness Testing; High Temperature Testing; Hydrostatic Testing; Impact Testing; Life (Durability); Low Temperature Testing; Plastic Properties; Radiation Detection and Measurement; Shear Testing; Static Testing; Statistical Analysis; Torsion; Wear Testing

—**Belt Conveyors**
BSI BS 4890-73. 1973 Test Methods for Mechanical Joints in Conveyor Belting. 14 pp.

—**Bush Chains**
BSI BS 7615-92. 1992 Motor Cycle Chains (ISO 10190: 1992) (E). 14 pp.
ISO 10190-92. Motor Cycle Chains—Characteristics and Test Methods First Edition. 11 pp.

—**Child Restraining Devices**
SAA AS 3629.3-91. Methods of Testing Child Restraints—Part 3: Dynamic Testing of Upper Anchorage Components (AS 3629: Part 3:1991/NZS 5466: Part 3:1991) Amdt 1 October 1992 Amdt 2 April 1993. 3 pp.
SNZ NZS 5466.1-91. Methods of Testing Child Restraints Part 1: Dynamic Testing Amend: 1, 1992 (NZS 5466.1:1991/AS 3629.1-1991). 10 pp.
SNZ NZS 5466.3-91. Methods of Testing Child Restraints Part 3: Dynamic Testing of Upper Anchorage Components Amend: 1, 1992; 2, 1993 (NZS 5466.3:1991/AS 3629.3-1991). 4 pp.

—**Elastomers**
ISO 2856-81. Elastomers—General Requirements for Dynamic Testing Second Edition. 17 pp.

—**Elastomers—Vulcanized**
BSI BS 903: Part A24-92. 1992 Physical Testing of Rubber Part A24: Guide to the Determination of Dynamic Properties of Rubbers. 16 pp.
BSI BS 903: Part A24-76. 1976 Methods of Testing Vulcanized Rubber Part A24: Measurement of Dynamic Moduli. 10 pp.
CNS K6744-83. Method of Test for Dynamic Properties of Vulcanized Rubber (Feb)(10016).
JIS K 6394-76. Testing Method for Dynamic Properties of Rubber Vulcanizates (R 1985). 17 pp.

—**Hydraulic Pressure Pumps**
JIS B 8661-89. Test Methods for Electronically Controlled Oil Hydraulic Pumps. 21 pp.

—**Roller Chains**
BSI BS 7615-92. 1992 Motor Cycle Chains (ISO 10190: 1992) (E). 14 pp.
ISO 10190-92. Motor Cycle Chains—Characteristics and Test Methods First Edition. 11 pp.

—**Test Specimens—Mounting**
CNS C6337-90. Basic Environmental Testing Procedures (Part 2: Tests, Mounting of Components, Equipment and Articles for Dynamic Tests Including Shock (Ea), Bump (Eb), Vibration (Fc and Fd) and Steady—State Acceleration (Ga) and Guidance (May)(12716).

—**Vulcanized Rubber**
BSI BS 903: Part A31-76. 1976 Amd 1 Methods of Testing Vulcanized Rubber Part A31: Determination of the Low-Frequency Dynamic Properties of Rubbers by Means of a Torsion Pendulum. 9 pp.

Dynamic Viscosity
See Also: Intrinsic Viscosity; Kinematic Viscosity; Rheological Properties; Saybolt Viscosity; Viscosity

—**Coal Tar Pitch**
ISO 8003-85. Carbonaceous Materials Used in the Production of Aluminium—Pitch for Electrodes—Measurement of Dynamic Viscosity First Edition. 6 pp.

—**Glass**
BSI BS 7034: Part 1-88. 1988 Viscosity and Viscometric Fixed Points of Glass Part 1: Recommendations for Use in the Determination of Viscosity and Viscometric Fixed Points. 12 pp.
BSI BS 7034: Part 2-88. 1988 Viscosity and Viscometric Fixed Points of Glass Part 2: Method for the Determination of Viscosity by Rotation Viscometers. 14 pp.

Dynamic Viscosity *(Cont.)*
—**Glass** *(Cont.)*
BSI BS 7034: Part 3-88. 1988 Viscosity and Viscometric Fixed Points of Glass Part 3: Method for the Determination of Viscosity by Fibre Elongation Viscometer (ISO 7884/3: 1987). 11 pp.
BSI BS 7034: Part 4-88. 1988 Viscosity and Viscometric Fixed Points of Glass Part 4: Method for the Determination of Viscosity by Beam Bending. 16 pp.
BSI BS 7034: Part 5-88. 1988 Viscosity and Viscometric Fixed Points of Glass Part 5: Method for the Determination of Working Point by Sinking Bar Viscometer. 8 pp.
ISO 7884 Pt 1-87. Glass—Viscosity and Viscometric Fixed Points—Part 1: Principles for Determining Viscosity and Viscometric Fixed Points First Edition. 11 pp.
ISO 7884 Pt 2-87. Glass—Viscosity and Viscometric Fixed Points—Part 2: Determination of Viscosity by Rotation Viscometers First Edition. 12 pp.
ISO 7884 Pt 3-87. Glass—Viscosity and Viscometric Fixed Points—Part 3: Determination of Viscosity by Fibre Elongation Viscometer First Edition. 11 pp.
ISO 7884 Pt 4-87. Glass—Viscosity and Viscometric Fixed Points—Part 4: Determination of Viscosity by Beam Bending First Edition. 15 pp.
ISO 7884 Pt 5-87. Glass—Viscosity and Viscometric Fixed Points—Part 5: Determination of Working Point by Sinking Bar Viscometer First Edition. 8 pp.

—**Liquid Fuels—Viscometers**
DIN ENGL 51561-78. Testing of Mineral Oils, Liquid Fuels and Related Liquids; Measurement of Viscosity Using the Vogel-Ossag Viscometer; Temperature Range: Approximately 10 to 150 Deg. C (Dec). 4 pp.
DIN ENGL 51569-78. Testing of Mineral Oils, Liquid Fuels and Related Liquids; Measurement of Viscosity Using the Vogel-Ossag Viscometer; Temperature Range:-55 to Approximately +10 Deg. C (Dec). 5 pp.

—**Liquids**
BSI BS 188-77. 1977 Determination of the Viscosity of Liquids. 18 pp.
CNS K6329-72. Method of Test for Viscosities of Transparent and Opaque Liquids (Kinematic and Dynamic Viscosities) (Oct)(3390).

—**Mineral Oils**
DIN ENGL 51561-78. Testing of Mineral Oils, Liquid Fuels and Related Liquids; Measurement of Viscosity Using the Vogel-Ossag Viscometer; Temperature Range: Approximately 10 to 150 Deg. C (Dec). 4 pp.
DIN ENGL 51569-78. Testing of Mineral Oils, Liquid Fuels and Related Liquids; Measurement of Viscosity Using the Vogel-Ossag Viscometer; Temperature Range:-55 to Approximately +10 Deg. C (Dec). 5 pp.

—**Paints**
ISO 2884-74. Paints and Varnishes—Determination of Viscosity at a High Rate of Shear First Edition. 4 pp.

—**Petroleum Products**
BSI BS 2000: Part 71-90. 1990 Petroleum and Its Products Part 71: Kinematic Viscosity of Transparent and Opaque Liquids and Calculation of Dynamic Viscosity. 16 pp.
ISO 3104-76. Petroleum Products—Transparent and Opaque Liquids—Determination of Kinematic Viscosity and Calculation of Dynamic Viscosity First Edition; (Erratum—Aug 1976) (Erratum—July 1978). 9 pp.

—**Potassium Silicates**
ISO 2123-72. Sodium and Potassium Silicates for Industrial Use—Determination of Dynamic Viscosity First Edition. 4 pp.

—**Sodium Silicates**
ISO 2123-72. Sodium and Potassium Silicates for Industrial Use—Determination of Dynamic Viscosity First Edition. 4 pp.

—**Varnishes**
ISO 2884-74. Paints and Varnishes—Determination of Viscosity at a High Rate of Shear First Edition. 4 pp.

—**Water**
ISO TR3666-77. Viscosity of Water First Edition. 2 pp.

Dynamite
See Also: Dinitrotoluene; Explosives; Trinitrotoluene
JIS K 4801-68. Dynamite.

—**Glycerol**
JIS K 3351-56. Dynamite Glycerine. 3 pp.

Dynamometers
See Also: Measuring Instruments

—**Automotive**
CNS D3107-82. Method of Dynamometer Test for Simulated Mountain Fade for Automobiles (Mar)(8564).
CNS D3108-82. Method of Braking Device Dynamometer Test for Passenger Car (Mar)(8565).
ISO 10521-92. Motor Vehicle Road Load—Determination Under Reference Atmospheric Conditions and Reproduction on Chassis Dynamometer First Edition. 22 pp.

—**Automotive—Exhaust Gases**
CNS D3077-87. Method of Test for Exhaust Pollution for Gasoline Engine Automobiles (Sep)(7895). 16 pp.

—**Buses (Vehicles)**
CNS D3116-82. Method of Braking Device Dynamometer Test for Trucks and Buses (Apr)(8680).

—**Motorcycles—Exhaust Gases**
CNS D3165-87. Method of Test for Exhaust Pollution for Motorcycles (Sep)(11386). 14 pp.

—**Motorcycles—Fuel Consumption**
CNS D3029-87. Method of Test for Fuel Consumption for Motorcycles (Sep)(3105). 13 pp.
ISO 11486-93. Two-Wheeled Motorcycles—Fuel Consumption Measurements—Chassis Dynamometer Setting by Coastdown Method First Edition. 17 pp.

—**Trucks**
CNS D3116-82. Method of Braking Device Dynamometer Test for Trucks and Buses (Apr)(8680).

Dysprosic Radar Repeaters
Use: Transmitters

D2b
Use: Domestic Digital Bus Systems

E-Mail
Use: Electronic Mail Systems

EADI
Use: Attitude Indicators

EAEC
Use: European Communities

Ear Muffs
See Also: Hearing Protectors; Protective Clothing
BSI BS 6344: Part 1-89. 1989 Industrial Hearing Protectors Part 1: Specification for Ear Muffs. 19 pp.

—**Insertion Loss—Quality Assurance**
BSI DD 192-90. 1990 Method for Measurement of Insertion Loss of Ear-Muff Type Hearing Protectors (Simplified Method for Quality Inspection Purposes). 11 pp.
ISO TR4869 Pt 3-89. Acoustics—Hearing Protectors—Part 3: Simplified Method for the Measurement of Insertion Loss of Ear-Muff Type Protectors for Quality Inspection Purposes First Edition. 11 pp.

—**Safety**
CEN PREN 352 (Part 1)-90. Hearing Protectors—Safety Requirements and Testing—Part 1: Ear Muffs. 33 pp.

—**Safety Helmets—Safety**
CEN PREN 352-3-92. Hearing Protectors—Safety Requirements and Testing—Part 3: Ear Muffs Attached to an Industrial Safety Helmet. 41 pp.

Ear Plugs (Hearing Protection)
Use For: Plugs (Hearing Protection)
See Also: Hearing Protectors
BSI BS 6344: Part 2-88. 1988 Industrial Hearing Protectors Part 2: Ear Plugs. 8 pp.

—**Safety**
CEN PREN 352 (Part 2)-90. Hearing Protectors—Safety Requirements and Testing—Part 2: Ear Plugs. 20 pp.

Ear Protectors
Use: Hearing Protectors

Early Mobile Data Services
Use For: EMDS *See Also:* Automatic Vehicle Location Systems; Mobile Satellite Communications; Radio Communications; Radiolocation Services; Satellite Communications

Early Mobile Data Services (Cont.)
—UHF
CCIR Report 1183-90. Technical and Operational Considerations for the Land Mobile-Satellite Service (LMSS)—Section 8F—Frequencies, Orbits and Systems. 32 pp.

Earphones
See Also: Artificial Ears; Audio Equipment; Electroacoustic Transducers; Headsets; Hearing Aids; Speakers

—Calibration—Artificial Ears
BSI BS 4669-71. 1971 An Artificial Ear of the Wide Band Type for the Calibration of Earphones Audiometry. 12 pp.
IEC 303-70. IEC Provisional Reference Coupler for the Calibration of Earphones Used in Audiometry First Edition. 12 pp.
IEC 318-70. An IEC Artificial Ear, of the Wide Band Type, for the Calibration of Earphones Used in Audiometry First Edition. 17 pp.
SAA AS 1591.5-87. Acoustics—Instrumentation for Audiometry—Part 5: Wide Band Artifical Ear. 7 pp.

—Calibration—Reference Couplers
BSI BS 4668-71. 1971 An Acoustic Coupler (IEC Reference Type) for the Calibration of Earphones Used in Audiometry. 11 pp.
SAA AS Z43.3-69. Instrumentation for Audiometry—See Also AS 1591—Part 3: Reference Coupler for the Calibration of Earphones Used in Audiometry. 10 pp.

—Headphones
MOD UK DSTAN 59-39-01. Standardization Status for Earphone Elements (Insets) for Headsets, Electrical and Headsets, Microphone Issue 2; Amendment 1. 9 pp.

—Hearing Aids
BSI BS 6083: Part 5-84. 1984 Amd 1 Hearing Aids Part 5: Specifications for Dimensions of the Nipple and Sealing Device for Insert Earphones (AMD 6182) October 31, 1990. 8 pp.
CENELEC HD 305-75. IEC Reference Coupler for the Measurement of Hearing Aids Using Earphones Coupled to the Ear by Means of Ear Inserts. 2 pp.
CENELEC HD 450.5-85. Hearing Aids Part 5: Nipples for Insert Earphones. 2 pp.
DIN ENGL 45602-57. Electric Hearing Aids; Earphone Pressure Knob and Earphone Pressure Ring; Mating Dimensions (July). 1 p.
IEC 118 Pt 5-83. Hearing Aids Part 5: Nipples for Insert Earphones First Edition. 11 pp.
IEC 126-73. IEC Reference Coupler for the Measurement of Hearing Aids Using Earphones Coupled to the Ear by Means of Ear Inserts Second Edition. 20 pp.
SAA AS 1088.5-87. Hearing Aids—Part 5: Nipples for Insert Earphones. 3 pp.
SAA AS 1089-71. Reference Coupler for the Measurement of the Electro-Acoustic Characteristics of Hearing Aid Earphones. 8 pp.

—Midget
CNS C7123-80. Midget Earphone (Aug)(6133).
JIS C 5508-78. Midget Earphones (R 1984). 11 pp.

—Occluded Ear Simulators
BSI BS 6310-82. 1982 Amd 1 Occluded-Ear Simulator for the Measurement of Earphones Coupled to the Ear by Ear Inserts (AMD 6233) May 31, 1990. 16 pp.
CENELEC HD 443 S1-88. Occluded-Ear Simulator for the Measurement of Earphones Coupled to the Ear by Ear Inserts. 3 pp.
IEC 711-81. Occluded-Ear Simulator for the Measurement of Earphones Coupled to the Ear by Ear Inserts First Edition. 28 pp.

Ears, Artificial
Use: Artificial Ears

Earth (Soil)
Use: Soils

Earth Augers
Use For: Earth Drills See Also: Augers; Boring Machines

—Specification Forms
CNS A4012-88. Standard Form of Specification of Earth Augers (Oct)(9554).
JIS A 8504-78. Standard Form of Specification of Earth Augers.
JIS A 8505-88. Standard Form of Specifications of Earth-Drill. 12 pp.

Earth Clamps
Use: Grounding Clamps

Earth Drills
Use: Earth Augers

Earth Electrodes
Use: Ground Electrodes

Earth Exploration Satellite Services
Use: Earth Exploration Satellites

Earth Exploration Satellites
Use For: Earth Exploration Satellite Services
See Also: Meteorological Satellites; Satellites
CCIR Report 535-4-90. Technical and Operational Considerations for the Earth Exploration-Satellite Service—Section 2F—Earth Exploration Satellites. 20 pp.
CCIR Report 693-3-90. Technical and Operational Considerations for the Earth Exploration-Satellite Service Preferred Frequency Bands for Active and Passive Microwave Sensors—Section 2F—Earth Exploration Satellites. 24 pp.
CCIR Report 1120-90. Method for Deriving Performance Criteria for the Earth Exploration-Satellite Service—Section 2F—Earth Exploration Satellites. 6 pp.
CCIR QUESTION 139/7-90. Radiocommunication Systems for Earth Exploration Satellites (Not Including Meteorological Satellites)—Questions Concerning Study Group 7—Science Services. 1 p.

—Active Detectors—Frequency Bands
CCIR RECMN 516-76. Frequency Bands for Active Sensors Used on Earth Exploration and Meteorological Satellites—Section 2F—Earth Exploration Satellites. 1 p.

—Bandwidths
CCIR RECMN 514-1-90. Telecommunication Links for Earth Exploration Satellites Frequencies, Bandwidths and Criteria for Protection from Interference—Section 2F—Earth Exploration Satellites. 2 pp.

—Coordination Thresholds
CCIR Report 1122-90. Methodology for Deriving Sharing Criteria and Coordination Thresholds for the Earth Exploration-Satellite Service—Section 2F—Earth Exploration Satellites. 24 pp.

—Data Collection
CCIR Report 538-4-90. Earth Exploration Satellites Satellites for Location of Platforms and for Data Collection—Section 2F—Earth Exploration Satellites. 11 pp.
CCIR QUESTION 142/7-90. Radiocommunications for Earth Exploration Satellites Data Collection and Position Location Systems—Questions Concerning Study Group 7—Science Services. 1 p.

—Electromagnetic Interference
CCIR RECMN 514-1-90. Telecommunication Links for Earth Exploration Satellites Frequencies, Bandwidths and Criteria for Protection from Interference—Section 2F—Earth Exploration Satellites. 2 pp.
CCIR Report 1122-90. Methodology for Deriving Sharing Criteria and Coordination Thresholds for the Earth Exploration-Satellite Service—Section 2F—Earth Exploration Satellites. 24 pp.
CCIR Report 1123-90. Methodology for Deriving Interference Criteria for the Earth Exploration-Satellite Service—Section 2F—Earth Exploration Satellites. 8 pp.

—Fixed Satellite Services—Frequency Band Sharing
CCIR QUESTION 67/4-90. Frequency Sharing Between the Fixed-Satellite Service and the Earth Exploration (Passive) and Space Research (Passive) Services—Questions of Study Group 4 Fixed-Satellite Service. 1 p.

—Fixed Satellites—Geostationary Satellites—Frequency Band Sharing
CCIR Report 540-1-82. Feasibility of Frequency Sharing Between an Earth Exploration-Satellite (EES) System and Fixed Satellite, Meteorological Satellite and Terrestrial Fixed and Mobile Services—Section 2F—Earth Exploration Satellites. 11 pp.

—Frequencies
CCIR RECMN 514-1-90. Telecommunication Links for Earth Exploration Satellites Frequencies, Bandwidths and Criteria for Protection from Interference—Section 2F—Earth Exploration Satellites. 2 pp.

—Frequency Band Sharing
CCIR Report 1122-90. Methodology for Deriving Sharing Criteria and Coordination Thresholds for the Earth Exploration-Satellite Service—Section 2F—Earth Exploration Satellites. 24 pp.

—Frequency Bands
CCIR Report 692-2-86. Preferred Frequency Bands and Power Flux-Density Considerations for Earth Exploration Satellites—Section 2F—Earth Exploration Satellites. 6 pp.

—Geostationary Satellites—Meteorological—Frequency Band Sharing
CCIR Report 540-1-82. Feasibility of Frequency Sharing Between an Earth Exploration-Satellite (EES) System and Fixed Satellite, Meteorological Satellite and Terrestrial Fixed and Mobile Services—Section 2F—Earth Exploration Satellites. 11 pp.

—Global Positioning Systems
CCIR QUESTION 142/7-90. Radiocommunications for Earth Exploration Satellites Data Collection and Position Location Systems—Questions Concerning Study Group 7—Science Services. 1 p.

—Hypothetical Reference Model
CCIR Report 1120-90. Method for Deriving Performance Criteria for the Earth Exploration-Satellite Service—Section 2F—Earth Exploration Satellites. 6 pp.

—Meteorological Services—Frequency Band Sharing
CCIR QUESTION 3/12-90. Sharing Between the Earth Exploration-Satellite Service or the Meteorological-Satellite Service on the One Hand and Other Space Services or the Metgl. Aids Service on the Other—Questions Concerning Study Group 12—Inter-Service Sharing and Compatibility. 1 p.

—Platforms
CCIR Report 538-4-90. Earth Exploration Satellites Satellites for Location of Platforms and for Data Collection—Section 2F—Earth Exploration Satellites. 11 pp.

—Power Flux Density
CCIR Report 692-2-86. Preferred Frequency Bands and Power Flux-Density Considerations for Earth Exploration Satellites—Section 2F—Earth Exploration Satellites. 6 pp.

—Radio Communications
CCIR QUESTION 138/7-90. Radiocommunication Systems for Earth Exploration Satellites, Including Meteorological Satellites—Questions Concerning Study Group 7—Science Services. 1 p.

—Radio Relay Systems—Frequency Band Sharing
CCIR Report 540-1-82. Feasibility of Frequency Sharing Between an Earth Exploration-Satellite (EES) System and Fixed Satellite, Meteorological Satellite and Terrestrial Fixed and Mobile Services—Section 2F—Earth Exploration Satellites. 11 pp.
CCIR QUESTION 113/9-90. Frequency Sharing Between Radio-Relay Systems and Systems of the Earth Exploration-Satellite Service and the Space Research Service—Questions of Study Group 9 Fixed Service. 1 p.

—Radio Relay Systems—Space Operation Services—2 GHz
CCIR Report 981-1-90. Sharing Considerations Near 2 GHz Between Systems in the Earth Exploration Satellite, Space Research, and Space Operation Services and Terrestrial Line-of-Sight Radio-Relay Systems—Section 2B—Topics of General Interest. 21 pp.

—Radiolocation Services—Radar—Frequency Band Sharing
CCIR Report 695-2-86. Feasibility of Frequency Sharing Between Spaceborne Radars and Terrestrial Radars in the Radiolocation Service—Section 2F—Earth Exploration Satellites. 13 pp.

—Satellite Communications—Frequency Band Sharing
CCIR QUESTION 3/12-90. Sharing Between the Earth Exploration-Satellite Service or the Meteorological-Satellite Service on the One Hand and Other Space Services or the Metgl. Aids Service on the Other—Questions Concerning Study Group 12—Inter-Service Sharing and Compatibility. 1 p.

—Sensors
CCIR QUESTION 140/7-90. Sensors Used by Earth Exploration Satellites Including Meteorological Satellites—Questions Concerning Study Group 7—Science Services. 1 p.

Earth Fault Protection
Use: Ground Fault Protection

Earth Fill
See Also: Foundations; Soils

INTERNATIONAL AND NON-U.S. NATIONAL STANDARDS
SUBJECT INDEX

Earth Fill (Cont.)
—Residential Buildings
SNZ NZS 4431-89. Code of Practice for Earth Fill for Residential Development Amend: 1, 1992. 20 pp.

Earth Handling Equipment
Use: Earthmoving Equipment

Earth Leakage Circuit Breakers
See Also: Circuit Breakers
MOD UK DSTAN 61-16-82. Differential Current Operated Earth-Leakage Circuit Breakers Issue 1. 23 pp.

—Voltage Operated
BSI BS 842-65. 1965 Amd 1 A.C. Voltage-Operated Earth-Leakage Circuit-Breakers. 29 pp.

Earth Moving Machinery
Use: Earthmoving Equipment

Earth Pressure
See Also: Loads (Forces); Pressure; Static Loads; Stresses

—Concrete Pipes
SAA AS 3725-89. Loads on Buried Concrete Pipes (This is a Joint Standard with SANZ NZS 3725). 16 pp.
SNZ NZS/AS 3725-89. Loads on Buried Concrete Pipes (This is a Joint Standard with SAA AS 3725). 16 pp.
SNZ NZS/AS 3725 Supplement 1-89. Loads on Buried Concrete Pipes. Commentary (This is a Joint Standard with SAA AS 3725). 25 pp.

Earth Resistance Meters
Use: Ground Resistance Meters

Earth Sciences
See Also: Geodesy; Geodynamics; Geography

—Classification
BSI BS 1000: (55)-82. 1982 Universal Decimal Classification (UDC). English Full Edition (55): Earth Sciences. Geology. Geophysics. 82 pp.
SNZ NZS/BS 1000 (55)-82. Universal Decimal Classification Earth Sciences. Geology. Geophysics. 84 pp.

Earth Stations
See Also: Satellite Communication Equipment; Satellite Communications; Terrestrial Stations
SAA AS 3516.3-91. Siting of Radiocommunications Facilities—Part 3: Fixed Location Satellite Earth Stations. 37 pp.

—Airborne—Coordination Contours
CCIR Report 773-78. Concept of Coordination and Protection Contours for Use in the Coordination of Mobile Earth Stations—Section 8H—Efficient Use of the Radio Spectrum Characteristics and Sharing of Frequency Resources. 6 pp.

—Airborne—Protection Contours
CCIR Report 773-78. Concept of Coordination and Protection Contours for Use in the Coordination of Mobile Earth Stations—Section 8H—Efficient Use of the Radio Spectrum Characteristics and Sharing of Frequency Resources. 6 pp.

—Antennas
CCIR Report 677-2-90. Radiation Patterns of Antennas for Space Research Earth Stations—Section 2A—Research in Space Technology. 5 pp.
DIN VDE 0855 Pt 12 (P)-88. Cabled Distribution Systems; Equipment for Receiving Aerial Systems with Satellite Receiving Equipment (Nov). 35 pp.

—Antennas—Coupling—Terrestrial Link Antennas
CCIR Report 709-1-82. Consideration of the Coupling Between an Earth-Station Antenna and a Terrestrial Link Antenna—Section 4/9B—Co-Ordination and Interference Calculations. 9 pp.

—Antennas—Cross Polarization
CCIR Report 1141-90. Polarization Discrimination in Interference Calculation—Section 4D1—Permissible Levels of Interference. 20 pp.

—Antennas—Fixed Satellite Services
CCIR Report 390-6-90. Earth-Station Antennas for the Fixed-Satellite Service—Section 4C—Earth Station and Baseband Characteristics —Earth Station Antennas —Maintenance of Earth Stations. 23 pp.
CCIR QUESTION 42/4-90. Characteristics of Antennas at Earth Stations in the Fixed-Satellite Service—Questions of Study Group 4 Fixed-Satellite Service. 1 p.

Earth Stations (Cont.)
—Antennas—Fixed Satellites—Radiation Diagrams—Design
CCIR Report 391-6-90. Radiation Diagrams of Antennas for Earth Stations in the Fixed-Satellite Service for Use in Interference Studies and for the Determination of a Design Objective—Section 4C—Earth Station and Baseband Characteristics—Earth Station Antennas-. 16 pp.
CCIR QUESTION 40/4-90. Reference Radiation Diagram of Antennas at Earth Stations in the Fixed-Satellite Service—Questions of Study Group 4 Fixed-Satellite Service. 1 p.

—Antennas—Gain to Temperature Ratio—Fixed Satellite Services
CCIR RECMN 733-92. Measurement of the G/T Ratio for Earth Stations Operating in the Fixed-Satellite Service—Section 4C—Earth Station and Baseband Characteristics—Earth Station Antennas—Maintenance of Earth Stations. 20 pp.

—Antennas—Geostationary Satellites—Design
CCIR RECMN 580-2-90. Radiation Diagrams for Use as Design Objectives for Antennas of Earth Stations Operating with Geostationary Satellites —Section 4C—Earth Station and Baseband Characteristics—Earth Station Antennas—Maintenance of Earth Stations. 3 pp.
CCIR RECMN 580-3-92. Radiation Diagrams for Use as Design Objectives for Antennas of Earth Stations Operating with Geostationary Satellites —Section 4C—Earth Station and Baseband Characteristics—Earth Station Antennas—Maintenance of Earth Stations. 3 pp.

—Antennas—Mobile Satellite Communications
CCIR Report 1047-1-90. Compact Antennas for Mobile Satellite Communication—Section 8I—Technical and Operating Characteristics of Mobile Satellite Services. 18 pp.
CCIR QUESTION 88-1/8-90. Propagation and Mobile Earth Station Antenna Characteristics for Mobile-Satellite Services—Questions Concerning Study Group 8 —Mobile, Radiodetermination, Amateur and Related Satellite Services. 1 p.

—Antennas—Noise Temperature—Fixed Satellite Services
CCIR Report 390-6-90. Earth-Station Antennas for the Fixed-Satellite Service—Section 4C—Earth Station and Baseband Characteristics —Earth Station Antennas —Maintenance of Earth Stations. 23 pp.

—Antennas—Performance Measurement
CCIR Report 998-1-90. Performance of Small Earth-Station Antennas for the Fixed-Satellite Service—Section 4C—Earth Station and Baseband Characteristics —Earth Station Antennas —Maintenance of Earth Stations. 20 pp.

—Antennas—Radiation
CCIR Report 390-6-90. Earth-Station Antennas for the Fixed-Satellite Service—Section 4C—Earth Station and Baseband Characteristics —Earth Station Antennas —Maintenance of Earth Stations. 23 pp.

—Antennas—Radiation Patterns—Fixed Satellite Services
CCIR RECMN 465-3-90. Reference Earth-Station Radiation Pattern for Use in Coordination and Interference Assessment in the Frequency Range from 2 to About 30 GHz—Section 4C—Earth Station and Baseband Characteristics—Earth Station Antennas—Maintenance of Earth Stations. 2 pp.
CCIR RECMN 465-4-92. Reference Earth-Station Radiation Pattern for Use in Coordination and Interference Assessment in the Frequency Range from 2 to About 30 GHz—Section 4C—Earth Station and Baseband Characteristics—Earth Station Antennas—Maintenance of Earth Stations. 2 pp.
CCIR RECMN 731-92. Reference Earth-Station Cross-Polarized Radiation Pattern for Use in Frequency Coordination and Interference Assessment in the Frequency Range from 2 to About 30 GHz—Section 4C—Earth Station and Baseband Characteristics—Earth Station Antennas-. 15 pp.

—Antennas—Radio Frequency Interference—Radiation Diagrams
CCIR Report 391-6-90. Radiation Diagrams of Antennas for Earth Stations in the Fixed-Satellite Service for Use in Interference Studies and for the Determination of a Design Objective—Section 4C—Earth Station and Baseband Characteristics—Earth Station Antennas-. 16 pp.
CCIR QUESTION 40/4-90. Reference Radiation Diagram of Antennas at Earth Stations in the Fixed-Satellite Service—Questions of Study Group 4 Fixed-Satellite Service. 1 p.

Earth Stations (Cont.)
—Antennas—Side Lobe Peak—Fixed Satellite Services
CCIR RECMN 732-92. Method for Statistical Processing of Earth-Station Antenna Side-Lobe Peaks—Section 4C—Earth Station and Baseband Characteristics—Earth Station Antennas—Maintenance of Earth Stations. 2 pp.

—Atmospheric Scattering—Radio Wave Propagation
CCIR Report 1010-86. Propagation Data for Bi-Directional Coordination of Earth Stations—Section 5G—Propagation Data Required for the Evaluation of Interference: Space and Terrestrial Systems. 5 pp.

—Bidirectional Coordination
CCIR Report 999-86. Determination of the Bidirectional Coordination Area—Section 4D2—Coordination Methods. 14 pp.

—Controllers
CEPT T/CCTS 2-85. Lignes Directrices Pour La Specification Des Equipements De Petites Stations Terriennes. 9 pp.

—Coordination Areas
CCIR RECMN 848-92. Determination of the Coordination Area of a Transmitting Earth Station Using the Same Frequency Band as Receiving Earth Stations in Bidirectionally Allocated Frequency Bands. 9 pp.
CCIR RECMN 850-92. Coordination Areas Using Predetermined Coordination Distances. 6 pp.

—Coordination Areas—Radio Wave Propagation
CCIR RECMN 620-86. Propagation Data Required for the Calculation of Coordination Distances—Section 5G—Propagation Data Required for the Evaluation of Interference: Space and Terrestrial Systems. 1 p.
CCIR Report 724-2-86. Propagation Data Required for the Evaluation of Coordination Distance in the Frequency Range 1-40 GHz—Section 5G—Propagation Data Required for the Evaluation of Interference: Space and Terrestrial Systems. 41 pp.
CCIR Report 1010-86. Propagation Data for Bi-Directional Coordination of Earth Stations—Section 5G—Propagation Data Required for the Evaluation of Interference: Space and Terrestrial Systems. 5 pp.
CCIR RECMN 620-1-92. Propagation Data Required for the Calculation of Coordination Distances in the Frequency Range 1-40 GHz. 9 pp.

—Coordination Distances
CCIR RECMN 850-92. Coordination Areas Using Predetermined Coordination Distances. 6 pp.

—Deep Space—Electromagnetic Interference
CCIR Report 844-1-90. Potential Interference Between Deep-Space Telecommunications and Some Other Services in Harmonically Related Bands—Section 2B—Topics of General Interest. 12 pp.

—Deep Space—Frequency Band Sharing
CCIR RECMN 578-82. Protection Criteria and Sharing Considerations Relating to Deep-Space Research—Section 2E—Space Research. 1 p.
CCIR QUESTION 114/7-90. Feasibility of Frequency Sharing Between Deep-Space Research Stations and Stations of Other Services—Questions Concerning Study Group 7 —Science Services. 1 p.

—Deep Space—Protection
CCIR RECMN 578-82. Protection Criteria and Sharing Considerations Relating to Deep-Space Research—Section 2E—Space Research. 1 p.

—Deep Space—Radiation—Aviation Personnel
CCIR Report 682-1-86. Probability of Hazards to Personnel Within Aircraft Due to Radiation from Deep-Space Earth Stations—Section 2A—Research in Space Technology. 2 pp.

—Deep Space—Space Stations—Frequency Bands
CCIR Report 849-2-90. Frequency Bands in the 40 to 120 GHz Range That Are Preferred for Deep-Space Research—Section 2E—Space Research. 9 pp.

—Digital Microwave Radios
BSI BS 7573: Sec 1.1-92. 1992 Methods of Measurement for Equipment Used in Digital Microwave Radio Transmission Systems Part 1: Measurements Common to Terrestrial Radio-Relay Systems and Satellite Earth Stations Section 1.1: General (IEC 835-1-1: 1990). 20 pp.

INDUSTRY STANDARDS

Earth Stations (Cont.)

—Digital Microwave Radios (Cont.)

BSI BS 7573: Sec 3.1-92. 1992 Methods of Measurement for Equipment Used in Digital Microwave Radio Transmission Systems Part 3: Measurements on Satellite Earth Stations Section 3.1: General (IEC 835-3-1: 1990). 6 pp.

BSI BS 7573: Sec 3.12-93. 1993 Methods of Measurement for Equipment Used in Digital Microwave Radio Transmission Systems Part 3: Measurements on Satellite Earth Stations Section 3.12: Overall System Performance (IEC 835-3-12: 1993) (S). 7 pp.

CENELEC EN 60835-3-1-92. Methods of Measurement for Equipment Used in Digital Microwave Radio Transmission Systems Part 3: Measurements on Satellite Earth Stations Section One: General; (IEC 835-3-1: 1990). 5 pp.

CENELEC EN 60835-3-1-92. Methods of Measurement for Equipment Used in Digital Microwave Radio Transmission Systems Part 3: Measurements on Satellite Earth Stations Section One: General (IEC 835-3-1:1990). 5 pp.

IEC 835 Pt 1-1-90. Methods of Measurement for Equipment Used in Digital Microwave Radio Transmission Systems Part 1: Measurements Common to Terrestrial Radio-Relay Systems and Satellite Earth Stations Section One—General First Edition; (CENELEC EN 60835-1-1: 1992). 37 pp.

IEC 835 Pt 3-1-90. Methods of Measurement for Equipment Used in Digital Microwave Radio Transmission Systems Part 3: Measurements on Satellite Earth Stations Section One—General First Edition; (CENELEC EN 60835-3-1: 1992). 9 pp.

IEC 835 Pt 3-12-93. Methods of Measurement for Equipment Used in Digital Microwave Radio Transmission Systems Part 3: Measurements on Satellite Earth Stations Section 12: Overall System Performance First Edition. 14 pp.

—Digital Microwave Radios—Adaptive Equalizers

IEC 835 Pt 2-8-93. Methods of Measurement for Equipment Used in Digital Microwave Radio Transmissions Systems Part 2: Measurements on Terrestrial Radio-Relay Systems Section 8: Adaptive Equalizer First Edition. 44 pp.

—Digital Microwave Radios—Data Terminal Equipment

IEC 835 Pt 3-9-93. Methods of Measurement for Equipment Used in Digital Microwave Radio Transmission Systems Part 3: Measurement on Satellite Earth Stations Section 9—Terminal Equipment SCP-PSK First Edition. 58 pp.

—Digital Microwave Radios—Digital Signal Processing

IEC 835 Pt 2-5-93. Methods of Measurement for Equipment Used in Digital Microwave Radio Transmission Systems Part 2: Measurements on Terrestrial Radio-Relay Systems Section 5: Digital Signal Processing Sub-System First Edition. 48 pp.

—Digital Microwave Radios—Low Noise Amplifiers

BSI BS 7573: Sec 3.4-93. 1993 Methods of Measurement for Equipment Used in Digital Microwave Radio Transmission Systems Part 3: Measurements on Satellite Earth Stations Section 3.4: Low Noise Amplifier (IEC 835-3-4: 1993) (S). 8 pp.

IEC 835 Pt 3-4-93. Methods of Measurement for Equipment Used in Digital Microwave Radio Transmission Systems Part 3: Measurements on Satellite Earth Stations Section 4: Low Noise Amplifier First Edition. 20 pp.

—Digital Microwave Radios—Relay Systems

BSI BS 7573: Sec 1.2-93. 1993 Methods of Measurement for Equipment Used in Digital Microwave Radio Transmission Systems Part 1: Measurements Common to Terrestrial Radio-Relay Systems and Satellite Earth Stations Section 1.2: Bacic Characteristics (IEC 835-1-2: 1992) (S). 21 pp.

BSI BS 7573: Sec 2.10-93. 1993 Methods of Measurement for Equipment Usedin Digital Microwave Radio Transmission Systems Part 2: Measurements on Terrestrial Radio-Relay Systems Section 2.10: Overall System Performance (IEC 835-2-10: 1992) (S). 15 pp.

BSI BS EN 60835-2-10-93. 1993 Methods of Measurement for Equipment Used in Digital Microwave Radio Transmission Systems Part 2: Measurements on Terrestrial Radio-Relay Systems Section 2.10: Overall System Performance (IEC 835-2-10: 1992) (S). 15 pp.

Earth Stations (Cont.)

—Digital Microwave Radios—Relay Systems (Cont.)

CENELEC EN 60835-1-1-92. Methods of Measurement for Equipment Used in Digital Microwave Radio Transmission Systems Part 1: Measurements Common to Terrestrial Radio-Relay Systems and Satellite Earth Stations Section One: General; (IEC 835-1-1: 1990). 5 pp.

CENELEC EN 60835-1-1-92. Methods of Measurement for Equipment Used in Digital Microwave Radio Transmission Systems Part 1: Measurements Common to Terrestrial Radio-Relay Systems and Satellite Earth Stations Section One: General (IEC 835-1-1:1990). 5 pp.

CENELEC EN 60835-2-1-92. Methods of Measurement for Equipment Used in Digital Microwave Radio Transmission Systems Part 2: Measurements on Terrestrial Radio-Relay Systems Section One: General (IEC 835-2-1:1990). 5 pp.

CENELEC EN 60835-2-10-93. Methods of Measurement for Equipment Used in Digital Microwave Radio Transmission Systems Part 2: Measurements on Terrestrial Radio-Relay Systems Section Ten—Overall System Performance (IEC 835-2-10:1992). 5 pp.

CENELEC EN 60835-2-10-93. Methods of Measurement for Equipment Used in Digital Microwave Radio Transmission Systems Part 2: Measurements on Terrestrial Radio-Relay Systems Section Ten—Overall System Performance (IEC 835-2-10: 1992). 9 pp.

IEC 835 Pt 1-2-92. Methods of Measurement for Equipment Used in Digital Microwave Radio Transmission Systems Part 1: Measurements Common to Terrestrial Radio-Relay Systems and Satellite Earth Stations Section 2: Basic Characteristics First Edition. 43 pp.

IEC 835 Pt 2-1-90. Methods of Measurement for Equipment Used in Digital Microwave Radio Transmission Systems Part 2: Measurements on Terrestrial Radio-Relay Systems Section One—General First Edition; (CENELEC EN 60835-2-1: 1992). 9 pp.

IEC 835 Pt 2-10-92. Methods of Measurement for Equipment Used in Digital Microwave Radio Transmission Systems Part 2: Measurements on Terrestrial Radio-Relay Systems Section Ten—Overall System Performance First Edition; (CENELEC EN 60835-2-10: 1993). 26 pp.

—Digital Microwave Radios—Relay Systems—Transmission

BSI BS 7573: Sec 1.3-93. 1993 Methods of Measurement for Equipment Used in Digital Microwave Radio Transmission Systems Part 1: Measurements Common to Terrestrial Radio-Relay Systems and Satellite Earth Stations Sect. 1.3: Transmission Characteristics (IEC 835-1-3: 1992) (S). 22 pp.

BSI BS 7573: Sec 1.4-93. 1993 Methods of Measurement for Equipment Used in Digital Microwave Radio Transmission Systems Part 1: Measurements Common to Terrestrial Radio-Relay Systems and Satellite Earth Stations Sec. 1.4: Transmission Performance (IEC 835-1-4: 1992) (S). 17 pp.

IEC 835 Pt 1-3-92. Methods of Measurement for Equipment Used in Digital Microwave Radio Transmission Systems Part 1: Measurements Common to Terrestrial Radio-Relay Systems and Satellite Earth Stations Section 3: Transmission Characteristics First Edition. 37 pp.

IEC 835 Pt 1-4-92. Methods of Measurement for Equipment Used in Digital Microwave Radio Transmission Systems Part 1: Measurements Common to Terrestrial Radio-Relay Systems and Satellite Earth Stations Section 4: Transmission Performance First Edition. 31 pp.

—Digital Microwave Radios—Transceivers

BSI BS 7573: Sec 2.4-93. 1993 Methods of Measurement for Equipment Used in Digital Microwave Radio Transmission Systems Part 2: Measurements on Terrestrial Radio-Relay Systems Section 2.4: Transmitter/Receiver Including Modulator/Demodulator (IEC 835-2-4: 1993) (S). 23 pp.

IEC 835 Pt 2-4-93. Methods of Measurement for Equipment Used in Digital Microwave Radio Transmission Systems Part 2: Measurements on Terrestrial Radio-Relay Systems Section 4: Transmitter/Receiver Including Modulator/Demodulator First Edition. 48 pp.

—Electrical Measurement

IEC 510 Pt 2-1-78. Methods of Measurement for Radio Equipment Used in Satellite Earth Stations Part 2: Measurements for Sub-Systems Section One—General Section Two—Antenna (Including Feed Network) First Edition. 61 pp.

Earth Stations (Cont.)

—Electromagnetic Interference—Terrestrial Stations

CCIR Report 448-5-90. Determination of the Interference Potential Between Earth Stations and Terrestrial Stations —Section 4/9B—Co-Ordination and Interference Calculations. 9 pp.

—Emergency Relief Operations

CCIR Report 554-4-90. Use of Small Earth Stations for Relief Operation in the Event of Natural Disasters and Similar Emergencies—Section 4C—Earth Station and Baseband Characteristics—Earth Station Antennas—Maintenance of Earth Stations. 9 pp.

—Emergency Relief Operations—Natural Disaster

CCIR Report 554-4-90. Use of Small Earth Stations for Relief Operation in the Event of Natural Disasters and Similar Emergencies—Section 4C—Earth Station and Baseband Characteristics—Earth Station Antennas—Maintenance of Earth Stations. 9 pp.

—Feeder Links—Fixed Satellite Services—Space Stations

CCIR QUESTION 30/4-86. Provision of Feeder Links Between Fixed-Earth Stations and Space Stations of Various Services in Frequency Bands Allocated to the Fixed-Satellite Service—Questions of Study Group 4 Fixed-Satellite Service. 2 pp.

—Fixed Satellite Services

CCIR Report 553-3-86. Operation and Maintenance of Earth Stations in the Fixed-Satellite Service—Section 4C—Earth Station and Baseband Characteristics—Earth Station Antennas—Maintenance of Earth Stations. 7 pp.

—Fixed Satellite Services—Coordinate Determination—Sharing

CCIR RECMN 359-5-82. Determination of the Co-Ordination Area of Earth Stations in the Fixed-Satellite Service Using the Same Frequency Bands as the Systems in the Fixed Terrestrial Service—Section 4/9B—Co-Ordination and Interference Calculations. 1 p.

—Fixed Satellite Services—Coordination Areas

CCIR Report 1163-90. Coordination Area of an Earth Station of the Fixed-Satellite Service Sharing the Same Frequency Band with the Radionavigation Service—Section 8D—Radiodetermination, Global Maritime Distress and Safety System and Related Subjects. 2 pp.

CCIR QUESTION 5/12-90. Coordination Area of an Earth Station of the Fixed-Satellite Service Sharing the Same Frequency Band with the Radionavigation Service—Questions Concerning Study Group 12—Inter-Service Sharing and Compatibility. 1 p.

—Fixed Satellite Services—Effective Isotropic Radiated Power

CCIR RECMN 524-3-90. Maximum Permissible Levels of off-Axis e.i.r.p. Density from Earth Stations in the Fixed-Satellite Service Transmitting in the 6 and 14 GHz Frequency Bands—Section 4D1—Permissible Levels of Interference. 2 pp.

CCIR RECMN 524-4-92. Maximum Permissible Levels of Off-Axis e.i.r.p. Density from Earth Stations in the Fixed-Satellite Service Transmitting in the 6 and 14 GHz Frequency Bands—Section 4D1—Permissible Levels of Interference. 10 pp.

—Fixed Satellite Services—Emergency Relief Operations

CCIR QUESTION 43/4-90. Use of Small Earth Stations in the Fixed-Satellite Service in the Event of Natural Disasters, Epidemics, Famines and Similar Emergencies for Warning and Relief Operations—Questions of Study Group 4 Fixed-Satellite Service. 1 p.

—Fixed Satellite Services—Feeder Links—Broadcasting Satellites

CCIR QUESTION 44/4-90. Use of Transportable Transmitting Earth Stations in the Fixed-Satellite Service Including Use for Feeder Links to Broadcasting Satellites—Questions of Study Group 4 Fixed-Satellite Service. 1 p.

—Fixed Satellite Services—Frequency Management

CCIR RECMN 731-92. Reference Earth-Station Cross-Polarized Radiation Pattern for Use in Frequency Coordination and Interference Assessment in the Frequency Range from 2 to About 30 GHz—Section 4C—Earth Station and Baseband Characteristics—Earth Station Antennas-. 15 pp.

INTERNATIONAL AND NON-U.S. NATIONAL STANDARDS
SUBJECT INDEX

Earth

Earth Stations *(Cont.)*

—Fixed Satellite Services—Geostationary Orbits—Interference

CCIR QUESTION 70/4-90. Protection of the Geostationary-Satellite Orbit Against Unacceptable Interference from Transmitting Earth Stations in the Fixed-Satellite Service at Frequencies Above 10 GHz—Questions of Study Group 4 Fixed-Satellite Service. 1 p.

—Fixed Satellite Services—Interference Cancellation Systems

CCIR RECMN 734-92. Application of Interference Cancellers in the Fixed-Satellite Service—Section 4C—Earth Station and Baseband Characteristics—Earth Station Antennas—Maintenance of Earth Stations. 10 pp.

CCIR QUESTION 58/4-90. Interference Reduction and Cancellation Techniques for the Earth Stations in the Fixed-Satellite Service—Questions of Study Group 4 Fixed-Satellite Service. 1 p.

—Fixed Satellite Services—Maintenance

CCIR Report 553-3-86. Operation and Maintenance of Earth Stations in the Fixed-Satellite Service—Section 4C—Earth Station and Baseband Characteristics—Earth Station Antennas—Maintenance of Earth Stations. 7 pp.

—Fixed Satellite Services—Radio Frequency Emissions

CCIR QUESTION 25-1/4-82. Unwanted Emissions Radiated from and Received by Earth Stations and Space Stations of the Fixed-Satellite Service—Questions of Study Group 4 Fixed-Satellite Service. 1 p.

—Fixed Satellite Services—Radio Frequency Interference

CCIR RECMN 731-92. Reference Earth-Station Cross-Polarized Radiation Pattern for Use in Frequency Coordination and Interference Assessment in the Frequency Range from 2 to About 30 GHz—Section 4C—Earth Station and Baseband Characteristics—Earth Station Antennas—. 15 pp.

CCIR QUESTION 58/4-90. Interference Reduction and Cancellation Techniques for the Earth Stations in the Fixed-Satellite Service—Questions of Study Group 4 Fixed-Satellite Service. 1 p.

—Fixed Satellite Services—Radio Navigation Services—Sharing

CCIR QUESTION 5/12-90. Coordination Area of an Earth Station of the Fixed-Satellite Service Sharing the Same Frequency Band with the Radionavigation Service—Questions Concerning Study Group 12—Inter-Service Sharing and Compatibility. 1 p.

—Fixed Satellite Services—Spurious Radiation

CCIR Report 713-1-82. Spurious Emissions from Earth Stations and Space Stations of the Fixed-Satellite Service—Section 4E—Frequency Sharing Between Networks of the Fixed-Satellite Service and Those of Other Space Radiocommunications Systems. 8 pp.

—Fixed Satellite Services—Terrestrial Stations—Frequency Sharing

CCIR QUESTION 57/4-90. Preferred Technical Characteristics and Selection of Sites for Earth Stations in the Fixed-Satellite Service to Facilitate Sharing with Terrestrial Services—Questions of Study Group 4 Fixed-Satellite Service. 1 p.

CCIR QUESTION 59/4-90. Preferred Technical Characteristics of Space Stations in the Fixed-Satellite Service to Facilitate Sharing with Terrestrial Services—Questions of Study Group 4 Fixed-Satellite Service. 1 p.

—Fixed Satellite Services—Terrestrial Stations—Interference

CCIR QUESTION 39/4-89. Technical Criteria to Be Used in the Board's Examinations of the Probability of Harmful Interference Required by Provisions Nos. 1354, 1506 and 1509 of the Radio Regulations—Questions of Study Group 4 Fixed-Satellite Service. 1 p.

—Fixed Satellite Services—Transportable—Transmitting

CCIR QUESTION 44/4-90. Use of Transportable Transmitting Earth Stations in the Fixed-Satellite Service Including Use for Feeder Links to Broadcasting Satellites—Questions of Study Group 4 Fixed-Satellite Service. 1 p.

Earth Stations *(Cont.)*

—Fixed Satellites—Broadcasting Satellite Services—Emissions

CCIR Report 712-1-82. Factors Concerning the Protection of Fixed-Satellite Earth Stations Operating in Adjacent Frequency Band Allocations Against Unwanted Emissions from Broadcasting Satellites Operating in Frequency Bands Around 12 GHz—Section 4E—Frequency Sharing Between Networks. 12 pp.

—Frequencies

CCIR RECMN 465-3-90. Reference Earth-Station Radiation Pattern for Use in Coordination and Interference Assessment in the Frequency Range from 2 to About 30 GHz—Section 4C—Earth Station and Baseband Characteristics—Earth Station Antennas—Maintenance of Earth Stations. 2 pp.

CCIR RECMN 465-4-92. Reference Earth-Station Radiation Pattern for Use in Coordination and Interference Assessment in the Frequency Range from 2 to About 30 GHz—Section 4C—Earth Station and Baseband Characteristics—Earth Station Antennas—Maintenance of Earth Stations. 2 pp.

—Frequency Band Sharing

CCIR RECMN 848-92. Determination of the Coordination Area of a Transmitting Earth Station Using the Same Frequency Band as Receiving Earth Stations in Bidirectionally Allocated Frequency Bands. 9 pp.

—Frequency Bands—Fixed Satellite Services

CCIR Report 386-3-82. Determination of the Power in Any 4 kHz Band Radiated Towards the Horizon by Earth Stations of the Fixed-Satellite Service Sharing Frequency Bands Below 15 GHz with the Terrestrial Services—Section 4/9A—Sharing Conditions. 4 pp.

—Frequency Demodulators—Measurement

IEC 510 Pt 2-6-92. Methods of Measurement for Radio Equipment Used in Satellite Earth Stations Part 2: Measurements for Sub-Systems Section Six: Frequency Demodulators First Edition. 49 pp.

—Frequency Division Multiplexers—Measurement

IEC 510 Pt 3-4-92. Methods of Measurement for Radio Equipment Used in Satellite Earth Stations Part 3: Methods of Measurement on Combinations of Sub-Systems Section Four: Measurements for Frequency Division Multiplex (f.d.m.) Transmission First Edition. 49 pp.

—Frequency Modulators—Measurement

IEC 510 Pt 2-5-92. Methods of Measurement for Radio Equipment Used in Satellite Earth Stations Part 2: Measurements for Sub-Systems Section Five: Frequency Modulators First Edition. 35 pp.

—Geostationary Space Stations—Coordination Areas

CCIR RECMN 847-92. Determination of the Coordination Area of an Earth Station Operating with a Geostationary Space Station and Using the Same Frequency Band as a System in a Terrestrial Service. 30 pp.

—INMARSAT

CCIR Report 918-1-90. Availability of Communications Circuits in the Maritime Mobile-Satellite Service—Section 8G—Availability, Performance Objectives and Interworking with Terrestrial Networks. 12 pp.

—Low Capacity—Fixed Satellite Services

CCIR Report 869-2-90. Low Capacity Earth Stations and Associated Satellite Systems in the Fixed-Satellite Service—Section 4C—Earth Station and Baseband Characteristics—Earth Station Antennas—Maintenance of Earth Stations. 15 pp.

CCIR QUESTION 23-1/4-86. Low Capacity Earth Stations and Associated Satellite Systems—Questions of Study Group 4 Fixed-Satellite Service. 1 p.

—Low Noise Amplifiers—Measurement

CENELEC HD 467.2.3 S1-90. Methods of Measurement for Radio Equipment Used in Satellite Earth Stations Part 2: Measurements for Sub-Systems Section Three—Low-Noise Amplifier. 3 pp.

IEC 510 Pt 2-3-89. Methods of Measurement for Radio Equipment Used in Satellite Earth Stations Part 2: Measurements for Sub-Systems Section Three—Low-Noise Amplifier First Edition. 15 pp.

—Meteorological Satellites

CCIR Report 1121-90. Performance, Interference and Sharing Criteria for Receiving Earth Stations in the Meteorological-Satellite Service Operating in the 1 670—1 710 MHz Band with Satellites in Low Earth Orbit—Section 2F—Earth Exploration Satellites. 12 pp.

Earth Stations *(Cont.)*

—Meteorological Satellites *(Cont.)*

CCIR Report 1124-90. Performance, Interference and Sharing Criteria for Receiving Earth Stations in the Meteorological-Satellite Service Operating in the 7 450—7 550 MHz Band with Satellites in Low Earth Orbit—Section 2F—Earth Exploration Satellites. 7 pp.

—Meteorological Satellites—Electromagnetic Interference

CCIR Report 1121-90. Performance, Interference and Sharing Criteria for Receiving Earth Stations in the Meteorological-Satellite Service Operating in the 1 670—1 710 MHz Band with Satellites in Low Earth Orbit—Section 2F—Earth Exploration Satellites. 12 pp.

CCIR Report 1124-90. Performance, Interference and Sharing Criteria for Receiving Earth Stations in the Meteorological-Satellite Service Operating in the 7 450—7 550 MHz Band with Satellites in Low Earth Orbit—Section 2F—Earth Exploration Satellites. 7 pp.

—Meteorological Satellites—Frequency Band Sharing

CCIR Report 1121-90. Performance, Interference and Sharing Criteria for Receiving Earth Stations in the Meteorological-Satellite Service Operating in the 1 670—1 710 MHz Band with Satellites in Low Earth Orbit—Section 2F—Earth Exploration Satellites. 12 pp.

CCIR Report 1124-90. Performance, Interference and Sharing Criteria for Receiving Earth Stations in the Meteorological-Satellite Service Operating in the 7 450—7 550 MHz Band with Satellites in Low Earth Orbit—Section 2F—Earth Exploration Satellites. 7 pp.

—Mobile Satellite Communications

CCIR Report 771-3-90. Considerations for the Mobile-Satellite Service—Section 8F—Frequencies, Orbits and Systems. 7 pp.

CENELEC PRETS 300 254-92. Satellite Earth Stations (SES); Land Mobile Earth Stations (LMESs) Operating in the 1,5/1,6/2,5 GHz Bands Providing Low Bit Rate Data Communications (LBRDCs). 24 pp.

CENELEC PRETS 300 255-92. Satellite Earth Stations (SES); Land Mobile Earth Stations (LMESs) Operating in the 11/12/14 GHz Bands Providing Low Bit Rate Data Communications (LBRDCs). 26 pp.

CENELEC PRETS 300 327-93. Satellite Earth Stations (SES); Satellite News Gathering (SNG) Transportable Earth Stations (TES) (13-14/11-12 GHz). 30 pp.

CEPT T/R 21-09 E-93. Euteltracs Service with Omnitracs Satellite Terminals in Europe. 1 p.

CEPT T/R 21-09 E-91. Euteltracs Service with OMNItracs Satellite Terminals in Europe. 2 pp.

CEPT T/R 31-02 E-93. Inmarsat-C Land Mobile-Satellite Service Terminals in Europe. 1 p.

CEPT T/R 31-02 E-91. Inmarsat-C Land Mobile-Satellite Service Terminals in Europe. 2 pp.

ETSI PRETS 300 254-92. Satellite Earth Stations (SES); Land Mobile Earth Stations (LMESs) Operating in the 1,5/1,6/2,5 GHz Bands Providing Low Bit Rate Data Communications (LBRDCs). 24 pp.

ETSI PRETS 300 255-92. Satellite Earth Stations (SES); Land Mobile Earth Stations (LMESs) Operating in the 11/12/14 GHz Bands Providing Low Bit Rate Data Communications (LBRDCs). 26 pp.

ETSI PRETS 300 327-93. Satellite Earth Stations (SES); Satellite News Gathering (SNG) Transportable Earth Stations (TES) (13-14/11-12 GHz). 30 pp.

—Mobile Satellite Communications—Control Facilities

CENELEC PRETS 300 282-92. Satellite Earth Stations (SES); Network Control Facilities (NCF) for Land Mobile Earth Stations (LMESs) Operating in the 1,5/1,6/2,5 GHz and 11/12/14 GHz Bands Providing Low Bit Rate Data Communications (LBRDCs). 7 pp.

ETSI PRETS 300 282-92. Satellite Earth Stations (SES); Network Control Facilities (NCF) for Land Mobile Earth Stations (LMESs) Operating in the 1, 5/1,6/2,5 GHz and 11/12/14 GHz Bands Providing Low Bit Rate Data Communications (LBRDCs). 7 pp.

—Mobile Satellite Communications—Coordination Contours

CCIR Report 773-78. Concept of Coordination and Protection Contours for Use in the Coordination of Mobile Earth Stations—Section 8H—Efficient Use of the Radio Spectrum Characteristics and Sharing of Frequency Resources. 6 pp.

INDUSTRY STANDARDS

Earth

Earth Stations (Cont.)

—Mobile Satellite Communications—PAD
CCITT RECMN X.351-89. Special Requirements to Be Met for Packet Assembly/Disassembly Facilities (PADs) Located at or in Association with Coast Earth Stations in the Public Mobile Satellite Service—Data Communication Networks—Interworking Between Networks, Mobile Data. 11 pp.

—Mobile Satellite Communications—Protection Contours
CCIR Report 773-78. Concept of Coordination and Protection Contours for Use in the Coordination of Mobile Earth Stations—Section 8H—Efficient Use of the Radio Spectrum Characteristics and Sharing of Frequency Resources. 6 pp.

—Mobile Satellite Communications—SHF
CCIR Report 1183-90. Technical and Operational Considerations for the Land Mobile-Satellite Service (LMSS)—Section 8F—Frequencies, Orbits and Systems. 32 pp.

—Mobile Satellite Communications—UHF
CCIR Report 1173-90. Technical and Operational Considerations for Aeronautical Mobile-Satellite Communications—Section 8F—Frequencies, Orbits and Systems. 40 pp.

—Mobile Stations (Communications)—Coordination Areas
CCIR QUESTION 4/12-90. Coordination Between an Earth Station and Mobile Stations in the Mobile Services—Questions Concerning Study Group 12—Inter-Service Sharing and Compatibility. 1 p.

—Nongeostationary Space Stations—Coordination Areas
CCIR RECMN 849-92. Determination of Coordination Area for Earth Stations Operating with Non-Geostationary Spacecraft in Bands Shared with Terrestrial Services. 10 pp.

—Power Amplifiers—Measurement
IEC 510 Pt 2-7-89. Methods of Measurement for Radio Equipment Used in Satellite Earth Stations Part 2: Measurements for Sub-Systems Section Seven—High Power Amplifier First Edition. 31 pp.

—Radio Communications—Terrestrial—Frequency Band Sharing
CCIR Report 386-3-82. Determination of the Power in Any 4 kHz Band Radiated Towards the Horizon by Earth Stations of the Fixed-Satellite Service Sharing Frequency Bands Below 15 GHz with the Terrestrial Services—Section 4/9A—Sharing Conditions. 4 pp.

—Radio Equipment—Measurement
CENELEC HD 467.1.2 S1-88. Methods of Measurement for Radio Equipment Used in Satellite Earth stations—Part 1: Measurements Common to Sub-Systems and Combinations of Sub-Systems. Section Two: Measurements in the r.f. Range. 3 pp.

IEC 510 Pt 1-75. Methods of Measurement for Radio Equipment Used in Satellite Earth Stations Part 1: General Sub-Clause 6.2: D.C. Source Conditions First Edition; (Supplement A-1980). 32 pp.

IEC 510 Pt 1-2-84. Methods of Measurement for Radio Equipment Used in Satellite Earth Stations Part 1: Measurements Common to Sub-Systems and Combinations of Sub-Systems Section Two: Measurements in the R.F. Range First Edition. 50 pp.

IEC 510 Pt 1-3-80. Methods of Measurement for Radio Equipment Used in Satellite Earth Stations Part 1: Measurements Common to Sub-Systems and Combinations of Sub-Systems Section Three—Measurements in the I.F. Range First Edition; (Amendment 1-1988). 48 pp.

IEC 510 Pt 1-4-86. Methods of Measurement for Radio Equipment Used in Satellite Earth Stations Part 1: Measurements Common to Sub-Systems and Combinations of Sub-Systems Section Four—Measurements in the Baseband First Edition. 31 pp.

IEC 510 Pt 1-5-88. Methods of Measurement for Radio Equipment Used in Satellite Earth Stations Part 1: Measurements Common to Sub-Systems and Combinations of Sub-Systems Section Five—Noise Temperature Measurements First Edition. 37 pp.

IEC 510 Pt 2-4-88. Methods of Measurement for Radio Equipment Used in Satellite Earth Stations Part 2: Measurements for Sub-Systems Section Four—up-and Down-Converters First Edition. 38 pp.

IEC 510 Pt 3-1-81. Methods of Measurement for Radio Equipment Used in Satellite Earth Stations Part 3: Methods of Measurement for Combinations of Sub-Systems Section One—General First Edition. 13 pp.

IEC 510 Pt 3-2-80. Methods of Measurement for Radio Equipment Used in Satellite Earth Stations Part 3: Methods of Measurement for Combinations of Sub-Systems Section Two: Measurement of the Figure of Merit (G/T) of the Receiving System in the 4 GHz to 6 GHz Range First Edition. 31 pp.

—Radio Frequency Interference
CCIR RECMN 465-3-90. Reference Earth-Station Radiation Pattern for Use in Coordination and Interference Assessment in the Frequency Range from 2 to About 30 GHz—Section 4C—Earth Station and Baseband Characteristics—Earth Station Antennas—Maintenance of Earth Stations. 2 pp.

CCIR RECMN 465-4-92. Reference Earth-Station Radiation Pattern for Use in Coordination and Interference Assessment in the Frequency Range from 2 to About 30 GHz—Section 4C—Earth Station and Baseband Characteristics—Earth Station Antennas—Maintenance of Earth Stations. 2 pp.

CEPT T/R 01-03-80. Controle Des Transmissions Par Satellite. 5 pp.

CEPT T/R 01-03 E-80. Monitoring of Satellite Transmissions. 6 pp.

—Radio Frequency Radiation—Safety
CCIR Report 543-1-86. Safety Aspects of Radio-Frequency Radiation from Space Research Earth Stations—Section 2A—Research in Space Technology. 4 pp.

CCIR QUESTION 52-2/1-86. Safety Aspects of Radio-Frequency Radiations from Earth Stations and Terrestrial Stations—Questions Concerning Study Group 1—Spectrum Management Techniques (Spectrum Engineering, Planning, Sharing, Monitoring and Utilization). 1 p.

—Radio Links—Data Relay Satellites—Spacecraft
CCIR QUESTION 117/7-90. Radio Links Between Earth Stations and Spacecraft by Means of Data Relay Satellites—Questions Concerning Study Group 7—Science Services. 1 p.

—Receivers—Space Communications—Coordinates—Frequency Band Sharing
CCIR Report 382-6-90. Determination of Co-Ordination Area—Section 4/9B—Co-Ordination and Interference Calculations. 63 pp.

—Receiving Antennas—Broadcasting Satellite Services
CCIR RECMN 652-86. 12 GHz Receiving Earth-Station Antenna and Satellite Transmitting Antenna Reference Patterns for the Broadcasting-Satellite Service—Section 10/11C—Technology. 6 pp.

CCIR RECMN 652-1-92. Reference Patterns for Earth-Station and Satellite Antennas for the Broadcasting-Satellite Service in the 12 GHz Band and for the Associated Feeder Links in the 14 GHz and 17 GHz Bands—Section 10/11C—Technology. 17 pp.

CCIR Report 810-3-90. Broadcasting-Satellite Service (Sound and Television) Reference Patterns and Technology for Transmitting and Receiving Antennas—Section 10/11C—Technology. 22 pp.

—Receiving Antennas—Noise Temperature
CCIR Report 868-1-86. Contributions to the Noise Temperature of an Earth-Station Receiving Antenna—Section 4C—Earth Station and Baseband Characteristics —Earth Station Antennas—Maintenance of Earth Stations. 4 pp.

—Receiving—Bidirectional Coordination
CCIR Report 999-86. Determination of the Bidirectional Coordination Area—Section 4D2—Coordination Methods. 14 pp.

CCIR RECMN 848-92. Determination of the Coordination Area of a Transmitting Earth Station Using the Same Frequency Band as Receiving Earth Stations in Bidirectionally Allocated Frequency Bands. 9 pp.

—Research—Interference
CCIR RECMN 509-1-90. Generalized Space Research Earth Station Antenna Radiation Pattern for Use in Interference Calculations, Including Coordination Procedures—Section 2A—Research in Space Technology. 1 p.

—Satellite News Gathering—Communication Circuits
CCIR RECMN 771-92. Auxiliary Coordination Satellite Circuits for SNG Terminals. 2 pp.

—Shipboard—Antennas
CCIR Report 922-1-86. Reference Radiation Pattern for Ship Earth-Station Antennas—Section 8I—Technical and Operating Characteristics of Mobile Satellite Services. 10 pp.

—Shipboard—Antennas—Multipath Fading
CCIR Report 1048-1-90. Fading Reduction Techniques Applicable to Ship Earth-Station Antennas—Section 8I—Technical and Operating Characteristics of Mobile Satellite Services. 13 pp.

—Shipboard—Antennas—Radiation
CCIR RECMN 694-90. Reference Radiation Pattern for Ship Earth Station Antennas—Section 8I—Technical and Operating Characteristics of Mobile Satellite Services. 2 pp.

CCIR RECMN 922-1-86. Reference Radiation Pattern for Ship Earth-Station Antennas—Section 8I—Technical and Operating Characteristics of Mobile Satellite Services. 10 pp.

—Shipboard—Digital
CCIR Report 921-2-90. System Aspects of Digital Ship Earth Stations—Section 8F—Frequencies, Orbits and Systems. 46 pp.

—Shipboard—INMARSAT
CCIR Report 921-2-90. System Aspects of Digital Ship Earth Stations—Section 8F—Frequencies, Orbits and Systems. 46 pp.

CCIR Report 918-1-90. Availability of Communications Circuits in the Maritime Mobile-Satellite Service—Section 8G—Availability, Performance Objectives and Interworking with Terrestrial Networks. 12 pp.

—Shipboard—Radio Frequency Interference
CCIR Report 764-2-86. Interference and Noise Problems for Maritime Mobile-Satellite Systems Using Frequencies in the Region of 1.5 and 1.6 GHz—Section 8G—Availability, Performance Objectives and Interworking with Terrestrial Networks. 7 pp.

—Space Research Services—Antennas—Radiation Patterns
CCIR QUESTION 127/7-90. Radiation Patterns and Side Lobe Characteristics of Large Antennas Used for Space Research Earth Stations and Radioastronomy—Questions Concerning Study Group 7—Science Services. 1 p.

—Space Research Services—Protection
CCIR RECMN 609-86. Protection Criteria for Telecommunication Links for Manned and Unmanned Near-Earth Research Satellites—Section 2E—Space Research. 1 p.

CCIR RECMN 609-1-92. Protection Criteria for Telecommunication Links for Manned and Unmanned Near-Earth Research Satellites—Section 2E—Space Research. 4 pp.

—Space Research Services—Radio Frequency Radiation—Equipment
CCIR QUESTION 125/7-90. Protection of Telecommunications Equipment from Radio-Frequency Radiation from Space Research Earth Stations—Questions Concerning Study Group 7—Science Services. 1 p.

—Space Research Services—Radio Frequency Radiation—Safety
CCIR QUESTION 124/7-90. Safety Aspects of Radio-Frequency Radiation from Space Research Stations—Questions Concerning Study Group 7—Science Services. 1 p.

—Space Services—Spurious Radiation
CCIR QUESTION 129/7-90. Spurious Emissions Radiated from and Received by Stations of Space Services—Questions Concerning Study Group 7—Science Services. 1 p.

—Spacecraft
CCIR Report 536-4-90. Telecommunication Requirements for Manned and Unmanned Deep-Space Research—Section 2E—Space Research. 20 pp.

—Telephony—Maintenance
CCITT RECMN M.1120-89. Functions, Maintenance Responsibilities and Maintenance Facilities of a Coast Earth Station for Telephony Services—Maintenance of International Telegraph, Phototelegraph and Leased Circuits—Maintenance of the International Public Telephone Network-. 3 pp.

—Television Receive Only
CENELEC PRETS 300 158-91. Satellite Earth Stations (SES); Television Receive-Only (TVRO) Satellite Earth Stations (DE/SES-4001). 56 pp.

CENELEC PRETS 300 158-92. Satellite Earth Stations (SES); Television Receive Only (TVRO) Satellite Earth Stations Operating in the 11/12 GHz FSS Bands. 56 pp.

INTERNATIONAL AND NON-U.S. NATIONAL STANDARDS
SUBJECT INDEX

Earthing

Earth Stations (Cont.)
—Television Receive Only (Cont.)
CENELEC ETS 300 158-92. Satellite Earth Stations (SES); Television Receive Only (TVRO-FSS) Satellite Earth Stations Operating in the 11/12 GHz FSS Bands. 57 pp.
CENELEC PRETS 300 249-92. Satellite Earth Stations (SES); Television Receive-Only (TVRO-BSS). 59 pp.
CENELEC PRETS 300 249-93. Satellite Earth Stations (SES); Television Receive-Only (TVRO-BSS). 57 pp.
ETSI ETS 300 158-92. Satellite Earth Stations (SES); Television Receive Only (TVRO-FSS) Satellite Earth Stations Operating in the 11/12 GHz FSS Bands. 57 pp.
ETSI PRETS 300 158-92. Satellite Earth Stations (SES); Television Receive Only (TVRO) Satellite Earth Stations Operating in the 11/12 GHz FSS Bands. 56 pp.
ETSI PRETS 300 158-91. Satellite Earth Stations (SES); Television Receive-Only (TVRO) Satellite Earth Stations (DE/SES-4001). 56 pp.
ETSI PRETS 300 249-93. Satellite Earth Stations (SES); Television Receive-Only (TVRO-BSS). 57 pp.
ETSI PRETS 300 249-92. Satellite Earth Stations (SES); Television Receive-Only (TVRO-BSS). 59 pp.

—Television Transmission—Measurement
IEC 510 Pt 3-3-88. Methods of Measurement for Radio Equipment Used in Satellite Earth Stations Part 3: Methods of Measurement for Combination of Sub-Systems Section Three—Measurements for Monochrome and Colour Television Transmission First Edition. 68 pp.

—Terrestrial Stations—Frequency Band Sharing
CCIR RECMN 847-92. Determination of the Coordination Area of an Earth Station Operating with a Geostationary Space Station and Using the Same Frequency Band as a System in a Terrestrial Service. 30 pp.
CCIR RECMN 849-92. Determination of Coordination Area for Earth Stations Operating with Non-Geostationary Spacecraft in Bands Shared with Terrestrial Services. 10 pp.

—Transmitters—Coordinate Determination
CCIR Report 382-6-90. Determination of Co-Ordination Area—Section 4/9B—Co-Ordination and Interference Calculations. 63 pp.

—Transmitting—Bidirectional Coordination
CCIR Report 999-86. Determination of the Bidirectional Coordination Area—Section 4D2—Coordination Methods. 14 pp.
CCIR RECMN 848-92. Determination of the Coordination Area of a Transmitting Earth Station Using the Same Frequency Band as Receiving Earth Stations in Bidirectionally Allocated Frequency Bands. 9 pp.

—Transportable—Transmitting—Outside Broadcasts
CCIR Report 1090-1-90. Use of Transportable Transmitting Earth Stations for the Transmission of Television Outside Broadcasts over Satellites—Section CMTT A—Television Transmission Standards and Performance Objectives. 6 pp.
CCIR QUESTION 32/CMTT-90. Use of Portable and Transportable Transmitting Earth Stations for Outside Television Broadcasts (Analogue) over Satellites—Questions Concerning the CMTT CCIR/CCITT Joint Study Group for Television and Sound Transmission. 1 p.

—Transportable—Transmitting—Outside Broadcasts—HDTV
CCIR QUESTION 74/CMTT-90. Use of Portable and Transportable Satellite Earth Stations for the Transmission of High Definition Television Outside Broadcasts over Satellites—Questions Concerning the CMTT CCIR/CCITT Joint Study Group for Television and Sound Transmission. 1 p.

—Transportable—Transmitting—Satellite News Gathering
CCIR QUESTION 33/CMTT-90. Use of Portable and Transportable Transmitting Earth Stations for News Gathering Via Satellites—Questions Concerning the CMTT CCIR/CCITT Joint Study Group for Television and Sound Transmission. 1 p.
CCIR QUESTION 75/CMTT-90. Use of Portable and Transportable Satellite Earth Stations for the Transmission of High Definition Television for News Gathering Via Satellite—Questions Concerning the CMTT CCIR/CCITT Joint Study Group for Television and Sound Transmission. 1 p.

Earth Stations (Cont.)
—Transportable—Transmitting—Satellite News Gathering—HDTV
CCIR QUESTION 75/CMTT-90. Use of Portable and Transportable Satellite Earth Stations for the Transmission of High Definition Television for News Gathering Via Satellite—Questions Concerning the CMTT CCIR/CCITT Joint Study Group for Television and Sound Transmission. 1 p.

—Unspecified Location—Geostationary Satellites—Sharing
CCIR Report 454-5-90. Method of Calculation for Determining if Coordination is Required Between Geostationary-Satellite Networks Sharing the Same Frequency Bands—Section 4D2—Coordination Methods. 39 pp.

—Very Small Aperture Terminals
CCIR RECMN 725-92. Technical Characteristics for Very Small Aperture Terminals (VSATs)—Section 4B1—Systems Aspects. 2 pp.
CENELEC PRETS 300 157-91. Satellite Earth Stations (SES); Receive-Only VSATs Used for Data Distribution (DE/SES-2001). 19 pp.
CENELEC PRETS 300 157-92. Satellite Earth Stations (SES); Receive-Only Very Small Aperture Terminals (VSATs) Used for Data Distribution Operating in the 11/12 GHz Frequency Bands. 18 pp.
CENELEC ETS 300 157-92. Satellite Earth Stations (SES); Receive-Only Very Small Aperture Terminals (VSATs) Used for Data Distribution Operating in the 11/12 GHz Frequency Bands. 18 pp.
CENELEC PRETS 300 159-91. Satellite Earth Stations (SES); Transmit/Receive VSATs Used for Data Communications (DE/SES-2002). 25 pp.
CENELEC PRETS 300 159-92. Satellite Earth Stations (SES); Transmit/Receive Very Small Aperture Terminals (VSATs) Used for Data Communications Operating in the Fixed Satellite Service (FSS) 11/12/14 GHz Frequency Bands. 24 pp.
CENELEC ETS 300 159-92. Satellite Earth Stations (SES); Transmit/Receive Very Small Aperture Terminals (VSATs) Used for Data Communications Operating in the Fixed Satellite Service (FSS) 11/12/14 GHz Frequency Bands. 23 pp.
CENELEC PRETS 300 160-91. Satellite Earth Stations (SES); Control and Monitoring Functions at a VSAT (DE/SES-3005). 13 pp.
CENELEC PRETS 300 160-92. Satellite Earth Stations (SES); Control and Monitoring Functions at a VSAT. 13 pp.
CENELEC ETS 300 160-92. Satellite Earth Stations (SES); Control and Monitoring Functions at a Very Small Aperture Terminal (VSAT). 13 pp.
CENELEC PRETS 300 161-91. Satellite Earth Stations (SES); Centralised Control and Monitoring Functions for VSAT Networks (DE/SES-3004). 9 pp.
CENELEC PRETS 300 161-92. Satellite Earth Stations (SES); Centralised Control and Monitoring Functions for VSAT Networks. 8 pp.
CENELEC ETS 300 161-92. Satellite Earth Stations (SES); Centralised Control and Monitoring Functions for VSAT Networks. 8 pp.
ETSI PRETS 300 157-91. Satellite Earth Stations (SES); Receive—Only VSATs Used for Data Distribution (DE/SES-2001).
ETSI PRETS 300 159-91. Satellite Earth Stations (SES); Transmit/Receive VSATs Used for Data Communications (DE/SES-2002).
ETSI PRETS 300 160-91. Satellite Earth Stations (SES); Control and Monitoring Functions at a VSAT (DE/SES-3005).
ETSI PRETS 300 161-91. Satellite Earth Stations (SES); Centralised Control and Monitoring Functions for VSAT Networks (DE/SES-3004).
ETSI ETS 300 157-92. Satellite Earth Stations (SES); Receive-Only Very Small Aperture Terminals (VSATs) Used for Data Distribution Operating in the 11/12 GHz Frequency Bands. 18 pp.
ETSI PRETS 300 157-92. Satellite Earth Stations (SES); Receive-Only Very Small Aperture Terminals (VSATs) Used for Data Distribution Operating in the 11/12 GHz Frequency Bands. 18 pp.
ETSI PRETS 300 157-91. Satellite Earth Stations (SES); Receive-Only VSATs Used for Data Distribution (DE/SES-2001). 19 pp.
ETSI ETS 300 159-92. Satellite Earth Stations (SES); Transmit/Receive Very Small Aperture Terminals (VSATs) Used for Data Communications Operating in the Fixed Satellite Service (FSS) 11/12/14 GHz Frequency Bands. 23 pp.
ETSI PRETS 300 159-92. Satellite Earth Stations (SES); Transmit/Receive Very Small Aperture Terminals (VSATs) Used for Data Communications Operating in the Fixed Satellite Service (FSS) 11/12/14 GHz Frequency Bands. 24 pp.
ETSI PRETS 300 159-91. Satellite Earth Stations (SES); Transmit/Receive VSATs Used for Data Communications (DE/SES-2002). 25 pp.

Earth Stations (Cont.)
—Very Small Aperture Terminals (Cont.)
ETSI ETS 300 160-92. Satellite Earth Stations (SES); Control and Monitoring Functions at a Very Small Aperture Terminal (VSAT). 13 pp.
ETSI PRETS 300 160-92. Satellite Earth Stations (SES); Control and Monitoring Functions at a VSAT. 13 pp.
ETSI PRETS 300 160-91. Satellite Earth Stations (SES); Control and Monitoring Functions at a VSAT (DE/SES-3005). 13 pp.
ETSI ETS 300 161-92. Satellite Earth Stations (SES); Centralised Control and Monitoring Functions for VSAT Networks. 8 pp.
ETSI PRETS 300 161-92. Satellite Earth Stations (SES); Centralised Control and Monitoring Functions for VSAT Networks. 8 pp.
ETSI PRETS 300 161-91. Satellite Earth Stations (SES); Centralised Control and Monitoring Functions for VSAT Networks (DE/SES-3004). 9 pp.

—Very Small Aperture Terminals—Cross Polarization
CCIR RECMN 727-92. Cross-Polarization Isolation from Very Small ApertureTerminals (VSATs)—Section 4B1—Systems Aspects. 1 p.

—Very Small Aperture Terminals—Effective Isotropic Radiated Power
CCIR RECMN 728-92. Maximum Permissible Level of Off-Axis e.i.r.p. Density from Very Small Aperture Terminals (VSATs)—Section 4B1—Systems Aspects. 6 pp.

—Very Small Aperture Terminals—Monitoring
CCIR RECMN 729-92. Control and Monitoring Function of Very Small Aperture Terminals (VSATs)—Section 4B1—Systems Aspects. 2 pp.

—Very Small Aperture Terminals—Spurious Radiation
CCIR RECMN 726-92. Maximum Permissible Level of Spurious Emissions from Very Small Aperture Terminals (VSATs)—Section 4B1—Systems Aspects. 2 pp.

Earth Surface
—Electrical Properties
CCIR Report 229-6-90. Electrical Characteristics of the Surface of the Earth—Section 5B—Effects of the Ground (Including Ground-Wave Propagation). 7 pp.
CCIR Report 879-1-86. Methods for Estimating Effective Electrical Characteristics of the Surface of the Earth—Section 5B—Effects of the Ground (Including Ground-Wave Propagation). 9 pp.

—Electrical Properties—Ground Wave Propagation
CCIR RECMN 527-2-90. Electrical Characteristics of the Surface of the Earth—Section 5B—Effects of the Ground (Including Ground-Wave Propagation). 3 pp.
CCIR RECMN 527-3-92. Electrical Characteristics of the Surface of the Earth. 5 pp.

—Radio Signals—Reflection
CCIR Report 1008-1-90. Reflection from the Surface of the Earth—Section 5B—Effects of the Ground (Including Ground-Wave Propagation). 8 pp.

—Radio Wave Propagation—Forecasting
CCIR QUESTION 9-1/5-90. Methods for Predicting Propagation over the Surface of the Earth—Questions Concerning Study Group 5—Radio Wave Propagation in Non-Ionized Media. 1 p.

Earth Terminals
Use: Ground Terminals

Earth Terminals, Satellite
Use: Earth Stations

Earth Testers
See Also: Testing Equipment
CNS C4170-80. Earth Testers (Feb)(5203).
JIS C 1304-76. Earth Testers. 11 pp.

Earthing (Electrical)
Use: Electrical Grounding

Earthing Conductors
Use: Grounding Conductors

Earthing Connectors
Use: Grounding Connectors

Earthing Devices
Use: Grounding Devices

INDUSTRY STANDARDS

Earthing

Earthing Reactors
Use: Grounding Reactors

Earthing/Short Circuiting Devices
Use: Short Circuiting/Grounding Devices

Earthing Switches
Use: Grounding Switches

Earthing Transformers
Use: Grounding Transformers

Earthmoving Equipment
See Also: Backhoes; Caterpillars (Vehicles); Construction Equipment; Dump Trucks; Excavating Equipment; Forestry Equipment; Front End Loaders; Graders; Hydraulic Excavating Equipment; Industrial Equipment; Loaders; Pipe Layers; Plate Compactors; Rollers (Compactors); Scrapers (Earthmoving Equipment); Tractors; Trench Excavators

BSI BS 6420-83. (WITHDRAWN) 1983 Measuring the Dimensions of Whole Machines with Their Equipment Used for Earth-Moving. 8 pp.
BSI BS 6911: Part 7-92. 1992 Testing Earth-Moving Machinery Part 7: Specification for Units of Measurement and Tolerances (ISO 9248: 1992). 7 pp.
ISO 6746 Pt 1-87. Earth-Moving Machinery—Definitions of Dimensions and Symbols—Part 1: Base Machine Second Edition. 12 pp.
ISO 6746 Pt 2-87. Earth-Moving Machinery—Definitions of Dimensions and Symbols—Part 2: Equipment Second Edition. 13 pp.
ISO 9248-92. Earth-Moving Machinery—Units for Dimensions, Performance and Capacities, and Their Measurement Accuracies First Edition. 5 pp.

—**Access Doors**
BSI BS 6112-83. (WITHDRAWN) 1983 Minimum Access Dimensions for Earth-Moving Machinery (Superseded by BS 6912: part 5: 1992). 6 pp.
BSI BS 6112-01. (WITHDRAWN) 1983 Amd 1 Minimum Access Dimensions for Earth-Moving Machinery (AMD 6877) May 1, 1992 (Superseded by BS 6912: Part 5: 1992). 11 pp.
BSI BS 6912: Part 5-92. 1992 Safety of Earth-Moving Machinery Part 5: Recommendations for Minimum Access Dimensions (ISO 2860: 1992). 9 pp.
CEN EN 22860-85. Earth-Moving Machinery Minimum Access Dimensions. 6 pp.
CNS A4014-88. Earth-Moving Machinery-Minimum Access Dimensions (Dec)(9945).
CNS A4015-88. Earth-Moving Machinery Access Systems (Dec)(9946).
CSA CAN/CSA-M2860-M91. Minimum Access Dimensions for Servicing Machines (EMM,FM) (ISO 2860-1983). 15 pp.
ISO 2860-92. Earth-Moving Machinery—Minimum Access Dimensions Fourth Edition. 7 pp.
ISO 2867-89. Earth-Moving Machinery—Access Systems Fourth Edition. 10 pp.
SAA AS 3868-91. Earth-Moving Machinery—Design Guide for Access Systems. 10 pp.

—**Access Fittings—Safety**
CNS A4015-88. Earth-Moving Machinery Access Systems (Dec)(9946).
ISO 2867-89. Earth-Moving Machinery—Access Systems Fourth Edition. 10 pp.

—**Backup Alarms—Acoustic Testing**
BSI BS 6912: Part 3-90. 1990 Safety of Earth Moving Machinery Part 3: Sound Test Method for Machine-Mounted Forward and Reverse Warning Alarm. 12 pp.
ISO 9533-89. Earth-Moving Machinery—Machine-Mounted Forward and Reverse Audible Warning Alarm—Sound Test Method First Edition. 8 pp.

—**Brakes**
BSI BS 6824-87. 1987 Performance and Test of Braking Systems for Wheeled Earth-Moving Machinery. 10 pp.
CNS A4017-88. Off-Highway Earth-Moving Machinery Performance Criteria for Brake Systems (Dec)(9948).
CSA CAN/CSA-M3450-M92. Braking Systems—Performance Requirements and Test Procedures—Wheeled Machines; (Gen Instr 1). 21 pp.
ISO 3450-85. Earth-Moving Machinery—Wheeled Machines—Performance Requirements and Test Procedures for Braking Systems Second Edition. 8 pp.

—**Center of Gravity**
BSI BS 3318-78. 1978 Earth-Moving Machinery. Method for Locating the Centre of Gravity. 9 pp.
CNS A4022-88. Earth-Moving Machinery Method for Locating the Center of Gravity (Dec)(9953).
ISO 5005-77. Earth-Moving Machinery—Method for Locating the Centre of Gravity First Edition. 8 pp.
JIS A 8915-82. Earth-Moving Machinery-Method for Locating the Centre of Gravity. 12 pp.

Earthmoving Equipment (Cont.)

—**Circuits—Identification Systems**
BSI BS 6913: Part 6-91. 1991 Operation and Maintenance of Earth-Moving Machinery Part 6: Specification for Identification and Marking of Electrical Wires and Cables (ISO 9247: 1990). 7 pp.
ISO 9247-90. Earth-Moving Machinery—Electrical Wires and Cables—Principles of Identification and Marking First Edition. 6 pp.

—**Diagnostic Testing**
BSI BS 5635-82. (WITHDRAWN) 1982 Service Instrumentation of Earth-Moving Machinery (Superseded by BS 6913: Part 4: 1990). 11 pp.
BSI BS 6913: Part 4-90. 1990 Operation and Maintenance of Earth-Moving Machinery Part 4: Recommendations for Service Instrumentation. 12 pp.
ISO 6012-89. Earth-Moving Machinery—Service Instrumentation Third Edition. 10 pp.

—**Drawbars**
BSI BS 6419-83. 1983 Measurement of Drawbar Pull of Earth-Moving Machinery. 11 pp.
ISO 7464-83. Earth-Moving Machinery—Method of Test for the Measurement of Drawbar Pull First Edition. 10 pp.

—**Ergonomics**
BSI BS 5538-83. 1983 Amd 2 Minimum Operator Space Envelope for Earth-Moving Machinery (AMD 6657) January 31, 1991. 15 pp.
BSI BS 6112-83. (WITHDRAWN) 1983 Minimum Access Dimensions for Earth-Moving Machinery (Superseded by BS 6912: part 5: 1992). 6 pp.
BSI BS 6112-01. (WITHDRAWN) 1983 Amd 1 Minimum Access Dimensions for Earth-Moving Machinery (AMD 6877) May 1, 1992 (Superseded by BS 6912: Part 5: 1992). 11 pp.
BSI BS 6124-87. 1987 Amd 1 Zones of Comfort and Reach for Controls in Earth-Moving Machinery (AMD 6420) December 21, 1990. 15 pp.
BSI BS 6912: Part 5-92. 1992 Safety of Earth-Moving Machinery Part 5: Recommendations for Minimum Access Dimensions (ISO 2860: 1992). 9 pp.
CEN EN 22860-85. Earth-Moving Machinery Minimum Access Dimensions. 6 pp.
CNS A4014-88. Earth-Moving Machinery-Minimum Access Dimensions (Dec)(9945).
CNS A4015-88. Earth-Moving Machinery Access Systems (Dec)(9946).
CNS A4016-88. Earth-Moving Machinery Human Physical Dimensions of Operators and Minimum Operator Space Envelope (Dec)(9947).
CSA CAN/CSA-M2860-M91. Minimum Access Dimensions for Servicing Machines (EMM,FM) (ISO 2860-1983). 15 pp.
CSA CAN/CSA-M3411-M91. Human Physical Dimensions of Operators—Minimum Operator Space Envelope—Machinery (EMM, AM, FM) (ISO 3411-1982). 19 pp.
CSA CAN/CSA-M6682-M91. Zones of Comfort and Reach for Controls—Machinery (EMM,FM) (Amd 1 Included); (ISO 6682-1986). 23 pp.
ISO 2860-92. Earth-Moving Machinery—Minimum Access Dimensions Fourth Edition. 7 pp.
ISO 2867-89. Earth-Moving Machinery—Access Systems Fourth Edition. 10 pp.
ISO 3411-82. Earth-Moving Machinery—Human Physical Dimensions of Operators and Minimum Operator Space Envelope Second Edition; (Amendment 1-1989). 13 pp.
ISO 6682-86. Earth-Moving Machinery—Zones of Comfort and Reach for Controls Second Edition; (Amendment 1-1989). 16 pp.

—**Falling Object Protective Structures**
BSI BS 5495-80. 1980 Amd 1 Laboratory Evaluations of Roll-Over and Falling-Object Protective Structures: The Deflection-Limiting Volume for Earth-Moving Machinery (ISO 3164: 1979) (AMD 3711) October 30, 1981. 9 pp.
BSI BS 5495-02. (WITHDRAWN) 1980 Amd 2 Laboratory Evaluations of Roll-Over and Falling-Object Protective Structures: The Deflection-Limiting Volume for Earth-Moving Machinery (ISO 3164: 1979) (AMD 6878) February 28, 1992 (Superseded by BS 6912: Part 7: 1992). 14 pp.
BSI BS 5526-85. (WITHDRAWN) 1985 Amd 1 Falling-Object Protective Structures on Earth Moving Machinery: Laboratory Test and Performance Requirements (AMD 6114) October 31, 1990 (Superseded by BS 6912: Part 6: 1992). 17 pp.
BSI BS 6912: Part 6-92. 1992 Safety of Earth-Moving Machinery Part 6: Specification for Falling-Object Protective Structures on Earth-Moving Machinery: Laboratory Tests and Performance Requirements (ISO 3449: 1992). 15 pp.
BSI BS 6912: Part 7-92. 1992 Safety of Earth-Moving Machinery Part 7: Specification for Laboratory Evaluations of Roll-Over and Falling-Object Protective Structures: the Deflection-Limiting Volume for Earth-Moving Machinery (ISO 3164: 1992). 7 pp.

Earthmoving Equipment (Cont.)

—**Falling Object Protective Structures (Cont.)**
CNS A4006-81. Deflection Limiting Volume of Earth-Moving Machinery-Roll-Over Protective Structure for Driver (Sep)(7850).
CNS A4024-88. Earth-Moving Machinery Laboratory Evaluations of Roll-Over and Falling-Object Protective Structures Specifications for the Deflection-Limiting Volume (Dec)(9955).
CNS A4025-88. Earth-Moving Machinery Falling-Object Protective Structures Laboratory Tests for Performance Requirements (Dec)(9956).
ISO 3164-92. Earth-Moving Machinery—Laboratory Evaluations of Roll-over and Falling-Object Protective Structures—Specifications for Deflection-Limiting Volume Fourth Edition; (Incorporating Amendment 1). 6 pp.
ISO 3449-92. Earth-Moving Machinery—Falling-Object Protective Structures—Laboratory Tests and Performance Requirements Fourth Edition. 12 pp.
JIS A 8920-90. Earth-Moving Machinery—Falling-Object Protective Structures—Laboratory Tests and Performance Requirements (ISO 3449-1984).

—**Fluid Power Equipment—Ports**
ISO 8925-89. Earth-Moving Machinery—Diagnostic Ports First Edition. 4 pp.

—**Fluid Power Equipment—Slope Limits—Static Testing**
BSI BS 6911: Part 8-93. 1993 Testing Earth-Moving Machinery Part 8: Determination of Slope Limits for Machine Fluid Systems Operation (Static Test Method) (ISO 10266: 1992). 8 pp.
ISO 10266-92. Earth-Moving Machinery—Determination of Slope Limits for Machine Fluid Systems Operation—Static Test Method First Edition. 6 pp.

—**Fuel Dispensing Equipment—Openings**
JIS A 8912-88. Earth-Moving Machinery—Dimensions of Fuel Filler Opening (ISO 3541-1985).

—**Fuel Tanks**
BSI BS 5279-76. 1976 Dimensions of Fuel Filler Openings of Earth-Moving Machinery. 4 pp.
CNS A4019-88. Earth-Moving Machinery Dimensions of Fuel Filler Opening (Dec)(9950).
ISO 3541-85. Earth-Moving Machinery—Dimensions of Fuel Filler Opening Second Edition. 6 pp.

—**Glossaries**
BSI BS 6913: Part 5-90. 1990 Operation and Maintenance of Earth-Moving Machinery Part 5: Diagnostic Ports. 4 pp.
BSI BS 6913: Part 7-91. 1991 Operation and Maintenance of Earth-Moving Machinery Part 7: Glossary for Machine Availability (ISO 8927: 1991). 24 pp.
BSI BS 6914: Part 1-88. 1988 Terminology (Including Definitions of Dimensions and Symbols) for Earth-Moving Machinery Part 1: Glossary of Terms for Basic Types of Earth-Moving Machinery. 4 pp.
BSI BS 6914: Part 2-88. 1988 Terminology (Including Definitions of Dimensions and Symbols) for Earth-Moving Machinery Part 2: Glossary of Terms for Base Machine. 12 pp.
BSI BS 6914: Part 3-88. 1988 Terminology (Including Definitions of Dimensions and Symbols) for Earth-Moving Machinery Part 3: Glossary of Terms for Equipment. 13 pp.
BSI BS 6914: Part 7-92. 1992 Terminology (Including Definitions of Dimensions and Symbols) for Earth-Moving Machinery Part 7: Vocabulary, Symbols and Units for Machine Performance (ISO 9245: 1991). 10 pp.
DIN ENGL 24095-83. Earth-Moving Machinery; Determination of Performance; Terminology, Units, Symbols (Jan). 4 pp.
ISO 6165-87. Earth-Moving Machinery—Basic Types—Vocabulary Second Edition. 4 pp.
ISO 6746 Pt 1-87. Earth-Moving Machinery—Definitions of Dimensions and Symbols—Part 1: Base Machine Second Edition. 12 pp.
ISO 8927-91. Earth-Moving Machinery—Machine Availability—Vocabulary First Edition. 25 pp.
ISO 9245-91. Earth-Moving Machinery—Machine Productivity—Vocabulary, Symbols and Units First Edition. 7 pp.
JIS A 8308-92. Earth-Moving Machinery—Basic Types—Vocabulary.

—**Grease Nipples**
BSI BS 6060-81. 1981 Nipple-Type Lubrication Fittings for Earth-Moving Machinery. 7 pp.
ISO 6392-80. Earth Moving Machinery—Lubrication Fittings—Nipple Type First Edition. 6 pp.

—**Ground Speed**
BSI BS 5982-87. 1987 Method for Determination of Ground Speed of Earth-Moving Machinery. 6 pp.
ISO 6014-86. Earth-Moving Machinery—Determination of Ground Speed Second Edition. 5 pp.

Earthmoving Equipment (Cont.)

—Guards (Protective)—Glossaries
BSI BS 5945-87. 1987 Guards and Shields for Earth-Moving Machinery. 8 pp.

CNS A4018-88. Earth-Moving Machinery Guards and Shields-Definitions and Specifications (Dec)(9949).

ISO 3457-86. Earth-Moving Machinery—Guards and Shields—Definitions and Specifications Third Edition. 7 pp.

JIS A 8307-91. Earth-Moving Machinery—Guards and Shields—Definitions and Specifications (ISO 3457-1986).

—Hand Signals
DIN ENGL 24081-78. Safety in the Organization of Work; Earth-Moving Machinery; Hand Signals (Nov). 4 pp.

—Hydraulic Tools—Movement Time
BSI BS 6198-81. (WITHDRAWN) 1981 Measurement of Tool Movement Time of Earth-Moving Machinery. 7 pp.

JIS A 8306-90. Earth-Moving Machinery—Test Method for Measurement of Tool Movement Time (ISO 5004-1987).

—Instrument Panels
BSI BS 5768-79. 1979 Operating Instrumentation for Earth-Moving Machinery. 7 pp.

ISO 6011-87. Earth-Moving Machinery—Operating Instrumentation Second Edition. 6 pp.

JIS A 8110-88. Construction Machinery—Service Instruments. 7 pp.

—Internal Combustion Engines—Net Power
BSI BS 6911: Part 4-90. 1990 Testing Earth-Moving Machinery Part 4: Method for the Evaluation of Engine Power Output. 24 pp.

ISO 9249-89. Earth-Moving Machinery—Engine Test Code—Net Power First Edition. 20 pp.

—Lift Arms—Supports
BSI BS 6912: Part 11-93. 1993 Safety of Earth-Moving Machinery Part 11: Requirements for Lift Arm Support Devices (ISO 10533: 1993) (R). 8 pp.

ISO 10533-93. Earth-Moving Machinery—Lift-Arm Support Devices First Edition. 5 pp.

—Lift Capacity
BSI BS 6911: Part 6-92. 1992 Testing Earth-Moving Machinery Part 6: Method for the Evaluation of Lift Capacity of Pipelayers and Wheel Tractors or Loaders Equipped with Side Boom (ISO 8813: 1992). 22 pp.

BSI BS 6911: Part 7-92. 1992 Testing Earth-Moving Machinery Part 7: Specification for Units of Measurement and Tolerances (ISO 9248: 1992). 7 pp.

ISO 8813-92. Earth-Moving Machinery—Lift Capacity of Pipelayers and Wheeled Tractors or Loaders Equipped with Side Boom First Edition. 19 pp.

ISO 9248-92. Earth-Moving Machinery—Units for Dimensions, Performance and Capacities, and Their Measurement Accuracies First Edition. 5 pp.

—Lighting
MOD UK DSTAN 62-7-81. Lights, Light Units and Reflectors for Service Vehicles Issue 2. 20 pp.

—Locks (Hardware)
BSI BS 6912: Part 8-92. 1992 Safety of Earth-Moving Machinery Part 8: Specification for Performance Requirements of an Articulated Frame Locking Device (ISO 10570: 1992). 6 pp.

ISO 10570-92. Earth-Moving Machinery—Articulated Frame Lock—Performance Requirements First Edition. 4 pp.

—Maintenance Personnel—Training
BSI BS 6574-85. 1985 Training of Mechanics Concerned with the Operation and Maintenance of Earth-Moving Machinery. 12 pp.

ISO 8152-84. Earth-Moving Machinery—Operation and Maintenance—Training of Mechanics First Edition. 10 pp.

—Manual Controls
BSI BS 6124-87. 1987 Amd 1 Zones of Comfort and Reach for Controls in Earth-Moving Machinery (AMD 6420) December 21, 1990. 15 pp.

CSA CAN/CSA-M6682-M91. Zones of Comfort and Reach for Controls—Machinery (EMM,FM) (Amd 1 Included); (ISO 6682-1986). 23 pp.

ISO 6682-86. Earth-Moving Machinery—Zones of Comfort and Reach for Controls Second Edition; (Amendment 1-1989). 16 pp.

—Manual Controls—Symbols
BSI BS 6309-82. (WITHDRAWN) 1982 Symbols for Operator Controls and for Controls Other Than Operator Controls for Use on Earth-Moving Machinery (Superseded by BS 6913: Part 8: 1992). 16 pp.

BSI BS 6913: Part 8-92. 1992 Operation and Maintenance of Earth-Moving Machinery Part 8: Specification for Common Symbols for Operator Controls and Other Displays (ISO 6405-1: 1991). 36 pp.

CSA CAN/CSA-M6405-91. Symbols—Operator Controls and Others—Machinery (EMM, FM) (ISO 6405-1982). 25 pp.

ISO 6405 Pt 1-91. Earth-Moving Machinery—Symbols for Operator Controls and Other Displays—Part 1: Common Symbols First Edition. 33 pp.

—Marine
CNS F3238-89. Scrapers for Marine Use (Jul)(12568).

—Mass
BSI BS 6300-82. 1982 Measuring the Masses of Whole Machines, Their Equipment and Components of Earth-Moving Machinery. 10 pp.

ISO 6016-82. Earth-Moving Machinery—Methods of Measuring the Masses of Whole Machines, Their Equipment and Components First Edition. 8 pp.

—Noise
BSI BS 6812: Part 1-87. 1987 Airborne Noise Emitted by Earth-Moving Machinery Part 1: Methods of Measurement of Exterior Noise in a Stationary Test Condition. 11 pp.

BSI BS 6812: Part 2-87. 1987 Airborne Noise Emitted by Earth-Moving Machinery Part 2: Method of Measurement at the Operators Position in a Stationary Test Condition. 8 pp.

BSI BS 6812: Part 3-91. 1991 Airborne Noise Emitted by Earth-Moving Machinery Part 3: Method of Measurement of Exterior Noise in Dynamic Test Conditions (ISO 6395: 1988). 19 pp.

ISO 6393-85. Acoustics—Measurement of Airborne Noise Emitted by Earth-Moving Machinery—Method for Determining Compliance with Limits for Exterior Noise—Stationary Test Condition First Edition; (SAE J 2102). 9 pp.

ISO 6394-85. Acoustics—Measurement of Airborne Noise Emitted by Earth-Moving Machinery—Operator's Position—Stationary Test Condition First Edition; (SAE J 2103). 6 pp.

ISO 6395-88. Acoustics—Measurement of Exterior Noise Emitted by Earth-Moving Machinery—Dynamic Test Conditions First Edition; (SAE J 2104). 16 pp.

ISO 6396-92. Acoustics—Measurement at the Operator's Position of Noise Emitted by Earth-Moving Machinery—Dynamic Test Conditions First Edition. 10 pp.

SAA AS 2012.1-90. Acoustics—Measurement of Airborne Noise Emitted by Earthmoving Machinery and Agricultural Tractors—Stationary Test Condition—Part 1: Determination of Compliance with Limits for Exterior Noise. 8 pp.

SAA AS 2012.2-90. Acoustics—Measurement of Airborne Noise Emitted by Earthmoving Machinery and Agricultural Tractors—Stationary Test Condition—Part 2: Operator's Position. 4 pp.

—Operators—Training
BSI BS 6264-82. 1982 Procedure for Operator Training for Earth-Moving Machinery. 11 pp.

ISO 7130-81. Earth-Moving Machinery—Guide to Procedure for Operator Training First Edition. 9 pp.

—Preservation
BSI BS 6197-85. 1985 Preservation and Storage of Earth-Moving Machinery. 8 pp.

ISO 6749-84. Earth-Moving Machinery—Preservation and Storage Second Edition. 7 pp.

—Protective Structures
SAA AS 2294-90. Earth-Moving Machinery—Protective Structures Amdt 1 March 1991 (In Professional Package 45). 34 pp.

—Roll Over Protective Structures
BSI BS 5495-80. 1980 Amd 1 Laboratory Evaluations of Roll-Over and Falling-Object Protective Structures: The Deflection-Limiting Volume for Earth-Moving Machinery (ISO 3164: 1979) (AMD 3711) October 30, 1981. 9 pp.

BSI BS 5495-02. (WITHDRAWN) 1980 Amd 2 Laboratory Evaluations of Roll-Over and Falling-Object Protective Structures: the Deflection-Limiting Volume for Earth-Moving Machinery (ISO 3164: 1979) (AMD 6878) February 28, 1992 (Superseded by BS 6912: Part 7: 1992). 14 pp.

BSI BS 5527: Part 1-87. 1987 Roll-Over Protective Structures on Earth-Moving Machinery: Laboratory Tests and Preformance Requirements Part 1: Crawler, Wheel Loaders and Tractors, Backhoe Loaders,Graders, Tractor Scrapers and Articulated Steel Dumpers. 19 pp.

BSI BS 6912: Part 7-92. 1992 Safety of Earth-Moving Machinery Part 7: Specification for Laboratory Evaluations of Roll-Over and Falling-Object Protective Structures: the Deflection-Limiting Volume for Earth-Moving Machinery (ISO 3164: 1992). 7 pp.

CNS A3126-81. Earth-Moving Machinery-Roll-Over Protective Structures for Driver Laboratory Tests and Performance Requirements (Sep)(7849).

CNS A4006-81. Deflection Limiting Volume of Earth-Moving Machinery-Roll-Over Protective Structure for Driver (Sep)(7850).

CNS A4024-88. Earth-Moving Machinery Laboratory Evaluations of Roll-Over and Falling-Object Protective Structures Specifications for the Deflection-Limiting Volume (Dec)(9955).

CSA B352-M1980. Rollover Protective Structures (ROPS) for Agricultural, Construction, Earthmoving, Forestry, Industrial, and Mining Machines; (Gen Instr 1). 76 pp.

ISO 3164-92. Earth-Moving Machinery—Laboratory Evaluations of Roll-over and Falling-Object Protective Structures—Specifications for Deflection-Limiting Volume Fourth Edition; (Incorporating Amendment 1). 6 pp.

ISO 3471 Pt 1-86. Earth-Moving Machinery—Roll-over Protective Structures—Laboratory Tests and Performance Requirements—Part 1: Crawler, Wheel Loaders and Tractors, Backhoe Loaders, Graders, Tractor Scrapers, Articulated Steer Dumpers First Edition. 18 pp.

JIS A 8910-89. Earth-Moving Machinery—Roll-Over Protective Structures—Laboratory Tests and Performance Requirements. 25 pp.

—Safety
CEN PREN 474-3-93. Earth-Moving Machinery—Safety—Part 3: Requirements for Loaders. 14 pp.

CEN PREN 474-4-93. Earth-Moving Machinery—Safety—Part 4: Requirements for Backhoe Loaders. 16 pp.

CEN PREN 474-5-93. Earth-Moving Machinery—Safety—Part 5: Requirements for Hydraulic Excavators. 16 pp.

CEN PREN 474-6-93. Earth-Moving Machinery—Safety—Part 6: Requirements for Dumpers and Site Carriers. 10 pp.

DIN ENGL 24092-85. Earth-Moving Machinery; Safety Requirements (Dec). 3 pp.

—Safety Belts
BSI BS 6218-81. 1981 Amd 1 Performance Requirements for Seat Belts and Seat Belt Anchorages for Earth-Moving Machinery Fitted with Roll-over Protective Structures (AMD 6358) June 28, 1991. 7 pp.

CNS A4005-88. Seat Belt Assemblies and Anchorage for Earth-Moving Machineries (Oct)(7848).

CSA CAN/CSA-M6683-92. Seat Belts and Anchorages—Machinery (EMM, FM) (ISO 6683-1981); (Gen Instr 1). 17 pp.

ISO 6683-81. Earth-Moving Machinery—Seat Belts and Seat Belt Anchorages First Edition; (Amendment 1-1990) (CAN/CSA-M6683-92). 8 pp.

JIS A 8911-79. Seat Belt Assemblies and Anchorage for Earth-Moving Machineries (R 1984). 25 pp.

—Seat Index Point
BSI BS 5631-78. 1978 Amd 2 Earth-Moving Machinery and Tractors and Machinery for Agriculture and Forestry—Seat Index Point (AMD 6164) December 31, 1989. 15 pp.

CSA CAN/CSA-M5353-M90. Seat Index Point (SIP)—Mobile Machines (EMM, AM, FM, MM) (ISO 5353-1978); (Gen Instr 1). 24 pp.

DIN ENGL EN 25353-92. Earth-Moving Machinery, and Tractors and Machinery for Agriculture and Forestry; Seat Index Point (ISO 5353: 1978, A1: 1981, A2: 1984) (Jan). 10 pp.

ISO 5353-78. Earth-Moving Machinery—Seat Index Point First Edition; (Amendment 1-1981) (Amendment 2-1984). 11 pp.

—Seats—Vibration
BSI BS 6294-82. 1982 Measurement of Body Vibration Transmitted to the Operator of Earth-Moving Machinery. 18 pp.

ISO 7096-82. Earth-Moving Machinery—Operator Seat—Transmitted Vibration First Edition. 16 pp.

—Shields—Glossaries
BSI BS 5945-87. 1987 Guards and Shields for Earth-Moving Machinery. 8 pp.

ISO 3457-86. Earth-Moving Machinery—Guards and Shields—Definitions and Specifications Third Edition. 7 pp.

JIS A 8307-91. Earth-Moving Machinery—Guards and Shields—Definitions and Specifications (ISO 3457-1986).

Earthmoving Equipment (Cont.)
—Starters—Key Locked
BSI BS 6912: Part 4-91. 1991 Safety of Earth-Moving Machinery Part 4: Specification for Key-Locked Starting Systems (ISO 10264: 1990). 6 pp.
ISO 10264-90. Earth-Moving Machinery—Key-Locked Starting Systems First Edition. 4 pp.

—Steering
BSI BS 6573-85. (WITHDRAWN) 1985 Steering Capability of Rubber Tyred Earth-Moving Machines (Superseded by BS 6912: Part 10: 1993). 14 pp.
BSI BS 6912: Part 10-93. 1993 Safety of Earth-Moving Machinery Part 10: Specification for Steering Capability of Rubber Tyred Earth-Moving Machines (ISO 5010: 1992) (Supersedes BS 6573: 1985). 12 pp.
ISO 5010-92. Earth-Moving Machinery—Rubber-Tyred Machines—Steering Requirements Second Edition. 10 pp.

—Stoppers
BSI BS 5796-87. 1987 Drain, Fill and Level Plugs for Earth-Moving Machinery. 8 pp.
ISO 6302-86. Earth-Moving Machinery—Drain, Fill and Level Plugs Second Edition. 6 pp.
JIS A 8913-91. Earth-Moving Machinery—Drain, Fill and Level Plugs.

—Storage
BSI BS 6197-85. 1985 Preservation and Storage of Earth-Moving Machinery. 8 pp.
ISO 6749-84. Earth-Moving Machinery—Preservation and Storage Second Edition. 7 pp.

—Symbols
BSI BS 6914: Part 2-88. 1988 Terminology (Including Definitions of Dimensions and Symbols) for Earth-Moving Machinery Part 2: Glossary of Terms for Base Machine. 12 pp.
BSI BS 6914: Part 3-88. 1988 Terminology (Including Definitions of Dimensions and Symbols) for Earth-Moving Machinery Part 3: Glossary of Terms for Equipment. 13 pp.
ISO 6746 Pt 1-87. Earth-Moving Machinery—Definitions of Dimensions and Symbols—Part 1: Base Machine Second Edition. 12 pp.
ISO 6746 Pt 2-87. Earth-Moving Machinery—Definitions of Dimensions and Symbols—Part 2: Equipment Second Edition. 13 pp.

—Technical Manuals
BSI BS 6228-85. 1985 Format and Content of Manuals on Operation and Maintenance for Earth-Moving Machinery. 20 pp.
CNS A4023-88. Earth-Moving Machinery Operation and Maintenance Format and Content of Manuals (Dec)(9954).
ISO 6750-84. Earth-Moving Machinery—Operation and Maintenance—Format and Content of Manuals Second Edition. 19 pp.

—Tire Valves
BSI BS AU 50: Pt 3: Sec 6-82. 1982 Tyres and Wheels Part 3: Valves Section 6: Dimensions for ISO Core Chamber No. 2 for Tyre Inflation Valves. 4 pp.
ISO 6762-82. Tyre Valves—ISO Core Chamber No. 2 Second Edition. 5 pp.

—Tires
BSI BS AU 50:Pt 1:SUBSEC 3.1-89. (WITHDRAWN) 1989 Tyres and Wheels Part 1: Tyres Section 3: Off-The-Road Tyres Subsection 3.1: Narrow and Wide Base Tyre Designations (Superseded by BS AU 50: Subsection 3.1a: 1991). 8 pp.
BSI BS AU 50:Pt1: SUBSEC 3.1A-91. 1991 Amd 1 Tyres and Wheels Part 1: Tyres Section 3: Off-the-Road Tyres Subsection 3.1a: Specification for Narrow and Wide Base Tyre Designations and Dimensions (E) (Spsds BS AU 50: Part 1: Subsec 3.1: 1989) (ISO 4250-1: 1988). 13 pp.
BSI BS AU 50:Pt 1:SUBSEC 3.2-89. (WITHDRAWN) 1989 Tyres and Wheels Part 1: Tyres Section 3: Off-The-Road Tyres Subsection 3.2: Narrow and Wide Base Tyre Loads and Inflation Pressures (Superseded by BS AU 50: Part 1: Subsection 3.2a: 1991). 12 pp.
BSI BS AU 50:Pt 1:SUBSEC3.2A-91. 1991 Tyres and Wheels Part 1: Tyres Section 3: Off-the-Road Tyres Subsection 3.2A: Specification for Narrow and Wide Base Tyre Loads and Inflation Pressures (ISO 4250-2: 1991). 15 pp.
CGSB 20-GP-5D-80. Tires, Pneumatic, Low Speed, Off Highway, Standard for. 23 pp.
CNS K4030-80. Dimensions of Tires for Industrial Vehicles and off the Road Service (Feb)(3673). 33 pp.
ISO 4250 Pt 1-88. Narrow and Wide Base off-Road Tyres and Rims—Part 1: Tyre Designation and Dimensions First Edition; (Amendment 1-1990). 16 pp.
ISO 4250 Pt 2-87. Narrow and Wide Base off-Road Tyres and Rims—Part 2: Loads and Inflation Pressures First Edition. 11 pp.

Earthmoving Equipment (Cont.)
—Tools
BSI BS 5485: Part 1-87. (WITHDRAWN) 1987 Service Tools Used With Earth-Moving Machinery Part 1: Guild to Common Maintenance and Adjustment Tools (Superseded by BS 6913: Part 1: 1989). 11 pp.
BSI BS 5485: Part 2-87. 1987 Service Tools Used with Earth-Moving Machinery Part 2: Guide to Common Repair Tools: Mechanical Pullers and Pushers. 11 pp.
BSI BS 6913: Part 1-89. 1989 Operation and Maintenance of Earth-Moving Machinery Part 1: Guide to Common Maintenance and Adjustment Service Tools. 11 pp.
CNS A4020-88. Earth-Moving Machinery Maintenance and Adjustment Tools (Dec)(9951).
ISO 4510 Pt 1-87. Earth-Moving Machinery—Service Tools—Part 1: Common Maintenance and Adjustment Tools Second Edition. 10 pp.
ISO 4510 Pt 2-86. Earth-Moving Machinery—Service Tools—Part 2: Common Repair Tools—Mechanical Pullers and Pushers First Edition. 9 pp.

—Turning Radius
BSI BS 6486-84. 1984 Measurement of Turning Dimensions of Wheeled Earthmoving Machinery. 11 pp.
ISO 7457-83. Earth-Moving Machinery—Measurement of Turning Dimensions of Wheeled Machines First Edition. 9 pp.

—Visibility
BSI BS 6911: Part 5-92. 1992 Testing Earth-Moving Machinery Part 5: Determination of Operator's Field of View (ISO 5006-1: 1991). 19 pp.
ISO 5006 Pt 1-91. Earth-Moving Machinery—Operator's Field of View —Part 1: Test Method First Edition. 16 pp.
ISO 5006 Pt 2-93. Earth-Moving Machinery—Operator's Field of View—Part 2: Evaluation Method First Edition. 10 pp.
ISO 5006 Pt 3-93. Earth-Moving Machinery—Operator's Field of View—Part 3: Criteria First Edition. 6 pp.

—Wheels
BSI BS AU 50:Pt 1:SUBSEC 3.1-89. (WITHDRAWN) 1989 Tyres and Wheels Part 1: Tyres Section 3: Off-The-Road Tyres Subsection 3.1: Narrow and Wide Base Tyre Designations (Superseded by BS AU 50: Subsection 3.1a: 1991). 8 pp.
BSI BS AU 50:Pt1: SUBSEC 3.1A-91. 1991 Amd 1 Tyres and Wheels Part 1: Tyres Section 3: Off-the-Road Tyres Subsection 3.1a: Specification for Narrow and Wide Base Tyre Designations and Dimensions (E) (Spsds BS AU 50: Part 1: Subsec 3.1: 1989) (ISO 4250-1: 1988). 13 pp.
ISO 4250 Pt 1-88. Narrow and Wide Base off-Road Tyres and Rims—Part 1: Tyre Designation and Dimensions First Edition; (Amendment 1-1990). 16 pp.

Earthquake Resistant Building Design
Use: Seismic Design

Earthquake Resistant Structures
See Also: Concrete Construction
SNZ NZS 3404: Part 2-92. Steel Structures Standard Part 2: Commentary to the Steel Structures Standard. 244 pp.

—Germany
DIN ENGL 4149 Pt 1-81. Buildings in German Earthquake Zones; Design Loads, Dimensioning, Design and Construction of Conventional Buildings (Apr). 15 pp.

Earthwork
Use For: Excavation; Site Preparation
See Also: Construction Sites; Dikes; Slopes

—Construction
BSI BS 8000: Part 1-89. 1989 Workmanship on Building Sites Part 1: Code of Practice for Excavation and Filling. 15 pp.

—Construction Contracts
DIN ENGL 18300-88. Tendering and Performance Stipulations in Con-tracts for Construction Works (VOB); Part C: General Technical Speci-fications in Contracts for Construction Works (ATV) Earthworks (Sept) (This Standard, Together with DIN 18299, Sept. 1988 Edition, Supersedes October 1979 Edition.). 7 pp.

—Glossaries
BSI BS 6100: SUBSEC 2.2.1-92. 1992 Glossary of Building and Civil Engineering Terms Part 2: Civil Engineering Section 2.2: Substructures. Earthworks. Foundations. Tunnels Subsection 2.2.1: Earthworks. 34 pp.

Earthwork (Cont.)
—Glossaries (Cont.)
BSI BS 6100: SUBSEC 2.2.1-90. 1990 Glossary of Building and Civil Engineering Terms Part 2: Civil Engineering Section 2.2: Substructures. Earthworks. Foundations. Tunnels Subsection 2.2.1: Earthworks. 29 pp.

Easy Chairs
Use: Armchairs

Eaves Trough Gutters
Use: Gutters—Eaves Troughs

Ebauches
Use: Watch Movements

Ebonite
Scope Note: For additional listings, see also specific products made from ebonite *Use For:* Hard Rubber
See Also: Vulcanization; Vulcanized Rubber

—Breaking Load
CNS K6450-80. Method of Test for Cross-Breaking Strength of Ebonite (Jan)(5140).

—Chlorine Content
DIN ENGL 53474-76. Testing of Plastics, Rubber and Elastomers; Determination of the Chlorine Content (Aug). 11 pp.

—Compression Testing
CNS K6449-80. Method of Test for Crushing Strength of Ebonite (Jan)(5139).
ISO 2474-72. Ebonite—Determination of Crushing Strength First Edition. 4 pp.

—Deflection Temperature
CNS K6616-80. Determination of Ester Value After Hot Formulation of Essential Oils of Geranium and Rose (Dec)(6682).

—Deflection Temperature—Bending Stress
BSI BS 2782:Pt1: METH 121A-B-91. 1991 Methods of Testing Plastics Part 1: Thermal Properties: Methods 121A and 121B: Determination of Temperature of Deflection of Plastics Under a Bending Stress (ISO 75: 1987) (V). 8 pp.
BSI BS 2782:Pt1: METH 121A-C-76. (WITHDRAWN) 1976 Methods of Testing Plastics Part 1: Thermal Properties: Method 121A-C: (A) Determ of Temp of Deflection Under a Bending Stress of 1.8 MPa of Plastics and Ebonite.(B) Determ of Temp of Deflec Under a Bending Stress of 0.45 MPa of Plastic & Ebonite. 4 pp.
CNS K6617-80. Plastics and Ebonite Determination of Temperature of Deflection Under Load (Dec)(6683).
ISO 75 Pt 2-93. Plastics—Determination of Temperature of Deflection Under Load—Part 2: Plastics and Ebonite First Edition; (Cancels and Replaces ISO 75:1987). 6 pp.

—Durometer Hardness Testing
BSI BS 2782:Pt3: METH 365B-92. 1992 Methods of Testing Plastics Part 3: Mechanical Properties Method 365B: Determination of Indentation Hardness by Means of a Durometer (Shore Hardness) (ISO 868: 1985) (V). 6 pp.
BSI BS 2782:Pt3: METH 365B-81. 1981 Methods of Testing Plastics Part 3: Mechanical Properties Method 365B: Determination of Indentation Hardness by Means of a Durometer (Shore Hardness). 4 pp.
ISO 868-85. Plastics and Ebonite—Determination of Indentation Hardness by Means of a Durometer (Shore Hardness) Second Edition. 5 pp.

—Elongation
CNS K6451-80. Method of Test for Tensile Strength and Elongation at Break of Ebonite (Jan)(5140).

—Indentation Hardness Testing
BSI BS 2782:Pt3: METH 365D-91. 1991 Methods of Testing Plastics Part 3: Mechanical Properties Method 365D: Determination of Hardness by the Ball Indentation Method (ISO 2039-1: 1987). 6 pp.
ISO 2039 Pt 1-93. Plastics—Determination of Hardness—Part 1: Ball Indentation Method Second Edition. 8 pp.

—Lead Content—Atomic Absorption Spectroscopy
DIN ENGL 53599 Pt 1-78. Testing of Rubber and Elastomers; Determination of the Lead Content; Determination by Atomic Absorption Spectroscopy for Lead Contents up to 1000 mg/kg (0.1%) (May). 4 pp.

Ebonite (Cont.)

—**Lead Content—Electrogravimetric Analysis**

DIN ENGL 53599 Pt 2-78. Testing of Rubber and Elastomers; Determination of the Lead Content; Electrogravimetric and Gravimetric Method for Lead Contents over 1000 mg/kg (0.1%) (May). 3 pp.

—**Lead Content—Gravimetric Analysis**

DIN ENGL 53599 Pt 2-78. Testing of Rubber and Elastomers; Determination of the Lead Content; Electrogravimetric and Gravimetric Method for Lead Contents over 1000 mg/kg (0.1%) (May). 3 pp.

—**Lead Content—Photometry**

DIN ENGL 53599 Pt 3-78. Testing of Rubber and Elastomers; Determination of the Lead Content; Photometric Method for Lead Contents up to 1000 mg/kg (0.1%) (May). 4 pp.

—**Sampling**

DIN ENGL 53551-70. Testing of Rubber and Ebonite; Sampling and Preparation of Specimens for Chemical Testings (July). 2 pp.

—**Sheets**

DIN ENGL 7712-75. Hard Rubber Sheets (Feb). 2 pp.

—**Sulfur Content**

DIN ENGL 53561-79. Testing of Rubber and Elastomers; Determination of Total Sulphur Content (Jan). 5 pp.

—**Tensile Testing**

CNS K6451-80. Method of Test for Tensile Strength and Elongation at Break of Ebonite (Jan)(5140).

—**Test Specimens**

DIN ENGL 53551-70. Testing of Rubber and Ebonite; Sampling and Preparation of Specimens for Chemical Testings (July). 2 pp.

—**Water Absorption**

BSI BS 903: Part A18-73. 1973 Methods of Testing Vulcanized Rubber Part A18: Determination of Equilibrium Water Vapour Absorption. 7 pp.

Ebulliometry

See Also: Boiling Points; Thermodynamic Properties

—**Soldering Fluxes**

ISO 9455 Pt 2-93. Soft Soldering Fluxes—Test Methods—Part 2: Determination of Non-Volatile Matter, Ebulliometric Method First Edition. 9 pp.

EC

Use: European Communities

EC Testing

Use: Electrolytic Corrosion

Eccentric Presses

See Also: Mechanical Presses

—**Bench Type—Open Front**

DIN ENGL 55170-61. Open-Front Bench Type Eccentric Presses; Sizes (Oct). 2 pp.

ECDIS

Use: Electronic Chart Display Systems

ECG Monitors

Use: Electrocardiograph Monitors

ECG Recorders

Use: Electrocardiograph Recorders

ECH

Use: Epichlorohydrin

Echo (Telecommunications)

See Also: Echo Cancellers; Echo Control Equipment; Echo Path Loss; Echo Suppressors; Electromagnetic Noise; End Delay

—**Bridges (Electrical)—Conference Calling**

CCITT RECMN G.172-89. Transmission Plan Aspects of International Conference Calls—General Characteristics of International Telephone Connections and Circuits (Study Groups XII and XV) 3 pp. 3 pp.

—**Fixed Satellite Services**

CCIR QUESTION 7-2/4-86. Baseband Transmission Variability, Delay, Echoes and Switching Discontinuities in Systems in the Fixed-Satellite Service—Questions of Study Group 4 Fixed-Satellite Service. 1 p.

Echo (Telecommunications) (Cont.)

—**Loudness Ratings**

CCITT RECMN G.111-89. Loudness Ratings (LRs) in an International Connection—General Characteristics of International Telephone Connections and Circuits (Study Groups XII and XV) 22 pp. 22 pp.

—**Telephone Circuits**

CCITT RECMN G.114-89. Mean One-Way Propagation Time—General Characteristics of International Telephone Connections and Circuits (Study Groups XII and XV) 11 pp. 11 pp.

CCITT RECMN G.131-89. Stability and Echo—General Characteristics of International Telephone Connections and Circuits (Study Groups XII and XV) 13 pp. 13 pp.

CCITT RECMN Q.42-89. Stability and Echo (Echo Suppressors)—General Recommendations on Telephone Switching and Signalling—Functions and Information Flows for Services in the ISDN—Supplements (Study Group XI) 1 pp. 1 p.

—**Telephone Circuits—Measurement**

CCITT RECMN G.114-89. Mean One-Way Propagation Time—General Characteristics of International Telephone Connections and Circuits (Study Groups XII and XV) 11 pp. 11 pp.

—**Telephone Circuits—Private Switched Networks**

CCITT RECMN G.171-89. Transmission Plan Aspects of Privately Operated Networks—General Characteristics of International Telephone Connections and Circuits (Study Groups XII and XV) 19 pp. 19 pp.

CENELEC PRETS 300 283-92. Business Telecommunications (BT); Planning of Loudness Rating and Echo Values for Private Networks Digitally Connected to the Public Network. 13 pp.

ETSI PRETS 300 283-92. Business Telecommunications (BT); Planning of Loudness Rating and Echo Values for Private Networks Digitally Connected to the Public Network. 13 pp.

—**Telephone Connections**

CCITT FASCICLE III.1. General Characteristics of International Telephone Connections and Circuits Recommendations G.101-G.181. 332 pp.

CCITT RECMN G.122-89. Influence of National Systems on Stability, Talker Echo, and Listener Echo in International Connections—General Characteristics of International Telephone Connections and Circuits (Study Groups XII and XV) 13 pp. 13 pp.

—**Voice Band Data Transmission**

CCITT RECMN G.113-89. Transmission Impairments—General Characteristics of International Telephone Connections and Circuits (Study Groups XII and XV) 22 pp. 22 pp.

Echo Cancellers

See Also: Communication Transmission Lines; Echo (Telecommunications); Echo Control Equipment; Echo Suppressors; Telecommunication Equipment; Telephone Equipment

CCITT RECMN G.114-89. Mean One-Way Propagation Time—General Characteristics of International Telephone Connections and Circuits (Study Groups XII and XV) 11 pp. 11 pp.

CCITT RECMN G.131-89. Stability and Echo—General Characteristics of International Telephone Connections and Circuits (Study Groups XII and XV) 13 pp. 13 pp.

CCITT RECMN G.165-89. Echo Cancellers—General Characteristics of International Telephone Connections and Circuits (Study Groups XII and XV) 23 pp. 23 pp.

—**Attenuation Distortion**

CCITT RECMN G.165-89. Echo Cancellers—General Characteristics of International Telephone Connections and Circuits (Study Groups XII and XV) 23 pp. 23 pp.

—**Comprehensive Testing**

CCITT RECMN M.660-89. Periodical In-Station Tests of Echo Suppressors Complying with Recommendations G.161 and G.164—General Maintenance Principles—Maintenance of International Transmission Systems and Telephone Circuits (Study Group IV) 5 pp. 5 pp.

—**Delay Distortion**

CCITT RECMN G.165-89. Echo Cancellers—General Characteristics of International Telephone Connections and Circuits (Study Groups XII and XV) 23 pp. 23 pp.

Echo Cancellers (Cont.)

—**Electrical Properties**

CCITT RECMN M.665-89. Testing of Echo Cancellers—General Maintenance Principles—Maintenance of International Transmission Systems and Telephone Circuits (Study Group IV) 2 pp. 2 pp.

—**Glossaries**

CCITT RECMN G.165-89. Echo Cancellers—General Characteristics of International Telephone Connections and Circuits (Study Groups XII and XV) 23 pp. 23 pp.

—**Group Delay**

CCITT RECMN G.165-89. Echo Cancellers—General Characteristics of International Telephone Connections and Circuits (Study Groups XII and XV) 23 pp. 23 pp.

—**International Switching Centers**

CCITT RECMN Q.115-89. Control of Echo Suppressors and Echo Cancellers by International Switching Centres—General Recommendations on Telephone Switching and Signalling—Functions and Information Flows for Services in the ISDN—Supplements (Study Group XI) 7 pp. 7 pp.

—**Maritime Mobile Satellite Communications**

CCIR RECMN 550-1-86. Use of Echo Suppressors in the Maritime Mobile-Satellite Service—Section 8I—Technical and Operating Characteristics of Mobile Satellite Services. 1 p.

—**Private Switched Networks**

CCITT RECMN G.171-89. Transmission Plan Aspects of Privately Operated Networks—General Characteristics of International Telephone Connections and Circuits (Study Groups XII and XV) 19 pp. 19 pp.

—**Release Time**

CCITT RECMN G.165-89. Echo Cancellers—General Characteristics of International Telephone Connections and Circuits (Study Groups XII and XV) 23 pp. 23 pp.

—**Return Loss**

CCITT RECMN G.165-89. Echo Cancellers—General Characteristics of International Telephone Connections and Circuits (Study Groups XII and XV) 23 pp. 23 pp.

—**Telephone Circuits—Electrical Properties**

CCITT RECMN M.665-89. Testing of Echo Cancellers—General Maintenance Principles—Maintenance of International Transmission Systems and Telephone Circuits (Study Group IV) 2 pp. 2 pp.

—**Testing Equipment**

CCITT RECMN O.27-89. In-Station Echo Canceller Test Equipment—Specifications for Measuring Equipment (Study Group IV) 7 pp. 7 pp.

—**Threshold Voltage**

CCITT RECMN G.165-89. Echo Cancellers—General Characteristics of International Telephone Connections and Circuits (Study Groups XII and XV) 23 pp. 23 pp.

—**Tones (Telephone Services)**

CCITT RECMN G.165-89. Echo Cancellers—General Characteristics of International Telephone Connections and Circuits (Study Groups XII and XV) 23 pp. 23 pp.

—**Transmission Loss**

CCITT RECMN G.122-89. Influence of National Systems on Stability, Talker Echo, and Listener Echo in International Connections—General Characteristics of International Telephone Connections and Circuits (Study Groups XII and XV) 13 pp. 13 pp.

Echo Chambers

Use: Reverberation Chambers

Echo Control Equipment

See Also: Echo (Telecommunications); Echo Cancellers; Echo Suppressors; Telecommunication Equipment

—**Disabling—Public Switched Telephone Networks**

CENELEC ETS 300 001 Chp 9-92. Attachments to Public Switched Telephone Network (PSTN); General Technical Requirements for Equipment Connected to an Analogue Subscriber Interface in the PSTN Chapter 9: Special Functions (Candidate NET 4). 198 pp.

Echo

Echo Control Equipment (Cont.)

—Disabling—Public Switched Telephone Networks (Cont.)

ETSI ETS 300 001 Chp 9-92. Attachments to Public Switched Telephone Network (PSTN); General Technical Requirements for Equipment Connected to an Analogue Subscriber Interface in the PSTN Chapter 9: Special Functions (Candidate NET 4). 198 pp.

—Maritime Mobile Satellite Communications—Automatic Services

CCITT RECMN G.473-89. Interconnection of a Maritime Mobile Satellite System with the International Automatic Switched Telephone Service; Transmission Aspects—International Analogue Carrier Systems (Study Group XV) 10 pp. 10 pp.

Echo Loss

Use: Echo Path Loss

Echo Path Loss

See Also: Data Transmission; Echo (Telecommunications); Return Loss (Data Transmission); Transmission Loss

—Digital Exchanges

CCITT RECMN G.142-89. Transmission Characteristics of Exchanges—General Characteristics of International Telephone Connections and Circuits (Study Groups XII and XV) 6 pp. 6 pp.

—Echo Suppressors

CCITT RECMN G.164-89. Echo Suppressors—General Characteristics of International Telephone Connections and Circuits (Study Groups XII and XV) 36 pp. 36 pp.

—Telephone Circuits

CCITT RECMN G.122-89. Influence of National Systems on Stability, Talker Echo, and Listener Echo in International Connections—General Characteristics of International Telephone Connections and Circuits (Study Groups XII and XV) 13 pp. 13 pp.
CCITT RECMN G.131-89. Stability and Echo—General Characteristics of International Telephone Connections and Circuits (Study Groups XII and XV) 13 pp. 13 pp.

—Telephone Circuits—Measurement

CCITT RECMN G.122-89. Influence of National Systems on Stability, Talker Echo, and Listener Echo in International Connections—General Characteristics of International Telephone Connections and Circuits (Study Groups XII and XV) 13 pp. 13 pp.

—Telephone Connections

CCITT RECMN G.122-89. Influence of National Systems on Stability, Talker Echo, and Listener Echo in International Connections—General Characteristics of International Telephone Connections and Circuits (Study Groups XII and XV) 13 pp. 13 pp.

—Telephones

CCITT FASCICLE III.1. General Characteristics of International Telephone Connections and Circuits Recommendations G.101-G.181. 332 pp.

Echo Sounders

Use: Sonic Depth Finders

Echo Suppressors

See Also: Communication Transmission Lines; Differential Sensitivity; Echo (Telecommunications); Echo Cancellers; Echo Control Equipment; End Delay; Telecommunication Equipment; Telephone Equipment

CCITT RECMN G.114-89. Mean One-Way Propagation Time—General Characteristics of International Telephone Connections and Circuits (Study Groups XII and XV) 11 pp. 11 pp.
CCITT RECMN G.131-89. Stability and Echo—General Characteristics of International Telephone Connections and Circuits (Study Groups XII and XV) 13 pp. 13 pp.
CCITT RECMN G.151-89. General Performance Objectives Applicable to All Modern International Circuits and National Extension Circuits—General Characteristics of International Telephone Connections and Circuits (Study Groups XII and XV) 6 pp. 6 pp.
CCITT RECMN G.164-89. Echo Suppressors—General Characteristics of International Telephone Connections and Circuits (Study Groups XII and XV) 36 pp. 36 pp.
CCITT RECMN Q.42-89. Stability and Echo (Echo Suppressors) —General Recommendations on Telephone Switching and Signalling—Functions and Information Flows for Services in the ISDN—Supplements (Study Group XI) 1 pp. 1 p.

Echo Suppressors (Cont.)

—Attenuation Distortion

CCITT RECMN G.164-89. Echo Suppressors—General Characteristics of International Telephone Connections and Circuits (Study Groups XII and XV) 36 pp. 36 pp.

—CCITT R2 Signaling Systems

CCITT RECMN Q.479-89. Signalling Procedures—Echo-Suppressor Control—Signalling Requirements—Specifications of Signalling Systems R1 and R2 (Study Group XI) 4 pp. 4 pp.

—Circuit Noise

CCITT RECMN G.164-89. Echo Suppressors—General Characteristics of International Telephone Connections and Circuits (Study Groups XII and XV) 36 pp. 36 pp.

—Comprehensive Testing

CCITT RECMN M.660-89. Periodical In-Station Tests of Echo Suppressors Complying with Recommendations G.161 and G.164—General Maintenance Principles—Maintenance of International Transmission Systems and Telephone Circuits (Study Group IV) 5 pp. 5 pp.

—Crosstalk

CCITT RECMN G.164-89. Echo Suppressors—General Characteristics of International Telephone Connections and Circuits (Study Groups XII and XV) 36 pp. 36 pp.

—Delay Distortion

CCITT RECMN G.164-89. Echo Suppressors—General Characteristics of International Telephone Connections and Circuits (Study Groups XII and XV) 36 pp. 36 pp.

—Echo Path Loss

CCITT RECMN G.164-89. Echo Suppressors—General Characteristics of International Telephone Connections and Circuits (Study Groups XII and XV) 36 pp. 36 pp.

—Electrical Impedance

CCITT RECMN G.164-89. Echo Suppressors—General Characteristics of International Telephone Connections and Circuits (Study Groups XII and XV) 36 pp. 36 pp.

—Electrical Measurement

CCITT RECMN G.164-89. Echo Suppressors—General Characteristics of International Telephone Connections and Circuits (Study Groups XII and XV) 36 pp. 36 pp.

—Glossaries

CCITT RECMN G.164-89. Echo Suppressors—General Characteristics of International Telephone Connections and Circuits (Study Groups XII and XV) 36 pp. 36 pp.

—Group Delay

CCITT RECMN G.164-89. Echo Suppressors—General Characteristics of International Telephone Connections and Circuits (Study Groups XII and XV) 36 pp. 36 pp.

—Harmonic Distortion

CCITT RECMN G.164-89. Echo Suppressors—General Characteristics of International Telephone Connections and Circuits (Study Groups XII and XV) 36 pp. 36 pp.

—Insertion Loss

CCITT RECMN G.164-89. Echo Suppressors—General Characteristics of International Telephone Connections and Circuits (Study Groups XII and XV) 36 pp. 36 pp.

—Intermodulation

CCITT RECMN G.164-89. Echo Suppressors—General Characteristics of International Telephone Connections and Circuits (Study Groups XII and XV) 36 pp. 36 pp.

—International Switching Centers

CCITT RECMN Q.115-89. Control of Echo Suppressors and Echo Cancellers by International Switching Centres—General Recommendations on Telephone Switching and Signalling—Functions and Information Flows for Services in the ISDN—Supplements (Study Group XI) 7 pp. 7 pp.

—Maritime Mobile Satellite Communications

CCIR RECMN 550-1-86. Use of Echo Suppressors in the Maritime Mobile-Satellite Service—Section 8I—Technical and Operating Characteristics of Mobile Satellite Services. 1 p.

Echo Suppressors (Cont.)

—Noise

CCITT RECMN G.164-89. Echo Suppressors—General Characteristics of International Telephone Connections and Circuits (Study Groups XII and XV) 36 pp. 36 pp.

—Private Switched Networks

CCITT RECMN G.171-89. Transmission Plan Aspects of Privately Operated Networks—General Characteristics of International Telephone Connections and Circuits (Study Groups XII and XV) 19 pp. 19 pp.

—Propagation Time

CCITT RECMN G.161-89. Echo-Suppressors Suitable for Circuits Having Either Short or Long Propagation Times—General Characteristics of International Telephone Connections and Circuits (Study Groups XII and XV) 1 pp. 1 p.
CCITT RECMN G.164-89. Echo Suppressors—General Characteristics of International Telephone Connections and Circuits (Study Groups XII and XV) 36 pp. 36 pp.

—Psophometric Power

CCITT RECMN G.164-89. Echo Suppressors—General Characteristics of International Telephone Connections and Circuits (Study Groups XII and XV) 36 pp. 36 pp.

—Radiotelephone Circuits

CCIR Report 1131-90. Adaptive HF Communications Systems—Section 3Ad—Operational Questions. 5 pp.

—Release Time

CCITT RECMN G.164-89. Echo Suppressors—General Characteristics of International Telephone Connections and Circuits (Study Groups XII and XV) 36 pp. 36 pp.

—Satellite Communication Equipment

CCITT RECMN G.151-89. General Performance Objectives Applicable to All Modern International Circuits and National Extension Circuits—General Characteristics of International Telephone Connections and Circuits (Study Groups XII and XV) 6 pp. 6 pp.

—Testing Equipment

CCITT RECMN O.25-89. Semiautomatic In-Circuit Echo Suppressor Testing System (ESTS)—Specifications for Measuring Equipment (Study Group IV) 7 pp. 7 pp.

—Threshold Voltage

CCITT RECMN G.164-89. Echo Suppressors—General Characteristics of International Telephone Connections and Circuits (Study Groups XII and XV) 36 pp. 36 pp.

—Tones (Telephone Services)

CCITT RECMN G.164-89. Echo Suppressors—General Characteristics of International Telephone Connections and Circuits (Study Groups XII and XV) 36 pp. 36 pp.

—Transient Response

CCITT RECMN G.164-89. Echo Suppressors—General Characteristics of International Telephone Connections and Circuits (Study Groups XII and XV) 36 pp. 36 pp.

—Transmission Loss

CCITT RECMN G.122-89. Influence of National Systems on Stability, Talker Echo, and Listener Echo in International Connections—General Characteristics of International Telephone Connections and Circuits (Study Groups XII and XV) 13 pp. 13 pp.

ECL

Use: Emitter Coupled Logic

Economics

See Also: Cost Analysis; Expenses

—Classification

BSI BS 1000: (33)-85. 1985 Universal Decimal Classification (UDC). English Full Edition (33): Economics. 122 pp.
SNZ NZS/BS 1000 (33)-85. Universal Decimal Classification Economics. 124 pp.

ECS

Use: European Communications Satellite

ECSC

Use: European Communities

EDD

Use: Envelope Delay Distortion

INTERNATIONAL AND NON-U.S. NATIONAL STANDARDS
SUBJECT INDEX

EDIFACT

EDDP Content Analysis
Use: Edifenphos Content Analysis

Eddy Current Scanners
Use: Eddy Current Testers

Eddy Current Testers
Use For: Eddy Current Scanners; Eddyscopes
See Also: Eddy Current Testing; Testing Equipment
JIS Z 2314-91. Test Methods for Performance Characteristics of Eddy Current Testing Instruments. 11 pp.

Eddy Current Testing
Use For: Foncault Current Testing *See Also:* Eddy Current Testers; Electromagnetic Testing; Physical Testing
CGSB CAN/CGSB-48.14-M86. Advanced Manual for: Eddy Current Test Method. 199 pp.
CNS Z8051-84. General Rules for Eddy Current Testing (Sep)(11050).

—**Airframes**
CAA LEAFLET 4-8 07.90. Eddy Current Methods. 13 pp.

—**Boilers**
BSI BS 1113: ENQ CASE 1-86. 1986 Design and Manufacture of Water-Tube Steam Generating Plant (Including Superheaters, Reheaters and Steel Tube Economizers) Enquiry Case 1: Eddy Current Testing in Lieu of Hydraulic Testing. 1 p.

—**Copper Pipes**
JIS H 0502-86. Method of Eddy Current Testing for Copper and Copper-Alloy Pipes and Tubes. 6 pp.

—**Glossaries**
BSI BS 3683: Part 5-65. 1965 Glossary of Terms Used in Non-Destructive Testing: Eddy Current Flaw Detection. 13 pp.
CNS Z8068-87. Glossary of Terms Related to Nondestructive Testing (Eddy Current Testing Terms) (Jan)(11823).

—**Metal Coatings—Thickness Measurement**
BSI BS 5411: Part 3-84. 1984 Methods of Test for Metallic and Related Coatings Part 3: Eddy Current Method for Measurement of Coating Thickness of Non-Conductive Coatings on Non-Magnetic Basis Metals. 7 pp.
ISO 2360-82. Non-Conductive Coatings on Non-Magnetic Basis Metals—Measurement of Coating Thickness—Eddy Current Method Second Edition. 5 pp.

—**Metal Tubes**
CEN PREN 2002 (Part 20)-92. Test Methods for Metallic Materials Part 20—Eddy Current Inspection of Circular Cross-Section Tubes. 15 pp.
SAA AS 2084-87. Non-Destructive Testing—Eddy Current Testing of Metal Tubes. 9 pp.

—**Metals**
CNS Z8062-85. Method of Eddy Current Test for Metal Bars, Rods and Wires (Oct)(11400).

—**Nonferrous Pipes**
BSI BS 3889: Part 2B-66. 1966 Non-Destructive Testing of Pipes and Tubes Part 2B: Eddy Current Testing of Non-Ferrous Tubes. 11 pp.

—**Pipes**
DIN ENGL 54141 Pt 1-82. Non-Destructive Testing; Eddy Current Testing of Pipes and Tubes; General Remarks on Testing Using Concentric Test Coils and the Single-Frequency Method (Oct). 3 pp.
DIN ENGL 54141 Pt 2-82. Non-Destructive Testing; Eddy Current Testing of Pipes and Tubes; Reference Method for the Determination and Calibration of the Properties of an Eddy Current Testing System Using Concentric Test Coils (Oct). 15 pp.
DIN ENGL 54141 Pt 3-87. Non-Destructive Testing; Eddy Current Testing of Pipes and Tubes; Procedure (Feb). 12 pp.

—**Ships**
MOD UK NES 729: Part 3-84. Requirements for Non-Destructive Examination Methods Part 3: Eddy Current Issue 1 (11.84). 17 pp.
MOD UK NES 729: Part 3-91. Requirements for Non—Destructive Examination Methods Part 3: Eddy Current Issue 2 (12.91). 23 pp.

—**Steel Pipes**
BSI BS 3889: Part 2A-86. 1986 Non-Destructive Testing of Pipes and Tubes Part 2A: Automatic Eddy Current Testing of Wrought Steel Tubes. 11 pp.
ISO 9304-89. Seamless and Welded (Except Submerged Arc-Welded) Steel Tubes for Pressure Purposes—Eddy Current Testing for the Detection of Imperfections First Edition. 11 pp.
JIS G 0583-78. Eddy Current Examination of Steel Pipes and Tubes. 10 pp.

Eddy Current Testing *(Cont.)*
—**Steel Pipes** *(Cont.)*
JIS Z 2315-91. Test Methods for Performance Characteristics of Eddy Current Flaw Detecting System. 10 pp.

—**Steels**
JIS G 0568-93. Method for Eddy Current Testing of Steel Products by Encircling Coil Technique. 11 pp.

—**Submarines**
MOD UK NES 729: Part 3-84. Requirements for Non-Destructive Examination Methods Part 3: Eddy Current Issue 1 (11.84). 17 pp.
MOD UK NES 729: Part 3-91. Requirements for Non—Destructive Examination Methods Part 3: Eddy Current Issue 2 (12.91). 23 pp.

—**Testing Personnel—Certification**
CGSB 48-GP-13M-79. Certification of Nondestructive Testing Personnel (Eddy Current Method); (Reprinted July 1984 Incorporating Amendment 1) (Amendment 5 Incorporates Amendments 2, 3 and 4 November 1990) (Amendment 8 Incorporates Amendment 7 November 1992). 149 pp.

—**Titanium Pipes**
JIS H 0515-92. Eddy Current Inspection of Titanium Pipes and Tubes. 10 pp.

Eddyscopes
Use: Eddy Current Testers

Edetate Disodium
Use For: Disodium Ethylenediaminetetraacetate
CNS K7206-66. Chemical Reagent (Disodium Ethylenediamine Tetraacetate) (Jun)(1706).

Edetic Acid Content Analysis
Use: EDTA Content Analysis

Edge Connectors
Use: Card Edge Connectors

Edge Preparation
Use: Welding—Edge Preparation

Edge Shears
Use: Shears (Hand Tools)—Edge

Edge Socket Connectors
Use: Card Edge Connectors

Edge Tear Testing
Use: Tear Strength

Edgeboard Connectors
Use: Card Edge Connectors

Edgecard Connectors
Use: Card Edge Connectors

Edging
See Also: Fabrics

—**Fabric—Tatami**
JIS L 3108-92. Edging Cloths for TATAMI.

Edgley EA 7 Optica Aircraft
See Also: Aircraft

—**Antenna Positions**
CAA. Edgley EA7 Optica (Approved Aerial Positions). 1 p.

EDI Messaging Services
Use For: Electronic Data Interchange Messaging Services *See Also:* EDIFACT; Electronic Mail Systems; Information Interchange; Message Handling Systems; Message Transfer Systems; Network Services (OSI)
CCITT RECMN F.435-91. Message Handling: Electronic Data Interchange Messaging Service (Study Group I) 50 pp. 50 pp.
CCITT RECMN X.435-91. Message Handling Systems: Electronic Data Interchange Messaging System (Study Group VII) 123 pp. 123 pp.
IEC DIS 13208-92. Information Technology—Text and Office Systems—Message Handling Systems—Electronic Data Interchange Messaging System ***CD-ROM ONLY***. 124 pp.
IEC DIS 13209-92. Information Technology—Text and Office Systems—Message Handling—Electronic Data Interchange Messaging Service ***CD-ROM ONLY***. 51 pp.
ISO DIS 13208-92. Information Technology—Text and Office Systems—Message Handling Systems—Electronic Data Interchange Messaging System ***CD-ROM ONLY***. 124 pp.

EDI Messaging Services *(Cont.)*
ISO DIS 13209-92. Information Technology—Text and Office Systems—Message Handling—Electronic Data Interchange Messaging Service ***CD-ROM ONLY***. 51 pp.
JTC1 DIS13208-92. Information Technology—Text and Office Systems—Message Handling Systems—Electronic Data Interchange Messaging System ***CD-ROM ONLY***. 124 pp.
JTC1 DIS13209-92. Information Technology—Text and Office Systems—Message Handling—Electronic Data Interchange Messaging Service ***CD-ROM ONLY***. 51 pp.

—**Glossaries**
CCITT RECMN F.435-91. Message Handling: Electronic Data Interchange Messaging Service (Study Group I) 50 pp. 50 pp.

—**Quality of Service**
CCITT RECMN F.435-91. Message Handling: Electronic Data Interchange Messaging Service (Study Group I) 50 pp. 50 pp.

Edible Oils
Use: Vegetable Oils

Edible Podded Peas
Use: Peas

EDIFACT
See Also: Data Transmission; EDI Messaging Services; Information Interchange; Open Systems Interconnection; Protocols
CSA CAN/CSA-Z243.310-91. UN/EDIFACT Invoice Message; (CGSB CAN/CGSB-200.26-91). 28 pp.
IEC DIS 13208-92. Information Technology—Text and Office Systems—Message Handling Systems—Electronic Data Interchange Messaging System ***CD-ROM ONLY***. 124 pp.
ISO DIS 13208-92. Information Technology—Text and Office Systems—Message Handling Systems—Electronic Data Interchange Messaging System ***CD-ROM ONLY***. 124 pp.
JTC1 DIS13208-92. Information Technology—Text and Office Systems—Message Handling Systems—Electronic Data Interchange Messaging System ***CD-ROM ONLY***. 124 pp.

—**Australian Port Codes**
SAA AS 3804-91. Electronic Data Interchange for Administration, Commerce and Transport (EDIFACT)—Australian Port Codes. 3 pp.

—**Business Forms**
SAA AS 3805-91. Electronic Data Interchange for Administration, Commerce and Transport (EDIFACT)—Forms Design—Basic Layout (ISO 8439:1990). 3 pp.

—**Invoices**
CSA CAN/CSA-Z243.301-91. UN/EDIFACT Message Design Guidelines; (CGSB CAN/CGSB-200.27-91). 40 pp.

—**Syntax Rules**
BSI BS EN 29735-92. 1992 Electronic Data Interchange for Administration, Commerce and Transport (EDIFACT)—Application Level Syntax Rules (ISO 9735: 1988). 30 pp.
CEN EN 29 735-90. Electronic Data Interchange for Administration, Commerce and Trade (EDIFACT)—Application Level Syntax Rules. 2 pp.
CEN EN 29735-92. Electronic Data Interchange for Administration, Commerce and Transport (EDIFACT)—Application Level Syntax Rules (ISO 9735: 1988) (Amended and Reprinted in 1990). 38 pp.
CGSB CAN/CGSB-200.10-89. Electronic Data Interchange for Administration, Commerce, and Transport (EDIFACT)—Application Level Syntax Rules (ISO 9735:1988) (CAN/CSA-Z243.300-89). 29 pp.
CSA CAN/CSA-Z243.300-89. Electronic Data Interchange for Administration, Commerce, and Transport (EDIFACT)—Application Level Syntax Rules (ISO 9735:1988). 29 pp.
DIN ENGL EN 29735-92. Electronic Data Interchange for Administration, Commerce and Transport (EDIFACT); Application Level Syntax Rules (ISO 9735: 1988, as Amended in 1990) (Aug). 25 pp.
ISO 9735-88. Electronic Data Interchange for Administration, Commerce and Transport (EDIFACT)—Application Level Syntax Rules First Edition; (Amended and Reprinted-1990) (Amendment 1-1992) (CEN EN 29735: 1992). 33 pp.
JTC1 9735-88. Electronic Data Interchange for Administration, Commerce and Transport (EDIFACT)—Application Level Syntax Rules First Edition; (Amended and Reprinted—1990) (Amendment 1-1992). 25 pp.

INDUSTRY STANDARDS

INTERNATIONAL AND NON-U.S. NATIONAL STANDARDS SUBJECT INDEX

EDIFACT (Cont.)
—Syntax Rules (Cont.)
OSI ISO DIS 9735-87. Electronic Data Interchange for Administration, Commerce and Trade (EDIFACT)—Application Level Syntax Rules. 24 pp.

SNZ NZS/ISO 9735-88. Electronic Data Interchange for Administration, Commerce and Transport (EDIFACT) TThis is a Joint Standard with SAA AS 3801). 19 pp.

Edifenphos Content Analysis
Use For: EDDP Content Analysis
—Waste Water
CNS K9093-82. Methods of Test for EDDP in Industrial Waste Water (Jun)(9002).

Editing
See Also: Copy Preparation; Proofreading
—Glossaries
CGSB 94-GP-2-73. Glossary of Editorial Terms in General Use in the Graphic Arts. 25 pp.
—Symbols
ISO 5776-83. Graphic Technology—Symbols for Text Correction First Edition. 7 pp.

Editing Systems
Use For: Text Entry Systems See Also: Computers; Data Processing; Software; Text Processing
—Cursor Control
IEC DIS 10741 Pt 1-92. Information Technology—User System Interfaces for Text and Office Systems—Dialogue Interaction—Part 1: Cursor Control for Text Editing ***CD-ROM ONLY***. 16 pp.

ISO DIS 10741 Pt 1-92. Information Technology—User System Interfaces for Text and Office Systems—Dialogue Interaction—Part 1: Cursor Control for Text Editing ***CD-ROM ONLY***. 16 pp.

—Glossaries
ISO DIS 2382 Pt 1.2-92. Information Technology—Vocabulary—Part 1: Fundamental Terms (Revision of ISO 2382-1:1984) ***CD-ROM ONLY***. 36 pp.

—Syntax Directed—SGML
BSI BS ISO/IEC TR 10037-91. 1991 Information Technology—SGML and Text-Entry Systems—Guidelines for SGML Syntax-Directed Editing Systems. 19 pp.

IEC TR10037-91. Information Technology—SGML and Text-Entry Systems—Guidelines for SGML Syntax-Directed Editing Systems First Edition. 16 pp.

ISO TR10037-91. Information Technology—SGML and Text-Entry Systems—Guidelines for SGML Syntax-Directed Editing Systems First Edition. 16 pp.

JTC1 TR10037-91. Information Technology—SGML and Text-Entry Systems—Guidelines for SGML Syntax-Directed Editing Systems First Edition. 16 pp.

SNZ NZS/AS 4110-93. Information Technology—SGML and Text-Entry Systems—Guidelines for SGML Syntax-Directed Editing Systems (ISO/IEC TR 10037:1991) (This is a Joint Standard with SAA AS 4110). 11 pp.

EDM Machines
Use: Electric Discharge Machine Tools

EDPM
Use: Ethylene Propylene Terpolymers

EDTA
Use For: Ethylenediaminetetraacetic Acid
CNS K7218-66. Chemical Reagent (Ethylenediamine Tetraacetic Acid) (Jun)(1718).
—Waste Water
DIN ENGL 38413 Pt 5-90. German Standard Methods for the Examination of Water, Waste Water and Sludge; Individual Constituents (Group P) Determination of Ethyl-enediaminetetraacetic Acid (EDTA) and Nitrolotriacetic Acid (NTA) by Polarography (P 5) (Oct). 8 pp.
—Water Pollution
DIN ENGL 38413 Pt 5-90. German Standard Methods for the Examination of Water, Waste Water and Sludge; Individual Constituents (Group P) Determination of Ethyl-enediaminetetraacetic Acid (EDTA) and Nitrolotriacetic Acid (NTA) by Polarography (P 5) (Oct). 8 pp.

EDTA Content Analysis
Use For: Edetic Acid Content Analysis; EDETIC ACID CONTENT ANALYSIS; Ethylenediaminetetraacetic Acid Content Analysis
—Cleaning Agents—Volumetric Analysis
BSI BS 3762: Sec 3.23-90. 1990 Analysis of Formulated Detergents Part 3: Quantitative Test Methods Section 3.23: Method for Determination of Chelating Agents Content. 6 pp.

ISO 4325-90. Soaps and Detergents—Determination of Chelating Agent Content—Titrimetric Method Second Edition. 4 pp.
—Soaps—Volumetric Analysis
BSI BS 1715: Sec 2.9-90. 1990 Analysis of Soaps Part 2: Quantitative Test Methods Section 2.9: Method for Determination of EDTA Content. 8 pp.
—Waste Water—Polarographic Analysis
DIN ENGL 38413 Pt 5-90. German Standard Methods for the Examination of Water, Waste Water and Sludge; Individual Constituents (Group P) Determination of Ethyl-enediaminetetraacetic Acid (EDTA) and Nitrolotriacetic Acid (NTA) by Polarography (P 5) (Oct). 8 pp.
—Water
BSI BS 2690 Part 15-74. 1974 Water Used in Industry Part 15: Free EDTA, Total Salts of EDTA, Polyacrylate and Polymethacrylate. 11 pp.
—Water Pollution—Polarographic Analysis
DIN ENGL 38413 Pt 5-90. German Standard Methods for the Examination of Water, Waste Water and Sludge; Individual Constituents (Group P) Determination of Ethyl-enediaminetetraacetic Acid (EDTA) and Nitrolotriacetic Acid (NTA) by Polarography (P 5) (Oct). 8 pp.

Education
Scope Note: For additional listings, use a more specific term See Also: Audiovisual Equipment; Ergonomics; Higher Education; Personnel Management; Vocational Education
—Classification
BSI BS 1000: (37)-75. 1975 Universal Decimal Classification (UDC). English Full Edition (37): Education. Teaching. Training. Leisure. 42 pp.

SNZ NZS/BS 1000 (37)-75. Universal Decimal Classification Education. Teaching. Training. Leisure. 44 pp.
—Standardization Courses
ISO Guide 9-76. Courses in Standardization First Edition. 8 pp.

Educational Equipment
Scope Note: See the subheading Educational under specific equipment

Educters
Use: Eductors

Eductors
Use For: Educters See Also: Ejectors (Pumps)
—Ships
MOD UK NES 327-81. Pumps and Eductors Issue 1 (11.81). 78 pp.
—Submarines
MOD UK NES 327-81. Pumps and Eductors Issue 1 (11.81). 78 pp.

EEC
Use: European Communities

EEG
Use: Electroencephalographs

EEIG
Use: European Economic Interest Groupings

Eels
See Also: Fish
—Frozen—Roasted
CNS N5121-80. Frozen Roasted Eel (May)(3900). 1 p.

Effective Capacitance
See Also: Capacitance; Electrical Measurement; Electrical Properties
—Communication Cables
DIN VDE 0472 Pt 504-83. Testing of Cables, Wires and Flexible Cords; Effective Capacitance (Apr). 5 pp.

Effective Capacitance (Cont.)
—Cords (Electric)
DIN VDE 0472 Pt 504-83. Testing of Cables, Wires and Flexible Cords; Effective Capacitance (Apr). 5 pp.
—Telephone Cables
CCITT RECMN G.611-89. Characteristics of Symmetric Cable Pairs for Analogue Transmission—Transmission Media Characteristics (Study Group XV) 5 pp. 5 pp.

Effective Capacity
Use: Effective Capacitance

Effective Isotropic Radiated Power
Use For: EIRP See Also: Antennas; Isotropic Radiated Power
—Fixed Services—Feeder Links—Broadcasting Satellite Services
CCIR Report 1006-86. Fixed Service e.i.r.p. Limits for the Protection of the Broadcasting-Satellite Feeder Links Around 18 GHz—Section 4/9A—Sharing Conditions. 2 pp.
—Off Axis—Earth Stations—Fixed Satellite Services
CCIR RECMN 524-3-90. Maximum Permissible Levels of off-Axis e.i.r.p. Density from Earth Stations in the Fixed-Satellite Service Transmitting in the 6 and 14 GHz Frequency Bands—Section 4D1—Permissible Levels of Interference. 2 pp.

CCIR RECMN 524-4-92. Maximum Permissible Levels of Off-Axis e.i.r.p. Density from Earth Stations in the Fixed-Satellite Service Transmitting in the 6 and 14 GHz Frequency Bands—Section 4D1—Permissible Levels of Interference. 10 pp.
—Off Axis—Very Small Aperture Terminals—Earth Stations
CCIR RECMN 728-92. Maximum Permissible Level of Off-Axis e.i.r.p. Density from Very Small Aperture Terminals (VSATs)—Section 4B1—Systems Aspects. 6 pp.
—Radio Relay Systems—Digital Satellites—Frequency Band Sharing
CCIR Report 790-1-82. E.i.r.p. and Power Limits for Terrestrial Radio-Relay Systems Sharing with Digital Satellite Systems in Bands Between 11 to 14 GHz and Around 30 GHz—Section 4/9A—Sharing Conditions. 9 pp.
—Radio Relay Systems—Fixed Satellites—Frequency Band Sharing
CCIR QUESTION 114/9-90. Maximum e.i.r.p. for Line-of-Sight Radio-Relay Transmitters Operating in Frequency Bands Shared with the Fixed-Satellite Service—Questions of Study Group 9 Fixed Service. 1 p.

Effervescent Bath Appliances
Use: Whirlpool Baths

Efficiency
Use For: Efficiency Testing See Also: Efficiency Factor; Qualification Testing; Testing
—Geostationary Satellite Orbit—Communications Satellites—Sharing
CCIR Report 453-5-90. Technical Factors Influencing the Efficiency of Use of the Geostationary-Satellite Orbit by Radiocommunication Satellites Sharing the Same Frequency Bands General Summary—Section 4D2—Coordination Methods. 36 pp.

CCIR QUESTION 48/4-90. Technical Factors Influencing the Efficiency of Use of the Geostationary-Satellite Orbit by Radiocommunication Sat. Networks Sharing Frequency Bands Allocated to the Fixed-Sat. Service—Questions of Study Group 4 Fixed-Satellite Service. 2 pp.
—Radio Spectra Utilization
CCIR Report 662-3-90. Definition of Spectrum Use and Efficiency—Section 1B—Spectrum Sharing and Planning Principles and Techniques. 24 pp.
—Radiotelephone Circuits
CCIR QUESTION 146/9-90. Improvements in the Performance and Efficiency of HF Radiotelephone Circuits—Questions of Study Group 9 Fixed Service. 1 p.
—Rotating Machines
IEC 34 Pt 2-72. Rotating Electrical Machines Part 2: Methods for Determining Losses and Efficiency of Rotating Electrical Machinery from Tests (Excluding Machines for Traction Vehicles) Measurement of Losses by the Calorimetric Method First Edition; (Supplement A-1974). 95 pp.

INTERNATIONAL AND NON-U.S. NATIONAL STANDARDS
SUBJECT INDEX

Efficiency *(Cont.)*
—**Rotating Machines** *(Cont.)*
SAA AS 1359.33-83. Rotating Electrical Machines—General Requirements—Part 33: Methods for Determining Losses and Efficiency. 29 pp.

Efficiency Factor
See Also: Efficiency; Telegraphy
—**Radiotelegraph Circuits**
CCIR QUESTION 144/9-90. Efficiency Factor and Telegraph Distortion on ARQ Circuits—Questions of Study Group 9 Fixed Service. 1 p.
—**Switching Circuits—Telex Communications**
CCIR Report 437-70. Operational Use of the Efficiency Factor—Section 3Cd—Performance of Digital Transmission Systems. 1 p.

Efficiency Testing
Use: Efficiency

Effluent Monitors
Use: Radiation Monitors—Effluents

Effluents
Scope Note: For additional listings, use a more specific term *See Also:* Contamination; Exhaust Gases; Filtration; Industrial Wastes; Radioactive Effluents; Sewage Treatment; Waste Disposal
—**Biochemical Oxygen Demand**
CPPA H.2-91. Determination of Biochemical Oxygen Demand. 5 pp.
—**Coal Preparation Plants**
JIS M 0201-74. Testing Method for Effluents from Coal Preparation Plant.
—**Fire—Corrosion**
IEC 695 Pt 5-1-93. Fire Hazard Testing Part 5: Assessment of Potential Corrosion Damage by Fire Effluent—Section 1: General Guidance. 25 pp.
—**Fire—Gas Analysis**
ISO TR9122 Pt 3-93. Toxicity Testing of Fire Effluents—Part 3: Methods for the Analysis of Gases and Vapours in Fire Effluents First Edition. 41 pp.
—**Fire—Toxicity**
BSI PD 6503: Part 1-90. 1990 Toxicity of Combustion Products Part 1: General (ISO/TR 9122-1: 1989). 25 pp.
ISO TR9122 Pt 1-89. Toxicity Testing of Fire Effluents—Part 1: General First Edition; (Corrected and Reprinted -1989). 23 pp.
ISO TR9122 Pt 2-90. Toxicity Testing of Fire Effluents—Part 2: Guidelines for Biological Assays to Determine the Acute Inhalation Toxicity of Fire Effluents (Basic Principles, Criteria and Methodology) First Edition. 18 pp.
ISO TR9122 Pt 3-93. Toxicity Testing of Fire Effluents—Part 3: Methods for the Analysis of Gases and Vapours in Fire Effluents First Edition. 41 pp.
ISO TR9122 Pt 5-93. Toxicity Testing of Fire Effluents—Part 5: Prediction of Toxic Effects of Fire Effluents First Edition. 21 pp.
—**Fire—Toxicity—Models**
ISO TR9122 Pt 4-93. Toxicity Testing of Fire Effluents—Part 4: the Fire Model (Furnaces and Combustion Apparatus Used in Small-Scale Testing) First Edition. 21 pp.
ISO TR9122 Pt 5-93. Toxicity Testing of Fire Effluents—Part 5: Prediction of Toxic Effects of Fire Effluents First Edition. 21 pp.
—**Fire—Toxicity—Ships**
MOD UK NES 713-85. Determination of the Toxicity Index of the Products of Combustion from Small Specimens of Materials Issue 3 (03.85). 16 pp.
—**Fire—Vapors**
ISO TR9122 Pt 3-93. Toxicity Testing of Fire Effluents—Part 3: Methods for the Analysis of Gases and Vapours in Fire Effluents First Edition. 41 pp.
—**Mills (Industrial Plants)—Solids Content**
CPPA H.1-91. Determination of Solids Content of Pulp and Paper Mill Effluents. 3 pp.
—**Photographic Processing—Hexacyanoferrate Content—Spectrometry**
ISO 7766 Pt 1-93. Photography—Processing Wastes—Analysis of Cyanides—Part 1: Determination of Hexacyanoferrate(II) and Hexacyanoferrate(III) by Spectrometry First Edition. 9 pp.

Effluents *(Cont.)*
—**Photographic Processing—Silver Content—AAS**
ISO 10348-93. Photography—Processing Wastes—Determination of Silver Content First Edition. 20 pp.
—**Photographic Processing—Silver Content—Potentiometric Analysis**
ISO 10348-93. Photography—Processing Wastes—Determination of Silver Content First Edition. 20 pp.
—**Pulp Mills—Colorimetry**
CPPA H.5-91. Colour of Pulp Mill Effluents. 3 pp.
—**Septic Tanks—Waste Disposal**
SAA AS 1547-73. Code of Practice for Disposal of Effluent from Small Septic Tanks (Incorporating Amdt 1). 20 pp.
—**Sulfate Content—Gravimetric Analysis**
BSI BS 6068: Sec 2.39-91. 1991 Water Quality Part 2: Physical, Chemical and Biochemical Methods Section 2.39: Method for the Determination of Sulphate Using Barium Chloride and Gravimetry (ISO 9280: 1990) (SUPERSEDES BS 2690: PART 6: 1968). 9 pp.
ISO 9280-90. Water Quality—Determination of Sulfate —Gravimetric Method Using Barium Chloride First Edition. 8 pp.
—**Suspended Particles—Filtration**
CEN PREN 871-92. Water Quality—Determination of Suspended Solids—Determination of Coarse Particles and Fibres Seperated on a 60 to 70 um Sieve. 5 pp.
—**Suspended Solids—Filtration**
CEN PREN 872-92. Water Quality—Determination of Suspended Solids—Determination of Suspended Solids by Glass Fibre Filters. 8 pp.

EFS
Use: European Automatic Freephone Service

EFTA
Use For: European Free Trade Association
—**Cooperation—CEN**
EURO MEMORANDUM 4-85. General Guidelines. 41 pp.
—**Cooperation—Cenelec**
EURO MEMORANDUM 4-85. General Guidelines. 41 pp.

Egg Albumin
See Also: Serum Albumin
CNS K7584-81. Albumin from Eggs (Aug)(7820).

Egg Products
Use: Eggs

Eggplants
See Also: Food; Vegetables
—**Grading**
CNS N1029-82. Grades and Packing of Eggplant (Jan)(2096). 3 pp.
—**Packaging**
CNS N1029-82. Grades and Packing of Eggplant (Jan)(2096). 3 pp.
—**Preserved**
CNS N5152-84. Brine-Cured Eggplant (Jul)(4840). 2 pp.

Eggs
Use For: Egg Products *See Also:* Dairy Products; Food; Poultry; Poultry Meat; Roe
CGSB 32.184M-88. Shell Eggs. 6 pp.
CNS N1033-90. Shell Chicken Eggs (Dec)(2100). 2 pp.
CNS N4013-91. Method of Test for Shell Eggs (Feb)(2101). 6 pp.
—**Albumins**
CNS K7584-81. Albumin from Eggs (Aug)(7820).
—**Alkalized**
CNS N5045-86. Alkalized Eggs (Thousand Year Eggs, Pidan) (Sep)(2173). 2 pp.
—**Alkalized—Inspection**
CNS N6039-86. Method of Test for Alkalized Eggs (Thousand Year Eggs, Pidan) (Sep)(2174). 3 pp.
—**Alkalized—Sampling**
CNS N6039-86. Method of Test for Alkalized Eggs (Thousand Year Eggs, Pidan) (Sep)(2174). 3 pp.

Eggs *(Cont.)*
—**Chemical Analysis**
SAA AS 1383-74. Methods for the Chemical Examination of Eggs and Egg Products (R 1993). 55 pp.
—**Dried—Albumins**
CNS N5098-71. Egg White Powder (Jul)(3282). 2 pp.
—**Duck**
CNS N1123-90. Shell Duck Eggs (Dec)(12826).
—**Frozen**
CNS N5078-80. Frozen Eggs (Jul)(2812). 3 pp.
—**Frozen—Acidity**
CNS N6050-80. Method of Test for Frozen Eggs (Jul)(2923). 3 pp.
—**Frozen—Ash Content**
CNS N6050-80. Method of Test for Frozen Eggs (Jul)(2923). 3 pp.
—**Frozen—Fat Content**
CNS N6050-80. Method of Test for Frozen Eggs (Jul)(2923). 3 pp.
—**Frozen—Fatty Acid Content**
CNS N6050-80. Method of Test for Frozen Eggs (Jul)(2923). 3 pp.
—**Frozen—Inspection**
CNS N6050-80. Method of Test for Frozen Eggs (Jul)(2923). 3 pp.
—**Frozen—Microbiological Analysis**
CNS N6050-80. Method of Test for Frozen Eggs (Jul)(2923). 3 pp.
—**Frozen—Nitrogen Content**
CNS N6050-80. Method of Test for Frozen Eggs (Jul)(2923). 3 pp.
—**Frozen—Protein Content**
CNS N6050-80. Method of Test for Frozen Eggs (Jul)(2923). 3 pp.
—**Frozen—Sampling**
CNS N6050-80. Method of Test for Frozen Eggs (Jul)(2923). 3 pp.
—**Importation**
EC COM(89) 9-89. Proposal for a Council Regulation (EEC) on Animal Health Conditions Governing Intra-Community Trade in and Imports from Third Countries of Poultry and Hatching Eggs. 48 pp.
EC 90/539/EEC-90. Council Directive on Animal Health Conditions Governing Intra-Community Trade in and Imports from Third Countries of, Poultry and Hatching eggs. 23 pp.
—**Inspection**
EC COM(87) 46-87. Proposal for a Council Directive on health Problems Affecting the Production and the Placing on the Market of Egg Products. 32 pp.
EC COM(88) 646-88. Amendment to the Proposal for a Council Directive on Health Problems Affecting the Production and the Placing on the Market of Egg Products (Presented by the Commission Pursuant to Article 149(3) of the EEC Treaty). 9 pp.
EC 89/437/EEC-89. Council Directive on Hygiene and Health Problems Affecting the Production and the Placing on the Market of EPg products. 14 pp.
—**Medicinal Products—Residues**
EC COM(93) 287-93. Proposal for a Council Regulation (EEC). 12 pp.
EC EEC/2377/90-90. Council Regulation Laying Down a Community Procedure for the Establishment of Maximum Residue Limits of Veterinary Medicinal Products in Foodstuffs of Animal Origin. 5 pp.
—**Microbiological Analysis**
SAA AS 1766.3.8-91. Methods for the Microbiological Examination of Food—Part 3: Examination of Specific Products—Part 3.8: Eggs and Egg Products. 3 pp.
—**Pesticide Residues**
EC 86/363/EEC-86. Council Directive on the Fixing of Maximum Levels for Pesticide Residues in and on Foodstuffs of Animal Origin. 5 pp.
EC 93/57/EEC-93. Council Directive Amending the Annexes to Directives 86/362/EEC and 86/363/EEC on the Fixing of Maximum Levels for Pesticide Residues in and on Cereals and Foodstuffs of Animal Origin Respectively. 5 pp.
—**Salted**
CNS N5044-80. Salted Duck Eggs (Jul)(2171). 3 pp.
—**Salted—Inspection**
CNS N6038-80. Method of Test for Salted Eggs (Jul)(2172). 2 pp.

INDUSTRY STANDARDS

INTERNATIONAL AND NON-U.S. NATIONAL STANDARDS
SUBJECT INDEX

Eggs (Cont.)
—Salted—Sampling
CNS N6038-80. Method of Test for Salted Eggs (Jul)(2172). 2 pp.

—Sampling
SAA AS 1918-92. Methods for Sampling Eggs and Egg Products. 13 pp.

EIRP
Use: Effective Isotropic Radiated Power

Ejector Cartridges
Use: PAD Cartridges

Ejector Drifts
—Taper Shanks
DIN ENGL 317-80. Ejector Drifts for Taper Shanks with Flat Tang (July). 1 p.

Ejector Pins
See Also: Ejector Sleeves; Pins

—Dies
BSI BS 7564: Part 5-92. 1992 Tools for Moulding Part 5: Ejector Pins with Cylindrical Head—Dimensions (ISO 6751: 1986). 6 pp.
BSI BS 7564: Part 6-92. 1992 Tools for Moulding Part 6: Flat Ejector Pins (ISO 8693: 1987). 8 pp.
BSI BS 7564: Part 7-92. 1992 Tools for Moulding Part 7: Shouldered Ejector Pins (ISO 8694: 1987). 7 pp.
CNS B2782-87. Ejector Rod of Die Casting Molds (Sep)(12069).
CNS B3371-80. Ejector Pins of Dies for Die Casting (Aug)(5997).
ISO 6751-86. Ejector Pins with Cylindrical Head—Dimensions Second Edition. 4 pp.
ISO 8693-87. Tools for Moulding—Flat Ejector Pins First Edition. 5 pp.
ISO 8694-87. Tools for Moulding—Shouldered Ejector Pins First Edition. 4 pp.
JIS B 5103-89. Ejector Pins of Dies for Die Casting. 8 pp.

—Molds (Casting)
BSI BS 7564: Part 5-92. 1992 Tools for Moulding Part 5: Ejector Pins with Cylindrical Head—Dimensions (ISO 6751: 1986). 6 pp.
BSI BS 7564: Part 6-92. 1992 Tools for Moulding Part 6: Flat Ejector Pins (ISO 8693: 1987). 8 pp.
BSI BS 7564: Part 7-92. 1992 Tools for Moulding Part 7: Shouldered Ejector Pins (ISO 8694: 1987). 7 pp.
CNS B2773-87. Ejector Guide Pins of Molds for Plastics (May)(11933).
CNS B3386-80. Ejector Pins of Molds for Plastics (Oct)(6547).
ISO 6751-86. Ejector Pins with Cylindrical Head—Dimensions Second Edition. 4 pp.
ISO 8693-87. Tools for Moulding—Flat Ejector Pins First Edition. 5 pp.
ISO 8694-87. Tools for Moulding—Shouldered Ejector Pins First Edition. 4 pp.
JIS B 5108-89. Ejector Pins of Moulds for Plastics. 7 pp.
JIS B 5114-89. Shouldered Ejector Pins of Moulds for Plastics. 8 pp.
JIS B 5121-89. Flat Ejector Pins of Moulds for Plastics. 7 pp.

Ejector Rods
Use: Ejector Pins

Ejector Sleeves
See Also: Ejector Pins

—Dies
BSI BS 7564: Part 9-92. 1992 Tools for Moulding Part 9: Ejector Sleeves with Cylindrical Head—Basic Series for General Purposes (ISO 8405: 1986). 7 pp.
ISO 8405-86. Ejector Sleeves with Cylindrical Head—Basic Series for General Purposes First Edition. 4 pp.

—Molds (Casting)
CNS B2771-87. Ejector Sleeves of Molds for Plastics (May)(11931).
JIS B 5117-89. Ejector Sleeves of Moulds for Plastics. 7 pp.

Ejectors (Pumps)
See Also: Eductors; Pumps

—Ships
MOD UK NES 327-81. Pumps and Eductors Issue 1 (11.81). 78 pp.

—Submarines
MOD UK NES 327-81. Pumps and Eductors Issue 1 (11.81). 78 pp.

EKG Monitors
Use: Electrocardiograph Monitors

EKG Recorders
Use: Electrocardiograph Recorders

Elapsed Time Indicators
See Also: Time Measuring Instruments
MOD UK DSTAN 59-61: 90/188-82. Semiconductor Device, Elapsed Time Indicator Issue 1. 16 pp.

Elastic Fabrics
Use For: Stretch Fabrics *See Also:* Fabrics; Spandex Fibers; Stretch Yarns
CNS L3103-89. Method of Test for Elastic Articles (Dec) (6842). 3 pp.
CNS L4119-89. Elastic Articles (Dec)(6841).

—Dimensional Stability
CGSB CAN/CGSB-4.2 NO.56.1-M87. Textile Test Methods Unidirectional Extension and Recovery Properties of Elastic Fabrics. 12 pp.

—Elastic Properties
CNS L3139-81. Testing Method for Stretch Properties of Stretch Woven Fabrics (Oct)(8039).

—Test Specimens
BSI BS 4952-92. 1992 Methods of Test for Elastic Fabrics. 13 pp.
BSI BS 4952-73. 1973 Methods of Test for Elastic Fabrics. 14 pp.

Elastic Modulus
Use: Modulus of Elasticity

Elastic Properties
Use For: Elasticity *See Also:* Modulus of Elasticity; Shear Modulus; Tensile Testing; Viscoelasticity

—Elastic Fabrics
CNS L3139-81. Testing Method for Stretch Properties of Stretch Woven Fabrics (Oct)(8039).

—Elastomers
DIN ENGL 53517-87. Testing of Rubber and Elastomers; Determination of Compression Set After Constant Strain (Apr) (Supersedes January 1972 Editions of DIN 53517 Parts 1 and 2). 4 pp.
DIN ENGL 53518-79. Testing of Rubber and Elastomers; Determination of Tension Set After Constant Deformation (Mar). 7 pp.

—Fabrics
BSI BS 4294-68. (WITHDRAWN) 1968 Methods of Test for the Stretch and Recovery Properties of Fabrics (Superseded by BS 4952: 1992). 8 pp.

—Fibers
BSI BS 4029-78. 1978 Determination of Tensile Elastic Recovery of Single Fibres and Filaments (Constant-Rate-Of-Extension Machines). 7 pp.
DIN ENGL 53835 Pt 2-81. Testing of Textiles; Determination of the Elastic Behaviour of Single and Plied Elastomeric Yarns by Repeated Application of Tensile Load Between Constant Extension Limits (Aug). 4 pp.

—Fish Paste—Frozen
CNS N6107-82. Method of Test for Frozen Minced Fish Meat (Nov)(4641). 2 pp.

—Joint Sealants
BSI BS EN 27389-91. 1991 Building Construction—Jointing Products—Determination of Elastic Recovery (ISO 7389: 1987). 10 pp.
CEN EN 27 389-90. Building Construction—Jointing Products—Determination of Elastic Recovery. 9 pp.
DIN ENGL EN 27389-91. Jointing Products in Building Construction; Determination of Elastic Recovery (ISO 7389:1987) (May) (Supersedes DIN 52458, April 1987 Edition). 6 pp.
ISO 7389-87. Building Construction—Jointing Products—Determination of Elastic Recovery Second Edition. 6 pp.

—Polyurethane Resins—Athletic Fields
CNS K6591-86. Method of Test for Polyurethane (PU) Athletic Installation Material (May)(6483). 6 pp.

—Rubber
DIN ENGL 53517-87. Testing of Rubber and Elastomers; Determination of Compression Set After Constant Strain (Apr) (Supersedes January 1972 Editions of DIN 53517 Parts 1 and 2). 4 pp.
DIN ENGL 53518-79. Testing of Rubber and Elastomers; Determination of Tension Set After Constant Deformation (Mar). 7 pp.

—Skis
ISO 5902-80. Alpine Skis—Determination of the Elastic Properties First Edition. 8 pp.

Elastic Properties (Cont.)
—Varnishes—Communication Cables
DIN VDE 0472 Pt 624-83. Testing of Cables, Wires and Flexible Cords; Testing of the Elasticity of the Varnished Coating (Jan). 4 pp.

—Vulcanized Rubber
BSI BS 903: Part A5-93. 1993 Physical Testing of Rubber Part A5: Method for Determination of Tension Set at Normal and High Temperatures (ISO 2285: 1988) (V). 10 pp.
BSI BS 903: Part A5-74. 1974 Amd 1 Methods of Testing Vulcanized Rubber Part A5: Determination of Tension Set. 11 pp.
ISO 2285-88. Rubber, Vulcanized or Thermoplastic—Determination of Tension Set at Normal and High Temperatures Third Edition. 7 pp.
SAA AS 1683.9-92. Methods of Test for Elastomers—Part 9: Rubber, Vulcanized or Thermoplastic—Determination of Tension Set at Normal and High Temperatures (ISO 2285:1988). 4 pp.

—Yarns
DIN ENGL 53835 Pt 3-81. Testing of Textiles; Determination of the Elastic Behaviour of Single and Plied Yarns by a Single Application of Tensile Load Between Constant Extension Limits (Aug). 7 pp.
DIN ENGL 53835 Pt 4-81. Testing of Textiles; Determination of the Elastic Behaviour of Single and Plied Yarns by a Single Application of Tensile Load Between Constant Force Limits (Aug). 6 pp.
DIN ENGL 53835 Pt 13-83. Testing of Textiles; Determination of the Elastic Behaviour of Textile Fabrics by a Single Application of Tensile Load Between Constant Extension Limits (Nov). 6 pp.

Elastic Waves
Use For: Acoustic Vibration *See Also:* Noise; Vibration

—Aircraft—Environmental Testing
BSI 3G 100:Pt 2: SUBSEC 3.14-73. 1973 General Requirements for Equipment for Use in Aircraft Part 2: All Equipment Section 3: Environmental Conditions Subsection 3.14: Acoustic Vibration. 10 pp.
ISO 2671-82. Environmental Tests for Aircraft Equipment—Part 3.4: Acoustic Vibration First Edition. 11 pp.

Elasticity
Use: Elastic Properties

Elastomer Coatings
See Also: Coatings

—Drying
CGSB 1-GP-71 METH 5.12-74. Methods of Testing Paints and Pigments Drying Time Drying Time Over Elastomers. 1 p.

Elastomer Sealants
See Also: Polysulfide Sealants; Polyurethane Sealants
CGSB CAN/CGSB-19.13-M87. Sealing Compound, One-Component, Elastomeric, Chemical Curing. 12 pp.

—Appliances—Gas—Safety
DIN ENGL 3535 Pt 2-83. Sealants for Gas Supply; Homogeneous Elastomers for Sealants for Domestic Appliances; Safety Requirements, Testing (Apr). 5 pp.

—Dairy Equipment
DIN ENGL 11483 Pt 2-84. Dairy Installations; Cleaning and Disinfection; Consideration of the Action on Sealing Materials (Feb). 2 pp.

—Drainpipes
DIN ENGL 4060-88. Elastomer Seals for Pipe Joints in Drains and Sewers; Requirements and Testing (Dec). 10 pp.

—Gas Pipelines
DIN ENGL 3535 Pt 3-86. Sealants for Gas Supply; Elastomer Sealants for Gas Supply Mains and Pipelines; Requirements and Testing (Apr). 5 pp.

—Plastic Pipes—PVC—Unplasticized
SAA AS 1462.17-88. Methods of Test for Unplasticized PVC (UPVC) Pipes and Fittings—Part 17: Method for Testing Pressure Pipe Joints with Elastomeric Seals. 4 pp.

—Sewer Pipes
DIN ENGL 4060-88. Elastomer Seals for Pipe Joints in Drains and Sewers; Requirements and Testing (Dec). 10 pp.

Elastomeric Adhesives
See Also: Adhesives; Elastomers

INTERNATIONAL AND NON-U.S. NATIONAL STANDARDS
SUBJECT INDEX
Elastomers

Elastomeric Adhesives (Cont.)
—**Waterproof—Quarry Tiles**
CGSB 71-GP-29M-79. Adhesive, Elastomeric, for Installation of Quarry Tiles, Standard for. 16 pp.

Elastomeric Films
Use: Elastomeric Sheets

Elastomeric Sheets
Use For: Elastomeric Films See Also: Elastomers; Sheet Rubber

—**Vapor Transmission—Electrolysis**
DIN ENGL 53122 Pt 2-82. Determination of Water Vapour Transmission (Density of Moisture Flow Rate) of Plastic Films, Elastomer Films, Paper, Board and Other Sheet Materials; Electrolysis Method (July). 4 pp.

—**Vapor Transmission—Gravimetric Analysis**
DIN ENGL 53122 Pt 1-74. Testing of Plastics Films, Elastomer Films, Paper, Board and Other Sheet Materials; Determination of Water Vapour Transmission Rate; Gravimetric Method (Nov). 4 pp.

Elastomeric Yarns
Use: Stretch Yarns

Elastomers
Scope Note: For additional listings, see also specific products made from elastomeric materials
See Also: Chlorobutyl Rubber; Elastomeric Adhesives; Elastomeric Sheets; Latex; Plastics; Polyamide Resins; Polymers; Polysulfide Polymers; Polyurethane Rubber; Rubber; Silicone Resins; Silicone Rubber; Synthetic Rubber; Thermoplastic Elastomers
MOD UK DSTAN 93-24-85. Method of Testing Elastomeric Materials Issue 2. 6 pp.
SAA AS 1683. Method of Test for Elastomers Complete Set in Binder.

—**Abrasion Testing**
DIN ENGL 53516-87. Testing of Rubber and Elastomers; Determination of Abrasion Resistance (June). 6 pp.

—**Acrylonitrile Content**
DIN ENGL 53538 Pt 2-81. Standard Reference Elastomers; Acrylonitrile-Butadiene Rubber (NBR) with Low Acrylonitrile Content for Characterizing Test Liquids and Greases Based on Mineral Oil (Sept). 3 pp.

—**Acrylonitrile Content—Kjeldahl Method**
DIN ENGL 53621 Pt 3-76. Testing of Rubber and Elastomers; Quantitative Determination of Polymers; Determination of the Acrylonitrile Content and Calculation of the Acrylonitrile-Butadiene Copolymer Content (July). 3 pp.

—**Adhesion Testing**
DIN ENGL 53539-79. Testing of Elastomers; Evaluation of Tear Propagation, Adhesion and Peel Tests (Sept). 2 pp.

—**Aging Testing**
DIN ENGL 53508-77. Testing of Elastomers; Accelerated Ageing (July). 4 pp.
DIN ENGL 53509 Pt 2-77. Testing of Rubber and Elastomers; Accelerated Test of Ageing in Elastomers by Exposure to Ozone; Determination of Ozone Concentration (Reference Method) (Mar). 3 pp.

—**Antidegradant Content—Thin Layer Chromatography**
DIN ENGL 53622 Pt 2-76. Testing of Rubber and Elastomers; Thin-Layer Chromatographic Analysis; Identification of Antidegradants in Rubber and Elastomers (June). 4 pp.

—**Antioxidant Content—Thin Layer Chromatography**
DIN ENGL 53622 Pt 2-76. Testing of Rubber and Elastomers; Thin-Layer Chromatographic Analysis; Identification of Antidegradants in Rubber and Elastomers (June). 4 pp.

—**Automotive**
JIS K 6403-81. Classification System for Elastomeric Materials for Automotive Applications.

—**Butyl Rubber Content**
DIN ENGL 53621 Pt 8-77. Testing of Rubber and Elastomers; Quantitative Determination of Polymers; Determination of the Content of Butyl Rubber and of Polyisobutylene (July). 4 pp.

Elastomers (Cont.)
—**Carbon Black Content—Pyrolysis**
DIN ENGL 53585-73. Determination of the Carbon Black Content in Rubber and Elastomers (July). 2 pp.

—**Chemical Resistance**
DIN ENGL 53521-87. Determination of the Behaviour of Rubber and Elastomers When Exposed to Fluids and Vapours (Nov). 14 pp.

—**Chlorobutyl Rubber Content**
DIN ENGL 53621 Pt 5-76. Testing of Rubber and Elastomers; Quantitative Determination of Polymers; Determination of the Chlorobutyl Rubber Content (July). 2 pp.

—**Chloroprene Rubber Content**
DIN ENGL 53621 Pt 2-74. Testing of Rubber and Elastomers; Quantitative Determination of Polymers; Determination of the Chloroprene Rubber Content (Nov). 2 pp.

—**Copper Content—Atomic Absorption Spectroscopy**
DIN ENGL 53569 Pt 2-72. Testing of Rubber and Elastomers; Determination of Copper Content; Determination by Atomic Absorption Spectroscopy (Nov). 7 pp.

—**Copper Content—Photometry**
DIN ENGL 53569 Pt 1-72. Testing of Rubber and Elastomers; Determination of Copper Content; Photometric Method (Nov). 4 pp.

—**Cracking (Fracturing)—Flexure**
DIN ENGL 53522 Pt 2-79. Testing of Rubber and Elastomers; Flexing Endurance Test; Determination of Resistance to Flex Cracking (Jan). 2 pp.

—**Cut Growth—Flexure**
DIN ENGL 53522 Pt 3-79. Testing of Rubber and Elastomers; Flexing Endurance Test; Determination of Resistance to Cut Growth (Jan). 2 pp.

—**Dynamic Testing**
ISO 2856-81. Elastomers—General Requirements for Dynamic Testing Second Edition. 17 pp.

—**Elastic Properties—Compression Testing**
DIN ENGL 53517-87. Testing of Rubber and Elastomers; Determination of Compression Set After Constant Strain (Apr) (Supersedes January 1972 Editions of DIN 53517 Parts 1 and 2). 4 pp.

—**Elastic Properties—Tensile Testing**
DIN ENGL 53518-79. Testing of Rubber and Elastomers; Determination of Tension Set After Constant Deformation (Mar). 7 pp.

—**Environmental Testing**
DIN ENGL 53387-89. Artificial Weathering and Ageing of Plastics and Elastomers by Exposure to Filtered Xenon Arc Radiation (Apr) (Supersedes June 1982 Edition and the April 1974 Edition of DIN 53389, Withdrawn in 1986). 8 pp.

—**Epichlorhydrin Content**
DIN ENGL 53621 Pt 6-76. Testing of Rubber and Elastomers; Quantitative Determination of Polymers; Determination of Polyepichlorhydrin Content and Epichlorhydrin-Ethylene Oxide Copolymer Content (May). 2 pp.

—**Flexure**
DIN ENGL 53522 Pt 1-79. Testing of Rubber and Elastomers; Flexing Endurance Test; Definitions, Apparatus, Preparation of Test Pieces (Jan). 2 pp.

—**Gas Permeability**
DIN ENGL 53536-85. Testing of Rubber and Elastomers; Determination of Permeability to Gases (Feb). 5 pp.

—**Lead Content—Atomic Absorption Spectroscopy**
DIN ENGL 53599 Pt 1-78. Testing of Rubber and Elastomers; Determination of the Lead Content; Determination by Atomic Absorption Spectroscopy for Lead Contents up to 1000 mg/kg (0.1%) (May). 4 pp.

—**Lead Content—Electrogravimetric Analysis**
DIN ENGL 53599 Pt 2-78. Testing of Rubber and Elastomers; Determination of the Lead Content; Electrogravimetric and Gravimetric Method for Lead Contents over 1000 mg/kg (0.1%) (May). 3 pp.

—**Lead Content—Gravimetric Analysis**
DIN ENGL 53599 Pt 2-78. Testing of Rubber and Elastomers; Determination of the Lead Content; Electrogravimetric and Gravimetric Method for Lead Contents over 1000 mg/kg (0.1%) (May). 3 pp.

Elastomers (Cont.)
—**Lead Content—Photometry**
DIN ENGL 53599 Pt 3-78. Testing of Rubber and Elastomers; Determination of the Lead Content; Photometric Method for Lead Contents up to 1000 mg/kg (0.1%) (May). 4 pp.

—**Light Testing**
DIN ENGL 53387-89. Artificial Weathering and Ageing of Plastics and Elastomers by Exposure to Filtered Xenon Arc Radiation (Apr) (Supersedes June 1982 Edition and the April 1974 Edition of DIN 53389, Withdrawn in 1986). 8 pp.

—**Manganese Content—Photometry**
DIN ENGL 53589 Pt 1-72. Testing of Rubber and Elastomers; Determination of Manganese Content; Photometric Method (Nov). 4 pp.

—**Oil Content**
DIN ENGL 53588-75. Testing of Rubber and Elastomers; Determination of Factice in Elastomers (Aug). 2 pp.

—**Peel Strength**
DIN ENGL 53539-79. Testing of Elastomers; Evaluation of Tear Propagation, Adhesion and Peel Tests (Sept). 2 pp.

—**Polyisobutylene Content**
DIN ENGL 53621 Pt 8-77. Testing of Rubber and Elastomers; Quantitative Determination of Polymers; Determination of the Content of Butyl Rubber and of Polyisobutylene (July). 4 pp.

—**Residue on Ignition Determination**
DIN ENGL 53568 Pt 1-74. Testing of Plastics, Rubber and Elastomers; Determination of Residue on Ignition Without Chemical Pretreatment of the Specimen (July). 2 pp.
DIN ENGL 53568 Pt 2-78. Testing of Rubber, Elastomers and Plastics; Determination of Residue on Ignition After Pretreatment of the Specimen with Acid (Sulphate Ash) (May). 2 pp.

—**Resilience**
DIN ENGL 53512-88. Determination of Rebound Resilience of Rubber (Dec). 4 pp.

—**Shore Hardness Testing**
DIN ENGL 53505-87. Testing of Rubber, Elastomers and Plastics; Shore Hardness Testing A and D (June). 4 pp.

—**Solar Water Heaters**
BSI BS 7431-91. 1991 Assessing Solar Water Heaters—Elastomeric Materials for Absorbers, Connecting Pipes and Fittings (ISO 9808: 1990). 8 pp.
ISO 9808-90. Solar Water Heaters—Elastomeric Materials for Absorbers, Connecting Pipes and Fittings—Method of Assessment First Edition. 6 pp.

—**Stabilizer Content—Thin Layer Chromatography**
DIN ENGL 53622 Pt 2-76. Testing of Rubber and Elastomers; Thin-Layer Chromatographic Analysis; Identification of Antidegradants in Rubber and Elastomers (June). 4 pp.

—**Styrene Content**
DIN ENGL 53621 Pt 7-77. Testing of Rubber and Elastomers; Quantitative Determination of Polymers; Determination of the Styrene Content and Calculation of the Content of Styrene-Butadiene Copolymer (July). 4 pp.

—**Sulfur Content**
DIN ENGL 53561-79. Testing of Rubber and Elastomers; Determination of Total Sulphur Content (Jan). 5 pp.

—**Tear Strength**
DIN ENGL 53507-83. Testing Rubber and Elastomers; Determination of the Tear Strength of Elastomers; Trouser Test Piece (Mar). 7 pp.
DIN ENGL 53515-90. Determination of Tear Strength of Rubber, Elastomers and Plastic Film Using Graves Angle Test Piece with Nick (Jan). 3 pp.
DIN ENGL 53539-79. Testing of Elastomers; Evaluation of Tear Propagation, Adhesion and Peel Tests (Sept). 2 pp.

—**Test Specimens**
DIN ENGL 53502-72. Testing of Elastomers and of Fabrics Coated with Elastomers; Test Specimens; Directions for Preparing (Aug). 3 pp.

—**Thin Layer Chromatography**
DIN ENGL 53622 Pt 1-75. Testing of Rubber and Elastomers; Thin Layer Chromatographic Analysis; Concepts and General Principles (Sept). 3 pp.

Elastomers

Elastomers (Cont.)

—Vinyl Acetate Content
DIN ENGL 53621 Pt 4-76. Testing of Rubber and Elastomers; Quantitative Determination of Polymers; Determination of the Vinyl Acetate Content and Calculation of the Ethylene-Vinyl Acetate Copolymer Content (Dec). 4 pp.

—Viscoelasticity—Vibration
DIN ENGL 53513-90. Determination of Viscoelastic Properties of Elastomers on Exposure to Forced Vibration at Non-Resonant Frequencies (Mar). 8 pp.

—Vulcanized—Storage
SAA AS CK15-69. Code of Recommended Practice for the Storage of Vulcanized Eoastomers. 7 pp.

—Vulcanizing Agent Content—Thin Layer Chromatography
DIN ENGL 53622 Pt 3-82. Testing of Rubber and Elastomers; Thin Layer Chromatographic Analysis; Detection of Vulcanizing Agents (June). 4 pp.

—Zinc Content—Complexometric Titrations
DIN ENGL 53581-74. Testing of Rubber and Elastomers; Determination of the Zinc Content (Mar). 4 pp.

—Zinc Content—Polarographic Analysis
DIN ENGL 53581-74. Testing of Rubber and Elastomers; Determination of the Zinc Content (Mar). 4 pp.

Elbows (Fittings)
Scope Note: See the subheading Elbow under specific equipment. For addition l listings, consult the following list *See Also:* Pipe Elbows

Elderly Persons
Use: Aging Persons

Electric Actuators
Use: Actuators

Electric Adapters
Use: Adapters (Electric)

Electric Appliances
Use: Appliances

Electric Arc Furnaces
Use For: Arc Furnaces *See Also:* Direct Arc Furnaces; Electroslag Remelting Furnaces; Furnaces; Submerged Arc Furnaces

—Firebricks
BSI BS 3056: Part 4-85. 1985 Sizes of Refractory Bricks Part 4: Bricks for Electric Arc Furnace Roofs. 7 pp.
BSI BS 3056: Part 7-86. 1986 Sizes of Refractory Bricks Part 7: Basic Bricks for Steel Making. 4 pp.
DIN ENGL 1082 Pt 3-88. Refractory Materials; Basic Arch Bricks and Refractory Bricks for Oxygen Steel-Making Converters and Arc Furnaces; Dimensions (Jan). 3 pp.
ISO 5019 Pt 4-88. Refractory Bricks—Dimensions—Part 4: Dome Bricks for Electric Arc Furnace Roofs Second Edition. 8 pp.

—Graphite Electrodes
CENELEC HD 564 S1-90. Nominal Dimensions of Cylindrical Machined Graphite Electrodes with Threaded Sockets and Nipples for Use in Electric Arc Furnaces. 10 pp.
IEC 239-87. Nominal Dimensions of Cylindrical Machined Graphite Electrodes with Threaded Sockets and Nipples for Use in Electric Arc Furnaces Second Edition. 22 pp.

—Heat Balance
JIS G 0703-77. Heat Balance of Arc Furnace.

—Safety
IEC 519 Pt 4-77. Safety in Electro-Heat Installations Part 4: Particular Requirements for Arc Furnace Installations First Edition. 18 pp.

Electric Arc Welding
Scope Note: See listing under Arc Welding

Electric Ballasts
Use: Ballasts (Electric)

Electric Batteries
Use: Batteries

Electric Bells
Use: Bells

Electric Blankets
See Also: Blankets; Heating Blankets; Heating Equipment

Electric Blankets (Cont.)
BSI BS 3999: Part 6-69. 1969 Amd 2 Methods of Measuring the Performance of Household Electrical Appliances Part 6: Electrically-Heated Blankets. 24 pp.
BSI BS 5129-82. 1982 Dimensions of Bed Blankets. 11 pp.
IEC 299-69. Measurement of the Performance Characteristics of Electric Blankets First Edition. 19 pp.
JIS C 9210-92. Electrically Heated Blankets. 37 pp.

—Flammability Testing
CSA C22.2 NO 101-M1984. Electrically Heated Bedding Appliances for Household Use (R 1992); (Gen Instr 1 Thru 4). 61 pp.
CSA 1169 Bull. Electrical Bulletin 1169 June 27, 1978 to C22.2 NO 101. 2 pp.

—Safety
BSI BS 3456: Sec A4-71. (WITHDRAWN) 1971 Amd 9 Safety of Household Electrical Appliances Part A: Heating and Cooking Appliances Section A4: Electrically Heated Blankets (Withdrawn, superseded by BS En 60967: 1991). 67 pp.
BSI BS EN 60967-91. 1991 Safety of Electrically Heated Blankets, Pads and Similar Flexible Heating Appliances for Household Use. 123 pp.
CENELEC EN 60 967-90. Safety of Electrically Heated Blankets, Pads and Similar Flexible Heating Appliances for Household Use. 13 pp.
CENELEC EN 60967-92. Safety of Electrically Heated Blankets, Pads and Similar Flexible Heating Appliances for Household Use. 3 pp.
CSA C22.2 NO 101-M1984. Electrically Heated Bedding Appliances for Household Use (R 1992); (Gen Instr 1 Thru 4). 61 pp.
CSA 1169 Bull. Electrical Bulletin 1169 June 27, 1978 to C22.2 NO 101. 2 pp.
IEC 335 Pt 17-74. Safety of Household and Similar Electrical Appliances Particular Requirements for Electrically Heated Blankets, Pads and Mattresses, Recm. First Edition. 95 pp.
IEC 967-88. Safety of Electrically Heated Blankets, Pads and Similar Flexible Heating Appliances for Household Use First Edition; (Amendment 1-1991). 260 pp.
SNZ NZCP 28-62. Code of Recommended Practice for the Repair of Electrically Heated Blankets for Domestic Use. 15 pp.
SNZ NZS 6317-88. Specification for Safety of Electric Blankets. 60 pp.

Electric Boilers
See Also: Boilers
CSA CAN/CSA-C22. 2 NO 165-92. Electric Boilers; (Gen Instr 1). 61 pp.

—Design
BSI BS 1894-92. 1992 Design and Manufacture of Electric Boilers of Welded Construction (Q). 97 pp.
BSI BS 1894-52. 1952 Electrode Boilers of Riveted Seamless, Welded and Cast Iron Construction for Water Heating and Steam Generating. 100 pp.

Electric Brakes
Use: Brakes

Electric Breakdown
Use: Dielectric Breakdown

Electric Cable Connectors
Use: Connectors

Electric Cables
Use: Cables (Electric)

Electric Chart Recorders
Use: Chart Recorders

Electric Circuits
Use: Circuits

Electric Clocks
See Also: Clocks

—Analog
CSA C22.2 NO 6-M1984. Electric Clocks; (Gen Instr 1). 35 pp.
CSA 1169 Bull. Electrical Bulletin 1169 June 27, 1978 to C22.2 NO 6. 2 pp.

—Digital
CSA C22.2 NO 6-M1984. Electric Clocks; (Gen Instr 1). 35 pp.
CSA 1169 Bull. Electrical Bulletin 1169 June 27, 1978 to C22.2 NO 6. 2 pp.

—Motors
CSA C22.2 NO 6-M1984. Electric Clocks; (Gen Instr 1). 35 pp.
CSA 1169 Bull. Electrical Bulletin 1169 June 27, 1978 to C22.2 NO 6. 2 pp.

Electric Clocks (Cont.)

—Power Transformers
CSA C22.2 NO 6-M1984. Electric Clocks; (Gen Instr 1). 35 pp.
CSA 1169 Bull. Electrical Bulletin 1169 June 27, 1978 to C22.2 NO 6. 2 pp.

—Quartz
CNS C6167-83. Method of Test for Electronic Clocks (Quartz) (Nov)(8942).

—Transformers—Electrical Insulation
CSA C22.2 NO 6-M1984. Electric Clocks; (Gen Instr 1). 35 pp.
CSA 1169 Bull. Electrical Bulletin 1169 June 27, 1978 to C22.2 NO 6. 2 pp.

Electric Coils
Use For: Coils (Electric) *See Also:* Choke Coils; Electrical Components; Filter Inductors; Fixed Inductors; Ignition Coils; Inductors; Magnetic Cores; Oscillator Coils; Power Inductors; Radio Frequency Chokes; Resistors; Saturable Reactors; Solenoids; Transformers; Windings
CNS C7021-71. Peaking Coil (Jun)(3260).

—Electromagnetic Testing Symbols
DIN ENGL 54140 Pt 3-89. Non-Destructive Testing; Electromagnetic Methods; Representation and General Characteristics of Test Coil Assemblies (Jan). 6 pp.

—Insulating Papers
CNS C4212-85. Coil Insulating Paper (Nov)(6045).
JIS C 2304-84. Coil Insulating Papers. 7 pp.

—Insulating Varnishes
JIS C 2353-83. Coil Impregnating Varnishes for Electrical Insulation. 9 pp.

—Laminates
MOD UK DSTAN 59-55-01. Laminations, Magnetic Issue 1; Amendment 1. 68 pp.

—Lead Wire
CSA C22.2 NO 116-1980. Coil-Lead Wires (R 1992); (Gen Instr 1 Thru 8). 52 pp.
CSA 1227 Bull. Electrical Bulletin 1227 May 16, 1979 to C22.2 NO 116. 9 pp.
CSA 1236 Bull. Electrical Bulletin 1236 September 12, 1979 to C22.2 NO 116. 9 pp.
CSA 1333 Bull. Electrical Bulletin 1333 August 17, 1981 to C22.2 NO 116. 9 pp.
CSA 1383 Bull. Electrical Bulletin 1383 November 29, 1982 to C22.2 NO 116. 9 pp.
CSA 1409 Bull. Electrical Bulletin 1409 July 28, 1983 to C22.2 NO 116. 9 pp.

—Magnetic Cores—Electromagnetic Interference
JIS C 2569-90. Ring Cores Made of Ferromagnetic Oxide. 12 pp.

—Radio Frequency—Noise Suppressors—Preferred Products List
CECC MUAHAG Vol 4 IS 3-90. Preferred Products List; Magnetic Components and Ferrite Materials (En, Fr, Ge). 46 pp.

—Radio Frequency—Printed Circuit Mount—Preferred Products List
CECC CECC MUAHAG Vol 4 IS 3-90. Preferred Products List; Magnetic Components and Ferrite Materials (En, Fr, Ge). 46 pp.

—Radio Frequency—Surface Mount—Preferred Products List
CECC CECC MUAHAG Vol 4 IS 3-90. Preferred Products List; Magnetic Components and Ferrite Materials (En, Fr, Ge). 46 pp.

Electric Commuter Cars
Use: Commuter Cars—Electric

Electric Condensers
Use: Capacitors

Electric Conductors
Scope Note: For additional listings, use a more specific term *Use For:* Conductors (Electric); Insulated Conductors; Sheathed Conductors *See Also:* Alarm Cables; Aluminum Conductors; Aluminum Wire; Arc Welding Cables; Armored Cables; Bare Conductors; Battery Electrolytes; Bonding Wire; Braided Wire; Bus Conductors; Busbars; Cable Penetration; Cables (Electric); Coaxial Cables; Communication Cables; Connectors; Control Cables (Electric); Copper Conductors; Copper Wire; Cords (Electric); Dielectrics; Drop Wire; Electric Coils; Electric Conduits; Electric Raceways; Electric Wire; Electric Wiring; Electrical Components;

Electric Conductors (Cont.)

See Also: (Cont.)
Electrical Insulation; Electrical Leads; Electrically Conductive Liquids; Elevator Cables; Flat Cables; Gas Filled Cables; Grounding Conductors; Heater Cords; Heating Cables; Instrumentation Cables; Iron Wire; Marine Cables (Electric); Neutral Cables; Oil Filled Cables; Phase Conductors; Power Cables; Power Line Hardware; Power Lines; Protective Conductors; Resistors; Rigid Conductors; Semiconductor Devices; Semiconductors; Service Entrance Cables; Service Entrance Equipment; Shielded Conductor Cables; Steel Conductors; Steel Wire; Stranded Conductors; Superconductors; Tinsel Cords; Transmission Lines; Windings; Wire Wrap Wire

DIN VDE 0295-92. Conductors for Cables and Flexible Cords for Power Installations (June) (Supersedes DIN VDE 0295/05.86). 16 pp.
ISO 2635-79. Aircraft—Conductors for General Purpose Aircraft Electrical Cables and Aerospace Applications—Dimensions and Characteristics First Edition. 4 pp.
SAA AS 1125-93. Conductors in Insulated Electric Cables and Flexible Cords. 14 pp.
SAA AS 1660. Methods of Test for Electric Cables, Cords and Conductors Complete Set in Binder.
SAA AS 1660.1-93. Methods of Test for Electric Cables, Cords and Conductors—Part 1: Conductors and Metallic Components. 6 pp.

—**Aerospace**
BSI G 231-83. (WITHDRAWN) 1983 Conductors for General Purpose Aircraft Electrical Cables and Aerospace Applications (Superseded by 2G 231: 1990). 6 pp.

—**Aircraft**
BSI G 231-83. (WITHDRAWN) 1983 Conductors for General Purpose Aircraft Electrical Cables and Aerospace Applications (Superseded by 2G 231: 1990). 6 pp.

—**Aircraft—Tensile Testing**
CEN PREN 3475 (Part 505)-92. Cables, Electrical, Aircraft Use Test Methods Part 505—Tensile Test on Conductors and Strands. 2 pp.

—**Color Coding**
CENELEC HD 324-76. Identification of Insulated and Bare Conductors by Colours. 2 pp.
CNS C1051-84. General Rules of Color Identification for Protective Conductors and Neutral Conductor and Terminal Marking for Apparatus (Dec)(5202).
IEC 446-89. Identification of Conductors by Colours or Numerals Second Edition. 18 pp.
JIS C 0602-84. General Rules of Colour Identification for Protective Conductor and Neutral Conductor and Terminal Marking for Apparatus. 5 pp.

—**Conductivity Testing**
CNS C3008-80. Testing Standard for Conductivity of Insulated Conductors (Jul)(686). 1 p.

—**Connectors**
CSA 1122A Bull. Electrical Bulletin 1122A January 10, 1979 to C22.2 NO 65. 3 pp.
CSA C22.2 NO 188-M1983. Splicing Wire and Cable Connectors (R 1989); (Gen Instr 1 Thru 2). 38 pp.
CSA 1122A Bull. Electrical Bulletin 1122A January 10, 1979 to C22.2 NO 188. 3 pp.
CSA 1122B Bull. Electrical Bulletin 1122B September 30, 1985 to C22.2 NO 188. 1 p.

—**Cotton Covered—Winding—Rectangular**
BSI BS 1791: Part 2-57. 1957 Amd 1 Cotton-Covered Copper Conductors Part 2: Rectangular Conductors. 19 pp.

—**Cranes**
CNS C4311-81. Contact Conductor with Insulating Cover for Crane (Jun)(7598).

—**Electrical Installations**
SAA AS 3000 Supp 1-91. Electrical Installations—Buildings, Structures and Premises—Supplement 1: Cable and Conductor Tables (Imperial Units) (Supplement to AS 3000—1991). 13 pp.

—**Electrical Properties**
SAA AS 1660.3-93. Electrical Tests. 22 pp.

—**Electronic Components**
CSA C22.2 NO 16-M1986. Insulated Conductors for Power-Operated Electronic Devices (R 1992); (Gen Instr 1 Thru 4). 64 pp.

—**Electronic Equipment**
CSA C22.2 NO 16-M1986. Insulated Conductors for Power-Operated Electronic Devices (R 1992); (Gen Instr 1 Thru 4). 64 pp.

—**Identification Systems**
IEC 391-72. Marking of Insulated Conductors, Recm. First Edition. 27 pp.

Electric Conductors (Cont.)

—**KU Values**
DIN VDE 0800 Pt 9-89. Telecommunications KU Values of Safety-Related Components and Insulation (May). 36 pp.

—**Overhead Power Lines**
SAA AS 1154.1-85. Insulator and Conductor Fittings for Overhead Power Lines—Part 1: Performance and General Requirements. 20 pp.
SAA AS 1154.2-85. Insulator and Conductor Fittings for Overhead Power Lines—Part 2: Dimensions. 53 pp.
SAA AS 1154.3-85. Insulator and Conductor Fittings for Overhead Power Lines—Part 3: Performance and General Requirements for Helical Fittings. 11 pp.

—**Reels**
SAA AS 3983-91. Metal Drums for Insulated Electric Cables and Bare Conductors (Supersedes AS C365.2—1970). 10 pp.

—**Resistance Testing**
CNS C3007-80. Testing Standard for D.C. Resistance of Insulated Conductors (Jul)(685). 2 pp.

—**Soldering Lugs**
CSA C22.2 NO 19-1935. Construction and Test of Soldering Lugs (R 1981); (Gen Instr 1). 16 pp.

—**Splicing**
CSA 1122A Bull. Electrical Bulletin 1122A January 10, 1979 to C22.2 NO 65. 3 pp.
CSA C22.2 NO 188-M1983. Splicing Wire and Cable Connectors (R 1989); (Gen Instr 1 Thru 2). 38 pp.
CSA 1122A Bull. Electrical Bulletin 1122A January 10, 1979 to C22.2 NO 188. 3 pp.

—**Symbols**
CNS C5113-81. Graphic Symbols for Conductor, Cable and Wiring (May)(7362).
SAA AS 1102.103-89. Graphical Symbols for Electrotechnology—Part 103: Conductors and Connecting Devices. 10 pp.

—**Symbols—Diagrams**
SAA AS 1102.103-89. Graphical Symbols for Electrotechnology—Part 103: Conductors and Connecting Devices. 10 pp.

—**Television Cables**
CSA C22.2 NO 16-M1986. Insulated Conductors for Power-Operated Electronic Devices (R 1992); (Gen Instr 1 Thru 4). 64 pp.

—**Washers (Fasteners)**
DIN ENGL 46288-74. Connecting Washers for Conductors (Aug). 3 pp.

Electric Conduit Boxes

Scope Note: For additional listings, see specific items or equipment *Use For:* Conduit Boxes
See Also: Electric Conduits; Electric Outlet Boxes; Electrical Enclosures; Junction Boxes

CNS C3066-79. Method of Test for Boxes for Unplasticized Polyvinyl Chloride (UPVC) Conduit (Sep)(4971). 3 pp.
CNS C4152-79. Boxes for Unplasticized Polyvinyl Chloride (UPVC) Conduit (Sep)(4970). 6 pp.
CNS C4234-82. Circular Surface Boxes for Rigid Steel Conduit (Dec)(6090).
CNS C4257-90. Boxes for Unplasticized Polyvinyl Chloride (UPVC) Conduit (Apr)(6113).
CSA CAN/CSA-C22. 2 NO 18-92. Outlet Boxes, Conduit Boxes, and Fittings; (Gen Instr 1 Thru 2). 118 pp.
CSA 1208 Bull. Electrical Bulletin 1208 January 3, 1979 to C22.2 NO 85. 1 p.
DIN VDE 0606 Pt 1-84. Connecting Material up to 660 V Installation Boxes to Accommodate Equipment and/or Connecting Terminals (VDE Specification) (Nov) (Partially Supersedes DIN 57606/VDE 0606/02.76 and DIN 57606b/VDE 0606b/02.80). 55 pp.
DIN VDE 0606 Pt 1 A1 (D)-85. Connecting Material up to 660 V; Installation Boxes for Accommodation of Equipment and/or Connecting Terminals; Amendment 1 (Nov). 3 pp.
DIN VDE 0606 Pt 1 A2 (D)-86. Connecting Material up to 660 V; Installation Boxes for Accommodation of Equipment and/or Connecting Terminals; Amendment 2 (Nov). 5 pp.
DIN VDE 0606 Pt 1 A3 (D)-88. Connecting Material up to 660 V; Installation Boxes for Accommodation of Equipment and/or Connecting Terminals; Amendment 3 (Dec). 3 pp.
JIS C 8336-91. Boxes for Rigid Metal Conduits. 28 pp.
JIS C 8435-88. Unplasticized Polyvinyl Chloride (UPVC) Boxes and Box Covers. 24 pp.

—**Concrete**
CNS C4232-82. Concrete Boxes for Rigid Steel Conduit (Dec)(6088).

Electric Conduit Boxes (Cont.)

—**Covers**
CNS C4233-82. Box Covers for Rigid Steel Conduit (Dec)(6089).

—**Electrical Enclosures—Fittings**
CSA CAN/CSA-C22. 2 NO 85-M89. Rigid PVC Boxes and Fittings; (Gen Instr 1 Thru 2). 37 pp.

—**Junction Boxes—Fittings**
CSA CAN/CSA-C22. 2 NO 85-M89. Rigid PVC Boxes and Fittings; (Gen Instr 1 Thru 2). 37 pp.

—**Switchgear**
CNS C4231-82. Switch Boxes for Rigid Steel Conduit (Dec)(6087).
CNS C4235-82. Surface Switch Boxes for Rigid Steel Conduit (Dec)(6091).

Electric Conduit Fittings

Scope Note: Includes components such as couplings, bends, and saddles for connecting electric conduits
Use For: Conduit Fittings (Electric)
See Also: Busway Fittings; Electric Conduits; Flexible Metal Conduit Fittings; Flexible Nonmetallic Conduit Fittings; Liquidtight Flexible Conduit Fittings; Rigid Metal Conduit Fittings; Rigid Nonmetallic Conduit Fittings

CENELEC EN 500 86-1-91. Conduit Systems for Electrical Installations Part 1: General Requirements. 42 pp.
CSA CAN/CSA-C22. 2 NO 18-92. Outlet Boxes, Conduit Boxes, and Fittings; (Gen Instr 1 Thru 2). 118 pp.
DIN ENGL 49020-59. Conduit for Electrical Wiring; Screwed Steel Conduit; Plain Conduit; Couplers (Jan). 2 pp.
DIN VDE 0605-82. Conduits and Fittings for Electrical Installations (Apr). 18 pp.
IEC 1035 Pt 1-90. Specification for Conduit Fittings for Electrical Installations Part 1: General Requirements First Edition. 49 pp.
SAA AS 2052-77. Metallic Conduits and Fittings. 28 pp.
SAA AS 2053-84. Non-Metallic Conduits and Fittings. 28 pp.
SNZ NZS 127-49. Steel Conduit and Fittings for Electrical Wiring Amend: 2A, 1967. 44 pp.
SNZ NZS 159: Part 1-54. Flexible Steel Conduit for Cable Protection and Flexible Steel Tubing to Enclose Flexible Drives Part 1: Flexible Steel Conduit and Adaptors for the Protection of Electric Cable. 16 pp.

—**Aluminum—Couplings**
CNS C4251-80. Couplings for Aluminium Conduit (Aug)(6107).

—**Aluminum—Elbows**
CNS C4252-80. Elbows for Aluminium Conduit (Aug)(6108).

—**Capping—Aircraft**
SBAC AS 8671 ISSUE 3. Capping (Cable Duct) 1.28 Inches Wide, Fire Resistant.
SBAC AS 8672 ISSUE 3. Capping (Cable Duct) 1.09 Inches Wide, Fire Resistant.
SBAC AS 8673 ISSUE 3. Capping (Cable Duct) .88 Inches Wide, Fire Resistant.
SBAC AS 8674 ISSUE 3. Capping (Cable Duct) .72 Inches Wide, Fire Resistant.

—**Caps—Packing Pieces—Aircraft**
SBAC AS 4817-4820 ISSUE 2. Cap for Packing Piece. 1 p.

—**Construction**
CSA C22.2 NO 0.5-1982. Threaded Conduit Entries (R 1992); (Gen Instr 1). 22 pp.

—**Insulated**
IEC 1035 Pt 2-2-93. Specification for Conduit Fittings for Electrical Installations—Part 2: Particular Specifications—Section 2: Conduit Fittings of Insulating Material First Edition. 27 pp.
IEC 1035 Pt 2-3-93. Specification for Conduit Fittings for Electrical Installations Part 2: Particular Specifications Section 3: Fittings for Flexible Conduits of Metal, Insulating or Composite Materials and for Pliable Conduits of Metal or Composite Materials First Edition. 27 pp.

—**PVC—Caps**
CNS C4260-90. Caps for Unplasticized Polyvinyl Chloride (UPVC) Conduit (Apr)(6116).
JIS C 8438-77. Caps for Unplasticized Polyvinyl Chloride (UPVC) Conduit. 7 pp.

—**PVC—Connectors**
CNS C4255-90. Connectors for Unplasticized Polyvinyl Chloride (UPVC) Conduit (Apr)(6111).
JIS C 8433-77. Connectors for Unplasticized Polyvinyl Chloride (UPVC) Conduit. 7 pp.

INTERNATIONAL AND NON-U.S. NATIONAL STANDARDS
SUBJECT INDEX
Electric

Electric Conduit Fittings (Cont.)
—PVC—Couplings
  CNS C4254-90. Couplings for Unplasticized Polyvinyl Chloride (UPVC) Conduit (Apr)(6110).
  JIS C 8432-77. Couplings for Unplasticized Polyvinyl Chloride (UPVC) Conduit. 7 pp.
—PVC—Elbows
  CNS C4256-90. Elbows for Unplasticized Polyvinyl Chloride (UPVC) Conduit (Apr)(6112).
  JIS C 8434-77. Normal Bends for Unplasticized Polyvinyl Chloride (UPVC) Conduit. 6 pp.
—PVC—Saddles
  CNS C4259-90. Saddles for Unplasticized Polyvinyl Chloride (UPVC) Conduit (Apr)(6115).
  JIS C 8437-77. Saddles for Unplasticized Polyvinyl Chloride (UPVC) Conduit. 5 pp.
—Screw Threads
  BSI BS 6053-81. 1981 Outside Diameters of Conduits for Electrical Installations and Threads for Conduits and Fittings. 8 pp.
  CENELEC HD 393-79. Outside Diameters of Conduits for Conduits for Electrical Installations and Threads for Conduits and Fittings. 2 pp.
  IEC 423-73. Outside Diameters of Conduits for Electrical Installations and Threads for Conduits and Fittings First Edition; (Supplement A-1978). 27 pp.
—Spacers—Aircraft
  SBAC AS 4821 ISSUE 2. Spacer (Cable Ducts). 1 p.
—Spring Clips—Aircraft
  SBAC AS 4799 ISSUE 2. Spring Clip (Cable Duct). 1 p.
  SBAC AS 4822-4825 ISSUE 2. Spring Clip (Cable Duct). 1 p.
—Thermoplastic
  BSI DD 200-91. 1991 Pliable Non-Rigid Conduit and Fittings for Direct Burial Underground (Type DB). 13 pp.

Electric Conduits
Use For: Cable Ducts; Flexible Tubing (Electrical); Tubing (Electrical) See Also: Busways; Electric Conduit Boxes; Electric Conduit Fittings; Electric Raceways; Electric Wire; Electrical Flexible Tubing; Electrical Metallic Tubing; Electrical Nonmetallic Tubing; Flexible Metal Conduits; Flexible Nonmetallic Conduits; Liquidtight Flexible Conduits; Rigid Metal Conduits; Rigid Nonmetallic Conduits; Wireways
  BSI BS 6099: Part 1-81. 1981 Amd 1 Conduits for Electrical Installations Part 1: General Requirements. 26 pp.
  CENELEC HD 394.1-79. Specification for Conduits for Electrical Installations Part 1 General Requirements. 2 pp.
  CENELEC EN 500 86-1-91. Conduit Systems for Electrical Installations Part 1: General Requirements. 42 pp.
  DIN VDE 0605-82. Conduits and Fittings for Electrical Installations (Apr). 18 pp.
  IEC 614 Pt 1-78. Specification for Conduits for Electrical Installations Part 1: General Requirements First Edition; (Amendment 2-1984, Incorporates Amendment 1). 43 pp.
  IEC 1084 Pt 1-91. Cable Trunking and Ducting Systems for Electrical Installations Part 1: General Requirements First Edition. 50 pp.
  SAA AS 2052-77. Metallic Conduits and Fittings. 28 pp.
  SAA AS 2053-84. Non-Metallic Conduits and Fittings. 28 pp.
  SNZ NZS 127-49. Steel Conduit and Fittings for Electrical Wiring Amend: 2A, 1967. 44 pp.
  SNZ NZS 159: Part 1-54. Flexible Steel Conduit for Cable Protection and Flexible Steel Tubing to Enclose Flexible Drives Part 1: Flexible Steel Conduit and Adaptors for the Protection of Electric Cable. 16 pp.
—Aerospace—Identification Systems
  NATO STANAG 3104 ED 6 AMD 2-88. Identification of Aircraft and Missile Pipeline and Electrical Conduits. 7 pp.
  NATO STANAG 3104 ED 6 AMD 3-88. Identification of Aircraft and Missile Pipeline and Electrical Conduits. 8 pp.
  NATO STANAG 3104 ED 6 AMD 4-88. Identification of Aircraft and Missile Pipeline and Electrical Conduits. 5 pp.
—Aircraft—Assembly
  SBAC AS 4800 ISSUE 4. Cable Duct (Typical Assembly). 1 p.
—Aircraft—Identification Systems
  NATO STANAG 3104 ED 6 AMD 2-88. Identification of Aircraft and Missile Pipeline and Electrical Conduits. 7 pp.

Electric Conduits (Cont.)
—Aircraft—Identification Systems (Cont.)
  NATO STANAG 3104 ED 6 AMD 3-88. Identification of Aircraft and Missile Pipeline and Electrical Conduits. 8 pp.
  NATO STANAG 3104 ED 6 AMD 4-88. Identification of Aircraft and Missile Pipeline and Electrical Conduits. 5 pp.
—Aircraft—Packing Pieces
  SBAC AS 4809-4812 ISSUE 2. Packing Piece—Cable Duct. 1 p.
  SBAC AS 4813-4816 ISSUE 2. Tee Piece for Packing. 1 p.
—Aluminum
  BSI BS 2898-70. 1970 Amd 1 Wrought Aluminium and Aluminium Alloys for Electrical Purposes. Bars, Extruded Round Tube and Sections. 32 pp.
  CNS C4250-80. Aluminium Conduit (Aug)(6106).
—Asbestos Packed
  SNZ NZS 159: Part 1-54. Flexible Steel Conduit for Cable Protection and Flexible Steel Tubing to Enclose Flexible Drives Part 1: Flexible Steel Conduit and Adaptors for the Protection of Electric Cable. 16 pp.
—Composite Materials
  IEC 614 Pt 2-6-92. Specification for Conduits for Electrical Installations Part 2: Particular Specifications for Conduits Section 6: Pliable Conduits of Metal or Composite Materials First Edition. 37 pp.
—Copper
  BSI BS 1977-76. 1976 High Conductivity Copper Tubes for Electrical Purposes. 10 pp.
—Dies
  BSI BS 1813-74. 1974 Dimensions of Conduit Dies, Diestocks and Guides. 13 pp.
—Electrical Enclosures—Construction
  CSA C22.2 NO 0.5-1982. Threaded Conduit Entries (R 1992); (Gen Instr 1). 22 pp.
—Electrical Insulation
  BSI BS 6099: Sec 2.2-82. 1982 Conduits for Electrical Installations Part 2: Particular Specifications Section 2.2: Rigid Plain Conduits of Insulating Materials. 15 pp.
  IEC 614 Pt 2-2-80. Specification for Conduits for Electrical Installations Part 2: Particular Specification for Rigid Plain Conduits Insulating Materials First Edition. 22 pp.
  IEC 614 Pt 2-3-90. Specification for Conduits for Electrical Installations Part 2: Particular Specifications for Conduits Section Three—Pliable Conduits of Insulating Material First Edition. 29 pp.
  IEC 614 Pt 2-4-85. Specification for Conduits for Electrical Installations Part 2: Particular Specifications for Conduits Section Four—Pliable Self-Recovering Conduits of Insulating Materials First Edition. 24 pp.
—Missiles—Identification Systems
  NATO STANAG 3104 ED 6 AMD 2-88. Identification of Aircraft and Missile Pipeline and Electrical Conduits. 7 pp.
  NATO STANAG 3104 ED 6 AMD 3-88. Identification of Aircraft and Missile Pipeline and Electrical Conduits. 8 pp.
  NATO STANAG 3104 ED 6 AMD 4-88. Identification of Aircraft and Missile Pipeline and Electrical Conduits. 5 pp.
—Outside Diameters
  BSI BS 6053-81. 1981 Outside Diameters of Conduits for Electrical Installations and Threads for Conduits and Fittings. 8 pp.
  CENELEC HD 393-79. Outside Diameters of Conduits for Electrical Installations and Threads for Conduits and Fittings. 2 pp.
  IEC 423-73. Outside Diameters of Conduits for Electrical Installations and Threads for Conduits and Fittings First Edition; (Supplement A-1978). 27 pp.
—Overpressure Testing—Hazardous Locations
  CSA 1273 Bull. Electrical Bulletin 1273 July 2, 1980 to C22.2 NO 25. 2 pp.
  CSA C22.2 NO 30-M1986. Explosion-Proof Enclosures for Use in Class I Hazardous Locations (R 1992); (Gen Instr 1 Thru 2). 41 pp.
  CSA 1273 Bull. Electrical Bulletin 1273 July 2, 1980 to C22.2 NO 30. 2 pp.
—Resinated Asbestos—Aircraft
  SBAC AS 4798 ISSUE 1. Cable Duct. 1 p.
  SBAC AS 4801-4804 ISSUE 3. Cable Duct. 1 p.

Electric Conduits (Cont.)
—Rubber Packed
  SNZ NZS 159: Part 1-54. Flexible Steel Conduit for Cable Protection and Flexible Steel Tubing to Enclose Flexible Drives Part 1: Flexible Steel Conduit and Adaptors for the Protection of Electric Cable. 16 pp.
—Thermoplastic
  BSI DD 200-91. 1991 Pliable Non-Rigid Conduit and Fittings for Direct Burial Underground (Type DB). 13 pp.
—Underground
  CENELEC PREN 50086-2-4-92. Conduit Systems for Electrical Installations Buried Underground. 9 pp.

Electric Connections
Use: Connectors

Electric Connectors
Use: Connectors

Electric Contactors
Use: Contactors

Electric Contacts
See Also: Breaker Points; Brushes (Machine Components); Chatter; Connectors; Contact Bounce; Crimp Contacts; Dry Reed Contacts; Electric Switches; Electric Terminals; Relay Contacts; Solder Contacts; Triaxial Contacts; Wiping Contacts
  CNS C7128-80. Small Contact Standard for Electrical Connectors (Aug)(6145).
  CSA C22.2 NO 65-93. Wire Connectors; (Gen Instr 1). 67 pp.
—Aerospace
  CEN PREN 2591 (Part B16)-92. Elements of Electrical and Optical Connection Test Methods Part B16—Engagement of Contacts. 4 pp.
—Aerospace—Bend Testing
  CEN PREN 2591 (Part D16)-92. Elements of Electrical and Optical Connection Test Methods Part D16—Contact Bending Strength. 4 pp.
—Aerospace—Connectors—Quality Assurance
  CEN PREN 3155-001-90. Electrical Contacts Used in Connection Devices Technical Specification. 34 pp.
  CEN PREN 3155-001-93. Aerospace Series Electrical Contacts Used in Elements of Connection Part 001—Technical Specification. 27 pp.
—Aerospace—Contact Force
  CEN PREN 2591 (Part D12)-92. Elements of Electrical and Optical Connection Test Methods Part D12—Contact Insertion and Extraction Forces. 4 pp.
—Aerospace—Protection Effectiveness (Scoop Proof)
  CEN PREN 2591 (Part E5)-92. Elements of Electrical and Optical Connection Test Methods Part E5—Contact Protection Effectiveness (Scoop-Proof). 4 pp.
—Aerospace—Stability Testing
  CEN PREN 2591 (Part D19)-92. Elements of Electrical and Optical Connection Test Methods Part D19—Stability of Male Contacts in Insert. 4 pp.
—Aircraft
  SBAC TS 301 ISSUE 2. SBAC Manufacturing Specification TS 301 for ESC Series of Electrical Contacts.
  SBAC ESC 30 ISSUE 2. High Temperature Electrical Contacts (Generally to Mil-C-39029). 7 pp.
—Atmospheric Corrosion Testing
  BSI BS 2011: Part 2.1KC-91. 1991 Environmental Testing Part 2.1: Tests Test Kc and Guidance. Sulphur Dioxide Test for Contacts and Connections. 21 pp.
  BSI BS 2011: Part 2.1KD-77. 1977 Basic Environmental Testing Procedures Part 2.1: Tests Part 2.1KD: Test Kd. Hydrogen Sulfide Test for Contacts and Connections. 8 pp.
  BSI BS 2011: Part 2.2KD-84. 1984 Basic Environmental Testing Procedures Part 2.2: Guidance Part 2.2KD: Test Kd. Guidance on Test Kd: Hydrogen Sulfide Test for Contacts and Connections. 12 pp.
  CENELEC HD 323.2.46 S1-88. Basic Environmental Testing Procedures—Part 2: Tests. Guidance to Test Kd: Hydrogen Sulphide Test for Contacts and Connections. 3 pp.
  IEC 68 Pt 2-42-82. Basic Environmental Testing Procedures Part 2: Tests Test Kc: Sulphur Dioxide Test for Contacts and Connections Second Edition. 18 pp.

INTERNATIONAL AND NON-U.S. NATIONAL STANDARDS
SUBJECT INDEX
Electric

Electric Contacts (Cont.)
—**Atmospheric Corrosion Testing** (Cont.)
IEC 68 Pt 2-43-76. Basic Environmental Testing Procedures Part 2: Tests Test Kd: Hydrogen Sulphide Test for Contacts and Connections First Edition. 16 pp.
IEC 68 Pt 2-46-82. Basic Environmental Testing Procedures Part 2: Tests Guidance to Test Kd: Hydrogen Sulphide Test for Contacts and Connections First Edition. 27 pp.
IEC 68 Pt 2-49-83. Basic Environmental Testing Procedures Part 2: Tests Guidance to Test Kc: Sulphur Dioxide Test for Contacts and Connections First Edition. 23 pp.
IEC 68 Pt 2-60 TTD-89. Environmental Testing Part 2: Tests—Test Ke: Corrosion Tests in Artificial Atmosphere at Very Low Concentration of Polluting Gas(es). 25 pp.
SAA AS 1099.2KC-81. Basic Environmental Testing Procedures for Electrotechnology—Part 2: Tests—Part 2Kc: Sulphur Dioxide Test for Contacts and Connections. 9 pp.
SAA AS 1099.2KD-81. Basic Environmental Testing Procedures for Electrotechnology—Part 2: Tests—Part 2Kd: Hydrogen Sulphide Test for Contacts and Connections. 6 pp.
SNZ IEC 68: Part 2-42-82. Basic Environmental Testing Procedures Part 2-42: Test Kc: Sulphur Dioxide Test for Controls and Connections. 14 pp.
SNZ IEC 68: Part 2-43-76. Basic Environmental Testing Procedures Part 2-43: Test Kd: Hydrogen Sulphide Test for Contacts and Connections. 14 pp.
SNZ IEC 68: Part 2-46-82. Basic Environmental Testing Procedures Part 2-46: Guidance to Test Kd: Hydrogen Sulphide Test for Contacts and Connections. 23 pp.
SNZ IEC 68: Part 2-49-83. Basic Environmental Testing Procedures Part 2-49: Guidance to Test Kc: Sulphur Dioxide for Contacts and Connections. 19 pp.

—**Communication Equipment**
CNS C4340-82. Electric Contact Material for Electric Communication Equipment and Apparatus (Jan)(8369).
JIS C 2509-65. Electric Contact Materials for Electric Communication Equipment and Apparatus. 7 pp.

—**Electromechanical Components**
SAA AS 1029.1-85. Low Voltage Contactors—Part 1: Electromechanical—up to and Including 1000 V a.c. and 1200 V d.c. 40 pp.
SNZ NZS/AS 1029. 1-85. Low Voltage Contactors Part 1: Electromechanical—up to and Including 1000 V a.c. and 1200 V d.c. (This is a Joint Standard with SAA AS 1029.1). 40 pp.

—**Extraction Tools—Aerospace**
CEN PREN 2591 (Part E6)-92. Elements of Electrical and Optical Connection Test Methods Part E6—Use of Tools. 4 pp.

—**Gages—Aerospace—Contact Force**
CEN PREN 2591 (Part D18)-92. Elements of Electrical and Optical Connection Test Methods Part D18—Gauge Insertion and Extraction Forces in a Female Contact. 4 pp.

—**Gages—Aerospace—Insertion Force**
CEN PREN 2591 (Part D18)-92. Elements of Electrical and Optical Connection Test Methods Part D18—Gauge Insertion and Extraction Forces in a Female Contact. 4 pp.

—**Glosssaries**
IEC 50 Group 30-57. International Electrotechnical Vocabulary Group 30: Electric Traction Second Edition. 98 pp.
SNZ IEC 50: 50(30)-57. International Electrotechnical Vocabulary 50(30): Electric Traction. 94 pp.

—**Hermetic Seals—Aerospace**
CEN PREN 2591 (Part C22)-92. Elements of Electrical and Optical Connection Test Methods Part C22—Hermeticity. 4 pp.

—**Insertion Tools—Aerospace**
CEN PREN 2591 (Part E6)-92. Elements of Electrical and Optical Connection Test Methods Part E6—Use of Tools. 4 pp.

—**Low Voltage**
SAA AS 1029.1-85. Low Voltage Contactors—Part 1: Electromechanical—up to and Including 1000 V a.c. and 1200 V d.c.. 40 pp.
SNZ NZS/AS 1029. 1-85. Low Voltage Contactors Part 1: Electromechanical—up to and Including 1000 V a.c. and 1200 V d.c. (This is a Joint Standard with SAA AS 1029.1). 40 pp.

Electric Contacts (Cont.)
—**Mechanical Testing**
BSI BS 5772: Part 8-85. 1985 Amd 1 Basic Testing Procedures and Measuring Methods for Electromechanical Components for Electronic Equip. Part 8: Connector Tests (Mechanical) and Mech. Tests on Contacts and Terminations (AMD 5076) May 31, 1990. 32 pp.
IEC 512 Pt 8-93. Electromechanical Components for Electronic Equipment; Basic Testing Procedures and Measuring Methods Part 8: Connector Tests (Mechanical) and Mechanical Tests on Contacts and Terminations Third Edition. 72 pp.

—**Quality Assurance**
BSI BS 9521-83. 1983 Amd 1 Removable Contacts of Assessed Quality for Electrical Connectors as Defined by BS 9520: Generic Data, Methods of Test and Rules for the Preparation of Detail Specifications. 37 pp.

—**Removable Walkout—Aerospace—Contact Force**
CEN PREN 2591 (Part D26)-92. Elements of Electrical and Optical Connection Test Methods Part D26—Contact Retention System Effectiveness (Removable Contact Walkout). 4 pp.

—**Screws**
CNS B2233-81. Cheese Head Countersunk Screws for Electric Contact Set (Apr)(4413). 3 pp.

—**Semiconductor Devices**
SAA AS 1029.2-82. Low Voltage Contactors—Part 2: Semiconductor (Solid State) (Up to and Including 1000 V a.c. and V d.c.). 30 pp.
SNZ NZS/AS 1029. 2-82. Low Voltage Contactors Part 2: Semiconductor (Solid State) (up to and Including 1000 V a.c and 1500 V d.c.) (This is a Joint Standard with SAA AS 1029.2). 30 pp.

—**Speech Circuits**
CCITT RECMN Q.30-89. Improving the Reliabilty of Contacts in Speech Circuits—General Recommendations on Telephone Switching and Signalling—Functions and Information Flows for Services in the ISDN—Supplements (Study Group XI) 1 pp. 1 p.

—**Symbols**
CNS C5115-81. Graphic Symbols for Contacts (May)(7364).

—**Telecommunication Equipment—Lubricants**
MOD UK DSTAN 68-7-01. Cleaning and Lubricating Product, Electrical Contact, ZX-33; Amendment 1. 8 pp.

—**Voltage Measurement**
CNS C3129-81. Method of Measuring Voltage Drop on Closed Contacts Where Over 100 Amperes Flow Through the Contacts (Jul)(7632).

Electric Control Gear
Use: Controlgear

Electric Controllers
Use: Controllers

Electric Converters
Use: Converters

Electric Cords
Use: Cords (Electric)

Electric Corona
See Also: Static Electricity

—**Connectors**
CNS C6254-86. Method of Test for Low Frequency (Below 3MHz) Electrical Connectors (TP-44 Corona Test) (May)(11571).

Electric Cutouts
See Also: Electrical Faults; Thermal Cutouts
CNS C3096-83. Method of Test for Low-Voltage Cut Outs (Oct)(6060).
CNS C4222-83. Low Voltage Cut Outs (Oct)(6059).
CNS C4357-82. High Voltage Cutout for Cubicle Type Power Receiving Unit (Jul)(9091).
JIS C 8301-88. Cut-out Switches. 12 pp.

—**Rolling Stock**
CNS E1052-85. Cut-Out Cocks for Railway Rolling Stock (Nov)(11410).
CNS E1054-85. Cut-Out Cocks with Side Vent for Railway Rolling Stock (Nov)(11412).
JIS E 4103-88. Cut-out Cocks for Railway Rolling Stock.

Electric Discharge Machine Tools
Use For: EDM Machines; Electric Spark Machines; Electrodischarge Machines; Electroerosive Machine; Spark Machines *See Also:* Cutting Tools; Drilling Equipment; Machine Tools
CNS B4066-86. Test Code for Performance and Accuracy of Electrical Discharge Machines and Numerically Controlled Electrical Discharge Machines (Feb)(11493).
JIS B 6360-83. Test Code for Performance and Accuracy of Wire Electrical Discharge Machines. 28 pp.
JIS B 6361-85. Test Code for Performance and Accuracy of Electrical Discharge Machines and Numerically Controlled Electrical Discharge Machines. 22 pp.

—**Acceptance Testing**
DIN ENGL 8662 Pt 1-88. Machine Tools; Cavity Sinking EDM Machines; Single Column Machines with Cross Slide Table; Acceptance Conditions (Aug). 6 pp.
DIN ENGL 8662 Pt 2-88. Machine Tools; Cavity Sinking EDM Machines; Single Column Machines with Fixed Table; Acceptance Conditions (Aug). 7 pp.
DIN ENGL 8662 Pt 3-88. Machine Tools; Cavity Sinking EDM Machines; Straight Sided Machines with Cross Slide Table; Acceptance Conditions (Aug). 8 pp.

—**Numerical Control**
CNS B4066-86. Test Code for Performance and Accuracy of Electrical Discharge Machines and Numerically Controlled Electrical Discharge Machines (Feb)(11493).
JIS B 6361-85. Test Code for Performance and Accuracy of Electrical Discharge Machines and Numerically Controlled Electrical Discharge Machines. 22 pp.

Electric Discharges
Use For: Partial Discharges
BSI BS 4828-85. 1985 Guide to Partial Discharge Measurements. 32 pp.
IEC 270-81. Partial Discharge Measurements Second Edition. 59 pp.
SAA AS 1018-85. Partial Discharge Measurements. 25 pp.

—**Cables (Electric)**
DIN VDE 0472 Pt 513-82. Testing of Cables, Wires and Flexible Cords; Partial Discharge Measurements (July). 28 pp.
IEC 840-88. Tests for Power Cables with Extruded Insulation for Rated Voltages Above 30 kV (Um = 36 kV) up to 150 kV (Um = 170 kV) First Edition; (Corrigendum—Nov 1988) (Amendment 2-1993, Including Amendment 1) (Corrigendum—July 1993). 71 pp.
IEC 885 Pt 2-87. Electrical Test Methods for Electric Cables Part 2: Partial Discharge Tests First Edition. 13 pp.
IEC 885 Pt 3-88. Electrical Test Methods for Electric Cables Part 3: Test Methods for Partial Discharge Measurements on Lengths of Extruded Power Cable First Edition. 56 pp.

—**Electrical Equipment**
DIN VDE 0110 Pt 20-90. Insulation Coordination for Electrical Equipment Within Low-Voltage Systems; Partial Discharge Tests Application Guide (Aug). 23 pp.

—**Instrument Transformers**
BSI BS 6184-81. 1981 Measuring Partial Discharges in Instrument Transformers. 7 pp.
IEC 44 Pt 4-80. Instrument Transformers Part 4: Measurement of Partial Discharges First Edition. 14 pp.

Electric Distribution Systems
Use: Power Distribution Systems

Electric Door Operators
Use: Door Openers

Electric Endurance Testing
See Also: Electrical Measurement; Endurance Testing

—**Digital Circuits**
BSI BS CECC 90100-86. (WITHDRAWN) 1986 Amd 2 Harmonized System of Quality Assessment for Electronic Components; Sectional Specification: Digital Monolithic Integrated Circuits (AMD 5954) March 28, 1991 (Renumbered as BS EN 190100: 1993). 69 pp.

—**Electric Switches**
CNS C6089-88. Method of Test for Electromechanical Switches (Electrical Endurance) (Dec)(6152).
CNS C6239-84. Method of Test for Electromechanical Switches (Logic (TTL) Level Endurance) (Nov)(11102).

INDUSTRY STANDARDS

Electric

Electric Endurance Testing (Cont.)

—Electric Switches (Cont.)
CNS C6240-84. Method of Test for Electromechanical Switches (Low Level Endurance) (Nov)(11103).

—Electrical Insulation
IEC 727 Pt 1-82. Evaluation of Electrical Endurance of Electrical Insulation Systems Part 1: General Considerations and Evaluation Procedures Based on Normal Distributions First Edition. 34 pp.

IEC 727 Pt 2-93. Evaluation of Electrical Endurance of Electrical Insulation Systems Part 2: Evaluation Procedures Based on Extreme-Value Distributions First Edition. 74 pp.

—Electronic Components
CNS C6248-85. Endurance (Electrical) Testing Method for Electronic Component (Apr)(11239).

JIS C 5036-75. Endurance (Electrical) Testing Method for Electronic Components. 4 pp.

Electric Energy Meters
See Also: Electric Measuring Instruments; Electric Power Meters; Electrical Measurement; Watt Hour Meters

EC 76/891/EEC-76. Council Directive on the Approximation of the Laws of the Member States Relating to Electrical Energy Meters. 19 pp.

—Testing Equipment
IEC 736-82. Testing Equipment for Electrical Energy Meters First Edition. 36 pp.

Electric Eyes
Use: Photoelectric Cells

Electric Fans
Use: Fans

Electric Fences
See Also: Fences

DIN VDE 0131-84. Construction and Operation of Electric Fence Equipment (Apr). 13 pp.

SAA AS 3014-91. Electrical Installations —Electrical Fences. 7 pp.

—Controllers
BSI BS 2632-80. (WITHDRAWN) 1980 Amd 1 Mains-Operated Electric Fence Controllers (Superseded by BS EN 61011: 1993). 36 pp.

CENELEC HD 92-74. Mains Operated Electric Fence Controllers. 5 pp.

SNZ NZS 6203: Part 1-87. Specification for Agricultural Electric Fencing (Safety Requirements Part 1: Alternating Current Mains-Operated Electric Fence Energizers Amend: 1, 1989; 2, 1991. 24 pp.

—Controllers—Battery Operated
BSI BS 6167-81. (WITHDRAWN) 1981 Amd 1 Battery-Operated Electric Fence Controllers Not Suitable for Connection to the Supply Mains (Supperseded by BS EN 61011-2: 1993). 28 pp.

BSI BS 6369-83. (WITHDRAWN) 1983 Amd 1 Battery-Operated Electric Fence Contollers Suitable for Connection to the Supply Mains (Superseded by BS EN 61011-1: 1993). 40 pp.

SNZ NZS 2210-68. Battery Operated Electric Fence Controllers Amend: 1, 1972. 16 pp.

—Controllers—Battery Operated—Safety
BSI BS EN 61011-1-93. 1993 Electric Fence Energizers Safety Requirements for Battery-Operated Electric Fence Energizers Suitable for Connection to the Supply Mains (Supersedes BS 6369: 1983) (E). 24 pp.

CENELEC PREN 61011-1-90. Electric Fence Energizers Part 1: Safety Requirements for Battery-Operated Electric Fence Energizers Suitable for Connection to the Supply Mains.

CENELEC EN 61011-1-92. Electric Fence Energizers Part 1: Safety Requirements for Battery-Operated Electric Fence Energizers Suitable for Connection to the Supply Mains. 8 pp.

CENELEC EN 61011-1-92. Electric Fence Energizers Safety Requirements for Battery-Operated Electric Fence Energizers Suitable for Connection to the Supply Mains. 10 pp.

CENELEC EN 61011-1/PRAA-92. AMD A Electric Fence Energizers Safety Requirements for Battery -Operated Electric Fence Energizers Suitable for Connection to the Supply Mains. 2 pp.

IEC 1011 Pt 1-89. Electric Fence Energizers Safety Requirements for Battery-Operated Electric Fence Energizers Suitable for Connection to the Supply Mains First Edition; (Amendment 2-1993, Incorporating Amendment 1). 45 pp.

IEC 1011 Pt 2-90. Electric Fence Energizers Safety Requirements for Battery-Operated Electric Fence Energizers Not for Connection to the Supply Mains First Edition; (Amendment 2-1993, Incorporating Amendment 1). 51 pp.

Electric Fences (Cont.)

—Controllers—Peak Discharge
CSA CAN/CSA-C22. 2 NO 103-M92. Electric Fence Controllers; (Gen Instr 1 Thru 2). 59 pp.

CSA 1169 Bull. Electrical Bulletin 1169 June 27, 1978 to C22.2 NO 103. 2 pp.

—Controllers—Safety
BSI BS EN 61011-93. 1993 Electric Fence Energizers Safety Requirements for Mains-Operated Electric Fence Energizers (Supersedes BS 2632: 1980) (E). 116 pp.

BSI BS EN 61011-2-93. 1993 Electric Fence Energizers Safety Requirements for Battery-Operated Electric Fence Energizers Not for Connection to the Supply Mains (Supersedes BS 6167: 1981) (E). 27 pp.

CENELEC PREN 61011-91. Electric Fence Energizers. Safety Requirements for Mains-Operated Electric Fence Energizers. 111 pp.

CENELEC EN 61011-92. Electric Fence Energizers Safety Requirements for Mains-Operated Electric Fence Energizers (IEC 1011: 1989, Modified). 107 pp.

CENELEC EN 61011/PRAA-92. AMD A Electric Fence Energizers—Safety Requirements for Mains-Operated Electric Fence Energizers. 2 pp.

CENELEC PREN 61011-2-91. Electric Fence Energizers Part 2: Safety Requirements for Battery-Operated Electric Fence Energizers Not for Connection to the Supply Mains. 27 pp.

CENELEC EN 61011-2-92. Electric Fence Energizers Part 2: Safety Requirements for Battery-Operated Electric Fence Energizers Not for Connection to the Supply Mains. 7 pp.

CENELEC EN 61011-2/PRAA-92. AMD A Electric Fence Energizers Safety Requirements for Battery -Operated Electric Fence Energizers Not for Connection to the Supply Mains. 2 pp.

IEC 1011-89. Electric Fence Energizers Safety Requirements for Mains-Operated Electric Fence Energizers First Edition; (Amendment 1-1991) (Corrigendum—June 1993) (Amendment 2-1993) (CENELEC EN 61011:1992). 268 pp.

—Controllers—Timed Sinusoidal
CSA CAN/CSA-C22. 2 NO 103-M92. Electric Fence Controllers; (Gen Instr 1 Thru 2). 59 pp.

CSA 1169 Bull. Electrical Bulletin 1169 June 27, 1978 to C22.2 NO 103. 2 pp.

Electric Field Strength
See Also: Magnetic Field Strength

IEC 833-87. Measurement of Power-Frequency Electric Fields First Edition. 48 pp.

SNZ NZS/IEC 833-87. Measurement of Power-Frequency Electric Fields. 46 pp.

—Antennas
CCIR Report 1117-90. Determination of the Electric Field Strength in the Near Field Zone of High-Power LF/MF Antennas—Section 1D—Spectrum Utilization and Applications. 12 pp.

Electric Filters
Scope Note: For additional listings, use a more specific term *See Also:* Band Reject Filters; Bandpass Filters; Cavity Filters; Ceramic Filters; Coaxial Filters; Crystal Filters; Dielectric Filters; Electromagnetic Interference Filters; Electronic Components; Filter/Surge Protectors; Line Filters (Electric); Low Pass Filters; Passive Filters; Pi Filters; Quartz Crystal Filters; Radio Frequency Filters; Resistors; Surface Acoustic Wave Filters; Through Connection Filters; Through Group Filters; Through Mastergroup Filters; Through Supergroup Filters; Through Supermastergroup Filters; Through 15 Supergroup Assembly Filters; White Noise Filters

BSI DD 158-87. 1987 Filters for Mains Signalling Systems. 12 pp.

MOD UK DSTAN 59-45: Part 1-85. Filters and Networks of Assessed Quality Part 1: General Requirements Issue 3. 15 pp.

MOD UK DSTAN 59-45: Part 80-85. Filters and Networks Part 80: Imported Filters and Networks Detail Specifications Issue 2. 6 pp.

MOD UK DSTAN 59-45: Part 90-85. Filters and Networks Part 90: Detail Specifications Issue 2. 7 pp.

—Quality Assurance
MOD UK DSTAN 59-45: Part 1-85. Filters and Networks of Assessed Quality Part 1: General Requirements Issue 3. 15 pp.

—Symbols
CNS C1143-4-89. Graphical Symbols for Use on Electrical and Electronic Equipment (Alarm, Signal, Filter) (Feb)(12491-4).

Electric Fittings
Scope Note: Use a more specific term

Electric Fittings (Cont.)
See: Busway Fittings; Busways; Cable Fittings; Copper Wire Fittings; Electric Conduit Fittings; Electric Outlet Boxes; Electric Raceway Fittings; Electrical Insulator Fittings; Power Line Hardware; Wireway Fittings; Wiring Devices

Electric Frying Pans
Use: Frying Utensils

Electric Fuses
Use: Fuses (Electric)

Electric Generating Stations
Use: Generating Stations

Electric Generators
Use: Generators

Electric Grills
Use: Grills (Appliances)—Electric

Electric Heating Elements
Use: Heating Elements

Electric Heating Pads
Use: Heating Pads—Electric

Electric Inlets
Use For: Inlets (Electrical) *See Also:* Connectors; Wiring Devices

CSA 895B Bull. Electrical Bulletin 895B May 12, 1976 to C22.2 NO 21. 2 pp.

CSA C22.2 NO 42-M1984. General Use Receptacles, Attachment Plugs, and Similar Wiring Devices; (Gen Instr 1 Thru 10). 101 pp.

CSA 895B Bull. Electrical Bulletin 895B May 12, 1976 to C22.2 NO 42. 2 pp.

CSA C22.2 NO 182.3-M1987. Special Use Attachment Plugs, Receptacles, and Connectors (R 1993); (Gen Instr 1 Thru 2). 41 pp.

—Appliances
BSI BS 3283: Part 1-80. 1980 Amd 2 Non-Reversible Connectors and Appliances Inlets for Portable Electrical Appliances(for Circuits up to 250 Volts) Part 1: 13A Connector and Appliance Inlet. 17 pp.

BSI BS 4491: Part 1-89. 1989 Amd 1 Appliance Couplers for Household and Similar General Purposes Part 1: Specification of General Requirements (AMD 6462) April 30, 1991. 123 pp.

BSI BS 4491: Sec 2.1-90. 1990 Appliance Couplers for Household and Similar General Purposes Part 2: Requirements for Individual Types of Coupler Section 2.1: Sewing Machine Couplers. 15 pp.

BSI BS EN 60320-2-2-92. 1992 Appliance Couplers for Household and Similar General Purposes Part 2: Specification for Individual Types of Coupler Section 2.2: Interconnection Couplers for Household and Similar Equipment. 37 pp.

CENELEC EN 60 320-87. Appliance Couplers for Household and Similar General Purposes Part 2: Sewing Machine Couplers. 15 pp.

CENELEC EN 60 320-87. Appliance Couplers for Household and Similar General Purposes. 115 pp.

CENELEC EN 60 320/A1-89. AMD 1 Appliance Couplers for Household and Similar General Purposes. 7 pp.

CENELEC EN 60 320 (Part 2.1)-87. Appliance Couplers for Household and Similar General Purposes Part 2: Sewing Machine Couplers. 15 pp.

CENELEC EN 60320-2-2-91. Appliance Couplers for Household and Similar General Purposes Part 2: Interconnection Couplers for Household and Similar Equipment. 7 pp.

DIN VDE 0627-86. Connectors and Plug-And-Socket Devices for Rated Voltages up to 1000 V a.c. and 1200 V d.c. and Rated Currents up to 500 A for Each Pole (June). 36 pp.

IEC 320-81. Appliance Couplers for Household and Similar General Purposes Second Edition; (Amendment 3-1987, Incorporating Amendment 1 and 2). 181 pp.

IEC 320 Pt 2-1-84. Appliance Couplers for Household and Similar General Purposes Part 2: Sewing Machine Couplers First Edition. 20 pp.

IEC 320 Pt 2-2-90. Appliance Couplers for Household and Similar General Purposes Part 2: Interconnection Couplers for Household and Similar Equipment First Edition; (Corrigendum—Aug 1990). 69 pp.

IEC 884 Pt 2-2-89. Plugs and Socket-Outlets for Household and Similar Purposes Part 2: Particular Requirements for Socket-Outlets for Appliances First Edition. 31 pp.

IEC 884 Pt 2-3-89. Plugs and Socket-Outlets for Household and Similar Purposes Part 2: Particular Requirements for Switched Socket-Outlets Without Interlock for Fixed Installations First Edition. 35 pp.

SNZ NZS 1801-63. Specification for Domestic Appliance Connectors and Appliance Inlet Sockets. 17 pp.

Electric Inlets (Cont.)

—Appliances—Three Wire
CSA CAN/CSA-C22. 2 NO 21-M90. Cord Sets and Power-Supply Cords; (Gen Instr 1 Thru 2). 74 pp.
CSA 895B Bull. Electrical Bulletin 895B May 12, 1976 to C22.2 NO 21. 2 pp.

—Electronic Equipment
CSA C22.2 NO 182.3-M1987. Special Use Attachment Plugs, Receptacles, and Connectors (R 1993); (Gen Instr 1 Thru 2). 41 pp.

—Locking
CSA C22.2 NO 182.2-M1987. Industrial Locking Type, Special Use Attachment Plugs, Receptacles, and Connectors (R 1993); (Gen Instr 1). 43 pp.

—Locking—Hospital Grade
CSA C22.2 NO 182.2-M1987. Industrial Locking Type, Special Use Attachment Plugs, Receptacles, and Connectors (R 1993); (Gen Instr 1). 43 pp.

—Medical Electrical Equipment—Three Wire
CSA CAN/CSA-C22. 2 NO 21-M90. Cord Sets and Power-Supply Cords; (Gen Instr 1 Thru 2). 74 pp.
CSA 895B Bull. Electrical Bulletin 895B May 12, 1976 to C22.2 NO 21. 2 pp.

—Office Machines—Three Wire
CSA CAN/CSA-C22. 2 NO 21-M90. Cord Sets and Power-Supply Cords; (Gen Instr 1 Thru 2). 74 pp.
CSA 895B Bull. Electrical Bulletin 895B May 12, 1976 to C22.2 NO 21. 2 pp.

Electric Insect Killers
Use: Insect Electrocution Devices

Electric Insulators
Use: Electrical Insulators

Electric Irons
Use: Irons (Electric)

Electric Lamps
Use: Lamps

Electric Lifts
Use: Elevators (Lifts)

Electric Lines
Use: Transmission Lines

Electric Locomotives
Use: Locomotives—Electric

Electric Measuring Instruments
Use For: Analog Electric Measuring Instruments; Digital Measuring Instruments *See Also:* Ammeters; Ampere Hour Meters; Chart Recorders; Coulombmeters; Electric Energy Meters; Electric Power Meters; Electrical Measurement; Electrical Properties; Electronic Measuring Instruments; Frequency Meters; Harmonic Analyzers; Measuring Instruments; Multimeters; Ohmmeters; Oscilloscopes; Power Factor Meters; Probes (Sensors); Radiation Measuring Instruments; Ratio Meters; Test Sets; Varhour Meters; Varmeters; Voltmeters; Volume Indicators; Volume Unit Meters; Watt Hour Meters; Wattmeters

CENELEC HD 326-76. Electrical Measuring Instruments Utilizing Radio-Active Sources. 5 pp.
CNS C1136-85. Tolerance of the Combined Indicator and the Driving Elements (Dec)(11443).
IEC 359-87. Expression of the Performance of Electrical and Electronic Measuring Equipment Second Edition; (Amendment 1-1991) 53 pp.
IEC 443-74. Stabilized Supply Apparatus for Measurement First Edition. 84 pp.
IEC 473-74. Dimensions for Panel Mounted Indicating and Recording Electrical Measuring Instruments First Edition; (Amendment 1-1979). 24 pp.
SAA AS 1024-71. Direct Recording Electrical Measuring Instruments and Their Accessories. 70 pp.
SAA AS 1042-73. Direct-Acting Indicating Electrical Measuring Instruments and Their Accessories. 71 pp.

—Biomechanical Effects (Human Body)
BSI DD ENV 28041-93. 1993 Human Response to Vibration Measuring Instrumentation (ISO 8041: 1990) (E). 32 pp.
CEN ENV 28041-93. Human Response to Vibration—Measuring Instrumentation (ISO 8041: 1990). 26 pp.
ISO 8041-90. Human Response to Vibration—Measuring Instrumentation First Edition; (Corrigendum 1-1993) (CEN EN 28041: 1993). 30 pp.

—Circuits—Hazardous Environments
BSI BS 6705-87. 1987 Electrical Measuring Instruments for Use on Intrinsically Safe Circuits in Coal Mines. 4 pp.

—Direct Acting—Indicating
BSI BS 89-77. (WITHDRAWN) 1977 Amd 1 Direct Acting Indicating Electrical Measuring Instruments and Their Accessories (Superseded by BS 89: Parts 1 to 9: 1990). 55 pp.
BSI BS 89: Part 1-90. 1990 Direct Acting Indicating Analogue Electrical Measuring Instruments and Their Accessories Part 1: Definitions and General Requirements Common to All Parts. 42 pp.
BSI BS 89: Part 7-90. 1990 Direct Acting Indicating Analogue Electrical Measuring Instruments and Their Accessories Part 7: Requirements for Multi-Function Instruments. 12 pp.
BSI BS 89: Part 8-90. 1990 Direct Acting Indicating Analogue Electrical Measuring Instruments and Their Accessories Part 8: Special Requirements for Accessories. 14 pp.
BSI BS 89: Part 9-90. 1990 Direct Acting Indicating Analogue Electrical Measuring Instruments and Their Accessories Part 9: Recommended Test Methods. 71 pp.
CENELEC HD 233-75. Recommendations for Direct Acting Indicating Electrical Measuring Instruments and Their Accessories (Superseded by EN 60 051 Series). 3 pp.
CENELEC EN 60 051-1-89. Direct Acting Indicating Analogue Electrical Measuring Instruments and Their Accessories Part 1: Definitions and General Requirements Common to All Parts (Supersedes HD 233). 5 pp.
CENELEC EN 60 051-7-89. Direct Acting Indicating Analogue Electrical Measuring Instruments and Their Accessories Part 7: Special Requirements for Multi-Function Instruments (Supersedes HD 233). 4 pp.
CENELEC EN 60 051-8-89. Direct Acting Indicating Analogue Electrical Measuring Instruments and Their Accessories Part 8: Special Requirements for Accessories (Supersedes HD 233). 4 pp.
CENELEC EN 60 051-9-89. Direct Acting Indicating Analogue Electrical Measuring Instruments and Their Accessories Part 9: Recommended Test Methods (Supersedes HD 233). 4 pp.
IEC 51 Pt 1-84. Direct Acting Indicating Analogue Electrical Measuring Instruments and Their Accessories Part 1: Definitions and General Requirements Common to All Parts Fourth Edition. 74 pp.
IEC 51 Pt 7-84. Direct Acting Indicating Analogue Electrical Measuring Instruments and Their Accessories Part 7: Special Requirements for Multi-Function Instruments Fourth Edition. 16 pp.
IEC 51 Pt 8-84. Direct Acting Indicating Analogue Electrical Measuring Instruments and Their Accessories Part 8: Special Requirements for Accessories Fourth Edition. 20 pp.
IEC 51 Pt 9-88. Direct Acting Indicating Analogue Electrical Measuring Instruments and Their Accessories Part 9: Recommended Test Methods Fourth Edition. 134 pp.

—Direct Acting—Recording
BSI BS 90-75. 1975 Direct Acting Electrical Recording Instruments and Their Accessories. 45 pp.
CENELEC HD 368-77. Direct Acting Recording Electrical Measuring Instruments and Their Accessories. 2 pp.
CNS C3193-84. Method of Test for Direct-Acting Electrical Recording Instruments (Jun)(10922).
CNS C4414-84. Direct-Acting Electrical Recording Instruments (Jun)(10921).
IEC 258-68. Direct Acting Recording Electrical Measuring Instruments and Their Accessories First Edition; (Amendment 1-1976). 130 pp.

—Electrical Codes
CSA 587 Bull. Electrical Bulletin 587 December 30, 1963 to C22.2 NO 115. 1 p.

—Environmental Testing
CNS C1107-82. Weather-Proof Performance of Electricity Meters (Sep)(9357).
CNS C3162-82. Method of Test for Weather-Proof Performance of Electricity Meters (Sep)(9358).
JIS C 1281-79. Weather-Proof Performance of Electricity Meters (R 1984). 14 pp.

—Glossaries
BSI BS 4727:Pt1: Group 04-86. 1986 Electrotechnical, Power, Telecommunication, Electronics, Lighting and Colour Terms Part 1: Terms Common to Power, Telecommunications and Electronics Group 04: Measurement Terminology. 34 pp.
SAA AS 1852.302-88. International Electrotechnical Vocabulary—Part 302: Electrical Measuring Instruments.

—Indirect Acting
CENELEC HD 301-75. Indirect Acting Electrical Measuring Instruments. 2 pp.

—Length Measurement
DIN ENGL 32876 Pt 1-86. Electrical Length Measurement with Analogue Data Acquisition; Concepts, Requirements, Testing (Apr). 6 pp.
DIN ENGL 32876 Pt 1 Suppl. 1-86. Electrical Length Measurement; General Information and Examples of Application (Apr). 4 pp.
JIS B 7450-89. Digital Position Readout. 18 pp.

—Pointers
DIN ENGL 43802 Pt 2-91. Line Scales and Pointers for Indicating Electrical Measuring Instruments; General Requirements (Jan) (the June 1964 Editions of DIN 43802 Parts 1, 5 and 6 Have Been Withdrawn on Publication of the Revised Versions of DIN 43802 Parts 2 to 4.). 4 pp.
DIN ENGL 43802 Pt 4-91. Line Scales and Pointers for Indicating Electrical Measuring Instruments; Scale Graduation and Numbering (Jan) (the June 1964 Edition of DIN 43802 Parts 1, 5 and 6 Have Been Withdrawn on Publication of the Revised Versions of DIN 43802 Parts 2 to 4.). 15 pp.

—Pointers—Design
DIN ENGL 43802 Pt 3-91. Line Scales and Pointers for Indicating Electrical Measuring Instruments; Designs and Dimensions (Jan) (the June 1964 Editions of DIN 43802 Parts 1, 5 and 6 Have Been Withdrawn on Publication of the Revised Versions of DIN 43802 Parts 2 to 4.). 5 pp.

—Radio Receivers
CNS C6209-83. Method of Test for Land Mobile Communication FM or PM Receiver, 25-947 MHz (Characteristics of the Measuring Equipment and Reference Sensitivity) (Dec)(10694).

—Recorders
BSI BS 7437-91. (WITHDRAWN) 1991 Electrical Measuring Instruments—X-Y Recorders (IEC 1028: 1991) (Renumbered as BS EN 61028: 1993). 34 pp.

—Safety
BSI BS 5458-77. 1977 Safety Requirements for Indicating and Recording Electrical Measuring Instruments and Their Accessories. 28 pp.
CSA CAN/CSA-C22. 2NO231SERM89. CSA Safety Requirements for Electrical and Electronic Measuring and Test Equipment; (Gen Instr 1). 133 pp.
CSA C22.2 NO 231 SP-M1990. Audit Checklist and Test Pack for Use with CAN/CSA-C22.2 No. 231 Series-M89 CSA Safety Requirements for Electrical and Electronic Measuring and Test Equipment. 27 pp.
DIN VDE 0410-76. VDE Specification for Electrical Measuring Instruments; Safety Requirements for Indicating and Recording Electrical Measuring Instruments and Their Accessories (Oct). 42 pp.
IEC 414-73. Safety Requirements for Indicating and Recording Electrical Measuring Instruments and Their Accessories First Edition. 52 pp.

—Scales (Indicators)
DIN ENGL 43802 Pt 2-91. Line Scales and Pointers for Indicating Electrical Measuring Instruments; General Requirements (Jan) (the June 1964 Editions of DIN 43802 Parts 1, 5 and 6 Have Been Withdrawn on Publication of the Revised Versions of DIN 43802 Parts 2 to 4.). 4 pp.
DIN ENGL 43802 Pt 4-91. Line Scales and Pointers for Indicating Electrical Measuring Instruments; Scale Graduation and Numbering (Jan) (the June 1964 Edition of DIN 43802 Parts 1, 5 and 6 Have Been Withdrawn on Publication of the Revised Versions of DIN 43802 Parts 2 to 4.). 15 pp.

—Scales (Indicators)—Design
DIN ENGL 43802 Pt 3-91. Line Scales and Pointers for Indicating Electrical Measuring Instruments; Designs and Dimensions (Jan) (the June 1964 Editions of DIN 43802 Parts 1, 5 and 6 Have Been Withdrawn on Publication of the Revised Versions of DIN 43802 Parts 2 to 4.). 5 pp.

—Spiral Springs
CNS B2605-81. Spiral Spring for Electrical Measuring Instruments (Sep)(7866).

—Symbols—Diagrams
BSI BS 3939: Part 8-85. 1985 Graphical Symbols for Electrical Power, Telecommunications and Electronics Diagrams Part 8: Measuring Instruments, Lamps and Signalling Devices (G). 15 pp.
IEC 617 Pt 8-83. Graphical Symbols for Diagrams Part 8: Measuring Instruments, Lamps and Signalling Devices First Edition. 26 pp.

Electric

Electric Measuring Instruments (Cont.)
—**Symbols—Diagrams** (Cont.)
SNZ IEC 617: Part 8-83. Graphical Symbols for Diagrams Part 8: Measuring Instruments, Lamps and Signalling Devices. 23 pp.

Electric Motors
Use: Motors

Electric Outlet Boxes
Use For: Outlet Boxes (Electric) *See Also:* Electric Conduit Boxes; Electric Outlet Covers; Electric Outlets; Electrical Enclosures; Junction Boxes; Wiring Devices

CNS C4230-82. Outlet Boxes for Rigid Steel Conduit (Dec)(6086).
CSA CAN/CSA-C22. 2 NO 18-92. Outlet Boxes, Conduit Boxes, and Fittings; (Gen Instr 1 Thru 2). 118 pp.
CSA C22.2 NO 42-M1984. General Use Receptacles, Attachment Plugs, and Similar Wiring Devices; (Gen Instr 1 Thru 10). 101 pp.
SNZ NZS 1201-70. Specification for Outlet Boxes for Use with Flush-Mounting Electrical Accessories Amend: 1971. 10 pp.

—**Fittings**
CSA CAN/CSA-C22. 2 NO 18-92. Outlet Boxes, Conduit Boxes, and Fittings; (Gen Instr 1 Thru 2). 118 pp.

—**Ships—Refrigeration Equipment**
CNS F5104-86. Socket-Outlet Boxes for Refrigerated Containers (Watertight Type) (Jun)(11616).
JIS F 8837-89. Watertight Socket-Outlet Boxes for Refrigerated Containers.

—**Switches**
IEC 670-89. General Requirements for Enclosures for Accessories for Household and Similar Fixed Electrical Installations Second Edition. 49 pp.

Electric Outlet Covers
See Also: Covers; Electric Outlet Boxes; Electric Outlets; Wiring Devices
CSA CAN/CSA-C22. 2 NO 18-92. Outlet Boxes, Conduit Boxes, and Fittings; (Gen Instr 1 Thru 2). 118 pp.

Electric Outlet Strips
Use For: Electronic Outlet Strips; Multiple Outlet Strips; Outlet Strips (Electric); Power Bars; Power Strips; Receptacle Strips *See Also:* Electric Outlets

—**Hospital Grade**
CSA C22.2 NO 0.4-M1982. Bonding and Grounding of Electrical Equipment (Protective Grounding) (R 1993); (Gen Instr 1). 19 pp.
CSA C22.2 NO 42-M1984. General Use Receptacles, Attachment Plugs, and Similar Wiring Devices; (Gen Instr 1 Thru 10). 101 pp.
CSA CAN/CSA-C22. 2 NO 49-92. Flexible Cords and Cables; (Gen Instr 1). 121 pp.
CSA C22.2 NO 125-M1984. Electromedical Equipment (R 1992); (Gen Instr 1 Thru 2). 68 pp.
CSA 1169 Bull. Electrical Bulletin 1169 June 27, 1978 to C22.2 NO 125. 2 pp.
CSA CAN/CSA-Z32.2-M89. Electrical Safety in Patient Care Areas; (Gen Instr 1). 64 pp.

Electric Outlets
Use For: Cable Outlets; Electric Receptacles; Outlets (Electric); Socket Outlets; Wall Socket Outlets
See Also: Connectors; Electric Outlet Boxes; Electric Outlet Covers; Electric Outlet Strips; Electric Plug Adapters; Electric Plug Receptacles; Electric Plugs; Jacks (Electric); Phone Jacks; Pin Grid Array Sockets; Sockets (Electric); Telephone Jacks; Wall Plates; Wiring Devices

BSI BS 196-61. 1961 Amd 4 Protected-Type Non-Reversible Plugs, Socket-Outlets, Cable-Couplers and Appliance Couplers, with Earthing Contacts for Single Phase a.c. Circuits up to 250 Volts. 64 pp.
BSI BS 546-50. 1950 Amd 7 Two-Pole and Earthing Pin-Plugs, Socket-Outlets and Socket-Outlet Adaptors for Circuits up to 250 Volts. 49 pp.
CNS C3045-88. Method of Test for Plugs and Receptacles for Domestic and Similar General Use (Mar)(3907).
CNS C4012-90. Plugs and Receptacles for Domestic and Similar General Use (Nov)(690).
CNS C4427-85. Hanging Receptacles and Plugs (Mar)(11214).
CSA C22.2 NO 42-M1984. General Use Receptacles, Attachment Plugs, and Similar Wiring Devices; (Gen Instr 1 Thru 10). 101 pp.
CSA C22.2 NO 182.3-M1987. Special Use Attachment Plugs, Receptacles, and Connectors (R 1993); (Gen Instr 1 Thru 2). 41 pp.

Electric Outlets (Cont.)
DIN VDE 0100 Pt 550-88. Erection of Power Installations with Nominal Voltages up to 1000 V; Selection and Erection of Electrical Equipment, Plug-And-Socket Devices, Switches and Installation Accessories (Apr) (Supersedes VDE 0100g/07.76 and VDE 0100/05.73). 9 pp.
DIN VDE 0620-84. Plugs and Socket Outlets up to 250 V, 25 A (Nov). 61 pp.
JIS C 7709-89. Lamp Caps and Holders. 126 pp.
SNZ NZS 1125-53. Specification for Round-Pin, Two-Pole and Earthing-Pin Plugs, Socket-Outlets and Socket-Outlet Adaptors for Use on Circuits up to 250 Volts and up to and Including 30 Amperes Amend: A, 1953; 4A, 1978; Supplement No. 1, 1963; 5A, 1984. 44 pp.
SNZ NZS 1989-65. Specification for Protected-Type Non-Reversible Plugs, Socket-Outlets, Cable-Couplers and Appliance-Couplers, with Earthing Contacts for Single Phase Alternating Current Circuits up to 250 Volts. 64 pp.
SNZ NZS/AS 3131-90. Approval and Test Specification—Plugs and Socket-Outlets for Use in Installation Wiring Systems (This is a Joint Standard with SAA AS 3131). 17 pp.
SNZ NZS 6216-89. Specification for Plugs and Socket-Outlets. 28 pp.
SNZ NZS 6217-89. Specification for Cord Extension Sockets. 16 pp.
SNZ NZS 6218-89. Specification for Plug Socket Adaptors. 16 pp.

—**Aerospace**
CEN PREN 3660-005-93. Aerospace Series Cable Outlet Accessories for Circular and Rectangular Electrical and Optical Connectors Part 005—Cable Outlet, Type A, 005, 90 Degree, Unsealed, with Clamp Strain Relief Product Standard. 9 pp.

—**Agricultural Equipment**
BSI BS 6777: Part 1-86. (WITHDRAWN) 1986 Electrical Connections for Agricultural Machinery Part 1: Double-Pole Connectors (Superseded by BS EN 24165: 1992). 3 pp.

—**Aircraft—Ground Support Equipment**
BSI 4G 173: Part 1-85. 1985 Connectors for Ground Electrical Supplies for Aircraft Part 1: Design, Performance and Test Requirements. 6 pp.
BSI 4G 173: Part 2-85. 1985 Connectors for Ground Electrical Supplies for Aircraft Part 2: Dimensions. 9 pp.
ISO 461 Pt 1-85. Aircraft—Connectors for Ground Electrical Supplies—Part 1: Design, Performance and Test Requirements First Edition. 6 pp.
ISO 461 Pt 2-85. Aircraft—Connectors for Ground Electrical Supplies—Part 2: Dimensions First Edition. 9 pp.

—**Aircraft Refueling Equipment**
BSI G 175-59. 1959 Amd 1 Aircraft Fuel Nozzle Grounding Plugs and Sockets. 2 pp.
ISO 46-73. Aircraft—Fuel Nozzle Grounding Plugs and Sockets First Edition. 3 pp.

—**Appliances**
IEC 884 Pt 2-2-89. Plugs and Socket-Outlets for Household and Similar Purposes Part 2: Particular Requirements for Socket-Outlets for Appliances First Edition. 31 pp.
IEC 884 Pt 2-3-89. Plugs and Socket-Outlets for Household and Similar Purposes Part 2: Particular Requirements for Switched Socket-Outlets Without Interlock for Fixed Installations First Edition. 35 pp.

—**Arc Welding Cables**
IEC 501-75. Safety Requirements for Arc Welding Equipment—Plugs, Socket-Outlets and Couplers for Welding Cables First Edition. 13 pp.

—**Automotive—Mounting**
BSI BS AU 201: Part 1-85. 1985 Relays and Flashers Part 1: Mounting and Positioning Dimensions of Male Tabs and Socket Apertures for Relays and Flashers for Road Vehicles. 5 pp.
ISO 7588-83. Road Vehicles—Relays and Flashers—Mounting and Positioning Dimensions of Male Tabs and Socket Apertures for Relays and Flashers First Edition. 5 pp.

—**Batteries**
MOD UK DSTAN 59-35: Pt 1:Sec 7-70. Plugs and Sockets, Electrical Pattern 112 for Use with Primary Batteries Issue 1. 17 pp.

—**Cords (Electric)**
CNS C4412-19-85. General-Purpose Nonlocking Plugs and Receptacles for Cord Sets (Feb)(10917-19).
CNS C4412-20-85. Specific-Purpose Locking Plugs and Receptacles for Cord Sets (Feb)(10917-20).
CNS C4412-21-85. Other Plugs and Receptacles for Cord Sets (Feb)(10917-21).

Electric Outlets (Cont.)
—**Electric Terminals**
CSA 943B Bull. Electrical Bulletin 943B June 25, 1976 to C22.2 NO 55. 2 pp.

—**Electronic Equipment**
CSA C22.2 NO 182.3-M1987. Special Use Attachment Plugs, Receptacles, and Connectors (R 1993); (Gen Instr 1 Thru 2). 41 pp.

—**Field—Power Distribution**
MOD UK DSTAN 59-4: Part 9-76. Plugs and Sockets, Electrical Pattern 124 for Power Distribution in the Field Issue 1. 18 pp.

—**Hospital Grade**
JIS T 1021-90. "Hospital Grade" Outlet-Sockets and Plugs. 22 pp.

—**Household**
CENELEC PRHD 535 S1-88. Plugs and Socket-Outlets for Household and Similar Purposes Part 1: General Requirements. 46 pp.
IEC 83-75. Plugs and Socket-Outlets for Domestic and Similar General Use Standards Second Edition; (Amendment 1-1979). 48 pp.
IEC 629-78. Standard Sheets for a Modular System (for Installation Accessories for Use in Domestic and Similar Installations) First Edition. 9 pp.
IEC 884 Pt 1-87. Plugs and Socket-Outlets for Household and Similar Purposes Part 1: General Requirements First Edition; (Amendment 1-1988) (Amendment 2-1991). 201 pp.
IEC 906 Pt 1-86. IEC System of Plugs and Socket-Outlets for Household and Similar Purposes Part 1:Plugs and Socket-Outlets 16 A 250 V A.C. First Edition. 35 pp.
IEC 906 Pt 2-92. IEC System of Plugs and Socket-Outlets for Household and Similar Purposes Part 2: Plugs and Socket-Outlets 15 A 125 V a.c. First Edition. 22 pp.
JIS C 8303-88. Plugs and Receptacles for Domestic and Similar General Use. 55 pp.
MOD UK DSTAN 59-4: Part 5-01. Plugs and Sockets, Electrical Part 5: Domestic Types, Including Sockets-Switch Issue 4; Corrigenda (Obsolescent). 19 pp.

—**Industrial**
BSI BS 4343-92. 1992 Plugs, Socket-Outlets and Couplers for Industrial Purposes Part 2: Dimensional Interchangeability Requirements for Pin and Contact-Tube Accessories of Harmonized Configurations (E). 74 pp.
BSI BS 4343-01. 1992 Amd 1 Plugs, Socket-Outlets and Couplers for Industrial Purposes Part 2: Dimensional Interchangeability Requirements for Pin and Contact-Tube Accessories of Harmonized Configurations (AMD 7890) August 15, 1993 (E). 75 pp.
BSI BS 4343-68. 1968 Amd 1 Industrial Plugs, Socket-Outlets and Couplers for a.c and d.c. Supplies. 139 pp.
BSI BS EN 60309-1-92. 1992 Plugs, Socket-Outlets and Couplers for Industrial Purposes Part 1: General Requirements. 67 pp.
BSI BS EN 60309-1-01. 1992 Amd 1 Plugs, Socket-Outlets and Couplers for Industrial Purposes Part 1: General Requirements (AMD 7889) August 15, 1993 (E). 68 pp.
BSI BS EN 60309-2-92. 1992 Plugs, Socket-Outlets and Couplers for Industrial Purposes Part 2: Dimensional Interchangeability Requirements for Pin and Contact-Tube Accessories of Harmonized Configurations (E). 74 pp.
BSI BS EN 60309-2-01. 1992 Amd 1 Plugs, Socket-Outlets and Couplers for Industrial Purposes Part 2: Dimensional Interchangeability Requirements for Pin and Contact-Tube Accessories of Harmonized Configurations (AMD 7890) August 15, 1993 (E). 75 pp.
CENELEC HD 196-75. Plugs, Sockets—Outlets and Coupless for Industrial Purposes. 7 pp.
CENELEC EN 60309-1-92. Plugs, Socket-Outlets and Couplers for Industrial Purposes Part 1: General Requirements. 8 pp.
CENELEC EN 60309-2-92. Plugs, Socket-Outlets and Couplers for Industrial Purposes Part 2: Dimensional Interchangeability Requirements for Pin and Contact-Tube Accessories. 7 pp.
CSA C22.2 NO 182.1-1988. Industrial Type, Special Use Attachment Plugs, Receptacles, and Connectors (R 1993); (Gen Instr 1 Thru 4). 43 pp.
DIN VDE 0623 & 0623A-77. Regulations for Plugs, Socket-Outlets, Couplers and Connectors for Industrial Purposes up to 200 A and up to 750 V (Mar). 68 pp.
IEC 309 Pt 1-88. Plugs, Socket-Outlets and Couplers for Industrial Purposes Part 1: General Requirements Second Edition; (Corrigendum—March 1992). 110 pp.

INTERNATIONAL AND NON-U.S. NATIONAL STANDARDS
SUBJECT INDEX

Electric

Electric Outlets *(Cont.)*
—Industrial *(Cont.)*
IEC 309 Pt 2-89. Plugs, Socket-Outlets and Couplers for Industrial Purposes Part 2: Dimensional Interchangeability Requirements for Pin and Contact-Tube Accessories Second Edition; (Corrigendum—April 1992). 95 pp.

—Locking
CSA C22.2 NO 182.2-M1987. Industrial Locking Type, Special Use Attachment Plugs, Receptacles, and Connectors (R 1993); (Gen Instr 1). 43 pp.

—Locking—Hospital Grade
CSA C22.2 NO 182.2-M1987. Industrial Locking Type, Special Use Attachment Plugs, Receptacles, and Connectors (R 1993); (Gen Instr 1). 43 pp.

—Mining Equipment
BSI BS 5125-90. 1990 50A, 650V Flameproof, Restrained and Bolted Plugs and Sockets for Use in Coal Mines. 21 pp.
BSI BS 5620-79. 1979 200A Flameproof Restrained and Bolted Plugs and Sockets for Voltages Not Exceeding 1100V, Primarily for Use in Mining. 16 pp.
SAA AS 1299-93. Electrical Equipment for Coal Mines—Flameproof Restrained Plugs and Receptacles (in Professional Package 32). 28 pp.

—Modular
IEC 629-78. Standard Sheets for a Modular System (for Installation Accessories for Use in Domestic and Similar Installations) First Edition. 9 pp.

—Plug In Units
MOD UK DEF-5251-A. General Requirements for Sockets for Electronic Valves and Plug-In Electronic Components (Withdrawn).

—Power Supplies
MOD UK DSTAN 59-4: Part 7-69. Plugs and Sockets, Electrical. Part 7: Orientations and Contact Allocations Reserved for Power Supplies (Withdrawn). 18 pp.

—Residual Current Devices
BSI BS 7288-90. 1990 Socket-Outlets Incorporating Residual Current Devices (S. R. C. D's). 30 pp.
BSI BS 7288-01. 1990 Amd 1 Socket-Outlets Incorporating Residual Current Devices (S. R. C. D's) (AMD 7469) January 15, 1993. 32 pp.

—Safety
BSI BS 1363-84. 1984 Amd 6 13 A Fused Plugs and Switched and Unswitched Socket-Outlets (AMD 6302) February 28, 1990. 87 pp.
DIN VDE 0620 A1-87. Plugs and Socket Outlets up to 250 V, 25 A: Amendment 1 (June). 9 pp.
DIN VDE 0620 A3 (D)-86. Plugs and Socket Outlets up to 380 V 25 A: Amendment 3 (Nov). 6 pp.
DIN VDE 0620 A5 (D)-87. Plugs and Socket Outlets up to 380 V 25 A: Amendment 5 (May). 7 pp.
DIN VDE 0620 A8 (D)-89. Plugs and Socket-Outlets up to 380 V 25 A: Amendment 8 (Aug) (Supersedes Draft DIN VDE 0620 A4/11.86). 6 pp.
IEC 501-75. Safety Requirements for Arc Welding Equipment—Plugs, Socket-Outlets and Couplers for Welding Cables First Edition. 13 pp.
SNZ NZS/AS 3105-93. Approval and Test Specification—Electrical Portable Outlet Devices (This is a Joint Standard with SAA AS 3105). 9 pp.

—Screwless
DIN VDE 0620 A6 (D)-88. Plugs and Socket-Outlets up to 380 V 25 A; Fixed Socket-Outlets with Screwless Terminals for 10/16A: Amendment 6 (June). 20 pp.

—Shavers
BSI BS 4573-70. 1970 Amd 2 Two-Pin Reversible Plugs and Shaver Socket-Outlets. 31 pp.

—Ships
CNS F5074-86. Marine Receptacles (Watertight Type) (Oct)(10565).
CNS F5075-86. Marine Receptacles (Non-Watertight Type) (Oct)(10566).
CNS F5076-86. Marine Plugs and Socket—Outlets (Watertight Type) (Oct)(10623).
JIS F 8833-90. Watertight Type Receptacles for Marine Use.
JIS F 8835-91. Marine Non-Watertight Type Receptacles.
JIS F 8836-89. Marine Watertight Type Plugs and Socket-Outlets.

—Ships—Refrigerated Containers
CNS F5105-86. Socket-Outlet Elements for Power Supply of Refrigerated Containers (Jun)(11617).

—Ships—Refrigerated Containers—Monitors
CNS F5106-86. Socket-Outlet Elements for Monitor of Refrigerated Containers (Jun)(11618).

Electric Outlets *(Cont.)*
—Straight Blade—Aerospace
CEN PREN 3660-004-93. Aerospace Series Cable Outlet Accessories for Circular and Rectangular Electrical and Optical Connectors Part 004—Cable Outlet, Type A, 004, Straight, Unsealed, with Clamp Strain Relief Product Standard. 9 pp.

—Symbols—Architectural Drawings
CNS C1100-87. Symbols of Lamps and Socket of Interior Wiring Diagram for Architectural Plans (Dec)(9111).

—Television Equipment
BSI BS 5550: SUBSEC 7.5.3-81. 1981 Cinematography Part 7: Production and Presentation Section 7.5: Film and Television Locating Lighting Subsection 7.5.3: Plugs and Socket Connectors. 35 pp.
BSI BS 5550: SUBSEC 7.5.4-81. 1981 Cinematography Part 7: Production and Presentation Section 7.5: Film and Television Location Lighting Subsection 7.5.4: Single Pole High Current Plugs and Socket Connectors. 23 pp.

—Watt Hour Meters
SAA AS 1284.4-91. Electricity Metering—Part 4: Socket Mounting System. 21 pp.

—X-Ray Equipment
BSI BS 6038-80. 1980 High-Voltage Cable Plug and Socket Connections for Medical X-Ray Equipment. 13 pp.
CENELEC HD 364 S2-83. High Voltage Cable Plug and Socket Connections for Medical X-Ray Equipment. 3 pp.
IEC 526-78. High-Voltage Cable Plug and Socket Connections for Medical X-Ray Equipment Second Edition. 21 pp.
JIS Z 4731-82. High Voltage Cable Plug and Socket for Medical X-Ray Equipment.

Electric Ovens
Use: Ovens—Electric

Electric Pilot Lights
Use: Pilot Lamps

Electric Plug Adapters
Use For: Electric Plug Adaptors; Socket Outlet Adapters *See Also:* Adapters (Electric); Connectors; Electric Outlets; Electric Plug Receptacles; Wiring Devices

BSI BS 546-50. 1950 Amd 7 Two-Pole and Earthing Pin-Plugs, Socket-Outlets and Socket-Outlet Adaptors for Circuits up to 250 Volts. 49 pp.
BSI BS 1363: Part 3-89. 1989 Amd 1 General Purpose Fuse Links for Domestic and Socket-Outlets and Boxes Part 3: Adaptors (AMD 6223) May 31, 1989. 57 pp.
CSA 895B Bull. Electrical Bulletin 895B May 12, 1976 to C22.2 NO 21. 2 pp.
CSA C22.2 NO 42-M1984. General Use Receptacles, Attachment Plugs, and Similar Wiring Devices; (Gen Instr 1 Thru 10). 101 pp.
CSA 895B Bull. Electrical Bulletin 895B May 12, 1976 to C22.2 NO 42. 2 pp.
DIN ENGL 1808-64. Adaptor Sockets for Cotter Retention (June). 2 pp.
DIN VDE 0620 A7 (D)-89. Plugs and Socket Outlets up to 389 V 25 A: Amendment 7: Socket-Adapters and Appliances with Integrated Plugs (June) (Supersedes Draft DIN VDE 0620 A2/07.86). 8 pp.
SNZ NZS 6218-89. Specification for Plug Socket Adaptors. 16 pp.

—Hospital Grade
CSA 895B Bull. Electrical Bulletin 895B May 12, 1976 to C22.2 NO 21. 2 pp.
CSA C22.2 NO 42-M1984. General Use Receptacles, Attachment Plugs, and Similar Wiring Devices; (Gen Instr 1 Thru 10). 101 pp.
CSA 895B Bull. Electrical Bulletin 895B May 12, 1976 to C22.2 NO 42. 2 pp.

—Safety
SNZ NZS/AS 3122-93. Approval and Test Specification—Socket-Outlet Adaptors (This is a Joint Standard with SAA AS 3122). 9 pp.

Electric Plug Adaptors
Use: Electric Plug Adapters

Electric Plug Receptacles
Use For: Receptacles (Electric Plugs)
See Also: Connectors; Electric Outlets; Electric Plug Adapters; Wiring Devices

CSA 895B Bull. Electrical Bulletin 895B May 12, 1976 to C22.2 NO 21. 2 pp.
CSA C22.2 NO 42-M1984. General Use Receptacles, Attachment Plugs, and Similar Wiring Devices; (Gen Instr 1 Thru 10). 101 pp.
CSA 895B Bull. Electrical Bulletin 895B May 12, 1976 to C22.2 NO 42. 2 pp.

Electric Plug Receptacles *(Cont.)*
CSA C22.2 NO 159-M1987. Attachment Plugs, Receptacles, and Similar Wiring Devices for Use in Hazardous Locations: Class I, Groups A, B, C, and D; Class II, Group G, in Coal or Coke Dust, and in Gaseous Mines (R 1993). 39 pp.
CSA C22.2 NO 182.3-M1987. Special Use Attachment Plugs, Receptacles, and Connectors (R 1993); (Gen Instr 1 Thru 2). 41 pp.
SNZ NZS 6219-89. Specification for Electrical Portable Outlet Devices. 16 pp.

—Cords (Electric)
CNS C4412-19-85. General-Purpose Nonlocking Plugs and Receptacles for Cord Sets (Feb)(10917-19).
CNS C4412-20-85. Specific-Purpose Locking Plugs and Receptacles for Cord Sets (Feb)(10917-20).
CNS C4412-21-85. Other Plugs and Receptacles for Cord Sets (Feb)(10917-21).

—Electronic Equipment
CSA C22.2 NO 182.3-M1987. Special Use Attachment Plugs, Receptacles, and Connectors (R 1993); (Gen Instr 1 Thru 2). 41 pp.

—Explosive Atmospheres—Mines
CSA C22.2 NO 159-M1987. Attachment Plugs, Receptacles, and Similar Wiring Devices for Use in Hazardous Locations: Class I, Groups A, B, C, and D; Class II, Group G, in Coal or Coke Dust, and in Gaseous Mines (R 1993). 39 pp.

—Hazardous Locations
CSA 895B Bull. Electrical Bulletin 895B May 12, 1976 to C22.2 NO 21. 2 pp.
CSA C22.2 NO 42-M1984. General Use Receptacles, Attachment Plugs, and Similar Wiring Devices; (Gen Instr 1 Thru 10). 101 pp.
CSA 895B Bull. Electrical Bulletin 895B May 12, 1976 to C22.2 NO 42. 2 pp.
CSA C22.2 NO 159-M1987. Attachment Plugs, Receptacles, and Similar Wiring Devices for Use in Hazardous Locations: Class I, Groups A, B, C, and D; Class II, Group G, in Coal or Coke Dust, and in Gaseous Mines (R 1993). 39 pp.

—Hospital Grade
CSA 895B Bull. Electrical Bulletin 895B May 12, 1976 to C22.2 NO 21. 2 pp.
CSA C22.2 NO 42-M1984. General Use Receptacles, Attachment Plugs, and Similar Wiring Devices; (Gen Instr 1 Thru 10). 101 pp.
CSA 895B Bull. Electrical Bulletin 895B May 12, 1976 to C22.2 NO 42. 2 pp.

—Industrial
CSA C22.2 NO 182.1-1988. Industrial Type, Special Use Attachment Plugs, Receptacles, and Connectors (R 1993); (Gen Instr 1 Thru 4). 43 pp.

—Locking
CSA C22.2 NO 182.2-M1987. Industrial Locking Type, Special Use Attachment Plugs, Receptacles, and Connectors (R 1993); (Gen Instr 1). 43 pp.

—Locking—Hospital Grade
CSA C22.2 NO 182.2-M1987. Industrial Locking Type, Special Use Attachment Plugs, Receptacles, and Connectors (R 1993); (Gen Instr 1). 43 pp.

Electric Plugs
Use For: Attachment Plugs; Cable Plugs; Plug Connectors *See Also:* Connectors; Electric Outlets; Phone Plugs; Telephone Plugs; Wiring Devices

BSI BS 196-61. 1961 Amd 4 Protected-Type Non-Reversible Plugs, Socket-Outlets, Cable-Couplers and Appliance Couplers, with Earthing Contacts for Single Phase a.c. Circuits up to 250 Volts. 64 pp.
BSI BS 546-50. 1950 Amd 7 Two-Pole and Earthing Pin-Plugs, Socket-Outlets and Socket-Outlet Adaptors for Circuits up to 250 Volts. 49 pp.
CNS C1056-85. Concentric Plugs and Jacks (Aug)(5428).
CNS C3045-88. Method of Test for Plugs and Receptacles for Domestic and Similar General Use (Mar)(3907).
CNS C4012-90. Plugs and Receptacles for Domestic and Similar General Use (Nov)(690).
CNS C4221-80. Separable Plug Body (Aug)(6058).
CNS C4412-19-85. General-Purpose Nonlocking Plugs and Receptacles for Cord Sets (Feb)(10917-19).
CNS C4412-20-85. Specific-Purpose Locking Plugs and Receptacles for Cord Sets (Feb)(10917-20).
CNS C4412-21-85. Other Plugs and Receptacles for Cord Sets (Feb)(10917-21).
CNS C4427-85. Hanging Receptacles and Plugs (Mar)(11214).
CSA 895B Bull. Electrical Bulletin 895B May 12, 1976 to C22.2 NO 21. 2 pp.
CSA C22.2 NO 42-M1984. General Use Receptacles, Attachment Plugs, and Similar Wiring Devices; (Gen Instr 1 Thru 10). 101 pp.
CSA 895B Bull. Electrical Bulletin 895B May 12, 1976 to C22.2 NO 42. 2 pp.

INDUSTRY STANDARDS

INTERNATIONAL AND NON-U.S. NATIONAL STANDARDS
SUBJECT INDEX

Electric

Electric Plugs (Cont.)

CSA C22.2 NO 182.3-M1987. Special Use Attachment Plugs, Receptacles, and Connectors (R 1993); (Gen Instr 1 Thru 2). 41 pp.

DIN VDE 0100 Pt 550-88. Erection of Power Installations with Nominal Voltages up to 1000 V; Selection and Erection of Electrical Equipment, Plug-And-Socket Devices, Switches and Installation Accessories (Apr) (Supersedes VDE 0100g/07.76 and VDE 0100/05.73). 9 pp.

DIN VDE 0278 Pt 6-88. Power Cable Accessories with Rated Voltages U up to 30 kV; Plug—In Type or Screw—Type Enclosed Cable Connections Uo/U Above 0.6/1 kV (Aug). 22 pp.

DIN VDE 0620-84. Plugs and Socket Outlets up to 250 V, 25 A (Nov). 61 pp.

JIS C 6560-79. Concentric Plugs and Jacks (R 1984). 20 pp.

SNZ NZS 1125-53. Specification for Round-Pin, Two-Pole and Earthing-Pin Plugs, Socket-Outlets and Socket-Outlet Adaptors for Use on Circuits up to 250 Volts and up to and Including 30 Amperes Amend: A, 1953; 4A, 1978; Supplement No. 1, 1963; 5A, 1984. 44 pp.

SNZ NZS 1989-65. Specification for Protected-Type Non-Reversible Plugs, Socket-Outlets, Cable-Couplers and Appliance-Couplers, with Earthing Contacts for Single Phase Alternating Current Circuits up to 250 Volts. 64 pp.

SNZ NZS/AS 3131-90. Approval and Test Specification—Plugs and Socket-Outlets for Use in Installation Wiring Systems (This is a Joint Standard with SAA AS 3131). 17 pp.

SNZ NZS 6216-89. Specification for Plugs and Socket-Outlets. 28 pp.

—**Agricultural Equipment**

BSI BS 6777: Part 1-86. (WITHDRAWN) 1986 Electrical Connections for Agricultural Machinery Part 1: Double-Pole Connectors (Superseded by BS EN 24165: 1992). 3 pp.

—**Aircraft—Ground Support Equipment**

BSI 4G 173: Part 1-85. 1985 Connectors for Ground Electrical Supplies for Aircraft Part 1: Design, Performance and Test Requirements. 6 pp.

BSI 4G 173: Part 2-85. 1985 Connectors for Ground Electrical Supplies for Aircraft Part 2: Dimensions. 9 pp.

ISO 461 Pt 1-85. Aircraft—Connectors for Ground Electrical Supplies—Part 1: Design, Performance and Test Requirements First Edition. 6 pp.

ISO 461 Pt 2-85. Aircraft—Connectors for Ground Electrical Supplies—Part 2: Dimensions First Edition. 9 pp.

—**Aircraft Refueling Equipment**

BSI G 175-59. 1959 Amd 1 Aircraft Fuel Nozzle Grounding Plugs and Sockets. 2 pp.

ISO 46-73. Aircraft—Fuel Nozzle Grounding Plugs and Sockets First Edition. 3 pp.

—**Appliances**

BSI BS 3283: Part 1-80. 1980 Amd 2 Non-Reversible Connectors and Appliances Inlets for Portable Electrical Appliances(for Circuits up to 250 Volts) Part 1: 13A Connector and Appliance Inlet. 17 pp.

BSI BS 4491: Part 1-89. 1989 Amd 1 Appliance Couplers for Household and Similar General Purposes Part 1: Specification of General Requirements (AMD 6462) April 30, 1991. 123 pp.

BSI BS 4491: Sec 2.1-90. 1990 Appliance Couplers for Household and Similar General Purposes Part 2: Requirements for Individual Types of Coupler Section 2.1: Sewing Machine Couplers. 15 pp.

BSI BS EN 60320-2-2-92. 1992 Appliance Couplers for Household and Similar General Purposes Part 2: Specification for Individual Types of Coupler Section 2.2: Interconnection Couplers for Household and Similar Equipment. 37 pp.

CENELEC EN 60 320-87. Appliance Couplers for Household and Similar General Purposes Part 2: Sewing Machine Couplers. 15 pp.

CENELEC EN 60 320-87. Appliance Couplers for Household and Similar General Purposes. 115 pp.

CENELEC EN 60 320/A1-89. AMD 1 Appliance Couplers for Household and Similar General Purposes. 7 pp.

CENELEC EN 60 320 (Part 2.1)-87. Appliance Couplers for Household and Similar General Purposes Part 2: Sewing Machine Couplers. 15 pp.

CENELEC EN 60320-2-2-91. Appliance Couplers for Household and Similar General Purposes Part 2: Interconnection Couplers for Household and Similar Equipment. 7 pp.

DIN VDE 0627-86. Connectors and Plug-And-Socket Devices for Rated Voltages up to 1000 V a.c. and 1200 V d.c. and Rated Currents up to 500 A for Each Pole (June). 36 pp.

IEC 320-81. Appliance Couplers for Household and Similar General Purposes Second Edition; (Amendment 3-1987, Incorporating Amendment 1 and 2). 181 pp.

Electric Plugs (Cont.)
—**Appliances (Cont.)**

IEC 320 Pt 2-1-84. Appliance Couplers for Household and Similar General Purposes Part 2: Sewing Machine Couplers First Edition. 20 pp.

IEC 320 Pt 2-2-90. Appliance Couplers for Household and Similar General Purposes Part 2: Interconnection Couplers for Household and Similar Equipment First Edition; (Corrigendum—Aug 1990). 69 pp.

IEC 884 Pt 2-2-89. Plugs and Socket-Outlets for Household and Similar Purposes Part 2: Particular Requirements for Socket-Outlets for Appliances First Edition. 31 pp.

IEC 884 Pt 2-3-89. Plugs and Socket-Outlets for Household and Similar Purposes Part 2: Particular Requirements for Switched Socket-Outlets Without Interlock for Fixed Installations First Edition. 35 pp.

SNZ NZS 1801-63. Specification for Domestic Appliance Connectors and Appliance Inlet Sockets. 17 pp.

—**Appliances—Heater Cords**

CSA C22.2 NO 57-M1985. Appliance Plugs for Heater Cord Sets (R 1993); (Gen Instr 1 Thru 3). 26 pp.

—**Arc Welding Cables**

IEC 501-75. Safety Requirements for Arc Welding Equipment—Plugs, Socket-Outlets and Couplers for Welding Cables First Edition. 13 pp.

—**Assemblies**

MOD UK DSTAN 59-95: Pt 1:Sec 1-74. Connector/Cable Assemblies Section 1: Assembly of Pattern 104, Plugs and Sockets, Electrical Issue 1 (Obsolescent). 21 pp.

—**Automotive**

BSI BS EN 24165-92. 1992 Road Vehicles—Electrical Connections—Double-Pole Connector (ISO 4165: 1979). 8 pp.

CEN EN 24165-91. Road Vehicles—Electrical Connections—Double Pole Connector. 3 pp.

ISO 4165-79. Road Vehicles—Electrical Connections—Double-Pole Connector First Edition. 3 pp.

—**Battery Locomotives**

CNS M2079-80. Plug Connection Devices for Explosion-Proof Type Battery Locomotives in Mines (Dec)(6703).

—**Electronic Equipment**

CSA C22.2 NO 182.3-M1987. Special Use Attachment Plugs, Receptacles, and Connectors (R 1993); (Gen Instr 1 Thru 2). 41 pp.

—**Explosive Atmospheres—Mines**

CSA C22.2 NO 159-M1987. Attachment Plugs, Receptacles, and Similar Wiring Devices for Use in Hazardous Locations: Class I, Groups A, B, C, and D; Class II, Group G, in Coal or Coke Dust, and in Gaseous Mines (R 1993). 39 pp.

—**Field—Power Distribution**

MOD UK DSTAN 59-4: Part 9-76. Plugs and Sockets, Electrical Pattern 124 for Power Distribution in the Field Issue 1. 18 pp.

—**Fluid Power Equipment**

BSI BS 6361-88. 1988 Three-Pin Plug Connectors for Electrically Controlled Fluid Power Equipment. 8 pp.

ISO 4400-85. Fluid Power Systems and Components—Three-Pin Electrical Plug Connector—Characteristics and Requirements Second Edition. 6 pp.

ISO 6952-89. Fluid Power Systems and Components—Two-Pin Electrical Plug Connector with Earth Contact—Characteristics and Requirements First Edition. 7 pp.

—**Hazardous Locations**

CSA C22.2 NO 159-M1987. Attachment Plugs, Receptacles, and Similar Wiring Devices for Use in Hazardous Locations: Class I, Groups A, B, C, and D; Class II, Group G, in Coal or Coke Dust, and in Gaseous Mines (R 1993). 39 pp.

—**Hospital Grade**

CSA 895B Bull. Electrical Bulletin 895B May 12, 1976 to C22.2 NO 21. 2 pp.

CSA C22.2 NO 42-M1984. General Use Receptacles, Attachment Plugs, and Similar Wiring Devices; (Gen Instr 1 Thru 10). 101 pp.

CSA 895B Bull. Electrical Bulletin 895B May 12, 1976 to C22.2 NO 42. 2 pp.

CSA CAN/CSA-C22. 2NO601.1-M90. Medical Electrical Equipment, Part 1: General Requirements for Safety; (Gen Instr 1). 240 pp.

JIS T 1021-90. "Hospital Grade" Outlet-Sockets and Plugs. 22 pp.

Electric Plugs (Cont.)
—**Household**

BSI BS EN 50075-91. 1991 Flat Non-Wirable Two-Pole Plugs, 2,5 A 250 V, with Cord, for the Connection of Class II-Equipment for Household and Similar Purposes. 26 pp.

CENELEC PRHD 535 S1-88. Plugs and Socket-Outlets for Household and Similar Purposes Part 1: General Requirements. 46 pp.

CENELEC EN 50075-90. Flat Non-Wirable Two-Pole Plugs, 2, 5 A 250 V, With Cord, for the Connection of Class II-Equipment for Household and Similar Purposes. 26 pp.

IEC 83-75. Plugs and Socket-Outlets for Domestic and Similar General Use Standards Second Edition; (Amendment 1-1979). 48 pp.

IEC 884 Pt 1-87. Plugs and Socket-Outlets for Household and Similar Purposes Part 1: General Requirements First Edition; (Amendment 1-1988) (Amendment 2-1991). 201 pp.

IEC 906 Pt 1-86. IEC System of Plugs and Socket-Outlets for Household and Similar Purposes Part 1:Plugs and Socket-Outlets 16 A 250 V A.C. First Edition. 35 pp.

IEC 906 Pt 2-92. IEC System of Plugs and Socket-Outlets for Household and Similar Purposes Part 2: Plugs and Socket-Outlets 15 A 125 V a.c. First Edition. 22 pp.

JIS C 8303-88. Plugs and Receptacles for Domestic and Similar General Use. 55 pp.

MOD UK DSTAN 59-4: Part 5-01. Plugs and Sockets, Electrical Part 5: Domestic Types, Including Sockets-Switch Issue 4; Corrigenda (Obsolescent). 19 pp.

—**Household—Fused**

IEC 884 Pt 2-1-87. Plugs and Socket-Outlets for Household and Similar Purposes Part 2: Particular Requirements for Fused Plugs First Edition. 19 pp.

—**Industrial**

BSI BS 4343-92. 1992 Plugs, Socket-Outlets and Couplers for Industrial Purposes Part 2: Dimensional Interchangeability Requirements for Pin and Contact-Tube Accessories of Harmonized Configurations (E). 74 pp.

BSI BS 4343-01. 1992 Amd 1 Plugs, Socket-Outlets and Couplers for Industrial Purposes Part 2: Dimensional Interchangeability Requirements for Pin and Contact-Tube Accessories of Harmonized Configurations (AMD 7890) August 15, 1993 (E). 75 pp.

BSI BS 4343-68. 1968 Amd 1 Industrial Plugs, Socket-Outlets and Couplers for a.c and d.c. Supplies. 139 pp.

BSI BS EN 60309-1-92. 1992 Plugs, Socket-Outlets and Couplers for Industrial Purposes Part 1: General Requirements. 67 pp.

BSI BS EN 60309-1-01. 1992 Amd 1 Plugs, Socket-Outlets and Couplers for Industrial Purposes Part 1: General Requirements (AMD 7889) August 15, 1993 (E). 68 pp.

BSI BS EN 60309-2-92. 1992 Plugs, Socket-Outlets and Couplers for Industrial Purposes Part 2: Dimensional Interchangeability Requirements for Pin and Contact-Tube Accessories of Harmonized Configurations (E). 74 pp.

BSI BS EN 60309-2-01. 1992 Amd 1 Plugs, Socket-Outlets and Couplers for Industrial Purposes Part 2: Dimensional Interchangeability Requirements for Pin and Contact-Tube Accessories of Harmonized Configurations (AMD 7890) August 15, 1993 (E). 75 pp.

CENELEC EN 60309-1-92. Plugs, Socket-Outlets and Couplers for Industrial Purposes Part 1: General Requirements. 8 pp.

CENELEC EN 60309-2-92. Plugs, Socket-Outlets and Couplers for Industrial Purposes Part 2: Dimensional Interchangeability Requirements for Pin and Contact-Tube Accessories. 7 pp.

CSA C22.2 NO 182.1-1988. Industrial Type, Special Use Attachment Plugs, Receptacles, and Connectors (R 1993); (Gen Instr 1 Thru 4). 43 pp.

DIN VDE 0623 & 0623A-77. Regulations for Plugs, Socket-Outlets, Couplers and Connectors for Industrial Purposes up to 200 A and up to 750 V (Mar). 68 pp.

IEC 309 Pt 1-88. Plugs, Socket-Outlets and Couplers for Industrial Purposes Part 1: General Requirements Second Edition; (Corrigendum—March 1992). 110 pp.

IEC 309 Pt 2-89. Plugs, Socket-Outlets and Couplers for Industrial Purposes Part 2: Dimensional Interchangeability Requirements for Pin and Contact-Tube Accessories Second Edition; (Corrigendum—April 1992). 95 pp.

—**Locking**

CSA C22.2 NO 182.2-M1987. Industrial Locking Type, Special Use Attachment Plugs, Receptacles, and Connectors (R 1993); (Gen Instr 1). 43 pp.

INTERNATIONAL AND NON-U.S. NATIONAL STANDARDS
SUBJECT INDEX

Electric

Electric Plugs (Cont.)

—**Locking—Hospital Grade**
CSA C22.2 NO 182.2-M1987. Industrial Locking Type, Special Use Attachment Plugs, Receptacles, and Connectors (R 1993); (Gen Instr 1). 43 pp.

—**Medical Electrical Equipment**
CSA CAN/CSA-C22. 2NO601.1-M90. Medical Electrical Equipment, Part 1: General Requirements for Safety; (Gen Instr 1). 240 pp.

—**Mining Equipment**
BSI BS 5125-90. 1990 50A, 650V Flameproof, Restrained and Bolted Plugs and Sockets for Use in Coal Mines. 21 pp.
BSI BS 5620-79. 1979 200A Flameproof Restrained and Bolted Plugs and Sockets for Voltages Not Exceeding 1100V, Primarily for Use in Mining. 16 pp.
SAA AS 1299-93. Electrical Equipment for Coal Mines—Flameproof Restrained Plugs and Receptacles (in Professional Package 32). 28 pp.

—**Polarized—Fluorescent Lighting**
CSA 1276 Bull. Electrical Bulletin 1276 July 16, 1980 to C22.2 NO 12. 2 pp.

—**Polarized—Incandescent Lighting**
CSA 1276 Bull. Electrical Bulletin 1276 July 16, 1980 to C22.2 NO 12. 2 pp.

—**Power Supplies**
MOD UK DSTAN 59-4: Part 7-69. Plugs and Sockets, Electrical. Part 7: Orientations and Contact Allocations Reserved for Power Supplies (Withdrawn). 18 pp.

—**Safety**
BSI BS 1363-84. 1984 Amd 6 13 A Fused Plugs and Switched and Unswitched Socket-Outlets (AMD 6302) February 28, 1990. 87 pp.
DIN VDE 0620 A1-87. Plugs and Socket Outlets up to 250 V, 25 A: Amendment 1 (June). 9 pp.
DIN VDE 0620 A3 (D)-86. Plugs and Socket Outlets up to 380 V 25 A: Amendment 3 (Nov). 6 pp.
DIN VDE 0620 A5 (D)-87. Plugs and Socket Outlets up to 380 V 25 A: Amendment 5 (May). 7 pp.
DIN VDE 0620 A8 (D)-89. Plugs and Socket-Outlets up to 380 V 25 A: Amendment 8 (Aug) (Supersedes Draft DIN VDE 0620 A4/11.86). 6 pp.
IEC 501-75. Safety Requirements for Arc Welding Equipment—Plugs, Socket-Outlets and Couplers for Welding Cables First Edition. 13 pp.

—**Shavers**
BSI BS 4573-70. 1970 Amd 2 Two-Pin Reversible Plugs and Shaver Socket-Outlets. 31 pp.

—**Ships**
CNS F5072-86. Marine Plugs (Non-Watertight Type) (Oct)(10563).
CNS F5073-86. Marine Plugs (Watertight Type) (Oct)(10564).
CNS F5076-86. Marine Plugs and Socket—Outlets (Watertight Type) (Oct)(10623).
JIS F 8831-84. Marine Non-Watertight Type Plugs.
JIS F 8832-84. Marine Watertight Type Plugs.
JIS F 8836-89. Marine Watertight Type Plugs and Socket-Outlets.

—**Storage Batteries—Aircraft**
NATO STANAG 3660 ED 2 AMD 4-77. Aircraft Main Battery Electric Plug. 13 pp.
NATO STANAG 3660 ED 2 AMD 6-77. Aircraft Main Battery Electric Plug. 14 pp.
NATO STANAG 3660 ED 2 AMD 7-77. Aircraft Main Battery Electric Plug. 15 pp.

—**Telecommunication Equipment**
BSI BS 6312-85. 1985 Plugs for Use with British Telecommunications Line Jack Units. 14 pp.

—**Television Equipment**
BSI BS 5550: SUBSEC 7.5.3-81. 1981 Cinematography Part 7: Production and Presentation Section 7.5: Film and Television Locating Lighting Subsection 7.5.3: Plugs and Socket Connectors. 35 pp.
BSI BS 5550: SUBSEC 7.5.4-81. 1981 Cinematography Part 7: Production and Presentation Section 7.5: Film and Television Location Lighting Subsection 7.5.4: Single Pole High Current Plugs and Socket Connectors. 23 pp.

—**Towed Vehicles**
BSI BS AU 194-84. 1984 Performance of Electrical Connections Between Towing Vehicles and Trailers. 4 pp.
BSI BS AU 195-84. (WITHDRAWN) 1984 Mounting Electrical Connections on Rear Cross Members of Towing Vehicles (Superseded by BS AU 195A: 1991). 4 pp.
BSI BS AU 195A-91. 1991 Mounting Electrical Connections on Rear Cross Members of Towing Vehicles (ISO 4009: 1989). 6 pp.

Electric Plugs (Cont.)

—**Towed Vehicles (Cont.)**
BSI BS AU 197-84. 1984 Electrical Connections Between Towing Vehicles and Trailers with 24V Equipment: Type 24N (Normal). 7 pp.
BSI BS AU 198-84. 1984 Electrical Connections Between Towing and Trailers with 24V Electrical Equipment: Type 24S (Supplementary). 7 pp.
ISO 4009-89. Towing Vehicles—Mounting of Electrical Connections on Rear Cross-Members Second Edition. 4 pp.
ISO 4091-92. Road Vehicles—Connectors for Electrical Connections Between Towing Vehicles and Trailers—Test Methods and Performance Requirements Second Edition. 10 pp.

—**Towed Vehicles—Interchangeability**
BSI BS AU 149A-80. 1980 Electrical Connections Between Towing Vehicles and Trailers with 6V or 12V Electrical Equipment: Type 12N (Normal). 6 pp.
BSI BS AU 177A-80. 1980 Amd 1 Electrical Connections Between Towing Vehicles and Trailers with 6V or 12V Electrical Equipment: Type 12S (Supplementary). 7 pp.
ISO 1185-75. Road Vehicles—Electrical Connections Between Towing Vehicles and Towed Vehicles with 24 V Electrical Equipment—Type 24 N (Normal) First Edition. 7 pp.
ISO 1724-80. Road Vehicles—Electrical Connections Between Towing Vehicles and Towed Vehicles with 6 or 12 V Electrical Equipment—Type 12 N (Normal) Second Edition. 6 pp.
ISO 3731-80. Road Vehicles—Electrical Connections Between Towing Vehicles and Trailers with 24 V Electrical Equipment—Type 24 S (Supplementary) Second Edition. 6 pp.
ISO 3732-82. Road Vehicles—Electrical Connections Between Towing Vehicles and Trailers with 6 or 12 V Electrical Equipment—Type 12 S (Supplementary) Second Edition. 6 pp.

—**X-Ray Equipment**
BSI BS 6038-80. 1980 High-Voltage Cable Plug and Socket Connections for Medical X-Ray Equipment. 13 pp.
CENELEC HD 364 S2-83. High Voltage Cable Plug and Socket Connections for Medical X-Ray Equipment. 3 pp.
IEC 526-78. High-Voltage Cable Plug and Socket Connections for Medical X-Ray Equipment Second Edition. 21 pp.
JIS Z 4731-82. High Voltage Cable Plug and Socket for Medical X-Ray Equipment.

Electric Power

Scope Note: Use a more specific term *See:* Air Switches; Busbars; Cables (Electric); Cabling Systems (Electric Power); Capacitors; Carrier Leaks; Decibels; Electric Outlets; Electric Plugs; Electrical Grounding; Electrical Measurement; Electrical Properties; Gain (Amplification); Loads (Electric); Loss Factor; Power Distribution Systems; Power Systems; Power Transmission; Psophometric Power; Regulators; Signal Levels; Substations (Electric); Transmission Lines

Electric Power Distribution
Use: Power Distribution Systems

Electric Power Generation
Use: Generating Stations

Electric Power Generation Equipment
Use: Generating Stations

Electric Power Meters
See Also: Demand Meters; Electric Energy Meters; Electric Measuring Instruments; Electrical Measurement; Multimeters; Phase Meters; Volume Indicators; Volume Unit Meters; Watt Hour Meters; Wattmeters
CNS C4429-85. General Rules for Electricity Meters (Dec)(11428).

—**Electrical Enclosures**
CNS K3089-90. Plastic Box for Low Voltage A.C. Power Meter (Sep)(11908).
CNS K6903-90. Method Test for Plastic Box for Low Voltage A.C. Power Meter (Sep)(11909). 6 pp.
SNZ NZS 6206-80. Specification for Domestic Electric Meter Boxes Amend: 1, 1988; 2, 1992. 7 pp.

—**Instrument Transformers—Error Analysis**
CNS C1134-85. Calculation Method for the Error of the Instrument Transformers for Metering Service (Dec)(11439).

Electric Power Plants
Use: Power Plants

Electric Power Stations
Use: Power Plants

Electric Power Substations
Use: Substations (Electric)

Electric Power System Disturbances
Use: Power Systems—Disturbances

Electric Power Systems
Use: Power Systems

Electric Power Transmission
Use: Power Transmission

Electric Primers

—**Arabic Gum**
CNS K1272-82. Arabic Gum (for Electric Primer Uses) (Aug)(9301).
CNS K6725-82. Method of Test for Arabic Gum (for Electric Primer Use) (Aug)(9302).

Electric Properties
Use: Electrical Properties

Electric Propulsion

—**Ships**
IEC 92 Pt 501-84. Electrical Installations in Ships Part 501: Special Features—Electric Propulsion Plant Third Edition. 32 pp.
JIS F 8073-86. Electrical Installations in Ships Part 501 Special Features—Electric Propulsion Plant (IEC 92-501-1894).

Electric Raceway Fittings
See Also: Busway Fittings; Electric Raceways

—**Floor—Concrete**
CSA C22.2 NO 79-1978. Cellular Metal and Cellular Concrete Floor Raceways and Fittings (R 1992); (Amd 1-4 January 1983) (Amd 5 October 1984) (Amd 6 July 1986) (Amd 7-8 December 1990) (Amd 9-20 August 1991). 43 pp.

—**Floor—Metal**
CSA C22.2 NO 79-1978. Cellular Metal and Cellular Concrete Floor Raceways and Fittings (R 1992); (Amd 1-4 January 1983) (Amd 5 October 1984) (Amd 6 July 1986) (Amd 7-8 December 1990) (Amd 9-20 August 1991). 43 pp.

—**Lighting**
CSA C22.2 NO 62-93. Surface Raceway Systems; (Gen Instr 1). 30 pp.

—**Surface—Metallic**
CSA C22.2 NO 62-93. Surface Raceway Systems; (Gen Instr 1). 30 pp.

—**Surface—Nonmetallic**
CSA C22.2 NO 62-93. Surface Raceway Systems; (Gen Instr 1). 30 pp.

—**Underfloor**
CSA C22.2 NO 80-1978. Underfloor Raceways and Fittings (R 1992); (Amd 1-5 January 1983) (Amd 6 October 1984) (Amd 7 July 1986) (Amd 8-13 August 1991). 27 pp.
DIN VDE 0634 Pt 2-87. Ducting and Trunking and Their Accessories for Underfloor Electrical Installations (Dec). 34 pp.

Electric Raceways
Use For: Cable Trunking; Raceways (Electric); Skirt Trunking; Trunking (Enclosures) *See Also:* Auxiliary Gutters; Bushings; Busways; Cable Trays; Cabling Systems (Electric Power); Electric Conduits; Electric Raceway Fittings; Electrical Enclosures; Electrical Metallic Tubing; Electrical Nonmetallic Tubing; Liquidtight Flexible Conduits; Power Distribution Systems; Pull Boxes; Trolley Busways; Wireways
DIN VDE 0604 Pt 3-86. Trunking on Mounted Walls and Ceilings for Electrical Installations; Skirting Board Ducts (May). 6 pp.
IEC 1084 Pt 1-91. Cable Trunking and Ducting Systems for Electrical Installations Part 1: General Requirements First Edition. 50 pp.

—**Ceiling**
DIN VDE 0604 Pt 1-86. Trunking Mounted on Walls and Ceilings for Electrical Installations; General Requirements. (May). 18 pp.
DIN VDE 0604 Pt 2-86. Trunking Mounted on Walls and Ceilings for Electrical Installations; Electrical Installation Ducts (May). 6 pp.

—**Concrete—Floors**
CSA C22.2 NO 79-1978. Cellular Metal and Cellular Concrete Floor Raceways and Fittings (R 1992); (Amd 1-4 January 1983) (Amd 5 October 1984) (Amd 6 July 1986) (Amd 7-8 December 1990) (Amd 9-20 August 1991). 43 pp.

INDUSTRY STANDARDS

INTERNATIONAL AND NON-U.S. NATIONAL STANDARDS
SUBJECT INDEX

Electric

Electric Raceways (Cont.)
—**Insulating Material**
BSI BS 4678: Part 4-82. 1982 Amd 1 Cable Trunking Part 4: Cable Trunking Made of Insulating Material. 12 pp.

—**Lighting**
CSA C22.2 NO 62-93. Surface Raceway Systems; (Gen Instr 1). 30 pp.

—**Lighting Poles**
BSI BS 5649: Part 5-82. 1982 Lighting Columns Part 5: Base Compartments and Cableways. 7 pp.
CEN EN 40-5-82. Lighting Columns—Part 5: Base Compartments and Cableways. 5 pp.

—**Metal—Floors**
CSA C22.2 NO 79-1978. Cellular Metal and Cellular Concrete Floor Raceways and Fittings (R 1992); (Amd 1-4 January 1983) (Amd 5 October 1984) (Amd 6 July 1986) (Amd 7-8 December 1990) (Amd 9-20 August 1991). 43 pp.

—**Steel**
BSI BS 4678: Part 1-71. 1971 Amd 3 Cable Trunking Part 1: Steel Surface Trunking. 19 pp.

—**Steel—Underfloor**
BSI BS 4678: Part 2-73. 1973 Amd 1 Cable Trunking Part 2: Steel Underfloor (Duct) Trunking. 8 pp.

—**Surface—Metallic**
CSA C22.2 NO 62-93. Surface Raceway Systems; (Gen Instr 1). 30 pp.

—**Surface—Nonmetallic**
CSA C22.2 NO 62-93. Surface Raceway Systems; (Gen Instr 1). 30 pp.

—**Surface—Plastic**
JIS C 8425-84. Plastic Surface Raceways for Interior Wiring. 8 pp.

—**Underfloor**
CSA C22.2 NO 80-1978. Underfloor Raceways and Fittings (R 1992); (Amd 1-5 January 1983) (Amd 6 October 1984) (Amd 7 July 1986) (Amd 8-13 August 1991). 27 pp.
DIN VDE 0634 Pt 2-87. Ducting and Trunking and Their Accessories for Underfloor Electrical Installations (Dec). 34 pp.

—**Wall**
DIN VDE 0604 Pt 1-86. Trunking Mounted on Walls and Ceilings for Electrical Installations; General Requirements. (May). 18 pp.
DIN VDE 0604 Pt 2-86. Trunking Mounted on Walls and Ceilings for Electrical Installations; Electrical Installation Ducts (May). 6 pp.

Electric Railroad Signals
Use: Railroad Signaling Systems—Electric

Electric Railroads
Use For: Electric Railways *See Also:* Electric Rolling Stock; Monorails; Railroad Traction Equipment; Rolling Stock; Sanders (Rolling Stock); Streetcars

—**Overhead Contact System**
IEC 913-88. Electric Traction Overhead Lines First Edition. 48 pp.

—**Overhead Contact System—Fittings**
JIS E 2002-89. Test Methods of Fittings for Overhead Contact System of Electric Railway. 17 pp.
JIS E 2205-89. Hanger Fittings for Overhead Contact System of Electric Railway. 17 pp.
JIS E 2206-89. Splicing Fittings for Overhead Contact System of Electric Railway. 18 pp.
JIS E 2207-89. Fittings for Feeder in Overhead Contact System of Electric Railway.
JIS E 2208-89. Anchor Fittings for Overhead Contact System of Electric Railway. 15 pp.

—**Overhead Contact System—Fittings—Glossaries**
JIS E 2001-89. Glossary of Terms Relating to Fittings for Overhead Contact System of Electric Railway.

—**Overhead Contact System—Insulators**
JIS E 2219-90. Section Insulators for Overhead Contact System of Electric Railway.
JIS E 2219-89. Section Insulators for Overhead Contact System of Electric Railway. 13 pp.
JIS E 2301-80. Insulator for Electric Overhead Line (R 1985). 19 pp.

—**Overhead Contact System—Power Supplies**
CSA CAN/CSA-C22. 3 NO 8-M91. Railway Electrification Guidelines; (Gen Instr 1 Thru 3). 103 pp.

Electric Railroads (Cont.)
—**Overhead Contact System—Sleeves**
JIS E 2220-89. Connecting Sleeves of Stranded Conductors for Overhead Contact System of Electric Railway. 18 pp.

—**Overhead Contact System—Trolley Poles**
JIS E 2202-89. Steady Arm for Overhead Contact System of Electric Railway. 17 pp.

—**Overhead Contact System—Trolley Retrievers**
JIS E 2201-89. Pull-off for Overhead Contact System of Electric Railway.

—**Overhead Contact System—Turnbuckles**
JIS E 2221-89. Wire Turnbuckles for Overhead Contact System of Electric Railway. 12 pp.

—**Overhead Contact System—Wire Clips**
JIS E 2209-89. Wire Clips for Overhead Contact System of Electric Railway. 12 pp.

Electric Railways
Use: Electric Railroads

Electric Ranges
Use: Ranges—Electric

Electric Reactors
See Also: Resistors; Saturable Reactors; Series Reactors; Shunt Reactors; Transformers; Voltage Dividers

BSI BS 7452-91. 1991 Separating Transformers, Autotransformers, Variable Transformers and Reactors (IEC 989: 1991). 130 pp.
CENELEC HD 539 S1-91. Reactors. 5 pp.
CENELEC HD 539 S2-92. Reactors. 13 pp.
CENELEC HD 539 S2-92. CORRIGENDUM Reactors. 1 p.
DIN VDE 0550 Pt 1 (D)-87. Separating Transformers, Autotransformers, Variable Transformers and Reactors. Identical with IEC 14D(CO)29 (Nov) (Partially Supersedes 0550 Part 1/12.69, Part 3/ 12.69 and Part 6/4.66) (Supersedes 0550 Part 4/ 4.66, Part 5/9.72 and 0552/5.69). 117 pp.
DIN VDE 0550 Pt 6-66. Specifications for Small Transformers; Part 6: Particular Specifications for Reactors (Line Reactors, Pre-Magnetised Reactors and RFI Reactors) (Apr) (Partially Superseded by Draft 0550 Part 1/11.87). 11 pp.
IEC 289-88. Reactors Second Edition. 92 pp.
IEC 989-91. Separating Transformers, Autotransformers, Variable Transformers and Reactors First Edition. 241 pp.

—**Electrical Enclosures**
BSI BS 6435-84. 1984 Unfilled Enclosures for the Dry Termination of HV Cables for Transformers and Reactors. 7 pp.

—**Glossaries**
IEC 50 Chap 421-90. International Electrotechnical Vocabulary Chapter 421: Power Transformers and Reactors First Edition. 67 pp.

—**Impulse Voltage Testing**
IEC 722-82. Guide to the Lightning Impulse and Switching Impulse Testing of Power Transformers and Reactors First Edition. 70 pp.

—**Insulating Oils—Gas Analysis**
BSI BS EN 61181-93. 1993 Impregnated Insulating Materials—Application of Dissolved Gas Analysis (DGA) to Factory Tests on Electrical Equipment (IEC 1181: 1993) (Q). 15 pp.
CENELEC PREN 61181-92. Impregnated Insulating Materials—Application of Dissolved Gas Analysis (DGA) to Factory Tests on Electrical Equipment. 13 pp.
IEC 1181-93. Impregnated Insulating Materials—Application of Dissolved Gas Analysis (DGA) to Factory Tests on Electrical Equipment First Edition. 21 pp.

—**Junction Boxes**
BSI BS 2562-79. 1979 Cable Boxes for Transformers and Reactors. 63 pp.

—**Railroad Traction Equipment**
CENELEC PREN 60310-93. Railway Applications—Traction Transformers and Reactors on Rolling Stock. 15 pp.

—**Semiconductor Converters**
BSI BS EN 60146-1-3-93. 1993 Semiconductor Convertors General Requirements and Line Commutated Convertors Part 1-3: Transformers and Reactors (IEC 146-1-3: 1991) (S). 19 pp.
CENELEC EN 60146-1-3-93. Semiconductor Convertors General Requirements and Line Commutated Convertors Part 1-3: Transformers and Reactors (IEC 146-1-3:1991). 5 pp.

Electric Reactors (Cont.)
—**Semiconductor Converters** (Cont.)
CENELEC EN 60146-1-3-93. Semiconductor Convertors General Requirements and Line Commutated Convertors Part 1-3: Transformers and Reactors (IEC 146-1-3:1991). 14 pp.
IEC 146 Pt 1-3-91. Semiconductor Convertors General Requirements and Line Commutated Convertors Part 1-3: Transformers and Reactors Third Edition; (CENELEC EN 60146-1-3: 1993). 28 pp.

—**Sound Level Measurement**
BSI BS 6056-91. 1991 Determination of Transformer and Reactor Sound Levels. 30 pp.
BSI BS 6056-81. 1981 Measurement of Transformer and Reactor Sound Levels. 22 pp.
CENELEC HD 399-79. Measurement of Transformer and Reactor Sound Level. 8 pp.
CENELEC HD 399 S2-91. Determination of Transformer and Reactor Sound Levels. 10 pp.
CENELEC EN 60551-92. Determination of Transformer and Reactor Sound Levels (IEC 551:1987) (Supersedes HD 399 S2:1991). 8 pp.
IEC 551-87. Determination of Transformer and Reactor Sound Levels Second Edition; (CENELEC EN 60551:1992). 44 pp.

Electric Receptacles
Use: Electric Outlets

Electric Relays
Use: Relays

Electric Resistance Furnaces
Use: Resistance Furnaces

Electric Resistance Strain Gages
Use: Strain Gages

Electric Resistors
Use: Resistors

Electric Rolling Stock
See Also: Commuter Cars; Electric Railroads; Electric Traction; Mine Cars; Monorails; Rolling Stock; Sanders (Rolling Stock)

IEC 1133-92. Electric Traction—Rolling Stock—Test Methods for Electric and Thermal/Electric Rolling Stock on Completion of Construction and Before Entry into Service First Edition. 90 pp.

—**Choppers (Circuits)**
JIS E 6201-91. General Rules of Choppers for Electric Rolling Stock.

—**Control Systems Equipment**
JIS E 5004-91. Control Equipment for Electric Rolling Stock—Test Methods.

—**Electric Wiring**
JIS E 6001-76. Wire Code for DC Electric Railcar.

—**Wire Code**
JIS E 6001-76. Wire Code for DC Electric Railcar.

Electric Screwdrivers
Use: Screwdrivers—Electric

Electric Shavers
Use: Shavers

Electric Shock
See Also: Electrical Grounding; Electrical Protection Equipment; Electrical Safety

—**Biomechanical Effects (Human Body)**
BSI PD 6519: Part 1-88. 1988 Guide to Effects of Current Passing Through the Human Body Part 1: General Aspects. 21 pp.
BSI PD 6519: Part 2-88. 1988 Guide to Effects of Current Passing Through the Human Body Part 2: Special Aspects. 24 pp.
IEC 479 Pt 1-84. Effects of Current Passing Through the Human Body Part 1: General Aspects Chapter 1: Electrical Impedance of the Human Body Chapter 2: Effects of Alternating Current in the Range of 15 Hz to 100 Hz Chapter 3: Effects of Direct Current Second Edition. 40 pp.
IEC 479 Pt 2-87. Effects of Current Passing Through the Human Body Part 2: Special Aspects Chapter 4: Effects of Alternating Current with Frequencies Above 100 Hz Chapter 5: Effects of Special Waveforms of Current Chapter 6: Effects of Unidirectional Single. 48 pp.
IEC 990-90. Methods of Measurement of Touch-Current and Protection Conductor Current First Edition; (Corrigendum—Aug 1991). 73 pp.
SAA AS 3859-91. Guide to the Effects of Current Passing Through the Human Body (IEC 479-1:1984, IEC 479-2:1987) (In Professional Package 47). 39 pp.

INTERNATIONAL AND NON-U.S. NATIONAL STANDARDS
SUBJECT INDEX

Electric

Electric Shock (Cont.)
—Electric Measuring Instruments
CNS C6035-79. Shocks Testing for Electrical Indicating Instruments (Jun)(4876).

—Fiber Optics
CNS C6259-86. Method of Test for Fiber Optic Devices (FOTP-14 Shock Test, Specified Pulse) (Sep)(11703).

—Medical Electrical Equipment
CSA CAN/CSA-C22. 2NO601.1-M90. Medical Electrical Equipment, Part 1: General Requirements for Safety; (Gen Instr 1). 240 pp.

—Protection
BSI BS 921-76. 1976 Rubber Mats for Electrical Purposes. 7 pp.
BSI BS 2754-76. 1976 Amd 1 Memorandum Construction of Electrical Equipment for Protection Against Electric Shock. 26 pp.
BSI BS 3042-71. 1971 Amd 3 Standard Test Fingers and Probes for Checking Protection Against Electrical, Mechanical and Thermal Hazard. 23 pp.
BSI BS 6907: Part 2-88. 1988 Electrical Installations for Open-Cast Mines and Quarries Part 2: General Recommendations for Protection Against Direct Contact and Electric Shock. 22 pp.
BSI BS CECC 31400-82. 1982 Fixed Ceramic Capacitors of Dielectric Class 1, for Electrical Shock Hazard Protection: Sectional Specification. 25 pp.
BSI BS CECC 31400-01. 1982 Amd 1 Sectional Specification: Fixed Ceramic Capacitors of Dielectric Class 1, for Electrical Shock Hazard Protection (AMD 7404) February 15, 1993. 28 pp.
BSI BS CECC 31401-88. 1988 Fixed Ceramic Dielectric Capacitors, Dielectric Class 1, for Electrical Shock Hazard Protection. 14 pp.
BSI BS CECC 31500-82. 1982 Fixed Ceramic Capacitors of Dielectric Class 2, for Electrical Shock Hazard Protection: Sectional Specification. 25 pp.
BSI BS CECC 31501-88. 1988 Fixed Ceramic Dielectric Capacitors, Dielectric Class 2, for Electrical Shock Hazard Protection. 14 pp.
BSI BS EN 50 059-91. 1991 Electrostatic Hand-Held Spraying Equipment for Non-Flammable Material for Painting and Finishing. 15 pp.
BSI PD 2754: Part 2-93. 1993 Classification of Electrical and Electronic Equipment with Regard to Protection Against Electric Shock Part 2: Guide to Requirements for Protection Against Electric Shock (IEC 536-2: 1992) (E). 22 pp.
BSI PD 6535-93. 1993 Common Aspects for Installation and Equipment for Protection Against Electric Shock (IEC 1440: 1992) (E). 15 pp.
CENELEC HD 366-77. Classification of Electrical and Electronic Equipment with Regard to Protection Against Electric Shock. 2 pp.
CENELEC HD 384.4.41-80. Electrical Installations of Buildings Part 4: Protection for Safety Chapter 41: Protection Against Electric Shock. 46 pp.
CENELEC HD 384.4.41/A1-80. AMD 1 Electrical Installations of Buildings Part 4: Protection for Safety Chapter 41: Protection Against Electric Shock. 2 pp.
CENELEC HD 384.4.41 S1/PRA1-93. AMD 1 Electrical Installations of Buildings Part 4: Protection for Safety Chapter 41: Protection Against Electric Shock. 13 pp.
CENELEC HD 384.4.47 S1-88. Electrical Installations of Buildings Part 4: Protection for Safety Chapter 47: Application of Protective Measures for Safety Section 470—General—Section 471—Measures of Protection Against Electric Shock. 5 pp.
CENELEC EN 50 059-90. Specification for Electrostatic Hand-Held Spraying Equipment for Non-Flammable Material for Painting and Finishing. 24 pp.
CSA Z259.4-M1979. Rubber Insulating Gloves and Mitts; (Erratum April 1980) (Erratum November 1980). 29 pp.
CSA Z259.5-M1979. Rubber Insulating Sleeves; (Erratum August 1980). 31 pp.
CSA Z259.6-M1981. Rubber Insulating Blankets. 16 pp.
CSA Z259.7-M1981. Rubber Insulating Line Hose; (Gen Instr 1). 18 pp.
CSA Z259.8-M1981. Rubber Insulating Covers; (Gen Instr 1). 18 pp.
DIN VDE 0100 Pt 410 (S)-83. Installation of Power Plant with Rated Voltages Not Exceeding 1000 V; Protective Measures; Protection Against Electric Shock (Nov) (Superseded by 0100 Part 460/10.88). 53 pp.
DIN VDE 0100 Pt 701-84. Installation of Power Plant with Rated Voltages up to 1000 V; Locations Containing a Bath or Shower (May). 19 pp.
DIN VDE 0106 Pt 1 A3 (D)-92. Protection Against Shock Current; Classification of Electrical & Electronic Equipment with Regard to Protection Against Shock: Amendment 3 (Jan). 4 pp.
DIN VDE 0106 Pt 100-83. Protection Against Electric Shock; Location of Control Elements in the Vicinity of Shock-Hazard Parts (Mar). 19 pp.

Electric Shock (Cont.)
—Protection (Cont.)
DIN VDE 0106 Pt 101-86. Protection Against Electric Shock; Basis Requirements for Protective Separation in Electrical Equipment (Nov). 13 pp.
DIN VDE 0106 Pt 101 A1 (D)-89. Protection Against Electric Shock; Basis Requirements for Protective Separation in Electrical Equipment; Amendment 1 (Feb). 4 pp.
DIN VDE 0470 Pt 1-84. Testing Equipment and Test Methods; Testing of Protection Against Contact; IEC Test Finger (Dec). 7 pp.
IEC 364 Pt 4-41-92. Electrical Installations of Buildings Part 4: Protection for Safety Chapter 41: Protection Against Electric Shock Third Edition. 53 pp.
IEC 364 Pt 4-47-81. Electrical Installations of Buildings Part 4: Protection for Safety Chapter 47: Application of Protective Measures for Safety Section 470: General Section 471: Measures of Protection Against Electric Shock First Edition. 12 pp.
IEC 536-76. Classification of Electrical and Electronic Equipment with Regard to Protection Against Electric Shock First Edition. 13 pp.
IEC 621 Pt 2-87. Electrical Installations for Outdoor Sites Under Heavy Conditions (Including Open-Cast Mines and Quarries) Part 2: General Protection Requirements Second Edition. 106 pp.
IEC 755-83. General Requirements for Residual Current Operated Protective Devices First Edition; (Amendment 1-1988) (Amendment 2-1992). 140 pp.
IEC 1140-92. Protection Against Electric Shock Common Aspects for Installation and Equipment First Edition. 29 pp.
SNZ NZS 1582-61. Specification for Standard Test Fingers for Checking Protection Against Electric Shock and Mechanical Hazard Amend: A, 1961; 1, 1968; 2, 1968 (Reconfirmed 1978). 8 pp.

—Protection—Ceramic Capacitors—Preferred Products List
CECC CECC MUAHAG Vol 1 IS 4-90. Preferred Products List; Capacitors (En, Fr, Ge). 64 pp.

—Protection—Classification
IEC 536 Pt 2-92. Classification of Electrical and Electronic Equipment with Regard to Protection Against Electric Shock Part 2: Guidelines to Requirements for Protection Against Electric Shock First Edition. 41 pp.

—Protection—Electrical Installations
IEC 364 Pt 4-481-93. Electrical Installations of Buildings Part 4: Protection for Safety Chapter 48: Choice of Protective Measures as a Function of External Influences Section 481—Selection of Measures for Protection Against Electric Shock in Relation to External Influences First Edition. 23 pp.

—Protection Equipment—Installation
IEC 1140-92. Protection Against Electric Shock Common Aspects for Installation and Equipment First Edition. 29 pp.

Electric Shock Testing
Use: Electric Shock

Electric Signs
Use: Signs

Electric Sockets
Use: Sockets (Electric)

Electric Spark Machines
Use: Electric Discharge Machine Tools

Electric Streetcars
Use: Streetcars

Electric Strength
Use: Dielectric Strength

Electric Supply Stations
Use: Power Systems

Electric Switches
Scope Note: For additional listings, use a more specific term *Use For:* Electromechanical Switches
See Also: Air Switches; Automatic Transfer Switches; Bypass Switches; Circuit Breakers; Control Switches; Control Systems Equipment; Cord Switches; Cryostats; Diode Switches; Disconnecting Switches; Dry Reed Switches; Electric Contacts; Electromagnetic Switches; Electronic Switches; Enclosed Switches; Fusible Switches; Grounding Switches; Interrupter Switches; Isolating Switches; Joystick Switches; Keyswitches; Knife Switches; Latching Relays; Lever Switches; Limit Switches; Load Break Switches; Matrix Switches; Membrane Switches; Multiposition Switches; Oil Switches; Pilot

Electric Switches (Cont.)
See Also: (Cont.)
Switches; Polarized Relays; Position Switches; Pressure Switches; Programming Switches; Pull Switches; Pushbutton Switches; Relays; Remote Control Switches; Rocker Switches; Rotary Switches; Selector Switches; Sensitive Switches; Slide Switches; Snap Switches; Substations (Electric); Switchgear; Switching Circuits; Switching Systems; Thermal Switches; Thermostatic Switches; Thermostats; Thumbwheel Switches; Time Delay Relays; Time Delay Switches; Time Switches; Toggle Switches; Transistors; Vacuum Switches; Voltage Regulators; Wall Plates; Wiring Devices

BSI BS EN 60947-3-92. 1992 Low-Voltage Switchgear and Controlgear Part 3: Switches, Disconnectors, Switch-Disconnectors and Fuse-Combination Units. 62 pp.
BSI BS EN 60947-3-01. 1992 Amd 1 Low-Voltage Switchgear and Controlgear Part 3: Switches, Disconnectors, Switch-Disconnectors and Fuse-Combination Units (AMD 7854) July 15, 1993 (E). 63 pp.
CENELEC EN 60947-3-92. Low-Voltage Switchgear and Controlgear Part 3: Switches, Disconnectors, Switch-Disconnectors and Fuse-Combination Units. 8 pp.
CENELEC EN 60947-3-92. CORRIGENDUM Low-Voltage Switchgear and Controlgear Part 3: Switches, Disconnectors, Switch Diconnectors and Fuse-Combination Units. 1 p.
CNS C4421-84. All-Cover Switch (Nov)(11092).
CNS C6083-88. Method of Test for Electromechanical Switches (General Rules) (Dec)(6146).
CSA C22.2 NO 55-M1986. Special Use Switches (R 1992); (Gen Instr 1 Thru 2). 34 pp.
CSA 1052 Bull. Electrical Bulletin 1052 April 28, 1976 to C22.2 NO 55. 1 p.
CSA 1169D Bull. Electrical Bulletin 1169D July 18, 1979 to C22.2 NO 55. 2 pp.
CSA C22.2 NO 111-M1986. General Use Switches (R 1992); (Gen Instr 1 Thru 2). 40 pp.
CSA 1052 Bull. Electrical Bulletin 1052 April 28, 1976 to C22.2 NO 111. 1 p.
CSA 1169D Bull. Electrical Bulletin 1169D July 18, 1979 to C22.2 NO 111. 2 pp.
DIN VDE 0100 Pt 550-88. Erection of Power Installations with Nominal Voltages up to 1000 V; Selection and Erection of Electrical Equipment, Plug-And-Socket Devices, Switches and Installation Accessories (Apr) (Supersedes VDE 0100g/07.76 and VDE 0100/05.73). 9 pp.
DIN VDE 0660 Pt 107 A1 (D)-91. Switchgear and Controlgear; Low Voltage Switchgear and Controlgear; Part 3: Switches, Disconnectors, Switch-Disconnectors and Fuse-Combination Units (Nov). 5 pp.
DIN VDE 0660 Pt 107 A2 (D)-92. Switchgear and Controlgear; Low-Voltage Switchgear and Controlgear; Part 3: Switches, Disconnectors and Fuse Combination Units (Feb). 5 pp.
IEC 947 Pt 3-90. Low-Voltage Switchgear and Controlgear Part 3: Switches, Disconnectors, Switch-Disconnectors and Fuse-Combination Units First Edition; (Corrigendum—Dec 1991). 115 pp.

—Actuators
CNS C6092-88. Method of Test for Electromechanical Switches (Actuator/Mounting Bushing Resistance) (Dec)(6155).

—Alarm Systems
BSI BS 4737: Sec 3.3-77. 1977 Intruder Alarm Systems in Buildings Part 3: Components Section 3.3: Protective Switches. 2 pp.

—Appliances
BSI BS 3676-63. (WITHDRAWN) 1963 Amd 7 Switches for Domestic and Similar Purposes (For Fixed or Portable Mounting) (Superseded by BS 3676: Part 1: 1989). 49 pp.
BSI BS EN 61058-1-92. 1992 Switches for Appliances Part 1: General Requirements (E). 108 pp.
CENELEC EN 61058-1-92. Switches for Appliances Part 1: General Requirements. 7 pp.
CENELEC EN 61058-1-92. Switches for Appliances Part 1: General Requirements. 9 pp.
CENELEC EN 61058-1-92. CORRIGENDUM Switches for Appliances Part 1: General Requirements (IEC 1058-1:1990). 1 p.
DIN VDE 0630-86. Switches for Appliances for a Rated Voltage Not Exceeding 500 V and a Rated Current Not Exceeding 63 A (Apr). 23 pp.
DIN VDE 0630 A4 (D)-89. Switches for Appliances for a Rated Voltage Not Exceeding 500V and a Rated Current Not Exceeding 63A; Amendment 4: No-Load Switches (Apr). 11 pp.
DIN VDE 0630 Pt 10-88. Switches for Appliances for a Rated Voltage Not Exceeding 500 V and a Rated Current Not Exceeding 63 A (Sept). 17 pp.
IEC 1058 Pt 1-90. Switches for Appliances Part 1: General Requirements First Edition; (Amendment 1-1993) (CENELEC EN 61058-1: 1992). 211 pp.

INTERNATIONAL AND NON-U.S. NATIONAL STANDARDS
SUBJECT INDEX

Electric

Electric Switches *(Cont.)*

—**Automotive—Backup Lights**
CNS D2056-87. Switches of Back-Up Lamps for Automobiles (Jan)(6324).
JIS D 5810-74. Back-up-Lamp Switches for Automobiles.

—**Automotive—Brake Lights**
CNS D2026-87. Mechanical Stop Lamp Switches for Automobiles (Aug)(5780).
CNS D2052-87. Switches of Hydraulic Stop Lamp for Automobiles (Jan)(6174).
JIS D 5808-76. Mechanical Stop Lamp Switches for Automobiles.
JIS D 5808-72. Mechanical Stop Lamp Switches for Automobiles (R 1976). 9 pp.

—**Automotive—Environmental Testing**
JIS D 0208-79. General Rules of Environmental Requirements for Switches of Automobiles (R 1983). 19 pp.

—**Automotive—Flasher Units**
JIS D 5811-74. Inspection of Hazard Warning Switches for Automobiles (R 1983). 10 pp.

—**Automotive—Parking Lights**
CNS D2053-87. Switches of Parking Lamps for Automobiles (Jan)(6175).

—**Automotive—Starters**
CNS D2024-87. Starter Switch for Automobiles (Aug)(5567).
JIS D 5806-79. Starter Switches for Automobiles.

—**Capacitance Measurement**
CNS C6085-88. Method of Test for Electromechanical Switches (Capacitance) (Dec)(6148).

—**Chatter**
CNS C6094-88. Method of Test for Electromechanical Switches (Monitoring Contact Chatter) (Dec)(6157).

—**Chromaticity**
CNS C6237-84. Method of Test for Electromechanical Switches (Chromaticity) (Nov)(11100).

—**Contact Bounce**
CNS C6235-84. Method of Test for Electromechanical Switches (Contact Bounce) (Nov)(11098).

—**Discharge Lamps**
CSA C22.2 NO 111-M1986. General Use Switches (R 1992); (Gen Instr 1 Thru 2). 40 pp.

—**Dust**
CNS C6095-88. Method of Test for Electromechanical Switches (Sand and Dust) (Dec)(6158).

—**Electric Endurance Testing**
CNS C6089-88. Method of Test for Electromechanical Switches (Electrical Endurance) (Dec)(6152).
CNS C6239-84. Method of Test for Electromechanical Switches (Logic (TTL) Level Endurance) (Nov)(11102).
CNS C6240-84. Method of Test for Electromechanical Switches (Low Level Endurance) (Nov)(11103).

—**Electrical Resistance**
CNS C6084-88. Method of Test for Electromechanical Switches (Switch Resistance) (Dec)(6147).

—**Endurance Testing**
CNS C6090-88. Method of Test for Electromechanical Switches (Mechanical Endurance) (Dec)(6153).

—**Explosion Proof**
CNS C4444-86. Explosion Proof Electric Switches (Dec)(117800).

—**Fluorescent Lighting**
CSA C22.2 NO 111-M1986. General Use Switches (R 1992); (Gen Instr 1 Thru 2). 40 pp.

—**Flush Mounting Boxes**
IEC 670-89. General Requirements for Enclosures for Accessories for Household and Similar Fixed Electrical Installations Second Edition. 49 pp.

—**High Temperature Testing**
CNS C6091-88. Method of Test for Electromechanical Switches (High/Low Temperature Operation) (Dec)(6154).

—**High Voltage**
BSI BS 5463-77. (WITHDRAWN) 1977 A.C. Switches of Rated Voltage Above 1 kV (Superseded by BS 5463: Part 1 and 2: 1991). 40 pp.
BSI BS 5463: Part 1-91. 1991 High-Voltage Switches Part 1: High-Voltage Switches for Rated Voltages Above 1kV and Less Than 52 kV. 34 pp.
BSI BS 5463: Part 2-91. 1991 High-Voltage Switches Part 2: High-Voltage Switches for Rated Voltages of 52 kV and Above. 52 pp.
CENELEC HD 355-76. High voltage Switches. 8 pp.

Electric Switches *(Cont.)*

—**High Voltage** *(Cont.)*
CENELEC HD 355.1 S2-91. High-Voltage Switches Part 1: High-Voltage Switches for Rated Voltages Above 1kV and Less Than 52 kv. 7 pp.
CENELEC HD 355.2 S2-91. High-Voltage Switches Part 2: High-Voltage Switches for Rated Voltages of 52 kV and Above. 6 pp.
IEC 265 Pt 1-83. High-Voltage Switches Part 1: High-Voltage Switches for Rated Voltages Above 1 kV and Less Than 52 kV Second Edition; (Amendment 1-1984) (Corrigendum—Oct 1986) (Corrigendum—Feb 1990). 66 pp.
IEC 265 Pt 2-88. High-Voltage Switches Part 2: High-Voltage Switches for Rated Voltages of 52 kV and Above First Edition; (Corrigendum—Feb 1990). 97 pp.
SAA AS 1025.1-84. High-Voltage a.c. Switchgear and Controlgear—Switches and Switch-Disconnectors—Part 1: For Rated Voltages Above 1 kV and Less Than 52 kV. 16 pp.
SAA AS 1025.2-89. High-Voltage a.c. Switchgear and Controlgear—Switches and Switch-Disconnectors—Part 2: For Rated Voltages of 52 kV and Above. 43 pp.

—**Household**
BSI BS 3676: Part 1-89. 1989 Switches for Household and Similar Fixed Electrical Installations Part 1: General Requirements. 47 pp.
CNS C4015-87. Small Switches for Indoor Use (Jul)(695). 14 pp.
DIN VDE 0632 Pt 1-91. Switches for Household and Similar Fixed-Electrical Installations General Requirements (Apr). 115 pp.
IEC 669 Pt 1-81. Switches for Household and Similar Fixed-Electrical Installations Part 1: General Requirements First Edition; (Amendment 1-1987) (Amendment 2-1992). 195 pp.
JIS C 8304-88. Small Switches for Indoor Use. 20 pp.
SNZ IEC 669: Part 1-81. Switches for Household and Similar Fixed-Electrical Installations Part 1: General Requirements. 112 pp.
SNZ NZS 2065-66. Specification for Switches for Domestic and Similar Purposes (for Fixed or Portable Mounting) Amend: A, 1966. 40 pp.

—**In Line**
SAA AS 3127-87. Approval and Test Specification—Cord-Line Switches Amdt 1,December 1987 Amdt 2, November 1989 (This is a Joint Standard with SANZ NZS 3127). 3 pp.
SNZ NZS/AS 3127-87. Approval and Test Specification—Cord-Line Switches Amend: 1, 1987; 2, 1989 (This is a Joint Standard with SAA AS 3127). 3 pp.

—**Insulation Resistance**
CNS C6092-88. Method of Test for Electromechanical Switches (Actuator/Mounting Bushing Resistance) (Dec)(6155).

—**Low Temperature Testing**
CNS C6091-88. Method of Test for Electromechanical Switches (High/Low Temperature Operation) (Dec)(6154).

—**Luminance**
CNS C6238-86. Method of Test for Electromechanical Switches—TP-3 Transmittancy (Luminance) (Dec)(11101).

—**Marine**
CNS F5028-86. Marine Small Switches (Non-Watertight Type) (Aug)(8583).
CNS F5029-86. Marine Small Switches (Watertight Type) (Aug)(8584).
CNS F5031-82. Unit Switches for Marine Use (Mar)(8586).
JIS F 8841-84. Marine Watertight Type Small Switches.
JIS F 8844-89. Unit Switches for Marine Use.

—**Mechanical Shock**
CNS C6093-88. Method of Test for Electromechanical Switches (Shock Test (Specified Rules) (Dec)(6156).

—**Modular**
IEC 629-78. Standard Sheets for a Modular System (for Installation Accessories for Use in Domestic and Similar Installations) First Edition. 9 pp.

—**Modules**
IEC 629-78. Standard Sheets for a Modular System (for Installation Accessories for Use in Domestic and Similar Installations) First Edition. 9 pp.

—**Motors**
BSI BS EN 60947-3-92. 1992 Low-Voltage Switchgear and Controlgear Part 3: Switches, Disconnectors, Switch-Disconnectors and Fuse-Combination Units. 62 pp.

Electric Switches *(Cont.)*

—**Motors** *(Cont.)*
BSI BS EN 60947-3-01. 1992 Amd 1 Low-Voltage Switchgear and Controlgear Part 3: Switches, Disconnectors, Switch-Disconnectors and Fuse-Combination Units (AMD 7854) July 15, 1993 (E). 63 pp.
CENELEC EN 60947-3-92. Low-Voltage Switchgear and Controlgear Part 3: Switches, Disconnectors, Switch-Disconnectors and Fuse-Combination Units. 8 pp.
CENELEC EN 60947-3-92. CORRIGENDUM Low-Voltage Switchgear and Controlgear Part 3: Switches, Disconnectors, Switch Diconnectors and Fuse-Combination Units. 1 pp.
CNS C3079-87. Method of Test for Small Switch of Single-Phase Motors (Jul)(5536).
CNS C4193-87. Small Switch for Single-Phase Motors (Mar)(5535). 5 pp.
DIN VDE 0660 Pt 107 A1 (D)-91. Switchgear and Controlgear; Low Voltage Switchgear and Controlgear; Part 3: Switches, Disconnectors, Switch-Disconnectors and Fuse-Combination Units (Nov). 5 pp.
DIN VDE 0660 Pt 107 A2 (D)-92. Switchgear and Controlgear; Low-Voltage Switchgear and Controlgear; Part 3: Switches, Disconnectors and Fuse Combination Units (Feb). 5 pp.
IEC 947 Pt 3-90. Low-Voltage Switchgear and Controlgear Part 3: Switches, Disconnectors, Switch-Disconnectors and Fuse-Combination Units First Edition; (Corrigendum—Dec 1991). 115 pp.
JIS C 4506-79. Small Switches for Single-Phase Motors.

—**Mountings—Mechanical Properties**
CNS C6086-88. Method of Test for Electromechanical Switches (Strength of Mounting Means) (Dec)(6149).

—**Overload Testing**
CNS C6088-88. Method of Test for Electromechanical Switches (Overload) (Dec)(6151).

—**Packaging**
MOD UK DSTAN 81-36-89. Packaging of Switches Issue 2. 15 pp.

—**Quality Assurance**
BSI BS CECC 96000-87. 1987 Electromechanical Switches: Generic Specification. 44 pp.
BSI BS EN 196000-93. 1993 Generic Specification: Electromechanical Switches. 54 pp.
CENELEC EN 196000-92. Generic Specification: Electromechanical Switches. 48 pp.

—**Reliability Assured**
MOD UK DSTAN 59-75: Part 1-77. Switches of Assessed Quality (Listed on EPIC Database) Part 1: General Requirements Issue 1. 9 pp.

—**Sands**
CNS C6095-88. Method of Test for Electromechanical Switches (Sand and Dust) (Dec)(6158).

—**Snap Action**
CSA 1052 Bull. Electrical Bulletin 1052 April 28, 1976 to C22.2 NO 55. 1 p.
CSA C22.2 NO 111-M1986. General Use Switches (R 1992); (Gen Instr 1 Thru 2). 40 pp.
CSA 1052 Bull. Electrical Bulletin 1052 April 28, 1976 to C22.2 NO 111. 1 p.

—**Soldering**
CNS C6236-84. Method of Test for Electromechanical Switches (Environmental Effects of Machine Soldering) (Nov)(11099).

—**Sound Activated**
CSA C22.2 NO 205-M1983. Signal Equipment (R 1992); (Gen Instr 1). 32 pp.

—**Street Lamps**
CNS C3093-80. Method of Test for Small Switches for Street Lamps (Aug)(6053).
CNS C4219-80. Small Switches for Street Lamps (Aug)(6052).

—**Symbols**
CNS C1060-83. Symbols for Sequential Control (Breakers and Switches) (Nov)(5528).
CNS C5116-81. Graphic Symbols for Switches and Contractors (May)(7365).

—**Symbols—Architectural Drawings**
CNS C1091-87. Symbols of Switch of Interior Wiring Diagram for Architectural Plans (Dec)(9102).

—**Terminals—Temperature Measurement**
CNS C6087-88. Method of Test for Electromechanical Switches (Terminal Temperature Rise) (Dec)(6150).

—**Uninterruptible Power Supplies**
IEC 146 Pt 5-88. Semiconductor Convertors Part 5: Switches for Uninterruptible Power Systems (UPS Switches) First Edition. 48 pp.

INDEX and DIRECTORY of

INTERNATIONAL AND NON-U.S. NATIONAL STANDARDS
SUBJECT INDEX

Electric

Electric Switchgear
Use: Switchgear

Electric Terminals
Use For: Cable Terminations *See Also:* Battery Terminals; Binding Posts; Connectors; Electric Contacts; Electrical Components; Feedthrough Insulators; Ferrules; Ground Terminals; Power Line Hardware; Soldering Lugs; Splices (Electrical); Standoffs; Terminal Blocks; Terminal Boards; Terminal Lugs; Testing Points; Wiring Space

BSI BS 7240-90. 1990 Pin Allocations for Microprocessor Systems Using the IEC 603.2 Connector. 8 pp.
CNS C4423-84. Non-Thread Terminals (Nov)(11094).
CNS C6015-90. Robustness of Termination Testing Procedures for Electronic Components (Dec)(3628).
CNS C6310-88. Method of Test for Terminal (Jun)(12340).
CSA C22.2 NO 65-93. Wire Connectors; (Gen Instr 1). 67 pp.
IEC 828-88. Pin Allocations for Microprocessor Systems Using the IEC 603-2 Connector First Edition. 11 pp.
JIS C 0051-86. Robustness of Terminations and Integral Mounting Devices for Electronic Components. 17 pp.
JTC1 828-88. Pin Allocations for Microprocessor Systems Using the IEC 603-2 Connector First Edition. 11 pp.
MOD UK DSTAN 59-3: Part 6-68. Terminals, Electrical Part 6: Terminals, Post Issue 1. 17 pp.
MOD UK DSTAN 59-3: Pt 6: Add. Terminals, Electrical Part 6: Terminals, Post Issue 1. 3 pp.
MOD UK DSTAN 59-40: Part 7-90. Inter-Circuit Terminations Part 7: Sealed Terminals (Leadthrough and Post) Sectional Specification Issue 1. 27 pp.
MOD UK DEF-5334-B-62. Boards, Stand-Off Insulators, and Terminals (Superseded by Def Stan 59-40: Part 2). 27 pp.

—**Aerospace—Busway Fittings—Impedance Measurement**
CEN PREN 2591 (Part G7)-92. Elements of Electrical and Optical Connection Test Methods Part G7—Measurement of Characteristic Impedance of a Bus or a Stub Terminator. 3 pp.

—**Aircraft**
BSI G 225-79. (OBSOLESCENT) 1979 Performance of Environment-Resistant Terminal Junction Modules with Removable Crimp-Type Contacts. 37 pp.
ISO 8668 Pt 1-86. Aircraft—Terminal Junction Systems—Part 1: Characteristics First Edition. 9 pp.
ISO 8668 Pt 2-86. Aircraft—Terminal Junction Systems—Part 2: Tests First Edition. 13 pp.
ISO 8668 Pt 5-92. Aircraft—Terminal Junction Systems—Part 5: Detail Specification for Type 3 System First Edition. 64 pp.

—**Aircraft—Crimp Contacts**
BSI G 184-G188-64. (OBSOLESCENT) 1964 Amd 1 Aluminium Terminal Ends and Inline Connectors for Hexagonal Crimping to Aircraft Aluminium Electric Cables. 17 pp.
BSI G 204-67. 1967 Amd 1 Copper Terminal Ends for Crimping to Electric Cables with Copper Conductors. 13 pp.
ISO 1965-73. Aluminium Terminal Ends for Crimping to Aircraft Aluminium Electrical Cables First Edition. 4 pp.
ISO 1966-73. Crimped Joints for Aircraft Electrical Cables First Edition. 9 pp.

—**Aircraft—Crimp Contacts—Thermocouples**
BSI 2G 215: Part 2-89. 1989 Extension Cables for Aircraft Nickel-Chromium and Nickel-Aluminium Thermocouple Part 2: Terminations. 6 pp.
BSI G 220-76. (WITHDRAWN) 1976 Amd 1 Crimped Terminations to Stranded Flexible Thermocouple Cables of the Nickel-Chromium and Nickel-Aluminium Types (Superseded by BS 2G 215: Part 2: 1989). 5 pp.
ISO 8056 Pt 2-88. Aircraft—Nickel-Chromium and Nickel-Aluminium Thermocouple Extension Cables—Part 2: Terminations—General Requirements and Tests First Edition. 7 pp.
ISO 8056 Pt 3-87. Aircraft—Nickel-Chromium and Nickel-Aluminium Thermocouple Extension Cables—Part 3: Crimp-Type Ring Terminal Ends—Dimensions First Edition. 4 pp.

—**Aircraft—Electrical Grounding—Bases**
SBAC AS 4639 ISSUE 1. Base Earth Terminal 4 B.A.. 1 p.
SBAC AS 4640 ISSUE 1. Base Earth Terminal 6 B.A.. 1 p.

—**Aircraft—Fasteners**
BSI G 203-67. 1967 Unified Screws with Captive Facing and Locking Washers. 4 pp.

Electric Terminals *(Cont.)*
—**Aircraft—Spark Plugs**
JIS W 4502-77. Spark Plug Terminals for Aircraft Piston Engines.

—**Automotive**
CNS D2064-90. Cable Terminals for Automobiles (Type LA) (Aug)(6902).
CNS D2065-90. Cable Terminals for Automobiles (Type LB) (Aug)(6903).
CNS D2066-90. Cable Terminals for Automobiles (Type LD) (Aug)(6904).
CNS D2067-90. Cable Terminals for Automobiles (Type LE) (Aug)(6905).
JIS D 5403-89. Cable Terminals for Automobiles.

—**Automotive—Cylindrical**
CNS D2075-88. Male Cylindric Terminals for Automobiles (CA Type) (Jan)(7024).
CNS D2076-88. Female Cylindric Terminals for Automobiles (CB Type) (Jan)(7025).
CNS D2077-88. Compound Cylindric Terminals for Automobiles (CW Type) (Jan)(7026).

—**Automotive—Distributors**
CNS D1066-87. Dimension of High Tension Connection for Ignition Coils and Distributors for Automobiles (Oct)(12123).
CNS D2022-2-86. High Tension Cable Connection for Ignition Coils and Distributors for Automobiles (Nov)(5561-2).
CNS D2022-3-86. Low Tension Cable Connections for Ignition Coils for Automobiles (Nov)(5561-3).
CNS D2073-90. High Tension Cable Terminals of Distributor for Automobiles (LA Type) (Aug)(6911).
ISO 3553 Pt 1-87. Road Vehicles—High-Tension Connections for Ignition Coils and Distributors—Part 1: Socket-Type First Edition. 6 pp.
ISO 3553 Pt 2-87. Road Vehicles—High-Tension Connections for Ignition Coils and Distributors—Part 2: Plug-Type First Edition. 4 pp.

—**Automotive—Flasher Units**
ISO TR8857-86. Road Vehicles—Flashers—Functional Allocation of Terminals First Edition. 3 pp.

—**Automotive—Ignition Coils**
BSI BS AU 167-78. (WITHDRAWN) 1978 Low Tension Cable Connections for Ignition Cells (Superseded by BS ISO 4024: 1992). 3 pp.
CNS D1066-87. Dimension of High Tension Connection for Ignition Coils and Distributors for Automobiles (Oct)(12123).
CNS D2022-2-86. High Tension Cable Connection for Ignition Coils and Distributors for Automobiles (Nov)(5561-2).
CNS D2022-3-86. Low Tension Cable Connections for Ignition Coils for Automobiles (Nov)(5561-3).
ISO 3553 Pt 1-87. Road Vehicles—High-Tension Connections for Ignition Coils and Distributors—Part 1: Socket-Type First Edition. 6 pp.
ISO 3553 Pt 2-87. Road Vehicles—High-Tension Connections for Ignition Coils and Distributors—Part 2: Plug-Type First Edition. 4 pp.
ISO 4024-92. Road Vehicles—Ignition Coils—Low-Tension Cable Connections Second Edition. 5 pp.

—**Automotive—Plate Shape**
CNS D2078-88. Plate Shape of Male Terminals for Automobiles (PA Type) (Jan)(7027).

—**Automotive—Relays**
BSI BS AU 201: Part 2-85. 1985 Relays and Flashers Part 2: Arrangement and Functional Attribution of Relay Terminals for Road Vehicles. 3 pp.
ISO 7880-84. Road Vehicles—Relays—Arrangement and Functional Attribution of Relay Terminals First Edition. 3 pp.
JIS D 5011-92. Relays for Automobiles—Arrangement and Functional Attribution of Arrayed Terminals and Dimension of Relays. 14 pp.

—**Automotive—Spark Plugs**
CNS D2074-90. High Tension Cable Terminals of Spark Plug for Automobiles (IB Type) (Aug)(7023).

—**Automotive—Storage Batteries**
CNS D2068-90. Storage Battery Cable Terminals for Automobiles (BA Type) (Aug)(6906).
CNS D2069-90. Storage Battery Cable Terminals for Automobiles (BB Type) (Aug)(6907).
CNS D2070-90. Storage Battery Cable Terminals for Automobiles (BC Type) (Aug)(6908).
CNS D2071-90. Storage Battery Cable Terminals for Automobiles (BD Type) (Aug)(6909).
CNS D2072-90. Storage Battery Cable Terminal Connectors for Automobiles (BE Type) (Aug)(6910).

—**Automotive—T Shape**
CNS D2079-88. T Shape Male Terminals of Cable Connection for Automobiles (TA Type) (Jan)(7028).

Electric Terminals *(Cont.)*
—**Combat Vehicles—Starter Batteries**
NATO STANAG 4015 Pt2 ED1 AMD1-65. Sizes and Positions of Starter Battery Terminal Posts for Wheeled Tactical Vehicles. 8 pp.
NATO STANAG 4015 Pt2 ED2 AMD1-00. Sizes and Positions of Starter Battery Terminal Posts for Wheeled Tactical Vehicles. 11 pp.

—**Compressed**
BSI BS 5372-89. 1989 Dimensions of Cable Terminations for 3-Core and 4-Core Polymeric Insulated Cables of Rated Voltages 600/1000V and 1900/3300V Having Aluminium Conductors. 8 pp.
BSI BS 5372-76. 1976 Cable Terminations for Electrical Equipment. 6 pp.
CNS C4184-90. Compression Terminals (Dec)(5517).
JIS C 2804-84. Compression Terminals. 25 pp.

—**Contactors—Identification Systems**
BSI BS 5582-78. 1978 Low-Voltage Switch Gear and Controlgear for Industrial Use. Terminal Marking and Distinctive Number for Auxiliary Contacts of Particular Contractors. 6 pp.
BSI BS 5583-78. 1978 Low-Voltage Switch-Gear and Controlgear for Industrial Use. Terminal Marking, Distinctive Number and Distinctive Letter for Particular Contractor Relays. 8 pp.

—**Controlgear**
BSI BS 5994-80. (WITHDRAWN) 1980 Industrial Low Voltage Switchgear and Control-Gear. Terminal Aperture Sizes for Unprepared Round Copper Conductors (Superseded by BS EN 60947-1: 1992). 7 pp.
BSI BS 6733-87. (WITHDRAWN) 1987 Low-Voltage Switchgear and Controlgear for Industrial Use. Requirements Applicable to Terminals Concerning Cross-Sections of Connectable Conductors (Superseded by BS EN 60947-1: 1992). 6 pp.

—**Controlgear—Identification Systems**
BSI BS 5472-77. 1977 Low Voltage Switchgear and Controlgear for Industrial Use. Terminal Marking and Distinctive Number. General Rules. 9 pp.
BSI BS 5581-78. 1978 Specification for Low-Voltage Switchgear and Controlgear for Industrial Use. Terminal Marking and Distinctive Number for Particular Control Switches. 6 pp.
BSI BS 6272-82. 1982 Low Voltage Switchgear and Controlgear for Industrial Use. Terminal Marking. Terminals for External Associated Electronic Circuit, Components and Contacts. 9 pp.

—**Copper Conductors**
DIN ENGL 46211-65. Stamped Cable Sockets for Copper Conductors (Mar). 4 pp.

—**Copper Conductors—Quick Connect—Flat Cables—Safety**
IEC 1210-93. Connecting Devices—Flat Quick-Connect Terminations for Electrical Copper Conductors—Safety Requirements First Edition. 50 pp.

—**Crimp Contacts**
CNS C4118-90. Non-Insulated Crimp-Style Terminals for Copper Conductors (Dec)(3434).
CNS C4118-80. Non-Insulated Crimp-Style Terminals for Copper Conductors (May)(3434). 15 pp.

—**High Voltage—Controlgear**
IEC 518-75. Dimensional Standardization of Terminals for High-Voltage Switchgear and Controlgear First Edition. 4 pp.

—**Hospital Grade**
JIS C 2808-79. Hospital Grade Earth Centers and Terminals (R 1989). 15 pp.

—**Identification Systems**
BSI BS 5559-91. 1991 Identification of Equipment Terminals and of Terminations of Certain Designated Conductors, Including General Rules for an Alphanumeric System. 14 pp.
CENELEC HD 241 S2-81. Identification of Apparatus Terminals and General Rules for a Uniform System of Terminal Marking, Using an Alpha-Numeric Notation. 2 pp.
CENELEC EN 60 445-90. Identification of Equipment Terminals and of Terminations of Certain Designated Conductors, Including General Rules for an Alphanumeric System. 3 pp.
CNS C1051-84. General Rules of Color Identification for Protective Conductors and Neutral Conductor and Terminal Marking for Apparatus (Dec)(5202).
IEC 445-88. Identification of Equipment Terminals and of Terminations of Certain Designated Conductors, Including General Rules for an Alphanumeric System Second Edition. 16 pp.
JIS C 0602-84. General Rules of Colour Identification for Protective Conductor and Neutral Conductor and Terminal Marking for Apparatus. 5 pp.

INDUSTRY STANDARDS

INTERNATIONAL AND NON-U.S. NATIONAL STANDARDS
SUBJECT INDEX

Electric

Electric Terminals *(Cont.)*

—**Identification Systems** *(Cont.)*
MOD UK DSTAN 61-7: Part 1-90. Identification of Electrical and Electronic Systems. Wiring and Components Part 1: Colour Coding and Marking for Polarity and Phase Distinction in Electrical Systems Issue 6. 11 pp.

—**Indexes (Documentation)**
MOD UK DSTAN 59-40: Part 90-92. Inter-Circuit Terminations Part 90: Detail Specifications Cross-Reference Index Issue 1. 7 pp.

—**Low Voltage—Contactors—Identification Systems**
CENELEC EN 50 012-77. Low Voltage Switchgear and Controlgear for Industrial use Terminal Marking and Distinctive Number for Auxiliary Contacts of Particular Contactors. 4 pp.

—**Low Voltage—Control Switches—Identification Systems**
CENELEC EN 50 013-77. Low Voltage Switchgear and Controlgear for Industrial use Terminal Marking and Distinctive Number for Particular Control Switches. 4 pp.

—**Low Voltage—Controlgear**
CENELEC EN 50 027-79. Industrial Voltage Switchgear and Controlgear: Terminal Aperture Sizes for Unprepared Round Copper Conductors. 5 pp.
CENELEC EN 50 051-85. Low-Voltage Switchgear and Controlgear for Industrial Use Requirements Applicable to Terminals Concerning Cross-Sections of Connectable Conductors. 4 pp.
DIN VDE 0660 Pt 12-82. Switchgear and Control Gear; Low Voltage Switchgear and Control Gear; Terminals for Protective Conductors (Sept). 7 pp.
DIN VDE 0660 Pt 99-82. Switchgear and Control Gear; Low-Voltage Switchgear and Control Gear; Connectable Conductor Cross-Sections (Sept). 5 pp.

—**Low Voltage—Controlgear—Identification Systems**
CENELEC EN 50 005-76. Low Voltage Switchgear and Controlgear for Industrial Use: Terminal Marking and Distinctive Number: General Rules. 7 pp.
CENELEC EN 50 042-80. Low Voltage Switchgear and Controlgear for Industrial Use: Terminal Marking: Terminals for External Associated Electronic Circuit Components and Contacts. 7 pp.

—**Low Voltage—Copper Conductors—Twist On**
BSI BS EN 60998-2-4-93. 1993 Connecting Devices for Low-Voltage Circuits for Household and Similar Purposes Part 2-4: Particular Requirements for Twist-on Connecting Devices (IEC 998-2-4: 1993) (E). 27 pp.
CENELEC EN 60998-2-4-93. Connecting Devices for Low-Voltage Circuits for Household and Similar Purposes Part 2-4: Particular Requirements for Twist-on Connecting Devices (IEC 998-2-4: 1993). 22 pp.
IEC 998 Pt 2-4-93. Connecting Devices for Low-Voltage Circuits for Household and Similar Purposes Part 2-4: Particular Requirements for Twist-on Connecting Devices First Edition; (Supersedes 685 Pt 2-4) (CENELEC EN 60998-2-4: 1993). 44 pp.

—**Low Voltage—Switchgear**
CENELEC EN 50 027-79. Industrial Voltage Switchgear and Controlgear: Terminal Aperture Sizes for Unprepared Round Copper Conductors. 5 pp.
CENELEC EN 50 051-85. Low-Voltage Switchgear and Controlgear for Industrial Use Requirements Applicable to Terminals Concerning Cross-Sections of Connectable Conductors. 4 pp.
DIN VDE 0660 Pt 99-82. Switchgear and Control Gear; Low-Voltage Switchgear and Control Gear; Connectable Conductor Cross-Sections (Sept). 5 pp.

—**Low Voltage—Switchgear—Identification Systems**
CENELEC EN 50 005-76. Low Voltage Switchgear and Controlgear for Industrial Use: Terminal Marking and Distinctive Number: General Rules. 7 pp.
CENELEC EN 50 042-80. Low Voltage Switchgear and Controlgear for Industrial Use: Terminal Marking: Terminals for External Associated Electronic Circuit Components and Contacts. 7 pp.

—**Marine**
CNS F5006-81. Small Size Terminal for Marine Use (Sep)(7908).
CNS F5111-86. Minimum Length Between the Cable Entrances and the Terminals of Electric Machinery and Equipment for Marine Use (Dec)(11794).
JIS F 8811-90. Small Size Terminals for Marine Use.

Electric Terminals *(Cont.)*

—**Mechanical Testing**
BSI BS 2011: Part 2.1U-84. 1984 Amd 1 Basic Environmental Testing Procedures Part 2.1: Tests Part 2.1U: Test U. Robustness of Terminations and Integral Mounting Devices (Amd 6106) November 30, 1990. 21 pp.
BSI BS 5772: Part 8-85. 1985 Amd 1 Basic Testing Procedures and Measuring Methods for Electromechanical Components for Electronic Equip. Part 8: Connector Tests (Mechanical) and Mech. Tests on Contacts and Terminations (AMD 5076) May 31, 1990. 32 pp.
CENELEC HD 323.2.21 S3-88. Basic Environmental Testing Procedures-Part 2: Tests. Test V: Robustness of Terminations and Integral Mounting Devices. 3 pp.
IEC 68 Pt 2-21-83. Basic Environmental Test-ing Procedures Part 2: Tests—Test U: Robust-ness of Terminations and Intergral Mounting Devices Fourth Edition; (Amendment 2-1991 Incorporating Amendment 1) (Corrigendum—Nov 1991) (Amendment 3-1992). 65 pp.
IEC 512 Pt 8-93. Electromechanical Components for Electronic Equipment; Basic Testing Procedures and Measuring Methods Part 8: Connector Tests (Mechanical) and Mechanical Tests on Contacts and Terminations Third Edition. 72 pp.
SAA AS 1099.2U-76. Basic Environmental Testing Procedures for Electrotechnology—Part 2: Tests—Part 2U: Robustness of Terminations. 8 pp.
SNZ IEC 68: Part 2-21-83. Basic Environmental Testing Procedures Part 2-21: Test U: Robustness of Terminations and Integral Mounting Devices Amend: 1. 36 pp.

—**Microprocessors**
BSI BS 7240-90. 1990 Pin Allocations for Microprocessor Systems Using the IEC 603.2 Connector. 8 pp.
IEC 828-88. Pin Allocations for Microprocessor Systems Using the IEC 603-2 Connector First Edition. 11 pp.
JTC1 828-88. Pin Allocations for Microprocessor Systems Using the IEC 603-2 Connector First Edition. 11 pp.

—**Pin**
MOD UK DSTAN 59-40: Part 1-90. Inter-Circuit Terminations Part 1: General Requirements and Test Methods for Qualification Approval Generic Specification Issue 2. 19 pp.

—**Power Lines**
DIN VDE 0220 Pts 1 & 1A-71. Specifications for Detachable Cable Terminals in Heavy-Current Cable Installations up to 1000 V (Nov). 22 pp.

—**Power Transformers—Identification Systems**
IEC 616-78. Terminal and Tapping Markings for Power Transformers, Rept. First Edition. 19 pp.

—**Quick Connect**
BSI BS 5057-92. 1992 Flat, Quick-Connect Terminations (IEC 760: 1989) (E). 34 pp.
BSI BS 5057-88. 1988 Flat Quick-Connect Terminations. 30 pp.
CNS C3185-84. Method of Test for Flat Quick-Connect Terminals (Jun)(10899).
CNS C4403-84. Flat Quick-Connect Terminals (Jun)(10898).
CSA C22.2 NO 153-M1981. Quick-Connect Terminals (R 1992); (Gen Instr 1 Thru 5). 32 pp.
IEC 760-89. Flat, Quick-Connect Terminations Second Edition; (Amendment 1-1993). 83 pp.
JIS C 2809-92. Flat Quick-Connect Terminals. 19 pp.
SAA AS 3169-89. Approval and Test Specifications—Flat, Quick-Connect Terminations (This is a Joint Standard with SANZ NZS 3169). 9 pp.
SNZ NZS/AS 3169-89. Approval and Test Specification—Flat, Quick-Connect Terminations (This is a Joint Standard with SAA AS 3169). 9 pp.

—**Quick Connect—Automotive**
BSI BS EN 28092: Part 2-92. 1992 Road Vehicles—Flat, Quick-Connect Terminations Part 2: Tests and Performance Requirements for Single Pole Connections (ISO 8092-2: 1988). 13 pp.
CEN EN 28092-2-91. Road Vehicles—Flat, Quick-Connect Terminations—Part 2: Test and Performance Requirements for Single Pole Connections. 13 pp.
DIN ENGL EN 28092 Pt 2-92. Flat, Quick-Connect Terminations for Road Vehicles; Tests and Performance Requirements for Single Pole Connections (ISO 8092-2: 1988) (Jan). 8 pp.
ISO 8092 Pt 2-88. Road Vehicles—Flat, Quick-Connect Terminations—Part 2: Tests and Performance Requirements for Single Pole Connections First Edition. 9 pp.

Electric Terminals *(Cont.)*

—**Railway Signals**
BSI BS 442-50. 1950 Terminals for Electrical Apparatus for Railway Signalling Purposes. 8 pp.
JIS E 3104-78. Connecting Terminals for Railway Signal.

—**Ring**
CSA 1410 Bull. Electrical Bulletin 1410 September 13, 1983 to C22.2 NO 65. 1 p.
CSA 1165B Bull. Electrical Bulletin 1165B June 25, 1984 to C22.2 NO 65. 5 pp.
CSA 1422 Bull. Electrical Bulletin 1422 July 6, 1984 to C22.2 NO 65. 3 pp.
CSA 1165C Bull. Electrical Bulletin 1165C December 4, 1984 to C22.2 NO 65. 2 pp.
CSA 1165D Bull. Electrical Bulletin 1165D April 23, 1986 to C22.2 NO 65. 2 pp.

—**Rotating Machines—Identification Systems**
BSI BS 822: Part 6-64. 1964 Amd 2 Terminal Markings for Electrical Machinery and Apparatus Part 6: Terminal Markings for Rotating Electrical Machinery. 33 pp.
BSI BS 4999: Part 108-87. 1987 Amd 1 General Requirements for Rotating Electrical Machines Part 108: Terminal Marking and Direction of Rotation (Amd 6528) November 30, 1990. 22 pp.
CENELEC HD 53.8 S3-85. Rotating Electrical Machines—Part 8: Terminal Markings and Direction of Rotation of Rotating Machines. 13 pp.
CENELEC HD 53.8 S3/A1-89. AMD 8 Rotating Electrical Machines—Part 8: Terminal Markings and Direction of Rotation of Rotating Machines. 4 pp.
DIN VDE 0530 Pt 8-87. Rotating Electrical Machines; Terminal Markings and Direction of Rotation (July) (Supersedes DIN 57 530 Part 8 and Partially Supersedes VDE 0530 Part 1). 35 pp.

—**Screw Type**
DIN ENGL 46207-71. Connections for Electrical Devices; Sleeve Terminals and Side Opening Sleeve Terminals; Principal Dimensions and Correlation (June). 2 pp.
MOD UK DSTAN 59-3: Part 2-01. Terminals, Electrical Part 2: Terminals, Screwed Head, and Terminals, Spring Head Issue 1; Amendment 2. 16 pp.

—**Screw Type—Clamping Points**
DIN VDE 0609 Pt 1-83. Screw Terminal Clamping Points for Connection of Copper Conductors up to 240 mm2; General Requirements (June). 15 pp.
DIN VDE 0609 Pt 1 A1 (D)-86. Screwed Terminal Clamping Points for Connection of Copper Conductors up to 240 mm2; General Requirements (Jan). 3 pp.
DIN VDE 0609 Pt 1 A2 (D)-87. Screw Terminal Clamping Points for Connection of Copper Conductors up to 240mm2; General Requirements (Aug) (Intended Amendment 2 to DIN 57609 Part 1/VDE 0609 Part 1/06.83 and Partial Amendment to Draft DIN VDE 0609 Part 1 A1/01.86). 3 pp.

—**Screwless**
IEC 685 Pt 2-1-80. Connecting Devices (Junction and/or Tapping) for Household and Similar Fixed Electrical Installations Part 2: Particular Requirements—Screwless Terminals for Connecting Copper Conductors Without Special Preparation First Edition. 26 pp.
MOD UK DSTAN 59-3: Part 2-01. Terminals, Electrical Part 2: Terminals, Screwed Head, and Terminals, Spring Head Issue 1; Amendment 2. 16 pp.

—**Screwless—Clamp Sites**
DIN VDE 0607-74. VDE Specification for Clamp Sites of Screwless Terminals for Connecting or Joining Copper Leads of 0.5 mm2 to 16 mm2 (Nov). 11 pp.

—**Screwless—Lampholders**
CSA 943B Bull. Electrical Bulletin 943B June 25, 1976 to C22.2 NO 55. 2 pp.

—**Screwless—Outlets**
CSA 943B Bull. Electrical Bulletin 943B June 25, 1976 to C22.2 NO 55. 2 pp.
DIN VDE 0620 A6 (D)-88. Plugs and Socket-Outlets up to 380 V 25 A; Fixed Socket-Outlets with Screwless Terminals for 10/16A: Amendment 6 (June). 20 pp.

—**Screwless—Switches**
CSA C22.2 NO 55-M1986. Special Use Switches (R 1992); (Gen Instr 1 Thru 2). 34 pp.
CSA 943B Bull. Electrical Bulletin 943B June 25, 1976 to C22.2 NO 55. 2 pp.

INTERNATIONAL AND NON-U.S. NATIONAL STANDARDS
SUBJECT INDEX

Electric

Electric Terminals (Cont.)
—Ships
- MOD UK NES 512: Part 10-01. Guide to Cables, Electrical and Associated Items Part 10: Terminations and Connectors Issue 2 (03.91); Amendment 1. 51 pp.
- MOD UK NES 514-81. Guide to Cable Entry, Termination and Junction Components for Equipment Issue 2 (01.81). 101 pp.
- MOD UK NES 514-92. Guide to Cable Entry, Termination and Junction Components for Equipment Issue 3 (04.92). 79 pp.

—Ships—Motor Starters—Identification Systems
- CNS F5120-87. Terminal Markings for Motor Starters in Ships (Jun)(11998).

—Spacers (Electric)
- CSA CAN/CSA-C22. 2 NO 0-M91. General Requirements—Canadian Electrical Code, Part II; (Gen Instr 1). 48 pp.

—Spade
- CSA 1410 Bull. Electrical Bulletin 1410 September 13, 1983 to C22.2 NO 65. 1 p.
- CSA 1165B Bull. Electrical Bulletin 1165B June 25, 1984 to C22.2 NO 65. 5 pp.
- CSA 1422 Bull. Electrical Bulletin 1422 July 6, 1984 to C22.2 NO 65. 3 pp.
- CSA 1165C Bull. Electrical Bulletin 1165C December 4, 1984 to C22.2 NO 65. 2 pp.
- CSA 1165D Bull. Electrical Bulletin 1165D April 23, 1986 to C22.2 NO 65. 2 pp.

—Submarines
- MOD UK NES 512: Part 10-01. Guide to Cables, Electrical and Associated Items Part 10: Terminations and Connectors Issue 2 (03.91); Amendment 1. 51 pp.

—Switchgear
- BSI BS 5994-80. (WITHDRAWN) 1980 Industrial Low Voltage Switchgear and Control-Gear. Terminal Aperture Sizes for Unprepared Round Copper Conductors (Superseded by BS EN 60947-1: 1992). 7 pp.
- BSI BS 6733-87. (WITHDRAWN) 1987 Low-Voltage Switchgear and Controlgear for Industrial Use. Requirements Applicable to Terminals Concerning Cross-Sections of Connectable Conductors (Superseded by BS EN 60947-1: 1992). 6 pp.
- IEC 518-75. Dimensional Standardization of Terminals for High-Voltage Switchgear and Controlgear First Edition. 4 pp.

—Switchgear—Identification Systems
- BSI BS 5472-77. 1977 Low Voltage Switchgear and Controlgear for Industrial Use. Terminal Marking and Distinctive Number. General Rules. 9 pp.
- BSI BS 5581-78. 1978 Specification for Low-Voltage Switchgear and Controlgear for Industrial Use. Terminal Marking and Distinctive Number for Particular Control Switches. 6 pp.
- BSI BS 6272-82. 1982 Low Voltage Switchgear and Controlgear for Industrial Use. Terminal Marking. Terminals for External Associated Electronic Circuit, Components and Contacts. 9 pp.

—Switchgear—Temperature
- BSI PD 6524-89. 1989 Guide to Specifying Permissible Temperature and Temperature Rise for Parts of Electrical Equipment, in Particular for Terminals. 81 pp.
- IEC 943-89. Guide for the Specification of Permissible Temperature and Temperature Rise for Parts of Electrical Equipment, in Particular for Terminals First Edition. 160 pp.

—Tabs—Quick Connect—Automotive
- BSI BS EN 28092: Part 1-92. 1992 Road Vehicles—Flat, Quick-Connect Terminations Part 1: Tabs for Single Pole Connections (ISO 8092-1: 1989). 11 pp.
- CEN EN 28092-1-91. Road Vehicles—Flat, Quick-Connect Terminations—Part 1: Tabs for Single Pole Connections. 3 pp.
- DIN ENGL EN 28092 Pt 1-92. Flat, Quick-Connect Terminations for Road Vehicles; Tabs for Single Pole Connections (ISO 8092-1: 1989) (Jan). 6 pp.
- ISO 8092 Pt 1-89. Road Vehicles—Flat, Quick-Connect Terminations—Part 1: Tabs for Single Pole Connections First Edition. 7 pp.

—Taper Pin
- MOD UK DSTAN 59-40: Part 8-91. Inter-Circuit Terminations Part 8: Terminal Blocks, Taper Socket and Terminals, Taper Pin Electrical Sectional Specification Codification Service Nomenclature: Terminals, Taper Receptacle, Electrical Issue 2. 48 pp.

Electric Terminals (Cont.)
—Taper Socket
- MOD UK DSTAN 59-40: Part 8-91. Inter-Circuit Terminations Part 8: Terminal Blocks, Taper Socket and Terminals, Taper Pin Electrical Sectional Specification Codification Service Nomenclature: Terminals, Taper Receptacle, Electrical Issue 2. 48 pp.

—Tappings—Identification Systems
- IEC 616-78. Terminal and Tapping Markings for Power Transformers, Rept. First Edition. 19 pp.

—Telecommunication Equipment
- CCITT RECMN L.9-89. Methods of Terminating Metallic Cable Conductors—Construction, Installation and Protection of Cable and Other Elements of Outside Plant (Study Group VI) 4 pp. 4 pp.

—Telephone Sets
- CNS C7168-84. Terminal for Flat-Oval Telephone Cord (May)(10871).

—Video Tape Recorders
- CNS C7092-85. Connection of Video Tape Recorder (Sep)(4956).
- JIS C 5573-74. Connection of Video Tape Recorder. 10 pp.

Electric Thermometers
See Also: Resistance Thermometers; Thermoelectric Thermometers; Thermometers
- JIS T 1140-89. Clinical Electrical Thermometers with Maximum Device.

—Fast Response
- DIN ENGL 43771-89. Measurement and Control; Electrical Temperature Sensors; Fast Response Thermometers (Dec). 4 pp.

—Thermowells
- DIN ENGL 43763-86. Measurement and Control; Electrical Temperature Sensors; Metal Protecting Tubes for Thermometers with Inserted Sensor Units (Mar). 12 pp.

Electric Toasters
Use: Toasters

Electric Tracing Equipment
See Also: Heat Tracing Systems

—Heating Cables—Pipes—Industrial
- CSA CAN/CSA-C22. 2NO130.1-M90. Heat-Tracing Cable Systems for Use in Industrial Locations; (Gen Instr 1). 32 pp.

—Heating Cables—Tanks (Containers)—Industrial
- CSA CAN/CSA-C22. 2NO130.1-M90. Heat-Tracing Cable Systems for Use in Industrial Locations; (Gen Instr 1). 32 pp.

Electric Traction
See Also: Electric Rolling Stock; Railroad Traction Equipment; Rolling Stock

—Glossaries
- BSI BS 4727:Pt2: Group 18-92. 1992 Electrotechnical, Power, Telecommunication, Electronics, Lighting and Colour Terms Part 2: Terms Particular to Power Engineering Group 18: Electric Traction (G). 71 pp.
- IEC 50 Chap 811-91. International Electrotechnical Vocabulary Chapter 811: Electric Traction First Edition. 242 pp.
- SAA AS 1852.30-70. International Electrotechnical Vocabulary—Part 30: Electric Traction Being IEC 50(30). 94 pp.

—Symbols
- SAA AS 1102.12-84. Graphical Symbols for Electrotechnology—Part 12: Electric Traction. 12 pp.

Electric Transducers
Use For: Electrical Transducers
See Also: Transducers

—Power Systems—Safety
- IEC 65-85. Safety Requirements for Mains Operated Electronic and Related Apparatus for Household and Similar General Use Fifth Edition; (Amendment 2-1989, Incorporating Amendment 1) (Amendment 3-1992). 210 pp.

Electric Transformers
Use: Transformers

Electric Trucks
See Also: Electric Vehicles; Forklifts; Ground Vehicles; Lift Trucks; Trucks

Electric Trucks (Cont.)
—Rotating Machines
- DIN VDE 0525-83. Rotating Electrical Machines for Industrial Trucks Driven by Electric Motors (Mar). 14 pp.

—Traction Batteries—Voltage Ratings
- ISO 1044-85. Industrial Trucks—Traction Batteries for Electric Trucks—Voltages Second Edition. 3 pp.

Electric Vehicles
See Also: Automatic Guided Vehicles; Automobiles; Electric Trucks; Forklifts; Ground Vehicles; Industrial Trucks; Lift Trucks; Motor Vehicles; Vehicles (Transportation)

—AC Motors
- IEC 349 Pt 2-93. Electric Traction—Rotating Electrical Machines for Rail and Road Vehicles—Part 2: Electronic Convertor-Fed Alternating Current Motors First Edition. 62 pp.

—Batteries
- BSI BS 7483-91. 1991 Lead-Acid Batteries for the Propulsion of Light Electric Vehicles. 8 pp.

—Battery Chargers
- IEC 718-92. Electrical Equipment for the Supply of Energy to Battery-Powered Road Vehicles Second Edition. 59 pp.

—Battery Powered—Electric Connectors
- BSI BS 3214-74. 1974 Amd 3 Locking Connectors for Battery Operated Vehicles, Excluding Industrial Trucks, (320 Ampere Rating) (AMD 6125) July 31, 1989. 14 pp.

—Connectors
- IEC 783-84. Wiring and Connectors for Electric Road Vehicles First Edition. 22 pp.

—Controllers
- IEC 786-84. Controllers for Electric Road Vehicles First Edition. 30 pp.

—Instrumentation
- IEC 784-84. Instrumentation for Electric Road Vehicles First Edition. 16 pp.

—Lead Acid Batteries
- BSI BS 2550-83. 1983 Amd 2 Lead-Acid Traction Batteries (AMD 4911) September 30, 1985. 14 pp.
- CENELEC HD 465.1-87. Lead—Acid Traction Batteries Part 1: General Requirements and Methods of Test. 3 pp.
- CNS D2180-84. Lead-Acid Traction Batteries (Aug)(10980).
- IEC 254 Pt 1-83. Lead-Acid Traction Batteries Part 1: General Requirements and Methods of Test Second Edition; (Amendment 1-1990). 29 pp.
- JIS D 5303-86. Lead-Acid Traction Batteries. 14 pp.

—Ohmic Resistors
- CENELEC HD 91-74. Rules for Ohmic Resistors Used in the Power Circuits of Electrically Powered Vehicles. 2 pp.
- IEC 322-70. Rules for Ohmic Resistors Used in the Power Circuits of Electrically Powered Vehicles First Edition. 21 pp.

—Power Resistors
- BSI BS 4846-72. 1972 Resistors for Traction Purposes. 13 pp.

—Rotating Machines
- BSI BS 173: Part 1-92. 1992 Electric Traction (Rail and Road Vehicles) Part 1: Specification for Rotating Electrical Machines (IEC 349: 1991) (E). 69 pp.
- BSI BS 173-80. (WITHDRWAN) 1980 Amd 1 Performance of Rotating Electrical Machines for Rail and Road Vehicles (AMD 4729) November 30, 1984 (Superseded by BS 173: Part 1: 1992 and BS 7571: 1992). 49 pp.
- CENELEC HD 225-74. Rules for Rotating Electrical Machines for Rail and Road Vehicles. 2 pp.
- DIN VDE 0535 Pt 1-79. Deviations from IEC 349 1st Edition 1971 (Dec). 17 pp.
- DIN VDE 0535 Pt 1 A1-83. Rules for Rotating Electrical Machines for Rail and Road Vehicles (July). 4 pp.
- IEC 349-91. Electric Traction Rotating Electrical Machines for Rail and Road Vehicles Second Edition. 136 pp.
- IEC 785-84. Rotating Machines for Electric Road Vehicles First Edition. 20 pp.

—Traction Batteries
- CENELEC HD 465.1-87. Lead—Acid Traction Batteries Part 1: General Requirements and Methods of Test. 3 pp.
- CNS D2180-84. Lead-Acid Traction Batteries (Aug)(10980).

INDUSTRY STANDARDS

Electric Vehicles (Cont.)

—Traction Batteries (Cont.)

IEC 254 Pt 1-83. Lead-Acid Traction Batteries Part 1: General Requirements and Methods of Test Second Edition; (Amendment 1-1990). 29 pp.

JIS D 5303-86. Lead-Acid Traction Batteries. 14 pp.

—Wiring

IEC 783-84. Wiring and Connectors for Electric Road Vehicles First Edition. 22 pp.

Electric Waves

Use: Electromagnetic Radiation

Electric Welding

See Also: Arc Welding; Electron Beam Welding; Electroslag Welding; Fusion Welding; Gas Metal Arc Welding; Gas Tungsten Arc Welding; Projection Welding; Resistance Welding; Seam Welding; Shielded Metal Arc Welding; Spot Welds; Stud Welding; Welding

—Glossaries

IEC 50 Chap 851-91. International Electrotechnical Vocabulary Chapter 851: Electric Welding First Edition. 41 pp.

Electric Wire

Scope Note: For additional listings, see the subheading Electric Wire under specific products or equipment *Use For:* Insulated Wire
See Also: Aluminum Conductors; Aluminum Wire; Bare Conductors; Busbars; Cables (Electric); Circuits; Color Coding; Communication Cables; Copper Conductors; Copper Wire; Drop Wire; Dumet Wire; Electric Conduits; Electric Wiring; Electrical Insulation; Harnesses (Electric); Hookup Wire; Iron Wire; Metal Products; Motor Lead Wire; Power Lines; Steel Conductors; Steel Wire; Stranded Conductors; Telephone Cables; Telephone Wire; Thermocouple Wire; Transmission Lines; Windings; Wire; Wire Wrap Wire

BSI BS 5733-79. 1979 Amd 3 General Requirements for Electrical Accessories. 20 pp.

CSA CAN/CSA-C22. 2 NO 0.3-92. Test Methods for Electrical Wires and Cables; (Gen Instr 1). 121 pp.

CSA 1334 Bull. Electrical Bulletin 1334 August 28, 1981 to C22.2 NO 16. 1 p.

CSA 1334 Bull. Electrical Bulletin 1334 August 28, 1981 to C22.2 NO 127. 1 p.

DIN VDE 0472 Suppl. 1-87. Testing of Cables, Wires and Flexible Cords; List of Standards in the DIN VDE 0472 Series (June). 18 pp.

DIN VDE 0472 Pt 1-87. Testing of Cables, Wires and Flexible Cords; General Requirements (June). 13 pp.

MOD UK DSTAN 61-12: Part 11-01. Wires, Cords, and Cables, Electrical-Metric Units Part 11: Imperial/Metric Cross Reference Index Issue 2; Amendment 1. 89 pp.

—Aging Testing

DIN VDE 0472 Pt 303-90. Testing of Cables, Wires and Flexible Cords; Ageing Procedures (May). 15 pp.

—Aircraft

MOD UK DSTAN 61-12: Part 21-01. Wires, Cords, and Cables, Electrical—Metric Units Part 21: Wires and Cables, Lightweight, Extruded Insulation, 135 Degrees C Tin Plated, and 150 Degrees C Silver Plated Conductors Issue 2; Amendment 2. 63 pp.

MOD UK DSTAN 61-12: Part 21-02. Wires, Cords, and Cables, Electrical-Metric Units Part 21: Wires and Cables, Lightweight, Extruded Insulation, 135 Degrees C Tin Plated, and 150 Degrees C Silver Plated Conductors Issue 2; Amendment 3 (Superseded in Part by Def Stan 61-12: Part 33). 64 pp.

MOD UK DSTAN 61-12: Part 24-01. Wires, Cords, and Cables, Electrical—Metric Units Part 24: Wires and Cables, Lightweight, Taped Insulation, 150 Degrees C Silver Plated and 260 Degrees C Nickel Plated Conductors Issue 1; Amendment 2 (Superseded by Def Stan 61-12: Part 33). 90 pp.

MOD UK DSTAN 61-12: Part 33-92. Wires Cords and Cables Electrical—Metric Units Part 33: Airframe Wires and Cables in Temperature Categories 135 Degrees C, 150 Degrees C, 200 Degrees C and 260 Degrees C Sectional Specification Issue 1 (Supersedes 61-12: Part 21 and Part 24). 40 pp.

—Aircraft—Crimp Contacts

BSI 5G 178: Part 1-93. 1993 Crimped Joints for Aircraft Electrical Cables and Wires Part 1: Specification for Design Requirements (Including Tests) for Components and Tools (S). 15 pp.

BSI 4G 178: Part 1-84. (WITHDRAWN) 1984 Amd 1 Crimped Joints for Aircraft Electrical Cables and Wires Part 1: Design Requirements (Including Tests) for Components and Tools (Superseded by 5G 178: Part 1: 1993). 11 pp.

Electric Wire (Cont.)

—Aircraft—Crimp Contacts (Cont.)

BSI 4G 178: Part 2-86. 1986 Crimped Joints for Aircraft Electrical Cables and Wires Part 2: Control of Crimping (Including User Control Tests). 10 pp.

—Aircraft—Heat Resistant

ISO 2032-73. Heat-Resisting Equipment Wires for Aircraft First Edition. 6 pp.

ISO 2436-73. Heat-Resisting Equipment Wires for Aircraft—Methods of Test First Edition. 4 pp.

—Aircraft—Sleeves—Identification Systems

BSI 3G 198: Part 1-89. 1989 Amd 1 Sleeves for Aircraft Electric Cables and Equipment Wires Part 1: Specification for Elastomeric Sleeves for Binding and Identification (AMD 6389) September 30, 1991. 17 pp.

—Appliances—Breakage

CNS C3170-83. Method of Test for Wires Breakage of Appliance (Jan)(9815).

—Bend Testing

DIN VDE 0472 Pt 610-85. Testing of Cables, Wires and Flexible Cords; Bending Test at Low Temperature (Jan). 6 pp.

—Breaking Length

DIN VDE 0472 Pt 626-83. Testing of Cables, Wires and Flexible Cords; Breaking Length (Jan). 5 pp.

—Cambric Insulated

CNS C2087-80. Cambric Insulated Wires (Dec)(6645). 12 pp.

JIS C 3403-78. Cambric Insulated Wires.

JIS C 3403-71. Cambric Insulated Wires. 14 pp.

—Capacitance Measurement

DIN VDE 0472 Pt 506-83. Testing of Cables, Wires and Flexible Cords Capacitative Unbalance (VDE Specification) (Apr). 5 pp.

—Continuity Testing

CNS C3006-80. Testing Standard for Electric Continuance of Insulated Wires (Jul)(684). 1 p.

—Cross Reference Index

MOD UK DSTAN 61-12: Part 90-89. Wires, Cords, and Cables of Assessed Quality Part 90: Detail Specifications Issue 1. 5 pp.

MOD UK DSTAN 61-12: Part 90-93. Wires, Cords and Cables Electrical Metric Units Part 90: Detail Specification Cross Reference Index Issue 2. 5 pp.

MOD UK DSTAN 61-12: Part 90-01. Wires, Cords and Cables Electrical Metric Units Part 90: Detail Specification Cross Reference Index Issue 2; Amendment 1. 6 pp.

—Dielectric Strength

CSA 1277 Bull. Electrical Bulletin 1277 July 17, 1980 to C22.2 NO 116. 2 pp.

CSA 1277 Bull. Electrical Bulletin 1277 July 17, 1980 to C22.2 NO 127. 2 pp.

DIN VDE 0472 Pt 509-86. Testing of Cables, Wires and Flexible Cords; Dielectric Strength on Cables, Wires and Cords for Telecommunications and Information Processing Systems (Oct). 11 pp.

—Elastomer Sheathed—Fire Resistant

MOD UK DSTAN 61-12: Part 27-87. Wires, Cords and Cables Electrical—Metric Units Part 27: Cables and Wires, Electrical (for Power Connections to Mobile Equipment) Issue 1. 34 pp.

—Electrical Codes

CSA 851 Bull. Electrical Bulletin 851 November 17, 1971 to C22.2 NO 38. 2 pp.

CSA 851 Bull. Electrical Bulletin 851 November 17, 1971 to C22.2 NO 127. 2 pp.

—Electrical Properties

IEC 885 Pt 1-87. Electrical Test Methods for Electric Cables Part 1: Electrical Tests for Cables, Cords and Wires for Voltages up to and Including 450/750 V First Edition. 13 pp.

—Electronic Equipment—Density

CNS C6096-80. Method of Test for Density of Fire Wire and Ribbon for Electronic Devices (Aug)(6316).

—Elongation

DIN VDE 0472 Pt 625-83. Testing of Cables and Insulated Conductors; Tensile and Elongation Characteristics of Strain Bearing Elements (Jan). 5 pp.

—Enameled—Impregnating Agents—Bonding Strength

IEC 1033-91. Test Methods for the Determination of Bond Strength of Impregnating Agents to an Enamelled Wire Substrate First Edition. 30 pp.

Electric Wire (Cont.)

—Enameled—Windings

CENELEC EN 60182-2-90. Basic Dimensions of Winding Wires Part 2: Maximum Overall Diameters of Enamelled Round Winding Wires (Supersedes HD 42.2 S2-1979). 8 pp.

—Ethylene Tetrafluoroethylene Insulated

MOD UK DSTAN 61-12: Part 29-01. Wires, Cords, and Cables, Electrical-Metric Units Part 29: Wires and Cables, Extruded Insulation, Types ETFE and ETFE Metsheath 135 Degrees C Tin Plated, and 150 Degrees C Silver Plated Conductors Issue 1; Amendment 2. 40 pp.

—Ethylene Tetrafluoroethylene Insulated—Data Processing Equipment

DIN VDE 0891 Pt 9-90. Use of Cables and Insulated Cords for Telecommunications Systems and Data Processing Systems; Special Guidelines for Equipment Wires with Solid and Stranded Conductors with Extended Temperature Range According to DIN VDE 0881 (May). 15 pp.

—Ethylene Tetrafluoroethylene Insulated—Telecommunications

DIN VDE 0891 Pt 9-90. Use of Cables and Insulated Cords for Telecommunications Systems and Data Processing Systems; Special Guidelines for Equipment Wires with Solid and Stranded Conductors with Extended Temperature Range According to DIN VDE 0881 (May). 15 pp.

—Fire Hazards

MOD UK DSTAN 61-12: Part 18-90. Wires Cords and Cables, Metric Units Part 18: Equipment Wires, Limited Fire Hazard Issue 3. 41 pp.

MOD UK DSTAN 61-12: Part 18-01. Wires Cords and Cables, Metric Units Part 18: Equipment Wires, Limited Fire Hazard Issue 3; Amendment 2. 43 pp.

MOD UK DSTAN 61-12: Part 18-02. Wires Cords and Cables, Metric Units Part 18: Equipment Wires, Limited Fire Hazard Issue 3; Amendment 3. 44 pp.

—Fire Testing

BSI BS 4066: Part 1-80. 1980 Tests on Electric Cables Under Fire Conditions Part 1: Method of Test on a Single Vertical Insulated Wire or Cable. 8 pp.

BSI BS 4066: Part 2-89. 1989 Tests on Electric Cables Under Fire Conditions Part 2: Method of Test on a Single Small Vertical Insulated Wire or Cable. 8 pp.

CENELEC HD 405.1-85. Test on Electric Cables Under Fire Conditions Part 1: Test on a Single Vertical Insulated Wire or Cable. 3 pp.

CENELEC HD 405.2 S1-91. Tests on Electric Cables Under Fire Conditions Part 2: Test on a Single Small Vertical Insulated Copper Wire or Cable. 7 pp.

DIN VDE 0472 Pt 804-89. Testing of Cables, Wires and Flexible Cords; Burning. 17 pp.

IEC 332 Pt 1-93. Tests on Electric Cables Under Fire Conditions Part 1: Test on a Single Vertical Insulated Wire or Cable Third Edition; (Corrigendum—May 1993). 23 pp.

IEC 332 Pt 3-92. Tests on Electric Cables Under Fire Conditions Part 3: Tests on Bunched Wires or Cables Second Edition. 62 pp.

—Fluorescent Lighting—1000 V

JIS C 3309-91. 1000 V Grade Insulated Wires for Fluorescent Lamps. 10 pp.

—Fluorinated Ethylene Propylene Insulated—Data Processing

DIN VDE 0891 Pt 9-90. Use of Cables and Insulated Cords for Telecommunications Systems and Data Processing Systems; Special Guidelines for Equipment Wires with Solid and Stranded Conductors with Extended Temperature Range According to DIN VDE 0881 (May). 15 pp.

—Fluorinated Ethylene Propylene Insulated—Telecommunications

DIN VDE 0891 Pt 9-90. Use of Cables and Insulated Cords for Telecommunications Systems and Data Processing Systems; Special Guidelines for Equipment Wires with Solid and Stranded Conductors with Extended Temperature Range According to DIN VDE 0881 (May). 15 pp.

—Fluorinated Polyhydrocarbon Insulated

BSI BS 6156-81. 1981 Amd 1 Low-Frequency Miniature Equipment Wires with Solid or Stranded Conductor, Fluorinated Polyhydrocarbon Type Insulation, Single. 19 pp.

IEC 673-80. Low-Frequency Miniature Equipment Wires with Solid or Stranded Conductor, Fluorinated Polyhydrocarbon Type Insulation, Single First Edition; (Amendment 3-1989, Incorporating Amendment 1 and 2). 44 pp.

INTERNATIONAL AND NON-U.S. NATIONAL STANDARDS
SUBJECT INDEX

Electric

Electric Wire (Cont.)

—**Glossaries**
DIN VDE 0289 Pt 4-88. Definitions for Cables, Wires and Flexible Cords for Power Installations Testing (Mar) (Together with DIN VDE 0289 Part 1 and DIN VDE 0289 Part 2 Supersedes DIN 57 289 Part 100/VDE 0289 Part 100). 6 pp.
DIN VDE 0289 Part 5-88. Definitions for Cables, Wires and Flexible Cords for Power Installations; Lengths (Mar). 3 pp.

—**Halogen Free**
DIN VDE 0472 Pt 815-89. Testing of Cables, Wires and Flexible Cords; Non-Halogen Verification (Mar). 14 pp.

—**Heat Resistant**
CNS C2157-83. High Temperature Insulated Electrical Wire (General Rules) (Feb)(10238).

—**High Temperature Testing**
DIN VDE 0472 Pt 632-87. Testing of Cables and Insulated Cords; Behaviour at High Temperatures (Oct) (Partially Supersedes VDE 0271/03.69 Which Was Withdrawn in December 1988). 4 pp.

—**Identification Systems**
CSA 1368 Bull. Electrical Bulletin 1368 April 7, 1982 to C22.2 NO 38. 2 pp.
CSA C22.2 NO 75-M1983. Thermoplastic Insulated Wires and Cables (R 1992); (Gen Instr 1 Thru 6). 48 pp.
CSA 1368 Bull. Electrical Bulletin 1368 April 7, 1982 to C22.2 NO 75. 2 pp.
CSA C22.2 NO 127-1988. Equipment Wires; (Gen Instr 1 Thru 3). 47 pp.
CSA 1368 Bull. Electrical Bulletin 1368 April 7, 1982 to C22.2 NO 127. 2 pp.

—**Identification Systems—Earthmoving Equipment**
BSI BS 6913: Part 6-91. 1991 Operation and Maintenance of Earth-Moving Machinery Part 6: Specification for Identification and Marking of Electrical Wires and Cables (ISO 9247: 1990). 7 pp.
ISO 9247-90. Earth-Moving Machinery—Electrical Wires and Cables—Principles of Identification and Marking First Edition. 6 pp.

—**Immersion Testing**
DIN VDE 0472 Pt 803-86. Testing of Cables, Wires and Flexible Cords; Oil Immersion (Apr). 7 pp.

—**Insulation Resistance Testing**
CSA 1433 Bull. Electrical Bulletin 1433 April 8, 1985 to C22.2 NO 38. 3 pp.

—**Mica Glass Insulated**
CSA 1315A Bull. Electrical Bulletin 1315A February 24, 1989 to C22.2 NO 210.2. 6 pp.

—**Multiconductor**
CSA C22.2 NO 127-1988. Equipment Wires; (Gen Instr 1 Thru 3). 47 pp.

—**Naval Ships**
MOD UK NES 512: Part 1-91. Guide to Cables, Electrical and Associated Items Part 1: General Issue 1 (10.91). 17 pp.
MOD UK NES 512: Part 6-91. Guide to Cables, Electrical and Associated Items Part 6: Cables, Electrical, Miniature and Sub—Miniature and Equipment Wires Issue 2 (05.91). 100 pp.
MOD UK NES 512: Part 6-01. Guide to Cables, Electrical and Associated Items Part 6: Cables, Electrical, Miniature and Sub-Miniature and Equipment Wires Issue 2 (05.91); Amendment 2. 116 pp.
MOD UK NES 512: Part 7-91. Guide to Cables, Electrical and Associated Items Part 7: Cables and Wires, Electrical, Miscellaneous for Special Services Issue 3 (04.91). 90 pp.
MOD UK NES 512: Part 7-01. Guide to Cables, Electrical and Associated Items Part 7: Cables and Wires, Electrical, Miscellaneous for Special Services Issue 3 (04.91); Amendment 1. 91 pp.
MOD UK NES 512: Part 7-93. Guide to Cables, Electrical and Associated Items Part 7: Cables and Wires, Electrical, Miscellaneous for Special Services Issue 4 (02.93). 84 pp.
MOD UK NES 523-88. Guide to the Installation of Thin-Wall Insulated, Limited Fire Hazard Equipment Wire and Electric Cable Issue 2 (05.88). 51 pp.
MOD UK NES 523-92. Guide to the Installation of Thin—Wall Insulated, Limited Fire Hazard Equipment Wire and Electric Cable Issue 3 (01.92). 52 pp.

—**Neon Tubes**
CNS C2073-87. Insulated Wires for Neon Tube (Oct)(6064). 4 pp.
CNS C3097-87. Testing Methods of Insulated Wires for Neon Tube (Oct)(6065). 3 pp.

Electric Wire (Cont.)

—**Neon Tubes (Cont.)**
JIS C 3308-91. Insulated Wires for Neon Tube. 10 pp.

—**Paper Covered**
BSI BS 4927: Part 1-74. 1974 Paper-, Cotton-, Silk-or Rayon-Covered Copper Conductors Part 1: Round Wire. 18 pp.

—**Plastic Insulated**
CNS C3004-80. General Rules for Testing Standard of Insulated Wires (Jul)(682). 1 p.
CNS C3011-88. Methods of Test for Plastic Insulated Wires and Cables (Mar)(689). 21 pp.
JIS C 3005-91. Testing Methods for Rubber or Plastic Insulated Wires and Cables. 39 pp.

—**Polyamide Protective Sleeves—Data Processing Equipment**
DIN VDE 0891 Pt 9-90. Use of Cables and Insulated Cords for Telecommunications Systems and Data Processing Systems; Special Guidelines for Equipment Wires with Solid and Stranded Conductors with Extended Temperature Range According to DIN VDE 0881 (May). 15 pp.

—**Polyamide Protective Sleeves—Telecommunication Equipment**
DIN VDE 0891 Pt 9-90. Use of Cables and Insulated Cords for Telecommunications Systems and Data Processing Systems; Special Guidelines for Equipment Wires with Solid and Stranded Conductors with Extended Temperature Range According to DIN VDE 0881 (May). 15 pp.

—**Polyethylene Insulated**
CSA 1433 Bull. Electrical Bulletin 1433 April 8, 1985 to C22.2 NO 38. 3 pp.
IEC 771-83. Calculation of Maximum Overall Diameter of Cables and Specification of Minimum Tensile Strength of Suspension Strand for Low-Frequency Cables with Polyolefin Insulation and Moisture Barrier Polyolefin Sheath First Edition. 13 pp.
MOD UK DSTAN 61-12: Part 6-01. Wires, Cords, and Cables, Electrical-Metric Units Part 6: Polyvinyl Chloride (PVC), Polyethylene, or Silicone Rubber Insulated Equipment Wires Issue 5; Amendment 1. 40 pp.

—**Polyethylene Insulated—Identification Systems**
CSA 1213A Bull. Electrical Bulletin 1213A October 25, 1979 to C22.2 NO 38. 3 pp.

—**Polyethylene Insulated—Weatherproof**
CNS C2035-87. Weather—Proof P.E. Insulated Wire (Jul)(1369). 5 pp.

—**Polyethylene Insulated—600 V**
CNS C2158-83. 600V Grade Polyethylene Insulated Wires (Jun)(10314).
JIS C 3326-92. 600 V Grade Polyethylene Insulated Wires. 10 pp.

—**Polytetrafluoroethylene Insulated**
CNS C2160-83. Polytetrafluoroethylene (Teflon) Insulated Electrical Wire (Oct)(10612).
MOD UK DSTAN 61-12: Part 8-01. Wires, Cords and Cables, Electrical—Metric Units Part 8: PTFE Insulated Equipment Wires Issue 2; Amendment 1. 20 pp.

—**Polytetrafluoroethylene Insulated—Data Processing Equipment**
DIN VDE 0891 Pt 9-90. Use of Cables and Insulated Cords for Telecommunications Systems and Data Processing Systems; Special Guidelines for Equipment Wires with Solid and Stranded Conductors with Extended Temperature Range According to DIN VDE 0881 (May). 15 pp.

—**Polytetrafluoroethylene Insulated—Polyamide Fiber Braided**
CNS C2162-83. Polytetrafluoroethylene (Teflon) Insulated Polyamide Fiber (Nylon) Braided Electrical Wire (Oct)(10614).

—**Polytetrafluoroethylene Insulated—Telecommunication Equipment**
DIN VDE 0891 Pt 9-90. Use of Cables and Insulated Cords for Telecommunications Systems and Data Processing Systems; Special Guidelines for Equipment Wires with Solid and Stranded Conductors with Extended Temperature Range According to DIN VDE 0881 (May). 15 pp.

—**PVC Asbestos Insulated—Panels—Marine—660 V**
CNS C2134-88. 660V PVC Asbestos Insulating Wire of Panel (660V-STW) for Marine Use (Jul)(9149).

Electric Wire (Cont.)

—**PVC Asbestos Insulated—Panels—Marine—660 V (Cont.)**
CNS C2135-88. 660V Single-Core Flexible PVC Asbestos Insulating Wire of Panel (660V-STWP) for Marine Use (Jul)(9150).

—**PVC Insulated**
JIS K 6723-83. Plasticized Polyvinyl Chloride Compounds. 14 pp.
MOD UK DSTAN 61-12: Part 6-01. Wires, Cords, and Cables, Electrical-Metric Units Part 6: Polyvinyl Chloride (PVC), Polyethylene, or Silicone Rubber Insulated Equipment Wires Issue 5; Amendment 1. 40 pp.

—**PVC Insulated—Data Processing Equipment**
DIN VDE 0812-88. Equipment Wires with Solid and Stranded Conductors for Telecommunication Systems and Data Processing Systems (Nov) (Replaces VDE 0812a/12.77). 25 pp.
DIN VDE 0891 Pt 9-90. Use of Cables and Insulated Cords for Telecommunications Systems and Data Processing Systems; Special Guidelines for Equipment Wires with Solid and Stranded Conductors with Extended Temperature Range According to DIN VDE 0881 (May). 15 pp.

—**PVC Insulated—Fluorescent Lighting**
CNS C2072-89. 1000 V Grade Insulated Wires for Fluorescent Lamps (Jul)(6063). 6 pp.

—**PVC Insulated—Heat Resistant**
CNS C2103-82. 600V Grade Heat Resistant Polyvinyl Chloride Insulated Wires (Jan)(8379).
JIS C 3317-87. 600 V Grade Heat-Resistant Polyvinyl Chloride Insulated Wires. 11 pp.

—**PVC Insulated—PVC Sheathed**
IEC 189 Pt 1-86. Low-Frequency Cables and Wires with PVC Insulation and PVC Sheath Part 1: General Test and Measuring Methods Second Edition; (Amendment 3-1992, Incorporating Amendment 1 and 2). 47 pp.
IEC 189 Pt 3-88. Low-Frequency Cables and Wires with PVC Insulation and PVC Sheath Part 3: Equipment Wires with Solid or Stranded Conductor, PVC Insulated, in Singles, Pairs and Triples Third Edition; (Amendment 1-1989). 52 pp.
IEC 189 Pt 4-80. Low-Frequency Cables and Wires with P.V.C. Insulation and P.V.C. Sheath Part 4: Distribution Wires with Solid Conductors, P.V.C. Insulated, in Pairs, Triples, Quads and Quintuples Second Edition; (Amendment 1-1989). 27 pp.
IEC 189 Pt 5-80. Low-Frequency Cables and Wires with P.V.C. Insulation and P.V.C. Sheath Part 5: Equipment Wires and Cables with Solid or Stranded Conductors, PVC Insulated, Screened, Single or One Pair Second Edition; (Amendment 2-1992, Incorporating Amd 1). 43 pp.
IEC 189 Pt 7-82. Low-Frequency Cables and Wires with P.V.C. Insulation and P.V.C. Sheath Part 7: Distribution Wires with Solid Conductors, P.V.C. Insulated, Polyamide Coated, in Singles, Pairs, Triples, Quads and Quintuples Second Edition (Amendment 1-1989). 32 pp.

—**PVC Insulated—PVC Sheathed—Telecommunication Equipment**
BSI BS 4808: Part 1-72. 1972 L.F. Cables and Wires with PVC Insulation and PVC Sheath for Telecommunication Part 1: General Requirements and Test. 9 pp.
BSI BS 4808: Part 2-72. 1972 L.F. Cables and Wires with PVC Insulation and PVC Sheath for Telecommunication Part 2: Equipment Wires with Solid or Stranded Conductors, Unscreened, Single. 8 pp.
BSI BS 4808: Part 3-72. 1972 L.F. Cables and Wires with PVC Insulation and PVC Sheath for Telecommunication Part 3: Cables and Equipment Wires, with Solid or Stranded Conductors, Screened, Single. 9 pp.
BSI BS 4808: Part 4-74. 1974 L.F. Cables and Wires with PVC Insulation and PVC Sheath for Telecommunication Part 4: Equipment Wires with Solid or Stranded Conductors, Unscreened in Pairs, Triples, Quads and Quintuples. 9 pp.
BSI BS 4808: Part 5-73. 1973 L.F. Cables and Wires with PVC Insulation and PVC Sheath for Telecommunication Part 5: Cables with Solid or Stranded Conductors, Screened and Sheathed, One Pair. 9 pp.

—**PVC Insulated—Telecommunication Equipment**
DIN VDE 0812-88. Equipment Wires with Solid and Stranded Conductors for Telecommunication Systems and Data Processing Systems (Nov) (Replaces VDE 0812a/12.77). 25 pp.

INDUSTRY STANDARDS

Electric Wire (Cont.)

—PVC Insulated—Telecommunication Equipment (Cont.)
DIN VDE 0891 Pt 9-90. Use of Cables and Insulated Cords for Telecommunications Systems and Data Processing Systems; Special Guidelines for Equipment Wires with Solid and Stranded Conductors with Extended Temperature Range According to DIN VDE 0881 (May). 15 pp.

—PVC Insulated—Weatherproof
CNS C2098-81. ACSR Conductor Polyvinyl Chloride Insulated Out-Door Weather-Proof Wires (Aug)(7782).
CNS C2105-87. Out-Door Weather Proof Polyvinyl Chloride Insulated Wires (Nov)(8559).
JIS C 3340-92. Outdoor Weatherproof Polyvinyl Chloride Insulated Wires. 9 pp.
JIS C 3370-87. ACSR Conductor Polyvinyl Chloride Insulated out-Door Weather-Proof Wires. 12 pp.

—PVC Insulated—600 V
CNS C2012-89. 600V Grade Polyvinyl Chloride Insulated Wires (Nov)(679).
CNS C2077-86. 600 V Polyvinyl Chloride Insulated Wire for Electrical Apparatus (Dec)(6070). 5 pp.
JIS C 3316-87. 600 V Grade Polyvinyl Chloride Insulated Wires for Electrical Apparatus. 9 pp.

—Quality Assurance
BSI BS 9530-88. (OBSOLESCENT) 1988 Cable Fitting Accessories of Assessed Quality for Circular Electrical Connectors: Generic Data, Methods of Test and Rules for the Preparation of Detail Specifications. 29 pp.
BSI BS 9550-89. (OBSOLESCENT) 1989 Capability Approval Procedures for d.c. and Low Frequency Connector Cable Assemblies and Wiring Harnesses: Generic Data. 24 pp.

—Rayon Covered
BSI BS 4927: Part 1-74. 1974 Paper-, Cotton-, Silk-or Rayon-Covered Copper Conductors Part 1: Round Wire. 18 pp.

—Reels
CNS C4308-81. Small Wooden Reel for Electric Wires (May)(7356).

—Rubber Insulated
CNS C3004-80. General Rules for Testing Standard of Insulated Wires (Jul)(682). 1 p.
CNS C3009-87. Method of Test for Rubber-Insulated Wires & Cables (Sep)(687).
CSA C22.2 NO 38-M1986. Thermoset Insulated Wires and Cables; (Gen Instr 1 Thru 3). 64 pp.
CSA 983B Bull. Electrical Bulletin 983B April 25, 1977 to C22.2 NO 38. 39 pp.
DIN VDE 0282 Pt 1-85. Rubber Cables, Wires and Flexible Cords for Power Installation; General Requirements (Apr). 16 pp.
JIS C 3005-91. Testing Methods for Rubber or Plastic Insulated Wires and Cables. 39 pp.

—Rubber Insulated—Appliances—600 V
CNS C2010-80. 600 V Grade Rubber Insulated Wire (Tentative) (Jul)(675). 4 pp.

—Rubber Insulated—Identification Systems
CSA C22.2 NO 38-M1986. Thermoset Insulated Wires and Cables; (Gen Instr 1 Thru 3). 64 pp.

—Safety Lamps—Mining Equipment
CNS M2009-71. Electric Sheathed Wires for Safety Lamps (Aug) (3175). 3 pp.

—Sheathed
CSA 1334 Bull. Electrical Bulletin 1334 August 28, 1981 to C22.2 NO 16. 1 p.

—Silicone—Heat Resistant
DIN VDE 0250 Pt 502-85. Cables, Wires and Flexible Cords for Power Installations; Heat-Resistant Silicone Wire (Mar) (Partial Replacement for VDE 0250c/08.75). 8 pp.

—Silicone Rubber Insulated
MOD UK DSTAN 61-12: Part 6-01. Wires, Cords, and Cables, Electrical-Metric Units Part 6: Polyvinyl Chloride (PVC), Polyethylene, or Silicone Rubber Insulated Equipment Wires Issue 5; Amendment 1. 40 pp.

—Silicone Rubber Insulated—Glass Fiber
CNS C2078-87. 600 V Grade Silicon Rubber Insulated Glass Fiber Braided Wires (Dec)(6071). 5 pp.
JIS C 3323-87. 600 V Grade Silicone Rubber Insulated Glass Fiber Braided Wires. 8 pp.

—Sleeves—Identification Systems
BSI BS 3858-92. 1992 Binding and Identification Sleeves for Use on Electric Cables and Wires. 15 pp.

—Sleeves—Identification Systems (Cont.)
BSI BS 3858-65. 1965 Amd 2 Binding and Identification Sleeves for Use on Electric Cables and Wires. 26 pp.

—Solderability
DIN VDE 0472 Pt 808-84. Testing of Cables, Wires and Flexible Cords; Tinning, Solderability and Soldering Shrinkage (VDE Specification) (Feb). 7 pp.

—Spool Holders
DIN ENGL 46399 Pt 1-82. Delivery Spools for Bare and Insulated Wires. Dimensions (Sept). 6 pp.

—Spools
DIN ENGL 46399 Pt 1-82. Delivery Spools for Bare and Insulated Wires. Dimensions (Sept). 6 pp.

—Submarines
MOD UK NES 512: Part 1-91. Guide to Cables, Electrical and Associated Items Part 1: General Issue 1 (10.91). 17 pp.
MOD UK NES 512: Part 6-91. Guide to Cables, Electrical and Associated Items Part 6: Cables, Electrical, Miniature and Sub-Miniature and Equipment Wires Issue 2 (05.91). 100 pp.
MOD UK NES 512: Part 6-01. Guide to Cables, Electrical and Associated Items Part 6: Cables, Electrical, Miniature and Sub-Miniature and Equipment Wires Issue 2 (05.91); Amendment 2. 116 pp.
MOD UK NES 512: Part 7-91. Guide to Cables, Electrical and Associated Items Part 7: Cables and Wires, Electrical, Miscellaneous for Special Services Issue 3 (04.91). 90 pp.
MOD UK NES 512: Part 7-01. Guide to Cables, Electrical and Associated Items Part 7: Cables and Wires, Electrical, Miscellaneous for Special Services Issue 3 (04.91); Amendment 1. 91 pp.
MOD UK NES 512: Part 7-93. Guide to Cables, Electrical and Associated Items Part 7: Cables and Wires, Electrical, Miscellaneous for Special Services Issue 4 (02.93). 84 pp.
MOD UK NES 523-88. Guide to the Installation of Thin-Wall Insulated, Limited Fire Hazard Equipment Wire and Electric Cable Issue 2 (05.88). 51 pp.
MOD UK NES 523-92. Guide to the Installation of Thin—Wall Insulated, Limited Fire Hazard Equipment Wire and Electric Cable Issue 3 (01.92). 52 pp.

—Substations (Electric)—6.6 kV
CNS C2082-89. 6.6 KV Insulated Wire for Cubicle Type Unit Substation (Jul)(6075). 4 pp.
CNS C3099-80. Method of Test for 6.6KV Insulated Wire for Cubicle Type Unit Substation (Aug)(6076).
JIS C 3611-91. Insulated Wires for Cubicle Type Unit Substation for 6.6 kV Receiving. 9 pp.

—Tear Strength
DIN VDE 0472 Pt 613-85. Testing of Cables, Wires and Flexible Cords; Tear Resistance (Nov) (Partial Replacement for VDE 0472/09.71 and VDE 0472d/12.77). 5 pp.

—Telecommunication Systems—Standards
CCITT FASCICLE I.2-88. Opinions and Resolutions Recommendations on the Organisation and Working Procedures of CCITT (Series A). 64 pp.

—Television Receivers—Fire Resistant
IEC 197-65. High-Voltage Connecting Wire with Flame Retarding Insulation for Use in Television Receivers First Edition. 15 pp.

—Temperature
CNS C1141-88. Temperature Limitation of Low Voltage Insulated Wires (Jul)(12357).

—Tensile Testing
DIN VDE 0472 Pt 625-83. Testing of Cables and Insulated Conductors; Tensile and Elongation Characteristics of Strain Bearing Elements (Jan). 5 pp.

—Testing Equipment
CSA CAN/CSA-C22. 2 NO 0.3-92. Test Methods for Electrical Wires and Cables; (Gen Instr 1). 121 pp.

—Thermoplastic Insulated
CSA 1334 Bull. Electrical Bulletin 1334 August 28, 1981 to C22.2 NO 16. 1 p.
CSA 632 Bull. Electrical Bulletin 632 March 29, 1966 to C22.2 NO 38. 2 pp.
CSA 671 Bull. Electrical Bulletin 671 March 16, 1967 to C22.2 NO 38. 3 pp.
CSA 730A Bull. Electrical Bulletin 730A May 31, 1974 to C22.2 NO 38. 2 pp.
CSA 983B Bull. Electrical Bulletin 983B April 25, 1977 to C22.2 NO 38. 39 pp.
CSA 983C Bull. Electrical Bulletin 983C November 14, 1978 to C22.2 NO 38. 3 pp.

—Thermoplastic Insulated (Cont.)
CSA 1368 Bull. Electrical Bulletin 1368 April 7, 1982 to C22.2 NO 38. 2 pp.
CSA 1403 Bull. Electrical Bulletin 1403 July 5, 1983 to C22.2 NO 38. 4 pp.
CSA 1316A Bull. Electrical Bulletin 1316A February 28, 1985 to C22.2 NO 38. 2 pp.
CSA 1240 Bull. Electrical Bulletin 1240 September 24, 1979 to C22.2 NO 51. 2 pp.
CSA 632 Bull. Electrical Bulletin 632 March 29, 1966 to C22.2 NO 75. 2 pp.
CSA 671 Bull. Electrical Bulletin 671 March 16, 1967 to C22.2 NO 75. 3 pp.
CSA 730A Bull. Electrical Bulletin 730A May 31, 1974 to C22.2 NO 75. 2 pp.
CSA 983B Bull. Electrical Bulletin 983B April 25, 1977 to C22.2 NO 75. 39 pp.
CSA 983C Bull. Electrical Bulletin 983C November 14, 1978 to C22.2 NO 75. 3 pp.
CSA 1240 Bull. Electrical Bulletin 1240 September 24, 1979 to C22.2 NO 75. 2 pp.
CSA 1334 Bull. Electrical Bulletin 1334 August 28, 1981 to C22.2 NO 75. 1 pp.
CSA 1368 Bull. Electrical Bulletin 1368 April 7, 1982 to C22.2 NO 75. 2 pp.
CSA 1403 Bull. Electrical Bulletin 1403 July 5, 1983 to C22.2 NO 75. 4 pp.
CSA 1316A Bull. Electrical Bulletin 1316A February 28, 1985 to C22.2 NO 75. 2 pp.
CSA C22.2 NO 127-1988. Equipment Wires; (Gen Instr 1 Thru 3). 47 pp.
CSA 1281 Bull. Electrical Bulletin 1281 August 13, 1980 to C22.2 NO 127. 1 p.
CSA 1364 Bull. Electrical Bulletin 1364 February 18, 1982 to C22.2 NO 127. 2 pp.

—Thermoplastic Insulated—600 V
CSA C22.2 NO 75-M1983. Thermoplastic Insulated Wires and Cables (R 1992); (Gen Instr 1 Thru 6). 48 pp.

—Thermoset Insulated
CSA 916 Bull. Electrical Bulletin 916 July 3, 1973 to C22.2 NO 16. 2 pp.
CSA 1334 Bull. Electrical Bulletin 1334 August 28, 1981 to C22.2 NO 16. 1 p.
CSA C22.2 NO 38-M1986. Thermoset Insulated Wires and Cables; (Gen Instr 1 Thru 3). 64 pp.
CSA 632 Bull. Electrical Bulletin 632 March 29, 1966 to C22.2 NO 38. 2 pp.
CSA 671 Bull. Electrical Bulletin 671 March 16, 1967 to C22.2 NO 38. 3 pp.
CSA 851 Bull. Electrical Bulletin 851 November 17, 1971 to C22.2 NO 38. 2 pp.
CSA 916 Bull. Electrical Bulletin 916 July 3, 1973 to C22.2 NO 38. 2 pp.
CSA 730A Bull. Electrical Bulletin 730A May 31, 1974 to C22.2 NO 38. 2 pp.
CSA 983B Bull. Electrical Bulletin 983B April 25, 1977 to C22.2 NO 38. 39 pp.
CSA 983C Bull. Electrical Bulletin 983C November 14, 1978 to C22.2 NO 38. 3 pp.
CSA 1334 Bull. Electrical Bulletin 1334 August 28, 1981 to C22.2 NO 38. 1 pp.
CSA 1368 Bull. Electrical Bulletin 1368 April 7, 1982 to C22.2 NO 38. 2 pp.
CSA 1316A Bull. Electrical Bulletin 1316A February 28, 1985 to C22.2 NO 38. 2 pp.
CSA 1316A Bull. Electrical Bulletin 1316A February 28, 1985 to C22.2 NO 51. 2 pp.
CSA C22.2 NO 52-M1989. Service-Entrance Cables; (Gen Instr 1 Thru 3). 34 pp.
CSA 1316A Bull. Electrical Bulletin 1316A February 28, 1985 to C22.2 NO 52. 2 pp.
CSA C22.2 NO 127-1988. Equipment Wires; (Gen Instr 1 Thru 3). 47 pp.
CSA 1316A Bull. Electrical Bulletin 1316A February 28, 1985 to C22.2 NO 131. 2 pp.

—Thermoset Insulated—Deep Well Pumps
CSA 1403 Bull. Electrical Bulletin 1403 July 5, 1983 to C22.2 NO 38. 4 pp.

—Thermoset Insulated—Identification Systems
CSA C22.2 NO 38-M1986. Thermoset Insulated Wires and Cables; (Gen Instr 1 Thru 3). 64 pp.

—Thermoset Insulated—Multiconductor
CSA C22.2 NO 38-M1986. Thermoset Insulated Wires and Cables; (Gen Instr 1 Thru 3). 64 pp.
CSA 983B Bull. Electrical Bulletin 983B April 25, 1977 to C22.2 NO 38. 39 pp.

—Thermoset Insulated—Submersible Pumps
CSA 1403 Bull. Electrical Bulletin 1403 July 5, 1983 to C22.2 NO 38. 4 pp.

—Tin Coatings (Made From Tin)
DIN VDE 0472 Pt 808-84. Testing of Cables, Wires and Flexible Cords; Tinning, Solderability and Soldering Shrinkage (VDE Specification) (Feb). 7 pp.

INTERNATIONAL AND NON-U.S. NATIONAL STANDARDS
SUBJECT INDEX

Electrical

Electric Wire (Cont.)
—**Varnished Cloth Insulated—Appliances**
CNS C2048-80. Varnish Cloth Insulated Wire (Jul)(2698). 15 pp.

—**Wax Coated Nylon**
CNS Z7203-87. Wax Coated Nylon Band for Electric Wire and Cable (Jan)(11822).

Electric Wiring
See Also: Cables (Electric); Electric Wire; Electrical Installations; Lacing Cords; Live Wire Working; Power Systems; Wiring Devices; Wiring Space

SAA AS 3000N-91. Notes on the Changes in the SAA Wiring Rules—10th Edition (AS 3000—1991) as Compared with the 9th Edition (AS 3000—1986) (Supersedes AS DOC 3000N —1986). 15 pp.

—**Aerospace Vehicles**
JIS W 2010-89. Wiring for Aerospace Vehicle.

—**Aircraft**
AECMA PREN2283-88. Testing of Aircraft Wiring. 9 pp.

—**Architectural Drawings**
SNZ NZS 5902: Part 4-76. Building and Civil Engineering Drawing Practice Part 4: Services—Electrical Amend: 1, 1984; 2, 1985 (Superseded bu NZS/AS 1100 Parts 101-501).

—**Architectural Drawings—Symbols**
CNS C1090-87. Graphical Symbols of Interior Wiring Diagram for Architectural Plans (Dec)(9101).
CNS C1091-87. Symbols of Switch of Interior Wiring Diagram for Architectural Plans (Dec)(9102).
CNS C1092-87. Symbols of Relays of Interior Wiring Diagram for Architectural Plans (Dec)(9103).
CNS C1093-87. Symbols of Meters of Interior Wiring Diagram for Architectural Plans (Dec)(9104).
CNS C1094-87. Symbols of Electric Machines of Interior Wiring Diagram for Architectural Plans (Dec)(9105).
CNS C1095-87. Symbols of Potential Transformers and Current Transformers of Interior Wiring Diagrams for Architectural Plans (Dec)(9106).
CNS C1096-87. Symbols of Connecting of Interior Wiring Diagrams for Architectural Plans (Dec)(9107).
CNS C1097-87. Symbols of Distribution of Interior Wiring Diagram for Architectural Plans (Dec)(9108).
CNS C1098-87. Symbols of Wiring of Interior Wiring Diagram for Architectural Plans (Dec)(9109).
CNS C1099-87. Symbols of Bus-Bar of Interior Wiring Diagram for Architectural Plans (Dec)(9110).
CNS C1100-87. Symbols of Lamps and Socket of Interior Wiring Diagram for Architectural Plans (Dec)(9111).
CNS C1101-87. Symbols of Telephone and Bell of Interior Wiring Diagram for Architectural Plans (Dec)(9112).
CNS C1102-87. Symbols of Fire Alarm System of Interior Wiring Diagrams for Architectural Plans (Dec)(9113).
CNS C1103-87. Symbols of Television and Broadcast Equipment of Interior Wiring Diagram for Architectural Plans (Dec)(9114).
CSA Z99.3-1979. Graphical Electrical Symbols for Architectural Plans (R 1989). 32 pp.
JIS C 0303-84. Graphical Symbols of Interior Wiring Diagram for Architectural Plans. 52 pp.

—**Buildings**
DIN ENGL 18015 Pt 3-90. Electrical Installations in Dwellings; Circuit Arrangement and Location of Electrical Equipment (July). 3 pp.
IEC 364 Pt 5-523-83. Electrical Installations of Buildings Part 5: Selection and Erection of Electrical Equipment Chapter 52: Wiring Systems Section 523—Current-Carrying Capacities First Edition. 72 pp.
SAA AS 3000-91. Electrical Installations —Buildings, Structures and Premises (Known as the SAA Wiring Rules) Amdt 1 April 1992 Amdt 2 January/February 1993 Amdt 3 May 1993 (in Professional Packages 17F, 21, 55) 68). 23 pp.
SAA AS 3000 DOC R/1-91. Electrical Installations —Buildings, Structures and Premises—Part R/1: Rulings to SAA Wiring Rules (AS 3000—1991) (Supersedes AS Doc 3000 R/1—1986, AS Doc 3000 R/2—1987, AS Doc 3000 R/3—1988, AS Doc 3000 R/4—1989, AS Doc 3000 R/5—1989, AS Doc 3000 R/6—1990 and. 61 pp.
SAA AS DOC 3000 R/7-91. Electrical Installations —Buildings, Structures and Premises—Part R/7: Rulings to SAA Wiring Rules (AS 3000—1986) Seventh Group (Superseded by AS 3000 Doc R/1—1991).

Electric Wiring (Cont.)
—**Buildings—Construction Contracts**
DIN ENGL 18382-88. Tendering and Performance Stipulations in Contracts for Construction Works (VOB); Part C: General Technical Specifications in Contracts for Construction Works (ATV); Electrical Installations in Buildings (Sept). 3 pp.

—**Chassis—Color Coding**
CNS C5072-80. Color Coding of Chassis Wiring (Aug)(6321).

—**Color Coding**
MOD UK DSTAN 61-7: Part 1-90. Identification of Electrical and Electronic Systems. Wiring and Components Part 1: Colour Coding and Marking for Polarity and Phase Distinction in Electrical Systems Issue 6. 11 pp.

—**Data Processing Equipment**
DIN VDE 0891 Pt 5-85. Use of Cables and Insulated Cords for Telecommunications and Information Processing Installations. Special Guidance on Wiring Cables and Cords According to DIN/VDE 0815 (Nov). 11 pp.

—**Electric Rolling Stock**
JIS E 6001-76. Wire Code for DC Electric Railcar.

—**Electric Vehicles**
IEC 783-84. Wiring and Connectors for Electric Road Vehicles First Edition. 22 pp.

—**Electrical Enclosures**
CSA C22.2 NO 0.12-M1985. Wiring Space and Wire Bending Space in Enclosures for Equipment Rated 750 V or Less (R 1992); (Gen Instr 1 Thru 2). 23 pp.

—**Fire Alarm and Extinguishing Equipment**
BSI DD 190-90. 1990 Performance of Alarm and Emergency Wiring Systems Under Fire Conditions (E). 15 pp.

—**Inductive Coordination—Outside**
CSA C22.3 NO 3-1954. Inductive Co-Ordination Definitions, Principles and Practices (R 1970). 22 pp.

—**Inductive Coordination—Outside—Glossaries**
CSA C22.3 NO 3-1954. Inductive Co-Ordination Definitions, Principles and Practices (R 1970). 22 pp.

—**Marine Engines—Control Consoles**
CNS F5026-82. Internal Wirings of Marine Engine Remote Control Consoles (Mar)(8581).
JIS F 0414-87. Internal Wirings and Pipings of Marine Engine Control Console.

—**Symbols**
CNS C5113-81. Graphic Symbols for Conductor, Cable and Wiring (May)(7362).

—**Telecommunication Systems**
DIN VDE 0891 Pt 5-85. Use of Cables and Insulated Cords for Telecommunications and Information Processing Installations. Special Guidance on Wiring Cables and Cords According to DIN/VDE 0815 (Nov). 11 pp.

Electrical Apparatus
Use: Electrical Equipment

Electrical Bonding
See Also: Bonding Conductors; Electrical Grounding

—**Aerospace Equipment**
JIS W 2009-78. Bonding, Electrical and Lightning Protection, for Aerospace Systems.

—**Aerostats**
CAA Chapter Q4-6 12.79. Electrical Bonding and Lightning Discharge Protection (Non-Rigid Airships). 5 pp.
CAA Chapter Q4-6 App 12.79. Primary Conductors (Non-Rigid Airships). 1 p.

—**Aircraft**
CAA Chapter K4-6 03.67. Electrical Bonding and Lightning Discharge Protection (Light Aeroplanes). 5 pp.
CAA Chapter P4-6. Electrical Bonding and Lightning Discharge Protection (Provisional Airworthiness Requirements for Civil Powered Lift Aircraft). 18 pp.
CAA LEAFLET 9-1 07.90. Bonding and Circuit Testing. 12 pp.
NATO STANAG 3659 ED 1 AMD 6-75. Bonding and in-Flight Lightning Protection for Aircraft. 17 pp.

Electrical Bonding (Cont.)
—**Aircraft (Cont.)**
NATO STANAG 3659 ED 1 AMD 7-75. Bonding and In-Flight Lightning Protection for Aircraft. 19 pp.

—**Aircraft—Radio Equipment**
CAA Chapter R4-5 04.74. Bonding and Lightning Discharge Protection (Radio). 1 p.

—**Buildings—Telecommunication**
CCITT RECMN K.27-91. Bonding Configurations and Earthing Inside a Telecommunication Building (Study Group V) 25 pp. 25 pp.
CENELEC PRETS 300 253-92. Equipment Engineering (EE); Earthing and Bonding of Telecommunication Equipment in Telecommunication Centres. 21 pp.
ETSI PRETS 300 253-92. Equipment Engineering (EE); Earthing and Bonding of Telecommunication Equipment in Telecommunication Centres. 21 pp.

—**Bushings**
CSA C22.2 NO 41-M1987. Grounding and Bonding Equipment (R 1993); (Gen Instr 1). 30 pp.

—**Electrical Enclosures—Hazardous Locations**
CSA C22.2 NO 25-1966. Enclosures for Use in Class II Groups E, F, and G Hazardous Locations (R 1992); (Gen Instr 1). 18 pp.
CSA 1273 Bull. Electrical Bulletin 1273 July 2, 1980 to C22.2 NO 25. 2 pp.
CSA 1273A Bull. Electrical Bulletin 1273A February 8, 1983 to C22.2 NO 25. 2 pp.
CSA C22.2 NO 30-M1986. Explosion-Proof Enclosures for Use in Class I Hazardous Locations (R 1992); (Gen Instr 1 Thru 2). 41 pp.

—**Electrical Equipment**
CSA C22.2 NO 0.4-M1982. Bonding and Grounding of Electrical Equipment (Protective Grounding) (R 1993); (Gen Instr 1). 19 pp.

—**Equipotential—Telecommunication Systems**
DIN VDE 0800 Pt 2-85. Telecommunications; Earthing and Equipotential Bonding (July). 84 pp.
DIN VDE 0800 Pt 2 A1 (D)-88. Telecommunications; Earthing and Equipotential Bonding; Amendment 1 (Sept). 29 pp.

—**Rotary Wing Aircraft**
CAA Chapter G4-6 11.75. Electrical Bonding and Lightning Discharge Protection (Rotocraft). 4 pp.
CAA Chapter G4-6 App #1 11.75. Electrical Bonding and Lightning Discharge Protection (Rotocraft). 5 pp.
CAA Chapter G4-6 App #2 11.75. Resistance and Continuity Measurement (Rotocraft). 3 pp.
CAA CAP524 SUB-Part D 12.86. (Rotorcraft). 31 pp.

—**Supersonic Transports**
CAA STANDARD NO. 8-6&AP 07.69. Electrical Bonding and Lightning Discharge Protection. 13 pp.

—**Telecommunication Equipment**
CENELEC PRETS 300 253-92. Equipment Engineering (EE); Earthing and Bonding of Telecommunication Equipment in Telecommunication Centres. 21 pp.
ETSI PRETS 300 253-92. Equipment Engineering (EE); Earthing and Bonding of Telecommunication Equipment in Telecommunication Centres. 21 pp.

—**Telecommunication Systems**
CCITT RECMN K.27-91. Bonding Configurations and Earthing Inside a Telecommunication Building (Study Group V) 25 pp. 25 pp.

Electrical Capacitance Liquid Level Gages
Use: Capacitance Level Indicators

Electrical Codes
See Also: Codes; Electrical Equipment; Electrical Installations; Electrical Safety

CSA C22.1-1990. Safety Standard for Electrical Installations. 568 pp.
CSA CE CODE Part I-1990. CE Code Handbook an Explanation of Rules of the CE Code, Part I (Errata March 1992). 1248 pp.
CSA C22.1EP-1990. What's New in 1990 Changes to CE Code Part I Explained. 143 pp.
CSA 1164 Bull. Electrical Bulletin 1164 April 13, 1978 to C22.2 NO 68. 2 pp.
CSA C22.10-1992. Canadian Electrical Code Part I (16th Edition) Plus Quebec Amendments. 584 pp.
SNZ NZMP 6002-87. Guide to New Zealand Electrical Requirements and Approvals. 84 pp.

INDUSTRY STANDARDS

Electrical
INTERNATIONAL AND NON-U.S. NATIONAL STANDARDS
SUBJECT INDEX

Electrical Codes *(Cont.)*

—Appliances

CSA C22.2 NO 1335.1-93. Portable Electrical Motor-Operated and Heating Appliances: General Requirements; (Gen Instr 1). 208 pp.
CSA C22.2 NO 1335.2.9-93. Portable Electrical Motor-Operated and Heating Appliances: Particular Requirements for Portable Electric Cooking Appliances; (Gen Instr 1). 40 pp.

—Data Processing Equipment

CSA CAN/CSA-C22. 2 NO 220-M91. Information Processing and Business Equipment; (Gen Instr 1). 119 pp.

—Heat Pumps

CSA CAN/CSA-C273.3-M91. Performance Standard for Split-System Central Air-Conditioners and Heat Pumps; (Gen Instr 1 Thru 3). 54 pp.
CSA C273.5-1980. Installation Requirements for Air-to-Air Heat Pumps (R 1991); (Amd 1-2 September 1980). 16 pp.

—Heaters

CSA C22.2 NO 1335.1-93. Portable Electrical Motor-Operated and Heating Appliances: General Requirements; (Gen Instr 1). 208 pp.
CSA C22.2 NO 1335.2.9-93. Portable Electrical Motor-Operated and Heating Appliances: Particular Requirements for Portable Electric Cooking Appliances; (Gen Instr 1). 40 pp.

—Vacuum Cleaners

CSA CAN/CSA-C22. 2 NO 243-M91. Vacuum Cleaners and Blower Cleaners; (UL 1017) (Gen Instr 1 Thru 3). 145 pp.

Electrical Components

Scope Note: For additional listings, use a more specific term ***Use For:*** Electrotechnical Components; Electrotechnical Equipment ***See Also:*** Capacitors; Connectors; Electric Coils; Electric Conductors; Electric Terminals; Electrical Engineering; Electrical Equipment; Electrical Installations; Electrodes; Electromechanical Components; Electron Tubes; Electronic Components; Electronic Equipment; Resistors
CNS Z5078-82. Preparation for Delivery of Electrical and Electronic Components (May)(6982).
IEC 750-83. Item Designation in Electrotechnology First Edition. 42 pp.

—Accelerated Testing

BSI BS 2011: Part 2.1GA-84. (WITHDRAWN) 1984 Amd 1 Basic Environmental Testing Procedures Part 2.1: Tests Part 2.1GA: Acceleration, Steady State (Superseded by BS EN 60068-2-7: 1993). 12 pp.
BSI BS 2011: Sec 4.1-83. (WITHDRAWN) 1983 Basic Environmental Testing Procedures Part 4: Miscellaneous Section 4.1: Specification for Mounting of Components, Equipment and Other Articles for Dynamic Tests (Renumbered as BS EN 60068-2-47: 1993). 15 pp.
BSI BS EN 60068-2-7-93. 1993 Environmental Testing Part 2: Tests Test Ga and Guidance: Acceleration, Steady State (IEC 68-2-7: 1983) (G). 16 pp.
BSI BS EN 60068-2-47-93. 1993 Environmental Testing Part 2: Tests Mounting of Components, Equipment and Other Articles for Dynamic Tests Including Shock (Ea), Bump (Eb), Vibration (Fc and Fd) and Steady-State Acceleration (Ga) and Guidance (IEC 68-2-47: 1982) (G). 23 pp.
CENELEC HD 323.2.47 S1-88. Basic Environmental Testing Procedures—Part 2: Tests. Mounting of Components. Equipment and Other Articles for Dynamic Tests Including Shock (Es). Bump (Eb). Vibration (Fc and Fd) and Steady-State Acceleration (Ga) and Guidance. 3 pp.
CENELEC EN 60068-2-7-93. Basic Environmental Testing Procedures Part 2: Tests Test Ga and Guidance: Acceleration, Steady State (IEC 68-2-7: 1983 + A1: 1986). 11 pp.
CENELEC EN 60068-2-47-93. Basic Environmental Testing Procedures Part 2: Tests Mounting of Components, Equipment and Other Articles for Dynamic Tests Including Shock (Ea), Bump (Eb), Vibration (Fc and Fd) and Steady-State Acceleration (Ga) and Guidance (IEC 68-2-47: 1982). 16 pp.
IEC 68 Pt 2-7-83. Basic Environmental Testing Procedures Part 2: Tests—Test Ga and Guidance: Acceleration, Steady State Second Edition; (Amendment 1-1986) (CENELEC EN 60068-2-7: 1993). 32 pp.
IEC 68 Pt 2-47-82. Basic Environmental Testing Procedures Part 2: Tests Mounting of Components, Equipment and Other Articles for Dynamic Tests Including Shock (Ea), Bump (Eb), Vibration (Fc and Fd) and Steady-State Acceleration (Ga) and Guidance First Edition (CENE EN 60068-2-47:93). 29 pp.
SAA AS 1099.2.7-90. Basic Environmental Testing Procedures for Electrotechnology—Part 2: Tests—Part 2.7: Test Ga—Acceleration, Steady State (IEC 68-2-7). 8 pp.

Electrical Components *(Cont.)*

—Accelerated Testing *(Cont.)*

SAA AS 1099.3.7-81. Basic Environmental Testing Procedures for Electrotechnology—Part 3: Background Information—Part 3.7: Section 7—Appraisal of the Problems of Accelerated Testing for Atmospheric Corrosion. 4 pp.
SNZ IEC 68: Part 2-7-83. Basic Environmental Testing Procedures Part 2-7: Test Ga and Guidance: Acceleration, Steady State. 23 pp.
SNZ IEC 68: Part 2-47-82. Basic Environmental Testing Procedures Part 2-47: Mounting of Components and Other Articles for Dynamic Tests Including Shock (Ea), Bump (Eb), Vibration (Fc and Fd) and Steady-State Acceleration (Ga) and Guidance. 25 pp.

—Atmospheric Corrosion Testing

IEC 68 Pt 2-60 TTD-89. Environmental Testing Part 2: Tests—Test Ke: Corrosion Tests in Artificial Atmosphere at Very Low Concentration of Polluting Gas(es). 25 pp.
SAA AS 1099.3.7-81. Basic Environmental Testing Procedures for Electrotechnology—Part 3: Background Information—Part 3.7: Section 7—Appraisal of the Problems of Accelerated Testing for Atmospheric Corrosion. 4 pp.

—Charts—Glossaries

SAA AS 1103.1-73. Diagrams, Charts and Tables for Electrotechnology—Part 1: Definitions and Classifications Reconfirmed 1982. 5 pp.
SNZ NZS/AS 1103. 1-73. Diagrams, Charts and Tables for Electrotechnology Part 1: Definitions and Classifications (This is a Joint Standard with SAA AS 1103.1). 5 pp.

—Circuit Diagrams

SAA AS 1103.4-92. Diagrams, Charts and Tables for Electrotechnology—Part 4: Guiding Principles for the Preparation of Circuit Diagrams (in Professional Package 56). 46 pp.
SAA AS 1103.6-81. Diagrams, Charts and Tables for Electrotechnology—Part 6: Preparation of Unit Wiring Diagrams and Tables. 14 pp.
SNZ NZS/AS 1103. 4-78. Diagrams, Charts and Tables for Electrotechnology Part 4: Guiding Principles for the Preparation of Circuit Diagrams (This is a Joint Standard with SAA AS 1103.4). 44 pp.
SNZ NZS/AS 1103. 6-81. Diagrams, Charts and Tables for Electrotechnology Part 6: Preparation of Unit Wiring Diagrams and Tables (This is a Joint Standard with SAA AS 1103.6). 14 pp.

—Damp Heat Testing

BSI BS 2011: Part 2.1CA-77. 1977 Amd 1 Basic Environmental Testing Procedures Part 2.1: Tests Part2.1Ca: Test Ca. Damp Heat, Steady State (AMD 5362) March 31, 1987. 4 pp.
BSI BS 2011: Part 2.1CB-90. 1990 Basic Environmental Testing Procedures Part 2.1: Tests Part 2.1Cb: Damp Heat, Steady State, Primarily for Equipment. 8 pp.
BSI BS 2011: Part 2.1DB-81. 1981 Amd 1 Basic Environmental Testing Procedures Part 2.1: Tests Part 2.1Db: Test Db and Guidance: Damp Heat, Cyclic (12 + 12-Hour Cycle). 13 pp.
BSI BS 2011: Part 2.1Z/AD-77. 1977 Basic Environmental Testing Procedures Part 2.1: Tests Part 2.1Z/AD: Test Z/AD. Composite Temperature/Humidity Cyclic Test. 12 pp.
BSI BS 2011: Pt 2.1Z/AMD-77. 1977 Basic Environmental Testing Procedures Part 2.1: Tests Part 2.1Z/AMD: Test Z/AMD. Combined Sequential Cold, Low Air Pressure and Damp Heat Test. 4 pp.
BSI BS 2011: Pt 2.2C & D-90. 1990 Basic Environmental Testing Procedures Part 2.2: Guidance Part 2.2C and D: Guidance for Damp Heat Tests Guidance. 20 pp.
CENELEC HD 323.2.28 S1-88. Basic Environmental Testing Procedures—Part 2: Tests. Guidance for Damp Heat Tests. 3 pp.
CENELEC HD 323.2.30 S3-88. Basic Environmental Testing Procedures—Part 2: Tests. Test Db and Guidance: Damp Heat, Cyclic (12 + 12-Hour Cycle). 3 pp.
CENELEC HD 323.2.38 S1-76. Basic Environmental Testing Procedures—Part 2: Tests. Test Z/AD: Composite Temperature/Humidity Cyclic Test. 2 pp.
CENELEC HD 323.2.39 S1-76. Basic Environmental Testing Procedures-Part 2: Tests. Test Z/Amd: Combined Sequential Cold, Low Air Pressure, and Damp Heat Test. 2 pp.
CENELEC HD 323.2.56 S1-90. Basic Environmental Testing Procedures Part 2: Tests Test Cb; Damp Heat, Steady State, Primarly for Equipment. 3 pp.
IEC 68 Pt 2-3-69. Basic Environmental Testing Procedures Part 2: Tests—Test Ca: Damp Heat, Steady State Third Edition; (Amendment 1-1984). 13 pp.

Electrical Components *(Cont.)*

—Damp Heat Testing *(Cont.)*

IEC 68 Pt 2-28-90. Environmental Testing Part 2: Tests—Guidance for Damp Heat Tests Third Edition. 37 pp.
IEC 68 Pt 2-30-80. Basic Environmental Testing Procedures Part 2: Tests—Test Db and Guidance: Damp Heat, Cyclic (12 + 12-Hour Cycle) Second Edition; (Amendment 1-1985). 25 pp.
IEC 68 Pt 2-38-74. Basic Environmental Testing Procedures Part 2: Tests—Test Z/AD: Composite Temperature/Humidity Cyclic Test First Edition. 23 pp.
IEC 68 Pt 2-39-76. Basic Environmental Testing Procedures Part 2: Tests—Test Z/AMD: Combined Sequential Cold, Low Air Pressure, and Damp Heat Test First Edition. 12 pp.
IEC 68 Pt 2-56-88. Environmental Testing Part 2: Tests—Test Cb: Damp Heat, Steady State, Primarily for Equipment First Edition. 19 pp.
JIS C 0022-87. Basic Environmental Testing Procedures Part 2: Tests, Test Ca: Damp Heat, Steady State. 7 pp.
JIS C 0032-90. Environmental Testing Test Cb: Damp Heat, Steady State, Primarily for Equipment. 9 pp.
SAA AS 1099.2DA-71. Basic Environmental Testing Procedures for Electrotechnology—Part 2: Tests—Part 2Da: Damp Heat, 24-Hour Cycle Reconfirmed 1985. 2 pp.
SAA AS 1099.2DB-82. Basic Environmental Testing Procedures for Electrotechnology—Part 2: Tests—Part 2Db: Damp Heat, Cyclic (12 + 12 Hours). 7 pp.
SAA AS 1099.2Z/A D-80. Basic Environmental Testing Procedures for Electrotechnology—Part 2: Tests—Part 2Z/AD: Composite Temperature/Humidity Cyclic Test Reconfirmed 1985. 8 pp.
SAA AS 1099.2Z/A MD-80. Basic Environmental Testing Procedures for Electrotechnology—Part 2: Tests—Part 2Z/AMD: Combined Sequential Cold, Low Air Pressure and Damp Heat Test Reconfirmed 1985. 2 pp.
SAA AS 1099.2Z/B M-80. Basic Environmental Testing Procedures for Electrotechnology—Part 2: Tests—Part 2Z/BM: Combined Dry Heat/Low Air Pressure Tests Reconfirmed 1985. 6 pp.
SAA AS 1099.2.3-90. Basic Environmental Testing Procedures for Electrotechnology—Part 2: Tests—Part 2.3: Test Ca—Damp Heat, Steady State (IEC 68-2-3:1969). 3 pp.
SAA AS 1099.3.3-82. Basic Environmental Testing Procedures for Electrotechnology—Part 3: Background Information—Part 3.3: Section 3-Guidance for Damp Heat Tests. 11 pp.
SNZ IEC 68: Part 2-3-69. Basic Environmental Testing Procedures Part 2-3: Test Ca: Damp Heat, Steady State. 9 pp.
SNZ IEC 68: Part 2-28-90. Basic Environmental Testing Procedures Part 2-28: Tests—Guidance for Damp Heat Tests. 33 pp.
SNZ IEC 68: Part 2-30-80. Basic Environmental Testing Procedures Part 2-30: Test Db and Guidance: Damp Heat, Cyclic (12 + 12-Hour Cycle) Amend: 1. 19 pp.
SNZ IEC 68: Part 2-38-74. Basic Environmental Testing Procedures Part 2-38: Test 2/AD: Composite Temperature/Humidity Cyclic Test. 20 pp.
SNZ IEC 68: Part 2-39-76. Basic Environmental Testing Procedures Part 2-39: Test 2/AMD: Combined Sequential Cold, Low Air Pressure, and Damp Heat Test. 8 pp.

—Design—Ships

MOD UK NES 501-88. General Requirements for the Design of Electrotechnical Equipment Issue 4 (11.88). 180 pp.

—Design—Submarines

MOD UK NES 501-88. General Requirements for the Design of Electrotechnical Equipment Issue 4 (11.88). 180 pp.

—Diagrams

BSI BS 5070: Part 2-88. 1988 Engineering Diagram Drawing Practice Part 2: Recommendations for Electrotechnology Diagrams. 72 pp.
CENELEC HD 246.1-75. Diagrams, charts, tables Part 1: Definitions and classification. 2 pp.
CENELEC HD 246.2-75. Diagrams, Charts, Tables Part 2: Item Designation. 2 pp.
CENELEC HD 246.3-75. Diagrams, Charts, Tables Part 3: General Recommendations for the Preparation of Diagrams. 2 pp.
CENELEC HD 246.5-76. Diagrams, Charts, Tables Part 5: Preparation of Interconnection Diagrams and Tables. 2 pp.
IEC 113 Pt 5-75. Diagrams, Charts, Tables Part 5: Preparation of Interconnection Diagrams and Tables First Edition. 19 pp.
SAA AS 1103.3-89. Diagrams, Charts and Tables for Electrotechology—Part 3: General Rules for the Preparation of Diagrams and Associated Documents. 8 pp.

896
INDEX and DIRECTORY of

Electrical Components *(Cont.)*

—Diagrams *(Cont.)*

SAA AS 1103.5-78. Diagrams, Charts and Tables for Electrotechnology—Part 5: Preparation of Interconnection Diagrams and Tables. 8 pp.

SNZ NZS/AS 1103. 3-89. Diagrams, Charts and Tables for Electrotechnology Part 3: General Rules for the Preparation of Diagrams and Associated Documents (This is a Joint Standard with SAA AS 1103.3). 8 pp.

SNZ NZS/AS 1103. 5-78. Diagrams, Charts and Tables for Electrotechnology Part 5: Preparation of Interconnection Diagrams and Tables (This is a Joint Standard with SAA AS 1103.5). 8 pp.

—Diagrams—Glossaries

SAA AS 1103.1-73. Diagrams, Charts and Tables for Electrotechnology—Part 1: Definitions and Classifications Reconfirmed 1982. 5 pp.

SNZ NZS/AS 1103. 1-73. Diagrams, Charts and Tables for Electrotechnology Part 1: Definitions and Classifications (This is a Joint Standard with SAA AS 1103.1). 5 pp.

—Diagrams—Units of Measurement

SAA AS 1103.7-84. Diagrams, Charts and Tables for Electrotechnology—Part 7: Representation of Values of Units of Physical Quantities. 2 pp.

SNZ NZS/AS 1103. 7-84. Diagrams, Charts and Tables for Electrotechnology Part 7: Representation of Values of Units of Physical Quantities (This is a Joint Standard with SAA AS 1103.7). 2 pp.

—Distribution Systems—Naval Ships

MOD UK NES 538-89. Guide to the Selection of Small Electrical Fittings and Components Issue 3 (08.89). 306 pp.

MOD UK NES 538-92. Guide to the Selection of Small Electrical Fittings and Components Issue 4 (10.92). 285 pp.

—Distribution Systems—Submarines

MOD UK NES 538-89. Guide to the Selection of Small Electrical Fittings and Components Issue 3 (08.89). 306 pp.

MOD UK NES 538-92. Guide to the Selection of Small Electrical Fittings and Components Issue 4 (10.92). 285 pp.

—Documents

IEC 1082 Pt 1-91. Preparation of Documents Used in Electrotechnology Part 1: General Requirements First Edition; (Supersedes 113 Pt 1 and 3). 160 pp.

—Dry Heat Testing

BSI BS 2011: Part 2.1B-77. (WITHDRAWN) 1977 Amd 1 Basic Environmental Testing Procedures Part 2.1: Tests Part 2.1B: Test B. Dry Heat (Superseded by BS EN 60068-2-2: 1993). 32 pp.

BSI BS 2011: Pt 2.1Z/BFC-84. 1984 Basic Environmental Testing Procedures Part 2.1: Tests Part 2.1Z/BFc: Test Z/BF Combined Dry Heat/ Vibration (Sinusoidal) Tests for Both Heat-Dissipating and Non-Heat-Dissipating Specimens. 13 pp.

BSI BS 2011: Part 2.1Z/BM-77. 1977 Amd 1 Basic Environmental Testing Procedures Part 2.1: Tests Part 2.1Z/BM: Test Z/BM. Combined Dry Heat/ Low Air Pressure Tests. 12 pp.

BSI BS 2011:Pt2. 2Z/AFC/BFC-86. 1986 Basic Environmental Testing Procedures Part 2.2: Guidance Part 2.2Z/AFc and Z/BFc: Test Z/AFc and Z/BFc. Guidance on Combined Temperature (Cold and Dry Heat) and Vibration (Sinusoidal) Tests. 8 pp.

BSI BS 2011: Part 3A & B-77. 1977 Amd 1 Basic Environmental Testing Procedures Part 3: Background Information Part 3A and B: Tests A (Cold) and Tests B (Dry Heat). 37 pp.

BSI BS 2011:Pt 3 A & B:Supp 1-80. 1980 Basic Environmental Testing Procedures Part 3: Background Information Part 3A and B: Tests A (Cold) and Tests B (Dry Heat): Supplement No. 1:. 5 pp.

BSI BS EN 60068-2-2-93. 1993 Environmental Testing Part 2: Tests Tests B. Dry Heat (IEC 68-2-2: 1974) (G). 39 pp.

CENELEC HD 323.2.2 S1-77. Basic Environmental Testing Procedures—Part 2: Tests. Test B: Dry Heat. 1 p.

CENELEC HD 323.2.41 S1-88. Basic Environmental Testing Procedures-Part 2: Tests. Test Z/BM: Combined Dry Heat/Low Air Pressure Tests. 3 pp.

CENELEC HD 323.2.51 S1-87. Basic Environmental Testing Procedures Part 2: Tests Z/BFc: Combined Dry Heat/Vibration (Sinusoidal) Tests for Both Heat-Dissipating and Non Heat-Dissipating Specimens. 3 pp.

CENELEC HD 323.3.1 S1-88. Basic Environmental Testing Procedures Part J: Background Information Section One—Cold and Dry Heat Tests. 2 pp.

Electrical Components *(Cont.)*

—Dry Heat Testing *(Cont.)*

CENELEC EN 60068-2-2-93. Basic Environmental Testing Procedures Part 2: Tests Tests B: Dry Heat (Includes Amendment A1: 1993) (Supersedes HD 323.2.2 S1: 1988) (IEC 68-2-2: 1974 + IEC 68-2-2A: 1976 + A1: 1993). 34 pp.

IEC 68 Pt 2-2-74. Basic Environmental Testing Procedures Part 2: Tests—Tests B: Dry Heat Fourth Edition; (Supplement A-1976) (Amendment 1-1993) (CENE EN 60068-2-2: 1993 Includes Amendment A1: 1993). 74 pp.

IEC 68 Pt 2-41-76. Basic Environmental Testing Procedures Part 2: Tests—Test Z/BM: Combined Dry Heat/ Low Air Pressure Tests First Edition; (Amendment 1-1983). 24 pp.

IEC 68 Pt 2-51-83. Basic Environmental Testing Procedures Part 2: Tests Tests Z/BFc: Combined Dry Heat/Vibration (Sinusoidal) Tests for Both Heat-Dissipating and Non-Heat-Dissipating Specimens First Edition. 25 pp.

IEC 68 Pt 2-53-84. Basic Environmental Testing Procedures Part 2: Tests Guidance to Tests Z/AFc and Z/BFc: Combined Temperature (Cold and Dry Heat) and Vibration (Sinusoidal) Tests First Edition. 14 pp.

IEC 68 Pt 3-1-74. Basic Environmental Testing Procedures Part 3: Background Information Section One—Cold and Dry Heat Tests First Edition; (Supplement A-1978) (Corrigenda—March 1980). 73 pp.

JIS C 0021-87. Basic Environmental Testing Procedures Part 2: Tests, Tests B: Dry Heat. 35 pp.

JIS C 0031-90. Basic Environmental Testing Procedures Part 2: Tests-Test Z/BM Combined Dry Heat/Low Air Pressure Tests. 10 pp.

SAA AS 1099.2BA-80. Basic Environmental Testing Procedures for Electrotechnology—Part 2: Tests—Part 2Ba: Dry Heat Test for Non-Heat-Dissipating Specimens with Sudden Change of Temperature.

SAA AS 1099.2BB-80. Basic Environmental Testing Procedures for Electrotechnology—Part 2: Tests—Part 2Bb: Dry Heat Test for Non-Heat-Dissipating Specimens with Gradual Change of Temperature.

SAA AS 1099.2BC-80. Basic Environmental Testing Procedures for Electrotechnology—Part 2: Tests—Part 2Bc: Dry Heat Test for Heat-Dissipating Specimens with Sudden Change of Temperature.

SAA AS 1099.2BD-80. Basic Environmental Testing Procedures for Electrotechnology—Part 2: Tests—Part 2Bd: Dry Heat Test for Heat-Dissipating Specimens with Gradual Change of Temperature (Bound Together). 24 pp.

SAA AS 1099.3.1-80. Basic Environmental Testing Procedures for Electrotechnology—Part 3: Background Information—Part 3.1: Section 1 Tests A and B-Cold and Dry Heat Tests. 32 pp.

SNZ IEC 68: Part 2-2-74. Basic Environmental Testing Procedures Part 2-2: Tests B: Dry Heat. 63 pp.

SNZ IEC 68: Part 2-2A-76. Basic Environmental Testing Procedures Part 2-2A: Tests B: Dry Heat. 2 pp.

SNZ IEC 68: Part 2-41-76. Basic Environmental Testing Procedures Part 2-41: Test Z/Bm: Combined Dry Heat/Low Air Pressure Tests Amend: 1. 19 pp.

SNZ IEC 68: Part 2-51-83. Basic Environmental Testing Procedures Part 2-51: Tests z/BFc: Combined Dry Heat/Vibration (Sinusoidal) Tests for Both Heat-Dissipating and Non-Heat -Dissipating Specimens. 21 pp.

SNZ IEC 68: Part 2-53-84. Basic Environmental Testing Procedures Part 2-53: Guidance to Tests Z/AFc and Z/BFc: Combined Temperature (Cold and Dry Heat) and Vibration (Sinusoidal) Tests. 10 pp.

SNZ IEC 68: Part 3-1-74. Basic Environmental Testing Procedures Part 3-1: Background Information. Section One —Cold and Dry Heat Tests. 59 pp.

SNZ IEC 68: Part 3-1A-78. Basic Environmental Testing Procedures Part 3-1A: Background Information. Section One —Cold and Dry Heat Tests. 7 pp.

—Electromagnetic Compatibility—Naval Ships

MOD UK NES 515-90. Guide to Electromagnetic Compatibility Issue 2 (09.90). 152 pp.

MOD UK NES 515-01. Guide to Electromagnetic Compatibility Issue 2 (09.90); Amendment 1. 157 pp.

—Electromagnetic Compatibility—Submarines

MOD UK NES 515-90. Guide to Electromagnetic Compatibility Issue 2 (09.90). 152 pp.

MOD UK NES 515-01. Guide to Electromagnetic Compatibility Issue 2 (09.90); Amendment 1. 157 pp.

Electrical Components *(Cont.)*

—Environmental Conditions—Classification

BSI BS 7527: Sec 2.2-91. 1991 Classification of Environmental Conditions Part 2: Environmental Conditions Appearing in Nature Section 2.2: Precipitation and Wind (IEC 721-2-2: 1988). 15 pp.

BSI BS 7527: Sec 2.3-91. 1991 Classification of Environmental Conditions Part 2: Environmental Conditions Appearing in Nature Section 2.3: Air Pressure (IEC 721-2-3: 1987). 7 pp.

BSI BS 7527: Sec 2.4-91. 1991 Classification of Environmental Conditions Part 2: Environmental Conditions Appearing in Nature Section 2.4: Solar Radiation and Temperature (IEC 721-2-4: 1987). 16 pp.

BSI BS 7527: Sec 2.7-91. 1991 Classification of Environmental Conditions Part 2: Environmental Conditions Appearing in Nature Section 2.7: Fauna and Flora (IEC 721-2-7: 1987). 10 pp.

BSI BS 7527: Sec 3.0-91. 1991 Classification of Environmental Conditions Part 3: Classification of Groups of Environmental Parameters and Their Severities Section 3.0: Introduction (IEC 721-3-0: 1984). 14 pp.

BSI BS 7527: Sec 3.1-91. (WITHDRAWN) 1991 Classification of Environmental Conditions Part 3: Classification of Groups of Environmental Parameters and Their Severities Section 3.1: Storage (IEC 721-3-1: 1987) (Renumbered as BS EN 60721-3-1: 1993). 18 pp.

BSI BS 7527: Sec 3.2-91. 1991 Classification of Environmental Conditions Part 3: Classification of Groups of Environmental Parameters and Their Severities Section 3.2: Transportation (IEC 721-3-2: 1985) (Renumbered as BS EN 60721-3-2: 1993) (G). 22 pp.

BSI BS 7527: Sec 3.3-91. (WITHDRAWN) 1991 Classification of Environmental Conditions Part 3: Classification of Groups of Environmental Parameters and Their Severities Section 3.3: Stationary Use at Weatherprotected Locations (IEC 721-3-3: 1987). 43 pp.

BSI BS 7527: Sec 3.4-91. (WITHDRAWN) 1991 Classification of Environmental Conditions Part 3: Classification of Groups of Environmental Parameters and Their Severities Section 3.4: Stationary Use at Non-Weatherprotected Locations (IEC 721-3-4: 1987) (Q). 24 pp.

BSI BS 7527: Sec 3.7-91. (WITHDRAWN) 1991 Classification of Environmental Conditions Part 3: Classification of Groups of Environmental Parameters and Their Severities Section 3.7: Portable and Non-Stationary Use (IEC 721-3-7: 1987) (Renumbered as BS EN 60721-3-7: 1993). 32 pp.

BSI BS EN 60721-3-1-93. 1993 Classification of Environmental Conditions Part 3: Classification of Groups of Environmental Parameters and Their Severities Section 3.1: Storage (IEC 721-3-1: 1987) (AMD 7909) August 15, 1993 (G). 28 pp.

BSI BS EN 60721-3-2-93. 1991 Amd 1 Classification of Environmental Conditions Part 3: Classification of Groups of Environmental Parameters and Their Severities Section 3.2: Transportation (AMD 7955) September 15, 1993 (G). 30 pp.

BSI BS EN 60721-3-3-93. 1993 Amd 1 Classification of Environmental Conditions Part 3: Classification of Groups of Environmental Parameters and Their Severities Section 3.3: Stationary Use at Weatherprotected Locations (IEC 721-3-3: 1987) (AMD 7931) August 15, 1993 (G). 54 pp.

BSI BS EN 60721-3-4-93. 1993 Amd 1 Classification of Environmental Conditions Part 3: Classification of Groups of Environmental Parameters and Their Severities Sec. 3.4: Stationary Use at Non-Weatherprotected Locations (IEC 721-3-4: 1987) (AMD 7929) August 15, 1993 (G). 29 pp.

BSI BS EN 60721-3-7-93. 1993 Amd 1 Classification of Environmental Conditions Part 3: Classification of Groups of Environmental Parameters and Their Severities Section 3.7: Portable and Non-Stationary Use (IEC 721-3-7: 1987) (AMD 7949) August 15, 1993 (G). 40 pp.

CENELEC HD 478.2.2 S1-90. Classification of Environmental Conditions Part 2: Environmental Conditions Appearing in Nature Precipitation and Wind. 3 pp.

CENELEC HD 478.2.7 S1-90. Classification of Environmental Conditions Part 2: Environmental Conditions Appearing in Nature—Fauna and Flora. 3 pp.

CENELEC EN 60721-3-1-93. Classification of Environmental Conditions Part 3: Classification of Groups of Environmental Parameters and Their Severities—Storage (IEC 721-3-1: 1987 + A1: 1991). 15 pp.

CENELEC EN 60721-3-2-93. Classification of Environmental Conditions Part 3: Classification of Groups of Environmental Parameters and Their Severities—Transportation (IEC 721-3-2: 1985 + A1: 1991). 19 pp.

INDUSTRY STANDARDS
897

Electrical
INTERNATIONAL AND NON-U.S. NATIONAL STANDARDS
SUBJECT INDEX

Electrical Components *(Cont.)*

—Environmental Conditions—Classification *(Cont.)*

CENELEC EN 60721-3-3-93. Classification of Environmental Conditions Part 3: Classification of Groups of Environmental Parameters and Their Severities Stationary Use at Weatherprotected Locations (IEC 721-3-3: 1987 + A1: 1991)(Supersedes HD 478.3.3 S1: 1989). 40 pp.

CENELEC EN 60721-3-4-93. Classification of Environmental Conditions Part 3: Classification of Groups of Environmental Parameters and Their Severities Stationary Use at Non-Weatherprotected Locations (IEC 721-3-4: 1987+A1:1991)(Supersedes HD 478.3.4 S1: 1989). 22 pp.

CENELEC EN 60721-3-7-93. Classification of Environmental Conditions Part 3: Classification of Groups of Environmental Parameters and Their Severities Portable and Non-Stationary Use (IEC 721-3-7: 1987 + A1: 1991) (Supersedes HD 478.3.7 S1: 1989). 30 pp.

IEC 721 Pt 2-2-88. Classification of Environmental Conditions Part 2: Environmental Conditions Appearing in Nature Precipitation and Wind First Edition. 24 pp.

IEC 721 Pt 2-3-87. Classification of Environmental Conditions Part 2: Environmental Conditions Appearing in Nature Air Pressure First Edition. 10 pp.

IEC 721 Pt 2-4-87. Classification of Environmental Conditions Part 2: Environmental Conditions Appearing in Nature Solar Radiation and Temperature First Edition; (Amendment 1-1988). 28 pp.

IEC 721 Pt 2-7-87. Classification of Environmental Conditions Part 2: Environmental Conditions Appearing in Nature Fauna and Flora First Edition. 14 pp.

IEC 721 Pt 3-0-84. Classification of Environmental Conditions Part 3: Classification of Groups of Environmental Parameters and Their Severities Introduction First Edition; (Corrigendum—Dec 1985) (Amendment 1-1987). 31 pp.

IEC 721 Pt 3-1-87. Classification of Environmental Conditions Part 3: Classification of Groups of Environmental Parameters and Their Severities Storage First Edition; (Corrigendum—March 1990) (Amendment 1-1991) (Amendment 2-1993). 49 pp.

IEC 721 Pt 3-2-85. Classification of Environmental Conditions Part 3: Classification of Groups of Environmental Parameters and Their Severities Transportation First Edition; (Amendment 1-1991) (Amendment 2-1993) (CENELEC EN 60721-3-2: 1993). 52 pp.

IEC 721 Pt 3-3-87. Classification of Environmental Conditions Part 3: Classification of Groups of Environmental Parameters and Their Severities Stationary Use at Weatherprotected Locations First Edition; (Corrigendum—April 1988). 93 pp.

IEC 721 Pt 3-4-87. Classification of Environmental Conditions Part 3: Classification of Groups of Environmental Parameters and Their Severities Stationary Use at Non-Weatherprotected Locations First Edition; (Corrigendum—March 1990). 62 pp.

IEC 721 Pt 3-7-87. Classification of Environmental Conditions Part 3: Classification of Groups of Environmental Parameters and Their Severities Portable and Non-Stationary Use First Edition; (Amendment 1-1991) (Amendment 2-1993) (CENE EN 60721-3-7:1993). 82 pp.

IEC 721 Pt 3-9-93. Classification of Environmental Conditions Part 3: Classification of Groups of Environmental Parameters and Their Severities Section 9: Microclimates Inside Products First Edition. 25 pp.

—Environmental Testing

BSI BS 2011: Part 1.1-89. 1989 Basic Environmental Testing Procedures Part 1.1: General and Guidance. 27 pp.

BSI BS 2011: Pt 2.1Z/ABDM-83. 1983 Basic Environmental Testing Procedures Part 2.1: Tests Part 2.1Z/ABDM: Test Z/ABDM. Climatic Sequence, Primarily Intended for Components. 6 pp.

BSI BS 2011: Sec 4.2-84. 1984 Basic Environmental Testing Procedures Part 4: Miscellaneous Section 4.2: Guidance on the Application of the Tests of BS 2011 to Simulate the Effects of Storage. 8 pp.

BSI BS 2011: Sec 4.3-91. (WITHDRAWN) 1991 Environmental Testing Part 4: Miscellaneous Section 4.3: Guide to Seismic Test Methods for Equipments (IEC 68-3-3: 1991) (Renumbered as BS EN 60068-3-3: 1993). 47 pp.

BSI BS 2011: Sec 5.1-91. 1991 Environmental Testing Part 5: Guide to Drafting of Test Methods Section 5.1: General Principles (IEC 68-5-1: 1991) (G). 14 pp.

BSI BS 2011: Sec 5.2-92. 1992 Environmental Testing Part 5: Guide to Drafting of Test Methods Section 5.2: Terms and Definitions (IEC 68-5-2: 1990) (G). 16 pp.

BSI PD 6521-88. 1988 Summaries of Test Methods for Environmental Testing of Electrotechnical Products. 54 pp.

Electrical Components *(Cont.)*

—Environmental Testing *(Cont.)*

CENELEC HD 323.1 S1-82. Basic Environment Testing Procedures Part 1: General and Guidance. 1 p.

CENELEC HD 323.1 S2-88. Basic Environmental Testing Procedures—Part 1: General and Guidance. 3 pp.

CENELEC HD 323.1 S2-91. Environmental Testing Part 1: General and Guidance. 4 pp.

CENELEC HD 323.2.48 S1-88. Basic Environmental Testing Procedures—Part 2: Tests. Guidance on the Application of the Tests of IEC Publication 68 to Simulate the Effects of Storage. 3 pp.

CENELEC HD 323.3.3 S1-91. Environmental Testing Part 3: Guidance—Seismic Test Methods for Equipments. 6 pp.

CENELEC HD 323.5.2 S1-91. Environmental Testing Part 5: Guide to Drafting of Test Methods Terms and Definitions. 6 pp.

IEC 68 Pt 1-88. Environmental Testing Part 1: General and Guidance Sixth Edition; (Amendment 1-1992). 63 pp.

IEC 68 Pt 2-18-89. Environmental Testing Part 2: Tests—Test R and Guidance: Water First Edition; (Corrigendum—May 1991) (Amendment 1-1993). 96 pp.

IEC 68 Pt 2-48-82. Basic Environmental Testing Procedures Part 2: Tests Guidance on the Application of the Tests of IEC Publication 68 to Simulate the Effects of Storage First Edition. 15 pp.

IEC 68 Pt 2-61-91. Environmental Testing Part 2: Test Methods Test Z/ABDM: Climatic Sequence First Edition. 37 pp.

IEC 68 Pt 4-87. Basic Environmental Testing Procedures Part 4: Information for Specification Writers—Test Summaries First Edition; (Amendment 1-1992). 137 pp.

IEC 68 Pt 5-1-91. Environmental Testing Part 5: Guide to Drafting of Test Methods —General Principles First Edition. 27 pp.

IEC 68 Pt 5-2-90. Environmental Testing Part 5: Guide to Drafting of Test Methods —Terms and Definitions First Edition. 31 pp.

MOD UK DSTAN 00-35: Part 3-90. Environmental Handbook for Defence Materiel Part 3: Environmental Testing Issue 1. 100 pp.

SAA AS 1099. Basic Environmental Testing Procedures for Electrotechnology (Complete Set in Binder).

SAA AS 1099.1-89. Basic Environmental Testing Procedures for Electrotechnology —Part 1: General. 24 pp.

SAA AS 1099.2. Basic Environmental Testing Procedures for Electrotechnology —Part 2: Tests.

SAA AS 1099.2.48-88. Basic Environmental Testing Procedures for Electrotechnology—Part 2: Tests—Part 2.48: Guidance on the Application of Tests A, B, and Ca to Simulate the Effects of Storage. 4 pp.

SNZ IEC 68: Part 1-82. Basic Environmental Testing Procedures Part 1: General and Guidance. 47 pp.

SNZ IEC 68: Part 2-48-82. Basic Environmental Testing Procedures Part 2-48: Guidance on the Application of the Tests of IEC Publication 68 to Simulate the Effects of Storage. 11 pp.

—Fire Hazards

BSI BS 6458: Part 1-90. 1990 Fire Hazard Testing for Electrotechnical Products Part 1: Glossary of Terms. 22 pp.

BSI BS 6458: Sec 2.1-84. 1984 Fire Hazard Testing for Electrotechnical Products Part 2: Methods of Test Section 2.1: Glow Wire Test. 12 pp.

BSI BS 6458: Sec 2.2-93. 1993 Fire Hazard Testing for Electrotechnical Products Part 2: Test Methods Section 2.2: Needle-Flame Test (IEC 695-2-2: 1991) (G). 15 pp.

BSI BS 6458: Sec 2.2-84. 1984 Fire Hazard Testing for Electrotechnical Products Part 2: Methods of Test Section 2.2: Needle-Flame Test. 10 pp.

BSI BS 6458: Sec 2.3-85. 1985 Fire Hazard Testing for Electrotechnical Products Part 2: Methods of Test Section 2.3: Bad-Connection Test with Heaters. 20 pp.

CENELEC HD 444.2.1-83. Fire Hazard Testing Part 2: Test Methods Glow-Wire Test and Guidance. 3 pp.

CENELEC HD 444.2.2-83. Fire Hazard Testing Part 2: Test Methods Meedle-Flame Test. 3 pp.

CENELEC HD 444.2.2 S2-92. Fire Hazard Testing Part 2: Test Methods Section 2—Needle-Flame Test (IEC 695-2-2:1991). 4 pp.

CENELEC HD 444.2.3-87. Fire Hazard Testing Part 2: Test Methods Bad-Connection Test with Heaters. 2 pp.

IEC 695 Pt 1-1-82. Fire Hazard Testing Part 1: Guidance for the Preparation of Requirements and Test Specifications for Assessing Fire Hazard of Electrotechnical Products General Guidance First Edition. 21 pp.

Electrical Components *(Cont.)*

—Fire Hazards *(Cont.)*

IEC 695 Pt 1-2-82. Fire Hazard Testing Part 1: Guidance for the Preparation of Requirements and Test Specifications for Assessing Fire Hazard of Electrotechnical Products Guidance for Electronic Components First Edition. 17 pp.

IEC 695 Pt 1-3-86. Fire Hazard Testing Part 1: Guidance for the Preparation of Requirements and Test Specifications for Assessing Fire Hazard of Electrotechnical Products Guidance for Use of Preselection Procedures First Edition. 19 pp.

IEC 695 Pt 2-1-91. Fire Hazard Testing Part 2: Test Methods Section 1: Glow-Wire Test and Guidance Second Edition. 28 pp.

IEC 695 Pt 2-2-91. Fire Hazard Testing Part 2: Test Methods Section 2—Needle-Flame Test Second Edition; (CENELEC HD 444.2.2 S2:1992). 21 pp.

IEC 695 Pt 2-3-84. Fire Hazard Testing Part 2: Test Methods Bad-Connection Test with Heaters First Edition. 33 pp.

IEC 695 Pt 3-1-82. Fire Hazard Testing Part 3: Examples of Fire Hazard Assessment Procedures and Interpretation of Results Combustion Characteristics and Survey of Test Methods for Their Determination First Edition. 19 pp.

IEC 695 Pt 5-1-93. Fire Hazard Testing Part 5: Assessment of Potential Corrosion Damage by Fire Effluent—Section 1: General Guidance. 25 pp.

JIS C 0060-85. Fire Hazard Testing Part 2: Test Methods Glow-Wire Test and Guidance. 13 pp.

JIS C 0061-85. Fire Hazard Testing Part 2: Test Methods Needle Flame Test. 10 pp.

JIS C 0062-87. Fire Hazard Testing Part 2: Test Methods Bad-Connection Test with Heaters. 17 pp.

—Flammability Testing

BSI BS 2011: Part 2.1PZ-70. 1970 Amd 1 Basic Environmental Testing Procedures Part 2.1: Tests Part 2.1PZ: Flammability. 5 pp.

—Gas Burners

CSA C22.2 NO 3-M1988. Electrical Features of Fuel-Burning Equipment (R 1993); (Gen Instr 1 Thru 2). 64 pp.

—Glossaries

IEC. Electricity, Electronics and Telecommunications Multilingual Dictionary Volume 1: English-French -Russian-German-Spanish-Dutch-Italian-Swedish-Polish Second Edition. 1944 pp.

IEC. Electricity, Electronics and Telecommunications Multilingual Dictionary Volume 2: English-French -Russian-German-Spanish-Dutch-Italian-Swedish-Polish Second Edition. 964 pp.

IEC. Electricity, Electronics and Telecommunications Multilingual Dictionary Volume 3: German, Dutch and Swedish Indexes Second Edition. 793 pp.

IEC. Electricity, Electronics and Telecommunications Multilingual Dictionary Volume 4: Spanish and Italian Indexes Second Edition. 540 pp.

IEC. Electricity, Electronics and Telecommunications Multilingual Dictionary Volume 5: Russian and Polish Indexes Second Edition. 578 pp.

IEC 50 Chap 00-79. General Index International Electrotechnical Vocabulary Second Edition. 1007 pp.

MOD UK DSTAN 05-17-80. Electrotechnical Terms and Graphical Symbols Issue 2. 39 pp.

SAA AS 1852.00-80. International Electrotechnical Vocabulary—Part 00: General Index Being IEC 50(00). 1011 pp.

SNZ IEC 50: 50(00)-79. International Electrotechnical Vocabulary 50(00): General Index. 1011 pp.

—Identification Systems

MOD UK DSTAN 61-7: Part 2-91. Identification of Electrical and Electronic Systems, Wiring and Components Part 2: Marking Code for Electrical and Electronic Equipment Issue 3. 24 pp.

—Immersion Testing

BSI BS 2011: Part 2.1XA-81. 1981 Basic Environmental Testing Procedures Part 2.1: Tests Part 2.1XA: Tests. Test XA and Guidance. Immersion in Cleaning Solvents (Renumbered as BS EN 60068-2-45: 1993). 10 pp.

BSI BS EN 60068-2-45-93. 1993 Amd 1 Environmental Testing Part 2.1: Tests Test XA and Guidance. Immersion in Cleaning Solvents (IEC 68-2-45: 1980) (AMD 7553) April 15, 1993. 19 pp.

CENELEC HD 323.2.45 S1-88. Basic Environmental Testing Procedures-Part 2: Tests. Test XA and Guidance: Immersion in Cleaning Solvents. 3 pp.

CENELEC EN 60068-2-45-92. Basic Safety Publication —Environmental Testing Part 2: Test Methods Test XA and Guidance: Immersion in Cleaning Solvents. 5 pp.

CENELEC EN 60068-2-45-92. Basic Environmental Testing Procedures Part 2: Tests Test XA and Guidance: Immersion in Cleaning Solvents. 5 pp.

Electrical Components *(Cont.)*

—Immersion Testing *(Cont.)*

CENELEC EN 60068-2-45-92. Basic Safety Publication —Environmental Testing Part 2: Test Methods Test XA and Guidance: Immersion in Cleaning Solvents (IEC 68-2-45: 1980 + Corrigendum 1981). 12 pp.

IEC 68 Pt 2-18-89. Environmental Testing Part 2: Tests—Test R and Guidance: Water First Edition; (Corrigendum—May 1991) (Amendment 1-1993). 96 pp.

IEC 68 Pt 2-45-80. Basic Environmental Testing Procedures Part 2: Tests Test XA and Guidance: Immersion in Cleaning Solvents First Edition; (Amendment 1-1993) (CENELEC EN 60068-2-45: 1992). 30 pp.

SNZ IEC 68: Part 2-45-80. Basic Environmental Testing Procedures Part 2-45: Test XA and Guidance: Immersion in Cleaning Solvents. 20 pp.

—Impact Testing

BSI BS 2011: Part 2.1EC-77. 1977 Amd 1 Basic Environmental Testing Procedures Part 2.1: Tests Part 2.1EC: Test Ec. Drop and Topple, Primarily for Equipment-Type Specimens (Renumbered as BS EN 68-2-31: 1969). 7 pp.

BSI BS 2011: Part 2.1ED-92. 1992 Amd 2 Environmental Testing Part 2.1: Tests Test Ed: Free Fall (IEC 68-2-32: 1975) (G). 13 pp.

BSI BS 2011: Part 2.1ED-77. 1977 Amd 1 Basic Environmental Testing Procedures Part 2.1: Tests Part 2.1Ed: Test Ed. Free Fall (Renumbered as BS EN 60068-2-32: 1993). 11 pp.

BSI BS 2011: Part 2.1EF-92. 1992 Environmental Testing Part 2.1: Tests Test Ef: Impact, Pendulum Hammer (IEC 68-2-62: 1991) (G). 15 pp.

BSI BS EN 60068-2-31-93. 1993 Amd 2 Environmental Testing Part 2.1: Tests Test Ec. Drop and Topple, Primarily for Equipment-Type Specimens (IEC 68-2-31: 1969) (AMD 7921) September 15, 1993 (G). 11 pp.

BSI BS EN 60068-2-32-93. 1993 Amd 1 Environmental Testing Part 2.1: Tests Test Ed. Free Fall (IEC 68-2-32: 1975) (AMD 7920) September 15, 1993 (G). 16 pp.

CENELEC HD 323.2.31 S1-88. Basic Environmental Testing Procedures—Part 2: Tests. Test Ec: Drop and Topple, Primarily for Equipment-Type Specimens (Superseded by EN 60068-2-31: 1993). 3 pp.

CENELEC HD 323.2.32 S1-88. Basic Environmental Testing Procedures—Part 2: Tests. Test Ed: Free Fall. 3 pp.

CENELEC EN 60068-2-31-93. Basic Environmental Testing Procedures Part 2: Tests Test Ec: Drop and Topple, Primarily for Equipment-Type Specimens (IEC 68-2-31: 1969 + A1: 1982) (Supersedes HD 323.2.31 S1: 1988). 8 pp.

CENELEC EN 60068-2-32-93. Basic Environmental Testing Procedures Part 2: Tests Test Ed: Free Fall (IEC 68-2-32: 1975 + A1: 1982 + A2: 1990) (Supersedes HD 323.2.32 S2: 1991). 12 pp.

IEC 68 Pt 2-31-69. Basic Environmental Testing Procedures Part 2: Tests—Test Ec: Drop and Topple, Primarily for Equipment-Type Specimens First Edition; (Amendment 1-1982) (CENELEC EN 60068-2-31: 1993). 13 pp.

IEC 68 Pt 2-32-75. Basic Environmental Testing Procedures Part 2: Tests Test Ed: Free Fall Second Edition; (Amendment 2-1990; Incorporating Amendment 1) (CENELEC EN 60068-2-32: 1993). 31 pp.

IEC 68 Pt 2-62-91. Environmental Testing Part 2: Test Methods Test Ef: Impact, Pendulum Hammer First Edition. 26 pp.

JIS C 0043-90. Basic Environmental Testing Procedures Part 2: Tests-Ec: Drop and Topple, Primarily for Equipment-Type Specimens. 10 pp.

JIS C 0044-90. Basic Environmental Testing Procedures Part 2: Tests-Test Ed: Free Fall. 10 pp.

SAA AS 1099.2ED-78. Basic Environmental Testing Procedures for Electrotechnology—Part 2: Tests—Part 2Ed: Free Fall Reconfirmed 1985. 3 pp.

SAA AS 1099.2.31-90. Basic Environmental Testing Procedures for Electrotechnology—Part 2: Tests—Part 2.31: Test Ec—Drop and Topple, Primarily for Equipment (IEC 68-2-31). 6 pp.

SNZ IEC 68: Part 2-31-69. Basic Environmental Testing Procedures Part 2-31: Test Ec: Drop and Topple, Primarily for Equipment-Type Specimens Amend: 1. 7 pp.

SNZ IEC 68: Part 2-32-75. Basic Environmental Testing Procedures Part 2-32: Test Ed: Free Fall Amend: 1. 12 pp.

—Leakage

BSI BS 2011: Part 2.1Q-81. 1981 Amd 1 Basic Environmental Testing Procedures Part 2.1: Tests Part 2.1Q: Test Q. Sealing (Amd 6093) November 30, 1990. 41 pp.

CENELEC HD 323.2.17 S2-87. Basic Environmental Testing Procedures—Part 2: Tests. Test Q: Sealing. 3 pp.

Electrical Components *(Cont.)*

—Leakage *(Cont.)*

CENELEC HD 323.2.17 S3-88. Basic Environmental Testing Procedures-Part 2: Tests. Test Q: Sealing. 3 pp.

CENELEC HD 323.2.17 S4-90. Basic Environmental Testing Procedures Part 2: Tests Test Q: Sealing. 3 pp.

IEC 68 Pt 2-17-78. Basic Environmental Testing Procedures Part 2: Tests—Test Q: Sealing Third Edition; (Amendment 4-1991 Incorporating Amendments 1 Thru 3). 118 pp.

JIS C 0026-89. Basic Environmental Testing Procedures Part 2: Tests, Test Q: Sealing. 34 pp.

SAA AS 1099.2Q-80. Basic Environmental Testing Procedures for Electrotechnology—Part 2: Tests—Part 2Q: Sealing. 20 pp.

SAA AS 1099.3.6-80. Basic Environmental Testing Procedures for Electrotechnology—Part 3: Background Information—Part 3.6: Section 6 Test Q—Sealing. 4 pp.

SNZ IEC 68: Part 2-17-78. Basic Environmental Testing Procedures Part 2-17: Test Q: Sealing Amend: 1. 71 pp.

—Liquid Penetrant Testing

BSI BS 2011: Part 2.1R-90. 1990 Basic Environmental Testing Procedures Part 2.1: Tests Part 2.1R: Test R and Guidance. Water. 43 pp.

—Logic Diagrams

SAA AS 1103.8-86. Diagrams, Charts and Tables for Electrotechnology—Part 8: Guiding Principles for the Preparation of Logic Diagrams. 26 pp.

SNZ NZS/AS 1103. 8-86. Diagrams, Charts and Tables for Electrotechnology Part 8: Guiding Principles for the Preparation of Logic Diagrams (This is a Joint Standard with SAA AS 1103.8). 26 pp.

—Low Pressure Testing

BSI BS 2011: Part 2.1M-84. 1984 Basic Environmental Testing Procedures Part 2.1 Tests Part 2.1M: Low Air Pressure. 7 pp.

BSI BS 2011: Part 2.1Z/AM-77. 1977 Amd 1 Basic Environmental Testing Procedures Part 2.1: Tests Part 2.1Z/AM: Test Z/AM. Combined Cold/Low Air Pressure Tests. 10 pp.

BSI BS 2011: Pt 2.1Z/AMD-77. 1977 Basic Environmental Testing Procedures Part 2.1: Tests Part 2.1Z/AMD: Test Z/AMD. Combined Sequential Cold, Low Air Pressure and Damp Heat Test. 4 pp.

BSI BS 2011: Part 2.1Z/BM-77. 1977 Amd 1 Basic Environmental Testing Procedures Part 2.1: Tests Part 2.1Z/BM: Test Z/BM. Combined Dry Heat/ Low Air Pressure Tests. 12 pp.

BSI BS 2011:Pt 3 Z/AM & Z/BM-77. 1977 Basic Environmental Testing Procedures Part 3: Background Information Part 3Z/AM and Z/BM: Test Z/AM and Z/BM. Combined Temperature/ Low Air Pressure Tests. 6 pp.

CENELEC HD 323.2.13-87. Basic Environmental Testing Procedures Part 2: Tests Test M: Low air Pressure. 3 pp.

CENELEC HD 323.2.39 S1-76. Basic Environmental Testing Procedures-Part 2: Tests. Test Z/Amd: Combined Sequential Cold, Low Air Pressure, and Damp Heat Test. 2 pp.

CENELEC HD 323.2.40 S1-88. Basic Environmental Testing Procedures—Part 2: Tests Test Z/AM: Combined Cold/Low Air Pressure Tests. 3 pp.

CENELEC HD 323.2.41 S1-88. Basic Environmental Testing Procedures-Part 2: Tests. Test Z/BM: Combined Dry Heat/Low Air Pressure Tests. 3 pp.

CENELEC HD 323.3.2 S1-88. Basic Environmental Testing Procedures-Part 3: Background Information. Section Two: Combined Temperature/Low Air Pressure Tests. 3 pp.

IEC 68 Pt 2-13-83. Basic Environmental Testing Procedures Part 2: Tests Test M: Low Air Pressure Fourth Edition. 17 pp.

IEC 68 Pt 2-39-76. Basic Environmental Testing Procedures Part 2: Tests—Test Z/AMD: Combined Sequential Cold, Low Air Pressure, and Damp Heat Test First Edition. 12 pp.

IEC 68 Pt 2-40-76. Basic Environmental Testing Procedures Part 2: Tests—Test Z/AM: Combined Cold/Low Air Pressure Tests First Edition; (Amendment 1-1983). 22 pp.

IEC 68 Pt 2-41-76. Basic Environmental Testing Procedures Part 2: Tests—Test Z/BM: Combined Dry Heat/ Low Air Pressure Tests First Edition; (Amendment 1-1983). 24 pp.

IEC 68 Pt 3-2-76. Basic Environmental Testing Procedures Part 3: Background Information Section Two-Combined Temperature/Low Air Pressure Tests First Edition. 14 pp.

JIS C 0029-89. Basic Environmental Testing Procedures Part 2: Tests, Test M: Low Air Pressure. 7 pp.

JIS C 0030-90. Basic Environmental Testing Procedures Part 2: Tests-Test Z/AM Combined Cold/Low Air Pressure Tests. 9 pp.

Electrical Components *(Cont.)*

—Low Pressure Testing *(Cont.)*

JIS C 0031-90. Basic Environmental Testing Procedures Part 2: Tests-Test Z/BM Combined Dry Heat/Low Air Pressure Tests. 10 pp.

SAA AS 1099.2Z/A M-80. Basic Environmental Testing Procedures for Electrotechnology—Part 2: Tests—Part 2Z/AM: Combined Cold/Low Air Pressure Tests Reconfirmed 1985. 6 pp.

SAA AS 1099.2Z/A MD-80. Basic Environmental Testing Procedures for Electrotechnology—Part 2: Tests—Part 2Z/AMD: Combined Sequential Cold, Low Air Pressure and Damp Heat Test Reconfirmed 1985. 2 pp.

SAA AS 1099.2Z/B M-80. Basic Environmental Testing Procedures for Electrotechnology—Part 2: Tests—Part 2Z/BM: Combined Dry Heat/Low Air Pressure Tests Reconfirmed 1985. 6 pp.

SAA AS 1099.2.13-90. Basic Environmental Testing Procedures for Electrotechnology—Part 2: Tests—Part 2.13: Test M—Low Air Pressure (IEC 68-2-13). 4 pp.

SAA AS 1099.3.2-80. Basic Environmental Testing Procedures for Electrotechnology—Part 3: Background Information—Part 3.2: Section 2—Combined Temperature/Low Air Pressure Tests. 4 pp.

SNZ IEC 68: Part 2-13-83. Basic Environmental Testing Procedures Part 2-13: Test M: Low Air Pressure. 13 pp.

SNZ IEC 68: Part 2-39-76. Basic Environmental Testing Procedures Part 2-39: Test 2/AMD: Combined Sequential Cold, Low Air Pressure, and Damp Heat Test. 8 pp.

SNZ IEC 68: Part 2-40-76. Basic Environmental Testing Procedures Part 2-40: Test 2/Am: Combined Cold/Low Air Pressure Tests Amend: 1. 17 pp.

SNZ IEC 68: Part 2-41-76. Basic Environmental Testing Procedures Part 2-41: Test Z/Bm: Combined Dry Heat/Low Air Pressure Tests Amend: 1. 19 pp.

SNZ IEC 68: Part 3-2-76. Basic Environmental Testing Procedures Part 3-2: Background Information. Section Two —Combined Temperature/ Low Air Pressure Tests. 12 pp.

—Low Temperature Testing

BSI BS 2011: Part 2.1A-90. (WITHDRAWN) 1990 Amd 2 Basic Environmental Testing Procedures Part 2.1: Tests Part 2.1A: Test A. Cold (Renumbered as BS EN 60068-2-1: 1993). 26 pp.

BSI BS EN 60068-2-1-93. 1993 Amd 1 Environmental Testing Part 2.1: Tests Tests A. Cold (AMD 7783) July 15, 1993 (G). 39 pp.

BSI BS 2011: Part 2.1Z/AD-77. 1977 Basic Environmental Testing Procedures Part 2.1: Tests Part 2.1Z/AD: Test Z/AD. Composite Temperature/Humidity Cyclic Test. 12 pp.

BSI BS 2011: Pt 2.1Z/AFC-84. 1984 Basic Environmental Testing Procedures Part 2.1: Tests: Part 2.1Z/AFc: Test Combined Cold/Vibration (Sinusoidal) Tests for Both Heat-Dissipating and Non-Heat-Dissipating Specimens. 12 pp.

BSI BS 2011: Part 2.1Z/AM-77. 1977 Amd 1 Basic Environmental Testing Procedures Part 2.1: Tests Part 2.1Z/AM: Test Z/AM. Combined Cold/Low Air Pressure Tests. 10 pp.

BSI BS 2011: Pt 2.1Z/AMD-77. 1977 Basic Environmental Testing Procedures Part 2.1: Tests Part 2.1Z/AMD: Test Z/AMD. Combined Sequential Cold, Low Air Pressure and Damp Heat Test. 4 pp.

BSI BS 2011:Pt2. 2Z/AFC/BFC-86. 1986 Basic Environmental Testing Procedures Part 2.2: Guidance Part 2.2Z/AFc and Z/BFc: Test Z/AFc and Z/BFc. Guidance on Combined Temperature (Cold and Dry Heat) and Vibration (Sinusoidal) Tests. 8 pp.

BSI BS 2011: Part 3A & B-77. 1977 Amd 1 Basic Environmental Testing Procedures Part 3: Background Information Part 3A and B: Tests A (Cold) and Tests B (Dry Heat). 37 pp.

BSI BS 2011:Pt 3 A & B:Supp 1-80. 1980 Basic Environmental Testing Procedures Part 3: Background Information Part 3A and B: Tests A (Cold) and Tests B (Dry Heat): Supplement No. 1:. 5 pp.

CENELEC HD 323.2.38 S1-76. Basic Environmental Testing Procedures—Part 2: Tests. Test Z/AD: Composite Temperature/Humidity Cyclic Test. 2 pp.

CENELEC HD 323.2.39 S1-76. Basic Environmental Testing Procedures-Part 2: Tests. Test Z/Amd: Combined Sequential Cold, Low Air Pressure, and Damp Heat Test. 2 pp.

CENELEC HD 323.2.40 S1-88. Basic Environmental Testing Procedures—Part 2: Tests Test Z/AM: Combined Cold/Low Air Pressure Tests. 3 pp.

CENELEC HD 323.3.1 S1-88. Basic Environmental Testing Procedures Part J: Background Information Section One—Cold and Dry Heat Tests. 2 pp.

CENELEC EN 60068-2-1-93. Environmental Testing Part 2: Tests Tests A: Cold (IEC 68-2-1:1990) (Supersedes HD 323.2.1 S2:1987). 5 pp.

INDUSTRY STANDARDS
899

Electrical
INTERNATIONAL AND NON-U.S. NATIONAL STANDARDS
SUBJECT INDEX

Electrical Components *(Cont.)*

—Low Temperature Testing *(Cont.)*

CENELEC EN 60068-2-1-93. Environmental Testing Part 2: Tests Tests A: Cold (IEC 68-2-1: 1990). 26 pp.

IEC 68 Pt 2-1-90. Environmental Testing Part 2: Tests—Tests A: Cold Fifth Edition; (Amendment 1-1993) (CENELEC EN 60068-2-1: 1993). 53 pp.

IEC 68 Pt 2-38-74. Basic Environmental Testing Procedures Part 2: Tests—Test Z/AD: Composite Temperature/Humidity Cyclic Test First Edition. 23 pp.

IEC 68 Pt 2-39-76. Basic Environmental Testing Procedures Part 2: Tests—Test Z/AMD: Combined Sequential Cold, Low Air Pressure, and Damp Heat Test First Edition. 12 pp.

IEC 68 Pt 2-40-76. Basic Environmental Testing Procedures Part 2: Tests—Test Z/AM: Combined Cold/Low Air Pressure Tests First Edition; (Amendment 1-1983). 22 pp.

IEC 68 Pt 2-50-83. Basic Environmental Testing Procedures Part 2: Tests Tests Z/AFc: Combined Cold/Vibration (Sinusoidal) Tests for Both Heat-Dissipating and Non-Heat Dissipating Specimens First Edition. 25 pp.

IEC 68 Pt 2-53-84. Basic Environmental Testing Procedures Part 2: Tests Guidance to Tests Z/AFc and Z/BFc: Combined Temperature (Cold and Dry Heat) and Vibration (Sinusoidal) Tests First Edition. 14 pp.

IEC 68 Pt 3-1-74. Basic Environmental Testing Procedures Part 3: Background Information Section One—Cold and Dry Heat Tests First Edition; (Supplement A-1978) (Corrigenda—March 1980). 73 pp.

JIS C 0020-87. Basic Environmental Testing Procedures Part 2: Tests, Tests A: Cold. 22 pp.

JIS C 0030-90. Basic Environmental Testing Procedures Part 2: Tests-Test Z/AM Combined Cold/Low Air Pressure Tests. 9 pp.

SAA AS 1099.2AA-80. Basic Environmental Testing Procedures for Electrotechnology—Part 2: Tests—Part 2Aa: Cold Test for Non-Heat-Dissipating Specimens with Gradual Change of Temperature.

SAA AS 1099.2AB-80. Basic Environmental Testing Procedures for Electrotechnology—Part 2: Tests—Part 2Ab: Cold Test for Non-Heat-Dissipating Specimens with Gradual Change of Temperature.

SAA AS 1099.2AD-80. Basic Environmental Testing Procedures for Electrotechnology—Part 2: Tests—Part 2Ad: Cold Test for Heat-Dissipating Specimens with Gradual Change of Temperature (Bound Together). 20 pp.

SAA AS 1099.2Z/A D-80. Basic Environmental Testing Procedures for Electrotechnology—Part 2: Tests—Part 2Z/AD: Composite Temperature/ Humidity Cyclic Test Reconfirmed 1985. 8 pp.

SAA AS 1099.2Z/A M-80. Basic Environmental Testing Procedures for Electrotechnology—Part 2: Tests—Part 2Z/AM: Combined Cold/Low Air Pressure Tests Reconfirmed 1985. 6 pp.

SAA AS 1099.2Z/A MD-80. Basic Environmental Testing Procedures for Electrotechnology—Part 2: Tests—Part 2Z/AMD: Combined Sequential Cold, Low Air Pressure and Damp Heat Test Reconfirmed 1985. 2 pp.

SAA AS 1099.3.1-80. Basic Environmental Testing Procedures for Electrotechnology—Part 3: Background Information—Part 3.1: Section 1 Tests A and B-Cold and Dry Heat Tests. 32 pp.

SNZ IEC 68: Part 2-1-74. Basic Environmental Testing Procedures Part 2-1: Tests A: Cold Amend: 1. 43 pp.

SNZ IEC 68: Part 2-1A-76. Basic Environmental Testing Procedures Part 2-1A: Tests A: Cold. 2 pp.

SNZ IEC 68: Part 2-38-74. Basic Environmental Testing Procedures Part 2-38: Test 2/AD: Composite Temperature/Humidity Cyclic Test. 20 pp.

SNZ IEC 68: Part 2-39-76. Basic Environmental Testing Procedures Part 2-39: Test 2/AMD: Combined Sequential Cold, Low Air Pressure, and Damp Heat Test. 8 pp.

SNZ IEC 68: Part 2-40-76. Basic Environmental Testing Procedures Part 2-40: Test 2/Am: Combined Cold/Low Air Pressure Tests Amend: 1. 17 pp.

SNZ IEC 68: Part 2-50-83. Basic Environmental Testing Procedures Part 2-50: Test 2/AFc: Combined Cold-Vibration (Sinusoidal) Tests for Both Heat Dissipating and Non-Heat-Dissipating Specimens. 21 pp.

SNZ IEC 68: Part 2-53-84. Basic Environmental Testing Procedures Part 2-53: Guidance to Tests Z/ AFc and Z/BFc: Combined Temperature (Cold and Dry Heat) and Vibration (Sinusoidal) Tests. 10 pp.

SNZ IEC 68: Part 3-1-74. Basic Environmental Testing Procedures Part 3-1: Background Information. Section One —Cold and Dry Heat Tests. 59 pp.

SNZ IEC 68: Part 3-1A-78. Basic Environmental Testing Procedures Part 3-1A: Background Information. Section One —Cold and Dry Heat Tests. 7 pp.

Electrical Components *(Cont.)*

—Mechanical Shock

BSI BS 2011: Part 2.1EA-88. (WITHDRAWN) 1988 Basic Environmental Testing Procedures Part 2.1: Tests Part 2.1EA: Test Ea. Shock (Renumbered as BS EN 60068-2-27: 1993). 31 pp.

BSI BS 2011: Part 2.1EB-87. (WITHDRAWN) 1987 Basic Environmental Testing Procedures Part 2.1: Tests Part 2.1EB: Test Eb. Bump (Renumbered as BS EN 60068-2-29: 1993). 16 pp.

BSI BS 2011: Part 2.1EE-88. (WITHDRAWN) 1988 Basic Environmental Testing Procedures Part 2.1: Tests Part 2.1EE: Bounce (Renumbered as BS EN 60068-2-55: 1987). 16 pp.

BSI BS 2011: Sec 4.1-83. (WITHDRAWN) 1983 Basic Environmental Testing Procedures Part 4: Miscellaneous Section 4.1: Specification for Mounting of Components, Equipment and Other Articles for Dynamic Tests (Renumbered as BS EN 60068-2-47: 1993). 15 pp.

BSI BS EN 60068-2-27-93. 1993 Amd 1 Environmental Testing Part 2: Tests Test Ea and Guidance: Shock (IEC 68-2-27: 1987) (G). 40 pp.

BSI BS EN 60068-2-29-93. 1993 Amd 1 Environmental Testing Part 2: Tests Test Eb and Guidance: Bump (IEC 68-2-29: 1987) (AMD 7827) June 15, 1993 (G). 25 pp.

BSI BS EN 60068-2-47-93. 1993 Environmental Testing Part 2: Tests Mounting of Components, Equipment and Other Articles for Dynamic Tests Including Shock (Ea), Bump (Eb), Vibration (Fc and Fd) and Steady-State Acceleration (Ga) and Guidance (IEC 68-2-47: 1982) (G). 23 pp.

BSI BS EN 60068-2-55-93. 1993 Environmental Testing Part 2: Tests Test E3 and Guidance: Bounce (IEC 68-2-55: 1987) (G). 25 pp.

CENELEC HD 323.2.27 S2-88. Basic Environmental Testing Procedures Part 2: Tests Test Ea and Guidance: Shock. 3 pp.

CENELEC HD 323.2.27 S2-89. Basic Environmental Testing Procedures Part 2: Tests. 3 pp.

CENELEC HD 323.2.29 S1-88. Basic Environmental Testing Procedures—Part 2: Tests. Test Eb: Bump. 3 pp.

CENELEC HD 323.2.29 S2-89. Basic Environmental Testing Procedures-Part 2: Tests. Test Eb and Guidance: Bump. 3 pp.

CENELEC HD 323.2.47 S1-88. Basic Environmental Testing Procedures—Part 2: Tests. Mounting of Components. Equipment and Other Articles for Dynamic Tests Including Shock (Es). Bump (Eb). Vibration (Fc and Fd) and Steady-State Acceleration (Ga) and Guidance. 3 pp.

CENELEC HD 323.2.55 S1-89. Basic Environmental Testing Procedures Part 2: Tests Test Ee and Guidance: Bounce. 3 pp.

CENELEC EN 60068-2-27-93. Basic Environmental Testing Procedures Part 2: Tests Test Ea and Guidance: Shock (IEC 68-2-27: 1987). 33 pp.

CENELEC EN 60068-2-29-93. Basic Environmental Testing Procedures Part 2: Tests Test Eb and Guidance: Bump (IEC 68-2-29: 1987 + Corrigendum). 19 pp.

CENELEC EN 60068-2-47-93. Basic Environmental Testing Procedures Part 2: Tests Mounting of Components, Equipment and Other Articles for Dynamic Tests Including Shock (Ea), Bump (Eb), Vibration (Fc and Fd) and Steady-State Acceleration (Ga) and Guidance (IEC 68-2-47: 1982). 16 pp.

CENELEC EN 60068-2-55-93. Basic Environmental Testing Procedures Part 2: Tests Test Ee and Guidance: Bounce (IEC 68-2-55: 1987). 18 pp.

IEC 68 Pt 2-27-87. Basic Environmental Testing Procedures Part 2: Tests—Test Ea and Guidance: Shock Third Edition (CENELEC EN 60068-2-27: 1993). 53 pp.

IEC 68 Pt 2-29-87. Basic Environmental Testing Procedures Part 2: Tests—Tests Eb and Guidance: Bump Second Edition; (Corrigendum—October 1987) (CENELEC EN 60068-2-29-1993). 34 pp.

IEC 68 Pt 2-47-82. Basic Environmental Testing Procedures Part 2: Tests Mounting of Components, Equipment and Other Articles for Dynamic Tests Including Shock (Ea), Bump (Eb), Vibration (Fc and Fd) and Steady-State Acceleration (Ga) and Guidance First Edition (CENE EN 60068-2-47:93). 29 pp.

IEC 68 Pt 2-55-87. Basic Environmental Testing Procedures Part 2: Tests Test Ee and Guidance: Bounce First Edition (CENELEC EN 60068-2-55: 1993). 32 pp.

JIS C 0041-87. Basic Environmental Testing Procedures Part 2: Tests, Test Ea: Shock. 45 pp.

JIS C 0042-87. Basic Environmental Testing Procedures Part 2: Tests, Test Eb: Bump. 29 pp.

SAA AS 1099.2.27-88. Basic Environmental Testing Procedures for Electrotechnology—Part 2: Tests—Part 2.27: Test Ea—Shock. 28 pp.

SAA AS 1099.2.29-90. Basic Environmental Testing Procedures for Electrotechnology—Part 2: Tests—Part 2.29: Test Eb—Bump and Guidance (IEC 68-2-29). 13 pp.

Electrical Components *(Cont.)*

—Mechanical Shock *(Cont.)*

SNZ IEC 68: Part 2-27-72. Basic Environmental Testing Procedures Part 2-27: Test Ea: Shock Amend: 1; 2. 37 pp.

SNZ IEC 68: Part 2-29-87. Basic Environmental Testing Procedures Part 2-29: Test Eb and Guidance: Bump. 30 pp.

SNZ IEC 68: Part 2-47-82. Basic Environmental Testing Procedures Part 2-47: Mounting of Components and Other Articles for Dynamic Tests Including Shock (Ea), Bump (Eb), Vibration (Fc and Fd) and Steady-State Acceleration (Ga) and Guidance. 25 pp.

—Medical Electrical Equipment

CSA CAN/CSA-C22. 2NO601.1-M90. Medical Electrical Equipment, Part 1: General Requirements for Safety; (Gen Instr 1). 240 pp.

—Mold Growth Testing

BSI BS 2011: Part 2.1J-89. 1989 Basic Environmental Testing Procedures Part 2.1: Tests Part 2.1J: Test J and Guidance. Mould Growth. 25 pp.

CENELEC HD 323.2.10 S2-87. Basic Environmental Testing Procedures—Part 2: Tests. Test J: Mould Growth. 3 pp.

CENELEC HD 323.2.10 S3-88. Basic Environmental Testing Procedures-Part 2: Tests Test J and Guidance: Mould Growth. 3 pp.

CENELEC HD 323.2.10 S3-92. Basic Environmental Testing Procedures Part 2: Tests Test J and Guidance: Mould Growth; (IEC 68-2-10:1988). 3 pp.

IEC 68 Pt 2-10-88. Basic Environmental Testing Procedures Part 2: Tests—Test J and Guidance: Mould Growth Fifth Edition; (Corrigendum—April 1989) (CENELEC HD 323.2.10 S3:1988). 54 pp.

SAA AS 1099.2.10-89. Basic Environmental Testing Procedures for Electrotechnology—Part 2: Tests—Part 2.10: Test J-Mould Growth. 23 pp.

SNZ IEC 68: Part 2-10-84. Basic Environmental Testing Procedures Part 2-10: Test J: Mould Growth. 17 pp.

—Naval Ships

MOD UK NES 507-91. Requirements for Materials and Finishes for Electrotechnical Equipment Issue 4 (03.91). 32 pp.

MOD UK NES 581: Part 1-90. Requirements for the Selection and Approval of Electrical and Electronic Components Part 1: Method of Selection and Approval Issue 2 (07.90). 21 pp.

MOD UK NES 581: Part 2-91. Requirements for the Selection and Approval of Electrical and Electronic Components Part 2: Standard, Approved and Acceptable Components Issue 5 (09.91). 199 pp.

MOD UK NES 581: Part 2-01. Requirements for the Selection and Approval of Electrical and Electronic Components Part 2: Standard, Approved and Acceptable Components Issue 5 (09.91); Amendment 1. 229 pp.

MOD UK NES 581: Part 3-91. Requirements for the Selection and Approval of Electrical and Electronic Components Part 3: Non—Standard Project Approved Components Issue 2 (12.91). 44 pp.

MOD UK NES 581: Part 3-01. Requirements for the Selection and Approval of Electrical and Electronic Components Part 3: Non-Standard Project Approved Components Issue 2 (12.91); Amendment 1. 85 pp.

MOD UK NES 609-88. Guide to the Determination of Creepage and Clearance Distances for Electrotechnical Equipment Issue 2 (10.88). 14 pp.

MOD UK NES 609-91. Guide to the Determination of Creepage and Clearance Distances for Electrotechnical Equipment Issue 3 (12.91). 14 pp.

MOD UK NES 694-01. Guide to the Preparation of Production Test Specifications for Electrotechnical Equipment Issue 3 (11.88); Amendment 1. 19 pp.

MOD UK NES 694-92. Guide to the Preparation of Production Test Specifications for Electro Technical Equipment Issue 4 (01.92). 21 pp.

—Oil Burners

CSA C22.2 NO 3-M1988. Electrical Features of Fuel-Burning Equipment (R 1993); (Gen Instr 1 Thru 2). 64 pp.

—Qualification Approvals

MOD UK DSTAN 05-14-01. Mutual Acceptance of Qualification Approvals for Electrical/Electronic Components Within NATO Countries Issue 2; Amendment 1. 19 pp.

NATO STANAG 4093 ED 3 AMD 2-76. Mutual Acceptance of Qualification Approvals for Electronic Parts. 14 pp.

NATO STANAG 4093 ED 4 AMD 0-89. Mutual Acceptance by NATO Member Countries of Qualification of Electronic and Electrical Components for Military Use. 24 pp.

NATO STANAG 4093 ED 4 AMD 0-93. Mutual Acceptance by NATO Member Countries of Qualification of Electronic and Electrical Components for Military Use. 23 pp.

INTERNATIONAL AND NON-U.S. NATIONAL STANDARDS
SUBJECT INDEX
Electrical

Electrical Components *(Cont.)*

—Qualified Products List

MOD UK DSTAN 59-59-91. Electrical/Electronic Components for Defence Use Services Qualified Products List Including Components of Assessed Quality Issue 6. 513 pp.
MOD UK DSTAN 59-59-01. Electrical/Electronic Components for Defence Use Services Qualified Products List Including Components of Assessed Quality Issue 6; Amendment 1. 560 pp.
MOD UK DSTAN 59-59-02. Electrical/Electronic Components for Defence Use Services Qualified Products List Including Components of Assessed Quality Issue 6; Amendment 2. 615 pp.
MOD UK DSTAN 59-59-03. Electrical/Electronic Components for Defence Use Services Qualified Products List Including Components of Assessed Quality Issue 6; Amendment 3. 601 pp.
MOD UK DSTAN 59-59-04. Electrical/Electronic Components for Defence Use Services Qualified Products List Including Components of Assessed Quality Issue 6; Corrigendum to Amendment 3. 602 pp.

—Quality Assurance

MOD UK DSTAN 61-12: 6112/33/003-93. Wires Cords and Cables Electrical—Metric Units Part 33: Airframe Wires and Cables in Temperature Categories 135 Degrees C, 150 Degrees C, 200 Degrees C and 260 Degrees C, Sectional Specification. 4 pp.

—Reliability

BSI DD 57-78. 1978 Development of Methods of Equipment Reliability Testing. 47 pp.

—Salt Spray Testing

BSI BS 2011: Part 2.1KA-82. 1982 Basic Environmental Testing Procedures Part 2.1: Tests Part 2.1KA: Tests Ka. Salt Mist. 7 pp.
BSI BS 2011: Part 2.1KB-87. 1987 Basic Environmental Testing Procedures Part 2.1: Tests Part 2.1KB: Test Kb. Salt Mist Cylic (Sodium Chloride Solution). 9 pp.
CENELEC HD 323.2.11-88. Basic Environmental Testing Procedures Part 2: Tests Test Ka: Salt Mist. 4 pp.
CENELEC HD 323.2.11 S1-85. Basic Environmental Testing Prcedures Part 2: Tests Test Ka: Salt Mist. 1 p.
CENELEC HD 323.2.52-87. Basic Environmental Testing Procedures Part 2: Tests Test kb: Salt Mist, Cylic (Sodium Chloride Solution). 3 pp.
IEC 68 Pt 2-11-81. Basic Environmental Testing Procedures Part 2: Tests Test Ka: Salt Mist Third Edition. 15 pp.
IEC 68 Pt 2-52-84. Basic Environmental Testing Procedures Part 2: Tests Test Kb: Salt Mist, Cyclic (Sodium Chloride Solution) First Edition. 19 pp.
JIS C 0023-89. Basic Environmental Testing Procedures Part 2: Tests-Test Ka: Salt Mist. 8 pp.
JIS C 0024-89. Basic Environmental Testing Procedures Part 2: Tests-Test Kb: Salt Mist, Cyclic (Sodium Chloride Solution). 8 pp.
SAA AS 1099.2KA-78. Basic Environmental Testing Procedures for Electrotechnology—Part 2: Tests—Part 2Ka: Salt Mist Reconfirmed 1985. 2 pp.
SNZ IEC 68: Part 2-11-81. Basic Environmental Testing Procedures Part 2-11: Test Ka: Salt Mist. 11 pp.
SNZ IEC 68: Part 2-52-84. Basic Environmental Testing Procedures Part 2-52: Test Kb: Salt Mist, Cyclic (Sodium Chloride Solution). 15 pp.

—Sealing

BSI BS 2011: Part 2.1Q-81. 1981 Amd 1 Basic Environmental Testing Procedures Part 2.1: Tests Part 2.1Q: Test Q. Sealing (Amd 6093) November 30, 1990. 41 pp.

—Signal Designations

IEC 1175-93. Designations for Signals and Connections First Edition. 74 pp.

—Solar Radiation

BSI BS 2011: Part 2.1SA-77. 1977 Basic Environmental Testing Procedures Part 2.1: Tests Part 2.1SA: Test Sa. Simulated Solar Radiation at Ground Level. 7 pp.
BSI BS 2011: Part 2.2SA-77. 1977 Amd 1 Basic Environmental Testing Procedures Part 2.2: Guidance Part 2.2Sa: Test Sa. Guidance for Solar Radiation Testing (Amd 6468) December 21, 1990. 24 pp.
BSI BS 7527: Sec 2.4-91. 1991 Classification of Environmental Conditions Part 2: Environmental Conditions Appearing in Nature Section 2.4: Solar Radiation and Temperature (IEC 721-2-4: 1987). 16 pp.
CENELEC HD 323.2.5 S1-76. Basic Environmental Testing Procedures—Part 2: Tests. Test SA: Simulated Solar Radiation at Ground Level. 1 p.
CENELEC HD 323.2.9 S2-87. Basic Environmental Testing Procedures—Part 2: Tests. Guidance for Solar Radiation Testing. 3 pp.

Electrical Components *(Cont.)*

—Solar Radiation *(Cont.)*

IEC 68 Pt 2-5-75. Basic Environmental Testing Procedures Part 2: Tests Test Sa: Simulated Solar Radiation at Ground Level First Edition. 13 pp.
IEC 68 Pt 2-9-75. Basic Environmental Testing Procedures Part 2: Tests—Guidance for Solar Radiation Testing First Edition; (Amendment 1-1984) (Corrigendum—Aug 1989). 42 pp.
IEC 721 Pt 2-4-87. Classification of Environmental Conditions Part 2: Environmental Conditions Appearing in Nature Solar Radiation and Temperature First Edition; (Amendment 1-1988). 28 pp.
SAA AS 1099.2SA-80. Basic Environmental Testing Procedures for Electrotechnology—Part 2: Tests—Part 2Sa: Simulated Solar Radiation. 4 pp.
SAA AS 1099.3.5-80. Basic Environmental Testing Procedures for Electrotechnology—Part 3: Background Information—Part 3.5: Section 5 Test Sa—Simulated Solar Radiation. 12 pp.
SNZ IEC 68: Part 2-5-75. Basic Environmental Testing Procedures Part 2-5: Test Sa: Simulated Solar Radiation at Ground Level. 11 pp.
SNZ IEC 68: Part 2-9-75. Basic Environmental Testing Procedures Part 2-9: Guidance for Solar Radiation Testing Amend: 1. 34 pp.

—Soldering

BSI BS 2011: Part 2.1T-81. 1981 Amd 1 Basic Environmental Testing Procedures Part 2.1: Tests Part 2.1T: Test T. Soldering (Amd 6092) November 30, 1990. 32 pp.
BSI BS 2011: Part 2.1TA-89. 1989 Basic Environmental Testing Procedures Part 2.1Ta: Soldering. Solderability Testing by the Wetting Balance Method. 14 pp.
BSI BS 2011: Part 2.2T-81. 1981 Basic Environmental Testing Procedures Part 2.2: Guidance Part 2.2T: Test T. Guidance to Soldering. 10 pp.
CENELEC HD 323.2.20 S2-87. Basic Environmental Testing Procedures—Part 2: Tests. Test T: Soldering. 3 pp.
CENELEC HD 323.2.20 S3-88. Basic Environmental Testing Procedures-Part 2: Tests Test T: Soldering. 3 pp.
CENELEC HD 323.2.20 S3-92. Basic Environmental Testing Procedures Part 2: Tests Test T: Soldering; (IEC 68-2-20:1979 + A2:1987). 3 pp.
CENELEC HD 323.2.44 S1-88. Basic Environmental Testing Procedures—Part 2: Tests. Guidance on Test T: Soldering. 3 pp.
CENELEC HD 323.2.54-87. Basic Environmental Testing Procedures Part 2: Tests Test TA: Soldering, Solderability Testing by the Wetting Balance Method. 3 pp.
CENELEC HD 323.2.54 S1-87. Basic Environmental Testing Procedures Part 2: Tests Test Ta: Soldering Solderability Testing by the Wetting Balance Method. 3 pp.
IEC 68 Pt 2-20-79. Basic Environmental Testing Procedures Part 2: Tests Test T: Soldering Fourth Edition; (Amendment 1-1986) (Amendment 2-1987) (CENELEC HD 323.2.20 S3:1988). 65 pp.
IEC 68 Pt 2-44-79. Basic Environmental Testing Procedures Part 2: Tests Tests—Guidance on Test T: Soldering First Edition. 22 pp.
IEC 68 Pt 2-54-85. Basic Environmental Testing Procedures Part 2: Tests Test Ta: Soldering Solderability Testing by the Wetting Balance Method First Edition. 31 pp.
JIS C 0053-90. Basic Environmental Testing Procedures Test Ta: Soldering Solderability Testing by the Wetting Balance Method. 15 pp.
SAA AS 1099.2T-80. Basic Environmental Testing Procedures for Electrotechnology—Part 2: Tests—Part 2T: Soldering. 18 pp.
SAA AS 1099.3.4-80. Basic Environmental Testing Procedures for Electrotechnology—Part 3: Background Information—Part 3.4: Section 4—Guidance on Test T: Soldering. 5 pp.
SNZ IEC 68: Part 2-20-79. Basic Environmental Testing Procedures Part 2-20: Test T: Soldering. 52 pp.
SNZ IEC 68: Part 2-44-79. Basic Environmental Testing Procedures Part 2-44: Guidance on Test T: Soldering. 19 pp.
SNZ IEC 68: Part 2-54-85. Basic Environmental Testing Procedures Part 2-54: Test Ta: Soldering. Solderability Testing by the Wetting Balance Method. 27 pp.

—Solid Fuel Burning Equipment

CSA C22.2 NO 3-M1988. Electrical Features of Fuel-Burning Equipment (R 1993); (Gen Instr 1 Thru 2). 64 pp.
CSA 1343 Bull. Electrical Bulletin 1343 November 3, 1981 to C22.2 NO 3. 4 pp.

—Solvent Resistance Testing

BSI BS 2011: Part 2.1R-90. 1990 Basic Environmental Testing Procedures Part 2.1: Tests Part 2.1R: Test R and Guidance. Water. 43 pp.

Electrical Components *(Cont.)*

—Steady State Performance

BSI BS 5760: Sec 10.3-93. 1993 Reliability of Systems, Equipment and Components Part 10: Guide to Reliability Testing Section 10.3: Compliance Test Procedures for Steady-State Availability (IEC 1070: 1991). 32 pp.
IEC 1070-91. Compliance Test Procedures for Steady-State Availability First Edition. 56 pp.

—Submarines

MOD UK NES 507-91. Requirements for Materials and Finishes for Electrotechnical Equipment Issue 4 (03.91). 32 pp.
MOD UK NES 581: Part 1-90. Requirements for the Selection and Approval of Electrical and Electronic Components Part 1: Method of Selection and Approval Issue 2 (07.90). 21 pp.
MOD UK NES 581: Part 2-91. Requirements for the Selection and Approval of Electrical and Electronic Components Part 2: Standard, Approved and Acceptable Components Issue 5 (09.91). 199 pp.
MOD UK NES 581: Part 2-01. Requirements for the Selection and Approval of Electrical and Electronic Components Part 2: Standard, Approved and Acceptable Components Issue 5 (09.91); Amendment 1. 229 pp.
MOD UK NES 581: Part 3-91. Requirements for the Selection and Approval of Electrical and Electronic Components Part 3: Non—Standard Project Approved Components Issue 2 (12.91). 44 pp.
MOD UK NES 581: Part 3-01. Requirements for the Selection and Approval of Electrical and Electronic Components Part 3: Non-Standard Project Approved Components Issue 2 (12.91); Amendment 1. 85 pp.
MOD UK NES 609-88. Guide to the Determination of Creepage and Clearance Distances for Electrotechnical Equipment Issue 2 (10.88). 14 pp.
MOD UK NES 609-91. Guide to the Determination of Creepage and Clearance Distances for Electrotechnical Equipment Issue 3 (12.91). 14 pp.

—Symbols

SAA AS 1102. Graphical Symbols for Electrotechnology.
SAA AS 1102.8-86. Graphical Symbols for Electrotechnology—Part 8: Symbols for Location Diagrams. 38 pp.

—Symbols—Diagrams

SAA AS 1102.101-89. Graphical Symbols for Electrotechnology—Part 101: General Information and General Index. 29 pp.
SAA AS 1102.102-89. Graphical Symbols for Electrotechnology—Part 102: Symbol Elements, Qualifying Symbols and Other Symbols Having General Application. 25 pp.

—Tables (Data)

SAA AS 1103.5-78. Diagrams, Charts and Tables for Electrotechnology—Part 5: Preparation of Interconnection Diagrams and Tables. 8 pp.
SAA AS 1103.6-81. Diagrams, Charts and Tables for Electrotechnology—Part 6: Preparation of Unit Wiring Diagrams and Tables. 14 pp.
SNZ NZS/AS 1103. 5-78. Diagrams, Charts and Tables for Electrotechnology Part 5: Preparation of Interconnection Diagrams and Tables (This is a Joint Standard with SAA AS 1103.5). 8 pp.
SNZ NZS/AS 1103. 6-81. Diagrams, Charts and Tables for Electrotechnology Part 6: Preparation of Unit Wiring Diagrams and Tables (This is a Joint Standard with SAA AS 1103.6). 14 pp.

—Tables (Data)—Glossaries

SAA AS 1103.1-73. Diagrams, Charts and Tables for Electrotechnology—Part 1: Definitions and Classifications Reconfirmed 1982. 5 pp.
SNZ NZS/AS 1103. 1-73. Diagrams, Charts and Tables for Electrotechnology Part 1: Definitions and Classifications (This is a Joint Standard with SAA AS 1103.1). 5 pp.

—Temperature Change Testing

BSI BS 2011: Part 2.1N-85. 1985 Amd 1 Basic Environmental Testing Procedures Part 2.1: Tests Part 2.1N: Test N. Change of Temperature (AMD 5435) March 31, 1987. 15 pp.
BSI BS 2011: Part 2.2N-77. 1977 Basic Environmental Testing Procedures Part 2.2: Guidance Part 2.2N: Test N. Guidance on Change of Temperature Tests. 8 pp.
BSI BS 2011:Pt 3 Z/AM & Z/BM-77. 1977 Basic Environmental Testing Procedures Part 3: Background Information Part 3Z/AM and Z/BM: Test Z/AM and Z/BM. Combined Temperature/Low Air Pressure Tests. 6 pp.
CENELEC HD 323.2.14 S2-87. Basic Environmental Testing Procedures—Part 2: Tests. Test N: Change of Temperature. 3 pp.
CENELEC HD 323.2.33 S1-88. Basic Environmental Testing Procedures-Part 2: Tests. Guidance on Change of Temperature Tests. 2 pp.

INDUSTRY STANDARDS
901

Electrical

INTERNATIONAL AND NON-U.S. NATIONAL STANDARDS
SUBJECT INDEX

Electrical Components *(Cont.)*

—Temperature Change Testing *(Cont.)*

CENELEC HD 323.2.33 S1-92. Basic Environmental Testing Procedures Part 2: Tests Guidance on Change of Temperature Tests; (IEC 68-2-33:1971 + A1:1978). 3 pp.

CENELEC HD 323.3.2 S1-88. Basic Environmental Testing Procedures-Part 3: Background Information. Section Two: Combined Temperature/Low Air Pressure Tests. 3 pp.

IEC 68 Pt 2-14-84. Basic Environmental Testing Procedures Part 2: Tests Test N: Change of Temperature Fifth Edition; (Amendment 1-1986). 38 pp.

IEC 68 Pt 2-33-71. Basic Environmental Testing Procedures Part 2: Tests—Guidance on Change of Temperature Tests First Edition; (Amendment 1-1978) (CENELEC HD 323.2.33 S1:1988). 18 pp.

IEC 68 Pt 3-2-76. Basic Environmental Testing Procedures Part 3: Background Information Section Two-Combined Temperature/Low Air Pressure Tests First Edition. 14 pp.

SAA AS 1099.2AA-80. Basic Environmental Testing Procedures for Electrotechnology—Part 2: Tests—Part 2Aa: Cold Test for Non-Heat-Dissipating Specimens with Gradual Change of Temperature.

SAA AS 1099.2AB-80. Basic Environmental Testing Procedures for Electrotechnology—Part 2: Tests—Part 2Ab: Cold Test for Non-Heat-Dissipating Specimens with Gradual Change of Temperature.

SAA AS 1099.2AD-80. Basic Environmental Testing Procedures for Electrotechnology—Part 2: Tests—Part 2Ad: Cold Test for Heat-Dissipating Specimens with Gradual Change of Temperature (Bound Together). 20 pp.

SAA AS 1099.2BA-80. Basic Environmental Testing Procedures for Electrotechnology—Part 2: Tests—Part 2Ba: Dry Heat Test for Non-Heat-Dissipating Specimens with Sudden Change of Temperature.

SAA AS 1099.2BB-80. Basic Environmental Testing Procedures for Electrotechnology—Part 2: Tests—Part 2Bb: Dry Heat Test for Non-Heat-Dissipating Specimens with Gradual Change of Temperature.

SAA AS 1099.2BC-80. Basic Environmental Testing Procedures for Electrotechnology—Part 2: Tests—Part 2Bc: Dry Heat Test for Heat-Dissipating Specimens with Sudden Change of Temperature.

SAA AS 1099.2N-73. Basic Environmental Testing Procedures for Electrotechnology—Part 2: Tests—Part 2N: Change of Temperature—Guidance on Change of Temperature Tests. 3 pp.

SAA AS 1099.2NA-73. Basic Environmental Testing Procedures for Electrotechnology—Part 2: Tests—Part 2Na: Rapid Change of Temperature, Two-Chamber Method. 2 pp.

SAA AS 1099.2NB-73. Basic Environmental Testing Procedures for Electrotechnology—Part 2: Tests—Part 2Nb: Change of Temperature, One-Chamber Method. 2 pp.

SAA AS 1099.2NC-73. Basic Environmental Testing Procedures for Electrotechnology—Part 2: Tests—Part 2Nc: Rapid Change of Temperature, Two-Water-Bath Method. 2 pp.

SAA AS 1099.3.2-80. Basic Environmental Testing Procedures for Electrotechnology—Part 3: Background Information—Part 3.2: Section 2—Combined Temperature/Low Air Pressure Tests. 4 pp.

SNZ IEC 68: Part 2-14-84. Basic Environmental Testing Procedures Part 2-14: Test N: Change of Temperature Amend: 1. 29 pp.

SNZ IEC 68: Part 2-33-71. Basic Environmental Testing Procedures Part 2-33: Guidance on Change of Temperature Tests Amend: 1. 13 pp.

SNZ IEC 68: Part 3-2-76. Basic Environmental Testing Procedures Part 3-2: Background Information. Section Two —Combined Temperature/Low Air Pressure Tests. 12 pp.

—Vibration

BSI BS 2011: Part 2.1EE-88. (WITHDRAWN) 1988 Basic Environmental Testing Procedures Part 2.1: Tests Part 2.1EE: Bounce (Renumbered as BS EN 60068-2-55: 1987). 16 pp.

BSI BS 2011: Part 2.1FC-83. 1983 Amd 2 Basic Environmental Testing Procedures Part 2.1: Tests Part 2.1Fc: Test Fc. Vibration (Sinusoidal). 32 pp.

BSI BS 2011: Part 2.1FD-73. 1973 Amd 2 Basic Environmental Testing Procedures Part 2.1: Tests Part 2.1FD: Random Vibration, Wide Band, General Requirements. 13 pp.

BSI BS 2011: Part 2.1FDA-73. 1973 Amd 2 Basic Environmental Testing Procedures Part 2.1: Tests Part 2.1FDA: Random Vibration, Wide Band. Reproducibility High. 24 pp.

BSI BS 2011: Part 2.1FDB-73. 1973 Amd 2 Basic Environmental Testing Procedures Part 2.1: Tests Part 2.1FDB: Random Vibration-Wide Band. Reproducibility Medium. 21 pp.

BSI BS 2011: Part 2.1FDC-73. 1973 Amd 2 Basic Environmental Testing Procedures Part 2.1: Tests Part 2.1FDC: Random Vibration-Wide Band. Reproducibility Low. 11 pp.

Electrical Components *(Cont.)*

—Vibration *(Cont.)*

BSI BS 2011: Part 2.1FE-91. (WITHDRAWN) 1991 Environmental Testing Part 2.1: Tests Test Fe and Guidance. Vibration (Sine-Beat Method) (IEC 68-2-59: 1990) (Renumbered as BS EN 60068-2-59: 1993). 31 pp.

BSI BS 2011: Part 2.1FF-89. (WITHDRAWN) 1989 Basic Environmental Testing Procedures Part 2.1: Tests Part 2.1Ff: Vibration-Time-History Method (Renumbered as BS EN 60068-2-57). 24 pp.

BSI BS 2011: Pt 2.1Z/AFC-84. 1984 Basic Environmental Testing Procedures Part 2.1: Tests: Part 2.1Z/AFc: Test Combined Cold/Vibration (Sinusoidal) Tests for Both Heat-Dissipating and Non-Heat-Dissipating Specimens. 12 pp.

BSI BS 2011: Pt 2.1Z/BFC-84. 1984 Basic Environmental Testing Procedures Part 2.1: Tests Part 2.1Z/BFc: Test Z/BF Combined Dry Heat/Vibration (Sinusoidal) Tests for Both Heat-Dissipating and Non-Heat-Dissipating Specimens. 13 pp.

BSI BS 2011:Pt2. 2Z/AFC/BFC-86. 1986 Basic Environmental Testing Procedures Part 2.2: Guidance Part 2.2Z/AFc and Z/BFc: Test Z/AFc and Z/BFc. Guidance on Combined Temperature (Cold and Dry Heat) and Vibration (Sinusoidal) Tests. 8 pp.

BSI BS 2011: Sec 4.1-83. (WITHDRAWN) 1983 Basic Environmental Testing Procedures Part 4: Miscellaneous Section 4.1: Specification for Mounting of Components, Equipment and Other Articles for Dynamic Tests (Renumbered as BS EN 60068-2-47: 1993). 15 pp.

BSI BS EN 60068-2-47-93. 1993 Environmental Testing Part 2: Tests Mounting of Components, Equipment and Other Articles for Dynamic Tests Including Shock (Ea), Bump (Eb), Vibration (Fc and Fd) and Steady-State Acceleration (Ga) and Guidance (IEC 68-2-47: 1982) (G). 23 pp.

BSI BS EN 60068-2-55-93. 1993 Environmental Testing Part 2: Tests Test E3 and Guidance: Bounce (IEC 68-2-55: 1987) (G). 25 pp.

BSI BS EN 60068-2-57-93. 1993 Amd 1 Environmental Testing Part 2: Tests Test Ff: Vibration—Time-History Method (IEC 68-2-57: 1989) (AMD 7834) June 15, 1993 (G). 33 pp.

BSI BS EN 60068-2-59-93. 1993 Amd 1 Environmental Testing Part 2: Test Methods Test Fe: Vibration—Sine Beat Method (AMD 7841) July 15, 1993 (G). 39 pp.

CENELEC HD 323.2.6 S2-88. Basic Environmental Testing Procedures—Part 2: Tests. Test Fc and Guidance: Vibration (Sinusoidal). 3 pp.

CENELEC HD 323.2.34 S1-88. Basic Environmental Testing Procedures Part 2: Tests Test Fd: Random Vibration Wide Band—General Requirements. 3 pp.

CENELEC HD 323.2.34 S1-88. Basic Environmental Testing Procedures—Part 2: Tests. Test Fd: Random Vibration Wide Band—General Requirements. 4 pp.

CENELEC HD 323.2.35 S1-88. Basic Environmental Testing Procedures Part 2: Tests Test Fda: Random Vibration Wide Band—Reproducibility High. 3 pp.

CENELEC HD 323.2.35 S1-88. Basic Environmental Testing Procedures—Part 2: Tests. Test Fda: Random Vibration Wide Band—Reproducibility High. 4 pp.

CENELEC HD 323.2.36 S1-88. Basic Environmental Testing Procedures Part 2: Tests Test Fdb: Random Vibration Wide Band—Reproducibility Medium. 3 pp.

CENELEC HD 323.2.36 S1-88. Basic Environmental Testing Procedures—Part 2: Tests. Test Fdb: Random Vibration Wide Band—Reproducibility Medium. 4 pp.

CENELEC HD 323.2.37 S1-88. Basic Environmental Testing Procedures Part 2: Tests Test Fdc: Random Vibration Wide Band—Reproducibility Low. 3 pp.

CENELEC HD 323.2.37 S1-88. Basic Environmental Testing Procedures—Part 2: Tests. Test Fdc: Random Vibration Wide Band—Reproducibility Low. 4 pp.

CENELEC HD 323.2.47 S1-88. Basic Environmental Testing Procedures—Part 2: Tests. Mounting of Components. Equipment and Other Articles for Dynamic Tests Including Shock (Es). Bump (Eb). Vibration (Fc and Fd) and Steady-State Acceleration (Ga) and Guidance. 3 pp.

CENELEC HD 323.2.55 S1-89. Basic Environmental Testing Procedures Part 2: Tests Test Ee and Guidance: Bounce. 3 pp.

CENELEC HD 323.2.57 S1-90. Basic Environmental Testing Procedures Part 2: Tests Test Ff: Vibration Time-history Method. 3 pp.

CENELEC HD 323.2.57 S1-92. Basic Environmental Testing Procedures Part 2: Tests Test Ff: Vibration—Time-History Method; (IEC 68-2-57: 1989). 3 pp.

CENELEC HD 323.2.59 S1-91. Basic Environmental Testing Procedures Part 2: Tests Test Fe: Vibration—Sine Beat Method. 6 pp.

Electrical Components *(Cont.)*

—Vibration *(Cont.)*

CENELEC EN 60068-2-47-93. Basic Environmental Testing Procedures Part 2: Tests Mounting of Components, Equipment and Other Articles for Dynamic Tests Including Shock (Ea), Bump (Eb), Vibration (Fc and Fd) and Steady-State Acceleration (Ga) and Guidance (IEC 68-2-47: 1982). 16 pp.

CENELEC EN 60068-2-55-93. Basic Environmental Testing Procedures Part 2: Tests Test Ee and Guidance: Bounce (IEC 68-2-55: 1987). 18 pp.

CENELEC EN 60068-2-57-93. Environmental Testing Part 2: Test Methods—Test Ff: Vibration Time-History Method (IEC 68-2-57: 1989). 27 pp.

CENELEC EN 60068-2-59-93. Environmental Testing Part 2: Test Methods Test Fe: Vibration—Sine Beat Method (IEC 68-2-59: 1990) (Supersedes HD 323.2.59 S1: 1991). 30 pp.

IEC 68 Pt 2-6-82. Basic Environmental Testing Procedures Part 2: Tests Test Fc and Guidance: Vibration (Sinusoidal) Fifth Edition; (Amendment 1-1983) (Amendment 2-1985). 69 pp.

IEC 68 Pt 2-34-73. Basic Environmental Testing Procedures Part 2: Tests Test Fd: Random Vibration Wide Band—General Requirements First Edition; (Amendment 1-1983). 41 pp.

IEC 68 Pt 2-35-73. Basic Environmental Testing Procedures Part 2: Tests Test Fda: Random Vibration Wide Band—Reproducibility High First Edition; (Amendment 1-1983). 53 pp.

IEC 68 Pt 2-36-73. Basic Environmental Testing Procedures Part 2: Tests—Test Fdb: Random Vibration Wide Band—Reproducibility Medium First Edition; (Amendment 1-1983). 53 pp.

IEC 68 Pt 2-37-73. Basic Environmental Testing Procedures Part 2: Tests—Test Fdc: Random Vibration Wide Band—Reproducibility Low First Edition; (Amendment 1-1983). 33 pp.

IEC 68 Pt 2-47-82. Basic Environmental Testing Procedures Part 2: Tests Mounting of Components, Equipment and Other Articles for Dynamic Tests Including Shock (Ea), Bump (Eb), Vibration (Fc and Fd) and Steady-State Acceleration (Ga) and Guidance First Edition (CENE EN 60068-2-47:93). 29 pp.

IEC 68 Pt 2-50-83. Basic Environmental Testing Procedures Part 2: Tests Tests Z/AFc: Combined Cold/Vibration (Sinusoidal) Tests for Both Heat-Dissipating and Non-Heat Dissipating Specimens First Edition. 25 pp.

IEC 68 Pt 2-51-83. Basic Environmental Testing Procedures Part 2: Tests Tests Z/BFc: Combined Dry Heat/Vibration (Sinusoidal) Tests for Both Heat-Dissipating and Non-Heat-Dissipating Specimens First Edition. 25 pp.

IEC 68 Pt 2-53-84. Basic Environmental Testing Procedures Part 2: Tests Guidance to Tests Z/AFc and Z/BFc: Combined Temperature (Cold and Dry Heat) and Vibration (Sinusoidal) Tests First Edition. 14 pp.

IEC 68 Pt 2-55-87. Basic Environmental Testing Procedures Part 2: Tests Test Ee and Guidance: Bounce First Edition (CENELEC EN 60068-2-55: 1993). 32 pp.

IEC 68 Pt 2-57-89. Environmental Testing Part 2: Test Methods Test Ff: Vibration—Time-History Method First Edition; (CENELEC HD 323.2.57 S1:1990) (CENELEC EN 60068-2-57: 1993). 51 pp.

IEC 68 Pt 2-59-90. Environmental Testing Part 2: Test Methods Test Fe: Vibration—Sine-Beat Method First Edition; (CENELEC EN 60068-2-59: 1993). 54 pp.

IEC 68 Pt 2-64-93. Environmental Testing Part 2: Test Methods Test Fh: Vibration, Broad-Band Random (Digital Control) and Guidance First Edition. 90 pp.

JIS C 0040-87. Basic Environmental Testing Procedures Part 2: Tests, Test Fc and Guidance: Vibration (Sinusoidal). 47 pp.

SAA AS 1099.2.6-88. Basic Environmental Testing Procedures for Electrotechnology—Part 2: Tests—Part 2.6: Test Fc—Vibration (Sinusoidal). 26 pp.

SNZ IEC 68: Part 2-6-82. Basic Environmental Testing Procedures Part 2-6: Test Fc and Guidance: Vibration (Sinusoidal). 59 pp.

SNZ IEC 68: Part 2-34-73. Basic Environmental Testing Procedures Part 2-34: Test Fd: Random Vibration Wide Band. General Requirements Amend: 1. 35 pp.

SNZ IEC 68: Part 2-35-73. Basic Environmental Testing Procedures Part 2-35: Test Fda: Random Vibration Wide Band. Reproducibility High Amend: 1. 47 pp.

SNZ IEC 68: Part 2-36-73. Basic Environmental Testing Procedures Part 2-36: Test Fdb: Random Vibration Wide Band. Reproducibility Medium Amend: 1. 47 pp.

SNZ IEC 68: Part 2-37-73. Basic Environmental Testing Procedures Part 2-37: Test Fdc: Random Vibration Wide Band. Reproducibility Low Amend: 1. 27 pp.

INTERNATIONAL AND NON-U.S. NATIONAL STANDARDS
SUBJECT INDEX
Electrical

Electrical Components *(Cont.)*
—Vibration *(Cont.)*
SNZ IEC 68: Part 2-47-82. Basic Environmental Testing Procedures Part 2-47: Mounting of Components and Other Articles for Dynamic Tests Including Shock (Ea), Bump (Eb), Vibration (Fc and Fd) and Steady-State Acceleration (Ga) and Guidance. 25 pp.

SNZ IEC 68: Part 2-50-83. Basic Environmental Testing Procedures Part 2-50: Test 2/AFc: Combined Cold-Vibration (Sinusoidal) Tests for Both Heat Dissipating and Non-Heat-Dissipating Specimens. 21 pp.

SNZ IEC 68: Part 2-51-83. Basic Environmental Testing Procedures Part 2-51: Tests z/BFc: Combined Dry Heat/Vibration (Sinusoidal) Tests for Both Heat-Dissipating and Non-Heat -Dissipating Specimens. 21 pp.

SNZ IEC 68: Part 2-53-84. Basic Environmental Testing Procedures Part 2-53: Guidance to Tests Z/AFc and Z/BFc: Combined Temperature (Cold and Dry Heat) and Vibration (Sinusoidal) Tests. 10 pp.

Electrical Conductance
Use: Electrical Resistance

Electrical Conductivity
Use: Electrical Resistivity

Electrical Copper Wire
Use: Copper Conductors

Electrical Cores
Use: Magnetic Cores

Electrical/Electronic Hardware
Scope Note: Use a more specific term *See:* Adapters (Electric); Bobbins (Electric); Cable Clamps; Communication Transmission Line Hardware; Electric Terminals; Electric Wire; Electrical Components; Electrical Equipment; Electrical Insulator Pins; Electrical Insulators; Electronic Component Packaging; Electronic Components; Electronic Equipment; Heat Sinks; Knobs; Magnetic Cores; Mounting Hardware; Power Line Hardware; Racks (Electrical); Shells (Electric); Sockets (Electric); Spacers (Electric); Spring Clips; Terminal Boards; Wiring Devices

Electrical Enclosures
Use For: Boxes (Electrical); Cases (Electrical); Enclosures (Electrical) *See Also:* Auxiliary Gutters; Cabinets (Electrical); Circuit Boxes; Connector Housings; Control Panels; Cutout Boxes; Degree of Protection (Electrical); Electric Conduit Boxes; Electric Outlet Boxes; Electric Raceways; Electrical Protection Equipment; Fuse Boxes; Instrument Housings; Junction Boxes; Mechanical Protection; Meter Sockets; Panel Boards (Electrical); Pull Boxes; Racks (Electrical); Shells (Electric)

BSI BS 4662-05. 1970 Amd 5 Boxes for the Enclosure of Electrical Accessories (AMD 6347) August 30, 1991. 42 pp.

BSI BS 5490-77. (WITHDRAWN) 1977 Amd 2 Classification of Degrees of Protection Provided by Enclosures (IEC 529: 1976) (AMD 4458) December 30, 1983 (Superseded by BS EN 60529: 1992). 23 pp.

BSI BS 6467: Part 1-85. 1985 Amd 1 Electrical Apparatus with Protection by Enclosure for Use in the Presence of Combustible Dusts Part 1: Specification for Apparatus (AMD 6457) November 30, 1990. 15 pp.

BSI BS 6467: Part 2-88. 1988 Electrical Apparatus with Protection by Enclosure for Use in the Presence of Combustible Dusts Part 2: Guide to Selection, Installation and Maintenance. 16 pp.

BSI BS EN 60529-92. 1992 Degrees of Protection Provided by Enclosures (IP Code). 50 pp.

BSI BS EN 60529-01. 1992 Amd 1 Degrees of Protection Provided by Enclosures (IP Code) (AMD 7643) July 15, 1993 (E). 51 pp.

CENELEC EN 60529-91. Degrees of Protection Provided by Enclosures (IP Code). 6 pp.

CENELEC EN 60529-92. CORRIGENDUM Degrees of Protection Provided by Enclosures (IP Code). 1 p.

DIN ENGL 40050-80. Degrees of Protection Provided by Enclosures; Protection of Electrical Equipment Against Contact, Foreign Bodies and Water (July). 6 pp.

IEC 529-89. Degrees of Protection Provided by Enclosures (IP Code) Second Edition. 76 pp.

MOD UK DSTAN 59-46-71. Cases, Equipment, Rack Mounting and Associated Lids, Panels, and Shelves-Suitable for Use with 19 Inch Racks Issue 1. 13 pp.

SAA AS 1939-90. Degrees of Protection Provided by Enclosures for Electrical Equipment (IP Code). 44 pp.

Electrical Enclosures *(Cont.)*
SAA AS 1939 Supp 1-90. Degrees of Protection Provided by Enclosures for Electrical Equipment (IP Code)—Supplement 1: Wallchart 1 (Supplement to AS 1939—1990).

SAA AS 1939 Supp 2-90. Degrees of Protection Provided by Enclosures for Electrical Equipment (IP Code)—Supplement 2: Wallchart 2 (Supplement to AS 1939—1990).

—Bonding—Hazardous Locations
CSA C22.2 NO 25-1966. Enclosures for Use in Class II Groups E, F, and G Hazardous Locations (R 1992); (Gen Instr 1). 18 pp.

CSA 1273 Bull. Electrical Bulletin 1273 July 2, 1980 to C22.2 NO 25. 2 pp.

CSA 1273A Bull. Electrical Bulletin 1273A February 8, 1983 to C22.2 NO 25. 2 pp.

CSA C22.2 NO 30-M1986. Explosion-Proof Enclosures for Use in Class I Hazardous Locations (R 1992); (Gen Instr 1 Thru 2). 41 pp.

—Classification
CENELEC HD 365 S3-85. Classification of Degrees of Protection Provided by Enclosures. 2 pp.

CENELEC HD 365 S3-88. Classification of Degrees of Protection Provided by Enclosures. 3 pp.

—Controlgear
BSI BS 5420-77. (WITHDRAWN) 1977 Degrees of Protection of Enclosures of Switchgear and Controlgear for Voltages up to and Including 1000 V a.C. and 1200 V d.c. (Superseded by BS EN 60947-1: 1992). 15 pp.

BSI BS 6878-88. 1988 Amd 1 High-Voltage Switchgear and Controlgear for Industrial Use. Cast Aluminium Alloy Enclosures for Gas-Filled High-Voltage Switchgear and Controlgear (EN 50 052: 1986) (AMD 6552) April 30, 1991). 21 pp.

BSI BS 7315-90. 1990 Wrought Aluminium and Aluminium Alloy Enclosures for Gas-Filled High Voltage Switchgear and Controlgear. 53 pp.

SAA AS 3132-91. Approval and Test Specification—Enclosures of Insulating Material for Switchgear and Controlgear Amdt 1 January/February 1993 (in Professional Package 28). 4 pp.

—Controlgear—High Voltage
BSI BS EN 50 068-91. 1991 Wrought Steel Enclosures for Gas-Filled High-Voltage Switchgear and Controlgear. 56 pp.

BSI BS EN 50069-91. 1991 Welded Composite Enclosures of Cast and Wrought Aluminium Alloys for Gas-Filled High-Voltage Switchgear and Controlgear. 18 pp.

CENELEC EN 50 052-86. Cast Aluminium Alloy Enclosures for Gas-Filled High-Voltage Switchgear and Control Gear. 16 pp.

CENELEC EN 50052/A1-90. AMD 1 Cast Aluminium Alloy Enclosures for Gas-Filled High-Voltage Switchgear and Control Gear. 6 pp.

CENELEC EN 50052 PRA2-93. AMD prA2 Cast Aluminium Alloy Enclosures for Gas-Filled High-Voltage Switchgear and Controlgear. 6 pp.

CENELEC EN 50 064-89. Wrought Aluminium and Aluminium Alloy Enclosures for Gas-Filled High-Voltage Switchgear and Controlgear. 82 pp.

CENELEC EN 50064 PRA1-93. AMD prA1 Wrought Aluminium and Aluminium Alloy Enclosures for Gas-Filled High-Voltage Switchgear and Controlgear. 3 pp.

CENELEC EN 50 068-91. Wrought Steel Enclosures for Gas-Filled High-Voltage Switchgear and Controlgear. 86 pp.

CENELEC EN 50068 PRA1-93. AMD prA1 Wrought Steel Enclosures for Gas-Filled High-Voltage Switchgear and Controlgear. 2 pp.

CENELEC EN 50 069-91. Welded Composite Enclosure of Cost and Wrought Aluminum for Gas-Filled High-Voltage Switchgear and Control Gear. 27 pp.

CENELEC EN 50069 PRA1-93. AMD prA1 Welded Composite Enclosures of Cast and Wrought Aluminium Alloys for Gas-Filled High-Voltage Switchgear and Controlgear. 3 pp.

—Covers—Screws
CSA 1208 Bull. Electrical Bulletin 1208 January 3, 1979 to C22.2 NO 85. 1 p.

—Dustproof
IEC 1241 Pt 1-1-93. Electrical Apparatus for Use in the Presence of Combustible Dust—Part 1: Electrical Apparatus Protected by Enclosures—Section 1: Specification for Apparatus First Edition. 47 pp.

IEC 1241 Pt 1-2-93. Electrical Apparatus for Use in the Presence of Combustible Dust—Part 1: Electrical Apparatus Protected by Enclosures—Section 2: Selection, Installation, and Maintenance First Edition. 39 pp.

Electrical Enclosures *(Cont.)*
—Dustproof—Explosion Proof
IEC 1241 Pt 1-1-93. Electrical Apparatus for Use in the Presence of Combustible Dust—Part 1: Electrical Apparatus Protected by Enclosures—Section 1: Specification for Apparatus First Edition. 47 pp.

—Dustproof—Installation
IEC 1241 Pt 1-2-93. Electrical Apparatus for Use in the Presence of Combustible Dust—Part 1: Electrical Apparatus Protected by Enclosures—Section 2: Selection, Installation, and Maintenance First Edition. 39 pp.

—Dustproof—Maintenance
IEC 1241 Pt 1-2-93. Electrical Apparatus for Use in the Presence of Combustible Dust—Part 1: Electrical Apparatus Protected by Enclosures—Section 2: Selection, Installation, and Maintenance First Edition. 39 pp.

—Electric Conduit Boxes—Fittings
CSA CAN/CSA-C22. 2 NO 85-M89. Rigid PVC Boxes and Fittings; (Gen Instr 1 Thru 2). 37 pp.

—Electric Conduits—Threaded—Construction
CSA C22.2 NO 0.5-1982. Threaded Conduit Entries (R 1992); (Gen Instr 1). 22 pp.

—Electric Power Meters
CNS K3089-90. Plastic Box for Low Voltage A.C. Power Meter (Sep)(11908).

CNS K6903-90. Method Test for Plastic Box for Low Voltage A.C. Power Meter (Sep)(11909). 6 pp.

SNZ NZS 6206-80. Specification for Domestic Electric Meter Boxes Amend: 1, 1988; 2, 1992. 7 pp.

—Electric Reactors
BSI BS 6435-84. 1984 Unfilled Enclosures for the Dry Termination of HV Cables for Transformers and Reactors. 7 pp.

—Explosion Proof
CSA 1273 Bull. Electrical Bulletin 1273 July 2, 1980 to C22.2 NO 25. 2 pp.

CSA 1273A Bull. Electrical Bulletin 1273A February 8, 1983 to C22.2 NO 25. 2 pp.

CSA C22.2 NO 30-M1986. Explosion-Proof Enclosures for Use in Class I Hazardous Locations (R 1992); (Gen Instr 1 Thru 2). 41 pp.

CSA 1273 Bull. Electrical Bulletin 1273 July 2, 1980 to C22.2 NO 30. 2 pp.

CSA 1273A Bull. Electrical Bulletin 1273A February 8, 1983 to C22.2 NO 30. 2 pp.

SAA AS 1076.3-77. Code of Practice for Selection, Instl. and Maint. of Elec. Ap and Assoc. Equipment for Use in Explosive Atmospheres (Other Than Mining Applications)—Pt. 3: Apparatus with Type of Protection 'd'—Flame-Proof Enclosure (Superseded by AS 2381.2—1992).

SAA AS 2380.2-91. Electrical Equipment for Explosive Atmospheres—Explosion-Protection Techniques—Part 2: Flameproof Enclosure d Amdt 1 July 1992 (in Professional Packages 40, 55) (Supersedes AS 2480—1986). 34 pp.

—Fiberglass Reinforced Plastics
CNS K3082-86. Glassfiber Reinforced Plastic Electric Enclosure (Aug)(11650). 7 pp.

CNS K6881-86. Method of Test for Glassfiber Reinforced Plastic Electric Enclosure (Aug)(11651). 9 pp.

—Grounding—Hazardous Locations
CSA C22.2 NO 25-1966. Enclosures for Use in Class II Groups E, F, and G Hazardous Locations (R 1992); (Gen Instr 1). 18 pp.

CSA 1273 Bull. Electrical Bulletin 1273 July 2, 1980 to C22.2 NO 25. 2 pp.

CSA 1273A Bull. Electrical Bulletin 1273A February 8, 1983 to C22.2 NO 25. 2 pp.

CSA C22.2 NO 30-M1986. Explosion-Proof Enclosures for Use in Class I Hazardous Locations (R 1992); (Gen Instr 1 Thru 2). 41 pp.

—Hazardous Locations
BSI BS 5345: Part 3-79. 1979 Amd 1 Installation and Maint-enance of Electrical Apparatus for Use in Potentially Explosive Atmospheres Part 3: Installation and Maintenance Requirements for Electrical Apparatus with Type of Protection 'd'. Flameproof Enclosure. 16 pp.

CSA C22.2 NO 25-1966. Enclosures for Use in Class II Groups E, F, and G Hazardous Locations (R 1992); (Gen Instr 1). 18 pp.

CSA 929 Bull. Electrical Bulletin 929 November 16, 1973 to C22.2 NO 25. 2 pp.

CSA 1310 Bull. Electrical Bulletin 1310 March 24, 1981 to C22.2 NO 25. 2 pp.

CSA 1273A Bull. Electrical Bulletin 1273A February 8, 1983 to C22.2 NO 25. 2 pp.

CSA C22.2 NO 30-M1986. Explosion-Proof Enclosures for Use in Class I Hazardous Locations (R 1992); (Gen Instr 1 Thru 2). 41 pp.

INTERNATIONAL AND NON-U.S. NATIONAL STANDARDS
SUBJECT INDEX

Electrical

Electrical Enclosures *(Cont.)*

—Indoor
CSA CAN/CSA-C22. 2 NO 94-M91. Special Purpose Enclosures; (Gen Instr 1). 45 pp.

—Indoor/Outdoor
CSA CAN/CSA-C22. 2 NO 94-M91. Special Purpose Enclosures; (Gen Instr 1). 45 pp.

—Junction Boxes—Fittings
CSA CAN/CSA-C22. 2 NO 85-M89. Rigid PVC Boxes and Fittings; (Gen Instr 1 Thru 2). 37 pp.

—Marine
CNS F5090-84. General Requirement for Degree of Protection and Inspection of Enclosures for Marine Electrical Apparatus (Apr)(10854).
JIS F 8007-89. General Requirements for Degree of Protection and Inspection of Enclosures for Marine Electrical Apparatus.

—Medical Electrical Equipment
CSA CAN/CSA-C22. 2NO601.1-M90. Medical Electrical Equipment, Part 1: General Requirements for Safety; (Gen Instr 1). 240 pp.

—Metal Sheets
CSA CAN/CSA-C22. 2 NO 0-M91. General Requirements—Canadian Electrical Code, Part II; (Gen Instr 1). 48 pp.

—Outdoor
CSA CAN/CSA-C22. 2 NO 94-M91. Special Purpose Enclosures; (Gen Instr 1). 45 pp.

—Panel Boards (Electrical)—Hazardous Locations
CSA C22.2 NO 30-M1986. Explosion-Proof Enclosures for Use in Class I Hazardous Locations (R 1992); (Gen Instr 1 Thru 2). 41 pp.

—Pressurized
BSI BS 5345: Part 5-83. 1983 Amd 1 Installation & Mainten-ance Electrical Appara-tus for Use in Potenti-ally Explosive Atmos-pheres Pt 5:Installation & Maintenance Require-ments for Electrical Apparatus Protected by Pressurization Including Continuous Dilution, and for Pressurized Rooms. 16 pp.
BSI BS 5501: Part 3-77. 1977 Amd 2 Electrical Apparatus for Potentially Explosive Atomspheres Part 3: Pressurized Apparatus 'p' (AMD 6437) March 30, 1990. 16 pp.

—Pressurized—Explosion Proof
SAA AS 1825-82. Electrical Equipment for Explosive Atmospheres—Pressurized Enclosures—Type of Protection P. 15 pp.

—Probes (Sensors)
BSI BS 3042-92. 1992 Test Probes to Verify Protection by Enclosures (IEC 1032: 1990) (G). 24 pp.
CENELEC HD 601 S1-91. Test Probes to Verify Protection by Enclosures. 6 pp.
IEC 1032-90. Test Probes to Verify Protection by Enclosures First Edition. (Corrigendum—Sept 1990). 41 pp.

—PVC
CSA CAN/CSA-C22. 2 NO 85-M89. Rigid PVC Boxes and Fittings; (Gen Instr 1 Thru 2). 37 pp.
CSA 1208 Bull. Electrical Bulletin 1208 January 3, 1979 to C22.2 NO 85. 1 p.

—Rotating Machines
BSI BS 4999: Part 105-88. 1988 General Requirements for Rotating Electrical Machines Part 105: Classification of Degrees of Protection Provided by Enclosures for Rotating Machinery. 25 pp.
CENELEC EN 60 034-86. Rotating Electrical Machines—Part 5: Classification of Degrees of Protection Provided by Enclosures for Rotating Machinery. 21 pp.
IEC 34 Pt 5-91. Rotating Electrical Machines Part 5: Classification of Degrees of Protection Provided by Enclosures of Rotating Electrical Machines (IP Code) Third Edition. 41 pp.
SAA AS 1359.20-80. Rotating Electrical Machines—General Requirements—Part 20: Classification of Types of Enclosure. 3 pp.

—Slides—Glossaries
IEC 916-88. Mechanical Structures for Electronic Equipment Terminology First Edition. 16 pp.

—Switchgear
BSI BS 5420-77. (WITHDRAWN) 1977 Degrees of Protection of Enclosures of Switchgear and Controlgear for Voltages up to and Including 1000 V a.C. and 1200 V d.c. (Superseded by BS EN 60947-1: 1992). 15 pp.

Electrical Enclosures *(Cont.)*

—Switchgear *(Cont.)*
BSI BS 6878-88. 1988 Amd 1 High-Voltage Switchgear and Controlgear for Industrial Use. Cast Aluminium Alloy Enclosures for Gas-Filled High-Voltage Switchgear and Controlgear (EN 50 052: 1986) (AMD 6552) April 30, 1991). 21 pp.
BSI BS 7315-90. 1990 Wrought Aluminium and Aluminium Alloy Enclosures for Gas-Filled High Voltage Switchgear and Controlgear. 53 pp.
SAA AS 3132-91. Approval and Test Specification—Enclosures of Insulating Material for Switchgear and Controlgear Amdt 1 January/February 1993 (in Professional Package 28). 4 pp.

—Switchgear—High Voltage
BSI BS EN 50 068-91. 1991 Wrought Steel Enclosures for Gas-Filled High-Voltage Switchgear and Controlgear. 56 pp.
BSI BS EN 50069-91. 1991 Welded Composite Enclosures of Cast and Wrought Aluminium Alloys for Gas-Filled High-Voltage Switchgear and Controlgear. 18 pp.
CENELEC EN 50 052-86. Cast Aluminium Alloy Enclosures for Gas-Filled High-Voltage Switchgear and Control Gear. 16 pp.
CENELEC EN 50052/A1-90. AMD 1 Cast Aluminium Alloy Enclosures for Gas-Filled High-Voltage Switchgear and Controlgear. 6 pp.
CENELEC EN 50052 PRA2-93. AMD prA2 Cast Aluminium Alloy Enclosures for Gas-Filled High-Voltage Switchgear and Controlgear. 6 pp.
CENELEC EN 50 064-89. Wrought Aluminium and Aluminium Alloy Enclosures for Gas-Filled High-Voltage Switchgear and Controlgear. 82 pp.
CENELEC EN 50064 PRA1-93. AMD prA1 Wrought Aluminium and Aluminium Alloy Enclosures for Gas-Filled High-Voltage Switchgear and Controlgear. 3 pp.
CENELEC EN 50 068-91. Wrought Steel Enclosures for Gas-Filled High-Voltage Switchgear and Controlgear. 86 pp.
CENELEC EN 50068 PRA1-93. AMD prA1 Wrought Steel Enclosures for Gas-Filled High-Voltage Switchgear and Controlgear. 2 pp.
CENELEC EN 50 069-91. Welded Composite Enclosure of Cost and Wrought Aluminum for Gas-Filled High-Voltage Switchgear and Control Gear. 27 pp.
CENELEC EN 50069 PRA1-93. AMD prA1 Welded Composite Enclosures of Cast and Wrought Aluminium Alloys for Gas-Filled High-Voltage Switchgear and Controlgear. 3 pp.

—Transformers
BSI BS 6435-84. 1984 Unfilled Enclosures for the Dry Termination of HV Cables for Transformers and Reactors. 7 pp.

—Wiring Space
CSA C22.2 NO 0.12-M1985. Wiring Space and Wire Bending Space in Enclosures for Equipment Rated 750 V or Less (R 1992); (Gen Instr 1 Thru 2). 23 pp.

—Wiring Space—Bending
CSA C22.2 NO 0.12-M1985. Wiring Space and Wire Bending Space in Enclosures for Equipment Rated 750 V or Less (R 1992); (Gen Instr 1 Thru 2). 23 pp.

Electrical Engineering
See Also: Electrical Components; Electronic Engineering; Mechanical Engineering

—Classification
BSI BS 1000: (621.3)-89. 1989 Universal Decimal Classification (UDC). English Full Edition (621.3): Electrical Engineering. 100 pp.
SNZ NZS/BS 1000 (621.3)-75. Universal Decimal Classification Electrical Engineering. 92 pp.

—Communications—Symbols
CNS C5001-82. Fundamental Symbols for Electric Communications Engineering (Jul)(413).

—Glossaries
IEC. Electricity, Electronics and Telecommunications Multilingual Dictionary Volume 1: English-French -Russian-German-Spanish-Dutch-Italian-Swedish-Polish Second Edition. 1944 pp.
IEC. Electricity, Electronics and Telecommunications Multilingual Dictionary Volume 2: English-French -Russian-German-Spanish-Dutch-Italian-Swedish-Polish Second Edition. 964 pp.
IEC. Electricity, Electronics and Telecommunications Multilingual Dictionary Volume 3: German, Dutch and Swedish Indexes Second Edition. 793 pp.
IEC. Electricity, Electronics and Telecommunications Multilingual Dictionary Volume 4: Spanish and Italian Indexes Second Edition. 540 pp.
IEC. Electricity, Electronics and Telecommunications Multilingual Dictionary Volume 5: Russian and Polish Indexes Second Edition. 578 pp.

Electrical Engineering *(Cont.)*

—Symbols
CNS C1017-82. Drawing Symbol of Electrical Engineering, Classification and Designation (Jul)(401).
CNS C1018-82. Fundamental Symbols for Electrical Power Engineering (Jul)(402).

—Yearbooks
IEC. IEC Yearbook 1993. 653 pp.

Electrical Equipment
Scope Note: For additional listings, see also specific types of equipment *See Also:* Communication Equipment; DC Equipment; Electrical Codes; Electrical Components; Electrical Installations; Electrical Protection Equipment; Electrical Safety; Electromagnetic Equipment; Electronic Equipment; Equalizers (Electrical); Essential Electrical Systems; Generating Stations; Interconnection Systems; Intervalometers; Magnetic Devices; Medical Electrical Equipment; Office Machines; Power Supplies; Radio Frequency Equipment; Stunning Equipment

CSA CAN/CSA-C22. 2 NO 0-M91. General Requirements—Canadian Electrical Code, Part II; (Gen Instr 1). 48 pp.
CSA 946 Bull. Electrical Bulletin 946 April 19, 1974 to C22.2 NO 0. 8 pp.
MOD UK DSTAN 59-36-90. Electronic and Associated Electrical Components for Defence Purposes Procedure for the Selection and Specification of Electronic and Associated Electrical Components for Use in Defence Equipment Issue 4. 25 pp.
SAA AS 3100-90. Approval and Test Specification—General Requirements for Electrical Equipment Amdt 2 January 1991 Amdt 3 March 1992 Amdt 4 October 1992 Amdt 5 Septemner 1993 (in Professional package 28). 1 p.
SNZ NZMP 6002-87. Guide to New Zealand Electrical Requirements and Approvals. 84 pp.

—AC—Symbols
CNS C1022-57. Symbols of A.C. Machines (Jan)(406) (R 1971).

—Aerospace—Glossaries
ISO 5843 Pt 1-85. Aerospace Construction—List of Equivalent Terms—Part 1: Aerospace Electrical Equipment First Edition. 29 pp.

—Aerostats
CAA 858 (Q). Section Q Non-Rigid Airships Electrical Supply, Systems and Equipment. 45 pp.

—Air Conditioning Equipment—Refrigerants
CSA C22.2 NO 140.3-M1987. Refrigerant-Containing Components for Use in Electrical Equipment (R 1993). 19 pp.

—Aircraft
AECMA PREN2282-90. Characteristics of Aircraft Electrical Supplies. 27 pp.
BSI BS EN 2282-92. 1992 Characteristics of Aircraft Electrical Supplies. 32 pp.
CAA 707 (G). Section G Rotorcraft Electrical Supply, Systems and Equipment (Blue Papers). 47 pp.
CAA Chapter J3-2 09.66. Equipment Including Cables. 2 pp.
CAA Chap G6-14 App #1 11.85. Installation (Rotorcraft). 3 pp.
CEN EN 2282-92. Characteristics of Aircraft Electrical Supplies. 28 pp.
JIS W 2011-78. Electrical Equipment, Aircraft, Selection and Installation. 4 pp.
JIS W 7201-83. Design and Installation of Aircraft Electrical Systems, General Specification for.

—Aircraft—Comprehensive Testing
CAA Chapter J5-1 06.53. Tests—General. 3 pp.
CAA Chapter J5-1 App 06.53. Tests. 3 pp.

—Aircraft—Construction
BSI 3G 100:Pt 4: Sec 1-73. 1973 General Requirements for Equipment for Use in Aircraft Part 4: Electrical Equipment Section 1: Construction and General. 4 pp.

—Aircraft—Cooling Systems
SBAC TS 16 ISSUE 2. Recommended Standard Method for the Presentation of Cooling Data for Aircraft Electrical Machines.

—Aircraft—Design—Domestic
CAA Chapter K6-1 3 App4 10.92. Design and Installation of Galleys and Domestic Equipment. 1 p.

—Aircraft—Design—Galleys
CAA Chapter K6-1 3 App4 10.92. Design and Installation of Galleys and Domestic Equipment. 1 p.

INDEX and DIRECTORY of

INTERNATIONAL AND NON-U.S. NATIONAL STANDARDS
SUBJECT INDEX
Electrical

Electrical Equipment (Cont.)

—Aircraft—Electromagnetic Compatibility—Test Methods
MOD UK DSTAN 59-41: Pt 3: Supp K-93. Electromagnetic Compatibility Part 3: Technical Requirements Test Methods and Limits Supplement K: Test Method DCS04 Imported Transient Susceptibility (Aircraft) Issue 1. 18 pp.

—Aircraft—Electromagnetic Interference
BSI 3G 100:Pt 4: Sec 2-80. (OBSOLESCENT) 1980 Amd 1 General Requirements for Equipment for Use in Aircraft Part 4: Electrical Equipment Section 2: Electromagnetic Interference at Radio and Audio Frequencies. 21 pp.
CAA 707 (G). Section G Rotorcraft Electrical Supply, Systems and Equipment (Blue Papers). 47 pp.
CAA Chap G6-14 App #2 11.85. Interference (Rotorcraft). 2 pp.
NATO STANAG 3516 ED 3 AMD 0-93. Electromagnetic Interference and Test Methods for Aircraft Electrical and Electronic Equipment. 51 pp.

—Aircraft—Environments
BSI 2G 100: Part 2-62. 1962 Amd 8 General Requirements for Electrical Equipment and Indicating Instruments for Aircraft Part 2: Environmental and Operating Conditions. 14 pp.

—Aircraft—Flight Testing
JIS W 7203-84. Flight Testing of Electric Systems in Aircraft, General Requirements for.

—Aircraft—Glossaries
BSI BS 185: Sec 10-70. 1970 Glossary of Aeronautical and Astronautical Terms Section 10: Auxiliary Services (Hydraulic, Pneumatic, Electrical, Air Conditioning and Refuelling). 7 pp.
JIS W 0107-87. Glossary of Terms for Aircraft Electrical and Lighting Systems.

—Aircraft—Grounding
CAA Chapter J3-3 06.53. Installation. 3 pp.

—Aircraft—High Voltage Testing
BSI 3G 100:Pt 4: SUBSEC 1.1-82. 1982 General Requirements for Equipment for Use in Aircraft Part 4: Electrical Equipment Section 1: Construction and General Subsection 1.1: Electrical Insulation Tests. 4 pp.
ISO 2678-85. Environmental Tests for Aircraft Equipment—Insulation Resistance and High Voltage Tests for Electrical Equipment Second Edition. 5 pp.

—Aircraft—Installation
CAA Chapter J3-3 06.53. Installation. 3 pp.
CAA Chapter J3-3 App 06.53. Installation Separation of Services and Prevention of Inadvertent Operation. 3 pp.
CAA Chapter K6-13 10.92. Utilisation and Installation of Electrically Operated Systems and Equipment. 10 pp.
CAA Chapter K6-1 3 App1 10.92. Installation. 2 pp.

—Aircraft—Installation—Cables (Electric)
CAA Chapter K6-1 3 App3 10.92. Cables and Cable Installation. 1 p.

—Aircraft—Installation—Domestic
CAA Chapter K6-1 3 App4 10.92. Design and Installation of Galleys and Domestic Equipment. 1 p.

—Aircraft—Installation—Galleys
CAA Chapter K6-1 3 App4 10.92. Design and Installation of Galleys and Domestic Equipment. 1 p.

—Aircraft—Insulation Resistance
BSI 3G 100:Pt 4: SUBSEC 1.1-82. 1982 General Requirements for Equipment for Use in Aircraft Part 4: Electrical Equipment Section 1: Construction and General Subsection 1.1: Electrical Insulation Tests. 4 pp.
ISO 2678-85. Environmental Tests for Aircraft Equipment—Insulation Resistance and High Voltage Tests for Electrical Equipment Second Edition. 5 pp.

—Aircraft—Interference
CAA Chapter K6-1 3 App2 10.92. Interference. 1 p.

—Aircraft—Operating Conditions
BSI 2G 100: Part 2-62. 1962 Amd 8 General Requirements for Electrical Equipment and Indicating Instruments for Aircraft Part 2: Environmental and Operating Conditions. 14 pp.

—Aircraft—Soldering
JIS W 7204-84. Method of Soldering for Aircraft Electrical and Electronic Equipments.

—Asbestos Cement Boards
BSI BS 3497-79. 1979 Unimpregnated Asbestos Cement Boards for Electrical Purposes. 14 pp.

Electrical Equipment (Cont.)

—Automotive—Direction of Rotation
CNS D1012-81. Designation of the Direction of Rotation of Electric Equipment for Automobiles (Jul)(7674).
JIS D 5001-72. Designation of the Direction of Rotation of Electric Equipment for Automobiles.

—Automotive—Glossaries
CNS D1024-85. Glossary of Terms Relating to Electric Equipments for Automobiles (Charging, Starting, Ignition, Preheating System) (Jul)(8572).
CNS D1025-85. Glossary of Terms Relating to Electric Equipments for Automobiles (Lamp System) (Jul)(8573).
CNS D1026-85. Glossary of Terms Relating to Electric Equipments for Automobiles (Windshield Wiper System) (Jul)(8574).
CNS D1027-85. Glossary of Terms Relating to Electric Equipments for Automobiles (Meter System) (Jul)(8575).
CNS D1028-85. Glossary of Terms Relating to Electric Equipments for Automobiles (Relay Braker and Flasher System) (Jul)(8576).
CNS D1029-85. Glossary of Terms Relating to Electric Equipments for Automobiles (Sound Signalling Device System) (Jul)(8577).
CNS D1030-85. Glossary of Terms Relating to Electric Equipments for Automobiles (Switch System) (Jul)(8578).
CNS D1031-85. Glossary of Terms Relating to Electric Equipments for Automobiles (Wiring System) (Jul)(8579).
JIS D 0103-82. Glossary of Terms Relating to Electric Equipments for Automobiles (R 1987). 117 pp.

—Automotive—Interference
BSI BS AU 243: Part 1-91. 1991 Methods of Test for Electrical Disturbance by Conduction and Coupling Part 1: General (ISO 7637-0: 1990). 7 pp.
BSI BS AU 243: Part 2-91. 1991 Methods of Test for Electrical Disturbance by Conduction and Coupling Part 2: Electrical Transient Conduction Along Supply Lines Only on Passenger Cars and Light Commercial Vehicles with 12 V Supply (ISO 7637-1: 1990). 17 pp.
BSI BS AU 243: Part 3-91. 1991 Methods of Test for Electrical Disturbance by Conduction and Coupling Part 3: Electrical Transient Conduction Along Supply Lines Only on Commercial Vehicles with 24 V Supply (ISO 7637-2: 1990). 17 pp.

—Automotive—Interference—Glossaries
ISO 7637 Pt 0-90. Road Vehicles—Electrical Disturbance by Conduction and Coupling—Part 0: Definitions and General First Edition. 6 pp.

—Automotive—Transient Voltage Measurement
ISO 7637 Pt 1-90. Road Vehicles—Electrical Disturbance by Conduction and Coupling—Part 1: Passenger Cars and Light Commercial Vehicles with Nominal 12 V Supply Voltage—Electrical Transient Conduction Along Supply Lines Only First Edition. 17 pp.
ISO 7637 Pt 2-90. Road Vehicles—Electrical Disturbance by Conduction and Coupling—Part 2: Commercial Vehicles with Nominal 24 V Supply Voltage—Electrical Transient Conduction Along Supply Lines Only First Edition. 15 pp.

—Automotive—Voltage Measurement
CNS D3053-87. Testing Voltages for Automotive Electric Equipments (Oct)(5565).
JIS D 5005-89. Nominal Voltages and Test Voltages for Automotive Electric Equipments. 5 pp.

—Ball Bearings
CNS B1161-80. Ball Bearings; Conrad Type, for Electrical Machines; Tolerances and Radial Clearance (May)(5506).

—Bonding
CSA C22.2 NO 0.4-M1982. Bonding and Grounding of Electrical Equipment (Protective Grounding) (R 1993); (Gen Instr 1). 19 pp.

—Ceramic Capacitors
BSI BS CECC 31400-82. 1982 Fixed Ceramic Capacitors of Dielectric Class 1, for Electrical Shock Hazard Protection: Sectional Specification. 25 pp.
BSI BS CECC 31400-01. 1982 Amd 1 Sectional Specification: Fixed Ceramic Capacitors of Dielectric Class 1, for Electrical Shock Hazard Protection (AMD 7404) February 15, 1993. 28 pp.

—Ceramics
CNS C3038-73. Method of Test for Vitrified Ceramic Materials for Electrical Applications (Aug)(3543).

—Certification—Glossaries
CSA Volume I. List of Certified Electrical Equipment Volume I Electrical Service Equipment, Wiring Hardware and Controls. 1127 pp.

Electrical Equipment (Cont.)

—Certification—Glossaries (Cont.)
CSA Volume II. List of Certified Electrical Equipment Volume II Electrically-Equipped Consumer, Commercial and Industrial Products. 1527 pp.
CSA Volume IV. List of Certified Electrical Equipment Volume IV Electrical Components and Accessories. 1087 pp.

—Color Coding
BSI BS 7645-93. 1993 Code for Designation of Colours (IEC 757: 1983) (H). 9 pp.
CENELEC HD 457-85. Code for Designation of Colours. 2 pp.
CNS C5076-80. Colors for Color Indentification and Coding (Dec)(6661).
IEC 757-83. Code for Designation of Colours First Edition. 10 pp.
SNZ IEC 757-83. Code for Designation of Colours. 7 pp.

—Comprehensive Testing
DIN VDE 0470-61. Rules for Testing Equipment and Testing Methods (Jan). 30 pp.

—Conducted Interference
DIN VDE 0847 Pt 1-81. Procedure for Measuring Electromagnetic Compatibility; Measurement of Conducted Interference (Nov). 24 pp.
DIN VDE 0847 Pt 2 (D)-87. Measuring Method for Evaluation of Electromagnetic Compatibility; Immunity from Conducted Disturbances (Oct). 86 pp.

—Construction
CSA CAN/CSA-C22. 2 NO 0-M91. General Requirements—Canadian Electrical Code, Part II; (Gen Instr 1). 48 pp.
CSA 946 Bull. Electrical Bulletin 946 April 19, 1974 to C22.2 NO 0. 8 pp.

—Corrosion Inhibitors
MOD UK TS 10151. Corrosion Preventive and Water Displacing Fluid for Use on Electrical Equipment PX-29.

—Distribution Systems—Naval Ships
MOD UK NES 539-81. Guide to the Design of Supply Systems for Portable Electrical Equipment Issue 1 (06.81). 29 pp.
MOD UK NES 539-93. Guide to the Design of Supply Systems for Portable Electrical Equipment Issue 2 (01.93). 26 pp.

—Distribution Systems—Submarines
MOD UK NES 539-81. Guide to the Design of Supply Systems for Portable Electrical Equipment Issue 1 (06.81). 29 pp.
MOD UK NES 539-93. Guide to the Design of Supply Systems for Portable Electrical Equipment Issue 2 (01.93). 26 pp.

—Double Insulated
CSA C22.2 NO 0.1-M1985. General Requirements for Double-Insulated Equipment; (Gen Instr 1). 19 pp.

—Double Insulated—Cords (Electric)
CSA C22.2 NO 0.1-M1985. General Requirements for Double-Insulated Equipment; (Gen Instr 1). 19 pp.

—Dust—Electrical Resistivity
IEC 1241 Pt 2-2-93. Electrical Apparatus for Use in the Presence of Combustible Dust—Part 2: Test Methods—Section 2: Method for Determining the Electrical Resistivity of Dust in Layers First Edition. 21 pp.

—Electric Discharges
DIN VDE 0110 Pt 20-90. Insulation Coordination for Electrical Equipment Within Low-Voltage Systems; Partial Discharge Tests Application Guide (Aug). 23 pp.

—Electric Shock—Classification
IEC 536 Pt 2-92. Classification of Electrical and Electronic Equipment with Regard to Protection Against Electric Shock Part 2: Guidelines to Requirements for Protection Against Electric Shock First Edition. 41 pp.

—Electrical Insulation—Coal Urines
SAA AS 1147.1-89. Electrical Equipment for Coal Mines—Insulating Materials—Part 1: Materials for Insulating Power Conducting Components. 7 pp.

—Electromagnetic Compatibility
MOD UK DSTAN 59-41: Part 1-88. Electromagnetic Compatibility Part 1: General Requirements Issue 4 (Superseded by Def Stan 59-41: Part 1: Section 1). 13 pp.
MOD UK DSTAN 59-41: Pt 1:Sec 1-93. Electromagnetic Compatibility Part 1: Introduction Section 1: General Requirements Issue 5. 13 pp.

INDUSTRY STANDARDS

Electrical
INTERNATIONAL AND NON-U.S. NATIONAL STANDARDS
SUBJECT INDEX

Electrical Equipment *(Cont.)*

—Electromagnetic Compatibility *(Cont.)*

MOD UK DSTAN 59-41: Pt 1:Sec 2-93. Electromagnetic Compatibility Part 1: Introduction Section 2: Guide to the Specification and Selection of EMC Requirements Issue 5. 45 pp.

MOD UK DSTAN 59-41: Part 2-88. Electromagnetic Compatibility Part 2: Management and Planning Procedures Issue 3. 10 pp.

—Electromagnetic Compatibility—Test Methods

MOD UK DSTAN 59-41: Part 3-01. Electromagnetic Compatibility Part 3: Technical Requirements Test Methods and Limits Issue 3; Amendment 4. 167 pp.

MOD UK DSTAN 59-41: Part 3-93. Electromagnetic Compatibility Part 3: Technical Requirements Test Methods and Limits Issue 4. 42 pp.

MOD UK DSTAN 59-41: Pt 3: Supp A-93. Electromagnetic Compatibility Part 3: Technical Requirements Test Methods and Limits Supplement A: Test Method DCE01 Conducted Emission on Power Lines 20Hz—150MHz Issue 1. 10 pp.

MOD UK DSTAN 59-41: Pt 3: Supp B-93. Electromagnetic Compatibility Part 3: Technical Requirements Test Methods and Limits Supplement B: Test Method DCE02 Conducted Emission on Control and Signal Lines 20Hz—150MHz Issue 1. 10 pp.

MOD UK DSTAN 59-41: Pt 3: Supp C-93. Electromagnetic Compatibility Part 3: Technical Requirements Test Methods and Limits Supplement C: Test Method DCE03 Conducted Emission on Exported Transients, Power Lines Issue 1. 12 pp.

MOD UK DSTAN 59-41: Pt 3: Supp D-93. Electromagnetic Compatibility Part 3: Technical Requirements Test Methods and Limits Supplement D: Test Method DRE01 Radiated Emissions E Field 14kHz—18GHz Issue 1. 14 pp.

MOD UK DSTAN 59-41: Pt 3: Supp E-93. Electromagnetic Compatibility Part 3: Technical Requirements Test Methods and Limits Supplement E: Test Method DRE02 H Field Radiation 20Hz—50kHz Issue 1. 10 pp.

MOD UK DSTAN 59-41: Pt 3: Supp F-93. Electromagnetic Compatibility Part 3: Technical Requirements Test Methods and Limits Supplement F: Test Method DRE03 Radiated Emmisions Installed Antenna 1MHz—76MHz Issue 1. 12 pp.

MOD UK DSTAN 59-41: Pt 3: Supp G-93. Electromagnetic Compatibility Part 3: Technical Requirements Test Methods and Limits Supplement G: Test Method DCS01 Conducted Susceptibility, Power Lines 20Hz—50kHz Issue 1. 11 pp.

MOD UK DSTAN 59-41: Pt 3: Supp H-93. Electromagnetic Compatibility Part 3: Technical Requirements Test Methods and Limits Supplement H: Test Method DCS02 Conducted Susceptibility, Power, Control and Signal Lines 50kHz—400MHz Issue 1. 20 pp.

MOD UK DSTAN 59-41: Pt 3: Supp J-93. Electromagnetic Compatibility Part 3: Technical Requirements Test Methods and Limits Supplement J: Test Method DCS03 Conducted Susceptibility, Control and Signal Lines 20Hz—50kHz Issue 1. 10 pp.

MOD UK DSTAN 59-41: Pt 3: Supp U-93. Electromagnetic Compatibility Part 3: Technical Requirements Test Methods and Limits Supplement U: Test Method DRS01 H Field Susceptibility, 20Hz—50kHz Issue 1. 11 pp.

MOD UK DSTAN 59-41: Pt 3: Supp V-93. Electromagnetic Compatibility Part 3: Technical Requirements Test Methods and Limits Supplement V: Test Method DRS02 Radiated Susceptibility E Field, 14kHz—18GHz Issue 1. 11 pp.

MOD UK DSTAN 59-41: Pt 3: Supp W-93. Electromagnetic Compatibility Part 3: Technical Requirements Test Methods and Limits Supplement W: Test Method DMFS01 Magnetostatic Field Susceptibility Issue 1. 8 pp.

—Electromagnetic Interference

BSI BS 613-77. 1977 Amd 2 Components and Filter Units for Electromagnetic Interference Suppression. 24 pp.

DIN VDE 0847 Pt 4 (D)-87. Procedures for Measurement of Electromagnetic Compatibility; Immunity Against Radiated Interference Variables (Jan). 35 pp.

—Electrostatic Sensitivity—Symbols

CNS Z5133-84. Symbol and Label for Electrostatic Sensitive Devices (Nov)(11144).

—Environmental Conditions—Ratings

IEC Guide 106-89. Guide for Specifying Environmental Conditions for Equipment Performance Rating First Edition. 24 pp.

—Explosion Proof

CNS C1038-83. General Rules for Explosion-Proof Construction of Electric Machineries and Apparatuses for General Use (Jan)(3376).

Electrical Equipment *(Cont.)*

—Explosion Proof *(Cont.)*

CNS C1108-83. Endurance Pressure Explosion-Proof Construction of Electric Machineries and Apparatuses of General Use (Jan)(9817).

CNS C1109-83. Into Oil Explosion-Proof Construction of Electric Machineries and Apparatuses of General Use (Jan)(9818).

CNS C1110-83. Inside Pressure Explosion-Proof Construction of Electric Machineries and Apparatuses of General Use (Jan)(9819).

CNS C1111-83. Increase Security Explosion-Proof Construction of Electric Machineries and Apparatuses of General Use (Jan)(9820).

CNS C1112-83. Substance Security Explosion-Proof Construction of Electric Machineries and Apparatuses of General Use (Jan)(9821).

CNS C1113-83. Loose Prevention Frame of General Rules for Explosion-Proof Construction of Electric Machineries and Apparatuses of General Use (Jan)(9822).

CNS C1114-83. Line Connection of Terminal Box to General Rules for Explosion-Proof Construction of Electric Machineries and Apparatuses of General Use (Jan)(9824).

CNS C1115-83. Outer Conductor and Terminal Box Connection of General Rules for Explosion-Proof Construction of Electric Machineries and Apparatuses of General Use (Jan)(9825).

CNS C4387-83. Terminal Box of General Rules for Explosion-Proof Construction of Electric Machineries and Apparatuses of General Use (Jan)(9823).

CNS F5082-84. General Requirements for Construction and Inspection of Electrical Explosion-Proof Apparatus for Marine Use (Feb)(10765).

—Explosive Atmospheres

BSI BS 229-57. 1957 Amd 9 Flameproof Enclosure of Electrical Apparatus. 56 pp.

BSI BS 4683: Part 1-71. 1971 Electrical Apparatus for Explosive Atmospheres Part 1: Classification of Maximum Surface Temperature. 8 pp.

BSI BS 4683: Part 2-71. 1971 Amd 3 Electrical Apparatus for Explosive Atmospheres Part 2: The Construction and Testing of Flameproof Enclosures of Electrical Apparatus Electrical Apparatus (AMD 6430) March 30, 1990. 50 pp.

BSI BS 4683: Part 4-73. 1973 Amd 2 Electrical Apparatus for Explosive Atmospheres Part 4: Type of Protection 'E'. 26 pp.

BSI BS 5345: Part 1-89. 1989 Amd 2 Code of Practice for the Selection, Install. and Maint. of Electrical Apparatus for Use in Potent. Explosive Atmospheres (Other Than Mining Applns. or Expl. Proc. and Man.) Part 1: General Recommendations (AMD 6333) January 31, 1991. 47 pp.

BSI BS 5345: Part 1-03. 1989 Amd 3 Selection, Installation and Maintenance of Electrical Apparatus for Use in Potentially Expolsive Atmospheres (Other Than Mining Applications or Explosives Processing and Manufacture) Part 1: General Recommendations (AMD 7871) Sept 15, 1993. 48 pp.

BSI BS 5345: Part 3-79. 1979 Amd 1 Installation and Maint-enance of Electrical Apparatus for Use in Potentially Explosive Atmospheres Part 3: Installation and Maintenance Requirements for Electrical Apparatus with Type of Protection 'd'. Flameproof Enclosure. 16 pp.

BSI BS 5345: Part 4-77. 1977 Installation and Maint-enance of Electrical Apparatus for Use in Potentially Explosive Atmospheres Part 4: Installation and Maint-enance Requirements for Electrical Apparatus with Type of Protection 'i'. Intrinsically Safe Elect. Appar. & Systems. 18 pp.

BSI BS 5345: Part 5-83. 1983 Amd 1 Installation & Mainten-ance Electrical Appara-tus for Use in Potenti-ally Explosive Atmos-pheres Pt 5:Installation & Maintenance Require-ments for Electrical Apparatus Protected by Pressurization Including Continuous Dilution, and for Pressurized Rooms. 16 pp.

BSI BS 5345: Part 6-78. 1978 Amd 1 Code of Practice for the Selection, and Maint. of Electrical Apparatus for Use in Potent. Explo. Atmosph. (Other Than Mining Applic. or Explos. Proc) Part 6: Recomm. for Type of Prot qqe' Incr Safety (AMD 5557) November 30, 1989. 17 pp.

BSI BS 5345: Part 7-79. 1979 Amd 1 Installation and Main-enance of Electrical Apparatus for Use in Potentially Explosive Atmospheres Part 7: Installation and Maintenance Requirements for Electrical Apparatus with Type of Protection N. 12 pp.

BSI BS 5345: Part 8-80. 1980 Installation and Maint-enance of Electrical Apparatus for Use in Potentially Explosive Atmospheres Part 8: Installation and Maintenance Requirements for Electrical Apparatus with Type of Protection 's'. Special Protection. 10 pp.

BSI BS 5501: Part 1-77. 1977 Amd 7 Electrical Apparatus for Potentially Explosive Atomspheres Part 1: General Requirements (AMD 6434) March 30, 1990. 42 pp.

Electrical Equipment *(Cont.)*

—Explosive Atmospheres *(Cont.)*

BSI BS 5501: Part 2-77. 1977 Amd 2 Electrical Apparatus for Potentially Explosive Atmospheres Part 2: Oil Immersion 'o' (AMD 6438) March 30, 1990. 10 pp.

BSI BS 5501: Part 3-77. 1977 Amd 2 Electrical Apparatus for Potentially Explosive Atomspheres Part 3: Pressurized Apparatus 'p' (AMD 6437) March 30, 1990. 16 pp.

BSI BS 5501: Part 4-77. 1977 Amd 2 Electrical Apparatus for Potentially Explosive Atmospheres Part 4: Powder Filling 'q' (AMD 6440) March 30, 1990. 22 pp.

BSI BS 5501: Part 5-77. 1977 Amd 7 Electrical Apparatus for Potentially Explosive Atmospheres Part 5: Flameproof Enclosure 'd' (AMD 6517) April 30, 1991. 69 pp.

BSI BS 5501: Part 6-77. 1977 Amd 8 Electrical Apparatus for Potentially Explosive Atomspheres Part 6: Increased Safety 'e' (AMD 6655) December 21, 1990. 73 pp.

BSI BS 5501: Part 7-77. 1977 Amd 12 Electrical Apparatus for Potentially Explosive Atmospheres Part 7: Intrinsic Safety 'i' (Q) (AMD 6436) March 30, 1990. 110 pp.

BSI BS 5501: Part 7-16. 1977 Amd 16 Electrical Apparatus for Potentially Explosive Atmospheres Part 7: Intrinsic Safety 'i'. 88 pp.

BSI BS 5501: Part 8-88. 1988 Amd 1 Electrical Apparatus for Potentially Explosive Atmospheres Part 8: Encapsulation 'qqm' (AMD 6439) March 30, 1990. 15 pp.

BSI BS 5501: Part 9-82. 1982 Amd 1 Electrical Apparatus for Potentially Explosive Atmospheres Part 9: Specification for Intrinsically Safe Electrical Systems 'i' (AMD 6435) March 30, 1990. 8 pp.

BSI BS 6941-88. 1988 Electrical Apparatus for Explosive Atmospheres with Type of Protection N. 23 pp.

BSI CP 1003: Part 3-67. 1967 Amd 1 Electrical Apparatus and Associated Equipment for Use in Explosive Atmospheres of Gas or Vapour Other Than Mining Applications Part 3: Division 2 Areas (AMD 5075) April 30, 1986. 15 pp.

BSI BS EN 50014-93. 1993 Electrical Apparatus for Potentially Explosive Atmospheres—General Requirements (E). 41 pp.

CENELEC EN 50 014-77. Electrical Apparatus for Potentially Explosive Atmospheres General Requirements. 72 pp.

CENELEC EN 50014-92. Electrical Apparatus for Potentially Explosive Atmospheres General Requirements. 35 pp.

CENELEC EN 50014-92. Electrical Apparatus for Potentially Explosive Atmospheres—General Requirements. 65 pp.

CENELEC EN 50 015-77. Electrical Apparatus for Potentially Explosive Atmospheres: Oil Immersion 'O'. 7 pp.

CENELEC EN 50 015/A1 (E)-79. AMD 1 Electrical Apparatus for Potentially Explosive Atmospheres: Oil Immersion 'O'. 4 pp.

CENELEC EN 50 016-77. Electrical Apparatus for Potentially Explosive Atmospheres: Pressurized Apparatus 'P'. 13 pp.

CENELEC EN 50 016/A1 (E)-79. AMD 1 Electrical Apparatus for Potentially Explosive Atmospheres: Pressurized Apparatus 'P'. 4 pp.

CENELEC EN 50 017-77. Electrical Apparatus for Potentially Explosive Atmospheres: Powder Filling 'Q'. 21 pp.

CENELEC EN 50 018-77. Electrical Apparatus for Potentially Explosive Atmospheres: Flameproof Enclosure 'D'. 54 pp.

CENELEC EN 50 018/A2-82. AMD 2 Electrical Apparatus for Potentially Explosive Atmospheres: Flameproof Enclosure 'D'.

CENELEC EN 50 018/A3-85. AMD 3 Electrical Apparatus for Potentially Explosive Atmospheres: Flameproof Enclosure 'D'.

CENELEC PREN 50018/PRAA-91. AMD prAA Electrical Apparatus for Potentially Explosive Atmospheres: Flameproof Enclosure 'D'. 71 pp.

CENELEC EN 50 019-85. Electrical Apparatus for Potentially Explosive Atmospheres: Increased Safety 'E'. 42 pp.

CENELEC EN 50 019/A4-89. AMD 4 Electrical Apparatus for Potentially Explosive Atmospheres: Increased Safety 'E'. 15 pp.

CENELEC EN 50 019/A5-90. AMD 5 Electrical Apparatus for Potentially Explosive Atmospheres: Increased Safety 'E'. 8 pp.

CENELEC EN 50 020-77. Electrical Apparatus for Potentially Explosive Atmospheres Intrinsic Safety 'I'. 71 pp.

CENELEC EN 50 020/A2-85. AMD 2 Electrical Apparatus for Potentially Explosive Atmospheres Intrinsic Safety 'I'. 3 pp.

CENELEC EN 50 020/A4-90. AMD 4 Electrical Apparatus for Potentially Explosive Atmospheres Intrinsic Safety 'I'. 14 pp.

906
INDEX and DIRECTORY of

INTERNATIONAL AND NON-U.S. NATIONAL STANDARDS
SUBJECT INDEX
Electrical

Electrical Equipment *(Cont.)*
—**Explosive Atmospheres** *(Cont.)*

CENELEC EN 50 020/A5-90. AMD 5 Electrical Apparatus for Potentially Explosive Atmospheres Intrinsic Safety 'I'. 2 pp.

CENELEC PREN 50021-92. Standard for Electrical Apparatus with Type of Protection 'n'. 86 pp.

CENELEC EN 50 028-87. Electrical Apparatus for Potentially Explosive Atmospheres. Encapsulation "M". 12 pp.

CENELEC EN 50 039-80. Electrical Apparatus for Potentially Explosive Atmospheres: Intrinsically Safe Electrical Systems 'I'. 5 pp.

CNS C3171-83. Method of Test for Electrical Apparatus for Use in Explosive Gas Atmospheres (Jan)(9826).

CNS M3009-85. Method of Test for Explosion-Proof Constructions of Electric Machineries and Apparatuses of Mines (Aug)(3378).

DIN VDE 0166-81. Electrical Installations and Equipment for Use in Atmospheres Potentially Endangered by Explosive Material (May). 28 pp.

DIN VDE 0170/0171 Pt 1 A102-88. Electrical Apparatus for Potentially Explosive Atmospheres; General Requirements; Amendment 102 (May). 4 pp.

DIN VDE 0170/0171 Pt 12 (D)-82. Electrical Equipment for Potentially Explosive Atmospheres; Requirements for Apparatus for Zone 0 (Nov). 26 pp.

EC 76/117/EEC-75. Council Directive on the Approximation of the Laws of the Member States Concerning Electrical Equipment for Use in Potentially Explosive Atmospheres. 4 pp.

EC 79/196/EEC-79. Council Directive on the Approximation of the Laws of the Member States Concerning Electrical Equipment for Use in Potentially Explosive Atmospheres Employing Certain Types of Protection. 3 pp.

EC 84/47/EEC-84. Commission Directive Adapting to Technical Progress Council Directive 79/196/EEC on the Approximation of the Laws of the Member States Concerning Electrical Equipment for Use in Potentially Explosive Atmospheres Employing Certain Types of Protection. 3 pp.

EC 88/35/EEC-87. Commission Directive Adapting to Technical Progress Council Directive 82/130/EEC on the Approximation of the Laws of the Member States Concerning Electrical Equipment for Use in Potentially Explosive Atmospheres in Mines Susceptible to Firedamp. 5 pp.

EC 88/571/EEC-88. Commission Directive Adapting to Technical Progress Council Directive 79/196/EEC on the Approximation of the Laws of the Member States Concerning Electrical Equipment for Use in Potentially Explosive Atmospheres Employing Certain Types of Protection. 1 p.

EC 91/269/EEC-91. Commission Directive Adapting to Technical Progress Council Directive 82/130/EEC on the Approximation of the Laws of the Member States Concerning Electrical Equipment for Use in Potentially Explosive Atmospheres in Mines Susceptible to Firedamp. 5 pp.

EC 82/490/EEC-82. Commission Recommendation Relating to the Certificates of Conformity Provided for in Council Directive 76/117/EEC on the Approximation of the Laws of the Member States Concerning Electrical Equipment for Use in Potentially-Explosive Atmospheres. 3 pp.

IEC 79 Pt 0-83. Electrical Apparatus for Explosive Gas Atmospheres Part 0: General Requirements Second Edition; (Amendment 2-1991, Incorporating Amendment 1). 71 pp.

IEC 79 Pt 1-90. Electrical Apparatus for Explosive Gas Atmospheres Part 1: Construction and Verification Test of Flameproof Enclosures of Electrical Apparatus Third Edition; (Amendment 1-1993). 131 pp.

IEC 79 Pt 2-83. Electrical Apparatus for Explosive Gas Atmospheres Part 2: Electrical Apparatus—Type of Protection "P" Third Edition. 56 pp.

IEC 79 Pt 3-90. Electrical Apparatus for Explosive Gas Atmospheres Part 3: Spark-Test Apparatus for Intrinsically-Safe Circuits Third Edition. 31 pp.

IEC 79 Pt 4-75. Electrical Apparatus for Explosive Gas Atmospheres Part 4: Method of Test for Ignition Temperature Second Edition. 22 pp.

IEC 79 Pt 5-67. Electrical Apparatus for Explosive Gas Atmospheres Part 5: Sand-Filled Apparatus First Edition; (Incorporating Supplement A-1969). 40 pp.

IEC 79 Pt 6-68. Electrical Apparatus for Explosive Gas Atmospheres Part 6: Oil-Immersed Apparatus First Edition. 13 pp.

IEC 79 Pt 7-90. Electrical Apparatus for Explosive Gas Atmospheres Part 7: Increased Safety "E" Second Edition; (Amendment 1-1991) (Amendment 2-1993). 112 pp.

IEC 79 Pt 10-86. Electrical Apparatus for Explosive Gas Atmospheres Part 10: Classification of Hazardous Areas Second Edition. 52 pp.

IEC 79 Pt 11-91. Electrical Apparatus for Explosive Gas Atmospheres Part 11: Intrinsic Safety "i" Third Edition. 105 pp.

Electrical Equipment *(Cont.)*
—**Explosive Atmospheres** *(Cont.)*

IEC 79 Pt 12-78. Electrical Apparatus for Explosive Gas Atmospheres Part 12: Classification of Mixtures of Gases or Vapours with Air According to Their Maximum Experimental Safe Gaps and Minimum Igniting Currents First Edition. 20 pp.

IEC 79 Pt 13-82. Electrical Apparatus for Explosive Gas Atmospheres Part 13: Construction and Use of Rooms or Buildings Protected by Pressurization First Edition. 20 pp.

IEC 79 Pt 14-84. Electrical Apparatus for Explosive Gas Atmospheres Part 14: Electrical Installations in Explosive Gas Atmospheres (Other Than Mines) First Edition. 46 pp.

IEC 79 Pt 15-87. Electrical Apparatus for Explosive Gas Atmospheres Part 15: Electrical Apparatus with Type of Protection "N" First Edition. 67 pp.

IEC 79 Pt 16-90. Electrical Apparatus for Explosive Gas Atmospheres Part 16: Artificial Ventilation for the Protection of Analyzer(s) Houses First Edition. 44 pp.

IEC 79 Pt 17-90. Electrical Apparatus for Explosive Gas Atmospheres Part 17: Recommendations for Inspection and Maintenance of Electrical Installations in Hazardous Areas (Other Than Mines) First Edition. 40 pp.

IEC 79 Pt 18-92. Electrical Apparatus for Explosive Gas Atmospheres Part 18: Encapsulation "m" First Edition. 52 pp.

JIS C 0901-83. Electrical Apparatus for Use in Gassy Coal Mines. 70 pp.

JIS C 0902-65. Testing Methods for Explosion Proof Construction of Electrical Machinery and Apparatus (for Coal Mines) (R 1971). 8 pp.

JIS C 0903-83. Electrical Apparatus for Explosive Atmospheres in General Industry. 65 pp.

JIS C 0904-83. Test Methods on Electrical Apparatus for Explosive Gas Atmospheres in General Industry. 26 pp.

JIS C 0905-83. Supplementary Requirements for Construction of Electrical Apparatus for Explosive Atmospheres in General Industry. 12 pp.

JIS C 4501-77. Explosion-Proof Type Plug Connection Devices for Coal Mines (R 1985). 18 pp.

SAA AS 1021-80. Protection by Purging of Electrical Equipment for Explosive Gas Atmospheres. 28 pp.

SAA AS 1039-86. Electrical Equipment in Coal Mines—Explosion Protected Distribution and Control Boxes for Voltages up to 3300 V a.c.. 15 pp.

SAA AS 1076. Code of Practice for Selection, Installation and Maintenance of Electrical Apparatus and Associated Equipment for Use in Explosive Atmospheres (Other than Mining Applications) NOTE: See Also AS 2381 Series.

SAA AS 1076.1-77. Code of Practice for Selection, Installation and Maintenance of Electrical Apparatus and Associated Equipment for Use in Explosive Atmospheres (Other Than Mining Applications)—Part 1: Basic Requirements Obsolescent January 1991. 88 pp.

SAA AS 1076.3-77. Code of Practice for Selection, Instl. and Maint. of Elec. Ap and Assoc. Equipment for Use in Explosive Atmospheres (Other Than Mining Applications)—Pt. 3: Apparatus with Type of Protection 'd'—Flame-Proof Enclosure (Superseded by AS 2381.2—1992).

SAA AS 1076.6-77. Code of Practice for Sel, Instl. and Maint. of Electrical Ap and Associated Equipment for Use in Explosive Atmospheres (Other Than Mining Applications)—Part 6: Apparatus with Type of Protection 'e'—Increased Safety (Superseded by AS 2381.6 —1993).

SAA AS 1076.7-77. Code of Practice for Selection, Installation and Maintenance of Electrical Apparatus and Associated Equipment for Use in Explosive Atmospheres (Other Than Mining Applications)—Part 7: Apparatus with Type of Protection 'n'—Non-Sparking Apparatus. 24 pp.

SAA AS 1076.8-77. Code of Practice for Selection, Installation and Maintenance of Electrical Apparatus and Associated Equipment for Use in Explosive Atmospheres (Other Than Mining Applications)—Part 8: Apparatus with Type of Protection 's'—Special Protection. 20 pp.

SAA AS 1826-83. Electrical Equipment for Explosive Atmospheres—Special Protection—Type of Protection S. 4 pp.

SAA AS 1915-92. Electrical Equipment for Explosive Atmospheres—Battery-Operated Vehicles (in Professional Package 40). 4 pp.

SAA AS 2380.9-91. Electrical Equipment for Explosive Atmospheres—Explosion-Protection Techniques—Part 9: Type of Protection n—Non-Sparking (in Professional Package 40) (Supersedes AS 2238—1982). 16 pp.

SAA AS 2381.1-91. Electrical Equipment for Explosive Atmospheres—Selection, Installation and Maintenance—Part 1: General Requirements Amdt 1 April 1992 (in Professional Packages 21, 23, 24, 40, 68). 2 pp.

Electrical Equipment *(Cont.)*
—**Explosive Atmospheres** *(Cont.)*

SNZ NZS 6109: Part 1-88. Electrical Systems of Dispensing Equipment for Explosive Atmospheres Part 1: Flammable Liquids Dispensing Equipment. 16 pp.

—**Explosive Atmospheres—Explosion Proof**

CNS M2005-85. General Rules for Explosion-Proof Construction of Electric Machineries and Apparatuses of Mines (Aug)(2867).

CNS M2005-2-85. Lamina Protected Explosion-Proof Construction of Electric Machineries and Apparatuses of Mines (Aug)(2867-2).

CNS M2005-3-85. Oil Immersed Explosion-Proof Construction of Electric Machineries and Apparatuses of Mines (Aug)(2867-3).

CNS M2005-4-85. Pressurized Explosion-Proof Construction of Electric Machineries and Apparatuses of Mines (Aug)(2867-4).

CNS M2005-7-85. Explosion-Proof Construction of Electric Machineries and Apparatuses (Aug)(2867-7).

CNS M2005-8-85. Loose Prevention Frame for Explosion-Proof Construction of Electric Machineries and Apparatuses of Mines (Aug)(2867-8).

CNS M2005-9-85. Terminals Box for Explosion-Proof Construction of Electric Machineries and Apparatuses of Mines (Aug)(2867-9).

CNS M2005-10-85. Line Connection of Terminal Box of Explosion-Proof Construction of Electric Machineries and Apparatuses of Mines (Aug)(2867-10).

CNS M2005-11-85. Outer Conductor and Terminal Box Connection for Explosion-Proof Construction of Electric Machineries and Apparatuses of Mines (Aug)(2867-11).

CNS M2005-12-85. Outer Conductor Connection for Explosion-Proof Construction of Electric Machineries and Apparatuses of Mines (Aug)(2867-12).

—**Explosive Atmospheres—Fire Resistance**

CNS M2005-1-85. Flameproof Endurance Pressure Explosion-Proof Construction of Electric Machineries and Apparatuses of Mines (Aug)(2867-1).

—**Explosive Atmospheres—Glossaries**

BSI BS 4727:Pt2: Group 13-91. 1991 Electrotechnical, Power, Telecommunication, Electronics, Lighting and Colour Terms Part 2: Terms Particular to Power Engineering Group 13: Electrical Apparatus for Explosive Atmospheres (IEC 50(426): 1990). 40 pp.

IEC 50 Chap 426-90. International Electrotechnical Vocabulary Chapter 426: Electrical Apparatus for Explosive Atmospheres First Edition. 44 pp.

—**Explosive Atmospheres—Safety**

CNS M2005-5-85. Increase Safety Explosion-Proof Construction of Electric Machineries and Apparatuses of Mines (Aug)(2867-5).

CNS M2005-6-85. Intrinsic Safety Explosion-Proof Construction of Electric Machineries and Apparatuses of Mines (Aug)(2867-6).

EC 76/117/EEC-75. Council Directive on the Approximation of the Laws of the Member States Concerning Electrical Equipment for Use in Potentially Explosive Atmospheres. 4 pp.

—**Explosive Atmospheres—Spray Painting Booths**

SNZ NZS/AS 4114 (INT)-93. Electrical Equipment for Explosive Atmospheres—Spray Painting Booths—Design, Construction, Installation and Maintenance (This is a Joint Standard with SAA AS 4114 (INT)). 15 pp.

—**Explosive Atmospheres—Ventilation**

SAA AS 1482-85. Electrical Equipment for Explosive Atmospheres—Protection by Ventilation—Type of Protection V. 17 pp.

—**Failure Rate**

IEC 605 Pt 6-86. Equipment Reliability Testing Part 6: Tests for the Validity of a Constant Failure Rate Assumption First Edition; (Amendment 1-1989). 23 pp.

IEC 605 Pt 7-78. Equipment Reliability Testing Part 7: Com-pliance Test Plans for Failure Rate and Mean Time Between Failures Assuming Constant Fail-ure Rate Clause 6—Pro-cedures for Design and Application of Time Ter-minated Test Plans First Edition; (Amendment 1-1990). 77 pp.

—**Fire Alarm and Extinguishing Equipment**

BSI BS 7273: Part 1-90. 1990 Operation of Fire Protection Measures Part 1: Electrical Actuation of Gaseous Total Flooding Extinguishing Systems. 12 pp.

BSI BS 7273: Part 2-92. 1992 Operation of Fire Protection Measures Part 2: Mechanical Actuation of Gaseous Total Flooding and Local Application Extinguishing Systems. 12 pp.

INDUSTRY STANDARDS

Electrical Equipment (Cont.)

—Firesafety
SAA HB37.2-93. Handbook of Australian Fire Standards—Part 2: Electrical Equipment (in Professional Package 44). 20 pp.

—Fuel Dispensing Equipment—Flammable Liquids
CSA C22.2 NO 22-M1986. Electrical Equipment for Flammable and Combustible Fuel Dispensers (R 1992); (Gen Instr 1). 19 pp.
CSA 1169 Bull. Electrical Bulletin 1169 June 27, 1978 to C22.2 NO 22. 2 pp.
SNZ NZS 6109: Part 1-88. Electrical Systems of Dispensing Equipment for Explosive Atmospheres Part 1: Flammable Liquids Dispensing Equipment. 16 pp.

—Fuel Dispensing Equipment—Liquefied Petroleum Gas
SNZ NZS 6109: Part 2-88. Electrical Systems of Dispensing Equipment for Explosive Atmospheres Part 2: Liquefied Petroleum Gas Dispensing Equipment. 16 pp.

—Furnaces
CENELEC PREN 50156-1-93. Electrical Equipment for Furnaces Part 1: Rules for Installation. 51 pp.
DIN VDE 0116-89. Electrical Equipment for Furnaces (Oct). 62 pp.

—Gas Processing Plants
CNS Z2073-88. Electrical Device for Liquefied Petroleum Gas Plant Use (Dec)(12478).

—Glossaries
BSI BS 4727:Pt1: Group 02-80. 1980 Amd 1 Electrotechnical, Power, Telecommunication, Electronics, Lighting and Colour Terms Part 1: Terms Common to Power, Telecommunicat. and Electronics Group 2: Electrical and Magnetic Devices (AMD 5813) May 31, 1989. 15 pp.
CSA CAN/CSA-C22. 2 NO 0-M91. General Requirements—Canadian Electrical Code, Part II; (Gen Instr 1). 48 pp.
CSA CAN/CSA-C22. 2 NO 68-92. Motor-Operated Appliances (Household and Commercial); (Gen Instr 1 Thru 2). 115 pp.
CSA CAN/CSA-C22. 2 NO 108-M89. Liquid Pumps; (Gen Instr 1 Thru 2). 68 pp.
IEC. Electricity, Electronics and Telecommunications Multilingual Dictionary Volume 1: English-French -Russian-German-Spanish-Dutch-Italian-Swedish-Polish Second Edition. 1944 pp.
IEC. Electricity, Electronics and Telecommunications Multilingual Dictionary Volume 2: English-French -Russian-German-Spanish-Dutch-Italian-Swedish-Polish Second Edition. 964 pp.
IEC. Electricity, Electronics and Telecommunications Multilingual Dictionary Volume 3: German, Dutch and Swedish Indexes Second Edition. 793 pp.
IEC. Electricity, Electronics and Telecommunications Multilingual Dictionary Volume 4: Spanish and Italian Indexes Second Edition. 540 pp.
IEC. Electricity, Electronics and Telecommunications Multilingual Dictionary Volume 5: Russian and Polish Indexes Second Edition. 578 pp.
IEC. Vocabulary of Fundamental Concepts. 237 pp.
IEC 50 Chap 151-78. International Electrotechnical Vocabulary Chapter 151: Electrical and Magnetic Devices; (Amendment 1-1987). 64 pp.
SAA AS 1852.151-88. International Electrotechnical Vocabulary—Part 151: Electrical and Magnetic Devices. 32 pp.
SNZ IEC 50: 50(151)-78. International Electrotechnical Vocabulary 50(151): Electrical and Magnetic Devices. 52 pp.

—Grounding
CSA C22.2 NO 0.4-M1982. Bonding and Grounding of Electrical Equipment (Protective Grounding) (R 1993); (Gen Instr 1). 19 pp.

—Identification Systems
CSA CAN/CSA-C22. 2 NO 0-M91. General Requirements—Canadian Electrical Code, Part II; (Gen Instr 1). 48 pp.
CSA 946 Bull. Electrical Bulletin 946 April 19, 1974 to C22.2 NO 0. 8 pp.

—Industrial
SAA AS 1543-85. Electrical Equipment of Industrial Machines. 26 pp.

—Industrial Machines
CENELEC HD 93.2-74. Part 2: Electrical Equipment of Machines Used in Large Series Production Lines (Superseded by EN 60204-1-1985). 3 pp.
DIN VDE 0113 Pt 1-86. Electrical Equipment of Industrial Machines; Part 1: General Requirements (Feb). 22 pp.
IEC 204 Pt 1-92. Electrical Equipment of Industrial Machines—Part 1: General Requirements Third Edition; (CENELEC EN 60204-1: 1992). 199 pp.

—Industrial Machines—Diagrams
IEC 204 Pt 2-84. Electrical Equipment of Industrial Machines Part 2: Item Designation and Examples of Drawings, Diagrams, Tables and Instructions (Appendices D and E of Publication 204-1) Second Edition. 67 pp.

—Insulated Wire
CNS C2066-85. Rubber Insulated Lead Wires for Electric Machinery and Apparatus (May)(5747). 15 pp.
JIS C 3315-87. Rubber Insulated Lead Wires for Electric Machinery and Apparatus. 20 pp.

—Insulating Oils—Gas Analysis
BSI BS 5574-78. (WITHDRAWN) 1978 Guide for the Sampling of Gases and of Oil from Oil-Filled Electrical Equipment and for the Analysis of Free and Dissolved Gases (Superseded by BS EN 60567: 1993). 32 pp.
BSI BS 5800-79. 1979 Guide for the Interpretation of the Analysis of Gases in Transformers and Other Oil-Filled Electrical Equipment in Service. 15 pp.
BSI BS EN 60567-93. 1993 Guide for the Sampling of Gases and of Oil from Oil-Filled Electrical Equipment and for the Analysis of Free and Dissolved Gases (IEC 567: 1992) (Supersedes BS 5574: 1978). 54 pp.
CENELEC HD 397-79. Interpretation of the Analysis of Gases in Transformers and Other Oil-Filled Electrical Equipment in Service. 1 p.
CENELEC EN 60567-92. Guide for the Sampling of Gases and of Oil From Oil-Filled Electrical Equipment and for the Analysis of Free and Dissolved Gases. 5 pp.
IEC 567-92. Guide for the Sampling of Gases and of Oil from Oil-Filled Electrical Equipment and for the Analysis of Free and Dissolved Gases Second Edition. 85 pp.
IEC 599-78. Interpretation of the Analysis of Gases in Transformers and Other Oil-Filled Electrical Equipment in Service First Edition. 28 pp.

—Insulation Coordination
CSA C22.2 NO 0.2-93. Insulation Coordination; (Gen Instr 1). 43 pp.
DIN VDE 0111 Pt 1-79. Insulation Co-Ordination to Equipment for Three-Phase a.c. Systems Above 1 kV Insulation Phase-To-Earth (VDE Specification) (Oct). 43 pp.
DIN VDE 0111 Pt 1 A1-86. Insulation Co-Ordination to Equipment for Three-Phase a.c. Systems 1 kV Insulation Phase-To-Earth: Amendment 1 (Sept). 5 pp.
DIN VDE 0111 Pt 2-83. Insulation Co-Ordination for Equipment in Three-Phase a.c. Systems Above 1 kV Phase-to-Insulation (VDE Specification) (Jan). 19 pp.
DIN VDE 0111 Pt 3-82. Insulation Co-Ordination for Equipment in Three-Phase a.c. Systems Above 1 kV Application Guide (VDE Guide) (Nov). 76 pp.

—Intrinsically Safe—Hazardous Locations
CSA CAN/CSA-C22. 2 NO 157-92. Intrinsically Safe and Non-Incendive Equipment for Use in Hazardous Locations; (Gen Instr 1). 61 pp.

—Labels—Adhesive
CSA CAN/CSA-C22. 2 NO0.15-M90. Adhesive Labels; (Gen Instr 1). 35 pp.

—Laminated Wood
IEC 1061 Pt 1-91. Specification for Non-Impregnated, Densified Laminated Wood for Electrical Purposes Part 1: Definitions, Designation and General Requirements First Edition. 19 pp.
IEC 1061 Pt 2-92. Specification for Non-Impregnated, Densified Laminated Wood for Electrical Purposes Part 2: Methods of Test First Edition. 33 pp.

—Land Service—Electromagnetic Compatibility—Test Methods
MOD UK DSTAN 59-41: Pt 3: Supp M-93. Electromagnetic Compatibility Part 3: Technical Requirements Test Methods and Limits Supplement M: Test Method DCS06 Imported Long Transient Susceptibility 28 Volt Systems (Land Service) Issue 1. 9 pp.
MOD UK DSTAN 59-41: Pt 3: Supp N-93. Electromagnetic Compatibility Part 3: Technical Requirements Test Methods and Limits Supplement N: Test Method DCS07 Imported Short Transient Susceptibility (Land Service) Issue 1. 10 pp.

—Leakage Paths
DIN VDE 0110 Pt 2-89. Insulation Co-Ordination for Equipment Within Low-Voltage Systems; Dimensioning of Clearances and Creepage Distances (Jan). 15 pp.

—Low Voltage Installations
IEC 1204-93. Low-Voltage Power Supply Devices, d.c. Output—Performance Characteristics and Safety Requirements First Edition. 59 pp.

—Machine Tools
CSA C22.2 NO 73-1953. Construction and Test of Electrically Equipped Machine Tools (R 1992). 19 pp.
CSA 257A Bull. Electrical Bulletin 257A June 24, 1975 to C22.2 NO 73. 2 pp.
CSA 1169 Bull. Electrical Bulletin 1169 June 27, 1978 to C22.2 NO 73. 2 pp.
JIS B 6015-89. Electrical Equipment of Machine Tools. 111 pp.

—Machinery—Safety
BSI BS EN 60204-1-93. 1993 Safety of Machinery—Electrical Equipment of Machines Part 1: Specification for General Requirements (F). 118 pp.
CENELEC EN 60 204-1-85. Electrical Equipment of Industrial Machines—Part 1: General Requirements; (Supersedes HD 93.2: 1974). 92 pp.
CENELEC EN 60 204-1/A1-88. AMD 1 Electrical Equipment of Industrial Machines—Part 1: General Requirements; (Supersedes HD 93.2: 1974). 4 pp.
CENELEC EN 60204-1-92. Safety of Machinery—Electrical Equipment of Machines Part 1: General Requirements; (IEC 204-1:1992, Modified). 105 pp.

—Maintainability
BSI BS 6548: Part 1-84. 1984 Maintainability of Equipment Part 1: Specifying and Contracting for Maintainability. 20 pp.
BSI BS 6548: Part 2-92. 1992 Maintainability of Equipment Part 2: Guide to Maintainability Studies During the Design Phase (IEC 706-2: 1990). 23 pp.
BSI BS 6548: Part 3-92. 1992 Maintainability of Equipment Part 3: Guide to Maintainability Verification, and the Collection, Analysis and Presentation of Maintainability Data (IEC 706-3: 1987) (F). 27 pp.
BSI BS 6548: Part 4-93. 1993 Maintainability of Equipment Part 4: Guide to the Planning of Maintenance and Maintenance Support (IEC 706-4: 1992) (F). 35 pp.
CSA CAN/CSA-Q632-90. Reliability and Maintainability Management Guidelines; (Gen Instr 1). 25 pp.
IEC 706 Pt 1-82. Guide on Maintainability of Equipment Part 1: Sections One, Two and Three Introduction, Requirements and Maintainability Programme First Edition. 42 pp.
IEC 706 Pt 2-90. Guide on Maintainability of Equipment Part 2: Section Five—Maintainability Studies During the Design Phase First Edition. 42 pp.
IEC 706 Pt 3-87. Guide on Maintainability of Equipment Part 3: Sections Six and Seven Verification and Collection, Analysis and Presentation of Data First Edition. 50 pp.

—Marine
CNS F5082-84. General Requirements for Construction and Inspection of Electrical Explosion-Proof Apparatus for Marine Use (Feb)(10765).
CNS F5089-84. General Requirement for Construction and Inspection of Marine Electrical Intrinsically Safe Apparatus (Apr)(10853).
JIS F 8004-88. General Requirements for Marine Electrical Flameproof Appararatus Flame Proof Apparatus for Marine Use.
JIS F 8005-80. General Requirements for Construction and Inspection of Marine Electrical Intrinsically Safe Apparatus.

—Marine—Electromagnetic Interference
BSI BS 7027-90. 1990 Limits and Methods of Measurement of the Immunity of Marine Electrical and Electronic Equipment to Conducted and Radiated Electromagnetic Interference. 14 pp.

—Marine—Phenolic Resins
MOD UK NES 2053-88. Specification for Sheets, Cellulose Paper and Cotton Fabric, Phenolic Synthetic Resin-Bonded, for Electrical Purposes Issue 1 (12.88). 10 pp.
MOD UK NES 2053-91. Specification for Sheets, Cellulose Paper and Cotton Fabric, Phenolic Synthetic Resin-Bonded, for Electrical Purposes Issue 2 (12.91). 11 pp.
MOD UK NES 2054-88. Specification for Tubes and Rods, Cellulose Paper and Cotton Fabric, Phenolic Synthetic Resin-Bonded, for Electrical Purposes Issue 1 (12.88). 12 pp.
MOD UK NES 2054-91. Specification for Tubes and Rods, Cellulose Paper and Cotton Fabric, Phenolic Synthetic Resin-Bonded, for Electrical Purposes Issue 2 (12.91). 11 pp.

—Mechanical Shock
CNS C3072-85. Method of Shock Test for Electric Machine and Equipment (Jun)(5311).
JIS C 0912-84. Shock Testing Procedure for Electric Machines and Equipment. 9 pp.

INTERNATIONAL AND NON-U.S. NATIONAL STANDARDS
SUBJECT INDEX
Electrical

Electrical Equipment (Cont.)

—Metal Sheets
CSA CAN/CSA-C22. 2 NO 0-M91. General Requirements—Canadian Electrical Code, Part II; (Gen Instr 1). 48 pp.

—Metal Working Equipment
CSA C22.2 NO 73-1953. Construction and Test of Electrically Equipped Machine Tools (R 1992). 19 pp.
CSA 1169 Bull. Electrical Bulletin 1169 June 27, 1978 to C22.2 NO 73. 2 pp.

—Mining Equipment
BSI BS 6704-87. 1987 Selection, Installation and Maintenance of Intrinsically Safe Electrical Equipment in Coal Mines. 23 pp.
BSI BS 6709-86. 1986 Interconnection of Electrical Apparatus, Constructed to Two or More British Standards, for Use in Mines Susceptible to Firedamp. 12 pp.
CSA CAN/CSA-M421-93. Use of Electricity in Mines; (Gen Instr 1). 67 pp.
SAA AS 1039-86. Electrical Equipment in Coal Mines—Explosion Protected Distribution and Control Boxes for Voltages up to 3300 V a.c.. 15 pp.

—Mining Equipment—Glossaries
BSI BS 3618: Sec 7-73. 1973 Glossary of Mining Terms Section 7: Electrical Engineering and Lighting. 11 pp.

—Mounting Rails
JIS C 2812-88. Mounting Rails for Devices. 12 pp.

—Name Plates—Adhesive
CSA CAN/CSA-C22. 2 NO0.15-M90. Adhesive Labels; (Gen Instr 1). 35 pp.

—Naval Ships
MOD UK NES 511-91. Requirements for Electrical Testing of Equipment Issue 3 (07.91). 22 pp.
MOD UK NES 519-88. Requirements for Electrical and Installations Fitted in Magazines Adjacent Compartments and Designated Danger Areas in Surface Ships Issue 2 (12.88). 20 pp.
MOD UK NES 519-93. Requirements and Safety Regulations for Electrical Equipment and Installations Fitted in Magazines, Submarine Weapon Stowage Compartments, Adjacent Compartments and Designated Danger Areas Issue 3 (01.93). 31 pp.
MOD UK NES 541-90. Guide to the Selection of Small Commercial Electrical Fittings and Components Issue 1 (11.90). 121 pp.
MOD UK NES 593-88. List of Approved Electrical Fittings for Use in Magazines Issue 2 (12.88). 24 pp.
MOD UK NES 593-01. List of Approved Electrical Fittings for Use in Magazines of Surface Ships Issue 3 (02.92); Amendment 1. 21 pp.

—Oxygen Measurement—Safety
CENELEC PREN 50104-92. Electrical Apparatus for the Detection and Measurement of Oxygen—Performance Requirements and Test Methods. 19 pp.

—Polymers
CSA CAN/CSA-C22. 2 NO 0.17-92. Evaluation of Properties of Polymeric Materials; (Gen Instr 1). 57 pp.

—Polymers—Classification
CSA C22.2 NO 0.11-M1985. Classification of Polymeric Compounds; (Gen Instr 1). 23 pp.

—Polymers—Flammability Testing
CSA C22.2 NO 0.6-M1982. Flammability Testing of Polymeric Materials; (Gen Instr 1 Thru 4). 32 pp.

—Qualification Approvals
MOD UK DSTAN 05-14-01. Mutual Acceptance of Qualification Approvals for Electrical/Electronic Components Within NATO Countries Issue 2; Amendment 1. 19 pp.

—Radio Communications—Electromagnetic Compatibility
CCIR QUESTION 81/1-90. Electromagnetic Compatibility Requirements Concerning Radiocommunication Services, in Particular Safety Services—Questions Concerning Study Group 1-Spectrum Management Techniques (Spectrum Engineering, Planning,Sharing, Monitoring and. 1 p.
CENELEC PRETS 300 279-92. Radio Equipment and Systems (RES); EMC Standard for Private Mobile Radio (PMR), Trunked and Ancillary Equipment (Analogue Speech and Combined Analogue Speech/Non-Speech Equipment). 115 pp.
ETSI PRETS 300 279-92. Radio Equipment and Systems (RES); EMC Standard for Private Mobile Radio (PMR), Trunked and Ancillary Equipment (Analogue Speech and Combined Analogue Speech/Non-Speech Equipment). 115 pp.

Electrical Equipment (Cont.)

—Radio Frequency Interference—Radio Communications
CCIR QUESTION 4-3/1-86. Limitation of Unwanted Radiation from Electrical Apparatus and Installations—Questions Concerning Study Group 1—Spectrum Management Techniques (Spectrum Engineering, Planning, Sharing, Monitoring and Utilization). 1 p.

—Refrigeration Equipment
CSA C22.2 NO 140.3-M1987. Refrigerant-Containing Components for Use in Electrical Equipment (R 1993). 19 pp.

—Reliability
BSI BS 5760: Sec 10.1-93. 1993 Reliability of Systems, Equipment and Components Part 10: Guide to Reliability Testing Section 10.1: General Requirements (IEC 605-1: 1978) (G). 38 pp.
CSA CAN/CSA-Q632-90. Reliability and Maintainability Management Guidelines; (Gen Instr 1). 25 pp.
IEC 605 Pt 1-78. Equipment Reliability Testing Part 1: General Requirements First Edition; (Amendment 1-1982). 67 pp.
IEC 605 Pt 3-1-86. Equipment Reliability Testing Part 3: Preferred Test Conditions Indoor Portable Equipment—Low Degree of Simulation First Edition. 22 pp.
IEC 605 Pt 3-2-86. Equipment Reliability Testing Part 3: Preferred Test Conditions Equipment for Stationary Use in Weatherprotected Locations—High Degree of Simulation First Edition. 24 pp.
IEC 605 Pt 3-3-92. Equipment Reliability Testing Part 3: Preferred Test Conditions Section 3: Test Cycle 3: Equipment for Stationary Use in Partially Weatherprotected Locations—Low Degree of Simulation First Edition. 33 pp.
IEC 605 Pt 3-4-92. Equipment Reliability Testing Part 3: Preferred Test Conditions Section 4: Test Cycle 4: Equipment for Portable and Non-Stationary Use—Low Degree of Stimulation First Edition. 37 pp.
IEC 605 Pt 4-86. Equipment Reliability Testing Part 4: Procedures for Determining Point Estimates and Confidence Limits for Equipment Reliability Determination Tests First Edition; (Amendment 1-1989). 61 pp.

—Repair Contracts
MOD UK DEFCON 112CU-90. Conditions of Repair Contracts for Naval Electrical, Electronic and Weapon Equipments 7/90. 8 pp.

—Repair—Explosive Atmospheres
SNZ NZS 6112-89. Code of Practice for the Repair of Electrical Apparatus for Use in Potentially Explosive Atmospheres. 52 pp.

—Robots
JIS B 8438-91. Industrial Robots—Electrical Equipment. 82 pp.

—Rotary Wing Aircraft
CAA Chapter G6-14 11.85. Utilisation and Installation of Electrically Operated Systems and Equipment (Rotocraft). 15 pp.
CAA CAP 524 SUB-Part F 12.86. Equipment (Rotorcraft). 17 pp.

—Safety
BSI BS EN 61010-1-93. 1993 Safety Requirements for Electrical Equipment for Measurement, Control, and Laboratory Use Part 1: General Requirements. 123 pp.
CENELEC EN 61010-1-93. Safety Requirements for Electrical Equipment for Measurement, Control and Laboratory Use Part 1: General Requirements (IEC 1010-1: 1990 + A1: 1992, Modified). 117 pp.
CSA CAN/CSA-C22. 2 NO 0-M91. General Requirements—Canadian Electrical Code, Part II; (Gen Instr 1). 48 pp.
CSA 946 Bull. Electrical Bulletin 946 April 19, 1974 to C22.2 NO 0. 8 pp.
IEC 1010 Pt 1-90. Safety Requirements for Electrical Equipment for Measurement, Control and Laboratory Use Part 1: General Requirements First Edition (CENELEC EN 61010-1: 1993). 190 pp.

—Safety—Standards Preparation
EURO MEMORANDUM 2-74. Preparation of Standards For Safety in the Design and Construction of Electrical Equipment. 5 pp.

—Service Units—Underfloor
DIN VDE 0634 Pt 1-87. Underfloor Electrical Installation Service Units (Sept). 29 pp.

Electrical Equipment (Cont.)

—Ships—Electromagnetic Compatibility—Test Methods
MOD UK DSTAN 59-41: Pt 3: Supp L-93. Electromagnetic Compatibility Part 3: Technical Requirements Test Methods and Limits Supplement L: Test Method DCS05 Externally Generated Transients (Ships) Issue 1. 11 pp.

—Ships—Electromagnetic Interference
CCIR RECMN 218-2-90. Prevention of Interference to Radio Reception on Board Ships —Section 8B—Maritime Mobile Service; Telegraphy and Related Subjects. 2 pp.
NATO STANAG 4435 ED 1 AMD 0-00. Electro-Magnetic Compatibility Testing Procedure and Requirements for Naval Electrical and Electronic Equipment (Surface Ship, Metallic Hull). 58 pp.
NATO STANAG 4435 ED 1 AMD 0-93. Electromagnetic Compatibility Testing Procedure and Requirements for Naval Electrical and Electronic Equipment (Surface Ship, Metallic Hull). 58 pp.
NATO STANAG 4436 ED 1 AMD 0-00. Electro-Magnetic Compatibility Testing Procedure and Requirements for Naval Electrical and Electronic Equipment (Surface Ship, Non-Metallic Hull). 56 pp.
NATO STANAG 4436 ED 1 AMD 0-93. Electro-Magnetic Compatibility Testing Procedure and Requirements for Naval Electrical and Electronic Equipment (Surface Ship, Non-Metallic Hull). 58 pp.

—Ships—Engineering Drawings—Symbols
CNS F5001-84. Graphical Symbols for Electrical Apparatus (Power) for Marine Engineering Drawings (May)(7784).
JIS F 8011-90. Graphical Symbols for Marine Engineering Apparatus (Power) for Marine Engineering Drawings.

—Ships—Environmental Testing
JIS F 0808-87. General Rules for Environmental Test of Marine Electrical Apparatus.

—Ships—Glossaries
JIS F 0031-89. Glossary of Terms for Shipbuilding (Electric Part).

—Ships—Inspection
MOD UK NES 696: Part 1-89. Final Electrical Inspections Part 1: Surface Ships Issue 2 (01.89). 29 pp.

—Ships—Name Plates
CNS F5110-86. Name Plate of Marine Electric Machinery and Equipment (Dec)(11793).

—Ships—Parts Lists
CNS F5108-86. Form of Spare Parts List of Electrical Apparatus for Marine Use (Dec)(11791).

—Ships—Spare Parts—Boxes (Containers)
CNS F5109-86. Spare-Part Boxes of Electrical Equipment for Marine Use (Dec)(11792).

—Ships—Vibration
CNS F5060-83. General Requirements for Vibration Test of Electrical Apparatus for Marine Use (May)(10255).
JIS F 8006-79. General Requirements for Vibration Test of Electrical Apparatus for Marine Use.

—Shock Protection
BSI BS 2754-76. 1976 Amd 1 Memorandum Construction of Electrical Equipment for Protection Against Electric Shock. 26 pp.
BSI PD 2754: Part 2-93. 1993 Classification of Electrical and Electronic Equipment with Regard to Protection Against Electric Shock Part 2: Guide to Requirements for Protection Against Electric Shock (IEC 536-2: 1992) (E). 22 pp.
BSI PD 6535-93. 1993 Common Aspects for Installation and Equipment for Protection Against Electric Shock (IEC 1440: 1992) (E). 15 pp.
CENELEC HD 366-77. Classification of Electrical and Electronic Equipment with Regard to Protection Against Electric Shock. 2 pp.
DIN VDE 0100 Pt 410 (S)-83. Installation of Power Plant with Rated Voltages Not Exceeding 1000 V; Protective Measures; Protection Against Electric Shock (Nov) (Superseded by 0100 Part 460/10.88). 53 pp.
DIN VDE 0100 Pt 701-84. Installation of Power Plant with Rated Voltages up to 1000 V; Locations Containing a Bath or Shower (May). 19 pp.
DIN VDE 0106 Pt 1 A3 (D)-92. Protection Against Shock Current; Classification of Electrical & Electronic Equipment with Regard to Protection Against Shock: Amendment 3 (Jan). 4 pp.
DIN VDE 0106 Pt 101-86. Protection Against Electric Shock; Basis Requirements for Protective Separation in Electrical Equipment (Nov). 13 pp.
DIN VDE 0106 Pt 101 A1 (D)-89. Protection Against Electric Shock; Basis Requirements for Protective Separation in Electrical Equipment; Amendment 1 (Feb). 4 pp.

INDUSTRY STANDARDS

Electrical — INTERNATIONAL AND NON-U.S. NATIONAL STANDARDS SUBJECT INDEX

Electrical Equipment (Cont.)

—Shock Protection (Cont.)
IEC 536-76. Classification of Electrical and Electronic Equipment with Regard to Protection Against Electric Shock First Edition. 13 pp.

—Silicon Steels
CNS C4304-81. Cold Rolled Silicon Steel Strip (Apr)(7215). 4 pp.

—Spas
SAA AS 3136-92. Approval and Test Specification—Electrical Equipment for Spa-Baths and Spa and Swimming Pools (AS 3136:1992/NZS 6232:1992) Amdt 1 June 1992 Amdt 2 September 1993 (in Professional Package 28). 10 pp.
SNZ NZS 6232-92. Approval and Test Specification—Electrical Equipment for Spa-Baths and Spa and Swimming Pools Amend: 1, 1992 (NZS 6232:1992/AS 3136-1992). 9 pp.

—Standard Voltages—Identification Systems
CSA CAN/CSA-C22. 2 NO 0-M91. General Requirements—Canadian Electrical Code, Part II; (Gen Instr 1). 48 pp.
CSA CAN3-C235-83. Preferred Voltage Levels for AC Systems, 0 to 50,000 V; (Gen Instr 1). 12 pp.

—Standards Preparation—Certification/Conformance
EURO MEMORANDUM 3-87. Implementation of the EEC Low Voltage Direction of 19 February 1973 with Respect to Reports, Certificates and Manufacturer's Declaration of Conformity. 50 pp.
EURO MEMORANDUM 4-84. Certification Organizations for Electrical Equipment Within Cenelec Countries Granting Marks of Conformity Related to Safety. 21 pp.
EURO MEMORANDUM 7-80. Guide to Certification of Products Not Fully Covered by Safety Standards Because of Technical Progress. 7 pp.

—Submarines
MOD UK NES 511-91. Requirements for Electrical Testing of Equipment Issue 3 (07.91). 22 pp.
MOD UK NES 519-88. Requirements for Electrical and Installations Fitted in Magazines Adjacent Compartments and Designated Danger Areas in Surface Ships Issue 2 (12.88). 20 pp.
MOD UK NES 519-93. Requirements and Safety Regulations for Electrical Equipment and Installations Fitted in Magazines, Submarine Weapon Stowage Compartments, Adjacent Compartments and Designated Danger Areas Issue 3 (01.93). 31 pp.
MOD UK NES 541-90. Guide to the Selection of Small Commercial Electrical Fittings and Components Issue 1 (11.90). 121 pp.

—Submarines—Electromagnetic Compatibility
NATO STANAG 4437 ED 1 AMD 0-00. (Draft) Electro-Magnetic Compatibility Testing Procedure and Requirements for Naval Electrical and Electronic Equipment (Submarines). 61 pp.

—Submarines—Inspection
MOD UK NES 696: Part 2-83. Final Electrical Inspections Part 2: Submarines Issue 1 (07.83). 12 pp.
MOD UK NES 696: Part 2-92. Final Electrical Inspections Part 2: Submarines Issue 2 (06.92). 34 pp.

—Sulfur Hexafluoride
BSI BS 5207-75. 1975 Sulphur Hexafluoride for Electrical Equipment. 24 pp.
BSI BS 5209-75. 1975 Testing of Sulphur Hexafluoride Taken from Electrical Equipment. 24 pp.
CNS C3156-82. Method of Test for Sulphur Hexafluoride (SF6) for Electrical Use (Jul)(9085).
IEC 480-74. Guide to the Checking of Sulphur Hexafluoride (SF6) Taken from Electrical Equipment First Edition. 44 pp.
JIS C 2131-76. Testing Methods of Sulphur Hexafluoride (SF6) for Electrical Uses.

—Swimming Pools
SAA AS 3136-92. Approval and Test Specification—Electrical Equipment for Spa-Baths and Spa and Swimming Pools (AS 3136:1992/NZS 6232:1992) Amdt 1 June 1992 Amdt 2 September 1993 (in Professional Package 28). 10 pp.
SNZ NZS 6232-92. Approval and Test Specification—Electrical Equipment for Spa-Baths and Spa and Swimming Pools Amend: 1, 1992 (NZS 6232:1992/AS 3136-1992). 9 pp.

—Symbols
BSI BS 6217-81. 1981 Amd 2 Graphical Symbols for Use on Electrical Equipment (IEC 417: 1973, 417A: 1974, 417B: 1975: 417C: 1977, 417D: 1978, 417E: 1980, 417F: 1982, 417G: 1985, 417H: 1987) (AMD 6126) May 31, 1991. 377 pp.
BSI BS 6217-03. 1981 Amd 3 Graphical Symbols for Use on Electrical Equipment (IEC 417: 1973, 417A: 1974, 417B: 1975: 417C: 1977, 417D: 1978, 417E: 1980, 417F: 1982, 417G: 1985, 417H: 1987, 417J: 1990) (AMD 7015) May 1, 1992. 467 pp.
BSI BS 6217-04. 1981 Amd 4 Graphical Symbols for Use on Electrical Equipment (IEC 417: 1973, 417A: 1974, 417B: 1975: 417C: 1977, 417D: 1978, 417E: 1980, 417F: 1982, 417G: 1985, 417H: 1987, 417J: 1990) (AMD 7875) August 15, 1993 (F). 481 pp.
BSI BS 7477-91. 1991 General Principles for the Creation of Graphical Symbols for Use on Equipment (IEC 416: 1988) (ISO 3461-1: 1988) (G). 15 pp.
CENELEC HD 243 S7-88. Graphical Symbols for Use on Equipment. Index, Survey and Compilation of the Single Sheets. 3 pp.
CENELEC HD 243 S8-89. Graphical Symbols for Use on Equipment. Index, Survey and Compilation of the Single Sheets. 3 pp.
CENELEC HD 243 S9-91. Graphical Symbols for Use on Equipment—Index, Survey and Compilation of the Single Sheets. 5 pp.
CENELEC HD 243 S10-93. Graphical Symbols for Use on Equipment—Index, Survey and Compilation of the Single Sheets (IEC 417:1973 + 417A: 1974 to 417K:1991). 4 pp.
CENELEC HD 245.1 S3-79. Letter Symbols to be Used in Electrical Technology —Part 1: General. 2 pp.
CENELEC HD 245.2-85. Letter Symbols to be Used in Electrical Technology Part 2: Telecommunications and Electronics. 2 pp.
CENELEC HD 245.3-76. Letter Symbols to be Used in Electrical Technology Part 3: Logarithmic Quantities and Units. 2 pp.
CENELEC HD 245.3 S2-91. Letter Symbols to Be Used in Electrical Technology Part 3: Logarithmic Quantities and Units. 5 pp.
CENELEC HD 571 S1-90. General Principles for the Creation of Graphical Symbols for Use on Equipment. 3 pp.
CNS C1064-83. Symbols for Sequential Control (General Uses) (Nov)(5532).
CNS C1065-83. Symbols for Sequential Control (Functions) (Nov)(5533).
CNS C1143-89. Graphical Symbols for Use on Electrical and Electronic Equipment (General Rules) (Feb)(12491).
CNS C1143-1-89. Graphical Symbols for Use on Electrical and Electronic Equipment (Source, Voltage) (Feb)(12491-1).
CNS C1143-2-89. Graphical Symbols for Use on Electrical and Electronic Equipment (General) (Feb)(12491-2).
CNS C1143-3-89. Graphical Symbols for Use on Electrical and Electronic Equipment (Movement, Action, Run, Adjustment) (Feb)(12491-3).
CNS C1143-4-89. Graphical Symbols for Use on Electrical and Electronic Equipment (Alarm, Signal, Filter) (Feb)(12491-4).
CNS C1143-5-89. Graphical Symbols for Use on Electrical and Electronic Equipment (Digital Transmission) (Feb)(12491-5).
CNS C1143-6-89. Graphical Symbols for Use on Electrical and Electronic Equipment (Telecommunication) (Feb)(12491-6).
CNS C5112-81. Fundamental Symbols for Electrical and Electronics (May)(7361).
CNS C5134-82. Additional Symbols for Electrical and Telecommunication Graphical Symbols (Apr)(8666).
IEC. Letter Symbols (Reprint—1984). 122 pp.
IEC 27 Pt 1-92. Letter Symbols to Be Used in Electrical Technology Part 1: General Sixth Edition; (Corrigendum April 1993). 115 pp.
IEC 27 Pt 2-72. Letter Symbols to Be Used in Electrical Technology Part 2: Telecommunications and Electronics First Edition; (Supplement A-1975) (Supplement B-1980). 83 pp.
IEC 27 Pt 3-89. Letter Symbols to Be Used in Electrical Technology Part 3: Logarithmic Quantities and Units Second Edition. 26 pp.
IEC 416-88. General Principles for the Creation of Graphical Symbols for Use on Equipment Second Edition. 26 pp.
IEC 417-73. Graphical Symbols for Use on Equipment First Ed.; (Supp A-1974), (Supp B-1975), (Supp C-1977) (Supplement D-1978) (Supplement E-1980) (Supplement F-1982) (Supplement G-1985) (Supplement H-1987) (Supplement J-1990) (Supplement K-1991) (Supplement L-1993). 461 pp.
IEC 617 Pt 1-85. Graphical Symbols for Diagrams Part 1: General Information, General Index Cross-Reference Tables First Edition. 92 pp.
IEC 617 Pt 2-83. Graphical Symbols for Diagrams Part 2: Symbol Elements, Qualifying Symbols and Other Symbols Having General Application First Edition. 35 pp.
ISO 3461 Pt 1-88. General Principles for the Creation of Graphical Symbols—Part 1: Graphical Symbols for Use on Equipment First Edition. 10 pp.
ISO 7000-89. Graphical Symbols for Use on Equipment—Index and Synopsis Second Edition. 78 pp.
JIS C 0301-90. Graphical Symbols for Electrical Apparatus.
SAA AS 1046.1-78. Letter Symbols for Use in Electrotechnology—Part 1: General. 48 pp.
SAA AS 1046.2-78. Letter Symbols for Use in Electrotechnology—Part 2: Telecommunications and Electronics. 32 pp.
SAA AS 1046.3-91. Letter Symbols for Use in Electrotechnology—Part 3: Logarithmic Quantities and Units (IEC 27-3:1989) (in Professional Package 56). 9 pp.
SAA AS 1104-78. Informative Symbols for Use on Electrical and Electronic Equipment. 24 pp.
SAA AS 1104S. Informative Symbols for Use on Electrical and Electronic Equipment—Individual Symbol Sheets (Complete Set in Binder).
SNZ IEC 416-88. General Principles for the Creation of Graphical Symbols for Use on Equipment. 23 pp.
SNZ IEC 617: Part 2-83. Graphical Symbols for Diagrams Part 2: Symbol Elements, Qualifying Symbols and Other Symbols Having General Application. 32 pp.

—Symbols—Diagrams
BSI BS 3939: Part 1-86. 1986 Graphical Symbols for Electrical Power, Telecommunications and Electronics Diagrams Part 1: General Information, General Index (G). 40 pp.
BSI BS 3939: Part 2-85. 1985 Graphical Symbols for Electrical Power, Telecommunications and Electronics Diagrams Part 2: Symbol Elements, Qualifying Symbols and Other Symbols Having General Application (G). 19 pp.
CNS C1094-87. Symbols of Electric Machines of Interior Wiring Diagram for Architectural Plans (Dec)(9105).

—Telecommunication Networks
CSA CAN/CSA-C22. 2 NO 225-M90. Telecommunication Equipment; (Gen Instr 1 Thru 3). 115 pp.

—Temperature
BSI PD 6524-89. 1989 Guide to Specifying Permissible Temperature and Temperature Rise for Parts of Electrical Equipment, in Particular for Terminals. 81 pp.
IEC 943-89. Guide for the Specification of Permissible Temperature and Temperature Rise for Parts of Electrical Equipment, in Particular for Terminals First Edition. 160 pp.

—Testing Equipment
DIN VDE 0104-89. Erection & Operation of Electrical Test Equipment (Oct). 46 pp.
MOD UK DSTAN 66-31-81. Basic Requirements and Tests for Proprietary Electronic and Electrical Test Equipment Issue 1. 20 pp.
MOD UK NES 43-88. Ancillary Support Equipment, Test Equipment and Tools Issue 2 (09.88). 45 pp.

—Vibration
CNS C3078-85. Method of Vibration Test for Electric Machines and Equipment (Jun)(5424).
JIS C 0911-84. Vibration Testing Procedure for Electric Machines and Equipment. 11 pp.

—Voltage Endurance Testing
CENELEC PREN 50093-91. Basic Immunity Standard for Voltage Dips Short Interruptions and Voltage Variations. 16 pp.

—Voltage Ratings—Identification Systems
EURO MEMORANDUM 14-89. Recommended Making of the Rated Voltages of Electrical Equipment. 6 pp.

—Water Resistance Testing
CNS C3152-82. Method for Test to Prove Protection Against Ingress of Water for Electrical Equipment (Jul)(9081).
JIS C 0920-82. Tests to Prove Protection Against Ingress of Water for Electrical Equipment. 9 pp.

—Wood Blocks
CNS C4215-80. Wood Blocks for Electric Devices (Aug)(6048).

Electrical Faults
Use For: Breakdown (Electrical); Faulting (Electrical)
See Also: Dielectric Breakdown; Electric Cutouts; Power Distribution Systems; Short Circuits

—Aircraft—Electrical Installations
CAA Chapter J1-3 09.66. Safety Precautions and Fault Conditions. 4 pp.

INTERNATIONAL AND NON-U.S. NATIONAL STANDARDS
SUBJECT INDEX
Electrical

Electrical Faults *(Cont.)*

—**Cable Insulation**
DIN VDE 0472 Pt 514-85. Testing of Cables, Wires and Flexible Cords; Faults on Insulation (May). 7 pp.

—**High Voltage Installations**
IEC 364 Pt 4-442-93. Electrical Installations of Buildings Part 4: Protection for Safety Chapter 44: Protection Against Overvoltages Section 442—Protection of Low-Voltage Installations Against Faults Between High-Voltage Systems and Earth First Edition. 45 pp.
IEC 919 Pt 2-91. Performance of High-Voltage d.c. (HVDC) Systems Part 2: Faults and Switching First Edition. 161 pp.

—**Switching—High Voltage Installations**
IEC 919 Pt 2-91. Performance of High-Voltage d.c. (HVDC) Systems Part 2: Faults and Switching First Edition. 161 pp.

Electrical Feedthrough
Use: Feedthrough Connectors

Electrical Flexible Tubing
Use For: Flexible Nonmetallic Tubing; Flexible Tubing (Electrical) *See Also:* Electric Conduits; Electrical Metallic Tubing
CSA CAN/CSA-C22. 2NO227.3-M91. Flexible Nonmetallic Tubing; (Gen Instr 1). 29 pp.

—**Cable Clamps**
CSA CAN/CSA-C22. 2 NO 18-92. Outlet Boxes, Conduit Boxes, and Fittings; (Gen Instr 1 Thru 2). 118 pp.

—**Connectors**
CSA CAN/CSA-C22. 2 NO 18-92. Outlet Boxes, Conduit Boxes, and Fittings; (Gen Instr 1 Thru 2). 118 pp.

—**Insulating**
CSA 1389A Bull. Electrical Bulletin 1389A January 16, 1989 to C22.2 NO 198.1. 2 pp.

—**Insulating—Extruded**
CSA C22.2 NO 198.1-M1986. Extruded Insulating Tubing (R 1992); (Gen Instr 1 Thru 4). 89 pp.

Electrical Grounding
Use For: Earthing (Electrical) *See Also:* Circuits; Electric Shock; Electrical Bonding; Electrical Protection Equipment; Ground Electrodes; Ground Fault Circuit Interrupters; Ground Fault Protection; Ground Relays; Ground Rods; Ground Terminals; Grounding Clamps; Grounding Conductors; Grounding Connectors; Grounding Devices; Grounding Reactors; Grounding Switches; Grounding Transformers; Power Systems; Protective Conductors; Surge Arresters; Transformers
BSI BS 4444-89. 1989 Guide to Electrical Earth Monitoring and Protective Conductor Proving. 12 pp.
BSI BS 4444-69. (WITHDRAWN) 1969 Amd 1 Guide to Electrical Earth Monitoring. 20 pp.
BSI BS 7430-91. 1991 Earthing. 85 pp.
BSI CP 1013-65. (WITHDRAWN) 1965 Earthing (Superseded by BS 7430: 1991). 130 pp.

—**Aircraft**
CAA 706 (K). Section K Light Aeroplanes Electrical Supply, Systems and Equipment (Blue Papers). 54 pp.
CAA Chapter K4-6 03.67. Electrical Bonding and Lightning Discharge Protection (Light Aeroplanes). 5 pp.
CAA Chapter K4-6 App 03.67. Electrical Bonding and Lightning Discharge Protection (Light Aeroplanes). 2 pp.

—**Aircraft—Bolts**
SBAC RS 682 ISSUE 1. Standard Servicing Earth Bolt (Ground Contact).

—**Aircraft—Channels**
SBAC AS 3114 ISSUE 1. Channel, Earthing Point 19 Amp. 1 p.
SBAC AS 3115 ISSUE 1. Channel, Earthing Point 37 Amp. 1 p.

—**Aircraft—Connectors**
NATO STANAG 3632 ED 3 AMD 2-86. Electrical Safety Connections for Aircraft and Ground Support Equipment. 12 pp.
NATO STANAG 3632 ED 4 AMD 0-91. Aircraft and Ground Support Equipment Electrical Connections for Static Grounding. 9 pp.
NATO STANAG 3632 ED 4 AMD 1-91. Aircraft and Ground Support Equipment Electrical Connections for Static Grounding. 10 pp.

Electrical Grounding *(Cont.)*

—**Aircraft—Electric Terminals—Bases**
SBAC AS 4639 ISSUE 1. Base Earth Terminal 4 B.A.. 1 p.
SBAC AS 4640 ISSUE 1. Base Earth Terminal 6 B.A.. 1 p.

—**Aircraft—Electrical Equipment**
CAA Chapter J3-3 06.53. Installation. 3 pp.

—**Aircraft—Electrical Installations**
CAA 707 (G). Section G Rotorcraft Electrical Supply, Systems and Equipment (Blue Papers). 47 pp.
CAA Chapter G6-13 11.85. Electrical Generation, Supply and Distribution (Rotorcraft). 11 pp.

—**Aircraft—Plates—Nuts**
SBAC AS 3116 ISSUE 1. Earthing Plate (For D.A. Nut 4 Or 2 B.A.). 1 p.

—**Aircraft Refueling Equipment**
BSI G 175-59. 1959 Amd 1 Aircraft Fuel Nozzle Grounding Plugs and Sockets. 2 pp.
ISO 46-73. Aircraft—Fuel Nozzle Grounding Plugs and Sockets First Edition. 3 pp.

—**Brooders**
CSA 1374 Bull. Electrical Bulletin 1374 June 28, 1982 to C22.2 NO 102. 1 p.

—**Buildings**
CENELEC HD 384.5.54 S1-88. Electrical Installatioans of Buildings—Part 5: Selection and Erection of Electrical Equipment. Chapter 54: Earthing Arrangements and Protective Conductors. 7 pp.
IEC 364 Pt 5-54-80. Electrical Installations of Buildings Part 5: Selection and Erection of Electrical Equipment Chapter 54: Earthing Arrangements and Protective Conductors First Edition; (Amendment 1-1982). 35 pp.

—**Buildings—Telecommunications**
CCITT RECMN K.27-91. Bonding Configurations and Earthing Inside a Telecommunication Building (Study Group V) 25 pp. 25 pp.
CENELEC PRETS 300 253-92. Equipment Engineering (EE); Earthing and Bonding of Telecommunication Equipment in Telecommunication Centres. 21 pp.
ETSI PRETS 300 253-92. Equipment Engineering (EE); Earthing and Bonding of Telecommunication Equipment in Telecommunication Centres. 21 pp.

—**Bushings**
CSA C22.2 NO 41-M1987. Grounding and Bonding Equipment (R 1993); (Gen Instr 1). 30 pp.

—**Data Processing Equipment**
IEC 364 Pt 7-707-84. Electrical Installations of Buildings Part 7: Requirements for Special Installations or Locations Section 707—Earthing Requirements for the Installation of Data Processing Equipment First Edition. 23 pp.

—**Dehumidifiers—Household**
CSA 649A Bull. Electrical Bulletin 649A January 10, 1967 to C22.2 NO 92. 1 p.

—**Electric Conductors—Color Coding**
CNS C1051-84. General Rules of Color Identification for Protective Conductors and Neutral Conductor and Terminal Marking for Apparatus (Dec)(5202).
JIS C 0602-84. General Rules of Colour Identification for Protective Conductor and Neutral Conductor and Terminal Marking for Apparatus. 5 pp.

—**Electrical Enclosures—Hazardous Locations**
CSA C22.2 NO 25-1966. Enclosures for Use in Class II Groups E, F, and G Hazardous Locations (R 1992); (Gen Instr 1). 18 pp.
CSA 1273 Bull. Electrical Bulletin 1273 July 2, 1980 to C22.2 NO 25. 2 pp.
CSA 1273A Bull. Electrical Bulletin 1273A February 8, 1983 to C22.2 NO 25. 2 pp.
CSA C22.2 NO 30-M1986. Explosion-Proof Enclosures for Use in Class I Hazardous Locations (R 1992); (Gen Instr 1 Thru 2). 41 pp.

—**Electrical Equipment**
CSA C22.2 NO 0.4-M1982. Bonding and Grounding of Electrical Equipment (Protective Grounding) (R 1993); (Gen Instr 1). 19 pp.

—**Electrical Equipment—Hospitals**
CNS C4268-80. Hospital Grade Earth Centers and Terminals (Dec)(6767).

—**Electrical Installations**
DIN VDE 0100 Pt 540-91. Installation of Power Plant with Rated Voltages Not Exceeding 1000 V; Selection and Insallation of Electrical Equipment; Earthing Arrangements, Protective Conductors and Equipotential Bonding Conductors (Nov). 37 pp.

Electrical Grounding *(Cont.)*

—**Electrical Installations** *(Cont.)*
DIN VDE 0100 Pt 540 A2 (D)-92. Erection of Power Installations with Nominal Voltages up to 1000 V; Selection and Erection of Equipment; Earthing Arrangements, Protective Conductors, Equipotential Bonding Conductors; Amendment 2 Identical with IEC 64 (Sec)570 (Jan). 8 pp.
DIN VDE 0141-89. Earthing Systems for Power Installations with Rated Voltages Above 1kV (July). 81 pp.

—**Enclosed Switches**
CSA 761 Bull. Electrical Bulletin 761 May 26, 1969 to C22.2 NO 4. 1 p.

—**Freezers—Household**
CSA 649A Bull. Electrical Bulletin 649A January 10, 1967 to C22.2 NO 92. 1 p.

—**Generating Stations**
CSA C22.3 NO 2-1975. General Grounding Requirements and Grounding Requirements for Electrical Supply Stations (R 1980). 12 pp.

—**Ground Support Equipment—Connectors**
NATO STANAG 3632 ED 3 AMD 2-86. Electrical Safety Connections for Aircraft and Ground Support Equipment. 12 pp.
NATO STANAG 3632 ED 4 AMD 0-91. Aircraft and Ground Support Equipment Electrical Connections for Static Grounding. 9 pp.
NATO STANAG 3632 ED 4 AMD 1-91. Aircraft and Ground Support Equipment Electrical Connections for Static Grounding. 10 pp.

—**Laboratory Equipment—Safety**
CSA C22.2 NO 151-M1986. Laboratory Equipment (R 1992); (Gen Instr 1 Thru 2). 61 pp.

—**Lighting Poles**
BSI BS 5649: Part 5-82. 1982 Lighting Columns Part 5: Base Compartments and Cableways. 7 pp.
CEN EN 40-5-82. Lighting Columns—Part 5: Base Compartments and Cableways. 5 pp.

—**Pipe Clamps**
BSI BS 951-86. 1986 Amd 1 Clamps for Earthing and Bonding Purposes (AMD 6635) April 30, 1991. 11 pp.

—**Power Plants**
CSA C22.3 NO 2-1975. General Grounding Requirements and Grounding Requirements for Electrical Supply Stations (R 1980). 12 pp.

—**Power Systems**
CSA C22.3 NO 2-1975. General Grounding Requirements and Grounding Requirements for Electrical Supply Stations (R 1980). 12 pp.
MOD UK DSTAN 61-5: Pt 3: Sec 1-01. Electrical Power Supply Systems Below 650 Volts Part 3: Distribution and Protection Requirements Section 1: Earthing Issue 1; Amendment 1. 24 pp.
MOD UK DSTAN 61-5: Pt 3: Sec 1-92. Electrical Power Supply Systems Below 650 Volts Part 3: Distribution and Protection Requirements Section 1: Earthing Issue 2. 22 pp.

—**Refrigeration Equipment—Household**
CSA 649A Bull. Electrical Bulletin 649A January 10, 1967 to C22.2 NO 92. 1 p.

—**Refrigerators—Household**
CSA 649A Bull. Electrical Bulletin 649A January 10, 1967 to C22.2 NO 92. 1 p.

—**Service Entrance Equipment**
CSA 761 Bull. Electrical Bulletin 761 May 26, 1969 to C22.2 NO 4. 1 p.

—**Substations (Electric)**
CSA C22.3 NO 2-1975. General Grounding Requirements and Grounding Requirements for Electrical Supply Stations (R 1980). 12 pp.

—**Telecommunication Equipment**
CENELEC PRETS 300 253-92. Equipment Engineering (EE); Earthing and Bonding of Telecommunication Equipment in Telecommunication Centres. 21 pp.
ETSI PRETS 300 253-92. Equipment Engineering (EE); Earthing and Bonding of Telecommunication Equipment in Telecommunication Centres. 21 pp.

—**Telecommunication Systems**
CCITT RECMN K.27-91. Bonding Configurations and Earthing Inside a Telecommunication Building (Study Group V) 25 pp. 25 pp.
DIN VDE 0800 Pt 2-85. Telecommunications; Earthing and Equipotential Bonding (July). 84 pp.
DIN VDE 0800 Pt 2 A1 (D)-88. Telecommunications; Earthing and Equipotential Bonding; Amendment 1 (Sept). 29 pp.

INDUSTRY STANDARDS

Electrical

Electrical Grounding *(Cont.)*
—Water Coolers—Household
CSA 649A Bull. Electrical Bulletin 649A January 10, 1967 to C22.2 NO 92. 1 p.

Electrical Heat Shrink Tubing
Use: Heat Shrink Tubing

Electrical Impedance
Use For: Admittance; Impedance
See Also: Characteristic Impedance; Electrical Properties; Electrical Resistance; Impedance Measurement; Input Impedance; Insulation Resistance; Nondestructive Testing; Ohmmeters; Surface Resistance; Transfer Impedance; Unbalance (Data Transmission)

IEC 725-81. Considerations on Reference Impedances for Use in Determining the Disturbance Characteristics of Household Appliances and Similar Electrical Equipment First Edition. 13 pp.

—Anodic Coatings
DIN ENGL 50949-84. Non-Destructive Testing of Anodic Oxidation Coatings on Pure Aluminium and Aluminium Alloys by Measurement of Admittance (Feb). 3 pp.

—Anodic Coatings—Sealing
BSI BS 6161: Part 6-84. 1984 Methods of Test for Anodic Oxidation Coatings on Aluminium and Its Alloys Part 6: Assessment of Sealing Quality by Measurement of Admittance or Impedance. 7 pp.

—Bipolar Transistors
CNS C6105-81. Method of Test for the Collector-Base Time Constant and for the Resistive Part of the Common-Emitter Input Impedance (Mar)(7012).

—Communication Terminal Equipment—Groups
CCITT RECMN G.232-89. 12-Channel Terminal Equipments—International Analogue Carrier Systems (Study Group XV) 13 pp. 13 pp.

—Compandors—Telephone Circuits
CCITT RECMN G.162-89. Characteristics of Compandors for Telephony—General Characteristics of International Telephone Connections and Circuits (Study Groups XII and XV) 8 pp. 8 pp.
CCITT RECMN G.166-89. Characteristics of Syllabic Compandors for Telephony on High Capacity Long Distance Systems—General Characteristics of International Telephone Connections and Circuits (Study Groups XII and XV) 7 pp. 7 pp.

—Echo Suppressors
CCITT RECMN G.164-89. Echo Suppressors—General Characteristics of International Telephone Connections and Circuits (Study Groups XII and XV) 36 pp. 36 pp.

—Frequency Division Multiplexers—Carrier Systems
CCITT RECMN G.371-89. FDM Carrier Systems for Submarine Cable—International Analogue Carrier Systems (Study Group XV) 3 pp. 3 pp.

—Oxide Coatings
DIN ENGL 50949-84. Non-Destructive Testing of Anodic Oxidation Coatings on Pure Aluminium and Aluminium Alloys by Measurement of Admittance (Feb). 3 pp.

—Speakers
CNS C5046-80. Loudspeaker, Dynamic, Magnetic Structures and Impedance (Jul)(5756).

—Telephone Cables
CCITT RECMN G.611-89. Characteristics of Symmetric Cable Pairs for Analogue Transmission—Transmission Media Characteristics (Study Group XV) 5 pp. 5 pp.
CCITT RECMN G.612-89. Characteristics of Symmetric Cable Pairs Designed for the Transmission of Systems with Bit Rates of the Order of 6 to 34 Mbit/s—Transmission Media Characteristics (Study Group XV) 4 pp. 4 pp.
CCITT RECMN G.613-89. Characteristics of Symmetric Cable Pairs Usable Wholly for the Transmission of Digital Systems with a Bit Rate of up to 2 Mbits—Transmission Media Characteristics (Study Group XV) 4 pp. 4 pp.
CCITT RECMN G.614-89. Characteristics of Symmetric Pair Star-Quad Cables Designed Earlier for Analogue Transmission Systems and Being Used Now for Digital System Transmission at Bit Rates of 6 to 34 Mbit/s—Transmission Media Characteristics (Study Group XV) 5 pp. 5 pp.
CCITT RECMN G.621-89. Characteristics of 0.7/2.9 mm Coaxial Cable Pairs—Transmission Media Characteristics (Study Group XV) 4 pp. 4 pp.

Electrical Impedance *(Cont.)*
—Telephone Cables *(Cont.)*
CCITT RECMN G.622-89. Characteristics of 1.2/4.4 mm Coaxial Cable Pairs—Transmission Media Characteristics (Study Group XV) 8 pp. 8 pp.
CCITT RECMN G.623-89. Characteristics of 2.6/9.5 mm Coaxial Cable Pairs—Transmission Media Characteristics (Study Group XV) 8 pp. 8 pp.
CCITT FASCICLE III.5-89. Digital Networks, Digital Sections and Digital Line Systems—Recommendations G.801—G.961. 292 pp.

—Telephone Repeaters
CCITT RECMN G.326-89. Typical Systems on Symmetric Cable Pairs ((12 + 12) Systems)—International Analogue Carrier Systems (Study Group XV) 3 pp. 3 pp.
CCITT RECMN G.611-89. Characteristics of Symmetric Cable Pairs for Analogue Transmission—Transmission Media Characteristics (Study Group XV) 5 pp. 5 pp.

—Telephone Repeaters—Carrier Systems
CCITT RECMN G.312-89. Intermediate Repeaters for Open-Wire Carrier Systems Conforming to Recommendation G.311—International Analogue Carrier Systems (Study Group XV) 2 pp. 2 pp.
CCITT RECMN G.322-89. General Characteristics Recommended for Systems on Symmetric Pair Cables —International Analogue Carrier Systems (Study Group XV) 8 pp. 8 pp.

—Telephone Repeaters—HF
CCITT RECMN G.334-89. 18 MHz Systems on Standardized 2.6/9.5 mm Coaxial Cable Pairs—International Analogue Carrier Systems (Study Group XV) 9 pp. 9 pp.
CCITT RECMN G.344-89. 6 MHz Systems on Standardized 1.2/4.4 mm Coaxial Cable Pairs—International Analogue Carrier Systems (Study Group XV) 4 pp. 4 pp.
CCITT RECMN G.345-89. 12 MHz Systems on Standardized 1.2/4.4 mm Coaxial Cable Pairs—International Analogue Carrier Systems (Study Group XV) 1 pp. 1 p.
CCITT RECMN G.346-89. 18 MHz Systems on Standardized 1.2/4.4 mm Coaxial Cable Pairs—International Analogue Carrier Systems (Study Group XV) 1 pp. 1 p.

—Telephone Repeaters—MF
CCITT RECMN G.341-89. 1.3 MHz Systems on Standardized 1.2/4.4 mm Coaxial Cable Pairs—International Analogue Carrier Systems (Study Group XV) 5 pp. 5 pp.

—Telephone Repeaters—MF/HF
CCITT RECMN G.332-89. 12 MHz Systems on Standardized 2.6/9.5 mm Coaxial Cable Pairs—International Analogue Carrier Systems (Study Group XV) 11 pp. 11 pp.

—Transformers—Radio Receivers
CNS C5012-89. Impedance and Ratios of Standard Q's for Medium Frequency Transformers of Transistor Radios (Jun)(3772).

Electrical Installations
See Also: Electric Wiring; Electrical Codes; Electrical Components; Electrical Equipment; Electronic Equipment; Emergency and Standby Power Supplies; High Voltage Installations; Low Voltage Installations; Power Systems; Service Entrance Equipment

CENELEC PREN 50110-93. Operation of Electrical Installations. 34 pp.
CNS C4001-80. Operation of Handles and Indication of Electric Installations (Aug)(39).
DIN VDE 0100 Suppl. 2-90. Erection of Power Installations with Nominal Voltages up to 1000 V; List of Relevant Standards (Mar). 18 pp.
DIN VDE 0100 Pt 100-82. Erection of Power Installations with Rated Voltages up to 1000 V; Scope—General Requirements (May). 4 pp.
DIN VDE 0100 Pt 100 (D)-85. Erection of Power Installations with Rated Voltages up to 1000 V; Scope—General Requirements (Feb). 7 pp.
DIN VDE 0100 Pt 510-87. Erection of Power Installations with Nominal Voltages up to 1000 V; Selection and Erection of Equipment; Common Rules (June). 18 pp.
DIN VDE 0100 Pt 520-85. Erection of Power Installations with Rated Voltages of up to 1000 V; Selection and Erection of Electrical Apparatus; Cables, Conductors and Busbars (Nov). 50 pp.
DIN VDE 0100 Pt 520 A1 (D)-86. Erection of Power Station Installations with Nominal Voltages up to 1000 Volts; Selection and Erection of Equipment Wiring Systems: Amendment 1. Identical with IEC 64 (Sec) 430 and 431 (Feb). 44 pp.

Electrical Installations *(Cont.)*
DIN VDE 0100 Pt 550-88. Erection of Power Installations with Nominal Voltages up to 1000 V; Selection and Erection of Electrical Equipment, Plug-And-Socket Devices, Switches and Installation Accessories (Apr) (Supersedes VDE 0100g/07.76 and VDE 0100/05.73). 9 pp.
DIN VDE 0100 Pt 600-87. Erection of Power Installations with Nominal Voltages up to 1000V (Nov). 42 pp.
DIN VDE 0101-89. Erection of Power Installations with Nominal Voltage Exceeding 1 kV (May). 97 pp.
IEC 1200 Pt 52-93. Electrical Installation Guide—Part 52: Selection and Erection of Electrical Equipment—Wiring Systems First Edition. 19 pp.
JIS C 0601-61. Operation of Handles and Indication of Electric Installations (R 1971). 12 pp.
SAA HB30-92. Electrical Installations Amdt 1 July 1993. 56 pp.

—Accident Prevention
DIN VDE 0105 Pt 1-83. Operation of Electrical Power Installations; General Requirements (July). 105 pp.
DIN VDE 0106 Pt 100-83. Protection Against Electric Shock; Location of Control Elements in the Vicinity of Shock-Hazard Parts (Mar). 19 pp.

—Aerostats
CAA 858 (Q). Section Q Non-Rigid Airships Electrical Supply, Systems and Equipment. 45 pp.

—Agricultural Buildings
BSI BS 5502: Part 25-91. 1991 Buildings and Structures for Agriculture Part 25: Code of Practice for Design and Installation of Services and Facilities. 28 pp.

—Aircraft
CAA 706 (K). Section K Light Aeroplanes Electrical Supply, Systems and Equipment (Blue Papers). 54 pp.
CAA 707 (G). Section G Rotorcraft Electrical Supply, Systems and Equipment (Blue Papers). 47 pp.
CAA. Explanatory Note 09.66. 3 pp.
CAA. Contents 09.66. 1 p.
CAA. Foreword 09.66. 2 pp.
CAA. Index 09.66. 2 pp.
CAA Chapter J1-1 09.66. General and Definitions—General. 3 pp.
CAA Chapter J1-2 09.66. Definitions. 6 pp.
CAA Chapter J1-2 App 09.66. Definitions. 4 pp.
CAA Chapter J1-3 09.66. Safety Precautions and Fault Conditions. 4 pp.
CAA Chapter J1-3 App 09.66. Safety Precautions and Fault Analysis. 2 pp.
CAA Chapter J2-1 09.66. Supply and Distribution. 5 pp.
CAA Chapter J2-1 App #1 09.66. Electrical Load Analyses. 13 pp.
CAA Chapter G6-13 11.85. Electrical Generation, Supply and Distribution (Rotocraft). 11 pp.
CAA Chapter G6-14 11.85. Utilisation and Installation of Electrically Operated Systems and Equipment (Rotocraft). 15 pp.
CAA Chap G6-14 App #1 11.85. Installation (Rotocraft). 3 pp.
CAA Chapter K6-13 10.92. Utilisation and Installation of Electrically Operated Systems and Equipment. 10 pp.
CAA Chapter K6-1 3 App1 10.92. Installation. 2 pp.
ISO 1540-84. Aerospace—Characteristics of Aircraft Electrical Systems Second Edition. 22 pp.

—Aircraft—Batteries
CAA Chapter K6-1 2 App4 10.92. Generator Switchgear and Battery Installation. 3 pp.

—Aircraft—Cables (Electric)
CAA Chapter K6-1 3 App3 10.92. Cables and Cable Installation. 1 p.
CEN PREN 2712-002-91. Single and Multicore Electrical Cables Screened and Jacketed for General Use Operating Temperatures Between -55 and +150 Degrees C Product Standard: General. 7 pp.
CEN PREN 2712-003-91. Single and Multicore Electrical Cables Screened (Spiral) and Jacketed for General Use Operating Temperatures Between -55 and +150 Degrees C Product Standard. 8 pp.
CEN PREN 2712-004-91. Single and Multicore Electrical Cables Screened (Braided) and Jacketed for General Use Operating Temperatures Between -55 and +150 Degrees C Product Standard. 8 pp.
CEN PREN 2712-005-91. Single and Multicore Electrical Cables Screened (Spiral) and Jacketed for General Use Operating Temperatures Between -55 and +150 Degrees C CO2 Laser Printable Product Standard. 8 pp.
CEN PREN 2712-007-91. Single and Multicore Electrical Cables Screened (Spiral) and Jacketed for General Use Operating Temperatures Between -55 and +150 Degrees C UV Laser Printable Product Standard. 8 pp.

Electrical Installations (Cont.)

—Aircraft—Cables (Electric) (Cont.)

CEN PREN 2712-008-91. Single and Multicore Electrical Cables Screened (Braided) and Jacketed for General Use Operating Temperatures Between -55 and +150 Degrees C UV Laser Printable Product Standard. 8 pp.

CEN PREN 2712-009-91. Single and Multicore Electrical Cables Screened (Spiral) and Jacketed for General Use Operating Temperatures Between -55 and +150 Degrees C YAG Laser Printable Product Standard. 8 pp.

CEN PREN 2712-010-91. Single and Multicore Electrical Cables Screened (Braided) and Jacketed for General Use Operating Temperatures Between -55 and +150 Degrees C YAG-X3 Laser Printable Product Standard. 8 pp.

CEN PREN 2713-002-91. Single and Multicore Electrical Cables Screened and Jacketed for General Use Operating Temperatures Between -55 and +200 Degrees C Product Standard: General. 7 pp.

CEN PREN 2713-003-91. Single and Multicore Electrical Cables Screened (Spiral) and Jacketed for General Use Operating Temperatures Between -55 and +200 Degrees C Product Standard. 8 pp.

CEN PREN 2713-004-91. Single and Multicore Electrical Cables Screened (Braided) and Jacketed for General Use Operating Temperatures Between -55 and +200 Degrees C Product Standard. 8 pp.

CEN PREN 2713-005-91. Single and Multicore Electrical Cables Screened (Spiral) and Jacketed for General Use Operating Temperatures Between -55 and +200 Degrees C CO2 Laser Printable Product Standard. 8 pp.

CEN PREN 2713-006-91. Single and Multicore Electrical Cables Screened (Braided) and Jacketed for General Use Operating Temperatures Between -55 and +200 Degrees C CO2 Laser Printable Product Standard. 8 pp.

CEN PREN 2713-008-91. Single and Multicore Electrical Cables Screened (Braided) and Jacketed for General Use Operating Temperatures Between -55 and +200 Degrees C UV Laser Printable Product Standard. 8 pp.

CEN PREN 2713-010-91. Single and Multicore Electrical Cables Screened (Braided) and Jacketed for General Use Operating Temperatures Between -55 and +200 Degrees C YAG-X3 Laser Printable Product Standard. 8 pp.

CEN PREN 2714-002-91. Single and Multicore Electrical Cables Screened and Jacketed for General Use Operating Temperatures Between -55 and +260 Degrees C Product Standard: General. 7 pp.

CEN PREN 2714-003-91. Single and Multicore Electrical Cables Screened (Spiral) and Jacketed for General Use Operating Temperatures Between -55 and +260 Degrees C Product Standard. 8 pp.

CEN PREN 2714-004-91. Single and Multicore Electrical Cables Screened (Braided) and Jacketed for General Use Operating Temperatures Between -55 and +260 Degrees C Product Standard. 8 pp.

CEN PREN 2714-005-91. Single and Multicore Electrical Cables Screened (Spiral) and Jacketed for General Use Operating Temperatures Between -55 and +260 Degrees C CO2 Laser Printable Product Standard. 8 pp.

CEN PREN 2714-006-91. Single and Multicore Electrical Cables Screened (Braided) and Jacketed for General Use Operating Temperatures Between -55 and +260 Degrees C CO2 Laser Printable Product Standard. 8 pp.

CEN PREN 2714-007-91. Single and Multicore Electrical Cables Screened (Spiral) and Jacketed for General Use Operating Temperatures Between -55 and +260 Degrees C UV Laser Printable Product Standard. 8 pp.

CEN PREN 2714-008-91. Single and Multicore Electrical Cables Screened (Braided) and Jacketed for General Use Operating Temperatures Between -55 and +260 Degrees C UV Laser Printable Product Standard. 8 pp.

CEN PREN 2714-009-91. Single and Multicore Electrical Cables Screened (Spiral) and Jacketed for General Use Operating Temperatures Between -55 and +260 Degrees C YAG-X3 Laser Printable Product Standard. 8 pp.

CEN PREN 2714-010-91. Single and Multicore Electrical Cables Screened (Braided) and Jacketed for General Use Operating Temperatures Between -55 and +260 Degrees C YAG-X3 Laser Printable Product Standard. 8 pp.

—Aircraft—Circuits

CAA Chapter J2-3 06.48. Circuit Control and Protection. 2 pp.

—Aircraft—Comprehensive Testing

CAA Chapter J5-1 06.53. Tests—General. 3 pp.
CAA Chapter J5-1 App 06.53. Tests. 3 pp.

—Aircraft—Design

NATO STANAG 3456 ED 4 AMD 4-76. Aircraft Electrical System Characteristics. 40 pp.
NATO STANAG 3456 ED 4 AMD 5-76. Aircraft Electrical System Characteristics. 42 pp.

—Aircraft—Electrical Faults

CAA Chapter J1-3 09.66. Safety Precautions and Fault Conditions. 4 pp.

—Aircraft—Flight Testing

CAA 706 (K). Section K Light Aeroplanes Electrical Supply, Systems and Equipment (Blue Papers). 54 pp.
CAA 707 (G). Section G Rotorcraft Electrical Supply, Systems and Equipment (Blue Papers). 47 pp.
CAA Chapter G6-13 11.85. Electrical Generation, Supply and Distribution (Rotorcraft). 11 pp.
CAA Chap G6-13 App #2 11.85. Ground and Flight Tests (Rotorcraft). 1 p.
CAA Chapter K6-1 2 App2 10.92. Ground and Flight Tests. 1 p.

—Aircraft—Glossaries

CAA Chapter J1-2 09.66. Definitions. 6 pp.
CAA Chapter J1-2 App 09.66. Definitions. 4 pp.

—Aircraft—Ground Testing

CAA 706 (K). Section K Light Aeroplanes Electrical Supply, Systems and Equipment (Blue Papers). 54 pp.
CAA 707 (G). Section G Rotorcraft Electrical Supply, Systems and Equipment (Blue Papers). 47 pp.
CAA Chapter G6-13 11.85. Electrical Generation, Supply and Distribution (Rotorcraft). 11 pp.
CAA Chap G6-13 App #2 11.85. Ground and Flight Tests (Rotorcraft). 1 p.
CAA Chapter K6-1 2 App2 10.92. Ground and Flight Tests. 1 p.

—Aircraft—Interference

CAA Chapter K6-1 3 App2 10.92. Interference. 1 p.

—Aircraft—Loads (Electric)

CAA Chapter K6-1 2 App3 10.92. Electrical Load Analyses. 3 pp.

—Aircraft—Loads (Electronic)

CAA 706 (K). Section K Light Aeroplanes Electrical Supply, Systems and Equipment (Blue Papers). 54 pp.
CAA 707 (G). Section G Rotorcraft Electrical Supply, Systems and Equipment (Blue Papers). 47 pp.
CAA Chapter J2-1 App #1 09.66. Electrical Load Analyses. 13 pp.
CAA Chapter G6-13 11.85. Electrical Generation, Supply and Distribution (Rotorcraft). 11 pp.
CAA Chap G6-13 App #3 11.85. Electrical Load Analyses (Rotorcraft). 3 pp.

—Aircraft—Reliability

CAA 706 (K). Section K Light Aeroplanes Electrical Supply, Systems and Equipment (Blue Papers). 54 pp.
CAA 707 (G). Section G Rotorcraft Electrical Supply, Systems and Equipment (Blue Papers). 47 pp.
CAA Chapter J2-1 App #3 09.66. System Reliability. 3 pp.
CAA Chapter G6-13 11.85. Electrical Generation, Supply and Distribution (Rotorcraft). 11 pp.
CAA Chap G6-13 App #1 11.85. Systems Reliability (Rotorcraft). 5 pp.
CAA Chapter K6-12 10.92. Electrical Generation, Supply and Distribution. 7 pp.
CAA Chapter K6-1 2 App1 10.92. Systems Reliability. 3 pp.

—Aircraft—Safety

CAA Chapter J1-3 09.66. Safety Precautions and Fault Conditions. 4 pp.

—Aircraft—Switchgear

CAA Chapter K6-1 2 App4 10.92. Generator Switchgear and Battery Installation. 3 pp.

—Aircraft—Switching Devices

CEN PREN 2349-90. Requirements and Test Procedures for Switching Devices. 55 pp.

—Boats (Marine)

CSA C22.2 NO 183.1-M1982. Alternating-Current (AC) Electrical Installations on Boats (R 1992); (Gen Instr 1). 27 pp.
CSA C22.2 NO 183.2-M1983. DC Electrical Installations on Boats (R 1992); (Gen Instr 1). 31 pp.
DIN VDE 0100 Pt 721-84. Erection of Power Installations with Rated Voltages up to 1000 V; Caravans, Boats and Yachts, as Well as Power Supply Thereof at Camping Sites and Berths (Apr). 10 pp.

—Buildings

BSI BS 4363-68. 1968 Amd 2 Distribution Units for Electricity Supplies for Construction and Building Sites. 26 pp.

—Buildings (Cont.)

CENELEC HD 384.1-78. Electrical Installations of Buildings Part 1: Scope. 2 pp.
CENELEC HD 384.3-85. Electrical Installations of Buildings Part 3: Assessment of General Characteristics. 21 pp.
CENELEC PRHD 384.3 S2-92. Electrical Installations of Buildings Part 3: Assessment of General Characteristics of Installations. 27 pp.
CENELEC HD 384.5.51-85. Electrical Installations of Buildings Part 5: Selection and Erection of Electrical Equipment Chapter 51: Common Rules. 9 pp.
CENELEC HD 384.5.54 S1-88. Electrical Installatioans of Buildings—Part 5: Selection and Erection of Electrical Equipment. Chapter 54: Earthing Arrangements and Protective Conductors. 7 pp.
CENELEC HD 384.5.523-91. Electrical Installations of Buildings Part 5: Selection and Erection of Electrical Equipment Chapter 52: Wiring Systems Section 523—Current-Carrying Capacities. 10 pp.
CENELEC HD 384.7.705 S1-91. Electrical Installations of Buildings Part 7: Requirements for Special Installations or Locations Section 705—Electrical Installations of Agricultural and Horticultural Premises. 5 pp.
CENELEC HD 384.7.706 S1-91. Electrical Installations of Buildings Part 7: Requirements for Special Installations or Locations Section 706—Restrictive Conducting Locations. 7 pp.
DIN ENGL 18015 Pt 3-90. Electrical Installations in Dwellings; Circuit Arrangement and Location of Electrical Equipment (July). 3 pp.
DIN VDE 0100 Pt 701-84. Installation of Power Plant with Rated Voltages up to 1000 V; Locations Containing a Bath or Shower (May). 19 pp.
DIN VDE 0108 Pt 1-89. Power Installations and Safety Power Supply in Communal Buildings; General (Oct) (VDE 0107/ 11.89 with DIN VDE 0108 Part 1 to Part 8 (Issues of 10.89) Supersedes DIN 57108/VDE 0108/12.79). 76 pp.
DIN VDE 0108 Pt 1 Suppl 1-89. Power Installations and Safety Power Supply in Communal Buildings; Building Regulations (Oct). 36 pp.
DIN VDE 0108 Pt 2-89. Power Installations and Safety Power Supply in Communal Buildings; Communal Facilities (Oct) (VDE 0107/11.89 with DIN VDE 0108 Part 1 to Part 8 (Issues of 10.89) Supersedes DIN 57 108/VDE 0108/12.79). 20 pp.
DIN VDE 0108 Pt 4-89. Power Installations and Safety Power Supply in Communal Buildings; High-Rise Buildings (Oct) (VDE 0107/11.89 with DIN VDE 0108 Part 1 to Part 8 (Issues of 10.89) Supersedes DIN 57108/VDE 0108/12.79). 5 pp.
DIN VDE 0108 Pt 7-89. Power Installations and Safety Power Supply in Communal Buildings; Working and Business Premises (Oct) (VDE 0107/ 11.89 with DIN VDE 0108 Part 1 to Part 8 (Issues of 10.89) Supersedes DIN 57108/VDE 0108/12.79). 9 pp.
IEC 364 Pt 1-92. Electrical Installations of Buildings Part 1: Scope, Object and Fundamental Principles Third Edition. 35 pp.
IEC 364 Pt 3-93. Electrical Installations of Buildings Part 3: Assessment of General Characteristics Second Edition. 63 pp.
IEC 364 Pt 5-51-79. Electrical Installations of Buildings Part 5: Selection and Erection of Electrical Equipment Chapter 51: Common Rules First Edition; (Amendment 1-1982) (Amendment 2-1993). 33 pp.
IEC 364 Pt 5-54-80. Electrical Installations of Buildings Part 5: Selection and Erection of Electrical Equipment Chapter 54: Earthing Arrangements and Protective Conductors First Edition; (Amendment 1-1982). 35 pp.
IEC 364 Pt 5-523-83. Electrical Installations of Buildings Part 5: Selection and Erection of Electrical Equipment Chapter 52: Wiring Systems Section 523—Current-Carrying Capacities First Edition. 72 pp.
IEC 364 Pt 7-701-84. Electrical Installations of Buildings Part 7: Requirements for Special Installations or Locations Section 701: Locations Containing a Bath Tub or Shower Basin First Edition. 20 pp.
IEC 364 Pt 7-704-89. Electrical Installations of Buildings Part 7: Requirements for Special Installations or Locations Section 704—Construction and Demolition Site Installations First Edition. 15 pp.
IEC 364 Pt 7-705-84. Electrical Installations of Buildings Part 7: Requirements for Special Installations or Locations Section 705—Electrical Installations of Agricultural and Horticultural Premises First Edition. 13 pp.
IEC 364 Pt 7-706-83. Electrical Installations of Buildings Part 7: Requirements for Special Installations or Locations Section 706—Restrictive Conducting Locations First Edition. 12 pp.

Electrical
INTERNATIONAL AND NON-U.S. NATIONAL STANDARDS
SUBJECT INDEX

Electrical Installations *(Cont.)*

—Buildings *(Cont.)*

SAA AS 1852.826-83. International Electrotechnical Vocabulary—Part 826: Electrical Installations of Buildings. 13 pp.

SAA AS 3000-91. Electrical Installations —Buildings, Structures and Premises (Known as the SAA Wiring Rules) Amdt 1 April 1992 Amdt 2 January/ February 1993 Amdt 3 May 1993 (in Professional Packages 17F, 21, 55) 68). 23 pp.

SAA AS 3000 DOC R/1-91. Electrical Installations —Buildings, Structures and Premises—Part R/1: Rulings to SAA Wiring Rules (AS 3000—1991) (Supersedes AS Doc 3000 R/1—1986, AS Doc 3000 R/2—1987, AS Doc 3000 R/3—1988, AS Doc 3000 R/4—1989, AS Doc 3000 R/5—1989, AS Doc 3000 R/6—1990 and. 61 pp.

SAA AS DOC 3000 R/7-91. Electrical Installations —Buildings, Structures and Premises—Part R/7: Rulings to SAA Wiring Rules (AS 3000—1986) Seventh Group (Superseded by AS 3000 Doc R/1—1991).

SNZ IEC 50: 50(826)-82. International Electrotechnical Vocabulary 50(826): Electrical Installations of Buildings. 24 pp.

—Buildings—Acceptance Testing

CENELEC HD 384.6.61 S1-92. Electrical Installations of Buildings Part 6: Verification Chapter 61: Initial Verification. 7 pp.

IEC 364 Pt 6-61-86. Electrical Installations of Buildings Part 6: Verification Chapter 61: Initial Verification First Edition; (Amendment 1-1993). 47 pp.

—Buildings—Construction Contracts

DIN ENGL 18382-88. Tendering and Performance Stipulations in Contracts for Construction Works (VOB); Part C: General Technical Specifications in Contracts for Construction Works (ATV); Electrical Installations in Buildings (Sept). 3 pp.

—Buildings—Construction Sites

CENELEC PRHD 384.7. 704 S1-91. Electrical Installations of Buildings Part 7: Requirements for Special Installations of Locations Section 704—Construction and Demolition Site Installations. 7 pp.

—Buildings—Demolition Sites

CENELEC PRHD 384.7. 704 S1-91. Electrical Installations of Buildings Part 7: Requirements for Special Installations of Locations Section 704—Construction and Demolition Site Installations. 7 pp.

—Buildings—Electric Shock Protection

IEC 364 Pt 4-481-93. Electrical Installations of Buildings Part 4: Protection for Safety Chapter 48: Choice of Protective Measures as a Function of External Influences Section 481—Selection of Measures for Protection Against Electric Shock in Relation to External Influences First Edition. 23 pp.

—Buildings—Fire Protection

IEC 364 Pt 4-482-82. Electrical Installations of Buildings Part 4: Protection for Safety Chapter 48: Choice of Protective Measures as a Function of External Influences Section 482-Protection Against Fire First Edition. 12 pp.

—Buildings—Safety

BSI PD 6535-93. 1993 Common Aspects for Installation and Equipment for Protection Against Electric Shock (IEC 1440: 1992) (E). 15 pp.

CENELEC HD 384.4.41-80. Electrical Installations of Buildings Part 4: Protection for Safety Chapter 41: Protection Against Electric Shock. 46 pp.

CENELEC HD 384.4.41/A1-80. AMD 1 Electrical Installations of Buildings Part 4: Protection for Safety Chapter 41: Protection Against Electric Shock. 2 pp.

CENELEC HD 384.4.41 S1/PRA1-93. AMD 1 Electrical Installations of Buildings Part 4: Protection for Safety Chapter 41: Protection Against Electric Shock. 13 pp.

CENELEC HD 384.4.42-85. Electrical Installations of Buildings Part 4: Protection for Safety Chapter 42: Protection Against Thermal Effects. 8 pp.

CENELEC HD 384.4.42 S1/A1-92. AMD 1 Electrical Installations of Buildings Part 4: Protection for Safety Chapter 42: Protection Against Thermal Effects. 5 pp.

CENELEC HD 384.4.43-78. Electrical Installations of Buildings Part 4: Protection for Safety Chapter 43: Protection Against Overcurrent. 19 pp.

CENELEC HD 384.4.45 S1-89. Electrical Installations of Buildings-Part 4: Protection for Safety. Chapter 45: Protection Against Undervoltage. 3 pp.

CENELEC HD 384.4.46 S1-87. Electrical Installatioans of Buildings—Part 4: Protection for Safety. Chapter 46: Isolation and Switching. 5 pp.

CENELEC HD 384.4.46 S1/A1-93. AMD 1 Electrical Installations of Buildings Part 4: Protection for Safety Chapter 46: Isolation and Switching. 5 pp.

Electrical Installations *(Cont.)*

—Buildings—Safety *(Cont.)*

CENELEC HD 384.4.47 S1-88. Electrical Installations of Buildings Part 4: Protection for Safety Chapter 47: Application of Protective Measures for Safety Section 470—General—Section 471—Measures of Protection Against Electric Shock. 5 pp.

CENELEC HD 384.4.473-78. Electrical Installations of Buildings Part 4: Protection for Safety Chapter 47: Application of Protective Measures Section 473: Protection Against Overcurrent. 2 pp.

CENELEC HD 384.5.56-85. Electrical Installations of Buildings Part 5: Selection and Erection of Electrical Equipment Chapter 56: Supplies for Safety Services. 8 pp.

IEC 364 Pt 4-41-92. Electrical Installations of Buildings Part 4: Protection for Safety Chapter 41: Protection Against Electric Shock Third Edition. 53 pp.

IEC 364 Pt 4-42-80. Electrical Installations of Buildings Part 4: Protection of Safety Chapter 42: Protection Against Thermal Effects First Edition. 12 pp.

IEC 364 Pt 4-43-77. Electrical Installations of Buildings Part 4: Protection for Safety Chapter 43: Protection Against Overcurrent First Edition. 14 pp.

IEC 364 Pt 4-45-84. Electrical Installations of Buildings Part 4: Protection for Safety Chapter 45: Protection Against Undervoltage First Edition. 9 pp.

IEC 364 Pt 4-46-81. Electrical Installations of Buildings Part 4: Protection for Safety Chapter 46: Isolation and Switching First Edition. 12 pp.

IEC 364 Pt 4-47-81. Electrical Installations of Buildings Part 4: Protection for Safety Chapter 47: Application of Protective Measures for Safety Section 470: General Section 471: Measures of Protection Against Electric Shock First Edition. 12 pp.

IEC 364 Pt 4-473-77. Electrical Installations of Buildings Part 4: Protection for Safety Chapter 47: Application of Protective Measures for Safety Section 473-Measures of Protection Against Overcurrent First Edition. 14 pp.

IEC 364 Pt 5-56-80. Electrical Installations of Buildings Part 5: Selection and Erection of Electrical Equipment Chapter 56: Safety Services First Edition. 12 pp.

—Buildings—Temporary

DIN VDE 0100 Pt 722-84. Erection of Power Installations with Rated Voltages up to 1000 V; Temporary Buildings, Vehicles for Traveling Exhibitions and Caravans (May). 25 pp.

DIN VDE 0108 Pt 8-89. Power Installations and Safety Power in Communal Buildings; Temporary Bldgs. Used in Communal Facilities, Stores and Shops, Exhibition Rooms, Public Houses and Restaurants (Oct) (VDE 0107/11.89 with DIN VDE 0108 Pt 1 to Pt 8(Issues of 10.89) Supersedes DIN 57108/VDE 0108/12.79). 11 pp.

—Cable Ties

CENELEC PREN 50146-93. Specification for Cable Ties for Electrical Installations. 17 pp.

—Cables (Electric)

DIN VDE 0100 Pt 730-86. Erection of Power Installations with Nominal Voltages up to 1000 V; Laying of Cables and Cords in Hollow Walls and in Buildings and Structures Made of Chiefly Combustible Building Materials According to DIN 4102 (Feb). 7 pp.

DIN VDE 0100 Pt 732-90. Erection of Power Installations with Nominal Voltages up to 1000 V; Cable Entries into Buildings in Public Cable Networks (Nov) (With DIN VDE 0211/12.85 Supersedes DIN 57 100 Part 732/VDE 0100 Part 732/03.83). 10 pp.

DIN VDE 0245 Pt 1 (D)-90. Cables & Cords for Electrical & Electronic Equipment in Power Installations; General Requirements (Oct). 15 pp.

DIN VDE 0298 Pt 1-82. Application of Cables and Flexible Cords in Power Installations; General Requirement for Cables with Rated Voltages Uo/U up to 18/30 kV (Nov). 29 pp.

DIN VDE 0619-87. Cable Entries for Cables and Cords; Requirements, Tests (Sept). 20 pp.

SAA AS 3000 Supp 1-91. Electrical Installations —Buildings, Structures and Premises—Supplement 1: Cable and Conductor Tables (Imperial Units) (Supplement to AS 3000—1991). 13 pp.

—Cables (Electric)—Length Measurement

DIN VDE 0100 Suppl. 5 (D)-88. Erection of Power Installations with Rated Voltages up to 1000 V; Maximum Permissible Lengths of Cables and Cords Taking into Consideration Protection Against Electric Shock in the Event of Fault, Short Circuit of Voltage Drop (Mar). 54 pp.

Electrical Installations *(Cont.)*

—Campgrounds

CENELEC HD 384.7.708 S1-92. Electrical Installations of Buildings Part 7: Requirements for Special Installations or Locations Section 708: Electrical Installations in Caravan Parks and Caravans. 12 pp.

DIN VDE 0100 Pt 708 (D)-90. Erection of Power Installations with Nominal Voltages up to 1000 V; Electrical Installations in Caravan Parks and Caravans; German Version prHD 384.7.708 (1989) (Mar). 27 pp.

DIN VDE 0100 Pt 721-84. Erection of Power Installations with Rated Voltages up to 1000 V; Caravans, Boats and Yachts, as Well as Power Supply Thereof at Camping Sites and Berths (Apr). 10 pp.

IEC 364 Pt 7-708-88. Electrical Installations of Buildings Part 7: Requirements for Special Installations or Locations—Section 708: Electrical Installations in Caravan Parks and Caravans First Edition; (Amendment 1-1993). 34 pp.

—Combat Vehicles—Standard Voltages

NATO STANAG 2601 ED 2 AMD 0-79. Standardization of Voltage of Electrical Systems in Tactical Vehicles. 9 pp.

NATO STANAG 2601 ED 3 AMD 0-00. Standardization of Electrical Systems in Tactical Vehicles. 5 pp.

—Construction—Controlgear—Fire Protection

IEC 364 Pt 5-53-86. Electrical Installations of Buildings Part 5: Selection and Erection of Electrical Equipment—Chapter 53: Switchgear and Controlgear First Edition; (Amendment 2-1992, Includes Amendment 1). 30 pp.

—Cords (Electric)

DIN VDE 0100 Pt 730-86. Erection of Power Installations with Nominal Voltages up to 1000 V; Laying of Cables and Cords in Hollow Walls and in Buildings and Structures Made of Chiefly Combustible Building Materials According to DIN 4102 (Feb). 7 pp.

DIN VDE 0245 Pt 1 (D)-90. Cables & Cords for Electrical & Electronic Equipment in Power Installations; General Requirements (Oct). 15 pp.

DIN VDE 0298 Pt 1-82. Application of Cables and Flexible Cords in Power Installations; General Requirement for Cables with Rated Voltages Uo/U up to 18/30 kV (Nov). 29 pp.

—Corrosion Prevention

DIN VDE 0150-83. Protection Against Corrosion Due to Stray Currents from DC Installations (Apr). 40 pp.

—Data Processing Equipment

IEC 364 Pt 7-707-84. Electrical Installations of Buildings Part 7: Requirements for Special Installations or Locations Section 707—Earthing Requirements for the Installation of Data Processing Equipment First Edition. 23 pp.

—Electric Conductors

SAA AS 3000 Supp 1-91. Electrical Installations —Buildings, Structures and Premises—Supplement 1: Cable and Conductor Tables (Imperial Units) (Supplement to AS 3000—1991). 13 pp.

—Electrical Codes

CSA C22.1-1990. Safety Standard for Electrical Installations. 568 pp.

CSA C22.1EP-1990. What's New in 1990 Changes to CE Code Part I Explained. 143 pp.

CSA C22.10-1992. Canadian Electrical Code Part I (16th Edition) Plus Quebec Amendments. 584 pp.

—Electrical Grounding

DIN VDE 0100 Pt 540-91. Installation of Power Plant with Rated Voltages Not Exceeding 1000 V; Selection and Insallation of Electrical Equipment; Earthing Arrangements, Protective Conductors and Equipotential Bonding Conductors (Nov). 37 pp.

DIN VDE 0100 Pt 540 A2 (D)-92. Erection of Power Installations with Nominal Voltages up to 1000 V; Selection and Erection of Equipment; Earthing Arrangements, Protective Conductors, Equipotential Bonding Conductors; Amendment 2 Identical with IEC 64 (Sec)570 (Jan). 8 pp.

DIN VDE 0141-89. Earthing Systems for Power Installations with Rated Voltages Above 1kV (July). 81 pp.

—Electromagnetic Compatibility

DIN VDE 0839 Pt 10 (D)-91. Electromagnetic Compatibility (EMC); Measurement of Immunity to Conducted Interference and Fields (Dec). 31 pp.

—Electronic Equipment

DIN VDE 0160-88. Electronic Equipment for Use in Electrical Power Installations and Their Assembly into Electrical Power Installations (May). 158 pp.

INTERNATIONAL AND NON-U.S. NATIONAL STANDARDS
SUBJECT INDEX
Electrical

Electrical Installations *(Cont.)*
—**Electronic Equipment** *(Cont.)*
DIN VDE 0160 (D)-90. Assembly of Electronic Equipment in Power Installations (Dec). 193 pp.
DIN VDE 0160 A1-89. Electronic Equipment for Use in Electrical Power Installations and Their Assembly into Electrical Power Installations: Amendment 1 (Apr). 11 pp.
DIN VDE 0160 A2 (D)-88. Electronic Equipment for Use in Electrical Power Installations and Their Assembly into Electrical Power Installations: Amendment 2 (Oct). 32 pp.

—**Electrostatic Precipitators**
DIN VDE 0146-80. Erection of Electrostatic-Precipitators (Electrostatic Filters) (VDE-Specification) (Mar). 18 pp.

—**Explosive Atmospheres**
DIN VDE 0105 Pt 7-87. Operation of Power Installations; Supplementary Requirements for Atmospheres Endangered by Potentially Explosive Material (Dec). 6 pp.
DIN VDE 0105 Pt 9-86. Operation of Power Installations Part 9: Supplementary Requirements for Potentially Explosive Atmospheres (May). 10 pp.
DIN VDE 0165-91. Erection of Electrical Installations in Potentially Explosive Atmospheres (Feb) (Supersedes DIN VDE 0165 A1/09.86). 65 pp.
DIN VDE 0166-81. Electrical Installations and Equipment for Use in Atmospheres Potentially Endangered by Explosive Material (May). 28 pp.
IEC 79 Pt 14-84. Electrical Apparatus for Explosive Gas Atmospheres Part 14: Electrical Installations in Explosive Gas Atmospheres (Other Than Mines) First Edition. 46 pp.
IEC 79 Pt 17-90. Electrical Apparatus for Explosive Gas Atmospheres Part 17: Recommendations for Inspection and Maintenance of Electrical Installations in Hazardous Areas (Other Than Mines) First Edition. 40 pp.

—**Fire Extinguishers**
CEN EN 3-2-78. Portable Fire Extinguishers—Part 2. 7 pp.
CEN PREN 3-2-93. Fight Against Fire—Portable Fire Extinguishers—Part 2: Retention of Charge, Dielectric Test, Compaction, Special Provision. 9 pp.

—**Fire Hazards**
DIN VDE 0100 Pt 720-83. Erection of Power Installations with Rated Voltages up to 1000 V; Locations Exposed to Fire Hazards (Mar). 15 pp.

—**Fire Protection**
DIN VDE 0100 Pt 420-91. Installation of Power Plant with Rated Voltages up to 1000 V; Protective Measures; Protection Against Thermal Effects (Nov) (Supersedes Din 57 100 Part 420 Refer to Transitional Period). 8 pp.

—**Firesafety**
CENELEC PRHD 384-4-48-92. Electrical Installations of Buildings Part 4: Protection for Safety Chapter 48: Choice of Protective Measures as a Function of External Influences Section 482: Protection Against Fire. 8 pp.

—**Furniture**
DIN VDE 0100 Pt 724-80. Erection of Power Installations with Rated Voltages up to 1000 V; Electrical Equipment in Furniture and Similar Fitments e.g. Curtain-Ledges, Decorative Coverings (June). 7 pp.

—**Gliders**
CAA JAR-22 SUBPART F. Equipment (Joint Airworthiness Requirements). 4 pp.

—**Glossaries**
CENELEC HD 384.2-86. International Electrotechnical Vocabulary Chapter 826: Electrical Installations of Buildings. 2 pp.
CENELEC HD 384.2 S1/A1-92. AMD 1 International Electrotechnical Vocabulary (IEV) Chapter 826: Electrical Installations of Buildings (IEC 50(826):1982/A1: 1990). 4 pp.
DIN VDE 0100 Pt 200 A1 (D)-88. Installation of Power Plant with Nominal Voltages up to 1000 V; Standard Terminology; Amendment 1 (Dec). 5 pp.
DIN VDE 0100 Pt 200 A2 (D)-89. Erection of Power Installations with Nominal Voltages up to 1000 V; Definitions; Amendment 2; Identical with IEC 64(CO)172 (Feb). 6 pp.
IEC. Electricity, Electronics and Telecommunications Multilingual Dictionary Volume 1: English-French-Russian-German-Spanish-Dutch-Italian-Swedish-Polish Second Edition. 1944 pp.
IEC. Electricity, Electronics and Telecommunications Multilingual Dictionary Volume 2: English-French-Russian-German-Spanish-Dutch-Italian-Swedish-Polish Second Edition. 964 pp.
IEC. Electricity, Electronics and Telecommunications Multilingual Dictionary Volume 3: German, Dutch and Swedish Indexes Second Edition. 793 pp.

Electrical Installations *(Cont.)*
—**Glossaries** *(Cont.)*
IEC. Electricity, Electronics and Telecommunications Multilingual Dictionary Volume 4: Spanish and Italian Indexes Second Edition. 540 pp.
IEC. Electricity, Electronics and Telecommunications Multilingual Dictionary Volume 5: Russian and Polish Indexes Second Edition. 578 pp.
IEC 50 Chap 826-82. International Electrotechnical Vocabulary Chapter 826: Electrical Installations of Buildings First Edition; (Amendment 1-1990) (CENELEC HD 384.2 S1/A1:1992). 48 pp.
IEC 364 Pt 2-21-93. Electrical Installations of Buildings Part 2: Definitions Chapter 21: Guide to General Terms First Edition. 19 pp.
SAA AS 1852.826-83. International Electrotechnical Vocabulary—Part 826: Electrical Installations of Buildings. 13 pp.
SNZ IEC 50: 50(826)-82. International Electrotechnical Vocabulary 50(826): Electrical Installations of Buildings. 24 pp.

—**Ground Effect Machines**
CAA Chapter B6-7 01.91. Electrical Systems. 8 pp.

—**Grounding Conductors**
CSA C22.2 NO 41-M1987. Grounding and Bonding Equipment (R 1993); (Gen Instr 1). 30 pp.

—**Harmonic Analysis—Safety**
CSA CAN/CSA-C22. 2 NO.16-M92. Measurement of Harmonic Currents; (Gen Instr 1 Thru 2). 26 pp.

—**Hoists**
DIN VDE 0100 Pt 726-90. Erection of Power Installations with Nominal Voltages up to 1000 V; Lifting and Hoisting Devices (Mar). 35 pp.

—**Hospitals**
DIN VDE 0107-89. Electrical Installations in Hospitals and Locations for Medical Use Outside Hospitals (Nov) (Supersedes DIN 57107 A1/VDE 0107 A1/11.82 with DIN VDE 0108 Part 1 to Part 8 (Issues of 10.89) Supersedes DIN 57108/VDE 0108/12.79). 55 pp.
DIN VDE 0107 Suppl 1-89. Electrical Installations in Hospitals and Locations for Medical Use Outside Hospitals; Extracts from Building and Industrial Safety Regulations (Nov). 64 pp.
JIS T 1022-82. Safety Requirements of Electrical Installations for Medically Used Rooms in Hospitals and Clinics. 19 pp.

—**Hospitals—Safety**
CEN PREN 793-92. Medical Electrical Equipment—Particular Requirements for Safety of Medical Supply Units. 43 pp.

—**Hotels**
DIN VDE 0108 Pt 5-89. Power Installations and Safety Power Supply in Communal Buildings; Restaurants (Oct) (VDE 0107/11.89 with DIN VDE 0108 Part 1 to Part 8 (Issues of 10.89) Supersedes DIN 57108/VDE 0108/12.79.). 7 pp.

—**Humid Locations**
DIN VDE 0100 Pt 737-90. Erection of Power Installations with Rated Voltages up to 1000 V; Humid and Wet Areas and Locations; Outdoor Installations (Nov). 9 pp.

—**Isolating Switches**
DIN VDE 0100 Pt 460-88. Erection of Power Installations with Nominal Voltages up to 1000 V; Protective Measures; Isolation and Switching (Oct) (Super-sedes VDE 0100/05.73 Sec 31 b)2, VDE 0100g/07. 76 Sec 31 a)1.2 and DIN 57100 Part 410/VDE 0100 Part 410/11.83, Sub-clause 9. 12 pp.
DIN VDE 0100 Pt 537-88. Erection of Power Installations with Nominal Voltages up to 1000 V; Selection and Erection of Electrical Equipment; Devices for Isolation and Switching (Oct) (Supersedes VDE 0100/05.73 Sec 31 b)1, VDE 0100g/07.76 Sec 31 a)1.1.1.3. With DIN VDE 0100 Part 460/10.88. 12 pp.

—**Lifting Equipment**
DIN VDE 0100 Pt 726-90. Erection of Power Installations with Nominal Voltages up to 1000 V; Lifting and Hoisting Devices (Mar). 35 pp.

—**Lighting**
DIN VDE 0100 Pt 559-83. Installation of Power Plant with Rated Voltages up to 1000 V; Luminaires and Lighting Equipment (Mar). 15 pp.

—**Low Voltage**
BSI BS EN 50065-1-92. 1992 Signalling on Low-Voltage Electrical Installations in the Frequency Range 3 kHz to 148.5 kHz Part 1: General Requirements, Frequency Bands and Electromagnetic Disturbances. 15 pp.

Electrical Installations *(Cont.)*
—**Low Voltage** *(Cont.)*
BSI BS EN 50065-1-01. 1992 Amd 1 Signalling on Low-Voltage Electrical Installations in the Frequency Range 3 kHz to 148.5 kHz Part 1: General Requirements, Frequency Bands and Electromagnetic Disturbances (AMD 7950) September 15, 1993 (S). 24 pp.
CENELEC EN 50 065-1-91. Signalling on Low-Voltage Electrical Installations in the Frequency Range 3/kHz to 148.5 kHz Part 1: General Requirements, Frequency Bands and Electromagnetic Disturbances. 23 pp.
CENELEC EN 50065-1/A1-92. AMD 1 Signalling on Low-Voltage Electrical Installations in the Frequency Range 3 khz to 148.5 khz Part 1: General Requirements, Frequency Bands and Electromagnetic Disturbances. 11 pp.

—**Low Voltage—Overvoltage Protection**
CENELEC PRHD 384.4. 442 S1-92. Electrical Installations of Buildings Part 4: Protection for Safety Chapter 44: Protection Against Overvoltages Section 442: Protection of Low-Voltage Installations Against Faults Between High Voltage Systems and Earth. 22 pp.
SAA AS 4070-92. Protection of Low-Voltage Electrical Installations and Equipment in MEN Systems from Transient Overvoltages. 3 pp.

—**Marinas**
DIN VDE 0100 Pt 721-84. Erection of Power Installations with Rated Voltages up to 1000 V; Caravans, Boats and Yachts, as Well as Power Supply Thereof at Camping Sites and Berths (Apr). 10 pp.

—**Medical Equipment**
DIN VDE 0107 (S)-81. Erection and Testing of Electrical Installations in Medically Used Rooms (June). 56 pp.

—**Medical Equipment—Safety**
BSI BS 5724: Sec 2.1: Supp 1-91. 1991 Medical Electrical Equipment Part 2: Particular Requirements for Safety Section 2.1: Medical Electron Accelerators in the Range 1 MeV to 50 MeV Supplement No. 1: Revised and Additional Text. 26 pp.

—**Mines**
BSI BS 6907: Part 2-88. 1988 Electrical Installations for Open-Cast Mines and Quarnes Part 2: General Recommendations for Protection Against Direct Contact and Electric Shock. 22 pp.
BSI BS 6907: Part 3-88. 1988 Electrical Installations for Open-Cast Mines and Quarnes Part 3: Recommendations for Equipment and Ancillaries. 12 pp.
BSI BS 6907: Part 4-88. 1988 Electrical Installations for Open-Cast Mines and Quarnes Part 4: Recommendations for Winning, Stadeing and Processing Machinery, Pumps and Low Signal Level and Communications Systems. 14 pp.
BSI BS 6907: Part 5-88. 1988 Electrical Installations for Open-Cast Mines and Quarnes Part 5: Recommendations for Operation. 8 pp.
DIN VDE 0105 Pt 11-90. Operation of Electrical Installations; Particularly Requirements for Underground Areas (Sept). 75 pp.
DIN VDE 0118 Pt 1-90. Erection of Electrical Installations in Mines; General Requirements (Sept) (with DIN VDE 0018 Part 2/09.90 and DIN VDE 0118 Part 3/09.90 Replacement for VDE 0118/02.70 and VDE 0118b/05.72). 105 pp.
DIN VDE 0118 Pt 2-90. Erection of Electrical Installations in Mines. Supplementary Requirements for Power Installations (Sept) (with DIN VDE 0118 Part 1 and DIN VDE 0118 Part 3 Replacement for VDE 0118 and VDE 0118 B See Transition Period.). 83 pp.
DIN VDE 0168-92. Erection of Electrical Installation in Open-Cast Mines, Quarries and Similar Works (Jan). 57 pp.

—**Mines—Glossaries**
BSI BS 6907: Part 1-88. 1988 Electrical Installations for Open-Cast Mines and Quarnes Part 1: Glossary. 10 pp.

—**Naval Ships**
MOD UK NES 1010-92. Weatherdeck Installation Requirements for Electromagnetic Environmental Control Issue 2 (10.92). 26 pp.
MOD UK NES 1027-87. Electromagnetic Compatibility Requirements for Installations in Ships Issue 1 (05.87). 32 pp.
MOD UK NES 502-90. Requirements for Electrical Installations Issue 4 (10.90). 75 pp.
MOD UK NES 502-01. Requirements for Electrical Installations Issue 4 (10.90); Amendment 1. 85 pp.
MOD UK NES 502-02. Requirements for Electrical Installations Issue 4 (10.90); Amendment 2. 86 pp.

INDUSTRY STANDARDS

Electrical Installations (Cont.)

—Naval Ships (Cont.)
MOD UK NES 529: Part 2-91. Nuclear Hardening Guide Part 2: Electrical Installation in Surface Ships Issue 1 (11.91). 35 pp.

—Naval Ships—Electromagnetic Compatibility
MOD UK NES 515-90. Guide to Electromagnetic Compatibility Issue 2 (09.90). 152 pp.

MOD UK NES 515-01. Guide to Electromagnetic Compatibility Issue 2 (09.90); Amendment 1. 157 pp.

—Office Furniture
BSI BS 6396-90. 1990 Electrical Systems in Office Furniture and Office Screens. 9 pp.

BSI BS 6396-01. 1990 Amd 1 Electrical Systems in Office Furniture and Office Screens (AMD 7170) July 15, 1992. 12 pp.

—Offshore Platforms
IEC 92 Pt 505-84. Electrical Installations in Ships Part 505: Special Features—Mobile Offshore Drilling Units Third Edition; (Amendment 1-1993). 37 pp.

—Operating Requirements
DIN VDE 0105 Pt 1-83. Operation of Electrical Power Installations; General Requirements (July). 105 pp.

—Operating Requirements—Explosive Atmospheres
DIN VDE 0105 Pt 7-87. Operation of Power Installations; Supplementary Requirements for Atmospheres Endangered by Potentially Explosive Material (Dec). 6 pp.

DIN VDE 0105 Pt 9-86. Operation of Power Installations Part 9: Supplementary Requirements for Potentially Explosive Atmospheres (May). 10 pp.

—Outdoor
BSI BS 6907: Part 2-88. 1988 Electrical Installations for Open-Cast Mines and Quarries Part 2: General Recommendations for Protection Against Direct Contact and Electric Shock. 22 pp.

BSI BS 6907: Part 3-88. 1988 Electrical Installations for Open-Cast Mines and Quarries Part 3: Recommendations for Equipment and Ancillaries. 12 pp.

BSI BS 6907: Part 4-88. 1988 Electrical Installations for Open-Cast Mines and Quarries Part 4: Recommendations for Winning, Stadeing and Processing Machinery, Pumps and Low Signal Level and Communications Systems. 14 pp.

BSI BS 6907: Part 5-88. 1988 Electrical Installations for Open-Cast Mines and Quarries Part 5: Recommendations for Operation. 8 pp.

DIN VDE 0100 Pt 737-90. Erection of Power Installations with Rated Voltages up to 1000 V; Humid and Wet Areas and Locations; Outdoor Installations (Nov). 9 pp.

IEC 621 Pt 1-87. Electrical Installations for Outdoor Sites Under Heavy Conditions (Including Open-Cast Mines and Quarries) Part 1: Scope and Definitions Second Edition. 26 pp.

IEC 621 Pt 2-87. Electrical Installations for Outdoor Sites Under Heavy Conditions (Including Open-Cast Mines and Quarries) Part 2: General Protection Requirements Second Edition. 106 pp.

IEC 621 Pt 3-79. Electrical Installations for Outdoor Sites Under Heavy Conditions (Including Open-Cast Mines and Quarries) Part 3: General Requirements for Equipment and Ancillaries First Edition; (Amendment 1-1986). 58 pp.

IEC 621 Pt 4-81. Electrical Installations for Outdoor Sites Under Heavy Conditions (Including Open-Cast Mines and Quarries) Part 4: Requirements for the Installation First Edition; (Amendment 1-1987). 60 pp.

IEC 621 Pt 5-87. Electrical Installations for Outdoor Sites Under Heavy Conditions (Including Open-Cast Mines and Quarries) Part 5: Operating Requirements First Edition. 26 pp.

—Outdoor—Glossaries
BSI BS 6907: Part 1-88. 1988 Electrical Installations for Open-Cast Mines and Quarries Part 1: Glossary. 10 pp.

—Outlets
DIN VDE 0100 Pt 550-88. Erection of Power Installations with Nominal Voltages up to 1000 V; Selection and Erection of Electrical Equipment, Plug-And-Socket Devices, Switches and Installation Accessories (Apr) (Supersedes VDE 0100g/07.76 and VDE 0100/05.73). 9 pp.

—Outpatient Clinics
DIN VDE 0107-89. Electrical Installations in Hospitals and Locations for Medical Use Outside Hospitals (Nov) (Supersedes DIN 57107 A1/VDE 0107 A1/11.82 with DIN VDE 0108 Part 1 to Part 8 (Issues of 10.89) Supersedes DIN 57108/VDE 0108/12.79). 55 pp.

DIN VDE 0107 Suppl 1-89. Electrical Installations in Hospitals and Locations for Medical Use Outside Hospitals; Extracts from Building and Industrial Safety Regulations (Nov). 64 pp.

—Overvoltage Protection
DIN VDE 0100 Pt 443 (D)-87. Erection of Power Installations with Nominal Voltages up to 1000 V; Protective Measures; Protection Against Overvoltages of Atmospheric Origin; Identical with IEC 64 (Central Office) 168 (Apr). 16 pp.

—Panel Meters
DIN VDE 0603-86. Small Distribution Boards and Meter Panels up to 250 V to Earth (Dec). 36 pp.

DIN VDE 0603 A1 (D)-89. Small Distribution Boards and Meter Panels up to 250 V to Earth; Amendment 1 (Feb). 7 pp.

—Partitions (Walls)
BSI BS 6396-90. 1990 Electrical Systems in Office Furniture and Office Screens. 9 pp.

BSI BS 6396-01. 1990 Amd 1 Electrical Systems in Office Furniture and Office Screens (AMD 7170) July 15, 1992. 12 pp.

—Planning
DIN VDE 0100 Pt 300-85. Erection of Power Installations with Rated Voltages up to 1000 V; General Data Concerning Planning of Electrical Installations (Nov). 12 pp.

—Quarries
BSI BS 6907: Part 2-88. 1988 Electrical Installations for Open-Cast Mines and Quarries Part 2: General Recommendations for Protection Against Direct Contact and Electric Shock. 22 pp.

BSI BS 6907: Part 3-88. 1988 Electrical Installations for Open-Cast Mines and Quarries Part 3: Recommendations for Equipment and Ancillaries. 12 pp.

BSI BS 6907: Part 4-88. 1988 Electrical Installations for Open-Cast Mines and Quarries Part 4: Recommendations for Winning, Stadeing and Processing Machinery, Pumps and Low Signal Level and Communications Systems. 14 pp.

BSI BS 6907: Part 5-88. 1988 Electrical Installations for Open-Cast Mines and Quarries Part 5: Recommendations for Operation. 8 pp.

—Quarries—Glossaries
BSI BS 6907: Part 1-88. 1988 Electrical Installations for Open-Cast Mines and Quarries Part 1: Glossary. 10 pp.

—Radio Communications—Radio Frequency Interference
CCIR QUESTION 4-3/1-86. Limitation of Unwanted Radiation from Electrical Apparatus and Installations—Questions Concerning Study Group 1—Spectrum Management Techniques (Spectrum Engineering, Planning, Sharing, Monitoring and Utilization). 1 p.

—Railroad Traction Equipment
DIN VDE 0115 Pt 2-82. Traction Systems; Particular Requirements for Vehicles and Their Equipment (June). 29 pp.

DIN VDE 0115 Pt 3-82. Traction Systems; Particular Requirements for Stationary Installations (June). 43 pp.

—Recreational Vehicle Parks
CENELEC HD 384.7.708 S1-92. Electrical Installations of Buildings Part 7: Requirements for Special Installations or Locations Section 708: Electrical Installations in Caravan Parks and Caravans. 12 pp.

DIN VDE 0100 Pt 708 (D)-90. Erection of Power Installations with Nominal Voltages up to 1000 V; Electrical Installations in Caravan Parks and Caravans; German Version prHD 384.7.708 (1989) (Mar). 27 pp.

DIN VDE 0100 Pt 721-84. Erection of Power Installations with Rated Voltages up to 1000 V; Caravans, Boats and Yachts, as Well as Power Supply Thereof at Camping Sites and Berths (Apr). 10 pp.

IEC 364 Pt 7-708-88. Electrical Installations of Buildings Part 7: Requirements for Special Installations or Locations—Section 708: Electrical Installations in Caravan Parks and Caravans First Edition; (Amendment 1-1993). 34 pp.

—Recreational Vehicles
CENELEC HD 23-87. Electrical Installations for Caravans with Supply Voltages up to 250 V. 3 pp.

CENELEC HD 384.7.708 S1-92. Electrical Installations of Buildings Part 7: Requirements for Special Installations or Locations Section 708: Electrical Installations in Caravan Parks and Caravans. 12 pp.

DIN VDE 0100 Pt 708 (D)-90. Erection of Power Installations with Nominal Voltages up to 1000 V; Electrical Installations in Caravan Parks and Caravans; German Version prHD 384.7.708 (1989) (Mar). 27 pp.

DIN VDE 0100 Pt 721-84. Erection of Power Installations with Rated Voltages up to 1000 V; Caravans, Boats and Yachts, as Well as Power Supply Thereof at Camping Sites and Berths (Apr). 10 pp.

DIN VDE 0100 Pt 722-84. Erection of Power Installations with Rated Voltages up to 1000 V; Temporary Buildings, Vehicles for Traveling Exhibitions and Caravans (May). 25 pp.

IEC 364 Pt 7-708-88. Electrical Installations of Buildings Part 7: Requirements for Special Installations or Locations—Section 708: Electrical Installations in Caravan Parks and Caravans First Edition; (Amendment 1-1993). 34 pp.

—Restaurants
DIN VDE 0108 Part 5-89. Power Installations and Safety Power Supply in Communal Buildings; Restaurants (Oct) (VDE 0107/11.89 with DIN VDE 0108 Part 1 to Part 8 (Issues of 10.89) Supersedes DIN 57108/VDE 0108/12.79.). 7 pp.

—Safety
SAA AS 4048.1 (INT)-92. Functional Safety of Electrical/Electronic/Programmable Electronic Systems—Generic Aspects—Part 1(Int): General Requirements (IEC 65A(Secretariat)123). 140 pp.

SNZ NZS 6200-88. Specification for General Requirements for Electrical Apparatus and Material. 64 pp.

—Saunas
CENELEC HD 384.7.703 S1-91. Electrical Installations of Buildings Part 7: Requirements for Special Installations or Locations Section 703- Locations Containing Sauna Heaters. 6 pp.

IEC 364 Pt 7-703-84. Electrical Installations of Buildings Part 7: Requirements for Special Installations or Locations Section 703—Locations Containing Sauna Heaters First Edition. 12 pp.

—Ships
IEC 92 Pt 101-80. Electrical Installations in Ships Part 101: Definitions and General Requirements Third Edition; (Amendment 1-1984) (Amendment 2-1987). 54 pp.

IEC 92 Pt 201-80. Electrical Installations in Ships Part 201: System Design—General Third Edition; (Amendment 5-1990, Incorporating Amendments 1 Thru 4). 89 pp.

IEC 92 Pt 401-80. Electrical Installations in Ships Part 401: Installation and Test of Completed Installation Third Edition; (Amendment 1-1987). 70 pp.

IEC 92 Pt 503-75. Electrical Installations in Ships Special Features A.C. Supply Systems with Voltages in the Range Above 1 kV up to and Including 11 kV First Edition. 21 pp.

IEC 533-77. Electromagnetic Compatibility of Electrical and Electronic Installations in Ships First Edition. 88 pp.

JIS F 8061-86. Electrical Installations in Ships Part 101 Definitions and General Requirements.

JIS F 8062-86. Electrical Installations in Ships Part 201 System Design—General (IEC 92-201-1980).

JIS F 8072-86. Electrical Installations in Ships Part 401 Installation and Test of Completed Installation (IEC 92-401-1980).

JIS F 8073-86. Electrical Installations in Ships Part 501 Special Features—Electric Propulsion Plant (IEC 92-501-1894).

JIS F 8076-86. Electrical Installations in Ships Part 504 Special Features—Control and Instrumentation (IEC 92-504-1974).

JIS F 8077-86. Electrical Installations in Ships Part 505 Special Features—Mobile Offshore Drilling Units (IEC 92-505-1984).

—Ships—Codes
CNS F5118-87. Device Function Numbers for Control Devices and Equipment of Electrical Installations in Ships (Jun)(11996).

—Ships—Glossaries
JIS F 8061-86. Electrical Installations in Ships Part 101 Definitions and General Requirements.

—Short Circuit Currents
DIN VDE 0103-88. Mechanical and Thermal Short-Circuit Protection of Electrical Power Installations (Apr). 65 pp.

INTERNATIONAL AND NON-U.S. NATIONAL STANDARDS
SUBJECT INDEX
Electrical

Electrical Installations *(Cont.)*

—**Skirting Board Ducts**
DIN VDE 0604 Pt 3-86. Trunking on Mounted Walls and Ceilings for Electrical Installations; Skirting Board Ducts (May). 6 pp.

—**Storage Batteries**
SAA AS 3011.2-92. Electrical Installations —Secondary Batteries Installed in Buildings—Part 2: Sealed Cells (in Professional Package 24). 7 pp.

—**Submarines**
MOD UK NES 1010-92. Weatherdeck Installation Requirements for Electromagnetic Environmental Control Issue 2 (10.92). 26 pp.
MOD UK NES 1027-87. Electromagnetic Compatibility Requirements for Installations in Ships Issue 1 (05.87). 32 pp.
MOD UK NES 502-90. Requirements for Electrical Installations Issue 4 (10.90). 75 pp.
MOD UK NES 502-01. Requirements for Electrical Installations Issue 4 (10.90); Amendment 1. 85 pp.
MOD UK NES 502-02. Requirements for Electrical Installations Issue 4 (10.90); Amendment 2. 86 pp.

—**Submarines—Electromagnetic Compatibility**
MOD UK NES 515-90. Guide to Electromagnetic Compatibility Issue 2 (09.90). 152 pp.
MOD UK NES 515-01. Guide to Electromagnetic Compatibility Issue 2 (09.90); Amendment 1. 157 pp.

—**Supersonic Transports**
CAA STANDARD NO. 8-0&AP 06.69. Electrical Systems—Definitions and General. 12 pp.
CAA STANDARD NO. 8-5 05.73. Electrical Systems—Tests. 4 pp.

—**Swimming Pools**
CENELEC HD 384.7.702 S1-91. Electrical Installations of Buildings Part 7: Requirements for Special Installations or Locations Section 702—Swimming Pools. 10 pp.
CENELEC HD 384.7.702-91. CORRIGENDUM Electrical Installations of Buildings Part 7: Requirements for Special Installations or Locations Section 702—Swimming Pools. 2 pp.
IEC 364 Pt 7-702-83. Electrical Installations of Buildings Part 7: Requirements for Special Installations or Locations Section 702—Swimming Pools First Edition. 16 pp.

—**Switchboards**
DIN VDE 0100 Pt 729-86. Erection of Power Installation with Nominal Voltages up to 1000 V; Installation and Connection of Switchgear and Distribution Boards (Nov) (Supersedes in Part VDE 0100/05.73). 11 pp.
DIN VDE 0603-86. Small Distribution Boards and Meter Panels up to 250 V to Earth (Dec). 36 pp.
DIN VDE 0603 A1 (D)-89. Small Distribution Boards and Meter Panels up to 250 V to Earth; Amendment 1 (Feb). 7 pp.

—**Switchboards—Safety**
DIN VDE 0100 Pt 736-83. Erection of Power Installations with Rated Voltages up to 1000 V; Low Voltage Circuits in High Voltage Switchboards (Nov). 6 pp.

—**Switches**
DIN VDE 0100 Pt 550-88. Erection of Power Installations with Nominal Voltages up to 1000 V; Selection and Erection of Electrical Equipment, Plug-And-Socket Devices, Switches and Installation Accessories (Apr) (Supersedes VDE 0100g/07.76 and VDE 0100/05.73). 9 pp.

—**Switchgear**
DIN VDE 0100 Pt 729-86. Erection of Power Installation with Nominal Voltages up to 1000 V; Installation and Connection of Switchgear and Distribution Boards (Nov) (Supersedes in Part VDE 0100/05.73). 11 pp.

—**Switchgear—Construction—Fire Protection**
IEC 364 Pt 5-53-86. Electrical Installations of Buildings Part 5: Selection and Erection of Electrical Equipment—Chapter 53: Switchgear and Controlgear First Edition; (Amendment 2-1992, Includes Amendment 1). 30 pp.

—**Switching Devices**
DIN VDE 0100 Pt 460-88. Erection of Power Installations with Nominal Voltages up to 1000 V; Protective Measures; Isolation and Switching (Oct) (Super-sedes VDE 0100/05.73 Sec 31 b)2, VDE 0100g/07. 76 Sec 31 a)1.2 and DIN 57100 Part 410/VDE 0100 Part 410/11.83, Sub-clause 9. 12 pp.

Electrical Installations *(Cont.)*

—**Switching Devices** *(Cont.)*
DIN VDE 0100 Pt 537-88. Erection of Power Installations with Nominal Voltages up to 1000 V; Selection and Erection of Electrical Equipment; Devices for Isolation and Switching (Oct) (Supersedes VDE 0100/05.73 Sec 31 b)1, VDE 0100g/07.76 Sec 31 a)1.1.1.3. With DIN VDE 0100 Part 460/10.88. 12 pp.

—**Tankers**
IEC 92 Pt 502-80. Electrical Installations in Ships Part 502: Special Features—Tankers Third Edition. 48 pp.
JIS F 8074-86. Electrical Installations in Ships Part 502 Special Features—Tankers (IEC 92-502-1980). 4 pp.

—**Telecommunication Systems—Radio Frequency Interference**
DIN VDE 0228 Pt 2-87. Measures to Be Taken Against Interference with Telecommunication Systems by Electric Power Installations: Part 2: Interference by Three-Phase Systems (Dec). 55 pp.
DIN VDE 0873 Pt 2 (D)-88. Measures Against Radio Interference from Electric Utility Plants and Electric Traction Systems; Radio Interference from Systems Below 10 kv and from Electric Trains (Oct) (Intended as a Replacement for DIN 57873 Part 2). 21 pp.

—**Telecommunications—Semiconductor Arrester Assemblies**
CCITT RECMN K.28-91. Characteristics of Semi-Conductor Arrester Assemblies for the Protection of Telecommunications Installations (Study Group V) 15 pp. 15 pp.

—**Telephone Cables—Ground Fault Protection**
CCITT RECMN K.8-89. Separation in the Soil Between Telecommunication Cables and Earthing System of Power Facilities—Protection Against Interference (Study Group V) 5 pp. 5 pp.

—**Telephone Cables—Railroad Tracks—Ground Fault Protection**
CCITT RECMN K.9-89. Protection of Telecommunication Staff and Plant Against a Large Earth Potential Due to a Neighbouring Electric Traction Line (Study Group V) 2 pp. 2 pp.

—**Thermal Protection**
DIN VDE 0100 Pt 420-91. Installation of Power Plant with Rated Voltages up to 1000 V; Protective Measures; Protection Against Thermal Effects (Nov) (Supersedes Din 57 100 Part 420 Refer to Transitional Period). 8 pp.

—**Voltage Bands**
CENELEC HD 193 S2-82. Voltage Bands for Electrical Installations of Buildings. 2 pp.
IEC 449-93. Voltage Bands for Electrical Installations of Buildings First Edition; (Amendment 1-1979). 13 pp.

—**Wet Locations**
DIN VDE 0100 Pt 737-90. Erection of Power Installations with Rated Voltages up to 1000 V; Humid and Wet Areas and Locations; Outdoor Installations (Nov). 9 pp.

Electrical Insulating Boards
Use: Electrical Insulating Papers

Electrical Insulating Bushings
Use For: Insulating Bushings (Electrical); Insulator Bushings *See Also:* Bushings; Electrical Insulation

—**Electric Conduits**
CNS C4239-86. Insulated Bushings for Rigid Steel Conduit (Oct)(6095).
CNS C4247-83. Insulated Bushings for Flexible Metal Conduit (Oct)(6103).

—**Electronic Equipment**
BSI BS 7039-89. 1989 Recommendations for Dimensions of Spindle Ends and Bushes, and for Dimensions of the Mounting of Single-Hole, Bush-Mounted Spindle-Operated Variable Capacitors, Variable Resistors and Potentio-meters for Electronic Equipment. 16 pp.
IEC 915-87. Capacitors and Resistors for Electronic Equipment Preferred Dimensions of Spindle Ends, Bushes and for the Mounting of Single-Hole, Bush-Mounted, Spindle-Operated Electronic Components First Edition. 22 pp.

—**Glass**
BSI BS 419-66. 1966 Varnished Fabrics and Tapes for Electrical Purposes. 36 pp.
CNS C4115-88. Varnished Glass Cloth (Oct)(3428).
CNS C4116-88. Silicon Varnished Glass Cloth (Oct)(3429).

Electrical Insulating Bushings *(Cont.)*

—**Glass** *(Cont.)*
CNS C4314-88. Silicons Rubber Coated Glass Cloth on Both Sides (Oct)(7602).
DIN ENGL 40632 Pt 1-67. Varnished Fibre Materials Used for Electrical Insulation; Varnished Glass Fabrics; Dimensions (Nov). 1 p.
IEC 394 Pt 3-2-88. Varnished Fabrics for Electrical Purposes Part 3: Specifications for Individual Materials Sheet 2: Glass-Fabric Based Varnished Fabrics with Epoxy, Polyurethane, Silicone, Polyester, Bituminous or Oleoresinous Varnish First Edition. 17 pp.
JIS C 2360-81. Varnished Glass Cloths. 8 pp.
JIS C 2366-81. Silicone Varnished Glass Cloths.
JIS C 2366-71. Silicone Varnished Glass Cloths. 5 pp.
JIS C 2367-81. Silicone Rubber Coated Glass Cloths on Both Sides.
JIS C 2367-71. Silicone Rubber Coated Glass Cloths on Both Sides. 5 pp.

—**Glass—Sleeving**
IEC 684 Pt 3-300-87. Specification for Flexible Insulating Sleeving Part 3: Specification Requirements for Individual Types of Sleeving Sheet 300: Glass Textile Fibre Sleeving, Braided, Uncoated First Edition. 11 pp.
IEC 684 Pt 3-403 TO 405-88. Specification for Flexible Insulating Sleeving Part 3: Specification Requirements for Individual Types of Sleeving Sheets 403 to 405: Glass Textile Sleeving with Acrylic Based Coating First Edition. 15 pp.
IEC 684 Pt 3-406 TO 408-88. Specification for Flexible Insulating Sleeving Part 3: Specification Requirements for Individual Types of Sleeving Sheets 406 to 408: Glass Textile Sleeving with PVC Based Coating First Edition. 15 pp.

—**Glossaries**
CNS C1105-82. Glossary of Terms for Insulator and Bushing (Sep)(9355).
JIS C 3803-77. Glossary of Terms for Insulator and Bushing.

—**Nylon**
BSI BS 419-66. 1966 Varnished Fabrics and Tapes for Electrical Purposes. 36 pp.

—**Polyethylene Terephthalate**
BSI BS 419-66. 1966 Varnished Fabrics and Tapes for Electrical Purposes. 36 pp.

—**Silk**
BSI BS 419-66. 1966 Varnished Fabrics and Tapes for Electrical Purposes. 36 pp.
CNS C4114-88. Varnished Silk (Oct)(3427).

—**Transformers**
BSI BS 7616-93. 1993 Bushings for Liquid Filled Transformers Above 1 kV and up to 36 kV (E). 18 pp.
CENELEC HD 506 S1-89. Bushings for Liquid Transformers Above 1kv up to 33kv. 12 pp.
CENELEC HD 506 S1/A1-92. AMD 1 Bushings for Liquid Transformers Above 1kv up to 33kv. 5 pp.
DIN ENGL 42531 Pt 1-68. Transformers; Bushings for Indoor and Outdoor Types; Insulation Classes 10 N to 30 N, 250 A; Assembly (Sept). 2 pp.
DIN ENGL 42531 Pt 2-68. Transformers; Bushings for Indoor and Outdoor Types; Insulation Classes 10 N to 30 N, 250 A; Single Parts (Sept). 3 pp.
DIN ENGL 42532 Pt 1-69. Indoor and Outdoor Transformer Bushings; Insulation Class 10 N to 30 N, 630 A; Assembly (Sept). 2 pp.
DIN ENGL 42532 Pt 2-69. Indoor and Outdoor Transformer Bushings; Insulation Class 10 N to 30 N, 630 A; Single Parts (Sept). 4 pp.
DIN ENGL 42533 Pt 1-69. Indoor and Outdoor Transformer Bushings; Insulation Classes 10 N to 30 N, 1000 to 3150 A; Assembly (Jan). 4 pp.
DIN ENGL 42533 Pt 2-69. Indoor and Outdoor Transformer Bushings; Insulation Class 10 N to 30 N, 1000 to 3150 A; Single Parts (Jan). 5 pp.
DIN ENGL 42538-69. Transformers; Fixing Arrangement for Bushings (Feb). 4 pp.
DIN ENGL 42539 Pt 1-68. Transformers; Bushings for Indoor and Outdoor Types; Insulation Class 3 N for 250 A to 3150 A; Assembly (Aug). 4 pp.
DIN ENGL 42539 Pt 2-68. Transformers; Bushings for Indoor and Outdoor Types; Insulation Class 3 N for 250 A to 3150 A; Single Parts (Aug). 3 pp.

Electrical Insulating Fabrics
Use For: Coated Fabrics (Electrical Insulation); Varnished Fabrics (Electrical Insulation)
See Also: Electrical Insulating Tapes; Electrical Insulation; Fabrics
BSI BS 5689: Part 1-79. 1979 Varnished Fabrics for Electrical Purposes Part 1: Definitions and General Requirements. 7 pp.
BSI BS 5689: Part 2-79. 1979 Varnished Fabrics for Electrical Purposes Part 2: Methods of Test. 16 pp.
CNS C3120-81. Method of Test for Varnished Cloth and Paper for Electrical Insulation (Jun)(7600).

INDUSTRY STANDARDS

INTERNATIONAL AND NON-U.S. NATIONAL STANDARDS
SUBJECT INDEX

Electrical

Electrical Insulating Fabrics (Cont.)
- DIN ENGL 40623 Pt 1-67. Varnished Fibre Materials Used for Electrical Insulation; Varnished Fabrics; Dimensions (Nov). 1 p.
- DIN ENGL 40623 Pt 2-67. Varnished Fibre Materials Used for Electrical Insulation; Varnished Fabrics; Technical Conditions of Delivery (Nov). 4 pp.
- IEC 394 Pt 1-72. Varnished Fabrics for Electrical Purposes Part 1: Definitions and General Requirements First Edition. 9 pp.
- IEC 394 Pt 2-72. Varnished Fabrics for Electrical Purposes Part 2: Methods of Test First Edition. 24 pp.
- JIS C 2120-92. Testing Method of Varnished Cloths and Tapes for Electrical Insulation. 18 pp.

—Cotton
- BSI BS 419-66. 1966 Varnished Fabrics and Tapes for Electrical Purposes. 36 pp.
- CNS C4113-88. Varnished Cotton Cloth (Oct)(3426).
- IEC 394 Pt 3-1-76. Varnished Fabrics for Electrical Purposes Part 3: Specifications for Individual Materials Sheet 1: Oleoresinous Varnish-Cotton Base, OR/C First Edition. 12 pp.
- JIS C 2340-81. Varnished Cotton Cloths. 7 pp.

—Glass
- BSI BS 419-66. 1966 Varnished Fabrics and Tapes for Electrical Purposes. 36 pp.
- CNS C4115-88. Varnished Glass Cloth (Oct)(3428).
- CNS C4116-88. Silicon Varnished Glass Cloth (Oct)(3429).
- CNS C4314-88. Silicons Rubber Coated Glass Cloth on Both Sides (Oct)(7602).
- DIN ENGL 40632 Pt 1-67. Varnished Fibre Materials Used for Electrical Insulation; Varnished Glass Fabrics; Dimensions (Nov). 1 p.
- DIN ENGL 40632 Pt 2-68. Varnished Fibre Materials Used for Electrical Insulation; Varnished Glass Fabric; Technical Conditions of Delivery (July). 5 pp.
- IEC 394 Pt 3-2-88. Varnished Fabrics for Electrical Purposes Part 3: Specifications for Individual Materials Sheet 2: Glass-Fabric Based Varnished Fabrics with Epoxy, Polyurethane, Silicone, Polyester, Bituminous or Oleoresinous Varnish First Edition. 17 pp.
- JIS C 2360-81. Varnished Glass Cloths. 8 pp.
- JIS C 2366-81. Silicone Varnished Glass Cloths.
- JIS C 2366-71. Silicone Varnished Glass Cloths. 5 pp.
- JIS C 2367-81. Silicone Rubber Coated Glass Cloths on Both Sides.
- JIS C 2367-71. Silicone Rubber Coated Glass Cloths on Both Sides. 5 pp.

—Glass—Sleeving
- IEC 684 Pt 3-300-87. Specification for Flexible Insulating Sleeving Part 3: Specification Requirements for Individual Types of Sleeving Sheet 300: Glass Textile Fibre Sleeving, Braided, Uncoated First Edition. 11 pp.
- IEC 684 Pt 3-403 TO 405-88. Specification for Flexible Insulating Sleeving Part 3: Specification Requirements for Individual Types of Sleeving Sheets 403 to 405: Glass Textile Sleeving with Acrylic Based Coating First Edition. 15 pp.
- IEC 684 Pt 3-406 TO 408-88. Specification for Flexible Insulating Sleeving Part 3: Specification Requirements for Individual Types of Sleeving Sheets 406 to 408: Glass Textile Sleeving with PVC Based Coating First Edition. 15 pp.

—Nylon
- BSI BS 419-66. 1966 Varnished Fabrics and Tapes for Electrical Purposes. 36 pp.

—Polyethylene Terephthalate
- BSI BS 419-66. 1966 Varnished Fabrics and Tapes for Electrical Purposes. 36 pp.

—Polyethylene Terephthalate—Sleeving
- IEC 684 Pt 3-320-87. Specification for Flexible Insulating Sleeving Part 3: Specification Requirements for Individual Types of Sleeving Sheet 320: Polyethylene Terephthalate Textile, Lightly Impregnated First Edition. 11 pp.

—Rayon
- CNS C4317-88. Varnished Rayon Cloth (Oct)(7605).
- JIS C 2401-71. Varnished Rayon Cloths. 6 pp.

—Silk
- BSI BS 419-66. 1966 Varnished Fabrics and Tapes for Electrical Purposes. 36 pp.
- CNS C4114-88. Varnished Silk (Oct)(3427).

—Sleeving
- DIN ENGL 40620 Pt 1-69. Sleevings Embodying Fabric Used for Electrical Insulation; Dimensions (May). 2 pp.
- DIN ENGL 40620 Pt 2-75. Textile Sleevings for Electrical Insulation; Test Methods (Sept). 5 pp.
- DIN ENGL 40620 Pt 3-75. Textile Sleevings for Electrical Insulation; Requirements (Sept). 2 pp.

Electrical Insulating Fabrics (Cont.)
—Terephthalate
- CNS C4315-89. Varnished Terephthalate Cloth (Jan)(7603).
- JIS C 2375-81. Varnished Terephthalate Cloths.
- JIS C 2375-71. Varnished Terephthalate Cloths. 6 pp.

Electrical Insulating Films
See Also: Electrical Insulation
- DIN ENGL 40634 Pt 3-71. Electrical Insulating Films; Dimensions (July). 2 pp.
- DIN VDE 0345-65. Specifications for Electrical Insulating Films (Jan). 37 pp.
- DIN VDE 0345A-69. Specification for Electrical Insulating Films: Amendment A (Aug). 14 pp.

—Comprehensive Testing
- DIN ENGL 40634 Pt 1-69. Electrical Insulating Films; Test Methods (Nov). 12 pp.

—Fluorinated Ethylene Propylene Resins
- IEC 674 Pt 3-7-92. Specification for Plastic Films for Electrical Purposes Part 3: Specifications for Individual Materials Sheet 7: Requirements for Fluoroethylene-Propylene (FEP) Films Used for Electrical Insulation First Edition. 24 pp.

—Physical Properties
- DIN ENGL 40634 Pt 2-69. Electrical Insulating Films; Material Properties; Classification (Nov). 5 pp.

—Polycarbonate Resins
- IEC 674 Pt 3-3-92. Specification for Plastic Films for Electrical Purposes Part 3: Specifications for Individual Materials Sheet 3: Requirements for Polycarbonate (PC) Films Used for Electrical Insulation First Edition. 28 pp.

—Polyester
- CNS C4186-85. Polyester Film Lamination for Electrical Insulation (Nov)(5519).
- CNS C4188-83. Polyester Films for Electrical Purpose (Jan)(5521).
- JIS C 2317-84. Polyester Film Lamination for Electrical Insulation. 9 pp.
- JIS C 2318-75. Polyester Films for Electrical Purposes (R 1978). 18 pp.

—Polyethylene Terephthalate
- IEC 674 Pt 3-2-92. Specification for Plastic Films for Electrical Purposes Part 3: Specifications for Individual Materials Sheet 2: Requirements for Balanced Biaxially Oriented Polyethylene Terephthalate (PET) Films Used for Electrical Insulation First Edition. 28 pp.

—Polypropylene
- JIS C 2330-78. Polypropylene Films for Electrical Purposes. 17 pp.

—Polytetrafluoroethylene
- CNS K3069-83. Polytetrafluoroethylene Sheets (Aug)(10511). 7 pp.
- JIS K 6888-77. Polytetrafluoroethylene Sheets (R 1980). 10 pp.

Electrical Insulating Liquids
See Also: Dielectrics; Electrical Insulating Oils; Electrical Insulation

—Aromatic Hydrocarbons
- BSI BS 6802-87. 1987 Unused Insulating Liquids Based on Synthetic Aromatic Hydrocarbons. 12 pp.
- CENELEC HD 497 S1-87. Specifications for Unused Insulating Liquids Based on Synthetic Aromatic Hydrocarbons. 3 pp.
- CENELEC HD 497 S1-88. Specifications for Unused Insulating Liquids Based on Synthetic Aromatic Hydrocarbons. 3 pp.
- CENELEC PREN 60867-92. Insulating Liquids—Specifications for Unused Liquids Based on Synthetic Aromatic Hydrocarbons. 15 pp.
- IEC 867-93. Insulating Liquids—Specifications for Unused Liquids Based on Synthetic Aromatic Hydrocarbons Second Edition. 31 pp.

—Breakdown Voltage
- IEC 897-87. Methods for the Determination of the Lightning Impulse Breakdown Voltage of Insulating Liquids First Edition. 30 pp.

—Calorific Value—Classification
- BSI BS EN 61100-93. 1993 Classification of Insulating Liquids According to Fire Point and Net Calorific Value (IEC 1100: 1992) (E). 12 pp.
- CENELEC EN 61100-92. Classification of Insulating Liquids According to Fire Point and Net Calorific Value. 5 pp.
- IEC 1100-92. Classification of Insulating Liquids According to Fire-Point and Net Calorific Value First Edition. 15 pp.

Electrical Insulating Liquids (Cont.)
—Classification
- CENELEC HD 618 S1-92. General Classification of Insulating Liquids. 5 pp.
- IEC 1039-90. General Classification of Insulating Liquids First Edition. 15 pp.

—Dielectric Dissipation Factor
- BSI BS 5737-79. 1979 Measurement of Relative Premittivity, Dielectric Dissipation Factor and d.c. Resistivity of Insulating Liquids. 19 pp.
- IEC 247-78. Measurement of Relative Permittivity, Dielectric Dissipation Factor and D.C. Resistivity of Insulating Liquids Second Edition. 35 pp.

—Electrical Resistivity
- BSI BS 5737-79. 1979 Measurement of Relative Premittivity, Dielectric Dissipation Factor and d.c. Resistivity of Insulating Liquids. 19 pp.
- IEC 247-78. Measurement of Relative Permittivity, Dielectric Dissipation Factor and D.C. Resistivity of Insulating Liquids Second Edition. 35 pp.

—Esters
- BSI BS EN 61099-92. 1992 Unsuded Synthetic Organic Esters for Electrical Purposes (IEC 1099: 1992) (F). 17 pp.
- CENELEC EN 61099-92. Specification for Unused Synthetic Organic Esters for Electrical Purposes. 5 pp.
- CENELEC EN 61099-92. Specification for Unused Synthetic Organic Esters for Electrical Purposes; (IEC 1099: 1992). 12 pp.
- IEC 1099-92. Specifications for Unused Synthetic Organic Esters for Electrical Purposes First Edition; (CENELEC EN 61099:1992). 25 pp.

—Fire Point—Classification
- BSI BS EN 61100-93. 1993 Classification of Insulating Liquids According to Fire Point and Net Calorific Value (IEC 1100: 1992) (E). 12 pp.
- CENELEC EN 61100-92. Classification of Insulating Liquids According to Fire Point and Net Calorific Value. 5 pp.
- IEC 1100-92. Classification of Insulating Liquids According to Fire-Point and Net Calorific Value First Edition. 15 pp.

—Flame Propagation
- CENELEC PREN 61197-92. Insulating Liquids—Test Method for the Linear Flame Propagation of Insulating Liquids Using a Glass-Fibre Tape. 13 pp.
- IEC 1197-93. Insulating Liquids—Linear Flame Propagation—Test Method Using a Glass-Fibre Tape First Edition. 24 pp.

—Gassing
- BSI BS 5797-86. 1986 Amd 1 Methods for Measurement of Gassing of Insulating Liquids Under Electrical Stress and Ionization. 19 pp.
- CENELEC HD 488 S1-88. Gassing of Insulating Liquids Under Electrical Stress and Ionization. 3 pp.
- IEC 628-85. Gassing of Insulating Liquids Under Electrical Stress and Ionization Second Edition; (Corrigendum—Oct 1986). 38 pp.

—Glossaries
- BSI BS 4727:Pt1: Group 10-91. 1991 Electrotechnical, Power, Telecommunication, Electronics, Lighting and Colour Terms Part 1: Terms Common to Power Telecommunications and Electronics Group 10: Insulating Solids, Liquids and Gases. 101 pp.
- IEC 50 Chap 212-90. International Electrotechnical Vocabulary Chapter 212: Insulating Solids, Liquids and Gases First Edition. 105 pp.

—Hydrocarbons—Oxidation Resistance
- BSI BS EN 61125-93. 1993 Unused Hydrocarbon-Based Insulating Liquids—Test Methods for Evaluating the Oxidation Stability (IEC 1125: 1992) (F). 32 pp.
- CENELEC HD 486 S1-88. Test Method for Evaluating the Oxidation Stability of Hydrocarbon Insulating Liquids. 3 pp.
- CENELEC EN 61125-93. Unused Hydrocarbon-Based Insulating Liquids Test Methods for Evaluating the Oxidation Stability (IEC 1125: 1992 + Corrigendum 1992). 27 pp.
- IEC 1125-92. Unused Hydrocarbon-Based Insulating Liquids—Test Methods for Evaluating the Oxidation Stability First Edition; (Corrigendum—September 1992) (Replaces 74, 474 and 813). 54 pp.

—Oxygen Index
- BSI BS EN 61144-93. 1993 Test Method for the Determination of Oxygen Index of Insulating Liquids (IEC 1144: 1992). 17 pp.
- CENELEC EN 61144-93. Test Method for the Determination of Oxygen Index of Insulation Liquids; (IEC 1144:1992). 4 pp.
- CENELEC EN 61144-93. Test Method for the Determination of Oxygen Index of Insulating Liquids (IEC 1144:1992). 12 pp.

INTERNATIONAL AND NON-U.S. NATIONAL STANDARDS
SUBJECT INDEX
Electrical

Electrical Insulating Liquids (Cont.)

—Oxygen Index (Cont.)
IEC 1144-92. Test Method for the Determination of Oxygen Index of Insulating Liquids First Edition; (CENELEC EN 61144:1993). 24 pp.

—Particle Density
IEC 970-89. Methods for Counting and Sizing Particles in Insulating Liquids First Edition. 38 pp.

—Particle Size Distribution
IEC 970-89. Methods for Counting and Sizing Particles in Insulating Liquids First Edition. 38 pp.

—Permittivity
BSI BS 5737-79. 1979 Measurement of Relative Premittivity, Dielectric Dissipation Factor and d.c. Resistivity of Insulating Liquids. 19 pp.
IEC 247-78. Measurement of Relative Permittivity, Dielectric Dissipation Factor and D.C. Resistivity of Insulating Liquids Second Edition. 35 pp.

—Polyisobutylene
CENELEC HD 582 S1-91. Specification for Unused Polybuthenes. 8 pp.
IEC 963-88. Specification for Unused Polybutenes First Edition. 21 pp.

—Resins
CNS C3204-88. Method of Test for Solventless Liquid Resins of Electrical Insulating (Apr)(12247).
JIS C 2105-92. Testing Methods of Solventless Liquid Resins for Electrical Insulation. 50 pp.

—Silicone
CENELEC HD 565 S1-93. Specifications for Silicone Liquids for Electrical Purposes (IEC 836:1988). 5 pp.
IEC 836-88. Specifications for Silicone Liquids for Electrical Purposes First Edition; (CENELEC HD 565 S1: 1993). 31 pp.

—Silicone—Transformers—Maintenance
BSI BS 7713-93. 1993 Guide for the Maintenance of Silicone Transformer Liquids (IEC 944: 1988) (E). 10 pp.
IEC 944-88. Guide for the Maintenance of Silicone Transformer Liquids First Edition. 16 pp.

—Transformers—Esters
IEC 1203-92. Synthetic Organic Esters for Electrical Purposes—Guide for Maintenance of Transformer Esters in Equipment First Ediiton. 19 pp.

—Water Content—Karl Fischer Method
CENELEC HD 487 S1-88. Determination of Water in Insulating Liquids by Automatic Coulometric Karl Fischer Titration. 3 pp.
IEC 814-85. Determination of Water in Insulating Liquids by Automatic Coulometric Karl Fischer Titration First Edition. 27 pp.

Electrical Insulating Materials
Use: Electrical Insulation

Electrical Insulating Mats
See Also: Electrical Insulation; Mats

—Elastomeric—Floor Coverings
IEC 1111-92. Matting of Insulating Material for Electrical Purposes First Edition. 62 pp.

Electrical Insulating Matting
Use: Electrical Insulating Mats

Electrical Insulating Oils
See Also: Askarels; Dielectrics; Electrical Insulating Liquids; Electrical Insulating Varnishes; Electrical Insulation; Mineral Oils; Oil Filled Transformers; Switchgear; Transformer Oils; Transformers
CNS C3025-87. Method of Test for Insulating Oil (Jul)(1327).
CNS C4035-87. Insulating Oil (for General Use) (Jul)(1326).
CNS C4093-87. Insulating Oil (for Cold Area) (Jul)(3070).
JIS C 2101-88. Testing Methods of Electrical Insulating Oils. 85 pp.
JIS C 2320-88. Electrical Insulating Oils. 11 pp.
SAA AS 1883-92. Guide to Maintenance and Supervision of Insulating Oils in Service (IEC 422:1989) (in Professional Package 57). 34 pp.

—Aging Testing
DIN ENGL 51554 Pt 2-78. Testing of Mineral Oils; Test of Susceptibility to Ageing According to Baader; Testing at 110 Deg. C (Sept). 3 pp.

—Antioxidant Content
BSI BS 5984-80. 1980 Detection and Determination of Specified Anti-Oxidant Additives in Insulating Oils. 15 pp.

Electrical Insulating Oils (Cont.)

—Antioxidant Content (Cont.)
CENELEC HD 415-80. Detection and Determination of Specified Anti-Oxydant Additives in Insulating Oils. 2 pp.
IEC 666-79. Detection and Determination of Specified Anti-Oxidant Additives in Insulating Oils First Edition. 27 pp.

—Aromatic Hydrocarbon Content
CENELEC HD 382-78. Determination of the Aromatic Hydorcarbon Content of New Mineral Insulating Oils. 2 pp.
IEC 590-77. Determination of the Aromatic Hydrocarbon Content of New Mineral Insulating Oils First Edition; (Corrigendum—July 1978). 21 pp.

—Dielectric Strength
BSI BS 5874-80. 1980 Amd 1 Determination of the Electric Strength of Insulating Oils. 10 pp.
IEC 156-63. Method for the Determination of the Electric Strength of Insulating Oils First Edition. 15 pp.

—Electric Reactors—Gas Analysis
BSI BS EN 61181-93. 1993 Impregnated Insulating Materials—Application of Dissolved Gas Analysis (DGA) to Factory Tests on Electrical Equipment (IEC 1181: 1993) (Q). 15 pp.
CENELEC PREN 61181-92. Impregnated Insulating Materials—Application of Dissolved Gas Analysis (DGA) to Factory Tests on Electrical Equipment. 13 pp.
IEC 1181-93. Impregnated Insulating Materials—Application of Dissolved Gas Analysis (DGA) to Factory Tests on Electrical Equipment First Edition. 21 pp.

—Furan Compound Content—Liquid Chromatography
CENELEC PREN 61198-92. Mineral Insulating Oils—Methods for the Determination of 2-Furfural and Related Compounds. 17 pp.
IEC 1198-93. Mineral Insulating Oils—Methods for the Determination of 2-Furfural and Related Compounds First Edition. 32 pp.

—Furfural Content—Liquid Chromatography
CENELEC PREN 61198-92. Mineral Insulating Oils—Methods for the Determination of 2-Furfural and Related Compounds. 17 pp.
IEC 1198-93. Mineral Insulating Oils—Methods for the Determination of 2-Furfural and Related Compounds First Edition. 32 pp.

—Gas Analysis
BSI BS 5574-78. (WITHDRAWN) 1978 Guide for the Sampling of Gases and of Oil from Oil-Filled Electrical Equipment and for the Analysis of Free and Dissolved Gases (Superseded by BS EN 60567: 1993). 32 pp.
BSI BS EN 60567-93. 1993 Guide for the Sampling of Gases and of Oil from Oil-Filled Electrical Equipment and for the Analysis of Free and Dissolved Gases (IEC 567: 1992) (Supersedes BS 5574: 1978). 54 pp.
CENELEC EN 60567-92. Guide for the Sampling of Gases and of Oil From Oil-Filled Electrical Equipment and for the Analysis of Free and Dissolved Gases. 5 pp.
IEC 567-92. Guide for the Sampling of Gases and of Oil from Oil-Filled Electrical Equipment and for the Analysis of Free and Dissolved Gases Second Edition. 85 pp.

—Maintenance
BSI BS 5730-79. 1979 Maintenance of Insulating Oil. 12 pp.

—Oil Filled Cables
BSI BS EN 60465-92. 1992 Unused Insulating Mineral Oils for Cables with Oil Ducts (IEC 465: 1988). 13 pp.
CENELEC EN 60465-90. Specification for Unused Insulating Mineral Oils For Cables with Oil Ducts. 4 pp.
IEC 465-88. Specification for Unused Insulating Mineral Oils for Cables with Oil Ducts Second Edition. 22 pp.

—Oxidation Resistance
BSI BS 2000: Part 307-83. (WITHDRAWN) 1983 Petroleum and Its Products Part 307: Oxidation Stability of Mineral Insulating Oil (Superseded by BS EN 61125: 1993). 4 pp.
BSI BS EN 61125-93. 1993 Unused Hydrocarbon-Based Insulating Liquids—Test Methods for Evaluating the Oxidation Stability (IEC 1125: 1992) (F). 32 pp.
CENELEC EN 61125-93. Unused Hydrocarbon-Based Insulating Liquids Test Methods for Evaluating the Oxidation Stability (IEC 1125: 1992 + Corrigendum 1992). 27 pp.

Electrical Insulating Oils (Cont.)

—Oxidation Resistance (Cont.)
IEC 1125-92. Unused Hydrocarbon-Based Insulating Liquids—Test Methods for Evaluating the Oxidation Stability First Edition; (Corrigendum—September 1992) (Replaces 74, 474 and 813). 54 pp.

—Polychlorinated Biphenyl Content—Gas Chromatography
IEC 997-89. Determination of Polychlorinated Biphenyls (PCBs) in Mineral Insulating Oils by Packed Column Gas Chromatography (GC) First Edition. 48 pp.

—Pour Point Depressants—Aging Testing
BSI BS EN 61065-93. 1993 Method for Evaluating the Low Temperature Flow Properties of Mineral Insulating Oils After Aging (IEC 1065: 1991). 14 pp.
CENELEC EN 61065-93. Method for Evaluating the Low Temperature Flow Properties of Mineral Insulating Oils After Ageing (IEC 1065:1991). 5 pp.
IEC 1065-91. Method for Evaluating the Low Temperature Flow Properties of Mineral Insulating Oils After Ageing First Edition; (CENELEC EN 61065:1993). 19 pp.

—Sampling
CNS C4039-83. Method of Sampling for Insulating Oil (Feb)(1371).

—Sulfur Content
BSI BS 5680-79. 1979 Detection of Corrosive Sulphur in Petroleum-Based Electrical Insulating Oils. 4 pp.
ISO 5662-78. Petroleum Products—Electrical Insulating Oils—Detection of Corrosive Sulphur First Edition. 4 pp.

—Sulfur Content—Silver Strip Test
DIN ENGL 51353-85. Testing of Insulating Oils; Detection of Corrosive Sulfur; Silver Strip Test (Dec). 3 pp.

—Switches
CSA C50-1976. Insulating Oil, Electrical for Transformers and Switches; (Erratum April 1976) (Amd 1-5 June 1979). 14 pp.

—Switchgear
BSI BS 148-84. 1984 Amd 1 Unused Mineral Insulating Oils for Transformers and Switchgear. 20 pp.
DIN VDE 0370 Pt 1-78. Insulating Oils; New Insulating Oils for Transformers and Switchgear (Dec). 21 pp.
DIN VDE 0370 Pt 2-78. Insulating Oils; Insulating Oils in Service in Transformers and Switchgear (Dec). 7 pp.
IEC 296-82. Specification for Unused Mineral Insulating Oils for Transformers and Switchgear Second Edition; (Amendment 1-1986). 27 pp.
SAA AS 1767-75. Insulating Oil for Transformers and Switchgear (Incorporating Amdt 1). 44 pp.

—Switchgear—Maintenance
IEC 422-89. Supervision and Maintenance Guide for Mineral Insulating Oils in Electrical Equipment Second Edition. 74 pp.

—Transformers
BSI BS 148-84. 1984 Amd 1 Unused Mineral Insulating Oils for Transformers and Switchgear. 20 pp.
CSA C50-1976. Insulating Oil, Electrical for Transformers and Switches; (Erratum April 1976) (Amd 1-5 June 1979). 14 pp.
DIN VDE 0370 Pt 1-78. Insulating Oils; New Insulating Oils for Transformers and Switchgear (Dec). 21 pp.
DIN VDE 0370 Pt 2-78. Insulating Oils; Insulating Oils in Service in Transformers and Switchgear (Dec). 7 pp.
IEC 296-82. Specification for Unused Mineral Insulating Oils for Transformers and Switchgear Second Edition; (Amendment 1-1986). 27 pp.
SAA AS 1767-75. Insulating Oil for Transformers and Switchgear (Incorporating Amdt 1). 44 pp.

—Transformers—Gas Analysis
BSI BS 5800-79. 1979 Guide for the Interpretation of the Analysis of Gases in Transformers and Other Oil-Filled Electrical Equipment in Service. 15 pp.
BSI BS EN 61181-93. 1993 Impregnated Insulating Materials—Application of Dissolved Gas Analysis (DGA) to Factory Tests on Electrical Equipment (IEC 1181: 1993) (Q). 15 pp.
CENELEC HD 397-79. Interpretation of the Analysis of Gases in Transformers and Other Oil-Filled Electrical Equipment in Service. 1 p.
CENELEC PREN 61181-92. Impregnated Insulating Materials—Application of Dissolved Gas Analysis (DGA) to Factory Tests on Electrical Equipment. 13 pp.

INDUSTRY STANDARDS

Electrical

INTERNATIONAL AND NON-U.S. NATIONAL STANDARDS
SUBJECT INDEX

Electrical Insulating Oils (Cont.)

—Transformers—Gas Analysis (Cont.)
IEC 599-78. Interpretation of the Analysis of Gases in Transformers and Other Oil-Filled Electrical Equipment in Service First Edition. 28 pp.
IEC 1181-93. Impregnated Insulating Materials—Application of Dissolved Gas Analysis (DGA) to Factory Tests on Electrical Equipment First Edition. 21 pp.

—Transformers—Maintenance
IEC 422-89. Supervision and Maintenance Guide for Mineral Insulating Oils in Electrical Equipment Second Edition. 74 pp.

—Water Content
BSI BS 6470-84. 1984 Determination of Water in Insulating Oils, and in Oil-Impregnated Paper and Pressboard. 24 pp.
BSI BS 6522-84. 1984 Determination of Percentage Water Saturation of Insulating Oil. 4 pp.
IEC 733-82. Determination of Water in Insulating Oils, and in Oil-Impregnated Paper and Pressboard First Edition. 43 pp.

Electrical Insulating Papers

Use For: Electrical Insulating Boards; Insulating Papers (Electrical) *See Also:* Capacitor Papers; Crepe Papers; Electrical Insulation; Papers; Waxed Papers

IEC 641 Pt 3-2-92. Specification for Pressboard and Presspaper for Electrical Purposes Part 3: Specifications for Individual Materials —Sheet 2: Requirements for Presspaper, Types P.2.1, P.4.1, P.4.2, P.4.3, P.6.1 and P.7.1 First Edition. 30 pp.

—Asbestos
BSI BS 3057-58. 1958 Amd 2 Untreated Asbestos Paper for Electrical Purposes. 29 pp.

—Cables (Electric)
CNS C4209-85. Insulating Papers for Electric Power Cables (Nov)(6042).
CNS C4210-85. Insulating Paper for Communication Cables (Nov)(6043).
JIS C 2307-84. Insulating Papers for Electric Power Cables. 7 pp.
JIS C 2308-84. Insulating Papers for Communication Cables. 7 pp.

—Cellulosic
BSI BS 5626: Part 1-79. 1979 Amd 1 Cellulosic Papers for Electrical Purposes Part 1: Definitions and General Requirements. 9 pp.
BSI BS 5626: Part 2-79. 1979 Amd 1 Cellulosic Papers for Electrical Purposes Part 2: Methods of Test. 35 pp.
BSI BS 5626: Sec 3.1-81. 1981 Cellulosic Papers for Electrical Purposes Part 3: Specifications for Individual Materials Section 3.1: General Purpose Paper. 11 pp.
DIN VDE 0311 Pt 20-88. Cellulosic Paper for Electrical Purposes; Test Methods (Cover Sheet, Modifications and Explanatory Notes Only) (Nov). 12 pp.
IEC 554 Pt 1-77. Specification for Cellulosic Papers for Electrical Purposes Part 1: Definitions and General Requirements First Edition; (Erratum—March 1979) (Amendment 1-1983). 19 pp.
IEC 554 Pt 2-77. Specification for Cellulosic Papers for Electrical Purposes Part 2: Methods of Test First Edition; (Amendment 1-1982) (Amendment 2-1984). 64 pp.
IEC 554 Pt 3-1-79. Specification for Cellulosic Papers for Electrical Purposes Part 3: Specifications for Individual Materials Sheet 1: General Purpose Electrical Paper First Edition. 22 pp.
IEC 554 Pt 3-5-84. Specification for Cellulosic Papers for Electrical Purposes Part 3: Specifications for Individual Materials Sheet 5: Special Papers First Edition. 39 pp.

—Coils
CNS C4212-85. Coil Insulating Paper (Nov)(6045).
JIS C 2304-84. Coil Insulating Papers. 7 pp.

—Glossaries
CNS C1069-80. Standard Definitions of Terms Relating to Electrical Insulating Papers (Aug)(6015).

—Laminated
BSI BS 5102-74. (WITHDRAWN) 1974 Amd 2 Phenolic Resin Bonded Paper Laminated Sheets for Electrical Applications (Superseded by BS 2572: 1990). 38 pp.
CNS C4186-85. Polyester Film Lamination for Electrical Insulation (Nov)(5519).
JIS C 2317-84. Polyester Film Lamination for Electrical Insulation. 9 pp.

—Mica
CNS C4328-88. Mica Splittings and Untreated Mica Papers for Electrical Insulation (Aug)(8356).

Electrical Insulating Papers (Cont.)

—Mica (Cont.)
CNS C4334-88. Mica with Paper on Both Sides (Sheets and Tapes) (Sep)(8363).
CNS C4335-88. Mica with Paper on One Side (Sheets and Tapes) (Sep)(8364).
DIN ENGL 40613-71. Products Based on Built-Up or Treated Mica Paper for Electrical Insulation; Insulating Laminate for Commutators; Dimensions (July). 3 pp.
DIN ENGL 40614-70. Products Based on Built-Up or Treated Mica Paper for Electrical Insulation; Laminated Wide and Small Strips; Dimensions (May). 2 pp.
IEC 371 Pt 3-2-91. Specification for Insulating Materials Based on Mica Part 3: Specifications for Individual Materials Sheet 2: Mica Paper First Edition. 26 pp.
IEC 371 Pt 3-4-92. Specification for Insulating Materials Based on Mica Part 3: Specifications for Individual Materials Sheet 4: Polyester Film-Backed Mica Paper with a B-Stage Epoxy Resin Binder First Edition. 26 pp.
IEC 371 Pt 3-5-92. Specification for Insulating Materials Based on Mica Part 3: Specifications for Individual Materials Sheet 5: Glass-Backed Mica Paper with an Epoxy Resin Binder for Post-Impregnation (VPI) First Edition. 26 pp.
IEC 371 Pt 3-6-92. Specification for Insulating Materials Based on Mica Part 3: Specifications for Individual Materials Sheet 6: Glass-Backed Mica Paper with a B-Stage Epoxy Resin Binder First Edition. 28 pp.
JIS C 2256-77. Mica with Papers on Both Sides (Sheets and Tapes) (R 1985). 9 pp.
JIS C 2257-77. Mica with Papers on One Side (Sheets and Tapes) (R 1985). 9 pp.
JIS C 2262-92. Post-Impregnatable Mica Paper Materials. 9 pp.
JIS C 2263-92. Resin Rich Mica Paper Materials. 9 pp.

—Noncellulosic
IEC 819 Pt 1-85. Specification for Non-Cellulosic Papers for Electrical Purposes Part 1: Definitions and General Requirements First Edition. 11 pp.

—Noncellulosic—Aramid Fibers
IEC 819 Pt 3-3-91. Specification for Non-Cellulosic Papers for Electrical Purposes Part 3: Specifications for Individual Materials Sheet 3: Unfilled Aramid (Aromatic Polyamide) Papers First Edition. 17 pp.

—Oil Impregnated—Water Content
BSI BS 6470-84. 1984 Determination of Water in Insulating Oils, and in Oil-Impregnated Paper and Pressboard. 24 pp.
IEC 733-82. Determination of Water in Insulating Oils, and in Oil-Impregnated Paper and Pressboard First Edition. 43 pp.

—Polymerization
IEC 450-74. Measurement of the Average Viscometric Degree of Polymerization of New and Aged Electrical Papers First Edition. 30 pp.

—Pressboard
BSI BS 231-75. 1975 Pressboard for Electrical Purposes. 19 pp.
BSI BS 3255-75. 1975 Presspaper for Electrical Purposes. 20 pp.
BSI BS 5354-76. 1976 Laminated Pressboard for Electrical Purposes. 15 pp.
BSI BS 5937: Part 2-80. 1980 Pressboard and Presspaper for Electrical Purposes Part 2: Methods of Test. 24 pp.
CENELEC HD 410.2-80. Specification for Press Board and Press Paper for Electric Purposes Part 2: Method of Test. 2 pp.
CNS C3092-85. Method of Test for Electrical Insulating Papers, Pressboard and Presspaper (Sep)(6041).
CNS C4211-85. Pressboard for Electrical Purpose (Nov)(6044).
DIN ENGL 7733-62. Laminated Products; Pressboard for Electrical Engineering; Types (June). 4 pp.
DIN ENGL 7734-58. Laminated Products; Pressboard for Electrical Engineering; Acceptance; Test Procedure (Oct). 9 pp.
DIN ENGL 7739 Pt 1-65. Laminated Products; Combined Pressboard for Electrical Insulation; Requirements; Testing (July). 5 pp.
DIN ENGL 7739 Pt 2-67. Laminated Products; Combined Pressboard for Electrical Insulation; Types (July). 3 pp.
DIN ENGL 40600-58. Laminated Products; Pressboard for Electrical Engineering; Sheets; Strips; Rolls; Tapes (Oct). 2 pp.
DIN VDE 0315-58. Rules for Methods of Testing Laminated Moulded Materials; Presspahn (Pressboard) for Electrical Engineering (Oct). 23 pp.
IEC 641 Pt 2-79. Specification for Pressboard and Presspaper for Electrical Purposes Part 2: Methods of Test First Edition; (Amendment 1-1993). 55 pp.

Electrical Insulating Papers (Cont.)

—Pressboard (Cont.)
IEC 641 Pt 3-1-92. Specification for Pressboard and Presspaper for Electrical Purposes Part 3: Specifications for Individual Materials —Sheet 1: Requirements for Pressboard, Types B.0.1, B.2.1, B.2.3, B.3.1, B.3.3, B.4.1, B.4.3, B.5.1, B.6.1 and B.7.1. 62 pp.
JIS C 2111-84. Testing Methods of Electrical Insulating Papers, Pressboard and Presspaper. 19 pp.
JIS C 2305-84. Pressboards for Electrical Purpose. 8 pp.

—Pressboard—Glossaries
BSI BS 5937: Part 1-80. 1980 Pressboard and Presspaper for Electrical Purposes Part 1: Definitions and General Requirements. 7 pp.
CENELEC HD 410.1-80. Specification for Pressboard and Presspaper for Electric Purposes Part 1: Definitions and General Requirements. 2 pp.
IEC 641 Pt 1-79. Specification for Pressboard and Presspaper for Electrical Purposes Part 1: Definitions and General Requirements First Edition; (Amendment 1-1993). 21 pp.

—Tissue
CNS C4079-85. Insulated Tissue Papers (Nov)(2898).
JIS C 2303-84. Insulated Tissue Papers. 7 pp.

—Varnished
CNS C4320-88. Varnished Paper (Oct)(7608).
DIN ENGL 40622 Pt 1-67. Varnished Fibre Materials Used for Electrical Insulation; Varnished Papers; Dimensions (Nov). 1 p.
DIN ENGL 40622 Pt 2-67. Varnished Fibre Materials Used for Electrical Insulation; Varnished Papers; Technical Conditions of Delivery (Nov). 4 pp.

—Vulcanized Fiber
CNS C4187-85. Vulcanized Fibre Sheet (May)(5520).
JIS C 2315-84. Vulcanized Fibre Sheet. 9 pp.

Electrical Insulating Tapes

Use For: Electrical Tapes (Insulating)
See Also: Adhesive Tapes; Electrical Insulating Fabrics; Electrical Insulation

JIS C 2120-92. Testing Method of Varnished Cloths and Tapes for Electrical Insulation. 18 pp.
SAA AS 1621.1-88. Pressure-Sensitive Adhesive Electrical Tapes—Part 1: General Requirements. 4 pp.
SAA AS 1621.2-88. Pressure-Sensitive Adhesive Electrical Tapes—Part 2: Methods of Test. 12 pp.

—Amalgamative
MOD UK TS 10215. Self Amalgamating Tape.

—Asbestos—Woven
BSI BS 1944-73. 1973 Amd 2 Woven Asbestos Tape for Electrical Insulating Purposes. 14 pp.

—Cellulosic Paper
IEC 454 Pt 3-4-78. Specification for Pressure-Sensitive Adhesive Tapes for Electrical Purposes Part 3: Specifications for Individual Materials Sheet 4: Requirements for Cellulosic Paper, Creped, with Thermosetting Adhesive First Edition. 15 pp.
IEC 454 Pt 3-5-80. Specification for Pressure-Sensitive Adhesive Tapes for Electrical Purposes Part 3: Specifications for Individual Materials Sheet 5: Requirements for Cellulosic Paper with Thermosetting Adhesive First Edition. 16 pp.
SAA AS 1621.3.4-88. Pressure-Sensitive Adhesive Electrical Tapes—Part 3: Specific Requirements—Part 3.4: Cellulosic Paper Tapes, Creped, with Thermosetting Adhesive. 5 pp.
SAA AS 1621.3.5-88. Pressure-Sensitive Adhesive Electrical Tapes—Part 3: Specific Requirements—Part 3.5: Cellulosic Paper Tapes with Thermosetting Adhesive. 5 pp.

—Cotton Fabric
BSI BS 633-70. 1970 Cotton Selvedge Tapes and Webbing for Electrical Purposes. 20 pp.

—Cotton Fabric—Varnished
BSI BS 419-66. 1966 Varnished Fabrics and Tapes for Electrical Purposes. 36 pp.
CNS C4078-88. Varnished Cotton Cloth Tape (Oct)(2897).
CNS C4111-88. Varnished Cotton Bias Tape (Oct)(3424).
JIS C 2341-81. Varnished Cotton Cloth Tapes. 8 pp.
JIS C 2342-81. Varnished Cotton Bias Tapes. 8 pp.

—Fabric
DIN ENGL 40633 Pt 2-67. Pressure-Sensitive Adhesive Tapes Used for Electrical Insulation; Fabric-Based Tapes; Types; Requirements; Testing (July). 4 pp.
DIN VDE 0340 Pt 2-67. Specifications for Self-Adhesive Insulating Tapes; Part 2: Fabric Tapes (July). 15 pp.

INDEX and DIRECTORY of

INTERNATIONAL AND NON-U.S. NATIONAL STANDARDS
SUBJECT INDEX
Electrical

Electrical Insulating Tapes (Cont.)
—**Fiberglass Reinforced**
CNS K3065-82. Glass Tape (Dec)(9719). 5 pp.
CNS K3066-82. Finished Glass Tapes (Dec)(9720). 6 pp.
JIS R 3422-85. Finished Glass Tapes. 9 pp.

—**Fiberglass Reinforced—Varnished**
MOD UK NES 2009-92. Specification for Varnished Woven Glass Fibre and Woven Polyethylene Terephthalate (Terylene) Tapes for Electrical Purposes Issue 1 (10.92). 10 pp.

—**Fiberglass Reinforced—Woven**
BSI BS 3779-85. 1985 Glass and Glass Polyester Fibre Woven Tapes for Electrical Purposes. 12 pp.

—**Glass Fabric**
IEC 454 Pt 3-8-86. Specification for Pressure-Sensitive Adhesive Tapes for Electrical Purposes Part 3: Specifications for Individual Materials Sheet 8: Requirements for Glass Fabric Tapes with Thermosetting Adhesive First Edition. 11 pp.

—**Glass Fabric—Resin Coated**
CNS C4459-88. Resin Preimpregnated Glass Binding Tape for Electrical Purpose (Oct)(12441).
JIS C 2412-75. Resin Preimpregnated Glass Binding Tapes for Electrical Purpose. 18 pp.

—**Glass Fabric—Varnished**
BSI BS 419-66. 1966 Varnished Fabrics and Tapes for Electrical Purposes. 36 pp.
CNS C4313-88. Silicons Varnished Glass Cloth Tape (Oct)(7601).
CNS C4321-88. Varnished Glass Cloth Tape (Oct)(7609).
CNS C4322-88. Varnished Glass Bias Tape (Oct)(7610).
JIS C 2361-81. Varnished Glass Cloth Tapes. 8 pp.
JIS C 2362-81. Varnished Glass Bias Tapes. 9 pp.
JIS C 2365-81. Silicone Varnished Glass Cloth Tapes.
JIS C 2365-71. Silicone Varnished Glass Cloth Tapes. 5 pp.

—**Nylon Fabric—Varnished**
BSI BS 419-66. 1966 Varnished Fabrics and Tapes for Electrical Purposes. 36 pp.

—**Polycarbonate Film**
IEC 454 Pt 3-6-84. Specification for Pressure-Sensitive Adhesive Tapes for Electrical Purposes Part 3: Specifications for Individual Materials Sheet 6: Requirements for Polycarbonate Film Tapes with Non-Thermosetting Adhesive First Edition. 11 pp.

—**Polyester Film**
IEC 454 Pt 3-2-81. Specification for Pressure-Sensitive Adhesive Tapes for Electrical Purposes Part 3: Specifications for Individual Materials Sheet 2: Requirements for Polyester Film Tapes (PETP) with Thermosetting Adhesive First Edition. 13 pp.
IEC 454 Pt 3-3-81. Specification for Pressure-Sensitive Adhesive Tapes for Electrical Purposes Part 3: Specifications for Individual Materials Sheet 3: Requirements for Polyester Film Tapes (PETP) with Non-Thermosetting Adhesive First Edition. 13 pp.
SAA AS 1621.3.2-88. Pressure-Sensitive Adhesive Electrical Tapes—Part 3: Specific Requirements—Part 3.2: Polyester Film Tapes (PETP) with Thermosetting Adhesive. 3 pp.
SAA AS 1621.3.3-88. Pressure-Sensitive Adhesive Electrical Tapes—Part 3: Specific Requirements—Part 3.3: Polyester Film Tapes (PETP) with Non-Thermosetting Adhesive. 3 pp.

—**Polyester—Pressure Sensitive**
CNS C4453-88. Pressure-Sensitive Polyester Adhesive Tapes for Electrical Insulation (Oct)(12435).
JIS C 2338-91. Pressure-Sensitive Adhesive Polyester Tapes for Electrical Insulation. 9 pp.

—**Polyester—Woven**
BSI BS 6551-85. 1985 Polyester Fibre Woven Tapes for Electrical Purposes. 10 pp.
CNS C4197-80. Polyester Woven Tapes for Electrical Insulation (Jul)(5746).

—**Polyethylene Terephthalate**
MOD UK NES 2009-92. Specification for Varnished Woven Glass Fibre and Woven Polyethylene Terephthalate (Terylene) Tapes for Electrical Purposes Issue 1 (10.92). 10 pp.

—**Polyethylene Terephthalate—Varnished**
BSI BS 419-66. 1966 Varnished Fabrics and Tapes for Electrical Purposes. 36 pp.

—**Polyimide Film**
IEC 454 Pt 3-7-84. Specification for Pressure-Sensitive Adhesive Tapes for Electrical Purposes Part 3: Specifications for Individual Materials Sheet 7: Requirements for Polyimide Film Tapes with Thermosetting Adhesive First Edition. 13 pp.

Electrical Insulating Tapes (Cont.)
—**Polytetrafluoroethylene Film**
JIS K 6885-79. Unsintered Polytetrafluoroethylene Tapes for Thread Sealing. 10 pp.
JIS K 6887-77. Polytetrafluoroethylene Tapes (R 1980). 12 pp.

—**Pressure Sensitive**
BSI BS 3924-78. 1978 Amd 1 Pressure-Sensitive Adhesive Tapes for Electrical Insulating Purposes. 21 pp.
CNS C3205-88. Method of Test for Pressure-Sensitive Adhesive Tapes for Electrical Insulation (Apr)(12248).
CNS C3206-88. Subsidiary Method of Test for Pressure-Sensitive Adhesive Tapes for Electrical Insulation (Apr)(12249).
DIN ENGL 40631-68. Pressure-Sensitive Adhesive Tapes Used for Electrical Insulation; Dimensions (Jan). 2 pp.
DIN ENGL 40633 Pt 3-70. Pressure-Sensitive Adhesive Tapes Used for Electrical Insulation; Tapes with Thermosetting Lamina of Adhesive; Types; Requirements; Testing (July). 9 pp.
DIN VDE 0340 Pt 1-75. VDE Specification for Pressure-Sensitive Adhesive Tapes Used for Electrical Insulation Plastic Tapes (May). 28 pp.
DIN VDE 0340 Pt 1 (D)-90. Pressure-Sensitive Adhesive Tapes for Electrical Purposes; General Requirements Identical to IEC 15C (CO) 278 (Dec). 16 pp.
IEC 454 Pt 1-92. Specifications for Pressure-Sensitive Adhesive Tapes for Electrical Purposes Part 1: General Requirements Second Edition. 26 pp.
IEC 454 Pt 2-74. Specification for Pressure-Sensitive Adhesive Tapes for Electrical Purposes Part 2: Methods of Test First Edition, (Supplement A-1978). 40 pp.
JIS C 2107-91. Testing Methods of Pressure-Sensitive Adhesive Tapes for Electrical Insulation. 19 pp.

—**PVC Film**
CNS C4049-82. Adhesive Polyvinyl Chloride Tape for Electric Insulation (Mar)(2064).
IEC 454 Pt 3-1-76. Specification for Pressure-Sensitive Adhesive Tapes for Electrical Purposes Part 3: Specifications for Individual Materials Sheet 1: Requirements for Plasticized Polyvinylchloride with Non-Thermosetting Adhesive First Edition. 12 pp.
JIS C 2336-91. Pressure-Sensitive Adhesive Polyvinyl Chloride Tapes for Electrical Insulation. 7 pp.
SAA AS 1621.3.1-88. Pressure-Sensitive Adhesive Electrical Tapes—Part 3: Specific Requirements—Part 3.1: Plasticized Polyvinylchloride Tapes with Non-Thermosetting Adhesive. 3 pp.

—**PVC Film—Cable Joints**
CSA C22.2 NO 197-M1983. PVC Insulating Tape (R 1992); (Gen Instr 1). 15 pp.

—**PVC Film—Splices (Electrical)**
CSA C22.2 NO 197-M1983. PVC Insulating Tape (R 1992); (Gen Instr 1). 15 pp.

—**Rayon Fabric—Varnished**
CNS C4318-88. Varnished Rayon Cloth Tape (Oct)(7606).
CNS C4319-88. Varnished Rayon Bias Tape (Oct)(7607).
JIS C 2403-71. Varnished Rayon Bias Tapes. 6 pp.

—**Rubber**
CNS C3021-69. Method of Test for Insulated Rubber Tape (Jun)(1144) (R 1973). 3 pp.
CNS C4026-69. Insulating Rubber Tape (Jun)(1143)(R 1973).

—**Silk Fabric—Varnished**
BSI BS 419-66. 1966 Varnished Fabrics and Tapes for Electrical Purposes. 36 pp.
CNS C4112-88. Varnished Silk Tape (Oct)(3425).

—**Terephthalate Fabric—Varnished**
CNS C4316-89. Varnished Terephthalate Cloth Tape (Jan)(7604).
JIS C 2376-81. Varnished Terephthalate Cloth Tapes.
JIS C 2376-71. Varnished Terephthalate Cloth Tapes. 6 pp.

—**Thermosetting**
DIN VDE 0340 Pt 3-70. Specifications for Self-Adhesive Insulating Tapes; Part 3: Tapes with Thermosetting Adhesive Layer (Aug). 34 pp.

—**Woven**
CNS C3155-82. Method of Test for Woven Tapes for Electrical Insulation (Jul)(9084).

Electrical Insulating Varnishes
See Also: Electrical Insulating Oils; Electrical Insulation; Varnishes
CNS C3027-88. Method of Test for Varnish of Electrical Insulation (Oct) (2604).
CNS C3033-85. Method of Test for Insulating Varnish (Aug)(3226).

Electrical Insulating Varnishes (Cont.)
CNS C4052-80. Insulated Varnish (Jul)(2355).
CNS C4455-88. Finished Varnishes for Electrical Insulation (Oct)(12437).
JIS C 2103-91. Testing Methods of Varnishes for Electrical Insulation. 33 pp.
MOD UK DSTAN 80-129-89. Varnish Electrical Insulating Brushing Type 1: Standard Type 2: Fungicidal Issue 1. 18 pp.
MOD UK DSTAN 80-129-93. Varnish Electrical Insulating Brushing Type 1: Standard Type 2: Fungicidal Issue 2. 18 pp.
SAA AS 1556-73. Air-Drying Insulating and Sealing Varnishes for Electrical Purposes Corrig.. 14 pp.
SAA AS 1557-73. Baking Insulating and Impregnating Varnishes for Electrical Purposes Corrig.. 16 pp.

—**Adhesive**
CNS C4458-88. Adhesive Varnish (Oct)(12440).

—**Bonding Strength—Wire Bundle Tests**
IEC 699-81. Test Method for the Evaluation of Bond Strength of Impregnating Varnishes by the Wire Bundle Test First Edition. 17 pp.

—**Coils**
CNS C4457-88. Coil Impregnating Varnishes for Electrical Insulation (Oct)(12439).
JIS C 2353-83. Coil Impregnating Varnishes for Electrical Insulation. 9 pp.

—**Drying**
MOD UK DEF-1053: METH NO. 38. Standard Methods of Testing Paint, Varnish, Lacquer and Related Products Method 38: Insulating Varnishes (Clear and Pigmented). Preparation of Specimens for Electrical Tests and for Assessment of Drying Properties (Withdrawn).

—**Electrical Properties**
MOD UK DEF-1053: METH NO. 38. Standard Methods of Testing Paint, Varnish, Lacquer and Related Products Method 38: Insulating Varnishes (Clear and Pigmented). Preparation of Specimens for Electrical Tests and for Assessment of Drying Properties (Withdrawn).

—**Electrical Resistivity**
MOD UK DEF-1053: METH NO. 45. Standard Methods of Testing Paint, Varnish, Lacquer and Related Products. Indices Method 45: Surface Resistivity of Insulating Varnishes (Clear and Pigmented) (Withdrawn).

—**Enameled Wires**
CNS C4456-88. Varnish for Enameled Wire (Oct)(12438).
JIS C 2351-88. Varnishes for Enameled Wires. 12 pp.

—**Finishing**
DIN ENGL 46449-70. Insulating Varnishes and Insulating Resin Materials for Electrical Purposes; Finishing Varnishes; Test Methods (Apr). 8 pp.
JIS C 2350-83. Finishing Varnishes for Electrical Insulation. 9 pp.

—**Flexibility**
MOD UK DEF-1053: METH NO. 40. Standard Methods of Testing Paint, Varnish, Lacquer and Related Products. Indices Method 40: The Effects of Heating on the Flexibility of Electrical Insulating Varnishes (Withdrawn).

—**Fungus Resistance**
MOD UK DSTAN 80-129-89. Varnish Electrical Insulating Brushing Type 1: Standard Type 2: Fungicidal Issue 1. 18 pp.
MOD UK DSTAN 80-129-93. Varnish Electrical Insulating Brushing Type 1: Standard Type 2: Fungicidal Issue 2. 18 pp.

—**Marine**
MOD UK NES 2058-89. Specification for Anti-Tracking Varnish NSN 8010-99-942-9331 for Application to Synthetic Resin-Bonded Materials for Electrical Purposes Issue 1 (05.89). 8 pp.
MOD UK NES 2058-93. Specification for Anti-Tracking Varnish NSN 8010-99-942-9331 for Application to Synthetic Resin-Bonded Materials for Electrical Purposes Issue 2 (04.93). 9 pp.

—**Phenolic Resins**
CNS C3203-88. Method of Test for Oil-Soluble 100% Phenolic Resins of Electrical Insulating Varnished (Apr)(12246).
JIS C 2104-75. Testing Methods of Oil-Soluble 100 % Phenolic Resins for Electrical Insulating Varnishes.

—**Printed Circuit Base Materials—Fungus Resistant**
MOD UK TS 478. Varnish, Fungicidal for Printed Circuit Boards (Withdrawn).

INDUSTRY STANDARDS

Electrical

Electrical Insulating Varnishes (Cont.)

—Silicone
CNS C3153-82. Method of Test for Silicone Varnish for Electrical Insulation (Jul)(9082).

—Sleeving
CNS C4454-88. Varnished Sleeving for Electrical Insulation (Oct)(12436).

—Solvents
BSI BS 5629: Part 1-78. 1978 Insulating Varnishes Containing Solvent Part 1: Definitions and General Requirements. 4 pp.
BSI BS 5629: Part 2-79. 1979 Insulating Varnishes Containing Solvent Part 2: Test Methods. 20 pp.
BSI BS 5629: Part 3-79. 1979 Insulating Varnishes Containing Solvent Part 3: Method for Specifying Requirements for Individual Materials. 12 pp.
IEC 464 Pt 1-76. Specification for Insulating Varnishes Containing Solvent Part 1: Definitions and General Requirements First Edition. 9 pp.
IEC 464 Pt 2-74. Specification for Insulating Varnishes Containing Solvent Part 2: Test Methods First Edition. 36 pp.
IEC 464 Pt 3-1-86. Specification for Insulating Varnishes Containing Solvent Part 3: Specifications for Individual Materials Sheet 1: Requirements for Cold Curing Finishing Varnishes First Edition. 11 pp.
IEC 464 Pt 3-2-89. Specification for Insulating Varnishes Containing Solvent Part 3: Specifications for Individual Materials Sheet 2: Requirements for Hot Curing Impregnating Varnishes First Edition. 27 pp.

—Thermodynamic Properties
CENELEC HD 570 S1-90. Test Procedure for Thermal Endurance of Insulating Varnishes—Electric Strength Method. 3 pp.
IEC 290-69. Evaluation of the Thermal Endurance of Electrical Insulating Varnishes by the Helical Coil Bond Test First Edition. 17 pp.
IEC 370-71. Test Procedure for Thermal Endurance of Insulating Varnishes—Electric Strength Method First Edition. 18 pp.
IEC 699-81. Test Method for the Evaluation of Bond Strength of Impregnating Varnishes by the Wire Bundle Test First Edition. 17 pp.

—Windings
MOD UK DEF-31-A-64. Varnish, Impregnating, for Electrical Purposes, Baking (Reprinted June, 1968). 5 pp.

Electrical Insulation

Use For: Electrical Insulating Materials
See Also: Blanket Insulation; Cable Insulation; Dielectrics; Electric Wire; Electrical Insulating Bushings; Electrical Insulating Fabrics; Electrical Insulating Films; Electrical Insulating Liquids; Electrical Insulating Mats; Electrical Insulating Oils; Electrical Insulating Papers; Electrical Insulating Tapes; Electrical Insulating Varnishes; Electrical Insulators; Electrical Properties; Fiberglass Ropes; Filling Compounds; Heat Shrink Tubing; Insulation Coordination; Insulation Resistance; Porcelain Insulators; Post Insulators; Protective Clothing; Sleeving; Treeing

CENELEC HD 437-84. Standard Conditions for Use Prior to and During the Testing of Solid Electrical Insulating Materials. 2 pp.
CNS C3028-68. Method of Test for Insulating Compound (Jan)(2802) (R 1973).
CNS C4050-83. Classification of Insulating Materials for Electrical Machineries and Apparatus (Feb)(2147).
CNS C4447-87. Electrical Insulating Compounds (Nov)(12151).
CSA CAN/CSA-C22. 2 NO 0-M91. General Requirements—Canadian Electrical Code, Part II; (Gen Instr 1). 48 pp.
IEC 212-71. Standard Conditions for Use Prior to and During the Testing of Solid Electrical Insulating Materials Second Edition. 17 pp.
IEC 505-75. Guide for the Evaluation and Identification of Insulation Systems of Electrical Equipment First Edition. 59 pp.
JIS C 2380-86. Electrical Insulating Compounds. 14 pp.
JIS C 4003-77. Classification of Materials for Insulation of Electrical Machinery and Apparatus. 15 pp.
MOD UK DSTAN 68-146-91. Insulating and Filling Compound (Chatterton's Type) for Electrical Purposes Issue 1. 11 pp.
MOD UK DEF-144-63. Insulating & Filling Compound (Chatterton's Type) for Electrical Purposes (Superseded by Def Stan 68-146). 3 pp.

Electrical Insulation (Cont.)

—Aging Testing
BSI BS 5691: Part 1-79. 1979 Determination of Thermal Endurance Properties of Electrical Insulating Materials Part 1: General Procedures for the Determination of Thermal Endurance Properties, Temperature Indices and Thermal Endurance Profiles. 19 pp.
BSI BS 5691: Part 2-79. 1979 Amd 1 Determination of Thermal Endurancre Properties of Electrical Insulating Materials Part 2: List of Materials and Available Tests. 11 pp.
CENELEC HD 611.1 S1-92. Guide for the Determination of Thermal Endurance Properties of Electrical Insulating Materials Part 1: General Guidelines for Ageing Procedures and Evaluation of Test Results. 6 pp.
IEC 216 Pt 1-90. Guide for the Determination of Thermal Endurance Properties of Electrical Insulating Materials Part 1: General Guidelines for Ageing Procedures and Evaluation of Test Results Fourth Edition. 55 pp.
IEC 610-78. Principal Aspects of Functional Evaluation of Electrical Insulation Systems: Aging Mechanisms and Diagnostic Procedures First Edition. 23 pp.
IEC 941-88. Mechanical Endurance Functional Tests for Electrical Insulation Systems First Edition. 24 pp.

—Appliances
CNS C1035-67. Insulating Compound (for Common Electric Appliance) (Oct)(2801) (R 1973). 3 pp.
CNS C3172-83. Classification and Test Methods for Insulation of Household Appliances and Equipments (Jun)(10315).
JIS C 0702-74. General Rules for Insulation Construction of Class II Electrical Appliances (R 1980). 18 pp.
JIS C 0703-80. Classification and Test Methods for Insulation of Household Appliances and Equipments. 16 pp.

—Arc Testing
DIN VDE 0303 Pt 5-90. Testing of Electrical Insulating Materials; Low-Voltage High-Current Arc Test (July) (Replaces DIN 53484/10.55X). 11 pp.

—Asbestos
BSI BS 7018-88. 1988 Guide to Use of Electrical Insulating Materials Containing Asbestos. 11 pp.
CNS R2176-85. Asbestos Yarns and Twisted Ropes (May) (11265).
CNS R2178-85. Asbestos Mill Boards (May)(11267).
JIS R 3450-79. Asbestos Yarns and Twisted Ropes. 10 pp.
JIS R 3454-79. Asbestos Mill Boards. 6 pp.

—Asbestos Cement
CNS C4352-82. Asbestos Cement Plate for Electric Insulation (Jul)(9086).
JIS C 2210-75. Asbestos Cement Plate for Electric Insulation (R 1983). 9 pp.

—Boards—Classification
SAA AS 1795.1-83. Sheets and Boards for Electrical Purposes—Part 1: Classification and General Requirements. 15 pp.

—Breakdown Voltage
DIN VDE 0303 Pt 2-74. Electrical Tests of Insulating Materials; Breakdown Voltage; Electric Strength (Nov). 23 pp.

—Building Equipment
CNS A3111-80. Method of Test for Electric Insulation of Equipment Units for Dwellings (Aug)(6302).
JIS A 1711-76. Method of Test for Electric Insulation of Equipment Unit for Dwellings (R 1983). 7 pp.

—Ceramics
BSI BS 6045: Part 1-81. 1981 Ceramic and Glass Electrical Insulating Materials Part 1: Definitions and Classification. 4 pp.
CENELEC HD 426.1 S1-82. Specification for Ceramic and Glass Insulating Materials—Part 1: Definitions and classification. 1 p.
CENELEC HD 426.2 S1-88. Specification for Ceramic and Glass Insulating Materials—Part 2: Methods of Test. 3 pp.
CENELEC HD 426.3 S1-88. Specification for Ceramic and Glass Insulating Materials-Part 3: Individual Materials. 3 pp.
IEC 672 Pt 1-80. Specification for Ceramic and Glass Insulating Materials Part 1: Definitions and Classification First Edition. 11 pp.
IEC 672 Pt 2-80. Specification for Ceramic and Glass Insulating Materials Part 2: Methods of Test First Edition; (Corrigendum—April 1990). 45 pp.
IEC 672 Pt 3-84. Specification for Ceramic and Glass Insulating Materials Part 3: Individual Materials First Edition. 17 pp.

Electrical Insulation (Cont.)

—Ceramics—Cracking (Fracturing)
BSI BS 7134: Sec 1.1-89. 1989 Testing of Engineering Ceramics Part 1: General and Textural Properties Section 1.1: Method for Determination of the Presence of Cracks and Other Defects by Dye Penetration Tests. 4 pp.

—Ceramics—Density
BSI BS 7134: Sec 1.2-89. 1989 Testing of Engineering Ceramics Part 1: General and Textural Properties Section 1.2: Methods of Determination of Density and Porosity. 10 pp.

—Ceramics—Porosity
BSI BS 7134: Sec 1.2-89. 1989 Testing of Engineering Ceramics Part 1: General and Textural Properties Section 1.2: Methods of Determination of Density and Porosity. 10 pp.

—Ceramics—Sampling
BSI BS 7134: Part 0-89. 1989 Testing of Engineering Ceramics Part 0: Introduction and Guide to Sampling. 7 pp.

—Chloroprene—Tubes—Heat Shrinkable
CNS C4380-82. Heat-Shrinkable Chloroprene Tubing for Electrical Insulation (Tubular Type) (Dec)(9690)

—Coal Urines
SAA AS 1147.1-89. Electrical Equipment for Coal Mines—Insulating Materials—Part 1: Materials for Insulating Power Conducting Components. 7 pp.

—Coatings—Interlaminar—Adhesion Testing
BSI BS 6404: Part 12-93. 1993 Magnetic Materials Part 12: Guide to Methods of Assessment of Temperature Capability of Interlaminar Insulation Coatings (IEC 404-12: 1992) (V). 16 pp.
IEC 404 Pt 12-92. Magnetic Materials Part 12: Guide to Methods of Assessment of Temperature Capability of Interlaminar Insulation Coatings First Edition. 28 pp.

—Coatings—Interlaminar—Compressibility Factors
BSI BS 6404: Part 12-93. 1993 Magnetic Materials Part 12: Guide to Methods of Assessment of Temperature Capability of Interlaminar Insulation Coatings (IEC 404-12: 1992) (V). 16 pp.
IEC 404 Pt 12-92. Magnetic Materials Part 12: Guide to Methods of Assessment of Temperature Capability of Interlaminar Insulation Coatings First Edition. 28 pp.

—Coatings—Interlaminar—Temperature Testing
BSI BS 6404: Part 12-93. 1993 Magnetic Materials Part 12: Guide to Methods of Assessment of Temperature Capability of Interlaminar Insulation Coatings (IEC 404-12: 1992) (V). 16 pp.
IEC 404 Pt 12-92. Magnetic Materials Part 12: Guide to Methods of Assessment of Temperature Capability of Interlaminar Insulation Coatings First Edition. 28 pp.

—Color Coding
BSI BS 6746C-69. 1969 Amd 1 Colour Chart for Insulation and Sheath of Electric Cables. 4 pp.
CENELEC HD 402 S2-85. Standard Colours for Thermoplastic Materials Used for the Insulation for Low-Frequency Cables and Wires. 2 pp.
IEC 304-82. Standard Colours for Insulation for Low-Frequency Cables and Wires Third Edition. 11 pp.

—Conduits
BSI BS 4607: Part 5-82. 1982 Amd 2 Non-Metallic Conduit Fittings for Electrical Installations Part 5: Rigid Conduit, Fittings and Components of Insulating Material. 20 pp.
BSI BS 6099: Sec 2.2-82. 1982 Conduits for Electrical Installations Part 2: Particular Specifications Section 2.2: Rigid Plain Conduits of Insulating Materials. 15 pp.
IEC 614 Pt 2-2-80. Specification for Conduits for Electrical Installations Part 2: Particular Specification for Rigid Plain Conduits Insulating Materials First Edition. 22 pp.
IEC 614 Pt 2-3-90. Specification for Conduits for Electrical Installations Part 2: Particular Specifications for Conduits Section Three—Pliable Conduits of Insulating Material First Edition. 29 pp.
IEC 614 Pt 2-4-85. Specification for Conduits for Electrical Installations Part 2: Particular Specifications for Conduits Section Four—Pliable Self-Recovering Conduits of Insulating Materials First Edition. 24 pp.
IEC 614 Pt 2-5-92. Specification for Conduits for Electrical Installations Part 2: Particular Specifications for Conduits Section 5: Flexible Conduits First Edition. 37 pp.

Electrical Insulation *(Cont.)*

—Data Processing Equipment

CSA CAN/CSA-C22. 2 NO 220-M91. Information Processing and Business Equipment; (Gen Instr 1). 119 pp.

—Data Processing Equipment—Capacitors

CSA CAN/CSA-C22. 2 NO 220-M91. Information Processing and Business Equipment; (Gen Instr 1). 119 pp.

—Dielectric Dissipation Factor

BSI BS 4542-70. 1970 Determination of Loss Tangent and Permittivity of Electrical Insulating Materials in Sheet Form (Lynch Method). 12 pp.
IEC 250-69. Recommended Methods for the Determination of the Permittivity and Dielectric Dissipation Factor of Electrical Insulating Materials at Power, Audio and Radio Frequencies Including Metre Wavelengths First Edition. 56 pp.
IEC 377 Pt 1-73. Recommended Methods for the Determination of the Dielectric Properties of Insulating Materials at Frequencies Above 300 MHz Part 1: General First Edition. 26 pp.
IEC 377 Pt 2-77. Methods for the Determination of the Dielectric Properties of Insulating Materials at Frequencies Above 300 MHz Part 2: Resonance Methods First Edition. 40 pp.

—Dielectric Properties

DIN ENGL 53483 Suppl. 1-55. Testing of Insulating Materials; Determination of the Dielectric Constant (Relative Permittivity) and of the Loss Factor; Measuring Equipment (Oct). 4 pp.
DIN ENGL 53483 Pt 1-69. Testing of Insulating Materials; Determination of Dielectric Properties; Definitions; General Information (July). 4 pp.
DIN ENGL 53483 Pt 2-70. Testing of Insulating Materials; Determination of Dielectric Properties; Testing at Standard Frequencies of 50 Hz, 1 kHz, 1 MHz (Mar). 8 pp.
DIN ENGL 53483 Pt 3-69. Testing of Insulating Materials; Determination of Dielectric Properties; Measuring Cells for Liquids for Frequencies up to 100 MHz (July). 3 pp.
DIN VDE 0303 Pt 2-74. Electrical Tests of Insulating Materials; Breakdown Voltage; Electric Strength (Nov). 23 pp.
DIN VDE 0303 Pt 4-69. Regulations for Electrical Tests on Insulating Materials; Part 4: Determination of Dielectric Properties (Dec). 38 pp.

—Dielectric Strength

CENELEC HD 559.1 S1-91. Methods of Test for Electric Strength of Solid Insulating Materials Part 1: Tests at Power Frequencies. 6 pp.
CENELEC HD 559.2 S1-91. Methods of Test for Electric Strength of Solid Insulating Materials—Part 2: Additional Requirements for Tests Using Direct Voltage. 6 pp.
CNS C3091-80. Method of Test for Electric Strength of Solid Insulating Materials (Aug)(6040).
IEC 243 Pt 1-88. Methods of Test for Electric Strength of Solid Insulating Materials Part 1: Tests at Power Frequencies First Edition; (Corrigendum—March 1989). 41 pp.
IEC 243 Pt 2-90. Methods of Test for Electric Strength of Solid Insulating Materials Part 2: Additional Requirements for Tests Using Direct Voltage First Edition. 17 pp.
JIS C 2110-75. Testing Methods for Electric Strength of Solid Insulating Materials. 13 pp.

—Elastomeric—Sleeving

MOD UK DSTAN 59-15: Part 1-90. Elastomeric Insulating Materials and Polytetrafluoroethylene (PTFE) Part 1: General Requirements, Tests and Test Methods for Qualification Approval Sleeving in Continuous Lengths Sleeves for Binding and/or Identification Purposes. 36 pp.

—Electric Endurance Testing

IEC 727 Pt 1-82. Evaluation of Electrical Endurance of Electrical Insulation Systems Part 1: General Considerations and Evaluation Procedures Based on Normal Distributions First Edition. 34 pp.
IEC 727 Pt 2-93. Evaluation of Electrical Endurance of Electrical Insulation Systems Part 2: Evaluation Procedures Based on Extreme-Value Distributions First Edition. 74 pp.

—Electrical Resistance

CNS C6111-81. Method of Test for D.C. Resistance or Conductance of Insulating Materials (May)(7367).

—Electrical Resistivity

BSI BS 6233-82. 1982 Volume Resistivity and Surface Resistivity of Solid Electrical Insulating Materials. 24 pp.
CENELEC HD 429-83. Methods of Test for Volume Resistivity and Surface Resistivity of Solid Electrical Insulating Materials. 2 pp.
IEC 93-80. Methods of Test for Volume Resistivity and Surface Resistivity of Solid Electrical Insulating Materials Second Edition. 44 pp.

Electrical Insulation *(Cont.)*

—Electrolytic Corrosion

BSI BS 5735-79. 1979 Test Methods for Determining Electrolytic Corrosion with Electric Insulating Materials. 24 pp.
DIN ENGL 53489-68. Testing of Electrical Insulating Materials; Assessing the Effect of Electrolytic Corrosion (Jan). 5 pp.
DIN VDE 0303 Pt 6-68. Regulations for Electrical Tests on Insulating Materials; Part 6: Determination of Electrolytic Corrosive Effects (Mar). 15 pp.
IEC 426-73. Test Methods for Determining Electrolytic Corrosion with Insulating Materials First Edition. 47 pp.

—Electrostatic Properties

DIN VDE 0303 Pt 8-75. VDE Specifications for Electrical Tests on Insulating Materials; Evaluation of the Electrostatic Behaviour (Apr). 18 pp.

—Endurance Testing

IEC 216 Pt 3-2-93. Guide for the Determination of Thermal Endurance Properties of Electrical Insulating Materials Part 3: In-structions for Calculating Thermal Endurance Characteristics Section 2—Calculations for Incomplete Data: Proof Test Results up to and Including the Median. 66 pp.
IEC 1026-91. Guidelines for Application of Analytical Test Methods for Thermal Endurance Testing of Electrical Insulating Materials First Edition. 67 pp.

—Environmental Testing

BSI BS 5604-86. 1986 Evaluating Resistance to Tracking and Erosion of Electrical Insulating Materials Used Under Severe Ambient Conditions. 12 pp.
CENELEC HD 380 S2-87. Test Methods for Evaluating Resistance to Tracking and Erosion of Electrical Insulating Materials Used Under Severe Ambient Conditions. 3 pp.
IEC 587-84. Test Methods for Evaluating Resistance to Tracking and Erosion of Electrical Insulating Materials Used Under Severe Ambient Conditions Second Edition. 23 pp.

—Epoxy Resins

BSI BS 3815-64. 1964 Epoxide Resin Casting Systems for Electrical Applications. 17 pp.
BSI BS 3816-89. 1989 Epoxy Resin Casting Systems Used for Electrical Insulating Applications at Power Frequencies. 4 pp.
CENELEC HD 307.3.1 S1-89. Specification for Solventless Polymerisable Resinous Compounds Used for Electrical Insulation-Part 3: Specifications for Individual Materials. Sheet one: Unfilled Epoxy Resinous Compounds. 3 pp.
CENELEC HD 307.3.2 S1-88. Specification for Solventless Polymerisable Resinous Compounds Used for Electrical Insulation-Part 3: Specifications for Individual Materials. Sheet Two: Quartz Filled Epoxy Resinous Compounds. 3 pp.
IEC 455 Pt 3-1-81. Specification for Solventless Polymerisable Resinous Compounds Used for Electrical Insulation Part 3: Specifications for Individual Materials Sheet 1: Unfilled Epoxy Resinous Compounds First Edition. 13 pp.
IEC 455 Pt 3-2-87. Specification for Solventless Polymerisable Resinous Compounds Used for Electrical Insulation Part 3: Specifications for Individual Materials Sheet 2: Quartz Filled Epoxy Resinous Compounds First Edition. 11 pp.
IEC 893 Pt 3-2-93. Specification for Industrial Rigid Laminated Sheets Based on Thermosetting Resins for Electrical Purposes Part 3: Specifications for Individual Materials —Sheet 2: Requirements for Rigid Laminated Sheets Based on Epoxide Resins First Edition. 28 pp.

—Epoxy Resins—Coatings

CENELEC HD 307.3.11 S1-90. Specification for Solventless Polymerisable Resinous Compounds Used for Electrical Insulation Part 3: Specification for Individual Materials Sheet 11: Epoxy Resin-Based Coating Powders. 3 pp.
IEC 455 Pt 3-11-88. Specification for Solventless Polymerisable Resinous Compounds Used for Electrical Insulation Part 3: Specifications for Individual Materials Sheet 11: Epoxy Resin-Based Coating Powders First Edition. 17 pp.

—Fatigue (Materials)

IEC 727 Pt 1-82. Evaluation of Electrical Endurance of Electrical Insulation Systems Part 1: General Considerations and Evaluation Procedures Based on Normal Distributions First Edition. 34 pp.

—Fire Testing

SAA AS 2420-87. Fire Test Methods for Solid Insulation Materials and Non-Metallic Enclosures Used in Electrical Equipment Amdt 1 January 1991 (In Professional Packages 28, 44). 29 pp.
SNZ AS 2420-87. Fire Test Methods for Solid Insulating Materials and Non-Metallic Enclosures Used in Electrical Equipment. 34 pp.

Electrical Insulation *(Cont.)*

—Flammability Testing

BSI BS 6334-83. 1983 Determination of the Flammability of Solid Electrical Insulating Materials When Exposed to an Igniting Source. 12 pp.
BSI BS 7447-91. 1991 Method of Test for Determination of the Ignitability of Soild Electrical Insulating Materials When Exposed to an Electrically Heated Wire Source. 10 pp.
CENELEC HD 441-83. Methods of Test for the Determination of the Flammability of Solid Electrical Insulating Materials When Exposed to an Igniting Source. 2 pp.
CENELEC HD 541 S1-91. Methods of Test for the Determination of the Ignitability of Solid Electrical Insulating Materials When Exposed to Electrically Heated Wire Sources. 7 pp.
DIN VDE 0304 Pt 3-85. Thermal Properties of Electrical Insulating Materials; Flammability Under the Action of Igniting Sources (National Foreword Only Corresponds to IEC 707—1981) (Sept). 6 pp.
IEC 707-81. Methods of Test for the Determination of the Flammability of Solid Electrical Insulating Materials When Exposed to an Igniting Source First Edition; (Amendment 1-1992). 33 pp.
IEC 829-88. Methods of Test for the Determination of the Ignitability of Solid Electrical Insulating Materials When Exposed to Electrically Heated Wire Sources First Edition. 26 pp.

—Flexible Materials

BSI BS 5785: Part 2-79. 1979 Combined Flexible Material for Electrical Insulation Part 2: Methods of Test. 15 pp.
BSI BS 6921-88. 1988 Method of Evaluation of the Thermal Endurance of Flexible Sheet Electrical Insulating Materials Using the Wrapped Tube Method. 7 pp.
CENELEC HD 480 S1-88. Test Method for Evaluating Thermal Endurance of Flexible Sheet Materials Using the Wrapped Tube Method. 3 pp.
IEC 626 Pt 1-79. Specification for Combined Flexible Materials for Electrical Insulation Part 1: Definitions and General Requirements First Edition. 12 pp.
IEC 626 Pt 2-78. Specification for Combined Flexible Materials for Electrical Insulation Part 2: Methods of Test First Edition. 26 pp.
IEC 626 Pt 3-88. Specification for Combined Flexible Materials for Electrical Insulation Part 3: Specifications for Individual Materials First Edition; (Corrigendum—Sept 1988). 38 pp.

—Flexible Materials—Endurance Testing

IEC 795-84. Test Method for Evaluating Thermal Endurance of Flexible Sheet Materials Using the Wrapped Tube Method First Edition. 13 pp.

—Fluorinated Ethylene Propylene—Tubing—Heat Shrinkable

CNS C4378-82. Heat-Shrinkable FEP Tubing for Electrical Insulation (Tubular Type) (Dec)(9688).

—Functional Testing

IEC 610-78. Principal Aspects of Functional Evaluation of Electrical Insulation Systems: Aging Mechanisms and Diagnostic Procedures First Edition. 23 pp.
IEC 791-84. Performance Evaluation of Insulation Systems Based on Service Experience and Functional Tests First Edition. 20 pp.
IEC 792 Pt 1-85. Multi-Factor Functional Testing of Electrical Insulation Systems Part 1: Test Procedures First Edition. 34 pp.
IEC 941-88. Mechanical Endurance Functional Tests for Electrical Insulation Systems First Edition. 24 pp.

—Functional Testing—Bibliographies

IEC 792 Pt 2-93. Multi-Factor Functional Testing of Electrical Insulation Systems Part 2: Bibliography First Edition. 53 pp.

—Glass

BSI BS 6045: Part 1-81. 1981 Ceramic and Glass Electrical Insulating Materials Part 1: Definitions and Classification. 4 pp.
CENELEC HD 426.1 S1-82. Specification for Ceramic and Glass Insulating Materials—Part 1: Definitions and classification. 1 p.
CENELEC HD 426.2 S1-88. Specification for Ceramic and Glass Insulating Materials—Part 2: Methods of Test. 3 pp.
CENELEC HD 426.3 S1-88. Specification for Ceramic and Glass Insulating Materials-Part 3: Individual Materials. 3 pp.
IEC 672 Pt 1-80. Specification for Ceramic and Glass Insulating Materials Part 1: Definitions and Classification First Edition. 11 pp.
IEC 672 Pt 2-80. Specification for Ceramic and Glass Insulating Materials Part 2: Methods of Test First Edition; (Corrigendum—April 1990). 45 pp.
IEC 672 Pt 3-84. Specification for Ceramic and Glass Insulating Materials Part 3: Individual Materials First Edition. 17 pp.

Electrical Insulation (Cont.)

—Glass—Sleeving
JIS C 2411-78. Silicone Varnished Glass Sleeving for Electrical Insulation. 11 pp.

—Glass Transition Temperature
BSI BS EN 61006-93. 1993 Methods of Test for Determination of the Glass Transition Temperature of Electrical Insulating Materials (IEC 1006: 1991) (F). 29 pp.
CENELEC EN 61006-93. Methods of Test for the Determination of the Glass Transition Temperature of Electrical Insulating Materials (IEC 1006: 1991). 24 pp.
IEC 1006-91. Methods of Test for the Determination of the Glass Transition Temperature of Electrical Insulating Materials First Edition; (CENELEC EN 61006:1993). 50 pp.

—Glossaries
BSI BS 4727:Pt1: Group 10-91. 1991 Electrotechnical, Power, Telecommunication, Electronics, Lighting and Colour Terms Part 1: Terms Common to Power Telecommunications and Electronics Group 10: Insulating Solids, Liquids and Gases. 101 pp.
BSI BS 6236: Part 1-82. 1982 Electrical Insulating Materials Based on Mica Part 1: General Requirements. 4 pp.
CENELEC HD 307.1 S2-81. Specification for Solventless Polymerizable Resinous Compounds Used for Electrical Insulation—Part 1: Definitions and General Requirements. Basis for Classification of Polymerisable Resinous Compounds. 2 pp.
CENELEC HD 352.1 S2-83. Specification for Insulating Materials Based on Mica Part 1: Definitions and General Requirements. 3 pp.
CENELEC HD 352.1 S2-89. Specification for Insulating Materials Based on Mica-Part 1: Definitions and General Requirements. 2 pp.
CENELEC HD 426.1 S1-82. Specification for Ceramic and Glass Insulating Materials—Part 1: Definitions and classification. 1 p.
CNS C1068-80. Standard Definitions of Terms Relating to Electric Insulation (Aug)(6014).
CNS C1071-80. Standard Definitions of Terms Relating to Hook-Up Wire Insulation (Aug)(6017).
CSA CAN/CSA-C22. 2 NO 0-M91. General Requirements—Canadian Electrical Code, Part II; (Gen Instr 1). 48 pp.
IEC 50 Chap 212-90. International Electrotechnical Vocabulary Chapter 212: Insulating Solids, Liquids and Gases First Edition. 105 pp.
IEC 371 Pt 1-80. Specification for Insulating Materials Based on Mica Part 1: Definitions and General Requirements Second Edition. 11 pp.

—Heat of Crystallization—Calorimetry
BSI BS EN 61074-93. 1993 Method of Test for Determination of Heats and Temperatures of Melting and Crystallization of Electrical Insulating Materials by Differential Scanning Calorimetry (IEC 1074: 1991) (F). 18 pp.
CENELEC EN 61074-93. Determination of Heats and Temperatures of Melting and Crystallization of Electrical Insulating Materials by Differential Scanning Calorimetry (IEC 1074: 1991). 13 pp.
IEC 1074-91. Determination of Heats and Temperatures of Melting and Crystallization of Electrical Insulating Materials by Differential Scanning Calorimetry First Edition; (CENELEC EN 61074:1993). 29 pp.

—Heat of Fusion—Calorimetry
BSI BS EN 61074-93. 1993 Method of Test for Determination of Heats and Temperatures of Melting and Crystallization of Electrical Insulating Materials by Differential Scanning Calorimetry (IEC 1074: 1991) (F). 18 pp.
CENELEC EN 61074-93. Determination of Heats and Temperatures of Melting and Crystallization of Electrical Insulating Materials by Differential Scanning Calorimetry (IEC 1074: 1991). 13 pp.
IEC 1074-91. Determination of Heats and Temperatures of Melting and Crystallization of Electrical Insulating Materials by Differential Scanning Calorimetry First Edition; (CENELEC EN 61074:1993). 29 pp.

—Hoses
CNS C4370-82. Poly Insulating Line Hose for High Voltage (Nov)(9576).

—Impact Testing
CNS K6724-82. Method of Test for Impact Resistance of Plastics and Electrical Insulating Materials (Aug)(9284).

—Impregnating Resins
DIN ENGL 46448 Pt 1-69. Insulating Varnishes and Insulating Resin Materials for Electrical Engineering; Impregnating Resin Materials; Test Methods (Jan). 12 pp.
DIN ENGL 46448 Pt 2-69. Insulating Varnishes and Insulating Resin Materials for Electrical Engineering; Impregnating Resin Materials; Types (Jan). 3 pp.

Electrical Insulation (Cont.)

—Impregnating Resins (Cont.)
DIN VDE 0360 Pt 2-69. Requirements for Electrical Insulating Varnishes and Resins; Part 2: Impregnating Resins (May). 38 pp.

—Impulse Voltage Testing
IEC 243 Pt 3-93. Methods of Test for Electric Strength of Solid Insulating Materials Part 3: Additional Requirements for Impulse Tests First Edition. 26 pp.

—Ions
BSI BS 5591-78. 1978 Amd 1 Determination of Ionic Impurities in Electrical Insulating Materials by Extraction with Liquids. 9 pp.
CENELEC HD 381-78. Methods of Test for the Determination of Ionic Impurities in Electric Insulation Materials by Extraction with Liquids. 2 pp.
IEC 589-77. Methods of Test for the Determination of Ionic Purities in Electrical Insulating Materials by Extraction with Liquids First Edition; (Corrigendum—June 1978). 17 pp.

—KU Values
DIN VDE 0800 Pt 9-89. Telecommunications KU Values of Safety-Related Components and Insulation (May). 36 pp.

—Melamine Resins
IEC 893 Pt 3-3-93. Specification for Industrial Rigid Laminated Sheets Based on Thermosetting Resins for Electrical Purposes Part 3: Specifications for Individual Materials —Sheet 3: Requirements for Rigid Laminated Sheets Based on Melamine Resins First Edition. 24 pp.

—Mica
BSI BS 4145-67. 1967 Glass Mica Boards for Electrical Purposes. 29 pp.
BSI BS 6236: Part 1-82. 1982 Electrical Insulating Materials Based on Mica Part 1: General Requirements. 4 pp.
BSI BS 6236: Part 2-88. 1988 Electrical Insulating Materials Based on Mica Part 2: Methods of Test. 24 pp.
CENELEC HD 352.1 S2-83. Specification for Insulating Materials Based on Mica Part 1: Definitions and General Requirements. 3 pp.
CENELEC HD 352.1 S2-89. Specification for Insulating Materials Based on Mica-Part 1: Definitions and General Requirements. 2 pp.
CENELEC HD 352.2-76. Specification for Insulating Materials Based on Built-up Mica or Treated Mica Paper Part 2: Methods of Test. 2 pp.
CENELEC HD 352.3.1 S2-89. Specification for Insulating Materials Based on Mica-Part 3: Specifications for Individual Materials. Sheet One: Commutator Separators and Materials. 3 pp.
CENELEC HD 352.3.3 S1-89. Specification for Insulating Materials Based on Mica-Part 3: Specifications for Individual Materials. Sheet Three: Specifications for Rigid Mica Materials for Heating Equipment. 3 pp.
CNS C3131-88. Method of Test for Electrical Insulating Mica Product (Aug)8355).
CNS C4329-88. Heat Formable Mica Plate (Sep)(8358).
CNS C4330-88. Mica Plates for Commutator Separators (Sep)(8359).
CNS C4331-88. Press Work Mica Plate (Sep)(8360).
CNS C4332-88. Heater Mica Plate (Sep)(8361).
CNS C4333-88. Flexible Mica (Sep)(8363).
CNS C4336-88. Mica with Glass Cloth on Both Sides (Sheets and Tapes) (Sep)(8365).
CNS C4337-88. Mica with Glass Cloth on One Side (Sheets and Tapes) (Sep)(8366).
CNS C4338-88. Mica with Film on Both Sides (Sheets and Tapes) (Sep)(8367).
CNS C4339-88. Mica with Film on One Side (Sheets and Tapes) (Sep)(8368).
IEC 371 Pt 1-80. Specification for Insulating Materials Based on Mica Part 1: Definitions and General Requirements Second Edition. 11 pp.
IEC 371 Pt 2-87. Specification for Insulating Materials Based on Mica Part 2: Methods of Test Second Edition. 49 pp.
IEC 371 Pt 3-1-84. Specification for Insulating Materials Based on Mica Part 3: Specifications for Individual Materials Sheet 1: Commutator Separators and Materials Second Edition. 19 pp.
IEC 371 Pt 3-3-83. Specification for Insulating Materials Based on Mica Part 3: Specifications for Individual Materials Sheet 3: Specification for Rigid Mica Materials for Heating Equipment First Edition. 13 pp.
JIS C 2116-92. Testing Methods for Electrical Insulating Mica Products. 24 pp.
JIS C 2220-92. Mica Paper. 8 pp.
JIS C 2250-92. General Rules for Electrical Insulating Mica Products. 23 pp.
JIS C 2251-92. Rigid Mica Materials for Heat Molding. 7 pp.

Electrical Insulation (Cont.)

—Mica (Cont.)
JIS C 2252-92. Rigid Mica Materials for Commutator Separators. 7 pp.
JIS C 2254-92. Rigid Mica Materials for Electric Heating Equipment. 7 pp.
JIS C 2255-92. Flexible Mica Materials. 12 pp.
JIS C 2260-77. Mica with Films on Both Sides (Sheets and Tapes) (R 1985). 7 pp.
JIS C 2261-77. Mica with Films on One Side (Sheets and Tapes) (R 1985). 7 pp.

—Mica—Designations
CNS C1080-88. General Rules for Electrical Insulating Mica Product (Aug)(8357).

—Mica—Marine
MOD UK NES 2059-89. Specification for Mica and Micanite for Electrical Purposes Issue 1 (07.89). 13 pp.
MOD UK NES 2059-92. Specification for Mica and Micanite for Electrical Purposes Issue 2 (10.92). 17 pp.

—Office Machines
CSA CAN/CSA-C22. 2 NO 220-M91. Information Processing and Business Equipment; (Gen Instr 1). 119 pp.

—Olefin Resins—Sleeving
IEC 684 Pt 3-209-87. Specification for Flexible Insulating Sleeving Part 3: Specification Requirements for Individual Types of Sleeving Sheet 209: Heat Shrinkable Sleeving, General Purpose, Flame Retarded Polyolefin Shrink Ratio 2:1 First Edition. 17 pp.

—Panels—Classification
SAA AS 1795.1-83. Sheets and Boards for Electrical Purposes—Part 1: Classification and General Requirements. 15 pp.

—Permittivity
BSI BS 4542-70. 1970 Determination of Loss Tangent and Permittivity of Electrical Insulating Materials in Sheet Form (Lynch Method). 12 pp.
DIN VDE 0303 Pt 4-69. Regulations for Electrical Tests on Insulating Materials; Part 4: Determination of Dielectric Properties (Dec). 38 pp.
IEC 250-69. Recommended Methods for the Determination of the Permittivity and Dielectric Dissipation Factor of Electrical Insulating Materials at Power, Audio and Radio Frequencies Including Metre Wavelengths First Edition. 56 pp.
IEC 377 Pt 1-73. Recommended Methods for the Determination of the Dielectric Properties of Insulating Materials at Frequencies Above 300 MHz Part 1: General First Edition. 26 pp.
IEC 377 Pt 2-77. Methods for the Determination of the Dielectric Properties of Insulating Materials at Frequencies Above 300 MHz Part 2: Resonance Methods First Edition. 40 pp.

—Phenolic Resins
IEC 893 Pt 3-4-93. Specification for Industrial Rigid Laminated Sheets Based on Thermosetting Resins for Electrical Purposes Part 3: Specifications for Individual Materials —Sheet 4: Requirements for Rigid Laminated Sheets Based on Phenolic Resins First Edition. 28 pp.

—Plastic—Dielectric Properties
SAA AS 1255.4-74. Methods of Test for Electrical Characteristics of Solid Plastics Insulating Materials—Part 4: Determination of the Permittivity and Dielectric Dissipation Factor at Power, Audio and Radio Frequencies up to 300 MHZ. 38 pp.

—Plastic—Dielectric Strength
SAA AS 1255.3-73. Methods of Test for Electrical Characteristics of Solid Plastics Insulating Materials—Part 3: Determination of Electric Strength at Power Frequencies. 19 pp.

—Plastic—Electrical Resistance
SAA AS 1255.2-73. Methods of Test for Electrical Characteristics of Solid Plastics Insulating Materials—Part 2: Determination of Insulation Resistance. 14 pp.

—Plastic—Electrical Resistivity
SAA AS 1255.1-73. Methods of Test for Electrical Characteristics of Solid Plastics Insulating Materials—Part 1: Determination of Volume and Surface Resistivities. 19 pp.

—Plastic Sheets
BSI BS 5961-80. 1980 Determination of Coefficients of Friction of Plastic Film and Sheeting for Use as Electrical Insulation. 11 pp.
BSI BS 7290: Part 1-90. 1990 Plastics Films for Electrical Purposes Part 1: Definitions and General Requirements. 8 pp.
BSI BS 7290: Part 2-90. 1990 Plastics Films for Electrical Purposes Part 2: Methods of Test. 32 pp.

INTERNATIONAL AND NON-U.S. NATIONAL STANDARDS
SUBJECT INDEX
Electrical

Electrical Insulation (Cont.)
—Plastic Sheets (Cont.)
IEC 648-79. Method of Test for Coefficients of Friction of Plastic Film and Sheeting for Use as Electrical Insulation First Edition. 21 pp.
IEC 674 Pt 1-80. Specification for Plastic Films for Electrical Purposes Part 1: Definitions and General Requirements First Edition. 13 pp.
IEC 674 Pt 2-88. Specification for Plastic Films for Electrical Purposes Part 2: Methods of Test First Edition. 58 pp.
IEC 674 Pt 3-4 TO 6-93. Specification for Plastic Films for Electrical Purposes Part 3: Specifications for Individual Materials Sheets 4 to 6: Requirements for Polyimide Films Used for Electrical Insulation First Edition. 36 pp.
JIS C 2151-90. Testing Methods of Plastic Films for Electrical Purposes. 32 pp.
SAA AS 1366.1-92. Rigid Cellular Plastics Sheets for Thermal Insulation—Part 1: Rigid Cellular Polyurethane (RC/PUR) Amdt 1 November 1992. 6 pp.

—Plastics—Test Specimens
DIN ENGL 50005-75. Testing of Plastics and Other Electrical Insulating Materials; Selection of Climates for Preconditioning, Conditioning and Testing of Test Specimens (July). 4 pp.

—Polyester Resins
BSI BS 5734: Part 1-90. 1990 Polyester Moulding Compounds for Electrical and Other Purposes Part 1: Methods of Test. 14 pp.
BSI BS 5734: Part 4-90. 1990 Polyester Moulding Compounds for Electrical and Other Purposes Part 4: Specification for Sheet Moulding Compounds for Electrical Purposes. 10 pp.
BSI BS 5734: Part 5-90. 1990 Polyester Moulding Compounds for Electrical and Other Purposes Part 5: Sheet Moulding Compounds for Mechanical Purposes. 9 pp.
CENELEC HD 307.3.5 S1-91. Specification for Solventless Polymerisable Resinous Compounds Used for Electrical Insulation Part 3: Specifications for Individual Materials Sheet 5: Unsaturated Polyester Impregnating Resins. 6 pp.
IEC 455 Pt 3-5-89. Specification for Solventless Polymerisable Resinous Compounds Used for Electrical Insulation Part 3: Specifications for Individual Materials Sheet 5: Unsaturated Polyester Impregnating Resins First Edition. 25 pp.
IEC 893 Pt 3-5-93. Specification for Industrial Rigid Laminated Sheets Based on Thermosetting Resins for Electrical Purposes Part 3: Specifications for Individual Materials —Sheet 5: Requirements for Rigid Laminated Sheets Based on Polyester Resins First Edition. 26 pp.

—Polyester Resins—Glass Laminated—Sheets
BSI BS 6673-86. 1986 Polyester Glass Mat Sheets for Electrical Purposes. 10 pp.

—Polyethylene
BSI BS 6469-90. (WITHDRAWN) 1990 Insulation and Sheaths of Electric Cables (Superseded by Various Sections of Parts 1-5 of BS 6469). 43 pp.

—Polyolefin—Tubes—Heat Shrinkable
CNS C4376-82. Heat-Shrinkable Polyolefin Tubing for Electrical Insulation (Tubular Type) (Dec)(9686).

—Polytetrafluoroethylene—Rods
JIS K 6889-77. Polytetrafluoroethylene Rods (R 1980). 11 pp.

—Polytetrafluoroethylene—Sleeving
MOD UK DSTAN 59-15: Part 1-90. Elastomeric Insulating Materials and Polytetrafluoroethylene (PTFE) Part 1: General Requirements, Tests and Test Methods for Qualification Approval Sleeving in Continuous Lengths Sleeves for Binding and/or Identification Purposes. 36 pp.

—Polytetrafluoroethylene—Tubes
JIS K 6890-77. Polytetrafluoroethylene Tubes (R 1980). 11 pp.

—Polytetrafluoroethylene—Tubes—Heat Shrinkable
CNS C4377-82. Heat-Shrinkable PTFE Tubing for Electrical Insulation (Tubular Type) (Dec)(9687).

—Polyurethane Resins
IEC 455 Pt 3-3-84. Specification for Solventless Polymerisable Resinous Compounds Used for Electrical Insulation Part 3: Specifications for Individual Materials Sheet 3: Unfilled Polyurethane Compounds First Edition. 13 pp.
IEC 455 Pt 3-4-84. Specification for Solventless Polymerisable Resinous Compounds Used for Electrical Insulation Part 3: Specifications for Individual Materials Sheet 4: Filled Polyurethane Compounds First Edition. 11 pp.

Electrical Insulation (Cont.)
—Polyvinylidene Fluoride—Tubes—Heat Shrinkable
CNS C4379-82. Heat-Shrinkable PVF2 Tubing for Electrical Insulation (Tubular Type) (Dec)(9689).

—Porcelain—Visual
CNS C3108-80. Permissible Limits of Visual Defects for Insulating Porcelains (Dec)(6773).
JIS C 3802-64. Permissible Limits of Visual Defects for Insulating Porcelains (R 1971). 12 pp.

—Power Transformers
BSI BS 171: Part 3-87. 1987 Power Transformers Part 3: Insulation Levels and Dielectric Tests. 35 pp.
CENELEC HD 398.3-86. Power Transformers Insulation Levels and Dielectric Tests. 10 pp.
IEC 76 Pt 3-80. Power Transformers Part 3: Insulation Levels and Dielectric Tests First Edition; (Amendment 1-1981). 81 pp.
IEC 76 Pt 3-1-87. Power Transformers Part 3: Insulation Levels and Dielectric Tests External Clearances in Air First Edition. 26 pp.
SAA AS 2374.3.0-82. Power Transformers—Part 3.0: Insulation Levels and Dielectric Tests (Redesignated by Amdt 1 September 1992 from AS 2374.3—1982). 1 p.
SAA AS 2374.3.1-92. Power Transformers—Part 3: Insulation Levels and Dielectric Tests—Part 3.1: External Clearances in Air. 13 pp.

—PVC
BSI BS 6469-90. (WITHDRAWN) 1990 Insulation and Sheaths of Electric Cables (Superseded by Various Sections of Parts 1-5 of BS 6469). 43 pp.
CNS K3001-89. Vinyl Chloride Resin Compounds (for Electrical Wires and Cables) (Nov)(1087). 2 pp.

—Radiation Effects
IEC 544 Pt 1-77. Guide for Determining the Effects of Ionizing Radiation on Insulating Materials Part 1: Radiation Interaction First Edition. 45 pp.
IEC 544 Pt 2-91. Guide for Determining the Effects of Ionizing Radiation on Insulating Materials Part 2: Procedures for Irradiation and Test Second Edition. 45 pp.
IEC 544 Pt 3-79. Guide for Determining the Effects of Ionizing Radiation on Insulating Materials Part 3: Test Procedures for Permanent Effects First Edition. 19 pp.
IEC 544 Pt 4-85. Guide for Determining the Effects of Ionizing Radiation on Insulating Materials Part 4: Classification System for Service in Radiation Environments First Edition. 19 pp.

—Rail Track Materials
BSI BS 456-61. 1961 Dimensions of Track-Circuit Insulation. 43 pp.

—Resins
CENELEC HD 307.1 S2-81. Specification for Solventless Polymerizable Resinous Compounds Used for Electrical Insulation—Part 1: Definitions and General Requirements. Basis for Classification of Polymerisable Resinous Compounds. 2 pp.
CENELEC HD 307.2 S1-89. Specification for Solventless Polymerisable Resinous Compounds Used for Electrical Insulation-Part 2: Methods of Test. 3 pp.
IEC 455 Pt 2-77. Specification for Solventless Polymerisable Resinous Compounds Used for Electrical Insulation Part 2: Methods of Test First Edition; (Amendment 1-1982). 44 pp.
JIS C 2105-92. Testing Methods of Solventless Liquid Resins for Electrical Insulation. 50 pp.

—Resins—Coatings
BSI BS 5917-80. 1980 Conformal Coating Material for Use on Printed Circuit Assemblies. 9 pp.
IEC 455 Pt 2-2-84. Specification for Solventless Polymerisable Resinous Compounds Used for Electrical Insulation Part 2: Methods of Test Test Methods for Coating Powders for Electrical Purposes First Edition. 43 pp.

—Resins—Glossaries
BSI BS 5664: Part 1-78. 1978 Amd 1 Solventless Polymersible Resinous Compounds Used for Electrical Insulation Part 1: Definitions and Requirements. 7 pp.
IEC 455 Pt 1-74. Specification for Solventless Polymerisable Resinous Compounds Used for Electrical Insulation Part 1: Definitions and General Requirements First Edition; (Supplement A-1980). 14 pp.

—Rods—Live Wire Working
CENELEC HD 496 S1-88. Insulating Foam-Filled Tubes and Solid Rods for Live Working. 4 pp.
IEC 855-85. Insulating Foam-Filled Tubes and Solid Rods for Live Working First Edition. 39 pp.

—Rolling Stock—Electrical Resistance
JIS E 4014-89. Test Methods for Insulation Resistance and Withstand Voltage of Railway Rolling Stock.

Electrical Insulation (Cont.)
—Rotating Machines
BSI BS 4999: Part 144-87. 1987 Amd 1 General Requirements for Rotating Electrical Machines Part 144: Specification for the Insulation of Bars and Coils of High Voltage Machines, Including Test Methods (AMD 6636) April 30, 1991. 8 pp.
CENELEC HD 345-76. Test of the Insulation of Bars and Coils of High Voltage Machines. 8 pp.
DIN VDE 0530 Pt 20 (D)-87. Rotating Electrical Machines; Functional Evaluation Insulation Systems for Rotating Electrical Machines, General Guidelines; Identical with IEC 2J (Central Office) 4 (Nov). 36 pp.
IEC 34 Pt 18-1-92. Functional Evaluation of Insulation Systems for Rotating Electrical Machines Part 1: General Guidelines First Edition; (Corrigendum—Aug 1992). 50 pp.

—Rotating Machines—Windings
DIN VDE 0530 Pt 21 (D)-88. Rotating Electrical Machines; Functional Evaluation of Insulation Systems for Rotating Electrical Machines; Test Procedures for Wire-Sound Windings, Thermal Evaluation and Classification; Identical with IEC 2J (Central Office) 5 (Oct). 57 pp.
DIN VDE 0530 Pt 22 (D)-88. Rotating Elect. Machines; Functional Eval. of Ins. Systems for Rotating Elect. Mach.; Test Proc. for Form-Wound Windings; Thermal Eval. and Class-ification of Insulation Systems Used in Machine up to and Including 50 MVA and 15 kV; Indenti-cal with IEC 2J (Central Office) 6 (Oct). 22 pp.
IEC 34 Pt 18-21-92. Rotating Electrical Machines Part 18: Functional Evaluation of Insulation Systems Section 21: Test Procedures for Wire-Wound Windings—Thermal Evaluation and Classification First Edition. 78 pp.
IEC 34 Pt 18-31-92. Rotating Electrical Machines Part 18: Functional Evaluation of Insulation Systems Section 31: Test Proce-dures for Form-Wound Windings—Thermal Eval-uation and Classifica-tion of Insulation Sys-tems Used in Machines up to and Including 50 MVA and 15 kV First Edition. 31 pp.

—Sheets—Dielectric Dissipation Factor
BSI BS 7663-93. 1993 Methods of Test for Determination of Permittivity and Dissipation Factor of Electrical Insulating Material in Sheet or Tubular Form (F). 18 pp.

—Sheets—Permittivity
BSI BS 7663-93. 1993 Methods of Test for Determination of Permittivity and Dissipation Factor of Electrical Insulating Material in Sheet or Tubular Form (F). 18 pp.

—Sheets—Silicone Resins—Glass Laminated
JIS C 2241-78. Silicone Resin Glass Laminated Sheets for Electrical Insulation. 13 pp.

—Silicone Resins
IEC 893 Pt 3-6-93. Specification for Industrial Rigid Laminated Sheets Based on Thermosetting Resins for Electrical Purposes Part 3: Specifications for Individual Materials —Sheet 6: Requirements for Rigid Laminated Sheets Based on Silicone Resins First Edition. 24 pp.
MOD UK DSTAN 68-69-90. Silicone Compound Electrical Insulating NATO Code S—736 Joint Service Designation XG—250 Issue 1. 30 pp.

—Silicone Resins—Glass Laminated
CNS C4353-82. Silicone Resin Glass Laminated Sheets for Electric Insulation (Jul)(9087).

—Silicone Sealants
CNS C3154-82. Method of Test for Electrical Silicone Rubber Compounds (Jul)(9083).
JIS C 2123-86. Testing Method for Electrical Silicone Rubber Compounds.

—Sleeving—Silicone Resins
CENELEC HD 523.3.121. 122 S1-90. Specification for Flexible Insulating Sleeving Part 3: Specification Requirement for Individual Types of Sleeving Sheets 121 and 122: Extruded Silicone Sleevings. 3 pp.
IEC 684 Pt 3-121 And 122-88. Specification for Flexible Insulating Sleeving Part 3: Specification Requirements for Individual Types of Sleeving Sheets 121 and 122: Extruded Silicone Sleeving First Edition. 17 pp.

—Test Specimens
DIN ENGL 50005-75. Testing of Plastics and Other Electrical Insulating Materials; Selection of Climates for Preconditioning, Conditioning and Testing of Test Specimens (July). 4 pp.

—Thermodynamic Properties
BSI BS 2757-86. 1986 Method for Determining the Thermal Classification of Electrical Insulation. 10 pp.

INDUSTRY STANDARDS

INTERNATIONAL AND NON-U.S. NATIONAL STANDARDS SUBJECT INDEX

Electrical

Electrical Insulation (Cont.)
—Thermodynamic Properties (Cont.)
BSI BS 5691: Part 1-79. 1979 Determination of Thermal Endurance Properties of Electrical Insulating Materials Part 1: General Procedures for the Determination of Thermal Endurance Properties, Temperature Indices and Thermal Endurance Profiles. 19 pp.
BSI BS 5691: Part 2-79. 1979 Amd 1 Determination of Thermal Endurancre Properties of Electrical Insulating Materials Part 2: List of Materials and Available Tests. 11 pp.
CENELEC HD 566 S1-90. Thermal Evaluation and Classification of Electrical Insulation. 4 pp.
CENELEC HD 611.2 S1-92. Guide for the Determination Thermal Endurance Properties of Electrical Insulating Materials Part 2: Choice of Test Criteria. 7 pp.
CENELEC HD 611.3.1 S1-92. Guide for the Determination Thermal Endurance Properties of Electrical Insulating Materials Part 3: Instructions for Calculating Thermal Endurance Characteristics Section One: Calculations Using Mean Values of Normally Distr. Complete Data. 5 pp.
CENELEC HD 611.4.1 S1-92. Guide for the Determination Thermal Endurance Properties of Electrical Insulating Materials Part 4: Ageing Ovens Section One: Single-Chamber Ovens. 5 pp.
CENELEC HD 611.5 S1-92. Guide for the Determination Thermal Endurance Properties of Electrical Insulating Materials Part 5: Guidelines for the Application of Thermal Endurance Characteristics. 5 pp.
CNS C1120-88. Surrounding Temperature Affect for Security Current of Wires Insulation (Jul)(9831).
DIN VDE 0304 Pts 1 & 2-59. Recommendations for Testing Solid Insulating Materials for Assessment of Their Thermal Stability; Part 1: Thermal Properties of Solid Insulating Materials; Part 2: Effects of Prolonged Action of Heat on the Properties of Insulating Materials (July). 18 pp.
DIN VDE 0304 Pt 3-85. Thermal Properties of Electrical Insulating Materials; Flammability Under the Action of Igniting Sources (National Foreword Only Corresponds to IEC 707—1981) (Sept). 6 pp.
IEC 85-84. Thermal Evaluation and Classification of Electrical Insulation Second Edition. 21 pp.
IEC 216 Pt 2-90. Guide for the Determination of Thermal Endurance Properties of Electrical Insulating Materials Part 2: Choice of Test Criteria Third Edition. 33 pp.
IEC 216 Pt 3-1-90. Guide for the Determination of Thermal Endurance Properties of Electrical Insulating Materials Part 3: In-structions for Calcula-ting Thermal Endurance Characteristics Section 1—Calculations Using Mean Values of Normally Distributed Complete Data First Edition. 50 pp.
IEC 216 Pt 4-1-90. Guide for the Determination of Thermal Endurance Properties of Electrical Insulating Materials Part 4: Ageing Ovens Section One: Single-Chamber Ovens Third Edition. 21 pp.
IEC 216 Pt 5-90. Guide for the Determination of Thermal Endurance Properties of Electrical Insulating Materials Part 5: Guidelines for the Application of Thermal Endurance Characteristics First Edition. 19 pp.
IEC 611-78. Guide for the Preparation of Test Procedures for Evaluating the Thermal Endurance of Electrical Insulation Systems First Edition. 18 pp.
SNZ NZS/IEC 85-84. Thermal Evaluation and Classification of Electrical Insulation. 17 pp.

—Thermosetting Resins
BSI BS 5734: Part 2-90. 1990 Polyester Moulding Compounds for Electrical and Other Purposes Part 2: Specification for Dough Moulding Compounds for Electrical Purposes. 9 pp.
IEC 893 Pt 1-87. Specification for Industrial Rigid Laminated Sheets Based on Thermosetting Resins for Electrical Purposes Part 1: Definitions, Designations and General Requirements First Edition. 13 pp.
IEC 893 Pt 2-92. Specification for Industrial Rigid Laminated Sheets Based on Thermosetting Resins for Electrical Purposes Part 2: Methods of Test First Edition. 38 pp.
IEC 893 Pt 3-1-92. Specification for Industrial Rigid Laminated Sheets Based on Thermosetting Resins for Electrical Purposes Part 3: Specifications for Individual Materials —Sheet 1: Types of Industrial Rigid Laminated Sheets First Edition. 20 pp.

—Tracking Resistance
BSI BS 5604-86. 1986 Evaluating Resistance to Tracking and Erosion of Electrical Insulating Materials Used Under Severe Ambient Conditions. 12 pp.
DIN VDE 0303 Pt 10-83. Testing Electrical Insulating Materials; Test Method for Evaluating Tracking Resistance at High Voltage (June). 13 pp.

Electrical Insulation (Cont.)
—Tracking Testing—Moist Conditions
BSI BS 5901-80. 1980 Determining the Comparative and the Proof Tracking Indices of Solid Insulating Materials Under Moist Conditions. 12 pp.
CENELEC HD 214 S2-80. Recommended Method for Determining the Comparative Tracking Index of Solid Insulating Material Under Moist Conditions. 2 pp.
IEC 112-79. Method for Determining the Comparative and the Proof Tracking Indices of Solid Insulating Materials Under Moist Conditions Third Edition. 20 pp.

—Treatment
BSI BS 2844-72. 1972 Memorandum on Conditioning of Solid Electrical Insulating Materials Prior to and During Testing. 13 pp.

—Treeing—Electrical Resistance
IEC 1072-91. Methods of Test for Evaluating the Resistance of Insulating Materials Against the Initiation of Electrical Trees First Edition. 55 pp.

—Trunking
BSI BS 4678: Part 4-82. 1982 Amd 1 Cable Trunking Part 4: Cable Trunking Made of Insulating Material. 12 pp.

—Tubes
CSA C22.2 NO 198.1-M1986. Extruded Insulating Tubing (R 1992); (Gen Instr 1 Thru 4). 89 pp.
JIS C 2133-92. Testing Methods of Tubing for Electrical Insulation. 26 pp.

—Tubes—Dielectric Dissipation Factor
BSI BS 7663-93. 1993 Methods of Test for Determination of Permittivity and Dissipation Factor of Electrical Insulating Material in Sheet or Tubular Form (F). 18 pp.

—Tubes—Live Wire Working
CENELEC HD 496 S1-88. Insulating Foam-Filled Tubes and Solid Rods for Live Working. 4 pp.
IEC 855-85. Insulating Foam-Filled Tubes and Solid Rods for Live Working First Edition. 39 pp.
IEC 1235-93. Live Working—Insulating Hollow Tubes for Electrical Purposes First Edition. 55 pp.

—Tubes—Permittivity
BSI BS 7663-93. 1993 Methods of Test for Determination of Permittivity and Dissipation Factor of Electrical Insulating Material in Sheet or Tubular Form (F). 18 pp.

—Varnishes
CNS C3027-1-88. Subsidiary Methods of Test for Varnish for Electrical Insulation (Oct)(2604-1).

—Voltage Endurance Testing
IEC 1251-93. Electrical Insulating Materials—A.C. Voltage Endurance Evaluation—Introduction First Edition. 56 pp.

—Vulcanized Fibers
BSI BS 6091: Part 1-81. 1981 Vulcanized Fibre for Electrical Purposes Part 1: Specification of General Requirements. 6 pp.
BSI BS 6091: Part 2-87. 1987 Vulcanized Fibre for Electrical Purposes Part 2: Methods of Test. 14 pp.
CENELEC HD 416.1-84. Specification for Vulcanized Fibre for Electrical Purposes Part 1: Definitions and General Requirements. 2 pp.
CENELEC HD 416.2 S1-89. Specification for Vulcanized Flore for Electrical Purposes-Part 2: Methods of Test. 3 pp.
CENELEC HD 416.3.1 S1-89. Specification for Vulcanized Fibre for Electrical Purposes Part 3: Specifications for Individual Materials. Sheet one: Flat Sheets. 3 pp.
IEC 667 Pt 1-80. Specification for Vulcanized Fibre for Electrical Purposes Part 1: Definitions and General Requirements First Edition. 13 pp.
IEC 667 Pt 2-82. Specification for Vulcanized Fibre for Electrical Purposes Part 2: Methods of Test First Edition; (Amendment 1-1986). 34 pp.
IEC 667 Pt 3-1-86. Specification for Vulcanized Fibre for Electrical Purposes Part 3: Specifications for Individual Materials Sheet 1: Flat Sheets First Edition. 15 pp.

—Withstand Voltage
DIN VDE 0303 Pt 2-74. Electrical Tests of Insulating Materials; Breakdown Voltage; Electric Strength (Nov). 23 pp.
SAA AS 1824.1-85. Insulation Coordination (Phase-to-Earth and Phase-to-Phase, Above 1 kV)—Part 1: Basic Principles, Standard Insulation Levels and Test Procedures. 16 pp.
SAA AS 1824.2-85. Insulation Coordination (Phase-to-Earth and Phase-to-Phase, Above 1 kV)—Part 2: Application Guide. 37 pp.

Electrical Insulator Fittings
See Also: Electrical Insulators

Electrical Insulator Fittings (Cont.)
—Metal—Overhead Power Lines
CNS C4269-86. Metal Fitting for Insulator (Jun)(6771).
JIS C 3703-75. Metal Fittings for Insulators (R 1978). 14 pp.

—Overhead Power Lines
BSI BS 3288: Part 1-73. 1973 Insulator and Conductor Fittings for Overhead Power Lines Part 1: Performance and General Requirements. 15 pp.
BSI BS 3288: Part 2-90. 1990 Insulator and Conductor Fittings for Overhead Power Lines Part 2: Range of Fittings. 87 pp.
DIN VDE 0212 Pt 55-89. Fittings for Overhead Lines and Switchgear; Insulation Behaviour of Fittings for Insulated Overhead Lines (Apr). 6 pp.
SNZ BS 3288: Part 1-73. Specification for Insulator and Conductor Fittings for Overhead Power Lines Part 1: Performance and General Requirements. 16 pp.

—Telegraph Lines
BSI BS 16-74. 1974 Amd 1 Telegraph Material (Insulators, Pole Fittings, etc). 55 pp.

Electrical Insulator Pins
See Also: Electrical Insulators; Pins; Power Line Hardware

—Resistance Welding Equipment
BSI BS 4215: Part 15-01. 1990 Amd 1 Resistance Spot Welding Electrodes, Electrode Holders and Ancillary Equipment Part 15: Specification for Insulated Pins for Use in Electrode Back-ups (AMD 7132) July 15, 1992. 8 pp.
DIN ENGL 44764-73. Parallel Pins Made of Insulating Material (Oct). 1 p.
ISO 9312-90. Resistance Welding Equipment—Insulated Pins for Use in Electrode Back-ups First Edition. 4 pp.

—Telephone Cables
CNS C3022-81. Method of Test for Pins for Telephone Line Insulators (May)(1149).
CNS C4027-81. Pins for Telephone Line Insulators (May)(1145).
CNS C4028-81. Number One of Pins for Telephone Line Insulators (May)(1146).
CNS C4029-81. Number Two of Pins for Telephone Line Insulators (May)(1147).
CNS C4030-81. Number Three of Pins for Telephone Line Insulators (May)(1148).
CNS C4309-81. Number Four of Pins for Telephone Line Insulators (May)(7359).

—Wood—Physical Properties
CSA O124-1957. Specification for the Physical Properties of Power and Communication Wood Insulator Pins (R 1972); (Erratum October 1964) (Rev August 1967). 13 pp.

Electrical Insulators
See Also: Cap and Pin Insulators; Electrical Insulation; Electrical Insulator Fittings; Electrical Insulator Pins; Insulator Strings; Pin Insulators; Porcelain Insulators; Post Insulators; Power Line Hardware; Standoffs; Strain Insulators; Suspension Insulators

CENELEC HD 329-76. Test on Hollow Insulators for Use in Electrical Equipment. 2 pp.
CNS C3107-80. Method of Test for Insulators (Dec)(6772).
CNS C4358-82. High Voltage Insulator for Cubicle Type Power Receiving Unit (Jul)(9092).
DIN VDE 0674-62. Rules for Insulating Bodies and Insulators to be Used in a.c. Equipment and Installations with Rated Voltages Above 1 kV (Oct). 34 pp.
IEC 233-74. Tests on Hollow Insulators for Use in Electrical Equipment Second Edition; (Amendment 1-1988). 30 pp.
IEC 438-73. Tests and Dimensions for High-Voltage D.C. Insulators First Edition. 16 pp.
MOD UK DSTAN 59-38: Part 1-71. Insulators, Electrical Part 1: Insulators, Bead Issue 1. 5 pp.

—Air Pollution
BSI BS EN 60507-93. 1993 Artificial Pollution Tests on High-Voltage Insulators to be Used on a.c. Systems (IEC 507: 1991) (Q). 38 pp.
CENELEC EN 60507-93. Artificial Pollution Tests on High-Voltage Insulators to Be Used on A.C. Systems (IEC 507: 1991). 33 pp.
IEC 507-91. Artificial Pollution Tests on High-Voltage Insulators to Be Used on a.c. Systems Second Edition (CENELEC EN 60507:1993). 59 pp.
IEC 815-86. Guide for the Selection of Insulators in Respect of Polluted Conditions First Edition. 42 pp.

—Ceramic
BSI BS EN 50062-92. 1992 Ceramic Pressurized Hollow Insulators for High-Voltage Switchgear and Controlgear. 18 pp.

INTERNATIONAL AND NON-U.S. NATIONAL STANDARDS
SUBJECT INDEX
Electrical

Electrical Insulators *(Cont.)*
—Ceramic *(Cont.)*
CENELEC EN 50062-91. Ceramic Pressurized Hollow Insulators for High-Voltage Switchgear and Controlgear. 25 pp.
JIS C 2141-92. Testing Methods of Ceramic Insulators for Electrical and Electronic Applications. 38 pp.

—Ceramic—Communication Transmission Lines
CNS C3016-70. Method of Test for Ceramic Insulators for Communication Line (Jan)(907).

—Ceramic/Glass
SAA AS 1372-74. Tests on Hollow Insulators of Ceramic Material or Glass for Use in Electrical Equipment for Voltages Above 1000 V. 16 pp.

—Ceramic/Glass—Acceptance Testing
BSI BS 4963-73. 1973 Tests on Hollow Insulators for Use in High Voltage Electrical Equipment. 16 pp.

—Ceramic/Glass—Overhead Contact Systems
IEC 383 Pt 1-93. Insulators for Overhead Lines with a Nominal Voltage Above 1 000 V Part 1: Ceramic or Glass Insulator Units for a.c. Systems— Definitions, Test Methods and Acceptance Criteria Fourth Edition. 115 pp.

—Ceramic/Glass—Overhead Power Lines
IEC 383 Pt 1-93. Insulators for Overhead Lines with a Nominal Voltage Above 1 000 V Part 1: Ceramic or Glass Insulator Units for a.c. Systems— Definitions, Test Methods and Acceptance Criteria Fourth Edition. 115 pp.

—Ceramic/Glass—Substations (Electric)
IEC 383 Pt 1-93. Insulators for Overhead Lines with a Nominal Voltage Above 1 000 V Part 1: Ceramic or Glass Insulator Units for a.c. Systems— Definitions, Test Methods and Acceptance Criteria Fourth Edition. 115 pp.

—Ceramic—Overhead Power Lines
BSI BS 137: Part 1-82. 1982 Amd 1 Insulators of Ceramic Material or Glass for Overhead Lines with a Nominal Voltage Greater Than 1000V Part 1: Methods of Test. 19 pp.
BSI BS 137: Part 2-73. 1973 Amd 1 Insulators of Ceramic Material or Glass for Overhead Lines with a Nominal Voltage Greater Than 1000V Part 2: Requirements (AMD 4380) October 31, 1983. 24 pp.
IEC 815-86. Guide for the Selection of Insulators in Respect of Polluted Conditions First Edition. 42 pp.
SNZ IEC 383-83. Test on Insulators of Ceramic Material or Glass for Overhead Lines with a Nominal Voltage Greater Than 1000 V. 65 pp.

—Ceramic—Overhead Power Lines— Electrical Properties
IEC 591-78. Sampling Rules and Acceptance Criteria When Applying Statistical Control Methods for Mech and Elmch Tests on Insulators of Ceramic Material or Glass for Overhead Lines with a Nominal Voltage Greater Than 1000 V First Edition. 27 pp.

—Ceramic—Overhead Power Lines— Mechanical Testing
IEC 591-78. Sampling Rules and Acceptance Criteria When Applying Statistical Control Methods for Mech and Elmch Tests on Insulators of Ceramic Material or Glass for Overhead Lines with a Nominal Voltage Greater Than 1000 V First Edition. 27 pp.

—Composite—Outdoor
DIN VDE 0441 Pt 1-85. Testing of Plastic Insulators for Alternating Voltages Above 1kV; Testing of Materials for Outdoor Insulators (July). 21 pp.
DIN VDE 0441 Pt 2-82. Testing Plastic Insulators for Systems with Operating a.c. Voltages Greater Than 1 kV; Tests on Outdoor Composite Insulators with Glass-Fibre Reinforced Core (VDE Specification) (Oct). 14 pp.

—Composite—Overhead Power Lines
IEC 1109-92. Composite Insulators for a.c. Overhead Lines with a Nominal Voltage Greater Than 1 000 V—Definitions, Test Methods and Acceptance Criteria First Edition. 76 pp.

—Electric Railroads—Overhead Contact Systems
JIS E 2219-90. Section Insulators for Overhead Contact System of Electric Railway.
JIS E 2219-89. Section Insulators for Overhead Contact System of Electric Railway. 13 pp.
JIS E 2301-80. Insulator for Electric Overhead Line (R 1985). 19 pp.

Electrical Insulators *(Cont.)*
—Glass—Overhead Power Lines
BSI BS 137: Part 1-82. 1982 Amd 1 Insulators of Ceramic Material or Glass for Overhead Lines with a Nominal Voltage Greater Than 1000V Part 1: Methods of Test. 19 pp.
BSI BS 137: Part 2-73. 1973 Amd 1 Insulators of Ceramic Material or Glass for Overhead Lines with a Nominal Voltage Greater Than 1000V Part 2: Requirements (AMD 4380) October 31, 1983. 24 pp.
IEC 815-86. Guide for the Selection of Insulators in Respect of Polluted Conditions First Edition. 42 pp.
SNZ IEC 383-83. Test on Insulators of Ceramic Material or Glass for Overhead Lines with a Nominal Voltage Greater Than 1000 V. 65 pp.

—Glass—Overhead Power Lines—Electrical Properties
IEC 591-78. Sampling Rules and Acceptance Criteria When Applying Statistical Control Methods for Mech and Elmch Tests on Insulators of Ceramic Material or Glass for Overhead Lines with a Nominal Voltage Greater Than 1000 V First Edition. 27 pp.

—Glass—Overhead Power Lines— Mechanical Testing
IEC 591-78. Sampling Rules and Acceptance Criteria When Applying Statistical Control Methods for Mech and Elmch Tests on Insulators of Ceramic Material or Glass for Overhead Lines with a Nominal Voltage Greater Than 1000 V First Edition. 27 pp.

—Glossaries
BSI BS 4727:Pt1: Group 12-91. 1991 Electrotechnical, Power, Telecommunication, Electronics, Lighting and Colour Terms Part 1: Terms Common to Power, Telecommunications and Electronics Group 12: Insulators (G) (IEC 50(471): 1984). 28 pp.
CNS C1105-82. Glossary of Terms for Insulator and Bushing (Sep)(9355).
IEC 50 Chap 471-84. International Electrotechnical Vocabulary Chapter 471: Insulators. 32 pp.
JIS C 3803-77. Glossary of Terms for Insulator and Bushing.
SAA AS 1852.471-88. International Electrotechnical Vocabulary—Part 471: Insulators. 11 pp.
SNZ IEC 50: 50(471)-84. International Electrotechnical Vocabulary 50(471): Insulators. 22 pp.

—Impulse Voltage Testing
IEC 506-75. Standard Impulse Tests on High-Voltage Insulators First Edition. 23 pp.

—Overhead Power Lines
SAA AS 1154.1-85. Insulator and Conductor Fittings for Overhead Power Lines—Part 1: Performance and General Requirements. 20 pp.
SAA AS 1154.2-85. Insulator and Conductor Fittings for Overhead Power Lines—Part 2: Dimensions. 53 pp.
SAA AS 1154.3-85. Insulator and Conductor Fittings for Overhead Power Lines—Part 3: Performance and General Requirements for Helical Fittings. 11 pp.

—Radio Frequency Interference
IEC 437-73. Radio Interference Test on High-Voltage Insulators First Edition. 17 pp.

—Switchgear—Indoor
CNS C4273-84. Indoor Switch and Bus Insulators (Apr)(6777).

—Telegraph Lines
BSI BS 16-74. 1974 Amd 1 Telegraph Material (Insulators, Pole Fittings, etc). 55 pp.

—Washing Systems
DIN VDE 0143-84. Live Line Washing Systems for Power Installations with Nominal Voltages Above 1 kV (Oct). 14 pp.

Electrical Leads
Use For: Lead Assemblies; Leads (Electrical)
See Also: Axial Leads; Electric Conductors; Electron Tubes; Electronic Component Packaging; Integrated Circuits; Radial Leads

—Aircraft
SBAC AS 46771 ISSUE 1. Lead, Electrical (for Primary Bonding Purposes).
SBAC AS 46772 ISSUE 3. Lead, Electrical (for Static Bonding Purposes) Unsuitable for Lighting Strike Purposes.
SBAC AGS 2100 ISSUE 1. High Temperature Bonding Leads.

—Cables
CENELEC HD 21.10 S1-93. Polyvinyl Chloride Insulated Cables of Rated Voltages up to and Including 450/750 V Part 10: Extensible Leads. 14 pp.

Electrical Leads *(Cont.)*
—Surface Acoustic Wave Resonators
IEC 1019 Pt 3-91. Surface Acoustic Wave (SAW) Resonators Part 3: Standard Outlines and Lead Connections First Edition. 25 pp.

Electrical Loads
Use: Loads (Electric)

Electrical Measurement
Scope Note: Includes documents on the measurement of electrical properties, quantities, and conditions
See Also: Ammeters; Amplitude Response; Capacitance Measurement; Current Noise; Effective Capacitance; Electric Endurance Testing; Electric Energy Meters; Electric Measuring Instruments; Electric Power Meters; Electrical Properties; Flowmeters; Frequency Meters; Frequency Response; Impedance Measurement; Ohmmeters; Oscilloscopes; Phase Measurement; Phase Meters; Phase Response; Power Factor Meters; Radiation Pyrometers; Recovery Time; Resistance Thermometers; Reverse Recovery Time; Selectivity (Electrical); Short Circuit Current Testing; Transient Voltage Measurement; Voltage Levels; Voltage Measurement; Voltmeters; Volume Indicators; Watt Hour Meters; Wattmeters; Winding Resistance Measurement

IEC 164-64. Recommendations in the Field of Quantities and Units Used in Electricity First Edition. 64 pp.

—Antennas
BSI BS 5640: Part 1-78. 1978 Aerials for the Reception of Sound and Television Broadcasting in the Frequency Range to 30 MHz to 1 GHz Part 1: Specification for Electrical and Mechanical Characteristics. 10 pp.
BSI BS 5640: Part 2-78. 1978 Aerials for the Reception of Sound and Television Broadcasting in the Frequency Range 30 MHz to 1 GHz Part 2: Methods of Measurement of Electrical Performance Parameters. 15 pp.
CENELEC HD 95.2-78. Aerials for the Reception of Sound and Television Broadcasting in the Frequency Range 30 MHz to 2 GHz Part 2: Methods of Measurement of Electrical Performance Parameters. 2 pp.
CENELEC EN 61114-1-93. Methods of Measurement on Receiving Antennas for Satellite Broadcast Transmission in the 12 GHz Band Part 1: Electrical Measurements on DBS Receiving Antennas; (IEC 1114-1: 1992). 5 pp.
IEC 489 Pt 8-84. Methods of Measurement for Radio Equipment Used in the Mobile Services Part 8: Methods of Measurement for Antennas First Edition. 31 pp.
IEC 510 Pt 2-1-78. Methods of Measurement for Radio Equipment Used in Satellite Earth Stations Part 2: Measurements for Sub-Systems Section One—General Section Two—Antenna (Including Feed Network) First Edition. 61 pp.
IEC 597 Pt 2-77. Aerials for the Reception of Sound and Television Broadcasting in the Frequency Range 30 MHz to 1 GHz Part 2: Methods of Measurement of Electrical Performance Parameters First Edition. 27 pp.
IEC 1114 Pt 1-92. Methods of Measurement on Receiving Antennas for Satellite Broadcast Transmissions in the 12 GHz Band First Edition; (CENELEC EN 61114-1: 1993). 72 pp.

—Broadcast Receivers—Satellite Communications
IEC 1079 Pt 3-93. Methods of Measurement on Receivers for Satellite Broadcast Transmissions in the 12 GHz Band Part 3: Electrical Measurements of Overall Performance of Receiver Systems Comprising an Outdoor Unit and a DBS Tuner Unit First Edition. 52 pp.

—Converters
IEC 510 Pt 2-4-88. Methods of Measurement for Radio Equipment Used in Satellite Earth Stations Part 2: Measurements for Sub-Systems Section Four—up-and Down-Converters First Edition. 38 pp.

—Crystal Resonators
IEC 483-76. Guide to Dynamic Measurements of Piezoelectric Ceramics with High Electromechanical Coupling First Edition. 37 pp.

—Data Communication—Interchange Circuits
CCITT RECMN X.26-89. Electrical Characteristics for Unbalanced Double-Current Interchange Circuits for General Use with Integrated Circuit Equipment in the Field of Data Communications —Data Communication Networks: Services and Facilities, Interfaces (Study Group VII) 1 pp. 1 p.

INDUSTRY STANDARDS

INTERNATIONAL AND NON-U.S. NATIONAL STANDARDS
SUBJECT INDEX

Electrical

Electrical Measurement *(Cont.)*

—Data Communication—Interchange Circuits *(Cont.)*
CCITT RECMN X.27-89. Electrical Characteristics for Balanced Double-Current Interchange Circuits for General Use with Integrated Cir. Equipment in the Field of Data Communications—Data Comm. Networks:Services and Facilities, Inter. (Study Group VII) 1 pp. 1 p.

—Data Links—Voice Frequency Telegraphy
CCITT RECMN M.830-89. Routine Measurements to Be Made on International Voice-Frequency Telegraph Links—Maintenance of International Telegraph, Phototelegraph and Leased Circuits—Maintenance of the International Public Telephone Network—Maintenance of Maritime. 1 p.

—Data Terminal Equipment—Interchange Circuits—Interfaces
CCITT RECMN V.10-89. Electrical Characteristics for Unbalanced Double-Current Interchange Circuits for General Use with Integrated Circuit Equipment in the Field of Data Communications—Data Communication Over the Telephone Network (Study Group XVII) 17 pp. 17 pp.
CCITT RECMN V.11-89. Electrical Characteristics for Balanced Double-Current Interchange Circuits for General Use with Integrated Circuit Equipment in the Field of Data Communications—Data Communication Over the Telephone Network (Study Group XVII) 12 pp. 12 pp.

—Decoders—MAC/Packet
IEC 1079 Pt 5-93. Methods of Measurement on Receivers for Satellite Broadcast Transmissions in the 12 GHz Band Part 5: Electrical Measurements on Decoder Units for MAC/Packet Systems First Edition. 156 pp.

—Digital Circuits
BSI BS CECC 90100-86. (WITHDRAWN) 1986 Amd 2 Harmonized System of Quality Assessment for Electronic Components: Sectional Specification: Digital Monolithic Integrated Circuits (AMD 5954) March 28, 1991 (Renumbered as BS EN 190100: 1993). 69 pp.

—Echo Suppressors
CCITT RECMN G.164-89. Echo Suppressors—General Characteristics of International Telephone Connections and Circuits (Study Groups XII and XV) 36 pp. 36 pp.

—Electron Tubes
BSI BS 4478-69. 1969 Methods of Measurement of Intermodulation Products in Electronic Tubes or Valves Intended for Use in Colour Television Transposers. 8 pp.
IEC 151 Pt 14-75. Measurements of the Electrical Properties of Electronic Tubes Part 14: Methods of Measurement of Radar and Oscilloscope Cathode-Ray Tubes Second Edition. 33 pp.
IEC 151 Pt 17-69. Measurements of the Electrical Properties of Electronic Tubes and Valves Part 17: Methods of Measurement of Gasfilled Tubes and Valves Second Edition. 51 pp.
IEC 151 Pt 19-69. Measurements of the Electrical Properties of Electronic Tubes and Valves Part 19: Methods of Measurement on Corona Stabilizers First Edition. 17 pp.
IEC 151 Pt 20-69. Measurements of the Electrical Properties of Electronic Tubes and Valves Part 20: Methods of Measurement of Thyratron Pulse Modulators First Edition. 32 pp.
IEC 151 Pt 22-70. Measurements of the Electrical Properties of Electronic Tubes and Valves Part 22: Methods of Measurement for Cold Cathode Counting and Indicator Tubes First Edition. 28 pp.
IEC 151 Pt 23-70. Measurements of the Electrical Properties of Electronic Tubes and Valves Part 23: Methods of Measurement of Vacuum Pulse Modulator Tubes and Valves First Edition. 23 pp.
IEC 151 Pt 24-71. Measurements of the Electrical Properties of Electronic Tubes Part 24: Methods of Measurement of Cathode-Ray Charge-Storage Tubes First Edition. 60 pp.
IEC 151 Pt 25-71. Measurements of the Electrical Properties of Electronic Tubes Part 25: Methods of Measurement of Geiger-Muller Counter Tubes First Edition. 34 pp.
IEC 151 Pt 26-71. Measurements of the Electrical Properties of Electronic Tubes Part 26: Methods of Measurement for Camera Tubes First Edition. 43 pp.
IEC 151 Pt 27-74. Measurements of the Electrical Properties of Electronic Tubes Part 27: Methods of Measurement for Intermodulation Products in Transmitting Tubes First Edition. 18 pp.

Electrical Measurement *(Cont.)*

—Glossaries
BSI BS 4727:Pt1: Group 04-86. 1986 Electrotechnical, Power, Telecommunication, Electronics, Lighting and Colour Terms Part 1: Terms Common to Power, Telecommunications and Electronics Group 04: Measurement Terminology. 34 pp.
IEC 235 Pt 1-72. Measurement of the Electrical Properties of Microwave Tubes Part 1: Terminology Second Edition; (Supplement A-1975). 91 pp.

—Magnetic Materials
BSI BS 6404: Part 2-85. 1985 Amd 1 Magnetic Materials Part 2: Methods of Measurement of Magnetic, Electrical and Physical Properties of Magnetic Sheet and Strip. 47 pp.
BSI BS 6404: Part 3-92. 1992 Magnetic Materials Part 3: Methods of Measurement of the Magnetic Properties of Magnetic Sheet and Strip by Means of a Single (IEC 404-3: 1992). 23 pp.
BSI BS 6404: Part 3-85. 1985 Magnetic Materials Part 3: Methods of Measurement of Specific Total Losses of Magnetic Sheet and Strip by Means of a Single Sheet Tester. 14 pp.
IEC 404 Pt 2-78. Magnetic Materials Part 2: Methods of Measurement of Magnetic, Electrical and Physical Properties of Magnetic Sheet and Strip Second Edition. 87 pp.
IEC 404 Pt 3-92. Magnetic Materials Part 3: Methods of Measurement of the Magnetic Properties of Magnetic Sheet and Strip by Means of a Single Sheet Tester Second Edition. 38 pp.

—Microwave Tubes
IEC 235 Pt 1-72. Measurement of the Electrical Properties of Microwave Tubes Part 1: Terminology Second Edition; (Supplement A-1975). 91 pp.
IEC 235 Pt 2-72. Measurement of the Electrical Properties of Microwave Tubes Part 2: General Measurements, Chapter VI: Methods of Measuring the Effects of Non-Linearity First Edition; (Supplement A-1974) (Supplement B-1975) (Supplement C-E:1976). 268 pp.
IEC 235 Pt 3-72. Measurement of the Electrical Properties of Microwave Tubes Part 3: Disk Seal Tubes First Edition. 23 pp.
IEC 235 Pt 4-72. Measurement of the Electrical Properties of Microwave Tubes Part 4: Magnetrons Clause 4: Voltage Tunable Magnetron First Edition; (Supplement A-1975). 27 pp.
IEC 235 Pt 5-72. Measurement of the Electrical Properties of Microwave Tubes Part 5: Low-Power Oscillator Klystrons First Edition. 15 pp.
IEC 235 Pt 6-72. Measurement of the Electrical Properties of Microwave Tubes Part 6: High-Power Klystrons First Edition. 21 pp.
IEC 235 Pt 7-72. Measurement of the Electrical Properties of Microwave Tubes Part 7: Gas-Filled Microwave Switching Devices First Edition. 35 pp.
IEC 235 Pt 8-72. Measurement of the Electrical Properties of Microwave Tubes Part 8: Backward-Wave Oscillator Tubes—"O" Type First Edition; (Supplement A-1974). 17 pp.
IEC 235 Pt 9-75. Measurement of the Electrical Properties of Microwave Tubes Part 9: Crossed-Field Amplifier Tubes First Edition. 23 pp.

—Picture Tubes
IEC 151 Pt 16-68. Measurements of the Electrical Properties of Electronic Tubes and Valves Part 16: Methods of Measurement for Television Picture Tubes First Edition. 19 pp.
IEC 151 Pt 28-78. Measurements of the Electrical Properties of Electronic Tubes Part 28: Methods of Measurement of Colour Television Picture Tubes First Edition. 24 pp.

—Quartz Crystal Units
IEC 1080-91. Guide to the Measurement of Equivalent Electrical Parameters of Quartz Crystal Units First Edition. 101 pp.

—Resistors
IEC 440-73. Method of Measurement of Non-Linearity in Resistors First Edition; (Corrigendum—Sept 1974). 20 pp.

—Semiconductor Devices—Temperature
CNS C5058-80. Standard Temperatures for Electrical Measurement and Rating Specification Semiconductor Devices (Jul)(5775).

—Signal Generators—Interchange Circuits
CCITT RECMN V.28-89. Electrical Characteristics for Unbalanced Double-Current Interchange Circuits—Data Communication over the Telephone Network (Study Group XVII) 5 pp. 5 pp.

Electrical Measurement *(Cont.)*

—Signal Generators—Interchange Circuits—Communication Interfaces
CCITT RECMN V.10-89. Electrical Characteristics for Unbalanced Double-Current Interchange Circuits for General Use with Integrated Circuit Equipment in the Field of Data Communications—Data Communication Over the Telephone Network (Study Group XVII) 17 pp. 17 pp.
CCITT RECMN V.11-89. Electrical Characteristics for Balanced Double-Current Interchange Circuits for General Use with Integrated Circuit Equipment in the Field of Data Communications—Data Communication Over the Telephone Network (Study Group XVII) 12 pp. 12 pp.

—Sound/Data Decoders—Satellite Communications
IEC 1079 Pt 4-93. Methods of Measurement on Receivers for Satellite Broadcast Transmissions in the 12 GHz Band Part 4: Electrical Measurements on Sound/Data Decoder Units for the Digital Subcarrier NTSC System. 43 pp.

—Telephone Cables—Glossaries
CCITT RECMN G.601-89. Terminology for Cables—Transmission Media Characteristics (Study Group XV) 6 pp. 6 pp.

—Telephone Circuits
CCITT RECMN G.613-89. Characteristics of Symmetric Cable Pairs Usable Wholly for the Transmission of Digital Systems with a Bit Rate of up to 2 Mbits—Transmission Media Characteristics (Study Group XV) 4 pp. 4 pp.

—Television Receivers
BSI BS 3549: Part 1-79. 1979 Amd 1 Measuring and Expressing the Performance of Television Receivers Part 1: General Considerations and Electrical Measurements Other Than Those at Audio-Frequencies. 142 pp.
BSI BS 3549: Part 2-81. 1981 Measuring and Expressing the Performance of Television Receivers Part 2: Electrical and Acoustic Measurements at Audio Frequencies. 30 pp.
CENELEC HD 567.6 S1-90. Recommended Methods of Measurement on Receivers for Television Broadcast Transmissions Part 6: Measurement Under Conditions Different from Broadcast Signal Standards. 3 pp.
CENELEC EN 60107-5-92. Recommended Methods of Measurement on Receivers for Television Broadcast Transmissions Part 5: Electrical Measurements on Multichannel Sound Television Receivers Using the NICAM Two-Channel Digital Sound-System. 5 pp.
IEC 107 Pt 1-77. Recommended Methods of Measurement on Receivers for Television Broadcast Transmissions Part 1: General Considerations Electrical Measurements Other Than Those at Audio-Frequencies Second Edition; (Amendment 1-1987). 257 pp.
IEC 107 Pt 2-80. Recommended Methods of Measurement on Receivers for Television Broadcast Transmissions Part 2: Electrical and Acoustic Measurements at Audio-Frequencies First Edition. 59 pp.
IEC 107 Pt 3-88. Recommended Methods of Measurement on Receivers for Television Broadcast Transmissions Part 3: Electrical Measurements on Multichannel Sound Television Receivers Using Subcarrier Systems First Edition. 55 pp.
IEC 107 Pt 4-88. Recommended Methods of Measurement on Receivers for Television Broadcast Transmissions Part 4: Electrical Measurements on Multichannel Sound Television Receivers Using the Two-Carrier FM-System First Edition. 35 pp.
IEC 107 Pt 5-92. Recommended Methods of Measurement on Receivers for Television Broadcast Transmissions Part 5: Electrical Measurements on Multichannel Sound Television Receivers Using the NICAM Two-Channel Digital Sound-System First Edition. 38 pp.
IEC 107 Pt 6-89. Recommended Methods of Measurement on Receivers for Television Broadcast Transmissions Part 6: Measurement Under Conditions Different from Broadcast Signal Standards First Edition. 85 pp.
SAA AS 1173.1-79. Recommended Methods of Measurement on Receivers for Television Broadcast Transmissions—Part 1: General Considerations—Electrical Measurements Other Than Those at Audio-Frequencies. 116 pp.
SAA AS 1173.2-86. Recommended Methods of Measurement on Receivers for Television Broadcast Transmissions—Part 2: Electrical and Acoustic Measurements at Audio-Frequencies. 22 pp.

INDEX and DIRECTORY of

INTERNATIONAL AND NON-U.S. NATIONAL STANDARDS
SUBJECT INDEX
Electrical

Electrical Measurement *(Cont.)*
—Transducers
BSI BS 6253: Part 1-82. (WITHDRAWN) 1982 Amd 2 Electrical Measuring Transducers for Converting a.c. Electrical Quantities into d.c. Electrical Quantities Part 1: Method for Specifying General Purpose Transducers (AMD 6225) March 28, 1991. 26 pp.
BSI BS EN 60688-92. 1992 Electrical Measuring Transducers for Converting a.c. Electrical Quantities to Analogue or Digital Signals (IEC 688: 1992) (Supersedes BS 6253: Parts 1 & 2: 1982). 42 pp.
CENELEC EN 60688-92. Electrical Measuring Transducers for Converting a.c. Electrical Quantities to Analogue or Digital Signals. 5 pp.
IEC 688-92. Electrical Measuring Transducers for Converting a.c. Electrical Quantities to Analogue or Digital Signals Second Edition. 74 pp.
MOD UK DSTAN 66-20-78. Transducers Electrical (Used in the Measurement of Electrical Quantities) Issue 1. 20 pp.
SAA AS 1384-73. Transducers for Electrical Measurements. 35 pp.

—Vacuum
IEC 206-66. Designation of the Quantities Characterizing the Magnetic and Electric Properties of Vacuum and a Substance First Edition. 5 pp.

—Video Tapes
BSI BS 6412-83. (WITHDRAWN) 1983 Amd 1 Measurement of Video Tape Properties (AMD 6315) August 31, 1990 (Superseded by BS EN 60735: 1992). 16 pp.
BSI BS EN 60735-92. 1992 Methods for Measurement of Video Tape Properties (S). 24 pp.
CENELEC HD 454-84. Measuring Methods for Video Tape Properties. 2 pp.
CENELEC EN 60735-91. Measuring Methods for Video Tape Properties. 6 pp.
IEC 735-91. Measuring Methods for Video Tape Properties Second Edition. 33 pp.

Electrical Medical Equipment
Use: Medical Electrical Equipment; Medical Equipment

Electrical Metallic Tubing
Use For: EMT *See Also:* Electric Conduits; Electric Raceways; Electrical Flexible Tubing

—Bends
CSA C22.2 NO 83-M1985. Electrical Metallic Tubing (R 1992); (Gen Instr 1). 19 pp.

—Couplings
CSA C22.2 NO 83-M1985. Electrical Metallic Tubing (R 1992); (Gen Instr 1). 19 pp.

—Elbows
CSA C22.2 NO 83-M1985. Electrical Metallic Tubing (R 1992); (Gen Instr 1). 19 pp.

—Fittings
CSA CAN/CSA-C22. 2 NO 18-92. Outlet Boxes, Conduit Boxes, and Fittings; (Gen Instr 1 Thru 2). 118 pp.

—Fittings—Threaded
CSA CAN/CSA-C22. 2 NO 18-92. Outlet Boxes, Conduit Boxes, and Fittings; (Gen Instr 1 Thru 2). 118 pp.

—Hangers (Fasteners)
CSA CAN/CSA-C22. 2 NO 18-92. Outlet Boxes, Conduit Boxes, and Fittings; (Gen Instr 1 Thru 2). 118 pp.

—Safety
CSA C22.2 NO 83-M1985. Electrical Metallic Tubing (R 1992); (Gen Instr 1). 19 pp.

—Straps (Fasteners)
CSA CAN/CSA-C22. 2 NO 18-92. Outlet Boxes, Conduit Boxes, and Fittings; (Gen Instr 1 Thru 2). 118 pp.

Electrical Nonmetallic Tubing
Use For: ENT *See Also:* Electric Conduits; Electric Raceways; Electrical Nonmetallic Tubing Fittings

—Corrugated
CSA C22.2 NO 227.1-93. Electrical Nonmetallic Tubing; (Gen Instr 1). 29 pp.

Electrical Nonmetallic Tubing Fittings
See Also: Electrical Nonmetallic Tubing; Tube Couplings
CSA C22.2 NO 227.1-93. Electrical Nonmetallic Tubing; (Gen Instr 1). 29 pp.
CSA CAN/CSA-C22. 2NO227.3-M91. Flexible Nonmetallic Tubing; (Gen Instr 1). 29 pp.

Electrical Nonmetallic Tubing Fittings *(Cont.)*
—Couplings
CSA C22.2 NO 227.1-93. Electrical Nonmetallic Tubing; (Gen Instr 1). 29 pp.

Electrical Power Feedthrough Connectors
Use: Feedthrough Connectors

Electrical Properties
Scope Note: For additional listings, use a more specific term *Use For:* Electrical Testing
See Also: Capacitance; Capillarity; Contact Resistance; Density; Dielectric Breakdown; Dielectric Dissipation Factor; Dielectric Loss; Dielectric Strength; Effective Capacitance; Electric Measuring Instruments; Electrical Impedance; Electrical Insulation; Electrical Measurement; Electrical Resistance; Electrical Resistivity; Electricity; Electrostatic Induction; Electrostatic Properties; Field Strength; Gain (Amplification); Impulse Voltage Testing; Insulation Resistance; Linearity; Mutual Inductance; Optical Properties; Permittivity; Photoconductivity; Photoelectric Emission; Reflection Factor; Solid State Physics; Steady State Performance; Surface Resistance; Transfer Impedance; Treeing; Voltage

—Airborne Radar
EUROCAE WG9/1-71 06.72. MPS for Airborne Secondary Surveillance Radar Transponder Apparatus (Amendment No. 1 Incorporated). 29 pp.

—Aircraft—Area Navigation Computing Instruments
EUROCAE ED-28 07.82. MPS for Airborne Area Navigation Computing Equipment Based on VOR and DME as Sensors. 31 pp.

—Aircraft Power
JIS W 7001-91. Aircraft Electrical Power Characteristics.

—Antennas
BSI BS 5640: Part 1-78. 1978 Aerials for the Reception of Sound and Television Broadcasting in the Frequency Range to 30 MHz to 1 GHz Part 1: Specification for Electrical and Mechanical Characteristics. 10 pp.
CENELEC HD 95.1-78. Aerials for the Reception of Sound and Television Broadcasting in the Frequency Range 30 MHz to 2 GHz Part 1: Electrical and Mechanical Characteristics. 2 pp.
IEC 597 Pt 1-77. Aerials for the Reception of Sound and Television Broadcasting in the Frequency Range 30 MHz to 1 GHz Part 1: Electrical and Mechanical Characteristics First Edition. 18 pp.

—Area Navigation Computing Instruments
EUROCAE ED-40 06.84. MPS for Airborne Computing Equipment for Area Navigation System Using Two DME as Sensors. 33 pp.

—Cables (Electric)
IEC 840-88. Tests for Power Cables with Extruded Insulation for Rated Voltages Above 30 kV (Um = 36 kV) up to 150 kV (Um = 170 kV) First Edition; (Corrigendum—Nov 1988) (Amendment 2-1993, Including Amendment 1) (Corrigendum—July 1993). 71 pp.
IEC 885 Pt 1-87. Electrical Test Methods for Electric Cables Part 1: Electrical Tests for Cables, Cords and Wires for Voltages up to and Including 450/750 V First Edition. 13 pp.
SAA AS 1660.3-93. Electrical Tests. 22 pp.

—Control Valves
IEC 730 Pt 2-8-92. Automatic Electrical Controls for Household and Similar Use Part 2: Particular Requirements for Electrically Operated Water Valves, Including Mechanical Requirements First Edition. 70 pp.

—Cords (Electric)
SAA AS 1660.3-93. Electrical Tests. 22 pp.

—Dead Reckoning Computers
EUROCAE WG7C/2-74 08.74. MPS for Airborne Automatic Dead Reckoning Computer Equipment Utilizing Aircraft Heading and Doppler Obtained Velocity Vector Data. 19 pp.

—Distance Measuring Equipment
EUROCAE ED-54 01.87. MOPR for DME/N and DME/P Operating Within the Radio Frequency Range 960 to 1215 MHz. 144 pp.

Electrical Properties *(Cont.)*
—Earth Surface
CCIR Report 229-6-90. Electrical Characteristics of the Surface of the Earth—Section 5B—Effects of the Ground (Including Ground-Wave Propagation). 7 pp.
CCIR Report 879-1-86. Methods for Estimating Effective Electrical Characteristics of the Surface of the Earth—Section 5B—Effects of the Ground (Including Ground-Wave Propagation). 9 pp.

—Earth Surface—Ground Wave Propagation
CCIR RECMN 527-2-90. Electrical Characteristics of the Surface of the Earth—Section 5B—Effects of the Ground (Including Ground-Wave Propagation). 3 pp.
CCIR RECMN 527-3-92. Electrical Characteristics of the Surface of the Earth. 5 pp.

—Echo Cancellers
CCITT RECMN M.665-89. Testing of Echo Cancellers—General Maintenance Principles—Maintenance of International Transmission Systems and Telephone Circuits (Study Group IV) 2 pp. 2 pp.

—Echo Cancellers—Telephone Circuits
CCITT RECMN M.665-89. Testing of Echo Cancellers—General Maintenance Principles—Maintenance of International Transmission Systems and Telephone Circuits (Study Group IV) 2 pp. 2 pp.

—Electric Conductors
SAA AS 1660.3-93. Electrical Tests. 22 pp.

—Flasher Units—Automotive
ISO 4082-81. Road Vehicles—Motor Vehicles—Flasher Units First Edition. 9 pp.

—Generator Sets
NATO STANAG 4135 ED 2 AMD 0-92. Electrical Characteristics of Rotating Alternating Current Generating Sets. 15 pp.

—Glide Path Receiving Instruments
EUROCAE ED-47A 01.88. MPS for Airborne ILS Receiving Equipment (Glide Path). 223 pp.

—Insulators
IEC 591-78. Sampling Rules and Acceptance Criteria When Applying Statistical Control Methods for Mech and Elmch Tests on Insulators of Ceramic Material or Glass for Overhead Lines with a Nominal Voltage Greater Than 1000 V First Edition. 27 pp.

—Localizers
EUROCAE ED-51 10.83. MPS for Airborne Automatic Direction Finding Equipment. 31 pp.

—Low Range Radio Altimeters
EUROCAE ED-30 03.80. MPS for Airborne Low Range Radio (Radar) Altimeter Equipment. 22 pp.

—Marker Beacons—Receiving Equipment
EUROCAE WG7/70 01.70. MPS for Airborne 75 MHz Marker Beacon Receiving Equipment. 26 pp.

—Microwave Landing Systems
EUROCAE ED-36 03.83. MOPR for Microwave Landing System (MLS) (Airborne Receiving Equipment). 69 pp.

—Radio Receivers
EUROCAE ED-23A 05.86. MPS for Airborne VHF Communications Equipment Operating in the Frequency Range 117.975-137.000 MHz. 125 pp.

—Signal Generators
IEC 716-81. Expression of the Properties of Signal Generators First Edition. 98 pp.

—Tape Recorders
BSI BS EN 61120-1-93. 1993 Digital Audio Tape Recorder Reel to Reel System, Using 6.3 mm Magnetic Tape, for Professional Use Part 1: General Requirements (IEC 1120-1: 1991) (S). 17 pp.
CENELEC EN 61120-1-93. Digital Audio Tape Recorder Reel to Reel System, Using 6,3 mm Magnetic Tape, for Professional Use Part 1: General Requirements (IEC 1120-1: 1991). 12 pp.
IEC 1120 Pt 1-91. Digital Audio Tape Recorder Reel to Reel System, Using 6,3 mm Magnetic Tape, for Professional Use Part 1: General Requirements First Edition; (CENELEC EN 61120-1: 1993). 26 pp.

—Transmitters
EUROCAE ED-23A 05.86. MPS for Airborne VHF Communications Equipment Operating in the Frequency Range 117.975-137.000 MHz. 125 pp.

INDUSTRY STANDARDS

Electrical

INTERNATIONAL AND NON-U.S. NATIONAL STANDARDS SUBJECT INDEX

Electrical Properties (Cont.)

—Transponders
EUROCAE WG9/1-71 06.72. MPS for Airborne Secondary Surveillance Radar Transponder Apparatus (Amendment No. 1 Incorporated). 29 pp.

—Ultrasonic Testing Equipment
JIS Z 2351-92. Method for Assessing the Electrical Characteristics of Ultrasonic Testing Instrument Using Pulse Echo Technique.

—Video Cassette Recorders
BSI BS 7486: Part 1-91. 1991 Helical-Scan Video Tape Cassette System Using 12,65 mm (0,5 in) Magnetic Tape on Type Beta Format—FM Audio Recording Part 1: 625 Line—50 Field Systems (S) (IEC 1053-1: 1991). 14 pp.

BSI BS 7487-91. (WITHDRAWN) 1991 Helical-Scan Video Tape Cassette System Using 12,65 mm (0,5 in) Magnetic Tape on Type VHS—FM Audio Recording (IEC 1054: 1991) (Renumbered as BS EN 61054: 1993). 12 pp.

BSI BS EN 61054-93. 1993 AMD 1 Helical-Scan Video Tape Cassette System Using 12,65 mm (0,5 in) Magnetic Tape on Type VHS—FM Audio Recording (IEC 1054: 1991) (AMD 7737) April 15, 1993 (S). 20 pp.

CENELEC EN 61054-93. Helical-Scan Video Tape Cassette System Using 12.65 mm (0.5 in) Magnetic Tape on Type VHS FM Audio Recording; (IEC 1054:1991). 5 pp.

CENELEC EN 61054-93. Helical-Scan Video Tape Cassette System Using 12,65 mm (0,5 in) Magnetic Tape on Type VHS—FM Audio Recording (IEC 1054: 1991). 10 pp.

IEC 1053 Pt 1-91. Helical-Scan Video Tape Cassette System Using 12,65 mm (0,5 in) Magnetic Tape on Type Beta Format—FM Audio Recording Part 1: 625 Line-50 Field Systems First Edition. 25 pp.

IEC 1053 Pt 2-91. Helical-Scan Video Tape Cassette System Using 12,65 mm (0,5 in) Magnetic Tape on Type Beta Format—FM Audio Recording Part 2: 525 Line-60 Field Systems First Edition. 23 pp.

IEC 1054-91. Helical-Scan Video Tape Cassette System Using 12,65 mm (0,5 in) Magnetic Tape on Type VHS—FM Audio Recording First Edition; (CENELEC EN 61054:1993). 23 pp.

—Windings
BSI BS 6811: Sec 2.5-92. 1992 Winding Wires Part 2: Methods of Test Section 2.5: Electrical Properties (IEC 851-5: 1988). 22 pp.

CENELEC HD 490.5 S1-88. Methods of Test for Winding Wires Part 5: Electrical Properties. 5 pp.

CENELEC HD 490.5 S2-91. Methods of Test for Winding Wires Part 5: Electrical Properties. 6 pp.

IEC 851 Pt 5-88. Method of Test for Winding Wires Part 5: Electrical Properties Second Edition. 53 pp.

—Xenon Arc Lamps
IEC 1127-92. High Pressure Xenon Short Arc Lamps—Dimensional, Electrical and Photometric Data and Cap Types First Edition. 52 pp.

Electrical Protection Equipment

See Also: Circuit Breakers; Circuit Protectors; Crowbars (Electrical); Degree of Protection (Electrical); Disconnecting Switches; Earth Leakage Circuit Breakers; Electric Shock; Electrical Enclosures; Electrical Equipment; Electrical Grounding; Electrical Safety; Electromagnetic Protection; Explosive Atmospheres; Ground Fault Protection; Ground Fault Protection Equipment; Ground Rods; Lightning Protection Equipment; Linemen's Equipment; Load Break Switches; Motor Protectors; Overcurrent Protection Equipment; Overvoltage Protection Equipment; Power Conditioning Equipment; Protective Equipment; Protective Relays; Residual Current Devices; Semiconductor Arrester Assemblies; Surge Arresters; Tripping Devices; Undervoltage Protectors

—Aerospace
JIS W 2009-78. Bonding, Electrical and Lightning Protection, for Aerospace Systems.

—Aircraft—Radio Equipment
CAA Chapter R4-3 04.74. Power Supplies and Circuit Protection (Radio). 2 pp.

—Boats
BSI BS 7489-91. 1991 Protection of Electrical Devices Used on Small Craft to Prevent Ignition of Surrounding Flammable Gases. 12 pp.

BSI BS EN 28846-93. 1993 Small Craft—Electrical Devices—Protection Against Ignition of Surrounding Flammable Gases (ISO 8846: 1990) (S). 14 pp.

ISO 8846-90. Small Craft—Electrical Devices—Protection Against Ignition of Surrounding Flammable Gases First Edition. 10 pp.

Electrical Protection Equipment (Cont.)

—Cables (Electric)
DIN VDE 0100 Pt 430-81. Installation of Power Plant with Rated Voltages up to 1000 V; Protection of Cables and Cords Against Undue Temperature Rise (June). 22 pp.

DIN VDE 0100 Pt 430 Suppl. 1-91. Erection of Power Installations with Nominal Voltages up to 1000V; Protection of Cables and Cords Against Overcurrent; Recommended Values for Current-Carrying Capacity Iz and the Allocation of Overcurrent. 17 pp.

—Cords (Electric)
DIN VDE 0100 Pt 430-81. Installation of Power Plant with Rated Voltages up to 1000 V; Protection of Cables and Cords Against Undue Temperature Rise (June). 22 pp.

DIN VDE 0100 Pt 430 Suppl. 1-91. Erection of Power Installations with Nominal Voltages up to 1000V; Protection of Cables and Cords Against Overcurrent; Recommended Values for Current-Carrying Capacity Iz and the Allocation of Overcurrent. 17 pp.

—Covers—Live Wire Working
IEC 1229-93. Rigid Protective Covers for Live Working on a.c. Installations First Edition. 66 pp.

—Current Transformers
IEC 44 Pt 6-92. Instrument Transformers Part 6: Requirements for Protective Current Transformers for Transient Performance First Edition. 90 pp.

—Electric Shock
IEC 1140-92. Protection Against Electric Shock Common Aspects for Installation and Equipment First Edition. 29 pp.

—Electrical Codes
IEC 364 Pt 4-443-90. Electrical Installations of Buildings Part 4: Protection for Safety Chapter 44: Protection Against Overvoltages Section 443—Protection Against Overvoltages of Atmospheric Origin or Due to Switching First Edition. 25 pp.

—Electrical Disturbance Testing
BSI BS 142: SUBSEC 1.4.1-90. 1990 Electrical Protection Relays Part 1: Information and Requirements for All Protection Relays Section 1.4: Electrical Disturbance Tests Subsection 1.4.1: 1 MHz Burst Disturbance Tests. 20 pp.

BSI BS 142: SUBSEC 1.4.4-93. 1993 Electrical Protection Relays Part 1: Information and Requirements for All Protection Relays Sec. 1.4: Specification for Electrical Disturbance Tests Subsection 1.4.4: Fast Transient Disturbance Test (IEC 255-22-4: 1992). 16 pp.

IEC 255 Pt 22-1-88. Electrical Disturbance Tests for Measuring Relays and Protection Equipment Part 1: 1 MHz Burst Disturbance Tests First Edition. 37 pp.

IEC 255 Pt 22-4-92. Electrical Relays Part 22: Electrical Disturbance Tests for Measuring Relays and Protection Equipment Section 4: Fast Transient Disturbance Test First Edition. 31 pp.

—Electromagnetic Interference
IEC 255 Pt 22-3-89. Electrical Relays Part 22: Electrical Disturbance Tests for Measuring Relays and Protection Equipment Section Three—Radiated Electromagnetic Field Disturbance Tests First Edition. 37 pp.

—Glossaries
IEC 50 Chap 448-87. International Electrotechnical Vocabulary Chapter 448: Power System Protection First Edition; (Corrigendum—October 1987). 41 pp.

SAA AS 1852.448-89. International Electrotechnical Vocabulary—Part 448: Power System Protection. 18 pp.

SNZ IEC 50: 50(448)-87. International Electrotechnical Vocabulary 50(448): Power System Protection. 32 pp.

—Hazardous Locations
CSA C22.2 NO 213-M1987. Non-Incendive Electrical Equipment for Use in Class I, Division 2 Hazardous Locations (R 1992). 30 pp.

—High Voltage Installations
IEC 919 Pt 2-91. Performance of High-Voltage d.c. (HVDC) Systems Part 2: Faults and Switching First Edition. 161 pp.

—Live Wire Working
IEC 1057-91. Aerial Devices with Insulating Boom Used for Live Working First Edition. 119 pp.

—Mats—Rubber
BSI BS 921-76. 1976 Rubber Mats for Electrical Purposes. 7 pp.

Electrical Protection Equipment (Cont.)

—Mining Equipment
SAA AS 2081.1-88. Electrical Equipment for Coal and Shale Mines—Electrical Protection Devices—Part 1: General Requirements. 7 pp.

SAA AS 2081.2-88. Electrical Equipment for Coal and Shale Mines—Electrical Protection Devices—Part 2: Earth-Continuity Monitoring Devices. 3 pp.

—Operating Rods—Live Wire Working
DIN VDE 0680 Pt 3-77. Personal Protective Equipment; Protective Devices and Apparatus for Working on Live Equipment up to 1000 V; Part 3: Operating Rods (Sept). 19 pp.

—Personal
JIS T 8010-79. Testing Method for Withstand Voltage of Personal Protective Equipment and Insulating Devices.

—Portable
DIN VDE 0661-88. Portable Protective Devices to Increase the Protection Level for 230 V a.c. Rated Voltage 16 A Rated Current and Rated Residual Current I Delta n Less Than or Equal to 30 mA (Apr). 38 pp.

—Ships
IEC 92 Pt 202-80. Electrical Installations in Ships Part 202: System Design Protection Third Edition; (Amendment 2-1989, Incorporating Amendment 1). 36 pp.

JIS F 8063-86. Electrical Installations in Ships Part 202 System Design—Protection (IEC 92-202-1980).

—Signal Devices—Telecommunications
BSI BS 7494: Part 1-91. 1991 Performance and Testing of Teleprotection Equipment of Power Systems Part 1: Specification for Narrow-Band Command Systems. 43 pp.

BSI BS 7494: Part 2-93. 1993 Performance and Testing of Teleprotection Equipment of Power Systems Part 2: Analogue Comparison System (IEC 834-2: 1993) (S). 40 pp.

CENELEC HD 543.1 S1-91. Performance and Testing of Teleprotection Equipment of Power Systems Part 1: Narrow-Band Command Systems. 5 pp.

IEC 834 Pt 1-88. Performance and Testing of Teleprotection Equipment of Power Systems Part 1: Narrow-Band Command Systems First Edition. 78 pp.

IEC 834 Pt 2-93. Performance and Testing of Teleprotection Equipment of Power Systems—Part 2: Analogue Comparison Systems First Edition. 81 pp.

—Symbols
CNS C1028-57. Symbols of Protective Devices (Jan)(412) (R 1971).

DIN ENGL 40713 Suppl. 3-75. Graphical Symbols; Protective Appliance Examples (Jan). 7 pp.

—Symbols—Diagrams
BSI BS 3939: Part 7-85. 1985 Graphical Symbols for Electrical Power, Telecommunications and Electronics Diagrams Part 7: Switchgear, Controlgear and Protective Devices (G). 32 pp.

IEC 617 Pt 7-83. Graphical Symbols for Diagrams Part 7: Switchgear, Controlgear and Protective Devices First Edition. 58 pp.

SAA AS 1102.107-89. Graphical Symbols for Electrotechnology—Part 107: Switchgear, Controlgear and Protective Devices. 40 pp.

SNZ IEC 617: Part 7-83. Graphical Symbols for Diagrams Part 7: Switchgear, Controlgear and Protective Devices. 56 pp.

—Telecommunication Networks
CSA CAN/CSA-C22. 2 NO 226-92. Protectors in Telecommunication Networks; (Gen Instr 1). 88 pp.

—Three Phase Motors
CNS C1140-88. Protecting Equipment and Distribution of Three-Phase 220v Motors for General Use (Jul)(12356).

—Tools
DIN VDE 0680 Pt 2-78. Personnel Protective Equipment; Protective Devices and Apparatus for Working on Live Equipment up to 1000 V; Part 2: Insulated Tools (Mar). 39 pp.

DIN VDE 0680 Pt 2 A2 (D)-87. Personal Protective Equipment, Protective Devices and Apparatus for Work on Electrically Energized Systems up to 1000 V; Insulated Tools: Amendment 2 (Jan). 16 pp.

—Vacuum Cleaners
CSA C22.2 NO 67-M1985. Vacuum Cleaners and Floor Polishers; (Gen Instr 1 Thru 4). 57 pp.

INTERNATIONAL AND NON-U.S. NATIONAL STANDARDS
SUBJECT INDEX

Electrical

Electrical Protection Equipment (Cont.)

—Vibration
BSI BS 142: SUBSEC 1.5.1-89. 1989 Electrical Protection Relays Part 1: Information and Requirements for All Protection Relays Section 1.5: Vibration, Shock, Bump and Seismic Testing Subsection 1.5.1: Vibration Tests (Sinusoidal). 14 pp.
IEC 255 Pt 21-1-88. Electrical Relays Part 21: Vibration, Shock, Bump and Seismic Tests on Measuring Relays and Protection Equipment Section One—Vibration Tests (Sinusoidal) First Edition. 26 pp.

Electrical Resistance
Use For: Electrical Conductance *See Also:* Contact Resistance; Dielectric Dissipation Factor; Electrical Impedance; Electrical Properties; Electrical Resistivity; Insulation Resistance; Surface Resistance; Tracking Resistance

—Alloy Wire—Temperature Coefficient
BSI BS 3467-62. 1962 Method of Test for Temperature Coefficient of Resistance of Alloy Wire for Precision Resistors. 13 pp.

—Cables (Electric)
DIN VDE 0472 Pt 501-83. Testing of Cables and Insulated Cords; Conductor Resistance (Apr). 7 pp.
DIN VDE 0472 Pt 510-84. Testing of Cables and Insulated Flexible Cords; Resistance to Direct Current (Sept). 6 pp.
DIN VDE 0472 Pt 512-85. Testing of Cables & Insulated Conductors; Resistance Between Protective Conductor and Conductive Layer (May). 5 pp.

—Cables (Electric)—Aircraft
CEN PREN 3475 (Part 301)-92. Cables, Electrical, Aircraft Use Test Methods Part 301—Electrical Resistance per Unit Length. 2 pp.

—Conveyor Belts
BSI BS EN 20284-93. 1993 Conveyor Belts—Electrical Conductivity—Specification and Method of Test (ISO 284: 1982) (Q). 10 pp.
CEN EN 20284-93. Conveyor Belts—Electrical Conductivity—Specification and Method of Test (ISO 284: 1982). 5 pp.
CSA CAN/CSA-M422-M87. Fire-Performance and Antistatic Requirements for Conveyor Belting; (Gen Instr 1). 26 pp.
ISO 284-82. Conveyor Belts—Electrical Conductivity—Specification and Method of Test Second Edition; (CEN EN 20284: 1993). 5 pp.
SAA AS 1334.9-82. Methods of Testing Conveyor and Elevator Belting—Part 9: Determination of Electrical Resistance of Conveyor Belting. 3 pp.

—Copper
CNS C1002-80. Resistance of Copper (Jul)(32).
CNS C1132-84. Resistance and Conductivity of Copper Materials for Electrical Purposes (Jun)(10913).
JIS C 3001-81. Resistance of Copper Materials for Electrical Purposes. 6 pp.

—Copper Conductors
IEC 28-25. International Standard of Resistance for Copper Second Edition. 9 pp.

—Copper Conductors—Aircraft
ISO 1967-74. Aircraft—Fire-Resisting Electrical Cables; Dimensions, Conductor Resistance and Mass First Edition. 3 pp.

—Cords
DIN VDE 0472 Pt 501-83. Testing of Cables and Insulated Cords; Conductor Resistance (Apr). 7 pp.
DIN VDE 0472 Pt 510-84. Testing of Cables and Insulated Flexible Cords; Resistance to Direct Current (Sept). 6 pp.

—Dust
JIS B 9915-89. Measuring Methods for Dust Resistivity (with Parallel Electrodes). 15 pp.

—Earth—Textile Floor Coverings
BSI BS 7078-89. 1989 Method for Determination of the Electrical Resistance to Earth of an Installed Textile Floor Covering. 4 pp.

—Electric Conductors
CNS C3007-80. Testing Standard for D.C. Resistance of Insulated Conductors (Jul)(685). 2 pp.

—Electric Switches
CNS C6084-88. Method of Test for Electromechanical Switches (Switch Resistance) (Dec)(6147).

—Electrical Insulation
CNS C6111-81. Method of Test for D.C. Resistance or Conductance of Insulating Materials (May)(7367).
JIS E 4014-89. Test Methods for Insulation Resistance and Withstand Voltage of Railway Rolling Stock.

Electrical Resistance (Cont.)

—Electrical Insulation—Plastics
SAA AS 1255.2-73. Methods of Test for Electrical Characteristics of Solid Plastics Insulating Materials—Part 2: Determination of Insulation Resistance. 14 pp.

—Floor Coverings
CEN PREN 1081-93. Resilient Floorcoverings—Determination of the Electrical Resistance. 9 pp.

—Hose Assemblies
BSI BS 5173: Part 4-77. (WITHDRAWN) 1977 Rubber and Plastics Hoses and Hose Assemblies Part 4: Electrical Tests (Superseded by BS 5173: Sections 104.1 & 104.2: 1991). 8 pp.
BSI BS 5173: Sec 104.1-91. (WITHDRAWN) 1991 Rubber and Plastics Hoses and Hose Assemblies Part 104: Electrical Tests Section 104.1: Measurement of Electrical Resistance of Hoses and Hose Assemblies (Superseded BS BS EN 28031: 1993). 9 pp.
BSI BS 5173: Sec 104.2-91. (WITHDRAWN) 1991 Rubber and Plastics Hoses and Hose Assemblies Part 104: Electrical Test Section 104.2: Measurement of Electrical Continuity and Discontinuity of Hoses and Hose Assemblies (Superseded by BS BS EN 28031: 1993). 6 pp.
BSI BS EN 28031-93. 1993 Ruber and Plastics Hoses and Hose Assemblies—Determination of Electrical Resistance (ISO 8031: 1987) (Supersedes BS 5173: Sections 104.1: 1991 & 104.2: 1990) (E). 11 pp.
CEN EN 28031-93. Rubber and Plasitics Hoses and Hose Assemblies—Determination of Electrical Resistance (ISO 8031: 1987). 6 pp.
ISO 8031-87. Rubber and Plastics Hoses and Hose Assemblies—Determination of Electrical Resistance First Edition; (CEN EN 28031:1993). 7 pp.

—Hoses
BSI BS 5173: Part 4-77. (WITHDRAWN) 1977 Rubber and Plastics Hoses and Hose Assemblies Part 4: Electrical Tests (Superseded by BS 5173: Sections 104.1 & 104.2: 1991). 8 pp.
BSI BS 5173: Sec 104.1-91. (WITHDRAWN) 1991 Rubber and Plastics Hoses and Hose Assemblies Part 104: Electrical Tests Section 104.1: Measurement of Electrical Resistance of Hoses and Hose Assemblies (Superseded BS BS EN 28031: 1993). 9 pp.
BSI BS EN 28031-93. 1993 Ruber and Plastics Hoses and Hose Assemblies—Determination of Electrical Resistance (ISO 8031: 1987) (Supersedes BS 5173: Sections 104.1: 1991 & 104.2: 1990) (E). 11 pp.
CEN EN 28031-93. Rubber and Plasitics Hoses and Hose Assemblies—Determination of Electrical Resistance (ISO 8031: 1987). 6 pp.
ISO 8031-87. Rubber and Plastics Hoses and Hose Assemblies—Determination of Electrical Resistance First Edition; (CEN EN 28031:1993). 7 pp.
SAA AS 1180.13A-83. Methods of Test for Hose Made from Elastomeric Materials—Part 13A: Determination of Electrical Resistance of Hose and Hose Components. 1 p.
SAA AS 1180.13B-87. Methods of Test for Hose Made from Elastomeric Materials—Part 13B: Determination of Electrical Resistance of Hose Assembly. 1 p.

—Magnetic Tapes
CNS C6098-80. Method of Test for Magnetic Tape Electrical Resistance Coating (Dec)(6664).

—Metals
BSI BS 115-54. 1954 Metallic Resistance Materials for Electrical Purposes. 15 pp.
BSI BS 3466-62. 1962 Methods of Test for Resistance Per Unit Length of Metallic Electrical Resistance Material. 11 pp.
CNS C3126-81. Method of Test for Change of Resistance with Temperature of Metallic Materials for Electrical Heating (Jul)(7629).
MOD UK NES 2061-89. Specification for Metallic Resistance Materials for Electrical Purposes Issue 1 (10.89). 7 pp.

—Metals—Temperature
CNS C6158-82. Method of Test for Electrical Resistance Temperature Characteristics of Metallic Resistance Materials (Apr)(8660).
JIS C 2526-79. Testing Method for Electrical Resistance-Temperature Characteristics of Metallic Resistance Materials.

—Metals—Thermoelectromotive Force
CNS C6159-82. Method of Test for Thermo-Electromotive Force of Metallic Resistance Materials (Apr)(8661).

—Nonmetallic Materials
DIN VDE 0303 Pt 3-83. Tests on Materials for Electrical Purposes; Measurement of the Electrical Resistance of Non-Metallic Materials (May). 41 pp.

Electrical Resistance (Cont.)

—Polymers
BSI BS 2050-78. 1978 Amd 2 Electrical Resistance of Conducting and Antistatic Products Made from Flexible Polymeric Material. 10 pp.

—Rubber
ISO 2878-87. Rubber, Vulcanized—Antistatic and Conductive Products—Determination of Electrical Resistance Second Edition. 8 pp.
ISO 2882-79. Rubber, Vulcanized—Antistatic and Conductive Products for Hospital Use—Electrical Resistance Limits First Edition. 4 pp.
ISO 2883-80. Rubber, Vulcanized—Antistatic and Conductive Products for Industrial Use—Electrical Resistance Limits First Edition. 4 pp.

—Telephone Cables
CCITT RECMN G.613-89. Characteristics of Symmetric Cable Pairs Usable Wholly for the Transmission of Digital Systems with a Bit Rate of up to 2 Mbits—Transmission Media Characteristics (Study Group XV) 4 pp. 4 pp.

—Treeing
IEC 1072-91. Methods of Test for Evaluating the Resistance of Insulating Materials Against the Initiation of Electrical Trees First Edition. 55 pp.

Electrical Resistivity
Use For: Electrical Conductivity; Surface Resistivity; Volume Resistivity *See Also:* Electrical Properties; Electrical Resistance; Insulation Resistance; Photoconductivity

—Aircraft Fuels
CNS K6965-89. Method of Test for Electrical Conductivity of Aviation and Distillate Fuels Containing a Static Dissipator Additive (Oct)(12616).
ISO 6297-83. Petroleum Products—Aviation and Distillate Fuels Containing a Static Dissipator Additive—Determination of Electrical Conductivity First Edition. 6 pp.

—Aluminum Alloys
BSI BS EN 2004-1-93. 1993 Test Methods for Aluminium and Aluminium Alloy Products Part 1: Determination of Electrical Conductivity of Wrought Aluminium Alloys (S). 11 pp.
CEN PREN 2004 (Part 1)-91. Test Methods for Aluminium and Aluminium Alloy Products Part 1—Determination of Electrical Conductivity of Wrought Aluminium Alloys. 7 pp.
CEN EN 2004-1-93. Aerospace Series Test Methods for Aluminium and Aluminium Alloy Products Part 1. Determination of Electrical Conductivity of Wrought Aluminium Alloys. 6 pp.

—Cable Insulation
DIN VDE 0472 Pt 502-91. Testing of Cables & Insulated Flexible Cords; Insulation Resistance and Volume Resistivity (Nov) (Supersedes DIN 57 472 Part 502). 15 pp.

—Conveyor Belts
BSI BS 490: Part 3-91. 1991 Conveyor and Elevator Belting Part 3: Specification for Flammability and Anti-Static Properties of Rubber and of Plastics Conveyor Belting of Textile Construction for General Use. 7 pp.
BSI BS 490: Sec 11.4-90. (WITHDRAWN) 1990 Conveyor and Elevator Belting Part 11: Methods of Test for Safety Section 11.4: Determination of Electrical Conductivity (Superseded by BS EN 20284: 1993). 7 pp.
CSA CAN/CSA-M422-M87. Fire-Performance and Antistatic Requirements for Conveyor Belting; (Gen Instr 1). 26 pp.
SNZ NZS/BS 490: Part 3-91. Conveyor and Elevator Belting Part 3: Specification for Flammability and Anti-Static Properties of Rubber and of Plastics Conveyor Belting of Textile Construction for General Use. 8 pp.
SNZ NZS/BS 490: Pt11:Sec11.4-90. Conveyor and Elevator Belting Part 11: Methods of Test for Safety Section 11.4: Determination of Electrical Conductivity. 6 pp.

—Dust
IEC 1241 Pt 2-2-93. Electrical Apparatus for Use in the Presence of Combustible Dust—Part 2: Test Methods—Section 2: Method for Determining the Electrical Resistivity of Dust in Layers First Edition. 21 pp.

—Electric Conductors
CNS C3008-80. Testing Standard for Conductivity of Insulated Conductors (Jul)(686). 1 p.

—Electrical Insulating Liquids
BSI BS 5737-79. 1979 Measurement of Relative Premittivity, Dielectric Dissipation Factor and d.c. Resistivity of Insulating Liquids. 19 pp.

INDUSTRY STANDARDS

Electrical Resistivity (Cont.)

—Electrical Insulating Varnishes
MOD UK DEF-1053: METH NO. 45. Standard Methods of Testing Paint, Varnish, Lacquer and Related Products. Indices Method 45: Surface Resistivity of Insulating Varnishes (Clear and Pigmented) (Withdrawn).

—Electrical Insulation
BSI BS 6233-82. 1982 Volume Resistivity and Surface Resistivity of Solid Electrical Insulating Materials. 24 pp.
CENELEC HD 429-83. Methods of Test for Volume Resistivity and Surface Resistivity of Solid Electrical Insulating Materials. 2 pp.
IEC 93-80. Methods of Test for Volume Resistivity and Surface Resistivity of Solid Electrical Insulating Materials Second Edition. 44 pp.
IEC 247-78. Measurement of Relative Permittivity, Dielectric Dissipation Factor and D.C. Resistivity of Insulating Liquids Second Edition. 35 pp.

—Electrical Insulation—Heat Shrinkable Tubing
CNS C3139-88. Method for Determining Resistivity of Heat Shrinkable Tubing for Electrical Insulation (Sep)(8786).

—Electrical Insulation—Plastics
SAA AS 1255.1-73. Methods of Test for Electrical Characteristics of Solid Plastics Insulating Materials—Part 1: Determination of Volume and Surface Resistivities. 19 pp.

—Extenders
BSI BS 3483: Part C5-74. 1974 Methods for Testing Pigments in Paints Part C5: Determination of Resistivity of Aqueous Extract. 3 pp.

—Fabrics
BSI BS 3266-81. 1981 Determination of Conductivity, pH, Water Soluble Matter, Chloride and Sulphate in Aqueous Extracts of Textile Materials. 9 pp.
BSI BS 6524-84. 1984 Determination of the Surface Resistivity of a Textile Fabric. 8 pp.

—Germanium
DIN ENGL 50431-88. Testing of Semiconductor Materials; Measurement of the Resistivity of Silicon or Germanium Single Crystals by Means of the Four Probe/Direct Current Method with Collinear Array (May) (ASTM F43-88). 5 pp.
JIS H 0601-62. Testing Methods of Resistivity for Germanium (R 1983). 6 pp.

—Germanium Wafers
DIN ENGL 50435-88. Testing of Semiconductor Materials; Determination of the Radial Resistivity Variation of Silicon or Germanium Slices by Means of the Four-Probe/Direct Current Method (May) (ASTM F81-89). 3 pp.

—Metals
BSI BS 115-54. 1954 Metallic Resistance Materials for Electrical Purposes. 15 pp.
CNS C6157-82. Method of Test for Conductor and Resistivity of Metallic Resistance Materials (Apr)(8659).
CNS H2050-88. Measuring Methods for Electrical Resistivity and Conductivity of Non-Ferrous Materials (Dec)(5129).
IEC 468-74. Method of Measurement of Resistivity of Metallic Materials First Edition. 33 pp.
JIS C 2525-79. Testing Method for Conductor-Resistance and Resistivity of Metallic Materials.
JIS H 0505-75. Measuring Methods for Electrical Resistivity and Conductivity of Non-Ferrous Materials.
MOD UK NES 2061-89. Specification for Metallic Resistance Materials for Electrical Purposes Issue 1 (10.89). 7 pp.

—Paperboard
BSI BS 2924: Part 2-92. 1992 Aqueous Extracts of Paper, Board and Pulp Part 2: Determination of Conductivity (ISO 6587: 1992) (Supersedes BS 2924: 1968). 7 pp.
BSI BS 2924: Part 2-83. 1983 Aqueous Extracts of Paper, Board and Pulp Part 2: Method for Determination of Conductivity. 7 pp.
ISO 6587-92. Paper, Board and Pulps—Determination of Conductivity of Aqueous Extracts Second Edition. 6 pp.

—Papers
BSI BS 2924: Part 2-92. 1992 Aqueous Extracts of Paper, Board and Pulp Part 2: Determination of Conductivity (ISO 6587: 1992) (Supersedes BS 2924: 1968). 7 pp.
BSI BS 2924: Part 2-83. 1983 Aqueous Extracts of Paper, Board and Pulp Part 2: Method for Determination of Conductivity. 7 pp.
ISO 6587-92. Paper, Board and Pulps—Determination of Conductivity of Aqueous Extracts Second Edition. 6 pp.

—Phenolic Resins
ISO 9944-90. Plastics—Phenolic Resins—Determination of Electrical Conductivity of Resin Extracts First Edition. 4 pp.

—Pigments
BSI BS 3483: Part C5-74. 1974 Methods for Testing Pigments in Paints Part C5: Determination of Resistivity of Aqueous Extract. 3 pp.
ISO 787 Pt XIV-73. General Methods of Test for Pigments—Part XIV: Determination of Resistivity of Aqueous Extract First Edition. 6 pp.

—Plastics
BSI BS 2044-84. 1984 Determination of Resistivity of Conductive and Antistatic Plastics and Rubbers (Laboratory Methods). 10 pp.
BSI BS 2782:Pt2: METH 231A-91. 1991 Methods of Testing Plastics Part 2: Electrical Properties Method 231A: Determination of Surface Resistivity (Incorporating BS 903: Part C1: 1991). 15 pp.
BSI BS 4618: Sec 2.3-75. 1975 Recommendations for the Presentation of Plastics Design Data Part 2: Electrical Properties Section 2.3: Volume Resistivity. 8 pp.
BSI BS 4618: Sec 2.4-75. 1975 Recommendations for the Presentation of Plastics Design Data Part 2: Electrical Properties Section 2.4: Surface Resistivity. 3 pp.
ISO 3915-81. Plastics—Measurement of Resistivity of Conductive Plastics First Edition. 6 pp.

—Pulps
BSI BS 2924: Part 2-92. 1992 Aqueous Extracts of Paper, Board and Pulp Part 2: Determination of Conductivity (ISO 6587: 1992) (Supersedes BS 2924: 1968). 7 pp.
BSI BS 2924: Part 2-83. 1983 Aqueous Extracts of Paper, Board and Pulp Part 2: Method for Determination of Conductivity. 7 pp.
ISO 6587-92. Paper, Board and Pulps—Determination of Conductivity of Aqueous Extracts Second Edition. 6 pp.

—Refractory Materials
BSI BS 6043: Sec 4.8-91. 1991 Methods of Sampling and Test for Carbonaceous Materials Used in Aluminium Manufacture Part 4: Cold Ramming Pastes Section 4.8: Determination of Electrical Resistivity of Baked Rammed Paste. 8 pp.

—Rubber
BSI BS 903: Part C1-91. 1991 Methods of Testing Vulcanized Rubber Part C1: Determination of Surface Resistivity Incorporating BS 2782: Part 2: Method 231A: 1991. 15 pp.
BSI BS 903: Part C1-82. 1982 Methods of Testing Vulcanized Rubber Part C1: Determination of Surface Resistivity Incorporating BS 2872: Part 2: Method 231A. 10 pp.
BSI BS 903: Part C2-82. 1982 Amd 1 Methods of Testing Vulcanized Rubber Part C2: Determination of Volume Resistivity (Incorporating BS 2782: Part 2: Method 230A). 10 pp.
BSI BS 2044-84. 1984 Determination of Resistivity of Conductive and Antistatic Plastics and Rubbers (Laboratory Methods). 10 pp.

—Silicon
DIN ENGL 50431-88. Testing of Semiconductor Materials; Measurement of the Resistivity of Silicon or Germanium Single Crystals by Means of the Four Probe/Direct Current Method with Collinear Array (May) (ASTM F43-88). 5 pp.
JIS H 0602-90. Testing Method of Resistivity for Silicon Crystals and Silicon Wafers with Four-Point Probe. 12 pp.

—Silicon Wafers
DIN ENGL 50435-88. Testing of Semiconductor Materials; Determination of the Radial Resistivity Variation of Silicon or Germanium Slices by Means of the Four-Probe/Direct Current Method (May) (ASTM F81-89). 3 pp.

—Soil Analysis
BSI BS 1377: Part 3-90. 1990 Methods of Test for Soils for Civil Engineering Purposes Part 3: Chemical and Electro-Chemical Tests. 45 pp.
SAA AS 1289.D4.1-80. Methods of Testing Soil for Engineering Purposes—Part D4.1: Soil Chemical Tests—Determination of the Electrical Resistivity of Sands and Granular Materials. 2 pp.

—Synchronous Belts
ISO 9563-90. Belt Drives—Electrical Conductivity of Antistatic Endless Synchronous Belts—Characteristics and Test Method First Edition. 7 pp.

—Textile Floor Coverings
BSI BS 6654-85. 1985 Method for Determination of the Electrical Resistivity of Textile Floor Coverings. 8 pp.

—Thermoelectric Materials
CNS C6109-81. Room Temperature Resistivity Measurements on Thermoelectric Materials (Apr)(7230).

—Water
BSI BS 6068: Sec 2.35-89. 1989 Water Quality Part 2: Physical, Chemical and Biochemical Methods Section 2.35: Method for the Determination of Electrical Conductivity. 8 pp.
ISO 7888-85. Water Quality—Determination of Electrical Conductivity First Edition. 8 pp.
JIS K 0552-88. Testing Methods for Electric Conductivity in Highly Purified Water. 9 pp.

Electrical Safety
See Also: Consumer Safety; Electric Shock; Electrical Codes; Electrical Equipment; Electrical Protection Equipment; Electricity; Live Wire Working; Mechanical Protection; Occupational Safety and Health

CSA C22.2 NO 0.8-M1986. Safety Functions Incorporating Electronic Technology (R 1992); (Gen Instr 1). 37 pp.
SNZ NZMP 6002-87. Guide to New Zealand Electrical Requirements and Approvals. 84 pp.
SNZ NZS 6200-88. Specification for General Requirements for Electrical Apparatus and Material. 64 pp.

—Air Cleaners
DIN VDE 0700 Pt 259-88. Safety of Household and Similar Electrical Appliances; Air Cleaning Appliances (July). 16 pp.

—Antennas
BSI BS 5373-77. 1977 Electrical Safety Requirements for Room Aerials. 7 pp.

—Appliances
CNS C4126-90. Safety Testing Standards for Household Electrical Appliances (Oct)(3766).
CNS C4126-81. Safety Testing Standards for Household Electrical Appliances (Jun)(3766). 3 pp.
CSA C22.2 NO 1335.1-93. Portable Electrical Motor-Operated and Heating Appliances: General Requirements; (Gen Instr 1). 208 pp.
DIN VDE 0700 Pt 1 A2-86. Safety of Household and Similar Electrical Appliances; General Requirements; Amendment 2 (Sept). 8 pp.
DIN VDE 0700 Pt 1 A3 (D)-81. Safety of Household and Similar Electrical Appliances; General Requirements (VDE Specification): Amendment 3 (July). 4 pp.
DIN VDE 0700 Pt 1 A7 (D)-82. Safety of Household and Similar Electrical Appliances; Part 1: General Requirements—Calibration of Spring-Operated Impact Test Apparatus: Amendment 7 (VDE Specification) (Feb). 10 pp.
DIN VDE 0700 Pt 1 A8 (D)-82. Safety of Household and Similar Electrical Appliances; General Requirements: Amendment 8 (VDE Specification) (Mar). 6 pp.
DIN VDE 0700 Pt 1 A11 (D)-83. (Amendment to IEC 335-1) Safety of Household and Similar Electrical Appliances; Part 1: General Requirements: Amendment 11 (VDE Specification) (DIN IEC 61(CO)330) (May). 9 pp.
DIN VDE 0700 Pt 1 A13 (D)-84. (Amendment to IEC 335-1) Safety of Household and Similar Electrical Appliances; Part 1: General Requirements: Amendment 13 (VDE Specification) (DIN IEC 61(CO)342) (Mar). 7 pp.
DIN VDE 0700 Pt 1 A14 (D)-84. (Amendment to IEC 335-1) Safety of Household and Similar Electrical Appliances; Part 1: General Requirements: Amendment 14 (VDE Specification) (DIN IEC 61(CO)346) (Mar). 5 pp.
DIN VDE 0700 Pt 1 A15 (D)-84. Safety of Household and Similar Electrical Appliances; General Requirements: Amendment 15 (VDE Specification) (Mar). 4 pp.
DIN VDE 0700 Pt 1 A17 (D)-84. Safety of Household and Similar Electrical Appliances; General Requirements: Amendment 17 (VDE Specification) (Sept). 6 pp.
DIN VDE 0700 Pt 1 A18 (D)-84. Safety of Household and Similar Electrical Appliances; General Requirements: Amendment 18 (VDE Specification) (Dec). 12 pp.
DIN VDE 0700 Pt 1 A19 (D)-85. Safety of Household and Similar Electrical Appliances; General Requirements: Amendment 19 (June). 8 pp.
DIN VDE 0700 Pt 1 A21 (D)-86. Safety of Household and Similar Electrical Appliances; General Requirements: Amendment 21; Identical to IEC 61(CO)458 (Sept). 7 pp.

Electrical Safety *(Cont.)*

—Appliances *(Cont.)*

DIN VDE 0700 Pt 1 A24 (D)-86. Safety of Household and Similar Electrical Appliances; General Requirements: Amendment 24; Identical to IEC 61(CO)468 (Sept). 7 pp.

DIN VDE 0700 Pt 1 A25 (D)-87. Safety of Household and Similar Electrical Appliances; General Requirements: Amendment 25 (Mar). 6 pp.

DIN VDE 0700 Pt 1 A26 (D)-87. Safety of Household and Similar Electrical Appliances; General Requirements: Amendment 26; Identical to IEC 61(CO)484 (Sept) (Supersedes Draft 0700 Part 1 A20/5.86). 18 pp.

DIN VDE 0700 Pt 1 A27 (D)-87. Safety of Household and Similar Electrical Appliances; General Requirements: Amendment 27 (Sept). 6 pp.

DIN VDE 0700 Pt 1 A28 (D)-87. Safety of Household and Similar Electrical Appliances; General Requirements: Amendment 28 (Oct). 5 pp.

DIN VDE 0700 Pt 1 A29 (D)-88. Safety of Household and Similar Electrical Appliances; General Requirements: Amendment 29 (Oct). 4 pp.

DIN VDE 0700 Pt 1 A30 (D)-89. Safety of Household and Similar Electrical Appliances; General Requirements: Amendment 30 (Mar). 5 pp.

IEC 65-85. Safety Requirements for Mains Operated Electronic and Related Apparatus for Household and Similar General Use Fifth Edition; (Amendment 2-1989, Incorporating Amendment 1) (Amendment 3-1992). 210 pp.

SNZ NZS 6300-92. Approval and Test Specification—General Requirements for Household and Similar Electrical Appliances Amend: 1, 1992 (NZS 6300:1992/AS 3300-1992). 89 pp.

—Batteries

BSI BS 6133-85. 1985 Amd 1 Code of Practice for Safe Operation of Lead-Acid Stationary Cells and Batteries. 19 pp.

IEC 1056 Pt 3-91. Portable Lead-Acid Cells and Batteries (Valve-Regulated Types) Part 3: Safety Recommendations for Use in Electric Appliances First Edition. 12 pp.

—Bedding—Electric

CSA C22.2 NO 101-M1984. Electrically Heated Bedding Appliances for Household Use (R 1992); (Gen Instr 1 Thru 4). 61 pp.

CSA 1169 Bull. Electrical Bulletin 1169 June 27, 1978 to C22.2 NO 101. 2 pp.

SNZ NZCP 28-62. Code of Recommended Practice for the Repair of Electrically Heated Blankets for Domestic Use. 15 pp.

—Buildings

BSI PD 6535-93. 1993 Common Aspects for Installation and Equipment for Protection Against Electric Shock (IEC 1440: 1992) (E). 15 pp.

CENELEC HD 384.4.41-80. Electrical Installations of Buildings Part 4: Protection for Safety Chapter 41: Protection Against Electric Shock. 46 pp.

CENELEC HD 384.4.41/A1-80. AMD 1 Electrical Installations of Buildings Part 4: Protection for Safety Chapter 41: Protection Against Electric Shock. 2 pp.

CENELEC HD 384.4.41 S1/PRA1-93. AMD 1 Electrical Installations of Buildings Part 4: Protection for Safety Chapter 41: Protection Against Electric Shock. 13 pp.

CENELEC HD 384.4.42-85. Electrical Installations of Buildings Part 4: Protection for Safety Chapter 42: Protection Against Thermal Effects. 8 pp.

CENELEC HD 384.4.42 S1/A1-92. AMD 1 Electrical Installations of Buildings Part 4: Protection for Safety Chapter 42: Protection Against Thermal Effects. 5 pp.

CENELEC HD 384.4.43-78. Electrical Installations of Buildings Part 4: Protection for Safety Chapter 43: Protection Against Overcurrent. 19 pp.

CENELEC HD 384.4.45 S1-89. Electrical Installations of Buildings-Part 4: Protection for Safety. Chapter 45: Protection Against Undervoltage. 3 pp.

CENELEC HD 384.4.46 S1-87. Electrical Installatioans of Buildings—Part 4: Protection for Safety. Chapter 46: Isolation and Switching. 5 pp.

CENELEC HD 384.4.46 S1/A1-93. AMD 1 Electrical Installations of Buildings Part 4: Protection for Safety Chapter 46: Isolation and Switching. 5 pp.

CENELEC HD 384.4.47 S1-88. Electrical Installations of Buildings Part 4: Protection for Safety Chapter 47: Application of Protective Measures for Safety Section 470—General—Section 471—Measures of Protection Against Electric Shock. 5 pp.

CENELEC HD 384.4.473-78. Electrical Installations of Buildings Part 4: Protection for Safety Chapter 47: Application of Protective Measures Section 473: Protection Against Overcurrent. 2 pp.

CENELEC HD 384.5.56-85. Electrical Installations of Buildings Part 5: Selection and Erection of Electrical Equipment Chapter 56: Supplies for Safety Services. 8 pp.

Electrical Safety *(Cont.)*

—Buildings *(Cont.)*

IEC 364 Pt 4-41-92. Electrical Installations of Buildings Part 4: Protection for Safety Chapter 41: Protection Against Electric Shock Third Edition. 53 pp.

IEC 364 Pt 4-42-80. Electrical Installations of Buildings Part 4: Protection of Safety Chapter 42: Protection Against Thermal Effects First Edition. 12 pp.

IEC 364 Pt 4-43-77. Electrical Installations of Buildings Part 4: Protection for Safety Chapter 43: Protection Against Overcurrent First Edition. 14 pp.

IEC 364 Pt 4-45-84. Electrical Installations of Buildings Part 4: Protection for Safety Chapter 45: Protection Against Undervoltage First Edition. 9 pp.

IEC 364 Pt 4-46-81. Electrical Installations of Buildings Part 4: Protection for Safety Chapter 46: Isolation and Switching First Edition. 12 pp.

IEC 364 Pt 4-47-81. Electrical Installations of Buildings Part 4: Protection for Safety Chapter 47: Application of Protective Measures for Safety Section 470: General Section 471: Measures of Protection Against Electric Shock First Edition. 12 pp.

IEC 364 Pt 4-473-77. Electrical Installations of Buildings Part 4: Protection for Safety Chapter 47: Application of Protective Measures for Safety Section 473-Measures of Protection Against Overcurrent First Edition. 14 pp.

IEC 364 Pt 5-56-80. Electrical Installations of Buildings Part 5: Selection and Erection of Electrical Equipment Chapter 56: Safety Services First Edition. 12 pp.

—Centrifuges

IEC 1010 Pt 2-020-92. Safety Requirements for Electrical Equipment for Measurement, Control, and Laboratory Use Part 2-020: Particular Requirements for Laboratory Centrifuges First Edition. 60 pp.

—Control Systems Equipment

BSI BS EN 61010-1-93. 1993 Safety Requirements for Electrical Equipment for Measurement, Control, and Laboratory Use Part 1: General Requirements. 123 pp.

CENELEC EN 61010-1-93. Safety Requirements for Electrical Equipment for Measurement, Control and Laboratory Use Part 1: General Requirements (IEC 1010-1: 1990 + A1: 1992, Modified). 117 pp.

CSA CAN/CSA-C22. 2NO1010.1-92. Safety Requirements for Electrical Equipment for Measurement, Control, and Laboratory Use, Part 1: General Requirements (IEC 1010-1:1990); (Gen Instr 1). 160 pp.

DIN VDE 0411 Pt 100-86. Measurement and Control; Safety Requirements for Electrically Operated Measuring, Control and Laboratory Equipment; General Requirements; Identical with IEC 66E (Sec) 22 (Aug). 137 pp.

IEC 1010 Pt 1-90. Safety Requirements for Electrical Equipment for Measurement, Control and Laboratory Use Part 1: General Requirements First Edition (CENELEC EN 61010-1: 1993). 190 pp.

—Controllers—Fences

BSI BS EN 61011-93. 1993 Electric Fence Energizers Safety Requirements for Mains-Operated Electric Fence Energizers (Supersedes BS 2632: 1980) (E). 116 pp.

BSI BS EN 61011-1-93. 1993 Electric Fence Energizers Safety Requirements for Battery-Operated Electric Fence Energizers Suitable for Connection to the Supply Mains (Supersedes BS 6369: 1983) (E). 24 pp.

BSI BS EN 61011-2-93. 1993 Electric Fence Energizers Safety Requirements for Battery-Operated Electric Fence Energizers Not for Connection to the Supply Mains (Supersedes BS 6167: 1981) (E). 27 pp.

CENELEC PREN 61011-91. Electric Fence Energizers. Safety Requirements for Mains-Operated Electric Fence Energizers. 111 pp.

CENELEC EN 61011-92. Electric Fence Energizers Safety Requirements for Mains-Operated Electric Fence Energizers (IEC 1011: 1989, Modified). 107 pp.

CENELEC EN 61011/PRAA-92. AMD A Electric Fence Energizers—Safety Requirements for Mains-Operated Electric Fence Energizers. 2 pp.

CENELEC PREN 61011-1-90. Electric Fence Energizers Part 1: Safety Requirements for Battery -Operated Electric Fence Energizers Suitable for Connection to the Supply Mains.

CENELEC EN 61011-1-92. Electric Fence Energizers Part 1: Safety Requirements for Battery-Operated Electric Fence Energizers Suitable for Connection to the Supply Mains. 8 pp.

CENELEC EN 61011-1-92. Electric Fence Energizers Safety Requirements for Battery-Operated Electric Fence Energizers Suitable for Connection to the Supply Mains. 10 pp.

Electrical Safety *(Cont.)*

—Controllers—Fences *(Cont.)*

CENELEC EN 61011-1/PRAA-92. AMD A Electric Fence Energizers Safety Requirements for Battery -Operated Electric Fence Energizers Suitable for Connection to the Supply Mains. 2 pp.

CENELEC PREN 61011-2-91. Electric Fence Energizers Part 2: Safety Requirements for Battery-Operated Electric Fence Energizers Not for Connection to the Supply Mains. 27 pp.

CENELEC EN 61011-2-92. Electric Fence Energizers Part 2: Safety Requirements for Battery-Operated Electric Fence Energizers Not for Connection to the Supply Mains. 7 pp.

CENELEC EN 61011-2/PRAA-92. AMD A Electric Fence Energizers Safety Requirements for Battery -Operated Electric Fence Energizers Not for Connection to the Supply Mains. 2 pp.

IEC 1011-89. Electric Fence Energizers Safety Requirements for Mains-Operated Electric Fence Energizers First Edition; (Amendment 1-1991) (Corrigendum—June 1993) (Amendment 2-1993) (CENELEC EN 61011:1992). 268 pp.

IEC 1011 Pt 1-89. Electric Fence Energizers Safety Requirements for Battery-Operated Electric Fence Energizers Suitable for Connection to the Supply Mains First Edition; (Amendment 2-1993, Incorporating Amendment 1). 45 pp.

IEC 1011 Pt 2-90. Electric Fence Energizers Safety Requirements for Battery-Operated Electric Fence Energizers Not for Connection to the Supply Mains First Edition; (Amendment 2-1993, Incorporating Amendment 1). 51 pp.

SNZ NZS 6203: Part 1-87. Specification for Agricultural Electric Fencing (Safety Requirements Part 1: Alternating Current Mains-Operated Electric Fence Energizers Amend: 1, 1989; 2, 1991. 24 pp.

—Converters

BSI BS EN 61046-92. 1992 D.C. or a.c. Supplied Electronic Step-Down Convertors for Filament Lamps. General and Safety Requirements. 31 pp.

BSI BS EN 61046-01. 1992 Amd 1 D.C. or a.c. Supplied Electronic Step-Down Convertors for Filament Lamps. General and Safety Requirements (IEC 1046: 1991) (AMD 7608) March 15, 1993. 40 pp.

CENELEC EN 61046-92. D.C. or a.c. Supplied Electronic Step-Down Convertors for Filament Lamps General and Safety Requirements. 2 pp.

CENELEC EN 61046/A1-92. AMD 1 D.C. or a.c Supplied Electronic Step-Down Convertors for Filament Lamps—General and Safety Requirements; (IEC 1046; 1991/A1: 1991). 5 pp.

IEC 1046-91. D.C. or a.c. Supplied Electronic Step-down Convertors for Filament Lamps General and Safety Requirements First Edition; (Amendment 1-1991) (CENELEC EN 61046/A1: 1992). 53 pp.

—Cooking Appliances

CSA C22.2 NO 1335.2.9-93. Portable Electrical Motor-Operated and Heating Appliances: Particular Requirements for Portable Electric Cooking Appliances; (Gen Instr 1). 40 pp.

—Data Processing Equipment

DIN VDE 0701 Pt 240-86. Repair, Modification and Inspection of Electrical Appliances; Safety Requirements for Data Processing Equipment and Office Machines (Apr). 10 pp.

DIN VDE 0701 Pt 240 (D)-90. Repair, Modification & Inspection of Electrical Appliances; Information Technology Equipment (Feb). 12 pp.

ECMA ECMA 166-92. Information Technology Equipment—Routine Electrical Safety Testing in Production. 7 pp.

SNZ NZS/AS 3260-93. Approval and Test Specification—Safety of Information Technology Equipment Including Electrical Business Equipment (This is a Joint Standard with SAA AS 3260). 175 pp.

—Decorative Lighting

SNZ NZS/AS 3152-92. Approval and Test Specification—Decorative Lighting Outfits AMENDMENT No. 1, 1993. 6 pp.

—Dehumidifiers

DIN VDE 0700 Pt 258-88. Safety of Household and Similar Electrical Appliances; Dehumidifiers (May). 24 pp.

—Dishwashers

BSI BS EN 60335-2-58-93. 1993 Safety of Household and Similar Electrical Appliances Part 2: Particular Requirements Section 2.58: Commercial Electric Dishwashing Machines (L). 33 pp.

DIN VDE 0700 Pt 231-84. Safety of Household & Similar Electrical Appliances; Commercial Dishwashers (VDE Specification) (Aug) (Supplements or Replaces Corresponding Sections in DIN 57700 Part 1/VDE 0700 Part 1). 34 pp.

INDUSTRY STANDARDS
933

INTERNATIONAL AND NON-U.S. NATIONAL STANDARDS
SUBJECT INDEX — Electrical

Electrical Safety (Cont.)

—**Earth Stations—Radio Frequency Radiation**
CCIR Report 543-1-86. Safety Aspects of Radio-Frequency Radiation from Space Research Earth Stations—Section 2A—Research in Space Technology. 4 pp.

—**Electric Measuring Instruments**
BSI BS 4743-79. (WITHDRAWN) 1979 Safety Requirements for Electronic Measuring Apparatus (Superseded by BS EN 61010-1: 1993). 61 pp.
BSI BS 5458-77. 1977 Safety Requirements for Indicating and Recording Electrical Measuring Instruments and Their Accessories. 28 pp.
CENELEC HD 401-80. Safety Requirements for Electronic Measuring Apparatus. 2 pp.
CSA CAN/CSA-C22. 2NO231SERM89. CSA Safety Requirements for Electrical and Electronic Measuring and Test Equipment; (Gen Instr 1). 133 pp.
CSA C22.2 NO 231 SP-M1990. Audit Checklist and Test Pack for Use with CAN/CSA-C22.2 No. 231 Series-M89 CSA Safety Requirements for Electrical and Electronic Measuring and Test Equipment. 27 pp.
DIN VDE 0410-76. VDE Specification for Electrical Measuring Instruments; Safety Requirements for Indicating and Recording Electrical Measuring Instruments and Their Accessories (Oct). 42 pp.
DIN VDE 0411 Pt 2-69. VDE Specifications for Electronic Measuring Instruments and Automatic Controls (Dec). 5 pp.
IEC 348-78. Safety Requirements for Electronic Measuring Apparatus Second Edition. 118 pp.
IEC 414-73. Safety Requirements for Indicating and Recording Electrical Measuring Instruments and Their Accessories First Edition. 52 pp.

—**Electric Outlets**
SNZ NZS/AS 3105-93. Approval and Test Specification—Electrical Portable Outlet Devices (This is a Joint Standard with SAA AS 3105). 9 pp.

—**Electric Plug Adapters**
SNZ NZS/AS 3122-93. Approval and Test Specification—Socket-Outlet Adaptors (This is a Joint Standard with SAA AS 3122). 9 pp.

—**Electric Terminals—Quick Connect**
IEC 1210-93. Connecting Devices—Flat Quick-Connect Terminations for Electrical Copper Conductors—Safety Requirements First Edition. 50 pp.

—**Electronic Measuring Instruments**
DIN VDE 0411 Pt 1-73. VDE Specifications for Electronic Measuring Instruments and Automatic Controls (Oct). 67 pp.
DIN VDE 0411 Pt 1A-80. Electronic Measuring Apparatus; Part Amendment to Part 1: Safety Requirements for Electronic Measuring Apparatus (Feb). 9 pp.

—**Electrosurgical Equipment**
BSI BS 5724: Sec 2.2-92. 1992 Medical Electrical Equipment Part 2: Particular Requirements for Safety Section 2.2: Specification for High Frequency Surgical Equipment (IEC 601-2-2: 1991). 44 pp.
BSI BS 5724: Sec 2.2-01. 1992 Amd 1 Medical Electrical Equipment Part 2: Particular Requirements for Safety Section 2.2: Specification for High Frequency Surgical Equipment (BS EN 60601-2-2: 1992) (AMD 7868) August 15, 1993 (N). 52 pp.
BSI BS 5724: Sec 2.2-83. 1983 Medical Electrical Equipment Part 2: Particular Requirements for Safety Section 2.2: High Frequency Surgical Equipment. 30 pp.
BSI BS EN 60601-2-2-93. 1993 Amd 1 Medical Electrical Equipment Part 2: Particular Requirements for Safety Section 2.2: Specification for High Frequency Surgical Equipment (IEC 601-2-2: 1991) (AMD 7868) August 15, 1993 (N). 52 pp.
CENELEC EN 60601-2-2-93. Medical Electrical Equipment Part 2: Particular Requirements for Safety Section 2.2 Specification for High Frequency Surgical Equipment (IEC 601-2-2: 1991). 39 pp.
CSA CAN/CSA-C22. 2N601.2.2-92. Medical Electrical Equipment Part 2: Particular Requirements for the Safety of High Frequency Surgical Equipment (IEC 601-2-2:1991); (Gen Instr 1). 56 pp.
IEC 601 Pt 2-2-91. Medical Electrical Equipment Part 2: Particular Requirements for the Safety of High Frequency Surgical Equipment Second Edition; (CAN/CSA-C22.2 No. 601.2.2-92) (CENELEC EN 60601-2-2: 1993). 78 pp.
SAA AS 3200.2.2-92. Approval and Test Spec.—Medical Electrical Equipment—Part 2: Particular Requirements for Safety—Part 2.2: High Frequency Surgical Equipment (IEC 601-2-2: 1991) (In Prof. Pkgs. 28, 17C) (Supersedes AS 3202—1989) (This is a Joint Standard with SNZ NZS 3200.2.2). 37 pp.

Electrical Safety (Cont.)

—**Electrosurgical Equipment** (Cont.)
SNZ NZS/AS 3200. 2.2-92. Approval and Test Specification—Medical Electrical Equipment Part 2.2: Particular Requirements for Safety—High Frequency Surgical Equipment (This is a Joint Standard with SAA AS 3200.2.2). 37 pp.

—**Elevating Platforms**
IEC 1057-91. Aerial Devices with Insulating Boom Used for Live Working First Edition. 119 pp.

—**Elevators (Lifts)**
CSA CAN/CSA-B44.1-M91. Elevator and Escalator Electrical Equipment (ASME A17.5-1991). 97 pp.

—**Escalators**
CSA CAN/CSA-B44.1-M91. Elevator and Escalator Electrical Equipment (ASME A17.5-1991). 97 pp.

—**Fans**
BSI BS 848: Part 5-86. 1986 Fans for General Purposes Part 5: Guide for Mechanical and Electrical Safety. 16 pp.
DIN VDE 0700 Pt 234 (D)-81. Safety of Household and Similar Electrical Appliances; Fans and Associated Regulators (Aug). 25 pp.

—**Floor Cleaning Equipment**
CSA C22.2 NO 10-1965. Electric Floor Surfacing and Cleaning Machines (Second Edition) (R 1992). 24 pp.
CSA 1169 Bull. Electrical Bulletin 1169 June 27, 1978 to C22.2 NO 10. 2 pp.
IEC 335 Pt 2-67-92. Safety of Household and Similar Electrical Appliances Part 2: Particular Requirements for Floor Treatment and Floor Cleaning Machines, for Industrial and Commercial Use First Edition. 34 pp.

—**Ground Support Equipment**
BSI 2G 219-83. 1983 General Requirements for Ground Support Electrical Supplies for Aircraft. 12 pp.
ISO 6858-82. Aircraft—Ground Support Electrical Supplies—General Requirements First Edition. 9 pp.
JIS W 7007-91. Aircraft—Ground Support Electrical Supplies—General Requirements (ISO 6858:1982).

—**Grounding Devices**
CSA C22.2 NO 151-M1986. Laboratory Equipment (R 1992); (Gen Instr 1 Thru 2). 61 pp.

—**Harmonic Analysis**
CSA CAN/CSA-C22. 2 NO0.16-M92. Measurement of Harmonic Currents; (Gen Instr 1 Thru 2). 26 pp.

—**Heaters**
BSI BS 7687: Sec 2.10-93. 1993 Safety Requirements for Electrical Equipment for Measurement, Control and Laboratory Use Part 2: Particular Requirements Section 2.10: Specification for Laboratory Equipment for the Heating of Materials (IEC 1010-2-010: 1992) (S). 21 pp.
CENELEC HD 283 S1-92. Safety of Household and Similar Electrical Appliances Particular Requirement for the Maximum Temperature Allowed for the Surfaces of Air-Outlet Grilles of Thermal Storage Room Heating Appliances. 9 pp.
CNS C4125-90. Safety Testing Code for Household Electric Heaters (Jan)(3765).
CSA C22.2 NO 1335.1-93. Portable Electrical Motor-Operated and Heating Appliances: General Requirements; (Gen Instr 1). 208 pp.
IEC 335 Pt 2-61-92. Safety of Household and Similar Electrical Appliances Part 2: Particular Requirements for Thermal Storage Room Heaters First Edition. 44 pp.
IEC 1010 Pt 2-010-92. Safety Requirements for Electrical Equipment for Measurement, Control, and Laboratory Use Part 2-010: Particular Requirements for Laboratory Equipment for the Heating of Materials First Edition. 38 pp.

—**Helmets**
CNS Z3015-88. Method of Test for Safety Helmets for Electrical Workers (Apr)(4599).

—**Laboratories**
SAA AS 2243.7-91. Safety in Laboratories—Part 7: Electrical Aspects (in Professional Packages 34C, 47). 22 pp.

—**Laboratory Equipment**
BSI BS 7687: Sec 2.10-93. 1993 Safety Requirements for Electrical Equipment for Measurement, Control and Laboratory Use Part 2: Particular Requirements Section 2.10: Specification for Laboratory Equipment for the Heating of Materials (IEC 1010-2-010: 1992) (S). 21 pp.
BSI BS EN 61010-1-93. 1993 Safety Requirements for Electrical Equipment for Measurement, Control, and Laboratory Use Part 1: General Requirements. 123 pp.

Electrical Safety (Cont.)

—**Laboratory Equipment** (Cont.)
CENELEC EN 61010-1-93. Safety Requirements for Electrical Equipment for Measurement, Control and Laboratory Use Part 1: General Requirements (IEC 1010-1: 1990 + A1: 1992, Modified). 117 pp.
CSA CAN/CSA-C22. 2NO1010.1-92. Safety Requirements for Electrical Equipment for Measurement, Control, and Laboratory Use, Part 1: General Requirements (IEC 1010-1:1990); (Gen Instr 1). 160 pp.
DIN VDE 0411 Pt 100-86. Measurement and Control; Safety Requirements for Electrically Operated Measuring, Control and Laboratory Equipment; General Requirements; Identical with IEC 66E (Sec) 22 (Aug). 137 pp.
IEC 1010 Pt 1-90. Safety Requirements for Electrical Equipment for Measurement, Control and Laboratory Use Part 1: General Requirements First Edition (CENELEC EN 61010-1: 1993). 190 pp.
IEC 1010 Pt 2-010-92. Safety Requirements for Electrical Equipment for Measurement, Control, and Laboratory Use Part 2-010: Particular Requirements for Laboratory Equipment for the Heating of Materials First Edition. 38 pp.
IEC 1010 Pt 2-020-92. Safety Requirements for Electrical Equipment for Measurement, Control, and Laboratory Use Part 2-020: Particular Requirements for Laboratory Centrifuges First Edition. 60 pp.

—**Ladders**
CNS Z2070-88. Insulated Ladder for Electric Work (Oct)(12452).

—**Lampholders**
BSI BS 5101: Part 1-75. (WITHDRAWN) 1975 Amd 5 Lamp Caps and Holders Together with Gauges for the Control of Interchangeability and Safety Part 1: Lamp Caps (IEC 61-1) (AMD 6519) January 31, 1991 (Superseded by BS EN 60061-2: 1993). 184 pp.
BSI BS 5101: Part 2-75. (WITHDRAWN) 1975 Amd 5 Lamp Caps and Holders Together with Gauges for the Control of Interchangeability and Safety Part 2: Lamp Holders (IEC 61-2) (AMD 6521) January 31, 1991 (Superseded by BS EN 60061-2: 1993). 137 pp.
BSI BS 5101: Part 3-75. (WITHDRAWN) 1975 Amd 5 Lamp Caps and Holders Together with Gauges for the Control of Interchangeability and Safety Part 3: Gauges (AMD 6520) January 31, 1991 (Superseded by BS EN 60061-3: 1993). 338 pp.
BSI BS 5101: Part 4-80. 1980 Amd 1 Lamp Caps and Holders Together with Gauges for the Control of Interchangeability and Safety Part 4: Lamp Caps, Lampholders and Gauges Used in the United Kingdom but Not Specified in Parts 1, 2 or 3. 18 pp.
BSI BS 5101: Part 5-90. (WITHDRAWN) 1990 Lamp Caps and Holders Together with Gauges for the Control of Interchangeability and Safety Part 5: Guidelines and General Information (Renumbered as BS EN 60061-4: 1992). 22 pp.
BSI BS 6972-88. 1988 General Requirements for Luminaire Supporting Couplers for Domestic, Light Industrial and Commercial Use. 22 pp.
BSI BS 7001-88. 1988 Interchangeability and Safety of a Standardized Luminaire Supporting Coupler. 14 pp.
BSI BS 7001-01. 1988 Amd 1 Interchangeability and Safety of a Standardized Luminaire Supporting Coupler (AMD 7625) June 15, 1993 (E). 23 pp.
BSI BS EN 60061-1-93. 1993 Lamp Caps and Holders Together with Gauges for the Control of Interchangeability and Safety Part 1: Lamp Caps (Supersedes BS 5101: part 1: 1975) (E). 202 pp.
BSI BS EN 60061-2-93. 1993 Lamp Caps and Holders Together with Gauges for the Control of Interchangebility and Safety Part 2: Lampholders (Supersedes BS 5101: part 2: 1975) (E). 158 pp.
BSI BS EN 60061-3-93. 1993 Lamp Caps and Holders Together with Gauges for the Control of Interchangeability and Safety Part 3: Gauges (E). 425 pp.
BSI BS EN 60061-4-92. 1992 Amd 1 Lamp Caps and Holders Together with Gauges for the Control of Interchangeability and Safety Part 4: Guidelines and General Information (AMD 7556) December 15, 1992. 24 pp.
CENELEC HD 65-82. Lamp Caps and Holders Together with Gauges for the Control of Interchangeability and Safety. 4 pp.
CENELEC HD 65.1 S1-88. Lamp Caps and Holders Together with Gauges for the Control of Interchangeability and Safety Part 1: Lamp Caps. 3 pp.
CENELEC HD 65.2-87. Lamp Caps and Holders Together with Gauges for the Control of Interchangeability and Safety Part 2: Lampholders. 3 pp.

INTERNATIONAL AND NON-U.S. NATIONAL STANDARDS
SUBJECT INDEX
Electrical

Electrical Safety *(Cont.)*

—Lampholders *(Cont.)*

CENELEC EN 60061-1-93. Lamp Caps and Holders Together with Gauges for the Control of Interchangeability and Safety Part 1: Lamp Caps (IEC 61-1: 1969 + Supplements A: 1970 to N: 1992, Modified) (Supersedes HD 65.1 S1: 1978). 197 pp.

CENELEC EN 60061-2-93. Lamp Caps and Holders Together with Gauges for the Control of Interchangeability and Safety Part 2: Lampholders (IEC 61-2: 1969 + Supplements A: 1970 to K: 1992, Modified). 152 pp.

CENELEC EN 60061-3-93. Lamp Caps and Holders Together with Gauges for the Control of Interchangeability and Safety Part 3: Gauges (IEC 61-3: 1969 + Supplements A: 1970 to M: 1992, Modified). 420 pp.

CENELEC EN 60061-4-92. Lamp Caps and Holders Together with Gauges for the Control of Interchangeability and Safety Part 4: Guidelines and General Information; (IEC 61-4: 1990, Mod). 4 pp.

CENELEC EN 60061-4-92. Lamp Caps and Holders Together with Gauges for the Control of Interchangeability and Safety Part 4: Guidelines and General Information. 13 pp.

IEC 61 Pt 1-69. Lamp Caps and Holders Together with Gauges for the Control of Interchangeability and Safety Part 1: Lamp Caps Third Edition; (Supplements A-J-1970-1980)(Supplement K-1983) (Supplement L-1987) (Supplement M-1989) (Supplement N-1992); (CENELEC EN60061-1:1993). 279 pp.

IEC 61 Pt 2-69. Lamp Caps and Holders Together with Gauges for the Control of Inter-changeability and Safety Part 2: Lampholders Third Edition; (Supplements A-F-1970-1980)(Supplement G-1983) (Supplement H-1987) (Supplement J-1989) (Supplement K-1992) (CENE EN 60061-2: 1993). 202 pp.

IEC 61 Pt 3-69. Lamp Caps and Holders Together with Gauges for the Control of Interchangeability and Safety Part 3: Gauges Third Edition; (Errata—Dec 1984) (Supplement K-1987) (Supplement L-1989) (Supplement M-1992) (CENE EN 60061-3: 1993). 506 pp.

IEC 61 Pt 4-90. Lamp Caps and Holders Together with Gauges for the Control of Interchangeability and Safety Part 4: Guidelines and General Information First Edition; (Supplement A-1992) (CENELEC EN 60061-4: 1992). 85 pp.

IEC 838 Pt 1-93. Miscellaneous Lampholders Part 1: General Requirements and Tests Second Edition; (Corrigendum—Aug 1993). 62 pp.

—Lasers

CENELEC HD 194-74. Requirements Concerning the Electrical Safety of Laser—Apparatus and Installations. 55 pp.

DIN VDE 0836-77. VDE Specification for the Electrical Safety of Laser Equipment and Installations (Feb). 65 pp.

IEC 820-86. Electrical Safety of Laser Equipment and Installations First Edition. 81 pp.

—Low Voltage Installations

IEC 1204-93. Low-Voltage Power Supply Devices, d.c. Output—Performance Characteristics and Safety Requirements First Edition. 59 pp.

—Measuring Instruments

BSI BS EN 61010-1-93. 1993 Safety Requirements for Electrical Equipment for Measurement, Control, and Laboratory Use Part 1: General Requirements. 123 pp.

CENELEC EN 61010-1-93. Safety Requirements for Electrical Equipment for Measurement, Control and Laboratory Use Part 1: General Requirements (IEC 1010-1: 1990 + A1: 1992, Modified). 117 pp.

CSA CAN/CSA-C22. 2NO1010.1-92. Safety Requirements for Electrical Equipment for Measurement, Control, and Laboratory Use, Part 1: General Requirements (IEC 1010-1:1990); (Gen Instr 1). 160 pp.

DIN VDE 0411 Pt 100-86. Measurement and Control; Safety Requirements for Electrically Operated Measuring, Control and Laboratory Equipment; General Requirements; Identical with IEC 66E (Sec) 22 (Aug). 137 pp.

IEC 1010 Pt 1-90. Safety Requirements for Electrical Equipment for Measurement, Control and Laboratory Use Part 1: General Requirements First Edition (CENELEC EN 61010-1: 1993). 190 pp.

—Medical Equipment

BSI BS 5724: Part 1-89. 1989 Medical Electrical Equipment Part 1: General Requirements for Safety. 212 pp.

BSI BS 5724: Part 1-01. 1989 Amd 1 Medical Electrical Equipment Part 1: General Safety Requirements (AMD 6715) December 21, 1990. 216 pp.

Electrical Safety *(Cont.)*

—Medical Equipment *(Cont.)*

BSI BS 5724: Sec 1.1-92. 1992 Medical Electrical Equipment Part 1: General Requirements for Safety Section 1.1: Collateral Standard: Safety Requirements for Medical Electrical Systems (IEC 601-1-1: 1992). 26 pp.

BSI BS 5724: Sec 1.2-93. 1993 Medical Electrical Equipment Part 1: General Requirements for Safety Section 1.2: Collateral Standard: Electromagnetic Compatibility—Requirements and Tests (ALSO KNOWN AS BS EN 60601-1-2: 1993) (N). 21 pp.

BSI BS 5724: Sec 2.1: Supp 1-91. 1991 Medical Electrical Equipment Part 2: Particular Requirements for Safety Section 2.1: Medical Electron Accelerators in the Range 1 MeV to 50 MeV Supplement No. 1: Revised and Additional Text. 26 pp.

BSI BS 5724: Sec 2.1-89. 1989 Medical Electrical Equipment Part 2: Particular Requirements for Safety Section 2.1: Specification for Medical Electron Accelerators in the Range 1 MeV to 50 MeV. 36 pp.

BSI BS 5724: Sec 2.3-92. 1992 Medical Electrical Equipment Part 2: Particular Requirements for Safety Section 2.3: Specification for Short-Wave Therapy Equipment (IEC 601-2-3: 1991) (N). 24 pp.

BSI BS 5724: Sec 2.3-01. 1992 Amd 1 Medical Electrical Equipment Part 2: Particular Requirements for Safety Section 2.3: Specification for Short-Wave Therapy Equipment (BS EN 60601-2-3: 1993) (AMD 7879) August 15, 1993 (N). 32 pp.

BSI BS 5724: Sec 2.3-83. 1983 Medical Electrical Equipment Part 2: Particular Requirements for Safety Section 2.3: Short Wave Therapy Equipment. 20 pp.

BSI BS 5724: Sec 2.5-85. 1985 Medical Electrical Equipment Part 2: Particular Requirements for Safety Section 2.5: Safety of Ultrasonic Therapy Equipment. 20 pp.

BSI BS 5724: Sec 2.5-01. 1985 Amd 1 Medical Electrical Equipment Part 2: Particular Requirements for Safety Section 2.5: Specification for Safety of Ultrasonic Therapy Equipment (IEC 601-2-5: 1984) (AMD 7016) July 15, 1992 (N). 21 pp.

BSI BS 5724: Sec 2.6-85. 1985 Medical Electrical Equipment Part 2: Particular Requirements for Safety Section 2.6: Microwave Therapy Equipment. 18 pp.

BSI BS 5724: Sec 2.6-01. 1985 Amd 1 Medical Electrical Equipment Part 2: Particular Requirements for Safety Section 2.6: Specification for Microwave Therapy Equipment (IEC 601-2-6: 1984) (AMD 7017) July 15, 1992. 20 pp.

BSI BS 5724: Sec 2.7-87. 1987 Medical Electrical Equipment Part 2: Particular Requirements for Safety Section 2.7: High Voltage Generators of Diagnostic X-Ray Generators. 62 pp.

BSI BS 5724: Sec 2.7-01. 1987 Amd 1 Medical Electrical Equipment Part 2: Particular Requirements for Safety Section 2.7: Spec for High-Voltage Generators of Diagnostic X-Ray Generators (IEC 601-2-7: 1987) (AMD 6988) April 1, 1992. 63 pp.

BSI BS 5724: Sec 2.8-87. 1987 Medical Electrical Equipment Part 2: Particular Requirements for Safety Section 2.8: Therapeutic X-Ray Generators (IEC 601-2-8: 1987). 37 pp.

BSI BS 5724: Sec 2.8-01. 1987 Amd 1 Medical Electrical Equipment Part 2: Particular Requirements for Safety Section 2.8: Specification for Therapeutic X-Ray Generators (IEC 601-2-8: 1987) (AMD 6998) February 28, 1992. 38 pp.

BSI BS 5724: Sec 2.9-01. 1988 Amd 1 Medical Electrical Equipment Part 2: Particular Requirements for Safety Section 2.9: Specification for Dosimeters Used in Radiotherapy with Electrically-Connected Radiation Detectors (IEC 601-2-9: 1987). 14 pp.

BSI BS 5724: Sec 2.10-88. 1988 Medical Electrical Equipment Part 2: Particular Requirements for Safety Section 2.10: Nerve and Muscle Stimulators. 14 pp.

BSI BS 5724: Sec 2.10-01. 1988 Amd 1 Medical Electrical Equipment Part 2: Particular Requirements for Safety Section 2.10: Specification for Nerve and Muscle Stimulators (IEC 601-2-10: 1987) (AMD 7018) July 15, 1992. 21 pp.

BSI BS 5724: Sec 2.11-89. 1989 Medical Electrical Equipment Part 2: Particular Requirements for Safety Section 2.11: Specification for Gamma Beam Therapy Equipment (IEC 601-2-11: 1987). 35 pp.

BSI BS 5724: Sec 2.11-01. 1989 Amd 1 Medical Electrical Equipment Part 2: Particular Requirements for Safety Section 2.11: Specification for Gamma Beam Therapy Equipment (IEC 601-2-11: 1987) (AMD 6804) November 29, 1991. 36 pp.

BSI BS 5724: Sec 2.11: Supp 1-89. 1989 Medical Electrical Equipment Part 2: Particular Requirements for Safety Section 2.11: Specification for Gamma Beam Therapy Equipment Supplement 1: Revised and Additional Text (IEC 601-2-11: 1988). 15 pp.

Electrical Safety *(Cont.)*

—Medical Equipment *(Cont.)*

BSI BS 5724: Sec 2.11: Supp 1-01. 1989 Amd 1 Medical Electrical Equipment Part 2: Particular Requirements Safety Section 2.11: Specification for Gamma Beam Therapy Equipment Supplement 1: Revised and Additional Text (IEC 601-2-11: 1988) (AMD 6805) November 29, 1991. 16 pp.

BSI BS 5724: Sec 2.11: Supp 2-93. 1993 Medical Electrical Equipment Part 2: Particular Requirements Safety Section 2.11: Specification for Gamma Beam Therapy Equipment Supplement 2: Methods of Test for Radiation Safety (IEC 601-2-11: Amendment 2: 1993) (N). 28 pp.

BSI BS 5724: Sec 2.12-90. 1990 Medical Electrical Equipment Part 2: Particular Requirements for Safety Section 2.12: Lung Ventilators. 20 pp.

BSI BS 5724: Sec 2.12-01. 1990 Amd 1 Medical Electrical Equipment Part 2: Particular Requirements for Safety Section 2.12: Specification for Lung Ventilators (IEC 601-2-12: 1988) (AMD 7446) January 15, 1993. 20 pp.

BSI BS 5724: Sec 2.13-90. 1990 Medical Electrical Equipment Part 2: Particular Requirements for Safety Section 2.13: Anaesthetic Machines. 24 pp.

BSI BS 5724: Sec 2.13-01. 1990 Amd 1 Medical Electrical Equipment Part 2: Particular Requirements for Safety Section 2.13: Specification for Anaesthetic Machines (IEC 601-2-13: 1989) (AMD 7447) January 15, 1993. 24 pp.

BSI BS 5724: Sec 2.14-89. 1989 Medical Electrical Equipment Part 2: Particular Requirements for Safety Section 2.14: Electroconvulsive Theraphy Equipment (N). 20 pp.

BSI BS 5724: Sec 2.14-01. 1989 Amd 1 Medical Electrical Equipment Part 2: Particular Requirements for Safety Section 2.14: Specification for Electroconvulsive Therapy Equipment (IEC 601-2-14: 1989) (AMD 6800) December 24, 1991. 21 pp.

BSI BS 5724: Sec 2.15-89. 1989 Medical Electrical Equipment Part 2: Particular Requirements for Safety Section 2.15: Specification for Capacitor Discharge X-Ray Generators (IEC 601-2-15: 1988). 39 pp.

BSI BS 5724: Sec 2.15-01. 1989 Amd 1 Medical Electrical Equipment Part 2: Particular Requirements for Safety Section 2.15: Specification for Capacitor Discharge X-Ray Generators (IEC 601-2-15: 1988) (AMD 6989) April 1, 1992. 42 pp.

BSI BS 5724: Sec 2.16-89. 1989 Medical Electrical Equipment Part 2: Particular Requirements for Safety Section 2.16: Specification for Haemodialysis Equipment (IEC 601-2-16: 1989). 31 pp.

BSI BS 5724: Sec 2.16-01. 1989 Amd 1 Medical Electrical Equipment Part 2: Particular Requirements for Safety Section 2.16: Specification for Haemodialysis Equipment (IEC 601-2-16: 1989) (AMD 6504) December 21, 1990. 32 pp.

BSI BS 5724: Sec 2.16-02. 1989 Amd 2 Medical Electrical Equipment Part 2: Particular Requirements for Safety Section 2.16: Specification for Haemodialysis Equipment (IEC 601-2-16: 1989) (AMD 6783) November 29, 1991. 33 pp.

BSI BS 5724: Sec 2.17-90. 1990 Medical Electrical Equipment Part 2: Particular Requirements for Safety Section 2.17: Remote Controlled Automatically Driven Gamma Ray After-loading Equipment. 31 pp.

BSI BS 5724: Sec 2.17-01. 1990 Amd 1 Medical Electrical Equipment Part 2: Particular Requirements for Safety Section 2.17: Specification for Remote-Controlled Automatically-Driven Gamma-Ray Afterloading Equipment (IEC 601-2-17: 1989) (AMD 7210) September 15, 1992. 33 pp.

BSI BS 5724: Sec 2.20-82. (WITHDRAWN) 1982 Medical Electrical Equipment Part 2: Particular Requirements for Safety Section 2.20: Nursing Incubators (Superseded by BS 5724: Section 2.119: 1991). 17 pp.

BSI BS 5724: Sec 2.21-83. (WITHDRAWN) 1983 Medical Electrical Equipment Part 2: Particular Requirements for Safety Section 2.21: Transport Incubators (Superseded by BS 5734: Section 2.120: 1991). 16 pp.

BSI BS 5724: Sec 2.22-87. 1987 Medical Electrical Equipment Part 2: Particular Requirements for Safety Section 2.22: General Operating Tables. 21 pp.

BSI BS 5724: Sec 2.23-89. 1989 Medical Electrical Equipment Part 2: Particular Requirements for Safety Section 2.23: Specification for Oxygen Concentrators (ISO 8359: 1988). 20 pp.

BSI BS 5724: Sec 2.119-91. 1991 Medical Electrical Equipment Part 2: Particular Requirements for Safety Section 2.119: Specification for Baby Incubators (N). 33 pp.

BSI BS 5724: Sec 2.119-01. 1991 Amd 1 Medical Electrical Equipment Part 2: Particular Requirements for Safety Section 2.119: Specification for Baby Incubators (IEC 601-2-19: 1990) (AMD 7319) December 15, 1992. 37 pp.

INDUSTRY STANDARDS
935

Electrical Safety *(Cont.)*

—Medical Equipment *(Cont.)*

BSI BS 5724: Sec 2.120-91. 1991 Medical Electrical Equipment Part 2: Particular Requirements for Safety Section 2.120: Specification for Transport Incubators (IEC 601-2-20: 1990). 35 pp.
BSI BS 5724: Sec 2.120-01. 1991 Amd 1 Medical Electrical Equipment Part 2: Particular Requirements for Safety Section 2.120: Specification for Transport Incubators (IEC 601-2-20: 1990) (AMD 7320) December 15, 1992. 39 pp.
BSI BS 5724: Sec 2.122-93. 1993 Medical Electrical Equipment Part 2: Particular Requirements for Safety Section 2.122: Specification for Diagnostic and Therapeutic Laser Equipment (ALSO KNOWN AS 60601-2-22: 1993). 27 pp.
BSI BS EN 60601-1-2-93. 1993 Medical Electrical Equipment Part 1: General Requirements for Safety Section 1.2: Collateral Standard: Electromagnetic Compatibility—Requirements and Tests (IEC 601-1-2: 1993) (N). 21 pp.
BSI BS EN 60601-2-3-93. 1993 Amd 1 Medical Electrical Equipment Part 2: Particular Requirements for Safety Section 2.3: Specification for Short-Wave Therapy Equipment (IEC 601-2-3: 1991) (AMD 7879) August 15, 1993 (N). 32 pp.
BSI BS EN 60601-2-22-93. 1993 Medical Electrical Equipment Part 2: Particular Requirements for Safety Section 2.122: Specification for Diagnostic and Therapeutic Laser Equipment (IEC 601-2-22: 1992). 27 pp.
CEN PREN 601-2-28-92. Medical Electrical Equipment Part 2: Particular Requirements for Safety Section 2.28: Specification for X-Ray Source Assemblies for Medical Diagnosis. 30 pp.
CEN PREN 793-92. Medical Electrical Equipment—Particular Requirements for Safety of Medical Supply Units. 43 pp.
CEN PREN 794-1-92. Medical Electrical Equipment—Particular Requirements for Safety of Lung Ventilators—Part 1: Lung Ventilators for Medical Use. 55 pp.
CENELEC HD 379-78. Area Exposure Product Meter. 2 pp.
CENELEC HD 395.1-79. Safety of Medical Electrical Equipment Part 1 General Requirements. 2 pp.
CENELEC HD 395.1 S2-88. Medical Electrical Equipment Part 1: General Requirements. 2 pp.
CENELEC HD 395.1 S2/A1-93. AMD 1 Medical Electrical Equipment Part 1: General Requirements for Safety. 4 pp.
CENELEC HD 395.2.2-85. Medical Electrical Equipment Part 2: Particular Requirements for the Safety of High Frequency Surgical Equipment. 2 pp.
CENELEC HD 395.2.3-85. Medical Electrical Equipment Part 2: Particular Requirements for the Safety of Short-Wave Therapy Equipment. 3 pp.
CENELEC HD 395.2.5-86. Medical Electrical Equipment Part 2: Particular Requirements for the Safety of Ultrasonic Therapy Equipment. 2 pp.
CENELEC HD 395.2.6-87. Medical Electrical Equipment Part 2: Particular Requirements for the Safety of Microwave Therapy Equipment. 2 pp.
CENELEC HD 395.2.10 S1-89. Medical Electrical Equipment-Part 2: Particular Requirements for the Safety of Nerve and Muscle Stimulators. 3 pp.
CENELEC HD 395.2.11 S1-89. Medical Electrical Equipment-Part 2: Particular Requirements for the Safety of Gamma Beam Therapy Equipment. 3 pp.
CENELEC HD 395.2.11 S2-90. Medical Electrical Equipment Part 2: Particular Requirements for the Safety of Gamma Beam Therapy Equipment. 3 pp.
CENELEC HD 395.2.11 S2-92. Medical Electrical Equipment Part 2: Particular Requirements for the Safety of Gamma Beam Therapy Equipment; (IEC 601-2-11:1987 + A1:1988). 3 pp.
CENELEC HD 395.2.14 S1-89. Medical Electrical Equipment part 2: Particular Requirements for the Safety of Electro Convulsive Therapy Equipment. 3 pp.
CENELEC HD 395.2.15 S1-89. Medical Electrical Equipment Part 2: Particular Requirements for the Safety of Capacitor Discharge X-Ray Generators. 5 pp.
CENELEC HD 395.2.17 S1-92. Medical Electrical Equipment Part 2: Particular Requirements for the Safety of Remote-Controlled Automatically-Driven Gamma-Ray Afterloading Equipment. 6 pp.
CENELEC HD 395.2.19 S1-92. Medical Electrical Equipment Part 2: Particular Requirements for Safety of Baby Incubators. 6 pp.
CENELEC HD 395.2.20 S1-92. Medical Electrical Equipment Part 2: Particular Requirements for Safety of Transport Incubators. 6 pp.
CENELEC EN 50 061-88. Safety of Implantable Cardiac Pacemakers. 27 pp.
CENELEC PREN 50115-93. Medical Electrical Equipment—Part 2: Particular Safety Requirements for Operating Tables. 26 pp.
CENELEC EN 60601-1-90. Medical Electrical Equipment Part 1: General Requirements for Safety; (IEC 601-1:1988). 415 pp.

Electrical Safety *(Cont.)*

—Medical Equipment *(Cont.)*

CENELEC EN 60601-1-2-93. Medical Electrical Equipment Part 1: General Requirements for Safety 2. Collateral Standard: Electromagnetic Compatibility—Requirements and Tests (IEC 601-1-2: 1993). 16 pp.
CENELEC EN 60601-2-3-93. Medical Electrical Equipment Part 2: Particular Requirements for the Safety of Short-Wave Therapy Equipment (IEC 601-2-3: 1991). 20 pp.
CENELEC EN 60601-2-22-92. Medical Electrical Equipment Part 2: Particular Requirements for the Safety of Diagnostic and Therapeutic Laser Equipment. 5 pp.
CENELEC EN 60601-2-22-92. Medical Electrical Equipment Part 2: Particular Requirements for the Safety of Diagnostic and Therapeutic Laser Equipment; (IEC 601-2-22: 1992). 22 pp.
CENELEC PREN 60601-2-27-92. Medical Electrical Equipment Part 2: Particular Requirements for Safety Section 2.126: Specification for Electroencephalographs. 28 pp.
CSA CAN/CSA-C22. 2NO601.1-M90. Medical Electrical Equipment, Part 1: General Requirements for Safety; (Gen Instr 1). 240 pp.
CSA CAN/CSA-C22. 2N601.2.1-92. Medical Electrical Equipment Part 2: Particular Requirements for the Safety of Medical Electron Accelerators in the Range 1 MeV to 50 MeV (IEC 601-2-1:1981); (Gen Instr 1). 85 pp.
CSA CAN/CSA-C22. 2N601.2.3-92. Medical Electrical Equipment Part 2: Particular Requirements for the Safety of Short-Wave Therapy Equipment (IEC 601-2-3:1991); (Gen Instr 1). 36 pp.
CSA CAN/CSA-C22. 2N601.2.4M90. Medical Electrical Equipment, Part 2: Particular Requirements for the Safety of Cardiac Defibrillators and Cardiac Defibrillator-Monitors; (Gen Instr 1). 78 pp.
CSA CAN/CSA-C22. 2N601.2.5-92. Medical Electrical Equipment Part 2: Particular Requirements for the Safety of Ultrasonic Therapy Equipment (IEC 601-2-5:1984); (Gen Instr 1). 44 pp.
CSA CAN/CSA-C22. 2N601.2.6-92. Medical Electrical Equipment Part 2: Particular Requirements for the Safety of Microwave Therapy Equipment (IEC 601-2-6:1984); (Gen Instr 1). 37 pp.
CSA CAN/CSA-C22. 2N601.2.7-92. Medical Electrical Equipment Part 2: Particular Requirements for the Safety of High-Voltage Generators of Diagnostic X-Ray Generators (IEC 601-2-7:1987); (Gen Instr 1). 79 pp.
CSA CAN/CSA-C22. 2N601.2.8-92. Medical Electrical Equipment Part 2: Particular Requirements for the Safety of Therapeutic X-Ray Generators (IEC 601-2-8:1987); (Gen Instr 1). 57 pp.
CSA CAN/CSA-C22. 2N601.2.9-92. Medical Electrical Equipment Part 2: Particular Requirements for the Safety of Dosimeters Used in Radiotherapy with Electrically-Connected Radiation Detectors (IEC 601-2-9:1987); (Gen Instr 1). 29 pp.
CSA CAN/CSA-C22. 2601.2.10-92. Medical Electrical Equipment Part 2: Particular Requirements for the Safety of Nerve and Muscle Stimulators (IEC 601-2-10:1987); (Gen Instr 1 Thru 2). 40 pp.
CSA CAN/CSA-C22. 2601.2.11-92. Medical Electrical Equipment Part 2: Particular Requirements for the Safety of Gamma Beam Therapy Equipment (IEC 601-2-11:1988); (Gen Instr 1). 60 pp.
CSA CAN/CSA-C22. 2601.2.14-92. Medical Electrical Equipment Part 2: Particular Requirements for the Safety of Electroconvulsive Therapy Equipment (IEC 601-2-14:1989); (Gen Instr 1). 39 pp.
CSA CAN/CSA-C22. 2601.2.15-92. Medical Electrical Equipment Part 2: Particular Requirements for the Safety of Capacitor Discharge X-Ray Generators (IEC 601-2-15:1988); (Gen Instr 1). 58 pp.
CSA CAN/CSA-C22. 2601.2.16-92. Medical Electrical Equipment Part 2: Particular Requirements for the Safety of Haemodialysis Equipment (IEC 601-2-16:1989); (Gen Instr 1). 49 pp.
CSA CAN/CSA-C22. 2601.2.18-92. Medical Electrical Equipment Part 2: Particular Requirements for the Safety of Endoscopic Equipment (IEC 601-2-18:1990); (Gen Instr 1). 36 pp.
CSA CAN/CSA-C22. 2601.2.19-92. Medical Electrical Equipment Part 2: Particular Requirements for the Safety of Baby Incubators (IEC 601-2-19:1990); (Gen Instr 1). 47 pp.
CSA CAN/CSA-C22. 2601.2.20-92. Medical Electrical Equipment Part 2: Particular Requirements for the Safety of Transport Incubators (IEC 601-2-20:1990); (Gen Instr 1). 50 pp.
CSA CAN/CSA-Z32.2-M89. Electrical Safety in Patient Care Areas; (Gen Instr 1). 64 pp.
CSA Z323.3.1-1982. Electrical Aids for Physically Disabled Persons. 17 pp.
DIN VDE 0107 (S)-81. Erection and Testing of Electrical Installations in Medically Used Rooms (June). 56 pp.
DIN VDE 0750 Pt 9 (D)-87. Medical Electrical Equipment; Implantable Cardiac Pacemakers; Requirements for Safety (June). 48 pp.

Electrical Safety *(Cont.)*

—Medical Equipment *(Cont.)*

DIN VDE 0750 Pt 217-87. Medical Electrical Equipment; Transport Incubators; Special Safety Regulations (Feb). 39 pp.
DIN VDE 0750 Pt 227 (D)-88. Medical, Electrical Equipment; Infant Radiant Warmers; Special Regulations for Safety (Oct). 26 pp.
IEC 513-76. Basic Aspects of the Safety Philosophy of Electrical Equipment Used in Medical Practice First Edition. 31 pp.
IEC 580-77. Area Exposure Product Meter First Edition. 46 pp.
IEC 601 Pt 1-88. Medical Electrical Equipment Part 1: General Requirements for Safety Second Edition; (Amendment 1-1991) (CENELEC EN 60601-1: 1990). 408 pp.
IEC 601 Pt 1-1-92. Medical Electrical Equipment Part 1: General Requirements for Safety 1. Collateral Standard: Safety Requirements for Medical Electrical Systems First Edition. 51 pp.
IEC 601 Pt 1-2-93. Medical Electrical Equipment Part 1: General Requirements for Safety 2. Collateral Standard: Electromagnetic Compatibility—Requirements and Tests First Edition (CENELEC EN 60601-1-2: 1993). 37 pp.
IEC 601 Pt 2-1-81. Safety of Medical Elec-trical Equipment Part 2: Particular Requirements for Medical Electron Ac-celerators in the Range 1 MeV to 50 MeV Section One: General Section Two: Radiation Safety for Equipment First Edi-tion; (Amendment 1-1984) (Amendment 2-1990) (CAN/CSA-C22.2 NO 601.2.1-92). 136 pp.
IEC 601 Pt 2-3-91. Medical Electrical Equipment Part 2: Particular Requirements for the Safety of Short-Wave Therapy Equipment Second Edition; (CAN/CSA-C22.2 No. 601.2.3-92) (CENELEC EN 60601-2-3: 1993). 43 pp.
IEC 601 Pt 2-4-83. Medical Electrical Equipment Part 2: Particular Requirements for the Safety of Cardiac Defibrillators and Cardiac Defibrillator-Monitors First Edition. 80 pp.
IEC 601 Pt 2-5-84. Medical Electrical Equipment Part 2: Particular Requirements for the Safety of Ultrasonic Therapy Equipment First Edition. 40 pp.
IEC 601 Pt 2-6-84. Medical Electrical Equipment Part 2: Particular Requirements for the Safety of Microwave Therapy Equipment First Edition. 38 pp.
IEC 601 Pt 2-7-87. Medical Electrical Equipment Part 2: Particular Requirements for the Safety of High-Voltage Generators of Diagnostic X-Ray Generators First Edition. 125 pp.
IEC 601 Pt 2-8-87. Medical Electrical Equipment Part 2: Particular Requirements for the Safety of Therapeutic X-Ray Generators First Edition; (CAN/CSA-C22.2 No.2.8-92). 76 pp.
IEC 601 Pt 2-9-87. Medical Electrical Equipment Part 2: Particular Requirements for the Safety of Dosimeters Used in Radiotherapy with Electrically-Connected Radiation Detectors First Edition; (CAN/CSA-C22.2 No. 601.2.9-92). 27 pp.
IEC 601 Pt 2-10-87. Medical Electrical Equipment Part 2: Particular Requirements for the Safety of Nerve and Muscle Stimulators First Edition. 39 pp.
IEC 601 Pt 2-11-87. Medical Electrical Equipment Part 2: Particular Requirements for the Safety of Gamma Beam Therapy Equipment First Edition; (Amendment 1-1988) (Amendment 2-1993) (CENELEC HD 395.2.11 S2:1990). 133 pp.
IEC 601 Pt 2-13-89. Medical Electrical Equipment Part 2: Particular Requirements for the Safety of Anaesthetic Machines First Edition. 43 pp.
IEC 601 Pt 2-14-89. Medical Electrical Equipment Part 2: Particular Requirements for the Safety of Electroconvulsive Therapy Equipment First Edition; (CAN/CSA-C22.2 No. 601.2.14-92). 39 pp.
IEC 601 Pt 2-15-88. Medical Electrical Equipment Part 2: Particular Requirements for the Safety of Capacitor Discharge X-Ray Generators First Edition. 81 pp.
IEC 601 Pt 2-16-89. Medical Electrical Equipment Part 2: Particular Requirements for Safety of Haemodialysis Equipment First Edition. 61 pp.
IEC 601 Pt 2-17-89. Medical Electrical Equipment Part 2: Particular Requirements for the Safety of Remote -Controlled Automatically-Driven Gamma-Ray Afterloading Equipment First Edition. 55 pp.
IEC 601 Pt 2-18-90. Medical Electrical Equipment Part 2: Particular Requirements for the Safety of Endoscopic Equipment First Edition. 37 pp.
IEC 601 Pt 2-19-90. Medical Electrical Equipment Part 2: Particular Requirements for Safety of Baby Incubators First Edition. 61 pp.
IEC 601 Pt 2-20-90. Medical Electrical Equipment Part 2: Particular Requirements for Safety of Transport Incubators First Edition. 65 pp.
IEC 601 Pt 2-22-92. Medical Electrical Equipment Part 2: Particular Requirements for the Safety of Diagnostic and Therapeutic Laser Equipment First Edition; (CENELEC EN 60601-2-22: 1992). 46 pp.
IEC 930-88. Guidelines for Administrative, Medical, and Nursing Staff Concerned with the Safe Use of Medical Electrical Equipment First Edition. 25 pp.

INTERNATIONAL AND NON-U.S. NATIONAL STANDARDS
SUBJECT INDEX Electrical

Electrical Safety *(Cont.)*

—Medical Equipment *(Cont.)*

JIS T 1001-92. General Requirements for Safety of Medical Electrical Equipment. 71 pp.
JIS T 1002-92. General Rules of Testing Methods for Safety of Medical Electrical Equipment. 39 pp.
JIS T 1022-82. Safety Requirements of Electrical Installations for Medically Used Rooms in Hospitals and Clinics. 19 pp.
SAA AS 3200.2.3-92. Approval and Test Specification—Medical Electrical Equipment—Part 2: Particular Requirements for Safety—Part 2.3: Short-Wave Therapy Equipment (IEC 601-2-3:1991) (Supersedes AS 3209—1986 (Which Will Remain Current)) (in Professional. 17 pp.
SAA AS 3200.2.5-92. Approval and Test Specification—Medical Electrical Equipment—Part 2: Particular Requirements for Safety—Part 2.5: Ultrasonic Therapy Equipment (IEC 601-2-5:1984) (Supersedes AS 3211—1986 (Which Will Remain Current)) (in Professional. 14 pp.
SAA AS 3200.2.6-92. Approval and Test Specification—Medical Electrical Equipment—Part 2: Particular Requirements for Safety—Part 2.6: Microwave Therapy Equipment (IEC 601-2-6:1984) (Supersedes AS 3210—1986 (Which Will Remain Current)) (in Professional. 11 pp.
SAA AS 3200.2.14-92. Approval and Test Specification—Medical Electrical Equipment—Part 2: Particular Requirements for Safety—Part 2.14: Electroconvulsive Therapy Equipment (IEC 601-2-14:1988) (in Prof. Pkgs. 17C,28) (This is a Joint Stand. with SNZ NZS 3200.2.14). 15 pp.
SAA AS 3200.2.16-92. Approval and Test Specification—Medical Electrical Equipment—Part 2: Particular Requirements for Safety—Part 2.16: Haemodialysis Equipment (Supersedes AS 3207—1981) (in Professional Packages 17C, 28) (This is a Joint Standard with SNZ NZS 3200.2.16). 27 pp.
SAA AS 3200.2.18-92. Approval and Test Specification—Medical Electrical Equipment—Part 2: Particular Requirements for Safety—Part 2.18: Endoscopic Equipment (IEC 601-2-18: 1990) (in Professional Packages 28, 17C) (This is a Joint Standard with SNZ NZS 3200.2.18). 13 pp.
SNZ NZS/AS 3200. 2-93. Approval and Test Specification—Medical Electrical Equipment Part 2: Particular Requirements for Safety (This is a Joint Standard with SAA AS 3200.2).
SNZ NZS/AS 3200. 2.22-93. Approval and Test Specification—Medical Electrical Equipment Part 2.22: Diagnostic and Therapeutic Laser Equipment (This is a Joint Standard with SAA AS 3200.2.22). 19 pp.
SNZ NZS 6150-90. Approval and Test Specification—Medical Electrical Equipment—General Requirements for Safety Amend: 1, 1992 (NZS 6150:1990/AS 3200.1-1990). 185 pp.

—Office Furniture Cabling Systems

CSA CAN/CSA-C22. 2 NO 14-M91. Industrial Control Equipment; (Gen Instr 1 Thru 4). 142 pp.

—Office Machines

BSI BS 7002-92. 1992 Safety of Information Technology Equipment, Including Electrical Business Equipment (S). 195 pp.
BSI BS 7002-89. 1989 Amd 1 Safety of Information Technology Equipment Including Electrical Business Equipment (AMD 6650) June 28, 1991. 140 pp.
BSI BS 7002-02. (WITHDRAWN) 1989 Amd 2 Safety of Information Technology Equipment Including Electrical Business Equipment (AMD 6974) January 31, 1992 (Superseded by BS EN 60950: 1992). 180 pp.
BSI BS EN 60950-92. 1992 Safety of Information Technology Equipment, Including Electrical Business Equipment (Supersedes BS 7002: 1989). 195 pp.
BSI BS EN 60950-01. 1992 Amd 1 Safety of Information Technology Equipment, Including Electrical Business Equipment (S). 212 pp.
CENELEC HD 372-78. Safety of Electrically Energized Office Machines. 4 pp.
CENELEC HD 372 S1-88. Safety of Electrically Energized Office Machines. 3 pp.
CENELEC PREN 50116-93. Information Technology Equipment—Routine Electrical Safety Testing in Production (Possible New British Standard). 4 pp.
CENELEC EN 60 950-88. Safety of Information Technology Equipment Including Electrical Business Equipment. 141 pp.
CENELEC EN 60950/A1-90. AMD 1 Safety of Information Technology Equipment Including Electrical Business Equipment. 5 pp.
CENELEC EN 60950/A2-91. AMD 2 Safety of Information Technology Equipment Including Electrical Business Equipment. 5 pp.
CENELEC EN 60950-92. Safety of Information Technology Equipment, Including Electrical Business Equipment. 15 pp.
CENELEC EN 60950-92. Safety of Information Technology Equipment, Including Electrical Business Equipment; (IEC 950: 1991, Modified). 184 pp.

Electrical Safety *(Cont.)*

—Office Machines *(Cont.)*

CENELEC EN 60950/PRAA-92. AMD A Safety of Information Technology Equipment Including Electrical Business Equipment. 4 pp.
CENELEC EN 60950/A1-93. AMD 1 Safety of Information Technology Equipment, Including Electrical Business Equipment (IEC 950: 1991/A1: 1992). 186 pp.
CSA CAN/CSA-C22. 2 NO 950-M89. Safety of Information Technology Equipment, Including Electrical Business Equipment; (Gen Instr 1) (Supersedes C22.2 No 220). 183 pp.
DIN VDE 0701 Pt 240-86. Repair, Modification and Inspection of Electrical Appliances; Safety Requirements for Data Processing Equipment and Office Machines (Apr). 10 pp.
DIN VDE 0701 Pt 240 (D)-90. Repair, Modification & Inspection of Electrical Appliances; Information Technology Equipment (Feb). 12 pp.
DIN VDE 0805 (D)-84. Deviations from IEC 435/ 2nd Ed (1983); DIN IEC 435 A1/VDE 0805 A1 Draft; Safety of Data Processing Equipment (Nov). 36 pp.
DIN VDE 0806-81. Deviations of IEC 380; Safety of Electrically Energized Office Machines (Aug). 14 pp.
ECMA ECMA-TR 28-85. Safety Verification (SAVE) Report—ECMA 57/IEC 435. 52 pp.
ECMA ECMA TR 39-92. Compliance Verification Report (Cover) ECMA 129—IEC 950—EN60950. 77 pp.
ECMA ECMA-TR 39-88. Compliance Verification (Cover) Report IEC 950. 63 pp.
IEC 950-91. Safety of Information Technology Equipment, Including Electrical Business Equipment Second Edition; (Amendment 1-1992) (Amendment 2-1993) (CENELEC EN 60950:1992) (CENELEC EN 60950/A1: 1993). 372 pp.
SNZ NZS 6661-89. Safety of Information Technology Equipment, Including Electrical Business Equipment Amend: A. 245 pp.

—Passenger Conveyors

CSA CAN/CSA-B44.1-M91. Elevator and Escalator Electrical Equipment (ASME A17.5-1991). 97 pp.

—Patient Monitoring Equipment

IEC 601 Pt 2-23-93. Medical Electrical Equipment Part 2: Particular Requirements for the Safety of Transcutaneous Partial Pressure Monitoring Equipment First Edition. 58 pp.

—Power Lines—Telephone Cables

CCITT RECMN K.19-89. Joint Use of Trenches and Tunnels for Telecommunication and Power Cables—Protection Against Interference (Study Group V) 2 pp. 2 pp.

—Power Systems

DIN VDE 0105 Pt 1-83. Operation of Electrical Power Installations; General Requirements (July). 105 pp.
IEC 65-85. Safety Requirements for Mains Operated Electronic and Related Apparatus for Household and Similar General Use Fifth Edition; (Amendment 2-1989, Incorporating Amendment 1) (Amendment 3-1992). 210 pp.

—Probes (Sensors)

IEC 1010 Pt 2-031-93. Safety Requirements for Electrical Equipment for Measurement, Control, and Laboratory Use Part 2-031: Particular Requirements for Hand-Held PROBE ASSEMBLIES for Electrical Measurement and Test First Edition. 48 pp.

—Process Control Equipment

CSA CAN/CSA-C22. 2NO1010.1-92. Safety Requirements for Electrical Equipment for Measurement, Control, and Laboratory Use, Part 1: General Requirements (IEC 1010-1:1990); (Gen Instr 1). 160 pp.

—Radiation Therapy Equipment

BSI BS 5724: Sec 2.129-93. 1993 Medical Electrical Equipment Part 2: Particular Requirements for Safety Section 2.129: Specification for Radiotherapy Simulators (IEC 601-2-29: 1993) (N). 27 pp.
IEC 601 Pt 2-29-93. Medical Electrical Equipment Part 2: Particular Requirements for the Safety of Radiotherapy Simulators First Edition. 55 pp.

—Robots

JIS B 8438-91. Industrial Robots—Electrical Equipment. 82 pp.

—Sockets (Electric)

SNZ NZS/AS 3120-93. Approval and Test Specification Extension Sockets (This is a Joint Standard with SAA AS 3120). 16 pp.

—Sprayers

BSI BS EN 50 059-91. 1991 Electrostatic Hand-Held Spraying Equipment for Non-Flammable Material for Painting and Finishing. 15 pp.

Electrical Safety *(Cont.)*

—Sprayers *(Cont.)*

CENELEC EN 50 059-90. Specification for Electrostatic Hand-Held Spraying Equipment for Non-Flammable Material for Painting and Finishing. 24 pp.

—Standards Preparation

EURO MEMORANDUM 2-74. Preparation of Standards For Safety in the Design and Construction of Electrical Equipment. 5 pp.
IEC Guide 104-84. Guide to the Drafting of Safety Standards, and the Role of Committees with Safety Pilot Functions and Safety Group Functions Second Edition; (Amendment 1-1989). 41 pp.

—Storage Water Heaters

DIN VDE 0700 Pt 21 A1 (D)-83. Safety of Household and Similar Electrical Appliances; Part 2: Particular Requirements for Water Heaters (Storage Heaters and Boilers); Amendment 1 (Apr). 5 pp.
DIN VDE 0700 Pt 21 A3 (D)-87. Safety of Household and Similar Electrical Appliances; Water Heaters (Storage Water Heaters and Hot Water Boilers); Amendment 3 to Draft DIN VDE 0700 Part 21 (Apr). 7 pp.

—Telecommunication Systems

BSI BS 6484-84. 1984 Amd 2 Electrical Safety Requirements for Independent Power Supply Units for Indirect Connection to Certain Telecommunications Network. 4 pp.
BSI BS EN 41003-93. 1993 Particular Safety Requirements for Equipment to Be Connected to Telecommunication Networks (S). 14 pp.
BSI BS EN 41003-91. 1991 Particular Safety Requirements for Equipment to Be Connected to Telecommunication Networks. 23 pp.
CENELEC ENV 41 003-88. Particular Safety Requirements for Equipment to be Connected to Telecommunications Network.
CENELEC PREN 41 003-90. Particular Safety Requirements for Equipment to be Connected to Telecommunication Networks. 31 pp.
CENELEC EN 41003-91. Particular Safety Requirements for Equipment to be Connected to Telecommunication Networks (Supersedes HD 270 S1). 24 pp.
CENELEC EN 41003-93. Particular Safety Requirements for Equipment to Be Connected to Telecommunication Networks. 11 pp.
DIN VDE 0800 Pt 1-89. Telecommunications; General Definitions, Requirements and Tests for the Safety of Installations and Equipment (May) (Together with VDE 0800 Part 4/03.86 and DIN VDE 0800 Part 10/05.89, Supersedes DIN 57800 Part 1/VDE 0800 Part 1/04.84.. 105 pp.
ECMA ECMA-TR 35-87. Particular Safety Requirements for Equipment to be Connected to Telecommunication Networks. 25 pp.
IEC Guide 105-85. Principles Concerning the Safety of Equipment Electrically Connected to a Telecommunications Network First Edition. 20 pp.

—Telephone Cables—Crossing

CCITT RECMN K.6-89. Precautions at Crossings —Protection Against Interference (Study Group V) 2 pp. 2 pp.

—Vending Machines

DIN VDE 0700 Pt 224-82. Safety of Household and Similar Electrical Appliances; Particular Requirements for Coin-Operated Automatic Machines and Control (Oct). 21 pp.

—Welding

SAA AS 1673.5-82. Methods for the Analysis of Iron Ores—Part 5: Determination of Loss on Ignition at 1 000 degrees Celsius. 3 pp.
SAA AS 1674.2-90. Safety in Welding and Allied Processes—Part 2: Electrical. 31 pp.

—X-Ray Equipment

BSI BS 5724: Sec 2.128-93. 1993 Medical Electrical Equipment Part 2: Particular Requirements for Safety Section 2.128: Specification for X-Ray Source Assemblies and X-Ray Tube Assemblies for Medical Diagnosis (ALSO KNOWN AS BS EN 60601-2-28: 1993) (N). 26 pp.
BSI BS EN 60601-2-28-93. 1993 Medical Electrical Equipment Part 2: Particular Requirements for Safety Section 2.128: Specification for X-Ray Source Assemblies and X-Ray Tube Assemblies for Medical Diagnosis (IEC 601-2-28: 1993) (N). 26 pp.
CENELEC EN 60601-2-28-93. Medical Electrical Equipment Part 2: Particular Requirements for the Safety of X-Ray Source Assemblies and X-Ray Tube Assemblies for Medical Diagnosis (IEC 601-2-28: 1993). 21 pp.
IEC 601 Pt 2-28-93. Medical Electrical Equipment Part 2: Particular Requirements for the Safety of X-ray Source Assemblies and X-ray Tube Assemblies for Medical Diagnosis First Edition; (Replaces 637) (CENELEC EN 60601-2-28: 1993). 45 pp.

INDUSTRY STANDARDS 937

Electrical

Electrical Shock
Use: Electric Shock

Electrical Steels
Scope Note: For additional listings, see also specific products made from electrical steels *See Also:* Alloy Steels; Low Alloy Steels; Steels
CNS C3117-81. Method of Test for Electrical Silicon Steel Sheets (Apr)(7214). 26 pp.

—**Sheets**
JIS C 2550-75. Methods of Test for Electrical Steel Sheets. 30 pp.
JIS C 2552-86. Non-Oriented Magnetic Steel Sheet and Strip. 11 pp.
JIS C 2553-86. Grain-Oriented Magnetic Steel Sheet and Strip. 9 pp.

—**Sheets—Cores**
JIS C 2555-92. Steel Sheets and Strip for Pole Core. 14 pp.

—**Sheets—Magnetic Measurement**
DIN ENGL 50462 Pt 4-79. Testing of Steel; Method for Determining the Magnetic Properties of Electrical Sheet and Strip in the 25 cm Epstein Square; Determination of Total Loss in an Alternating Magnetic Field Using the Bridge Method (July). 5 pp.
DIN ENGL 50462 Pt 5-79. Testing of Steel; Method for Determination of the Magnetic Properties of Electrical Steel Sheet and Strip in the 25 cm Epstein Square; Deter-mination of the Commuta-tion Values of the Magnetic Polarization and the Remanent Induc-tion in a Continuous Magnetic Field (June). 3 pp.

—**Strips**
JIS C 2552-86. Non-Oriented Magnetic Steel Sheet and Strip. 11 pp.
JIS C 2553-86. Grain-Oriented Magnetic Steel Sheet and Strip. 9 pp.

—**Strips—Cores**
JIS C 2555-92. Steel Sheets and Strip for Pole Core. 14 pp.

—**Strips—Magnetic Measurement**
DIN ENGL 50462 Pt 4-79. Testing of Steel; Method for Determining the Magnetic Properties of Electrical Sheet and Strip in the 25 cm Epstein Square; Determination of Total Loss in an Alternating Magnetic Field Using the Bridge Method (July). 5 pp.
DIN ENGL 50462 Pt 5-79. Testing of Steel; Method for Determination of the Magnetic Properties of Electrical Steel Sheet and Strip in the 25 cm Epstein Square; Deter-mination of the Commuta-tion Values of the Magnetic Polarization and the Remanent Induc-tion in a Continuous Magnetic Field (June). 3 pp.

Electrical Strain Gages
Use: Strain Gages

Electrical Systems
Use: Electrical Installations

Electrical Systems (Avionics)
Use: Avionics

Electrical Tapes (Insulating)
Use: Electrical Insulating Tapes

Electrical Testing
Use: Electrical Properties

Electrical Transducers
Use: Electric Transducers

Electrical Tubing
Use: Electric Conduits

Electrically Conductive Liquids
See Also: Battery Electrolytes; Electric Conductors

—**Electromagnetic Flowmeters**
BSI BS 5792: Part 1-93. 1993 Measurement of Conductive Liquid Flow in Closed Conduits Part 1: Method Using Eloectromagnetic Flowmeters (ISO 6817: 1992) (Supersedes BS 5792: 1980). 24 pp.
ISO 6817-92. Measurement of Conductive Liquid Flow in Closed Conduits—Method Using Electromagnetic Flowmeters First Edition. 22 pp.

Electrically Powered Vehicles
Use: Electric Vehicles

Electricity
Scope Note: For additional listings, use a more specific term *See Also:* Electrical Properties; Electrical Safety; Electromagnetism; Physics; Power Transmission

Electricity (Cont.)

—**Glossaries**
IEC. Electricity, Electronics and Telecommunications Multilingual Dictionary Volume 1: English-French -Russian-German-Spanish-Dutch-Italian-Swedish-Polish Second Edition. 1944 pp.
IEC. Electricity, Electronics and Telecommunications Multilingual Dictionary Volume 2: English-French -Russian-German-Spanish-Dutch-Italian-Swedish-Polish Second Edition. 964 pp.
IEC. Electricity, Electronics and Telecommunications Multilingual Dictionary Volume 3: German, Dutch and Swedish Indexes Second Edition. 793 pp.
IEC. Electricity, Electronics and Telecommunications Multilingual Dictionary Volume 4: Spanish and Italian Indexes Second Edition. 540 pp.
IEC. Electricity, Electronics and Telecommunications Multilingual Dictionary Volume 5: Russian and Polish Indexes Second Edition. 578 pp.
IEC 50 Chap 301,302,303-83. Advance Edition of International Electrotechnical Vocabulary Chapter 301: General Terms on Measurements in Electricity Chapter 302: Electrical Measuring Instruments Chapter 303: Electronic Measuring Instruments. 119 pp.
SAA AS 1852.301-88. International Electrotechnical Vocabulary—Part 301: General Terms on Measurements in Electricity Bound Together with Parts 301-303.
SNZ IEC 50: 50(301,302,303)-83. International Electrotechnical Vocabulary 50(301, 302, 303): General Terms on Measurements in Electricity. Electrical Measuring Instruments. Electronic Measuring Instruments. 107 pp.

—**Symbols**
CNS C1016-65. Letter Symbols of Electricity (Apr)(400) (R 1971).

—**Tariffs—Glossaries**
BSI BS 4727:Pt2: Group 07-73. 1973 Electrotechnical, Power, Telecommunication, Electronics, Lighting and Colour Terms Part 2: Terms Particular to Power Engineering Group 07: Tariffs Terminology. 18 pp.
IEC 50 Chap 691-73. International Electrotechnical Vocabulary Chapter 691: Tariffs for Electricity First Edition. 50 pp.
SAA AS 1852.691-75. International Electrotechnical Vocabulary—Part 691: Tariffs for Electricity Being IEC 50(691). 51 pp.
SNZ IEC 50: 50(691)-73. International Electrotechnical Vocabulary 50(691): Tariffs for Electricity. 44 pp.

—**Units of Measurement**
BSI BS 5775: Part 5-93. 1993 Quantities, Units and Symbols Part 5: Electricity and Magnestism (ISO 31-5: 1992) (G). 36 pp.
BSI BS 5775: Part 5-80. 1980 Amd 1 Specification for Quantities, Units and Symbols Part 5: Electricity and Magnetism (AMD 5850) July 31, 1989. 28 pp.
CNS Z7196-5-85. Quantities and Units of Electricity and Magnetism (Oct)(11296-5).
ISO 31 Pt 5-92. Quantities and Units—Part 5: Electricity and Magnetism Second Edition; (Corrected and Reprinted —1993). 35 pp.

Electroacoustic Measuring Instruments
Use: Acoustic Measuring Instruments

Electroacoustic Transducers
Use For: Acoustic Transducers; Sound Transducers
See Also: Bells; Earphones; Headphones; Hydrophones; Microphones; Piezoelectric Transducers; Sound Recording and Reproduction Equipment; Speakers; Telephones; Transducers

—**Glossaries**
JIS Z 8107-84. Glossary of Acoustical Terms (Transducers and Instruments).

Electroacoustics
See Also: Acoustics; Audio Amplifiers; Tape Recorders

—**Glossaries**
BSI BS 4727:Pt3: Group 08-85. 1985 Electrotechnical, Power, Telecommunication, Electronics, Lighting and Colour Terms Part 3: Terms Particular to Telecommunications and Electronics Group 08: Acoustics and Electroacoustics Terminology. 36 pp.
SAA AS 1852.801-88. International Electrotechnical Vocabulary—Part 801: Acoustics and Electro-Acoustics. 116 pp.

—**Hearing Aids**
BSI BS 6083: Part 0-84. (WITHDRAWN) 1984 AMD 1 Hearing Aids Part 0: Methods for Measurement of Electroacoustical Characteristics (AMD 6179) October 31, 1990 (Renumbered as BS EN 60118-0: 1993). 27 pp.

Electroacoustics (Cont.)

—**Hearing Aids (Cont.)**
BSI BS 6083: Part 3-91. 1991 Amd 1 Hearing Aids Part 3: Methods for Measurement of Electroacoustical Characteristics of Hearing Aid Equipment Not Entirely Worn on the Listener (AMD 6398) May 31 1991. 11 pp.
BSI BS EN 60118-0-93. 1993 Amd 2 Hearing Aids Part 0: Methods for Measurement of Electroacoustical Characteristics (IEC 118-0: 1983) (AMD 7836) August 15, 1993 (N). 32 pp.
CENELEC HD 450.0-84. Hearing Aids Part 0: Measurement of Electroacoustical Characteristics. 3 pp.
CENELEC EN 60118-0-93. Hearing Aids Part 0: Measurement of Electroacoustical Characteristics (IEC 118-0:1983) (Supersedes HD 450.0 S1:1984). 5 pp.
CENELEC EN 60118-0-93. Hearing Aids Part 0: Measurement of Electroacoustical Characteristics (IEC 118-0: 1983) (Supersedes HD 450.0 S1: 1984). 25 pp.
IEC 118 Pt 0-83. Hearing Aids Part 0: Measurement of Electroacoustical Characteristics Second Edition; (CENELEC EN 60118-0: 1993). 54 pp.
SAA AS 1088.0-87. Hearing Aids—Part 0: Measurement of Electroacoustical Characteristics. 21 pp.

—**Monitor Headphones—Sound Broadcasting**
CCIR RECMN 708-90. Determination of the Electro-Acoustical Properties of Studio Monitor Headphones—Section 10C—Audio-Frequency Characteristics of Sound-Broadcasting Signals. 4 pp.
CCIR QUESTION 83/10-90. Determination of the Electro-Acoustical Properties of Studio Monitor Headphones—Questions Concerning Study Group 10—Broadcasting Service (Sound). 1 p.

—**Video Telephones**
CENELEC PRI-ETS 300 302-1-93. Integrated Services Digital Network (ISDN); Videotelephony Teleservice Part 1: Electroacoustic Characteristics for Handset Terminals When Using PCM Encoding. 12 pp.
ETSI PRI-ETS 300 302-1-93. Integrated Services Digital Network (ISDN); Videotelephony Teleservice Part 1: Electroacoustic Characteristics for Handset Terminals When Using PCM Encoding. 12 pp.

Electroanalysis
Use: Electrolytic Analysis

Electrocardiograph Monitors
Use For: ECG Monitors; EKG Monitors
See Also: Electrocardiographs; Medical Electrical Equipment; Medical Equipment; Monitors
CSA Z303.2.1-M1983. Physiological Monitors—Electrocardiographic; (Gen Instr 1). 31 pp.
JIS T 1304-85. ECG Monitors. 31 pp.

—**Wireless**
JIS T 1307-85. Wireless ECG Monitors.

Electrocardiograph Recorders
Use For: ECG Recorders; EKG Recorders
See Also: Electrocardiographs; Medical Electrical Equipment; Medical Equipment
JIS T 1117-88. Ambulatory ECG Recording System (Holter System). 16 pp.

—**Modems—Data Transmission**
CCITT RECMN V.16-89. Medical Analogue Data Transmission Modems—Data Communication over the Telephone Network (Study Group XVII) 7 pp. 7 pp.

Electrocardiographs
See Also: Electrocardiograph Monitors; Electrocardiograph Recorders; Medical Electrical Equipment; Medical Equipment
CSA Z303.1-M1982. Diagnostic Electrocardiographs: Minimum Performance Requirements. 32 pp.
JIS T 1202-84. Electrocardiographs. 27 pp.
SAA AS T52-71. Direct-Writing Electrocardiographs (Including Recording Chart). 20 pp.

—**Safety**
BSI BS 5724: Sec 2.125-93. 1993 Medical Electrical Equipment Part 2: Particular Requirements for Safety Section 2.125: Specification for Electrocardiographs (IEC 601-2-25: 1993). 31 pp.
IEC 601 Pt 2-25-93. Medical Electrical Equipment Part 2: Particular Requirements for the Safety of Electrocardiographs First Edition. 63 pp.

Electrochemical Analysis
See Also: Electrochemistry; Electrogravimetric Analysis; Electrolytic Analysis

INTERNATIONAL AND NON-U.S. NATIONAL STANDARDS
SUBJECT INDEX

Electrochemical Analysis (Cont.)
—Olefins
ISO 8173-86. Light Olefins for Industrial Use—Determination of Molecular Oxygen in Gaseous Phase—Electrochemical Method with a Membrane-Covered Cell First Edition. 7 pp.

—Water
BSI BS 6068: Sec 2.15-86. (WITHDRAWN) 1986 Water Quality Part 2: Physical, Chemical and Bio-Chemical Methods Section 2.15: Determination of Dissolved Oxygen: Electrochemical Probe Method (Superseded by BS EN 25814: 1992). 8 pp.

BSI BS EN 25814-92. 1992 Method for Determination of Dissolved Oxygen in Water: Electrochemical Probe Method (ISO 5814) (1990) (Supersedes BS 6068: Section 2.15: 1986). 15 pp.

CEN EN 25814-92. Water Quality—Determination of Dissolved Oxygen—Electrochemical Probe Method. 10 pp.

DIN ENGL EN 25814-92. Water Quality; Determination of Dissolved Oxygen by the Electrochemical Probe Method (ISO 5814:1990) (Nov) (Supersedes DIN 38408 Part 22, November 1986 Edition). 12 pp.

IEC 746 Pt 4-92. Expression of Performance of Electrochemical Analyzers Part 4: Dissolved Oxygen in Water Measured by Membrane Covered Amperometric Sensors First Edition. 59 pp.

ISO 5814-90. Water Quality—Determination of Dissolved Oxygen—Electrochemical Probe Method Second Edition. 11 pp.

Electrochemical Analyzers
See Also: Analyzers; Electrochemical Probes; Laboratory Equipment

BSI BS 6438: Part 1-84. 1984 Electrochemical Analyzers Part 1: Method of Specifying Performance Common to All Analyzers. 22 pp.

IEC 746 Pt 1-82. Expression of Performance of Electrochemical Analyzers Part 1: General First Edition. 46 pp.

—Electrode Potentials
IEC 746 Pt 5-92. Expression of Performance of Electrochemical Analyzers Part 5: Oxidation-Reduction Potential or Redox Potential First Edition. 35 pp.

—Electrolytic Conductivity
BSI BS 6438: Part 3-86. 1986 Electrochemical Analyzers Part 3: Method for Specifying Performance of Electrolytic Conductivity Analyzers. 23 pp.

IEC 746 Pt 3-85. Expression of Performance of Electrochemical Analyzers Part 3: Electrolytic Conductivity First Edition. 38 pp.

—PH
BSI BS 6438: Part 2-84. 1984 Electrochemical Analyzers Part 2: Method for Specifying Performance of pH Value Analyzers. 22 pp.

IEC 746 Pt 2-82. Expression of Performance of Electrochemical Analyzers Part 2: pH Value First Edition. 44 pp.

—Waste Water—Dissolved Oxygen Content—Amperometry
IEC 746 Pt 4-92. Expression of Performance of Electrochemical Analyzers Part 4: Dissolved Oxygen in Water Measured by Membrane Covered Amperometric Sensors First Edition. 59 pp.

—Water—Dissolved Oxygen Content—Amperometry
IEC 746 Pt 4-92. Expression of Performance of Electrochemical Analyzers Part 4: Dissolved Oxygen in Water Measured by Membrane Covered Amperometric Sensors First Edition. 59 pp.

Electrochemical Corrosion
See Also: Corrosion; Electrolytic Corrosion

DIN ENGL 50927-85. Planning and Application of Electrochemical Corrosion Protection of Internal Surfaces of Apparatus, Containers and Tubes (Internal Protection) (Aug). 12 pp.

—Metals—Underground Structures
CSA C22.3 NO 4-1974. Control of Electrochemical Corrosion of Underground Metallic Structures (R 1980). 27 pp.

—Stainless Steels
CNS G2268-88. Method of Electrochemical Reactivation Ratio Measurement for Stainless Steels (Apr)(12258).

JIS G 0580-86. Method of Electrochemical Potentiokinetic Reactivation Ratio Measurement for Stainless Steels. 7 pp.

Electrochemical Corrosion Testing
Use: Electrochemical Corrosion

Electrochemical Polishing
Use: Electropolishing

Electrochemical Probes
See Also: Electrochemical Analyzers; Probes (Sensors)

—Surface Water—Fluoride Content
BSI BS 6068: Sec 2.41-93. 1993 Water Quality Part 2: Physical, Chemical and Biochemical Methods Section 2.41: Determination of Fluoride: Electrochemical Probe Method for Potable and Lightly Polluted Water (ISO 10359-1: 1992). 11 pp.

ISO 10359 Pt 1-92. Water Quality—Determination of Fluoride—Part 1: Electrochemical Probe Method for Potable and Lightly Polluted Water First Edition. 10 pp.

—Waste Water—Fluoride Content
BSI BS 6068: Sec 2.41-93. 1993 Water Quality Part 2: Physical, Chemical and Biochemical Methods Section 2.41: Determination of Fluoride: Electrochemical Probe Method for Potable and Lightly Polluted Water (ISO 10359-1: 1992). 11 pp.

ISO 10359 Pt 1-92. Water Quality—Determination of Fluoride—Part 1: Electrochemical Probe Method for Potable and Lightly Polluted Water First Edition. 10 pp.

—Water—Fluoride Content
BSI BS 6068: Sec 2.41-93. 1993 Water Quality Part 2: Physical, Chemical and Biochemical Methods Section 2.41: Determination of Fluoride: Electrochemical Probe Method for Potable and Lightly Polluted Water (ISO 10359-1: 1992). 11 pp.

ISO 10359 Pt 1-92. Water Quality—Determination of Fluoride—Part 1: Electrochemical Probe Method for Potable and Lightly Polluted Water First Edition. 10 pp.

Electrochemistry
See Also: Corrosion; Electrochemical Analysis

—Glossaries
CNS K0045-89. Technical Terms for Analytical Chemistry (Electro Chemistry Division) (Aug)(12587).

IEC 50 Chap 111-02-84. Advance Edition of the International Electrotechnical Vocabulary Chapter 111: Physics and Chemistry Section 111-02—Electrochemical Concepts. 31 pp.

JIS K 0213-91. Technical Terms for Analytical Chemistry (Electrochemistry Division).

SAA AS 1852.111. 02-88. International Electrotechnical Vocabulary—Part 111.02: Physics and Chemistry—Electrochemical Concepts. 10 pp.

SNZ IEC 50: 50(111-02)-84. International Electrotechnical Vocabulary 50(111-02): Physics and Chemistry Section 111-02: Electrochemical Concepts. 21 pp.

Electroconvulsive Therapy Equipment
See Also: Medical Electrical Equipment; Medical Equipment

—Safety
BSI BS 5724: Sec 2.14-89. 1989 Medical Electrical Equipment Part 2: Particular Requirements for Safety Section 2.14: Electroconvulsive Theraphy Equipment (N). 20 pp.

BSI BS 5724: Sec 2.14-01. 1989 Amd 1 Medical Electrical Equipment Part 2: Particular Requirements for Safety Section 2.14: Specification for Electroconvulsive Therapy Equipment (IEC 601-2-14: 1989) (AMD 6800) December 24, 1991. 21 pp.

CSA CAN/CSA-C22. 2601.2.14-92. Medical Electrical Equipment Part 2: Particular Requirements for the Safety of Electroconvulsive Therapy Equipment (IEC 601-2-14:1989); (Gen Instr 1). 39 pp.

IEC 601 Pt 2-14-89. Medical Electrical Equipment Part 2: Particular Requirements for the Safety of Electroconvulsive Therapy Equipment First Edition; (CAN/CSA-C22.2 No. 601.2.14-92). 39 pp.

SAA AS 3200.2.14-92. Approval and Test Specification—Medical Electrical Equipment—Part 2: Particular Requirements for Safety—Part 2.14: Electroconvulsive Therapy Equipment (IEC 601-2-14:1988) (in Prof. Pkgs. 17C,28) (This is a Joint Stand. with SNZ NZS 3200.2.14). 15 pp.

Electrode Fittings
—Receptacles—Gas Discharge Tubes
CSA C22.2 NO 34-M1987. Electrode Receptacles, Fittings, and Connectors for Gas Tubes (R 1993); (Gen Instr 1 Thru 2). 23 pp.

Electrode Holders
See Also: Holders; Welding Equipment

—Metal Arc Welding
CNS C4154-79. Welding Electrode Holder (Nov)(5021). 4 pp.

Electrode Holders (Cont.)
—Metal Arc Welding (Cont.)
DIN ENGL 8569 Pt 1-74. Rod Electrode Holders for Metal Arc Welding; Sizes; Requirements; Testing (Oct). 3 pp.

—Resistance Spot Welding
BSI BS EN 28430-1-92. 1992 Resistance Spot Welding—Electrode Holders Part 1: Taper Fixing 1:10 (ISO 8430-1: 1988). 10 pp.

BSI BS EN 28430-2-92. 1992 Resistance Spot Welding—Electrode Holders Part 2: Morse Taper Fixing (ISO 8430-2: 1988). 10 pp.

BSI BS EN 28430-3-92. 1992 Resistance Spot Welding—Electrode Holders Part 3: Parallel Shank Fixing for End Thrust (ISO 8430-3: 1988). 10 pp.

CEN EN 28430-1-92. Resistance Spot Welding—Electrode Holders—Part 1: Taper Fixing 1:10. 5 pp.

CEN EN 28430-2-92. Resistance Spot Welding—Electrode Holders—Part 2: Morse Taper Fixing. 5 pp.

CEN EN 28430-3-92. Resistance Spot Welding—Electrode Holders—Part 3: Parallel Shank Fixing for End Thrust. 5 pp.

DIN ENGL EN 28430 Pt 1-92. Resistance Spot Welding; Electrode Holders; Taper Fixing 1:10 (ISO 8430-1:1988) (July) (Supersedes DIN 44768 Part 1, January 1980 Edition). 7 pp.

DIN ENGL EN 28430 Pt 2-92. Resistance Spot Welding; Electrode Holders; Morse Taper Fixing (ISO 8430-2:1988) (July) (Supersedes DIN 44768 Part 2, January 1980 Edition). 7 pp.

DIN ENGL EN 28430 Pt 3-92. Resistance Spot Welding; Electrode Holders; Parallel Shank Fixing for End Thrust (ISO 8430-3:1988) (July) (Supersedes DIN 44768 Part 3, January 1980 Edition). 7 pp.

ISO 8430 Pt 1-88. Resistance Spot Welding—Electrode Holders—Part 1: Taper Fixing 1: 10 First Edition. 6 pp.

ISO 8430 Pt 2-88. Resistance Spot Welding—Electrode Holders—Part 2: Morse Taper Fixing First Edition. 6 pp.

ISO 8430 Pt 3-88. Resistance Spot Welding—Electrode Holders—Part 3: Parallel Shank Fixing for End Thrust First Edition; (Corrigendum 1-1990). 7 pp.

Electrode Potentials
Use For: Oxidation Reduction Potentials; Redox Potential

—Electrochemical Analyzers
IEC 746 Pt 5-92. Expression of Performance of Electrochemical Analyzers Part 5: Oxidation-Reduction Potential or Redox Potential First Edition. 35 pp.

—Soil Analysis
BSI BS 1377: Part 3-90. 1990 Methods of Test for Soils for Civil Engineering Purposes Part 3: Chemical and Electro-Chemical Tests. 45 pp.

—Waste Water
DIN ENGL 38404 Pt 6-84. German Standard Methods for the Examination of Water, Waste Water and Sludge; Physical and Physico-Chemical Parameters; (Group C); Determination of the Oxidation Reduction (Redox) Potential (C 6) (May). 5 pp.

—Water
DIN ENGL 38404 Pt 6-84. German Standard Methods for the Examination of Water, Waste Water and Sludge; Physical and Physico-Chemical Parameters; (Group C); Determination of the Oxidation Reduction (Redox) Potential (C 6) (May). 5 pp.

Electrodeposited Coatings
Scope Note: Includes electrodeposited processes and materials *Use For:* Electrodeposition; Electrolytic Deposition *See Also:* Coatings; Electroless Coatings; Electrophoresis; Electroplated Coatings; Metal Coatings (Made From Metal); Nonmetallic Coatings

—Cadmium
MOD UK DTD-904C-63. Cadmium Plating (Reprinted December 1966, Incorporating Amendment No. 1) (Superseded by Def Stan 03-19). 5 pp.

—Cadmium—Aircraft
AECMA PREN2133-81. Cadmium Plating of Steels with Maximum Specified Tensile Strength Less Than or Equal to 1450 MPa and Copper and Copper Alloys (C7/SC4/D). 14 pp.

JIS W 1114-84. Plating Cadmium (Electrodeposited for Aircraft).

—Chromium
BSI BS EN 10 202-90. 1990 Cold Reduced Electrolytic Chromium/Chromium Steel Coated Steel. 24 pp.

INDUSTRY STANDARDS

INTERNATIONAL AND NON-U.S. NATIONAL STANDARDS
SUBJECT INDEX

Electrodeposited

Electrodeposited Coatings (Cont.)
—Chromium (Cont.)
- CEN PREN 10 170-87. Single Cold Reduced Electrolytic Chromium/Chromium Oxide Coated Steel: Sheet. 27 pp.
- CEN EN 10 202-89. Cold Reduced Electrolytic Chromium/Chromium Oxide Coated Steel. 34 pp.
- CNS G2240-85. Method of Test for Chromium Plated Tin Free Steel (Apr)(11242).
- CNS G3226-85. Chromium Plated Tin Free Steel (Apr)(11241).
- DIN ENGL EN 10202-90. Cold Reduced Electrolytic Chromium/Chromium Oxide Coated Steel; English Version of DIN EN 10202 (Mar). 17 pp.
- ISO 8110 Pt 1-88. Single Cold-Reduced Electrolytic Chromium/Chromium Oxide-Coated Steel—Part 1: Sheets First Edition. 18 pp.
- ISO 8111 Pt 1-88. Double Cold-Reduced Electrolytic Chromium/Chromium Oxide-Coated Steel—Part 1: Sheets First Edition. 20 pp.
- JIS G 3315-87. Chromium Plated Tin Free Steel. 29 pp.
- MOD UK DSTAN 03-14-79. Electro-Deposition of Chromium for Engineering Purposes Issue 1. 22 pp.
- MOD UK DSTAN 03-31-93. Guide to the Use of Chromium Plating for Engineering Purposes Issue 1. 14 pp.
- MOD UK DG-13. Design Aspects of Chromium Plating for Engineering Purposes (Superseded by Def Stan 03-31).

—Chromium—Aerospace
- AECMA PREN2132-82. Electrodeposition of Chromium for Engineering Purposes. 12 pp.
- MOD UK DTD-943-74. Electrodeposited Cobalt/Chromium Carbide Composite Coatings. 2 pp.

—Chromium—Aircraft
- JIS W 1113-84. Chromium Plating (Electrodeposited for Aircraft).

—Chromium Oxide
- CEN PREN 10 170-87. Single Cold Reduced Electrolytic Chromium/Chromium Oxide Coated Steel: Sheet. 27 pp.
- ISO 8110 Pt 1-88. Single Cold-Reduced Electrolytic Chromium/Chromium Oxide-Coated Steel—Part 1: Sheets First Edition. 18 pp.
- ISO 8111 Pt 1-88. Double Cold-Reduced Electrolytic Chromium/Chromium Oxide-Coated Steel—Part 1: Sheets First Edition. 20 pp.

—Cobalt—Aerospace
- MOD UK DTD-943-74. Electrodeposited Cobalt/Chromium Carbide Composite Coatings. 2 pp.

—Epoxy Primers—Steel
- MOD UK TS 10245. Powder Coating, Priming for Phosphate Steel (Withdrawn).

—Glossaries
- ISO 2080-81. Electroplating and Related Processes—Vocabulary Second Edition. 59 pp.

—Gold
- ISO 4523-85. Metallic Coatings—Electrodeposited Gold and Gold Alloy Coatings for Engineering Purposes First Edition. 8 pp.
- MOD UK DSTAN 03-17-79. Electro-Deposition of Gold Issue 1. 21 pp.
- MOD UK DSTAN 03-17-92. Electro—Deposition of Gold Issue 2. 22 pp.

—Gold—Adhesion Testing
- BSI BS 6670: Part 5-86. 1986 Methods of Test for Electroplated Gold and Gold Alloy Coatings Part 5: Adhesion Tests. 4 pp.
- ISO 4524 Pt 5-85. Metallic Coatings—Test Methods for Electrodeposited Gold and Gold Alloy Coatings—Part 5: Adhesion Tests First Edition. 3 pp.

—Gold—Environmental Testing
- BSI BS 6670: Part 2-86. 1986 Methods of Test for Electroplated Gold and Gold Alloy Coatings Part 2: Environmental Tests. 7 pp.
- ISO 4524 Pt 2-85. Metallic Coatings—Test Methods for Electrodeposited Gold and Gold Alloy Coatings—Part 2: Environmental Tests First Edition. 5 pp.

—Gold—Gold Content
- BSI BS 6670: Part 4-86. 1986 Methods of Test for Electroplated Gold and Gold Alloy Coatings Part 4: Determination of Gold Content. 4 pp.
- ISO 4524 Pt 4-85. Metallic Coatings—Test Methods for Electrodeposited Gold and Gold Alloy Coatings—Part 4: Determination of Gold Content First Edition. 4 pp.

—Gold—Porosity
- BSI BS 6670: Part 3-86. 1986 Methods of Test for Electroplated Gold and Gold Alloy Coatings Part 3: Electrographic Tests for Porosity. 7 pp.

Electrodeposited Coatings (Cont.)
—Gold—Porosity (Cont.)
- ISO 4524 Pt 3-85. Metallic Coatings—Test Methods for Electrodeposited Gold and Gold Alloy Coatings—Part 3: Electrographic Tests for Porosity First Edition; (Corrected and Reprinted -1986). 5 pp.

—Gold—Residual Salt Content
- BSI BS 6670: Part 6-89. 1989 Methods of Test for Electroplated Gold and Gold Alloy Coatings Part 6: Detection of Residual Salts. 4 pp.
- ISO 4524 Pt 6-88. Metallic Coatings—Test Methods for Electrodeposited Gold and Gold Alloy Coatings—Part 6: Determination of the Presence of Residual Salts First Edition. 4 pp.

—Gold—Thickness Measurement
- BSI BS 6670: Part 1-86. 1986 Methods of Test for Electroplated Gold and Gold Alloy Coatings Part 1: Determination of Coating Thickness. 8 pp.
- ISO 4524 Pt 1-85. Metallic Coatings—Test Methods for Electrodeposited Gold and Gold Alloy Coatings—Part 1: Determination of Coating Thickness First Edition. 7 pp.

—Infrared Reflective
- MOD UK TS 10278. Powder Coating, IRR Matt, Stoving (Superseded by Def Stan 80-122).

—Metal—Adhesion Testing
- BSI BS 5411: Part 10-81. 1981 Methods of Test for Metallic and Related Coatings Part 10: Review of Methods Available for Testing Adhesion of Electro-Deposited and Chemically Deposited Metallic Coatings on Metallic Substrates. 10 pp.
- ISO 2819-80. Metallic Coatings on Metallic Substrates—Electrodeposited and Chemically Deposited Coatings—Review of Methods Available for Testing Adhesion Second Edition. 8 pp.

—Metal—Sampling
- BSI BS 6041-81. 1981 Amd 1 Method of Sampling of Electrodeposited Metallic Coatings and Related Finishes: Procedures for Inspections by Attributes. 15 pp.
- ISO 4519-80. Electrodeposited Metallic Coatings and Related Finishes—Sampling Procedures for Inspection by Attributes First Edition; (Erratum—Dec 1981). 13 pp.

—Nickel
- BSI BS EN 248-92. 1992 Sanitary Taps: General Technical Specifications for Electrodeposited Nickel Chrome Coatings. 10 pp.
- CEN EN 248-89. Sanitary Taps: General Technical Specifications for Electrodeposited Nickel Chrome Coatings. 5 pp.
- ISO 1456-88. Metallic Coatings—Electrodeposited Coatings of Nickel Plus Chromium and of Copper Plus Nickel Plus Chromium Second Edition; (Supersedes 1457). 19 pp.
- ISO 1458-88. Metallic Coatings—Electrodeposited Coatings of Nickel Second Edition. 16 pp.
- MOD UK DSTAN 03-27-87. Electrodeposition of Nickel for Engineering Purposes Issue 1. 21 pp.

—Nickel—Thickness
- BSI BS 5411: Part 9-84. 1984 Methods of Test for Metallic and Related Coatings Part 9: Measurement of Coating Thickness of Electrodeposited Nickel Coatings on Magnetic and Non-Magnetic Substrates Magnetic Method. 8 pp.
- ISO 2361-82. Electrodeposited Nickel Coatings on Magnetic and Non-Magnetic Substrates—Measurement of Coating Thickness—Magnetic Method Second Edition. 6 pp.

—Polyester Primers—Steel
- MOD UK TS 10245. Powder Coating, Priming for Phosphate Steel (Withdrawn).

—Silver
- ISO 4521-85. Metallic Coatings—Electrodeposited Silver and Silver Alloy Coatings for Engineering Purposes First Edition. 8 pp.
- MOD UK DSTAN 03-9-75. Electro-Deposition of Silver Issue 1. 11 pp.
- MOD UK DSTAN 03-9-92. Electro—Deposition of Silver Issue 2. 13 pp.

—Silver—Adhesion Testing
- BSI BS 6669: Part 2-86. 1986 Methods of Test for Electroplated Silver and Silver Alloy Coatings Part 2: Adhesion Tests. 7 pp.
- ISO 4522 Pt 2-85. Metallic Coatings—Test Methods for Electrodeposited Silver and Silver Alloy Coatings—Part 2: Adhesion Tests First Edition. 5 pp.

—Silver—Residual Salt Content
- BSI BS 6669: Part 3-89. 1989 Methods of Test for Electroplated Silver and Silver Alloy Coatings Part 3: Detection of Residual Salts. 4 pp.

Electrodeposited Coatings (Cont.)
—Silver—Residual Salt Content (Cont.)
- ISO 4522 Pt 3-88. Metallic Coatings—Test Methods for Electrodeposited Silver and Silver Alloy Coatings—Part 3: Determination of the Presence of Residual Salts First Edition. 4 pp.

—Silver—Thickness Measurement
- BSI BS 6669: Part 1-86. 1986 Methods of Test for Electroplated Silver and Silver Alloy Coatings Part 1: Determination of Coating Thickness. 6 pp.
- ISO 4522 Pt 1-85. Metallic Coatings—Test Methods for Electrodeposited Silver and Silver Alloy Coatings—Part 1: Determination of Coating Thickness First Edition. 4 pp.

—Tin
- BSI BS EN 10203-91. 1991 Cold Reduced Electrolytic Tinplate (Supersedes BS 2920: 1973) (V). 24 pp.
- CEN PREN10203-88. Cold Reduced Rlectrolytic Tinplate. 35 pp.
- CEN EN 10203-91. Cold Reduced Electrolytic Tinplate. 36 pp.
- DIN ENGL EN 10203-91. Cold Reduced Electrolytic Tinplate (Aug) (Supersedes Parts of DIN 1616, October 1984 Edition). 18 pp.
- ISO 1111 Pt 1-83. Single Cold-Reduced Tinplate and Single Cold-Reduced Blackplate—Part 1: Electrolytic and Hot-Dipped Tinplate Sheet and Blackplate Sheet First Edition. 16 pp.
- ISO 1111 Pt 2-83. Single Cold-Reduced Tinplate and Single Cold-Reduced Blackplate—Part 2: Electrolytic Tinplate Coil and Blackplate Coil for Subsequent Cutting into Sheet Form Second Edition. 17 pp.
- ISO 4977 Pt 1-84. Double Cold-Reduced Electrolytic Tinplate—Part 1: Sheet First Edition. 16 pp.
- ISO 4977 Pt 2-88. Double Cold-Reduced Electrolytic Tinplate—Part 2: Coil for Subsequent Cutting into Sheets First Edition. 19 pp.
- ISO 5950-91. Continuous Electrolytic Tin-Coated Cold-Reduced Carbon Steel Sheet of Commercial and Drawing Qualities Second Edition; DoD Adopted. 14 pp.
- MOD UK DSTAN 03-15-79. Electro-Deposition of Tin-Lead Alloy for Soldering Purposes Issue 1. 16 pp.
- MOD UK DSTAN 03-15-92. Electro—Deposition of Tin—Lead Alloy for Soldering Purposes Issue 2. 16 pp.
- MOD UK DTD-927A-56. Tin-Zinc Alloy Plating (Reprinted October 1962). 5 pp.

—Zinc
- BSI BS 6687-86. 1986 Amd 1 Electrolytically Zinc Coated Steel Flat Rolled Products. 11 pp.
- CNS G2191-83. Method of Test for Electrolytic Zinc-Coated Steel Sheets and Strips (Sep)(10569).
- CNS G3211-83. Electrolytic Zinc-Coated Steel Sheets and Strips (Sep)(10568).
- DIN ENGL 17163-88. Steel Flat Products; Electrolytically Zinc Coated Cold Rolled Strip and Sheet; Technical Delivery Conditions (Mar). 6 pp.
- ISO 5002-82. Hot-Rolled and Cold-Reduced Electrolytic Zinc-Coated Carbon Steel Sheet of Commercial and Drawing Qualities First Edition; DoD Adopted. 14 pp.
- JIS G 3313-90. Electrolytic Zinc-Coated Steel Sheets and Coils. 77 pp.

Electrodeposition
Use: Electrodeposited Coatings

Electrodes
Scope Note: For additional listings, use a more specific term *See Also:* Anodes; Bare Electrodes; Carbon Electrodes; Coated Electrodes; Deflecting Electrodes; Electrical Components; Electrochemistry; Electrode Fittings; Electrode Holders; Electrolysis; Ground Electrodes; Ion Selective Electrodes; pH Electrodes; Reference Electrodes; Sphere Gaps; Welding Electrodes; Welding Rods

—Coke
- BSI BS 6043: Sec 2.1-85. 1985 Methods of Sampling and Test for Carbonaceous Materials Used in Aluminium Manufacture Part 2: Electrode Coke Section 2.1: Sampling. 8 pp.
- ISO 6375-80. Carbonaceous Materials for the Production of Aluminium—Cokes for Electrodes—Sampling First Edition. 7 pp.

—Fillet Welds
- JIS Z 3181-73. Method of Test for Fillet Weld of Covered Electrode. 4 pp.

—Numbering
- CENELEC HD 145-74. Numbering of Electrodes and Designation of Units in Electronic Tubes and Valves. 2 pp.
- CNS C5093-81. Numbering of Electrodes in Multiple Electrode Semiconductor Devices and Designation of Units in Multiple Unit Semiconductor Devices (Mar)(7011).

INTERNATIONAL AND NON-U.S. NATIONAL STANDARDS
SUBJECT INDEX

Electrodes (Cont.)
—Numbering (Cont.)
IEC 135-61. Numbering of Electrodes and Designation of Units in Electronic Tubes and Valves First Edition. 9 pp.

—Pitch
BSI BS 6043: Sec 1.4-81. 1981 Methods of Sampling and Test for Carbonaceous Materials Used in Aluminium Manufacture Part 1: Electrode Pitch Section 1.4: Determination of Content of Toluene-Insoluble Material. 4 pp.

BSI BS 6043: Sec 1.5-83. 1983 Methods of Sampling and Test for Carbonaceous Materials Used in Aluminium Manufacture Part 1: Electrode Pitch Section 1.5: Determination of Content of Quinoline-Insoluble Material. 7 pp.

ISO 6376-80. Carbonaceous Materials for the Production of Aluminium—Pitch for Electrodes—Determination of Content of Toluene-Insoluble Material First Edition. 4 pp.

ISO 6791-81. Carbonaceous Materials for the Production of Aluminium—Pitch for Electrodes—Determination of Content of Quinoline-Insoluble Material First Edition. 5 pp.

—Pitch—Sulfur Content Analysis
ISO 9055-88. Carbonaceous Materials for the Production of Aluminium—Pitch for Electrodes—Determination of Sulfur Content by the Bomb Method First Edition. 7 pp.

Electrodischarge Machines
Use: Electric Discharge Machine Tools

Electroencephalographs
Use For: EEG *See Also:* Medical Electrical Equipment; Medical Equipment
JIS T 1203-85. Electroencephalographs.

—Photic Stimulators
JIS T 1213-84. Photic Stimulators for Electroencephalographs.

—Safety
CENELEC PREN 60601-2-27-92. Medical Electrical Equipment Part 2: Particular Requirements for Safety Section 2.126: Specification for Electroencephalographs. 28 pp.

Electroerosive Machine
Use: Electric Discharge Machine Tools

Electroexplosive Devices
Use: Initiators (Explosives)

Electroflux Distribution Signals
Use: Signaling Connection Control Part

Electrogas Welding
Use: Gas Metal Arc Welding

Electrogravimetric Analysis
See Also: Electrochemical Analysis

—Nickel Castings
JIS H 1272-88. Methods for Determination of Copper in Nickel and Nickel Alloy Castings. 8 pp.

—Ores
CNS M3074-81. Method for Determination of Copper in Ores (Electrogravimetric Method) (Oct)(8042).

Electroheating Installations
Use: Heating Plants

Electrohydraulic Proportional Flow Control Valves
Use: Proportional Valves—Electrohydraulic

Electroless Coatings
Use For: Autocatalytic Coatings; Electroless Plating
See Also: Coatings; Electrodeposited Coatings; Electroplated Coatings; Metal Coatings (Made From Metal); Surface Finishing

—Aerospace—Codes
DIN ENGL LN 9368 Pt 6-83. Aerospace; Designation of Surface Treatments; Code Numbers for Methods of Electroless Produced Metallic Coatings (Apr). 2 pp.

—Metal
BSI BS 4479: Part 2-90. 1990 Design of Articles That Are to Be Coated Part 2: Recommendations for Electroplated and Autocatalytic Coatings. 15 pp.

—Nickel
CNS H3147-87. Autocatalytic Nickel Deposition on Metals for Engineering Use (Sep)(12090).
JIS H 8645-89. Autocatalytic Nickel—Phosphorus Coatings on Metals. 19 pp.

Electroless Plating
Use: Electroless Coatings

Electroluminescent Displays
See Also: Display Devices

—Preferred Products List
CECC CECC MUAHAG Vol 10 IS 2-92. Preferred Products List; OPTO Electronic Devices (En, Fr, Ge). 48 pp.

Electrolysers
Use: Electrolyzers

Electrolysis
See Also: Corrosion

—Elastomeric Sheets
DIN ENGL 53122 Pt 2-82. Determination of Water Vapour Transmission (Density of Moisture Flow Rate) of Plastic Films, Elastomer Films, Paper, Board and Other Sheet Materials; Electrolysis Method (July). 4 pp.

—Paperboard
DIN ENGL 53122 Pt 2-82. Determination of Water Vapour Transmission (Density of Moisture Flow Rate) of Plastic Films, Elastomer Films, Paper, Board and Other Sheet Materials; Electrolysis Method (July). 4 pp.

—Papers
DIN ENGL 53122 Pt 2-82. Determination of Water Vapour Transmission (Density of Moisture Flow Rate) of Plastic Films, Elastomer Films, Paper, Board and Other Sheet Materials; Electrolysis Method (July). 4 pp.

—Plastic Sheets
DIN ENGL 53122 Pt 2-82. Determination of Water Vapour Transmission (Density of Moisture Flow Rate) of Plastic Films, Elastomer Films, Paper, Board and Other Sheet Materials; Electrolysis Method (July). 4 pp.

Electrolytes
Scope Note: Use a more specific term *See:* Battery Electrolytes; Electrically Conductive Liquids

Electrolytic Analysis
Use For: Electroanalysis *See Also:* Chemical Analysis; Electrochemical Analysis; Electrometric Analysis; Quantitative Analysis

—Aluminum
ISO 796-73. Aluminium Alloys—Determination of Copper—Electrolytic Method First Edition. 7 pp.

—Aluminum Alloys
ISO 796-73. Aluminium Alloys—Determination of Copper—Electrolytic Method First Edition. 7 pp.

—Chlorine
ISO 2202-72. Liquid Chlorine for Industrial Use—Determination of Water Content Using an Electrolytic Analyser First Edition; (Addendum 1-1975). 12 pp.

—Copper
ISO 1553-76. Unalloyed Copper Containing Not Less Than 99,90 % of Copper—Determination of Copper Content—Electrolytic Method First Edition. 6 pp.
SAA AS K208.1-70. Methods for the Analysis of Unalloyed Copper—Part 1: Method for the Electrolytic Determination of Copper in Unalloyed Copper Containing Not Less Than 99.85 Percent Copper Amdt 1 April 1971. 7 pp.

—Copper Alloys
BSI BS 1748: Parts 1-5-61. 1961 Methods for the Analysis of Copper Alloys Parts 1-5: Determination of Copper, Lead, Iron, Aluminium and Nickel in Copper Alloys. 23 pp.
BSI BS 1748: Parts 11-12-64. 1964 Methods for the Analysis of Copper Alloys Parts 11-12: The Determination of Copper and Lead in Leaded Bronze Alloys. 12 pp.
BSI BS 1825-52. 1952 Determination of Cadmium in Copper-Cadmium Alloys (Electrolytic Method). 8 pp.
ISO 1554-76. Wrought and Cast Copper Alloys—Determination of Copper Content—Electrolytic Method First Edition. 4 pp.
SAA AS 1515.4-78. Copper Alloys—Part 4: Method for the Electrolytic Determination of Copper in Wrought and Cast Copper Alloys Reconfirmed 1989. 10 pp.

—Iron Ores
BSI BS 7020: Part 19-88. 1988 Analysis of Iron Ores Part 19: Method for the Determination of Fluorine Content: Ion-Selective Electrode Method. 12 pp.

Electrolytic Analysis (Cont.)
—Iron Ores (Cont.)
ISO 4694-87. Iron Ores—Determination of Fluorine Content—Ion-Selective Electrode Method First Edition. 11 pp.

—Water
CNS R2180-85. Asbestos Cloth Diaphragms for Electrolysis of Water (May)(11269).
JIS R 3456-79. Asbestos Cloth Diaphragms for Electrolysis of Water. 8 pp.

—White Metals
BSI BS 3338: Part 15-65. 1965 Methods for the Sampling and Analysis of Tin and Tin Alloys Part 15: Determination of Copper and Lead in White Metal Bearing Alloys (Electrodeposition Method). 8 pp.

—Zinc Alloys
BSI BS 3630: Part 13-72. 1972 Methods for the Sampling and Analysis of Zinc and Zinc Alloys Part 13: Copper in Zinc Alloys (Alloy B) (Electrolytic Method). 8 pp.
ISO 1976-75. Zinc Alloys—Determination of Copper Content—Electrolytic Method First Edition. 4 pp.

Electrolytic Capacitors
Scope Note: For additional listings, use a more specific term *See Also:* Aluminum Capacitors; Aluminum Chip Capacitors; Capacitors; Fixed Capacitors; Tantalum Capacitors; Tantalum Chip Capacitors
CNS C5009-84. General Rules for Electrolytic Capacitors (Oct)(3695).
JIS C 5140-89. General Rules of Fixed Electrolytic Capacitors for Use in Electronic Equipment. 46 pp.

—Motors
CNS C6193-82. Method of Test for Electrolytic Capacitors for Motor Starting (Nov)(9584).
CNS C7029-85. Electrolytic Capacitors for Motor Starting (Jan)(4328).
JIS C 4905-91. Electrolytic Capacitors for Motor Starting. 12 pp.

—Reliability Assured
CNS C5033-86. General Rules for Reliability Assured Electrolytic Capacitors (Jan)(4906).

Electrolytic Conductivity
—Electrochemical Analyzers
BSI BS 6438: Part 3-86. 1986 Electrochemical Analyzers Part 3: Method for Specifying Performance of Electrolytic Conductivity Analyzers. 23 pp.
IEC 746 Pt 3-85. Expression of Performance of Electrochemical Analyzers Part 3: Electrolytic Conductivity First Edition. 38 pp.

Electrolytic Copper
Use For: Cathode Copper *See Also:* Copper
CNS H3001-83. Electrolytic Cathode Copper (Jun)(6).
DIN ENGL 1708-73. Copper; Cathodes and Refinery Shapes (Jan). 4 pp.
JIS H 2121-61. Electrolytic Cathode Copper. 3 pp.

—Antimony Content—Atomic Absorption Spectrophotometry
BSI BS 7317: Part 3-90. 1990 Methods for Analysis of High Purity Copper Cathode Cu-CATH-1 Part 3: Method for Determination of Antimony, Arsenic, Bismuth, Selenium, Tellurium and Tin by Hydride Generation and Atomic Absorption Spectrophotometry. 12 pp.
BSI BS 7317: Part 4-90. 1990 Methods for Analysis of High Purity Copper Cathode Cu-CATH-1 Part 4: Method for Determination of Antimony, Arsenic, Bismuth, Lead, Selenium, Tellurium and Tin by Electrothermal Atomiza-tion Atomic Absorption Spectrophotometry. 9 pp.
BSI DD 95: Part 3-84. (WITHDRAWN) 1984 Analysis of Higher Purity Copper Cathode Cu-CATH-1 Part 3: Methods for Determination of Antimony, Arsenic, Bismuth, Selenium, Tellurium and Tin by Hydride Generation and Atomic Absorption Spectrophotometry. 14 pp.
BSI DD 95: Part 4-86. (WITHDRAWN) 1986 Analysis of Higher Purity Copper Cathode CU-CATH-1 Part 4: Method for Determination of Antimony, Arsenic, Bismuth, Lead, Selenium, Tellurium and Tin by Electrothermal Atomization Atomic. 9 pp.

—Antimony Content—Hydride Generation
BSI BS 7317: Part 3-90. 1990 Methods for Analysis of High Purity Copper Cathode Cu-CATH-1 Part 3: Method for Determination of Antimony, Arsenic, Bismuth, Selenium, Tellurium and Tin by Hydride Generation and Atomic Absorption Spectrophotometry. 12 pp.

INDUSTRY STANDARDS

Electrolytic Copper *(Cont.)*

—Antimony Content—Hydride Generation *(Cont.)*

BSI DD 95: Part 3-84. (WITHDRAWN) 1984 Analysis of Higher Purity Copper Cathode Cu-CATH-1 Part 3: Methods for Determination of Antimony, Arsenic, Bismuth, Selenium, Tellurium and Tin by Hydride Generation and Atomic Absorption Spectrophotometry. 14 pp.

—Arsenic Content—Atomic Absorption Spectrophotometry

BSI BS 7317: Part 3-90. 1990 Methods for Analysis of High Purity Copper Cathode Cu-CATH-1 Part 3: Method for Determination of Antimony, Arsenic, Bismuth, Selenium, Tellurium and Tin by Hydride Generation and Atomic Absorption Spectrophotometry. 12 pp.

BSI BS 7317: Part 4-90. 1990 Methods for Analysis of High Purity Copper Cathode Cu-CATH-1 Part 4: Method for Determination of Antimony, Arsenic, Bismuth, Lead, Selenium, Tellurium and Tin by Electrothermal Atomiza-tion Atomic Absorption Spectrophotometry. 9 pp.

BSI DD 95: Part 3-84. (WITHDRAWN) 1984 Analysis of Higher Purity Copper Cathode Cu-CATH-1 Part 3: Methods for Determination of Antimony, Arsenic, Bismuth, Selenium, Tellurium and Tin by Hydride Generation and Atomic Absorption Spectrophotometry. 14 pp.

BSI DD 95: Part 4-86. (WITHDRAWN) 1986 Analysis of Higher Purity Copper Cathode CU-CATH-1 Part 4: Method for Determination of Antimony, Arsenic, Bismuth, Lead, Selenium, Tellurium and Tin by Electrothermal Atomization Atomic. 9 pp.

—Arsenic Content—Hydride Generation

BSI BS 7317: Part 3-90. 1990 Methods for Analysis of High Purity Copper Cathode Cu-CATH-1 Part 3: Method for Determination of Antimony, Arsenic, Bismuth, Selenium, Tellurium and Tin by Hydride Generation and Atomic Absorption Spectrophotmetry. 12 pp.

BSI DD 95: Part 3-84. (WITHDRAWN) 1984 Analysis of Higher Purity Copper Cathode Cu-CATH-1 Part 3: Methods for Determination of Antimony, Arsenic, Bismuth, Selenium, Tellurium and Tin by Hydride Generation and Atomic Absorption Spectrophotometry. 14 pp.

—Billets—Oxygen Free

CNS H3119-83. Oxygen-Free Electrolytic Copper Billets and Cakes (Jun)(10346).

—Bismuth Content—Atomic Absorption Spectrophotometry

BSI BS 7317: Part 3-90. 1990 Methods for Analysis of High Purity Copper Cathode Cu-CATH-1 Part 3: Method for Determination of Antimony, Arsenic, Bismuth, Selenium, Tellurium and Tin by Hydride Generation and Atomic Absorption Spectrophotometry. 12 pp.

BSI BS 7317: Part 4-90. 1990 Methods for Analysis of High Purity Copper Cathode Cu-CATH-1 Part 4: Method for Determination of Antimony, Arsenic, Bismuth, Lead, Selenium, Tellurium and Tin by Electrothermal Atomiza-tion Atomic Absorption Spectrophotometry. 9 pp.

BSI DD 95: Part 3-84. (WITHDRAWN) 1984 Analysis of Higher Purity Copper Cathode Cu-CATH-1 Part 3: Methods for Determination of Antimony, Arsenic, Bismuth, Selenium, Tellurium and Tin by Hydride Generation and Atomic Absorption Spectrophotometry. 14 pp.

BSI DD 95: Part 4-86. (WITHDRAWN) 1986 Analysis of Higher Purity Copper Cathode CU-CATH-1 Part 4: Method for Determination of Antimony, Arsenic, Bismuth, Lead, Selenium, Tellurium and Tin by Electrothermal Atomization Atomic. 9 pp.

—Bismuth Content—Hydride Generation

BSI BS 7317: Part 3-90. 1990 Methods for Analysis of High Purity Copper Cathode Cu-CATH-1 Part 3: Method for Determination of Antimony, Arsenic, Bismuth, Selenium, Tellurium and Tin by Hydride Generation and Atomic Absorption Spectrophotometry. 12 pp.

BSI DD 95: Part 3-84. (WITHDRAWN) 1984 Analysis of Higher Purity Copper Cathode Cu-CATH-1 Part 3: Methods for Determination of Antimony, Arsenic, Bismuth, Selenium, Tellurium and Tin by Hydride Generation and Atomic Absorption Spectrophotometry. 14 pp.

Electrolytic Copper *(Cont.)*

—Cadmium Content—Atomic Absorption Spectrophotometry

BSI BS 7317: Part 1-90. 1990 Methods for Analysis of High Purity Copper Cathode Cu-CATH-1 Part 1: Method for Determination of Cadmium Manganese and Silver (Screening Procedure for Chromium, Cobalt, Iron, Nickel and Zinc) by Atomic Absorption. 9 pp.

BSI DD 95: Part 1-84. (WITHDRAWN) 1984 Amd 1 Analysis of Higher Purity Copper Cathode Cu-CATH-1 Part 1: Method for Determination of Cadmium, Manganese and Silver (Screening Procedures for Chromium, Cobalt, Iron, Nickel and Zinc) by Atomic Absorption. 10 pp.

—Cakes—Oxygen Free

CNS H3119-83. Oxygen-Free Electrolytic Copper Billets and Cakes (Jun)(10346).

—Chemical Analysis

CNS H2010-88. Method of Determining Chemical Components of Electrolytic Cathode Copper (Dec)(273).

JIS H 1101-90. Methods for Chemical Analysis of Electrolytic Cathode Copper. 64 pp.

—Chromium Content—Atomic Absorption Spectrophotometry

BSI BS 7317: Part 1-90. 1990 Methods for Analysis of High Purity Copper Cathode Cu-CATH-1 Part 1: Method for Determination of Cadmium Manganese and Silver (Screening Procedure for Chromium, Cobalt, Iron, Nickel and Zinc) by Atomic Absorption. 9 pp.

BSI BS 7317: Part 2-90. 1990 Methods for Analysis of High Purity Copper Cathode Cu-CATH-1 Part 2: Method for Determination of Chromium, Cobalt, Iron, Nickel and Zinc by Discrete Volume Nebulization Atomic Absorption Spectrophotometry. 9 pp.

BSI DD 95: Part 2-84. (WITHDRAWN) 1984 Amd 1 Analysis of Higher Purity Copper Cathode Cu-CATH-1 Part 2: Method for Determination of Chromium, Cobalt, Iron, Nickel, and Zinc by Discrete Volume Nebulization Atomic Absorption Spectrophotometry. 10 pp.

—Cobalt Content—Atomic Absorption Spectrophotometry

BSI BS 7317: Part 1-90. 1990 Methods for Analysis of High Purity Copper Cathode Cu-CATH-1 Part 1: Method for Determination of Cadmium Manganese and Silver (Screening Procedure for Chromium, Cobalt, Iron, Nickel and Zinc) by Atomic Absorption. 9 pp.

BSI BS 7317: Part 2-90. 1990 Methods for Analysis of High Purity Copper Cathode Cu-CATH-1 Part 2: Method for Determination of Chromium, Cobalt, Iron, Nickel and Zinc by Discrete Volume Nebulization Atomic Absorption Spectrophotometry. 9 pp.

BSI DD 95: Part 2-84. (WITHDRAWN) 1984 Amd 1 Analysis of Higher Purity Copper Cathode Cu-CATH-1 Part 2: Method for Determination of Chromium, Cobalt, Iron, Nickel, and Zinc by Discrete Volume Nebulization Atomic Absorption Spectrophotmetry. 10 pp.

—Iron Content—Atomic Absorption Spectrophotometry

BSI BS 7317: Part 1-90. 1990 Methods for Analysis of High Purity Copper Cathode Cu-CATH-1 Part 1: Method for Determination of Cadmium Manganese and Silver (Screening Procedure for Chromium, Cobalt, Iron, Nickel and Zinc) by Atomic Absorption. 9 pp.

BSI BS 7317: Part 2-90. 1990 Methods for Analysis of High Purity Copper Cathode Cu-CATH-1 Part 2: Method for Determination of Chromium, Cobalt, Iron, Nickel and Zinc by Discrete Volume Nebulization Atomic Absorption Spectrophotometry. 9 pp.

BSI DD 95: Part 2-84. (WITHDRAWN) 1984 Amd 1 Analysis of Higher Purity Copper Cathode Cu-CATH-1 Part 2: Method for Determination of Chromium, Cobalt, Iron, Nickel, and Zinc by Discrete Volume Nebulization Atomic Absorption Spectrophotmetry. 10 pp.

—Lead Content—Atomic Absorption Spectrophotometry

BSI BS 7317: Part 7-90. 1990 Methods for Analysis of High Purity Copper Cathode Cu-CATH-1 Part 7: Method for Determination of Lead by Lanthanum Hydroxide Separation and Atomic Absorption Spectrophotometry. 9 pp.

BSI DD 95: Part 4-86. (WITHDRAWN) 1986 Analysis of Higher Purity Copper Cathode CU-CATH-1 Part 4: Method for Determination of Antimony, Arsenic, Bismuth, Lead, Selenium, Tellurium and Tin by Electrothermal Atomization Atomic. 9 pp.

Electrolytic Copper *(Cont.)*

—Lead Content—Atomic Absorption Spectrophotometry *(Cont.)*

BSI DD 95: Part 7-86. (WITHDRAWN) 1986 Analysis of Higher Purity Copper Cathode Cu-CATH-1 Part 7: Method for Determination of Lead by Lanthanum Hydroxide Separation and Atomic Absorption Spectrophotometry (Superseded by BS 7317: Part 7: 1990). 8 pp.

—Lead Content—Lanthanum Hydroxide Separation

BSI BS 7317: Part 7-90. 1990 Methods for Analysis of High Purity Copper Cathode Cu-CATH-1 Part 7: Method for Determination of Lead by Lanthanum Hydroxide Separation and Atomic Absorption Spectrophotometry. 9 pp.

BSI 7348-90. 1990 Electrical Apparatus for the Detection of Combustible Gases in Domestic Premises (Q). 11 pp.

BSI 7348-01. 1990 Amd 1 Electrical Apparatus for the Detection of Combustible Gases in Domestic Premises (AMD 6869) October 31, 1991. 12 pp.

BSI DD 95: Part 7-86. (WITHDRAWN) 1986 Analysis of Higher Purity Copper Cathode Cu-CATH-1 Part 7: Method for Determination of Lead by Lanthanum Hydroxide Separation and Atomic Absorption Spectrophotometry (Superseded by BS 7317: Part 7: 1990). 8 pp.

—Manganese Content—Atomic Absorption Spectrophotometry

BSI BS 7317: Part 1-90. 1990 Methods for Analysis of High Purity Copper Cathode Cu-CATH-1 Part 1: Method for Determination of Cadmium Manganese and Silver (Screening Procedure for Chromium, Cobalt, Iron, Nickel and Zinc) by Atomic Absorption. 9 pp.

BSI DD 95: Part 1-84. (WITHDRAWN) 1984 Amd 1 Analysis of Higher Purity Copper Cathode Cu-CATH-1 Part 1: Method for Determination of Cadmium, Manganese and Silver (Screening Procedures for Chromium, Cobalt, Iron, Nickel and Zinc) by Atomic Absorption. 10 pp.

—Nickel Content—Atomic Absorption Spectrophotometry

BSI BS 7317: Part 1-90. 1990 Methods for Analysis of High Purity Copper Cathode Cu-CATH-1 Part 1: Method for Determination of Cadmium Manganese and Silver (Screening Procedure for Chromium, Cobalt, Iron, Nickel and Zinc) by Atomic Absorption. 9 pp.

BSI BS 7317: Part 2-90. 1990 Methods for Analysis of High Purity Copper Cathode Cu-CATH-1 Part 2: Method for Determination of Chromium, Cobalt, Iron, Nickel and Zinc by Discrete Volume Nebulization Atomic Absorption Spectrophotometry. 9 pp.

BSI DD 95: Part 2-84. (WITHDRAWN) 1984 Amd 1 Analysis of Higher Purity Copper Cathode Cu-CATH-1 Part 2: Method for Determination of Chromium, Cobalt, Iron, Nickel, and Zinc by Discrete Volume Nebulization Atomic Absorption Spectrophotmetry. 10 pp.

—Phosphorus Content—Spectrophotometry

BSI BS 7317: Part 6-90. 1990 Methods for Analysis of High Purity Copper Cathode Cu-CATH-1 Part 6: Method for Determination of Phosphorus and Silicon by Spectrophotometry. 9 pp.

BSI DD 95: Part 6-87. (WITHDRAWN) 1987 Analysis of Higher Purity Copper Cathode Cu-CATH-1 Part 6: Methods for Determination of Phosphorous and Silicon by Spectrophotometry (Superseded by BS 7317: Part 6: 1990). 9 pp.

—Selenium Content—Atomic Absorption Spectrophotometry

BSI BS 7317: Part 3-90. 1990 Methods for Analysis of High Purity Copper Cathode Cu-CATH-1 Part 3: Method for Determination of Antimony, Arsenic, Bismuth, Selenium, Tellurium and Tin by Hydride Generation and Atomic Absorption Spectrophotometry. 12 pp.

BSI BS 7317: Part 4-90. 1990 Methods for Analysis of High Purity Copper Cathode Cu-CATH-1 Part 4: Method for Determination of Antimony, Arsenic, Bismuth, Lead, Selenium, Tellurium and Tin by Electrothermal Atomiza-tion Atomic Absorption Spectrophotometry. 9 pp.

BSI DD 95: Part 3-84. (WITHDRAWN) 1984 Analysis of Higher Purity Copper Cathode Cu-CATH-1 Part 3: Methods for Determination of Antimony, Arsenic, Bismuth, Selenium, Tellurium and Tin by Hydride Generation and Atomic Absorption Spectrophotometry. 14 pp.

BSI DD 95: Part 4-86. (WITHDRAWN) 1986 Analysis of Higher Purity Copper Cathode CU-CATH-1 Part 4: Method for Determination of Antimony, Arsenic, Bismuth, Lead, Selenium, Tellurium and Tin by Electrothermal Atomization Atomic. 9 pp.

Electrolytic Copper *(Cont.)*

—Selenium Content—Hydride Generation
BSI BS 7317: Part 3-90. 1990 Methods for Analysis of High Purity Copper Cathode Cu-CATH-1 Part 3: Method for Determination of Antimony, Arsenic, Bismuth, Selenium, Tellurium and Tin by Hydride Generation and Atomic Absorption Spectrophotometry. 12 pp.
BSI DD 95: Part 3-84. (WITHDRAWN) 1984 Analysis of Higher Purity Copper Cathode Cu-CATH-1 Part 3: Methods for Determination of Antimony, Arsenic, Bismuth, Selenium, Tellurium and Tin by Hydride Generation and Atomic Absorption Spectrophotometry. 14 pp.

—Silicon Content—Spectrophotometry
BSI BS 7317: Part 6-90. 1990 Methods for Analysis of High Purity Copper Cathode Cu-CATH-1 Part 6: Method for Determination of Phosphorus and Silicon by Spectrophotometry. 9 pp.
BSI DD 95: Part 6-87. (WITHDRAWN) 1987 Analysis of Higher Purity Copper Cathode Cu-CATH-1 Part 6: Methods for Determination of Phosphorous and Silicon by Spectrophotometry (Superseded by BS 7317: Part 6: 1990). 9 pp.

—Silver Content—Atomic Absorption Spectrophotometry
BSI BS 7317: Part 1-90. 1990 Methods for Analysis of High Purity Copper Cathode Cu-CATH-1 Part 1: Method for Determination of Cadmium Manganese and Silver (Screening Procedure for Chromium, Cobalt, Iron, Nickel and Zinc) by Atomic Absorption. 9 pp.
BSI DD 95: Part 1-84. (WITHDRAWN) 1984 Amd 1 Analysis of Higher Purity Copper Cathode Cu-CATH-1 Part 1: Method for Determination of Cadmium, Manganese and Silver (Screening Procedures for Chromium, Cobalt, Iron, Nickel and Zinc) by Atomic Absorption. 10 pp.

—Spectrochemical Analysis
JIS H 1103-76. Methods for Emission Spectrochemical Analysis of Electrolytic Cathode Copper.

—Sulfur Content—Hydrogen Sulfide Evolution
BSI BS 7317: Part 5-90. 1990 Methods for Analysis of High Purity Copper Cathode Cu-CATH-1 Part 5: Method for Determination of Sulphur by Hydrogen Sulphide Evolution and Spectrophotometry. 11 pp.
BSI DD 95: Part 5-86. (WITHDRAWN) 1986 Analysis of Higher Purity Copper Cathode Cu-CATH-1 Part 5: Method for the Determination of Sulphur by Hydrogen Sulphide Evolution and Spectrophotometry (Superseded by BS 7317: Part 5: 1990). 10 pp.

—Sulfur Content—Spectrophotometry
BSI BS 7317: Part 5-90. 1990 Methods for Analysis of High Purity Copper Cathode Cu-CATH-1 Part 5: Method for Determination of Sulphur by Hydrogen Sulphide Evolution and Spectrophotometry. 11 pp.
BSI DD 95: Part 5-86. (WITHDRAWN) 1986 Analysis of Higher Purity Copper Cathode Cu-CATH-1 Part 5: Method for the Determination of Sulphur by Hydrogen Sulphide Evolution and Spectrophotometry (Superseded by BS 7317: Part 5: 1990). 10 pp.

—Tellurium Content—Atomic Absorption Spectrophotometry
BSI BS 7317: Part 3-90. 1990 Methods for Analysis of High Purity Copper Cathode Cu-CATH-1 Part 3: Method for Determination of Antimony, Arsenic, Bismuth, Selenium, Tellurium and Tin by Hydride Generation and Atomic Absorption Spectrophotometry. 12 pp.
BSI BS 7317: Part 4-90. 1990 Methods for Analysis of High Purity Copper Cathode Cu-CATH-1 Part 4: Method for Determination of Antimony, Arsenic, Bismuth, Lead, Selenium, Tellurium and Tin by Electrothermal Atomiza-tion Atomic Absorption Spectrophotometry. 9 pp.
BSI DD 95: Part 3-84. (WITHDRAWN) 1984 Analysis of Higher Purity Copper Cathode Cu-CATH-1 Part 3: Methods for Determination of Antimony, Arsenic, Bismuth, Selenium, Tellurium and Tin by Hydride Generation and Atomic Absorption Spectrophotometry. 14 pp.
BSI DD 95: Part 4-86. (WITHDRAWN) 1986 Analysis of Higher Purity Copper Cathode CU-CATH-1 Part 4: Method for Determination of Antimony, Arsenic, Bismuth, Lead, Selenium, Tellurium and Tin by Electrothermal Atomization Atomic. 9 pp.

—Tellurium Content—Hydride Generation
BSI DD 95: Part 3-84. (WITHDRAWN) 1984 Analysis of Higher Purity Copper Cathode Cu-CATH-1 Part 3: Methods for Determination of Antimony, Arsenic, Bismuth, Selenium, Tellurium and Tin by Hydride Generation and Atomic Absorption Spectrophotometry. 14 pp.

Electrolytic Copper *(Cont.)*

—Tin Content—Atomic Absorption Spectrophotometry
BSI BS 7317: Part 3-90. 1990 Methods for Analysis of High Purity Copper Cathode Cu-CATH-1 Part 3: Method for Determination of Antimony, Arsenic, Bismuth, Selenium, Tellurium and Tin by Hydride Generation and Atomic Absorption Spectrophotometry. 12 pp.
BSI BS 7317: Part 4-90. 1990 Methods for Analysis of High Purity Copper Cathode Cu-CATH-1 Part 4: Method for Determination of Antimony, Arsenic, Bismuth, Lead, Selenium, Tellurium and Tin by Electrothermal Atomiza-tion Atomic Absorption Spectrophotometry. 9 pp.
BSI DD 95: Part 3-84. (WITHDRAWN) 1984 Analysis of Higher Purity Copper Cathode Cu-CATH-1 Part 3: Methods for Determination of Antimony, Arsenic, Bismuth, Selenium, Tellurium and Tin by Hydride Generation and Atomic Absorption Spectrophotometry. 14 pp.
BSI DD 95: Part 4-86. (WITHDRAWN) 1986 Analysis of Higher Purity Copper Cathode CU-CATH-1 Part 4: Method for Determination of Antimony, Arsenic, Bismuth, Lead, Selenium, Tellurium and Tin by Electrothermal Atomization Atomic. 9 pp.

—Tin Content—Hydride Generation
BSI BS 7317: Part 3-90. 1990 Methods for Analysis of High Purity Copper Cathode Cu-CATH-1 Part 3: Method for Determination of Antimony, Arsenic, Bismuth, Selenium, Tellurium and Tin by Hydride Generation and Atomic Absorption Spectrophotometry. 12 pp.
BSI DD 95: Part 3-84. (WITHDRAWN) 1984 Analysis of Higher Purity Copper Cathode Cu-CATH-1 Part 3: Methods for Determination of Antimony, Arsenic, Bismuth, Selenium, Tellurium and Tin by Hydride Generation and Atomic Absorption Spectrophotometry. 14 pp.

—Zinc Content—Atomic Absorption Spectrophotometry
BSI BS 7317: Part 1-90. 1990 Methods for Analysis of High Purity Copper Cathode Cu-CATH-1 Part 1: Method for Determination of Cadmium Manganese and Silver (Screening Procedure for Chromium, Cobalt, Iron, Nickel and Zinc) by Atomic Absorption. 9 pp.
BSI BS 7317: Part 2-90. 1990 Methods for Analysis of High Purity Copper Cathode Cu-CATH-1 Part 2: Method for Determination of Chromium, Cobalt, Iron, Nickel and Zinc by Discrete Volume Nebulization Atomic Absorption Spectrophotometry. 9 pp.
BSI DD 95: Part 2-84. (WITHDRAWN) 1984 Amd 1 Analysis of Higher Purity Copper Cathode Cu-CATH-1 Part 2: Method for Determination of Chromium, Cobalt, Iron, Nickel, and Zinc by Discrete Volume Nebulization Atomic Absorption Spectrophotmetry. 10 pp.

Electrolytic Corrosion

See Also: Corrosion; Electrochemical Corrosion

—Chromium Coatings
ISO 4539-80. Electrodeposited Chromium Coatings—Electrolytic Corrosion Testing (EC Test) First Edition. 8 pp.

—Electrical Insulation
BSI BS 5735-79. 1979 Test Methods for Determining Electrolytic Corrosion with Electric Insulating Materials. 24 pp.
DIN ENGL 53489-68. Testing of Electrical Insulating Materials; Assessing the Effect of Electrolytic Corrosion (Jan). 5 pp.
DIN VDE 0303 Pt 6-68. Regulations for Electrical Tests on Insulating Materials; Part 6: Determination of Electrolytic Corrosive Effects (Mar). 15 pp.
IEC 426-73. Test Methods for Determining Electrolytic Corrosion with Insulating Materials First Edition. 47 pp.

—Metals
DIN ENGL 50929 Pt 1-85. Corrosion of Metals; Probability of Corrosion of Metallic Materials When Subject to Corrosion from the Outside; General (Sept). 5 pp.

Electrolytic Corrosion Testing
Use: Electrolytic Corrosion

Electrolytic Deposition
Use: Electrodeposited Coatings

Electrolytic Etching
See Also: Etching

—Austenitic Stainless Steels
JIS G 0571-80. Method of 10 Per Cent Oxalic Acid Etch Test for Stainless Steels. 10 pp.

Electrolytic Polishing
Use: Electropolishing

Electrolyzers
Use For: Electrolysers

—Cleanliness—Submarines
MOD UK NES 372-83. Electrolyser System Cleanliness Issue 1 (04.83). 47 pp.
MOD UK NES 372: Part 1-88. Oxygen Clean Standards (OX) Part 1: HP Electrolysers Issue 2 (08.88). 60 pp.
MOD UK NES 372: Part 2-88. Oxygen Clean Standards (OX) Part 2—LP Electrolysers Issue 2 (12.88). 52 pp.

Electromagnetic Brakes
Use: Brakes

Electromagnetic Clutches
See Also: Clutches; Electromagnetic Equipment
JIS B 1403-76. Electromagnetic Multiple Disc Clutches (Wet Type) (R 1985). 9 pp.

Electromagnetic Compatibility
Use For: EMC; Immunity to Interference (Electromagnetic Compatibility)
See Also: Electromagnetic Environment; Electromagnetic Fields; Electromagnetic Radiation; Harmonic Analysis; Line Impedance Stabilization Networks; Power Density; Radio Frequency Interference

BSI BS EN 50081-1-92. 1992 Electromagnetic Compatibility—Generic Emission Standard Part 1: Residential, Commercial and Light Industry. 15 pp.
BSI BS EN 50082-1-92. 1992 Electromagnetic Compatibility—Generic Immunity Standard Part 1: Residential, Commercial and Light Industry. 19 pp.
CENELEC PREN 50 081-1-90. Electromagnetic Compatibility—Generic Emission Standard Generic Standard Class: Domestic, Commercial and Light Industry. 7 pp.
CENELEC EN 50081-1-92. Electromagnetic Compatibility—Generic Emission Standard Part 1: Residential, Commercial and Light Industry. 11 pp.
CENELEC EN 50081-1-92. Electromagnetic Compatibility—Generic Emission Standard Part 1: Residential, Commercial and Light Industry. 13 pp.
CENELEC PREN 50 082-1-90. Electromagnetic Compatibility—Generic Immunity Standard Generic Standard Class: Domestic, Commercial and Light Industry. 11 pp.
CENELEC EN 50082-1-92. Electromagnetic Compatibility Generic Immunity Standard Part 1: Residential, Commercial and Light Industry. 14 pp.
CENELEC EN 50082-1-92. Electromagnetic Compatibility—Generic Immunity Standard Part 1: Residential, Commercial and Light Industry. 16 pp.
CENELEC EN 50082-1-92. CORRIGENDUM Electromagnetic Compatibility—Generic Immunity Standard Part 1: Residential, Commercial and Light Industry. 1 p.
EC 89/336/EEC-89. Council Directive on the Approximation of the Laws of the Member States Relating to Electromagnetic Compatability (NEW APP) (for Rel Stds See COMMUN/89/336). 8 pp.
EC COMMUN/89/336-89. Commission Communication in the Framework of the Implementation of the 'New Approach' Directives. 2 pp.
EC 92/31/EEC-92. Council Directive Amending Directive 89/336/EEC on the Approximation of the Laws of the Member States Relating to Electromagnetic Compatibility (NEW APP). 1 p.
EC 93/68/EEC-93. Council Directive Amending Directives 87/404/EEC (Simple Pressure Vessels), 88/378/EEC (Safety of Toys), 89/106/EEC (Construction Products), 89/336/EEC (Electromagnetic Compatibility), 89/392/EEC (Machinery), 89/686/ EEC (Personal Protective. 22 pp.
IEC 1000 Pt 2-3-92. Electromagnetic Compatibility (EMC) Part 2: Environment Section 3: Description of the Environment—Radiated and Non-Network-Frequency-Related Conducted Phenomena First Edition. 138 pp.
IEC 1000 Pt 4-1-92. Electromagnetic Compatibility (EMC) Part 4: Testing and Measurement Techniques Section 1: Overview of Immunity Tests Basic EMC Publication First Edition. 144 pp.
IEC Guide 107-89. Electromagnetic Compatibility. Guide to the Drafting of Electromagnetic Compatibility Publications. 28 pp.
NATO STANAG 3731 ED 2 AMD 2-90. Bibliography on Electromagnetic Compatibility (EMC). 25 pp.

—Aerospace Equipment
JIS W 7005-84. Electromagnetic Compatibility Requirements for Aerospace Systems.

INTERNATIONAL AND NON-U.S. NATIONAL STANDARDS SUBJECT INDEX

Electromagnetic Compatibility *(Cont.)*

—Aircraft
NATO STANAG 3614 ED 2 AMD 0-83. Electromagnetic Compatibility (EMC) of Aircraft Systems. 18 pp.

—Aircraft Equipment
MOD UK AVP 118-76. Guide to Electromagnetic Compatibility in Aircraft Systems. 5 pp.

MOD UK AVP 118: Chapter 7-76. Installation of Equipment and Systems. 56 pp.

MOD UK AVP 118: Chapter 8-81. Equipment Design for Electromagnetic Compatibility. 34 pp.

NATO STANAG 3516 ED 2 AMD 4-77. Electromagnatic Compatibility for Aircraft Electrical and Electronic Equipment. 47 pp.

NATO STANAG 3614 ED 3 AMD 0-89. Electromagnetic Compatibility (EMC) of Aircraft Systems. 9 pp.

—Avionics—Glossaries
NATO STANAG 3968 ED 1 AMD 1-85. NATO Glossary of Electromagnetic Terms and Definitions. 11 pp.

NATO STANAG 3968 ED 2 AMD 0-92. NATO Glossary of Electromagnetic Terminology. 7 pp.

NATO STANAG 3968 ED 2 AMD 1-92. NATO Glossary of Electromagnetic Terminology. 8 pp.

—Bibliographies
NATO STANAG 3731 ED 1 AMD 1-88. Bibliography on Electromagnetic Compatibility (EMC). 22 pp.

NATO STANAG 3731 ED 2 AMD 0-90. Bibliography on Electromagnetic Compatibility (EMC). 22 pp.

NATO STANAG 3731 ED 2 AMD 1-90. Bibliography on Electromagnetic Compatibility (EMC). 23 pp.

—Certification
EC COM(87) 527-87. Proposal for a Council Directive on the Approximation of the Laws of the Member States Relating to Electromagnetic Compatibliity. 30 pp.

EC COM(88) 548-88. Amendment to the Proposal for a Council Directive on the Approximation of the Laws of the Member States Relating to Electromagnetic Compatibility. 8 pp.

—Couplings
DIN VDE 0846 Pt 12 (D)-90. Measuring Apparatus for Assessment of Electromagnetic Compatibility; Coupling Equipment (Jan). 18 pp.

—Damped Oscillatory Magnetic Field
IEC 1000 Pt 4-10-93. Electromagnetic Compatibility (EMC)—Part 4: Testing and Measurement Techniques—Section 10: Damped Oscillatory Magnetic Field Immunity Test Basic EMC Publication First Edition. 63 pp.

—Decoders—Emergency Locator Transmitters
CCIR Report 1101-90. Preliminary Studies of Radio Interference Limits for MAC/Packet Decoders—Section 1D—Spectrum Utilization and Applications. 27 pp.

—Decoders—Emergency Position Indicating Radio Beacons
CCIR Report 1101-90. Preliminary Studies of Radio Interference Limits for MAC/Packet Decoders—Section 1D—Spectrum Utilization and Applications. 27 pp.

—Electrical Components—Naval Ships
MOD UK NES 515-90. Guide to Electromagnetic Compatibility Issue 2 (09.90). 152 pp.

MOD UK NES 515-01. Guide to Electromagnetic Compatibility Issue 2 (09.90); Amendment 1. 157 pp.

—Electrical Components—Submarines
MOD UK NES 515-90. Guide to Electromagnetic Compatibility Issue 2 (09.90). 152 pp.

MOD UK NES 515-01. Guide to Electromagnetic Compatibility Issue 2 (09.90); Amendment 1. 157 pp.

—Electrical Equipment
DIN VDE 0847 Pt 1-81. Procedure for Measuring Electromagnetic Compatibility; Measurement of Conducted Interference (Nov). 24 pp.

DIN VDE 0847 Pt 2 (D)-87. Measuring Method for Evaluation of Electromagnetic Compatibility; Immunity from Conducted Disturbances (Oct). 86 pp.

DIN VDE 0847 Pt 4 (D)-87. Procedures for Measurement of Electromagnetic Compatibility; Immunity Against Radiated Interference Variables (Jan). 35 pp.

Electromagnetic Compatibility *(Cont.)*

—Electrical Equipment *(Cont.)*
MOD UK DSTAN 59-41: Part 1-88. Electromagnetic Compatibility Part 1: General Requirements Issue 4 (Superseded by Def Stan 59-41: Part 1: Section 1). 13 pp.

MOD UK DSTAN 59-41: Pt 1:Sec 1-93. Electromagnetic Compatibility Part 1: Introduction Section 1: General Requirements Issue 5. 13 pp.

MOD UK DSTAN 59-41: Pt 1:Sec 2-93. Electromagnetic Compatibility Part 1: Introduction Section 2: Guide to the Specification and Selection of EMC Requirements Issue 5. 45 pp.

MOD UK DSTAN 59-41: Part 2-88. Electromagnetic Compatibility Part 2: Management and Planning Procedures Issue 3. 10 pp.

MOD UK DSTAN 59-41: Part 3-01. Electromagnetic Compatibility Part 3: Technical Requirements Test Methods and Limits Issue 3; Amendment 4. 167 pp.

MOD UK DSTAN 59-41: Part 3-93. Electromagnetic Compatibility Part 3: Technical Requirements Test Methods and Limits Issue 4. 42 pp.

MOD UK DSTAN 59-41: Pt 3: Supp A-93. Electromagnetic Compatibility Part 3: Technical Requirements Test Methods and Limits Supplement A: Test Method DCE01 Conducted Emission on Power Lines 20Hz—150MHz Issue 1. 10 pp.

MOD UK DSTAN 59-41: Pt 3: Supp B-93. Electromagnetic Compatibility Part 3: Technical Requirements Test Methods and Limits Supplement B: Test Method DCE02 Conducted Emission on Control and Signal Lines 20Hz—150MHz Issue 1. 10 pp.

MOD UK DSTAN 59-41: Pt 3: Supp C-93. Electromagnetic Compatibility Part 3: Technical Requirements Test Methods and Limits Supplement C: Test Method DCE03 Conducted Emission on Exported Transients, Power Lines Issue 1. 12 pp.

MOD UK DSTAN 59-41: Pt 3: Supp D-93. Electromagnetic Compatibility Part 3: Technical Requirements Test Methods and Limits Supplement D: Test Method DRE01 Radiated Emissions E Field 14kHz—18GHz Issue 1. 14 pp.

MOD UK DSTAN 59-41: Pt 3: Supp E-93. Electromagnetic Compatibility Part 3: Technical Requirements Test Methods and Limits Supplement E: Test Method DRE02 H Field Radiation 20Hz—50kHz Issue 1. 10 pp.

MOD UK DSTAN 59-41: Pt 3: Supp F-93. Electromagnetic Compatibility Part 3: Technical Requirements Test Methods and Limits Supplement F: Test Method DRE03 Radiated Emmisions Installed Antenna 1MHz—76MHz Issue 1. 12 pp.

MOD UK DSTAN 59-41: Pt 3: Supp G-93. Electromagnetic Compatibility Part 3: Technical Requirements Test Methods and Limits Supplement G: Test Method DCS01 Conducted Susceptibility, Power Lines 20Hz—50kHz Issue 1. 11 pp.

MOD UK DSTAN 59-41: Pt 3: Supp H-93. Electromagnetic Compatibility Part 3: Technical Requirements Test Methods and Limits Supplement H: Test Method DCS02 Conducted Susceptibility, Power, Control and Signal Lines 50kHz—400MHz Issue 1. 20 pp.

MOD UK DSTAN 59-41: Pt 3: Supp J-93. Electromagnetic Compatibility Part 3: Technical Requirements Test Methods and Limits Supplement J: Test Method DCS03 Conducted Susceptibility, Control and Signal Lines 20Hz—50kHz Issue 1. 10 pp.

MOD UK DSTAN 59-41: Pt 3: Supp K-93. Electromagnetic Compatibility Part 3: Technical Requirements Test Methods and Limits Supplement K: Test Method DCS04 Imported Transient Susceptibility (Aircraft) Issue 1. 18 pp.

MOD UK DSTAN 59-41: Pt 3: Supp L-93. Electromagnetic Compatibility Part 3: Technical Requirements Test Methods and Limits Supplement L: Test Method DCS05 Externally Generated Transients (Ships) Issue 1. 11 pp.

MOD UK DSTAN 59-41: Pt 3: Supp M-93. Electromagnetic Compatibility Part 3: Technical Requirements Test Methods and Limits Supplement M: Test Method DCS06 Imported Long Transient Susceptibility 28 Volt Systems (Land Service) Issue 1. 9 pp.

MOD UK DSTAN 59-41: Pt 3: Supp N-93. Electromagnetic Compatibility Part 3: Technical Requirements Test Methods and Limits Supplement N: Test Method DCS07 Imported Short Transient Susceptibility (Land Service) Issue 1. 10 pp.

MOD UK DSTAN 59-41: Pt 3: Supp U-93. Electromagnetic Compatibility Part 3: Technical Requirements Test Methods and Limits Supplement U: Test Method DRS01 H Field Susceptibility, 20Hz—50kHz Issue 1. 11 pp.

MOD UK DSTAN 59-41: Pt 3: Supp V-93. Electromagnetic Compatibility Part 3: Technical Requirements Test Methods and Limits Supplement V: Test Method DRS02 Radiated Susceptibility E Field, 14kHz—18GHz Issue 1. 11 pp.

Electromagnetic Compatibility *(Cont.)*

—Electrical Equipment *(Cont.)*
MOD UK DSTAN 59-41: Pt 3: Supp W-93. Electromagnetic Compatibility Part 3: Technical Requirements Test Methods and Limits Supplement W: Test Method DMFS01 Magnetostatic Field Susceptibility Issue 1. 8 pp.

—Electrical Equipment—Submarines
NATO STANAG 4437 ED 1 AMD 0-00. (Draft) Electro-Magnetic Compatibility Testing Procedure and Requirements for Naval Electrical and Electronic Equipment (Submarines). 61 pp.

—Electrical Installations—Conducted Interference
DIN VDE 0839 Pt 10 (D)-91. Electromagnetic Compatibility (EMC); Measurement of Immunity to Conducted Interference and Fields (Dec). 31 pp.

—Electrical Installations—Naval Ships
MOD UK NES 1010-92. Weatherdeck Installation Requirements for Electromagnetic Environmental Control Issue 2 (10.92). 26 pp.

MOD UK NES 1027-87. Electromagnetic Compatibility Requirements for Installations in Ships Issue 1 (05.87). 32 pp.

MOD UK NES 515-01. Guide to Electromagnetic Compatibility Issue 2 (09.90); Amendment 1. 157 pp.

—Electrical Installations—Ships
IEC 533-77. Electromagnetic Compatibility of Electrical and Electronic Installations in Ships First Edition. 88 pp.

—Electrical Installations—Submarines
MOD UK NES 1010-92. Weatherdeck Installation Requirements for Electromagnetic Environmental Control Issue 2 (10.92). 26 pp.

MOD UK NES 1027-87. Electromagnetic Compatibility Requirements for Installations in Ships Issue 1 (05.87). 32 pp.

MOD UK NES 515-01. Guide to Electromagnetic Compatibility Issue 2 (09.90); Amendment 1. 157 pp.

—Electronic Equipment
MOD UK DSTAN 59-41: Part 1-88. Electromagnetic Compatibility Part 1: General Requirements Issue 4 (Superseded by Def Stan 59-41: Part 1: Section 1). 13 pp.

MOD UK DSTAN 59-41: Pt 1:Sec 1-93. Electromagnetic Compatibility Part 1: Introduction Section 1: General Requirements Issue 5. 13 pp.

MOD UK DSTAN 59-41: Pt 1:Sec 2-93. Electromagnetic Compatibility Part 1: Introduction Section 2: Guide to the Specification and Selection of EMC Requirements Issue 5. 45 pp.

MOD UK DSTAN 59-41: Part 2-88. Electromagnetic Compatibility Part 2: Management and Planning Procedures Issue 3. 10 pp.

MOD UK DSTAN 59-41: Part 3-01. Electromagnetic Compatibility Part 3: Technical Requirements Test Methods and Limits Issue 3; Amendment 4. 167 pp.

MOD UK DSTAN 59-41: Part 3-93. Electromagnetic Compatibility Part 3: Technical Requirements Test Methods and Limits Issue 4. 42 pp.

MOD UK DSTAN 59-41: Pt 3: Supp A-93. Electromagnetic Compatibility Part 3: Technical Requirements Test Methods and Limits Supplement A: Test Method DCE01 Conducted Emission on Power Lines 20Hz—150MHz Issue 1. 10 pp.

MOD UK DSTAN 59-41: Pt 3: Supp B-93. Electromagnetic Compatibility Part 3: Technical Requirements Test Methods and Limits Supplement B: Test Method DCE02 Conducted Emission on Control and Signal Lines 20Hz—150MHz Issue 1. 10 pp.

MOD UK DSTAN 59-41: Pt 3: Supp C-93. Electromagnetic Compatibility Part 3: Technical Requirements Test Methods and Limits Supplement C: Test Method DCE03 Conducted Emission on Exported Transients, Power Lines Issue 1. 12 pp.

MOD UK DSTAN 59-41: Pt 3: Supp D-93. Electromagnetic Compatibility Part 3: Technical Requirements Test Methods and Limits Supplement D: Test Method DRE01 Radiated Emissions E Field 14kHz—18GHz Issue 1. 14 pp.

MOD UK DSTAN 59-41: Pt 3: Supp E-93. Electromagnetic Compatibility Part 3: Technical Requirements Test Methods and Limits Supplement E: Test Method DRE02 H Field Radiation 20Hz—50kHz Issue 1. 10 pp.

MOD UK DSTAN 59-41: Pt 3: Supp F-93. Electromagnetic Compatibility Part 3: Technical Requirements Test Methods and Limits Supplement F: Test Method DRE03 Radiated Emmisions Installed Antenna 1MHz—76MHz Issue 1. 12 pp.

MOD UK DSTAN 59-41: Pt 3: Supp G-93. Electromagnetic Compatibility Part 3: Technical Requirements Test Methods and Limits Supplement G: Test Method DCS01 Conducted Susceptibility, Power Lines 20Hz—50kHz Issue 1. 11 pp.

Electromagnetic Compatibility (Cont.)
—Electronic Equipment (Cont.)

MOD UK DSTAN 59-41: Pt 3: Supp H-93. Electromagnetic Compatibility Part 3: Technical Requirements Test Methods and Limits Supplement H: Test Method DCS02 Conducted Susceptibility, Power, Control and Signal Lines 50kHz—400MHz Issue 1. 20 pp.

MOD UK DSTAN 59-41: Pt 3: Supp J-93. Electromagnetic Compatibility Part 3: Technical Requirements Test Methods and Limits Supplement J: Test Method DCS03 Conducted Susceptibility, Control and Signal Lines 20Hz—50kHz Issue 1. 10 pp.

MOD UK DSTAN 59-41: Pt 3: Supp K-93. Electromagnetic Compatibility Part 3: Technical Requirements Test Methods and Limits Supplement K: Test Method DCS04 Imported Transient Susceptibility (Aircraft) Issue 1. 18 pp.

MOD UK DSTAN 59-41: Pt 3: Supp L-93. Electromagnetic Compatibility Part 3: Technical Requirements Test Methods and Limits Supplement L: Test Method DCS05 Externally Generated Transients (Ships) Issue 1. 11 pp.

MOD UK DSTAN 59-41: Pt 3: Supp M-93. Electromagnetic Compatibility Part 3: Technical Requirements Test Methods and Limits Supplement M: Test Method DCS06 Imported Long Transient Susceptibility 28 Volt Systems (Land Service) Issue 1. 9 pp.

MOD UK DSTAN 59-41: Pt 3: Supp N-93. Electromagnetic Compatibility Part 3: Technical Requirements Test Methods and Limits Supplement N: Test Method DCS07 Imported Short Transient Susceptibility (Land Service) Issue 1. 10 pp.

MOD UK DSTAN 59-41: Pt 3: Supp U-93. Electromagnetic Compatibility Part 3: Technical Requirements Test Methods and Limits Supplement U: Test Method DRS01 H Field Susceptibility, 20Hz—50kHz Issue 1. 11 pp.

MOD UK DSTAN 59-41: Pt 3: Supp V-93. Electromagnetic Compatibility Part 3: Technical Requirements Test Methods and Limits Supplement V: Test Method DRS02 Radiated Susceptibility E Field, 14kHz—18GHz Issue 1. 11 pp.

MOD UK DSTAN 59-41: Pt 3: Supp W-93. Electromagnetic Compatibility Part 3: Technical Requirements Test Methods and Limits Supplement W: Test Method DMFS01 Magnetostatic Field Susceptibility Issue 1. 8 pp.

—Electronic Equipment—Submarines

NATO STANAG 4437 ED 1 AMD 0-00. (Draft) Electro-Magnetic Compatibility Testing Procedure and Requirements for Naval Electrical and Electronic Equipment (Submarines). 61 pp.

—European Digital Cordless Telecommunications

CENELEC PRETS 300 329-93. Radio Equipment and Systems (RES); Electro-Magnetic Compatibility (EMC) for Digital European Cordless Telecommunications (DECT) Equipment. 24 pp.

ETSI PRETS 300 329-93. Radio Equipment and Systems (RES); Electro-Magnetic Compatibility (EMC) for Digital European Cordless Telecommunications (DECT) Equipment. 24 pp.

—Generator Sets

MOD UK DSTAN 61-5: Pt 2: Sec 5-92. Electrical Power Supply Systems Below 650 Volts Part 2: Ground Generating Set Characteristics Section 5: Electromagnetic Compatibility Requirements Issue 1. 10 pp.

—Glossaries

IEC 50 Chap 161-90. International Electrotechnical Vocabulary Chapter 161: Electromagnetic Compatibility First Edition. 75 pp.

—Integrated Services Digital Networks—Interfaces

CENELEC PREN 50096-91. Integrated Services Digital Network (ISDN) Equipment with ISDN User-Network Interface at Basic and Primary Rate—EMC Requirements. 34 pp.

CENELEC PRETS 300 126-90. Integrated Services Digital Network (ISDN) Equipment with ISDN Interface at Basic and Primary Rate EMC Requirements (DE/EE-4001). 24 pp.

—Measuring Instruments

DIN VDE 0846 Pt 13 (D)-90. Measuring Apparatus for Assessment of Electromagnetic Compatibility; Measuring Ancillaries (Jan). 14 pp.

—Medical Electrical Equipment

BSI BS 5724: Sec 1.2-93. 1993 Medical Electrical Equipment Part 1: General Requirements for Safety Section 1.2: Collateral Standard: Electromagnetic Compatibility—Requirements and Tests (ALSO KNOWN AS BS EN 60601-1-2: 1993) (N). 21 pp.

Electromagnetic Compatibility (Cont.)
—Medical Electrical Equipment (Cont.)

BSI BS EN 60601-1-2-93. 1993 Medical Electrical Equipment Part 1: General Requirements for Safety Section 1.2: Collateral Standard: Electromagnetic Compatibility—Requirements and Tests (IEC 601-1-2: 1993) (N). 21 pp.

CENELEC EN 60601-1-2-93. Medical Electrical Equipment Part 1: General Requirements for Safety 2. Collateral Standard: Electromagnetic Compatibility—Requirements and Tests (IEC 601-1-2: 1993). 16 pp.

IEC 601 Pt 1-2-93. Medical Electrical Equipment Part 1: General Requirements for Safety 2. Collateral Standard: Electromagnetic Compatibility—Requirements and Tests First Edition (CENELEC EN 60601-1-2: 1993). 37 pp.

—Mobile Radio Equipment

CCIR Report 1103-90. Setting Limits on EMC Parameters for Mobile Service Communication Equipment Based on an Analysis of the Interference Environment—Section 1A—Spectrum Engineering and Computer-Aided Principles and Techniques. 6 pp.

CENELEC PRETS 300 342-93. Radio Equipment and Systems (RES); Electro-Magnetic Compatibility (EMC) for European Digital Cellular Telecommunications System (GSM) Mobile Radio and Ancillary Equipment. 25 pp.

ETSI PRETS 300 342-93. Radio Equipment and Systems (RES); Electro-Magnetic Compatibility (EMC) for European Digital Cellular Telecommunications System (GSM) Mobile Radio and Ancillary Equipment. 25 pp.

MOD UK DSTAN 58-6-91. EMC of Mobile Communications Installations Issue 1. 45 pp.

MOD UK DSTAN 58-6-01. EMC of Mobile Communications Installations Issue 1; Amendment 1. 46 pp.

—Open Site Testing

MOD UK DSTAN 59-41: Part 4-88. Electromagnetic Compatibility Part 4: Open Site Testing Issue 2. 15 pp.

—Power Amplifiers

DIN VDE 0846 Pt 14-90. Measuring Equipment for Assessment of Electromagnetic Compatibility Power Amplifiers (Jan). 12 pp.

—Power Frequency Magnetic Field

IEC 1000 Pt 4-8-93. Electromagnetic Compatibility (EMC)—Part 4: Testing and Measurement Techniques—Section 8: Power Frequency Magnetic Field Immunity Test Basic EMC Publication First Edition. 63 pp.

—Power Systems—Disturbances

CENELEC ENV 61000-2-2-93. Electromagnetic Compatibility (EMC) Part 2: Environment Section 2: Compatibility Levels for Low-Frequency Conducted Disturbances and Signalling in Public Low-Voltage Power Supply Systems (IEC 1000-2-2: 1990, Modified). 12 pp.

IEC 1000 Pt 2-2-90. Electromagnetic Compatibility (EMC) Part 2: Environment Section 2: Compatibility Levels for Low-Frequency Conducted Disturbances and Signalling in Public Low-Voltage Power Supply Systems First Edition (CENELEC ENV 61000-2-2: 1993). 26 pp.

—Process Control Equipment

BSI BS 6667: Part 1-85. 1985 Electromagnetic Compatibility for Industrial-Process Measurement and Control Equipment Part 1: General Introduction. 7 pp.

CENELEC HD 481.1-87. Electromagnetic Compatibility for Industrial-Process Measurement and Control Equipment Part 1: General Introduction. 3 pp.

IEC 801 Pt 1-84. Electromagnetic Compatibility for Industrial-Process Measurement and Control Equipment Part 1: General Introduction First Edition (CENELEC EN 60801-2: 1993). 12 pp.

—Process Control Equipment—Electrostatic Discharge

BSI BS 6667: Part 2-85. (WITHDRAWN) 1985 Electromagnetic Compatibility for Industrial-Process Measurement and Control Equipment Part 2: Method of Evaluating Susceptibility to Electrostatic Discharge (Superseded by BS EN 60801-2: 1993). 22 pp.

BSI BS EN 60801-2-93. 1993 Electromagnetic Compatibility for Industrial-Process Measurement and Control Equipment Part 2: Electrostatic Discharge Requirements (IEC 801-2: 1991) (F). 37 pp.

CENELEC HD 481.2-87. Electromagnetic Compatibility for Industrial-Process Measurement and Control Equipment Part 2: Electrostatic Discharge Requirements. 3 pp.

Electromagnetic Compatibility (Cont.)
—Process Control Equipment—Electrostatic Discharge (Cont.)

CENELEC EN 60801-2-93. Electromagnetic Compatibility for Industrial-Process Measurement and Control Equipment Part 2: Electrostatic Discharge Requirements (IEC 801-2:1991) (Supersedes HD 481.2 S1:1987). 5 pp.

CENELEC EN 60801-2-93. Electromagnetic Compatibility for Industrial-Process Measurement and Control Equipment Part 2: Electrostatic Discharge Requirements (IEC 801-2:1991) (Supersedes HD 481.2 S1:1987). 32 pp.

IEC 801 Pt 2-91. Electromagnetic Compatibility for Industrial-Process Measurement and Control Equipment Part 2: Electrostatic Discharge Requirements Second Edition; (CENELEC EN 60801-2: 1993). 56 pp.

—Process Control Equipment—Radiated Electromagnetic Field

BSI BS 6667: Part 3-85. 1985 Electromagnetic Compatibility for Industrial-Process Measurement and Control Equipment Part 3: Method of Evaluating Susceptibility to Radiated Electromagnetic Energy. 26 pp.

CENELEC HD 481.3 S1-86. Electromagnetic Compatibility for Industrial-Process Measurement and Control Equipment—Part 3: Radiated Electromagnetic Field Requirements. 1 p.

IEC 801 Pt 3-84. Electromagnetic Compatibility for Industrial-Process Measurement and Control Equipment Part 3: Radiated Electromagnetic Field Requirements First Edition. 44 pp.

—Pulse Magnetic Field

IEC 1000 Pt 4-9-93. Electromagnetic Compatibility (EMC)—Part 4: Testing and Measurement Techniques—Section 9: Pulse Magnetic Field Immunity Test Basic EMC Publication First Edition. 61 pp.

—Radio Communications—Electrical Equipment

CCIR QUESTION 81/1-90. Electromagnetic Compatibility Requirements Concerning Radiocommunication Services, in Particular Safety Services—Questions Concerning Study Group 1-Spectrum Management Techniques (Spectrum Engineering, Planning,Sharing, Monitoring and. 1 p.

CENELEC PRETS 300 279-92. Radio Equipment and Systems (RES); EMC Standard for Private Mobile Radio (PMR), Trunked and Ancillary Equipment (Analogue Speech and Combined Analogue Speech/Non-Speech Equipment). 115 pp.

ETSI PRETS 300 279-92. Radio Equipment and Systems (RES); EMC Standard for Private Mobile Radio (PMR), Trunked and Ancillary Equipment (Analogue Speech and Combined Analogue Speech/Non-Speech Equipment). 115 pp.

—Radio Equipment

CENELEC PRETS 300 339-93. Radio Equipment and Systems (RES); Generic Electro-Magnetic Compatibility (EMC) for Radio Equipments. 111 pp.

ETSI PRETS 300 339-93. Radio Equipment and Systems (RES); Generic Electro-Magnetic Compatibility (EMC) for Radio Equipments. 111 pp.

—Radio Paging Systems—Receivers

CCIR Report 1099-90. Procedure for Analyzing the Electromagnetic Compatibility of Frequency-Hopping Spread Spectrum Radios—Section 1A—Spectrum Engineering and Computer-Aided Principles and Techniques. 13 pp.

CENELEC PRETS 300 340-93. Radio Equipment and Systems (RES); Electro-Magnetic Compatibility (EMC) for European Radio Message System (ERMES) Paging Receivers. 22 pp.

ETSI PRETS 300 340-93. Radio Equipment and Systems (RES); Electro-Magnetic Compatibility (EMC) for European Radio Message System (ERMES) Paging Receivers. 22 pp.

—Signal Generators

DIN VDE 0846 Pt 11 (D)-90. Measuring Apparatus for Judgement of Electromagnetic Compatibility; Signal Generators for Testing (Jan). 38 pp.

—Speech Transmission Systems—Radio

CCIR Report 526-1-78. Measures of Voice Transmission Performance Applicable for Electromagnetic Compatibility Analysis—Section 1A—Spectrum Engineering and Computer-Aided Principles and Techniques. 8 pp.

—Spread Spectrum Modulation

CCIR Report 975-86. Application of Spread-Spectrum Techniques to Radiolocation and Radio Compatibility—Section 1D—Spectrum Utilization and Applications. 4 pp.

Electromagnetic Compatibility *(Cont.)*

—Subscriber Lines
DIN VDE 0878 Pt 200 (D)-87. Radio Interference Suppression of Telecommunication Systems and Equipment (May). 72 pp.

—Telecommunication Systems
CCIR Report 833-82. Multi-Function Telecommunication Systems—Section 1D—Spectrum Utilization and Applications. 1 p.
CCITT RECMN K.27-91. Bonding Configurations and Earthing Inside a Telecommunication Building (Study Group V) 25 pp. 25 pp.

Electromagnetic Contactors
Use: Magnetic Contactors

Electromagnetic Counters
CNS C4381-83. Electro Magnetic Counters (Jan)(9809).
JIS C 4555-80. Electro Magnetic Counters. 24 pp.

Electromagnetic Couplings
See Also: Couplings

—Earth Stations—Antennas—Terrestrial Link Antennas
CCIR Report 709-1-82. Consideration of the Coupling Between an Earth-Station Antenna and a Terrestrial Link Antenna—Section 4/9B—Co-Ordination and Interference Calculations. 9 pp.

Electromagnetic Energy
Use: Electromagnetic Radiation

Electromagnetic Environment
See Also: Electromagnetic Compatibility; Electromagnetic Fields; Power Density
IEC 1000 Pt 2-3-92. Electromagnetic Compatibility (EMC) Part 2: Environment Section 3: Description of the Environment—Radiated and Non-Network-Frequency-Related Conducted Phenomena First Edition. 138 pp.

—Avionics—Glossaries
NATO STANAG 3968 ED 1 AMD 1-85. NATO Glossary of Electromagnetic Terms and Definitions. 11 pp.
NATO STANAG 3968 ED 2 AMD 0-92. NATO Glossary of Electromagnetic Terminology. 7 pp.
NATO STANAG 3968 ED 2 AMD 1-92. NATO Glossary of Electromagnetic Terminology. 8 pp.

—Naval Operations—Radar
NATO STANAG 1307 ED 1 AMD 2-86. Maximum NATO Naval Operational Electromagnetic Environment Produced by Radio and Radar. 8 pp.
NATO STANAG 1307 ED 1 AMD 3-86. Maximum NATO Naval Operational Electromagnetic Environment Produced by Radio and Radar. 10 pp.

—Naval Operations—Radio Equipment
NATO STANAG 1307 ED 1 AMD 2-86. Maximum NATO Naval Operational Electromagnetic Environment Produced by Radio and Radar. 8 pp.
NATO STANAG 1307 ED 1 AMD 3-86. Maximum NATO Naval Operational Electromagnetic Environment Produced by Radio and Radar. 10 pp.

—Power Systems—Disturbances
BSI BS 7484-91. 1991 Electromagnetic Environment for Low-Frequency Conducted Disturbances and Signalling in Public Power Supply Systems (IEC 1000-2-1: 1990). 30 pp.
BSI DD ENV 61000-2-2-93. 1993 Electromagnetic Compatibility (EMC) Part 2: Environment Section 2: Compatibility Levels for Low-Frequency Conducted Disturbances and Signalling in Public Low-Voltage Power Supply Systems (S). 18 pp.
CENELEC PREN 61000-2-2-92. Electromagnetic Compatibility (EMC) Part 2: Environment Section 2: Compatibility Levels for Low-Frequency Conducted Disturbances and Signalling in Public Low-Voltage Power Supply Systems. 12 pp.
IEC 1000 Pt 2-1-90. Electromagnetic Compatibility (EMC) Part 2: Environment Section 1: Description of the Environment—Electromagnetic Environment for Low-Frequency Conducted Disturbances and Signalling in Public Power Supply Systems First Edition. 51 pp.

—Power Systems—Glossaries
IEC 1000 Pt 1-1-92. Electromagnetic Compatibility (EMC) Part 1: General Section 1: Application and Interpretation of Fundamental Definitions and Terms First Edition. 62 pp.

Electromagnetic Equipment
See Also: Electrical Equipment; Electromagnetic Clutches; Electromagnetic Flowmeters; Electromagnetic Interference Filters; Electromagnetic Interference Measuring Instruments; Electromagnetic Relays; Electromagnetic Switches; Power Density
DIN VDE 0580 AD-79. Electromagnetic Devices Partial Amendment D (VDE Specification) (Sept). 10 pp.

—Covermeters—Reinforced Concrete
BSI BS 1881: Part 204-88. 1988 Amd 1 Methods of Testing Concrete Part 204: Recommendations on the Use of Electromagnetic Covermeters (AMD 6201) December 22, 1989. 13 pp.

—Glossaries
SAA AS 1852.901A-78. International Electrotechnical Vocabulary—Part 901A: Magnetism—Terms and Definitions Relating Non-Reciprocal Electromagnetic Components Being IEC 50(901A). 23 pp.
SNZ IEC 50: 50(901A)-75. International Electrotechnical Vocabulary 50(901A): Magnetism. First Supplement: Section 901-905. Terms and Definitions Relating to Non-Reciprocal Electromagnetic Components. 19 pp.

Electromagnetic Fields
See Also: Electromagnetic Compatibility; Electromagnetic Environment; Field Strength

—Automotive—Generation
ISO TR10305-92. Road Vehicles—Generation of Standard EM Fields for Calibration of Power Density Meters from 20 kHz to 1 000 MHz First Edition. 48 pp.

—Hazards
DIN VDE 0848 Pt 1-82. Hazards from Electromagnetic Fields; Methods for Measurement and Calculation (Feb). 24 pp.

—Safety
DIN VDE 0848 Pt 2-84. Hazards by Electromagnetic Fields; Protection of Persons in the Frequency Range from 10 kHz to 3000 GHz (July). 14 pp.
DIN VDE 0848 Pt 4-89. Safety at Electromagnetic Fields; Limits of Field Strengths for the Protection of Persons in the Frequency Range from 0 to 30 kHz (Oct) (Partially Supersedes DIN 57848 Part 2/VDE 0848 Part 2/07.84). 16 pp.
DIN VDE 0848 Pt 4 A1 (D). Safety at Electromagnetic Fields; Limits of Field Strengths for the Protection of Persons in the Frequency Range from 0 to 30 kHz; Amendment No. 1. 35 pp.

Electromagnetic Flowmeters
See Also: Electromagnetic Equipment; Flowmeters
BSI BS 5792-80. (WITHDRAWN) 1980 Amd 1 Electromagnetic Flow-Meters (Superseded by BS 5792: Part 1: 1993). 15 pp.
BSI BS 7526-91. 1991 Methods of Evaluating the Performance of Electromagnetic Flowmeters (ISO 9104: 1991). 25 pp.
ISO 9104-91. Measurement of Fluid Flow in Closed Conduits—Methods of Evaluating the Performance of Electromagnetic Flow-Meters for Liquids First Edition. 23 pp.
JIS B 7554-84. Electromagnetic Flowmeters. 17 pp.
JIS Z 8764-80. Method of Flow Measurement by Electromagnetic Flowmeters. 17 pp.

—Electrically Conductive Liquids
BSI BS 5792: Part 1-93. 1993 Measurement of Conductive Liquid Flow in Closed Conduits Part 1: Method Using Eloectromagnetic Flowmeters (ISO 6817: 1992) (Supersedes BS 5792: 1980). 24 pp.
ISO 6817-92. Measurement of Conductive Liquid Flow in Closed Conduits—Method Using Electromagnetic Flowmeters First Edition. 22 pp.

Electromagnetic Interference
Use For: Conducted Interference; EMI
See Also: Crosstalk; Electromagnetic Intrusion; Electromagnetic Noise; Electromagnetic Radiation; Environmental Conditions; Interference Cancellation Systems; Jamming; Multipath Distortion; Precipitation Static; Radio Frequency Interference
IEC 478 Pt 3-89. Stabilized Power Supplies, D.C. Output Part 3: Reference Levels and Measurement of Conducted Electromagnetic Interference (EMI) Second Edition. 20 pp.
IEC 478 Pt 5-93. Stabilized Power Supplies, d.c. Output Part 5: Measurement of the Magnetic Component of the Reactive Near Field First Ediiton. 23 pp.

—Aircraft
BSI 3G 100:Pt 4: Sec 2-80. (OBSOLESCENT) 1980 Amd 1 General Requirements for Equipment for Use in Aircraft Part 4: Electrical Equipment Section 2: Electromagnetic Interference at Radio and Audio Frequencies. 21 pp.
CAA 706 (K). Section K Light Aeroplanes Electrical Supply, Systems and Equipment (Blue Papers). 54 pp.
CAA 707 (G). Section G Rotorcraft Electrical Supply, Systems and Equipment (Blue Papers). 47 pp.
CAA Chap G6-14 App #2 11.85. Interference (Rotorcraft). 2 pp.
MOD UK AVP 118: Chapter 1-76. Nature of Electromagnetic Interference in Aircraft. 4 pp.
MOD UK AVP 118: Chapter 2-76. Methods of Measurement. 9 pp.
MOD UK AVP 118: Chapter 3-76. Types of Interference. 3 pp.
MOD UK AVP 118: Chapter 4-76. Propagation of Interference. 4 pp.
MOD UK AVP 118: Chapter 5-76. Effects of Interference. 5 pp.
MOD UK AVP 118: Chapter 6-76. Control of Interference. 19 pp.

—Aircraft—Compasses
CAA Chapter J4-1 App 09.66. Interference Avoidance of Magnetic Interference with the Compass. 1 p.

—Aircraft—Electric Connectors
SBAC ESC 65 ISSUE 3. Backshell, Assembly, Straight, RFI/EMI Cone Grounding, Anti-Decoupling. (MS 3155 Connector Interface MoD). 7 pp.
SBAC ESC 66 ISSUE 3. Backshell, Assembly, 90 Degrees RFI/EMI Cone Grounding, Anti Decoupling. (MS 3155 Connector Interface MoD). 7 pp.
SBAC ESC 67 ISSUE 3. Backshell, Assembly, 45 Degrees RFI/EMI Cone Grounding, Anti—Decoupling. (MS 3155 Connector Interface MoD). 7 pp.
SBAC ESC 72 ISSUE 1. Backshell, Assembly, Straight, RFI/EMI Cone Grounding, Anti-Decoupling. (MS 3155 Connector Interface MoD). 7 pp.
SBAC ESC 73 ISSUE 1. Backshell, Assembly, 90 Degrees RFI/EMI Cone Grounding, Anti Decoupling. (MS 3155 Connector Interface MoD). 7 pp.
SBAC ESC 74 ISSUE 1. Backshell, Assembly, 45 Degrees RFI/EMI Cone Grounding, Anti—Decoupling. (MS 3155 Connector Interface MoD). 7 pp.

—Aircraft—Electrical Equipment
CAA Chapter J4-1 09.66. Interference—General. 2 pp.
NATO STANAG 3516 ED 3 AMD 0-93. Electromagnetic Interference and Test Methods for Aircraft Electrical and Electronic Equipment. 51 pp.

—Aircraft—Electronic Equipment
NATO STANAG 3516 ED 3 AMD 0-93. Electromagnetic Interference and Test Methods for Aircraft Electrical and Electronic Equipment. 51 pp.

—Automotive—Electrical Equipment
BSI BS AU 243: Part 1-91. 1991 Methods of Test for Electrical Disturbance by Conduction and Coupling Part 1: General (ISO 7637-0: 1990). 7 pp.
BSI BS AU 243: Part 2-91. 1991 Methods of Test for Electrical Disturbance by Conduction and Coupling Part 2: Electrical Transient Conduction Along Supply Lines Only on Passenger Cars and Light Commercial Vehicles with 12 V Supply (ISO 7637-1: 1990). 17 pp.
BSI BS AU 243: Part 3-91. 1991 Methods of Test for Electrical Disturbance by Conduction and Coupling Part 3: Electrical Transient Conduction Along Supply Lines Only on Commercial Vehicles with 24 V Supply (ISO 7637-2: 1990). 17 pp.

—Automotive—Electrical Equipment—Glossaries
ISO 7637 Pt 0-90. Road Vehicles—Electrical Disturbance by Conduction and Coupling—Part 0: Definitions and General First Edition. 6 pp.

—Broadcast Receivers
CSA CAN/CSA-C108.9-M91. Sound and Television Broadcasting Receivers and Associated Equipment—Limits and Methods of Measurement of Immunity Characteristics; (Gen Instr 1). 64 pp.

—Cabled Distribution Systems
CENELEC PREN 50083-2-91. Cable Distribution Systems ENC Requirements Radiation and Immunity Performance Requirements. 72 pp.

—Communication Cables
CCITT RECMN K.29-92. Coordinated Protection Schemes for Telecommunication Cables Below Ground (Study Group V) 7 pp. 7 pp.

—Data Processing Equipment
CENELEC PREN 55 101-4-90. Immunity Requirements for Information Technology Equipment—Part 4: Conducted Immunity (See PREN 55024-4). 14 pp.

INTERNATIONAL AND NON-U.S. NATIONAL STANDARDS
SUBJECT INDEX
Electromagnetic

Electromagnetic Interference *(Cont.)*

—**Earth Exploration Satellites**
 CCIR RECMN 514-1-90. Telecommunication Links for Earth Exploration Satellites Frequencies, Bandwidths and Criteria for Protection from Interference—Section 2F—Earth Exploration Satellites. 2 pp.
 CCIR Report 1122-90. Methodology for Deriving Sharing Criteria and Coordination Thresholds for the Earth Exploration-Satellite Service—Section 2F—Earth Exploration Satellites. 24 pp.
 CCIR Report 1123-90. Methodology for Deriving Interference Criteria for the Earth Exploration-Satellite Service—Section 2F—Earth Exploration Satellites. 8 pp.

—**Earth Stations**
 CCIR Report 844-1-90. Potential Interference Between Deep-Space Telecommunications and Some Other Services in Harmonically Related Bands—Section 2B—Topics of General Interest. 12 pp.

—**Earth Stations—Meteorological Satellites**
 CCIR Report 1121-90. Performance, Interference and Sharing Criteria for Receiving Earth Stations in the Meteorological-Satellite Service Operating in the 1 670–1 710 MHz Band with Satellites in Low Earth Orbit—Section 2F—Earth Exploration Satellites. 12 pp.
 CCIR Report 1124-90. Performance, Interference and Sharing Criteria for Receiving Earth Stations in the Meteorological-Satellite Service Operating in the 7 450–7 550 MHz Band with Satellites in Low Earth Orbit—Section 2F—Earth Exploration Satellites. 7 pp.

—**Earth Stations—Research**
 CCIR RECMN 509-1-90. Generalized Space Research Earth Station Antenna Radiation Pattern for Use in Interference Calculations, Including Coordination Procedures—Section 2A—Research in Space Technology. 1 p.

—**Earth Stations—Terrestrial Stations**
 CCIR Report 448-5-90. Determination of the Interference Potential Between Earth Stations and Terrestrial Stations —Section 4/9B—Co-Ordination and Interference Calculations. 9 pp.

—**Electrical Equipment**
 DIN VDE 0839 Pt 10 (D)-91. Electromagnetic Compatibility (EMC); Measurement of Immunity to Conducted Interference and Fields (Dec). 31 pp.
 NATO STANAG 4435 ED 1 AMD 0-00. Electro-Magnetic Compatibility Testing Procedure and Requirements for Naval Electrical and Electronic Equipment (Surface Ship, Metallic Hull). 58 pp.
 NATO STANAG 4436 ED 1 AMD 0-00. Electro-Magnetic Compatibility Testing Procedure and Requirements for Naval Electrical and Electronic Equipment (Surface Ship, Non-Metallic Hull). 56 pp.

—**Electrical Protection Equipment**
 IEC 255 Pt 22-3-89. Electrical Relays Part 22: Electrical Disturbance Tests for Measuring Relays and Protection Equipment Section Three—Radiated Electromagnetic Field Disturbance Tests First Edition. 37 pp.

—**Fixed Satellite Services**
 CCIR Report 388-6-90. Methods for Determining the Effects of Interference on the Performance and the Availability of Terrestrial Radio-Relay Systems and Systems in the Fixed-Satellite Service—Section 4/9B—Co-Ordination and Interference Calculations. 40 pp.

—**Fixed Satellite Services—Frequencies**
 CCIR Report 710-3-90. Interference Allocations in Systems Operating at Frequencies Greater Than 10 GHz in the Fixed-Satellite Service—Section 4D1—Permissible Levels of Interference. 11 pp.

—**Geostationary Satellites**
 CCIR Report 844-1-90. Potential Interference Between Deep-Space Telecommunications and Some Other Services in Harmonically Related Bands—Section 2B—Topics of General Interest. 12 pp.

—**Glossaries**
 DIN VDE 0870 Pt 1-84. Electromagnetic Interference (EMI); Terms (July). 16 pp.
 SAA AS 1852.902-76. International Electrotechnical Vocabulary—Part 902: Electromagnetic Interference. 8 pp.

—**Infrared Radiation—Sound Transmission**
 IEC 1147-93. Uses of Infra-Red Transmission and the Prevention or Control of Interference Between Systems First Edition. 43 pp.

Electromagnetic Interference *(Cont.)*

—**ISM Equipment**
 CSA CAN/CSA-C108.6-M91. Limits and Methods of Measurement of Electromagnetic Disturbance Characteristics of Industrial, Scientific and Medical (ISM) Radio-Frequency Equipment (CISPR 11:1990); (Gen Instr 1). 39 pp.
 SAA AS 2064-90. Electromagnetic Interference—Industrial, Scientific and Medical (ISM) Radio-frequency Equipment—Limits and Methods of Measurement (Superseded (in Part) by AS/NZS 2064.1—1992 and AS/NZS 2064.2—1992).

—**Marine**
 BSI BS 1597-85. 1985 Limits and Methods of Measurement of Electromagnetic Interference Generated by Marine Equipment and Installations. 12 pp.

—**Marine—Electrical Equipment**
 BSI BS 7027-90. 1990 Limits and Methods of Measurement of the Immunity of Marine Electrical and Electronic Equipment to Conducted and Radiated Electromagnetic Interference. 14 pp.

—**Marine—Electronic Equipment**
 BSI BS 7027-90. 1990 Limits and Methods of Measurement of the Immunity of Marine Electrical and Electronic Equipment to Conducted and Radiated Electromagnetic Interference. 14 pp.

—**Maritime Mobile Services**
 CCIR RECMN 218-2-90. Prevention of Interference to Radio Reception on Board Ships —Section 8B—Maritime Mobile Service; Telegraphy and Related Subjects. 2 pp.

—**Measuring Relays**
 IEC 255 Pt 22-3-89. Electrical Relays Part 22: Electrical Disturbance Tests for Measuring Relays and Protection Equipment Section Three—Radiated Electromagnetic Field Disturbance Tests First Edition. 37 pp.

—**Mobile Radio Equipment**
 CCIR Report 1103-90. Setting Limits on EMC Parameters for Mobile Service Communication Equipment Based on an Analysis of the Interference Environment—Section 1A—Spectrum Engineering and Computer-Aided Principles and Techniques. 6 pp.

—**Motor Boats**
 IEC CISPR 12-90. Limits and Methods of Measurement of Radio Interference Characteristics of Vehicles, Motor Boats, and Spark-Ignited Engine-Driven Devices Third Edition. 81 pp.
 SAA AS/NZS 2557-92. Limits and Methods of Measurement of Radio Interference Characteristics of Vehicles, Motor Boats and Spark-Ignited Engine-Driven Devices (IEC/CISPR 12:1990) (Supersedes AS 2557–1982). 40 pp.
 SNZ NZS/AS 2557-92. Limits and Methods of Measurement of Radio Interference Characteristics of Vehicles, Motor Boats, and Spark-Ignited Engine-Driven Devices (This is a Joint Standard with SAA AS 2557). 40 pp.

—**Passive Microwave Detectors—Satellites**
 CCIR Report 987-86. Interference to Spaceborne Remote Passive Microwave Sensors from Active Services in Adjacent and Sub-Harmonic Bands—Section 2F—Earth Exploration Satellites. 7 pp.

—**Power Distribution Systems—Telephone Cables**
 CCITT RECMN K.3-89. Interference Caused by Audio-Frequency Signals Injected into a Power Distribution Network—Protection Against Interference (Study Group V) 1 pp. 1 p.

—**Power Systems—Repeaters**
 CCITT RECMN K.2-89. Protection of Repeater Power-Feeding Systems Against Interference from Neighbouring Electricity Lines—Protection Against Interference (Study Group V) 1 pp. 1 p.

—**Radio Receivers**
 CNS C6272-86. Method of Test for Electromagnetic Interference of Frequency Modulation Broadcast Receivers (Dec)(11783).

—**Radio Relay Systems**
 CCIR Report 388-6-90. Methods for Determining the Effects of Interference on the Performance and the Availability of Terrestrial Radio-Relay Systems and Systems in the Fixed-Satellite Service—Section 4/9B—Co-Ordination and Interference Calculations. 40 pp.

—**Reporting—Warning Systems**
 NATO STANAG 6004 ED 3 AMD 0-92. Meaconing, Intrusion Jamming and Interference Report. 5 pp.

Electromagnetic Interference *(Cont.)*

—**Satellites**
 CCIR Report 1137-90. Stochastic Approach in the Evaluation of Interference Between Satellite Networks—Section 4D2—Coordination Methods. 30 pp.

—**Satellites—Statistical Analysis**
 CCIR Report 1137-90. Stochastic Approach in the Evaluation of Interference Between Satellite Networks—Section 4D2—Coordination Methods. 30 pp.

—**Scattered Radiation—Optical Communication Links**
 CCIR Report 667-78. Interference of Optical Communication Links Due to Direct and Scattered Radiation—Section 1D—Spectrum Utilization and Applications. 6 pp.

—**Ships—Antennas**
 CCIR RECMN 218-2-90. Prevention of Interference to Radio Reception on Board Ships —Section 8B—Maritime Mobile Service; Telegraphy and Related Subjects. 2 pp.

—**Ships—Electrical Equipment**
 CCIR RECMN 218-2-90. Prevention of Interference to Radio Reception on Board Ships —Section 8B—Maritime Mobile Service; Telegraphy and Related Subjects. 2 pp.
 NATO STANAG 4435 ED 1 AMD 0-00. Electro-Magnetic Compatibility Testing Procedure and Requirements for Naval Electrical and Electronic Equipment (Surface Ship, Metallic Hull). 58 pp.
 NATO STANAG 4435 ED 1 AMD 0-93. Electromagnetic Compatibility Testing Procedure and Requirements for Naval Electrical and Electronic Equipment (Surface Ship, Metallic Hull). 58 pp.
 NATO STANAG 4436 ED 1 AMD 0-00. Electro-Magnetic Compatibility Testing Procedure and Requirements for Naval Electrical and Electronic Equipment (Surface Ship, Non-Metallic Hull). 56 pp.
 NATO STANAG 4436 ED 1 AMD 0-93. Electro-Magnetic Compatibility Testing Procedure and Requirements for Naval Electrical and Electronic Equipment (Surface Ship, Non-Metallic Hull). 58 pp.

—**Ships—Electronic Equipment**
 NATO STANAG 4435 ED 1 AMD 0-00. Electro-Magnetic Compatibility Testing Procedure and Requirements for Naval Electrical and Electronic Equipment (Surface Ship, Metallic Hull). 58 pp.
 NATO STANAG 4435 ED 1 AMD 0-93. Electromagnetic Compatibility Testing Procedure and Requirements for Naval Electrical and Electronic Equipment (Surface Ship, Metallic Hull). 58 pp.
 NATO STANAG 4436 ED 1 AMD 0-00. Electro-Magnetic Compatibility Testing Procedure and Requirements for Naval Electrical and Electronic Equipment (Surface Ship, Non-Metallic Hull). 56 pp.
 NATO STANAG 4436 ED 1 AMD 0-93. Electro-Magnetic Compatibility Testing Procedure and Requirements for Naval Electrical and Electronic Equipment (Surface Ship, Non-Metallic Hull). 58 pp.

—**Spark Ignition Engines**
 IEC CISPR 12-90. Limits and Methods of Measurement of Radio Interference Characteristics of Vehicles, Motor Boats, and Spark-Ignited Engine-Driven Devices Third Edition. 81 pp.
 SAA AS/NZS 2557-92. Limits and Methods of Measurement of Radio Interference Characteristics of Vehicles, Motor Boats and Spark-Ignited Engine-Driven Devices (IEC/CISPR 12:1990) (Supersedes AS 2557–1982). 40 pp.
 SNZ NZS/AS 2557-92. Limits and Methods of Measurement of Radio Interference Characteristics of Vehicles, Motor Boats, and Spark-Ignited Engine-Driven Devices (This is a Joint Standard with SAA AS 2557). 40 pp.

—**Standard Frequency Satellite Services**
 CCIR RECMN 376-1-66. Avoidance of External Interference with Emissions of the Standard-Frequency Service in the Bands Allocated to That Service—Section 7B—Specifications for the Standard-Frequency and Time-Signal Services. 1 p.
 CCIR RECMN 537-78. Reduction of Mutual Interference Between Emissions of the Standard-Frequency and Time-Signal Service on the Allocated Frequencies in Bands 6 and 7—Section 7C—Systems for Dissemination and Comparison. 1 p.

—**Supersonic Transports**
 CAA STANDARD NO. 8-2 05.73. Electrical Systems—Circuit Control and Protection. 5 pp.
 CAA STANDARD NO. 8-4 06.69. Electrical Systems—Interference. 3 pp.

—**Suppression—Appliances**
 BSI BS 613-77. 1977 Amd 2 Components and Filter Units for Electromagnetic Interference Suppression. 24 pp.

INDUSTRY STANDARDS

Electromagnetic

Electromagnetic Interference (Cont.)

—Suppression—Electrical Equipment
BSI BS 613-77. 1977 Amd 2 Components and Filter Units for Electromagnetic Interference Suppression. 24 pp.

—Suppression—Fixed Inductors—Quality Assurance
CENELEC PREN 138 000-93. Generic Specification: Fixed Inductors for Electromagnetic Interference Suppression. 28 pp.
CENELEC PREN 138 100-93. Sectional Specification: Fixed Inductors for Electromagnetic Interference Suppression —Inductors for Which Safety Tests are Required. 32 pp.
CENELEC PREN 138 101-93. Blank Detail Specification: Fixed Inductors for Electromagnetic Interference Suppression —Inductors for Which Safety Tests are Required. Assessment Level D. 13 pp.

—Suppression—Magnetic Cores
JIS C 2569-90. Ring Cores Made of Ferromagnetic Oxide. 12 pp.

—Telecommunication Systems
BSI BS 6839: Part 1-87. (WITHDRAWN) 1987 Amd 1 Mains Signalling Equipment Part 1: Communications and Interference Limits and Measurements (Superseded by BS EN 50065-1: 1992). 13 pp.

—Telephone Cables
CCITT RECMN K.4-89. Disturbance to Signalling—Protection Against Interference (Study Group V) 2 pp. 2 pp.

—Telephone Cables—Installation—Construction
CCITT Volume 1X. Protection Against Interference—Series K Recommendations Construction, Installation and Protection of Cable and Other Elements of Outside Plant—Series L Recommendations. 174 pp.

—Television Broadcasting
CNS C5216-87. Electromagnetic Interference of Color (Monochrome) Television Broadcast Transmissions (Jul)(12009).
CNS C6289-87. Method of Test for Electromagnetic Interference of Color (Monochrome) Television Broadcast Transmissions (Jul)(12008).

—Television Systems—Frequency Modulation—Fixed Satellite—Sharing
CCIR Report 449-1-74. Measured Interference into Frequency-Modulation Television Systems Using Frequencies Shared Within Systems in the Fixed-Satellite Service or Between These Systems and Terrestrial Systems—Section 4/9B—Co-Ordination and Inter-ference Calculations. 7 pp.

—Time Signal Services
CCIR RECMN 376-1-66. Avoidance of External Interference with Emissions of the Standard-Frequency Service in the Bands Allocated to That Service—Section 7B—Specifications for the Standard-Frequency and Time-Signal Services. 1 p.
CCIR RECMN 537-78. Reduction of Mutual Interference Between Emissions of the Standard-Frequency and Time-Signal Service on the Allocated Frequencies in Bands 6 and 7—Section 7C—Systems for Dissemination and Comparison. 1 p.

—Vehicles
IEC CISPR 12-90. Limits and Methods of Measurement of Radio Interference Characteristics of Vehicles, Motor Boats, and Spark-Ignited Engine-Driven Devices Third Edition. 81 pp.
SAA AS/NZS 2557-92. Limits and Methods of Measurement of Radio Interference Characteristics of Vehicles, Motor Boats and Spark-Ignited Engine-Driven Devices (IEC/CISPR 12:1990) (Supersedes AS 2557–1982). 40 pp.
SNZ NZS/AS 2557-92. Limits and Methods of Measurement of Radio Interference Characteristics of Vehicles, Motor Boats, and Spark-Ignited Engine-Driven Devices (This is a Joint Standard with SAA AS 2557). 40 pp.

Electromagnetic Interference Capacitors
Use: Filter Capacitors

Electromagnetic Interference Filters
Use For: EMI Filters *See Also:* Electromagnetic Equipment; Electromagnetic Radiation; Filter Capacitors; Line Filters (Electric); Optical Filters
CSA C22.2 NO 8-M1986. Electromagnetic Interference (EMI) Filters (R 1992); (Gen Instr 1 Thru 2). 43 pp.
JIS C 5870-92. General Rules of Interference Filters. 8 pp.

Electromagnetic Interference Filters (Cont.)
JIS C 5871-92. Test Methods of Interference Filter. 9 pp.

—Capacitors
CECC CECC 32 400 ISSUE 1-92. Sectional Specification: Fixed Capacitors for Electromagnetic Interference Suppression and Connection to the Supply Mains; (Assessment Level D) (En, Ge) Erratum (En, Ge). 116 pp.

—Passive—Ceramic—Tubular—Military—Preferred Products List
CECC CECC MUAHAG Vol 11 IS 2-92. Preferred Products List; Filters (En, Fr, Ge). 88 pp.

—Passive—Disk—Military—Preferred Products List
CECC CECC MUAHAG Vol 11 IS 2-92. Preferred Products List; Filters (En, Fr, Ge). 88 pp.

—Passive—Military—Preferred Products List
CECC CECC MUAHAG Vol 11 IS 2-92. Preferred Products List; Filters (En, Fr, Ge). 88 pp.

—Passive—Plastic—Tubular—Military—Preferred Products List
CECC CECC MUAHAG Vol 11 IS 2-92. Preferred Products List; Filters (En, Fr, Ge). 88 pp.

—Passive—Power Lines—Military—Preferred Products List
CECC CECC MUAHAG Vol 11 IS 2-92. Preferred Products List; Filters (En, Fr, Ge). 88 pp.

—Passive—Printed Circuit Mount—Military—Preferred Products List
CECC CECC MUAHAG Vol 11 IS 2-92. Preferred Products List; Filters (En, Fr, Ge). 88 pp.

—Power Transmission Lines
MOD UK DSTAN 59-45: 90/005-83. Power Line Interference Suppression Filters Issue 1. 17 pp.
MOD UK DSTAN 59-45: 90/006-83. Power Line Interference Suppression Filters Issue 1. 17 pp.

Electromagnetic Interference Measuring Instruments
See Also: Electromagnetic Equipment; Electromagnetic Radiation; Measuring Instruments
CSA C108.1.1-1977. Electromagnetic Interference Measuring Instrument—C.I.S.P.R. Type; (Amd 1-2 September 1980). 17 pp.
CSA C108.1.2-M1981. Electromagnetic Interference Measuring Instrument—ANSI Type; (Gen Instr 1). 14 pp.

Electromagnetic Intrusion
See Also: Electromagnetic Interference; Electronic Warfare

—Reporting—Warning Systems
NATO STANAG 6004 ED 3 AMD 0-92. Meaconing, Intrusion Jamming and Interference Report. 5 pp.

Electromagnetic Noise
See Also: Crosstalk; Echo (Telecommunications); Electromagnetic Interference; Electromagnetic Radiation; Hum (Electrical); Impulse Noise; Interference Cancellation Systems; Noise (Spurious Signals); Noise Temperature; Radio Frequency Interference; Radio Receivers; White Noise

—Mobile Radio Services—Protection
CCIR Report 926-82. Factors That Should Be Considered When Establishing Protection Criteria for Aeronautical Safety Services—Section 8K—Aeronautical Mobile Service (Terrestrial). 5 pp.

—Mobile Satellite Communications
CCIR Report 764-2-86. Interference and Noise Problems for Maritime Mobile-Satellite Systems Using Frequencies in the Region of 1.5 and 1.6 GHz—Section 8G—Availability, Performance Objectives and Interworking with Terrestrial Networks. 7 pp.

—Power Transmission Lines
CSA CAN3-C108 .3.1-M84. Limits and Measurement Methods of Electro-Magnetic Noise from AC Power Systems, 0.15–30 MHz. 25 pp.

—Transistors
CNS C6100-80. Measurement of Transistor Noise Figure at HF and VHF (Dec)(6800).
CNS C6101-80. Standard for the Measurement of Transistor Noise Figure at Frequencies up to 20KHz by Sinusoidal Signal-Generator Method (Dec)(6801).

Electromagnetic Noise (Cont.)
—Voltage Regulator Diodes
CNS C5054-80. Voltage Regulator Diode Noise Voltage Measurement (Jul)(5764).

Electromagnetic Properties
Scope Note: For additional listings, use a more specific term *See Also:* Chromaticity; Color; Electromagnetic Radiation; Optical Activity; Optical Properties; Photoconductivity; Photoelectric Emission; Power Density; Reflectance; Refractive Index; Transmittance

—Video Tapes
BSI BS 6412-83. (WITHDRAWN) 1983 Amd 1 Measurement of Video Tape Properties (AMD 6315) August 31, 1990 (Superseded by BS EN 60735: 1992). 16 pp.
BSI BS EN 60735-92. 1992 Methods for Measurement of Video Tape Properties (S). 24 pp.
CENELEC HD 454-84. Measuring Methods for Video Tape Properties. 2 pp.
CENELEC EN 60735-91. Measuring Methods for Video Tape Properties. 6 pp.
IEC 735-91. Measuring Methods for Video Tape Properties Second Edition. 33 pp.

Electromagnetic Protection
See Also: Electrical Protection Equipment; Lightning Protection; Overload Protection; Overvoltage Protection

—Radio Stations
CCIR Report 861-82. Protection of Radio Stations Against Lightning and Other Electromagnetic Disturbances—Section 3Ad—Operational Questions. 17 pp.
CCIR QUESTION 153/9-90. Protection of Radio Stations Against Lightning and Other Electromagnetic Disturbances—Questions of Study Group 9 Fixed Service. 1 p.

Electromagnetic Radiation
Use For: Electric Waves; Electromagnetic Energy; Electromagnetic Spectra; Electromagnetic Waves *See Also:* Antennas; Diffraction; Electromagnetic Compatibility; Electromagnetic Fields; Electromagnetic Interference; Electromagnetic Interference Filters; Electromagnetic Interference Measuring Instruments; Electromagnetic Noise; Electromagnetic Properties; Electromagnetic Relays; Electromagnetic Switches; Electromagnetism; Gamma Rays; Infrared Radiation; Laser Radiation; Light (Visible Radiation); Radio Frequency Radiation; Radio Wave Propagation; Radiometry; Reflection; Signal Intensity; Solar Radiation; Ultraviolet Radiation; White Light; X-Rays
CNS Z7196-6-85. Quantities and Units of Light and Related Electromagnetic Radiations (Oct)(11296-6).

—Avionics—Glossaries
NATO STANAG 3968 ED 1 AMD 1-85. NATO Glossary of Electromagnetic Terms and Definitions. 11 pp.
NATO STANAG 3968 ED 2 AMD 0-92. NATO Glossary of Electromagnetic Terminology. 7 pp.
NATO STANAG 3968 ED 2 AMD 1-92. NATO Glossary of Electromagnetic Terminology. 8 pp.

—Data Processing Equipment
CSA C108.8-M1983. Electromagnetic Emissions from Data Processing Equipment and Electronic Office Machines (R 1989); (Amd 1-5 September 1983). 20 pp.

—Flammable Atmospheres
BSI BS 6656-91. 1991 Prevention of Inadvertent Ignition of Flammable Atmospheres by Radio-Frequency Radiation. 55 pp.
BSI BS 6656-86. 1986 Guide to Prevention of Inadvertent Ignition of Flammable Atmospheres by Radio-Frequency Radiation. 52 pp.

—Initiators (Explosives)
BSI BS 6657-91. 1991 Prevention of Inadvertent Initiation of Electro-Explosive Devices by Radio-Frequency Radiation. 33 pp.
BSI BS 6657-86. 1986 Guide to Prevention of Inadvertent Initiation of Electro-Explosive Devices by Radio-Frequency Radiation. 32 pp.

—Initiators (Explosives)—Munitions
NATO STANAG 4234 ED 1 AMD 0-00. (DRAFT) Electromagnetic Radiation (Radio Frequency)—200 khz to 40 GHz Environment—Affecting the Design of Material for Use by NATO Forces. 9 pp.
NATO STANAG 4234 ED 1 AMD 0-92. Electromagnetic Radiation (Radio Frequency)—200 kHz to 40 GHz Environment—Affecting the Design of Materiel for Use by NATO Forces. 10 pp.

Electromagnetic Radiation *(Cont.)*
—Initiators (Explosives)—Munitions *(Cont.)*
NATO STANAG 4324 ED 1 AMD 0-91. Electromagnetic Radiation (Radio Frequency) Test Information to Determine the Safety and Suitability for Service of Electro-Explosive Devices and Associated Electronic Systems in Munitions and Weapon Systems. 11 pp.

—Initiators (Explosives)—Weapons Systems
NATO STANAG 4234 ED 1 AMD 0-00. (DRAFT) Electromagnetic Radiation (Radio Frequency)—200 khz to 40 GHz Environment—Affecting the Design of Material for Use by NATO Forces. 9 pp.

NATO STANAG 4234 ED 1 AMD 0-92. Electromagnetic Radiation (Radio Frequency)—200 kHz to 40 GHz Environment—Affecting the Design of Materiel for Use by NATO Forces. 10 pp.

NATO STANAG 4324 ED 1 AMD 0-91. Electromagnetic Radiation (Radio Frequency) Test Information to Determine the Safety and Suitability for Service of Electro-Explosive Devices and Associated Electronic Systems in Munitions and Weapon Systems. 11 pp.

—Measurement
SNZ NZS 6609: Part 2-90. Radiofrequency Radiation Part 2: Principles and Methods of Measurement—300 kHz to 100 GHz Amend: A, 1990. 36 pp.

—Office Machines
CSA C108.8-M1983. Electromagnetic Emissions from Data Processing Equipment and Electronic Office Machines (R 1989); (Amd 1-5 September 1983). 20 pp.

—Safety Levels
CNS C5063-80. Safety Level of Electromagnetic Radiation with Respect to Personnel (Aug)(6141).

SNZ NZS 6609: Part 1-90. Radiofrequency Radiation Part 1: Maximum Exposure Levels—100 kHz to 300 GHz Amend: A, 1990. 20 pp.

—Sound Broadcasting—Glossaries
CCIR RECMN 561-2-86. Definitions of Radiation in LF, MF and HF Broadcasting Bands—Section 10A-1—Amplitude-Modulation Sound Broadcasting in Bands 5 (LF), 6 (MF) and 7 (HF). 3 pp.

—Telecommunication Systems
CCIR Report 666-78. System for Telecommunications Using the Electromagnetic Spectrum Above 3000 GHz—Section 1D—Spectrum Utilization and Applications. 9 pp.

CCIR Report 680-2-90. Techniques for Telecommunication by Means of Electromagnetic Waves in the Infra-Red and Visible Regions of the Spectrum—Section 2B—Topics of General Interest. 16 pp.

—Units of Measurement
BSI BS 5775: Part 6-93. 1993 Quantities, Units and Symbols Part 6: Light and Related Electromagnetic Radiations (ISO 31-6: 1992) (G). 27 pp.

BSI BS 5775: Part 6-82. 1982 Amd 1 Specification for Quantities, Units and Symbols Part 6: Light and Related Electromagnetic Radiations (AMD 5851) July 31, 1989. 17 pp.

ISO 31 Pt 6-92. Quantities and Units—Part 6: Light and Related Electromagnetic Radiations Third Edition. 26 pp.

Electromagnetic Relays
See Also: Electromagnetic Equipment; Electromagnetic Radiation; Relays

CNS C3179-83. Method of Test for Hinge Type Electromagnetic Relays (Oct)(10601).

CNS C4395-83. Hinge Type Electromagnetic Relays (Oct)(10600).

JIS C 4530-87. Hinge Type Electromagnetic Relays. 28 pp.

—Aerospace
CEN PREN 2548-001-89. Electromagentic Relays 10A Plug-in Type, Two and Four Pole. 6 pp.

CEN PREN 2548-003-89. Electromagnetic Relays 10A Plug-in Type, Two and Four Pole Without Mounting Hardware Dimensions Detail Standard. 5 pp.

CEN PREN 2548-004-89. Electromagnetic Relays 10A Plug-In Type Two and Four Pole with Imperial Mounting Hardware Dimensions Detail Specification. 5 pp.

CEN PREN 2548-005-89. Electromagnetic Relays 10A Plug-In Type Two and Four Pole with Metric Mounting Hardware Dimensions Detail Specification. 5 pp.

CEN PREN 3203-001-90. Electromagnetic Relays 5A Plug-in Type, Two and Four Pole Technical Specification. 6 pp.

CEN PREN 3203-003-90. Electromagnetic Relays 5A Plug-in Type, Two and Four Pole Without Mounting Hardware—Dimensions—Product Standards. 5 pp.

Electromagnetic Relays *(Cont.)*
—Aerospace *(Cont.)*
CEN PREN 3204-001-90. Electromagnetic Relays 25A Plug-in Type, One and Three Pole Technical Specification. 7 pp.

CEN PREN 3204-003-90. Electromagnetic Relays 25A Plug-in Type, One and Three Pole Without Mounting Hardware—Dimensions—Product Standard. 5 pp.

CEN PREN 3204-004-90. Electromagnetic Relays 25 A Plug-in Type, One and Three Pole with Unified Mounting Hardware—Dimensions—Product Standard. 5 pp.

CEN PREN 3204-005-90. Electromagnetic Relays 25A Plug-in Type, One and Three Pole with Metric Mounting Hardware—Dimensions—Product Standard. 5 pp.

—Aerospace—Bases
CEN PREN 2593-001-89. Bases for 10A Plug-In Relays, Two and Four Poles. 6 pp.

CEN PREN 2593-003-89. Base with Metric Hardware Non-Captive for Electromagentic Relays Two and Four Poles to EN 2548-003 Dimensions Detail Specification. 5 pp.

CEN PREN 2593-005-89. Base with Metric Hardware Non-Captive for Electromagnetic Relays Two and Four Poles to EN 2548-005 Dimensions Detail Specification. 5 pp.

CENELEC PREN 2593-004-89. Base with Imperial Hardware Non-Captive for Electromagnetic Relays Two and Four Poles to EN 2548-004—Dimensions—Specification-. 5 pp.

—Aircraft
ISO 2315-80. Aircraft—Two-and Four-Pole Sealed Electromagnetic Relays, 2 A and 3 A—Clearance and Fixing Dimensions Second Edition. 10 pp.

—Control Circuits
JIS C 5440-80. General Rules for Reliability Assured Lowpower Electromagnetic Relays for Industrial Control Circuits.

JIS C 5442-86. Test Methods of Low Power Electromagnetic Relays for Industrial Control Circuits. 70 pp.

Electromagnetic Spectra
Use: Electromagnetic Radiation

Electromagnetic Switches
See Also: Electric Switches; Electromagnetic Equipment; Electromagnetic Radiation

CNS C3149-82. Method for Test for A.C. Electromagnetic Switches (May)(8796).

CNS C4084-82. A. C. Electromagnetic Switches (May)(2930). 11 pp.

JIS C 8325-83. AC Electromagnetic Switches. 26 pp.

—Remote Control
IEC 669 Pt 2-2-84. Switches for Household and Similar Fixed-Electrical Installations Part 2: Particular Requirements for Electromagnetic Remote-Control Switches (R.C.S.) First Edition. 21 pp.

Electromagnetic Testing
See Also: Eddy Current Testing

—Electric Coils—Symbols
DIN ENGL 54140 Pt 3-89. Non-Destructive Testing; Electromagnetic Methods; Representation and General Characteristics of Test Coil Assemblies (Jan). 6 pp.

—Steel Pipes
CEN PREN 29302-91. Seamless and Welded (Except Submerged Arc-Welded) Steel Tubes for Pressure Purposes—Electromagnetic Testing for Verification of Hydraulic Leak—Tightness. 9 pp.

ISO 9302-89. Seamless and Welded (Except Submerged Arc-Welded) Steel Tubes for Pressure Purposes—Electromagnetic Testing for Verification of Hydraulic Leak-Tightness First Edition. 8 pp.

Electromagnetic Warfare
Use: Electronic Warfare

Electromagnetic Waves
Use: Electromagnetic Radiation

Electromagnetism
See Also: Electricity; Electromagnetic Radiation

—Glossaries
BSI BS 4727:Pt1: Group 01-83. 1983 Amd 1 Electrotechnical, Power, Telecommunication, Electronics, Lighting and Colour Terms Part 1: Terms Common to Power, Telecommunications and Electronics Group 01: Fundamental Terminology. 56 pp.

IEC. Vocabulary of Fundamental Concepts. 237 pp.

IEC 50 Chap 121-78. International Electrotechnical Vocabulary Chapter 121: Electromagnetism. 61 pp.

Electromagnetism *(Cont.)*
—Glossaries *(Cont.)*
SAA AS 1852.121-79. International Electrotechnical Vocabulary—Part 121: Electromagnetism Being IEC 50(121). 59 pp.

SNZ IEC 50: 50(121)-78. International Electrotechnical Vocabulary 50(121): Electromagnetism. 51 pp.

Electromechanical Components
Scope Note: For additional listings, see also specific types of equipment *See Also:* Electrical Components; Mechanical Components

BSI BS 5772: Part 1-85. 1985 Basic Testing Procedures and Measuring Methods for Electromechanical Components for Electronic Equipment Part 1: General. 12 pp.

BSI BS CECC 00009-82. 1982 Basic Testing Procedures and Measuring Methods for Electromechanical Components: Basic Specification. 30 pp.

—Accelerated Testing
BSI BS 5772: Part 4-79. 1979 Basic Testing Procedures and Measuring Methods for Electromechancial Components for Electronic Equipment Part 4: Dynamic Stress Tests. 8 pp.

IEC 512 Pt 4-76. Electromechanical Components for Electronic Equipment; Basic Testing Procedures and Measuring Methods Part 4: Dynamic Stress Tests First Edition. 17 pp.

—Cable Clamp Testing
BSI BS 5772: Part 9-93. 1993 Electromechanical Components for Electronic Equipment: Basic Testing Procedures and Measuring Methods Part 9: Miscellaneous Tests (IEC 512-9: 1992) (S). 20 pp.

BSI BS 5772: Part 9-79. 1979 Basic Testing Procedures and Measuring Methods for Electromechanical Components for Electronic Equipment Part 9: Cable Clamping Tests, Explosion Hazard Tests, r.f Resistance Test, Capacitance Tests Sheilding and Filtering. 12 pp.

IEC 512 Pt 9-92. Electromechanical Components for Electronic Equipment; Basic Testing Procedures and Measuring Methods Part 9: Miscellaneous Tests Second Edition. 42 pp.

—Capacitance Measurement
BSI BS 5772: Part 9-93. 1993 Electromechanical Components for Electronic Equipment: Basic Testing Procedures and Measuring Methods Part 9: Miscellaneous Tests (IEC 512-9: 1992) (S). 20 pp.

BSI BS 5772: Part 9-79. 1979 Basic Testing Procedures and Measuring Methods for Electromechanical Components for Electronic Equipment Part 9: Cable Clamping Tests, Explosion Hazard Tests, r.f Resistance Test, Capacitance Tests Sheilding and Filtering. 12 pp.

IEC 512 Pt 9-92. Electromechanical Components for Electronic Equipment; Basic Testing Procedures and Measuring Methods Part 9: Miscellaneous Tests Second Edition. 42 pp.

—Comprehensive Testing
CECC CECC 00 009 ISSUE 1-80. Basic Specification: Basic Testing Procedures and Measuring Methods for Electromechanical Components (En, Fr, Ge). 48 pp.

IEC 512 Pt 1-84. Electromechanical Components for Electronic Equipment; Basic Testing Procedures and Measuring Methods Part 1: General Second Edition; (Amendment 1-1988). 29 pp.

—Connectors—Mechanical Testing
IEC 512 Pt 8-93. Electromechanical Components for Electronic Equipment; Basic Testing Procedures and Measuring Methods Part 8: Connector Tests (Mechanical) and Mechanical Tests on Contacts and Terminations Third Edition. 72 pp.

—Contact Resistance
BSI BS 5772: Part 2-79. 1979 Basic Testing Procedures and Measuring Methods for Electromechanical Components for Electronic Equipment Part 2: General Examination, Electrical Continuity and Contact Resistance Tests, Insulation Tests and Voltage Stress Tests. 16 pp.

BSI BS 5772: Pt 2: Supp 1-81. 1981 Electromechanical Components for Electronic Equipment: Basic Testing Procedures Part 2: General Examination, Electrical Continuity and Contract Resistance Tests, Insulation Tests and Voltage Stress Supp 1: 1981 Test 4c. 4 pp.

IEC 512 Pt 2-85. Electromechanical Components for Electronic Equipment; Basic Testing Procedures and Measuring Methods Part 2: General Examination, Electrical Continuity and Contact Resistance Tests, Insulation Tests and Voltage Stress Tests Second Edition. 39 pp.

—Contacts
SAA AS 1029.1-85. Low Voltage Contactors—Part 1: Electromechanical—up to and Including 1000 V a.c. and 1200 V d.c. 40 pp.

Electromechanical Components (Cont.)

—Contacts (Cont.)
SNZ NZS/AS 1029. 1-85. Low Voltage Contactors Part 1: Electromechanical—up to and Including 1000 V a.c. and 1200 V d.c. (This is a Joint Standard with SAA AS 1029.1). 40 pp.

—Contacts—Mechanical Testing
IEC 512 Pt 8-93. Electromechanical Components for Electronic Equipment; Basic Testing Procedures and Measuring Methods Part 8: Connector Tests (Mechanical) and Mechanical Tests on Contacts and Terminations Third Edition. 72 pp.

—Continuity Testing
BSI BS 5772: Part 2-79. 1979 Basic Testing Procedures and Measuring Methods for Electromechanical Components for Electronic Equipment Part 2: General Examination, Electrical Continuity and Contact Resistance Tests, Insulation Tests and Voltage Stress Tests. 16 pp.

BSI BS 5772: Pt 2: Supp 1-81. 1981 Electromechanical Components for Electronic Equipment: Basic Testing Procedures Part 2: General Examination, Electrical Continuity and Contract Resistance Tests, Insulation Tests and Voltage Stress Supp 1: 1981 Test 4c. 4 pp.

IEC 512 Pt 2-85. Electromechanical Components for Electronic Equipment; Basic Testing Procedures and Measuring Methods Part 2: General Examination, Electrical Continuity and Contact Resistance Tests, Insulation Tests and Voltage Stress Tests Second Edition. 39 pp.

—Current Carrying Capacity
BSI BS 5772: Part 3-79. 1979 Basic Testing Procedures and Measuring Methods for Electromechanical Components for Electronic Methods Part 3: Current Carrying Capacity Tests. 8 pp.

IEC 512 Pt 3-76. Electromechanical Components for Electronic Equipment; Basic Testing Procedures and Measuring Methods Part 3: Current-Carrying Capacity Tests First Edition. 16 pp.

—Endurance Testing
BSI BS 5772: Part 5-79. 1979 Basic Testing Procedures and Measuring Methods for Electromechanical Components for Electronic Equipment Part 5: Impact Tests (Free Components), Static Load Tests (Fixed Components), Endurance Tests and Overload Tests (Switches). 16 pp.

BSI BS 5772: Pt 5: Supp 1-81. 1981 Basic Testing Procedures and Measuring Methods for Electromechanical Components for Electronic Equip. Part 5: Impact Tests (Free Components), Static Load Tests (FixedComponents) Endurance Test and Over-Load Tests (Switches) Supp 1: Tests 7b and 10a. 6 pp.

BSI BS 5772: Pt 5: Supp 2-82. 1982 Basic Testing Procedures and Measuring Methods for Electromechanical Components for Electronic Equip. Part 5: Impact Tests (Free Components), Static Load Tests (FixedComp.), Enduranceand Overload Tests (Switches) Supp. 2:Test 10c. 6 pp.

—Environmental Testing
BSI BS 5772: Part 6-84. 1984 Basic Testing Procedures and Measuring Methods for Electromechanical Components for Electronic Equipment Part 6: Climatic Tests and Soldering Tests. 26 pp.

BSI BS 5772: Pt 6: Supp 1-81. (WITHDRAWN) 1981 Basic Testing Procedures and Electromechanical Components for Electronic Equipment Part 6: Climatic Tests and Soldering Tests Supplement 1: Test 11b (Superseded by BS 5772: Part 6: 1984). 4 pp.

IEC 512 Pt 6-84. Electromechanical Components for Electronic Equipment; Basic Testing Procedures and Measuring Methods Part 6: Climatic Tests and Soldering Tests Second Edition. 57 pp.

—Error Control
CCITT RECMN V.40-89. Error Indication with Electromechanical Equipment—Data Communication over the Telephone Network (Study Group XVII) 1 pp. 1 p.

—Examination (Quality Assurance)
BSI BS 5772: Part 2-79. 1979 Basic Testing Procedures and Measuring Methods for Electromechanical Components for Electronic Equipment Part 2: General Examination, Electrical Continuity and Contact Resistance Tests, Insulation Tests and Voltage Stress Tests. 16 pp.

BSI BS 5772: Pt 2: Supp 1-81. 1981 Electromechanical Components for Electronic Equipment: Basic Testing Procedures Part 2: General Examination, Electrical Continuity and Contract Resistance Tests, Insulation Tests and Voltage Stress Supp 1: 1981 Test 4c. 4 pp.

Electromechanical Components (Cont.)

—Examination (Quality Assurance) (Cont.)
IEC 512 Pt 2-85. Electromechanical Components for Electronic Equipment; Basic Testing Procedures and Measuring Methods Part 2: General Examination, Electrical Continuity and Contact Resistance Tests, Insulation Tests and Voltage Stress Tests Second Edition. 39 pp.

—Explosion Hazard Testing
BSI BS 5772: Part 9-93. 1993 Electromechanical Components for Electronic Equipment: Basic Testing Procedures and Measuring Methods Part 9: Miscellaneous Tests (IEC 512-9: 1992) (S). 20 pp.

BSI BS 5772: Part 9-79. 1979 Basic Testing Procedures and Measuring Methods for Electromechanical Components for Electronic Equipment Part 9: Cable Clamping Tests, Explosion Hazard Tests, r.f Resistance Test, Capacitance Tests Sheilding and Filtering. 12 pp.

IEC 512 Pt 9-92. Electromechanical Components for Electronic Equipment; Basic Testing Procedures and Measuring Methods Part 9: Miscellaneous Tests Second Edition. 42 pp.

—Fatigue Testing
BSI BS 5772: Part 5-93. 1993 Electromechanical Components for Electronic Equipment: Testing Procedures and Measuring Methods Part 5: Impact Tests (Free Components), Static Load Tests (Fixed Components), Endurance Tests and Overload Tests (IEC 512-5: 1992) (S). 24 pp.

IEC 512 Pt 5-92. Electromechanical Components for Electronic Equipment; Basic Testing Procedures and Measuring Methods Part 5: Impact Tests (Free Components), Static Load Tests (Fixed Components), Endurance Tests and Overload Tests Second Edition. 50 pp.

—Flammability Testing
BSI BS 5772: Part 9-93. 1993 Electromechanical Components for Electronic Equipment: Basic Testing Procedures and Measuring Methods Part 9: Miscellaneous Tests (IEC 512-9: 1992) (S). 20 pp.

BSI BS 5772: Part 9-79. 1979 Basic Testing Procedures and Measuring Methods for Electromechanical Components for Electronic Equipment Part 9: Cable Clamping Tests, Explosion Hazard Tests, r.f Resistance Test, Capacitance Tests Sheilding and Filtering. 12 pp.

IEC 512 Pt 9-92. Electromechanical Components for Electronic Equipment; Basic Testing Procedures and Measuring Methods Part 9: Miscellaneous Tests Second Edition. 42 pp.

—Glossaries
BSI BS 4727:Pt1: Group 13-91. 1991 Electrotechnical, Power, Telecommunication, Electronics, Lighting and Colour Terms Part 1: Terms Common to Power, Telecommunications and Electronics Group 13: Electromechanical Components for Electronic Equipment (G) (IEC 50(581): 1978). 92 pp.

IEC 50 Chap 581-78. Advance Edition of the International Electrotechnical Vocabulary Chapter 581: Electromechanical Components for Electronic Equipment. 95 pp.

SAA AS 1852.581-79. International Electrotechnical Vocabulary—Part 581: Electromechanical Components for Electronic Equipment Being IEC 50(581). 93 pp.

SNZ IEC 50: 50(581)-78. International Electrotechnical Vocabulary 50(581): Electromechanical Components for Electronic Equipment. 85 pp.

—Hydroelectric Power Plants
IEC 1116-92. Electromechanical Equipment Guide for Small Hydroelectric Installations First Edition. 112 pp.

—Impact Testing
BSI BS 5772: Part 5-93. 1993 Electromechanical Components for Electronic Equipment: Testing Procedures and Measuring Methods Part 5: Impact Tests (Free Components), Static Load Tests (Fixed Components), Endurance Tests and Overload Tests (IEC 512-5: 1992) (S). 24 pp.

BSI BS 5772: Part 5-79. 1979 Basic Testing Procedures and Measuring Methods for Electromechanical Components for Electronic Equipment Part 5: Impact Tests (Free Components), Static Load Tests (Fixed Components), Endurance Tests and Overload Tests (Switches). 16 pp.

BSI BS 5772: Pt 5: Supp 1-81. 1981 Basic Testing Procedures and Measuring Methods for Electromechanical Components for Electronic Equip. Part 5: Impact Tests (Free Components), Static Load Tests (FixedComponents) Endurance Test and Over-Load Tests (Switches) Supp 1: Tests 7b and 10a. 6 pp.

Electromechanical Components (Cont.)

—Impact Testing (Cont.)
BSI BS 5772: Pt 5: Supp 2-82. 1982 Basic Testing Procedures and Measuring Methods for Electromechanical Components for Electronic Equip. Part 5: Impact Tests (Free Components), Static Load Tests (FixedComp.), Enduranceand Overload Tests (Switches) Supp. 2:Test 10c. 6 pp.

IEC 512 Pt 5-92. Electromechanical Components for Electronic Equipment; Basic Testing Procedures and Measuring Methods Part 5: Impact Tests (Free Components), Static Load Tests (Fixed Components), Endurance Tests and Overload Tests Second Edition. 50 pp.

—Insulation Resistance
BSI BS 5772: Part 2-79. 1979 Basic Testing Procedures and Measuring Methods for Electromechanical Components for Electronic Equipment Part 2: General Examination, Electrical Continuity and Contact Resistance Tests, Insulation Tests and Voltage Stress Tests. 16 pp.

BSI BS 5772: Pt 2: Supp 1-81. 1981 Electromechanical Components for Electronic Equipment: Basic Testing Procedures Part 2: General Examination, Electrical Continuity and Contract Resistance Tests, Insulation Tests and Voltage Stress Supp 1: 1981 Test 4c. 4 pp.

IEC 512 Pt 2-85. Electromechanical Components for Electronic Equipment; Basic Testing Procedures and Measuring Methods Part 2: General Examination, Electrical Continuity and Contact Resistance Tests, Insulation Tests and Voltage Stress Tests Second Edition. 39 pp.

—Leakage
IEC 512 Pt 7-93. Electromechanical Components for Electronic Equipment; Basic Testing Procedures and Measuring Methods Part 7: Mechanical Operating Tests and Sealing Tests Third Edition. 40 pp.

—Mechanical Shock
BSI BS 5772: Part 4-79. 1979 Basic Testing Procedures and Measuring Methods for Electromechanical Components for Electronic Equipment Part 4: Dynamic Stress Tests. 8 pp.

IEC 512 Pt 4-76. Electromechanical Components for Electronic Equipment; Basic Testing Procedures and Measuring Methods Part 4: Dynamic Stress Tests First Edition. 17 pp.

—Mechanical Testing
BSI BS 5772: Part 7-81. 1981 Basic Testing Procedures and Measuring Methods for Electromechanical Components for Electronic Equipment Part 7: Mechanical Operating Tests and Sealing Tests. 12 pp.

BSI BS 5772: Part 8-85. 1985 Amd 1 Basic Testing Procedures and Measuring Methods for Electromechanical Components for Electronic Equip. Part 8: Connector Tests (Mechanical) and Mech. Tests on Contacts and Terminations (AMD 5076) May 31, 1990. 32 pp.

IEC 512 Pt 7-93. Electromechanical Components for Electronic Equipment; Basic Testing Procedures and Measuring Methods Part 7: Mechanical Operating Tests and Sealing Tests Third Edition. 40 pp.

IEC 512 Pt 8-93. Electromechanical Components for Electronic Equipment; Basic Testing Procedures and Measuring Methods Part 8: Connector Tests (Mechanical) and Mechanical Tests on Contacts and Terminations Third Edition. 72 pp.

—Overload Testing
BSI BS 5772: Part 5-93. 1993 Electromechanical Components for Electronic Equipment: Testing Procedures and Measuring Methods Part 5: Impact Tests (Free Components), Static Load Tests (Fixed Components), Endurance Tests and Overload Tests (IEC 512-5: 1992) (S). 24 pp.

BSI BS 5772: Part 5-79. 1979 Basic Testing Procedures and Measuring Methods for Electromechanical Components for Electronic Equipment Part 5: Impact Tests (Free Components), Static Load Tests (Fixed Components), Endurance Tests and Overload Tests (Switches). 16 pp.

BSI BS 5772: Pt 5: Supp 1-81. 1981 Basic Testing Procedures and Measuring Methods for Electromechanical Components for Electronic Equip. Part 5: Impact Tests (Free Components), Static Load Tests (FixedComponents) Endurance Test and Over-Load Tests (Switches) Supp 1: Tests 7b and 10a. 6 pp.

BSI BS 5772: Pt 5: Supp 2-82. 1982 Basic Testing Procedures and Measuring Methods for Electromechanical Components for Electronic Equip. Part 5: Impact Tests (Free Components), Static Load Tests (FixedComp.), Enduranceand Overload Tests (Switches) Supp. 2:Test 10c. 6 pp.

INTERNATIONAL AND NON-U.S. NATIONAL STANDARDS
SUBJECT INDEX

Electromechanical Components (Cont.)

—Overload Testing (Cont.)
IEC 512 Pt 5-92. Electromechanical Components for Electronic Equipment; Basic Testing Procedures and Measuring Methods Part 5: Impact Tests (Free Components), Static Load Tests (Fixed Components), Endurance Tests and Overload Tests Second Edition. 50 pp.

—Radio Frequency Resistance
BSI BS 5772: Part 9-93. 1993 Electromechanical Components for Electronic Equipment: Basic Testing Procedures and Measuring Methods Part 9: Miscellaneous Tests (IEC 512-9: 1992) (S). 20 pp.
BSI BS 5772: Part 9-79. 1979 Basic Testing Procedures and Measuring Methods for Electromechanical Components for Electronic Equipment Part 9: Cable Clamping Tests, Explosion Hazard Tests, r.f Resistance Test, Capacitance Tests Sheilding and Filtering. 12 pp.
IEC 512 Pt 9-92. Electromechanical Components for Electronic Equipment; Basic Testing Procedures and Measuring Methods Part 9: Miscellaneous Tests Second Edition. 42 pp.

—Sealing
BSI BS 5772: Part 7-81. 1981 Basic Testing Procedures and Measuring Methods for Electromechanical Components for Electronic Equipment Part 7: Mechanical Operating Tests and Sealing Tests. 12 pp.
IEC 512 Pt 7-93. Electromechanical Components for Electronic Equipment; Basic Testing Procedures and Measuring Methods Part 7: Mechanical Operating Tests and Sealing Tests Third Edition. 40 pp.

—Shielding Testing
BSI BS 5772: Part 9-93. 1993 Electromechanical Components for Electronic Equipment: Basic Testing Procedures and Measuring Methods Part 9: Miscellaneous Tests (IEC 512-9: 1992) (S). 20 pp.
BSI BS 5772: Part 9-79. 1979 Basic Testing Procedures and Measuring Methods for Electromechanical Components for Electronic Equipment Part 9: Cable Clamping Tests, Explosion Hazard Tests, r.f Resistance Test, Capacitance Tests Sheilding and Filtering. 12 pp.
IEC 512 Pt 9-92. Electromechanical Components for Electronic Equipment; Basic Testing Procedures and Measuring Methods Part 9: Miscellaneous Tests Second Edition. 42 pp.

—Soldering
IEC 512 Pt 6-84. Electromechanical Components for Electronic Equipment; Basic Testing Procedures and Measuring Methods Part 6: Climatic Tests and Soldering Tests Second Edition. 57 pp.

—Static Loads
BSI BS 5772: Part 5-93. 1993 Electromechanical Components for Electronic Equipment: Testing Procedures and Measuring Methods Part 5: Impact Tests (Free Components), Static Load Tests (Fixed Components), Endurance Tests and Overload Tests (IEC 512-5: 1992) (S). 24 pp.
BSI BS 5772: Part 5-79. 1979 Basic Testing Procedures and Measuring Methods for Electromechanical Components for Electronic Equipment Part 5: Impact Tests (Free Components), Static Load Tests (Fixed Components), Endurance Tests and Overload Tests (Switches). 16 pp.
BSI BS 5772: Pt 5: Supp 1-81. 1981 Basic Testing Procedures and Measuring Methods for Electromechanical Components for Electronic Equip. Part 5: Impact Tests (Free Components), Static Load Tests (FixedComponents) Endurance Test and Over-Load Tests (Switches) Supp 1: Tests 7b and 10a. 6 pp.
BSI BS 5772: Pt 5: Supp 2-82. 1982 Basic Testing Procedures and Measuring Methods for Electromechanical Components for Electronic Equip. Part 5: Impact Tests (Free Components), Static Load Tests (FixedComp.), Enduranceand Overload Tests (Switches) Supp. 2:Test 10c. 6 pp.
IEC 512 Pt 5-92. Electromechanical Components for Electronic Equipment; Basic Testing Procedures and Measuring Methods Part 5: Impact Tests (Free Components), Static Load Tests (Fixed Components), Endurance Tests and Overload Tests Second Edition. 50 pp.

—Switches
IECQ QC 960000-91. Electromechanical Switches for Use in Electronic Equipment Part 1: Generic Specification (IEC 1020-1 ED 1). 78 pp.

—Switches—Quality Assurance
BSI BS QC 960000-92. 1992 Electromechanical Switches for Use in Electronic Equipment Generic Specification (IEC 1020-1: 1991). 39 pp.
IEC 1020 Pt 1-91. Electromechanical Switches for Use in Electronic Equipment Part 1: Generic Specification; (IECQ QC 960000). 78 pp.

Electromechanical Components (Cont.)

—Terminals—Mechanical Testing
IEC 512 Pt 8-93. Electromechanical Components for Electronic Equipment; Basic Testing Procedures and Measuring Methods Part 8: Connector Tests (Mechanical) and Mechanical Tests on Contacts and Terminations Third Edition. 72 pp.

—Vibration
BSI BS 5772: Part 4-79. 1979 Basic Testing Procedures and Measuring Methods for Electromechancial Components for Electronic Equipment Part 4: Dynamic Stress Tests. 8 pp.
IEC 512 Pt 4-76. Electromechanical Components for Electronic Equipment; Basic Testing Procedures and Measuring Methods Part 4: Dynamic Stress Tests First Edition. 17 pp.

—Voltage Stress
BSI BS 5772: Part 2-79. 1979 Basic Testing Procedures and Measuring Methods for Electromechanical Components for Electronic Equipment Part 2: General Examination, Electrical Continuity and Contact Resistance Tests, Insulation Tests and Voltage Stress Tests. 16 pp.
BSI BS 5772: Pt 2: Supp 1-81. 1981 Electromechanical Components for Electronic Equipment: Basic Testing Procedures Part 2: General Examination, Electrical Continuity and Contract Resistance Tests, Insulation Tests and Voltage Stress Supp 1: 1981 Test 4c. 4 pp.
IEC 512 Pt 2-85. Electromechanical Components for Electronic Equipment; Basic Testing Procedures and Measuring Methods Part 2: General Examination, Electrical Continuity and Contact Resistance Tests, Insulation Tests and Voltage Stress Tests Second Edition. 39 pp.

Electromechanical Flip Flops
Use: Bistable Relays

Electromechanical Plotters
Use: Vector Plotters

Electromechanical Relays
Use: Relays

Electromechanical Switches
Use: Electric Switches

Electromechanical Timers
See Also: Time Measuring Instruments; Timers
CNS C3125-84. Method of Test for Motor Driven Timer (Jun)(7628).
CNS C4326-84. Motor Driven Timer (Jun)(7627).

Electromedical Equipment
Use: Medical Electrical Equipment; Medical Equipment

Electrometric Analysis
See Also: Amperometric Titration; Coulometric Analysis; Electrolytic Analysis; Polarographic Analysis; Potentiometric Analysis

—Aluminum Fluoride
CNS M3121-83. Method for Determining Water Content in Natural and Artificial Cryolite and Aluminium for Industrial Use (Electrometric Method) (Mar)(10120).
ISO 3392-76. Cryolite, Natural and Artificial, and Aluminium Fluoride for Industrial Use—Determination of Water Content—Electrometric Method First Edition. 7 pp.

—Cryolite
CNS M3121-83. Method for Determining Water Content in Natural and Artificial Cryolite and Aluminium for Industrial Use (Electrometric Method) (Mar)(10120).
ISO 3392-76. Cryolite, Natural and Artificial, and Aluminium Fluoride for Industrial Use—Determination of Water Content—Electrometric Method First Edition. 7 pp.

—Hydrocarbons
CNS K6522-80. Method of Test for Bromine Index of Petroleum Hydrocarbons by Electrometric Titration (Jul)(5839).

—Iron
BSI BS EN 10 071-91. 1991 Chemical Analysis of Ferrous Materials Determination of Manganese in Steels and Irons Electrometric Titration Method. 9 pp.
CEN EN 10 071-89. Chemical Analysis of Ferrous Materials Determination of Manganese in Steels and Irons Electrometric Titration Method. 5 pp.

Electrometric Analysis (Cont.)

—Iron (Cont.)
DIN ENGL EN 10071-90. Chemical Analysis of Ferrous Materials; Determination of Manganese in Steel and Iron; Electrometric Titration Method (Apr). 6 pp.

—Liquefied Petroleum Gas
BSI BS 2000: Part 272-93. 1993 Methods of Test for Petroleum and Its Products Part 272: Determination of Mercaptan Sulphur and Hydrogen Sulphide Content of LPG—Electrometric Titration Method (W). 7 pp.
BSI BS 2000: Part 272-87. 1987 Petroleum and Its Products Part 272: Determination of Mercaptan Sulphur and Hydrogen Sulphide Content of Liquefied Petroleum Gases by Electometric Titration. 11 pp.

—Olefins
BSI BS 5089-80. 1980 Amd 1 Determination of Bromine Number of Petroleum Distillates and Commercial Aliphatic Olefins by Electrometric Method. 16 pp.
CNS K6523-80. Method of Test for Bromine Number of Petroleum Distillates and Commerical Aliphatic Olefins by Electrometric Titration (Jul)(5840).
ISO 3839-78. Petroleum Distillates and Commercial Aliphatic Olefins—Determination of Bromine Number—Electrometric Method First Edition; (Erratum—Nov 1979) (Erratum—June 1980). 16 pp.

—Petroleum Products
BSI BS 5089-80. 1980 Amd 1 Determination of Bromine Number of Petroleum Distillates and Commercial Aliphatic Olefins by Electrometric Method. 16 pp.
CNS K6523-80. Method of Test for Bromine Number of Petroleum Distillates and Commerical Aliphatic Olefins by Electrometric Titration (Jul)(5840).
ISO 3839-78. Petroleum Distillates and Commercial Aliphatic Olefins—Determination of Bromine Number—Electrometric Method First Edition; (Erratum—Nov 1979) (Erratum—June 1980). 16 pp.
JIS K 2605-80. Testing Method for Bromine Number of Petroleum Products (Electrometric Titration Method).

—Soils
SNZ NZS 4402: Pt 3:TEST 3.3.1-86. Methods of Testing Soils for Civil Engineering Purposes Part 3: Soil Chemical Tests. Section 3.3: Determination of the pH Value. Test 3.3.1: Standard Method (Electrometric). 2 pp.

—Steels
BSI BS EN 10 071-91. 1991 Chemical Analysis of Ferrous Materials Determination of Manganese in Steels and Irons Electrometric Titration Method. 9 pp.
CEN EN 10 071-89. Chemical Analysis of Ferrous Materials Determination of Manganese in Steels and Irons Electrometric Titration Method. 5 pp.
DIN ENGL EN 10071-90. Chemical Analysis of Ferrous Materials; Determination of Manganese in Steel and Iron; Electrometric Titration Method (Apr). 6 pp.

—Water
BSI BS 2690: Part 109-84. 1984 Water Used in Industry Part 109: Alkalinity, Acidity, pH Value and Carbon Dioxide. 8 pp.

Electrometric Methods
Use: Electrometric Analysis

Electrometric Titration
Use: Electrometric Analysis

Electromyographs
See Also: Measuring Instruments; Medical Electrical Equipment; Medical Equipment
JIS T 1150-86. Electromyographs.

Electron Accelerators
See Also: Accelerators (Mechanical); Medical Electrical Equipment; Medical Equipment
BSI BS 5724: Sec 3.1-90. 1990 Medical Electrical Equipment Part 3: Particular Requirements for Performance Section 3.1: Methods of Declaring Functional Performance Characteristics of Medical Electron Accelerators in the Range 1 MeV to 50 MeV. 78 pp.
BSI BS 5724: Sec 3.1-01. 1990 Amd 1 Medical Electrical Equipment Part 3: Particular Requirements for Performance Section 3.1: Mehtods of Declaring Functional Performance Characteristics of Medical Electron Accelerators in the Range 1MeV to 50 MeV. 81 pp.

INDUSTRY STANDARDS

INTERNATIONAL AND NON-U.S. NATIONAL STANDARDS
SUBJECT INDEX

Electron

Electron Accelerators (Cont.)
BSI BS 5724: Sec 3.1: Supp 1-90. 1990 Medical Electrical Equipment Part 3: Particular Requirements for Performance Section 3.1: Methods of Declaring Functional Performance Characteristics of Medical Electron Accelerators in the Range 1 MeV to 50 MeV. 59 pp.
CENELEC HD 583 S1-91. Medical Electrical Equipment Medical Electron Accelerators Functional Performance Characteristics. 6 pp.
IEC 976-89. Medical Electrical Equipment Medical Electron Accelerators—Functional Performance Characteristics First Edition. 151 pp.
IEC 977-89. Medical Electrical Equipment Medical Electron Accelerators in the Range 1 MeV to 50 MeV—Guidelines for Functional Performance Characteristics First Edition. 113 pp.
JIS Z 4705-85. Medical Electron Accelerators.

—Medical—Safety
BSI BS 5724: Sec 2.1-89. 1989 Medical Electrical Equipment Part 2: Particular Requirements for Safety Section 2.1: Specification for Medical Electron Accelerators in the Range 1 MeV to 50 MeV. 36 pp.
CSA CAN/CSA-C22. 2N601.2.1-92. Medical Electrical Equipment Part 2: Particular Requirements for the Safety of Medical Electron Accelerators in the Range 1 MeV to 50 MeV (IEC 601-2-1:1981); (Gen Instr 1). 85 pp.
IEC 601 Pt 2-1-81. Safety of Medical Elec-trical Equipment Part 2: Particular Requirements for Medical Electron Ac-celerators in the Range 1 MeV to 50 MeV Section One: General Section Two: Radiation Safety for Equipment First Edi-tion; (Amendment 1-1984) (Amendment 2-1990) (CAN/CSA-C22.2 NO 601.2.1-92). 136 pp.

Electron Beam Welding
See Also: Electric Welding; Fusion Welding; Welding

—Quality Assurance
DIN ENGL 8563 Pt 11-87. Quality Assurance of Welding Operations; Electron Beam Welded Joints in Steel; Specification of Quality Classes (Oct). 4 pp.

Electron Beam Welding Machines
See Also: Arc Welding; Welding Equipment

—Acceptance Testing
DIN ENGL 32505 Pt 1-87. Acceptance Inspection of Electron Beam Welding Machines; Principles and Acceptance Conditions (Oct). 4 pp.
DIN ENGL 32505 Pt 2-88. Acceptance Inspection of Electron Beam Welding Machines; Measurement of Accelerating Voltage (Aug). 3 pp.
DIN ENGL 32505 Pt 3-87. Acceptance Inspection of Electron Beam Welding Machines; Measurement of Beam Current (Oct). 3 pp.
DIN ENGL 32505 Pt 4-87. Acceptance Inspection of Electron Beam Welding Machines; Measurement of Welding Speed (Oct). 3 pp.
DIN ENGL 32505 Pt 5-89. Acceptance Inspection of Electron Beam Welding Machines; Measurement of Tracking Accuracy (Apr). 6 pp.
DIN ENGL 32505 Pt 6-87. Acceptance Inspection of Electron Beam Welding Machines; Measurement of Stability of Focal Spot Position (Oct). 3 pp.

Electron Dosimeters
Use: Dosimeters—Beta Radiation

Electron Emission Current Measurement
Use: Emission Current Measurement

Electron Guns
See Also: Electron Tubes

—Heating Plants
BSI BS 7689-93. 1993 Test Methods for Electroheating Installations with Electron Guns (IEC 703: 1981) (F). 11 pp.
CENELEC HD 440-83. Test Methods for Electroheating Installations with Electron Guns. 2 pp.
IEC 519 Pt 7-83. Safety in Electroheat Installations Part 7: Particular Requirements for Installations with Electron Guns First Edition. 18 pp.
IEC 703-81. Test Methods for Electroheating Installations with Electron Guns First Edition. 18 pp.

—Safety
IEC 519 Pt 7-83. Safety in Electroheat Installations Part 7: Particular Requirements for Installations with Electron Guns First Edition. 18 pp.

Electron Irradiation
See Also: Irradiation

Electron Irradiation (Cont.)
—Sterilization—Medical
CEN PREN 552-91. Sterilization of Medical Devices—Validation and Routine Control of Sterilization by Irradiation—Requirements. 21 pp.
CEN PREN 553-91. Sterilization of Medical Devices—Validation and Routine Control of Sterilization by Irradiation—Guidance. 22 pp.

Electron Microscopy
Use For: Scanning Electron Microscopy
See Also: Microscopy

—Metal Coatings (Made From Metal)
BSI BS 5411: Part 16-89. 1989 Methods of Test for Metallic and Related Coatings Part 16: Scanning Electron Microscope Method for Measurement of Local Thickness of Coatings by Examination of Cross-Sections. 8 pp.
ISO 9220-88. Metallic Coatings—Measurement of Coating Thickness—Scanning Electron Microscope Method First Edition. 8 pp.

—Semiconductor Dice
BSI BS CECC 00013-85. 1985 Scanning Electron Microscope Inspection of Semiconductor Dice. Basic Specification. 25 pp.
CECC CECC 00 013 ISSUE 1-84. Basic Specification: Scanning Electron Microscope Inspection of Semiconductor Dice (En, Fr, Ge). 67 pp.

—Water
JIS K 0554-90. Testing Methods for Concentration of Fine Particles in Highly Purified Water. 21 pp.

Electron Tube Holders
See Also: Electron Tubes; Holders
MOD UK DEF-5251-58. Holders, Valve. Wiring Jigs (Reprinted November 1965). 72 pp.

Electron Tube Shields
See Also: Electron Tubes
IEC 288 Pt 1-69. Tube and Valve Shields Part 1: General Requirements and Methods of Test First Edition; (Amendment 1-1972). 26 pp.
IEC 288 Pt 2-69. Tube and Valve Shields Part 2: Specification Sheets for Shields for Tubes and Valves and Dimensions of Testing Devices and Gauges for Shields First Edition. 35 pp.

—Gages
IEC 288 Pt 2-69. Tube and Valve Shields Part 2: Specification Sheets for Shields for Tubes and Valves and Dimensions of Testing Devices and Gauges for Shields First Edition. 35 pp.

—Testing Equipment
IEC 288 Pt 2-69. Tube and Valve Shields Part 2: Specification Sheets for Shields for Tubes and Valves and Dimensions of Testing Devices and Gauges for Shields First Edition. 35 pp.

Electron Tubes
Scope Note: For additional listings, use a more specific term *Use For:* Electronic Tubes; Electronic Valves (Electron Tubes); Valves, Electronic
See Also: Antitransmit Receive Tubes; Backward Wave Tubes; Camera Tubes; Cathode Ray Storage Tubes; Cathode Ray Tubes; Circuits; Cold Cathode Counter Tubes; Cold Cathode Tubes; Corona Stabilizer Tubes; Counter Tubes; Crossed-Field Amplifier Tubes; Density Modulated Tubes; Disk Seal Tubes; Electrical Components; Electrical Leads; Electron Guns; Electron Tube Holders; Electron Tube Shields; Flashtubes; Gas Discharge Tubes; Gas Filled Tubes; Geiger Mueller Tubes; Gyrotrons; Ignitrons; Image Tubes; Indicator Tubes; Klystrons; Magnetrons; Microwave Tubes; Monitor Diode Tubes; Monoscopes; Neon Tubes; Photomultiplier Tubes; Phototubes; Picture Tubes; Planar Ceramic Tubes; Power Tetrodes; Power Tubes; Proportional Counter Tubes; Protector Tubes; Pulse Modulator Tubes; Receiving Tubes; Rectifier Tubes; Rectifiers; Space Charge Controlled Tubes; Switching Tubes; Thermionic Tubes; Thyratrons; Transmit Receive Tubes; Transmitting Tubes; Traveling Wave Tubes; Trigger Tubes; Triodes; Vidicons; Voltage Reference Tubes; Voltage Regulator Tubes; X-Ray Tubes
BSI BS 448: Part 1-81. 1981 Dimensions of Electronic Tubes and Valves Part 1: IEC Dimensions. 208 pp.
BSI BS 448: Part 1-01. 1981 Amd 1 Dimensions of Electronic Tubes and Valves Part 1: IEC Dimensions (IEC 67: 1966, IEC 67A: 1966, IEC 67B: 1969, IEC 67C: 1970, IEC 67D: 1977) (AMD 5998) September 15, 1992 (S). 212 pp.
BSI BS 448: Part 2-81. 1981 Dimensions of Electronic Tubes and Valves Part 2: Dimensions of United Kingdom Electronic Tubes. 70 pp.
BSI BS 448: Part 2-01. 1981 Amd 1 Dimensions of Electronic Tubes and Valves Part 2: Specification for Dimensions of United Kingdom Electronic Tubes (AMD 5999) September 15, 1992 (S). 73 pp.

Electron Tubes (Cont.)
BSI CP 1005-62. 1962 Use of Electronic Valves. 72 pp.
CENELEC HD 146 S4-88. Dimensions of Electronic Tubes and Valves. 3 pp.
IEC 67-66. Dimensions of Electronic Tubes and Valves Second Edition; (Supplement A-1967) (Supplement B-1969) (Supplement C-1970) (Supplement D-1977) (Supplement E-1986). 223 pp.
MOD UK DSTAN 59-60: Part 1-01. Valves, Electronic (Electronic Tubes) (Listed on EPIC Database) Part 1: General Requirements Issue 2; Amendment 3. 48 pp.
MOD UK DSTAN 59-60: 90/062-01. Valve Electronic, R.F. Input Tube Issue 1; Amendment 4. 19 pp.
SNZ NZSR 15-65. Use of Electronic Valves. 71 pp.

—Amplification Factor Measurement
IEC 151 Pt 12-66. Measurements of the Electrical Properties of Electronic Tubes and Valves Part 12: Methods of Measuring Electrode Resistance, Transconductance, Amplification Factor, Conversion Resistance and Conversion Transconductance First Edition. 28 pp.

—Audiofrequency Measurement
IEC 151 Pt 10-66. Measurements of the Electrical Properties of Electronic Tubes and Valves Part 10: Methods of Measurement of Audio-Frequency Output Power and Distortion First Edition. 11 pp.

—Bases
JIS C 7008-64. Designation System of Bases and Caps for Electron Tubes.

—Capacitance Measurement
CENELEC HD 148 S2-77. Methods for the Measurement of Direct Interelectrode Capacitances of Electronic Tubes and Valves. 2 pp.
CNS C5045-80. Rating Values of Interelement Capacitance for Electron Tubes (Jul)(5752).
IEC 100-62. Methods for the Measurement of Direct Interelectrode Capacitances of Electronic Tubes and Valves Second Edition; (Amendment 1-1969). 58 pp.

—Caps
JIS C 7008-64. Designation System of Bases and Caps for Electron Tubes.

—Cathode Heating and Heater Warm-up Times
IEC 151 Pt 8-66. Measurements of the Electrical Properties of Electronic Tubes and Valves Part 8: Measurement of Cathode Heating Time and Heater Warm-up Time First Edition; (Corrigendum—July 1967). 13 pp.

—Cathode Interface Impedance Measurement
IEC 151 Pt 9-66. Measurements of the Electrical Properties of Electronic Tubes and Valves Part 9: Methods of Measuring the Cathode-Interface Impedance First Edition. 39 pp.

—Conversion Resistance Measurement
IEC 151 Pt 12-66. Measurements of the Electrical Properties of Electronic Tubes and Valves Part 12: Methods of Measuring Electrode Resistance, Transconductance, Amplification Factor, Conversion Resistance and Conversion Transconductance First Edition. 28 pp.

—Conversion Transconductance Measurement
IEC 151 Pt 12-66. Measurements of the Electrical Properties of Electronic Tubes and Valves Part 12: Methods of Measuring Electrode Resistance, Transconductance, Amplification Factor, Conversion Resistance and Conversion Transconductance First Edition. 28 pp.

—Copper Materials
BSI BS 3839-78. 1978 Oxygen-Free High-Conductivity Copper for Electronic Tubes and Semi-Conductor Devices. 8 pp.
JIS H 1202-82. Method for Chemical Analysis of Oxygen-Free Copper Products for Electron Tubes (R 1988). 54 pp.
JIS H 3510-92. Oxygen Free Copper Sheet, Plate, Strip, Seamless Pipe and Tube, Rod, Bar and Wire for Electron Devices. 19 pp.
JIS H 4522-60. Seamless Nickel Tube for Cathode of Vacuum Tube (R 1964). 6 pp.

—Cross Modulation Measurement
IEC 151 Pt 21-69. Measurements of the Electrical Properties of Electronic Tubes and Valves Part 21: Methods of Measurement of Cross-Modulation in Electronic Tubes and Valves First Edition. 14 pp.

—Defense Contracts
MOD UK DEFCON 112F-90. Basic Set of Conditions of Contract—Electronic Valves, Semi-Conductors Etc 5/90. 2 pp.

INTERNATIONAL AND NON-U.S. NATIONAL STANDARDS
SUBJECT INDEX

Electronic

Electron Tubes *(Cont.)*

—Electrical Measurement

IEC 151 Pt 0-66. Measurements of the Electrical Properties of Electronic Tubes and Valves Part 0: Precautions Relating to Methods of Measurement of Electronic Tubes and Valves First Edition. 9 pp.

—Electrical Measurement—Intermodulation Products—Transposers

BSI BS 4478-69. 1969 Methods of Measurement of Intermodulation Products in Electronic Tubes or Valves Intended for Use in Colour Television Transposers. 8 pp.

—Electrode Current Measurement

IEC 151 Pt 1-63. Measurements of the Electrical Properties of Electronic Tubes and Valves Part 1: Measurement of Electrode Current First Edition. 5 pp.

IEC 151 Pt 15-67. Measurements of the Electrical Properties of Electronic Tubes and Valves Part 15: Methods of Measurement of Spurious and Unwanted Electrode Currents First Edition. 34 pp.

—Electrode Resistance Measurement

IEC 151 Pt 12-66. Measurements of the Electrical Properties of Electronic Tubes and Valves Part 12: Methods of Measuring Electrode Resistance, Transconductance, Amplification Factor, Conversion Resistance and Conversion Transconductance First Edition. 28 pp.

—Emission Current Measurement

IEC 151 Pt 13-66. Measurements of the Electrical Properties of Electronic Tubes and Valves Part 13: Methods of Measurement of Emission Current from Hot Cathodes for High-Vacuum Electronic Tubes and Valves First Edition. 11 pp.

—Envelopes—Ceramic

BSI BS 4789-72. 1972 Ceramic Components for Use in Envelopes for Electronic Tubes. 17 pp.

—Filament Current Measurement

IEC 151 Pt 2-63. Measurements of the Electrical Properties of Electronic Tubes and Valves Part 2: Measurement of Heater or Filament Current First Edition. 5 pp.

—Glossaries

BSI BS 4727:Pt1: Group 06-73. 1973 Electrotechnical, Power, Telecommunication, Electronics, Lighting and Colour Terms Part 1: Terms Common to Power, Telecommunications and Electronics Group 06: Electronic Tube Terminology. 33 pp.

IEC 50 Chap 531-74. International Electrotechnical Vocabulary Chapter 531: Electronic Tubes First Edition. 168 pp.

JIS C 7102-73. Glossary of Terms for Electron Tubes.

SAA AS 1852.531-75. International Electrotechnical Vocabulary—Part 531: Electronic Tubes Being IEC 50(531). 167 pp.

SNZ IEC 50: 50(531)-74. International Electrotechnical Vocabulary 50(531): Electronic Tubes. 157 pp.

—Indexes (Documentation)

MOD UK DSTAN 59-60: Part 90-01. Valves, Electronic (Electronic Tubes) (Listed on EPIC Database) Part 90: Detail Specifications Issue 2; Amendment 3 Corrigendum. 25 pp.

—Input and Output Admittance Measurement

BSI BS 4319: Part 1-68. 1968 Methods of Measurement of the Electrical Properties of Electronic Tubes and Valves (Excluding Microwave Devices) Part 1: Essential Conditions for Measuring Equivalent Input and Output Admittances. 8 pp.

IEC 151 Pt 3-63. Measurements of the Electrical Properties of Electronic Tubes and Valves Part 3: Measurement of Equivalent Input and Output Admittances First Edition. 9 pp.

—Ionizing Radiation—Detection and Measurement

IEC 562-76. Measurements of Incidental Ionizing Radiation from Electronic Tubes First Edition. 21 pp.

—Mechanical Shock

IEC 151 Pt 6-65. Measurements of the Electrical Properties of Electronic Tubes and Valves Part 6: Methods of Application of Mechanical Shock (Impulse) Excitation to Electronic Tubes and Valves First Edition. 15 pp.

—Nickel Materials

BSI BS 3504-62. 1962 Amd 1 Magnesium-Activated Nickel for Cathodes of Electronic Tubes and Valves. 12 pp.

BSI BS 3727: Part 1-66. 1966 Methods for the Analysis of Nickel for Use in Electronic Tubes and Valves Part 1: Determination of Aluminium (Photometric Method). 8 pp.

Electron Tubes *(Cont.)*

—Nickel Materials *(Cont.)*

BSI BS 3727: Part 2-68. 1968 Methods for the Analysis of Nickel for Use in Electronic Tubes and Valves Part 2: Determination of Boron (Photometric Method). 8 pp.

BSI BS 3727: Part 3-66. 1966 Amd 1 Methods for the Analysis of Nickel for Use in Electronic Tubes and Valves Part 3: Determination of Carbon (AMD 6007) July 31, 1989. 20 pp.

BSI BS 3727: Part 4-64. 1964 Methods for the Analysis of Nickel for Use in Electronic Tubes and Valves Part 4: Determination of Chromium (Photometric Method). 8 pp.

BSI BS 3727: Part 5-64. 1964 Methods for the Analysis of Nickel for Use in Electronic Tubes and Valves Part 5: Determination of Cobalt (Photometric Method). 9 pp.

BSI BS 3727: Part 6-64. 1964 Methods for the Analysis of Nickel for Use in Electronic Tubes and Valves Part 6: Determination of Copper (Photometric Method). 8 pp.

BSI BS 3727: Part 7-64. 1964 Methods for the Analysis of Nickel for Use in Electronic Tubes and Valves Part 7: Determination of Iron (Photometric Method). 9 pp.

BSI BS 3727: Part 8-64. 1964 Methods for the Analysis of Nickel for Use in Electronic Tubes and Valves Part 8: Determination of Manganese (Photometric Method). 8 pp.

BSI BS 3727: Part 9-67. 1967 Methods for the Analysis of Nickel for Use in Electronic Tubes and Valves Part 9: Determination of Magnesium (Photometric Method). 8 pp.

BSI BS 3727: Part 10-64. 1964 Methods for the Analysis of Nickel for Use in Electronic Tubes and Valves Part 10: Determination of Silicon, 0.020-025 Per Cent (Photometric Method). 9 pp.

BSI BS 3727: Part 11-65. 1965 Methods for the Analysis of Nickel for Use in Electronic Tubes and Valves Part 11: Determination of Silicon, 0.001-0.020 Per Cent (Photometric Method). 9 pp.

BSI BS 3727: Part 12-70. 1970 Amd 1 Methods for the Analysis of Nickel for Use in Electronic Tubes and Valves Part 12: Determination of Sulphur (Combustion Method) (AMD 6005) July 31, 1970. 13 pp.

BSI BS 3727: Part 13-65. 1965 Methods for the Analysis of Nickel for Use in Electronic Tubes and Valves Part 13: Determination of Titanium (Photometric Method). 9 pp.

BSI BS 3727: Part 14-64. 1964 Methods for the Analysis of Nickel for Use in Electronic Tubes and Valves Part 14: Determination of Tungsten (Gravimetric Method). 9 pp.

BSI BS 3727: Part 15-68. 1968 Methods for the Analysis of Nickel for Use in Electronic Tubes and Valves Part 15: Determination of Zinc (Photometric Method). 8 pp.

BSI BS 3727: Part 16-66. 1966 Methods for the Analysis of Nickel for Use in Electronic Tubes and Valves Part 16: Determination of Combined and Free Magnesium. 9 pp.

BSI BS 3727: Part 20-66. 1966 Methods for the Analysis of Nickel for Use in Electronic Tubes and Valves Part 20: Spectrographic Method. 13 pp.

BSI BS 3727: Part 21-66. 1966 Methods for the Analysis of Nickel for Use in Electronic Tubes and Valves Part 21: Determination of Magnesium (Atomic Absorption Method). 8 pp.

BSI BS 3727: Part 22-66. 1966 Methods for the Analysis of Nickel for Use in Electronic Tubes and Valves Part 22: Determination of Zinc (Atomic Absorption Method). 8 pp.

JIS H 1421-79. General Rules for Chemical Analysis of Nickel Materials for Electron Tubes. 6 pp.

JIS H 1422-79. Methods for Determination of Silicon in Nickel Materials for Electron Tubes (R 1984). 11 pp.

JIS H 1423-79. Methods for Determination of Magnesium in Nickel Materials for Electron Tubes (R 1984). 8 pp.

JIS H 1424-79. Methods for Determination of Copper in Nickel Materials for Electron Tubes.

JIS H 1425-79. Methods for Determination of Iron in Nickel Materials for Electron Tubes.

JIS H 1426-79. Methods for Determination of Manganese in Nickel Materials for Electron Tubes.

JIS H 1427-79. Methods for Determination of Carbon in Nickel Materials for Electron Tubes.

JIS H 1428-79. Methods for Determination of Sulfur in Nickel Materials for Electron Tubes.

JIS H 1429-79. Methods for Determination of Titanium in Nickel Materials for Electron Tubes.

JIS H 1430-79. Methods for Determination of Tungsten in Nickel Materials for Electron Tubes.

JIS H 1431-79. Methods for Determination of Cobalt in Nickel Materials for Electron Tubes.

JIS H 4501-90. Nickel Sheets and Strips for Electronic Tube. 9 pp.

JIS H 4502-90. Nickel Sheets and Strips for Cathode of Electronic Tube. 10 pp.

JIS H 4511-90. Nickel Bars and Wires for Electronic Tube. 8 pp.

Electron Tubes *(Cont.)*

—Noise

IEC 151 Pt 4-63. Measurements of the Electrical Properties of Electronic Tubes and Valves Part 4: Methods of Measuring Noise Factor First Edition. 13 pp.

IEC 151 Pt 5-64. Measurements of the Electrical Properties of Electronic Tubes and Valves Part 5: Methods of Measuring Hiss and Hum First Edition. 13 pp.

IEC 151 Pt 7-64. Measurements of the Electrical Properties of Electronic Tubes and Valves Part 7: Measurement of Equivalent Noise Resistance First Edition. 13 pp.

IEC 151 Pt 18-68. Measurements of the Electrical Properties of Electronic Tubes and Valves Part 18: Methods of Measurement of Noises Due to Mechanical or Acoustic Excitations First Edition. 20 pp.

—Numbering

CENELEC HD 145-74. Numbering of Electrodes and Designation of Units in Electronic Tubes and Valves. 2 pp.

IEC 135-61. Numbering of Electrodes and Designation of Units in Electronic Tubes and Valves First Edition. 9 pp.

—Packaging

MOD UK DSTAN 81-6-01. Packaging of Valves, Electronic (Electronic Tubes) Semiconductor Devices and Integrated Circuits Issue 3; Amendment 1. 19 pp.

—Radio Frequency Measurement

IEC 151 Pt 11-66. Measurements of the Electrical Properties of Electronic Tubes and Valves Part 11: Methods of Measurement of Radio-Frequency Output Power First Edition. 13 pp.

—Rating Systems

IEC 134-61. Rating Systems for Electronic Tubes and Valves and Analogous Semiconductor Devices First Edition. 9 pp.

—Spark Gaps

MOD UK DSTAN 59-60: 90/009-01. Valve, Electronic, Travelling—Wave Limiter and Power Amplifier, X-Band Issue 1; Amendment 2. 10 pp.

—Symbols

CNS C5005-82. Symbols of Electronic Tube (Jul)(417).

CNS C5094-81. Graphical Symbols for Electronic and Ionic Tubes (Mar)(7014).

—Symbols—Diagrams

BSI BS 3939: Part 5-85. 1985 Graphical Symbols for Electrical Power, Telecommunications and Electronics Diagrams Part 5: Semiconductors and Electron Tubes (G). 28 pp.

IEC 617 Pt 5-83. Graphical Symbols for Diagrams Part 5: Semiconductors and Electron Tubes First Edition. 47 pp.

SAA AS 1102.105-89. Graphical Symbols for Electrotechnology—Part 105: Semiconductors and Electron Tubes. 31 pp.

SNZ IEC 617: Part 5-83. Graphical Symbols for Diagrams Part 5: Semiconductors and Electron Tubes. 44 pp.

—Transconductance Measurement

IEC 151 Pt 12-66. Measurements of the Electrical Properties of Electronic Tubes and Valves Part 12: Methods of Measuring Electrode Resistance, Transconductance, Amplification Factor, Conversion Resistance and Conversion Transconductance First Edition. 28 pp.

—Tube Sockets

IEC 149 Pt 1-63. Sockets and Accessories for Electronic Plug-in Devices Part 1: General Requirements and Methods of Test First Edition; (Amendment 1-1970) (Amendment 2-1972). 51 pp.

IEC 149 Pt 2-65. Sockets and Accessories for Electronic Plug-in Devices Part 2: Specification Sheets for Sockets and Dimensions of Wiring Jigs and Pin Straighteners First Edition; (Supplement A-H:1972) (Supplement J-L:1976). 257 pp.

JIS C 7006-75. Sockets for Electronic Tubes. 27 pp.

MOD UK DEF-5251-A. General Requirements for Sockets for Electronic Valves and Plug-In Electronic Components (Withdrawn).

MOD UK DEF-5251-A-2. Sockets, Electronic Valve, Pattern VS-B4D (Withdrawn).

MOD UK DEF-5251-A-10. Detail Specification for Sockets, Electronic Valve, Pattern VS-B7G (Withdrawn).

MOD UK DEF-5251-A-14. Detail Specification for Sockets, Electronic Valve, Pattern VS-B9A (Withdrawn).

Electronic Alarms

Use: Alarm Systems

INTERNATIONAL AND NON-U.S. NATIONAL STANDARDS
SUBJECT INDEX
Electronic

Electronic Attitude Direction Indicators
Use: Attitude Indicators

Electronic Blackboards
Use: Telewriters

Electronic Chart Display Systems
Use For: Chart Display Systems, Electronic; ECDIS
See Also: Charts; Display Devices; Navigational Aids; Navigational Charts; Video Display Terminals

CCIR Report 1165-90. Transmission of Digital Data for the Updating of Electronic Chart Display Systems (ECDIS)—Section 8D—Radiodetermination, Global Maritime Distress and Safety System and Related Subjects. 15 pp.

CCIR RECMN 826-92. Transmission of Information for Updating Electronic Chart Display and Information Systems (ECDIS). 2 pp.

—Data Transmission—Updates
CCIR Report 1165-90. Transmission of Digital Data for the Updating of Electronic Chart Display Systems (ECDIS)—Section 8D—Radiodetermination, Global Maritime Distress and Safety System and Related Subjects. 15 pp.

CCIR QUESTION 98/8-90. Transmission of Digital Data for the Updating of Electronic Chart Display Systems (ECDIS)—Questions Concerning Study Group 8—Mobile, Radiodetermination, Amateur and Related Satellite Services. 1 p.

CCIR RECMN 826-92. Transmission of Information for Updating Electronic Chart Display and Information Systems (ECDIS). 2 pp.

Electronic Component Packaging
See Also: DIP Packages; Electrical Leads; Integrated Circuits; Packaging; Semiconductor/Integrated Circuit Packaging; Surface Mount Technology

MOD UK DSTAN 81-77-01. Packaging of Small Electronic and/or Mechanical Assemblies Issue 1; Amendment 1. 17 pp.

—Axial Leads
BSI BS 6062: Part 1-81. 1981 Packaging of Electronic Components for Automatic Handling Part 1: Specification for Tape Packaging of Components with Axial Leads on Continous Tapes. 8 pp.

CENELEC HD 143.1-83. Packaging of Components for Automatic Handling Part 1: Tape Packaging of Components with Axial Leads on Continuous Tapes. 2 pp.

CNS Z5077-81. Reel Packaging of Components with Axial Leads (Jan)(6981).

IEC 286 Pt 1-80. Packaging of Components for Automatic Handling Part 1: Tape Packaging of Components with Axial Leads on Continuous Tapes First Edition. 14 pp.

JIS C 0805-89. Packaging of Electronic Components on Continuous Tapes (Components with Axial Type and Radial Wire Lead Terminals). 20 pp.

—Leadless Components
BSI BS 6062: Part 3-88. 1988 Packaging of Electronic Components for Automatic Handling Part 3: Packaging of Leadless Components on Continous Tapes. 10 pp.

BSI BS 6062: Part 3-93. 1993 Packaging of Electronic Components for Automatic Handling Part 3: Specification for Packaging of Leadless Components on Continuous Tapes (IEC 286-3: 1991) (S). 19 pp.

CENELEC HD 143.3 S1-88. Packaging of Components for Automatic Handling Part 3: Packaging of Leadless Component;s on Continuous Tapes. 3 pp.

CENELEC HD 143.3 S2-92. Packaging of Components for Automatic Handling Part 3: Packaging of Leadless Components on Continuous Tapes (IEC 286-3:1991). 5 pp.

IEC 286 Pt 3-91. Packaging of Components for Automatic Handling Part 3: Packaging of Leadless Components on Continuous Tapes Second Edition (CENELEC HD 143.3 S2:1992). 33 pp.

—Radial Leads
CNS C5187-84. Lead Taping of Components in Hybrid Radial Lead Configuration for Automatic Insertion (Nov)(11104).

CNS C5188-84. Lead Taping of Components in the Radial Configuration for Automatic Insertion (Nov)(11105).

JIS C 0805-89. Packaging of Electronic Components on Continuous Tapes (Components with Axial Type and Radial Wire Lead Terminals). 20 pp.

—Surface Mounting
JIS C 0806-90. Packaging of Electronic Components on Continuous Tapes (Surface Mounting Devices). 23 pp.

Electronic Component Packaging (Cont.)
—Unidirectional Leads
BSI BS 6062: Part 2-88. 1988 Packaging of Electronic Components for Automatic Handling Part 2: Tape Packaging of Components with Unidirectional Leads on Continuous Tapes. 12 pp.

CENELEC HD 143.2-87. Packaging of Components for Automatic Handling Part 2: Tape Packaging of Components with Unidirectional Leads on Continuous Tapes. 2 pp.

IEC 286 Pt 2-85. Packaging of Components for Automatic Handling Part 2: Tape Packaging of Components with Unidirectional Leads on Continuous Tapes First Edition. 27 pp.

Electronic Components
Scope Note: For additional listings, use a more specific term *See Also:* Attenuators; CECC; Diodes; Discrete Components; Electric Filters; Electrical Components; Electronic Equipment; FEN; Inductors; Integrated Circuits; Magnetic Cores; Printed Circuit Board Components; Printed Circuit Board Manufacturing; Printed Circuit Boards; Semiconductor Devices; Transistors

CECC EN 100 014-91. Basic Specification: CECC Assessed Process Average Procedure (60% Confidence Limit) (Supersedes CECC 00 014 Issue 1: 1986). 7 pp.

CNS C5204-89. General Rules for Electronic Components (Jun)(11475).

CNS Z5078-82. Preparation for Delivery of Electrical and Electronic Components (May)(6982).

IEC Guide 103-80. Guide on Dimensional Co-Ordination First Edition. 44 pp.

JIS C 5001-87. General Rules for Electronic Components. 14 pp.

—Acceptance Testing
MOD UK DSTAN NRD 636 1.93-01. Schedule of Selected Electronic Components and Processes Undergoing Approval Tests Issue 50; Amendment 1. 59 pp.

—Cargo Transportation
CNS Z6027-80. Shipping Test for Electronic Components (Excluding Electronic Tubes) (May)(5638).

—Color Coding
CNS C5138-82. Color Code for Electronic Parts (Jul)(9079).

JIS C 0801-64. Color Code for Electronic Parts (R 1970). 4 pp.

—Damp Heat Testing
CNS C6010-89. Damp Heat (Steady State) Testing Procedures for Electronic Components (Jun)(3623).

CNS C6012-85. Damp Heat (Cyclic) Testing Procedures for Electronic Components (Apr)(3625). 5 pp.

—Defense Contracts—Production Permits
MOD UK DEFCON 17A-90. Use of Alternative Electronic Components in the Production/Repair/Modification of Defence Equipment 10/90. 1 p.

—Dry Heat Testing
CNS C6021-89. Dry Heat Testing Procedures for Electronic Components (Jul)(3634).

—Electric Conductors
CSA C22.2 NO 16-M1986. Insulated Conductors for Power-Operated Electronic Devices (R 1992); (Gen Instr 1 Thru 4). 64 pp.

—Electric Endurance Testing
CNS C6248-85. Endurance (Electrical) Testing Method for Electronic Component (Apr)(11239).

JIS C 5036-75. Endurance (Electrical) Testing Method for Electronic Components. 4 pp.

—Electrostatic Discharge—Protection
BSI BS CECC 00015: Pt 1-91. (WITHDRAWN) 1991 Basic Specification: Protection of Electrostatic Sensitive Devices Part 1: General Requirements (Renumbered BS EN 100015-1: 1992) (T). 62 pp.

BSI BS EN 100015-1-92. 1991 Amd 1 Basic Specification: Protection of Electrostatic Sensitive Devices Part 1: General Requirements (AMD 7443) November 15, 1992. 65 pp.

CECC CECC 00 015/I ISSUE 1-91. Basic Specification: Protection of Electrostatic Sensitive Devices Part 1: General Requirements (En, Fr, Ge) ERRATUM (En, Fr, Ge). 184 pp.

CENELEC EN 100015-1-92. Basic Specification: Protection of Electrostatic Sensitive Devices Part 1: General Requirements. 59 pp.

CENELEC PREN 100 015-2-92. Basic Specification: Protection of Electrostatic Sensitive Devices Part 2: Requirements for Low Humidity Conditions. 5 pp.

Electronic Components (Cont.)
—Electrostatic Discharge—Protection (Cont.)
CENELEC PREN 100 015-3-93. Basic Specification: Protection of Electrostatic Sensitive Devices. Part 3: Requirements for Clean Rooms. 7 pp.

CENELEC PREN 100 015-4-92. Protection of Electrostatic Sensitive Devices Part 4: Requirements for High Voltage Environments. 7 pp.

—Endurance Testing
CNS C6249-85. Endurance (Mechanical) Testing Method for Electronic Components (Apr)(11240).

JIS C 5037-75. Endurance (Mechanical) Testing Method for Electronic Components. 4 pp.

—Engineering Drawings
BSI BS 6943-88. 1988 Classification of Shapes of Electronic Components for Placement on Printed Wiring Boards. 49 pp.

—Environmental Engineering
JIS C 5002-68. Environmental Requirements for Electronic Components (R 1974). 4 pp.

—Environmental Testing
CECC CECC 00 006 ISSUE 2-81. Basic Specification: Environmental Test Procedures (En, Fr, Ge). 61 pp.

CNS C6009-89. General Rules for Basic Environmental Testing Procedures for Electronic Components (Jun)(3622).

—Failure Rate
CNS C5205-86. General Test Procedure of Failure Rate for Electronic Components (Jan)(11476).

JIS C 5003-74. General Test Procedure of Failure Rate for Electronic Components (R 1982); (Erratum). 13 pp.

—Fire Wire and Ribbon—Density
CNS C6096-80. Method of Test for Density of Fire Wire and Ribbon for Electronic Devices (Aug)(6316).

—Flammability Testing
CNS C6126-81. Flammability Tests for Electronic Components (Jul)(7650).

—Glossaries
IEC. Electricity, Electronics and Telecommunications Multilingual Dictionary Volume 1: English-French-Russian-German-Spanish-Dutch-Italian-Swedish-Polish Second Edition. 1944 pp.

IEC. Electricity, Electronics and Telecommunications Multilingual Dictionary Volume 2: English-French-Russian-German-Spanish-Dutch-Italian-Swedish-Polish Second Edition. 964 pp.

IEC. Electricity, Electronics and Telecommunications Multilingual Dictionary Volume 3: German, Dutch and Swedish Indexes Second Edition. 793 pp.

IEC. Electricity, Electronics and Telecommunications Multilingual Dictionary Volume 4: Spanish and Italian Indexes Second Edition. 540 pp.

IEC. Electricity, Electronics and Telecommunications Multilingual Dictionary Volume 5: Russian and Polish Indexes Second Edition. 578 pp.

—Identification Systems
MOD UK DSTAN 61-7: Part 2-91. Identification of Electrical and Electronic Systems, Wiring and Components Part 2: Marking Code for Electrical and Electronic Equipment Issue 3. 24 pp.

—Identification Systems—Electrostatic Sensitivity
CNS Z5133-84. Symbol and Label for Electrostatic Sensitive Devices (Nov)(11144).

—Immersion Testing
CNS C6246-85. Sealing (Immersion Cyclic) Testing Method for Electronic Components (Apr)(11237).

JIS C 0052-86. Testing Method of Resistance to Solvents (Immersion in Cleaning Solvents). 12 pp.

JIS C 5032-75. Sealing (Immersion Cyclic) Testing Method for Electronic Components. 4 pp.

—Inspection
IEC 419-73. Guide for the Inclusion of Lot-by-Lot and Periodic Inspection Procedures in Specifications for Electronic Components (or Parts) First Edition. 11 pp.

—Inspection by Attributes
CECC CECC 00 007 ISSUE 2-78. Basic Specification: Sampling Plans and Procedures for Inspection by Attributes (En, Fr, Ge). 15 pp.

—Insulation Resistance
CNS C6022-85. Insulation Resistance Testing Method for Electronic Components (Apr)(3635).

—Low Pressure Testing
CNS C6245-85. Low Air Pressure Testing Method for Electronic Components (Apr)(11236).

INTERNATIONAL AND NON-U.S. NATIONAL STANDARDS
SUBJECT INDEX
Electronic

Electronic Components (Cont.)

—**Low Temperature Testing**
CNS C6242-89. Cold Testing Procedures for Electronic Components (Jul)(11233).

—**Low Temperature Testing—Storage**
CNS C6244-85. Storage (Low Temperature) Testing Method for Electronic Components (Apr)(11235).

—**Mechanical Shock**
CNS C6243-90. Mechanical Shock Testing Method for Electronic Components (May)(11234).

—**Modular**
CECC CECC 60 000 ISSUE 1-92. Generic Specification: Modular Electronic Units (En, Fr). 44 pp.

—**Modular—Quality Assurance**
BSI BS EN 160000-93. 1993 Amd 1 Generic Specification: Modular Electronic Units (AMD 7860) August 15, 1993 (T). 29 pp.
CENELEC EN 160000-93. Generic Specification: Modular Electronic Units. 23 pp.

—**Mountings**
BSI BS 6103-86. 1986 Dimensions for the Mounting of Single-Hole, Bush-Mounted Spindle Operated Electronic Components. 24 pp.
CENELEC HD 391 S2-81. Dimensions for the Mounting of Single-Hole Bush-Mounted, Spindle-Operated Electronic Components.
CENELEC HD 391 S3-88. Dimensions for the Mounting of Single-Hole, Bush-Mounted, Spindle-Operated Electronic Components. 4 pp.
IEC 620-84. Dimensions for the Mounting of Single-Hole, Bush-Mounted, Spindle-Operated Electronic Components Second Edition. 31 pp.
IEC 915-87. Capacitors and Resistors for Electronic Equipment Preferred Dimensions of Spindle Ends, Bushes and for the Mounting of Single-Hole, Bush-Mounted, Spindle-Operated Electronic Components First Edition. 22 pp.
JIS C 0051-86. Robustness of Terminations and Integral Mounting Devices for Electronic Components. 17 pp.

—**Naval Ships**
MOD UK NES 581: Part 1-90. Requirements for the Selection and Approval of Electrical and Electronic Components Part 1: Method of Selection and Approval Issue 2 (07.90). 21 pp.
MOD UK NES 581: Part 2-91. Requirements for the Selection and Approval of Electrical and Electronic Components Part 2: Standard, Approved and Acceptable Components Issue 5 (09.91). 199 pp.
MOD UK NES 581: Part 2-01. Requirements for the Selection and Approval of Electrical and Electronic Components Part 2: Standard, Approved and Acceptable Components Issue 5 (09.91); Amendment 1. 229 pp.
MOD UK NES 581: Part 3-91. Requirements for the Selection and Approval of Electrical and Electronic Components Part 3: Non—Standard Project Approved Components Issue 2 (12.91). 44 pp.
MOD UK NES 581: Part 3-01. Requirements for the Selection and Approval of Electrical and Electronic Components Issue 2 (12.91); Amendment 1. 85 pp.

—**Printed Circuits**
IEC 321-70. Guidance for the Design and Use of Components Intended for Mounting on Boards with Printed Wiring and Printed Circuits First Edition; (Amendment 1-1975). 39 pp.

—**Procurement**
MOD UK DSTAN 05-37-89. Policy for the Procurement of Electronic Components Issue 2. 5 pp.

—**Qualification Approvals**
MOD UK DSTAN 05-14-01. Mutual Acceptance of Qualification Approvals for Electrical/Electronic Components Within NATO Countries Issue 2; Amendment 1. 19 pp.
NATO STANAG 4093 ED 3 AMD 2-76. Mutual Acceptance of Qualification Approvals for Electronic Parts. 14 pp.
NATO STANAG 4093 ED 4 AMD 0-89. Mutual Acceptance by NATO Member Countries of Qualification of Electronic and Electrical Components for Military Use. 24 pp.
NATO STANAG 4093 ED 4 AMD 0-93. Mutual Acceptance by NATO Member Countries of Qualification of Electronic and Electrical Components for Military Use. 23 pp.

—**Qualified Products List**
BSI PD 9002:2-92. 1992 BS 9000—IECQ Qualified Products List (T). 241 pp.
BSI PD 9002:2-91. 1991 BS 9000—IECQ Qualified Products List. 261 pp.
BSI PD 9002:3-92. 1992 BS 9000—IECQ Qualified Products List. 231 pp.

Electronic Components (Cont.)

—**Qualified Products List (Cont.)**
MOD UK DSTAN 59-59-91. Electrical/Electronic Components for Defence Use Services Qualified Products List Including Components of Assessed Quality Issue 6. 513 pp.
MOD UK DSTAN 59-59-01. Electrical/Electronic Components for Defence Use Services Qualified Products List Including Components of Assessed Quality Issue 6; Amendment 1. 560 pp.
MOD UK DSTAN 59-59-02. Electrical/Electronic Components for Defence Use Services Qualified Products List Including Components of Assessed Quality Issue 6; Amendment 2. 615 pp.
MOD UK DSTAN 59-59-03. Electrical/Electronic Components for Defence Use Services Qualified Products List Including Components of Assessed Quality Issue 6; Amendment 3. 601 pp.
MOD UK DSTAN 59-59-04. Electrical/Electronic Components for Defence Use Services Qualified Products List Including Components of Assessed Quality Issue 6; Corrigendum to Amendment 3. 602 pp.
QSS QC 001005-92. IECQ Qualified Products List ISSUE 2. 93 pp.

—**Quality Assurance**
BSI BS 9000: Part 1-89. (OBSOLESCENT) 1989 General Requirements for Electronic Components of Assessed Quality Part 1: General Procedures (Superseded by BSCECC 00111: Part 3:1991). 18 pp.
BSI BS 9000: Part 4-91. (OBSOLESCENT) 1991 General Requirements for a System for Electronic Components of Assessed Quality Part 4: Specification of Procedures for the Approval of an Organization (Superseded by BS CECC 00114: Part 1: 1991). 23 pp.
BSI BS 9000: Part 4-89. 1989 General Requirements for Electronic Components of Assessed Quality Part 4: Procedures for the Approval of an Organization. 19 pp.
BSI BS 9000: Part 6-89. (OBSOLESCENT)1989 General Requirements for Electronic Components of Assessed Quality Part 6: Capability Approval Procedures (Superseded by BS CECC 00114: Part 3: 1991). 19 pp.
BSI BS 9000: Part 7-89. (OBSOLESCENT) 1989 General Requirements for Electronic Components of Assessed Quality Part 7: Lot Formation, Release Procedures and Certified Test Records (Superseded by BS CECC 00114: Part 2: 1991). 13 pp.
BSI BS 9000: Sec 8.1-89. (WITHDRAWN) 1989 General Requirements for Electronic Components of Assessed Quality Part 8: Drafting Reqd. Section 8.1: Generic, Sectional and Blank Detail Spec. Produced Under Qualification Approval (Superseded by BS CECC 00111: Part 3: 1991). 20 pp.
BSI BS 9000: Sec 8.2-89. (WITHDRAWN) 1989 General Requirements for Electronic Components of Assessed Quality Part 8: Drafting Requirements: Section 8.2: Generic and Sectional Spec. Produced Under Capability Approval (Superseded by BS CECC 00111: Part 3: 1991). 11 pp.
BSI BS 9003-77. (WITHDRAWN) 1977 Amd 1 Requirements for the Manufacture of Electronic Components of Assessed Quality Intended for Long Life Applications. 15 pp.
CENELEC EN 100 014-91. Basic Specification: CECC Assessed Process Average Process Average Procedure (60% Confidence Limit). 7 pp.
CENELEC PREN 160 200-1-93. Sectional Specification: Microwave Modular Electronic Units of Assessed Quality. Part 1: Capability Approval Procedure. 46 pp.
CENELEC PREN 160 201-93. Blank Detail Specification: Microwave Modular Electronic Units of Assessed Quality (Capability Approval). 15 pp.
MOD UK DSTAN N/RD 636 5.91. Schedule of Electronic Components Undergoing Approval Tests to BS9000 /CECC or Defence Specifications Issue 45; Amendment 2. 67 pp.
MOD UK DSTAN N/RD 636 9.91. Schedule of Electronic Components Undergoing Approval Tests to BS9000 Issue 46. 50 pp.
MOD UK DSTAN N/RD 636 1.92. Schedule of Electronic Components Undergoing Approval Tests to BS9000 /CECC or Defence Specifications Issue 47. 49 pp.
MOD UK DSTAN NRD 636 5.92. Schedule of Electronic Components Undergoing Approval Tests to BS9000 /CECC or Defence Specifications Issue 48. 51 pp.
MOD UK DSTAN NRD 636 5.92-01. Schedule of Electronic Components Undergoing Approval Tests to BS9000 /CECC or Defence Specifications Issue 48; Amendment 1. 67 pp.
MOD UK DSTAN NRD 636 9.92-01. Schedule of Selected Electronic Components and Processes Undergoing Approval Tests Issue 49; Addendum. 62 pp.
MOD UK DSTAN NRD 636 1.93. Schedule of Selected Electronic Components and Processes Undergoing Approval Tests Issue 50. 58 pp.

Electronic Components (Cont.)

—**Quality Assurance (Cont.)**
MOD UK DSTAN NRD 636 5.93. Schedule of Selected Electronic Components and Processes Undergoing Approval Tests or Newly Approved Issue 51. 61 pp.
MOD UK DSTAN NRD 636 5.93-01. Schedule of Selected Electronic Components and Processes Undergoing Approval Tests or Newly Approved Issue 51; Amendment 1. 47 pp.
MOD UK DSTAN 61-12: 6112/33/003-93. Wires Cords and Cables Electrical—Metric Units Part 33: Airframe Wires and Cables in Temperature Categories 135 Degrees C, 150 Degrees C, 200 Degrees C and 260 Degrees C, Sectional Specification. 4 pp.
MOD UK DSTAN 66-31-81. Basic Requirements and Tests for Proprietary Electronic and Electrical Test Equipment Issue 1. 20 pp.
MOD UK DSTAN 66-31-91. Basic Requirements and Tests for Proprietary Electronic and Electrical Test Equipment Issue 2. 25 pp.

—**Quality Assurance—Calibration Systems—CECC Procedures**
CECC CECC 00 011 ISSUE 1-82. Basic Specification: Calibration Requirements for the CECC System (En, Fr, Ge). 24 pp.

—**Quality Assurance—CECC Procedures**
BSI BS 9000: Part 2-91. 1991 General Requirements for a System for Electronic Components of Assessed Quality Part 2: Specification for the National Implementation of the CECC System. 8 pp.
BSI BS 9000: Part 2-83. 1983 Amd 1 General Requirements for Electronic Components of Assessed Quality Part 2: National Implementation of CECC Basic Rules and Rules of Procedure. 8 pp.
BSI BS CECC 00014-88. (WITHDRAWN) 1988 CECC Assessed Process Average Procedure (60 Confidence Limit): Basic Specification (Renumbered as BS EN 100014: 1992). 9 pp.
BSI BS CECC 00104-91. 1991 Rule of Procedure 4 CECC Working Groups General Rules and Additional Rules. 16 pp.
BSI BS CECC 00104-01. 1991 Amd 1 Rule of Procedure 4 CECC Working Groups General Rules and Additional Rules (AMD 7732) May 15, 1993 (T). 22 pp.
BSI BS CECC 00108-91. 1991 Rule of Procedure 8 Attestation of Conformity. 20 pp.
BSI BS CECC 00109-91. 1991 Rule of Procedure 9 Certified Test Records. 9 pp.
BSI BS CECC 00111: Pt 1-91. 1991 Rule of Procedure 11 Specifications Part 1: General Regulations for CECC Specifications. 21 pp.
BSI BS CECC 00111: Pt 1-01. 1991 Amd 1 Rule of Procedure 11 Specifications Part 1: General Regulations for CECC Specifications (AMD 7333) January 15, 1993. 23 pp.
BSI BS CECC 00111: Pt 1-02. 1991 Amd 2 Rule of Procedure 11 Specifications Part 1: General Regulations for CECC Specifications (AMD 7733) May 15, 1993 (T). 41 pp.
BSI BS CECC 00111: Pt 2-91. 1991 Rule of Procedure 11 Specifications Part 2: Regulations for CECC Specifications for Components of Enhanced Assessment of Quality. 10 pp.
BSI BS CECC 00111: Pt 3-91. 1991 Rule of Procedure 11 Specifications Part 3: Regulations for CECC Specifications for Components for General and Professional (Civil and Military) Usage (Excluding Detail Specifications). 31 pp.
BSI BS CECC 00111: Pt 3-01. 1991 Amd 1 Rule of Procedure 11 Specifications Part 3: Regulations for CECC Specifications for Components for General and Professional (Civil and Military) Usage (Excluding Detail Specifications) (AMD 7329) January 15, 1993 (T). 40 pp.
BSI BS CECC 00111: Pt 4-91. 1991 Rule of Procedure 11 Specifications Part 4: Regulations for CECC Detail Specifications. 37 pp.
BSI BS CECC 00111: Pt 4-01. 1991 Amd 1 Rule of Procedure 11 Specifications Part 4: Regulations for CECC Detail Specifications (AMD 7334) January 15, 1993 (T). 44 pp.
BSI BS CECC 00111: Pt 4-02. 1991 Amd 2 Rule of Procedure 11 Specifications Part 4: Regulations for CECC Detail Specifications (AMD 7734) May 15, 1993 (T). 48 pp.
BSI BS CECC 00111: Pt 5-91. 1991 Rule of Procedure 11 Specifications Part 5: Preparation of Specifications Under the Single Originator Procedure. 17 pp.
BSI BS CECC 00114: Pt 0-92. 1992 Rule of Procedure 14 Quality Assessment Procedures Part 0: An Introduction to the Types of Approval Available Under the CECC System. 11 pp.
BSI BS CECC 00114: Pt 1-93. 1993 Rule of Procedure 14 Quality Assessment Procedures Part 1: Approval of Manufacturers and Other Organizations (T). 37 pp.

INTERNATIONAL AND NON-U.S. NATIONAL STANDARDS
SUBJECT INDEX
Electronic

Electronic Components *(Cont.)*
—Quality Assurance—CECC Procedures *(Cont.)*
BSI BS CECC 00114: Pt 1-01. 1993 Amd 1 Rule of Procedure 14 Quality Assessment Procedures Part 1: Approval of Manufacturers and Other Organizations (AMD 7782) July 15, 1993 (T). 41 pp.

BSI BS CECC 00114: Pt 1-91. 1991 Amd 1 Rule of Procedure 14. Quality Assessment Procedures Part 1: Approval of Manufacturers and Other Organizations. 28 pp.

BSI BS CECC 00114: Pt 1-02. 1991 Amd 2 Rule of Procedure 14 Quality Assessment Procedures Part 1: Approval of Manufacturers and Other Organizations (AMD 7184) July 15, 1992 (T). 30 pp.

BSI BS CECC 00114: Pt 2-93. 1993 Rule of Procedure 14 Quality Assessment Procedures Part 2: Qualification Approval of Electronic Components (T). 34 pp.

BSI BS CECC 00114: Pt 2-91. 1991 Rule of Procedure 14 Quality Assessment Procedures Part 2: Qualification Approval of Electronic Components (T). 33 pp.

BSI BS CECC 00114: Pt 3-93. 1993 Rule of Procedure 14 Quality Assessment Procedures Part 3: Capability Approval of an Electronic Component Manufacturing Activity (T). 50 pp.

BSI BS CECC 00114: Pt 3-91. 1991 Rule of Procedure 14 Quality Assessment Procedures Part 3: Capability Approval of an Electronic Component Manufacturing Activity. 33 pp.

BSI BS CECC 00114: Pt 3-01. 1991 Amd 1 Rule of Procedure 14 Quality Assessment Procedures Part 3: Capability Approval of an Electronic Component Manufacturing Activity (AMD 7185) December 15, 1992 (T). 68 pp.

BSI BS CECC 00114: Pt 4-92. 1992 Rule of Procedure 14 Quality Assessment Procedures Part 4: Procedure for Enhanced Assessment of Quality. 12 pp.

BSI BS CECC 00114: Pt 5-93. 1993 Rule of Procedure 14 Quality Assessment Procedures Part 5: Process Approval of Specialist Contractors Within the Electronic Components Industry (T). 45 pp.

BSI BS CECC 00802-91. 1991 Guidance Document: CECC Standard Method for the Specification of Surface Mounting Components (SMDs) of Assessed Quality. 22 pp.

BSI BS CECC 60000-92. 1992 Generic Specification: Modular Electronic Units (Renumbered as BS EN 160000: 1993) (T). 26 pp.

BSI BS CECC 200000-93. 1993 Requirements for Process Assessment Schedules for Process Approval (T). 34 pp.

BSI BS EN 100014-92. 1992 Amd 1 Basic Specification: CECC Assessed Process Average Procedure (60 Percent Confidence Limit) (AMD 7195) October 15, 1992. 13 pp.

BSI PD 9004-89. 1989 Amd 1 BS 9000 CECC & IECQ UK Administrative Guide February 1991. 147 pp.

CECC CECC 00 500 ISSUE 3. CECC System for Electronic Components of Assessed Quality Introduction to the System (En). 4 pp.

—Quality Assurance—IECQ Procedures
BSI BS 9000: Part 3-91. 1991 General Requirements for a System for Electronic Components of Assessed Quality Part 3: Specification for the National Implementation of the IECQ System. 8 pp.

BSI BS 9000: Part 3-87. 1987 General Requirements for Electronic Components of Assessed Quality Part 3: National Implementation of IECQ Basic Rules and Rules of Procedure. 11 pp.

BSI BS QC 001002-91. 1991 Rules of Procedure of the IEC Quality Assessment System for Electronic Components (IECQ). 91 pp.

BSI BS QC 001002-01. 1991 Amd 1 Rules of Procedure of the IEC Quality Assessment System for Electronic Components (IECQ) (AMD 7711) June 15, 1993 (T). 94 pp.

BSI PD 9004-89. 1989 Amd 1 BS 9000 CECC & IECQ UK Administrative Guide February 1991. 147 pp.

IECQ QC 001001-86. Basic Rules of the IEC Quality Assessment System for Electronic Components (IECQ); (Amendment 1-1992). 50 pp.

IECQ QC 001002-86. Rules of Procedure of the IEC Quality Assessment System for Electronic Components (IECQ); (Amendment 1-1992). 184 pp.

IECQ QC 001003-88. Guidance Documents Documents—Guides; (Amendment 1-1990). 93 pp.

IECQ QC 001005-93. Register of Firms, Products and Services Approved Under the IECQ System, Including ISO 9000. 99 pp.

QSS QC 001001-86. Basic Rules of the IEC Quality Assessment System for Electronic Components (IECQ) (Amendment 1-1992). 49 pp.

—Quality Assurance—IECQ Procedures—Specifications
IECQ QC 001004 ISSUE 1-93. Specifications List Information Current to 1992-12. 63 pp.

Electronic Components *(Cont.)*
—Quality Assurance—IECQ Specifications List
QSS QC 001004-92. IEC Quality Assessment System for Electronic Components (IECQ) Specifications List Information Current to 1992-06. 61 pp.

—Quality Assurance—PPM Approach
BSI BS CECC 00800-88. 1988 Code of Practice for the Use of the ppm Approach in Association with the CECC System (T). 13 pp.

CECC CECC 00 800 ISSUE 1-86. Code of Practice on the Use of the PPM Approach in Association with the CECC System (En, Fr, Ge). 30 pp.

—Quality Assurance—Radiography
BSI BS CECC 00012-85. 1985 Radiographic Inspection of Electronic Components: Basic Specification. 13 pp.

—Quality Assurance—Specifications Preparation
BSI BS CECC 00111: Pt 1-91. 1991 Rule of Procedure 11 Specifications Part 1: General Regulations for CECC Specifications. 21 pp.

BSI BS CECC 00111: Pt 1-01. 1991 Amd 1 Rule of Procedure 11 Specifications Part 1: General Regulations for CECC Specifications (AMD 7333) January 15, 1993. 23 pp.

BSI BS CECC 00111: Pt 1-02. 1991 Amd 2 Rule of Procedure 11 Specifications Part 1: General Regulations for CECC Specifications (AMD 7733) May 15, 1993 (T). 41 pp.

BSI BS CECC 00111: Pt 2-91. 1991 Rule of Procedure 11 Specifications Part 2: Regulations for CECC Specifications for Components of Enhanced Assessment of Quality. 10 pp.

BSI BS CECC 00111: Pt 3-91. 1991 Rule of Procedure 11 Specifications Part 3: Regulations for CECC Specifications for Components for General and Professional (Civil and Military) Usage (Excluding Detail Specifications). 31 pp.

BSI BS CECC 00111: Pt 3-01. 1991 Amd 1 Rule of Procedure 11 Specifications Part 3: Regulations for CECC Specifications for Components for General and Professional (Civil and Military) Usage (Excluding Detail Specifications) (AMD 7329) January 15, 1993 (T). 40 pp.

BSI BS CECC 00111: Pt 4-91. 1991 Rule of Procedure 11 Specifications Part 4: Regulations for CECC Detail Specifications. 37 pp.

BSI BS CECC 00111: Pt 4-01. 1991 Amd 1 Rule of Procedure 11 Specifications Part 4: Regulations for CECC Detail Specifications (AMD 7334) January 15, 1993 (T). 44 pp.

BSI BS CECC 00111: Pt 4-02. 1991 Amd 2 Rule of Procedure 11 Specifications Part 4: Regulations for CECC Detail Specifications (AMD 7734) May 15, 1993 (T). 48 pp.

BSI BS CECC 00111: Pt 5-91. 1991 Rule of Procedure 11 Specifications Part 5: Preparation of Specifications Under the Single Originator Procedure. 17 pp.

CECC CECC 00 400 ISSUE 3-86. Handbook for the Production of CECC Documents (En) ERRATUM (En, Fr, Ge) AMD 1 (En, Fr, Ge) AMD 2 (En, Fr, Ge) AMD 4 (En, Fr, Ge) AMD 5 (En, Fr, Ge) AMD 6 (En, Fr, Ge) AMD 7 (En, Fr, Ge) AMD 8 (En, Fr, Ge). 103 pp.

IEC Guide 102-89. Electronic Components Specification Structures for Quality Assessment (Qualification Approval and Capability Approval) Third Edition (IECQ Guide 102). 64 pp.

IECQ Guide 102-89. Electronic Components: Specification Structures for Quality Assessment (Qualification Approval and Capability Approval) (IEC GUIDE 102 ED 3). 64 pp.

QSS Guide 102-89. Electronic Components Specification Structures for Quality Assessment (Qualification Approval and Capability Approval) Third Edition. 64 pp.

—Radiography
CECC CECC 00 012 ISSUE 1-85. Basic Specification: Radiographic Inspection of Electronic Components (En, Fr, Ge). 30 pp.

CENELEC PREN 100 012-93. Basic Specification: X-Ray Inspection of Electronic Components. 10 pp.

—Radioscopy
CENELEC PREN 100 012-93. Basic Specification: X-Ray Inspection of Electronic Components. 10 pp.

—Reliability
BSI BS 4200: Part 3-79. (WITHDRAWN) 1979 Guide on the Reliability of Electronic Equipment and Parts Used Therein Part 3: Presentation of Reliability Data on Electronic Components (or Parts). 20 pp.

CECC CECC 00 801 ISSUE 1-90. Preliminary Guidance Document: Pi-Q Factors of CECC Approved Components for Use in Reliability Predictions (En, Fr, Ge). 33 pp.

Electronic Components *(Cont.)*
—Reliability *(Cont.)*
IEC 409-81. Guide for the Inclusion of Reliability Clauses into Specifications for Components (or Parts) for Electronic Equipment Second Edition. 32 pp.

—Reliability Assured
CNS C5029-86. General Rules for Reliability Assured Electronic Components (Jan)(4901).

JIS C 5700-74. General Rules for Reliability Assured Electronic Components (R 1982). 17 pp.

—Safety
CSA C22.2 NO 0.8-M1986. Safety Functions Incorporating Electronic Technology (R 1992); (Gen Instr 1). 37 pp.

—Salt Spray Testing
CNS C6014-85. Salt Mist Testing Method for Electronic Components (Apr)(3627).

—Sampling
CECC CECC 00 007 ISSUE 2-78. Basic Specification: Sampling Plans and Procedures for Inspection by Attributes (En, Fr, Ge). 15 pp.

—Sealing
CNS C6018-85. Sealing Test Methods for Electronic Components (Apr)(3631). 2 pp.

CNS C6246-85. Sealing (Immersion Cyclic) Testing Method for Electronic Components (Apr)(11237).

JIS C 5032-75. Sealing (Immersion Cyclic) Testing Method for Electronic Components. 4 pp.

—Solderability
CNS C6019-85. Solderability Testing Method for Electronic Components (Apr)(3632). 4 pp.

—Soldering—Heat Resistance
CNS C6247-85. Testing Method of Resistance to Soldering Heat for Electronic Components (Apr)(11238).

—Spindles
CENELEC HD 363-77. Dimensions of Spindle Ends for Manually Operated Electronic Components. 2 pp.

IEC 390-72. Dimensions of Spindle Ends for Manually Operated Electronic Components First Edition; (Supplement A-1976) (Amendment 1-1976). 50 pp.

—Submarines
MOD UK NES 581: Part 1-90. Requirements for the Selection and Approval of Electrical and Electronic Components Part 1: Method of Selection and Approval Issue 2 (07.90). 21 pp.

MOD UK NES 581: Part 2-91. Requirements for the Selection and Approval of Electrical and Electronic Components Part 2: Standard, Approved and Acceptable Components Issue 5 (09.91). 199 pp.

MOD UK NES 581: Part 2-01. Requirements for the Selection and Approval of Electrical and Electronic Components Part 2: Standard, Approved and Acceptable Components Issue 5 (09.91); Amendment 1. 229 pp.

MOD UK NES 581: Part 3-91. Requirements for the Selection and Approval of Electrical and Electronic Components Part 3: Non—Standard Project Approved Components Issue 2 (12.91). 44 pp.

MOD UK NES 581: Part 3-01. Requirements for the Selection and Approval of Electrical and Electronic Components Part 3: Non-Standard Project Approved Components Issue 2 (12.91); Amendment 1. 85 pp.

—Telephone Transmission
CCITT RECMN G.231-89. Arrangement of Carrier Equipment—International Analogue Carrier Systems (Study Group XV) 2 pp. 2 pp.

—Temperature Change Testing
CNS C6020-89. Change of Temperature Testing Procedures for Electronic Components (Oct)(3633).

—Terminations
CNS C6015-90. Robustness of Termination Testing Procedures for Electronic Components (Dec)(3628).

JIS C 0051-86. Robustness of Terminations and Integral Mounting Devices for Electronic Components. 17 pp.

—Vibration
CNS C6016-89. Vibration Testing Procedures for Electronic Components (Jun)(3629).

Electronic Control Equipment
See Also: Control Systems Equipment; Electronic Equipment

—Aircraft Engines
CAA JAR-E Section 3. Advisory Material Joint (Joint Airworthiness Requirements). 23 pp.

INTERNATIONAL AND NON-U.S. NATIONAL STANDARDS
SUBJECT INDEX
Electronic

Electronic Control Equipment *(Cont.)*
—**Aircraft Propellers**
CAA JAR-E Section 3. Advisory Material Joint (Joint Airworthiness Requirements). 23 pp.

Electronic Counters
JIS C 4556-87. Electronic Preset Counters. 21 pp.

Electronic Data Interchange Messaging Services
Use: EDI Messaging Services

Electronic Detectors
Use: Multimeters

Electronic Displays
Use: Display Devices

Electronic Engineering
See Also: Electrical Engineering

—**Glossaries**
IEC. Electricity, Electronics and Telecommunications Multilingual Dictionary Volume 1: English-French-Russian-German-Spanish-Dutch-Italian-Swedish-Polish Second Edition. 1944 pp.
IEC. Electricity, Electronics and Telecommunications Multilingual Dictionary Volume 2: English-French-Russian-German-Spanish-Dutch-Italian-Swedish-Polish Second Edition. 964 pp.
IEC. Electricity, Electronics and Telecommunications Multilingual Dictionary Volume 3: German, Dutch and Swedish Indexes Second Edition. 793 pp.
IEC. Electricity, Electronics and Telecommunications Multilingual Dictionary Volume 4: Spanish and Italian Indexes Second Edition. 540 pp.
IEC. Electricity, Electronics and Telecommunications Multilingual Dictionary Volume 5: Russian and Polish Indexes Second Edition. 578 pp.

—**Yearbooks**
IEC. IEC Yearbook 1993. 653 pp.

Electronic Equipment
Scope Note: For additional listings, see specific types of equipment *See Also:* Avionics; Chassis; Compandors; Computer Equipment; Electrical Components; Electrical Equipment; Electrical Installations; Electronic Components; Electronic Control Equipment; Electronic Measuring Instruments; Electronic Switches; Line Amplifiers; Linearity; Loads (Electric); Optoelectronic Devices; Plug In Units; Power Electronics; Signal Limiters; Time Bases

CSA CAN/CSA-C22. 2 NO 1-M90. Radio, Television, and Electronic Apparatus; (Gen Instr 1 Thru 3). 146 pp.
CSA 1423A Bull. Electrical Bulletin 1423A January 2, 1985 to C22.2 NO 1. 8 pp.
CSA 1434 Bull. Electrical Bulletin 1434 April 1, 1985 to C22.2 NO 1. 2 pp.
IEC Guide 103-80. Guide on Dimensional Co-Ordination First Edition. 44 pp.
MOD UK DSTAN 58-95: Part 90-85. Electronic Assemblies Part 90: Detail Specifications Issue 1. 8 pp.

—**Aircraft—Electromagnetic Compatibility—Test Methods**
MOD UK DSTAN 59-41: Pt 3: Supp K-93. Electromagnetic Compatibility Part 3: Technical Requirements Test Methods and Limits Supplement K: Test Method DCS04 Imported Transient Susceptibility (Aircraft) Issue 1. 18 pp.

—**Aircraft—Electromagnetic Interference**
NATO STANAG 3516 ED 3 AMD 0-93. Electromagnetic Interference and Test Methods for Aircraft Electrical and Electronic Equipment. 51 pp.

—**Aircraft—Procurement**
MOD UK DSTAN 05-123-01. Technical Procedures for the Procurement of Aircraft, Weapon and Electronic Systems Issue 1; Amendment 55. 20 pp.
MOD UK DSTAN 05-123-02. Technical Procedures for the Procurement of Aircraft, Weapon and Electronic Systems Issue 1; Amendment 57. 23 pp.
MOD UK DSTAN 05-123-03. Technical Procedures for the Procurement of Aircraft, Weapon and Electronic Systems Issue 1; Amendment 58. 25 pp.
MOD UK DSTAN 05-123-04. Technical Procedures for the Procurement of Aircraft, Weapon and Electronic Systems Issue 1; Amendment 59. 25 pp.

—**Aluminum Capacitors**
BSI BS QC 300301-93. 1993 Blank Detail Specification Fixed Capacitors for Use in Electronic Equipment. Part 4: Blank Detail Specification: Aluminium Electrolytic Capacitors with Non-Solid Electrolyte Assessment Level E. 16 pp.

Electronic Equipment *(Cont.)*
—**Aluminum Capacitors** *(Cont.)*
BSI BS CECC 30301-93. 1993 Blank Detail Specification Aluminium Electrolytic Capacitors with Non-Solid Electrolyte (Assessment Level E) (T). 23 pp.
BSI BS CECC 30301-77. 1977 Aluminium Electrolytic Capacitors with Non-Solid Electrolyte: Blank Detail Specification. 9 pp.
BSI BS CECC 30301 024-81. 1981 Amd 1 Aluminium Electrolytic Capacitors with Non-Solid Electrolyte. Long-Life Grade: Detail Specification: Full Plus Additional Assessment Level. 28 pp.

—**Armored Vehicles**
MOD UK DSTAN 00-32: Part 1-85. Systematic Approach to Vehicle Electronics (SAVE) Part 1: General Requirements Issue 1. 7 pp.

—**Automotive—Diagnostic Testing**
BSI BS AU 206: Part 4-86. 1986 Diagnostic Systems for Road Vehicles Part 4: General Requirements for Testing Electronic Systems. 5 pp.
ISO 8093-85. Road Vehicles—Diagnostic Testing of Electronic Systems First Edition. 5 pp.
ISO 9141-89. Road Vehicles—Diagnostic Systems—Requirements for Interchange of Digital Information First Edition. 16 pp.

—**Automotive—Environmental Testing**
CNS D1036-82. General Rules of Environmental Testing Method for Automotive Electronic Equipment (Nov)(9589).

—**Automotive—Glossaries**
CNS D1065-85. Glossary of Terms Relating to Electric Equipment for Automobiles (Electronic System) (Jul)(11301).

—**Cargo Transportation**
CNS Z6026-80. Shipping Test for Consumer Electronic Products (May)(5637).

—**Ceramic Capacitors**
BSI BS CECC 30700-91. 1991 Fixed Capacitors of Ceramic Dielectric Class 2: Sectional Specification. 39 pp.
BSI BS CECC 30700-01. 1991 Amd 1 Sectional Specification: Fixed Capacitors of Ceramic Dielectric, Class 2 (AMD 7401) February 15, 1993. 45 pp.
BSI BS CECC 31100-81. 1981 Fixed Ceramic Dielectric Capacitors of Barrier Layer Type (Dielectric Class 3): Sectional Specification. 28 pp.
BSI BS CECC 31100-01. 1981 Amd 1 Sectional Specification: Fixed Ceramic Dielectric Capacitors of Barrier Layer Type (Dielectric Class 3) (AMD March 15, 1993. 31 pp.
BSI BS CECC 31400-82. 1982 Fixed Ceramic Capacitors of Dielectric Class 1, for Electrical Shock Hazard Protection: Sectional Specification. 25 pp.
BSI BS CECC 31400-01. 1982 Amd 1 Sectional Specification: Fixed Ceramic Capacitors of Dielectric Class 1, for Electrical Shock Hazard Protection (AMD 7404) February 15, 1993. 28 pp.
IEC 384 Pt 8-88. Fixed Capacitors for Use in Electronic Equipment Part 8: Sectional Specification: Fixed Capacitors of Ceramic Dielectric Second Edition; (Amendment 1-1993) (IECQ QC 300600). 91 pp.
IEC 384 Pt 9-88. Fixed Capacitors for Use in Electronic Equipment Part 9: Sectional Specification: Fixed Capacitors of Ceramic Dielectric, Class 2 Second Edition (IECQ QC 300700). 67 pp.
IEC 384 Pt 9-1-88. Fixed Capacitors for Use in Electronic Equipment Part 9: Blank Detail Specification: Fixed Capacitors of Ceramic Dielectric, Class 2 Assessment Level E First Edition (IECQ QC 300701). 33 pp.
IECQ QC 300701-88. Fixed Capacitors for Use in Electronic Equipment Part 9: Blank Detail Specification: Fixed Capacitors of Ceramic Dielectric, Class 2 Assessment Level E (IEC 384-9-1 ED 1). 33 pp.

—**Ceramic Chip Capacitors**
BSI BS QC 301901-92. 1992 Fixed Capacitors for Use in Electronic Equipment. Part 10: Blank Detail Specification: Fixed Multilayer Ceramic Chip Capacitors Assessment Level E. 12 pp.

—**Ceramics—Bulk Density**
CNS C6068-80. Method of Test for Apparent Density of Ceramics for Electron Device and Semiconductor Application (Jul)(5751). 3 pp.

—**Connectors**
BSI BS CECC 75201-92. 1992 Blank Detail Specification Interim Example Detail Specification/ Blank Detail Specification Circular Connectors for Frequencies Below 3 MHz. 4 pp.

Electronic Equipment *(Cont.)*
—**Connectors** *(Cont.)*
IEC 807 Pt 7-91. Rectangular Connectors for Frequencies Below 3 MHz Part 7: Detail Specification for a Range of Connectors with Polarized Guides or Jackscrews and Size 16 (13 A) Round Contacts—Removable Crimp Contact Types with Closed Crimp Barrels, Rear Insertion/ Front Release, with. 112 pp.

—**Connectors—Television Receivers**
JIS C 5590-88. Digital Interconnection of Television Receivers and Electronic Equipments (8 Pin). 9 pp.
JIS C 5591-88. Analog Interconnection of Television Receivers and Electronic Equipments (21 Pin). 12 pp.

—**Defense Contracts**
MOD UK DEFCON 112N-89. Basic Set of Conditions of Contract—Military Electronics 5/89. 1 p.

—**Design**
MOD UK DSTAN 00-10-86. General Design and Manufacturing Requirements for Service Electronic Equipment Issue 3. 474 pp.
MOD UK DSTAN 00-10-01. General Design and Manufacturing Requirements for Service Electronic Equipment Issue 3; Amendment 1. 475 pp.
MOD UK DSTAN 00-10-02. General Design and Manufacturing Requirements for Service Electronic Equipment Issue 3; Amendment 2. 476 pp.
MOD UK DSTAN 58-95: Part 1-01. Electronic Assemblies Part 1: General Requirements Issue 1; Amendment 1. 18 pp.
MOD UK DSTAN 59-36-90. Electronic and Associated Electrical Components for Defence Purposes Procedure for the Selection and Specification of Electronic and Associated Electrical Components for Use in Defence Equipment Issue 4. 25 pp.

—**Design—Maintainability**
CSA CAN/CSA-Q633-90. Reliability, Availability, and Maintainability Design Guide for Electronic Products; (Gen Instr 1). 40 pp.

—**Design—Reliability**
CSA CAN/CSA-Q633-90. Reliability, Availability, and Maintainability Design Guide for Electronic Products; (Gen Instr 1). 40 pp.

—**DIP Switches**
BSI BS QC 960500-92. 1992 Electromechanical Switches for Use in Electronic Equipment Sectional Specification for in-Line Package Switches (IEC 1020-3: 1991). 16 pp.
BSI BS QC 960501-92. 1992 Electromechanical Switches for Use in Electronic Equipment Blank Detail Specification for in-Line Package Switches (IEC 1020-3-1: 1991). 10 pp.
IEC 1020 Pt 3-91. Electromechanical Switches for Use in Electronic Equipment Part 3: Sectional Specification for In-Line Package Switches First Edition; (IECQ QC 960500). 34 pp.
IEC 1020 Pt 3-1-91. Electromechanical Switches for Use in Electronic Equipment Part 3: Sectional Specification for In-Line Package Switches Section 1—Blank Detail Specification; First Edition; (IECQ QC 960501). 26 pp.

—**Electric Coils**
IEC 1007-90. Transformers and Inductors for Use in Electronic and Telecommunication Equipment—Measuring Methods and Test Procedures First Edition; (Amendment 1-1993). 113 pp.

—**Electric Conductors**
CSA C22.2 NO 16-M1986. Insulated Conductors for Power-Operated Electronic Devices (R 1992); (Gen Instr 1 Thru 4). 64 pp.

—**Electric Inlets**
CSA C22.2 NO 182.3-M1987. Special Use Attachment Plugs, Receptacles, and Connectors (R 1993); (Gen Instr 1 Thru 2). 41 pp.

—**Electric Outlets**
CSA C22.2 NO 182.3-M1987. Special Use Attachment Plugs, Receptacles, and Connectors (R 1993); (Gen Instr 1 Thru 2). 41 pp.

—**Electric Plug Receptacles**
CSA C22.2 NO 182.3-M1987. Special Use Attachment Plugs, Receptacles, and Connectors (R 1993); (Gen Instr 1 Thru 2). 41 pp.

—**Electric Plugs**
CSA C22.2 NO 182.3-M1987. Special Use Attachment Plugs, Receptacles, and Connectors (R 1993); (Gen Instr 1 Thru 2). 41 pp.

Electronic Equipment (Cont.)

—Electric Shock—Classification

IEC 536 Pt 2-92. Classification of Electrical and Electronic Equipment with Regard to Protection Against Electric Shock Part 2: Guidelines to Requirements for Protection Against Electric Shock First Edition. 41 pp.

—Electrical Installations

DIN VDE 0160-88. Electronic Equipment for Use in Electrical Power Installations and Their Assembly into Electrical Power Installations (May). 158 pp.

DIN VDE 0160 (D)-90. Assembly of Electronic Equipment in Power Installations (Dec). 193 pp.

DIN VDE 0160 A1-89. Electronic Equipment for Use in Electrical Power Installations and Their Assembly into Electrical Power Installations: Amendment 1 (Apr). 11 pp.

DIN VDE 0160 A2 (D)-88. Electronic Equipment for Use in Electrical Power Installations and Their Assembly into Electrical Power Installations: Amendment 2 (Oct). 32 pp.

—Electromagnetic Compatibility

MOD UK DSTAN 59-41: Part 1-88. Electromagnetic Compatibility Part 1: General Requirements Issue 4 (Superseded by Def Stan 59-41: Part 1: Section 1). 13 pp.

MOD UK DSTAN 59-41: Pt 1:Sec 1-93. Electromagnetic Compatibility Part 1: Introduction Section 1: General Requirements Issue 5. 13 pp.

MOD UK DSTAN 59-41: Pt 1:Sec 2-93. Electromagnetic Compatibility Part 1: Introduction Section 2: Guide to the Specification and Selection of EMC Requirements Issue 5. 45 pp.

MOD UK DSTAN 59-41: Part 2-88. Electromagnetic Compatibility Part 2: Management and Planning Procedures Issue 3. 10 pp.

—Electromagnetic Compatibility—Test Methods

MOD UK DSTAN 59-41: Part 3-01. Electromagnetic Compatibility Part 3: Technical Requirements Test Methods and Limits Issue 3; Amendment 4. 167 pp.

MOD UK DSTAN 59-41: Part 3-93. Electromagnetic Compatibility Part 3: Technical Requirements Test Methods and Limits Issue 4. 42 pp.

MOD UK DSTAN 59-41: Pt 3: Supp A-93. Electromagnetic Compatibility Part 3: Technical Requirements Test Methods and Limits Supplement A: Test Method DCE01 Conducted Emission on Power Lines 20Hz—150MHz Issue 1. 10 pp.

MOD UK DSTAN 59-41: Pt 3: Supp B-93. Electromagnetic Compatibility Part 3: Technical Requirements Test Methods and Limits Supplement B: Test Method DCE02 Conducted Emission on Control and Signal Lines 20Hz—150MHz Issue 1. 10 pp.

MOD UK DSTAN 59-41: Pt 3: Supp C-93. Electromagnetic Compatibility Part 3: Technical Requirements Test Methods and Limits Supplement C: Test Method DCE03 Conducted Emission on Exported Transients, Power Lines Issue 1. 12 pp.

MOD UK DSTAN 59-41: Pt 3: Supp D-93. Electromagnetic Compatibility Part 3: Technical Requirements Test Methods and Limits Supplement D: Test Method DRE01 Radiated Emissions E Field 14kHz—18GHz Issue 1. 14 pp.

MOD UK DSTAN 59-41: Pt 3: Supp E-93. Electromagnetic Compatibility Part 3: Technical Requirements Test Methods and Limits Supplement E: Test Method DRE02 H Field Radiation 20Hz—50kHz Issue 1. 10 pp.

MOD UK DSTAN 59-41: Pt 3: Supp F-93. Electromagnetic Compatibility Part 3: Technical Requirements Test Methods and Limits Supplement F: Test Method DRE03 Radiated Emmisions Installed Antenna 1MHz—76MHz Issue 1. 12 pp.

MOD UK DSTAN 59-41: Pt 3: Supp G-93. Electromagnetic Compatibility Part 3: Technical Requirements Test Methods and Limits Supplement G: Test Method DCS01 Conducted Susceptibility, Power Lines 20Hz—50kHz Issue 1. 11 pp.

MOD UK DSTAN 59-41: Pt 3: Supp H-93. Electromagnetic Compatibility Part 3: Technical Requirements Test Methods and Limits Supplement H: Test Method DCS02 Conducted Susceptibility, Power, Control and Signal Lines 50kHz—400MHz Issue 1. 20 pp.

MOD UK DSTAN 59-41: Pt 3: Supp J-93. Electromagnetic Compatibility Part 3: Technical Requirements Test Methods and Limits Supplement J: Test Method DCS03 Conducted Susceptibility, Control and Signal Lines 20Hz—50kHz Issue 1. 10 pp.

MOD UK DSTAN 59-41: Pt 3: Supp U-93. Electromagnetic Compatibility Part 3: Technical Requirements Test Methods and Limits Supplement U: Test Method DRS01 H Field Susceptibility, 20Hz—50kHz Issue 1. 11 pp.

MOD UK DSTAN 59-41: Pt 3: Supp V-93. Electromagnetic Compatibility Part 3: Technical Requirements Test Methods and Limits Supplement V: Test Method DRS02 Radiated Susceptibility E Field, 14kHz—18GHz Issue 1. 11 pp.

MOD UK DSTAN 59-41: Pt 3: Supp W-93. Electromagnetic Compatibility Part 3: Technical Requirements Test Methods and Limits Supplement W: Test Method DMFS01 Magnetostatic Field Susceptibility Issue 1. 8 pp.

—Engineering Drawings—Symbols

CNS B1001-10-89. Engineering Drawing Symbols for Electronic Power and Electronic System (Mar)(3-10).

—Environmental Engineering

CNS C5211-86. Environmental Requirements for Electronic Equipments (Nov)(11762).

CNS C5218-88. Environmental Requirements for Electronic Components (Jan)(12211).

JIS C 6007-69. Environmental Requirements for Electronic Equipments.

—Film Capacitors

IEC 384 Pt 12-88. Fixed Capacitors for Use in Electronic Equipment Part 12: Sectional Specification: Fixed Polycarbonate Film Dielectric Metal Foil D.C. Capacitors Second Edition (IECQ QC 301700). 59 pp.

IEC 384 Pt 13-1-91. Fixed Capacitors for Use in Electronic Equipment Part 13: Blank Detail Specification: Fixed Polypropylene Film Dielectric Metal Foil d.c. Capacitors. Assessment Level E First Edition (IECQ QC 301801). 33 pp.

—Film Resistors

BSI BS QC 390100-92. 1992 Fixed Film Resistor Networks for Use in Electronic Equipment Sectional Specification for Film Resistor Networks of Assessed Quality on the Basis of the Capability Approval Procedure (IEC 1045-2: 1991). 54 pp.

IEC 1045 Pt 2-91. Fixed Film Resistor Networks for Use in Electronic Equipment Part 2: Sectional Specification for Film Resistor Networks of Assessed Quality on the Basis of the Capability Approval Procedure First Edition (IECQ QC 390100). 109 pp.

IEC 1045 Pt 2-1-91. Fixed Film Resistor Networks for Use in Electronic Equipment Part 2: Blank Detail Specification for Film Resistor Networks of Assessed Quality on the Basis of the Capability Approval Procedure Assessment Level E First Edition (IECQ QC 390101). 35 pp.

IECQ QC 390101-91. Fixed Film Resistor Networks for Use in Electronic Equipment Part 2: Blank Detail Specification for Film Resistor Networks of Assessed Quality on the Basis of the Capability Approval Procedure Assessment Level E (IEC 1045-2-1 ED 1). 35 pp.

—Fire Wire and Ribbon—Density

CNS C6096-80. Method of Test for Density of Fire Wire and Ribbon for Electronic Devices (Aug)(6316).

—Fixed Resistors

IEC 115 Pt 1-82. Fixed Resistors for Use in Electronic Equipment Part 1: Generic Specification Second Edition; (Amendment 2-1987, Incorporating Amendment 1) (Amendment 3-1989) (Amendment 4-1993). 128 pp.

—Fixed Resistors—Precision

BSI BS QC 400302-92. 1992 Fixed Precision Resistors Assessment Level F (IEC 115-5-2: 1992). 13 pp.

IEC 115 Pt 5-2-92. Fixed Resistors for Use in Electronic Equipment Part 5: Blank Detail Specification: Fixed Precision Resistors Assessment Level F First Edition (IECQ QC 400302). 33 pp.

—Glossaries

BSI BS 4727:Pt1: Group 01-83. 1983 Amd 1 Electrotechnical, Power, Telecommunication, Electronics, Lighting and Colour Terms Part 1: Terms Common to Power, Telecommunications and Electronics Group 01: Fundamental Terminology. 56 pp.

BSI BS 4727:Pt3: Group 01-71. 1971 Electrotechnical, Power, Telecommunication, Electronics, Lighting and Colour Terms Part 3: Terms Particular to Telecommunications and Electronics Group 01: General Telecommunications and Electronics Terminology. 42 pp.

IEC. Electricity, Electronics and Telecommunications Multilingual Dictionary Volume 1: English-French-Russian-German-Spanish-Dutch-Italian-Swedish-Polish Second Edition. 1944 pp.

IEC. Electricity, Electronics and Telecommunications Multilingual Dictionary Volume 2: English-French-Russian-German-Spanish-Dutch-Italian-Swedish-Polish Second Edition. 964 pp.

IEC. Electricity, Electronics and Telecommunications Multilingual Dictionary Volume 3: German, Dutch and Swedish Indexes Second Edition. 793 pp.

IEC. Electricity, Electronics and Telecommunications Multilingual Dictionary Volume 4: Spanish and Italian Indexes Second Edition. 540 pp.

IEC. Electricity, Electronics and Telecommunications Multilingual Dictionary Volume 5: Russian and Polish Indexes Second Edition. 578 pp.

IEC 50 Chap 551-82. International Electrotechnical Vocabulary Chapter 551: Power Electronics. 75 pp.

JIS C 5600-77. Glossary of Terms Used in Electronics and Telecommunications (Basic Terms).

JIS C 5602-86. Glossary of Passive Components for Electronic Equipment; (Erratum). 87 pp.

SAA AS 1852.551-83. International Electrotechnical Vocabulary—Part 551: Power Electronics. 42 pp.

SNZ IEC 50: 50(551)-82. International Electrotechnical Vocabulary 50(551): Power Electronics. 65 pp.

—Household

SAA AS 3250-90. Approval and Test Specification—Mains Operated Electronic Equipment and Related Equipment for Household and Similar General Use (This is a Joint Standard with SANZ NZS 3250). 59 pp.

SNZ NZS/AS 3250-90. Approval and Test Specification—Mains Operated Electronic and Related Equipment for Household and Similar General Use (This is a Joint Standard with SAA AS 3250). 59 pp.

—Identification Systems

MOD UK DSTAN 61-7: Part 2-91. Identification of Electrical and Electronic Systems, Wiring and Components Part 2: Marking Code for Electrical and Electronic Equipment Issue 3. 24 pp.

—Identification Systems—Electrostatic Sensitivity

CNS Z5133-84. Symbol and Label for Electrostatic Sensitive Devices (Nov)(11144).

—Immersion Testing

JIS C 0052-86. Testing Method of Resistance to Solvents (Immersion in Cleaning Solvents). 12 pp.

—Indexes (Documentation)

MOD UK DSTAN 59-20-01. Index to NATO Electronic Parts Recommendations (NEPRs) and NATO Electronic Technical Recommendations (NETRs) Issue 3; Amendment 1. 22 pp.

—Industrial Machines

BSI BS 2771: Part 1-86. 1986 Electronic Equipment for Industrial Machines Part 1: General Requirements. 94 pp.

—Inspection by Attributes

BSI BS E9007-75. 1975 Basic Specification: Sampling Plans and Procedures for Inspection by Attributes. 6 pp.

—Land Service—Electromagnetic Compatibility—Test Methods

MOD UK DSTAN 59-41: Pt 3: Supp M-93. Electromagnetic Compatibility Part 3: Technical Requirements Test Methods and Limits Supplement M: Test Method DCS06 Imported Long Transient Susceptibility 28 Volt Systems (Land Service) Issue 1. 9 pp.

MOD UK DSTAN 59-41: Pt 3: Supp N-93. Electromagnetic Compatibility Part 3: Technical Requirements Test Methods and Limits Supplement N: Test Method DCS07 Imported Short Transient Susceptibility (Land Service) Issue 1. 10 pp.

—Liquid Nitrogen

MOD UK DSTAN 68-94-90. Nitrogen, Liquid: for Aircraft, Ship Systems and General Purpose Useage Issue 2. 7 pp.

MOD UK DSTAN 68-94-01. Nitrogen, Liquid: for Aircraft, Ship Systems and General Purpose Useage Issue 2; Amendment 1. 8 pp.

MOD UK DSTAN 68-94-02. Nitrogen, Liquid: for Aircraft, Ship Systems and General Purpose Useage Issue 2; Amendment 2. 9 pp.

—Marine—Electromagnetic Interference

BSI BS 7027-90. 1990 Limits and Methods of Measurement of the Immunity of Marine Electrical and Electronic Equipment to Conducted and Radiated Electromagnetic Interference. 14 pp.

—Mechanical Shock

ISO 8569-89. Mechanical Vibration—Shock-and-Vibration-Sensitive Electronic Equipment—Methods of Measurement and Reporting Data of Shock and Vibration Effects in Buildings First Edition. 13 pp.

INTERNATIONAL AND NON-U.S. NATIONAL STANDARDS
SUBJECT INDEX

Electronic Equipment (Cont.)
—Mechanical Structures—Glossaries
CENELEC HD 550 S1-89. Mechanical Structures for Electronic Equipment—Terminology. 3 pp.
CNS C5233-90. Mechanical Structures for Electronic Equipment Terminology (May)(12713).

—Modular
BSI BS EN 60917-92. 1992 Modular Order for the Development of Mechanical Structures for Electronic Equipment Practices (IEC 917: 1988). 18 pp.
BSI BS EN 60917-0-92. 1992 Modular Order for the Development of Mechanical Structures for Electronic Equipment Practices Part 0: Guide for the Users of IEC Publication 917 (IEC 917-0: 1989). 24 pp.
CENELEC EN 60 917-90. Modular Order for the Development of Mechanical Structures for Electronic Equipment Practices. 4 pp.
CENELEC EN 60917-0-92. Modular Order for the Development of Mechanical Structures for Electronic Equipment Practices Part 0: Guide for the Users of IEC Publication 917;. 5 pp.
CENELEC EN 60917-0-92. Modular Order for the Development of Mechanical Structures for Electronic Equipment Practices Part 0: Guide for the Users of IEC Publication 917; (IEC 917-0:1989). 7 pp.
IEC 917-88. Modular Order for the Development of Mechanical Structures for Electronic Equipment Practices First Edition; (Amendment 1-1993). 38 pp.
IEC 917 Pt 0-89. Modular Order for the Development of Mechanical Structures for Electronic Equipment Practices Part 0: Guide for the Users of IEC Publication 917 First Edition; (CENELEC EN 60917-0: 1992). 41 pp.

—Modular—Mechanical Structures
IEC 917 Pt 2-92. Modular Order for the Development of Mechanical Structures for Electronic Equipment Practices Part 2: Sectional Specification—Interface Co-Ordination Dimensions for the 25 mm Equipment Practice First Edition. 40 pp.

—Modules
BSI BS EN 60917-92. 1992 Modular Order for the Development of Mechanical Structures for Electronic Equipment Practices (IEC 917: 1988). 18 pp.
BSI BS EN 60917-0-92. 1992 Modular Order for the Development of Mechanical Structures for Electronic Equipment Practices Part 0: Guide for the Users of IEC Publication 917 (IEC 917-0: 1989). 24 pp.
CENELEC EN 60 917-90. Modular Order for the Development of Mechanical Structures for Electronic Equipment Practices. 4 pp.
CENELEC EN 60917-0-92. Modular Order for the Development of Mechanical Structures for Electronic Equipment Practices Part 0: Guide for the Users of IEC Publication 917;. 5 pp.
CENELEC EN 60917-0-92. Modular Order for the Development of Mechanical Structures for Electronic Equipment Practices Part 0: Guide for the Users of IEC Publication 917; (IEC 917-0:1989). 7 pp.
IEC 917-88. Modular Order for the Development of Mechanical Structures for Electronic Equipment Practices First Edition; (Amendment 1-1993). 38 pp.
IEC 917 Pt 0-89. Modular Order for the Development of Mechanical Structures for Electronic Equipment Practices Part 0: Guide for the Users of IEC Publication 917 First Edition; (CENELEC EN 60917-0: 1992). 41 pp.

—Molybdenum Materials
CNS H2051-85. General Rules for Test of Tungsten and Molybdenum Materials for Lighting and Electronic Equipments (Nov)(6337).
CNS H3090-85. Molybdenum Wires for Lighting and Electronic Equipments (Nov)(6343).
CNS H3091-85. Molybdenum Rods for Lighting and Electronic Equipments (Nov)(6344).
CNS H3092-85. Molybdenum Sheets for Lighting and Electronic Equipments (Nov)(6345).
JIS H 4460-84. General Rules for Test of Tungsten and Molybdenum Materials for Lighting and Electronic Equipments. 9 pp.
JIS H 4481-89. Molybdenum Wires for Lighting and Electronic Equipments. 10 pp.
JIS H 4482-89. Molybdenum Rods for Lighting and Electronic Equipments. 7 pp.
JIS H 4483-84. Molybdenum Sheets for Lighting and Electronic Equipments. 7 pp.

—Mountings
BSI BS 7039-89. 1989 Recommendations for Dimensions of Spindle Ends and Bushes, and for Dimensions for the Mounting of Single-Hole, Bush-Mounted Spindle-Operated Variable Capacitors, Variable Resistors and Potentio-meters for Electronic Equipment. 16 pp.

—Nitrogen
MOD UK DSTAN 16-19-93. Nitrogen, Compressed, Oil-Free: for All Applications Issue 2. 9 pp.
MOD UK DSTAN 68-95-90. Nitrogen, Compressed, High Purity Oil—Free: for All Applications Issue 2. 8 pp.

Electronic Equipment (Cont.)
—Nitrogen (Cont.)
MOD UK DSTAN 68-95-01. Nitrogen Gas, Compressed, High Purity Oil-Free: for All Applications Issue 2; Amendment 2. 10 pp.

—Nitrogen—Fuel Hoses
MOD UK DSTAN 16-19-93. Nitrogen, Compressed, Oil-Free: for All Applications Issue 2. 9 pp.
NATO STANAG 3990 ED 1 AMD 1-90. Characteristics of High Quality Oil-Free Gaseous Nitrogen, Supply Pressure and Hoses, for Aircraft, Weapons and Electronic Equipment Servicing. 6 pp.

—Nonwirewound Resistors
BSI BS QC 400102-92. 1992 Fixed Low-Power Non-Wirewound Resistors Assessment Level F (IEC 115-2-2: 1992). 13 pp.
BSI BS CECC 40102-81. 1981 Amd 1 Fixed Low Power Non-Wirewound Resistors: Blank Detail Specification: Assessment Level M. 16 pp.
BSI BS CECC 40104-92. 1992 Fixed Low Power Non-Wirewound Resistors of Enhanced Quality. 25 pp.

—Panels—Packaging
MOD UK DSTAN 81-65-01. Packaging of Panels, Electronic Circuit Issue 2; Amendment 1. 17 pp.

—Pliers
CNS B3190-87. Pliers for Radio (Apr)(2693). 3 pp.

—Post Design Services
MOD UK DSTAN 05-125-01. Technical Procedures for Post Design Services (Electronics) Issue 2; Amendment 1. 7 pp.
MOD UK DSTAN 05-125: Chapter 1-01. Outline of PDS Organisation Issue 2; Amendment 1. 3 pp.
MOD UK DSTAN 05-125: Chapter 2-01. Contracts and Administration Issue 2; Amendment 1. 5 pp.
MOD UK DSTAN 05-125: Chapter 4-01. Systems and Equipment Issue 2; Amendment 1. 4 pp.
MOD UK DSTAN 05-125: Chapter 6-01. Fault Investigation Procedure Issue 2; Amendment 1. 10 pp.
MOD UK DSTAN 05-125: Chapter 8-01. Forms: Description of Use Issue 2; Amendment 1. 112 pp.
MOD UK DSTAN 05-125: Appendix A-01. List of Addresses Issue 2; Amendment 1. 5 pp.
MOD UK DSTAN 05-125: Appendix B-01. Supply Source of Publications Issue 2; Amendment 1. 2 pp.
MOD UK DSTAN 05-125: Appendix C-01. List of Abbreviations Issue 2; Amendment 1. 5 pp.
MOD UK DSTAN 05-125: Index-01. Alphabetical Index Issue 2; Amendment 1. 9 pp.

—Power Capacitors
IEC 1071 Pt 1-91. Power Electronic Capacitors Part 1: General First Edition. 70 pp.

—Power Resistors
BSI BS 9940: Part 02.0-83. 1983 Fixed Resistors for Use in Electronic Equipment Part 02.0: Sectional Specification: Fixed Power Resistors. 17 pp.
BSI BS QC 400202-92. 1992 Fixed Power Resistors Assessment Level F (IEC 115-4-2: 1992). 12 pp.
IEC 115 Pt 4-82. Fixed Resistors for Use in Electronic Equipment Part 4: Sectional Specification: Fixed Power Resistors Second Edition; (Amendment 1-1993) (IECQ QC 400200). 44 pp.
IEC 115 Pt 4-1-83. Fixed Resistors for Use in Electronic Equipment Part 4: Blank Detail Specification: Fixed Power Resistors Assessment Level E First Edition; (Amendment 1-1993) (IECQ QC 400201). 34 pp.
IEC 115 Pt 4-2-92. Fixed Resistors for Use in Electronic Equipment Part 4: Blank Detail Specification: Fixed Power Resistors Assessment Level F First Edition (IECQ QC 400202). 31 pp.
IEC 115 Pt 4-3-93. Fixed Resistors for Use in Electronic Equipment Part 4: Blank Detail Specification: Fixed Power Resistors, Heat-Sink Types Assessment Level H First Edition (IECQ QC 400203). 35 pp.

—Power Systems—Safety
IEC 65-85. Safety Requirements for Mains Operated Electronic and Related Apparatus for Household and Similar General Use Fifth Edition; (Amendment 2-1989, Incorporating Amendment 1) (Amendment 3-1992). 210 pp.

—Printed Circuits
SAA AS 1560-89. Guidance for the Mounting of Discrete Electronic Components on Printed Boards (IEC 321:1970). 16 pp.

—Procurement
SNZ NZMP 6652-78. Code of Practice for the Procurement of Electronic Equipment Amend: 1, 1982. 34 pp.

Electronic Equipment (Cont.)
—Programmable—Safety
SAA AS 4048.1 (INT)-92. Functional Safety of Electrical/Electronic/Programmable Electronic Systems—Generic Aspects—Part 1(Int): General Requirements (IEC 65A(Secretariat)123). 140 pp.

—Pushbutton Switches
BSI BS QC 960400-92. 1992 Electromechanical Switches for Use in Electronic Equipment Sectional Specification for Pushbutton Switches (IEC 1020-5: 1991). 22 pp.
BSI BS QC 960401-92. 1992 Blank Detail Specification Electromechanical Switches for Use in Electronic Equipment Blank Detail Specification for Pushbutton Switches (IEC 1020-5-1: 1991). 10 pp.
IEC 1020 Pt 5-91. Electromechanical Switches for Use in Electronic Equipment Part 5: Sectional Specification for Pushbutton Switches; First Edition; (IECQ QC 960400). 44 pp.
IEC 1020 Pt 5-1-91. Electromechanical Switches for Use in Electronic Equipment Part 5: Sectional Specification for Pushbutton Switches Section 1—Blank Detail Specification; First Edition; (IECQ QC 960401). 26 pp.

—Qualified Products List
BSI PD 9002:2-92. 1992 BS 9000—IECQ Qualified Products List (T). 241 pp.
BSI PD 9002:2-91. 1991 BS 9000—IECQ Qualified Products List. 261 pp.
BSI PD 9002:3-92. 1992 BS 9000—IECQ Qualified Products List. 231 pp.

—Quality Assurance
BSI BS E9007-75. 1975 Basic Specification: Sampling Plans and Procedures for Inspection by Attributes. 6 pp.
MOD UK DSTAN 05-62: Part 14-81. Guidance on Quality Assurance Procedures Part 14: Guide to Quality Assurance Arrangements for Installation of Ground Based Equipments Incorporating Electronic Systems Issue 1. 4 pp.

—Quality Assurance—Defense Contracts
MOD UK DEFCON 17-90. Selection of Electronic and Assiciated Electrical Components for Defence Equipment 10/90. 1 p.
MOD UK DEFCON 112BS-90. Conditions of Contract for Electronic Equipment Procured to CECC/BS9000 Procedures 9/90. 1 p.

—Ratings—Environmental Conditions
IEC Guide 106-89. Guide for Specifying Environmental Conditions for Equipment Performance Rating First Edition. 24 pp.

—Rectangular Connectors
BSI BS CECC 75301-92. 1992 Interim Example Detail Specification/Blank Detail Specification: Rectangular Connectors for Frequencies Below 3 MHz. 4 pp.
MOD UK DSTAN 59-35: Pt 3:Sec 12-83. Connectors, Electrical Part 3: Connectors for d.c. and Low Frequency Applications Section 12: Pattern 9523 F0013 List of Items Conforming to BS 9523 F0013 Issue 1. 17 pp.

—Reliability
BSI BS 4200: Part 1-67. (WITHDRAWN) 1967 Guide on the Reliability of Electronic Equipment and Parts Used Therein Part 1: Introduction. 5 pp.
BSI BS 4200: Part 4-71. (WITHDRAWN) 1971 Guide on the Reliability of Electronic Equipment and Parts Used Therein Part 4: Collection of Reliability, Availability and Maintainability Data from Field Performance of Electronic Items. 9 pp.
BSI BS 4200: Part 5-68. (WITHDRAWN) 1968 Guide on the Reliability of Electronic Equipment and Parts Used Therein Part 5: Reliability Programmes for Equipment (Superseded by BS 5760: Part 1: 1985). 5 pp.
BSI BS 4200: Part 6-71. (WITHDRAWN) 1971 Guide on the Reliability of Electronic Equipment and Parts Used Therein Part 6: Feedback of Reliability Information on Equipment (Superseded by BS 5760: Part 1: 1985). 8 pp.
BSI BS 4200: Part 7-82. (WITHDRAWN) 1982 Guide on the Reliability of Electronic Equipment and Parts Used Therein Part 7: The Inclusion of of Reliability Clauses in Specifications for Components or Parts for Electronic Equipment (Superseded by BS 5760: Part 4: 1986). 16 pp.
IEC 300 Pt 3-1-91. Dependability Management Part 3: Application Guide Section 1: Analysis Techniques for Dependability: Guide on Methodology First Edition. 43 pp.
IEC 319-78. Presentation of Reliability Data on Electronic Components (or Parts) Second Edition. 37 pp.
IEC 362-71. Guide for the Collection of Reliability, Availability, and Maintainability Data from Field Performance of Electronic Items First Edition. 21 pp.

INTERNATIONAL AND NON-U.S. NATIONAL STANDARDS SUBJECT INDEX

Electronic

Electronic Equipment (Cont.)

—**Reliability—Glossaries**
BSI BS 4200: Part 2-74. (WITHDRAWN) 1974 Amd 2 Guide on the Reliability of Electronic Equipment and Parts Used Therein Part 2: Terminology (AMD 5376) June 30, 1988 (Superseded by BS 4778: Sections 3.1 & 3.2: 1991). 16 pp.

—**Repair Contracts**
MOD UK DEFCON 112CU-90. Conditions of Repair Contracts for Naval Electrical, Electronic and Weapon Equipments 7/90. 8 pp.

—**Rolling Stock**
JIS E 5006-88. General Rules for Tests of Electronic Equipment Used on Railway Rolling Stock.

—**Rotary Switches**
BSI BS 9563-79. (OBSOLESCENT) 1979 Rotary (Manual) Switches of Assessed Quality: Generic Data and Methods of Test: General Rules for the Preparation of Detail Specifications (Superseded by BS CECC 96100). 20 pp.
BSI BS QC 960100-92. 1992 Electromechanical Switches for Use in Electronic Equipment Sectional Specification for Rotary Switches (IEC 1020-2: 1991). 16 pp.
BSI BS QC 960101-92. 1992 Electromechanical Switches for Use in Electronic Equipment Blank Detail Specification for Rotary Switches (IEC 1020-2-1: 1991). 10 pp.
BSI BS CECC 96100-88. 1988 Sectional Specification Including Blank Detail Specification: Rotary Switches. 45 pp.
CENELEC PREN 196 103-92. Blank Detail Specification: Rotary Switches Assessment Level Y. 26 pp.
CENELEC PREN 196 110-92. Sectional Specification: Rotary Switches Capability Approval. 31 pp.
IEC 1020 Pt 2-91. Electromechanical Switches for Use in Electronic Equipment Part 2: Sectional Specification for Rotary Switches; First Edition; (IECQ QC 960100). 34 pp.
IEC 1020 Pt 2-1-91. Electromechanical Switches for Use in Electronic Equipment Part 2: Sectional Specification for Rotary Switches Section 1—Blank Detail Specification; First Edition; (IECQ QC 960101). 26 pp.
JIS C 6437-89. Rotary Switches for Electronic Equipment. 42 pp.

—**Safety**
BSI BS 415-90. 1990 Safety Requirements for Mains-Operated Electronic and Related Apparatus for Household and Similar General Use. 80 pp.
CENELEC HD 195 S4-85. Safety Requirements for Mains Operated Electronic and Related Apparatus for Household and Similar General Use. 4 pp.
CENELEC HD 195 S5-88. Safety Requirements for Mains Operated Electronic and Related Apparatus for Household and Similar General Use. 4 pp.
CENELEC HD 195 S6-89. Safety Requirements for Mains Operated Electronic and Related Apparatus for Household and Similar General Use. 19 pp.
CSA C22.2 NO 0.8-M1986. Safety Functions Incorporating Electronic Technology (R 1992); (Gen Instr 1). 37 pp.
SAA AS 4048.1 (INT)-92. Functional Safety of Electrical/Electronic/Programmable Electronic Systems—Generic Aspects—Part 1(Int): General Requirements (IEC 65A(Secretariat)123). 140 pp.

—**Sampling**
BSI BS E9007-75. 1975 Basic Specification: Sampling Plans and Procedures for Inspection by Attributes. 6 pp.

—**Screening Testing**
BSI BS 4200: Part 8-76. (WITHDRAWN) 1976 Guide on the Reliability of Electronic Equipment and Parts Used Therein Part 8: The Screening (Sorting) of Electronic Equipment and Parts. 8 pp.

—**Secure Voice Communications—Radio Telephones**
CCITT RECMN G.464-89. Principles of the Devices Used to Achieve Privacy in Radiotelephone Conversations—International Analogue Carrier Systems (Study Group XV) 2 pp. 2 pp.

—**Sensitive Switches**
BSI BS QC 960300-92. 1992 Electromechanical Switches for Use in Electronic Equipment Sectional Specification for Sensitive Switches (IEC 1020-6: 1991). 16 pp.
BSI BS QC 960301-92. 1992 Blank Detail Specification Electromechanical Switches for Use in Electronic Equipment Blank Detail Specification for Sensitive Switches (IEC 1020-6-1: 1991). 10 pp.
IEC 1020 Pt 6-91. Electromechanical Switches for Use in Electronic Equipment Part 6: Sectional Specification for Sensitive Switches First Edition; (IECQ QC 960300). 34 pp.

Electronic Equipment (Cont.)

—**Sensitive Switches** (Cont.)
IEC 1020 Pt 6-1-91. Electromechanical Switches for Use in Electronic Equipment Part 6: Sectional Specification for Sensitive Switches Section 1—Blank Detail Specification; First Edition; (IECQ QC 960301). 26 pp.

—**Ships—Electromagnetic Compatibility—Test Methods**
MOD UK DSTAN 59-41: Pt 3: Supp L-93. Electromagnetic Compatibility Part 3: Technical Requirements Test Methods and Limits Supplement L: Test Method DCS05 Externally Generated Transients (Ships) Issue 1. 11 pp.

—**Ships—Electromagnetic Interference**
NATO STANAG 4435 ED 1 AMD 0-00. Electro-Magnetic Compatibility Testing Procedure and Requirements for Naval Electrical and Electronic Equipment (Surface Ship, Metallic Hull). 58 pp.
NATO STANAG 4435 ED 1 AMD 0-93. Electromagnetic Compatibility Testing Procedure and Requirements for Naval Electrical and Electronic Equipment (Surface Ship, Metallic Hull). 58 pp.
NATO STANAG 4436 ED 1 AMD 0-00. Electro-Magnetic Compatibility Testing Procedure and Requirements for Naval Electrical and Electronic Equipment (Surface Ship, Non-Metallic Hull). 56 pp.
NATO STANAG 4436 ED 1 AMD 0-93. Electro-Magnetic Compatibility Testing Procedure and Requirements for Naval Electrical and Electronic Equipment (Surface Ship, Non-Metallic Hull). 58 pp.

—**Shock Protection**
BSI BS 2754-76. 1976 Amd 1 Memorandum Construction of Electrical Equipment for Protection Against Electric Shock. 26 pp.
BSI PD 2754: Part 2-93. 1993 Classification of Electrical and Electronic Equipment with Regard to Protection Against Electric Shock Part 2: Guide to Requirements for Protection Against Electric Shock (IEC 536-2: 1992) (E). 22 pp.
BSI PD 6535-93. 1993 Common Aspects for Installation and Equipment for Protection Against Electric Shock (IEC 1440: 1992) (E). 15 pp.
CENELEC HD 366-77. Classification of Electrical and Electronic Equipment with Regard to Protection Against Electric Shock. 2 pp.
DIN VDE 0106 Pt 1 A3 (D)-92. Protection Against Shock Current; Classification of Electrical & Electronic Equipment with Regard to Protection Against Shock: Amendment 3 (Jan). 4 pp.
IEC 536-76. Classification of Electrical and Electronic Equipment with Regard to Protection Against Electric Shock First Edition. 13 pp.

—**Spacing**
CSA 1388 Bull. Electrical Bulletin 1388 December 22, 1982 to C22.2 NO 1. 3 pp.

—**Spindles**
BSI BS 7039-89. 1989 Recommendations for Dimensions of Spindle Ends and Bushes, and for Dimensions for the Mounting of Single-Hole, Bush-Mounted Spindle-Operated Variable Capacitors, Variable Resistors and Potentio-meters for Electronic Equipment. 16 pp.
IEC 915-87. Capacitors and Resistors for Electronic Equipment Preferred Dimensions of Spindle Ends, Bushes and for the Mounting of Single-Hole, Bush-Mounted, Spindle-Operated Electronic Components First Edition. 22 pp.

—**Standard Voltages—Identification Systems**
CSA CAN/CSA-C22. 2 NO 0-M91. General Requirements—Canadian Electrical Code, Part II; (Gen Instr 1). 48 pp.
CSA CAN3-C235-83. Preferred Voltage Levels for AC Systems, 0 to 50,000 V; (Gen Instr 1). 12 pp.

—**Submarines—Electromagnetic Compatibility**
NATO STANAG 4437 ED 1 AMD 0-00. (Draft) Electro-Magnetic Compatibility Testing Procedure and Requirements for Naval Electrical and Electronic Equipment (Submarines). 61 pp.

—**Symbols**
CENELEC HD 245.2-85. Letter Symbols to be Used in Electrical Technology Part 2: Telecommunications and Electronics. 2 pp.
CNS C1143-89. Graphical Symbols for Use on Electrical and Electronic Equipment (General Rules) (Feb)(12491).
CNS C1143-1-89. Graphical Symbols for Use on Electrical and Electronic Equipment (Source, Voltage) (Feb)(12491-1).
CNS C1143-2-89. Graphical Symbols for Use on Electrical and Electronic Equipment (General) (Feb)(12491-2).
CNS C1143-3-89. Graphical Symbols for Use on Electrical and Electronic Equipment (Movement, Action, Run, Adjustment) (Feb)(12491-3).

Electronic Equipment (Cont.)

—**Symbols** (Cont.)
CNS C1143-4-89. Graphical Symbols for Use on Electrical and Electronic Equipment (Alarm, Signal, Filter) (Feb)(12491-4).
CNS C1143-5-89. Graphical Symbols for Use on Electrical and Electronic Equipment (Digital Transmission) (Feb)(12491-5).
CNS C1143-6-89. Graphical Symbols for Use on Electrical and Electronic Equipment (Telecommunication) (Feb)(12491-6).
CNS C5112-81. Fundamental Symbols for Electrical and Electronics (May)(7361).
IEC 27 Pt 2-72. Letter Symbols to Be Used in Electrical Technology Part 2: Telecommunications and Electronics First Edition; (Supplement A-1975) (Supplement B-1980). 83 pp.
SAA AS 1046.2-78. Letter Symbols for Use in Electrotechnology—Part 2: Telecommunications and Electronics. 32 pp.
SAA AS 1104-78. Informative Symbols for Use on Electrical and Electronic Equipment. 24 pp.
SAA AS 1104S. Informative Symbols for Use on Electrical and Electronic Equipment—Individual Symbol Sheets (Complete Set in Binder).

—**Symbols—Electrostatic Sensitivity**
CNS Z5133-84. Symbol and Label for Electrostatic Sensitive Devices (Nov)(11144).

—**Testability**
MOD UK DSTAN 00-13-01. Guide to the Achievement of Testability in Electronic and Allied Equipment Issue 2; Amendment 1. 45 pp.

—**Testing Equipment**
MOD UK DSTAN 66-31-81. Basic Requirements and Tests for Proprietary Electronic and Electrical Test Equipment Issue 1. 20 pp.

—**Toggle Switches**
BSI BS QC 960200-92. 1992 Electromechanical Switches for Use in Electronic Equipment Sectional Specification for Lever (Toggle) Switches (IEC 1020-4: 1991). 15 pp.
BSI BS QC 960201-92. 1992 Blank Detail Specification Electromechanical Switches for Use in Electronic Equipment Blank Detail Specifcation for Lever (Toggle) Switches (IEC 1020-4-1: 1991). 10 pp.
IEC 1020 Pt 4-91. Electromechanical Switches for Use in Electronic Equipment Part 4: Sectional Specification for Lever (Toggle) Switches; First Edition; (IECQ QC 960200). 30 pp.
IEC 1020 Pt 4-1-91. Electromechanical Switches for Use in Electronic Equipment Part 4: Sectional Specification for Lever (Toggle) Switches Section 1—Blank Detail Specification; First Edition; (IECQ QC 960201). 26 pp.

—**Transformers**
IEC 1007-90. Transformers and Inductors for Use in Electronic and Telecommunication Equipment—Measuring Methods and Test Procedures First Edition; (Amendment 1-1993). 113 pp.

—**Tungsten Materials**
CNS H2051-85. General Rules for Test of Tungsten and Molybdenum Materials for Lighting and Electronic Equipments (Nov)(6337).
CNS H3085-85. Tungsten Wires for Lighting and Electronic Equipments (Nov)(6338).
CNS H3086-85. Tungsten Rods for Lighting and Electronic Equipments (Nov)(6339).
CNS H3088-85. Thoriated Tungsten Wires and Rods for Lighting and Electronic Equipments (Nov)(6341).
JIS H 4460-84. General Rules for Test of Tungsten and Molybdenum Materials for Lighting and Electronic Equipments. 9 pp.
JIS H 4461-89. Tungsten Wires for Lighting and Electronic Equipments. 11 pp.
JIS H 4462-89. Tungsten Rods for Lighting and Electronic Equipments. 8 pp.
JIS H 4463-84. Thoriated Tungsten Wires and Rods for Lighting and Electronic Equipments. 8 pp.

—**Tungsten—Molybdenum Alloy Materials**
CNS H3089-85. Tungsten-Molybdenum Alloy Wires for Lighting and Electronic Equipments (Nov)(6342).
JIS H 4471-89. Tungsten-Molybdenum Alloy Wires for Lighting and Electronic Equipments. 8 pp.

—**Varistors—Surge Suppression**
IEC 1051 Pt 2-91. Varistors for Use in Electronic Equipment Part 2: Sectional Specification for Surge Suppression Varistors First Edition (IECQ QC 420100). 39 pp.
IEC 1051 Pt 2-1-91. Varistors for Use in Electronic Equipment Part 2: Blank Detail Specification for Silicon Carbide Surge Suppression Varistors—Assessment Level E First Edition (IECQ QC 420101). 31 pp.

INTERNATIONAL AND NON-U.S. NATIONAL STANDARDS
SUBJECT INDEX

Electronic Equipment *(Cont.)*
—**Vibration**
ISO 8569-89. Mechanical Vibration—Shock-and-Vibration-Sensitive Electronic Equipment—Methods of Measurement and Reporting Data of Shock and Vibration Effects in Buildings First Edition. 13 pp.

—**Voltage Endurance Testing**
CENELEC PREN 50093-91. Basic Immunity Standard for Voltage Dips Short Interruptions and Voltage Variations. 16 pp.
CNS C6013-85. Voltage Endurance Testing Method for Electronic Components (Apr)(3626). 3 pp.

—**Wire Nippers**
CNS B3148-84. Cutting Nippers (Oct)(1133). 5 pp.

Electronic Explosive Devices
Use: Initiators (Explosives)

Electronic Flash Equipment
Use: Photographic Flash Equipment

Electronic Funds Transfer
See Also: Banking

—**Algorithms**
SAA AS 2805.5.1-92. Electronic Funds Transfer—Requirements for Interfaces—Part 5: Ciphers—Part 5.1: Data Encipherment Algorithm 1 (DEA 1) (Supersedes AS 2805.5—1985 (in Part)). 10 pp.
SAA AS 2805.5.2-92. Electronic Funds Transfer—Requirements for Interfaces—Part 5: Ciphers—Part 5.2: Modes of Operation for an n-Bit Block Cipher Algorithm (Supersedes AS 2805.5—1985 (in Part)). 12 pp.
SAA AS 2805.5.3-92. Electronic Funds Transfer—Requirements for Interfaces—Part 5: Ciphers—Part 5.3: Data Encipherment Algorithm 2 (DEA 2). 5 pp.

—**Key Management**
SAA AS 2805.6.5. 1-92. Electronic Funds Transfer—Requirements for Interfaces—Part 6: Key Management—Part 6.5.1: TCU Initialization—Principles. 4 pp.
SAA AS 2805.6.5. 3-92. Electronic Funds Transfer—Requirements for Interfaces—Part 6: Key Management—Part 6.5.3: TCU Initialization—Asymmetric. 6 pp.

—**Privacy**
SAA AS 2805.9-91. Electrical Funds Transfer—Requirements for Interfaces—Part 9: Privacy of Communications. 7 pp.

Electronic Mail Systems
Use For: E-Mail; Mailboxes, Electronic
See Also: EDI Messaging Services; FAX Communications; FAX Machines; Telecommunication Equipment

—**Glossaries**
IEC DIS 2382 Pt 27-92. Information Technology—Vocabulary—Part 27: Office Automation ***CD-ROM ONLY***. 22 pp.
ISO DIS 2382 Pt 27-92. Information Technology—Vocabulary—Part 27: Office Automation ***CD-ROM ONLY***. 22 pp.
JTC1 DIS2382 Pt 27-92. Information Technology—Vocabulary—Part 27: Office Automation ***CD-ROM ONLY***. 22 pp.

—**Open Systems Interconnection**
BSI BS 7521-92. 1992 Electronic Messaging Enabling the Application of BS ISO/IEC 10021 in the United Kingdom. 29 pp.
BSI BS ISO/IEC 10021-1-90. 1990 Amd 1 Information Technology—Text Communication—Message-Oriented Text Interchange Systems (MOTIS)—Part 1: System and Service Overview (AMD 6846) September 30, 1991. 77 pp.
BSI BS ISO/IEC 10021-1-02. 1990 Amd 2 Information Technology—Text Communication—Message-Oriented Text Interchange Systems (MOTIS)—Part 1: System and Service Overview (AMD 7139) July 15, 1992 (S). 78 pp.
BSI BS ISO/IEC 10021-1-03. 1990 Amd 3 Information Technology—Text Communication—Message-Oriented Text Interchange Systems (MOTIS)—Part 1: System and Service Overview (AMD 7271) August 15, 1992 (Technical Corr 3) (S). 80 pp.
BSI BS ISO/IEC 10021-1-04. 1990 Amd 4 Information Technology—Text Communication—Message-Oriented Text Interchange Systems (MOTIS)—Part 1: System and Service Overview (AMD 7540) December 15, 1992 (Technical Corr 4) (S). 82 pp.

Electronic Mail Systems *(Cont.)*
—**Open Systems Interconnection** *(Cont.)*
BSI BS ISO/IEC 10021-1-05. 1990 Amd 5 Information Technology—Text Communication—Message-Oriented Text Interchange Systems (MOTIS)—Part 1: System and Service Overview (AMD 7682) March 15, 1993 (Technical Corr 5) (S). 85 pp.
BSI BS ISO/IEC 10021-2-90. 1990 Amd 1 Information Technology—Text Communication—Message-Oriented Text Interchange Systems (MOTIS)—Part 2: Overall Architecture (AMD 6847) September 30, 1991. 101 pp.
BSI BS ISO/IEC 10021-2-02. 1990 Amd 2 Information Technology—Text Communication—Message-Oriented Text Interchange Systems (MOTIS)—Part 2: Overall Architecture (AMD 7140) July 15, 1992 (Technical Corr 2) (S). 103 pp.
BSI BS ISO/IEC 10021-2-03. 1990 Amd 3 Information Technology—Text Communication—Message-Oriented Text Interchange Systems (MOTIS)—Part 2: Overall Architecture (AMD 7272) August 15, 1992 (Technical Corr 3) (S). 106 pp.
BSI BS ISO/IEC 10021-2-04. 1990 Amd 4 Information Technology—Text Communication—Message-Oriented Text Interchange Systems (MOTIS)—Part 2: Overall Architecture (AMD 7541) December 15, 1992 (Technical Corr 4). 113 pp.
BSI BS ISO/IEC 10021-3-90. 1990 Information Technology—Text Communication—Message-Oriented Text Interchange Systems (MOTIS)—Part 3: Abstract Service Definition Conventions (S). 41 pp.
BSI BS ISO/IEC 10021-3-01. 1990 Amd 1 Information Technology—Text Communication—Message-Oriented Text Interchange Systems (MOTIS)—Part 3: Abstract Service Definition Conventions (AMD 7273) August 15, 1992 (Technical Corr 1). 43 pp.
BSI BS ISO/IEC 10021-4-90. 1990 Amd 1 Information Technology—Text Communication—Message-Oriented Text Interchange Systems (MOTIS)—Part 4: Message Transfer System: Abstract Service Definition and Procedures (AMD 6848) September 30, 1991. 204 pp.
BSI BS ISO/IEC 10021-4-02. 1990 Amd 2 Information Technology—Text Communication—Message-Oriented Text Interchange Systems (MOTIS)—Part 4: Message Transfer System: Abstract Service Definition and Procedures (AMD 7141) July 15, 1992 (S). 212 pp.
BSI BS ISO/IEC 10021-4-03. 1990 Amd 3 Information Technology—Text Communication—Message-Oriented Text Interchange Systems (MOTIS)-Part 4: Message Transfer Sys: Abstract Service Definition and Procedures (AMD 7274) August 15, 1992 (Technical Corr 3) (S). 219 pp.
BSI BS ISO/IEC 10021-4-04. 1990 Amd 4 Information Technology—Text Communication—Message-Oriented Text Interchange Systems (MOTIS)—Part 4: Message Transfer Sys: Abstract Service Definition and Procedures (AMD 7542) December 15, 1992 (Technical Corr 4) (S). 235 pp.
BSI BS ISO/IEC 10021-4-05. 1990 Amd 5 Information Technology—Text Communication—Message-Oriented Text Interchange Systems (MOTIS)-Part 4: Message Transfer Sys: Abstract Service Definition and Procedures (AMD 7683) March 15, 1993 (Technical Corr 5) (S). 242 pp.
BSI BS ISO/IEC 10021-5-90. 1990 Amd 1 Information Technology—Text Communication—Message-Oriented Text Interchange Systems (MOTIS)—Part 5: Message Store: Abstract Service Definition (AMD 6849) September 30, 1991. 117 pp.
BSI BS ISO/IEC 10021-5-02. 1990 Amd 2 Information Technology—Text Communication—Message-Oriented Text Interchange Systems (MOTIS)—Part 5: Message Store: Abstract Service Definition (AMD 7142) July 15, 1992 (Technical Corr 2) (S). 124 pp.
BSI BS ISO/IEC 10021-5-03. 1990 Amd 3 Information Technology—Text Communication—Message-Oriented Text Interchange Systems (MOTIS)—Part 5: Message Store: Abstract Service Definition (AMD 7275) August 15, 1992 (Technical Corr 3) (S). 138 pp.
BSI BS ISO/IEC 10021-5-04. 1990 Amd 4 Information Technology—Text Communication—Message-Oriented Text Interchange Systems (MOTIS)—Part 5: Message Store: Abstract Service Definition (AMD 7543) December 15, 1992 (Technical Corr 4) (S). 146 pp.
BSI BS ISO/IEC 10021-5-05. 1990 Amd 5 Information Technology—Text Communication—Message-Oriented Text Interchange Systems (MOTIS)—Part 5: Message Store: Abstract Service Definition (AMD 7684) March 15, 1993 (Technical Corr 5) (S). 151 pp.
BSI BS ISO/IEC 10021-6-90. 1990 Information Technology—Text Communication—Message-Oriented Text Interchange Systems (MOTIS)—Part 6: Protocol Specifications (S). 61 pp.

Electronic Mail Systems *(Cont.)*
—**Open Systems Interconnection** *(Cont.)*
BSI BS ISO/IEC 10021-6-01. 1990 Amd 1 Information Technology—Text Communication—Message-Oriented Text Interchange Systems (MOTIS)—Part 6: Protocol Specifications (AMD 6850) September 30, 1991. 65 pp.
BSI BS ISO/IEC 10021-6-02. 1990 Amd 2 Information Technology—Text Communication—Message-Oriented Text Interchange Systems (MOTIS)—Part 6: Protocol Specifications (AMD 7143) July 15, 1992 (Technical Corr 2) (S). 68 pp.
BSI BS ISO/IEC 10021-6-03. 1990 Amd 3 Information Technology—Text Communication—Message-Oriented Text Interchange Systems (MOTIS)—Part 6: Protocol Specifications (AMD 7276) August 15, 1992 (Technical Corr 3) (S). 71 pp.
BSI BS ISO/IEC 10021-6-04. 1990 Amd 4 Information Technology—Text Communication—Message-Oriented Text Interchange Systems (MOTIS)—Part 6: Protocol Specifications (AMD 7544) December 15, 1992 (Technical Corr 4) (S). 74 pp.
BSI BS ISO/IEC 10021-6-05. 1990 Amd 5 Information Technology—Text Communication—Message-Oriented Text Interchange Systems (MOTIS)—Part 6: Protocol Specifications (AMD 7685) March 15 1993 (Technical Corr 5) (S). 77 pp.
BSI BS ISO/IEC 10021-7-90. 1990 Amd 1 Information Technology—Text Communication—Message-Oriented Text Interchange Systems (MOTIS)—Part 7: Interpersonal Messaging System (AMD 6851) September 30, 1991. 122 pp.
BSI BS ISO/IEC 10021-7-02. 1990 Amd 2 Information Technology—Text Communication—Message-Oriented Text Interchange Systems (MOTIS)—Part 7: Interpersonal Messaging System (AMD 7144) July 15, 1992 (Technical Corr 2) (S). 130 pp.
BSI BS ISO/IEC 10021-7-03. 1990 Amd 3 Information Technology—Text Communication—Message-Oriented Text Interchange Systems (MOTIS)—Part 7: Interpersonal Messaging System (AMD 7277) August 15, 1992 (Technical Corr 3) (S). 142 pp.
BSI BS ISO/IEC 10021-7-04. 1990 Amd 4 Information Technology—Text Communication—Message-Oriented Text Interchange Systems (MOTIS)—Part 7: Interpersonal Messaging System (AMD 7545) December 15, 1992 (Technical Corr 4) (S). 148 pp.
BSI BS ISO/IEC 10021-7-05. 1990 Amd 5 Information Technology—Text Communication—Message-Oriented Text Interchange Systems (MOTIS)—Part 7: Interpersonal Messaging System (AMD 7686) March 15, 1993 (Technical Corr 5) (S). 152 pp.
IEC 10021 Pt 1-90. Information Technology—Text Communication—Message-Oriented Text Interchange Systems (MOTIS)—Part 1: System and Service Overview First Edition; (Corrigendum 1-2:1991) (Corrigendum 3-5:1992) (Amendment 1-1993). 91 pp.
IEC 10021 Pt 2-90. Information Technology—Text Communication—Message-Oriented Text Interchange Systems (MOTIS)—Part 2: Overall Architecture First Edition; (Corrigendum 1-2:1991) (Corrigendum 3-4:1992). 106 pp.
IEC 10021 Pt 2 Draft AMD 1. Information Technology—Text Communication—Message-Oriented Text Interchange Systems (MOTIS)—Part 2: Overall Architecture Amendment 1: Representation of O/R Addresses for Human Exchange; (1992) ***CD-ROM ONLY***. 8 pp.
IEC 10021 Pt 2 Draft AMD 2. Information Technology—Text Communication—Message-Oriented Text Interchange Systems (MOTIS)—Part 2: Overall Architecture Amendment 2: Minor Enhancements; (1992) ***CD-ROM ONLY***. 7 pp.
IEC 10021 Pt 3-90. Information Technology—Text Communication—Message-Oriented Text Interchange Systems (MOTIS)—Part 3: Abstract Service Definition Conventions First Edition; (Corrigendum 1-1992). 40 pp.
IEC 10021 Pt 4-90. Information Technology—Text Communication—Message-Oriented Text Interchange Systems (MOTIS)—Part 4: Message Transfer System: Abstract Service Definition and Procedures First Edition; (Corrigendum 1-2:1991) (Corrigendum 3-5:1992). 232 pp.
IEC 10021 Pt 4 Draft AMD 1. Information Technology—Text Communication—Message-Oriented Text Interchange Systems (MOTIS)—Part 4: Message Transfer System: Abstract Service Definition and Procedures Amendment 1: Minor Enhancements; (1992) ***CD-ROM ONLY***. 5 pp.
IEC 10021 Pt 7 Draft AMD 1. Information Technology—Text Communication—Message-Oriented Text Interchange Systems (MOTIS)—Part 7: Interpersonal Messaging System Amendment 1: Minor Enhancements; (1992) ***CD-ROM ONLY***. 11 pp.

INDUSTRY STANDARDS

Electronic
INTERNATIONAL AND NON-U.S. NATIONAL STANDARDS
SUBJECT INDEX

Electronic Mail Systems *(Cont.)*

—Open Systems Interconnection *(Cont.)*

IEC 10021 Pt 7 Draft AMD1.2. Information Technology—Text Communication—Message-Oriented Text Interchange Systems (MOTIS)—Part 7: Interpersonal Messaging System Amendment 1: Minor Enhancements; (1993) ***CD-ROM ONLY***. 16 pp.

ISO 10021 Pt 1-90. Information Technology—Text Communication—Message-Oriented Text Interchange Systems (MOTIS)—Part 1: System and Service Overview First Edition; (Corrigendum 1-2:1991) (Corrigendum 3-5:1992) (Amendment 1-1993). 13 pp.

ISO 10021 Pt 2-90. Information Technology—Text Communication—Message-Oriented Text Interchange Systems (MOTIS)—Part 2: Overall Architecture First Edition; (Corrigendum 1-2:1991) (Corrigendum 3-4:1992). 106 pp.

ISO 10021 Pt 2 Draft AMD 1. Information Technology—Text Communication—Message-Oriented Text Interchange Systems (MOTIS)—Part 2: Overall Architecture Amendment 1: Representation of O/R Addresses for Human Exchange; (1992) ***CD-ROM ONLY***. 8 pp.

ISO 10021 Pt 2 Draft AMD 2. Information Technology—Text Communication—Message-Oriented Text Interchange Systems (MOTIS)—Part 2: Overall Architecture Amendment 2: Minor Enhancements; (1992) ***CD-ROM ONLY***. 7 pp.

ISO 10021 Pt 3-90. Information Technology—Text Communication—Message-Oriented Text Interchange Systems (MOTIS)—Part 3: Abstract Service Definition Conventions First Edition; (Corrigendum 1-1992). 40 pp.

ISO 10021 Pt 4-90. Information Technology—Text Communication—Message-Oriented Text Interchange Systems (MOTIS)—Part 4: Message Transfer System: Abstract Service Definition and Procedures First Edition; (Corrigendum 1-2:1991) (Corrigendum 3-5:1992). 232 pp.

ISO 10021 Pt 4 Draft AMD 1. Information Technology—Text Communication—Message-Oriented Text Interchange Systems (MOTIS)—Part 4: Message Transfer System: Abstract Service Definition and Procedures Amendment 1: Minor Enhancements; (1992) ***CD-ROM ONLY***. 5 pp.

ISO 10021 Pt 5-90. Information Technology—Text Communication—Message-Oriented Text Interchange Systems (MOTIS)—Part 5: Message Store: Abstract Service Definition First Edition; (Corrigendum 1-2:1991) (Corrigendum 3-5:1992) (Amendment 1-1992). 191 pp.

ISO 10021 Pt 6-90. Information Technology—Text Communication—Message-Oriented Text Interchange Systems (MOTIS)—Part 6: Protocol Specifications First Edition; (Corrigendum 1-2:1991) (Corrigendum 3-5:1992) (Amendment 1-1993). 12 pp.

ISO 10021 Pt 7-90. Information Technology—Text Communication—Message-Oriented Text Interchange Systems (MOTIS)—Part 7: Interpersonal Messaging System First Edition; (Corrigendum 1-2:1991) (Corrigendum 3-5:1992) (Amendment 3-1993). 79 pp.

ISO 10021 Pt 7 Draft AMD 1. Information Technology—Text Communication—Message-Oriented Text Interchange Systems (MOTIS)—Part 7: Interpersonal Messaging System Amendment 1: Minor Enhancements; (1992) ***CD-ROM ONLY***. 11 pp.

ISO 10021 Pt 7 Draft AMD1.2. Information Technology—Text Communication—Message-Oriented Text Interchange Systems (MOTIS)—Part 7: Interpersonal Messaging System Amendment 1: Minor Enhancements; (1993) ***CD-ROM ONLY***. 16 pp.

JTC1 10021 Pt 2-90. Information Technology—Text Communication—Message-Oriented Text Interchange Systems (MOTIS)—Part 2: Overall Architecture First Edition; (Corrigendum 1-2:1991) (Corrigendum 3-4:1992). 106 pp.

JTC1 10021 Pt2 Draft AMD 1. Information Technology—Text Communication—Message-Oriented Text Interchange Systems (MOTIS)—Part 2: Overall Architecture Amendment 1: Representation of O/R Addresses for Human Exchange; (1992) ***CD-ROM ONLY***. 8 pp.

JTC1 10021 Pt2 Draft AMD 2. Information Technology—Text Communication—Message-Oriented Text Interchange Systems (MOTIS)—Part 2: Overall Architecture Amendment 2: Minor Enhancements; (1992) ***CD-ROM ONLY***. 7 pp.

JTC1 10021 Pt 3-90. Information Technology—Text Communication—Message-Oriented Text Interchange Systems (MOTIS)—Part 3: Abstract Service Definition Conventions First Edition; (Corrigendum 1-1992). 40 pp.

JTC1 10021 Pt 4-90. Information Technology—Text Communication—Message-Oriented Text Interchange Systems (MOTIS)—Part 4: Message Transfer System: Abstract Service Definition and Procedures First Edition; (Corrigendum 1-2:1991) (Corrigendum 3-5:1992). 226 pp.

Electronic Mail Systems *(Cont.)*

—Open Systems Interconnection *(Cont.)*

JTC1 10021 Pt4 Draft AMD 1. Information Technology—Text Communication—Message-Oriented Text Interchange Systems (MOTIS)—Part 4: Message Transfer System: Abstract Service Definition and Procedures Amendment 1: Minor Enhancements; (1992) ***CD-ROM ONLY***. 5 pp.

JTC1 10021 Pt7 Draft AMD 1. Information Technology—Text Communication—Message-Oriented Text Interchange Systems (MOTIS)—Part 7: Interpersonal Messaging System Amendment 1: Minor Enhancements; (1992) ***CD-ROM ONLY***. 11 pp.

JTC1 10021 Pt7 Draft AMD1.2. Information Technology—Text Communication—Message-Oriented Text Interchange Systems (MOTIS)—Part 7: Interpersonal Messaging System Amendment 1: Minor Enhancements; (1993) ***CD-ROM ONLY***. 16 pp.

OSI ISO DIS 8505-86. Information Processing—Text Communication—Functional Description and Service Specification for Message Oriented Text Interchange Systems. 56 pp.

OSI ISO DIS 8883-86. Information Processing—Text Communication—Message Oriented Text Interchange System, Message Transfer Sublayer, Message Interchange Service and Message Transfer Protocol. 75 pp.

OSI ISO DIS 8883-86. Information Processing-Text Communication-Message Oriented Text Interchange System, Message Transfer Sublayer, Message Interchange Service and Message Transfer Protocol. 75 pp.

OSI ISO DIS 9065-87. Information Processing—Text Communication—Message Oriented Text Interchange System User Agent Sublayer—Interpersonal Messaging User Agent—Message Interchange Formats and Protocols. 53 pp.

OSI ISO DIS 9065-86. Information Processing—Text Communication—Message Oriented Text Interchange System User Agent Sublayer—Inter-Personal Messaging User Agent—Message Interchange Formats and Protocols. 52 pp.

OSI ISO/IEC 10021-1-90. Information Technology—Text Communication—Message-Oriented Text Interchange Systems (MOTIS) Part 1: System and Service Overview. 71 pp.

OSI ISO/IEC 10021-2-90. Information Technology—Text Communication—Message-Oriented Text Interchange Systems (MOTIS)—Part 2: Overall Architecture. 95 pp.

OSI ISO/IEC 10021-3-90. Information Technology—Text Communication—Message-Oriented Text Interchange Systems (MOTIS) Part 3: Abstract Service Definition Conventions. 39 pp.

OSI ISO/IEC 10021-4-90. Information Technology—Text Communication—Message-Oriented Text Interchange Systems (MOTIS) Part 4: Message Transfer System: Abstract Service Definition and Procedures. 192 pp.

OSI ISO/IEC 10021-5-90. Information Technology—Text Communication—Message-Oriented Text Interchange Systems (MOTIS) Part 5: Message Store: Abstract Service Definition. 109 pp.

OSI ISO/IEC 10021-6-90. Information Technology—Text Communication—Message-Oriented Text Interchange Systems (MOTIS)—Part 6: Protocol Specifications. 59 pp.

OSI ISO/IEC 10021-7-90. Information Technology—Text Communication—Message-Oriented Text Interchange Systems (MOTIS)—Part 7: Interpersonal Messaging System. 117 pp.

SAA AS 4033.1-92. Information Technology—Text Communication—Message-Oriented Text Interchange Systems—Part 1: System and Service Overview (ISO/IEC 10021-1:1990) (in Professional Package 26A). 68 pp.

SAA AS 4033.2-92. Information Technology—Text Communication—Message-Oriented Text Interchange Systems—Part 2: Overall Architecture (ISO/IEC 10021-2:1990) (in Professional Package 26A). 84 pp.

SAA AS 4033.3-92. Information Technology—Text Communication—Message-Oriented Text Interchange Systems—Part 3: Abstract Service Definition Conventions (ISO/IEC 10021-3:1990) (in Professional Package 26A). 32 pp.

SAA AS 4033.4-92. Information Technology—Text Communication—Message-Oriented Text Interchange Systems—Part 4: Message Transfer System—Abstract Service Definition and Procedures (ISO/IEC 10021-4:1990) (in Professional Package 26A). 191 pp.

SAA AS 4033.5-92. Information Technology—Text Communication—Message-Oriented Text Interchange Systems—Part 5: Message Store—Abstract Service Definition (ISO/IEC 10021-5:1990) (in Professional Package 26A). 119 pp.

SAA AS 4033.6-92. Information Technology—Text Communication—Message-Oriented Text Interchange Systems—Part 6: Protocol Specifications (ISO/IEC 10021-6:1990) (in Professional Package 26A). 52 pp.

Electronic Mail Systems *(Cont.)*

—Open Systems Interconnection *(Cont.)*

SAA AS 4033.7-92. Information Technology—Text Communication—Message-Oriented Text Interchange Systems—Part 7: Interpersonal Messaging System (ISO/IEC 10021-7:1990) (in Professional Package 26A). 121 pp.

—Telex Communications

CCITT RECMN F.74-89. Operational Provisions Relating to Mailbox Devices Connected to the Telex Network—Telegraph and Mobile Services Operations and Quality of Service (Study Group I) 3 pp. 3 pp.

—Telex Communications—Answerback

CCITT RECMN F.74-92. Intermediate Storage Devices Accessed from the International Telex Service Using Single Stage Selection—Answerback Format (Study Group I) 5 pp. 5 pp.

CCITT RECMN F.74-89. Operational Provisions Relating to Mailbox Devices Connected to the Telex Network—Telegraph and Mobile Services Operations and Quality of Service (Study Group I) 3 pp. 3 pp.

—Telex Communications—Answerback—Overflow Traffic

CCITT RECMN F.74-89. Operational Provisions Relating to Mailbox Devices Connected to the Telex Network—Telegraph and Mobile Services Operations and Quality of Service (Study Group I) 3 pp. 3 pp.

—Telex Communications—Answering Time

CCITT RECMN F.74-89. Operational Provisions Relating to Mailbox Devices Connected to the Telex Network—Telegraph and Mobile Services Operations and Quality of Service (Study Group I) 3 pp. 3 pp.

Electronic Measuring Instruments

See Also: Electronic Equipment

BSI BS 4308-79. (OBSOLESCENT) 1979 Amd 1 Documentation to Be Supplied with Electronic Measuring Apparatus (AMD 6202) December 21, 1990. 20 pp.

BSI BS 4889-90. 1990 Method for Specifying the Performance of Electrical and Electronic Measuring Equipment. 27 pp.

BSI BS 4889-73. (WITHDRAWN) 1973 Method for Specifying the Performance of Electronic Measuring Equipment. 16 pp.

CENELEC HD 312-77. Documentation to Be Supplied with Electronic Measuring Apparatus. 2 pp.

IEC 278-68. Documentation to be Supplied with Electronic Measuring Apparatus First Edition; (Supplement A-1974). 39 pp.

IEC 359-87. Expression of the Performance of Electrical and Electronic Measuring Equipment Second Edition; (Amendment 1-1991). 53 pp.

—Digital Display

BSI BS 7194-90. 1990 Direct Current and Low Frequency Electronic Measuring Instruments with a Digital Display. 25 pp.

BSI BS 7194-01. 1990 Amd 1 Direct-Current and Low-Frequency Electronic Measuring Instruments with a Digital Display (AMD 6892) May 1, 1992. 25 pp.

—Glossaries

CNS C5217-87. Glossary of Terms in Electronic Measuring Apparatus (Sep)(12080).

JIS C 1002-75. Glossary of Terms Used in Electronic Measuring Apparatus.

SAA AS 1852.303-88. International Electrotechnical Vocabulary—Part 303: Electronic Measuring Instruments.

—Interfaces

BSI BS 6146: Part 1-81. 1981 Interface System for Programmable Measuring Instruments (Byte Serial, Bit Parallel) Part 1: Specification for Functional, Elect-rical and Mechanical Requirements, System Applications and Require -ments for Designer and User. 112 pp.

BSI BS 6146: Part 2-81. (WITHDRAWN) 1981 Interface System for Programmable Measuring Instruments (Byte Serial, Bit Parallel) Part 2: Guide for Code and Format Conventions (Superseded by Revision of IEC 625-2). 27 pp.

CENELEC HD 414.1-80. An Interface System for Programmable Measuring Instruments (Byte Serial, Bit Parallel) Part 1: Functional Specifications, Electrical Specifications System Application and Requirements for the Designer and User. 2 pp.

CENELEC HD 414.2-83. An Interface System for Programmable Measuring Instruments (Byte Serial, Bit Parrallel) Part 2: Code and Format Conventions. 2 pp.

IEC 625 Pt 1-79. Interface System for Programmable Measuring Instruments (Byte, Serial, Bit Parallel) Part 1: Functional Specifications, Electri-cal, Specifications, Mechanical Specifica-tions, System Applica-tions and Requirements for the Designer and User First Edition. 226 pp.

INTERNATIONAL AND NON-U.S. NATIONAL STANDARDS
SUBJECT INDEX

Electronic Measuring Instruments (Cont.)
—**Interfaces** (Cont.)
- IEC 625 Pt 2-80. An Interface System for Programmable Measuring Instruments (Byte Serial, Bit Parallel) Part 2: Code and Format Conventions First Edition. 54 pp.
- JIS C 1901-87. Interface System for Programmable Measuring Instruments. 146 pp.

—**Levels**
- DIN ENGL 2276 Pt 2-86. Inclination Measuring Systems; Electronic Inclination Measuring Systems; Types and Requirements (June). 2 pp.

—**Safety**
- BSI BS 4743-79. (WITHDRAWN) 1979 Safety Requirements for Electronic Measuring Apparatus (Superseded by BS EN 61010-1: 1993). 61 pp.
- CENELEC HD 401-80. Safety Requirements for Electronic Measuring Apparatus. 2 pp.
- CSA CAN/CSA-C22. 2NO231SERM89. CSA Safety Requirements for Electrical and Electronic Measuring and Test Equipment; (Gen Instr 1). 133 pp.
- CSA C22.2 NO 231 SP-M1990. Audit Checklist and Test Pack for Use with CAN/CSA-C22.2 No. 231 Series-M89 CSA Safety Requirements for Electrical and Electronic Measuring and Test Equipment. 27 pp.
- DIN VDE 0411 Pt 1-73. VDE Specifications for Electronic Measuring Instruments and Automatic Controls (Oct). 67 pp.
- DIN VDE 0411 Pt 1A-80. Electronic Measuring Apparatus; Part Amendment to Part 1: Safety Requirements for Electronic Measuring Apparatus (Feb). 9 pp.
- DIN VDE 0411 Pt 2-69. VDE Specifications for Electronic Measuring Instruments and Automatic Controls (Dec). 5 pp.
- IEC 348-78. Safety Requirements for Electronic Measuring Apparatus Second Edition. 118 pp.

Electronic Messaging
Use: Message Handling Systems

Electronic News Gathering
Use For: ENG (Electronic News Gathering)
See Also: Electronic News Gathering Equipment

—**Television Program Exchange**
- CCIR RECMN 715-90. International Exchange of ENG Recordings—Section 10/11G—Exchange of Recorded Television Programmes. 7 pp.
- CCIR Report 803-2-90. International Exchange of ENG Recordings for Television News Programmes—Section 10/11G—Exchange of Recorded Television Programmes. 1 p.

—**Television Recording**
- CCIR RECMN 715-90. International Exchange of ENG Recordings—Section 10/11G—Exchange of Recorded Television Programmes. 7 pp.
- CCIR QUESTION 106/11-90. Television Recordings on Magnetic Tape for Electronic News Gathering—Questions Concerning Study Group 11—Broadcasting Service (Television). 1 p.

Electronic News Gathering Equipment
Use For: ENG Equipment *See Also:* Electronic News Gathering; Television Equipment

—**Interfaces—Video Tape Recorders**
- BSI BS EN 60933-3-93. 1993 Audio, Video and Audiovisual Systems—Interconnections and Matching Values Part 3: Specification for Interface for the Interconnection of ENG Cameras and Portable VTRs Using Non-Composite Signals, for 625 Line/50 Field Systems. 14 pp.
- CENELEC EN 60933-3-92. Audio, Video and Audiovisual Systems Interconnections and Matching Values Part 3: Interface for the Interconnection of ENG Cameras and Portable VTRs Using Non-Composite Signals, for 625 Line/50 Field Systems. 4 pp.
- CENELEC EN 60933-3-92. Audio, Video and Audiovisual Systems Interconnections and Matching Values Part 3: Interface for the Interconnection of ENG Cameras and Portable VTRs Using Non-Composite Signals, for 625 Line/50 Field Systems; (IEC 933-3: 1992). 9 pp.
- IEC 933 Pt 3-92. Audio, Video and Audiovisual Systems—Interconnections and Matching Values Part 3: Interface for the Interconnection of ENG Cameras and Portable VTRs Using Non-Composite Signals, for 625 Line/50 Field Systems First Edition; (CENELEC EN 60933-3:1992). 23 pp.

Electronic Outlet Strips
Use: Electric Outlet Strips

Electronic Prepress Systems
See Also: Graphic Arts; Photographic Processing Equipment; Printing; Registers (Alignment)

—**Color—Device Exchange Format—Processing Systems**
- OSI ISO/DIS 10758-90. Device Exchange Format for the Online Transfer of Colour Proofs from Electronic Prepress Systems to Direct Digital Colour Proofing Systems. 45 pp.

—**Color—Information Interchange—User Exchange Format**
- BSI BS 7607-92. 1992 Graphic Technology—Prepress Digital Data Exchange—Colour Picture Data on Magnetic Tape (ISO 10755: 1992) (H). 15 pp.
- ISO 10755-92. Graphic Technology—Prepress Digital Data Exchange—Colour Picture Data on Magnetic Tape First Edition; (ANSI IT8.1-1988). 14 pp.
- JIS X 0651-91. User Exchange Format for the Exchange of Color Picture Data Between Electronic Prepress Systems Via Magnetic Tape. 24 pp.
- JTC1 10755-92. Graphic Technology—Prepress Digital Data Exchange—Colour Picture Data on Magnetic Tape First Edition; (ANSI IT8.1-1988). 14 pp.
- OSI ISO/DIS 10755-90. User Exchange Format (UEF00) for the Exchange of Colour Picture Data Between Electronic Prepress Systems Via Magnetic Tape (DDE500). 18 pp.

—**Information Interchange—User Exchange Format**
- OSI ISO/DIS 10756-90. User Exchange Format (UEF01) for the Exchange of Line of Data Between Electronic Prepress Systems Via Magnetic Tape (DDE500). 21 pp.
- OSI ISO/DIS 10757-90. User Exchange Format (UEF02) for the Exchange of Geometric Information Between Electronic Prepress Systems Via Magnetic Tape (DDE500). 48 pp.

—**Monochrome—Information Interchange—User Exchange Format**
- OSI ISO/DIS 10759-90. User Exchange Format (UEF03) for the Exchange of Monochrome Image Data Between Electronic Prepress Systems Via Magnetic Tape (DDE500). 21 pp.

Electronic Publications
See Also: Documents

—**Statistics**
- BSI BS ISO 9707-91. 1991 Information and Documentation—Statistics on the Production and Distribution of Books, Newspapers, Periodicals and Electronic Publications. 19 pp.
- ISO 9707-91. Information and Documentation—Statistics on the Production and Distribution of Books, Newspapers, Periodicals and Electronic Publications First Edition. 17 pp.

Electronic Sockets
Use: Sockets (Electric)

Electronic Switches
Scope Note: For additional listings, use a more specific term *See Also:* Analog Switches; Crosspoint Switches; Electric Switches; Electronic Equipment; Programming Switches; Silicon Bilateral Switches; Silicon Controlled Switches; Solid State Switches; Switches
- DIN VDE 0632 AQ-79. Switches up to 750 V 63 A; Particular Requirements for Electronic Actuating Switches up to 250 V a.c. and 16 A: Amendment Q (Apr). 32 pp.
- DIN VDE 0632 AR-79. Switches up to 750 V 63 A; Particular Requirements for Electronic Switches up to 250 V a.c. and 16 A: Amendment R (Apr). 31 pp.
- IEC 669 Pt 2-1-84. Switches for Household and Similar Fixed-Electrical Installations Part 2: Particular Requirements for Electronic Switches First Edition; (Corrigendum—July 1985). 53 pp.
- IECQ QC 960000-91. Electromechanical Switches for Use in Electronic Equipment Part 1: Generic Specification (IEC 1020-1 ED 1). 78 pp.
- JIS C 5441-85. Testing Methods of Switches for Electronic Equipment. 52 pp.

—**Appliances**
- DIN VDE 0630 Pt 12-88. Switches for Appliances for a Rated Voltage Not Exceeding 500 V and a Rated Current Not Exceeding 63 A; Electronic Switches (Sept). 27 pp.

—**Quality Assurance**
- BSI BS QC 960000-92. 1992 Electromechanical Switches for Use in Electronic Equipment Generic Specification (IEC 1020-1: 1991). 39 pp.
- IEC 1020 Pt 1-91. Electromechanical Switches for Use in Electronic Equipment Part 1: Generic Specification; (IECQ QC 960000). 78 pp.

Electronic Switches (Cont.)
—**Remote Control—Household**
- DIN VDE 0637 Pt 2-86. Remote Control Switches and Switches Incorporating a Time Delay Device for Domestic and Similar Fixed Electrical Installations; Magnetic Remote Control Switches; Electronic Switches (Feb). 14 pp.

—**Rotary—Quality Assurance**
- BSI BS 9563-79. (OBSOLESCENT) 1979 Rotary (Manual) Switches of Assessed Quality: Generic Data and Methods of Test: General Rules for the Preparation of Detail Specifications (Superseded by BS CECC 96100). 20 pp.

—**Time Delay—Household**
- DIN VDE 0637 Pt 2-86. Remote Control Switches and Switches Incorporating a Time Delay Device for Domestic and Similar Fixed Electrical Installations; Magnetic Remote Control Switches; Electronic Switches (Feb). 14 pp.

Electronic Transducers
See Also: Piezoelectric Transducers; Transducers
- JIS C 1111-89. Electrical Measuring Transducers for Converting A.C. Electrical Quantities into D.C. Electrical Quantities. 34 pp.

—**Angular—Preferred Products List**
- CECC CECC MUAHAG Vol 13 IS 1-91. Preferred Products List; Servo Components (En, Fr, Ge). 27 pp.

—**Angular—Synchros—Preferred Products List**
- CECC CECC MUAHAG Vol 13 IS 1-91. Preferred Products List; Servo Components (En, Fr, Ge). 27 pp.

—**Electrical Measurement**
- BSI BS 6253: Part 1-82. (WITHDRAWN) 1982 Amd 2 Electrical Measuring Transducers for Converting a.c. Electrical Quantities into d.c. Electrical Quantities Part 1: Method for Specifying General Purpose Transducers (AMD 6225) March 28, 1991. 26 pp.
- BSI BS EN 60688-92. 1992 Electrical Measuring Transducers for Converting a.c. Electrical Quantities to Analogue or Digital Signals (IEC 688: 1992) (Supersedes BS 6253: Parts 1 & 2: 1982). 42 pp.
- CENELEC EN 60688-92. Electrical Measuring Transducers for Converting a.c. Electrical Quantities to Analogue or Digital Signals. 5 pp.
- IEC 688-92. Electrical Measuring Transducers for Converting a.c. Electrical Quantities to Analogue or Digital Signals Second Edition. 74 pp.
- MOD UK DSTAN 66-20-78. Transducers Electrical (Used in the Measurement of Electrical Quantities) Issue 1. 20 pp.

—**Outdoor**
- BSI BS 6253: Part 2-89. (WITHDRAWN) 1989 Electrical Measuring Transducers for Converting a.c. Electrical Quantities into d.c. Electrical Quantities Part 2: Method for Specifying Transducers for Outdoor Use and Other Severe Environments. 18 pp.
- IEC 688 Pt 2-88. Electrical Measuring Transducers for Converting A.C. Electrical Quantities into D.C. Electrical Quantities Part 2: Transducers for Outdoor Use and Other Severe Environments First Edition. 36 pp.

Electronic Transformers
See Also: Transformers
- BSI BS 6600: Part 1-87. 1987 Outline Dimensions of Transformers and Inductors for Use in Telecommunication Electronic Equipment Part 1: The Outline Dimensions of Transformers and Inductors Using YE1-1 Laminations. 14 pp.
- BSI BS 6600: Part 10-85. 1985 Outline Dimensions of Transformers and Inductors for Use in Telecommunication Electronic Equipment Part 10: Outline Dimensions of Transformers and Inductors Using the Q Range of C-Cores. 19 pp.

—**Intermediate Frequency**
- CNS C6241-85. Method of Test for High Frequency Inductors and Intermediate Frequency Transformers for Electronic Equipment (Jan)(11183).
- JIS C 5320-87. General Rules of High Frequency Inductors and Intermediate Frequency Transformers for Electronic Equipment. 16 pp.

—**Laminates**
- BSI BS 2857-76. 1976 Nickel-Iron Transformer and Choke Laminations. 30 pp.
- BSI BS 6554: Part 1-84. 1984 Laminations for Transformers and Inductors for Use in Telecommunications and Electronic Equipment Part 1: Characteristics and Electrical Testing. 24 pp.

Electronic Transformers *(Cont.)*
—Laminates *(Cont.)*
BSI BS 6600: Part 1-87. 1987 Outline Dimensions of Transformers and Inductors for Use in Telecommunication Electronic Equipment Part 1: The Outline Dimensions of Transformers and Inductors Using YE1-1 Laminations. 14 pp.
BSI BS 7382: Part 1-90. 1990 Laminated Core Packages for Transformers and Inductors Used in Telecommunication and Electronic Equipment Part 1: Dimensions. 11 pp.
IEC 740-82. Laminations for Transformers and Inductors for Use in Telecommunication and Electronic Equipment First Edition; (Amendment 1-1991). 57 pp.
IEC 852 Pt 1-86. Outline Dimensions of Transformers and Inductors for Use in Telecommunication and Electronic Equipment Part 1: Transformers and Inductors Using YEI-1 Laminations First Edition. 22 pp.
IEC 852 Pt 3-92. Outline Dimensions of Transformers and Inductors for Use in Telecommunication and Electronic Equipment Part 3: Transformers and Inductors Using YUI-1 Laminations First Edition. 21 pp.
IEC 1021 Pt 1-90. Laminated Core Packages for Transformers and Inductors Used in Telecommunication and Electronic Equipment Part 1: Dimensions First Edition. 17 pp.

Electronic Tubes
Use: Electron Tubes

Electronic Valve Shields
Use: Electron Tube Shields

Electronic Valves
Use: Electron Tubes

Electronic Valves (Electron Tubes)
Use: Electron Tubes

Electronic Warfare
Use For: Electromagnetic Warfare
See Also: Electromagnetic Intrusion; Jamming; Military Operations; Warfare
—Airborne Operations
NATO STANAG 3873 ED 3 AMD 0-90. Electronic Warfare (EW) in Air Operations—ATP-44(A). 5 pp.
—Land Battles—Allied Tactical Publications
NATO STANAG 6010 ED 1 AMD 1-89. Electronic Warfare in the Land Battle—ATP-51. 4 pp.

Electronystagmographs
Use For: ENG
JIS T 1160-85. Electronystagmographs.

Electrophoresis
Use For: Cataphoresis *See Also:* Electrodeposited Coatings
—Wool Fibers
ISO 2915-75. Wool—Determination of Cysteic Acid Content of Wool Hydrolysates by Paper Electrophoresis and Colorimetry First Edition. 7 pp.

Electroplated Coatings
Scope Note: Includes electroplating processes and materials *Use For:* Cathodic Coatings, Electroplated; Chemical Plating; Electroplating
See Also: Electrodeposited Coatings; Electroless Coatings; Metal Cleaning; Metal Coatings (Made From Metal); Metal Finishing; Metallizing
JIS H 8630-87. Electroplated Coatings on Plastics Materials for Decorative Purposes. 21 pp.
—Aerospace—Codes
DIN ENGL LN 9368 Pt 5-83. Aerospace; Designation of Surface Treatments; Code Numbers for Cathodic Treatments (Apr). 3 pp.
—Automotive
CNS D1009-81. General Rules of Electroplating for Automobile Parts (Jul)(7669).
—Bicycles
CNS B7071-75. Standard for Surface Treatment of Bicycle Parts (Dec)(3884). 3 pp.
—Bicycles—Bend Testing
CNS B7072-75. Method of Test for Surface Treatment of Bicycle Parts (Dec)(3885). 2 pp.
—Bicycles—Light Testing
CNS B7072-75. Method of Test for Surface Treatment of Bicycle Parts (Dec)(3885). 2 pp.
—Bicycles—Salt Spray Testing
CNS B7072-75. Method of Test for Surface Treatment of Bicycle Parts (Dec)(3885). 2 pp.

Electroplated Coatings *(Cont.)*
—Cadmium
BSI BS 1706-90. 1990 Amd 1 Electroplated Coatings of Zinc and Cadmium on Iron and Steel (AMD 6731) May 31, 1991. 17 pp.
BSI BS 2868-68. 1968 Cadmium Anodes and Cadmium Oxide for Electroplating. 8 pp.
BSI BS 3382: Parts 1 & 2-61. 1961 Amd 2 Electroplated Coatings on Threaded Components Part 1: Cadmium on Steel Components Part 2: Zinc on Steel Components. 34 pp.
BSI BS 6338-82. 1982 Chromate Conversion Coatings on Electroplated Zinc and Cadmium Coatings. 4 pp.
BSI BS 7371: Part 3-93. 1993 Coatings on Metal Fasteners Part 3: Specification for Electroplated Zinc and Cadmium Coatings (F). 14 pp.
CNS H3080-87. Electroplated Coatings of Cadmium on Iron and Steel (Sep)(4828).
ISO 2082-86. Metallic Coatings—Electroplated Coatings of Cadmium on Iron or Steel Second Edition. 7 pp.
ISO 4520-81. Chromate Conversion Coatings on Electroplated Zinc and Cadmium Coatings First Edition. 4 pp.
JIS H 8611-93. Electroplated Coatings of Cadmium on Steel. 8 pp.
JIS H 8625-93. Chromate Conversion Coatings on Electroplated Zinc and Cadmium Coatings. 14 pp.
SAA AS K132.1-73. Electroplated Coatings on Threaded Components —Part 1: Cadmium on Steel. 23 pp.
SAA AS 1790-84. Electroplated Coatings—Cadmium on Iron or Steel (R 1993). 5 pp.
SAA AS 1791-86. Chromate Conversion Coatings—Zinc and Cadmium (R 1993). 5 pp.
—Cadmium—Aerospace
CEN PREN 2133-92. Cadmium Plating of Steels with Maximum Specified Tensile Strength Equal to or Less Than 1450 MPa Copper, Copper Alloys and Nickel Alloys. 16 pp.
—Cadmium—Impact Testing
CNS H2048-79. Dropping Test for Electroplated Coatings of Cadmium (Apr)(4830).
—Chromium
BSI BS 1224-70. 1970 Amd 1 Electroplated Coatings of Nickel and Chromium. 29 pp.
BSI BS 3382: Parts 3 & 4-65. 1965 Amd 1 Electroplated Coatings on Threaded Components Part 3: Nickel or Nickel Plus Chromium on Steel Components Part 4: Nickel or Nickel Plus Chromium on Copper and Copper Alloy (Including Brass) Components. 33 pp.
BSI BS 4601-70. 1970 Electroplated Coatings of Nickel Plus Chromium on Plastics Materials. 20 pp.
BSI BS 4641-86. 1986 Electroplated Coatings of Chromium for Engineering Purposes. 14 pp.
CNS H3060-87. Electroplated Coatings of Nickel and Chromium (Sep)(4157).
CNS H3106-86. Operation Standard for Chromium Plating on Zinc Alloys (Feb)(8508).
CNS H3109-82. Recommended Practice for Copper-Nickel-Chromium Plating (Jun)(8977).
CNS H3136-87. Electroplated Coatings of Copper Nickel Chromium on Plastics (Oct)(11391).
CNS H3154-90. Electroplated Coatings of Chromium for Engineering Use (Mar)(12685).
ISO 1456-88. Metallic Coatings—Electrodeposited Coatings of Nickel Plus Chromium and of Copper Plus Nickel Plus Chromium Second Edition; (Supersedes 1457). 19 pp.
ISO 4525-85. Metallic Coatings—Electroplated Coatings of Nickel Plus Chromium on Plastics Materials First Edition. 7 pp.
ISO 6158-84. Metallic Coatings—Electroplated Coatings of Chromium for Engineering Purposes First Edition. 9 pp.
JIS H 8615-93. Electroplated Coatings of Chromium for Engineering Purposes. 12 pp.
JIS H 8617-91. Electroplated Coatings of Nickel and Chromium. 17 pp.
JIS H 9121-53. Standard Operation of Nickel Plating and Chromium Plating. 9 pp.
JIS H 9125-65. Operation Standard for Chromium Plating on Zinc Alloys (R 1983). 9 pp.
SAA AS 1192-82. Electroplated Coatings—Nickel and Chromium (R 1993). 9 pp.
SAA AS 1406-82. Electroplated Coatings—Nickel and Chromium on Plastics for Decorative Applications (R 1993). 7 pp.
—Copper
BSI BS 4493-69. 1969 Copper Salts for Electroplating. 25 pp.
CNS H3060-87. Electroplated Coatings of Nickel and Chromium (Sep)(4157).
CNS H3106-86. Operation Standard for Chromium Plating on Zinc Alloys (Feb)(8508).
CNS H3109-82. Recommended Practice for Copper-Nickel-Chromium Plating (Jun)(8977).
CNS H3137-87. Electroplated Coatings of Copper (Oct)(11392).

Electroplated Coatings *(Cont.)*
—Copper *(Cont.)*
DIN ENGL 50967-91. Electrodeposited Coatings of Nickel Plus Chromium and Copper Plus Nickel Plus Chromium (Jan). 8 pp.
JIS H 8617-91. Electroplated Coatings of Nickel and Chromium. 17 pp.
JIS H 9125-65. Operation Standard for Chromium Plating on Zinc Alloys (R 1983). 9 pp.
—Copper—Aerospace
CEN PREN 3765-92. Electrolytic Silver Plating for General Use Steels, Heat Resisting Alloys, Copper and Copper Alloys. 16 pp.
—Copper Alloys—Aerospace
CEN PREN 3765-92. Electrolytic Silver Plating for General Use Steels, Heat Resisting Alloys, Copper and Copper Alloys. 16 pp.
—Corrosion Testing
JIS H 8502-88. Methods of Corrosion Resistance Test for Metallic Coatings. 32 pp.
—Electroplating Equipment
CNS H3104-87. Equipment and Apparatus for Electroplating (Aug)(8505).
JIS H 9122-77. Equipment and Apparatus for Electroplating (R 1986). 7 pp.
—Engineering Drawings—Designations
CNS H1007-87. Drawing Indications for Electroplated Coatings (Aug)(7796).
DIN ENGL 50960 Pt 2-86. Electroplated and Chemical Coatings; Indications on Drawings (Feb) (Together with DIN 50960 Part 1, February 1986 Edition, Supersedes DIN 50960, June 1963 Edition Withdrawn in July 1976). 6 pp.
—Fasteners
CNS B1056-80. Electroplated Coating of Bolts, Screws and Nuts (May)(4240).
CNS B7085-78. Method of Test for Electroplated Coatings of Bolts, Screws and Nuts (Mar)(4241).
—Glossaries
CNS H1009-90. Glossary of Terms Used in Electroplating (Oct) (8589).
ISO 2080-81. Electroplating and Related Processes—Vocabulary Second Edition. 59 pp.
JIS H 0400-82. Glossary of Terms Used in Electroplating. 28 pp.
SAA AS K178-69. Glossary of Terms Used in Electroplating. 31 pp.
—Gold
BSI BS 4292: Part 1-89. 1989 Method for Specifying Electroplated Coatings of Gold and Gold Alloys Part 1: Gold and Gold Alloys for Engineering Purposes. 12 pp.
BSI BS 4292: Part 2-89. 1989 Method for Specifying Electroplated Coatings of Gold and Gold Alloys Part 2: Gold and Gold Alloys for Decorative and Protective Purposes. 10 pp.
BSI BS 5658-79. 1979 Gold Potassium Cyanide for Electro-Plating. 12 pp.
CNS H3076-87. Electroplated Coatings of Gold for Engineering Use (Aug)(4824).
CNS H3144-87. Electroplated Coatings of Gold for Decorative Use (Aug)(12048).
JIS H 8620-93. Electroplated Coatings of Gold and Gold Alloy for Engineering Purposes. 19 pp.
JIS H 8622-93. Electroplated Coatings of Gold and Gold Alloy for Decorative Purposes. 14 pp.
SAA AS 1901-93. Electroplated Coatings—Gold and Gold Alloys (Supersedes AS 1902—1976). 9 pp.
SAA AS 1902-76. Electroplated Coatings of Gold and Gold Alloys for Decorative Applications (Superseded by AS 1901—1993).
—Heat Resistant Alloys—Aerospace
CEN PREN 3765-92. Electrolytic Silver Plating for General Use Steels, Heat Resisting Alloys, Copper and Copper Alloys. 16 pp.
—Heat Resistant Steels—Aerospace
CEN PREN 3765-92. Electrolytic Silver Plating for General Use Steels, Heat Resisting Alloys, Copper and Copper Alloys. 16 pp.
—Lead
CNS H3146-87. Electroplated Coatings of Lead and Lead-Tin Alloys on Iron and Steel (Sep)(12089).
—Metal
BSI BS 4479: Part 2-90. 1990 Design of Articles That Are to Be Coated Part 2: Recommendations for Electroplated and Autocatalytic Coatings. 15 pp.
SAA AS CK13-66. Code of Recommended Practice for Preparation of Metal Surfaces for Electroplating. 32 pp.
—Nickel
BSI BS 558 & 564-70. 1970 Nickel Anodes, Anode Nickel and Nickel Salts for Electroplating. 28 pp.

Electroplated Coatings (Cont.)

—Nickel (Cont.)
BSI BS 1224-70. 1970 Amd 1 Electroplated Coatings of Nickel and Chromium. 29 pp.
BSI BS 3382: Parts 3 & 4-65. 1965 Amd 1 Electroplated Coatings on Threaded Components Part 3: Nickel or Nickel Plus Chromium on Steel Components Part 4: Nickel or Nickel Plus Chromium on Copper and Copper Alloy (Including Brass) Components. 33 pp.
BSI BS 3597-84. 1984 Electroplated Coatings of 65/35 Tin/Nickel Alloy. 10 pp.
BSI BS 4601-70. 1970 Electroplated Coatings of Nickel Plus Chromium on Plastics Materials. 20 pp.
BSI BS 4758-86. 1986 Method for Specifying Electroplated Coatings of Nickel for Engineering Purposes. 12 pp.
CNS H3060-87. Electroplated Coatings of Nickel and Chromium (Sep)(4157).
CNS H3106-86. Operation Standard for Chromium Plating on Zinc Alloys (Feb)(8508).
CNS H3109-82. Recommended Practice for Copper-Nickel-Chromium Plating (Jun)(8977).
CNS H3136-87. Electroplated Coatings of Copper Nickel Chromium on Plastics (Oct)(11391).
DIN ENGL 50967-91. Electrodeposited Coatings of Nickel Plus Chromium and Copper Plus Nickel Plus Chromium (Jan). 8 pp.
DIN ENGL 50968-91. Electrodeposited Coatings of Nickel and Nickel Plus Copper (Jan). 5 pp.
ISO 4525-85. Metallic Coatings—Electroplated Coatings of Nickel Plus Chromium on Plastics Materials First Edition. 7 pp.
ISO 4526-84. Metallic Coatings—Electroplated Coatings of Nickel for Engineering Purposes First Edition. 9 pp.
JIS H 8617-91. Electroplated Coatings of Nickel and Chromium. 17 pp.
JIS H 9121-53. Standard Operation of Nickel Plating and Chromium Plating. 9 pp.
JIS H 9125-65. Operation Standard for Chromium Plating on Zinc Alloys (R 1983). 9 pp.
MOD UK DSTAN 03-10-87. Electrodeposition of Nickel and Chromium Issue 2. 21 pp.
SAA AS 1192-82. Electroplated Coatings—Nickel and Chromium (R 1993). 9 pp.
SAA AS 1406-82. Electroplated Coatings—Nickel and Chromium on Plastics for Decorative Applications (R 1993). 7 pp.
SNZ NZS/BS 4758-86. Method for Specifying Electroplated Coatings of Nickel for Engineering Purposes. 12 pp.

—Nickel—Autocatalytic
DIN ENGL 50966-88. Electroplated Coatings; Autocatalytic Nickel-Phosphorus Coatings on Metal in Technical Applications (May). 5 pp.

—Nickel Sulfates
CNS K1166-74. Nickel Salt for Electro-Plating (Tentative) (Mar)(3700).

—Palladium
CNS H3149-87. Electroplated Coatings of Palladium for Engineering Use (Sep)(12092).

—Potassium Cyanides
BSI BS 622-67. 1967 Amd 1 Potassium and Sodium Cyanides for Electroplating. 13 pp.

—Rhodium
CNS H3148-87. Electroplated Coatings of Rhodium for Engineering Use (Sep)(12091).

—Silver
BSI BS 1561-66. 1966 Silver Anodes and Silver Salts for Electroplating. 12 pp.
BSI BS 2816-89. 1989 Methods for Specifying Electroplated Coatings of Silver and Silver Alloys for Engineering Purposes. 12 pp.
BSI BS 3382: Parts 5 & 6-67. 1967 Electroplated Coatings on Threaded Components Part 5: Tin on Copper and Copper Alloy (Including Brass) Components Part 6: Silveron Copper and Copper Alloy (Including Brass) Components. 33 pp.
BSI BS 4290-89. 1989 Electroplated Coatings of Silver and Silver Alloys for Decorative and Protective Purposes. 10 pp.
CNS C3036-72. General Rules for Silver-Plating (Jun)(3307).
CNS H3077-87. Electroplated Coatings of Silver for Decorative Use (Aug)(4825).
CNS H3135-87. Electroplated Coatings of Silver for Engineering Use (Aug)(11307).
JIS H 8621-93. Electroplated Coatings of Silver for Engineering Purposes. 14 pp.
JIS H 8623-93. Electroplated Coatings of Silver for Decorative Purposes. 12 pp.
SAA AS 1856-91. Electroplated Coatings—Silver (Supersedes AS 1857—1976). 6 pp.

—Silver—Cutlery
CNS S1185-87. Electroplated Coatings of Silver for Cutlery, Flat-Ware and Hollow-Ware (Oct)(11316).

—Silver—Fasteners—Aerospace
CEN PREN 2786-90. Electrolytic Silver Plating of Fasteners. 4 pp.
CEN PREN 2786-93. Aerospace Series Electrolytic Silver Plating of Fasteners. 4 pp.

—Silver—Tableware
CNS S1185-87. Electroplated Coatings of Silver for Cutlery, Flat-Ware and Hollow-Ware (Oct)(11316).

—Sodium Cyanide
BSI BS 622-67. 1967 Amd 1 Potassium and Sodium Cyanides for Electroplating. 13 pp.

—Symbols
JIS H 0404-88. Graphical Symbol for Electroplated Coating. 11 pp.

—Technical Documents—Designations
DIN ENGL 50960 Pt 1-86. Electroplated and Chemical Coatings; Designation and Specification in Technical Documents (Feb) (Together with DIN 50960 Part 2, February 1986 Edition, Supersedes DIN 50960, June 1963 Edition, Withdrawn in July 1976). 6 pp.

—Threaded Components
SAA AS 1897-76. Electroplated Coatings on Threaded Components (Metric Coarse Series). 40 pp.

—Tin
BSI BS 1468-67. 1967 Amd 1 Tin Anodes and Tin Salts for Electroplating. 19 pp.
BSI BS 1872-84. 1984 Electroplated Coatings of Tin. 12 pp.
BSI BS 3382: Parts 5 & 6-67. 1967 Electroplated Coatings on Threaded Components Part 5: Tin on Copper and Copper Alloy (Including Brass) Components Part 6: Silveron Copper and Copper Alloy (Including Brass) Components. 33 pp.
BSI BS 3597-84. 1984 Electroplated Coatings of 65/35 Tin/Nickel Alloy. 10 pp.
BSI BS 6137-82. 1982 Electroplated Coatings of Tin/Lead Alloys. 12 pp.
CNS H3078-87. Electroplated Coatings of Tin (Aug)(4826).
ISO 2093-86. Electroplated Coatings of Tin—Specification and Test Methods Second Edition. 10 pp.
ISO 2179-86. Electroplated Coatings of Tin-Nickel Alloy—Specification and Test Methods Second Edition. 8 pp.
ISO 7587-86. Electroplated Coatings of Tin-Lead Alloys—Specification and Test Methods First Edition. 13 pp.
JIS H 8619-93. Electroplated Coatings of Tin. 8 pp.
JIS H 8624-90. Electroplated Coatings of Tin-Lead Alloys. 8 pp.
MOD UK DSTAN 03-8-73. Electro-Deposition of Tin Issue 1. 23 pp.
MOD UK DSTAN 03-8-92. Electro—Deposition of Tin Issue 2. 19 pp.
SAA AS K141-66. Electroplated Coatings of Tin Being BS 1872:1964 Endorsed Subject to Australian Amendment Amdt 1 July 1966. 13 pp.

—Wear Testing
JIS H 8503-89. Methods of Wear Resistance Test for Metallic Coatings. 27 pp.

—Zinc
BSI BS 1706-90. 1990 Amd 1 Electroplated Coatings of Zinc and Cadmium on Iron and Steel (AMD 6731) May 31, 1991. 17 pp.
BSI BS 3382: Parts 1 & 2-61. 1961 Amd 2 Electroplated Coatings on Threaded Components Part 1: Cadmium on Steel Components Part 2: Zinc on Steel Components. 34 pp.
BSI BS 6338-82. 1982 Chromate Conversion Coatings on Electroplated Zinc and Cadmium Coatings. 4 pp.
BSI BS 7371: Part 3-93. 1993 Coatings on Metal Fasteners Part 3: Specification for Electroplated Zinc and Cadmium Coatings (F). 14 pp.
CNS H3079-87. Electroplated Coatings of Zinc on Iron and Steel (Sep)(4827).
DIN ENGL 50961-87. Electroplated Coatings; Zinc and Cadmium Coatings on Iron and Steel; Chromate Treatment of Zinc and Cadmium Coatings (June) (Supersedes DIN 50941, May 1978 Edition, and DIN 50962, April 1976 Edition). 6 pp.
ISO 2081-86. Metallic Coatings—Electroplated Coatings of Zinc on Iron or Steel Second Edition. 7 pp.
ISO 4520-81. Chromate Conversion Coatings on Electroplated Zinc and Cadmium Coatings First Edition. 4 pp.
JIS H 8610-91. Electroplated Coatings of Zinc on Iron or Steel. 8 pp.
JIS H 8625-93. Chromate Conversion Coatings on Electroplated Zinc and Cadmium Coatings. 14 pp.
SAA AS K132.2-73. Electroplated Coatings on Threaded Components —Part 2: Zinc on Steel. 23 pp.
SAA AS 1789-84. Electroplated Coatings—Zinc on Iron or Steel (R 1993). 4 pp.
SAA AS 1791-86. Chromate Conversion Coatings—Zinc and Cadmium (R 1993). 5 pp.

—Zinc—Impact Testing
CNS H2047-79. Dropping Test for Electroplated Coatings of Zinc (Apr)(4829).

Electroplating
Use: Electroplated Coatings

Electropolishing
Use For: Electrochemical Polishing; Electrolytic Polishing *See Also:* Metal Finishing

—Aerospace—Heat Resistant Alloys
CEN PREN 3769-90. Electrolytic Polishing of Steels and Heat Resisting Alloys. 6 pp.

—Aerospace—Steels
CEN PREN 3769-90. Electrolytic Polishing of Steels and Heat Resisting Alloys. 6 pp.

Electroreflectance Mesometers
Use: Cathode Ray Tubes

Electroretinographs
See Also: Medical Electrical Equipment; Medical Equipment
JIS T 1161-86. Electroretinographs.

Electrosensitive Safety Systems
Use For: ESSS *See Also:* Safety

—Machinery
BSI BS 6491: Part 1-84. 1984 Amd 1 Electro-Sensitive Safety Systems for Industrial Machines Part 1: General Requirements (AMD 5185) June 30, 1987. 24 pp.
BSI BS 6491: Part 2-87. 1987 Electro-Sensitive Safety Systems for Industrial Machines Part 2: Specifications for Particular Requirements for an Electro-Sensitive Safety System Incorporating a Photoelectric Sensing Unit. 10 pp.
CENELEC PREN 50100-1-92. Safety of Machinery: Electro-Sensitive Protective Devices Part 1: Specification for General Requirements. 85 pp.
SAA AS 4024.2 (INT)-92. Safeguarding of Machinery—Part 2(Int): Presence Sensing Systems (Expires 17 August, 1994) (in Professional Package 47). 107 pp.

Electroslag Remelting Furnaces
See Also: Electric Arc Furnaces; Furnaces
CENELEC HD 470 S1-87. Test Methods for Electro-Slag Remelting Furnaces. 3 pp.
IEC 779-83. Test Methods for Electro-Slag Remelting Furnaces First Edition. 19 pp.

—Safety
IEC 519 Pt 8-83. Safety in Electroheat Installations Part 8: Particular Requirements for Electro-Slag Remelting Furnaces First Edition. 18 pp.

Electroslag Welding
See Also: Electric Welding; Resistance Welding; Welding

—Consumable Guide Method—Structural Steels
JIS Z 3606-77. Recommended Practice for Consumable Nozzle Electroslag Welding (R 1980). 12 pp.

Electroslag Welding Equipment
See Also: Welding Equipment
DIN VDE 0544 Pt 2 (D)-78. Welding Equipment and Assemblies for Arc Welding and Allied Processes; Assemblies (VDE Specification) (Oct). 20 pp.

Electrostatic Air Cleaners
Use: Electrostatic Precipitators

Electrostatic Behavior
Use: Electrostatic Properties

Electrostatic Charge
See Also: Electrostatic Discharge; Electrostatic Properties
IEC 1087-91. Guide for Evaluating the Discharges from a Charged Surface First Edition. 27 pp.

—Fabrics
CNS L3148-82. Method of Test for Electrostatic Propensity of Woven and Knitted Fabrics (Jan)(8312).
JIS L 1094-88. Testing Methods for Electrostatic Propensity of Woven and Knitted Fabrics. 25 pp.

Electrostatic

Electrostatic Charge (Cont.)

—Floor Coverings—Hazardous Environments
DIN ENGL 51953-75. Testing of Organic Floor Coverings; Testing of the Ability of Floor Coverings in Explosion-Hazarded Rooms to Dissipate Electrostatic Charges (Aug). 2 pp.

—Initiators (Explosives)—Munitions
NATO STANAG 4235 ED 1 AMD 0-00. (DRAFT) Electrostatic Environment Conditions Affecting the Design of Material for Use by NATO Forces. 5 pp.
NATO STANAG 4235 ED 1 AMD 0-93. Electrostatic Environmental Conditions Affecting the Design of Materiel for Use by NATO Forces. 9 pp.

—Initiators (Explosives)—Weapons Systems
NATO STANAG 4235 ED 1 AMD 0-00. (DRAFT) Electrostatic Environment Conditions Affecting the Design of Material for Use by NATO Forces. 5 pp.
NATO STANAG 4235 ED 1 AMD 0-93. Electrostatic Environmental Conditions Affecting the Design of Materiel for Use by NATO Forces. 9 pp.

—Liquid Fuel Transfer—Safety
NATO STANAG 3682 ED 4 AMD 0-90. Electrostatic Safety Connection Procedures for Liquid Fuel Loading/ Unloading Operations During Ground Transfer. 7 pp.
NATO STANAG 3682 ED 4 AMD 1-90. Electrostatic Safety Connection Procedures for Liquid Fuel Loading/ Unloading Operations During Ground Transfer. 7 pp.
NATO STANAG 3682 ED 4 AMD 2-90. Electrostatic Safety Connection Procedures for Liquid Fuel Loading/ Unloading Operations During Ground Transfer. 8 pp.

Electrostatic Discharge
See Also: Electrostatic Charge
IEC 1087-91. Guide for Evaluating the Discharges from a Charged Surface First Edition. 27 pp.

—Data Processing Equipment
CENELEC PREN 55024-2-89. Immunity Requirements for Information Technology Equipment Part 2: Electrostatic Discharge Requirements (See EN 55101-2). 44 pp.
CENELEC PREN 55024-3-89. Immunity Characteristics of Information Technology Equipment Part 3: Immunity to Radiated Radio Frequency Disturbances (See EN 55101-3). 20 pp.
CENELEC EN 55101-2-89. (Draft) Immunity Requirements for Information Technology Equipment Part 2: Electrostatic Discharge Requirements (See PREN 55024-2). 44 pp.
CENELEC EN 55101-3-89. (Draft) Immunity Characteristics of Information Technology Equipment. Part 3: Immunity to Radiated Radio Frequency Disturbances (See PREN 55024-3). 20 pp.

—Electronic Components—Protection
BSI BS CECC 00015: Pt 1-91. (WITHDRAWN) 1991 Basic Specification: Protection of Electrostatic Sensitive Devices Part 1: General Requirements (Renumbered BS EN 100015-1: 1992) (T). 62 pp.
BSI BS EN 100015-1-92. 1991 Amd 1 Basic Specification: Protection of Electrostatic Sensitive Devices Part 1: General Requirements (AMD 7443) November 15, 1992. 65 pp.
CECC CECC 00 015/I ISSUE 1-91. Basic Specification: Protection of Electrostatic Sensitive Devices Part I: General Requirements (En, Fr, Ge) ERRATUM (En, Fr, Ge). 184 pp.
CENELEC EN 100015-1-92. Basic Specification: Protection of Electrostatic Sensitive Devices Part 1: General Requirements. 59 pp.
CENELEC PREN 100 015-2-92. Basic Specification: Protection of Electrostatic Sensitive Devices Part 2: Requirements for Low Humidity Conditions. 5 pp.
CENELEC PREN 100 015-3-93. Basic Specification: Protection of Electrostatic Sensitive Devices. Part 3: Requirements for Clean Rooms. 7 pp.
CENELEC PREN 100 015-4-92. Protection of Electrostatic Sensitive Devices Part 4: Requirements for High Voltage Environments. 7 pp.

—Telecommunication Equipment—Protection
DIN VDE 0845 Pt 1-87. Protection of Telecommunication Installations Against Lightning, Electrostatic Discharges and Overvoltages from Power Installations; Measures Against Overvoltages (Oct) (Partially Supersedes DIN 57845/VDE 0845/ 04.76, DIN 57 845 A1/VDE 0845A1/04.81). 79 pp.

Electrostatic Dust Samplers
Use: Dust

Electrostatic Induction
Use For: Induction, Electrostatic See Also: Electrical Properties; Electrostatic Properties

Electrostatic Induction (Cont.)

—Circuit Noise—Power Lines—Telephone Networks
CCITT RECMN G.123-89. Circuit Noise in National Networks—General Characteristics of International Telephone Connections and Circuits (Study Groups XII and XV) 6 pp. 6 pp.

Electrostatic Microphones
Use: Capacitor Microphones

Electrostatic Precipitators
Use For: Electrostatic Air Cleaners See Also: Air Cleaners; Precipitators; Separators (Mechanical)

—Commercial
CSA C22.2 NO 187-M1986. Electrostatic Air Cleaners (R 1992); (Gen Instr 1). 34 pp.

—Electrical Installations
DIN VDE 0146-80. Erection of Electrostatic-Precipitators (Electrostatic Filters) (VDE-Specification) (Mar). 18 pp.

—Residential
CSA C22.2 NO 187-M1986. Electrostatic Air Cleaners (R 1992); (Gen Instr 1). 34 pp.

Electrostatic Properties
See Also: Electrical Properties; Electrostatic Charge; Electrostatic Induction

—Carpets
JIS L 1023-92. Testing Methods for Several Characteristics of Textile Floor Coverings. 26 pp.

—Electrical Insulation
DIN VDE 0303 Pt 8-75. VDE Specifications for Electrical Tests on Insulating Materials; Evaluation of the Electrostatic Behaviour (Apr). 18 pp.

—Plastics
DIN VDE 0303 Pt 8-75. VDE Specifications for Electrical Tests on Insulating Materials; Evaluation of the Electrostatic Behaviour (Apr). 18 pp.

—Rubber
DIN VDE 0303 Pt 8-75. VDE Specifications for Electrical Tests on Insulating Materials; Evaluation of the Electrostatic Behaviour (Apr). 18 pp.

Electrostatic Spray Guns
See Also: Electrostatic Spraying Equipment; Spray Guns
BSI BS 6742: Part 1-87. 1987 Amd 1 Electrostatic Painting and Finishing Equipment Using Flammable Materials Part 1: Hand-Held Spray Guns and Associated Apparatus (AMD 6424) May 31, 1990. 11 pp.
BSI BS 6742: Part 2-87. 1987 Electrostatic Painting and Finishing Equipment Using Flammable Materials Part 2: Selection, Installation and Use of Hand-Held Paint Spray Guns (Energy Limit of 3.24mJ) and Associated Apparatus. 10 pp.
BSI BS EN 50 059-91. 1991 Electrostatic Hand-Held Spraying Equipment for Non-Flammable Material for Painting and Finishing. 15 pp.
CENELEC EN 50 059-90. Specification for Electrostatic Hand-Held Spraying Equipment for Non-Flammable Material for Painting and Finishing. 24 pp.

—Explosive Atmospheres
BSI BS 6742: Part 3-90. 1990 Electrostatic Painting and Finishing Equipment Using Flammable Materials Part 3: Selection, Installation and Use of Hand Held Powder Spray Guns (Energy Limit of 5 mJ) and Their Associated Apparatus. 12 pp.
BSI BS 6742: Part 4-90. 1990 Electrostatic Painting and Finishing Equipment Using Flammable Materials Part 4: Selection, Installation and Use of Hand Held Flock Spray Guns (Energy Limit of 0.24 mJ or 5 mJ) and Their Associated Apparatus. 12 pp.
CENELEC EN 50 050-86. Electrical Apparatus for Potentially Explosive Atmospheres: Electrostatic Hand-Held Spraying Equipment. 7 pp.
CENELEC EN 50 053-87. Requirements for the Selection, Installation and Use of Electrostatic Spraying Equipment for Flammable Materials—Part 1: Hand-held Electrostatic Paint Spray Guns with an Energy Limit of 0,24 mj and Their Associated Apparatus. 8 pp.
CENELEC EN 50 053-2-89. Requirements for the Selection, Installation and Use of Electrostatic Spraying Equipment for Flammable Materials Part 2: Hand-Held Electrostatic Powder Spray Guns with an Energy Limit of 5 mJ and Their Associated Apparatus. 18 pp.

Electrostatic Spray Guns (Cont.)

—Explosive Atmospheres (Cont.)
CENELEC EN 50 053-3-89. Requirements for the Selection, Installation and Use of Electrostatic Spraying Equipment for Flammable Materials Part 3: Hand-Held Electrostatic Flock Spray Guns with an Energy Limit of 0.24 mJ or 5 mJ and Their Associated Apparatus. 19 pp.

Electrostatic Spraying Equipment
See Also: Electrostatic Spray Guns
BSI BS EN 50 059-91. 1991 Electrostatic Hand-Held Spraying Equipment for Non-Flammable Material for Painting and Finishing. 15 pp.
CENELEC EN 50 059-90. Specification for Electrostatic Hand-Held Spraying Equipment for Non-Flammable Material for Painting and Finishing. 24 pp.

—Explosive Atmospheres
BSI BS 6742: Part 3-90. 1990 Electrostatic Painting and Finishing Equipment Using Flammable Materials Part 3: Selection, Installation and Use of Hand Held Powder Spray Guns (Energy Limit of 5 mJ) and Their Associated Apparatus. 12 pp.
BSI BS 6742: Part 4-90. 1990 Electrostatic Painting and Finishing Equipment Using Flammable Materials Part 4: Selection, Installation and Use of Hand Held Flock Spray Guns (Energy Limit of 0.24 mJ or 5 mJ) and Their Associated Apparatus. 12 pp.
CENELEC EN 50 050-86. Electrical Apparatus for Potentially Explosive Atmospheres: Electrostatic Hand-Held Spraying Equipment. 7 pp.
CENELEC EN 50 053-87. Requirements for the Selection, Installation and Use of Electrostatic Spraying Equipment for Flammable Materials—Part 1: Hand-held Electrostatic Paint Spray Guns with an Energy Limit of 0,24 mj and Their Associated Apparatus. 8 pp.
CENELEC EN 50 053-2-89. Requirements for the Selection, Installation and Use of Electrostatic Spraying Equipment for Flammable Materials Part 2: Hand-Held Electrostatic Powder Spray Guns with an Energy Limit of 5 mJ and Their Associated Apparatus. 18 pp.
CENELEC EN 50 053-3-89. Requirements for the Selection, Installation and Use of Electrostatic Spraying Equipment for Flammable Materials Part 3: Hand-Held Electrostatic Flock Spray Guns with an Energy Limit of 0.24 mJ or 5 mJ and Their Associated Apparatus. 19 pp.

Electrosurgical Equipment
See Also: Dental Equipment; Medical Electrical Equipment; Medical Equipment; Surgical Instruments

—High Frequency—Safety
BSI BS 5724: Sec 2.2-92. 1992 Medical Electrical Equipment Part 2: Particular Requirements for Safety Section 2.2: Specification for High Frequency Surgical Equipment (IEC 601-2-2: 1991). 44 pp.
BSI BS 5724: Sec 2.2-01. 1992 Amd 1 Medical Electrical Equipment Part 2: Particular Requirements for Safety Section 2.2: Specification for High Frequency Surgical Equipment (BS EN 60601-2-2: 1992) (AMD 7868) August 15, 1993 (N). 52 pp.
BSI BS 5724: Sec 2.2-83. 1983 Medical Electrical Equipment Part 2: Particular Requirements for Safety Section 2.2: High Frequency Surgical Equipment. 30 pp.
BSI BS EN 60601-2-2-93. 1993 Amd 1 Medical Electrical Equipment Part 2: Particular Requirements for Safety Section 2.2: Specification for High Frequency Surgical Equipment (IEC 601-2-2: 1991) (AMD 7868) August 15, 1993 (N). 52 pp.
CENELEC EN 60601-2-2-93. Medical Electrical Equipment Part 2: Particular Requirements for Safety Section 2.2 Specification for High Frequency Surgical Equipment (IEC 601-2-2: 1991). 39 pp.
CSA CAN/CSA-C22. 2N601.2.2-92. Medical Electrical Equipment Part 2: Particular Requirements for the Safety of High Frequency Surgical Equipment (IEC 601-2-2:1991); (Gen Instr 1). 56 pp.
IEC 601 Pt 2-2-91. Medical Electrical Equipment Part 2: Particular Requirements for the Safety of High Frequency Surgical Equipment Second Edition; (CAN/CSA-C22.2 No. 601.2.2-92) (CENELEC EN 60601-2-2: 1993). 78 pp.
SAA AS 3200.2.2-92. Approval and Test Spec.— Medical Electrical Equipment—Part 2: Particular Requirements for Safety—Part 2.2: High Frequency Surgical Equipment (IEC 601-2-2: 1991) (In Prof. Pkgs. 28, 17C) (Supersedes AS 3202—1989) (This is a Joint Standard with SNZ NZS 3200.2.2). 37 pp.
SNZ NZS/AS 3200. 2.2-92. Approval and Test Specification—Medical Electrical Equipment Part 2.2: Particular Requirements for Safety—High Frequency Surgical Equipment (This is a Joint Standard with SAA AS 3200.2.2). 37 pp.

Electrotechnical Components
Use: Electrical Components

INTERNATIONAL AND NON-U.S. NATIONAL STANDARDS
SUBJECT INDEX
Elevators

Electrotechnical Equipment
Use: Electrical Components

Electrotherapy Equipment
See Also: Diathermy Machines; Medical Electrical Equipment; Medical Equipment; Muscle Stimulators; Nerve Stimulators

JIS C 6310-86. Low Frequency Therapy Equipment. 12 pp.

—**Safety**

CENELEC HD 395.2.14 S1-89. Medical Electrical Equipment part 2: Particular Requirements for the Safety of Electro Convulsive Therapy Equipment. 3 pp.

Electrothermal Atomic Absorption Spectrophotometry
Use: Atomic Absorption Spectrophotometry

Electrothermal Atomization Atomic Absorption Spectrometry
Use: Atomic Absorption Spectrometry

Electrothermal Wave Eliminators
Use: Degreasers (Cleaning Agents)

Element Error Rate
—**Telegraph Distortion**

CCITT RECMN R.2-89. Element Error Rate—Telegraph Transmission (Study Group IX) 1 pp. 1 p.

—**Telegraph Distortion—Units of Measurement**

CCITT RECMN R.4-89. Methods for the Separate Measurements of the Degrees of Various Types of Telegraph Distortion—Telegraph Transmission (Study Group IX) 1 pp. 1 p.

—**Telegraph Distortion—50 Bauds**

CCITT RECMN R.54-89. Conventional Degree of Distortion Tolerable for Standardized Start-Stop 50-Baud Systems—Telegraph Transmission (Study Group IX) 2 pp. 2 pp.

CCITT RECMN R.55-89. Conventional Degree of Distortion—Telegraph Transmission (Study Group IX) 1 pp. 1 p.

Elevating Platforms
Use: Platforms—Elevating

Elevator Belts
See Also: Belts (Machinery); Conveyor Belts

—**Glossaries**

SAA AS 4035-92. Conveyor and Elebator Belting—Glossary of Terms (Supersedes AS B255—1969). 6 pp.

Elevator Cables
See Also: Cables (Electric); Cables (Mechanical); Electric Conductors

—**Control—300 V**

CSA CAN/CSA-C22. 2 NO 49-92. Flexible Cords and Cables; (Gen Instr 1). 121 pp.

—**Control—600 V**

CSA CAN/CSA-C22. 2 NO 49-92. Flexible Cords and Cables; (Gen Instr 1). 121 pp.

—**PVC Insulated**

CENELEC HD 359-76. Flat Polyvinylchloride Sheated Cables for Lifts and Similar Applications. 8 pp.

CENELEC HD 359 S2-90. Flat Polyvinylchloride Sheathed Lift Cables. 17 pp.

CENELEC HD 359.2-79. Flat Polyvinyl Chloride Sheathed Flexible Cables for Lifts and Similar Applications. 1 p.

IEC 227 Pt 6-85. Polyvinyl Chloride Insulated Cables of Rated Voltages up to and Including 450/750 V Part 6: Lift Cables and Cables for Flexible Connections Second Edition. 25 pp.

—**Rubber Insulated**

CENELEC HD 360-76. Rubber Insulated Lift Cables for Normal Use. 9 pp.

CENELEC HD 360 S2-90. Circular Rubber Insulated Lift Cables for Normal Use. 17 pp.

CENELEC HD 360 S2/A1-91. AMD 1 Circular Rubber Insulated Lift Cables for Normal Use. 6 pp.

CENELEC HD 360.2-79. Rubber Insulated Lift Cables for Normal Use. 3 pp.

DIN VDE 0282 Pt 807-89. Rubber Cabies, Wires and Flexible Cords for Power Installation; Rubber-Insulated Lift Cables 05RT and 05RN (June). 12 pp.

Elevator Cables (Cont.)
—**Rubber Insulated (Cont.)**

DIN VDE 0282 Pt 808-89. Rubber Cables, Wires and Flexible Cords for Power Installation; Rubber-Insulated Lift Cables 07RT and 07RN (June). 11 pp.

IEC 245 Pt 5-80. Rubber Insulated Cables of Rated Voltages up to and Including 450/750 V Part 5: Lift Cables First Edition; (Amendment 1-1985). 19 pp.

—**Signal Circuits—300 V**

CSA CAN/CSA-C22. 2 NO 49-92. Flexible Cords and Cables; (Gen Instr 1). 121 pp.

—**Signal Circuits—600 V**

CSA CAN/CSA-C22. 2 NO 49-92. Flexible Cords and Cables; (Gen Instr 1). 121 pp.

Elevators (Conveyors)
See Also: Bucket Elevators; Conveyors; Elevators (Lifts)

—**Safety**

BSI BS 5667: Part 13-79. 1979 Continuous Mechanical Handling Equipment. Safety Requirements Part 13: Unit Loads: Arm Elevators and Push Bar Conveyors. 3 pp.

BSI BS 5667: Part 18-79. 1979 Continuous Mechanical Handling Equipment. Safety Requirements Part 18: Conveyors and Elevators with Chain-Elements: Examples for Guarding of Nip Points. 26 pp.

ISO TR5046-77. Continuous Mechanical Handling Equipment—Safety Code for Conveyors and Elevators with Chain-Elements—Examples for Guarding of Nip Points First Edition. 23 pp.

Elevators (Grain)
Use: Grain Elevators

Elevators (Lifts)
Use For: Electric Lifts; Hydraulic Lifts; Lifting Platforms; Lifts (Elevators); Paternosters
See Also: Elevators (Conveyors); Escalators; Hoists; Machine Rooms; Passenger Conveyors; Service Lifts; Stairlifts; Wheelchair Lifts; Winches

BSI BS 2655: Part 8-71. 1971 Amd 2 Lifts, Escalators, Passenger Conveyors and Paternosters Part 8: Modernization or Reconstruction of Lifts, Escalators and Paternosters. 29 pp.

BSI BS 5655: Part 5-89. 1989 Lifts and Service Lifts Part 5: Dimensions of Standard Lift Arrangements. 30 pp.

BSI BS 5655: Part 5-81. (WITHDRAWN) 1981 Amd 1 Lifts and Service Lifts Part 5: Specifications of Standard Electric Lift Arrangements. 24 pp.

BSI BS 5655: Part 11-89. 1989 Lifts and Service Lifts Part 11: Recommendations for the Installation of New, and the Modernization of, Electric Lifts in Existing Buildings. 15 pp.

BSI BS 5655: Part 12-89. 1989 Amd 1 Lifts and Service Lifts Part 12: Recommendations for the Installation of New, and the Modernization of, Hydraulic Lifts in Existing Buildings (AMD 6762) September 30, 1991. 17 pp.

CNS B1337-88. Structure Standard for Elevators (Jan)(10594).

CNS B4065-88. Hydraulic Elevator (Jan)(11380).

CNS R7042-88. Testing Standard for Elevator, Escalator and Dumbwaiter (Jan)(2866).

EC 84/529/EEC-84. Council Directive on the Approximation of the Laws of the Member States Relating to Electrically Operated Lifts. 9 pp.

EC 86/312/EEC-86. Commission Directive Adapting to Technical Progress Council Directive 84/529/EEC on the Approximation of the Laws of the Member States Relating to Electrically Operated Lifts. 2 pp.

ISO 4190 Pt 1-90. Lift Installation—Part 1: Lifts of Classes I, II and III Second Edition; (Incorporating Addendum 2). 15 pp.

ISO 4190 Pt 2-82. Passenger Lifts and Service Lifts—Part 2: Lifts of Class IV First Edition. 6 pp.

SAA AS 1418.10-87. Cranes (Including Hoists and Winches) (Known as the SAA Crane Code)—Part 10: Elevating Work Platforms Amdt 1 August 1989. 27 pp.

SAA AS 1735. Lifts, Escalators, and Moving Walks (Known as the SAA Lift Code) Complete Set in Binder.

SAA AS 1735.1-86. Lifts, Escalators, and Moving Walks (Known as the SAA Lift Code)—Part 1: General Requirements Amdt 1 July 1989. 15 pp.

SAA AS 1735.2-93. Lifts, Escalators, and Moving Walks—Part 2: Passenger and Goods Lifts—Electric Amdt 1 August 1986 Amdt 2 May 1987 Amdt 3 July 1989 Amdt 4 July 1990 (in Professional Packages 55, 62, 63, 64, 65, 66, 67, 68, 69). 147 pp.

SAA AS 1735.3-86. Lifts, Escalators, and Moving Walks (Known as the SAA Lift Code)—Part 3: Passenger and Goods Lifts—Electrohydraulic Amdt 1 July 1989. 24 pp.

Elevators (Lifts) (Cont.)

SNZ NZS/AS 1735. 8-86. Lifts, Escalators and Moving Walks (SAA Lift Code) Part 8: Inclined Lifts Amend: 1 (This is a Joint Standard with SAA AS 1735.8). 20 pp.

—**Alarm Systems**

CEN PREN 627-92. Specification for Data Logging, Alarm Reporting and Remote Monitoring of Lifts, Escalators and Passenger Conveyors. 10 pp.

—**Automobiles**

BSI BS AU 161-73. (WITHDRAWN) 1973 Vehicle Lifts (Superseded by BS AU 161: Part 1A: 1983 and BS AU 161: Part 2: 1989). 7 pp.

CNS B1338-87. Size of Car and Hoistway of Elevators (May)(10595).

CNS D4004-77. Lifts for Automobiles (Apr)(4077). 9 pp.

JIS A 4301-83. Size of Car and Hoistway of Elevators. 15 pp.

JIS D 8107-77. Lifts for Automobiles (R 1983). 12 pp.

—**Automobiles—Mechanical Properties**

CNS D4005-77. Automobile Series Specification for Vehicle Lifts (Apr)(4078). 5 pp.

—**Automobiles—Mobile**

BSI BS AU 161: Part 2-89. 1989 Vehicle Lifts Part 2: Mobile Lifts. 15 pp.

—**Automobiles—Safety**

BSI BS 6109: Part 2-89. 1989 Tail Lifts, Mobile Lifts and Ramps Associated with Vehicles Part 2: Code of Practice for Passenger Lifts and Ramps. 17 pp.

BSI BS AU 161: Part 1A-83. 1983 Amd 1 Vehicle Lifts Part 1A: Fixed Lifts. 19 pp.

CNS D4005-77. Automobile Series Specification for Vehicle Lifts (Apr)(4078). 5 pp.

—**Cables (Electric)**

SAA AS 1979-93. Electric Cables—Lifts—Flexible Travelling (in Professional Package 55). 9 pp.

—**Comprehensive Testing**

BSI BS 5655: Part 10-86. 1986 Amd 1 Lifts and Service Lifts Part 10: Testing and Inspection of Electric and Hydraulic Lifts (AMD 6002) May 31, 1989. 48 pp.

SAA AS 1735.10-86. Lifts, Escalators, and Moving Walks (Known as the SAA Lift Code)-Part 10: Tests. 11 pp.

—**Construction—Safety**

CEN EN 81-1-85. Safety Rules for the Construction and Installation of Lifts and Service Lifts—Part 1: Electric Lifts. 67 pp.

CEN EN 81-2-87. Safety Rules for the Construction and Installation of Lifts and Service Lifts Part 2 Hydraulic Lifts. 73 pp.

—**Contactors—Endurance Testing**

CSA CAN/CSA-C22. 2 NO 14-M91. Industrial Control Equipment; (Gen Instr 1 Thru 4). 142 pp.

—**Control Systems Equipment**

BSI BS 5655: Part 7-83. 1983 Amd 1 Lifts and Service Lifts Part 7: Manual Control Devices, Indicators and Additional Fittings. 10 pp.

ISO 4190 Pt 5-87. Lifts and Service Lifts (USA: Elevators and Dumbwaiters)—Part 5: Control Devices, Signals and Additional Fittings Second Edition. 10 pp.

—**Control Systems Equipment—Handicapped Persons**

CSA CAN/CSA-B355-M86. Elevating Devices for the Handicapped. 35 pp.

CSA CAN/CSA-B613-M87. Elevating Devices for the Handicapped in Private Residences; (Gen Instr 1). 30 pp.

—**Eyebolts—Lift Suspension**

BSI BS 5655: Part 8-83. 1983 Lifts and Service Lifts Part 8: Specification for Eyebolts for Lift Suspension. 8 pp.

—**Fire Doors**

SAA AS 1735.11-86. Lifts, Escalators, and Moving Walks (Known as the SAA Lift Code)—Part 11: Fire-Rated Landing Doors. 6 pp.

SNZ NZS/AS 1735. 11-86. Lifts, Escalators and Moving Walks (SAA Lift Code) Part 11: Fire-Rated Landing Doors (This is a Joint Standard with SAA AS 1735.11). 6 pp.

—**Fire Fighting**

BSI BS 5588: Part 5-91. 1991 Fire Precautions in the Design, Construction and Use of Buildings Part 5: Code of Practice for Firefighting Stairs and Lifts. 43 pp.

BSI BS 5588: Part 5-86. 1986 Fire Precautions in the Design and Construction of Buildings Part 5: Code of Practice for Firefighting Stairs and Lifts. 24 pp.

INDUSTRY STANDARDS

INTERNATIONAL AND NON-U.S. NATIONAL STANDARDS
SUBJECT INDEX

Elevators

Elevators (Lifts) (Cont.)

—Fire Fighting (Cont.)
BSI BS 5588: Part 5-01. 1991 Amd 1 Fire Precautions in the Design, Construction and Use of Buildings Part 5: Code of Practice for Firefighting Stairs and Lifts (AMD 7196) June 15, 1992. 48 pp.

—Glossaries
BSI BS 3810: Part 2-65. 1965 Amd 1 Glossary of Terms Used in Materials Handling Part 2: Terms Used in Connection with Conveyors and Elevators (Excluding Pneumatic and Hydraulic Handling). 29 pp.

BSI BS 3810: Part 8-75. 1975 Glossary of Terms Used in Materials Handling Part 8: Terms Used in Connection with Lifts, Lifting Platforms and Inclinded Haulage. 15 pp.

BSI BS 6100: SUBSEC 3.2.2-84. 1984 Glossary of Building and Civil Engineering Terms Part 3: Services Section 3.2: Internal Communication and Transport Subsection 3.2.2: Internal Transport. 7 pp.

BSI BS 6100: SUBSEC 3.2.2-01. 1984 Amd 1 Building and Civil Engineering Terms Part 3: Services Section 3.2: Internal Communication and Transport Subsection 3.2.2: Internal Transport (AMD 7258) August 15, 1992. 8 pp.

SNZ NZS 2000: Part 2-66. Glossary of Terms Used in Materials Handling Part 2: Terms Used in Connection with Conveyors and Elevators (Excluding Pneumatic and Hydraulic Handling) (Reconfirmed 1978). 28 pp.

—Guide Rails
BSI BS 5655: Part 9-85. 1985 Amd 2 Lifts and Service Lifts Part 9: Guide Rails. 22 pp.

ISO 7465-83. Passenger Lifts and Service Lifts—Guide Rails for Lifts and Counterweights—T-Type First Edition. 11 pp.

—Handicapped Accessible Equipment
SAA AS 1735.12-86. Lifts, Escalators, and Moving Walks (Known as the SAA Lift Code)—Part 12: Facilities for Persons with Disabilities Amdt 1 December 1991. 29 pp.

—Handicapped Persons
SAA AS 1735.13-86. Lifts, Escalators, and Moving Walks (Known as the SAA Lift Code)—Part 13: Lifts for Persons with Limited Mobility—Manually Powered Amdt 1 July 1989. 6 pp.

SAA AS 1735.14-90. Lifts, Escalators, and Moving Walks (Known as the SAA Lift Code)—Part 14: Lifts for People with Limited Mobility—Restricted Use—Low-Rise Platforms. 7 pp.

SAA AS 1735.15-90. Lifts, Escalators, and Moving Walks (Known as the SAA Lift Code)—Lifts for People with Limited Mobility—Restricted Use—Non-Automatically Controlled. 17 pp.

—Handicapped Persons—Safety
BSI BS 6440-83. 1983 Powered Lifting Platforms for Use by Disabled Persons. 16 pp.

—Inclined
SAA AS 1735.8-86. Lifts, Escalators, and Moving Walks (Known as the SAA Lift Code)—Part 8: Inclined Lifts Amdt 1 July 1989. 20 pp.

—Inspection
BSI BS 5655: Part 10-86. 1986 Amd 1 Lifts and Service Lifts Part 10: Testing and Inspection of Electric and Hydraulic Lifts (AMD 6002) May 31, 1989. 48 pp.

JIS A 4302-92. Inspection Standard of Elevator, Escalator and Dumbwaiter. 23 pp.

—Installation
BSI BS 5655: Part 6-90. 1990 Lifts and Service Lifts Part 6: Code of Practice for Selection and Installation. 48 pp.

—Installation—Safety
CEN EN 81-1-85. Safety Rules for the Construction and Installation of Lifts and Service Lifts—Part 1: Electric Lifts. 67 pp.

CEN EN 81-2-87. Safety Rules for the Construction and Installation of Lifts and Service Lifts Part 2 Hydraulic Lifts. 73 pp.

—Insulated Cables
CENELEC HD 359-76. Flat Polyvinylchloride Sheated Cables for Lifts and Similar Applications. 8 pp.

CENELEC HD 359 S2-90. Flat Polyvinylchloride Sheathed Lift Cables. 17 pp.

CENELEC HD 359.2-79. Flat Polyvinyl Chloride Sheathed Flexible Cables for Lifts and Similar Applications. 1 p.

CENELEC HD 360-76. Rubber Insulated Lift Cables for Normal Use. 9 pp.

CENELEC HD 360 S2-90. Circular Rubber Insulated Lift Cables for Normal Use. 17 pp.

CENELEC HD 360 S2/A1-91. AMD 1 Circular Rubber Insulated Lift Cables for Normal Use. 6 pp.

CENELEC HD 360.2-79. Rubber Insulated Lift Cables for Normal Use. 3 pp.

Elevators (Lifts) (Cont.)

—Insulated Cables (Cont.)
CNS C2109-90. Travelling Cables for Elevators (Mar)(8936).

CNS C3150-90. Method for Test for Traveling Cables for Elevators (Mar)(8937).

DIN VDE 0282 Pt 807-89. Rubber Cables, Wires and Flexible Cords for Power Installation; Rubber-Insulated Lift Cables 05RT and 05RN (June). 12 pp.

DIN VDE 0282 Pt 808-89. Rubber Cables, Wires and Flexible Cords for Power Installation; Rubber-Insulated Lift Cables 07RT and 07RN (June). 11 pp.

IEC 227 Pt 6-85. Polyvinyl Chloride Insulated Cables of Rated Voltages up to and Including 450/750 V Part 6: Lift Cables and Cables for Flexible Connections Second Edition. 25 pp.

IEC 245 Pt 5-80. Rubber Insulated Cables of Rated Voltages up to and Including 450/750 V Part 5: Lift Cables First Edition; (Amendment 1-1985). 19 pp.

JIS C 3408-87. Travelling Cables for Elevators. 22 pp.

—Monitors
BSI DD 176-88. 1988 Draft for Development Data Logging and Remote Monitoring Equipment for Lifts, Escalators and Passenger Conveyors. 7 pp.

CEN PREN 627-92. Specification for Data Logging, Alarm Reporting and Remote Monitoring of Lifts, Escalators and Passenger Conveyors. 10 pp.

—Residential Buildings
ISO 4190 Pt 6-84. Lifts and Service Lifts (USA: Elevators and Dumbwaiters)—Part 6: Passenger Lifts to Be Installed in Residential Buildings—Planning and Selection First Edition. 10 pp.

—Safety
BSI BS 5655: Part 1-86. 1986 Amd 1 Lifts and Service Lifts Part 1: Safety Rules for the Construction and Installation of Electric Lifts (AMD 5840) September 29, 1989. 71 pp.

BSI BS 5655: Part 2-88. 1988 Amd 1 Lifts and Service Lifts Part 2: Hydraulic Lifts (AMD 6220) April 28, 1989. 77 pp.

BSI BS 7255-89. 1989 Amd 1 Code of Practice for Safe Working on Lifts (AMD 6375) April 30, 1991. 30 pp.

BSI PD 6500-86. 1986 Explanatory Supplement to BS 5655 'Lifts and Service Lifts' Part 1: 'Safety Rules for the Construction and Installation of Electric Lifts' (EN 81 Part 1). 21 pp.

BSI DD 197-90. 1990 Recommendations for Vandal Resistant Lifts. 11 pp.

CSA CAN/CSA-B44-M90. Safety Code for Elevators; (Gen Instr 1 Thru 3) (Supplement 1 1992). 415 pp.

CSA CAN/CSA-B44.1-M91. Elevator and Escalator Electrical Equipment (ASME A17.5-1991). 97 pp.

EC COM(93) 240-93. Amended Proposal for a Council Directive on the Approximation of the Laws of the Member States Relating to Lifts. 13 pp.

ISO TR11071 Pt 1-90. Comparison of Worldwide Lift Safety Standards—Part 1: Electric Lifts (Elevators) First Edition. 56 pp.

—Shafts
CNS B1338-87. Size of Car and Hoistway of Elevators (May)(10595).

JIS A 4301-83. Size of Car and Hoistway of Elevators. 15 pp.

—Ships
MOD UK NES 113: Part 4-89. Requirements for Mechanical Handling Part 4: Lifts Issue 2 (11.89). 26 pp.

—Ships—Safety
ISO 8383-85. Lifts on Ships—Specific Requirements First Edition; (Corrected and Reprinted -1991). 6 pp.

—Wire Rope
BSI BS 302: Part 4-87. 1987 Stranded Steel Wire Ropes Part 4: Ropes for Lifts. 12 pp.

CSA G387-M1981. Steel Wire Rope for Elevators (R 1992); (Gen Instr 1). 20 pp.

ISO 4101-83. Drawn Steel Wire for Elevator Ropes—Specifications First Edition. 5 pp.

ISO 4344-83. Steel Wire Ropes for Lifts First Edition. 6 pp.

SNZ NZS/BS 302: Part 4-87. Stranded Steel Wire Ropes Part 4: Specification for Ropes for Lifts. 12 pp.

Elevators (Surgical)
Use For: Surgical Elevators *See Also:* Surgical Instruments

—Dental
CNS T3036-81. Dental Elevators (Jul)(7756).

JIS T 5407-79. Dental Elevators (R 1985). 6 pp.

Elevators (Surgical) (Cont.)

—Periosteal
CNS T1031-80. Periosteal Elevators (Aug)(6290).

JIS T 2607-53. Periosteal Elevators.

Ells (Pipe)
Use: Pipe Elbows

Elmendorf Testers
Use: Tear Testers—Elmendorf

Elongation
Use For: Elongation at Break; Elongation Values
See Also: Deformation; Expansion; Extensibility; Tensile Testing

—Adhesive Tapes
SAA AS 1635.6.1-74. Methods of Testing Pressure Sensitive Adhesive Tape—Part 6.1: Elongation. 1 p.

—Aggregates
BSI BS 812: Part 105.2-90. 1990 Testing Aggregates Part 105.2: Elongation Index of Coarse Aggregate. 9 pp.

—Cable Insulation
BSI BS 6469: Sec 4.2-92. 1992 Insulating and Sheathing Materials of Electric Cables Part 4: Methods of Test Specific to Polyethylene and Polypropylene Compounds Section 4.2: Elongation at Break After Pre-Conditioning—Wrapping Test After Pre-Conditioning-Wrapping. 24 pp.

CENELEC HD 505.4.2 S1-92. Common Test Methods for Insulating and Sheathing Materials of Electric Cables Part 4: Methods Specific to Polyethylene and Polypropylene Compounds Section Two—Elongation at Break After Pre-Conditioning—Wrapping Test After Thermal Ageing in Air—Measurement of Mass. 6 pp.

CNS C6233-84. Method of Test for Fiber Optic Devices (Cable Jacket Elongation) (Jun)(10933).

DIN VDE 0472 Pt 602-83. Testing of Cables and Insulated Cords; Tensile Strength and Elongation at Break (Aug). 13 pp.

DIN VDE 0472 Pt 616-84. Testing of Cables, Wires and Flexible Cords. Elongation Test at Low Temperature (Oct). 7 pp.

IEC 811 Pt 4-2-90. Common Test Methods for Insulating and Sheathing Materials of Electric Cables Part 4: Methods Specific to Polyethylene and Polypropylene Compounds Section Two—Elongation at Break After Pre-Conditioning—Wrapping Test After Pre-Conditioning—Wrapping Test After. 38 pp.

—Cables (Electric)
DIN VDE 0472 Pt 625-83. Testing of Cables and Insulated Conductors; Tensile and Elongation Characteristics of Strain Bearing Elements (Jan). 5 pp.

—Cellular Plastics
BSI BS 4443: Part 1-88. 1988 Amd 1 Methods of Test for Flexible Cellular Materials Part 1: Methods 1 to 6. 16 pp.

—Coated Fabrics
BSI BS 3424: Part 4-82. 1982 Testing Coated Fabrics Part 4. Method 6. Method for Determination of Breaking Strength and Elongation at Break. 4 pp.

BSI BS 3424: Part 21-93. 1993 Testing Coated Fabrics Part Method for Determination of Elongation and Tension Set. 10 pp.

BSI BS 3424: Part 21-87. 1987 Testing Coated Fabrics Part Method for Determination of Elongation and Tension Set. 6 pp.

ISO 1421-77. Fabrics Coated with Rubber or Plastics—Determination of Breaking Strength and Elongation at Break First Edition. 4 pp.

SAA AS 1441.9-73. Methods of Test for Coated Fabrics—Part 9: Method for Determination of Elongation and Stretch-Set.

—Condoms
CEN PREN 600-8-91. Latex Rubber Condoms—Part 8: Determination of Force and Elongation at Break. 7 pp.

—Conveyor Belts
BSI BS 490: Sec 10.2-83. 1983 Conveyor and Elevator Belting Part 10: Testing for Physical Properties Section 10.2: Method for Determination of Full Thickness Tensile Strength and Elongation of Rubber and Plastics Conveyor Belting of Textile Construction. 4 pp.

BSI BS 490: Sec 10.5-84. 1984 Conveyor and Elevator Belting Part 10: Testing for Physical Properties Section 10.5: Method for Determination of Tensile Strength and Elongation at Break of Rubber Covers. 2 pp.

ISO 283-90. Conveyor Belts—Full Thickness Tensile Strength and Elongation—Specifications and Method of Test Second Edition. 11 pp.

INDEX and DIRECTORY of

Elongation (Cont.)

—Conveyor Belts (Cont.)
- ISO 7622 Pt 1-84. Steel Cord Conveyor Belts—Longitudinal Traction Test—Part 1: Measurement of Elongation First Edition. 4 pp.
- SNZ NZS/BS 490: Pt10:Sec10.2-83. Conveyor and Elevator Belting Part 10: Testing and Physical Properties Section 10.2: Method for Determination of Full Thickness Tensile Strength and Elongation of Rubber and Plastics Conveyor Belting of Textile Construction. 3 pp.
- SNZ NZS/BS 490: Pt10:Sec10.5-84. Conveyor and Elevator Belting Part 10: Testing and Physical Properties Section 10.5: Method for Determination of Tensile Strength and Elongation at Break of Rubber Covers. 2 pp.

—Copper
- BSI DD 79-81. 1981 Method for Performing the Spiral Elongation Test on High Conductivity Copper. 7 pp.
- ISO TR4745-78. High Conductivity Copper—Spiral Elongation Test First Edition. 5 pp.

—Copper Conductors
- DIN VDE 0472 Pt 623-83. Testing of Cables, Wires and Flexible Cords; Elongation at Break of the Copper Conductor (Jan). 4 pp.

—Cordage
- CGSB CAN/CGSB-4.2 NO.10-M87. Textile Test Methods Elongation. 9 pp.

—Ebonite
- CNS K6451-80. Method of Test for Tensile Strength and Elongation at Break of Ebonite (Jan)(5140).

—Fabrics
- BSI BS 2576-86. 1986 Amd 1 Method for Determination of Breaking Strength and Elongation (Strip Method) of Woven Fabrics (AMD 6199) September 29, 1989. 11 pp.
- BSI BS 7304: Part 3-90. (WITHDRAWN) 1990 Methods of Test for Nonwovens Part 3: Determination of Tensile Strength and Elongation (Renumbered as BS EN 29073-3: 1992). 8 pp.
- CGSB CAN/CGSB-4.2 NO.9.1-M90. Textile Test Methods Breaking Strength of Fabrics—Strip Method—Constant-Time-to-Break Principle. 11 pp.
- CGSB CAN/CGSB-4.2 NO.9.3-M90. Textile Test Methods Breaking Strength of High-Strength Fabrics—Constant-Time-to-Break Principle. 13 pp.
- CGSB CAN/CGSB-4.2 NO.10-M87. Textile Test Methods Elongation. 9 pp.
- ISO 5081-77. Textiles—Woven Fabrics—Determination of Breaking Strength and Elongation (Strip Method) First Edition. 10 pp.

—Fibers
- ISO 5079-77. Textiles—Man-Made Fibres—Determination of Breaking Strength and Elongation of Individual Fibres First Edition. 6 pp.

—Fishing Nets
- BSI BS 5414-76. 1976 Determination of Elongation of Netting Yarns for Fishing Nets. 7 pp.
- ISO 3790-76. Fishing Nets—Determination of Elongation of Netting Yarns First Edition. 6 pp.

—Geotextiles
- CGSB CAN/CGSB-148. 1 NO.7.3-92. Methods of Testing Geotextiles and Geomembranes Grab Tensile Test for Geotextiles. 9 pp.

—Glass
- BSI BS 7034: Part 3-88. 1988 Viscosity and Viscometric Fixed Points of Glass Part 3: Method for the Determination of Viscosity by Fibre Elongation Viscometer (ISO 7884/3: 1987). 11 pp.
- ISO 7884 Pt 3-87. Glass—Viscosity and Viscometric Fixed Points—Part 3: Determination of Viscosity by Fibre Elongation Viscometer First Edition. 11 pp.

—Glass Fabrics
- ISO 4606-79. Textile Glass—Woven Fabric—Determination of Tensile Breaking Force and Breaking Elongation by the Strip Method First Edition. 7 pp.

—Insulated Wire
- DIN VDE 0472 Pt 625-83. Testing of Cables and Insulated Conductors; Tensile and Elongation Characteristics of Strain Bearing Elements (Jan). 5 pp.

—Leather
- CNS K6125-81. Method of Test for Elongation Rate of Leather (Aug)(1281).
- ISO 3376-76. Leather—Determination of Tensile Strength and Elongation First Edition. 5 pp.

Elongation (Cont.)

—Nonwoven Fabrics
- BSI BS EN 29073-3-92. 1992 Amd 1 Methods of Test for Nonwovens Part 3: Determination of Tensile Strength and Elongation (ISO 9073-3: 1989) (AMD 7010) November 15, 1992. 14 pp.
- CEN EN 29073-3-92. Textiles—Test Methods for Nonwovens—Part 3: Determination of Tensile Strength and Elongation. 4 pp.
- DIN ENGL EN 29073 Pt 3-92. Textiles; Test Methods for Nonwovens; Determination of Tensile Strength and Elongation (ISO 9073-3: 1989) (Aug). 5 pp.
- ISO 9073 Pt 3-89. Textiles—Test Methods for Nonwovens—Part 3: Determination of Tensile Strength and Elongation First Edition. 6 pp.

—Paperboard
- BSI BS 4415: Part 2-86. 1986 Tensile Properties of Paper and Board Part 2: Constant Rate of Elongation Method. 10 pp.
- ISO 1924 Pt 2-85. Paper and Board—Determination of Tensile Properties—Part 2: Constant Rate of Elongation Method First Edition. 8 pp.

—Papers
- BSI BS 4415: Part 2-86. 1986 Tensile Properties of Paper and Board Part 2: Constant Rate of Elongation Method. 10 pp.
- ISO 1924 Pt 2-85. Paper and Board—Determination of Tensile Properties—Part 2: Constant Rate of Elongation Method First Edition. 8 pp.

—Plastic Gutters
- BSI BS 2782:Pt11: METH 1110-89. 1989 Methods of Testing Plastics Part 11: Thermoplastics Pipes, Fittings and Valves Method 1110: Tensile Properties of Dumb-Bell Specimens from PVC Gutter Profiles or Pipes for Non-Pressure Applications. 5 pp.

—Plastic Pipes
- BSI BS 2782:Pt11: METH 1110-89. 1989 Methods of Testing Plastics Part 11: Thermoplastics Pipes, Fittings and Valves Method 1110: Tensile Properties of Dumb-Bell Specimens from PVC Gutter Profiles or Pipes for Non-Pressure Applications. 5 pp.
- CNS K6139-60. Method of Test of Soft Polyvinyl Chloride Pipes (Apr)(1297)(R 1968). 3 pp.

—Plastic Sheets
- CNS K6160-62. Method of Test for Soft Polyvinyl Chloride Sheet and Film (Apr)(1441) (R 1973). 6 pp.

—Plastics
- BSI BS 2782:Pt3: METH 320A-F-76. 1976 Methods of Testing Plastics Part 3: Mechanical Properties Method 320A-F: Determination of Tensile Strength, Elongation and Elastic Modulus. 11 pp.
- BSI BS 2782:Pt3: METH 326A-C-77. 1977 Methods of Testing Plastics Part 3: Mechanical Properties Method 326A-C: Determination of Tensile Strength and Elongation of Plastics Films. 4 pp.
- BSI BS 2782:Pt3: METH 327A-93. 1993 Methods of Testing Plastics Part 3: Mechanical Properties Method 327A: Determination of Tensile Strength and Elongation at Break of Polytetrafluoroetheylene (PTFE) Products. 8 pp.
- BSI BS 2782:Pt3: METH 327A-82. 1982 Methods of Testing Plastics Part 3: Mechanical Properties Method 327A: Determination of Tensile Strength and Elongation at Break of Polytetrafluoroetheylene (PTFE) Products. 5 pp.

—Polyurethane Resins—Athletic Fields
- CNS K6591-86. Method of Test for Polyurethane (PU) Athletic Installation Material (May)(6483). 6 pp.

—Rubber
- DIN ENGL 53504-85. Testing of Vulcanized Rubber; Determination of Tensile Strength at Break, Tensile Strength, Elongation at Break and Stress Values in a Tensile Test (Mar). 5 pp.

—Thread (Textiles)
- CGSB CAN/CGSB-4.2 NO.10-M87. Textile Test Methods Elongation. 9 pp.

—Vulcanized Rubber
- BSI BS 903: Part A52-86. 1986 Methods of Testing Vulcanized Rubber Part A52: Determination of Aging Characteristics by Measurement of Stress at a Given Elongation. 8 pp.
- CNS K6354-85. Method of Test for Low Elongation Stress of Vulcanized Rubber (Dec)(3563). 6 pp.
- ISO 6914-85. Rubber, Vulcanized—Determination of Ageing Characteristics by Measurement of Stress at a Given Elongation First Edition. 6 pp.

—Yarns
- CGSB CAN/CGSB-4.2 NO.10-M87. Textile Test Methods Elongation. 9 pp.

Elongation (Cont.)

—Yarns—Glass Fabrics
- ISO 3341-84. Textile Glass—Yarns—Determination of Breaking Force and Breaking Elongation Second Edition. 7 pp.

Elongation at Break
Use: Elongation

Elongation Index
Use: Elongation

Elongation Testing
Use: Elongation

Elongation Values
Use: Elongation

ELT
Use: Crash Locator Beacons

Elutriation

—Particle Size Distribution—Powders
- BSI BS 3406: Part 3-63. 1963 Amd 1 Determination of Particle Size Distribution Part 3: Air Elutriation Methods (AMD 4537) April 30, 1984. 35 pp.
- CEN PREN 3640-88. Test Methods Heat Resisting Alloys Atomized Powder Procedure for the Inspection of Loose Powder Cleaners by Water Elutriation. 7 pp.
- CEN PREN 3640-92. Test Methods Heat Resisting Alloys Atomized Powder Procedure for the Inspection of Loose Powder Cleanness by Water Elutriation. 8 pp.

EM Fields
Use: Electromagnetic Fields

Embarkation

—Amphibious Operations
- NATO STANAG 1195 ED 3 AMD 0-92. Amphibious Embarkation—ATP-39(A). 4 pp.

Embarkation Ladders
Use: Ladders

Embraer Bandeirante EMB-P1 Aircraft
See Also: Aircraft

—Foreign Airworthiness Directives
- CAA. Embraer Bandeirante EMB-110K1, P1 and P2 (Foreign Airworthiness Directives). 30 pp.

Embraer Bandeirante EMB-P2 Aircraft
See Also: Aircraft

—Foreign Airworthiness Directives
- CAA. Embraer Bandeirante EMB-110K1, P1 and P2 (Foreign Airworthiness Directives). 30 pp.

Embraer Bandeirante EMB-110 Aircraft
See Also: Aircraft

—Accidents
- CAA. Embraer EMB—110 Bandeirante (World Airline Accident Summary). 1 p.

—Foreign Airworthiness Directives
- CAA. Embraer Bandeirante EMB—110 Series Aircraft. 21 pp.

Embraer Bandeirante EMB-110K1 Aircraft
See Also: Aircraft

—Foreign Airworthiness Directives
- CAA. Embraer Bandeirante EMB-110K1, P1 and P2 (Foreign Airworthiness Directives). 30 pp.

Embraer Brasilia EMB-120 Aircraft
See Also: Aircraft

—Antenna Positions
- CAA. Embraer 120 Brasilia. 1 p.

—Certification
- CAA. Embraer EMB-120 and-120 RT Brasilia. 12 pp.

Embraer Brasilia EMB-120RT Aircraft
See Also: Aircraft

—Certification
- CAA. Embraer EMB-120 and-120 RT Brasilia. 12 pp.

Embraer

INTERNATIONAL AND NON-U.S. NATIONAL STANDARDS
SUBJECT INDEX

Embraer EMB-110 Aircraft
See Also: Aircraft
—**Data Sheets**
CAA FA20 ISSUE 2. Embraer EMB-110. 4 pp.

Embraer EMB-110 Bandeirante Aircraft
See Also: Aircraft
—**Accidents**
CAA. Embraer EMB-110 Bandeirante. 1 p.

Embraer EMB-120 Aircraft
See Also: Aircraft
—**Data Sheets**
CAA FA37 ISSUE 1. Embraer, EMB-120. 2 pp.
—**Foreign Airworthiness Directives**
CAA. Embraer EMB—120 Series Aircraft. 18 pp.

Embraer EMB-120 Brasilia Aircraft
See Also: Aircraft
—**Accidents**
CAA. Embraer EMB-120 Brasitia (World Airline Accident Summary). 1 p.

Embraer EMB-121 Aircraft
See Also: Aircraft
—**Data Sheets**
CAA FA22 ISSUE 1. Embraer EMB-121. 3 pp.

Embraer Xingu EMB-121 Aircraft
See Also: Aircraft
—**Certification**
CAA. Embraer Xingu EMB-121. 6 pp.

Embraer 110 P1 Aircraft
See Also: Aircraft
—**Antenna Positions**
CAA. Embraer 110P1 and 110P2 (Approved Aerial Positions). 1 p.

Embraer 110 P2 Aircraft
See Also: Aircraft
—**Antenna Positions**
CAA. Embraer 110P1 and 110P2 (Approved Aerial Positions). 1 p.

Embraer 121 Aircraft
See Also: Aircraft
—**Antenna Positions**
CAA. Embraer 121 (Approved Aerial Positions). 1 p.

Embrittlement, Hydrogen
Use: Hydrogen Embrittlement

EMC
Use: Electromagnetic Compatibility

EMDS
Use: Early Mobile Data Services

Emeraude (Piel) Aircraft
See Also: Aircraft
—**Foreign Airworthiness Directives**
CAA. Piel Emeraude Series Aircraft (Foreign Airworthiness Directives). 4 pp.

Emeraude (Rousseau) Aircraft
See Also: Aircraft
—**Foreign Airworthiness Directives**
CAA. Piel Emeraude Series Aircraft (Foreign Airworthiness Directives). 4 pp.

Emeraude (Scintex) Aircraft
See Also: Aircraft
—**Foreign Airworthiness Directives**
CAA. Piel Emeraude Series Aircraft (Foreign Airworthiness Directives). 4 pp.

Emergency Alighting
Use: Aircraft Landings

Emergency and Backup Power Supplies
Use: Emergency and Standby Power Supplies

Emergency and Standby Power Supplies
Use For: Automatic Standby Power Plants; Backup Power Supplies; Emergency Electrical Installations; Emergency Generating Stations *See Also:* Automatic Transfer Switches; Auxiliary Power Units; Electrical Installations; Emergency Planning; Emergency Relief Operations; Essential Electrical Systems; Power Supplies; Uninterruptible Power Supplies
DIN VDE 0100 Pt 728-90. Erection of Power Installations with Rated Voltages up to 1000 V; Standby Power Supply Installations (Mar). 11 pp.

—**Automatic Transfer Switches**
BSI BS EN 60947-6-1-92. 1992 Low-Voltage Switchgear and Controlgear Part 6: Multiple Function Equipment Section 1: Automatic Transfer Switching Equipment. 46 pp.
BSI BS EN 60947-6-1-01. 1992 Amd 1 Low-Voltage Switchgear and Controlgear Part 6: Multiple Function Equipment Section 1: Automatic Transfer Switching Equipment (AMD 7851) July 15, 1993 (E). 47 pp.
CENELEC EN 60947-6-1-91. Low-Voltage Switchgear and Controlgear Part 6: Multiple Function Equipment Section One—Automatic Transfer Switching Equipment. 6 pp.
IEC 947 Pt 6-1-89. Low-Voltage Switchgear and Controlgear Part 6: Multiple Function Equipment Section One—Automatic Transfer Switching Equipment First Edition. 81 pp.

—**Buildings**
CSA CAN/CSA-C282-M89. Emergency Electrical Power Supply for Buildings; (Gen Instr 1). 36 pp.
DIN VDE 0108 Pt 1-89. Power Installations and Safety Power Supply in Communal Buildings; General (Oct) (VDE 0107/ 11.89 with DIN VDE 0108 Part 1 to Part 8 (Issues of 10.89) Supersedes DIN 57108/VDE 0108/12.79). 76 pp.
DIN VDE 0108 Pt 1 Suppl 1-89. Power Installations and Safety Power Supply in Communal Buildings; Building Regulations (Oct). 36 pp.
DIN VDE 0108 Pt 2-89. Power Installations and Safety Power Supply in Communal Buildings; Communal Facilities (Oct) (VDE 0107/11.89 with DIN VDE 0108 Part 1 to Part 8 (Issues of 10.89) Supersedes DIN 57 108/VDE 0108/12.79). 20 pp.
DIN VDE 0108 Pt 4-89. Power Installations and Safety Power Supply in Communal Buildings; High-Rise Buildings (Oct) (VDE 0107/11.89 with DIN VDE 0108 Part 1 to Part 8 (Issues of 10.89) Supersedes DIN 57108/VDE 0108/12.79). 5 pp.
DIN VDE 0108 Pt 7-89. Power Installations and Safety Power Supply in Communal Buildings; Working and Business Premises (Oct) (VDE 0107/ 11.89 with DIN VDE 0108 Part 1 to Part 8 (Issues of 10.89) Supersedes DIN 57108/VDE 0108/12.79). 9 pp.
DIN VDE 0108 Pt 8-89. Power Installations and Safety Power in Communal Buildings; Temporary Bldgs. Used in Communal Facilities, Stores and Shops, Exhibition Rooms, Public Houses and Restaurants (Oct) (VDE 0107/11.89 with DIN VDE 0108 Pt 1 to Pt 8(Issues of 10.89) Supersedes DIN 57108/VDE 0108/12.79). 11 pp.
SNZ NZS 6104-81. Specification for Emergency Electricity Supply in Buildings (Reconfirmed 1991). 27 pp.

—**Exhibition Facilities**
DIN VDE 0108 Pt 3-89. Power Installations and Safety Power Supply in Communal Buildings; Stores and Shops and Exhibition Rooms (Oct) (VDE 0107/ 11.89 with DIN VDE 0108 Part 1 to Part 8 (Issues of 10.89) Supersedes DIN 57108/VDE 0108/12.79). 9 pp.
DIN VDE 0108 Pt 8-89. Power Installations and Safety Power in Communal Buildings; Temporary Bldgs. Used in Communal Facilities, Stores and Shops, Exhibition Rooms, Public Houses and Restaurants (Oct) (VDE 0107/11.89 with DIN VDE 0108 Pt 1 to Pt 8(Issues of 10.89) Supersedes DIN 57108/VDE 0108/12.79). 11 pp.

—**Fire Fighting**
CNS Z2054-83. Cubicle Emergency Power Source for Fire Fighting Use (Nov)(10673).
CNS Z3023-83. Method of Test for Fire Fighting Emergency Power Source (Apr)(10204).

—**Ground Effect Machines**
CAA Chapter B6-7 01.91. Electrical Systems. 8 pp.
CAA Chapter B6-7 App 08.83. Electrical Systems. 1 p.

—**Gyrohorizons**
CAA NOTICE #81 ISSUE 1. Emergency Power Supply for Electrically Operated Gyroscopic Bank and Pitch Indicators (Artificial Horizons) (Airworthiness Notices). 5 pp.

Emergency and Standby Power Supplies *(Cont.)*
—**Hotels**
DIN VDE 0108 Pt 5-89. Power Installations and Safety Power Supply in Communal Buildings; Restaurants (Oct) (VDE 0107/11.89 with DIN VDE 0108 Part 1 to Part 8 (Issues of 10.89) Supersedes DIN 57108/VDE 0108/12.79.). 7 pp.

—**Nuclear Power Plants**
CSA CAN/CSA-N290.5-M90. Requirements for the Support Power Systems of CANDU Nuclear Power Plants; (Gen Instr 1). 39 pp.

—**Parking Facilities**
DIN VDE 0108 Pt 6-89. Power Installations and Safety Power Supply in Communal Buildings; Enclosed Car-Parks (Oct) (VDE 0107/11.89 with DIN VDE 0108 Part 1 to Part 8 (Issues of 10.89) Supersedes DIN 57108/VDE 0108/12.79.). 7 pp.

—**Restaurants**
DIN VDE 0108 Pt 5-89. Power Installations and Safety Power Supply in Communal Buildings; Restaurants (Oct) (VDE 0107/11.89 with DIN VDE 0108 Part 1 to Part 8 (Issues of 10.89) Supersedes DIN 57108/VDE 0108/12.79.). 7 pp.
DIN VDE 0108 Pt 8-89. Power Installations and Safety Power in Communal Buildings; Temporary Bldgs. Used in Communal Facilities, Stores and Shops, Exhibition Rooms, Public Houses and Restaurants (Oct) (VDE 0107/11.89 with DIN VDE 0108 Pt 1 to Pt 8(Issues of 10.89) Supersedes DIN 57108/VDE 0108/12.79). 11 pp.

—**Retail Stores**
DIN VDE 0108 Pt 3-89. Power Installations and Safety Power Supply in Communal Buildings; Stores and Shops and Exhibition Rooms (Oct) (VDE 0107/ 11.89 with DIN VDE 0108 Part 1 to Part 8 (Issues of 10.89) Supersedes DIN 57108/VDE 0108/12.79). 9 pp.
DIN VDE 0108 Pt 8-89. Power Installations and Safety Power in Communal Buildings; Temporary Bldgs. Used in Communal Facilities, Stores and Shops, Exhibition Rooms, Public Houses and Restaurants (Oct) (VDE 0107/11.89 with DIN VDE 0108 Pt 1 to Pt 8(Issues of 10.89) Supersedes DIN 57108/VDE 0108/12.79). 11 pp.

—**Switchboards**
CNS Z2056-84. Main and Branch Distribution Switchboards for Emergency Power Source (Jul)(10977).

—**Telecommunication Networks**
CCIR Report 992-86. HF Network for Relief Operations and Related Equipment Characteristics—Section 3Ad—Operational Questions. 5 pp.

—**Telecommunication Systems**
CCIR Report 992-86. HF Network for Relief Operations and Related Equipment Characteristics—Section 3Ad—Operational Questions. 5 pp.

Emergency Call Services
See Also: Service Telecommunications; Telephone Services
CCITT RECMN E.140-89. Principles for the Operation of International Telephone Services—Telephone Network and ISDN—Operation, Numbering, Routing and Mobile Service (Study Group II) 3 pp. 3 pp.

—**Symbols**
CCITT RECMN E.121-89. Pictograms and Symbols to Assist Users of the Telephone Service—Telephone Network and ISDN—Operation, Numbering, Routing and Mobile Service (Study Group II) 11 pp. 11 pp.

Emergency Electrical Installations
Use: Emergency and Standby Power Supplies

Emergency Exits
See Also: Aircraft Safety; Doors; Windows
—**Aircraft—Break-In Points**
CAA Spec. NO. 7 ISSUE 2.
—**Aircraft Cabins**
CAA Chapter K4-3 04.74. Compartment Design and Safety Provisions (Light Aeroplanes). 3 pp.
—**Aircraft—Signs**
NATO STANAG 3230 ED 4 AMD 4-84. Emergency Markings on Aircraft. 9 pp.
NATO STANAG 3230 ED 4 AMD 5-84. Emergency Markings on Aircraft. 10 pp.
—**Fasteners**
BSI BS 5725: Part 1-81. 1981 Emergency Exit Devices Part 1: Panic Bolts and Panic Latches Mechanically Operated by a Horizontal Push-Bar. 16 pp.

INTERNATIONAL AND NON-U.S. NATIONAL STANDARDS
SUBJECT INDEX
Emergency

Emergency Exits (Cont.)
—**Ground Effect Machines**
CAA Chapter B5-6 01.91. Exits. 4 pp.
CAA Chapter B5-6 App 11.72. Exits. 3 pp.

—**Rotary Wing Aircraft**
CAA Chapter G4-2 12.80. Flight Crew Compartment Design (Rotocraft). 3 pp.
CAA CAP 524 SUB-Part D 12.86. Design and Construction (Rotorcraft). 23 pp.

Emergency Flasher Units
Use: Flasher Units

Emergency Generating Stations
Use: Emergency and Standby Power Supplies

Emergency Lighting
Use For: Battery Powered Emergency Lighting; Standby Lighting *See Also:* Beacon Lights; Emergency Planning; Lighting
BSI BS 4533: Sec 102.22-90. 1990 Luminaires Part 102: Particular Requirements Section 102.22: Luminaires for Emergency Lighting. 29 pp.
BSI BS 5266: Part 1-88. 1988 Emergency Lighting Part 1: Code of Practice for the Emergency Lighting of Premises Other Than Cinemas and Certain Other Specified Premises Used for Entertainment. 24 pp.
BSI BS 5266: Part 1-75. 1975 Emergency Lighting Part 1: Code of Practice for the Emergency Lighting of Premises Other Than Cinemas and Certain Other Specified Premises Used for Entertainment. 23 pp.
CENELEC EN 60598-2-22-90. Luminaires Part 2: Particular Requirements Section Twenty-Two: Luminaires for Emergency Lighting. 13 pp.
CNS C4348-85. Automatic Lights for Emergent Service (Dec)(8802).
IEC 598 Pt 2-22-90. Luminaires Part 2: Particular Requirements Section Twenty-Two—Luminaires for Emergency Lighting Second Edition. 41 pp.

—**Aircraft**
CAA NOTICE #42 ISSUE 1. Internal Emergency Lighting Systems (Airworthiness Notices). 2 pp.
CAA NOTICE #56 ISSUE 4. Floor Proximity Emergency Escape Path Marking (Airworthiness Notices). 6 pp.

—**Aircraft Cabins**
NATO STANAG 3870 ED 1 AMD 2-86. Emergency Escape/Evacuation Lighting. 10 pp.

—**Buildings**
DIN ENGL 5035 Pt 5-87. Artificial Lighting of Interiors; Emergency Lighting (Dec). 7 pp.
SNZ NZS 6742-71. Code of Practice for Emergency Lighting in Buildings. 18 pp.

—**Construction**
CSA C22.2 NO 141-M1985. Unit Equipment for Emergency Lighting (R 1992); (Gen Instr 1). 31 pp.
CSA 902B Bull. Electrical Bulletin 902B September 27, 1978 to C22.2 NO 141. 2 pp.

—**Contactors**
BSI BS 764-90. 1990 Automatic Change-Over Contactors for Emergency Lighting Systems. 12 pp.

—**Exhibition Facilities**
DIN VDE 0108 Pt 3-89. Power Installations and Safety Power Supply in Communal Buildings; Stores and Shops and Exhibition Rooms (Oct) (VDE 0107/11.89 with DIN VDE 0108 Part 1 to Part 8 (Issues of 10.89) Supersedes DIN 57108/VDE 0108/12.79). 9 pp.

—**Flammability Testing**
CSA C22.2 NO 141-M1985. Unit Equipment for Emergency Lighting (R 1992); (Gen Instr 1). 31 pp.
CSA 902B Bull. Electrical Bulletin 902B September 27, 1978 to C22.2 NO 141. 2 pp.

—**Ground Effect Machines**
CAA Chapter B5-6 01.91. Exits. 4 pp.
CAA Chapter B6-7 01.91. Electrical Systems. 8 pp.
CAA Chapter B6-7 App 08.83. Electrical Systems. 1 p.

—**Indoor**
CSA C22.2 NO 141-M1985. Unit Equipment for Emergency Lighting (R 1992); (Gen Instr 1). 31 pp.
CSA 902B Bull. Electrical Bulletin 902B September 27, 1978 to C22.2 NO 141. 2 pp.

—**Parking Facilities**
DIN VDE 0108 Pt 6-89. Power Installations and Safety Power Supply in Communal Buildings; Enclosed Car-Parks (Oct) (VDE 0107/11.89 with DIN VDE 0108 Part 1 to Part 8 (Issues of 10.89) Supersedes DIN 57108/VDE 0108/12.79.). 7 pp.

Emergency Lighting (Cont.)
—**Photometry**
BSI BS 5225: Part 3-82. 1982 Photometric Data for Luminaires Part 3: Method of Photometric Measurement of Battery-Operated Emergency Lighting Luminaires. 11 pp.

—**Power Relays**
BSI BS 5266: Part 3-81. 1981 Emergency Lighting Part 3: Small Power Relays (Electromagnetic) for Emergency Lighting Applications up to and Including 32 A. 10 pp.

—**Restaurants**
DIN VDE 0108 Pt 5-89. Power Installations and Safety Power Supply in Communal Buildings; Restaurants (Oct) (VDE 0107/11.89 with DIN VDE 0108 Part 1 to Part 8 (Issues of 10.89) Supersedes DIN 57108/VDE 0108/12.79.). 7 pp.

—**Retail Stores**
DIN VDE 0108 Pt 3-89. Power Installations and Safety Power Supply in Communal Buildings; Stores and Shops and Exhibition Rooms (Oct) (VDE 0107/11.89 with DIN VDE 0108 Part 1 to Part 8 (Issues of 10.89) Supersedes DIN 57108/VDE 0108/12.79). 9 pp.

Emergency Locator Transmitters
—**Aircraft—Rescue Equipment**
NATO STANAG 3650 ED 2 AMD 2-88. Essential Sar Location Equipment and Associated Characteristics (Aircraft). 5 pp.

—**Electromagnetic Compatibility**
CCIR Report 1101-90. Preliminary Studies of Radio Interference Limits for MAC/Packet Decoders—Section 1D—Spectrum Utilization and Applications. 27 pp.

Emergency Medical Services
See Also: Emergency Relief Operations

—**Interdepartmental Procurement**
NATO STANAG 2127 ED 3 AMD 1-89. Medical, Surgical and Dental Instruments, Equipment and Supplies. 10 pp.

—**Military—Distribution Systems**
NATO STANAG 2105 ED 5 AMD 1-86. NATO Table of Military Equivalents (AMedP-1(D)). 8 pp.

—**Military—Training**
NATO STANAG 2122 ED 2 AMD 4-75. Medical Training in First-Aid, Basic Hygiene and Emergency Care. 12 pp.
NATO STANAG 2122 ED 2 AMD 5-75. Medical Training in First-Aid, Basic Hygiene and Emergency Care. 12 pp.

—**War Surgery**
NATO STANAG 2068 ED 4 AMD 1-86. Emergency War Surgery. 5 pp.
NATO STANAG 2068 ED 4 AMD 3-86. Emergency War Surgery. 5 pp.

Emergency Planning
See Also: Emergency and Standby Power Supplies; Emergency Lighting; Emergency Relief Operations

—**Aircraft**
NATO STANAG 3465 ED 4 AMD 10-73. Safety, Emergency and Signalling Procedures for Military Air Movement—Fixed Wing Aircraft. 15 pp.

—**Airports**
CAA Part 8 CAP 168. Rescue and Fire Fighting Services. 29 pp.
CAA Part 9 CAP 168. Medical Services. 4 pp.
ICAO 9137 Part 7. Airport Services Manual Part 7 Airport Emergency Planning First Edition—1980. 76 pp.

—**Airports—Training**
CAA Part 8 CAP 168. Rescue and Fire Fighting Services. 29 pp.

—**Hazardous Materials—Aircraft**
ICAO 9481. Emergency Response Guidance for Aircraft Incidents Involving Dangerous Goods 1993-1994 Edition. 78 pp.

—**Industrial**
CSA CAN/CSA-Z731-M91. Emergency Planning for Industry; (Gen Instr 1). 64 pp.

—**War Burials**
NATO STANAG 2070 ED 3 AMD 3-74. Emergency War Burial Procedures. 13 pp.

Emergency Position Indicating Radio Beacons
Use For: EPIRB *See Also:* Crash Locator Beacons

Emergency Position Indicating Radio Beacons (Cont.)
CCIR RECMN 439-3-82. Emergency Position-Indicating Radio Beacons Operating at the Frequency 2182 kHz—Section 8D—Radiodetermination, Global Maritime Distress and Safety System and Related Subjects. 2 pp.
CENELEC PRETS 300 066-90. Float-Free Maritime Satellite Emergency Position-Indicating Radio Beacons (EPIBRs) Operating on 406.025 MHz Technical Characteristics and Methods of Measurement. 41 pp.
CENELEC PRETS 300 066-92. Radio Equipment and Systems (RES); Float-Free Maritime Satellite; Emergency Position Indicating Radio Beacons (EPIRBs) Operating on 406,025 MHz; Technical Characteristics and Methods of Measurement. 36 pp.
CENELEC ETS 300 066-92. Radio Equipment and Systems (RES); Float-Free Maritime Satellite Emergency Position Indicating Radio Beacons (EPIRBs) Operating on 406,025 MHz; Technical Characteristics and Methods of Measurement. 36 pp.
ETSI PRETS 300 066-90. Float-Free Maritime Satellite Emergency Position—Indicating Radio Beacons (EPIRBs) Operating on 406.025 MHz Technical Characteristics and Methods of Measurement. 41 pp.
ETSI ETS 300 066-92. Radio Equipment and Systems (RES); Float-Free Maritime Satellite Emergency Position Indicating Radio Beacons (EPIRBs) Operating on 406,025 MHz; Technical Characteristics and Methods of Measurement. 36 pp.
ETSI PRETS 300 066-92. Radio Equipment and Systems (RES); Float-Free Maritime Satellite; Emergency Position Indicating Radio Beacons (EPIRBs) Operating on 406,025 MHz; Technical Characteristics and Methods of Measurement. 36 pp.
ETSI PRETS 300 066-90. Float-Free Maritime Satellite Emergency Position-Indicating Radio Beacons (EPIBRs) Operating on 406.025 MHz Technical Characteristics and Methods of Measurement. 41 pp.
NATO STANAG 3281 ED 5 AMD 6-78. Personal Locator Beacons. 10 pp.

—**Carrier Frequencies**
CCIR RECMN 690-90. Transmission Characteristics of Emergency Position-Indicating Radio Beacons (EPIRBs) Operating on Carrier Frequencies of 121.5 MHz and 243 MHz—Section 8D—Radiodetermination, Global Maritime Distress and Safety System and Related Subjects. 1 p.

—**Electromagnetic Compatibility**
CCIR Report 1101-90. Preliminary Studies of Radio Interference Limits for MAC/Packet Decoders—Section 1D—Spectrum Utilization and Applications. 27 pp.

—**Identification Signals**
CCIR Report 749-3-90. Future Use and Characteristics of Emergency Position-Indicating Radio Beacons in the Mobile and Mobile-Satellite Services—Section 8D—Radiodetermination, Global Maritime Distress and Safety System and Related Subjects. 7 pp.

—**Maritime Mobile Services**
CCIR Report 749-3-90. Future Use and Characteristics of Emergency Position-Indicating Radio Beacons in the Mobile and Mobile-Satellite Services—Section 8D—Radiodetermination, Global Maritime Distress and Safety System and Related Subjects. 7 pp.
CCIR QUESTION 31-4/8-90. Future Use and Characteristics of Emergency Position-Indicating Radio Beacons in the Mobile Service—Questions Concerning Study Group 8—Mobile, Radiodetermination, Amateur and Related Satellite Services. 1 p.

—**Maritime Mobile Services—VHF**
CCIR Report 749-3-90. Future Use and Characteristics of Emergency Position-Indicating Radio Beacons in the Mobile and Mobile-Satellite Services—Section 8D—Radiodetermination, Global Maritime Distress and Safety System and Related Subjects. 7 pp.

—**Mobile Satellite Communications**
CCIR Report 749-3-90. Future Use and Characteristics of Emergency Position-Indicating Radio Beacons in the Mobile and Mobile-Satellite Services—Section 8D—Radiodetermination, Global Maritime Distress and Safety System and Related Subjects. 7 pp.
CCIR QUESTION 31-4/8-90. Future Use and Characteristics of Emergency Position-Indicating Radio Beacons in the Mobile Service—Questions Concerning Study Group 8—Mobile, Radiodetermination, Amateur and Related Satellite Services. 1 p.
CCIR QUESTION 90/8-88. Technical and Operating Characteristics of Systems Providing Radiocommunication Using Satellite Techniques for Distress and Safety Operations—Questions Concerning Study Group 8 —Mobile, Radiodetermination, Amateur and Related Satellite Services. 1 p.

INDUSTRY STANDARDS

INTERNATIONAL AND NON-U.S. NATIONAL STANDARDS
SUBJECT INDEX

Emergency

Emergency Position Indicating Radio Beacons *(Cont.)*

—**Mobile Satellite Communications—Interference—Monitoring**

CCIR Report 979-1-90. Monitoring Techniques for Detecting and Locating Sources of Interference Likely to Disturb the Satellite Emergency Position-Indicating Radio Beacon System of the Mobile-Satellite (Earth-to-Space) Service in the Band 406-406.1 MHz—Section 1C—Spectrum Monitoring Tech. 5 pp.

CCIR QUESTION 63-1/1-86. Monitoring of the Band 406-406.1 MHz—Questions Concerning Study Group 1—Spectrum Management Techniques (Spectrum Engineering, Planning, Sharing, Monitoring and Utilization). 1 p.

—**Modulation**

CCIR Report 749-3-90. Future Use and Characteristics of Emergency Position-Indicating Radio Beacons in the Mobile and Mobile-Satellite Services—Section 8D—Radiodetermination, Global Maritime Distress and Safety System and Related Subjects. 7 pp.

—**Radio Spectra—Measurement**

CCIR Report 1162-90. Power Distribution Requirements for EPIRBs Operating on the Frequencies of 121.5 MHz and 243 MHz—Section 8D—Radiodetermination, Global Maritime Distress and Safety System and Related Subjects. 12 pp.

—**SHF—Global Maritime Distress and Safety Systems**

CCIR Report 1036-1-90. Frequencies for Homing and Locating in the Global Maritime Distress and Safety System (GMDSS)—Section 8D—Radiodetermination, Global Maritime Distress and Safety System and Related Subjects. 23 pp.

—**Transmission—Radio Signals**

CENELEC PRETS 300 152-91. Radio Equipment and Systems Maritime Emergency Position Indicating Radio Beacons (EPIRBs) Intended for the Use on the Frequency 121.5 MHz or the Frequencies 121.5 MHz and 243 MHz for Homing Purposed Only Technical Characteristics and Methods of Measurement. 17 pp.

CENELEC ETS 300 152-91. Radio Equipment and Systems; Maritime Emergency Position Indicating Radio Beacons (EPIRBs) Intended for Use on the Frequency 121,5 MHz or the Frequencies 121,5 MHz and 243 MHz for Homing Purposes Only Technical Characteristics and Methods of Measurement. 19 pp.

ETSI PRETS 300 152-91. Final Draft—Radio Equipment and Systems—Maritime Emergency Position Indicating Radio Beacons (EPIRBs) Intended for Use on the Frequency 121.5 MHz or the Frequencies 121.5 MHz and 243 MHz for Homing Purposes Only—Technical Characteristics.

ETSI ETS 300 152-91. Radio Equipment and Systems; Maritime Emergency Position Indicating Radio Beacons (EPIRBs) Intended for Use on the Frequency 121.5 MHz or the Frequencies 121,5 MHz and 243 MHz for Homing Purposes Only Technical Characteristics and Methods of Measurement. 19 pp.

—**UHF—Encoded Data Transmission**

CCIR RECMN 633-1-90. Transmission Characteristics of a Satellite Emergency Position-Indicating Radiobeacon (Satellite EPIRB) System Operating Through a Low Polar-Orbiting Satellite System in the 406 MHz Band—Section 8J—Technical and Operating Characteristics of. 17 pp.

—**UHF—Geostationary Satellites**

CCIR RECMN 632-1-90. Transmission Characteristics of a Satellite Emergency Position-Indicating Radiobeacon (Satellite EPIRB) System Operating Through Geostationary Satellites in the 1.6 GHz Band—Section 8J—Technical and Operating Characteristics of Radiocommunications. 7 pp.

—**UHF—Global Maritime Distress and Safety Systems**

CCIR Report 1033-1-90. VHF Radiotelephone Systems with Automatic Facilities for the Maritime Mobile Service—Section 8C—Maritime Mobile Service; Telephony and Related Subjects. 6 pp.

—**UHF—INMARSAT**

CCIR RECMN 632-1-90. Transmission Characteristics of a Satellite Emergency Position-Indicating Radiobeacon (Satellite EPIRB) System Operating Through Geostationary Satellites in the 1.6 GHz Band—Section 8J—Technical and Operating Characteristics of Radiocommunications. 7 pp.

Emergency Position Indicating Radio Beacons *(Cont.)*

—**UHF—INMARSAT—Duty Cycles**

CCIR RECMN 632-1-90. Transmission Characteristics of a Satellite Emergency Position-Indicating Radiobeacon (Satellite EPIRB) System Operating Through Geostationary Satellites in the 1.6 GHz Band—Section 8J—Technical and Operating Characteristics of Radiocommunications. 7 pp.

—**UHF—INMARSAT—Geostationary Satellites**

CCIR Report 1184-90. Pre-Operational Demonstrations of the 1.6 GHz Satellite EPIRB System Using the Inmarsat Geostationary Space Segment—Section 8J—Technical and Operating Characteristics of Radiocommunications Using Satellite Distress and Safety Operation and. 8 pp.

—**UHF—Mobile Satellite Communications**

CCIR Report 761-3-90. Technical and Operating Characteristics of Distress Systems in the Mobile-Satellite Service —Section 8J—Technical and Operating Characteristics of Radiocommunications Using Satellite Distress and Safety Operation and of Radio Determination Satellite Services. 19 pp.

CCIR Report 1045-1-90. Satellite EPIRB Coordinated Trials Programme and Pre-Operational Demonstrations Using the Inmarsat Geostationary Space Segment Operating in the 1.6 GHz Band—Section 8J—Technical and Operating Characteristics of Radiocommunications. 12 pp.

—**UHF—Search and Rescue Satellite Aided Tracking**

CCIR RECMN 633-1-90. Transmission Characteristics of a Satellite Emergency Position-Indicating Radiobeacon (Satellite EPIRB) System Operating Through a Low Polar-Orbiting Satellite System in the 406 MHz Band—Section 8J—Technical and Operating Characteristics of. 17 pp.

CCIR Report 1042-86. Compatibility Between Satellite EPIRBs Using the Band 406-406.1 MHz and Other Radio Services Using Adjacent Bands—Section 8D—Radiodetermination, Global Maritime Distress and Safety System and Related Subjects. 8 pp.

CCIR Report 761-3-90. Technical and Operating Characteristics of Distress Systems in the Mobile-Satellite Service —Section 8J—Technical and Operating Characteristics of Radiocommunications Using Satellite Distress and Safety Operation and of Radio Determination Satellite Services. 19 pp.

CCIR Report 919-2-90. Performance of a Low-Altitude, Polar-Orbiting Satellite EPIRB System—Section 8J—Technical and Operating Characteristics of Radiocommunications Using Satellite Distress and Safety Operation and of Radio Determination Satellite Services. 38 pp.

—**VHF—Global Maritime Distress and Safety Systems**

CCIR Report 1036-1-90. Frequencies for Homing and Locating in the Global Maritime Distress and Safety System (GMDSS)—Section 8D—Radiodetermination, Global Maritime Distress and Safety System and Related Subjects. 23 pp.

—**VHF—Mobile Satellite Communications**

CCIR Report 761-3-90. Technical and Operating Characteristics of Distress Systems in the Mobile-Satellite Service —Section 8J—Technical and Operating Characteristics of Radiocommunications Using Satellite Distress and Safety Operation and of Radio Determination Satellite Services. 19 pp.

—**VHF—Search and Rescue Satellite Aided Tracking**

CCIR Report 1162-90. Power Distribution Requirements for EPIRBs Operating on the Frequencies of 121.5 MHz and 243 MHz—Section 8D—Radiodetermination, Global Maritime Distress and Safety System and Related Subjects. 12 pp.

CCIR Report 919-2-90. Performance of a Low-Altitude, Polar-Orbiting Satellite EPIRB System—Section 8J—Technical and Operating Characteristics of Radiocommunications Using Satellite Distress and Safety Operation and of Radio Determination Satellite Services. 38 pp.

—**VHF—SELCAL Systems**

CCIR RECMN 693-90. Technical Characteristics of VHF Emergency Position-Indicating Radio Beacons Using Digital Selective Calling (DSC VHF EPIRB)—Section 8D—Radiodetermination, Global Maritime Distress and Safety System and Related Subjects. 2 pp.

CCIR Report 749-3-90. Future Use and Characteristics of Emergency Position-Indicating Radio Beacons in the Mobile and Mobile-Satellite Services—Section 8D—Radiodetermination, Global Maritime Distress and Safety System and Related Subjects. 7 pp.

Emergency Position Indicating Radio Beacons *(Cont.)*

—**VHF—SELCAL Systems—Message Formats**

CCIR RECMN 693-90. Technical Characteristics of VHF Emergency Position-Indicating Radio Beacons Using Digital Selective Calling (DSC VHF EPIRB)—Section 8D—Radiodetermination, Global Maritime Distress and Safety System and Related Subjects. 2 pp.

Emergency Power Supplies
Use: Emergency and Standby Power Supplies

Emergency Power Systems
Use: Emergency and Standby Power Supplies

Emergency Procedures
Use: Emergency Planning

Emergency Rations
See Also: Emergency Relief Operations; Food

—**Marine Safety Equipment**

CNS N5124-76. Emergency Rations for Life Boats and Life Rafts (Tentative) (Sep)(4015). 2 pp.

—**Nutritional—Interchangeable**

NATO STANAG 2937 ED 2 AMD 1-87. Survival, Emergency and Individual Combat Rations—Nutritional Values and Packaging. 8 pp.

Emergency Relief Operations
See Also: Emergency and Standby Power Supplies; Emergency Medical Services; Emergency Planning; Emergency Rations

—**Earth Stations**

CCIR Report 554-4-90. Use of Small Earth Stations for Relief Operation in the Event of Natural Disasters and Similar Emergencies—Section 4C—Earth Station and Baseband Characteristics—Earth Station Antennas—Maintenance of Earth Stations. 9 pp.

CCIR QUESTION 43/4-90. Use of Small Earth Stations in the Fixed-Satellite Service in the Event of Natural Disasters, Epidemics, Famines and Similar Emergencies for Warning and Relief Operations—Questions of Study Group 4 Fixed-Satellite Service. 1 p.

—**Earth Stations—Natural Disaster**

CCIR Report 554-4-90. Use of Small Earth Stations for Relief Operation in the Event of Natural Disasters and Similar Emergencies—Section 4C—Earth Station and Baseband Characteristics—Earth Station Antennas—Maintenance of Earth Stations. 9 pp.

CCIR QUESTION 43/4-90. Use of Small Earth Stations in the Fixed-Satellite Service in the Event of Natural Disasters, Epidemics, Famines and Similar Emergencies for Warning and Relief Operations—Questions of Study Group 4 Fixed-Satellite Service. 1 p.

—**Fixed Satellite Services—Fixed Services—Frequency Band Sharing**

CCIR QUESTION 65/4-90. Frequency Sharing Between Systems in the Fixed-Satellite Service and the Fixed Service in the Case of Relief Operations and Other Temporary Applications—Questions of Study Group 4 Fixed-Satellite Service. 1 p.

—**Radio Equipment**

CCIR Decision 63-2-89. Use of Automatically Controlled Radio Systems Below 30 MHz to Provide Service to Sparsely Populated Areas and for Emergency Relief Operations—Annex to Volume III—Fixed Service at Frequencies Below About 30 MHz. 2 pp.

—**Radio Equipment—Transportable—Fixed Services**

CCIR QUESTION 148/9-90. Transportable Fixed Service Radiocommunication Equipment for Relief Operations—Questions of Study Group 9 Fixed Service. 1 p.

—**Radio Links—Transportable—Radio Relay Systems**

CCIR Report 615-1-82. Transportable Fixed Radiocommunications Equipment for Relief Operations—Section 9E1—Line-of-Sight Radio-Relay Systems. 4 pp.

—**Radio Relay Systems—Transportable**

CCIR QUESTION 121/9-90. Transportable Fixed-Service Radiocommunication Equipment for Relief Operations—Questions of Study Group 9 Fixed Service. 1 p.

INDEX and DIRECTORY of

INTERNATIONAL AND NON-U.S. NATIONAL STANDARDS SUBJECT INDEX

Emergency Survival Kits
Use: Survival Kits

Emergency Water Supplies
Use For: EWS

—Indicator Plates
BSI BS 3251-76. 1976 Amd 1 Indicator Plates for Fire Hydrants and Emergency Water Supplies (AMD 6736) September 30, 1991. 14 pp.

Emery Cloth
Use: Emery Paper

Emery Paper
Use For: Emery Cloth *See Also:* Abrasive Sheets; Abrasives; Papers
SAA AS B292-69. Emery Cloth and Glass Paper. 12 pp.

EMI
Use: Electromagnetic Interference

EMI Filters
Use: Electromagnetic Interference Filters

Emission Control Devices (Engines)
Use: Exhaust Emission Control Equipment

Emission Control Equipment
Use: Exhaust Emission Control Equipment

Emission Current Measurement
—Electron Tubes
IEC 151 Pt 13-66. Measurements of the Electrical Properties of Electronic Tubes and Valves Part 13: Methods of Measurement of Emission Current from Hot Cathodes for High-Vacuum Electronic Tubes and Valves First Edition. 11 pp.

Emission Spectrochemical Analysis
Use: Spectrochemical Analysis

Emission Spectrography
Use: Spectrography

Emission Spectroscopy
Use For: Flame Emission Spectroscopy; Optical Emission Spectral Analysis; Plasma Emission Spectroscopy
CNS K0030-88. General Rules for Emission Spectroscopic Analysis (Sep)(12416).

—Carbon Steels
CNS G2230-84. Emission-Spectroscopic Analysis for Carbon Steel and Low Alloy Steel (Oct)(11071).
JIS G 1252-75. Emission-Spectroscopic Analysis for Carbon Steel and Low Alloy Steel. 8 pp.

—Cast Iron
CNS G2183-84. Emission-Spectroscopic Analysis for Pig Iron and Cast Iron (Oct)(10503).
JIS G 1251-76. Emission-Spectroscopic Analysis for Pig Iron and Cast Iron (R 1984). 8 pp.

—Hydrofluoric Acid
DIN ENGL 50451 Pt 2-90. Determination of Cobalt, Chromium, Copper, Iron and Nickel as Impurities in Hydrofluoric Acid for Use in Semiconductor Technology by Plasma-Induced Emission Spectrometry (Oct). 2 pp.

—Iron
CNS G1021-84. General Rules on Emission-Spectroscopic Analysis for Iron and Steel (Oct)(9705).
JIS G 1202-75. General Rules on Emission-Spectroscopic Analysis for Iron and Steel (R 1978). 23 pp.

—Low Alloy Steels
CNS G2230-84. Emission-Spectroscopic Analysis for Carbon Steel and Low Alloy Steel (Oct)(11071).
JIS G 1252-75. Emission-Spectroscopic Analysis for Carbon Steel and Low Alloy Steel. 8 pp.

—Pig Iron
CNS G2183-84. Emission-Spectroscopic Analysis for Pig Iron and Cast Iron (Oct)(10503).
JIS G 1251-76. Emission-Spectroscopic Analysis for Pig Iron and Cast Iron (R 1984). 8 pp.

—Steels
CNS G1021-84. General Rules on Emission-Spectroscopic Analysis for Iron and Steel (Oct)(9705).
JIS G 1202-75. General Rules on Emission-Spectroscopic Analysis for Iron and Steel (R 1978). 23 pp.

Emissions
See Also: Exhaust Gases

Emissions (Cont.)
—Aircraft
CAA Chapter M4-1 05.86. General (Emissions Certification). 1 p.

—Aircraft—Certification
CAA 847 (M). Section M Emissions Certification. 57 pp.
CAA. Contents. 1 p.
CAA. Foreword 05.86. 3 pp.
CAA Chapter M1-1 05.86. General (Emissions Certification). 1 p.
CAA Chapter M2-1 05.86. Fuel Venting Certification (Emissions Certification). 1 p.
CAA Chapter M2-2 05.86. Engine Emissions Certification (Emissions Certification). 1 p.
CAA Chapter M3-1 05.86. Emission Standards for Turbofan and Turbojet Engines (Emissions Certification). 1 p.

—Aircraft Engines
CAA Chapter M4-2 05.86. Test Conditions (Emissions Certification). 2 pp.
CAA Chapter M4-2 App 05.86. Specification for Fuel to Be Used in Aircraft Turbine Engine Emmission Testing (Emissions Certification). 1 p.

—Aircraft Engines—Modifications
CAA Chapter M5-1 12.86. Modifications (Emissions Certification). 1 p.

—Aircraft—Glossaries
CAA Chapter M1-2 05.86. Definitions and Symbols (Emissions Certification). 2 pp.
CAA Chapter M1-2 App 05.86. Measurement of Reference Pressure Ratio (Emissions Certification). 1 p.

—Aircraft—Modifications
CAA Chapter M5-1 12.86. Modifications (Emissions Certification). 1 p.

—Aircraft—Symbols
CAA Chapter M1-2 05.86. Definitions and Symbols (Emissions Certification). 2 pp.

—Motor Vehicles—Air Pollution
EC COM(93) 277-93. Re-Examined Proposal for a Council Directive Amending Directive 70/220/EEC on the Approximation of the Laws of the Member States Relating to Measures to Be Taken Against Air Pollution by Emmissions from Motor Vehicles. 14 pp.
EC 91/441/EEC-91. Council Directive Amending Directive 70/220/EEC on the Approximation of the Laws of the Member States Relating to be Taken Against Air Pollution by Emissions from Motor Vehicles. 106 pp.

Emissivity
See Also: Optical Density; Optical Properties; Temperature; Thermodynamic Properties

—Building Components—Infrared Thermometers
CNS A3269-87. Simplified Test Method for Emissivity by Infrared Radio Meter (Sep)(12066).
JIS A 1423-83. Simplified Test Method for Emissivity by Infrared Radio Meter. 8 pp.

Emitter Coupled Logic
Use For: ECL *See Also:* Logic Levels

—Connectors—Electronic Modules—Nuclear Instruments
IEC 912-87. ECL (Emitter Coupled Logic) Front Panel Interconnections in Counter Logic First Edition. 11 pp.

Emitters (Irrigation Equipment)
Use: Irrigation Equipment (Agricultural)—Emitters

EMS
Use: European Monetary System

EMT
Use: Electrical Metallic Tubing

Emulsification
See Also: Demulsification; Dispersants

—Dry Cleaning Solvents
ISO 6837-82. Surface Active Agents—Water Dispersing Power in Dry Cleaning Solvents First Edition. 7 pp.

—Hydraulic Fluids
MOD UK DSTAN 05-50: Part 11-01. Methods for Testing Fuels, Lubricants and Associated Products Part 11: Emulsification Properties of Oils and Hydraulic Fluid Issue 1; Amendment 1. 9 pp.
MOD UK DSTAN 91-35-01. Hydraulic Fluid Petroleum: Emulsifying Joint Service Designation: OX-30 Issue 1; Amendment 2. 15 pp.

Emulsification (Cont.)
—Lubricating Oils
BSI BS 2000: Part 19-93. 1993 Methods of Test for Petroleum and Its Products Part 19: Determination of Demulsibility Characteristics of Lubricating Oil. 4 pp.
BSI BS 2000: Part 19-90. 1990 Petroleum and Its Products Part 19: Demulsification Number of Lubricating Oil. 6 pp.
DIN ENGL 51599-75. Testing of Lubricating Oils; Determination of Demulsification Capacity According to the Stirring Method (Oct). 3 pp.
JIS K 2520-91. Petroleum Products—Lubricating Oils—Determination of Demulsibility Characteristics. 17 pp.

Emulsified Asphalt Coatings (Made From Asphalt)
Use: Asphalt Coatings (Made From Asphalt)

Emulsified Asphalts
Use For: Asphalt Emulsions
CNS K5016-83. Emulsified Asphalt (Jul)(1304). 4 pp.
JIS K 2208-80. Emulsified Asphalt (R 1985). 39 pp.

—Charged Particles
CNS K6781-83. Method of Test for Particle Charge of Emulsified Asphalts (Jun)(10364).

—Classification
CNS K6797-83. Method of Test for Classification of Rapid Setting Cationic Emulsified Asphalt (Jul)(10458).

—Coating Ability
CNS K6787-83. Method of Test for Coating Ability and Water Resistance of Emulsified Asphalts (Jun)(10370).

—Demulsification
CNS K6780-83. Method of Test for Demulsibility of Emulsified Asphalts (Jun)(10363). 2 pp.

—Environmental Testing
CNS K6788-83. Method of Test for Storage Stability of Asphalt Emulsion (Jun)(10371).

—Freezing
CNS K6786-83. Method of Test for Freezing of Emulsified Asphalts (Jun)(10369).

—Miscibility
CNS K6785-83. Method of Test for Miscibility with Water of Emulsified Asphalts (Jun)(10368).

—Mixing
CNS K6783-83. Method of Test for Cement Mixing of Emulsified Asphalts (Jun)(10366).

—Oils—Distillation Methods
CNS K6794-83. Method of Test for Identification of Oil Distillate Obtained from Emulsified Asphalts by Micro-Distillation (Jul)(10455).

—Residue Content—Distillation Methods
CNS K6793-83. Method of Test for Residue by Distillation of Emulsified Asphalts (Jul)(10454).
CNS K6796-83. Method of Test for Examination of Residue Obtained from Distillation or Evaporation of Emulsified Asphalts (Jul)(10457).

—Residue Content—Evaporation
CNS K6795-83. Method of Test for Residue by Evaporation of Emulsified Asphalts (Jul)(10456).
CNS K6796-83. Method of Test for Examination of Residue Obtained from Distillation or Evaporation of Emulsified Asphalts (Jul)(10457).

—Settling
CNS K6782-83. Method of Test for Settlement of Emulsified Asphalts (Jun)(10365).

—Sieve Analysis
CNS K6784-83. Method of Test for Sieve Test of Emulsified Asphalts (Jun)(10367).

—Water Resistance Testing
CNS K6787-83. Method of Test for Coating Ability and Water Resistance of Emulsified Asphalts (Jun)(10370).

Emulsified Paints
Use: Paints

Emulsifiers
Use: Emulsifying Agents

Emulsifying Agents
Use For: Emulsifiers *See Also:* Demulsification; Dispersants; Emulsions; Surfactants

INDUSTRY STANDARDS

Emulsifying Agents (Cont.)

—Food

EC 74/329/EEC-74. Council Directive on the Approximation of the Laws of the Member States Relating to Emulsifiers, Stabilizers, Thickeners and Gelling Agents for Use in Foodstuffs. 7 pp.

EC 78/612/EEC-78. Council Directive Amending for the First Time Directive 74/329/EEC on the Approximation of the Laws of the Member States Relating to Emulsifiers, Stabilizers, Thickeners and Gelling Agents for Use in Foodstuffs. 4 pp.

EC 80/597/EEC-80. Council Directive Amending for the Second Time Directive 74/329/EEC on the Approximation of the Laws of the Member States Relating to Emulsifiers, Stabilizers, Thickeners and Gelling Agents for Use in Foodstuffs. 3 pp.

EC 85/6/EEC-84. Council Directive Amending for the Third Time Directive 74/329/EEC on the Approximation of the Laws of the Member States Relating to Emulsifiers, Stabilizers, Thickeners and Gelling Agents for Use in Foodstuffs. 1 p.

EC 86/102/EEC-86. Council Directive Amending for the Fourth Time Directive 74/329/EEC on the Approximation of the Laws of the Member States Relating to Emulsifiers, Stabilizers, Thickeners and Gelling Agents for Use in Foodstuffs. 2 pp.

EC 89/393/EEC-89. Council Directive Amending for the Fifth Time Directive 74/329/EEC on the Approximation of the Laws of the Member States Relating to Emulsifiers, Stabilizers, Thickeners and Gelling Agents for Use in Foodstuffs. 1 p.

EC 90/612/EEC-90. Commission Directive Amending Council Directive 78/663/EEC Laying down Specific Purity Criteria for Emulsifiers, Stabilizers, Thickeners and Gelling Agents for Use in Foodstuffs. 2 pp.

—Food—Purity

EC 78/663/EEC-78. Council Directive Laying Down Specific Criteria of Purity for Emulsifiers, Stabilizers, Thickeners and Gelling Agents for Use in Foodstuffs. 23 pp.

EC 82/504/EEC-82. Council Directive Amending Directive 78/663/EEC Laying down Specific Criteria of Purity for Emulsifiers, Stabilizers, Thickeners and Gelling Agents for Use in Foodstuffs. 3 pp.

—Solvents

MOD UK DSTAN 68-188-93. Emulsifier for Solvent Cleaner Issue 1. 16 pp.

MOD UK TS 354A. Emulsifier for Solvent Cleaner (Superseded by Def Stan 68-188).

—Solvents—Flash Point

MOD UK M 9536/68. Examination of Emulsifier for Solvent Cleaner (Superseded by Def Stan 68-188).

—Solvents—Miscibility

MOD UK M 9536/68. Examination of Emulsifier for Solvent Cleaner (Superseded by Def Stan 68-188).

—Solvents—PH

MOD UK M 9536/68. Examination of Emulsifier for Solvent Cleaner (Superseded by Def Stan 68-188).

—Solvents—Visual Inspection

MOD UK M 9536/68. Examination of Emulsifier for Solvent Cleaner (Superseded by Def Stan 68-188).

Emulsifying Oils
Use: Soluble Oils

Emulsion Paints
Use: Paints

Emulsions
See Also: Demulsification; Emulsifying Agents; Vinyl Acrylic Emulsions

—Cleaning Agents—Acidity

CGSB 31-GP-0A METH 51.1-57. Methods of Testing Corrosion-Prevention Materials and Processes Free Alkali or Free Acid. 1 p.

—Cleaning Agents—Alkalinity

CGSB 31-GP-0A METH 51.1-57. Methods of Testing Corrosion-Prevention Materials and Processes Free Alkali or Free Acid. 1 p.

—Cleaning Agents—Corrosion Testing

CGSB 31-GP-0A METH 11.5-57. Methods of Testing Corrosion-Prevention Materials and Processes Corrosion (Emulsion Cleaners); (Amended September 1966) (Re-Edited March 1982). 1 p.

—Cleaning Agents—Sodium Chromate Content

CGSB 31-GP-0A METH 52.1-57. Methods of Testing Corrosion-Prevention Materials and Processes Determination of Sodium Chromate. 2 pp.

Emulsions (Cont.)

—Solvents—Stability Testing

CGSB 31-GP-0A METH 9.6-57. Methods of Testing Corrosion-Prevention Materials and Processes Stability of Emulsions. 1 p.

—Stability Testing

CGSB 25-GP-1M METH 5.1-84. Methods of Sampling and Testing Waxes and Polishes Sediment and Stability of Emulsion. 1 p.

CGSB 25-GP-1M METH 5.2-84. Methods of Sampling and Testing Waxes and Polishes Stability. 1 p.

Enamel Paints
See Also: Acrylic Enamel Paints; Alkyd Enamel Paints; Aminoalkyd Enamel Paints; Cashew Resin Enamel Paints; Enamels; Lacquers; Lead Silicochromate Alkyd Enamel Paints; Organic Coatings; Paints; Phthalic Resin Enamel Paints; PVC Enamel Paints; Solvents; Varnishes

—Automotive—Refinishing

CGSB 1-GP-191M-80. Enamel, Automotive Refinishing, Fast Drying, Gloss, Standard for; (Amendment 1 Mar 1981). 12 pp.

—Concrete Floors—Interior

CGSB CAN/CGSB-1.66-M89. Interior Enamel for Concrete Floors. 10 pp.

—Concrete Floors—Interior/Exterior

CGSB CAN/CGSB-1.73-M91. Exterior and Interior Enamel for Floors. 8 pp.

—Consistency

CGSB 1-GP-71 METH 4.5-74. Methods of Testing Paints and Pigments Consistency Stormer Consistency. 2 pp.

—Conversion Coatings—Salt Spray Testing

CGSB 31-GP-0A METH 14.7A-62. Methods of Testing Corrosion-Prevention Materials and Processes Salt Spray Resistance of Painted Parts; (Supersedes 14.7 15 March 1957) (Amended September 1974) (Re-Edited March 1982). 1 p.

—Gloss—Exterior

SAA AS 3730.6-91. Guide to the Properties of Paints for Buildings—Part 6: Solvent-Borne—Exterior—Full Gloss Enamel (Supersedes SAA TR1.6: 2309—1982). 6 pp.

—Gloss—Interior

SAA AS 3730.11-91. Guide to the Properties of Paints for Buildings—Part 11: Solvent-Borne —Interior— Full Gloss Enamel. 4 pp.

—Gloss—Interior/Exterior

MOD UK DSTAN 80-59-90. Paint System, Hard Gloss Enamel Paint, Finishing, Hard Gloss Enamel Paint, Undercoat, for Paint, Finishing, Hard Gloss Enamel Types: Brushing Spraying Issue 3. 19 pp.

MOD UK DSTAN 80-59-01. Paint System, Hard Gloss Enamel Paint, Finishing, Hard Gloss Enamel Paint, Undercoat, for Paint, Finishing, Hard Gloss Enamel Types: Brushing Spraying Issue 3; Amendment 1. 22 pp.

SAA AS 4025.1-92. Paints for Equipment Including Ships—Part 1: Solvent-Borne—Interior and Exterior—Full Gloss Enamel (in Professional Package 30A) (Supersedes AS K126—1964). 7 pp.

—Laboratory Ware—Chemical Resistance

BSI BS 3996-78. 1978 Colour Coding for One-Mark and Graduated Pipettes (Including Requirements for the Service Performance of the Colour Coding Enamels). 7 pp.

ISO 4794-82. Laboratory Glassware—Methods for Assessing Chemical Resistance of Enamels Used for Colour Coding and Colour Marking First Edition. 4 pp.

—Metal—Heat Resistant—Interior/Exterior

CGSB CAN/CGSB-1.76-M91. Interior and Exterior Heat-Resistant Enamel. 8 pp.

—Metal—Rapid Dry—Flat

CGSB CAN/CGSB-1.114-M91. Quick-Drying Flat Enamel. 8 pp.

—Rapid Dry

CNS K2045-89. Lustless Enamel (Quick-Drying) (Jan)(2627). 2 pp.

CNS K6826-89. Method of Test for Lustless Enamel (Quick Drying) (Jan)(11539).

—Rust Inhibitive—Semigloss

CNS K2046-86. Rust Inhibiting Enamel (Semi-Gloss) (Apr)(2628). 2 pp.

CNS K6827-86. Method of Test for Rust Inhibiting Enamel (Semi-Gloss) (Apr)(11540).

Enamel Paints (Cont.)

—Undercoatings—Fiberboard

CGSB 1-GP-71 METH 134.9-78. Methods of Testing Paints and Pigments Applicability and Appearance Enamel Undercoats on Fiberboard Panels; (Amended September 1980) (Re-Edited September 1982). 1 p.

—Undercoatings—Gloss—Interior/Exterior

MOD UK DSTAN 80-59-01. Paint System, Hard Gloss Enamel Paint, Finishing, Hard Gloss Enamel Paint, Undercoat, for Paint, Finishing, Hard Gloss Enamel Types: Brushing Spraying Issue 3; Amendment 1. 22 pp.

—Undercoatings—Interior

CGSB CAN/CGSB-1.38-M91. Interior Enamel Undercoater. 8 pp.

—Undercoatings—Wood—Semigloss

CGSB 1-GP-71 METH 134.10-78. Methods of Testing Paints and Pigments Applicability and Appearance Enamel Undercoats, Topcoated. 1 p.

—Wallboard

CGSB 1-GP-71 METH 134.7-78. Methods of Testing Paints and Pigments Applicability and Appearance Paints and Enamels on Sealed Boards, Brush and Roller Application. 1 p.

—Wallboard—Flat

CGSB 1-GP-71 METH 134.1-78. Methods of Testing Paints and Pigments Applicability and Appearance Enamels on Sealed Boards. 1 p.

—Wallboard—Gloss

CGSB 1-GP-71 METH 134.1-78. Methods of Testing Paints and Pigments Applicability and Appearance Enamels on Sealed Boards. 1 p.

—Wallboard—Hiding Power

CGSB 1-GP-71 METH 134.1-78. Methods of Testing Paints and Pigments Applicability and Appearance Enamels on Sealed Boards. 1 p.

CGSB 1-GP-71 METH 134.7-78. Methods of Testing Paints and Pigments Applicability and Appearance Paints and Enamels on Sealed Boards, Brush and Roller Application. 1 p.

—Wallboard—Primer Sealers

CGSB 1-GP-71 METH 134.12-78. Methods of Testing Paints and Pigments Applicability and Appearance Primer-Sealers, Brush and Roller Application; (Amended September 1980) (Re-Edited September 1982). 2 pp.

—Wallboard—Semigloss

CGSB 1-GP-71 METH 134.1-78. Methods of Testing Paints and Pigments Applicability and Appearance Enamels on Sealed Boards. 1 p.

—Wooden Floors

CGSB 1-GP-71 METH 134.5-75. Methods of Testing Paints and Pigments Applicability and Appearance Enamels for Wooden Floors. 1 p.

Enameled Wire
Scope Note: See the subheading Enameled under specific types of wire

Enamels
Scope Note: For additional listings, use a more specific term *See Also:* Baking Enamels; Coatings; Enamel Paints; Finishes; Lacquer Enamels; Nonmagnetic Coatings; Nonmetallic Coatings; Paints; Vitreous Enamels

CNS K2011-84. Enamel (Feb)(606). 2 pp.

—Aluminum—Self Lifting Resistance

CGSB 1-GP-71 METH 133.1-79. Methods of Testing Paints and Pigments Behavior Towards Undercoats and Self-Lifting Finishes on Primed Aluminum; (Amended September 1980) (Re-Edited September 1982). 1 p.

—Aluminum—Undercoatings

CGSB 1-GP-71 METH 133.1-79. Methods of Testing Paints and Pigments Behavior Towards Undercoats and Self-Lifting Finishes on Primed Aluminum; (Amended September 1980) (Re-Edited September 1982). 1 p.

—Bleeding

CGSB 1-GP-71 METH 37.3-80. Methods of Testing Paints and Pigments Bleeding Resistance. 1 p.

—Corrosion Testing

DIN ENGL 51174-88. Corrosion Testing of Enamel in Closed Systems (May). 4 pp.

—Density

CGSB 1-GP-71 METH 2.1-78. Methods of Testing Paints and Pigments Density (Weight Per Gallon). 2 pp.

INTERNATIONAL AND NON-U.S. NATIONAL STANDARDS
SUBJECT INDEX

Enamels *(Cont.)*

—Drying
CGSB 1-GP-71 METH 5.1-78. Methods of Testing Paints and Pigments Drying Time General Method. 2 pp.

—Fire Resistant
SAA AS K179-69. Semi-Gloss Enamel—Low Fire Hazard Type for Non-Combustible Surfaces (Primarily for Service Use). 15 pp.

—Floors—Concrete—Brushed
CGSB 1-GP-71 METH 134.6-79. Methods of Testing Paints and Pigments Applicability and Appearance Enamels for Concrete Floors. 1 p.

—Floors—Concrete—Roller Applied
CGSB 1-GP-71 METH 134.6-79. Methods of Testing Paints and Pigments Applicability and Appearance Enamels for Concrete Floors. 1 p.

—Gasoline Resistant
SAA AS K126-64. Full Gloss Enamel, Oil and Petrol Resistant (Superseded by AS 4025.1—1992).
SAA AS K143-62. Semi-Gloss Enamel, Oil and Petrol Resistant (Superseded by AS 4025.3—1993).

—Glass—Acid Resistance Testing
CGSB 1-GP-71 METH 105.2-79. Methods of Testing Paints and Pigments Acid Resistance Enamel Films on Glass Panels. 1 p.

—Impact Testing
BSI BS 1344: Part 21-93. 1993 Methods of Testing Vitreous Enamel Finishes Part 21: Determination of the Resistance of Vitreous Enamelled Articles to Impact: Pistol Test (ISO 4532: 1991) (F). 11 pp.
ISO 4532-91. Vitreous and Porcelain Enamels—Determination of the Resistance of Enamelled Articles to Impact—Pistol Test First Edition. 8 pp.

—Metals
CNS R1009-79. General Rules for Enamel Ware (Nov)(890). 2 pp.
CNS R3025-83. Method of Test for Enamel Ware (Jun)(903). 3 pp.
JIS K 5980-86. Baking Enamel Film on Metal Substrate. 12 pp.

—Metals—Radiation Detection and Measurement
CNS R3114-83. Method of Test for Radiation of Enamel (and Cloisonne) Products (Jun)(10379). 1 p.

—Nonvolatile Matter Content
CGSB 1-GP-71 METH 17.1-80. Methods of Testing Paints and Pigments Volatile Matter Paints, Enamels and Lacquers. 1 p.
CNS K6030-72. Method of Test for Enamel (Jun)(627). 7 pp.

—Oil Resistant
SAA AS K126-64. Full Gloss Enamel, Oil and Petrol Resistant (Superseded by AS 4025.1—1992).
SAA AS K143-62. Semi-Gloss Enamel, Oil and Petrol Resistant (Superseded by AS 4025.3—1993).

—Particle Content
CNS K6030-72. Method of Test for Enamel (Jun)(627). 7 pp.

—Physical Properties
CNS K6030-72. Method of Test for Enamel (Jun)(627). 7 pp.

—Pigment Content
CNS K6030-72. Method of Test for Enamel (Jun)(627). 7 pp.

—Sagging
CGSB 1-GP-71 METH 32.1-81. Methods of Testing Paints and Pigments Sagging. 1 p.

—Steel—Brushed
CGSB 1-GP-71 METH 134.4-78. Methods of Testing Paints and Pigments Applicability and Appearance Enamels on Bare Steel. 1 p.

—Steel—Brushed—Flat
CGSB 1-GP-71 METH 134.3-78. Methods of Testing Paints and Pigments Applicability and Appearance Enamels on Primed Steel, Brush and Roller Application. 1 p.

—Steel—Brushed—Gloss
CGSB 1-GP-71 METH 134.3-78. Methods of Testing Paints and Pigments Applicability and Appearance Enamels on Primed Steel, Brush and Roller Application. 1 p.

—Steel—Brushed—Hiding Power
CGSB 1-GP-71 METH 134.3-78. Methods of Testing Paints and Pigments Applicability and Appearance Enamels on Primed Steel, Brush and Roller Application. 1 p.

Enamels *(Cont.)*

—Steel—Brushed—Hiding Power *(Cont.)*
CGSB 1-GP-71 METH 134.4-78. Methods of Testing Paints and Pigments Applicability and Appearance Enamels on Bare Steel. 1 p.

—Steel—Brushed—Semigloss
CGSB 1-GP-71 METH 134.3-78. Methods of Testing Paints and Pigments Applicability and Appearance Enamels on Primed Steel, Brush and Roller Application. 1 p.

—Steel—Flat—Environmental Testing
CGSB 1-GP-71 METH 143.4-78. Methods of Testing Paints and Pigments Long-Term Outdoor Performance Test for One Coat Styrenated Flat Alkyd Enamel on Steel. 1 p.

—Steel—Heat Resistance
CGSB 1-GP-71 METH 112.2-79. Methods of Testing Paints and Pigments Heat Resistance Enamel. 1 p.

—Steel—Roller Applied—Flat
CGSB 1-GP-71 METH 134.3-78. Methods of Testing Paints and Pigments Applicability and Appearance Enamels on Primed Steel, Brush and Roller Application. 1 p.

—Steel—Roller Applied—Gloss
CGSB 1-GP-71 METH 134.3-78. Methods of Testing Paints and Pigments Applicability and Appearance Enamels on Primed Steel, Brush and Roller Application. 1 p.

—Steel—Roller Applied—Hiding Power
CGSB 1-GP-71 METH 134.3-78. Methods of Testing Paints and Pigments Applicability and Appearance Enamels on Primed Steel, Brush and Roller Application. 1 p.

—Steel—Roller Applied—Semigloss
CGSB 1-GP-71 METH 134.3-78. Methods of Testing Paints and Pigments Applicability and Appearance Enamels on Primed Steel, Brush and Roller Application. 1 p.

—Undercoatings—Gasoline Resistant
SAA AS K127-64. Undercoat for Oil and Petrol Resistant Enamels (Superseded by AS 4025.2—1993).

—Undercoatings—Oil Resistant
SAA AS K127-64. Undercoat for Oil and Petrol Resistant Enamels (Superseded by AS 4025.2—1993).

—Volatile Matter Content
CGSB 1-GP-71 METH 17.1-80. Methods of Testing Paints and Pigments Volatile Matter Paints, Enamels and Lacquers. 1 p.

—Water Content
CNS K6030-72. Method of Test for Enamel (Jun)(627). 7 pp.

—Water Resistance Testing
CNS K6030-72. Method of Test for Enamel (Jun)(627). 7 pp.

—Wood Sealants
CGSB 1-GP-71 METH 123.2-79. Methods of Testing Paints and Pigments Sealing Properties; (Amended September 1980) (Re-Edited September 1982). 1 p.

Encapsulating
See Also: Sealants

—Explosives—Silicones
MOD UK TS 10275. RTV Silicone Two-Part Encapsulating System—Type QX.

Encipherment, Data
Use: Data Encryption

Enclosed Gas Streams
Use: Flue Gases

Enclosed Switches
See Also: Electric Switches; Vacuum Switches

—Air
CSA CAN/CSA-C22. 2 NO 4-M89. Enclosed Switches; (Gen Instr 1 Thru 3). 78 pp.
CSA 655 Bull. Electrical Bulletin 655 October 12, 1966 to C22.2 NO 4. 4 pp.

—Electrical Codes
CSA 1165 Bull. Electrical Bulletin 1165 April 12, 1978 to C22.2 NO 4. 2 pp.

—Electrical Grounding
CSA 761 Bull. Electrical Bulletin 761 May 26, 1969 to C22.2 NO 4. 1 p.

Enclosed Switches *(Cont.)*

—Low Voltage
CNS C3074-83. Method of Test for Enclosed Switches for Low Voltage (Jun)(5317).
CNS C4173-83. Enclosed Switches for Low Voltage (Jun)(5316).
JIS C 8326-88. Low-Voltage Enclosed Switches (AC 600 V and Below). 26 pp.
JIS C 8362-88. Enclosed Switches (AC 250 V and Below). 16 pp.

—Sensitive
CNS C3197-85. Method of Test for Enclosed Sensitive Switches (Jan)(11181).
CNS C4426-85. Enclosed Sensitive Switches (Jan)(11180).
JIS C 4508-90. Enclosed Sensitive Switches. 45 pp.

Enclosures (Electrical)
Use: Electrical Enclosures

Enclosures (Test Chambers)
Use: Test Chambers

Encoded Data Transmission
See Also: Codes; Data Codes; Data Transmission; Encoded Information Types

—Emergency Position Indicating Radio Beacons
CCIR RECMN 633-1-90. Transmission Characteristics of a Satellite Emergency Position-Indicating Radiobeacon (Satellite EPIRB) System Operating Through a Low Polar-Orbiting Satellite System in the 406 MHz Band—Section 8J—Technical and Operating Characteristics of. 17 pp.

—Maritime Mobile Services—Accounting
CCITT RECMN D.91-91. Transmission in Encoded Form of Maritime Telecommunications Accounting Information (Study Group III) 13 pp. 13 pp.
CCITT RECMN D.91-89. Transmission in Encoded Form of Maritime Telecommunications Accounting Information—General Tariff Principles—Charging and Accounting in International Telecommunications Services (Study Group III) 8 pp. 8 pp.

—Reverse Charging—Accounting
CCITT RECMN D.176 (REV 1)-92. Transmission in Encoded Form of Telephone Reversed Charge Billing and Accounting Information (Study Group III) 9 pp (Same as Recmn E.276). 9 pp.
CCITT RECMN D.176-89. Transmission in Encoded Form of Telephone Reversed Charge Billing and Accounting Information—General Tariff Principles—Charging and Accounting in International Telecommunications Services (Study Group III) 6 pp. 6 pp.
CCITT RECMN E.276-89. Transmission in Encoded Form of Telephone Reversed Charge Billing and Accounting Information—Telephone Network and ISDN—Operation, Numbering, Routing and Mobile Service (Study Group II) 1 pp (Same as Recmn D.176). 1 p.

—Reverse Charging—Tariffs
CCITT RECMN D.176 (REV 1)-92. Transmission in Encoded Form of Telephone Reversed Charge Billing and Accounting Information (Study Group III) 9 pp (Same as Recmn E.276). 9 pp.
CCITT RECMN D.176-89. Transmission in Encoded Form of Telephone Reversed Charge Billing and Accounting Information—General Tariff Principles—Charging and Accounting in International Telecommunications Services (Study Group III) 6 pp. 6 pp.
CCITT RECMN E.276-89. Transmission in Encoded Form of Telephone Reversed Charge Billing and Accounting Information—Telephone Network and ISDN—Operation, Numbering, Routing and Mobile Service (Study Group II) 1 pp (Same as Recmn D.176). 1 p.

—Telegraphy—Accounting
CCITT RECMN D.190-89. Transmission in Encoded Form of Monthly International Accounting Information—General Tariff Principles—Charging and Accounting in International Telecommunications Services (Study Group III) 6 pp (Same as Recmn E.275). 6 pp.
CCITT RECMN E.275-89. Transmission in Encoded Form of Monthly International Accounting Information—Telephone Network and ISDN—Operation, Numbering, Routing and Mobile Service (Study Group II) 1 pp (Same as Recmn D.190). 1 p.

—Telephone Services—Accounting
CCITT RECMN D.190-89. Transmission in Encoded Form of Monthly International Accounting Information—General Tariff Principles—Charging and Accounting in International Telecommunications Services (Study Group III) 6 pp (Same as Recmn E.275). 6 pp.

INDUSTRY STANDARDS

Encoded Data Transmission (Cont.)
—Telephone Services—Accounting (Cont.)
CCITT RECMN E.275-89. Transmission in Encoded Form of Monthly International Accounting Information—Telephone Network and ISDN—Operation, Numbering, Routing and Mobile Service (Study Group II) 1 pp (Same as Recmn D.190). 1 p.

—Telex Communications—Accounting
CCITT RECMN D.190-89. Transmission in Encoded Form of Monthly International Accounting Information—General Tariff Principles—Charging and Accounting in International Telecommunications Services (Study Group III) 6 pp (Same as Recmn E.275). 6 pp.

CCITT RECMN E.275-89. Transmission in Encoded Form of Monthly International Accounting Information—Telephone Network and ISDN—Operation, Numbering, Routing and Mobile Service (Study Group II) 1 pp (Same as Recmn D.190). 1 p.

Encoded Information Types
See Also: Codes; Encoded Data Transmission; FAX Communications; Interpersonal Messaging Systems; Message Handling Systems; Teletex Communications

CCITT RECMN F.400-92. Message Handling Services: Message Handling System and Service Overview (Study Group I) 82 pp (Same as Recmn X.400). 82 pp.

CCITT RECMN F.400-89. Message Handling System and Service Overview—Message Handling and Directory Services—Operations and Definition of Service (Study Group I) 73 pp. 73 pp.

CCITT RECMN F.415-89. Message Handling Services: Intercommunication with Public Physical Delivery Services—Message Handling and Directory Services—Operations and Definition of Service (Study Group I) 15 pp (Erratum in Recmn F.410). 15 pp.

CCITT RECMN X.400-93. Message Handling Services: Message Handling System and Service Overview (Study Group VII) 82 pp (Same as Recmn F.400). 82 pp.

—Interpersonal Messaging Services
CCITT RECMN F.420-92. Message Handling Services: Public Interpersonal Messaging Service (Study Group I) 16 pp. 16 pp.

CCITT RECMN F.420-89. Message Handling Services: the Public Interpersonal Messaging Service—Message Handling and Directory Services—Operations and Definition of Service (Study Group I) 15 pp. 15 pp.

—Message Handling Systems—Conversion Rules
CCITT RECMN X.408-89. Message Handling Systems: Encoded Information Type Conversion Rules—Data Communication Networks—Message Handling Systems (Study Group VII) 44 pp. 44 pp.

Encoder/Decoders
Use For: CODEC; Codecs; Coder/Decoders
See Also: Decoders; Encoders

—Adaptive Delta Pulse Code Modulation—16 kbit/s
CCITT RECMN G.726-90. 40, 32, 24, 16 kbit/s Adaptive Differential Pulse Code Modulation (ADPCM) (Study Group XV) 60 pp (Replaces Recmn G.721 and G.723). 60 pp.

—Adaptive Delta Pulse Code Modulation—2-Bit
CCITT RECMN G.727-90. 5-. 4-, 3-amd 2-Bits Sample Embedded Adaptive Differential Pulse Code Modulation (ADPCM) (Study Group XV) 58 pp. 58 pp.

—Adaptive Delta Pulse Code Modulation—24 kbit/s
CCITT RECMN G.726-90. 40, 32, 24, 16 kbit/s Adaptive Differential Pulse Code Modulation (ADPCM) (Study Group XV) 60 pp (Replaces Recmn G.721 and G.723). 60 pp.

—Adaptive Delta Pulse Code Modulation—3-Bit
CCITT RECMN G.727-90. 5-. 4-, 3-amd 2-Bits Sample Embedded Adaptive Differential Pulse Code Modulation (ADPCM) (Study Group XV) 58 pp. 58 pp.

—Adaptive Delta Pulse Code Modulation—32 kbit/s
CCITT RECMN G.721-89. 32 kbit/s Adaptive Differential Pulse Code Modulation (ADPCM)—General Aspects of Digital Transmission Systems; Terminal Equipments (Study Groups XV and XVIII) 38 pp (Replaced by Recmn G.726). 38 pp.

Encoder/Decoders (Cont.)
—Adaptive Delta Pulse Code Modulation—32 kbit/s (Cont.)
CCITT RECMN G.726-90. 40, 32, 24, 16 kbit/s Adaptive Differential Pulse Code Modulation (ADPCM) (Study Group XV) 60 pp (Replaces Recmn G.721 and G.723). 60 pp.

—Adaptive Delta Pulse Code Modulation—4-Bit
CCITT RECMN G.727-90. 5-. 4-, 3-amd 2-Bits Sample Embedded Adaptive Differential Pulse Code Modulation (ADPCM) (Study Group XV) 58 pp. 58 pp.

—Adaptive Delta Pulse Code Modulation—40 kbit/s
CCITT RECMN G.726-90. 40, 32, 24, 16 kbit/s Adaptive Differential Pulse Code Modulation (ADPCM) (Study Group XV) 60 pp (Replaces Recmn G.721 and G.723). 60 pp.

—Adaptive Delta Pulse Code Modulation—5-Bit
CCITT RECMN G.727-90. 5-. 4-, 3-amd 2-Bits Sample Embedded Adaptive Differential Pulse Code Modulation (ADPCM) (Study Group XV) 58 pp. 58 pp.

—Adaptive Delta Pulse Code Modulation—64 kbit/s
CCITT RECMN G.722-89. 7 kHz Audio-Coding Within 64 kBit/s—General Aspects of Digital Transmission Systems; Terminal Equipments (Study Groups XV and XVIII) 73 pp. 73 pp.

—Audio
BSI BS 7133-89. 1989 Amd 1 Audio Recording—PCM Encoder/Decoder System (AMD 6718) August 30, 1991. 23 pp.

CENELEC HD 544 S1-89. Audio Recording—PCM Encoder/Decoder System. 3 pp.

IEC 841-88. Audio Recording—PCM Encoder/Decoder System First Edition. 37 pp.

—Audiovisual Services—1920 kbit/s
CCITT RECMN H.261-90. Video Codec for Audiovisual Services at p x 64 kBit/s (Study Group XV) 30 pp. 30 pp.

—Audiovisual Services—384 kbit/s
CCITT RECMN H.261-89. Codec for Audiovisual Services at n x 384 kBit/s Transmission of Non-Telephone Signals—Transmission of Sound-Programme and Television Signals (Study Group XV) 9 pp. 9 pp.

—Audiovisual Services—64 kbit/s
CCITT RECMN G.725-89. System Aspects for the Use of the 7 kHz Audio Codec Within 64 kbit/s—General Aspects of Digital Transmission Systems; Terminal Equipments (Study Groups XV and XVIII) 11 pp. 11 pp.

—Frequency Division Multiplexers
CCITT RECMN G.795-89. Characteristics of Codecs for FDM Assemblies—General Aspects of Digital Transmission Systems; Terminal Equipments (Study Groups XV and XVIII) 4 pp. 4 pp.

—Frequency Division Multiplexers—Transmultiplexers—(DIV) Systems
CCITT FASCICLE III.4-89. General Aspects of Digital Transmission Systems Terminal Equipments—Recommendations G.700-G.795. 621 pp.

—Frequency Division Multiplexers—Transmultiplexers—(DOV) Systems
CCITT FASCICLE III.4-89. General Aspects of Digital Transmission Systems Terminal Equipments—Recommendations G.700-G.795. 621 pp.

—Low Delay Code Excited Linear Prediction—16 kbits
CCITT RECMN G.728-92. Coding of Speech at 16 kBit/s Using Low-Delay Code Excited Linear Prediction (Study Group XV) 65 pp. 65 pp.

—Pulse Code Modulation—Sound Recording
CENELEC HD 544 S1-89. Audio Recording—PCM Encoder/Decoder System. 3 pp.

—Pulse Code Modulation—Television Systems
BSI BS 7133-89. 1989 Amd 1 Audio Recording—PCM Encoder/Decoder System (AMD 6718) August 30, 1991. 23 pp.

IEC 841-88. Audio Recording—PCM Encoder/Decoder System First Edition. 37 pp.

Encoder/Decoders (Cont.)
—Pulse Code Modulation—2 Wire Interfaces (Voice Band)
CCITT RECMN G.715-89. Separate Performance Characteristics for the Encoding and Decoding Side of PCM Channels Applicable to 2-Wire Interfaces—General Aspects of Transmission Systems; Terminal Equipments (Study Groups XV and XVIII) 15 pp (Replaced by RecmnG.712). 15 pp.

—Pulse Code Modulation—4 Wire Interfaces (Voice Band)
CCITT RECMN G.714-89. Separate Performance Characteristics for the Encoding and Decoding Sides of PCM Channels Applicable to 4-Wire Voice-Frequency Interfaces—General Aspects of Digital Transmission Systems; Terminal Equipments (Study Groups XV and XVIII) 13 pp. 13 pp.

—Sound Recording and Reproduction Equipment—320 kbit/s
CCITT RECMN J.43-89. Characteristics of Equipment for the Coding of Analogue High Quality Sound Programme Signals for Transmission on 320 KBit/s Channels—Line Transmission of Non-Telephone Signals—Transmission of Sound-Programme Television Signals (Study Group XV) 10 pp. 10 pp.

CCITT RECMN J.44-89. Characteristics of Equipment for the Coding of Analogue Medium Quality Sound-Programme Signals for Transmission on 320 kBit/s Channels—Line Transmission of Non-Telephone Signals—Transmission of Sound-Programme and Television Signals (Study Group XV) 2 pp. 2 pp.

—Sound Recording and Reproduction Equipment—384 kbit/s
CCITT RECMN J.41-89. Characteristics of Equipment for the Coding of Analogue High Quality Sound Programme Signals for Transmission on 384 KBit/s Channels—Line Transmission of Non-Telephone Signals—Transmission of Sound-Programme and Television Signals (Study Group XV) 12 pp. 12 pp.

CCITT RECMN J.42-89. Characteristics of Equipment for the Coding of Analogue Medium Quality Sound-Programme Signals for Transmission on 384-kBit/s Channels—Line Transmission of Non-Telephone Signals—Transmission of Sound-Programme and Television Signals—(Study Group XV) 4 pp. 4 pp.

—Speech Transmission Systems—Military—Interoperability
NATO STANAG 4380 ED 1 AMD 0-88. (DRAFT) Technical Standards for Analogue-Digital Conversion of Voice Signals—16 KBPS CVSD. 10 pp.

—Television Equipment—Test Pictures
CCIR Report 1213-90. Test Pictures and Sequences for Subjective Assessments of Digital Codecs—Section 11F—Digital Methods of Transmitting Television Information. 6 pp.

CCIR RECMN 802-92. Test Pictures and Sequences for Subjective Assessments of Digital Codecs Conveying Signals Produced According to Recommendation 601. 5 pp.

—Television Equipment—Test Sequences
CCIR Report 1213-90. Test Pictures and Sequences for Subjective Assessments of Digital Codecs—Section 11F—Digital Methods of Transmitting Television Information. 6 pp.

CCIR RECMN 802-92. Test Pictures and Sequences for Subjective Assessments of Digital Codecs Conveying Signals Produced According to Recommendation 601. 5 pp.

—Video Telephone Services—Frame Structures
CCITT RECMN H.130-89. Frame Structures for Use in the International Interconnection of Digital Codecs for Videoconferencing or Visual Telephony—Line Transmission of Non-Telephone Signals—Transmission of Sound-Programme and Television Signals (Study Group XV) 16 pp. 16 pp.

—Videoconferencing Services—Data Transmission
CCITT RECMN H.120-89. Codecs for Videoconferencing Using Primary Digital Group Transmission—Line Transmission of Non-Telephone Signals—Transmission of Sound-Programme and Television Signals (Study Group XV) 59 pp. 59 pp.

—Videoconferencing Services—Frame Structures
CCITT RECMN H.130-89. Frame Structures for Use in the International Interconnection of Digital Codecs for Videoconferencing or Visual Telephony—Line Transmission of Non-Telephone Signals—Transmission of Sound-Programme and Television Signals (Study Group XV) 16 pp. 16 pp.

INTERNATIONAL AND NON-U.S. NATIONAL STANDARDS
SUBJECT INDEX
Encoding

Encoders

See Also: Data Communication Equipment; Encoder/Decoders; Priority Interrupt Controllers; Speech Encoding

MOD UK DSTAN 59-27: Part 4-01. Precision Instrument, Rotating, Servocomponents Part 4: Encoders, Shaft Angle to Digital Issue 1; Amendment 1. 35 pp.

—Angular—Preferred Products List

CECC CECC MUAHAG Vol 13 IS 1-91. Preferred Products List; Servo Components (En, Fr, Ge). 27 pp.

—Pulse Code Modulation Measuring Instruments

CCITT RECMN O.133-89. Equipment for Measuring the Performance of PCM Encoders and Decoders—Specifications for Measuring Equipment (Study Group IV) 26 pp. 26 pp.

Encoding

Use For: Binary Encoding; Coding (Data Conversion); Data Encoding ***See Also:*** Data Encryption; Data Processing; Digital Audio Coding

—Acoustic Signals—Sound Broadcasting

CCIR RECMN 646-86. Source Encoding for Digital Sound Signals in Broadcasting Studios—Section 10C—Audio-Frequency Characteristics of Sound-Broadcasting Signals. 1 p.

CCIR RECMN 646-1-92. Source Encoding for Digital Sound Signals in Broadcasting Studios—Section 10C—Audio-Frequency Characteristics of Sound-Broadcasting Signals. 1 p.

CCIR QUESTION 87/10-90. Digital Coding in Broadcasting for Emission of Sound Signals—Questions Concerning Study Group 10—Broadcasting Service (Sound). 1 p.

—Audio Systems

IEC 11172 Pt 3-93. Information Technology—Coding of Moving Pictures and Associated Audio for Digital Storage Media at up to About 1,5 Mbit/s—Part 3: Audio First Edition. 157 pp.

ISO 11172 Pt 3-93. Information Technology—Coding of Moving Pictures and Associated Audio for Digital Storage Media at up to About 1,5 Mbit/s—Part 3: Audio First Edition. 157 pp.

JTC1 11172 Pt 3-93. Information Technology—Coding of Moving Pictures and Associated Audio for Digital Storage Media at up to About 1,5 Mbit/s—Part 3: Audio First Edition. 157 pp.

—Audio/Video Systems

IEC 11172 Pt 1-93. Information Technology—Coding of Moving Pictures and Associated Audio for Digital Storage Media at up to About 1,5 Mbit/s—Part 1: Systems First Edition. 60 pp.

ISO 11172 Pt 1-93. Information Technology—Coding of Moving Pictures and Associated Audio for Digital Storage Media at up to About 1,5 Mbit/s—Part 1: Systems First Edition. 60 pp.

JTC1 11172 Pt 1-93. Information Technology—Coding of Moving Pictures and Associated Audio for Digital Storage Media at up to About 1,5 Mbit/s—Part 1: Systems First Edition. 60 pp.

—Computer Graphics Interface

BSI BS ISO/IEC 9637-2-92. 1992 Information Technology—Computer Graphics—Interfacing Techniques for Dialogues with Graphical Devices (CGI)—Data Stream Binding—Part 2: Binary Encoding (S). 80 pp.

IEC DIS 9637 Pt 1-92. Information Technology—Interface Techniques for Dialogues with Graphical Devices—CGI Data Stream Encoding—Part 1: Character Encoding ***CD-ROM ONLY***. 69 pp.

IEC 9637 Pt 2-92. Information Technology—Computer Graphics—Interfacing Techniques for Dialogues with Graphical Devices (CGI)—Data Stream Binding—Part 2: Binary Encoding First Edition. 78 pp.

ISO 9637 Pt 2-92. Information Technology—Computer Graphics—Interfacing Techniques for Dialogues with Graphical Devices (CGI)—Data Stream Binding—Part 2: Binary Encoding First Edition. 78 pp.

JTC1 9637-92. Information Technology—Computer Graphics—Interfacing Techniques for Dialogues with Graphical Devices (CGI)—Data Stream Binding—Part 2: Binary Encoding First Edition. 78 pp.

OSI ISO IEC DIS 9637-2-91. Information Technology—Computer Graphics—Interfacing Techniques for Dialogues with Graphical Devices (CGI)—Data Stream Binding—Part 2: Binary Encoding.

—Computer Graphics Metafile

BSI BS 6945: Part 2-88. (WITHDRAWN) 1988 Amd 1 Computer Graphics: Metafile for the Storage and Transfer of Picture Description Information (CGM) Part 2: Character Encoding (ISO 8632-2: 1987) (AMD 6354) August 30, 1992 (Renumbered as BS EN 28632-2: 1992). 88 pp.

Encoding *(Cont.)*

—Computer Graphics Metafile *(Cont.)*

BSI BS 6945: Part 3-88. 1988 Computer Graphics: Metafile for the Storage and Transfer of Picture Description Information Part 3: Binary Encoding. 54 pp.

BSI BS 6945: Part 3-01. (WITHDRAWN) 1988 Amd 1 Computer Graphics: Metafile for the Storage and Transfer of Picture Description Information (CGM) Part 3: Binary Encoding (AMD 6355) November 29, 1991 (Renumbered as BS EN 28632-3: 1992). 71 pp.

BSI BS 6945: Part 4-88. (WITHDRAWN) 1988 Amd 1 Computer Graphics: Metafile for the Storage and Transfer of Picture Description Information (CGM) Part 4: Clear Text Encoding (ISO 8632-4: 1987) (AMD 6356) August 30, 1991 (Renumbered as BS EN 28632-4: 1992). 36 pp.

BSI BS EN 28632-2-92. 1992 Computer Graphics: Metafile for the Storage and Transfer of Picture Description Information (CGM) Part 2: Character Encoding (AMD 7373) November 15, 1992. 93 pp.

BSI BS EN 28632-3-92. 1992 Computer Graphics: Metafile for the Storage and Transfer of Picture Description Information (CGM) Part 3: Binary Encoding (AMD 7374) November 15, 1992. 77 pp.

BSI BS EN 28632-4-92. 1992 Computer Graphics: Metafile for the Storage and Transfer of Picture Description Information (CGM) Part 4: Clear Text Encoding (AMD 7375) November 15, 1992. 61 pp.

CEN EN 28632-2-92. Information Processing Systems—Computer Graphics—Metafile for the Storage and Transfer of Picture Description Information—Part 2: Character Encoding. 89 pp.

CEN EN 28632-3-92. Information Processing Systems—Computer Graphics—Metafile for the Storage and Transfer of Picture Description Information—Part 3: Binary Encoding. 56 pp.

CEN EN 28632-4-92. Information Processing Systems—Computer Graphics—Metafile for the Storage and Transfer of Picture Description Information—Part 4: Clear Text Encoding. 60 pp.

CEN EN 28632-4-92. Information Processing Systems—Computer Graphics—Metafile for the Storage and Transfer of Picture Description Information Part 4: Clear Text Encoding. 56 pp.

IEC 8632 Pt 2-92. Information Technology—Computer Graphics—Metafile for the Storage and Transfer of Picture Description Information—Part 2: Character Encoding Second Edition. 98 pp.

IEC 8632 Pt 2 Draft AMD 1. Information Technology—Computer Graphics—Metafile for the Storage and Transfer of Picture Description Information—Part 2: Character Encoding; (1993) ***CD-ROM ONLY***. 6 pp.

IEC 8632 Pt 3-92. Information Technology—Computer Graphics—Metafile for the Storage and Transfer of Picture Description Information—Part 3: Binary Encoding Second Edition. 79 pp.

IEC 8632 Pt 3 Draft AMD 1. Information Technology—Computer Graphics—Metafile for the Storage and Transfer of Picture Description Information—Part 3: Binary Encoding; (1993) ***CD-ROM ONLY***. 6 pp.

IEC 8632 Pt 4-92. Information Technology—Computer Graphics—Metafile for the Storage and Transfer of Picture Description Information—Part 4: Clear Text Encoding Second Edition. 63 pp.

IEC 8632 Pt 4 Draft AMD 1. Information Technology—Computer Graphics—Metafile for the Storage and Transfer of Picture Description Information—Part 4: Clear Text Encoding; (1993) ***CD-ROM ONLY***. 6 pp.

ISO 8632 Pt 2-92. Information Technology—Computer Graphics—Metafile for the Storage and Transfer of Picture Description Information—Part 2: Character Encoding Second Edition; (CEN EN 28632-2:1992). 98 pp.

ISO 8632 Pt 2 Draft AMD 1. Information Technology—Computer Graphics—Metafile for the Storage and Transfer of Picture Description Information—Part 2: Character Encoding; (1993) ***CD-ROM ONLY***. 6 pp.

ISO 8632 Pt 3-92. Information Technology—Computer Graphics—Metafile for the Storage and Transfer of Picture Description Information—Part 3: Binary Encoding Second Edition. 79 pp.

ISO 8632 Pt 3 Draft AMD 1. Information Technology—Computer Graphics—Metafile for the Storage and Transfer of Picture Description Information—Part 3: Binary Encoding; (1993) ***CD-ROM ONLY***. 6 pp.

ISO 8632 Pt 4-92. Information Technology—Computer Graphics—Metafile for the Storage and Transfer of Picture Description Information—Part 4: Clear Text Encoding Second Edition. 63 pp.

ISO 8632 Pt 4 Draft AMD 1. Information Technology—Computer Graphics—Metafile for the Storage and Transfer of Picture Description Information—Part 4: Clear Text Encoding; (1993) ***CD-ROM ONLY***. 6 pp.

Encoding *(Cont.)*

—Computer Graphics Metafile *(Cont.)*

JTC1 8632-92. Information Technology—Computer Graphics—Metafile for the Storage and Transfer of Picture Description Information—Part 2: Character Encoding Second Edition. 98 pp.

JTC1 8632 Pt2 Draft AMD 1. Information Technology—Computer Graphics—Metafile for the Storage and Transfer of Picture Description Information—Part 2: Character Encoding; (1993) ***CD-ROM ONLY***. 6 pp.

JTC1 8632-92. Information Technology—Computer Graphics—Metafile for the Storage and Transfer of Picture Description Information—Part 3: Binary Encoding Second Edition. 79 pp.

JTC1 8632 Pt3 Draft AMD 1. Information Technology—Computer Graphics—Metafile for the Storage and Transfer of Picture Description Information—Part 3: Binary Encoding; (1993) ***CD-ROM ONLY***. 6 pp.

JTC1 8632-92. Information Technology—Computer Graphics—Metafile for the Storage and Transfer of Picture Description Information—Part 4: Clear Text Encoding Second Edition. 63 pp.

JTC1 8632 Pt4 Draft AMD 1. Information Technology—Computer Graphics—Metafile for the Storage and Transfer of Picture Description Information—Part 4: Clear Text Encoding; (1993) ***CD-ROM ONLY***. 6 pp.

OSI ISO 8632-2-90. Information Processing Systems—for Computer Graphics—Metafile for the Storage and Transfer of Picture Description Information—Part 2: Character Encoding—Amendment 1. 14 pp.

OSI ISO 8632-2 DAM 3-90. Information Processing Systems—Computer Graphics—Metafile for the Storage and Transfer of Picture Description Information—Part 2: Character Encoding AMENDMENT 3. 28 pp.

OSI ISO 8632-2 PDAD 1-87. Information Processing Systems; Computer Graphics; Metafile for the Storage and Transfer of Picture Descriptioninformation-Part 2: Character Encoding, Proposed Draft Addendum 1 to ISO 8632-2.

OSI ISO 8632-3-87. Information Processing Systems—Computer Graphics—Metafile for the Storage and Transfer of Picture Description Information—Part 3—Binary Encoding (with Amendment 1). 74 pp.

OSI ISO 8632-3/DAM 3-91. Information Processing Systems—Computer Graphics—Metafile for the Storage and Transfer of Picture Description Information—Part 3—Binary Encoding AMENDMENT 3. 22 pp.

OSI ISO 8632-3 PDAD 1-87. Information Processing Systems; Computer Graphics; Metafile for the Storage and Transfer of Picture Description Information Part 3: Binary Encoding, Proposed Draft Addendum 1 to ISO 8632-3.

OSI ISO 8632-4-87. Information Processing Systems—Computer Graphics—Metafile for the Storage and Transfer of Picture Description Information—Part 4: Clear Text Encoding. 59 pp.

OSI ISO 8632-4 DAM 3-91. Information Processing Systems—Computer Graphics—Metafile for the Storage and Transfer of Picture Description Information—Part 4: Clear Text Encoding—Amendment 3. 18 pp.

OSI ISO 8632-4 PDAD 1-87. Information Processing Systems; Computer Graphics; Metafile for the Storage and Transfer of Picture Description Information Part 4: Clear Text Encoding, Proposed Draft Addendum 1 to ISO 8632—4.

SNZ NZS/ISO 8632. 2-87. Computer Graphics—Metafiles for the Storage and Transfer of Picture Description Information Part 2: Character Encoding. 60 pp.

SNZ NZS/ISO 8632. 3-87. Computer Graphics—Metafiles for the Storage and Transfer of Picture Description Information Part 3: Binary Encoding. 50 pp.

SNZ NZS/ISO 8632. 4-87. Computer Graphics—Metafiles for the Storage and Transfer of Picture Description Information Part 4: Clear Text Encoding. 33 pp.

—Digital Television—Television Studios

CCIR RECMN 601-2-90. Encoding Parameters of Digital Television for Studios—Section 11F—Digital Methods of Transmitting Television Information. 10 pp.

CCIR RECMN 601-3-92. Encoding Parameters of Digital Television for Studios—Section 11F—Digital Methods of Transmitting Television Information. 11 pp.

—Image Processing

IEC DIS 10918 Pt 1-92. Information Technology—Digital Compression and Coding of Continuous-Tone Still Images—Part 1: Requirements and Guidelines ***CD-ROM ONLY***. 203 pp.

IEC DIS 10918 Pt 2-93. Information Technology—Digital Compression and Coding of Continuous-Tone Still Images—Part 2: Compliance Testing ***CD-ROM ONLY***. 77 pp.

INDUSTRY STANDARDS
977

Encoding (Cont.)

—Image Processing (Cont.)

ISO DIS 10918 Pt 1-92. Information Technology—Digital Compression and Coding of Continuous-Tone Still Images—Part 1: Requirements and Guidelines ***CD-ROM ONLY***. 203 pp.

ISO DIS 10918 Pt 2-93. Information Technology—Digital Compression and Coding of Continuous-Tone Still Images—Part 2: Compliance Testing ***CD-ROM ONLY***. 77 pp.

JTC1 DIS10918 Pt 1-92. Information Technology—Digital Compression and Coding of Continuous-Tone Still Images—Part 1: Requirements and Guidelines ***CD-ROM ONLY***. 203 pp.

JTC1 DIS10918 Pt 2-93. Information Technology—Digital Compression and Coding of Continuous-Tone Still Images—Part 2: Compliance Testing ***CD-ROM ONLY***. 77 pp.

—Open Systems Interconnection

BSI BS 6963-88. (WITHDRAWN) 1988 Open Systems Interconnection: Specification for Basic Encoding Rules for Abstract Syntax Notation One (ASN1) (Superseded by BS ISO/IEC 8825: 1990). 19 pp.

BSI BS ISO/IEC 8825-90. 1990 Information Technology—Open Systems Interconnection—Specification of Basic Encoding Rules for Abstract Syntax Notation One (ASN.1). 26 pp.

CCITT RECMN X.209-89. Specification of Basic Encoding Rules for Abstract Syntax Notation One (ASN.1)—Data Communication Networks—Open Systems Interconnection (OSI) Model and Notation, Service Definition (Study Group VII) 21 pp. 21 pp.

CSA CAN/CSA-Z243.162-89. Information Processing Systems—Open Systems Interconnection—Specification of Basic Encoding Rules for Abstract Syntax Notation One (ASN.1) (ISO 8825-1987). 30 pp.

IEC 8825-90. Information Technology—Open Systems Interconnection—Specification of Basic Encoding Rules for Abstract Syntax Notation One (ASN.1) Second Edition. 23 pp.

IEC DIS 8825 Pt 1-92. Information Technology—Open Systems Interconnection—Specification of ASN.1 Encoding Rules—Part 1: Basic Encoding Rules (BER) ***CD-ROM ONLY***. 43 pp.

IEC DIS 8825 Pt 3-92. Information Technology—Open Systems Interconnection—Specification of ASN.1 Encoding Rules—Part 3: Distinguished Canonical Encoding Rules ***CD-ROM ONLY***. 21 pp.

ISO 8825-90. Information Technology—Open Systems Interconnection—Specification of Basic Encoding Rules for Abstract Syntax Notation One (ASN.1) Second Edition. 23 pp.

ISO DIS 8825 Pt 1-92. Information Technology—Open Systems Interconnection—Specification of ASN.1 Encoding Rules—Part 1: Basic Encoding Rules (BER) ***CD-ROM ONLY***. 43 pp.

ISO DIS 8825 Pt 3-92. Information Technology—Open Systems Interconnection—Specification of ASN.1 Encoding Rules—Part 3: Distinguished Canonical Encoding Rules ***CD-ROM ONLY***. 21 pp.

JIS X 5604-90. Information Processing Systems—Open Systems Interconnection—Specification of Basic Encoding Rules for Abstract Syntax Notation One (ASN.1) (ISO 8825:1987).

JTC1 DIS8825 Pt 1-92. Information Technology—Open Systems Interconnection—Specification of ASN.1 Encoding Rules—Part 1: Basic Encoding Rules (BER) ***CD-ROM ONLY***. 43 pp.

JTC1 DIS8825 Pt 3-92. Information Technology—Open Systems Interconnection—Specification of ASN.1 Encoding Rules—Part 3: Distinguished Canonical Encoding Rules ***CD-ROM ONLY***. 21 pp.

OSI ISO 8825 PDAD 1-87. Information Processing Systems—Open Systems Interconnection—Abstract Syntax Notation One (ASN.1)—Proposed Draft Addendum 1: Extensions to ASN.1 Basic Encoding Rules. 10 pp.

OSI ISO/IEC 8825-90. Information Technology—Open Systems Interconnection—Specification of Basic Encoding Rules for Abstract Syntax Notation One (ASN.1). 23 pp.

OSI ISO/IEC 8825 DAD 1-88. Information Processing Systems—Open Systems Interconnection—Specification of Basic Encoding Rules for Abstract Syntax Notation One (ASN.1) Addendum 1: ASN.1 Extensions. 8 pp.

OSI ISO 8825-87. Information Processing Systems—Open Systems Interconnection—Specification of Basic Encoding Rules for Abstract Syntax Notation One (ASN.1). 20 pp.

SAA AS 3626-91. Information Technology—Open Systems Interconnection—Specification of Basic Encoding Rules for Abstract Syntax Notation One (ASN.1) (ISO/IEC 8825:1990) (in Professional Packages 26A, 26C). 17 pp.

SNZ NZS/ISO-IEC 8825-90. Information Technology—Open Systems Interconnection—Specification of Basic Encoding Rules for Abstract Syntax Notation One (ASN.1). 17 pp.

Encoding (Cont.)

—Television Signals

CCIR Report 629-4-90. Digital Coding of Colour Television Signals—Section 11F—Digital Methods of Transmitting Television Information. 11 pp.

CCIR Report 962-2-90. Filtering, Sampling and Multiplexing for Digital Encoding of Colour Television Signals—Section 11F—Digital Methods of Transmitting Television Information. 10 pp.

CCIR QUESTION 25-3/11-90. Standards for Television Systems Using Digital Modulation—Questions Concerning Study Group 11—Broadcasting Service (Television). 1 p.

CCIR QUESTION 60/11-90. Digital Encoding of Colour Television Signals—Questions Concerning Study Group 11—Broadcasting Service (Television). 1 p.

—Television Signals—Bit Rate Reduction

CCIR QUESTION 81/11-90. Reduction in the Bit Rate in the Digital Coding of Colour Television Signals—Questions Concerning Study Group 11—Broadcasting Service (Television). 1 p.

—Television Signals—Enhancements

CCIR RECMN 796-92. Parameters for Enhanced Compatible Coding Systems Based on 625-Line PAL and SECAM Television Systems. 5 pp.

—Television Signals—Filtering

CCIR QUESTION 61/11-90. Filtering and Sampling for Digital Encoding of Colour Television Signals—Questions Concerning Study Group 11—Broadcasting Service (Television). 1 p.

—Television Signals—Sampling

CCIR QUESTION 61/11-90. Filtering and Sampling for Digital Encoding of Colour Television Signals—Questions Concerning Study Group 11—Broadcasting Service (Television). 1 p.

—Video Systems

IEC 11172 Pt 2-93. Information Technology—Coding of Moving Pictures and Associated Audio for Digital Storage Media at up to About 1,5 Mbit/s—Part 2: Video First Edition. 122 pp.

ISO 11172 Pt 2-93. Information Technology—Coding of Moving Pictures and Associated Audio for Digital Storage Media at up to About 1,5 Mbit/s—Part 2: Video First Edition. 122 pp.

JTC1 11172 Pt 2-93. Information Technology—Coding of Moving Pictures and Associated Audio for Digital Storage Media at up to About 1,5 Mbit/s—Part 2: Video First Edition. 122 pp.

Encryption, Data

Use: Data Encryption

End Delay

Use For: Round Trip End Delay *See Also:* Delay; Echo (Telecommunications); Echo Suppressors; Time

CCITT RECMN G.164-89. Echo Suppressors—General Characteristics of International Telephone Connections and Circuits (Study Groups XII and XV) 36 pp. 36 pp.

End Mills

See Also: Machine Tools; Milling Cutters; Milling Machines; Shell End Mills

CNS B1034-88. Hand of Cut and Helix for Plain and End Milling Cutters (Oct)(3595).

—Carbide

DIN ENGL 8011-63. Carbide Tips for Reamers, Countersinks, Counterbores and End Mills (Sept). 1 p.

—Helical—Parallel Shanks—Hard Metal

ISO 10145 Pt 1-93. End Mills with Brazed Helical Hardmetal Tips—Part 1: Dimensions of End Mills with Parallel Shank First Edition. 5 pp.

—Helical—Taper Shanks—Hard Metal

ISO 10145 Pt 2-93. End Mills with Brazed Helical Hardmetal Tips—Part 2: Dimensions of End Mills with 7/24 Taper Shank First Edition. 5 pp.

—Life Testing

ISO 8688 Pt 2-89. Tool Life Testing in Milling—Part 2: End Milling First Edition. 31 pp.

—Morse Taper Shanks

BSI BS 4193: Part 10-82. 1982 Hardmetal Insert Tooling Part 10: Mills with Indexable Inserts and Morse Taper Shank. 4 pp.

DIN ENGL 845 Pt 2-90. Morse Taper Shank End Mills; Technical Delivery Conditions (Apr). 6 pp.

ISO 1641 Pt II-78. End Mills and Slot Drills—Part II: Milling Cutters with Morse Taper Shanks First Edition. 5 pp.

ISO 6262 Pt 2-82. End Mills with Indexable Inserts—Part 2: End Mills with Morse Taper Shank First Edition. 3 pp.

End Mills (Cont.)

—Parallel Shanks

BSI BS 4193: Part 9-82. 1982 Hardmetal Insert Tooling Part 9: End Mills with Indexable Inserts and Flatted Parallel Shank. 4 pp.

DIN ENGL 844 Pt 1-89. Parallel Shank End Mills; Dimensions (Apr). 3 pp.

DIN ENGL 844 Pt 2-90. Parallel Shank End Mills; Technical Delivery Conditions (Apr). 7 pp.

DIN ENGL 6527-91. Solid Hardmetal End Mills with Stepped Parallel Shank; Dimensions (Feb). 3 pp.

DIN ENGL 6528-91. Solid Hardmetal End Mills with Continuous Parallel Shank; Dimensions (Feb). 3 pp.

DIN ENGL 6529-92. Parallel Shank Solid Hardmetal End Mills; Technical Delivery Conditions (Feb). 8 pp.

DIN ENGL 6535-92. Parallel Shanks for Hardmetal Twist Drills and End Mills; Dimensions (Feb). 4 pp.

ISO 1641 Pt I-78. End Mills and Slot Drills—Part I: Milling Cutters with Parellel Shanks First Edition. 5 pp.

ISO 6262 Pt 1-82. End Mills with Indexable Inserts—Part 1: End Mills with Flatted Parallel Shank First Edition. 3 pp.

JIS B 4114-83. Parallel Shank End Mills Brazed with Cemented Carbide Tips. 17 pp.

JIS B 4211-88. End Mills with Parallel Shanks. 12 pp.

—Screwed Shanks

BSI BS 122: Part 4-80. 1980 Amd 1 Milling Cutters Part 4: Screwed Shank End Mills and Slot Drills. 20 pp.

—Square

JIS B 4116-87. Solid Carbide Square End Mills. 13 pp.

—Straight Shanks

CNS B3233-74. End Mills with Straight Shank (Jan)(3608). 3 pp.

CNS B3235-74. Two-Flute End Mills with Straight Shank (Jan)(3610). 3 pp.

—Taper Shanks

CNS B3234-74. End Mills with Taper Shank (Jan)(3609). 4 pp.

CNS B3238-74. Two-Flute End Mills with Taper Shank (Jan)(3613). 5 pp.

DIN ENGL 2328 Pt 2-90. End Mills with 7/24 Taper Shank; Technical Delivery Conditions (Apr). 6 pp.

ISO 1641 Pt III-78. End Mills and Slot Drills—Part III: Milling Cutters with 7/24 Taper Shanks First Edition. 5 pp.

JIS B 4212-88. End Mills with Taper Shanks. 10 pp.

End of Address

Use For: EOA *See Also:* Address Codes; Codes; Control Codes

—Direct Dialing In Services—Integrated Services Digital Networks

CCITT RECMN E.164-91. Numbering Plan for the ISDN Era (Study Group II) 19 pp (Same as Recmn I.331). 19 pp.

CCITT RECMN E.164-89. Numbering Plan for the ISDN Era—Telephone Network and ISDN—Operation, Numbering, Routing and Mobile Service (Study Group II) 6 pp (Same as Recmn I.331). 6 pp.

End of Message Codes

See Also: Codes; Data Communication; Data Transmission

—Telegraph Repeaters

CCITT RECMN F.30-89. Use of Various Sequences of Combinations for Special Purposes—Telegraph and Mobile Services Operations and Quality of Service (Study Group I) 2 pp. 2 pp.

CCITT RECMN F.31-89. Telegram Retransmission System—Telegraph and Mobile Services Operations and Quality of Service (Study Group I) 12 pp. 12 pp.

—Telegraphy—Message Switching

CCITT RECMN F.35-89. Provisions Applying to the Operation of an International Public Automatic Message Switching Service for Equipments Utilizing the International Telegraph Alphabet No. 2—Telegraph and Mobile Services Operations and Quality of Service (Study Group I) 7 pp. 7 pp.

—Telex Communications—Message Switching

CCITT RECMN F.35-89. Provisions Applying to the Operation of an International Public Automatic Message Switching Service for Equipments Utilizing the International Telegraph Alphabet No. 2—Telegraph and Mobile Services Operations and Quality of Service (Study Group I) 7 pp. 7 pp.

INTERNATIONAL AND NON-U.S. NATIONAL STANDARDS
SUBJECT INDEX
Endurance

End of Message Codes *(Cont.)*
—Telex Communications—Store and Forward Units
CCITT RECMN F.72-89. International Telex Store and Forward—General Principles and Operational Aspects—Telegraph and Mobile Services Operations and Quality of Service (Study Group I) 18 pp. 18 pp.

End of Pulsing Signals
See Also: Signaling Systems

—Interregister Signaling—Termination—CCITT R2 Signaling Systems
CCITT RECMN Q.473-89. Signalling Procedures—Use of End-of-Pulsing Signal I-15 in International Working—Specifications of Signalling Systems R1 and R2 (Study Group XI) 2 pp. 2 pp.

—Register Signaling—CCITT No. 5 Signaling Systems
CCITT RECMN Q.152-89. End-of-Pulsing Conditions—Register Arrangements Concerning ST (End-of-Pulsing) Signal—Specifications of Signalling Systems Nos. 4 and 5 (Study Group XI) 2 pp. 2 pp.

—Register Signaling—CCITT R1 Signaling Systems
CCITT RECMN Q.321-89. End-of-Pulsing Conditions—Register Arrangements Concerning ST Signal—Specifications of Signalling Systems R1 and R2 (Study Group XI) 1 pp. 1 p.

End of Transmission Codes
See Also: Codes; Data Transmission

—Telex Communications—Store and Forward Units
CCITT RECMN F.72-89. International Telex Store and Forward—General Principles and Operational Aspects—Telegraph and Mobile Services Operations and Quality of Service (Study Group I) 18 pp. 18 pp.

End Quenching Testing
Use: Jominy Testing

Endoscopes
See Also: Gastroscopes; Medical Instruments
JIS T 1553-84. Endoscopic Apparatus for Medical Use. 13 pp.

—Lamps
BSI BS 1929-53. (OBSOLESCENT) 1953 Screw Threads and External Dimensions of Lamps for Endoscopic and Diagnostic Instruments. 8 pp.

—Safety
CSA CAN/CSA-C22. 2601.2.18-92. Medical Electrical Equipment Part 2: Particular Requirements for the Safety of Endoscopic Equipment (IEC 601-2-18:1990); (Gen Instr 1). 36 pp.
IEC 601 Pt 2-18-90. Medical Electrical Equipment Part 2: Particular Requirements for the Safety of Endoscopic Equipment First Edition. 37 pp.
SAA AS 3200.2.18-92. Approval and Test Specification—Medical Electrical Equipment—Part 2: Particular Requirements for Safety—Part 2.18: Endoscopic Equipment (IEC 601-2-18: 1990) (in Professional Packages 28, 17C) (This is a Joint Standard with SNZ NZS 3200.2.18). 13 pp.

Endotracheal Tubes
Use: Tracheal Tubes

Enduracidin Content Analysis
Use For: Enramycin Content Analysis

—Animal Feed
CNS N4126-84. Method of Test for Feed Additives: Determination of Enramycin (Nov)(11132).

Endurance Testing
Scope Note: For additional listings, use a more specific term *See Also:* Electric Endurance Testing; Fatigue Testing; Folding Endurance

—Appliances
JIS C 9101-87. General Rules of Endurance Test Methods of Household and Similar Electrical Appliances. 23 pp.

—Cable Couplers
CEN PREN 2591 (Part CG1)-92. Elements of Electrical and Optical Connection Test Methods Part CG1—Endurance Test. 4 pp.

—Cables (Electric)—Aircraft
CEN PREN 3475 (Part 410)-92. Cables, Electrical, Aircraft Use Test Methods Part 410—Thermal Endurance. 2 pp.

Endurance Testing *(Cont.)*
—Carburetors
CNS D3103-89. Endurance Test Code of Carburetors for Automobiles (Jan)(8260).

—Connectors
CEN PREN 2591 (Part D6)-92. Elements of Electrical and Optical Connection Test Methods Part D6—Mechanical Endurance. 4 pp.
SBAC TS 363 ISSUE 1. Test for Electrical Connectors Mechanical Endurance.

—Control Systems Equipment—Elevator Contactors
CSA CAN/CSA-C22. 2 NO 14-M91. Industrial Control Equipment; (Gen Instr 1 Thru 4). 142 pp.

—Electric Switches
CNS C6090-88. Method of Test for Electromechanical Switches (Mechanical Endurance) (Dec)(6153).

—Electrical Insulation
IEC 216 Pt 3-2-93. Guide for the Determination of Thermal Endurance Properties of Electrical Insulating Materials Part 3: In-structions for Calculating Thermal Endurance Characteristics Section 2—Calculations for Incomplete Data: Proof Test Results up to and Including the Median. 66 pp.
IEC 795-84. Test Method for Evaluating Thermal Endurance of Flexible Sheet Materials Using the Wrapped Tube Method First Edition. 13 pp.
IEC 1026-91. Guidelines for Application of Analytical Test Methods for Thermal Endurance Testing of Electrical Insulating Materials First Edition. 67 pp.

—Electromechanical Components
BSI BS 5772: Part 5-93. 1993 Electromechanical Components for Electronic Equipment: Testing Procedures and Measuring Methods Part 5: Impact Tests (Free Components), Static Load Tests (Fixed Components), Endurance Tests and Overload Tests (IEC 512-5: 1992) (S). 24 pp.
BSI BS 5772: Part 5-79. 1979 Basic Testing Procedures and Measuring Methods for Electromechanical Components for Electronic Equipment Part 5: Impact Tests (Free Components), Static Load Tests (Fixed Components), Endurance Tests and Overload Tests (Switches). 16 pp.
BSI BS 5772: Pt 5: Supp 1-81. 1981 Basic Testing Procedures and Measuring Methods for Electromechanical Components for Electronic Equip. Part 5: Impact Tests (Free Components), Static Load Tests (FixedComponents) Endurance Test and Over-Load Tests (Switches) Supp 1: Tests 7b and 10a. 6 pp.
BSI BS 5772: Pt 5: Supp 2-82. 1982 Basic Testing Procedures and Measuring Methods for Electromechanical Components for Electronic Equip. Part 5: Impact Tests (Free Components), Static Load Tests (FixedComp.), Enduranceand Overload Tests (Switches) Supp. 2:Test 10c. 6 pp.
IEC 512 Pt 5-92. Electromechanical Components for Electronic Equipment; Basic Testing Procedures and Measuring Methods Part 5: Impact Tests (Free Components), Static Load Tests (Fixed Components), Endurance Tests and Overload Tests Second Edition. 50 pp.

—Electronic Components
CNS C6249-85. Endurance (Mechanical) Testing Method for Electronic Components (Apr)(11240).
JIS C 5037-75. Endurance (Mechanical) Testing Method for Electronic Components. 4 pp.

—Faucets
BSI BS 5412 & 5413: Part 4-76. 1976 Amd 3 Performance of Draw-Off Taps with Metal Bodies for Water Services and and with Plastic Bodies for Water Services Part 4: Mechanical and Endurance Characteristics. 12 pp.

—Field Effect Transistors
CNS C6058-88. Environmental Testing Methods and Endurance Testing Methods for Discrete Semiconductor Devices (Continuous Operation Test of Field-Effect Transistor) (Sep)(5542).
CNS C6060-88. Environmental Testing Methods and Endurance Testing Methods for Discrete Semiconductor Devices (Intermittent Operation Test of Field-Effect Transistor) (Sep)(5544).
CNS C6062-88. Environmental Testing Methods and Endurance Testing Methods of Discrete Semiconductor Devices (Reverse Bias Test of Field-Effect Transistor Under High Temperature) (Sep)(5546).

—Hinges
BSI BS 7352-90. 1990 Strength and Durability Performance of Metal Hinges for Side Hanging Applications and Dimensional Requirements for Template Drilled Hinges. 24 pp.

Endurance Testing *(Cont.)*
—Hip Prostheses
BSI BS 7251: Part 5-90. 1990 Orthopaedic Joint Prostheses Part 5: Method for Determination of Endurance Properties of Stemmed Femoral Components of Hip Joint Prostheses with Application of Torsion. 11 pp.
BSI BS 7251: Part 6-90. 1990 Orthopaedic Joint Prostheses Part 6: Method for Determination of Endurance Properties of Stemmed Femoral Components of Hip Joint Prostheses Without Application of Torsion. 12 pp.
BSI BS 7251: Part 10-92. 1992 Orthopaedic Joint Prostheses Part 10: Method of Determination of Endurance Properties of the Head and Neck Region of Stemmed Femoral Components of Hip Joint Prostheses (ISO 7206-6: 1992). 11 pp.
BSI BS 7251: Part 11-93. 1993 Orthopaedic Joint Prostheses Part 11: Specification for Endurance of Stemmed Femoral Components Without Application of Torsion (ISO 7206-7: 1993) (N). 7 pp.
BSI DD 91-86. (WITHDRAWN) 1986 Method for Determination of Endurance Properties of Stemmed Femoral Components of Hip Joint Prostheses (Superseded by BS 7251: Part 5: 1990). 11 pp.
ISO 7206 Pt 3-88. Implants for Surgery—Partial and Total Hip Joint Prostheses—Part 3: Determination of Endurance Properties of Stemmed Femoral Components Without Application of Torsion First Edition. 14 pp.
ISO 7206 Pt 4-89. Implants for Surgery—Partial and Total Hip Joint Prostheses—Part 4: Determination of Endurance Properties of Stemmed Femoral Components with Application of Torsion First Edition. 12 pp.
ISO 7206 Pt 6-92. Implants for Surgery—Partial and Total Hip Joint Prostheses—Part 6: Determination of Endurance Properties of Head and Neck Region of Stemmed Femoral Components First Edition. 10 pp.
ISO 7206 Pt 7-93. Implants for Surgery—Partial and Total Hip Joint Prostheses—Part 7: Endurance Performance of Stemmed Femoral Components Without Application of Torsion First Edition. 6 pp.

—Integrated Circuits
JIS C 7022-79. Environmental Testing Methods and Endurance Testing Methods for Semiconductor Integrated Circuits (R 1984). 50 pp.

—Photovoltaic Cells
JIS C 8917-89. Environmental and Endurance Test Methods for Crystalline Solar Cell Modules. 41 pp.

—Plastics
BSI BS 4618: Sec 4.6-74. 1974 Recommendations for the Presentation of Plastics Design Data Part 4: Enviromental and Chemical Effects Section 4.6: The Thermal Endurance of Plastics. 10 pp.

—Relay Contacts
BSI BS 5992: Part 4-83. 1983 Electrical Relays Part 4: Contact Loads Preferred Values Used in Endurance Tests for Electrical Relay Contacts. 14 pp.
BSI BS 5992: Part 5-83. 1983 Electrical Relays Part 5: Test Equipment Used in Endurance Tests for Electrical Relay Contacts. 9 pp.
IEC 255 Pt 14-81. Electrical Relays Part 14: Endurance Test for Electrical Relay Contacts—Preferred Values for Contact Loads First Edition. 30 pp.
IEC 255 Pt 15-81. Electrical Relays Part 15: Endurance Tests for Electrical Relay Contacts—Specification for the Characteristics of Test Equipment First Edition. 16 pp.

—Semiconductor Devices
CNS C6041-83. Environmental Testing Methods and Endurance Testing Method for Discrete Semiconductor Devices (General Rules) (Apr)(5066).
CNS C6042-88. Environmental Testing Methods and Endurance Testing Methods for Discrete Semiconductor Devices (Test of Soft Solders for Heat Endurance) (Jun)(5067).
JIS C 7021-77. Environmental Testing Methods and Endurance Testing Methods for Discrete Semiconductor Devices (R 1980). 72 pp.

—Shunt Power Capacitors
BSI BS 7264: Part 2-90. 1990 Shunt Capacitors for a.c. Power Systems Having a Rated Voltage Above 660V Part 2: Guide to Endurance Testing. 17 pp.
BSI BS 7264: Part 2-01. 1990 Amd 1 Shunt Capacitors for a.c. Power Systems Having a Rated Voltage Above 1000V Part 2: Guide to Endurance Testing (AMD 7530) February 15, 1993. 26 pp.
CENELEC HD 525.2 S1-89. Shunt Capacitors for a.c. Power Systems Having a Rated Voltage Above 660v Part 2: Endurance Testing. 6 pp.
CENELEC HD 525.2 S1/A1-91. AMD 1 Shunt Capacitors for A.C. Power Systems Having a Rated Voltage Above 660V Part 2: Endurance Testing. 6 pp.

INDUSTRY STANDARDS

Endurance Testing *(Cont.)*
—**Shunt Power Capacitors** *(Cont.)*
IEC 871 Pt 2-87. Shunt Capacitors for a.c. Power Systems Having a Rated Voltage Above 660 V Part 2: Endurance Testing First Edition; (Amendment 1-1991). 33 pp.

—**Surge Arresters**
IECQ PQC 77-88. Measuring and Testing Methods for Surge Protective Devices. 39 pp.

Energy Absorbers
Use: Fall Arresting Devices—Energy Absorbers

Energy Conservation
See Also: Energy Consumption; Thermal Insulation
SAA AS 3596-92. Energy Management Programs—Guidelines for Definition and Analysis of Energy and Cost Savings. 18 pp.

—**Agricultural Buildings**
BSI BS 5502: Sec 1.4-86. 1986 Design of Buildings and Structures for Agriculture Part 1: General Considerations Section 1.4: Energy. 4 pp.

—**Buildings**
SNZ NZS 4220-82. Code of Practice for Energy Conservation in Non-Residential Buildings. 30 pp.

—**Houses**
BSI BS 8211: Part 1-88. 1988 Energy Efficiency in Housing Part 1: Code of Practice for Energy Efficient Refurbishment of Housing. 17 pp.

Energy Consumption
See Also: Energy Conservation; Fuel Consumption

—**Boilers**
CSA B212-93. Seasonal Energy Utilization Efficiencies of Oil-Fired Furnaces and Boilers; (Gen Instr 1). 72 pp.

—**Broadcast Transmitters—Sound Broadcasting**
CCIR Report 1060-1-90. Energy Saving Methods in Amplitude Modulation Broadcasting and Their Influence on Reception Quality—Section 10A-1—Amplitude-Modulation Sound Broadcasting in Bands 5 (LF), 6 (MF) and 7 (HF). 4 pp.

—**Buildings**
BSI BS 8207-85. 1985 Code of Practice for Energy Efficiency in Buildings. 39 pp.
BSI BS 8207: Supplement-85. 1985 Code of Practice for Engery Efficiency in Buildings: Supplement: Energy Design Guide. 40 pp.

—**Clothes Dryers**
CSA CAN/CSA-C361-92. Test Method for Measuring Energy Consumption and Drum Volume of Electrically Heated Household Tumble-Type Clothes Dryers; (Gen Instr 1). 25 pp.
ISO 9398 Pt 2-93. Specifications for Industrial Laundry Machines—Definitions and Testing of Capacity and Consumption Characteristics—Part 2: Batch Drying Tumblers First Edition. 7 pp.

—**Dishwashers**
CENELEC HD 378-78. Methods to be Used for Measuring Energy Consumption of Electric Automatic Dishwashers for Cold Water Supply Only, for Household Use and for the Purpose of Informing Consumers of It. 5 pp.
CSA CAN/CSA-C373-92. Energy Consumption Test Methods for Household Dishwashers; (Gen Instr 1). 21 pp.

—**Food Preparation Equipment**
CEN PREN 203-2-92. Gas Heated Catering Equipment—Part 2: Rational Use of Energy. 10 pp.

—**Freezers**
BSI BS EN 153-90. 1990 Methods of Measuring the Energy Consumption of Electric Mains Operated Household Refrigerators, Frozen Food Storage Cabinets, Food Freezers and Their Combinations, Together with Associated Characteristics. 8 pp.
CEN EN 153-82. Methods of Measuring the Energy Consumption of Electric Mains Operated Household Refrigerators, Frozen Food Storage Cabinets, Food Freezers and Their Combinations, Together with Associated Characteristics. 7 pp.
CEN EN 153-90. Methods of Measuring the Energy Consumption of Electric Mains Operated Household Refrigerators, Frozen Food Storage Cabinets, Food Freezers and Their Combinations Together with Associated Characteristics. 7 pp.
CSA CAN/CSA-C300-M89. Capacity Measurement and Energy Consumption Test Methods for Refrigerators, Combination Refrigerator-Freezers, and Household Freezers; (Gen Instr 1). 90 pp.

Energy Consumption *(Cont.)*
—**Freezers** *(Cont.)*
CSA CAN/CSA-C300-M91. Capacity Measurement and Energy Consumption Test Methods for Refrigerators, Combination Refrigerator-Freezers, and Freezers; (Gen Instr 1). 95 pp.
SNZ NZS 6205: Part 1-89. Energy Labelling of Appliances Part 1: Refrigerators, Refrigerator/Freezers and Freezers—Specification for Appliance Energy Rating Label (NZS 6205.1:1989/AS 2575.1:1989). 12 pp.
SNZ NZS 6205: Part 2-89. Energy Labelling of Appliances Part 2: Refrigerators, Refrigerator/Freezers and Freezers—Determination of Energy Consumption and Efficiency Rating (NZS 6205.2:1989/AS 2575.2:1989). 36 pp.

—**Gas Meters**
SNZ NZS 5259-91. Gas Metering. Gas Meter Acceptance Testing, Meter Installation and Calculation of Energy from Measured Volume Amend: 1, 1993. 61 pp.

—**Irons (Electric)—Flatwork**
ISO 9398 Pt 1-93. Specifications for Industrial Laundry Machines—Definitions and Testing of Capacity and Consumption Characteristics—Part 1: Flatwork Ironing Machines First Edition. 8 pp.

—**Microwave Ovens**
CSA CAN/CSA-C388-M89. Energy Consumption Test Methods for Household Microwave Ovens; (Gen Instr 1). 20 pp.

—**Oil Furnaces**
CSA B212-93. Seasonal Energy Utilization Efficiencies of Oil-Fired Furnaces and Boilers; (Gen Instr 1). 72 pp.

—**Ranges**
CENELEC HD 376 S2-84. Methods to be Used for Measuring Energy Consumption of Electric Ovens for Household Use and for the Purpose of Informing Consumers of It. 8 pp.
CSA CAN/CSA-C358-M89. Energy Consumption Test Methods for Household Electric Ranges; (Gen Instr 1). 32 pp.

—**Refrigerated Display Cases**
CEN PREN 441-9-91. Commercial Refrigerated Cabinets—Part 9: Electrical Energy Consumption Test. 6 pp.
DIN ENGL 8954 Pt 6-74. Open Refrigerated Display Cabinets; Electrical Energy Consumption Testing (Feb). 1 p.

—**Refrigerator/Freezers**
CSA CAN/CSA-C300-M91. Capacity Measurement and Energy Consumption Test Methods for Refrigerators, Combination Refrigerator-Freezers, and Freezers; (Gen Instr 1). 95 pp.

—**Refrigerators**
BSI BS EN 153-90. 1990 Methods of Measuring the Energy Consumption of Electric Mains Operated Household Refrigerators, Frozen Food Storage Cabinets, Food Freezers and Their Combinations, Together with Associated Characteristics. 8 pp.
CEN EN 153-82. Methods of Measuring the Energy Consumption of Electric Mains Operated Household Refrigerators, Frozen Food Storage Cabinets, Food Freezers and Their Combinations, Together with Associated Characteristics. 7 pp.
CEN EN 153-90. Methods of Measuring the Energy Consumption of Electric Mains Operated Household Refrigerators, Frozen Food Storage Cabinets, Food Freezers and Their Combinations Together with Associated Characteristics. 7 pp.
CSA CAN/CSA-C300-M89. Capacity Measurement and Energy Consumption Test Methods for Refrigerators, Combination Refrigerator-Freezers, and Household Freezers; (Gen Instr 1). 90 pp.
CSA CAN/CSA-C300-M91. Capacity Measurement and Energy Consumption Test Methods for Refrigerators, Combination Refrigerator-Freezers, and Freezers; (Gen Instr 1). 95 pp.
SNZ NZS 6205: Part 1-89. Energy Labelling of Appliances Part 1: Refrigerators, Refrigerator/Freezers and Freezers—Specification for Appliance Energy Rating Label (NZS 6205.1:1989/AS 2575.1:1989). 12 pp.
SNZ NZS 6205: Part 2-89. Energy Labelling of Appliances Part 2: Refrigerators, Refrigerator/Freezers and Freezers—Determination of Energy Consumption and Efficiency Rating (NZS 6205.2:1989/AS 2575.2:1989). 36 pp.

—**Residential Buildings—Thermal Insulation**
CEN PREN 832-92. Thermal Performance of Buildings—Calculation of Energy Use for Heating—Residential Buildings. 55 pp.
ISO 9164-89. Thermal Insulation—Calculation of Space Heating Requirements for Residential Buildings First Edition. 29 pp.

Energy Consumption *(Cont.)*
—**Storage Water Heaters**
CENELEC HD 500 S1-88. Methods to be Used for Measuring Energy Consumption of Thermal Storage Water Heaters and for the Purpose of Informing Consumers of it. 5 pp.
SAA AS 1056.4-90. Storage Water Heaters—Part 4: Calculations of Energy Consumption. 6 pp.

—**Three Phase Motors**
CSA C390-93. Energy Efficiency Test Methods for Three-Phase Induction Motors; (Gen Instr 1). 51 pp.

—**Washing Machines**
CSA CAN/CSA-C360-92. Test Method for Measuring Energy Consumption and Capacity of Automatic Household Clothes Washers; (Gen Instr 1). 32 pp.
ISO 9398 Pt 4-93. Specifications for Industrial Laundry Machines—Definitions and Testing of Capacity and Consumption Characteristics—Part 4: Washers-Extractors First Edition. 9 pp.

—**Washing Tunnels**
ISO 9398 Pt 3-93. Specifications for Industrial Laundry Machines—Definitions and Testing of Capacity and Consumption Characteristics—Part 3: Washing Tunnels First Edition. 7 pp.

—**Water Extractors**
ISO 9398 Pt 4-93. Specifications for Industrial Laundry Machines—Definitions and Testing of Capacity and Consumption Characteristics—Part 4: Washers-Extractors First Edition. 9 pp.

—**Wheelchairs**
ISO 7176 Pt 4-88. Wheelchairs—Part 4: Determination of Energy Consumption of Electric Wheelchairs First Edition. 6 pp.
SAA AS 3696.4-92. Wheelchairs—Part 4: Determination of Energy Consumption of Electric Wheelchairs (ISO 7176-4:1988) (in Professional Package 17G). 3 pp.

Energy Efficiency
Use: Energy Consumption

Energy Management
See Also: Environmental Engineering; Power Distribution Systems
CSA C22.2 NO 205-M1983. Signal Equipment (R 1992); (Gen Instr 1). 32 pp.
JIS Z 9204-91. General Rules for Energy Evaluation Method by Available Energy. 13 pp.

—**Classification**
BSI BS 1000: (620)-90. 1990 Universal Decimal Classification (UDC). English Full Edition (620): Materials Testing. Commercial Materials. Power Stations. Economics of Energy. 28 pp.
SNZ NZS/BS 1000 (620)-73. Universal Decimal Classification Materials Testing. Commercial Materials. Power Stations. Economics of Energy. 32 pp.

—**Glossaries**
JIS Z 9211-82. Technical Terms Used in Energy Management.
JIS Z 9212-83. Technical Terms Used in Energy Management (No. 2).

Energy Regulators (Controllers)
Use: Power Regulators (Controllers)

Energy Sources
See Also: Fuels; Nuclear Power; Solar Radiation

—**Broadcasting—Developing Countries**
CCIR QUESTION 54/10-86. Broadcasting System Using Unconventional Energy Sources—Questions Concerning Study Group 10—Broadcasting Service (Sound). 1 p.

—**Civil Engineering—Glossaries**
BSI BS 6100: Sec 3.1-86. 1986 Glossary of Building and Civil Engineering Terms Part 3: Services Section 3.1: Energy Sources and Distribution. 24 pp.
BSI BS 6100: Sec 3.1-01. 1986 Amd 1 Glossary of Building and Civil Engineering Terms Part 3: Services Section 3.1: Energy Sources and Distribution (AMD 7256) August 15, 1992. 28 pp.

—**Construction Engineering—Glossaries**
BSI BS 6100: Sec 3.1-86. 1986 Glossary of Building and Civil Engineering Terms Part 3: Services Section 3.1: Energy Sources and Distribution. 24 pp.
BSI BS 6100: Sec 3.1-01. 1986 Amd 1 Glossary of Building and Civil Engineering Terms Part 3: Services Section 3.1: Energy Sources and Distribution (AMD 7256) August 15, 1992. 28 pp.

Energy Transmission
See Also: Power Transmission

INTERNATIONAL AND NON-U.S. NATIONAL STANDARDS
SUBJECT INDEX
Engineering

Energy Transmission *(Cont.)*
—Microwave Beam—Free Space
CCIR Report 679-2-86. Characteristics and Effects of Radio Techniques for the Transmission of Energy from Space—Section 2B—Topics of General Interest. 7 pp.

—Microwave Beam—Satellites
CCIR Report 679-2-86. Characteristics and Effects of Radio Techniques for the Transmission of Energy from Space—Section 2B—Topics of General Interest. 7 pp.

—Radio Transmission
CCIR QUESTION 130/7-90. Characteristics and Effects of Radio Techniques for the Transmission of Energy—Questions Concerning Study Group 7—Science Services. 1 p.

ENG
Use: Electronystagmographs

ENG (Electronic News Gathering)
Use: Electronic News Gathering

ENG Equipment
Use: Electronic News Gathering Equipment

Engine Blocks
Use: Cylinder Blocks

Engine Coolants
See Also: Coolants

—Aircraft—Hose Fittings
BSI 2C 4-73. 1973 Amd 2 Coupling Dimensions for Aero-Engine Refrigerant Pressure Replenishment Connections. 9 pp.

—Alkalinity
CNS K6481-80. Method of Test for Reserve Alkalinity of Engine Antifreezes Antirusts and Coolants (May)(5584).

—Ash Content
CNS K6484-80. Method of Test for Ash Content of Engine Antifreezes, Antirusts and Coolants (May)(5587).

—Corrosion Inhibitors
BSI BS 4959-74. 1974 Recommendations for Corrosion and Scale Prevention in Engine Cooling Water Systems. 19 pp.
BSI BS 6580-92. 1992 Corrosion Inhibiting, Engine Coolant Concentrate ('Antifreeze'). 10 pp.
BSI BS 6580-85. 1985 Corrosion Inhibiting, Engine Coolant Concentrate. 8 pp.
JIS K 2408-90. Corrosion Inhibitors for Engine Coolant. 22 pp.

—Ethylene Glycol
CGSB CAN/CGSB-3.890-M83. Coolant Concentrate, Automotive Engine. 9 pp.
MOD UK DSTAN 68-108-91. Ethanediol (Ethylene Glycol) Issue 1. 16 pp.
MOD UK DSTAN 68-108-93. Ethanediol (Ethylene Glycol) Joint Service Designation AL-20 Issue 2. 16 pp.

—Scale Prevention
BSI BS 4959-74. 1974 Recommendations for Corrosion and Scale Prevention in Engine Cooling Water Systems. 19 pp.

Engine Cylinders
See Also: Combustion Chambers; Cylinder Blocks; Cylinder Heads; Cylinders (Containers); Exhaust Valves; Inlet Valves; Internal Combustion Engines; Pistons
CNS D2136-87. Cylinder for Small Internal Combustion Engines of Land Use (Jul)(8953).

—Aircraft—Corrosion Inhibitors—Lubricating Oils
MOD UK DTD-791C-72. Corrosion Preventive Oil, Aircraft Piston Engine: Static Preservation, Upper Cylinder NATO Code Number: C-613 Joint Service Designation: PX-13. 6 pp.

—Aircraft—Numbering
BSI 2M 41-77. 1977 Methods of Numbering Propulsion Units and Components and Describing Their Direction of Rotation. 6 pp.
ISO 482-77. Aircraft—Propulsion Units and Components—Methods of Numbering and Describing Direction of Rotation First Edition. 6 pp.

—Automotive—Linings
CNS D2088-85. Cylinder Liners for Automobile Engines (Cast Iron) (Jun)(7136). 11 pp.
JIS D 3103-89. Cylinder Liners for Automotive Engines. 19 pp.

Engine Cylinders *(Cont.)*
—Forgings—Aluminum Alloys—Aerospace
MOD UK DTD-324B-71. Forging Stock and Forgings for Engine Cylinders and Pistons of Aluminium—Silicon—Magnesium—Copper—Nickel Alloy (Solution Treated and Precipitation Treated) (Si 11.5, Mg 1.1, Cu 1.0, Ni 1.0). 2 pp.

—Gasoline Engines—Pressure—Testing Equipment
CNS B6074-82. Test Device for Cylinder Pressure of Gasoline Engine (Oct)(9466).

—Gasoline Engines—Testing Equipment
CNS B7209-82. Method of Test for Test Device for Cylinder Pressure of Gasoline Engine (Oct)(9467).

—Internal Combustion Engines—Automotive
CNS D2136-87. Cylinder for Small Internal Combustion Engines of Land Use (Jul)(8953).

—Internal Combustion Piston Engines—Designations
BSI BS 5673-79. (WITHDRAWN) 1979 Designation of the Cylinders of Reciprocating Internal Combustion Engines (Superseded by BS 5672: 1991). 7 pp.

Engine Generators
Use: Generator Sets

Engine Heaters
See Also: Glow Plugs; Heaters
CSA CAN/CSA-C22. 2 NO 191-M89. Engine Heaters and Battery Warmers; (Gen Instr 1 Thru 3). 54 pp.

—Automotive
CSA CAN/CSA-C22. 2 NO 191-M89. Engine Heaters and Battery Warmers; (Gen Instr 1 Thru 3). 54 pp.

—Trucks
CSA CAN/CSA-C22. 2 NO 191-M89. Engine Heaters and Battery Warmers; (Gen Instr 1 Thru 3). 54 pp.

Engine Lathes
See Also: Lathes; Machine Tools
CNS B1098-79. Specified Items of Machine Tools Engine Lathe (Jan)(4670-1). 2 pp.

Engine Noses
Use: Nose Cones

Engine Oils
Use: Motor Oils

Engine Starters
Use For: Starter Circuits; Starting Motors
See Also: Internal Combustion Engines; Motor Starters; PAD Cartridges; Starters

—Aircraft
NATO STANAG 3368 ED 2 AMD 7-68. Internal Aircraft Engine Starting Systems. 9 pp.
NATO STANAG 3368 ED 2 AMD 8-68. Internal Aircraft Engine Starting Systems. 10 pp.

—Aircraft—Air—Connectors
ISO 2026-74. Aircraft—Connections for Starting Engines by Air First Edition. 6 pp.
NATO STANAG 3372 ED 5 AMD 2-88. Low Pressure Air and Associated Electrical Connections for Aircraft Engine Starting. 10 pp.
NATO STANAG 3372 ED 5 AMD 3-88. Low Pressure Air and Associated Electrical Connections for Aircraft Engine Starting. 11 pp.

—Aircraft—Firesafety
CAA NOTICE #33 ISSUE 3. Unprotected Starter Circuits in Aircraft Not Exceeding 12500 lb (Airworthiness Notices). 2 pp.

—Aircraft—Flanges
SBAC RS 310 (V). 5.750 Inch P.C.DIA. Starter Mounting Flange.
SBAC RS 311 (V). 7 Inch P.C.DIA. Starter Mounting Flange.

—Automobiles
CNS D1006-87. Rated Output of Starting Motors for Automobiles (Oct)(5781).
CNS D3044-87. Mounting Dimensions of Starting Motors for Automobiles (Oct)(5321).
JIS D 5203-76. Mounting Dimensions of Starting Motors for Automobiles.

—Automobiles—Inspection
CNS D3064-87. Method of Test for Starting Motors of Automobiles (Oct)(6171).
JIS D 1607-89. Test Methods of Starter Motors for Automobiles.

Engine Starters *(Cont.)*
—Internal Combustion Piston Engines—Glossaries
ISO 7967 Pt 8-90. Reciprocating Internal Combustion Engines—Vocabulary of Components and Systems—Part 8: Starting Systems First Edition. 13 pp.

Engine Timing Lights
Use: Timing Lamps

Engine Valves
See Also: Combustion Chambers; Cylinder Heads; Internal Combustion Engines; Valves

—Internal Combustion Piston Engines—Glossaries
BSI BS 7016: Part 3-88. 1988 Components and Systems of Reciprocating Internal Combustion Engines Part 3: Glossary of Terms of Valves, Camshaft Drive and Actuating Mechanisms. 19 pp.
ISO 7967 Pt 3-87. Reciprocating Internal Combustion Engines—Vocabulary of Components and Systems—Part 3: Valves, Camshaft Drive and Actuating Mechanisms First Edition. 19 pp.

—Motorcycles
CNS D3129-82. Method of Test for Engine Valves for Motorcycles (Jun)(8958).

Engineering
Scope Note: Use a more specific term
See: Aeronautical Engineering; Civil Engineering; Control Engineering; Electrical Engineering; Electronic Engineering; Engineering Drawings; Environmental Engineering; Ergonomics; Mechanical Engineering; Military Engineering; Nuclear Engineering; Sanitary Engineering; Structural Engineering; Traffic Engineering (Telecommunications); Water Supply Engineering

Engineering Drawings
Use For: Mechanical Drawings; Sections (Drawings); Technical Drawings *See Also:* Architectural Drawings; Axonometric Projections; Drafting; Drawings; Normal Projections; Symbols; Title Blocks; Views
BSI BS 308: Part 1-84. 1984 Amd 3 Engineering Drawing Practice Part 1: Recommendations for General Principles (AMD 6710) April 30, 1991. 55 pp.
BSI BS 308: Part 2-85. 1985 Amd 1 Engineering Drawing Practice Part 2: Recommendations for Dimensioning and Tolerancing of Size (AMD 6711) April 30, 1991. 55 pp.
BSI BS 308: Part 2-02. 1985 Amd 2 Engineering Drawing Practice Part 2: Recommendations for Dimensioning and Tolerancing of Size (AMD 7157) September 15, 1992. 64 pp.
BSI BS 2856-73. 1973 Precise Conversion of Inch and Metric Sizes on Engineering Drawings. 17 pp.
BSI BS 5070: Part 1-88. 1988 Engineering Diagram Drawing Practice Part 1: Recommendations for General Principles. 15 pp.
BSI PP 7305-91. 1991 Geometrical Tolerancing Solutions to Excercises in Geometrical Tolerancing (G). 49 pp.
CNS B1001-4-88. Engineering Drawing Geometrical Tolerancing (Dec)(3-4).
CSA CAN3-B78.1-M83. Technical Drawings—General Principles; (R 1990) (Erratum August 1983). 60 pp.
DIN ENGL 6 Pt 2-86. Technical Drawings; Representation in Normal Projection; Sections (Dec). 8 pp.
ISO 5458-87. Technical Drawings—Geometrical Tolerancing—Positional Tolerancing First Edition. 15 pp.
ISO TR5460-85. Technical Drawings—Geometrical Tolerancing—Tolerancing of Form, Orientation, Location and Run-out—Verification Principles and Methods—Guidelines First Edition. 71 pp.
ISO 8015-85. Technical Drawings—Fundamental Tolerancing Principle First Edition. 7 pp.
JIS B 0001-85. Technical Drawing for Mechanical Engineering. 81 pp.
JIS B 0024-88. Technical Drawings-Fundamental Tolerancing Principle; (ISO 8015-1985). 10 pp.
JIS B 0025-91. Technical Drawings—Geometrical Tolerancing —Positional Tolerancing (ISO 5458-1987).
JIS Z 8311-84. Technical Drawing—Sizes and Format of Drawings. 13 pp.
JIS Z 8316-84. Technical Drawing—Presentation of Views and Sections. 30 pp.
NATO ASTANP-2 V1 ED2 AMD0-82. NATO Guide to Variances in Drawing Practice. 149 pp.
SAA AS 1100. Technical Drawing See Also Under Handbooks —HB1, HB3, HB6.
SAA AS 1100.101-92. Technical Drawing—Part 101: General Principles (in Professional Package 56). 229 pp.

INDUSTRY STANDARDS

Engineering Drawings (Cont.)

SAA AS 1100.201-92. Technical Drawing—Part 201: Mechanical Engineering Drawing (in Professional Package 56). 71 pp.
SAA AS 1100.401-84. Technical Drawing—Part 401: Engineering Survey and Engineering Survey Design Drawing Amdt 1 May 1984. 57 pp.
SAA AS 1100.501-85. Technical Drawing—Part 501: Structural Engineering Drawing. 18 pp.
SAA AS 1100.501 Supp 1-86. Technical Drawing 501: Structural Engineering Drawing—Supplement 1: Structural Engineering Drawings (Supplement to AS 1100. 501—1985). 17 pp.
SNZ NZS/BS 308: Part 1-84. Engineering Drawing Practice Part 1: Recommendations for General Principles (Withdrawn—Superseded by NZS/AS 1100 Parts 101-501).
SNZ NZS/BS 308: Part 2-85. Engineering Drawing Practice Part 2: Recommendations for Dimensioning and Tolerancing of Size (Withdrawn—Superseded by NZS/AS 1100 Parts 101-501).
SNZ NZS/AS 1100: Part 101-92. Technical Drawing Part 101: General Principles. 229 pp.
SNZ NZS/AS 1100: Part 201-92. Technical Drawing Part 201: Mechanical Engineering Drawing. 71 pp.
SNZ NZS/AS 1100: Part 401-84. Technical Drawing Part 401: Engineering Survey and Engineering Survey Design Drawing Amend: 1. 57 pp.
SNZ NZS/AS 1100: Part 501-85. Technical Drawing Part 501: Structural Engineering Drawing. 18 pp.
SNZ NZS/AS 1100: Pt501: Suppl-86. Technical Drawing Part 501: Structural Engineering Drawing. 17 pp.
SNZ BS 5070-74. Drawing Practice for Engineering Diagrams. 65 pp.
SNZ BS 5070: Part 1-88. Engineering Diagram Drawing Practice Part 1: Recommendations for General Principles. 16 pp.
SNZ NZMP 5905-81. Standard Technical Drawing Practice for School Certificate Courses in New Zealand. 31 pp.
SNZ NZMP 5906-84. Standard Technical Drawing Practice for Senior Courses in Schools and Technical Institutes. 74 pp.

—Abbreviations
CSA Z85-1983. ABBR Abbreviations for Scientific and Engineering Terms. 280 pp.

—Aerospace
EURO CTI/CO/84/15 166-86. Drawing System for European Aerospace Manufacturers. 44 pp.

—Aerospace—Hardware
SBAC AGS 1 (V). Procedure for the Establishment of Standards for Proprietary Items.

—Aerospace—Identification Systems
AECMA PREN2851-89. Marking or Parts and Assemblies Indications on Drawings. 7 pp.
BSI BS EN 2851-92. 1992 Marking of Parts and Assemblies Other Than Engines—Indications on Drawings. 12 pp.
CEN EN 2851-92. Marking of Parts and Assemblies Other Than Engines—Indications on Drawings. 8 pp.

—Aerospace Rivets
AECMA PREN2544-83. Representation of Rivets on Drawings for Aerospace Equipment. 7 pp.
BSI BS EN 2544-90. 1990 Representation of Rivets on Drawings for Aerospace Equipment. 10 pp.
CEN EN 2544-87. Representation of Rivets on Drawings for Aerospace Equipment. 7 pp.

—Air Handling Equipment—Symbols
ISO 4067 Pt 1-84. Technical Drawings—Installations—Part 1: Graphical Symbols for Plumbing, Heating, Ventilation and Ducting First Edition. 12 pp.

—Air Terminal Devices
ISO 6412 Pt 3-93. Technical Drawings—Simplified Representation of Pipelines—Part 3: Terminal Features of Ventilation and Drainage Systems First Edition. 8 pp.

—Aircraft
CAA LEAFLET 2-1 07.90. Engineering Drawings. 17 pp.

—Aircraft—Definitions
SBAC RS 622 ISSUE 4. Definition of Drawing State.

—Aircraft Engines
MOD UK AVP 115: Part 3-01. Requirements and Definitions for Drawings, Drawing Introduction Sheet and Design Changes for Equipment Under the Design Control of DG/ENG (PE); Amendment 24 (Superseded by Def Stan 05-124). 89 pp.

—Aircraft—Part Numbering
SBAC TS 66 ISSUE 3. Draughting and Part Numbering Procedures for S.B.A.C. Drawings. 15 pp.
SBAC TS 88 ISSUE 1. Standard Drawing Numbering System and Drawing Format.

Engineering Drawings (Cont.)

—Aircraft—Threaded Components
SBAC RS 726 (V). Removal of Incomplete, Threads.

—Antifriction Bearings—Symbols
ISO 8826 Pt 1-89. Technical Drawings—Rolling Bearings—Part 1: General Simplified Representation First Edition. 6 pp.

—Automatic Control Equipment—Symbols
ISO TR8545-84. Technical Drawings—Installations—Graphical Symbols for Automatic Control First Edition. 6 pp.

—Automotive—Test Dummies—Safety Belts
SAA AS E35/TD-73. Working Drawing for Test Dummy for Seat Belt Assemblies 1 Drawing 1065MM x 1565 MM.

—Aviation Facilities
SAA AS 1100.401 Supp 1-84. Technical Drawing—Part 401: Engineering Survey and Engineering Survey Design Drawing—Supplement 1: Aviation Facilities (Supplement to AS 1100. 401—1984). 7 pp.
SNZ NZS/AS 1100: Pt401: Suppl-84. Technical Drawing Part 401: Engineering Survey and Engineering Survey Design Drawing Supplement 1: Aviation Facilities. 7 pp.

—Brake Pipes
CNS D1032-82. Simplified Drawing Methods for Brake Pipes of Automobiles (Sep)(9385).

—CAD
DIN ENGL 6774 Pt 10-84. Technical Drawings; Principles for the Computer Aided Preparation of Drawings (Dec). 2 pp.
ISO TR10127-90. Computer-Aided Design (CAD) Technique—Use of Computers for the Preparation of Construction Drawings First Edition. 20 pp.
JIS B 3402-89. Technical Drawing—Drawing by CAD. 12 pp.
JTC1 TR10127-90. Computer-Aided Design (CAD) Technique—Use of Computers for the Preparation of Construction Drawings First Editon. 20 pp.
OSI ISO TR 10127-90. Computer-Aided Design (CAD) Technique—Use of Computers for the Preparation of Construction Drawings. 20 pp.

—Center Holes
ISO 6411-82. Technical Drawings—Simplified Representation of Centre Holes First Edition. 7 pp.

—Coatings—Designations
DIN ENGL 50960 Pt 2-86. Electroplated and Chemical Coatings; Indications on Drawings (Feb) (Together with DIN 50960 Part 1, February 1986 Edition, Supersedes DIN 50960, June 1963 Edition Withdrawn in July 1976). 6 pp.

—Codes—Glossaries
DIN ENGL 199 Pt 3-78. Terms in Drawings and Parts Lists; Processing of Parts Lists, Terms in Code Systems (Aug). 6 pp.

—Cones
ISO 3040-90. Technical Drawings—Dimensioning and Tolerancing—Cones Second Edition. 10 pp.

—Datum
ISO 5459-81. Technical Drawings—Geometrical Tolerancing—Datums and Datum-Systems for Geometrical Tolerances First Edition. 19 pp.

—Diagrams
CNS B1001-11-89. Engineering Drawing Representation of Diagram (Mar)(3-11).

—Die Forgings
CNS B1315-83. Steel Die Forgings Directive Items for Forging Drawings (Jun)(10306).

—Dimensioning
CNS B1001-1-88. Engineering Drawing Dimensioning (Dec)(3-1).
CSA CAN/CSA-B78.2-M91. Dimensioning and Tolerancing of Technical Drawings. 147 pp.
DIN ENGL 406 Pt 1-77. Dimensioning in Drawings; Kinds (Apr). 3 pp.
DIN ENGL 406 Pt 2-81. Dimensioning on Drawings; Rules (Aug). 20 pp.
DIN ENGL 406 Pt 3-75. Dimensioning in Drawings; Dimensioning by Co-Ordinates (July). 8 pp.
ISO 129-85. Technical Drawings—Dimensioning—General Principles, Definitions, Methods of Execution and Special Indications First Edition; (Supersedes 2595). 15 pp.
ISO 10579-93. Technical Drawings—Dimensioning and Tolerancing—Non-Rigid Parts First Edition. 7 pp.
JIS Z 8317-84. Technical Drawing—Dimensioning. 36 pp.

Engineering Drawings (Cont.)

—Drainage Systems—Terminals
ISO 6412 Pt 3-93. Technical Drawings—Simplified Representation of Pipelines—Part 3: Terminal Features of Ventilation and Drainage Systems First Edition. 8 pp.

—Drainpipes—Symbols
ISO 4067 Pt 6-85. Technical Drawings—Installations—Part 6: Graphical Symbols for Supply Water and Drainage Systems in the Ground First Edition. 7 pp.

—Drawing Inks
ISO 9957 Pt 1-92. Fluid Draughting Media—Part 1: Water-Based India Ink—Requirements and Test Conditions First Edition. 9 pp.

—Drawing Paper
ISO 5457-80. Technical Drawings—Sizes and Layout of Drawing Sheets First Edition. 8 pp.
ISO 9431-90. Construction Drawings—Spaces for Drawing and for Text, and Title Blocks on Drawing Sheets First Edition. 8 pp.

—Ducts—Symbols
ISO 4067 Pt 1-84. Technical Drawings—Installations—Part 1: Graphical Symbols for Plumbing, Heating, Ventilation and Ducting First Edition. 12 pp.
SAA AS 1101.5-84. Graphical Symbols for General Engineering—Part 5: Piping, Ducting and Mechanical Services for Buildings. 18 pp.

—Duplicating Masters
DIN ENGL 6774 Pt 4-82. Technical Drawings; Rules for Preparation; Drawn Masters for Printing Purposes (Apr). 5 pp.

—Electronic Equipment—Symbols
CNS B1001-10-89. Engineering Drawing Symbols for Electronic Power and Electronic System (Mar)(3-10).

—Electroplated Coatings—Designations
CNS H1007-87. Drawing Indications for Electroplated Coatings (Aug)(7796).
DIN ENGL 50960 Pt 2-86. Electroplated and Chemical Coatings; Indications on Drawings (Feb) (Together with DIN 50960 Part 1, February 1986 Edition, Supersedes DIN 50960, June 1963 Edition Withdrawn in July 1976). 6 pp.

—Ferroalloys—Heat Treatment
DIN ENGL 6773 Pt 2-77. Heat-Treatment of Ferrous Metals; Heat-Treated Parts; Representation and Indications in Drawings; Hardening, Hardening and Tempering, Quenching and Tempering (May). 4 pp.
DIN ENGL 6773 Pt 3-76. Heat Treatment of Ferrous Metals; Heat-Treated Parts; Representation and Indications in Drawings; Surface Layer Hardening (Nov). 8 pp.
DIN ENGL 6773 Pt 4-77. Heat Treatment of Ferrous Metals; Heat-Treated Parts; Representation and Indications in Drawings; Case Hardening (May). 7 pp.
DIN ENGL 6773 Pt 5-77. Heat Treatment of Ferrous Metals; Heat-Treated Parts; Representation and Indications in Drawings; Nitriding (May). 5 pp.

—Fire Protection—Symbols
BSI BS 1635-90. 1990 Graphical Symbols and Abbreviations for Fire Protection Drawings. 27 pp.

—Folding
DIN ENGL 824-81. Technical Drawings; Folding to Filing Size (Mar). 4 pp.

—Format
JIS Z 8311-84. Technical Drawing—Sizes and Format of Drawings. 13 pp.

—Gas Distribution
SAA AS 1100.401 Supp 2-84. Technical Drawing—Part 401: Engineering Survey and Engineering Survey Design Drawing—Supplement 2: Gas Distribution (Supplement to AS 1100. 401—1984). 3 pp.
SNZ NZS/AS 1100: Pt401: Supp2-84. Technical Drawing Part 401: Engineering Survey and Engineering Survey Design Drawing Supplement 2: Gas Distribution. 3 pp.

—Gear Teeth
DIN ENGL 3966 Pt 1-78. Information on Gear Teeth in Drawings; Information on Involute Teeth for Cylindrical Gears (Aug). 8 pp.
DIN ENGL 3966 Pt 2-78. Information on Gear Teeth in Drawings; Information on Straight Bevel Gear Teeth (Aug). 6 pp.

—Gears
DIN ENGL 58405 Pt 3-72. Spur Gear Drives for Fine Mechanics; Indication in Drawings, Examples for Calculation (May). 19 pp.

INTERNATIONAL AND NON-U.S. NATIONAL STANDARDS
SUBJECT INDEX

Engineering

Engineering Drawings (Cont.)

—**Gears** (Cont.)
ISO 1341-76. Straight Bevel Gears—Information to Be Given to the Manufacturer by the Purchaser in Order to Obtain the Gear Required First Edition. 4 pp.

—**Gears—Symbols**
ISO 2203-73. Technical Drawings—Conventional Representation of Gears First Edition. 8 pp.

—**Geometrical Tolerances**
BSI BS 308: Part 3-90. 1990 Engineering Drawing Practice Part 3: Recommendations for Geometrical Tolerancing (G). 87 pp.
BSI BS 308: Part 3-01. 1990 Amd 1 Engineering Drawing Practice Part 3: Recommendations for Geometrical Tolerancing (AMD 6973) June 15, 1992 (G). 93 pp.
BSI BS 308: Part 3-02. 1990 Amd 2 Engineering Drawing Practice Part 3: Recommendations for Geometrical Tolerancing (AMD 7421) January 15, 1993. 94 pp.
CSA CAN/CSA-B78.2-M91. Dimensioning and Tolerancing of Technical Drawings. 147 pp.
ISO 7083-83. Technical Drawings—Symbols for Geometrical Tolerancing—Proportions and Dimensions First Edition. 11 pp.
SNZ NZS/BS 308: Part 3-90. Engineering Drawing Practice Part 3: Recommendations for Geometrical Tolerancing (Withdrawn—Superseded by NZS/AS 1100 Parts 101-501).

—**Geometrical Tolerances—Symbols**
BSI BS 308: Part 3-90. 1990 Engineering Drawing Practice Part 3: Recommendations for Geometrical Tolerancing (G). 87 pp.
BSI BS 308: Part 3-01. 1990 Amd 1 Engineering Drawing Practice Part 3: Recommendations for Geometrical Tolerancing (AMD 6973) June 15, 1992 (G). 93 pp.
BSI BS 308: Part 3-02. 1990 Amd 2 Engineering Drawing Practice Part 3: Recommendations for Geometrical Tolerancing (AMD 7421) January 15, 1993. 94 pp.
ISO 1101-83. Technical Drawings—Geometrical Tolerancing—Tolerancing of Form, Orientation, Location and Run-out—Generalities, Definitions, Symbols, Indications on Drawings First Edition. 29 pp.
ISO 7083-83. Technical Drawings—Symbols for Geometrical Tolerancing—Proportions and Dimensions First Edition. 11 pp.
SNZ NZS/BS 308: Part 3-90. Engineering Drawing Practice Part 3: Recommendations for Geometrical Tolerancing (Withdrawn—Superseded by NZS/AS 1100 Parts 101-501).

—**Glassware**
ISO 6414-82. Technical Drawings for Glassware First Edition. 8 pp.

—**Glossaries**
BSI BS 6100: SUBSEC 1.5.7-88. 1988 Glossary of Building and Civil Engineering Terms Part 1: General and Miscellaneous Section 1.5: Operations: Associated Plant and Equipment Subsection 1.5.7: Drawings. 7 pp.
BSI BS 6100: SUBSEC 1.5.7-01. 1988 Amd 1 Glossary of Building and Civil Engineering Terms Part 1: General and Miscellaneous Section 1.5: Operations; Associated Plant and Equipment Subsection 1.5.7: Drawings (AMD 7245) August 15, 1992. 8 pp.
DIN ENGL 199 Pt 1-84. Terminology Associated with Technical Drawings and Item Lists; Technical Drawings (May). 6 pp.
DIN ENGL 199 Pt 2-77. Terms in Drawings and Parts Lists; Parts Lists (Dec). 12 pp.
DIN ENGL 199 Pt 4-81. Terminology Associated with Technical Drawings and Parts Lists; Amendments (Oct). 6 pp.
DIN ENGL 4764-82. Surfaces of Components Used in Mechanical Engineering and Light Engineering; Terminology According to Stress Conditions (June). 4 pp.
ISO 1101-83. Technical Drawings—Geometrical Tolerancing—Tolerancing of Form, Orientation, Location and Run-out—Generalities, Definitions, Symbols, Indications on Drawings First Edition. 29 pp.
ISO 10209 Pt 1-92. Technical Product Documentation—Vocabulary—Part 1: Terms Relating to Technical Drawings: General and Types of Drawings First Edition; (Replaces 5457). 11 pp.
JIS B 3410-90. Glossary of Terms Used in Plotters Technical Drawing—Numerically Controlled Drafting Machines.

—**Heating Equipment—Symbols**
ISO 4067 Pt 1-84. Technical Drawings—Installations—Part 1: Graphical Symbols for Plumbing, Heating, Ventilation and Ducting First Edition. 12 pp.

Engineering Drawings (Cont.)

—**Hydraulic Systems—Symbols**
CNS B1001-9-81. Engineering Drawing Symbols for Hydraulic and Pneumatic Power System (Jul)(3-9).
SAA AS 1101.1-93. Graphical Symbols for General Engineering—Part 1: Hydraulic and Pneumatic Systems (in Profesional Package 56A). 35 pp.

—**Industrial Plants—Pipelines**
DIN ENGL 2408 Pt 1-82. Process Plant Piping; Project and Construction Engineering Documents; Terminology (May). 6 pp.

—**Integrated Circuits**
BSI BS 3934-65. (WITHDRAWN) 1965 (Includes Addenda 1,2,3, 4, and 5) Dimensions of Semiconductor Devices (Includes Addenda 1,2, and 3) (Superseded by BS 3934: Parts 1 to 6 and Part 10: 1992). 270 pp.

—**Item Lists**
ISO 6433-81. Technical Drawings—Item References First Edition. 4 pp.
ISO 7573-83. Technical Drawings—Item Lists First Edition. 4 pp.

—**Laboratory Ware**
BSI BS 2774-83. 1983 Drawing Conventions for Laboratory Glass Apparatus. 10 pp.

—**Lamps**
IEC 1126-92. Procedure for Use in the Preparation of Maximum Lamp Outlines First Edition. 84 pp.

—**Lettering**
ISO 3098 Pt I-74. Technical Drawings—Lettering—Part I: Currently Used Characters First Edition. 8 pp.
JIS Z 8313-84. Technical Drawing—Lettering. 11 pp.

—**Lettering—Cyrillic Characters**
ISO 3098 Pt 4-84. Technical Drawings—Lettering—Part 4: Cyrillic Characters First Edition. 7 pp.

—**Lettering—Greek Characters**
ISO 3098 Pt 2-84. Technical Drawings—Lettering—Part 2: Greek Characters First Edition. 7 pp.

—**Lettering—Roman Characters**
ISO 3098 Pt 3-87. Technical Drawings—Lettering—Part 3: Diacritical and Particular Marks for the Latin Alphabet First Edition. 8 pp.

—**Line Conventions**
JIS Z 8312-84. Technical Drawing—Line Conventions. 14 pp.

—**Lines**
DIN ENGL 15 Pt 1-84. Technical Drawings; Lines; Principles (June). 3 pp.
DIN ENGL 15 Pt 2-84. Technical Drawings; Lines; General Application (June). 5 pp.

—**Locking Nuts—Aerospace**
AECMA PREN3065-88. Installation Holes for Self-Locking Serrated Shank Nuts Design Specifications. 5 pp.

—**Manufacturing Spigots**
DIN ENGL 6785-82. Manufacturing Spigots on Parts Turned on Lathes; Indications on Drawings (May). 3 pp.

—**Marine—Electrical Equipment**
CNS F5001-84. Graphical Symbols for Electrical Apparatus (Power) for Marine Engineering Drawings (May)(7784).
CNS F5002-84. Graphical Symbols for Electrical Apparatus (Lighting Fittings and Accessories) for Marine Engineering Drawings (May)(7785).
CNS F5003-84. Graphical Symbols for Electrical Apparatus (Communication) for Marine Engineering Drawings (May)(7786).
JIS F 8011-90. Graphical Symbols for Electrical Apparatus (Power) for Marine Engineering Drawings.
JIS F 8012-90. Graphical Symbols for Electrical Apparatus for Marine Use Engineering Drawings—Lighting Fittings and Accessories.
JIS F 8013-91. Graphical Symbols for Electrical Apparatus for Marine Use Engineering Drawings—Communication, Instrumentation, Navigation and Radio.

—**Marine—Hulls**
JIS F 0201-83. General Rules for Automatic Drawing of Basic Hull Construction Plans.

—**Microfilming**
BSI BS 4210: Part 1-77. 1977 35 mm Microcopying of Technical Drawings Part 1: Operating Procedures. 11 pp.
BSI BS 4210: Part 2-77. 1977 35 mm Microcopying of Technical Drawings Part 2: Photographic Requirements for Silver Film. 7 pp.

Engineering Drawings (Cont.)

—**Microfilming** (Cont.)
BSI BS 4210: Part 3-77. 1977 35 mm Microcopying of Technical Drawings Part 3: Unitized Microfilm Carriers. 8 pp.
BSI BS 5536-88. 1988 Preparation of Technical Drawings and Diagrams for Microfilming. 10 pp.
BSI BS 6342-83. 1983 105 mm Microcopying of Technical Drawings (Single Image, A6 Size). 8 pp.
CGSB CAN/CGSB-72.8-M80. Examination and Preparation of Drawings to Be Microfilmed. 20 pp.
CGSB CAN/CGSB-72.9-M81. Operating Procedures for Microfilming Drawings. 11 pp.
DIN ENGL 19051 Pt 3 Suppl. 1-81. Tests for Reprographic Use; DIN-Test Assembly (Test Table) for Testing the Microfilming of Technical Drawings; Neutral Density Cards for Use in Practice (Apr). 3 pp.
ISO 3272 Pt 1-83. Microfilming of Technical Drawings and Other Drawing Office Documents—Part 1: Operating Procedures First Edition. 6 pp.
ISO 3272 Pt III-75. Microcopying of Technical Drawings and Other Drawing Office Documents—Part III: Unitized 35 mm Microfilm Carriers First Edition. 6 pp.
ISO 6428-82. Technical Drawings—Requirements for Microcopying First Edition. 5 pp.
JIS Z 6004-89. Operating Procedures for Microfilming of Technical Drawings on 35 mm Microfilm. 11 pp.
JIS Z 6005-89. Quality Requirements for Processed 35 mm Silver Gelatin Microfilm of Technical Drawings. 7 pp.
SAA AS 1203-72. Microfilming of Engineering Documents (35 mm) Amdt 1 April 1973 Amdt 2 March 1979. 20 pp.

—**Microfilming—Aerospace**
AECMA PREN2484-84. Microfilming of Drawings—Aperture Card for 35 mm Microfilm (CO/01A). 8 pp.
BSI BS EN 2484-89. 1989 Microfilming of Drawings: Aperture Card for 35mm Microfilm. 12 pp.
CEN EN 2484-89. Microfilming of Drawings Aperture Card for 35 mm Microfilm. 11 pp.

—**Microfilming—Aperture Cards—Aerospace**
JIS Z 6006-89. Aperture Card for 35 mm Microfilm of Technical Drawings. 13 pp.

—**Microfilming—Quality Assurance**
CGSB CAN/CGSB-72.12-M81. Microcopying of Technical Drawings—Quality Criteria and Control. 9 pp.
ISO 3272 Pt 2-78. Microfilming of Technical Drawings and Other Drawing Office Documents—Part 2: Quality Criteria and Control First Edition; (Amendment 1-1980). 7 pp.

—**Naval Ships**
MOD UK NES 722-91. Requirements for the Preparation Identification and Management of Drawings Issue 3 (03.91). 33 pp.
MOD UK NES 722-93. Requirements for the Preparation Identification and Management of Drawings Issue 4 (06.93). 39 pp.

—**Optical Equipment**
BSI BS 4301-91. 1991 Preparation of Drawings for Optical Elements and Systems (G). 23 pp.

—**Orthographical Projection**
CNS B1001-88. Engineering Drawing General Principle of Presentation (Dec)(3).
ISO 128-82. Technical Drawings—General Principles of Presentation First Edition. 18 pp.

—**Pipelines**
DIN ENGL 2425 Pt 3-80. Plans for Public Utilities, Water Resources and Long-Distance Lines; Plans for Long-Distance Pipelines; Technical Regulation of the DVGW (German Gas and Water Engineers Association) (May). 14 pp.
DIN ENGL 2425 Pt 4-80. Plans for Public Utilities, Water Resources and Long-Distance Lines; Sewer Network Drawings of Public Sewerage Systems (May). 7 pp.
DIN ENGL 2425 Pt 5-83. Plans for Public Supplies, Water Engineering and Long-Distance Pipelines; Maps and Plans for Water Engineering (Oct). 28 pp.
DIN ENGL 2425 Pt 6-82. Plans for Public Utilities, Water Resources Management and for Long-Distance Pipelines; Maps and Plans for Water Engineering, Flood and Coastal Protection (Feb). 13 pp.
ISO 6412 Pt 1-89. Technical Drawings—Simplified Representation of Pipelines—Part 1: General Rules and Orthogonal Representation First Edition. 13 pp.

—**Pipelines—Symbols**
BSI BS 1553: Part 1-77. 1977 Specification for Graphical Symbols for General Engineering Part 1: Piping Systems and Plant. 36 pp.
CNS B1001-8-89. Engineering Drawing Piping (Mar)(3-8).

INDUSTRY STANDARDS

Engineering Drawings (Cont.)

—Pipelines—Symbols (Cont.)
SAA AS 1101.5-84. Graphical Symbols for General Engineering—Part 5: Piping, Ducting and Mechanical Services for Buildings. 18 pp.

—Plotters
ISO 9959 Pt 1-92. Numerically Controlled Draughting Machines—Drawing Test for the Evaluation of Performance—Part 1: Vector Plotters First Edition. 17 pp.

—Plumbing—Symbols
ISO 4067 Pt 1-84. Technical Drawings—Installations—Part 1: Graphical Symbols for Plumbing, Heating, Ventilation and Ducting First Edition. 12 pp.

—Pneumatic Systems—Symbols
CNS B1001-9-81. Engineering Drawing Symbols for Hydraulic and Pneumatic Power System (Jul)(3-9).
SAA AS 1101.1-93. Graphical Symbols for General Engineering—Part 1: Hydraulic and Pneumatic Systems (in Profesional Package 56A). 35 pp.

—Power Systems—Symbols
CNS B1001-10-89. Engineering Drawing Symbols for Electronic Power and Electronic System (Mar)(3-10).

—Practices
BSI BS 1192: Part 2-87. 1987 Constructon Drawing Practice Part 2: Recommendations for Architectural and Engineering Drawings. 58 pp.
BSI PP 7308-86. 1986 Engineering Drawing Practice for Schools and Colleges. 46 pp.
JIS A 0101-77. Drawing Office Practice for Civil Engineering Works.
JIS Z 8310-84. Technical Drawing—General Code of Drawing Practice. 11 pp.

—Printed Circuit Boards—Components
BSI BS 6943-88. 1988 Classification of Shapes of Electronic Components for Placement on Printed Wiring Boards. 49 pp.

—Profiles
ISO 1660-87. Technical Drawings—Dimensioning and Tolerancing of Profiles Second Edition. 8 pp.

—Projection
JIS Z 8315-84. Technical Drawing—Projection Methods. 9 pp.

—Projection—Glossaries
DIN ENGL 5 Pt 10-86. Technical Drawings; Projections; Terminology (Dec). 13 pp.

—Radio Frequency Connectors
BSI BS 9210 N006: Pt 2-78. 1978 Detail Specification for Radio Frequency Connect-ors (Series SMA), 50 ohms, Unsealed, Solder-ed, Centre Contact, Clamp Outer for Flexible Cables, Solder Outer for Semi-Rigid Cables: Part 2: Control Drawings, Mating Face Details and Gauge Information. 6 pp.
BSI BS 9210 N009: Pt 2-78. 1978 Detail Specification for Electronic Components of Assessed Quality, Soldered Part 2: Control Drawings, Mating Face Details and Gauge Information. 6 pp.

—Radio Frequency Connectors—Quality Assurance
BSI BS 9210 N001: Pt 2-82. 1982 Radio Frequency Connector (Type BNC), Sealed, Soldered, Captive Content, 50 ohms Part 2: Control Drawing, Mating Face Details and Gauge Information. 6 pp.

—Reprography
DIN ENGL 6774 Pt 1-86. Technical Drawings; Rules for the Preparation of Drawings for Reproduction Purposes (Dec). 3 pp.

—Reprography—Film Slides
DIN ENGL 6774 Pt 3-82. Technical Drawings; Rules for Preparation; Drawn Originals for Slides (June). 14 pp.

—Riveting—Symbols
CNS B1001-5-88. Engineering Drawing Symbols for Riveting (Dec)(3-5).

—Roads
SAA AS 1100.401 Supp 4-84. Technical Drawing—Part 401: Engineering Survey and Engineering Survey Design Drawing—Supplement 4: Roads (Supplement to AS 1100. 401—1984) Amdt 1 May 1984. 11 pp.
SNZ NZS/AS 1100: Pt401: Supp4-84. Technical Drawing Part 401: Engineering Survey and Engineering Survey Design Drawing Supplement 4: Roads Amend: 1. 11 pp.

—Rotating Machines—Naval Ships
MOD UK NES 628-89. Requirements for Preparation of Drawings for Rotating Electrical Machinery Issue 3 (05.89). 60 pp.

Engineering Drawings (Cont.)

—Rotating Machines—Naval Ships (Cont.)
MOD UK NES 628-93. Requirements for Preparation of Drawings for Rotating Electrical Machinery Issue 4 (02.93). 67 pp.

—Rotating Machines—Submarines
MOD UK NES 628-89. Requirements for Preparation of Drawings for Rotating Electrical Machinery Issue 3 (05.89). 60 pp.
MOD UK NES 628-93. Requirements for Preparation of Drawings for Rotating Electrical Machinery Issue 4 (02.93). 67 pp.

—Scale (Ratio)
ISO 5455-79. Technical Drawings—Scales First Edition. 3 pp.
JIS Z 8314-84. Technical Drawing—Scales. 5 pp.

—Screw Threads
ISO 6410 Pt 1-93. Technical Drawings—Screw Threads and Threaded Parts—Part 1: General Conventions First Edition. 11 pp.
ISO 6410 Pt 3-93. Technical Drawings—Screw Threads and Threaded Parts—Part 3: Simplified Representation First Edition. 7 pp.

—Seals—Symbols
ISO 9222 Pt 1-89. Technical Drawings—Seals for Dynamic Application—Part 1: General Simplified Representation First Edition; (Corrected and Reprinted -1989). 6 pp.
ISO 9222 Pt 2-89. Technical Drawings—Seals for Dynamic Application—Part 2: Detailed Simplified Representation First Edition. 15 pp.

—Semiconductor Devices
BSI BS 3934-65. (WITHDRAWN) 1965 (Includes Addenda 1,2,3, 4, and 5) Dimensions of Semiconductor Devices (Includes Addenda 1,2, and 3) (Superseded by BS 3934: Parts 1 to 6 and Part 10: 1992). 270 pp.

—Serrations—Symbols
ISO 6413-88. Technical Drawings—Representation of Splines and Serrations First Edition. 8 pp.

—Sewage Systems
SAA AS 1100.401 Supp 3-84. Technical Drawing—Part 401: Engineering Survey and Engineering Survey Design Drawing—Supplement 3: Sewerage and Water (Supplement to AS 1100. 401—1984). 6 pp.
SNZ NZS/AS 1100: Pt401: Supp3-84. Technical Drawing Part 401: Engineering Survey and Engineering Survey Design Drawing Supplement 3: Sewerage and Water Supply. 6 pp.

—Splines—Symbols
ISO 6413-88. Technical Drawings—Representation of Splines and Serrations First Edition. 8 pp.

—Springs (Elastic)—Symbols
ISO 2162-73. Technical Drawings—Representation of Springs First Edition. 6 pp.

—Steel Construction
CNS B1001-7-88. Engineering Drawing Symbols for Steel Construction (Dec)(3-7).
ISO 5261-81. Engineering Drawings for Structural Metal Work First Edition. 12 pp.

—Steel Construction—Symbols
CNS B1001-7-88. Engineering Drawing Symbols for Steel Construction (Dec)(3-7).

—Surface Texture
CNS B1001-3-88. Engineering Drawing Method of Indicating Surface Texture on Drawing (Dec)(3-3).
ISO 1302-92. Technical Drawings—Method of Indicating Surface Texture Third Edition. 19 pp.

—Symbols
BSI BS 308: Part 3-90. 1990 Engineering Drawing Practice Part 3: Recommendations for Geometrical Tolerancing (G). 87 pp.
BSI BS 308: Part 3-01. 1990 Amd 1 Engineering Drawing Practice Part 3: Recommendations for Geometrical Tolerancing (AMD 6973) June 15, 1992 (G). 93 pp.
BSI BS 308: Part 3-02. 1990 Amd 2 Engineering Drawing Practice Part 3: Recommendations for Geometrical Tolerancing (AMD 7421) January 15, 1993. 94 pp.
CSA Z85-1983. ABBR Abbreviations for Scientific and Engineering Terms. 280 pp.
ISO 3461 Pt 2-87. General Principles for the Creation of Graphical Symbols—Part 2: Graphical Symbols for Use in Technical Product Documentation First Edition. 7 pp.
ISO 7083-83. Technical Drawings—Symbols for Geometrical Tolerancing—Proportions and Dimensions First Edition. 11 pp.
JIS Z 8313-84. Technical Drawing—Lettering. 11 pp.

Engineering Drawings (Cont.)

—Symbols (Cont.)
SNZ NZS/BS 308: Part 3-90. Engineering Drawing Practice Part 3: Recommendations for Geometrical Tolerancing (Withdrawn—Superseded by NZS/AS 1100 Parts 101-501).

—Thread Inserts
ISO 6410 Pt 2-93. Technical Drawings—Screw Threads and Threaded Parts—Part 2: Screw Thread Inserts First Edition. 9 pp.

—Threaded Fasteners—Scale (Ratio)
BSI PD 7300-79. 1979 Nuts and Bolts: Recommended Drawing Ratios for Schools and Colleges. Teachers' Notes. 21 pp.

—Tolerancing
ISO 406-87. Technical Drawings—Tolerancing of Linear and Angular Dimensions Second Edition. 7 pp.
ISO 2692-88. Technical Drawings—Geometrical Tolerancing—Maximum Material Principle Amendment 1: Least Material Requirement First Edition; (Amendment 1-1992). 35 pp.
ISO 10578-92. Technical Drawings—Tolerancing of Orientation and Location —Projected Tolerance Zone First Edition. 10 pp.
ISO 10579-93. Technical Drawings—Dimensioning and Tolerancing—Non-Rigid Parts First Edition. 7 pp.
JIS Z 8318-84. Technical Drawing—Tolerancing of Linear and Angular Dimensions. 13 pp.

—Water Supply—Symbols
DIN ENGL 1080 Pt 7-79. Terms, Symbols and Units Used in Civil Engineering; Water Engineering (Mar). 4 pp.
ISO 4067 Pt 6-85. Technical Drawings—Installations—Part 6: Graphical Symbols for Supply Water and Drainage Systems in the Ground First Edition. 7 pp.

—Welding
DIN ENGL 8570 Pt 1-87. General Tolerances for Welded Structures; Linear and Angular Dimensions (Oct). 4 pp.
DIN ENGL 8570 Pt 3-87. General Tolerances for Welded Structures; Geometrical Tolerances (Oct). 4 pp.

—Welding—Aerospace
AECMA PREN2574-88. Welds Information on Drawings. 2 pp.
BSI BS EN 2574-90. 1990 Welds. Information on Drawings. 6 pp.
CEN EN 2574-90. Welds Information on Drawings. 3 pp.

—Welding—Symbols
CNS B1001-6-89. Engineering Drawing (Symbols for Welding) (Jan)(3-6).
SAA AS 1101.3-87. Graphical Symbols for General Engineering—Part 3: Welding and Non-Destructive Examination (This ia A Joint Standard with SANZ NZS 1101:Part3:1987). 80 pp.
SNZ NZS/AS 1101: Part 3-87. Graphical Symbols for Engineering Part 3: Welding and Non-Destructive Examination (This is a Joint Standard with SAA AS 1101:Part 3). 80 pp.

—Woodworking
CNS O1037-88. Technical Drawings for Woodworking—General Principles (Dec)(11666).
CNS O1037-1-88. Technical Drawings for Woodworking—Dimensioning (Dec)(11666-1).
CNS O1037-4-88. Technical Drawings for Woodworking—Tolerances and Fits (Dec)(11666-4).
CNS O1037-5-88. Technical Drawings for Woodworking—Woodworking for Windows and Doors (Dec)(11666-5).
CNS O1037-6-88. Technical Drawings for Woodworking—Examples of Woodworking Drawings (Dec)(11666-6).

—Woodworking—Surface Texture
CNS O1037-3-88. Technical Drawings for Woodworking—Surface Texture on Wood Drawings (Dec)(11666-3).

—Woodworking—Symbols
CNS O1037-2-88. Technical Drawings for Woodworking—Symbols for Woodworking Drawings (Dec)(11666-2).

—Workpieces—Edge Conditions
DIN ENGL 6784-82. Edges of Workpieces; Concepts; Indications on Drawings (Feb). 8 pp.

Engineering Equipment
Use: Equipment

Engineering Reports

—Military Operations
NATO STANAG 2096 ED 4 AMD 1-88. Reporting Engineer Information in the Field. 19 pp.

—Publication—Format
CNS Z7029-69. Format for Publication of Engineering and Scientific Reports (Tentative) (Jul)(3030).

Engines

Scope Note: For additional listings, use a more specific term *See Also:* Aircraft Engines; Automotive Engines; Combustion Chambers; Cylinder Blocks; Dental Engines; Diesel Engines; Engine Heaters; Engine Starters; Engine Valves; Exhaust Systems; Flywheels; Fuel Consumption; Gas Turbine Engines; Gasoline Engines; Helicopter Engines; Ignition Systems; Internal Combustion Engines; Internal Combustion Piston Engines; Marine Engines; Model Engines; Motors; Multifuel Engines; Outboard Engines; Piston Engines; Prime Movers; Propulsion Systems; Ramjet Engines; Spark Ignition Engines; Steam Engines; Turbine Engines; Turbines; Turbofan Engines; Turbojet Engines; Turboprop Engines; Turboshaft Engines

—Classification
BSI BS 1000: (621.1/22)-84. 1984 Universal Decimal Classificaton (UDC). English Full Edition (621.1/22): UDC 621.1 Heat Engines in General. Steam Power. Steam Engines. Boilers UDC 621.22 Water Power. Hydraulic Energy. 22 pp.
BSI BS 1000: (621.4)-76. 1976 Universal Decimal Classification (UDC). English Full Edition (621.4): Heat Engines (Other Than Steam Engines). 19 pp.
SNZ NZS/BS 1000 (621.1/.22)-84. Universal Decimal Classification Heat Engines in General. Steam Power. Steam Engines. Boilers. Water Power. Hydraulic Energy. 24 pp.
SNZ NZS/BS 1000 (621.4)-76. Universal Decimal Classification Heat Engines (Other Than Steam Engines). 20 pp.

English Flutes
Use: Recorders (Musical)

Engravers
Use: Engraving Equipment

Engraving

—Arabic Characters
JIS Z 8905-76. Standard Type of Letters Used in Mechanical Engraving (Arabic Figures and Roman Types).

—Chinese Characters
JIS Z 8903-84. Standard Type of Letters Used in Mechanical Engraving (Joyo Kanji, Common-Use Chinese Characters) (R 1989). 43 pp.

—Japanese Characters
JIS Z 8904-76. Standard Type of Letters Used in Mechanical Engraving (Katakana Characters) (R 1989). 8 pp.
JIS Z 8906-77. Standard Type of Letters Used in Mechanical Engraving (Hiragana Characters) (R 1988). 8 pp.

—Roman Characters
JIS Z 8905-76. Standard Type of Letters Used in Mechanical Engraving (Arabic Figures and Roman Types).

Engraving Equipment
Use For: Engraving Machines *See Also:* Machining

—Classification
BSI BS 1000: (681.8/9)-72. 1972 Amd 1 Universal Decimal Classification (UDC). English Full Edition (681.8/.9): Technical Acoustics. Musical Instruments. Engraving and Sculpting Machines. 26 pp.
SNZ NZS/BS 1000 (681.8/.9)-72. Universal Decimal Classification Technical Acoustics. Musical Instruments. Engraving and Sculpting Machines Amend: 1. 24 pp.

—Electric—Hand Held
CSA CAN/CSA-C22. 2 NO 68-92. Motor-Operated Appliances (Household and Commercial); (Gen Instr 1 Thru 2). 115 pp.
MOD UK DSTAN 51-19: Part 2-92. Hand Tools, Powered Part 2: Electric Issue 1. 31 pp.

—Templates—Template Grooves
DIN ENGL 55076-73. Machine Tools; Engraving Machines; Templates and Template Grooves; Cross Sections (Sept). 2 pp.

Engraving Machines
Use: Engraving Equipment

Enhanced Quality Television Systems
See Also: High Definition Television Equipment; Television Systems
CCIR Decision 60-3-89. Digital Television Standards—Annex to Volume XI-1—Broadcasting Service (Television). 4 pp.
CCIR QUESTION 42-1/11-90. Enhanced Television—Questions Concerning Study Group 11—Broadcasting Service (Television). 1 p.

—Aspect Ratio—Normal
CCIR Report 1077-1-90. Enhanced 4:3 Aspect Ratio Television Systems—Section 11A—Characteristics of Systems for Monochrome and Colour Television. 9 pp.

—Audio Systems
CCIR QUESTION 79/10-90. Suitable Sound Systems to Accompany High-Definition Television and Enhanced Television Systems—Questions Concerning Study Group 10—Broadcasting Service (Sound). 1 p.

—Broadcasting
CCIR QUESTION 79/11-90. Terrestrial Emission of Enhanced Television—Questions Concerning Study Group 11—Broadcasting Service (Television). 1 p.

Enlargers (Photography)
Use For: Photoenlargers; Photographic Enlargers
JIS B 7177-80. Photographic Enlargers. 19 pp.

—Household
CSA C22.2 NO 118-1959. Construction and Test of Picture Machines and Appliances (R 1992); (Erratum April 1959) (Rev 1-5 March 1969). 21 pp.
CSA 649J Bull. Electrical Bulletin 649J November 17, 1975 to C22.2 NO 118. 1 p.
CSA 1169 Bull. Electrical Bulletin 1169 June 27, 1978 to C22.2 NO 118. 2 pp.

—Portable
CSA C22.2 NO 118-1959. Construction and Test of Picture Machines and Appliances (R 1992); (Erratum April 1959) (Rev 1-5 March 1969). 21 pp.
CSA 1169 Bull. Electrical Bulletin 1169 June 27, 1978 to C22.2 NO 118. 2 pp.

Enlarging
Use: Expansion

Enramycin Content Analysis
Use: Enduracidin Content Analysis

Enrichment

—Enterobacteriaceae Content Analysis
BSI BS 5763: Part 15-91. 1991 Methods for Microbiological Examination of Food and Animal Feeding Stuffs Part 15: Detection of Enterobacteriaceae with Pre-Enrichment (W) (ISO 8523: 1991). 14 pp.
ISO 8523-91. Microbiology—General Guidance for the Detection of Enterobacteriaceae with Pre-Enrichment First Edition. 11 pp.

—Water
BSI BS 6068: Sec 4.6-89. 1989 Water Quality Part 4: Microbiological Methods Section 4.6: Detection and Enumeration of Pseudomonas Aeruginosa by Enrichment in Liquid Medium (ISO 8360-1: 1988). 8 pp.
BSI BS 6068: Sec 4.8-93. 1993 Water Quality Detection and Enumeration of the Spores of Sulfite-Reducing Anaerobes (Clostridia) Part 1: Method by Enrichment in a Liquid Medium (ISO 6461/1: 1986) (N). 10 pp.
BSI BS EN 26461-1-93. 1993 Water Quality Detection and Enumeration of the Spores of Sulfite-Reducing Anaerobes (Clostridia) Part 1: Method by Enrichment in a Liquid Medium (ISO 6461/1: 1986) (N). 10 pp.
CEN EN 26461-1-93. Water Quality—Detection and Enumeration of the Spores of Sulfite-Reducing Anaerobes (Clostridia)—Part 1: Method by Enrichment in a Liquid Medium (ISO 6461-1: 1986). 5 pp.
ISO 6461 Pt 1-86. Water Quality—Detection and Enumeration of the Spores of Sulfite-Reducing Anaerobes (Clostridia)—Part 1: Method by Enrichment in a Liquid Medium First Edition; (CEN EN 26461-1: 1993). 5 pp.
ISO 7899 Pt 1-84. Water Quality—Detection and Enumeration of Faecal Streptococci—Part 1: Method by Enrichment in a Liquid Medium First Edition. 5 pp.
ISO 8360 Pt 1-88. Water Quality—Detection and Enumeration of Pseudomonas Aeruginosa—Part 1: Method by Enrichment in Liquid Medium First Edtion. 9 pp.

Enstrom F 28A Aircraft
See Also: Aircraft

—Antenna Positions
CAA. Enstrom F28A Series and 280 Series (Approved Aerial Positions). 1 p.

Enstrom Helicopters
See Also: Helicopters

—Foreign Airworthiness Directives
CAA. Enstrom Series Helicopters (Foreign Airworthiness Directives). 6 pp.

Enstrom 280 Aircraft
See Also: Aircraft

—Antenna Positions
CAA. Enstrom F28A Series and 280 Series (Approved Aerial Positions). 1 p.

ENT
Use: Electrical Nonmetallic Tubing

Enteral Feeding Kits
Use: Feeding Devices (Patient)

Entercoccus
Use: Streptococcus

Enterobacteriaceae
See Also: Bacteria; Coliform Bacteria; Escherichia Coli; Salmonella; Shigella

—Food
ISO 7402-93. Microbiology—General Guidance for the Enumeration of Enterobacteriaceae Without Resuscitation—MPN Technique and Colony-Count Technique Second Edition. 14 pp.

—Meat
ISO 5552-79. Meat and Meat Products—Detection and Enumeration of Enterobacteriaceae (Reference Methods) First Edition; (Erratum—June 1979). 9 pp.

Enterobacteriaceae Content Analysis

—Food—Enrichment
BSI BS 5763: Part 15-91. 1991 Methods for Microbiological Examination of Food and Animal Feeding Stuffs Part 15: Detection of Enterobacteriaceae with Pre-Enrichment (W) (ISO 8523: 1991). 14 pp.
ISO 8523-91. Microbiology—General Guidance for the Detection of Enterobacteriaceae with Pre-Enrichment First Edition. 11 pp.

—Meat
DIN ENGL 10164 Pt 1-86. Microbiological Examination of Meat and Meat Products; Determination of Enterobacteriaceae by the Spatula Method (Reference Method) (Aug). 3 pp.
DIN ENGL 10164 Pt 2-86. Microbiological Examination of Meat and Meat Products; Determination of Enterobacteriaceae by the Drop Method (Aug). 3 pp.
ISO 5552-79. Meat and Meat Products—Detection and Enumeration of Enterobacteriaceae (Reference Methods) First Edition; (Erratum—June 1979). 9 pp.

Enterococcus
Use: Streptococcus

Entertainment Equipment
See Also: Audio Equipment; Television Equipment; Video Equipment

—Portable—Connectors
IEC 130 Pt 10-71. Connectors for Frequencies Below 3 MHz Part 10: Connectors for Coupling an External Low-Voltage Power Supply to Portable Entertainment Equipment First Edition. 21 pp.

Entrainment

—Fats
BSI BS 684: Sec 1.16-81. 1981 Amd 1 Methods of Analysis of Fats and Fatty Oils Part 1: Physical Methods Section 1.16: Determination of Water Content: Entrainment Method. 5 pp.
ISO 934-80. Animal and Vegetable Fats and Oils—Determination of Water Content—Entrainment Method First Edition. 4 pp.

—Oils
BSI BS 684: Sec 1.16-81. 1981 Amd 1 Methods of Analysis of Fats and Fatty Oils Part 1: Physical Methods Section 1.16: Determination of Water Content: Entrainment Method. 5 pp.
ISO 934-80. Animal and Vegetable Fats and Oils—Determination of Water Content—Entrainment Method First Edition. 4 pp.

Entrainment (Cont.)
—Seasonings
BSI BS 4585: Part 2-82. 1982 Methods of Test for Spices and Condiments Part 2: Determination of Moisture Content (Entrainment Method). 8 pp.
CNS N6151-81. Method of Test for Spices and Condiments—Determination of Moisture Content—Entrainment Method (Feb)(7070). 3 pp.
ISO 939-80. Spices and Condiments—Determination of Moisture Content—Entrainment Method First Edition. 6 pp.

Entrenching Tools
See Also: Hand Tools; Tools
NATO STANAG 4365 ED 1 AMD 0-00. (Draft) Design Criteria for a Lightweight Hand Entrenching Tool and Carrier. 5 pp.
NATO STANAG 4365 ED 1 AMD 0-92. Design Criteria for a Lightweight Hand Entrenching Tool and Carrier. 7 pp.

Entry Control Systems
Use: Access Control Systems

Envelope Delay Distortion
Use For: EDD; Group Delay Distortion
See Also: Delay Distortion; Distortion (Electrical); Group Delay

—Communication Terminal Equipment—Groups
CCITT RECMN G.232-89. 12-Channel Terminal Equipments—International Analogue Carrier Systems (Study Group XV) 13 pp. 13 pp.

—Frequency Translation Equipment
CCITT RECMN G.233-89. Recommendations Concerning Translating Equipments—International Analogue Carrier Systems (Study Group XV) 10 pp. 10 pp.

—Maritime Mobile Satellite Communications—Automatic Services
CCITT RECMN G.473-89. Interconnection of a Maritime Mobile Satellite System with the International Automatic Switched Telephone Service; Transmission Aspects—International Analogue Carrier Systems (Study Group XV) 10 pp. 10 pp.

—Telephone Circuits
CCITT FASCICLE III.1. General Characteristics of International Telephone Connections and Circuits Recommendations G.101-G.181. 332 pp.

—Telephone Circuits—Public Switched Telephone Networks
CCITT RECMN G.113-89. Transmission Impairments—General Characteristics of International Telephone Connections and Circuits (Study Groups XII and XV) 22 pp. 22 pp.

—Telephone Connections
CCITT RECMN G.113-89. Transmission Impairments—General Characteristics of International Telephone Connections and Circuits (Study Groups XII and XV) 22 pp. 22 pp.

—Telephone Exchanges—Public Switched Telephone Networks
CCITT RECMN G.113-89. Transmission Impairments—General Characteristics of International Telephone Connections and Circuits (Study Groups XII and XV) 22 pp. 22 pp.

—Through Group Filters
CCITT RECMN G.242-89. Through-Connection of Groups, Supergroups, Etc.—Analogue Carrier Systems (Study Group XV) 7 pp. 7 pp.

—Through Supergroup Filters
CCITT RECMN G.242-89. Through-Connection of Groups, Supergroups, Etc.—Analogue Carrier Systems (Study Group XV) 7 pp. 7 pp.

—Voice Band Data Transmission
CCITT RECMN G.113-89. Transmission Impairments—General Characteristics of International Telephone Connections and Circuits (Study Groups XII and XV) 22 pp. 22 pp.

Envelope Display Distortion
—Telephone Circuits
CCITT RECMN G.133-89. Group-Delay Distortion—General Characteristics of International Telephone Connections and Circuits (Study Groups XII and XV) 1 pp. 1 pp.

Envelopes
See Also: Stationery

Envelopes (Cont.)
BSI BS 1133: SUBSEC 7.2-86. 1986 Packaging Code Section 7: Paper and Board Wrappers, Bags and Containers Section 7.1: Wrapping Papers Subsection 7.1: Bags and Envelopes. 12 pp.
BSI BS 1133: SUBSEC 7.2-01. 1986 Amd 1 Packaging Code Section 7: Paper and Board Wrappers, Bags and Containers Subsection 7.2: Bags and Envelopes (AMD 7215) August 15, 1992 (H). 13 pp.
CNS P1004-89. Envelopes for Official and General Uses (Mar)(3403). 2 pp.
JIS S 5502-88. Envelopes and Pockets. 20 pp.
SAA AS P2-69. Mailing Envelopes. 11 pp.

—Cancellation Area
ISO 415-75. Envelopes, Postcards and Similar Articles—Cancellation Area First Edition. 3 pp.

—Glossaries
BSI BS 6456-83. 1983 Correspondence Envelopes. 7 pp.
ISO 6924-83. Correspondence Envelopes —Vocabulary First Edition. 11 pp.

—Kraft Paper
CGSB 9-GP-31MA-87. Paper, Golden Kraft, for Envelopes. 8 pp.

—Radiographic Film
BSI BS 3043-73. 1973 Amd 1 Storage Envelopes for Processed X-Ray Films for Medical Radiography (AMD 1662) February 14, 1975. 9 pp.

—Sizes
BSI BS 4264-87. 1987 Amd 1 Envelopes for Commercial, Official and Professional Use. 10 pp.
ISO 269-85. Correspondence Envelopes—Designation and Sizes Second Edition. 4 pp.

—Waterproof
MOD UK DSTAN 81-17-70. Envelopes, Packing Documents, Waterproof, Tongued and Slotted Issue 1. 7 pp.

Environmental Conditions
See Also: Ambient Temperature; Atmospheric Pressure; Controlled Atmospheres; Electromagnetic Interference; Environmental Engineering; Environmental Protection; Ergonomics; Humidity; Noise; Pressure; Temperature; Thermal Environments
BSI DD 93-84. (WITHDRAWN) 1984 Methods for Assessing Exposure to Wind Driven Rain (Superseded by BS 8104: 1992). 61 pp.
CENELEC HD 478.3.0 S1-86. Classification of Environmental Conditions-Part 3: Classification of Groups of Environmental Parameters and Their Severities-Introduction. 1 p.

—Aircraft
BSI 3G 100:Pt 2: SUBSEC 3.9-72. 1972 General Requirements for Equipment for Use in Aircraft Part 2: All Equipment Section 3: Environmental Conditions Subsection 3.9: Ice Formation. 3 pp.
BSI 3G 100:Pt 2: SUBSEC 3.10-74. 1974 General Requirements for Equipment for Use in Aircraft Part 2: All Equipment Section 3: Environmental Conditions Subsection 3.10: Impact Icing. 12 pp.
BSI 2G 229-87. (WITHDRAWN) 1987 Environmental Conditions and Test Procedures for Airborne Equipment (Superseded by BS 36 229: 1993). 4 pp.
BSI 3G 229-93. 1993 Environmental Conditions and Test Procedures for Airborne Equipment (ISO 7137: 1992) (S). 6 pp.
CAA STANDARD NO. 1-0 &AP03.76. General Definitions and Including Atmospheric Conditions. 42 pp.
EUROCAE ED-14C/RTCA DO160C 12.89. Environmental Conditions and Test Procedures for Airborne Equipment. 480 pp.
ISO 7137-92. Aircraft—Environmental Conditions and Test Procedures for Airborne Equipment Third Edition. 8 pp.

—Aircraft Equipment—Explosion Proof
BSI 3G 100:Pt 2: SUBSEC 3.5-72. 1972 General Requirements for Equipment for Use in Aircraft Part 2: All Equipment Section 3: Environmental Conditions Subsection 3.5: Explosion-Proofness. 12 pp.

—Aircraft Equipment—Fluid Contamination
BSI 3G 100:Pt 2: SUBSEC 3.12-91. 1991 General Requirements for Equipment for Use on Aircraft Part 2: All Equipment Section 3: Environmental Conditions Subsection 3.12: Fluid Contamination. 11 pp.

—Aircraft Equipment—Glossaries
ISO 5843 Pt 5-90. Aerospace—List of Equivalent Terms—Part 5: Environmental and Operating Conditions for Aircraft Equipment First Edition. 39 pp.

Environmental Conditions (Cont.)
—Aircraft Equipment—Mold Growth
BSI 3G 100:Pt 2: SUBSEC 3.3-72. 1972 Amd 1 General Requirements for Equipment for Use in Aircraft Part 2: All Equipment Section 3: Environmental Conditions Subsection 3.3: Mould Growth. 2 pp.

—Aircraft—Glossaries
JIS W 0110-87. Glossary of Terms for Aircraft Environmental Control.

—Computer Equipment
BSI BS 7083-89. 1989 Amd 1 Recommendations for the Accommodation and Operating Environment of Computer Equipment (AMD 6624) December 21, 1990. 17 pp.

—Computer Equipment—Ships
JIS F 0416-91. Commercial Computing Apparatus and Their Peripherals—Environmental Conditions in Ships.

—Computers
BSI BS 7083-89. 1989 Amd 1 Recommendations for the Accommodation and Operating Environment of Computer Equipment (AMD 6624) December 21, 1990. 17 pp.

—Data Terminal Equipment
CEPT T/CS 34-06-81. Conditions D'Environnement De L'Equipement Terminal Installe Chez L'Abonne. 2 pp.

—Discrete Components—Industrial
ISO TR10450-91. Industrial Automation Systems and Integration—Operating Conditions for Discrete Part Manufacturing—Equipment in Industrial Environments First Edition. 13 pp.

—Electrical Components
BSI BS 7527: Part 1-91. 1991 Classification of Environmental Conditions Part 1: Environmental Parameters and Their Severities (IEC 721-1: 1990). 23 pp.
BSI BS 7527: Sec 3.0-91. 1991 Classification of Environmental Conditions Part 3: Classification of Groups of Environmental Parameters and Their Severities Section 3.0: Introduction (IEC 721-3-0: 1984). 14 pp.
BSI BS 7527: Sec 3.4-91. (WITHDRAWN) 1991 Classification of Environmental Conditions Part 3: Classification of Groups of Environmental Parameters and Their Severities Section 3.4: Stationary Use at Non-Weatherprotected Locations (IEC 721-3-4: 1987) (Q). 24 pp.
BSI BS 7527: Sec 3.7-91. (WITHDRAWN) 1991 Classification of Environmental Conditions Part 3: Classification of Groups of Environmental Parameters and Their Severities Section 3.7: Portable and Non-Stationary Use (IEC 721-3-7: 1987) (Renumbered as BS EN 60721-3-7: 1993). 32 pp.
BSI BS EN 60721-3-4-93. 1993 Amd 1 Classification of Environmental Conditions Part 3: Classification of Groups of Environmental Parameters and Their Severities Sec. 3.4: Stationary Use at Non-Weatherprotected Locations (IEC 721-3-4: 1987) (AMD 7929) August 15, 1993 (G). 29 pp.
BSI BS EN 60721-3-7-93. 1993 Amd 1 Classification of Environmental Conditions Part 3: Classification of Groups of Environmental Parameters and Their Severities Section 3.7: Portable and Non-Stationary Use (IEC 721-3-7: 1987) (AMD 7949) August 15, 1993 (G). 40 pp.
CENELEC HD 478.1 S1-92. Classification of Environmental Conditions Part 1: Environmental Parameters and Their Severities. 5 pp.
CENELEC HD 478.3.0 S2-89. Classification of Environmental Conditions-Part 3: Classification of Groups of Environmental Parameters and Their Severities. Introduction. 3 pp.
CENELEC HD 478.3.4 S1-89. Classification of Environmental Conditions Part 3: Classification of Groups of Environmental Parameters and Their Severities Stationary Use at Non-Weatherprotected Locations (Superseded by EN 60721-3-4: 1993). 3 pp.
CENELEC HD 478.3.7 S1-89. Classification of Environmental Conditions Part 3: Classification of Groups of Environmental Parameters and Their Severities Portable and Non-Stationary Use (Superseded by EN 60721-3-7: 1993). 3 pp.
CENELEC EN 60721-3-1-93. Classification of Environmental Conditions Part 3: Classification of Groups of Environmental Parameters and Their Severities—Storage (IEC 721-3-1: 1987 + A1: 1991). 15 pp.
CENELEC EN 60721-3-4-93. Classification of Environmental Conditions Part 3: Classification of Groups of Environmental Parameters and Their Severities Stationary Use at Non-Weatherprotected Locations (IEC 721-3-4: 1987+A1:1991)(Supersedes HD 478.3.4 S1: 1989). 22 pp.

INTERNATIONAL AND NON-U.S. NATIONAL STANDARDS
SUBJECT INDEX
Environmental

Environmental Conditions *(Cont.)*

—**Electrical Components** *(Cont.)*

CENELEC EN 60721-3-7-93. Classification of Environmental Conditions Part 3: Classification of Groups of Environmental Parameters and Their Severities Portable and Non-Stationary Use (IEC 721-3-7: 1987 + A1: 1991) (Supersedes HD 478.3.7 S1: 1989). 30 pp.

IEC 721 Pt 1-90. Classification of Environmental Conditions Part 1: Environmental Parameters and Their Severities Second Edition; (Amendment 1-1992). 48 pp.

IEC 721 Pt 3-0-84. Classification of Environmental Conditions Part 3: Classification of Groups of Environmental Parameters and Their Severities Introduction First Edition; (Corrigendum—Dec 1985) (Amendment 1-1987). 31 pp.

IEC 721 Pt 3-4-87. Classification of Environmental Conditions Part 3: Classification of Groups of Environmental Parameters and Their Severities Stationary Use at Non-Weatherprotected Locations First Edition; (Corrigendum—March 1990). 62 pp.

IEC 721 Pt 3-7-87. Classification of Environmental Conditions Part 3: Classification of Groups of Environmental Parameters and Their Severities Portable and Non-Stationary Use First Edition; (Amendment 1-1991) (Amendment 2-1993) (CENE EN 60721-3-7:1993). 82 pp.

—**Electrical Components—Air Pressure**

BSI BS 7527: Sec 2.3-91. 1991 Classification of Environmental Conditions Part 2: Environmental Conditions Appearing in Nature Section 2.3: Air Pressure (IEC 721-2-3: 1987). 7 pp.

CENELEC HD 478.2.3 S1-90. Classification of Environmental Conditions Part 2: Environmental Conditions Appearing in Nature—Air Pressure. 3 pp.

IEC 721 Pt 2-3-87. Classification of Environmental Conditions Part 2: Environmental Conditions Appearing in Nature Air Pressure First Edition. 10 pp.

—**Electrical Components—Dust and Sand**

BSI BS 7527: Sec 2.5-91. 1991 Classification of Environmental Conditions Part 2: Environmental Conditions Appearing in Nature Section 2.5: Dust, Sand, Salt Mist (IEC 721-2-5: 1991). 21 pp.

IEC 721 Pt 2-5-91. Classification of Environmental Conditions Part 2: Environmental Conditions Appearing in Nature Section 5: Dust, Sand, Salt Mist First Edition. 38 pp.

—**Electrical Components—Earthquakes—Vibration**

BSI BS 7527: Sec 2.6-91. 1991 Classification of Environmental Conditions Part 2: Environmental Conditions Appearing in Nature Section 2.6: Earthquake Vibration and Shock (IEC 721-2-6: 1990). 17 pp.

CENELEC HD 478.2.6 S1-92. Classification of Environmental Conditions Part 2: Environmental Conditions Appearing in Nature Earthquake Vibration and Shock (IEC 721-2-6:1990). 5 pp.

IEC 721 Pt 2-6-90. Classification of Environmental Conditions Part 2: Environmental Conditions Appearing in Nature—Earthquake Vibration and Shock First Edition; (CENELEC HD 478.2.6 S1:1992). 25 pp.

—**Electrical Components—Ground Vehicle Installations**

BSI BS 7527: Sec 3.5-91. 1991 Classification of Environmental Conditions Part 3: Classification of Groups of Environmental Parameters and Their Severities Section 3.5: Ground Vehicle Installations (IEC 721-3-5: 1985) (Renumbered as BS EN 60721-3-5: 1993) (Q). 28 pp.

BSI BS EN 60721-3-5-93. 1993 Amd 1 Classification of Environmental Conditions Part 3: Classification of Groups of Environmental Parameters and Their Severities Section 3.5: Ground Vehicle Installations (AMD September 15, 1993 (G). 43 pp.

CENELEC EN 60721-3-5-93. Classification of Environmental Conditions Part 3: Classification of Groups of Environmental Parameters and Their Severities Ground Vehicle Installations (IEC 721-3-5: 1985 + A1: 1991). 25 pp.

IEC 721 Pt 3-5-85. Classification of Environmental Conditions Part 3: Classification of Groups of Environmental Parameters and Their Severities Ground Vehicle Installations First Edition; (Amendment 1-1991) (CENELEC EN 60721-3-5: 1993). 57 pp.

—**Electrical Components—Humidity**

BSI BS 7527: Sec 2.1-91. 1991 Classification of Environmental Conditions Part 2: Environmental Conditions Appearing in Nature Section 2.1: Temperature and Humidity (IEC 721-2-1: 1982). 31 pp.

Environmental Conditions *(Cont.)*

—**Electrical Components—Humidity** *(Cont.)*

IEC 721 Pt 2-1-82. Classification of Environmental Conditions Part 2: Environmental Conditions Appearing in Nature Temperature and Humidity First Edition; (Amendment 1-1987). 46 pp.

—**Electrical Components—Microclimatic Conditions**

IEC 721 Pt 3-9-93. Classification of Environmental Conditions Part 3: Classification of Groups of Environmental Parameters and Their Severities Section 9: Microclimates Inside Products First Edition. 25 pp.

—**Electrical Components—Plants and Animals**

BSI BS 7527: Sec 2.7-91. 1991 Classification of Environmental Conditions Part 2: Environmental Conditions Appearing in Nature Section 2.7: Fauna and Flora (IEC 721-2-7: 1987). 10 pp.

IEC 721 Pt 2-7-87. Classification of Environmental Conditions Part 2: Environmental Conditions Appearing in Nature Fauna and Flora First Edition. 14 pp.

—**Electrical Components—Precipitation and Wind**

BSI BS 7527: Sec 2.2-91. 1991 Classification of Environmental Conditions Part 2: Environmental Conditions Appearing in Nature Section 2.2: Precipitation and Wind (IEC 721-2-2: 1988). 15 pp.

IEC 721 Pt 2-2-88. Classification of Environmental Conditions Part 2: Environmental Conditions Appearing in Nature Precipitation and Wind First Edition. 24 pp.

—**Electrical Components—Salt Mist**

BSI BS 7527: Sec 2.5-91. 1991 Classification of Environmental Conditions Part 2: Environmental Conditions Appearing in Nature Section 2.5: Dust, Sand, Salt Mist (IEC 721-2-5: 1991). 21 pp.

IEC 721 Pt 2-5-91. Classification of Environmental Conditions Part 2: Environmental Conditions Appearing in Nature Section 5: Dust, Sand, Salt Mist First Edition. 38 pp.

—**Electrical Components—Ships**

BSI BS 7527: Sec 3.6-91. (WITHDRAWN) 1991 Classification of Environmental Conditions Part 3: Classification of Groups of Environmental Parameters and Their Severities Section 3.6: Ship Environment (IEC 721-3-6: 1987) (Renumbered as BS EN 60721-3-6: 1993). 16 pp.

BSI BS EN 60721-3-6-93. 1993 Amd 1 Classification of Environmental Conditions Part 3: Classification of Groups of Environmental Parameters and Their Severities Section 3.6: Ship Environment (IEC 721-3-6: 1987) (AMD 7948) August 15, 1993 (G). 21 pp.

CENELEC HD 478.3.6 S1-89. Classification of Environmental Conditions Part 3: Classification of Groups of Environmental Parameters and Their Severities Ship Environment (Superseded by EN 60721-3-6: 1993). 3 pp.

CENELEC EN 60721-3-6-93. Classification of Environmental Conditions Part 3: Classification of Groups of Environmental Parameters and Their Severities Ship Environment (IEC 721-3-6: 1987 + A1: 1991) (Supersedes HD 478.3.6 S1: 1989). 13 pp.

IEC 721 Pt 3-6-87. Classification of Environmental Conditions Part 3: Classification of Groups of Environmental Parameters and Their Severities Ship Environment First Edition; (Corrigendum 1987) (Amendment 1-1991) (CENE EN 60721-3-6:1993). 32 pp.

—**Electrical Components—Solar Radiation—Temperature**

BSI BS 7527: Sec 2.4-91. 1991 Classification of Environmental Conditions Part 2: Environmental Conditions Appearing in Nature Section 2.4: Solar Radiation and Temperature (IEC 721-2-4: 1987). 16 pp.

IEC 721 Pt 2-4-87. Classification of Environmental Conditions Part 2: Environmental Conditions Appearing in Nature Solar Radiation and Temperature First Edition; (Amendment 1-1988). 28 pp.

—**Electrical Components—Storage**

BSI BS 7527: Sec 3.1-91. (WITHDRAWN) 1991 Classification of Environmental Conditions Part 3: Classification of Groups of Environmental Parameters and Their Severities Section 3.1: Storage (IEC 721-3-1: 1987) (Renumbered as BS EN 60721-3-1: 1993). 18 pp.

BSI BS EN 60721-3-1-93. 1993 Classification of Environmental Conditions Part 3: Classification of Groups of Environmental Parameters and Their Severities Section 3.1: Storage (IEC 721-3-1: 1987) (AMD 7909) August 15, 1993 (G). 28 pp.

Environmental Conditions *(Cont.)*

—**Electrical Components—Storage** *(Cont.)*

CENELEC HD 478.3.1 S1-90. Classification of Environmental Conditions Part 3: Classification of Groups of Environmental Parameters and Their Severities—Storage. 3 pp.

—**Electrical Components—Temperature**

BSI BS 7527: Sec 2.1-91. 1991 Classification of Environmental Conditions Part 2: Environmental Conditions Appearing in Nature Section 2.1: Temperature and Humidity (IEC 721-2-1: 1982). 31 pp.

IEC 721 Pt 2-1-82. Classification of Environmental Conditions Part 2: Environmental Conditions Appearing in Nature Temperature and Humidity First Edition; (Amendment 1-1987). 46 pp.

—**Electrical Components—Transportation Conditions**

BSI BS 7527: Sec 3.2-91. 1991 Classification of Environmental Conditions Part 3: Classification of Groups of Environmental Parameters and Their Severities Section 3.2: Transportation (IEC 721-3-2: 1985) (Renumbered as BS EN 60721-3-2: 1993) (G). 22 pp.

BSI BS EN 60721-3-2-93. 1991 Amd 1 Classification of Environmental Conditions Part 3: Classification of Groups of Environmental Parameters and Their Severities Section 3.2: Transportation (AMD 7955) September 15, 1993 (G). 30 pp.

CENELEC EN 60721-3-2-93. Classification of Environmental Conditions Part 3: Classification of Groups of Environmental Parameters and Their Severities—Transportation (IEC 721-3-2: 1985 + A1: 1991). 19 pp.

IEC 721 Pt 3-2-85. Classification of Environmental Conditions Part 3: Classification of Groups of Environmental Parameters and Their Severities Transportation First Edition; (Amendment 1-1991) (Amendment 2-1993) (CENELEC EN 60721-3-2: 1993). 52 pp.

—**Electrical Components—Weather Protected Locations**

BSI BS 7527: Sec 3.3-91. (WITHDRAWN) 1991 Classification of Environmental Conditions Part 3: Classification of Groups of Environmental Parameters and Their Severities Section 3.3: Stationary Use at Weatherprotected Locations (IEC 721-3-3: 1987). 43 pp.

BSI BS EN 60721-3-3-93. 1993 Amd 1 Classification of Environmental Conditions Part 3: Classification of Groups of Environmental Parameters and Their Severities Section 3.3: Stationary Use at Weatherprotected Locations (IEC 721-3-3: 1987) (AMD 7931) August 15, 1993 (G). 54 pp.

CENELEC HD 478.3.3 S1-89. Classification of Environmental Conditions Part 3: Classification of Groups of Environmental Parameters and Their Severities Stationary Use at Weatherprotected Locations (Superseded by EN 60721-3-3: 1993). 3 pp.

CENELEC EN 60721-3-3-93. Classification of Environmental Conditions Part 3: Classification of Groups of Environmental Parameters and Their Severities Stationary Use at Weatherprotected Locations (IEC 721-3-3: 1987 + A1: 1991)(Supersedes HD 478.3.3 S1: 1989). 40 pp.

IEC 721 Pt 3-3-87. Classification of Environmental Conditions Part 3: Classification of Groups of Environmental Parameters and Their Severities Stationary Use at Weatherprotected Locations First Edition; (Corrigendum—April 1988). 93 pp.

—**Electrical Equipment**

IEC Guide 106-89. Guide for Specifying Environmental Conditions for Equipment Performance Rating First Edition. 24 pp.

—**Electrical Equipment—Aircraft**

BSI 2G 100: Part 2-62. 1962 Amd 8 General Requirements for Electrical Equipment and Indicating Instruments for Aircraft Part 2: Environmental and Operating Conditions. 14 pp.

—**Electronic Equipment**

IEC Guide 106-89. Guide for Specifying Environmental Conditions for Equipment Performance Rating First Edition. 24 pp.

—**Gamma Ray Equipment—Airborne**

IEC 1134-92. Airborne Instrumentation for Measurement of Terrestrial Gamma Radiation First Edition. 57 pp.

—**Indicating Instruments—Aircraft**

BSI 2G 100: Part 2-62. 1962 Amd 8 General Requirements for Electrical Equipment and Indicating Instruments for Aircraft Part 2: Environmental and Operating Conditions. 14 pp.

INDUSTRY STANDARDS

Environmental
INTERNATIONAL AND NON-U.S. NATIONAL STANDARDS
SUBJECT INDEX

Environmental Conditions *(Cont.)*

—Machine Tools

BSI BS 6751-86. 1986 Protection of Machine Tools Intended for Use in Extreme Environmental Conditions. 14 pp.

—Medical Electrical Equipment

CSA CAN/CSA-C22. 2NO601.1-M90. Medical Electrical Equipment, Part 1: General Requirements for Safety; (Gen Instr 1). 240 pp.

—Medical Electrical Equipment—Storage

CSA CAN/CSA-C22. 2NO601.1-M90. Medical Electrical Equipment, Part 1: General Requirements for Safety; (Gen Instr 1). 240 pp.

—Medical Electrical Equipment—Transport

CSA CAN/CSA-C22. 2NO601.1-M90. Medical Electrical Equipment, Part 1: General Requirements for Safety; (Gen Instr 1). 240 pp.

—Offshore Drilling

CSA CAN/CSA-S471-92. General Requirements, Design Criteria, the Environment, and Loads; (Gen Instr 1). 93 pp.

CSA S471.1-1992 SP. Commentary to CSA Standard CAN/CSA-S471-92, General Requirements, Design Criteria, the Environment, and Loads; (Gen Instr 1). 27 pp.

—Remote Control Equipment

BSI BS 7404: Sec 2.1-91. 1991 Telecontrol Equipment and Systems Part 2: Operating Conditions Section 2.1: Specification for Classes of Environmental Conditions and Power Supplies. 28 pp.

IEC 870 Pt 2-1-87. Telecontrol Equipment and Systems Part 2: Operating Conditions Section One—Environmental Conditions and Power Supplies First Edition. 44 pp.

—Telecommunication Equipment

CENELEC PRETS 300 019-A-90. Equipment Engineering: Environmental Conditions and Environmental Tests for Telecommunications Equipment Part A—Introduction and Terminology. 43 pp.

CENELEC PRETS 300 019-B-90. Equipment Engineering Environmental Conditions and Environmental Tests for Telecommunications Equipment Part B—Classification of Environmental Conditions (T/TR 02-12). 105 pp.

CENELEC PRETS 300 019-1-0-92. Equipment Engineering (EE); Environmental Conditions and Environmental Tests for Telecommunications Equipment Part 1-0: Classification of Environmental Conditions Introduction. 10 pp.

CENELEC PRETS 300 019-1-7-91. Equipment Engineering (EE); Environmental Conditions and Environmental Tests for Telecommunications Equipment Part 1-7: Classification of Environmental Conditions Portable and Non-Stationary Use. 17 pp.

CENELEC ETS 300 019-1-7-92. Equipment Engineering (EE); Environmental Conditions and Environmental Tests for Telecommunications Equipment Part 1-7: Classification of Environmental Conditions Portable and Non-Stationary Use. 24 pp.

CENELEC PRETS 300 019-2-7-92. Equipment Engineering (EE); Environmental Conditions and Environmental Tests for Telecommunications Equipment Part 2-7: Specification of Environmental Tests T 7.1 to T 7.3E Portable and Non-Stationary Use. 18 pp.

CEPT T/L 02-12 E-88. Environmental Conditions and Environmental Tests for Telecommunications Equipment. 24 pp.

CEPT T/TR 02-03-86. Conditions D'Environnement Pour Equipements Destines Auxcentres De Telecommunication. 11 pp.

CEPT T/TR 02-03 E-86. Environmental Conditions for the Equipment to be Installed in Telecommunication Centres. 11 pp.

CEPT T/TR 02-06-86. Conditions D'Environement Pour Equipements De Telecommunication Installes a L'Air Libre Et Dans Des Emplacements Situes a L'Abri Des Intemperies (a L'Exception Des Centres De Telecommunications Avec Regulation De La Temperature Et De L'Humidite De L'Air. 20 pp.

CEPT T/TR 02-06 E-86. Environmental Conditions for Telecommunications Equipment at Weather-Protected Locations (Excluding Telecommunication Centres with Temperature and Humidity Controls) and in the Open Air. 20 pp.

CEPT T/TR 02-09-86. Classification Des Conditions D'Environnement Pour Des Instalations De Telecommunication a Poste Fixe Et a L'Abri Des Intemperies Categorie D'Environnement 3.1-3.5. 15 pp.

CEPT T/TR 02-09 E-86. Classification of Environmental Conditions for Stationary Telecommunications Installations at Weather-Protected Locations Environmental Classes 3.1-3.5. 15 pp.

Environmental Conditions *(Cont.)*

—Telecommunication Equipment *(Cont.)*

ETSI PRETS 300 019-A-90. Equipment Engineering; Environmental Conditions and Environmental Tests for Telecommunications Equipment—Part A—Introduction and Terminology (T/TR 02-12). 43 pp.

ETSI PRETS 300 019-B-90. Equipment Engineering. Environmental Conditions and Environmental Tests for Telecommunications Equipment—Part B—Classification of Environmental Conditions (T/TR 02-12). 105 pp.

ETSI PRETS 300 019-1-4-91. Equipment Engineering (EE); Environmental Conditions and Environmental Tests for Telecommunications Equipment Part 1-4: Classification of Environmental Conditions. Stationary use at Non-Weatherprotected Locations. 13 pp.

ETSI PRETS 300 019-1-7-91. Equipment Engineering (EE); Environmental Conditions and Environmental Tests for Telecommunications Equipment Part 1-7: Classification of Environmental Conditions. Portable and Non-Stationary use. 16 pp.

ETSI PRETS 300 019-A-90. Equipment Engineering: Environmental Conditions and Environmental Tests for Telecommunications Equipment Part A—Introduction and Terminology. 43 pp.

ETSI PRETS 300 019-B-90. Equipment Engineering Environmental Conditions and Environmental Tests for Telecommunications Equipment Part B—Classification of Environmental Conditions (T/TR 02-12). 105 pp.

ETSI PRETS 300 019-1-0-92. Equipment Engineering (EE); Environmental Conditions and Environmental Tests for Telecommunications Equipment Part 1-0: Classification of Environmental Conditions Introduction. 10 pp.

ETSI PRETS 300 019-1-4-91. Equipment Engineering (EE); Environmental Conditions and Environmental Test for Telecommunications Equipment Part 1-4: Classification of Environmental Conditions Stationary Use at Non-Weatherprotected Locations. 14 pp.

ETSI ETS 300 019-1-7-92. Equipment Engineering (EE); Environmental Conditions and Environmental Tests for Telecommunications Equipment Part 1-7: Classification of Environmental Conditions Portable and Non-Stationary Use. 24 pp.

ETSI PRETS 300 019-1-7-91. Equipment Engineering (EE); Environmental Conditions and Environmental Tests for Telecommunications Equipment Part 1-7: Classification of Environmental Conditions Portable and Non-Stationary Use. 17 pp.

ETSI PRETS 300 019-2-7-92. Equipment Engineering (EE); Environmental Conditions and Environmental Tests for Telecommunications Equipment Part 2-7: Specification of Environmental Tests T 7.1 to T 7.3E Portable and Non-Stationary Use. 18 pp.

—Telecommunication Equipment—Ground Vehicle Installations

CENELEC PRETS 300 019-1-5-91. Equipment Engineering (EE); Environmental Conditions and Environmental Tests for Telecommunications Equipment Part 1-5: Classification of Environmental Conditions Ground Vehicle Installations. 16 pp.

CENELEC ETS 300 019-1-5-92. Equipment Engineering (EE); Environmental Conditions and Environmental Tests for Telecommunications Equipment Part 1-5: Classification of Environmental Conditions Ground Vehicle Installations. 16 pp.

CENELEC PRETS 300 019-2-5-92. Equipment Engineering (EE); Environmental Conditions and Environmental Tests for Telecommunications Equipment Part 2-5: Specification of Environmental Tests T 5.1 and T 5.2 Ground Vehicle Installations. 15 pp.

ETSI PRETS 300 019-1-5-91. Equipment Engineering (EE); Environmental Conditions and Environmental Tests for Telecommunications Equipment Part 1-5: Classification of Environmental Conditions. Ground Vehicle Installations. 15 pp.

ETSI ETS 300 019-1-5-92. Equipment Engineering (EE); Environmental Conditions and Environmental Tests for Telecommunications Equipment Part 1-5: Classification of Environmental Conditions Ground Vehicle Installations. 16 pp.

ETSI PRETS 300 019-1-5-91. Equipment Engineering (EE); Environmental Conditions and Environmental Tests for Telecommunications Equipment Part 1-5: Classification of Environmental Conditions Ground Vehicle Installations. 16 pp.

ETSI PRETS 300 019-2-5-92. Equipment Engineering (EE); Environmental Conditions and Environmental Tests for Telecommunications Equipment Part 2-5: Specification of Environmental Tests T 5.1 and T 5.2 Ground Vehicle Installations. 15 pp.

—Telecommunication Equipment—Installation

CCITT RECMN K.10-89. Unbalance About Earth of Telecommunication Installations—Protection Against Interference (Study Group V) 8 pp. 8 pp.

Environmental Conditions *(Cont.)*

—Telecommunication Equipment—Nonweather Protected Locations

CENELEC PRETS 300 019-1-4-91. Equipment Engineering (EE); Environmental Conditions and Environmental Test for Telecommunications Equipment Part 1-4: Classification of Environmental Conditions Stationary Use at Non-Weatherprotected Locations. 14 pp.

CENELEC ETS 300 019-1-4-92. Introductory Element Equipment Engineering (EE); Environmental Conditions and Environmental Test for Telecommunications Equipment Part 1-4: Classification of Environmental Conditions Stationary Use at Non-Weatherprotected Locations. 14 pp.

CENELEC PRETS 300 019-2-4-92. Equipment Engineering (EE); Environmental Conditions and Environmental Tests for Telecommunications Equipment Part 2-4: Specification of Environmental Tests T 4.1 and T 4.1E Stationary Use at Non-Weatherprotected Locations. 13 pp.

ETSI ETS 300 019-1-4-92. Introductory Element Equipment Engineering (EE); Environmental Conditions and Environmental Test for Telecommunications Equipment Part 1-4: Classification of Environmental Conditions Stationary Use at Non-Weatherprotected Locations. 14 pp.

ETSI PRETS 300 019-2-4-92. Equipment Engineering (EE); Environmental Conditions and Environmental Tests for Telecommunications Equipment Part 2-4: Specification of Environmental Tests T 4.1 and T 4.1E Stationary Use at Non-Weatherprotected Locations. 13 pp.

—Telecommunication Equipment—Ships

CENELEC PRETS 300 019-1-6-91. Equipment Engineering (EE); Environmental Conditions and Environmental Tests for Telecommunications Equipment Part 1-6: Classification of Environmental Conditions Ship Environments. 13 pp.

CENELEC ETS 300 019-1-6-92. Equipment Engineering (EE); Environmental Conditions and Environmental Tests for Telecommunications Equipment Part 1-6: Classification of Environmental Conditions Ship Environments. 14 pp.

CENELEC PRETS 300 019-2-6-92. Equipment Engineering (EE); Environmental Conditions and Environmental Tests for Telecommunications Equipment Part 2-6: Specification of Environmental Tests T 6.1 to T 6.3 Ship Environment. 19 pp.

ETSI PRETS 300 019-1-6-91. Equipment Engineering (EE); Environmental Conditions and Environmental Tests for Telecommunications Equipment Part 1-6: Classification of Environmental Conditions. Ship Environments. 13 pp.

ETSI ETS 300 019-1-6-92. Equipment Engineering (EE); Environmental Conditions and Environmental Tests for Telecommunications Equipment Part 1-6: Classification of Environmental Conditions Ship Environments. 14 pp.

ETSI PRETS 300 019-1-6-91. Equipment Engineering (EE); Environmental Conditions and Environmental Tests for Telecommunications Equipment Part 1-6: Classification of Environmental Conditions Ship Environments. 13 pp.

ETSI PRETS 300 019-2-6-92. Equipment Engineering (EE); Environmental Conditions and Environmental Tests for Telecommunications Equipment Part 2-6: Specification of Environmental Tests T 6.1 to T 6.3 Ship Environment. 19 pp.

—Telecommunication Equipment—Storage

CENELEC PRETS 300 019-1-1-91. Equipment Engineering (EE); Environmental Conditions and Environmental Tests for Telecommunications Equipment Part 1-1: Classification of Environmental Conditions Storage. 15 pp.

CENELEC ETS 300 019-1-1-92. Equipment Engineering (EE); Environmental Conditions and Environmental Tests for Telecommunications Equipment Part 1-1: Classification of Environmental Conditions Storage. 15 pp.

CENELEC PRETS 300 019-2-1-92. Equipment Engineering (EE); Environmental Conditions and Environmental Tests for Telecommunications Equipment Part 2-1: Specification of Environmental Tests T 1.1 to T 1.3 Storage. 19 pp.

ETSI PRETS 300 019-1-1-91. Equipment Engineering (EE); Environmental Conditions and Environmental Tests for Telecommunications Equipment—Part 1-1: Classification of Environmental Conditions Storage.

ETSI ETS 300 019-1-1-92. Equipment Engineering (EE); Environmental Conditions and Environmental Tests for Telecommunications Equipment Part 1-1: Classification of Environmental Conditions Storage. 15 pp.

ETSI PRETS 300 019-1-1-91. Equipment Engineering (EE); Environmental Conditions and Environmental Tests for Telecommunications Equipment Part 1-1: Classification of Environmental Conditions Storage. 15 pp.

988
INDEX and DIRECTORY of

INTERNATIONAL AND NON-U.S. NATIONAL STANDARDS
SUBJECT INDEX
Environmental

Environmental Conditions (Cont.)
—Telecommunication Equipment—Storage (Cont.)
ETSI PRETS 300 019-2-1-92. Equipment Engineering (EE); Environmental Conditions and Environmental Tests for Telecommunications Equipment Part 2-1: Specification of Environmental Tests T 1.1 to T 1.3 Storage. 19 pp.

—Telecommunication Equipment—Transportation
CENELEC PRETS 300 019-1-2-91. Equipment Engineering (EE); Environmental Conditions and Environmental Tests for Telecommunications Equipment Part 1-2: Classification of Environmental Conditions Transportation. 15 pp.

CENELEC ETS 300 019-1-2-92. Equipment Engineering (EE); Environmental Conditions and Environmental Tests for Telecommunications Equipment Part 1-2: Classification of Environmental Conditions Transportation. 16 pp.

CENELEC PRETS 300 019-2-2-92. Equipment Engineering (EE); Environmental Conditions and Environmental Tests for Telecommunications Equipment Part 2-2: Specification of Environmental Tests T 2.1 to T 2.3 Transportation. 19 pp.

ETSI PRETS 300 019-1-2-91. Equipment Engineering (EE); Environmental Conditions and Environmental Tests for Telecommunications Equipment Part 1-2: Classification of Environmental Conditions Transportation. 14 pp.

ETSI ETS 300 019-1-2-92. Equipment Engineering (EE); Environmental Conditions and Environmental Tests for Telecommunications Equipment Part 1-2: Classification of Environmental Conditions Transportation. 16 pp.

ETSI PRETS 300 019-1-2-91. Equipment Engineering (EE); Environmental Conditions and Environmental Tests for Telecommunications Equipment Part 1-2: Classification of Environmental Conditions Transportation. 15 pp.

ETSI PRETS 300 019-2-2-92. Equipment Engineering (EE); Environmental Conditions and Environmental Tests for Telecommunications Equipment Part 2-2: Specification of Environmental Tests T 2.1 to T 2.3 Transportation. 19 pp.

—Telecommunication Equipment—Weather Protected Locations
CENELEC PRETS 300 019-1-3-91. Equipment Engineering (EE); Environmental Conditions and Environmental Tests for Telecommunications Equipment Part 1-3: Classification of Environmental Conditions Stationary Use at Weatherprotected Locations. 20 pp.

CENELEC ETS 300 019-1-3-92. Equipment Engineering (EE); Environmental Conditions and Environmental Tests for Telecommunications Equipment Part 1-3: Classification of Environmental Conditions Stationary Use at Weatherprotected Locations. 20 pp.

CENELEC PRETS 300 019-2-3-92. Equipment Engineering (EE); Environmental Conditions and Environmental Tests for Telecommunications Equipment Part 2-3: Specification of Environmental Tests T 3.1 to T 3.5 Stationary Use at Weatherprotected Locations. 24 pp.

ETSI PRETS 300 019-1-3-91. Equipment Engineering (EE); Environmental Conditions and Environmental Tests for Telecommunications Equipment Part 1-3: Classification of Environmental Conditions. Stationary use at Weatherprotected Locations. 19 pp.

ETSI ETS 300 019-1-3-92. Equipment Engineering (EE); Environmental Conditions and Environmental Tests for Telecommunications Equipment Part 1-3: Classification of Environmental Conditions Stationary Use at Weatherprotected Locations. 20 pp.

ETSI PRETS 300 019-1-3-91. Equipment Engineering (EE); Environmental Conditions and Environmental Tests for Telecommunications Equipment Part 1-3: Classification of Environmental Conditions Stationary Use at Weatherprotected Locations. 20 pp.

ETSI PRETS 300 019-2-3-92. Equipment Engineering (EE); Environmental Conditions and Environmental Tests for Telecommunications Equipment Part 2-3: Specification of Environmental Tests T 3.1 to T 3.5 Stationary Use at Weatherprotected Locations. 24 pp.

—Workplace—Diffusion Samplers
DIN ENGL 33403 Pt 1-84. Climate at Workplaces and in Working Environments; Basic Principles for Determining Climates (Apr). 7 pp.

Environmental Engineering
See Also: Air Conditioners; Building Automation Systems; Cooling Systems; Energy Management; Environmental Conditions; Environmental Protection; Ergonomics; Heating Equipment; Ventilation Equipment; Waste Disposal

Environmental Engineering (Cont.)
—Defense Equipment
MOD UK DSTAN 00-25: Part 5-92. Human Factors for Designers of Equipment Part 5: Stresses and Hazards Issue 1. 61 pp.

—Electronic Equipment
CNS C5211-86. Environmental Requirements for Electronic Equipments (Nov)(11762).

CNS C5218-88. Environmental Requirements for Electronic Components (Jan)(12211).

JIS C 5002-68. Environmental Requirements for Electronic Components (R 1974). 4 pp.

JIS C 6007-69. Environmental Requirements for Electronic Equipments.

—Generator Sets
MOD UK DSTAN 61-5: Pt 2: Sec 4-92. Electrical Power Supply Systems Below 650 Volts Part 2: Ground Generating Set Characteristics Section 4: Environmental Requirements Issue 1. 18 pp.

—Materiel
MOD UK DSTAN 00-35: Part 1-86. Environmental Handbook for Defence Materiel Part 1: General Requirements Issue 1. 28 pp.

MOD UK DSTAN 00-35: Part 2-87. Environmental Handbook for Defence Materiel Part 2: Specification of Service Environments Issue 1. 22 pp.

MOD UK DSTAN 00-35: Part 4-86. Environmental Handbook for Defence Materiel Part 4: Natural Environments Issue 1. 334 pp.

—Materiel—Fires
MOD UK DSTAN 00-28-81. Fire Environmental Conditions Affecting Design Criteria of Military Materiel Issue 1. 15 pp.

—Public Health—Glossaries
BSI BS 6100: Sec 2.7-92. 1992 Glossary of Building and Civil Engineering Terms Part 2: Civil Engineering Section 2.7: Public Health. Environmental Engineering. 26 pp.

BSI BS 6100: Sec 2.7-88. 1988 Glossary of Building and Civil Engineering Terms Part 2: Civil Engineering Section 2.7: Public Health. Environmental Engineering. 22 pp.

—Telecommunication Equipment
CENELEC ETR 035-92. Equipment Engineering (EE); Environmental Engineering Guidance and Terminology. 36 pp.

ETSI ETR 035-92. Equipment Engineering (EE); Environmental Engineering Guidance and Terminology. 36 pp.

—Telecommunication Equipment—Glossaries
CENELEC ETR 035-92. Equipment Engineering (EE); Environmental Engineering Guidance and Terminology. 36 pp.

ETSI ETR 035-92. Equipment Engineering (EE); Environmental Engineering Guidance and Terminology. 36 pp.

Environmental Noise
Use: Noise

Environmental Protection
See Also: Environmental Conditions; Environmental Engineering

EURO 1991 Aug. Pollution Knows No Frontiers. 9 pp.

—Air Transportation
ICAO 9592. Committee on Aviation Environmental Protection Second Meeting Montreal, 2-13 December 1991. 191 pp.

JIS W 2017-84. Environmental Control, Environmental Protection, and Engine Bleed Air Systems, Aircraft, General Specification for.

—Chemical Products—Safety Data Sheets
DIN ENGL 52900-83. DIN Safety Data Sheet for Chemical Substances and Preparations; Form and Instructions on How to Fill in the Form (Feb). 10 pp.

—Classification
BSI BS 1000: (5/50)-83. 1983 Universal Decimal Classification (UDC). English Full Edition (5/50): Exact Sciences in General. Nature Study. Environmental Sciences. 16 pp.

SNZ NZS/BS 1000 (5/50)-83. Universal Decimal Classification Exact Sciences in General. Nature Study. Conservation. Environmental Sciences. 16 pp.

—European Communities
EURO 1990-4. European Community and Environmental Protection. 11 pp.

EURO 1991 Aug. Pollution Knows No Frontiers. 9 pp.

Environmental Protection (Cont.)
—European Communities—Germany—Transitional Measures
EC 90/660/EEC-90. Council Directive on the Transitional Measures Applicable in Germany with Regard to Certain Community Provisions Relating to the Protection of the Environment, in Connection with the Internal Market. 2 pp.

Environmental Requirements
Use: Environmental Engineering

Environmental Stress Cracking
Use For: ESC *See Also:* Cracking (Fracturing); Environmental Testing; Stresses

—Cable Insulation
DIN VDE 0472 Pt 810-85. Testing of Cables, Wires and Flexible Cords; Resistance to Environmental Stress Cracking (Sept). 11 pp.

—Pipe Fittings
BSI BS 2782:Pt11: METH 1109A-89. 1989 Methods of Testing Plastics Part 11: Thermoplastics Pipes, Fittings and Valves Method 1109A: Resistance to Environmental Stress Cracking of Polyethylene Pipes and Fittings for Non-Pressure Applications. 4 pp.

—Plastic Pipes
BSI BS 2782:Pt11: METH 1109A-89. 1989 Methods of Testing Plastics Part 11: Thermoplastics Pipes, Fittings and Valves Method 1109A: Resistance to Environmental Stress Cracking of Polyethylene Pipes and Fittings for Non-Pressure Applications. 4 pp.

ISO 8796-89. Polyethylene (PE) 25 Pipes for Irrigation Laterals—Susceptibility to Environmental Stress-Cracking Induced by Insert-Type Fittings—Test Method and Specification First Edition. 6 pp.

—Plastics
BSI BS 2782:Pt8: METH 831A-B-93. 1993 Methods of Testing Plastics Part 8: Other Properties Methods 831A to 831B: Determination of Environmental Stress Cracking (ESC) by the Ball or Pin Impression Method (ISO 4600: 1992). 12 pp.

BSI BS 2782:Pt8: METH 832A-91. 1991 Methods of Testing Plastics Part 8: Other Properties Method 832A: Determination of Resistance to Environmental Stress Cracking (ESC) by the Bent Strip Method (ISO 4599: 1986). 11 pp.

BSI BS 2782:Pt8: METH 833A-C-93. 1993 Methods of Testing Plastics Part 8: Other Properties Methods 833A to 833C: Determination of Environmental Stress Cracking (ESC) by the Constant Tensile Stress Method (ISO 6252: 1992). 14 pp.

BSI BS 4618: SUBSEC 1.3.3-76. 1976 Recommendations for the Presentation of Plastics Design Data Part 1: Mechanical Properties Section 1.3: Strength Subsection 1.3.3: Environmental Stress Cracking. 2 pp.

ISO 4599-86. Plastics—Determination of Resistance to Environmental Stress Cracking (ESC)—Bent Strip Method First Edition. 8 pp.

ISO 4600-92. Plastics—Determination of Environmental Stress Cracking (ESC)—Ball or Pin Impression Method Second Edition. 11 pp.

ISO 6252-92. Plastics—Determination of Environmental Stress Cracking (ESC)—Constant-Tensile-Stress Method Second Edition. 12 pp.

Environmental Systems (HVAC)
Use: Heating, Ventilating and Air Conditioning Equipment

Environmental Testing
Use For: Accelerated Weathering Testing; Ambient Conditions Testing; Artificial Weathering Tests; Climatic Testing; Outdoor Exposure Testing; Storage Testing; Weather Resistance Testing; Weathering Testing; Weathering Tests *See Also:* Ambient Temperature; Corrosion Testing; Damp Heat Testing; Daylight Testing; Dry Heat Testing; Environmental Stress Cracking; Environmental Testing Equipment; Freeze Thaw Testing; High Temperature Testing; Immersion Testing; Low Pressure Testing; Low Temperature Testing; Materials Testing; Physical Testing; Random Vibration; Salt Spray Testing; Temperature Change Testing; Test Chambers; Vibration

JIS C 0010-85. Basic Environmental Testing Procedures (General and Guidance). 26 pp.

—Adhesives
CNS K6178-75. Method of Test for Urea Resin Adhesives (May)(2233). 4 pp.

CNS K6442-80. Method of Test for Storage Stability of Adhesives (Jan)(5132).

JIS K 6860-74. General Recommended Practices for Atmospheric Exposure of Adhesive Bonds (R 1989). 8 pp.

INDUSTRY STANDARDS

INTERNATIONAL AND NON-U.S. NATIONAL STANDARDS SUBJECT INDEX

Environmental

Environmental Testing (Cont.)

—Aerospace Equipment

JIS W 0801-83. Environmental Test Methods for Aerospace Equipment.

—Aircraft

CAA JAR-1 Section 2. Interpretations (Joint Airworthiness Requirements). 1 p.
ISO 2671-82. Environmental Tests for Aircraft Equipment—Part 3.4: Acoustic Vibration First Edition. 11 pp.
NATO STANAG 3518 ED 4 AMD 0-90. Environmental Test Methods (Withdrawn 93-06). 5 pp.

—Aircraft—Airborne Radar

EUROCAE WG9/1-71 06.72. MPS for Airborne Secondary Surveillance Radar Transponder Apparatus (Amendment No. 1 Incorporated). 29 pp.
EUROCAE ED-38 06.83. MPS for Airborne Weather, Ground Mapping and Assisted Approach Pulse Radars. 82 pp.

—Aircraft—Altitude Measuring Instruments

EUROCAE ED-26 03.79. MPS for Airborne Altitude Measurement and Coding Systems. 57 pp.
EUROCAE ED-43 10.83. MOPR for the SSR Transpander and the Altitude Measurement and Coding System. 96 pp.

—Aircraft—Area Navigation Computing Instruments

EUROCAE ED-28 07.82. MPS for Airborne Area Navigation Computing Equipment Based on VOR and DME as Sensors. 31 pp.
EUROCAE ED-40 06.84. MPS for Airborne Computing Equipment for Area Navigation System Using Two DME as Sensors. 33 pp.

—Aircraft—Area Navigation Systems

EUROCAE ED-27 10.79. MOPR for Airborne Area Navigation Systems, Based on VOR and DME as Sensors. 63 pp.

—Aircraft—Audio Amplifiers

EUROCAE ED-18 07.85. Audio Systems Characteristics and MPS Aircraft Microphones (Except Carbon), Aircraft Headsets, Handsets and Loudspeakers, Aircraft Audio Selector Panels and Amplifiers. 108 pp.

—Aircraft—Audio Selector Panels

EUROCAE ED-18 07.85. Audio Systems Characteristics and MPS Aircraft Microphones (Except Carbon), Aircraft Headsets, Handsets and Loudspeakers, Aircraft Audio Selector Panels and Amplifiers. 108 pp.

—Aircraft—Dead Reckoning Computers

EUROCAE WG7C/2-74 08.74. MPS for Airborne Automatic Dead Reckoning Computer Equipment Utilizing Aircraft Heading and Doppler Obtained Velocity Vector Data. 19 pp.

—Aircraft—Distance Measuring Equipment

EUROCAE ED-54 01.87. MOPR for DME/N and DME/P Operating Within the Radio Frequency Range 960 to 1215 MHz. 144 pp.
EUROCAE ED-57 12.86. MPS for Distance Measuring Equipment (DME/N and DME/P). 96 pp.

—Aircraft Equipment

BSI 3G 100: Part 0-80. 1980 Amd 1 General Requirements for Equipment for Use in Aircraft Part 0: Introduction (AMD 5366) April 30, 1987. 5 pp.
BSI 3G 100:Pt 2: SUBSEC 3.0-72. 1972 Amd 1 General Requirements for Equipment for Use in Aircraft Part 2: All Equipment Section 3: Environmental Conditions Subsection 3.0: Standard Test Requirements. 4 pp.
BSI 3G 100:Pt 2: SUBSEC 3.2-70. 1970 Amd 2 General Requirements for Equipment for Use in Aircraft Part 2: All Equipment Section 3: Environmental Conditions Subsection 3.2: Temperature-Pressure Requirements. 15 pp.
BSI 3G 100:Pt 2: SUBSEC 3.4-72. 1972 General Requirements for Equipment for Use in Aircraft Part 2: All Equipment Section 3: Environmental Conditions Subsection 3.4: Differential Pressure Requirements. 6 pp.
BSI 3G 100:Pt 2: SUBSEC 3.6-72. 1972 Amd 1 General Requirements for Equipment for Use in Aircraft Part 2: All Equipment Section 3: Environmental Conditions Subsection 3.6: Acceleration Requirements. 7 pp.
BSI 3G 100:Pt 2: SUBSEC 3.8-72. 1972 Amd 1 General Requirements for Equipment for Use in Aircraft Part 2: All Equipment Section 3: Environmental Conditions Subsection 3.8: Salt Mist. 3 pp.

Environmental Testing (Cont.)

—Aircraft Equipment (Cont.)

BSI 3G 100:Pt 2: SUBSEC 3.14-73. 1973 General Requirements for Equipment for Use in Aircraft Part 2: All Equipment Section 3: Environmental Conditions Subsection 3.14: Acoustic Vibration. 10 pp.
BSI 2G 229-87. (WITHDRAWN) 1987 Environmental Conditions and Test Procedures for Airborne Equipment (Superseded by BS 36 229: 1993). 4 pp.
BSI 3G 229-93. 1993 Environmental Conditions and Test Procedures for Airborne Equipment (ISO 7137: 1992) (S). 6 pp.
EUROCAE ED-14C/RTCA DO160C 12.89. Environmental Conditions and Test Procedures for Airborne Equipment. 480 pp.
ISO 7137-92. Aircraft—Environmental Conditions and Test Procedures for Airborne Equipment Third Edition. 8 pp.
JIS W 7002-81. Environmental Conditions and Test Procedures for Airborne Equipment and Instruments.
MOD UK DTD-1085C-61. Climatic Testing of Airborne Instruments and Equipment. 3 pp.
MOD UK DTD-1085C-01. Climatic Testing of Airborne Instruments and Equipment; Amendment 1. 4 pp.

—Aircraft—Flowmeters

EUROCAE ED-42 10.83. MPS for a Fuel Flowmeter to Aircraft Standards. 24 pp.

—Aircraft—Fuel Gages

EUROCAE ED-41 12.83. MPS for Airborne Fuel Quantity Gauging Systems. 30 pp.

—Aircraft—Fuel Tanks—Float Gages

EUROCAE ED-41 12.83. MPS for Airborne Fuel Quantity Gauging Systems. 30 pp.

—Aircraft Fuels—Indicators

EUROCAE ED-42 10.83. MPS for a Fuel Flowmeter to Aircraft Standards. 24 pp.

—Aircraft Fuels—Transmitters

EUROCAE ED-42 10.83. MPS for a Fuel Flowmeter to Aircraft Standards. 24 pp.

—Aircraft—Glide Path Receiving Instruments

EUROCAE ED-47A 01.88. MPS for Airborne ILS Receiving Equipment (Glide Path). 223 pp.

—Aircraft—Handsets

EUROCAE ED-18 07.85. Audio Systems Characteristics and MPS Aircraft Microphones (Except Carbon), Aircraft Headsets, Handsets and Loudspeakers, Aircraft Audio Selector Panels and Amplifiers. 108 pp.

—Aircraft—Headphones

EUROCAE ED-18 07.85. Audio Systems Characteristics and MPS Aircraft Microphones (Except Carbon), Aircraft Headsets, Handsets and Loudspeakers, Aircraft Audio Selector Panels and Amplifiers. 108 pp.

—Aircraft—L Band Avionics

EUROCAE ED-25 07.76. MPS for Experimental AEROSAT L-Band Avionics. 75 pp.

—Aircraft—Localizers

EUROCAE ED-51 10.83. MPS for Airborne Automatic Direction Finding Equipment. 31 pp.

—Aircraft—Loudspeakers

EUROCAE ED-18 07.85. Audio Systems Characteristics and MPS Aircraft Microphones (Except Carbon), Aircraft Headsets, Handsets and Loudspeakers, Aircraft Audio Selector Panels and Amplifiers. 108 pp.

—Aircraft—Low Range Radio Altimeters

EUROCAE ED-30 03.80. MPS for Airborne Low Range Radio (Radar) Altimeter Equipment. 22 pp.

—Aircraft—Marker Beacons—Receiving Equipment

EUROCAE WG7/70 01.70. MPS for Airborne 75 MHz Marker Beacon Receiving Equipment. 26 pp.

—Aircraft—Microphones

EUROCAE ED-18 07.85. Audio Systems Characteristics and MPS Aircraft Microphones (Except Carbon), Aircraft Headsets, Handsets and Loudspeakers, Aircraft Audio Selector Panels and Amplifiers. 108 pp.

—Aircraft—Microwave Landing Systems

EUROCAE ED-36 03.83. MOPR for Microwave Landing System (MLS) (Airborne Receiving Equipment). 69 pp.

Environmental Testing (Cont.)

—Aircraft—Omega Navigation System

EUROCAE ED-29 10.77. MPS for Airborne Omega Navigation Equipment. 45 pp.

—Aircraft—Pipe Couplings—Swaged

SBAC RS 736 ISSUE 1. British Metric Swaged Pipe Coupling (AS 43020 and AS 43090 Series) Summary of Tests.
SBAC RS 740 ISSUE 1. British Metric Swaged Pipe Coupling (AS 43020 and AS 43090 Series) Designers's Guide.

—Aircraft—Radio Receivers

EUROCAE ED-23A 05.86. MPS for Airborne VHF Communications Equipment Operating in the Frequency Range 117.975-137.000 MHz. 125 pp.

—Aircraft—Stores Management Systems

NATO STANAG 3557 ED 1 AMD 1-00. Standard NATO Safety Tests for Airborne Pyrotechnic Stores. 16 pp.

—Aircraft—Transmitters

EUROCAE ED-23A 05.86. MPS for Airborne VHF Communications Equipment Operating in the Frequency Range 117.975-137.000 MHz. 125 pp.

—Aircraft—Transponders

EUROCAE WG9/1-71 06.72. MPS for Airborne Secondary Surveillance Radar Transponder Apparatus (Amendment No. 1 Incorporated). 29 pp.
EUROCAE ED-38 06.83. MPS for Airborne Weather, Ground Mapping and Assisted Approach Pulse Radars. 82 pp.
EUROCAE ED-43 10.83. MOPR for the SSR Transpander and the Altitude Measurement and Coding System. 96 pp.

—Alarm Systems

DIN VDE 0830 Pt 13 (D)-87. Alarm Systems; General Requirements; Environmental Testing for Alarm Systems (Aug). 69 pp.
IEC 839 Pt 1-3-88. Alarm Systems Part 1: General Requirements Section Three—Environmental Testing First Edition. 90 pp.

—Alkyd Primers

CGSB 1-GP-71 METH 143.9-78. Methods of Testing Paints and Pigments Long-Term Outdoor Performance Test for Alkyd Primer on Wood. 1 p.

—Aluminum Coatings (On Aluminum)

CGSB 1-GP-71 METH 122.8-79. Methods of Testing Paints and Pigments Accelerated Weathering Test for One Coat of Air-Drying Material on Steel or Aluminum Panels. 1 p.
CGSB 1-GP-71 METH 122.10-79. Methods of Testing Paints and Pigments Accelerated Weathering Test for Aircraft Finishes on Primed Anodized Aluminum Panels. 1 p.
ISO TR11728-93. Anodized Aluminium and Aluminium Alloys—Accelerated Test of Weather Fastness of Coloured Anodic Oxide Coatings Using Cyclic Artificial Light and Pollution Gas First Edition. 9 pp.

—Antennas

CENELEC HD 95.3-84. Aerials for the Reception of Sound and Television Broadcasting in the Frequency Range 30 MHz to 1GHz Part 3: Methods of Measurements of Mechanical Properties, Vibration and Environmental Tests. 2 pp.
CENELEC HD 95.3 S1-89. Aerials for the Reception of Sound and Television Broadcasting in the Frequency Range 30 MHz to 1 GHz-Part 3: Methods of Measurement of Mechanical Properties, Vibration and Environmental Tests. 3 pp.
IEC 597 Pt 3-83. Aerials for the Reception of Sound and Television Broadcasting in the Frequency Range 30 MHz to 1 GHz Part 3: Methods of Measurement of Mechanical Properties, Vibration and Environmental Tests First Edition. 23 pp.

—Artificial—Lacquers

SAA AS 1580.483. 1-92. Paints and Related Materials—Methods of Test—Part 483.1: Resistance to Artificial Weathering (Carbon-Arc Type Instruments) (in Professional Packages 30, 39). 8 pp.
SNZ NZS/AS 1580. 483.1-92. Methods of Test for Paints and Related Materials Part 483.1: Resistance to Artificial Weathering (Carbon-Arc Type Instruments) (This is a Joint Standard with SAA AS 1580.483.1). 8 pp.

—Artificial—Paints

SAA AS 1580.483. 1-92. Paints and Related Materials—Methods of Test—Part 483.1: Resistance to Artificial Weathering (Carbon-Arc Type Instruments) (in Professional Packages 30, 39). 8 pp.
SNZ NZS/AS 1580. 483.1-92. Methods of Test for Paints and Related Materials Part 483.1: Resistance to Artificial Weathering (Carbon-Arc Type Instruments) (This is a Joint Standard with SAA AS 1580.483.1). 8 pp.

INTERNATIONAL AND NON-U.S. NATIONAL STANDARDS
SUBJECT INDEX
Environmental

Environmental Testing (Cont.)

—Artificial—Varnishes
SAA AS 1580.483. 1-92. Paints and Related Materials—Methods of Test—Part 483.1: Resistance to Artificial Weathering (Carbon-Arc Type Instruments) (in Professional Packages 30, 39). 8 pp.

SNZ NZS/AS 1580. 483.1-92. Methods of Test for Paints and Related Materials Part 483.1: Resistance to Artificial Weathering (Carbon-Arc Type Instruments) (This is a Joint Standard with SAA AS 1580.483.1). 8 pp.

—Asphalts
CNS K6788-83. Method of Test for Storage Stability of Asphalt Emulsion (Jun)(10371).

—Athletic Fields
BSI BS 7044: Sec 2.4-89. 1989 Artificial Sports Surfaces Part 2: Methods of Test Section 2.4: Methods for Determination of Environmental Resistance. 8 pp.

—Automotive
CNS D1011-81. General Rules of Weatherability for Automobile Parts (Jul)(7671). 33 pp.

JIS D 0205-87. Test Method of Weatherability for Automotive Parts. 52 pp.

—Avionics
JIS W 7004-81. Testing, Environmental, Airborne Electronic and Associated Equipment.

—Cable Insulation
BSI BS 6469: Sec 4.1-92. 1992 Insulation and Sheathing Materials of Electric Cables Part 4: Methods of Test Specific to Polyethylene and Polypropylene Compounds Section 4.1: Resistance to Environmental Stess Cracking—Wrapping Test After Thermal Ageing in. 27 pp.

CENELEC HD 395.2.16 S1-89. Medical Electrical Equipment Part 2: Particular Requirements for the Safety of Heamodialysis Equipment. 3 pp.

CENELEC HD 505.4.1 S1-88. Test Methods—Insulating/Sheathing Materials of Electric Cables Part 4: Polyethylene/Polyphlylene Compounds Section One—Resistance —Environmental Stress Cracking—Wrapping Test Thermal Ageing in Air—Measurement of Melt Flow Index—Carbon Black. 3 pp.

CENELEC HD 505.4.1 S2-90. Common Test Methods-Insulating/Sheathing Materials Electric Cables Part 4: Methods Polyethylene/Polypropylene Compounds Sec 1-Resistance Environmental Stress-Cracking/Wrapping Test After Thermal Ageing in Air-Measurement of Melt Flow Index-Carbon Black. 3 pp.

IEC 811 Pt 4-1-85. Common Test Methods for Insulating and Sheathing Materials of Electric Cables Part 4: Methods Specific to Polyethylene and Polypropylene Compounds Section One—Resistance to Environmental Stress Cracking—Wrapping Test After Thermal Ageing in Air—Measurement of the. 52 pp.

—Cargo Nets
ISO TR8647-90. Environmental Degradation of Textiles Used in Air Cargo Restraint Equipment First Edition. 21 pp.

—Coatings
CGSB 1-GP-71 METH 122.2-74. Methods of Testing Paints and Pigments Accelerated Weathering General Procedure. 2 pp.

CGSB 1-GP-71 METH 143.1-80. Methods of Testing Paints and Pigments Long-Term Outdoor Performance Selection, Exposure and Examination of Panels. 2 pp.

—Coatings—Test Panels
CGSB 1-GP-71 METH 122.1-74. Methods of Testing Paints and Pigments Accelerated Weathering Selection, Exposure and Examination of Panels. 1 p.

—Connectors
BSI BS 2011: Part 2.1KC-91. 1991 Environmental Testing Part 2.1: Tests Test Kc and Guidance. Sulphur Dioxide Test for Contacts and Connections. 21 pp.

BSI BS 2011: Part 2.1KD-77. 1977 Basic Environmental Testing Procedures Part 2.1: Tests Part 2.1KD: Test Kd. Hydrogen Sulfide Test for Contacts and Connections. 8 pp.

BSI BS 2011: Part 2.2KD-84. 1984 Basic Environmental Testing Procedures Part 2.2: Guidance Part 2.2KD: Test Kd. Guidance on Test Kd: Hydrogen Sulfide Test for Contacts and Connections. 12 pp.

CEN PREN 2591 (Part C2)-92. Elements of Electrical and Optical Connection Test Methods Part C2—Climatic Sequence. 6 pp.

CENELEC HD 323.2.46 S1-88. Basic Environmental Testing Procedures—Part 2: Tests. Guidance to Test Kd: Hydrogen Sulphide Test for Contacts and Connections. 3 pp.

—Connectors (Cont.)
IEC 68 Pt 2-42-82. Basic Environmental Testing Procedures Part 2: Tests Test Kc: Sulphur Dioxide Test for Contacts and Connections Second Edition. 18 pp.

IEC 68 Pt 2-43-76. Basic Environmental Testing Procedures Part 2: Tests Test Kd: Hydrogen Sulphide Test for Contacts and Connections First Edition. 16 pp.

IEC 68 Pt 2-46-82. Basic Environmental Testing Procedures Part 2: Tests Guidance to Test Kd: Hydrogen Sulphide Test for Contacts and Connections First Edition. 27 pp.

IEC 68 Pt 2-49-83. Basic Environmental Testing Procedures Part 2: Tests Guidance to Test Kc: Sulphur Dioxide Test for Contacts and Connections First Edition. 23 pp.

SAA AS 1099.2KC-81. Basic Environmental Testing Procedures for Electrotechnology—Part 2: Tests—Part 2Kc: Sulphur Dioxide Test for Contacts and Connections. 9 pp.

SAA AS 1099.2KD-81. Basic Environmental Testing Procedures for Electrotechnology—Part 2: Tests—Part 2Kd: Hydrogen Sulphide Test for Contacts and Connections. 6 pp.

SBAC TS 331 ISSUE 2. Test for Electrical Connectors Climatic Tests.

SBAC TS 337 ISSUE 1. Test for Electrical Connectors Environmental Sealing.

SNZ IEC 68: Part 2-42-82. Basic Environmental Testing Procedures Part 2-42: Test Kc: Sulphur Dioxide Test for Controls and Connections. 14 pp.

SNZ IEC 68: Part 2-43-76. Basic Environmental Testing Procedures Part 2-43: Test Kd: Hydrogen Sulphide Test for Contacts and Connections. 14 pp.

SNZ IEC 68: Part 2-46-82. Basic Environmental Testing Procedures Part 2-46: Guidance to Test Kd: Hydrogen Sulphide Test for Contacts and Connections. 23 pp.

SNZ IEC 68: Part 2-49-83. Basic Environmental Testing Procedures Part 2-49: Guidance to Test Kc: Sulphur Dioxide for Contacts and Connections. 19 pp.

—Construction Materials
CNS A3156-82. Method of Test for Change in Properties of Plastics Building Materials Resulting from Out-Door Exposure (Jun)(8908).

CNS A3157-82. Method of Test for Out-Door Exposure of Plastics Building Materials (Jun)(8909).

CNS A3158-82. Method of Test for Accelerated Artificial Exposure of Plastics Building Materials (Jun)(8910).

JIS A 1410-68. Recommended Practice for Out-Door Exposure of Plastics Building Materials (R 1982). 7 pp.

JIS A 1411-68. Standard Method of Test for Change in Properties of Plastics Building Materials Resulting from Out-Door Exposure (R 1982). 8 pp.

JIS A 1415-77. Recommended Practice for Accelerated Artificial Exposure of Plastics Building Materials (R 1982). 18 pp.

—Corrosion Inhibitors
CGSB 31-GP-0A METH 16.1-57. Methods of Testing Corrosion-Prevention Materials and Processes Accelerated Weathering; (Re-Edited April 1982). 2 pp.

CGSB 31-GP-0A METH 17.1-57. Methods of Testing Corrosion-Prevention Materials and Processes Shed Storage. 2 pp.

—Doors
CEN EN 79-77. Methods of Testing Doors: Behaviour of Door Leaves Placed Between Two Different Climates. 4 pp.

CEN PREN 1121-2-93. Behaviour Between Two Different Climates—Test Method—Part 2: Doors. 14 pp.

ISO 6445-89. Doors and Doorsets—Test of Behaviour Between Two Different Climates First Edition. 6 pp.

ISO 8273-85. Doors and Doorsets—Standard Atmospheres for Testing the Performance of Doors and Doorsets Placed Between Different Climates First Edition. 4 pp.

—Elastomers
DIN ENGL 53386-82. Testing of Plastics and Elastomers; Exposure to Natural Weathering (June). 7 pp.

DIN ENGL 53387-89. Artificial Weathering and Ageing of Plastics and Elastomers by Exposure to Filtered Xenon Arc Radiation (Apr) (Supersedes June 1982 Edition and the April 1974 Edition of DIN 53389, Withdrawn in 1986). 8 pp.

—Electric Contacts
BSI BS 2011: Part 2.1KC-91. 1991 Environmental Testing Part 2.1: Tests Test Kc and Guidance. Sulphur Dioxide Test for Contacts and Connections. 21 pp.

—Electric Contacts (Cont.)
BSI BS 2011: Part 2.1KD-77. 1977 Basic Environmental Testing Procedures Part 2.1: Tests Part 2.1KD: Test Kd. Hydrogen Sulfide Test for Contacts and Connections. 8 pp.

BSI BS 2011: Part 2.2KD-84. 1984 Basic Environmental Testing Procedures Part 2.2: Guidance Part 2.2KD: Test Kd. Guidance on Test Kd: Hydrogen Sulfide Test for Contacts and Connections. 12 pp.

CENELEC HD 323.2.46 S1-88. Basic Environmental Testing Procedures—Part 2: Tests. Guidance to Test Kd: Hydrogen Sulphide Test for Contacts and Connections. 3 pp.

IEC 68 Pt 2-42-82. Basic Environmental Testing Procedures Part 2: Tests Test Kc: Sulphur Dioxide Test for Contacts and Connections Second Edition. 18 pp.

IEC 68 Pt 2-43-76. Basic Environmental Testing Procedures Part 2: Tests Test Kd: Hydrogen Sulphide Test for Contacts and Connections First Edition. 16 pp.

IEC 68 Pt 2-46-82. Basic Environmental Testing Procedures Part 2: Tests Guidance to Test Kd: Hydrogen Sulphide Test for Contacts and Connections First Edition. 27 pp.

IEC 68 Pt 2-49-83. Basic Environmental Testing Procedures Part 2: Tests Guidance to Test Kc: Sulphur Dioxide Test for Contacts and Connections First Edition. 23 pp.

SAA AS 1099.2KC-81. Basic Environmental Testing Procedures for Electrotechnology—Part 2: Tests—Part 2Kc: Sulphur Dioxide Test for Contacts and Connections. 9 pp.

SAA AS 1099.2KD-81. Basic Environmental Testing Procedures for Electrotechnology—Part 2: Tests—Part 2Kd: Hydrogen Sulphide Test for Contacts and Connections. 6 pp.

SNZ IEC 68: Part 2-42-82. Basic Environmental Testing Procedures Part 2-42: Test Kc: Sulphur Dioxide Test for Controls and Connections. 14 pp.

SNZ IEC 68: Part 2-43-76. Basic Environmental Testing Procedures Part 2-43: Test Kd: Hydrogen Sulphide Test for Contacts and Connections. 14 pp.

SNZ IEC 68: Part 2-46-82. Basic Environmental Testing Procedures Part 2-46: Guidance to Test Kd: Hydrogen Sulphide Test for Contacts and Connections. 23 pp.

SNZ IEC 68: Part 2-49-83. Basic Environmental Testing Procedures Part 2-49: Guidance to Test Kc: Sulphur Dioxide for Contacts and Connections. 19 pp.

—Electric Measuring Instruments
CNS C1107-82. Weather-Proof Performance of Electricity Meters (Sep)(9357).

CNS C3162-82. Method of Test for Weather-Proof Performance of Electricity Meters (Sep)(9358).

JIS C 1281-79. Weather-Proof Performance of Electricity Meters (R 1984). 14 pp.

—Electric Switches
JIS D 0208-79. General Rules of Environmental Requirements for Switches of Automobiles (R 1983). 19 pp.

—Electrical Components
BSI BS 2011: Part 1.1-89. 1989 Basic Environmental Testing Procedures Part 1.1: General and Guidance. 27 pp.

BSI BS 2011: Pt 2.1Z/ABDM-83. 1983 Basic Environmental Testing Procedures Part 2.1: Tests Part 2.1Z/ABDM: Test Z/ABDM. Climatic Sequence, Primarily Intended for Components. 6 pp.

BSI BS 2011: Sec 4.2-84. 1984 Basic Environmental Testing Procedures Part 4: Miscellaneous Section 4.2: Guidance on the Application of the Tests of BS 2011 to Simulate the Effects of Storage. 8 pp.

BSI BS 2011: Sec 4.3-91. (WITHDRAWN) 1991 Environmental Testing Part 4: Miscellaneous Section 4.3: Guide to Seismic Test Methods for Equipments (IEC 68-3-3: 1991) (Renumbered as BS EN 60068-3-3: 1993). 47 pp.

BSI BS 2011: Sec 5.1-91. 1991 Environmental Testing Part 5: Guide to Drafting of Test Methods Section 5.1: General Principles (IEC 68-5-1: 1991) (G). 14 pp.

BSI BS 2011: Sec 5.2-92. 1992 Environmental Testing Part 5: Guide to Drafting of Test Methods Section 5.2: Terms and Definitions (IEC 68-5-2: 1990) (G). 16 pp.

BSI PD 6521-88. 1988 Summaries of Test Methods for Environmental Testing of Electrotechnical Products. 54 pp.

CENELEC HD 323.1 S1-82. Basic Environment Testing Procedures Part 1: General and Guidance. 1 p.

CENELEC HD 323.1 S2-88. Basic Environmental Testing Procedures—Part 1: General and Guidance. 3 pp.

INDUSTRY STANDARDS

INTERNATIONAL AND NON-U.S. NATIONAL STANDARDS
SUBJECT INDEX

Environmental

Environmental Testing (Cont.)
—Electrical Components (Cont.)
- CENELEC HD 323.2.48 S1-88. Basic Environmental Testing Procedures—Part 2: Tests. Guidance on the Application of the Tests of IEC Publication 68 to Simulate the Effects of Storage. 3 pp.
- CENELEC HD 323.3.3 S1-91. Environmental Testing Part 3: Guidance—Seismic Test Methods for Equipments. 6 pp.
- CENELEC HD 323.5.2 S1-91. Environmental Testing Part 5: Guide to Drafting of Test Methods Terms and Definitions. 6 pp.
- IEC 68 Pt 1-88. Environmental Testing Part 1: General and Guidance Sixth Edition; (Amendment 1-1992). 63 pp.
- IEC 68 Pt 2-18-89. Environmental Testing Part 2: Tests—Test R and Guidance: Water First Edition; (Corrigendum—May 1991) (Amendment 1-1993). 96 pp.
- IEC 68 Pt 2-48-82. Basic Environmental Testing Procedures Part 2: Tests Guidance on the Application of the Tests of IEC Publication 68 to Simulate the Effects of Storage First Edition. 15 pp.
- IEC 68 Pt 2-61-91. Environmental Testing Part 2: Test Methods Test Z/ABDM: Climatic Sequence First Edition. 37 pp.
- IEC 68 Pt 4-87. Basic Environmental Testing Procedures Part 4: Information for Specification Writers—Test Summaries First Edition; (Amendment 1-1992). 137 pp.
- IEC 68 Pt 5-1-91. Environmental Testing Part 5: Guide to Drafting of Test Methods —General Principles First Edition. 27 pp.
- IEC 68 Pt 5-2-90. Environmental Testing Part 5: Guide to Drafting of Test Methods —Terms and Definitions First Edition. 31 pp.
- MOD UK DSTAN 00-35: Part 3-90. Environmental Handbook for Defence Materiel Part 3: Environmental Testing Issue 1. 100 pp.
- SAA AS 1099. Basic Environmental Testing Procedures for Electrotechnology (Complete Set in Binder).
- SAA AS 1099.1-89. Basic Environmental Testing Procedures for Electrotechnology —Part 1: General. 24 pp.
- SAA AS 1099.2. Basic Environmental Testing Procedures for Electrotechnology —Part 2: Tests.
- SAA AS 1099.2.48-88. Basic Environmental Testing Procedures for Electrotechnology—Part 2: Tests—Part 2.48: Guidance on the Application of Tests A, B, and Ca to Simulate the Effects of Storage. 4 pp.
- SNZ IEC 68: Part 1-82. Basic Environmental Testing Procedures Part 1: General and Guidance. 47 pp.
- SNZ IEC 68: Part 2-48-82. Basic Environmental Testing Procedures Part 2-48: Guidance on the Application of the Tests of IEC Publication 68 to Simulate the Effects of Storage. 11 pp.

—Electrical Equipment
- JIS C 0010-85. Basic Environmental Testing Procedures (General and Guidance). 26 pp.

—Electrical Insulation
- BSI BS 5604-86. 1986 Evaluating Resistance to Tracking and Erosion of Electrical Insulating Materials Used Under Severe Ambient Conditions. 12 pp.
- CENELEC HD 380 S2-87. Test Methods for Evaluating Resistance to Tracking and Erosion of Electrical Insulating Materials Used Under Severe Ambient Conditions. 8 pp.
- IEC 587-84. Test Methods for Evaluating Resistance to Tracking and Erosion of Electrical Insulating Materials Used Under Severe Ambient Conditions Second Edition. 23 pp.

—Electromechanical Components
- BSI BS 5772: Part 6-84. 1984 Basic Testing Procedures and Measuring Methods for Electromechanical Components for Electronic Equipment Part 6: Climatic Tests and Soldering Tests. 26 pp.
- BSI BS 5772: Pt 6: Supp 1-81. (WITHDRAWN) 1981 Basic Testing Procedures and Electromechanical Components for Electronic Equipment Part 6: Climatic Tests and Soldering Tests Supplement 1: Test 11b (Superseded by BS 5772: Part 6: 1984). 4 pp.
- IEC 512 Pt 6-84. Electromechanical Components for Electronic Equipment; Basic Testing Procedures and Measuring Methods Part 6: Climatic Tests and Soldering Tests Second Edition. 57 pp.

—Electronic Components
- CECC CECC 00 006 ISSUE 2-81. Basic Specification: Environmental Test Procedures (En, Fr, Ge). 61 pp.

—Electronic Equipment
- CNS C6009-89. General Rules for Basic Environmental Testing Procedures for Electronic Components (Jun)(3622).
- CNS D1036-82. General Rules of Environmental Testing Method for Automotive Electronic Equipment (Nov)(9589).

Environmental Testing (Cont.)
—Enamel Paints
- CGSB 1-GP-71 METH 143.4-78. Methods of Testing Paints and Pigments Long-Term Outdoor Performance Test for One Coat Styrenated Flat Alkyd Enamel on Steel. 1 p.
- CNS K6031-87. Method of Test for Lacquer Enamel (May) (628). 6 pp.

—Fabrics
- BSI BS 2087: Part 1-92. 1992 Preservative Textile Treatments Part 1: Specification for Treatments. 17 pp.
- BSI BS 2087: Part 1-81. 1981 Amd 1 Preservative Treatments for Textiles Part 1: Specification for Treatments (L). 15 pp.
- BSI BS 2087: Part 2-92. 1992 Preservative Textile Treatments Part 2: Methods of Test. 23 pp.
- BSI BS 2087: Part 2-81. 1981 Amd 1 Preservative Treatments for Textiles Part 2: Methods of Test. 23 pp.
- CGSB CAN/CGSB-4.2 NO.35.1-M90. Textile Test Methods Colourfastness to Burnt Gas Fumes. 11 pp.
- CGSB CAN/CGSB-4.2 NO.69-M91. Textile Test Methods Weather Resistance—Xenon Arc Radiation. 10 pp.
- ISO 105 Pt G-78. Textiles—Tests for Colour Fastness Part G: Colour Fastness to Atmospheric Contaminants First Edition. 20 pp.
- ISO 105 Pt G04-89. Textiles—Tests for Colour Fastness Part G04: Colour Fastness to Oxides of Nitrogen in the Atmosphere at High Humidities First Edition. 7 pp.

—Field Effect Transistors
- CNS C6058-88. Environmental Testing Methods and Endurance Testing Methods for Discrete Semiconductor Devices (Continuous Operation Test of Field-Effect Transistor) (Sep)(5542).
- CNS C6060-88. Environmental Testing Methods and Endurance Testing Methods for Discrete Semiconductor Devices (Intermittent Operation Test of Field-Effect Transistor) (Sep)(5544).
- CNS C6062-88. Environmental Testing Methods and Endurance Testing Methods of Discrete Semiconductor Devices (Reverse Bias Test of Field-Effect Transistor Under High Temperature) (Sep)(5546).

—Glass
- DIN ENGL 52344-84. Testing of Glass; Testing the Effect of Alternating Atmosphere on Multilayer Insulating Glass (May). 4 pp.

—Gold Coatings (Made From Gold)
- BSI BS 6670: Part 2-86. 1986 Methods of Test for Electroplated Gold and Gold Alloy Coatings Part 2: Environmental Tests. 7 pp.
- ISO 4524 Pt 2-85. Metallic Coatings—Test Methods for Electrodeposited Gold and Gold Alloy Coatings—Part 2: Environmental Tests First Edition. 5 pp.

—Ground Support Equipment
- MOD UK DTD-1086C-60. Climatic Testing of Ground Instruments and Equipment (Reprinted April 1962). 3 pp.

—Hoses
- BSI BS 5173: Part 6-77. (WITHDRAWN) 1977 Amd 1 Rubber and Plastics Hoses and Hose Assemblies Part 6: Environmental Tests (AMD 5188) July 31, 1986 (Superseded by BS 5173: Sections 106.1: 1989, 106.2: 1990, 106.3: 1986, 106.4: 1989). 9 pp.

—Industrial Materials
- JIS Z 2381-87. Recommended Practice for Weathering Test. 33 pp.

—Integrated Circuits
- JIS C 7022-79. Environmental Testing Methods and Endurance Testing Methods for Semiconductor Integrated Circuits (R 1984). 50 pp.

—Lacquers
- SAA AS 1580.457. 1-88. Paints and Related Materials—Methods of Test—Part 457.1: Resistance to Natural Weathering. 6 pp.
- SNZ NZS/AS 1580. 457.1-88. Methods of Test for Paints and Related Materials Part 457.1: Resistance to Outdoor Weathering (This is a Joint Standard with SAA AS 1580.457.1). 6 pp.

—Leakage—Electrical Components
- BSI BS 2011: Part 2.1Q-81. 1981 Amd 1 Basic Environmental Testing Procedures Part 2.1: Tests Part 2.1Q: Test Q. Sealing (Amd 6093) November 30, 1990. 41 pp.
- CENELEC HD 323.2.17 S2-87. Basic Environmental Testing Procedures—Part 2: Tests. Test Q: Sealing. 3 pp.
- CENELEC HD 323.2.17 S3-88. Basic Environmental Testing Procedures-Part 2: Tests. Test Q: Sealing. 3 pp.

Environmental Testing (Cont.)
—Leakage—Electrical Components (Cont.)
- CENELEC HD 323.2.17 S4-90. Basic Environmental Testing Procedures Part 2: Tests Test Q: Sealing. 3 pp.
- IEC 68 Pt 2-17-78. Basic Environmental Testing Procedures Part 2: Tests—Test Q: Sealing Third Edition; (Amendment 4-1991 Incorporating Amendments 1 Thru 3). 118 pp.
- JIS C 0026-89. Basic Environmental Testing Procedures Part 2: Tests, Test Q: Sealing. 34 pp.
- SNZ IEC 68: Part 2-17-78. Basic Environmental Testing Procedures Part 2-17: Test Q: Sealing Amend: 1. 71 pp.

—Marine Instruments
- CNS F1009-84. General Rules for Environmental Tests of Control and Instrumentation Equipment for Marine Use (Apr)(10856).
- JIS F 0807-89. General Rules for Environmental Tests of Control and Instrumentation Equipment for Marine Use.

—Masonry Coatings
- CGSB 1-GP-71 METH 122.13-79. Methods of Testing Paints and Pigments Accelerated Weathering Test for Masonry Coatings. 1 p.
- CGSB 1-GP-71 METH 143.11-78. Methods of Testing Paints and Pigments Long-Term Outdoor Performance Masonry Finishes on Mortar Panels. 2 pp.

—Materiel
- MOD UK DSTAN 07-55: Part 1-01. Environmental Testing of Service Materiel Part 1: General Requirements Issue 1. Amendment 3. 40 pp.
- MOD UK DSTAN 07-55: Pt 2:Sec 2-01. Environmental Testing of Service Materiel Part 2: Tests Section 2: Climatic Issue 1; Amendment 4. 125 pp.
- MOD UK DSTAN 07-55: Pt 2:Sec 2-75. Environmental Testing of Service Materiel Part 2: Tests Section 2: Climatic Issue 1 (Reprinted November, 1984 Incorporating Amendments 1-4). 113 pp.
- MOD UK DSTAN 07-55: Pt 2:Sec 3-01. Environmental Testing of Service Materiel Part 2: Tests Section 3: Chemical and Biological Attack Issue 2; Amendment 2. 18 pp.
- NATO STANAG 4370 ED 1 AMD 0-00. Enviromental Testing. 7 pp.

—Materiel Packaging
- MOD UK DSTAN 81-41: Part 3-91. Packaging of Defence Materiel Part 3: Environmental Testing Issue 3. 88 pp.

—Measuring Instruments— Telecommunication Equipment
- CCITT RECMN O.3-89. Climatic Conditions and Relevant Tests for Measuring Equipment—Specifications for Measuring Equipment (Study Group IV) 3 pp. 3 pp.

—Medical Electrical Equipment
- CSA CAN/CSA-C22. 2NO601.1-M90. Medical Electrical Equipment, Part 1: General Requirements for Safety; (Gen Instr 1). 240 pp.

—Medical Electrical Equipment—Abnormal Operations
- CSA CAN/CSA-C22. 2NO601.1-M90. Medical Electrical Equipment, Part 1: General Requirements for Safety; (Gen Instr 1). 240 pp.

—Medical Electrical Equipment—Fault Conditions
- CSA CAN/CSA-C22. 2NO601.1-M90. Medical Electrical Equipment, Part 1: General Requirements for Safety; (Gen Instr 1). 240 pp.

—Metal Coatings (On Metal)
- JIS Z 0304-74. Outdoor Exposure Test for Protected Metals. 12 pp.

—Munitions—Air Launched
- NATO STANAG 4325 ED 1 AMD 0-92. Environmental and Safety Tests for the Appraisal of Air-Launched Munitions. 27 pp.

—Munitions—Surface Launched
- NATO STANAG 4337 ED 1 AMD 0-91. Environmental and Safety Tests for the Appraisal of Surface-Launched Munitions (SLM) (Withdrawn 93-06). 30 pp.

—Paints
- BSI BS 3900: Part F2-73. 1973 Amd 1 Methods of Test for Paints Group F: Durability Tests on Paint Films Part F2: Determination of Resistance to Humidity Under Condensation Conditions. 3 pp.

INTERNATIONAL AND NON-U.S. NATIONAL STANDARDS
SUBJECT INDEX
Environmental

Environmental Testing *(Cont.)*

—Paints *(Cont.)*

BSI BS 3900: Part F3-71. (OBSOLESCENT) 1971 Amd 1 Methods of Test for Paints Group F: Durability Tests on Paint Films Part F3: Resistance to Artificial Weathering (Enclosed Carbon Arc). 4 pp.
BSI BS 3900: Part F5-72. 1972 Methods of Test for Paints Group F: Durability Tests on Paint Films Part F5: Determination of Light Fastness of Paints for Interior Use (Exposed to Artificial Light Sources). 5 pp.
BSI BS 3900: Part F6-76. 1976 Methods of Test for Paints Group F: Durability Tests on Paint Films Part F6: Notes for Guidance on the Conduct of Natural Weathering Tests. 7 pp.
BSI BS 3900: Part F8-93. 1993 Methods of Test for Paints Part F8: Determination of Resistance to Humid Atmospheres Containing Sulfur Dioxide (ISO 3231: 1993) (W). 9 pp.
BSI BS 3900: Part F8-76. 1976 Methods of Test for Paints Part F: Durability Tests on Paint Films Part F8: Determination of Resistance to Humid Atmospheres Containing Sulphur Dioxide. 3 pp.
BSI BS 3900: Part F9-82. 1982 Methods of Test for Paints Group F: Durability Tests on Paint Films: Part F9: Determination of Resistance to Humidity (Continuous Condensation). 5 pp.
BSI BS 3900: Part G7-76. 1976 Methods of Test for Paints Group G: Environmental Tests on Paint Films (Including Tests for Resistance to Corrosion and Chemicals) Part G7: Determination of the Effect of Heat. 2 pp.
BSI BS AU 148: Part 12-69. 1969 Methods of Test for Motor Vehicle Paints Part 12: Resistance to Accelerated Weathering. 8 pp.
BSI BS AU 148: Part 14-69. 1969 Methods of Test for Motor Vehicle Paints Part 14: Resistance to Outdoor Exposure. 2 pp.
BSI BS EN 23270-91. 1991 Temperatures and Humidities for Conditioning and Testing Paints, Varnishes and Their Raw Materials (ISO 3270: 1984). 9 pp.
CEN EN 23270-91. Paints and Varnishes and Their Raw Materials—Temperatures and Humidities for Conditioning and Testing. 5 pp.
CGSB 1-GP-71 METH 122.2-74. Methods of Testing Paints and Pigments Accelerated Weathering General Procedure. 2 pp.
CGSB 1-GP-71 METH 122.4-79. Methods of Testing Paints and Pigments Accelerated Weathering Tests for House Paints on Wooden Panels. 1 p.
CGSB 1-GP-71 METH 143.1-80. Methods of Testing Paints and Pigments Long-Term Outdoor Performance Selection, Exposure and Examination of Panels. 2 pp.
CNS K6143-87. Method of Test for Traffic Paints (May)(1334). 19 pp.
DIN ENGL 53231-91. Artificial Weathering and Ageing of Paint Coatings by Exposure to Filtered Xenon Arc Radiation (Apr). 9 pp.
DIN ENGL EN 23270-91. Paints, Varnishes and Their Raw Materials; Temperatures and Humidities for Conditioning and Testing (ISO 3270:1984) (Sept). 5 pp.
ISO 2809-76. Paints and Varnishes—Determination of Light Fastness of Paints for Interior Use First Edition. 5 pp.
ISO 2810-74. Paints and Varnishes—Notes for Guidance on the Conduct of Natural Weathering Tests First Edition. 9 pp.
ISO 3231-93. Paints and Varnishes—Determination of Resistance to Humid Atmospheres Containing Sulphur Dioxide Second Edition. 8 pp.
ISO 3248-75. Paints and Varnishes—Determination of the Effect of Heat First Edition. 4 pp.
ISO 3270-84. Paints and Varnishes and Their Raw Materials—Temperatures and Humidities for Conditioning and Testing Third Edition. 4 pp.
ISO 6270-80. Paints and Varnishes—Determination of Resistance to Humidity (Continuous Condensation) First Edition. 5 pp.
SAA AS 1580.457. 1-88. Paints and Related Materials—Methods of Test—Part 457.1: Resistance to Natural Weathering. 6 pp.
SAA AS 1580.481. 0-91. Paints and Related Materials—Methods of Test—Part 481: Coatings—Part 481.0: Guide to Assessing Paint Systems Exposed to Weathering Conditions (in Professional Packages 30, 39). 7 pp.
SAA AS 1580.481. 1.1-91. Paints and Related Materials—Methods of Test—Part 481: Coatings—Part 481.1: Assessment of Individual Defects of Exposed Films —Part 481.1.1: Exposed to Weathering—General Appearance (Supersedes AS 1580.481.1—1975 (in Part)) (in Professional Packages 30, 39). 2 pp.
SNZ NZS/AS 1580. 457.1-88. Methods of Test for Paints and Related Materials Part 457.1: Resistance to Outdoor Weathering (This is a Joint Standard with SAA AS 1580.457.1). 6 pp.
SNZ NZS/AS 1580. 481.0-91. Methods of Test for Paints and Related Materials Part 480.1: Coatings—Guide to Assessing Paint Systems Exposed to Weathering Conditions (This is a Joint Standard with SAA AS 1580.481.0). 7 pp.
SNZ NZS/AS 1580. 481.1.1-91. Methods of Test for Paints and Related Materials Part 481.1.1: Coatings—Exposed to Weathering—General Appearance (This is a Joint Standard with SAA AS 1580.481.1.1). 2 pp.

—Paints—Algae

SAA AS 1580.481. 1.13-92. Paints and Related Materials—Methods of Test—Pt. 481: Coatings —Pt. 481.1: Assmt. of Indiv. Defects of Exposed Films—Pt. 481. 1.13: Exposedto Weathering—Degree to Fungal or Algal Growth Amdt 1 January/February 1993 (in Professional Packages 30, 39). 5 pp.

—Paints—Fungi

SAA AS 1580.481. 1.13-92. Paints and Related Materials—Methods of Test—Pt. 481: Coatings —Pt. 481.1: Assmt. of Indiv. Defects of Exposed Films—Pt. 481. 1.13: Exposedto Weathering—Degree to Fungal or Algal Growth Amdt 1 January/February 1993 (in Professional Packages 30, 39). 5 pp.

—Paints—Test Panels

CGSB 1-GP-71 METH 122.1-74. Methods of Testing Paints and Pigments Accelerated Weathering Selection, Exposure and Examination of Panels. 1 p.

—Photovoltaic Cells

JIS C 8917-89. Environmental and Endurance Test Methods for Crystalline Solar Cell Modules. 41 pp.

—Pipelines—Plastic

CEN PREN 1056-93. Plastics Piping and Ducting Systems—Plastics Pipes and Fittings—Test Method for Resistance to Direct (Natural) Weathering. 10 pp.

—Plastic Sheets

CNS K6309-71. Method of Test for Polymethylmethacrylate Corrugated Sheet (Weathering Test) (Apr)(3280). 1 p.

—Plastics

BSI BS 2782:Pt5: METH 540A-77. 1977 Methods of Testing Plastics Part 5: Optical and Colour Properties, Weathering Method 540A: Determination of Resistance to Change upon Exposure Under Glass to Daylight. 12 pp.
BSI BS 2782:Pt5: METH 540C-88. 1988 Methods of Testing Plastics Part 5: Optical and Colour Properties Method 540C: Determination of Ultraviolet Radiation Intensity Using Polysulphane Film. 8 pp.
BSI BS 2782:Pt5: METH 550A-81. 1981 Methods of Testing Plastics Part 5: Optical and Colour Properties, Weathering Method 550A: Methods of Exposure to Natural Weathering. 10 pp.
BSI BS 2782:Pt5: METH 551A-88. 1988 Methods of Testing Plastics Part 5: Optical and Colour Properties, Weathering Method 551A: Determination of the Effects of Exposure to Damp Heat, Water Spray and Salt Mist. 10 pp.
BSI BS 2782:Pt5: METH 552A-81. 1981 Methods of Testing Plastics: Part 5: Optical and Colour Properties, Weathering: Method 552A: Determination of Changes in Colour Variations in Prop After Exposure to Daylight Under Glass, Natural Weathering or Artificial Light. 10 pp.
BSI BS 4618: Sec 4.2-72. 1972 Recommendations for the Presentation of Plastics Design Data Part 4: Enviromental and Chemical Effects Section 4.2: Resistance to Natural Weathering. 11 pp.
BSI BS 4618: Sec 4.4-73. 1973 Recommendations for the Presentation of Plastics Design Data Part 4: Enviromental and Chemical Effects Section 4.4: The Effects on Plastics of Marine Exposure. 7 pp.
BSI BS 4618: Sec 4.5-74. 1974 Recommendations for the Presentation of Plastics Design Data Part 4: Enviromental and Chemical Effects Section 4.5: The Effects on Plastics of Soil Burial and Biological Attack. 3 pp.
DIN ENGL 53384-89. Artificial Weathering and Ageing of Plastics by Exposure to Laboratory UV Radiation Sources (Apr). 5 pp.
DIN ENGL 53387-89. Artificial Weathering and Ageing of Plastics and Elastomers by Exposure to Filtered Xenon Arc Radiation (Apr) (Supersedes June 1982 Edition and the April 1974 Edition of DIN 53389, Withdrawn in 1986). 8 pp.
ISO 4582-80. Plastics—Determination of Changes in Colour and Variations in Properties After Exposure to Daylight Under Glass, Natural Weathering or Artificial Light First Edition. 8 pp.
ISO 4607-78. Plastics—Methods of Exposure to Natural Weathering First Edition. 9 pp.
ISO 4611-87. Plastics—Determination of the Effects of Exposure to Damp Heat, Water Spray and Salt Mist Second Edition. 10 pp.

Environmental Testing *(Cont.)*

—Plastics *(Cont.)*

JIS K 7219-86. Recommended Practice for Outdoor Exposure of Plastics. 23 pp.
SAA AS 1745.1-89. Outdoor Weathering of Plastics in the Australian Environment —Part 1: Commercial Products. 20 pp.
SAA AS 1745.2-89. Outdoor Weathering of Plastics in the Australian Environment —Part 2: Guide for Design Purposes. 12 pp.

—Plastics—Australia

SAA AS CK24 Part 1-72. Code of Practice for Outdoor Weathering of Plastics in the Australian Environment Part 1: Products. 30 pp.

—Polyvinyl Acetate Adhesives

BSI BS 3544-62. 1962 Amd 1 Methods of Test for Polyvinyl Acetate Adhesives for Wood (AMD PD 5936) November 3, 1966 (W). 17 pp.
BSI BS 3544-02. 1962 Amd 2 Methods of Test for Polyvinyl Acetate Adhesives for Wood (AMD 6873) January 31, 1992 (W). 18 pp.

—Powder Coatings

BSI BS 3900: Part J4-87. 1987 Methods of Test for Paints Group J: Testing of Coating Powders Part J4: Determination of the Storage Stability of Coating Powders. 4 pp.

—Precious Metals

IEC 68 Pt 2-42-82. Basic Environmental Testing Procedures Part 2: Tests Test Kc: Sulphur Dioxide Test for Contacts and Connections Second Edition. 18 pp.
IEC 68 Pt 2-49-83. Basic Environmental Testing Procedures Part 2: Tests Guidance to Test Kc: Sulphur Dioxide Test for Contacts and Connections First Edition. 23 pp.
SAA AS 1099.2KC-81. Basic Environmental Testing Procedures for Electrotechnology—Part 2: Tests—Part 2Kc: Sulphur Dioxide Test for Contacts and Connections. 9 pp.
SNZ IEC 68: Part 2-42-82. Basic Environmental Testing Procedures Part 2-42: Test Kc: Sulphur Dioxide Test for Controls and Connections. 14 pp.
SNZ IEC 68: Part 2-49-83. Basic Environmental Testing Procedures Part 2-49: Guidance to Test Kc: Sulphur Dioxide for Contacts and Connections. 19 pp.

—Primers

CGSB 1-GP-71 METH 122.11-79. Methods of Testing Paints and Pigments Accelerated Weathering Test for Wood Primer on Sealed Wooden Panels. 1 p.
CGSB 1-GP-71 METH 122.12-79. Methods of Testing Paints and Pigments Accelerated Weathering Test for Wood Primer on Sealed and Topcoated Wooden Panels. 1 p.

—PVC—Windows

CEN PREN 513-91. Unplasticised Polyvinylchloride (PVC-U) Profiles for the Construction of Windows—Determination of the Resistance to Artificial Weathering. 16 pp.
CEN PREN 513-93. Unplasticised Polyvinylchloride (PVC-U) Profiles for the Construction of Windows—Determination of the Resistance to Artificial Weathering. 14 pp.

—Rubber

DIN ENGL 53386-82. Testing of Plastics and Elastomers; Exposure to Natural Weathering (June). 7 pp.

—Sealants

CGSB CAN2-19.0-M77METH17.1-78. Methods of Testing Putty, Caulking and Sealing Compounds Accelerated Weathering General Procedure. 2 pp.
ISO 11431-93. Building Construction—Sealants—Determination of Adhesion/Cohesion Properties After Exposure to Artificial Light Through Glass First Edition. 7 pp.

—Semiconductor Devices

BSI BS 6493: Part 3-85. 1985 Semiconductor Devices Part 3: Mechanical and Climatic Test Methods (IEC 749: 1984). 30 pp.
BSI BS 6493: Part 3-01. 1985 Amd 1 Semiconductor Devices Part 3: Mechanical and Climatic Test Methods (IEC 749: 1984) (AMD 7121) August 15, 1992. 54 pp.
CNS C6041-83. Environmental Testing Methods and Endurance Testing Method for Discrete Semiconductor Devices (General Rules) (Apr)(5066).
CNS C6042-88. Environmental Testing Methods and Endurance Testing Methods for Discrete Semiconductor Devices (Test of Soft Solders for Heat Endurance) (Jun)(5067).
IEC 749-84. Semiconductor Devices Mechanical and Climatic Test Methods First Edition; (Amendment 1-1991) (Amendment 2-1993). 130 pp.
JIS C 7021-77. Environmental Testing Methods and Endurance Testing Methods for Discrete Semiconductor Devices (R 1980). 72 pp.

INDUSTRY STANDARDS
993

Environmental Testing (Cont.)

—Ships—Electrical Equipment
JIS F 0808-87. General Rules for Environmental Test of Marine Electrical Apparatus.

—Steel Coatings
CGSB 1-GP-71 METH 122.6-79. Methods of Testing Paints and Pigments Accelerated Weathering Test for Two Coats of Air-Drying Material on Steel Panels. 1 p.
CGSB 1-GP-71 METH 122.7-80. Methods of Testing Paints and Pigments Accelerated Weathering Test for Two Coats of Baking Material on Steel Panels. 1 p.
CGSB 1-GP-71 METH 122.8-79. Methods of Testing Paints and Pigments Accelerated Weathering Test for One Coat of Air-Drying Material on Steel or Aluminum Panels. 1 p.
CGSB 1-GP-71 METH 122.9-79. Methods of Testing Paints and Pigments Accelerated Weathering Test for One Coat Material on Primed Steel Panels. 1 p.
CGSB 1-GP-71 METH 143.2-78. Methods of Testing Paints and Pigments Long-Term Outdoor Performance Test for One Air-Drying Coat on Primed Steel. 1 p.
CGSB 1-GP-71 METH 143.3-78. Methods of Testing Paints and Pigments Long-Term Outdoor Performance Test for One Baked Coat of Material on Primed Steel. 1 p.
CGSB 1-GP-71 METH 143.6-78. Methods of Testing Paints and Pigments Long-Term Outdoor Performance Test for Two Air-Drying Coats of Material on Steel. 1 p.

—Surge Arresters
IECQ PQC 77-88. Measuring and Testing Methods for Surge Protective Devices. 39 pp.

—Telecommunication Equipment
CENELEC PRETS 300 019-2-0-92. Equipment Engineering (EE); Environmental Conditions and Environmental Tests for Telecommunications Equipment Part 2-0: Specification of Environmental Tests Introduction. 16 pp.
ETSI PRETS 300 019-2-0-92. Equipment Engineering (EE); Environmental Conditions and Environmental Tests for Telecommunications Equipment Part 2-0: Specification of Environmental Tests Introduction. 16 pp.

—Telecommunication Systems
CENELEC PRETS 300 019-A-90. Equipment Engineering: Environmental Conditions and Environmental Tests for Telecommunications Equipment Part A—Introduction and Terminology. 43 pp.
CEPT T/L 02-12 E-88. Environmental Conditions and Environmental Tests for Telecommunications Equipment. 24 pp.
ETSI PRETS 300 019-A-90. Equipment Engineering; Environmental Conditions and Environmental Tests for Telecommunications Equipment—Part A—Introduction and Terminology (T/TR 02-12). 43 pp.
ETSI PRETS 300 019-A-90. Equipment Engineering: Environmental Conditions and Environmental Tests for Telecommunications Equipment Part A—Introduction and Terminology. 43 pp.

—Tie Downs—Fabrics
ISO TR8647-90. Environmental Degradation of Textiles Used in Air Cargo Restraint Equipment First Edition. 21 pp.

—Unit Loads
ISO 10531-92. Packaging—Complete, Filled Transport Packages—Stability Testing of Unit Loads First Edition. 10 pp.

—Varnishes
BSI BS 3900: Part F2-73. 1973 Amd 1 Methods of Test for Paints Group F: Durability Tests on Paint Films Part F2: Determination of Resistance to Humidity Under Condensation Conditions. 3 pp.
BSI BS 3900: Part F8-93. 1993 Methods of Test for Paints Part F8: Determination of Resistance to Humid Atmospheres Containing Sulfur Dioxide (ISO 3231: 1993) (W). 9 pp.
BSI BS 3900: Part F8-76. 1976 Methods of Test for Paints Part F: Durability Tests on Paint Films Part F8: Determination of Resistance to Humid Atmospheres Containing Sulphur Dioxide. 3 pp.
BSI BS 3900: Part G7-76. 1976 Methods of Test for Paints Group G: Environmental Tests on Paint Films (Including Tests for Resistance to Corrosion and Chemicals) Part G7: Determination of the Effect of Heat. 2 pp.
BSI BS EN 23270-91. 1991 Temperatures and Humidities for Conditioning and Testing Paints, Varnishes and Their Raw Materials (ISO 3270: 1984). 9 pp.
CEN EN 23270-91. Paints and Varnishes and Their Raw Materials—Temperatures and Humidities for Conditioning and Testing. 5 pp.

Environmental Testing (Cont.)

—Varnishes (Cont.)
DIN ENGL 53231-91. Artificial Weathering and Ageing of Paint Coatings by Exposure to Filtered Xenon Arc Radiation (Apr). 9 pp.
DIN ENGL EN 23270-91. Paints, Varnishes and Their Raw Materials; Temperatures and Humidities for Conditioning and Testing (ISO 3270:1984) (Sept). 5 pp.
ISO 2809-76. Paints and Varnishes—Determination of Light Fastness of Paints for Interior Use First Edition. 5 pp.
ISO 2810-74. Paints and Varnishes—Notes for Guidance on the Conduct of Natural Weathering Tests First Edition. 9 pp.
ISO 3231-93. Paints and Varnishes—Determination of Resistance to Humid Atmospheres Containing Sulphur Dioxide Second Edition. 8 pp.
ISO 3248-75. Paints and Varnishes—Determination of the Effect of Heat First Edition. 4 pp.
ISO 3270-84. Paints and Varnishes and Their Raw Materials—Temperatures and Humidities for Conditioning and Testing Third Edition. 4 pp.
ISO 6270-80. Paints and Varnishes—Determination of Resistance to Humidity (Continuous Condensation) First Edition. 5 pp.
SAA AS 1580.457. 1-88. Paints and Related Materials—Methods of Test—Part 457.1: Resistance to Natural Weathering. 6 pp.
SNZ NZS/AS 1580. 457.1-88. Methods of Test for Paints and Related Materials Part 457.1: Resistance to Outdoor Weathering (This is a Joint Standard with SAA AS 1580.457.1). 6 pp.

—Vulcanized Rubber
BSI BS 903: Part A53-89. 1989 Methods of Testing Vulcanized Rubber Part A53: Methods for Exposure to Natural Weathering, for Use in Determining Resistance to Weathering. 12 pp.
BSI BS 903: Part A55-89. 1989 Methods of Testing Vulcanized Rubber Part A55: Methods for Assessment of Changes in Properties After Exposure to Natural Weathering or Artificial Light, for Use in Determining Resistance to Weathering (ISO 4665/1: 1985) (V). 8 pp.
ISO 4665 Pt 1-85. Rubber, Vulcanized—Resistance to Weathering—Part 1: Assessment of Changes in Properties After Exposure to Natural Weathering or Artificial Light First Edition. 7 pp.
ISO 4665 Pt 2-85. Rubber, Vulcanized—Resistance to Weathering—Part 2: Methods of Exposure to Natural Weathering First Edition. 10 pp.

—Wheelchairs
ISO 7176 Pt 9-88. Wheelchairs—Part 9: Climatic Tests for Electric Wheelchairs First Edition. 4 pp.

—Wood Coatings
CGSB 1-GP-71 METH 122.19-79. Methods of Testing Paints and Pigments Accelerated Weathering Test for Two Coats of Pigmented Stain on Wood Panels. 1 p.

—Wood Coatings—Test Panels
CGSB 1-GP-71 METH 97.1-74. Methods of Testing Paints and Pigments Preparation of Wooden Panels General Conditions. 2 pp.
CGSB 1-GP-71 METH 143.8-80. Methods of Testing Paints and Pigments Long-Term Outdoor Performance Preparation of Wooden Panels for Exposure of Exterior Finishes. 1 p.

—Wood Stains
CGSB 1-GP-71 METH 143.12-74. Methods of Testing Paints and Pigments Long-Term Outdoor Performance Two Coats of Exterior Stain on Wood. 1 p.

Environmental Testing Equipment

See Also: Environmental Testing; Testing Equipment; Vibration Testers

—Biomechanical Effects (Human Body)
BSI DD ENV 28041-93. 1993 Human Response to Vibration Measuring Instrumentation (ISO 8041: 1990) (E). 32 pp.
CEN ENV 28041-93. Human Response to Vibration—Measuring Instrumentation (ISO 8041: 1990). 26 pp.
ISO 8041-90. Human Response to Vibration—Measuring Instrumentation First Edition; (Corrigendum 1-1993) (CEN EN 28041: 1993). 30 pp.

—Light Exposure—Calibration
JIS K 7200-86. Standard Reference Test Specimens for Calibrating a Laboratory Light Exposure Apparatus. 15 pp.

—Light Exposure—Water Exposure
CNS B6088-85. Light-and-Water-Exposure Apparatus (Enclosed Carbon-Arc Type) (Apr)(11230).
CNS B6089-85. Light-and-Water-Exposure Apparatus (Open-Flame Sunshine Carbon-Arc Type) (Apr)(11231).

Environmental Testing Equipment (Cont.)

—Light Exposure—Water Exposure (Cont.)
CNS B6090-85. Light-Exposure and Light-and-Water-Exposure Apparatus (Xenon-Arc Lamp Type) (Apr)(11232).
JIS B 7753-88. Light-and Water-Exposure Apparatus (Open-Flame Sunshine Carbon-Arc Type). 26 pp.
JIS B 7754-91. Light-Exposure and Light-and-Water-Exposure Apparatus (Xenon-Arc Lamp Type). 31 pp.

—Paints
BSI BS 3900: Part F3:Add1-78. 1978 Methods of Test for Paints Group F: Durability Tests on Paint Films Part F3: Resistance to Artificial Weathering (Enclosed Carbon Arc) Addendum 1: Notes for Guidance on the Operation of Artificial Weathering Apparatus. 2 pp.

Environments
Use: Environmental Conditions

Enzymatic Activity
See Also: Amylase Activity; Phosphatase Activity; Urease Activity

—Dairy Products
DIN ENGL 10326-86. Determination of Sucrose and Glucose Content in Milk Products and Ice Cream; Enzymatic Method (Feb). 5 pp.

—Dairy Products—Spectrometry
BSI BS 1743: Part 8-87. 1987 Analysis of Dried Milk and Dried Milk Products Part 8: of the Lactic Acid and Lactates Content of Dried Milk. 6 pp.
ISO 8069-86. Dried Milk—Determination of Lactic Acid and Lactates Content—Enzymatic Method First Edition; (Corrected and Reprinted -1988). 6 pp.

—Fruit Juices
CNS N6222-91. Method of Test for Fruit and Vegetable Juices and Drinks (Jun)(12633). 2 pp.

—Tobacco
ISO 8451-91. Tobacco—Determination of Starch Content—Enzymatic Method First Edition. 8 pp.

—Vegetable Juices
CNS N6222-91. Method of Test for Fruit and Vegetable Juices and Drinks (Jun)(12633). 2 pp.

Enzyme Assay

—Dairy Products
DIN ENGL 10326-86. Determination of Sucrose and Glucose Content in Milk Products and Ice Cream; Enzymatic Method (Feb). 5 pp.

—Ice Cream
DIN ENGL 10326-86. Determination of Sucrose and Glucose Content in Milk Products and Ice Cream; Enzymatic Method (Feb). 5 pp.

Enzymes
Scope Note: Use a more specific term *See:* Amylase; Cholinesterase; Glutathione; Lipase; Trypsin

EOA
Use: End of Address

Eosin Y, Yellowish
Use: Eosin Yellowish-(YS)

Eosin Yellowish-(YS)
Use For: Tetrabromofluorscein Sodium
See Also: Dyes; Fluorescein
CNS K7208-66. Chemical Reagent (Eosin) (Tetrabromofluorescein Soluble) (Jun)(1708).
JIS K 8651-88. Eosin Y (Sodium Tetrabromofluorescein).

EP Lubricants
Use: Extreme Pressure Lubricants

EP 9 Aircraft
See Also: Aircraft

—Antenna Positions
CAA. EP9 (Approved Aerial Positions). 1 p.

EPDM
Use: Ethylene Propylene Terpolymers

Epichlorohydrin
Use For: 1-Chloro-2,3-Epoxy Propane; Chloromethyloxirane; ECH *See Also:* Epoxy Compounds; Hazardous Materials
CNS K7567-88. Chemical Reagent (Chloromethyloxirane) (Epichlorohydrin) (Oct)(7731).

INTERNATIONAL AND NON-U.S. NATIONAL STANDARDS
SUBJECT INDEX
Epitaxial

Epichlorohydrin *(Cont.)*
—**Highway Transportation—Emergency Procedures**
SAA AS 1678.6.0. 007-83. Emergency Procedure Guides—Transport—Part 6.0.007: Epichlorhydrin (ECH or 1-Chloro-2, 3-Epoxypropane).

Epichlorohydrin Content Analysis
Use For: Polyepichlorohydrin Content Analysis
—**Elastomers**
DIN ENGL 53621 Pt 6-76. Testing of Rubber and Elastomers; Quantitative Determination of Polymers; Determination of Polyepichlorohydrin Content and Epichlorhydrin-Ethylene Oxide Copolymer Content (May). 2 pp.
—**Rubber**
DIN ENGL 53621 Pt 6-76. Testing of Rubber and Elastomers; Quantitative Determination of Polymers; Determination of Polyepichlorohydrin Content and Epichlorhydrin-Ethylene Oxide Copolymer Content (May). 2 pp.

EPIRB
Use: Emergency Position Indicating Radio Beacons

Epitaxial Transistors
See Also: Transistors

—**Darlington—NPN—LF**
CECC CECC 50 003-025. BS CECC 50 003-025 Issue 1; Case Rated Low Frequency Amplification Darlington Bipolar Transistors (En). 10 pp.
CECC CECC 50 003-026. BS CECC 50 003-026 Issue 1; Case Rated Low Frequency Amplification Darlington Bipolar Transistors (En). 10 pp.
CECC CECC 50 003-036 ISSUE 1-80. BS CECC 50 003-036; Case Rated Low Frequency Amplification Bipolar Darlington Transistor (En). 8 pp.

—**Darlington—NPN/PNP—LF**
CECC CECC 50 003-030 ISSUE 1-79. BS CECC 50 003-030; Case Rated Low Frequencey Amplification Darlington Bipolar Transistors (En). 13 pp.

—**Darlington—PNP—LF**
CECC CECC 50 003-027. BS CECC 50 003-027 Issue 1; Case Rated Low Frequency Amplification Darlington Bipolar Transistor (En). 10 pp.
CECC CECC 50 003-028. BS CECC 50 003-028 Issue 1; Case Rated Low Frequency Amplification Darlington Bipolar Transistor (En). 10 pp.
CECC CECC 50 003-029. UTE C 86 613-029 Edition 2; Case Rated Bipolar Transistors for Low Frequency Amplification (Fr). 6 pp.
CECC CECC 50 003-037 ISSUE 1-80. BS CECC 50 003-037; Case Rated Low Frequency Amplification Bipolar Darlington Transistor (En). 8 pp.

—**Silicon—NPN**
CECC CECC 50 002-186 ISSUE 1-82. BS CECC 50 002-186 Issue 1; Ambient Rated Amplifier Transistors (En). 12 pp.

—**Silicon—NPN—Dual**
MOD UK DSTAN 59-61: 90/110-01. Semiconductor Device Dual Transistor Issue 2; Amendment 1. 15 pp.

—**Silicon—NPN—HF**
CECC CECC 50 002-095. BS CECC 50 002-095 Issue 1; Ambient Rated Dual High Frequency Amplification Bipolar Transistors (En). 10 pp.
CECC CECC 50 002-134. BS CECC 50 002-134 Issue 1; Ambient Rated High Frequency Amplification Bipolar Transistor (En). 8 pp.

—**Silicon—NPN—LF**
CECC CECC 50 002-007 ISSUE 1-77. BS CECC 50 002-007; Ambient-Rated Bipolar Transistors for Low Frequency Amplification (En). 14 pp.
CECC CECC 50 002-008 ISSUE 1-77. BS CECC 50 002-008; Ambient-Rated Bipolar Transistors for Low Frequency Amplification (En). 13 pp.
CECC CECC 50 002-087. BS CECC 50 002-087 Issue 1; Ambient Rated Low Frequency Amplification Bipolar Transistors (En). 9 pp.
CECC CECC 50 002-130. BS CECC 50 002-130 Issue 1; Ambient Rated Dual Low Frequency Amplification Bipolar Transistor (En). 11 pp.
CECC CECC 50 002-170 ISSUE 1-80. BS CECC 50 002-170 Issue 1; Ambient Rated Low Frequency Amplification Bipolar Transistors (En). 9 pp.
CECC CECC 50 002-189 ISSUE 1-81. BS CECC 50 002-189 Issue 1; Ambient Rated Low Frequency Amplification Bipolar Transistor (En). 8 pp.
CECC CECC 50 002-289-291 IS 1. NL-CECC 50 002-289 to 291 Issue 1; NPN Silicon Ambient-Rated Planar Epitaxial Bipolar Low Frequency Amplifier in Hermetically Sealed Metal Case; BC107 CECC 50 002-289 BC108 CECC 50 002-290 BC109 CECC 50 002-291 (En). 3 pp.

Epitaxial Transistors *(Cont.)*
—**Silicon—NPN—LF** *(Cont.)*
CECC CECC 50 002-292-294 IS 1. NL-CECC 50 002-292 to 294 Issue 1; NPN Silicon Ambient-Rated Planar Epitaxial Bipolar Low Frequency Amplifier in Hermetically Sealed Metal Case; BCY58 CECC 50 002-292 BCY59 CECC 50 002-293 BCY65E CECC 50 002-294 (En). 2 pp.
CECC CECC 50 002-302 ISSUE 1. NL CECC 50 002-302 Issue 1; NPN, Silicon, Ambient-Rated Planar Epitaxial Bipolar Low Frequency Amplifier in Hermetically Sealed Metal Case; CV10806 (En). 1 p.
CECC CECC 50 003-012 ISSUE 2-83. BS CECC 50 003-012; Case Rated Low Frequency Amplification Bipolar Transistor (En). 9 pp.
CECC CECC 50 003-014 ISSUE 2-81. BS CECC 50 003-014; Case Rated Medium Power Low Frequency Amplification Bipolar Transistor (En). 10 pp.
CECC CECC 50 003-015 ISSUE 1-79. BS CECC 50 003-015; Case Rated Low Frequency Amplification Bipolar Transistor (En). 9 pp.
CECC CECC 50 003-018. BS CECC 50 003-018 Issue 1; Case Rated Low Frequency Amplification Bipolar Transistor (En). 13 pp.
CECC CECC 50 003-020 ISSUE 2-81. BS CECC 50 003-020; Case Rated Low Frequency Amplification Bipolar Transistor (En). 8 pp.
CECC CECC 50 003-023. BS CECC 50 003-023 Issue 1; Case Rated Low Frequency Amplification Bipolar Transistors (En). 9 pp.
CECC CECC 50 003-031 ISSUE 2-81. BS CECC 50 003-031; Case Rated Low Frequency Amplification Bipolar Transistor (En). 9 pp.
CECC CECC 50 003-032 ISSUE 2-81. BS CECC 50 003-032; Case Rated Low Frequency Amplification Bipolar Transistor (En). 9 pp.
CECC CECC 50 003-033 ISSUE 2-81. BS CECC 50 003-033; Case Rated Low Frequency Amplification Bipolar Transistor (En). 9 pp.
CECC CECC 50 003-034 ISSUE 2-81. BS CECC 50 003-034; Case Rated Low Frequency Amplification Bipolar Transistor (En). 11 pp.

—**Silicon—NPN—LF/HF**
CECC CECC 50 002-088. BS CECC 50 002-088 Issue 1; Ambient Rated Bipolar Transistor for Low and High Frequency Amplification (En). 8 pp.

—**Silicon—NPN—Power—Switching**
CECC CECC 50 004-121 ISSUE 1-82. BS CECC 50 004-121; Case Rated Bipolar Switching Transistors (En). 14 pp.

—**Silicon—NPN—Power—VHF**
CECC CECC 50 007-001 ISSUE 1-78. BS CECC 50 007-001; NPN Silicon Planar Epitaxial Transistor Metal and Ceramic Case with Epoxy Seal (En). 16 pp.

—**Silicon—NPN—Switching**
CECC CECC 50 004-004 ISSUE 1. BS CECC 50 004-004 Ambient Rated High Speed Saturated Switching Transistor (En). 11 pp.
CECC CECC 50 004-033 ISSUE 1-81. BS CECC 50 004-033; Case Rated Switching Bipolar Transistors (En). 11 pp.
CECC CECC 50 004-034 ISSUE 1-81. BS CECC 50 004-034; Case Rated Switching Bipolar Transistor (En). 11 pp.
CECC CECC 50 004-035 ISSUE 1-81. BS CECC 50 004-035; Case Rated Switching Bipolar Transistor (En). 11 pp.
CECC CECC 50 004-038 ISSUE 2-81. BS CECC 50 004-038; Ambient Rated High Voltage, High Current Switching Bipolar Transistor (En). 9 pp.
CECC CECC 50 004-039 ISSUE 2-81. BS CECC 50 004-039; Case Rated Switching Bipolar Transistor (En). 10 pp.
CECC CECC 50 004-040 ISSUE 2-84. BS CECC 50 004-040; Case Rated Switching Bipolar Transistor (En). 9 pp.
CECC CECC 50 004-041 ISSUE 1-81. BS CECC 50 004-041; Case Rated Switching Bipolar Transistors (En). 13 pp.
CECC CECC 50 004-048 ISSUE 1-78. BS CECC 50 004-048; Ambient Rated Switching Bipolar Transistors (En). 18 pp.
CECC CECC 50 004-065 ISSUE 1-81. BS CECC 50 004-065; Case Rated Switching Bipolar Transistors (En). 13 pp.
CECC CECC 50 004-066 ISSUE 1-79. BS CECC 50 004-066; Ambient Rated Switching Bipolar Transistor (En). 9 pp.
CECC CECC 50 004-088 ISSUE 1-82. BS CECC 50 004-088; Case Rated Bipolar Switching Transistor (En). 10 pp.
CECC CECC 50 004-147 ISSUE 1-84. BS CECC 50 004-147; Ambient Rated Switching Bipolar Transistor (En). 11 pp.

—**Silicon—PNP—Dual**
MOD UK DSTAN 59-61: 90/115-72. Semiconductor Devices, Transistor Issue 1. 16 pp.

Epitaxial Transistors *(Cont.)*
—**Silicon—PNP—LF**
CECC CECC 50 002-006. BS E9372 F069 Issue 1; Ambient Rated Bipolar Transistors for Low Frequency Amplification (En). 15 pp.
CECC CECC 50 002-007 ISSUE 1-77. BS CECC 50 002-007; Ambient-Rated Bipolar Transistors for Low Frequency Amplification (En). 14 pp.
CECC CECC 50 002-075. BS CECC 50 002-075 Issue 1; Ambient Rated Bipolar Transistors for Low Frequency Low Level, Low Noise Amplification; PNP Epitaxial Silicon Planar Transistor in Metal Can, Glass-Metal Sealed (En). 13 pp.
CECC CECC 50 002-084. BS CECC 50 002-084 Issue 1; Ambient Rated Bipolar Transistors for Low Frequency Amplification Bipolar Transistors (En). 9 pp.
CECC CECC 50 002-093. BS CECC 50 002-093 Issue 1; Ambient Rated Dual Low Frequency Amplification Bipolar Transistor (En). 10 pp.
CECC CECC 50 002-094. BS CECC 50 002-094 Issue 1; Ambient Rated Dual Low Frequency Amplification Bipolar Transistor (En). 10 pp.
CECC CECC 50 002-096. BS CECC 50 002-096 Issue 1; Ambient Rated Dual Low Frequency Amplification Bipolar Transistor (En). 10 pp.
CECC CECC 50 002-107. BS CECC 50 002-107 Issue 1; Ambient Rated Bipolar Transistors for High Voltage, Low Frequency Applications (En). 16 pp.
CECC CECC 50 002-169 ED 1-85. CEI CECC 50 002-169 Edition 1; Ambient-Rated Bipolar Transistors for Low and High Frequency Amplification (En). 10 pp.
CECC CECC 50 002-169 ISSUE 1-80. BS CECC 50 002-169 Issue 1; Ambient Rated Low Frequency Amplification Bipolar Transistors (En). 9 pp.
CECC CECC 50 002-295-297 IS 1. NL-CECC 50 002-295 to 297 Issue 1; PNP, Silicon, Ambient-Rated Planar Epitaxial Bipolar Low Frequency Amplifier in Hermetically Sealed Metal Case; BCY70 CECC 50 002-295 BCY71 CECC 50 002-296 BCY72 CECC 50 002-297 (En). 2 pp.
CECC CECC 50 002-298 ISSUE 1. NL-CECC 50 002-298 Issue 1; PNP, Silicon Ambient-Rated Planar Epitaxial Bipolar Low Frequency Amplifier in Hermetically Sealed Metal Case; 2N2904 2N2905 2N2904A 2N2905A (En). 2 pp.
CECC CECC 50 002-299 ISSUE 1. NL CECC 50 002-299 Issue 1; PNP, Silicon, Ambient-Rated Planar Epitaxial Bipolar Low Frequency Amplifier in Hermetically Sealed Metal Case; 2N2906 2N2907 2N2906A 2N2907A (En). 2 pp.
CECC CECC 50 002-300 ISSUE 2. NL CECC 50 002-300 Issue 2; PNP, Silicon Ambient-Rated Planar Epitaxial Bipolar Low Frequency Amplifier in Hermetically Sealed Metal Case; CV8669 CV7670 CV7671 CV7672 (En). 2 pp.
CECC CECC 50 002-301 ISSUE 1. NL CECC 50 002-301 Issue 1; PNP, Silicon Ambient-Rated Planar Epitaxial Bipolar Low Frequency Amplifier in Hermetically Sealed Metal Case; CV7673 CV7674 CV7675 CV7676 (En). 2 pp.
CECC CECC 50 003-013 ISSUE 2-81. BS CECC 50 003-013; Case Rated Medium Power Low Frequency Amplification Bipolar Transistors (En). 10 pp.
CECC CECC 50 003-016 ISSUE 2-81. BS CECC 50 003-016; Case Rated Low Frequency Amplification Bipolar Transistor (En). 8 pp.
CECC CECC 50 003-019. BS CECC 50 003-019 Issue 1; Case Rated Low Frequency Amplification Bipolar Transistor (En). 12 pp.
CECC CECC 50 003-021 ISSUE 2-82. BS CECC 50 003-021; Case Rated Low Frequency Amplification Bipolar Transistors (En). 10 pp.

—**Silicon—PNP—LF/HF**
CECC CECC 50 002-012 ISSUE 2-80. BS CECC 50 002-012; Ambient Rated Transistor for High and Low Frequency Amplification (En) AMD 1 (En). 9 pp.
CECC CECC 50 002-013 ISSUE 2-80. BS CECC 50 002-013; Ambient Rated Transistor for High and Low Frequency Amplification (En) AMD 1 (En). 9 pp.
CECC CECC 50 002-014 ISSUE 2-80. BS CECC 50 002-014; Ambient Rated Transistor for High and Low Frequency Amplification (En) AMD 1 (En). 9 pp.
CECC CECC 50 002-085. BS CECC 50 002-085 Issue 1; Ambient Rated Bipolar Transistors for Low and High Frequency Amplification (En). 8 pp.

—**Silicon—PNP—Switching**
CECC CECC 50 004-037 ISSUE 2-81. BS CECC 50 004-037; Case Rated Switching Bipolar Transistor (En). 11 pp.
CECC CECC 50 004-065 ISSUE 1-81. BS CECC 50 004-065; Case Rated Switching Bipolar Transistors (En). 13 pp.
CECC CECC 50 004-133 ISSUE 1-83. BS CECC 50 004-133; Case Rated Bipolar Switching Transistor (En). 12 pp.

EPN Content Analysis
—Waste Water
CNS K9091-82. Method of Test for Parathion, Methyl-Parathion and EPN in Industrial Waste Water (Jun)(9000).

Epoxide Adhesives
Use: Epoxy Adhesives

Epoxide Resins
Use: Epoxy Resins

Epoxides
Use: Epoxy Compounds

Epoxy Adhesives
See Also: Adhesives
MOD UK DSTAN 80-105-85. Adhesive, Epoxide, Two-Part (Solventless), Hot Curing No 4 Issue 1 (Withdrawn). 10 pp.
MOD UK DSTAN 80-148-91. Adhesive, Epoxide, General Purpose Issue 1. 12 pp.

—Ceramic Tile
DIN ENGL 18156 Pt 4-84. Materials Used for the Application of Ceramic Tiling by the Thin Bed Method; Epoxy Resin Adhesives (Dec). 3 pp.
DIN ENGL 18157 Pt 3-86. Application of Ceramic Tiling by the Thin Bed Method Using Epoxy Resin Adhesives (Apr). 4 pp.

—Injection—Building Maintenance
CNS A2151-83. Epoxy Injection Adhesives for Repairing in Buildings (Apr)(10141). 2 pp.
CNS A3181-83. Method of Test for Epoxy Injection Adhesives for Repairing in Buildings (Apr)(10142).
JIS A 6024-92. Epoxy Injection Adhesives for Repairing in Buildings. 13 pp.

—Metal to Metal
MOD UK DSTAN 80-177-93. Adhesive, Epoxide, Solution Hot Curing, No 1, Type QX Issue 1. 22 pp.
MOD UK TS 344B. Adhesive, Epoxide, Solution, Hot Curing, No.1 Type QX (Superseded by Def Stan 80-177).
MOD UK TS 400C. Adhesive, Expoxide Solution, Hot Curing, No. 2, LF Quality (Withdrawn).
MOD UK TS 466C. Adhesive (Two-Part), Epoxide Solution, Hot Curing No.3 Type QX.
MOD UK TS 467C. Adhesive (Three-Part), Epoxide Solventless, Hot Curing No.2 Type QX (Dormant).
MOD UK TS 489. Adhesive, Epoxide, Solventless, Hot Curing No. 3 LF Quality (Withdrawn).

—Waterproof—Quarry Tiles
CGSB 71-GP-30M-79. Adhesive, Epoxy and Modified Mortar Systems, for Installation of Quarry Tiles, Standard for. 15 pp.

Epoxy Coatings
Use For: Epoxy Paints *See Also:* Bonded Coatings; Coatings; Epoxy Primers; Epoxy Resins; Nonmetallic Coatings
CNS K2089-87. Epoxy Resin Paint (May)(4938). 3 pp.
CNS K6693-87. Method of Test for Epoxy Resin Paint (May)(8012).

—Aircraft—Exterior
MOD UK DTD-5555A-65. Exterior Glossy Finishing Schemes (Cold Curing Epoxide Type) Scheme I, Scheme II and Scheme III (Reprinted May 1975). 7 pp.
MOD UK DTD-5567A-64. Interior and Exterior Protective Finishing Scheme (Cold Curing Epoxide Type) (Reprinted July 1969, Incorporating Amendment No. 1). 8 pp.

—Aircraft—Interior
MOD UK DTD-5567A-64. Interior and Exterior Protective Finishing Scheme (Cold Curing Epoxide Type) (Reprinted July 1969, Incorporating Amendment No. 1). 8 pp.

—Amphibious Vehicles
MOD UK DSTAN 80-163-93. Paint System Defence Equipment Heavy Duty (Two-Pack) Paint, Priming, Defence Equipment Heavy Duty, Epoxide Paint, Undercoat, Defence Equipment Epoxide/Pitch Paint, Finishing, Defence Equipment Heavy Duty, Epoxide Issue 1. 30 pp.

—Concretes
CGSB 1-GP-146M-79. Coating, Epoxy, Cold Cured, Gloss, Standard for; (Amendment 1 Feb 1981) (QPL Apr 1986). 14 pp.
CGSB CAN/CGSB-1.153-M90. High Build, Gloss, Epoxy Coating. 13 pp.

—Decks—Nonskid
CGSB 1-GP-192MA-85. Deck Coating, Nonslip, Epoxy. 14 pp.

—Exterior
MOD UK DTD-5567A-64. Interior and Exterior Protective Finishing Scheme (Cold Curing Epoxide Type) (Reprinted July 1969, Incorporating Amendment No. 1). 8 pp.

—Iron Pipes
JIS G 5528-84. Epoxy-Powder Coating for Interior of Ductile Iron Pipes and Fittings. 26 pp.

—Low Temperature Curing—Concrete
CGSB CAN/CGSB-1.207-M91. Low Temperature Curing Epoxy Coating. 11 pp.

—Low Temperature Curing—Corrosion Resistant Steel
CGSB CAN/CGSB-1.207-M91. Low Temperature Curing Epoxy Coating. 11 pp.

—Marine
MOD UK TS 10222. Paint, Epoxy, Solvent Free (2 Pack), Type: Brushing (Superseded by Def Stan 80-179).

—Materiel
MOD UK DSTAN 80-163-93. Paint System Defence Equipment Heavy Duty (Two-Pack) Paint, Priming, Defence Equipment Heavy Duty, Epoxide Paint, Undercoat, Defence Equipment Epoxide/Pitch Paint, Finishing, Defence Equipment Heavy Duty, Epoxide Issue 1. 30 pp.

—Metal
CGSB 1-GP-146M-79. Coating, Epoxy, Cold Cured, Gloss, Standard for; (Amendment 1 Feb 1981) (QPL Apr 1986). 14 pp.
CGSB CAN/CGSB-1.153-M90. High Build, Gloss, Epoxy Coating. 13 pp.
MOD UK DSTAN 80-93-83. Paint, Epoxide, Stoving Issue 1. 9 pp.
MOD UK TS 10095C. Powder Coating, Stoving, Matt; Semi-Gloss, Glossy (Superseded by Def Stan 80-169).

—Metal—Marine
MOD UK TS 463B. System Defence Equipment Heavy Duty Paint (Two-Pack) (Superseded by Def Stan 80-163).

—Pipe Fittings
JIS G 5528-84. Epoxy-Powder Coating for Interior of Ductile Iron Pipes and Fittings. 26 pp.

—Plaster
CGSB 1-GP-146M-79. Coating, Epoxy, Cold Cured, Gloss, Standard for; (Amendment 1 Feb 1981) (QPL Apr 1986). 14 pp.
CGSB CAN/CGSB-1.153-M90. High Build, Gloss, Epoxy Coating. 13 pp.

—Powder
MOD UK DSTAN 80-169-93. Powder Coating, Stoving Types: Matt, Semi-Gloss and Glossy Issue 1. 17 pp.

—Powder—Iron Pipes
JIS G 5528-84. Epoxy-Powder Coating for Interior of Ductile Iron Pipes and Fittings. 26 pp.

—Roller Applied
CGSB 1-GP-71 METH 134.17-78. Methods of Testing Paints and Pigments Applicability and Appearance Two Component Coatings, Roller Application. 1 p.

—Roller Applied—Hiding Power
CGSB 1-GP-71 METH 134.17-78. Methods of Testing Paints and Pigments Applicability and Appearance Two Component Coatings, Roller Application. 1 p.

—Self Priming—Corrosion Inhibitive—Concrete
CGSB CAN/CGSB-1.207-M91. Low Temperature Curing Epoxy Coating. 11 pp.

—Self Priming—Corrosion Resistant Steel
CGSB CAN/CGSB-1.207-M91. Low Temperature Curing Epoxy Coating. 11 pp.

—Ships
MOD UK DSTAN 80-112-85. Paint System Coal Tar Epoxy Two-Pack Issue 1. 12 pp.
MOD UK DSTAN 80-179-93. Paint, Epoxy, Solvent Free, Two-Pack Issue 1. 21 pp.
MOD UK NES 756-88. Application of Coal Tar Epoxy Paints Issue 3 (02.88). 21 pp.

—Steel
JIS K 5551-91. Epoxy Resin Paint. 13 pp.
MOD UK DSTAN 80-112-85. Paint System Coal Tar Epoxy Two-Pack Issue 1. 12 pp.

—Steel Bars
BSI BS 7295: Part 2-90. 1990 Fusion Bonded Epoxy Coated Carbon Steel Bars for the Reinforcement of Concrete Part 2: Coatings (R). 9 pp.
BSI BS 7295: Part 2-01. 1990 Amd 1 Fusion Bonded Epoxy Coated Carbon Steel Bars for the Reinforcement of Concrete Part 2: Specification for Coatings (AMD 6956) February 28, 1992. 10 pp.

—Steel—Marine
CGSB 1-GP-193MA-83. Coating, High-Build Epoxy, Marine, Standard for. 15 pp.

—Steel Pipes
CSA CAN/CSA-Z245.20-M92. External Fusion Bond Epoxy Coating for Steel Pipe; (Gen Instr 1 Thru 2). 87 pp.
SAA AS 3862-91. External Fusion-Bonded Epoxy Coating for Steel Pipes Amdt 1 September 1992. 47 pp.

—Submarines
MOD UK DSTAN 80-179-93. Paint, Epoxy, Solvent Free, Two-Pack Issue 1. 21 pp.
MOD UK NES 756-88. Application of Coal Tar Epoxy Paints Issue 3 (02.88). 21 pp.

—Thinners
CGSB CAN/CGSB-1.197-92. Thinner for Epoxy Coatings. 7 pp.
MOD UK DSTAN 80-38-85. Thinners for: Paint Epoxy Two-Pack, Cellulose Nitrate Paints, Dopes and Lacquers Issue 2. 8 pp.

—Wood
CGSB 1-GP-146M-79. Coating, Epoxy, Cold Cured, Gloss, Standard for; (Amendment 1 Feb 1981) (QPL Apr 1986). 14 pp.
CGSB CAN/CGSB-1.153-M90. High Build, Gloss, Epoxy Coating. 13 pp.

Epoxy Compounds
Use For: Epoxides *See Also:* Epichlorohydrin; Epoxy Equivalent; Ethers; Ethylene Oxide; Propylene Oxide
MOD UK NES 2074-01. Specification for Epoxy Syntactic Foam Issue 2 (09.92); Amendment 1. 23 pp.
MOD UK TS 10272. Epoxide System for Fibre-Reinforced Composites.

—Epoxy Equivalent—Volumetric Analysis
BSI BS 2782:Pt4: METH 433C-D-79. 1979 Methods of Testing Plastics Part 4: Chemical Properties Methods 433C and 433D: Determination of Epoxide Equivalent of Epoxide Compounds. 4 pp.
ISO 3001-78. Plastics—Epoxide Compounds—Determination of Epoxide Equivalent Second Edition. 5 pp.

Epoxy Equivalent
See Also: Chemical Analysis; Epoxy Compounds

—Epoxy Compounds—Volumetric Analysis
BSI BS 2782:Pt4: METH 433C-D-79. 1979 Methods of Testing Plastics Part 4: Chemical Properties Methods 433C and 433D: Determination of Epoxide Equivalent of Epoxide Compounds. 4 pp.
ISO 3001-78. Plastics—Epoxide Compounds—Determination of Epoxide Equivalent Second Edition. 5 pp.

—Epoxy Resins—Indicator Titration Method
JIS K 7236-86. Testing Methods for Epoxy Equivalent of Epoxide Resins. 9 pp.

—Epoxy Resins—Potentiometric Analysis
JIS K 7236-86. Testing Methods for Epoxy Equivalent of Epoxide Resins. 9 pp.

Epoxy Laminates
See Also: Epoxy Resins; Laminates
BSI BS 2572-90. 1990 Phenolic Laminated Sheet and Epoxy Cotton Fabric Laminated Sheet (V). 21 pp.
BSI BS 2572-01. 1990 Amd 1 Phenolic Laminated Sheet and Epoxy Cotton Fabric Laminated Sheet (AMD 6966) February 28, 1992 (V). 22 pp.
MOD UK TS 10272. Epoxide System for Fibre-Reinforced Composites.

—Printed Circuit Base Materials
BSI BS 4584: Part 3-72. (WITHDRAWN) 1972 Amd 1 Metal-Clad Base Materials for Printed Circuits Part 3: Epoxide Woven Glass Fabric Copper-Clad Laminated Sheet, Flame Retardant Grade: EP-GC-Cu-3 (S) (Superseded by BS 4584: Section 102.5: 1990). 18 pp.
BSI BS 4584: Part 11-77. (WITHDRAWN) 1977 Metal-Clad Base Materials For Printed Circuits Part 11: Bonding Sheet Material for Use in the Fabrication of Multi-layer Printed Boards. EP-GC-11 (Superseded by BS 4584: Section 103.1: 1990). 13 pp.

INTERNATIONAL AND NON-U.S. NATIONAL STANDARDS
SUBJECT INDEX
Epoxy

Epoxy Laminates *(Cont.)*

—Printed Circuit Base Materials *(Cont.)*

BSI BS 4584: Sec 102.3-90. 1990 Metal-Clad Base Materials for Printed Circuits Part 102: Copper-Clad Base Materials Section 102.3: Epoxide Cellulose Paper Copper-Clad Laminated Sheet of Defined Flammability (Vertical Burning Test). 12 pp.

BSI BS 4584: Sec 102.4-90. 1990 Metal-Clad Base Materials for Printed Circuits Part 102: Copper-Clad Base Materials Section 102.4: Epoxide Woven Glass Fabric Copper-Clad Laminated Sheet, General Purpose Grade. 12 pp.

BSI BS 4584: Sec 102.5-90. 1990 Metal-Clad Base Materials for Printed Circuits Part 102: Copper-Clad Base Materials Section 102.5: Epoxide Woven Glass Fabric Copper-Clad Laminated Sheet of Defined Flammability (Vertical Burning Test). 14 pp.

BSI BS 4584: Sec 102.9-90. 1990 Metal-Clad Base Materials for Printed Circuits Part 102: Copper-Clad Base Materials Section 102.9: Epoxide Cellulose Paper Core, Epoxide Glass Cloth Surf. Copper-Clad Lamin. Sheet of Defined Flamm. (Vertical Burning Test). 12 pp.

BSI BS 4584: Sec 102.9-01. 1990 Amd 1 Metal-Clad Base Materials for Printed Wiring Boards Part 102: Copper-Clad Base Materials Section 102.9: Specification for Epoxide Cellulose Paper Core, Epoxide Glass Cloth Surfaces Copper-Clad Laminated Sheet of. 15 pp.

BSI BS 4584: Sec 102.10-90. 1990 Metal-Clad Base Materials for Printed Circuits Part 102: Copper-Clad Base Materials Section 102.10: Epoxide Non-Woven/Woven Glass Reinforced Copper-Clad Laminated Sheet of Defined Flammability (Vertical Burning Test). 12 pp.

BSI BS 4584:Sec 102.10-01. 1990 Amd 1 Metal-Clad Base Materials for Printed Wiring Boards Part 102: Copper-Clad Base Materials Section 102.10: Specification for Epoxide Non-Woven/Woven Glass Reinforced Copper-Clad Laminated Sheet of Defined Flammability. 15 pp.

BSI BS 4584: Sec 102.11-90. 1990 Metal-Clad Base Materials for Printed Circuits Part 102: Copper-Clad Base Materials Section 102. 11: Thin Epoxide Woven Glass Fabric Copper-Clad Laminated Sheet, General Purpose Grade, for Use in the Fabrication of Multilayer Prtd Boards. 10 pp.

BSI BS 4584:Sec 102.11-01. 1990 Amd 1 Metal-Clad Base Materials for Printed Wiring Boards Part 102: Copper-Clad Base Materials Section 102.11: Specification for Thin Epoxide Woven Glass Fabric Copper-Clad Laminated Sheet, General Purpose Grade, for Use. 13 pp.

BSI BS 4584: Sec 102.12-90. 1990 Metal-Clad Base Materials for Printed Circuits Part 102: Copper-Clad Base Materials Section 102. 12: Thin Epoxide Woven Glass Fabric Copper-Clad Lamin. Sheet of Defined Flammability, for Use in the Fabrication of Multilayer Prtd Boards. 10 pp.

BSI BS 4584:Sec 102.12-01. 1990 Amd 1 Metal-Clad Base Materials for Printed Wiring Boards Part 102: Copper-Clad Base Materials Section 102.12: Specification for Thin Epoxide Woven Glass Fabric Copper-Clad Laminated Sheet of Defined Flammability,. 12 pp.

BSI BS 4584: Sec 102.18-92. 1992 Metal-Clad Base Materials for Printed Wiring Boards Part 102: Copper-Clad Base Materials Section 102.18: Specification for Bismaleimide/Triazine Modified Epoxide Woven Glass Fabric Copper-Clad Laminated Sheet of. 19 pp.

BSI BS 4584: Sec 102.19-92. 1992 Metal-Clad Base Materials for Printed Wiring Boards Part 102: Copper-Clad Base Materials Section 102. 19: Spec. for Thin Bismaleimide/Triazine Modified Epoxide Glass Fabric Copper-Clad Laminated Sheets of Defined Flammability for. 16 pp.

CENELEC HD 313.2.3 S1-89. Base Materials for Printed Circuits-Part 2: Specifications Specification No 3: Epoxide Cellulose Paper Copper-Clad Laminated Sheet of Defined Flammability (Vertical Burning Test). 3 pp.

CENELEC HD 313.2.3 S2-90. Base Materials for Printed Circuits Part 2: Specifications Specification No.3: Epoxide Cellulose Paper Copper-Clad Laminated Sheet of Defined Flammability (Vertical Burning Test). 3 pp.

CENELEC HD 313.2.4 S2-90. Base Materials for Printed Circuits Part 2: Specifications Specification No.4: Epoxide Woven Glass Fabric Copper-Clad Laminated Sheet, General Purpose Grade. 3 pp.

CENELEC HD 313.2.5 S2-90. Base Materials for Printed Circuits Part 2: Specifications Specification No.5: Epoxide Woven Glass Fabric Copper-Clad Laminated Sheet of Defined Flammability (Vertical Burning Test). 3 pp.

CENELEC HD 313.2.9 S2-90. Base Materials for Printed Circuits Part 2: Specifications Specification No.9: Epoxide Cellulose Paper Core, Epoxide Glass Cloth Surfaces Copper-Clad Laminated Sheet of Defined Flammability (Vertical Burning Test). 4 pp.

Epoxy Laminates *(Cont.)*

—Printed Circuit Base Materials *(Cont.)*

CENELEC HD 313.2.9 S3-91. Base Materials for Printed Circuits Part 2: Specifications Specification No. 9: Epoxide Cellulose Paper Core, Epoxide Glass Cloth Surfaces Copper-Clad Laminated Sheet of Defined Flammability (Vertical Burning Test). 6 pp.

CENELEC HD 313.2.10 S1-89. Base Materials for Printed Circuits-Part 2: Specifications Specification No 10: Epoxide Non-Woven/Woven Glass Reinforced Copper-Clad Laminated Sheet of Defined Flammability (Vertical Burning Test). 3 pp.

CENELEC HD 313.2.10 S2-90. Base Materials for Printed Circuits Part 2: Specifications Specification No. 10: Epoxide Non-Woven Glass Reinforced Copper-Clad Laminated Sheet of Defined Flammability (Vertical Burning Test). 3 pp.

CENELEC HD 313.2.10 S3-91. Base Materials for Printed Circuits Part 2: Specifications Specification No. 10: Epoxide Non-Woven/Woven Glass Reinforced Copper-Clad Laminated Sheet of Defined Flammability (Vertical Burning Test). 6 pp.

CENELEC HD 313.2.11 S1-89. Base Materials for Printed Circuits-Part 2: Specifications Specification No 11: Thin Epoxide Woven Glass Fabric Copper-Clad Laminated Sheet, General Purpose Grade for Use in the Fabrication of Multilayer Printed Boards. 3 pp.

CENELEC HD 313.2.11 S2-90. Base Materials for Printed Circuits Part 2: Specifications Specification No. 11: Thin Epoxide Woven Glass Fabric Copper-Clad Laminated Sheet, General Purpose Grade, for Use in the Fabrication of Multilayer Printed Boards. 3 pp.

CENELEC HD 313.2.12 S1-89. Base Materials for Printed Circuits-Part 2: Specifications Specification No 12: Thin Epoxide Woven Glass Fabric Copper-Clad Laminated Sheet of Defined Flammability for Use in the Fabrication of Multilayer Printed Boards. 3 pp.

CENELEC HD 313.2.12 S2-90. Base Materials for Printed Circuits Part 2: Specifications Specification No. 12: Thin Epoxide Woven Glass Fabric Copper-Clad Laminated Sheet of Defined Flammability, for Use in the Fabrication of Multilayer Printed Boards. 3 pp.

IEC 249 Pt 2-3-87. Base Materials for Printed Circuits Part 2: Specifications Specification No. 3: Epoxide Cellulose Paper Copper-Clad Laminated Sheet of Defined Flammability (Vertical Burning Test) Second Edition; (Amendment 2-1993, Incorporating Amendment 1). 37 pp.

IEC 249 Pt 2-4-87. Base Materials for Printed Circuits Part 2: Specifications Specification No. 4: Epoxide Woven Glass Fabric Copper-Clad Laminated Sheet, General Purpose Grade Second Edition; (Amendment 3-1993, Incorporating Amendment 1 and 2). 39 pp.

IEC 249 Pt 2-5-87. Base Materials for Printed Circuits Part 2: Specifications Specification No. 5: Epoxide Woven Glass Fabric Copper-Clad Laminated Sheet of Defined Flammability (Vertical Burning Test) Second Edition; (Amd. 3-1993, Incorporating Amendment 1 and 2). 43 pp.

IEC 249 Pt 2-9-87. Base Materials for Printed Circuits Part 2: Specifications Specification No. 9: Epoxide Cellulose Paper Core, Epoxide Glass Cloth Surfaces Copper-Clad Laminated Sheet of Defined Flammability (Vertical Burning Test) First Ed.; (Amd 3-1993, Incorp. Amds 1 and 2). 38 pp.

IEC 249 Pt 2-10-87. Base Materials for Printed Circuits Part 2: Specifications Spec. No. 10: Epoxide Non-Woven/Woven Glass Reinforced Copper-Clad Laminated Sheet of Defined Flammability (Vertical Burning Test) First Ed.; (Amendment 3-1993, Incorporating Amendment 1 and 2). 39 pp.

IEC 249 Pt 2-11-87. Base Materials for Printed Circuits Part 2: Specifications Spec. No. 11: Thin Epoxide Woven Glass Fabric Copper-Clad Laminated Sheet, General Purpose Grade, for Use in the Fabrication of Multilayer Printed Boards First Edition (Amendment 2-1993, Incorporating Amd 1). 33 pp.

IEC 249 Pt 2-12-87. Base Materials for Printed Circuits Part 2: Specifications Spec. No. 12: Thin Epoxide Woven Glass Fabric Copper-Clad Laminated Sheet of Defined Flammability, for Use in the Fabrication of Multi-layer Printed Boards Second Ed.; (Amd 2-1993, Incorporating Amd 1). 35 pp.

IEC 249 Pt 2-18-92. Base Materials for Printed Circuits Part 2: Specifications Specification No. 18: Bismaleimide/Triazine Modified Epoxide Woven Glass Fabric Copper-Clad Laminated Sheet of Defined Flammability (Vertical Burning Test) First Edition; (Amendment 1-1993). 41 pp.

IEC 249 Pt 2-19-92. Base Materials for Printed Circuits Part 2: Specifications Specification No. 19: Thin Bismaleimide/Triazine Modified Epoxide Woven Glass Fabric Copper-Clad Laminated Sheet of Defined Flammability for Use in the Fabrication of Multilayer Printed. 37 pp.

Epoxy Laminates *(Cont.)*

—Printed Circuit Base Materials *(Cont.)*

MOD UK DSTAN 59-50-01. Requirements for Plastics Sheet, Laminated Copper Clad, Epoxide Resin Bonded, Woven Glass Fabric Base-Fire-Retardant (Metal Clad Base Materials for Printed Circuits) Issue 1; Amendment 1. 18 pp.

Epoxy Paints

Use: Epoxy Coatings

Epoxy Primers

See Also: Epoxy Coatings; Primers (Coatings)

CNS K2086-86. Epoxy Resin Non-Zinc Primer (May)(4935). 2 pp.

CNS K2087-86. Epoxy Resin Zinc Primer (May)(4936). 2 pp.

CNS K6852-86. Method of Test for Epoxy Resin Non-Zinc Primer (May)(11582).

CNS K6853-86. Method of Test for Epoxy Resin Zinc Primer (May)(11583).

—Aluminum

CNS K2202-88. Epoxy Resin Aluminum Tri-Polyphosphate Anti-Corrosive Primer (Apr)(12268). 2 pp.

—Aluminum—Aerospace—Hydraulic Fluid Resistant

MOD UK DTD-5640-90. Paint, Priming for Aircraft, Epoxide Cold Curing, Improved Adhesion, Two Pack. 9 pp.

—Aluminum Alloys—Aerospace—Hydraulic Fluid Resistant

MOD UK DTD-5640-90. Paint, Priming for Aircraft, Epoxide Cold Curing, Improved Adhesion, Two Pack. 9 pp.

—Aluminum—Tripolyphosphate Content

CNS K6953-88. Method of Test for Epoxy Resin Aluminum Tri-Polyphosphate Anticorrosive Primer (Apr)(12269). 3 pp.

—Amphibious Vehicles

MOD UK DSTAN 80-163-93. Paint System Defence Equipment Heavy Duty (Two-Pack) Paint, Priming, Defence Equipment Heavy Duty, Epoxide Paint, Undercoat, Defence Equipment Epoxide/Pitch Paint, Finishing, Defence Equipment Heavy Duty, Epoxide Issue 1. 30 pp.

—Electrodeposited

MOD UK TS 10245. Powder Coating, Priming for Phosphate Steel (Withdrawn).

—Materiel

MOD UK DSTAN 80-163-93. Paint System Defence Equipment Heavy Duty (Two-Pack) Paint, Priming, Defence Equipment Heavy Duty, Epoxide Paint, Undercoat, Defence Equipment Epoxide/Pitch Paint, Finishing, Defence Equipment Heavy Duty, Epoxide Issue 1. 30 pp.

—Metal

CGSB CAN/CGSB-1.165-M89. Cold Curing Epoxy Primer; (Amendment 1 May 1991). 13 pp.

Epoxy Resin Coatings

Use: Epoxy Coatings

Epoxy Resins

See Also: Adhesives; Epoxy Coatings; Epoxy Laminates; Masonry Cements; Resins; Thermosetting Resins

ISO 3673 Pt 1-80. Plastics—Epoxide Resins—Part 1: Designation First Edition. 4 pp.

JIS K 7121-87. Testing Methods for Transition Temperatures of Plastics. 12 pp.

JIS K 7231-86. General Rules for Testing Methods of Epoxide Resins and Hardeners. 6 pp.

MOD UK DSTAN 80-157-93. Epoxide Resin Putty System Issue 1. 13 pp.

—Accelerating Agents

ISO 4597 Pt 1-83. Plastics—Hardeners and Accelerators for Epoxide Resins—Part 1: Designation First Edition. 4 pp.

—Aerospace—Primers

BSI X 33-91. 1991 Two Component Epoxy Primer for Aerospace Purposes. 10 pp.

BSI X 33-01. 1991 Amd 1 Two Component Epoxy Primer for Aerospace Purposes (AMD 7164) May 1, 1992. 12 pp.

—Chlorine Content

BSI BS 2782:Pt4: METH 433B-79. 1979 Methods of Testing Plastics Part 4: Chemical Properties Method 433B: Determination of Easily Saponifiable Chlorine in Epoxide Resins and Related Materials. 4 pp.

INDUSTRY STANDARDS
997

Epoxy Resins (Cont.)

—Chlorine Content (Cont.)
ISO 4583-78. Plastics—Epoxide Resins and Related Materials—Determination of Easily Saponifiable Chlorine First Edition. 5 pp.

—Chlorine Content—Combustion
ISO 4615-79. Plastics—Unsaturated Polyesters and Epoxide Resins—Determination of Total Chlorine Content First Edition. 7 pp.

—Chlorine Content—Potentiometric Analysis
BSI BS 2782:Pt4: METH 433A-79. 1979 Methods of Testing Plastics Part 4: Chemical Properties Method 433A: Determination of Inorganic Chlorine in Epoxide Resins and Glycidyl Esters. 4 pp.

ISO 4573-78. Plastics—Epoxide Resins and Glycidyl Esters—Determination of Inorganic Chlorine First Edition. 4 pp.

—Crystallization
ISO 4895-87. Plastics—Liquid Epoxide Resins—Determination of Tendency to Crystallize First Edition. 4 pp.

—Density
JIS K 7232-86. Testing Methods for Specific Gravity of Epoxide Resins and Hardeners. 16 pp.

—Designations
JIS K 7238-91. Designation of Epoxide Resins. 6 pp.

—Electrical Insulation
BSI BS 3815-64. 1964 Epoxide Resin Casting Systems for Electrical Applications. 17 pp.

BSI BS 3816-89. 1989 Epoxy Resin Casting Systems Used for Electrical Insulating Applications at Power Frequencies. 4 pp.

CENELEC HD 307.3.11 S1-90. Specification for Solventless Polymerisable Resinous Compounds Used for Electrical Insulation Part 3: Specification for Individual Materials Sheet 11: Epoxy Resin-Based Coating Powders. 3 pp.

IEC 455 Pt 3-11-88. Specification for Solventless Polymerisable Resinous Compounds Used for Electrical Insulation Part 3: Specifications for Individual Materials Sheet 11: Epoxy Resin-Based Coating Powders First Edition. 17 pp.

IEC 893 Pt 3-2-93. Specification for Industrial Rigid Laminated Sheets Based on Thermosetting Resins for Electrical Purposes Part 3: Specifications for Individual Materials —Sheet 2: Requirements for Rigid Laminated Sheets Based on Epoxide Resins First Edition. 28 pp.

—Epoxy Equivalent—Indicator Titration Method
JIS K 7236-86. Testing Methods for Epoxy Equivalent of Epoxide Resins. 9 pp.

—Epoxy Equivalent—Potentiometric Analysis
JIS K 7236-86. Testing Methods for Epoxy Equivalent of Epoxide Resins. 9 pp.

—Floors
CGSB 81-GP-4M-77. Flooring, Seamless, Decorative Epoxy, Troweled Finish, Standard for. 12 pp.

CGSB 81-GP-5M-78. Flooring, Seamless Epoxy, Broadcast Quartz, Standard for. 10 pp.

CGSB 81-GP-6M-77. Flooring, Seamless, Epoxy Terrazzo, Standard for. 11 pp.

—Foucry Testing
CGSB 1-GP-71 METH 75.3-82. Methods of Testing Paints and Pigments Foucry (Nitric Acid) Test for Epoxy Resins. 1 p.

—Gel Time
BSI BS 2782:Pt8: METH 835C-80. 1980 Methods of Testing Plastics Part 8: Other Properties Method 835C: Determination of Gelation Time of Polyester and Epoxy Resins Using a Gel Timer. 4 pp.

—Hardeners
ISO 4597 Pt 1-83. Plastics—Hardeners and Accelerators for Epoxide Resins—Part 1: Designation First Edition. 4 pp.

JIS K 7231-86. General Rules for Testing Methods of Epoxide Resins and Hardeners. 6 pp.

—Hardeners—Acidity
JIS K 7239-91. Determination of Free Acid in Acid Anhydride-Based Hardeners for Epoxide Resins. 8 pp.

—Hardeners—Density
JIS K 7232-86. Testing Methods for Specific Gravity of Epoxide Resins and Hardeners. 16 pp.

—Hardeners—Viscosity
JIS K 7233-86. Testing Methods for Viscosity of Epoxide Resins and Hardeners. 13 pp.

Epoxy Resins (Cont.)

—Nonvolatile Matter Content—Heat Loss Method
JIS K 7235-86. Testing Methods for Non-Volatile Matter in Solvent-Diluted Epoxide Resins. 5 pp.

—Paints—Binders
ISO 7142-84. Binders for Paints and Varnishes—Epoxy Resins—General Methods of Test First Edition. 6 pp.

—Quartz Filled—Electrical Insulation
IEC 455 Pt 3-2-87. Specification for Solventless Polymerisable Resinous Compounds Used for Electrical Insulation Part 3: Specifications for Individual Materials Sheet 2: Quartz Filled Epoxy Resinous Compounds First Edition. 11 pp.

—Shrinkage
BSI BS 2782:Pt6: METH 644A-79. 1979 Methods of Testing Plastics Part 6: Dimensional Properties Method 644A: Determination of Total Volume Shrinkage of Polyester and Epoxide Casting Resins. 4 pp.

ISO 3521-76. Plastics—Polyester and Epoxy Casting Resins—Determination of Total Volume Shrinkage First Edition. 4 pp.

—Softening Points
JIS K 7234-86. Testing Methods for Softening Point of Epoxide Resins. 13 pp.

—Swann Spot Testing
CGSB 1-GP-71 METH 75.4-82. Methods of Testing Paints and Pigments Swann Spot Test for Epoxy Resins. 1 p.

—Unfilled—Electrical Insulation
IEC 455 Pt 3-1-81. Specification for Solventless Polymerisable Resinous Compounds Used for Electrical Insulation Part 3: Specifications for Individual Materials Sheet 1: Unfilled Epoxy Resinous Compounds First Edition. 13 pp.

—Varnishes—Binders
ISO 7142-84. Binders for Paints and Varnishes—Epoxy Resins—General Methods of Test First Edition. 6 pp.

—Viscosity
JIS K 7233-86. Testing Methods for Viscosity of Epoxide Resins and Hardeners. 13 pp.

Epoxy Varnishes
See Also: Varnishes

—Baking—Steel—Ammunition
MOD UK DSTAN 80-175-93. Varnish, Ammunition, Stoving, Type QX Class A Expoxy Ester B Silicone Alkyd C Tung Oil Phenolic Types: Brushing Spraying Dipping Issue 1. 16 pp.

—Metal
MOD UK DSTAN 80-35-01. Varnish, Epoxide, Stoving, Type QX Type: Spraying Issue 1; Amendment 1. 10 pp.

Epoxyethame
Use: Ethylene Oxide

EPROM
Use: EPROMs

EPROMs
Use For: EPROM; Erasable Programmable Read Only Memories; Rewritable Programmable Read Only Memories *See Also:* Memory Circuits; PROMs; ROMs

MOD UK DSTAN 59-62: Part 7-83. Microcircuits Electronic (Integrated Circuits) (Listed on EPIC Database) Part 7: Memories/Microprocessors Issue 1. 14 pp.

—CMOS—UV Erasable—Preferred Products List
CECC CECC MUAHAG Vol 7 IS 8-92. Preferred Products List; Active Microcircuits (En, Fe, Ge). 89 pp.

—Data Storage Devices—Information Interchange
ECMA ECMA 167-92. Volume and File Structure of Write-Once and Rewritable Media Using Non-Sequential Recording for Information Interchange. 120 pp.

IEC DIS 13346-93. Information Technology—Volume and File Structure of Write-Once and Rewritable Media Using Non-Sequential Recording for Information Interchange ***CD-ROM ONLY***. 123 pp.

EPROMs (Cont.)

—Data Storage Devices—Information Interchange (Cont.)
ISO DIS 13346-93. Information Technology—Volume and File Structure of Write-Once and Rewritable Media Using Non-Sequential Recording for Information Interchange ***CD-ROM ONLY***. 123 pp.

JTC1 DIS13346-93. Information Technology—Volume and File Structure of Write-Once and Rewritable Media Using Non-Sequential Recording for Information Interchange ***CD-ROM ONLY***. 123 pp.

—Preferred Products List
CECC CECC MUAHAG Vol 7 IS 8-92. Preferred Products List; Active Microcircuits (En, Fe, Ge). 89 pp.

—Quality Assurance
BSI BS CECC 90113-87. 1987 Amd 1 MOS Ultra-Violet Light Erasable Electrically Programmable Read Only Memories Silicon Monolithic Circuits (AMD 5997) November 30, 1990. 29 pp.

Equality Comparators
See Also: Comparators; Magnitude Comparators

—8-Bit—Preferred Products List
CECC CECC MUAHAG Vol 7 IS 8-92. Preferred Products List; Active Microcircuits (En, Fe, Ge). 89 pp.

Equalizers (Electrical)
See Also: Attenuators; Deemphasis Networks; Electrical Equipment

—Telephone Circuits—Modems
CCITT RECMN V.27-89. 4800 Bits per Second Modem with Manual Equalizer Standardized for Use on Leased Telephone-Type Circuits—Data Communication over the Telephone Network (Study Group XVII) 6 pp. 6 pp.

CCITT RECMN V.27 BIS-89. 4800/2400 Bits per Second Modem with Automatic Equalizer Standardized for Use on Leased Telephone-Type Circuits—Data Communication over the Telephone Network (Study Group XVII) 12 pp. 12 pp.

CCITT RECMN V.27 TER-89. 4800/2400 Bits per Second Modem Standardized for Use in the General Switched Telephone Network—Data Communication over the Telephone Network (Study Group XVII) 13 pp. 13 pp.

Equilibration
Use: Balancing

Equilibrium Moisture Content Analysis
See Also: Moisture Content Analysis

—Coal
JIS M 8803-76. Methods for Test of Equilibrium Moisture of Coal at 96 to 97 Percent Relative Humidity and 30 Degrees C.

Equines
See Also: Livestock

—Breeding Stock
EC 90/427/EEC-90. Council Directive on the Zootechnical and Genealogical Conditions Governing Intra-Community Trade in Equidae. 5 pp.

—Competitions
EC 90/428/EEC-90. Council Directive on Trade in Equidae Intended for Competitions and Laying Down the Conditions for Participation Therein. 2 pp.

—Embryo Collection
EC 90/427/EEC-90. Council Directive on the Zootechnical and Genealogical Conditions Governing Intra-Community Trade in Equidae. 5 pp.

—Health and Veterinary Inspection
EC 90/425/EEC-90. Council Directive Concerning Veterinary and Zootechnical Checks Applicable in Intra-Community Trade in Certain Live Animals and Products with a View to the Completion of the Internal Market. 13 pp.

EC 91/628/EEC-91. Council Directive on the Protection of Animals During Transport and Amending Directives 90/425/EEC and 91/496/EEC. 12 pp.

—Health Inspection—Importation
EC 90/426/EEC-90. Council Directive on Animal Health Conditions Governing the Movement and Unipart from Third Countries of Equidae. 13 pp.

INTERNATIONAL AND NON-U.S. NATIONAL STANDARDS
SUBJECT INDEX

Equines *(Cont.)*
—Ova Collection
EC 90/427/EEC-90. Council Directive on the Zootechnical and Genealogical Conditions Governing Intra-Community Trade in Equidae. 5 pp.

—Semen Collection
EC 90/427/EEC-90. Council Directive on the Zootechnical and Genealogical Conditions Governing Intra-Community Trade in Equidae. 5 pp.

—Veterinary Inspection—Importation
EC 90/426/EEC-90. Council Directive on Animal Health Conditions Governing the Movement and Unipart from Third Countries of Equidae. 13 pp.

Equipment
Scope Note: For additional listings, use a more specific term *Use For:* Engineering Equipment; Technical Equipment *See Also:* Agricultural Equipment; Air Conditioning Equipment; Air Handling Equipment; Audio Equipment; Audiovisual Equipment; Chemical Equipment; Chemical Process Equipment; Communication Equipment; Construction Equipment; Control Systems Equipment; Data Processing Equipment; Dental Equipment; Distillation Equipment; Electrical Equipment; Entertainment Equipment; Fishing Equipment; Forestry Equipment; Ground Support Equipment; Heating Equipment; Industrial Equipment; Mechanical Equipment; Medical Equipment; Mining Equipment; Radar Equipment; Radio Equipment; Rice Milling Equipment; Ultrasonic Equipment; Weeding Equipment; Welding Equipment

—Coding
NATO STANAG 4438 ED 1 AMD 0-00. Codification of Equipment—Uniform System of Dissemination of Data Association with NATO Stock Numbers. 5 pp.

—Land Service—Commercial Literature
MOD UK DEFCON 4A-90. Technical Contract Conditions for Supply of Commercially Produced Literature for Land Service Equipment 7/90. 11 pp.

—Mine Countermeasures
NATO STANAG 1202 ED 1 AMD 2-80. Stanag 1202 MW (Edition No. 2)—NATO Mine Countermeasures Vehicles and Equipment (AMP-3)(A) (Navy) Volume I. 5 pp.
NATO STANAG 1202 ED 2 AMD 3-80. NATO Mine Countermeasures Vehicles and Equipment (AMP-3)(A)(NAVY) Volume I. 5 pp.

—Safety
DIN VDE 31000 Pt 2-87. General Guide for Designing Technical Equipment to Satisfy Safety Requirements; Safety Technology Concepts; Basic Concepts (Dec). 9 pp.

—Symbols
BSI BS 7324-90. 1990 Guide to Graphical Symbols for Use on Equipment. 73 pp.
CNS C1142-89. General Principles for the Formulation of Graphical Symbols for Use on Equipment (Feb)(12490).

—Technical Contract Conditions
MOD UK DEFCON 4A-90. Technical Contract Conditions for Supply of Commercially Produced Literature for Land Service Equipment 7/90. 11 pp.

—Testing Equipment
MOD UK NES 43-88. Ancillary Support Equipment, Test Equipment and Tools Issue 2 (09.88). 45 pp.

Equivalence Gates
Use: Exclusive NOR Gates

Erasable Programmable Read Only Memories
Use: EPROMs

Erasers
Use For: Rubbers (Erasers)

—Plastic
CNS S1106-80. Plastic Erasers (Dec)(6856). 4 pp.
JIS S 6050-88. Plastic Erasers. 8 pp.

—Rubber
CGSB 53-GP-67M-78. Eraser, Rubber, Standard for. 9 pp.
CNS S1127-81. Eraser Rubbers (Jul)(7753).
CNS S2016-81. Method of Test for Eraser Rubbers (Jul)(965).
JIS S 6004-88. Rubber Erasers. 10 pp.

Ercoupe 415 Forney FIA Aircraft
See Also: Aircraft

Ercoupe 415 Forney FIA Aircraft *(Cont.)*
—Antenna Positions
CAA. Ercoupe 415 (Approved Aerial Positions). 1 p.

ERG
Use: Electroretinographs

Ergonomics
Use For: Human Factors Engineering
See Also: Anthropometric Characteristics; Biomechanical Effects (Human Body); Education; Environmental Conditions; Environmental Engineering; Manual Controls; Organization and Methods; Systems Engineering; Test Dummies; Work Measurement

BSI PP 7317-87. 1987 Ergonomics-Standards and Guidelines for Designers. 216 pp.
ICAO Circular 238. Human Factors Digest No. 6 Ergonomics—1992. 48 pp.
MOD UK DSTAN 00-25: Part 1-01. Human Factors for Designers of Equipment Part 1: Introduction Issue 2; Amendment 1 Corrections. 8 pp.
MOD UK DSTAN 00-25: Part 2-85. Human Factors for Designers of Equipment Part 2: Body Size Issue 1. 59 pp.
MOD UK DSTAN 00-25: Part 3-01. Human Factors for Designers of Equipment Part 3: Body Strength and Stamina Issue 1; Corrigendum. 49 pp.
MOD UK DSTAN 00-25: Part 8-89. Human Factors for Designers of Equipment Part 8: Auditory Information Issue 1. 37 pp.
MOD UK DSTAN 00-25: Part 12-89. Human Factors for Designers of Equipment Part 12: Systems Issue 1. 70 pp.

—Actuators—Machinery
CEN PREN 894-1-92. Safety of Machinery—Ergonomics Requirements for the Design of Displays and Control Actuators—Part 1: Human Interactions with Displays and Control Actuators. 19 pp.
CEN PREN 894-3-92. Safety of Machinery—Ergonomics Requirements for the Design of Displays and Control Actuators—Part 3: Control Actuators. 37 pp.
IEC 447-93. Man-Machine Interface (MMI)—Actuating Principles Second Edition. 41 pp.

—Aging Persons
BSI BS 4467-91. 1991 Guide to Dimensions in Designing for Elderly People. 20 pp.

—Agricultural Buildings
BSI BS 5502: Sec 1.6-86. (WITHDRAWN) 1986 Design of Buildings and Structures for Agriculture Part 1: General Considerations Section 1.6: Human and Animal Welfare (Superseded by BS 5502: Part 20: 1990 and BS 5502: Part 80:1990). 7 pp.
BSI BS 5502: Sec 3.5-78. (WITHDRAWN) 1978 Amd 1 Design of Buildings and Structures for Agriculture Part 3: Appendices: Legislation, Technical Data and References Section 3.5: Reference Data: Environment (Superseded by BS 5502: Parts 20, 80 and 82: 1990). 4 pp.

—Agricultural Equipment
BSI BS 6055-81. 1981 Methods for Measurement of Whole-Body Vibration of the Operators of Agricultural Wheeled Tractors and Machinery. 14 pp.
ISO 5007-90. Agricultural Wheeled Tractors—Operator's Seat—Laboratory Measurement of Transmitted Vibration First Edition. 17 pp.
ISO 5008-79. Agricultural Wheeled Tractors and Field Machinery—Measurement of Whole-Body Vibration of the Operator First Edition; (Erratum—June 1980). 14 pp.

—Air Traffic Control
ICAO Circular 241. Human Factors Digest No. 8 Human Factors in Air Traffic Control—1993. 48 pp.

—Aircraft Safety
ICAO Circular 216. Human Factors Digest No. 1 Fundamental Human Factors Concepts—1989. 36 pp.
ICAO Circular 217. Human Factors Digest No. 2 Flight Crew Training: Cockpit Resource Management (CRM) and Line-Oriented Flight Training (LOFT)—1989. 67 pp.
ICAO Circular 227. Human Factors Digest No. 3 Training of Operational Personnel in Human Factors—1991. 58 pp.
ICAO Circular 229. Human Factors Digest No. 4 Proceedings of the ICAO Human Factors Seminar—1990. 628 pp.
ICAO Circular 240. Human Factors Digest No. 7 Investigation of Human Factors in Accidents and Incidents—1993. 61 pp.

Ergonomics *(Cont.)*
—Body Dimensions
BSI BS 7231: Part 1-90. 1990 Amd 1 Body Measurements of Boys and Girls from Birth up to 16.9 Years Part 1: Information in the Form of Tables (G) (AMD 6749) September 30, 1991. 260 pp.
BSI BS 7231: Part 2-90. 1990 Body Measurements of Boys and Girls from Birth up to 16.9 Years Part 2: Recommendations of Body Dimensions for Children. 15 pp.
CEN PREN 979-92. Basic List of Definitions of Human Body Dimensions for Technical Design. 26 pp.

—Body Dimensions—Access Doors
DIN ENGL 33402 Pt 4-86. Human Body Dimensions; Principles of Dimensioning Passages and Accesses (Oct). 6 pp.

—Body Dimensions—Walkways
DIN ENGL 33402 Pt 4-86. Human Body Dimensions; Principles of Dimensioning Passages and Accesses (Oct). 6 pp.

—Buildings—Acoustics
BSI BS 7643: Part 3-93. 1993 Building Construction—Expression of Users' Requirements Part 3: Acoustical Requirements (ISO 6242-3: 1992). 12 pp.
ISO 6242 Pt 3-92. Building Construction—Expression of Users' Requirements—Part 3: Acoustical Requirements First Edition. 9 pp.

—Buildings—Air Pollution
BSI BS 7643: Part 2-93. 1993 Building Construction—Expression of Users' Requirements Part 2: Air Purity Requirements (ISO 6242-2: 1992). 9 pp.
ISO 6242 Pt 2-92. Building Construction—Expression of Users' Requirements—Part 2: Air Purity Requirements First Edition. 8 pp.

—Buildings—Thermal Environments
BSI BS 7643: Part 1-93. 1993 Building Construction—Expression of Users' Requirements Part 1: Thermal Requirements (ISO 6242-1: 1992). 11 pp.
ISO 6242 Pt 1-92. Building Construction—Expression of Users' Requirements—Part 1: Thermal Requirements First Edition. 9 pp.

—Chairs
BSI BS 3044-90. 1990 Ergonomics Principles in the Design and Selection of Office Furniture. 24 pp.

—Construction Equipment
JIS A 8301-86. Construction Machinery-Minimum Access Dimensions. 9 pp.

—Control Devices
MOD UK DSTAN 00-25: Part 10-92. Human Factors for Designers of Equipment Part 10: Controls Issue 1. 84 pp.

—Control Rooms
DIN ENGL 33414 Pt 1-85. Ergonomic Design of Control Rooms; Seated Work Stations; Terms and Definitions, Principles, Dimensions (Apr). 16 pp.

—Control Rooms—Nuclear Power Plants
IEC 964-89. Design for Control Rooms of Nuclear Power Plants First Edition. 133 pp.
IEC 1227-93. Nuclear Power Plants—Control Rooms—Operator Controls First Edition. 35 pp.

—Defense Equipment
MOD UK DSTAN 00-25: Part 5-92. Human Factors for Designers of Equipment Part 5: Stresses and Hazards Issue 1. 61 pp.

—Desks
DIN ENGL 4549-82. Office Furniture; Desks, Machine Operator's Desks and Visual Display Unit (VDU) Desks; Dimensions (Nov). 3 pp.

—Direction of Movement
DIN ENGL 33417-87. Description of Position, Orientation and Direction of Movement of Objects (Aug). 4 pp.

—Display Devices—Aircraft Cabins
NATO STANAG 3705 ED 3 AMD 0-92. Human Engineering Design Criteria for Controls and Displays in Aircrew Stations. 21 pp.
NATO STANAG 3705 ED 3 AMD 1-92. Human Engineering Design Criteria for Controls and Displays in Aircrew Stations. 22 pp.

—Display Devices—Machinery
CEN PREN 894-1-92. Safety of Machinery—Ergonomics Requirements for the Design of Displays and Control Actuators—Part 1: Human Interactions with Displays and Control Actuators. 19 pp.
CEN PREN 894-2-92. Safety of Machinery—Ergonomics Requirements for the Design of Displays and Control Actuators—Part 2: Displays. 24 pp.

INDUSTRY STANDARDS

INTERNATIONAL AND NON-U.S. NATIONAL STANDARDS
SUBJECT INDEX

Ergonomics

Ergonomics (Cont.)

—Earthmoving Equipment
BSI BS 6112-83. (WITHDRAWN) 1983 Minimum Access Dimensions for Earth-Moving Machinery (Superseded by BS 6912: part 5: 1992). 6 pp.
BSI BS 6112-01. (WITHDRAWN) 1983 Amd 1 Minimum Access Dimensions for Earth-Moving Machinery (AMD 6877) May 1, 1992 (Superseded by BS 6912: Part 5: 1992). 11 pp.
BSI BS 6912: Part 5-92. 1992 Safety of Earth-Moving Machinery Part 5: Recommendations for Minimum Access Dimensions (ISO 2860: 1992). 9 pp.
CEN EN 22860-85. Earth-Moving Machinery Minimum Access Dimensions. 6 pp.
CNS A4014-88. Earth-Moving Machinery-Minimum Access Dimensions (Dec)(9945).
CNS A4015-88. Earth-Moving Machinery Access Systems (Dec)(9946).
CSA CAN/CSA-M2860-M91. Minimum Access Dimensions for Servicing Machines (EMM,FM) (ISO 2860-1983). 15 pp.
CSA CAN/CSA-M3411-M91. Human Physical Dimensions of Operators—Minimum Operator Space Envelope—Machinery (EMM, AM, FM) (ISO 3411-1982). 19 pp.
CSA CAN/CSA-M6682-M91. Zones of Comfort and Reach for Controls—Machinery (EMM,FM) (Amd 1 Included); (ISO 6682-1986). 23 pp.
ISO 2860-92. Earth-Moving Machinery—Minimum Access Dimensions Fourth Edition. 7 pp.
ISO 2867-89. Earth-Moving Machinery—Access Systems Fourth Edition. 10 pp.
ISO 3411-82. Earth-Moving Machinery—Human Physical Dimensions of Operators and Minimum Operator Space Envelope Second Edition; (Amendment 1-1989). 13 pp.

—Environmental Conditions—Workplace
DIN ENGL 33403 Pt 1-84. Climate at Workplaces and in Working Environments; Basic Principles for Determining Climates (Apr). 7 pp.

—Flight Crew Systems
NATO STANAG 3994 ED 1 AMD 0-91. Application of Human Engineering to Advanced Aircrew Systems. 10 pp.
NATO STANAG 3994 ED 1 AMD 1-91. Application of Human Engineering to Advanced Aircrew Systems. 11 pp.

—Flight Decks—Automation
ICAO Circular 234. Human Factors Digest No. 5 Operational Implications of Automation in Advanced Technology Flight Decks—1992. 46 pp.

—Forestry Equipment
CSA CAN/CSA-M6682-M91. Zones of Comfort and Reach for Controls—Machinery (EMM,FM) (Amd 1 Included); (ISO 6682-1986). 23 pp.

—Hand Wheels
DIN ENGL 33411 Pt 3-86. Human Physical Strength; Maximum Static Action Moments Applied by Male Operators When Actuating Hand-Wheels (Dec). 5 pp.

—Indicating Instruments
DIN ENGL 33413 Pt 1-84. Ergonomic Aspects of Indicating Devices; Types, Observation Tasks, Suitability (June). 4 pp.

—Industrial Plants
SAA AS 1837-76. Code of Practice for Application of Ergonomics to Factory and Office Work. 32 pp.

—Kitchens
ISO 3055-85. Kitchen Equipment—Coordinating Sizes Second Edition. 18 pp.

—Lighting
MOD UK DSTAN 00-25: Part 6-86. Human Factors for Designers of Equipment Part 6: Vision and Lighting Issue 1. 74 pp.

—Machinery
BSI BS EN 349-93. 1993 Safety of Machinery Minimum Gaps to Avoid Crushing of Parts of the Human Body. 10 pp.
CEN PREN 349-90. Safety of Machinery—Minimum Distances to Avoid Crushing of Parts of the Human Body. 7 pp.
CEN EN 349-93. Safety of Machinery—Minimum Gaps to Avoid Crushing of Parts of the Human Body. 5 pp.

—Machinery—Vibration Reduction
CEN PREN 1030-1-93. Hand-Arm Vibration—Guidelines for Vibration Hazards Reduction—Part 1: Engineering Methods by Design of Machinery. 10 pp.

—Maintainability
MOD UK DSTAN 00-25: Part 11-88. Human Factors for Designers of Equipment Part 11: Design for Maintainability Issue 1. 18 pp.

Ergonomics (Cont.)

—Manual Controls—Aircraft Cabins
NATO STANAG 3705 ED 3 AMD 0-92. Human Engineering Design Criteria for Controls and Displays in Aircrew Stations. 21 pp.
NATO STANAG 3705 ED 3 AMD 1-92. Human Engineering Design Criteria for Controls and Displays in Aircrew Stations. 22 pp.

—Manual Controls—Automotive
ISO TR9511-91. Road Vehicles—Driver Hand-Control Reach—in-Vehicle Checking Procedure First Edition. 7 pp.

—Mental Workload—Glossaries
ISO 10075-91. Ergonomic Principles Related to Mental Work-Load—General Terms and Definitions First Edition. 9 pp.

—Office Systems
CSA CAN/CSA-Z412-M89. Office Ergonomics. 118 pp.
SAA AS 1837-76. Code of Practice for Application of Ergonomics to Factory and Office Work. 32 pp.

—Orientation
DIN ENGL 33417-87. Description of Position, Orientation and Direction of Movement of Objects (Aug). 4 pp.

—Pay Telephones
CCITT RECMN E.120-89. Instructions for Users of the International Telephone Service—Telephone Network and ISDN—Operation, Numbering, Routing and Mobile Service (Study Group II) 7 pp.

—Pictograms—Multiple Index Approach
CENELEC ETR 070-93. Human Factors (HF); the Multiple Index Approach (MIA) for the Evaluation of Pictograms. 21 pp.
ETSI ETR 070-93. Human Factors (HF); the Multiple Index Approach (MIA) for the Evaluation of Pictograms. 21 pp.

—Position
DIN ENGL 33417-87. Description of Position, Orientation and Direction of Movement of Objects (Aug). 4 pp.

—Speech Transmission Systems
MOD UK DSTAN 00-25: Part 9-91. Human Factors for Designers of Equipment Part 9: Voice Communication Issue 1. 45 pp.

—Static Action Forces
DIN ENGL 33411 Pt 4-87. Human Physical Strength; Maximum Static Action Forces (Isodynes) (May). 20 pp.

—Tables (Furniture)
BSI BS 3044-90. 1990 Ergonomics Principles in the Design and Selection of Office Furniture. 24 pp.

—Telecommunication Equipment
CENELEC ETR 029-91. Human Factors (HF); Access to Telecommunications for People with Special Needs Recommendations for Improving and Adapting Telecommunication Terminals and Services for People with Impairments. 21 pp.
CENELEC ETR 039-92. Human Factors (HF); Human Factors Standards for Telecommunications Applications. 30 pp.
ETSI ETR 029-91. Human Factors (HF); Access to Telecommunications for People with Special Needs Recommendations for Improving and Adapting Telecommunication Terminals and Services for People with Impairments. 21 pp.
ETSI ETR 039-92. Human Factors (HF); Human Factors Standards for Telecommunications Applications. 30 pp.

—Telecommunication Services
CENELEC ETR 029-91. Human Factors (HF); Access to Telecommunications for People with Special Needs Recommendations for Improving and Adapting Telecommunication Terminals and Services for People with Impairments. 21 pp.
CENELEC ETR 039-92. Human Factors (HF); Human Factors Standards for Telecommunications Applications. 30 pp.
ETSI ETR 029-91. Human Factors (HF); Access to Telecommunications for People with Special Needs Recommendations for Improving and Adapting Telecommunication Terminals and Services for People with Impairments. 21 pp.
ETSI ETR 039-92. Human Factors (HF); Human Factors Standards for Telecommunications Applications. 30 pp.

—Telephones
CENELEC ETR 051-92. Human Factors (HF); Usability Checklist for Telephones Basic Requirements. 15 pp.

Ergonomics (Cont.)

—Telephones (Cont.)
ETSI ETR 051-92. Human Factors (HF); Usability Checklist for Telephones Basic Requirements. 15 pp.

—Thermal Environments
ISO 7243-89. Hot Environments—Estimation of the Heat Stress on Working Man, Based on the WBGT-Index (Wet Bulb Globe Temperature) Second Edition. 13 pp.
ISO 7730-84. Moderate Thermal Environments—Determination of the PMV and PPD Indices and Specification of the Conditions for Thermal Comfort First Edition. 22 pp.
ISO 7933-89. Hot Environments—Analytical Determination and Interpretation of Thermal Stress Using Calculation of Required Sweat Rate First Edition. 24 pp.
ISO 8996-90. Ergonomics—Determination of Metabolic Heat Production First Edition. 21 pp.
SNZ NZS/ISO 7243-89. Hot Environments. Estimation of the Heat Stress on Working Man, Based on the WBGT-Index (Wet Bulb Globe Temperature). 9 pp.

—Tractors
ISO 4252-92. Agricultural Tractors—Operator's Workplace, Access and Exit—Dimensions Second Edition. 8 pp.

—Video Display Terminals
BSI BS 7179: Part 1-90. (WITHDRAWN) 1990 Ergonomics of Design and Use of Visual Display Terminals (VDTs) in Offices Part 1: Introduction (Superseded by BS EN 29241-2: 1993). 12 pp.
BSI BS 7179: Part 2-90. (WITHDRAWN) 1990 Ergonomics of Design and Use of Visual Display Terminals (VDTs) in Offices Part 2: Recommendations for the Design of Office VDT Tasks (Superseded by BE EN 29241-3: 1993). 9 pp.
BSI BS 7179: Part 3-90. (WITHDRAWN) 1990 Ergonomics of Design and Use of Visual Display Terminals (VDTs) in Offices Part 3: Specification for Visual Displays (Superseded by BS EN 29241-3: 1993). 28 pp.
BSI BS 7179: Part 5-90. 1990 Ergonomics of Design and Use of Visual Display Terminals (VDTs) in Offices Part 5: VDT Work Stations. 11 pp.
BSI BS 7179: Part 6-90. 1990 Ergonomics of Design and Use of Visual Display Terminals (VDTs) in Offices Part 6: Code of Practice for the Design of VDT Work Environments. 10 pp.
BSI BS EN 29241-1-93. 1993 Ergonomic Requirements for Office Work with Visual Display Terminals (VDTs) Part 1: General Introduction (ISO 9241-1: 1992) (S). 14 pp.
BSI BS EN 29241-2-93. 1993 Ergonomic Requirements for Office Work with Visual Display Terminals (VDTs) Part 2: Guidance on Task Requirements (ISO 9241-2: 1992) (S). 11 pp.
CEN EN 29241-2-93. Ergonomic Requirements for Office Work with Visual Display Terminals (VDTs)—Part 2: Guidance on Task Requirements (ISO 9241-2: 1992). 6 pp.
CEPT T/SF 40 E-85. Human Factor Aspects of Visual Display Terminals for Telecommunication Services. 11 pp.
DIN ENGL 66234 Pt 1-80. VDU Work Stations; Geometrical Design of Characters (Mar). 3 pp.
DIN ENGL 66234 Pt 2-83. VDU Work Stations; Perceptibility of Characters on Screens (May). 4 pp.
DIN ENGL 66234 Pt 3-81. VDU Work Stations; Grouping and Formatting of Data (Mar). 2 pp.
DIN ENGL 66234 Pt 5-81. VDU Work Stations; Coding of Information (Mar). 7 pp.
DIN ENGL 66234 Pt 8-88. VDU Work Stations; Principles of Ergonomic Dialogue Design (Feb). 6 pp.
ECMA ECMA 110-85. Ergonomics-Requirements for Monochromatic Visual Display Devices. 57 pp.
ECMA ECMA 126-87. Ergonomics-Requirements for Colour Visual Display Devices. 21 pp.
ECMA ECMA 136-89. Ergonomics—Requirements for Non-CRT Visual Display Units. 17 pp.
ECMA ECMA-TR 22-84. Ergonomics Recommendations for VDU Work Places. 32 pp.
ISO 9241 Pt 1-92. Ergonomic Requirements for Office Work with Visual Display Terminals (VDTs)—Part 1: General Introduction First Edition. 11 pp.
ISO 9241 Pt 2-92. Ergonomic Requirements for Office Work with Visual Display Terminals (VDTs)—Part 2: Guidance on Task Requirements First Edition (CEN EN 29241-2: 1993). 8 pp.
JTC1 9241 Pt 1-92. Ergonomic Requirements for Office Work with Visual Display Terminals (VDTs)—Part 1: General Introduction First Edition. 11 pp.
JTC1 9241 Pt 2-92. Ergonomic Requirements for Office Work with Visual Display Terminals (VDTs)—Part 2: Guidance on Task Requirements First Edition. 8 pp.
OSI DIS 9241-3.2-90. Visual Display Terminals (VDTS) Used for Office Task—Ergonomic Requirements—Part 3: Visual Display Requirements. 42 pp.

INTERNATIONAL AND NON-U.S. NATIONAL STANDARDS SUBJECT INDEX

Ergonomics (Cont.)
—Video Display Terminals—Image Quality
BSI BS EN 29241-3-93. 1993 Ergonomic Requirements for Office Work with Visual Display Terminals (VDTs) Part 3: Visual Display Requirements (ISO 9241-3: 1992) (S). 35 pp.
CEN EN 29241-3-93. Ergonomic Requirements for Office Work with Visual Display Terminals (VDTs)—Part 3: Visual Display Requirements (ISO 9241-3:1992). 30 pp.
ISO 9241 Pt 3-92. Ergonomic Requirements for Office Work with Visual Display Terminals (VDTs)—Part 3: Visual Display Requirements First Edition; (CEN EN 29241-3:1993). 36 pp.
JTC1 9241 Pt 3-92. Ergonomic Requirements for Office Work with Visual Display Terminals (VDTs)—Part 3: Visual Display Requirements First Edition. 36 pp.

—Video Display Terminals—Keyboards
BSI BS 7179: Part 4-90. 1990 Ergonomics of Design and Use of Visual Display Terminals (VDTs) in Offices Part 4: Keyboards. 16 pp.

—Video Display Terminals—Lighting
DIN ENGL 5035 Pt 7-88. Artificial Lighting of Interiors; Lighting of Rooms with VDU Work Stations or VDU Assisted Workplaces (Sept). 8 pp.

—Work Systems
BSI DD 202-91. 1991 Ergonomic Principles in the Design of Work Systems (ISO 6385: 1981). 12 pp.
ISO 6385-81. Ergonomic Principles in the Design of Work Systems First Edition. 6 pp.
MOD UK DSTAN 00-25: Part 4-91. Human Factors for Designers of Equipment Part 4: Workplace Design Issue 1. 104 pp.

—Work Systems—Lighting
ISO 8995-89. Principles of Visual Ergonomics—the Lighting of Indoor Work Systems First Edition. 31 pp.

Erichsen Cup Testing
Use: Ball Cup Testing

Erichsen Cupping Testers
Use: Cupping Testers—Erichsen

Eriochrome Black T (TM)
Use: CI Mordant Black 11

ERMES
Use: Radio Paging Systems

Erosion
See Also: Cavitation Corrosion; Water Erosion

—Electrical Insulation
BSI BS 5604-86. 1986 Evaluating Resistance to Tracking and Erosion of Electrical Insulating Materials Used Under Severe Ambient Conditions. 12 pp.
CENELEC HD 380 S2-87. Test Methods for Evaluating Resistance to Tracking and Erosion of Electrical Insulating Materials Used Under Severe Ambient Conditions. 3 pp.
IEC 587-84. Test Methods for Evaluating Resistance to Tracking and Erosion of Electrical Insulating Materials Used Under Severe Ambient Conditions Second Edition. 23 pp.

Error Analysis
See Also: Character Error Rate; Experimental Design; Reliability; Tolerances

—Discharge Measurement—Open Channel Flow
BSI BS 3680: Part 3F-86. 1986 Measurement of Liquid Flow in Open Channels Part 3F: Stream Flow Measurement: Collection and Processing of Data for Determination of Errors in Measurement. 24 pp.
CGSB CAN/CGSB-157.6-M91. Liquid Flow Measurement in Open Channels—Velocity-Area Methods—Collection and Processing of Data for Determination of Errors in Measurement (ISO 1088:1985). 27 pp.
ISO 1088-85. Liquid Flow Measurement in Open Channels—Velocity-Area Methods—Collection and Processing of Data for Determination of Errors in Measurement Second Edition. 24 pp.
ISO 5168-78. Measurement of Fluid Flow—Estimation of Uncertainty of a Flow-Rate Measurement First Edition. 29 pp.
ISO TR7178-83. Liquid Flow Measurement in Open Channels—Velocity-Area Methods—Investigation of Total Error First Edition. 27 pp.

—Flow Measurement
BSI BS 5844-80. 1980 Measurement of Fluid Flow: Estimation of Uncertainty of a Flow-Rate Measurement. 28 pp.

Error Analysis (Cont.)
—Instrument Transformers
CNS C1133-85. Tolerance of Allowable Error of Instrument Transformer (Dec)(11436).
CNS C1134-85. Calculation Method for the Error of the Instrument Transformers for Metering Service (Dec)(11439).

—Length Measurement—Glossaries
DIN ENGL 2257 Pt 2-74. Definitions of Length Verification Practice; Measurement Errors and Uncertainties (Aug). 7 pp.

Error Control
See Also: Automatic Request-Repeat Equipment; Automatic Request-Repeat Mode; Character Error Rate; Cyclic Redundancy Check; Error Correction; Error Performance; Error Recovery (Telecommunications); Failure (Quality Control); Frequency Error; Pulse Communications

—Data Broadcasting
CCIR Report 1210-90. Error-Protection Strategies for Data Broadcasting Services—Section 11B—Ancillary Television Services. 28 pp.

—Data Circuit Terminating Equipment
CCITT RECMN V.41-89. Code-Independent Error-Control System—Data Communication over the Telephone Network (Study Group XVII) 10 pp. 10 pp.

—Data Terminal Equipment
CCITT RECMN V.41-89. Code-Independent Error-Control System—Data Communication over the Telephone Network (Study Group XVII) 10 pp. 10 pp.

—Digital Satellite Dedicated Networks
CCIR Report 1134-90. Digital Satellite Dedicated Networks—Section 4B1—Systems Aspects. 20 pp.

—Digital Transmission—Radio Circuits
CCIR Report 435-70. Error Statistics and Error Control in Digital Transmission Over Operating Radio Circuits—Section 3Cd—Performance of Digital Transmission Systems. 4 pp.

—Directory Services
CCITT RECMN F.500-92. International Public Directory Services (Study Group I) 39 pp. 39 pp.
CCITT RECMN F.500-89. International Public Directory Services—Message Handling Services—Operations and Definition of Service (Study Group I) 32 pp. 32 pp.

—Electromechanical Components
CCITT RECMN V.40-89. Error Indication with Electromechanical Equipment—Data Communication over the Telephone Network (Study Group XVII) 1 pp. 1 p.

—Message Transfer Systems
CCITT RECMN F.410-89. Message Handling Services: Message Transfer Service—Message Handling and Directory Services—Operations and Definition of Service (Study Group I) 10 pp. 10 pp.

—Public Data Networks—Data Link Layers
CCITT RECMN X.141-89. General Principles for the Detection and Correction of Errors in Public Data Networks—Data Communication Networks—Transmission, Signalling and Switching, Network Aspects, Maintenance and Administrative Arrangements (Study Group VII) 8 pp. 8 pp.

—Signaling Links—CCITT No. 6 Signaling Systems
CCITT RECMN Q.277-89. 6.7 Error Control—Specifications of Signalling System No. 6 (Study Group XI) 3 pp. 3 pp.

—Telex Communications—Store and Forward Units
CCITT RECMN F.72-89. International Telex Store and Forward—General Principles and Operational Aspects—Telegraph and Mobile Services Operations and Quality of Service (Study Group I) 18 pp. 18 pp.

Error Correction
See Also: Automatic Error Correction; Error Control; Error Performance; Failure (Quality Control); Forward Error Correction; Transmission Systems

—Data Circuit Terminating Equipment
CCITT RECMN V.42 BIS-90. Data Compression Procedures for Data Circuit Terminating Equipment (DCE) Using Error Correction Procedures (Study Group XVII) 30 pp. 30 pp.

Error Correction (Cont.)
—Data Circuit Terminating Equipment—Asynchronous to Synchronous
CCITT RECMN V.42-89. Error-Correcting Procedures for DCEs Using Asynchronous-to-Synchronous Conversion—Data Communication over the Telephone Network (Study Group XVII) 75 pp. 75 pp.

—Public Data Networks—Data Link Layers
CCITT RECMN X.141-89. General Principles for the Detection and Correction of Errors in Public Data Networks—Data Communication Networks—Transmission, Signalling and Switching, Network Aspects, Maintenance and Administrative Arrangements (Study Group VII) 8 pp. 8 pp.

—Radiotelegraphy—Maritime Mobile Services
CCIR RECMN 476-4-86. Direct-Printing Telegraph Equipment in the Maritime Mobile Service—Section 8B—Maritime Mobile Service; Telegraphy and Related Subjects. 10 pp.

—Telegraphy—Radio Circuits
CCIR RECMN 342-2-70. Automatic Error-Correcting System for Telegraph Signals Transmitted Over Radio Circuits—Section 3Ca—Radiotelegraph Circuits. 9 pp.

—Voice Band Data Transmission
CCITT RECMN G.113-89. Transmission Impairments—General Characteristics of International Telephone Connections and Circuits (Study Groups XII and XV) 22 pp. 22 pp.

Error Performance
See Also: Error Control; Error Correction; Performance Measurement; Performance Parameters (Telecommunications)

—Digital Transmission—Fixed Satellite Services
CCIR Report 1139-90. General System and Performance Aspects of Digital Transmission in the Fixed-Satellite Service—Section 4B1—Systems Aspects. 19 pp.

—Hypothetical Reference Digital Path—Fixed Satellite Services—ISDN
CCIR RECMN 614-1-90. Allowable Error Performance for a Hypothetical Reference Digital Path in the Fixed-Satellite Service Operating Below 15 GHz When Forming Part of an International Connection in an Integrated Services Digital Network—Section 4B2—Perform-ance and Availability. 2 pp.
CCIR RECMN 614-2-92. Allowable Error Performance for a Hypothetical Reference Digital Path in the Fixed-Satellite Service Operating Below 15 GHz When Forming Part of an International Connection in an Integrated Services Digital Network—Section 4B2—Perform-ance and Availability. 42 pp.

Error Protection
Use: Error Control

Error Recovery (Telecommunications)
Use For: Failure Recovery (Telecommunications)
See Also: Error Control; Failure (Quality Control)

—Message Transfer Systems
CCITT RECMN F.202-89. Interworking Between the Telex Service and the Teletex Service—General Procedures and Operational Requirements for the International Interconnection of Telex/Teletex Conversion Facilities—Telematic, Data Transmission and Teleconference Services—Operations and. 4 pp.

Erthrocytometers
Use For: Red Cell Sedimentation Measuring Instruments; Wintrobe Tubes
BSI BS 4316-68. 1968 Amd 1 Apparatus for Measurement of Packed Red Cell Volume. 14 pp.

Erucic Acid
See Also: Erucic Acid Content Analysis

—Rapeseeds
BSI BS 7073: Part 2-89. 1989 Oilseeds Part 2: Rape (Colza) Seeds with High Erucic Acid Content. 4 pp.
BSI BS 7073: Part 3-89. 1989 Oilseeds Part 3: Rape (Colza) Seeds with Low Erucic Acid Content-Specification. 4 pp.
ISO 7556-87. Rape (Colza) Seeds with High Erucic Acid Content—Specification First Edition. 4 pp.

Erucic Acid Content Analysis
See Also: Erucic Acid

Erucic Acid Content Analysis (Cont.)

—Fats
BSI BS 684: Sec 2.41-87. 1987 Methods of Analysis of Fats and Fatty Oils Part 2: Other Methods Section 2.41: Determination of Erucic Acid. 6 pp.
EC 76/621/EEC-76. Council Directive Relating to the Fixing of the Maximum Level of Erucic Acid in Oils and Fats Intended as Such for Human Consumption and in Foodstuffs Containing Added Oils or Fats. 3 pp.
EC 80/891/EEC-80. Commission Directive Relating to the Community Method of Analysis for Determining the Erucic Acid Content in Oils and Fats Intended to be Used as Such for Human Consumption and Foodstuffs Containing Added Oils or Fats. 7 pp.
ISO 8209-86. Animal and Vegetable Fats and Oils—Determination of Erucic Acid First Edition. 6 pp.

—Oils
BSI BS 684: Sec 2.41-87. 1987 Methods of Analysis of Fats and Fatty Oils Part 2: Other Methods Section 2.41: Determination of Erucic Acid. 6 pp.
CNS N6098-77. Method of Test for Edible Vegetable Oils (Determination of Euric Acid) (Apr)(4091). 1 p.
EC 76/621/EEC-76. Council Directive Relating to the Fixing of the Maximum Level of Erucic Acid in Oils and Fats Intended as Such for Human Consumption and in Foodstuffs Containing Added Oils or Fats. 3 pp.
EC 80/891/EEC-80. Commission Directive Relating to the Community Method of Analysis for Determining the Erucic Acid Content in Oils and Fats Intended to be Used as Such for Human Consumption and Foodstuffs Containing Added Oils or Fats. 7 pp.
ISO 8209-86. Animal and Vegetable Fats and Oils—Determination of Erucic Acid First Edition. 6 pp.

Erythrocyte Sedimentation Rate Measuring Instruments
Use: Erythrocytometers

Erythrocytometers
Use For: Erythrocyte Sedimentation Rate Measuring Instruments; Red Cell Sedimentation Measuring Instruments; Westergren Inbes; Westergren Tubes
See Also: Medical Electrical Equipment; Medical Equipment
BSI BS 2554-87. 1987 Westergren Tubes and Support for the Measure-ment of Erythrocyte Sedimentation Rate. 14 pp.

—Supports
BSI BS 2554-87. 1987 Westergren Tubes and Support for the Measure-ment of Erythrocyte Sedimentation Rate. 14 pp.

Erythromycin Content Analysis

—Animal Feed
CNS N4103-82. Method of Test for Erythromycin in Feed (Oct)(9533).

ESC
Use: Environmental Stress Cracking

Escalators
See Also: Elevators (Lifts); Hoists; Passenger Conveyors; Stairways
BSI BS 2655: Part 8-71. 1971 Amd 2 Lifts, Escalators, Passenger Conveyors and Paternosters Part 8: Modernization or Reconstruction of Lifts, Escalators and Paternosters. 29 pp.
CNS B1358-90. Structure for Escalators (Jan)(12651).
CNS B7042-88. Testing Standard for Elevator, Escalator and Dumbwaiter (Jan)(2866).
SAA AS 1735. Lifts, Escalators, and Moving Walks (Known as the SAA Lift Code) Complete Set in Binder.
SAA AS 1735.1-86. Lifts,Escalators, and Moving Walks (Known as the SAA Lift Code)—Part 1: General Requirements Amdt 1 July 1989. 15 pp.
SAA AS 1735.5-86. Lifts, Escalators, and Moving Walks (Known as the SAA Lift Code)—Part 5: Escalators Amdt 1 July 1989. 17 pp.
SNZ NZS/AS 1735. 8-86. Lifts, Escalators and Moving Walks (SAA Lift Code) Part 8: Inclined Lifts Amend: 1 (This is a Joint Standard with SAA AS 1735.8). 20 pp.

—Alarm Systems
CEN PREN 627-92. Specification for Data Logging, Alarm Reporting and Remote Monitoring of Lifts, Escalators and Passenger Conveyors. 10 pp.

—Construction—Safety
CEN EN 115-83. Safety Rules for the Construction and Installation of Escalators and Passenger Conveyors. 30 pp.
CEN PREN 115-92. Safety Rules for the Construction and Installation of Escalators and Passenger Conveyors. 72 pp.

Escalators (Cont.)

—Glossaries
BSI BS 6100: SUBSEC 3.2.2-84. 1984 Glossary of Building and Civil Engineering Terms Part 3: Services Section 3.2: Internal Communication and Transport Subsection 3.2.2: Internal Transport. 7 pp.
BSI BS 6100: SUBSEC 3.2.2-01. 1984 Amd 1 Building and Civil Engineering Terms Part 3: Services Section 3.2: Internal Communication and Transport Subsection 3.2.2: Internal Transport (AMD 7258) August 15, 1992. 8 pp.

—Handicapped Accessible Equipment
SAA AS 1735.12-86. Lifts, Escalators, and Moving Walks (Known as the SAA Lift Code)—Part 12: Facilities for Persons with Disabilities Amdt 1 December 1991. 29 pp.

—Inspection
JIS A 4302-92. Inspection Standard of Elevator, Escalator and Dumbwaiter. 23 pp.

—Installation—Safety
CEN EN 115-83. Safety Rules for the Construction and Installation of Escalators and Passenger Conveyors. 30 pp.
CEN PREN 115-92. Safety Rules for the Construction and Installation of Escalators and Passenger Conveyors. 72 pp.

—Monitors
BSI DD 176-88. 1988 Draft for Development Data Logging and Remote Monitoring Equipment for Lifts, Escalators and Passenger Conveyors. 7 pp.
CEN PREN 627-92. Specification for Data Logging, Alarm Reporting and Remote Monitoring of Lifts, Escalators and Passenger Conveyors. 10 pp.

—Safety
BSI BS 5656-83. 1983 Safety Rules for the Construction and Installation of Escalators and Passenger Conveyors. 32 pp.
CSA CAN/CSA-B44.1-M91. Elevator and Escalator Electrical Equipment (ASME A17.5-1991). 97 pp.

Escape Breathing Apparatus
Use: Breathing Apparatus

Escape Characters
ISO 2375-85. Data Processing—Procedure for Registration of Escape Sequences Third Edition. 10 pp.
JTC1 2375-85. Data Processing—Procedure for Registration of Escape Sequences Third Edition. 10 pp.
OSI ISO 2375-85. Data Processing—Procedure for Registration of Escape Sequences. 10 pp.

—Numbering Plans—Integrated Services Digital Networks
CCITT RECMN E.166-89. Numbering Plan Interworking in the ISDN Era—Telephone Network and ISDN—Operation, Numbering, Routing and Mobile Service (Study Group II) 14 pp (Same as Recmn X.122). 14 pp.

Escape Chutes
Use: Evacuation Equipment

Escape Codes
Use: Escape Characters

Escape Slides
Use: Evacuation Equipment

Escape Valves
Use: Safety Valves

Escherichia Coli
See Also: Bacteria; Coliform Bacteria; Enterobacteriaceae
CNS N6211-89. Method of Test for Microbiology—Test of Enteropathogenic Escherichia Coli (Jun)(12539). 12 pp.

—Dairy Products
BSI BS 4285: Sec 3.8-88. 1988 Microbiological Examination for Dairy Purposes Part 3: Methods for Detection and/or Enumeration of Specific Groups of Microorganisms Section 3.8: Enumeration of Presumptive Esherichia Coli. 6 pp.
SAA AS 1095.3.10-84. Microbiological Methods for the Dairy Industry—Part 3: Methods of Examination for Specific Groups of Microorganisms—Part 3.10: Escherichia Coli—Direct Plate Method. 3 pp.

—Food
BSI BS 5763: Part 8-85. 1985 Microbiological Examination of Food and Animal Feeding Stuffs Part 8: Enumeration of Presumptive Escherichia Coli. 12 pp.

Escherichia Coli (Cont.)

—Food (Cont.)
BSI BS 5763: Part 13-89. 1989 Microbiological Examination of Food and Animal Feeding Stuffs Part 13: Enumeration of Escherichia Coli: Colony Count Technique Using Membranes. 8 pp.
CNS N6192-88. Methods of Test for Food Microbiology-Test of Escherichia Coli (Nov)(10951). 7 pp.
ISO 7251-84. Microbiology—General Guidance for Enumeration of Presumptive Escherichia Coli—Most Probable Number Technique First Edition. 11 pp.
SAA AS 1766.2.3-92. Food Microbiology—Part 2: Examination for Specific Organisms—Part 2.3: Coliforms and Escherichia Coli (Supersedes AS 1095.3.1—1987). 7 pp.
SAA AS 1766.2.12-84. Method for the Microbiological Examination of Food—Part 2: Examination for Specific Organisms—Part 2.12: Escherichia Coli—Direct Plate Method. 3 pp.

—Meat
ISO 3811-79. Meat and Meat Products—Detection and Enumeration of Presumptive Coliform Bacteria and Presumptive Escherichia Coli (Reference Method) First Edition. 7 pp.
ISO 6391-88. Meat and Meat Products—Enumeration of Escherichia Coli—Colony Count Technique at 44 Degrees C Using Membranes First Edition. 9 pp.

—Water
DIN ENGL 38411 Pt 6-91. German Standard Methods for the Examination of Water, Waste Water and Sludge; Microbiological Methods (Group K); Determination of Escherichia Coli and Coliform Organisms (K 6) (June). 6 pp.
ISO 9308 Pt 1-90. Water Quality—Detection and Enumeration of Coliform Organisms, Thermotolerant Coliform Organisms and Presumptive Escherichia Coli—Part 1: Membrane Filtration Method First Edition. 14 pp.
ISO 9308 Pt 2-90. Water Quality—Detection and Enumeration of Coliform Organisms, Thermotolerant Coliform Organisms and Presumptive Escherichia Coli—Part 2: Multiple Tube (Most Probable Number) Method First Edition. 13 pp.
SAA AS 1095.4.1. 4-81. Microbiological Methods for the Dairy Industry—Part 4: Methods for the Examination of Water and Air—Part 4.1.4: Microbiological Examination of Water—Escherichia Coli by Multiple Tube Dilution. 3 pp.
SAA AS 1095.4.1. 6-81. Microbiological Methods for the Dairy Industry—Part 4: Methods for the Examination of Water and Air—Part 4.1.6: Microbiological Examination of Water—Escherichia Coli by Membrane Filtration. 4 pp.

Eschka Method
CNS K7210-66. Chemical Reagent (Eschka's Mixture) (Jun)(1710).

—Coal
BSI BS 1016: Part 8-77. 1977 Amd 1 Methods for Analysis and Testing of Coal and Coke Part 8: Chlorine in Coal and Coke (AMD 3472) December 31, 1980. 9 pp.
BSI BS 1016: Part 8-02. 1977 Amd 2 Methods for Analysis and Testing of Coal and Coke Part 8: Chlorine in Coal and Coke (AMD 6984) May 1, 1992. 12 pp.
BSI BS 1016: SUB SEC 106.4.1-93. 1993 Methods for Analysis and Testing of Coal and Coke Part 106: Ultimate Analysis of Coal and Coke Section 106.4: Determination of Total Sulfur Content Subsection 106.4.1: Eschka Method (W). 11 pp.
CNS M3145-84. Method for Determination of Total Sulfur of Coal and Coke (Eschka Method) (Mar)(10826).
ISO 334-92. Solid Mineral Fuels—Determination of Total Sulfur—Eschka Method Second Edition. 9 pp.
SAA AS 1038.6.3. 1-86. Methods for the Analysis and Testing of Coal and Coke—Part 6.3.1: Ultimate Analysis of Higher Rank Coal—Determination of Total Sulphur (Eschka Method) (R 1992). 2 pp.
SAA AS 1038.8.1-92. Coal and Coke—Analysis and Testing—Part 8.1: Coal and Coke—Chlorine —Eschka Method (Supersedes AS 1038.8—1980 (in Part)) Which Will Remain Current) (in Professional Package 32). 2 pp.

—Coke
BSI BS 1016: Part 8-77. 1977 Amd 1 Methods for Analysis and Testing of Coal and Coke Part 8: Chlorine in Coal and Coke (AMD 3472) December 31, 1980. 9 pp.
BSI BS 1016: Part 8-02. 1977 Amd 2 Methods for Analysis and Testing of Coal and Coke Part 8: Chlorine in Coal and Coke (AMD 6984) May 1, 1992. 12 pp.
BSI BS 1016: SUB SEC 106.4.1-93. 1993 Methods for Analysis and Testing of Coal and Coke Part 106: Ultimate Analysis of Coal and Coke Section 106.4: Determination of Total Sulfur Content Subsection 106.4.1: Eschka Method (W). 11 pp.

INTERNATIONAL AND NON-U.S. NATIONAL STANDARDS
SUBJECT INDEX

Ester

Eschka Method (Cont.)
—Coke (Cont.)
CNS M3145-84. Method for Determination of Total Sulfur of Coal and Coke (Eschka Method) (Mar)(10826).
ISO 334-92. Solid Mineral Fuels—Determination of Total Sulfur—Eschka Method Second Edition. 9 pp.
SAA AS 1038.8.1-92. Coal and Coke—Analysis and Testing—Part 8.1: Coal and Coke—Chlorine—Eschka Method (Supersedes AS 1038.8—1980 (in Part)) Which Will Remain Current) (in Professional Package 32). 2 pp.

—Solid Fuels
ISO 587-81. Solid Mineral Fuels—Determination of Chlorine Using Eschka Mixture Second Edition. 5 pp.

Essential Electrical Systems
See Also: Electrical Equipment; Emergency and Standby Power Supplies; Medical Electrical Equipment; Medical Equipment; Power Systems
DIN VDE 0100 Pt 560-84. Installation of Power Plant with Rated Voltages of up to 1000 V; Selection and Installation Equipment; Supplies for Safety Services (Nov). 11 pp.

—Medical Equipment
CSA CAN/CSA-Z32.4-M86. Essential Electrical Systems for Hospitals; (Gen Instr 1). 38 pp.

Essential Oil Content Analysis
See Also: Essential Oils; Oil Content Analysis

—Beverages
CNS N6175-82. Method of Test for Beverage: Determination of Essential Oil (Sep)(9433). 1 p.

—Citrus Fruits
ISO 1955-82. Citrus Fruits and Derived Products—Determination of Essential Oils Content (Reference Method) First Edition. 5 pp.

—Perfumes
CNS S2062-81. Method of Test for Essential Oil Content in Perfume (for Cosmetic) (Nov)(8161).

Essential Oils
Scope Note: For additional listings, use a more specific term Use For: Synthetic Essential Oils
See Also: Anise Oil; Artemisia Oil; Bay Oil; Bergamot Oil; Bitter Orange Oil; Calamus Oil; Caraway Oil; Cardamom Oil; Cassia Oil; Cedarwood Oil; Celery Seed Oil; Cinnamon Leaf Oil; Citronella Oil; Citrus Oil; Coriander Oil; Cubeb Oil; Essential Oil Content Analysis; Eucalyptus Oil; Flavorings; Geranium Oil; Grapefruit Oil; Hyssop Oil; Ilang-Ilang Oil; Juniper Berry Oil; Lavender Oil; Lemon Oil; Lemongrass Oil; Lime Oil; Mace Oil; Mandarin Oil; Neroli Oil; Nutmeg Oil; Orange Oil; Palmarosa Oil; Parsley Oil; Patchouli Oil; Pennyroyal Oil; Pepper Oil; Peppermint Oil; Petitgrain Oil; Pimenta Oil; Rose Oil; Rosemary Oil; Rosewood Oil; Sage Oil; Sandalwood Oil; Sassafras Oil; Spearmint Oil; Spices; Spike Oil; Thyme Oil; Turpentine; Vetiver Oil

—Alcohol Content
CNS K6611-80. Estimation of Free Alcohols Content by Determination of Ester Value After Acetylation of Essential Oil (Dec)(6677).
ISO 1241-80. Essential Oils—Determination of Ester Value After Acetylation and Evaluation of Free Alcohols and Total Alcohols Content First Edition. 5 pp.
ISO 3794-76. Essential Oils (Containing Tertiary Alcohols)—Estimation of Free Alcohols Content by Determination of Ester Value After Acetylation First Edition. 4 pp.
ISO 4096-78. Essential Oils (Containing Tertiary Alcohols)—Evaluation of Free Alcohols Content by Determination of Ester Value After Cold Formylation First Edition. 4 pp.

—Alcohol Content—Acetylation
CNS K6614-80. Estimation of Free Alcohols Content by Determination of Ester Value After Acetylation of Essential Oils (Containing Tertiary Alcohols) (Dec)(6680).
CNS K6615-80. Estimation of Primary and Secondary Free Alcohols Content of Essential Oil by Acetylation in Pyridine (Dec)(6681).
ISO 3793-76. Essential Oils—Estimation of Primary and Secondary Free Alcohols Content by Acetylation in Pyridine First Edition. 4 pp.

—Carbonyl Compound Content
CNS K6612-80. Determination of Carbonyl Compounds Content of Essential Oils by Free Hydroxylamine Method (Dec)(6678).
CNS K6613-80. Determination of Carbonyl Compounds Content of Essential Oils by Hydroxylammonium Chloride Method (Dec)(6679).

Essential Oils (Cont.)
—Carbonyl Compound Content (Cont.)
ISO 1279-84. Essential Oils—Determination of Carbonyl Value—Hydroxylammonium Chloride Method Second Edition. 5 pp.

—Carbonyl Value
ISO 1271-83. Essential Oils—Determination of Carbonyl Value—Free Hydroxylamine Method Second Edition; (Corrected and Reprinted -1985). 5 pp.

—Density
CNS K6598-80. Determination of the Density and Relative Density of Essential Oils (Oct)(6571).

—Density—Pycnometric Analysis
ISO 279-81. Essential Oils—Determination of Relative Density at 20 Degree C (Reference Method) First Edition. 4 pp.

—Ester Value
CNS K6602-80. Determination of Ester Value of Essential Oils (Oct)(6575).
ISO 709-80. Essential Oils—Determination of Ester Value First Edition. 4 pp.
ISO 4096-78. Essential Oils (Containing Tertiary Alcohols)—Evaluation of Free Alcohols Content by Determination of Ester Value After Cold Formylation First Edition. 4 pp.
ISO 7660-83. Essential Oils—Determination of Ester Value of Oils Containing Difficult-to-Saponify Esters First Edition. 4 pp.

—Ester Value—Acetylation
CNS K6611-80. Estimation of Free Alcohols Content by Determination of Ester Value After Acetylation of Essential Oil (Dec)(6677).
CNS K6614-80. Estimation of Free Alcohols Content by Determination of Ester Value After Acetylation of Essential Oils (Containing Tertiary Alcohols) (Dec)(6680).
ISO 1241-80. Essential Oils—Determination of Ester Value After Acetylation and Evaluation of Free Alcohols and Total Alcohols Content First Edition. 5 pp.
ISO 3794-76. Essential Oils (Containing Tertiary Alcohols)—Estimation of Free Alcohols Content by Determination of Ester Value After Acetylation First Edition. 4 pp.

—Eucalyptol Content
CNS K6605-80. Determination of Cineole Content of Essential Oils (Oct)(6578).
ISO 1202-81. Essential Oils—Determination of 1,8-Cineole Content First Edition. 5 pp.

—Freezing Points
CNS K6604-80. Determination of Freezing Point of Essential Oils (Oct)(6577).
ISO 1041-73. Essential Oils—Determination of Freezing Point First Edition. 4 pp.

—Gas Chromatography
ISO 7359-85. Essential Oils—Analysis by Gas Chromatography on Packed Columns—General Method First Edition. 9 pp.
ISO 7609-85. Essential Oils—Analysis by Gas Chromatography on Capillary Columns—General Method First Edition. 9 pp.

—Glossaries
ISO 3218-76. Essential Oils—Principles of Nomenclature First Edition. 4 pp.
ISO 4720-92. Essential Oils—Nomenclature First Edition. 13 pp.

—Identification Systems
CNS Z5066-80. Labelling and Marking Containers for Essential Oils (Oct)(6640).
ISO R211-61. Essential Oils-Labelling and Marking Containers First Edition. 3 pp.

—Liquid Chromatography
ISO 8432-87. Essential Oils—Analysis by High Performance Liquid Chromatography—General Method First Edition. 7 pp.

—Miscibility
ISO 875-81. Essential Oils—Evaluation of Miscibility in Ethanol First Edition. 5 pp.

—Optical Activity
CNS K6601-80. Determination of the Optical Rotation of Essential Oils (Oct)(6574).
ISO 592-81. Essential Oils—Determination of Optical Rotation First Edition. 4 pp.

—Packaging
CNS Z5065-80. Packing for Essential Oils (Oct)(6639).
ISO R210-61. Essential Oils-Packing First Edition. 3 pp.

Essential Oils (Cont.)
—Phenol Content
CNS K6607-80. Determination of Phenols Content of Essential Oils (Oct)(6580).
ISO 1272-73. Essential Oils—Determination of Phenols Content First Edition. 4 pp.

—Refractive Index
CNS K6599-80. Determination of Refractive Index of Essential Oils (Oct)(6572).
ISO 280-76. Essential Oils—Determination of Refractive Index First Edition. 4 pp.

—Residue Content—Distillation Methods
CNS K6703-81. Essential Oils Determination of Residue from Distillation Under Reduced Pressure (Nov)(8137).
ISO 5991-79. Essential Oils—Determination of Residue from Distillation Under Reduced Pressure First Edition. 5 pp.

—Residue Content—Evaporation
CNS K6702-81. Essential Oils Quantitative Evaluation of Residue on Evaporation (Nov)(8136).
ISO 4715-78. Essential Oils—Quantitative Evaluation of Residue on Evaporation First Edition. 4 pp.

—Sampling
CNS K6597-80. Essential Oils Sampling (Oct)(6570).
CNS K6600-80. Preparation of Test Sample of Essential Oils (Oct)(6573).
ISO 212-73. Essential Oils—Sampling First Edition. 5 pp.
ISO 356-77. Essential Oils—Preparation of Test Sample First Edition. 3 pp.

—Saponification Number
CNS K6606-80. Determination of the Acid Value of Essential Oils (Oct)(6579).
ISO 1242-73. Essential Oils—Determination of the Acid Value First Edition. 4 pp.

—Solubility
CNS K6603-80. Determination of Solubility of Essential Oils in Ethanol (Oct)(6576).

ESSS
Use: Electrosensitive Safety Systems

Ester Content Analysis
See Also: Esters

—n-Butyl Acetate
BSI BS 551-90. 1990 Butyl Acetate for Industrial Use. 8 pp.

—Adipates—Volumetric Analysis
ISO 2527-74. Adipate Esters for Industrial Use—Determination of Ester Content—Volumetric Method First Edition. 5 pp.

—Amyl Acetate
BSI BS 552-70. 1970 Amyl Acetate. 12 pp.
BSI BS 552-01. 1970 Amd 1 Amyl Acetate (AMD 7218) September 15, 1992. 14 pp.

—Ethanols—Volumetric Analysis
BSI BS 6392: Part 8-83. 1983 Amd 1 Testing of Ethanol for Industrial Use Part 8: Method for Determination of Esters Content. 5 pp.
ISO 1388 Pt 9-81. Ethanol for Industrial Use—Methods of Test—Part 9: Determination of Esters Content—Titrimetric Method After Saponification First Edition. 5 pp.

—Ethyl Acetate
BSI BS 553-90. 1990 Ethyl Acetate for Industrial Use. 8 pp.

—Isopropyl Acetate
BSI BS 1834-68. 1968 Isopropyl Acetate. 11 pp.
BSI BS 1834-01. 1968 Amd 1 Isopropyl Acetate (AMD 7219) September 15, 1992. 13 pp.

—Peppermint Oil
CNS K6102-72. Method of Test for Crude and De-Mentholized Peppermint Oil (Jun)(1094). 2 pp.

—Phthalate Esters—Volumetric Analysis
ISO 1385 Pt V-77. Phthalate Esters for Industrial Use—Methods of Test—Part V: Determination of Ester Content—Titrimetric Method After Saponification First Edition. 4 pp.

Ester Gum
See Also: Rosins
CNS K2169-87. Ester Gum (Nov)(12162). 1 p.
CNS K6931-87. Method of Test for Ester Gum (Nov)(12163).
JIS K 5903-78. Ester Gum.

—Copal
MOD UK CS 1669. Copal Ester Gum, LF Quality (Withdrawn).

INDUSTRY STANDARDS

Ester Value

Use For: Rosin Esters

—Beeswax
MOD UK M 2012/71. Examination of Beeswax GS and Beeswax, Lead Free.

—Chemical Products
JIS K 0070-92. Test Methods for Acid Value, Saponification Value, Ester Value, Iodine Value, Hydroxyl Value and Unsaponifiable Matter of Chemical Products. 20 pp.

—Essential Oils
CNS K6602-80. Determination of Ester Value of Essential Oils (Oct)(6575).
CNS K6616-80. Determination of Ester Value After Hot Formulation of Essential Oils of Geranium and Rose (Dec)(6682).
ISO 709-80. Essential Oils—Determination of Ester Value First Edition. 4 pp.
ISO 4096-78. Essential Oils (Containing Tertiary Alcohols)—Evaluation of Free Alcohols Content by Determination of Ester Value After Cold Formylation First Edition. 4 pp.
ISO 7660-83. Essential Oils—Determination of Ester Value of Oils Containing Difficult-to-Saponify Esters First Edition. 4 pp.

—Essential Oils—Acetylation
CNS K6611-80. Estimation of Free Alcohols Content by Determination of Ester Value After Acetylation of Essential Oil (Dec)(6677).
CNS K6614-80. Estimation of Free Alcohols Content by Determination of Ester Value After Acetylation of Essential Oils (Containing Tertiary Alcohols) (Dec)(6680).
ISO 1241-80. Essential Oils—Determination of Ester Value After Acetylation and Evaluation of Free Alcohols and Total Alcohols Content First Edition. 5 pp.
ISO 3794-76. Essential Oils (Containing Tertiary Alcohols)—Estimation of Free Alcohols Content by Determination of Ester Value After Acetylation First Edition. 4 pp.

—Geranium Oil
CNS K6616-80. Determination of Ester Value After Hot Formulation of Essential Oils of Geranium and Rose (Dec)(6682).

Esterification

See Also: Acetylation

—Polyoxyalkylene Resins
ISO 6796-81. Polyglycols for Industrial Use—Determination of Hydroxyl Number—Phthalic Anhydride Esterification Method First Edition. 5 pp.

Esters

Use For: Synthetic Esters *See Also:* Acetates; Adipates; Adipic Esters; Alginates; Aluminum Acetate; Ammonium Acetate; Ammonium Oxalate; Ammonium Tartrate; Antimony Potassium Tartrate; n-Butyl Acetate; Dimethyl Sulfate; Ester Content Analysis; Ethyl Acetate; Ethyl Oleate; Fats; Glycidyl Esters; Lead Salicylate; Lead 2-Ethylhexanoate; Methacrylates; Methyl Esters; 3-Methylbutyl Acetate; Oxalates; Phthalate Esters; Phthalic Esters; Potassium Biphthalate; Sodium Citrate; Sodium Formate; Sodium Oxalate; Sodium Salicylate; Triaryl Phosphate Esters; Uranyl Acetate; Urethanes; Vinyl Acetate; Zinc Acetate

—Boiling Points
ISO 4626-80. Volatile Organic Liquids—Determination of Boiling Range of Organic Solvents Used as Raw Materials First Edition. 14 pp.

—Electrical Insulating Liquids
BSI BS EN 61099-92. 1992 Unsuded Synthetic Organic Esters for Electrical Purposes (IEC 1099: 1992) (F). 17 pp.
CENELEC EN 61099-92. Specification for Unused Synthetic Organic Esters for Electrical Purposes. 5 pp.
CENELEC EN 61099-92. Specification for Unused Synthetic Organic Esters for Electrical Purposes; (IEC 1099: 1992). 12 pp.
IEC 1099-92. Specifications for Unused Synthetic Organic Esters for Electrical Purposes First Edition; (CENELEC EN 61099:1992). 25 pp.

—Electrical Insulating Liquids—Transformers
IEC 1203-92. Synthetic Organic Esters for Electrical Purposes—Guide for Maintenance of Transformer Esters in Equipment First Ediiton. 19 pp.

Estimates

Scope Note: Use a more specific term *See:* Cost Estimates

Estrogen Content Analysis

—Cosmetics—Hygiene
CNS S2086-82. Methods of Hygienic Test for Cosmetics — Estroigenic Hormones (Oct)(9540).

Etch Primers

Use: Wash Primers

Etched Crystals

See Also: Etching

—Germanium—Etch Pit Density
JIS H 0610-66. Method of Measurement of Etch Pit Density of Germanium Crystal (R 1983). 7 pp.

—Silicon—Etch Pit Density
JIS H 0609-71. Methods of Measurement of Etch Pit Density and Lineage for Silicon Crystals (R 1983). 9 pp.

—Silicon—Lineage
JIS H 0609-71. Methods of Measurement of Etch Pit Density and Lineage for Silicon Crystals (R 1983). 9 pp.

Etching

See Also: Corrosion; Electrolytic Etching; Etched Crystals

—Metals
BSI M 37-70. 1970 Method for the Etch Inspection of Metallic Material and Components. 9 pp.
CGSB 31-GP-0A METH 40.1-57. Methods of Testing Corrosion-Prevention Materials and Processes Etching Action of Metal Conditioners. 1 p.

—Silica Coatings—Etch Rate—Optical Method
DIN ENGL 50453 Pt 2-90. Spectrophotometric Determination of Etch Rate of Mixtures for Etching Silicon Dioxide Coatings for Use in Semiconductor Technology (Oct). 2 pp.

—Silicon—Crystal Defects
DIN ENGL 50434-86. Testing of Materials for Semiconductor Technology; Detection of Crystal Defects in Monocrystalline Silicon Using Etching Techniques on (111) and (100) Surfaces (Feb). 9 pp.

—Silicon Wafers—Etch Rate—Gravimetric Analysis
DIN ENGL 50453 Pt 1-90. Gravimetric Determination of Etch Rate of Mixtures for Etching Silicon Single Crystals for Use in Semiconductor Technology (Oct). 3 pp.

Ethanal

Use: Acetaldehyde

Ethanediol

Use: Ethylene Glycol

Ethanol

Use For: Anhydrous Alcohol; Ethanol Absolute; Ethyl Alcohol *See Also:* Alcohols; Ethanol Content Analysis; Ethanolamine; Hazardous Materials
BSI BS 3591-85. 1985 Industrial Methylated Spirits. 7 pp.
CNS K7028-61. Chemical Reagent (Alcohol, 95 Percent) (Ethyl Alcohol) (Dec)(1528).
CNS K7029-61. Chemical Reagent (Anhydrous Alcohol, Anhydrous Ethanol) (Dec)(1529).
JIS K 8101-87. Ethanol (99.5) (Ethyl Alcohol (99.5)).
JIS K 8102-87. Ethanol (95) (Ethyl Alcohol (95)).

—Acidity
BSI BS 6392: Part 1-83. 1983 Amd 1 Testing of Ethanol for Industrial Use Part 1: Method for Detection of Alkalinity or Determination of Acidity to Phenolphthalein. 5 pp.
ISO 1388 Pt 2-81. Ethanol for Industrial Use—Methods of Test—Part 2: Detection of Alkalinity or Determination of Acidity to Phenolphthalein First Edition. 4 pp.

—Aldehyde Content—Colorimetric Analysis
BSI BS 6392: Part 4-83. 1983 Amd 1 Testing of Ethanol for Industrial Use Part 4: Method for Determination of Aldehydes Content. 5 pp.
ISO 1388 Pt 5-81. Ethanol for Industrial Use—Methods of Test—Part 5: Determination of Aldehydes Content—Visual Colorimetric Method First Edition. 5 pp.

—Alkalinity
BSI BS 6392: Part 1-83. 1983 Amd 1 Testing of Ethanol for Industrial Use Part 1: Method for Detection of Alkalinity or Determination of Acidity to Phenolphthalein. 5 pp.
ISO 1388 Pt 2-81. Ethanol for Industrial Use—Methods of Test—Part 2: Detection of Alkalinity or Determination of Acidity to Phenolphthalein First Edition. 4 pp.

Ethanol (Cont.)

—Carbinol Content—Colorimetric Analysis
ISO 1388 Pt 8-81. Ethanol for Industrial Use—Methods of Test—Part 8: Determination of Methanol Content (Methanol Contents Between 0,10 and 1,50 % (V/V))—Visual Colorimetric Method First Edition. 5 pp.

—Carbinol Content—Photometry
ISO 1388 Pt 7-81. Ethanol for Industrial Use—Methods of Test—Part 7: Determination of Methanol Content (Methanol Contents Between 0,01 and 0,20 % (V/V))—Photometric Method First Edition. 6 pp.

—Carbonyl Compound Content—Photometry
BSI BS 6392: Part 2-83. 1983 Amd 1 Testing of Ethanol for Industrial Use Part 2: Method for Determination of Carbonyl Compounds Content Present in Small Amounts (Photometric Method). 5 pp.
ISO 1388 Pt 3-81. Ethanol for Industrial Use—Methods of Test—Part 3: Estimation of Content of Carbonyl Compounds Present in Small Amounts—Photometric Method First Edition. 5 pp.

—Carbonyl Compound Content—Volumetric Analysis
BSI BS 6392: Part 3-83. 1983 Amd 1 Testing of Ethanol for Industrial Use Part 3: Method for Determination of Carbonyl Compounds Content Present in Moderate Amounts (Titrimetric Method). 5 pp.
ISO 1388 Pt 4-81. Ethanol for Industrial Use—Methods of Test—Part 4: Estimation of Content of Carbonyl Compounds Present in Moderate Amounts—Titrimetric Method First Edition. 5 pp.

—Denaturant Content
CNS K6713-82. Method of Test for Volatile Denaturants in Alcohol (Feb)(8525).

—Ester Content—Volumetric Analysis
BSI BS 6392: Part 8-83. 1983 Amd 1 Testing of Ethanol for Industrial Use Part 8: Method for Determination of Esters Content. 5 pp.
ISO 1388 Pt 9-81. Ethanol for Industrial Use—Methods of Test—Part 9: Determination of Esters Content—Titrimetric Method After Saponification First Edition. 5 pp.

—Furfural Content
ISO 1388 Pt 11-81. Ethanol for Industrial Use—Methods of Test—Part 11: Test for Detection of Furfural First Edition. 4 pp.

—Hydrocarbon Content—Distillation Methods
ISO 1388 Pt 10-81. Ethanol for Industrial Use—Methods of Test—Part 10: Estimation of Hydrocarbons Content—Distillation Method First Edition. 7 pp.

—Industrial
BSI BS 507-85. 1985 Ethanol for Industrial Use. 7 pp.
BSI BS 6392: Part 0-83. 1983 Amd 1 Testing of Ethanol for Industrial Use Part 0: General Introduction. 5 pp.
ISO 1388 Pt 1-81. Ethanol for Industrial Use—Methods of Test—Part 1: General First Edition. 4 pp.
SAA AS K121-68. Ethanol Being BS 507:1966 Endorsed Without Amendment. 12 pp.

—Methanol Content—Colorimetric Analysis
BSI BS 6392: Part 7-83. 1983 Amd 1 Testing of Ethanol for Industrial Use Part 7: Method for Determination of Methanol Content (0.10% (V/V) to 1.50% (V/V)) (Visual Colorimetric Method). 5 pp.

—Methanol Content—Photometry
BSI BS 6392: Part 6-83. 1983 Amd 1 Testing of Ethanol for Industrial Use Part 6: Method for Determination of Methanol Content (0.01% (V/V) to 0.20% (V/V) (Photometric Method). 7 pp.

—Miscibility
BSI BS 6392: Part 5-83. 1983 Amd 1 Testing of Ethanol for Industrial Use Part 5: Method of Test for Miscibility with Water. 4 pp.
ISO 1388 Pt 6-81. Ethanol for Industrial Use—Methods of Test—Part 6: Test for Miscibility with Water First Edition. 4 pp.

—Permanganate Time
BSI BS 6392: Part 9-83. 1983 Amd 1 Testing of Ethanol for Industrial Use Part 9: Method for Determination of Permanganate Time. 5 pp.
BSI BS 6392: Part 9-02. 1983 Amd 2 Testing of Ethanol for Industrial Use Part 9: Method for Determination of Permanganate Time (AMD 7764) June 15, 1993 (W). 6 pp.

INTERNATIONAL AND NON-U.S. NATIONAL STANDARDS
SUBJECT INDEX

Ethyl

Ethanol (Cont.)
—Permanganate Time (Cont.)
ISO 1388 Pt 12-81. Ethanol for Industrial Use—Methods of Test—Part 12: Determination of Permanganate Time First Edition. 5 pp.

Ethanol Absolute
Use: Ethanol

Ethanol Content Analysis
See Also: Ethanol

—Beverages—Gas Chromatography
CNS N6181-82. Method of Test for Beverage—Determination of Ethanol Content (May)(10292). 1 p.

—Fruits
ISO 2448-73. Fruit and Vegetable Products—Determination of Ethanol First Edition. 5 pp.

—Hydrometers
BSI BS 5470-77. 1977 Glass Alcohol Hydrometers Not Incorporating a Thermometer. 7 pp.

—Methanols
CNS K6263-80. Method of Test for Methyl Alcohol (Methanol) (Feb)(2790). 7 pp.

—Perfumes
CNS S2063-81. Method of Test for Methanol and Ethanol Content in Perfume (for Cosmetic (Nov)(8162).

—Vegetables
ISO 2448-73. Fruit and Vegetable Products—Determination of Ethanol First Edition. 5 pp.

Ethanolamine
Use For: Aminoethanol; 2 Aminoethanol; Monoethanolamine *See Also:* Ethanol
CNS K1144-71. Ethanolamine (Industrial Grade) (Jan)(3164).
CNS K6298-71. Method of Test for Ethanolamine (Apr)(3234).
CNS K7211-66. Chemical Reagent (Ethanolamine) (Jun)(1711).
JIS K 8109-88. 2-Aminoethauol (Monoethanolamine).

Ethene
Use: Ethylene

Ethernet (BTN)
Use For: Carrier Sense Multiple Access/CD (Ethernet); CSMA/CD (Ethernet); IEEE 802.3
See Also: Local Area Networks
BSI BS 6531: Part 1-84. (WITHDRAWN) 1984 10 Mbps Slotted Ring Local Area Network Part 1: Specification for the Coding of Bits and Structure of Slots and Mini-Packets (Superseded by BS ISO 8802-7: 1991). 9 pp.
BSI BS 7246-90. 1990 Guide for Local Area Networks CSMA/DC 10 Mbits/s Baseband Planning and Installation. 44 pp.
BSI BS 8802/3-89. (WITHDRAWN) 1989 Carrier Sense Multiple Access with Collision Detection (CSMA/CD) Access Method and Physical Layer Specifications (Superseded by BS ISO/IEC 8802-3: 1992). 180 pp.
BSI BS ISO/IEC 8802-3-92. 1992 Information Technology—Local and Metropolitan Area Networks—Part 3: Carrier Sense Multiple Access with Collision Detection (CSMA/CD) Access Method and Physical Layer Specifications. 314 pp.
BSI DD ENV 41801-3-92. 1992 Information Systems Interconnection—Relaying the Connectionless-Mode Network Service—Part 3: ISO 8802-3 CSMA/CD Subnetwork Dependent Media Dependent Requirements. 18 pp.
BSI DD ENV 41802-3-92. 1992 Information Systems Interconnection—X.25 Protocol Relaying Part 3: ISO 8802-3 CSMA/CS Subnetwork Dependent Media Dependent Requirements. 13 pp.
CEN ENV 41 101-86. Information Systems Interconnection: Local Area Networks; Provision of the OSI Connection-Mode Transport Service Using Connectionless-Mode Network Service on a CSMA/CD Single LAN. 16 pp.
CEN ENV 41102-86. Information Systems Interconnection: Local Area Networks; Provision of the OSI Connection-Mode Transport Service and the OSI Connectionless-Mode Network Service on a CSMA/CO LAN in a Single or Multiple LAN Configuration. 18 pp.
CEN ENV 41801-3-92. Information Systems Interconnection—Relaying the Connectionless-Mode Network Service—Part 3: ISO 8802-3 CSMA/CD Subnetwork Dependent Media Dependent Requirements. 26 pp.
CEN ENV 41802-3-92. Information Systems Interconnection—X.25 Protocol Relaying—Part 3: ISO 8802-3 CSMA/CD Subnetwork Dependent Media Dependent Requirements. 10 pp.

Ethernet (BTN) (Cont.)
ECMA ECMA 80-84. Local Area Networks (CSMA/CD Baseband) Coaxial Cable System. 18 pp.
ECMA ECMA 81-84. Local Area Networks (CSMA/CD Baseband) Physical Layer. 17 pp.
ECMA ECMA 82-84. Local Area Networks (CSMA/CD Baseband) Link Layer. 40 pp.
ECMA ECMA-TR 26-90. Planning and Installation Guide for CSMA/CD 10 Mbit/s Baseband LAN Coaxial Cable Systems. 64 pp.
IEC 907-89. Local Area Networks CSMA/CD 10 Mbit/s Baseband Planning and Installation Guide First Edition. 86 pp.
JTC1 907-89. Local Area Networks CSMA/CD 10 Mbit/s Baseband Planning and Installation Guide First Edition. 86 pp.
OSI ISO 8802-3 DAM 4-90. Part 3: Carrier Sense Multiple Access with Collision Detection (CSMA/CD) Access Method and Physical Layer Specifications Amendment 4: Physical Signalling, Medium Attachment and Baseband Medium Specifications, StarLAN, Type 1 Bases. 62 pp.
OSI ISO 8802-3 DAM 3-90. Information Processing Systems—Local Area Networks—Part 3: Carrier Sense Multiple Access with Collision Detection. Amendment 3: Broadband Medium Attachment Unit and Broadband Medium Specification, Type10 BROAD 36. 43 pp.
OSI ISO DIS 8802/3 PDAD-87. Local Area Networks—Part 3: CSMA/CD—Addendum Broadband Medium Attachment Unit and Broadband Medium Specifications, Type IOBROAD36.
OSI ISO IEC 8802-3 DAD 5-89. Information Processing Systems—Local Area Networks Part 3: Carrier Sense Multiple Access with Collision Detection Addendum 5: Medium Attachment Unit and Baseband Medium Specification for a Vendor Independent Fibre Optic Inter Repeater Link (FOIRL-Standard). 24 pp.
OSI ISO 8802-3 DAD 2-87. Information Processing Systems—Local Area Networks Part 3: Carrier Sense Multiple Access with Collision Detection Addendum 2: Repeater Set and Repeater Unit Specification for Use with 10 Base 5 and 10 Base 2 Networks. 19 pp.
OSI ISO 8802-3 DAD 1-87. Information Processing—Local Area Networks PT3: Carrier Sense Multiple Access with Collision Detection Addendum 1: Medium Attachment Unit and Baseband Medium Specifications for Type 10 Base 2. 36 pp.
OSI ISO DIS 8802-3 PDAD-87. Local Area Networks—Part 3: CSMA/CD—Addendum Broadband Medium Attachment Unit and Broadband Medium Specifications, Type IOBROAD36. 230 pp.
OSI ISO/IEC 8802-3-90. Information Processing Systems—Local Area Networks—Part 3: Carrier Sense Multiple Access with Collision Detection (CSMA/CD) Access Method and Physical Layer Specifications (with Correction Sheet 09-90). 217 pp.
OSI ISO DIS 8802-3-87. Information Processing Systems—Local Area Networks Part 3: Carrier Sense Multiple Access with Collision Detection—Access Method and Physical Layer Specifications. 5 pp.
OSI ISO 8802-3-89. Information Processing Systems—Local Area Networks Part 3: Carrier Sense Multiple Access with Collision Detection (CSMA/CD) Access Method and Physical Layer Specifications. 178 pp.
OSI ISO 8802 3 DAD 1-87. Information Processing—Local Area Networks PT3: Carrier Sense Multiple Access with Collision Detection Addendum 1: Medium Attachment Unit and Baseband Medium Specifications for Type 10 Base 2.
OSI ISO 8802 3 DAD 2-87. Information Processing Systems-Local Area Networks Part 3: Carrier Sense Multiple Access with Collision Detection Addendum 2: Repeater Set and Repeater Unit Specification Foruse with 10 Base 5 and 10 Base 2 Networks.
OSI ISO/IEC DISP 10608-2-90. Information Technology—International Standardized Profile TAnnnn—Connection-Mode Transport Service over Connectionless-Mode Network Service—Part 2: TA51 Profile Including Subnetwork-Dependent Requirements for CSMS/CD Local Area Networks (LANs). 37 pp.

Ethers
See Also: Anisole; Epoxy Compounds; Ethyl Ether; gamma-Glycidoxypropyltrimethoxysilane; Isopropyl Ether; Polyethers; Vanillin
CNS K1222-80. Ether (Feb)(5228).
CNS K6457-80. Method of Test for Ether (Feb)(5241).
CNS K7212-66. Chemical Reagent (Ether) (Jun)(1712).
CNS K7213-66. Chemical Reagent (Ether, Absolute) (Jun)(1713).

—Boiling Points
ISO 4626-80. Volatile Organic Liquids—Determination of Boiling Range of Organic Solvents Used as Raw Materials First Edition. 14 pp.

Ethnology
—Classification
BSI BS 1000: (39)-70. 1970 Universal Decimal Classification (UDC). English Full Edition (39): Ethnology. Ethnography. 16 pp.
SNZ NZS/BS 1000 (39)-70. Universal Decimal Classification Ethnology. Ethnography. 16 pp.

Ethopabate Content Analysis
—Animal Feed
CNS N4088-82. Method of Test for Feed Additives: Determination of Ethopabate (Feb)(8534).

2-Ethoxyethanol Acetate
—Aerospace
MOD UK DTD-5589-67. 2-Ethoxyethanol Acetate (Urethane Grade). 2 pp.

2-Ethoxyethanol
Use: Ethylene Glycol Monoethyl Ether

Ethoxylated Alcohol Sulfates
—Unsulfated Matter Content
BSI BS 6829: Sec 5.3-89. 1989 Analysis of Surface Active Agents (Raw Materials) Part 5: Ethoxylated Alcohol and Alkylphenol Sulphates Section 5.3: Method for Determination of Unsulphated Matter Content. 4 pp.
ISO 8799-88. Surface Active Agents—Sulfated Ethoxylated Alcohols and Alkylphenols—Determination of Content of Unsulfated Matter First Edition. 5 pp.

Ethoxylated Alkylphenol Sulfates
—Unsulfated Matter Content
BSI BS 4607: Part 1-84. 1984 Amd 1 Non-Metallic Conduit Fittings for Electrical Installations Part 1: Fittings and Components of Insulating Material. 22 pp.
BSI BS 6829: Sec 5.3-89. 1989 Analysis of Surface Active Agents (Raw Materials) Part 5: Ethoxylated Alcohol and Alkylphenol Sulphates Section 5.3: Method for Determination of Unsulphated Matter Content. 4 pp.
ISO 8799-88. Surface Active Agents—Sulfated Ethoxylated Alcohols and Alkylphenols—Determination of Content of Unsulfated Matter First Edition. 5 pp.

Ethoxyquin Content Analysis
—Animal Feed
CNS N4095-82. Method of Test for Feed Additives: Determination of Ethoxyguin (Jun)(9025).

Ethyl Acetate
Use For: Acetic Ester *See Also:* Acetates; Esters; Hazardous Materials
BSI BS 553-90. 1990 Ethyl Acetate for Industrial Use. 8 pp.
CNS K1117-78. Ethyl Acetate (May)(2616).
CNS K7214-66. Chemical Reagent (Ethyl Acetate) (Jun)(1714).
JIS K 1511-78. Testing Method for Acetic Ester.
JIS K 1513-57. Ethyl Acetate.
JIS K 8361-86. Ethyl Acetate.

—Storage and Handling—Information Cards
SAA AS 2508.3.00 2-82. Safe Storage and Handling Information Cards for Hazardous Materials—Part 3.002: Acetates (Ethyl Acetate, Propyl Acetates, Butyl Acetates, Amyl Acetate) Double Sided Card.

Ethyl Acetate Content Analysis
—Ink Removers
MOD UK M 2003/59. Examination of Solvent, Cleaning.

—Paint Removers
MOD UK M 2003/59. Examination of Solvent, Cleaning.

Ethyl Acetoacetate
Use For: Ethyl 3-Oxobutanate
CNS K7602-88. Chemical Reagent (Ethyl 3-Oxobutanate) (Oct)(8521).
JIS K 8031-80. Ethyl 3-Oxobutanate.

Ethyl Acrylate
See Also: Hazardous Materials
CNS K1251-80. Ethyl Acrylate (Jul)(5830).

Ethyl Alcohol
Use: Ethanol

Ethyl Aldehyde
Use: Acetaldehyde

INDUSTRY STANDARDS

INTERNATIONAL AND NON-U.S. NATIONAL STANDARDS
SUBJECT INDEX

Ethyl

Ethyl Aminobenzoate Content Analysis
—Cosmetics
CNS S2097-83. Methods of Hygienic Test for Cosmetics Ethyl p-Aminoibenzoate and Ethyl O-Aminobenzoate (Jan)(9914).

Ethyl Benzene
Use: Ethylbenzene

Ethyl Bromide
CNS K7215-66. Chemical Reagent (Ethyl Bromide) (Jun)(1715).

Ethyl Carbamate
Use: Urethanes

Ethyl Cellulose
—Aircraft
MOD UK DTD-426-40. Ethyl Cellulose (Reprinted June 1962). 2 pp.
—Dipped Coatings
MOD UK CS 2486B. Protective PX 15.
—Inhibitive Coatings
MOD UK CS 2724. Ethyl Cellulose, High Viscosity.
—Lacquers
MOD UK TS 491A. Ethyl Cellulose for Lacquers, LF Quality (Withdrawn).

Ethyl Cellulose Coatings
See Also: Coatings
MOD UK CS 2486B. Protective PX 15.

Ethyl Cellulose Lacquers
See Also: Coatings; Lacquers
MOD UK TS 491A. Ethyl Cellulose for Lacquers, LF Quality (Withdrawn).
—Weapons
MOD UK TS 492A. Lacquer, Ethyl Cellulose, LF Quality (Withdrawn).

Ethyl Centralite
—Propellants—Stability Testing
NATO STANAG 4117 ED 2 AMD 4-73. Stability Test Procedures and Requirements for Propellants Stabilized with Diphenylamine, Ethyl Centralite or Mixtures of Both. 44 pp.

Ethyl Chloride
Use For: Chloroethane; Chloroethanes
BSI BS 5598: Part 7-80. 1980 Methods of Sampling and Test for the Halogenated Hydrocarbons Part 7: Methods of Test for Methyl Chloride and Ethyl Chloride. 10 pp.
ISO 5787-79. Methyl Chloride and Ethyl Chloride for Industrial Use—Methods of Test First Edition. 8 pp.

Ethyl Cyanoacetate
CNS K7216-66. Chemical Reagent (Ethyl Cyanoacetate) (Jun)(1716).
JIS K 8449-61. Ethyl Cyanoacetate.

Ethyl Cyanoacrylate Adhesives
Use: Cyanoacrylate Adhesives

Ethyl Ether
Use For: Diethyl Ether *See Also:* Ethers
BSI BS 579-57. 1957 Diethyl Ether (Technical). 12 pp.
JIS K 1506-79. Ethyl Ether for Industrial Use.
JIS K 1506-57. Ethyl Ether for Industrial Use. 4 pp.
JIS K 8103-87. Diethyl Ether.

Ethyl Iodide
Use For: Iodoethane
CNS K7220-66. Chemical Reagent (Ethyl Iodide) (Jun)(1720).
JIS K 8911-81. Iodoethane.

Ethyl Isoamyl Ketone
Use For: Isoamyl Ethyl Ketone *See Also:* Ketones
ISO 2500-74. isoAmyl Ethyl Ketone for Industrial Use—List of Methods of Test First Edition. 4 pp.
—Acidity—Volumetric Analysis
ISO 2887-73. secButyl Alcohol, Methyl Ethyl Ketone, isoButyl Methyl Ketone, isoAmyl Ethyl Ketone, Diacetone Alcohol, and Hexylene Glycol for Industrial Use—Determination of Acidity to Phenolphthalein—Volumetric Method First Edition. 4 pp.

Ethyl Mercaptans
CNS K7554-88. Chemical Reagent (Ethyl Mercaptan) (Oct)(7712).

Ethyl Methyl Ketone
Use: Methyl Ethyl Ketone

Ethyl Methyl Ketone Peroxide
Use: Methyl Ethyl Ketone Peroxide

Ethyl Oleate
See Also: Esters; Oleic Acid
MOD UK DSTAN 68-70-92. Ethyl Oleate, Type QX Issue 1. 9 pp.

N-Ethyl-m-Toluidine
JIS K 4109-84. Aminobenzenes (Aniline,o-Toluidin,p-Toluinene, m-Toluidine, N-Ethyl-m-Toluidine).

Ethyl 3-Oxobutanate
Use: Ethyl Acetoacetate

N-Ethylaniline
See Also: Aniline
CNS K2182-87. N-Ethyle N-Ethylaniline (Dec)(12198). 5 pp.
JIS K 4112-81. N-Substituted Anilines (N, N-Dimethylaniline.N, N-Diethylaniline. N-Ethylaniline. N-Methylaniline. N-Benzyl-N-Ethylaniline).

Ethylbenzene
Use For: Ethyl Benzene *See Also:* Hazardous Materials
—Explosives
MOD UK TS 10287. Ethylbenzene.

Ethylene
Use For: Ethene; Olefiant Gas *See Also:* Aliphatic Hydrocarbons; Hexanitrostilbene; Hydrocarbons; Olefins
—Acetone Content—Gas Chromatography
ISO 8174-86. Ethylene and Propylene for Industrial Use—Determination of Acetone, Acetonitrile, Propan-2-ol and Methanol—Gas Chromatographic Method First Edition. 8 pp.
—Acetonitrile Content—Gas Chromatography
ISO 8174-86. Ethylene and Propylene for Industrial Use—Determination of Acetone, Acetonitrile, Propan-2-ol and Methanol—Gas Chromatographic Method First Edition. 8 pp.
—Acetylene Content
CNS K6397-76. Method of Test for Trace Acetylene and Carbon Dioxide in Ethylene (Dec)(4054).
—Carbon Dioxide Content
CNS K6397-76. Method of Test for Trace Acetylene and Carbon Dioxide in Ethylene (Dec)(4054).
—Carbon Dioxide Content—Gas Chromatography
ISO 6381-81. Ethylene and Propylene for Industrial Use—Determination of Traces of Carbon Monoxide and Carbon Dioxide—Gas Chromatographic Method First Edition. 8 pp.
—Carbon Monoxide Content—Gas Chromatography
ISO 6381-81. Ethylene and Propylene for Industrial Use—Determination of Traces of Carbon Monoxide and Carbon Dioxide—Gas Chromatographic Method First Edition. 8 pp.
—Hydrocarbon Content—Gas Chromatography
ISO 6379-81. Ethylene for Industrial Use—Determination of Hydrocarbon Impurities—Gas Chromatographic Method First Edition. 6 pp.
—Hydrogen Content
CNS K6388-76. Method of Test for Ethylene, Other Olefins, Inerts and Hydrogen in High Purity Ethylene (Jun)(3947).
—Impurities Content
JIS K 1800-78. Analytical Methods for Trace of Impurities in High Purity Ethylene.
—Isopropyl Alcohol Content—Gas Chromatography
ISO 8174-86. Ethylene and Propylene for Industrial Use—Determination of Acetone, Acetonitrile, Propan-2-ol and Methanol—Gas Chromatographic Method First Edition. 8 pp.
—Methanol Content—Gas Chromatography
ISO 8174-86. Ethylene and Propylene for Industrial Use—Determination of Acetone, Acetonitrile, Propan-2-ol and Methanol—Gas Chromatographic Method First Edition. 8 pp.

Ethylene (Cont.)
—Moisture Content
CNS K6389-76. Method of Test for Moisture in Ethylene (Jun)(3948).
—Olefin Content
CNS K6388-76. Method of Test for Ethylene, Other Olefins, Inerts and Hydrogen in High Purity Ethylene (Jun)(3947).
—Rare Gases Content
CNS K6388-76. Method of Test for Ethylene, Other Olefins, Inerts and Hydrogen in High Purity Ethylene (Jun)(3947).
—Sampling
ISO 7382-86. Ethylene for Industrial Use—Sampling in the Liquid and the Gaseous Phase First Edition. 8 pp.
—Sampling—Polar Compounds Content
ISO 8916-88. Ethylene for Industrial Use—Determination of Traces of Polar Compounds—Preparation of Condensate Samples by Low-Temperature Scrubbing Technique First Edition. 9 pp.
—Storage and Handling—Information Cards
SAA AS 2508.3.00 3-92. Safe Storage and Handling Information Cards for Hazardous Materials—Part 3.013: Ethylene (Ethene) (in Professional Package 38). 2 pp.
—Sulfur Content
CNS K6398-76. Method of Test for Trace Quantities of Total Sulfur in Ethylene (Dec)(4055).

Ethylene Chlorohydrin
CNS K7580-88. Chemical Reagent (Ethylene Chlorohydrin) (2-Chloroethanol) (Oct)(7815).
JIS K 8106-78. Ethylene Chlorohydrin, 2-Chloroethanol.

Ethylene Copolymer Bitumen Coatings
See Also: Bitumens; Coatings
—Roofing
DIN ENGL 16729-84. Ethylene Copolymer Bitumen (ECB) Roofing Felt and Waterproofing Sheet; Requirements (Sept). 4 pp.

Ethylene Copolymers
See Also: Copolymers; Polyethylene
ISO 1872 Pt 1-86. Plastics—Polyethylene (PE) and Ethylene Copolymer Thermoplastics—Part 1: Designation First Edition. 8 pp.
—Cellular Plastics
CNS K6271-72. Method of Test for Polyethylene Resin (Jun)(2940). 6 pp.
ISO 7214-85. Cellular Plastics—Polyethylene—Methods of Test First Edition. 5 pp.

Ethylene Dichloride
Use For: sym-Dichloroethane; 1,2-Dichloroethane
See Also: Hazardous Materials
CNS K1156-72. Dichloroethane (Jan)(3335).
CNS K6316-72. Method of Test for Dichloroethane (Jan)(3336).
CNS K7644-82. Chemical Reagent (1,2-Dichloroethane) (Sep)(9406).
JIS K 8465-80. 1,2-Dichloroethane.
MOD UK TS 452B. Ethylene Dichloride 1, Dichloroethane (Withdrawn).

Ethylene Glycol
Use For: Ethanediol
CNS K1228-80. Ethylene Glycol (Feb)(5234).
CNS K6399-81. Methods of Test for Ethylene Glycol (Oct)(4088).
CNS K7219-66. Chemical Reagent (Ethylene Glycol) (Jun)(1719).
JIS K 1527-78. Ethylene Glycol (R 1983). 13 pp.
JIS K 8105-79. Ethylene Glycol.
—Antifreezes
MOD UK DSTAN 68-127-89. Antifreeze, Inhibited Ethanediol NATO Code S-757 Joint Service Designation AL-39 Issue 1. 14 pp.
—Engine Coolants
CGSB CAN/CGSB-3.890-M83. Coolant Concentrate, Automotive Engine. 9 pp.
MOD UK DSTAN 68-108-91. Ethanediol (Ethylene Glycol) Issue 1. 16 pp.
MOD UK DSTAN 68-108-93. Ethanediol (Ethylene Glycol) Joint Service Designation AL-20 Issue 2. 16 pp.
—Heat Transfer Fluids
CGSB 3-GP-855M-79. Ethylene Glycol, Uninhibited, Standard for. 7 pp.

Ethylene Glycol Content Analysis
—Motor Oils
DIN ENGL 51375 Pt 2-86. Testing of Lubricants; Determination of 1,2-Ethanediol Content of Unused Motor Oils by Gas Chromatography (Mar). 3 pp.

Ethylene Glycol Ether
CNS K1216-79. Acetate Ester of Ethylene Glycol Ether (95% Grade) (Jun)(4885).

Ethylene Glycol Monobutyl Ether
Use For: 2-Butoxyethanol
CNS K1208-78. Ethylene Glycol Monobutyl Ether (Industrial Grade) (Oct)(4639).
—Decarbonizing—Aircraft Engines
MOD UK TS 10127. 2-Butoxyethanol, Technical.

Ethylene Glycol Monoethyl Ether
Use For: 2-Ethoxyethanol
BSI BS 2713-66. 1966 2-Ethoxyethanol (Ethylene Glycol Monoethyl Ether). 8 pp.
BSI BS 2713-01. 1966 Amd 1 2-Ethoxyethanol (Ethylene Glycol Monoethyl Ether) (AMD 7628) May 15, 1993 (W). 9 pp.
CNS K1207-78. 2-Ethoxy Ethanol (Industrial Grade) (Oct)(4638).

Ethylene Glycol Monomethyl Ether
—Adhesives
MOD UK CS 3006. Ethylene Glycol Monom Methyl Ether (Withdrawn).
—Antiicing Additives
CAA NOTICE #23 ISSUE 1. Fuel Additives—Health Hazards (Airworthiness Notices). 2 pp.
CGSB CAN/CGSB-3.526-M87. Icing Inhibitor for Aviation Fuels. 9 pp.
MOD UK DERD 2451-80. Fuel System Icing Inhibitor Issue 3. 6 pp.

Ethylene Oxide
Use For: Epoxyethame See Also: Epoxy Compounds; Hazardous Materials
JIS K 1526-78. Ethylene Oxide.
JIS K 1526-61. Ethylene Oxide. 11 pp.
—Highway Transportation—Emergency Procedures
SAA AS 1678.2.1. 005-85. Emergency Procedure Guide—Transport—Part 2.1.005: Ethylene Oxide.
—Medical Equipment—Biological Hazards
CEN PREN 30993-7-93. Biological Testing of Medical and Dental Materials and Devices—Part 7: Ethylene Oxide Sterilization Residuals (ISO/DIS 10993-7). 63 pp.
—Plants (Botany)—Prohibited Use
EC 86/355/EEC-86. Council Directive Amending Directive 79/117/EEC Prohibiting the Placing on the Market and Use of Plant Protection Products Containing Certain Active Substances. 2 pp.
—Sterilizers—Medical
CEN PREN 550-91. Sterilization of Medical Devices—Validation and Routine Control of Ethylene Oxide Sterilization—Requirements. 19 pp.
CEN PREN 551-91. Sterilization of Medical Devices—Validation and Routine Control of Ethylene Oxide Sterilization—Guidance. 30 pp.
SAA AS 1714-90. Sterilizers—12/88 Ethylene Oxide—Hospital Use. 29 pp.
—Sterilizers—Medical—Aerator Cabinets
SAA AS 1862-76. Aeration Cabinets (for Use with Ethylene Oxide Sterilizers). 12 pp.
—Surfactants—Iodometry
BSI BS 6829: Sec 4.3-90. 1990 Analysis of Surface Active Agents (Raw Materials) Part 4: Ethylene Oxide Adducts Section 4.3: Methods for Determination of Oxyethylene Groups. 8 pp.
ISO 2270-89. Non-Ionic Surface Active Agents—Polyethoxylated Derivatives—Iodometric Determination of Oxyethylene Groups Second Edition. 8 pp.
—Surfactants—Nonionic—Cloud Point
BSI BS 6829: Sec 4.1-91. 1991 Analysis of Surface Active Agents (Raw Materials) Part 4: Ethylene Oxide Adducts Section 4.1: Methods for Determination of Cloud Temperature (Cloud Point). 11 pp.
BSI BS 6829: Sec 4.1-88. 1988 Analysis of Surface Active Agents (Raw Materials) Part 4: Ethylene Oxide Adducts Section 4.1: Methods for Determination of Cloud Temperature (Cloud Point). 7 pp.
ISO 1065-91. Non-Ionic Surface-Active Agents Obtained from Ethylene Oxide and Mixed Non-Ionic Surface-Active Agents—Determination of Cloud Point Second Edition. 9 pp.

Ethylene Plastics
Use: Polyethylene

Ethylene Propylene Copolymers
Use For: Ethylene Propylene Rubber
See Also: Copolymers
BSI BS 6014-91. 1991 Ethylene Propylene Rubber Compounds. 10 pp.
MOD UK DSTAN 93-43-92. Ethylene Propylene, Rubber Compounds, Type QX Issue 1. 12 pp.
—Aerospace
CEN PREN2428-89. Ethylen Propylen—Rubber (EPM/EPDM) Hardness 50 IRHD. 5 pp.
CEN PREN2429-89. Ethylen Propylen—Rubber (EPM/EPDM) Hardness 60 IRHD. 5 pp.
CEN PREN2430-89. Ethylen Propylen—Rubber (EPM/EPDM) Hardness 70 IRHD. 5 pp.
CEN PREN 2431-89. Ethylene Propylene Rubber (EPM/EPDM) Hardness 80 IRHD. 5 pp.
CEN PREN 3705-89. Ethylene—Propylene Rubber (EPM/EPDM) Hardness 80 IRHD for Hydraulic Circuits. 6 pp.
MOD UK DTD-5596A. Vulcanized Ehtylene-Propylene Rubbers (General Purpose). 7 pp.
—Aircraft
SBAC TS 79 ISSUE 2. General Purpose Grade of Ethylene Propylene Rubber.
—Fluid Resistant—Aerospace
MOD UK DTD-5597A. Vulcanized Ethylene-Propylene Rubbers (Fluid Resistant). 8 pp.
—Respirators
MOD UK DSTAN 93-70-93. Ethylene Propylene Diene Rubber Black, 91 IRHD Issue 1. 10 pp.
—Test Specimens
MOD UK DSTAN 93-52: Part 3-88. Vulcanized Rubbers for Use as Standard Test Materials Part 3: Vulcanized Ethylene-Propylene Rubber Sheet, Standard EP/1 Issue 1. 9 pp.
—Thermogravimetric Analysis
ISO 9924 Pt 1-93. Rubber and Rubber Products—Determination of the Composition of Vulcanizates and Uncured Compounds by Thermogravimetry—Part 1: Butadiene, Ethylene-Propylene Copolymer and Terpolymer, Isobutene-Isoprene, Isoprene and Styrene-Butadiene. 8 pp.

Ethylene Propylene Rubber
Use: Ethylene Propylene Copolymers

Ethylene Propylene Terpolymers
Use For: Propylene Diene Rubber
BSI BS 6063-92. 1992 Method for Evaluation of General Purpose Ethylene-Propylene-Diene Rubber (EPDM) (ISO 4097: 1991). 12 pp.
BSI BS 6063-81. 1981 Methods of Test for Non-Oil Extended Raw General Purpose Ethylene-Propylene-Diene Rubbers (EPDM). 8 pp.
ISO 4097-91. Rubber, Ethylene-Propylene-Diene (EPDM)—General Purpose Types—Evaluation Procedure Third Edition. 10 pp.
—Fenders—Naval Ships
MOD UK NES 170-93. Requirements for EPDM Rubber Circular and 'D' Section Marine Fenders Issue 1 (05.93). 27 pp.
—Standard Elastomer
DIN ENGL 53538 Pt 5-88. Standard Reference Elastomers; Ethylene Propylene Diene Rubber (EPDM), Peroxide Cured, for Characterizing Service Fluids with Respect to Their Action on EPDM (Oct). 3 pp.

Ethylene Vinyl Acetate Copolymers
See Also: Polyolefins; Thermoplastic Resins
CNS K3067-83. Ethylene Vinyl Acetate Resin (Jan)(9896). 2 pp.
—Molding Materials—Test Specimens
DIN ENGL 16778 Pt 2-88. Plastics Moulding Materials; Ethylene Vinyl Acetate (E/VAC) Moulding Materials; Preparation of Specimens and Determination of Their Properties (Apr). 7 pp.
—Moldings
MOD UK TS 70002. Crosslinked Expanded Ethylene-Vinyl Acetate Sheets and Mouldings.
—Sheets
MOD UK TS 70002. Crosslinked Expanded Ethylene-Vinyl Acetate Sheets and Mouldings.
—Vinyl Acetate Content
CNS K6738-83. Method of Test for Vinyl Acetate Content in EVA Resin (Jan)(9897).
CNS K6739-83. Method of Test for Vinyl Acetate Content in EVA Resin (Saponification Method) (Jan)(9898).

Ethylenediamine
CNS K7217-66. Chemical Reagent (Ethylenediamine) (Jun)(1717).

Ethylenediaminetetraacetic Acid
Use: EDTA

Ethylenediaminetetraacetic Acid Content Analysis
Use: EDTA Content Analysis

Ethylenediaminetetraacetic Acid Disodium Salt
JIS K 8107-76. Ethylenediaminetetra-acetic Acid Disodium Salt.

2-Ethylhexan-1-01
BSI BS 1835-91. 1991 2-Ethylhexan-1-ol for Industrial Use. 6 pp.

2-Ethylhexyl Acrylate
CNS K1253-80. 2-Ethylhexyl Acrylate (Jul)(5832).

2-Ethylhexyl Alcohol
Use For: Octanol; Octyl Alcohol See Also: Alcohols
CNS K1177-86. Octanols (for Industrial Use) (Jun)(3923).
CNS K6394-86. Method of Test for Octanols (for Industrial Use) (Jun)(4012).
JIS K 1525-60. Octyl Alcohol (Octanol).

N-Ethylpiperidine
CNS K7221-66. Chemical Reagent (N-Ethylpiperidine) (Jun)(1721).

ETSI
Use For: European Telecommunications Standards Institute See Also: European Standards; Standards
—Bulletins
EURO 1992-05 Bulletin-92. Bulletin of the European Standards Organizations. 19 pp.
EURO 1992-06 Bulletin-92. Bulletin of the European Standards Organizations. 24 pp.
EURO 1992-07 Bulletin-92. Bulletin of the European Standards Organizations. 26 pp.
—Scientific Reports
CENELEC ETR 049-92. Advanced Testing Methods (ATM); State of Research in the Area of Formal Test Specification Methods. 54 pp.
ETSI ETR 049-92. Advanced Testing Methods (ATM); State of Research in the Area of Formal Test Specification Methods. 54 pp.

Eubacteriales
Scope Note: Use a more specific term
See: Anthracis; Bacillus Cereus; Bacteria; Clostridium; Coliform Bacteria; Enterobacteriaceae; Escherichia Coli; Listeria Monocytogenes; Salmonella; Shigella; Staphylococcus; Staphylococcus Aureus; Streptococcus

Eucalyptol Content Analysis
Use For: Cineol Content Analysis
—Essential Oils
CNS K6605-80. Determination of Cineole Content of Essential Oils (Oct)(6578).
ISO 1202-81. Essential Oils—Determination of 1,8-Cineole Content First Edition. 5 pp.

Eucalyptus Oil
Use For: Myrtaceae Oil See Also: Essential Oils
CNS K5068-80. Oil of Eucalyptus Globulus (Aug)(6347).
CNS K5078-80. Oil of Australian Eucalyptus (80 to 85 Percent Cineole Content) (Sep)(6449).
ISO 770-80. Oil of Eucalyptus Globulus First Edition. 4 pp.
ISO 3044-74. Oil of Eucalyptus Citriodora First Edition. 4 pp.
ISO 3065-74. Oil of Australian Eucalyptus, 80 to 85 % Cineole Content First Edition. 4 pp.
ISO 4732-83. Rectified Oil of Eucalyptus Globulus Labillardiere, Portugal First Edition. 4 pp.

Eugenol
See Also: Clove Oil
—Dental Materials
BSI BS 4284-81. 1981 Dental Impression Pastes (Zinc Oxide Type). 8 pp.
BSI BS 7214-89. 1989 Dental Zinc Oxide/Eugenol Cements and Zinc Oxide Non-Eugenol Cements (ISO 3107: 1988). 8 pp.
BSI BS 7214-01. 1989 Amd 1 Dental Zinc Oxide/Eugenol Cements and Zinc Oxide Non-Eugenol Cements (ISO 3107: 1988) (AMD 7039) May 1, 1992. 19 pp.

Eugenol (Cont.)
—Dental Materials (Cont.)
CEN EN 23107-91. Dental Zinc Oxide/Eugenol Cements and Zinc Oxide Non-Eugenol Cements. 3 pp.
ISO 3107-88. Dental Zinc Oxide/Eugenol Cements and Zinc Oxide Non-Eugenol Cements Second Edition; (Supersedes 3106). 13 pp.

Eumycophyta
Use: Fungi

Euratom
Use: European Communities

Eurobonds
See Also: Securities
ISO 8109-90. Banking and Related Financial Services—Securities—Format of Eurobonds Second Edition. 14 pp.
JTC1 8109-90. Banking and Related Financial Services—Securities—Format of Eurobonds Second Edition. 14 pp.
OSI ISO DIS 8109-89. Banking and Related Financial Services—Securities—Format of Eurobonds. 18 pp.

Eurocard Connectors
Use: DIN Connectors

Eurocopter France AS 332L2 Super Puma Helicopters
Use: AS 332L2 Super Puma Helicopters

European Association of Aerospace Manufacturers
Use: AECMA

European Atomic Energy Community
Use: European Communities

European Automatic Freephone Service
Use For: EFS *See Also:* Freephone Services; Telephone Services
CEPT T/TPH 40 E-87. European Automatic Freephone Service (EFS) (Revised Edinburgh 1988). 4 pp.

—Integrated Services Digital Networks
CENELEC PRETS 300 208-92. Integrated Services Digital Network (ISDN); Freephone (FPH) Supplementary Service Service Description. 16 pp.
CENELEC PRETS 300 209-92. Integrated Services Digital Network (ISDN); Freephone (FPH) Supplementary Service Functional Capabilities and Information Flows. 61 pp.
CENELEC PRETS 300 210-92. Integrated Services Digital Network (ISDN); Freephone (FPH) Supplementary Service Digital Subscriber Signalling System No. One (DSS1) Protocol. 22 pp.
ETSI PRETS 300 208-92. Integrated Services Digital Network (ISDN); Freephone (FPH) Supplementary Service Service Description. 16 pp.
ETSI PRETS 300 209-92. Integrated Services Digital Network (ISDN); Freephone (FPH) Supplementary Service Functional Capabilities and Information Flows. 61 pp.
ETSI PRETS 300 210-92. Integrated Services Digital Network (ISDN); Freephone (FPH) Supplementary Service Digital Subscriber Signalling System No. One (DSS1) Protocol. 22 pp.

European Broadcasting Union
See Also: Broadcasting

—International Leased Circuits
CEPT T/TPH 6-86. Location Permanente Et Temporaire De Circuits Internationaux De Telecommunications Aux Membres De L'UER Et a Leur Utilisation Par Des Tiers Et Par Des Administrations Des Telecommunications. 4 pp.
CEPT T/TPH 6 E-90. Permanent and Occasional Lease of International Telecommunications Circuits to the EBU and Its Members and Their Use by Third Parties and Telecommunications Administrations. 3 pp.

European Coal and Steel Community
Use: European Communities

European Committee for Electrotechnical Standardization
Use: Cenelec

European Committee for Standardization
Use: CEN

European Communications Satellite
Use For: ECS *See Also:* European Telecommunications Satellite Organization; Satellite Communications

—Telephone Networks—Signaling Systems
CEPT T/CS 42-05-85. Essais De Signalisation Pour Le Systeme De Signalisation R2 Via ECS. 5 pp.
CEPT T/CS 42-05 E-85. System R2 Signalling Tests Via the European Communications Satellite (ECS). 5 pp.
CEPT T/CS 42-09-85. Surveillance De Circuit Et De Signalisation Via Le Satellite ECS. 7 pp.
CEPT T/CS 42-09 E-85. Signalling and Circuit Supervision Via the European Communications Satellite (ECS). 7 pp.

European Communities
Use For: Common Market; EAEC; EC; ECSC; EEC; Euratom; European Atomic Energy Community; European Coal and Steel Community; European Economic Community; SEM; Single European Market
See Also: CEC (Commission of the European Communities); European Monetary System; European Standards; International Agreements
EC COM(92) 519-92. Commission Proposal for a Council Regulation Amending the Council Regulation (EEC, Euratom) No 1552/89 Implementing Decision 88/376/EEC, Euratom on the System of the Communities' Own Resources. 6 pp.
EC COM(92) 383-92. Commission of the European Communities Seventh Report of the Commission to the Council and the European Parliament Concerning the Implementation of the White Paper on the Completion of the Internal Market. 141 pp.
EC 92/481/EEC-92. Council Decision on the Adoption of an Action Plan for the Exchange Between Member State Administrations of National Officials Who Are Engaged in the Implementation of Community Legislation Required to Achieve the Internal Market. 3 pp.
EURO 1985 14 June. Completing the Internal Market White Paper from the Commission to the European Council. 95 pp.
EURO 1986 26 May. Completing the Internal Market First Report from the Commission to the Council and the European Parliament (COM (86)300). 50 pp.
EURO 1987 11 May. Completing the Internal Market Second Report from the Commission to the Council and the European Parliament (COM (87)203). 73 pp.
EURO 1988 21 March. Completing the Internal Market Third Report from the Commission to the Council and the European Parliament (COM(88)134). 92 pp.
EURO 1989 20 June. Completing the Internal Market Fourth Progress Report of the Commission to the Council and the European Parliament (COM (89)311). 89 pp.
EURO 1988 XXIIND. XXIInd General Report on the Activities of the European Communities 1988. 460 pp.
EURO 1988 May. European Economy. 196 pp.
EURO 1988 Feb. Europe Without Frontiers—Completing the Internal Market. 70 pp.
EURO 1989-16. Europe—Our future; The Institutions of the European Community. 15 pp.
EURO 1991 Jan. Working for the Regions. 9 pp.
EURO 1991 May. Economic and Monetary Union. 9 pp.
EURO 1991-4. Trans-European Networks for a Community Without Frontiers. 12 pp.
EURO 1991 Jan. Freedom of Movement. 9 pp.
EURO 1991 May. Social Challenge. 9 pp.
EURO 1991 May. What is the EMS. 8 pp.
EURO 1991 Aug. Young People's Europe. 8 pp.
EURO 1991 JUN. Education and Training. 9 pp.
EURO 1989-01. The Single Market Progress on Commission White Paper. 36 pp.
EURO 1989-02. The Single Market Progress on Commission White Paper. 54 pp.
EURO 1989-03. The Single Market Progress on Commission White Paper. 36 pp.
EURO 1989-11. Single Market Progress on Commission White Paper. 57 pp.
EURO 1990-02. The Single Market Progress on Commission White Paper. 59 pp.
EURO 1990-04. Single Market Progress on Commission White Paper. 56 pp.
EURO 1990-06. Single Market Progress on Commission White Paper. 60 pp.
EURO 1990-09. Single Market Progress on Commission White Paper. 52 pp.
EURO 1990-12. Single Market Progress on Commission White Paper. 55 pp.
EURO 1991-3. The Single Market Progress on Commission White Paper. 60 pp.
EURO 1991-07. Single Market Progress on Commission White Paper. 61 pp.
EURO 1991-11. Single Market Progress on Commission White Paper. 56 pp.

European Communities (Cont.)
EURO 1992-02. Single Market Progress on Commission White Paper. 57 pp.
EURO 1992-04. Single Market Progress on Commission White Paper. 64 pp.
EURO 1992-08. Single Market Progress on Commission White Paper. 60 pp.

—Aerospace Industry
EURO EC/88/20420/E,F-89. Panorama of EC Industry—Aerospace Industry. 32 pp.
EURO AECMA/23396/92-92. European Aeronautical Industry Towards the 21st Century. 34 pp.

—Agriculture
EURO 1990-1. Community's Agricultural Policy on the Threshold of the 1990's. 11 pp.

—Air Transportation
EURO JLJ/FMR/23179-92. AECMA Views on European Air Transport Policy. 8 pp.

—Aircraft—Trade Balance
EURO ECO/4690/F & E And 4675/-75. Demand Prospects for Civil Transport Aircraft Currency Balance of European Airlines Purchases and European Industry Sales. 41 pp.

—Bosnia Herzegovina—Importation
EC EEC/343/92-92. Commission Regulation on the Definition of the Concept of Originating Products and Methods of Administrative Cooperation Applicable to Imports into the Community of Products Originating in the Republics of Croatia and Slovenia and the Yugoslav Republics of. 64 pp.

—Catalogs
EURO 1989-18. Catalogue 1979-89. 9 pp.
EURO 1990-15. Catalogue 1979-90. 10 pp.

—Commerce
EC 86/653/EEC-86. Council Directive on the Coordination of the Laws of the Member States Relating to Self-Employed Commercial Agents. 5 pp.
EC EEC 1901/85-85. Council Regulation Amending Regulation (EEC) No 222/77 on Community Transit. 3 pp.
EC EEC/1674/87-87. Council Regulations Amending Regulation 222/ 77 on Community Transit. 2 pp.
EC EEC/474/90-90. Council Regulation Amending, with a View to Abolishing Lodgement of the Transit Advice Note on Crossing an Internal Frontier of the Community, Regulation (EEC) No 222/77 on Community Transit. 2 pp.
EC EEC/2726/90-90. Council Regulation on Community Transit. 10 pp.
EC EEC/3330/91-91. Council Regulation on the Statistics Relating to the Trading of Goods Between Member States. 11 pp.

—Company Law
EURO 1989 October. Company Law in the European Community. 11 pp.

—Consumer Protection
EURO 1990-14. European Community and Consumer Protection. 12 pp.

—Copyrights
EURO 1989-17. Patents, Trade Marks and Copyright in the European Community. 9 pp.

—Croatia—Importation
EC EEC/343/92-92. Commission Regulation on the Definition of the Concept of Originating Products and Methods of Administrative Cooperation Applicable to Imports into the Community of Products Originating in the Republics of Croatia and Slovenia and the Yugoslav Republics of. 64 pp.

—Developing Countries—Cooperation
EURO 1991 Mar. ECU. 7 pp.

—Eastern Europe—Economic Relations
EURO 1991 Aug. Community and Its Eastern Neighbours. 9 pp.

—Environmental Protection
EURO 1990-4. European Community and Environmental Protection. 11 pp.
EURO 1991 Aug. Pollution Knows No Frontiers. 9 pp.

—Environmental Protection—Germany—Transitional Measures
EC 90/660/EEC-90. Council Directive on the Transitional Measures Applicable in Germany with Regard to Certain Community Provisions Relating to the Protection of the Environment, in Connection with the Internal Market. 2 pp.

—Fisheries
EURO 1991-3. Common Fisheries Policy. 12 pp.

INTERNATIONAL AND NON-U.S. NATIONAL STANDARDS
SUBJECT INDEX
European

European Communities *(Cont.)*

—Food—Committees
EC 80/1073/EEC-80. Commission Decision Establishing a New Statute of the Advisory Committee on Foodstuffs. 4 pp.

—Germany—Transitional Measures
EC 90/650/EEC-90. Council Directive on Transitional Measures Applicable in Germany in the Context of the Harmonization of Technical Rules for Certain Products. 2 pp.
EC 90/657/EEC-90. Council Directive on Transitional Measures Applicable in Germany in the Context of the Harmonization of Technical Rules. 8 pp.

—Higher Education
EURO 1990-5. Education and Training-the Approach to 1992. 12 pp.
EURO 1991 JUN. Education and Training. 9 pp.

—Higher Education—Diplomas
EC 89/48/EEC-88. Council Directive on a General System for the Recognition of Higher-Education Diplomas Awarded on Completion of Professional Education and Training of at Least Three Years' Duration. 8 pp.
EURO 1989 Oct. European Community and Recognition of Diplomas for Professional Purposes. 11 pp.

—Industrial Products—Conformance
EC COM(93) 144-93. Amended Proposal for a Council Directive Amending Council Directives 87/404/EEC (Simple Pressure Vessels), 88/378/EEC (Safety of Toys), 89/106/EEC (Construction Products), 89/336/EEC (Electromagnetic Compatibility), 89/392/EEC (Machinery),. 4 pp.
EC 90/683/EEC-90. Council Decision Concerning the Modules for the Various Phases of the Conformity Assessment Procedures Which are Intended to Be Used in the Technical Harmonization Directive. 15 pp.

—Industrial Products—Identification Systems—Conformance
EC COM(93) 144-93. Amended Proposal for a Council Directive Amending Council Directives 87/404/EEC (Simple Pressure Vessels), 88/378/EEC (Safety of Toys), 89/106/EEC (Construction Products), 89/336/EEC (Electromagnetic Compatibility), 89/392/EEC (Machinery),. 4 pp.
EC COM(92) 499-92. Modification of the Proposal for a Council Regulation (EEC) Concerning the Affixing and Use of the CE Mark of Conformity on Industrial Products in a Proposal for a Council Directive and a Proposal for a Council Decision. 23 pp.

—Industrial Products—Importation—Conformance
EC COM(92) 466-92. Commission Proposal for a Council Regulation (EEC) on Checks for Conformity with the Rules on Product Safety in the Case of Products Imported from Third Countries. 4 pp.

—Industrial Products—Trade Barriers
EURO 1990. Catalogue of Community Legal Acts and Other Texts Relating to the Elimination of Technical Barriers to Trade for Industrial Products. 179 pp.

—Information Systems
EURO 1989 Nov. Esprit: Key to the Technological Awakening of Europe. 11 pp.

—Information Systems—Standards Preparation
EURO STANDARDIZAT ION-FACTST 0-88. Foreword. 2 pp.
EURO STANDARDIZAT ION-FACTST 1-88. The Challenge of Standardization. 3 pp.
EURO STANDARDIZAT ION-FACTST 2. Networks, Communication and Interworking. 3 pp.
EURO STANDARDIZAT ION-FACTST 3-88. Community Standardization Policy. 3 pp.
EURO STANDARDIZAT ION-FACTST 4-88. Standards Bodies: the Changing Scene. 3 pp.
EURO STANDARDIZAT ION-FACTST 5-88. European Standards. 3 pp.
EURO STANDARDIZAT ION-FACTST 6. Conformance Testing and Certification Policy. 2 pp.
EURO STANDARDIZAT ION-FACTST 7-88. The CTS Programme. 3 pp.
EURO STANDARDIZAT ION-FACTST 8-88. Public Procurement. 2 pp.
EURO STANDARDIZAT ION-FACTST 9-88. International Cooperation. 1 p.
EURO STANDARDIZAT ION-REF ST 1-88. Glossary. 3 pp.
EURO STANDARDIZAT ION-REF ST 2-88. Milestones in Community Standardization policy. 1 p.

European Communities *(Cont.)*

—Information Systems—Standards Preparation *(Cont.)*
EURO STANDARDIZAT ION-REF ST 3-88. List of Standardization Organizations and Technical Bodies. 1 p.

—Internal Market
EC COM(93) 256-93. Communication from the Commission to the Council and the European Parliament Reinforcing the Effectiveness of the Internal Market. 63 pp.
EC COM(93) 261-93. Second Progress Report on the Internal Energy Market. 24 pp.

—Invitation for Bids
MOD UK DEFCON 47M-91. Invitation to Tender—Special Notices and Instructions—EC Works Directive 9/91. 1 p.

—Legislation—Directories
EURO 1988 11TH EDITION. Directory of Community Legislation in Force and Other Acts of the Community Institutions—Volume 11. 165 pp.
EURO 1989 13TH EDITION. Directory of Community Legislation in Force and Other Acts of the Community Institutions-Volume 1. 813 pp.
EURO 1989 14TH ED VOLUME 1. Directory of Community Legislation in Force and Other Acts of the Community Institutions—Volume 1. 752 pp.
EURO 1989 13TH ED. Directory of Community Legislation in Force and Other Acts of Community Institutions-Volume II. 192 pp.
EURO 1989 14TH ED VOLUME II. Directory of Community Legislation in Force and Other Acts of the Community Institutions—Volume II. 178 pp.
EURO 1990-1 Volume I. Directory of Community Legislation in Force and Other Acts of the Community Institutions-Volume I. 761 pp.
EURO 1991-1 Volume I. Directory of Community Legislation in Force and Other Acts of the Community Institutions Volume I. 853 pp.
EURO 1992-1 Volume I. Directory of Community Legislation in Force and Other Acts of the Community Institutions Volume I. 847 pp.
EURO 1992-1 Volume I. Directory of Community Legislation in Force and Other Acts of the Community Institutions Volume I. 847 pp.
EURO 1993-1 Volume I. Directory of Community Legislation in Force and Other Acts of the Community Institutions Volume I Analytical Register. 859 pp.
EURO 1990-2 Volume II. Directory of Community Legislation in Force and Other Acts of the Community Institutions-Volume II. 181 pp.
EURO 1991-2 Volume II. Directory of Community Legislation in Force and Other Acts of the Community Institutions Volume II. 198 pp.
EURO 1992-2 Volume II. Directory of Community Legislation in Force and Other Acts of the Community Institutions Volume II. 195 pp.
EURO 1992-2 Volume II. Directory of Community Legislation in Force and Other Acts of the Community Institutions Volume II. 197 pp.
EURO 1993-2 Volume II. Directory of Community Legislation in Force and Other Acts of the Community Institutions Volume II Chronological Index Alphabetical Index. 198 pp.

—Legislation—Indexes (Documentation)
EURO 1989 13TH ED. Directory of Community Legislation in Force and Other Acts of the Community Institutions-Volume II. 192 pp.
EURO 1989 14TH ED VOLUME II. Directory of Community Legislation in Force and Other Acts of the Community Institutions—Volume II. 178 pp.
EURO 1990-2 Volume II. Directory of Community Legislation in Force and Other Acts of the Community Institutions-Volume II. 181 pp.
EURO 1991-2 Volume II. Directory of Community Legislation in Force and Other Acts of the Community Institutions Volume II. 198 pp.
EURO 1992-2 Volume II. Directory of Community Legislation in Force and Other Acts of the Community Institutions Volume II. 195 pp.
EURO 1992-2 Volume II. Directory of Community Legislation in Force and Other Acts of the Community Institutions Volume II. 197 pp.
EURO 1993-2 Volume II. Directory of Community Legislation in Force and Other Acts of the Community Institutions Volume II Chronological Index Alphabetical Index. 198 pp.

—Macedonia—Importation
EC EEC/343/92-92. Commission Regulation on the Definition of the Concept of Originating Products and Methods of Administrative Cooperation Applicable to Imports into the Community of Products Originating in the Republics of Croatia and Slovenia and the Yugoslav Republics of. 64 pp.

—Marine Transportation
EC COM(85) 90-85. Progress Towards a Common Transport Policy—Maritime Transport. 126 pp.

European Communities *(Cont.)*

—Occupational Safety and Health
EURO 1990-3. Health and Safety at Work in the European Community. 11 pp.

—Official Journal—Bids
MOD UK DEFCON 49-89. Tender Notice for the Official Journal—EEC 12/89. 2 pp.
MOD UK DEFCON 49A-89. Request for Certificate of Posting DEFCON 49 12/89. 1 p.

—Patents
EURO 1989-17. Patents, Trade Marks and Copyright in the European Community. 9 pp.

—Pesticides—Committees
EC 78/436/EEC-78. Commission Decision Establishing a Scientific Committee for Pesticides. 2 pp.
EC 86/105/EEC-86. Commission Decision Amending Decisions 76/791/EEC, 78/436/EEC and 81/651/EEC with Respect to the Number of Members of the Scientific Committees. 1 p.

—Roads—Safety
EC COM(93) 246-93. Communication from the Commission to the Council for an Action Programme on Road Safety. 56 pp.

—Rural Development
EURO 1991-5. European Community and Rural Development. 10 pp.

—Science and Technology
EURO 1991 Mar. Science and Technology in Europe. 9 pp.

—Slovenia—Importation
EC EEC/343/92-92. Commission Regulation on the Definition of the Concept of Originating Products and Methods of Administrative Cooperation Applicable to Imports into the Community of Products Originating in the Republics of Croatia and Slovenia and the Yugoslav Republics of. 64 pp.

—Standards Preparation
EC 83/189/EEC-83. Council Directive Laying Down a Procedure for the Provision of Information in the Field of Technical Standards and Regulations (for Rel Stds See TEST/83/189). 5 pp.
EC TEST/83/189-83. Information Procedure—Technical Regulations. 4 pp.
EURO 1988 Dec. New Community Standards Policy. 185 pp.
EURO 1990. Green Paper on Standards Development (COM (90) 456). 63 pp.

—Structural Funds
EURO 1990-7/8. New Structural Policies of the European Community. 15 pp.
EURO 1991 Jan. Working for the Regions. 9 pp.
EURO 1991-5. European Community and Rural Development. 10 pp.

—Taxes—Business Corporations
EC 90/435/EEC-90. Council Directive on the Common System of Taxation Applicable in the Case of Parent Companies and Subsidiaries of Different Member States. 4 pp.

—Telecommunication Equipment
EURO 1988 September. Telecommunications in Europe. 257 pp.

—Thailand—Trade
EC 91/478/EEC-91. Council Decision Concerning the Provisional Application of the Agreed Minute Amending the Agreement Between the European Economic Community and the Kingdom of Thailand on Trade in Textile Products. 3 pp.

—Tourism
EURO 1990-2. 1990, European Tourism Year. 11 pp.

—Trademarks
EURO 1989-17. Patents, Trade Marks and Copyright in the European Community. 9 pp.

—Veterinary Medicine—Committees
EC 86/105/EEC-86. Commission Decision Amending Decisions 76/791/EEC, 78/436/EEC and 81/651/EEC with Respect to the Number of Members of the Scientific Committees. 1 p.

—Veterinary Medicine—Exchange Program
EC 93/88/EEC-92. Commission Decision Fixing the Community Financial Contribution to the Implementation of a Second Programme for the Exchange of Officials Competent for Veterinary Matters. 3 pp.

—Vocational Education
EC 85/368/EEC-85. Council Decision on the Comparability of Vocational Training Qualifications Between the Member States of the European Community. 4 pp.

INDUSTRY STANDARDS

European

European Communities (Cont.)
—**Vocational Education** (Cont.)
EURO 1990-5. Education and Training-the Approach to 1992. 12 pp.
EURO 1991 JUN. Education and Training. 9 pp.

—**Wastepapers**
CEN PREN 643-92. List of European Standard Qualities of Waste Paper. 9 pp.

—**Workers' Rights**
EURO 1990-6. Community Charter of Fundamental Social Rights for Workers. 12 pp.

European Digital Cordless Telecommunications
Use For: DECT; Digital European Cordless Telecommunications *See Also:* Telecommunication Systems

CENELEC PRETS 300 175-1-91. Radio Equipment and Systems Digital European Cordless Telecommunications Common Interface Part 1: Overview (DE/RES 3001-1). 30 pp.
CENELEC PRETS 300 175-1-92. Radio Equipment and Systems (RES); Digital European Cordless Telecommunications (DECT) Common Interface Part 1: Overview. 29 pp.
CENELEC ETS 300 175-1-92. Radio Equipment and Systems (RES); Digital European Cordless Telecommunications (DECT) Common Interface Part 1: Overview. 28 pp.
CENELEC ETR 015-91. Radio Equipment and Systems Digital European Cordless Telecommunications (DECT) Reference Document. 42 pp.
CENELEC ETR 042-92. Radio Equipment and Systems; Digital European Cordless Telecommunications (DECT) a Guide to DECT Features That Influence the Traffic Capacity and the Maintenance of High Radio Link Transmission Quality, Including the Results of Simulations. 52 pp.
CEPT T/SF 66 E-90. Services and Facilities Aspects of Digital European Cordless Telecommunications (DECT). 5 pp.
ETSI ETS 300 175-1-92. Radio Equipment and Systems (RES); Digital European Cordless Telecommunications (DECT) Common Interface Part 1: Overview. 28 pp.
ETSI PRETS 300 175-1-92. Radio Equipment and Systems (RES); Digital European Cordless Telecommunications (DECT) Common Interface Part 1: Overview. 29 pp.
ETSI PRETS 300 175-1-91. Radio Equipment and Systems Digital European Cordless Telecommunications Common Interface Part 1: Overview (DE/RES 3001-1). 30 pp.
ETSI ETR 015-91. Radio Equipment and Systems Digital European Cordless Telecommunications (DECT) Reference Document. 42 pp.
ETSI ETR 042-92. Radio Equipment and Systems (RES); Digital European Cordless Telecommunications (DECT) a Guide to DECT Features That Influence the Traffic Capacity and the Maintenance of High Radio Link Transmission Quality, Including the Results of Simulations. 52 pp.

—**Addressing Systems**
CENELEC PRETS 300 175-6-91. Radio Equipment and Systems Digital European Cordless Telecommunications Common Interface Part 6: Identities and Addressing (DE/RES 3001-6). 40 pp.
CENELEC PRETS 300 175-6-92. Radio Equipment and Systems (RES); Digital European Cordless Telecommunications (DECT) Common Interface Part 6: Identities and Addressing. 130 pp.
CENELEC ETS 300 175-6-92. Radio Equipment and Systems (RES); Digital European Cordless Telecommunications (DECT) Common Interface Part 6: Identities and Addressing. 40 pp.
CENELEC PRI-ETS 300 176-91. Radio Equipment and Systems Digital European Cordless Telecommunications Approval Test Specification (DI/RES 3002). 109 pp.
CENELEC PRI-ETS 300 176-92. Radio Equipment and Systems (RES); Digital European Cordless Telecommunications (DECT) Approval Test Specification. 117 pp.
CENELEC I-ETS 300 176-92. Radio Equipment and Systems (RES); Digital European Cordless Telecommunications (DECT) Approval Test Specification. 118 pp.
ETSI ETS 300 175-6-92. Radio Equipment and Systems (RES); Digital European Cordless Telecommunications (DECT) Common Interface Part 6: Identities and Addressing. 40 pp.
ETSI PRETS 300 175-6-92. Radio Equipment and Systems (RES); Digital European Cordless Telecommunications (DECT) Common Interface Part 6: Identities and Addressing. 130 pp.
ETSI PRETS 300 175-6-91. Radio Equipment and Systems Digital European Cordless Telecommunications Common Interface Part 6: Identities and Addressing (DE/RES 3001-6). 40 pp.

European Digital Cordless Telecommunications (Cont.)
—**Addressing Systems** (Cont.)
ETSI I-ETS 300 176-92. Radio Equipment and Systems (RES); Digital European Cordless Telecommunications (DECT) Approval Test Specification. 118 pp.
ETSI PRI-ETS 300 176-92. Radio Equipment and Systems (RES); Digital European Cordless Telecommunications (DECT) Approval Test Specification. 117 pp.
ETSI PRI-ETS 300 176-91. Radio Equipment and Systems Digital European Cordless Telecommunications Approval Test Specification (DI/RES 3002). 109 pp.

—**Data Link Layer (OSI)**
CENELEC PRETS 300 175-4-91. Radio Equipment and Systems Digital European Cordless Telecommunications Common Interface Part 4: Data Link Control Layer (DE/RES 3001-4). 137 pp.
CENELEC PRETS 300 175-4-92. Radio Equipment and Systems (RES); Digital European Cordless Telecommunications (DECT) Common Interface Part 4: Data Link Control Layer. 43 pp.
CENELEC ETS 300 175-4-92. Radio Equipment and Systems (RES); Digital European Cordless Telecommunications (DECT) Common Interface Part 4: Data Link Control Layer. 126 pp.
ETSI ETS 300 175-4-92. Radio Equipment and Systems (RES); Digital European Cordless Telecommunications (DECT) Common Interface Part 4: Data Link Control Layer. 126 pp.
ETSI PRETS 300 175-4-92. Radio Equipment and Systems (RES); Digital European Cordless Telecommunications (DECT) Common Interface Part 4: Data Link Control Layer. 43 pp.
ETSI PRETS 300 175-4-91. Radio Equipment and Systems Digital European Cordless Telecommunications Common Interface Part 4: Data Link Control Layer (DE/RES 3001-4). 137 pp.

—**Electromagnetic Compatibility**
CENELEC PRETS 300 329-93. Radio Equipment and Systems (RES); Electro-Magnetic Compatibility (EMC) for Digital European Cordless Telecommunications (DECT) Equipment. 24 pp.
ETSI PRETS 300 329-93. Radio Equipment and Systems (RES); Electro-Magnetic Compatibility (EMC) for Digital European Cordless Telecommunications (DECT) Equipment. 24 pp.

—**Fixed Termination—PICS**
CENELEC PRETS 300 323-6-93. Radio Equipment and Systems (RES); Digital European Cordless Telecommunications (DECT) Public Access Profile (PAP) Test Specification Part 6: FT PICS Proforma. 161 pp.
ETSI PRETS 300 323-6-93. Radio Equipment and Systems (RES); Digital European Cordless Telecommunications (DECT) Public Access Profile (PAP) Test Specification Part 6: FT PICS Proforma. 161 pp.

—**Fixed Termination—PIXIT**
CENELEC PRETS 300 323-7-93. Radio Equipment and Systems (RES); Digital European Cordless Telecommunications (DECT) Public Access Profile (PAP) Test Specification Part 7: FT PIXIT Proforma. 46 pp.
ETSI PRETS 300 323-7-93. Radio Equipment and Systems (RES); Digital European Cordless Telecommunications (DECT) Public Access Profile (PAP) Test Specification Part 7: FT PIXIT Proforma. 46 pp.

—**Frequency Bands**
CEPT T/R 22-02 E-89. Frequency Band to Be Designated for the European Digital-Cordless Telecommunication System (DECT). 1 p.
EC 91/287/EEC-91. Council Directive on the Frequency Band to be Designated for the Coordinated Introduction of Digital European Cordless Telecommunications (DECT) into the Community. 2 pp.

—**Information Security**
CENELEC PRETS 300 175-7-91. Radio Equipment and Systems Digital European Cordless Telecommunications Common Interface Part 7: Security Features (DE/RES 3001-7). 99 pp.
CENELEC PRETS 300 175-7-92. Radio Equipment and Systems (RES); Digital European Cordless Telecommunications (DECT) Common Interface Part 7: Security Features. 102 pp.
CENELEC ETS 300 175-7-92. Radio Equipment and Systems (RES); Digital European Cordless Telecommunications (DECT) Common Interface Part 7: Security Features. 104 pp.
ETSI ETS 300 175-7-92. Radio Equipment and Systems (RES); Digital European Cordless Telecommunications (DECT) Common Interface Part 7: Security Features. 104 pp.

European Digital Cordless Telecommunications (Cont.)
—**Information Security** (Cont.)
ETSI PRETS 300 175-7-92. Radio Equipment and Systems (RES); Digital European Cordless Telecommunications (DECT) Common Interface Part 7: Security Features. 102 pp.
ETSI PRETS 300 175-7-91. Radio Equipment and Systems Digital European Cordless Telecommunications Common Interface Part 7: Security Features (DE/RES 3001-7). 99 pp.

—**Interfaces**
CENELEC ETR 043-92. Radio Equipment and Systems (RES); Digital European Cordless Telecommunications (DECT) Common Interface Services and Facilities Requirements Specification. 94 pp.
ETSI ETR 043-92. Radio Equipment and Systems (RES); Digital European Cordless Telecommunications (DECT) Common Interface Services and Facilities Requirements Specification. 94 pp.

—**Medium Access Control Layer (OSI)**
CENELEC PRETS 300 175-3-91. Radio Equipment and Systems Digital European Cordless Telecommunications Common Interface Part 3: Medium Access Control Layer (DE/RES 3001-3). 176 pp.
CENELEC PRETS 300 175-3-92. Radio Equipment and Systems (RES); Digital European Cordless Telecommunications (DECT) Common Interface Part 3: Medium Access Control Layer. 194 pp.
CENELEC ETS 300 175-3-92. Radio Equipment and Systems (RES); Digital European Cordless Telecommunications (DECT) Common Interface Part 3: Medium Access Control Layer. 197 pp.
ETSI PRETS 300 175-3-92. Radio Equipment and Systems (RES); Digital European Cordless Telecommunications (DECT) Common Interface Part 3: Medium Access Control Layer. 194 pp.
ETSI PRETS 300 175-3-91. Radio Equipment and Systems Digital European Cordless Telecommunications Common Interface Part 3: Medium Access Control Layer (DE/RES 3001-3). 176 pp.

—**Network Layer (OSI)**
CENELEC PRETS 300 175-5-91. Radio Equipment and Systems Digital European Cordless Telecommunications Common Interface Part 5: Network Layer (DE/RES 3001-5). 224 pp.
CENELEC PRETS 300 175-5-92. Radio Equipment and Systems (RES); Digital European Cordless Telecommunications (DECT) Common Interface Part 5: Network Layer. 246 pp.
CENELEC ETS 300 175-5-92. Radio Equipment and Systems (RES); Digital European Cordless Telecommunications (DECT) Common Interface Part 5: Network Layer. 240 pp.
ETSI ETS 300 175-5-92. Radio Equipment and Systems (RES); Digital European Cordless Telecommunications (DECT) Common Interface Part 5: Network Layer. 240 pp.
ETSI PRETS 300 175-5-92. Radio Equipment and Systems (RES); Digital European Cordless Telecommunications (DECT) Common Interface Part 5: Network Layer. 246 pp.
ETSI PRETS 300 175-5-91. Radio Equipment and Systems Digital European Cordless Telecommunications Common Interface Part 5: Network Layer (DE/RES 3001-5). 224 pp.

—**Physical Layer (OSI)**
CENELEC PRETS 300 175-2-91. Radio Equipment and Systems Digital European Cordless Telecommunications Common Interface Part 2: Physical Layer (DE/RES 3001-2). 36 pp.
CENELEC PRETS 300 175-2-92. Radio Equipment and Systems (RES); Digital European Cordless Telecommunications (DECT) Common Interface Part 2: Physical Layer. 38 pp.
CENELEC ETS 300 175-2-92. Radio Equipment and Systems (RES); Digital European Cordless Telecommunications (DECT) Common Interface Part 2: Physical Layer. 37 pp.
ETSI ETS 300 175-2-92. Radio Equipment and Systems (RES); Digital European Cordless Telecommunications (DECT) Common Interface Part 2: Physical Layer. 37 pp.
ETSI PRETS 300 175-2-92. Radio Equipment and Systems (RES); Digital European Cordless Telecommunications (DECT) Common Interface Part 2: Physical Layer. 38 pp.
ETSI PRETS 300 175-2-91. Radio Equipment and Systems Digital European Cordless Telecommunications Common Interface Part 2: Physical Layer (DE/RES 3001-2). 36 pp.
ETSI ETS 300 175-3-92. Radio Equipment and Systems (RES); Digital European Cordless Telecommunications (DECT) Common Interface Part 3: Medium Access Control Layer. 197 pp.

INTERNATIONAL AND NON-U.S. NATIONAL STANDARDS
SUBJECT INDEX

European

European Digital Cordless Telecommunications (Cont.)

—Portable Termination—PICS

CENELEC PRETS 300 323-3-93. Radio Equipment and Systems (RES); Digital European Cordless Telecommunications (DECT) Public Access Profile (PAP) Test Specification Part 3: PT PICS Proforma. 158 pp.

ETSI PRETS 300 323-3-93. Radio Equipment and Systems (RES); Digital European Cordless Telecommunications (DECT) Public Access Profile (PAP) Test Specification Part 3: PT PICS Proforma. 158 pp.

—Portable Termination—PIXIT

CENELEC PRETS 300 323-4-93. Radio Equipment and Systems (RES); Digital European Cordless Telecommunications (DECT) Public Access Profile (PAP) Test Specification Part 4: PT PIXIT Proforma. 52 pp.

ETSI PRETS 300 323-4-93. Radio Equipment and Systems (RES); Digital European Cordless Telecommunications (DECT) Public Access Profile (PAP) Test Specification Part 4: PT PIXIT Proforma. 52 pp.

—Public Access Profile

CENELEC PRETS 300 175-9-91. Radio Equipment and Systems Digital European Cordless Telecommunications Common Interface Part 9: Public Access Profile (DE/RES 3001-9). 68 pp.

CENELEC PRETS 300 175-9-92. Radio Equipment and Systems (RES); Digital European Cordless Telecommunications (DECT) Common Interface Part 9: Public Access Profile. 71 pp.

CENELEC ETS 300 175-9-92. Radio Equipment and Systems (RES); Digital European Cordless Telecommunications (DECT) Common Interface Part 9: Public Access Profile. 69 pp.

CENELEC PRETS 300 323-1-93. Radio Equipment and Systems (RES); Digital European Cordless Telecommunications (DECT) Public Access Profile (PAP) Test Specification Part 1: Overview. 107 pp.

ETSI ETS 300 175-9-92. Radio Equipment and Systems (RES); Digital European Cordless Telecommunications (DECT) Common Interface Part 9: Public Access Profile. 69 pp.

ETSI PRETS 300 175-9-92. Radio Equipment and Systems (RES); Digital European Cordless Telecommunications (DECT) Common Interface Part 9: Public Access Profile. 71 pp.

ETSI PRETS 300 175-9-91. Radio Equipment and Systems Digital European Cordless Telecommunications Common Interface Part 9: Public Access Profile (DE/RES 3001-9). 68 pp.

ETSI PRETS 300 323-1-93. Radio Equipment and Systems (RES); Digital European Cordless Telecommunications (DECT) Public Access Profile (PAP) Test Specification Part 1: Overview. 107 pp.

—Speech Encoding

CENELEC PRETS 300 175-8-91. Radio Equipment and Systems Digital European Cordless Telecommunications Common Interface Part 8: Speech Coding and Transmission (DE/RES 3001-8). 37 pp.

CENELEC PRETS 300 175-8-92. Radio Equipment and Systems (RES); Digital European Cordless Telecommunications (DECT) Common Interface Part 8: Speech Coding and Transmission. 40 pp.

CENELEC ETS 300 175-8-92. Radio Equipment and Systems (RES); Digital European Cordless Telecommunications (DECT) Common Interface Part 8: Speech Coding and Transmission. 37 pp.

ETSI ETS 300 175-8-92. Radio Equipment and Systems (RES); Digital European Cordless Telecommunications (DECT) Common Interface Part 8: Speech Coding and Transmission. 37 pp.

ETSI PRETS 300 175-8-92. Radio Equipment and Systems (RES); Digital European Cordless Telecommunications (DECT) Common Interface Part 8: Speech Coding and Transmission. 40 pp.

ETSI PRETS 300 175-8-91. Radio Equipment and Systems Digital European Cordless Telecommunications Common Interface Part 8: Speech Coding and Transmission (DE/RES 3001-8). 37 pp.

—Speech Transmission Systems

CENELEC PRETS 300 175-8-91. Radio Equipment and Systems Digital European Cordless Telecommunications Common Interface Part 8: Speech Coding and Transmission (DE/RES 3001-8). 37 pp.

CENELEC PRETS 300 175-8-92. Radio Equipment and Systems (RES); Digital European Cordless Telecommunications (DECT) Common Interface Part 8: Speech Coding and Transmission. 40 pp.

CENELEC ETS 300 175-8-92. Radio Equipment and Systems (RES); Digital European Cordless Telecommunications (DECT) Common Interface Part 8: Speech Coding and Transmission. 37 pp.

European Digital Cordless Telecommunications (Cont.)

—Speech Transmission Systems (Cont.)

ETSI ETS 300 175-8-92. Radio Equipment and Systems (RES); Digital European Cordless Telecommunications (DECT) Common Interface Part 8: Speech Coding and Transmission. 37 pp.

ETSI PRETS 300 175-8-92. Radio Equipment and Systems (RES); Digital European Cordless Telecommunications (DECT) Common Interface Part 8: Speech Coding and Transmission. 40 pp.

ETSI PRETS 300 175-8-91. Radio Equipment and Systems Digital European Cordless Telecommunications Common Interface Part 8: Speech Coding and Transmission (DE/RES 3001-8). 37 pp.

—Speech Transmission Systems—Interworking

CENELEC ETR 041-92. Transmission and Multiplexing (TM); Digital European Cordless Telecommunication (DECT) Transmission Aspects 3, 1 kHz Telephony Interworking with Other Networks. 27 pp.

ETSI ETR 041-92. Transmission and Multiplexing (TM); Digital European Cordless Telecommunication (DECT) Transmission Aspects 3,1 kHz Telephony Interworking with Other Networks. 27 pp.

European Economic Community
Use: European Communities

European Economic Interest Groupings
Use For: EEIG

EC EEC/2137/85-85. Council Regulation on the European Economic Interest Grouping EEIG. 9 pp.

European Free Trade Association
Use: EFTA

European Monetary System
Use For: EMS *See Also:* Currencies; European Communities

—Currencies

EURO 1991 Mar. Community and the Third World. 9 pp.

European Radio Messaging Systems
Use: Radio Paging Systems

European Standards
See Also: AECMA; CECC; CEN; Cenelec; CEPT; ETSI; European Communities; Standards

EC 83/189/EEC-83. Council Directive Laying Down a Procedure for the Provision of Information in the Field of Technical Standards and Regulations (for Rel Stds See TEST/83/189). 5 pp.

EC TEST/83/189-83. Information Procedure—Technical Regulations. 4 pp.

EC 88/182/EEC-88. Council Directive Amending Directive 83/189/EEC Laying Down a Procedure for the Provision of Information in the Field of Technical Standards and Regulations. 2 pp.

EURO 1988 CS. Common Standards for Enterprises. 74 pp.

EURO 1988 Dec. New Community Standards Policy. 185 pp.

EURO 1990. Green Paper on Standards Development (COM (90) 456). 63 pp.

—Bulletins

EURO 1992-05 Bulletin-92. Bulletin of the European Standards Organizations. 19 pp.

EURO 1992-06 Bulletin-92. Bulletin of the European Standards Organizations. 24 pp.

EURO 1992-07 Bulletin-92. Bulletin of the European Standards Organizations. 26 pp.

—CEN

EC SM 5. Public Procurement. 4 pp.

EURO MEMORANDUM 1-88. Status of the European Standard. 8 pp.

EURO MEMORANDUM 2-77. Consumer Interests and the Preparation of Standards. 7 pp.

EURO MEMORANDUM 5-89. Trade Unions and the Preparation of European Standards. 8 pp.

EURO 1988 Part 2. CEN/Cenelec Internal Regulations Part 2 Common Rules for Standards work. 53 pp.

EURO 1989 ED 13. CEN N525: Edition 13. National Implementation of European Standards. 337 pp.

EURO 1990 ED 14. CEN N525: Edition 14. National Implementation of European Standards. 509 pp.

EURO 1990 Part 2. CEN/Cenelec Internal Regulations Part 2: Common Rules for Standard Work. 60 pp.

EURO 1993 CEN STANDARDS. CEN Standards for Access to the European Market. 382 pp.

EURO 1983 Jan. Rules for Presentation of European Standards. 58 pp.

European Standards (Cont.)

—CEN (Cont.)

EURO 1988 CS. Common Standards for Enterprises. 74 pp.

—CEN—Certification Systems

EURO 1983 Part 3. CEN Internal Regulations Part 3: Directives for Certification Work Edition 1. 30 pp.

EURO 1989 March. European Framework for Testing and Certification. 27 pp.

—CEN—Construction Materials

EURO 1989 June. The European Harmonization of Construction Products. 213 pp.

—CEN—Medical Equipment

EURO 1988 Dec 12-14. Workshop of the European Standardization of Medical Devices. 611 pp.

—Cenelec

CENELEC ETR 010-93. ISDN Standards Management (ISM); the ETSI Basic Guide on the European Integrated Services Digital Network. 41 pp.

CENELEC ETR 076-93. Integrated Services Digital Network (ISDN); Standards Guide. 88 pp.

ETSI ETR 010-93. ISDN Standards Management (ISM); the ETSI Basic Guide on the European Integrated Services Digital Network. 41 pp.

ETSI ETR 076-93. Integrated Services Digital Network (ISDN); Standards Guide. 88 pp.

EURO MEMORANDUM 1-88. Status of the European Standard. 8 pp.

EURO MEMORANDUM 2-77. Consumer Interests and the Preparation of Standards. 7 pp.

EURO MEMORANDUM 5-89. Trade Unions and the Preparation of European Standards. 8 pp.

EURO 1988 Part 2. CEN/Cenelec Internal Regulations Part 2 Common Rules for Standards work. 53 pp.

EURO 1990 Part 2. CEN/Cenelec Internal Regulations Part 2: Common Rules for Standard Work. 60 pp.

EURO MEMORANDUM 1-74. Interrelation Between Regulations and Standards. 13 pp.

EURO MEMORANDUM 15-89. Establishment of Technical Specification. 11 pp.

EURO 1983 Jan. Rules for Presentation of European Standards. 58 pp.

EURO 1984 Dec. Cenelec EN/HD Rules. 55 pp.

EURO 1989-01-30. National Implementation of Approved Documents. 636 pp.

EURO 1988 CS. Common Standards for Enterprises. 74 pp.

EURO 1989. Electrotechnical Standards for the European Economic Space. 34 pp.

—Cenelec—Certification Systems

EURO 1989 March. European Framework for Testing and Certification. 27 pp.

EURO MEMORANDUM 6-88. Adoption of New Standards as a Basis for the Approval of Products in the Cenelec Member Countries. 13 pp.

EURO MEMORANDUM 11-82. Elements for the Assessment of Reciprocity in the Field of Certification. 16 pp.

EURO MEMORANDUM 13-84. Cenelec Certification Agreement (CCA). 48 pp.

—Cenelec—Electrical Equipment—Safety

EURO MEMORANDUM 2-74. Preparation of Standards For Safety in the Design and Construction of Electrical Equipment. 5 pp.

—Cenelec—Low Voltage Directive

EURO MEMORANDUM 3-87. Implementation of the EEC Low Voltage Direction of 19 February 1973 with Respect to Reports, Certificates and Manufacturer's Declaration of Conformity. 50 pp.

EURO MEMORANDUM 4-84. Certification Organizations for Electrical Equipment Within Cenelec Countries Granting Marks of Conformity Related to Safety. 21 pp.

EURO MEMORANDUM 5-78. Policy Paper on National Deviations to Harmonisation Documents with Particular Reference to the Low Voltage Directive 73/23/ EEC of the European Economic Community. 13 pp.

EURO MEMORANDUM 10-82. Publication of Cenelec Results in the Field of the Low Voltage Directive in the Form of European Standards. 13 pp.

—Cenelec—Medical Equipment

EURO 1988 Dec 12-14. Workshop of the European Standardization of Medical Devices. 611 pp.

—Consumer Participation

EC 88/41/EEC-87. Council Recommendation on the Involvement and Improvement of Consumer Participation in Standardization. 1 p.

EURO MEMORANDUM 2-77. Consumer Interests and the Preparation of Standards. 7 pp.

INTERNATIONAL AND NON-U.S. NATIONAL STANDARDS
SUBJECT INDEX

European

European Standards *(Cont.)*

—**Information Systems**
EURO STANDARDIZAT ION-FACTST 3-88. Community Standardization Policy. 3 pp.
EURO STANDARDIZAT ION-FACTST 4-88. Standards Bodies: the Changing Scene. 3 pp.
EURO STANDARDIZAT ION-FACTST 5-88. European Standards. 3 pp.

—**Intellectual Property Rights**
EURO MEMORANDUM 8-92. Standardization and Intellectual Property Rights (IPR). 20 pp.

—**Labor Union Participation**
EURO MEMORANDUM 5-89. Trade Unions and the Preparation of European Standards. 8 pp.

—**Reviews**
EURO 1989-3 REVIEW. Ongoing Activities in European Standards. 37 pp.
EURO 1989-4 REVIEW. Ongoing Activities in European Standards. 31 pp.
EURO 1989-5 REVIEW. Ongoing Activities in European Standards. 37 pp.
EURO 1989-6 REVIEW. Ongoing Activities in European Standards. 39 pp.
EURO 1989-7 REVIEW. Ongoing Activities in European Standards. 30 pp.
EURO 1989-9 REVIEW. Ongoing Activities in European Standards. 29 pp.
EURO 1989-10 REVIEW. Ongoing Activities in European Standards. 26 pp.
EURO 1989-11 REVIEW. Ongoing Activities in European Standards. 30 pp.
EURO 1989-12 REVIEW. Ongoing Activities in European Standards. 26 pp.
EURO 1990-01 REVIEW. Ongoing Activities in European Standards. 16 pp.
EURO 1990-02 REVIEW. Ongoing Activities in European Standards. 28 pp.
EURO 1990-03 REVIEW. Ongoing Activities in European Standards. 26 pp.
EURO 1990-04 REVIEW. Ongoing Activities in European Standards. 27 pp.
EURO 1990-05 REVIEW. Ongoing Activities in European Standards. 25 pp.
EURO 1990-06 REVIEW. Ongoing Activities in European Standards. 22 pp.
EURO 1990-07 REVIEW. Ongoing Activities in European Standards. 23 pp.
EURO 1990-09 REVIEW. Ongoing Activities in European Standards. 49 pp.
EURO 1990-10 REVIEW. Ongoing Activities in European Standards. 31 pp.
EURO 1990-11 REVIEW. Ongoing Activities in European Standards. 24 pp.
EURO 1990-12 REVIEW. Ongoing Activities in European Standards. 31 pp.
EURO 1991-01 REVIEW. Ongoing Activities in European Standards. 33 pp.
EURO 1991-02 REVIEW. Ongoing Activities in European Standards. 33 pp.
EURO 1991-03 REVIEW. Ongoing Activities in European Standards. 36 pp.
EURO 1991-04 REVIEW. Ongoing Activities in European Standards. 39 pp.
EURO 1991-05 REVIEW. Ongoing Activities in European Standards. 40 pp.
EURO 1991-06 REVIEW. Ongoing Activities in European Standards. 32 pp.
EURO 1991-07 REVIEW. Ongoing Activities in European Standards. 45 pp.
EURO 1991-09 REVIEW. Ongoing Activities in European Standards. 58 pp.
EURO 1991-10 REVIEW. Ongoing Activities in European Standards. 31 pp.
EURO 1991-11 REVIEW. Ongoing Activities in European Standards. 48 pp.
EURO 1991-12 REVIEW. Ongoing Activities in European Standards. 38 pp.
EURO 1992-01 REVIEW. Ongoing Activities in European Standards. 44 pp.
EURO 1992-02 REVIEW. Ongoing Activities in European Standards. 44 pp.
EURO 1992-03 REVIEW. Ongoing Activities in European Standards. 51 pp.

European Telecommunications Satellite Organization

Use For: EUTELSAT *See Also:* European Communications Satellite; Satellite Communications; Telecommunication Equipment

—**Business Services**
CEPT T/PGT 15-87. Principes Generaux De Tarification Et De Comptabilite Relatifs Aux Services Numeriques Internationaux D'Affairesofferts Via Les Systemes Multiservices Par Satellites (SMS) D'EUTELSAT. 5 pp.

European Telecommunications Standards Institute

Use: ETSI

European Telephone Networks

See Also: Telephone Networks; Telephone Systems

—**Announcements (Telephone Services)—Glossaries**
CEPT T/CS 28-02-81. Denomination Et Signification Des Tonalites Et Designations Generales Pour Announces Parlees. 3 pp.
CEPT T/CS 28-02 E-86. Names and Meanings of Tones and General Designation for Verbal Announcements. 4 pp.

—**Signaling Systems**
CEPT T/CS 14-01-83. Utilisation Du Systeme De Signalisation R 2 Dans Les Reseaux Telephoniques Europeens. 1 p.
CEPT T/CS 14-01 E-86. Use of Signalling System R2 in European Telephone Networks. 1 p.

—**Signaling Systems—Format**
CEPT T/CS 01-02-82. Mise En Forme Des Specifications Relatives a La Signalisation. 1 p.
CEPT T/CS 01-02 E-82. Format of Signalling System Specifications. 1 p.

—**Tones (Telephone Services)—Glossaries**
CEPT T/CS 28-02-81. Denomination Et Signification Des Tonalites Et Designations Generales Pour Announces Parlees. 3 pp.
CEPT T/CS 28-02 E-86. Names and Meanings of Tones and General Designation for Verbal Announcements. 4 pp.

Eurotrans Committee Mandates
Use: CEN

EUTELSAT
Use: European Telecommunications Satellite Organization

Euthanasia Cabinets
Use: Veterinary Equipment—Euthanasia Cabinets

EVA
Use: Ethylene Vinyl Acetate Copolymers

Evacuation Equipment
Use For: Escape Chutes; Escape Slides
See Also: Aircraft Equipment

—**Aircraft**
CAA Chapter P6-6. Life Rafts and Escape Chutes (Provisional Airworthiness Requirements for Civil Powered Lift Aircraft). 4 pp.
CAA LEAFLET 14-18. Air Cruisers Company Evacuation Systems. 1 p.
CAA Spec. NO. 6 ISSUE 2.

—**Helicopters**
CAA NOTICE #27 ISSUE 2. Helicopter Emergency Escape Facilities. 4 pp.

—**Rotary Wing Aircraft**
CAA Chapter G6-6 12.80. Life Rafts and Escape Chutes/Slides (Rotocraft). 2 pp.

Evaluation
See Also: Figure of Merit; Inspection; Military Intelligence; Quality Assurance; Quality Control

—**Aeronautical Navigation Charts**
NATO STANAG 2215 ED 5 AMD 1-89. Evaluation of Land Maps, Aeronautical Charts and Digital Topographic Data. 29 pp.

—**Maps**
NATO STANAG 2215 ED 5 AMD 1-89. Evaluation of Land Maps, Aeronautical Charts and Digital Topographic Data. 29 pp.

—**Topography**
NATO STANAG 2215 ED 5 AMD 1-89. Evaluation of Land Maps, Aeronautical Charts and Digital Topographic Data. 29 pp.

Evaporated Milk
See Also: Beverages; Condensed Milk; Dairy Products; Milk

CGSB 32.166M-89. Milk, Condensed or Evaporated (Supercedes 32-GP-167M October 1978, 32-GP-174M October 1978). 8 pp.
CNS N5057-80. Evaporated Milk (Aug)(2342). 2 pp.

—**Fat Content—Gravimetric Analysis**
BSI BS 1742: Part 3-90. 1990 Methods for Chemical Analysis of Condensed Milks Part 3: Determination of Fat Content. 12 pp.
ISO 1737-85. Evaporated Milk and Sweetened Condensed Milk—Determination of Fat Content—Gravimetric Method (Reference Method) Second Edition. 10 pp.

Evaporated Milk *(Cont.)*

—**Solids Content**
BSI BS 1741: Part 2-90. 1990 Methods for Chemical Analysis of Liquid Milk and Cream Part 2: Determination of Total Solids Content of Liquid Milk, Cream and Unsweetened Condensed Milk. 7 pp.
ISO 6731-89. Milk, Cream and Evaporated Milk—Determination of Total Solids Content (Reference Method) First Edition. 6 pp.

—**Test Specimens**
CNS N6056-80. Method of Test for Milk and Milk Products (General Rules) (Aug)(3440). 2 pp.

Evaporation Loss
Use For: Weight Loss (Evaporation)

—**Cable Insulation**
DIN VDE 0472 Pt 612-83. Testing of Cables and Insulated Cords; Weight Loss by Evaporation (Aug). 8 pp.

—**Chemical Products**
CNS K0019-82. Method for Determining Loss and Residue of Chemical Products (May)(8838).
JIS K 0067-92. Test Methods for Loss and Residue of Chemical Products. 18 pp.

—**Corrosion Inhibitors**
CGSB 31-GP-0A METH 2.3-62. Methods of Testing Corrosion-Prevention Materials and Processes Loss by Evaporation. 1 p.

—**Greases**
CNS K6769-83. Method of Test for Evaporation Loss of Lubricating Greases and Oils (May)(10267).
CNS K6770-83. Method of Test for Evaporation Loss of Lubricating Greases Over Wide-Temperature Range (May)(10268).

—**Lubricating Oils**
CNS K6769-83. Method of Test for Evaporation Loss of Lubricating Greases and Oils (May)(10267).
DIN ENGL 51581-83. Testing of Lubricants; Determination of Evaporation Loss of Lubricating Oils; (Noack Test) (Sept). 4 pp.

—**Plasticizers—Coated Fabrics**
SAA AS 1441.8-73. Methods of Test for Coated Fabrics—Part 8: Method for Determination of Plasticizer Evaporation.

—**Scouring Powders**
MOD UK M 9504/59. Determination of Powder, Cleaning.

Evaporation Residue Analysis
See Also: Residue Content Analysis

—**Ammonia**
BSI BS 4431-89. 1989 Methods of Sampling and Test for Liquefied Anhydrous Ammonia. 16 pp.
ISO 4276-78. Anhydrous Ammonia for Industrial Use—Evaluation of Residue on Evaporation—Gravimetric Method First Edition. 5 pp.
ISO 7109-85. Ammonia Solution for Industrial Use—Determination of Residue After Evaporation at 105 Degrees Celsius—Gravimetric Method First Edition. 3 pp.

—**Ammonia Liquor**
MOD UK M 2006/60. Examination of Ammonia Liquor.

—**Ammonium Hydroxide**
BSI BS 4651: Part 3-87. 1987 Ammonia Solution Part 3: Method for Determination of Residue on Evaporation. 3 pp.
BSI BS 4651: Part 4-88. 1988 Ammonia Solution Part 4: Methods for Determination of Residue on Heating at 850 Degrees Celsius. 4 pp.

—**Aromatic Hydrocarbons**
ISO 5277-81. Aromatic Hydrocarbons—Determination of Residue on Evaporation of Products Having Boiling Points up to 150 Degrees Celsius First Edition. 4 pp.

—**Chemical Products**
CNS K0019-82. Method for Determining Loss and Residue of Chemical Products (May)(8838).
JIS K 0067-92. Test Methods for Loss and Residue of Chemical Products. 18 pp.

—**Coatings**
CGSB 1-GP-71 METH 20.1-78. Methods of Testing Paints and Pigments Residue on Evaporation. 1 p.

—**Detergents**
MOD UK M 9521/65. Examination of Detergent Solution for Plastic Food Utensils.

INDEX and DIRECTORY of

INTERNATIONAL AND NON-U.S. NATIONAL STANDARDS
SUBJECT INDEX

Evaporation Residue Analysis (Cont.)
—**Emulsified Asphalts**
CNS K6795-83. Method of Test for Residue by Evaporation of Emulsified Asphalts (Jul)(10456).
CNS K6796-83. Method of Test for Examination of Residue Obtained from Distillation or Evaporation of Emulsified Asphalts (Jul)(10457).

—**Essential Oils**
CNS K6702-81. Essential Oils Quantitative Evaluation of Residue on Evaporation (Nov)(8130).
ISO 4715-78. Essential Oils—Quantitative Evaluation of Residue on Evaporation First Edition. 4 pp.

—**Fluorohydrocarbons**
BSI BS 5598: Part 8-80. 1980 Methods of Sampling and Test for Halogenated Hydrocarbons Part 8: Determination of Non-Volatile Residue in Fluorinated Hydrocarbons. 4 pp.
ISO 5789-79. Fluorinated Hydrocarbons for Industrial Use—Determination of Non-Volatile Residue First Edition. 4 pp.

—**Halogenated Hydrocarbons**
BSI BS 5598: Part 3-79. 1979 Methods of Sampling and Test for Halogenated Hydrocarbons Part 3: Determination of Residue on Evaporation. 4 pp.
CNS K6562-80. Method of Test for Evaporation Residue of Liquid Halogenated Hydrocarbons (Aug)(6207).
ISO 2210-72. Liquid Halogenated Hydrocarbons for Industrial Use—Determination of Residue on Evaporation First Edition. 4 pp.

—**Hydrocarbons**
DIN ENGL 51613-66. Testing of Liquefied Petroleum Gases; Determination of Elementary Sulphur and Residue on Evaporation (Apr). 3 pp.

—**Liquefied Petroleum Gas**
DIN ENGL 51613-66. Testing of Liquefied Petroleum Gases; Determination of Elementary Sulphur and Residue on Evaporation (Apr). 3 pp.

—**Naphthalene**
CNS K6650-81. Method of Test for Evaporation Residue of Naphthalene (Mar)(7169).

—**Phenolic Resins**
BSI BS 2782:Pt4: METH 451M-91. 1991 Methods of Testing Plastics Part 4: Chemical Properties Method 451M: Determination of Non-Volatile Material in Liquid Phenolic Resins (ISO 8618: 1987) (V). 5 pp.
ISO 8618-87. Plastics—Liquid Phenolic Resins—Conventional Determination of Non-Volatile Matter First Edition. 4 pp.

—**Phenols**
ISO 1897 Pt 10-82. Phenol, o-Cresol, m-Cresol, p-Cresol, Cresylic Acid and Xylenols for Industrial Use—Methods of Test—Part 10: Determination of Dry Residue After Evaporation on a Water Bath (Excluding Cresylic Acid and Xylenols) First Edition. 4 pp.

—**Propane**
CNS K6252-67. Method of Test for Evaporation Residue of Commercial Propane Gas (F.P. of Mercury Method) (Aug)(2753)(R 1973).

—**Refrigeration Oils**
DIN ENGL 51590 Pt 1-85. Testing of Lubricants; Determination of the R 12 Insolubles in Refrigerator Oils Using the-30 Degrees Celsius Precipitation Method (Sept). 3 pp.

—**Titanium Tetrachloride**
MOD UK M 888/72. Examination of Titanium Tetrachloride (Withdrawn).

—**Volatile Fluids**
ISO 759-81. Volatile Organic Liquids for Industrial Use—Determination of Dry Residue After Evaporation on a Water Bath—General Method First Edition. 4 pp.

—**Waste Water**
CNS K9005-81. Methods of Test for Industrial Waste Water Solid Matter and Evaporation Residue (Apr)(3793).
DIN ENGL 38409 Pt 1-87. German Standard Methods for the Examination of Water, Waste Water and Sludge; Parameters Characterizing Effects and Substances (Group H); Determination of Total Dry Residue, Filtrate Dry Residue and Residue on Ignition (H 1) (Jan). 4 pp.

—**Water**
DIN ENGL 38409 Pt 1-87. German Standard Methods for the Examination of Water, Waste Water and Sludge; Parameters Characterizing Effects and Substances (Group H); Determination of Total Dry Residue, Filtrate Dry Residue and Residue on Ignition (H 1) (Jan). 4 pp.

Evaporation Residue Analysis (Cont.)
—**Water Baths**
BSI BS 4524-83. 1983 Determination of Residue on Evaporation on a Water Bath. 4 pp.

Evaporative Coolers
See Also: Coolers; Cooling Systems; Cooling Towers; Water Coolers

—**Motor Operated**
CSA C22.2 NO 104-M1983. Humidifiers and Evaporative Coolers; (Gen Instr 1 Thru 2). 37 pp.

—**Motor Operated—Cooling Systems**
CSA C22.2 NO 104-M1983. Humidifiers and Evaporative Coolers; (Gen Instr 1 Thru 2). 37 pp.

—**Motor Operated—Heating Elements**
CSA C22.2 NO 104-M1983. Humidifiers and Evaporative Coolers; (Gen Instr 1 Thru 2). 37 pp.

—**Motor Operated—Heating Equipment**
CSA C22.2 NO 104-M1983. Humidifiers and Evaporative Coolers; (Gen Instr 1 Thru 2). 37 pp.

—**Motor Operated—In Duct**
CSA C22.2 NO 104-M1983. Humidifiers and Evaporative Coolers; (Gen Instr 1 Thru 2). 37 pp.

—**Motor Operated—Residential**
CSA C22.2 NO 104-M1983. Humidifiers and Evaporative Coolers; (Gen Instr 1 Thru 2). 37 pp.

Evaporative Emission Control Systems
Use: Exhaust Emission Control Equipment

Evaporators
Scope Note: Use a more specific term
See: Compressor/Evaporators

Evra Propellers
Use: Aircraft Propellers—Evra

EWS
Use: Emergency Water Supplies

Excavating Equipment
Scope Note: For additional listings, use a more specific term Use For: Excavators
See Also: Backhoes; Boring Machines; Bulldozers; Caterpillars (Vehicles); Construction Equipment; Dredging Equipment; Earthmoving Equipment; Front End Loaders; Graders; Hydraulic Excavating Equipment; Mining Equipment; Pits (Excavations); Scrapers (Earthmoving Equipment); Shaft Sinking; Tractors; Trench Excavators; Tunneling (Excavations)
BSI BS 1761-51. 1951 Single Bucket Excavators of the Crawler Mounted Friction-Driven Type. 60 pp.

—**Cable Operated—Glossaries**
DIN ENGL 24080-79. Earth-Moving Machinery; Hydraulic Excavators; Cable-Operated, Excavators; Terms (Mar). 5 pp.

—**Manual Controls**
BSI BS 5528-81. 1981 Amd 1 Operator's Controls on Excavators Used for Earth Moving. 5 pp.
CNS A4021-88. Earth-Moving Machinery Excavator-Operator's Controls (Dec)(9952).
ISO 4557-82. Earth-Moving Machinery—Excavators—Operator's Controls Third Edition. 4 pp.
JIS A 8405-90. Excavators—Operator's Controls. 16 pp.

—**Noise—Limitation**
EC COM(93) 154-93. Commission Proposal for a Council Directive to Amend Council Directive 86/662/EEC on the Limitation of Noise Emitted by Earth-Moving Machinery (93/C 157/11). 4 pp.
EC 86/662/EEC-86. Council Directive on the Limitation of Noise Emited by Hydraulic Excavators, Rope-Operated Excavators, Dozers, Loaders and Excavator-Loaders. 11 pp.

—**Safety**
CEN PREN 474-5-93. Earth-Moving Machinery—Safety—Part 5: Requirements for Hydraulic Excavators. 16 pp.

Excavation
Use: Earthwork

Excavators
Use: Excavating Equipment

Exchange Call Set Up Delay
See Also: Delay; Seizing (Telecommunications); Telephone Exchanges; Time

Exchange Call Set Up Delay (Cont.)
—**Grade of Service—Digital Exchanges**
CCITT RECMN E.543-89. Grades of Service in Digital International Telephone Exchanges—Telephone Network and ISDN—Quality of Service, Network Management and Traffic Engineering (Study Group II) 4 pp. 4 pp.

Exchange Capacity Measurement
Use For: Ion Exchange Capacity Measurement
See Also: Capacity Measurement

—**Anion Exchangers**
DIN ENGL 54402-85. Testing of Ion Exchangers; Determination of the Total Capacity of Anion Exchangers (July). 9 pp.

—**Soils—Cation**
DIN ENGL 19684 Pt 8-77. Methods of Soil Analysis for Water Management for Agricultural Purposes; Chemical Laboratory Tests; Determination of Exchange Capacity of a Soil and of the Amount of Exchangeable Cations (Feb). 3 pp.

Exchange Control Systems
—**Network Management Systems**
CCITT RECMN E.412-89. Network Management Controls—Telephone Network and ISDN—Quality of Service, Network Management and Traffic Engineering (Study Group II) 9 pp. 9 pp.

Exchangers (Ion)
Use: Ion Exchangers

Exchanges (Telephone)
Use: Telephone Exchanges

Excise Taxes
See Also: Taxes; Value-Added Tax

—**Alcoholic Beverages**
EC COM(85) 150-85. Proposal for a Council Directive Laying Down Certain Rules on Indirect Taxes Which Affect the Consumption of Alcoholic Drinks. 10 pp.
EC COM(85) 151-85. Proposal for a Council Directive Concerning the Harmonization of Excise Duties on Fortified Wine and Similar Products. 24 pp.
EC COM(87) 328-87. Proposal for a Council Directive Concerning Approximation of the Rates of Excise Duty on Alcoholic Beverages and on the Alcohol Contained in Other Products. 18 pp.
EC 92/12/EEC-92. Council Directive on the General Arrangements for Products Subject to Excise Duty and on the Holding, Movement and Monitoring of Such Products. 13 pp.

—**Beer**
EC COM(87) 328-87. Proposal for a Council Directive Concerning Approximation of the Rates of Excise Duty on Alcoholic Beverages and on the Alcohol Contained in Other Products. 18 pp.

—**Cigarettes**
EC COM(87) 325/2-87. Proposal for a Council Directive on the Approximation of Taxes on Cigarettes. 23 pp.

—**Exemptions**
EC 85/348/EEC-85. Council Directive Amending Directive 69/169/EEC on the Harmonization of Provisions Laid Down by Law, Regulation or Administrative Action Relating to Exemption from Turnover Tax and Excise Duty on Imports in International Travel. 3 pp.

—**Exemptions—Non-Commercial Goods**
EC 85/349/EEC-85. Council Directive Amending Directive 74/651/EEC on the Tax Reliefs to be Allowed on the Importation of Goods in Small Consignments of a Non-Commercial Character Within the Community. 2 pp.

—**Mineral Oils**
EC COM(87) 327/2-87. Proposal for a Council Directive on the Approximation of the Rates of Excise Duty on Mineral Oils. 15 pp.
EC 92/12/EEC-92. Council Directive on the General Arrangements for Products Subject to Excise Duty and on the Holding, Movement and Monitoring of Such Products. 13 pp.

—**Rate System**
EC COM(87) 324-87. Proposal for a Council Directive Instituting a Process of Convergence of Rates of Value-Added Tax and Excise Duties. 12 pp.

—**Tobacco**
EC COM(87) 326/2-87. Proposal for a Council Directive on the Approximation of Taxes on Manufactured Tobacco Other Than Cigarettes. 23 pp.

INDUSTRY STANDARDS

Excise Taxes (Cont.)
—Tobacco (Cont.)
EC 80/1275/EEC-80. Council Directive Amending Directive 72/464/EEC on Taxes Other than Turnover Taxes Which Affect the Consumption of Manufactured Tobacco. 1 p.
EC 92/12/EEC-92. Council Directive on the General Arrangements for Products Subject to Excise Duty and on the Holding, Movement and Monitoring of Such Products. 13 pp.

—Wines
EC COM(87) 328-87. Proposal for a Council Directive Concerning Approximation of the Rates of Excise Duty on Alcoholic Beverages and on the Alcohol Contained in Other Products. 18 pp.

Excitation Systems
Use: Exciters

Exciter Coils
Use: Exciters

Exciter Lamps
See Also: Lamps

—Motion Picture Projectors
BSI BS 5550: SUBSEC 7.1.1-80. 1980 Amd 3 Cinematography Part 7: Production and Presentation Section 7.1: Light Sources and Lighting Subsection 7.1.1: Exciter Lamps. 20 pp.

Exciters
Use For: Excitation Systems See Also: Antennas; Oscillators; Starters; Transmitters

—Synchronous Machines—Glossaries
BSI BS 4999: Sec 116.1-92. 1992 Rotating Electrical Machines Part 116: Excitation Systems for Synchronous Machines Sec 116.1: Definitions (IEC 34-16-1: 1991). 14 pp.
CENELEC HD 53.16.1 S1-91. Rotating Electrical Machines Part 16: Excitation Systems for Synchronous Machines Chapter 1: Definitions. 5 pp.
IEC 34 Pt 16-1-91. Rotating Electrical Machines Part 16: Excitation Systems for Synchronous Machines Chapter 1: Definitions First Edition; (Corrigendum—Apr 1992). 24 pp.

—Synchronous Machines—Power System Stability—Models
IEC 34 Pt 16-2-91. Rotating Electrical Machines Part 16: Excitation Systems for Synchronous Machines Chapter 2: Models for Power System Studies First Edition. 71 pp.

Exclusive NOR Gates
Use For: Equivalence Gates; KNOR
See Also: Digital Circuits; Exclusive OR NOR Gates; Gate Circuits
CECC CECC 90 104-052 ISSUE 2. NL CECC 90 104-052 Issue 2; Digital Integrated Circuits in Accordance with FS 90 104; HEC/HEF 4077B; Quadruple Exclusive-NOR Gate (En). 6 pp.
CECC CECC 90 109-723 ISSUE 1-87. Digital Integrated Circuits in Accordance with FS 90 109; 54 HC 7266; 74 HC 7266; Exclusive NOR Gate (En, Fr). 6 pp.
CECC CECC 90 109-753 ISSUE 1-87. Digital Integrated Circuits in Accordance with FS 90 109; 54 HC 266, 74 HC 266; Exclusive NOR-Gate with Open Drain Outputs (En, Fr). 5 pp.

—Quad
CECC CECC 90 104-146 ISSUE 1-81. BS CECC 90 104-146; Silicon Complementary MOS with (B) Buffered Outputs and Cavity Packaging (En). 23 pp.
CECC CECC 90 104-146 ISSUE 1-86. CEI CECC 90 104-146; Silicon Complementary MOS with (B) Buffered Outputs Cavity and Non Cavity Packaging (En). 2 pp.

—2-Input
CECC CECC 90 103-182 ISSUE 1-82. BS 90 103-182; Quadruple 2-Input Exclusive NOR Gates with Open Collector Outputs (En). 16 pp.

—2-Input—Preferred Products List
CECC CECC MUAHAG Vol 7 IS 8-92. Preferred Products List; Active Microcircuits (En, Fe, Ge). 89 pp.

Exclusive OR Gates
Use For: XOR Gates See Also: Digital Circuits; Exclusive OR NOR Gates; Gate Circuits
CECC CECC 90 104-187 ISSUE 1-81. BS CECC 90 104-187 Issue 1; Quad Exclusive OR Gate (En). 22 pp.
CECC CECC 90 104-187 ISSUE 1-86. CEI CECC 90 104-187 Issue 1; Quad Exclusive OR Gate (En). 2 pp.

Exclusive OR Gates (Cont.)
CECC CECC 90 109-620 ISSUE 1-86. Digital Integrated Circuits in Accordance with FS 90 109; 54 HC 86, 74 HC 86; Exclusive OR Gates (En, Fr). 5 pp.
CECC CECC 90 109-710 ISSUE 1-87. Digital Integrated Circuits in Accordance with FS 90 109; 54 HCT 86, 74 HCT 86; Exclusive OR Gates (En, Fr). 6 pp.
CECC CECC 90 109-717 ISSUE 1-87. Digital Integrated Circuits in Accordance with FS 90 109; 54 HC 386, 74 HC 386; Exclusive OR-Gates (En, Fr). 6 pp.

—Preferred Products List
CECC CECC MUAHAG Vol 7 IS 8-92. Preferred Products List; Active Microcircuits (En, Fe, Ge). 89 pp.

—2-Input
CECC CECC 90 101-028 ISSUE 1-87. UTE C 86-213/B 48; Digital Integrated Circuits in Accordance with FS 90 101; 54/64/74 86; Quad 2-Input Exclusive OR Gate (En). 11 pp.
CECC CECC 90 101-040 ISSUE 1-87. UTE C 86-213/B 60; Digital Integrated Circuits in Accordance with FS 90 101; 54/64/74 136; Quad 2-Input Exclusive OR Gate with Open Collector Outputs (En). 18 pp.
CECC CECC 90 102-022 ISSUE 1-81. BS CECC 90 102-022; Quadruple 2-Input Exclusive OR Gates (En) AMD 1 (En). 18 pp.
CECC CECC 90 103-040 ISSUE 1-81. BS CECC 90 103-040; Quadruple 2-Input Exclusive OR Gates (En) AMD 1 (En). 17 pp.
CECC CECC 90 103-127 ISSUE 1-81. BS CECC 90 103-127; Quad 2-Input Exclusive OR Gate Circuits (En) AMD 1 (En). 18 pp.
CECC CECC 90 104-046 ISSUE 2. NL CECC 90 104-046 Issue 2; Digital Integrated Circuits in Accordance with FS 90 104; HEC/HEF 4070B; Quadruple Exclusive-OR Gate (En). 6 pp.
CECC CECC 90 106-060 ISSUE 1-85. UTE C 86-217; Digital Integrated Circuits in Accordance with FS 90 106; 54 ALS 86, 74 ALS 86; Exclusive OR Gate (En, Fr) ADD 3 (En, Fr). 15 pp.
CECC CECC 90 107-011 ISSUE 2-89. UTE C 86-218 ADD 2/FA 11; Digital Integrated Circuits in Accordance with FS 90 107; 54/74 F 86; Exclusive OR-Gate (En, Fr) ADD 3 (En, Fr). 9 pp.

—2-Input—Preferred Products List
CECC CECC MUAHAG Vol 7 IS 8-92. Preferred Products List; Active Microcircuits (En, Fe, Ge). 89 pp.

—4-Input
CECC CECC 90 104-125 ISSUE 1-81. BS CECC 90 104-125; Silicon Complementary MOS with (B) Buffered Outputs and Cavity Packaging (En). 23 pp.
CECC CECC 90 104-125 ISSUE 1-86. CEI CECC 90 104-125; Silicon Complementary MOS with (B) Buffered Outputs Cavity and Non Cavity Packaging (En). 2 pp.

Exclusive OR NOR Gates
See Also: Exclusive NOR Gates; Exclusive OR Gates; Gate Circuits

—Quad
CECC CECC 90 102-049 ISSUE 1-82. BS CECC 90 102-049; Quadruple Exclusive OR/NOR Gates (En). 18 pp.

Exercise Bicycles
See Also: Bicycles; Cycles; Exercise Equipment; Sports Equipment
CNS B7200-82. Method of Test for Exerciser Bike (Sep)(9339).

—Stationary—Safety
CEN PREN 957-5-92. Stationary Training Equipment—Part 5: Additional Specific Safety Requirements and Test Methods for Pedal Crank Training Equipment (Type P). 10 pp.

Exercise Books
Use: Notebooks

Exercise Equipment
Use For: Fitness Equipment See Also: Exercise Bicycles; Sports Equipment

—Chest Expanders
CNS S1221-90. Expander (Sep)(12791). 2 pp.

—Motor Operated
CSA CAN/CSA-C22. 2 NO 68-92. Motor-Operated Appliances (Household and Commercial); (Gen Instr 1 Thru 2). 115 pp.
CSA 1169 Bull. Electrical Bulletin 1169 June 27, 1978 to C22.2 NO 68. 2 pp.

Exercise Equipment (Cont.)
—Stationary—Benches—Safety
CEN PREN 957-4-92. Stationary Training Equipment—Part 4: Additional Specific Safety Requirements and Test Methods for Strength Training Benches (Type B). 5 pp.

—Stationary—Safety
CEN PREN 957-1-92. Stationary Training Equipment—Part 1: General Safety Requirements and Test Methods. 9 pp.
CEN PREN 957-2-92. Stationary Training Equipment—Part 2: Additional Specific Safety Requirements and Test Methods for Domestic Strength Training Equipment (Type H). 5 pp.
CEN PREN 957-3-92. Stationary Training Equipment—Part 3: Additional Specific Safety Requirements and Test Methods for Institutional Strength Training Equipment (Type S). 7 pp.

Exhaust Blowers
Use For: Exhausters See Also: Blowers (Ventilation)

—Acceptance Testing
BSI BS 1571: Part 1-87. 1987 Testing of Positive Displacement Compressors and Exhausters Part 1: Methods for Acceptance Testing. 64 pp.
BSI BS 1571: Part 2-75. 1975 Amd 3 Testing of Positive Displacement Compressors and Exhausters Part 2: Methods for Simplified Acceptance Testing for Air Compressors and Exhausters (AMD 5576) August 28, 1987. 22 pp.
BSI BS 1571: Part 2-04. 1975 Amd 4 Testing of Positive Displacement Compressors and Exhausters Part 2: Methods for Simplified Acceptance Testing for Air Compressors and Exhausters (AMD 6879) February 28, 1992. 23 pp.
SNZ NZS/BS 1571: Part 1-87. Testing of Positive Displacement Compressors and Exhausters Part 1: Methods for Acceptance Testing. 64 pp.
SNZ NZS/BS 1571: Part 2-75. Testing of Positive Displacement Compressors and Exhausters Part 2: Simplified Acceptance Tests for Air Compressors and Exhausters Amend: 3; 4, 1992. 24 pp.

—Turbo
ISO 5389-92. Turbocompressors—Performance Test Code First Edition. 177 pp.

Exhaust Emission Control Equipment
Use For: Emission Control Devices (Engines); Emission Control Equipment; Evaporative Emission Control Systems See Also: Air Pollution; Catalytic Converters; Exhaust Gases

—Automotive
CNS D1017-81. Test Code of Exhaust Emission Control Devices for Automobiles (Oct)(7980). 12 pp.
CNS D3173-87. Method of Test for Evaporative Emission Control System for Gasoline Engine Automobiles (Sep)(11534). 10 pp.

Exhaust Gases
Use For: Waste Gases See Also: Air Monitoring; Air Pollution; Diluents; Effluents; Emissions; Exhaust Emission Control Equipment; Exhaust Pipes; Exhaust Systems; Flue Gases; Fumes; Gases; Gasohol; Industrial Wastes; Odors; Rocket Exhaust; Vapors; Waste Disposal

—Acrolein Content
JIS K 0089-83. Methods for Determination of Acrolein in Exhaust Gas. 12 pp.

—Ammonia Content
CNS K9099-82. Methods of Test for Ammonia in Exhaust Gas (Jul)(9186).
JIS K 0099-83. Methods for Determination of Ammonia in Exhaust Gas. 22 pp.

—Automotive
CNS D3097-89. Exhaust Gas Measurement of Carburetors for Automobiles (Jan)(8254).
EC COM(92) 290-92. Amendment to the Proposal for a Council Directive Amending Directive 77/143/EEC on the Approximation of the Laws of the Member States Relating to Roadworthiness Tests for Motor Vehicles and Their Trailers (Emissions). 2 pp.
EC 92/55/EEC-92. Council Directive Amending Directive 77/143/EEC on the Approximation of the Laws of the Member States Relating to Roadworthiness Tests for Motor Vehicles and Their Trailers (Exhaust Emissions). 4 pp.

—Benzene Content
CNS K9108-90. Method for Determination of Benzene in Exhaust Gas (Mar)(12688).
JIS K 0088-83. Methods for Determination of Benzene in Exhaust Gas. 13 pp.

—Bromine Content
JIS K 0085-83. Methods for Determination of Bromine in Exhaust Gas. 13 pp.

INTERNATIONAL AND NON-U.S. NATIONAL STANDARDS
SUBJECT INDEX

Exhaust

Exhaust Gases (Cont.)

—Carbon Dioxide Content—Automotive
  JIS D 1030-76. Analytical Procedure for Continuous Measurement of Carbon Monoxide, Carbon Dioxide and Hydrocarbon in Automobile Exhaust Gas.

—Carbon Disulfide Content
  CNS K9111-90. Method of Test for Carbon Disulfide in Exhaust Gas (Dec)(12825).
  JIS K 0091-83. Methods for Determination of Carbon Disulfide in Exhaust Gas. 12 pp.

—Carbon Monoxide Content
  CNS K9109-90. Method of Test for Carbon Monoxide in Exhaust Gas (Dec)(12824).
  JIS K 0098-88. Methods for Determination of Carbon Monoxide in Flue Gas. 9 pp.

—Carbon Monoxide Content—Automotive
  CNS D3050-80. Measuring Methods of Carbon Monoxide in Automobile Exhaust Gas (Idling) (May)(5562).
  ISO 3929-76. Road Vehicles—Determination of Exhaust Carbon Monoxide Concentration at Idle Speed First Edition. 4 pp.
  JIS D 1028-83. Measuring Method of Carbon Monoxide in Automobile Exhaust Gas (Idling). 5 pp.
  JIS D 1030-76. Analytical Procedure for Continuous Measurement of Carbon Monoxide, Carbon Dioxide and Hydrocarbon in Automobile Exhaust Gas.
  JIS D 8006-83. Simplified Carbonmonoxide Testers for Automobiles. 9 pp.
  SNZ NZS 5429-82. Code of Practice for In-Field Testing of Road Vehicles for Exhaust Carbon Monoxide and Hydrocarbons Concentration at Idle Speed. 8 pp.

—Chemical Analysis
  CNS K9018-76. Composition Analysis for Waste Gases of Combustion Equipments (Dec)(4049).

—Chlorine Content
  CNS K9062-81. Methods of Test for Chlorine in Exhaust Gas (O-Tolidine Method) (Mar)(7054).
  JIS K 0106-82. Methods for Determination of Chlorine in Exhaust Gas. 14 pp.

—Concentration (Composition)—Automotive
  CNS D3185-89. Method of Test for Exhausts Concentration for Gasoline Engine Automobiles (Jul)(12567).

—Diesel Engines
  CNS D3174-88. Method of Test for Exhaust Smoke Under Noload and Rapid Acceleration for Diesel Engine Automobiles (Sep)(11644). 10 pp.
  CNS D3175-88. Method of Test for Exhaust Smoke Under Full Load and Steady Speed for Diesel Engine Automobiles (Sep)(11645). 9 pp.
  EC 88/77/EEC-87. Council Directive on the Approximation of the Laws of the Member States Relating to the Measures to be Taken Against the Emission of Gaseous Pollutants from Diesel Engines for Use in Vehicles. 29 pp.
  EC 91/542/EEC-91. Council Directive Amending Directive 88/77/EEC on the Approximation of the Laws of the Member States Relating to the Measures to be Taken Against the Emission of Gaseous Pollutants from Diesel Engines for Use in Vehicles. 19 pp.

—Diesel Engines—Tractors
  EC 77/537/EEC-77. Council Directive on the Approximation of the Laws of the Member States Relating to the Measures to Be Taken Against the Emission of Pollutants from Diesel Engines for Use in Wheeled Agricultural or Forestry Tractors. 22 pp.

—Fiber Content
  BSI BS 6069: Sec 4.2-91. 1991 Characterization of Air Quality Part 4: Stationary Source Emissions Section 4.2: Method for the Determination of Asbestos Plant Emissions by Fibre Count Measurement. 24 pp.

—Fluorine Compound Content
  CNS K9069-81. Method of Test for Fluorine Compounds in Exhaust Gas (Aug)(7810).
  JIS K 0105-82. Methods for Determination of Fluorine Compounds in Exhaust Gas. 25 pp.

—Gasoline Engines—Dynamometers
  CNS D3077-87. Method of Test for Exhaust Pollution for Gasoline Engine Automobiles (Sep)(7895). 16 pp.

—Hydrocarbon Content—Automotive
  CNS D3173-87. Method of Test for Evaporative Emission Control System for Gasoline Engine Automobiles (Sep)(11534). 10 pp.
  JIS D 1030-76. Analytical Procedure for Continuous Measurement of Carbon Monoxide, Carbon Dioxide and Hydrocarbon in Automobile Exhaust Gas.

Exhaust Gases (Cont.)

—Hydrocarbon Content—Automotive (Cont.)
  SNZ NZS 5429-82. Code of Practice for In-Field Testing of Road Vehicles for Exhaust Carbon Monoxide and Hydrocarbons Concentration at Idle Speed. 8 pp.

—Hydrogen Chloride Content
  CNS K9063-81. Method of Test for Hydrogen Chloride in Exhaust Gas (Mar)(7055).
  JIS K 0107-82. Methods for Determination of Hydrogen Chloride in Exhaust Gas. 15 pp.

—Hydrogen Cyanide Content
  JIS K 0109-82. Methods for Determination of Hydrogen Cyanide in Exhaust Gas. 23 pp.

—Hydrogen Sulfide Content
  CNS K9070-81. Method of Test for Hydrogen Sulfide in Exhaust Gas (Aug)(7811).
  JIS K 0108-83. Analytical Methods for Determination of Hydrogen Sulfide in Exhaust Gas. 23 pp.

—Internal Combustion Piston Engines
  CEN PREN 28178-2-93. Reciprocating Internal Combustion Engines—Exhaust Emission Measurement Part 2: Measurement of Gaseous and Particulate Emissions at Site. 42 pp.
  CEN PREN 28178-4-93. Reciprocating Internal Combustion Engines—Exhaust Emission Measurement Part 4: Test Cycles for Different Engine Applications. 26 pp.

—Motorcycles
  CNS D3184-88. Method of Test for Exhaust Smoke Under No Load and Acceleration Conditions for Motorcycles (May)(12299).

—Motorcycles—Dynamometers
  CNS D3165-87. Method of Test for Exhaust Pollution for Motorcycles (Sep)(11386). 14 pp.

—Nitrogen Dioxide Content
  CNS K9022-77. Method of Test for Nitrogen Dioxide in Exhaust Gases (Saltzman Method) (Apr)(4084).

—Nitrogen Oxide Content
  CNS K9023-77. Method of Test for Oxides of Nitrogen in Exhaust Gases (PDS Phenoldisulfonic Acid Method) (Apr)(4085).

—Opacity—Diesel Engines
  ISO 3173-74. Road Vehicles—Apparatus for Measurement of the Opacity of Exhaust Gas from Diesel Engines Operating Under Steady State Conditions First Edition. 29 pp.
  ISO TR4011-76. Road Vehicles—Apparatus for Measurement of the Opacity of Exhaust Gas from Diesel Engines First Edition. 17 pp.
  ISO 7644-88. Road Vehicles—Measurement of Opacity of Exhaust Gas from Compression-Ignition (Diesel) Engines—Lug-Down Test First Edition. 6 pp.
  ISO 7645-88. Road Vehicles—Measurement of Opacity of Exhaust Gas from Compression-Ignition (Diesel) Engines—Steady Single-Speed Test First Edition. 4 pp.

—Phenol Content
  CNS K9106-90. Method for Determination of Phenols in Exhaust Gas (Mar)(12686).
  JIS K 0086-83. Methods of Determination of Phenols in Exhaust Gas. 15 pp.

—Phosgene Content
  JIS K 0090-83. Method of Determination of Phosgene in Exhaust Gas. 7 pp.

—Pyridine Content
  CNS K9107-90. Method for Determination of Pyridine in Exhaust Gas (Mar)(12687).
  JIS K 0087-83. Methods of Determination of Pyridine in Exhaust Gas. 9 pp.

—Sampling
  DIN ENGL 51853-79. Testing of Fuel Gases, Protective Gases and Exhaust Gases; Sampling (Jan). 10 pp.

—Smoke Content—Aircraft Engines
  CAA Chapter M4-1 05.86. General (Emissions Certification). 1 p.
  CAA Chapter M4-3 05.86. Smoke Emission Evaluation (Emissions Certification). 5 pp.
  CAA Chapter M4-6 05.86. Test Schedules—Subsonic Engines (Emissions Certification). 1 p.
  CAA Chapter M4-7 05.86. Test Schedules—Supersonic Engines (Emissions Certification). 1 p.
  CAA Chapter M4-8 05.86. Calculation of Smoke Numbers (Emissions Certification). 1 p.

—Smoke Content—Diesel Engines
  CNS D4009-83. Reflection Type Smokemeters for Measuring Smoke Concentration of Exhaust Gas for Diesel Automobiles (Jan)(9845).

Exhaust Gases (Cont.)

—Smoke Content—Diesel Engines (Cont.)
  ISO TR9310-87. Road Vehicles—Smoke Measurement of Compression-Ignition (Diesel) Engines—Survey of Short in-Service Tests First Edition. 18 pp.
  JIS D 1101-85. Diesel Engine Smoke Measurement. 5 pp.
  JIS D 8004-71. Reflection Type Smokemeters for Measuring Carbon Concentration of Exhaust Smoke for Diesel Automobiles. 12 pp.

—Smoke Content—Tractors
  ISO 789 Pt 4-86. Agricultural Tractors—Test Procedures—Part 4: Measurement of Exhaust Smoke Second Edition. 9 pp.

—Sulfur Dioxide Content
  BSI BS 6069: Sec 4.1-90. 1990 Characterization of Air Quality Part 4: Stationary Source Emissions Section 4.1: Method for the Determination of the Mass Concentration of Sulphur Dioxide: Hydrogen Peroxide/Barium Perchlorate/Thorin Method. 10 pp.
  ISO 7934-89. Stationary Source Emissions—Determination of the Mass Concentration of Sulfur Dioxide—Hydrogen Peroxide/Barium Perchlorate/Thorin Method First Edition. 9 pp.

—Sulfur Dioxide Content—Automatic Test Equipment
  BSI BS 6069: Sec 4.4-93. 1993 Characterization of Air Quality Part 4: Stationary Source Emissions Section 4.4: Determination of the Mass Concentration of Sulfur Dioxide—Performance Characteristics of Auto-mated Measuring Methods (ISO 7935: 1992). 16 pp.
  ISO 7935-92. Stationary Source Emissions—Determination of the Mass Concentration of Sulphur Dioxide—Performance Characteristics of Automated Measuring Methods First Edition. 16 pp.

—Thiol Content
  JIS K 0092-83. Methods for Determination of Mercaptan in Exhaust Gas. 15 pp.

—Thiol Content—Spectrophotometry
  CNS K9098-82. Method of Test for Mercaptan in Exhaust Gas — Spectrophotometric Method (Jul)(9185).

Exhaust Hoods

Scope Note: Use a more specific term See: Range Hoods; Ventilation Equipment

Exhaust Pipes

See Also: Exhaust Gases; Exhaust Systems; Internal Combustion Engines; Internal Combustion Piston Engines; Pipes; Piston Engines; Vents

—Marine—Flanges
  BSI BS MA 9-70. 1970 Amd 3 Flanges, Bolting and Gaskets for Exhaust Gas Piping for Diesel Engines and Boiler Uptakes. 10 pp.
  JIS F 7805-76. Basic Dimensions of Steel Flanges for Marine Exhaust Gas Pipe.

—Motorcycles—Packings (Seals)
  CNS D2140-88. Exhaust Pipe Packing for Motorcycle (Jul)(8957).

—Vitreous Enameled
  JIS A 5533-71. Porcelain Enameled Exhaust Pipes.

Exhaust Systems

Use For: Air Exhaust Equipment; Mechanical Exhaust Systems See Also: Air Cleaners; Air Pollution; Blowers (Ventilation); Chimneys; Cooling Systems; Ducts; Engines; Exhaust Gases; Exhaust Pipes; Flues; Ventilation Equipment; Waste Disposal
  CSA CAN/CSA-C260-M90. Rating the Performance of Residential Mechanical Ventilating Equipment; (Gen Instr 1). 50 pp.

—Aerostats—Engines
  CAA Chapter Q5-6 12.79. Exhaust Systems (Non-Rigid Airships). 2 pp.

—Aircraft Engines
  CAA Chapter K5-6 04.74. Exhaust Systems (Light Aeroplanes). 2 pp.
  CAA Chapter P5-6. Compressed Air and Exhaust Systems (Provisional Airworthiness Requirements for Civil Powered Lift Aircraft). 2 pp.

—Automotive
  EC 92/97/EEC-92. Council Directive Amending Directive 70/157/EEC on the Approximation of the Laws of the Member States Relating to the Permissible Sound Level and the Exhaust System of Motor Vehicles. 31 pp.

INDUSTRY STANDARDS

Exhaust

INTERNATIONAL AND NON-U.S. NATIONAL STANDARDS
SUBJECT INDEX

Exhaust Systems *(Cont.)*
—Heaters—Aircraft—Repair/Maintenance
CAA NOTICE #41 ISSUE 8. Maintenance of Cockpit and Cabin Combustion Heaters and Their Associated Exhaust Systems (Airworthiness Notices). 2 pp.

—Inspection
JIS A 4303-76. Inspection Standards of Smoke Exhaust Equipment.

—Internal Combustion Piston Engines—Glossaries
BSI BS 7016: Part 4-88. 1988 Components and Systems of Reciprocating Internal Combustion Engines Part 4: Glossary of Terms for Pressure Charging and Air/Exhaust Gas Ducting Systems. 17 pp.
ISO 7967 Pt 4-88. Reciprocating Internal Combustion Engines—Vocabulary of Components and Systems—Part 4: Pressure Charging and Air/Exhaust Ducting Systems First Edition. 18 pp.

—Mopeds
BSI BS AU 193A-90. 1990 Replacement Motor Cycle and Moped Exhaust Systems. 11 pp.

—Motorcycles
BSI BS AU 193A-90. 1990 Replacement Motor Cycle and Moped Exhaust Systems. 11 pp.
EC 78/1015/EEC-78. Council Directive on the Approximation of the Laws of the Member States on the Permissible Sound Level and Exhaust System of Motorcycles. 10 pp.

—Residential Buildings
CSA CAN/CSA-F326-M91. Residential Mechanical Ventilation Systems; (Gen Instr 1). 97 pp.

—Rotary Wing Aircraft
CAA Chapter G5-6 06.76. Exhaust Systems (Rotocraft). 2 pp.
CAA CAP 524 SUB-Part E 12.86. Power-Plant (Rotorcraft). 27 pp.

Exhaust Valves
See Also: Engine Cylinders; Inlet Valves; Valves

—Motorcycles
CNS D2190-88. Intake and Exhaust Valves for Motorcycles (Jul)(12374).

Exhausters
Use: Exhaust Blowers

Exhibition Facilities
—Electrical Installations
DIN VDE 0108 Pt 3-89. Power Installations and Safety Power Supply in Communal Buildings; Stores and Shops and Exhibition Rooms (Oct) (VDE 0107/11.89 with DIN VDE 0108 Part 1 to Part 8 (Issues of 10.89) Supersedes DIN 57108/VDE 0108/12.79). 9 pp.
DIN VDE 0108 Pt 8-89. Power Installations and Safety Power in Communal Buildings; Temporary Bldgs. Used in Communal Facilities, Stores and Shops, Exhibition Rooms, Public Houses and Restaurants (Oct) (VDE 0107/11.89 with DIN VDE 0108 Pt 1 to Pt 8(Issues of 10.89) Supersedes DIN 57108/VDE 0108/12.79). 11 pp.

—Emergency and Standby Power Supplies
DIN VDE 0108 Pt 8-89. Power Installations and Safety Power in Communal Buildings; Temporary Bldgs. Used in Communal Facilities, Stores and Shops, Exhibition Rooms, Public Houses and Restaurants (Oct) (VDE 0107/11.89 with DIN VDE 0108 Pt 1 to Pt 8(Issues of 10.89) Supersedes DIN 57108/VDE 0108/12.79). 11 pp.

Exhibitions
—Construction Contracts
MOD UK DEFCON 112W-90. Conditions of Contract Governing the Construction Erection and Dismantling of Exhibitions, Exhibition Stands and Standfitting 4/90. 4 pp.

Exits (Aircraft)
Use: Aircraft Doors

Expanded Metals
Use For: Expanded Steels *See Also:* Metals
CNS G3246-90. Expanded Steels (Jun)(12728).
JIS G 3351-87. Expanded Metals. 10 pp.

Expanded Polystyrene
Use: Polystyrene

Expanded Rubber
Use: Foam Rubber

Expanded Steels
Use: Expanded Metals

Expanders (Tools)
See Also: Pipe Cutters; Tools
CNS S2150-90. Method of Test for Expander (Sep)(12792). 3 pp.

Expanding
Use: Expansion

Expansion
Use For: Enlarging *See Also:* Elongation; Swelling; Thermal Expansion

—Bricks
SAA AS 1226.5-84. Methods of Sampling and Testing Clay Building Bricks—Part 5: Method for Determining Characteristic Expansion. 4 pp.

—Cements
CNS R3044-85. Method of Test for Autoclave Expansion of Portland Cement (May) (1258).

—Ceramic Tiles
BSI BS 6431: Part 21-84. 1984 Ceramic Floor and Tiles Part 21: Method for Determination of Moisture Expansion Using Boiling Water. Unglazed Tiles. 7 pp.
BSI BS 6431: Part 21-01. 1984 Amd 1 Ceramic Floor and Tiles Part 21: Method for Determination of Moisture Expansion Using Boiling Water. Unglazed Tiles (AMD 7110) July 15, 1992. 10 pp.
CEN EN 155-84. Ceramic Tiles: Determination of Moisture Expansion Using Boiling Water; Unglazed Tiles. 5 pp.
CEN EN 155-91. Ceramic Tiles—Determination of Moisture Expansion Using Boiling Water—Unglazed Tiles. 4 pp.

—Ceramics
CNS R3056-69. Method of Test for Moisture Expansion of Filed Fine Ceramic and Porcelain Products (Jan)(2892)(R 1975).

—Concretes
BSI BS 1881: Part 5-70. 1970 Amd 2 Methods of Testing Concrete Part 5: Methods of Testing Hardened Concrete for Other Than Strength (AMD 6267) March 30, 1990. 27 pp.

—Fabrics
BSI BS 4768-72. 1972 Amd 1 Determination of the Bursting Strength and Bursting Distension of Fabrics (AMD 6340) June 28, 1991. 9 pp.

—Hoses
BSI BS 5173: Sec 102.2-85. (WITHDRAWN) 1985 Amd 1 Rubber and Plastics Hoses and Hose Assemblies Part 102: Hydraulic Pressure Tests Section 102.2: Determination of Volumetric Expansion (AMD 5965) May 31, 1989 (Renumbered as BS EN 26801: 1993). 7 pp.
BSI BS EN 26801-93. 1993 Amd 2 Rubber and Plastics Hoses—Determination of Volumetric Expansion (AMD 7520) April 15, 1993. 14 pp.
CEN EN 26801-93. Rubber and Plastics Hoses—Determination of Volumetric Expansion (ISO 6801: 1983). 6 pp.
ISO 6801-83. Rubber or Plastics Hoses—Determination of Volumetric Expansion First Edition (CEN EN 26801: 1993). 5 pp.

—Mortars
CEN PREN 480 (Part 3)-91. Admixtures for Concrete, Mortar and Grout—Test Methods—Part 3: Determination of Shrinkage and Expansion. 10 pp.
CNS R3111-82. Method of Test for Potential Expansion of Portland Cement Mortars Exposed to Sulfate (Dec)(9746).
DIN ENGL 52450-85. Testing of Inorganic Non-Metallic Building Materials; Determination of Shrinkage and Expansion on Small Specimens (Aug). 8 pp.

—Porcelain
CNS R3056-69. Method of Test for Moisture Expansion of Filed Fine Ceramic and Porcelain Products (Jan)(2892)(R 1975).

Expansion Bolts
Use: Anchor Bolts

Expansion Compensators, Bellows
Use: Bellows Expansion Joints

Expansion Joints
See Also: Bellows; Bellows Expansion Joints; Joints; Pipe Joints

—Cast Iron—Ships
CNS F3056-80. Ships' Cast Iron Pipe Sleeve Type Expansion Joints (Aug)(6330).

Expansion Joints *(Cont.)*
—Cast Iron—Ships *(Cont.)*
CNS F3057-80. Ships' Cast Steel Pipe Sleeve Type Expansion Joint (Sep)(6437).
JIS F 3016-90. Ships' Cast Iron Pipe Sleeve Type Expansion Joints.
JIS F 3016-71. Ships' Cast Iron Pipe Sleeve Type Expansion Joints. 6 pp.

—Cork
ISO 3867-82. Agglomerated Cork Material of Expansion Joints for Construction and Building—Test Methods First Edition. 7 pp.

—Cork—Packaging
ISO 3869-81. Agglomerated Cork—Filler Material of Expansion Joints for Construction and Buildings—Characteristics, Sampling and Packing First Edition. 3 pp.

—Pressure Vessels
CNS B5089-83. Construction of Pressure Vessels (6) (Expansion Joints) (Jan)(9793).

—Steel
JIS F 3017-90. Ships' Cast Steel Pipe Sleeve Type Expansion Joints.
JIS F 3017-71. Ships' Cast Steel Pipe Sleeve Type Expansion Joints. 6 pp.

Expansion Plugs
See Also: Plug Valves; Safety Plugs

—Automotive
CNS D2036-89. Expansion Plugs for Automobiles (Jul)(5792).

—Drums (Containers)—Steel
JIS Z 1604-84. Plugs and Flanges for Steel Drums. 12 pp.

—Drums (Containers)—Steel—Oil
CNS Z5016-86. Plugs and Flanges for Steel Drums for Oil (Apr)(1322).

Expansion Ratio (Transmission Gain)
See Also: Compression Ratio (Transmission Gain); Signal Levels

—Compandors—Telephone Circuits
CCITT RECMN G.162-89. Characteristics of Compandors for Telephony—General Characteristics of International Telephone Connections and Circuits (Study Groups XII and XV) 8 pp. 8 pp.
CCITT RECMN G.166-89. Characteristics of Syllabic Compandors for Telephony on High Capacity Long Distance Systems—General Characteristics of International Telephone Connections and Circuits (Study Groups XII and XV) 7 pp. 7 pp.

Expansion Shields
MOD UK DSTAN 53-74-75. Shield, Expansion (Masonry Plugs, Machine Bolt Type, and Woodscrew Type) Issue 1. 13 pp.

Expansion Tanks
Use For: Expansion Vessels *See Also:* Tanks (Containers)

—Cooling Systems
BSI BS 7074: Part 3-89. 1989 Application, Selection and Installation of Expansion Vessels and Ancillary Equipment for Sealed Water Systems Part 3: Code of Practice for Chilled and Condenser Systems. 20 pp.

—Water Heaters
BSI BS 4814-90. 1990 Expansion Vessels Using an Internal Diaphragm, for Sealed Hot Water Heating Systems. 12 pp.
BSI BS 6144-90. 1990 Expansion Vessels Using an Internal Diaphragm, for Unvented Hot Water Supply Systems. 13 pp.
BSI BS 7074: Part 1-89. 1989 Application, Selection and Installation of Expansion Vessels and Ancillary Equipment for Sealed Water Systems Part 1: Code of Practice for Domestic Heating and Hot Water Supply. 12 pp.
BSI BS 7074: Part 2-89. 1989 Application, Selection and Installation of Expansion Vessels and Ancillary Equipment for Sealed Water Systems Part 2: Code of Practice for Low and Medium Temperature Hot Water Heating Systems. 24 pp.

Expansion Thermometers
See Also: Thermometers
BSI BS 1041: Sec 2.1-85. 1985 Temperature Measurement Part 2: Expansion Thermometers Section 2.1: Guide to Selection and Use of Liquid in Thermometers. 20 pp.
BSI BS 1041: Sec 2.2-89. 1989 Temperature Measurement Part 2: Expansion Thermometers Section 2.2: Guide to Selection and Use of Dial-Type Expansion Thermometers. 12 pp.

INTERNATIONAL AND NON-U.S. NATIONAL STANDARDS
SUBJECT INDEX

Expansion Thermometers *(Cont.)*
BSI BS 5235-75. 1975 Amd 1 Dial-Type Expansion Thermometers. 19 pp.

Expansion Valves
See Also: Gas Valves; Refrigeration Equipment; Refrigeration Valves; Valves

—**Storage Water Heaters**
BSI BS 6283: Part 1-91. 1991 Safety and Control Devices for Use in Hot Water Systems Part 1: Specification for Expansion Valves for Pressures up to and Including 10 Bar. 14 pp.
BSI BS 6283: Part 1-82. 1982 Safety Devices for Use in Hot Water Systems Part 1: Specification for Expansion Valves for Pressures up to and Including 10 Bar. 10 pp.

—**Thermostatic**
JIS B 8619-84. Methods of Testing for Performance of Thermostatic Refrigerant Expansion Valves.

Expansion Vessels
Use: Expansion Tanks

Expenses
Use For: Costs *See Also:* Accounting; Construction Contracts; Construction Costs; Cost Analysis; Cost Estimates; Economics; Prices

—**Circuit Groups**
CCITT RECMN E.522-89. Number of Circuits in a High-Usage Group—Telephone Network and ISDN—Quality of Service, Network Management and Traffic Engineering (Study Group II) 10 pp. 10 pp.

—**Circuit Groups—Alternative Routing**
CCITT RECMN E.522-89. Number of Circuits in a High-Usage Group—Telephone Network and ISDN—Quality of Service, Network Management and Traffic Engineering (Study Group II) 10 pp. 10 pp.

—**Interruption of Service—Telecommunication Networks**
CCITT RECMN E.862 (REV 1)-92. Dependability Planning of Telecommunication Networks (Study Group II) 16 pp. 16 pp.
CCITT RECMN E.862-89. Dependability Planning of Telecommunication Networks—Telephone Network and ISDN—Quality of Service, Network Management and Traffic Engineering (Study Group II) 12 pp. 12 pp.

—**Project Management**
BSI BS 6046: Part 4-92. 1992 Use of Network Techniques in Project Management Part 4: Guide to Resource Analysis and Cost Control (G). 34 pp.
BSI BS 6046: Part 4-81. 1981 Use of Network Techniques in Project Management Part 4: Guide to Resource Analysis and Cost Control. 21 pp.

—**Satellite Communications—Regional**
CCITT FASCICLE II.1-88. General Tariff Principles—Charging and Accounting in International Telecommunications Services Series D Recommendations. 371 pp.

—**Supergroups—Regional**
CCITT FASCICLE II.1-88. General Tariff Principles—Charging and Accounting in International Telecommunications Services Series D Recommendations. 371 pp.

—**Telecommunication Networks**
CCITT RECMN E.862 (REV 1)-92. Dependability Planning of Telecommunication Networks (Study Group II) 16 pp. 16 pp.
CCITT RECMN E.862-89. Dependability Planning of Telecommunication Networks—Telephone Network and ISDN—Quality of Service, Network Management and Traffic Engineering (Study Group II) 12 pp. 12 pp.

—**Telegraphy**
CCITT RECMN D.40 (REV 1)-92. General Tariff Principles Applicable to Telegrams Exchanged in the International Public Telegram Service (Study Group III) 7 pp. 7 pp.
CCITT RECMN D.40-89. General Tariff Principles Applicable to Telegrams Exchanged in the International Public Telegram Service—General Tariff Principles—Charging and Accounting in International Telecommunications Services (Study Group III) 4 pp. 4 pp.

—**Telephone Circuits—Europe**
CCITT FASCICLE II.1-88. General Tariff Principles—Charging and Accounting in International Telecommunications Services Series D Recommendations. 371 pp.

Expenses *(Cont.)*
—**Telephone Circuits—Mediterranean Basin**
CCITT FASCICLE II.1-88. General Tariff Principles—Charging and Accounting in International Telecommunications Services Series D Recommendations. 371 pp.

—**Telephone Circuits—Regional**
CCITT FASCICLE II.1-88. General Tariff Principles—Charging and Accounting in International Telecommunications Services Series D Recommendations. 371 pp.

—**Telephone Circuits—Tables (Data)**
CCITT FASCICLE II.1-88. General Tariff Principles—Charging and Accounting in International Telecommunications Services Series D Recommendations. 371 pp.

—**Telephone Services—Europe**
CCITT FASCICLE II.1-88. General Tariff Principles—Charging and Accounting in International Telecommunications Services Series D Recommendations. 371 pp.

—**Telephone Services—Handbooks**
CCITT FASCICLE II.1-88. General Tariff Principles—Charging and Accounting in International Telecommunications Services Series D Recommendations. 371 pp.

—**Telephone Services—Mediterranean Basin**
CCITT FASCICLE II.1-88. General Tariff Principles—Charging and Accounting in International Telecommunications Services Series D Recommendations. 371 pp.

—**Telephone Services—Regional**
CCITT FASCICLE II.1-88. General Tariff Principles—Charging and Accounting in International Telecommunications Services Series D Recommendations. 371 pp.

—**Telephone Services—Tables (Data)**
CCITT FASCICLE II.1-88. General Tariff Principles—Charging and Accounting in International Telecommunications Services Series D Recommendations. 371 pp.

—**Telex Communications—Europe**
CCITT FASCICLE II.1-88. General Tariff Principles—Charging and Accounting in International Telecommunications Services Series D Recommendations. 371 pp.

—**Telex Communications—Handbooks**
CCITT FASCICLE II.1-88. General Tariff Principles—Charging and Accounting in International Telecommunications Services Series D Recommendations. 371 pp.

—**Telex Communications—Mediterranean Basin**
CCITT FASCICLE II.1-88. General Tariff Principles—Charging and Accounting in International Telecommunications Services Series D Recommendations. 371 pp.

—**Telex Communications—Regional**
CCITT FASCICLE II.1-88. General Tariff Principles—Charging and Accounting in International Telecommunications Services Series D Recommendations. 371 pp.

—**Telex Communications—Tables (Data)**
CCITT FASCICLE II.1-88. General Tariff Principles—Charging and Accounting in International Telecommunications Services Series D Recommendations. 371 pp.

Experimental Certificates
—**Air Cushion Vehicles**
CAA Chapter A3-2 04.79. Experimental Certs. 3 pp.
CAA Chapter A3-2 App 08.81. Experimental Certs. 2 pp.

Experimental Design
See Also: Error Analysis; Laboratories; Statistical Analysis

—**Glossaries**
BSI BS 5532: Part 3-86. 1986 Statistical Terminology Part 3: Glossary of Terms Relating to the Design of Experiments. 40 pp.
ISO 3534 Pt 3-85. Statistics—Vocabulary and Symbols—Part 3: Design of Experiments First Edition. 42 pp.

—**Symbols**
BSI BS 5532: Part 3-86. 1986 Statistical Terminology Part 3: Glossary of Terms Relating to the Design of Experiments. 40 pp.

Experimental Design *(Cont.)*
—**Symbols** *(Cont.)*
ISO 3534 Pt 3-85. Statistics—Vocabulary and Symbols—Part 3: Design of Experiments First Edition. 42 pp.

Explosion Indices
See Also: Explosions

—**Dusts—Explosion Protection Systems**
BSI BS 6713: Part 1-86. 1986 Explosion Protection Systems Part 1: Method for Determination of Explosion Indices of Combustible Dusts in Air. 8 pp.
BSI BS 6713: Part 1-01. 1986 Amd 1 Explosion Protection Systems Part 1: Method for Determination of Explosion Indices of Combustible Dusts in Air (ISO 6184/1: 1985) (AMD 7083) February 28, 1992. 13 pp.
DIN ENGL EN 26184 Pt 1-91. Explosion Protection Systems; Determination of Explosion Indices of Combustible Dusts in Air (ISO 6184-1: 1985) (June). 8 pp.
ISO 6184 Pt 1-85. Explosion Protection Systems—Part 1: Determination of Explosion Indices of Combustible Dusts in Air First Edition. 7 pp.

—**Fuels—Explosion Protection Systems**
BSI BS 6713: Part 3-86. 1986 Explosion Protection Systems Part 3: Method for Determination of Explosion Indices of Fuel/Air Mixtures Other Than Dust/Air and Gas/Air Mixtures. 7 pp.
BSI BS 6713: Part 3-01. 1986 Amd 1 Explosion Protection Systems Part 3: Method for Determination of Explosion Indices of Fuel/Air Mixtures Other Than Dust/Air and Gas/Air Mixtures (ISO 6184/3: 1985) (AMD 7085) February 28, 1992. 12 pp.
DIN ENGL EN 26184 Pt 3-91. Explosion Protection Systems; Determination of Explosion Indices of Fuel/Air Mixtures Other Than Dust/Air and Gas/Air Mixtures (ISO 6184-3: 1985) (June). 6 pp.
ISO 6184 Pt 3-85. Explosion Protection Systems—Part 3: Determination of Explosion Indices of Fuel/Air Mixtures Other Than Dust/Air and Gas/Air Mixtures First Edition. 6 pp.

—**Gases—Explosion Protection Systems**
BSI BS 6713: Part 2-86. 1986 Explosion Protection Systems Part 2: Method for Determination of Explosion Indices of Combustible Gases in Air. 8 pp.
BSI BS 6713: Part 2-01. 1986 Amd 1 Explosion Protection Systems Part 2: Method for Determination of Explosion Indices of Combustible Gases in Air (ISO 6184/2: 1985) (AMD 7084) February 28, 1992. 13 pp.
DIN ENGL EN 26184 Pt 2-91. Explosion Protection Systems; Determination of Explosion Indices of Combustible Gases in Air (ISO 6184-2: 1985) (June). 8 pp.
ISO 6184 Pt 2-85. Explosion Protection Systems—Part 2: Determination of Explosion Indices of Combustible Gases in Air First Edition. 7 pp.

Explosion Prevention
Use: Explosions

Explosion Protection Systems
See Also: Explosions
CEN PREN 1127-1-93. Safety of Machinery—Fire and Explosions—Part 1: Explosion Prevention and Protection. 50 pp.

—**Dusts—Explosion Indices**
BSI BS 6713: Part 1-86. 1986 Explosion Protection Systems Part 1: Method for Determination of Explosion Indices of Combustible Dusts in Air. 8 pp.
BSI BS 6713: Part 1-01. 1986 Amd 1 Explosion Protection Systems Part 1: Method for Determination of Explosion Indices of Combustible Dusts in Air (ISO 6184/1: 1985) (AMD 7083) February 28, 1992. 13 pp.
DIN ENGL EN 26184 Pt 1-91. Explosion Protection Systems; Determination of Explosion Indices of Combustible Dusts in Air (ISO 6184-1: 1985) (June). 8 pp.
ISO 6184 Pt 1-85. Explosion Protection Systems—Part 1: Determination of Explosion Indices of Combustible Dusts in Air First Edition. 7 pp.

—**Efficacy**
BSI BS 6713: Part 4-86. 1986 Explosion Protecton Systems Part 4: Method for Determination of Efficacy of Explosion Suppression Systems. 7 pp.
BSI BS 6713: Part 4-01. 1986 Amd 1 Explosion Protecton Systems Part 4: Method for Determination of Efficacy of Explosion Suppression Systems (ISO 6184/4: 1985) (AMD 7086) February 28, 1992. 12 pp.
ISO 6184 Pt 4-85. Explosion Protection Systems—Part 4: Determination of Efficacy of Explosion Suppression Systems First Edition. 6 pp.

INTERNATIONAL AND NON-U.S. NATIONAL STANDARDS SUBJECT INDEX

Explosion

Explosion Protection Systems *(Cont.)*

—Fuels—Explosion Indices

BSI BS 6713: Part 3-86. 1986 Explosion Protection Systems Part 3: Method for Determination of Explosion Indices of Fuel/Air Mixtures Other Than Dust/Air and Gas/Air Mixtures. 7 pp.

BSI BS 6713: Part 3-01. 1986 Amd 1 Explosion Protection Systems Part 3: Method for Determination of Explosion Indices of Fuel/Air Mixtures Other Than Dust/Air and Gas/Air Mixtures (ISO 6184/3: 1985) (AMD 7085) February 28, 1992. 12 pp.

DIN ENGL EN 26184 Pt 3-91. Explosion Protection Systems; Determination of Explosion Indices of Fuel/Air Mixtures Other Than Dust/Air and Gas/Air Mixtures (ISO 6184-3: 1985) (June). 6 pp.

DIN ENGL EN 26184 Pt 4-91. Explosion Protection Systems; Determination of Efficacy of Explosion Suppression Systems (ISO 6184-4: 1985) (June). 7 pp.

ISO 6184 Pt 3-85. Explosion Protection Systems—Part 3: Determination of Explosion Indices of Fuel/Air Mixtures Other Than Dust/Air and Gas/Air Mixtures First Edition. 6 pp.

—Gases—Explosion Indices

BSI BS 6713: Part 2-86. 1986 Explosion Protection Systems Part 2: Method for Determination of Explosion Indices of Combustible Gases in Air. 8 pp.

BSI BS 6713: Part 2-01. 1986 Amd 1 Explosion Protection Systems Part 2: Method for Determination of Explosion Indices of Combustible Gases in Air (ISO 6184/2: 1985) (AMD 7084) February 28, 1992. 13 pp.

DIN ENGL EN 26184 Pt 2-91. Explosion Protection Systems; Determination of Explosion Indices of Combustible Gases in Air (ISO 6184-2: 1985) (June). 8 pp.

ISO 6184 Pt 2-85. Explosion Protection Systems—Part 2: Determination of Explosion Indices of Combustible Gases in Air First Edition. 7 pp.

Explosions

See Also: Combustion; Demolition; Dust Explosions; Explosion Indices; Explosion Protection Systems; Explosive Ordnance Disposal; Explosives; Fires; Flame Propagation; Overpressure; Sound Pressure; Warning Systems

—Glossaries

BSI BS 4422: Part 7-88. 1988 Glossary of Terms Associated with Fire Part 7: Explosion Detection and Suppression Means. 3 pp.

ISO 8421 Pt 7-87. Fire Protection—Vocabulary—Part 7: Explosion Detection and Suppression Means First Edition. 6 pp.

—Machinery

CEN PREN 1127-1-93. Safety of Machinery—Fire and Explosions—Part 1: Explosion Prevention and Protection. 50 pp.

Explosive Actuated Tools

Use: Powder Actuated Tools

Explosive Atmospheres

Use For: Firedamp Hazards *See Also:* Electrical Protection Equipment; Explosive Gases; Hazardous Environments; Hazardous Locations

—Cable Glands

SAA AS 1828-84. Electrical Equipment for Explosive Atmospheres—Cable Glands Amdt 1 November 1985. 11 pp.

—Cap Lamps

BSI BS EN 50033-91. 1991 Construction and Testing of Miners' Caplamps in Relation to the Risk of Explosion, for Mines Susceptible to Firedamp. 13 pp.

BSI BS EN 50033-01. 1991 Amd 1 Construction and Testing of Miners' Caplamps in Relation to the Risk of Explosion, for Mines Susceptible to Firedamp (AMD 7877) July 15, 1993 (Supersedes BS 6881: 1988) (Q). 14 pp.

CENELEC EN 50 033-86. Electrical Apparatus for Potentially Explosive Atmospheres: Caplamps for Mines Susceptible to Firedamp. 7 pp.

—Classification

BSI BS 5345: Part 2-83. 1983 Amd 1 Code of Practice for the Selection, Install. and Maint. of Electrical Apparatus for Use in Potent. Explosive Atmospheres (Other Than Mining Applic. or Explo. Proc.) Part 2: Classif. of Hazardous Areas (AMD 5754) July 31, 1989. 24 pp.

SNZ NZS 6101: Part 1-88. Classification of Hazardous Areas Part 1: Flammable Gas Vapour Atmospheres. 24 pp.

SNZ NZS 6101: Part 2-90. Classification of Hazardous Areas Part 2: Combustible Dusts. 24 pp.

Explosive Atmospheres *(Cont.)*

—Classification *(Cont.)*

SNZ NZS 6101: Part 3-91. Classification of Hazardous Areas Part 3: Specific Occupancies (Flammable Gas and Vapour Atmospheres). 88 pp.

—Dust

SNZ NZS 6101: Part 2-90. Classification of Hazardous Areas Part 2: Combustible Dusts. 24 pp.

—Electric Plug Receptacles—Mining Equipment

CSA C22.2 NO 159-M1987. Attachment Plugs, Receptacles, and Similar Wiring Devices for Use in Hazardous Locations: Class I, Groups A, B, C, and D; Class II, Group G, in Coal or Coke Dust, and in Gaseous Mines (R 1993). 39 pp.

—Electric Plugs—Mining Equipment

CSA C22.2 NO 159-M1987. Attachment Plugs, Receptacles, and Similar Wiring Devices for Use in Hazardous Locations: Class I, Groups A, B, C, and D; Class II, Group G, in Coal or Coke Dust, and in Gaseous Mines (R 1993). 39 pp.

—Electrical Enclosures—Mining Equipment

CSA 1273 Bull. Electrical Bulletin 1273 July 2, 1980 to C22.2 NO 25. 2 pp.

CSA 1273A Bull. Electrical Bulletin 1273A February 8, 1983 to C22.2 NO 25. 2 pp.

CSA C22.2 NO 30-M1986. Explosion-Proof Enclosures for Use in Class I Hazardous Locations (R 1992); (Gen Instr 1 Thru 2). 41 pp.

CSA 1273 Bull. Electrical Bulletin 1273 July 2, 1980 to C22.2 NO 30. 2 pp.

—Electrical Equipment

BSI BS 229-57. 1957 Amd 9 Flameproof Enclosure of Electrical Apparatus. 56 pp.

BSI BS 4683: Part 1-71. 1971 Electrical Apparatus for Explosive Atmospheres Part 1: Classification of Maximum Surface Temperature. 8 pp.

BSI BS 4683: Part 2-71. 1971 Amd 3 Electrical Apparatus for Explosive Atmospheres Part 2: The Construction and Testing of Flameproof Enclosures of Electrical Apparatus Electrical Apparatus (AMD 6430) March 30, 1990. 50 pp.

BSI BS 4683: Part 4-73. 1973 Amd 2 Electrical Apparatus for Explosive Atmospheres Part 4: Type of Protection 'E'. 26 pp.

BSI BS 4727:Pt2: Group 13-91. 1991 Electrotechnical, Power, Telecommunication, Electronics, Lighting and Colour Terms Part 2: Terms Particular to Power Engineering Group 13: Electrical Apparatus for Explosive Atmospheres (IEC 50(426): 1990). 40 pp.

BSI BS 5345: Part 1-89. 1989 Amd 2 Code of Practice for the Selection, Install. and Maint. of Electrical Apparatus for Use in Potent. Explosive Atmospheres (Other Than Mining Applns. or Expl. Proc. and Man.) Part 1: General Recommendations (AMD 6333) January 31, 1991. 47 pp.

BSI BS 5345: Part 1-03. 1989 Amd 3 Selection, Installation and Maintenance of Electrical Apparatus for Use in Potentially Expolsive Atmospheres (Other Than Mining Applications or Explosives Processing and Manufacture) Part 1: General Recommendations (AMD 7871) Sept 15, 1993. 48 pp.

BSI BS 5345: Part 2-83. 1983 Amd 1 Code of Practice for the Selection, Install. and Maint. of Electrical Apparatus for Use in Potent. Explosive Atmospheres (Other Than Mining Applic. or Explo. Proc.) Part 2: Classif. of Hazardous Areas (AMD 5754) July 31, 1989. 24 pp.

BSI BS 5345: Part 4-77. 1977 Installation and Maint-enance of Electrical Apparatus for Use in Potentially Explosive Atmospheres Part 4: Installation and Maint-enance Requirements for Electrical Apparatus with Type of Protection 'i'. Intrinsically Safe Elect. Appar. & Systems. 18 pp.

BSI BS 5345: Part 6-78. 1978 Amd 1 Code of Practice for the Selection, and Maint. of Electrical Apparatus for Use in Potent. Explo. Atmosph. (Other Than Mining Applic. or Explos. Proc) Part 6: Recomm. for Type of Prot qqe' Incr Safety (AMD 5557) November 30, 1989. 17 pp.

BSI BS 5345: Part 7-79. 1979 Amd 1 Installation and Main-enance of Electrical Apparatus for Use in Potentially Explosive Atmospheres Part 7: Installation and Maintenance Requirements for Electrical Apparatus with Type of Protection N. 12 pp.

BSI BS 5345: Part 8-80. 1980 Installation and Maint-enance of Electrical Apparatus for Use in Potentially Explosive Atmospheres Part 8: Installation and Maintenance Requirements for Electrical Apparatus with Type of Protection 's'. Special Protection. 10 pp.

BSI BS 5501: Part 1-77. 1977 Amd 7 Electrical Apparatus for Potentially Explosive Atomspheres Part 1: General Requirements (AMD 6434) March 30, 1990. 42 pp.

Explosive Atmospheres *(Cont.)*

—Electrical Equipment *(Cont.)*

BSI BS 5501: Part 2-77. 1977 Amd 2 Electrical Apparatus for Potentially Explosive Atmospheres Part 2: Oil Immersion 'o' (AMD 6438) March 30, 1990. 10 pp.

BSI BS 5501: Part 3-77. 1977 Amd 2 Electrical Apparatus for Potentially Explosive Atomspheres Part 3: Pressurized Apparatus 'p' (AMD 6437) March 30, 1990. 16 pp.

BSI BS 5501: Part 4-77. 1977 Amd 2 Electrical Apparatus for Potentially Explosive Atmospheres Part 4: Powder Filling 'q' (AMD 6440) March 30, 1990. 22 pp.

BSI BS 5501: Part 5-77. 1977 Amd 7 Electrical Apparatus for Potentially Explosive Atmospheres Part 5: Flameproof Enclosure 'd' (AMD 6517) April 30, 1991. 69 pp.

BSI BS 5501: Part 6-77. 1977 Amd 8 Electrical Apparatus for Potentially Explosive Atomspheres Part 6: Increased Safety 'e' (AMD 6655) December 21, 1990. 73 pp.

BSI BS 5501: Part 7-77. 1977 Amd 12 Electrical Apparatus for Potentially Explosive Atmospheres Part 7: Intrinsic Safety 'i' (Q) (AMD 6436) March 30, 1990. 110 pp.

BSI BS 5501: Part 7-16. 1977 Amd 16 Electrical Apparatus for Potentially Explosive Atmospheres Part 7: Intrinsic Safety 'i'. 88 pp.

BSI BS 5501: Part 8-88. 1988 Amd 1 Electrical Apparatus for Potentially Explosive Atomspheres Part 8: Encapsulation 'qqm' (AMD 6439) March 30, 1990. 15 pp.

BSI BS 5501: Part 9-82. 1982 Amd 1 Electrical Apparatus for Potentially Explosive Atmospheres Part 9: Specification for Intrinsically Safe Electrical Systems 'i' (AMD 6435) March 30, 1990. 8 pp.

BSI BS 6941-88. 1988 Electrical Apparatus for Explosive Atmospheres with Type of Protection N. 23 pp.

BSI CP 1003: Part 3-67. 1967 Amd 1 Electrical Apparatus and Associated Equipment for Use in Explosive Atmospheres of Gas or Vapour Other Than Mining Applications Part 3: Division 2 Areas (AMD 5075) April 30, 1986. 15 pp.

BSI BS EN 50014-93. 1993 Electrical Apparatus for Potentially Explosive Atmospheres—General Requirements (E). 41 pp.

CENELEC EN 50 014-77. Electrical Apparatus for Potentially Explosive Atmospheres General Requirements. 72 pp.

CENELEC EN 50014-92. Electrical Apparatus for Potentially Explosive Atmospheres General Requirements. 35 pp.

CENELEC EN 50014-92. Electrical Apparatus for Potentially Explosive Atmospheres—General Requirements. 65 pp.

CENELEC EN 50 015-77. Electrical Apparatus for Potentially Explosive Atmospheres: Oil Immersion 'O'. 7 pp.

CENELEC EN 50 015/A1 (E)-79. AMD 1 Electrical Apparatus for Potentially Explosive Atmospheres: Oil Immersion 'O'. 4 pp.

CENELEC EN 50 016-77. Electrical Apparatus for Potentially Explosive Atmospheres: Pressurized Apparatus 'P'. 13 pp.

CENELEC EN 50 016/A1 (E)-79. AMD 1 Electrical Apparatus for Potentially Explosive Atmospheres: Pressurized Apparatus 'P'. 4 pp.

CENELEC EN 50 017-77. Electrical Apparatus for Potentially Explosive Atmospheres: Powder Filling 'Q'. 21 pp.

CENELEC EN 50 018-77. Electrical Apparatus for Potentially Explosive Atmospheres: Flameproof Enclosure 'D'. 54 pp.

CENELEC EN 50 018/A2-82. AMD 2 Electrical Apparatus for Potentially Explosive Atmospheres: Flameproof Enclosure 'D'.

CENELEC EN 50 018/A3-85. AMD 3 Electrical Apparatus for Potentially Explosive Atmospheres: Flameproof Enclosure 'D'.

CENELEC PREN 50018/PRAA-91. AMD prAA Electrical Apparatus for Potentially Explosive Atmospheres: Flameproof Enclosure 'D'. 71 pp.

CENELEC EN 50 019-85. Electrical Apparatus for Potentially Explosive Atmospheres: Increased Safety 'E'. 42 pp.

CENELEC EN 50 019/A4-89. AMD 4 Electrical Apparatus for Potentially Explosive Atmospheres: Increased Safety 'E'. 15 pp.

CENELEC EN 50 019/A5-90. AMD 5 Electrical Apparatus for Potentially Explosive Atmospheres: Increased Safety 'E'. 8 pp.

CENELEC EN 50 020-77. Electrical Apparatus for Potentially Explosive Atmospheres Intrinsic Safety 'I'. 71 pp.

CENELEC EN 50 020/A2-85. AMD 2 Electrical Apparatus for Potentially Explosive Atmospheres Intrinsic Safety 'I'. 3 pp.

CENELEC EN 50 020/A3-90. AMD 3 Electrical Apparatus for Potentially Explosive Atmospheres Intrinsic Safety 'I'. 8 pp.

INTERNATIONAL AND NON-U.S. NATIONAL STANDARDS
SUBJECT INDEX
Explosive

Explosive Atmospheres *(Cont.)*

—Electrical Equipment *(Cont.)*

CENELEC EN 50 020/A4-90. AMD 4 Electrical Apparatus for Potentially Explosive Atmospheres Intrinsic Safety 'I'. 14 pp.
CENELEC EN 50 020/A5-90. AMD 5 Electrical Apparatus for Potentially Explosive Atmospheres Intrinsic Safety 'I'. 2 pp.
CENELEC EN 50020-92. CORRIGENDUM Electrical Apparatus for Potentially Explosive Atmospheres Intrinsic Safety 'I'. 1 p.
CENELEC PREN 50021-92. Standard for Electrical Apparatus with Type of Protection 'n'. 86 pp.
CENELEC EN 50 028-87. Electrical Apparatus for Potentially Explosive Atmosphers. Encapsulation "M". 12 pp.
CENELEC EN 50 039-80. Electrical Apparatus for Potentially Explosive Atmospheres: Intrinsically Safe Electrical Systems 'I'. 5 pp.
CNS C3171-83. Method of Test for Electrical Apparatus for Use in Explosive Gas Atmospheres (Jan)(9826).
CNS F5082-84. General Requirements for Construction and Inspection of Electrical Explosion-Proof Apparatus for Marine Use (Feb)(10765).
CNS M2005-85. General Rules for Explosion-Proof Construction of Electric Machineries and Apparatuses of Mines (Aug)(2867).
CNS M2005-1-85. Flameproof Endurance Pressure Explosion-Proof Construction of Electric Machineries and Apparatuses of Mines (Aug)(2867-1).
CNS M2005-2-85. Lamina Protected Explosion-Proof Construction of Electric Machineries and Apparatuses of Mines (Aug)(2867-2).
CNS M2005-3-85. Oil Immersed Explosion-Proof Construction of Electric Machineries and Apparatuses of Mines (Aug)(2867-3).
CNS M2005-4-85. Pressurized Explosion-Proof Construction of Electric Machineries and Apparatuses of Mines (Aug)(2867-4).
CNS M2005-7-85. Explosion-Proof Construction of Electric Machineries and Apparatuses (Aug)(2867-7).
CNS M2005-8-85. Loose Prevention Frame for Explosion-Proof Construction of Electric Machineries and Apparatuses of Mines (Aug)(2867-8).
CNS M2005-9-85. Terminals Box for Explosion-Proof Construction of Electric Machineries and Apparatuses of Mines (Aug)(2867-9).
CNS M2005-10-85. Line Connection of Terminal Box of Explosion-Proof Construction of Electric Machineries and Apparatuses of Mines (Aug)(2867-10).
CNS M2005-11-85. Outer Conductor and Terminal Box Connection for Explosion-Proof Construction of Electric Machineries and Apparatuses of Mines (Aug)(2867-11).
CNS M2005-12-85. Outer Conductor Connection for Explosion-Proof Construction of Electric Machineries and Apparatuses of Mines (Aug)(2867-12).
CNS M3009-85. Method of Test for Explosion-Proof Constructions of Electric Machineries and Apparatuses of Mines (Aug)(3378).
CSA CAN/CSA-M421-93. Use of Electricity in Mines; (Gen Instr 1). 67 pp.
DIN VDE 0166-81. Electrical Installations and Equipment for Use in Atmospheres Potentially Endangered by Explosive Material (May). 28 pp.
DIN VDE 0170/0171 Pt 1 A102-88. Electrical Apparatus for Potentially Explosive Atmospheres; General Requirements; Amendment 102 (May). 4 pp.
DIN VDE 0170/0171 Pt 12 (D)-82. Electrical Equipment for Potentially Explosive Atmospheres; Requirements for Apparatus for Zone 0 (Nov). 26 pp.
EC 76/117/EEC-75. Council Directive on the Approximation of the Laws of the Member States Concerning Electrical Equipment for Use in Potentially Explosive Atmospheres. 4 pp.
EC 79/196/EEC-79. Council Directive on the Approximation of the Laws of the Member States Concerning Electrical Equipment for Use in Potentially Explosive Atmospheres Employing Certain Types of Protection. 3 pp.
EC 84/47/EEC-84. Commission Directive Adapting to Technical Progress Council Directive 79/196/EEC on the Approximation of the Laws of the Member States Concerning Electrical Equipment for Use in Potentially Explosive Atmospheres Employing Certain Types of Protection. 3 pp.
EC 88/571/EEC-88. Commission Directive Adapting to Technical Progress Council Directive 79/196/EEC on the Approximation of the Laws of the Member States Concerning Electrical Equipment for Use in Potentially Explosive Atmospheres Employing Certain Types of Protection. 1 p.
EC 82/490/EEC-82. Commission Recommendation Relating to the Certificates of Conformity Provided for in Council Directive 76/117/EEC on the Approximation of the Laws of the Member States Concerning Electrical Equipment for Use in Potentially-Explosive Atmospheres. 3 pp.

Explosive Atmospheres *(Cont.)*

—Electrical Equipment *(Cont.)*

IEC 50 Chap 426-90. International Electrotechnical Vocabulary Chapter 426: Electrical Apparatus for Explosive Atmospheres First Edition. 44 pp.
IEC 79 Pt 0-83. Electrical Apparatus for Explosive Gas Atmospheres Part 0: General Requirements Second Edition; (Amendment 2-1991, Incorporating Amendment 1). 71 pp.
IEC 79 Pt 1-90. Electrical Apparatus for Explosive Gas Atmospheres Part 1: Construction and Verification Test of Flameproof Enclosures of Electrical Apparatus Third Edition; (Amendment 1-1993). 131 pp.
IEC 79 Pt 2-83. Electrical Apparatus for Explosive Gas Atmospheres Part 2: Electrical Apparatus—Type of Protection "P" Third Edition. 56 pp.
IEC 79 Pt 3-90. Electrical Apparatus for Explosive Gas Atmospheres Part 3: Spark-Test Apparatus for Intrinsically-Safe Circuits Third Edition. 31 pp.
IEC 79 Pt 4-75. Electrical Apparatus for Explosive Gas Atmospheres Part 4: Method of Test for Ignition Temperature Second Edition. 22 pp.
IEC 79 Pt 5-67. Electrical Apparatus for Explosive Gas Atmospheres Part 5: Sand-Filled Apparatus First Edition; (Incorporating Supplement A-1969). 40 pp.
IEC 79 Pt 6-68. Electrical Apparatus for Explosive Gas Atmospheres Part 6: Oil-Immersed Apparatus First Edition. 13 pp.
IEC 79 Pt 7-90. Electrical Apparatus for Explosive Gas Atmospheres Part 7: Increased Safety "E" Second Edition; (Amendment 1-1991) (Amendment 2-1993). 112 pp.
IEC 79 Pt 10-86. Electrical Apparatus for Explosive Gas Atmospheres Part 10: Classification of Hazardous Areas Second Edition. 52 pp.
IEC 79 Pt 11-91. Electrical Apparatus for Explosive Gas Atmospheres Part 11: Intrinsic Safety "i" Third Edition. 105 pp.
IEC 79 Pt 12-78. Electrical Apparatus for Explosive Gas Atmospheres Part 12: Classification of Mixtures of Gases or Vapours with Air According to Their Maximum Experimental Safe Gaps and Minimum Igniting Currents First Edition. 20 pp.
IEC 79 Pt 13-82. Electrical Apparatus for Explosive Gas Atmospheres Part 13: Construction and Use of Rooms or Buildings Protected by Pressurization First Edition. 20 pp.
IEC 79 Pt 14-84. Electrical Apparatus for Explosive Gas Atmospheres Part 14: Electrical Installations in Explosive Gas Atmospheres (Other Than Mines) First Edition. 46 pp.
IEC 79 Pt 15-87. Electrical Apparatus for Explosive Gas Atmospheres Part 15: Electrical Apparatus with Type of Protection "N" First Edition. 67 pp.
IEC 79 Pt 16-90. Electrical Apparatus for Explosive Gas Atmospheres Part 16: Artificial Ventilation for the Protection of Analyzer(s) Houses First Edition. 44 pp.
IEC 79 Pt 17-90. Electrical Apparatus for Explosive Gas Atmospheres Part 17: Recommendations for Inspection and Maintenance of Electrical Installations in Hazardous Areas (Other Than Mines) First Edition. 40 pp.
IEC 79 Pt 18-92. Electrical Apparatus for Explosive Gas Atmospheres Part 18: Encapsulation "m" First Edition. 52 pp.
JIS C 0901-83. Electrical Apparatus for Use in Gassy Coal Mines. 70 pp.
JIS C 0902-65. Testing Methods for Explosion Proof Construction of Electrical Machinery and Apparatus (for Coal Mines) (R 1971). 8 pp.
JIS C 0903-83. Electrical Apparatus for Explosive Atmospheres in General Industry. 65 pp.
JIS C 0904-83. Test Methods on Electrical Apparatus for Explosive Gas Atmospheres in General Industry. 26 pp.
JIS C 0905-83. Supplementary Requirements for Construction of Electrical Apparatus for Explosive Atmospheres in General Industry. 12 pp.
JIS C 4501-77. Explosion-Proof Type Plug Connection Devices for Coal Mines (R 1985). 18 pp.
JIS F 8004-88. General Requirements for Marine Electrical Flameproof Appararatus Flame Proof Apparatus for Marine Use.
SAA AS 1021-80. Protection by Purging of Electrical Equipment for Explosive Gas Atmospheres. 28 pp.
SAA AS 1039-86. Electrical Equipment in Coal Mines—Explosion Protected Distribution and Control Boxes for Voltages up to 3300 V a.c.. 15 pp.
SAA AS 1076. Code of Practice for Selection, Installation and Maintenance of Electrical Apparatus and Associated Equipment for Use in Explosive Atmospheres (Other than Mining Applications) NOTE: See Also AS 2381 Series.
SAA AS 1076.1-77. Code of Practice for Selection, Installation and Maintenance of Electrical Apparatus and Associated Equipment for Use in Explosive Atmospheres (Other Than Mining Applications)—Part 1: Basic Requirements Obsolescent January 1991. 88 pp.

Explosive Atmospheres *(Cont.)*

—Electrical Equipment *(Cont.)*

SAA AS 1076.3-77. Code of Practice for Selection, Instl. and Maint. of Elec. Ap and Assoc. Equipment for Use in Explosive Atmospheres (Other Than Mining Applications)—Pt. 3: Apparatus with Type of Protection 'd'—Flame-Proof Enclosure (Superseded by AS 2381.2—1992).
SAA AS 1076.6-77. Code of Practice for Sel, Instl. and Maint. of Electrical Ap and Associated Equipment for Use in Explosive Atmospheres (Other Than Mining Applications)—Part 6: Apparatus with Type of Protection 'e'—Increased Safety (Superseded by AS 2381.6 —1993).
SAA AS 1076.7-77. Code of Practice for Selection, Installation and Maintenance of Electrical Apparatus and Associated Equipment for Use in Explosive Atmospheres (Other Than Mining Applications)—Part 7: Apparatus with Type of Protection 'n'—Non-Sparking Apparatus. 24 pp.
SAA AS 1076.8-77. Code of Practice for Selection, Installation and Maintenance of Electrical Apparatus and Associated Equipment for Use in Explosive Atmospheres (Other Than Mining Applications)—Part 8: Apparatus with Type of Protection 's'—Special Protection. 20 pp.
SAA AS 1826-83. Electrical Equipment for Explosive Atmospheres—Special Protection—Type of Protection S. 4 pp.
SAA AS 1915-92. Electrical Equipment for Explosive Atmospheres—Battery-Operated Vehicles (in Professional Package 40). 4 pp.
SAA AS 2380.2-91. Electrical Equipment for Explosive Atmospheres—Explosion-Protection Techniques—Part 2: Flameproof Enclosure d Amdt 1 July 1992 (in Professional Packages 40, 55) (Supersedes AS 2480—1986). 34 pp.
SAA AS 2380.9-91. Electrical Equipment for Explosive Atmospheres—Explosion-Protection Techniques—Part 9: Type of Protection n—Non-Sparking (in Professional Package 40) (Supersedes AS 2238—1982). 16 pp.
SAA AS 2381.1-91. Electrical Equipment for Explosive Atmospheres—Selection, Installation and Maintenance—Part 1: General Requirements Amdt 1 April 1992 (in Professional Packages 21, 23, 24, 40, 68). 2 pp.
SNZ NZS 6109: Part 1-88. Electrical Systems of Dispensing Equipment for Explosive Atmospheres Part 1: Flammable Liquids Dispensing Equipment. 16 pp.

—Electrical Equipment—Mining Equipment

EC 88/35/EEC-87. Commission Directive Adapting to Technical Progress Council Directive 82/130/EEC on the Approximation of the Laws of the Member States Concerning Electrical Equipment for Use in Potentially Explosive Atmospheres in Mines Susceptible to Firedamp. 5 pp.
EC 91/269/EEC-91. Commission Directive Adapting to Technical Progress Council Directive 82/130/EEC on the Approximation of the Laws of the Member States Concerning Electrical Equipment for Use in Potentially Explosive Atmospheres in Mines Susceptible to Firedamp. 5 pp.

—Electrical Equipment—Repair

SNZ NZS 6112-89. Code of Practice for the Repair of Electrical Apparatus for Use in Potentially Explosive Atmospheres. 52 pp.

—Electrical Equipment—Safety

CNS M2005-5-85. Increase Safety Explosion-Proof Construction of Electric Machineries and Apparatuses of Mines (Aug)(2867-5).
CNS M2005-6-85. Intrinsic Safety Explosion-Proof Construction of Electric Machineries and Apparatuses of Mines (Aug)(2867-6).
EC 76/117/EEC-75. Council Directive on the Approximation of the Laws of the Member States Concerning Electrical Equipment for Use in Potentially Explosive Atmospheres. 4 pp.

—Electrical Equipment—Spray Painting Booths

SNZ NZS/AS 4114 (INT)-93. Electrical Equipment for Explosive Atmospheres—Spray Painting Booths—Design, Construction, Installation and Maintenance (This is a Joint Standard with SAA AS 4114 (INT)). 15 pp.

—Electrical Equipment—Ventilation

SAA AS 1482-85. Electrical Equipment for Explosive Atmospheres—Protection by Ventilation—Type of Protection V. 17 pp.

—Electrical Installations

DIN VDE 0105 Pt 7-87. Operation of Power Installations; Supplementary Requirements for Atmospheres Endangered by Potentially Explosive Material (Dec). 6 pp.
DIN VDE 0105 Pt 9-86. Operation of Power Installations Part 9: Supplementary Requirements for Potentially Explosive Atmospheres (May). 10 pp.

INDUSTRY STANDARDS
1019

INTERNATIONAL AND NON-U.S. NATIONAL STANDARDS SUBJECT INDEX

Explosive

Explosive Atmospheres (Cont.)

—Electrical Installations (Cont.)
DIN VDE 0165-91. Erection of Electrical Installations in Potentially Explosive Atmospheres (Feb) (Supersedes DIN VDE 0165 A1/09.86). 65 pp.
DIN VDE 0166-81. Electrical Installations and Equipment for Use in Atmospheres Potentially Endangered by Explosive Material (May). 28 pp.

—Electrostatic Spraying Equipment
CENELEC EN 50 050-86. Electrical Apparatus for Potentially Explosive Atmospheres: Electrostatic Hand-Held Spraying Equipment. 7 pp.
CENELEC EN 50 053-87. Requirements for the Selection, Installation and Use of Electrostatic Spraying Equipment for Flammable Materials—Part 1: Hand-held Electrostatic Paint Spray Guns with an Energy Limit of 0,24 mj and Their Associated Apparatus. 8 pp.
CENELEC EN 50 053-2-89. Requirements for the Selection, Installation and Use of Electrostatic Spraying Equipment for Flammable Materials Part 2: Hand-Held Electrostatic Powder Spray Guns with an Energy Limit of 5 mJ and Their Associated Apparatus. 18 pp.
CENELEC EN 50 053-3-89. Requirements for the Selection, Installation and Use of Electrostatic Spraying Equipment for Flammable Materials Part 3: Hand-Held Electrostatic Flock Spray Guns with an Energy Limit of 0.24 mJ or 5 mJ and Their Associated Apparatus. 19 pp.

—Fittings
BSI BS 889-65. (WITHDRAWN) 1965 Amd 3 Flameproof Electric Lighting Fittings (AMD 5050) May 30, 1986. 65 pp.

—Flashlights
CNS F5084-85. Explosion-Proof Portable Electric Lamps for Marine Use (Dry Cells and Batteries Type) (Sep)(10767).
JIS F 8425-90. Flameproof Portable Electric Lamps for Marine Use—Dry Cells and Batteries Type.
JIS F 8425-62. Explosion-Proof Flash Lights for Marine Use (Dry Battery Type) (R 1968). 8 pp.

—Instrumentation
SAA AS 1076.13-77. Code of Practice for Selection, Installation and Maintenance of Electrical Apparatus and Associated Equipment for Use in Explosive Atmospheres (Other Than Mining Applications)—Part 13: Installation and Maintenance Requirements for Instrumentation. 36 pp.

—Lamps—Portable
BSI BS 889-65. (WITHDRAWN) 1965 Amd 3 Flameproof Electric Lighting Fittings (AMD 5050) May 30, 1986. 65 pp.
JIS C 8004-85. Hand Lanterns for Explosive Atmospheres. 10 pp.

—Lighting
CNS F5033-86. Marine Control Switches for Explosion Proof (Aug)(8588).
CNS F5083-84. Explosion-Proof Bulkhead Lights for Marine Use (Feb)(10766).
JIS C 8001-91. Luminaires for Explosive Atmospheres. 28 pp.
JIS F 8846-90. Control Switches of Flameproof Light for Marine Use.

—Wiring Devices—Mining Equipment
CSA C22.2 NO 159-M1987. Attachment Plugs, Receptacles, and Similar Wiring Devices for Use in Hazardous Locations: Class I, Groups A, B, C, and D; Class II, Group G, in Coal or Coke Dust, and in Gaseous Mines (R 1993). 39 pp.

Explosive Bonded Clad Metals
Use: Clad Metals

Explosive Gases
Use For: Flammable Gases *See Also:* Explosive Atmospheres; Gases; Vinyl Chloride

—Classification
IEC 79 Pt 12-78. Electrical Apparatus for Explosive Gas Atmospheres Part 12: Classification of Mixtures of Gases or Vapours with Air According to Their Maximum Experimental Safe Gaps and Minimum Igniting Currents First Edition. 20 pp.
SNZ NZS 6101: Part 1-88. Classification of Hazardous Areas Part 1: Flammable Gas Vapour Atmospheres. 24 pp.
SNZ NZS 6101: Part 3-91. Classification of Hazardous Areas Part 3: Specific Occupancies (Flammable Gas and Vapour Atmospheres). 88 pp.

—Explosion Limits
DIN ENGL 51649 Pt 1-86. Determination of Explosion Limits of Gases and Gas/Air Mixtures (Dec). 7 pp.

Explosive Gases (Cont.)

—Ignition Temperature
DIN ENGL 51794-78. Testing of Mineral Oil Hydrocarbons; Determination of Ignition Temperature (Jan). 5 pp.
IEC 79 Pt 4-75. Electrical Apparatus for Explosive Gas Atmospheres Part 4: Method of Test for Ignition Temperature Second Edition. 22 pp.

—Plastics—High Temperature Testing
ISO 871-80. Plastics—Determination of Temperature of Evolution of Flammable Gases (Decomposition Temperature) from a Small Sample of Pulverized Material First Edition. 5 pp.

Explosive Ordnance Disposal
See Also: Explosions; Explosive Ordnance Reconnaissance; Explosives; Ordnance; Safety
NATO STANAG 2143 ED 2 AMD 1-87. Explosive Ordnance Reconnaissance/Explosive Ordnance Disposal (EOR/EOD). 17 pp.
NATO STANAG 2143 ED 3 AMD 0-91. Explosive Ordnance Reconnaissance/Explosive Ordnance Disposal. 14 pp.
NATO STANAG 2370 ED 1 AMD 1-87. Principles of Improvised Explosive Device Disposal—AEODP-3(A). 5 pp.

—Information Centers
NATO STANAG 2834 ED 1 AMD 7-75. The Operation of the Explosive Ordnance Disposal Technical Information Centre (EODTIC). 10 pp.

—Munitions
NATO STANAG 2369 ED 1 AMD 0-87. Identification and Disposal of Surface and Air Munitions-AEODP-2. 5 pp.
NATO STANAG 2369 ED 1 AMD 1-87. Identification and Disposal of Surface and Air Munitions-AEODP-2. 5 pp.
NATO STANAG 2369 ED 1 AMD 2-87. Identification and Disposal of Surface and Air Munitions-AEODP-2. 5 pp.

—Procedures
NATO STANAG 2377 ED 1 AMD 1-87. Procedure for the Management of an EOD Incident for Use When Working with Other Agencies. 15 pp.

—Radiography
NATO STANAG 2371 ED 1 AMD 0-90. EOD Radiography—AEODP-4. 5 pp.

—Recovery—Fixed Installations
NATO STANAG 2391 ED 1 AMD 0-87. Explosive Ordnance Disposal Recovery Operations on Fixed Installations—AEODP-5. 5 pp.
NATO STANAG 2391 ED 1 AMD 1-87. Explosive Ordnance Disposal Recovery Operations on Fixed Installations—AEODP-5. 5 pp.

—Underwater
NATO STANAG 2884 ED 2 AMD 0-86. Underwater Munition Disposal Procedures—AEODP-1. 5 pp.
NATO STANAG 2884 ED 2 AMD 2-86. Underwater Munition Disposal Procedures—AEODP-1. 5 pp.
NATO STANAG 2884 ED 2 AMD 3-86. Underwater Munition Disposal Procedures—AEODP-1. 5 pp.

Explosive Ordnance Reconnaissance
See Also: Explosive Ordnance Disposal; Explosives; Ordnance; Reconnaissance
NATO STANAG 2143 ED 2 AMD 1-87. Explosive Ordnance Reconnaissance/Explosive Ordnance Disposal (EOR/EOD). 17 pp.
NATO STANAG 2143 ED 3 AMD 0-91. Explosive Ordnance Reconnaissance/Explosive Ordnance Disposal. 14 pp.

Explosives
See Also: Ammonium Nitrate Fuel Oil Explosives; Black Powder; Blasters; Blasting Caps; Bombs (Ordnance); Burning Rate; Carlit; Cartridges (Explosives); Cyclonite; Demolition; Detonating Cords; Detonators; Dinitrotoluene; Drilling Equipment; Dynamite; Explosions; Explosive Ordnance Disposal; Explosive Ordnance Reconnaissance; Fires; Fireworks; Gelatin (Explosives); Gunpowder; Hazardous Materials; Hexanitrostilbene; High Explosives; Incendiary Mixtures; Initiators (Explosives); Mines (Ordnance); Munitions; Nitroglycerin; Nitroguanidene; Octogene; Ordnance; Picric Acid; Pyrotechnics; Sheathed Explosives; Smokeless Powder; Tetryl; Trinitrotoluene
CNS M3007-88. Method of Test for Industrial Explosives (Jun)(2767).
JIS K 4804-68. Ammon Explosives and TNT Explosives.
JIS K 4809-68. Analytical Methods of Explosives.
JIS K 4810-79. Testing Methods of Explosives.
JIS K 4811-68. Testing Methods for Permissible Explosives.
JIS K 4819-68. Permissible Explosive.

Explosives (Cont.)

—Acrylonitrile Copolymers
MOD UK TS 10285. Carboxy Terminated Acrylonitrile/Butadiene Reactive Liquid Polymer Type 2, QX.

—Aluminum Magnesium Alloys
MOD UK DSTAN 68-157-93. Magnesium-Aluminium Alloy 50/50 Powdered, Sizes 120 and 200 Issue 1. 12 pp.

—Aluminum Powders—Quality Assurance
NATO STANAG 4300 ED 1 AMD 0-00. (DRAFT) Test Procedures for Assessing the Quality of Aluminium Powder, for Use in Explosive Formulation, for Deliveries from One NATO Nation to Another. 13 pp.
NATO STANAG 4300 ED 1 AMD 0-93. Test Procedures for Assessing the Quality of Aluminium Powder, for Use in Explosive Formulation, for Deliveries from One NATO Nation to Another. 13 pp.

—Ammonium Nitrate—Delivery Systems
NATO STANAG 4024 ED 2 AMD 2-62. Specification for Ammonium Nitrate Crystal and Prill-Type A and Type B (for Use in Explosives) for Deliveries from One NATO Nation to Another. 9 pp.

—Barium Acetate
MOD UK DSTAN 68-41-91. Barium Acetate Anhydrous and Monohydrate Issue 2. 8 pp.

—Barium Nitrates
MOD UK DSTAN 68-74-01. Barium Nitrate Grades 1 and 1A Issue 1; Amendment 1. 8 pp.

—Cartridge Cases
MOD UK TS 10002B. Pulp Kraft, Bleached for Combustible Charge Components. Type 1 and Type 2.

—Charcoal
MOD UK DSTAN 91-31-89. Charcoal, Wood for Explosives and Pyrotechnics Types: A, D, F, G and J Issue 2. 9 pp.

—Classification
BSI BS 1000: (662)-83. 1983 Universal Decimal Classification (UDC). English Full Edition (662): Explosives. Fuels. 38 pp.
SNZ NZS/BS 1000 (662)-83. Universal Decimal Classification Explosives. Fuels. 40 pp.

—Data Sheets
MOD UK DSTAN 13-129-92. Requirements for Explosives Hazard Data Sheets for MOD Use Issue 1. 17 pp.

—Design—Chemical Compatibility
NATO STANAG 4147 ED 1 AMD 0-82. Chemical Compatibility of Ammunition Components with Explosives and Propellants (Non Nuclear Applications). 19 pp.
NATO STANAG 4147 ED 1 AMD 0-92. Chemical Compatibility of Ammunition Components with Explosives and Propellants (Non Nuclear Applications). 17 pp.

—Dyes
MOD UK DSTAN 68-58-01. Dyestuffs for Explosives Applications Issue 1; Amendment 3. 34 pp.

—Ethylbenzene
MOD UK TS 10287. Ethylbenzene.

—Glossaries
JIS K 4800-85. Glossary of Explosive Terms.

—Inhibitive Coatings—Ethyl Cellulose
MOD UK CS 2724. Ethyl Cellulose, High Viscosity.

—Lead Dioxide
MOD UK DSTAN 68-78-84. Lead Dioxide Issue 1. 6 pp.

—Lead Pipes
MOD UK DSTAN 47-24-87. Piping, Lead (for Explosive Filled Cord) Issue 2. 8 pp.

—Methyl Cellulose
MOD UK DSTAN 68-79-84. Methyl Cellulose Issue 1. 6 pp.

—Nitric Acid
MOD UK DSTAN 68-50-84. Nitric Acid, Special Issue 1. 9 pp.

—Nitrocellulose—Safety
NATO STANAG 4178 ED 1 AMD 0-91. Test Procedures for Assessing the Quality of Deliveries of Nitrocellulose from One NATO Nation to Another. 44 pp.

Explosives *(Cont.)*

—Noise
MOD UK DSTAN 00-27-85. Acceptable Limits for Exposure to Impulse Noise from Military Weapons, Explosives and Pyrotechnics Issue 1. 10 pp.

—Packaging—Shipping Containers
CGSB CAN/CGSB-43.151-M90. Packing of Explosives, Class 1, for Transportation; (Corrigendum—Jan 1991). 75 pp.

—Pentaerythritol Dioleate
MOD UK DSTAN 68-103-88. Pentaerythritol Dioleate, LF Quality Issue 1. 8 pp.

—Phthalocyanine Blue
MOD UK TS 10085A. Copper Phthalocyanine LF Quality.

—Plastic—Identification Systems
ICAO 9571. Convention on the Marking of Plastic Explosives for the Purpose of Detection, Done at Montreal on 1 March 1991. 46 pp.

—Polyester Resins
MOD UK TS 10300. Polyester Resin, Type QX.

—Polyethylene
MOD UK DSTAN 93-9-84. Polyethylene, Type QX, Grades 70 and 200 Issue 2. 8 pp.

—Polyisobutylene
MOD UK DSTAN 68-23-81. Polyisobutene (Low Molecular Weight) Polyisobutene (Low Molecular Weight, LF Quality) and Polyisobutene (Medium Molecular Weight, LF Quality, Types 1, 2 and 3) Issue 2. 10 pp.

—Potassium Nitrate
MOD UK DSTAN 68-63-90. Potassium Nitrate Issue 2. 9 pp.

—Potassium Perchlorate
MOD UK DSTAN 68-80-01. Potassium Perchlorate Issue 1; Amendment 1. 10 pp.

—Powdered—Mining
CNS M2003-88. Powdered Permissible Explosives for Coal Mines (for Taiwan Area) (Jun)(2765).
CNS M2090-88. Plyrock Powdered Permissible Explosives for Coal Mines (for Taiwan Area) (Sep)(12422).

—PRH-1
MOD UK DSTAN 13-501-87. PRH-1 Issue 1. 10 pp.

—Qualification Testing
NATO STANAG 4170 ED 1 AMD 1-85. Principles and Methodology for the Qualification of Explosive Materials for Military Use. 19 pp.

—Quality Assurance
MOD UK DSTAN 05-61: Part 17-92. Quality Assurance Procedural Requirements Part 17: Proof of Explosive Operated Devices That Discharge Projectiles Issue 1. 8 pp.

—Quality Assurance—Material Compatibility
MOD UK QTR 16/MQAD. Testing of Materials Which Are Required to Be Compatible with Explosives Issue 1 (Superseded by Def Stan 13-102).
MOD UK DSTAN 13-102-01. Testing of Materials Which Are Required to Be Compatible with Explosives Issue 1; Amendment 1. 10 pp.

—Reactivity
CGSB 1-GP-71 METH 87.1-82. Methods of Testing Paints and Pigments Reactivity (Compatibility with Explosives) Test. 5 pp.

—Rosins—Sealing Materials
MOD UK DSTAN 68-87-88. Rosin, Type QX Issue 1. 18 pp.

—Safety
BSI BS 5607-88. 1988 Safe Use of Explosives in the Construction Industry. 36 pp.
BSI BS 5607-78. 1978 Code of Practice for the Safe Use of Exolosives in the Construction Industry. 30 pp.
SNZ NZS 4403-76. Code of Practice for the Storage, Handling and Use of Explosives (Explosives Code). 68 pp.

—Silica
MOD UK TS 10270. Silica Powder, Microfine, Hydrophobic.

—Silicon Powder
MOD UK DSTAN 68-113-89. Silicon, Powdered Issue 1. 8 pp.

Explosives *(Cont.)*

—Silicones—Encapsulating
MOD UK TS 10275. RTV Silicone Two-Part Encapsulating System—Type QX.

—Sodium Sulfate Anhydrous
CNS K1255-80. Anhydrous Sodium Sulfate for Use in Explosives (Jul)(5842). 1 p.
CNS K6525-80. Method of Test for Anhydrous Sodium Sulfate Used in Explosives (Jul)(5843).

—Storage
SNZ NZS 4403-76. Code of Practice for the Storage, Handling and Use of Explosives (Explosives Code). 68 pp.

—Storage—Classification
NATO STANAG 4123 ED 3 AMD 0-00. (DRAFT) Determination of the Classification of Military Ammunition and Explosives. 9 pp.

—Storage—Quality Assurance
MOD UK DSTAN 05-61: Part 10-01. Quality Assurance Procedural Requirements Part 10: Full Trials Certification Required for Certain Explosive Related Stores Ordered for Trial Purposes Issue 1; Amendment 1. 6 pp.
MOD UK DSTAN 05-61: Part 10-92. Quality Assurance Procedural Requirements Part 10: Full Trials Certification Required for Explosive Related Items Ordered for Trial Purposes Issue 2. 12 pp.

—TATB
MOD UK DSTAN 13-500-87. TATB Type A Issue 1. 11 pp.

—Transportation—Classification
NATO STANAG 4123 ED 3 AMD 0-00. (DRAFT) Determination of the Classification of Military Ammunition and Explosives. 9 pp.

—Transportation—Safety
NATO STANAG 2890 ED 2 AMD 2-88. Regulations for Safety in the Transport of Ammunition and Explosives by Military Train. 10 pp.

—Waxes
MOD UK TS 10265. Wax Special No. 4.5.
MOD UK TS 10269. Wax Special No.10 Type QX (Withdrawn).

Exponential Atmospheres
Use: Isothermal Atmospheres

Export Documents

—Declaration Forms
EC EEC 1900/85-85. Council Regulation Introducing Community Export and Import Declaration Forms. 2 pp.

Exposure Meters
Use For: Photographic Exposure Meters
See Also: Light Meters; Photographic Flash Equipment; Radiation Measuring Instruments
ISO 2720-74. Photography—General Purpose Photographic Exposure Meters (Photoelectric Type)—Guide to Product Specification First Edition. 10 pp.
ISO 10157-91. Photography—Flash Exposure Meter—Requirements First Edition; (Corrected and Reprinted -1991). 8 pp.

Exposure Rate Meters
Use For: Rate Meters *See Also:* Kerma Rate Meters; Radiation Measuring Instruments; Radiation Meters
IEC 846-89. Beta, X and Gamma Radiation Dose Equivalent and Dose Equivalent Rate Meters for Use in Radiation Protection First Edition. 81 pp.
ISO 4071-78. Exposure Meters and Dosimeters—General Methods for Testing First Edition. 32 pp.

—Calibration
BSI BS 5869-80. 1980 Specification for X and Y Reference Radiations for Calibrating Dosemeters and Dose Ratemeters and for Determining Their Response as a Function of Photon Energy. 36 pp.
BSI BS 6689-86. 1986 Reference Beta Radiations for Calibrating Dosemeters and Doseratemeters and for Determining Their Response as a Function of Beta Radiation Energy. 12 pp.
CNS J2092-83. Methods of Calibration for Radiation Exposure Meters (Sep)(10570). 9 pp.
ISO 4037-79. X and Gamma Reference Radiations for Calibrating Dosemeters and Dose Ratemeters and for Determining Their Response as a Function of Photon Energy First Edition; (Addendum 1-1983) (Amendment 1-1983) (Addendum 2-1989). 61 pp.
ISO 6980-84. Reference Beta Radiations for Calibrating Dosemeters and Doseratemeters and for Determining Their Response as a Function of Beta Radiation Energy First Edition. 11 pp.

Exposure Rate Meters *(Cont.)*

—Calibration *(Cont.)*
JIS Z 4511-91. Methods of Calibration for Exposure Meters and Dose-Equivalent Meters. 27 pp.

—Installed
BSI BS 5566-78. 1978 Recommendation for Installed Exposure Rate Meters, Warning Assemblies and Monitors for X or Gamma Radiation of Energy Between 80 keV and 3 MeV. 21 pp.
IEC 532-92. Radiation Protection Instrumentation—Installed Dose Ratemeters, Warning Assemblies and Monitors—X and Gamma Radiation of Energy Between 50 keV and 7 MeV Second Edition. 81 pp.

—Medical
CENELEC HD 379-78. Area Exposure Product Meter. 2 pp.
IEC 580-77. Area Exposure Product Meter First Edition. 46 pp.

—Personal
JIS Z 4312-90. Personal Alarm Dosemeters for X-Rays and Gamma-Rays. 13 pp.

—Pocket—Direct Reading
BSI BS 3385-73. 1973 Direct Reading Pocket Type Electroscope Exposure Meters. 7 pp.
ISO 1758-76. Direct-Reading Electroscope-Type Pocket Exposure Meters First Edition. 4 pp.

—Pocket—Indirect Reading
ISO 1759-76. Indirect-Reading Capacitor-Type Pocket Exposure Meters and Accessory Electrometers First Edition. 4 pp.

—Portable
IEC 395-72. Portable X or Gamma Radiation Exposure Rate Meters and Monitors for Use in Radiological Protection First Edition. 45 pp.
IEC 463-74. Low Energy X or Gamma Radiation Portable Exposure Rate Meters and Monitors for Use in Radiological Protection First Edition. 46 pp.
IEC 1005-90. Portable Neutron Ambient Dose Equivalent Ratemeters for Use in Radiation Protection First Edition. 60 pp.
JIS Z 4333-90. Portable Photon Ambient Dose Equivalent Ratemeters for Radiation Protection. 17 pp.

Exposure Suits
Use For: Immersion Suits *See Also:* Chemical Resistant Clothing; Clothing; Heat Resistant Clothing; Hoods (Headgear)

—Helicopters
CAA Spec. NO. 19.
CGSB CAN/CGSB-65.17-M88. Helicopter Passenger Transportation Suit Systems. 35 pp.

—Marine
CGSB CAN/CGSB-65.16-M89. Marine Abandonment Immersion Suit Systems; (Amendment 1 June 1989). 49 pp.
CGSB CAN/CGSB-65.21-M89. Marine Anti-Exposure Work Suit Systems; (Amendment 1 May 1990). 33 pp.
CNS F1014-87. Method of Test for Immersion Suit for Marine Use (Feb)(11832).
CNS F4018-87. Immersion Suit for Marine Use (Feb)(11831).

Expulsion Fuses
See Also: Fuses (Electric)

—High Voltage
BSI BS 2692: Part 2-56. 1956 Amd 6 Fuses for Voltages Exceeding 1000 V a.c. Part 2: Expulsion Fuses. 69 pp.
IEC 282 Pt 2-70. High-Voltage Fuses Part 2: Expulsion and Similar Fuses First Edition; (Amendment 1-1978). 62 pp.
SAA AS 1033.1-90. High Voltage Fuses (For Rated Voltages Exceeding 1000 V—Part 1: Explusion Type. 71 pp.

—High Voltage—Short Circuit Power Factor
IEC 282 Pt 3-76. High-Voltage Fuses Part 3: Determination of Short-Circuit Power Factor for Testing Current-Limiting Fuses and Expulsion and Similar Fuses First Edition. 10 pp.

Extended Pascal
Use: Pascal

Extenders
See Also: Additives; Paint Extenders; Solvents

—Acidity
BSI BS 3483: Part C3-82. 1982 Methods for Testing Pigments in Paints Part C3: Determination of Acidity or Alkalinity of the Aqueous Extract. 4 pp.

Extenders (Cont.)

—Acidity (Cont.)
ISO 787 Pt 4-81. General Methods of Test for Pigments and Extenders—Part 4: Determination of Acidity or Alkalinity of the Aqueous Extract First Edition. 5 pp.

—Alkalinity
BSI BS 3483: Part C3-82. 1982 Methods for Testing Pigments in Paints Part C3: Determination of Acidity or Alkalinity of the Aqueous Extract. 4 pp.
ISO 787 Pt 4-81. General Methods of Test for Pigments and Extenders—Part 4: Determination of Acidity or Alkalinity of the Aqueous Extract First Edition. 5 pp.

—Bulk Density—Tamping
BSI BS 3483: Part B10-82. 1982 Methods for Testing Pigments in Paints Part B10: Determination of Tamped Volume and Apparent Density After Tamping. 6 pp.
ISO 787 Pt 11-81. General Methods of Test for Pigments and Extenders—Part 11: Determination of Tamped Volume and Apparent Density After Tamping First Edition. 6 pp.

—Density—Centrifuging
BSI BS 3483: Part B9-80. 1980 Methods for Testing Pigments in Paints Part B9: Determination of Density Relative to Water (Using a Centrifuge). 4 pp.

—Density—Pycnometric Analysis
BSI BS 3483: Part B8-93. 1993 Methods for Testing Pigments for Paints Part B8: Determination of Density (Pyknometer Method) (ISO 787-10: 1993) (W). 9 pp.
BSI BS 3483: Part B8-82. 1982 Methods for Testing Pigments in Paints Part B8: Determination of Density (Pyknometer Method) 4 Degrees Centigrate. 6 pp.
ISO 787 Pt 10-93. General Methods of Test for Pigments and Extenders—Part 10: Determination of Density—Pyknometer Method Second Edition. 9 pp.
ISO 787 Pt 23-79. General Methods of Test for Pigments and Extenders—Part 23: Determination of Density (Using a Centrifuge to Remove Entrained Air) First Edition. 5 pp.

—Dispersion
BSI BS 3483: Part D1-91. 1991 Testing Pigments for Paints Part D1: Method of Dispersion for Assessment of Dispersion Characteristics (Introduction) (ISO 8780-1: 1990). 6 pp.
ISO 8780 Pt 1-90. Pigments and Extenders—Methods of Dispersion for Assessment of Dispersion Characteristics—Part 1: Introduction First Edition. 4 pp.

—Dispersion—Automatic Mullers
BSI BS 3483: Part D5-91. 1991 Testing Pigments for Paints Part D5: Method of Dispersion for Assessment of Dispersion Characteristics Using an Automatic Muller (ISO 8780-5: 1990). 8 pp.
ISO 8780 Pt 5-90. Pigments and Extenders—Methods of Dispersion for Assessment of Dispersion Characteristics—Part 5: Dispersion Using an Automatic Muller First Edition. 7 pp.

—Dispersion—Bead Mills
BSI BS 3483: Part D4-91. 1991 Testing Pigments for Paints Part D4: Method of Dispersion for Assessment of Dispersion Characteristics Using a Bead Mill (ISO 8780-4: 1990). 9 pp.
ISO 8780 Pt 4-90. Pigments and Extenders—Methods of Dispersion for Assessment of Dispersion Characteristics—Part 4: Dispersion Using a Bead Mill First Edition. 8 pp.

—Dispersion—Impeller Mills
BSI BS 3483: Part D3-91. 1991 Testing Pigments for Paints Part D3: Method of Dispersion for Assessment of Dispersion Characteristics Using a High-Speed Impeller Mill (ISO 8780-3: 1990). 9 pp.
ISO 8780 Pt 3-90. Pigments and Extenders—Methods of Dispersion for Assessment of Dispersion Characteristics—Part 3: Dispersion Using a High-Speed Impeller Mill First Edition. 8 pp.

—Dispersion—Oscillatory Shaking Machines
BSI BS 3483: Part D2-91. 1991 Testing Pigments for Paints Part D2: Method of Dispersion for Assessment of Dispersion Characteristics Using an Oscillatory Shaking Machine (ISO 8780-2: 1990). 9 pp.
ISO 8780 Pt 2-90. Pigments and Extenders—Methods of Dispersion for Assessment of Dispersion Characteristics—Part 2: Dispersion Using an Oscillatory Shaking Machine First Edition. 8 pp.

—Dispersion—Triple Roll Mills
BSI BS 3483: Part D6-91. 1991 Testing Pigments for Paints Part D6: Method of Dispersion for Assessment of Dispersion Characteristics Using a Triple-Roll Mill (ISO 8780-6: 1990). 8 pp.

Extenders (Cont.)

—Dispersion—Triple Roll Mills (Cont.)
ISO 8780 Pt 6-90. Pigments and Extenders—Methods of Dispersion for Assessment of Dispersion Characteristics—Part 6: Dispersion Using a Triple-Roll Mill First Edition. 7 pp.

—Electrical Resistivity—Extraction Analysis
BSI BS 3483: Part C5-74. 1974 Methods for Testing Pigments in Paints Part C5: Determination of Resistivity of Aqueous Extract. 3 pp.

—Oil Absorption
BSI BS 3483: Part B7-82. 1982 Methods for Testing Pigments in Paints Part B7: Determination of Oil Absorption Value. 4 pp.
ISO 787 Pt 5-80. General Methods of Test for Pigments and Extenders—Part 5: Determination of Oil Absorption Value First Edition. 5 pp.

—pH
BSI BS 3483: Part C4-82. 1982 Methods for Testing Pigments in Paints Part C4: Determination of pH Value of an Aqueous Suspension. 4 pp.
ISO 787 Pt 9-81. General Methods of Test for Pigments and Extenders—Part 9: Determination of pH Value of an Aqueous Suspension First Edition. 5 pp.

—Residue Content—Sieve Analysis
BSI BS 3483: Part B3-82. 1982 Methods for Testing Pigments in Paints Part B3: Determination of Residue on Sieve (Water Method, Using a Manual Procedure). 4 pp.
ISO 787 Pt 7-81. General Methods of Test for Pigments and Extenders—Part 7: Determination of Residue on Sieve—Water Method—Manual Procedure First Edition. 6 pp.
ISO 787 Pt 18-83. General Methods of Test for Pigments and Extenders—Part 18: Determination of Residue on Sieve—Mechanical Flushing Procedure Second Edition. 7 pp.

—Soluble Matter Content—Extraction Analysis
BSI BS 3483: Part C1-80. 1980 Methods for Testing Pigments in Paints Part C1: Determination of Matter Soluble in Water (Hot Extraction Method). 4 pp.
BSI BS 3483: Part C2-80. 1980 Methods for Testing Pigments in Paints Part C2: Determination of Matter Soluble in Water (Cold Extraction Method). 4 pp.
ISO 787 Pt 3-79. General Methods of Test for Pigments and Extenders—Part 3: Determination of Matter Soluble in Water—Hot Extraction Method First Edition. 4 pp.
ISO 787 Pt 8-79. General Methods of Test for Pigments and Extenders—Part 8: Determination of Matter Soluble in Water—Cold Extraction Method First Edition. 4 pp.

—Volatile Matter Content—High Temperature Testing
BSI BS 3483: Part B6-74. 1974 Methods for Testing Pigments for Paints Part B6: Determination of Matter Volatile at 105 Degrees Centigrade. 4 pp.
ISO 787 Pt 2-81. General Methods of Test for Pigments and Extenders—Part 2: Determination of Matter Volatile at 105 Degrees Celsius First Edition. 5 pp.

Extenders (Logic Circuits)
See Also: Gate Circuits
MOD UK DSTAN 59-62: 90/055-69. Integrated Circuit Issue 1. 2 pp.

Extensibility
See Also: Deflection; Deformation; Elongation; Plastic Properties

—Bituminous Sealants
DIN ENGL 1996 Pt 19-84. Testing of Asphalt; Determination of Extensibility and Adhesion Using a Rabe Joint Model (May). 6 pp.

—Joint Sealants
DIN ENGL 1996 Pt 19-84. Testing of Asphalt; Determination of Extensibility and Adhesion Using a Rabe Joint Model (May). 6 pp.

Extensible Papers
See Also: Packaging Materials; Papers

—Kraft
JIS P 3412-76. Kraft Extensible Paper.

Extension Cords
See Also: Cords (Electric); Power Supply Cords
SAA AS 3199-82. Approval and Test Specification for Cord Extension Sets Amdt 1 April 1985 (This is a Joint Standard with SANZ NZS 3199). 8 pp.

Extension Cords (Cont.)
SNZ NZS/AS 3199-82. Approval and Test Specification for Cord Extension Sets Amend: 1 (This is a Joint Standard with SAA AS 3199). 8 pp.

—Sockets—Safety
SNZ NZS/AS 3120-93. Approval and Test Specification Extension Sockets (This is a Joint Standard with SAA AS 3120). 16 pp.

Extension Ladders
See Also: Ladders

—Aluminum
BSI BS 2037-90. 1990 Amd 1 Portable Aluminium Ladders, Steps, Trestles and Lightweight Stagings (AMD 6394) September 30, 1991. 21 pp.
BSI BS 2037-02. 1990 Amd 2 Portable Aluminium Ladders, Steps, Trestles and Lightweight Stagings (AMD 7895) September 15, 1993 (R). 22 pp.
BSI BS 2037-84. (WITHDRAWN) 1984 Amd 1 Portable Aluminium Ladders, Steps, Trestles and Lightweight Stagings. 22 pp.

—Wood
BSI BS 1129-90. 1990 Portable Timber Ladders, Steps, Trestles and Lightweight Stagings. 36 pp.
BSI BS 1129-01. 1990 Amd 1 Portable Timber Ladders, Steps, Trestles and Lightweight Stagings (AMD 7414) January 15, 1993. 38 pp.
BSI BS 1129-02. 1990 Amd 2 Portable Timber Ladders, Steps, Trestles and Lightweight Stagings (AMD 7896) September 15, 1993 (R). 39 pp.

Extension Springs
Use For: Tension Springs See Also: Springs (Elastic)

—Helical
CNS B1268-81. Calculation and Design for Helical Tension Springs Round Wire or Round Bar (Sep)(7856).
CNS B1270-81. Calculation and Design for Helical Tensional Springs Wire or Rod (Sep)(7858).
CNS B2595-81. Quality Specification for Cold Coiled Helical Tension Springs Made of Round Metal Wire. (Jun)(7595). 15 pp.
DIN ENGL 2099 Pt 2-73. Helical Springs Made of Round Wire; Data for Tension Springs; Printed Form (Nov). 3 pp.
JIS B 2704-87. Design of Helical Compression and Extension Springs.
JIS B 2704-78. Design of Helical Compression and Extension Springs. 30 pp.
JIS B 2708-87. Cold Coiled Helical Tension Springs. 8 pp.

—Helical—Design
BSI BS 1726: Part 2-88. 1988 Amd 1 Coil Springs Part 2: Guide for the Design of Helical Extension Springs (AMD 6110) October 31, 1989. 18 pp.

—Helical—Quality Assurance
DIN ENGL 2097-73. Helical Springs Made of Round Wire; Quality Specifications for Cold Coiled Tension Springs (May). 10 pp.

—Testing Equipment
JIS B 7738-84. Testing Machines for Helical Compression and Extension Springs. 11 pp.

Extensions (Tools)
Use For: Tool Extensions See Also: Tools

—Bars
CNS B3269-85. Assembly Tools for Screws and Nuts Bar (Apr)(4190).
CNS B3426-85. Assembly Tools for Screws and Nuts Bar (for Family Use) (Apr)(8347).

—Driving Squares
DIN ENGL 377-69. Extensions for Tools with Driving Square (Aug). 2 pp.

Extensometers
See Also: Dimensional Measuring Instruments; Strain Gages
BSI BS 3846-70. 1970 Calibration and Grading of Extensometers for Testing of Metals. 25 pp.
SAA AS 1545-76. Methods for the Calibration and Grading of Extensometers Reconfirmed 1989. 32 pp.

—Metals
CEN PREN 10 002-4-90. Metallic Materials—Verification of Extensometres Used in Uniaxial Testing. 8 pp.
ISO 9513-89. Metallic Materials—Verification of Extensometers Used in Uniaxial Testing First Edition; (Corrected and Reprinted -1989). 8 pp.

—Metals—Tensile Testing
JIS B 7741-91. Extensometers Used in Metallic Material Tensile Testing. 14 pp.

INTERNATIONAL AND NON-U.S. NATIONAL STANDARDS
SUBJECT INDEX

Extensometers *(Cont.)*
—Thermal Expansion
DIN ENGL 51045 Pt 1-89. Determination of the Thermal Expansion of Solids; Principles (Sept). 8 pp.

Exterior Paints
Scope Note: See the subheading Exterior under the specific type of paint or item being painted

External Tab Washers
Use: Tab Washers—External

External Vibrators
Use: Concrete Vibrators—External

Extra 230 Aircraft
See Also: Aircraft

—Foreign Airworthiness Directives
CAA. Extra EA-230 Series Aircraft. 2 pp.

Extra 300 Aircraft
See Also: Aircraft

—Foreign Airworthiness Directives
CAA. Extra EA-300 Series Aircraft. 1 p.

Extracting Tools
Scope Note: Use a more specific term *See:* Hand Tools; Tools

Extraction Analysis
Scope Note: For extraction analysis of specific products or materials, see the subheading Extraction Analysis under the specific material or product. For additional references, consult the following list *See:* Centrifuging; Distillation Methods; Leachate Extraction; Solvent Extraction

Extraction Solvents
Use: Solvents

Extraction Tools
See Also: Insertion Tools; Nail Pullers; Screw Extractors

—Electric Contacts
CEN PREN 2591 (Part E6)-92. Elements of Electrical and Optical Connection Test Methods Part E6—Use of Tools. 4 pp.

Extraction Zones
See Also: Air Drop Operations; Drop Zones

—Tactical Air Transport Operations
NATO STANAG 3570 ED 3 AMD 4-86. Drop Zones and Extraction Zones—Criteria and Markings. 26 pp.
NATO STANAG 3570 ED 3 AMD 5-86. Drop Zones and Extraction Zones—Criteria and Markings. 27 pp.

—Tactical Air Transport Operations—Identification Systems
NATO STANAG 3570 ED 3 AMD 4-86. Drop Zones and Extraction Zones—Criteria and Markings. 26 pp.
NATO STANAG 3570 ED 3 AMD 5-86. Drop Zones and Extraction Zones—Criteria and Markings. 27 pp.

Extractors, Comedo
Use: Comedo Extractors

Extractors (Glassware)
Use For: Soxhlet Extractors *See Also:* Laboratory Ware
BSI BS 2071-89. 1989 Soxhlet Extractors (H). 12 pp.

—Assembly
CNS R2136-86. Assembly of Soshlet's Glass Extractor for Chemical Analysis (Mar) (7312).

Extractors (Laundry)
Use: Water Extractors

Extractors (Tools)
Use: Extraction Tools

Extraneous Matter Content Analysis
Use: Impurities Content Analysis

Extraterrestrial Intelligence
See Also: Astronomy; Space Communications; Space Research Services

Extraterrestrial Intelligence *(Cont.)*
—Radio Communications
CCIR Report 700-2-90. Radiocommunication Aspects of Systems to Search for Extra-Terrestrial Intelligence (SETI)—Section 2G—Radioastronomy and Radar Astronomy. 7 pp.
CCIR QUESTION 150/7-90. Radiocommunication Requirements for Systems to Search for Extraterrestrial Intelligence—Questions Concerning Study Group 7—Science Services. 1 p.

Extraterrestrial Plasmas
Use: Plasmas (Physics)—Extraterrestrial

Extreme Pressure Lubricants
Use For: EP Lubricants *See Also:* Greases; Lubricants; Lubricating Oils; Mineral Oils
CGSB 3-GP-380MA-90. Extreme Pressure Industrial Gear Type Lubricating Oil (ISO Viscosity Grade 150). 8 pp.

—Aging Testing
DIN ENGL 51586-77. Testing of Lubricants; Determination of the Ageing Properties of Lubricating Oils for High Pressure Load (Aug). 4 pp.

—Aircraft
MOD UK DSTAN 91-53-01. Grease, Aircraft: Synthetic, Extreme Pressure NATO Code No: G-354 Joint Service Designation: XG-287 Issue 1; Amendment 1. 14 pp.

—Loads (Forces)—Four Ball Testers
DIN ENGL 51350 Pt 4-84. Testing of Lubricants; Testing by the Shell Four-Ball Tester; Determination of Welding Load of Consistent Lubricants (Jan). 3 pp.

Extrudability
—Joint Sealants
BSI BS EN 28394-91. 1991 Building Construction—Jointing Products—Determination of Extrudability of One-Component Sealants (ISO 8394: 1988). 9 pp.
BSI BS EN 29048-91. 1991 Building Construction—Jointing Products—Determination of Extrudability of Sealants Using Standardized Apparatus (ISO 9048: 1987). 11 pp.
CEN EN 28 394-90. Building Construction—Jointing Products—Determination of Extrudability of One-Component Sealants. 6 pp.
CEN EN 29 048-90. Building Construction—Jointing Products—Determination of Extrudability of Sealants Using Standardized Apparatus. 9 pp.
DIN ENGL EN 28394-91. Jointing Products in Building Construction; Determination of Extrudability of One-Component Sealants (ISO 8394:1988) (May). 5 pp.
DIN ENGL EN 29048-91. Jointing Products in Building Construction; Determination of Extrudability of Sealants Using Standardized Apparatus (ISO 9048:1987) (May) (Supersedes DIN 52456, May 1976 Edition). 7 pp.
ISO 8394-88. Building Construction—Jointing Products—Determination of Extrudability of One-Component Sealants First Edition. 4 pp.
ISO 9048-87. Building Construction—Jointing Products—Determination of Extrudability of Sealants Using Standardized Apparatus First Edition. 7 pp.

—Sealants
CGSB CAN2-19.0-M77METH 3.1-78. Methods of Testing Putty, Caulking and Sealing Compounds Extrudability of Prepackaged Cartridges of Single Component Sealing Compounds. 1 p.
CGSB CAN2-19.0-M77METH 3.2-78. Methods of Testing Putty, Caulking and Sealing Compounds Extrudability of Bulk Packaged Single and Multicomponent Sealing Compounds. 1 p.

Extruders
See Also: Extrusion Compounds; Plastics Working Machinery; Presses

—Plastics—Safety
CEN PREN 1114-1-93. Rubber and Plastics Machines—Safety—Extruders and Extrusion Lines—Requirements for the Design and Construction—Part 1: Extruders. 57 pp.

—Rubber—Safety
CEN PREN 1114-1-93. Rubber and Plastics Machines—Safety—Extruders and Extrusion Lines—Requirements for the Design and Construction—Part 1: Extruders. 57 pp.

Extrusion Compounds
See Also: Extruders; Molding Materials

Extrusion Compounds *(Cont.)*
—Aluminum—Aerospace—Ultrasonic Testing
CEN PREN2004-02-75. Test Methods for Aluminium and Aluminium Alloy Products—Part 2—Ultrasonic Testing of Plates Forgings and Extrusions (C5/21A). 7 pp.

—Aluminum—Test Specimens
CGSB CAN2-19.0-M77METH12.2-78. Methods of Testing Putty, Caulking and Sealing Compounds Preparation of Plain Aluminum Panels and Extrusions. 1 p.

—Doors
CGSB 41-GP-19MA-78. Rigid Vinyl Extrusions for Windows and Doors, Standard for (R 1984). 13 pp.
CGSB 41-GP-20M-76. Extrusions, Vinyl, Nonrigid, for Windows and Doors, Standard for (R 1983). 9 pp.

—Polycarbonate Resins
MOD UK DSTAN 93-61-90. Method for Specifying Polycarbonate (PC) Moulding and Extrusion Materials Issue 1. 13 pp.
MOD UK TS 423D. Polycarbonate Moulding and Extrusion Material, Type QX.
MOD UK TS 437B. Polycarbonate Moulding and Extrusion Material (Withdrawn).

—Polyethylene
BSI BS 3412-92. 1992 General Purpose Polyethylene Materials for Moulding and Extrusion (V). 18 pp.
BSI BS 3412-76. 1976 Amd 2 Polyethylene Materials for Moulding and Extrusion (AMD 4972) August 30, 1985. 19 pp.
MOD UK DSTAN 93-36-01. Polyethylene Materials for Moulding and Extrusion, Type QX Issue 1; Amendment 1. 10 pp.
MOD UK TS 479B. Polyethylene Moulding and Extrusion Material, Black, for Drinking Water Applications.

—Polymethyl Methacrylate
ISO 8257 Pt 1-87. Plastics—Poly(Methyl Methacrylate) (PMMA) Moulding and Extrusion Materials—Part 1: Designation First Edition. 7 pp.

—Polymethyl Methacrylate—Test Specimens
ISO 8257 Pt 2-90. Plastics—Poly (Methyl Methacrylate) (PMMA) Moulding and Extrusion Materials—Part 2: Preparation of Test Specimens and Determination of Properties First Edition. 8 pp.

—Polytetrafluoroethylene
BSI BS 6564: Sec 1.1-89. 1989 Polytetrafluoroethylene (PTFE) Materials and Products Part 1: Polytetrafluoroethylene Powders for Moulding and Extrusion Section 1.1: Specification. 8 pp.
BSI BS 6564: Sec 1.2-89. 1989 Polytetrafluoroethylene (PTFE) Materials and Products Part 1: Polytetrafluoroethylene Powders for Moulding and Extrusion Section 1.2: Method of Specifying. 23 pp.
BSI BS 6564: Sec 1.2-01. 1989 Amd 1 Polytetrafluoroethylene (PTFE) Materials and Products Part 1: Polytetrafluoroethylene Powders for Moulding and Extrusion Section 1.2: Method of Specifying (AMD 7125) May 1, 1992. 24 pp.
SAA AS 1198-73. Polytetrafluoroethylene (PTFE) Extruded Rod and Tube Reconfirmed 1986. 11 pp.

—Sealants—Aerospace
CEN PREN 3094-89. Sealants Test Method Determination of Application Time. 6 pp.

—Vulcanized Rubber—Marine
MOD UK NES 2006: Part 2-01. Specification for Ozone-Resistant Natural Rubber Part 2: Extrusions Issue 1 (07.84); Amendment 1. 12 pp.

Extrusions
Scope Note: For extrusions of specific products or materials, see the subheading Extrusions under the specific material or product. For additional references consult the following list *See:* Construction Materials; Extrudability; Extruders; Extrusion Compounds; Moldings; Rods; Tubes

Exudation Resistance
—Packaging Papers
JIS Z 0220-60. Testing Method for Exudation Resistance of Tarpaulin Paper for Wrappings.

Eye Charts
Use: Visual Acuity Charts

Eye Ends
Use: Eye Fasteners

Eye Fasteners
Use For: Eye Ends *See Also:* Fasteners

—Aircraft
SBAC AS 2900 ISSUE 2. Eye End. 1 p.
SBAC AS 6540 ISSUE 1(I). Eye End.

INDUSTRY STANDARDS

INTERNATIONAL AND NON-U.S. NATIONAL STANDARDS
SUBJECT INDEX

Eye

Eye Glasses
Use: Eyeglasses

Eye Protectors
See Also: Goggles; Industrial Equipment; Protective Clothing; Sunglasses; Visors

BSI BS 2092-87. 1987 Eye-Protectors for Industrial and Non-Industrial Uses. 24 pp.
CEN PREN 166-89. Personal Eye-Protection: Specifications. 33 pp.
CNS Z2010-65. Safety Goggles (for General Use) (Jan)(2397)(R 1972).
CSA CAN/CSA-Z94.3-92. Industrial Eye and Face Protectors; (Gen Instr 1). 64 pp.
ISO 4849-81. Personal Eye-Protectors—Specifications First Edition. 7 pp.
ISO 4854-81. Personal Eye-Protectors—Optical Test Methods First Edition. 20 pp.
ISO 4855-81. Personal Eye-Protectors—Non-Optical Test Methods First Edition. 12 pp.
ISO 4856-82. Personal Eye-Protectors—Synoptic Tables of Requirements for Oculars and Eye-Protectors First Edition. 6 pp.
JIS T 8148-89. Goggle Type Eye Protectors for Industry. 10 pp.
SAA AS 1336-82. Recommended Practices for Eye Protection in the Industrial Environment. 16 pp.
SAA AS/NZS 1337-92. Eye Protectors for Industrial Applications (in Professional Packages 32, 36, 47) (Supersedes AS 1337—1984 and NZS 5816:1986). 40 pp.
SNZ NZS/AS 1337-92. Eye Protectors for Industrial Applications (This is a Joint Standard with SAA AS 1337). 40 pp.

—Automotive

BSI BS 4110-79. 1979 Amd 4 Eye-Protectors for Vehicle Users (AMD 5898) April 28, 1989. 17 pp.
SAA AS 1609-81. Eye Protectors for Motor Cyclists and Racing Car Drivers Amdt 1, August 1982. 12 pp.

—Deformation

CEN PREN 168-89. Personal Eye Protection: Non-Optical Test Methods. 24 pp.

—Glossaries

BSI BS 6967-88. 1988 Glossary of Terms for Personal Eye-Protection. 8 pp.
CEN EN 165-86. Personal Eye-Protection: Vocabulary. 6 pp.
CEN PREN 165-91. Personal Eye-Protection—Vocabulary. 41 pp.
ISO 4007-77. Personal Eye-Protectors—Vocabulary First Edition. 7 pp.

—Helmets—Motorcycle

CEN PREN 173-91. Personal Eye-Protection—Visors for Motorcycle Helmets. 12 pp.

—Impact Testing

CEN PREN 168-89. Personal Eye Protection: Non-Optical Test Methods. 24 pp.

—Infrared Filters

BSI BS EN 171-92. 1992 Infra-Red Filters Used in Personal Eye-Protection Equipment. 9 pp.
CEN EN 171-86. Personal Eye-Protection: Infrared Filters; Transmittance Requirements and Recommended Use. 5 pp.
CEN EN 171-92. Personal Eye-Protection—Infra Red Filters—Transmittance Requirements and Recommended Use. 4 pp.
CEN PREN 172-89. Personal Eye Protection: Filters for General and Industrial Use. 25 pp.
CNS Z2031-87. Eye Protectors for Radiations (Jul)(7174).
DIN ENGL 4647 Pt 3-77. Glasses for Eye-Protection Equipment; Protective Filters Against Infra-Red Radiation (Protective IR Filters) (Feb). 4 pp.
ISO 4852-78. Personal Eye-Protectors—Infra-Red Filters—Utilisation and Transmittance Requirements First Edition. 6 pp.
JIS T 8141-80. Eye Protectors for Radiations. 19 pp.
SAA AS 1338.3-81. Filters for Eye Protectors—Part 3: Filters for Protection Against Infrared Radiation (Superseded by AS/NZS 1338.3:1992).
SAA AS/NZS 1338. 3-92. Filters for Eye Protectors—Part 3: Filters for Protection Against Infra-Red Radiation (in Professional Packages 32, 36, 47) (Supersedes AS 1338.3—1981) 7 pp.
SNZ NZS/AS 1338. 3-92. Filters for Eye Protectors Part 3: Filters for Protection Against Infra-Red Radiation (This is a Joint Standard with SAA AS 1338.2). 7 pp.

—Laser Filters

ISO 6161-81. Personal Eye-Protectors—Filters and Eye-Protectors Against Laser Radiation First Edition; (Amendment Slip-1982). 7 pp.

—Lenses

DIN ENGL 4646 Pt 1-83. Lenses for Eye Protectors; Principles, Requirements, Dimensions, Marking (Mar). 9 pp.

Eye Protectors *(Cont.)*

—Lenses—Antifogging Tests

DIN ENGL 4646 Pt 8-86. Lenses for Eye Protectors; Antifogging Test (Sept). 2 pp.

—Lenses—Heat Treated Glass

CNS Z2033-81. Eye Protector with Tempered Glass Lenses (Mar)(7176).
JIS T 8146-71. Eye Protector with Tempered Glass Lenses. 6 pp.

—Lenses—Plastic

CNS Z2034-81. Eye Protector with Plastic Lenses (Mar)(7177).
JIS T 8147-75. Eye Protector with Plastic Lenses. 11 pp.

—Maintenance

BSI BS 7028-88. 1988 Selection, Use and Maintenance of Eye-Protection for Industrial and Other Uses. 24 pp.

—Mechanical Properties

CEN PREN 168-89. Personal Eye Protection: Non-Optical Test Methods. 24 pp.

—Motor Vehicle Operators

JIS T 8137-82. Eye Protectors for Vehicular Users.

—Motorcycles

SAA AS 1609-81. Eye Protectors for Motor Cyclists and Racing Car Drivers Amdt 1, August 1982. 12 pp.

—Optical Properties

CEN PREN 167-89. Personal Eye-Protection: Optical Test Methods. 22 pp.
DIN ENGL 4646 Pt 2-75. Glasses for Eye-Protection Equipment; Optical Test Methods (Dec). 6 pp.

—Racquet Sports

CSA P400-M1982P. Racquet Sports Eye Protection; (Amd 1-2 June 1986). 17 pp.
SAA AS/NZS 4066-92. Eye Protectors for Racquet Sports. 13 pp.
SNZ NZS/AS 4066-92. Eye Protectors for Racquet Sports (This is a Joint Standard with SAA AS 4066). 13 pp.

—Radiation

BSI BS 1542-82. 1982 Equipment for Eye, Face and Neck Protection Against Non-Ionizing Radiation Arising During Welding and Similar Operations. 8 pp.
BSI BS 2724-87. 1987 Amd 1 Sun Glare Eye Protectors for General Use. 20 pp.
CEN PREN 172-89. Personal Eye Protection: Filters for General and Industrial Use. 25 pp.
CNS Z2031-87. Eye Protectors for Radiations (Jul)(7174).
JIS T 8141-80. Eye Protectors for Radiations. 19 pp.

—Skiing

CEN PREN 174-93. Specification for Personal Eye Protection—Ski Goggles for Downhill Skiing. 11 pp.

—Ultraviolet Filters

BSI BS EN 170-92. 1992 Ultraviolet Filters Used in Personal Eye-Protection Equipment. 9 pp.
CEN EN 170-86. Personal Eye-Protection: Ultraviolet Filters; Transmittance Requirements and Recommended Use. 5 pp.
CEN EN 170-92. Personal Eye-Protection—Ultraviolet Filters—Transmittance Requirements and Recommended Use. 4 pp.
CEN PREN 172-89. Personal Eye Protection: Filters for General and Industrial Use. 25 pp.
CNS Z2031-87. Eye Protectors for Radiations (Jul)(7174).
DIN ENGL 4647 Pt 2-77. Glasses for Eye-Protection Equipment; Protective Filters Against Ultra-Violet Radiation (Protective UV Filters) (Sept). 3 pp.
ISO 4851-79. Personal Eye-Protectors—Ultra-Violet Filters—Utilisation and Transmittance Requirements First Edition. 4 pp.
JIS T 8141-80. Eye Protectors for Radiations. 19 pp.
SAA AS 1338.2-81. Filters for Eye Protectors—Part 2: Filters for Protection Against Ultraviolet Radiation (Superseded by AS/NZS 1338.2:1992).
SAA AS/NZS 1338. 2-92. Filters for Eye Protectors—Part 2: Filters for Protection Against Ultraviolet Radiation (in Professional Packages 32, 36, 47) (Supersedes AS 1338.2—1981 and NZS 5817.2:1983). 5 pp.
SNZ NZS/AS 1338. 2-92. Filters for Eye Protectors Part 2: Filters for Protection Against Ultraviolet Radiation (This is a Joint Standard with SAA AS 1338.2). 5 pp.

—Welding

BSI BS 679-89. 1989 Filters, Cover Lenses and Backing Lenses for Use During Welding and Similar Operations. 14 pp.

Eye Protectors *(Cont.)*

—Welding *(Cont.)*

BSI BS 1542-82. 1982 Equipment for Eye, Face and Neck Protection Against Non-Ionizing Radiation Arising During Welding and Similar Operations. 8 pp.
BSI BS EN 169-92. 1992 Filters for Personal Eye-Protection Equipment Used in Welding and Similar Operations. 12 pp.
CEN EN 169-86. Personal Eye-Protection: Filters for Welding and Related Techniques; Transmittance Requirements and Recommended Use. 7 pp.
CEN EN 169-92. Personal Eye-Protection—Filters for Welding and Related Techniques—Transmittance Requirements and Recommended Utilisation. 7 pp.
CEN PREN 379-90. Personal Eye Protection—Welding Filters with Variable Shade or Dual Shade. 11 pp.
CNS Z2032-87. Shields and Helmets for Welders (Jul)(7175).
DIN ENGL 4647 Pt 1-77. Glasses for Eye-Protection Equipment; Protective Filters for Use in Welding (Feb). 5 pp.
DIN ENGL 4647 Pt 6-77. Glasses for Eye-Protection Equipment; Cover Glasses (Feb). 2 pp.
ISO 4850-79. Personal Eye-Protectors—Filters for Welding and Related Techniques—Filters—Utilisation and Transmittance Requirements First Edition. 7 pp.
JIS T 8142-89. Helmet Type and Handshield Type Protectors for Welders. 12 pp.
SAA AS 1338.1-81. Filters for Eye Protectors—Part 1: Filters for Protection Against Radiation Generated in Welding and Allied Operations (Superseded by AS/NZS 1338.1:1992).
SNZ NZS/AS 1338. 1-92. Filters for Eye Protectors Part 1: Filters for Protection Against Radiation Generated in Welding and Allied Operations (This is a Joint Standard with SAA AS 1338.1). 21 pp.

—Welding—Filters

SAA AS/NZS 1338. 1-92. Filters for Eye Protectors—Part 1: Filters for Protection Against Radiation Generated in Welding and Allied Operations (in Professional Packages 32, 36, 47) (Supersedes AS 1338.1—1981 and NZS 5817.1: 1983). 21 pp.

Eye Screws
Use For: Ring Head Screws *See Also:* Screws
CNS A2010-89. Screws, Ring Headed (Jul)(860). 2 pp.

Eyebolts
See Also: Bolts; Hooks; Lifting Tackle; Ringbolts

BSI BS 4278-84. 1984 Eyebolts for Lifting Purposes. 14 pp.
CNS B2280-81. Eye Bolts (Apr)(4490).
CNS B2281-81. Lifting Eye Bolts (Apr)(4491).
CNS B2282-81. Eye Bolts with Small Eyes (M20) (Apr)(4492).
DIN ENGL 444-83. Eyebolts (Apr). 5 pp.
DIN ENGL 580-72. Lifting Eye Bolts (Mar). 3 pp.
ISO 3266-84. Eyebolts for General Lifting Purposes First Edition. 9 pp.
JIS B 1168-75. Eyebolts. 13 pp.
MOD UK DSTAN 53-24-01. Eyebolts and Bolts, Ring for Lifting Purposes Issue 1; Amendment 1. 10 pp.
SNZ NZS/BS 4278-84. Specification for Eyebolts for Lifting Purposes. 16 pp.

—Aircraft

BSI SP 114-55. 1955 Amd 3 Eye Bolts (Unified Threads) for Aircraft (AMD 3350) July 31, 1980. 8 pp.

—Aircraft—Rudder Bars

SBAC AS 2187 ISSUE 5. Eye bolt—(Rudder Bar). 1 p.

—Aircraft—Stirrups—Fuel Filters

SBAC AS 3244 ISSUE 1. Stirrup Eyebolt—Filter Fuel Type 'A'. 1 p.
SBAC AS 3269 ISSUE 1. Stirrup Eyebolt—Filter Fuel Type 'B'. 1 p.

—Lift Suspension—Elevators (Lifts)

BSI BS 5655: Part 8-83. 1983 Lifts and Service Lifts Part 8: Specification for Eyebolts for Lift Suspension. 8 pp.

—Swivels

DIN ENGL 82006-71. Oval Eyes with Screwed Shank for Swivels and Turnbuckles (Apr). 3 pp.
DIN ENGL 82010-71. Round Eyes with Screwed Shank for Swivels and Turnbuckles (Apr). 3 pp.

—Turnbuckles

DIN ENGL 82006-71. Oval Eyes with Screwed Shank for Swivels and Turnbuckles (Apr). 3 pp.
DIN ENGL 82010-71. Round Eyes with Screwed Shank for Swivels and Turnbuckles (Apr). 3 pp.

—Turnbuckles—Ships

JIS F 7020-89. Marine Turnbuckles with Eye Bolts.

INTERNATIONAL AND NON-U.S. NATIONAL STANDARDS
SUBJECT INDEX

Eyeglasses
Use For: Spectacles *See Also:* Contact Lenses; Optical Equipment; Sunglasses
BSI BS 3199-72. 1972 Method for Measurement of Spectacles, Including a Glossary of Terms. 25 pp.
BSI BS 5043-01. (OBSOLESCENT) 1973 Amd 1 Book Holders, and Prismatic Spectacles for Use as Reading Aids in Hospitals and the Home (AMD 6895) February 28, 1992. 14 pp.
BSI BS 7394: Part 1-91. 1991 Complete Spectacles Part 1: Ready-to-Wear Near-Vision Spectacles. 10 pp.
BSI BS 7394: Part 1-01. 1991 Amd 1 Complete Spectacles Part 1: Specification for Ready-to-Wear Near-Vision Spectacles (AMD 7048) June 15, 1992. 11 pp.
SAA AS 1067.2-90. Sunglasses and Fashion Spectacles—Part 2: Performance Requirements. 5 pp.
SNZ NZS 5835-84. Sunglasses and Fashion Spectacles—Non-Prescription Types Amend: A, 1984. 20 pp.

—**Antifogging Agents**
JIS S 4030-83. Methods of Antifogging Agent Test for Spectacle Lens.

—**Frames**
BSI BS 6625-85. 1985 Amd 1 Spectacle Frames (AMD 6577) January 31, 1990 (N). 13 pp.
BSI BS 6625: Part 1-85. 1985 Amd 2 Spectacle Frames Part 1: Specification for General Requirements (AMD 7052) February 28, 1992 (N). 15 pp.
DIN ENGL 58199-89. Spectacle Frames; Requirements and Testing (June). 5 pp.
JIS B 7281-90. Measuring System for Spectacle Frames. 7 pp.
SAA AS 2228.2-92. Spectacles—Part 2: Spectacle Frames. 19 pp.

—**Frames—Glossaries**
BSI BS 3521: Parts 1 & 2-62. 1962 Amd 1 Glossary of Terms Relating to Ophthalmic Lenses and Spectacle Frames Part 1: Glossary Relating to Ophthalmic Lenses Part 2: Glossary Relating to Spectacle Frames. 75 pp.
BSI BS 3521: Part 2-91. 1991 Terms Relating to Ophthalmic Optics and Spectacle Frames Part 2: Glossary of Terms Relating to Spectacle Frames. 36 pp.
ISO 7998-84. Optics and Optical Instruments—Spectacle Frames—Vocabulary and Lists of Equivalent Terms First Edition. 12 pp.
JIS B 7280-87. Glossary of Terms on Spectacle Frames.

—**Frames—Identification Systems**
BSI BS 6625: Part 2-92. 1992 Spectacle Frames Part 2: Specification for Marking (ISO 9456: 1991). 8 pp.
ISO 9456-91. Optics and Optical Instruments—Ophthalmic Optics—Marking of Spectacle Frames First Edition. 6 pp.
JIS B 7282-92. Spectacle Frames—Marking. 5 pp.

—**Frames—Machine Screws/Nuts**
JIS B 1119-86. Machine Screws and Nuts for Spectacle Frames.

—**Frames—Measurement**
BSI BS 3199-92. 1992 Measuring System for Spectacle Frames (ISO 8624: 1991). 7 pp.
BSI BS 3199-72. 1972 Method for Measurement of Spectacles, Including a Glossary of Terms. 25 pp.
ISO 8624-91. Optics and Optical Instruments—Ophthalmic Optics—Measuring System for Spectacle Frames First Edition. 6 pp.

—**Frames—Screw Threads**
BSI BS 3172-87. 1987 Screw Threads for Spectacle Frames. 4 pp.

—**Glossaries**
BSI BS 3199-72. 1972 Method for Measurement of Spectacles, Including a Glossary of Terms. 25 pp.

—**Lenses**
CNS T4018-88. Method of Test for Spectacle Lenses (Oct)(12449).
SAA AS 2228.1-92. Spectacles—Part 1: Spectacle Lenses (Supersedes AS 2228—1978). 23 pp.

—**Lenses—Blanks**
BSI BS 2738: Part 4-92. 1992 Spectacle Lenses Part 4: Specification for Single-Vision and Multifocal Semi-Finished Lens Blanks (ISO 10322: Part 1: 1991). 12 pp.
BSI BS 2738: Part 5-92. 1992 Spectacle Lenses Part 5: Specification for Progressive Power Semi-Finished Lens Blanks (ISO 10322: Part 2: 1991). 12 pp.
ISO 10322 Pt 1-91. Ophthalmic Optics—Semi-Finished Lens Blanks—Part 1: Specifications for Single-Vision and Multifocal Lens Blanks First Edition. 9 pp.
ISO 10322 Pt 2-91. Ophthalmic Optics—Semi-Finished Lens Blanks—Part 2: Specifications for Progressive Power Lens Blanks First Edition. 9 pp.

Eyeglasses (Cont.)
—**Lenses—Glossaries**
CNS T2037-88. Glossary of Terms Used in Spectacke Lenses (Oct)(12448).

—**Lenses—Single Vision—Geometric Requirements**
ISO 8980 Pt 1-92. Ophthalmic Optics—Finished Single-Vision Corrective Lenses—Part 1: General Requirements First Edition. 7 pp.

—**Lenses—Single Vision—Mechanical Properties**
ISO 8980 Pt 1-92. Ophthalmic Optics—Finished Single-Vision Corrective Lenses—Part 1: General Requirements First Edition. 7 pp.

—**Lenses—Single Vision—Optical Properties**
ISO 8980 Pt 1-92. Ophthalmic Optics—Finished Single-Vision Corrective Lenses—Part 1: General Requirements First Edition. 7 pp.

—**Prescriptions**
NATO STANAG 2346 ED 1 AMD 5-75. Standard Method of Writing Prescriptions for Spectacles. 7 pp.

—**Safety**
SAA AS 1067.1-90. Sunglasses and Fashion Spectacles—Part 1: Safety Requirements Amdt 1 December 1990 Amdt 2 July 1993. 15 pp.

—**Scales (Indicators)—Tabo Graduated Dial**
DIN ENGL 58201-60. Tabo Graduated Dial Scale; Particulars; Application (Sept). 2 pp.

—**Templates**
DIN ENGL 5345-77. Formers for Spectacle Lenses; Types; Principal Dimensions; Inscription (Apr). 2 pp.

Eyelets (Fasteners)
BSI BS 3102-59. 1959 Brass Eyelets and Washers for General Purposes. 34 pp.
JIS S 9021-71. Eyelets.
MOD UK DSTAN 53-22-01. Eyelets, Metallic, and Grommets, Metallic Issue 2; Amendment 1. 11 pp.

—**Aircraft**
SBAC AGS 232 (V). Eyelets (Aluminium and Brass).

—**Aircraft—Drainage Holes**
SBAC AGS 840 ISSUE 4. Drainage Hole Eyelets.
SBAC AGS 889 ISSUE 4. Drainage Hole Eyelets (Marine Type).

—**Canvas Sheets**
CNS Z7094-81. Eyelets for Canvas (Jul)(7771).

—**Paperboard—Labels**
CNS Z7095-81. Eyelets for Cardboard Labels (Jul)(7772).

Eyenuts
See Also: Nuts (Fasteners)
CNS B2363-84. Lifting Eye Nuts (Jul)(4730).
DIN ENGL 582-71. Lifting Eye Nuts (Apr). 2 pp.
JIS B 1169-75. Eyenuts. 10 pp.

Eyepiece Micrometers
See Also: Micrometers

—**Microscopes**
JIS B 7149-67. 1/10 mm Eyepiece Micrometers for Microscope (R 1973). 3 pp.

Eyepieces
Scope Note: Use a more specific term *See:* Filar Micrometer Eyepieces; Microscope Eyepieces

Eyeplates
See Also: Lifting Equipment

—**Ships**
BSI BS MA 10-71. 1971 Oval Eyeplates. 3 pp.
CNS F3039-80. Ships' Eye Plates (May)(5572).
CNS F3076-80. Ships' Wire Rope Stay Eye Plate (Dec)(6828).
ISO 8146-85. Shipbuilding and Marine Structures—Oval Eyeplates First Edition. 5 pp.
JIS F 1015-87. Yachts' U-Type Bow Eyes.
JIS F 3410-90. Ships' Eye Plates.
JIS F 3410-74. Ships' Eye Plates. 5 pp.
JIS F 3415-74. Ships' Wire Rope Stay Eye Plates. 4 pp.

—**Ships—Chains**
CNS F3081-80. Ships' Eye Plates for Chainlet (Dec)(6833).
JIS F 3908-75. Ships' Eye Plates for Chainlet. 4 pp.

Eyes (Tools)
Scope Note: See the subheading Eyes under the specific product

F Coaxial Connectors
Use For: F Connectors *See Also:* Coaxial Connectors

— **75 Ohms**
IEC 169 Pt 24-91. Radio-Frequency Connectors Part 24: Radio-Frequency Coaxial Connectors with Screw Coupling, Typically for Use in 75 ohm Cable Distribution Systems (Type F) First Edition. 18 pp.

F Connectors
Use: F Coaxial Connectors

Fabric Brighteners
Use: Fluorescent Brighteners

Fabric Coatings
Use: Fabric Finishes

Fabric Finishes
See Also: Coatings; Dyes; Fabrics
JIS L 1063-85. Testing Methods for Migration of Dyestuffs and Finishing Agents on Woven Fabrics and Knitted Fabrics.

—**Solvents**
CGSB CAN/CGSB-4.2 NO.30.1-M89. Textile Test Methods Effect of Solvents on the Permanence of Textile Finishes. 7 pp.

Fabric Finishing Equipment
See Also: Fabrics

—**Laundering Testing**
BSI BS 6246: Part 5-91. 1991 Industrial Laundry Machinery Part 5: Methods for the Assessment of the Effect of Shaped Garment Finishing Machinery on Textiles. 32 pp.

Fabric Laminates
Use: Laminated Fabrics

Fabric Sizings
See Also: Bleaching Agents; Fabrics; Lubricants; Sizing (Surface Treatment); Sizing Compounds; Waxes
BSI BS 4032-78. 1978 Determination of Certain Water-Or Alkali-Soluble Additives in Cellulosic or Synthetic Fibres, Yarns and Fabrics or Yarns and Fabrics Made from Blends of Such Fibres. 4 pp.

Fabricating Services
Use: Fabrication Services

Fabrication Services
Use For: Fabricating Services *See Also:* Construction
ISO 3834-78. Welding—Factors to Be Considered When Assessing Firms Using Welding as a Prime Means of Fabrication First Edition. 4 pp.

—**Acrylic Panels/Shapings—Aerospace**
MOD UK DTD-925D-69. Fabrication of Acrylic Panels and Shapings. 3 pp.

—**Rotary Wing Aircraft**
CAA 235 (G). Section G Rotorcraft Miscellaneous Amendments Derived from Section D, "Aeroplanes" (Blue Papers). 6 pp.
CAA Chapter G4-1 01.75. Design and Construction—General (Rotocraft). 4 pp.

—**Ships**
MOD UK NES 706-01. Welding and Fabrication of Ship Structure Issue 2 (03.83); Amendment 5. 197 pp.

—**Steels—Marine**
MOD UK NES 770: Part 1-01. Welding and Fabrication of Q1N, HY80 and QT35 Steel Part 1: General Issue 1 (06.89); Amendment 1. 44 pp.

Fabrics
Scope Note: For additional listings, see also specific products made from fabrics *Use For:* Cloth; Knitted Fabrics; Textile Fabrics; Textile Materials; Textiles; Woven Fabrics *See Also:* Abrasive Cloth; Acetate Fabrics; Aerospace Fabrics; Asbestos Fabrics; Awnings; Bed Sheets; Bedspreads; Bonded Fabrics; Burlaps; Canvas; Carbon Fabrics; Carpets; Charcoal Fabrics; Coated Fabrics; Coatings; Corduroy; Cotton/Cellulose Blend Fabrics; Cotton Fabrics; Cotton/Polyester Blend Fabrics; Draperies; Durable Press Fabrics; Dyes; Edging; Elastic Fabrics; Electrical Insulating Fabrics; Fabric Finishes; Fabric Finishing Equipment; Fabric Sizings; Felts; Fibers; Fleece; Flocked Fabrics; Foam Laminated Fabrics; Fused Fabrics; Geotextiles; Glass Fabrics; Haircloth; Jute; Laces (Fabrics); Laminated Fabrics; Laminates;

Fabrics

Fabrics (Cont.)

See Also: (Cont.)
Linen; Nets; Nonwoven Fabrics; Nylon Fabrics; Papers; Parachute Fabrics; Pile Fabrics; Polyester Fabrics; Polypropylene Fabrics; Poplin; Ramie; Rayon; Roofing Fabrics; Satin; Serge; Silica Fabrics; Silk; Tablecloths; Taffeta; Tarpaulins; Textile Industry; Textile Preservatives; Ticking; Towel Fabrics; Twill; Upholstery; Viscose Fabrics; Weaving; Webbing; Wool/Cotton Blend Fabrics; Wool Fabrics; Wool/Polyester Blend Fabrics; Yarns

BSI BS 5441-88. 1988 Amd 1 Methods of Test for Knitted Fabrics (AMD 6338) June 28, 1991. 19 pp.
BSI PD 6527-90. 1990 Methods of Test for Textiles (Other Than Those Methods Already Published as British Standards). 55 pp.
BSI PD 6537: Part 1-93. 1992 Variability of Methods of Test for Textiles and Reports of Inter-Laboratory Trials on Certain Methods Part 1: Phase 1: BS 2576, BS 4303 and ISO 9290 (L). 8 pp.
BSI PP 787-88. 1988 Standards and Textiles. 49 pp.
CNS L3080-87. Method of Test for Nonwoven Fabrics (Sep)(5610).
CNS L4065-77. Mixed Fabrics (Finished) (Jun)(3853). 3 pp.
CNS L4155-86. Pile Knitted Fabrics (Aug)(11662).
ISO 2959-73. Textiles—Woven Fabric Descriptions First Edition. 3 pp.
JIS L 1018-90. Testing Methods for Knitted Fabrics. 59 pp.

—Abrasion Testing
BSI BS 5690-91. 1991 Abrasion Resistance of Fabrics. 13 pp.
SAA AS 2001.2.25-90. Methods of Test for Textiles—Part 2: Physical Tests—Part 2.25: Determination of Flat Abrasion Resistance of Textile Fabrics (Martindale Abrasion Method). 6 pp.
SAA AS 2001.2.26-90. Methods of Test for Textiles—Part 2: Physical Tests—Part 2.26: Determination of Abrasion Resistance of Textile Fabrics (Flexing and Abrasion Method). 6 pp.
SAA AS 2001.2.27-90. Methods of Test for Textiles—Part 2: Physical Tests—Part 2.27: Determination of Abrasion Resistance of Textile Fabrics (Inflated Diaphragm Method). 5 pp.
SAA AS 2001.2.28-92. Methods of Test for Textiles—Part 2: Physical Tests—Part 2.28: Determination of Abrasion Resistance of Textile Fabrics (Rotary Platform, Double-Head Method). 8 pp.

—Aerostats—Tear Resistant
CAA 862 (Q). Section Q Non-Rigid Airships—Airship Fabric Tear Resistance. 2 pp.

—Air Filters
JIS Z 8908-89. Filter Fabrics for Dust Collection. 12 pp.

—Aircraft—Doping
BSI X 26-66. 1966 Doping and Finishing Schemes for Fabric Covered Aircraft. 7 pp.
CAA LEAFLET 2-9 07.90. Doping. 10 pp.

—Aircraft Equipment
CAA LEAFLET 2-8 07.90. Fabric Covering. 15 pp.

—Aircraft—Finishing Compounds
BSI X 26-66. 1966 Doping and Finishing Schemes for Fabric Covered Aircraft. 7 pp.

—Aircraft—Flame Resistant
CAA NOTICE #58 ISSUE 4. Flame Resistant Furnishing Materials (Airworthiness Notices). 2 pp.

—Aircraft—Flammability Testing
SBAC TS 80 ISSUE 1. Recommendations for the Control of Flammability Testing of Non Metallic Materials and Textiles for Aeronautical Purposes.

—Aqueous Solutions—pH
BSI BS 3266-81. 1981 Determination of Conductivity, pH, Water Soluble Matter, Chloride and Sulphate in Aqueous Extracts of Textile Materials. 9 pp.
CGSB CAN/CGSB-4.2 NO.74-M91. Textile Test Methods Textiles—Determination of pH of the Aqueous Extract (ISO 3071:1980). 8 pp.
ISO 3071-80. Determination of pH of the Aqueous Extract Second Edition. 5 pp.
SAA AS 2001.3.1-77. Methods of Test for Textiles—Part 3: Chemical Tests—Part 3.1: Determination of pH of Aqueous Extract. 2 pp.

—Ash Content
BSI BS 6646-85. 1985 Amd 1 Method for Determination of Ash of Textiles (AMD 5720) July 29, 1988. 6 pp.
CNS L3166-82. Determination of the Contents of Ash and Sulfur in Textile Materials (Apr)(8709).

—Asphalt Roofing
JIS A 6012-77. Woven Fabrics Asphalt Roofings.

Fabrics (Cont.)

—Bacteria
JIS L 1902-90. Testing Method for Antibacterial of Textiles. 13 pp.

—Bagging
JIS L 1061-87. Testing Methods Bagging of Woven and Knitted Fabrics.

—Bast Fibers
CNS L3038-77. Hard and Bast Fabrics (Unfinished) (Jun)(2273).
CNS L4066-77. Hard and Bast Fabrics (Jun)(3855).

—Breaking Load
BSI BS 2576-86. 1986 Amd 1 Method for Determination of Breaking Strength and Elongation (Strip Method) of Woven Fabrics (AMD 6199) September 29, 1989. 11 pp.
CGSB CAN/CGSB-4.2 NO.9.1-M90. Textile Test Methods Breaking Strength of Fabrics—Strip Method—Constant-Time-to-Break Principle. 11 pp.
CGSB CAN/CGSB-4.2 NO.9.2-M90. Textile Test Methods Breaking Strength of Fabrics—Grab Method—Constant-Time-to-Break Principle. 10 pp.
CGSB CAN/CGSB-4.2 NO.9.3-M90. Textile Test Methods Breaking Strength of High-Strength Fabrics—Constant-Time-to-Break Principle. 13 pp.
ISO 5081-77. Textiles—Woven Fabrics—Determination of Breaking Strength and Elongation (Strip Method) First Edition. 10 pp.
ISO 5082-82. Textiles—Woven Fabrics—Determination of Breaking Strength—Grab Method First Edition. 9 pp.
SAA AS 2001.2.3-88. Methods of Test for Textiles—Part 2: Physical Tests—Part 2.3: Determination of Breaking Force and Extension of Textile Fabrics. 4 pp.
SAA AS 2001.2.7-87. Methods of Test for Textiles—Part 2: Physical Tests—Part 2.7: Determination of Breaking Force and Extension of Yarns. 7 pp.
SAA AS 2001.2.20-86. Methods of Test for Textiles—Part 2: Physical Tests—Part 2.20: Determination of Seam Breaking Force. 3 pp.
SAA AS 2001.2.21-89. Methods of Test for Textiles—Part 2: Physical Tests—Part 2.21: Determination of Seam Opening Due to the Application of Force in the Transverse Direction. 5 pp.
SAA AS 2001.2.22-86. Methods of Test for Textiles—Part 2: Physical Tests—Part 2.22: Determination of Yarn Slippage in Woven Fabrics at a Standard Stitched Seam. 4 pp.

—Burning Behavior
CEN PREN 1103-93. Textiles—Burning Behaviour—Fabrics for Apparel—Detailed Procedure to Determine the Burning Behaviour of Fabrics for Apparel. 4 pp.

—Bursting Strength
BSI BS 3424: Part 6-82. 1982 Testing Coated Fabrics Part 6: Methods 8A and 8B. Methods for Determination of Bursting Strength. 6 pp.
BSI BS 4768-72. 1972 Amd 1 Determination of the Bursting Strength and Bursting Distension of Fabrics (AMD 6340) June 28, 1991. 9 pp.
CGSB CAN/CGSB-4.2 NO.11.1-M88. Textile Test Methods Bursting Strength—Diaphragm Pressure Test. 9 pp.
CGSB CAN/CGSB-4.2 NO.11.2-M89. Textile Test Methods Bursting Strength—Ball Burst Test. 9 pp.
CNS L3082-80. Method of Testing for Bursting Strength of Fabrics (May)(5613). 3 pp.
ISO 2960-74. Textiles—Determination of Bursting Strength and Bursting Distension—Diaphragm Method First Edition. 4 pp.
SAA AS 2001.2.4-90. Methods of Test for Textiles—Part 2: Physical Tests—Part 2.4: Determination of Bursting Pressure of Textile Fabrics—Hydraulic Diaphragm Method. 7 pp.
SAA AS 2001.2.19-88. Methods of Test for Textiles—Part 2: Physical Tests—Part 2.19: Determination of Bursting Force of Textile Fabrics—Ball Burst Method. 2 pp.

—Chemical Analysis
ISO 1833-77. Textiles—Binary Fibre Mixtures—Quantitative Chemical Analysis Second Edition; (Amendment 1-1980). 20 pp.

—Chloride Content—Gravimetric Analysis
BSI BS 3266-81. 1981 Determination of Conductivity, pH, Water Soluble Matter, Chloride and Sulphate in Aqueous Extracts of Textile Materials. 9 pp.

—Chloride Content—Volumetric Analysis
BSI BS 3266-81. 1981 Determination of Conductivity, pH, Water Soluble Matter, Chloride and Sulphate in Aqueous Extracts of Textile Materials. 9 pp.

—Chromaticity
CGSB CAN/CGSB-4.2 NO.64-M91. Textile Test Methods Chromatic Transference Scale. 9 pp.

Fabrics (Cont.)

—Civil Engineering
CNS A2183-85. Nonwoven Fabric Used in Civil Engineering (Apr)(11228).

—Classification
BSI BS 1000: (677)-81. 1981 Amd 1 Universal Decimal Classification (UDC). English Full Edition (677): Textile Industry. 69 pp.
SNZ NZS/BS 1000 (677)-81. Universal Decimal Classification Textile Industry Amend: 1. 68 pp.

—Colorfastness Testing
BSI BS 1006-90. 1990 Methods of Test for Colour Fastness of Textiles and Leather (L). 234 pp.
BSI BS 1006-01. 1990 Amd 1 Methods of Test for Colour Fastness of Textiles and Leather (AMD 7284) October 15, 1992 (L). 247 pp.
BSI BS 1006-02. 1990 Amd 2 Methods of Test for Colour Fastness of Textiles and Leather (AMD 7201) January 15, 1993 (L). 248 pp.
BSI BS EN 20105-A01-93. 1993 Textiles—Tests for Colour Fastness Part A01: General Principles of Testing (ISO 105-A01: 1989) (L). 15 pp.
BSI BS EN 20105-A02-93. 1993 Textiles—Tests for Colour Fastness Part A02: Grey Scale for Assessing Change in Colour (ISO 105-A02: 1987) (L). 9 pp.
BSI BS EN 20105-A03-93. 1993 Textiles—Tests for Colour Fastness Part A03: Grey Scale for Assessing Staining (ISO 105-A03: 1987) (L). 9 pp.
BSI BS EN 20105-B02-93. 1993 Textiles—Tests for Colour Fastness Part B02: Colour Fastness to Artificial Light (Xenon Arc Fading Lamp Test) (ISO 105-B02: 1988) (L). 18 pp.
BSI BS EN 20105-C06-93. 1993 Textiles—Tests for Colour Fastness Part C06: Colour Fastness to Domestic and Commercial Laundering (ISO 105-C06: 1987) (L). 11 pp.
CEN EN 20105-A01-92. Textiles—Tests for Colour Fastness Part A01: General Principles of Testing (ISO 105-AO1: 1989). 7 pp.
CEN EN 20105-A03-92. Textiles—Tests for Colour Fastness Part A03: Grey Scale for Assessing Staining (ISO 105-AO3: 1987). 4 pp.
CEN EN 20105-B02-92. Textiles—Tests for Colour Fastness Part B02. Colour Fastness to Artificial Light: Xenon Arc Fading Lamp Test (ISO 105-B02: 1988). 13 pp.
CEN EN 20105-C06-92. Textiles—Tests for Colour Fastness—Part C06: Colour Fastness to Domestic and Commercial Laundering (ISO 105-C06: 1987). 8 pp.
CGSB CAN/CGSB-4.2 NO.18.1-M90. Textile Test Methods Colourfastness to Artificial Light: Carbon-Arc Radiation. 20 pp.
CGSB CAN/CGSB-4.2 NO.18.2-M90. Textile Test Methods Textiles—Tests for Colourfastness—Part B01: Colourfastness to Light: Daylight (ISO 105-B01:1988). 20 pp.
CGSB CAN/CGSB-4.2 NO.18.3-M90. Textile Test Methods Textiles—Tests for Colourfastness—Part B02: Colourfastness to Artificial Light: Xenon Arc Fading Lamp Test (ISO 105-B02:1988). 28 pp.
CGSB CAN/CGSB-4.2 NO.19.1-M90. Textile Test Methods Colourfastness to Washing—Accelerated Test—Launder-Ometer; (Amendment 1 June 1992). 12 pp.
CGSB CAN/CGSB-4.2 NO.20-M89. Textile Test Methods Colourfastness to Water. 9 pp.
CGSB CAN/CGSB-4.2 NO.21-M89. Textile Test Methods Colourfastness to Sea Water. 9 pp.
CGSB CAN/CGSB-4.2 NO.22-M90. Textile Test Methods Colourfastness to Rubbing (Crocking). 11 pp.
CGSB CAN/CGSB-4.2 NO.23-M90. Textile Test Methods Colourfastness to Perspiration. 9 pp.
CGSB CAN/CGSB-4.2 NO.24-M91. Textile Test Methods Colourfastness and Dimensional Change in Commercial Laundering. 7 pp.
CGSB CAN/CGSB-4.2 NO.29.1-M89. Textile Test Methods Colourfastness to Dry Cleaning Solvent. 9 pp.
CGSB CAN/CGSB-4.2 NO.31-M89. Textile Test Methods Textiles—Tests for Colourfastness—Part X11: Colourfastness to Hot Pressing (ISO 105/X11-1987). 11 pp.
CGSB CAN/CGSB-4.2 NO.35.1-M90. Textile Test Methods Colourfastness to Burnt Gas Fumes. 11 pp.
CGSB CAN/CGSB-4.2 NO.46-M90. Textile Test Methods Textiles—Tests for Colourfastness—Part A02: Grey Scale for Assessing Change in Colour (ISO 105-A02:1987). 10 pp.
CGSB CAN/CGSB-4.2 NO.47-M90. Textile Test Methods Textiles—Tests for Colourfastness—Part A03: Grey Scale for Assessing Staining (ISO 105-A03:1987). 10 pp.
CGSB CAN/CGSB-4.2 NO.52.1-M90. Textile Test Methods Colourfastness to Chlorinated Water. 9 pp.
CGSB CAN/CGSB-4.2 NO.52.2-M91. Textile Test Methods Textiles—Tests for Colourfastness—Part E03: Colourfastness to Chlorinated Water (Swimming-Pool Water) (ISO 105-E03:1987). 7 pp.

INTERNATIONAL AND NON-U.S. NATIONAL STANDARDS
SUBJECT INDEX
Fabrics

Fabrics *(Cont.)*

—Colorfastness Testing *(Cont.)*

CGSB CAN/CGSB-4.2 NO.55-M90. Textile Test Methods Loss in Strength and Colour Change of Fabrics Due to Retained Chlorine. 13 pp.
CGSB CAN/CGSB-4.2 NO.58-M90. Textile Test Methods Colourfastness and Dimensional Change in Domestic Laundering of Textiles. 18 pp.
CGSB CAN/CGSB-4.2 NO.68-92. Textile Test Methods Textiles—Tests for Colourfastness—Part J01: Measurement of Colour and Colour Differences (ISO 105-J01:1989). 8 pp.
CNS L1007-82. Undyed Cloth for Staining of Color Fastness Test (Feb)(3841).
CNS L3026-82. Method of Test for Colour Fastness to Sunlight and Daylight (Jun)(1493).
CNS L3027-83. Method of Test for Colour Fastness to Washing (Mar)(1494). 6 pp.
CNS L3028-83. Method of Test for Colour Fastness to Hot Water (Mar)(1495).
CNS L3029-83. Method of Test for Colour Fastness to Perspiration (Mar)(1496).
CNS L3030-83. Method of Test for Colour Fastness to Water (Mar)(1497).
CNS L3031-83. Method of Test for Colour Fastness to Bleaching with Hypochlorite (Mar)(1498).
CNS L3032-83. Method of Test for Colour Fastness to Rubbing (Mar)(1499). 2 pp.
CNS L3074-82. Method of Test for Colour Fastness to Carbon Arc Lamp Light (Jun)(3845). 4 pp.
CNS L3075-82. Method of Test for Colour Fastness to Xenon Arc Lamp Light (Jun)(3846).
CNS L3151-82. Method of Test for Colour Fastness to Sea Water (Jan)(8430).
CNS L3152-82. Method of Test for Colour Fastness to Organic Solvents (Jan) (8431).
CNS L3153-82. Method of Test for Colour Fastness to Rubbing with Organic Solvents (Jan)(8432).
CNS L3154-82. Method of Test for Colour Fastness to Steaming (Jan)(8433).
CNS L3155-82. Method of Test for Colour Fastness to Dry Cleaning (Jan)(8434).
CNS L3156-83. Method of Test for Colour Fastness to Hot Pressing (Jun)(8532).
CNS L3157-82. Method of Test for Colour Fastness to Sublimation in Storage (Feb)(8533).
CNS L3158-82. Method of Test for Colour Fastness to Mercerizing (Mar)(8616).
CNS L3159-82. Method of Test for Colour Fastness to Soda Boiling (Mar)(8617).
CNS L3161-82. Method of Test for Colour Fastness to Bleaching with Peroxide (Mar)(8619).
CNS L3162-82. Method of Test for Colour Fastness to Stoving (Mar)(8620).
CNS L3163-82. Method of Test for Colour Fastness to Bleaching Sodium Chlorite (Mar)(8621).
CNS L3168-82. Method of Test for Colour Fastness to Resin Finish (Apr)(8711).
CNS L3169-82. Method of Test for Colour Fastness to Vulcanizing with Sulphur Monochloride (Apr)(8712).
CNS L3170-82. Method of Test for Colour Fastness to Steam Vulcanizing (Apr) (8713).
CNS L3171-82. Method of Test for Colour Fastness to Vulcanizing with Hot Air (Apr)(8714).
CNS L3172-82. Method of Test for Colour Fastness to Formaldehyde (Jun)(9015).
CNS L3173-82. Method of Test for Colour Fastness to Nitrogen Oxide (Jun) (9016).
CNS L3174-82. Method of Test for Colour Fastness to Chlorinated Water (Jun) (9017).
CNS L3175-82. Method of Test for Colour Fastness to Water Spotting (Jun) (9018).
CNS L3176-82. Method of Test for Colour Fastness to Acid Spotting (Jun) (9019).
CNS L3177-82. Method of Test for Colour Fastness to Alkali Spotting (Jun) (9020).
CNS L3178-82. Method of Test for Colour Fastness to Dry Heating (Jun)(9021).
CNS L3179-82. Method of Test for Colour Fastness to High Temperature Steaming (Jun)(9022).
CNS L3180-82. Method of Test for Colour Fastness to Pleating (Jun)(9023).
CNS L3183-82. Method of Test for Colour Fastness to Chlorination (Aug)(9306).
CNS L3184-82. Method of Test for Colour Fastness to Decatizing (Aug)(9307).
CNS L3186-82. Method of Test for Colour Fastness to Alkali Milling (Aug) (9309).
CNS L3187-82. Method of Test for Colour Fastness to Carbonizing with Aluminium Chloride (Aug)(9310).
CNS L3188-82. Method of Test for Colour Fastness to Carbonizing with Sulphuric Acid (Aug) (9311).
CNS L3189-82. Method of Test for Colour Fastness to Metals in the Dyebath: Chromium Salt (Aug) (9312).
CNS L3190-82. Method of Test for Colour Fastness to Metals in the Dyebath: Iron and Copper (Aug) (9313).
CNS L3192-83. Method of Test for Colour Fastness to Light of Fluorescent Whitening Agents and Fluorescent Whitened Textiles (Jan)(9901).

Fabrics *(Cont.)*

—Colorfastness Testing *(Cont.)*

CNS L3193-83. Method of Test for Colour Fastness to Light and Perspiration (Jan)(9902).
CNS L3194-83. Method of Test for Detection and Colour Fastness to Photochromism (Jan) (9903).
CNS L4152-84. Weft Knitted Fabrics (Finished) (Jul)(10970). 2 pp.
ISO 105 Pt A01-89. Textiles—Tests for Colour Fastness—Part A01: General Principles of Testing Fourth Edition; (Corrected and Reprinted -1990) (CEN EN 20105-A01:1992). 10 pp.
ISO 105 Pt A02-93. Textiles—Tests for Colour Fastness—Part A02: Grey Scale for Assessing Change in Colour Fourth Edition. 5 pp.
ISO 105 Pt A03-93. Textiles—Tests for Colour Fastness—Part A03: Grey Scale for Assessing Staining Fourth Edition. 5 pp.
ISO 105 Pt A04-89. Textiles—Tests for Colour Fastness—Part A04: Method for the Instrumental Assessment of the Degree of Staining of Adjacent Fabrics First Edition. 4 pp.
ISO 105 Pt B01-89. Textiles—Tests for Colour Fastness—Part B01: Colour Fastness to Light: Daylight Fourth Edition. 10 pp.
ISO 105 Pt B02-88. Textiles—Tests for Colour Fastness—Part B02: Colour Fastness to Artifical Light: Xenon Arc Fading Lamp Test Third Edition; (CEN EN 20105-B02:1992). 14 pp.
ISO 105 Pt B03-88. Textiles—Tests for Colour Fastness—Part B03: Colour Fastness to Weathering: Outdoor Exposure Third Edition. 8 pp.
ISO 105 Pt B04-88. Textiles—Tests for Colour Fastness—Part B04: Colour Fastness to Weathering: Xenon Arc Third Edition. 8 pp.
ISO 105 Pt B05-88. Textiles—Tests for Colour Fastness—Part B05: Detection and Assessment of Photochromism Third Edition. 7 pp.
ISO 105 Pt B06-92. Textiles—Tests for Colour Fastness—Part B06: Colour Fastness to Artificial Light at High Temperatures: Xenon Arc Fading Lamp Test First Edition. 15 pp.
ISO 105 Pt C01-89. Textiles—Tests for Colour Fastness—Part C01: Colour Fastness to Washing: Test 1 Fourth Edition; (CEN EN 20105-C01:1992). 6 pp.
ISO 105 Pt C02-89. Textiles—Tests for Colour Fastness—Part C02: Colour Fastness to Washing: Test 2 Fourth Edition. 6 pp.
ISO 105 Pt C03-89. Textiles—Tests for Colour Fastness—Part C03: Colour Fastness to Washing: Test 3 Fourth Edition; (CEN EN 20105-C03:1992). 6 pp.
ISO 105 Pt C04-89. Textiles—Tests for Colour Fastness—Part C04: Colour Fastness to Washing: Test 4 Fourth Edition; (CEN EN 20105-C04:1992). 6 pp.
ISO 105 Pt C05-89. Textiles—Tests for Colour Fastness—Part C05: Colour Fastness to Washing: Test 5 Fourth Edition; (CEN EN 20105-C05:1992). 6 pp.
ISO 105 Pt C06-87. Textiles—Tests for Colour Fastness—Part C06: Colour Fastness to Domestic and Commercial Laundering Second Edition; (CEN EN 20105-C06:1992). 7 pp.
ISO 105 Pt D01-87. Textiles—Tests for Colour Fastness—Part D01: Colour Fastness to Dry Cleaning Third Edition. 4 pp.
ISO 105 Pt D02-87. Textiles—Tests for Colour Fastness—Part D02: Colour Fastness to Rubbing: Organic Solvents Third Edition. 4 pp.
ISO 105 Pt E01-89. Textiles—Tests for Colour Fastness—Part E01: Colour Fastness to Water Third Edition. 5 pp.
ISO 105 Pt E02-89. Textiles—Tests for Colour Fastness—Part E02: Colour Fastness to Sea Water Third Edition. 5 pp.
ISO 105 Pt E03-87. Textiles—Tests for Colour Fastness—Part E03: Colour Fastness to Chlorinated Water (Swimming-Bath Water) Second Edition. 4 pp.
ISO 105 Pt E04-89. Textiles—Tests for Colour Fastness—Part E04: Colour Fastness to Perspiration Third Edition. 6 pp.
ISO 105 Pt E05-89. Textiles—Tests for Colour Fastness—Part E05: Colour Fastness to Spotting: Acid Third Edition. 4 pp.
ISO 105 Pt E06-89. Textiles—Tests for Colour Fastness—Part E06: Colour Fastness to Spotting: Alkali Third Edition. 4 pp.
ISO 105 Pt E07-89. Textiles—Tests for Colour Fastness—Part E07: Colour Fastness to Spotting: Water Third Edition. 4 pp.
ISO 105 Pt E08-87. Textiles—Tests for Colour Fastness—Part E08: Colour Fastness to Water: Hot Water Second Edition. 4 pp.
ISO 105 Pt E09-89. Textiles—Tests for Colour Fastness—Part E09: Colour Fastness to Potting Third Edition. 4 pp.
ISO 105 Pt E10-87. Textiles—Tests for Colour Fastness—Part E10: Colour Fastness to Decatizing Second Edition. 4 pp.

Fabrics *(Cont.)*

—Colorfastness Testing *(Cont.)*

ISO 105 Pt E11-87. Textiles—Tests for Colour Fastness—Part E11: Colour Fastness to Steaming Second Edition. 4 pp.
ISO 105 Pt E12-89. Textiles—Tests for Colour Fastness—Part E12: Colour Fastness to Milling: Alkaline Milling Third Edition. 6 pp.
ISO 105 Pt E13-87. Textiles—Tests for Colour Fastness—Part E13: Colour Fastness to Acid-Felting: Severe Second Edition. 4 pp.
ISO 105 Pt E14-87. Textiles—Tests for Colour Fastness—Part E14: Colour Fastness to Acid-Felting: Mild Second Edition. 4 pp.
ISO 105 Pt F-85. Textiles—Tests for Colour Fastness—Part F: Standard Adjacent Fabrics Third Edition. 24 pp.
ISO 105 Pt F10-89. Textiles—Tests for Colour Fastness—Part F10: Specification for Adjacent Fabric: Multifibre First Edition. 7 pp.
ISO 105 Pt G-78. Textiles—Tests for Colour Fastness Part G: Colour Fastness to Atmospheric Contaminants First Edition. 20 pp.
ISO 105 Pt G04-89. Textiles—Tests for Colour Fastness Part G04: Colour Fastness to Oxides of Nitrogen in the Atmosphere at High Humidities First Edition. 7 pp.
ISO 105 Pt J01-89. Textiles—Tests for Colour Fastness—Part J01: Measurement of Colour and Colour Differences Third Edition; (CAN/CGSB-4.2 No.68-92). 4 pp.
ISO 105 Pt J02-87. Textiles—Tests for Colour Fastness—Part J02: Method for the Instrumental Assessment of Whiteness First Edition. 4 pp.
ISO 105 Pt N-78. Textiles—Tests for Colour Fastness Part N: Colour Fastness to Bleaching Agencies First Edition. 15 pp.
ISO 105 Pt P-78. Textiles—Tests for Colour Fastness Part P: Colour Fastness to Heat Treatments First Edition. 11 pp.
ISO 105 Pt S-78. Textiles—Tests for Colour Fastness Part S: Colour Fastness to Vulcanizing First Edition. 11 pp.
ISO 105 Pt X01-87. Textiles—Tests for Colour Fastness—Part X01: Colour Fastness to Carbonizing: Aluminium Chloride Third Edition. 4 pp.
ISO 105 Pt X11-87. Textiles—Tests for Colour Fastness—Part X11: Colour Fastness to Hot Pressing Third Edition. 4 pp.
ISO 105 Pt X13-87. Textiles—Tests for Colour Fastness—Part X13: Colour Fastness of Wool Dyes to Processes Using Chemical Means for Creasing, Pleating and Setting Third Edition. 6 pp.
ISO 105 Pt Z-78. Textiles—Tests for Colour Fastness Part Z: Colorant Characteristics First Edition. 8 pp.
JIS L 0801-78. General Principles of Testing Methods for Colour Fastness (R 1983). 14 pp.
JIS L 0803-92. Standard Adjacent Fabrics for Staining of Colour Fastness Test. 10 pp.
JIS L 0841-92. Testing Methods for Colour Fastness to Daylight. 12 pp.
JIS L 0842-88. Testing Methods for Colour Fastness to Carbon Arc Lamp Light. 6 pp.
JIS L 0843-88. Testing Methods for Colour Fastness to Xenon Arc Lamp Light. 9 pp.
JIS L 0844-86. Testing Methods for Colour Fastness to Washing and Laundering. 14 pp.
JIS L 0845-75. Testing Method for Colour Fastness to Hot Water (R 1983). 6 pp.
JIS L 0846-92. Testing Method for Colour Fastness to Water. 6 pp.
JIS L 0847-75. Testing Method for Colour Fastness to Sea Water (R 1983). 6 pp.
JIS L 0848-78. Testing Method for Colour Fastness to Perspiration. 10 pp.
JIS L 0849-71. Testing Method for Colour Fastness to Rubbing (R 1978). 6 pp.
JIS L 0850-75. Testing Method for Colour Fastness to Hot Pressing (R 1978). 9 pp.
JIS L 0851-75. Testing Method for Colour Fastness to Acid Spotting. 4 pp.
JIS L 0852-67. Testing Method for Colour Fastness to Alkali Spotting. 4 pp.
JIS L 0853-75. Testing Method for Colour Fastness to Water Spotting. 4 pp.
JIS L 0854-75. Testing Method for Colour Fastness to Sublimation in Storage (R 1983). 5 pp.
JIS L 0855-92. Testing Methods for Colour Fastness to Nitrogen Oxides. 10 pp.
JIS L 0856-83. Testing Methods for Colour Fastness to Bleaching with Hypochlorite. 7 pp.
JIS L 0857-75. Testing Method for Colour Fastness to Bleaching with Peroxide. 5 pp.
JIS L 0858-75. Testing Method for Colour Fastness to Stoving (R 1983). 6 pp.
JIS L 0859-75. Testing Method for Colour Fastness to Bleaching with Sodium Chlorite (R 1988). 7 pp.
JIS L 0860-74. Testing Method for Colour Fastness to Dry Cleaning. 4 pp.
JIS L 0861-75. Testing Method for Colour Fastness to Organic Solvents (R 1983). 5 pp.
JIS L 0862-75. Testing Method for Colour Fastness to Rubbing with Organic Solvents (R 1983). 7 pp.

INDUSTRY STANDARDS
1027

Fabrics

Fabrics (Cont.)

—Colorfastness Testing (Cont.)

JIS L 0863-92. Testing Method for Colour Fastness to Mercerizing. 6 pp.
JIS L 0864-76. Testing Method for Colour Fastness to Soda Boiling.
JIS L 0866-76. Testing Method for Colour Fastness to Carbonizing with Aluminium Chloride.
JIS L 0867-76. Testing Method for Colour Fastness to Carbonizing with Sulphuric Acid.
JIS L 0868-75. Testing Method for Colour Fastness to Formaldehyde. 5 pp.
JIS L 0869-77. Testing Method for Colour Fastness to Steaming.
JIS L 0870-75. Testing Method for Colour Fastness to Metals in the Dyebath: Chromium Salt.
JIS L 0871-75. Testing Method for Colour Fastness to Metals in the Dyebath: Iron and Copper.
JIS L 0873-74. Testing Method for Colour Fastness to Chlorination.
JIS L 0874-75. Testing Method for Colour Fastness to Decatizing (R 1983). 8 pp.
JIS L 0876-70. Testing Method for Colour Fastness to Alkali Milling.
JIS L 0878-75. Testing Method for Colour Fastness to High Temperature Steaming (R 1983). 9 pp.
JIS L 0879-75. Testing Method for Colour Fastness to Dry Heating (R 1983). 6 pp.
JIS L 0880-77. Testing Method for Colour Fastness to Pleating.
JIS L 0881-71. Testing Method for Colour Fastness to Vulcanizing with Sulphur Monochloride.
JIS L 0882-75. Testing Method for Colour Fastness to Steam Vulcanizing.
JIS L 0883-71. Testing Method for Colour Fastness to Vulcanizing with Hot Air.
JIS L 0884-83. Testing Methods for Colour Fastness to Chlorinated Water. 6 pp.
JIS L 0885-92. Testing Method for Colour Fastness to Resin Finish. 7 pp.
JIS L 0886-92. Testing Methods for the Detection and Assessment of Photochromism. 7 pp.
JIS L 0887-75. Testing Method for Colour Fastness to Light of Fluorescent Whitening Agents and Fluorescent Whitened Textiles.
JIS L 0888-88. Testing Method for Colour Fastness to Light and Perspiration.
JIS L 0889-87. Testing Methods for Colour Fastness to Bleaching with Sodium Percarbonate. 10 pp.
JIS Z 8715-91. Whiteness of Near-White Opaque Materials—Specifying Method. 8 pp.
SAA AS 2001.4. Methods of Test for Textiles—Part 4: Colourfastness Tests Subset in Binder.
SAA AS 2001.4.1-80. Methods of Test for Textiles—Part 4: Colourfastness Tests—Part 4.1: Definitions and General Requirements. 4 pp.
SAA AS 2001.4.2-82. Methods of Test for Textiles—Part 4: Colourfastness Tests—Part 4.2: Determination of Colourfastness to Daylight. 3 pp.
SAA AS 2001.4.3-81. Methods of Test for Textiles—Part 4: Colourfastness Tests—Part 4.3: Determination of Colourfastness to Rubbing. 2 pp.
SAA AS 2001.4.4-81. Methods of Test for Textiles—Part 4: Colourfastness Tests—Part 4.4: Determination of Colourfastness to Water Spotting. 1 p.
SAA AS 2001.4.5-81. Methods of Test for Textiles—Part 4: Colourfastness Tests—Part 4.5: Determination of Colourfastness to Chlorinated Swimming Pool Water. 4 pp.
SAA AS 2001.4.6-90. Methods of Test for Textiles—Part 4: Colourfastness Tests—Part 4.6: Determination of Colourfastness to Hot Pressing. 3 pp.
SAA AS 2001.4.7-81. Methods of Test for Textiles—Part 4: Colourfastness Tests—Part 4.7: Determination of Colourfastness to Soda Boiling. 2 pp.
SAA AS 2001.4.8-79. Methods of Test for Textiles—Part 4: Colourfastness Tests—Part 4.8: Determination of Colourfastness to Water Reconfirmed 1986. 4 pp.
SAA AS 2001.4.9-81. Methods of Test for Textiles—Part 4: Colourfastness Tests—Part 4.9: Determination of Colourfastness to Alkali Spotting. 1 p.
SAA AS 2001.4.10-87. Methods of Test for Textiles—Part 4: Colourfastness Tests—Part 4.10: Determination of Colourfastness to Alkaline Milling. 2 pp.
SAA AS 2001.4.11-82. Methods of Test for Textiles—Part 4: Colourfastness Tests—Part 4.11: Determination of Colourfastness to Bleaching with Sodium Hypochlorite. 3 pp.
SAA AS 2001.4.12-81. Methods of Test for Textiles—Part 4: Colourfastness Tests—Part 4.12: Determination of Colourfastness to Acid Spotting. 2 pp.
SAA AS 2001.4.13-87. Methods of Test for Textiles—Part 4: Colourfastness Tests—Part 4.13: Determination of Colourfastness to Acid Milling. 2 pp.

Fabrics (Cont.)

—Colorfastness Testing (Cont.)

SAA AS 2001.4.14-80. Methods of Test for Textiles—Part 4: Colourfastness Tests—Part 4.14: Determination of Colourfastness to Seawater. 4 pp.
SAA AS 2001.4.15-87. Methods of Test for Textiles—Part 4: Colourfastness Tests—Part 4.15: Determination of Colourfastness to Washing. 5 pp.
SAA AS 2001.4.16-81. Methods of Test for Textiles—Part 4: Colourfastness Tests—Part 4.16: Determination of Colourfastness to Dry Cleaning Solvents. 2 pp.
SAA AS 2001.4.17-80. Methods of Test for Textiles—Part 4: Colourfastness Tests—Part 4.17: Determination of Colourfastness to Perspiration. 4 pp.
SAA AS 2001.4.21-79. Methods of Test for Textiles—Part 4: Colourfastness Tests—Part 4.21: Determination of Colourfastness to Light Using an Artificial Light Source (Mercury Vapour—Tungsten Filament—Internally Phosphor-Coated Lamp) Reconfirmed 1986. 4 pp.
SAA AS 2001.4.22-91. Methods of Test for Textiles—Part 4: Colourfastness Tests—Part 4.22: Determination of Colour Change Due to Flat Abrasion (Frosting) of Textile Fabrics (Screen Wire Method). 5 pp.
SNZ NZS/BS 1006-78. Methods of Test for Colour Fastness of Textiles and Leather Amend: 7. 212 pp.

—Colorfastness Testing Equipment

CNS L2020-75. Perspiration Testing Machine for Color Fastness Test (Aug)(3842).
CNS L2021-75. Rubbing Testing Machine for Color Fastness Test (Aug)(3843).
CNS L2027-81. Apparatus for Testing of Color Fastness to Washing (Nov)(8149).
CNS L2053-82. Apparatus for Testing of Colour Fastness to Carbon Arc Lamp Light (Jun)(9024).
JIS L 0821-83. Apparatus for Testing of Colour Fastness to Washing. 5 pp.
JIS L 0822-92. Apparatus for Testing of Colour Fastness to Perspiration. 5 pp.
JIS L 0823-71. Apparatus for Testing of Colour Fastness to Rubbing.
JIS L 0824-88. Apparatus for Testing of Colour Fastness to Carbon Arc Lamp Light. 9 pp.

—Colorfastness Testing—Laundering

BSI BS EN 20105-C01-93. 1993 Textiles—Tests for Colour Fastness Part C01: Colour Fastness to Washing: Test 1 (ISO 105-C01: 1989) (L). 10 pp.
BSI BS EN 20105-C02-93. 1993 Textiles—Tests for Colour Fastness Part C02: Colour Fastness to Washing: Test 2 (ISO 105-C02: 1989) (L). 10 pp.
BSI BS EN 20105-C03-93. 1993 Textiles—Tests for Colour Fastness Part C03: Colour Fastness to Washing: Test 3 (ISO 105-C03: 1989) (L). 10 pp.
BSI BS EN 20105-C04-93. 1993 Textiles—Tests for Colour Fastness Part C04: Colour Fastness to Washing: Test 4 (ISO 105-C04: 1989) (L). 10 pp.
BSI BS EN 20105-C05-93. 1993 Textiles—Tests for Colour Fastness Part C05: Colour Fastness to Washing: Test 5 (ISO 105-C05: 1989) (L). 10 pp.
CEN EN 20105-C01-92. Textiles—Tests for Colour Fastness Part C01: Colour Fastness to Washing: Test 1 (ISO 105-C01: 1989). 7 pp.
CEN EN 20105-C03-92. Textiles—Tests for Colour Fastness Part C03: Colour Fastness to Washing: Test 3 (ISO 105-C03: 1989). 7 pp.
CEN EN 20105-C04-92. Textiles—Tests for Colour Fastness—Part C04: Colour Fastness to Washing: Test 4 (ISO 105-C04: 1989). 7 pp.
CEN EN 20105-C05-92. Textiles—Tests for Colour Fastness—Part C05: Colour Fastness to Washing: Test 5 (ISO 105-C05: 1989). 5 pp.
DIN ENGL EN 20105 Pt C01-93. Tests for Colour Fastness of Textiles; Colour Fastness to Washing: Test 1 (ISO 105-C01: 1989) (Mar). 6 pp.
DIN ENGL EN 20105 Pt C02-93. Tests for Colour Fastness of Textiles; Colour Fastness to Washing: Test 2 (ISO 105-C02: 1989) (Mar). 6 pp.
DIN ENGL EN 20105 Pt C03-93. Tests for Colour Fastness of Textiles; Colour Fastness to Washing: Test 3 (ISO 105-C03: 1989) (Mar). 6 pp.
DIN ENGL EN 20105 Pt C04-93. Tests for Colour Fastness of Textiles; Colour Fastness to Washing: Test 4 (ISO 105-C04: 1989) (Mar). 6 pp.
DIN ENGL EN 20105 Pt C05-93. Tests for Colour Fastness of Textiles; Colour Fastness to Washing: Test 5 (ISO 105-C05: 1989) (Mar). 6 pp.
ISO 105 Pt C01-89. Textiles—Tests for Colour Fastness—Part C01: Colour Fastness to Washing: Test 1 Fourth Edition; (CEN EN 20105-C01:1992). 6 pp.

—Colorimetry—Light Sources

CGSB CAN/CGSB-4.2 NO.41-M91. Textile Test Methods Standard Light Sources for Colour Matching of Textiles. 6 pp.

—Comprehensive Testing

JIS L 1041-83. Testing Methods for Resin Finished Textiles. 36 pp.

Fabrics (Cont.)

—Comprehensive Testing (Cont.)

JIS L 1096-90. Testing Methods for Woven Fabrics. 89 pp.
SAA AS 2001. Methods of Test for Textiles Complete Set in Binder.

—Conditioning

SAA AS 2001.1-84. Methods of Test for Textiles—Part 1: Conditioning Procedures. 7 pp.

—Copper Content—Atomic Absorption Spectrophotometry

CGSB CAN/CGSB-4.2 NO.42-M91. Textile Test Methods Copper Content of Textiles. 6 pp.

—Courses

CGSB CAN/CGSB-4.2 NO.7-M88. Textile Test Methods Knitted Fabric Count—Wales and Courses Per Centimetre. 7 pp.
SAA AS 2001.2.6-81. Methods of Test for Textiles—Part 2: Physical Tests—Part 2.6: Determination of the Number of Wales and Courses per Unit Length in Knitted Fabric. 3 pp.

—Creasing

BSI BS 3086-72. (WITHDRAWN) 1972 Method for Determination of Recovery from Creasing of Textile Fabrics by Measuring the Angle of Recovery (Superseded by BS EN 22313: 1992). 11 pp.
BSI BS EN 22313-92. 1992 Textile Fabrics—Determination of the Recovery from Creasing of a Horizontally Folded Specimen by Measuring the Angle of Recovery (ISO 2313: 1972) (Supersedes BS 3086: 1972). 11 pp.
CEN EN 22313-92. Textile Fabrics—Determination of the Recovery from Creasing of a Horizontally Folded Specimen by Measuring the Angle of Recovery. 6 pp.
CGSB CAN/CGSB-4.2 NO.45-M88. Textile Test Methods Textile Fabrics—Determination of the Recovery from Creasing of a Horizontally Folded Specimen by Measuring the Angle-of-Recovery. 10 pp.
CNS L3085-80. Method of Test for Recovery from Creasing of Fabrics (May)(5616).
DIN ENGL 53890-72. Testing of Textiles; Determination of the Crease Recovery Angle of Area-Measured Textiles; Method Using an Air-Dry Specimen with Horizontal Fold and Erected Free Limb (Jan). 6 pp.
DIN ENGL EN 22313-92. Textiles; Determination of the Recovery from Creasing of a Horizontally Folded Specimen by Measuring the Angle of Recovery (ISO 2313:1972) (Aug) (Supersedes DIN 53891 Part 1, February 1978 Edition). 7 pp.
ISO 2313-72. Textile Fabrics—Determination of the Recovery from Creasing of a Horizontally Folded Specimen by Measuring the Angle of Recovery First Edition. 6 pp.
JIS L 1059-92. Testing Methods for Crease Recovery of Woven Fabrics.

—Creasing—Laundering Testing

CGSB CAN/CGSB-4.2 NO.59.3-M88. Textile Test Methods Appearance After Repeated Domestic Launderings—Pressed-in Creases. 11 pp.
CNS L3149-82. Method of Test for Crease Evaluation of Woven and Knitted Fabrics After Laundering (Jan)(8313).

—Defects

CNS L3076-77. Testing Method for the Woven Fabric Defects (Jun)(3854). 4 pp.
CNS L3086-80. Method of Test for Nonwoven Fabrics Defects (May)(5618).
CNS L3209-84. Method of Test for the Weft Knitted Fabric Defects (Jul)(10968).
CNS L4152-84. Weft Knitted Fabrics (Finished) (Jul)(10970). 2 pp.

—Defects—Glossaries

BSI BS 7342-90. 1990 Glossary of Terms for Defects in Woven Fabrics. 28 pp.
BSI BS 7343-90. 1990 Glossary of Terms for Defects in Knitted Fabrics. 22 pp.
ISO 8498-90. Woven Fabrics—Description of Defects—Vocabulary First Edition. 27 pp.
ISO 8499-90. Knitted Fabrics—Description of Defects—Vocabulary First Edition. 22 pp.
SAA AS 1083-71. Glossary of Terms for Defects in Woven and Knitted Textile Piece Goods Corrig. Reconfirmed 1986. 23 pp.

—Defense Contracts

MOD UK DEFCON 112B-91. Conditions of Contract for Clothing and Textiles 2/91. 7 pp.

—Dimensional Stability

BSI BS 2819-90. 1990 Determination of Bow, Skew and Lengthway Distortion in Woven and Knitted Fabrics. 8 pp.
BSI BS 4323-79. 1979 Determination of Dimensional Change of Fabrics Induced by Free Steam. 7 pp.

INTERNATIONAL AND NON-U.S. NATIONAL STANDARDS
SUBJECT INDEX

Fabrics (Cont.)
—Dimensional Stability (Cont.)
BSI BS 4736-85. 1985 Method for Determination of Dimensional Changes of Fabrics Induced by Cold-Water Immersion. 7 pp.
BSI BS 4931-86. 1986 Preparation, Marking and Measuring of Textile Fabrics, Garments and Fabric Assemblies in Tests for Assessing Dimensional Change. 8 pp.
BSI BS 4961: Part 1-80. 1980 Determination of Dimensional Stability of Textiles to Dry Cleaning in Tetrachloroethylene Part 1: Machine Method. 7 pp.
BSI BS 5807-87. 1987 Determination of Dimensional Change of Textiles in Domestic Washing and Drying. 4 pp.
BSI BS 7552: Part 2-92. 1992 Effect of Dry Heat on Fabrics Under Low Pressure Part 2: Method for Determination of Dimensional Change in Fabrics Exposed to Dry Heat (ISO 9866-2: 1991). 8 pp.
CGSB CAN/CGSB-4.2 NO.24-M91. Textile Test Methods Colourfastness and Dimensional Change in Commercial Laundering. 7 pp.
CGSB CAN/CGSB-4.2 NO.24.1-M88. Textile Test Methods Dimensional Change in Washing of Woven Fabrics —Accelerated Method. 9 pp.
CGSB CAN/CGSB-4.2 NO.24.2-M91. Textile Test Methods Dimensional Change in Commercial Type Laundering of Textiles (Washwheel). 13 pp.
CGSB CAN/CGSB-4.2 NO.25.1-M90. Textile Test Methods Dimensional Change in Wetting. 9 pp.
CGSB CAN/CGSB-4.2 NO.30-M90. Textile Test Methods Dimensional Change in Dry Cleaning. 9 pp.
CGSB CAN/CGSB-4.2 NO.58-M90. Textile Test Methods Colourfastness and Dimensional Change in Domestic Laundering of Textiles. 18 pp.
CGSB CAN/CGSB-4.2 NO.67-M90. Textile Test Methods Dimensional Change and Appearance After Laundering of Coated, Bonded, Laminated and Fused Fabrics. 11 pp.
CNS L3083-80. Method of Test for Dimensional Change of Fabrics Induced by Free-Stream (May)(5614).
ISO 675-79. Textiles—Woven Fabrics—Determination of Dimensional Change on Commercial Laundering Near the Boiling Point First Edition; (Corrigendum 1-1990). 7 pp.
ISO 3005-78. Textiles—Determination of Dimensional Change of Fabrics Induced by Free-Steam Second Edition. 6 pp.
ISO 3175-79. Textiles—Determination of Dimensional Change on Dry Cleaning in Perchloroethylene—Machine Method Second Edition. 6 pp.
ISO 3759-84. Textiles—Preparation, Marking and Measuring of Fabric Specimens and Garments in Tests for Determination of Dimensional Change Second Edition. 8 pp.
ISO 5077-84. Textiles—Determination of Dimensional Change in Washing and Drying First Edition. 5 pp.
ISO 7771-85. Textiles—Determination of Dimensional Changes of Fabrics Induced by Cold-Water Immersion First Edition. 6 pp.
ISO 9866 Pt 1-91. Textiles—Effect of Dry Heat on Fabrics Under Low Pressure—Part 1: Procedure for Dry-Heat Treatment of Fabrics First Edition. 5 pp.
ISO 9866 Pt 2-91. Textiles—Effect of Dry Heat on Fabrics Under Low Pressure—Part 2: Determination of Dimensional Change in Fabrics Exposed to Dry Heat First Edition. 6 pp.
SAA AS 1287.2-72. Methods for the Determination of Dimensional Change in Fabrics—Part 2: Relaxation in Steam Reconfirmed 1989. 7 pp.
SAA AS 1287.4-74. Methods for the Determination of Dimensional Change in Fabrics—Part 4: Dimensional Stability to Heat. 6 pp.
SAA AS 2001.5.1-87. Methods of Test for Textiles—Part 5: Dimensional Change—Part 5.1: General Requirements. 7 pp.
SAA AS 2001.5.3-85. Methods of Test for Textiles—Part 5: Dimensional Change—Part 5.3: Determination of Relaxation in Aqueous Solution. 2 pp.
SAA AS 2001.5.4-87. Methods of Test for Textiles—Part 5: Dimensional Change—Part 5.4: Determination of Dimensional Change in Laundering of Textile Fabrics and Garments—Automatic Machine Method (in Professional Packages 62, 69).
SAA AS 2001.5.5-87. Method of Test for Textiles—Part 5: Dimensional Change—Part 5.5: Determination of Dimensional Change in Laundering of Textile Fabrics and Garments—Cubex Machine Method. 2 pp.
SAA AS 2001.5.7-86. Method of Test for Textiles—Part 5: Dimensional Change—Part 5.7: Determination of Dimensional Change on Dry Cleaning in Perchloroethylene Excluding Finishing—Machine Method. 3 pp.

—Displacement Resistance
DIN ENGL 53934-86. Testing of Textiles; Determination of the Resistance of Fabrics to Displacement (July). 4 pp.

—Drapability
BSI BS 5058-73. 1973 Methods for the Assessment of Drape of Fabrics. 9 pp.

—Dry Cleaning
BSI BS 5742-89. 1989 Textile Labels Requiring to Be Washed and/or Dry Cleaned. 8 pp.

—Dry Heat Testing
BSI BS 7552: Part 1-92. 1992 Effect of Dry Heat on Fabrics Under Low Pressure Part 1: Method for Dry Heat Treatment of Fabrics (ISO 9866-1: 1991). 8 pp.
BSI BS 7552: Part 2-92. 1992 Effect of Dry Heat on Fabrics Under Low Pressure Part 2: Method for Determination of Dimensional Change in Fabrics Exposed to Dry Heat (ISO 9866-2: 1991). 8 pp.
ISO 9866 Pt 1-91. Textiles—Effect of Dry Heat on Fabrics Under Low Pressure—Part 1: Procedure for Dry-Heat Treatment of Fabrics First Edition. 5 pp.
ISO 9866 Pt 2-91. Textiles—Effect of Dry Heat on Fabrics Under Low Pressure—Part 2: Determination of Dimensional Change in Fabrics Exposed to Dry Heat First Edition. 6 pp.

—Elastic Properties
BSI BS 4294-68. (WITHDRAWN) 1968 Methods of Test for the Stretch and Recovery Properties of Fabrics (Superseded by BS 4952: 1992). 8 pp.
CNS L3139-81. Testing Method for Stretch Properties of Stretch Woven Fabrics (Oct)(8039).

—Electrostatic Charge
CNS L3148-82. Method of Test for Electrostatic Propensity of Woven and Knitted Fabrics (Jan)(8312).
JIS L 1094-88. Testing Methods for Electrostatic Propensity of Woven and Knitted Fabrics. 25 pp.

—Elongation
BSI BS 2576-86. 1986 Amd 1 Method for Determination of Breaking Strength and Elongation (Strip Method) of Woven Fabrics (AMD 6199) September 29, 1989. 11 pp.
BSI BS 7304: Part 3-90. (WITHDRAWN) 1990 Methods of Test for Nonwovens Part 3: Determination of Tensile Strength and Elongation (Renumbered as BS EN 29073-3: 1992). 8 pp.
CGSB CAN/CGSB-4.2 NO.9.1-M90. Textile Test Methods Breaking Strength of Fabrics—Strip Method—Constant-Time-to-Break Principle. 11 pp.
CGSB CAN/CGSB-4.2 NO.9.3-M90. Textile Test Methods Breaking Strength of High-Strength Fabrics—Constant-Time-to-Break Principle. 13 pp.
CGSB CAN/CGSB-4.2 NO.10-M87. Textile Test Methods Elongation. 9 pp.
ISO 5081-77. Textiles—Woven Fabrics—Determination of Breaking Strength and Elongation (Strip Method) First Edition. 10 pp.

—Environmental Testing
BSI BS 2087: Part 1-92. 1992 Preservative Textile Treatments Part 1: Specification for Treatments. 17 pp.
BSI BS 2087: Part 1-81. 1981 Amd 1 Preservative Treatments for Textiles Part 1: Specification for Treatments (L). 15 pp.
BSI BS 2087: Part 2-92. 1992 Preservative Textile Treatments Part 2: Methods of Test. 23 pp.
BSI BS 2087: Part 2-81. 1981 Amd 1 Preservative Treatments for Textiles Part 2: Methods of Test. 23 pp.
CGSB CAN/CGSB-4.2 NO.69-M91. Textile Test Methods Weather Resistance—Xenon Arc Radiation. 10 pp.

—Expansion
BSI BS 4768-72. 1972 Amd 1 Determination of the Bursting Strength and Bursting Distension of Fabrics (AMD 6340) June 28, 1991. 9 pp.

—Extraction Analysis
CNS L3165-82. Determination of the Contents of Extractable Matter in Textile Materials (Organic Solvents Method) (Apr) (8708).

—Faults
BSI BS 6395-83. 1983 Numerical Designation of Fabric Faults by Visual Inspection. 5 pp.

—Fire Resistant—Fire Testing
CGSB CAN/CGSB-4.2 NO.27.10-M91. Textile Test Methods Flame Resistance—Vertically Oriented Textile Fabric or Fabric Assembly Test. 11 pp.

—Fire Retardants
CNS Z2016-71. Fire-Proof Agents for Paper, Cloth and Canvas (Tentative) (Jul)(3078).

—Fire Retardants—Removal
CGSB CAN/CGSB-4.2 NO.30.3-M87. Textile Test Methods Procedure for Removal of Flame-Retardant Treatments from Textile Products. 8 pp.

—Flammability Testing
BSI BS 4569-83. 1983 Ignitability (Surface Flash) of Pile Fabrics and Assemblies Having Pile on the Surface. 7 pp.
BSI BS 5438-89. 1989 Amd 1 Methods of Test for Flammability of Textile Fabrics When Subjected to a Small Igniting Flame Applied to the Face or Bottom Edge of Vertically Oriented Specimens (AMD 6509) June 29, 1990. 25 pp.
BSI BS 5438-76. 1976 Amd 1 Methods of Test for Flammability of Vertically Oriented Textile Fabrics and Fabric Assemblies Subjected to a Small Igniting Flame. 19 pp.
BSI BS 5651-89. 1989 Cleansing and Wetting Procedures for Use in the Assessment of the Effect of Cleansing and Wetting on the Flammability of Textile Fabrics and Fabric Assemblies. 12 pp.
BSI BS 5722-91. 1991 Flammability Performance of Fabrics and Fabric Combinations Used in Nightwear Garments. 12 pp.
BSI BS 5867: Part 2-80. 1980 Amd 1 Fabrics for Curtains and Drapes Part 2: Flammability Requirements. 8 pp.
BSI PD 2777-77. 1977 Fabric Flammability Burning Accidents and the Relevance of BS 5438. 14 pp.
CEN PREN 532-91. Clothing for Protection Against Heat and Flame—Method of Test for Limited Flame Spread. 14 pp.
CEN PREN 1101-93. Textiles and Textile Products—Burning Behaviour—Curtains and Drapes—Detailed Procedure to Determine the Ignitability of Vertically Oriented Specimens (Small Flame). 4 pp.
CEN PREN 1102-93. Textiles and Textile Products—Burning Behaviour—Curtains and Drapes—Detailed Procedure to Determine the Flame Spread of Vertically Oriented Specimens. 6 pp.
CGSB CAN/CGSB-4.2 NO.27-M88. Textile Test Methods Burning Behaviour—Selection of Methods. 9 pp.
CGSB CAN/CGSB-4.2 NO.27.1-M87. Textile Test Methods Flame Resistance—Vertical Burning Test. 11 pp.
CGSB CAN/CGSB-4.2 NO.27.2-M87. Textile Test Methods Flame Resistance—Surface Burning Test. 10 pp.
CGSB CAN/CGSB-4.2 NO.27.3-M86. Textile Test Methods Textile Fabrics—Burning Behaviour—Measurement of Flame Spread Properties of Vertically Oriented Specimens (ISO 6941-1984). 27 pp.
CGSB CAN/CGSB-4.2 NO.27.4-M86. Textile Test Methods Textile Fabrics—Burning Behaviour—Determination of Ease of Ignition of Vertically Oriented Specimens (ISO 6940-1984). 29 pp.
CGSB CAN/CGSB-4.2 NO.27.5-M87. Textile Test Methods Flame-Resistance—45 Degree Angle Test—One Second Flame Impingement. 12 pp.
CNS L3196-83. Method of Test for Flammability of Clothes — Test for the Degree of Charring Extent (May)(10285).
CNS S2148-90. Method of Test for Tent—Flammability of Wall and Top Materials (Mar)(12693). 4 pp.
DIN ENGL 54336-86. Testing of Textiles; Determination of Burning Behaviour of Vertically Oriented Specimens with Ignition at Lower Edge (Nov). 6 pp.
DIN ENGL 66083-86. Characteristic Values for the Burning Behaviour of Textile Products; Textile Fabrics for Industrial and Protective Clothing (Nov). 2 pp.
ISO 6940-84. Textile Fabrics—Burning Behaviour—Determination of Ease of Ignition of Vertically Oriented Specimens First Edition; (Amendment 1-1993). 16 pp.
ISO 6941-84. Textile Fabrics—Burning Behaviour—Measurement of Flame Spread Properties of Vertically Oriented Specimens First Edition; (Amendment 1-1992). 16 pp.
ISO 10047-93. Textiles—Determination of Surface Burning Time of Fabrics First Edition. 10 pp.
JIS A 1323-84. Flame Retardant Testing Method for Spark Droplets of Welding and Gas Cutting on Fabric Sheets in Construction Works. 9 pp.
JIS L 1091-92. Testing Methods for Flammability of Textiles. 23 pp.
SAA AS 1176.1-82. Textiles—Methods of Test for Combustion Properties—Part 1: The Determination of Ease of Ignition of Certain Textile Materials in a Horizontal Plane. 7 pp.
SAA AS 1176.1D-72. Textiles—Methods of Test for Combustion Properties—Part 1D: Drawings of Apparatus for Australian Standard Method for the Determination of Ease of Ignition of Textiles Corrig. Reconfirmed 1986.
SAA AS 1176.2D-72. Textiles—Methods of Test for Combustion Properties—Part 2D: Drawings of Apparatus for Australian Standard Method for the Determination of Rate of Burning and Heat Output of Textiles from Which Clothing May Be Made Reconfirmed 1986.

INTERNATIONAL AND NON-U.S. NATIONAL STANDARDS
SUBJECT INDEX

Fabrics

Fabrics (Cont.)

—**Flammability Testing** (Cont.)
SAA AS 1176.2-82. Textiles—Methods of Test for Combustion Properties—Part 2: The Determination of Burning Time of Textile Materials. 7 pp.
SAA AS 1176.3-82. Textiles—Methods of Test for Combustion Properties—Part 3: Determination of Surface Burning Properties of Certain Textile Materials. 10 pp.
SNZ NZS 8703-77. Methods of Test for Combustion Characteristics of Textile Materials Amend: A, 1977. 24 pp.

—**Flammability Testing—Glossaries**
BSI BS 6373-85. 1985 Glossary of Terms Relating to Burning Behaviour of Textiles and Textile Products. 11 pp.
ISO 4880-84. Burning Behaviour of Textiles and Textile Products—Vocabulary First Edition; (Amendment 1-1992) (Amendment 2-1993). 21 pp.

—**Fluorescent Whitening Agents**
CNS L1017-83. Scale for Assessing the Intensity of Fluorescent Whitening Effect (Jan)(9899).
CNS L1018-83. Scale for Assessing the Fastness of Fluorescent Whitened Textiles (Jan)(9900).
JIS L 0806-71. Scale for Assessing the Fastness of Fluorescent Whitened Textiles.
JIS L 0807-71. Scale for Assessing the Intensity of Fluorescent Whitening Effect.
JIS L 1064-85. Identification of Fluorescent Brightening Agents Classes on Textiles.

—**Formaldehyde Content**
BSI BS 6806: Part 1-87. 1987 Amd 1 Formaldehyde in Textiles Part 1: Method for Determination of Total Formaldehyde. 9 pp.
BSI BS 6806: Part 2-87. 1987 Amd 1 Formaldehyde in Textiles Part 2: Method for Determination of 'Free' Formaldehyde. 9 pp.
BSI BS 6806: Part 3-87. 1987 Amd 1 Formaldehyde in Textiles Part 3: Method for Determination of Released Formaldehyde. 9 pp.
CGSB CAN/CGSB-4.2 NO.63.2-M91. Textile Test Methods Formaldehyde in Textiles—Part 2: Method for Determination of "Free" Formaldehyde (BS 6806:Part 2:1987). 10 pp.
CGSB CAN/CGSB-4.2 NO.63.3-M91. Textile Test Methods Formaldehyde in Textiles—Part 3: Method for Determination of Released Formaldehyde (BS 6806:Part 3:1987); (Including Amendment 1 January 1988). 10 pp.

—**Friction**
JIS L 1056-87. Testing Methods for Sliding Frictional Melting of Woven and Knitted Fabrics. 11 pp.

—**Frosting**
JIS L 1901-88. Testing Methods for Frosting Due to Yarn Reversing of Woven and Knitted Fabrics. 11 pp.

—**Fungus Resistance Testing**
CGSB CAN/CGSB-4.2 NO.28.1-M91. Textile Test Methods Resistance to Micro-Organisms—Fungus Damage Test—Pure Culture—Quantitative. 8 pp.
CGSB CAN/CGSB-4.2 NO.28.2-M91. Textile Test Methods Resistance to Micro-Organisms—Surface-Growing Fungus Test—Pure Culture. 7 pp.
CGSB CAN/CGSB-4.2 NO.28.4-M91. Textile Test Methods Resistance to Micro-Organisms—Fungus Damage Test—Pure Culture—Qualitative. 7 pp.
SAA AS 1157.2-72. Methods of Testing Materials for Resistance to Fungal Growth—Part 2: Resistance of Textiles to Fungal Growth. 9 pp.

—**Gas Permeability**
BSI BS 5636-90. 1990 Amd 1 Determination of Permeability of Fabrics to Air (AMD 6341) June 28, 1991. 6 pp.
CGSB CAN/CGSB-4.2 NO.36-M89. Textile Test Methods Air Permeability. 8 pp.
CNS L3081-80. Method of Test for Air Permeability of Fabrics (May)(5612).
DIN ENGL 53887-86. Testing of Textiles; Determination of Air Permeability of Textile Fabrics (Aug). 4 pp.

—**Glossaries**
BSI BS 6189-89. 1989 Terms Relating to Fabrics and Associated Fibres, Yarns and Processes. 28 pp.
BSI BS 7089-89. (WITHDRAWN) 1989 Definition for Nonwovens (Superseded by BS EN 29092: 1992). 3 pp.
JIS L 0206-76. Glossary of Terms Used in Textile Industry (Fabrics).
JIS L 0208-92. Glossary of Terms Used in Textile Industry—Testing.
JIS L 0212-84. Glossary of Terms Used in Textile End Product.
JIS L 0213-83. Glossary of Terms Used in Textile Sundry Goods.
MOD UK DL-14. Textile Fabrics and Threads.
SNZ NZS 8701-71. Definitions of Commonly Used Terms Relating to Textiles Amend: 1, 1976. 10 pp.

Fabrics (Cont.)

—**Government Supplied Property—Order Forms**
MOD UK DEFCON 198-91. Clothing and Textile Contracts Issues of Government Property on Embodiment Loan Terms 2/91. 2 pp.
MOD UK DEFCON 198A-91. Clothing and Textile Contracts Issues of Government Property on Prepayment or Repayment Terms 2/91. 2 pp.
MOD UK DEFCON 198B-80. Materials to Be Issued to the Contractor 9/80. 1 p.

—**Grading**
CNS L3053-65. Inspecting Standard of Narrow-Width Textiles (for Export) (Apr)(2462)(R 1973). 3 pp.
CNS L3068-81. Mixed Fabrics (Unfinished) (May)(3394). 2 pp.
CNS L4151-84. Weft Knitted Fabrics (Unfinished) (Jul) (10969). 2 pp.
CNS L4152-84. Weft Knitted Fabrics (Finished) (Jul)(10970). 2 pp.

—**Greases**
CNS L4010-62. Grease for Textile Products (May)(1442) (R 1973). 1 p.

—**Household—Labels**
SAA AS 1957-87. Care Labelling of Clothing, Household Textiles, Furnishings, Upholstered Furniture, Bedding, Piece Goods and Yarns. 7 pp.

—**Identification Systems**
BSI BS 2747-86. 1986 Amd 1 Code of Practice for Textile Care Labelling (AMD 5529) October 31, 1986. 19 pp.
CGSB CAN/CGSB-86.1-M91. Care Labelling of Textiles. 18 pp.
CNS L1016-81. Care Labelling of Textile Goods (Nov)(8148). 10 pp.
CNS L3068-81. Mixed Fabrics (Unfinished) (May)(3394). 2 pp.
CNS L4151-84. Weft Knitted Fabrics (Unfinished) (Jul) (10969). 2 pp.
CNS L4152-84. Weft Knitted Fabrics (Finished) (Jul)(10970). 2 pp.
ISO 3758-91. Textiles—Care Labelling Code Using Symbols First Edition. 10 pp.
JIS L 0217-76. Care Labelling of Textile Goods.
SNZ NZS 8721-88. Care Labelling of Clothing, Household Textiles, Furnishings, Upholstered Furniture, Bedding, Piece Goods and Yarns Amend: A, 1988. 7 pp.
SNZ NZS 8722-88. Care Labelling—Guide to the Selection of Correct Care Labelling Instructions from NZS 8721 Amend: A, 1988. 15 pp.

—**Inspection**
CNS L3053-65. Inspecting Standard of Narrow-Width Textiles (for Export) (Apr)(2462)(R 1973). 3 pp.

—**Inspection—Glossaries**
JIS L 0220-92. Glossary of Terms Used in Textile Industry—Inspection.

—**Invitation for Bids**
MOD UK DEFCON 47D-86. Invitation to Tender for Clothing and Textile Contracts 8/86. 2 pp.
MOD UK DEFCON 47D-91. Invitation to Tender for Clothing and Textile Contracts 9/91. 2 pp.

—**Ironing**
BSI BS 7305-90. 1990 Method for Determination of the Maximum Safe Ironing Temperature of Fabrics. 4 pp.
CGSB CAN/CGSB-4.2 NO.31-M89. Textile Test Methods Textiles—Tests for Colourfastness—Part X11: Colourfastness to Hot Pressing (ISO 105/X11-1987). 9 pp.
CGSB CAN/CGSB-4.2 NO.33-M86. Textile Test Methods Pressing—Selection of Methods. 6 pp.
CGSB CAN/CGSB-4.2 NO.33.1-M86. Textile Test Methods Method of Pressing Ironing. 8 pp.
CGSB CAN/CGSB-4.2 NO.33.4-M86. Textile Test Methods Method of Pressing Tension Pressing. 11 pp.
CGSB CAN/CGSB-4.2 NO.57-M90. Textile Test Methods Determination of Maximum Safe Ironing Temperature. 9 pp.
CNS L3156-83. Method of Test for Colour Fastness to Hot Pressing (Jun)(8532).
ISO 105 Pt X11-87. Textiles—Tests for Colour Fastness—Part X11: Colour Fastness to Hot Pressing Third Edition. 4 pp.
ISO 105 Pt X13-87. Textiles—Tests for Colour Fastness—Part X13: Colour Fastness of Wool Dyes to Processes Using Chemical Means for Creasing, Pleating and Setting Third Edition. 6 pp.
JIS L 0850-75. Testing Method for Colour Fastness to Hot Pressing (R 1978). 9 pp.

—**Ironing—Shrinkage**
CNS L3225-88. Method of Test for Shrinkage Percentage by Ironing of Woven and Knitted Fabrics (Dec)(12474).

Fabrics (Cont.)

—**Jute**
BSI BS 7144-89. 1989 Flax and Jute Fabrics for Industrial Use. 16 pp.
MOD UK DSTAN 83-25-01. Cloths, Jute, and Mixtures Containing Jute Issue 2; Amendment 1. 13 pp.

—**Jute—Oil Content**
BSI BS 3845-90. 1990 Determination of Added Oil Content of Jute Yarn, Rove and Fabric. 8 pp.

—**Laundering—Commercial**
CGSB CAN/CGSB-4.2 NO.34-M89. Textile Test Methods Standard Commercial Laundering Procedures. 13 pp.

—**Laundering Testing**
BSI BS 4923-91. 1991 Methods for Individual Domestic Washing and Drying for Use in Textile Testing. 17 pp.
BSI BS 5742-89. 1989 Textile Labels Requiring to Be Washed and/or Dry Cleaned. 8 pp.
BSI BS 6246: Part 1-87. 1987 Industrial Laundry Machinery Part 1: Methods for the Assessment of the Effect of Rotary Washing Machines on Textiles. 26 pp.
BSI BS 6246: Part 2-83. 1983 Industrial Laundry Machinery Part 2: Methods for the Assessment of the Effect of Extracting Machines on Textiles. 20 pp.
BSI BS 6246: Part 3-85. 1985 Industrial Laundry Machinery Part 3: Methods for the Assessment of the Effect of Flatwork Ironing Machines on Textiles. 18 pp.
BSI BS 6246: Part 4-83. 1983 Industrial Laundry Machinery Part 4: Methods for the Assessment of the Effect of Batch Drying Tumblers on Textiles. 11 pp.
BSI BS 6246: Part 5-91. 1991 Industrial Laundry Machinery Part 5: Methods for the Assessment of the Effect of Shaped Garment Finishing Machinery on Textiles. 32 pp.
CGSB CAN/CGSB-4.2 NO.24-M91. Textile Test Methods Colourfastness and Dimensional Change in Commercial Laundering. 7 pp.
CGSB CAN/CGSB-4.2 NO.24.1-M88. Textile Test Methods Dimensional Change in Washing of Woven Fabrics —Accelerated Method. 9 pp.
CGSB CAN/CGSB-4.2 NO.24.2-M91. Textile Test Methods Dimensional Change in Commercial Type Laundering of Textiles (Washwheel). 13 pp.
CGSB CAN/CGSB-4.2 NO.58-M90. Textile Test Methods Colourfastness and Dimensional Change in Domestic Laundering of Textiles. 18 pp.
CNS L3218-85. Method of Test for Soil Release of Fabric (Jul)(11309).
ISO 6330-84. Textiles—Domestic Washing and Drying Procedures for Textile Testing First Edition. 8 pp.

—**Length Measurement**
BSI BS 1931-68. 1968 Determination of Length of Woven or Knitted Fabrics When Relaxed at Zero Tension. 9 pp.
CGSB CAN/CGSB-4.2 NO.4.2-M87. Textile Test Methods Textiles Fabrics—Measurement of Length of Pieces. 11 pp.
CNS L3089-80. Measurement of Length of Fabrics (Aug)(6244).
ISO 3933-76. Textiles—Woven Fabrics—Measurement of Length of Pieces First Edition. 5 pp.
SAA AS 1587-73. Methods for Measurements of Textile Fabrics—Length, Width, Thickness, Mass Per Unit Length and Mass Per Unit Area Corrig. Amdt 1 June 1987. 12 pp.

—**Life Rafts—Ships**
ISO TR6065-91. Shipbuilding and Marine Structures—Inflatable Liferafts—Materials First Edition. 15 pp.

—**Linings**
BSI BS 4973: Part 1-73. 1973 Interlinings Part 1: Nonwoven Sew-In Interlinings. 9 pp.
BSI BS 4973: Part 4-90. 1990 Interlining Part 4: Woven Sew-in Interlinings. 10 pp.
JIS L 1085-92. Testing Methods for Nonwoven Interlining Fabrics. 15 pp.

—**Linings—Defects**
CNS L4076-83. Nonwoven Interlining Fabric (Jun)(5617). 2 pp.

—**Linings—Fusible**
BSI BS 4973: Part 2-73. 1973 Interlinings Part 2: Nonwoven Fusible Interlinings. 9 pp.
BSI BS 4973: Part 3-76. 1976 Interlining Part 3: Woven and Knitted Fusible Interlinings. 4 pp.
JIS L 1086-83. Testing Methods for Fusible Interlining Fabrics.

—**Linings—Grading**
CNS L4076-83. Nonwoven Interlining Fabric (Jun)(5617). 2 pp.

INTERNATIONAL AND NON-U.S. NATIONAL STANDARDS
SUBJECT INDEX

Fabrics

Fabrics (Cont.)

—Linings—Identification Systems
CNS L4076-83. Nonwoven Interlining Fabric (Jun)(5617). 2 pp.

—Linings—Quality Assurance
CNS L4035-66. Colored Lining Fabrics (for Winter Clothes) (Tentative) (Sep)(2564) (R 1973). 1 p.
CNS L4036-66. Colored Lining Fabrics, for Clothes and Raincoats (Tentative) (Sep)(2565)(R1973). 2 pp.
CNS L4040-66. Lining Fabrics for Woolen and Worsted Clothes (Tentative) (Sep)(2639) (R 1973). 2 pp.

—Linting
BSI BS 6909-88. 1988 Method for Generation and Counting of the Airborne Linting Propensity of Fabrics in the Dry. 10 pp.

—Low Fire Danger
SNZ NZS 8704-89. Low Fire Danger Fabrics for Domestic Apparel. 12 pp.
SNZ NZS 8705-89. Children's Night Clothes Having Low Fire Danger. 26 pp.

—Mass
BSI BS 2471-78. 1978 Methods of Test for Textiles. Woven Fabrics. Determination of Mass Per Unit Length and Mass Per Unit Area. 8 pp.
BSI BS 2866-84. 1984 Methods for Determination of the Mass of Warp and Weft per Unit Area of Fabric. 4 pp.
BSI BS 7304: Part 1-90. (WITHDRAWN) 1990 Methods of Test for Nonwovens Part 1: Determination of Mass per Unit Area (Renumbered as BS EN 29073-1: 1992). 8 pp.
CGSB CAN/CGSB-4.2 NO.5.1-M90. Textile Test Methods Unit Mass of Fabrics. 9 pp.
CNS L3087-80. Determination of Mass of Fabrics (Aug)(6242).
ISO 3801-77. Textiles—Woven Fabrics—Determination of Mass per Unit Length and Mass per Unit Area First Edition. 6 pp.
ISO 7211 Pt 6-84. Textiles—Woven Fabrics—Construction—Methods of Analysis—Part 6: Determination of the Mass of Warp and Weft per Unit Area of Fabric First Edition. 4 pp.
SAA AS 1587-73. Methods for Measurements of Textile Fabrics—Length, Width, Thickness, Mass Per Unit Length and Mass Per Unit Area Corrig. Amdt 1 June 1987. 12 pp.
SAA AS 2001.2.13-87. Methods of Test for Textiles—Part 2: Physical Tests—Part 2.13: Determination of Mass per Unit Area and Mass per Unit Length of Fabrics. 2 pp.

—Mass—Glossaries
BSI BS 1051-92. 1992 Glossary of Terms Relating to the Mass Determination of Textiles (ISO 6348: 1980). 6 pp.
BSI BS 1051-81. 1981 Glossary of Terms Relating to the Conditioning, of Textiles Testing and Mass Determination. 7 pp.
CGSB CAN/CGSB-4. 175-M91 Pt2. Textiles—Determination of Mass—Vocabulary Part 2 (ISO 6348:1980). 6 pp.
ISO 6348-80. Textiles—Determination of Mass—Vocabulary First Edition. 3 pp.

—Measurement Precision
CGSB CAN/CGSB-4.2 NO.1-M87. Textile Test Methods Precision and Accuracy of Measurements. 16 pp.

—Medical—Flammability Testing
CGSB CAN/CGSB-4.162-M80. Hospital Textiles—Flammability Performance Requirements. 34 pp.

—Metal Content
BSI BS 6463: Part 3-87. 1987 Quicklime, Hydrated Lime and Natural Calcium Carbonate Part 3: Methods for Determination of Metals in Textiles. 13 pp.
BSI BS 6648-86. 1986 Wet Ashing of Textile Materials for Subsequent Determination of Metal Content. 4 pp.
BSI BS 6648-01. 1986 Amd 1 Methods for Wet Ashing of Textile Materials for Subsequent Determination of Metal Content (AMD 7007) July 15, 1992. 5 pp.

—Metric System
SAA AS 1128-72. Preferred Metric Units for Textiles Reconfirmed 1986 (This is a Joint Standard with SANZ NZS 1128). 6 pp.
SNZ NZS/AS 1128-72. Preferred Metric Units for Textiles (Reconfirmed 1986) (This is a Joint Standard with SAA AS 1128). 6 pp.

—Microbiological Resistance Testing
BSI BS 2087: Part 1-92. 1992 Preservative Textile Treatments Part 1: Specification for Treatments. 17 pp.

Fabrics (Cont.)

—Microbiological Resistance Testing (Cont.)
BSI BS 2087: Part 1-81. 1981 Amd 1 Preservative Treatments for Textiles Part 1: Specification for Treatments (L). 15 pp.
BSI BS 6085-92. 1992 Methods for Determination of the Resistance of Textiles to Microbiological Deterioration. 16 pp.
BSI BS 6085-81. 1981 Methods of Test for Determination of the Resistance of Textiles to Microbiological Deterioration. 8 pp.

—Microorganism Resistance Testing
CGSB CAN/CGSB-4.2 NO.28.3-M91. Textile Test Methods Resistance to Micro-Organisms—Soil Burial Test. 7 pp.

—Moisture Content
CGSB CAN/CGSB-4.2 NO.3-M88. Textile Test Methods Determination of Moisture in Textiles. 9 pp.

—Molding Materials—Volatile Matter Content
ISO 9782-93. Plastics—Reinforced Moulding Compounds and Prepregs—Determination of Apparent Volatile-Matter Content First Edition. 6 pp.

—Nightwear—Children's
SNZ NZS 8705-89. Children's Night Clothes Having Low Fire Danger. 26 pp.

—Nonfibrous Materials—Quantitative Analysis
CGSB CAN/CGSB-4.2 NO.15-M88. Textile Test Methods Nonfibrous Materials on Textiles. 11 pp.

—Nonfibrous Materials—Removal
CGSB CAN/CGSB-4.2 NO.15-M88. Textile Test Methods Nonfibrous Materials on Textiles. 11 pp.

—Oil Resistance Testing
CNS L3217-85. Method of Test for Oil Repellency of Fabric (Jul)(11308).

—Packaging Papers
MOD UK TS 403. Paper, Textile Wrapping.

—Permeability
SAA AS 2001.2.34-90. Methods of Test for Textiles—Part 2: Physical Tests—Part 2.34: Determination of Permeability of Fabrics. 6 pp.

—Pest Resistance
BSI BS 2087: Part 1-92. 1992 Preservative Textile Treatments Part 1: Specification for Treatments. 17 pp.
BSI BS 2087: Part 1-81. 1981 Amd 1 Preservative Treatments for Textiles Part 1: Specification for Treatments (L). 15 pp.
BSI BS 2087: Part 2-92. 1992 Preservative Textile Treatments for Textiles Part 2: Methods of Test. 23 pp.
BSI BS 2087: Part 2-81. 1981 Amd 1 Preservative Treatments for Textiles Part 2: Methods of Test. 23 pp.
BSI BS 4797-78. 1978 Method of Test for Textiles. Determination of Resistance to Certain Insect Pests. 8 pp.
CGSB CAN/CGSB-4.2 NO.38-M89. Textile Test Methods Resistance to Insect Pests. 9 pp.
ISO 3998-77. Textiles—Determination of Resistance to Certain Insect Pests First Edition. 7 pp.
SAA AS 2001.6.1-80. Methods of Test for Textiles—Part 6: Miscellaneous Tests—Part 6.1: Determination of the Resistance of Textiles to Certain Insect Pests. 8 pp.

—Physical Testing
JIS L 0105-92. General Principles of Physical Testing Methods for Textiles.

—Pile Retention
JIS L 1075-83. Testing Methods for Pile Retention of Woven and Knitted Fabrics.

—Pilling
BSI BS 5811-86. 1986 Determination of the Resistance to Pilling of Woven Fabrics (Pill Testing Box Method). 8 pp.
CGSB CAN/CGSB-4.2 NO.51.1-M87. Textile Test Methods Resistance to Pilling, Rotating Box Method. 16 pp.
CGSB CAN/CGSB-4.2 NO.51.2-M87. Textile Test Methods Resistance to Pilling—Random Tumble Pilling Tester. 13 pp.
CNS L3140-81. Testing Method for Pilling of Woven Fabrics and Knitted Fabrics (Oct)(8040).
JIS L 1076-92. Testing Methods for Pilling of Woven Fabrics and Knitted Fabrics. 26 pp.

—Pleating
JIS L 1060-87. Testing Methods for Pleating of Woven and Knitted Fabrics.

Fabrics (Cont.)

—Polyacrylonitrile
CNS L4157-88. Polyacrylonitrile Woven Fabrics (Finished) (Oct)(12443).

—Porosity Measurement
BSI BS 3321-86. 1986 Method for Measurement of the Equivalent Pore Size of Fabric (Bubble Pressure Test). 8 pp.

—Radiation Decontamination
ISO 9271-92. Decontamination of Radioactively Contaminated Surfaces—Testing of Decontamination Agents for Textiles First Edition. 21 pp.

—Resistivity
BSI BS 3266-81. 1981 Determination of Conductivity, pH, Water Soluble Matter, Chloride and Sulphate in Aqueous Extracts of Textile Materials. 9 pp.
BSI BS 6524-84. 1984 Determination of the Surface Resistivity of a Textile Fabric. 8 pp.

—Seam Slippage
CGSB CAN/CGSB-4.2 NO.32.1-M91. Textile Test Methods Resistance of Woven Fabrics to Seam Slippage. 8 pp.

—Ships
MOD UK NES 129-84. Requirements for Soft Furnishings and Fabrics Issue 2 (04.84). 28 pp.

—Shrinkage
CNS L3138-83. Method of Test for Shrinkage Percentage of Woven Fabrics (Mar) (8038).
CNS L3225-88. Method of Test for Shrinkage Percentage by Ironing of Woven and Knitted Fabrics (Dec)(12474).
JIS L 1042-92. Testing Methods for Shrinkage Percentage of Woven Fabrics. 18 pp.
JIS L 1057-92. Testing Methods for Shrinkage Percentage by Ironing of Woven and Knitted Fabrics. 7 pp.

—Sizing Compounds
BSI BS 4032-78. 1978 Determination of Certain Water-Or Alkali-Soluble Additives in Cellulosic or Synthetic Fibres, Yarns and Fabrics or Yarns and Fabrics Made from Blends of Such Fibres. 4 pp.

—Slippage—Sewing Machines
BSI BS 5948: Part 3-85. 1985 Household Sewing Machines Part 3: Method for Determination of Creep of One Ply of Material over Another. 7 pp.
ISO 4818-84. Household Sewing Machines—Determination of Creep of One Ply of Material over Another First Edition. 6 pp.

—Snag Resistance
CGSB CAN/CGSB-4.2 NO.60-M89. Textile Test Methods Resistance to Snagging—Mace Test. 17 pp.
CNS L3226-88. Method of Test for Sang of Woven Fabrics and Knitted Fabrics (Dec)(12475).
JIS L 1058-83. Testing Methods for Snag of Woven Fabrics and Knitted Fabrics. 25 pp.

—Soil Resistance Testing
CGSB CAN/CGSB-4.2 NO.28.3-M91. Textile Test Methods Resistance to Micro-Organisms—Soil Burial Test. 7 pp.
CNS L3218-85. Method of Test for Soil Release of Fabric (Jul)(11309).
CNS L3219-85. Method of Test for Soil Redeposition, Resistance to Launder-Ometer (Jul) (11310).
CNS L3220-85. Method of Test for Soil Redeposition, Resistance to Terg-O-Tometer (Jul) (11311).

—Soluble Matter Content
BSI BS 3266-81. 1981 Determination of Conductivity, pH, Water Soluble Matter, Chloride and Sulphate in Aqueous Extracts of Textile Materials. 9 pp.
CGSB CAN/CGSB-4.2 NO.40-M91. Textile Test Methods Solvent-Extractable Material. 7 pp.

—Solvent Content
SAA AS 2001.3.4-81. Methods of Test for Textiles—Part 3: Chemical Tests—Part 3.4: Determination of Solvent-Soluble Matter. 3 pp.

—Staining
CNS L1006-82. Grey Scale for Assessing Staining (Feb)(3840).
JIS L 0805-83. Grey Scale for Assessing Staining. 7 pp.

—Standard Atmospheres
BSI BS EN 20139-92. 1992 Textiles—Standard Atmospheres for Conditioning and Testing (ISO 139: 1973). 8 pp.
CEN EN 20139-92. Textiles—Standard Atmospheres for Conditioning and Testing. 3 pp.
CGSB CAN/CGSB-4.2 NO.2-M88. Textile Test Methods Conditioning Textile Materials for Testing. 9 pp.

INDUSTRY STANDARDS

Fabrics

Fabrics (Cont.)

—Standard Atmospheres (Cont.)
- DIN ENGL 53802 (S)-79. Testing of Textiles; Conditioning of Samples to the Standard Atmosphere (July) Superseded by EN 20139 September 1992 Edition. 2 pp.
- DIN ENGL EN 20139-92. Textiles; Standard Atmospheres for Conditioning and Testing (ISO 139: 1973) (Supersedes DIN 53802, July 1979 Edition). 4 pp.
- ISO 139-73. Textiles—Standard Atmospheres for Conditioning and Testing First Edition. 3 pp.

—Statistical Analysis
- DIN ENGL 53804 Pt 1-81. Statistical Interpretation; Measureable (Continuous) Characteristics (Sept) (Supersedes DIN 53804/January 1961 Edition). 15 pp.

—Steaming
- CGSB CAN/CGSB-4.2 NO.33-M86. Textile Test Methods Pressing—Selection of Methods. 6 pp.
- CGSB CAN/CGSB-4.2 NO.33.2-M86. Textile Test Methods Method of Pressing Steam Pressing. 7 pp.
- CGSB CAN/CGSB-4.2 NO.33.3-M86. Textile Test Methods Method of Pressing Steaming. 6 pp.

—Steaming—Colorfastness Testing
- CNS L3154-82. Method of Test for Colour Fastness to Steaming (Jan)(8433).
- CNS L3179-82. Method of Test for Colour Fastness to High Temperature Steaming (Jun)(9022).
- JIS L 0869-77. Testing Method for Colour Fastness to Steaming.
- JIS L 0878-75. Testing Method for Colour Fastness to High Temperature Steaming (R 1983). 9 pp.

—Steaming—Dimensional Stability
- CGSB CAN/CGSB-4.2 NO.25.2-M89. Textile Test Methods Dimensional Change of Textile Fabrics to Open-Head Steaming. 8 pp.

—Stiffness Testing
- BSI BS 3356-90. 1990 Amd 1 Determination of Bending Length and Flexural Rigidity of Fabrics (AMD 6337) June 28, 1991. 9 pp.
- CNS L3084-80. Method of Test for Stiffness of Fabrics (May)(5615).
- SAA AS 2001.2.9-77. Methods of Test for Textiles—Part 2: Physical Tests—Part 2.9: Determination of Stiffness of Cloth Reconfirmed 1986. 4 pp.

—Strength Loss
- CGSB CAN/CGSB-4.2 NO.55-M90. Textile Test Methods Loss in Strength and Colour Change of Fabrics Due to Retained Chlorine. 13 pp.

—Stripping—Hydrochloric Acid
- DIN ENGL 54278 Pt 2-78. Testing of Textiles; Coatings and Attendant Materials; Determination of Condensation Products Which Can Be Stripped by Means of Hydrochloric Acid (Feb). 2 pp.

—Submarines
- MOD UK NES 129-84. Requirements for Soft Furnishings and Fabrics Issue 2 (04.84). 28 pp.

—Sulfate Content—Gravimetric Analysis
- BSI BS 3266-81. 1981 Determination of Conductivity, pH, Water Soluble Matter, Chloride and Sulphate in Aqueous Extracts of Textile Materials. 9 pp.

—Sulfur Content
- CNS L3166-82. Determination of the Contents of Ash and Sulfur in Textile Materials (Apr)(8709).

—Tear Strength
- BSI BS 4253-67. 1967 Amd 1 Ballistic Tear Testing of Woven Fabrics. 13 pp.
- BSI BS 4303-68. 1968 Method for the Determination of the Resistance to Tearing of Woven Fabrics by the Wing-Rip Technique. 12 pp.
- BSI BS 7304: Part 4-90. (WITHDRAWN) 1990 Methods of Test for Nonwovens Part 4: Determination of Tear Resistance (Renumbered as BS EN 29073-4: 1992). 8 pp.
- BSI BS EN 29073-4: 1992 Amd 1 Methods of Test for Nonwovens Part 4: Determination of Tear Resistance (ISO 9073-4: 1989) (AMD 7011) November 15, 1992. 13 pp.
- CGSB CAN/CGSB-4.2 NO.12.1-M90. Textile Test Methods Tearing Strength—Single-Rip Method. 10 pp.
- CGSB CAN/CGSB-4.2 NO.12.2-M90. Textile Test Methods Tearing Strength—Trapezoid Method. 10 pp.
- CGSB CAN/CGSB-4.2-M77METH12.3-84. Textile Test Methods Tearing Strength (Elmendorf Ballistic Method) (R 1991). 11 pp.
- DIN ENGL 53859 Pt 1-79. Testing of Textiles; Tear Growth Testing of Textile Fabrics; Tongue Tear Growth Test (Jan). 4 pp.
- DIN ENGL 53859 Pt 2-79. Testing of Textiles; Tear Growth Testing of Textile Fabrics; Leg Tear Growth Test (Jan). 3 pp.

Fabrics (Cont.)

—Tear Strength (Cont.)
- ISO 9073 Pt 4-89. Textiles—Test Methods for Nonwovens—Part 4: Determination of Tear Resistance First Edition. 6 pp.
- ISO 9290-90. Textiles—Woven Fabrics—Determination of Tear Resistance by the Falling Pendulum Method First Edition. 8 pp.
- SAA AS 2001.2.8-85. Methods of Test for Textiles—Part 2: Physical Tests—Part 2.8: Determination of Tear Resistance of Woven Textile Fabrics by the Falling Pendulum (Elmendorf) Apparatus. 6 pp.
- SAA AS 2001.2.10-86. Methods of Test for Textiles—Part 2: Physical Tests—Part 2.10: Determinaiton of the Tear Resistance of Woven Textile Fabrics by the Wing-Rip Method. 2 pp.

—Tensile Testing
- BSI BS 7304: Part 3-90. (WITHDRAWN) 1990 Methods of Test for Nonwovens Part 3: Determination of Tensile Strength and Elongation (Renumbered as BS EN 29073-3: 1992). 8 pp.

—Test Specimens
- BSI BS 4658-78. 1978 Amd 1 Methods of Test for Textiles. Preparation of Laboratory Test Samples and Test Specimens for Chemical Testing. 8 pp.
- CGSB CAN/CGSB-4.2 NO.1-M87. Textile Test Methods Precision and Accuracy of Measurements. 16 pp.
- CGSB CAN/CGSB-4.2 NO.2-M88. Textile Test Methods Conditioning Textile Materials for Testing. 9 pp.
- CNS L3099-80. Preparation, Marking and Measuring of Fabric Specimens in Tests for Determination of Dimensional Change (Dec)(6685).
- ISO 5089-77. Textiles—Preparation of Laboratory Test Samples and Test Specimens for Chemical Testing First Edition. 5 pp.

—Thermal Resistance
- BSI BS 4745-90. 1990 Determination of Thermal Resistance of Textiles. 12 pp.
- ISO 5085 Pt 1-89. Textiles—Determination of Thermal Resistance—Part 1: Low Thermal Resistance First Edition. 11 pp.
- ISO 5085 Pt 2-90. Textiles—Determination of Thermal Resistance—Part 2: High Thermal Resistance First Edition. 9 pp.

—Thickness Measurement
- BSI BS 2544-87. 1987 Amd 2 Determination of Thickness of Textile Materials (AMD 6342) May 31, 1991. 10 pp.
- BSI BS 7304: Part 2-90. (WITHDRAWN) 1990 Methods of Test for Nonwovens Part 2: Determination of Thickness (Renumbered as BS EN 29073-2: 1992). 11 pp.
- BSI BS EN 29073-2-92. 1992 Amd 1 Methods of Test for Nonwovens Part 2: Determination of Thickness (ISO 9073-2: 1989) (AMD 7009) November 15, 1992 (L). 17 pp.
- CGSB CAN/CGSB-4.2 NO.37-M87. Textile Test Methods Fabric Thickness. 8 pp.
- CNS L3090-80. Measurement of Thickness of Fabrics (Aug)(6245).
- DIN ENGL 53855 Pt 2 (S)-82. Testing of Textiles; Determination of the Thickness of Textile Fabrics with a Density Not Exceeding 0,05 g/cm3 (Oct) (Superseded by DIN EN 29073 Part 2). 4 pp.
- DIN ENGL 53855 Pt 3-79. Testing of Textiles; Determination of Thickness of Textile Fabrics; Floor Coverings (Jan). 2 pp.
- ISO 5084-77. Textiles—Determination of Thickness of Woven and Knitted Fabrics (Other Than Textile Floor Coverings) First Edition. 4 pp.
- ISO 9073 Pt 2-89. Textiles—Test Methods for Nonwovens—Part 2: Determination of Thickness First Edition. 7 pp.
- SAA AS 1587-73. Methods for Measurements of Textile Fabrics—Length, Width, Thickness, Mass Per Unit Length and Mass Per Unit Area Corrig. Amdt 1 June 1987. 12 pp.
- SAA AS 2001.2.15-89. Methods of Test for Textiles—Part 2: Physical Tests—Part 2.15: Determination of Thickness of Textile Fabrics. 7 pp.

—Thread Counts
- CEN PREN 1049-2-93. Textiles—Woven Fabrics—Construction—Methods of Analysis—Part 2: Determination of Number of Threads per Unit Length. 10 pp.
- CGSB CAN/CGSB-4.2 NO.6-M89. Textile Test Methods Textiles—Woven Fabrics—Construction—Methods of Analysis—Part 2: Determination of Number of Threads Per Unit Length (ISO 7211/2-1984). 20 pp.
- ISO 7211 Pt 2-84. Textiles—Woven Fabrics—Construction—Methods of Analysis—Part 2: Determination of Number of Threads per Unit Length First Edition. 8 pp.

Fabrics (Cont.)

—Thread Counts (Cont.)
- SAA AS 2001.2.5-91. Methods of Test for Textiles—Part 2: Physical Tests—Part 2.5: Determination of the Number of Threads per Unit Length in Woven Fabric. 4 pp.

—Toys—Colorfastness Testing
- CNS Z8016-9-86. Method of Test for Toy Safety (Testing for Color Fastness of Textile Materials to Perspiration) (Jun)(4798-9). 2 pp.

—Toys—Cracking (Fracturing)
- CNS Z8016-8-86. Method of Test for Toy Safety (Testing for Wet/Dry Crocking of Textile Materials) (Jun)(4798-8). 2 pp.

—Uniforms—Linings
- BSI BS 4560-90. 1990 Amd 1 Fabrics for Linings in Uniform Clothing. 13 pp.

—Upholstery
- BSI BS 2543-91. 1991 Woven and Knitted Fabrics for Upholstery. 15 pp.
- BSI BS 2543-01. 1991 Amd 1 Woven and Knitted Fabrics for Upholstery (AMD 6771) April 1, 1992. 16 pp.
- BSI BS 2543-02. 1991 Amd 2 Woven and Knitted Fabrics for Upholstery (AMD 7646) May 15, 1993. 17 pp.

—Wales
- CGSB CAN/CGSB-4.2 NO.7-M88. Textile Test Methods Knitted Fabric Count—Wales and Courses Per Centimetre. 7 pp.
- SAA AS 2001.2.6-81. Methods of Test for Textiles—Part 2: Physical Tests—Part 2.6: Determination of the Number of Wales and Courses per Unit Length in Knitted Fabric. 3 pp.

—Water Resistance Testing
- BSI BS 2823-82. (WITHDRAWN) 1982 Resistance of Fabrics to Penetration by Water (Hydrostatic Head Test) (Renumbered as BS EN 20811: 1992). 7 pp.
- BSI BS 3449-90. 1990 Method for Resistance of Fabrics to Water Absorption (Static Immersion Test). 4 pp.
- BSI BS 3702-82. (WITHDRAWN) 1982 Determination of Resistance of Textile Fabrics to Surface Wetting (Spray Test) (Renumbered as BS EN 20811). 6 pp.
- BSI BS 5066-74. 1974 Method of Test for the Resistance of Fabrics to an Artificial Shower. 14 pp.
- BSI BS 7209-90. 1990 Water Vapour Permeable Apparel Fabrics. 14 pp.
- BSI BS EN 20811-92. 1992 Amd 1 Textiles—Determination of Resistance to Water Penetration—Hydrostatic Pressure Test (ISO 811: 1981) (AMD 7006) November 15, 1992 (Supersedes BS 2823: 1982). 12 pp.
- BSI BS EN 24920-92. 1992 Amd 1 Textiles—Determination of Resistance to Surface Wetting (Spray Test) of Fabrics (ISO 4920: 1981) (AMD 7004) November 15, 1992. 12 pp.
- CEN EN 20811-92. Textiles—Determination of Resistance to Water Penetration—Hydrostatic Pressure Test. 5 pp.
- CEN EN 24920-92. Textiles—Determination of Resistance to Surface Wetting (Spray Test) of Fabrics. 5 pp.
- CGSB CAN/CGSB-4.2 NO.26.1-M88. Textile Test Methods Water Resistance—Static Head Penetration Test. 9 pp.
- CGSB CAN/CGSB-4.2 NO.26.3-M88. Textile Test Methods Water Resistance—Variable Head Penetration Test. 10 pp.
- CGSB CAN/CGSB-4.2 NO.26.4-M88. Textile Test Methods Water Resistance—Bundesmann Test. 10 pp.
- CGSB CAN/CGSB-4.2 NO.49-M91. Textile Test Methods Resistance of Materials to Water Vapour Diffusion. 10 pp.
- DIN ENGL EN 20811-92. Determination of Resistance of Textile Fabrics to Water Penetration; Hydrostatic Pressure Test (ISO 811: 1981) (Aug). 6 pp.
- DIN ENGL EN 24920-92. Determination of Resistance to Surface Wetting of Textile Fabrics (Spray Test); (ISO 4920: 1981) (Aug). 6 pp.
- ISO 811-81. Textile Fabrics—Determination of Resistance to Water Penetration—Hydrostatic Pressure Test First Edition. 5 pp.
- ISO 4920-81. Textile Fabrics—Determination of Resistance to Surface Wetting (Spray Test) First Edition. 5 pp.
- ISO 9865-91. Textiles—Determination of Water Repellency of Fabrics by the Bundesmann Rain-Shower Test First Edition; (Corrected and Reprinted -1992). 7 pp.
- SAA AS 2001.2.16-87. Methods of Test for Textiles—Part 2: Physical Tests—Part 2.16: Determination of Water Repellency of Textile Surfaces—Spray Rating Test. 3 pp.

INTERNATIONAL AND NON-U.S. NATIONAL STANDARDS
SUBJECT INDEX

Fabrics *(Cont.)*
—**Water Resistance Testing** *(Cont.)*
SAA AS 2001.2.17-87. Methods of Test for Textiles—Part 2: Physical Tests—Part 2.17: Determination of Resistance of Fabrics to Water Penetration—Hydrostatic Pressure Test. 3 pp.
SAA AS 2001.2.18-87. Methods of Test for Textiles—Part 2: Physical Tests—Part 2.18: Determination of Resistance of Fabrics to Water Penetration—Cone Test. 2 pp.
SAA AS 2001.2.24-90. Methods of Test for Textiles—Part 2: Physical Tests—Part 2.24: Determination of Resistance of Textile Fabrics to Water Vapour Diffusion—Control Dish Method. 7 pp.

—**Wettability**
BSI BS 4554-70. 1970 Method of Test for Wettability of Textile Fabrics. 17 pp.
CGSB CAN/CGSB-4.2 NO.26.2-M87. Textile Test Methods Water Resistance—Surface Wetting Test—Spray Test. 10 pp.

—**Width Measurement**
BSI BS 1930-90. 1990 Method for Determination of Width of Woven or Knitted Fabrics When Relaxed at Zero Tension. 8 pp.
CGSB CAN/CGSB-4.2 NO.4.1-M87. Textile Test Methods Textile Fabrics—Measurement of Width of Pieces. 11 pp.
CNS L3088-80. Measurement of Width of Fabrics (Aug)(6243).
ISO 3932-76. Textiles—Woven Fabrics—Measurement of Width of Pieces First Edition. 6 pp.
SAA AS 1587-73. Methods for Measurements of Textile Fabrics—Length, Width, Thickness, Mass Per Unit Length and Mass Per Unit Area Corrig. Amdt 1 June 1987. 12 pp.
SAA AS 2001.2.12-87. Methods of Test for Textiles—Part 2: Physical Tests—Part 2.12: Determination of Width of Fabrics. 2 pp.

—**Wrappers—Sterilizations**
CSA CAN/CSA-Z314.10-M90. Selection, Use, Maintenance, and Laundry of Reusable Textile Wrappers for Sterilization Products in Health Care Facilities; (Gen Instr 1). 35 pp.

—**Wrinkling—Visual Inspection**
ISO 9867-91. Textiles—Evaluation of the Wrinkle Recovery of Fabrics—Appearance Method First Edition. 8 pp.

—**Yarn Slippage**
BSI BS 3320-88. 1988 Amd 1 Method for the Determination of Slippage Resistance of Yarns in Woven Fabrics: Seam Method (AMD 6261) May 31, 1990. 9 pp.
JIS L 1062-87. Testing Methods for Distortion and Slippage of Yarn in Woven Fabrics.

Facades
See Also: Walls
ISO 7361-86. Performance Standards in Building—Presentation of Performance Levels of Facades Made of Same-Source Components First Edition. 12 pp.

—**Acoustical Insulation**
BSI BS 2750: Part 5-80. 1980 Measurement of Sound Insulation in Buildings and of Building Elements Part 5: Field Measurements of Airborne Sound Insulation of Facade Elements and Facades. 12 pp.
ISO 140 Pt V-78. Acoustics—Measurement of Sound Insulation in Buildings and of Building Elements—Part V: Field Measurements of Airborne Sound Insulation of Facade Elements and Facades First Edition. 10 pp.
SNZ NZS/ISO 140: Part 5-78. Acoustics Part 5: Field Measurements of Airborne Sound Insulation of Facade Elements and Facades. 8 pp.

—**Wind Resistance**
ISO 7895-87. Facades Made of Components—Tests for Resistance to Positive and Negative Static Pressure Generated by Wind First Edition. 10 pp.

Face Bend Testing
Use: Bend Testing

Face Masks
Use: Face Shields

Face Paints
See Also: Paints

—**Camouflage**
MOD UK DSTAN 68-92-01. Paint, Face, Camouflage, Brown Issue 1; Amendment 1. 13 pp.

Face Plates
Use: Wall Plates

Face Plates (Identification Marking)
Use: Name Plates

Face Protectors
Use: Face Shields

Face Shields
Use For: Face Masks; Face Protectors
See Also: Breathing Apparatus; Helmets; Protective Clothing; Sports Equipment; Visors
BSI BS 4532-69. 1969 Amd 1 Snorkels and Face Masks. 13 pp.
BSI BS 6016-80. (WITHDRAWN) 1980 Filtering Facepiece Dust Respirators (Superseded by BS EN 149: 1992). 16 pp.
BSI DD 97: Part 14-87. (WITHDRAWN) 1987 Respiratory Protective Equipment Part 14: Power Assisted Particle Filtering Devices Incorporating Full Face Masks, Half Masks and Quarter Masks (Superseded by BS EN 147: 1992). 14 pp.
CNS Z2019-87. Safety Face Shield (Jul)(3504).

—**Breathing Apparatus—Aerospace**
MOD UK DTD-5614A-82. Silicone Rubber Facepieces for Oxygen Masks. 3 pp.

—**Diving Equipment**
CNS S1214-89. Diving Mask (Feb)(12497). 2 pp.
CNS S2141-89. Method of Test for Diving Mask (Feb)(12498). 3 pp.

—**Hockey**
CSA CAN/CSA-Z262.2-M90. Face Protectors and Visors for Ice Hockey Players; (Gen Instr 1). 44 pp.

—**Industrial**
CSA CAN/CSA-Z94.3-92. Industrial Eye and Face Protectors; (Gen Instr 1). 64 pp.

—**Lacrosse**
CSA CAN/CSA-Z262.2-M90. Face Protectors and Visors for Ice Hockey Players; (Gen Instr 1). 44 pp.

—**Police**
CSA CAN/CSA-Z611-M86. Police Riot Helmets and Faceshield Protection; (Gen Instr 1). 28 pp.

Face Wrenches
See Also: Wrenches

—**Lockrings**
CNS B3325-80. Assembly Tools for Screws and Nuts for Slotted Lock Rings (Jan)(5104).

—**Pin Type**
CNS B3326-80. Assembly Tools for Screws and Nuts Face Wrench (Jan)(5105).

Faceplates (Electrical)
Use: Wall Plates

Faceplates (Identification Marking)
Use: Name Plates

Faceplates (Machine Tool Components)
—**Bayonet Type**
BSI BS 4442: Part 2-69. (WITHDRAWN) 1969 Amd 1 Lathe Spindle Noses and Faceplates Part 2:. 13 pp.
ISO 702 Pt III-75. Machine Tools—Spindle Noses and Face Plates—Sizes for Interchangeability—Part III: Bayonet Type First Edition. 10 pp.

—**Camlock Type**
BSI BS 4442: Part 1-69. 1969 Lathe Spindle Noses and Faceplates Part 1: Types A and Camlock. 19 pp.
DIN ENGL 55029-80. Machine Tools; Spindle Noses and Face Plates with Centering Taper; Camlock Type, Accessories; Dimensions (Mar). 10 pp.
ISO 702 Pt II-75. Machine Tools—Spindle Noses and Face Plates—Sizes for Interchangeability—Part II: Camlock Type First Edition. 16 pp.

—**Lathes**
CNS B3398-83. Spindle Noses and Face Plates of Latches (May)(6876).
JIS B 6109-82. Spindle Noses and Face Plates of Lathes. 19 pp.

—**Type A**
BSI BS 4442: Part 1-69. 1969 Lathe Spindle Noses and Faceplates Part 1: Types A and Camlock. 19 pp.
ISO 702 Pt I-75. Machine Tools—Spindle Noses and Face Plates—Sizes for Interchangeability—Part I: Type A First Edition. 7 pp.

Facial Tissues
See Also: Paper Products; Papers
CGSB CAN/CGSB-9.4-92. Facial Tissue. 7 pp.
CNS P2054-89. Facial Tissue (Dec)(4150). 2 pp.
JIS S 3104-85. Facial Tissues.

Facing Bricks
See Also: Bricks

—**Calcium Silicate**
DIN ENGL 106 Pt 2-80. Sandlime Bricks and Blocks; Facing Bricks and Hard-Burnt Facing Bricks (Nov). 6 pp.

—**Frost Resistance—Freeze Thaw Testing**
DIN ENGL 52252 Pt 1-86. Testing the Frost Resistance of Facing Bricks and Clinker Blocks; Freezing of Single Bricks on All Sides (Dec). 4 pp.
DIN ENGL 52252 Pt 2-86. Testing the Frost Resistance of Facing Bricks and Clinker Blocks; Freezing of Bricks Arranged in Test Blocks (Dec). 5 pp.
DIN ENGL 52252 Pt 3-86. Testing the Frost Resistance of Facing Bricks and Clinker Blocks; Freezing of Test Walls on One Side (Dec). 5 pp.

Facsimile Communications
Use: FAX Communications

Facsimile Equipment
Use: FAX Machines

Facsimile Modules
Use: FAX Modules

Facsimile Packet Assembly/Disassembly Facility
Use: Packet Assembly/Disassembly Devices—FAX Machines

Facsimile Transmission
Use: FAX Communications

Factories
Use: Industrial Plants

Fading
See Also: Attenuation; Multipath Fading; Radio Communications; Radio Waves

—**Feeder Links—Broadcasting Satellite Services**
CCIR RECMN 794-92. Techniques for Minimizing the Impact on the Overall BSS System Performance Due to Rain Along the Feeder-Link Path. 11 pp.

—**Radio Communications**
CCIR Report 266-7-90. Ionospheric Propagation and Noise Characteristics Pertinent to Terrestrial Radiocommunication Systems Design and Service Planning (Fading)—Section 6E—Ionospheric Propagation Prediction at Frequencies Between About 1.6 and 30 MHz. 19 pp.
CCIR QUESTION 35/6-90. Variations of Ionospheric Propagation Characteristics and Fading—Questions Concerning Study Group 6—Radio Wave Propagation in Ionized Media. 1 p.

—**Radio Equipment**
CCIR RECMN 339-6-86. Bandwidths, Signal-to-Noise Ratios and Fading Allowances in Complete Systems—Section 3Aa—Technical Characteristics. 3 pp.

—**Rebroadcasting—Single Sideband Reception—Tropical Regions**
CCIR QUESTION 66/10-90. Single-Sideband Reception for Reduction of Fading Effects for Re-Broadcast Applications Within the Tropical Zone—Questions Concerning Study Group 10—Broadcasting Service (Sound). 1 p.

—**Sound Broadcasting**
CCIR RECMN 411-4-90. Fading Allowances in HF Broadcasting—Section 10A-1—Amplitude-Modulation Sound Broadcasting in Bands 5 (LF), 6 (MF) and 7 (HF). 1 p.

—**Sound Broadcasting—Tropical Regions**
CCIR Report 304-3-90. Fading Characteristics for Sound Broadcasting in the Tropical Zone—Section 10A-2—Sound Broadcasting in the Tropical Zone. 18 pp.
CCIR QUESTION 43/6-90. Fading Characteristics of Sound Broadcasting in the Tropical Zone—Questions Concerning Study Group 6—Radio Wave Propagation in Ionized Media. 1 p.

Failed Calls
Use: Unsuccessful Calls

Failure (Quality Control)
Use For: Failure Mode and Effects
See Also: Corrosion; Cracking (Fracturing); Error Control; Error Correction; Error Recovery (Telecommunications); Failure Rate; Fault Tree Analysis; Quality Assurance; Screening Testing; Short Circuits; Wear

Failure (Quality Control) (Cont.)

BSI BS 5760: Part 5-91. 1991 Reliability of Systems, Equipment and Components Part 5: Guide to Failure Modes, Effects and Criticality Analysis (FMEA and FMECA) (G). 47 pp.
CENELEC HD 485-87. Analysis Techniques for System Reliability—Procedure for Failure Mode and Effects Analysis (FMEA). 3 pp.
CENELEC HD 485 S1-87. Analysis Techniques for System Reliability—Procedure for Failure Mode and Effects Analysis (FMEA). 4 pp.
IEC 812-85. Analysis Techniques for System Reliability—Procedure for Failure Mode and Effects Analysis (FMEA) First Edition. 44 pp.

—Aircraft—Altitude Measuring Instruments
EUROCAE ED-26 03.79. MPS for Airborne Altitude Measurement and Coding Systems. 57 pp.

—Aircraft—Area Navigation Systems
EUROCAE ED-27 10.79. MOPR for Airborne Area Navigation Systems, Based on VOR and DME as Sensors. 63 pp.
EUROCAE ED-39 06.84. MOPR for Airborne Area Navigation Systems, Based on Two DME as Sensors. 67 pp.

—Aircraft—Automatic Pilots
CAA 741 (K). Section K Light Aeroplanes Autopilots and Flight Directors (Blue Papers). 15 pp.
CAA Chapter K6-4 App #1 10.92. Failure Characteristics (Light Aeroplanes). 2 pp.

—Aircraft—Circuit Breakers
CAA NOTICE #87 ISSUE 1. (This Notice Gives Details of a Mandatory Action) Failure of Mechanical Products Inc. Circuit Breakers (Airworthiness Notices). 2 pp.

—Aircraft—Omega Navigation System
EUROCAE ED-29 10.77. MPS for Airborne Omega Navigation Equipment. 45 pp.

—Ammunition
NATO STANAG 2940 ED 2 AMD 2-85. Reporting of Major Ammunition Malfunctions. 14 pp.
NATO STANAG 2940 ED 2 AMD 3-85. Reporting of Major Ammunition Malfunctions. 16 pp.

—Common Channel Signaling Systems
CCITT RECMN E.413-89. International Network Management—Planning —Telephone Network and ISDN—Quality of Service, Network Management and Traffic Engineering (Study Group II) 5 pp. 5 pp.

—International Exchanges
CCITT RECMN E.550-89. Grade-of-Service and New Performance Criteria Under Failure Conditions in International Telephone Exchanges—Telephone Network and ISDN—Quality of Service, Network Management and Traffic Engineering (Study Group II) 8 pp. 8 pp.

—Network Management Systems
CCITT RECMN E.413-89. International Network Management—Planning —Telephone Network and ISDN—Quality of Service, Network Management and Traffic Engineering (Study Group II) 5 pp. 5 pp.

—Rotary Wing Aircraft
CAA CAP524 SUB-Part D 12.86. (Rotorcraft). 31 pp.

—Supersonic Transports
CAA STANDARD NO. 1-1 & App. Airworthiness Objectives and System Analysis. 6 pp.

—Telecommunication Equipment
CCITT FASCICLE II.3-88. Telephone Network and ISDN Quality of Service, Network Management and Traffic Engineering. Recommendations E.401-E.880. 368 pp.

—Telecommunication Equipment—Tables (Data)
CCITT FASCICLE II.3-88. Telephone Network and ISDN Quality of Service, Network Management and Traffic Engineering. Recommendations E.401-E.880. 368 pp.

—Telecommunication Networks
CCITT RECMN E.862 (REV 1)-92. Dependability Planning of Telecommunication Networks (Study Group II) 16 pp. 16 pp.
CCITT RECMN E.862-89. Dependability Planning of Telecommunication Networks—Telephone Network and ISDN—Quality of Service, Network Management and Traffic Engineering (Study Group II) 12 pp. 12 pp.

—Telephone Networks
CCITT RECMN M.1530-92. Network Maintenance Information (Study Group IV) 8 pp. 8 pp.

Failure Mode and Effects
Use: Failure (Quality Control)

Failure Rate
See Also: Failure (Quality Control); Reliability

—Electrical Equipment
IEC 605 Pt 6-86. Equipment Reliability Testing Part 6: Tests for the Validity of a Constant Failure Rate Assumption First Edition; (Amendment 1-1989). 23 pp.
IEC 605 Pt 7-78. Equipment Reliability Testing Part 7: Compliance Test Plans for Failure Rate and Mean Time Between Failures Assuming Constant Fail-ure Rate Clause 6—Pro-cedures for Design and Application of Time Ter-minated Test Plans First Edition; (Amendment 1-1990). 77 pp.

—Electronic Components
CNS C5205-86. General Test Procedure of Failure Rate for Electronic Components (Jan)(11476).
JIS C 5003-74. General Test Procedure of Failure Rate for Electronic Components (R 1982); (Erratum). 13 pp.

—Glossaries
CNS B8006-1-85. Glossary of Terms for Reliability (Terms of Failure) (Oct)(11381-1).

—Resistors
JIS C 5721-91. Fixed Metal Film Resistors for Use in Electronic Equipment—Form 05, Grade A. 36 pp.

Failure Rate Testing
Use: Failure Rate

Failure Recovery (Telecommunications)
Use: Error Recovery (Telecommunications)

Fairchild Argus Aircraft
See Also: Aircraft; Fairchild 24R-46A Argus III Aircraft

—Foreign Airworthiness Directives
CAA. Fairchild Argus Series Aircraft. 1 p.
CAA. Fairchild SA227 Series Aircraft. 1 p.

Fairchild F-27 Aircraft
See Also: Aircraft

—Accidents
CAA. Fokker/Fairchild F-27 Series (World Airline Accident Summary). 2 pp.
CAA. Fokker/Fairchild F-27 Series. 2 pp.

Fairchild-Hiller FH-227 Aircraft
See Also: Aircraft

—Accidents
CAA. Fairchild Hiller FH-227 (World Airline Accident Summary). 1 p.
CAA. Fairchild Hiller FH-227. 1 p.

Fairchild SA227-AC Metro III Aircraft
See Also: Aircraft

—Certification
CAA. Fairchild SA227-AC Metro III. 19 pp.

—Data Sheets
CAA FA43 ISSUE 1. Fairchild Aircraft, Model SA227-AC, Metro 111. 2 pp.

Fairchild/Swearingen Merlin III B Aircraft
See Also: Aircraft

—Data Sheets
CAA FA25 ISSUE 1. Fairchild Swearingen Merlin IIIB. 2 pp.

Fairchild 24R-46A Argus III Aircraft
See Also: Aircraft; Fairchild Argus Aircraft

—Antenna Positions
CAA. Fairchild 24R-46A Argus III (Approved Aerial Positions). 1 p.

Fairey Reed Propellers
Use: Aircraft Propellers—Fairey Reed

Fairings
—Ships
MOD UK NES 789-88. Requirements for the Application of Structural Fairing and Filling Materials Issue 2 (05.88). 19 pp.

—Submarines
MOD UK NES 789-88. Requirements for the Application of Structural Fairing and Filling Materials Issue 2 (05.88). 19 pp.

Fairleads
See Also: Mooring Hardware; Ships
BSI BS MA 19-73. 1973 Amd 1 Panama-Type Fairleads. 7 pp.
BSI BS MA 21-73. 1973 Deck-End Roller Fairleads. 8 pp.
BSI BS MA 22-73. 1973 Cast Roller Fairleads. 11 pp.
BSI BS MA 23-73. 1973 Multi-Angle Roller Fairleads. 5 pp.
CNS F3109-81. Small Size Fairleads (May)(7374).
CNS F3122-81. Universal Fairleaders (Jul)(7694).
CNS F3125-81. Fairleads (Aug)(7787).
ISO 4127 Pt 1-79. Shipbuilding—Inland Navigation—Fairleads—Part 1: Two-Lip Fairleads First Edition. 4 pp.
ISO 4127 Pt 2-80. Shipbuilding—Inland Vessels—Fairleads—Part 2: Roller Fairleads First Edition. 5 pp.
JIS F 1011-84. Yacht's Fairleads.
JIS F 2014-87. Fairleads.
JIS F 2014-69. Fair-Leads. 10 pp.
JIS F 2021-76. Small Size Fairleads. 17 pp.
JIS F 2026-80. Fairleads with Horizontal Rollers.
JIS F 2026-76. Fairleads with Horizontal Rollers. 10 pp.

—Aircraft
SBAC AS 2823 ISSUE 3. Fairlead. 1 p.

Falco F8L Aircraft
See Also: Aircraft

—Antenna Positions
CAA. Falco F8L (Approved Aerial Positions). 1 p.

—Foreign Airworthiness Directives
CAA. Falco F8L Series (Foreign Airworthiness Directives). 1 p.

Fall Arresting Devices
See Also: Safety; Safety Belts
BSI BS EN 341-93. 1993 Personal Protective Equipment Against Falls From a Height—Descender Devices (N). 10 pp.
BSI BS EN 363-93. 1993 Personal Protective Equipment Against Falls from a Height—Fall Arrest Systems (N). 14 pp.
BSI BS EN 364-93. 1993 Personal Protective Equipment Against Falls from a Height—Test Methods (N). 21 pp.
CEN PREN 341-90. Personal Equipment for Protection Against Falls —Personal Equipment for Work Positioning and Prevention of Falls from a Height—Rescue Equipment—Descender Devices. 8 pp.
CEN EN 341-92. Personal Protective Equipment Against Falls from a Height—Descender Devices. 5 pp.
CEN PREN 363-90. Personal Fall Arresting Systems: Systems. 6 pp.
CEN EN 363-92. Personal Protective Equipment Against Falls from a Height—Fall Arrest Systems. 7 pp.
DIN ENGL EN 364-93. Personal Protective Equipment Against Fall from a Height; Test Methods (Feb). 26 pp.
SAA AS 1891.3-92. Industrial Fall-Arrest Systems and Devices—Part 3: Fall-Arrest Devices (in Professional Package 47). 16 pp.

—Connectors
BSI BS EN 362-93. 1993 Personal Protective Equipment Against Falls from a Height—Connectors (Supersedes BS 1397: 1979 & BS 5062: Part 1: 1985) (N). 10 pp.
CEN PREN 362-90. Personal Fall Arresting Systems: Connectors. 4 pp.
CEN EN 362-92. Personal Protective Equipment Against Falls from a Height—Connectors. 4 pp.

—Energy Absorbers
BSI BS EN 355-93. 1993 Personal Protective Equipment Against Falls From a Height—Energy Absorbers (N). 10 pp.
CEN PREN 355-90. Personal Fall Arresting Systems: Energy Absorbers. 6 pp.
CEN EN 355-92. Personal Protective Equipment Against Falls from a Height—Energy Absorbers. 4 pp.

—Guided—Anchorage Line
BSI BS EN 353-1-93. 1993 Personal Protective Equipment Against Falls from a Height: Guided Type Fall Arresters Part 1: Specification for Guided Type Fall Arresters on a Rigid Anchorage Line (Supersedes bs 5062: Part 1: 1985) (N). 11 pp.

Fall Arresting Devices (Cont.)

—Guided—Anchorage Line (Cont.)
BSI BS EN 353-2-93. 1993 Personal Protective Equipment Against Falls from a Height: Guided Type Fall Arresters Part 2: Specification for Guided Type Fall Arresters on a Flexible Anchorage Line (Supersedes BS 5062: Part 1: 1985) (N). 11 pp.
CEN EN 353-1-92. Personal Protective Equipment Against Falls from a Height—Guided Type Fall Arresters on a Rigid Anchorage Line. 5 pp.
CEN EN 353-2-92. Personal Protective Equipment Against Falls from a Height—Guided Type Fall Arresters on a Flexible Anchorage Line. 5 pp.

—Identification Systems
BSI BS EN 365-93. 1993 Personal Protective Equipment Against Falls from a Height—General Requirements for Instructions for Use and for Marking (Supersedes BS 1397: 1979 & BS 5062: Part 1 & 2: 1985) (N). 9 pp.
CEN PREN 365-90. Protection Against Falls from Height: Requirements for General Instructions for Use. 5 pp.
CEN EN 365-92. Personal Protective Equipment Against Falls from a Height—General Requirements for Instructions for Use and for Marking. 3 pp.

—Instructional Materials
ESI BS EN 365-93. 1993 Personal Protective Equipment Against Falls from a Height—General Requirements for Instructions for Use and for Marking (Supersedes BS 1397: 1979 & BS 5062: Part 1 & 2: 1985) (N). 9 pp.
CEN PREN 365-90. Protection Against Falls from Height: Requirements for General Instructions for Use. 5 pp.
CEN EN 365-92. Personal Protective Equipment Against Falls from a Height—General Requirements for Instructions for Use and for Marking. 3 pp.

—Mountaineering Equipment—Safety
CEN PREN 958-92. Mountaineering Equipment —Fall Arrestor System—Safety Requirements and Test Method. 5 pp.

—Self Locking—Retractable
BSI BS EN 360-93. 1993 Personal Protective Equipment Against Falls from a Height—Retractable Type Fall Arresters (Supersedes BS 5062: Part 1: 1985) (N). 10 pp.
CEN PREN 360-90. Personal Fall Arresting Systems: Self-Locking Retractable Type Fall Arrestors. 8 pp.
CEN EN 360-92. Personal Protective Equipment Against Falls from a Height—Retractable Type Fall Arresters. 4 pp.

—Self Locking—Sliding
CEN PREN 353-90. Personal Fall Arresting Systems: Self-Locking Sliding Type Fall Arrestors. 10 pp.

—Shock Absorbers
CSA CAN/CSA-Z259.11-M92. Shock Absorbers for Personal Fall Arrest Systems; (Gen Instr 1). 21 pp.

Fall Arresting Safety Belts
Use: Safety Belts

Fall Arresting Safety Lines
Use: Safety Lines

Fall Arrestors
Use: Fall Arresting Devices

Falling Object Protective Structures
Use For: FOPS See Also: Protective Equipment; Safety

—Construction Equipment
EC 86/296/EEC-86. Council Directive on the Approximation of the Laws of the Member States Relating to Falling-Object Protective Structures (FOPS) for Certain Construction Plant. 9 pp.

—Earthmoving Equipment
BSI BS 5526-85. (WITHDRAWN) 1985 Amd 1 Falling-Object Protective Structures on Earth Moving Machinery: Laboratory Test and Performance Requirements (AMD 6114) October 31, 1990 (Superseded by BS 6912: Part 6: 1992). 17 pp.
BSI BS 6912: Part 6-92. 1992 Safety of Earth-Moving Machinery Part 6: Specification for Falling-Object Protective Structures on Earth-Moving Machinery: Laboratory Tests and Performance Requirements (ISO 3449: 1992). 15 pp.
CNS A4025-88. Earth-Moving Machinery Falling-Object Protective Structures Laboratory Tests for Performance Requirements (Dec)(9956).
ISO 3449-92. Earth-Moving Machinery—Falling-Object Protective Structures—Laboratory Tests and Performance Requirements Fourth Edition. 12 pp.
JIS A 8920-90. Earth-Moving Machinery—Falling-Object Protective Structures—Laboratory Tests and Performance Requirements (ISO 3449-1984).

Falling Object Protective Structures (Cont.)

—Earthmoving Equipment—Deflection
BSI BS 5495-80. 1980 Amd 1 Laboratory Evaluations of Roll-Over and Falling-Object Protective Structures: The Deflection-Limiting Volume for Earth-Moving Machinery (ISO 3164: 1979) (AMD 3711) October 30, 1981. 9 pp.
BSI BS 5495-02. (WITHDRAWN) 1980 Amd 2 Laboratory Evaluations of Roll-Over and Falling-Object Protective Structures: the Deflection-Limiting Volume for Earth-Moving Machinery (ISO 3164: 1979) (AMD 6878) February 28, 1992 (Superseded by BS 6912: Part 7: 1992). 14 pp.
BSI BS 6912: Part 7-92. 1992 Safety of Earth-Moving Machinery Part 7: Specification for Laboratory Evaluations of Roll-Over and Falling-Object Protective Structures: the Deflection-Limiting Volume for Earth-Moving Machinery (ISO 3164: 1992). 7 pp.
CNS A4006-81. Deflection Limiting Volume of Earth-Moving Machinery-Roll-Over Protective Structure for Driver (Sep)(7850).
CNS A4024-88. Earth-Moving Machinery Laboratory Evaluations of Roll-Over and Falling-Object Protective Structures Specifications for the Deflection-Limiting Volume (Dec)(9955).
ISO 3164-92. Earth-Moving Machinery—Laboratory Evaluations of Roll-over and Falling-Object Protective Structures—Specifications for Deflection-Limiting Volume Fourth Edition; (Incorporating Amendment 1). 6 pp.

—Forestry Equipment
ISO 8083-89. Machinery for Forestry—Falling-Object Protective Structures—Laboratory Tests and Performance Requirements First Edition. 8 pp.

—Forklifts
JIS D 6021-84. Overhead Guards for Fork Lift Trucks.

—Lift Trucks
BSI BS 5933-80. 1980 Overhead Guards for High-Lift Rider Trucks. 4 pp.
ISO 6055-79. High-Lift Rider Trucks—Overhead Guards—Specification and Testing First Edition. 4 pp.

Falling Weight Testing
Use: Impact Testing

False Celings
Use: Ceilings

False Teeth
Use: Dentures

Falsework
See Also: Construction Equipment; Temporary Structures
BSI BS 5975-82. 1982 Amd 2 Code of Practice for Falsework (AMD 5415) December 31, 1986. 87 pp.
CSA S269.1-1975. Falsework for Construction Purposes. 40 pp.
DIN ENGL 4421-82. Falsework; Calculation, Design and Construction (Aug). 20 pp.

—Floor Centers—Mechanical Testing
BSI BS 5507: Part 1-77. 1977 Amd 1 Methods of Test for False Work Equipment Part 1: Floor Centres (AMD 5360) May 31, 1989. 18 pp.

—Props
BSI BS 5507: Part 3-82. 1982 Methods of Test for False Work Equipment Part 3: Props. 7 pp.

—Tubular Steel
CENELEC HD 1039-90. Steel Tubes for Falsework and Working Scaffolds—Requirements, Tests. 11 pp.

Family Practice
Use For: General Practice

—Training
EC 86/457/EEC-86. Council Directive on Specific Training in General Medical Practice. 4 pp.

Fan Beams
See Also: Radio Communications; Radio Transmission; Satellite Communications; Spot Beams

—Multiple—Mobile Satellite Communications
CCIR Report 1172-90. Intersystem Frequency Sharing and Reuse in the Mobile-Satellite Services Operating at Mid to High Portions of Band 9—Section 8H—Efficient Use of the Radio Spectrum Characteristics and Sharing of Frequency Resources. 25 pp.

Fan Coil Units
See Also: Fans; Heat Exchangers; Heating, Ventilating and Air Conditioning Equipment
CSA CAN/CSA-C22. 2 NO 236-M90. Heating and Cooling Equipment; (Gen Instr 1) (ANSI/UL 1995). 136 pp.
JIS A 4008-87. Fancoil Units. 35 pp.

—Air Flow
BSI BS 4856: Part 1-72. 1972 Methods of Testing and Rating Fan Coil Units; Unit Heaters and Unit Coolers Part 1: Thermal and Volumetric Performance for Heating Duties: Without Additional Ducting. 26 pp.
SNZ NZS 4856: Part 1-72. Methods for Testing and Rating Fan Coil Units, Unit Heaters and Unit Coolers Part 1: Thermal and Volumetric Performance for Heating Duties; Without Additional Ducting. 24 pp.

—Noise
BSI BS 4856: Part 4-78. 1978 Methods of Testing and Rating Fan Coil Units; Unit Heaters and Unit Coolers Part 4: Acoustic Performance; Without Additional Ducting. 12 pp.
BSI BS 4856: Part 5-79. 1979 Methods of Testing and Rating Fan Coil Units; Unit Heaters and Unit Coolers Part 5: Acoustic Performance; with Ducting. 16 pp.

—Thermal Measurement
BSI BS 4856: Part 1-72. 1972 Methods of Testing and Rating Fan Coil Units; Unit Heaters and Unit Coolers Part 1: Thermal and Volumetric Performance for Heating Duties: Without Additional Ducting. 26 pp.
SNZ NZS/BS 4856: Part 1-72. Methods for Testing and Rating Fan Coil Units, Unit Heaters and Unit Coolers Part 1: Thermal and Volumetric Performance for Heating Duties; Without Additional Ducting. 24 pp.

Fan Convectors
Use: Convectors

Fan Jet Falcon Aircraft
See Also: Aircraft

—Data Sheets
CAA FA12 ISSUE 6. Fan Jet Falcon Series E and F. 3 pp.

Fan Jet Falcon 20 Series D Aircraft
See Also: Aircraft

—Antenna Positions
CAA. Falcon 20 Series E, F and D (Approved Aerial Positions). 1 p.

—Certification
CAA. Avion Marcel Dassault Fan Jet Falcon (Falcon 20) Basic, Series D, Series E and Series F. 13 pp.

Fan Jet Falcon 20 Series E Aircraft
See Also: Aircraft

—Antenna Positions
CAA. Falcon 20 Series E, F and D (Approved Aerial Positions). 1 p.

—Certification
CAA. Avion Marcel Dassault Fan Jet Falcon (Falcon 20) Basic, Series D, Series E and Series F. 13 pp.

Fan Jet Falcon 20 Series F Aircraft
See Also: Aircraft

—Certification
CAA. Avion Marcel Dassault Fan Jet Falcon (Falcon 20) Basic, Series D, Series E and Series F. 13 pp.

Fans
Use For: Circulating Fans; Electric Fans; Turbo Fans
See Also: Air Handling Equipment; Axial Fans; Blowers (Ventilation); Ceiling Fans; Centrifugal Fans; Cooling Systems; Fan Coil Units; Jet Fans; Materials Handling Equipment; Radiators (Heating); Refrigeration Equipment; Rotary Compressors; Rotary Pumps; Ventilation Equipment
CNS B4046-81. Turbo-Fans and Blowers (Aug)(7778).
CNS B7165-81. Methods of Test for Turbo-Fans and Blowers (Aug)(7779).
CNS C4008-86. Electric Fan (Desk and Wall Mounting Type) (Nov)(547). 9 pp.
CNS C4021-86. Floor Fan (Jan)(1028).
CNS C4046-86. Ventilating Fan (Jan)(2060). 6 pp.
CNS C4047-86. Stand Fan (Jan)(2061). 6 pp.
CNS C4371-86. Box Electric Fan (Jan)(9578). 7 pp.
DIN ENGL 24163 Pt 1-85. Fans; Performance Testing; Standard Characteristics (Jan). 14 pp.
DIN ENGL 24166-89. Fans; Technical Delivery Conditions (Jan). 12 pp.
JIS B 8330-81. Testing Methods for Turbo-Fans and Blowers. 36 pp.

INTERNATIONAL AND NON-U.S. NATIONAL STANDARDS
SUBJECT INDEX

Fans

Fans (Cont.)
JIS C 9601-90. Electric Fans. 29 pp.
JIS C 9603-88. Ventilating Fans. 27 pp.

—**Acceptance Testing**
SAA AS/NZS 3302-92. Approval and Test Specification—Particular Requirements for Electric Fans (Supersedes AS 3302—1988) Amdt 2 September 1993 (in Professional Package 28). 9 pp.
SNZ NZS/AS 3302-92. Approval and Test Specification—Particular Requirements for Electric Fans. 9 pp.

—**Aircraft—Design**
CAA Chapter P4-9. Rotor, Fan, Propeller and Transmission Systems (Provisional Airworthiness Requirements for Civil Powered Lift Aircraft). 3 pp.

—**Aircraft—Loads (Forces)**
CAA Chapter P3-4. Engine, Fan, Rotor, Propeller and Transmission System Loads (Provisional Airworthiness Requirements for Civil Powered Lift Aircraft). 10 pp.

—**Appliances**
DIN VDE 0730 Pt 2 F-81. Electric Motor-Operated Appliances for Domestic and Similar Purposes; Particular Requirements for Fans (June). 8 pp.

—**Attic**
CSA C22.2 NO 113-M1984. Fans and Ventilators (R 1993); (Gen Instr 1 Thru 6). 51 pp.
CSA 1169 Bull. Electrical Bulletin 1169 June 27, 1978 to C22.2 NO 113. 2 pp.

—**Automotive**
CNS D2014-86. Electric Fans for Automobiles (Jun)(4596).

—**Automotive—Overload Protection**
CNS D2014-1-87. Over Current Protector of Electric Fans for Automobiles (Mar)(4596-1).

—**Bathrooms**
CNS C4394-86. Ventilating Fan for Bath Room Use (Nov)(10597). 6 pp.

—**Boats—Air Flow**
ISO 9097-91. Small Craft—Electric Fans First Edition. 6 pp.

—**Canopy**
CSA C22.2 NO 113-M1984. Fans and Ventilators (R 1993); (Gen Instr 1 Thru 6). 51 pp.
CSA 1169 Bull. Electrical Bulletin 1169 June 27, 1978 to C22.2 NO 113. 2 pp.

—**Comprehensive Testing**
BSI BS 848: Part 1-80. 1980 Amd 2 Fans for General Purposes Part 1: Methods of Testing Performance. 86 pp.
BSI BS 848: Part 1-03. 1980 Amd 3 Fans for General Purposes Part 1: Methods of Testing Performance (AMD 7330) November 15, 1992 (Q). 87 pp.

—**Desk**
CSA C22.2 NO 113-M1984. Fans and Ventilators (R 1993); (Gen Instr 1 Thru 6). 51 pp.
CSA 1169 Bull. Electrical Bulletin 1169 June 27, 1978 to C22.2 NO 113. 2 pp.

—**Flanges**
BSI BS 6339-83. 1983 Dimensions of Circular Flanges for General Purpose Industrial Fans. 7 pp.
ISO 6580-81. General Purpose Industrial Fans—Circular Flanges—Dimensions First Edition. 4 pp.

—**Glossaries**
JIS B 0132-84. Glossary of Terms for Fans, Blowers and Compressors.

—**Hassock**
CSA C22.2 NO 113-M1984. Fans and Ventilators (R 1993); (Gen Instr 1 Thru 6). 51 pp.
CSA 1169 Bull. Electrical Bulletin 1169 June 27, 1978 to C22.2 NO 113. 2 pp.

—**Heaters**
CSA CAN/CSA-C22. 2 NO 236-M90. Heating and Cooling Equipment; (Gen Instr 1) (ANSI/UL 1995). 136 pp.

—**Household**
SAA AS 2071-84. Performance of Household Electrical Appliances—Circulating Fans. 19 pp.

—**Household—Hood**
CSA C22.2 NO 113-M1984. Fans and Ventilators (R 1993); (Gen Instr 1 Thru 6). 51 pp.
CSA 1169 Bull. Electrical Bulletin 1169 June 27, 1978 to C22.2 NO 113. 2 pp.

—**Lampholders**
CSA C22.2 NO 113-M1984. Fans and Ventilators (R 1993); (Gen Instr 1 Thru 6). 51 pp.

Fans (Cont.)
—**Lampholders** (Cont.)
CSA 1169 Bull. Electrical Bulletin 1169 June 27, 1978 to C22.2 NO 113. 2 pp.

—**Noise**
BSI BS 848: Part 2-85. 1985 Fans for General Purposes Part 2: Methods of Noise for Testing. 54 pp.
CNS Z8024-82. Method of Noise Level Measurement for Fans Blowers and Compressors (Apr)(8753).
CNS Z8025-82. Method of Noise Spectrum Measurement for Fans Blowers and Compressors (Apr)(8754).
DIN ENGL 45635 Pt 38-86. Measurement of Noise Emitted by Machines; Airborne Noise Emission; Enveloping Surface Method, Reverberation Room Method and In-Duct Method; Fans (Apr). 10 pp.

—**Pedestal**
CSA C22.2 NO 113-M1984. Fans and Ventilators (R 1993); (Gen Instr 1 Thru 6). 51 pp.
CSA 1169 Bull. Electrical Bulletin 1169 June 27, 1978 to C22.2 NO 113. 2 pp.

—**Power Transformers—Blade Diameters**
DIN ENGL 42565-51. Power Transformers; Fans; Blade Diameter 350 mm, for Open-Air Installation (Sept). 1 p.

—**Regulators**
BSI BS 5060-87. 1987 Performance and Construction of Electric Circulating Fans and Regulators. 15 pp.
BSI BS 5285-75. 1975 Amd 1 Performance of A.C. Electric Ventilating Fans and Regulators for Non-Industrial Use. 8 pp.
IEC 665-80. A.C. Electric Ventilating Fans and Regulators for Household and Similar Purposes First Edition. 56 pp.
IEC 879-86. Performance and Construction of Electric Circulating Fans and Regulators First Edition; (Corrigendum—Jan 1992). 28 pp.

—**Regulators—Safety**
BSI BS 3456: Sec 102.342-88. 1988 Safety of Household Electrical Appliances Part 102: Particular Requirements Section 102.342: Electric Fans and Regulators. 16 pp.
CENELEC HD 280 S1-86. Safety of Household and Similar Electrical Appliances: Particular Requirements for Electric Fans and Regulations. 6 pp.
DIN VDE 0700 Pt 234 (D)-81. Safety of Household and Similar Electrical Appliances; Fans and Associated Regulators (Aug). 25 pp.
IEC 342 Pt 1-81. Safety Requirements for Electric Fans and Regulators Part 1: Fans and Regulators for Household and Similar Purposes Second Edition; (Amendment 1-1982). 30 pp.

—**Regulators—Ships—Safety**
CENELEC HD 280.2 S1-90. Safety Requirements for Electric Fans and Regulators Part 2: Fans and Regulators for Use in Ships. 3 pp.
IEC 342 Pt 2-82. Safety Requirements for Electric Fans and Regulators Part 2: Fans and Regulators for Use in Ships Second Edition. 34 pp.

—**Safety**
BSI BS 848: Part 5-86. 1986 Fans for General Purposes Part 5: Guide for Mechanical and Electrical Safety. 16 pp.
DIN ENGL 24167 Pt 1-82. Fans; Protection Against Accidental Contact with Fan Impellers; Safety Requirements (Sept). 4 pp.

—**Sound Power**
ISO 5136-90. Acoustics—Determination of Sound Power Radiated into a Duct by Fans—in-Duct Method First Edition; (Corrigendum—Sept 1993). 28 pp.

—**Sound Pressure**
JIS B 8346-91. Fans, Blowers and Compressors—Determination of A-Weighted Sound Pressure Level. 60 pp.

—**Suitcase**
CSA C22.2 NO 113-M1984. Fans and Ventilators (R 1993); (Gen Instr 1 Thru 6). 51 pp.
CSA 1169 Bull. Electrical Bulletin 1169 June 27, 1978 to C22.2 NO 113. 2 pp.

—**Test Airways**
DIN ENGL 24163 Pt 2-85. Fans; Performance Testing; Standardized Test Airways (Jan). 35 pp.
DIN ENGL 24163 Pt 3-85. Fans; Performance Testing of Small Fans Using Standardized Test Airways (Jan). 19 pp.

—**Utility**
CSA C22.2 NO 113-M1984. Fans and Ventilators (R 1993); (Gen Instr 1 Thru 6). 51 pp.
CSA 1169 Bull. Electrical Bulletin 1169 June 27, 1978 to C22.2 NO 113. 2 pp.

Fans (Cont.)
—**Vibration**
BSI BS 848: Part 6-89. 1989 Fans for General Purposes Part 6: Method of Measurement of Fan Vibration. 23 pp.

—**Volumetric Analysis**
BSI BS 6583-85. 1985 Methods for Volumetric Testing for Rating of Fan Sections in Central Station Air Handling Units (Including Guidance on Rating). 12 pp.

—**Wall Insert**
CSA C22.2 NO 113-M1984. Fans and Ventilators (R 1993); (Gen Instr 1 Thru 6). 51 pp.
CSA 1169 Bull. Electrical Bulletin 1169 June 27, 1978 to C22.2 NO 113. 2 pp.

—**Window**
CSA C22.2 NO 113-M1984. Fans and Ventilators (R 1993); (Gen Instr 1 Thru 6). 51 pp.
CSA 1169 Bull. Electrical Bulletin 1169 June 27, 1978 to C22.2 NO 113. 2 pp.

FANS (ICAO)
Use: ICAO Special Committee on Future Air Navigation Systems

Farm Crops
Use: Agricultural Products

Farm Produce
Use: Agricultural Products

Farms
See Also: Agricultural Products; Animal Husbandry; Dairies; Livestock Equipment

—**Modernization**
EC 77/390/EEC-77. Council Directive Amending Directive 72/159/EEC on the Modernization of Farms. 1 p.

Fascia
See Also: Construction Equipment

—**Aluminum**
CGSB CAN/CGSB-93.2-M91. Prefinished Aluminum Siding, Soffits and Fascia, for Residential Use. 20 pp.

—**Aluminum—Installation**
CGSB CAN/CGSB-93.5-92. Installation of Metal Residential Siding, Soffits and Fascia. 7 pp.

—**Galvanized Steel**
CGSB CAN/CGSB-93.4-92. Galvanized Steel and Aluminum-Zinc Alloy Coated Steel Siding, Soffits and Fascia, Prefinished, Residential. 8 pp.

—**Galvanized Steel—Installation**
CGSB CAN/CGSB-93.5-92. Installation of Metal Residential Siding, Soffits and Fascia. 7 pp.

—**Pine Wood**
SAA AS 1496-73. Preservative Treated Radiata Pine Fascia Boards and Barge Boards Amdt 1 April 1977.

—**PVC**
CGSB 41-GP-24MA-83. Siding, Soffits and Fascia, Rigid Vinyl, Standard for. 14 pp.
CGSB CAN/CGSB-41.33-M87. Installation of Rigid Vinyl Residential Siding, Soffits and Fascia. 10 pp.

—**Softwoods**
SAA AS 1787-75. Preservative Treated Fascia Boards and Barge Boards Milled from Australian-Grown Conifers (Softwoods) (Excluding Radiata Pine and Cypress Pine).

—**Wood—Moisture Content**
SNZ NZS 3617-79. Specification for Profiles of Weatherboards, Fascia Boards, and Flooring. 15 pp.

—**Wood—Sizes**
SNZ NZS 3617-79. Specification for Profiles of Weatherboards, Fascia Boards, and Flooring. 15 pp.

Fast Recovery Rectifier Diodes
Use: Fast Recovery Rectifiers

Fast Recovery Rectifiers
Use For: Fast Recovery Rectifier Diodes
See Also: Diodes; Rectifier Diodes; Rectifiers; Switching Diodes
CECC CECC 50 008-006. Ambient Rated Fast Recovery Rectifier (En). 10 pp.
CECC CECC 50 008-007. Ambient Rated Fast Recovery Rectifier (En). 10 pp.
CECC CECC 50 008-008. Ambient Rated Fast Recovery Rectifier (En). 10 pp.

INTERNATIONAL AND NON-U.S. NATIONAL STANDARDS
SUBJECT INDEX
Fasteners

Fast Recovery Rectifiers (Cont.)
CECC CECC 50 008-009. Ambient Rated Fast Recovery Rectifier (En). 10 pp.
CECC CECC 50 008-010. Ambient Rated Fast Recovery Rectifier (En). 10 pp.
CECC CECC 50 008-011 ISSUE 2-81. BS CECC 50 008-011; Ambient Rated, Fast Recovery Low Power Rectifier Diodes (En). 9 pp.
CECC CECC 50 009-014 ISSUE 2-80. BS CECC 50 009-014; Fast Recovery Case Rated Silicon Rectifier Diode Primarily Intended for High Frequency Switching (En). 15 pp.

Fast Select Services
See Also: Data Transmission; Packet Switched Data Transmission Services; Virtual Call Services
—Packet Switched Public Data Networks—Accounting
CCITT RECMN D.21-89. Special Tariff Principles for Short Transaction Transmissions on the International Packet Switched Public Data Networks Using the Fast Select Facility with Restriction—General Tariff Principles—Charging and Accounting in International Telecommunications. 1 p.
—Packet Switched Public Data Networks—Tariffs
CCITT RECMN D.21-89. Special Tariff Principles for Short Transaction Transmissions on the International Packet Switched Public Data Networks Using the Fast Select Facility with Restriction—General Tariff Principles—Charging and Accounting in International Telecommunications. 1 p.

Fast Sky Blue
See Also: Pigments
—Pigments
CNS K2222-88. Fast Sky Blue (May)(12310).
JIS K 5243-71. Fast Sky Blue.

Fast Yellow G
See Also: Pigments
—Pigments
CNS K2141-83. Fast Yellow G (Jun)(10362). 3 pp.
JIS K 5216-71. Fast Yellow G. 5 pp.

Fast Yellow 10G
See Also: Pigments
—Pigments
CNS K2213-88. Fast Yellow 10 G (May)(12301).
JIS K 5217-71. Fast Yellow 10 G. 5 pp.

FASTBUS Systems
See Also: Data Acquisition Systems; Data Buses; Nuclear Instrumentation
IEC 935-90. IEC 935 FASTBUS Modular High Speed Data Acquisition System First Edition. 443 pp.
IEC 1052-91. IEC 1052 FASTBUS STANDARD ROUTINES Standard Routines for Use with FASTBUS Data Acquisition System First Edition. 239 pp.

Fasteners
Scope Note: For additional listings, use a more specific term *Use For:* Mechanical Fasteners
See Also: Adhesives; Aerospace Bolts; Aerospace Fasteners; Aerospace Nuts; Aerospace Rivets; Aerospace Screws; Aerospace Washers; Aircraft Bolts; Aircraft Fasteners; Aircraft Nuts; Aircraft Rivets; Aircraft Screws; Aircraft Washers; Anchor Nuts; Bands; Blanks (Fasteners); Bolts; Brackets; Buttons (Fasteners); Captive Fasteners; Casement Fasteners; Chains; Chucks; Clamping Nuts; Clamps; Clevis; Clips; Conical Seats; Connectors; Cotter Pins; Counterbore Nuts; Couplings; Dowels; Eye Fasteners; Grommet Nuts; Grommets; Hardware; Hasps; Hinges; Holders; Hooks; Indexable Inserts; Inserts (Fasteners); Jo-Bolts; Joints; Kingpins; Latches (Fasteners); Lock Tooth Washers; Lock Washers; Locking Fasteners; Locking Screws; Locknuts; Lockplates; Locks (Security); Lockwire; Lockwire Holes; Loop Clamps; Lugs (Fasteners); Machine Screws; Nails (Fasteners); Nuts (Fasteners); Panel Fasteners; Paper Clips; Paper Fasteners; Phillips Screws; Pins; Piston Pins; Plate Nuts; Quick Release Fasteners; Retaining Rings; Rivets; Roofing Fasteners; Ropes; Round Head Bolts; Rubber Bands; Safety Pins; Screws; Sems; Shackles; Shank Nuts; Sleeves (Fittings); Soldering Lugs; Solderless Connections; Spacers (Mechanical); Spikes (Fasteners); Splines; Staplers; Staples; Straps (Fasteners); Structural Members; Studs (Fasteners); Taper Pins; Tenons; Terminal Lugs; Thimbles; Thread Inserts; Threaded Fasteners; Triangle Nuts; Truss Clips
See Also: Wire; Wire Nails; Wire Ropes; Zippers
CNS Z5055-77. Fasteners (Oct)(4187).

Fasteners (Cont.)
DIN ENGL 267 Pt 1-82. Fasteners; Technical Delivery Conditions; General Requirements (Aug). 3 pp.
—Abbreviations
DIN ENGL 918-79. Fasteners; Terminology; Spelling of Terms; Abbreviations (Sept). 4 pp.
—Acceptance Testing
BSI BS 6587-85. 1985 Methods of Acceptance Inspection for Fasteners. 10 pp.
ISO 3269-88. Fasteners—Acceptance Inspection Second Edition. 11 pp.
JIS B 1091-91. Fasteners—Acceptance Inspection (ISO 3269-1988).
—Aircraft
CAA LEAFLET 3-5 07.90. Standard Fasteners of American Manufacture. 26 pp.
—Automotive—Disk Wheels
DIN ENGL 74361 Pt 2-82. Disc Wheels for Motor Vehicles and Trailers; Fasteners for Stud Centring (Nov). 5 pp.
DIN ENGL 74361 Pt 3-79. Disc Wheels for Motor Vehicles and Trailers; Dimensions and Fasteners for Attachment with Centring on Wheel Bore (Nov). 4 pp.
—Belt
CNS B2113-61. Belt Fastener (Jun)(1362)(R 1973).
—Busways—Corrosion Prevention
CSA C22.2 NO 27-1973. Busways; (Amd 1 June 1987). 19 pp.
CSA 1091 Bull. Electrical Bulletin 1091 February 8, 1977 to C22.2 NO 27. 1 p.
CSA 1169 Bull. Electrical Bulletin 1169 June 27, 1978 to C22.2 NO 27. 2 pp.
—Busways—Plating Protection
CSA C22.2 NO 27-1973. Busways; (Amd 1 June 1987). 19 pp.
CSA 1091 Bull. Electrical Bulletin 1091 February 8, 1977 to C22.2 NO 27. 1 p.
CSA 1169 Bull. Electrical Bulletin 1169 June 27, 1978 to C22.2 NO 27. 2 pp.
—Classification
BSI BS 1000: (621.8)-76. 1976 Amd 1 Universal Decimal Classification (UDC). English Full Edition (621.8): Mechanical Power Transmission. Machine Elements. Gearing. Materials Handling. Mechanical Attachments, Fixing, Fasteners. Lubrication. 30 pp.
SNZ NZS/BS 1000 (621.8)-76. Universal Decimal Classification Mechanical Power Transmission. Machine Elements. Gearing. Materials Handling. Mechanical Attachments, Fixing, Fasteners. Lubrication Amend: 1. 28 pp.
—Construction Materials
BSI BS 1494: Part 1-64. 1964 Amd 1 Fixing Accessories for Building Purposes Part 1: Fixings for Sheet, Roof and Wall Coverings. 27 pp.
—Documents—Standards Organizations—Numbering Systems
DIN ENGL 918 Suppl. 2-91. Fasteners; Synopsis of Available ISO Standards and DIN Standards (Oct). 28 pp.
—Glossaries
BSI BS 6040-81. 1981 Nomenclature for Bolts, Screws, Nuts and Accessories. 58 pp.
CNS B1050-78. General Terms of Fasteners (Oct)(4218).
CNS B1074-78. Terms of Fasteners for Portions of Head (Oct)(4530).
CNS B1075-78. Terms of Fasteners for Portions of Neck (Oct)(4531).
CNS B1076-78. Terms of Fasteners for Portions of End (Point) (Oct)(4532).
CNS B1077-78. Terms of Fasteners for Portions of Fit (Oct)(4533).
CNS B1086-78. Terms Relating to the Gauge of Fasteners (Oct)(4542).
CNS B1087-78. Terms of Fasteners for Portions of Appearance (Oct)(4543).
DIN ENGL 918 Suppl. 1-87. Fasteners; Representation of Standardized Fasteners and Their Nomenclature (Nov). 17 pp.
JIS B 0101-74. Glossary of Terms Relating to Fasteners (R 1983). 116 pp.
—Identification Systems
ISO 8991-86. Designation System for Fasteners First Edition. 4 pp.
—Manufacturing—Glossaries
CNS B1083-78. Terms Relating to the Manufacture of Fasteners (Oct)(4539).
CNS B1084-78. Terms Relating to the Machinery of Manufacturing Fasteners (Oct)(4540).

Fasteners (Cont.)
—Manufacturing—Glossaries (Cont.)
CNS B1085-78. Terms Relating to the Working Tools of Manufacturing Fasteners (Oct)(4541).
—Masonry
CEN PREN 845-1-92. Specification for Ancillary Components for Masonry—Part 1: Ties, Straps, Hangers, Brackets and Support Angles. 26 pp.
CSA CAN3-A370-M84. Connectors for Masonry; (Gen Instr 1 Thru 2). 59 pp.
—Masonry—Flexural Strength
CEN PREN 846-10-92. Methods of Test for Ancillary Components for Masonry—Part 10: Determination of Flexural Resistance and Stiffness of Brackets and Support Angles. 9 pp.
—Masonry—Stiffness Testing
CEN PREN 846-10-92. Methods of Test for Ancillary Components for Masonry—Part 10: Determination of Flexural Resistance and Stiffness of Brackets and Support Angles. 9 pp.
—Powder Actuated Tools
SAA AS 1873-78. Explosive-Powered Hand-Held Fastening Tools, Fasteners and Explosive Charges. 24 pp.
—Snap—Aerospace
MOD UK DSTAN 53-47-73. Fasteners—Snap Including Spring, Nylon/Metal Rivets, and Locking Snap Fasteners Issue 2. 10 pp.
—Snap—Spring—Fixed Oval Eye
CNS Z7085-81. Spring Snap With Fixed Round Eye and Oval Eye (Jul)(7762).
—Snap—Spring—Fixed Round Eye
CNS Z7085-81. Spring Snap With Fixed Round Eye and Oval Eye (Jul)(7762).
—Snap—Spring—Fixed Square Eye
CNS Z7084-81. Spring Snap With Fixed Square Eye (Jul)(7761).
—Snap—Trigger—Fixed Round Eye
CNS Z7086-81. Trigger Snap With Fixed Square Eye and Round Eye (Jul)(7763).
—Snap—Trigger—Fixed Square Eye
CNS Z7086-81. Trigger Snap With Fixed Square Eye and Round Eye (Jul)(7763).
—Steel Construction
BSI BS 3294: Part 1-60. (WITHDRAWN) 1960 Amd 6 The Use of High Strength Friction Grip Bolts in Structural Steelwork Part 1: General Grade Bolts. 15 pp.
BSI BS 4395: Part 1-69. 1969 Amd 2 High Strength Friction Grip Bolts and Associated Nuts and Washers for Structural Engineering Part 1: General Grade. 36 pp.
BSI BS 4395: Part 2-69. 1969 Amd 2 High Strength Friction Grip Bolts and Associated Nuts and Washers for Structural Engineering Part 2: Higher Grade Bolts and Nuts and General Grade Washers. 36 pp.
BSI BS 4604: Part 2-70. 1970 Amd 2 The Use of High Strength Friction Grip Bolts in Structural Steelwork. Metric Series Part 2: Higher Grade (Parallel Shank). 17 pp.
BSI BS 4604: Part 3-73. (WITHDRAWN) 1973 The Use of High Strength Friction Grip Bolts in Structural Steelwork. Metric Series Part 3: Higher Grade (Waisted Shank). 16 pp.
CNS B2123-87. Hexagon Head Bolts for Steel Structures (Oct)(3124).
CNS B2211-87. Hexagon Bolts with Large Widths Across Flats for Steel Structures (Joints with High-Tensile, Prestressed Bolts) (Oct)(4366).
CNS B2397-80. Washers for Steel Structure (Jan)(5112).
DIN ENGL 6914-89. High-Strength Hexagon Head Bolts with Large Widths Across Flats for Structural Steel Bolting (Oct). 6 pp.
DIN ENGL 6915-89. High-Strength Hexagon Nuts with Large Widths Across Flats for Structural Steel Bolting (Oct). 3 pp.
DIN ENGL 6916-89. Round Washers for High-Strength Structural Steel Bolting (Oct). 2 pp.
DIN ENGL 6917-89. Square Taper Washers for High-Strength Structural Bolting of Steel 1 Sections (Oct). 3 pp.
DIN ENGL 7969-89. Slotted Countersunk Head Screws for Structural Steel Bolting for Supply with or Without Nut (Oct). 5 pp.
DIN ENGL 7989-74. Plain Washers for Steel Construction (July). 1 p.
DIN ENGL 7990-89. Hexagon Head Bolts for Structural Steel Bolting for Supply with Nut (Oct). 6 pp.

INDUSTRY STANDARDS

INTERNATIONAL AND NON-U.S. NATIONAL STANDARDS
SUBJECT INDEX

Fasteners

Fasteners (Cont.)
—Steel Construction (Cont.)
ISO 4775-84. Hexagon Nuts for High-Strength Structural Bolting with Large Width Across Flats—Product Grade B—Property Classes 8 and 10 First Edition. 8 pp.

ISO 7411-84. Hexagon Bolts for High-Strength Structural Bolting with Large Width Across Flats (Thread Lengths According to ISO 888)—Product Grade C—Property Classes 8.8 and 10.9 First Edition. 9 pp.

ISO 7412-84. Hexagon Bolts for High-Strength Structural Bolting with Large Width Across Flats (Short Thread Length)—Product Grade C—Property Classes 8.8 and 10.9 First Edition. 9 pp.

SNZ NZS/BS 4395: Part 1-69. Specification for High Strength Friction Grip Bolts and Associated Nuts and Washers for Structural Engineering Part 1: General Grade. 36 pp.

SNZ NZS/BS 4395: Part 2-69. Specification for High Strength Friction Grip Bolts and Associated Nuts and Washers for Structural Engineering Part 2: Higher Grade Bolts and Nuts and General Grade Washers. 36 pp.

—Symbols
JIS B 0143-85. Symbols and Designations of Dimensions for Threaded Fasteners. 16 pp.

—Wood
BSI BS 6948-89. 1989 Methods of Test for Mechanically Fastened Joints in Timber and Wood-Based Materials. 39 pp.

BSI BS 6948-01. 1989 Amd 1 Methods of Test for Mechanically Fastened Joints in Timber and Wood-Based Materials (AMD 6916) February 28, 1992. 40 pp.

SAA AS 1649-74. Methods for the Determination of Basic Working Loads for Metal Fasteners for Timber. 32 pp.

SNZ NZS/AS 1649-74. Methods for the Determination of Basic Working Loads for Metal Fasteners for Timber (This is a Joint Standard with SAA AS 1649). 32 pp.

—Zinc Coatings (Made From Zinc)
BSI BS 7371: Part 11-93. 1993 Coatings on Metal Fasteners Part 11: Specification for Zinc Flake Non-Electrolytically Applied Cured Coatings (F). 11 pp.

Fastening Tools, Powder Actuated
Use: Powder Actuated Tools

Fat Content Analysis
Use For: Free Fat Content Analysis *See Also:* Oil Content Analysis; Saponification Number; Solid Fat Content Analysis; Vegetable Fat Content Analysis

—Animal Feed
CNS N4024-4-86. Method of Test for Feeds: Determination of Crude Fat (Aug)(2770-4).

—Baby Formula—Gravimetric Analysis
BSI BS 7142: Part 1-89. 1989 Analysis of Milk-Based Products Part 1: Determination of Fat Content of Milk-Based Infant Foods by the Rose-Gottlieb Gravimetric Method. 12 pp.

BSI BS 7142: Part 2-89. 1989 Analysis of Milk-Based Products Part 2: Determination of Fat Content of Milk-Based Infant Foods by the Weibull-Berntrop Gravimetric Method. 8 pp.

ISO 8262 Pt 1-87. Milk Products and Milk-Based Foods—Determination of Fat Content by the Weibull-Berntrop Gravimetric Method (Reference Method)—Part 1: Infant Foods First Edition. 8 pp.

ISO 8381-87. Milk-Based Infant Foods—Determination of Fat Content—Rose-Gottlieb Gravimetric Method (Reference Method) First Edition. 12 pp.

—Butter
BSI BS 5086: Part 2-84. 1984 Analysis of Butter Part 2: Methods for Determination of Water, Solids-Non-Fat and Fat Contents (Reference Method). 4 pp.

CNS N6072-86. Method of Test for Butter (Apr)(3528). 3 pp.

ISO 3727-77. Butter—Determination of Water, Solids-Not-Fat and Fat Contents on the Same Test Portion (Reference Method) First Edition. 5 pp.

SAA AS 2300.7.2-91. Methods of Chemical and Physical Testing for the Dairying Industry—Part 7: Butter—Part 7.2: Determination of Fat and Curd (Supersedes AS 1739—1975 (in Part)). 2 pp.

—Butter—Gravimetric Analysis
BSI BS 1743: Part 11-86. 1986 Analysis of Dried Milk and Dried Milk Products Part 11: Determination of Fat Content (Gravimetric Reference Method). 10 pp.

ISO 1736-85. Dried Milk, Dried Whey, Dried Buttermilk and Dried Butter Serum—Determination of Fat Content—Gravimetric Method (Reference Method) Second Edition. 10 pp.

Fat Content Analysis (Cont.)
—Caseins—Gravimetric Analysis
BSI BS 6248: Part 10-86. 1986 Amd 1 Caseins and Cadeinates Part 10: Method for Determination of Fats (Gravimetric Reference Method) (AMD 5868) July 29, 1988. 11 pp.

ISO 5543-86. Caseins and Caseinates—Determination of Fat Content—Gravimetric Method (Reference Method) First Edition (Corrected and Reprinted -1987). 10 pp.

—Cereal Products
ISO 7302-82. Cereals and Cereal Products—Determination of Total Fat Content First Edition. 5 pp.

—Cereals
ISO 7302-82. Cereals and Cereal Products—Determination of Total Fat Content First Edition. 5 pp.

—Cheeses
CNS N6066-72. Methods of Test for Milk and Milk Products (Detection of Iso-Fat) (Oct)(3450). 1 p.

SNZ NZS 501: Part 2-71. Gerber Method for the Determination of Fat in Milk and Milk Products Part 2: Methods Amend: A, 1982. 28 pp.

—Cheeses—Gravimetric Analysis
BSI BS 770: Part 3-89. 1989 Chemical Analysis of Cheese Part 3: Determination of Fat Content (Reference Method). 12 pp.

ISO 1735-87. Cheese and Processed Cheese Products—Determination of Fat Content—Gravimetric Method (Reference Method) Second Edition. 11 pp.

ISO 1854-87. Whey Cheese—Determination of Fat Content—Gravimetric Method (Reference Method) Second Edition. 11 pp.

—Cheeses—Van Gulik Method
ISO 3432-75. Cheese—Determination of Fat Content—Butyrometer for Van Gulik Method First Edition. 6 pp.

ISO 3433-75. Cheese—Determination of Fat Content—Van Gulik Method First Edition. 6 pp.

—Condensed Milk
CNS N6066-72. Methods of Test for Milk and Milk Products (Detection of Iso-Fat) (Oct)(3450). 1 p.

EC 79/1067/EEC-79. First Commission Directive Laying Down Community Methods of Analysis for Testing Certain Partly or Wholly Dehydrated Preserved Milk for Human Consumption. 24 pp.

—Condensed Milk—Gravimetric Analysis
ISO 1737-85. Evaporated Milk and Sweetened Condensed Milk—Determination of Fat Content—Gravimetric Method (Reference Method) Second Edition. 10 pp.

—Cotton Fibers
BSI BS 3477-62. 1962 Amd 1 Determination of Oils, Fats and Waxes in Cotton. 8 pp.

—Cream
SNZ NZS 501: Part 2-71. Gerber Method for the Determination of Fat in Milk and Milk Products Part 2: Methods Amend: A, 1982. 28 pp.

—Cream—Gravimetric Analysis
BSI BS 1741: Part 4-87. 1987 Methods for Chemical Analysis of Liquid Milk and Cream Part 4: Determination of Fat Content of Cream. 12 pp.

ISO 2450-85. Cream—Determination of Fat Content—Gravimetric Method (Reference Method) Second Edition. 10 pp.

—Dairy Products
SNZ NZS 501: Part 2-71. Gerber Method for the Determination of Fat in Milk and Milk Products Part 2: Methods Amend: A, 1982. 28 pp.

—Dairy Products—Gravimetric Analysis
BSI BS 5522-77. 1977 Milk and Milk Products. Determination of Fat Content. Mojonnier-Type Fat Extraction Flasks. 6 pp.

ISO 3889-77. Milk and Milk Products—Determination of Fat Content—Mojonnier-Type Fat Extraction Flasks First Edition. 5 pp.

ISO 8262 Pt 3-87. Milk Products and Milk-Based Foods—Determination of Fat Content by the Weibull-Berntrop Gravimetric Method (Reference Method)—Part 3: Special Cases First Edition. 8 pp.

—Dairy Products—Volumetric Analysis
BSI BS 696: Part 1-89. 1989 Gerber Method for the Determination of Fat in Milk and Milk Products Part 1: Apparatus. 23 pp.

BSI BS 696: Part 2-89. 1989 Gerber Method for the Determination of Fat in Milk and Milk Products Part 2: Methods. 12 pp.

CNS N6060-86. Method of Test for Milk and Milk Products (Determination of Fat Content) (Oct)(3444). 2 pp.

Fat Content Analysis (Cont.)
—Eggs—Frozen
CNS N6050-80. Method of Test for Frozen Eggs (Jul)(2923). 3 pp.

—Elastomers
DIN ENGL 53588-75. Testing of Rubber and Elastomers; Determination of Factice in Elastomers (Aug). 2 pp.

—Evaporated Milk—Gravimetric Analysis
ISO 1737-85. Evaporated Milk and Sweetened Condensed Milk—Determination of Fat Content—Gravimetric Method (Reference Method) Second Edition. 10 pp.

—Fish Meal
CNS N4015-78. Method of Test for Fish Meal and Paste (Mar)(2245). 2 pp.

—Fish Paste
CNS N4015-78. Method of Test for Fish Meal and Paste (Mar)(2245). 2 pp.

—Flours (Food)
CNS N6028-73. Method of Test for Soybean Powder (Nov)(1418). 3 pp.

—Food
CNS N6117-84. Method of Test for Crude Fat in Food (Jan)(5036). 4 pp.

ISO 7302-82. Cereals and Cereal Products—Determination of Total Fat Content First Edition. 5 pp.

—Ice Cream
CNS N6143-86. Method of Test for Ice Cream (Jan)(6509). 2 pp.

—Ice Cream—Gravimetric Analysis
BSI BS 2472: Part 3-89. 1989 Methods for Chemical Analysis of Ice Cream Part 3: Determination of Fat Content. 12 pp.

BSI BS 7142: Part 3-89. 1989 Analysis of Milk-Based Products Part 3: Determination of Fat Content of Milk-Based Edible Ices and Ice Mixes by the Weibull-Berntrop Gravimetric Method. 10 pp.

CNS N6202-86. Method of Test for Fat Content in Ice Cream Products-Rose-Gottlieb Gravimetric Method (Oct)(11736). 3 pp.

ISO 7328-84. Milk-Based Edible Ices and Ice-Mixes—Determination of Fat Content—Rose-Gottlieb Gravimetric Method (Reference Method) First Edition. 10 pp.

ISO 8262 Pt 2-87. Milk Products and Milk-Based Foods—Determination of Fat Content by the Weibull-Berntrop Gravimetric Method (Reference Method)—Part 2: Edible Ices and Ice-Mixes First Edition. 8 pp.

—Mayonnaise
CNS N6095-76. Method of Test for Mayonnaise (Jun)(3964). 2 pp.

—Meat
BSI BS 4401: Part 4-70. 1970 Amd 1 Methods of Test for Meat and Meat Products Part 4: Determination of Total Fat Content (AMD 6557) December 21, 1990. 14 pp.

BSI BS 4401: Part 5-70. 1970 Methods of Test for Meat and Meat Products Part 5: Determination of Free Fat Content. 8 pp.

CNS N6138-80. Method of Test for Meat and Meat Products—Determination of Total Fat Content (Aug)(6392). 3 pp.

CNS N6139-80. Method of Test for Meat and Meat Products—Determination of Free Fat Content (Aug)(6393). 3 pp.

ISO 1443-73. Meat and Meat Products—Determination of Total Fat Content First Edition. 4 pp.

ISO 1444-73. Meat and Meat Products—Determination of Free Fat Content First Edition. 4 pp.

—Meat—Dried
CNS N6036-87. Method of Test for Fried and Dried Shredded Meat (Jul)(2168). 2 pp.

—Milk
SNZ NZS 501: Part 2-71. Gerber Method for the Determination of Fat in Milk and Milk Products Part 2: Methods Amend: A, 1982. 28 pp.

—Milk—Gravimetric Analysis
BSI BS 1741: Part 3-87. 1987 Methods for Chemical Analysis of Liquid Milk and Cream Part 3: Determination of Fat Content of Liquid Milk. 12 pp.

BSI BS 1742: Part 3-90. 1990 Methods for Chemical Analysis of Condensed Milks Part 3: Determination of Fat Content. 12 pp.

BSI BS 1743: Part 11-86. 1986 Analysis of Dried Milk and Dried Milk Products Part 11: Determination of Fat Content (Gravimetric Reference Method). 10 pp.

INTERNATIONAL AND NON-U.S. NATIONAL STANDARDS
SUBJECT INDEX

Fatigue

Fat Content Analysis *(Cont.)*
—Milk—Gravimetric Analysis *(Cont.)*
BSI BS 5522-77. 1977 Milk and Milk Products. Determination of Fat Content. Mojonnier-Type Fat Extraction Flasks. 6 pp.
BSI BS 7142: Part 4-89. 1989 Analysis of Milk-Based Products Part 4: Determination of Fat Content of Liquid, Concentrated and Dried Milk Products by the Weibull-Berntrop Gravimetric Method. 8 pp.
ISO 1211-84. Milk—Determination of Fat Content—Gravimetric Method (Reference Method) First Edition. 10 pp.
ISO 1736-85. Dried Milk, Dried Whey, Dried Buttermilk and Dried Butter Serum—Determination of Fat Content—Gravimetric Method (Reference Method) Second Edition. 10 pp.
ISO 3889-77. Milk and Milk Products—Determination of Fat Content—Mojonnier-Type Fat Extraction Flasks First Edition. 5 pp.
ISO 7208-84. Skimmed Milk, Whey and Buttermilk—Determination of Fat Content—Gravimetric Method (Reference Method) First Edition. 9 pp.

—Milk—Volumetric Analysis
BSI BS 696: Part 1-89. 1989 Gerber Method for the Determination of Fat in Milk and Milk Products Part 1: Apparatus. 23 pp.
BSI BS 696: Part 2-89. 1989 Gerber Method for the Determination of Fat in Milk and Milk Products Part 2: Methods. 12 pp.
CNS N6060-86. Method of Test for Milk and Milk Products (Determination of Fat Content) (Oct)(3444). 2 pp.
ISO 488-83. Milk—Determination of Fat Content—Gerber Butyrometers First Edition. 12 pp.
ISO 2446-76. Milk—Determination of Fat Content (Routine Method) First Edition. 9 pp.

—Powdered Milk
EC 79/1067/EEC-79. First Commission Directive Laying Down Community Methods of Analysis for Testing Certain Partly or Wholly Dehydrated Preserved Milk for Human Consumption. 24 pp.

—Sausages
CNS N6032-87. Method of Test for Chinese Sausages (Jun)(2103). 2 pp.

—Soaps—Volumetric Analysis
BSI BS 1715: Sec 2.1-89. 1989 Analysis of Soaps Part 2: Quantitative Test Methods Section 2.1: Method for Determination of Total Alkali Content and Total Fatty Matter Content (ISO 685-1975). 7 pp.
ISO 685-75. Analysis of Soaps—Determination of Total Alkali Content and Total Fatty Matter Content First Edition. 5 pp.
MOD UK M 9532/67. Examination of Soap, Saddle, Glycerine (No Information) (Withdrawn).

—Starches
ISO 3947-77. Starches, Native or Modified—Determination of Total Fat Content First Edition. 4 pp.

—Whey
SNZ NZS 501: Part 2-71. Gerber Method for the Determination of Fat in Milk and Milk Products Part 2: Methods Amend: A, 1982. 28 pp.

—Whey—Gravimetric Analysis
BSI BS 1743: Part 11-86. 1986 Analysis of Dried Milk and Dried Milk Products Part 11: Determination of Fat Content (Gravimetric Reference Method). 10 pp.
ISO 1736-85. Dried Milk, Dried Whey, Dried Buttermilk and Dried Butter Serum—Determination of Fat Content—Gravimetric Method (Reference Method) Second Edition. 10 pp.
ISO 7208-84. Skimmed Milk, Whey and Buttermilk—Determination of Fat Content—Gravimetric Method (Reference Method) First Edition. 9 pp.

—Wool Fabrics
BSI BS 3582-81. 1981 Amd 1 Methods of Test for Determination of Oils, Fats and Waxes in Wool. 5 pp.

Fatigue (Materials)
Use For: Fatigue Properties; Strain Fatigue; Structural Fatigue *See Also:* Plastic Properties; Vibration
MOD UK DSTAN 03-21-83. Mechanical Methods for the Inducement of Compressive Surface Residual Stresses Issue 1. 39 pp.

—Adhesives
ISO 9664-93. Adhesives—Test Methods for Fatigue Properties of Structural Adhesives in Tensile Shear First Edition. 13 pp.

—Aircraft
CAA 799 (K). Semi-Aerobatic Aeroplanes (Blue Papers). 25 pp.

Fatigue (Materials) *(Cont.)*
—Aircraft *(Cont.)*
CAA NOTICE #89 ISSUE 2. Continuing Structural Integrity of Transport Aeroplanes (Airworthiness Notices). 5 pp.
CAA Chapter K3-1 App 10.92. Acceptable Fatigue Design Criteria (Light Aeroplanes). 2 pp.
CAA Chapter P3-9. Flutter Preventions and Structural Stiffness of Aerodynamic Surfaces (Provisional Airworthiness Requirements for Civil Powered Lift Aircraft). 3 pp.
CAA STANDARD NO. 4-2 08.69. Fatigue Strength. 4 pp.

—Aircraft Engines
CAA NOTICE #44 ISSUE 5. Gas Turbine Engine Parts Subject to Retirement or Ultimate (Scrap) Lives (Airworthiness Notices). 3 pp.

—Aircraft Equipment
CAA NOTICE #62 ISSUE 3. Fatigue Lives (Airworthiness Notices). 2 pp.

—Aircraft—Sonic
JIS W 0615-85. Airplane Strength and Rigidity, Sonic, Fatigue.

—Bridges (Structures)
BSI BS 5400: Part 10-80. 1980 Steel, Concrete and Composite Bridges Part 10: Code of Practice for Fatigue. 55 pp.
BSI BS 5400: Part 10C-80. 1980 Steel, Concrete and Composite Bridges Part 10C: Charts for Classification of Details for Fatigue. 2 pp.

—Rotary Wing Aircraft
CAA 816 (G). Section G Rotorcraft Engine and Transmission Systems Gearbox Fatigue Tests (Blue Papers). 4 pp.
CAA Chapter G3-4 06.66. Engine and Transmission Systems (Rotocraft). 1 p.
CAA CAP524 SUB-Part C 12.86. (Rotorcraft). 13 pp.

—Steels—Metal Arc Welding
CSA W59-1989. Welded Steel Construction (Metal Arc Welding); (Gen Instr 1 Thru 2). 211 pp.
CSA W59-M1989. Steel Fixed Offshore Structures Welded Steel Construction (Metal Arc Welding); (Gen Instr 1) (Supplement 1 1989) (Gen Instr 2 to Supplement 1 1989). 241 pp.
SAA AS 1554.5-89. Structural Steel Welding (Known as the SAA Structural Steel Welding Code)—Part 5: Welding of Steel Structures Subject to High Levels of Fatigue Loading. 44 pp.
SNZ NZS 4701-81. Metal-Arc Welding of Steel Structures. 52 pp.

—Turbine Engines—Ground Effect Machines
CAA Chapter B7-3 App 04.79. Engine Testing—Type Tests for Gas Turbine Shaft Power Engines. 7 pp.

Fatigue Properties
Use: Fatigue (Materials)

Fatigue Testing
Use For: Axial Load Fatigue Testing; Direct Stress Fatigue Testing; Plane Bending Flexural Fatigue Testing; Rotating Bar Bending Fatigue Testing; Rotating Bending Fatigue Testing; Torsional Stress Fatigue Testing *See Also:* Bend Testing; Compression Testing; Creep Properties; Ductility; Dynamic Testing; Endurance Testing; Folding Endurance; High Temperature Testing; Impact Testing; Low Temperature Testing; Radiation Detection and Measurement; Shear Testing; Stress (Testing); Tensile Testing; Torsion; Wear Testing

—Belt Drives
ISO 5287-85. Narrow V-Belt Drives for the Automotive Industry—Fatigue Test Second Edition. 7 pp.

—Cellular Plastics
BSI BS 4443: Part 5-80. 1980 Methods of Test for Flexible Cellular Materials Part 5: Method 13, Test for Dynamic Fatigue by Constant Load Pounding. 8 pp.
ISO 3385-89. Flexible Cellular Polymeric Materials—Determination of Fatigue by Constant-Load Pounding Third Edition. 7 pp.

—Chairs
BSI BS 4875: Part 1-85. 1985 Strength and Stability of Furniture Part 1: Methods for Determination of Strength of Chairs and Stools. 20 pp.
BSI BS 4875: Part 3-85. 1985 Strength and Stability of Furniture Part 3: Methods for Determination of Strength of Settees. 20 pp.
BSI BS 4875: Part 5-85. 1985 Strength and Stability of Furniture Part 5: Methods for Determination of Strength of Tables and Trolleys. 12 pp.

Fatigue Testing *(Cont.)*
—Chairs *(Cont.)*
BSI BS 4875: Part 7-85. 1985 Strength and Stability of Furniture Part 7: Methods for Determination of Strength of Storage Furniture. 20 pp.
BSI BS 5873: Part 3-85. 1985 Amd 1 Educational Furniture Part 3: Specification for Strength and Stability of Tables for Educational Institutions (AMD 6325) May 31, 1991. 16 pp.
ISO 7173-89. Furniture—Chairs and Stools—Determination of Strength and Durability First Edition. 23 pp.

—Fasteners
ISO 3800 Pt 1-77. Threaded Fasteners—Axial Load Fatigue Testing—Part 1: Test Methods First Edition. 9 pp.
JIS B 1081-91. Axial Load Fatigue Testing for Threaded Fasteners. 24 pp.

—Fluid Power Equipment
BSI BS 7268-90. 1990 Method for Determination of Fatigue Pressure Rating of Metal Pressure Containing Envelopes in Hydraulic Fluid Power Systems. 12 pp.

—Footwear
BSI BS 5131: Sec 4.9-91. 1991 Methods of Test for Footwear and Footwear Materials Part 4: Other Components Section 4.9: Fatigue Resistance of Heels of Ladies' Shoes. 7 pp.
BSI BS 5131: Sec 5.7-78. 1978 Methods of Test for Footwear and Footwear Materials Part 5: Testing of Complete Footwear Section 5.7: Fatigue Test for Rigid Units and Shoe Bottoms. 5 pp.

—Hydraulic Equipment—Tubes
ISO 8574-90. Aerospace—Hydraulic System Tubing—Qualification Tests First Edition. 8 pp.

—Metals
BSI BS 3518: Part 1-93. 1993 Methods of Fatigue Testing Part 1: Guide to General Principles (G). 33 pp.
BSI BS 3518: Part 1-62. 1962 Methods of Fatigue Testing Part 1: General Principles. 28 pp.
BSI BS 3518: Part 2-62. 1962 Amd 1 Methods of Fatigue Testing Part 2: Rotating Bending Fatigue Tests. 15 pp.
BSI BS 3518: Part 3-63. 1963 Methods of Fatigue Testing Part 3: Direct Stress Fatigue Tests. 12 pp.
BSI BS 3518: Part 4-63. (WITHDRAWN) 1963 Methods of Fatigue Testing Part 4: Torsional Stress Fatigue Tests. 12 pp.
BSI BS 6835-88. 1988 Amd 1 Determination of the Rate of Fatigue Crack Growth in Metallic Materials. 30 pp.
BSI BS 7270-90. 1990 Method for Constant Amplitude Strain Controlled Fatigue Testing. 21 pp.
CNS G2057-79. General Rules for Fatigue Testing of Metals (Aug)(4958). 8 pp.
CNS G2078-81. Method of Rotating Bending Fatigue Testing of Metals (May)(7375). 6 pp.
CNS G2079-81. Method of Plane Bending Fatigue Testing of Metal Plates (May)(7376). 6 pp.
DIN ENGL 50113-82. Testing of Metals; Rotating Bar Bending Fatigue Test (Mar). 4 pp.
DIN ENGL 50142-82. Testing of Metallic Materials; Flat Bending Fatigue Test (Mar). 3 pp.
ISO 1099-75. Metals—Axial Load Fatigue Testing First Edition. 7 pp.
JIS Z 2273-78. General Rules for Fatigue Testing of Metals. 15 pp.
JIS Z 2274-78. Method of Rotating Bending Fatigue Testing of Metals. 12 pp.
JIS Z 2275-78. Method of Plane Bending Fatigue Testing of Metal Plates. 11 pp.

—Plastics
BSI BS 4618: SUBSEC 1.3.1-75. 1975 Recommendations for the Presentation of Plastics Design Data Part 1: Mechanical Properties Section 1.3: Strength Subsection 1.3.1: Static Fatigue Failure Caused by a Constant Force. 6 pp.
CNS K6715-82. Method of Test for Flexural Fatigue of Rigid Plastics by Plane Bending (Feb)(8527).
JIS K 7118-80. General Rules for Testing Fatigue of Rigid Plastics. 13 pp.
JIS K 7119-72. Testing Method of Flexural Fatigue of Rigid Plastics by Plane Bending. 5 pp.

—Sintered Metals
BSI BS 5600: Sec 3.10-79. 1979 Powder Metallurgical Materials and Products Part 3: Methods of Testing Sintered Metal Materials and Products, Excluding Hardmetals Section 3.10: Fatigues Test Pieces. 6 pp.
ISO 3928-77. Sintered Metal Materials, Excluding Hardmetals—Fatigue Test Pieces First Edition. 5 pp.

—Skis
ISO 6266-80. Alpine Skis—Determination of Fatigue Indexes—Cyclic Loading Test First Edition. 6 pp.
ISO 7798-84. Cross-Country Skis—Determination of Fatigue Indexes—Cyclic Loading Test First Edition. 7 pp.

INDUSTRY STANDARDS

Fatigue Testing (Cont.)

—Statistical Analysis
BSI BS 3518: Part 5-66. 1966 Methods of Fatigue Testing Part 5: Guide to the Application of Statistics. 57 pp.

—Steels
ISO 1352-77. Steel—Torsional Stress Fatigue Testing First Edition. 6 pp.

—Stools
BSI BS 4875: Part 1-85. 1985 Strength and Stability of Furniture Part 1: Methods for Determination of Strength of Chairs and Stools. 20 pp.
ISO 7173-89. Furniture—Chairs and Stools—Determination of Strength and Durability First Edition. 23 pp.

—Structural Steels
BSI BS 7608-93. 1993 Fatigue Design and Assessment of Steel Structures. 82 pp.

—Valves
BSI BS 2782:Pt11: METH 1132-92. 1992 Thermoplastics Valves—Fatigue Strength—Test Method (ISO 8659: 1989). 9 pp.
BSI BS EN 28659-92. 1992 Thermoplastics Valves—Fatigue Strength—Test Method (ISO 8659: 1989). 9 pp.
CEN EN 28659-90. Thermoplastics Valves—Fatigue Strength—Test Method. 4 pp.
DIN ENGL EN 28659-91. Thermoplastics Valves; Testing of Fatigue Strength (ISO 8659:1989) (June). 5 pp.
ISO 8659-89. Thermoplastics Valves—Fatigue Strength—Test Method First Edition. 6 pp.

—Welded Joints
JIS Z 3103-87. Method of Repeated Tension Fatigue Testing for Fusion Welded Joints. 14 pp.
JIS Z 3138-89. Method of Fatigue Testing for Spot Welded Joint. 17 pp.

—Wheelchairs
SAA AS 3696.8 (INT)-91. Wheelchairs—Part 8(Int): Static, Impact and Fatigue Strength Tests (Expires 9 December 1993). 72 pp.

Fatigue Testing Equipment

See Also: Testing Equipment
DIN ENGL 51228-88. Materials Testing Machines; Fatigue Testing Machines; General Requirements (Sept). 10 pp.

—Calibration
BSI DD 2-71. 1971 Dynamic Force Calibration of Axial Load Fatigue Testing Machines by Means of a Strain Gauge Technique. 11 pp.
ISO 4965-79. Axial Load Fatigue Testing Machines—Dynamic Force Calibration—Strain Gauge Technique First Edition. 9 pp.

Fats

Scope Note: For additional listings, use a more specific term *Use For:* Crude Fats
See Also: Animal Fats; Cocoa Butter; Esters; Fatty Acids; Greases; Shortening; Vegetable Fats; Vegetable Oils
BSI BS 684: Sec 2.4-76. 1976 Methods of Analysis of Fats and Fatty Oils Part 2: Other Methods Section 2.4: Calculation of Total Fatty Matter. 2 pp.
CNS N6074-83. Methods of Test for Edible Oils and Fats (General Rules) (Mar)(3639). 2 pp.
EC EEC/356/92-92. Council Regulation Amending Regulation No 136/66/EEC on the Establishment of a Common Organization of the Market in Oils and Fats. 2 pp.

—Acetyl Number
BSI BS 684: Sec 2.9-77. 1977 Methods of Analysis of Fats and Fatty Oils Part 2: Other Methods Section 2.9: Determination of the Hydroxyl Value and the Acetyl Value. 4 pp.

—Acidity
BSI BS 684: Sec 2.10-88. 1988 Amd 1 Methods of Analysis of Fats and Fatty Oils Part 2: Other Methods Section 2.10: Determination of Acidity (AMD 6050) May 31, 1990. 7 pp.
CNS N6082-82. Methods of Test for Edible Oils and Fats (Determination of Acid Value) (Jan)(3647). 1 p.

—Aflatoxin Content
CNS N6097-79. Method of Test for Edible Fats and Oils (Determination of Aflatoxin) (Apr)(4090). 3 pp.

—Alkalinity—Volumetric Analysis
BSI BS 684: Sec 2.5-89. 1989 Amd 1 Methods of Analysis of Fats and Fatty Oils Part 2: Other Methods Section 2.5: Determination of Alkalinity (AMD 6550) November 30, 1990. 5 pp.

Fats (Cont.)

—Bleaching
CNS K6762-83. Method of Bleaching Test for Fats and Oils (Mar)(10097).
JIS K 3501-62. Method of Bleaching Test for Fats and Oils.
JIS K 3502-62. Method of Alkali Refining of Fats and Oils for Bleaching Test.

—Classification
BSI BS 1000: (665)-79. 1979 Universal Decimal Classification (UDC). English Full Edition (665): Oils. Fats. Waxes. Adhesives. Gums. Resins. 34 pp.
SNZ NZS/BS 1000 (665)-79. Universal Decimal Classification Oils. Fats. Waxes. Adhesives. Gums. Resins. 36 pp.

—Cloud Point
BSI BS 684: Sec 1.5-87. 1987 Amd 1 Methods of Analysis of Fats and Fatty Oils Part 1: Physical Methods Section 1.5: Determination of Cloud Point. 7 pp.
CNS N6086-82. Methods of Test for Edible Oils and Fats (Cloud Point Test) (Jan)(3651). 1 p.

—Cold Testing
CNS N6087-82. Methods of Test for Edible Oils and Fats (Cold Test) (Jan)(3652). 1 p.

—Color Testing—Lovibond Tintometer
CNS K6764-83. Method of Test for Colour of Fats and Oils by Means of Lovibond Tintometer (Mar)(10099).
CNS N6076-82. Methods of Test for Edible Oils and Fats (Color Test) (Jan)(3641). 1 p.
JIS K 3503-62. Testing Method of Colour of Fats and Oils by Means of Lovibond Tintometer.

—Crystallization
BSI BS 684: Sec 1.13-76. 1976 Methods of Analysis of Fats and Fatty Oils Part 1: Physical Methods Section 1.13: Determination of Cooling Curve. 4 pp.

—Density
CNS N6079-82. Methods of Test for Edible Oils and Fats (Determination of Specific Gravity) (Jan)(3644). 1 p.

—Dilatometry
BSI BS 684: Sec 1.12-90. 1990 Methods of Analysis of Fats and Fatty Oils Part 1: Physical Methods Section 1.12: Determination of Dilation. 13 pp.

—Dropping Point
BSI BS 684: Sec 1.4-76. 1976 Methods of Analysis of Fats and Fatty Oils Part 1: Physical Methods Section 1.4: Determination of Flow and Drop Points. 8 pp.

—Drying
BSI BS 684: Sec 1.7-76. 1976 Methods of Analysis of Fats and Fatty Oils Part 1: Physical Methods Section 1.7: Determination of Surface-Drying Time. 2 pp.

—Erucic Acid Content
EC 76/621/EEC-76. Council Directive Relating to the Fixing of the Maximum Level of Erucic Acid in Oils and Fats Intended as Such for Human Consumption and in Foodstuffs Containing Added Oils or Fats. 3 pp.
EC 80/891/EEC-80. Commission Directive Relating to the Community Method of Analysis for Determining the Erucic Acid Content in Oils and Fats Intended to be Used as Such for Human Consumption and Foodstuffs Containing Added Oils or Fats. 7 pp.

—Esterification
BSI BS 684: Sec 2.34-80. 1980 Amd 1 Methods of Analysis of Fats and Fatty Oils Part 2: Other Methods Section 2.34: Preparation of Methyl Esters of Fatty Acids. 9 pp.

—Fatty Acid Content
BSI BS 684: Sec 2.11-76. 1976 Methods of Analysis of Fats and Fatty Oils Part 2: Other Methods Section 2.11: Determination of Volatile Acids (Reichert, Polenske and Kirschner Values). 4 pp.
CNS N6110-82. Method of Test for Edible Oils and Fats Determination of Total Fatty Acids (Jan)(4842). 1 p.

—Flow Rate
BSI BS 684: Sec 1.4-76. 1976 Methods of Analysis of Fats and Fatty Oils Part 1: Physical Methods Section 1.4: Determination of Flow and Drop Points. 8 pp.

—Hydroxyl Value
BSI BS 684: Sec 2.9-77. 1977 Methods of Analysis of Fats and Fatty Oils Part 2: Other Methods Section 2.9: Determination of the Hydroxyl Value and the Acetyl Value. 4 pp.

Fats (Cont.)

—Impurities—Solvent Extraction
CNS N6078-82. Methods of Test for Edible Oils and Fats (Determination of Insoluble Impurities) (Jan)(3643). 2 pp.

—Iodine Number
CNS N6081-86. Methods of Test for Edible Oils and Fats (Determination of Iodine Value) (Nov)(3646). 3 pp.

—Melting Points
CNS N6109-82. Method of Test for Edible Oils and Fats Determination of Melting Point (Jan)(4841). 1 p.

—Melting Points—Open Tube Method
CNS N6162-82. Methods of Test for Edible Oils and Fats—Determination of Melting Point (Open-Tube Method) (Jan)(8452). 1 p.

—Moisture Content—Dehydration
CNS N6077-82. Methods of Test for Edible Oils and Fats (Determination of Moisture and Volatile Matter) (Jan)(3642). 1 p.

—Neutral Oil Content
BSI BS 684: Sec 2.8-77. 1977 Methods of Analysis of Fats and Fatty Oils Part 2: Other Methods Section 2.8: Determination of Total Neutral Oil. 4 pp.

—Penetration Resistance
BSI BS 684: Sec 1.11-76. 1976 Methods of Analysis of Fats and Fatty Oils Part 1: Physical Methods Section 1.11: Determination of Penetration Value. 4 pp.

—Peroxide Number
CNS N6085-82. Methods of Test for Edible Oils and Fats (Determination of Peroxide Value) (Jan)(3650). 1 p.

—Polyethylene Content
BSI BS 684: Sec 2.26-86. 1986 Methods of Analysis of Fats and Fatty Oils Part 2: Other Methods Section 2.26: Determination of Polyethylene-Type Plastics. 6 pp.

—Rancidity Index—Kreis Test
BSI BS 684: Sec 2.32-79. 1979 Methods of Analysis of Fats and Fatty Oils Part 2: Other Methods Section 2.32: Determination of Rancidity Index. 3 pp.

—Refining
CNS K6763-83. Method of Alkali Refining of Fats and Oils (Mar)(10098).

—Refractive Index
CNS N6080-86. Methods of Test for Edible Oils and Fats (Determination of Refractive Index) (Nov)(3645). 1 p.

—Sampling
CNS K6726-82. Sampling of Crude Fats and Oil (Aug)(9304).

—Saponifiable Matter Content
BSI BS 2000: Part 284-93. 1993 Methods of Test for Petroleum and Its Products Part 284: Determination of Saponifiable and Unsaponifiable Matter in Oils, Fats and Waxes (W). 6 pp.
BSI BS 2000: Part 284-88. 1988 Petroleum and Its Products Part 284: Saponifiable and Unsaponifiable Matter in Oils, Fats and Waxes. 8 pp.

—Saponification Number
BSI BS 684: Sec 2.6-89. 1989 Methods of Analysis of Fats and Fatty Oils Part 2: Other Methods Section 2.6: Determination of Saponification Value. 4 pp.
CNS K6765-83. Method of Test for Acid Value of Fats and Oils (Mar)(10100).
CNS N6082-82. Methods of Test for Edible Oils and Fats (Determination of Acid Value) (Jan)(3647). 1 p.
CNS N6083-82. Methods of Test for Edible Oils and Fats (Determination of Saponification Value) (Jan)(3648). 1 p.
DIN ENGL 53402-90. Determination of Acid Value (Sept). 3 pp.
JIS K 3504-62. Testing Method for Acid Value of Fats and Oils.

—Stability Testing
CNS N6088-85. Methods of Test for Edible Oils and Fats (AOM Method for Stability Testing) (Jan)(3653). 2 pp.

—Unsaponifiable Matter Content
BSI BS 684: Sec 2.7-89. 1989 Methods of Analysis of Fats and Fatty Oils Section 2.7: Determination of Unsaponifiable Matter. 8 pp.

INTERNATIONAL AND NON-U.S. NATIONAL STANDARDS
SUBJECT INDEX

Fatty

Fats *(Cont.)*
—Unsaponifiable Matter Content *(Cont.)*
BSI BS 2000: Part 284-93. 1993 Methods of Test for Petroleum and Its Products Part 284: Determination of Saponifiable and Unsaponifiable Matter in Oils, Fats and Waxes (W). 6 pp.
BSI BS 2000: Part 284-88. 1988 Petroleum and Its Products Part 284: Saponifiable and Unsaponifiable Matter in Oils, Fats and Waxes. 8 pp.
CNS N6084-86. Methods of Test for Edible Oils and Fats (Determination of Unsaponifiable Matter) (Nov)(3649). 1 p.

—Volatile Matter Content—Dehydration
CNS N6077-82. Methods of Test for Edible Oils and Fats (Determination of Moisture and Volatile Matter) (Jan)(3642). 1 p.

—Water Content—Karl Fischer Method
BSI BS 684: Sec 2.1-76. 1976 Methods of Analysis of Fats and Fatty Oils Part 2: Other Methods Section 2.1: Determination of Water by the (Karl Fischer). 6 pp.

Fatty Acid Content Analysis
See Also: Fatty Acids

—Binders (Materials)—Gas Chromatography
DIN ENGL 55957-83. Binders for Paints and Varnishes; Determination of Fatty Acids by Gas Chromatography (Oct). 4 pp.

—Butter
BSI BS 5086: Part 6-85. 1985 Analysis of Butter Part 6: Method for Determination of Free Free Fatty Acids in the Fat (Reference Method). 4 pp.

—Cereal Products
BSI BS 4317: Part 17-87. 1987 Cereals and Pulses Part 17: Determination of Fat Acidity. 6 pp.
ISO 7305-86. Milled Cereal Products—Determination of Fat Acidity First Edition. 5 pp.

—Coatings
CNS K6630-81. Method of Test for Fatty Acids Used in Protective Coatings (Mar)(7039).

—Eggs—Frozen
CNS N6050-80. Method of Test for Frozen Eggs (Jul)(2923). 3 pp.

—Fats
BSI BS 684: Sec 2.11-76. 1976 Methods of Analysis of Fats and Fatty Oils Part 2: Other Methods Section 2.11: Determination of Volatile Acids (Reichert, Polenske and Kirschner Values). 4 pp.
BSI BS 684: Sec 2.12-84. 1984 Methods of Analysis of Fats and Fatty Oils Part 2: Other Methods Section 2.12: Determination of Oxidised Fatty Acids. 4 pp.
BSI BS 684: Sec 2.39-86. 1986 Methods of Analysis of Fats and Fatty Oils Part 2: Other Methods Section 2.39: Determination of the Composition of Fatty Acids in the 2-Position. 8 pp.
BSI BS 684: Sec 2.43-88. 1988 Methods of Analysis of Fats and Fatty Oils Part 2: Other Methods Section 2.43: Determination of Polyunsaturated Fatty Acids with a Cis, Cis 1,4-Diene Structure. 8 pp.
CNS N6110-82. Method of Test for Edible Oils and Fats Determination of Total Fatty Acids (Jan)(4842). 1 p.
ISO 6800-85. Animal and Vegetable Fats and Oils—Determination of the Composition of Fatty Acids in the 2-Position First Edition. 7 pp.
ISO 7847-87. Animal and Vegetable Fats and Oils—Determination of Polyunsaturated Fatty Acids with a Cis,Cis 1,4-diene Structure First Edition. 8 pp.

—Fish Liver Oil
CNS K6114-63. Method of Test for Fish-Liver Oil (Crude) (Jul)(1223) (R 1973). 2 pp.

—Fruits
ISO 6632-81. Fruits, Vegetables and Derived Products—Determination of Volatile Acidity First Edition. 11 pp.

—Latex
BSI BS 6057: Sec 3.6-93. 1993 Rubber Latices Part 3: Methods of Test Section 3.6: Determination of Volatile Fatty Acid Number of Natural Rubber Latices (ISO 506: 1992). 8 pp.
BSI BS 6057: Sec 3.6-87. 1987 Rubber Latices Part 3: Methods of Test Section 3.6: Determination of Volatile Fatty Acid Number. 6 pp.
ISO 506-92. Rubber Latex, Natural, Concentrate—Determination of Volatile Fatty Acid Number Third Edition. 7 pp.
SAA AS 1683.6-74. Methods of Test for Elastomers—Part 6: Volatile Fatty Acid Number of Natural Rubber Latex Reconfirmed 1989.

Fatty Acid Content Analysis *(Cont.)*
—Oils
BSI BS 684: Sec 2.12-84. 1984 Methods of Analysis of Fats and Fatty Oils Part 2: Other Methods Section 2.12: Determination of Oxidised Fatty Acids. 4 pp.
BSI BS 684: Sec 2.39-86. 1986 Methods of Analysis of Fats and Fatty Oils Part 2: Other Methods Section 2.39: Determination of the Composition of Fatty Acids in the 2-Position. 8 pp.

—Organic Coatings—Qualitative Analysis
CNS K6804-14-84. Method of Test for Organic Coating (Chemical Analysis) — Qualitative Test of Fatty Acid in Solvent Solubles (Jul)(10880-14).

—Papers
MOD UK DSTAN 13-10-68. Paper, Wrapping, Unglazed, and Paper, Wrapping, Unglazed, Lead Free Issue 1. 9 pp.
MOD UK M 6512/61. Examination of Paper Wrapping, Unglazed and Paper, Wrapping, Unglazed, LF Quality (Superseded by Def Stan 13-10).

—Reagents
CNS K7620-82. Chemical Reagent (Fatty Oil, Fatty Acid and Higher Alcohol Determination) (Jun)(8992).

—Rosins
CNS K6469-80. Method of Test for Fatty Acids Content of Tall Oil Rosin (Mar)(5354).

—Soaps
BSI BS 1715: Sec 2.4-89. 1989 Analysis of Soaps Part 2: Quantitative Test Methods Section 2.4: Method for Determination of Free Fatty Acids Content. 4 pp.
CGSB 2-GP-11M METH 3.1-83. Methods of Testing and Analysis of Soaps and Detergents Fatty and Rosin Acids, Anhydrous Soap and Acid Number. 2 pp.

—Vegetable Oils
CNS N6110-82. Method of Test for Edible Oils and Fats Determination of Total Fatty Acids (Jan)(4842). 1 p.

—Vegetables
ISO 6632-81. Fruits, Vegetables and Derived Products—Determination of Volatile Acidity First Edition. 11 pp.

Fatty Acids
Scope Note: For additional listings, use a more specific term *Use For:* Polymerized Fatty Acids
See Also: Acetic Acid; n-Capric Acid; n-Caproic Acid; Fats; Fatty Acid Content Analysis; Oleic Acid; Stearic Acid; Tallow Fatty Acids
BSI BS 6947-88. 1988 Acid Oils and Recovered Fatty Acid Distillates. 4 pp.
CNS K1058-61. Fatty Acid (Jun)(1385)(R 1973).
CNS K6151-69. Method of Test for Fatty Acids (Jan)(1386)(R 1973).
CNS K6643-81. Method of Test for Polymerized Fatty Acid (Mar)(7162).
JIS K 3331-86. Hardened Oil and Fatty Acids for Industrial Use.

—Ash Content
CNS K6631-81. Method of Test for Ash in Drying Oils and Fatty Acid (Mar)(7040).

—Coconuts
CNS K1261-81. Distilled Coconut Fatty Acids (May)(7429).

—Colorimetry
CNS K6647-81. Method of Test for Color After Heating of Fatty Acid (Mar)(7166).

—Corn
CNS K1262-81. Distilled Corn Fatty Acids (May)(7430).

—Cottonseed
CNS K1264-81. Fractionated and Distilled Cottonseed Fatty Acids (May)(7432).

—Esterification
ISO 5509-78. Animal and Vegetable Fats and Oils—Preparation of Methyl Esters of Fatty Acids First Edition; (Erratum—Sept 1981). 9 pp.

—Hydroxyl Value
CNS K6648-81. Method of Test for Hydroxyl Value of Fatty Oils and Acid (Mar)(7167).

—Iodine Number
CNS K6646-81. Method of Test for Iodine Value of Drying Oils and Fatty Acid (Mar)(7165).

Fatty Acids *(Cont.)*
—Linseed
CNS K1265-81. Distilled Linseed Fatty Acids (May)(7433).

—Ricinoleic Acid
CNS K1260-81. Dehydrated Castor Acids (May)(5428).

—Rosin Acid Content
CNS K6465-80. Method of Test for Rosin Acids in Fatty Acids (Mar)(5350).

—Sampling
CNS K6660-81. Method of Sampling Liquid Oils, Fatty Acids and Polymerized Fatty Acids Commonly Used in Paints, Varnishes and Related Materials (Apr)(7261).

—Saponification Number
CNS K6632-81. Method of Test for Acid Value of Fatty Acids and Polymerized Fatty Acids (Mar)(7041).
CNS K6633-81. Method of Test for Saponification Value of Drying Oils, Fatty Acids, and Polymerized Fatty Acids (Mar)(7042).
DIN ENGL 53402-90. Determination of Acid Value (Sept). 3 pp.

—Soaps—Solidification Points
CGSB 2-GP-11M METH 8.1-83. Methods of Testing and Analysis of Soaps and Detergents Titer. 3 pp.

—Solidification Points
BSI BS 684: Sec 1.6-88. 1988 Methods of Analysis of Fats and Fatty Oils Part 1: Physical Methods Section 1.6: Determination of Titre. 6 pp.
CNS K6645-81. Method of Test for Titer of Fatty Acid (Mar)(7164).
ISO 935-88. Animal and Vegetable Fats and Oils—Determination of Titre First Edition. 6 pp.

—Soy Bean
CNS K1263-81. Distilled Soybean Fatty Acids (May)(7431).

—Tall Oil
CNS K1259-81. Tall Oil Fatty Acids (May)(7427).

—Unsaponifiable Matter Content
CNS K6634-81. Method of Test for Unsaponifiable Matter in Drying Oils, Fatty Acids, and Polymerized Fatty Acids (Mar)(7043).

Fatty Amines
—Amine Content
CNS K6690-81. Total, Primary, Secondary and Tertiary Amine Value of Fatty Amines by Alternative Indicator Method (Sep)(7914).
CNS K6692-81. Calculation of Percent of Primary, Secondary and Tertiary Amines in Fatty Amines (Sep)(7916).

—Amine Content—Potentiometric Analysis
CNS K6689-81. Total, Primary, Secondary and Tertiary Amine Values of Fatty Amines, Amidoamines and Diamines by Referee Potentiometric Method (Sep)(7913).

—Ethoxylated
ISO 6384-81. Surface Active Agents—Technical Ethoxylated Fatty Amines—Methods of Analysis First Edition. 9 pp.

—Iodine Number
CNS K6682-81. Method of Test for Iodine Value of Fatty Amines, Amidoamines, and Diamines (Jul)(7726).

Fatty Amines Content Analysis
—Water—Spectrophotometry
BSI BS 2690: Part 117-83. 1983 Water Used in Industry Part 117: Long-Chain Fatty Amines: Spectrophotometric Method. 4 pp.

Fatty Nitrogen Compounds
Use: Nitrogen Organic Compounds

Fatty Oil Content Analysis
—Reagents
CNS K7620-82. Chemical Reagent (Fatty Oil, Fatty Acid and Higher Alcohol Determination) (Jun)(8992).

Fatty Oils
See Also: Oils

—Hydroxyl Value
CNS K6648-81. Method of Test for Hydroxyl Value of Fatty Oils and Acid (Mar)(7167).

INDUSTRY STANDARDS

Fatty Quaternary Ammonium Chlorides
Use: Quaternary Ammonium Chloride

Faucets
Use For: Mixer Taps; Taps (Water); Water Faucets
See Also: Cocks; Gate Valves; Globe Valves; Hydraulic Valves; Plumbing Fixtures; Valves

BSI BS 1010: Part 2-73. 1973 Amd 4 Draw-off Taps and Stopvalves for Water Services (Screwdown Pattern) Part 2: Draw-off Taps and Ground Stopvalves. 71 pp.
CNS B2616-84. One Way Faucet (Vertical Type) (Jun)(8085). 2 pp.
CNS B2618-84. One Way Faucet (Long Shank Type) (Jun)(8087). 2 pp.
CNS B2619-84. Combination Faucet (Jun)(8088). 2 pp.
CNS B5004-84. One Way Faucet (Screw Type) (Jun)(711). 2 pp.
JIS B 2061-91. Faucets and Ball Taps. 52 pp.

—**Chemical Resistant**
BSI BS 5412 & 5413: Part 5-76. 1976 Amd 3 Performance of Draw-Off Taps with Metal Bodies for Water Services and with Plastic Bodies for Water Services Part 5: Physio-Chemical Characteristics: Materials, Coatings. 15 pp.

—**Design**
BSI BS 5412 & 5413: Part 1-76. 1976 Amd 4 Performance of Draw-Off Taps with Metal Bodies for Water Services and with Plastic Bodies for Water Services Part 1: Dimensional and Design Characteristics. 22 pp.

—**Endurance Testing**
BSI BS 5412 & 5413: Part 4-76. 1976 Amd 3 Performance of Draw-Off Taps with Metal Bodies for Water Services and and with Plastic Bodies for Water Services Part 4: Mechanical and Endurance Characteristics. 12 pp.

—**Flow Rate**
BSI BS EN 200-92. 1992 Sanitary Tapware: General Technical Specifications for Single Taps and Mixer Taps (Nominal Size 1/2) PN 10: Minimum Flow Pressure of 0.05 MPa (0.5 Bar). 33 pp.
CEN EN 200-89. Sanitary Tapware; General Technical Specifications for Single Taps and Mixer Taps (Nominal Size 1/2) PN 10 Minimum Flow Pressure of 0.05 MPa (0.5 Bar). 46 pp.

—**Flow Regulators**
BSI BS EN 246-92. 1992 Sanitary Tapware: General Specifications for Flow Rate Regulators. 18 pp.
CEN EN 246-89. Sanitary Tapware; General Specifications for Flow Rate Regulator. 17 pp.
DIN ENGL EN 246-90. Sanitary Taps; General Specifications for Flow Rate Regulators (Jan). 12 pp.

—**Hydraulics**
BSI BS 5412 & 5413: Part 3-76. 1976 Amd 2 Performance of Draw-Off Taps with Metal Bodies for Water Services and with Plastic Bodies for Water Services Part 3: Hydraulic Characteristics. 10 pp.

—**Mechanical Testing**
BSI BS 5412 & 5413: Part 4-76. 1976 Amd 3 Performance of Draw-Off Taps with Metal Bodies for Water Services and and with Plastic Bodies for Water Services Part 4: Mechanical and Endurance Characteristics. 12 pp.

—**Microorganism Resistant**
BSI BS 5412 & 5413: Part 5-76. 1976 Amd 3 Performance of Draw-Off Taps with Metal Bodies for Water Services and with Plastic Bodies for Water Services Part 5: Physio-Chemical Characteristics: Materials, Coatings. 15 pp.

—**Mixing**
CEN PREN 817-92. Mechanical Mixers—High Pressure. 44 pp.

—**Nickel Coatings**
BSI BS EN 248-92. 1992 Sanitary Taps: General Technical Specifications for Electrodeposited Nickel Chrome Coatings. 10 pp.
CEN EN 248-89. Sanitary Taps: General Technical Specifications for Electrodeposited Nickel Chrome Coatings. 5 pp.
DIN ENGL EN 248-90. Sanitary Taps; General Technical Specifications for Electrodeposited Nickel Chrome Coatings (Jan). 4 pp.

—**Noise**
BSI BS 6864: Part 2-87. 1987 Laboratory Tests on Noise Emission from Appliances and Equipment Intended for Use in Water Supply Installations Part 2: Method for Mounting and Operating Draw-Offs Taps. 8 pp.

Faucets (Cont.)
—**Noise** (Cont.)
ISO 3822 Pt 2-84. Acoustics—Laboratory Tests on Noise Emission from Appliances and Equipment Used in Water Supply Installations—Part 2: Mounting and Operating Conditions for Draw-off Taps First Edition. 6 pp.

—**Pressure Resistance**
BSI BS 5412 & 5413: Part 2-76. 1976 Amd 2 Performance of Draw-Off Taps with Metal Bodies for Water Services and with Plastic Bodies for Water Services Part 2: Water Tightness and Pressure Resistance Characteristics. 7 pp.

—**Sanitary**
CEN PREN 816-92. Automatic Shut-Off Valves for Sanitary Installations PN 10. 21 pp.

—**Seat Washers**
BSI BS 3457-73. 1973 Materials for Water Tap and Stopvalve Seat Washers. 7 pp.

—**Showers**
SNZ NZS 4611-82. Non-Thermostatic Shower Mixing Valves Amend: 1, 1990. 4 pp.

—**Spray**
BSI BS 5388-76. 1976 Spray Taps. 10 pp.
BSI BS 5779-79. 1979 Spray Mixing Taps. 14 pp.

—**Tempering Valves**
SNZ NZS 4617-89. Tempering (3-Port Mixing) Valves. 16 pp.

—**Watertightness**
BSI BS 5412 & 5413: Part 2-76. 1976 Amd 2 Performance of Draw-Off Taps with Metal Bodies for Water Services and with Plastic Bodies for Water Services Part 2: Water Tightness and Pressure Resistance Characteristics. 7 pp.

Fault Currents
Scope Note: Use a more specific term *See:* Leakage Currents; Short Circuit Current Testing

Fault Report Point
See Also: Maintenance Organizations (Telecommunications)

—**Telephone Circuits**
CCITT RECMN M.715-89. Fault Report Point (Circuit)—General Maintenance Principles—Maintenance of International Transmission Systems and Telephone Circuits (Study Group IV) 2 pp. 2 pp.

—**Telephone Networks**
CCITT RECMN M.716-89. Fault Report Point (Network)—General Maintenance Principles—Maintenance of International Transmission Systems and Telephone Circuits (Study Group IV) 3 pp. 3 pp.

Fault Tree Analysis
Use For: FTA

BSI BS 5760: Part 7-91. 1991 Reliability of Systems, Equipment and Components Part 7: Guide to Fault Tree Analysis (G) (IEC 1025: 1990). 24 pp.
DIN ENGL 25424 Pt 1-81. Fault Tree Analysis; Method and Graphical Symbols (Sept). 8 pp.

—**Symbols**
CENELEC HD 617 S1-92. Fault Tree Analysis (FTA). 5 pp.
DIN ENGL 25424 Pt 1-81. Fault Tree Analysis; Method and Graphical Symbols (Sept). 8 pp.
IEC 1025-90. Fault Tree Analysis (FTA) First Edition. 42 pp.

Faulting (Electrical)
Use: Electrical Faults

Faults (Telecommunications)
—**Data Transmission Equipment**
CCITT RECMN M.1355-89. Maintenance of International Data Transmission Systems Operating in the Range 2.4 to 14.4 kBit/s—Maintenance of International Telegraph, Phototelegraph and Leased Circuits—Maintenance of the International Public Telephone Network-. 2 pp.
CCITT RECMN M.1375-89. Maintenance of International Data Transmission Systems Operating at 48 kBit/s and Above—Maintenance of International Telegraph, Phototelegraph and Leased Circuits—Maintenance of the International Public Telephone Network-. 4 pp.

Faults (Telecommunications) (Cont.)
—**Telephone Circuits**
CCITT RECMN M.130-89. Operational Procedures in Locating and Clearing Transmission Faults—General Maintenance Principles—Maintenance of International Transmission Systems and Telephone Circuits (Study Group IV) 9 pp. 9 pp.

—**Telephone Networks**
CCITT RECMN M.1530-92. Network Maintenance Information (Study Group IV) 8 pp. 8 pp.

FAX Communications
Use For: International Public Facsimile Services; Phototelegram Services; Phototelegraph Transmission; Telefax Communications *See Also:* Bureaufax Communications; Datafax Communications; Electronic Mail Systems; Encoded Information Types; FAX Machines; Index of Cooperation; Information Interchange; Judder; Message Handling Systems; Phasing (Data Transmission); Phototelegraphy; Scanning Tracks; Teleinformatic Services; Telephone Services; Teleservices

CCITT RECMN A.21-89. Collaboration with Other International Organizations on CCITT-Defined Telematic Services—Terminal Equipment and Protocols for Telematic Services (Study Group VIII) 2 pp. 2 pp.
CCITT RECMN F.1-89. Operational Provisions for the International Public Telegram Service—Telegraph and Mobile Services Operations and Quality of Service (Study Group I) 54 pp. 54 pp.
CCITT RECMN F.160-89. General Operational Provisions for the International Public Facsimile Services—Telematic, Data Transmission and Teleconference Services —Operations and Quality of Service (Study Group I) 4 pp. 4 pp.
CCITT RECMN F.190-92. Operational Provisions for the International Facsimile Service Between Public Bureaus and Subscriber Stations and vice versa (Bureaufax—telefax and vice versa) (Study Group I) 5 pp. 5 pp.
CCITT RECMN F.190-89. Operational Provisions for the International Facsimile Service Between Public Bureaus and Subscriber Stations & Vice Versa (Bureaufax—Telefax and Vice-Versa)—Telematic, Data Transmission and Teleconference Serv.—Operations & Quality of Serv. (Study Gr. I) 3 pp. 3 pp.
CEPT T/SF 16-80. Definition D'un Service Public De Telecopie Utilisant Le Reseau Telephonique Public. 6 pp.
CEPT T/TG 18 E-89. Operational Aspects of an International Telefax Service. 2 pp.

—**Accounting**
CCITT RECMN D.71 (REV 1)-92. General Tariff Principles for the Public Facsimile Service Between Subscriber Stations (Telefax Service) (Study Group III) 5 pp. 5 pp.
CCITT RECMN D.71-89. General Tariff Principles for the Public Facsimile Between Subscriber Stations (Telefax Service)—General Tariff Principles—Charging and Accounting in International Telecommunications Services (Study Group III) 2 pp. 2 pp.

—**Address Codes**
CCITT RECMN F.1-89. Operational Provisions for the International Public Telegram Service—Telegraph and Mobile Services Operations and Quality of Service (Study Group I) 54 pp. 54 pp.

—**Bureaufax Communications—Interworking**
CCITT RECMN D.73 (REV 1)-92. General Tariff and Intl Accounting Principles for Interworking Between the International Bureaufax and Telefax Services (Study Group III) 5 pp. 5 pp.
CCITT RECMN D.73-89. General Tariff and Intl Accounting Principles for Interworking Between the International Bureaufax and Telefax Services— General Tariff Principles—Charging and Accounting in International Telecommunications Services (Study Group III) 2 pp. 2 pp.
CCITT RECMN F.190-92. Operational Provisions for the International Facsimile Service Between Public Bureaus and Subscriber Stations and vice versa (Bureaufax—telefax and vice versa) (Study Group I) 5 pp. 5 pp.
CCITT RECMN F.190-89. Operational Provisions for the International Facsimile Service Between Public Bureaus and Subscriber Stations & Vice Versa (Bureaufax—Telefax and Vice-Versa)—Telematic, Data Transmission and Teleconference Serv.—Operations & Quality of Serv. (Study Gr. I) 3 pp. 3 pp.

—**CCITT Group 1—Telephone Networks**
CCITT RECMN T.30-88. Procedures for Document Facsimile Transmission in the General Switched Telephone Network—Terminal Equipment and Protocols for Telematic Services (Study Group VIII) 98 pp. 98 pp.

INTERNATIONAL AND NON-U.S. NATIONAL STANDARDS
SUBJECT INDEX

FAX

FAX Communications *(Cont.)*

—CCITT Group 1—Telephone Networks *(Cont.)*

CCITT RECMN T.30-88. Procedures for Document Facsimile Transmission in the General Switched Telephone Network—Terminal Equipment and Protocols for Telematic Services (Study Group VIII) 98 pp (Revision (1990)). 101 pp.

—CCITT Group 2

CEPT T/SF 12-81. Services Et Facilities Pour Un Telecopieur (Groupe 2) Destine a Un Marche a Usage Faible, Moyen Et Intensif. 3 pp.

—CCITT Group 2—Public Networks

CCITT RECMN F.180-89. General Operational Provisions for the International Public Facsimile Service Between Subscribers' Stations (Telefax)—Telematic, Data Transmission and Teleconference Services —Operations and Quality of Service (Study Group I) 5 pp. 5 pp.

—CCITT Group 2—Public Switched Telephone Networks

CCITT RECMN F.182-89. Operational Provisions for the International Public Facsimile Service Between Subscriber Stations with Groups 2 and 3 Facsimile Machines (Telefax 2 and Telefax 3)—Telematic, Data Transmission and Teleconference Services —Operations and Quality of Service. 2 pp.

—CCITT Group 2—Telephone Networks

CCITT RECMN T.30-88. Procedures for Document Facsimile Transmission in the General Switched Telephone Network—Terminal Equipment and Protocols for Telematic Services (Study Group VIII) 98 pp. 98 pp.

CCITT RECMN T.30-88. Procedures for Document Facsimile Transmission in the General Switched Telephone Network—Terminal Equipment and Protocols for Telematic Services (Study Group VIII) 98 pp (Revision (1990)). 101 pp.

—CCITT Group 3

CENELEC PRI-ETS 300 071-92. European Digital Cellular Telecommunications System (Phase 1); Technical Realization of Facsimile Group 3 Non-Transparent. 57 pp.

CEPT T/SF 60 E-87. Definition of a Public Facsimile Service Between Users Using CCITT Recommendations for Group 3 and Group 4. 3 pp.

ETSI PRI-ETS 300 071-92. European Digital Cellular Telecommunications System (Phase 1); Technical Realization of Facsimile Group 3 Non-Transparent. 57 pp.

—CCITT Group 3—Public Land Mobile Networks

CENELEC PRI-ETS 300 070-92. European Digital Cellular Telecommunications System (Phase 1); Technical Realization of Facsimile Group 3 Transparent. 42 pp.

CENELEC GSM 03.45-93. See PRI-ETS 300 070. 40 pp.

ETSI PRI-ETS 300 070-92. European Digital Cellular Telecommunications System (Phase 1); Technical Realization of Facsimile Group 3 Transparent. 42 pp.

ETSI GSM 03.45-93. See PRI-ETS 300 070. 40 pp.

—CCITT Group 3—Public Networks

CCITT RECMN F.180-89. General Operational Provisions for the International Public Facsimile Service Between Subscribers' Stations (Telefax)—Telematic, Data Transmission and Teleconference Services —Operations and Quality of Service (Study Group I) 5 pp. 5 pp.

—CCITT Group 3—Public Switched Telephone Networks

CCITT RECMN F.182-89. Operational Provisions for the International Public Facsimile Service Between Subscriber Stations with Groups 2 and 3 Facsimile Machines (Telefax 2 and Telefax 3)—Telematic, Data Transmission and Teleconference Services —Operations and Quality of Service. 2 pp.

—CCITT Group 3—Telephone Networks

CCITT RECMN T.30-88. Procedures for Document Facsimile Transmission in the General Switched Telephone Network—Terminal Equipment and Protocols for Telematic Services (Study Group VIII) 98 pp. 98 pp.

CCITT RECMN T.30-88. Procedures for Document Facsimile Transmission in the General Switched Telephone Network—Terminal Equipment and Protocols for Telematic Services (Study Group VIII) 98 pp (Revision (1990)). 101 pp.

FAX Communications *(Cont.)*

—CCITT Group 4

CCITT RECMN F.184-89. Operational Provisions for the International Public Facsimile Service Between Subscriber Stations with Group 4 Facsimile Machines (Telefax 4)—Telematic, Data Transmission and Teleconference Services—Operations and Quality of Service (Study Group I) 9 pp. 9 pp.

CEPT T/SF 60 E-87. Definition of a Public Facsimile Service Between Users Using CCITT Recommendations for Group 3 and Group 4. 3 pp.

—CCITT Group 4—Circuit Switched Public Data Networks

CCITT RECMN F.184-89. Operational Provisions for the International Public Facsimile Service Between Subscriber Stations with Group 4 Facsimile Machines (Telefax 4)—Telematic, Data Transmission and Teleconference Services—Operations and Quality of Service (Study Group I) 9 pp. 9 pp.

—CCITT Group 4—Codes

CCITT RECMN F.184-89. Operational Provisions for the International Public Facsimile Service Between Subscriber Stations with Group 4 Facsimile Machines (Telefax 4)—Telematic, Data Transmission and Teleconference Services—Operations and Quality of Service (Study Group I) 9 pp. 9 pp.

—CCITT Group 4—Document Formats

CCITT RECMN F.184-89. Operational Provisions for the International Public Facsimile Service Between Subscriber Stations with Group 4 Facsimile Machines (Telefax 4)—Telematic, Data Transmission and Teleconference Services—Operations and Quality of Service (Study Group I) 9 pp. 9 pp.

—CCITT Group 4—Integrated Services Digital Networks

CCITT RECMN F.184-89. Operational Provisions for the International Public Facsimile Service Between Subscriber Stations with Group 4 Facsimile Machines (Telefax 4)—Telematic, Data Transmission and Teleconference Services—Operations and Quality of Service (Study Group I) 9 pp. 9 pp.

CENELEC PRETS 300 280-92. Terminal Equipment (TE); Facsimile Group 4 Class 1 Equipment on the Integrated Services Digital Network (ISDN) Terminal Testing. 44 pp.

ETSI PRETS 300 280-92. Terminal Equipment (TE); Facsimile Group 4 Class 1 Equipment on the Integrated Services Digital Network (ISDN) Terminal Testing. 44 pp.

—CCITT Group 4—Numbering Plans

CCITT RECMN F.184-89. Operational Provisions for the International Public Facsimile Service Between Subscriber Stations with Group 4 Facsimile Machines (Telefax 4)—Telematic, Data Transmission and Teleconference Services—Operations and Quality of Service (Study Group I) 9 pp. 9 pp.

—CCITT Group 4—Office Document Interchange Format

CCITT RECMN T.503-91. Document Application Profile for the Interchange of Group 4 Facsimile Documents (Study Group VIII) 11 pp. 11 pp.

CCITT RECMN T.503-89. Document Application Profile for the Interchange of Group 4 Facsimile Documents—Terminal Equipment and Protocols for Telematic Services (Study Group VIII) 9 pp. 9 pp.

—CCITT Group 4—Packet Switched Public Data Networks

CCITT RECMN F.184-89. Operational Provisions for the International Public Facsimile Service Between Subscriber Stations with Group 4 Facsimile Machines (Telefax 4)—Telematic, Data Transmission and Teleconference Services—Operations and Quality of Service (Study Group I) 9 pp. 9 pp.

—CCITT Group 4—Public Networks

CCITT RECMN F.180-89. General Operational Provisions for the International Public Facsimile Service Between Subscribers' Stations (Telefax)—Telematic, Data Transmission and Teleconference Services —Operations and Quality of Service (Study Group I) 5 pp. 5 pp.

—CCITT Group 4—Public Switched Telephone Networks

CCITT RECMN F.184-89. Operational Provisions for the International Public Facsimile Service Between Subscriber Stations with Group 4 Facsimile Machines (Telefax 4)—Telematic, Data Transmission and Teleconference Services—Operations and Quality of Service (Study Group I) 9 pp. 9 pp.

FAX Communications *(Cont.)*

—CCITT Group 4—Quality of Service

CCITT RECMN F.184-89. Operational Provisions for the International Public Facsimile Service Between Subscriber Stations with Group 4 Facsimile Machines (Telefax 4)—Telematic, Data Transmission and Teleconference Services—Operations and Quality of Service (Study Group I) 9 pp. 9 pp.

—CCITT Group 4—Teletex Communications—Interworking

CCITT RECMN F.184-89. Operational Provisions for the International Public Facsimile Service Between Subscriber Stations with Group 4 Facsimile Machines (Telefax 4)—Telematic, Data Transmission and Teleconference Services—Operations and Quality of Service (Study Group I) 9 pp. 9 pp.

—CCITT Group 4—Teletex—Control Procedures

CCITT RECMN T.62-89. Control Procedures for Teletex and Group 4 Facsimile Services—Terminal Equipment and Protocols for Telematic Services (Study Group VIII) 180 pp. 180 pp.

—CCITT Group 4—Teletex—Control Procedures—X.215

CCITT RECMN T.62 BIS-89. Control Procedures for Teletex and G4 Facsimile Services Based on Recommendations X.215 and X.225—Terminal Equipment and Protocols for Telematic Services (Study Group VIII) 32 pp. 32 pp.

—CCITT Group 4—Teletex—Control Procedures—X.225

CCITT RECMN T.62 BIS-89. Control Procedures for Teletex and G4 Facsimile Services Based on Recommendations X.215 and X.225—Terminal Equipment and Protocols for Telematic Services (Study Group VIII) 32 pp. 32 pp.

—CCITT Group 4—Teletex—Mixed Mode

CCITT RECMN T.561-89. Terminal Characteristics for Mixed Mode of Operation MM—Terminal Equipment and Protocols for Telematic Services (Study Group VIII) 12 pp. 12 pp.

—Directories

CCITT RECMN F.180-89. General Operational Provisions for the International Public Facsimile Service Between Subscribers' Stations (Telefax)—Telematic, Data Transmission and Teleconference Services —Operations and Quality of Service (Study Group I) 5 pp. 5 pp.

—Directories—Accounting

CCITT RECMN F.180-89. General Operational Provisions for the International Public Facsimile Service Between Subscribers' Stations (Telefax)—Telematic, Data Transmission and Teleconference Services —Operations and Quality of Service (Study Group I) 5 pp. 5 pp.

—Glossaries

BSI BS 4727:Pt3: Group 03-92. 1992 Glossary of Electrotechnical, Power, Telecommunication, Electronics, Lighting and Colour Terms Part 3: Terms Particular to Telecommunications and Electronics Group 03: Telegraphy, Facsimile and Data Communication (G). 77 pp.

BSI BS 4727:Pt3: Group 03-71. 1971 Electrotechnical, Power, Telecommunication, Electronics, Lighting and Colour Terms Part 3: Terms Particular to Telecommunications and Electronics Group 03: Telegraphy, Including Facsimile, Terminology. 26 pp.

—Integrated Services Digital Networks

CCITT RECMN I.241. 3-88. Telefax 4—Integrated Services Digital Network (ISDN)—General Structure and Service Capabilities (Study Group XVIII) 8 pp. 8 pp.

—International Leased Circuits

CCITT RECMN M.1015-89. Types of Transmission on Leased Circuits—Maintenance of International Telegraph, Phototelegraph and Leased Circuits—Maintenance of the International Public Telephone Network—Maintenance of Maritime Satellite and Data Transmission Systems. 3 pp.

—International Numbers (Telephone)—Notation

CCITT RECMN E.123-89. Notation for National and International Telephone Numbers—Telephone Network and ISDN—Operation, Numbering, Routing and Mobile Service (Study Group II) 5 pp. 5 pp.

INDUSTRY STANDARDS

FAX Communications (Cont.)

—Interpersonal Messaging Services—Intercommunication
CCITT RECMN F.423-92. Message Handling Services: Intercommunication Between the Interpersonal Messaging Service and the Telefax Service (Study Group I) 8 pp. 8 pp.

—Interworking
CENELEC ETR 030-92. Network Aspects (NA); Interworking Aspects of MoU-ISDN Priority 1 + 2 Services. 20 pp.
ETSI ETR 030-92. Network Aspects (NA); Interworking Aspects of MoU-ISDN Priority 1 + 2 Services. 20 pp.

—Leased Circuits
CCITT RECMN H.43-89. Document Facsimile Transmissions on Leased Telephone-Type Circuits—Line Transmission of Non-Telephone Signals—Transmission of Sound-Programme and Television Signals (Study Group XV) 1 pp. 1 pp.
CCITT RECMN T.10-89. Document Facsimile Transmissions on Leased Telephone-Type Circuits—Terminal Equipment and Protocols for Telematic Services (Study Group VIII) 2 pp. 2 pp.

—Maritime Mobile Services—Radiotelephony
CCIR QUESTION 14-1/8-78. Direct-Printing and Other Data Signals Using Voice-Frequency Techniques on VHF Radiotelephony Channels in the Maritime Mobile Service—Questions Concerning Study Group 8 —Mobile, Radiodetermination, Amateur and Related Satellite Services. 1 p.

—Metallic Circuits—Maritime Mobile Satellite Communications
CCIR Report 588-1-78. Black and White Facsimile Transmissions over Combined Metallic and Radio Circuits in the Maritime Mobile Service and in the Maritime Mobile-Satellite Service —Section 8B—Maritime Mobile Service: Telegraphy and Related Subjects. 3 pp.

—Metallic Circuits—Maritime Mobile Services
CCIR Report 588-1-78. Black and White Facsimile Transmissions over Combined Metallic and Radio Circuits in the Maritime Mobile Service and in the Maritime Mobile-Satellite Service —Section 8B—Maritime Mobile Service: Telegraphy and Related Subjects. 3 pp.

—Modems—Telephone Networks
CCITT RECMN V.17-91. 2-Wire Modem for Facsimile Applications with Rates up to 14 400 bit/s (Study Group XVII) 14 pp. 14 pp.

—National Significant Numbers—Notation
CCITT RECMN E.123-89. Notation for National and International Telephone Numbers—Telephone Network and ISDN—Operation, Numbering, Routing and Mobile Service (Study Group II) 5 pp. 5 pp.

—Protocols
CCITT FASCICLE VII.3. Terminal Equipment and Protocols for Telematic Services—Recommendations T.0—T.63. 494 pp.
CCITT FASCICLE VII.6. Terminal Equipment and Protocols for Telematic Services—Recommendations T.400—T.418. 482 pp.
CCITT FASCICLE VII.7. Terminal Equipment and Protocols for Telematic Services—Recommendations T.431—T.564. 329 pp.

—Public Networks
CCITT RECMN F.180-89. General Operational Provisions for the International Public Facsimile Service Between Subscribers' Stations (Telefax)—Telematic, Data Transmission and Teleconference Services —Operations and Quality of Service (Study Group I) 5 pp. 5 pp.

—Quality of Service
CCITT RECMN F.160-89. General Operational Provisions for the International Public Facsimile Services—Telematic, Data Transmission and Teleconference Services —Operations and Quality of Service (Study Group I) 4 pp. 4 pp.

—Radio Circuits—Maritime Mobile Satellite Communications
CCIR Report 588-1-78. Black and White Facsimile Transmissions over Combined Metallic and Radio Circuits in the Maritime Mobile Service and in the Maritime Mobile-Satellite Service —Section 8B—Maritime Mobile Service: Telegraphy and Related Subjects. 3 pp.

FAX Communications (Cont.)

—Radio Circuits—Maritime Mobile Services
CCIR Report 588-1-78. Black and White Facsimile Transmissions over Combined Metallic and Radio Circuits in the Maritime Mobile Service and in the Maritime Mobile-Satellite Service —Section 8B—Maritime Mobile Service: Telegraphy and Related Subjects. 3 pp.

—Radio Circuits—Meteorological Charts
CCIR RECMN 343-1-66. Facsimile Transmission of Meteorological Charts Over Radio Circuits—Section 3Cc—Phototelegraphy (Facsimile). 1 p.

—Service Messages
CCITT RECMN D.71 (REV 1)-92. General Tariff Principles for the Public Facsimile Service Between Subscriber Stations (Telefax Service) (Study Group III) 5 pp. 5 pp.
CCITT RECMN D.71-89. General Tariff Principles for the Public Facsimile Between Subscriber Stations (Telefax Service)—General Tariff Principles—Charging and Accounting in International Telecommunications Services (Study Group III) 2 pp. 2 pp.
CCITT RECMN F.160-89. General Operational Provisions for the International Public Facsimile Services—Telematic, Data Transmission and Teleconference Services —Operations and Quality of Service (Study Group I) 4 pp. 4 pp.

—Service Messages—Public Networks
CCITT RECMN F.180-89. General Operational Provisions for the International Public Facsimile Service Between Subscribers' Stations (Telefax)—Telematic, Data Transmission and Teleconference Services —Operations and Quality of Service (Study Group I) 5 pp. 5 pp.

—Store and Forward Mode
CCITT RECMN F.162-92. Service and Operational Requirements of Store-and Forward Facsimile Service (Study Group I) 13 pp. 13 pp.

—Store and Forward Mode—Address Codes
CCITT RECMN F.162-89. Operational Requirements of an International Store-and-Forward Facsimile Switching Service (Comfax)—Telematic, Data Transmission and Teleconference Services —Operations and Quality of Service—(Study Group I) 7 pp. 7 pp.

—Store and Forward Mode—Closed User Groups
CCITT RECMN F.162-89. Operational Requirements of an International Store-and-Forward Facsimile Switching Service (Comfax)—Telematic, Data Transmission and Teleconference Services —Operations and Quality of Service—(Study Group I) 7 pp. 7 pp.

—Store and Forward Mode—Message Handling Systems
CCITT RECMN F.162-89. Operational Requirements of an International Store-and-Forward Facsimile Switching Service (Comfax)—Telematic, Data Transmission and Teleconference Services —Operations and Quality of Service—(Study Group I) 7 pp. 7 pp.

—Store and Forward Mode—Message Switching
CCITT RECMN F.162-89. Operational Requirements of an International Store-and-Forward Facsimile Switching Service (Comfax)—Telematic, Data Transmission and Teleconference Services —Operations and Quality of Service—(Study Group I) 7 pp. 7 pp.

—Store and Forward Mode—Multiple Address Calls
CCITT RECMN F.162-89. Operational Requirements of an International Store-and-Forward Facsimile Switching Service (Comfax)—Telematic, Data Transmission and Teleconference Services —Operations and Quality of Service—(Study Group I) 7 pp. 7 pp.

—Store and Forward Mode—Quality of Service
CCITT RECMN F.162-89. Operational Requirements of an International Store-and-Forward Facsimile Switching Service (Comfax)—Telematic, Data Transmission and Teleconference Services —Operations and Quality of Service—(Study Group I) 7 pp. 7 pp.

FAX Communications (Cont.)

—Symbols
CCITT RECMN E.121-89. Pictograms and Symbols to Assist Users of the Telephone Service—Telephone Network and ISDN—Operation, Numbering, Routing and Mobile Service (Study Group II) 11 pp. 11 pp.
CEPT T/TG 23 E-89. Pictogram Designation Facsimile (Telefax, Bureaufax). 1 p.

—Tariffs
CCITT RECMN D.71 (REV 1)-92. General Tariff Principles for the Public Facsimile Service Between Subscriber Stations (Telefax Service) (Study Group III) 5 pp. 5 pp.
CCITT RECMN D.71-89. General Tariff Principles for the Public Facsimile Between Subscriber Stations (Telefax Service)—General Tariff Principles—Charging and Accounting in International Telecommunications Services (Study Group III) 2 pp. 2 pp.

—Tariffs—Refunds
CCITT RECMN D.71 (REV 1)-92. General Tariff Principles for the Public Facsimile Service Between Subscriber Stations (Telefax Service) (Study Group III) 5 pp. 5 pp.
CCITT RECMN D.71-89. General Tariff Principles for the Public Facsimile Between Subscriber Stations (Telefax Service)—General Tariff Principles—Charging and Accounting in International Telecommunications Services (Study Group III) 2 pp. 2 pp.

—Telephone Circuits
CCITT RECMN T.10-89. Document Facsimile Transmissions on Leased Telephone-Type Circuits—Terminal Equipment and Protocols for Telematic Services (Study Group VIII) 2 pp. 2 pp.

—Telephone Networks
CCITT RECMN T.10 BIS-89. Document Facsimile Transmissions in the General Switched Telephone Network—Terminal Equipment and Protocols for Telematic Services (Study Group VIII) 1 pp. 1 pp.

—Test Charts
CCITT RECMN T.20-89. Standardized Test Chart for Facsimile Transmissions—Terminal Equipment and Protocols for Telematic Services (Study Group VIII) 4 pp. 4 pp.
CCITT RECMN T.21-89. Standardized Test Charts for Document Facsimile Transmissions—Terminal Equipment and Protocols for Telematic Services (Study Group VIII) 6 pp. 6 pp.

—Trademark Rights—Exceptions
CCITT FASCICLE I.2-88. Opinions and Resolutions Recommendations on the Organisation and Working Procedures of CCITT (Series A). 64 pp.
CCITT FASCICLE II.4-88. Telegraph and Mobile Service—Operations and Quality of Service Recommendations F.1—F.140. 293 pp.
CCITT FASCICLE II.5-88. Telematic, Data Transmission and Teleconference Service—Operations and Quality of Service Recommendations F.160—F.353, F.600, F.601, F.710—F.730. 154 pp.
CCITT FASCICLE II.6-88. Message Handling and Directory Services—Operations and Definition of Service Recommendations F.400—F.422, F.500. 184 pp.

FAX Equipment
Use: FAX Machines

FAX Machines
Use For: Document Facsimile Terminals; Facsimile Equipment; FAX Equipment; FAX Terminals; Phototelegraph Systems; Telefax Equipment
See Also: Electronic Mail Systems; FAX Communications; FAX Modules; Index of Cooperation; Judder; Phasing (Data Transmission); Scanning Tracks; Telecommunication Equipment; Teleinformatic Services

CCITT FASCICLE VII.3. Terminal Equipment and Protocols for Telematic Services—Recommendations T.0—T.63. 494 pp.
CCITT FASCICLE VII.6. Terminal Equipment and Protocols for Telematic Services—Recommendations T.400—T.418. 482 pp.
CCITT FASCICLE VII.7. Terminal Equipment and Protocols for Telematic Services—Recommendations T.431—T.564. 329 pp.
CEPT T/SF 18-81. Mire Normalisee D'Abonne Pour Utilisation Avec Un Terminal TELEFAX. 2 pp.
OSI ISO DP 9071-1-87. Text and Office Systems—Basic and Optional Requirements—Part 1: Facsimile Equipment. 12 pp.

—CCITT Group 1
CCITT RECMN T.2-89. Standardization of Group 1 Facsimile Apparatus for Document Transmission—Terminal Equipment and Protocols for Telematic Services (Study Group VIII) 3 pp. 3 pp.

INTERNATIONAL AND NON-U.S. NATIONAL STANDARDS
SUBJECT INDEX

FAX

FAX Machines *(Cont.)*

—CCITT Group 1—Index of Cooperation
CCITT RECMN T.2-89. Standardization of Group 1 Facsimile Apparatus for Document Transmission—Terminal Equipment and Protocols for Telematic Services (Study Group VIII) 3 pp. 3 pp.

—CCITT Group 1—Public Networks—Classification
CCITT RECMN T.0-89. Classification of Facsimile Apparatus for Document Transmission over the Public Networks—Terminal Equipment and Protocols for Telematic Services (Study Group VIII) 4 pp. 4 pp.

—CCITT Group 1—Scanning Tracks
CCITT RECMN T.2-89. Standardization of Group 1 Facsimile Apparatus for Document Transmission—Terminal Equipment and Protocols for Telematic Services (Study Group VIII) 3 pp. 3 pp.

—CCITT Group 1—Telephone Transmission
CCITT RECMN T.2-89. Standardization of Group 1 Facsimile Apparatus for Document Transmission—Terminal Equipment and Protocols for Telematic Services (Study Group VIII) 3 pp. 3 pp.

—CCITT Group 2
CEPT T/CD 05-01 E-86. Specification of a CCITT Group 2 Facsimile Terminal for the Low Usage Market. 10 pp.

CEPT T/CD 05-02-81. Specifications D'un Telecopieur Du Grooupe 2 Du CCITT Pour Moyen a Fort Trafic. 11 pp.

CEPT T/CD 05-02 E-81. Specification of a CCITT Group 2 Document Facsimile Terminal for the Medium to High Usage Market. 10 pp.

CEPT T/SF 12-81. Services Et Facilities Pour Un Telecopieur (Groupe 2) Destine a Un Marche a Usage Faible, Moyen Et Intensif. 3 pp.

—CCITT Group 2—PSTN
CCITT RECMN F.182-89. Operational Provisions for the International Public Facsimile Service Between Subscriber Stations with Groups 2 and 3 Facsimile Machines (Telefax 2 and Telefax 3)—Telematic, Data Transmission and. Teleconference Services—Operations and Quality of Service. 2 pp.

—CCITT Group 2—Public Networks
CCITT RECMN F.180-89. General Operational Provisions for the International Public Facsimile Service Between Subscribers' Stations (Telefax)—Telematic, Data Transmission and Teleconference Services —Operations and Quality of Service (Study Group I) 5 pp. 5 pp.

—CCITT Group 2—Public Networks—Classification
CCITT RECMN T.0-89. Classification of Facsimile Apparatus for Document Transmission over the Public Networks—Terminal Equipment and Protocols for Telematic Services (Study Group VIII) 4 pp. 4 pp.

—CCITT Group 2—Telephone Transmission
CCITT RECMN T.3-89. Standardization of Group 2 Facsimile Apparatus for Document Transmission—Terminal Equipment and Protocols for Telematic Services (Study Group VIII) 7 pp. 7 pp.

—CCITT Group 2—Terminal Identification
CCITT RECMN F.182-89. Operational Provisions for the International Public Facsimile Service Between Subscriber Stations with Groups 2 and 3 Facsimile Machines (Telefax 2 and Telefax 3)—Telematic, Data Transmission and Teleconference Services—Operations and Quality of Service. 2 pp.

—CCITT Group 3
CENELEC PRETS 300 242-92. Terminal Equipment (TE); Group 3 Facsimile Equipment. 43 pp.

CENELEC ETS 300 242-92. Terminal Equipment (TE); Group 3 Facsimile Equipment. 43 pp.

CEPT T/CD 05-03-81. Specifications D'un Telecopieur Du Groupe 3 Du CCITT. 10 pp.

CEPT T/CD 05-03 E-81. Specification of a CCITT Group 3 Document Facsimile Terminal. 9 pp.

CEPT T/SF 21-82. Caracteristiques Applicables Aux Telecopieurs De Groupe 3. 3 pp.

ETSI ETS 300 242-92. Terminal Equipment (TE); Group 3 Facsimile Equipment. 43 pp.

ETSI PRETS 300 242-92. Terminal Equipment (TE); Group 3 Facsimile Equipment. 43 pp.

—CCITT Group 3—Codes—Procedures
CCITT RECMN T.35-89. Procedure for the Allocation of CCITT Members' Codes—Terminal Equipment and Protocols for Telematic Services (Study Group VIII) 5 pp. 5 pp.

FAX Machines *(Cont.)*

—CCITT Group 3—Interpersonal Messaging Services
CCITT RECMN F.420-92. Message Handling Services: Public Interpersonal Messaging Service (Study Group I) 16 pp. 16 pp.

CCITT RECMN F.420-89. Message Handling Services: the Public Interpersonal Messaging Service—Message Handling and Directory Services—Operations and Definition of Service (Study Group I) 15 pp. 15 pp.

—CCITT Group 3—PSTN
CCITT RECMN F.182-89. Operational Provisions for the International Public Facsimile Service Between Subscriber Stations with Groups 2 and 3 Facsimile Machines (Telefax 2 and Telefax 3)—Telematic, Data Transmission and Teleconference Services—Operations and Quality of Service. 2 pp.

—CCITT Group 3—PSTN—Telex Communications—Transfer of Messages
CCITT RECMN F.87-91. Operational Principles for the Transfer of Messages from Terminals on the Telex Network to Group 3 Facsimile Terminals Connected to the Public Switched Telephone Network (Study Group I) 12 pp. 12 pp.

—CCITT Group 3—Public Networks
CCITT RECMN F.180-89. General Operational Provisions for the International Public Facsimile Service Between Subscribers' Stations (Telefax)—Telematic, Data Transmission and Teleconference Services —Operations and Quality of Service (Study Group I) 5 pp. 5 pp.

—CCITT Group 3—Public Networks—Classification
CCITT RECMN T.0-89. Classification of Facsimile Apparatus for Document Transmission over the Public Networks—Terminal Equipment and Protocols for Telematic Services (Study Group VIII) 4 pp. 4 pp.

—CCITT Group 3—Telephone Transmission
CCITT RECMN T.4-88. Standardization of Group 3 Facsimile Apparatus for Document Transmission—Terminal Equipment and Protocols for Telematic Services (Study Group VIII) 27 pp. 27 pp.

CCITT RECMN T.4-88. Standardization of Group 3 Facsimile Apparatus for Document Transmission—Terminal Equipment and Protocols for Telematic Services (Study Group VIII) 27 pp (Revision (1990)). 29 pp.

—CCITT Group 3—Terminal Identification
CCITT RECMN F.182-89. Operational Provisions for the International Public Facsimile Service Between Subscriber Stations with Groups 2 and 3 Facsimile Machines (Telefax 2 and Telefax 3)—Telematic, Data Transmission and Teleconference Services —Operations and Quality of Service. 2 pp.

—CCITT Group 4
CCITT RECMN F.184-89. Operational Provisions for the International Public Facsimile Service Between Subscriber Stations with Group 4 Facsimile Machines (Telefax 4)—Telematic, Data Transmission and Teleconference Services—Operations and Quality of Service (Study Group I) 9 pp. 9 pp.

CENELEC PRETS 300 087-91. Integrated Services Digital Network (ISDN); Facsimile Group 4 Class 1 Equipment on the ISDN Functional Specification of the Equipment (T/TE 05.09). 12 pp.

CENELEC PRETS 300 088-91. Integrated Services Digital Network (ISDN); Facsimile Group 4 Class 1 Equipment on the ISDN General and Service Aspects (T/TE 05-06). 10 pp.

CENELEC PRETS 300 112-90. Integrated Services Digital Network (ISDN): Facsimile Group 4 Class 1 Equipment on the ISDN End-to-End Protocols (T/TE 05-07). 25 pp.

CENELEC PRETS 300 112-90. Integrated Services Digital Network (ISDN); Facsimile Group 4 Class 1 Equipment on the ISDN End-to-End Protocols (T/TE 05-07). 48 pp.

CENELEC PRETS 300 120-90. Integrated Services Digital Network (ISDN). 15 pp.

CENELEC PRETS 300 120-92. Integrated Services Digital Network (ISDN); Service Requirements for Telefax Group 4. 16 pp.

CENELEC ETS 300 120-92. Integrated Services Digital Network (ISDN); Service Requirements for Telefax Group 4. 16 pp.

CENELEC PRETS 300 155-91. Integrated Services Digital Network (ISDN); Facsimile Group 4, Class 1 Equipment on the ISDN End-to-End Protocols Tests (T/TE 05-08). 201 pp.

CENELEC PRETS 300 280-92. Terminal Equipment (TE); Facsimile Group 4 Class 1 Equipment on the Integrated Services Digital Network (ISDN) Terminal Testing. 44 pp.

FAX Machines *(Cont.)*

—CCITT Group 4 *(Cont.)*
CEPT T/SF 32-84. Exigences Pour Les Transmissions De Telecopies Sur Les Reseaux De Telecommunications a L'Aide De Telecopieurs Du Groupe 4. 6 pp.

ETSI PRETS 300 087-91. Integrated Services Digital Network (ISDN); Facsimile Group 4 Class 1 Equipment on the ISDN-Functional Specification of the Equipment (T/TE 05-09).

ETSI PRETS 300 088-91. Integrated Services Digital Network (ISDN); Facsimile Group 4 Class 1 Equipment on the ISDN—General and Service Aspects (T/TE 05-6).

ETSI PRETS 300 112-90. Integrated Services Digital Network (ISDN); Facsimile Group 4 Class 1 Equipment on the ISDN—End-To-End Protocols (T/TE 05-07). 48 pp.

ETSI PRETS 300 120-90. Integrated Services Digital Network (ISDN) Telefax G4 (T/NAI (90) 02). 14 pp.

ETSI PRETS 300 155-91. Integrated Services Digital Network (ISDN); Facsimile Group 4, Class 1 Equipment on the ISDN—End-to-End Protocols Test (T/TE 05-08). 201 pp.

ETSI PRETS 300 087-91. Integrated Services Digital Network (ISDN); Facsimile Group 4 Class 1 Equipment on the ISDN Functional Specification of the Equipment (T/TE 05.09). 12 pp.

ETSI PRETS 300 088-91. Integrated Services Digital Network (ISDN); Facsimile Group 4 Class 1 Equipment on the ISDN General and Service Aspects (T/TE 05-06). 10 pp.

ETSI PRETS 300 112-90. Integrated Services Digital Network (ISDN); Facsimile Group 4 Class 1 Equipment on the ISDN End-to-End Protocols (T/TE 05-07). 48 pp.

ETSI PRETS 300 112-90. Integrated Services Digital Network (ISDN): Facsimile Group 4 Class 1 Equipment on the ISDN End-to-End Protocols (T/TE 05-07). 25 pp.

ETSI ETS 300 120-92. Integrated Services Digital Network (ISDN); Service Requirements for Telefax Group 4. 16 pp.

ETSI PRETS 300 120-92. Integrated Services Digital Network (ISDN); Service Requirements for Telefax Group 4. 16 pp.

ETSI PRETS 300 120-90. Integrated Services Digital Network (ISDN). 15 pp.

ETSI PRETS 300 155-91. Integrated Services Digital Network (ISDN); Facsimile Group 4, Class 1 Equipment on the ISDN End-to-End Protocols Tests (T/TE 05-08). 201 pp.

ETSI PRETS 300 280-92. Terminal Equipment (TE); Facsimile Group 4 Class 1 Equipment on the Integrated Services Digital Network (ISDN) Terminal Testing. 44 pp.

—CCITT Group 4—Character Imaging Devices—Interworking
CCITT RECMN T.351-89. Imaging Process of Character Information on Facsimile Apparatus—Terminal Equipment and Protocols for Telematic Services (Study Group VIII) 4 pp. 4 pp.

—CCITT Group 4—Codes
CCITT RECMN T.6-89. Facsimile Coding Schemes and Coding Control Functions for Group 4 Facsimile Apparatus—Terminal Equipment and Protocols for Telematic Services (Study Group VIII) 10 pp. 10 pp.

—CCITT Group 4—Codes—Procedures
CCITT RECMN T.35-89. Procedure for the Allocation of CCITT Members' Codes—Terminal Equipment and Protocols for Telematic Services (Study Group VIII) 5 pp. 5 pp.

—CCITT Group 4—Integrated Services Digital Networks
CCITT RECMN T.563-91. Terminal Characteristics for Group 4 Facsimile Apparatus (Study Group VIII) 25 pp. 25 pp.

CCITT RECMN T.563-89. Terminal Characteristics for Group 4 Facsimile Apparatus—Terminal Equipment and Protocols for Telematic Services (Study Group VIII) 19 pp. 19 pp.

—CCITT Group 4—Integrated Services Digital Networks—Interworking
CCITT RECMN F.353-89. Provision of Telematic and Data Transmission Services on Integrated Services Digital Network (ISDN)—Telematic, Data Transmission and Teleconference Services—Operations and Quality of Service (Study Group I) 5 pp. 5 pp.

—CCITT Group 4—Interpersonal Messaging Services
CCITT RECMN F.420-92. Message Handling Services: Public Interpersonal Messaging Service (Study Group I) 16 pp. 16 pp.

INDUSTRY STANDARDS

INTERNATIONAL AND NON-U.S. NATIONAL STANDARDS
SUBJECT INDEX

FAX

FAX Machines (Cont.)

—CCITT Group 4—Interpersonal Messaging Services (Cont.)
CCITT RECMN F.420-89. Message Handling Services: the Public Interpersonal Messaging Service—Message Handling and Directory Services—Operations and Definition of Service (Study Group I) 15 pp. 15 pp.

—CCITT Group 4—PSTN
CCITT RECMN T.563-91. Terminal Characteristics for Group 4 Facsimile Apparatus (Study Group VIII) 25 pp. 25 pp.
CCITT RECMN T.563-89. Terminal Characteristics for Group 4 Facsimile Apparatus—Terminal Equipment and Protocols for Telematic Services (Study Group VIII) 19 pp. 19 pp.

—CCITT Group 4—Public Data Networks
CCITT RECMN T.563-91. Terminal Characteristics for Group 4 Facsimile Apparatus (Study Group VIII) 25 pp. 25 pp.
CCITT RECMN T.563-89. Terminal Characteristics for Group 4 Facsimile Apparatus—Terminal Equipment and Protocols for Telematic Services (Study Group VIII) 19 pp. 19 pp.

—CCITT Group 4—Public Networks
CCITT RECMN F.180-89. General Operational Provisions for the International Public Facsimile Service Between Subscribers' Stations (Telefax)—Telematic, Data Transmission and Teleconference Services—Operations and Quality of Service (Study Group I) 5 pp. 5 pp.

—CCITT Group 4—Public Networks—Classification
CCITT RECMN T.0-89. Classification of Facsimile Apparatus for Document Transmission over the Public Networks—Terminal Equipment and Protocols for Telematic Services (Study Group VIII) 4 pp. 4 pp.

—CCITT Group 4—Quality of Service
CCITT RECMN F.184-89. Operational Provisions for the International Public Facsimile Service Between Subscriber Stations with Group 4 Facsimile Machines (Telefax 4)—Telematic, Data Transmission and Teleconference Services—Operations and Quality of Service (Study Group I) 9 pp. 9 pp.

—CCITT Group 4—Teleinformatic—Transport Services (OSI)
CCITT RECMN T.70-89. Network-Independent Basic Transport Service for the Telematic Services—Terminal Equipment and Protocols for Telematic Services (Study Group VIII) 53 pp. 53 pp.

—CCITT Group 4—Teletex Communications—Interworking
CCITT RECMN F.184-89. Operational Provisions for the International Public Facsimile Service Between Subscriber Stations with Group 4 Facsimile Machines (Telefax 4)—Telematic, Data Transmission and Teleconference Services—Operations and Quality of Service (Study Group I) 9 pp. 9 pp.

—CCITT Group 4—Terminal Identification
CCITT RECMN F.184-89. Operational Provisions for the International Public Facsimile Service Between Subscriber Stations with Group 4 Facsimile Machines (Telefax 4)—Telematic, Data Transmission and Teleconference Services—Operations and Quality of Service (Study Group I) 9 pp. 9 pp.

—Digital—Military Communications—Interoperability
NATO STANAG 5000 ED 2 AMD 1-86. Interoperability of Tactical Digital Facsimile Equipment. 72 pp.

—Glossaries
CCITT RECMN T.0-89. Classification of Facsimile Apparatus for Document Transmission over the Public Networks—Terminal Equipment and Protocols for Telematic Services (Study Group VIII) 4 pp. 4 pp.
IEC 50 Chap 721-91. International Electrotechnical Vocabulary Chapter 721: Telegraphy, Facsimile and Data Communication First Edition. 320 pp.

—Integrated Services Digital Networks
CCITT RECMN I.333-89. Terminal Selection in ISDN—Integrated Services Digital Network (ISDN)—Overall Network Aspects and Functions, ISDN User-Network Interfaces (Study Group XVIII) 18 pp. 18 pp.

FAX Machines (Cont.)

—Interfaces—Data Circuit Terminating Equipment—Public Data Network
CCITT RECMN X.38-92. G3 Facsimile Equipment/DCE Interface for G3 Facsimile Equipment Accessing the Facsimile Packet Assembly/Disassembly Facility (FPAD) in a Public Data Network Situated in the Same Country (Study Group VII) 49 pp. 49 pp.

—Marine—Meteorological Charts
BSI BS 7381-90. 1990 Marine Facsimile Receivers for Meteorological Charts. 12 pp.
ISO 9876-89. Shipbuilding—Marine Facsimile Receivers for Meteorological Charts First Edition. 8 pp.
JIS F 9601-91. Marine Facsimile Receivers for Meteorological Charts.

—Public Data Networks
CCITT RECMN T.0-89. Classification of Facsimile Apparatus for Document Transmission over the Public Networks—Terminal Equipment and Protocols for Telematic Services (Study Group VIII) 4 pp. 4 pp.

—Public Networks
CCITT RECMN T.0-89. Classification of Facsimile Apparatus for Document Transmission over the Public Networks—Terminal Equipment and Protocols for Telematic Services (Study Group VIII) 4 pp. 4 pp.

—Public Telephone Networks
CCITT RECMN T.0-89. Classification of Facsimile Apparatus for Document Transmission over the Public Networks—Terminal Equipment and Protocols for Telematic Services (Study Group VIII) 4 pp. 4 pp.

—Radio Circuits—Metallic
CCIR RECMN 344-2-70. Standardization of Phototelegraph Systems for Use on Combined Radio and Metallic Circuits—Section 3Cc—Phototelegraphy (Facsimile). 2 pp.

—Reproduction Papers
CNS P2003-84. Simili Paper (Oct)(1315). 2 pp.

—Shipping Containers
ISO 9711 Pt 2-90. Freight Containers—Information Related to Containers on Board Vessels—Part 2: Telex Data Transmission First Edition. 11 pp.

—Specification Forms
OSI ISO DIS 9064-1. Information Processing—Text and Office Systems—Minimum Information to be Included in Specification Sheets.

—Store and Forward Mode
CCITT RECMN F.87-91. Operational Principles for the Transfer of Messages from Terminals on the Telex Network to Group 3 Facsimile Terminals Connected to the Public Switched Telephone Network (Study Group I) 12 pp. 12 pp.

—Store and Forward Units—Interfaces
CCITT RECMN F.163-92. Operational Requirements of the Interconnection of Facsimile Store-and-Forward Units (Study Group I) 10 pp. 10 pp.

—Terminal Identification
CCITT RECMN F.160-89. General Operational Provisions for the International Public Facsimile Services—Telematic, Data Transmission and Teleconference Services —Operations and Quality of Service (Study Group I) 4 pp. 4 pp.

—Test Charts
JIS B 9525-87. Monochrome Test Chart for Group 4 Facsimile Apparatus. 15 pp.
OSI ISO DP 9063-1-86. DP 9063/1 Revised—Information Processing—Text and Office Systems—Test Charts and Test Patterns—Part 1: Facsimile Equipment (Group 2 and Group 3). 9 pp.

—Text Patterns
OSI ISO DP 9063-2-86. DP 9063/2 Revised—Information Processing—Text and Office Systems—Test Charts and Text Patterns—Part 2: Telefax Equipment.

FAX Modules
Use For: Facsimile Modules
See Also: Demodulators; FAX Machines

—Digital Circuit Multiplication Equipment
CCITT RECMN G.766-92. Facsimile Demodulation/Remodulation for DCME (Study Group XV) 115 pp. 115 pp.

FAX Services
Use: FAX Communications

FAX Terminals
Use: FAX Machines

FCC (Flat Conductor Cables)
Use: Flat Cables

FD
Use: Frequency Distance

FDDI
Use: Fiber Distributed Data Interface

FDM
Use: Frequency Division Multiplexers

Fe-Alloys
Use: Ferroalloys

Feather Meal
See Also: Animal Feed; Bone Meal; Fish Meal

—Animal Feed
CNS N2020-84. Feather Meal (Feed Feeding) (Dec)(2595). 2 pp.

—Animal Feed—Protein Content
CNS N4023-78. Method of Test for Digestible Protein in the Feather Meal for Feeding (Mar)(2713). 1 p.

Featheredge Files
Use: Files (Tools)—Featheredge

Feathers
See Also: Animal Feed; Down; Plumage; Quilts
CGSB CAN/CGSB-139.2-M86. Determination of Composition of Mixtures of Feathers and Down, by Manual Sorting. 9 pp.
CGSB 139-GP-4M-84. Determination of Composition of Mixtures of Feathers and Down, by Mechanical Sorting, Standard for. 11 pp.
CNS N1009-85. Semi-Refined Waterfowl Feathers and Down (May)(820). 4 pp.
CNS N1034-85. Refined Waterfowl Feather and Down (May)(2119). 4 pp.
CNS N1083-85. Chicken Feathers (May)(2880). 4 pp.
CNS N4011-85. Method of Test for Waterfowl Feathers and Down (May)(1483). 3 pp.

—Decoration
CNS N1069-85. Feathers of Decoration (May)(2570). 5 pp.
CNS N4026-85. Method of Test for Feathers of Chickens and Decoration (May)(3026).

—Fertilizers
CNS N3049-88. Steamed Feather Fertilizer (Jun)(11915).

—Filling Power Index
BSI BS 5334-76. 1976 Amd 1 Method of Test for Filling Power Index of Down and Feathers. 5 pp.

—Glossaries
CGSB CAN/CGSB-139.1-M86. Feathers and Down—Terminology and Characteristics. 19 pp.
JIS L 0216-90. Glossary of Terms Used in Feathers. 7 pp.

—Paddings
JIS L 1903-90. Testing Methods for Feathers. 18 pp.
SAA AS 2479-87. Down and/or Feather Filling Materials and Filled Products (This is a Joint Standard with SANZ NZS 2479). 48 pp.
SNZ NZS/AS 2479-87. Down and/or Feather Filling Materials and Filled Products (This is a Joint Standard with SAA AS 2479). 48 pp.

—Paddings—Glossaries
BSI PD 6522-88. 1988 Terms Relevant to the Feather Industry for Fillings for Bedding, Upholstery and Other Domestic Articles. 7 pp.

FEC
Use: Forward Error Correction

Feed Pumps
Use For: Boiler Feed Pumps *See Also:* Centrifugal Pumps; Fuel Pumps; Pumps; Reciprocating Pumps
JIS B 8303-90. Testing Methods for Boiler Feed Pumps. 10 pp.

Feeder Busways
Use: Busways

Feeder Links
See Also: Downlinks; Satellite Links; Uplinks

INTERNATIONAL AND NON-U.S. NATIONAL STANDARDS
SUBJECT INDEX

Feldspars

Feeder Links (Cont.)

—Broadcasting Satellite Services—Fixed Services—EIRP
CCIR Report 1006-86. Fixed Service e.i.r.p. Limits for the Protection of the Broadcasting-Satellite Feeder Links Around 18 GHz—Section 4/9A—Sharing Conditions. 2 pp.

—Broadcasting Satellite Services—Frequency Assignment
CCIR Report 952-2-90. Technical Characteristics of Feeder Links to Broadcasting Satellites Elements Required for the Establishment of Plans of Frequency Assignments and Orbital Positions for the Broadcasting-Satellite Service and the Associated Feeder Links—Sharing in the Feeder-. 56 pp.
CCIR QUESTION 86/11-90. Frequencies for the Feeder Links to a Broadcasting Satellite (Sound and Television)—Questions Concerning Study Group 11—Broadcasting Service (Television). 2 pp.
CCIR QUESTION 98/11-90. Technical Characteristics of and Planning Methods for Feeder Links to Broadcasting Satellites (Television) in the 12 GHz Band in Regions 1 and 3—Questions Concerning Study Group 11—Broadcasting Service (Television). 1 p.

—Broadcasting Satellite Services—Orbits
CCIR Report 952-2-90. Technical Characteristics of Feeder Links to Broadcasting Satellites Elements Required for the Establishment of Plans of Frequency Assignments and Orbital Positions for the Broadcasting-Satellite Service and the Associated Feeder Links—Sharing in the Feeder-. 56 pp.

—Broadcasting Satellite Services—Radio Frequency Interference
CCIR RECMN 795-92. Techniques for Alleviating Mutual Interference Between Feeder Links to the BSS. 3 pp.

—Broadcasting Satellite Services—Rain Fading
CCIR RECMN 794-92. Techniques for Minimizing the Impact on the Overall BSS System Performance Due to Rain Along the Feeder-Link Path. 11 pp.

—Broadcasting Satellite Services—Sound-Program Signals
CCIR QUESTION 59/CMTT-90. Transmission of Digital Sound-Programme Signals over Broadcasting Satellite Feeder Links—Questions Concerning the CMTT CCIR/CCITT Joint Study Group for Television and Sound Transmission. 1 p.

—Downlinks—Noise—Broadcasting Satellite Services
CCIR RECMN 793-92. Partitioning of Noise Between Feeder Links for the Broadcasting-Satellite Service (BSS) and BSS Down Links. 5 pp.

—Earth Stations—Space Stations
CCIR QUESTION 30/4-86. Provision of Feeder Links Between Fixed-Earth Stations and Space Stations of Various Services in Frequency Bands Allocated to the Fixed-Satellite Service—Questions of Study Group 4 Fixed-Satellite Service. 2 pp.

—Fixed Satellite Services—Mobile Satellite Communications
CCIR QUESTION 55/4-90. Feeder Links in the Fixed-Satellite Service Used for the Connection of Satellites in Various Mobile-Satellite Services—Questions of Study Group 4 Fixed-Satellite Service. 1 p.

—Mobile Satellites—Hypothetical Reference Digital Paths
CCIR RECMN 827-92. Hypothetical Reference Digital Path for Systems in the Mobile-Satellite Service Using Feeder Links. 2 pp.

—Space Stations—Broadcasting Satellite Services
CCIR Report 561-4-90. Feeder Links to Space Stations in the Broadcasting-Satellite Service—Section 4E—Frequency Sharing Between Networks of the Fixed-Satellite Service and Those of Other Space Radiocommunications Systems. 13 pp.
CCIR QUESTION 54/4-90. Feeder Links for the Space Stations in the Broadcasting-Satellite Service—Questions of Study Group 4 Fixed-Satellite Service. 1 p.

—Space Stations—Frequency Band Sharing
CCIR Report 1076-86. Considerations Affecting the Accommodation of Spacecraft Service Functions (TTC) Within the Broadcasting-Satellite and Feeder-Link Service Bands—Section 10/11E—Sharing. 14 pp.

Feeders (Mechanical)
Scope Note: Use a more specific term
See: Agricultural Equipment; Conveyors; Drum Feeders; Filling Machines; Livestock Equipment; Materials Handling Equipment; Mixers; Unloaders; Vibrating Feeders; Water Treatment

Feeding Bottles (Babies)
Use For: Baby Bottles *See Also:* Bottles; Feeding Devices (Patient)

—Nipples—Nitrosamines Content—Extraction Analysis
BSI BS 7115: Part 1-88. 1988 Determination of Volatile Nitrosamines in Rubber Teats for Feeding Bottles and Babies Dummies Part 1: Dichloromethane Extraction Method. 8 pp.
BSI BS 7115: Part 2-88. 1988 Determination of Volatile Nitrosamines in Rubber Teats for Feeding Bottles and Babies Dummies Part 2: Artificial Saliva Extraction Method. 8 pp.

Feeding Devices (Patient)
Use For: Enteral Feeding Kits *See Also:* Feeding Bottles (Babies); Medical Equipment

—Bottles
CNS T2030-85. Feeding Bottles (Aug)(11348).
CNS T4016-85. Method of Test for Feeding Bottles (Aug)(11349).
JIS T 9112-75. Feeding Bottles.

—Naso-Enteric
BSI BS 6314-90. 1990 Sterile, Single Use, Naso-Enteric Feeding Tubes and Ancillary Equipment. 24 pp.

Feeds
Use: Animal Feed

Feedthrough (Electrical)
Use: Feedthrough Connectors

Feedthrough Connectors
Use For: Bulkhead Connectors; Bulkhead Penetrators; Electrical Feedthrough; Electrical Power Feedthrough Connectors; Feedthrough (Electrical)
See Also: Connectors; Feedthrough Insulators

—Coaxial—MCX—50 Ohms—Receptacles
CECC CECC 22 221-804 ISSUE 1-86. Detail Specification: Radio Frequency Coaxial Connectors; Series MCX (En, Fr, Ge). 33 pp.

—Coaxial—MCX—50 Ohms—Right Angle—Receptacles
CECC CECC 22 221-805 ISSUE 1-86. Detail Specification: Radio Frequency Coaxial Connectors; Series MCX (En, Fr, Ge). 33 pp.

—Coaxial—SMA—50 Ohms—Panel Seals—Receptacles
CECC CECC 22 111-814 ISSUE 1-85. Detail Specification: Radio Frequency Coaxial Connectors; Series SMA (En, Fr, Ge). 24 pp.

—Coaxial—SMA—50 Ohms—Receptacles
CECC CECC 22 111-813 ISSUE 1-85. Detail Specification: Radio Frequensy Coaxial Connectors; Series SMA (En, Fr, Ge). 24 pp.

Feedthrough Insulators
See Also: Electric Terminals; Feedthrough Connectors
MOD UK DSTAN 59-3: Part 1-01. Terminals, Electrical Part 1: Terminals, Leadthrough (Parts of This Document Have Been Superseded) Issue 1; Amendment 2. 12 pp.

Feedwater
Use For: Boiler Feedwater *See Also:* Boiler Water; Industrial Water; Water
CNS B1312-89. Boiler Code of Boiler Feed Water and Boiler Water (Apr)(10231).
JIS B 8224-86. Testing Methods for Boiler Feed Water and Boiler Water. 93 pp.

—Naval Ships
NATO STANAG 1010 ED 4 AMD 4-82. Specifications of Water to be Transferred to Ships of NATO Navies Intended for Use in Boiler Feedwater Systems. 13 pp.

Feedwater Tanks
See Also: Tanks (Containers)

—Paints (Aluminum)—Marine—Interior/Exterior
CGSB CAN/CGSB-1.93-92. Aluminum Marine Paint. 8 pp.

—Ships
MOD UK NES 344-87. Feed Steam and Drain Systems in Ships Fitted with Auxiliary Boilers Issue 2 (11.87). 59 pp.

Feedwater Treatment
See Also: Boiler Water Treatment; Distillation Methods; Water Treatment
JIS B 8223-89. Water Conditioning for Boiler Feed Water and Boiler Water. 15 pp.

—Marine
BSI BS 1170-83. 1983 Amd 1 Treatment of Water for Marine Boilers. 62 pp.
MOD UK DSTAN 68-18-71. Evaporator Feed Treatment and Descaling Chemicals for Shipboard Use Issue 1. 8 pp.

Feeler Gages
Use For: Feeler Stock *See Also:* Gages; Gap Gages; Maintenance Equipment
BSI BS 957: Part 1-41. 1941 Amd 3 Feeler Gauges Part 1: Inch Units. 10 pp.
BSI BS 957: Part 2-69. 1969 Feeler Gauges Part 2: Metric Units. 8 pp.
CNS B6032-82. Feeler Gauges (Jan)(4755).
CNS B6075-82. Feeler Gauges for Spark Plug (Oct)(9468).
DIN ENGL 2275-77. Feeler Gauges (Sept). 2 pp.
ISO R1938-71. ISO System of Limits and Fits Part II: Inspection of Plain Workpieces First Edition. 28 pp.
JIS B 7524-92. Feeler Gauges. 10 pp.
SNZ NZS 376: Part 1-49. Feeler Gauges Part 1: Inch Units. 8 pp.
SNZ NZS/ISO R1938-71. ISO System of Limits and Fits. Part 2 Inspection of Plain Workpieces. 28 pp.

—Spark Plugs
CNS B7210-82. Method of Test for Feeler Gauges for Spark Plugs (Oct)(9469).

Feeler Stock
Use: Feeler Gages

Fehling's Solution A
Use: Cupric Sulfate

Feldspars

—Aluminum Oxide Content—Gravimetric Analysis
CNS M3084-82. Method for Determination of Aluminium Oxide in Feldspar (Oxine Gravimetric Method) (Jan)(8439).

—Aluminum Oxide Content—Volumetric Analysis
CNS M3085-82. Method for Determination of Aluminium Oxide in Feldspar (EDTA-Zinc Back Titration Method) (Jan)(8440).

—Chemical Analysis
CNS M3080-82. General Rules for Chemical Analysis of Feldspar (Jan)(8435).
JIS M 8853-76. Methods for Chemical Analysis of Feldspar (R 1990). 23 pp.

—Ignition Loss
CNS M3081-82. Method for Determination of Ignition Loss in Feldspar (Jan)(8436).

—Iron Oxide Content—Absorptiometric Analysis
CNS M3086-82. Method for Determination of Iron Oxide in Feldspar (O-Phenanthroline Absorptiometric Method) (Jan)(8441).

—Potassium Oxide Content—Absorptiometric Analysis
CNS M3093-82. Method for Determination of Potassium Oxide in Feldspar (Atomic Absorptiometric Method) (Jan)(8448).

—Potassium Oxide Content—Gravimetric Analysis
CNS M3091-82. Method for Determination of Potassium Oxide in Feldspar (Gravimetric Method) (Jan)(8446).

—Potassium Oxide Content—Photometry
CNS M3092-82. Method for Determination of Potassium Oxide in Feldspar (Flame Photometric Method) (Jan)(8447).

—Silica Content—Absorptiometric Analysis
CNS M3082-82. Method for Determination of Silicon Dioxide in Feldspar (Dehydration Gravimetric and Absorptiometric Method) (Jan)(8437).
CNS M3083-82. Method for Determination of Silicon Dioxide in Feldspar (Flocculation Gravimetric and Absorptiometric Method) (Jan)(8438).

INDUSTRY STANDARDS

Feldspars

Feldspars (Cont.)
—Silica Content—Gravimetric Analysis
CNS M3082-82. Method for Determination of Silicon Dioxide in Feldspar (Dehydration Gravimetric and Absorptiometric Method) (Jan)(8437).
CNS M3083-82. Method for Determination of Silicon Dioxide in Feldspar (Flocculation Gravimetric and Absorptiometric Method) (Jan)(8438).

—Sodium Oxide Content—Absorptiometric Analysis
CNS M3090-82. Method for Determination of Sodium Oxide in Feldspar (Atomic Absorptiometric Method) (Jan)(8445).

—Sodium Oxide Content—Gravimetric Analysis
CNS M3088-82. Method for Determination of Sodium Oxide in Feldspar (Gravimetric Method) (Jan)(8443).

—Sodium Oxide Content—Photometry
CNS M3089-82. Method for Determination of Sodium Oxide in Feldspar (Flame Photometric Method) (Jan)(8444).

—Titanium Oxide Content—Absorptiometric Analysis
CNS M3087-82. Method for Determination of Titanium Oxide in Feldspar (Diantipyrylmethane Absorptiometric Method) (Jan)(8442).

Felts
See Also: Fabrics; Hair Felts; Nonwoven Fabrics; Roofing Felts; Wool; Wool Fabrics; Wool Felts
CNS L3062-80. Method of Test for Felt (Aug)(2689).
CNS L4044-80. Felt (Aug)(2688).
JIS L 3204-85. Recovered Fiber Felts.
MOD UK DSTAN 83-70: Part 1-80. Cloths, Felt, Synthetic Fibres Part 1: Needleloom, for General Purposes Issue 1. 11 pp.

—Ammunition
CGSB 4-GP-34MA-89. Felt, Ammunition Filling. 18 pp.

—Carpets—Underlays
CGSB 4-GP-36M-78. Carpet Underlay, Fibre Type, Standard for. 9 pp.

—Ceramic Fiber
BSI BS 7225: Sec 3.1-90. 1990 Classification of Refractories Part 3: Section 3.1: Blankets, Mats, Felts and Paper. 8 pp.

—Jute
CNS L3181-82. Method of Test for Jute Felts (Jul)(9188).
CNS L4144-82. Tute Felts (Jul)(9187).
JIS L 3203-84. Jute Felts (R 1990). 8 pp.

—Packaging
BSI BS 7200-89. 1989 Synthetic-Fibre Needlefelts. 8 pp.
CGSB 4-GP-35MA-88. Felt Sheet Low-Density. 11 pp.
CGSB 4-GP-37MA-88. Felt Sheet, High-Density. 10 pp.

—Polyester—Filter Elements
MOD UK DSTAN 83-85-93. Polyester Felt Issue 1. 11 pp.

—Rings—Bearing Housings
CNS B1289-82. Felt Ring Groove; for Rolling Bearing Housing (Sep)(9336).
CNS B2699-82. Felt Ring for Rolling Bearing Housing (Sep)(9337).
DIN ENGL 5419-59. Felt Rings; Felt Strips; Ring Grooves for Rolling Bearing Housings (Sept). 3 pp.

—Rings—Roller Bearings
CNS B2518-80. Roller Bearings for Electric Traction, Felt Ring and Felt Ring Groover (Aug)(6312).

—Strips—Bearing Housings
CNS B1289-82. Felt Ring Groove; for Rolling Bearing Housing (Sep)(9336).
CNS B2700-82. Felt Strip for Rolling Bearing Housing (Sep)(9338).
DIN ENGL 5419-59. Felt Rings; Felt Strips; Ring Grooves for Rolling Bearing Housings (Sept). 3 pp.

FEN
Use For: Association for the Promotion of Electrotechnical Standardization; Association pour Promotion de la Normalisation Electrotechnique ; Foerderverein fuer Elektrotechnische Normung
See Also: Electronic Components

—Internal Regulations
CECC CECC FEN ISSUE 2-91. Internal Regulations (En, Fr, Ge). 24 pp.

Fence Posts
See Also: Fences; Poles (Supports)

—Concrete
SAA AS N36-64. Concrete Fence Posts for Rural Use. 22 pp.

—Wood
SAA AS O70-58. Sawn House Stumps, Sole Plates, Fence Posts and Struts from South-Eastern Australian Hardwoods. 7 pp.

Fences
See Also: Chain Link Fences; Electric Fences; Fence Posts; Palisade Fences; Picket Fences; Wire Fences

—Aluminum Alloy
JIS A 6513-83. Fences and Gates with Metals. 21 pp.

—Close Boarded
BSI BS 1722: Part 5-86. 1986 Fences Part 5: Close Boarded Fences. 24 pp.

—Concrete
CNS A2223-88. Precast Reinforced Components for Concrete Fences (Jun)(12330).
CNS A3276-88. Method of Test for Precast Reinforced Components for Concrete Fences (Jun)(12331).
JIS A 5409-82. Precast Reinforced Components for Concrete Fences. 12 pp.
SNZ NZS 3153-74. Specification for Concrete Posts and Other Concrete Units for Wire Fencing in Rural Areas (Reconfirmed 1985). 16 pp.

—Continuous Bar—Steel
BSI BS 1722: Part 8-92. 1992 Fences Part 8: Specification for Mild Steel (Low Carbon Steel) Continuous Bar Fences. 13 pp.
BSI BS 1722: Part 8-78. 1978 Fences Part 8: Mild Steel (Low Carbon Steel) Continuous Bar Fences. 7 pp.

—Lap Boarded Panel
BSI BS 1722: Part 11-86. 1986 Fences Part 11: Woven Wood and Lap Boarded Panel Fences. 22 pp.

—Steel
BSI BS 1722: Part 9-92. 1992 Fences Part 9: Specification for Mild Steel (Low Carbon Steel) Fences with Round or Square Verticals and Flat Horizontals. 16 pp.
BSI BS 1722: Part 9-79. 1979 Fences Part 9: Mild Steel (Low Carbon Steel) Fences with Round or Square Verticals and Flat Posts and Horizontals. 7 pp.
JIS A 6513-83. Fences and Gates with Metals. 21 pp.

—Steel—Panel
BSI BS 1722: Part 14-92. 1992 Fences Part 14: Specification for Open Mesh Steel Panel Fences. 24 pp.

—Steel—Powder Coatings
BSI BS 1722: Part 16-92. 1992 Fences Part 16: Specification for Organic Powder Coatings to Be Used as a Plastics Finish to Components and Mesh (R). 13 pp.

—Steel—Residential Buildings
CNS A2134-82. Steel Fences for Dwellings (May)(8760).
CNS A3152-82. Method of Test for Steel Fences for Dwellings (May)(8761).

—Steel Wire—Galvanized
SAA AS 2423-91. Galvanized Wire Fencing Products. 13 pp.

—Swimming Pools
SAA AS 1926-86. Fences and Gates for Private Swimming Pools Amdt 1 March 1987 (Superseded by AS 1926.1 —1993 but will Remain Avalible Superseded). 10 pp.

—Wood
BSI BS 1722: Part 7-86. 1986 Fences Part 7: Wooden Post and Rail Fences. 16 pp.
SNZ NZS 3607-89. Specification for Round and Part-Round Timber Fence Posts. 8 pp.

—Wood—Woven
BSI BS 1722: Part 11-86. 1986 Fences Part 11: Woven Wood and Lap Boarded Panel Fences. 22 pp.

Fenders (Marine)
Use: Marine Fenders

Fenestration Materials
Use: Glazing Materials

Fenitrothion
See Also: Pesticides

Fenitrothion (Cont.)
SAA AS 1870.17D-79. Standard for Development—Pesticides for Agricultural Use—Part 17D: Fenitrothion. 19 pp.

Fennel Seed
See Also: Spices
ISO 7927 Pt 1-87. Fennel Seed, Whole or Ground (Powdered)—Part 1: Bitter Fennel Seed (Foeniculum Vulgare P. Miller Var. Vulgare)—Specification First Edition. 7 pp.

Fenugreek Seed
See Also: Spices
BSI BS 7087: Part 7-90. 1990 Herbs and Spices Ready for Food Use Part 7: Dried Fenugreek (Whole or Ground). 9 pp.
ISO 6575-82. Fenugreek, Whole or Ground (Powdered)—Specification First Edition. 5 pp.

FEP
Use: Fluorinated Ethylene Propylene Resins

Fermentation
—Classification
BSI BS 1000: (663)-85. 1985 Universal Decimal Classification (UDC). English Full Edition (663): Industrial Microbiology. Industrial Mycology. Fermentation Industry. Beverage Industry. Stimulant Industry. 30 pp.
SNZ NZS/BS 1000 (663)-85. Universal Decimal Classification Industrial Microbiology. Industrial Mycology. Fermentation Industry. Beverage Industry. Stimulant Industry.. 32 pp.

—Vinegars
CNS N6025-83. Method of Test for Vinegar (Jul)(1071). 1 p.

Fermented Milk
Use: Yogurt

Ferric Aluminum Oxide Content Analysis
—Sodium Sulfates
CNS K6085-74. Methods of Test for Sodium Sulfate of Industrial Grade (Oct)(1004). 5 pp.

Ferric Ammonium Citrate
Use: Ammonium Ferric Citrate

Ferric Ammonium Sulfate
Use: Ammonium Ferric Sulfate

Ferric Chloride
Use For: Iron (III) Chloride *See Also:* Ferric Chloride Sulfate; Ferric Sulfate
CNS K7223-66. Chemical Reagent (Ferric Chloride) (Jun)(1723).
CNS K7224-66. Chemical Reagent (Ferric Chloride, Low in Phosphorus) (Jun)(1723-1).
JIS K 1447-56. Ferric Chloride Solution.
JIS K 8142-76. Ferric Chloride.

—Water Treatment
CEN PREN 888-92. Iron(III) Chloride Used for Water Intended for Human Consumption. 14 pp.
DIN ENGL 19602-87. Iron(III) Salts for Water Treatment; Technical Delivery Conditions (Feb). 4 pp.

Ferric Chloride Sulfate
Use For: Iron (III) Chloric Sulfate *See Also:* Ferric Chloride; Ferric Sulfate

—Water Treatment
CEN PREN 891-92. Iron(III) Chloride Sulfate Used for Water Intended for Human Consumption. 11 pp.

Ferric Citrate
Use For: Iron Citrate *See Also:* Reagents
CNS K7225-66. Chemical Reagent (Ferric Citrate) (Jun)(1724).

Ferric Nitrate
See Also: Reagents
CNS K7226-66. Chemical Reagent (Ferric Nitrate) (Jun)(1725).
JIS K 8559-81. Iron (III) Nitrate Enneahydrate.

Ferric Oxide
Use For: Ferric Oxide, Red; Haematite; Hematite; Red Iron Trioxide
MOD UK TS 652. Haematite, Ground Fine.

—Ferrites
JIS K 1462-81. Iron (III) Oxide for Ferrite. 42 pp.

Ferric Oxide, Red
Use: Ferric Oxide

INTERNATIONAL AND NON-U.S. NATIONAL STANDARDS
SUBJECT INDEX
Ferritic

Ferric Oxide Content Analysis
—Aluminum Sulfate
CNS K6162-79. Method of Test for Aluminium Sulfate (Industrial Use) (Jun)(2073). 3 pp.
—Calcium Carbonates
CNS K6721-82. Method of Test for Calcium Carbonate for Rubber (Jun)(9004). 4 pp.
—Organic Coatings—Quantitative Analysis
CNS K6804-25-89. Method of Test for Organic Coating (Chemical Analysis) — Quantative Test of Ferric Oxide in Solvent Insolubles (Jan)(10880-25).

Ferric Sulfate
Use For: Iron (III) Sulfate *See Also:* Aluminum Iron Sulfate; Ferric Chloride; Ferric Chloride Sulfate; Reagents
CNS K7227-66. Chemical Reagent (Ferric Sulfate) (Jun)(1726).
JIS K 8981-76. Ferric Sulfate.
—Water Treatment
CEN PREN 890-92. Iron(III) Sulfate Used for Water Intended for Human Consumption. 12 pp.
DIN ENGL 19602-87. Iron(III) Salts for Water Treatment; Technical Delivery Conditions (Feb). 4 pp.

Ferris Wheels
See Also: Amusement Rides
—Inspection
CNS A3162-88. Inspection Standard of Amusements (Ferris Wheel) (Dec)(8915).
JIS A 1714-78. Inspection Standard of Amusements (Ferris Wheel).

Ferrite Circulators
See Also: Ferrites; Microwave Circulators
—Isolating—Radio Frequency
MOD UK DSTAN 59-85-01. Isolators, Radio Frequency (Reflection) and Circulators, Radio Frequency Issue 1; Amendment 1. 37 pp.
MOD UK DSTAN 59-85: Part 90-81. Isolators, Radio Frequency (Reflection) and Circulators, Radio Frequency Part 90: Detail Specifications Issue 1. 3 pp.
MOD UK DSTAN 59-85: 90/005-01. Isolators, Radio Frequency, Reflection Issue 1; Amendment 1. 14 pp.
MOD UK DSTAN 59-85: 90/006-01. Isolator, Radio Frequency, Reflection Issue 1; Amendment 1. 12 pp.
MOD UK DSTAN 59-85: 90/011-82. Gain Equalisers, Passive Issue 1. 13 pp.

Ferrite Content Analysis
—Austenitic Stainless Steels
JIS Z 3119-88. Methods of Measurement for Ferrite Content in Austenitic Stainless Steel Deposited Metal. 11 pp.
—Manganese
CNS G2097-81. Method of Chemical Analysis for Silicon, Phosphorus, Sulphur and Ferrite in Manganese Metal (Oct)(8005).
—Silicon—Absorptiometric Analysis
CNS G2136-82. Method of Chemical Analysis for Ferrite in Metallic Silicon (Sulfosalicylic Acid Absorptimetric Method) (Oct)(9495).
CNS G2137-82. Method of Chemical Analysis for Ferrite in Metallic Silicon (Atomic Absorptimetric Method) (Oct)(9496).
—Silicon—Volumetric Analysis
CNS G2135-82. Method of Chemical Analysis for Ferrite in Metallic Silicon (EDTA Titration Method) (Oct)(9494).
—Weld Metal
DIN ENGL 32514 Pt 1-90. Determination of Ferrite Number of Austenitic Weld Metal; Measurement Method (June). 2 pp.
JIS Z 3119-88. Methods of Measurement for Ferrite Content in Austenitic Stainless Steel Deposited Metal. 11 pp.

Ferrite Cores
Use For: Ferromagnetic Oxide Cores
See Also: Magnetic Cores; Pot Cores
CECC CECC 25 100-031. DIN 45 970 Teil 114; Magnetic Oxide Cores for Inductor Applications (Ge). 5 pp.
CECC CECC 25 100-032. DIN 45 970 Teil 115; Magnetic Oxide Cores for Inductor Applications (Ge). 5 pp.
CECC CECC 25 100-033. DIN 45 970 Teil 116; Magnetic Oxide Cores for Inductor Applications (Ge). 5 pp.
CECC CECC 25 100-034. DIN 45 970 Teil 117; Magnetic Oxide Cores for Inductor Applications (Ge). 5 pp.

Ferrite Cores (Cont.)
CECC CECC 25 100-035. DIN 45 970 Teil 118; Magnetic Oxide Cores for Inductor Applications (Ge). 5 pp.
CECC CECC 25 100-036. DIN 45 970 Teil 119; Magnetic Oxide Cores for Inductor Applications (Ge). 5 pp.
CECC CECC 25 100-037. DIN 45 970 Teil 1110; Magnetic Oxide Cores for Inductor Applications (Ge). 5 pp.
CECC CECC 25 100-038. DIN 45 970 Teil 1111; Magnetic Oxide Cores for Inductor Applications (Ge). 5 pp.
CECC CECC 25 100-042. DIN 45 970 Teil 1115; Magnetic Oxide Cores for Inductor Applications (Ge). 5 pp.
CECC CECC 25 100-043. DIN 45 970 Teil 1116; Magnetic Oxide Cores for Inductor Applications (Ge). 5 pp.
CNS C1121-83. General Rules for Method of Test for Ferrite Cores (Jun)(10320).
CNS C3177-83. Method of Test for Ferrite Cores for Memory Devices (Jun)(10322).
CNS C4342-82. Insert Core Made Ferromagnetic Oxide (Jan)(8371).
—Inductors
IECQ QC 250101/JP 0001-88. Detail Specification for Electronic Components; Ferrite Cores; Assessment Level: A. 11 pp.
IECQ QC 250101/JP 0002-88. Detail Specification for Electronic Components; Ferrite Cores; Assessment Level: A. 11 pp.
—RM Type
BSI BS 6555-84. 1984 Dimensions of Square Cores (RM-Cores) Made of Magnetic Oxides and Associated Parts. 17 pp.
CECC CECC 25 100-028. DIN 45 970 Teil 111; Magnetic Oxide Cores for Inductor Applications (Ge). 5 pp.
CECC CECC 25 100-029. DIN 45 970 Teil 112; Magnetic Oxide Cores for Inductor Applications (Ge). 5 pp.
CECC CECC 25 100-030. DIN 45 970 Teil 113; Magnetic Oxide Cores for Inductor Applications (Ge). 5 pp.
CECC CECC 25 100-039. DIN 45 970 Teil 1112; Magnetic Oxide Cores for Inductor Applications (Ge). 5 pp.
CECC CECC 25 100-040. DIN 45 970 Teil 1113; Magnetic Oxide Cores for Inductor Applications (Ge). 5 pp.
CECC CECC 25 100-041. DIN 45 970 Teil 1114; Magnetic Oxide Cores for Inductor Applications (Ge). 5 pp.
IEC 431-83. Dimensions of Square Cores (RM-Cores) Made of Magnetic Oxides and Associated Parts Second Edition. 29 pp.
—RM Type—Inductors
CECC CECC 25 100-019. NEN CECC 25 100-019 Edition 1; Magnetic Oxide Core Intended for Inductor Applications (En). 8 pp.
IECQ QC 250101/JP 0003-88. Detail Specification for Electronic Components: Ferrite Cores; Assessment Level: A. 11 pp.
IECQ QC 250101/JP 0004-88. Detail Specification for Electronic Components; Ferrite Cores; Assessment Level: A. 11 pp.
—RM Type—Transformers
IECQ QC 250101/JP 0003-88. Detail Specification for Electronic Components; Ferrite Cores; Assessment Level: A. 11 pp.
IECQ QC 250101/JP 0004-88. Detail Specification for Electronic Components; Ferrite Cores; Assessment Level: A. 11 pp.
—Transformers
IECQ QC 250101/JP 0001-88. Detail Specification for Electronic Components; Ferrite Cores; Assessment Level: A. 11 pp.
IECQ QC 250101/JP 0002-88. Detail Specification for Electronic Components; Ferrite Cores; Assessment Level: A. 11 pp.

Ferrite Isolators
Use For: Waveguide Isolators *See Also:* Ferrites
—Radio Frequency
MOD UK DSTAN 59-85-01. Isolators, Radio Frequency (Reflection) and Circulators, Radio Frequency Issue 1; Amendment 1. 37 pp.
MOD UK DSTAN 59-85: Part 90-81. Isolators, Radio Frequency (Reflection) and Circulators, Radio Frequency Part 90: Detail Specifications Issue 1. 3 pp.
MOD UK DSTAN 59-85: 90/005-01. Isolators, Radio Frequency, Reflection Issue 1; Amendment 1. 14 pp.
MOD UK DSTAN 59-85: 90/006-01. Isolator, Radio Frequency, Reflection Issue 1; Amendment 1. 12 pp.
MOD UK DSTAN 59-85: 90/011-82. Gain Equalisers, Passive Issue 1. 13 pp.

Ferrite Phase Shifters
Use: Microwave Phase Shifters

Ferrites
See Also: Ferrite Circulators; Ferrite Isolators; Gyromagnetic Materials; Magnetic Materials
BSI BS 6786-86. 1986 Limits for Physical Imperfections of Parts Made from Magnetic Oxides. 4 pp.
IEC 424-73. Guide to the Specification of Limits for Physical Imperfections of Parts Made from Magnetic Oxides First Edition. 9 pp.
—Ferric Oxide
JIS K 1462-81. Iron (III) Oxide for Ferrite. 42 pp.
—Magnetic Measurement
IEC 401-93. Ferrite Materials—Guide on the Format of Data Appearing in Manufacturers' Catalogues of Transformer and Inductor Cores Second Edition. 20 pp.
IEC 556-82. Measuring Methods for Properties of Gyromagnetic Materials Intended for Application at Microwave Frequencies First Edition. 106 pp.
—Microwave
IEC 392-72. Guide for the Drafting of Specifications for Microwave Ferrites First Edition. 9 pp.
IEC 556-82. Measuring Methods for Properties of Gyromagnetic Materials Intended for Application at Microwave Frequencies First Edition. 106 pp.
—Microwave—Preferred Products List
CECC CECC MUAHAG Vol 12 IS 1-90. Preferred Products List; Microwave Components (En, Fr, Ge). 76 pp.
—Physical Properties
CNS C3174-83. Method of Test for Fundamental Properties of Soft Ferrites (Jun)(10318).
JIS C 2561-81. Measuring Methods for Fundamental Properties of Soft Ferrites. 27 pp.

Ferritic Stainless Steels
See Also: Alloy Steels; Austenitic Stainless Steels; Corrosion Resistant Steels; Ferroalloys; Martensitic Stainless Steels; Stainless Steels; Steels
—Bars—Seamless
BSI BS 6258-88. 1988 Hollow Steel Bars for Machining. 14 pp.
BSI BS 6258-01. 1988 Amd 1 Hollow Steel Bars for Machining (AMD 7278) September 15, 1992. 15 pp.
—Chemical Analysis
CEN PREN 10088-1-93. Stainless Steels—Part 1: List of Stainless Steels. 19 pp.
—Forgings
BSI BS 1503-89. 1989 Amd 1 Steel Forgings for Pressure Purposes (AMD 6739) September 30, 1991. 35 pp.
BSI BS 1503-02. 1989 Amd 2 Steel Forgings for Pressure Purposes (AMD 7744) July 15, 1993 (Q). 36 pp.
SNZ NZS/BS 1503-89. Steel Forgings for Pressure Purposes Amend: 1, 1991. 32 pp.
—Physical Properties
CEN PREN 10088-1-93. Stainless Steels—Part 1: List of Stainless Steels. 19 pp.

Ferritic Steel Castings
See Also: Castings; Ferroalloy Castings; Steel Castings
DIN ENGL 17445-84. Stainless Steel Castings; Technical Delivery Conditions (Nov). 12 pp.
—High Temperature
DIN ENGL 17245-87. Ferritic Steel Castings with Elevated Temperature Properties; Technical Delivery Conditions (Dec). 16 pp.

Ferritic Steels
Scope Note: For additional listings, see also specific products made from ferritic steels
See Also: Ferroalloys; Steels
—Arc Welding
CEN PREN 1011-93. Recommendations for Arc Welding of Ferritic Steels. 67 pp.
—Flats—Ultrasonic Testing
BSI BS 5996-93. 1993 Acceptance Levels for Internal Imperfections in Steel Plate, Strip and Wide Flats, Based on Ultrasonic Testing (SUPERSEDES BS 4336: 1980) (V). 13 pp.
—Grain Size Analysis
BSI BS 4490-89. 1989 Amd 1 Methods for Micrographic Determination of the Grain Size of Steel (AMD 6597) January 31, 1991. 36 pp.
CNS G2178-83. Method of Ferrite Grain Size Test for Steel (Jul)(10437).

INDUSTRY STANDARDS

Ferritic Steels (Cont.)

—Grain Size Analysis (Cont.)
DIN ENGL 50601-85. Metallographic Examination; Determination of the Ferritic or Austenitic Grain Size of Steel and Ferrous Materials (Aug). 13 pp.
ISO 643-83. Steels—Micrographic Determination of the Ferritic or Austenitic Grain Size First Edition. 31 pp.
JIS G 0552-77. Methods of Ferrite Grain Size Test for Steel. 15 pp.

—Impact Testing
SAA AS 1663-74. Method for Dropweight Test for Nil-Ductility Transition Temperature of Ferritic Steels Reconfirmed 1989. 24 pp.

—Plates—Ultrasonic Testing
BSI BS 5996-93. 1993 Acceptance Levels for Internal Imperfections in Steel Plate, Strip and Wide Flats, Based on Ultrasonic Testing (SUPERSEDES BS 4336: 1980) (V). 13 pp.
BSI BS 5996-80. 1980 Amd 2 Ultrasonic Testing and Specifying Quality Grades of Ferritic Steel Plate. 10 pp.

—Shear Testing
SAA AS 1330-74. Method for the Dropweight Tear Test of Ferritic Steels Reconfirmed 1989. 12 pp.

—Strips—Ultrasonic Testing
BSI BS 5996-93. 1993 Acceptance Levels for Internal Imperfections in Steel Plate, Strip and Wide Flats, Based on Ultrasonic Testing (SUPERSEDES BS 4336: 1980) (V). 13 pp.

—Tubes—Color Coding
ISO 9095-90. Steel Tubes—Continuous Character Marking and Colour Coding for Material Identification First Edition. 9 pp.

—Tubes—Identification Marking
ISO 9095-90. Steel Tubes—Continuous Character Marking and Colour Coding for Material Identification First Edition. 9 pp.

—Welding
BSI BS 1821-82. 1982 Amd 1 Class 1 Oxyacetylene Welding of Ferritic Steel Pipework for Carrying Fluids. 21 pp.
BSI BS 2633-87. 1987 Amd 1 Class 1 Arc Welding of Ferritic Steel Pipework for Carrying Fluids (Amd 5798) February 29, 1988. 56 pp.
BSI BS 2633-02. 1987 Amd 2 Class I Arc Welding of Ferritic Steel Pipework for Carrying Fluids (Amd 6969) July 15, 1992. 60 pp.

Ferroalloy Castings
See Also: Carbon Steel Castings; Castings; Chromium Molybdenum Steel Castings; Ductile Iron Castings; Ferritic Steel Castings; Iron Castings; Stainless Steel Castings; Steel Castings
JIS G 5511-91. Low Thermal Expansive Fe-Alloy Castings. 9 pp.

Ferroalloy Coatings (On Ferroalloys)
See Also: Coatings
CGSB 31-GP-0A METH 61.2-62. Methods of Testing Corrosion-Prevention Materials and Processes Oxalic Acid Spot Test. 1 p.
MOD UK DEF-2331-A-61. Protective PX-1, Dyed Protective PX-1, Undyed (Consolidated Edition August 1976, Incorporating Amendments Nos. 1 and 2). 3 pp.
MOD UK DEF-2331-A-01. Protective PX-1, Dyed Protective PX-1, Undyed (Consolidated Edition August 1976 Incorporating Amendments Nos. 1 and 2); Amendment 3. 5 pp.

—Oxide—Hot Caustic Solutions
SAA AS CK3-47. Code of Recommended Practice for Oxide Coating of Ferrous Metals in Hot Caustic Solutions. 6 pp.

—Phosphate
ISO 9717-90. Phosphate Conversion Coatings for Metals—Method of Specifying Requirements First Edition. 19 pp.

Ferroalloys
Scope Note: For additional listings, see also specific products made from ferroalloys *Use For:* Ferrous Alloys; Ferrous Metals; Iron Alloys *See Also:* Alloy Steels; Alloys; Austenitic Stainless Steels; Bearing Alloys; Carbon Steels; Cast Iron; Chromium Iron Alloys; Chromium Molybdenum Steels; Chromium Steels; Corrosion Resistant Steels; Drill Steels; Ferritic Stainless Steels; Ferritic Steels; Ferroboron; Ferrochromium; Ferromanganese; Ferromolybdenum; Ferronickel; Ferroniobium; Ferrophosphorus; Ferrosilicochromium; Ferrosilicomanganese; Ferrosilicon; Ferrotitanium; Ferrotungsten; Ferrovanadium; Free Machining Steels; Gray Iron; Heat Resistant Steels; High Alloy Steels; High

Ferroalloys (Cont.)
See Also: (Cont.)
Strength Steels; Iron; Jominy Testing; Killed Steels; Low Alloy Steels; Malleable Cast Iron; Manganese Steels; Maraging Steels; Martensitic Stainless Steels; Nickel Chromium Molybdenum Steels; Nickel Chromium Steels; Nickel Cobalt Molybdenum Steels; Nickel Steels; Pig Iron; Reinforcing Steels; Silicon Steels; Stainless Steels; Steel Castings; Steels; Structural Steels; Tool Steels; White Iron

—Case Hardening—Metal Products—Engineering Drawings
DIN ENGL 6773 Pt 4-77. Heat Treatment of Ferrous Metals; Heat-Treated Parts; Representation and Indications in Drawings; Case Hardening (May). 7 pp.

—Chemical Analysis
BSI BS 6200: Part 1-91. 1991 Sampling and Analysis of Iron, Steel and Other Ferrous Metals Part 1: Introduction and Contents. 12 pp.
BSI BS 6200: Sec 3.0-91. 1991 Sampling and Analysis of Iron, Steel and Other Ferrous Metals Part 3: Methods of Analysis Section 3.0: Summary of Methods. 10 pp.
CNS G2003-62. Sampling Method of Chemical Test for FE-Alloys (Dec)(269) (R 1973).
CNS G2051-86. General Rules of Chemical Analysis for Ferroalloys (Jul)(4761). 5 pp.
JIS G 1301-87. General Rules for Chemical Analysis of Ferroalloys. 9 pp.

—Corrosion Inhibitors—Enclosed Systems
MOD UK TS 10035A. PX 25 Corrosion Preventive Oil: Enclosed Ferrous Metal Systems, Contact and Volatile Corrosion Inhibited (Withdrawn).

—Corrosion Inhibitors—Humidity
CGSB 31-GP-0A METH 13.1-57. Methods of Testing Corrosion-Prevention Materials and Processes Humidity Test. 9 pp.

—Corrosion Inhibitors—Oils
MOD UK TS 10035A. PX 25 Corrosion Preventive Oil: Enclosed Ferrous Metal Systems, Contact and Volatile Corrosion Inhibited (Withdrawn).

—Glossaries
ISO 8954 Pt 1-90. Ferroalloys—Vocabulary—Part 1: Materials First Edition. 11 pp.

—Hardening—Metal Products—Engineering Drawings
DIN ENGL 6773 Pt 2-77. Heat-Treatment of Ferrous Metals; Heat-Treated Parts; Representation and Indications in Drawings; Hardening, Hardening and Tempering, Quenching and Tempering (May). 4 pp.

—Heat Treated—Bars—Aerospace
BSI HR 51-73. 1973 Amd 2 Heat-Resisting Alloy Billets, Bars, Forgings and Parts (Solution Treated at 980 Degrees C) (Ni 25.5, Cr 14.7, Ti 2, Mn 1.5, Mo 1.2, Si 0.7, V 0.3, Fe Remainder). 5 pp.
BSI HR 52-73. 1973 Amd 2 Heat-Resisting Alloy Billets, Bars, Forgings and Parts (Ni 25.5, Cr 14.7, Ti 1.8, Mn 1.5, Mo 1.2, V 0.3, Fe Remainder). 5 pp.

—Heat Treated—Billets—Aerospace
BSI HR 51-73. 1973 Amd 2 Heat-Resisting Alloy Billets, Bars, Forgings and Parts (Solution Treated at 980 Degrees C) (Ni 25.5, Cr 14.7, Ti 2, Mn 1.5, Mo 1.2, Si 0.7, V 0.3, Fe Remainder). 5 pp.
BSI HR 52-73. 1973 Amd 2 Heat-Resisting Alloy Billets, Bars, Forgings and Parts (Ni 25.5, Cr 14.7, Ti 1.8, Mn 1.5, Mo 1.2, V 0.3, Fe Remainder). 5 pp.

—Heat Treated—Forgings—Aerospace
BSI HR 51-73. 1973 Amd 2 Heat-Resisting Alloy Billets, Bars, Forgings and Parts (Solution Treated at 980 Degrees C) (Ni 25.5, Cr 14.7, Ti 2, Mn 1.5, Mo 1.2, Si 0.7, V 0.3, Fe Remainder). 5 pp.
BSI HR 52-73. 1973 Amd 2 Heat-Resisting Alloy Billets, Bars, Forgings and Parts (Ni 25.5, Cr 14.7, Ti 1.8, Mn 1.5, Mo 1.2, V 0.3, Fe Remainder). 5 pp.

—Heat Treatment—Engineering Drawings
DIN ENGL 6773 Pt 2-77. Heat-Treatment of Ferrous Metals; Heat-Treated Parts; Representation and Indications in Drawings; Hardening, Hardening and Tempering, Quenching and Tempering (May). 4 pp.
DIN ENGL 6773 Pt 3-76. Heat Treatment of Ferrous Metals; Heat-Treated Parts; Representation and Indications in Drawings; Surface Layer Hardening (Nov). 8 pp.
DIN ENGL 6773 Pt 4-77. Heat Treatment of Ferrous Metals; Heat-Treated Parts; Representation and Indications in Drawings; Case Hardening (May). 7 pp.

Ferroalloys (Cont.)
—Heat Treatment—Engineering Drawings (Cont.)
DIN ENGL 6773 Pt 5-77. Heat Treatment of Ferrous Metals; Heat-Treated Parts; Representation and Indications in Drawings; Nitriding (May). 5 pp.

—Hot Dip Coatings
SAA AS 1650-89. Hot-Dipped Galvanized Coatings on Ferrous Articles (This is a Joint Standard with SANZ NZS 1650). 20 pp.
SNZ NZS/AS 1650-89. Hot-Dipped Galvanized Coatings on Ferrous Articles (This is a Joint Standard with SAA AS 1650). 20 pp.

—Nitriding—Metal Products—Engineering Drawings
DIN ENGL 6773 Pt 5-77. Heat Treatment of Ferrous Metals; Heat-Treated Parts; Representation and Indications in Drawings; Nitriding (May). 5 pp.

—Quench Hardening—Metal Products—Engineering Drawings
DIN ENGL 6773 Pt 2-77. Heat-Treatment of Ferrous Metals; Heat-Treated Parts; Representation and Indications in Drawings; Hardening, Hardening and Tempering, Quenching and Tempering (May). 4 pp.

—Sampling
BSI BS 1837-70. (WITHDRAWN) 1970 Methods for the Sampling of Iron, Steel, Permanent Magnet Alloys and Ferro-Alloys (Superseded by BS 6200: Section 2.2: 1993). 20 pp.
BSI BS 6200: Part 1-91. 1991 Sampling and Analysis of Iron, Steel and Other Ferrous Metals Part 1: Introduction and Contents. 12 pp.
CNS G2190-86. General Rules for Sampling of Ferroalloys (Jul)(10567).
CNS G2267-86. Method of Sampling for Size Analysis of Ferroalloys and Size Determination (Jul) (11632).
ISO 3713-87. Ferroalloys—Sampling and Preparation of Samples—General Rules First Edition. 19 pp.
ISO 4551-87. Ferroalloys—Sampling and Sieve Analysis First Edition; (Corrected and Reprinted -1988). 14 pp.
ISO 7087-84. Ferroalloys—Experimental Methods for the Evaluation of the Quality Variation and Methods for Checking the Precision of Sampling First Edition. 16 pp.
ISO 7347-87. Ferroalloys—Experimental Methods for Checking the Bias of Sampling and Sample Preparation First Edition. 15 pp.
ISO 7373-87. Ferroalloys—Experimental Methods for Checking the Precision of Sample Division First Edition; (Corrected and Reprinted -1988). 10 pp.
JIS G 1501-85. General Rules for Sampling of Ferroalloys. 34 pp.
JIS G 1641-85. Methods of Sampling for Size Analysis of Ferroalloys and Size Determination. 21 pp.

—Sampling—Chemical Analysis
CNS G2003-62. Sampling Method of Chemical Test for FE-Alloys (Dec)(269) (R 1973).
CNS G2264-86. Method of Sampling for Chemical Analysis of Ferroalloys Part 1 (Jul)(11629).
CNS G2265-86. Method of Sampling for Chemical Analysis of Ferroalloys Part 2 (Jul)(11630).
CNS G2266-86. Method of Sampling for Chemical Analysis of Ferroalloys Part 3 (Jul)(11631).

—Sampling—Glossaries
ISO 8954 Pt 2-90. Ferroalloys—Vocabulary—Part 2: Sampling and Sample First Edition. 11 pp.

—Sieve Analysis
CNS G2267-86. Method of Sampling for Size Analysis of Ferroalloys and Size Determination (Jul) (11632).
ISO 4551-87. Ferroalloys—Sampling and Sieve Analysis First Edition; (Corrected and Reprinted -1988). 14 pp.
JIS G 1641-85. Methods of Sampling for Size Analysis of Ferroalloys and Size Determination. 21 pp.

—Sieve Analysis—Glossaries
ISO 8954 Pt 3-90. Ferroalloys—Vocabulary—Part 3: Sieve Analysis First Edition. 11 pp.

—Surface Hardening—Metal Products—Engineering Drawings
DIN ENGL 6773 Pt 3-76. Heat Treatment of Ferrous Metals; Heat-Treated Parts; Representation and Indications in Drawings; Surface Layer Hardening (Nov). 8 pp.

—Tempering—Metal Products—Engineering Drawings
DIN ENGL 6773 Pt 2-77. Heat-Treatment of Ferrous Metals; Heat-Treated Parts; Representation and Indications in Drawings; Hardening, Hardening and Tempering, Quenching and Tempering (May). 4 pp.

INTERNATIONAL AND NON-U.S. NATIONAL STANDARDS
SUBJECT INDEX

Ferroalloys (Cont.)

—Tin Coatings—Lead Content—Quantitative Analysis
BSI BS 6534-84. 1984 Quantitative Determination of Lead in Tin Coatings. 8 pp.

—Wrought—Chemical Analysis
DIN ENGL 17745-73. Wrought Alloys of Nickel and Iron; Composition (Jan). 3 pp.

—X-Ray Fluorescence Spectrometry
CNS G2263-86. Method for Fluorescent X-Ray Analysis of Ferroalloys (May)(11576).
JIS G 1351-87. Method for X-Ray Fluorescence Spectrometric Analysis of Ferroalloys. 8 pp.

Ferroboron
See Also: Ferroalloys
CNS G3112-84. Ferroboron (Jun)(5121).
DIN ENGL 17567-70. Ferroboron; Technical Conditions of Delivery (Jan). 2 pp.
JIS G 2318-86. Ferroboron. 6 pp.

—Boron Content—Volumetric Analysis
BSI BS 6200: SUBSEC 3.5.2-91. 1991 Sampling and Analysis of Iron, Steel and Other Ferrous Metals Part 3: Methods of Analysis Section 3.5: Determination of Boron Subsection 3.5.2: Ferroboron: Volumetric Method. 7 pp.

—Chemical Analysis
CNS G2095-81. Method of Chemical Analysis for Ferroboron (Oct)(8003).
JIS G 1327-77. Methods for Chemical Analysis of Ferroboron (R 1988). 15 pp.

—Sampling
BSI BS 6200: Sec 2.2-93. 1993 Sampling and Analysis of Iron, Steel and Other Ferrous Metals Part 2: Sampling and Sample Preparation Section 2.2: Methods for Ferroalloys and Other Alloying Additives (V). 20 pp.
CNS G2217-84. Method of Sampling for Ferroboron (Mar)(10815).
JIS G 1603-85. Methods of Sampling for Chemical Analysis of Ferroalloys (Part 3 Ferrophosphorus, Manganese Metal, Silicon Metal, Chromium Metal, Calcium Silicon and Ferroboron). 10 pp.

Ferrochromium
See Also: Ferroalloys
CNS G3075-84. Ferrochromium (Jun)(3475).
DIN ENGL 17565-68. Ferro-Chromium, Ferro-Chromium-Silicon and Chromium; Technical Conditions of Delivery (Dec). 2 pp.
ISO 5448-81. Ferrochromium—Specification and Conditions of Delivery First Edition. 10 pp.
ISO 5449-80. Ferrosilicochromium—Specification and Conditions of Delivery First Edition. 5 pp.
JIS G 2303-86. Ferrochromium. 8 pp.

—Aluminum Content—Volumetric Analysis
BSI BS 6200: SUBSEC 3.1.1-91. 1991 Sampling and Analysis of Iron, Steel and Other Ferrous Metals Part 3: Methods of Analysis Section 3.1: Determination of Aluminium Subsection 3.1.1: Steel, Cast Iron, Low Carbon Ferro-Chromium Metal: Volumetric Method. 11 pp.

—Carbon Content
CNS G2185-83. Method of Chemical Analysis for Carbon in Ferrochromium (Aug)(10505).

—Chemical Analysis
JIS G 1313-89. Methods for Chemical Analysis of Ferrochromium. 23 pp.

—Chromium Content
CNS G2184-83. Method of Chemical Analysis for Chromium in Ferrochromium (Aug)(10504).
SAA AS 3587.3-91. Ferroalloys—Chemical Analysis—Part 3: Determination of Chromium Content of Ferrochromium and Ferrosilicochromium. 4 pp.

—Chromium Content—Potentiometric Analysis
BSI BS 6200: SUB SEC 3.10.4-85. 1985 Sampling and Analysis of Iron, Steel and Other Ferrous Metals Part 3: Methods of Analysis Section 3.10: Determination of Chromium Subsection 3.10.4: Ferrochromium and Ferrosilicochromium: Potentiometric Method. 6 pp.
ISO 4140-79. Ferrochromium and Ferrosilicochromium—Determination of Chromium Content—Potentiometric Method First Edition. 5 pp.

—Chromium Content—Volumetric Analysis
BSI BS 6200: SUB SEC 3.10.5-92. 1992 Sampling and Analysis of Iron, Steel and Other Ferrous Metals Part 3: Methods of Analysis Section 3.10: Determination of Chromium Subsection 3.10.5: Ferrochromium: Volumetric Method. 8 pp.

Ferrochromium (Cont.)

—Nitrogen Content
CNS G2189-83. Method of Chemical Analysis for Nitrogen in Ferrochromium (Aug)(10509).

—Particle Size
DIN ENGL 17599-85. Ferroalloys; Particle Size Ranges (June). 4 pp.

—Phosphorus Content
CNS G2187-83. Method of Chemical Analysis for Phosphorus in Ferrochromium (Aug)(10507).

—Phosphorus Content—Spectrophotometry
BSI BS 6200: SUB SEC 3.24.2-92. 1992 Sampling and Analysis of Iron, Steel and Other Ferrous Metals Part 3: Methods of Analysis Sec 3.24: Determination of Phosphorus Subsection 3.24.2: Ferrochromium, Ferromanganese and Ferromolybdenum:Spectrophotometric Method. 7 pp.

—Sampling
BSI BS 6200: Sec 2.2-93. 1993 Sampling and Analysis of Iron, Steel and Other Ferrous Metals Part 2: Sampling and Sample Preparation Section 2.2: Methods for Ferroalloys and Other Alloying Additives (V). 20 pp.
CNS G2199-84. Method of Sampling for High Carbon Ferrochromium (Jan)(10748).
CNS G2200-84. Method of Sampling for Medium Carbon Ferrochromium (Jan)(10749).
CNS G2201-84. Method of Sampling for Low Carbon Ferrochromium (Jan)(10750).

—Sampling—Chemical Analysis
ISO 4552 Pt 1-87. Ferroalloys—Sampling and Sample Preparation for Chemical Analysis—Part 1: Ferrochromium, Ferrosilicochromium, Ferrosilicon, Ferrosilicomanganese, Ferromanganese First Edition. 11 pp.
JIS G 1601-85. Methods of Sampling for Chemical Analysis of Ferroalloys (Part 1 Ferromanganese, Ferrosilicon, Ferrochromium, Ferrosilicomanganese and Ferrosilicochromium). 12 pp.

—Silicon Content
CNS G2186-83. Method of Chemical Analysis for Silicon in Ferrochromium (Aug)(10506).

—Sulfur Content
CNS G2188-83. Method of Chemical Analysis for Sulfur in Ferrochromium (Aug)(10508).

Ferromagnetic Materials
See Also: Magnetic Materials

—Leakage Flux
DIN ENGL 54136 Pt 1-88. Non-Destructive Testing; Magnetic Leakage Flux Testing by Scanning with Probes; Principles (Oct). 4 pp.

—Magnetic Particle Inspection
CNS G3245-90. Method of Test for Magnetic Particle Testing of Ferromagnetic Materials and Classification of Magnetic Particle Indication (Jan)(12657).
SAA AS 1171-76. Methods for Magnetic Particle Testing of Ferromagnetic Products and Components. 20 pp.

Ferromagnetic Oxide
Scope Note: For additional listings, see specific materials or products See Also: Magnetic Materials

—Tubes/Pins/Rods
IEC 220-66. Dimensions of Tubes, Pins and Rods of Ferromagnetic Oxides First Edition. 7 pp.

Ferromagnetic Oxide Cores
Use: Ferrite Cores

Ferromagnetic Pipes
See Also: Pipes

—Eddy Current Testing
DIN ENGL 54141 Pt 1-82. Non-Destructive Testing; Eddy Current Testing of Pipes and Tubes; General Remarks on Testing Using Concentric Test Coils and the Single-Frequency Method (Oct). 3 pp.

Ferromanganese
See Also: Ferroalloys
CNS G3035-83. Ferromanganese (Nov)(2150).
DIN ENGL 17564-68. Ferromanganese, Ferromanganese-Silicon and Manganese; Technical Conditions of Delivery (Dec). 3 pp.
ISO 5446-80. Ferromanganese—Specification and Conditions of Delivery First Edition. 6 pp.
JIS G 2301-86. Ferromanganese. 7 pp.

—Chemical Analysis
CNS G2016-63. Method of Chemical Analysis for Ferromanganese (Jul)(2151)(R 1976).
JIS G 1311-87. Methods for Chemical Analysis of Ferromanganese. 32 pp.

Ferromanganese (Cont.)

—Manganese Content—Potentiometric Analysis
BSI BS 6200: SUB SEC 3.18.3-85. 1985 Sampling and Analysis of Iron, Steel and Other Ferrous Metals Part 3: Methods of Analysis Section 3.18: Determination of Manganese Subsection 3.18.3: Ferromanganese and Ferrosilicomanganese: Potentiometric Method. 6 pp.
BSI BS 6200: SUB SEC 3.18.3-01. 1985 Amd 1 Sampling and Analysis of Iron, Steel and Other Ferrous Metals Part 3: Methods of Analysis Sec 3.18: Determination of Manganese Subsection 3.18.3: Ferromanganese and Ferrosilicomanganese: Potentiometric Method (ISO 4159: 1978). 11 pp.
CEN EN 24 159-89. Determination of Manganese Content Potentiometric Method. 3 pp.
DIN ENGL EN 24159-90. Determination of Manganese Content of Ferromanganese and Ferro-Silicomanganese; by the Potentiometric Method (ISO 4159: 1978) (Apr). 6 pp.
ISO 4159-78. Ferromanganese and Ferrosilicomanganese—Determination of Manganese Content—Potentiometric Method First Edition. 5 pp.

—Particle Size
DIN ENGL 17599-85. Ferroalloys; Particle Size Ranges (June). 4 pp.

—Phosphorus Content—Spectrophotometry
BSI BS 6200: SUB SEC 3.24.2-92. 1992 Sampling and Analysis of Iron, Steel and Other Ferrous Metals Part 3: Methods of Analysis Sec 3.24: Determination of Phosphorus Subsection 3.24.2: Ferrochromium, Ferromanganese and Ferromolybdenum:Spectrophotometric Method. 7 pp.

—Sampling
BSI BS 6200: Sec 2.2-93. 1993 Sampling and Analysis of Iron, Steel and Other Ferrous Metals Part 2: Sampling and Sample Preparation Section 2.2: Methods for Ferroalloys and Other Alloying Additives (V). 20 pp.
CNS G2193-83. Method of Sampling for Ferromanganese (Nov)(10655).
JIS G 1601-85. Methods of Sampling for Chemical Analysis of Ferroalloys (Part 1 Ferromanganese, Ferrosilicon, Ferrochromium, Ferrosilicomanganese and Ferrosilicochromium). 12 pp.

—Sampling—Chemical Analysis
ISO 4552 Pt 1-87. Ferroalloys—Sampling and Sample Preparation for Chemical Analysis—Part 1: Ferrochromium, Ferrosilicochromium, Ferrosilicon, Ferrosilicomanganese, Ferromanganese First Edition. 11 pp.

Ferromolybdenum
See Also: Ferroalloys; Heat Resistant Alloys; Refractory Materials
CNS G3084-84. Ferromolybdenum (Jun) (3826).
DIN ENGL 17561-65. Ferromolybdenum; Technical Conditions of Delivery (Dec). 2 pp.
ISO 5452-80. Ferromolybdenum—Specification and Conditions of Delivery First Edition. 5 pp.
JIS G 2307-86. Ferromolybdenum. 6 pp.

—Chemical Analysis
CNS G2118-82. Methods of Chemical Analysis for Ferromolybdenum (Jan)(8402).
JIS G 1317-82. Methods for Chemical Analysis of Ferromolybdenum. 40 pp.

—Molybdenum Content—Gravimetric Analysis
BSI BS 6200: SUB SEC 3.19.2-92. 1992 Sampling and Analysis of Iron, Steel and Other Ferrous Metals Part 3: Methods of Analysis Section 3.19: Determination of Molybdenum Subsection 3.19.2: Ferromolybdenum: Gravimetric Method. 7 pp.
ISO 4173-80. Ferromolybdenum—Determination of Molybdenum Content—Gravimetric Method First Edition. 5 pp.

—Particle Size
DIN ENGL 17599-85. Ferroalloys; Particle Size Ranges (June). 4 pp.

—Phosphorus Content—Spectrophotometry
BSI BS 6200: SUB SEC 3.24.2-92. 1992 Sampling and Analysis of Iron, Steel and Other Ferrous Metals Part 3: Methods of Analysis Sec 3.24: Determination of Phosphorus Subsection 3.24.2: Ferrochromium, Ferromanganese and Ferromolybdenum:Spectrophotometric Method. 7 pp.

—Sampling
BSI BS 6200: Sec 2.2-93. 1993 Sampling and Analysis of Iron, Steel and Other Ferrous Metals Part 2: Sampling and Sample Preparation Section 2.2: Methods for Ferroalloys and Other Alloying Additives (V). 20 pp.
CNS G2203-84. Method of Sampling for Ferromolybdenum (Jan) (10752).

Ferromolybdenum (Cont.)

—Sampling (Cont.)

JIS G 1602-85. Methods of Sampling for Chemical Analysis of Ferroalloys (Part 2 Ferrotungsten, Ferromolybdenum, Ferrovanadium, Ferrotitanium and Ferroniobium). 9 pp.

—Sampling—Chemical Analysis

ISO 4552 Pt 2-87. Ferroalloys—Sampling and Sample Preparation for Chemical Analysis—Part 2: Ferrotitanium, Ferromolybdenum, Ferrotungsten, Ferroniobium, Ferrovanadium First Edition. 10 pp.

Ferronickel

See Also: Ferroalloys; Nickel Alloys

DIN ENGL 17568-70. Ferronickel; Technical Conditions of Delivery (Jan). 2 pp.
JIS G 2316-86. Ferronickel. 6 pp.

—Carbon Content—Infrared Analysis

BSI BS 6783: Part 5-86. 1986 Sampling and Analysis of Nickel, Ferronickel and Nickel Alloys Part 5: Method for Determining of Carbon in Nickel Ferronickel and Nickel Alloys by Infra-Red Absorption After Induction Furnace Combustion. 11 pp.
ISO 7524-85. Nickel, Ferronickel and Nickel Alloys—Determination of Carbon Content—Infra-Red Absorption Method After Induction Furnace Combustion First Edition. 10 pp.

—Chemical Analysis

JIS G 1326-87. Methods of Chemical Analysis of Ferronickel. 65 pp.

—Cobalt Content—Atomic Absorption Spectrometry

BSI BS 6783: Part 3-86. 1986 Sampling and Analysis of Nickel, Ferronickel and Nickel Alloys Part 3: Method for Determination of Cobalt in Ferronickel by Flame Atomic Absorption Spectrometry. 6 pp.
BSI BS 6783: Part 3-01. 1986 Amd 1 Sampling and Analysis of Nickel, Ferronickel and Nickel Alloys Part 3: Method for Determination of Cobalt in Ferronickel by Flame Atomic Absorption Spectrometry (ISO 7520: 1985) (AMD 6992) February 28, 1992 (V). 10 pp.
CEN EN 27520-91. Ferronickel—Determination of Cobalt Content—Flame Atomic Absorption Spectrometric Method. 3 pp.
DIN ENGL EN 27520-92. Determination of Cobalt Content of Ferronickel; Flame Atomic Absorption Spectrometric Method (ISO 7520: 1985) (Feb). 6 pp.
DIN ENGL EN 27526-92. Determination of Sulfur Content of Nickel, Ferronickel and Nickel Alloys; Infrared Absorption Method After Induction Furnace Combustion (ISO 7526: 1985) (Feb). 10 pp.
ISO 7520-85. Ferronickel—Determination of Cobalt Content—Flame Atomic Absorption Spectrometric Method First Edition. 5 pp.

—Ingots

BSI BS EN 26501-92. 1992 Ferronickel—Specification and Delivery Requirements (ISO 6501: 1988) (V). 10 pp.
BSI BS EN 26501-01. 1992 Amd 1 Ferronickel—Specification and Delivery Requirements (ISO 6501: 1988) (AMD 7390) September 15, 1992. 11 pp.
CEN EN 26501-92. Ferronickel—Specification and Delivery Requirements. 5 pp.
DIN ENGL EN 26501-92. Ferronickel; Specification and Delivery Requirements (ISO 6501: 1988) (July). 6 pp.
ISO 6501-88. Ferronickel—Specification and Delivery Requirements First Edition. 6 pp.

—Ingots—Chemical Analysis—Sampling

BSI BS EN 28050-92. 1992 Ferronickel Ingots or Pieces—Sampling for Analysis (ISO 8050: 1988). 23 pp.
CEN EN 28050-92. Ferronickel Ingots or Pieces—Sampling for Analysis. 18 pp.
DIN ENGL EN 28050-92. Ferronickel Ingots or Pieces; Sampling for Analysis (ISO 8050: 1988) (July). 19 pp.
ISO 8050-88. Ferronickel Ingots or Pieces—Sampling for Analysis First Edition. 19 pp.

—Nickel Content—Gravimetric Analysis

BSI BS 6783: Part 2-86. 1986 Sampling and Analysis of Nickel, Ferronickel and Nickel Alloys Part 2: Method for Determination of Nickel in Ferronickel (Dimethylglyoxime Gravimetric Method). 10 pp.
BSI BS 6783: Part 2-01. 1986 Amd 1 Sampling and Analysis of Nickel, Ferronickel and Nickel Alloys Part 2: Method for Determination of Nickel in Ferronickel (Dimethylglyoxime Gravimetric Method) (ISO 6352: 1985) (AMD 6991) February 28, 1992 (V). 12 pp.
CEN EN 26352-91. Ferronickel—Determination of Nickel Content—Dimenthylglyoxime Gravimetric Method. 3 pp.

Ferronickel (Cont.)

—Nickel Content—Gravimetric Analysis (Cont.)

DIN ENGL EN 26352-92. Determination of Nickel Content of Ferronickel; Gravimetric Method Using Dimethylglyoxime (ISO 6352: 1985) (Feb). 9 pp.
ISO 6352-85. Ferronickel—Determination of Nickel Content—Dimethylglyoxime Gravimetric Method First Edition. 8 pp.

—Phosphorus Content—Molecular Absorption Spectrophotometry

BSI BS 6783: Part 13-92. 1992 Sampling and Analysis of Nickel, Ferronickel and Nickel Alloys Part 13: Method for the Detn. of Phosphorus in Nickel, Ferronickel and Nickel Alloys by Phosphovanadomolybdate Molecular Absorption Spectrometry (ISO 11400: 1992). 11 pp.
ISO 11400-92. Nickel, Ferronickel and Nickel Alloys—Determination of Phosphorus Content—Phosphovanadomolybdate Molecular Absorption Spectrometric Method First Edition. 8 pp.

—Sampling

JIS G 1604-85. Methods of Sampling for Chemical Analysis of Ferroalloys (Part 4 Ferronickel). 9 pp.

—Shot

BSI BS EN 26501-92. 1992 Ferronickel—Specification and Delivery Requirements (ISO 6501: 1988) (V). 10 pp.
BSI BS EN 26501-01. 1992 Amd 1 Ferronickel—Specification and Delivery Requirements (ISO 6501: 1988) (AMD 7390) September 15, 1992. 11 pp.
CEN EN 26501-92. Ferronickel—Specification and Delivery Requirements. 5 pp.
DIN ENGL EN 26501-92. Ferronickel; Specification and Delivery Requirements (ISO 6501: 1988) (July). 6 pp.
ISO 6501-88. Ferronickel—Specification and Delivery Requirements First Edition. 6 pp.

—Shot—Chemical Analysis—Sampling

BSI BS EN 28049-92. 1992 Ferronickel Shot—Sampling for Analysis (ISO 8049: 1988). 26 pp.
CEN EN 28049-92. Ferronickel Shot—Sampling for Analysis. 22 pp.
DIN ENGL EN 28049-92. Ferronickel Shot; Sampling for Analysis (ISO 8049: 1988) (July). 22 pp.
ISO 8049-88. Ferronickel Shot—Sampling for Analysis First Edition. 22 pp.

—Silicon Content—Gravimetric Analysis

BSI BS 6783: Part 9-86. 1986 Sampling and Analysis of Nickel, Ferronickel and Nickel Alloys Part 9: Method for Determination of Silicon in Ferronickel (Gravimetric Method). 4 pp.
BSI BS 6783: Part 9-01. 1986 Amd 1 Sampling and Analysis of Nickel, Ferronickel and Nickel Alloys Part 9: Method for Determination of Silicon in Ferronickel (Gravimetric Method) (ISO 8343: 1985) (AMD 6995) April 1, 1992. 7 pp.
CEN EN 28343-91. Ferronickel—Determination of Silicon Content—Gravimetric Method. 3 pp.
DIN ENGL EN 28343-92. Determination of Silicon Content of Ferronickel; Gravimetric Method (ISO 8343: 1985) (Feb). 5 pp.
ISO 8343-85. Ferronickel—Determination of Silicon Content—Gravimetric Method First Edition. 4 pp.

—Sulfur Content—Infrared Analysis

BSI BS 6783: Part 7-86. 1986 Sampling and Analysis of Nickel, Ferronickel and Nickel Alloys Part 7: Method for Determining of Sulphur in Nickel, Ferronickel and Nickel Alloy by Infra-Red Absorption After Induction Furnace Combustion. 11 pp.
BSI BS 6783: Part 7-01. 1986 Amd 1 Sampling and Analysis of Nickel, Ferronickel and Nickel Alloys Part 7: Method for Determination of Sulphur in Nickel, Ferronickel and Nickel Alloys by Infra-Red Absorption After Induction Furnace Combustion (ISO 7526: 1985). 14 pp.
CEN EN 27526-91. Nickel, Ferronickel and Nickel Alloys—Determination of Sulfur Content—Infra-Red Absorption Method After Induction Furnace Combustion. 3 pp.
ISO 7526-85. Nickel, Ferronickel and Nickel Alloys—Determination of Sulfur Content—Infra-Red Absorption Method After Induction Furnace Combustion First Edition. 9 pp.

—Sulfur Content—Volumetric Analysis

BSI BS 6783: Part 8-86. 1986 Sampling and Analysis of Nickel, Ferronickel and Nickel Alloys Part 8: Method for Determination of Sulphur in Nickel, Ferronickel and Nickel Alloys by Iodimetric After Induction Furnace Combustion. 10 pp.
BSI BS 6783: Part 8-01. 1986 Amd 1 Sampling and Analysis of Nickel, Ferronickel and Nickel Alloys Part 8: Method for Determination of Sulphur in Nickel, Ferronickel and Nickel Alloys by Iodimetric Titration Method After Induction Furnace Combustion (ISO 7527: 1985). 14 pp.

Ferronickel (Cont.)

—Sulfur Content—Volumetric Analysis (Cont.)

CEN EN 27527-91. Nickel, Ferronickel and Nickel Alloys—Determination of Sulfur Content—Iodimetric Titration Method After Induction Furnace Combustion. 3 pp.
DIN ENGL EN 27527-92. Determination of Sulfur Content of Nickel, Ferronickel and Nickel Alloys; Iodimetric Titration Method After Induction Furnace Combustion (ISO 7527: 1985) (Feb). 10 pp.
ISO 7527-85. Nickel, Ferronickel and Nickel Alloys—Determination of Sulfur Content—Iodimetric Titration Method After Induction Furnace Combustion First Edition. 9 pp.

Ferroniobium

See Also: Ferroalloys

CNS G3114-84. Ferroniobium (Jun)(5123).
ISO 5453-80. Ferroniobium—Specification and Conditions of Delivery First Edition. 5 pp.
JIS G 2319-86. Ferroniobium. 6 pp.

—Chemical Analysis

CNS G2103-81. Method of Chemical Analysis for Ferroniobium (Nov)(8122).
JIS G 1328-82. Methods for Chemical Analysis of Ferroniobium. 76 pp.

—Niobium Content—Gravimetric Analysis

ISO TR7955-82. Ferroniobium—Determination of Niobium Content—Gravimetric Method First Edition. 14 pp.

—Sampling

BSI BS 6200: Sec 2.2-93. 1993 Sampling and Analysis of Iron, Steel and Other Ferrous Metals Part 2: Sampling and Sample Preparation Section 2.2: Methods for Ferroalloys and Other Alloying Additives (V). 20 pp.
CNS G2218-84. Method of Sampling for Ferroniobium (Mar) (10816).
JIS G 1602-85. Methods of Sampling for Chemical Analysis of Ferroalloys (Part 2 Ferrotungsten, Ferromolybdenum, Ferrovanadium, Ferrotitanium and Ferroniobium). 9 pp.

—Sampling—Chemical Analysis

ISO 4552 Pt 2-87. Ferroalloys—Sampling and Sample Preparation for Chemical Analysis—Part 2: Ferrotitanium, Ferromolybdenum, Ferrotungsten, Ferroniobium, Ferrovanadium First Edition. 10 pp.

Ferrophosphorus

See Also: Ferroalloys

CNS G3040-84. Ferrophosphorus (Jun) (2520).
JIS G 2310-86. Ferrophosphorus. 6 pp.

—Chemical Analysis

CNS G2086-81. Method of Chemical Analysis for Ferro Phosphorous (Aug)(7793).

—Phosphorus Content—Gravimetric Analysis

JIS G 1320-68. Method for Chemical Analysis of Ferrophosphorus (R 1983). 6 pp.

—Sampling

CNS G2206-84. Method of Sampling for Ferrophosphorus (Jan) (10755).
JIS G 1603-85. Methods of Sampling for Chemical Analysis of Ferroalloys (Part 3 Ferrophosphorus, Manganese Metal, Silicon Metal, Chromium Metal, Calcium Silicon and Ferroboron). 10 pp.

Ferroresonant Power Transformers

Use: Ferroresonant Transformers

Ferroresonant Transformers

See Also: Transformers

CSA C22.2 NO 66-1988. Speciality Transformers; (Gen Instr 1). 53 pp.
CSA 1169 Bull. Electrical Bulletin 1169 June 27, 1978 to C22.2 NO 66. 2 pp.
CSA 1440 Bull. Electrical Bulletin 1440 June 23, 1986 to C22.2 NO 66. 2 pp.

— 601-999 V

CSA C22.2 NO 66-1988. Speciality Transformers; (Gen Instr 1). 53 pp.
CSA 1169 Bull. Electrical Bulletin 1169 June 27, 1978 to C22.2 NO 66. 2 pp.
CSA 1440 Bull. Electrical Bulletin 1440 June 23, 1986 to C22.2 NO 66. 2 pp.

Ferrosilicochromium

See Also: Ferroalloys

DIN ENGL 17565-68. Ferro-Chromium, Ferro-Chromium-Silicon and Chromium; Technical Conditions of Delivery (Dec). 2 pp.

INTERNATIONAL AND NON-U.S. NATIONAL STANDARDS
SUBJECT INDEX

Ferrosilicochromium (Cont.)
ISO 5449-80. Ferrosilicochromium—Specification and Conditions of Delivery First Edition. 5 pp.

—Carbon Content
CNS G2175-83. Method of Chemical Analysis for Carbon in Ferrosilico-Chromium (May)(10261).

—Chemical Analysis
JIS G 1325-89. Methods for Chemical Analysis of Ferrosilicochromium. 27 pp.

—Chromium Content
CNS G2174-83. Method of Chemical Analysis for Chromium in Ferrosilico-Chromium (May)(10260).
SAA AS 3587.3-91. Ferroalloys—Chemical Analysis—Part 3: Determination of Chromium Content of Ferrochromium and Ferrosilicochromium. 4 pp.

—Chromium Content—Potentiometric Analysis
BSI BS 6200: SUB SEC 3.10.4-85. 1985 Sampling and Analysis of Iron, Steel and Other Ferrous Metals Part 3: Methods of Analysis Section 3.10: Determination of Chromium Subsection 3.10.4: Ferrochromium and Ferrosilicochromium: Potentiometric Method. 6 pp.
ISO 4140-79. Ferrochromium and Ferrosilicochromium—Determination of Chromium Content—Potentiometric Method First Edition. 5 pp.

—Particle Size
DIN ENGL 17599-85. Ferroalloys; Particle Size Ranges (June). 4 pp.

—Phosphorus Content
CNS G2176-83. Method of Chemical Analysis for Phosphorus in Ferrosilico-Chromium (May)(10262).

—Sampling
BSI BS 6200: Sec 2.2-93. 1993 Sampling and Analysis of Iron, Steel and Other Ferrous Metals Part 2: Sampling and Sample Preparation Section 2.2: Methods for Ferroalloys and Other Alloying Additives (V). 20 pp.
JIS G 1601-85. Methods of Sampling for Chemical Analysis of Ferroalloys (Part 1 Ferromanganese, Ferrosilicon, Ferrochromium, Ferrosilicomanganese and Ferrosilicochromium). 12 pp.

—Sampling—Chemical Analysis
ISO 4552 Pt 1-87. Ferroalloys—Sampling and Sample Preparation for Chemical Analysis—Part 1: Ferrochromium, Ferrosilicochromium, Ferrosilicon, Ferrosilicomanganese, Ferromanganese First Edition. 11 pp.

—Silicon Content
CNS G2173-83. Method of Chemical Analysis for Silicon in Ferrosilico-Chromium (May)(10259).

—Silicon Content—Gravimetric Analysis
BSI BS 6200: SUB SEC 3.26.5-85. 1985 Sampling and Analysis of Iron, Steel and Other Ferrous Metals: Part 3: Methods of Analysis: Section 3.26: Determination of Silicon: Subsection 3.26.5: Ferrosilicon, Ferrosilicomanganese and Ferrosilicochromium: Gravimetric Method. 6 pp.
ISO 4158-78. Ferrosilicon, Ferrosilicomanganese and Ferrosilicochromium—Determination of Silicon Content—Gravimetric Method First Edition. 5 pp.

Ferrosilicomanganese
See Also: Ferroalloys
DIN ENGL 17564-68. Ferromanganese, Ferromanganese-Silicon and Manganese; Technical Conditions of Delivery (Dec). 3 pp.
ISO 5447-80. Ferrosilicomanganese—Specification and Conditions of Delivery First Edition. 5 pp.

—Manganese Content—Potentiometric Analysis
BSI BS 6200: SUB SEC 3.18.3-85. 1985 Sampling and Analysis of Iron, Steel and Other Ferrous Metals Part 3: Methods of Analysis Section 3.18: Determination of Manganese Subsection 3.18.3: Ferromanganese and Ferrosilicomanganese: Potentiometric Method. 6 pp.
BSI BS 6200: SUB SEC 3.18.3-01. 1985 Amd 1 Sampling and Analysis of Iron, Steel and Other Ferrous Metals Part 3: Methods of Analysis Sec 3.18: Determination of Manganese Subsection 3.18.3: Ferromanganese and Ferrosilicomanganese: Potentiometric Method (ISO 4159: 1978). 11 pp.
CEN EN 24 159-89. Determination of Manganese Content Potentiometric Method. 3 pp.
DIN ENGL EN 24159-90. Determination of Manganese Content of Ferromanganese and Ferro-Silicomanganese; by the Potentiometric Method (ISO 4159: 1978) (Apr). 6 pp.
ISO 4159-78. Ferromanganese and Ferrosilicomanganese—Determination of Manganese Content—Potentiometric Method First Edition. 5 pp.

—Particle Size
DIN ENGL 17599-85. Ferroalloys; Particle Size Ranges (June). 4 pp.

—Sampling
BSI BS 6200: Sec 2.2-93. 1993 Sampling and Analysis of Iron, Steel and Other Ferrous Metals Part 2: Sampling and Sample Preparation Section 2.2: Methods for Ferroalloys and Other Alloying Additives (V). 20 pp.
JIS G 1601-85. Methods of Sampling for Chemical Analysis of Ferroalloys (Part 1 Ferromanganese, Ferrosilicon, Ferrochromium, Ferrosilicomanganese and Ferrosilicochromium). 12 pp.

—Sampling—Chemical Analysis
ISO 4552 Pt 1-87. Ferroalloys—Sampling and Sample Preparation for Chemical Analysis—Part 1: Ferrochromium, Ferrosilicochromium, Ferrosilicon, Ferrosilicomanganese, Ferromanganese First Edition. 11 pp.

—Silicon Content—Gravimetric Analysis
BSI BS 6200: SUB SEC 3.26.5-85. 1985 Sampling and Analysis of Iron, Steel and Other Ferrous Metals: Part 3: Methods of Analysis: Section 3.26: Determination of Silicon: Subsection 3.26.5: Ferrosilicon, Ferrosilicomanganese and Ferrosilicochromium: Gravimetric Method. 6 pp.
ISO 4158-78. Ferrosilicon, Ferrosilicomanganese and Ferrosilicochromium—Determination of Silicon Content—Gravimetric Method First Edition. 5 pp.

Ferrosilicon
See Also: Ferroalloys
CNS G3034-83. Ferrosilicon (Nov)(2148).
DIN ENGL 17560-65. Ferrosilicon and Silicon; Technical Conditions of Delivery (Dec). 2 pp.
ISO 5445-80. Ferrosilicon—Specification and Conditions of Delivery First Edition. 6 pp.
JIS G 2302-86. Ferrosilicon. 6 pp.

—Aluminum Content—Atomic Absorption Spectrometry
BSI BS 6200: SUBSEC 3.1.5-85. 1985 Sampling and Analysis of Iron, Steel and Other Ferrous Metals Part 3: Methods of Analysis Section 3.1: Determination of Aluminium Subsection 3.1.5: Ferro-Silicon: Atomic Absorption Spectrometric Method. 6 pp.
ISO 4139-79. Ferrosilicon—Determination of Aluminium Content—Flame Atomic Absorption Spectrometric Method First Edition. 5 pp.

—Chemical Analysis
CNS G2015-80. Method of Chemical Analysis for Ferrosilicon (Jan)(2149).
JIS G 1312-89. Methods for Chemical Analysis of Ferrosilicon. 36 pp.

—Particle Size
DIN ENGL 17599-85. Ferroalloys; Particle Size Ranges (June). 4 pp.

—Sampling
BSI BS 6200: Sec 2.2-93. 1993 Sampling and Analysis of Iron, Steel and Other Ferrous Metals Part 2: Sampling and Sample Preparation Section 2.2: Methods for Ferroalloys and Other Alloying Additives (V). 20 pp.
CNS G2192-83. Method of Sampling for Ferrosilicon (Nov)(10654).
JIS G 1601-85. Methods of Sampling for Chemical Analysis of Ferroalloys (Part 1 Ferromanganese, Ferrosilicon, Ferrochromium, Ferrosilicomanganese and Ferrosilicochromium). 12 pp.

—Sampling—Chemical Analysis
ISO 4552 Pt 1-87. Ferroalloys—Sampling and Sample Preparation for Chemical Analysis—Part 1: Ferrochromium, Ferrosilicochromium, Ferrosilicon, Ferrosilicomanganese, Ferromanganese First Edition. 11 pp.

—Silicon Content—Gravimetric Analysis
BSI BS 6200: SUB SEC 3.26.5-85. 1985 Sampling and Analysis of Iron, Steel and Other Ferrous Metals: Part 3: Methods of Analysis: Section 3.26: Determination of Silicon: Subsection 3.26.5: Ferrosilicon, Ferrosilicomanganese and Ferrosilicochromium: Gravimetric Method. 6 pp.
ISO 4158-78. Ferrosilicon, Ferrosilicomanganese and Ferrosilicochromium—Determination of Silicon Content—Gravimetric Method First Edition. 5 pp.

Ferrotitanium
See Also: Ferroalloys
CNS G3113-84. Ferrotitanium (Jun) (5122).
DIN ENGL 17566-68. Ferrotitanium; Technical Conditions of Delivery (Dec). 2 pp.
ISO 5454-80. Ferrotitanium—Specification and Conditions of Delivery First Edition. 5 pp.
JIS G 2309-86. Ferrotitanium. 6 pp.

Ferrotitanium (Cont.)
—Chemical Analysis
CNS G2093-81. Method of Chemical Analysis for Ferrotitanium (Oct)(8001).
JIS G 1319-82. Methods for Chemical Analysis of Ferrotitanium. 44 pp.

—Particle Size
DIN ENGL 17599-85. Ferroalloys; Particle Size Ranges (June). 4 pp.

—Sampling
BSI BS 6200: Sec 2.2-93. 1993 Sampling and Analysis of Iron, Steel and Other Ferrous Metals Part 2: Sampling and Sample Preparation Section 2.2: Methods for Ferroalloys and Other Alloying Additives (V). 20 pp.
CNS G2205-84. Method of Sampling for Ferrotitanium (Jan) (10754).
JIS G 1602-85. Methods of Sampling for Chemical Analysis of Ferroalloys (Part 2 Ferrotungsten, Ferromolybdenum, Ferrovanadium, Ferrotitanium and Ferroniobium). 9 pp.

—Sampling—Chemical Analysis
ISO 4552 Pt 2-87. Ferroalloys—Sampling and Sample Preparation for Chemical Analysis—Part 2: Ferrotitanium, Ferromolybdenum, Ferrotungsten, Ferroniobium, Ferrovanadium First Edition. 10 pp.

—Titanium Content—Volumetric Analysis
BSI BS 6200: SUB SEC 3.32.4-85. 1985 Sampling and Analysis of Iron, Steel and Other Ferrous Metals Part 3: Methods of Analysis Section 3.32: Determination of Titanium Subsection 3.32.4: Ferrotitanium: Volumetric Method. 8 pp.
ISO 7692-83. Ferrotitanium—Determination of Titanium Content—Titrimetric Method First Edition. 7 pp.

Ferrotungsten
Use For: Ferrowolfram See Also: Ferroalloys; Heat Resistant Alloys; Refractory Materials
CNS G3083-84. Ferrotungsten (Jun)(3824).
DIN ENGL 17562-65. Ferrotungsten; Technical Conditions of Delivery (Dec). 2 pp.
ISO 5450-80. Ferrotungsten—Specification and Conditions of Delivery First Edition. 5 pp.
JIS G 2306-86. Ferrotungsten. 6 pp.

—Antimony Content
CNS G2156-83. Methods of Chemical Analysis for Antimony in Ferrotungsten (Jan)(9867).

—Arsenic Content
CNS G2158-83. Methods of Chemical Analysis for Arsenic in Ferrotungsten (Jan)(9869).

—Bismuth Content
CNS G2155-83. Methods of Chemical Analysis for Bismuth in Ferrotungsten (Jan)(9866).

—Carbon Content
CNS G2149-83. Methods of Chemical Analysis for Carbon in Ferrotungsten (Jan)(9860).

—Chemical Analysis
JIS G 1316-82. Methods for Chemical Analysis Ferrotungsten. 73 pp.

—Copper Content
CNS G2154-83. Methods of Chemical Analysis for Copper in Ferrotungsten (Jan)(9865).

—Manganese Content
CNS G2151-83. Methods of Chemical Analysis for Manganese in Ferrotungsten (Jan)(9862).

—Particle Size
DIN ENGL 17599-85. Ferroalloys; Particle Size Ranges (June). 4 pp.

—Phosphorus Content
CNS G2152-83. Methods of Chemical Analysis for Phosphorus in Ferrotungsten (Jan)(9863).

—Sampling
BSI BS 6200: Sec 2.2-93. 1993 Sampling and Analysis of Iron, Steel and Other Ferrous Metals Part 2: Sampling and Sample Preparation Section 2.2: Methods for Ferroalloys and Other Alloying Additives (V). 20 pp.
CNS G2202-84. Method of Sampling for Ferrotungsten (Jan) (10751).
JIS G 1602-85. Methods of Sampling for Chemical Analysis of Ferroalloys (Part 2 Ferrotungsten, Ferromolybdenum, Ferrovanadium, Ferrotitanium and Ferroniobium). 9 pp.

—Sampling—Chemical Analysis
ISO 4552 Pt 2-87. Ferroalloys—Sampling and Sample Preparation for Chemical Analysis—Part 2: Ferrotitanium, Ferromolybdenum, Ferrotungsten, Ferroniobium, Ferrovanadium First Edition. 10 pp.

Ferrotungsten (Cont.)

—Silicon Content
CNS G2150-83. Methods of Chemical Analysis for Silicon in Ferrotungsten (Jan)(9861).

—Sulfur Content
CNS G2153-83. Methods of Chemical Analysis for Sulfur in Ferrotungsten (Jan)(9864).

—Tin Content
CNS G2157-83. Methods of Chemical Analysis for Tin in Ferrotungsten (Jan)(9868).

—Tin Content—Volumetric Analysis
BSI BS 6200: SUB SEC 3.31.2-92. 1992 Sampling and Analysis of Iron, Steel and Other Ferrous Metals Part 3: Methods of Analysis Section 3.31: Determination of Tin Subsection 3.31.2: Ferrotungsten and Tungsten Metal: Volumetric Method. 7 pp.

—Tungsten Content
CNS G2148-83. Methods of Chemical Analysis for Tungsten in Ferrotungsten (Jan)(9859).

—Tungsten Content—Gravimetric Analysis
BSI BS 6200: SUB SEC 3.33.3-85. 1985 Sampling and Analysis of Iron, Steel and Other Ferrous Metals Part 3: Methods of Analysis Section 3.33: Determination of Tungsten Subsection 3.33.3: Ferrotungsten: Gravimetric Method. 6 pp.
ISO 7693-84. Ferrotungsten—Determination of Tungsten Content—Cinchonine Gravimetric Method First Edition. 6 pp.

Ferrous Alloys
Use: Ferroalloys

Ferrous Ammonium Sulfate
Use: Ammonium Ferrous Sulfate

Ferrous Chloride
Use For: Iron (II) Chloride **See Also:** Reagents
CNS K7229-66. Chemical Reagent (Ferrous Chloride) (Jun)(1728).
JIS K 8137-81. Iron (II) Chloride n Hydrate.

Ferrous Metals
Use: Ferroalloys

Ferrous Oxide Content Analysis

—Iron Ores
CNS M3018-80. Method for Determination of Ferrous Oxide in Iron Ores (Mar)(5380).
JIS M 8213-83. Method for Determination of Ferrous Oxide in Iron Ores.
JIS M 8213-71. Method for Determination of Ferrous Oxide in Iron Ores. 6 pp.

Ferrous Sulfate
Use For: Iron (II) Sulfate **See Also:** Reagents; Sulfates
CNS K7230-66. Chemical Reagent (Ferrous Sulfate) (Jun)(1729).
JIS K 8978-76. Ferrous Sulfate.

—Fertilizers
CNS N3095-88. Ferrous Sulfate, Fertilizer Grade (Jul)(12029).

—Water Treatment
CEN PREN 889-92. Iron(II) Sulfate Used for Water Intended for Human Consumption. 13 pp.
DIN ENGL 19609-87. Iron(II) Sulfate for Water Treatment; Technical Delivery Conditions (Feb). 3 pp.

Ferrous Sulfides
Use For: Iron Sulfides **See Also:** Pyrite
CNS K7231-66. Chemical Reagent (Ferrous Sulfide) (Jun)(1730).
CNS M1011-66. Iron Sulfide for Industrial Use (Jun)(2642) (R 1973).
JIS K 8948-80. Iron (II) Sulfide.

Ferrovanadium
See Also: Ferroalloys
CNS G3085-84. Ferrovanadium (Jun)(3827).
DIN ENGL 17563-65. Ferrovanadium; Technical Conditions of Delivery (Dec). 2 pp.
ISO 5451-80. Ferrovanadium—Specification and Conditions of Delivery First Edition. 5 pp.
JIS G 2308-86. Ferrovanadium. 6 pp.

—Chemical Analysis
CNS G2094-81. Method of Chemical Analysis for Ferrovandium (Oct)(8002).
JIS G 1318-82. Methods for Chemical Analysis of Ferrovanadium. 60 pp.

—Particle Size
DIN ENGL 17599-85. Ferroalloys; Particle Size Ranges (June). 4 pp.

Ferrovanadium (Cont.)

—Sampling
BSI BS 6200: Sec 2.2-93. 1993 Sampling and Analysis of Iron, Steel and Other Ferrous Metals Part 2: Sampling and Sample Preparation Section 2.2: Methods for Ferroalloys and Other Alloying Additives (V). 20 pp.
CNS G2204-84. Method of Sampling for Ferrovanadium (Jan) (10753).
JIS G 1602-85. Methods of Sampling for Chemical Analysis of Ferroalloys (Part 2 Ferrotungsten, Ferromolybdenum, Ferrovanadium, Ferrotitanium and Ferroniobium). 9 pp.

—Sampling—Chemical Analysis
ISO 4552 Pt 2-87. Ferroalloys—Sampling and Sample Preparation for Chemical Analysis—Part 2: Ferrotitanium, Ferromolybdenum, Ferrotungsten, Ferroniobium, Ferrovanadium First Edition. 10 pp.

—Vanadium Content—Potentiometric Analysis
BSI BS 6200: SUB SEC 3.34.4-85. 1985 Sampling and Analysis of Iron, Steel and Other Ferrous Metals Part 3: Methods of Analysis Section 3.34: Determination of Vanadium Subsection 3.34.4: Ferrovanadium: Potentiometric Method. 8 pp.
ISO 6467-80. Ferrovanadium—Determination of Vanadium Content—Potentiometric Method First Edition. 5 pp.

Ferrowolfram
Use: Ferrotungsten

Ferroxyl Testing

—Surface Finishing
CNS H2042-87. Method of Ferroxyl Test for Surface Finishing (Sep)(4160).

Ferruginous Manganese Ores
Use: Manganese Ores—Ferruginous

Ferrules
See Also: Electric Terminals; Flues; Pipe Fittings

—Aircraft
SBAC AS 15836-747 (V). Ferrule, Blank.

—Aircraft—Welding
SBAC AS 15720-743 (V). Ferrule, Welding, Pipe Fitting (for Inch Pipes).
SBAC AS 15766-769 ISSUE 3. Ferrule, Welding, Pipe Fitting—Assembly.
SBAC AS 26000-049 (V). Ferrule, Welding, Pipe Fitting (for Metric Pipes). 1 p.
SBAC AS 28013-018 ISSUE 1. Ferrule, Welding, Pipe Fitting—Assembly (For Metric Pipes).
SBAC AS 52710-719 ISSUE 1. Ferrule, Welding, Pipe, Fitting (Steel).
SBAC AS 52740-749 ISSUE 2. Ferrule, Welding, Pipe, Fitting (Titanium).

—Cables (Electric)
MOD UK DSTAN 59-23-70. Ferrules, Screened Cable Issue 1. 18 pp.

—Cords (Electric)—Aerospace
BSI 2SP 168-169-85. 1985 Ferrules and Assembly Wires for Braided Rubber Cord Assemblies for Aerospace Use. 4 pp.

—F-SMA Connectors—Preferred Products List
CECC CECC MUAHAG Vol 3B IS 4-91. Preferred Products List; Connectors; R.F. and Fibre Optics (En, Fr, Ge). 65 pp.

—Fluid Power Equipment—Titanium Alloy—Aircraft
AECMA PREN3269-90. Pipe Coupling 8 Degrees 30' in Titanium Alloy Blind Ferrule with Dynamic Beam Seal End. 5 pp.

—Fluid Power Equipment—Welded—Titanium Alloy—Aircraft
AECMA PREN3243-89. Pipe Coupling 8 Degrees 30' in Titanium Alloy Ferrule, Welded, with Dynamic Beam Seal End. 5 pp.
CEN PREN3561-90. Pipe Coupling 8 Degrees 30 in Titanium Alloy Ferrule with Dynamic Beam Seal End, Welded and Reduced at Pipe End. 6 pp.

—Hose Fittings—Aircraft
SBAC AGS 1233 (V). Ferrule Re-Usable End Fittings for Non-Fire Resistant Hose D.T.D. (R.D.I.) 3951.

—Steel Wire Rope—Fittings
BSI BS 5281-75. 1975 Amd 1 Ferrule Secured Eye Terminations for Wire Ropes. 8 pp.

Ferrules (Cont.)

—Vacuum Bottles—Aircraft
SBAC AS 3364 ISSUE 1. Ferrule—Vacuum Flask Stowage. 1 p.

Ferrying
See Also: Transportation

—Loads (Forces)—Classification
NATO STANAG 2021 ED 4 AMD 0-87. Computation of Bridge, Ferry, Raft and Vehicle Classifications. 31 pp.
NATO STANAG 2021 ED 5 AMD 0-90. Military Computation of Bridge, Ferry, Raft and Vehicle Classifications. 39 pp.

Fertilizer Distributors
Use: Fertilizers

Fertilizers
Use For: Fertilizer Distributors
See Also: Agricultural Chemicals; Ashes; Bone Meal; Dicyandiamide; Oil Meal; Soil Conditioners; Superphosphates; Urea
CNS N3017-88. Compound Fertilizer, Solid (Jun)(3076).
CNS N3062-88. Mixed Organic Fertilizer (Jun)(11928).
CNS N3100-88. Fritted Trace—Element (FTE) Fertilizer (Jul)(12034).
CNS N3102-88. Compound Fertilizer with Secondary & Micro-Nutrients, Solid (Jun)(12095).
CNS N3104-88. Compound Fertilizer with Organic Matters, Solid (Jun)(12097).
CNS N3105-88. Compound Fertilizer with Organic Matters and Secondary & Micro-Nutrients, Solid (Jun)(12098).
CNS N3106-88. Coated Compound Fertilizer (Jun)(12099).
CNS N3107-88. Horticultural Home-Use Compound Fertilizer (Jun)(12100).
CNS N3108-88. Multiple Secondary & Micro-Nutrients Fertilizer, Solid (Jun)(12101).
CNS N4137-88. Method of Test for Fertilizers: General Rules (Apr)(11463).
EC 76/116/EEC-75. Council Directive on the Approximation of the Laws of the Member States Relating to Fertilizers. 24 pp.

—Ammoniacal Nitrogen Content
CNS N4027-70. Method of Test for Compound Fertilizer (May)(3077). 6 pp.

—Ammoniacal Nitrogen Content—Volumetric Analysis
BSI BS 5551: SUBSEC 4.1.2-82. 1982 Fertilizers Part 4: Chemical Analysis Section 4.1: Determination of Nitrogen Subsection 4.1.2: Determination of Ammoniacal Nitrogen, Titrimetric Method After Distillations. 7 pp.
BSI BS 5551: SUBSEC 4.1.4-84. 1984 Fertilizers Part 4: Chemical Analysis Sec 4 Determination of Nitrogen Subsection 4.1.4: Method for Determination of Ammoniacal Nitrogen Content in the Presence of Other Substance Which Rel Ammonia When Treated with Sodium Hydroxide (Titrimetric Method). 6 pp.
ISO 5314-81. Fertilizers—Determination of Ammoniacal Nitrogen Content—Titrimetric Method After Distillation First Edition. 6 pp.
ISO 7408-83. Fertilizers—Determination of Ammoniacal Nitrogen Content in the Presence of Other Substances Which Release Ammonia When Treated with Sodium Hydroxide—Titrimetric Method First Edition. 7 pp.

—Ammonium Chloride
CNS N3002-88. Ammonium Chloride, Fertilizer Grade (Aug)(53).

—Ammonium Humate
CNS N3063-88. Ammonium Humate Fertilizer (Aug)(11950).

—Ammonium Nitrate
CNS N3003-88. Ammonium Nitrate, Fertilizer Grade (Jul)(55).
EC 80/876/EEC-80. Council Directive on the Approximation of the Laws of the Member States Relating to Straight Ammonium Nitrate Fertilizers of High Nitrogen Content. 5 pp.

—Ammonium Nitrate—Oil Retention
BSI BS 5551: Sec 3.4-87. 1987 Fertilizers Part 3: Physical Properties Section 3.4: Method for Determination of Oil Retention of High Nitrogen Content, Straight Ammonium Nitrate Fertilizers. 4 pp.
ISO 5313-86. High Nitrogen Content, Straight Ammonium Nitrate Fertilizers—Determination of Oil Retention First Edition. 4 pp.

—Ammonium Sulfate
CNS N3004-88. Ammonium Sulfate, Fertilizer Grade (Aug)(262).

INTERNATIONAL AND NON-U.S. NATIONAL STANDARDS
SUBJECT INDEX

Fertilizers

Fertilizers *(Cont.)*

—**Angle of Repose**
BSI BS 5551: Sec 3.6-89. 1989 Fertilizers Part 3: Physical Properties Section 3.6: Method for Determination of Static Angle of Repose of Solid Fertilizers. 6 pp.
ISO 8398-89. Solid Fertilizers—Measurement of Static Angle of Repose First Edition. 6 pp.

—**Arsenic Content**
CNS N4028-86. Method of Test for Hazardous Components in Compound Fertilizer (Jan)(3080). 4 pp.

—**Bean Curd**
CNS N3055-88. Dried Bean-Curd Residue Fertilizer (Jun)(11921).

—**Biuret Content**
CNS N4028-86. Method of Test for Hazardous Components in Compound Fertilizer (Jan)(3080). 4 pp.

—**Bone Meal**
CNS N3016-88. Steamed Bone Meal for Fertilizer (Aug)(2596).
CNS N3046-88. Flesh Meal Fertilizer (Jul)(11912).
CNS N3047-88. Flesh-Bone Meal Fertilizer (Jun)(11913).
CNS N3048-88. Raw Bone Meal Fertilizer (Jun)(11914).

—**Boric Acid**
CNS N3040-88. Boric Acid, Fertilizer Grade (Jul)(11863).

—**Boron Content**
EC COM(88) 562-88. Proposal for a Council Directive on the Approximation of the Laws of the Member States in Respect of the Trace Elements Boron, Cobalt, Copper, Iron, Manganese, Molybdenum and Zinc Contained in Fertilizers. 20 pp.
EC 89/530/EEC-89. Council Directive Supplementing and Amending Directive 76/116/EEC in Respect of the Trace Elements Boron, Cobalt, Copper Iron, Manganese, Molybdenum and Zinc Contained in Fertilizers. 9 pp.

—**Boron Inorganic Compound**
CNS N3041-88. Fused Boron Fertilizer (Jul)(11864).
CNS N3042-88. Processed Boron Fertilizer (Jul)(11865).

—**Bulk Density**
BSI BS 5551: Sec 3.1-93. 1993 Fertilizers Part 3: Physical Properties Section 3.1: Method for Determination of Bulk Density (Loose) (ISO 3944: 1992). 7 pp.
BSI BS 5551: Sec 3.1-81. 1981 Fertilizers Part 3: Physical Properties Section 3.1: Method for the Determination of Bulk Density (Loose). 4 pp.
BSI BS 5551: Sec 3.2-93. 1993 Fertilizers Part 3: Physical Properties Section 3.2: Method for Determination of Bulk Density (Tapped) (ISO 5311: 1992). 10 pp.
BSI BS 5551: Sec 3.2-83. 1983 Fertilizers Part 3: Physical Properties Section 3.2: Method for the Determination of Bulk Density (Tapped). 5 pp.
BSI BS 5551: Sec 3.3-92. 1992 Fertilizers Part 3: Physical Properties Section 3.3: Method for Determination of Bulk Density (Loose) of Fine-Grained Fertilizers (ISO 7837: 1992). 7 pp.
BSI BS 5551: Sec 3.3-84. 1984 Fertilizers Part 3: Physical Properties Section 3.3: Method for Determination of Bulk Density (Loose) of Fine-Grained Fertilizers. 5 pp.
ISO 3944-92. Fertilizers—Determination of Bulk Density (Loose) Third Edition. 7 pp.
ISO 5311-92. Fertilizers—Determination of Bulk Density (Tapped) Third Edition. 8 pp.
ISO 7837-92. Fertilizers—Determination of Bulk Density (Loose) of Fine-Grained Fertilizers Second Edition. 7 pp.

—**Byproducts—Animal**
CNS N3060-88. Animal by-Product Fertilizer (Jun)(11926).

—**Byproducts—Fish**
CNS N3059-88. Processed Fish Scrap Fertilizer (Jun)(11925).

—**Byproducts—Plant**
CNS N3061-88. Plant by-Product Fertilizer (Jun)(11927).

—**Calcium Ammonium Nitrate**
CNS N3066-88. Calcium Ammonium Nitrate Fertilizer (Jun)(11953).

—**Calcium Carbonates**
CNS N3024-88. Calcium Carbonate, Fertilizer Grade (Aug)(11847).

Fertilizers *(Cont.)*

—**Calcium Content**
EC COM(87) 646-87. Proposal for a Council Directive Supplementing and Amending Directive 76/116/EEC in Respect of the Calcium, Magnesium, Sodium and Sulphur Content of Fertilizers. 11 pp.
EC 89/284/EEC-89. Council Directive Supplementing and Amending Directive 76/116/EEC in Respect of the Calcium, Magnesium, Sodium and Sulphur Content of Fertilizers. 5 pp.
EC 89/519/EEC-89. Commission Directive Supplementing and Amending Directive 77/535/EEC on the Approximation of the Laws of the Member States Relating to Methods of Sampling and Analysis for Fertilizers. 18 pp.

—**Calcium Cyanamide**
CNS N3005-88. Calcium Cyanamide, Fertilizer Grade (Aug)(266).

—**Calcium Hydroxides**
CNS N3023-88. Slacked Lime, Fertilizer Grade (Aug)(11846).

—**Calcium Nitrate**
CNS N3065-88. Calcium Nitrate, Fertilizer Grade (Jul)(11952).

—**Calcium Oxides**
CNS N3022-88. Quick Lime, Fertilizer Grade (Aug)(11845).
CNS N3026-88. Mixed Lime, Fertilizer (Jul)(11849).
CNS N3027-88. By-Product Lime Fertilizer (Jul)(11850).

—**Classification**
BSI BS 5551: Sec 1.1-84. 1984 Fertilizers Section 1.1: Terminology and Labelling Section 1.1: Classification. 10 pp.
ISO 7851-83. Fertilizers and Soil Conditioners—Classification First Edition. 11 pp.

—**Cobalt Content**
EC COM(88) 562-88. Proposal for a Council Directive on the Approximation of the Laws of the Member States in Respect of the Trace Elements Boron, Cobalt, Copper, Iron, Manganese, Molybdenum and Zinc Contained in Fertilizers. 20 pp.
EC 89/530/EEC-89. Council Directive Supplementing and Amending Directive 76/116/EEC in Respect of the Trace Elements Boron, Cobalt, Copper Iron, Manganese, Molybdenum and Zinc Contained in Fertilizers. 9 pp.

—**Compost**
CNS N3020-85. Garbage Compost, Fertilizer Grade (Dec)(3960).

—**Copper**
CNS N3097-88. Chelated Copper Fertilizer (Jul)(12031).

—**Copper Content**
EC COM(88) 562-88. Proposal for a Council Directive on the Approximation of the Laws of the Member States in Respect of the Trace Elements Boron, Cobalt, Copper, Iron, Manganese, Molybdenum and Zinc Contained in Fertilizers. 20 pp.
EC 89/530/EEC-89. Council Directive Supplementing and Amending Directive 76/116/EEC in Respect of the Trace Elements Boron, Cobalt, Copper Iron, Manganese, Molybdenum and Zinc Contained in Fertilizers. 9 pp.

—**Cupric Sulfates**
CNS N3096-88. Copper Sulfate, Fertilizer Grade (Jul)(12030).

—**Feathers**
CNS N3049-88. Steamed Feather Fertilizer (Jun)(11915).

—**Ferrous Sulfate**
CNS N3095-88. Ferrous Sulfate, Fertilizer Grade (Jul)(12029).

—**Fish Meal**
CNS N3045-88. Fish Meal Fertilizer (Jul)(11911).

—**Glossaries**
BSI BS 5551: Sec 1.2-86. 1986 Fertilizers Part 1: Terminology and Labelling Section 1.2: Glossary of Terms. 18 pp.
ISO 8157-84. Fertilizers and Soil Conditioners—Vocabulary First Edition. 19 pp.

—**Ground Shell**
CNS N3025-88. Ground Shell, Fertilizer Grade (Jul)(11848).

—**Guanylurea Sulfate**
CNS N3071-88. Guanylurea Sulfate Fertilizer (Jul)(11958).

Fertilizers *(Cont.)*

—**Identification Systems**
BSI BS 5551: Sec 1.3-84. 1984 Fertilizers Part 1: Terminology and Labelling Section 1.3: Marking of Containers (Packages, Drums, Etc) and Labels. 4 pp.
ISO 7409-84. Fertilizers—Marking—Presentation and Declarations First Edition. 4 pp.

—**Iron**
CNS N3094-88. Chelated Iron Fertilizer (Jul)(12028).

—**Iron Content**
EC COM(88) 562-88. Proposal for a Council Directive on the Approximation of the Laws of the Member States in Respect of the Trace Elements Boron, Cobalt, Copper, Iron, Manganese, Molybdenum and Zinc Contained in Fertilizers. 20 pp.
EC 89/530/EEC-89. Council Directive Supplementing and Amending Directive 76/116/EEC in Respect of the Trace Elements Boron, Cobalt, Copper Iron, Manganese, Molybdenum and Zinc Contained in Fertilizers. 9 pp.

—**Leather**
CNS N3050-88. Steamed Leather Waste Fertilizer (Jun)(11916).

—**Liquid**
CNS N3101-88. Compound Fertilizer, Solid (Jun)(12094).
CNS N3103-88. Compound Fertilizer with Secondary & Micro-Nutrients, Fluid (Jun)(12096).
CNS N3109-88. Multiple Secondary & Micro-Nutrients Fertilizer, Fluid (Jun)(12102).
EC 88/183/EEC-88. Council Directive Amending Directive 76/116/EEC in Respect of Fluid Fertilizers. 7 pp.

—**Liquid—Sampling**
CNS N4057-80. Method of Test for Fertilizers—Sampling (Sep)(6500). 4 pp.

—**Liquid—Urea Ammonium Nitrate**
CNS N3076-88. Urea-Ammonium Nitrate Fertilizer, Fluid (Jul)(11963).

—**Magnesium**
CNS N3030-88. Processed Magnesium Fertilizer (Aug)(11853).
CNS N3031-88. Magnesium Humate, Fertilizer Grade (Jul)(11854).
CNS N3032-88. Magnesium Lignosulfonate Fertilizer (Aug)(11855).
CNS N3033-88. By-Product Magnesium Fertilizer (Aug)(11856).
CNS N3034-88. Mixed Magnesium Fertilizer (Aug)(11857).

—**Magnesium Content**
EC COM(87) 646-87. Proposal for a Council Directive Supplementing and Amending Directive 76/116/EEC in Respect of the Calcium, Magnesium, Sodium and Sulphur Content of Fertilizers. 11 pp.
EC 89/284/EEC-89. Council Directive Supplementing and Amending Directive 76/116/EEC in Respect of the Calcium, Magnesium, Sodium and Sulphur Content of Fertilizers. 5 pp.
EC 89/519/EEC-89. Commission Directive Supplementing and Amending Directive 77/535/EEC on the Approximation of the Laws of the Member States Relating to Methods of Sampling and Analysis for Fertilizers. 18 pp.

—**Magnesium Hydroxides**
CNS N3029-88. Magnesium Hydroxide, Fertilizer Grade (Aug)(11852).

—**Magnesium Oxide**
CNS N3112-89. Magnesium Oxide: Fertilizer Grade (Jul)(12572).

—**Magnesium Sulfates**
CNS N3028-88. Magnesium Sulfate, Fertilizer Grade (Aug)(11851).

—**Manganese**
CNS N3036-88. Processed Manganese Fertilizer (Jul)(11859).
CNS N3037-88. Manganese Sleg Fertilizer (Jul)(11860).
CNS N3038-88. Mixed Manganese Fertilizer (Jul)(11861).
CNS N3099-88. Chelated Manganese Fertilizer (Jul)(12033).

—**Manganese Content**
CNS N4137-2-85. Method of Test for Fertilizers: Determination of Manganese Content (Dec)(11463-2).
EC COM(88) 562-88. Proposal for a Council Directive on the Approximation of the Laws of the Member States in Respect of the Trace Elements Boron, Cobalt, Copper, Iron, Manganese, Molybdenum and Zinc Contained in Fertilizers. 20 pp.

INDUSTRY STANDARDS

INTERNATIONAL AND NON-U.S. NATIONAL STANDARDS
SUBJECT INDEX

Fertilizers

Fertilizers (Cont.)

—Manganese Content (Cont.)
EC 89/530/EEC-89. Council Directive Supplementing and Amending Directive 76/116/EEC in Respect of the Trace Elements Boron, Cobalt, Copper Iron, Manganese, Molybdenum and Zinc Contained in Fertilizers. 9 pp.

—Manganese Sulfate
CNS N3035-88. Manganese Sulfate, Fertilizer Grade (Jul)(11858).

—Manure
CNS N3056-88. Nitrogenous Guano Fertilizer (Jun)(11922).
CNS N3058-88. Processes Poultry Manure (Jun)(11924).

—Microbe
CNS N3057-88. Dried Microbe Fertilizer (Jun)(11923).

—Moisture Content
CNS N4027-70. Method of Test for Compound Fertilizer (May)(3077). 6 pp.
CNS N4058-80. Method of Test for Fertilizers—Determination of Moisture Content (Sep) (6501). 2 pp.

—Molybdenum Content
EC COM(88) 562-88. Proposal for a Council Directive on the Approximation of the Laws of the Member States in Respect of the Trace Elements Boron, Cobalt, Copper, Iron, Manganese, Molybdenum and Zinc Contained in Fertilizers. 20 pp.
EC 89/530/EEC-89. Council Directive Supplementing and Amending Directive 76/116/EEC in Respect of the Trace Elements Boron, Cobalt, Copper Iron, Manganese, Molybdenum and Zinc Contained in Fertilizers. 9 pp.

—Nitrate Nitrogen Content
CNS N4027-70. Method of Test for Compound Fertilizer (May)(3077). 6 pp.

—Nitrate Nitrogen Content—Gravimetric Analysis
BSI BS 5551: SUBSEC 4.1.1-81. 1981 Fertilizers Part 4: Chemical Analysis Section 4.1: Determination of Nitrogen Subsection 4.1.1: Nitron Gravimetric Method for Determination of Nitrate Nitrogen. 5 pp.
ISO 4176-81. Fertilizers—Determination of Nitrate Nitrogen Content—Nitron Gravimetric Method First Edition. 5 pp.

—Nitrates—Liquid
CNS N3015-88. Mixed Nitrate Fertilizer, Fluid (Aug)(1436).

—Nitrogen
CNS N3073-88. By-Product Nitrogen Fertilizer (Jul)(11960).
CNS N3074-88. Mixed Nitrogen Fertilizer (Jul)(11961).

—Nitrogen Content
CNS N4027-70. Method of Test for Compound Fertilizer (May)(3077). 6 pp.
CNS N4085-85. Method of Test for Fertilizers: Determination of Nitrogen Content (Dec)(8449).

—Nitrogen Content—Volumetric Analysis
BSI BS 5551: SUBSEC 4.1.3-86. 1986 Fertilizers Part 4: Chemical Analysis Section 4.1: Determi-nation of Nitrogen Subsection 4.1.3: Method for Determination of Total Nitrogen Content (Titrimetric Method After Distillation). 7 pp.
ISO 5315-84. Fertilizers—Determination of Total Nitrogen Content—Titrimetric Method After Distillation First Edition. 7 pp.

—Nitrogen—Liquid
CNS N3075-88. By-Product Nitrogen Fertilizer, Fluid (Jul)(11962).

—Nitrous Acid Content
CNS N4028-86. Method of Test for Hazardous Components in Compound Fertilizer (Jan)(3080). 4 pp.

—Oil Meal
CNS N3051-88. Soybean Meal Fertilizer (Jun)(11917).
CNS N3052-88. Peanut Meal Fertilizer (Jun)(11918).
CNS N3053-88. Linseed Meal Fertilizer (Jun)(11919).

—Organic Compounds—Liquid
CNS N3110-88. Compound Fertilizer with Organic Matters, Fluid (Jun)(12347).
CNS N3111-88. Compound Fertilizer with Organic Matters and Secondary & Micro-Nutrients, Fluid (Jun)(12348).

—Oxamide
CNS N3072-88. Oxamide Fertilizer (Jul)(11959).

Fertilizers (Cont.)

—Phosphate Content
CNS N4027-70. Method of Test for Compound Fertilizer (May)(3077). 6 pp.

—Phosphate Content—Extraction Analysis
BSI BS 5551: SUBSEC 4.2.1-78. 1978 Fertilizers Part 4: Chemical Analysis Section 4.2: Determination of Phosphorous Subsection 4.2.1: Extraction of Water-Soluble Phosphates. 4 pp.
BSI BS 5551: SUBSEC 4.2.2-87. 1987 Fertilizers Part 4: Chemical Analysis Section 4.2: Determination of Phosphorous Subsection 4.2.2: Extraction of Phosphates Soluble in Mineral Acids. 4 pp.
ISO 5316-77. Fertilizers—Extraction of Water-Soluble Phosphates First Edition. 3 pp.
ISO 7497-84. Fertilizers—Extraction of Phosphates Soluble in Mineral Acids First Edition. 4 pp.

—Phosphates
CNS N3007-88. Fused Phosphate Fertilizer (Jul)(428).
CNS N3085-88. Humate-Phosphate (Jul)(12019).
CNS N3086-88. Calcined Phosphate Fertilizer (Jul)(12020).
CNS N3087-88. Processed Phosphate Fertilizer (Jul)(12021).
CNS N3088-88. By-Product Phosphate Fertilizer (Jul)(12022).
CNS N3089-88. Concentrated Superphosphate, Fertilizer Grade (Jul)(12023).
CNS N3090-88. Mixed Phosphate Fertilizer (Jul)(12024).

—Phosphorus Content
CNS N4086-82. Method of Test for Fertilizers—Determination of Phosphorus Content (Jan) (8450). 8 pp.

—Phosphorus Content—Gravimetric Analysis
BSI BS 5551: SUBSEC 4.2.3-87. 1987 Fertilizers Part 4: Chemical Analysis Section 4.2: Determination of Phosphorous Subsection 4.2.3: Method for Determination of Phosphorous Content (Quinoline Molybdophosphate Gravimetric Method). 4 pp.
ISO 6598-85. Fertilizers—Determination of Phosphorus Content—Quinoline Phosphomolybdate Gravimetric Method First Edition. 5 pp.

—Potassium
CNS N3079-88. Crude Potassium Salts, Fertilizer Grade (Jul)(11966).
CNS N3080-88. Amended Bittern Potassium Fertilizer (Jul)(11967).
CNS N3081-88. Potassium Humate Fertilizer (Aug)(11968).
CNS N3082-88. Potassium Silicata, Fertilizer Grade (Jul)(11969).
CNS N3083-88. By-Product Potassium Fertilizer (Jul)(11970).
CNS N3084-88. Mixed Potassium Fertilizer (Jul)(12018).

—Potassium Bicarbonate
CNS N3078-88. Potassium Bicarbonate, Fertilizer Grade (Jul)(11965).

—Potassium Chloride
CNS N3019-88. Potassium Chloride, Fertilizer Grade (Aug)(3705).

—Potassium Content
BSI BS 5551: SUBSEC 4.3.1-83. 1983 Fertilizers Part 4: Chemical Analysis Section 4.3: Determination of Potassium Subsection 4.3.1: Preparation of the Test Solution for Determination of Water-Soluble Potassium Content. 3 pp.
BSI BS 5551: SUBSEC 4.3.2-83. 1983 Fertilizers Part 4: Chemical Analysis Section 4.3: Determination of Potassium Subsection 4.3.2: Preparation of the Test Solution for Determination of Acid-Soluble Potassium Content. 3 pp.
CNS N4087-82. Method of Test for Fertilizers—Determination of Potassium Content (Jan) (8451). 2 pp.
ISO 5317-83. Fertilizers—Determination of Water-Soluble Potassium Content—Preparation of the Test Solution First Edition. 3 pp.
ISO 7407-83. Fertilizers—Determination of Acid-Soluble Potassium Content—Preparation of the Test Solution First Edition. 3 pp.

—Potassium Content—Gravimetric Analysis
BSI BS 5551: SUBSEC 4.3.3-84. 1984 Amd 1 Fertilizers Part 4: Chemical Analysis Section 4.3: Determination of Potassium Subsection 4.3.3: Potassium Tetaphenylborate Gravimetric Methods. 6 pp.
ISO 5318-83. Fertilizers—Determination of Potassium Content—Potassium Tetraphenylborate Gravimetric Method (Reference Method) First Edition. 5 pp.

Fertilizers (Cont.)

—Potassium Content—Volumetric Analysis
BSI BS 5551: SUBSEC 4.3.4-88. 1988 Fertilizers Part 4: Chemical Analysis Section 4: Determination of Potassium Subsection 4.3.4: Determination of Potassium Content (Titrimetric Method). 7 pp.
ISO 5310-86. Fertilizers—Determination of Potassium Content—Titrimetric Method First Edition. 6 pp.

—Potassium Magnesium Sulfate
CNS N3077-88. Potassium Magnesium Sulfate Fertilizer (Jul)(11964).

—Potassium Oxide Content
CNS N4027-70. Method of Test for Compound Fertilizer (May)(3077). 6 pp.

—Potassium Sulfate
CNS N3018-88. Potassium Sulfate, Fertilizer Grade (Aug)(3436).

—Reduction
BSI BS 5551: Sec 2.5-89. 1989 Fertilizers Part 2: Sampling Section 2.5: Method for Reduction of Samples of Solid Fertilizers. 8 pp.
ISO 7742-88. Solid Fertilizers—Reduction of Samples First Edition. 8 pp.

—Rice Bran
CNS N3054-88. Rice Bran Meal Fertilizer (Jun)(11920).

—Sampling
BSI BS 5551: Sec 2.2-84. 1984 Fertilizers Part 2: Sampling Section 2.2: Presentation of Sampling Reports. 10 pp.
BSI BS 5551: Sec 2.3-78. 1978 Fertilizers Part 2: Sampling Section 2.3: Sampling from a Conveyor by Stopping the Belt (Reference Method) (ISO 3963: 1977). 3 pp.
BSI BS 5551: Sec 2.3-01. 1978 Amd 1 Fertilizers Part 2: Sampling Section 2.3: Sampling from a Conveyor by Stopping the Belt (Reference Method) (ISO 3963: 1977) (AMD 6962) February 28, 1992 (W). 5 pp.
BSI BS 5551: Sec 2.4-83. 1983 Fertilizers Part 2: Sampling Section 2.4: Practical Arrangements for Final Samples. 3 pp.
BSI BS 5551: Sec 2.7-88. 1988 Fertilizers Part 2: Sampling Section 2.7: Recommendations for Minimum Mass of Increment of a Solid Fertilizer to Be Taken to Be Representative of the Total Sampling Unit. 8 pp.
BSI BS 5551: Sec 2.9-91. 1991 Fertilizers Part 2: Sampling Section 2.9: Sampling Plan for the Evaluation of a Large Delivery of Solid Fertilizer (ISO 8634: 1991) (W). 15 pp.
CNS N4027-70. Method of Test for Compound Fertilizer (May)(3077). 6 pp.
CNS N4057-80. Method of Test for Fertilizers—Sampling (Sep)(6500). 4 pp.
EC 77/535/EEC-77. Commission Directive on the Approximation of the Laws of the Member States Relating to Methods of Sampling and Analysis for Fertilizers. 90 pp.
ISO 3963-77. Fertilizers—Sampling from a Conveyor by Stopping the Belt First Edition. 4 pp.
ISO 5306-83. Fertilizers—Presentation of Sampling Reports First Edition. 10 pp.
ISO 7410-83. Fertilizers and Soil Conditioners—Final Samples—Practical Arrangements First Edition. 3 pp.
ISO TR7553-87. Fertilizers—Sampling—Minimum Mass of Increment to Be Taken to Be Representative of the Total Sampling Unit First Edition. 7 pp.

—Sieve Analysis
BSI BS 5551: Sec 3.5-88. 1988 Fertilizers Part 3: Physical Properties Section 3.5: Method for Determination of Particle Size Distribution by Test Sieving. 5 pp.
ISO 8397-88. Solid Fertilizers and Soil Conditioners—Test Sieving Second Edition. 6 pp.

—Silicate Slag
CNS N3044-88. Silicate Slag Fertilizer (Jul)(11910).

—Siliceous Limestone
CNS N3043-88. Siliceous Limestone, Fertilizer Grade (Aug)(11866).

—Silicon Content
CNS N4137-1-85. Method of Test for Fertilizers: Determination of Silicon Content (Dec)(11463-1).

—Sodium Borate
CNS N3039-88. Borax, Fertilizer Grade (Jul)(11862).

—Sodium Content
EC COM(87) 646-87. Proposal for a Council Directive Supplementing and Amending Directive 76/116/EEC in Respect of the Calcium, Magnesium, Sodium and Sulphur Content of Fertilizers. 11 pp.

INTERNATIONAL AND NON-U.S. NATIONAL STANDARDS
SUBJECT INDEX

Fertilizers (Cont.)
—Sodium Content (Cont.)
EC 89/284/EEC-89. Council Directive Supplementing and Amending Directive 76/116/EEC in Respect of the Calcium, Magnesium, Sodium and Sulphur Content of Fertilizers. 5 pp.
EC 89/519/EEC-89. Commission Directive Supplementing and Amending Directive 77/535/EEC on the Approximation of the Laws of the Member States Relating to Methods of Sampling and Analysis for Fertilizers. 18 pp.

—Sodium Molybdate
CNS N3098-88. Sodium Molybdate, Fertilizer Grade (Jul)(12032).

—Sodium Nitrate
CNS N3064-88. Sodium Nitrate, Fertilizer Grade (Jul)(11951).

—Solid—Moisture Content—Gravimetric Analysis
BSI BS 5551: SUBSEC 4.4.1-93. 1993 Fertilizers Part 4: Chemical Analysis Section 4.4: Determination of Water Subsection 4.4.1: Gravimetric Method by Drying Under Reduced Pressure (ISO 8189: 1992). 6 pp.
BSI BS 5551: SUBSEC 4.4.2-93. 1993 Fertilizers Part 4: Chemical Analysis Section 4.4: Determination of Water Subsection 4.4.2: Gravimetric Method by Drying at 105 Degrees C (ISO 8190: 1992). 6 pp.
ISO 8189-92. Solid Fertilizers—Determination of Moisture Content—Gravimetric Method by Drying Under Reduced Pressure First Edition. 6 pp.
ISO 8190-92. Solid Fertilizers—Determination of Moisture Content—Gravimetric Method by Drying at (105 Plus or Minus 2) Degrees C First Edition. 6 pp.

—Solid—Nitrogen Content—Gravimetric Analysis
BSI BS 5551: SUBSEC 4.1.5-93. 1993 Fertilizers Part 4: Chemical Analysis Sec 4.1: Determination of Nitrogen Subsection 4.1.5: Method for Determination of Urea Nitrogen Content (Gravimetric Method Using Xanthydrol) (ISO 8603: 1993) (W). 8 pp.
ISO 8603-93. Solid Fertilizers—Determination of Urea Nitrogen Content—Gravimetric Method Using Xanthydrol First Edition. 7 pp.

—Solid—Samplers
ISO 5308-92. Solid Fertilizers—Method of Checking the Performance of Mechanical Devices for Sampling of Product Moving in Bulk First Edition. 6 pp.

—Solid—Sampling
BSI BS 5551: Sec 2.1-93. 1993 Fertilizers Part 2: Sampling Section 2.1: Simple Sampling Method for Small Lots of Solid Fertilizer (ISO 8633: 1992). 10 pp.
BSI BS 5551: Sec 2.5-89. 1989 Fertilizers Part 2: Sampling Section 2.5: Method for Reduction of Samples of Solid Fertilizers. 8 pp.
BSI BS 5551: Sec 2.6-92. 1992 Fertilizers Part 2: Sampling Section 2.6: Guide to Derivation of a Sampling Plan for the Evaluation of a Large Delivery of Solid Fertilizer (ISO TR 5307: 1991). 31 pp.
BSI BS 5551: Sec 2.10-93. 1993 Fertilizers Part 2: Sampling Section 2.10: Method of Checking the Performance of Mechanical Devices for Sampling of Solid Fertilizer Moving in Bulk (ISO 5308: 1992). 6 pp.
ISO TR5307-91. Solid Fertilizers—Derivation of a Sampling Plan for the Evaluation of a Large Delivery First Edition. 31 pp.
ISO 7742-88. Solid Fertilizers—Reduction of Samples First Edition. 8 pp.
ISO 8633-92. Solid Fertilizers—Simple Sampling Method for Small Lots First Edition. 8 pp.
ISO 8634-91. Solid Fertilizers—Sampling Plan for the Evaluation of a Large Delivery First Edition. 14 pp.

—Solid—Sulfate Content—Gravimetric Analysis
BSI BS 5551: Sec 4.5-93. 1993 Fertilizers Part 4: Chemical Analysis Section 4.5: Method for Determination of Mineral-Acid-Soluble Sulfate Content of Solid Fertilizers (ISO 10084: 1992). 6 pp.
ISO 10084-92. Solid Fertilizers—Determination of Mineral -Acid-Soluble Sulfate Content—Gravimetric Method First Edition. 6 pp.

—Sulfamic Acid Content
CNS N4028-86. Method of Test for Hazardous Components in Compound Fertilizer (Jan)(3080). 4 pp.

Fertilizers (Cont.)
—Sulfur Content
EC COM(87) 646-87. Proposal for a Council Directive Supplementing and Amending Directive 76/116/EEC in Respect of the Calcium, Magnesium, Sodium and Sulphur Content of Fertilizers. 11 pp.
EC 89/284/EEC-89. Council Directive Supplementing and Amending Directive 76/116/EEC in Respect of the Calcium, Magnesium, Sodium and Sulphur Content of Fertilizers. 5 pp.
EC 89/519/EEC-89. Commission Directive Supplementing and Amending Directive 77/535/EEC on the Approximation of the Laws of the Member States Relating to Methods of Sampling and Analysis for Fertilizers. 18 pp.

—Sulfuric Acid Content
CNS N4028-86. Method of Test for Hazardous Components in Compound Fertilizer (Jan)(3080). 4 pp.

—Test Specimens
BSI BS 5551: Sec 2.8-91. 1991 Fertilizers Part 2: Sampling Section 2.8: Methods for Preparation of Samples of Solid Fertilizer for Chemical and Physical Analysis (ISO 8358: 1991). 8 pp.

—Thiocyanic Acid Content
CNS N4028-86. Method of Test for Hazardous Components in Compound Fertilizer (Jan)(3080). 4 pp.

—Trace Element Content
EC 93/1/EEC-93. Commission Directive Amending Directive 77/535/EEC on the Approximation of the Laws of the Member States Relating to Methods of Sampling and Analysis for Fertilizers (Analysis Methods for Trace Elements). 20 pp.

—Urea
CNS N3013-88. Urea, Fertilizer Grade (Jul)(1309). 1 p.
CNS N3067-88. Crotonylidene Diurea Fertilizer (Jul)(11954).
CNS N3068-88. Isobutylidene Diurea Fertilizer (Jul)(11955).
CNS N3069-88. Coated Urea Fertilizer (Jul)(11956).
CNS N3070-88. Ureaform Fertilizer (Jul)(11957).

—Zinc
CNS N3092-88. Chelated Zinc Fertilizer (Jul)(12026).

—Zinc Content
EC COM(88) 562-88. Proposal for a Council Directive on the Approximation of the Laws of the Member States in Respect of the Trace Elements Boron, Cobalt, Copper, Iron, Manganese, Molybdenum and Zinc Contained in Fertilizers. 20 pp.
EC 89/530/EEC-89. Council Directive Supplementing and Amending Directive 76/116/EEC in Respect of the Trace Elements Boron, Cobalt, Copper Iron, Manganese, Molybdenum and Zinc Contained in Fertilizers. 9 pp.

—Zinc Oxide
CNS N3093-88. Zinc Oxide, Fertilizer Grade (Jul)(12027).

—Zinc Sulfate
CNS N3091-88. Zinc Sulfate, Fertilizer Grade (Jul)(12025).

Fertilizing Equipment
See Also: Agricultural Equipment
BSI BS 6483: Part 1-84. 1984 Fertilizer Distributors Part 1: Methods of Evaluating Performance Under Laboratory and Field Conditions. 19 pp.
BSI BS 6483: Part 2-85. 1985 Fertilizer Distributors Part 2: Method of Test for Fertilizer Distributors in Lines. 16 pp.
BSI BS 6550-85. 1985 Recommendation for Working Widths of Equipment for Sowing, Planting, Distributing Fertilizers and Spraying. 3 pp.
ISO 5690 Pt 1-85. Equipment for Distributing Fertilizers—Test Methods—Part 1: Full Width Fertilizer Distributors Second Edition. 17 pp.
ISO 5690 Pt 2-84. Equipment for Distributing Fertilizers—Test Methods—Part 2: Fertilizer Distributors in Lines First Edition. 16 pp.
ISO 6720-89. Agricultural Machinery—Equipment for Sowing, Planting, Distributing Fertilizers and Spraying—Recommended Working Widths Second Edition. 4 pp.

—Safety
DIN ENGL 11001 Pt 6-80. Agricultural Machines and Tractors; Implements for Soil Cultivation, Sowing, Plant Treatment and Fertilizing; Special Technical Safety Requirements and Testing (Aug). 4 pp.
ISO 4254 Pt 9-92. Tractors and Machinery for Agriculture and Forestry—Technical Means for Ensuring Safety—Part 9: Equipment for Sowing, Planting and Distributing Fertilizers First Edition. 7 pp.

Fetal Heart Detectors
Use: Fetal Monitors—Cardiac

Fetal Monitors
See Also: Medical Equipment; Patient Monitoring Equipment
JIS T 1303-84. Fetal Monitors.

—Cardiac—Ultrasonic
IEC 1206-93. Ultrasonics—Continuous-Wave Doppler Systems—Test Procedures First Edition. 65 pp.

FETs
Use For: Field Effect Transistors; Field Transistors; Unipolar Transistors See Also: GaAs FETs; JFETs; MOSFETs; Semiconductor Devices; Transistors
BSI BS 6493: Sec 1.4-92. 1992 Semiconductor Devices Part 1: Discrete Devices Section 1.4: Recommendations for Microwave Diodes and Transistors (IEC 747-4: 1991). 83 pp.
BSI BS 6493: Sec 1.8-85. 1985 Semiconductor Devices Part 1: Discrete Devices Section 1.8: Recommendations for Field-Effect Transistors. 46 pp.
BSI BS 6493: Sec 1.8-01. 1985 Amd 1 Semiconductor Devices: Discrete Devices and Integrated Circuits Part 1: Discrete Devices Section 1.8: Recommendations for Field-Effect Transistors (IEC 747-8: 1984) (AMD 7212) September 15, 1992. 58 pp.
IEC 747 Pt 4-91. Semiconductor Devices Discrete Devices Part 4: Microwave Diodes and Transistors First Edition. 163 pp.
IEC 747 Pt 8-84. Semiconductor Devices Discrete Devices Part 8: Field-Effect Transistors First Edition; (Amendment 1-1991). 112 pp.
MOD UK DSTAN 59-61: 90/047-71. Transistor Issue 1. 13 pp.
MOD UK DSTAN 59-61: 90/153-73. Semiconductor Device—Transistors Issue 1. 14 pp.

—Amplifiers—Preferred Products List
CECC CECC MUAHAG Vol 9 IS 3-90. Preferred Products List; Semiconductors (En, Fr, Ge) AMD 1 (En, Fr, Ge). 51 pp.

—Chopper
MOD UK DSTAN 59-61: Part 6-01. Semiconductor Devices (Listed on EPIC Database) Part 6: Field Effect Transistors Issue 1; Amendment 1. 28 pp.

—Constant Operation
CNS C6058-88. Environmental Testing Methods and Endurance Testing Methods for Discrete Semiconductor Devices (Continuous Operation Test of Field-Effect Transistor) (Sep)(5542).

—Intermittent Operation Testing
CNS C6060-88. Environmental Testing Methods and Endurance Testing Methods for Discrete Semiconductor Devices (Intermittent Operation Test of Field-Effect Transistor) (Sep)(5544).

—Microwave—Quality Assurance
BSI BS 9352-80. (WITHDRAWN) 1980 Rules for the Preparation of Detail Specifications for Semiconductor Devices of Assessed Quality: Field Effect Transistors for Microwave Applications. 10 pp.

—N-Channel—Enhancement Type—Switching
CECC CECC 50 012-040 ISSUE 1-87. BS CECC 50 012-040; N-Channel, Silicon, Case-Rated Quad Insulated Gate Enhancement MOSFET; Hermetically Sealed Dual-in-Line Encapsulation; High Speed Switching Application (En). 11 pp.
CECC CECC 50 012-041 ISSUE 1-87. BS CECC 50 012-041; 2 N-Channel, 2 P-Channel, Silicon, Case-Rated Quad Mospower Array; Quad Insulated Gate Enhancement FET; Hermetically Sealed Dual-in-Line; High Speed Switching Application (En). 15 pp.
CECC CECC 50 012-042 ISSUE 2-87. N-Channel Enhancement MOSFET Ambient-Rated, Low Frequency, High Voltage Transistor, Switching Applications; Plastic Encapsulation (En). 9 pp.
CECC CECC 50 012-051 ISSUE 1-88. NL CECC 50 012-051; Single Gate Field-Effect-Transistor (En). 8 pp.
CECC CECC 50 012-066 ISSUE 1-89. Mosfet Transistor; Enhancement Mode /Repetitive Avalanche; (En, Fr, Ge). 14 pp.
CECC CECC 50 012-067 ISSUE 1-89. Mosfet Transistor; Enhancement Mode /Repetitive Avalanche; (En, Ge, Ne). 14 pp.

—N-Channel—Enhancement Type—Switching—LF
CECC CECC 50 012-027 ISSUE 2-85. BS CECC 50 012-027; Single Gate Field-Effect Transistor, N-Channel, Enhanced Gate (En). 19 pp.

FETs (Cont.)

—N-Channel—Enhancement Type—Switching—LF (Cont.)

CECC CECC 50 012-028 ISSUE 2-85. BS CECC 50 012-028; Single Gate Field Effect Transistor, N Channel, Enhanced Gate (En). 19 pp.

CECC CECC 50 012-029 ISSUE 2-85. BS CECC 50 012-029; Single Gate Field Effect Transistor, N Channel, Enhanced Gate (En). 19 pp.

CECC CECC 50 012-043 ISSUE 2-87. N-Channel Enhancement MOSFET Ambient-Rated, Low Frequency, High Voltage Transistor, Switching Application; Plastic Encapsulation (En). 9 pp.

CECC CECC 50 012-056 ISSUE 1-91. BS CECC 50 012-056; Single Gate Field Effect Transisitor, N Channel, Enhanced Gate (En). 19 pp.

CECC CECC 50 012-061 ISSUE 1-91. BS CECC 50 012-061; Single Gate Field Effect Transistor, N Channel, Enhanced Gate; Type Numbers IRFY130(M), IRFY430(M) (En). 19 pp.

CECC CECC 50 012-062 ISSUE 1A-91. BS CECC 50 012-062; Single Gate Field Effect Transistor, N Channel, Enhanced Gate; Type Numbers IRFY044(M), IRFY140(M), IRFY240(M), IRFY340(M), IRFY440(M) (En). 19 pp.

—P-Channel—Enhancement Type—Switching

CECC CECC 50 012-041 ISSUE 1-87. BS CECC 50 012-041; 2 N-Channel, 2 P-Channel, Silicon, Case-Rated Quad Mospower Array; Quad Insulated Gate Enhancement FET; Hermetically Sealed Dual-in-Line; High Speed Switching Application (En). 15 pp.

CECC CECC 50 012-052 ISSUE 1-88. NL CECC 50 012-052; Single Gate Field-Effect-Transistor (En). 8 pp.

CECC CECC 50 012-053 ISSUE 1-87. BS CECC 50 012-053; P-Channel, Case-Rated Quad Insulated Gate Enhancement FET in Hermetically Sealed Dual-in-Line Encapsulation; High Speed Switching Application (En). 12 pp.

—P-Channel—Enhancement Type—Switching—LF

CECC CECC 50 012-036 ISSUE 1-90. BS CECC 50 012-036; Single Gate Field Effect Transistor, P Channel, Enhanced Gate (En). 21 pp.

CECC CECC 50 012-037 ISSUE D-89. BS CECC 50 012-037; Single Gate Field Effect Transistor, P Channel, Enhanced Gate (En). 21 pp.

CECC CECC 50 012-057 ISSUE 1-91. BS CECC 50 012-057; Single Gate Field Effect Transistor, P Channel, Enhanced Gate (En). 19 pp.

CECC CECC 50 012-063 ISSUE 1-90. BS CECC 50 012-063; Single Gate Field Effect Transistor, P Channel, Enhanced Gate; Type Numbers IRFY9120(M). 19 pp.

CECC CECC 50 012-064 ISSUE 1-90. BS CECC 50 012-064; Single Gate Field Effect Transistor, P Channel, Enhanced Gate; Type Numbers: IRF Y9130(M) (En). 19 pp.

CECC CECC 50 012-065 ISSUE 1-91. BS CECC 50 012-65; Single Gate Field Effect Transistor, P Channel, Enhanced Gate; Type Numbers IYFR140(M), IYFR240(M). 19 pp.

—Planar

MOD UK DSTAN 59-61: 90/094-01. Semiconductor Devices, Transistor Issue 1; Amendment 1. 23 pp.

MOD UK DSTAN 59-61: 90/153-75. Semiconductor Device—Transistors Issue 1. 14 pp.

—Power Amplifiers

IECQ QC 750106-93. Semiconductor Devices Discrete Devices Part 8: Field-Effect Transistors Section Two—Blank Detail Specification for Field-Effect Transistors for Case-Rated Power Amplifier Applications (IEC 747-8-2 ED 1). 41 pp.

—Power Amplifiers—Quality Assurance

BSI BS QC 750106-93. 1993 Blank Detail Specification Field-Effect Transistors for Case-Rated Power Amplifier Applications (IEC 747-8-2: 1993) (T). 18 pp.

IEC 747 Pt 8-2-93. Semiconductor Devices Discrete Devices Part 8: Field-Effect Transistors Section Two—Blank Detail Specification for Field-Effect Transistors for Case-Rated Power Amplifier Applications First Edition (IECQ QC 750106). 41 pp.

—Preferred Products List

CECC CECC MUAHAG Vol 9 IS 3-90. Preferred Products List; Semiconductors (En, Fr, Ge) AMD 1 (En, Fr, Ge). 51 pp.

—Reliability Assured

CNS C7121-86. Reliability Assured Field—Effect Transistors (Oct)(6131).

JIS C 7214-78. Reliability Assured Field-Effect Transistors.

—Reverse Bias Testing—High Temperature

CNS C6062-88. Environmental Testing Methods and Endurance Testing Methods of Discrete Semiconductor Devices (Reverse Bias Test of Field-Effect Transistor Under High Temperature) (Sep)(5546).

—Single Gate

IEC 747 Pt 8-1-87. Semiconductor Devices Discrete Devices Part 8: Field-Effect Transistors Section One—Blank Detail Specification for Single-Gate Field-Effect Transistors, up to 5 W and 1 GHz First Edition (IECQ QC 750112). 37 pp.

IECQ QC 750112-87. Semiconductor Devices Discrete Devices Part 8: Field-Effect Transistors Section One—Blank Detail Specification for Single-Gate Field-Effect Transistors, up to 5W and 1 GHz (IEC 747-8-1 ED 1). 37 pp.

MOD UK DSTAN 59-61: 80/036-85. Transistor, Field Effect Issue 1. 33 pp.

MOD UK DSTAN 59-61: 80/037-85. Transistor, Field Effect Issue 1. 33 pp.

MOD UK DSTAN 59-61: 80/038-85. Transistor, Field Effect Issue 1. 33 pp.

MOD UK DSTAN 59-61: 80/039-85. Transistor, Field Effect Issue 1. 33 pp.

MOD UK DSTAN 59-61: 80/040-85. Transistor, Field Effect Issue 1. 33 pp.

MOD UK DSTAN 59-61: 80/041-85. Transistor, Field Effect Issue 1. 33 pp.

MOD UK DSTAN 59-61: 90/094-01. Semiconductor Devices, Transistor Issue 1; Amendment 1. 23 pp.

—Single Gate—Quality Assurance

BSI BS QC 750112-88. 1988 Semiconductor Devices, Discrete Devices, Blank Detail Specification. 17 pp.

BSI BS CECC 50012-78. (WITHDRAWN) 1978 Amd 1 Single Gate Field-Effect Transistors: Blank Detail Specification (Renumbered as BS EN 150012: 1993). 67 pp.

BSI BS EN 150012-93. 1993 Amd 2 Blank Detail Specification: Single Gate Field-Effect Transistors (AMD 7593) February 1993 (T). 71 pp.

CENELEC EN 150 012-91. Blank Detail Specification: Single Gate Field-Effect Transistors. 62 pp.

—Switching

MOD UK DSTAN 59-61: Part 6-01. Semiconductor Devices (Listed on EPIC Database) Part 6: Field Effect Transistors Issue 1; Amendment 1. 28 pp.

MOD UK DSTAN 59-61: 90/099-72. Semiconductor Devices, Transistors Issue 1. 22 pp.

—Switching—Preferred Products List

CECC CECC MUAHAG Vol 9 IS 3-90. Preferred Products List; Semiconductors (En, Fr, Ge) AMD 1 (En, Fr, Ge). 51 pp.

Fever Thermometers
Use: Clinical Thermometers

FFA AS 202 Bravo Aircraft
See Also: Aircraft

—Antenna Positions
CAA. FFA AS202/184A Bravo. 1 p.

—Foreign Airworthiness Directives
CAA. FFA AS 202 Bravo Series Aircraft (Foreign Airworthiness Directives). 1 p.

Fiber Building Board
Use: Fiberboard

Fiber Cans
Use: Cans—Fiber

Fiber Cement Pipes
See Also: Pipes

—Drain
CEN PREN 588-1-91. Fibre-Cement Pipes for Sewer and Drains—Part 1: Pipes, Joints and Fittings for Gravity Systems. 39 pp.

—Sewer
CEN PREN 588-1-91. Fibre-Cement Pipes for Sewer and Drains—Part 1: Pipes, Joints and Fittings for Gravity Systems. 39 pp.

Fiber Composites
See Also: Fibers

—Aerospace—Mechanical Properties
CEN PREN 3783-92. Fibre Composite Materials Normalisation of Fibre Dominated Mechanical Properties. 7 pp.

—Quantitative Analysis
JIS L 1030-90. Testing Methods for Quantitative Analysis of Fibre Mixtures. 83 pp.

Fiber Content Analysis
See Also: Aramid Fiber Content Analysis; Carbon Fiber Content Analysis; Fibers; Glass Fiber Content Analysis; Inorganic Fiber Content Analysis

—Animal Feed
CNS N4024-8-86. Method of Test for Feeds: Determination of Crude Fiber (Aug)(2770-8).

—Exhaust Gases
BSI BS 6069: Sec 4.2-91. 1991 Characterization of Air Quality Part 4: Stationary Source Emissions Section 4.2: Method for the Determination of Asbestos Plant Emissions by Fibre Count Measurement. 24 pp.

—Flours (Food)
CNS N6028-73. Method of Test for Soybean Powder (Nov)(1418). 3 pp.

—Food
BSI BS 6215: Part 1-81. 1981 Amd 1 Agricultural Food Products Part 1: Determination of Crude Fibre Content (General Method). 12 pp.

BSI BS 6215: Part 1-02. 1981 Amd 2 Methods of Test for Agricultural Food Products Part 1: Determination of Crude Fibre Content (General Method) (ISO 5498: 1981) (AMD 7726) June 15, 1993 (W). 13 pp.

CNS N6118-79. Determination of Crude Fiber Content for Foodstuffs (Nov)(5037). 3 pp.

ISO 5498-81. Agricultural Food Products—Determination of Crude Fibre Content—General Method First Edition. 10 pp.

—Food—Scharrer Method
BSI BS 6215: Part 2-81. 1981 Amd 1 Agricultural Food Products Part 2: Determination of Crude Fibre Content (Modified Scharrer Method). 8 pp.

ISO 6541-81. Agricultural Food Products—Determination of Crude Fibre Content—Modified Scharrer Method First Edition. 6 pp.

Fiber Crops
See Also: Plants (Botany)

—Seeds
EC 88/380/EEC-88. Council Directive Amending Directives 66/400/EEC, 66/401/EEC, 66/ 402/EEC, 66/403/EEC, 69/ 208/EEC, 70/457/EEC AND 70/458/EEC on the Marketing of Beet Seed, Fodder Plant Seed, Cereal Seed, Seed Potatoes, Seed of Oil and Fibre Plants and Vegetable Seed. 18 pp.

Fiber Distributed Data Interface
Use For: FDDI; Fibre Distributed Data Interface
See Also: Fiber Optic Communication Equipment; Fiber Optic Equipment; Local Area Networks; Open Systems Interconnection

—Data Link Layer—Media Access Control
OSI ISO DIS 9314-2-88. Information Processing Systems-Fibre Distribution Data Interface (FDDI) Part 2: Media Access Control (MAC). 48 pp.

—Hybrid Ring Control
OSI ISO IEC DIS 9314-5-87. Information Processing Systems—Fibre Distributed Data Interface (FDDI)—Part 5: Hybrid Ring Control (HRC).

—Hybrid Rings
IEC DIS 9314 Pt 5-91. Information Processing Systems—Fibre Distributed Data Interface (FDDI)—Part 5: Hybrid Ring Control (HRC) ***CD-ROM ONLY***. 129 pp.

ISO DIS 9314 Pt 5-91. Information Processing Systems—Fibre Distributed Data Interface (FDDI)—Part 5: Hybrid Ring Control (HRC) ***CD-ROM ONLY***. 129 pp.

JTC1 DIS9314 Pt 5-91. Information Processing Systems—Fibre Distributed Data Interface (FDDI)—Part 5: Hybrid Ring Control (HRC) ***CD-ROM ONLY***. 129 pp.

—Network Services
IEC DISP 10608 Pt 6-93. Information Technology—International Standardized Profile TAnnnn—Connection-Mode Transport Service over Connectionless-Mode Network Service—Part 6: Definition of Profile TA54 for Operation over an FDDI LAN Subnetwork ***CD-ROM ONLY***. 20 pp.

ISO DISP 10608 Pt 6-93. Information Technology—International Standardized Profile TAnnnn—Connection-Mode Transport Service over Connectionless-Mode Network Service—Part 6: Definition of Profile TA54 for Operation over an FDDI LAN Subnetwork ***CD-ROM ONLY***. 20 pp.

INTERNATIONAL AND NON-U.S. NATIONAL STANDARDS
SUBJECT INDEX
Fiber

Fiber Distributed Data Interface (Cont.)
—Network Services (Cont.)
JTC1 DISP10608 Pt 6-93. Information Technology—International Standardized Profile TAnnnn—Connection-Mode Transport Service over Connectionless-Mode Network Service—Part 6: Definition of Profile TA54 for Operation over an FDDI LAN Subnetwork ***CD-ROM ONLY***. 20 pp.

—Physical Layer
BSI BS ISO/IEC 9314-3-90. 1990 Information Processing Systems—Fibre Distributed Data Interface (FDDI)—Part 3: Physical Layer Medium Dependent (PMD). 58 pp.
IEC 9314 Pt 3-90. Information Processing Systems—Fibre Distributed Data Interface (FDDI)—Part 3: Physical Layer Medium Dependent (PMD) First Edition. 55 pp.
ISO 9314 Pt 1-89. Information Processing Systems—Fibre Distributed Data Interface (FDDI)—Part 1: Token Ring Physical Layer Protocol (PHY) First Edition. 37 pp.
ISO 9314 Pt 3-90. Information Processing Systems—Fibre Distributed Data Interface (FDDI)—Part 3: Physical Layer Medium Dependent (PMD) First Edition. 55 pp.
JTC1 9314 Pt 1-89. Information Processing Systems—Fibre Distributed Data Interface (FDDI)—Part 1: Token Ring Physical Layer Protocol (PHY) First Edition. 37 pp.
JTC1 9314-90. Information Processing Systems—Fibre Distributed Data Interface (FDDI)—Part 3: Physical Layer Medium Dependent (PMD) First Edition. 55 pp.
OSI ISO DIS 9314-1-88. Information Processing Systems-Fibre Distributed Data Interface (FDDI) Part 1: Physical Layer Protocol (PHY). 32 pp.
OSI ISO/IEC 9314-3-90. Information Processing Systems—Fibre Distributed Data Interface (FDDI)—Patr 3: Physical Layer Medium Dependent (PMD). 72 pp.
OSI ISO DP 9314-3-87. Interconnection of Equipment Fibre Distributed Data Interface Physical Layer Medium Dependent (PMD). 52 pp.
SAA AS 3734.1-91. Information Systems—Equipment—Fibre Distributed Data Interface (FDDI)—Part 1: Token Ring PHYsical Layer Protocol (PHY) (ISO/IEC 9314-2:1989) (In Professional Packages 26A, 26C). 67 pp.
SAA AS 3734.3-91. Information Systems—Equipment—Fibre Distributed Data Interface (FDDI)—Part 3: Physical Layer Medium Dependent (PMD) (ISO/IEC 9314-3:1990). 47 pp.
SNZ NZS/ISO/IEC 9314.3-90. Information Systems—Equipment—Fibre Distributed Data Interface (FDDI) Part 3: Physical Layer Medium Dependent (PMD). 47 pp.

—Physical Layer—Token Rings
BSI BS 7233: Part 1-90. 1990 Fibre Distributed Data Interface (FDDI) Part 1: Token Ring Physical Layer Protocol (PHY). 36 pp.

—Token Ring—Media Access Control
SAA AS 3734.2-91. Information Systems—Equipment—Fibre Distributed Data Interface (FDDI)—Part 2: Token Ring Media Access Control (MAC) (ISO/IEC 9314-2: 1989) (In Professional Packages 26A, 26C). 67 pp.

—Token Rings
BSI BS 7233: Part 1-90. 1990 Fibre Distributed Data Interface (FDDI) Part 1: Token Ring Physical Layer Protocol (PHY). 36 pp.

—Token Rings—Media Access Control
BSI BS ISO 9314-2-89. 1989 Information Processing Systems—Fibre Distributed Data Interface (FDDI)—Part 2: Token Ring Media Access Control (MAC) (S). 76 pp.
ISO 9314 Pt 2-89. Information Processing Systems—Fibre Distributed Data Interface (FDDI)—Part 2: Token Ring Media Access Control (MAC) First Edition. 73 pp.
JTC1 9314 Pt 2-89. Information Processing Systems—Fibre Distributed Data Interface (FDDI)—Part 2: Token Ring Media Access Control (MAC) First Edition. 73 pp.

—Transport Services
IEC DISP 10608 Pt 6-93. Information Technology—International Standardized Profile TAnnnn—Connection-Mode Transport Service over Connectionless-Mode Network Service—Part 6: Definition of Profile TA54 for Operation over an FDDI LAN Subnetwork ***CD-ROM ONLY***. 20 pp.

Fiber Distributed Data Interface (Cont.)
—Transport Services (Cont.)
IEC DISP 10608 Pt 14-93. Information Technology—International Standardized Profile TAnnnn—Connection-Mode Transport Service over Connectionless-Mode Network Service—Part 14: MAC, PHY and PMD Sublayer Dependent and Stn. Mgmt. Reqmts. over an FDDI LAN Subnetwork ***CD-ROM ONLY***. 38 pp.
ISO DISP 10608 Pt 6-93. Information Technology—International Standardized Profile TAnnnn—Connection-Mode Transport Service over Connectionless-Mode Network Service—Part 6: Definition of Profile TA54 for Operation over an FDDI LAN Subnetwork ***CD-ROM ONLY***. 20 pp.
ISO DISP 10608 Pt 14-93. Information Technology—International Standardized Profile TAnnnn—Connection-Mode Transport Service over Connectionless-Mode Network Service—Part 14: MAC, PHY and PMD Sublayer Dependent and Stn. Mgmt. Reqmts. over an FDDI LAN Subnetwork ***CD-ROM ONLY***. 38 pp.
JTC1 DISP10608 Pt 6-93. Information Technology—International Standardized Profile TAnnnn—Connection-Mode Transport Service over Connectionless-Mode Network Service—Part 6: Definition of Profile TA54 for Operation over an FDDI LAN Subnetwork ***CD-ROM ONLY***. 20 pp.
JTC1 DISP10608 Pt 14-93. Information Technology—International Standardized Profile TAnnnn—Connection-Mode Transport Service over Connectionless-Mode Network Service—Part 14: MAC, PHY and PMD Sublayer Dependent and Stn. Mgmt. Reqmts. over an FDDI LAN Subnetwork ***CD-RON ONLY***. 38 pp.

Fiber Optic Build Outs
Use: Fiber Optic Cable Assemblies

Fiber Optic Cable Assemblies
Use For: Fiber Optic Build Outs; Fiberoptic Cable Assemblies *See Also:* Cable Assemblies (Electric); Fiber Optic Cables; Fiber Optic Equipment
MOD UK DSTAN 60-1: Part 4-82. Fibre Optics Part 4: Cable Assemblies, Fibre Optic Issue 1. 23 pp.

Fiber Optic Cables
Use For: Fibrelight Cables *See Also:* Communication Cables; Fiber Optic Cable Assemblies; Fiber Optic Connectors; Fiber Optic Equipment; Fiber Optic Splices
BSI BS 6558: Part 1-85. 1985 Optical Fibres and Cables Part 1: General Requirements. 58 pp.
CECC EN 187 000-92. Generic Specification: Optical Fibre Cables (Supersedes CECC 87 000 Issue 1: 1991). 57 pp.
CNS C6316-88. Method of Test for Fiber Optic Devices (FOTP-95 Optical Fibers and Cables Absolute Optical Power Test) (Jul)(12363).
IEC 794 Pt 2-89. Optical Fibre Cables Part 2: Product Specifications Second Edition. 32 pp.
JIS C 6830-91. Optical Fiber Cords. 11 pp.
MOD UK DSTAN 60-1-89. Cables, Fibre Optic Issue 2. 52 pp.

—Aerial—Cable Joints
CCITT RECMN L.12-92. Optical Fibre Joints (Study Group VI) 10 pp. 10 pp.
CCITT RECMN L.13-92. Sheath Joints and Organizers of Optical Fibre Cables in the Outside Plant (Study Group VI) 6 pp. 6 pp.

—Aerial—Insulation
CCITT RECMN L.10-89. Optical Fibre Cables for Duct, Tunnel, Aerial and Buried Application—Construction, Installation and Protection of Cable and Other Elements of Outside Plant (Study Group VI) 8 pp. 8 pp.

—Bend Testing
CNS C6276-86. Method of Test for Fiber Optic Devices (FOTP-37 Cable Bend Test Low and High Temperature) (Dec)(11788).
CNS C6297-87. Method of Test for Fiber Optic Devices (FOTP-33 Fiber Optic Cable Tensile Loading and Bending Test) (Sep)(12083).
CNS C6315-88. Method of Test for Fiber Optic Devices (FOTP-91 Fiber Optic Cable Twist-Bend Test) (Jul)(12362).

—Bend Testing—Attenuation
DIN VDE 0472 Pt 232-85. Testing of Cables, Wires and Flexible Cords; Attenuation Changes in Optical Fibre Cables During Bending (Jan). 4 pp.

—Cable Joints—Telecommunication Equipment
CCITT RECMN L.12-92. Optical Fibre Joints (Study Group VI) 10 pp. 10 pp.

Fiber Optic Cables (Cont.)
—Cable Joints—Telecommunication Equipment (Cont.)
CCITT RECMN L.13-92. Sheath Joints and Organizers of Optical Fibre Cables in the Outside Plant (Study Group VI) 6 pp. 6 pp.

—Compression Testing
CNS C6318-88. Method of Test for Fiber Optic Devices (FOTP-41 Compressive Loading Resistance of Fiber Optic Cable) (Jul)(12366).

—Conductive
CSA C22.2 NO 232-M1988. Optical Fiber Cables. 16 pp.
IEC 794 Pt 1-93. Optical Fibre Cables Part 1: Generic Specification Third Edition. 86 pp.

—Data Processing Equipment—Glossaries
DIN VDE 0888 Pt 1-88. Optical Fibres for Telecommunication and Data Processing Systems Definitions (June) (Supersedes DIN 57888 Part 1). 18 pp.

—Digital Line Systems
CCITT RECMN G.950-89. General Considerations on Digital Line Systems—Digital Networks, Digital Sections and Digital Line Systems (Study Groups XV and XVIII) 2 pp. 2 pp.
CCITT RECMN G.955-89. Digital Line Systems Based on the 1544 kBit/s Hierarchy on Optical Fibre Cables—Digital Networks, Digital Sections and Digital Line Systems (Study Groups XV and XVIII) 8 pp. 8 pp.
CCITT RECMN G.956-89. Digital Line Systems Based on the 2048 kBit/s Hierarchy on Optical Fibre Cables—Digital Networks, Digital Sections and Digital Line Systems (Study Groups XV and XVIII) 8 pp. 8 pp.

—Digital Line Systems—Synchronous Digital Hierarchy
CCITT RECMN G.958-90. Digital Line Systems Based on the Synchronous Digital Hierarchy for Use on Optical Fibre Cables (Study Group XV) 28 pp. 28 pp.

—Direct Burial—Cable Joints
CCITT RECMN L.12-92. Optical Fibre Joints (Study Group VI) 10 pp. 10 pp.
CCITT RECMN L.13-92. Sheath Joints and Organizers of Optical Fibre Cables in the Outside Plant (Study Group VI) 6 pp. 6 pp.

—Direct Burial—Insulation
CCITT RECMN L.10-89. Optical Fibre Cables for Duct, Tunnel, Aerial and Buried Application—Construction, Installation and Protection of Cable and Other Elements of Outside Plant (Study Group VI) 8 pp. 8 pp.

—Drip Testing
CNS C6282-87. Method of Test for Fiber Optic Devices (FOTP-81 Compound Flow (Drip) Test for Filled Fiber Optic Cable) (Apr)(11904).

—Elongation
CCITT RECMN L.14-92. Measurement Method to Determine the Tensile Performance of Optical Fibre Cables Under Load (Study Group VI) 8 pp. 8 pp.

—Environmental—Cable Joints
CCITT RECMN L.12-92. Optical Fibre Joints (Study Group VI) 10 pp. 10 pp.
CCITT RECMN L.13-92. Sheath Joints and Organizers of Optical Fibre Cables in the Outside Plant (Study Group VI) 6 pp. 6 pp.

—Filled—Liquid Penetrant
CNS C6301-87. Method of Test for Fiber Optic Devices (FOTP-82 Fluid Penetration Test for Filled Fiber Optic Cable) (Sep)(12087).

—Flammability Testing
CNS C6285-87. Method of Test for Fiber Optic Devices (FOTP-99 Gas Flame Test for Special Purpose Fiber Optic Cable) (Apr)(11907).

—Gas Leakage
CNS C6329-89. Method of Test for Fiber Optic Devices (FOTP-100 Gas Leakage Test for Gas—Blocked Fiber Optic Cables) (Jul)(12560).

—Glossaries
CCITT RECMN G.651-89. Characteristics of a 50/125 Micrometer Multimode Graded Index Optical Fibre Cable—Transmission Media Characteristics (Study Group XV) 30 pp. 30 pp.
CCITT RECMN G.652-89. Characteristics of a Single-Mode Optical Fibre Cable—Transmission Media Characteristics (Study Group XV) 34 pp. 34 pp.

Fiber Optic Cables (Cont.)

—Immersion Testing

CNS C6251-85. Method of Test for Fiber Optic Devices (Fiber Optic Cable Fluid Immersion Test) (Aug)(11333).

—Impact Testing

CNS C6223-84. Method of Test for Fiber Optic Devices (Impact Testing of Cable Assemblies) (May)(10878).

—Inspection

CNS C6295-87. Method of Test for Fiber Optic Devices (FOTP-13 Visual and Mechanical Inspection of Fibers, Cables, Connectors Other Fiber Optic Devices) (Sep)(12081).

—Insulation

CCITT RECMN L.10-89. Optical Fibre Cables for Duct, Tunnel, Aerial and Buried Application—Construction, Installation and Protection of Cable and Other Elements of Outside Plant (Study Group VI) 8 pp. 8 pp.

—Insulation—Adhesion Testing

CNS C6284-87. Method of Test for Fiber Optic Devices (FOTP-84 Jacket Self-Adhesion (Blocking) Test for Fiber Optic Cable) (Apr)(11906).

—Insulation—Elongation

CNS C6233-84. Method of Test for Fiber Optic Devices (Cable Jacket Elongation) (Jun)(10933).

—Insulation—Joints

CCITT RECMN L.13-92. Sheath Joints and Organizers of Optical Fibre Cables in the Outside Plant (Study Group VI) 6 pp. 6 pp.

—Insulation—Shrinkage

CNS C6232-84. Method of Test for Fiber Optic Devices (Cable Jacket Shrinkage) (Jun)(10932).

—Knot Testing

CNS C6252-88. Method of Test for Fiber Optic Devices (FOTP-87 Fiber Optic Cable Knot Test) (Jul)(11334).

—Lightning Protection

CCITT RECMN K.25-91. Lightning Protection of Optical Fibre Cables—Protection Against Interference (Study Group V) 4 pp. 4 pp.

—Low Temperature Testing

CNS C6317-88. Method of Test for Fiber Optic Devices (FOTP-98 Fiber Optic Cable External Freezing Test) (Jul)(12364).

—Marine

JIS F 8550-91. Marine Optical Transmission Apparatus - Criteria for Selection of Optical Fiber Cable and Optical Connector.

—Multimode

CCITT RECMN G.651-89. Characteristics of a 50/125 Micrometer Multimode Graded Index Optical Fibre Cable—Transmission Media Characteristics (Study Group XV) 30 pp. 30 pp.

—Multimode—Amplitude Response

CCITT RECMN G.651-89. Characteristics of a 50/125 Micrometer Multimode Graded Index Optical Fibre Cable—Transmission Media Characteristics (Study Group XV) 30 pp. 30 pp.

—Multimode—Attenuation

CCITT RECMN G.651-89. Characteristics of a 50/125 Micrometer Multimode Graded Index Optical Fibre Cable—Transmission Media Characteristics (Study Group XV) 30 pp. 30 pp.

—Multimode—Chromatic Dispersion

CCITT RECMN G.651-89. Characteristics of a 50/125 Micrometer Multimode Graded Index Optical Fibre Cable—Transmission Media Characteristics (Study Group XV) 30 pp. 30 pp.

—Multimode Distortion—Bandwidth

CCITT RECMN G.651-89. Characteristics of a 50/125 Micrometer Multimode Graded Index Optical Fibre Cable—Transmission Media Characteristics (Study Group XV) 30 pp. 30 pp.

—Multimode—Factory Lengths

CCITT RECMN G.651-89. Characteristics of a 50/125 Micrometer Multimode Graded Index Optical Fibre Cable—Transmission Media Characteristics (Study Group XV) 30 pp. 30 pp.

—Multimode—Frequency Response—Basebands

CCITT RECMN G.651-89. Characteristics of a 50/125 Micrometer Multimode Graded Index Optical Fibre Cable—Transmission Media Characteristics (Study Group XV) 30 pp. 30 pp.

Fiber Optic Cables (Cont.)

—Multimode—Mechanical Properties

JIS C 6861-91. Test Methods for Mechanical Characteristics of All Plastic Multimode Optical Fibers. 13 pp.

—Multimode—Optical Measurement

CCITT RECMN G.651-89. Characteristics of a 50/125 Micrometer Multimode Graded Index Optical Fibre Cable—Transmission Media Characteristics (Study Group XV) 30 pp. 30 pp.

—Multimode—Phase Response

CCITT RECMN G.651-89. Characteristics of a 50/125 Micrometer Multimode Graded Index Optical Fibre Cable—Transmission Media Characteristics (Study Group XV) 30 pp. 30 pp.

—Multimode—Plastic

JIS C 6836-92. All Plastic Multimode Optical Fiber Cords. 9 pp.

—Nonconductive

CSA C22.2 NO 232-M1988. Optical Fiber Cables. 16 pp.

—Quality Assurance

BSI BS 9230-86. 1986 Connectors of Assessed Quality for Optical Fibres and Cables: Generic Data and Methods of Test. 44 pp.

BSI BS 9230-01. 1986 Amd 1 Connectors of Assessed Quality for Optical Fibres and Cables: Generic Data and Methods of Test (AMD 6331) October 31, 1991. 64 pp.

CENELEC EN 187 000-92. Generic Specification: Optical Fibre Cables (Supersedes CECC 87 000 Issue 1: 1991). 57 pp.

MOD UK DSTAN 60-1: Part 0-92. Cables, Fibre Optic Part 0: General Requirements and Test Methods for Qualification Approval and Capability Approval Generic Specification Issue 1. 65 pp.

—Single Mode

CCITT RECMN G.652-89. Characteristics of a Single-Mode Optical Fibre Cable—Transmission Media Characteristics (Study Group XV) 34 pp. 34 pp.

—Single Mode—Aerial

CENELEC PRI-ETS 300 227-92. Transmission and Multiplexing (TM); CCITT Recommendation G.652-Type Single-Mode Optical Fibre. 15 pp.

CENELEC PRI-ETS 300 227-93. Transmission and Multiplexing (TM); CCITT Recommendation G.652-Type Single-Mode Optical Fibre. 14 pp.

CENELEC I-ETS 300 227-93. Transmission and Multiplexing (TM); ITU-T Recommendation G.652-Type Single-Mode Optical Fibre. 14 pp.

CENELEC PRI-ETS 300 229-92. Transmission and Multiplexing (TM); Single-Mode Optical Fibre Cables to be Used for Aerial Application. 20 pp.

CENELEC PRI-ETS 300 229-93. Transmission and Multiplexing (TM); Single-Mode Optical Fibre Cables to Be Used for Aerial Application. 19 pp.

CENELEC I-ETS 300 229-93. Transmission and Multiplexing (TM); Single-Mode Optical Fibre Cables to Be Used for Aerial Application. 19 pp.

ETSI I-ETS 300 227-93. Transmission and Multiplexing (TM); ITU-T Recommendation G.652-Type Single-Mode Optical Fibre. 14 pp.

ETSI PRI-ETS 300 227-93. Transmission and Multiplexing (TM); CCITT Recommendation G.652-Type Single-Mode Optical Fibre. 14 pp.

ETSI PRI-ETS 300 227-92. Transmission and Multiplexing (TM); CCITT Recommendation G.652-Type Single-Mode Optical Fibre. 15 pp.

ETSI I-ETS 300 229-93. Transmission and Multiplexing (TM); Single-Mode Optical Fibre Cables to Be Used for Aerial Application. 19 pp.

ETSI PRI-ETS 300 229-93. Transmission and Multiplexing (TM); Single-Mode Optical Fibre Cables to Be Used for Aerial Application. 19 pp.

ETSI PRI-ETS 300 229-92. Transmission and Multiplexing (TM); Single-Mode Optical Fibre Cables to be Used for Aerial Application. 20 pp.

—Single Mode—Attenuation

CCITT RECMN G.652-89. Characteristics of a Single-Mode Optical Fibre Cable—Transmission Media Characteristics (Study Group XV) 34 pp. 34 pp.

—Single Mode—Buried

CENELEC PRI-ETS 300 226-92. Transmission and Multiplexing (TM); Single-Mode Optical Fibre Cables to Be Used in Ducts and for Directly Buried Application. 19 pp.

CENELEC PRI-ETS 300 226-93. Transmission and Multiplexing (TM); Single-Mode Optical Fibre Cables to be Used in Ducts and for Directly Buried Application. 18 pp.

CENELEC I-ETS 300 226-93. Transmission and Multiplexing (TM); Single-Mode Optical Fibre Cables to Be Used in Ducts and for Directly Buried Application. 18 pp.

Fiber Optic Cables (Cont.)

—Single Mode—Buried (Cont.)

CENELEC PRI-ETS 300 227-92. Transmission and Multiplexing (TM); CCITT Recommendation G.652-Type Single-Mode Optical Fibre. 15 pp.

CENELEC PRI-ETS 300 227-93. Transmission and Multiplexing (TM); CCITT Recommendation G.652-Type Single-Mode Optical Fibre. 14 pp.

CENELEC I-ETS 300 227-93. Transmission and Multiplexing (TM); ITU-T Recommendation G.652-Type Single-Mode Optical Fibre. 14 pp.

ETSI I-ETS 300 226-93. Transmission and Multiplexing (TM); Single-Mode Optical Fibre Cables to Be Used in Ducts and for Directly Buried Application. 18 pp.

ETSI PRI-ETS 300 226-93. Transmission and Multiplexing (TM); Single-Mode Optical Fibre Cables to be Used in Ducts and for Directly Buried Application. 18 pp.

ETSI PRI-ETS 300 226-92. Transmission and Multiplexing (TM); Single-Mode Optical Fibre Cables to Be Used in Ducts and for Directly Buried Application. 19 pp.

ETSI I-ETS 300 227-93. Transmission and Multiplexing (TM); ITU-T Recommendation G.652-Type Single-Mode Optical Fibre. 14 pp.

ETSI PRI-ETS 300 227-93. Transmission and Multiplexing (TM); CCITT Recommendation G.652-Type Single-Mode Optical Fibre. 14 pp.

ETSI PRI-ETS 300 227-92. Transmission and Multiplexing (TM); CCITT Recommendation G.652-Type Single-Mode Optical Fibre. 15 pp.

—Single Mode—Chromatic Dispersion

CCITT RECMN G.652-89. Characteristics of a Single-Mode Optical Fibre Cable—Transmission Media Characteristics (Study Group XV) 34 pp. 34 pp.

—Single Mode—Dispersion Shifted

CCITT RECMN G.653-89. Characteristics of a Dispersion-Shifted Single-Mode Optical Fibre Cable—Transmission Media Characteristics (Study Group XV) 5 pp. 5 pp.

CENELEC PRI-ETS 300 228-92. Transmission and Multiplexing (TM); CCITT Recommendation G.653-Type Dispersion Shifted Single-Mode Optical Fibre. 8 pp.

CENELEC PRI-ETS 300 228-93. Transmission and Multiplexing (TM); CCITT Recommendation G.653-Type Dispersion Shifted Single-Mode Optical Fibre. 7 pp.

CENELEC I-ETS 300 228-93. Transmission and Multiplexing (TM); ITU-T Recommendation G.653-Type Dispersion Shifted Single-Mode Optical Fibre. 7 pp.

ETSI I-ETS 300 228-93. Transmission and Multiplexing (TM); ITU-T Recommendation G.653-Type Dispersion Shifted Single-Mode Optical Fibre. 7 pp.

ETSI PRI-ETS 300 228-93. Transmission and Multiplexing (TM); CCITT Recommendation G.653-Type Dispersion Shifted Single-Mode Optical Fibre. 7 pp.

ETSI PRI-ETS 300 228-92. Transmission and Multiplexing (TM); CCITT Recommendation G.653-Type Dispersion Shifted Single-Mode Optical Fibre. 8 pp.

—Single Mode—Dispersion Shifted—Attenuation

CCITT RECMN G.653-89. Characteristics of a Dispersion-Shifted Single-Mode Optical Fibre Cable—Transmission Media Characteristics (Study Group XV) 5 pp. 5 pp.

—Single Mode—Dispersion Shifted—Chromatic Dispersion

CCITT RECMN G.653-89. Characteristics of a Dispersion-Shifted Single-Mode Optical Fibre Cable—Transmission Media Characteristics (Study Group XV) 5 pp. 5 pp.

—Single Mode—Dispersion Shifted—Factory Lengths

CCITT RECMN G.653-89. Characteristics of a Dispersion-Shifted Single-Mode Optical Fibre Cable—Transmission Media Characteristics (Study Group XV) 5 pp. 5 pp.

—Single Mode—Dispersion Shifted—Optical Measurement

CCITT RECMN G.653-89. Characteristics of a Dispersion-Shifted Single-Mode Optical Fibre Cable—Transmission Media Characteristics (Study Group XV) 5 pp. 5 pp.

—Single Mode—Ducts

CENELEC PRI-ETS 300 227-92. Transmission and Multiplexing (TM); CCITT Recommendation G.652-Type Single-Mode Optical Fibre. 15 pp.

CENELEC PRI-ETS 300 227-93. Transmission and Multiplexing (TM); CCITT Recommendation G.652-Type Single-Mode Optical Fibre. 14 pp.

INTERNATIONAL AND NON-U.S. NATIONAL STANDARDS SUBJECT INDEX

Fiber

Fiber Optic Cables (Cont.)

—Single Mode—Ducts (Cont.)

CENELEC I-ETS 300 227-93. Transmission and Multiplexing (TM); ITU-T Recommendation G.652-Type Single-Mode Optical Fibre. 14 pp.

ETSI I-ETS 300 227-93. Transmission and Multiplexing (TM); ITU-T Recommendation G.652-Type Single-Mode Optical Fibre. 14 pp.

ETSI PRI-ETS 300 227-93. Transmission and Multiplexing (TM); CCITT Recommendation G.652-Type Single-Mode Optical Fibre. 14 pp.

ETSI PRI-ETS 300 227-92. Transmission and Multiplexing (TM); CCITT Recommendation G.652-Type Single-Mode Optical Fibre. 15 pp.

—Single Mode—Loss Minimized

CCITT RECMN G.654-89. Characteristics of a 1550 mm Wavelength Loss-Minimized Single-Mode Optical Fibre Cable—Transmission Media Characteristics (Study Group XV) 3 pp. 3 pp.

—Single Mode—Loss Minimized—Attenuation

CCITT RECMN G.654-89. Characteristics of a 1550 mm Wavelength Loss-Minimized Single-Mode Optical Fibre Cable—Transmission Media Characteristics (Study Group XV) 3 pp. 3 pp.

—Single Mode—Loss Minimized—Chromatic Dispersion

CCITT RECMN G.654-89. Characteristics of a 1550 mm Wavelength Loss-Minimized Single-Mode Optical Fibre Cable—Transmission Media Characteristics (Study Group XV) 3 pp. 3 pp.

—Single Mode—Loss Minimized—Factory Lengths

CCITT RECMN G.654-89. Characteristics of a 1550 mm Wavelength Loss-Minimized Single-Mode Optical Fibre Cable—Transmission Media Characteristics (Study Group XV) 3 pp. 3 pp.

—Single Mode—Loss Minimized—Optical Measurement

CCITT RECMN G.654-89. Characteristics of a 1550 mm Wavelength Loss-Minimized Single-Mode Optical Fibre Cable—Transmission Media Characteristics (Study Group XV) 3 pp. 3 pp.

—Single Mode—Optical Measurement

CCITT RECMN G.652-89. Characteristics of a Single-Mode Optical Fibre Cable—Transmission Media Characteristics (Study Group XV) 34 pp. 34 pp.

—Splicing

IECQ QC 850100-93. Splices for Optical Fibres and Cables Part 2: Sectional Specification: Splice Organizers and Closures for Optical Fibres and Cables (IEC 1073-2 ED 1). 129 pp.

—Splicing—Fusion

IEC 1073 Pt 3-93. Splices for Optical Fibres and Cables Part 3: Sectional Specification—Fusion Splices for Optical Fibres and Cables First Edition (IECQ QC 850200). 81 pp.

IECQ QC 850200-93. Splices for Optical Fibres and Cables Part 3: Sectional Specification: Fusion Splices for Optical Fibres and Cables (IEC 1073-3 ED 1). 81 pp.

—Splicing—Quality Assurance

IEC 1073 Pt 2-93. Splices for Optical Fibres and Cables Part 2: Sectional Specification—Splice Organizers and Closures for Optical Fibres and Cables First Edition (IECQ QC 850100). 128 pp.

—Stuffing Tubes—Compression Testing

CNS C6328-89. Method of Test for Fiber Optic Devices (FOTP-94 Fiber Cable Stuffing Tube Compression) (Jul)(12559).

—Surgical

BSI BS 4824-72. 1972 Fibrelight Cables and Fittings for Surgical Equipment. 12 pp.

—Surgical—Fittings

BSI BS 4824-72. 1972 Fibrelight Cables and Fittings for Surgical Equipment. 12 pp.

—Telecommunication Systems

IEC 794 Pt 1-93. Optical Fibre Cables Part 1: Generic Specification Third Edition. 86 pp.

—Telecommunication Systems—Glossaries

DIN VDE 0888 Pt 1-88. Optical Fibres for Telecommunication and Data Processing Systems Definitions (June) (Supersedes DIN 57888 Part 1). 18 pp.

—Temperature Change Testing—Attenuation

DIN VDE 0472 Pt 233-89. Testing of Cables, Wires and Flexible Cords; Attenuation of Changes in Optical Fibre Cables During Temperature Cycling (Mar). 6 pp.

Fiber Optic Cables (Cont.)

—Tensile Testing

CCITT RECMN L.14-92. Measurement Method to Determine the Tensile Performance of Optical Fibre Cables Under Load (Study Group VI) 8 pp. 8 pp.

CNS C6297-87. Method of Test for Fiber Optic Devices (FOTP-33 Fiber Optic Cable Tensile Loading and Bending Test) (Sep)(12083).

—Tensile Testing—Attenuation

CCITT RECMN L.14-92. Measurement Method to Determine the Tensile Performance of Optical Fibre Cables Under Load (Study Group VI) 8 pp. 8 pp.

DIN VDE 0472 Pt 231-89. Testing of Cables, Wires and Flexible Cords; Attenuation Changes in Optical Fibre Cables During Tensile Stress (Mar). 5 pp.

—Twisting

CNS C6298-87. Method of Test for Fiber Optic Devices (FOTP-36 Twist Test for Fiber Optic Cable Assemblies) (Sep)(12084).

CNS C6314-88. Method of Test for Fiber Optic Devices (FOTP-85 Fiber Optic Cable Twist Test) (Jul)(12361).

CNS C6315-88. Method of Test for Fiber Optic Devices (FOTP-91 Fiber Optic Cable Twist-Bend Test) (Jul)(12362).

—Underwater—Cable Joints

CCITT RECMN L.12-92. Optical Fibre Joints (Study Group VI) 10 pp. 10 pp.

CCITT RECMN L.13-92. Sheath Joints and Organizers of Optical Fibre Cables in the Outside Plant (Study Group VI) 6 pp. 6 pp.

—Wicking

CNS C6231-84. Method of Test for Fiber Optic Devices (Cable Wicking Test) (Jun)(10931).

Fiber Optic Communication Equipment

See Also: Fiber Distributed Data Interface; Fiber Optic Equipment

—Data Terminal Equipment

JIS C 6110-90. General Rules of Transmitting and/or Receiving Modules for Low Speed Fiber Optic Transmission. 16 pp.

JIS C 6111-90. Test Methods of Transmitting and/or Receiving Modules for Low Speed Fiber Optic Transmission. 27 pp.

—Data Transmission

CEN PREN 3758-92. Simplex High Speed Data Transmission System. 58 pp.

CEN PREN 3910-90. High Speed Data Transmission Under Stanag 3838 or Fiber Optic Equivalent Control. 100 pp.

CEN PREN 3910-92. High Speed Data Transmission Under STANAG 3838 or Fiber Optic Equivalent Control. 93 pp.

—Data Transmission—Avionics

MOD UK DSTAN 00-18: Pt 1:Sec 5-91. Avionic Data Transmission Interface Systems Part 1: Guide to Interface Systems Section 5: Guide to the Fibre Optic Interface Standardization Issue 2. 156 pp.

MOD UK DSTAN 00-18: Pt 2:Supp A-90. Avionic Data Transmission Interface Systems Part 2: Serial, Time Division, Command/Response Multiplex Data Bus Supplement A: Fibre Optic Supplement for a Point to Point Link Issue 1. 15 pp.

MOD UK DSTAN 00-18: Pt 2:Supp B-90. Avionic Data Transmission Interface Systems Part 2: Serial, Time Division, Command/Response Multiplex Data Bus Supplement B: Fibre Optic Supplement for a Single Transmissive Star Issue 1. 15 pp.

MOD UK DSTAN 00-18: Pt 2:Supp C-90. Avionic Data Transmission Interface Systems Part 2: Serial, Time Division, Command/Response Multiplex Data Bus Supplement C: Fibre Optic Supplement for a Single Reflective Star Issue 1. 15 pp.

MOD UK DSTAN 00-18: Pt 2:Supp D-90. Avionic Data Transmission Interface Systems Part 2: Serial, Time Division, Command/Response Multiplex Data Bus Supplement D: Fibre Optic Supplement for a Multi-Local Transmissive Star Issue 1. 20 pp.

—Glossaries

BSI BS 4727:Pt3: Group 13-92. 1992 Electrotechnical, Power, Telecommunication, Electronics, Lighting and Colour Terms Part 3: Terms Particular to Telecommunications and Electronics Group 13: Optical Fibre Communication (IEC 50(731): 1991) (G). 174 pp.

IEC 50 Chap 731-91. International Electrotechnical Vocabulary Chapter 731: Optical Fibre Communication First Edition; (Corrigendum—March and July—1992). 188 pp.

—Safety

IEC 825 Pt 2-93. Safety of Laser Products Part 2: Safety of Optical Fibre Communication Systems First Edition. 35 pp.

Fiber Optic Connectors

Use For: Optical Waveguide Connectors
See Also: Connectors; Fiber Optic Cables; Fiber Optic Equipment; Fiber Optic Splices; Star Couplers; Waveguides

CECC CECC 86 000 ISSUE 1-92. Generic Specification: Connector Sets for Optical Fibres and Cables; Part 1: Requirements, Test Methods and Qualification Approval Procedures (En, Fr, Ge). 374 pp.

CECC CECC 86 001 ISSUE 1-92. Blank Detail Specification: Connectors for Optical Fibres and Cables; Environmental Category I (En, Fr, Ge). 29 pp.

CECC CECC 86 002 ISSUE 1-92. Blank Detail Specification: Connectors for Optical Fibres and Cables; Environmental Category II (En, Fr, Ge). 29 pp.

CECC CECC 86 003 ISSUE 1-92. Blank Detail Specification: Connectors for Optical Fibres and Cables; Environmental Category III (En, Fr, Ge). 33 pp.

CECC CECC 86 004 ISSUE 1-92. Blank Detail Specification: Connectors for Optical Fibres and Cables; Environmental Category IV (En, Fr, Ge). 33 pp.

CECC EN 186 210-92. Sectional Specification: Connector Sets for Optical Fibres and Cables; Type CF08. 14 pp.

IEC 874 Pt 0-88. Connectors for Optical Fibres and Cables Part 0: Guide for the Construction of Sectional Specifications First Edition. 48 pp.

IEC 874 Pt 2-93. Connectors for Optical Fibres and Cables Part 2: Sectional Specification for Fibre Optic Connector—Type F-SMA Second Edition (IECQ QC 210100). 26 pp.

IEC 874 Pt 3-93. Connectors for Optical Fibres and Cables Part 3: Sectional Specification for Fibre Optic Connector CFO3 Second Edition (IECQ QC 210300). 22 pp.

IEC 874 Pt 4-93. Connectors for Optical Fibres and Cables Part 4: Sectional Specification for Fibre Optic Connector—Type CFO4 Second Edition; (IECQ QC 210500). 22 pp.

IEC 874 Pt 5-93. Connectors for Optical Fibres and Cables Part 5: Sectional Specification for Fibre Optic Connector—Type BAM Second Edition; (IECQ QC 210400). 34 pp.

IEC 874 Pt 6-93. Connectors for Optical Fibres and Cables Part 6: Sectional Specification for Fibre Optic Connector—Type LSA Second Edition (IECQ QC 210200). 34 pp.

IEC 874 Pt 7-93. Connectors for Optical Fibres and Cables Part 7: Sectional Specification for Fibre Optic Connector—Type FC Second Edition; (IECQ QC 210700). 149 pp.

IEC 874 Pt 8-93. Connectors for Optical Fibres and Cables Part 8: Sectional Specification for Fibre Optic Connector—Type D Second Edition; (IECQ QC 210600). 81 pp.

IEC 874 Pt 9-93. Connectors for Optical Fibres and Cables Part 9: Sectional Specification for Fibre Optic Connector—Type OF-2 Second Edition; (IECQ QC 210800). 28 pp.

IECQ QC 210000-93. Connectors for Optical Fibers and Cables Part 1: Generic Specification; (Corrigendum—April 1993) (IEC 874-1 ED 3). 273 pp.

IECQ QC 210100-93. Connectors for Optical Fibres and Cables Part 2: Sectional Specification for Fibre Optic Connector—Type F-SMA (IEC 874-2 ED 2). 26 pp.

IECQ QC 210101/US 0001-89. Fiber Optics for Use in Electronic Equipment: Detail Specification: Fiber Optic Connectors Type F—SMA; (Amendment 1-1990). 25 pp.

IECQ QC 210200-93. Connectors for Optical Fibres and Cables Part 6: Sectional Specification for Fibre Optic Connector—Type LSA (IEC 874-6 ED 2). 35 pp.

IECQ QC 210300-93. Connectors for Optical Fibres and Cables Part 3: Sectional Specification for Fibre Optic Connector—Type CFO3 (IEC 874-3 ED 2). 23 pp.

IECQ QC 210400-90. Connectors for Optical Fibres and Cables Part 5: Sectional Specification Fibre Optic Connector Type BAM (IEC 874-5 ED 2). 87 pp.

IECQ QC 210500-93. Connectors for Optical Fibres and Cables Part 4: Sectional Specification for Fibre Optic Connector—Type CFO4 (IEC 874-4 ED 2). 23 pp.

IECQ QC 210600-90. Connectors for Optical Fibres and Cables Part 8: Sectional Specification Fibre Optic Connector Type D (IEC 874-8 ED 1). 81 pp.

IECQ QC 210700-90. Connectors for Optical Fibres and Cables Part 7: Sectional Specification Fibre Optic Connector Type FC (IEC 874-7 ED 1). 149 pp.

IECQ QC 210800-93. Connectors for Optical Fibres and Cables Part 9: Sectional Specification for Fibre Optic Connector-Type OF-2 (IEC 874-9 ED 2). 28 pp.

INDUSTRY STANDARDS

Fiber Optic Connectors (Cont.)

IECQ QC 211200-92. Connectors for Optical Fibres and Cables Part 10: Sectional Specification Fibre Optic Connector Type BFOC/2,5 (IEC 874-10 ED 1). 23 pp.

IECQ QC 211700-93. Connectors for Optical Fibres and Cables Part 13: Sectional Specification for Fibre Optic Connector—Type CF 08 (IEC 874-13 ED 1). 23 pp.

IECQ QC 211800-93. Connectors for Optical Fibres and Cables Part 14: Sectional Specification for Fibre Optic Connector—Type SC (IEC 874-14 ED 1). 25 pp.

JIS C 5961-90. Test Methods of Connectors for Optical Fiber Cords. 49 pp.

JIS C 5962-90. General Rules of Connectors for Optical Fiber Cords. 15 pp.

JIS C 5970-87. F01 Type Connectors for Optical Fiber Cords. 33 pp.

JIS C 5971-87. F02 Type Connectors for Optical Fiber Cords. 29 pp.

JIS C 5972-87. F03 Type Connectors for Optical Fiber Cords. 22 pp.

JIS C 5973-90. F04 Type Connectors for Optical Fiber Cords. 34 pp.

JIS C 5974-90. F05 Type Connectors for Optical Fiber Cords. 29 pp.

JIS C 5975-90. F06 Type Connectors for Optical Fiber Cords. 33 pp.

JIS C 5976-90. F07 Type Connectors for Optical Fiber Cords. 26 pp.

JIS C 5977-90. F08 Type Connectors for Optical Fiber Cords. 26 pp.

JIS C 5978-90. F09 Type Connectors for Optical Fiber Cords. 22 pp.

JIS C 5979-90. F10 Type Connectors for Optical Fiber Cords. 20 pp.

—**BFOC**

IEC 874 Pt 10-92. Connectors for Optical Fibres and Cables Part 10: Sectional Specification Fibre Optic Connector Type BFOC/2,5 First Edition; (IECQ QC 211200). 23 pp.

—**Cable Flexing Testing**

CNS C6116-81. Method of Test for Fiber Optic Devices (Cable Flexing for Fiber Optic Connectors) (Jul)(7638).

—**Cable Retention Testing**

CNS C6118-81. Method of Test for Fiber Optic Devices (Cable Retention Test Procedure for Fiber Optic Cable Interconnecting Devices) (Jul)(7640).

—**CF**

IEC 874 Pt 13-93. Connectors for Optical Fibres and Cables Part 13: Sectional Specification for Fibre Optic Connector—Type CF 08 First Edition (IECQ QC 211700). 23 pp.

—**Compression Testing**

CNS C6275-86. Method of Test for Fiber Optic Devices (FOTP-26 Crush Resistance of Fiber Optic Cable Interconnecting Devices) (Dec)(11787).

CNS C6283-87. Method of Test for Fiber Optic Devices (FOTP-83 Cable to Interconnecting Device Axial Compressive Loading) (Apr)(11905).

—**F-SMA**

CECC CECC 86 104-801 ISSUE 1-93. Connector Sets for Optical Fibres and Cables; Type F-SMA (En, Fr, Ge). 28 pp.

—**F-SMA—Crimp—Plugs—Preferred Products List**

CECC CECC MUAHAG Vol 3B IS 4-91. Preferred Products List; Connectors; R.F. and Fibre Optics (En, Fr, Ge). 65 pp.

—**F-SMA—Crimp—Receptacles—Preferred Products List**

CECC CECC MUAHAG Vol 3B IS 4-91. Preferred Products List; Connectors; R.F. and Fibre Optics (En, Fr, Ge). 65 pp.

—**Fire Walls—Flammability Testing**

CNS C6333-89. Method of Test for Fiber Optic Devices (FOTP-172 Flame Resistance of Firewall Connector) (Jul)(12564).

—**Glossaries**

CNS C5077-86. Fiber Optic Connector Terminology (Oct)(6662).

DIN VDE 0888 Pt 1-88. Optical Fibres for Telecommunication and Data Processing Systems Definitions (June) (Supersedes DIN 57888 Part 1). 18 pp.

—**Hermaphroditic—Preferred Products List**

CECC CECC MUAHAG Vol 3B IS 4-91. Preferred Products List; Connectors; R.F. and Fibre Optics (En, Fr, Ge). 65 pp.

Fiber Optic Connectors (Cont.)

—**Humidity**

CNS C6117-81. Method of Test for Fiber Optic Devices (Humidity Test Procedure for Fiber Optic Connecting Devices) (Jul)(7639).

—**Insertion Loss**

CNS C6119-81. Method of Test for Fiber Optic Devices (Test Procedure for Fiber Optics Bundle Connector Insertion Loss) (Jul)(7641).

CNS C6121-87. Method of Test for Fiber Optic Devices (FOTP—34 Interconnection Device Insertion Loss) (Jun)(7643).

—**Inspection**

CNS C6295-87. Method of Test for Fiber Optic Devices (FOTP-13 Visual and Mechanical Inspection of Fibers, Cables, Connectors Other Fiber Optic Devices) (Sep)(12081).

—**Marine**

JIS F 8550-91. Marine Optical Transmission Apparatus—Criteria for Selection of Optical Fiber Cable and Optical Connector.

—**Multimode—Preferred Products List**

CECC CECC MUAHAG Vol 3B IS 4-91. Preferred Products List; Connectors; R.F. and Fibre Optics (En, Fr, Ge). 65 pp.

—**Multimode—Terminal—Military—Preferred Products List**

CECC CECC MUAHAG Vol 3B IS 4-91. Preferred Products List; Connectors; R.F. and Fibre Optics (En, Fr, Ge). 65 pp.

—**Multimode—Terminal—Telecommunications—Preferred Products List**

CECC CECC MUAHAG Vol 3B IS 4-91. Preferred Products List; Connectors; R.F. and Fibre Optics (En, Fr, Ge). 65 pp.

—**OCCA**

IEC 874 Pt 11-93. Connectors for Optical Fibres and Cables Part 11: Sectional Specification for Fibre Optic Connector—Type OCCA-PC First Edition. 24 pp.

IEC 874 Pt 12-93. Connectors for Optical Fibres and Cables Part 12: Sectional Specification for Fibre Optic Connector—Type OCCA-BU First Edition. 24 pp.

—**Preferred Products List**

CECC CECC MUAHAG Vol 3B IS 4-91. Preferred Products List; Connectors; R.F. and Fibre Optics (En, Fr, Ge). 65 pp.

—**Printed Circuit Mount—Crimp—Preferred Products List**

CECC CECC MUAHAG Vol 3B IS 4-91. Preferred Products List; Connectors; R.F. and Fibre Optics (En, Fr, Ge). 65 pp.

—**Quality Assurance**

CENELEC EN 186 210-92. Sectional Specification: Connector Sets for Optical Fibres and Cables Type CF08. 14 pp.

IEC 874 Pt 1-93. Connectors for Optical Fibres and Cables Part 1: Generic Specification Third Edition; (Corrigendum—April 1993) (IECQ QC 210000). 273 pp.

IEC 874 Pt 12-93. Connectors for Optical Fibres and Cables Part 12: Sectional Specification for Fibre Optic Connector—Type OCCA-BU First Edition. 24 pp.

MOD UK DSTAN 60-2-89. Connectors of Assessed Quality for Optical Fibres and Cables Issue 1. 20 pp.

—**Safety**

IEC 874 Pt 11-93. Connectors for Optical Fibres and Cables Part 11: Sectional Specification for Fibre Optic Connector—Type OCCA-PC First Edition. 24 pp.

—**SC**

IEC 874 Pt 14-93. Connectors for Optical Fibres and Cables Part 14: Sectional Specification for Fibre Optic Connector—Type SC First Edition (IECQ QC 211800). 25 pp.

Fiber Optic Equipment

Scope Note: For additional listings, use a more specific term *See Also:* Chromatic Dispersion; Communication Equipment; Fiber Distributed Data Interface; Fiber Optic Cable Assemblies; Fiber Optic Cables; Fiber Optic Communication Equipment; Fiber Optic Connectors; Fiber Optic Light Sources; Fiber Optic Modems; Fiber Optic Multiplexers; Fiber Optic Receivers; Fiber Optic Splices; Fiber Optic Switches; Fiber Optic Transmitters; Optical Fiber Pigtails; Optical Fibers; Optical Glass; Star Couplers; Stuffing

Fiber Optic Equipment (Cont.)

See Also: (Cont.)
Tubes

CNS C6115-81. Method of Test for Fiber Optic Devices (General Rules) (Jul)(7637).

CNS C6299-87. Method of Test for Fiber Optic Devices (FOTP-54 Mode Scrambler Launch Requirements for Information Transmission Capacity Measurements) (Sep)(12085).

CNS C6312-88. Method of Test for Fiber Optic Devices (FOTP-66 Measuring Relative Abrasion Resistance of Optical Waveguide Coating and Buffers) (Jul)(12359).

—**Acceleration Testing**

CNS C6219-87. Method of Test for Fiber Optic Devices (FOTP-18 Acceleration Test) (Sep)(10874).

—**Aging Testing**

CNS C6222-84. Method of Test for Fiber Optic Devices (Maintenance Aging) (May)(10877).

CNS C6330-89. Method of Test for Fiber Optic Devices (FOTP-101 Accelerated Oxygen Aging) (Jul)(12561).

—**Attenuation**

CNS C6324-88. Method of Test for Fiber Optic Devices (FOTP-53 Attenuation by Substitution Measurement for Multimode Graded-Index Optical Fibers or Fiber Assemblies Used in Long Length Communications System) (Jul)(12372).

—**Attenuators**

IEC 869 Pt 1-88. Fibre Optic Attenuators Part 1: Generic Specification First Edition (IECQ QC 8000000). 80 pp.

IECQ QC 800000-88. Fibre Optic Attenuators Part 1: Generic Specification (IEC 869-1 ED 1). 81 pp.

JIS C 5920-88. General Rules of Optical Attenuator. 10 pp.

—**Branching Devices**

IEC 875 Pt 1-92. Fibre Optic Branching Devices Part 1: Generic Specification Second Edition (IECQ QC 810000). 150 pp.

IECQ QC 810000-92. Fibre Optic Branching Devices Part 1: Generic Specification (IEC 875-1 ED 2). 151 pp.

JIS C 5910-87. General Rules of Fiber Optic Branching Devices. 15 pp.

—**Crosstalk**

CNS C6286-87. Method of Test for Fiber Optic Devices (FOTP—42 Optical Crosstalk in Fiber Optic Components) (Jun)(11992).

—**Diameters**

CNS C6322-88. Method of Test for Fiber Optic Devices (FOTP-48 On-Line Diameter Measurement) (Jul)(12370).

—**Diode Lasers**

JIS C 5940-89. General Rules of Laser Diodes for Fiber Optic Transmission. 14 pp.

JIS C 5941-89. Test Methods of Laser Diodes for Fiber Optic Transmission. 18 pp.

—**Discontinuity**

CNS C6229-84. Method of Test for Fiber Optic Devices (Circuit Discontinuities Test) (Jun)(10929).

—**Dust**

CNS C6230-84. Method of Test for Fiber Optic Devices (Dust (Fine Sand) Test) (Jun)(10930).

—**Electric Shock**

CNS C6259-86. Method of Test for Fiber Optic Devices (FOTP-14 Shock Test, Specified Pulse) (Sep)(11703).

—**Geometry—Microscopy**

CNS C6321-88. Method of Test for Fiber Optic Devices (FOTP-45 Microscopic Method for Measuring Fiber Geometry) (Jul)(12369).

—**Immersion Testing**

CNS C6225-84. Method of Test for Fiber Optic Devices (Fluid Immersion Test) (Jun)(10925).

CNS C6260-86. Method of Test for Fiber Optic Devices (FOTP-15 Altitude Immersion Test) (Sep)(11704).

CNS C6313-88. Method of Test for Fiber Optic Devices (FOTP-75 Fluid Immersion Test) (Jul)(12360).

—**Impact Testing**

CNS C6217-87. Method of Test for Fiber Optic Devices (FOTP—2 Impact Test) (Jun)(10872).

—**Inspection**

CNS C6295-87. Method of Test for Fiber Optic Devices (FOTP-13 Visual and Mechanical Inspection of Fibers, Cables, Connectors Other Fiber Optic Devices) (Sep)(12081).

INTERNATIONAL AND NON-U.S. NATIONAL STANDARDS
SUBJECT INDEX

Fiber Optic Equipment *(Cont.)*
—**Life (Durability)**
  CNS C6220-87. Method of Test for Fiber Optic Devices (FOTP—4 Temperature Life Test) (Jun)(10875).
—**Light Testing**
  CNS C6250-85. Method of Test for Fiber Optic Devices (Ambient Light Susceptibility) (Aug)(11332).
  CNS C6307-87. Method of Test for Fiber Optic Devices (FOTP-22 Ambient Light Susceptibility) (Dec)(12171).
—**Naval Ships**
  MOD UK NES 1009: Part 1-92. Requirements for the Application of Fibre Optics in Ships and Submarines Part 1: Fibre Optic Components Issue 1 (02.92). 172 pp.
  MOD UK NES 1009: Part 2-92. Requirements for the Application of Fibre Optics in Ships and Submarines Part 2: Fibre Optic System Design Issue 1 (04.92). 68 pp.
  MOD UK NES 1009: Part 3-92. Requirements for the Application of Fibre Optics in Ships and Submarines Part 3: Installation, Maintenance and Repair of Fibre Optic Systems Issue 1 (08.92). 92 pp.
—**Passive Devices**
  JIS C 5900-87. General Rules of Passive Device for Fiber Optic Transmission.
  JIS C 5901-87. Test Methods for Passive Devices for Fiber Optic Transmission. 15 pp.
—**Radiation Detection and Measurement**
  CNS C6262-86. Method of Test for Fiber Optic Devices (FOTP-47 Output Far-Field Radiation Pattern Measurement) (Sep)(11706).
  CNS C6319-88. Method of Test for Fiber Optic Devices (FOTP-43 Output Near-Fiber Radiation Pattern Measurement) (Jul)(12367).
  CNS C6323-88. Method of Test for Fiber Optic Devices (FOTP-49 Nuclear Radiation Effects in Fiber Optic Components Measurement) (Jul)(12371).
—**Refractive Index**
  CNS C6228-84. Method of Test for Fiber Optic Devices (Refractive Index Profile Transverse Interference Method) (Jun)(10928).
  CNS C6320-88. Method of Test for Fiber Optic Devices (FOTP-44 Refractive Index Profile, Refracted Ray Method) (Jul)(12368).
—**Safety**
  IEC 1218-93. Fibre Optics—Safety Guide First Edition. 36 pp.
—**Salt Spray Testing**
  CNS C6306-87. Method of Test for Fiber Optic Devices (FOTP-16 Salt Spray Corrosion Test) (Dec)(12170).
—**Seals—Leakage**
  CNS C6120-87. Method of Test for Fiber Optic Devices (FOTP-23 Air Leakage Testing of Fiber Optic Component Seals) (Apr)(7642).
—**Submarines**
  MOD UK NES 1009: Part 1-92. Requirements for the Application of Fibre Optics in Ships and Submarines Part 1: Fibre Optic Components Issue 1 (02.92). 172 pp.
  MOD UK NES 1009: Part 2-92. Requirements for the Application of Fibre Optics in Ships and Submarines Part 2: Fibre Optic System Design Issue 1 (04.92). 68 pp.
  MOD UK NES 1009: Part 3-92. Requirements for the Application of Fibre Optics in Ships and Submarines Part 3: Installation, Maintenance and Repair of Fibre Optic Systems Issue 1 (08.92). 92 pp.
—**Temperature Change Testing**
  CNS C6218-84. Method of Test for Fiber Optic Devices (Temperature Cycling, Thermal Shock) (May)(10873).
—**Tensile Testing**
  CNS C6224-84. Method of Test for Fiber Optic Devices (Tensile Proof Test) (Jun)(10924).
—**Thermal Shock**
  CNS C6218-84. Method of Test for Fiber Optic Devices (Temperature Cycling, Thermal Shock) (May)(10873).
—**Torsion**
  CNS C6311-88. Method of Test for Fiber Optic Devices (FOTP-63 Torsion Test) (Jul)(12358).
—**Transmittance**
  CNS C6261-86. Method of Test for Fiber Optic Devices (FOTP-20 Change of Optical Transmittance) (Sep)(11705).

Fiber Optic Equipment *(Cont.)*
—**Vibration**
  CNS C6221-84. Method of Test for Fiber Optic Devices (Vibration Test) (May)(10876).

Fiber Optic Light Sources
See Also: Fiber Optic Equipment; Light Sources
  BSI BS CECC 20006-88. 1988 Pin-Photodiodes for Fibre Optic Applications. 12 pp.
  JIS C 5950-89. General Rules of Light Emitting Diodes for Fiber Optic Transmission. 12 pp.
  JIS C 5951-89. Test Methods of Light Emitting Diodes for Fiber Optic Transmission. 16 pp.
  JIS C 5990-88. General Rules of Photodiodes for Fiber Optic Transmission. 13 pp.
  JIS C 5991-88. Test Methods of Photodiodes for Fiber Optic Transmission. 14 pp.
—**Lasers**
  JIS C 5940-89. General Rules of Laser Diodes for Fiber Optic Transmission. 14 pp.
  JIS C 5941-89. Test Methods of Laser Diodes for Fiber Optic Transmission. 18 pp.

Fiber Optic Modems
See Also: Fiber Optic Equipment; Modems
  IEC 875 Pt 3-92. Fibre Optic Branching Devices Part 3: Sectional Specification—Wavelength Selective Branching Devices Second Edition (IECQ QC 810200). 82 pp.
  IECQ QC 810200-92. Fibre Optic Branching Devices Part 3: Sectional Specification—Wavelength Selective Branching Devices (IEC 875-3 ED 2). 83 pp.

Fiber Optic Multiplexers
See Also: Fiber Optic Equipment
  IEC 875 Pt 3-92. Fibre Optic Branching Devices Part 3: Sectional Specification—Wavelength Selective Branching Devices Second Edition (IECQ QC 810200). 82 pp.
  IECQ QC 810200-92. Fibre Optic Branching Devices Part 3: Sectional Specification—Wavelength Selective Branching Devices (IEC 875-3 ED 2). 83 pp.

Fiber Optic Receivers
See Also: Fiber Optic Equipment; Receivers
  MOD UK DSTAN 59-61: 90/246-88. Fibre Optic Receiver Module (VX9548) Issue 1. 21 pp.
  MOD UK DSTAN 59-61: 90/247-88. Fibre Optic Receiver Module (VX9457) Issue 1. 20 pp.

Fiber Optic Splicers
Use: Fiber Optic Splices

Fiber Optic Splices
Use For: Fiber Optic Splicers *See Also:* Fiber Optic Cables; Fiber Optic Connectors; Fiber Optic Equipment; Splices (Electrical)
—**Fusion**
  IEC 1073 Pt 3-93. Splices for Optical Fibres and Cables Part 3: Sectional Specification—Fusion Splices for Optical Fibres and Cables First Edition (IECQ QC 850200). 81 pp.
  IECQ QC 850200-93. Splices for Optical Fibres and Cables Part 3: Sectional Specification: Fusion Splices for Optical Fibres and Cables (IEC 1073-3 ED 1). 81 pp.

Fiber Optic Switches
See Also: Fiber Optic Equipment
  IEC 876 Pt 1-86. Fibre Optic Switches Part 1: Generic Specification First Edition. 72 pp.
  JIS C 5930-88. General Rules of Optical Switches. 17 pp.
  JIS C 5931-88. Test Methods for Optical Switches. 9 pp.

Fiber Optic Systems
Use: Fiber Optic Equipment

Fiber Optic Systems and Equipment
Use: Fiber Optic Equipment

Fiber Optic Transmitters
See Also: Fiber Optic Equipment; Transmitters
  MOD UK DSTAN 59-61: 90/248-88. Fibre Optic Transmitter Module (VX9431) Issue 1. 21 pp.

Fiber Optics
Use: Fiber Optic Equipment

Fiber Reinforced Plastics
Use: Reinforced Plastics

Fiber Ropes
Use: Ropes

Fiber Waveguides
Use: Optical Waveguides

Fiberboard
Scope Note: For additional listings, see also specific products made from fiberboards *Use For:* Fiber Building Board; Fibre Building Board
See Also: Boxboard; Building Board; Corrugated Board; Corrugated Fiberboard; Hardboard; Paperboard; Particle Board; Pressboard; Wallboard
  BSI BS EN 324-2-93. 1993 Wood-Based Panels—Determination of Dimensions of Boards Part 2: Determination of Squareness and Edge Straightness (R). 10 pp.
  CEN PREN 324 (Part 2)-89. Wood-Based Panels—Determination of Dimensions of Boards—Part 2: Determination of Squareness and Edge Straightness. 7 pp.
  CEN EN 324-2-93. Wood-Based Panels—Determination of Dimensions of Boards—Part 2: Determination of Squareness and Edge Straightness. 7 pp.
  CEN PREN 622-1-92. Fibreboards—Specifications—Part 1: General Requirements. 6 pp.
  CEN PREN 622-2-93. Fibreboards—Specifications—Part 2: Requirements for General Purpose Boards for Use in Dry Conditions. 10 pp.
  CEN PREN 622-4-93. Fibreboards—Specifications—Part 4: Requirements for General Purpose Boards for Use in Humid Conditions. 10 pp.
  CNS A2225-88. Dressed Hard Fibreboards (Jul)(12349).
  CNS A3278-88. Method of Test for Dressed Hard Fiber Boards (Jul)(12350).
  CNS O1025-83. Hard Fibreboards (Jan)(9907). 3 pp.
  CNS O1026-83. Semihard Fibreboards (Jan)(9909). 2 pp.
  CNS O1030-83. Dressed Hard Fibreboard for Exterior Use (Jul)(10469). 4 pp.
  CNS O2052-83. Method of Test for Hard Fibreboards (Jan)(9908).
  CNS O2053-83. Method of Test for Semihard Fibreboards (Jan)(9910).
  CNS O2055-83. Method of Test for Dressed Hard Fibreboard for Exterior Use (Jul)(10470). 4 pp.
  ISO 2695-76. Fibre Building Boards—Hard and Medium Boards for General Purposes—Quality Specifications—Appearance, Shape and Dimensional Tolerances First Edition. 3 pp.
  JIS A 5903-87. Dressed Hard Fibreboards. 13 pp.
  JIS A 5906-83. Medium Density Fibreboards. 21 pp.
—**Acoustic Insulation**
  CNS A2038-86. Organic Fiber Sound Proof Boards (Dec)(2777). 1 p.
  CNS A3058-86. Method of Test for Organic Fiber Soundproof Boards (Dec)(2778). 3 pp.
  CNS O1029-83. Insulation Fiberboard for Acoustic Use (Jul)(10468). 8 pp.
  JIS A 6304-72. Soft Fibreboards for Acoustic Use.
—**Ammunition**
  MOD UK DEF-127-A-01. Board for Discs or Wads; Amendment 1. 5 pp.
—**Bagasse—Bituminous—Concretes**
  CNS A2033-63. Bituminous Bagasse Boards for Concrete (Sep)(2179) (R 1973). 1 p.
—**Bend Testing**
  BSI BS EN 310-93. 1993 Wood-Base Panels—Determination of Modulus of Elasticity in Bending and of Bending Strength (R). 12 pp.
  CEN PREN 310-90. Wood—Based Panels—Determination of Modulus of Elasticity—in Beding and Bending Strength. 7 pp.
  CEN EN 310-93. Wood-Based Panels—Determination of Modulus of Elasticity in Bending and of Bending Strength. 8 pp.
  CNS A3064-86. Method of Bending Test for Boards for Buildings (Feb)(3904). 2 pp.
  ISO 768-72. Fibre Building Boards—Determination of Bending Strength First Edition. 5 pp.
  JIS A 1408-77. Method of Bending Test for Boards of Buildings.
  JIS A 1408-64. Method of Bending Test for Boards of Buildings (R 1974). 6 pp.
—**Boxes (Containers)**
  CGSB CAN/CGSB-43.162-M91. Fibreboard Boxes and Partitions for Transport of Food Aid. 18 pp.
  SAA AS 1048-82. International Fibreboard Box Code. 32 pp.
—**Boxes (Containers)—Compression Testing**
  SAA AS 1301.800S-87. Methods of Test for Pulp and Paper (Metric Units)—Part 800s: Compression Resistance of Fibreboard Boxes (Cases) (This is a Joint Standard with SANZ NZS 1301). 2 pp.
  SNZ NZS/AS 1301. 800S-87. Methods of Test for Pulp and Paper Compression Resistance of Fibreboard Boxes (Cases) (This is a Joint Standard with SAA AS 1301.800S). 2 pp.

INDUSTRY STANDARDS

Fiberboard (Cont.)

—Boxes (Containers)—Waterproof Adhesives—Adhesion Testing
CPPA D.10U-77. Ply Separation of Combined Container Board. 1 p.

—Burst Testers
CPPA M-18-84. Perkins Model 'AH' Mullen Tester. 1 p.

—Bursting Strength
CPPA D.19P-90. Bursting Strength of Board. 2 pp.

—Cellulose—Lighting
CSA C22.2 NO 9-M1989. Luminaires; (Gen Instr 1) (Supercedes C22.2 No 97-1969). 83 pp.
CSA 696 Bull. Electrical Bulletin 696 September 21, 1967 to C22.2 NO 9. 1 p.
CSA 1169 Bull. Electrical Bulletin 1169 June 27, 1978 to C22.2 NO 9. 2 pp.

—Classification
BSI BS 1142-89. 1989 Fibre Building Boards. 42 pp.
BSI BS 1142-01. 1989 Amd 1 Fibre Building Boards (AMD 6970) April 1, 1992 (Supersedes BS 1142: Part 1: 1971). 43 pp.
BSI BS 1142-02. 1989 Amd 2 Fibre Building Boards (AMD 7776) July 15, 1993 (R). 44 pp.
BSI BS EN 316-93. 1993 Wood Fibreboards—Definition, Classification and Symbols (R). 7 pp.
CEN PREN 316-90. Wood Fibreboards—Definition, Classification and Marking. 5 pp.
CEN EN 316-93. Wood Fibreboards—Definition, Classification and Symbols. 4 pp.
ISO 818-75. Fibre Building Boards—Definition—Classification First Edition. 3 pp.
SNZ BS 1142-89. Specification for Fibre Building Boards. 40 pp.

—Contact Force
BSI BS EN 320-93. 1993 Fibreboards—Determination of Resistance to Axial Withdrawal of Screws (R). 11 pp.
CEN PREN 320-90. Fibreboards—Determination of Resistance to Axial Withdrawal of Screws. 6 pp.
CEN EN 320-93. Fibreboards—Determination of Resistance to Axial Withdrawal of Screws. 6 pp.

—Creep Properties
CEN PREN 1156-93. Wood-Based Panels—Determination of Duration of Load and Creep Factors. 11 pp.

—Cyclic Testing
BSI BS EN 321-93. 1993 Fibreboards—Cyclic Tests in Humid Conditions (R). 10 pp.
CEN PREN 321-90. Fibreboards—Wet Cyclic Test. 6 pp.
CEN EN 321-93. Fibreboards—Cyclic Tests in Humid Conditions. 5 pp.

—Density
BSI BS EN 323-93. 1993 Wood-Based Panels—Determination of Density (R). 10 pp.
CEN PREN 323-89. Wood-Based Panels—Determination of Density of Test Pieces. 7 pp.
CEN EN 323-93. Wood-Based Panels—Determination of Density. 5 pp.
ISO 9427-89. Wood-Based Panels—Determination of Density First Edition. 4 pp.

—Drums (Containers)
BSI BS 1133: SUBSEC 7.4-89. 1989 Packaging Code Section 7: Paper and Board Wrappers, Bags and Containers Subsection 7.4: Fibreboard Drums. 16 pp.
BSI BS 1596-92. 1992 Fibreboard Drums (H). 11 pp.
BSI BS 1596-74. 1974 Fibreboard Drums for Shipment of Goods Overseas. 9 pp.

—Footwear
BSI BS 5131: Sec 4.2-90. 1990 Methods of Test for Footwear and Footwear Materials Part 4: Other Components Section 4.2: Flexing Index of Fibreboard. 6 pp.
BSI BS 5131: Sec 4.3-90. 1990 Methods of Test for Footwear and Footwear Materials Part 4: Other Components Section 4.3: Resistance of Fibreboard to Stitch-Tear. 6 pp.
BSI BS 5131: Sec 4.5-90. 1990 Methods of Test for Footwear and Footwear Materials Part 4: Other Components Section 4.5: Tensile Strength of Fibreboard. 7 pp.
BSI BS 5131: Sec 4.12-90. 1990 Methods of Test for Footwear and Footwear Materials Part 4: Other Components Section 4.12: Fastness of Fibreboard Finishes to Rubbing in the Presence of Water and Perspiration. 10 pp.
BSI BS 5131: Sec 4.13-76. 1976 Methods of Test for Footwear and Footwear Materials Part 4: Other Components Section 4.13: Transverse Tensile Strength of Fibreboard. 2 pp.

Fiberboard (Cont.)

—Glazed—Ammunition
MOD UK DEF-43-A-67. Glazed Board and Glazed Board, Lead Free. 5 pp.

—Glossaries
JIS Z 0104-90. Glossary of Terms Used in Corrugated Fibreboard Industry. 30 pp.

—Humidity
BSI BS EN 318-93. 1993 Fibreboards—Determination of Dimensional Changes Associated with Changes in Relative Humidity (R). 11 pp.
CEN PREN 318-90. Fibreboards—Determination of Dimensional Changes Associated with Changes in Relative Humidity. 7 pp.
CEN EN 318-93. Fibreboards—Determination of Dimensional Changes Associated with Changes in Relative Humidity. 8 pp.

—Identification Systems
CEN PREN 316-90. Wood Fibreboards—Definition, Classification and Marking. 5 pp.

—Impact Testing
CNS A3178-83. Method of Impact Test for Boards of Buildings (Feb)(9961).
JIS A 1421-81. Method of Impact Test for Boards of Buildings. 9 pp.

—Insulating
CNS O1027-83. Heat Insulation Fiberboards (Jan)(9911). 4 pp.
CNS O2054-83. Method of Test for Heat Insulation Fiberboards (Jan)(9912).
CSA CAN/CSA-A247-M86. Insulating Fibreboard; (Gen Instr 1). 13 pp.
JIS A 5905-79. Insulation Fibreboards. 12 pp.

—Insulating—Nails (Fasteners)
CNS A2083-86. Nails for Sheathing Insulation Fiberboards (Dec)(6534). 2 pp.

—Length Measurement
BSI BS EN 324-1-93. 1993 Wood-Based Panels—Determination of Dimensions of Boards Part 1: Determination of Thickness, Width and Length (R). 10 pp.
CEN PREN 324 (Part 1)-89. Wood-Based Panels—Determination of Dimensions of Boards—Part 1: Determination of Thickness, Width and Length. 5 pp.
CEN EN 324-1-93. Wood-Based Panels—Determination of Dimensions of Boards Part 1: Determination of Thickness, Width and Length. 5 pp.
ISO 9426 Pt 1-89. Wood-Based Panels—Determination of Dimensions—Part 1: Determination of Thickness, Width and Length First Edition. 4 pp.

—Loads (Forces)
CEN PREN 622-3-93. Fibreboards—Specifications—Part 3: Requirements for Loadbearing Boards for Use in Dry Conditions. 10 pp.
CEN PREN 622-5-93. Fibreboards—Specifications—Part 5: Requirements for Loadbearing Boards for Use in Humid Conditions. 10 pp.
CEN PREN 1156-93. Wood-Based Panels—Determination of Duration of Load and Creep Factors. 11 pp.

—Magnesium Carbonate
CNS A2156-88. Magnesium Carbonate Boards (Nov)(10406).
CNS A3188-88. Method of Test for Magnesium Carbonate Boards (Nov)(10407).
JIS A 6701-83. Magnesium Carbonate Boards. 12 pp.

—Moisture Content
BSI BS EN 322-93. 1993 Wood-Based Panels—Determination of Moisture Content (R). 9 pp.
CEN PREN 322-89. Wood-Based Panels—Determination of Moisture Content. 6 pp.
CEN EN 322-93. Wood-Based Panels—Determination of Moisture Content. 6 pp.
CEN PREN 1087-1-93. Particleboards and Fibreboards—Moisture Resistance—Part 1: Boil Test. 7 pp.
ISO 9425-89. Wood-Based Panels—Determination of Moisture Content First Edition; (Supersedes 767, 823 and 3806). 4 pp.

—Paints—Alkali Resistance Testing
CGSB 1-GP-71 METH 106.3-78. Methods of Testing Paints and Pigments Alkali Resistance Emulsion Paints. 1 p.

—Plastic Veneer—Glowing Cigarette Test
DIN ENGL 51961-84. Testing of Plastics Surfaces; Behaviour on Exposure to Glowing Cigarettes (Aug). 2 pp.

Fiberboard (Cont.)

—Quality Assurance
DIN ENGL 68750-58. Wood Fibre Building Boards; Porous and Hard Wood Fibre Building Boards; Quality Conditions (Apr). 2 pp.

—Sand Content
ISO 3340-76. Fibre Building Boards—Determination of Sand Content First Edition. 5 pp.

—Stability Testing
ISO 3729-76. Fibre Building Boards—Determination of Surface Stability First Edition. 5 pp.

—Surface Absorption
BSI BS EN 382-1-93. 1993 Fibreboards—Determination of Surface Absorption Part 1: Test Method for Dry Process Fibreboards (R). 10 pp.
CEN PREN 382 (Part 1)-90. Fibreboards—Determination of Surface Absorption—Part 1: Toluene Test. 5 pp.
CEN EN 382-1-93. Fibreboards—Determination of Surface Absorption—Part 1: Test Method for Dry Process Fibreboards. 6 pp.

—Surface Roughness
ISO 3546-76. Fibre Building Boards—Determination of Surface Finish (Roughness) First Edition. 5 pp.

—Symbols
BSI BS EN 316-93. 1993 Wood Fibreboards—Definition, Classification and Symbols (R). 7 pp.
CEN EN 316-93. Wood Fibreboards—Definition, Classification and Symbols. 4 pp.

—Tensile Testing
BSI BS EN 319-93. 1993 Particleboards and Fibreboards—Determination of Tensile Strength Perpendicular to the Plane of the Board (R). 11 pp.
CEN PREN 319-90. Particleboards and Fibreboards-Determination of Transverse Tensile Strength. 6 pp.
CEN EN 319-93. Particleboards and Fibreboards—Determination of Tensile Strength Perpendicular to the Plane of the Board. 6 pp.

—Test Pieces
BSI BS EN 325-93. 1993 Wood-Based Panels—Determination of Dimensions of Test Pieces (R). 9 pp.
CEN PREN 325-89. Wood-Based Panels—Determination of Dimensions of Test Pieces. 5 pp.
CEN EN 325-93. Wood-Based Panels—Determination of Dimensions of Test Pieces. 4 pp.
CEN PREN 326 (Part 1)-90. Wood-Based Panels—Sampling, Cutting and Inspection—Part 1: Sampling and Cutting of Test Pieces and Evaluation of Test Results. 11 pp.
ISO 9424-89. Wood-Based Panels—Determination of Dimensions of Test Pieces First Edition. 4 pp.

—Thickness Measurement
BSI BS EN 317-93. 1993 Particleboards and Fibreboards—Determination of Swelling in Thickness After Immersion in Water (R). 10 pp.
BSI BS EN 324-1-93. 1993 Wood-Based Panels—Determination of Dimensions of Boards Part 1: Determination of Thickness, Width and Length (R). 10 pp.
CEN PREN 317-90. Particleboards and Fibreboards—Determination of Swelling in Thickness After Immersion in Water. 5 pp.
CEN EN 317-93. Particleboards and Fibreboards—Determination of Swelling in Thickness After Immersion in Water. 5 pp.
CEN PREN 324 (Part 1)-89. Wood-Based Panels—Determination of Dimensions of Boards—Part 1: Determination of Thickness, Width and Length. 5 pp.
CEN EN 324-1-93. Wood-Based Panels—Determination of Dimensions of Boards Part 1: Determination of Thickness, Width and Length. 5 pp.
ISO 9426 Pt 1-89. Wood-Based Panels—Determination of Dimensions—Part 1: Determination of Thickness, Width and Length First Edition. 4 pp.

—Water Absorption
ISO 769-72. Fibre Building Boards—Hard and Medium Boards—Determination of Water Absorption and of Swelling in Thickness After Immersion in Water First Edition. 4 pp.
ISO 2696-76. Fibre Building Boards—Hard and Medium Boards for General Purposes—Quality Specifications—Water Absorption and Swelling in Thickness First Edition. 3 pp.
SAA AS 1301.442S-91. Methods of Test for Pulp and Paper—Part 442S: Water Absorption of Solid Fibreboard (Total Immersion) (This is a Joint Standard with SANZ NZS 1301.442S). 2 pp.
SNZ NZS/AS 1301. 442S-91. Methods of Test for Pulp and Paper Water Absorption of Solid Fibre-Board (Total Immersion) (This is a Joint Standard with SAA AS 1301.442S). 2 pp.

INTERNATIONAL AND NON-U.S. NATIONAL STANDARDS
SUBJECT INDEX
Fiberglass

Fiberboard (Cont.)
—Weight Measurement
BSI BS 5881-80. 1980 Determination of Grammage of Single Layers of Solid Fibreboard. 4 pp.
ISO 5638-78. Solid Fibreboard—Determination of Grammage of Single Layers First Edition. 4 pp.

—Width Measurement
BSI BS EN 324-1-93. 1993 Wood-Based Panels—Determination of Dimensions of Boards Part 1: Determination of Thickness, Width and Length (R). 10 pp.
CEN PREN 324 (Part 1)-89. Wood-Based Panels—Determination of Dimensions of Boards—Part 1: Determination of Thickness, Width and Length. 5 pp.
CEN EN 324-1-93. Wood-Based Panels—Determination of Dimensions of Boards Part 1: Determination of Thickness, Width and Length. 5 pp.
ISO 9426 Pt 1-89. Wood-Based Panels—Determination of Dimensions—Part 1: Determination of Thickness, Width and Length First Edition. 4 pp.

Fiberglass
Scope Note: For products made from fiberglass, see the specific products *See Also:* Fiberglass Mats; Fiberglass Ropes; Rovings

—Rovings
BSI BS 3691-90. 1990 E Glass Fibre Rovings for Reinforcement of Polyester and Epoxy Resin Systems. 12 pp.
CNS K3054-81. Fiber Glass Roving (May)(7399). 2 pp.
CNS K3056-81. Fiber Glass Woven Roving (May)(7401). 2 pp.

Fiberglass Insulation
See Also: Acoustical Insulation; Thermal Insulation

—Rope—Marine
MOD UK NES 801: Part 1-89. Requirements for Insulation Material Part 1: Glass Fibre Products Glass Rope for Thermal Insulation Issue 2 (04.89). 11 pp.

—Webbing—Marine
MOD UK NES 801: Part 3-89. Requirements for Insulation Material Part 3: Glass Fibre Products Glass Webbing for Pipe Hangers Issue 2 (04.89). 11 pp.

Fiberglass Mats
See Also: Ceramic Fiber Mats; Fiberglass; Mats

—Pipe Insulation
CNS K3058-82. Glass Fiber Mat for Steel Pipe Over Wrapping (Jan)(8425). 1 p.

Fiberglass Reinforced Cements
Use For: Glass Fibre Reinforced Cements
See Also: Cements
BSI BS 6432-84. 1984 Determining Properties of Glass Fibre Reinforced Cement Material. 12 pp.

Fiberglass Reinforced Plastics
Scope Note: For additional listings, see specific products made from fiberglass reinforced plastics
See Also: Laminated Plastics; Plastics; Sheet Molding Compounds
CNS K6975-90. General Rules for Method of Test for Glass Fiber Reinforced Plastics (Sep)(12776).
JIS K 7011-89. Glass Fiber Reinforced Plastics for Structural Use. 10 pp.
JIS K 7051-87. General Rules for Testing Methods of Glass Fiber Reinforced Plastics. 7 pp.

—Aerospace—Conditioning
CEN PREN 2743-90. Reinforced Plastics Standard Procedures for Conditioning Prior to Testing. 6 pp.
CEN PREN 2743-91. Reinforced Plastics Standard Procedures for Conditioning Prior to Testing. 6 pp.

—Aerospace—Flexural Strength
AECMA PREN2746-90. Glass Fibre Reinforced Plastics Determination of Flexural Properties Three Point Bend Method. 9 pp.

—Aerospace—Shear Testing
AECMA PREN2377-85. Glass Fibre Reinforced Plastics—Test Method—Determination of Apparent Interlaminar Shear Strength. 8 pp.
BSI BS EN 2377-89. 1989 Glass Fibre Reinforced Plastics. Test Method. Determination of Apparent Interlaminar Shear Strength. 12 pp.
CEN PREN 2377-87. Glass Fibre Reinforced Plastics; Test Method Determination of Apparent Interlaminar Shear Strength. 9 pp.
CEN EN 2377-89. Glass Fibre Reinforced Plastics Test Method Determination of Apparent Interlaminar Shear Strength. 11 pp.

Fiberglass Reinforced Plastics (Cont.)
—Aerospace—Tensile Testing
AECMA PREN2747-90. Glass Fibre Reinforced Plastics Determination of Tensile Properties. 16 pp.

—Alkali Content
CNS K6662-81. Method of Test for Fiber Glass Products (May)(7397). 9 pp.

—Bend Testing
JIS K 7055-87. Testing Method for Flexural Properties of Glass Fiber Reinforced Plastics. 12 pp.

—Binders—Mats
ISO 2558-74. Textile Glass Chopped-Strand Mats for Reinforcement of Plastics—Determination of Time of Dissolution of the Binder in Styrene First Edition. 5 pp.

—Boats (Marine)—Construction
MOD UK NES 166-01. Requirements for Glass Woven Roving Fabrics for Ship Structures Issue 1 (11.83); Amendment 1. 32 pp.
MOD UK NES 166-93. Requirements for Glass Reinforcing Fabrics for Ships, Boats and Craft Structures Issue 2 (02.93). 53 pp.

—Boats (Marine)—Maintenance
MOD UK NES 742: Part 2-88. Preservation of Small Craft and Boats Part 2: GRP Construction Issue 3 (11.88). 13 pp.
MOD UK NES 742: Part 2-91. Requirements for the Preservation of Small Craft and Boats Part 2: G R P Construction Issue 4 (12.91). 14 pp.
MOD UK NES 752-81. GRP Ships and Boats Maintenance, Survey and Repair Issue 2 (04.81). 48 pp.

—Chemical Resistance
DIN ENGL 53393-76. Testing of Textile Glass-Reinforced Plastics; Behaviour to the Effect of Chemicals (Mar). 7 pp.

—Comprehensive Testing
ISO 4899-82. Textile Glass Reinforced Thermosetting Plastics—Properties and Test Methods First Edition. 5 pp.

—Compression Testing
CNS K6980-90. Method of Test for Compressive Properties of Glass Fiber Reinforced Plastics (Sep)(12781).
JIS K 7056-87. Testing Method for Compressive Properties of Glass Fiber Reinforced Plastics. 11 pp.

—Epoxy Resins—Glass Rovings
BSI BS 3691-90. 1990 E Glass Fibre Rovings for Reinforcement of Polyester and Epoxy Resin Systems. 12 pp.

—Fiber Content
CNS K6976-90. Method of Test for Fiber Content of Glass Fiber Reinforced Plastics (Sep)(12777).
JIS K 7052-87. Measuring Method for Fiber Content of Glass Fiber Reinforced Plastics. 8 pp.

—Flexural Strength
BSI BS 2782:Pt10: METH 1005-77. 1977 Methods of Testing Plastics Part 10: Glass Reinforced Plastics Method 1005: Determination of Flexural Properties. Three Point Method. 8 pp.
CEN EN 63-77. Glass Reinforced Plastics: Determination of Flexural Properties; Three Point Method. 6 pp.
JIS K 7055-87. Testing Method for Flexural Properties of Glass Fiber Reinforced Plastics. 12 pp.
JIS K 7074-88. Testing Methods for Flexural Properties of Carbon Fiber Reinforced Plastics. 11 pp.

—Glass Fabrics
BSI BS 3396: Part 1-91. 1991 Woven Glass Fibre Fabrics for Plastics Reinforcement Part 1: Specification for Loom-State Fabrics. 12 pp.
BSI BS 3396: Part 1-82. 1982 Woven Glass Fibre Fabrics for Plastics Reinforcement Part 1: Loom-State Fabrics. 10 pp.
BSI BS 3396: Part 2-87. 1987 Woven Glass Fibre Fabrics for Plastics Reinforcement Part 2: Desized Fabrics. 4 pp.
BSI BS 3396: Part 3-87. 1987 Woven Glass Fibre Fabrics for Plastics Reinforcement Part 3: Finished Fabrics for Use with Polyester Resin Systems. 8 pp.
DIN ENGL 61854 Pt 1-87. Textile Glass; Woven Glass Fabrics for Plastics Reinforcement; Woven Glass Filament Fabric and Woven Roving; Technical Delivery Conditions (Apr). 7 pp.
DIN ENGL 61854 Pt 2-87. Textile Glass; Woven Glass Fabrics for Plastics Reinforcement; Woven Glass Filament Fabric and Woven Roving; Types (Apr). 4 pp.

Fiberglass Reinforced Plastics (Cont.)
—Glass Fabrics (Cont.)
DIN ENGL 61855 Pt 1-87. Textile Glass; Glass Roving for Plastics Reinforcement; Technical Delivery Conditions (Apr) (Together with DIN 61855 Part 2, April 1987 Edition, Supersedes DIN 61855, August 1975 Edition). 6 pp.

—Glass Mats
BSI BS 3496-89. 1989 E Glass Fibre Chopped Strand Mat for Reinforcement of Polyester and Other Liquid Laminating Systems. 12 pp.
CNS K3053-81. Fiber Glass Chopped Strand Mats (May)(7398). 3 pp.
DIN ENGL 61853 Pt 1-87. Textile Glass; Textile Glass Mats for Plastics Reinforcement; Technical Delivery Conditions (Apr) (Together with DIN 61853 Part 2, April 1987 Edition, Supersedes DIN 61853, June 1975 Edition). 7 pp.
JIS R 3411-91. Textile Glass Chopped Strand Mats. 8 pp.

—Glass Mats—Classification
DIN ENGL 61853 Pt 2-87. Textile Glass; Textile Glass Mats for Plastics Reinforcement; Classification and Application (Apr) (Together with DIN 61853 Part 1, April 1987 Edition, Supersedes DIN 61853, June 1975 Edition). 10 pp.

—Glass Strands
CNS K3057-81. Fiber Glass Chopped Strands (May)(7402). 2 pp.
JIS R 3419-84. Glass Chopped Strands. 7 pp.

—Hardness Testing
BSI BS 2782:Pt10: METH 1001-77. 1977 Methods of Testing Plastics Part 10: Glass Reinforced Plastics Method 1001: Measurement of Hardness by Means of a Barcol Impressor. 7 pp.
CEN EN 59-77. Glass Reinforced Plastics: Measurement of Hardness by Means of a Barcol Impressor. 5 pp.
CNS K6984-90. Method of Test for Barcol Hardness of Glass Fiber Reinforced Plastics (Sep)(12785).
JIS K 7060-87. Testing Method for Barcol Hardness of Glass Fiber Reinforced Plastics. 8 pp.

—Ignition Loss
BSI BS 2782:Pt10: METH 1002-77. 1977 Amd 1 Methods of Testing Plastics Part 10: Glass Reinforced Plastics Method 1002: Determination of Loss on Ignition. 8 pp.
CEN EN 60-77. Glass Reinforced Plastics: Determination of the Loss on Ignition. 6 pp.
ISO 1172-75. Textile Glass Reinforced Plastics—Determination of Loss on Ignition First Edition. 5 pp.

—Impact Testing
JIS K 7061-92. Testing Method for Charpy Impact Strength of Glass Fiber Reinforced Plastics. 15 pp.
JIS K 7062-92. Testing Method of Izod Impact Strength of Glass Fiber Reinforced Plastics. 14 pp.

—Laminated—Aircraft
MOD UK DTD-933A-64. Glass Fabric Reinforced Polyester Laminates for Aircraft Structures and Airborne Radomes. 6 pp.

—Laminated—Compression Testing
ISO 8515-91. Textile-Glass-Reinforced Plastics—Determination of Compressive Properties in the Direction Parallel to the Plane of Lamination First Edition. 11 pp.

—Laminated—Panels
BSI BS 2782:Pt9: METH 920A-C-77. 1977 Methods of Testing Plastics Part 9: Sampling and Test Specimen Preparation Methods 920A to 920C: Preparation of Glass Fibre Reinforced, Resin Bonded, Low-Pressure Laminated Plates or Panels for Test Purposes. 8 pp.
ISO 1268-74. Plastics—Preparation of Glass Fibre Reinforced, Resin Bonded, Low-Pressure Laminated Plates or Panels for Test Purposes First Edition. 8 pp.

—Laminated—Plates
BSI BS 2782:Pt9: METH 920A-C-77. 1977 Methods of Testing Plastics Part 9: Sampling and Test Specimen Preparation Methods 920A to 920C: Preparation of Glass Fibre Reinforced, Resin Bonded, Low-Pressure Laminated Plates or Panels for Test Purposes. 8 pp.
ISO 1268-74. Plastics—Preparation of Glass Fibre Reinforced, Resin Bonded, Low-Pressure Laminated Plates or Panels for Test Purposes First Edition. 8 pp.

—Laminated—Quality Assurance
MOD UK QTM 32. Manufacture of Fibre Reinforced Laminated Components Issue 1 (Superseded by Def Stan 93-51).

INDUSTRY STANDARDS

INTERNATIONAL AND NON-U.S. NATIONAL STANDARDS
SUBJECT INDEX

Fiberglass

Fiberglass Reinforced Plastics (Cont.)

—**Laminated—Quality Assurance** (Cont.)
MOD UK DSTAN 93-51-88. Quality Assurance Requirements for the Manufacture of Fibre Reinforced Plastics Laminated Components Issue 1. 12 pp.

—**Laminated—Shear Strength**
CNS K6981-90. Method of Test for Apparent Interlaminar Shear Strength of Glass Fiber Reinforced Plastics (Sep)(12782).

—**Naval Ships—Construction**
MOD UK NES 166-01. Requirements for Glass Woven Roving Fabrics for Ship Structures Issue 1 (11.83); Amendment 1. 32 pp.
MOD UK NES 166-93. Requirements for Glass Reinforcing Fabrics for Ships, Boats and Craft Structures Issue 2 (02.93). 53 pp.

—**Naval Ships—Design**
JIS F 0303-92. Ships Made of Fiberglass Reinforced Plastics—Design of Earth.

—**Naval Ships—Maintenance**
MOD UK NES 752-81. GRP Ships and Boats Maintenance, Survey and Repair Issue 2 (04.81). 48 pp.

—**Physical Properties**
CNS K6662-81. Method of Test for Fiber Glass Products (May)(7397). 9 pp.

—**Plates—Test Specimens**
ISO 9353-91. Glass-Reinforced Plastics—Preparation of Plates with Unidirectional Reinforcements by Bag Moulding First Edition. 9 pp.

—**Polyester Resins**
BSI BS 3532-90. 1990 Method of Specifying Unsaturated Polyester Resin Systems. 11 pp.
BSI BS 3532-01. 1990 Amd 1 Method of Specifying Unsaturated Polyester Resin Systems (AMD 7344) November 15, 1992. 12 pp.
JIS K 6919-92. Liquid Unsaturated Polyester Resins for Fiber Reinforced Plastics. 14 pp.

—**Polyester Resins—Glass Fabrics**
BSI BS 3396: Part 3-87. 1987 Woven Glass Fibre Fabrics for Plastics Reinforcement Part 3: Finished Fabrics for Use with Polyester Resin Systems. 8 pp.
BSI BS 3749-91. 1991 E Glass Fibre Woven Roving Fabrics for the Reinforcement of Polyester and Epoxy Resin Systems. 12 pp.
BSI BS 3749-74. 1974 Woven Roving Fabrics of E Glass Fibre for the Reinforcement of Polyester Resin Systems. 8 pp.

—**Polyester Resins—Glass Rovings**
BSI BS 3691-90. 1990 E Glass Fibre Rovings for Reinforcement of Polyester and Epoxy Resin Systems. 12 pp.

—**Polyester Resins—Sheets—Corrugated**
BSI BS 4154: Part 1-85. 1985 Corrugated Plastics Translucent Sheets Made from Thermo-Setting Polyester Resin (Glass Fibre Reinforced) Part 1: Material and Performance Requirements. 16 pp.
BSI BS 4154: Part 2-85. 1985 Corrugated Plastics Translucent Sheets Made from Thermo-Setting Polyester Resin (Glass Fibre Reinforced) Part 2: Profiles and Dimensions. 4 pp.

—**Polyurethane Coatings—Interior/Exterior**
CGSB 1-GP-180MA-82. Coating, Polyurethane, Two-Package, General Purpose, Standard for. 15 pp.

—**Resin Content—Glass Fabrics**
BSI BS 2782:Pt10: METH 1006-78. 1978 Methods of Testing Plastics Part 10: Glass Reinforced Plastics Method 1006: Determination of Volatile Matter and Resin Content of Synthetic Resin-Impregnated Textile Glass Fibre. 2 pp.

—**Rods—Compression Testing**
ISO 3605-87. Textile Glass—Rovings—Determination of Compressive Strength of Rod Composites Second Edition. 8 pp.

—**Rods—Flexural Strength**
ISO 3597-77. Textile Glass Reinforced Plastics—Composites in the Form of Rods Made from Textile Glass Rovings—Determination of Flexural (Cross-Breaking) Strength First Edition. 5 pp.

—**Rovings**
CNS K3054-81. Fiber Glass Roving (May)(7399). 2 pp.
CNS K3056-81. Fiber Glass Woven Roving (May)(7401). 2 pp.
DIN ENGL 61854 Pt 2-87. Textile Glass; Woven Glass Fabrics for Plastics Reinforcement; Woven Glass Filament Fabric and Woven Roving; Types (Apr). 4 pp.
JIS R 3412-91. Textile Glass Rovings. 7 pp.

Fiberglass Reinforced Plastics (Cont.)

—**Rovings** (Cont.)
JIS R 3417-84. Woven Roving Glass Fabrics. 7 pp.

—**Rovings—Classification**
DIN ENGL 61855 Pt 2-87. Textile Glass; Glass Roving for Plastics Reinforcement; Classification and Application (Apr) (Together with DIN 61855 Part 1, April 1987 Edition, Supersedes DIN 61855, August 1975 Edition). 11 pp.

—**Shear Testing**
CNS K6982-90. Method of Test of Transverse Shear Strength of Glass Fiber Reinforced Plastics (Sep)(12783).
CNS K6983-90. Method of Test for in-Plane Shear Properties of Glass Reinforced Plastics (Sep)(12784).
ISO 4585-89. Textile Glass Reinforced Plastics—Determination of Apparent Interlaminar Shear Properties by Short-Beam Test First Edition. 6 pp.
JIS K 7057-87. Testing Method for Apparent Interlaminar Shear Strength of Glass Fiber Reinforced Plastics. 9 pp.
JIS K 7058-87. Testing Method for Transverse Shear Strength of Glass Fiber Reinforced Plastics. 9 pp.
JIS K 7059-87. Testing Method for In-Plane Shear Properties of Glass Fiber Reinforced Plastics. 9 pp.

—**Sheets—Corrugated**
BSI BS 4154: Part 1-85. 1985 Corrugated Plastics Translucent Sheets Made from Thermo-Setting Polyester Resin (Glass Fibre Reinforced) Part 1: Material and Performance Requirements. 16 pp.
BSI BS 4154: Part 2-85. 1985 Corrugated Plastics Translucent Sheets Made from Thermo-Setting Polyester Resin (Glass Fibre Reinforced) Part 2: Profiles and Dimensions. 4 pp.
CNS K3034-77. Glassfiber Reinforced Unsaturated Polyester Corrugated Sheets (Apr)(4089). 4 pp.
JIS A 5701-75. Glassfiber Reinforced Plastic Corrugated Sheets (R 1978). 11 pp.

—**Standard Atmospheres**
BSI BS 2782:Pt10: METH 1004-77. 1977 Methods of Testing Plastics Part 10: Glass Reinforced Plastics Method 1004: Standard Atmospheres for Conditioning and Testing. 6 pp.
CEN EN 62-77. Glass Reinforced Plastics: Standard Atmospheres for Conditioning and Testing. 4 pp.

—**Tensile Testing**
BSI BS 2782:Pt10: METH 1003-77. 1977 Methods of Testing Plastics Part 10: Glass Reinforced Plastics Method 1003: Determination of Tensile Properties. 12 pp.
CEN EN 61-77. Glass Reinforced Plastics: Determination of Tensile Properties. 10 pp.
CNS K6978-90. Method of Test for Tensile Properties of Glass Fiber Reinforced Plastics (Sep)(12779). 8 pp.
ISO 3268-78. Plastics—Glass-Reinforced Materials—Determination of Tensile Properties First Edition. 10 pp.
JIS K 7054-87. Testing Method for Tensile Properites of Glass Fiber Reinforced Plastics. 13 pp.
JIS K 7073-88. Testing Method for Tensile Properties of Carbon Fibre Reinforced Plastics. 11 pp.

—**Transverse Strain**
CNS K6982-90. Method of Test of Transverse Shear Strength of Glass Fiber Reinforced Plastics (Sep)(12783).
CNS K6983-90. Method of Test for in-Plane Shear Properties of Glass Reinforced Plastics (Sep)(12784).

—**Void Content**
CNS K6977-90. Method of Test for Void Content of Glass Fiber Reinforced Plastics (Sep)(12778).
ISO 7822-90. Textile Glass Reinforced Plastics—Determination of Void Content—Loss on Ignition, Mechanical Disintegration and Statistical Counting Methods First Edition. 11 pp.
JIS K 7053-87. Measuring Method for Void Content of Glass Fiber Reinforced Plastics. 7 pp.

—**Volatile Matter Content—Glass Fabrics**
BSI BS 2782:Pt10: METH 1006-78. 1978 Methods of Testing Plastics Part 10: Glass Reinforced Plastics Method 1006: Determination of Volatile Matter and Resin Content of Synthetic Resin-Impregnated Textile Glass Fibre. 2 pp.

Fiberglass Ropes
See Also: Electrical Insulation; Fiberglass; Ropes

—**Marine**
MOD UK NES 801: Part 1-89. Requirements for Insulation Material Part 1: Glass Fibre Products Glass Rope for Thermal Insulation Issue 2 (04.89). 11 pp.

Fiberglass Tapes
See Also: Tapes
CNS K3065-82. Glass Tape (Dec)(9719). 5 pp.

Fiberglass Tapes (Cont.)
CNS K3066-82. Finished Glass Tapes (Dec)(9720). 6 pp.
IEC 1067 Pt 1-91. Specification for Glass and Glass Polyester Fibre Woven Tapes Part 1: Definitions, Classification and General Requirements First Edition. 21 pp.
IEC 1067 Pt 2-92. Specification for Glass and Glass Polyester Fibre Woven Tapes Part 2: Methods of Test First Edition. 24 pp.
JIS R 3422-85. Finished Glass Tapes. 9 pp.

—**Aircraft**
MOD UK DTD-5546-59. Woven Glass Fibre Tape and Webbing ("E" Glass). 5 pp.

Fiberoptic Cable Assemblies
Use: Fiber Optic Cable Assemblies

Fibers
Scope Note: For additional listings, use a more specific term *Use For:* Natural Fibers; Textile Fibers
See Also: Acetate Fibers; Acrylic Fibers; Aramid Fibers; Asbestos; Batting (Fibers); Bristles; Carbon Fibers; Cellulose Fibers; Ceramic Fibers; Chlorofibers; Cordage; Cotton Fibers; Fabrics; Fiber Composites; Fiber Content Analysis; Flax; Glass Wool; Hair (Animal); Hemp; Linen; Linters; Mineral Wool; Modacrylic Fibers; Monofilaments; Nylon Fibers; Paddings; Papers; Polyamide Fibers; Polyester Fibers; Polypropylene Fibers; Pulps; PVC Fibers; Ramie; Rayon; Rayon Fibers; Rock Wool; Rovings; Silk; Sisal; Slivers; Spandex Fibers; Staple Fibers; Synthetic Fibers; Triacetate Fibers; Twist Balance Index; Viscose Fibers; Vulcanized Fibers; Wood Fibers; Wool; Wool Fibers; Yarns

—**Colorfastness Testing—Acid-Felting**
ISO 105 Pt E13-87. Textiles—Tests for Colour Fastness—Part E13: Colour Fastness to Acid-Felting: Severe Second Edition. 4 pp.
ISO 105 Pt E14-87. Textiles—Tests for Colour Fastness—Part E14: Colour Fastness to Acid-Felting: Mild Second Edition. 4 pp.

—**Colorfastness Testing—Alkaline Milling**
ISO 105 Pt E12-89. Textiles—Tests for Colour Fastness—Part E12: Colour Fastness to Milling: Alkaline Milling Third Edition. 6 pp.

—**Colorfastness Testing—Bleaching**
ISO 105 Pt N-78. Textiles—Tests for Colour Fastness Part N: Colour Fastness to Bleaching Agencies First Edition. 15 pp.

—**Colorfastness Testing—Carbonizing**
ISO 105 Pt X01-87. Textiles—Tests for Colour Fastness—Part X01: Colour Fastness to Carbonizing: Aluminium Chloride Third Edition. 4 pp.

—**Colorfastness Testing—Dry Cleaning**
ISO 105 Pt D01-87. Textiles—Tests for Colour Fastness—Part D01: Colour Fastness to Dry Cleaning Third Edition. 4 pp.

—**Colorfastness Testing—Laundering**
BSI BS EN 20105-C06-93. 1993 Textiles—Tests for Colour Fastness Part C06: Colour Fastness to Domestic and Commercial Laundering (ISO 105-C06: 1987) (L). 11 pp.
CEN EN 20105-C06-92. Textiles—Tests for Colour Fastness—Part C06: Colour Fastness to Domestic and Commercial Laundering (ISO 105-C06: 1987). 8 pp.
CGSB CAN/CGSB-4.2 NO.19.1-M90. Textile Test Methods Colourfastness to Washing—Accelerated Test—Launder-Ometer; (Amendment 1 June 1992). 12 pp.
ISO 105 Pt C06-87. Textiles—Tests for Colour Fastness—Part C06: Colour Fastness to Domestic and Commercial Laundering Second Edition; (CEN EN 20105-C06:1992). 7 pp.

—**Colorfastness Testing—Standard Adjacent Fabrics**
ISO 105 Pt F-85. Textiles—Tests for Colour Fastness—Part F: Standard Adjacent Fabrics Third Edition. 24 pp.

—**Colorfastness Testing—Steaming**
ISO 105 Pt E11-87. Textiles—Tests for Colour Fastness—Part E11: Colour Fastness to Steaming Second Edition. 4 pp.

—**Colorfastness Testing—Water**
CGSB CAN/CGSB-4.2 NO.20-M89. Textile Test Methods Colourfastness to Water. 9 pp.

—**Diameters**
SAA AS 2001.2.1-88. Methods of Test for Textiles—Part 2: Physical Tests—Part 2.1: Determination of Mean Fibre Diameter of Textile Fibres by Measurement of Projected Images. 6 pp.

INTERNATIONAL AND NON-U.S. NATIONAL STANDARDS
SUBJECT INDEX
Fibers

Fibers *(Cont.)*

—Elastic Properties

BSI BS 4029-78. 1978 Determination of Tensile Elastic Recovery of Single Fibres and Filaments (Constant-Rate-Of-Extension Machines). 7 pp.

—Fungus Resistance Testing

CNS L3063-75. Method of Test for Mildew Resistance of Fiber Products (May) (2690).

—Glossaries

BSI BS 4815-90. 1990 Glossary of Generic Names for Man-Made Fibres. 13 pp.
BSI BS 6189-89. 1989 Terms Relating to Fabrics and Associated Fibres, Yarns and Processes. 28 pp.
CGSB CAN/CGSB-4.157-M91. Generic Names for Man-Made Fibres. 14 pp.
CGSB CAN/CGSB-4. 175-M91 Pt4. Textiles—Morphology of Fibres and Yarns—Vocabulary Part 4 (ISO 8159:1987). 9 pp.
CNS L1001-71. Glossary of Terms Used for Fibers (Textile Materials) (Jan)(2336).
ISO 6938-84. Textiles—Natural Fibres—Generic Names and Definitions First Edition. 8 pp.
ISO 8159-87. Textiles—Morphology of Fibres and Yarns—Vocabulary First Edition. 6 pp.
JIS L 0204-72. Glossary of Terms Used in Fibre.

—Length Measurement

CNS L3104-81. Textile Fibers Determination of Length by Measuring Individual Fibers (Jan)(6934).

—Length Measurement—Comb Sorter Method

BSI BS 4044-89. 1989 Determination of Fibre Length by Comb Sorter Diagram. 12 pp.

—Linear Density

BSI BS 947-70. 1970 Amd 1 Universal System for Designating Linear Density of Textiles (Tex System). 16 pp.
BSI BS 2016-73. 1973 Determination of the Linear Density of Textile Fibres: Gravimetric Methods. 10 pp.
CEN PREN 21973-2-93. Textile Fibres—Determination of Linear Density—Gravimetric Method and Vibroscope Method. 15 pp.
ISO 1144-73. Textiles—Universal System for Designating Linear Density (Tex System) First Edition. 9 pp.
ISO 1973-76. Textile Fibres—Determination of Linear Density—Gravimetric Method First Edition. 4 pp.
JIS L 0101-78. Tex System to Designate Linear Density of Fibres, Yarn Intermediates, Yarns and Other Textile Materials.

—Mass

BSI BS 4784: Part 1-88. 1988 Methods for Determination of Commerical Mass of Consignments of Textiles Part 1: Mass Determination and Calculations. 20 pp.
CGSB CAN/CGSB-4.2 NO.72.1-M91. Textile Test Methods Textiles—Fibres and Yarns—Determination of Commercial Mass of Consignments—Part 1: Mass Determination and Calculations (ISO 6741-1:1989). 17 pp.
ISO 6741 Pt 1-89. Textiles—Fibres and Yarns—Determination of Commercial Mass of Consignments Part 1: Mass Determination and Calculations Second Edition. 14 pp.
ISO TR6741 Pt 4-87. Textiles—Fibres and Yarns—Determination of Commercial Mass of Consignments—Part 4: Values Used for the Commercial Allowances and the Commercial Moisture Regains First Edition. 6 pp.

—Mass—Test Specimens

BSI BS 4784: Part 2-88. 1988 Methods for Determination of Commercial Mass of Consignments of Textiles Part 2: Methods for Obtaining Laboratory Samples. 8 pp.
BSI BS 4784: Part 3-88. 1988 Methods for Determination of Commercial Mass of Consignments of Textiles Part 3: Specimen Cleaning Procedures. 7 pp.
CGSB CAN/CGSB-4.2 NO.72.2-M91. Textile Test Methods Textiles—Fibres and Yarns—Determination of Commercial Mass of Consignments—Part 2: Methods for Obtaining Laboratory Samples (ISO 6741-2:1987). 11 pp.
CGSB CAN/CGSB-4.2 NO.72.3-M91. Textile Test Methods Textiles—Fibres and Yarns—Determination of Commercial Mass of Consignments—Part 3: Specimen Cleaning Procedures (ISO 6741-3:1987). 9 pp.
ISO 6741 Pt 2-87. Textiles—Fibres and Yarns—Determination of Commercial Mass of Consignments—Part 2: Methods for Obtaining Laboratory Samples First Edition. 8 pp.
ISO 6741 Pt 3-87. Textiles—Fibres and Yarns—Determination of Commercial Mass of Consignments—Part 3: Specimen Cleaning Procedures First Edition. 7 pp.

Fibers *(Cont.)*

—Metal Content

BSI BS 6810-87. 1987 Methods for Determination of Metals in Textiles. 32 pp.

—Mixtures—Quantitative Analysis

BSI BS 4407-88. 1988 Amd 1 Quantitative Analysis of Fibre Mixtures (AMD 6297) September 28, 1990. 30 pp.
CGSB 4-GP-176M-89. Quantitative Analysis of Fibre Mixtures—Flotation Method. 10 pp.
CGSB CAN/CGSB-4.2 NO.0-M88. Textile Test Methods Table of Contents, Moisture Regain Values, SI Units Used in CAN/CGSB-4.2-M and Fibre, Yarn, Fabric, Garment and Carpet Properties. 43 pp.
CGSB CAN/CGSB-4.2 NO.14-M88. Textile Test Methods Quantitative Analysis of Fibre Mixtures—Introduction. 7 pp.
CGSB CAN/CGSB-4.2 NO.14.1-M88. Textile Test Methods Quantitative Analysis of Fibre Mixtures—Binary Mixtures Containing Wool —70% Sulfuric Acid Method. 12 pp.
CGSB CAN/CGSB-4.2 NO.14.5-M91. Textile Test Methods Quantitative Analysis of Fibre Mixtures—Binary Mixtures Containing Acrylic or Modacrylic Fibres—Butyrolactone Method. 9 pp.
CGSB CAN/CGSB-4.2 NO.14.7-M88. Textile Test Methods Quantitative Analysis of Fibre Mixtures—Binary Mixtures Containing Acetate Fibres—Acetone Method. 12 pp.
CGSB CAN/CGSB-4.2 NO.14.8-M88. Textile Test Methods Quantitative Analysis of Fibre Mixtures—Binary Mixtures Containing Triacetate Fibres—Dichloromethane Method. 12 pp.
CGSB CAN/CGSB-4.2 NO.14.9-M89. Textile Test Methods Quantitative Analysis of Fibre Mixtures—Microscopic Count Method. 12 pp.
CGSB CAN/CGSB-4.2 NO.14.11-M89. Textile Test Methods Quantitative Analysis of Fibre Mixtures—Binary Mixtures Containing Secondary and Triacetate Fibres—70% Acetone Method. 11 pp.
CGSB CAN/CGSB-4.2 NO.14.12-M88. Textile Test Methods Quantitative Analysis of Fibre Mixtures—Binary Mixtures Containing Nylon 6 or Nylon 6,6—80% Formic Acid Method. 12 pp.
CGSB CAN/CGSB-4.2 NO.14.13-M89. Textile Test Methods Quantitative Analysis of Fibre Mixtures—Binary Mixtures Containing Certain Spandex (Elastane) Fibres—Boiling Cyclohexanone Method. 13 pp.
CGSB CAN/CGSB-4.2 NO.14.14-M89. Textile Test Methods Quantitative Analysis of Fibre Mixtures—Binary Mixtures Containing Polyvinyl Chloride Fibres—Carbon Disulphide/Acetone Method. 11 pp.
CGSB CAN/CGSB-4.2 NO.14.15-M89. Textile Test Methods Quantitative Analysis of Fibre Mixtures—Binary Mixtures Containing Polypropylene Fibres—Xylene Method. 10 pp.
CGSB CAN/CGSB-4.2 NO.14.16-M88. Textile Test Methods Quantitative Analysis of Fibre Mixtures—Binary Mixtures Containing Acetate and Polyvinyl Chloride Fibres—Glacial Acetic Acid Method. 11 pp.
CGSB CAN/CGSB-4.2 NO.14.17-M89. Textile Test Methods Quantitative Analysis of Fibre Mixtures—Manual Separation. 10 pp.
CGSB CAN/CGSB-4.2 NO.14.18-M91. Textile Test Methods Quantitative Analysis of Multi-Fibre Blends. 11 pp.
CNS L3050-87. Method of Test for Quantitative Analysis of Fiber Mixtures (Oct)(2339).
EC 73/44/EEC-73. Council Directive on the Approximation of the Laws of the Member States Relating to the Quantitative Analysis of Ternary Fibre Mixtures. 19 pp.
EC 87/142/EEC-87. Commission Recommendation on Certain Methods for the Removal of Non-Fibrous Matter Prior to Quantitative Analysis of Fibre Mixtures. 4 pp.
EC 87/185/EEC-87. Commission Recommendation on Quantitative Methods of Analysis for the Identification of Acrylic and Modacrylic Fibres, Chlorofibres and Trivinyl Fibres. 6 pp.
ISO 5088-76. Textiles—Ternary Fibre Mixtures—Quantitative Analysis First Edition. 16 pp.
ISO TR5090-77. Textiles—Methods for the Removal of Non-Fibrous Matter Prior to Quantitative Analysis of Fibre Mixtures First Edition. 7 pp.
SAA AS 2001.7-83. Methods of Test for Textiles—Part 7: Quantitative Analysis of Fibre Mixtures Subset in Binder. 1 p.
SAA AS 2001.7.2-83. Methods of Test for Textiles—Part 7: Quantitative Analysis of Fibre Mixtures—Part 7.2: General Requirements. 10 pp.
SAA AS 2001.7.3-83. Methods of Test for Textiles—Part 7: Quantitative Analysis of Fibre Mixtures—Part 7.3: Manual Separation. 2 pp.
SAA AS 2001.7.4-83. Methods of Test for Textiles—Part 7: Quantitative Analysis of Fibre Mixtures—Part 7.4: Binary Mixtures of Acetate and Certain Other Fibres (Method Using Acetone). 2 pp.

Fibers *(Cont.)*

—Mixtures—Quantitative Analysis *(Cont.)*

SAA AS 2001.7.5-83. Methods of Test for Textiles—Part 7: Quantitative Analysis of Fibre Mixtures—Part 7.5: Binary Mixtures of Certain Protein Fibres (Wool, Animal Hair, Silk or Protein) and Certain Other Fibres (Method Using Alkaline Sodium Hypochlorite). 3 pp.
SAA AS 2001.7.6-83. Methods of Test for Textiles—Part 7: Quantitative Analysis of Fibre Mixtures—Part 7.6: Binary Mixtures of Viscose, Cupro or Certain Types of Polynosic (Modal) and Cotton Fibres (Method Using Formic Acid and Zinc Chloride). 2 pp.
SAA AS 2001.7.7-83. Methods of Test for Textiles—Part 7: Quantitative Analysis of Fibre Mixtures—Part 7.7: Binary Mixtures of Nylon 6 or Nylon 6.6 and Certain Other Fibres (Method Using Formic Acid 80 Percent m/m). 2 pp.
SAA AS 2001.7.8-83. Methods of Test for Textiles—Part 7: Quantitative Analysis of Fibre Mixtures—Part 7.8: Binary Mixtures of Acetate and Triacetate Fibres (Method Using Benzyl Alcohol). 2 pp.
SAA AS 2001.7.9-83. Methods of Test for Textiles—Part 7: Quantitative Analysis of Fibre Mixtures—Part 7.9: Binary Mixtures of Triacetate and Certain Other Fibres (Method Using Dichloromethane). 2 pp.
SAA AS 2001.7.10-83. Methods of Test for Textiles—Part 7: Quantitative Analysis of Fibre Mixtures—Part 7.10: Binary Mixtures of Certain Cellulose Fibres with Polyester (Method Using Sulphuric Acid 75 Percent m/m). 2 pp.
SAA AS 2001.7.11-83. Methods of Test for Textiles—Part 7: Quantitative Analysis of Fibre Mixtures—Part 7.11: Binary Mixtures of Acrylics, Certain Modacrylics or Certain Chlorofibres (Method Using Dimethyl Formamide). 2 pp.
SAA AS 2001.7.12-83. Methods of Test for Textiles—Part 7: Quantitative Analysis of Fibre Mixtures—Part 7.12: Binary Mixtures of Certain Chlorofibres and Certain Other Fibres (Method Using Carbon Disulphide/Acetone 55.5/44.5). 2 pp.
SAA AS 2001.7.13-83. Methods of Test for Textiles—Part 7: Quantitative Analysis of Fibre Mixtures—Part 7.13: Binary Mixtures of Acetate and Certain Chlorofibres (Method Using Glacial Acetic Acid). 2 pp.
SAA AS 2001.7.14-83. Methods of Test for Textiles—Part 7: Quantitative Analysis of Fibre Mixtures—Part 7.14: Binary Mixtures of Silk and Wool or Hair Fibres (Method Using Sulphuric Acid 75 Percent m/m). 2 pp.
SAA AS 2001.7.15-83. Methods of Test for Textiles—Part 7: Quantitative Analysis of Fibre Mixtures—Part 7.15: Binary Mixtures of Certain Cellulose Fibres and Wool or Hair (Method Using Sulphuric Acid 70 Percent m/m). 2 pp.
SAA AS 2001.7.16-83. Methods of Test for Textiles—Part 7: Quantitative Analysis of Fibre Mixtures—Part 7.16: Binary Mixtures of Jute and Certain Animal Fibres (Method of Determining the Nitrogen Content). 2 pp.
SAA AS 2001.7.18-83. Methods of Test for Textiles—Part 7: Quantitative Analysis of Fibre Mixtures—Part 7.18: Binary Mixtures of Polyolefin and Certain Other Fibres (Method Using Xylene). 2 pp.
SAA AS 2001.7.19-90. Methods of Test for Textiles—Part 7: Quantitative Analysis of Fibre Mixtures—Part 7.19: Binary Mixtures of Chlorofibres (Homopolymers of Vinyl Chloride) and Certain Other Fibres (Method Using Concentrated Sulfuric Acid). 2 pp.
SAA AS 2001.7.20-83. Methods of Test for Textiles—Part 7: Quantitative Analysis of Fibre Mixtures—Part 7.20: Binary Mixtures of Inorganic Based, Man-Made Fibres and Certain Fibres (Method Based on Ignition at 625 Degrees Celsius). 2 pp.

—Mixtures—Quantitative Analysis—Test Specimens

EC 73/44/EEC-73. Council Directive on the Approximation of the Laws of the Member States Relating to the Quantitative Analysis of Ternary Fibre Mixtures. 19 pp.
SAA AS 2001.7.1-83. Methods of Test for Textiles—Part 7: Quantitative Analysis of Fibre Mixtures—Part 7.1: Test Samples and Test Specimens. 3 pp.

—Sampling

BSI BS 2545-65. 1965 Methods of Fibre Sampling for Testing. 20 pp.
ISO 1130-75. Textile Fibres—Some Methods of Sampling for Testing First Edition; (Amendment Slip-1976). 12 pp.

—Size—Material Determination (Textiles)

BSI BS 4032-78. 1978 Determination of Certain Water-Or Alkali-Soluble Additives in Cellulosic or Synthetic Fibres, Yarns and Fabrics or Yarns and Fabrics Made from Blends of Such Fibres. 4 pp.

INDUSTRY STANDARDS
1067

Fibers (Cont.)

—Tensile Testing
BSI BS 3411-71. 1971 Determination of the Tensile Properties of Individual Textile Fibres. 12 pp.
CNS L3143-82. Method of Test for Tensile Strength of Fibers (Jan)(8306).
JIS L 1069-78. Tensile Strength Tests of Fibres.

—Twist—Designations
CGSB CAN/CGSB-4.2 NO.8.3-M89. Textile Test Methods Textiles—Designation of the Direction of Twist in Yarns and Related Products (ISO 2-1973). 8 pp.
ISO 2-73. Textiles—Designation of the Direction of Twist in Yarns and Related Products First Edition. 3 pp.

—Webbing—Aerospace
BSI 2F 111-73. (WITHDRAWN) 1973 Amd 1 Flax Harness Webbing (Superseded by BS 3F 111: 1991). 5 pp.
BSI 3F 111-91. 1991 Flax Harness Webbing for Aerospace Purposes. 6 pp.
BSI 3F 111-01. 1991 Amd 1 Flax Harness Webbing for Aerospace Purposes (AMD 7222) August 15, 1992 (S). 7 pp.

—Wire Rope—Cores
BSI BS 525-91. 1991 Fibre Cores for Wire Ropes. 8 pp.
BSI BS 5053-85. 1985 Cordage and Webbing Slings and for Fibre Cores for Wire Ropes. 22 pp.
SNZ NZS/BS 5053-85. Methods of Test for Cordage and Webbing Slings and for Fibre Cores for Wire Ropes. 24 pp.

Fibre Building Board
Use: Fiberboard

Fibre Cement Pipes
Use: Fiber Cement Pipes

Fibre Distributed Data Interface
Use: Fiber Distributed Data Interface

Fibre Optic Cables
Use: Fiber Optic Cables

Fibre Optics
Use: Fiber Optic Equipment

Fibreglass Reinforced Plastics
Use: Fiberglass Reinforced Plastics

Fibrelight Cables
Use: Fiber Optic Cables

Fiducial Points

—Automotive—Glossaries
BSI BS AU 173-79. 1979 Definitions for a Three-Dimensional Reference System and Fiducial Marks. 4 pp.
ISO 4130-78. Road Vehicles—Three-Dimensional Reference System and Fiducial Marks—Definitions First Edition. 4 pp.

Field Artillery
See Also: Artillery

—Scatterable Mines
NATO STANAG 2963 ED 1 AMD 0-92. Coordination of Field Artillery Delivered Scatterable Mines. 23 pp.

Field Buses
Use: Data Buses

Field Capacity
See Also: Moisture Content Analysis; Soil Analysis
CNS A3271-88. Method of Test for Field Vane Shear Test in Cohesive Soil (May)(12282).

Field Data Collection
Use: Data Collection

Field Effect Transistors
Use: FETs

Field Hospitals
See Also: Casualties; Hospitals; Medical Services

—Lighting—Surgical Equipment
NATO STANAG 2978 ED 1 AMD 1-83. Essential Performance Characteristics of Field Surgical Lights. 5 pp.

—Medical Electrical Equipment
NATO STANAG 2979 ED 1 AMD 2-83. Essential Characteristics of Electro-Surgical Apparatus. 6 pp.

Field Hospitals (Cont.)

—Military Operations—Camouflage
NATO STANAG 2931 ED 1 AMD 3-84. Orders for the Camouflage of the Red Cross and Red Crescent on Land in Tactical Operations. 10 pp.

Field Intensity
Use: Field Strength

Field of Vision
Use: Visual Field

Field Peas
Use: Peas

Field Strength
Use For: Field Intensity *See Also:* Electric Field Strength; Electromagnetic Fields; Magnetic Field Strength; Orientation; Radiation Patterns

—Aeronautical Mobile Services
CCIR RECMN 441-1-82. Signal-to-Interference Ratios and Minimum Field Strengths Required in the Aeronautical Mobile (R) Service Above 30 MHz—Section 8K—Aeronautical Mobile Service (Terrestrial). 1 p.

—Broadcasting
CCIR Report 228-3-86. Measurement of Field Strength for VHF (Metric) and UHF (Decimetric) Broadcast Services, Including Television—Section 5D—Aspects Relative to the Terrestrial Broadcasting and Mobile Services. 8 pp.
CCIR Report 1149-90. HF Field Strength Measurements Specifications for a Field Strength Measurement Campaign Intended for Future Improvements in Prediction Methods, Particularly Those Used for HF Broadcasting—Section 6G—Ionospheric Propagation Measurements. 17 pp.

—Diffraction—Ground Wave Propagation
CCIR RECMN 526-1-82. Propagation by Diffraction—Section 5B—Effects of the Ground (Including Ground-Wave Propagation). 1 p.
CCIR RECMN 526-2-92. Propagation by Diffraction. 16 pp.

—Frequency Curves—Ground Wave Propagation
CCIR RECMN 368-6-90. Ground-Wave Propagation Curves for Frequencies Between 10 kHz and 30 MHz—Section 5B—Effects of the Ground (Including Ground-Wave Propagation). 16 pp.
CCIR RECMN 368-7-92. Ground-Wave Propagation Curves for Frequencies Between 10 kHz and 30 MHz. 54 pp.

—Frequency Curves—Radio Wave Propagation—Broadcasting
CCIR RECMN 370-5-86. VHF and UHF Propagation Curves for the Frequency Range from 30 MHz to 1000 MHz. Broadcasting Services—Section 5D—Aspects Relative to the Terrestrial Broadcasting and Mobile Services. 28 pp.
CCIR Report 239-7-90. Propagation Statistics Required for Broadcasting Services Using the Frequency Range 30 to 1000 MHz—Section 5D—Aspects Relative to the Terrestrial Broadcasting and Mobile Services. 19 pp.

—Frequency Curves—Radio Wave Propagation—Mobile Radio Services
CCIR Report 567-4-90. Propagation Data and Prediction Methods for the Terrestrial Land Mobile Service Using the Frequency Range 30 MHz to 3 GHz—Section 5D—Aspects Relative to the Terrestrial Broadcasting and Mobile Services. 32 pp.

—Ground Wave Propagation
CCIR RECMN 527-2-90. Electrical Characteristics of the Surface of the Earth—Section 5B—Effects of the Ground (Including Ground-Wave Propagation). 3 pp.
CCIR RECMN 527-3-92. Electrical Characteristics of the Surface of the Earth. 5 pp.
CCIR RECMN 832-92. World Atlas of Ground Conductivities. 48 pp.

—Ionospheric Propagation
CCIR RECMN 684-90. Prediction of Field Strength at Frequencies Below About 500 khz—Section 6D—Ionospheric Propagation Prediction at Frequencies Below About 1.6 MHz. 1 p.
CCIR RESOLUTION 111-90. HF Field-Strength Measurement Campaign—Volume VI—Propagation in Ionized Media. 1 p.
CCIR OPINION 69-82. Field-Strength Measurements for Frequencies Below About 1.7 MHz—Volume VI—Propagation in Ionized Media. 1 p.
CCIR Report 895-2-90. Radio Propagation and Circuit Performance at Frequencies Below About 30 khz—Section 6D—Ionospheric Propagation Prediction at Frequencies Below About 1.6 MHz. 29 pp.

Field Strength (Cont.)

—Ionospheric Propagation (Cont.)
CCIR Report 1149-90. HF Field Strength Measurements Specifications for a Field Strength Measurement Campaign Intended for Future Improvements in Prediction Methods, Particularly Those Used for HF Broadcasting—Section 6G—Ionospheric Propagation Measurements. 17 pp.
CCIR Decision 84-89. Potential for an HF Field-Strength Measurement Campaign to Provide a Significant Measurement Data Bank in Time for Preparations for a WARC HFBC in 1993—Annex to Volume VI—Propagation in Ionized Media. 1 p.

—Land Mobile Services
CCIR Report 358-5-86. Protection Ratios and Minimum Field Strengths Required in the Mobile Services—Section 8A—Land Mobile Service and Related Subjects. 11 pp.

—Maritime Mobile Services
CCIR Report 358-5-86. Protection Ratios and Minimum Field Strengths Required in the Mobile Services—Section 8A—Land Mobile Service and Related Subjects. 11 pp.

—Mobile Radio Services
CCIR QUESTION 1-1/8-86. Signal-to-Interference Protection Ratios and Minimum Field Strengths Required in the Mobile Services—Questions Concerning Study Group 8—Mobile, Radiodetermination, Amateur and Related Satellite Services. 1 p.

—Radio Communications—HF
CCIR RECMN 845-92. HF Field-Strength Measurement. 29 pp.

—Radio Frequency Interference
DIN VDE 0877 Pt 2-85. Measurement of Radio Interferences; Measurement of Radio Interference Field Strength (Feb). 23 pp.

—Radio Frequency Interference—Motor Vehicles
DIN VDE 0879 Pt 1-79. Radio Interference Suppression of Vehicles, Vehicle Equipment and Internal Combustion Engines; Remote Interference Suppression of Vehicles and Units with Internal Combustion Engines (June). 29 pp.

—Radio Frequency Interference—Sound Broadcasting
CCIR Report 945-2-90. Methods for the Assessment of Multiple Interference—Section 10B—Frequency-Modulation Sound Broadcasting in Bands 8 (VHF) and 9 (UHF). 22 pp.

—Radio Frequency Interference—Television Services
CCIR RECMN 417-3-86. Minimum Field Strengths for Which Protection May Be Sought in Planning a Television Service—Section 11E—Planning of Television Networks, Protection Ratios Television Receivers and Antennas. 1 p.
CCIR RECMN 417-4-92. Minimum Field Strengths for Which Protection May Be Sought in Planning a Television Service—Section 11E—Planning of Television Networks, Protection Ratios Television Receivers and Antennas. 2 pp.

—Radio Spectra—Monitoring
CCIR RECMN 378-4-86. Field-Strength Measurements at Monitoring Stations and Expeditious Methods for Making These Measurements—Section 1C—Spectrum Monitoring Techniques. 2 pp.
CCIR RECMN 378-5-92. Field-Strength Measurements at Monitoring Stations—Section 1C—Spectrum Monitoring Techniques. 3 pp.
CCIR Report 272-5-86. Frequency Measurements at Monitoring Stations—Section 1C—Spectrum Monitoring Techniques. 6 pp.
CCIR Report 273-7-90. Field-Strength Measurements at Monitoring Stations—Section 1C—Spectrum Monitoring Techniques. 2 pp.
CCIR Report 668-3-90. Automatic Monitoring and Measurement in the Radio-Frequency Spectrum—Section 1C—Spectrum Monitoring Techniques. 17 pp.
CCIR QUESTION 24-1/1-82. Field-Strength Measurements at Monitoring Stations and Expeditious Methods for Making These Measurements—Questions Concerning Study Group 1—Spectrum Management Techniques (Spectrum Engineering, Planning, Sharing, Monitoring and Utilization). 1 p.

—Radio Wave Propagation
CCIR Report 227-3-82. General Methods of Measuring the Field Strength and Related Parameters (See Vol. V, Dubrovnik, 1986)—Section 5A—Texts of General Interest. 1 p.

INTERNATIONAL AND NON-U.S. NATIONAL STANDARDS
SUBJECT INDEX
Filament

Field Strength (Cont.)
—Sky Waves—Broadcasting
CCIR RECMN 435-7-92. Sky-Wave Field-Strength Prediction Method for the Broadcasting Service in the Frequency Range 150 to 1 600 kHz Section 6D—Ionospheric Propagation Prediction at Frequencies Below About 1.6 MHz. 25 pp.

—Sky Waves—Ionospheric Propagation
CCIR RECMN 435-6-90. Prediction of Sky-Wave Field Strength Between 150 and 1600 khz—Section 6D—Ionospheric Propagation Prediction at Frequencies Below About 1.6 MHz. 24 pp.
CCIR RECMN 683-90. Sky-Wave Field Strength Prediction Method for Propagation to Aircraft at About 500 khz—Section 6D—Ionospheric Propagation Prediction at Frequencies Below About 1.6 MHz. 7 pp.
CCIR RECMN 533-2-90. Estimating Sky-Wave Field Strength at Frequencies Between 2 and 30 MHz—Section 6E—Ionospheric Propagation Prediction at Frequencies Between About 1.6 and 30 MHz. 1 p.
CCIR RECMN 533-3-92. CCIR HF Propagation Prediction Method Section 6E—Ionospheric Propagation Prediction at Frequencies Between About 1.6 and 30 MHz. 15 pp.
CCIR Report 265-7-90. Sky-Wave Propagation and Circuit Performance at Frequencies Between About 30 khz and 500 khz—Section 6D—Ionospheric Propagation Prediction at Frequencies Below About 1.6 MHz. 18 pp.
CCIR Report 431-5-90. Analysis of Sky-Wave Propagation Measurements for the Frequency Range 150 to 1600 khz—Section 6D—Ionospheric Propagation Prediction at Frequencies Below About 1.6 MHz. 21 pp.
CCIR Report 432-2-86. Accuracy of Predictions of Sky-Wave Field Strength in Bands 5 (LF) and 6 (MF)—Section 6D—Ionospheric Propagation Prediction at Frequencies Below About 1.6 MHz. 6 pp.
CCIR Report 575-4-90. Methods for Predicting Sky-Wave Field Strengths at Frequencies Between 150 khz and 1705 khz—Section 6D—Ionospheric Propagation Prediction at Frequencies Below About 1.6 MHz. 18 pp.
CCIR Report 252-2-70. CCIR Interim Method for Estimating Sky-Wave Field Strength and Transmission Loss at Frequencies Between the Approximate Limits of 2 and 30 MHz—Section 6E—Ionospheric Propagation Prediction at Frequencies Between About 1.6 and 30 MHz. 1 p.
CCIR Report 252-2 Supplement-80. Second CCIR Computer-Based Interim Method for Estimating Sky-Wave Field Strength and Transmission Loss at Frequencies Between 2 and 30 MHz—Section 6E—Ionospheric Propagation Prediction at Frequencies Between About 1.6 and 30 MHz. 44 pp.
CCIR Report 252-2 Supp-82. Second CCIR Computer-Based Interim Method for Estimating Sky-Wave Field Strength and Transmission Loss Frequencies Between 2 and 30 MHz—Section 6E—Ionospheric Propagation Prediction at Frequencies Between About 1.6 and 30 MHz. 1 p.
CCIR Report 729-3-90. Developments in the Estimation of Sky-Wave Field Strength and Transmission Loss at Frequencies Above 1.5 MHz—Section 6E—Ionospheric Propagation Prediction at Frequencies Between About 1.6 and 30 MHz. 8 pp.
CCIR Decision 6-7-89. Sky-Wave Field Strength and Transmission Loss at Frequencies Above 1.6 MHz—Annex to Volume VI—Propagation in Ionized Media. 2 pp.
CCIR Decision 9-5-89. Radio Propagation and Circuit Performance at Frequencies Below About 500 khz—Annex to Volume VI—Propagation in Ionized Media. 2 pp.
CCIR Decision 57-2-89. Sky-Wave Propagation at Frequencies Between 150 and 1700 khz—Annex to Volume VI—Propagation in Ionized Media. 2 pp.

—Sky Waves—Radio Wave Propagation
CCIR QUESTION 31-2/6-90. Sky-Wave Field Strength and Circuit Performance at Frequencies Below About 1.7 MHz—Questions Concerning Study Group 6—Radio Wave Propagation in Ionized Media. 1 p.

—Sound Broadcasting
CCIR Report 516-4-90. Field Strength Resulting from Several Electromagnetic Fields—Section 10A-1—Amplitude-Modulation Sound Broadcasting in Bands 5 (LF), 6 (MF) and 7 (HF). 5 pp.
CCIR QUESTION 56/10-90. Minimum Usable Field Strength in LF, MF and HF Broadcasting—Questions Concerning Study Group 10—Broadcasting Service (Sound). 1 p.

—Sound Broadcasting—FM—VHF
CCIR RECMN 412-5-90. Planning Standards for FM Sound Broadcasting at VHF—Section 10B—Frequency-Modulation Sound Broadcasting in Bands 8 (VHF) and 9 (UHF). 6 pp.

Field Strength (Cont.)
—Sound Broadcasting—VHF
CCIR QUESTION 101/10-90. Minimum Usable Field Strength for Sound Broadcasting in Band 8 (VHF)—Questions Concerning Study Group 10—Broadcasting Service (Sound). 1 p.

—Sporadic E Propagation
CCIR RECMN 534-3-90. Method for Calculating Sporadic-E Field Strength—Section 6F—Ionospheric Propagation Prediction and Applications at Frequencies Above About 30 MHz. 24 pp.

Field Telephones
See Also: Telephones
NATO STANAG 4393 ED 1 AMD 0-00. (DRAFT) Military Characteristics for 2 Wire Analogue Ring-Down Field Telephone Sets. 8 pp.
NATO STANAG 5004 ED 2 AMD 4-77. Military Characteristics for Field Telephone Sets (Minimum Standard). 8 pp.

Field Transistors
Use: FETs

Fifes
See Also: Musical Instruments
JIS S 8512-84. Fifes.

Fifth Wheels
See Also: Hitches
SAA AS 1773-90. Articulated Vehicles—Fifth Wheel Assemblies (Supersedes AS 1772—1975).
SNZ NZS 5450-89. Coupling Devices for Articulated Vehicles—Fifth Wheel Assemblies. 24 pp.

—Coupling Pins
BSI BS AU 1B-77. (WITHDRAWN) 1977 Dimensions of 50 mm Fifth Wheel King Pin of Semi-Trailers (Superseded by BS AU 1C: 1989). 4 pp.
BSI BS AU 1C-89. 1989 Dimensions of '50' Fifth Wheel King Pin for Semi-Trailers. 4 pp.
BSI BS AU 2A-70. 1970 Dimensions of 3 1/2-4 1/2 Inch Diameter Fifth Wheel King Pin for Use with Extra Heavy Duty Semi-Trailers. 2 pp.
CNS D2081-86. Shape and Dimensions of Fifth Wheel Coupling Pins for Semi-Trailers (Aug)(7030).
ISO 337-81. Road Vehicles—50 Semi-Trailer Fifth Wheel Coupling Pin—Basic and Mounting/Interchangeability Dimensions Second Edition; (Corrigendum 1-1990). 5 pp.
ISO 4086-82. Road Vehicles—90 Semi-Trailer Fifth Wheel Coupling Pin—Basic and Mounting/Interchangeability Dimensions Second Edition. 4 pp.
JIS D 6602-85. Shapes and Dimensions of Fifth Wheel Coupling Pins for Semi-Trailers. 9 pp.

—Coupling Pins—Mounting
ISO 337-81. Road Vehicles—50 Semi-Trailer Fifth Wheel Coupling Pin—Basic and Mounting/Interchangeability Dimensions Second Edition; (Corrigendum 1-1990). 5 pp.

—Couplings—Mechanical Properties
BSI BS AU 235: Part 1-89. 1989 Fifth Wheel Couplings for Commercial Vehicles Part 1: Test Conditions and Strength Requirements. 6 pp.
BSI BS AU 235: Part 2-91. 1991 Fifth Wheel Couplings for Commercial Vehicles Part 2: Method of Testing Strength of Coupling Pins (ISO 8716: 1988). 6 pp.
ISO 8716-88. Commercial Road Vehicles—Fifth Wheel Coupling Pins—Strength Tests First Edition. 4 pp.
ISO 8717-88. Commercial Road Vehicles—Fifth Wheel Couplings—Strength Tests First Edition. 6 pp.

—Kingpins
NATO STANAG 4009 Pt1 ED1 AMD4-88. Fifth Wheel Attachments (Part I—NATO Kingpin). 6 pp.
SNZ NZS 5451-89. Coupling Devices for Articulated Vehicles—Fifth Wheel Kingpins. 12 pp.

—Kingpins—Semitrailers
NATO STANAG 4009 Pt2 ED1 AMD3-58. Fifth Wheel Attachments (Part II) Location and Height of Fifth Wheel Attachments on Tractors and Semi-Trailers, Class I. 7 pp.
NATO STANAG 4009 Pt3 ED1 AMD7-59. Fifth Wheel Attachments (Part III—Location and Height of Fifth Wheel Attachments on Tractors and Semi-Trailers, Class II). 18 pp.
NATO STANAG 4009 Pt4 ED1 AMD5-60. Fifth Wheel Kingpin (for Semi-Trailers of a Gross Weight from 22 to 72 Tonnes). 12 pp.

—Kingpins—Tractor Trucks
NATO STANAG 4009 Pt2 ED1 AMD3-58. Fifth Wheel Attachments (Part II) Location and Height of Fifth Wheel Attachments on Tractors and Semi-Trailers, Class I. 7 pp.

Fifth Wheels (Cont.)
—Kingpins—Tractor Trucks (Cont.)
NATO STANAG 4009 Pt3 ED1 AMD7-59. Fifth Wheel Attachments (Part III—Location and Height of Fifth Wheel Attachments on Tractors and Semi-Trailers, Class II). 18 pp.

—Mounting
ISO 3842-84. Road Vehicles—Fifth Wheel Mounting Second Edition. 4 pp.
JIS D 4011-84. Fifth Wheel Mounting for Semitrailer Tractors.
SAA AS 1771-87. Installation of Fifth Wheel and Turntable Assemblies. 11 pp.

—Positions
BSI BS AU 3-70. 1970 Clearance Dimensions and Fifth Wheel Positions for Tractors and Semi-Trailers. 6 pp.

Fighting Vehicles
Use: Combat Vehicles

Figure of Merit
See Also: Evaluation; Performance Measurement

—Broadcast Receivers—Broadcasting Satellite Services
CCIR RECMN 790-92. Characteristics of Receiving Equipment and Calculation of Receiver Figure-of-Merit (G/T) for the Broadcasting-Satellite Service. 3 pp.

Figured Glass
See Also: Glass; Plate Glass; Sheet Glass
CNS R2050-84. Figured Glasses (Aug) (2441).
JIS R 3203-81. Figured Glasses. 7 pp.

—Wired—Heat Absorbing
JIS R 3208-87. Heat Absorbing Glass. 8 pp.

Figures
Use: Diagrams

Filament Lamps
Use: Incandescent Lighting

Filament Yarns
See Also: Spun Yarns; Textured Yarns; Yarns
CNS L3013-67. Method of Test for Viscose Rayon Filament (Mar)(1225)(R 1973).
JIS L 1013-92. Testing Methods for Man-Made Filament Yarns.
JIS L 1013-69. Testing Method for Rayon Filament Yarn (R 1975). 23 pp.

—Crimping
DIN ENGL 53840 Pt 1-83. Testing of Textiles; Determination of Parameters for the Crimp of Textured Filament Yarns; Filament Yarns with a Linear Density of up to 500 dtex (Nov). 6 pp.
DIN ENGL 53840 Pt 2-83. Testing of Textiles; Determination of Parameters for the Crimp of Textured Filament Yarns; Filament Yarns with a Linear Density Exceeding 500 dtex (Nov). 5 pp.

—Glossaries
CGSB CAN/CGSB-4. 175-M91 Pt5. Textiles—Textured Filament Yarns—Vocabulary Part 5 (ISO 8160:1987). 9 pp.
ISO 8160-87. Textiles—Textured Filament Yarns—Vocabulary First Edition. 4 pp.

—Polyethylene Fibers—Tatami
JIS L 2502-78. Polyethylene Continuous Filament Yarns for TATAMI.

—Rayon
CNS L3145-82. Method of Test for Filament Rayon Twist Yarn (Jan)(8308).

—Rayon—Comprehensive Testing
CNS L3013-67. Method of Test for Viscose Rayon Filament (Mar)(1225)(R 1973).
JIS L 1013-69. Testing Method for Rayon Filament Yarn (R 1975). 23 pp.

—Synthetic—Stretch
JIS L 1090-92. Testing Methods for Synthetic Filament Stretch Yarns. 20 pp.

—Tensile Testing
CNS L3144-82. Method of Test for Tensile Strength of Filament Yarns (Jan) (8307).

—Textured—Glossaries
ISO 10132-93. Textiles—Textured Filament Yarn—Definitions First Edition. 7 pp.

—Vinylidene Chloride Resins
JIS L 1035-78. Testing Methods for Twisted Polyvinyliden Chloride Filament Yarns.

INDUSTRY STANDARDS

Filar Micrometer Eyepieces
—Microscopes
JIS B 7151-79. Filar Micrometer Eyepieces.
JIS B 7151-67. Filar Micrometer Eyepieces. 7 pp.
JIS B 7152-91. Biological Microscope Objectives and Eyepieces—Methods of Measurement of Performance. 24 pp.

File Folders
See Also: Stationery
BSI BS 1467-72. 1972 Amd 1 Folders and Files. 10 pp.
CGSB CAN/CGSB-53.30-M89. File Folders. 10 pp.
CGSB CAN/CGSB-53.112-M89. Folder, File, Suspension. 11 pp.
CNS S1187-85. Office Files (Flat Files) (Aug)(11344).
CNS S1188-85. Office Files (Folder and Guide) (Aug)(11345).
JIS S 5505-83. Office Files (Flat Files).
JIS S 5506-77. Office Files (Folder and Guide).

—Kraft Paper
CGSB 9-GP-32MA-87. Paper, Kraft, for File Folders. 8 pp.

—Pocket Style—Plastic
CGSB CAN/CGSB-53.145-M89. File Control and Records Storage Pocket. 10 pp.

—Sizes
ISO 623-74. Paper and Board—Folders and Files—Sizes First Edition. 3 pp.

—Wallet Style
CGSB CAN/CGSB-53.111-M89. Wallet Envelope. 8 pp.
CGSB CAN/CGSB-53.122-M89. Wallet Filing Jacket. 8 pp.

File Guides
Use: Guide Cards

File Transfer, Access and Management
Use: Application Layer (OSI)—File Transfer, Access, and Management

Files (Data)
—Holes
SAA AS P5-69. Punching Patterns for Round Holes Used in Files and Loose Leaf Binders Reconfirmed 1986. 4 pp.

Files (Tools)
Use For: Set Files *See Also:* Bone Files; Hand Tools; Knife Files; Rasps (Tools); Saw Files; Tools; Woodworking Equipment
BSI BS 498: Part 1-60. 1960 Files and Rasps Part 1: Rasps and Engineers' Files. 41 pp.
BSI BS 498-90. 1990 Rasps and Engineers' Files. 24 pp.
CGSB 39-GP-30B-75. Files and Rasps, Hand, American Pattern, Standard for. 29 pp.
CGSB 39-GP-53-62. Files; Hand, Swiss Pattern, Specification for. 40 pp.
CNS B3157-78. Files (General) (Mar)(1185). 2 pp.
CNS B7028-78. Method of Test for Files (Mar)(1214).
ISO 234 Pt 1-83. Files and Rasps—Part 1: Dimensions First Edition. 8 pp.
ISO 234 Pt 2-82. Files and Rasps—Part 2: Characteristics of Cut First Edition. 4 pp.
JIS B 4703-66. Files (R 1978). 8 pp.
JIS B 4704-64. Set Files (R 1972). 8 pp.
MOD UK DSTAN 51-11: Part 5-74. Hand Tools, General Purpose Part 5: File, Engineer's Issue 1. 19 pp.
MOD UK DSTAN 51-11: Part 6-76. Hand Tools, General Purpose Part 6: Files, Precision Issue 1. 14 pp.
MOD UK DEF-1098-A-63. Engineers' Files. 5 pp.

—Featheredge
CNS B3171-78. Feather-Edge File for Smith (Mar)(1199). 1 p.

—Flat
CNS B3159-78. Flat File for Carpenter with Wide Flat End (Mar)(1187). 1 p.
CNS B3160-78. Flat File for Carpenter with Sharp Flat End (Mar)(1188). 1 p.
CNS B3164-78. Flat File for Tinman (Mar)(1192). 1 p.
CNS B3166-78. Flat File for Smith with Wide Flat End (Mar)(1194). 1 p.
CNS B3167-78. Flat File for Smith with Sharp Flat End (Mar)(1195). 1 p.
CNS B3174-78. Special Flat File for Smith (Mar)(1202). 1 p.
CNS B3176-78. Heavy Flat File for Smith (Mar)(1204). 1 p.

—Half Round
CNS B3161-78. Half-Round File for Carpenter with Wide Flat End (Mar)(1189). 1 p.
CNS B3162-78. Half-Round File for Carpenter with Thin Sharp End (Mar)(1190). 1 p.
CNS B3165-78. Half-Round File for Tinman (Mar)(1193). 1 p.
CNS B3168-78. Half-Round File for Smith (Mar)(1196). 1 p.
CNS B3177-78. Heavy Half-Round File for Smith (Mar)(1205). 1 p.

—Handles
CNS B3111-54. Wooden Handle for File (Jul)(329).

—Rotary—Safety
CEN PREN 792-9-92. Handheld Non-Electric Power Tools—Safety Requirements—Part 9: Die Grinders. 9 pp.

—Round
CNS B3163-78. Round File for Carpenter (Mar)(1191). 1 p.
CNS B3169-78. Round File for Smith (Mar)(1197). 1 p.
CNS B3178-78. Heavy Round File for Smith (Mar)(1206). 1 p.

—Square
CNS B3170-78. Square File for Smith (Mar)(1198). 1 p.
CNS B3175-78. Special Square File for Smith (Mar)(1203). 1 p.
CNS B3179-78. Heavy Square File for Smith (Mar)(1207). 1 p.

—Teeth
CNS B3158-78. Density of File Teeth (Mar)(1186). 1 p.

—Triangular
CNS B3172-78. Triangular File for Smith (Mar)(1200). 1 p.
CNS B3180-78. Heavy Triangular File for Smith (Mar)(1208). 1 p.

Filiform Corrosion Testing
See Also: Corrosion Testing

—Steel—Paints and Varnishes
BSI BS 3900: Part F13-86. 1986 Methods of Test for Paints Group F: Durability Tests on Paint Films Part F13: Filiform Corrosion Test on Steel. 7 pp.
ISO 4623-84. Paints and Varnishes—Filiform Corrosion Test on Steel First Edition. 7 pp.

Filing Cabinets
See Also: Cabinets (Furniture); Card Cabinets; Furniture; Office Furniture
BSI BS 4438-69. 1969 Amd 1 Filing Cabinets and Suspended Filing Pockets. 13 pp.
CGSB CAN/CGSB-44.1-92. Vertical Filing Cabinets, Steel. 11 pp.
CNS S1074-84. Office Steel Filing Cabinets (Feb)(3118).
JIS S 1033-91. Office Furniture—Steel Filing Cabinets. 18 pp.

—Filing Pockets
BSI BS 4438-69. 1969 Amd 1 Filing Cabinets and Suspended Filing Pockets. 13 pp.

—Fire Resistant
CNS S1069-83. Steel Fire Resistive Safes (Jun)(2998).
JIS S 1037-89. Fire-Resistive Containers. 25 pp.

—Lateral—Steel
CGSB CAN/CGSB-44.2-M89. Lateral Filing Cabinets, Steel; (Amendment 1 September 1992). 16 pp.

—Metal—Index Cards
CGSB 44.16A-93. Metal Filing Cabinet for Index Cards. 9 pp.

Filing Systems
BSI BS 6478-84. 1984 Filing Bibliographic Information in Libraries and Documentation. 11 pp.
ISO 7154-83. Documentation—Bibliographic Filing Principles First Edition. 10 pp.
ISO TR8393-85. Documentation—ISO Bibliographic Filing Rules (International Standard Bibliographic Filing Rules)—Exemplification of Bibliographic Filing Principles in a Model Set of Rules First Edition. 26 pp.

—Alphanumeric
BSI BS 1749-85. 1985 Alphabetical Arrangement and the Filing Order of Numbers and Symbols. 16 pp.
CSA Z243.4.1-1992P. Canadian Alphanumeric Ordering Standard for Character Sets of CSA Standard CAN/CSA-Z243.4; (Gen Instr 1). 44 pp.

—Control Characters
ISO 6630-86. Documentation—Bibliographic Control Characters First Edition. 8 pp.
JTC1 6630-86. Documentation—Bibliographic Control Characters First Edition. 8 pp.
OSI ISO 6630-86. Documentation—Bibliographic Control Characters. 8 pp.

Fill Caps
Use: Filler Caps

Filler Caps
Use For: Fill Caps *See Also:* Caps (Lids); Filling Connectors; Filling Machines; Pressure Caps

—Aircraft
SBAC AS 1965 ISSUE 6. Filler Cap. 1 p.
SBAC AS 2398 ISSUE 5. Filler Cap Assembly. 1 p.
SBAC AS 2404 ISSUE 4. Washer (Filler Cap). 1 p.
SBAC AS 2511 ISSUE 2. Filler Cap. 1 p.
SBAC AS 2591 ISSUE 4. Filler Cap Base. 1 p.
SBAC AS 2593 ISSUE 5. Filler Cap. 1 p.
SBAC AS 3298 ISSUE 2. Filler Cap. 1 p.
SBAC AS 4644 ISSUE 2. Filler Cap. 1 p.
SBAC AS 6207 ISSUE 1. Filler Cap. 1 p.
SBAC AS 6264-6267 ISSUE 1. Filler Cap, Small Plastic. 1 p.
SBAC AS 6320 ISSUE 5. Filler Cap Base with Vent. 1 p.
SBAC AS 6321 ISSUE 3. Filler Cap Base Without Vent. 1 p.
SBAC AS 6349 ISSUE 1. Filler Ca
SBAC AS 6358 ISSUE 1. Filler Ca
SBAC AS 6744 ISSUE 2. Filler Cap Base with Vent.
SBAC AS 6745 ISSUE 1. Filler Cap Base Without Vent.
SBAC AS 6747 ISSUE 1. Filler Ca

—Aircraft—Air Valves
SBAC AS 3167 ISSUE 2. Assembly of Filler Cap—Air Valve Type. 1 p.
SBAC AS 3168 ISSUE 1. Air Valve for Filler Cap.AS.1964. 1 p.

—Aircraft—Aluminum Alloy
SBAC AS 2400 (V). Filler Cap. 1 p.

—Aircraft—Arms
SBAC AS 2402 (V). Arm Filler Cap. 1 p.

—Aircraft—Arms—Stampings
SBAC AS 2512 (V). Stamping for Arm (Filler Cap). 2 pp.

—Aircraft—Assembly
SBAC AS 2510 ISSUE 5. Filler Cap Assembly 2 Inch Diameter. 1 p.
SBAC AS 4641 ISSUE 1(I). Assembly of Filler Cap Mk.II with Vent. 1 p.
SBAC AS 4642 ISSUE 1(I). Assembly of Filler Cap Mk.II—Without Vent. 11 pp.
SBAC AS 6318 ISSUE 2(I). Assembly of Filler Cap with Vent. 1 p.
SBAC AS 6319 ISSUE 2(I). Assembly of Filler Cap Without Vent. 1 p.
SBAC AS 6348 ISSUE 1. Filler Cap Assembly.
SBAC AS 6357 ISSUE 1. Filler Cap Assembly (2 Inch Diameter).
SBAC AS 6541 ISSUE 2(I). Assembly of Filler Cap Mk.II with Vent.
SBAC AS 6542 ISSUE 2(I). Assembly of Filler Cap Mk.II—Without Vent.
SBAC AS 6742 ISSUE 2. Assembly of Filler Cap Mk.II with Vent.
SBAC AS 6743 ISSUE 2. Assembly of Filler Cap Mk.II Without Vent.

—Aircraft—Axis Pins
SBAC AS 3037 ISSUE 4. Axis Pin for Catch (Flush Filler Cap). 1 p.
SBAC AS 6749 ISSUE 1. Axis Pin for Catch, (Flush Filler Cap).

—Aircraft—Captive Nuts
SBAC AS 6352 ISSUE 2. Captive, Nut (Filler Cap).

—Aircraft—Clips
SBAC AS 2813 ISSUE 2. Cable Clip (Filler Cap). 1 p.

—Aircraft—Dipstick Assemblies
SBAC AS 3104-3105 ISSUE 1. Assembly of Cap and Dip Stick. 1 p.

—Aircraft—Fuel
NATO STANAG 3294 ED 3 AMD 5-73. Aircraft Fuel Caps and Fuel Cap Access Covers. 8 pp.

—Aircraft—Grub Screws
SBAC AS 6750 ISSUE 2. Grub Screw.

—Aircraft—Handles
SBAC AS 2409 ISSUE 7. Knob (Filler Cap). 1 p.

INTERNATIONAL AND NON-U.S. NATIONAL STANDARDS
SUBJECT INDEX

Filler Caps (Cont.)

—Aircraft—Levers
SBAC AS 2594 ISSUE 7. Filler Cap Locking Lever. 1 p.
SBAC AS 6748 ISSUE 2. Filler Cap Lever.

—Aircraft—Lids
SBAC AS 4643 ISSUE 2(I). Lid Assembly—Flush Filler Cap Mk.II. 1 p.
SBAC AS 6746 ISSUE 1. Lid Assembly—Flush Filler Cap Mk.II..

—Aircraft—Nuts (Fasteners)
SBAC AS 2408 ISSUE 4. Captive Nut (Filler Cap). 1 p.

—Aircraft—Pivot Screws
SBAC AS 6350 ISSUE 1. Pivot, Screw (Filler Cap).

—Aircraft—Plates (Supports)
SBAC AS 2410 ISSUE 4. Retaining Plate (Filler Cap). 1 p.

—Aircraft—Reinforcing Rings—Split
SBAC AS 4521 ISSUE 1. Split Reinforcing, Ring Used on Filler Cap AS 2398. 1 p.
SBAC AS 4522 ISSUE 1. Split Reinforcing, Ring Used on Filler Cap AS 2510. 1 p.

—Aircraft—Retaining Cables
SBAC AS 4649 ISSUE 2. Retaining Cable (Filler Cap). 1 p.

—Aircraft—Screws
SBAC AS 2406 ISSUE 3. Pivot Screw (Filler Cap). 1 p.
SBAC AS 2407 ISSUE 3. Screw (Filler Cap). 1 p.
SBAC AS 6351 ISSUE 1. Screw (Filler Cap).

—Aircraft—Sealing Rings
SBAC AS 5364 ISSUE 6. Filler Sealing Ring. 1 p.

—Aircraft—Seatings
SBAC AS 2399 (V). Seating—Filler Cap. 1 p.
SBAC AS 2405 ISSUE 4. Washer—Seating (Filler Cap). 1 p.
SBAC AS 2513 (V). Seating for Filler Cap. 2 pp.
SBAC AS 6354 ISSUE 1. Seating for Filler Ca

—Aircraft—Seatings—Inserts
SBAC AS 2514 ISSUE 3. Insert for Seating (Filler Cap). 1 p.
SBAC AS 6359 ISSUE 1. Insert for Seating, (Filler Cap).

—Aircraft—Sub-Assemblies
SBAC AS 2592 ISSUE 8. Filler Cap—(Sub-Assembly). 1 p.

—Aircraft—Washers (Fasteners)
SBAC AS 2405 ISSUE 4. Washer—Seating (Filler Cap). 1 p.
SBAC AS 2517 ISSUE 2. Washer for 2 Inch Filler Cap. 1 p.
SBAC AS 3031-3036 ISSUE 1. Packing Washer (Flush Filler Cap). 1 p.

—Automotive—Fuel
CNS D2043-89. Shape & Dimension of Fuel Filler Necks and Covers for Automobiles (Jul)(6159).
JIS D 2501-60. Shape and Dimension of Fuel Filler Necks and Covers for Automobiles.

Filler Connectors
Use: Filling Connectors

Filler Joists
Use: Joists—Filler

Filler Metal
See Also: Brazing; Brazing Alloys; Coated Electrodes; Metals; Soldering; Weld Metal; Weldability; Welded Joints; Welded Pipes; Welding; Welding Electrodes; Welding Rods; Weldments
DIN ENGL 8571-81. Filler Metals and Auxiliary Metals for Metal Welding; Concepts; Classification (July). 5 pp.

—Braze Welding
SAA AS 1167.1-93. Welding and Brazing—Filler Metals—Part 1: Filler Metal for Brazing and Braze Welding (in Professional Package 36). 7 pp.

—Cracking (Fracturing)
DIN ENGL 50129-73. Testing of Metallic Materials; Testing of Welding Filler Metals for Liability to Cracking (Oct). 4 pp.

—Fusion Welding—Aluminum
DIN ENGL 1732 Pt 1-88. Filler Metals for Welding Aluminium and Aluminium Alloys; Composition, Application and Technical Delivery Conditions (June). 8 pp.

Filler Metal (Cont.)

—Fusion Welding—Aluminum Alloys
DIN ENGL 1732 Pt 1-88. Filler Metals for Welding Aluminium and Aluminium Alloys; Composition, Application and Technical Delivery Conditions (June). 8 pp.

—Fusion Welding—Aluminum Alloys—Test Specimens
DIN ENGL 1732 Pt 3-88. Filler Metals for Welding Aluminium and Aluminium Alloys; Sample and Specimen Preparation, Mechanical Properties of All-Weld Metal (June). 2 pp.

—Fusion Welding—Aluminum—Test Specimens
DIN ENGL 1732 Pt 3-88. Filler Metals for Welding Aluminium and Aluminium Alloys; Sample and Specimen Preparation, Mechanical Properties of All-Weld Metal (June). 2 pp.

—Fusion Welding—Copper
DIN ENGL 1733 Pt 1-88. Filler Metals for Welding Copper and Copper Alloys; Composition, Application and Technical Delivery Conditions (June). 8 pp.

—Fusion Welding—Copper Alloys
DIN ENGL 1733 Pt 1-88. Filler Metals for Welding Copper and Copper Alloys; Composition, Application and Technical Delivery Conditions (June). 8 pp.

—Fusion Welding—Titanium Alloys—Test Specimens
DIN ENGL 1737 Pt 2-88. Filler Metals for Welding Titanium and Titanium-Palladium Alloys; Sample and Specimen Preparation, Mechanical Properties of All-Weld Metal (June). 2 pp.

—Fusion Welding—Titanium—Test Specimens
DIN ENGL 1737 Pt 2-88. Filler Metals for Welding Titanium and Titanium-Palladium Alloys; Sample and Specimen Preparation, Mechanical Properties of All-Weld Metal (June). 2 pp.

—Gas Shielded Arc Welding
CSA W48.4-M1980. Solid Mild Steel Filler Metals for Gas Shielded Arc Welding; (Gen Instr 1 Thru 2). 27 pp.
DIN ENGL 1737 Pt 1-84. Filler Metals for Welding Titanium and Titanium-Palladium Alloys; Chemical Composition, Technical Delivery Conditions (June). 3 pp.
DIN ENGL 8556 Pt 1-86. Filler Metals for Welding Stainless and Heat-Resisting Steels; Designation, Technical Delivery Conditions (May). 10 pp.
DIN ENGL 8559 Pt 1-84. Filler Metals for Gas-Shielded Arc Welding; Wire Electrodes, Filler Wires, Solid Rods and Solid Wires for Gas-Shielded Arc Welding of Unalloyed and Alloyed Steels (July). 8 pp.
DIN ENGL 8575 Pt 1-84. Filler Metals for Arc Welding of Creep-Resisting Steels; Classification, Designation, Technical Delivery Conditions (Apr). 11 pp.

—Gas Welding
BSI BS 1453-72. 1972 Amd 1 Filler Materials for Gas Welding. 25 pp.
BSI BS EN 20544-91. 1991 Sizes for Filler Metals for Manual Welding (F) (ISO 544: 1989). 9 pp.
CEN PREN 20 544-90. Filler Materials for Manual Welding—Size Requirements. 3 pp.
CEN EN 20544-91. Filler Materials for Manual Welding—Size Requirements. 3 pp.
DIN ENGL EN 20544-91. Filler Materials for Manual Welding; Size Requirements (ISO 544: 1989) (Dec). 4 pp.
ISO 544-89. Filler Materials for Manual Welding—Size Requirements Second Edition; (Supersedes 545, 546 and 547). 4 pp.

—Mechanical Properties
DIN ENGL 1732 Pt 3-88. Filler Metals for Welding Aluminium and Aluminium Alloys; Sample and Specimen Preparation, Mechanical Properties of All-Weld Metal (June). 2 pp.
DIN ENGL 1736 Pt 2-85. Welding Filler Metals for Nickel and Nickel Alloys; Sample, Test Pieces, Mechanical Properties (Aug). 4 pp.
DIN ENGL 1737 Pt 2-88. Filler Metals for Welding Titanium and Titanium-Palladium Alloys; Sample and Specimen Preparation, Mechanical Properties of All-Weld Metal (June). 2 pp.

—Nickel
DIN ENGL 1736 Pt 1-85. Welding Filler Metals for Nickel and Nickel Alloys; Composition, Application and Technical Delivery Conditions (Aug). 9 pp.

Filler Metal (Cont.)

—Nickel Alloys
DIN ENGL 1736 Pt 1-85. Welding Filler Metals for Nickel and Nickel Alloys; Composition, Application and Technical Delivery Conditions (Aug). 9 pp.

—Shear Testing
ISO 5187-85. Welding and Allied Processes—Assemblies Made with Soft Solders and Brazing Filler Metals—Mechanical Test Methods First Edition. 13 pp.

—Submerged Arc Welding
DIN ENGL 8557 Pt 1-81. Filler Metals for Submerged Arc Welding; Joint Welding of Unalloyed and Alloy Steels; Designation; Technical Delivery Conditions (Apr). 5 pp.

—Symbols
ISO 3677-92. Filler Metals for Soft Soldering, Brazing and Braze Welding—Designation Second Edition. 5 pp.

—Tensile Testing
ISO 5187-85. Welding and Allied Processes—Assemblies Made with Soft Solders and Brazing Filler Metals—Mechanical Test Methods First Edition. 13 pp.

—Titanium
DIN ENGL 1737 Pt 1-84. Filler Metals for Welding Titanium and Titanium-Palladium Alloys; Chemical Composition, Technical Delivery Conditions (June). 3 pp.

—Welding
CEN PREN 759-92. Technical Delivery Conditions for Steel Welding Filler Metals—Including Type of Product, Dimensions, Tolerances and Marking. 11 pp.
SAA AS 1167.2-86. Welding and Brazing—Filler Metals—Part 2: Filler Metal for Welding. 12 pp.

Filler Necks
Scope Note: For specific types of filler necks, see the specific equipment

—Aircraft
SBAC AS 2481 ISSUE 2. Filler Neck. 1 p.

Filler Rods
Use: Welding Rods

Filler Wire
Use: Welding Electrodes

Fillers
Scope Note: See the subheading Fillers under the specific product See Also: Furnish; Joint Fillers; Plastic Wood

Fillers (Papermaking)
See Also: Papermaking Equipment; Sizing (Surface Treatment)
CPPA J.18P-90. Analysis of Fillers and Pigments. 3 pp.

—Chemical Properties
CPPA J.2U-77. Physical & Chemical Properties of Mineral Fillers & Pigments. 1 p.

—Colorimetry
CPPA J.3U-77. Colour of Mineral Fillers and Pigments. 1 p.

—Opacity
CPPA J.4U-77. Opacifying Power of Mineral Fillers and Pigments. 2 pp.

—Physical Properties
CPPA J.2U-77. Physical & Chemical Properties of Mineral Fillers & Pigments. 1 p.

Fillet Bars (Printing)
Use: Printing Presses—Fillet Bars

Fillet Welds
See Also: Joints; Welded Joints; Welding
ISO R617-67. Calculation of Rectangular Symmetrical Fillet Welds Statically Loaded in Such a Way That the Transverse Section is Not Under Any Normal Stress (Parallel Stress is Equal to 0) First Edition. 6 pp.

—Bend Testing
JIS Z 3134-65. Method of Bend Test for T Type Fillet Welded Joint (R 1980). 6 pp.
JIS Z 3135-71. Method of Soundness Test for Fillet Welds. 5 pp.

—Break Testing
JIS Z 3133-70. Method of Break Test for Fillet Welds (R 1979). 5 pp.

Fillet Welds (Cont.)

—Cracking (Fracturing)
JIS Z 3153-93. Method of T-Joint Weld Cracking Test. 6 pp.

—Electrodes
JIS Z 3181-73. Method of Test for Fillet Weld of Covered Electrode. 4 pp.

—Shear Testing
JIS Z 3132-76. Method of Shear Test for Side Fillet Welded Joint. 6 pp.

—Test Specimens
DIN ENGL 50127-76. Testing of Steel; Bendover Test Specimen, Wedge Test Specimen, Notched Pipe Tension Test Specimen for Assessment of Fusion-Welded Fillet or Butt Joints (Aug). 4 pp.

Fillets
See Also: Welding
CNS B3025-47. Fillet and Chamfer (Mar)(149)(R1973).

Filling Compounds
See Also: Electrical Insulation
MOD UK DSTAN 68-146-91. Insulating and Filling Compound (Chatterton's Type) for Electrical Purposes Issue 1. 11 pp.
MOD UK DEF-144-63. Insulating & Filling Compound (Chatterton's Type) for Electrical Purposes (Superseded by Def Stan 68-146). 3 pp.

—Bituminous—Electrical Equipment
BSI BS 1858-73. 1973 Amd 1 Bitumen Based Filing Compounds for Electrical Purposes. 24 pp.

—Cables (Electric)
BSI BS 6469: Sec 5.1-92. 1992 Insulating and Sheathing Materials of Electric Cables Part 5: Methods of Test Specific to Filling Compounds Section 5.1: Drop-Point—Separation of Oil—Lower Temperature Brittleness—Total Acid Number—Absence of Corrosive Components. 24 pp.
CENELEC HD 505.5.1 S1-92. Common Test Methods for Insulating and Sheathing Materials of Electric Cables Part 5: Methods Specific to Filling Compounds Section One: Drop-Point—Separation of Oil Lower Temperature Brittleness—Total Acid Number Absence of Corrosive Components—Permittivity at 23. 6 pp.
IEC 811 Pt 5-1-90. Common Test Methods for Insulating and Sheathing Materials of Electric Cables Part 5: Methods Specific to Filling Compounds Section One—Drop-Point—Separation of Oil—Lower Temperature Brittleness—Total Acid Number—Absence of Corrosive Components-. 36 pp.

—Cables (Electric)—Accessories
DIN VDE 0291 Pt 1-72. Specification for Filling Compounds for Cable Accessories and for Boiling Out Compounds; Filling Compounds to Be Run in Hot, Cold Pressed Compounds, Cold-Cast Compounds and Boiling out Compounds (Feb). 24 pp.
DIN VDE 0291 Pt 2-79. Compounds for Use in Cable Accessories; Casting Resinous Compounds Before Cure and in the Cured State (Nov). 45 pp.

—Naval Ships
MOD UK NES 2060-89. Specification for Filling Compound NSN 8030-99-224-7760 for Rubber Filled Multiple Glands Issue 1 (06.89). 6 pp.
MOD UK NES 2060-92. Specification for Filling Compound NSN 8030-99-224-7760 for Rubber Filled Multiple Glands Issue 2 (11.92). 8 pp.
MOD UK NES 2074-01. Specification for Epoxy Syntactic Foam Issue 2 (09.92); Amendment 1. 23 pp.
MOD UK TS 10272. Epoxide System for Fibre-Reinforced Composites.
MOD UK NES 789-88. Requirements for the Application of Structural Fairing and Filling Materials Issue 2 (05.88). 19 pp.

—Pipes—Bending
MOD UK TS 10142. Composition, Filling, for Pipe Bending (Withdrawn).

—Submarines
MOD UK NES 2060-89. Specification for Filling Compound NSN 8030-99-224-7760 for Rubber Filled Multiple Glands Issue 1 (06.89). 6 pp.
MOD UK NES 2060-92. Specification for Filling Compound NSN 8030-99-224-7760 for Rubber Filled Multiple Glands Issue 2 (11.92). 8 pp.
MOD UK NES 2074-01. Specification for Epoxy Syntactic Foam Issue 2 (09.92); Amendment 1. 23 pp.
MOD UK TS 10272. Epoxide System for Fibre-Reinforced Composites.
MOD UK NES 789-88. Requirements for the Application of Structural Fairing and Filling Materials Issue 2 (05.88). 19 pp.

Filling Connectors
See Also: Couplings; Filler Caps; Pipe Couplings; Tube Couplings

—Aircraft—Fuel Dispensing Equipment
ISO 45-90. Aircraft—Pressure Refuelling Connections Second Edition. 7 pp.

—Water Storage Tanks—Ships
BSI BS MA 85-79. 1979 Shipbuilding: Filling Connection for Drinking Water Tanks on Ships. 6 pp.
ISO 5620 Pt 1-92. Shipbuilding and Marine Structures—Filling Connection for Drinking Water Tanks—Part 1: General Requirements First Edition. 7 pp.
ISO 5620 Pt 2-92. Shipbuilding and Marine Structures—Filling Connection for Drinking Water Tanks—Part 2: Components First Edition. 9 pp.

Filling Machines
See Also: Filler Caps; Packaging Machines
JIS B 7604-87. Automatic Gravimetric Filling Machines. 12 pp.

Filling Stations
Use: Service Stations

Fillings (Upholstery)
Use: Paddings

Fillister Head Screws
See Also: Screws

—Phillips
CNS B2293-81. Cross Recessed Raised Fillister Head Screws (Apr)(4559). 4 pp.

Film Adhesives
See Also: Adhesives

—Glass
JIS A 5759-88. Adhesive Films for Glazings. 19 pp.

—Labels
JIS Z 1529-91. Pressure Sensitive Adhesive Films for Printing. 11 pp.

—Structural—Aerospace—Water Absorption
CEN PREN 2667 (Part 6)-89. Test Method for the Determination of Water Absorption of Structural Foam Film Adhesives. 8 pp.

Film and Foil Capacitors
Use: Film Capacitors

Film Badges
Use: Photographic Dosimeters

Film Capacitors
Use For: Film and Foil Capacitors; Paper-Plastic Film Capacitors; Plastic Film Capacitors
See Also: Capacitors; Fixed Capacitors; Metallized Film Capacitors; Variable Capacitors
BSI BS 2131-65. 1965 Fixed Capacitors for Direct Current Using Impregnated Paper or Paper/Plastics Film Dielectric. 24 pp.
CNS C5022-79. General Rules of Paper and Plastic Film Capacitors for Electronic Equipment (Jan)(4706).
CNS C7026-73. Fixed Plastic Film Capacitors for Direct Current (Nov)(3664). 7 pp.
CNS C7064-85. AC Mains Supply Plastic Film Capacitors (Jan)(4592).
CNS C7072-90. Plastic Film Capacitors (Characteristic M) for Direct Current (Oct)(4708).
CNS C7072-79. Plastic Film Capacitors (Characteristic M) for Direct Current (Jan)(4708). 21 pp.
CNS C7073-79. Plastic Film Capacitors (Characteristic S) for Direct Current (Jan)(4709). 18 pp.
IEC 80-64. Fixed Capacitors for Direct Current Using Impregnated Paper or Paper/Plastic Film Dielectric Second Edition. 43 pp.
JIS C 4901-84. Low-Voltage Power Capacitors. 31 pp.
JIS C 5113-91. Fixed Polyester Film Capacitors for Use in Electronic Equipment. 28 pp.
JIS C 5114-91. Fixed Polystyrene Film Capacitors for Use in Electronic Equipment. 64 pp.
JIS C 5152-80. AC Mains Supply Plastic Film Capacitors (R 1985). 16 pp.
MOD UK DSTAN 59-44: Pt 4:Sec 2-01. Capacitors, Fixed, of Assessed Quality (Listed on EPIC Database) Part 4: Capacitors, Fixed, Polystyrene Dielectric Section 2: List of Items Conforming to BS 9074 N007 Issue 1; Corrigendum. 18 pp.
SNZ NZS 1601-66. Specification for Fixed Capacitors for Direct Current Using Impregnated Paper or Paper/Plastics Film Dielectric. 24 pp.

—Electric Equipment
JIS C 4908-88. Capacitors for Electrical Apparatus. 26 pp.

Film Capacitors (Cont.)

—Electronic Equipment
JIS C 5111-88. General Rules of Fixed Paper and Plastic Film Capacitors for Use in Electronic Equipment. 47 pp.
JIS C 5150-80. General Rules of AC Mains Supply Capacitors for Electronic Equipment (R 1985). 19 pp.

—Polycarbonate/Foil
IEC 384 Pt 12-88. Fixed Capacitors for Use in Electronic Equipment Part 12: Sectional Specification: Fixed Polycarbonate Film Dielectric Metal Foil D.C. Capacitors Second Edition (IECQ QC 301700). 59 pp.

—Polycarbonate/Foil—Cylindrical—Preferred Products List
CECC CECC MUAHAG Vol 1 IS 4-90. Preferred Products List; Capacitors (En, Fr, Ge). 64 pp.

—Polycarbonate/Foil—Rectangular—Preferred Products List
CECC CECC MUAHAG Vol 1 IS 4-90. Preferred Products List; Capacitors (En, Fr, Ge). 64 pp.

—Polycarbonate—Quality Assurance
BSI BS 9070: Sec 6-05. 1971 Amd 5 Fixed Capacitors of Assessed Quality: Generic Data and Methods of Test Section 6: Polycarbonate Dielectric Capacitors and Polyethylene Terephthalate Dielectric Capacitors for d.c. Use and Polypropylene Dielectric Capacitors. 17 pp.
BSI BS E9076: Part 1-76. 1976 Amd 1 Sectional Specification for Polycarbonate Dielectric Capacitors and Polyethylene Terephthalate Dielectric Capacitors for d.c. Use Part 1: Polyester Film Capacitors with Metal Foil Electrodes. 30 pp.

—Polycarbonate—Reliability Assured
MOD UK DSTAN 59-44: Pt 6:Sec 2-78. Capacitors, Fixed, of Assessed Quality (Listed on EPIC Database) Part 6: Capacitors, Fixed, Polycarbonate and Polyethylene Terephthalate Dielectrics Section 2: List of Items Conforming to BS 9076 F008 Issue 1. 19 pp.

—Polyethylene/Foil—Cylindrical—Preferred Products List
CECC CECC MUAHAG Vol 1 IS 4-90. Preferred Products List; Capacitors (En, Fr, Ge). 64 pp.

—Polyethylene/Foil—Rectangular—Preferred Products List
CECC CECC MUAHAG Vol 1 IS 4-90. Preferred Products List; Capacitors (En, Fr, Ge). 64 pp.

—Polyethylene Terephthalate/Foil
CECC CECC 30 100 ISSUE 2-85. Sectional Specification: Fixed Polyethylene-Terephthalate Film Dielectric Metal Foil Capacitors for Direct Current (En, Fr, Ge) AMD 1 (En, Fr, Ge) AMD 2 (En, Fr, Ge) AMD 3 (En, Fr, Ge) AMD 4 (En, Fr, Ge) AMD 6 and 7 (En, Fr, Ge). 106 pp.
CECC CECC 30 101 ISSUE 2-85. Blank Detail Specification: Fixed Polyethylene-Terephthalate Film Dielectric Metal Foil Capacitors for Direct Current (En, Fr, Ge) AMD 1 (En, Fr, Ge) AMD 2 (En, Fr, Ge). 45 pp.
IEC 384 Pt 2-82. Fixed Capacitors for Use in Electronic Equipment Part 2: Blank Detail Specification: Fixed Metallized Polyethylene-Terephthalate Film Dielectric D.C. Capacitors Assessment Level E Second Edition (Amendment 1-1987) (Amendment 2-1992) (IECQ QC 300400). 66 pp.
IEC 384 Pt 11-88. Fixed Capacitors for Use in Electronic Equipment Part 11: Sectional Specification: Fixed Polyethylene-Terephthalate Film Dielectric Metal Foil D.C. Capacitors Second Edition (IECQ QC 300100). 59 pp.
IEC 384 Pt 11-1-88. Fixed Capacitors for Use in Electronic Equipment Part 11: Blank Detail Specification: Fixed Polyethylene-Terephthalate Film Dielectric Metal Foil D.C. Capacitors Assessment Level E First Edition (IECQ QC 300101). 31 pp.
IECQ QC 300100-88. Fixed Capacitors for Use in Electronic Equipment Part 11: Sectional Specification: Fixed Polyethylene-Terephthalate Film Dielectric Metal Foil d.c. Capacitors (IEC 384-11 ED 2). 59 pp.
IECQ QC 300101-88. Fixed Capacitors for Use in Electronic Equipment Part 11: Blank Detail Specification: Fixed Polyethylene-Terephthalate Film Dielectric Metal Foil d.c. Capacitors Assessment Level E (IEC 384-11-1 ED 1). 31 pp.
IECQ QC 300101/CN 0001-89. Detail Specification for Electronic Components Fixed Polyethylene-Terephthalate Film Dielectric Metal Foil d.c. Capacitors, Type CL12 Assessment Level E. 12 pp.

Film Capacitors (Cont.)

—Polyethylene Terephthalate/Foil (Cont.)

IECQ QC 300101/CN 0002-92. Detail Specification for Electronic Components Fixed Polyethylene-Terephthalate Film Dielectric Metal Foil d.c. Capacitors, Type CL 11 Assessment Level E. 12 pp.

—Polyethylene Terephthalate/Foil—Cylindrical

CECC CECC 30 101-002. BS CECC 30 101-002 Issue 1; Polyethylene Terephthalte Dielectric (En) 10 pp.

—Polyethylene Terephthalate/Foil—Preferred Products List

CECC CECC MUAHAG Vol 1 IS 4-90. Preferred Products List; Capacitors (En, Fr, Ge). 64 pp.

—Polyethylene Terephthalate/Foil—Rectangular

CECC CECC 30 101-007. BS CECC 30 101-007 Issue 1; Terephthalate Film Dielectric Metal Foil d.c. Capacitor (En). 13 pp.

—Polyethylene Terephthalate/Foil—Rectangular—Printed Circuit Mount

CECC CECC 30 101-005. DIN 45 910 Teil 252; Polyethylene Terephthalate Film Dielectric Metal Foil Capacitors DC 63 to 400V, General-Purpose Grade, Rectangular Shape, Insulated, Radial Terminations for Printed Circuits (Ge). 8 pp.

CECC CECC 30 101-801 ISSUE 1-90. Detail Specification: Fixed Polyethylene-Terephthalate Film Dielectric Metal Foil Capacitors for Direct Current (En, Fr, Ge). 16 pp.

—Polyethylene Terephthalate—Quality Assurance

BSI BS 9070: Sec 6-05. 1971 Amd 5 Fixed Capacitors of Assessed Quality: Generic Data and Methods of Test Section 6: Polycarbonate Dielectric Capacitors and Polyethylene Terephthalate Dielectric Capacitors for d.c. Use and Polypropylene Dielectric Capacitors. 17 pp.

BSI BS E9076: Part 1-76. 1976 Amd 1 Sectional Specification for Polycarbonate Dielectric Capacitors and Polyethylene Terephthalate Dielectric Capacitors for d.c. Use Part 1: Polyester Film Capacitors with Metal Foil Electrodes. 30 pp.

—Polyethylene Terephthalate—Reliability Assured

MOD UK DSTAN 59-44: Pt 6:Sec 1-78. Capacitors, Fixed, of Assessed Quality (Listed on EPIC Database) Part 6: Capacitors, Fixed, Polycarbonate and Polyethylene Terephthalate Dielectrics Section 1: Sectional Requirements and Index to Sections Issue 1. 6 pp.

—Polypropylene/Foil

BSI BS QC 301701-92. 1992 Fixed Capacitors for Use in Electronic Equipment Blank Detail Specification Fixed Polycarbonate Film Dielectric Metal Foil d.c. Capacitors Assessment Level E (IEC 384-12-1: 1988) (T). 14 pp.

BSI BS QC 301800-91. 1991 Fixed Capacitors for Use in Electronic Equipment. Sectional Specification for Fixed Polypropylene Film Dielectric Metal Foil d.c. Capacitors (IEC 384-13: 1991). 33 pp.

BSI BS QC 301801-93. 1993 Fixed Capacitors for Use in Electronic Equipment Blank Detail Specification Fixed Polypropylene Film Dielectric Metal Foil d.c. Capacitors Assessment Level E (IEC 384-13-1: 1991). 14 pp.

CECC CECC 31 800 ISSUE 1-85. Sectional Specification: Fixed Polypropylene Film Dielectric Metal Foil d.c. Capacitors (En, Fr, Ge) AMD 1 (En, Fr, Ge) AMD 3 (En, Fr, Ge) AMD 4 (En, Fr, Ge) AMD 5 (En, Fr, Ge) AMD 6 (En, Fr, Ge) AMD 7 (En, Fr, Ge) AMD 8 & 9 (En, Fr, Ge). 111 pp.

CECC CECC 31 801 ISSUE 1-86. Blank Detail Specification: Fixed Polypropylene Film Dielectric Metal Foil d.c. Capacitors (En, Fr, Ge) AMD 1 (En, Fr, Ge) AMD 2 (En, Fr, Ge) AMD 3 (En, Fr, Ge). 45 pp.

IEC 384 Pt 13-1-91. Fixed Capacitors for Use in Electronic Equipment Part 13: Blank Detail Specification: Fixed Polypropylene Film Dielectric Metal Foil d.c. Capacitors. Assessment Level E First Edition (IECQ QC 301801). 33 pp.

—Polypropylene/Foil—Cylindrical—Preferred Products List

CECC CECC MUAHAG Vol 1 IS 4-90. Preferred Products List; Capacitors (En, Fr, Ge). 64 pp.

—Polypropylene/Foil—Cylindrical—Printed Circuit Mount

CECC CECC 31 801-002. BS CECC 31 801-002 Issue 1; Fixed Polypropylene Film Dielectric Metal Foil d.c. Capacitor (En). 14 pp.

CECC CECC 31 801-005. DIN 45910 Teil 274; Polypropylene Film Dielectric Metal Foil Capacitors dc 63 to 630 V, General-Purpose Grade, Cylindrical Shape, Insulated, Radial Terminations for Printed Circuits, Climatic Category 40/085/21, Stability Class 1 (Ge). 8 pp.

—Polypropylene/Foil—Preferred Products List

CECC CECC MUAHAG Vol 1 IS 4-90. Preferred Products List; Capacitors (En, Fr, Ge). 64 pp.

—Polypropylene/Foil—Rectangular—Preferred Products List

CECC CECC MUAHAG Vol 1 IS 4-90. Preferred Products List; Capacitors (En, Fr, Ge). 64 pp.

—Polypropylene/Foil—Rectangular—Printed Circuit Mount

CECC CECC 31 801-003. DIN 45910 Teil 272; Polypropylene Film Dielectric Metal Foil Capacitors dc 63 V, Long-Life Grade, Rectangular Shape, Insulated, Radial Terminations for Printed Circuits, Climatic Category 55/085/56, Stability Class 1 (Ge). 8 pp.

CECC CECC 31 801-004. DIN 45910 Teil 273; Polypropylene Film Dielectric Metal Foil Capacitors dc 63 to 100 V, Long-Life Grade, Rectangular Shape, Insulated, Radial Terminations for Printed Circuits, Climatic Category 55/085/56, Stability Class 2 (Ge). 9 pp.

—Polypropylene—Quality Assurance

BSI BS 9070: Sec 6-05. 1971 Amd 5 Fixed Capacitors of Assessed Quality: Generic Data and Methods of Test Section 6: Polycarbonate Dielectric Capacitors and Polyethylene Terephthalate Dielectric Capacitors for d.c. Use and Polypropylene Dielectric Capacitors. 17 pp.

IEC 384 Pt 12-1-88. Fixed Capacitors for Use in Electronic Equipment Part 12: Blank Detail Specification: Fixed Polycarbonate Film Dielectric Metal Foil D.C. Capacitors Assessment Level E First Edition (IECQ QC 301701). 31 pp.

IEC 384 Pt 13-91. Fixed Capacitors for Use in Electronic Equipment Part 13: Sectional Specification: Fixed Polypropylene Film Dielectric Metal Foil d.c. Capacitors Second Edition (IECQ QC 301800). 65 pp.

—Polystyrene

JIS C 5114-91. Fixed Polystyrene Film Capacitors for Use in Electronic Equipment. 64 pp.

—Polystyrene/Foil

CECC CECC 30 901 ISSUE 2-85. Blank Detail Specification: Fixed Polystyrene Film Dielectric Metal Foil d.c. Capacitors (En, Fr, Ge) AMD 1 (En, Fr, Ge) AMD 2 (En, Fr, Ge). 42 pp.

IECQ QC 300900-91. Fixed Capacitors for Use in Electronic Equipment: Part 7: Sectional Specification: Fixed Polystyrene Film Dielectric Metal Foil d.c. Capacitors (IEC 384-7 ED 2). 72 pp.

IECQ QC 300901-91. Fixed Capacitors for Use in Electronic Equipment Part 7: Blank Detail Specification for Fixed Polystyrene Film Dielectric Metal Foil d.c. Capacitors Assessment Level E (IEC 384-7-1 ED 1). 35 pp.

—Polystyrene/Foil—Molded—Rectangular—Printed Circuit Mount

CECC CECC 30 901-002. BS CECC 30 901-002 Issue 1; Fixed Capacitor Polystyrene Dielectric Extended Foil (En). 10 pp.

CECC CECC 30 901-003. BS CECC 30 901-003 Issue 1; Fixed Capacitor Polystyrene Film Dielectric Extended Foil (En) AMD 1 (En). 13 pp.

—Polystyrene/Foil—Preferred Products List

CECC CECC MUAHAG Vol 1 IS 4-90. Preferred Products List; Capacitors (En, Fr, Ge). 64 pp.

—Polystyrene/Foil—Quality Assurance

IEC 384 Pt 7-91. Fixed Capacitors for Use in Electronic Equipment Part 7: Sectional Specification: Fixed Polystyrene Film Dielectric Metal Foil d.c. Capacitors Second Edition (IECQ QC 300900). 72 pp.

IEC 384 Pt 7-1-91. Fixed Capacitors for Use in Electronic Equipment Part 7: Blank Detail Specification for Fixed Polystyrene Film Dielectric Metal Foil d.c. Capacitors Assessment Level E First Edition (IECQ QC 300901). 35 pp.

—Polystyrene/Foil—Rectangular—Preferred Products List

CECC CECC MUAHAG Vol 1 IS 4-90. Preferred Products List; Capacitors (En, Fr, Ge). 64 pp.

—Polystyrene—Molded—Rectangular—Printed Circuit Mount

CECC CECC 30 901-008. CEI CECC 30 901-008 Issue 1; Fixed Polystyrene Film Dielectric d.c. Capacitors (En). 9 pp.

—Polystyrene—Quality Assurance

BSI BS 9070: Sec 4-71. 1971 Amd 3 Fixed Capacitors of Assessed Quality: Generic Data and Methods of Test Section 4: Polystyrene Dielectric Capacitors. 14 pp.

BSI BS 9070: Sec 4-04. 1971 Amd 4 Fixed Capacitors of Assessed Quality: Generic Data and Methods of Test Section 4: Polystyrene Dielectric Capacitors (AMD 7395) March 15, 1993 (T). 15 pp.

BSI BS 9074 N001-74. 1974 Amd 1 Detail Specification for Fixed Polystyrene Foil Dielectric Capacitors with Insulated Tubular Case and Axial Terminations. Full Assessment. 13 pp.

BSI BS 9074 N001-02. 1974 Amd 2 Detail Specification for Fixed Polystyrene Foil Dielectric Capacitors with Insulated Tubular Case and Axial Terminations. Full Assessment Level (AMD 7410) January 15, 1993. 14 pp.

BSI BS 9074 N002-74. 1974 Amd 4 Detail Specification for Fixed Polystyrene Foil Dielectric Capacitors with Insulated Tubular Case and Axial Terminations. Capacitance Range 5pF to 500nF. Full Assessment (AMD 4101) November 30, 1982. 16 pp.

BSI BS 9074 N002-05. 1974 Amd 5 Detail Specification for Fixed Polystyrene Foil Dielectric Capacitors. Tubular Insulated Case, Axial Terminations Full Assessment Level (AMD 7411) January 15, 1993. 17 pp.

BSI BS 9074 N007-81. 1981 Amd 1 Detail Specification for Fixed Capacity Poly-styrene Film Dielectric Extended Foil. Typical Construction: Rectangular Non-Metallic Case. Unidirectional Terminations. Full Assessment Plus Additional Requirements. 16 pp.

BSI BS 9074 N007-02. 1981 Amd 2 Detail Specification for Fixed Polystyrene Film Dielectric Capacitors, Extended Foil. Rectangular Non-Metallic Case, Unidirectional Terminations. Full Plus Additional Assessment Level (AMD 7412) January 15, 1993. 17 pp.

BSI BS QC 300900-91. 1991 Fixed Capacitors for Use in Electronic Equipment. Sectional Specification: Fixed Polystyrene Film Dielectric Metal Foil d.c. Capacitors (IEC 384-7: 1991). 36 pp.

—Polystyrene—Rectangular—Printed Circuit Mount

CECC CECC 30 901-001-85. Fixed Polystyrene Film Dielectric d.c. Capacitors (En). 13 pp.

—Quality Assurance

BSI BS CECC 31600-91. 1991 Harmonized System of Quality Assessment for Electronic Components: Sectional Specification: Fixed Capacitors with Paper and Paper/Plastics Dielectrics (T). 42 pp.

BSI BS CECC 31601-93. 1993 Fixed Capacitors for Use in Electronic Equipment Blank Detail Specification: Fixed Capacitors with Paper or Paper/Plastic Dielectrics. 23 pp.

—Reliability Assured

CNS C5031-86. General Rules for Reliability Assured Fixed Paper and Plastic Film Capacitors (Jan)(4903).

CNS C7192-87. Reliability Assured Plastic Film Capacitors (Characteristic M) (Apr)(11898).

CNS C7193-87. Reliability Assured Plastic Film Capacitors (Characteristic S) (Apr)(11899).

MOD UK DSTAN 59-44: Pt 4:Sec 1-81. Capacitors, Fixed, of Assessed Quality (Listed on EPIC Database) Part 4: Capacitors, Fixed, Polystyrene Dielectric Section 1: Sectional Requirements and Index to Sections Issue 1. 6 pp.

MOD UK DSTAN 59-44: Pt 4:Sec 4-01. Capacitors, Fixed, of Assessed Quality (Listed on EPIC Database) Part 4: Capacitors, Fixed, Polystyrene Dielectric Section 4: List of Items Conforming to BS 9074 F0004 Issue 1; Corrigendum. 27 pp.

—Teflon—Tubular—Microwave—Preferred Products List

CECC CECC MUAHAG Vol 1 IS 4-90. Preferred Products List; Capacitors (En, Fr, Ge). 64 pp.

—Teflon—Tubular—Preferred Products List

CECC CECC MUAHAG Vol 1 IS 4-90. Preferred Products List; Capacitors (En, Fr, Ge). 64 pp.

—Tuning

IEC 415 Pt 1-73. Plastic Film Dielectric Rotary Variable Tuning Capacitors: Grade 2 Part 1: General Requirements for Tests and Measuring Methods First Edition. 35 pp.

JIS C 6465-87. Variable Tuning Capacitors Type A (Plastic Film Dielectric) for Use in Radio Receiver. 34 pp.

Film
INTERNATIONAL AND NON-U.S. NATIONAL STANDARDS
SUBJECT INDEX

Film Changers, X-Ray
Use: X-Ray Film Changers

Film Editors
—Portable—Construction
CSA C22.2 NO 118-1959. Construction and Test of Picture Machines and Appliances (R 1992); (Erratum April 1959) (Rev 1-5 March 1969). 21 pp.
CSA 1169 Bull. Electrical Bulletin 1169 June 27, 1978 to C22.2 NO 118. 2 pp.
—Portable—Testing
CSA C22.2 NO 118-1959. Construction and Test of Picture Machines and Appliances (R 1992); (Erratum April 1959) (Rev 1-5 March 1969). 21 pp.
CSA 1169 Bull. Electrical Bulletin 1169 June 27, 1978 to C22.2 NO 118. 2 pp.
—Power Supply Cords
CSA C22.2 NO 118-1959. Construction and Test of Picture Machines and Appliances (R 1992); (Erratum April 1959) (Rev 1-5 March 1969). 21 pp.
CSA 925 Bull. Electrical Bulletin 925 October 25, 1973 to C22.2 NO 118. 2 pp.
CSA 1169 Bull. Electrical Bulletin 1169 June 27, 1978 to C22.2 NO 118. 2 pp.

Film/Foil Capacitors
Use: Film Capacitors

Film Integrated Circuits
See Also: Integrated Circuits; Monolithic Integrated Circuits
BSI BS QC 763000-90. 1990 Film and Hybrid Film Integrated Circuits. Generic Specification (Document Number Changed to BS QC 760000: 1990 by Amendment 1). 45 pp.
CECC CECC 63 000 ISSUE 1-84. Generic Specification: Film and Hybrid Integrated Circuits (En, Fr, Ge) AMD 1 (En, Fr, Ge) AMD 2 (En, Fr, Ge) AMD 3 (En, Fr, Ge) AMD 4 (En, Fr, Ge) AMD 5 (En, Fr, Ge). 230 pp.
CECC CECC 63 100 ISSUE 1-84. Sectional Specification: Film and Hybrid Integrated Circuits (En, Fr, Ge) AMD 1 (En, Fr, Ge) AMD 2 (En, Fr, Ge) AMD 3 (En, Fr, Ge). 67 pp.
CECC CECC 63 200 ISSUE 1-84. Sectional Specification: Film and Hybrid Integrated Circuits; (Capability Approval) (En, Fr, Ge) AMD 1 (En, Fr, Ge) AMD 2 (En, Fr, Ge) AMD 3 (En, Fr, Ge) AMD 4 (En, Fr, Ge). 275 pp.
CECC CECC 63 201 ISSUE 1-84. Blank Detail Specification: Film and Hybrid Integrated Circuits; (Capability Approval) (En, Fr, Ge) AMD 1 (En, Fr, Ge) AMD 2 (En, Fr, Ge). 36 pp.
IEC 748 Pt 20-88. Semiconductor Devices Integrated Circuits Part 20: Generic Specification for Film Integrated Circuits and Hybrid Film Integrated Circuits First Edition (IECQ QC 760000). 87 pp.
IECQ QC 760000-88. Semiconductor Devices Integrated Circuits Part 20: Generic Specification for Film Integrated Circuits and Hybrid Film Integrated Circuits (IEC 748-20 ED 1). 87 pp.
IECQ QC 760100-91. Semiconductor Devices Integrated Circuits Part 21: Sectional Specification for Film Integrated Circuits and Hybrid Film Integrated Circuits on the Basis of Qualification Approval Procedure (IEC 748-21 ED 1). 36 pp.
IECQ QC 760101-91. Semiconductor Devices—Integrated Circuits Part 21: Section One: Blank Detail Specification for Film Integrated Circuits and Hybrid Film Integrated Circuits on the Basis of Qualification Approval Procedure (IEC 748-21-1 ED 1). 24 pp.
IECQ QC 760200-92. Semiconductor Devices Integrated Circuits Part 22: Sectional Specification for Film Integrated Circuits and Hybrid Film Integrated Circuits on the Basis of the Capability Approval Procedures (IEC 748-22 ED 1). 116 pp.
IECQ QC 760201-91. Semiconductor Devices Integrated Circuits Part 22: Section One: Blank Detail Specification for Film Integrated Circuits and Hybrid Film Integrated Circuits on the Basis of the Capability Approval Procedures (IEC 748-22-1 ED 1). 30 pp.
—Hybrid
BSI BS QC 763000-90. 1990 Film and Hybrid Film Integrated Circuits. Generic Specification (Document Number Changed to BS QC 760000: 1990 by Amendment 1). 45 pp.
IEC 748 Pt 20-88. Semiconductor Devices Integrated Circuits Part 20: Generic Specification for Film Integrated Circuits and Hybrid Film Integrated Circuits First Edition (IECQ QC 760000). 87 pp.
IECQ QC 760000-88. Semiconductor Devices Integrated Circuits Part 20: Generic Specification for Film Integrated Circuits and Hybrid Film Integrated Circuits (IEC 748-20 ED 1). 87 pp.
IECQ QC 760100-91. Semiconductor Devices Integrated Circuits Part 21: Sectional Specification for Film Integrated Circuits and Hybrid Film Integrated Circuits on the Basis of Qualification Approval Procedure (IEC 748-21 ED 1). 36 pp.

Film Integrated Circuits *(Cont.)*
—Hybrid *(Cont.)*
IECQ QC 760101-91. Semiconductor Devices—Integrated Circuits Part 21: Section One: Blank Detail Specification for Film Integrated Circuits and Hybrid Film Integrated Circuits on the Basis of Qualification Approval Procedure (IEC 748-21-1 ED 1). 24 pp.
IECQ QC 760200-92. Semiconductor Devices Integrated Circuits Part 22: Sectional Specification for Film Integrated Circuits and Hybrid Film Integrated Circuits on the Basis of the Capability Approval Procedures (IEC 748-22 ED 1). 116 pp.
IECQ QC 760201-91. Semiconductor Devices Integrated Circuits Part 22: Section One: Blank Detail Specification for Film Integrated Circuits and Hybrid Film Integrated Circuits on the Basis of the Capability Approval Procedures (IEC 748-22-1 ED 1). 30 pp.
—Hybrid—Quality Assurance
BSI BS QC 760100-91. 1991 Semiconductor Devices. Integrated Circuits. Sectional Specification for Film and Hybrid Integrated Circuits: Qualification Approval (IEC 748-21: 1991) (IEC 748-21: 1991). 19 pp.
BSI BS QC 760101-92. 1992 Film and Hybird Integrated Circuits: Qualification Approval Procedure. 10 pp.
BSI BS QC 760200-92. 1992 Semiconductor Devices Integrated Circuits. Sectional Specification for Film Integrated Circuits and Hybrid Film Integrated Circuits: Capability Approval (IEC 748-22: 1992). 59 pp.
BSI BS QC 760201-92. 1992 Semiconductor Devices. Integrated Circuits Blank Detail Specification Film and Hybrid Integrated Circuits: Capability Approval (IEC 748-22-1: 1991). 12 pp.
IEC 748 Pt 21-91. Semiconductor Devices Integrated Circuits Part 21: Sectional Specification for Film Integrated Circuits and Hybrid Film Intergrated Circuits on the Basis of Qualification Approval Procedure First Edition (IECQ QC 760100). 36 pp.
IEC 748 Pt 21-1-91. Semiconductor Devices—Integrated Circuits Part 21: Section One: Blank Detail Specification for Film Integrated Circuits and Hybrid Film Integrated Circuits on the Basis of Qualification Approval Procedure First Edition (IECQ QC 760101). 24 pp.
IEC 748 Pt 22-92. Semiconductor Devices Integrated Circuits Part 22: Sectional Specification for Film Integrated Circuits and Hybrid Film Integrated Circuits on the Basis of the Capability Approval Procedures First Edition (IECQ QC 760200). 116 pp.
IEC 748 Pt 22-1-91. Semiconductor Devices Integrated Circuits Part 22: Section One: Blank Detail Specification for Film Integrated Circuits and Hybrid Film Integrated Circuits on the Basis of the Capability Approval Procedures First Edition (IECQ QC 760201). 30 pp.
—Quality Assurance
BSI BS QC 760000-90. 1990 Amd 1 Film and Hybrid Film Integrated Circuits. General Specification (IEC 748-20: 1988) (AMD 6754) September 1991. 46 pp.
BSI BS QC 760100-91. 1991 Semiconductor Devices. Integrated Circuits. Sectional Specification for Film and Hybrid Integrated Circuits: Qualification Approval (IEC 748-21: 1991) (IEC 748-21: 1991). 19 pp.
BSI BS QC 760101-92. 1992 Film and Hybird Integrated Circuits: Qualification Approval Procedure. 10 pp.
BSI BS QC 760200-92. 1992 Semiconductor Devices Integrated Circuits. Sectional Specification for Film Integrated Circuits and Hybrid Film Integrated Circuits: Capability Approval (IEC 748-22: 1992). 59 pp.
BSI BS QC 760201-92. 1992 Semiconductor Devices. Integrated Circuits Blank Detail Specification Film and Hybrid Integrated Circuits: Capability Approval (IEC 748-22-1: 1991). 12 pp.
BSI BS CECC 63000-90. 1990 Film and Hybrid Integrated Circuits: Generic Specification. 76 pp.
BSI BS CECC 63100-85. 1985 Amd 3 Harmonized System of Quality Assessment for Electronic Components: Sectional Specification: Film and Hybrid Integrated Circuits (T) (AMD 6621) April 30, 1991. 36 pp.
BSI BS CECC 63101-85. 1985 Amd 1 Film and Hybrid Integrated Circuits: Blank Detail Specification (AMD 6595) July 31, 1990. 13 pp.
BSI BS CECC 63200-85. 1985 Amd 4 Harmonized System of Quality Assessment for Electronic Components: Sectional Specification: Film and Hybrid Integrated Circuits (Capability Approval) (AMD 6623) April 30, 1991. 90 pp.
BSI BS CECC 63201-85. 1985 Amd 2 Film and Hybrid Integrated Circuits: Blank Detail Specification (Capability Approval) (AMD 6583) July 31, 1990. 18 pp.

Film Integrated Circuits *(Cont.)*
—Quality Assurance *(Cont.)*
IEC 748 Pt 21-91. Semiconductor Devices Integrated Circuits Part 21: Sectional Specification for Film Integrated Circuits and Hybrid Film Intergrated Circuits on the Basis of Qualification Approval Procedure First Edition (IECQ QC 760100). 36 pp.
IEC 748 Pt 21-1-91. Semiconductor Devices—Integrated Circuits Part 21: Section One: Blank Detail Specification for Film Integrated Circuits and Hybrid Film Integrated Circuits on the Basis of Qualification Approval Procedure First Edition (IECQ QC 760101). 24 pp.
IEC 748 Pt 22-92. Semiconductor Devices Integrated Circuits Part 22: Sectional Specification for Film Integrated Circuits and Hybrid Film Integrated Circuits on the Basis of the Capability Approval Procedures First Edition (IECQ QC 760200). 116 pp.
IEC 748 Pt 22-1-91. Semiconductor Devices Integrated Circuits Part 22: Section One: Blank Detail Specification for Film Integrated Circuits and Hybrid Film Integrated Circuits on the Basis of the Capability Approval Procedures First Edition (IECQ QC 760201). 30 pp.

Film/Paper Capacitors
Use: Film Capacitors

Film Processing Equipment
Use: Photographic Processing Equipment

Film Resistors
See Also: Fixed Resistors; Metal Film Resistors; Resistors; Thick Film Resistor Networks; Thin Film Resistors; Variable Resistors
BSI BS QC 390000-92. 1992 Fixed Film Resistor Networks for Use in Electronic Equipment Generic Specification (IEC 1045-1: 1991). 48 pp.
BSI BS QC 390101-92. 1992 Film Resistor Networks of Assessed Quality on the Basis of Capability Approval Procedure Assessment Level E (IEC 1045-2-1: 1991). 15 pp.
IEC 1045 Pt 1-91. Fixed Film Resistor Networks for Use in Electronic Equipment Part 1: Generic Specification First Edition (IECQ QC 390000). 97 pp.
IECQ QC 390000-91. Fixed Film Resistor Networks for Use in Electronic Equipment Part 1: Generic Specification (IEC 1045-1 ED 1). 97 pp.
IECQ QC 390100-91. Fixed Film Resistor Networks for Use in Electronic Equipment Part 2: Sectional Specification for Film Resistor Networks of Assessed Quality on the Basis of the Capability Approval Procedure (IEC 1045-2 ED 1). 109 pp.
—Power
IECQ QC 400101/US 0002-87. Resistors for Use in Electronic Equipment: Detail Specification: Fixed Low-Power Non-Wirewound Resistors, Insulated, Standard Film. 12 pp.
—Power—Printed Circuit Mount
CECC CECC 40 101-020. CEI-CECC 40 101-020 Edition 1; Fixed Low Power Non-Wirewound Insulated Resistors; Stability Plus or Minus (1% + 0,05 Ohm) (En). 7 pp.
CECC CECC 40 101-021. CEI-CECC 40 101-021; Fixed Low Power Non-Wirewound Insulated Resistors; Stability Plus or Minus (2% + 0,1 Ohm) (En). 7 pp.
CECC CECC 40 101-026. CEI-CECC 40 101-026; Fixed Low Power Non-Wirewound Insulated Resistors; Stability Plus or Minus (1% + 0.05 Ohm) (En). 7 pp.
—Precision
CNS C6156-82. Fixed Film Resistors and Semiprecision (Feb)(8479).
—Quality Assurance
BSI BS QC 390100-92. 1992 Fixed Film Resistor Networks for Use in Electronic Equipment Sectional Specification for Film Resistor Networks of Assessed Quality on the Basis of the Capability Approval Procedure (IEC 1045-2: 1991). 54 pp.
IEC 1045 Pt 2-91. Fixed Film Resistor Networks for Use in Electronic Equipment Part 2: Sectional Specification for Film Resistor Networks of Assessed Quality on the Basis of the Capability Approval Procedure First Edition (IECQ QC 390100). 109 pp.
IEC 1045 Pt 2-1-91. Fixed Film Resistor Networks for Use in Electronic Equipment Part 2: Blank Detail Specification for Film Resistor Networks of Assessed Quality on the Basis of the Capability Approval Procedure Assessment Level E First Edition (IECQ QC 390101). 35 pp.
IECQ QC 390101-91. Fixed Film Resistor Networks for Use in Electronic Equipment Part 2: Blank Detail Specification for Film Resistor Networks of Assessed Quality on the Basis of the Capability Approval Procedure Assessment Level E (IEC 1045-2-1 ED 1). 35 pp.

INTERNATIONAL AND NON-U.S. NATIONAL STANDARDS
SUBJECT INDEX

Film Slides
Use: Slides (Photographic)

Film Strips
Use: Filmstrips

Films
Scope Note: Use a more specific term *See:* Diazo Films; Drafting Films; Film Adhesives; Microfilm; Motion Picture Films; Packaging Films; Photographic Films; Photomechanical Films; Polypropylene Films; Radiation Monitoring Films; Radiographic Films; Slides (Photographic); Training Films; Transparencies

Filmslides
Use: Slides (Photographic)

Filmstrip Projectors
Use For: Still Projectors *See Also:* Filmstrips; Projectors
BSI BS 1915-68. 1968 Amd 1 Still Projectors. 28 pp.
BSI BS 4120-68. 1968 Amd 1 Methods of Measurement of Performance of Still Projectors. 12 pp.

—Frame Masks
ISO 9559-89. Photography—35 mm Filmstrip Projectors—Specifications for Frame Masks First Edition. 4 pp.

—Illuminance
ISO 8341-89. Photography—Slide Projectors and Filmstrip Projectors—Illumination Test First Edition; (ANSI IT7.201-1991). 6 pp.

—Portable—Construction
CSA C22.2 NO 118-1959. Construction and Test of Picture Machines and Appliances (R 1992); (Erratum April 1959) (Rev 1-5 March 1969). 21 pp.
CSA 1169 Bull. Electrical Bulletin 1169 June 27, 1978 to C22.2 NO 118. 2 pp.

—Portable—Testing
CSA C22.2 NO 118-1959. Construction and Test of Picture Machines and Appliances (R 1992); (Erratum April 1959) (Rev 1-5 March 1969). 21 pp.
CSA 1169 Bull. Electrical Bulletin 1169 June 27, 1978 to C22.2 NO 118. 2 pp.

—Power Supply Cords
CSA C22.2 NO 118-1959. Construction and Test of Picture Machines and Appliances (R 1992); (Erratum April 1959) (Rev 1-5 March 1969). 21 pp.
CSA 925 Bull. Electrical Bulletin 925 October 25, 1973 to C22.2 NO 118. 2 pp.
CSA 1169 Bull. Electrical Bulletin 1169 June 27, 1978 to C22.2 NO 118. 2 pp.

—Safety
BSI BS 3456: Sec 3.14-80. 1980 1972-1981 Edition. Specification for Safety of Household Electrical Appliances Part 3: Particular Specifications Section 3.14: Projectors. 52 pp.

Filmstrips
Use For: Film Strips *See Also:* Audiovisual Equipment; Filmstrip Projectors; Photographic Films; Projectors; Transparencies
BSI BS 1917-68. 1968 Slides and Film Strips. 12 pp.

— 35mm
CNS Z9016-80. 35mm Filmstrips Dimensions and Formats (Jan)(5157).

—Containers
BSI BS 2698-71. 1971 Containers and Notes for Filmstrips. 12 pp.

—Formats
CNS Z9016-80. 35mm Filmstrips Dimensions and Formats (Jan)(5157).

—Identification Systems
BSI BS 2698-71. 1971 Containers and Notes for Filmstrips. 12 pp.

Filter Capacitors
Use For: Electromagnetic Interference Capacitors; Interference Suppressor Capacitors; Noise Filter Capacitors; Noise Suppression Capacitors; Radio Frequency Interference Capacitors; Suppression Capacitors *See Also:* Capacitors; Electromagnetic Interference Filters; Filters; Fixed Capacitors; Noise Suppressors; Power Supplies; Switching Power Supplies
BSI BS 6201: Part 3-82. 1982 Fixed Capacitors for Use in Electronic Equipment Part 3: Fixed Capacitors for Radio Interference Suppression. Selection of Methods of Test and General Requirements. 31 pp.
CNS C7180-86. Paper Dielectric Capacitors for AC and DC, by-Pass Radio-Interference Reduction (Jul)(11625).

Filter Capacitors (Cont.)
DIN VDE 0565 Pt 1-79. Radio Interference Suppression Devices; Radio Interference Suppression Capacitors (Dec). 66 pp.
DIN VDE 0565 Pt 1 A1-84. Radio Interference Suppression Devices; Radio Interference Suppression Capacitors: Amendment 1 (June). 2 pp.
IEC 940-88. Guidance Information on the Application of Capacitors, Resistors, Inductors, and Complete Filter Units for Radio Interference Suppression First Edition. 21 pp.
SAA AS 1541.14-83. Fixed Capacitors for Use in Electronic Equipment—Part 14: Fixed Capacitors for Radio Interference Suppression. 28 pp.

—Electromagnetic Interference
CECC CECC 32 400 ISSUE 1-92. Sectional Specification: Fixed Capacitors for Electromagnetic Interference Suppression and Connection to the Supply Mains; (Assessment Level D) (En, Ge) Erratum (En, Ge). 116 pp.
CECC CECC 32 401 ISSUE 1-92. Blank Detail Specification: Fixed Capacitors for Electromagnetic Interference Suppression and Connection to the Supply Mains (Assessment Level D) (En, Ge). 28 pp.

—Power Supplies
IECQ QC 302400-93. Fixed Capacitors for Use in Electronic Equipment Part 14: Sectional Specification: Fixed Capacitors for Electromagnetic Interference Suppression Connection to the Supply Mains (IEC 384-14 ED 2). 117 pp.
IECQ QC 302401-93. Fixed Capacitors for Use In Electronic Equipment Part 14: Blank Detail Specification: Fixed Capacitors for Electromagnetic Interference Suppression Connection to the Supply Mains Assessment Level D (IEC 384-14-1 ED 1). 33 pp.

—Power Supplies—Quality Assurance
IEC 384 Pt 14-93. Fixed Capacitors for Use in Electronic Equipment Part 14: Sectional Specification: Fixed Capacitors for Electromagnetic Interference Suppression and Connection to the Supply Mains Second Edition (IECQ QC 302400). 117 pp.
IEC 384 Pt 14-1-93. Fixed Capacitors for Use in Electronic Equipment Part 14: Blank Detail Specification: Fixed Capacitors for Electromagnetic Interference Suppression and Connection to the Supply Mains Assessment Level D First Edition (IECQ QC 302401). 33 pp.

—Quality Assurance
IEC 384 Pt 14-93. Fixed Capacitors for Use in Electronic Equipment Part 14: Sectional Specification: Fixed Capacitors for Electromagnetic Interference Suppression and Connection to the Supply Mains Second Edition (IECQ QC 302400). 117 pp.

Filter Crystal Units
See Also: Crystal Filters; Quartz Crystal Units

—Resonance Frequency
IEC 283-68. Methods for the Measurement of Frequency and Equivalent Resistance of Unwanted Resonances of Filter Crystal Units First Edition. 22 pp.

—Resonant Resistance
IEC 283-68. Methods for the Measurement of Frequency and Equivalent Resistance of Unwanted Resonances of Filter Crystal Units First Edition. 22 pp.

Filter Elements
See Also: Filters

—Air Filters—Agricultural Equipment
DIN ENGL 71459-87. Air Filter Elements for Commercial Vehicles; Dimensions (May). 6 pp.

—Air Filters—Automotive
DIN ENGL 71459-87. Air Filter Elements for Commercial Vehicles; Dimensions (May). 6 pp.
ISO 8027-84. Road Vehicles—Air Filter Elements for Passenger Cars—Types P and R—Dimensions First Edition. 6 pp.

—Air Filters—Compressors
DIN ENGL 24189-86. Testing of Air Cleaners for Internal Combustion Engines and Compressors; Test Methods (Jan). 10 pp.

—Air Filters—Construction Equipment
DIN ENGL 71459-87. Air Filter Elements for Commercial Vehicles; Dimensions (May). 6 pp.

—Felts
MOD UK DSTAN 83-85-93. Polyester Felt Issue 1. 11 pp.

Filter Flasks
Use: Flasks—Filtering

Filter Funnels
Use: Funnels—Filter

Filter Heads
See Also: Filters

—Aircraft—Bearings
SBAC AS 3312 ISSUE 3. Bearing for Filter Head. 1 p.

Filter Inductors
See Also: Electric Coils; Inductors
CNS C7130-80. Power Filter Inductors for Electronic Equipment (Sep)(6434).

Filter Laminations
—Aircraft
SBAC AS 3317 ISSUE 3. Filter Lamination. 1 p.

Filter Packs
—Aircraft—Baffle Plates
SBAC AS 3318 ISSUE 3. Baffle Plate Filter Pack. 1 p.

—Aircraft—Plates (Supports)
SBAC AS 3321 ISSUE 3. Top, Plate—Filter Pack. 1 p.

—Aircraft—Rings
SBAC AS 3319 ISSUE 3. Bottom, Ring—Filter Pack. 1 p.
SBAC AS 3322 ISSUE 3. Stabilizing, Ring—Filter Pack. 1 p.

Filter Paper
See Also: Laboratory Equipment; Papers
BSI BS 6410-91. 1991 Filter Papers. 39 pp.
MOD UK DEF-128-01. Paper, Filter; Amendment 1. 4 pp.

—Chemical Analysis
CNS P2058-87. Filter Paper for Chemical Analysis (May)(5038). 2 pp.
CNS P3084-87. Method of Test for Filter Paper for Chemical Analysis (May)(11972).
JIS P 3801-76. Filter Paper (for Chemical Analysis) (R 1985). 11 pp.

—Glass Fabrics
MOD UK TS 10066B. Paper, Filter, Glass Fibre (Water Repellant Treated).

—Spacers
MOD UK TS 10147. Base Material for Spacer Card.

Filter Pipe Fittings
See Also: Filter Pipes; Pipe Fittings

—Well—Flanged—Steel
DIN ENGL 4922 Pt 3-75. Steel Filter Pipes for Drilled Wells; Flanged Connection, NW 500 to NW 1000 (Nominal Diameter 500 to 1000) (Dec). 2 pp.

—Well—Screwed—Steel
DIN ENGL 4922 Pt 2-81. Steel Filter Pipes for Drilled Wells with Screwed Connection DN 100 to DN 500 (Apr). 4 pp.

Filter Pipe Joints
See Also: Filter Pipes; Pipe Joints

—Well—Butt Strap—Steel
DIN ENGL 4922 Pt 1-78. Steel Filter Pipes for Drilled Wells with Slot Perforation and Fishing; (Butt Strap Joint) (Feb). 4 pp.

Filter Pipes
See Also: Filter Pipe Fittings; Filter Pipe Joints; Pipes

—PVC—Unplasticized—Water Wells
DIN ENGL 4925 Pt 1-90. DN 40 to DN 100 Threaded Unplasticized Polyvinyl Chloride (PVC-U) Water Well Filter Pipes and Casings (Nov). 4 pp.
DIN ENGL 4925 Pt 2-90. DN 125 to DN 200 Threaded Unplasticized Polyvinyl Chloride (PVC-U) Water Well Filter Pipes and Casings (Nov). 5 pp.
DIN ENGL 4925 Pt 3-90. DN 250 to DN 400 Threaded Unplasticized Polyvinyl Chloride (PVC-U) Water Well Filter Pipes and Casings (Nov). 5 pp.

—Steel—Well
DIN ENGL 4920-83. DN 32 to DN 100 Threaded Steel Filter Pipes with Slot Perforation for Drilled Wells and Driven Wells (July). 3 pp.
DIN ENGL 4922 Pt 1-78. Steel Filter Pipes for Drilled Wells with Slot Perforation and Fishing; (Butt Strap Joint) (Feb). 4 pp.

Filter

Filter Pipes (Cont.)
—Steel—Well (Cont.)
DIN ENGL 4922 Pt 2-81. Steel Filter Pipes for Drilled Wells with Screwed Connection DN 100 to DN 500 (Apr). 4 pp.
DIN ENGL 4922 Pt 3-75. Steel Filter Pipes for Drilled Wells; Flanged Connection, NW 500 to NW 1000 (Nominal Diameter 500 to 1000) (Dec). 2 pp.

Filter Sand
See Also: Filters; Sand Filters; Sands
—Sand Filters
DIN ENGL 19623-78. Filter Sands and Filter Gravels for Water Purification Filters; Technical Conditions of Delivery (Jan). 5 pp.
—Well Filters
DIN ENGL 4924-72. Filter Sands and Filter Gravels for Well Filters (Feb). 1 p.

Filter/Strainers
See Also: Filters; Strainers; Sump Strainers
—Irrigation Equipment (Agricultural)
ISO 9912 Pt 2-92. Agricultural Irrigation Equipment—Filters—Part 2: Strainer-Type Filters First Edition. 9 pp.
—Self Cleaning—Irrigation Equipment (Agricultural)
ISO 9912 Pt 3-92. Agricultural Irrigation Equipment—Filters—Part 3: Automatic Self-Cleaning Strainer-Type Filters First Edition. 9 pp.

Filter/Surge Protectors
See Also: Electric Filters; Surge Arresters; Surge Suppressors
CSA C22.2 NO 8-M1986. Electromagnetic Interference (EMI) Filters (R 1992); (Gen Instr 1 Thru 2). 43 pp.

Filtering Capacitors
Use: Filter Capacitors

Filters
See Also: Air Filters; Bag Filters; Bandpass Filters; Cavity Filters; Cigarette Filters; Color Filters; Dielectric Filters; Electric Filters; Electromagnetic Interference Filters; Filter Capacitors; Filter Elements; Filter Heads; Filter Laminations; Filter Packs; Filter Sand; Filter/Strainers; Fuel Filters; Fuel Separators/Filters; Hydraulic Fluid Filters; Infrared Filters; Interference Screens; Laboratory Filters; Line Traps; Membrane Filters; Microwave Filters; Oil Filters; Ophthalmic Filters; Optical Filters; Passive Filters; Photographic Filters; Precoat Filters; Quartz Crystal Filters; Radio Frequency Filters; Sand Filters; Screens (Filters); Strainers; Surface Acoustic Wave Filters; Trickling Filters; Ultraviolet Filters; Water Filters; Waveguide Filters; Well Filters; White Noise Filters
—Refrigeration Equipment
CSA C22.2 NO 140.3-M1987. Refrigerant-Containing Components for Use in Electrical Equipment (R 1993). 19 pp.

Filters (Light)
Use: Optical Filters

Filtration
Scope Note: For additional listings, use a more specific term *See Also:* Biochemical Oxygen Demand; Concentrators; Effluents; Sewage Filtration; Sewage Treatment; Ultrafiltration; Water Treatment
—Effluents
CEN PREN 871-92. Water Quality—Determination of Suspended Solids—Determination of Coarse Particles and Fibres Seperated on a 60 to 70 um Sieve. 5 pp.
CEN PREN 872-92. Water Quality—Determination of Suspended Solids—Determination of Suspended Solids by Glass Fibre Filters. 8 pp.
—Fuel Oils
ISO 10307 Pt 1-93. Petroleum Products—Total Sediment in Residual Fuel Oils—Part 1: Determination by Hot Filtration First Edition. 9 pp.
—Sludge
DIN ENGL 38409 Pt 2-87. German Standard Methods for the Examination of Water, Waste Water and Sludge; Parameters Characterizing Effects and Substances (Group H); Determination of Filterable Matter and the Residue on Ignition (H 2) (Mar). 5 pp.
DIN ENGL 38411 Pt 5-83. German Standard Methods for the Examination of Water, Waste Water and Sludge; Microbiological Methods (Group K); Determination of Reproductive Germs Using the Membrane Filter Method (K 5) (Feb). 3 pp.

Filtration (Cont.)
—Sludge (Cont.)
DIN ENGL 38411 Pt 8-82. German Standard Methods for the Analysis of Water, Waste Water and Sludge; Microbiological Methods (Group K); Detection of Pseudomonas Aeruginosa (K 8) (May). 5 pp.
—Waste Water
DIN ENGL 38409 Pt 2-87. German Standard Methods for the Examination of Water, Waste Water and Sludge; Parameters Characterizing Effects and Substances (Group H); Determination of Filterable Matter and the Residue on Ignition (H 2) (Mar). 5 pp.
DIN ENGL 38411 Pt 5-83. German Standard Methods for the Examination of Water, Waste Water and Sludge; Microbiological Methods (Group K); Determination of Reproductive Germs Using the Membrane Filter Method (K 5) (Feb). 3 pp.
DIN ENGL 38411 Pt 8-82. German Standard Methods for the Analysis of Water, Waste Water and Sludge; Microbiological Methods (Group K); Detection of Pseudomonas Aeruginosa (K 8) (May). 5 pp.
—Water
BSI BS 2690: Part 104-83. 1983 Water Used in Industry Part 104: Silica: Reactive, Total and Suspended. 6 pp.
BSI BS 6068: Sec 4.7-89. 1989 Water Quality Part 4: Microbiological Methods Section 4.7: Detection and Enumeration of Pseudomonas Aeruginosa Membrane Filtration (ISO 8360-2: 1988). 8 pp.
BSI BS 6068: Sec 4.9-93. 1993 Water Quality Detection and Enumeration of the Spores of Sulfite-Reducing Anaerobes (Clostridia) Part 2: Method by Membrane Filtration (ISO 6461/2: 1986) (N). 10 pp.
BSI BS EN 26461-2-93. 1993 Water Quality Detection and Enumeration of the Spores of Sulfite-Reducing Anaerobes (Clostridia) Part 2: Method by Membrane Filtration (ISO 6461/2: 1986) (N). 10 pp.
CEN PREN 870-92. Water Quality—Determination of Suspended Solids—Determination by 0.4—0.45 um Pore Width Filters. 7 pp.
CEN PREN 871-92. Water Quality—Determination of Suspended Solids—Determination of Coarse Particles and Fibres Seperated on a 60 to 70 um Sieve. 5 pp.
CEN PREN 872-92. Water Quality—Determination of Suspended Solids—Determination of Suspended Solids by Glass Fibre Filters. 8 pp.
CEN EN 26461-2-93. Water Quality—Detection and Enumeration of the Spores of Sulfite-Reducing Anaerobes (Clostridia)—Part 2: Method by Membrane Filtration (ISO 6461/2: 1986). 5 pp.
DIN ENGL 38409 Pt 2-87. German Standard Methods for the Examination of Water, Waste Water and Sludge; Parameters Characterizing Effects and Substances (Group H); Determination of Filterable Matter and the Residue on Ignition (H 2) (Mar). 5 pp.
DIN ENGL 38411 Pt 5-83. German Standard Methods for the Examination of Water, Waste Water and Sludge; Microbiological Methods (Group K); Determination of Reproductive Germs Using the Membrane Filter Method (K 5) (Feb). 3 pp.
DIN ENGL 38411 Pt 8-82. German Standard Methods for the Analysis of Water, Waste Water and Sludge; Microbiological Methods (Group K); Detection of Pseudomonas Aeruginosa (K 8) (May). 5 pp.
ISO 6461 Pt 2-86. Water Quality—Detection and Enumeration of the Spores of Sulfite-Reducing Anaerobes (Clostridia)—Part 2: Method by Membrane Filtration First Edition; (CEN EN 26461-2: 1993). 5 pp.
ISO 7899 Pt 2-84. Water Quality—Detection and Enumeration of Faecal Streptococci—Part 2: Method by Membrane Filtration First Edition. 6 pp.
ISO 8360 Pt 2-88. Water Quality—Detection and Enumeration of Pseudomonas Aeruginosa—Part 2: Membrane Filtration Method First Edition. 9 pp.

FIMS
Use For: Form Interface Management System
See Also: Programming Languages
IEC DIS 11730-93. Information Technology—Programming Languages—Form Interface Management System (FIMS) ***CD-ROM ONLY***. 490 pp.
ISO DIS 11730-93. Information Technology—Programming Languages—Form Interface Management System (FIMS) ***CD-ROM ONLY***. 490 pp.
JTC1 DIS11730-93. Information Technology—Programming Languages—Form Interface Management System (FIMS) ***CD-ROM ONLY***. 490 pp.

FIN Stabilizers
Use: Stabilizers (Fluid Dynamics)

Financial Disclosure
See Also: Accounting
—Quarterly Report Forms—Defense Contracts
MOD UK DEFCON 136-75. Quarterly Financial Report 11/75. 1 p.

Financial Institutions
See Also: Banking; Credit Institutions; Currencies
—Annual Accounts—Information Disclosure
EC 86/635/EEC-86. Council Directive on the Annual Accounts and Consolidated Accounts of Banks and Other Financial Institutions. 17 pp.
EC 89/117/EEC-89. Council Directive on the Obligations of Branches Established in a Member State of Credit Institutions and Financial Institutions Having their Head Offices Outside That Member State Regarding the Publication of Annual Accounting Documents. 3 pp.
—Consolidated Accounts—Information Disclosure
EC 86/635/EEC-86. Council Directive on the Annual Accounts and Consolidated Accounts of Banks and Other Financial Institutions. 17 pp.
—Data Transmission—Sign-On Authentication
ISO 11131-92. Banking and Related Financial Services—Sign-on Authentication First Edition. 15 pp.

Financial Transaction Cards
Use: Bank Cards

Fine Papers
See Also: Art Papers; Drawing Papers; Manifold Papers; Papers; Printing Papers; Writing Papers
—Curl (Materials)
CNS P3026-89. Method of Test for Degree of Curl and Sizing of Paper (Curl Test) (Aug)(3318). 2 pp.
—Sizing (Surface Treatment)
CNS P3026-89. Method of Test for Degree of Curl and Sizing of Paper (Curl Test) (Aug)(3318). 2 pp.
—Water Resistance Testing
CNS P3026-89. Method of Test for Degree of Curl and Sizing of Paper (Curl Test) (Aug)(3318). 2 pp.

Finger Cots
See Also: Gloves; Medical Equipment
CNS T2007-80. Finger Sack for Medical Use (Oct)(6628).
JIS T 9108-55. Finger Sack for Medical Use.

Finger Joints
Use: Interlocking Joints

Finish Washers
See Also: Washers (Fasteners)
—Automotive
CNS D2168-83. Finishing Washers for Automobiles (Feb)(9988).

Finishes
Scope Note: For additional listings, use a more specific term *See Also:* Baking Finishes; Bleached Shellac; Coatings; Corrosion; Dopes; Enamel Paints; Enamels; Finishing Compounds; Flat Finishes; Lacquers; Matte Finishes; Metallizing; Nitrocellulose Finishes; Organic Coatings; Paints; Polishes; Semigloss; Shellacs; Sizing (Surface Treatment); Sprayed Coatings; Surface Properties; Varnishes; Veneers; Vitreous Enamels; Waxes
—Aluminum—Adhesion Testing
CGSB 1-GP-71 METH 135.6-78. Methods of Testing Paints and Pigments Adhesion Adhesion of Finishes to Primed Anodized Aluminum. 1 p.
—Baked—Salt Spray Testing
CGSB 1-GP-71 METH 129.4-79. Methods of Testing Paints and Pigments Salt-Spray Corrosion Resistance Corrosion Resistance of Baked Finishes. 1 p.
—Cabinets (Electrical)—Wooden
CNS C6107-81. Finish Test of Wooden Cabinets for Electronic Equipments (Apr)(7228).
—Furniture—Gloss
BSI BS 3962: Part 1-80. 1980 Methods of Test for Clear Finishes for Wooden Furniture Part 1: Assessment of Low Angle Glare by Measurements of Specular Gloss at 85 Degrees. 7 pp.

Finishes (Cont.)

—Glass Containers—Heights
ISO 9009-91. Glass Containers—Height and Non-Parallelism of Finish with Reference to Container Base—Test Methods First Edition. 7 pp.

—Glossaries
BSI BS 6100: SUBSEC 1.3.7-91. 1991 Glossary of Building and Civil Engineering Terms Part 1: General and Miscellaneous Section 1.3: Parts of Construction Works Subsection 1.3.7: Finishes. 21 pp.
BSI BS 6100: SUBSEC 1.3.7-01. 1991 Amd 1 Glossary of Building and Civil Engineering Terms Part 1: General and Miscellaneous Section 1.3: Parts of Construction Works Subsection 1.3.7: Finishes (AMD 7237) August 15, 1992. 22 pp.

—Hardware
CGSB CAN/CGSB-69.34-M90. Materials and Finishes (ANSI/BHMA A156.18-1984). 15 pp.

—Metal—Corrosion Inhibitive
MOD UK DSTAN 03-30: Part 2-91. Treatments for the Protection of Metal Parts of Service Stores and Equipment Against Corrosion Part 2: Schedule of Metallic and Allied Finishes for Iron and Steel and Non-Ferrous Metals and Alloys Issue 1. 161 pp.
MOD UK DG-8: Pt 1: Sec I. Defence Guide: Treatments for the Protection of Metal Parts of Service Stores and Equipments Against Corrosion Part 1: Schedule of Protective Coatings Section 1: Metallic and Allied Finishes for Iron and Steel Issue 2 (Superseded by Def Stan 03-30: Part 2).
MOD UK DG-8: Pt 1: Sec II. Defence Guide: Treatments for the Protection of Metal Parts of Service Stores and Equipments Against Corrosion Part 1: Schedule of Protective Coatings Section 2: Metallic and Allied Finishes for Non-Ferrous Metals and Alloys Issue 2.

—School Furniture
BSI BS 5873: Part 1-80. 1980 Amd 1 Educational Furniture Part 1: Functional Dimensions, Identification and Finish of Chairs and Tables for Educational Institutions. 13 pp.

—Scrub Resistance
CGSB 1-GP-71 METH 125.2-79. Methods of Testing Paints and Pigments Scrubbability Emulsion Finishes. 1 p.

—Vacuum Lug—Glass Containers
ISO 9100-92. Wide-Mouth Glass Containers—Vacuum Lug Finishes—Dimensions First Edition. 8 pp.

—Walls—Interior—Scrub Resistance
CGSB 1-GP-71 METH 125.4-79. Methods of Testing Paints and Pigments Scrubbability Interior Flat Wall Finishes, Emulsion and Alkyd Types. 1 p.

Finishing Agents
Use: Finishing Compounds

Finishing Cements
See Also: Cements
CGSB CAN/CGSB-51.12-M86. Cement, Thermal Insulating and Finishing. 11 pp.

—Thermal Insulation
BSI BS 3958: Part 6-72. 1972 Amd 1 Thermal Insulating Materials Part 6: Finishing Materials; Hard Setting Composition, Self-Setting Cement and Gypsum Plaster. 19 pp.

Finishing Compounds

—Aeronautical—Metals
BSI X 27-66. (WITHDRAWN) 1966 Cellulose Finishing Scheme for Aeronautical Purposes (Superseded by 2X 27: 1991). 5 pp.
BSI X 29-66. 1966 High Gloss Finishing Scheme for Aeronautical Purposes. 6 pp.
BSI X 30-66. (WITHDRAWN) 1966 Light-Weight Aluminium Finishing Scheme (Cellulose Base) for Aeronautical Purposes (Superseded by 2X 27: 1991). 5 pp.

—Aerospace—Metals
BSI 2X 27-91. 1991 Cellulose Finishing Scheme for Aerospace Purposes. 9 pp.

—Aircraft—Fabrics
BSI X 26-66. 1966 Doping and Finishing Schemes for Fabric Covered Aircraft. 7 pp.

—Fabrics
JIS L 1063-85. Testing Methods for Migration of Dyestuffs and Finishing Agents on Woven Fabrics and Knitted Fabrics.

—Leather—Adhesion Testing
CNS K6818-85. Method of Test for Adhesion of Finish to Leather (Dec)(11455).

Finishing Compounds (Cont.)

—Leather—Adhesion Testing (Cont.)
JIS K 6555-79. Testing Method for Adhesion of Finish to Leathers.

—Leather—Flexural Strength
JIS K 6545-70. Testing Method for Flexing Endurance of Light Leathers and Their Surface Finishes.

—Rubber
CNS K4022-70. Finishing Agent for Rubber Goods (Tentative) (Sep)(3141). 2 pp.

—Synthetic Fibers
JIS K 3361-79. Method of Testing for Finishing Agent of Higher Alcohols for Synthetic Fibers (R 1984). 15 pp.

—Wood—Latex
SAA AS 3730.16-91. Guide to the Properties of Paints for Buildings—Part 16: Latex—Timber Finish—Exterior. 5 pp.

Finishing Machinery
See Also: Construction Equipment; Honing Machinery; Lapping Machines; Machine Tools; Textile Machinery

—Asphalt
CNS A4029-85. Method of Test for Asphalt Finishers (Feb)(11193).
JIS A 8702-85. Test Code of Asphalt Finishers. 29 pp.

—Asphalt—Specification Forms
CNS A4028-85. Standard Form of Specification of Asphalt Finishers (Feb)(11192).
JIS A 8701-85. Standard Form of Specifications of Asphalt Finishers. 15 pp.

—Glossaries
CNS B1230-80. Glossary of Terms Relating to Parts of Surface-Finishing Machine (Dec)(6754).
CNS B1233-80. Glossary of Terms Relating to Parts of Finishing Machine (Dec)(6757).

—Textiles—Designations
BSI BS ISO 10459-92. 1992 Textile Machinery—Dyeing and Finishing Machines—Designation of Operating Ranges of Component Parts (L). 7 pp.
ISO 10459-92. Textile Machinery—Dyeing and Finishing Machines—Designation of Operating Ranges of Component Parts First Edition. 5 pp.

—Textiles—Mechanical Guides
JIS L 7508-92. Pneumatic Cloth Guiders for Dyeing and Finishing.

—Textiles—Nominal Speeds
BSI BS ISO 10457-91. 1991 Textile Machinery—Dyeing and Finishing Machines—Nominal Speeds (L). 7 pp.
ISO 10457-91. Textile Machinery—Dyeing and Finishing Machines—Nominal Speeds First Edition. 4 pp.

Finishing Paints
Use: Paints

Finnpipettes
Use: Pipettes

Fire Alarm and Extinguishing Equipment
Scope Note: Includes automated fire control systems
Use For: Automatic Sprinklers (Fire Protection); Fire Alarm Systems; Fire Detection Systems
See Also: Alarm Cables; Alarm Systems; Dry Pipe Valves; Fire Alarm Signals; Fire Alarm Switches; Fire Extinguishers; Fire Extinguishing Agents; Fire Fighting Equipment; Fire Protection; Fire Sprinkler Equipment; Fires; Firesafety; Flame Detectors; Heat Detectors; Smoke Detectors; Warning Systems

BSI BS 5306: Part 0-86. 1986 Amd 2 Fire Extinguishing Installations and Equipment on Premises Part 0: Guide for the Selection of Installed Systems and Other Fire Equipment (AMD 6653) July 31, 1991. 44 pp.
CEN EN 54-1-76. Components of Automatic Fire Detection Systems—Part 1: Introduction. 7 pp.
CEN PR EN 54-1-87. Components for Automatic Fire Detection Systems Part 1: Introduction. 6 pp.
CNS Z2040-86. Fire Alarm Equipments General Rules (Oct)(8873).
CNS Z2041-86. Detector of Fire Alarm Equipment (Oct)(8874).
DIN VDE 0833 Pt 1-89. Alarm Systems for Fire, Intrusion and Hold-Up; General Requirements (Jan). 25 pp.
DIN VDE 0833 Pt 2-82. Alarm Systems for Fire, Burglary and Hold-Ups; Requirements for Fire Alarm Systems (FAS) (VDE Specification) (Aug). 8 pp.

Fire Alarm and Extinguishing Equipment (Cont.)

ISO 6182 Pt 1-93. Fire Protection—Automatic Sprinkler Systems—Part 1: Requirements and Test Methods for Sprinklers First Edition. 37 pp.
SNZ NZS 4512-81. Automatic Fire Alarm Systems in Buildings Amend: 1, 1987; 2, 1990. 28 pp.
SNZ NZS 4561-73. Specification for Manual Fire Alarm Systems for Use in Buildings. 19 pp.

—Agricultural Buildings
BSI BS 5502: Part 52-91. 1991 Buildings and Structures for Agriculture Part 52: Code of Practice for Design of Alarm Systems and Emergency Ventilation for Livestock Housing. 14 pp.

—Automatic Control Equipment
SAA AS 1603.4-87. Automatic Fire Detection and Alarm Systems—Part 4: Control and Indicating Equipment Amdt 1 June 1988 Amdt 2 October 1989. 32 pp.

—Carbon Dioxide
BSI BS 5306: Part 4-86. 1986 Fire Extinguishing Installations and Equipment on Premises Part 4: Carbon Dioxide Systems. 42 pp.

—Design
BSI BS 5839: Part 1-88. 1988 Amd 1 Fire Detection and Alarm Systems for Buildings Part 1: Code of Practice for System Design, Installation and Servicing (AMD 6317) January 31, 1991. 72 pp.
BSI BS 5839: Part 1-02. 1988 Amd 2 Fire Detection and Alarm Systems for Buildings Part 1: Code of Practice for System Design, Installation and Servicing (AMD 6874) May 1, 1992. 75 pp.
BSI PD 6531-92. 1992 Queries and Interpretations on BS 5839: Parts 1 and 4 (as Amended). 13 pp.
SAA AS 1670-86. Automatic Fire Detection and Alarm Systems—System Design, Installation, and Commissioning Amdt 1 January 1987 Amdt 2 March 1988 Amdt 3 June 1988. 38 pp.

—Dry Chemical
BSI BS 5306: Part 7-88. 1988 Fire Extinguishing Installations and Equipment on Premises Part 7: Powder Systems. 35 pp.
CEN PREN 615-91. Fire Protection—Fire Extinguishing Media—Powder. 14 pp.

—Dry Pipe Valves
ISO 6182 Pt 3-93. Fire Protection—Automatic Sprinkler Systems—Part 3: Requirements and Test Methods for Dry Pipe Valves First Edition. 18 pp.

—Dry Pipe Valves—Quick Opening
ISO 6182 Pt 4-93. Fire Protection—Automatic Sprinkler Systems—Part 4: Requirements and Test Methods for Quick-Opening Devices First Edition. 11 pp.

—Electric Wire
BSI DD 190-90. 1990 Performance of Alarm and Emergency Wiring Systems Under Fire Conditions (E). 15 pp.

—Electric Wiring—Architectural Drawings—Symbols
CNS C1102-87. Symbols of Fire Alarm System of Interior Wiring Diagrams for Architectural Plans (Dec)(9113).

—Electrical Equipment
BSI BS 7273: Part 1-90. 1990 Operation of Fire Protection Measures Part 1: Electrical Actuation of Gaseous Total Flooding Extinguishing Systems. 12 pp.
BSI BS 7273: Part 2-92. 1992 Operation of Fire Protection Measures Part 2: Mechanical Actuation of Gaseous Total Flooding and Local Application Extinguishing Systems. 12 pp.

—Foam
BSI BS 5306: Sec 6.1-88. 1988 Fire Extinguishing Installations and Equipment on Premises Part 6: Foam Systems Section 6.1: Low Expansion Foam Systems. 32 pp.
BSI BS 5306: Sec 6.2-89. 1989 Fire Extinguishing Installations and Equipment on Premises Part 6: Foam Systems Section 6.2: Medium and High Expansion Foam Systems. 26 pp.

—Glossaries
BSI BS 4422: Part 3-90. 1990 Glossary of Terms Associated with Fire Part 3: Fire Detection and Alarm. 10 pp.
BSI BS 4422: Part 3-72. (WITHDRAWN) 1972 Glossary of Terms Associated with Fire Part 3: Means of Escape. 9 pp.
ISO 7240 Pt 1-88. Fire Detection and Alarm Systems—Part 1: General and Definitions First Edition. 9 pp.
ISO 8421 Pt 3-89. Fire Protection—Vocabulary—Part 3: Fire Detection and Alarm First Edition. 11 pp.

Fire Alarm and Extinguishing Equipment (Cont.)

—Ground Effect Machines
CAA Chapter B6-10 08.83. Fire Extinguishing Systems. 2 pp.
CAA Chapter B6-10 App 08.83. Fire Detection and Extinguishing Systems. 2 pp.

—Halon
BSI BS 5306: Sec 5.1-92. 1992 Fire Extinguishing Installations and Equipment on Premises Part 5: Halon Systems Section 5.1: Specification for Halon 1301 Total Flooding Systems. 66 pp.
BSI BS 5306: Sec 5.1-82. 1982 Amd 1 Fire Extinguishing Installations and Equipment on Premises Part 5: Halon Systems Section 5.1: Halon 1301 Total Flooding Systems. 48 pp.
BSI BS 5306: Sec 5.2-84. 1984 Fire Extinguishing Installations and Equipment on Premises Part 5: Halon Systems Section 5.2: Halon 1211 Total Flooding Systems. 40 pp.
SAA AS 1851.11-91. Maintenance of Fire Protection Equipment—Part 11: Halon 1301 Total Flooding Systems (in Professional Packages 21, 30, 44). 7 pp.

—Indicating Instruments
BSI BS 5839: Part 4-88. 1988 Amd 1 Fire Detection and Alarm Systems for Buildings Part 4: Specification for Control and Indicating Equipment (AMD 6654) December 21, 1990. 24 pp.
BSI BS 5839: Part 4-02. 1988 Amd 2 Fire Detection and Alarm Systems for Buildings Part 4: Specification for Control and Indicating Equipment (AMD 6875) May 1, 1992 (Supersedes BS 3116: Part 4: 1974). 26 pp.
SAA AS 1603.4-87. Automatic Fire Detection and Alarm Systems—Part 4: Control and Indicating Equipment Amdt 1 June 1988 Amdt 2 October 1989. 32 pp.
SAA AS 4050(INT)-92. Fire Detection and Fire Alarm Systems—Firefighters' Control and Indicating Facilities (in Professional Package 44). 2 pp.

—Installation
BSI BS 5839: Part 1-88. 1988 Amd 1 Fire Detection and Alarm Systems for Buildings Part 1: Code of Practice for System Design, Installation and Servicing (AMD 6317) January 31, 1991. 72 pp.
BSI BS 5839: Part 1-02. 1988 Amd 2 Fire Detection and Alarm Systems for Buildings Part 1: Code of Practice for System Design, Installation and Servicing (AMD 6874) May 1, 1992. 75 pp.
BSI PD 6531-92. 1992 Queries and Interpretations on BS 5839: Parts 1 and 4 (as Amended). 13 pp.
SAA AS 1670-86. Automatic Fire Detection and Alarm Systems—System Design, Installation, and Commissioning Amdt 1 January 1987 Amdt 2 March 1988 Amdt 3 June 1988. 38 pp.

—Maintenance
SAA AS 1851.8-87. Maintenance of Fire Protection Equipment—Part 8: Automatic Fire Detection and Alarm Systems. 7 pp.
SAA AS 1851.8 Supp 1-90. Maintenance of Fire Protection Equipment—Part 8: Automatic Fire Detection and Alarm Systems—Supplement 1: System Certificates and Maintenance Records (Supplement to AS 1851.8 —1987). 78 pp.
SAA AS 1851.10-89. Maintenance of Fire Protection Equipment—Part 10: Emergency Warning and Intercommunication Systems. 2 pp.

—Manual Call Points
SAA AS 1603.5-91. Automatic Fire Detection and Alarm Systems—Part 5: Manual Call Points (In Professional Packages 30,44). 7 pp.

—Manual Controls
BSI BS 5839: Part 4-88. 1988 Amd 1 Fire Detection and Alarm Systems for Buildings Part 4: Specification for Control and Indicating Equipment (AMD 6654) December 21, 1990. 24 pp.
BSI BS 5839: Part 4-02. 1988 Amd 2 Fire Detection and Alarm Systems for Buildings Part 4: Specification for Control and Indicating Equipment (AMD 6875) May 1, 1992 (Supersedes BS 3116: Part 4: 1974). 26 pp.
BSI PD 6531-92. 1992 Queries and Interpretations on BS 5839: Parts 1 and 4 (as Amended). 13 pp.
SAA AS 4050(INT)-92. Fire Detection and Fire Alarm Systems—Firefighters' Control and Indicating Facilities (in Professional Package 44). 2 pp.

—Paints-Alkyd Enamel
CGSB CAN/CGSB-1.88-92. Gloss Alkyd Enamel, Air Drying and Baking. 9 pp.

—Paints-Alkyd Enamel—Exterior
CGSB CAN/CGSB-1.59-M89. Alkyd, Exterior Gloss Enamel. 11 pp.

Fire Alarm and Extinguishing Equipment (Cont.)

—Power Supplies
CEN PREN 54 (Part 4)-89. Components of Automatic Fire Detection Systems; Part 4: Power Supply Equipment. 23 pp.

—Receivers
CNS Z2044-86. Receiver of Fire Alarm (Oct)(8877).
CNS Z3028-86. Method of Test for Receiver of Fire Alarm Equipment (Oct)(11039).

—Repeaters
CNS Z2042-86. Repeater of Fire Alarm Equipment (Oct)(8875).
CNS Z3027-86. Method of Test for Repeater of Fire Alarm Equipment (Oct)(11038).

—Retarding Chambers
ISO 6182 Pt 2-93. Fire Protection—Automatic Sprinkler Systems—Part 2: Requirements and Test Methods for Wet Alarm Valves, Retard Chambers and Water Motor Alarms First Edition. 20 pp.

—Sensitivity
BSI BS 5445: Part 9-84. 1984 Components of Automatic Fire Detection Systems Part 9: Methods of Test of Sensitivity in Fire. 14 pp.
CEN EN 54-9-82. Components of Automatic Fire Fetection Systems Part 9 Fire Sensitivity Test. 10 pp.

—Ships
MOD UK NES 602-91. Guide to the Installation of Thorn Security Automatic Fire Detection Systems Issue 2 (03.91). 338 pp.
MOD UK NES 603-90. Guide to Policy, Design and Installation of Fire Detection Systems in Ships Issue 1 (04.90). 74 pp.

—Submarines
MOD UK NES 602-91. Guide to the Installation of Thorn Security Automatic Fire Detection Systems Issue 2 (03.91). 338 pp.
MOD UK NES 603-90. Guide to Policy, Design and Installation of Fire Detection Systems in Ships Issue 1 (04.90). 74 pp.

—Tanks (Combat Vehicles)
NATO STANAG 4317 ED 1 AMD 0-00. (DRAFT) Specification of Common Characteristics for Fire Detection and Fighting System for Main Battle Tanks. 33 pp.
NATO STANAG 4317 ED 1 AMD 0-92. Specification of Common Characteristics for Fire Detection and Fire Fighting Systems for Future Main Battle Tanks. 32 pp.

—Water Motor Alarms
ISO 6182 Pt 2-93. Fire Protection—Automatic Sprinkler Systems—Part 2: Requirements and Test Methods for Wet Alarm Valves, Retard Chambers and Water Motor Alarms First Edition. 20 pp.

Fire Alarm Cables
Use: Alarm Cables

Fire Alarm Call Points
Use: Fire Alarm Switches

Fire Alarm Signals
Scope Note: Includes fire signaling devices, e.g., bells, horns, etc.; for automated fire alarm and control systems, see the listings under Fire Alarm and Extinguishing Equipment *See Also:* Alarm Signals; Alarm Valves; Fire Alarm and Extinguishing Equipment; Signal Devices
BSI BS 5839: Part 4-88. 1988 Amd 1 Fire Detection and Alarm Systems for Buildings Part 4: Specification for Control and Indicating Equipment (AMD 6654) December 21, 1990. 24 pp.
BSI BS 5839: Part 4-02. 1988 Amd 2 Fire Detection and Alarm Systems for Buildings Part 4: Specification for Control and Indicating Equipment (AMD 6875) May 1, 1992 (Supersedes BS 3116: Part 4: 1974). 26 pp.
BSI PD 6531-92. 1992 Queries and Interpretations on BS 5839: Parts 1 and 4 (as Amended). 13 pp.
CNS Z2043-86. Signal Generator of Fire Alarm (Oct)(8876).
CNS Z3026-86. Method of Test for Detector & Signal Generator of Fire Alarm Equipment (Oct)(11037).
SAA AS 1603.6-87. Automatic Fire Detection and Alarm Systems—Part 6: Fire Alarm Bells. 9 pp.
SAA AS 4050(INT)-92. Fire Detection and Fire Alarm Systems—Firefighters' Control and Indicating Facilities (in Professional Package 44). 2 pp.

Fire Alarm Stations
Use: Fire Alarm Switches

Fire Alarm Switches
Use For: Call Points; Fire Alarm Call Points; Fire Alarm Stations; Manual Call Points *See Also:* Fire Alarm and Extinguishing Equipment; Switches
BSI BS 5839: Part 2-83. 1983 Amd 2 Fire Detection and Alarm Systems in Buildings Part 2: Specifications for Manual Call Points. 16 pp.
CEN PREN54-11-89. Components of Automatic Fire Detection Systems Part 11: Manual Call Points. 23 pp.
CEN PREN 54-11-91. Components of Automatic Fire Detection Systems: Part 11: Manual Call Points. 23 pp.

Fire Alarm Systems
Use: Fire Alarm and Extinguishing Equipment

Fire Barriers
See Also: Barriers; Fire Dampers; Fire Doors; Fire Protection; Fire Walls

—Fire Testing
DIN ENGL 4102 Pt 5-77. Fire Behaviour of Building Materials and Building Components; Fire Barriers, Barriers in Lift Wells and Glazings Resistant Against Fire; Definitions, Requirements and Tests (Sept) (Superseded in Part by DIN 4102 Part 13). 8 pp.

—Naval Ships
MOD UK NES 703-01. Thermal and Acoustic Insulation of Hull and Machinery Issue 1 (01.81); Corrigendum. 118 pp.
MOD UK NES 703-81. Thermal and Acoustic Insulation of Hull and Machinery Issue 1 (01.81) (Amendment 1 Incorporated). 120 pp.

—Submarines
MOD UK NES 703-01. Thermal and Acoustic Insulation of Hull and Machinery Issue 1 (01.81); Corrigendum. 118 pp.
MOD UK NES 703-81. Thermal and Acoustic Insulation of Hull and Machinery Issue 1 (01.81) (Amendment 1 Incorporated). 120 pp.

Fire Blankets
Use For: Blankets (Fire) *See Also:* Blankets; Fire Fighting Equipment
BSI BS 6575-85. 1985 Fire Blankets. 17 pp.
SNZ NZS 4516-91. Fire Blankets Amend: A, 1991. 8 pp.

Fire Bricks
Use: Firebricks

Fire Codes
See Also: Standards
SAA HB37.1-93. Handbook of Australian Fire Standards—Part 1: Fire—General (in Professional Package 44). 17 pp.

—Model Bylaws
SNZ NZS 9231-71. Model Bylaw for Fire Prevention Amend: 1, 1973; 2, 1979. 14 pp.

Fire Dampers
See Also: Dampers; Ducts; Fire Barriers
SAA AS 1682.1-90. Fire Dampers—Part 1: Specification. 9 pp.
SAA AS 1682.2-90. Fire Dampers—Part 2: Installation. 7 pp.

—Fire Testing
DIN ENGL 4102 Pt 6-77. Fire Behaviour of Building Materials and Building Components; Ventilation Ducts; Definitions, Requirements and Tests (Sept). 5 pp.

—Ships
JIS F 2422-88. Ship's Fire Dampers.

—Smoke Leakage
JIS A 1314-92. Smoke-Proof Test Method for Fire Damper. 9 pp.

Fire Detection Systems
Use: Fire Alarm and Extinguishing Equipment

Fire Doors
See Also: Doors; Fire Barriers; Fire Walls; Firesafety; Smoke Curtains
BSI PD 6512: Part 1-85. (WITHDRAWN) 1985 Use of Elements of Structural Protection with Particular Reference to the Recommendations Given in BS 5588 Fire Precautions in the Design and Construction of Buildings Part 1: Guide to Fire Doors. 10 pp.
SAA AS 1905.1-90. Components for the Protection of Openings in Fire-Resistant Walls—Part 1: Fire-Resistant Doorsets. 30 pp.

INTERNATIONAL AND NON-U.S. NATIONAL STANDARDS
SUBJECT INDEX

Fire Doors (Cont.)
SAA AS 1905.1 Supp 1-88. Components for the Protection of Openings in Fire-Resistant Walls—Part 1: Fire-Resistant Doorsets—Supplement 1: Logbook for the Maintenance of Fire-Resistant Doorsets (Supplement to AS 1905. 1—1984). 21 pp.
SNZ NZS 4232: Part 1-88. Performance Criteria for Fire Resisting Closures Part 1: Internal and External Fire Doorsets Amend: 1, 1991. 38 pp.

—**Assemblies**
BSI BS 8214-90. 1990 Code of Practice for Fire Door Assemblies with Non-Metallic Leaves. 37 pp.
BSI BS 8214-01. 1990 Amd 1 Code of Practice for Fire Door Assemblies with Non-Metallic Leaves (AMD 7438) November 15, 1992. 38 pp.

—**Elevators (Lifts)**
SAA AS 1735.11-86. Lifts, Escalators, and Moving Walks (Known as the SAA Lift Code)—Part 11: Fire-Rated Landing Doors. 6 pp.
SNZ NZS/AS 1735. 11-86. Lifts, Escalators and Moving Walks (SAA Lift Code) Part 11: Fire-Rated Landing Doors (This is a Joint Standard with SAA AS 1735.11). 6 pp.

—**Fire Testing**
CNS A3223-85. Method of Fire-Protecting Test for Fire Door of Buildings (Apr)(11227).
JIS A 1311-75. Method of Fire Protecting Test of Fire Door for Buildings. 10 pp.

—**Maintenance**
SAA AS 1851.7-84. Maintenance of Fire Protection Equipment—Part 7: Fire-Resistant Doorsets. 6 pp.

—**Rolling**
JIS A 4705-91. Components of Rolling Fire Door for Buildings. 21 pp.
SAA AS 1905.2-89. Components for the Protection of Openings in Fire-Resistant Walls—Part 2: Fire-Resistant Roller Shutters. 8 pp.

—**Rolling—Inspection**
JIS A 1313-84. Inspection Standard of Rolling Fire Doors. 9 pp.

—**Ships—Steel**
JIS F 2337-87. Ships' Fire Retarding Steel Doors.

—**Smoke Control**
BSI BS 476: Sec 31.1-83. 1983 Fire Tests on Building Materials and Structures Part 31: Methods for Measuring Smoke Penetration Through Doorsets and Shutter Assemblies Section 31.1: Method of Measurement Under Ambient Temperature Conditions. 11 pp.
DIN ENGL 18095 Pt 1-88. Smoke Control Doors; Concepts and Requirements (Oct). 6 pp.
ISO 5925 Pt 1-81. Fire Tests—Evaluation of Performance of Smoke Control Door Assemblies—Part 1: Ambient Temperature Test First Edition. 8 pp.

—**Smoke Control—Leakage**
DIN ENGL 18095 Pt 2-91. Smoke Control Doors; Type Testing for Durability and Leakage (Mar). 8 pp.

—**Smoke Control—Life (Durability)**
DIN ENGL 18095 Pt 2-91. Smoke Control Doors; Type Testing for Durability and Leakage (Mar). 8 pp.

—**Spring Hinges**
CNS A2116-81. Adjustable Non-Load-Bearing Spring Hinge for Fire Protection Doors (Oct)(7936).
DIN ENGL 18262-69. Adjustable Non-Load-Bearing Spring Hinge for Fire Protecting Doors (May). 2 pp.

Fire Extinguishers
Scope Note: Includes mobile and portable equipment; for automated fire control systems see the listings under Fire Alarm and Extinguishing Equipment.
See Also: Fire Alarm and Extinguishing Equipment; Fire Extinguishing Agents; Fire Fighting Equipment; Fire Protection
BSI BS 5306: Part 3-85. 1985 Amd 1 Fire Extinguishing Installations and Equipment on Premises Part 3: Code of Practice for Selection, Installation and Maintenance of Portable Fire Extinguishers. 26 pp.
BSI BS 5306: Part 3-02. 1985 Amd 2 Fire Extinguishing Installations and Equipment on Premises Part 3: Code of Practice for Selection, Installation and Maintenance of Portable Fire Extinguishers (AMD 7739) September 15, 1993 (R). 32 pp.
BSI BS 5423-87. 1987 Amd 1 Portable Fire Extinguishers (AMD 5989) May 31, 1989. 38 pp.
CEN EN 3 (Part 1)-75. Fire Fighting Portable Fire Extinguishers—Part 1. 10 pp.
CEN EN 3 (Part 1) A1-86. AMD 1 Fire Fighting Portable Fire Extinguishers—Part 1. 4 pp.

Fire Extinguishers (Cont.)
CEN PREN 3-1-93. Fight Against Fire—Portable Fire Extinguishers—Part 1: Description, Duration of Operation, A and B Fire Test. 15 pp.
CEN EN 3-4-84. Portable Fire Extinguishers—Part 4. 6 pp.
CEN PREN 3-4-93. Fight Against Fire—Portable Fire Extinguishers—Part 4: Charges, Minimum Fire Performance. 7 pp.
CEN PREN 3-6-93. Fight Against Fire—Portable Fire Extinguishers—Part 6: Provisions for the Attestation of Conformity of Portable Fire Extinguishers to Parts 1 to 5 of EN 3. 30 pp.
CNS Z2057-85. Aerosol Type Mini Fire Extinguisher (Jul)(10978).
MOD UK DSTAN 42-20-01. Extinguisher, Fire, Vaporizing Liquid, (Bromochlorodifluoro-methane (BCF) 45 Kilogram Capacity, Trolley Type-Metric Issue 1; Amendment 1. 11 pp.
MOD UK DSTAN 42-32-85. Extinguisher, Fire, Vaporizing Liquid, Bromochlorodifluoro-methane (BCF) 25 Kilogram Capacity, Trolley Type-Metric Issue 1. 13 pp.
SAA MP26-82. Portable Fire Extinguisher Selection Chart. 2 pp.
SAA AS 1841.1-92. Portable Fire Extinguish-ers—Pt. 1: General Requirements (Supersedes AS 1841—1985, AS 1842—1985, AS1844—1985, AS 1845—1985, AS 1846—1985, AS1847—1985, AS 1848—1985 (All in Part) and AS 1849—1976 (in Full)) (in Prof. Packages 30, 44) Amdt 1 January/February 1993. 28 pp.
SAA AS 1841.2-92. Portable Fire Extinguishers—Part 2: Water Type (Supersedes AS 1841—1985 and AS 1842—1985 (Both in Part)) Amdt 1 January/February 1993 (in Professional Packages 30, 44). 3 pp.
SAA AS 1841.3-92. Portable Fire Extinguishers—Part 3: Wet Chemical Type (Supersedes AS 1842—1985 (in Part)) Amdt 1 January/February 1993 (in Professional Packages 30, 44). 3 pp.
SNZ NZS 4506-78. Specification for Portable Fire Extinguishers of the Water, Foam and Dry Powder Types Amend: 1, 1979. 16 pp.

—**Aircraft**
CAA NOTICE #60 ISSUE 1. Cabin and Toilet Fire Protection (Airworthiness Notices). 3 pp.

—**Aircraft Engine**
MOD UK DSTAN 42-25-01. Extinguisher, Fire, Vaporizing Liquid (Bromochlorodifluoro-methane (BCF) 12 Kilogram Capacity, Skid Type-Metric Issue 1; Amendment 1. 13 pp.

—**Back Pack**
CGSB 28-GP-18M-83. Extinguisher, Fire, Water, Back-Pack, Standard for. 18 pp.

—**Bombs**
CNS Z2004-87. Fire Extinguishing Bombs (Feb)(1408).

—**Bucket Pumps**
SNZ NZS 4511-79. Specification for Bucket Pump Fire Extinguishers. 7 pp.

—**Carbon Dioxide**
MOD UK DSTAN 42-15-01. Extinguishers, Fire, Carbon Dioxide 10 Kilogram Capacity, Trolley Type—Metric Issue 2; Amendment 1. 14 pp.
MOD UK DSTAN 42-17-01. Extinguishers, Fire, Carbon Dioxide 2 Kilogram and 5 Kilogram Capacity-Metric Issue 2; Amendment 3 (Withdrawn). 15 pp.
SAA AS 1841.6-92. Portable Fire Extinguishers—Part 6: Carbon Dioxide Type (Supersedes AS 1847—1985 (in Part)) Amdt 1 January/February 1993 (in Professional Packages 30, 44). 4 pp.
SNZ NZS 4508-79. Portable Carbon Dioxide Fire Extinguishers Amend: 1, 1980; 2, 1991. 11 pp.

—**Carbon Dioxide—Design**
ISO 6183-90. Fire Protection Equipment—Carbon Dioxide Extinguishing Systems for Use on Premises—Design and Installation First Edition. 27 pp.

—**Carbon Dioxide—Destroyers**
CNS Z2060-84. Capacity Valves Safety Devices and Destroyer for Carbon Dioxide Fire Extinguisher ETC (Dec)(11176).

—**Carbon Dioxide—Hoses**
CNS Z2061-84. Hoses, Nozzles, Nozzle Valves and Hoses Reel for Movable Carbon Dioxide Fire Extinguisher ETC (Dec)(11177).

—**Carbon Dioxide—Installation**
ISO 6183-90. Fire Protection Equipment—Carbon Dioxide Extinguishing Systems for Use on Premises—Design and Installation First Edition. 27 pp.

Fire Extinguishers (Cont.)
—**Carbon Dioxide—Nozzles**
CNS Z2061-84. Hoses, Nozzles, Nozzle Valves and Hoses Reel for Movable Carbon Dioxide Fire Extinguisher ETC (Dec)(11177).

—**Carbon Dioxide—Nozzles—Valves**
CNS Z2061-84. Hoses, Nozzles, Nozzle Valves and Hoses Reel for Movable Carbon Dioxide Fire Extinguisher ETC (Dec)(11177).

—**Carbon Dioxide—Safety Valves**
CNS Z2060-84. Capacity Valves Safety Devices and Destroyer for Carbon Dioxide Fire Extinguisher ETC (Dec)(11176).

—**Classification**
SAA AS 1850-81. Portable Fire Extinguishers—Classification, Rating and Fire Testing Amdt 1 April 1982 Amdt 2 July 1989. 11 pp.
SNZ NZS 4507-78. Specification for Fire Testing and Rating of Portable Fire Extinguishers. 8 pp.

—**Compression Testing**
CEN EN 3-2-78. Portable Fire Extinguishers—Part 2. 7 pp.
CEN PREN 3-2-93. Fight Against Fire—Portable Fire Extinguishers—Part 2: Retention of Charge, Dielectric Test, Compaction, Special Provision. 9 pp.
CEN EN 3-5-84. Portable Fire Extinguishers—Part 5: Complementary Requirements and Tests. 17 pp.
CEN PREN 3-5-93. Fight Against Fire—Portable Fire Extinguishers—Part 5: Complementary Requirements and Tests. 23 pp.

—**Corrosion Testing**
CEN EN 3-5-84. Portable Fire Extinguishers—Part 5: Complementary Requirements and Tests. 17 pp.
CEN PREN 3-5-93. Fight Against Fire—Portable Fire Extinguishers—Part 5: Complementary Requirements and Tests. 23 pp.

—**Dielectric Testing**
CEN EN 3-2-78. Portable Fire Extinguishers—Part 2. 7 pp.
CEN PREN 3-2-93. Fight Against Fire—Portable Fire Extinguishers—Part 2: Retention of Charge, Dielectric Test, Compaction, Special Provision. 9 pp.

—**Disposable**
BSI BS 6165-92. 1992 Small Disposable Fire Extinguishers of the Aerosol Type. 19 pp.
BSI BS 6165-81. 1981 Amd 1 Small Disposable Fire Extinguishers of the Aerosol Type. 19 pp.

—**Dry Chemical**
MOD UK DSTAN 42-6-01. Extinguisher, Fire, Dry Powder 25 lb (11 kg) Capacity Issue 2; Amendment 1. 9 pp.

—**Dry Powder**
CNS Z2021-87. Compressed Dry Chemical Powder Extinguishers, Portable and Wheeled Types (Mar)(3638).
MOD UK DSTAN 42-23-01. Extinguisher, Fire, Dry Powder, 150 lb Capacity, Trolley Type Issue 1; Amendment 1. 12 pp.
MOD UK DSTAN 42-36-92. Extinguisher, Fire, ABC Dry Powder, Stored Pressure, 2kg Capacity Issue 1. 9 pp.

—**Fire Testing**
SAA AS 1850-81. Portable Fire Extinguishers—Classification, Rating and Fire Testing Amdt 1 April 1982 Amdt 2 July 1989. 11 pp.
SNZ NZS 4507-78. Specification for Fire Testing and Rating of Portable Fire Extinguishers. 8 pp.

—**Foam**
CNS Z2001-84. Portable and Wheeled Types of Foam Fire Extinguisher (Dec)(441).
SAA AS 1841.4-92. Portable Fire Extinguishers—Part 4: Foam Type (Supersedes AS 1844—1985 and AS 1845—1985 (Both in Part)) Amdt 1 January/February 1993 (in Professional Packages 30, 44). 3 pp.

—**Foam—Aqueous Film Forming**
MOD UK DSTAN 42-33-93. Extinguishers, Fire AFFF, 90 Litre (20 Gallon) Capacity Trolley Mounted Issue 1. 16 pp.

—**Foam Inlets**
BSI BS 5306: Part 1-76. 1976 Amd 2 Fire Extinguishing Installations and Equipment on Premises Part 1: Hydrant Systems, Hose Reels and Foam Inlets (Formerly CP 402.101) (AMD 5756) February 29, 1988. 24 pp.

—**Glossaries**
ISO 8421 Pt 4-90. Fire Protection—Vocabulary—Part 4: Fire Extinction Equipment First Edition. 18 pp.

INDUSTRY STANDARDS

Fire Extinguishers (Cont.)

—Halon
BSI BS 6643: Part 1-85. 1985 Amd 2 Recharging Fire Extinguishers Part 1: Procedure and Materials. 8 pp.
BSI BS 6643: Part 1-03. 1985 Amd 3 Recharging Fire Extinguishers (Manufactured to BS 5423 'Specification for Portable Fire Extinguishers') Part 1: Specification for Procedure and Materials (AMD 7740) September 15, 1993 (N). 10 pp.
BSI BS 7327: Part 1-90. 1990 Small Fixed Fire Extinguishers Part 1: Small Fixed Halon Fire Extinguishers. 20 pp.
SAA AS 3689.2-91. Automatic Fire Extinguishing Systems Using Halogenated Hydrocarbons—Part 2: Mechanical Components Amdt 1 June 1991 (in Professional Package 44). 30 pp.
SNZ NZS 4551-74. Specification for Portable Fire Extinguishers of the Halogenated Hydrocarbon Type. 20 pp.

—Hose Reels
CNS Z2061-84. Hoses, Nozzles, Nozzle Valves and Hoses Reel for Movable Carbon Dioxide Fire Extinguisher ETC (Dec)(11177).

—Identification Systems
NATO STANAG 3995 ED 1 AMD 2-89. Marking of Portable Fire Extinguishers. 9 pp.
NATO STANAG 3995 ED 1 AMD 3-89. Marking of Portable Fire Extinguishers. 10 pp.
SNZ NZS 5807-80. Code of Practice for Industrial Identification by Colour, Wording or Other Coding Amend: 1, 1983; 2, 1988. 20 pp.

—Impact Testing
CEN EN 3-5-84. Portable Fire Extinguishers—Part 5: Complementary Requirements and Tests. 17 pp.
CEN PREN 3-5-93. Fight Against Fire—Portable Fire Extinguishers—Part 5: Complementary Requirements and Tests. 23 pp.

—Liquid Carbon Dioxide
CNS Z2003-84. Portable and Wheeled Types Liquid Carbon Dioxide Extinguisher (Dec)(1387).

—Maintenance
SAA AS 1851.1-89. Maintenance of Fire Protection Equipment—Part 1: Portable Fire Extinguishers Amdt 1 August 1990. 18 pp.

—Nacelles—Aircraft Engine—Doors
BSI C 6-66. 1966 Aircraft Engine Nacelle Fire Extinguisher Doors. 1 p.
ISO 1021-80. Aircraft—Engine Nacelle Fire Extinguisher Apertures and Doors First Edition. 3 pp.

—Powder
SAA AS 1841.5-92. Portable Fire Extinguishers—Part 5: Powder Type (Supersedes AS 1846—1985 (in Part)) Amdt 1 January/February 1993 (in Professional Packages 30, 44). 3 pp.

—Powder—Refill Charges
BSI BS 6643: Part 2-85. 1985 Amd 1 Recharging Fire Extinguishers Part 2: Powder Refill Charges. 9 pp.
BSI BS 6643: Part 2-02. 1985 Amd 4 Recharging Fire Extinguishers (Manufactured to BS 5423 'Specification for Portable Fire Extinguishers') Part 2: Specification for Powder Refill Charges (AMD 7741) September 15, 1993 (N). 11 pp.

—Pressure Measurement
CEN PREN 3-3-92. Portable Fire Extinguishers—Construction, Resistance to Pressure, Mechanical Tests. 23 pp.

—Rising Mains
SNZ NZS 4510-78. Code of Practice for Riser Mains for Fire Service Use. 12 pp.

—Vaporizing Liquid
SAA AS 1841.7-92. Portable Fire Extinguishers—Part 7: Vapourizing-Liquid Type (Supersedes AS 1848—1985 (in Part)) Amdt 1 January/February 1993 (in Professional Packages 30, 44). 3 pp.

—Water
SAA AS 1841.2-92. Portable Fire Extinguishers—Part 2: Water Type (Supersedes AS 1841—1985 and AS 1842—1985 (Both in Part)) Amdt 1 January/February 1993 (in Professional Packages 30, 44). 3 pp.

—Wet Chemical
SAA AS 1841.3-92. Portable Fire Extinguishers—Part 3: Wet Chemical Type (Supersedes AS 1842—1985 (in Part)) Amdt 1 January/February 1993 (in Professional Packages 30, 44). 3 pp.

Fire Extinguishing Agents
Use For: Fire Extinguishing Media *See Also:* Fire Alarm and Extinguishing Equipment; Fire Extinguishers; Fire Fighting; Fire Fighting Equipment; Fire Protection; Fires

—Airports
CAA Part 8 CAP 168. Rescue and Fire Fighting Services. 29 pp.

—Bromochlorodifluoromethane
MOD UK DSTAN 42-25-01. Extinguisher, Fire, Vaporizing Liquid (Bromochlorodifluoro-methane (BCF)) 12 Kilogram Capacity, Skid Type-Metric Issue 1; Amendment 1. 13 pp.

—Carbon Dioxide
BSI BS 6535: Part 1-90. 1990 Fire Extinguishing Media Part 1: Carbon Dioxide. 18 pp.
ISO 5923-89. Fire Protection—Fire Extinguishing Media—Carbon Dioxide Second Edition. 15 pp.
MOD UK DSTAN 68-3-01. Carbon Dioxide, Technical (55 Gramme) (Charge, Carbon Dioxide) Issue 2; Amendment 2. 11 pp.
MOD UK DSTAN 68-4-01. Carbon Dioxide, Technical, 227 Gramme (Charge, Carbon Dioxide) Issue 2; Amendment 1. 9 pp.
MOD UK DSTAN 68-65-01. Carbon Dioxide, Technical 74 Gram (Charge, Carbon Dioxide) Issue 1; Amendment 1. 10 pp.

—Dry Chemical
BSI BS 6535: Part 3-89. 1989 Fire Extinguishing Media Part 3: Specification for Powder. 16 pp.
CNS Z3005-72. Method of Test for Dry Powder for Fire Extinguishing (Jun)(3392).
ISO 7202-87. Fire Protection—Fire Extinguishing Media—Powder First Edition; (Corrected and Reprinted -1988). 15 pp.
MOD UK DEF-1420-63. Dry Powder, Fire Extinguishing, Foam Compatible. 3 pp.

—Foam
CGSB 28-GP-28M-83. Foam, Liquid Concentrate, Fire Extinguishing, Standard for. 22 pp.
CGSB CAN/CGSB-28.74-M90. Liquid Concentrate, Firefighting, Aqueous Film-Forming Foam (AFFF). 30 pp.
CNS Z2018-74. Chemical for Fire Extinguishing (for Foam Extinguishers) (Oct)(3239).
MOD UK DSTAN 42-21-01. Foam Liquid, Fire Extinguishing (Protein Type) Issue 1; Amendment 1. 17 pp.
MOD UK DSTAN 42-22-01. Foam Liquid, Fire Extinguishing (Fluoroprotein Type) Issue 1; Amendment 1. 17 pp.
MOD UK DSTAN 42-24-01. Foam Liquid, Fire Extinguishing (Fluorochemical Type) Issue 1; Amendment 3. 20 pp.
MOD UK DSTAN 42-30-90. Charges, Fire Extinguisher, Foam Issue 2. 8 pp.

—Glossaries
ISO 8421 Pt 4-90. Fire Protection—Vocabulary—Part 4: Fire Extinction Equipment First Edition. 18 pp.

—Halogenated Hydrocarbons
BSI BS 6535: Part 2-84. (WITHDRAWN) 1984 Fire Extinguishing Media Part 2: Specification for Halogenated Hydrocarbons (ISO 7201: 1982) (Superseded by BS 6535: Section 2.1: 1990). 8 pp.
BSI BS 6535: Sec 2.1-90. 1990 Fire Extinguishing Media Part 2: Halons Section 2.1: Halon 1211 and Halon 1301. 8 pp.
BSI BS 6535: Sec 2.2-89. 1989 Fire Extinguishing Media Part 2: Halons Section 2.2: Code of Practice for Safe Handling and Transfer Procedures. 8 pp.
ISO 7201 Pt 1-89. Fire Protection—Fire Extinguishing Media—Halogenated Hydrocarbons—Part 1: Specifications for Halon 1211 and Halon 1301 Second Edition. 7 pp.

—Halogenated Hydrocarbons—Safety
BSI BS 6535: Sec 2.2-89. 1989 Fire Extinguishing Media Part 2: Halons Section 2.2: Code of Practice for Safe Handling and Transfer Procedures. 8 pp.
ISO 7201 Pt 2-91. Fire Extinguishing Media—Halogenated Hydrocarbons—Part 2: Code of Practice for Safe Handling and Transfer Procedures of Halon 1211 and Halon 1301 First Edition. 8 pp.

Fire Extinguishing Media
Use: Fire Extinguishing Agents

Fire Fighting
Use For: Firefighting *See Also:* Fire Extinguishing Agents; Fire Protection; Fires

—Airports
CAA Part 8 CAP 168. Rescue and Fire Fighting Services. 29 pp.
ICAO 9137 Part 1. Airport Services Manual Part 1 Rescue and Fire Fighting Third Edition—1990. 223 pp.

—Airports—Identification Systems
NATO STANAG 3712 ED 2 AMD 4-83. Airport Rescue and Fire Fighting Services—Identification Categories. 6 pp.
NATO STANAG 3712 ED 2 AMD 5-83. Airport Rescue and Fire Fighting Services—Identification Categories. 7 pp.

—Airports—Training
CAA Part 8 CAP 168. Rescue and Fire Fighting Services. 29 pp.

—Evaluation Guides—Military
NATO STANAG 3929 ED 1 AMD 3-88. Evaluation Guide for NATO Crash/Fire/Rescue Services. 13 pp.
NATO STANAG 3929 ED 1 AMD 4-88. Evaluation Guide for NATO Crash/Fire/Rescue Services. 12 pp.

—Glossaries
ISO 8421 Pt 8-90. Fire Protection—Vocabulary—Part 8: Terms Specific to Fire-Fighting, Rescue Services and Handling Hazardous Materials First Edition. 27 pp.

—Heliports—Identification Systems
NATO STANAG 3861 ED 2 AMD 3-83. Heliport Rescue and Fire-Fighting Services—Identification Categories. 6 pp.

Fire Fighting Clothing
See Also: Clothing; Fire Fighting Equipment; Flame Retardant Clothing; Heat Resistant Clothing; Protective Clothing

CEN PREN 469-91. Protective Clothing for Firefighters. 9 pp.
CGSB CAN/CGSB-155.1-M88. Firefighters' Protective Clothing for Protection Against Heat and Flame; (Amendment 1 Nov 1990). 22 pp.
DIN ENGL 32761-83. Clothing for Protection Against Flame for Short Periods; Safety Requirements, Testing (Apr) (Supersedes DIN 4845, July 1963 Edition). 6 pp.
DIN ENGL 32764 Pt 1-87. Clothing for Protection Against Radiant Heat; Type WS (Heavy Duty) Protective Clothing; Safety Requirements, Testing (Feb). 4 pp.
ISO 2801-73. Clothing for Protection Against Heat and Fire—General Recommendations for Users and for Those in Charge of Such Users First Edition. 6 pp.

—Nuclear, Biological, and Chemical Warfare
NATO STANAG 7049 ED 1 AMD 0-93. Personal Protective Equipment Requirements Together with Standard Operating Procedures for CFR Operations in an NBC Environment. 6 pp.

—Protective Gloves
CEN PREN 659-92. Firefighters' Gloves—Protection Against Heat and Flames. 6 pp.

Fire Fighting Equipment
Scope Note: For additional listings, see specific types of equipment used in fire fighting *See Also:* Fire Alarm and Extinguishing Equipment; Fire Blankets; Fire Extinguishers; Fire Extinguishing Agents; Fire Fighting Clothing; Fire Hoses; Fire Hydrant Boxes; Fire Hydrants; Fire Protection; Fire Pumps; Fire Sprinkler Equipment; Fires; Stirrup Pumps

—Axes
BSI BS 2957-58. 1958 Fireman's Axe with Ash Handle. 11 pp.
BSI BS 3054-59. 1959 Fireman's Axe with Rubber Insulated Handle. 12 pp.

—Batteries
CNS Z2050-83. Batteries Device for Fire Fighting Emergency Use (Apr)(10205).

—Buildings—Elevators (Lifts)
BSI BS 5588: Part 5-91. 1991 Fire Precautions in the Design, Construction and Use of Buildings Part 5: Code of Practice for Firefighting Stairs and Lifts. 43 pp.
BSI BS 5588: Part 5-86. 1986 Fire Precautions in the Design and Construction of Buildings Part 5: Code of Practice for Firefighting Stairs and Lifts. 24 pp.
BSI BS 5588: Part 5-01. 1991 Amd 1 Fire Precautions in the Design, Construction and Use of Buildings Part 5: Code of Practice for Firefighting Stairs and Lifts (AMD 7196) June 15, 1992. 48 pp.

—Buildings—Installation
SNZ NZS 4503-74. Code of Practice for the Distribution, Installation and Maintenance of Hand Operated Fire Fighting Equipment for Use in Buildings. 16 pp.

INTERNATIONAL AND NON-U.S. NATIONAL STANDARDS
SUBJECT INDEX

Fire

Fire Fighting Equipment (Cont.)

—Buildings—Stairways
BSI BS 5588: Part 5-91. 1991 Fire Precautions in the Design, Construction and Use of Buildings Part 5: Code of Practice for Firefighting Stairs and Lifts. 43 pp.

BSI BS 5588: Part 5-86. 1986 Fire Precautions in the Design and Construction of Buildings Part 5: Code of Practice for Firefighting Stairs and Lifts. 24 pp.

BSI BS 5588: Part 5-01. 1991 Amd 1 Fire Precautions in the Design, Construction and Use of Buildings Part 5: Code of Practice for Firefighting Stairs and Lifts (AMD 7196) June 15, 1992. 48 pp.

—Connections—Enclosures
SNZ NZS 4521-74. Specification for Boxes for Fire Brigade Connections Amend: 1. 8 pp.

—Control Valves
CNS Z2055-87. Control Valve for Fire Fighting Use (Sep)(10763).

—Emergency and Standby Power Supplies
CNS Z2054-83. Cubicle Emergency Power Source for Fire Fighting Use (Nov)(10673).

CNS Z3023-83. Method of Test for Fire Fighting Emergency Power Source (Apr)(10204).

—Ground Effect Machines
CAA Chapter B6-10 08.83. Fire Extinguishing Systems. 2 pp.

CAA Chapter B6-10 App 08.83. Fire Detection and Extinguishing Systems. 2 pp.

—Maintenance
SNZ NZS 4503-74. Code of Practice for the Distribution, Installation and Maintenance of Hand Operated Fire Fighting Equipment for Use in Buildings. 16 pp.

—Merchant Ships
NATO STANAG 1169 ED 1 AMD 3-81. Firefighting Equipment and Principles for Harmonization of Present and Future Equipment and Materials. 16 pp.

—Naval Ships
NATO STANAG 1169 ED 1 AMD 3-81. Firefighting Equipment and Principles for Harmonization of Present and Future Equipment and Materials. 16 pp.

—Paints (Alkyd Enamel)
CGSB CAN/CGSB-1.88-92. Gloss Alkyd Enamel, Air Drying and Baking. 9 pp.

—Paints (Alkyd Enamel)—Exterior
CGSB CAN/CGSB-1.59-M89. Alkyd, Exterior Gloss Enamel. 11 pp.

—Ropes
BSI BS 3367-80. 1980 Fire Brigade and Industrial Ropes and Rescue Lines. 12 pp.

—Safety Nets
BSI BS 3566-82. 1982 Lightweight Salvage Sheets for Fire Service Use. 8 pp.

—Ships—Axes
CNS F3085-83. Fire Axes for Ships (Mar)(6915).
JIS F 3610-89. Ships' Fire Axes.

—Ships—Carbon Dioxide Cylinders
BSI BS 5396-76. 1976 Seamless Steel Co2 Containers for Fixed Fire-Fighting Installations on Ships. 4 pp.

ISO 3500-90. Seamless Steel CO2 Cylinders for Fixed Fire-Fighting Installations on Ships Second Edition. 4 pp.

—Ships—Pressure
ISO 3935-77. Shipbuilding—Inland Navigation—Fire-Fighting Water System—Pressures First Edition. 3 pp.

—Ships—Symbols
CNS F4008-81. Symbols for Life-Saving Appliance and Fire Fighting Appliance of Ships (Oct)(7983). 11 pp.

JIS F 0051-83. Symbols for Lifesaving Appliance and Fire Fighting of Ships.

—Symbols
ISO 6790-86. Equipment for Fire Protection and Fire Fighting—Graphical Symbols for Fire Protection Plans—Specification First Edition. 13 pp.

—Wrecking Bars
CNS B3413-84. Crowbars (for Fire Fighting Use) (Dec)(7344).

Fire Guards
Use For: Spark Guards *See Also:* Fire Screens

—Heaters
BSI BS 1945-71. 1971 Amd 2 Fireguards for Heating Appliances (Gas, Electric and Oil-Burning). 17 pp.

Fire Guards (Cont.)

—Heaters (Cont.)
BSI BS 6539-84. 1984 Amd 1 Fireguards for Use with Solid Fuel Appliances. 13 pp.

BSI BS 6778-86. 1986 Fireguards for Use with Portable Free-Standing or Wall-Mounted Heating Appliances. 11 pp.

BSI PD 6516-87. 1987 Guarding Fires and Heating Appliances. 8 pp.

SNZ NZS/BS 6539-84. Specification for Fireguards for Use with Solid Fuel Appliances Amend: 1. 12 pp.

Fire Hazard Testing
Use: Fire Hazards

Fire Hazards
Use For: Fire Hazard Testing *See Also:* Fire Protection; Firesafety; Flammable Atmospheres; Flammable Materials

BSI BS 6336-82. 1982 Development and Presentation of Fire Tests and Their Use in Hazard Assessment. 12 pp.

BSI BS EN 60695-2-4/0-93. 1993 Fire Hazard Testing Part 2: Test Methods Section 2.4/0 Sheet 0: Diffusion Type and Premixed Type Flame Test Methods (IEC 695-2-4/0: 1991) (G). 22 pp.

BSI BS EN 60695-2-4/1-93. 1993 Fire Hazard Testing Part 2: Test Methods Section 2.4/1 Sheet 1: 1 kW Nominal Pre-Mixed Test Flame and Guidance (IEC 695-2-4/1: 1991) (G). 23 pp.

CENELEC EN 60695-2-4/0-93. Fire Hazard Testing Part 2: Test Methods Section 4/Sheet 0: Diffusion Type and Premixed Type Flame Test Methods (IEC 695-2-4/0: 1991). 17 pp.

CENELEC EN 60695-2-4/1-93. Fire Hazard Testing Part 2: Test Methods Section 4/Sheet 1: 1 kW Nominal Pre-Mixed Test Flame and Guidance (IEC 695-2-4/1: 1991). 18 pp.

IEC 695 Pt 2-4/0-91. Fire Hazard Testing Part 2: Test Methods—Section 4/Sheet 0: Diffusion Type and Premixed Type Flame Test Methods First Edition; (CENELEC EN 60695-2-4/0: 1993). 31 pp.

IEC 695 Pt 2-4/1-91. Fire Hazard Testing Part 2: Test Methods Section 4/Sheet 1: 1 kW Nominal Pre-Mixed Test Flame and Guidance First Edition; (CENELEC EN 60695-2-4/1: 1993). 27 pp.

—Aircraft—Technical Manuals
NATO STANAG 3896 ED 2 AMD 0-90. Illustrated Information on Fire Hazards and Rescue Features for NATO Aircraft—AEP-11. 4 pp.

NATO STANAG 3896 ED 2 AMD 1-90. Illustrated Information on Fire Hazards and Rescue Features for NATO Aircraft—AEP-11. 4 pp.

—Cables (Electric)
MOD UK DSTAN 61-12: Part 9-90. Wires, Cords and Cables Electrical—Metric Units Part 9: Cables, Radio Frequency Including Limited Fire Hazard (LFH) Variants Issue 3. 36 pp.

MOD UK DSTAN 61-12: Part 9-93. Wires, Cords and Cables Electrical—Metric Units Part 9: Cables, Radio Frequency Including Limited Fire Hazard (LFH) Variants Issue 4. 49 pp.

MOD UK DSTAN 61-12: Part 18-90. Wires Cords and Cables, Metric Units Part 18: Equipment Wires, Limited Fire Hazard Issue 3. 41 pp.

MOD UK DSTAN 61-12: Part 18-01. Wires Cords and Cables, Metric Units Part 18: Equipment Wires, Limited Fire Hazard Issue 3; Amendment 2. 43 pp.

MOD UK DSTAN 61-12: Part 18-02. Wires Cords and Cables, Metric Units Part 18: Equipment Wires, Limited Fire Hazard Issue 3; Amendment 3. 44 pp.

MOD UK DSTAN 61-12: Part 25-01. Wires, Cords, and Cables, Electrical Metric Units Part 25: Cables, Electrical Limited Fire Hazard, up to Size 2.5mm2 Cross-Sectional Area Issue 1; Amendment 1. 45 pp.

MOD UK DSTAN 61-12: Part 25-92. Wires, Cords and Cables Electrical Metric Units Part 25: Cables Electrical Limited Fire Hazard, up to Conductor Size 2.5sq mm Cross Sectional Area Issue 2. 43 pp.

MOD UK DSTAN 61-12: Part 25-02. Wires, Cords and Cables Electrical Metric Units Part 25: Cables Electrical Limited Fire Hazard, up to Conductor Size 2.5sq mm Cross Sectional Area Issue 2; Amendment 1. 44 pp.

MOD UK DSTAN 61-12: Part 25-03. Wires, Cords and Cables Electrical Metric Units Part 25: Cables Electrical Limited Fire Hazard, up to Conductor Size 2.5sq mm Cross Sectional Area Issue 2; Amendment 2. 45 pp.

MOD UK DSTAN 61-12: Part 25-04. Wires, Cords and Cables Electrical Metric Units Part 25: Cables Electrical Limited Fire Hazard, up to Conductor Size 2.5sq mm Cross Sectional Area Issue 2; Amendment 3. 46 pp.

—Clothing—Design
ISO TR9240-92. Textiles—Design of Apparel for Reduced Fire Hazard First Edition. 12 pp.

Fire Hazards (Cont.)

—Construction
SAA AS 3959-91. Construction of Buildings in Bushfire-Prone Areas (in Professional Package 44). 10 pp.

—Electric Wire
MOD UK DSTAN 61-12: Part 18-90. Wires Cords and Cables, Metric Units Part 18: Equipment Wires, Limited Fire Hazard Issue 3. 41 pp.

MOD UK DSTAN 61-12: Part 18-01. Wires Cords and Cables, Metric Units Part 18: Equipment Wires, Limited Fire Hazard Issue 3; Amendment 2. 43 pp.

MOD UK DSTAN 61-12: Part 18-02. Wires Cords and Cables, Metric Units Part 18: Equipment Wires, Limited Fire Hazard Issue 3; Amendment 3. 44 pp.

—Electrical Components
BSI BS 6458: Part 1-90. 1990 Fire Hazard Testing for Electrotechnical Products Part 1: Glossary of Terms. 22 pp.

BSI BS 6458: Sec 2.1-84. 1984 Fire Hazard Testing for Electrotechnical Products Part 2: Methods of Test Section 2.1: Glow Wire Test. 12 pp.

BSI BS 6458: Sec 2.2-93. 1993 Fire Hazard Testing for Electrotechnical Products Part 2: Methods of Test Section 2.2: Needle-Flame Test (IEC 695-2-2: 1991) (G). 15 pp.

BSI BS 6458: Sec 2.2-84. 1984 Fire Hazard Testing for Electrotechnical Products Part 2: Methods of Test Section 2.2: Needle-Flame Test. 10 pp.

BSI BS 6458: Sec 2.3-85. 1985 Fire Hazard Testing for Electrotechnical Products Part 2: Methods of Test Section 2.3: Bad-Connection Test with Heaters. 20 pp.

CENELEC HD 444.2.1-83. Fire Hazard Testing Part 2: Test Methods Glow-Wire Test and Guidance. 3 pp.

CENELEC HD 444.2.2-83. Fire Hazard Testing Part 2: Test Methods Meedle-Flame Test. 3 pp.

CENELEC HD 444.2.2 S2-92. Fire Hazard Testing Part 2: Test Methods Section 2—Needle-Flame Test (IEC 695-2:2:1991). 4 pp.

CENELEC HD 444.2.3-87. Fire Hazard Testing Part 2: Test Methods Bad-Connection Test with Heaters. 2 pp.

IEC 695 Pt 1-1-82. Fire Hazard Testing Part 1: Guidance for the Preparation of Requirements and Test Specifications for Assessing Fire Hazard of Electrotechnical Products General Guidance First Edition. 21 pp.

IEC 695 Pt 1-2-82. Fire Hazard Testing Part 1: Guidance for the Preparation of Requirements and Test Specifications for Assessing Fire Hazard of Electrotechnical Products Guidance for Electronic Components First Edition. 17 pp.

IEC 695 Pt 1-3-86. Fire Hazard Testing Part 1: Guidance for the Preparation of Requirements and Test Specifications for Assessing Fire Hazard of Electrotechnical Products Guidance for Use of Preselection Procedures First Edition. 19 pp.

IEC 695 Pt 2-1-91. Fire Hazard Testing Part 2: Test Methods Section 1: Glow-Wire Test and Guidance Second Edition. 28 pp.

IEC 695 Pt 2-2-91. Fire Hazard Testing Part 2: Test Methods Section 2—Needle-Flame Test Second Edition; (CENELEC HD 444.2.2 S2:1992). 21 pp.

IEC 695 Pt 2-3-84. Fire Hazard Testing Part 2: Test Methods Bad-Connection Test with Heaters First Edition. 33 pp.

IEC 695 Pt 3-1-82. Fire Hazard Testing Part 3: Examples of Fire Hazard Assessment Procedures and Interpretation of Results Combustion Characteristics and Survey of Test Methods for Their Determination First Edition. 19 pp.

IEC 695 Pt 5-1-93. Fire Hazard Testing Part 5: Assessment of Potential Corrosion Damage by Fire Effluent—Section 1: General Guidance. 25 pp.

JIS C 0060-85. Fire Hazard Testing Part 2: Test Methods Glow-Wire Test and Guidance. 13 pp.

JIS C 0061-85. Fire Hazard Testing Part 2: Test Methods Needle Flame Test. 10 pp.

JIS C 0062-87. Fire Hazard Testing Part 2: Test Methods Bad-Connection Test with Heaters. 17 pp.

—Electrical Installations
DIN VDE 0100 Pt 720-83. Erection of Power Installations with Rated Voltages up to 1000 V; Locations Exposed to Fire Hazards (Mar). 15 pp.

—Glossaries
IEC 695 Pt 4-89. Fire Hazard Testing Part 4: Terminology Concerning Fire Tests First Edition. 37 pp.

—Munitions—Liquid Fuel Fires
NATO STANAG 4240 ED 1 AMD 0-91. Liquid Fuel Fire Tests for Munitions. 20 pp.

—Pressure Vessels
BSI BS 5500: ENQ CASE 99-93. 1993 Enquiry Case Case 99: Vessels Under Fire Exposure (Q). 1 p.

INDUSTRY STANDARDS

INTERNATIONAL AND NON-U.S. NATIONAL STANDARDS
SUBJECT INDEX — Fire

Fire Hazards *(Cont.)*
—Weapons Systems—Liquid Fuel Fires
- NATO STANAG 4240 ED 1 AMD 0-91. Liquid Fuel Fire Tests for Munitions. 20 pp.

Fire Hose Assemblies
See Also: Fire Hoses; Hose Assemblies
- BSI BS 3165-86. 1986 Rubber and Plastics Suction Hoses and Hose Assemblies for Fire-Fighting Purposes. 12 pp.
- BSI BS 6391-83. 1983 Amd 1 Non-Percolating Layflat Delivery Hoses and Hose Assemblies for Fire Fighting Purposes (AMD 5209) September 30, 1986. 14 pp.
- MOD UK DSTAN 42-31-90. Reel and Hose Assembly, Fire Fighting Issue 2. 11 pp.

Fire Hose Couplings
Use: Hose Couplings—Fire

Fire Hoses
See Also: Fire Fighting Equipment; Fire Hose Assemblies; Fire Hydrants; Hoses
- BSI BS 3165-86. 1986 Rubber and Plastics Suction Hoses and Hose Assemblies for Fire-Fighting Purposes. 12 pp.
- BSI BS 3169-86. 1986 First Aid Reel Hoses for Fire Fighting Purposes. 8 pp.
- BSI BS 6391-83. 1983 Amd 1 Non-Percolating Layflat Delivery Hoses and Hose Assemblies for Fire Fighting Purposes (AMD 5209) September 30, 1986. 14 pp.
- CEN PREN 694-92. Semi-Rigid Reel Hose for First Aid Fixed Installations. 13 pp.
- ISO 4642-78. Rubber Products—Hoses, Non-Collapsible, for Fire-Fighting Service First Edition. 6 pp.

—Fire Extinguishers
- CNS Z2061-84. Hoses, Nozzles, Nozzle Valves and Hoses Reel for Movable Carbon Dioxide Fire Extinguisher ETC (Dec)(11177).

—Hose Reels
- BSI BS 5274-85. 1985 Fire Hose Reels (Water) for Fixed Installations. 11 pp.
- BSI BS 5306: Part 1-76. 1976 Amd 2 Fire Extinguishing Installations and Equipment on Premises Part 1: Hydrant Systems, Hose Reels and Foam Inlets (Formerly CP 402.101) (AMD 5756) February 29, 1988. 24 pp.
- CEN PREN 671-1-92. Fixed Firefighting Systems—Hose Systems—Part 1: Hose Reels with Semi-Rigid Hose. 21 pp.
- MOD UK DSTAN 42-31-90. Reel and Hose Assembly, Fire Fighting Issue 2. 11 pp.
- SAA AS 1221-91. Fire Hose Reels Amdt 1 August 1992 (in Professional Packages 20, 21, 30, 41, 44, 62-69). 20 pp.
- SNZ NZS 4504-81. Specification for Fire Hose Reels Amend: 2, 1991. 16 pp.

—Hose Reels—Fire Extinguishers
- CNS Z2061-84. Hoses, Nozzles, Nozzle Valves and Hoses Reel for Movable Carbon Dioxide Fire Extinguisher ETC (Dec)(11177).

—Hose Reels—Fixed
- BSI BS 3169-86. 1986 First Aid Reel Hoses for Fire Fighting Purposes. 8 pp.
- CEN PREN 694-92. Semi-Rigid Reel Hose for First Aid Fixed Installations. 13 pp.

—Maintenance
- SAA AS 1851.9-88. Maintenance of Fire Protection Equipment—Part 9: Delivery Layflat Fire Hose. 4 pp.

—Reels—Maintenance
- SAA AS 1851.2-88. Maintenance of Fire Protection Equipment—Part 2: Fire Hose Reels. 1 p.

Fire Hydrant Boxes
See Also: Boxes (Containers); Fire Fighting Equipment

—Covers
- BSI BS 750-84. 1984 Underground Fire Hydrants and Surface Box Frames and Covers. 15 pp.
- BSI BS 750-01. 1984 Amd 1 Underground Fire Hydrants and Surface Box Frames and Covers (AMD 7658) June 15, 1993 (R). 17 pp.
- SNZ NZS/BS 750-84. Specification for Underground Fire Hydrants and Surface Box Frames and Covers. 16 pp.

—Frames
- BSI BS 750-84. 1984 Underground Fire Hydrants and Surface Box Frames and Covers. 15 pp.
- BSI BS 750-01. 1984 Amd 1 Underground Fire Hydrants and Surface Box Frames and Covers (AMD 7658) June 15, 1993 (R). 17 pp.

Fire Hydrant Boxes *(Cont.)*
—Frames *(Cont.)*
- SNZ NZS/BS 750-84. Specification for Underground Fire Hydrants and Surface Box Frames and Covers. 16 pp.

Fire Hydrants
Use For: Dry Risers; Wet Risers *See Also:* Fire Fighting Equipment; Fire Hoses; Fire Hydrant Boxes; Fire Protection; Rising Mains
- BSI BS 5306: Part 1-76. 1976 Amd 2 Fire Extinguishing Installations and Equipment on Premises Part 1: Hydrant Systems, Hose Reels and Foam Inlets (Formerly CP 402.101) (AMD 5756) February 29, 1988. 24 pp.

—Boxes
- BSI BS 5041: Part 3-75. 1975 Amd 1 Fire Hydrant Systems Equipment Part 3: Inlet Breechings for Dry Riser Inlets. 9 pp.
- BSI BS 5041: Part 4-75. 1975 Amd 1 Fire Hydrant Systems Equipment Part 4: Boxes for Landing Valves for Dry Risers. 10 pp.
- BSI BS 5041: Part 5-74. 1974 Amd 1 Fire Hydrant Systems Equipment Part 5: Boxes for Foam Inlets and Dry Risers. 14 pp.

—Design
- SAA AS 2419.1-91. Fire Hydrant Installations—Part 1: System Design, Installation and Commissioning (in Professional Packages 20, 21, 44, 62#69). 42 pp.

—Diaphragm Valves
- BSI BS 5041: Part 1-87. 1987 Amd 1 Fire Hydrant System Equipment Part 1: Landing Valves for Wet Risers (AMD 5912) September 30, 1988. 15 pp.

—Fire Extinguishers
- BSI BS 5306: Part 1-76. 1976 Amd 2 Fire Extinguishing Installations and Equipment on Premises Part 1: Hydrant Systems, Hose Reels and Foam Inlets (Formerly CP 402.101) (AMD 5756) February 29, 1988. 24 pp.

—Gate Valves
- BSI BS 5041: Part 2-87. 1987 Amd 1 Fire Hydrant Systems Equipment Part 2: Landing Valves for Dry Risers (AMD 5776) July 29, 1988. 12 pp.
- BSI BS 5041: Part 2-76. 1976 Fire Hydrant Systems Equipment Part 2: Landing Valves for Dry Risers. 7 pp.

—Globe Valves
- BSI BS 5041: Part 1-87. 1987 Amd 1 Fire Hydrant System Equipment Part 1: Landing Valves for Wet Risers (AMD 5912) September 30, 1988. 15 pp.

—Identification Systems
- SNZ NZS 4501-72. Code of Practice for the Location Marking of Fire Hydrants Amend: 1, 1980. 8 pp.

—Indicator Plates
- BSI BS 3251-76. 1976 Amd 1 Indicator Plates for Fire Hydrants and Emergency Water Supplies (AMD 6736) September 30, 1991. 14 pp.

—Installation
- SAA AS 2419.1-91. Fire Hydrant Installations—Part 1: System Design, Installation and Commissioning (in Professional Packages 20, 21, 44, 62#69). 42 pp.

—Maintenance
- SAA AS 1851.4-92. Maintenance of Fire Protection Equipment—Part 4: Fire Hydrant Installations (in Professional Packages 21, 30, 44, 65, 69). 10 pp.

—Stop Valves
- BSI BS 5041: Part 1-75. 1975 Fire Hydrant Systems Equipment Part 1: Landing Valves for Wet Risers. 9 pp.

—Underground
- BSI BS 750-84. 1984 Underground Fire Hydrants and Surface Box Frames and Covers. 15 pp.
- BSI BS 750-01. 1984 Amd 1 Underground Fire Hydrants and Surface Box Frames and Covers (AMD 7658) June 15, 1993 (R). 17 pp.
- SNZ NZS/BS 750-84. Specification for Underground Fire Hydrants and Surface Box Frames and Covers. 16 pp.

—Valves
- SAA AS 2419.2-91. Fire Hydrant Installations—Part 2: Fire Hydrant Valves (in Professional Packages 20, 21, 44, 62#69). 12 pp.
- SAA AS 3952-91. Water Supply—DN 80 Spring Hydrant Valve for General Purposes Amdt 1 June 1993 (in Professional Packages 44, 61). 11 pp.

—Wall
- CEN PREN 671-2-92. Fixed Firefighting Systems—Hose Systems—Part 2: Wall Hydrant with Lay-Flat Hose. 17 pp.

Fire Point
See Also: Flash Point

—Electrical Insulating Liquids
- BSI BS EN 61100-93. 1993 Classification of Insulating Liquids According to Fire Point and Net Calorific Value (IEC 1100: 1992) (E). 12 pp.
- CENELEC EN 61100-92. Classification of Insulating Liquids According to Fire Point and Net Calorific Value. 5 pp.
- IEC 1100-92. Classification of Insulating Liquids According to Fire-Point and Net Calorific Value First Edition. 15 pp.

—Petroleum Products
- BSI BS 2000: Part 35-93. 1993 Methods of Test for Petroleum and Its Products Part 35: Determination of Open Flash and Fire Point—Pensky-Martens Method (Supersedes BS 4688: 1991). 3 pp.
- BSI BS 2000: Part 35-82. 1982 Petroleum and Its Products Part 35: Flash Point (Open) and Fire Point of Petroleum Products by the Pensky-Martens Apparatus. 4 pp.
- BSI BS 4689-80. 1980 Amd 1 Determination of Flash and Fire Points of Petroleum Products: Cleveland Open Cup Method (AMD 6229) February 28, 1990 (W). 8 pp.
- CNS K6377-90. Determination of Flash Fire Points for Petroleum Products with Cleveland Open Cup (Aug)(3775).
- ISO 2592-73. Petroleum Products—Determination of Flash and Fire Points—Cleveland Open Cup Method First Edition. 6 pp.

Fire Propagation
Use: Flame Propagation

Fire Protection
See Also: Breathing Apparatus; Diving Equipment; Fire Alarm and Extinguishing Equipment; Fire Barriers; Fire Extinguishers; Fire Extinguishing Agents; Fire Fighting; Fire Fighting Equipment; Fire Hazards; Fire Hydrants; Fire Pumps; Fire Testing; Fire Walls; Fires; Firesafety; Flame Arresters; Flashback Arresters; Flood Protection; Gas Detectors; Protective Clothing; Safety Lamps; Thermal Insulation; Warning Systems

—Aerostats
- CAA Chapter Q5-8 12.79. Fire Precautions (Non-Rigid Airships). 3 pp.
- CAA Chapter Q5-8 App 12.79. Fire Precautions (Non-Rigid Airships). 1 p.

—Agricultural Buildings
- BSI BS 5502: Sec 1.3-86. (WITHDRAWN) 1986 Design of Buildings and Structures for Agriculture Part 1: General Considerations Section 1.3: Fire Protection (R) (Superseded by BS 5588: Part 1: 1990). 6 pp.
- BSI BS 5502: Part 23-90. 1990 Design of Buildings and Structures for Agriculture Part 23: Code of Practice for Fire Precautions. 16 pp.

—Aircraft
- CAA Chapter G5-8 App 06.76. Fire Precautions (Rotorcraft). 2 pp.

—Aircraft Cabins
- CAA Chapter K4-3 04.74. Compartment Design and Safety Provisions (Light Aeroplanes). 3 pp.
- CAA Chapter Q4-3 12.79. Compartment Design and Safety Provisions (Non-Rigid Airships). 5 pp.

—Aircraft Compartments
- CAA Chapter K4-3 04.74. Compartment Design and Safety Provisions (Light Aeroplanes). 3 pp.
- CAA Chapter Q4-3 12.79. Compartment Design and Safety Provisions (Non-Rigid Airships). 5 pp.
- ISO 2685-92. Aircraft—Environmental Conditions and Test Procedures for Airborne Equipment—Resistance to Fire in Designated Fire Zones First Edition. 33 pp.

—Aircraft Engines
- CAA Chapter K5-8 04.74. Fire Precautions (Light Aeroplanes). 3 pp.
- CAA Chapter P5-8. Fire Precautions (Provisional Airworthiness Requirements for Civil Powered Lift Aircraft). 15 pp.
- CAA Chapter Q5-8 12.79. Fire Precautions (Non-Rigid Airships). 3 pp.

—Aircraft Engines—Rotary Wing
- CAA CAP 524 SUB-Part E 12.86. Power-Plant (Rotorcraft). 27 pp.

—Aircraft Engines—Supersonic Transports
- CAA STANDARD NO. 6-3 03.76. Power Plant Installations: Fire Precautions. 11 pp.

—Aircraft Equipment
- CAA Chapter K5-8 04.74. Fire Precautions (Light Aeroplanes). 3 pp.

INTERNATIONAL AND NON-U.S. NATIONAL STANDARDS
SUBJECT INDEX

Fire

Fire Protection (Cont.)
—**Aircraft Equipment** (Cont.)
CAA Chapter K5-8 App 04.74. Fire Precautions (Light Aeroplanes). 1 p.

—**Aircraft—Polyester Fabrics**
MOD UK DTD-5624-81. Fibrous Polyester Material for Use as a Flame and Fire Suppressant in Aircraft Dry Bays. 9 pp.

—**Aircraft—Toilet Facilities**
CAA NOTICE #60 ISSUE 1. Cabin and Toilet Fire Protection (Airworthiness Notices). 3 pp.

—**Buildings**
BSI BS 5588: Part 1-90. 1990 Fire Precautions in the Design, and Construction of Buildings Part 1: Code of Practice for Residential Buildings. 64 pp.
BSI BS 5588: Part 1-01. 1990 Amd 1 Fire Precautions in the Design, and Construction of Buildings Part 1: Code of Practice for Residential Buildings (AMD 7840) September 15, 1993 (R). 71 pp.
BSI BS 5588: Part 2-85. (WITHDRAWN) 1985 Amd 1 Fire Precautions in the Design and Construction of Buildings Part 2: Code of Practice for Shops. 62 pp.
BSI BS 5588: Part 2-03. 1985 Amd 3 Fire Precautions in the Design and Construction of Buildings Part 2: Code of Practice for Shops (AMD 6478) August 31, 1990. 62 pp.
BSI BS 5588: Part 3-83. 1983 Amd 3 Fire Precautions in the Design and Construction of Buildings Part 3: Code of Practice for Office Buildings (AMD 6160) October 31, 1989. 61 pp.
BSI BS 5588: Part 6-91. 1991 Fire Precautions in the Design, Construction and Use of Buildings Part 6: Code of Practice for Places of Assembly. 67 pp.
BSI BS 5588: Part 10-91. 1991 Fire Precautions in the Design, Construction and Use of Buildings Part 10: Code of Practice for Shopping Complexes. 96 pp.

—**Buildings—Automatic Release Mechanisms**
BSI BS 5839: Part 3-88. 1988 Fire Detection and Alarm Systems in Buildings Part 3: Automatic Release Mechanisms for Certain Fire Protection Equipment. 16 pp.

—**Buildings—Ducts**
BSI BS 5588: Part 9-89. 1989 Fire Precautions in the Design and Construction of Buildings Part 9: Code of Practice for Ventilation and Air Conditioning Ductwork. 25 pp.

—**Buildings—Electrical Installations**
DIN VDE 0100 Pt 420-91. Installation of Power Plant with Rated Voltages up to 1000 V; Protective Measures; Protection Against Thermal Effects (Nov) (Supersedes Din 57 100 Part 420 Refer to Transitional Period). 8 pp.
IEC 364 Pt 4-482-82. Electrical Installations of Buildings Part 4: Protection for Safety Chapter 48: Choice of Protective Measures as a Function of External Influences Section 482-Protection Against Fire First Edition. 12 pp.

—**Buildings—Handicapped Persons**
BSI BS 5588: Part 8-88. 1988 Fire Precautions in the Design and Construction of Buildings Part 8: Code of Practice for Means of Escape for Disabled People. 24 pp.

—**Buildings—Heat Control**
DIN ENGL 18232 Pt 1-81. Structural Fire Protection; Smoke and Heat Control Installations; Terminology and Application (Sept). 3 pp.

—**Buildings—Smoke Control**
DIN ENGL 18232 Pt 1-81. Structural Fire Protection; Smoke and Heat Control Installations; Terminology and Application (Sept). 3 pp.
SAA AS 1668.1-91. Use of Mechanical Ventilation and Air-Conditioning in Buildings—Part 1: Fire and Smoke Control (Supersedes SAA MP47.C1—1980) (in Professional Packages 20, 21, 44, 50, 62-69). 53 pp.

—**Buildings—Smoke Vents**
DIN ENGL 18232 Pt 2-89. Structural Fire Protection in Industrial Buildings; Smoke and Heat Control Installations Design, Construction, Performance and Installation of Smoke Vents (Nov). 5 pp.
DIN ENGL 18232 Pt 3-84. Structural Fire Protection in Industrial Buildings; Smoke and Heat Control Installations; Smoke Vents; Testing (Sept). 7 pp.

—**Cement Mortars—Plastering**
CNS A3198-83. Method of Plastering of Cement Mortar for Fire Protection (Dec)(10679).
JIS A 7801-63. Method of Plastering of Cement Mortar for Fire Protection.

Fire Protection (Cont.)
—**Chemical Plants**
BSI BS 5908-90. 1990 Code of Practice for Fire Precautions in the Chemical and Allied Industries. 86 pp.

—**Construction Materials—Sprayed Coatings**
BSI BS 8202: Part 1-87. 1987 Coatings for Fire Protection of Building Elements Part 1: Code of Practice for the Selection and Installation of Sprayed Mineral Coatings. 20 pp.

—**Data Processing Equipment**
BSI BS 6266-92. 1992 Fire Protection for Electronic Data Processing Installations. 35 pp.
BSI BS 6266-82. 1982 Fire Protection for Electronic Data Processing Installations (Formerly CP 95). 16 pp.

—**Electrical Installations—Controlgear—Construction**
IEC 364 Pt 5-53-86. Electrical Installations of Buildings Part 5: Selection and Erection of Electrical Equipment—Chapter 53: Switchgear and Controlgear First Edition; (Amendment 2-1992, Includes Amendment 1). 30 pp.

—**Electrical Installations—Switchgear—Construction**
IEC 364 Pt 5-53-86. Electrical Installations of Buildings Part 5: Selection and Erection of Electrical Equipment—Chapter 53: Switchgear and Controlgear First Edition; (Amendment 2-1992, Includes Amendment 1). 30 pp.

—**Engineering Drawings—Symbols**
BSI BS 1635-90. 1990 Graphical Symbols and Abbreviations for Fire Protection Drawings. 27 pp.

—**Glossaries**
BSI BS 4422: Part 3-90. 1990 Glossary of Terms Associated with Fire Part 3: Fire Detection and Alarm. 10 pp.
BSI BS 4422: Part 3-72. (WITHDRAWN) 1972 Glossary of Terms Associated with Fire Part 3: Means of Escape. 9 pp.
BSI BS 4422: Part 4-75. 1975 Glossary of Terms Associated with Fire Part 4: Fire Protection Equipment. 13 pp.
BSI BS 4422: Part 5-89. 1989 Glossary of Terms Associated with Fire Part 5: Smoke Control. 4 pp.
BSI BS 4422: Part 5-76. (WITHDRAWN) 1976 Glossary of Terms Associated with Fire Part 5: Miscellaneous Terms. 10 pp.
BSI BS 4422: Part 6-88. 1988 Glossary of Terms Associated with Fire Part 6: Evacuation and Means of Escape. 4 pp.
BSI BS 4422: Part 7-88. 1988 Glossary of Terms Associated with Fire Part 7: Explosion Detection and Suppression Means. 3 pp.
ISO 8421 Pt 2-87. Fire Protection—Vocabulary—Part 2: Structural Fire Protection First Edition. 7 pp.
ISO 8421 Pt 3-89. Fire Protection—Vocabulary—Part 3: Fire Detection and Alarm First Edition. 11 pp.
ISO 8421 Pt 6-87. Fire Protection—Vocabulary—Part 6: Evacuation and Means of Escape First Edition. 8 pp.
ISO 8421 Pt 7-87. Fire Protection—Vocabulary—Part 7: Explosion Detection and Suppression Means First Edition. 6 pp.

—**Ground Effect Machines**
CAA Chapter B5-7 01.91. Fire Precautions. 6 pp.

—**Houses**
BSI BS 5588: Sec 1.1-84. (WITHDRAWN) 1984 Amd 1 Fire Precautions in the Design and Construction of Buildings Part 1: Resid Buildings Section 1.1: Code of Practice for Single-Family Dwelling Houses (AMD 5714) July 31, 1987 (Superseded by BS 5588: Part 1: 1990). 12 pp.
BSI CP 3: Ch IV: Part 1-71. 1971 Code of Basic Data for the Design of Buildings Chapter IV: Precautions Against Fire Part 1: Flats and Maisonettes (in Blocks over Two Storeys). 63 pp.

—**Internal Combustion Engines**
BSI BS 6327-82. 1982 Fire Protection of Reciprocating Internal Combustion Engines. 4 pp.
ISO 6826-82. Reciprocating Internal Combustion Engines—Fire Protection First Edition. 5 pp.

—**Nuclear Power Plants**
CSA CAN/CSA-N293-M87. Fire Protection for CANDU Nuclear Power Plants; (Gen Instr 1). 36 pp.

—**Rolling Stock**
BSI BS 6853-87. 1987 Code of Practice for Fire Precautions in the Design and Construction of Railway Passenger Rolling Stock. 29 pp.

Fire Protection (Cont.)
—**Rotary Wing Aircraft**
CAA CAP 524 SUB-Part D 12.86. Design and Construction (Rotorcraft). 23 pp.

—**Sheathing Board—Rock Wool**
CNS A2194-86. Rock Wool Sheathing Boards (Sep)(11701).
CNS A3243-86. Method of Test for Rock Wool Sheathing Boards (Sep)(11702).
JIS A 5451-80. Rock Wool Sheathing Boards. 12 pp.

—**Ships**
SAA AS 1266-86. Fire Control Plans for Ships. 13 pp.

—**Ships—Glossaries**
BSI BS 4422: Part 9-90. 1990 Glossary of Terms Associated with Fire Part 9: Marine Terms. 4 pp.

—**Smoke Control—Glossaries**
ISO 8421 Pt 5-88. Fire Protection—Vocabulary—Part 5: Smoke Control First Edition. 7 pp.

—**Symbols**
ISO 6790-86. Equipment for Fire Protection and Fire Fighting—Graphical Symbols for Fire Protection Plans—Specification First Edition. 13 pp.
SAA HB20-91. Graphical Symbols for Fire Protection Drawings (in Professional Package 44). 33 pp.

—**Water Supply Installations—DVGW Codes**
DIN ENGL 1988 Pt 6-88. Drinking Water Supply Systems; Fire Fighting Installations (DVGW Code of Practice) (Dec). 8 pp.

Fire Pumps
See Also: Fire Fighting Equipment; Fire Protection; Pumps
CNS Z2046-82. General Rules of Fire Pumps (Jul)(9192).
MOD UK DSTAN 42-29-01. Pump, Fire, Lightweight, Engine-Driven 1200 Litres/Minute Capacity (at 7 Bar) Issue 1; Amendment 2. 14 pp.

—**Carts**
CNS Z2048-82. Hand-Cart for Fire Pumps (Jul)(9194).
CNS Z2049-82. Portable-Cart for Fire Pumps (Jul)(9195).

—**Centrifugal—Stationary**
CNS B4052-86. Stationary Centrifugal Fire Pump (Mar)(8917). 4 pp.
CNS B4053-86. Prime Mover of Stationary Centrifugal Fire Pump (Mar)(8918).
CNS B4054-86. Auxiliaries of Stationary Centrifugal Fire Pump (Mar)(8919).

—**Portable**
SAA AS 1687-91. Knapsack Spray Pumps for Firefighting Amdt 1 August 1991 (in Professional Package 44).

—**Stirrup**
MOD UK DSTAN 42-5-01. Pump, Reciprocating, Hand Operated (Stirrup Pump) Issue 2; Amendment 1. 10 pp.

—**Vehicles**
CNS Z2047-86. Vehicles for Fire Pumps (Mar)(9193).

Fire Resistance Testing
Use: Fire Testing

Fire Resistant Clothing
Use: Flame Retardant Clothing

Fire Retardant Coatings
Use: Fire Retardant Paints

Fire Retardant Paints
Use For: Fire Retardant Coatings See Also: Fire Retardants; Intumescent Coatings; Paints
CNS K2146-86. Fire-Retardant Paints for Buildings (Oct)(11728). 2 pp.
JIS K 5661-70. Fire-Retardant Paints for Buildings (R 1983); (Erratum). 9 pp.
MOD UK DSTAN 80-78-79. Paint, Finishing, Fire-Retardant, White and Tinted White Type: Brushing Issue 1. 10 pp.

—**Alkyd Enamel—Marine**
CGSB 1-GP-52M-77. Enamel, Alkyd, Marine, Interior, Fire Retardant, Standard for; (Amendment 2 September 1984 Incorporates Amendment 1). 12 pp.

Fire Retardant Testing
Use: Flammability Testing

Fire Retardants
Use For: Flame Retardants See Also: Additives; Fire Retardant Paints; Flame Retardant Clothing

INDUSTRY STANDARDS

INTERNATIONAL AND NON-U.S. NATIONAL STANDARDS SUBJECT INDEX

Fire

Fire Retardants (Cont.)

—**Ammonium Dihydrogen Orthophosphate**
MOD UK CS 3049A. Ammonium Dihydrogen Orthophosphate (Withdrawn).

—**Canvas**
CNS Z2016-71. Fire-Proof Agents for Paper, Cloth and Canvas (Tentative) (Jul)(3078).

—**Fabrics**
CNS Z2016-71. Fire-Proof Agents for Paper, Cloth and Canvas (Tentative) (Jul)(3078).

—**Fabrics—Removal**
CGSB CAN/CGSB-4.2 NO.30.3-M87. Textile Test Methods Procedure for Removal of Flame-Retardant Treatments from Textile Products. 8 pp.

—**Papers**
CNS Z2016-71. Fire-Proof Agents for Paper, Cloth and Canvas (Tentative) (Jul)(3078).

—**Textile Floor Coverings—Removal**
CGSB CAN/CGSB-4.2 NO.30.2-M90. Textile Test Methods Procedure for the Removal of Non-Permanent Flame-Retardant Treatments on Textile Floor Coverings. 8 pp.

—**Timber**
CNS Z2017-71. Fire-Proof Agents for Timber (Tentative) (Jul)(3079).

Fire Safety
Use: Firesafety

Fire Screens
See Also: Fire Guards
BSI BS 3248-86. 1986 Sparkguards for Use with Solid Fuel Appliances. 8 pp.

Fire Sprinkler Equipment
Scope Note: Includes sprinkler hardware and parts; for automated fire control, see the listings under Fire Alarm and Extinguishing Equipment
See Also: Alarm Valves; Dry Pipe Valves; Fire Alarm and Extinguishing Equipment; Fire Fighting Equipment
BSI BS 5306: Part 0-86. 1986 Amd 2 Fire Extinguishing Installations and Equipment on Premises Part 0: Guide for the Selection of Installed Systems and Other Fire Equipment (AMD 6653) July 31, 1991. 44 pp.
BSI BS 5306: Part 2-90. 1990 Fire Extinguishing Installations and Equipment on Premises Part 2: Specification for Sprinkler Systems. 160 pp.
BSI BS 5306: Part 2-79. (WITHDRAWN) 1979 Amd 4 Code of Practice for Fire Extinguishing Installations and Equipment on Premises Part 2: Sprinkler Systems. 29 pp.
CNS Z2062-85. Automatic Sprinkler (Closed Type) (Apr)(11254).
CNS Z3029-85. Method of Test for Automatic Sprinkler (Closed Type) (Apr)(11255).
SNZ NZS 4541-87. Automatic Fire Sprinkler Systems Amend: 1; 2, 1991. 166 pp.

—**Maintenance**
SAA AS 1851.3-85. Maintenance of Fire Protection Equipment—Part 3: Automatic Fire Sprinkler Systems. 14 pp.

—**Residential**
SNZ NZS 4515-90. Residential Fire Sprinkler Systems. 64 pp.

Fire Testing
Use For: Fire Resistance Testing; Flame Testing
See Also: Burning Quality; Burning Rate; Fire Protection; Firesafety; Flame Propagation; Flammability Testing; Materials Testing; Physical Testing
BSI BS EN 60695-2-4/0-93. 1993 Fire Hazard Testing Part 2: Test Methods Section 2.4/0 Sheet 0: Diffusion Type and Premixed Type Flame Test Methods (IEC 695-2-4/0: 1991) (G). 22 pp.
BSI BS EN 60695-2-4/1-93. 1993 Fire Hazard Testing Part 2: Test Methods Section 2.4/1 Sheet 1: 1 kW Nominal Pre-Mixed Test Flame and Guidance (IEC 695-2-4/1: 1991) (G). 23 pp.
CENELEC EN 60695-2-4/0-93. Fire Hazard Testing Part 2: Test Methods Section 4/Sheet 0: Diffusion Type and Premixed Type Flame Test Methods (IEC 695-2-4/0: 1991). 17 pp.
CENELEC EN 60695-2-4/1-93. Fire Hazard Testing Part 2: Test Methods Section 4/Sheet 1: 1 kW Nominal Pre-Mixed Test Flame and Guidance (IEC 695-2-4/1: 1991). 18 pp.
DIN ENGL V 18230 Pt 1 (P)-87. Structural Fire Protection in Industrial Buildings; Required Fire Resistance Time Determined By Calculation (Sept). 16 pp.

Fire Testing (Cont.)
DIN ENGL 18230 Pt 2-87. Structural Fire Protection in Industrial Buildings; Determination of the Combustion Factor (Sept). 8 pp.
IEC 695 Pt 2-4/0-91. Fire Hazard Testing Part 2: Test Methods—Section 4/Sheet 0: Diffusion Type and Premixed Type Flame Test Methods First Edition; (CENELEC EN 60695-2-4/0: 1993). 31 pp.
IEC 695 Pt 2-4/1-91. Fire Hazard Testing Part 2: Test Methods Section 4/Sheet 1: 1 kW Nominal Pre-Mixed Test Flame and Guidance First Edition; (CENELEC EN 60695-2-4/1: 1993). 27 pp.

—**Aircraft**
BSI 3G 100:Pt 2: SUBSEC 3.13-73. 1973 General Requirements for Equipment for Use in Aircraft Part 2: All Equipment Section 3: Environmental Conditions Subsection 3.13: Resistance to Fire in Designated Fire Zones. 20 pp.

—**Aircraft Cabins**
CAA Spec. NO. 8 FIRST Draft.

—**Aircraft Compartments**
CAA Spec. NO. 8 FIRST Draft.
ISO 2685-92. Aircraft—Environmental Conditions and Test Procedures for Airborne Equipment—Resistance to Fire in Designated Fire Zones First Edition. 33 pp.

—**Aircraft Equipment**
ISO 2685-92. Aircraft—Environmental Conditions and Test Procedures for Airborne Equipment—Resistance to Fire in Designated Fire Zones First Edition. 33 pp.

—**Aircraft—Swaged Pipe Couplings**
SBAC RS 736 ISSUE 1. British Metric Swaged Pipe Coupling (AS 43020 and AS 43090 Series) Summary of Tests.
SBAC RS 740 ISSUE 1. British Metric Swaged Pipe Coupling (AS 43020 and AS 43090 Series) Designers's Guide.

—**Belt Conveyors**
BSI BS 490: Sec 11.2-91. 1991 Conveyor and Elevator Belting Part 11: Methods of Test for Safety Section 11.2: Low Energy and High Energy Propane Gallery Tests. 12 pp.
SNZ NZS/BS 490: Pt11:Sec11.2-91. Conveyor and Elevator Belting Part 11: Methods of Test for Safety Section 11.2: Low Energy and High Energy Propane Gallery Tests. 12 pp.

—**Buildings**
BSI BS 476: Part 1-53. 1953Amd 3 Fire Tests on Building Materials and Structures Part 1: (AMD 686) February 18, 1971. 27 pp.
BSI BS 476: Part 32-89. 1989 Fire Tests on Building Materials and Structures Part 32: Guide to Full Scale Fire Tests Within Buildings. 16 pp.
CNS A3185-83. Method of Fire Test for Wooden Structural Parts of Buildings (Apr) (10148).
JIS A 1301-75. Method of Fire Test for Wooden Structural Parts of Buildings. 8 pp.
JIS A 1302-75. Method of Fire Test for Noncombustible Structural Parts of Buildings. 9 pp.
JIS A 1304-75. Method of Fire Resistance Test for Structural Parts of Buildings. 11 pp.
JIS A 1706-76. Method of Safety Test for Combustion of Equipment Units for Dwellings.
SNZ NZS/BS 476: Part 8-72. Fire Tests on Building Materials and Structures Part 8: Test Methods and Criteria for the Fire Resistance of Elements of Building Construction Amend: 3. 17 pp.

—**Cables (Electric)**
BSI BS 4066: Part 1-80. 1980 Tests on Electric Cables Under Fire Conditions Part 1: Method of Test on a Single Vertical Insulated Wire or Cable. 8 pp.
BSI BS 4066: Part 2-89. 1989 Tests on Electric Cables Under Fire Conditions Part 2: Method of Test on a Single Small Vertical Insulated Wire or Cable. 8 pp.
BSI BS 4066: Part 3-86. 1986 Tests on Electric Cables Under Fire Conditions Part 3: Method for Classification of Flame Propagation Characteristics of Bunched Cables. 14 pp.
BSI BS 6387-83. 1983 Amd 2 Performance Requirements for Cables Required to Maintain Circuit Integrity Under Fire Conditions (AMD 5615) July 29, 1988. 26 pp.
BSI BS 7629-93. 1993 Thermosetting Insulated Cables with Limited Circuit Integrity When Affected by Fire (V). 17 pp.
CENELEC HD 405.1-85. Test on Electric Cables Under Fire Conditions Part 1: Test on a Single Vertical Insulated Wire or Cable. 3 pp.
CENELEC HD 405.2 S1-91. Tests on Electric Cables Under Fire Conditions Part 2: Test on a Single Small Vertical Insulated Copper Wire or Cable. 7 pp.
DIN ENGL 4102 Pt 9-90. Fire Behaviour of Building Materials and Elements; Seals for Cable Penetrations; Concepts, Requirements and Testing (May). 11 pp.

Fire Testing (Cont.)

—**Cables (Electric) (Cont.)**
DIN VDE 0472 Pt 804-89. Testing of Cables, Wires and Flexible Cords; Burning. 17 pp.
DIN VDE 0472 Pt 814-91. Testing of Cables and Flexible Cords; Fire Resisting Characteristics of Electric Cables (Jan). 10 pp.
IEC 332 Pt 1-93. Tests on Electric Cables Under Fire Conditions Part 1: Test on a Single Vertical Insulated Wire or Cable Third Edition; (Corrigendum—May 1993). 23 pp.
IEC 332 Pt 2-89. Tests on Electric Cables Under Fire Conditions Part 2: Test on a Single Small Vertical Insulated Copper Wire or Cable First Edition. 13 pp.
IEC 332 Pt 3-92. Tests on Electric Cables Under Fire Conditions Part 3: Tests on Bunched Wires or Cables Second Edition. 62 pp.

—**Cabling Systems (Electric Power)—Construction Materials**
DIN ENGL 4102 Pt 12-91. Fire Behaviour of Building Materials and Elements; Fire Resistance of Electrical Cable Systems; Requirements and Testing (Jan). 6 pp.

—**Cellular Materials**
BSI BS 4735-74. 1974 Laboratory Methods of Test for Assessment of the Horizontal Burning Characteristics of Specimens No Larger Than 150 mm X 50 mm X 13 mm (Nominal) of Cellular Plastics and Cellular Rubber Materials When Subjected to a Small Flame. 12 pp.
BSI BS 4735-01. 1974 Amd 1 Laboratory Method of Test for Assessment of the Horizontal Burning Char. of Specimens No Larger Than150 mm X 50 mm X 13 mm (Nominal) of Cellular Plastics and Cellular Rubber Materials When Subjected to a Small Flame (AMD 7687) July 15, 1993 (V). 13 pp.

—**Ceramics**
CNS R3054-69. Method of Test for Fire-Resistance of Dry Fired and Pressed Fine Ceramic and Porcelain Specimens at Room Temperature (Jan)(2890) (R 1975).

—**Chimneys**
ISO 4736-79. Fire Tests—Small Chimneys—Testing at Elevated Temperatures First Edition. 8 pp.

—**Coatings**
SNZ NZMP 9-89. Fire Properties of Building Materials and Elements of Structure. 207 pp.

—**Connectors**
CEN PREN 2591 (Part C18)-92. Elements of Electrical and Optical Connection Test Methods Part C18—Fire-Resistance. 6 pp.
CEN PREN 2591-FC18-93. Aerospace Series Elements of Electrical and Optical Connection Test Methods Part FC18—Optical Elements Fire Resistance. 3 pp.
SBAC TS 370 ISSUE 1. Test for Electrical Connectors Fire Resistance.

—**Construction Materials**
BSI BS 476: Part 6-89. 1989 Amd 1 Fire Tests on Building Materials and Structures Part 6: Method of Test for Fire Propagation for Products. 16 pp.
BSI BS 476: Part 6-81. 1981 Amd 1 Fire Tests on Building Materials and Structures Part 6: Method of Test for Fire Propagation for Products (AMD 4329) August 31, 1983. 20 pp.
BSI BS 476: Part 7-01. 1987 Amd 1 Fire Tests on Building Materials and Structures Part 7: Method for Classification of the Surface Spread of Flame of Products (AMD 6249) January 31, 1990. 27 pp.
BSI BS 476: Part 7-02. 1987 Amd 2 Fire Tests on Building Materials and Structures Part 7: Method for Classification of the Surface Spread of Flame of Products (AMD 7030) January 30, 1992. 28 pp.
BSI BS 476: Part 7-03. 1987 Amd 3 Fire Tests on Building Materials and Structures Part 7: Method for Classification of the Surface Spread of Flame of Products (AMD 7612) April 15, 1993 (R). 28 pp.
BSI BS 476: Part 7-71. 1971 Fire Tests on Building Materials and Structures Part 7: Surface Spread of Flame Tests for Materials. 18 pp.
BSI BS 476: Part 10-83. 1983 Fire Tests on Building Materials and Structures Part 10: Guide to the Principles and Application of Fire Testing. 12 pp.
BSI BS 476: Part 11-82. 1982 Fire Tests on Building Materials and Structures Part 11: Method for Assessing the Heat Emission from Building Materials. 18 pp.
BSI BS 476: Part 12-91. 1991 Fire Tests on Building Materials and Structures Part 12: Method of Test for Ignitability of Products by Direct Flame Impingement. 20 pp.
BSI BS 476: Part 13-87. 1987 Amd 1 Fire Tests on Building Materials and Structures Part 13: Method of Measuring the Ignitability of Products Subjected to Thermal Irradiance. 37 pp.

INTERNATIONAL AND NON-U.S. NATIONAL STANDARDS
SUBJECT INDEX — Fire

Fire Testing (Cont.)
—Construction Materials (Cont.)

BSI BS 476: Part 20-87. 1987 Amd 1 Fire Tests on Building Materials and Structures Part 20: Method for Determination of the Fire Resistance of Elements of Construction (General Principles) (AMD 6487) April 30, 1990. 44 pp.

BSI BS 476: Part 21-87. 1987 Fire Tests on Building Materials and Structures Part 21: Methods for Determination of the Fire Resistance of Loadbearing Elements of Construction. 19 pp.

BSI BS 476: Part 22-87. 1987 Fire Tests on Building Materials and Structures Part 22: Methods for Determination of the Fire Resistance of Non-Loadbearing Elements of Construction. 22 pp.

BSI BS 476: Part 23-87. 1987 Fire Tests on Building Materials and Structures Part 23: Methods for Determination of the Contribution of Components to the Fire Resistance of a Structure. 20 pp.

BSI BS 476: Part 24-87. 1987 Fire Tests on Building Materials and Structures Part 24: Method for Determination of the Fire Resistance of Ventilation Ducts. 24 pp.

BSI BS 476: Part 32-89. 1989 Fire Tests on Building Materials and Structures Part 32: Guide to Full Scale Fire Tests Within Buildings. 16 pp.

BSI PD 6496-81. 1981 Amd 1 Comparison Between the Technical Requirements of BS 576 Part 8: 1972 with Other Relevant International Standards and Documents on Fire Resistance Tests. 38 pp.

BSI PD 6520-88. 1988 Guide to Fire test Methods for Building Materials and Elements of Construction. 14 pp.

CNS A3113-80. Method of Test for Incombustibility of Internal Finish Material of Buildings (Oct)(6532).

CNS A3125-81. Method of Test for Incombustibility of Thin Materials for Buildings (Jul)(7614).

CNS Z3011-73. Fire Test of Building Construction and Materials (Nov)(3581).

DIN ENGL 4102 Suppl. 1-81. Fire Behaviour of Building Materials and Building Components; Tables of Contents (May). 5 pp.

DIN ENGL 4102 Pt 1-81. Fire Behaviour of Building Materials and Building Components; Building Materials; Concepts, Requirements and Tests (May) (Superseded in Parts by DIN 4102 Part 15). 16 pp.

DIN ENGL 4102 Pt 2-77. Fire Behaviour of Building Materials and Building Components; Building Components; Definitions, Requirements and Tests (Sept). 11 pp.

DIN ENGL 4102 Pt 4-81. Fire Behaviour of Building Materials and Building Components; Summary and Use of Classified Building Materials, Building Components and Special Building Components (Mar). 108 pp.

DIN ENGL 4102 Pt 8-86. Fire Behaviour of Building Materials and Components; Small Scale Test Furnace (May). 4 pp.

DIN ENGL 4102 Pt 15-90. Fire Behaviour of Building Materials and Elements; 'Brandschacht' (May) (Supersedes Parts of DIN 4102 Part 1, May 1981 Edition). 15 pp.

ISO 834-75. Fire-Resistance Tests—Elements of Building Construction First Edition; (Amendment 1-1979) (Amendment 2-1980). 22 pp.

ISO 1182-90. Fire Tests—Building Materials—Non-Combustibility Test Third Edition. 22 pp.

ISO TR3814-89. Tests for Measuring "Reaction-to-Fire" of Building Materials—Their Development and Application Second Edition; (Replaces TR6585). 17 pp.

ISO 5657-86. Fire Tests—Reaction to Fire—Ignitability of Building Products First Edition. 36 pp.

ISO TR5924-89. Fire Tests—Reaction to Fire—Smoke Generated by Building Products (Dual-Chamber Test) First Edition. 38 pp.

ISO TR6167-84. Fire-Resistance Tests—Contribution Made by Suspended Ceilings to the Protection of Steel Beams in Floor and Roof Assemblies First Edition. 13 pp.

ISO TR10158-91. Principles and Rationale Underlying Calculation Methods in Relation to Fire Resistance of Structural Elements First Edition. 27 pp.

JIS A 1321-75. Testing Method for Incombustibility of Internal Finish Material and Procedure of Buildings. 11 pp.

JIS A 1322-66. Testing Method for Incombustibility of Thin Materials for Buildings (R 1982). 8 pp.

SAA AS 1530.4-90. Methods for Fire Tests on Building Materials, Components and Structures—Part 4: Fire-Resistance Test of Elements of Building Construction (This is a Joint Standard with SNZ NZS 1530.4). 27 pp.

SNZ NZMP 9-89. Fire Properties of Building Materials and Elements of Structure. 207 pp.

SNZ NZS/BS 476: Part 20-87. Fire Tests on Building Materials and Structures Part 20: Method for Determination of the Fire Resistance of Elements of Construction (General Principles). 40 pp.

SNZ NZS/BS 476: Part 21-87. Fire Tests on Building Materials and Structures Part 21: Methods for Determination of the Fire Resistance of Loadbearing Elements of Construction. 20 pp.

SNZ NZS/BS 476: Part 22-87. Fire Tests on Building Materials and Structures Part 22: Methods for Determination of the Fire Resistance of Non-Loadbearing Elements of Construction. 24 pp.

SNZ NZS/ISO 834-75. Fire-Resistance Tests. Elements of Building Construction Amend: 1; 2. 16 pp.

SNZ NZS/AS 1530. 4-90. Methods for Fire Tests on Building Materials, Components and Structures Part 4: Fire-Resistance Test of Elements of Building Construction (This is a Joint Standard with SAA AS 1530.4). 27 pp.

—Construction Materials—Heat Release

BSI BS 476: Part 15-93. 1993 Fire Tests on Building Materials and Structures Part 15: Method of Measuring the Rate of Heat Release of Products (ISO 5660-1: 1993) (R). 38 pp.

ISO 5660 Pt 1-93. Fire Tests—Reaction to Fire—Part 1: Rate of Heat Release from Building Products (Cone Calorimeter Method) First Edition. 36 pp.

—Construction Materials—Smoke Density

BSI BS 6401-83. 1983 Measurement, in the Laboratory, of the Specific Optical Density of Smoke Generated by Materials. 32 pp.

JIS A 1306-83. Measuring Method of Smoke Density Using Light Extinction Method. 7 pp.

—Construction Materials—Surface Burning Characteristics

CNS Z3013-74. Method of Test for Surface Burning Characteristics of Building Materials (Mar)(3690).

—Cords (Electric)

DIN VDE 0472 Pt 804-89. Testing of Cables, Wires and Flexible Cords; Burning. 17 pp.

—Doors

CNS Z3007-72. Fire Tests of Door Assemblies (Oct)(3407). 6 pp.

ISO 3008-76. Fire-Resistance Tests—Door and Shutter Assemblies First Edition; (Amendment Slip-1976) (Amendment Slip-1977) (Erratum—March 1982) (Amendment 1-1984). 18 pp.

—Doors—Smoke Control

BSI BS 476: Sec 31.1-83. 1983 Fire Tests on Building Materials and Structures Part 31: Methods for Measuring Smoke Penetration Through Doorsets and Shutter Assemblies Section 31.1: Method of Measurement Under Ambient Temperature Conditions. 11 pp.

ISO 5925 Pt 1-81. Fire Tests—Evaluation of Performance of Smoke Control Door Assemblies—Part 1: Ambient Temperature Test First Edition. 8 pp.

—Ducts

DIN ENGL 4102 Pt 6-77. Fire Behaviour of Building Materials and Building Components; Ventilation Ducts; Definitions, Requirements and Tests (Sept). 5 pp.

DIN ENGL 4102 Pt 11-85. Fire Behaviour of Building Materials and Building Components; Pipe Encasements, Pipe Bushings, Service Shafts and Ducts, and Barriers Across Inspection Openings; Terminology, Requirements and Testing (Dec). 12 pp.

ISO 6944-85. Fire Resistance Tests—Ventilation Ducts First Edition. 24 pp.

—Electric Wire

BSI BS 4066: Part 1-80. 1980 Tests on Electric Cables Under Fire Conditions Part 1: Method of Test on a Single Vertical Insulated Wire or Cable. 8 pp.

BSI BS 4066: Part 2-89. 1989 Tests on Electric Cables Under Fire Conditions Part 2: Test on a Single Small Vertical Insulated Wire or Cable. 8 pp.

CENELEC HD 405.1-85. Test on Electric Cables Under Fire Conditions Part 1: Test on a Single Vertical Insulated Wire or Cable. 3 pp.

CENELEC HD 405.2 S1-91. Test on Electric Cables Under Fire Conditions Part 2: Test on a Single Small Vertical Insulated Copper Wire or Cable. 7 pp.

DIN VDE 0472 Pt 804-89. Testing of Cables, Wires and Flexible Cords; Burning. 17 pp.

IEC 332 Pt 1-93. Tests on Electric Cables Under Fire Conditions Part 1: Test on a Single Vertical Insulated Wire or Cable Third Edition; (Corrigendum—May 1993). 23 pp.

IEC 332 Pt 2-89. Tests on Electric Cables Under Fire Conditions Part 2: Test on a Single Small Vertical Insulated Copper Wire or Cable First Edition. 13 pp.

IEC 332 Pt 3-92. Tests on Electric Cables Under Fire Conditions Part 3: Tests on Bunched Wires or Cables Second Edition. 62 pp.

—Electrical Insulation

SAA AS 2420-87. Fire Test Methods for Solid Insulation Materials and Non-Metallic Enclosures Used in Electrical Equipment Amdt 1 January 1991 (In Professional Packages 28, 44). 29 pp.

SNZ AS 2420-87. Fire Test Methods for Solid Insulating Materials and Non-Metallic Enclosures Used in Electrical Equipment. 34 pp.

—Fabrics

DIN ENGL 66083-86. Characteristic Values for the Burning Behaviour of Textile Products; Textile Fabrics for Industrial and Protective Clothing (Nov). 2 pp.

—Fabrics—Fire Resistant

CGSB CAN/CGSB-4.2 NO.27.10-M91. Textile Test Methods Flame Resistance—Vertically Oriented Textile Fabric or Fabric Assembly Test. 11 pp.

—Fire Alarms

CEN EN 54-9-82. Components of Automatic Fire Fetection Systems Part 9 Fire Sensitivity Test. 10 pp.

—Fire Barriers

DIN ENGL 4102 Pt 5-77. Fire Behaviour of Building Materials and Building Components; Fire Barriers, Barriers in Lift Wells and Glazings Resistant Against Fire; Definitions, Requirements and Tests (Sept) (Superseded in Part by DIN 4102 Part 13). 8 pp.

—Fire Dampers

DIN ENGL 4102 Pt 6-77. Fire Behaviour of Building Materials and Building Components; Ventilation Ducts; Definitions, Requirements and Tests (Sept). 5 pp.

—Fire Dampers—Smoke Leakage

JIS A 1314-92. Smoke-Proof Test Method for Fire Damper. 9 pp.

—Fire Doors

CNS A3223-85. Method of Fire-Protecting Test for Fire Door of Buildings (Apr)(11227).

JIS A 1311-75. Method of Fire Protecting Test of Fire Door for Buildings. 10 pp.

—Fire Extinguishers

SAA AS 1850-81. Portable Fire Extinguishers—Classification, Rating and Fire Testing Amdt 1 April 1982 Amdt 2 July 1989. 11 pp.

SNZ NZS 4507-78. Specification for Fire Testing and Rating of Portable Fire Extinguishers. 8 pp.

—Floor Coverings

DIN ENGL 4102 Pt 14-90. Fire Behaviour of Building Materials and Elements; Determination of the Burning Behavior of Floor Covering Systems Using a Radiant Heat Source (May). 7 pp.

DIN ENGL 54332-75. Testing of Textiles; Determination of the Burning Behaviour of Textile Floor Coverings (Feb). 7 pp.

—Floors

BSI BS 5268: Sec 4.2-90. 1990 Structural Use of Timber Part 4: Fire Resistance of Timber Structures Section 4.2: Recommendations for Calculating Fire Resistance of Timber Stud Walls and Joisted Floor Constructions. 28 pp.

—Glass

BSI PD 6512: Part 3-87. 1987 Use of Elements of Structural Protection with Particular Reference to the Recommendations Given in BS 5588 Fire Precautions in the Design and Construction of Buildings Part 3: To the Fire Performance of Glass. 10 pp.

CEN PREN 357-90. Glass in Building—Glazed Assemblies Containing Fire Resistant Transparent or Translucent Glass, for Use in Buildings. 7 pp.

—Glazing Systems

DIN ENGL 4102 Pt 5-77. Fire Behaviour of Building Materials and Building Components; Fire Barriers, Barriers in Lift Wells and Glazings Resistant Against Fire; Definitions, Requirements and Tests (Sept) (Superseded in Part by DIN 4102 Part 13). 8 pp.

DIN ENGL 4102 Pt 13-90. Fire Behaviour of Building Materials and Elements; Fire Resistant Glazing; Concepts, Requirements and Testing (May) (Supersedes Parts of DIN 4102 Part 5, September 1977 Edition). 4 pp.

ISO 3009-76. Fire-Resistance Tests—Glazed Elements First Edition; (Amendment Slip-1977) (Amendment 1-1984). 13 pp.

—Glossaries

ISO 3261-75. Fire Tests—Vocabulary First Edition. 11 pp.

ISO Guide 52-90. Glossary of Fire Terms and Definitions First Edition. 33 pp.

INDUSTRY STANDARDS

Fire

Fire Testing (Cont.)

—Materiel
MOD UK DSTAN 07-55: Pt 2:Sec 6-01. Environmental Testing of Service Materiel Part 2: Tests Section 6: Fire and Explosion Issue 1; Amendment 1. 89 pp.

—Pipes
DIN ENGL 4102 Pt 11-85. Fire Behaviour of Building Materials and Building Components; Pipe Encasements, Pipe Bushings, Service Shafts and Ducts, and Barriers Across Inspection Openings; Terminology, Requirements and Testing (Dec). 12 pp.

—Plastics—Ignition Sources
ISO TR10353-92. Plastics—Survey of Ignition Sources Used for National and International Fire Tests First Edition. 34 pp.

—Plywood
CNS O2049-82. Method of Test for Flameproof of Fireproof Plywood (Apr)(8736). 5 pp.

—Polystyrene
BSI BS 6203-91. 1991 Guide to Fire Characteristics and Fire Performance of Expanded Polystyrene Materials Used in Building Applications. 33 pp.

—Porcelain
CNS R3054-69. Method of Test for Fire-Resistance of Dry Fired and Pressed Fine Ceramic and Porcelain Specimens at Room Temperature (Jan)(2890) (R 1975).

—Radiometers
BSI BS 6809-87. 1987 Calibration of Radiometers for Use in Fire Testing. 16 pp.

—Roofing
BSI BS 476: Part 3-58. 1958 Fire Tests on Building Materials and Structures Part 3: External Fire Exposure Roof Tests. 12 pp.
BSI BS 476: Part 3-75. 1975 Fire Tests on Building Materials and Structures Part 3: External Fire Exposure Roof Test. 14 pp.
CNS A3184-83. Method of Fire Test for Roofs of Buildings (Apr)(10147).
DIN ENGL 4102 Pt 7-87. Fire Behaviour of Building Materials and Building Components; Roof Coverings; Terminology, Requirements and Testing (Mar). 4 pp.
JIS A 1312-59. Method of Fire Test for Roof of Building (R 1973). 10 pp.

—Rooms
BSI BS 476: Part 33-93. 1993 Fire Tests on Building Materials and Structures Part 33: Full-Scale Room Test for Surface Products (ISO 9705: 1993) (R). 38 pp.
ISO 9705-93. Fire Tests—Full-Scale Room Test for Surface Products First Edition. 36 pp.

—Safety Glass—Automotive
ISO 3917-92. Road Vehicles—Safety Glazing Materials—Test Methods for Resistance to Radiation, High Temperature, Humidity, Fire and Simulated Weathering Second Edition. 8 pp.

—Shafts (Buildings)
DIN ENGL 4102 Pt 11-85. Fire Behaviour of Building Materials and Building Components; Pipe Encasements, Pipe Bushings, Service Shafts and Ducts, and Barriers Across Inspection Openings; Terminology, Requirements and Testing (Dec). 12 pp.

—Shutters (Blinds)
CNS A3072-85. Method of Test for Fire-Proof Shutters for Buildings (Apr)(4213). 2 pp.
ISO 3008-76. Fire-Resistance Tests—Door and Shutter Assemblies First Edition; (Amendment Slip-1976) (Amendment Slip-1977) (Erratum—March 1982) (Amendment 1-1984). 18 pp.

—Shutters (Blinds)—Smoke Control
BSI BS 476: Sec 31.1-83. 1983 Fire Tests on Building Materials and Structures Part 31: Methods for Measuring Smoke Penetration Through Doorsets and Shutter Assemblies Section 31.1: Method of Measurement Under Ambient Temperature Conditions. 11 pp.
ISO 5925 Pt 1-81. Fire Tests—Evaluation of Performance of Smoke Control Door Assemblies—Part 1: Ambient Temperature Test First Edition. 8 pp.

—Structural Design
ISO TR3956-75. Principles of Structural Fire-Engineering Design with Special Regard to the Connection Between Real Fire Exposure and the Heating Conditions of the Standard Fire-Resistance Test (ISO 834) First Edition. 8 pp.

Fire Testing (Cont.)

—Structural Members
CNS A3304-89. Method of Fire Test for Noncombustible Structural Parts of Buildings (May)(12513).
CNS A3305-89. Method of Fire Resistance Test for Structural Parts of Buildings (May) (12514).

—Surfacing Materials
BSI BS 7188-89. 1989 Amd 2 Methods of Test for Impact Absorbing Playground Surfaces (AMD 6855) September 30, 1991. 21 pp.

—Valves
BSI BS 6755: Part 2-87. 1987 Amd 2 Testing of Valves Part 2: Specification for Fire Type-Testing Requirements (AMD 6712) June 28, 1991. 20 pp.
ISO 10497-92. Testing of Valves—Fire Type-Testing Requirements First Edition. 18 pp.

—Walls
DIN ENGL 4102 Pt 3-77. Fire Behaviour of Building Materials and Building Components; Fire Walls and Non-Load-Bearing External Walls; Definitions, Requirements and Tests (Sept). 7 pp.

—Walls—Wood Frame
BSI BS 5268: Sec 4.2-90. 1990 Structural Use of Timber Part 4: Fire Resistance of Timber Structures Section 4.2: Recommendations for Calculating Fire Resistance of Timber Stud Walls and Joisted Floor Constructions. 28 pp.

—Windows
CNS Z3008-72. Fire Tests of Window Assemblies (Oct)(3466).

—Wooden Structures
BSI BS 5268: Sec 4.1-78. 1978 Amd 2 Structural Use of Timber Part 4: Fire Resistance of Timber Structures Section 4.1: Recommendations for Calculating Fire Resistance of Timber Members (AMD 6192) March 31, 1990. 13 pp.

Fire Tube Boilers
Use For: Shell Boilers *See Also:* Boilers; Waste Heat Boilers; Watertube Boilers
BSI BS 2790: CASE 20-87. (WITHDRAWN) 1987 Design and Manufacture of Shell Boilers of Welded Construction Enquiry Cases: Case 20: Accumulation Test. 1 p.
ISO 5730-92. Stationary Shell Boilers of Welded Construction (Other Than Water-Tube Boilers) First Edition. 214 pp.

—Construction
BSI BS 2790-92. 1992 Design and Manufacture of Shell Boilers of Welded Construction (Q). 151 pp.
BSI BS 2790-01. 1992 Amd 1 Design and Manufacture of Shell Boilers of Welded Construction (AMD 7437) November 15, 1992 (Q). 172 pp.
BSI BS 2790-89. 1989 Amd 3 Design and Manufacture of Shell Boilers of Welded Construction (AMD 6760) October 31, 1991 (Q). 149 pp.

—Design
BSI BS 2790-92. 1992 Design and Manufacture of Shell Boilers of Welded Construction (Q). 151 pp.
BSI BS 2790-01. 1992 Amd 1 Design and Manufacture of Shell Boilers of Welded Construction (AMD 7437) November 15, 1992 (Q). 172 pp.
BSI BS 2790-89. 1989 Amd 3 Design and Manufacture of Shell Boilers of Welded Construction (AMD 6760) October 31, 1991 (Q). 149 pp.

—Design Temperature
BSI BS 2790: CASE 26-93. 1993 Design and Manufacture of Shell Boilers of Welded Construction Enquiry Cases: Case 26: Incorporation of Design Temperature in Boiler Marking. 1 p.

—Plates—Temperature Measurement
BSI BS 2790: CASE 21-88. (WITHDRAWN) 1988 Design and Manufacture of Shell Boilers of Welded Construction Enquiry Cases: Case 21: Tube Plate Metal Temperatures. 1 p.

—Safety
BSI BS 2790: CASE 22-88. (WITHDRAWN) 1988 Design and Manufacture of Shell Boilers of Welded Construction Enquiry Cases: Case 22: Application of Safety Factor to Boilers. 1 p.

—Stresses
BSI BS 2790: CASE 15-90. 1990 Design and Manufacture of Shell Boilers of Welded Construction Enquiry Cases: Enquiry Case 15: Allowable Stresses in Shells Under Hydraulic Test Conditions at Boiler Supports. 1 p.

—Ultrasonic Testing
BSI BS 2790: CASE 23-88. 1988 Design and Manufacture of Shell Boilers of Welded Construction Enquiry Cases: Case 23: Ultrasonic Examination of Welds. 1 p.

Fire Walls
See Also: Fire Barriers; Fire Doors; Fire Protection; Walls

—Fiber Optic Connectors—Flammability Testing
CNS C6333-89. Method of Test for Fiber Optic Devices (FOTP-172 Flame Resistance of Firewall Connector) (Jul)(12564).

—Fire Testing
DIN ENGL 4102 Pt 3-77. Fire Behaviour of Building Materials and Building Components; Fire Walls and Non-Load-Bearing External Walls; Definitions, Requirements and Tests (Sept). 7 pp.

Fire Windows
See Also: Firesafety
SNZ NZS 4232: Part 2-88. Performance Criteria for Fire Resisting Closures Part 2: Fire Resisting Glazing Systems. 36 pp.

Firebacks
See Also: Chimneys; Fireplaces; Flues
BSI BS 1251-87. 1987 Open-Fireplace Components. 12 pp.
BSI BS 1251-70. 1970 Amd 2 Open Fireplace Components (AMD 1651) January 31, 1975. 28 pp.

Firebricks
Use For: Fire Bricks; Fireclay Bricks; High Alumina Bricks; Refractory Bricks *See Also:* Bricks; Ceramics; Combustion Chambers; Fireclay Refractories; Magnesia Bricks; Refractory Materials; Silica Bricks; Silicate Refractories
BSI BS 1902: Sec 3.11-83. 1983 Amd 1 Methods for Testing Refractory Materials Part 3: General and Textural Properties Section 3.11: Measurement of Dimensions and Shape of Refractory Bricks and Blocks (Methods 1902-311) (AMD 6290). 11 pp.
BSI BS 3056: Part 1-85. 1985 Sizes of Refractory Bricks Part 1: Multi-Purpose Bricks (Supersedes BS 3056: 1973). 23 pp.
BSI BS 3056: Part 2-85. 1985 Sizes of Refractory Bricks Part 2: Bricks for Use in Glass-Melting Furnaces. 32 pp.
BSI BS 3056: Part 2-01. 1985 Amd 1 Sizes of Refractory Bricks Part 2: Specification for Bricks for Use in Glass-Melting Furnaces (AMD 6917) May 1, 1992 (F). 33 pp.
BSI BS 3056: Part 3-86. 1986 Amd 1 Sizes of Refractory Bricks Part 3: Bricks for Rotary Cement Kilns. 17 pp.
BSI BS 3056: Part 8-87. 1987 Size of Refractory Bricks Part 8: Bricks for Ladles. 10 pp.
CGSB CAN/CGSB-10.1-92. Fireclay Refractory Brick for Stationary Boiler Service. 7 pp.
CNS R1006-86. Shapes and Dimensions of Refractory Bricks (Jul)(612).
CNS R1011-76. Standard Classification of Insulating Fire Brick (Dec)(1047).
CNS R1019-86. Shape and Dimension of Refractory Bricks for Rotary Kiln (Jul)(11636).
CNS R1020-86. Shape and Dimension of Refractory Bricks for Cupola (Jul)(11637).
CNS R2047-65. Fire Bricks for Coal Fired Steam Boilers (Jan)(2353)(R 1973).
CNS R2049-65. Fire Clay Bricks (Used for Metal Melting) (Jan)(2394)(R 1973). 4 pp.
CNS R2076-73. Fire Brick for Electric Melting Furnace of Steel (Aug)(3588).
CNS R3140-86. Method for Measuring Dimensions of Refractory Bricks (Jul)(11639).
ISO 5417-86. Refractory Bricks for Use in Rotary Kilns—Dimensions First Edition. 4 pp.
JIS R 2101-83. Shape and Dimension of Refractory Bricks. 7 pp.
JIS R 2102-83. Shape and Dimension of Refractory Bricks for Open-Hearth Furnace. 6 pp.
JIS R 2103-83. Shape and Dimension of Refractory Bricks for Rotary Kiln. 9 pp.
JIS R 2104-83. Shape and Dimension of Refractory Bricks for Cupola. 6 pp.
JIS R 2202-75. Methods of Measuring Dimensions of Refractory Bricks (R 1986). 5 pp.
JIS R 2304-76. Fireclay Brick. 7 pp.
JIS R 2305-76. High Alumina Brick (R 1986). 7 pp.
JIS R 2402-75. Specifications of Fireclay Bricks for Ladle Lining. 7 pp.
JIS R 2611-92. Insulating Fire Bricks. 7 pp.
SAA AS 1617.1-93. Refractory Bricks and Shapes—Part 1: Fireclay (in Professional Package 32) (Supersedes AS 1617—1974). 4 pp.
SAA AS 1619-74. Methods for Determining the Dimensions of Refractory Bricks Bound with AS 1618. 5 pp.
SNZ NZS 2174-67. Classification and Methods of Sampling and Testing of Insulating Refractory Bricks. 24 pp.

—Acid Resistant
JIS R 1536-91. Acid and Heat Resisting Bricks for Chemical Industry. 6 pp.

INTERNATIONAL AND NON-U.S. NATIONAL STANDARDS
SUBJECT INDEX

Fireclay

Firebricks (Cont.)

—Arch
CNS R2096-79. Dimensions of Refractory Arch Bricks (Jan)(4721).
DIN ENGL 1082 Pt 1-88. Refractory Materials; Refractory Arch Bricks; Dimensions (Jan). 3 pp.
DIN ENGL 1082 Pt 3-88. Refractory Materials; Basic Arch Bricks and Refractory Bricks for Oxygen Steel-Making Converters and Arc Furnaces; Dimensions (Jan). 3 pp.
DIN ENGL 1082 Pt 4-89. Refractory Materials; Refractory Arch Bricks for Use in Rotary Kilns; Dimensions (Feb). 2 pp.
ISO 5019 Pt 2-84. Refractory Bricks—Dimensions—Part 2: Arch Bricks First Edition. 4 pp.

—Basic
BSI BS 3056: Part 7-92. 1992 Sizes of Refractory Bricks Part 7: Specification for Basic Bricks for Steel Making. 8 pp.
BSI BS 3056: Part 7-86. 1986 Sizes of Refractory Bricks Part 7: Basic Bricks for Steel Making. 4 pp.
ISO 5019 Pt 6-84. Refractory Bricks—Dimensions—Part 6: Basic Bricks for Oxygen Steel-Making Converters First Edition. 5 pp.

—Boilers
CGSB CAN/CGSB-10.1-92. Fireclay Refractory Brick for Stationary Boiler Service. 7 pp.

—Boilers—Marine
CGSB CAN/CGSB-10.4-92. Fireclay Refractory Brick for Marine Oil-Fired Boilers. 7 pp.

—Chemical Analysis
JIS R 2212-91. Chemical Analysis of Refractory Bricks and Refractory Mortars. 109 pp.

—Classification
CNS R1007-76. Standard Classification of Fireclay Brick (Dec)(625).
CNS R2046-76. Standard Classification of High-Aluminum Brick (Dec)(2352).
SNZ NZS 2174-67. Classification and Methods of Sampling and Testing of Insulating Refractory Bricks. 24 pp.

—Compression Testing
CNS R3010-78. Method of Test for Cold Crushing Strength and Modulus of Rupture of Refractory Brick and Shapes at Room Temperature (Nov)(616).
CNS R3143-86. Method of Test for Crushing Strength of Insulating Fire Bricks (Oct)(11740).
JIS R 2206-91. Test Method for Cold Crushing Strength of Refractory Bricks. 6 pp.
JIS R 2615-85. Testing Method for Crushing Strength of Insulating Fire Bricks. 6 pp.

—Defects
BSI BS 1902: Sec 3.13-85. 1985 Amd 1 Methods for Testing Refractory Materials Part 3: General and Textural Prop. Sec 3.13: Mst. of Corner and Edge Defects and Other Sur. Imperf. of Refractory Bricks and Blocks (Att-ributive Prop.) (Method 1902-313) (AMD 6343) June 28, 1991. 10 pp.

—Density
CNS R3008-78. Method of Test for Size and Bulk Density of Refractory Brick and Insulating Firebrick (Nov)(614).
CNS R3013-86. Method of Test for Apparant Porosity Water Absorption and Specific Gravity of Refractory Bricks (Oct)(619).
CNS R3141-86. Method of Test for Specific Gravity and True Porosity of Insulating Fire Bricks (Oct)(11738).
JIS R 2205-92. Testing Method for Apparent Porosity, Water Absorption and Specific Gravity of Refractory Bricks. 8 pp.
JIS R 2614-85. Testing Method for Specific Gravity and True Porosity of Insulating Fire Bricks. 6 pp.

—Dome
BSI BS 3056: Part 4-85. 1985 Sizes of Refractory Bricks Part 4: Bricks for Electric Arc Furnace Roofs. 7 pp.
ISO 5019 Pt 4-88. Refractory Bricks—Dimensions—Part 4: Dome Bricks for Electric Arc Furnace Roofs Second Edition. 8 pp.

—Furnaces—Marine
CGSB CAN/CGSB-10.5-92. Insulating Refractory Brick. 7 pp.

—Furnaces—Waterproof Coatings
MOD UK DSTAN 80-173-93. Waterproofing Compound for Refractory Furnace Linings Type: Brushing/Spraying Issue 1. 18 pp.
MOD UK TS 10192A. Waterproofing Compound for Refractory Furnace Linings. Type: Brushing/Spraying (Superseded by Def Stan 80-1730.

—Gas Permeability
JIS R 2115-87. Testing Method for Permeability to Gases of Refractory Bricks. 8 pp.

Firebricks (Cont.)

—Identification Systems
ISO 9205-88. Refractory Bricks for Use in Rotary Kilns—Hot-Face Identification Marking First Edition. 7 pp.

—Inspection
CNS R3139-86. Method of Inspection for Dimension of Refractory Bricks (Jul)(11638).
JIS R 2150-83. Method of Inspection for Dimension of Refractory Bricks. 5 pp.

—Linear Change
SAA AS R31 METHOD 15-66. Methods for the Physical Testing of Refractories and Refractory Materials Method 15: The Determination of Permanent Linear Change on Reheating of Insulating Bricks. 2 pp.

—Measurement
BSI BS 1902: Sec 3.2-81. 1981 Methods for Testing Refractory Materials Part 3: General and Textural Properties Section 3.2: Measurement of the Dimensions of Specimens for Testing (Methods 1902-302). 4 pp.

—Modulus of Rupture
CNS R3010-78. Method of Test for Cold Crushing Strength and Modulus of Rupture of Refractory Brick and Shapes at Room Temperature (Nov)(616).
CNS R3142-86. Method of Test for Modulus of Rupture of Insulating Fire Bricks (Oct)(11739).
JIS R 2213-78. Test Method for Modulus of Rupture of Refractory Bricks (R 1986). 6 pp.
JIS R 2619-81. Testing Method for Modulus of Rupture of Insulating Fire Bricks. 7 pp.

—Porosity
CNS R3013-86. Method of Test for Apparant Porosity Water Absorption and Specific Gravity of Refractory Bricks (Oct)(619).
CNS R3141-86. Method of Test for Specific Gravity and True Porosity of Insulating Fire Bricks (Oct)(11738).
JIS R 2205-92. Testing Method for Apparent Porosity, Water Absorption and Specific Gravity of Refractory Bricks. 8 pp.
JIS R 2614-85. Testing Method for Specific Gravity and True Porosity of Insulating Fire Bricks. 6 pp.

—Rectangular
DIN ENGL 1081-88. Refractory Materials; Rectangular Refractory Bricks; Dimensions (Jan). 2 pp.
DIN ENGL 1082 Pt 3-88. Refractory Materials; Basic Arch Bricks and Refractory Bricks for Oxygen Steel-Making Converters and Arc Furnaces; Dimensions (Jan). 3 pp.
ISO 5019 Pt 1-84. Refractory Bricks—Dimensions—Part 1: Rectangular Bricks First Edition. 4 pp.

—Rectangular—Checker
ISO 5019 Pt 3-84. Refractory Bricks—Dimensions—Part 3: Rectangular Checker Bricks for Regenerative Furnaces First Edition. 3 pp.

—Refractoriness
CNS R3011-85. Method of Test for Refractoriness of Refractory Brick (Nov)(617).
CNS R3012-78. Method of Test for Reheat Change of Refractory Brick (Nov)(618).
CNS R3015-65. Method of Test for Softening of Refractory Bricks Under Load (Sep)(621)(R 1973).
JIS R 2204-91. Testing Method for Refractoriness of Refractory Brick. 6 pp.
JIS R 2209-91. Testing Method for Refractoriness Under Load of Refractory Bricks. 8 pp.

—Refractory Mortars—Furnaces
CGSB CAN/CGSB-10.3-92. Air-Setting Refractory Mortar. 6 pp.
CGSB CAN/CGSB-10.6-92. Heat-Setting Refractory Mortar. 7 pp.

—Sampling
CNS R3007-56. Method of Sampling for Refractory Bricks (Apr) (613)(R 1973).
SNZ NZS 2174-67. Classification and Methods of Sampling and Testing of Insulating Refractory Bricks. 24 pp.

—Shrinkage
CNS R3144-86. Method of Test for Thermal Expansion and Shrinkage of Insulating Fire Bricks (Oct)(11741).
CNS R3145-86. Method of Test for Reheat Shrinkage of Insulating Fire Bricks (Oct)(11742).
JIS R 2613-85. Testing Method for Reheat Shrinkage of Insulating Fire Bricks. 6 pp.
JIS R 2617-85. Testing Method for Thermal Expansion and Shrinkage of Insulating Fire Bricks. 6 pp.

Firebricks (Cont.)

—Size
CNS R3008-78. Method of Test for Size and Bulk Density of Refractory Brick and Insulating Firebrick (Nov)(614).
SAA AS 1618-74. Preferred Sizes for Refractory Bricks Bound with AS 1619. 10 pp.

—Skewbacks
DIN ENGL 1082 Pt 2-88. Refractory Materials; Refractory Skewbacks; Dimensions (Jan). 2 pp.
ISO 5019 Pt 5-84. Refractory Bricks—Dimensions—Part 5: Skewbacks First Edition. 4 pp.

—Temperature Change Testing—Spalling
CNS R3016-56. Method of Spalling Test for Refractory Bricks Under Sudden Change of Temperature (Apr)(622) (R 1973).

—Thermal Conductivity
CNS R3157-87. Method of Test for Thermal Conductivity of Insulating Fire Bricks by Water Circulation (Sep)(12110).
JIS R 2616-79. Testing Method for Thermal Conductivity of Insulating Fire Bricks by Water Circulation.
JIS R 2618-92. Testing Method for Thermal Conductivity of Insulating Fire Bricks by Hot Wire. 8 pp.

—Thermal Expansion
CNS R3144-86. Method of Test for Thermal Expansion and Shrinkage of Insulating Fire Bricks (Oct)(11741).
JIS R 2207-76. Test Method for the Rate of Linear Change of Refractory Brick on Heating (R 1986). 6 pp.
JIS R 2208-76. Test Method for Permanent Linear Change of Refractory Brick on Reheating (R 1986). 6 pp.
JIS R 2617-85. Testing Method for Thermal Expansion and Shrinkage of Insulating Fire Bricks. 6 pp.

—Thermal Shock Resistance
DIN ENGL 51068 Pt 1-76. Testing of Ceramic Materials; Determination of Resistance to Thermal Shock; Water Quenching Method for Refractory Bricks (May). 3 pp.

—Wall
BSI BS 3056: Part 5-85. (WITHDRAWN) 1985 Sizes of Refractory Bricks Part 5: Bricks for Blast Furnace Walls. 3 pp.

—Warpage
CNS R3009-56. Method of Test for Warpage of Refractory Bricks (Apr)(615) (R 1973).
JIS R 2203-75. Method of Measuring Warpage of Refractory Bricks (R 1983). 5 pp.

—Water Absorption
CNS R3013-86. Method of Test for Apparant Porosity Water Absorption and Specific Gravity of Refractory Bricks (Oct)(619).
JIS R 2205-92. Testing Method for Apparent Porosity, Water Absorption and Specific Gravity of Refractory Bricks. 8 pp.

—Water Resistance Testing
BSI BS 1902: Sec 3.14-87. 1987 Methods for Testing Refractory Materials Part 3: General and Textured Properties Section 3.14: Determination of Hydration Tendency. 8 pp.

—X-Ray Fluorescence Spectrometry
JIS R 2216-87. Method for X-Ray Fluorescence Spectrometric Analysis of Refractory Bricks and Refractory Mortars. 13 pp.

Fireclay Bricks
Use: Firebricks

Fireclay Refractories
See Also: Ceramics; Firebricks; Refractory Materials

—Castable
JIS R 2541-76. High Alumina and Fire Clay Castable Refractories (R 1986). 6 pp.

—Castable—Classification
CNS R1015-76. Standard Classification of Fireclay and High-Alumina Plastics and Ramming Mixes (Jun)(3971).

—Plastic
CGSB CAN/CGSB-10.7-93. Fireclay and Alumina Refractory Plastic. 7 pp.
JIS R 2561-76. High Alumina and Fire Clay Plastic Refractories (R 1986). 6 pp.

—Plastic—Classification
CNS R1015-76. Standard Classification of Fireclay and High-Alumina Plastics and Ramming Mixes (Jun)(3971).

INDUSTRY STANDARDS

Fireclay Refractories (Cont.)

—Plastic—Coefficient of Linear Variation
JIS R 2576-76. Method of Test for Permanent of Linear Change on High Alumina and Fireclay Plastic Refractories.

—Plastic—Compression Testing
CNS R3153-87. Method of Test for Crushing Strength and Modulus of Rupture of High Alumina and Fireclay Plastic Refractories (May) (11973).

JIS R 2575-92. Testing Method for Crushing Strength and Modulus of Rupture of High Alumina and Fireclay Plastic Refractories. 10 pp.

—Plastic—Modulus of Rupture
CNS R3153-87. Method of Test for Crushing Strength and Modulus of Rupture of High Alumina and Fireclay Plastic Refractories (May) (11973).

JIS R 2575-92. Testing Method for Crushing Strength and Modulus of Rupture of High Alumina and Fireclay Plastic Refractories. 10 pp.

—Plastic—Moisture Content
CNS R3133-85. Method of Test for Moisture Content of Fireclay Plastic Refractories (Nov) (11427).

—Plastic—Rate of Linear Change
JIS R 2577-81. Testing Method for the Rate of Linear Change of High Alumina and Fire Clay Plastic Refractories.

—Plastic—Refractoriness
CNS R3128-85. Method of Test for Refractoriness of Fireclay Plastic Refractories (Nov) (11422).

JIS R 2573-91. Testing Method for Refractoriness of Fireclay Plastic Refractories. 6 pp.

—Plastic—Sampling
CNS R3132-85. Method of Sampling for Fireclay Plastic Refractories (Nov) (11426).

JIS R 2571-85. Sampling Method for High Aluminous Plastic Refractories and Fireclay Plastic Refractories. 4 pp.

—Plastic—Water Content
JIS R 2572-85. Testing Method for Water Content of High Aluminous Plastic Refractories and Fireclay Plastic Refractories. 5 pp.

—Plastic—Workability
CNS R3158-87. Method of Test for Workability Index of Fireclay Plastic Refractories (Sep) (12111).

JIS R 2574-91. Testing Method for Workability Index of Fireclay Plastic. 11 pp.

—Refractoriness
ISO 528-83. Refractory Products—Determination of Pyrometric Cone Equivalent (Refractoriness) First Edition. 7 pp.

Firedamp Hazards
Use: Explosive Atmospheres

Firefighters' Helmets
Use: Helmets—Firefighters'

Firefighters' Jackets
Use: Jackets—Firefighters'

Firefighters' Uniforms
Use: Dress Uniforms—Firefighters'

Firefighting
Use: Fire Fighting

Firelighters
Use: Lighters

Fireplace Entrainments
Use: Fireplace Surrounds

Fireplace Screens
Use: Fire Screens

Fireplace Surrounds
Use For: Fireplace Entrainments
See Also: Fireplaces
BSI BS 1251-87. 1987 Open-Fireplace Components. 12 pp.

BSI BS 1251-70. 1970 Amd 2 Open Fireplace Components (AMD 1651) January 31, 1975. 28 pp.

Fireplaces
Use For: Hearths *See Also:* Firebacks; Fireplace Surrounds; Furnaces; Heating Equipment; Solid Fuel Burning Equipment; Space Heaters
BSI BS 1251-87. 1987 Open-Fireplace Components. 12 pp.

BSI BS 1251-70. 1970 Amd 2 Open Fireplace Components (AMD 1651) January 31, 1975. 28 pp.

Fireplaces (Cont.)
CSA CAN/CSA-B415.1-92. Performance Testing of Solid-Fuel-Burning Stoves, Inserts, and Low-Burn-Rate Factory-Built Fireplaces; (Gen Instr 1 Thru 2). 89 pp.

SAA HB33-92. Domestic Open Fireplaces. 10 pp.

—Inserts—Decorative—Gas
CEN PREN 509-91. Decorative Fuel Effect Gas Appliances. 88 pp.

—Installation
BSI CP 403-74. 1974 Installation of Domestic Heating and Cooking Appliances Burning Solid Fuel. 37 pp.

—Masonry
CSA CAN/CSA-A405-M87. Design and Construction of Masonry Chimneys and Fireplaces; (Gen Instr 1). 40 pp.

Fires
See Also: Combustion; Explosions; Explosives; Fire Alarm and Extinguishing Equipment; Fire Extinguishing Agents; Fire Fighting; Fire Fighting Equipment; Fire Protection; Firesafety
BSI BS 6336-82. 1982 Development and Presentation of Fire Tests and Their Use in Hazard Assessment. 12 pp.

ISO 3941-77. Classification of Fires First Edition. 3 pp.

—Classification
BSI BS 4547-72. (WITHDRAWN) 1972 Classification of Fires (Superseded by BS 4533: Section 102.57: 1990). 6 pp.

BSI BS EN 2-92. 1992 Classification of Fires (Supersedes BS 4547: 1972). 7 pp.

CEN EN 2-72. Classification of Fires. 4 pp.

CEN EN 2-92. Classification of Fires. 3 pp.

CNS Z1021-77. Classification for Fire Disaster (Jun)(3658).

DIN ENGL EN 2-93. Classification of Fires (Jan). 3 pp.

—Data Acquisition Systems
ISO TR7248-85. Fire Data—Collection and Presentation System First Edition. 15 pp.

—Effluents—Corrosion
IEC 695 Pt 5-1-93. Fire Hazard Testing Part 5: Assessment of Potential Corrosion Damage by Fire Effluent—Section 1: General Guidance. 25 pp.

—Effluents—Smoke Point—Ships
MOD UK NES 711-81. Determination of the Smoke Index of the Products of Combustion from Small Specimens of Materials Issue 2 (01.81). 21 pp.

—Effluents—Toxicity
BSI PD 6503: Part 1-90. 1990 Toxicity of Combustion Products Part 1: General (ISO/TR 9122-1: 1989). 25 pp.

ISO TR9122 Pt 1-89. Toxicity Testing of Fire Effluents—Part 1: General First Edition; (Corrected and Reprinted -1989). 23 pp.

ISO TR9122 Pt 2-90. Toxicity Testing of Fire Effluents—Part 2: Guidelines for Biological Assays to Determine the Acute Inhalation Toxicity of Fire Effluents (Basic Principles, Criteria and Methodology) First Edition. 18 pp.

ISO TR9122 Pt 3-93. Toxicity Testing of Fire Effluents—Part 3: Methods for the Analysis of Gases and Vapours in Fire Effluents First Edition. 41 pp.

ISO TR9122 Pt 5-93. Toxicity Testing of Fire Effluents—Part 5: Prediction of Toxic Effects of Fire Effluents First Edition. 21 pp.

—Effluents—Toxicity—Models
ISO TR9122 Pt 4-93. Toxicity Testing of Fire Effluents—Part 4: the Fire Model (Furnaces and Combustion Apparatus Used in Small-Scale Testing) First Edition. 21 pp.

ISO TR9122 Pt 5-93. Toxicity Testing of Fire Effluents—Part 5: Prediction of Toxic Effects of Fire Effluents First Edition. 21 pp.

—Effluents—Toxicity—Ships
MOD UK NES 713-85. Determination of the Toxicity Index of the Products of Combustion from Small Specimens of Materials Issue 3 (03.85). 16 pp.

—Glossaries
BSI BS 4422: Part 1-87. 1987 Glossary of Terms Associated with Fire Part 1: General Terms and Phenomena of Fire. 6 pp.

BSI BS 4422: Part 2-90. 1990 Glossary of Terms Associated with Fire Part 2: Structural Fire Protection. 4 pp.

BSI BS 4422: Part 2-71. (WITHDRAWN) 1971 Glossary of Terms Associated with Fire Part 2: Building Materials and Structures. 9 pp.

ISO 8421 Pt 1-87. Fire Protection—Vocabulary—Part 1: General Terms and Phenomena of Fire First Edition. 8 pp.

Fires (Cont.)

—Glossaries (Cont.)
ISO Guide 52-90. Glossary of Fire Terms and Definitions First Edition. 33 pp.

—Incident Reporting
SAA AS 2577.1-92. Australian Fire Incident Reporting System—Part 1: Description and Implementation (Supersedes AS 2577—1983 (in Part)) (in Professional Package 44). 44 pp.

—Incident Reporting—Classification
SAA AS 2577.2-92. Australian Fire Incident Reporting System—Part 2: Classification and Coding (Supersedes AS 2577—1983 (in Part)) (in Professional Package 44). 81 pp.

—Motor Vehicles
SAA AS 1678.0.0. 001-83. Emergency procedure Guide—Transport—Part 0.0.001: Vehicle Fire.

—Toxicity
BSI PD 6503: Part 2-88. 1988 Amd 1 Toxicity of Combustion Products Part 2: Guide to the Relevance of Small-Scale Tests for Measuring the Toxicity of Combustion Products of Materials (AMD 6030) September 30, 1988. 20 pp.

BSI DD 180-89. 1989 Guide for the Assessment of Toxic Hazards in Fire in Buildings and Transport. 21 pp.

ISO TR6543-79. Development of Tests for Measuring Toxic Hazards in Fire First Edition. 9 pp.

Firesafety
See Also: Accident Prevention; Fire Alarm and Extinguishing Equipment; Fire Doors; Fire Hazards; Fire Protection; Fire Testing; Fire Windows; Fires
SAA HB37.1-93. Handbook of Australian Fire Standards—Part 1: Fire—General (in Professional Package 44). 17 pp.

—Anesthesia Breathing Systems
SAA AS 1169-82. Minimizing of Combustion Hazards Arising from the Medical Use of Flammable Anaesthetic Agents. 23 pp.

—Atriums—Codes of Practice
SNZ NZS 4238-91. Code of Practice for Fire Safety in Atrium Buildings. 16 pp.

—Buildings—Model Bylaws
SNZ NZS 9232-91. Model Bylaw for Precautions Against Fire and Panic in Cinemas, Theatres and Places of Assembly. 20 pp.

—Cable Insulation—Naval
MOD UK NES 518. Limited Fire Hazard Sheathing for Electric Cables Issue 3 (02.89) (Superseded by Def Stan 61-12: Part 31).

—Cargo—Marine Transportation
EC EEC/4058/86-86. Council Regulation Concerning Coordinated Action to Safeguard Fire Access to Cargos in Ocean Trades. 3 pp.

—Electrical Equipment
SAA HB37.2-93. Handbook of Australian Fire Standards—Part 2: Electrical Equipment (in Professional Package 44). 20 pp.

—Electrical Installations
CENELEC PRHD 384-4-48-92. Electrical Installations of Buildings Part 4: Protection for Safety Chapter 48: Choice of Protective Measures as a Function of External Influences Section 482: Protection Against Fire. 8 pp.

—Halongenide
CNS Z2063-87. Safety Code of Halongenide for Fire Fighting (Mar)(11885).

—Hospitals
SNZ NZS 4208-73. Code of Practice for the Design of Hospitals with Respect to Fire Amend: 1, 1976. 20 pp.

—Hotels
EC 86/666/EEC-86. Council Recommendation on Fire Safety in Existing Hotels. 9 pp.

—Incinerators
SAA AS 1875-76. Domestic Incinerators (Fire Safety). 8 pp.

—Lithium Batteries
MOD UK DSTAN 61-19-89. Storage, Handling and Disposal of Lithium Batteries Issue 1. 15 pp.

—Meat Processing Facilities
SNZ NZS 4216-83. Code of Practice for Design of Meatworks Complexes for Fire Safety (Reconfirmed 1988). 26 pp.

INTERNATIONAL AND NON-U.S. NATIONAL STANDARDS
SUBJECT INDEX

Firesafety (Cont.)
—**Medical Electrical Equipment**
CSA CAN/CSA-C22. 2NO601.1-M90. Medical Electrical Equipment, Part 1: General Requirements for Safety; (Gen Instr 1). 240 pp.

—**Plastics**
SAA HB37.3-93. Handbook of Australian Fire Standards—Part 3: Plastics and Rubber—Materials and Products (in Professional Package 44). 16 pp.

—**Rubber**
SAA HB37.3-93. Handbook of Australian Fire Standards—Part 3: Plastics and Rubber—Materials and Products (in Professional Package 44). 16 pp.

—**Ships**
MOD UK NES 802: Part 3-89. Requirements for Acoustic and Thermal Insulatin Material Part 3: Mineral Wool Products Rock Wool Fire Protection Slabs Issue 2 (04.89). 13 pp.

—**Signs**
BSI BS 5499: Part 1-90. 1990 Fire Safety Signs, Notices and Graphic Symbols Part 1: Fire Safety Signs (N). 27 pp.
BSI BS 5499: Part 1-01. 1990 Amd 1 Fire Safety Signs, Notices and Graphic Symbols Part 1: Specification for Fire Safety Signs (AMD 7444) January 15, 1993. 28 pp.
BSI BS 5499: Part 2-86. 1986 Fire Safety Signs, Notices and Graphic Symbols Part 2: Self-Luminous Fire Safety Signs. 10 pp.
BSI BS 5499: Part 3-90. 1990 Fire Safety Signs, Notices and Graphic Symbols Part 3: Internally Illuminated Fire Safety Signs (Supersedes BS 2560: 1978). 8 pp.
ISO 6309-87. Fire Protection—Safety Signs First Edition. 10 pp.

—**Structural Design**
ISO TR3956-75. Principles of Structural Fire-Engineering Design with Special Regard to the Connection Between Real Fire Exposure and the Heating Conditions of the Standard Fire-Resistance Test (ISO 834) First Edition. 8 pp.

—**Submarines**
MOD UK NES 802: Part 3-89. Requirements for Acoustic and Thermal Insulatin Material Part 3: Mineral Wool Products Rock Wool Fire Protection Slabs Issue 2 (04.89). 13 pp.

—**Welding**
SAA AS 1674.1-90. Safety in Welding and Allied Processes—Part 1: Fire Precautions. 11 pp.

Fireworks
See Also: Explosives; Pyrotechnics
BSI BS 7114: Part 2-88. 1988 Fireworks Part 2: Specification for Fireworks. 20 pp.

—**Classification**
BSI BS 7114: Part 1-88. 1988 Fireworks Part 1: Classification of Fireworks. 12 pp.

—**Comprehensive Testing**
BSI BS 7114: Part 3-88. 1988 Fireworks Part 3: Methods of Test for Fireworks. 28 pp.

Firm Joint Calipers
CGSB CAN/CGSB-39.20-M88. Jointed Calipers. 16 pp.

Firmware
See Also: Computer Equipment; Software
MOD UK DSTAN 00-22-91. Identification and Marking of Programmable Items Issue 2. 21 pp.

First Aid
See Also: First Aid Equipment; Medicine

—**Airports—Training**
CAA Part 9 CAP 168. Medical Services. 4 pp.

—**Material—Chemical Injuries**
NATO STANAG 2871 ED 3 AMD 1-89. First-Aid Material for Chemical Injuries. 10 pp.

—**Military—Training**
NATO STANAG 2122 ED 2 AMD 4-75. Medical Training in First-Aid, Basic Hygiene and Emergency Care. 12 pp.
NATO STANAG 2122 ED 2 AMD 5-75. Medical Training in First-Aid, Basic Hygiene and Emergency Care. 12 pp.

—**Nuclear, Biological and Chemical Warfare**
NATO STANAG 2358 ED 2 AMD 1-89. First Aid and Hygiene Training in NBC Operations. 11 pp.

First Aid Equipment
See Also: First Aid; Medical Electrical Equipment; Medical Equipment

—**Fire Trucks**
SNZ NZS 5807-80. Code of Practice for Industrial Identification by Colour, Wording or Other Coding Amend: 1, 1983; 2, 1988. 20 pp.

—**Industrial**
SNZ NZS 2274-69. Specification for Industrial First-Aid Cabinets Amend: 1, 1978. 8 pp.

—**Military Operations**
NATO STANAG 2126 ED 5 AMD 4-83. First-Aid Kits and Emergency Medical Care Kits. 12 pp.

—**Military Operations—Aircraft**
NATO STANAG 3746 ED 1 AMD 4-75. Medical First Aid Equipment in Aircraft. 9 pp.
NATO STANAG 3746 ED 1 AMD 6-75. Medical First Aid Equipment in Aircraft. 9 pp.

First Aid Kits
NATO STANAG 2126 ED 5 AMD 3-83. First-Aid Kits and Emergency Medical Care Kits. 12 pp.

—**Life Rafts**
NATO STANAG 1185 ED 1 AMD 8-77. Minimum Essential Medical and Survival Equipment for Ship Life Rafts Including Guidelines for Survival at Sea. 10 pp.
NATO STANAG 1185 ED 1 AMD 9-77. Minimum Essential Medical and Survival Equipment for Ship Life Rafts Including Guidelines for Survival at Sea. 12 pp.

Fish
Use For: Fish Products *See Also:* Anchovies; Aquatic Organisms; Bluefish; Bonito; Crustacea; Eels; Fish Liver Oil; Fish Paste; Fisheries; Food; Herrings; Mackerel; Mollusks; Roe; Sardines; Scad; Shellfish
CNS N5199-84. Chilled Fish (Dec)(9636). 2 pp.
CNS N6178-82. Method of Test for Chilled Fish (Nov)(9637). 2 pp.

—**Canned**
CGSB 32.143M-91. Canned Fish. 11 pp.
CNS N5015-89. Canned Fished (Feb)(1229). 5 pp.
CNS N5138-91. Canned Tuna and Bonito (Feb)(4456). 5 pp.

—**Canned—Bleached Dark Meat Content**
CNS N6104-78. Method of Determination of Bleached Dark Meat in Canned Fish (Jul)(4455). 1 p.

—**Cans**
BSI BS 6992: Part 4-89. (OBSOLESCENT) 1989 Capacities and Related Cross Sections of Open-Top Cans for Food and Drinks Part 4: Recommendations for Cans for Fish and Other Fishery Products. 8 pp.
JIS Z 1608-63. Metal Fish Containers and Freezing Pans.

—**Cold—Food Services—Handling**
CGSB 32.72M-91. Handling, Packaging and Labelling of Meat, Poultry and Fish for Food Services. 10 pp.

—**Cold—Food Services—Identification Systems**
CGSB 32.72M-91. Handling, Packaging and Labelling of Meat, Poultry and Fish for Food Services. 10 pp.

—**Cold—Food Services—Packaging**
CGSB 32.72M-91. Handling, Packaging and Labelling of Meat, Poultry and Fish for Food Services. 10 pp.

—**Dried—Condiments**
CNS N5190-81. Flavored Condiments (Oct)(8055). 3 pp.

—**Dried—Larval**
CNS N5175-81. Dehydrated Larval Fish, Chilled (Packaged) (Jan)(6945). 2 pp.

—**Dried—Larval—Fluorescent Agents**
CNS N6152-81. Method of Test for Dehydrated Larval Fish, Chilled (Aug)(7839). 2 pp.

—**Dried—Larval—Hydrogen Peroxide Content**
CNS N6152-81. Method of Test for Dehydrated Larval Fish, Chilled (Aug)(7839). 2 pp.

—**Dried—Larval—Microbiological Analysis**
CNS N6152-81. Method of Test for Dehydrated Larval Fish, Chilled (Aug)(7839). 2 pp.

—**Dried—Larval—Moisture Content**
CNS N6152-81. Method of Test for Dehydrated Larval Fish, Chilled (Aug)(7839). 2 pp.

—**Dried—Larval—Salt Content**
CNS N6152-81. Method of Test for Dehydrated Larval Fish, Chilled (Aug)(7839). 2 pp.

Fish (Cont.)
—**Dried—Larval—Sampling**
CNS N6152-81. Method of Test for Dehydrated Larval Fish, Chilled (Aug)(7839). 2 pp.

—**Dried—Larval—Sensory Analysis**
CNS N6152-81. Method of Test for Dehydrated Larval Fish, Chilled (Aug)(7839). 2 pp.

—**Dried—Larval—Sodium Borate Content**
CNS N6152-81. Method of Test for Dehydrated Larval Fish, Chilled (Aug)(7839). 2 pp.

—**Dried—Larval—Temperature Measurement**
CNS N6152-81. Method of Test for Dehydrated Larval Fish, Chilled (Aug)(7839). 2 pp.

—**Dried—Larval—Visual Inspection**
CNS N6152-81. Method of Test for Dehydrated Larval Fish, Chilled (Aug)(7839). 2 pp.

—**Food Services—Cold Storage**
CGSB 32.72M-91. Handling, Packaging and Labelling of Meat, Poultry and Fish for Food Services. 10 pp.

—**Food Services—Refrigerated Transportation**
CGSB 32.72M-91. Handling, Packaging and Labelling of Meat, Poultry and Fish for Food Services. 10 pp.

—**Food Services—Refrigeration Equipment**
CGSB 32.72M-91. Handling, Packaging and Labelling of Meat, Poultry and Fish for Food Services. 10 pp.

—**Freezing Pans**
JIS Z 1608-63. Metal Fish Containers and Freezing Pans.

—**Fresh Water—Lethal Toxicity Testing**
BSI BS 6068: Sec 5.2-85. 1985 Water Quality: Part 5: Methods for Biological Testing: Section 5.2: Determination of the Acute Lethal Toxicity of Substances to a Freshwater Fish (Brachydanio Rerio Hamilton-Buchanan (Teleostei,Cyprinidae)): Static Method. 12 pp.
BSI BS 6068: Sec 5.3-85. 1985 Water Quality: Part 5: Methods for Biological Testing: Section 5.3: Determination of the Acute Lethal Toxicity of Substances to a Freshwater Fish (Brachydanio Rerio Hamilton-Buchanan (Teleostei,Cyprinidae)): Semi-Static Method. 12 pp.
BSI BS 6068: Sec 5.4-85. 1985 Water Quality: Part 5: Methods for Biological Testing: Section 5.4: Determination of the Acute Lethal Toxicity of Substances to a Freshwater Fish (Brachydanio Rerio Hamilton-Buchanan (Teleostei,Cyprinidae)): Flow Through Method. 13 pp.
ISO 7346 Pt 1-84. Water Quality—Determination of the Acute Lethal Toxicity of Substances to a Freshwater Fish (Brachydanio Rerio Hamilton-Buchanan (Teleostei, Cyprinidae))—Part 1: Static Method First Edition. 11 pp.
ISO 7346 Pt 2-84. Water Quality—Determination of the Acute Lethal Toxicity of Substances to a Freshwater Fish (Brachydanio Rerio Hamilton-Buchanan (Teleostei, Cyprinidae))—Part 2: Semi-Static Method First Edition. 11 pp.
ISO 7346 Pt 3-84. Water Quality—Determination of the Acute Lethal Toxicity of Substances to a Freshwater Fish (Brachydanio Rerio Hamilton-Buchanan (Teleostei, Cyprinidae))—Part 3: Flow-Through Method First Edition. 12 pp.

—**Frozen**
CGSB 32.141M-91. Fresh, Chilled or Frozen Fish and Fish Products. 13 pp.
CNS N5113-84. Frozen Fish (Dec)(3732). 2 pp.

—**Frozen—Food Services—Handling**
CGSB 32.72M-91. Handling, Packaging and Labelling of Meat, Poultry and Fish for Food Services. 10 pp.

—**Frozen—Food Services—Identification Systems**
CGSB 32.72M-91. Handling, Packaging and Labelling of Meat, Poultry and Fish for Food Services. 10 pp.

—**Frozen—Food Services—Packaging**
CGSB 32.72M-91. Handling, Packaging and Labelling of Meat, Poultry and Fish for Food Services. 10 pp.

—**Frozen—Inspection**
CNS N6029-84. Method of Test for Frozen Fishes (Aug)(1451). 4 pp.

—**Frozen—Microbiological Analysis**
CNS N6029-84. Method of Test for Frozen Fishes (Aug)(1451). 4 pp.

—**Frozen—Minced**
CNS N5140-82. Frozen Minced Fish Meat (Nov)(4640). 2 pp.

INDUSTRY STANDARDS

Fish (Cont.)

- **Frozen—Nitrogen Content**
  CNS N6029-84. Method of Test for Frozen Fishes (Aug)(1451). 4 pp.
- **Frozen—PH**
  CNS N6029-84. Method of Test for Frozen Fishes (Aug)(1451). 4 pp.
- **Frozen—Sampling**
  CNS N6029-84. Method of Test for Frozen Fishes (Aug)(1451). 4 pp.
- **Frozen—Temperature Measurement**
  CNS N6029-84. Method of Test for Frozen Fishes (Aug)(1451). 4 pp.
- **Frozen—Trimethylamine Content**
  CNS N6029-84. Method of Test for Frozen Fishes (Aug)(1451). 4 pp.
- **Frozen—Weight Measurement**
  CNS N6029-84. Method of Test for Frozen Fishes (Aug)(1451). 4 pp.
- **Health Inspection**
  EC 91/67/EEC-91. Council Directive Concerning the Animal Health Conditions Governing the Placing on the Market of Aquaculture Animals and Products. 21 pp.
  EC 91/493/EEC-91. Council Directive Laying Down the Health Conditions for the Production and the Placing on the Market of Fishery Products. 21 pp.
- **Inspection—Nematodes**
  EC COM(88) 47-88. Proposal for a Council Regulation (EEC) Laying Down Health Conditions for the Marketing of Fish and Fish Products Concerning Nematodes. 13 pp.
- **Microbiological Analysis**
  SAA AS 1766.3.5-83. Methods for the Microbiological Examination of Food—Part 3: Examination of Specific Products—Part 3.5: Molluscs, Crustaceans and Fish, and Products Thereof. 4 pp.
- **Pickled**
  CGSB 32.142M-91. Preserved Fish (Salted, Dried, Smoked or Pickled). 9 pp.
- **Prices**
  EC EEC/3931/86-86. Council Regulation Fixing the Guide Prices for the Fishery Products Listed in Annexe II to Regulation (EEC) No 3796/81 for the 1987 Fishing Year. 2 pp.
- **Processing**
  SNZ NZS 8402-74. Code of Practice for the Handling, Processing and Distribution of Fish. 35 pp.
- **Salted**
  CGSB 32.142M-91. Preserved Fish (Salted, Dried, Smoked or Pickled). 9 pp.
- **Shredded**
  CNS N5189-81. Shred Fish (Oct)(8053). 2 pp.
- **Shredded—Chemical Analysis**
  CNS N6160-81. Method of Test for Shred Fish (Oct)(8054). 2 pp.
- **Shredded—Sampling**
  CNS N6160-81. Method of Test for Shred Fish (Oct)(8054). 2 pp.
- **Smoked**
  CGSB 32.142M-91. Preserved Fish (Salted, Dried, Smoked or Pickled). 9 pp.
- **Waste Water—Toxicity**
  DIN ENGL 38412 Pt 31-89. German Standard Methods for the Examination of Water, Waste Water and Sludge; Bio-Assays (Group L); Determining the Tolerance of Fish to the Toxicity of Waste Water by Way of a Dilution (Mar). 4 pp.

Fish Bone Meal
Use: Fish Meal

Fish Glue
See Also: Adhesives
- **Sampling**
  BSI BS 647-81. 1981 Sampling and Testing Glues (Bone, Skin and Fish Glues). 22 pp.

Fish Liver Oil
See Also: Fish; Oils
  CNS N2041-86. Refined Fish Liver Oil (for Feeding) (Oct)(3924). 1 p.
  CNS N4140-86. Method of Test for Refined Fish Liver Oil (Oct)(11735).

Fish Liver Oil (Cont.)
- **Fatty Acid Content**
  CNS K6114-63. Method of Test for Fish-Liver Oil (Crude) (Jul)(1223) (R 1973). 2 pp.
- **Impurities Content**
  CNS K6114-63. Method of Test for Fish-Liver Oil (Crude) (Jul)(1223) (R 1973). 2 pp.
- **Moisture Content**
  CNS K6114-63. Method of Test for Fish-Liver Oil (Crude) (Jul)(1223) (R 1973). 2 pp.
- **Vitamin A Content**
  CNS K6114-63. Method of Test for Fish-Liver Oil (Crude) (Jul)(1223) (R 1973). 2 pp.

Fish Meal
See Also: Animal Feed; Bone Meal; Feather Meal; Oil Meal
- **Animal Feed**
  CNS N2004-84. Fish Meal (for Feeding) (Oct)(2243). 2 pp.
  CNS N2034-86. Fish Soluble Dried (for Feeding) (Mar)(3494). 1 p.
  CNS N2036-86. Fish Bone Meal (for Feeding) (Mar)(3708). 1 p.
  CNS N2039-86. Mixed Fish Soluble Meal (for Feeding) (Mar) (3730). 2 pp.
- **Fat Content**
  CNS N4015-78. Method of Test for Fish Meal and Paste (Mar)(2245). 2 pp.
- **Fertilizers**
  CNS N3045-88. Fish Meal Fertilizer (Jul)(11911).
- **Glossaries**
  ISO 7088-81. Fish-Meal—Vocabulary First Edition. 5 pp.
- **Insoluble Matter Content**
  CNS N4015-78. Method of Test for Fish Meal and Paste (Mar)(2245). 2 pp.
- **Moisture Content**
  CNS N4015-78. Method of Test for Fish Meal and Paste (Mar)(2245). 2 pp.
- **Phosphoric Acid Content**
  CNS N4015-78. Method of Test for Fish Meal and Paste (Mar)(2245). 2 pp.
- **Protein Content**
  CNS N4015-78. Method of Test for Fish Meal and Paste (Mar)(2245). 2 pp.
- **Salt Content**
  CNS N4015-78. Method of Test for Fish Meal and Paste (Mar)(2245). 2 pp.
- **Sieve Analysis**
  CNS N4015-78. Method of Test for Fish Meal and Paste (Mar)(2245). 2 pp.

Fish Paste
See Also: Animal Feed; Fish
- **Animal Feed**
  CNS N2005-84. Fish Solubles, Condensed (for Feeding) (Dec) (2244). 2 pp.
- **Fat Content**
  CNS N4015-78. Method of Test for Fish Meal and Paste (Mar)(2245). 2 pp.
- **Frozen—Elastic Properties**
  CNS N6107-82. Method of Test for Frozen Minced Fish Meat (Nov)(4641). 2 pp.
- **Frozen—Impurities Content**
  CNS N6107-82. Method of Test for Frozen Minced Fish Meat (Nov)(4641). 2 pp.
- **Frozen—Microbiological Analysis**
  CNS N6107-82. Method of Test for Frozen Minced Fish Meat (Nov)(4641). 2 pp.
- **Frozen—Moisture Content**
  CNS N6107-82. Method of Test for Frozen Minced Fish Meat (Nov)(4641). 2 pp.
- **Frozen—Salt Content**
  CNS N6107-82. Method of Test for Frozen Minced Fish Meat (Nov)(4641). 2 pp.
- **Frozen—Temperature Measurement**
  CNS N6107-82. Method of Test for Frozen Minced Fish Meat (Nov)(4641). 2 pp.
- **Insoluble Matter Content**
  CNS N4015-78. Method of Test for Fish Meal and Paste (Mar)(2245). 2 pp.

Fish Paste (Cont.)
- **Moisture Content**
  CNS N4015-78. Method of Test for Fish Meal and Paste (Mar)(2245). 2 pp.
- **Phosphoric Acid Content**
  CNS N4015-78. Method of Test for Fish Meal and Paste (Mar)(2245). 2 pp.
- **Polychlorinated Biphenyl Content**
  CNS N4067-81. Method of Test for PCB in Fish Solubles (For Feeding) (Jan)(6942). 1 p.
- **Protein Content**
  CNS N4015-78. Method of Test for Fish Meal and Paste (Mar)(2245). 2 pp.
- **Salt Content**
  CNS N4015-78. Method of Test for Fish Meal and Paste (Mar)(2245). 2 pp.
- **Sieve Analysis**
  CNS N4015-78. Method of Test for Fish Meal and Paste (Mar)(2245). 2 pp.

Fish Plates
See Also: Railroad Equipment
  BSI BS 47-59. (WITHDRAWN) 1959 Steel Fishplates for Bullhead and Flat Bottom Railway Rails (Superseded by BS 47: Part 1: 1991). 53 pp.
  BSI BS 47: Part 1-91. 1991 Fishplates for Railway Rails Part 1: Specification for Rolled Steel Fishplates. 39 pp.
  CNS E1006-74. Fish Plate for Regular Steel Rail (Oct)(2787).
  CNS E1007-74. Fish Plate for Light Steel Rail (Oct)(2788).
  DIN ENGL 5901 Pt 2-65. Rails up to 20 kg/m; Flat Fishplates; Angle Fishplates (Mar). 1 p.
  ISO 6305 Pt 1-81. Railway Components—Technical Delivery Requirements—Part 1: Rolled Steel Fishplates First Edition. 12 pp.
  JIS E 1102-92. Fish Plates for Rails. 19 pp.
  JIS E 1104-78. Fishplates for Light Rails.
  JIS E 1104-63. Fish Plates for Light Rails. 10 pp.
  JIS E 1116-66. Compromise Fish Plates for Rails.
  SAA AS 1085.2-93. Railway Permanent Way Material—Part 2: Fishplates (Supersedes AS 1085—1986). 11 pp.
- **Fasteners**
  BSI BS 64-92. 1992 Normal and High Strength Steel Bolts and Nuts for Railway Rail Fishplates (Q). 16 pp.
  BSI BS 64-46. 1946 Amd 2 Steel Fishbolts and Nuts for Railway Rails. 16 pp.
  CNS B2350-83. Fish Bolt with Round Head and Oval Neck (Mar)(4693).
  CNS B2351-79. Fish Bolt with Square Head (Jan)(4694).
  CNS E1009-81. Spring Washers for Fish Plate (Mar)(7033).
  ISO 6305 Pt 4-85. Railway Components—Technical Delivery Requirements—Part 4: Untreated Steel Nuts and Bolts and High-Strength Nuts and Bolts for Fish-Plates and Fastenings First Edition. 10 pp.
  JIS E 1107-88. Fish Bolts and Nuts.
  JIS E 1107-65. Fishbolts and Nuts (R 1971). 6 pp.
  JIS E 1113-88. Heat-Treated Fish Bolts and Nuts. 7 pp.
  JIS E 1114-78. Fish Bolts and Nuts for N Type Rails and 60kg Rails.
  JIS E 1114-65. Fishbolts and Nuts for N Rails. 5 pp.
  JIS E 1115-78. Spring Washers for Fish Plates.
  SAA AS 1085.4-88. Railway Permanent Way Material—Part 4: Fishbolts and Nuts. 7 pp.
  SAA AS 1085.7-78. Railway Permanent Way Material—Spring Washers. 8 pp.
- **Mining Equipment**
  BSI BS 248-69. 1969 Light Rails and Fishplates for Use in Mines. 23 pp.
  CNS M2023-80. I Section Steels and Fish Plates for Mine (Feb)(5251).
  JIS M 2504-77. I Section Steels and Fish Plates for Mine. 12 pp.

Fish Powder
Use: Fish Meal

Fish Products
Use: Fish

Fish Soluble Meal
Use: Fish Meal

Fish Solubles
Use: Fish Paste

Fishbolts
Use: Bolts—Fish Plates

INTERNATIONAL AND NON-U.S. NATIONAL STANDARDS
SUBJECT INDEX

Fisheries
See Also: Fish; Fishing
—European Communities
EURO 1991-3. Common Fisheries Policy. 12 pp.

Fishing
See Also: Fisheries; Fishing Permits
—Classification
BSI BS 1000: (636/639)-81. 1981 Universal Decimal Classification (UDC). English Full Edition (636/639): Animal Breeding. Animal Produce. Hunting. Fishing. 56 pp.
SNZ NZS/BS 1000 (636/639)-81. Universal Decimal Classification Animal Breeding. Animal Produce. Hunting. Fishing. 56 pp.

Fishing Equipment
See Also: Fishing Nets; Fishlines
—Needles
CNS S1167-83. Fishing Needles (Aug)(10515).
JIS S 7001-77. Fishing Needles.
—Steel Wire
CNS G3024-89. Steel Wire for Fishing (Sep)(1180)(R 1972).

Fishing Lines
Use: Fishlines

Fishing Nets
See Also: Fishing Equipment; Yarns
—Breaking Load
BSI BS 4650-70. 1970 Amd 1 Determination of the Mesh Breaking Load of Netting for Fishing. 12 pp.
BSI BS 4674-71. 1971 Amd 1 Determination of the Breaking Load and Knot Breaking Load of Netting Yarns for Fishing Nets. 12 pp.
ISO 1805-73. Fishing Nets—Determination of Breaking Load and Knot Breaking Load of Netting Yarns First Edition. 6 pp.
ISO 1806-73. Fishing Nets—Determination of Mesh Breaking Load of Netting First Edition. 6 pp.
—Classification
BSI BS 5172-75. 1975 Fishing Nets: Description and Designation of Knotted Netting. 9 pp.
ISO 1530-73. Fishing Nets—Description and Designation of Knotted Netting First Edition. 5 pp.
—Classification—Tex System
BSI BS 4406-69. 1969 Amd 1 Designation in the Tex System of Netting Yarns for Fishing Nets. 7 pp.
ISO 858-73. Fishing Nets—Designation of Netting Yarns in the Tex System First Edition. 4 pp.
—Cotton Spun Yarn
JIS L 2102-61. Cotton Twist Yarn for Fishing Nets and Implements. 6 pp.
—Cutting
ISO 1532-73. Fishing Nets—Cutting Knotted Netting to Shape ("Tapering") First Edition. 7 pp.
ISO 2075-72. Cutting Netting to Shape—Determination of the Cutting Rate First Edition. 7 pp.
—Dimensional Stability—Immersion
BSI BS 5171-74. 1974 Netting Yarns: Determination of Change in Length After Immersion in Water. 8 pp.
ISO 3090-74. Netting Yarns—Determination of Change in Length After Immersion in Water First Edition. 4 pp.
—Elongation
BSI BS 5414-76. 1976 Determination of Elongation of Netting Yarns for Fishing Nets. 7 pp.
ISO 3790-76. Fishing Nets—Determination of Elongation of Netting Yarns First Edition. 6 pp.
—Filament Yarns—Polyvinylidene Chloride
CNS L3206-84. Method of Test for Twisted Polyvinylidene Chloride and Polyvinyl Chloride Filament Yarns for Fishing Nets (Mar) (10818).
—Filament Yarns—PVC
CNS L3206-84. Method of Test for Twisted Polyvinylidene Chloride and Polyvinyl Chloride Filament Yarns for Fishing Nets (Mar) (10818).
—Glossaries
BSI BS 4440-74. 1974 Glossary of Basic Terms for Fishing Nets. 12 pp.
CGSB CAN/CGSB-55.1-M85. Fishing Yarns, Netting and Nets. 57 pp.
ISO 1107-74. Fishing Nets—Netting—Basic Terms and Definitions First Edition. 5 pp.
—Hanging—Glossaries
ISO 1531-73. Fishing Nets—Hanging of Netting—Basic Terms and Definitions First Edition. 3 pp.

Fishing Nets (Cont.)
—Mounting/Joining—Glossaries
BSI BS 5398-76. 1976 Classification of Methods for Mounting and Joining of Fishing Nets. 8 pp.
ISO 3660-76. Fishing Nets—Mounting and Joining of Netting—Terms and Illustrations First Edition. 12 pp.
—Nylon Fibers
CNS L3044-84. Method of Test for Twisted Nylon Filament of Fishing Nets (Feb)(2279).
JIS L 1034-78. Testing Method for Twisted Nylon Filament of Fishing Nets.
—Synthetic Fibers
CNS L3207-84. Method of Test for Synthetic Fiber Fishing Net (Mar)(10819).
JIS L 1043-92. Methods of Testing Synthetic Fibre Nettings for Fishing Net.
—Twine—Cotton
CNS L3033-63. Testing Standard for Cotton Fishing Net Twine (Mar)(1500)(R 1973).
—Vinylon Spun Yarn
CNS L3205-84. Method of Test for Twisted Spun Vinylon Yarn for Fishing Nets (Mar)(10817).
JIS L 1033-78. Testing Methods for Twisted Spun Vinylon Yarn for Fishing Nets.

Fishing Permits
See Also: Fishing
—Wartime
NATO STANAG 1080 ED 2 AMD 2-79. North Atlantic Treaty Organization Fishing Permit. 46 pp.
NATO STANAG 1080 ED 2 AMD 3-79. North Atlantic Treaty Organization Fishing Permit. 47 pp.

Fishing Tools
See Also: Drilling Equipment; Tools
—Tire Valves
CNS D4011-83. Fishing Tool for Pneumatic Tire Valve for Automobiles (May)(10249).

Fishlines
See Also: Fishing Equipment
—Cotton Spun Yarn
JIS L 2102-61. Cotton Twist Yarn for Fishing Nets and Implements. 6 pp.

Fissile Materials
Use: Fissionable Materials

Fission Reactors
Use: Nuclear Reactors

Fissionable Material Content Analysis
—Plutonium Nitrate—Gamma Ray Spectrometry
CNS J2047-82. Method of Test for Gamma-Emitting Fission Products, Uranium and Thorium in Nuclear-Grade Plutonium Nitrate Solutions by Gamma Ray Spectroscope (Mar)(8594).

Fissionable Materials
Use For: Fissile Materials See Also: Radioactive Materials
—Criticality
BSI BS 3598-70. 1970 Recommendations for Criticality Safety in Handling and Processing Fissile Materials. 12 pp.
ISO 1709-75. Nuclear Energy—Fissile Materials—Principles of Criticality Safety in Handling and Processing First Edition. 5 pp.
—Radiation Detection and Measurement
CNS J2083-83. Method for Determining Fission Product Activity in UF6 Counting Method (Jan)(9882).

Fistball Equipment
Use: Volleyball Equipment

FITES
Use For: Forward Interworking Telephone Events
See Also: Signaling Systems
CCITT RECMN Q.601-89. General—Interworking of Signalling Systems (Study Group XI) 2 pp. 2 pp.
CCITT RECMN Q.603-89. Events—Interworking of Signalling Systems (Study Group XI) 1 pp. 1 p.
CCITT RECMN Q.608-89. Miscellaneous Interworking Aspects—Interworking of Signalling Systems (Study Group XI) 24 pp. 24 pp.

Fitness Equipment
Use: Exercise Equipment

Fits
Use For: Standard Fits See Also: Interference Fits; Tolerances
BSI BS 1916: Part 2-53. 1953 Limits and Fits for Engineering Part 2: Guide to the Selection of Fits in BS 1916 Part 1. 53 pp.
BSI BS 1916: Part 3-63. 1963 Amd 2 Limits and Fits for Engineering Part 3: Recommendations for Tolerances, Limits, and Fits for Large Diameters. 36 pp.
BSI BS 4500: Sec 1.1-90. (WITHDRAWN) 1990 ISO Limits and Fits Part 1: General, Tolerances and Deviations Section 1.1: Bases of Tolerances, Deviations and Fits (Renumbered as BS EN 20286-1: 1993). 35 pp.
BSI BS 4500A-70. 1970 Data Sheet: Selected ISO Fits, Hole Basis. 3 pp.
BSI BS 4500B-70. 1970 Amd 1 Data Sheet: Selected ISO Fits, Shaft Basis. 3 pp.
BSI BS EN 20286-1-93. 1993 Amd 1 ISO System of Limits and Fits Part 1: Bases of Tolerances, Deviations and Fits (ISO 286-1: 1988) (AMD 7630) August 15, 1993 (H). 38 pp.
CEN EN 20286-1-93. ISO System of Limits and Fits—Part 1: Bases of Tolerances, Deviations and Fits (ISO 286-1: 1988). 33 pp.
CNS B1001-2-88. Engineering Drawing (Conventional Representation of Common Machine Elements) (Dec)(3-2).
CNS B1002-1-87. Limits and Fits (General, Tolerance and Deviation) (Jun)(4-1).
CNS B1002-2-87. Limits and Fits (Inspection of Plain Workpieces) (Jul)(4-2).
CSA B97.3-1970. Standard Fits for Mating Parts, Inch Sizes (R 1992); (Erratum June 1983). 38 pp.
CSA B97.3-M1982. Tolerances and Standard Fits for Mating Parts, Metric Sizes (R 1992). 50 pp.
DIN ENGL 7157-66. Recommended Selection of Fits; Tolerance Zones; Allowances; Fit Tolerances (Jan). 5 pp.
ISO 286 Pt 1-88. ISO System of Limits and Fits—Part 1: Bases of Tolerances, Deviations and Fits First Edition (CEN EN 20286-1: 1993). 34 pp.
JIS B 0401-86. System of Limits and Fits. 42 pp.
SAA AS 1654-74. Limits and Fits for Engineering Reconfirmed 1987. 68 pp.
SAA AS 1654 Supp 1-75. Limits and Fits for Engineering—Imperial Units.
SNZ NZS/ISO 286: Part 1-88. ISO System of Limits and Fits Part 1: Bases of Tolerances, Deviations and Fits. 30 pp.
SNZ NZS/BS 1916: Part 2-53. Limits and Fits for Engineering Part 2: Guide to the Selection of Fits in BS 1916:Part 1. 52 pp.
SNZ NZS/BS 1916: Part 3-63. Limits and Fits for Engineering Part 3: Recommendations for Tolerances, Limits, and Fits for Large Diameters Amend: 1; 2. 40 pp.

Fittings
See Also: Adapters (Fittings); Banjo Fittings; Bordeaux Connections; Busway Fittings; Cable Fittings; Connectors; Copper Wire Fittings; Corner Fittings; Couplings; Duct Fittings; Electric Conduit Fittings; Electric Outlet Boxes; Electric Raceway Fittings; Electrical Insulator Fittings; Electrode Fittings; Fasteners; Hardware; Hose Fittings; Indexable Inserts; Inserts (Fasteners); Joints; Pipe Fittings; Pipe Flanges; Power Line Hardware; Sockets (Fittings); Stuffing Tubes; Tie Downs; Timber Connectors; Tube Fittings; Vacuum Fittings; Valve Fittings; Welding; Wireway Fittings
—Plastic—Creep Rupture Strength
CNS K6640-81. Method of Test for Short-Time Rupture Strength of Plastic Rope, Tubing and Fittings (Mar)(7051).
—Ships—Steam Whistles
JIS F 2704-67. Fittings for Steam Whistle. 5 pp.

Five Way Valves
Use: Directional Control Valves

Fixed Capacitors
Scope Note: For additional listings, see specific types of capacitors See Also: Aluminum Capacitors; Aluminum Chip Capacitors; Capacitors; Ceramic Capacitors; Ceramic Chip Capacitors; Electrolytic Capacitors; Film Capacitors; Filter Capacitors; Glass Capacitors; Metallized Film Capacitors; Mica Capacitors; Paper Capacitors; Tantalum Capacitors; Tantalum Chip Capacitors
BSI BS QC 300000-83. 1983 Amd 1 Specification for Harmonized System of Quality Assessment for Electronic Components. Fixed Capacitors for Use in Electronic Equipment. Generic Specification (IEC 384-1: 1982) (AMD 5955) August 15, 1992 (T). 69 pp.

Fixed Capacitors (Cont.)

CECC CECC 30 000 ISSUE 3-83. Generic Specification: Fixed Capacitors (En, Fr, Ge) AMD 1 (En, Fr, Ge) AMD 2 (En, Fr, Ge) AMD 3 (En, Fr, Ge) AMD 4 (En, Fr, Ge) AMD 5 (En, Fr, Ge) AMD 6 (En, Fr, Ge) AMD 7 (En, Fr, Ge) AMD 8 (En, Fr, Ge). 151 pp.
CENELEC PREN 137 000-92. Generic Specification: Fixed a.c. Capacitors for Use with Motors or in Discharge Lamp Circuits. 39 pp.
CNS C6007-86. Method of Test for Fixed Capacitors for Electronic Equipment (Jan)(3432).
IEC 384 Pt 1-82. Fixed Capacitors for Use in Electronic Equipment Part 1: Generic Specification Second Edition; (Amendment 2-1987, Incorporating Amendment 1) (Amendment 3-1989) (Amendment 4-1992) (IECQ QC 300000). 138 pp.
IECQ QC 300000-82. Fixed Capacitors for Use in Electronic Equipment Part 1: Generic Specification; (Amendment 2-1987, Incorporating Amendment 1) (Amendment 3-1989) (Amendment 4-1992) (IEC 384-1 ED 2). 139 pp.
SAA AS C343-64. Fixed Ceramic Dielectric Capacitors (Temperature Compensating) Type 1 Corrig.. 23 pp.
SAA AS 1456-73. Fixed Capacitors for Direct Current—Polyester Film Dielectric. 27 pp.
SAA AS 1541.7-83. Fixed Capacitors for Use in Electronic Equipment—Part 7: Fixed Polystyrene Film Dielectric Direct Current Capacitors. 16 pp.

—Glossaries

SAA AS 1541.1-80. Fixed Capacitors for Use in Electronic Equipment—Part 1: Terminology and Methods of Test. 24 pp.

—Quality Assurance

BSI BS 9070: Sec 1 & 2-69. 1969 Amd 12 Fixed Capacitors of Assessed Quality: Generic Data and Methods of Test Section 1: Principles and Mandatory Procedures. Section 2: General Rules for Drafting Detail Specifications. 51 pp.
BSI BS 9070: Sec 1 & 2-13. 1969 Amd 13 Fixed Capacitors of Assessed Quality: Generic Data and Methods of Test Section 1: Principles and Mandatory Procedures Section 2: General Rules for Drafting Detail Specifications (AMD 7393) March 15, 1993. 52 pp.
BSI BS 9075 N001-03. 1981 Amd 3 Detail Specification for Fixed Ceramic Dielectric Capacitors, Type 1 and Type 2; Rectangular Monolithic Chips. Full Assessment Level (AMD 7511) June 15, 1993 (T). 24 pp.
BSI BS QC 300600-91. 1991 Fixed Capacitors for Use in Electronic Equipment. Sectional Specification for Fixed Capacitors of Ceramic Dielectric, Class 1 (IEC 384-8: 1988). 48 pp.
BSI BS QC 300601-93. 1993 Fixed Capacitors for Use in Electronic Equipment Blank Detail Specification Fixed Capacitors of Ceramic Dielectric, Class 1 Assessment Level E (IEC 384-8-1: 1988). 15 pp.
BSI BS QC 300700-92. 1992 Fixed Capacitors for Use in Electronic Equipment. Sectional Specification for Fixed Capacitors of Ceramic Dielectric, Class 2 (IEC 384-9: 1988) (T). 35 pp.
BSI BS QC 300701-93. 1993 Fixed Capacitors for Use in Electronic Equipment Blank Detail Specification Fixed Capacitors of Ceramic Dielectric, Class 2 Assessment Level E (IEC 384-9-1: 1988). 16 pp.
BSI BS CECC 30000-92. 1992 Generic Specification: Fixed Capacitors. 50 pp.
BSI BS CECC 30000-84. 1984 Fixed Capacitors: Generic Specification. 43 pp.
BSI BS CECC 30702-93. 1993 Blank Detail Specification: Fixed Capacitors of Ceramic Dielectric, Class 2 (Assessment Level P. Telecom Level). 17 pp.
MOD UK DSTAN 59-44: Part 1-01. Capacitors, Fixed, of Assessed Quality (Listed on EPIC Database) Part 1: General Requirements Issue 2; Amendment 1. 29 pp.
MOD UK DSTAN 59-44: 90/003-85. Capacitors, Fixed, High Voltage Issue 2. 21 pp.
SAA AS 1455-73. Fixed Ceramic Dielectric Capacitors—Type 2. 24 pp.
SAA AS 1541.1-80. Fixed Capacitors for Use in Electronic Equipment—Part 1: Terminology and Methods of Test. 24 pp.

Fixed Carbon Content Analysis

See Also: Carbon Content Analysis

—Briquets

JIS M 8812-84. Methods for Proximate Analysis of Coal and Coke. 28 pp.

—Charcoal

JIS M 8812-84. Methods for Proximate Analysis of Coal and Coke. 28 pp.

—Coal

JIS M 8812-84. Methods for Proximate Analysis of Coal and Coke. 28 pp.

Fixed Carbon Content Analysis (Cont.)

—Coke

JIS M 8812-84. Methods for Proximate Analysis of Coal and Coke. 28 pp.

Fixed Daily Measurement Hour

See Also: Average Daily Peak Hour; Busy Hour; Fixed Daily Measurement Period; Time Consistent Busy Hour; Traffic Units (Telecommunications); Units of Measurement

CCITT RECMN E.500 (REV 1)-92. Traffic Intensity Measurement Principles (Study Group II) 17 pp. 17 pp.
CCITT RECMN E.500-89. Traffic Intensity Measurement Principles —Telephone Network and ISDN—Quality of Service, Network Management and Traffic Engineering (Study Group II) 12 pp. 12 pp.

Fixed Daily Measurement Period

See Also: Average Daily Peak Hour; Busy Hour; Fixed Daily Measurement Hour; Telecommunication Circuits; Telecommunication Traffic; Time; Time Consistent Busy Hour; Units of Measurement

CCITT RECMN E.500 (REV 1)-92. Traffic Intensity Measurement Principles (Study Group II) 17 pp. 17 pp.
CCITT RECMN E.500-89. Traffic Intensity Measurement Principles —Telephone Network and ISDN—Quality of Service, Network Management and Traffic Engineering (Study Group II) 12 pp. 12 pp.

Fixed Inductors

See Also: Electric Coils; Inductors

—Electromagnetic Interference Suppression—Quality Assurance

CENELEC PREN 138 000-93. Generic Specification: Fixed Inductors for Electromagnetic Interference Suppression. 28 pp.
CENELEC PREN 138 100-93. Sectional Specification: Fixed Inductors for Electromagnetic Interference Suppression —Inductors for Which Safety Tests are Required. 32 pp.
CENELEC PREN 138 101-93. Blank Detail Specification: Fixed Inductors for Electromagnetic Interference Suppression —Inductors for Which Safety Tests are Required. Assessment Level D. 13 pp.

—Radio Frequency

CECC CECC 29 000 ISSUE 1-92. Generic Specification: Fixed RF Wound Inductors (En, Ge). 52 pp.
IEC 938 Pt 1-88. Fixed Inductors for Radio Interference Suppression Part 1: Generic Specification First Edition. 59 pp.
IEC 938 Pt 2-88. Fixed Inductors for Radio Interference Suppression Part 2: Sectional Specification Selection of Methods of Test and General Requirements First Edition. 49 pp.

—Radio Frequency—Quality Assurance

BSI BS 9750-90. 1990 Fixed Radio Frequency Inductors of Assessed Quality: Generic Data and Methods of Test. 35 pp.
BSI BS 9751-79. 1979 Amd 2 Blank Detail Specification for Fixed Insulated (Unshielded) r.f. Inductors at the Full Assessment. 18 pp.
BSI BS 9752-89. 1989 Fixed Insulated (Unshielded) r.f. Inductors Suitable for Surface Mounting at the Full Assessment Level. 40 pp.
CENELEC PREN 129 200-93. Sectional Specification: Fixed Inductors for Radio Frequency Circuits. 32 pp.
CENELEC PREN 129 201-93. Blank Detail Specification: Fixed Inductors for Radio Frequency Circuits. Assessment Level E. 16 pp.
CENELEC PREN 129 202-93. Blank Detail Specification: Fixed Inductors for Radio Frequency Circuits. Assessment Level P. 16 pp.

—Radio Frequency—Reliability Assured

MOD UK DSTAN 59-78: Part 1-82. Inductors, Fixed, RF of Assessed Quality (Listed on EPIC Database) Part 1: General Requirements Issue 1. 14 pp.
MOD UK DSTAN 59-78: Part 2-82. Inductors of Assessed Quality (Listed on EPIC Database) Part 2: Inductors, Fixed, Radio Frequency Issue 1. 19 pp.

—Wirewound—Quality Assurance

CENELEC PREN 129 102-93. Blank Detail Specification: Wirewound Surface Mounting Inductors. Assessment Level P. 14 pp.

Fixed Networks

Use: Fixed Terminal Networks

Fixed Power Resistors

Use: Power Resistors

Fixed Price Contracts

See Also: Defense Contracts; Incentive Contracts; Progress Payments

MOD UK DEFCON 127-80. Price Fixing Conditions for Contracts of Lesser Value 12/80. 2 pp.

Fixed Resistors

Scope Note: For additional listings, see specific types of resistors *See Also:* Carbon Film Resistors; Carbon Resistors; Ceramic Resistors; Chip Resistors; Film Resistors; Metal Film Resistors; Metal Oxide Resistors; Power Resistors; Precision Resistors; Resistors; Thermistors; Thin Film Resistors; Variable Resistors; Wirewound Resistors

BSI BS 9940: Part 0-83. (WITHDRAWN) 1983 Amd 1 Fixed Resistors for Use in Electronic Equipment part 0: Generic Specification. Prescribes Terms, Definitions, Methods of Test and Inspection Requirements for Use in Sectional and Detail Specifications for Fixed Resistors of Assessed. 39 pp.
BSI BS QC 400000-90. 1990 Fixed Resistors for Use in Electronic Equipment: Generic Specification. 52 pp.
CECC CECC 40 000 ISSUE 2-79. Generic Specification: Fixed Resistors (En, Fr, Ge) AMD 1—5 (En, Fr, Ge). 253 pp.
CENELEC PREN 140 000-92. Generic Specification: Fixed Resistors. 79 pp.
CNS C5023-85. General Rules of Fixed Resistors for Use in Electronic Equipment (Aug)(4711).
CNS C6037-79. Method of Test for Fixed Resistors of Electronic Equipment (Jul)(4899).
CNS C6156-82. Fixed Film Resistors and Semiprecision (Feb)(8479).
IEC 115 Pt 1-82. Fixed Resistors for Use in Electronic Equipment Part 1: Generic Specification Second Edition; (Amendment 2-1987, Incorporating Amendment 1) (Amendment 3-1989) (Amendment 4-1993). 128 pp.
JIS C 5201-88. General Rules of Resistors for Use in Electronic Equipment. 32 pp.
JIS C 5202-90. Test Methods of Fixed Resistors for Electronic Equipment. 106 pp.
MOD UK DSTAN 59-29-01. Selection of Resistors, Fixed, Non-Wirewound, for Logistics Support Purposes Issue 2; Corrigenda (Superseded in Part by Def Stan 59-30: Part 0). 9 pp.
MOD UK DSTAN 59-30: Part 1-01. Resistors, Fixed, of Assessed Quality (Listed on EPIC Database) Part 1: Resistors, Fixed, Non-Wirewound, Insulated, Medium Stability—Multiple Rating Full Assessment List of Items Conforming to BS CECC 40101-019 Issue 4; Amendment 1. 44 pp.
MOD UK DSTAN 59-30: Part 2-01. Resistors, Fixed, of Assessed Quality (Listed on EPIC Database) Part 2: Resistors, Fixed, Non-Wirewound, Insulated, Load Life Stability 0.5 Per Cent Maximum Full Assessment List of Items Conforming to BS CECC 40101-004 Issue 2; Amendment 2. 53 pp.
MOD UK DEF-5115-64. General Requirements for Resistors, Fixed (Reprinted October, 1965 Incorporating Amendment No. 1) (Superseded by Def Stan 59-30: Part 4). 16 pp.
SAA AS 1352.1-91. Fixed Resistors for Use in Electronic Equipment—Part 1: Terms and Methods of Test—Generic Specification (QC400000). 49 pp.

—Color Coding

CNS C1087-82. Colour Code for Miniature Fixed Resistor (Jul)(9080).
JIS C 0802-90. Colour Code for Fixed Resistors for Electronic Equipment. 8 pp.

—Current Noise

BSI BS 4119-67. 1967 Method of Measurement of Current Noise Generated in Fixed Resistors. 14 pp.
IEC 195-65. Method of Measurement of Current Noise Generated in Fixed Resistors First Edition. 27 pp.

—Glossaries

SAA AS 1352.1-91. Fixed Resistors for Use in Electronic Equipment—Part 1: Terms and Methods of Test—Generic Specification (QC400000). 49 pp.

—Packaging

MOD UK DSTAN 81-37-89. Packaging of Resistors Issue 2. 20 pp.

—Quality Assurance

BSI BS 9110-69. (WITHDRAWN) 1969 Amd 2 Fixed Resistors of Assessed Quality: Generic Data and Methods of Test (Superseded by CECC 40 100). 38 pp.
BSI BS 9940: Part 04.01-84. 1984 Fixed Resistors for Use in Electronic Equipment Part 04.1: Fixed Resistor Networks with Individually Measurable Resistors All of Equal Value and Equal Dissipation: Blank Detail Specification: Assessment Level E. 10 pp.

INTERNATIONAL AND NON-U.S. NATIONAL STANDARDS
SUBJECT INDEX

Fixed

Fixed Resistors (Cont.)
—Quality Assurance (Cont.)
BSI BS 9940: Part 04.02-84. 1984 Fixed Resistors for Use in Electronic Equipment Part 04.2: Fixed Resistor Networks with Individually Measurable Resistors of Either Different Resistance Values or Different Rated Dissipators: Blank Detail Spec. Assessment Level E. 10 pp.

BSI BS 9940: Part 05.01-85. 1985 Fixed Resistors for Use in Electronic Equipment Part 05.01: Fixed Resistor Networks in Which Not All Resistors Are Individually Measurable: Blank Detail Specification: Assessment Level E. 15 pp.

BSI BS CECC 40000-80. 1980 Amd 5 Fixed Resistors: Generic Specification (AMD 6254) August 31, 1990 (See Also BS E9110: 1974). 85 pp.

—Reliability Assured
CNS C5219-88. General Rules for Reliability Assured Fixed Resistors (May)(12287).

MOD UK DSTAN 59-30: Part 0-91. Resistors of Assessed Quality, Selection and Procurement Part 0: General Guidance and Requirements Issue 2. 31 pp.

MOD UK DSTAN 59-30: 80/004-90. Fixed, Low Power Chip Resistors Issue 1. 15 pp.

—Surge Arresters
MOD UK DSTAN 59-30: 90/003-87. Fixed Resistors Issue 1. 13 pp.

Fixed Satellite Services
See Also: Broadcasting Satellite Services; Communications Satellites; Fixed Services (Radio Communications); Geostationary Satellites; Satellite Communications; Satellites; Space Communications

CCIR Volume IV-1 TOC-90. Table of Contents. 2 pp.

CCIR Volume IV-1 NUM IND TXTS-90. Numerical Index of Texts. 1 p.

CCIR Volume IV-1 REF & INTRO-90. Terms of Reference of Study Group 4 and Introduction by the Chairman of Study Group 4. 8 pp.

CCIR Volume IV-1 Annex TOC-90. Table of Contents. 5 pp.

CCIR Volume IV-1 ANX NUM INDX-90. Numerical Index of Texts. 1 p.

CCIR Volume IV-1 ANX TXT DEL-90. Index of Texts Deleted. 1 p.

CCIR Volume IV&IX TOC-90. Table of Contents. 1 p.

CCIR Volume IV&IX NUM INDX TXT-90. Numerical Index of Texts. 1 p.

CCIR Volume IV&IX TOC-90. Table of Contents. 2 pp.

CCIR Volume IV&IX NUM IND TXTS-90. Numerical Index of Texts. 1 p.

CCIR Volume XV-4 TOC-90. Table of Contents.

—Analog Transmission—Hypothetical Reference Circuits
CCIR RECMN 352-4-82. Hypothetical Reference Circuit for Systems Using Analogue Transmission in the Fixed-Satellite Service—Section 4B2—Performance and Availability. 2 pp.

—Antennas—Earth Stations—Gain to Temperature Ratio
CCIR RECMN 733-92. Measurement of the G/T Ratio for Earth Stations Operating in the Fixed-Satellite Service—Section 4C—Earth Station and Baseband Characteristics—Earth Station Antennas—Maintenance of Earth Stations. 20 pp.

—Antennas—Earth Stations—Radiation Diagrams—Design
CCIR Report 391-6-90. Radiation Diagrams of Antennas for Earth Stations in the Fixed-Satellite Service for Use in Interference Studies and for the Determination of a Design Objective—Section 4C—Earth Station and Baseband Characteristics—Earth Station Antennas-. 16 pp.

CCIR QUESTION 40/4-90. Reference Radiation Diagram of Antennas at Earth Stations in the Fixed-Satellite Service—Questions of Study Group 4 Fixed-Satellite Service. 1 p.

—Antennas—Earth Stations—Radiation Patterns
CCIR RECMN 465-3-90. Reference Earth-Station Radiation Pattern for Use in Coordination and Interference Assessment in the Frequency Range from 2 to About 30 GHz—Section 4C—Earth Station and Baseband Characteristics—Earth Station Antennas—Maintenance of Earth Stations. 2 pp.

CCIR RECMN 465-4-92. Reference Earth-Station Radiation Pattern for Use in Coordination and Interference Assessment in the Frequency Range from 2 to About 30 GHz—Section 4C—Earth Station and Baseband Characteristics—Earth Station Antennas—Maintenance of Earth Stations. 2 pp.

Fixed Satellite Services (Cont.)
—Antennas—Earth Stations—Radiation Patterns (Cont.)
CCIR RECMN 731-92. Reference Earth-Station Cross-Polarized Radiation Pattern for Use in Frequency Coordination and Interference Assessment in the Frequency Range from 2 to About 30 GHz—Section 4C—Earth Station and Baseband Characteristics—Earth Station Antennas-. 15 pp.

—Antennas—Earth Stations—Side Lobe Peak
CCIR RECMN 732-92. Method for Statistical Processing of Earth-Station Antenna Side-Lobe Peaks—Section 4C—Earth Station and Baseband Characteristics—Earth Station Antennas—Maintenance of Earth Stations. 2 pp.

—Antennas—Satellites
CCIR Report 558-4-90. Satellite Antenna Patterns in the Fixed-Satellite Service—Section 4D3—Spacecraft Station Keeping—Satellite Antenna Radiation Pattern—Pointing Accuracy. 57 pp.

—Antennas—Satellites—Radiation
CCIR QUESTION 41/4-90. Radiation Characteristics of Satellite Antennas in the Fixed-Satellite Service—Questions of Study Group 4 Fixed-Satellite Service. 1 p.

—Broadcasting Satellite Services—Frequency Band Sharing
CCIR Report 631-4-90. Frequency Sharing Between the Broadcasting-Satellite Service (Sound and Television) and Terrestrial Services—Section 10/11E—Sharing. 49 pp.

CCIR Report 809-3-90. Inter-Regional Sharing of the 11.7 to 12.75 GHz Frequency Band Between the Broadcasting-Satellite Service and the Fixed-Satellite Service—Section 10/11E—Sharing. 6 pp.

CCIR QUESTION 88/11-90. Broadcasting-Satellite Service (Television) Criteria to Be Applied for Frequency Sharing Between the Broadcasting-Satellite Service and the Terrestrial and Space Services in the Frequency Range 2500 MHz to 2690 MHz—Questions Concerning. 1 p.

CCIR QUESTION 66/4-90. Frequency Sharing Between the Fixed-Satellite Service and the Broadcasting-Satellite Service—Questions of Study Group 4 Fixed-Satellite Service. 1 p.

—Broadcasting Satellite Services—Radio Frequency Interference
CCIR Report 873-2-90. Analysis of the Interference from the Broadcast-ing-Satellite Service of One Region into the Fixed-Satellite Service of Another Region Around 12 GHz—Section 4E—Frequency Sharing Between Networks of the Fixed-Satellite Service and Those of Other Space Radiocommunications Sys.. 7 pp.

—Carrier Systems—Angle Modulation
CCIR RECMN 446-2-78. Carrier Energy Dispersal for Systems Employing Angle Modulation by Analogue Signals or Digital Modulation in the Fixed-Satellite Service—Section 4C—Earth Station and Baseband Characteristics —Earth Station Antennas —Maintenance of Earth Stations. 1 p.

CCIR RECMN 446-3-92. Carrier Energy Dispersal for Systems Employing Angle Modulation by Analogue Signals or Digital Modulation in the Fixed-Satellite Service—Section 4C—Earth Station and Baseband Characteristics —Earth Station Antennas —Maintenance of Earth Stations. 20 pp.

—Carrier Systems—Angle Modulation—Power Density
CCIR Report 792-3-90. Calculation of the Maximum Power Density Averaged over 4 kHz of an Angle-Modulated Carrier—Section 4/9B—Co-Ordination and Interference Calculations. 6 pp.

—Carrier Systems—Digital Modulation
CCIR RECMN 446-2-78. Carrier Energy Dispersal for Systems Employing Angle Modulation by Analogue Signals or Digital Modulation in the Fixed-Satellite Service—Section 4C—Earth Station and Baseband Characteristics —Earth Station Antennas —Maintenance of Earth Stations. 1 p.

CCIR RECMN 446-3-92. Carrier Energy Dispersal for Systems Employing Angle Modulation by Analogue Signals or Digital Modulation in the Fixed-Satellite Service—Section 4C—Earth Station and Baseband Characteristics —Earth Station Antennas —Maintenance of Earth Stations. 20 pp.

—Carrier to Interference Ratio
CCIR RECMN 741-92. Carrier-to-Interference Calculations Between Networks in the Fixed-Satellite Service—Section 4D2—Coordination Methods. 1 p.

Fixed Satellite Services (Cont.)
—CCIR Handbooks
CCIR Decision 64-1-89. Updating of the Handbook on Satellite Communications (Fixed-Satellite Service)—Annex to Volume IV-1—Fixed-Satellite Service. 1 p.

—Communications Satellites—Frequency Band Sharing
CCIR QUESTION 48/4-90. Technical Factors Influencing the Efficiency of Use of the Geostationary-Satellite Orbit by Radiocommunication Sat. Networks Sharing Frequency Bands Allocated to the Fixed-Sat. Service—Questions of Study Group 4 Fixed-Satellite Service. 2 pp.

—Coordination
CCIR RECMN 737-92. Relationship of Technical Coordination Methods Within the Fixed-Satellite Service—Section 4D2—Coordination Methods. 2 pp.

CCIR RECMN 740-92. Technical Coordination Methods for Fixed-Satellite Networks—Section 4D2—Coordination Methods. 27 pp.

CCIR QUESTION 49/4-90. Technical Coordination Methods for Systems in the Fixed-Satellite Service—Questions of Study Group 4 Fixed-Satellite Service. 1 p.

—Degradation of Service
CCIR RECMN 496-2-86. Limits of Power Flux-Density of Radionavigation Transmitters to Protect Space Station Receivers in the Fixed-Satellite Service in the 14 GHz Band—Section 8D—Radiodetermination, Global Maritime Distress and Safety System and Related Subjects. 1 p.

CCIR RECMN 496-3-92. Limits of Power Flux-Density of Radionavigation Transmitters to Protect Space Station Receivers in the Fixed-Satellite Service in the 14 GHz Band—Section 8D—Radiodetermination, Global Maritime Distress and Safety System and Related Subjects. 2 pp.

—Design—Radio Wave Propagation
CCIR QUESTION 16-1/5-90. Propagation Data and Prediction Methods for Fixed-Satellite and Broadcasting-Satellite Services—Questions Concerning Study Group 5—Radio Wave Propagation in Non-Ionized Media. 2 pp.

—Digital Modulation—Energy Dispersal
CCIR Report 384-6-90. Energy Dispersal in the Fixed-Satellite Service—Section 4C—Earth Station and Baseband Characteristics—Earth Station Antennas—Maintenance of Earth Stations. 28 pp.

—Digital Modulation—Radio Frequency Interference
CCIR QUESTION 50/4-90. Interference Criteria and Calculation Methods for Networks in the Fixed-Satellite Service Using Digital Modulation—Questions of Study Group 4 Fixed-Satellite Service. 1 p.

—Digital Transmission
CCIR QUESTION 52/4-90. Characteristics for International Digital Transmission Links in the Fixed-Satellite Service—Questions of Study Group 4 Fixed-Satellite Service. 1 p.

—Digital Transmission—Error Performance
CCIR Report 1139-90. General System and Performance Aspects of Digital Transmission in the Fixed-Satellite Service—Section 4B1—Systems Aspects. 19 pp.

—Digital Transmission—Radio Frequency Interference
CCIR Report 867-2-90. Maximum Permissible Interference in Single-Channel-per-Carrier and Intermediate Rate Digital Transmissions in Networks of the Fixed-Satellite Service—Section 4D1—Permissible Levels of Interference. 23 pp.

—Doppler Shift
CCIR Report 214-4-86. Effects of Doppler Frequency-Shifts and Switching Discontinuities in the Fixed-Satellite Service—Section 4B2—Performance and Availability. 8 pp.

—Doppler Shift—Compensation
CCIR RECMN 730-92. Compensation of the Effects of Switching Discontinuities for Voice Band Data and of Doppler Frequency-Shifts in the Fixed-Satellite Service—Section 4B2—Performance and Availability. 12 pp.

—Earth Exploration Satellites—Frequency Band Sharing
CCIR QUESTION 67/4-90. Frequency Sharing Between the Fixed-Satellite Service and the Earth Exploration-Satellite (Passive) and Space Research (Passive) Services—Questions of Study Group 4 Fixed-Satellite Service. 1 p.

INDUSTRY STANDARDS

INTERNATIONAL AND NON-U.S. NATIONAL STANDARDS
SUBJECT INDEX

Fixed

Fixed Satellite Services (Cont.)
—Earth Stations

CCIR Report 553-3-86. Operation and Maintenance of Earth Stations in the Fixed-Satellite Service—Section 4C—Earth Station and Baseband Characteristics—Earth Station Antennas—Maintenance of Earth Stations. 7 pp.

CCIR Report 869-2-90. Low Capacity Earth Stations and Associated Satellite Systems in the Fixed-Satellite Service—Section 4C—Earth Station and Baseband Characteristics—Earth Station Antennas—Maintenance of Earth Stations. 15 pp.

CCIR QUESTION 23-1/4-86. Low Capacity Earth Stations and Associated Satellite Systems—Questions of Study Group 4 Fixed-Satellite Service. 1 p.

CCIR QUESTION 44/4-90. Use of Transportable Transmitting Earth Stations in the Fixed-Satellite Service Including Use for Feeder Links to Broadcasting Satellites—Questions of Study Group 4 Fixed-Satellite Service. 1 p.

—Earth Stations—Antennas

CCIR Report 390-6-90. Earth-Station Antennas for the Fixed-Satellite Service—Section 4C—Earth Station and Baseband Characteristics —Earth Station Antennas —Maintenance of Earth Stations. 23 pp.

CCIR QUESTION 42/4-90. Characteristics of Antennas at Earth Stations in the Fixed-Satellite Service—Questions of Study Group 4 Fixed-Satellite Service. 1 p.

—Earth Stations—Antennas—Noise Temperature

CCIR Report 390-6-90. Earth-Station Antennas for the Fixed-Satellite Service—Section 4C—Earth Station and Baseband Characteristics —Earth Station Antennas —Maintenance of Earth Stations. 23 pp.

—Earth Stations—Broadcasting Satellite Services—Emissions

CCIR Report 712-1-82. Factors Concerning the Protection of Fixed-Satellite Earth Stations Operating in Adjacent Frequency Band Allocations Against Unwanted Emissions from Broadcasting Satellites Operating in Frequency Bands Around 12 GHz—Section 4E—Frequency Sharing Between Networks. 12 pp.

—Earth Stations—Coordinate Determination—Frequency Band Sharing

CCIR RECMN 359-5-82. Determination of the Co-Ordination Area of Earth Stations in the Fixed-Satellite Service Using the Same Frequency Bands as the Systems in the Fixed Terrestrial Service—Section 4/9B—Co-Ordination and Interference Calculations. 1 p.

—Earth Stations—Coordination Areas

CCIR Report 1163-90. Coordination Area of an Earth Station of the Fixed-Satellite Service Sharing the Same Frequency Band with the Radionavigation Service—Section 8D—Radiodetermination, Global Maritime Distress and Safety System and Related Subjects. 2 pp.

CCIR QUESTION 5/12-90. Coordination Area of an Earth Station of the Fixed-Satellite Service Sharing the Same Frequency Band with the Radionavigation Service—Questions Concerning Study Group 12—Inter-Service Sharing and Compatibility. 1 p.

—Earth Stations—Effective Isotropic Radiated Power

CCIR RECMN 524-3-90. Maximum Permissible Levels of off-Axis e.i.r.p. Density from Earth Stations in the Fixed-Satellite Service Transmitting in the 6 and 14 GHz Frequency Bands—Section 4D1—Permissible Levels of Interference. 2 pp.

CCIR RECMN 524-4-90. Maximum Permissible Levels of Off-Axis e.i.r.p. Density from Earth Stations in the Fixed-Satellite Service Transmitting in the 6 and 14 GHz Frequency Bands—Section 4D1—Permissible Levels of Interference. 10 pp.

—Earth Stations—Emergency Relief Operations

CCIR QUESTION 43/4-90. Use of Small Earth Stations in the Fixed-Satellite Service in the Event of Natural Disasters, Epidemics, Famines and Similar Emergencies for Warning and Relief Operations—Questions of Study Group 4 Fixed-Satellite Service. 1 p.

—Earth Stations—Feeder Links—Broadcasting Satellites

CCIR QUESTION 44/4-90. Use of Transportable Transmitting Earth Stations in the Fixed-Satellite Service Including Use for Feeder Links to Broadcasting Satellites—Questions of Study Group 4 Fixed-Satellite Service. 1 p.

Fixed Satellite Services (Cont.)
—Earth Stations—Feeder Links—Space Stations

CCIR QUESTION 30/4-86. Provision of Feeder Links Between Fixed-Earth Stations and Space Stations of Various Services in Frequency Bands Allocated to the Fixed-Satellite Service—Questions of Study Group 4 Fixed-Satellite Service. 2 pp.

—Earth Stations—Frequency Bands

CCIR Report 386-3-82. Determination of the Power in Any 4 kHz Band Radiated Towards the Horizon by Earth Stations of the Fixed-Satellite Service Sharing Frequency Bands Below 15 GHz with the Terrestrial Services—Section 4/9A—Sharing Conditions. 4 pp.

—Earth Stations—Geostationary Orbits—Interference

CCIR QUESTION 70/4-90. Protection of the Geostationary-Satellite Orbit Against Unacceptable Interference from Transmitting Earth Stations in the Fixed-Satellite Service at Frequencies Above 10 GHz—Questions of Study Group 4 Fixed-Satellite Service. 1 p.

—Earth Stations—Interference Cancellation Systems

CCIR RECMN 734-92. Application of Interference Cancellers in the Fixed-Satellite Service—Section 4C—Earth Station and Baseband Characteristics—Earth Station Antennas—Maintenance of Earth Stations. 10 pp.

CCIR QUESTION 58/4-90. Interference Reduction and Cancellation Techniques for the Earth Stations in the Fixed-Satellite Service—Questions of Study Group 4 Fixed-Satellite Service. 1 p.

—Earth Stations—Maintenance

CCIR Report 553-3-86. Operation and Maintenance of Earth Stations in the Fixed-Satellite Service—Section 4C—Earth Station and Baseband Characteristics—Earth Station Antennas—Maintenance of Earth Stations. 7 pp.

—Earth Stations—Radio Frequency Emissions

CCIR QUESTION 25-1/4-82. Unwanted Emissions Radiated from and Received by Earth Stations and Space Stations of the Fixed-Satellite Service—Questions of Study Group 4 Fixed-Satellite Service. 1 p.

—Earth Stations—Radio Frequency Interference

CCIR QUESTION 58/4-90. Interference Reduction and Cancellation Techniques for the Earth Stations in the Fixed-Satellite Service—Questions of Study Group 4 Fixed-Satellite Service. 1 p.

—Earth Stations—Terrestrial Stations—Frequency Band Sharing

CCIR QUESTION 57/4-90. Preferred Technical Characteristics and Selection of Sites for Earth Stations in the Fixed-Satellite Service to Facilitate Sharing with Terrestrial Services—Questions of Study Group 4 Fixed-Satellite Service. 1 p.

CCIR QUESTION 59/4-90. Preferred Technical Characteristics of Space Stations in the Fixed-Satellite Service to Facilitate Sharing with Terrestrial Services—Questions of Study Group 4 Fixed-Satellite Service. 1 p.

—Earth Stations—Terrestrial Stations—Radio Frequency Interference

CCIR QUESTION 39/4-89. Technical Criteria to Be Used in the Board's Examinations of the Probability of Harmful Interference Required by Provisions Nos. 1354, 1506 and 1509 of the Radio Regulations—Questions of Study Group 4 Fixed-Satellite Service. 1 p.

—Echo

CCIR QUESTION 7-2/4-86. Baseband Transmission Variability, Delay, Echoes and Switching Discontinuities in Systems in the Fixed-Satellite Service—Questions of Study Group 4 Fixed-Satellite Service. 1 p.

—Electromagnetic Interference

CCIR Report 388-6-90. Methods for Determining the Effects of Interference on the Performance and the Availability of Terrestrial Radio-Relay Systems and Systems in the Fixed-Satellite Service—Section 4/9B—Co-Ordination and Interference Calculations. 40 pp.

—Energy Dispersal

CCIR Report 384-6-90. Energy Dispersal in the Fixed-Satellite Service—Section 4C—Earth Station and Baseband Characteristics—Earth Station Antennas—Maintenance of Earth Stations. 28 pp.

Fixed Satellite Services (Cont.)
—Feeder Links—Mobile Satellite Communications

CCIR QUESTION 55/4-90. Feeder Links in the Fixed-Satellite Service Used for the Connection of Satellites in Various Mobile-Satellite Services—Questions of Study Group 4 Fixed-Satellite Service. 1 p.

—Fixed Services (Radio Communications)—Frequency Band Sharing

CCIR Report 876-82. Frequency Sharing Between Systems in the Fixed-Satellite Service and the Fixed Service in Frequency Bands Above 40 GHz—Section 4/9A—Sharing Conditions. 7 pp.

CCIR Report 1005-86. Frequency Sharing Between Systems of the Fixed Service and Systems of the Fixed-Satellite Service Comprising Forward Band Working (FBW) Networks and Reverse Band Working (RBW) Networks—Section 4/9A—Sharing Conditions. 7 pp.

CCIR Report 1143-90. Frequency Sharing Between the Fixed-Satellite Service and the Fixed Service Under Provisions of RR Article 14 in Region 2—Section 4/9A—Sharing Conditions. 3 pp.

CCIR QUESTION 61/4-90. Criteria for Frequency Sharing Between the Fixed Service and the Fixed-Satellite Service in Bidirectionally Allocated Frequency Bands—Questions of Study Group 4 Fixed-Satellite Service. 1 p.

CCIR QUESTION 62/4-90. Frequency Sharing of the Fixed-Satellite Service and the Inter-Satellite Service with the Fixed Service Under Provisions of RR Article 14—Questions of Study Group 4 Fixed-Satellite Service. 1 p.

CCIR QUESTION 115/9-90. Criteria for Frequency Sharing Between the Fixed Service and the Fixed-Satellite Service in Bidirectionally Allocated Frequency Bands—Questions of Study Group 9 Fixed Service. 1 p.

—Fixed Services—Emergency Relief Operations—Frequency Band Sharing

CCIR QUESTION 65/4-90. Frequency Sharing Between Systems in the Fixed-Satellite Service and the Fixed Service in the Case of Relief Operations and Other Temporary Applications—Questions of Study Group 4 Fixed-Satellite Service. 1 p.

—Fixed Services—Tropospheric Propagation—Frequency Band Sharing

CCIR QUESTION 19-1/5-90. Propagation Factors Affecting Frequency-Sharing Between Fixed-Satellite Service and Fixed and Mobile Terrestrial Services—Questions Concerning Study Group 5—Radio Wave Propagation in Non-Ionized Media. 2 pp.

—Flux Density—Frequency Band Sharing

CCIR RECMN 674-90. Power Flux-Density Values to Facilitate the Application of Article 14 for FSS in Relation to the Fixed-Satellite Service in the 11.7-12.2 GHz Band in Region 2—Section 4/9A—Sharing Conditions. 1 p.

—Frequencies—Electromagnetic Interference

CCIR Report 710-3-90. Interference Allocations in Systems Operating at Frequencies Greater Than 10 GHz in the Fixed-Satellite Service—Section 4D1—Permissible Levels of Interference. 11 pp.

—Frequency Band Sharing

CCIR Report 455-5-90. Frequency Sharing Between Networks of the Fixed-Satellite Service—Section 4D1—Permissible Levels of Interference. 43 pp.

—Frequency Band Sharing—10 GHz+

CCIR Report 552-4-90. Use of Frequency Bands Above 10 GHz in the Fixed-Satellite Service—Section 4B1—Systems Aspects. 34 pp.

CCIR QUESTION 47/4-90. Use of Frequency Bands Above 10 GHz in the Fixed-Satellite Service—Questions of Study Group 4 Fixed-Satellite Service. 1 p.

—Frequency Bands—10 GHz+

CCIR Report 552-4-90. Use of Frequency Bands Above 10 GHz in the Fixed-Satellite Service—Section 4B1—Systems Aspects. 34 pp.

CCIR QUESTION 47/4-90. Use of Frequency Bands Above 10 GHz in the Fixed-Satellite Service—Questions of Study Group 4 Fixed-Satellite Service. 1 p.

—Geostationary—Coordination—Frequency Band Sharing

CCIR RECMN 738-92. Procedure for Determining if Coordination is Required Between Geostationary-Satellite Networks Sharing the Same Frequency Bands—Section 4D2—Coordination Methods. 12 pp.

INDEX and DIRECTORY of

INTERNATIONAL AND NON-U.S. NATIONAL STANDARDS
SUBJECT INDEX

Fixed Satellite Services (Cont.)

—Geostationary—Coordination—Frequency Band Sharing (Cont.)

CCIR RECMN 739-92. Additional Methods for Determining if Detailed Coordination is Necessary Between Geostationary-Satellite Networks in the Fixed-Satellite Service Sharing the Same Frequency Bands—Section 4D2—Coordination Methods. 10 pp.

—Geostationary—Earth Exploration—Frequency Band Sharing

CCIR Report 540-1-82. Feasibility of Frequency Sharing Between an Earth Exploration-Satellite (EES) System and Fixed Satellite, Meteorological Satellite and Terrestrial Fixed and Mobile Services—Section 2F—Earth Exploration Satellites. 11 pp.

—Geostationary—Modulation—Television Channels—Interference

CCIR RECMN 483-1-78. Maximum Permissible Level of Interference in a Television Channel of a Geostationary-Satellite Network in the Fixed-Satellite Service Employing Frequency Modulation, Caused by Other Networks of This Service—Section 4D1—Permissible Levels of Interference. 1 p.

CCIR RECMN 483-2-92. Maximum Permissible Level of Interference in a Television Channel of a Geostationary-Satellite Network in the Fixed-Satellite Service Employing Frequency Modulation, Caused by Other Networks of This Service—Section 4D1—Permissible Levels of Interference. 2 pp.

—Geostationary—Multiplexing—Telephone Channels—Radio Interference

CCIR RECMN 466-5-90. Maximum Permissible Level of Interference in a Telephone Channel of a Geostationary-Satellite Network in the Fixed-Satellite Service Employing Frequency Modulation with Frequency-Division Multiplex, Caused by Other Networks of this Service—Section 4D1-. 3 pp.

CCIR RECMN 466-6-92. Maximum Permissible Level of Interference in a Telephone Channel of a Geostationary-Satellite Network in the Fixed-Satellite Service Employing Frequency Modulation with Frequency-Division Multiplex, Caused by Other Networks of this Service—Section 4D1-. 3 pp.

—Geostationary Orbits—Fixed Services—Frequency Sharing

CCIR Report 1142-90. Frequency Sharing Between the Fixed Service and the Fixed-Satellite Service Using Satellites in Slightly Inclined Geostationary Orbits—Section 4/9A—Sharing Conditions. 29 pp.

CCIR QUESTION 64/4-90. Preferred Technical Characteristics of Fixed-Satellite Service Networks Using Satellites in Slightly Inclined Geostationary Orbits to Facilitate Sharing with the Fixed Service—Questions of Study Group 4 Fixed-Satellite Service. 1 p.

CCIR QUESTION 117/9-90. Criteria for Frequency Sharing Between the Fixed Service and FSS Networks Using Satellites in Slightly Inclined Geostationary Orbits—Questions of Study Group 9 Fixed Service. 1 p.

—Geostationary Orbits—Frequency Management

CCIR RECMN 743-92. Coordination of Satellite Networks Using Slightly Inclined Geostationary-Satellite Orbits and Between Such Networks and Satellite Networks Using Non-Inclined GSO Satellites—Section 4D2—Coordination Methods. 14 pp.

—Geostationary Orbits—Inclined

CCIR Report 1138-90. Intra-Service Implications of Using Slightly Inclined Geostationary Orbits for Fixed Satellite Service Networks Operational, Sharing and Coordination Considerations—Section 4D2—Coordination Methods. 22 pp.

CCIR QUESTION 51/4-90. Operation, Frequency Sharing and Coordination of Fixed Satellite Service Networks Using Satellites in Slightly Inclined Geostationary Orbits—Questions of Study Group 4 Fixed-Satellite Service. 1 p.

—Geostationary Orbits—Inclined—Frequency Band Sharing

CCIR Report 1138-90. Intra-Service Implications of Using Slightly Inclined Geostationary Orbits for Fixed Satellite Service Networks Operational, Sharing and Coordination Considerations—Section 4D2—Coordination Methods. 22 pp.

CCIR QUESTION 51/4-90. Operation, Frequency Sharing and Coordination of Fixed Satellite Service Networks Using Satellites in Slightly Inclined Geostationary Orbits—Questions of Study Group 4 Fixed-Satellite Service. 1 p.

Fixed Satellite Services (Cont.)

—Geostationary Orbits—Inclined—Interference

CCIR QUESTION 137/7-90. Effects on Study Group 2 Services of the Use of Inclined (near-Geostationary) Satellite Orbits by the Fixed-Satellite Service—Questions Concerning Study Group 7—Science Services. 1 p.

—Geostationary Orbits—Physical Interference

CCIR QUESTION 34/4-86. Physical Interference in the Geostationary-Satellite Orbit—Questions of Study Group 4 Fixed-Satellite Service. 1 p.

—Geostationary Orbits—Radio Astronomy—Frequency Sharing

CCIR QUESTION 137/7-90. Effects on Study Group 2 Services of the Use of Inclined (near-Geostationary) Satellite Orbits by the Fixed-Satellite Service—Questions Concerning Study Group 7—Science Services. 1 p.

—Geostationary Orbits—Space Stations—Spot Beams

CCIR QUESTION 36/4-87. Use of Steerable Spot Beams by Space Stations in the Geostationary-Satellite Orbit—Questions of Study Group 4 Fixed-Satellite Service. 1 p.

—Geostationary—Pulse Code Modulation—Telephony—Radio Interference

CCIR RECMN 523-3-90. Maximum Permissible Levels of Interference in a Geostationary-Satellite Network in the Fixed-Satellite Service Using 8-Bit PCM Encoded Telephony, Caused by Other Networks of this Service—Section 4D1—Permissible Levels of Interference. 3 pp.

CCIR RECMN 523-4-92. Maximum Permissible Levels of Interference in a Geostationary-Satellite Network in the Fixed-Satellite Service Using 8-Bit PCM Encoded Telephony, Caused by Other Networks of this Service—Section 4D1—Permissible Levels of Interference. 3 pp.

—Geostationary—Radio Spectra Utilization

CCIR RECMN 742-92. Spectrum Utilization Methodologies—Section 4D2—Coordination Methods. 5 pp.

—Geostationary Satellites—Frequency Bands—Station Keeping

CCIR RECMN 484-2-82. Station-Keeping in Longitude of Geostationary Satellites Using Frequency Bands Allocated to the Fixed-Satellite Service—Section 4D3—Spacecraft Station-Keeping—Satellite Antenna Radiation Pattern—Pointing Accuracy. 1 p.

CCIR RECMN 484-3-92. Station-Keeping in Longitude of Geostationary Satellites in the Fixed Satellite Service—Section 4D3—Spacecraft Station-Keeping—Satellite Antenna Radiation Pattern—Pointing Accuracy. 2 pp.

—Geostationary Satellites—Polarization Discrimination

CCIR RECMN 736-92. Estimation of Polarization Discrimination in the Interference Calculations Between Geostationary-Satellite Networks in the Fixed-Satellite Service—Section 4D1—Permissible Levels of Interference. 7 pp.

—Geostationary Satellites—Radio Frequency Interference

CCIR QUESTION 39/4-89. Technical Criteria to Be Used in the Board's Examinations of the Probability of Harmful Interference Required by Provisions Nos. 1354, 1506 and 1509 of the Radio Regulations—Questions of Study Group 4 Fixed-Satellite Service. 1 p.

—Geostationary Satellites—Station Keeping

CCIR Report 556-4-90. Factors Affecting Station-Keeping of Geostationary Satellites of the Fixed-Satellite Service—Section 4D3—Spacecraft Station Keeping—Satellite Antenna Radiation Pattern—Pointing Accuracy. 4 pp.

—Hypothetical Reference Circuits

CCIR Report 706-2-86. Availability of Circuits in the Fixed-Satellite Service—Section 4B2—Performance and Availability. 5 pp.

—Hypothetical Reference Circuits—Interruption of Services

CCIR QUESTION 45/4-90. Interruptions to Traffic on Digital Paths or Circuits in the Fixed-Satellite Service—Questions of Study Group 4 Fixed-Satellite Service. 2 pp.

Fixed Satellite Services (Cont.)

—Hypothetical Reference Circuits—Noise Level

CCIR RECMN 354-2-74. Video Bandwidth and Permissible Noise Level in the Hypothetical Reference Circuit for the Fixed-Satellite Service—Section 4B2—Performance and Availability. 1 p.

—Hypothetical Reference Circuits—Noise Power

CCIR RECMN 353-6-90. Allowable Noise Power in the Hypothetical Reference Circuit for Frequency-Division Multiplex Telephony in the Fixed-Satellite Service—Section 4B2—Performance and Availability. 2 pp.

CCIR RECMN 353-7-92. Allowable Noise Power in the Hypothetical Reference Circuit for Frequency-Division Multiplex Telephony in the Fixed-Satellite Service—Section 4B2—Performance and Availability. 7 pp.

—Hypothetical Reference Circuits—Sound Transmission

CCIR RECMN 502-2-82. Hypothetical Reference Circuits for Sound-Programme Transmissions Terrestrial Systems and Systems in the Fixed-Satellite Service—Section CMTT C—Transmission Standards and Performance Objectives for Sound-Programme Channels. 3 pp.

—Hypothetical Reference Circuits—Telephony

CCIR RECMN 579-1-86. Availability Objectives for a Hypothetical Reference Circuit and a Hypothetical Reference Digital Path When Used for Telephony Using Pulse-Code Modulation, or as Part of an Integrated Services Digital Network Hypothetical Reference Connection, in the. 2 pp.

CCIR RECMN 579-2-92. Availability Objectives for a Hypothetical Reference Circuit and a Hypothetical Reference Digital Path When Used for Telephony Using Pulse-Code Modulation, or as Part of an Integrated Services Digital Network Hypothetical Reference Connection, in the. 6 pp.

—Hypothetical Reference Circuits—Video Bandwidth

CCIR RECMN 354-2-74. Video Bandwidth and Permissible Noise Level in the Hypothetical Reference Circuit for the Fixed-Satellite Service—Section 4B2—Performance and Availability. 1 p.

—Hypothetical Reference Digital Path

CCIR Report 706-2-86. Availability of Circuits in the Fixed-Satellite Service—Section 4B2—Performance and Availability. 5 pp.

—Hypothetical Reference Digital Path—Digital Transmission

CCIR RECMN 521-2-86. Hypothetical Reference Digital Path for Systems Using Digital Transmission in the Fixed-Satellite Service—Section 4B2—Performance and Availability. 2 pp.

—Hypothetical Reference Digital Path—Interruption of Services

CCIR QUESTION 45/4-90. Interruptions to Traffic on Digital Paths or Circuits in the Fixed-Satellite Service—Questions of Study Group 4 Fixed-Satellite Service. 2 pp.

—Hypothetical Reference Digital Path—ISDN

CCIR Report 997-1-90. Characteristics of a Fixed-Satellite Service Hypothetical Reference Digital Path Forming Part of an Integrated Services Digital Network —Section 4B2—Performance and Availability. 33 pp.

—Hypothetical Reference Digital Path—Telephony

CCIR RECMN 579-1-86. Availability Objectives for a Hypothetical Reference Circuit and a Hypothetical Reference Digital Path When Used for Telephony Using Pulse-Code Modulation, or as Part of an Integrated Services Digital Network Hypothetical Reference Connection, in the. 2 pp.

CCIR RECMN 579-2-92. Availability Objectives for a Hypothetical Reference Circuit and a Hypothetical Reference Digital Path When Used for Telephony Using Pulse-Code Modulation, or as Part of an Integrated Services Digital Network Hypothetical Reference Connection, in the. 6 pp.

—Interference Calculations

CCIR Report 455-5-90. Frequency Sharing Between Networks of the Fixed-Satellite Service—Section 4D1—Permissible Levels of Interference. 43 pp.

INDUSTRY STANDARDS

INTERNATIONAL AND NON-U.S. NATIONAL STANDARDS
SUBJECT INDEX

Fixed

Fixed Satellite Services *(Cont.)*

—**Interference—Frequency Band Sharing—Digital Radio Systems**

CCIR Report 877-1-86. Interference Criteria for Digital Radio-Relay Systems Sharing Frequency Bands with the Fixed-Satellite Service—Section 4/9A—Sharing Conditions. 6 pp.

—**Interference—Television Systems—Frequency Modulation—Sharing**

CCIR Report 449-1-74. Measured Interference into Frequency-Modulation Television Systems Using Frequencies Shared Within Systems in the Fixed-Satellite Service or Between These Systems and Terrestrial Systems—Section 4/9B—Co-Ordination and Inter-ference Calculations. 7 pp.

—**Intersatellite Services—Frequency Band Sharing**

CCIR QUESTION 56/4-90. Frequency Sharing Between the Inter-Satellite Service When Used for Links of the Fixed-Satellite Service and Terrestrial Radiocommunication Services—Questions of Study Group 4 Fixed-Satellite Service. 1 p.

—**ISDN**

CCIR RESOLUTION 103-90. Liasion and Joint Studies with the CCITT in Mobile and Fixed-Satellite Services and High-Definition Television in ISDNs—Volume XIV—Administrative Texts of the CCIR. 2 pp.

—**ISDN—Hypothetical Reference Circuits**

CCIR RECMN 579-1-86. Availability Objectives for a Hypothetical Reference Circuit and a Hypothetical Reference Digital Path When Used for Telephony Using Pulse-Code Modulation, or as Part of an Integrated Services Digital Network Hypothetical Reference Connection, in the. 2 pp.

CCIR RECMN 579-2-92. Availability Objectives for a Hypothetical Reference Circuit and a Hypothetical Reference Digital Path When Used for Telephony Using Pulse-Code Modulation, or as Part of an Integrated Services Digital Network Hypothetical Reference Connection, in the. 6 pp.

—**ISDN—Hypothetical Reference Digital Path**

CCIR RECMN 579-1-86. Availability Objectives for a Hypothetical Reference Circuit and a Hypothetical Reference Digital Path When Used for Telephony Using Pulse-Code Modulation, or as Part of an Integrated Services Digital Network Hypothetical Reference Connection, in the. 2 pp.

CCIR RECMN 579-2-92. Availability Objectives for a Hypothetical Reference Circuit and a Hypothetical Reference Digital Path When Used for Telephony Using Pulse-Code Modulation, or as Part of an Integrated Services Digital Network Hypothetical Reference Connection, in the. 6 pp.

—**ISDN—Hypothetical Reference Digital Path—Error Performance**

CCIR RECMN 614-1-90. Allowable Error Performance for a Hypothetical Reference Digital Path in the Fixed-Satellite Service Operating Below 15 GHz When Forming Part of an International Connection in an Integrated Services Digital Network —Section 4B2—Perform-ance and Availability. 2 pp.

CCIR RECMN 614-2-92. Allowable Error Performance for a Hypothetical Reference Digital Path in the Fixed-Satellite Service Operating Below 15 GHz When Forming Part of an International Connection in an Integrated Services Digital Network —Section 4B2—Perform-ance and Availability. 42 pp.

—**ISDN—Hypothetical Reference Digital Paths—Interference**

CCIR RECMN 735-92. Maximum Permissible Levels of Interference in a Geostationary-Satellite Network for an HRDP When Forming Part of the ISDN in the Fixed-Satellite Service Caused by Other Networks of This Service Below 15 GHz—Section 4D1—Permissible Levels of Interference. 5 pp.

—**Meteorological Satellites—Frequency Band Sharing**

CCIR Report 1125-90. Frequency Sharing Between Systems in the Fixed-Satellite Service and the Meteorological-Satellite Service Operating with Satellites in Low Earth Orbit in the Band 7 450–7 550 MHz—Section 2F—Earth Exploration Satellites. 10 pp.

Fixed Satellite Services *(Cont.)*

—**Mobile Radio Communications—Tropospheric Propagation—Sharing**

CCIR QUESTION 19-1/5-90. Propagation Factors Affecting Frequency-Sharing Between Fixed-Satellite Service and Fixed and Mobile Terrestrial Services—Questions Concerning Study Group 5—Radio Wave Propagation in Non-Ionized Media. 2 pp.

—**Mobile Radio Services—Fixed Services—Frequency Band Sharing**

CCIR Report 540-1-82. Feasibility of Frequency Sharing Between an Earth Exploration-Satellite (EES) System and Fixed Satellite, Meteorological Satellite and Terrestrial Fixed and Mobile Services—Section 2F—Earth Exploration Satellites. 11 pp.

—**Multiple Access**

CCIR QUESTION 46/4-90. Preferred Multiple-Access Characteristics in the Fixed-Satellite Service—Questions of Study Group 4 Fixed-Satellite Service. 1 p.

—**Passive Detectors—Frequency Band Sharing**

CCIR Report 850-1-86. Frequency Sharing by Passive Sensors with the Fixed, Mobile Except Aeronautical Mobile, and Fixed-Satellite Services in the Band 18.6-18.8 GHz Minimum Restrictions to Other Services in Order to Ensure Satisfactory Operations of Passive Sensors—Section 2F—Earth. 16 pp.

—**Pre-Emphasis—Frequency Modulation—Telephony**

CCIR RECMN 464-1-82. Pre-Emphasis Characteristics for Frequency-Modulation Systems for Frequency-Division Multiplex Telephony in the Fixed-Satellite Service—Sec-tion 4C—Earth Station and Baseband Character-istics—Earth Station Antennas—Maintenance of Earth Stations. 4 pp.

CCIR RECMN 464-2-92. Pre-Emphasis Characteristics for Frequency-Modulation Systems for Frequency-Division Multiplex Telephony in the Fixed-Satellite Service—Sec 4C—Earth Station and Baseband Character-istics—Earth Station Antennas—Maintenance of Earth Stations. 5 pp.

CCIR Report 212-3-74. Use of Pre-Emphasis in Frequency-Modulation Systems for Frequency Division Multiplex Telephony and Television in the Fixed-Satellite Service—Section 4C—Earth Station and Baseband Characteristics—Earth Station Antennas —Maintenance of Earth Stations. 4 pp.

—**Pre-Emphasis—Television Transmission**

CCIR Report 212-3-74. Use of Pre-Emphasis in Frequency-Modulation Systems for Frequency Division Multiplex Telephony and Television in the Fixed-Satellite Service—Section 4C—Earth Station and Baseband Characteristics—Earth Station Antennas —Maintenance of Earth Stations. 4 pp.

—**Pulse Code Modulation—Hypothetical Reference Digital Path**

CCIR RECMN 522-3-90. Allowable Bit Error Ratios at the Output of the Hypothetical Reference Digital Path for Systems in the Fixed-Satellite Service Using Pulse-Code Modulation for Telephony —Section 4B2—Performance and Availability. 2 pp.

CCIR RECMN 522-4-92. Allowable Bit Error Ratios at the Output of the Hypothetical Reference Digital Path for Systems in the Fixed-Satellite Service Using Pulse-Code Modulation for Telephony —Section 4B2—Performance and Availability. 8 pp.

—**Radio Communications—Frequency Band Sharing**

CCIR RECMN 355-3-82. Frequency Sharing Between Systems in the Fixed-Satellite Service and Terrestrial Radio Services in the Same Frequency Bands—Section 4/9A—Sharing Conditions. 2 pp.

CCIR Report 209-5-86. Frequency Sharing Between Systems in the Fixed-Satellite Service and Terrestrial Radio Services—Section 4/9A—Sharing Conditions. 8 pp.

CCIR QUESTION 63/4-90. Frequency Sharing of the Fixed-Satellite Service and the Inter-Satellite Service with Terrestrial Radio Services Other Than the Fixed Service Under Provisions of RR Article 14—Questions of Study Group 4 Fixed-Satellite Service. 1 p.

—**Radio Frequency Interference**

CCIR RECMN 766-92. Methods for Determining the Effects of Interference on the Performance and the Availability of Terrestrial Radio-Relay Systems and Systems in the Fixed-Satellite Service Section 4/9B—Co-ordination and Interference Calculations. 38 pp.

Fixed Satellite Services *(Cont.)*

—**Radio Frequency Interference—Radio Relay Systems**

CCIR RECMN 615-86. Maximum Allowable Values of Interference from the Fixed-Satellite Service into Terrestrial Radio-Relay Systems Which May Form Part of an ISDN and Share the Same Frequency Band Below 15 GHz—Section 4/9A—Sharing Conditions. 1 p.

—**Radio Links—Radio Frequency Interference**

CCIR RECMN 558-2-86. Maximum Allowable Values of Interference from Terrestrial Radio Links to Systems in the Fixed-Satellite Service Employing 8-Bit PCM Encoded Telephony and Sharing the Same Frequency Bands—Section 4/9A—Sharing Conditions. 2 pp.

—**Radio Relay Systems—Flux Density**

CCIR RECMN 358-3-82. Maximum Permissible Values of Power Flux-Density at the Surface of the Earth Produced by Satellites in the Fixed-Satellite Service Using the Same Frequency Bands Above 1 GHz as Line-of-Sight Radio-Relay Systems—Section 4/9A—Sharing Conditions. 2 pp.

—**Radio Relay Systems—Frequency Band Sharing**

CCIR RECMN 355-3-82. Frequency Sharing Between Systems in the Fixed-Satellite Service and Terrestrial Radio Services in the Same Frequency Bands—Section 4/9A—Sharing Conditions. 2 pp.

CCIR RECMN 355-4-92. Frequency Sharing Between Systems in the Fixed-Satellite Service and Radio-Relay Systems in the Same Frequency Bands—Section 4/9A—Sharing Conditions. 6 pp.

CCIR RECMN 358-3-82. Maximum Permissible Values of Power Flux-Density at the Surface of the Earth Produced by Satellites in the Fixed-Satellite Service Using the Same Frequency Bands Above 1 GHz as Line-of-Sight Radio-Relay Systems—Section 4/9A—Sharing Conditions. 2 pp.

CCIR Report 209-5-86. Frequency Sharing Between Systems in the Fixed-Satellite Service and Terrestrial Radio Services—Section 4/9A—Sharing Conditions. 8 pp.

CCIR QUESTION 109/9-90. Criteria for Frequency Sharing Between Radio-Relay Systems and Systems in the Fixed-Satellite Service—Questions of Study Group 9 Fixed Service. 1 p.

—**Radio Relay Systems—Isotropic Radiated Power—Sharing**

CCIR RECMN 406-6-90. Maximum Equivalent Isotropically Radiated Power of Line-of-Sight Radio-Relay System Transmitters Operating in the Frequency Bands Shared with the Fixed-Satellite Service—Section 4/9A—Sharing Conditions. 3 pp.

CCIR RECMN 406-7-92. Maximum Equivalent Isotropically Radiated Power of Line-of-Sight Radio-Relay System Transmitters Operating in the Frequency Bands Shared with the Fixed-Satellite Service—Section 4/9A—Sharing Conditions. 2 pp.

CCIR QUESTION 114/9-90. Maximum e.i.r.p. for Line-of-Sight Radio-Relay Transmitters Operating in Frequency Bands Shared with the Fixed-Satellite Service—Questions of Study Group 9 Fixed Service. 1 p.

—**Radio Relay Systems—Telephone Channels—Interference**

CCIR RECMN 356-4-78. Maximum Allowable Values of Interference from Line-of-Sight Radio-Relay Systems in a Telephone Channel of a System in the Fixed-Satellite Svc Employing Frequency Modulation, When the Same Frequency Bands are Shared by Both Systems—Section 4/9A—Sharing Conditions. 4 pp.

CCIR RECMN 357-3-78. Maximum Allowable Values of Interference in a Telephone Channel of an Analogue Angle-Modulated Radio-Relay System Sharing the Same Frequency Bands as Systems in the Fixed-Satellite Service—Section 4/9A—Sharing Conditions. 4 pp.

—**Satellite Communications—Frequency Band Sharing**

CCIR QUESTION 68/4-90. Frequency Sharing of the Fixed-Satellite Service and the Inter-Satellite Service with Other Space Radio Services Under Provisions of RR Article 14—Questions of Study Group 4 Fixed-Satellite Service. 1 p.

—**Satellite Links**

CCIR Report 451-3-82. Factors Affecting the System Design and the Selection of Frequencies for Inter-Satellite Links of the Fixed-Satellite Service—Section 4B1—Systems Aspects. 13 pp.

INDEX and DIRECTORY of

INTERNATIONAL AND NON-U.S. NATIONAL STANDARDS
SUBJECT INDEX

Fixed Satellite Services (Cont.)

—Satellite Links—Frequency Band Sharing

CCIR Report 874-82. Frequency Sharing Between the Inter-Satellite Service When Used by the Fixed-Satellite Service and Other Space Services—Section 4E—Frequency Sharing Between Networks of the Fixed-Satellite Service and Those of Other Space Radiocommunications Systems. 2 pp.

—Satellite Links—Hypothetical Reference Chains

CCIR QUESTION 29/CMTT-90. Analogue Television Reference Chains for Terrestrial and Fixed-Satellite Links—Questions Concerning the CMTT CCIR/CCITT Joint Study Group for Television and Sound Transmission. 1 p.

—Satellites—Antennas—Radiation Pattern—Design

CCIR RECMN 672-90. Satellite Antenna Radiation Pattern for Use as a Design Objective in the Fixed-Satellite Service—Section 4D3—Spacecraft Station-Keeping—Satellite Antenna Radiation Pattern—Pointing Accuracy. 3 pp.

CCIR RECMN 672-1-92. Satellite Antenna Radiation Pattern for Use as a Design Objective in the Fixed-Satellite Service Employing Geostationary Satellites—Section 4D3—Spacecraft Station-Keeping—Satellite Antenna Radiation Pattern—Pointing Accuracy. 24 pp.

—Satellites—Flexible Positioning

CCIR RECMN 670-90. Flexibility in the Positioning of Satellites as a Design Objective—Section 4D3—Spacecraft Station-Keeping—Satellite Antenna Radiation Pattern—Pointing Accuracy. 1 p.

CCIR RECMN 670-1-92. Flexibility in the Positioning of Satellites as a Design Objective—Section 4D3—Spacecraft Station-Keeping—Satellite Antenna Radiation Pattern—Pointing Accuracy. 8 pp.

—Single Channel Per Carrier—Radio Frequency Interference

CCIR Report 867-2-90. Maximum Permissible Interference in Single-Channel-per-Carrier and Intermediate Rate Digital Transmissions in Networks of the Fixed-Satellite Service—Section 4D1—Permissible Levels of Interference. 23 pp.

—Space Research Services—Frequency Band Sharing

CCIR QUESTION 67/4-90. Frequency Sharing Between the Fixed-Satellite Service and the Earth Exploration-Satellite (Passive) and Space Research (Passive) Services—Questions of Study Group 4 Fixed-Satellite Service. 1 p.

—Space Stations—Fixed Services—Frequency Band Sharing

CCIR QUESTION 116/9-90. Sharing Criteria for Protecting the Fixed Service from Emissions of Space Stations in the Fixed-Satellite Service in Shared Frequency Bands—Questions of Study Group 9 Fixed Service. 1 p.

—Space Stations—Frequency Band Sharing

CCIR Report 560-2-82. Sharing Criteria for the Protection of Space Stations in the Fixed-Satellite Service Receiving in the Band 14 to 14.4 GHz—Section 4E—Frequency Sharing Between Networks of the Fixed-Satellite Service and Those of Other Space Radiocommunications Systems. 3 pp.

—Space Stations—Orbits—Antenna Beams—Intersection

CCIR RECMN 765-92. Intersection of Radio-Relay Antenna Beams with Orbits Used by Space Stations in the Fixed-Satellite Service Section 4/9A—Sharing Conditions. 18 pp.

CCIR Report 393-4-90. Intersections of Radio-Relay Antenna Beams with Orbits Used by Space Stations in the Fixed-Satellite Service—Section 4/9A—Sharing Conditions. 34 pp.

—Space Stations—Radio Frequency Emissions

CCIR QUESTION 25-1/4-82. Unwanted Emissions Radiated from and Received by Earth Stations and Space Stations of the Fixed-Satellite Service—Questions of Study Group 4 Fixed-Satellite Service. 1 p.

—Space Stations—Radio Relay Transmitters—Interference

CCIR QUESTION 60/4-90. Sharing Criteria for Protecting Receiving Space Stations in the Fixed-Satellite Service Against Interference from Line-of-Sight Radio-Relay Transmitters Operating in Shared Frequency Bands—Questions of Study Group 4 Fixed-Satellite Service. 1 p.

Fixed Satellite Services (Cont.)

—Space Stations—Satellite Links

CCIR Report 451-3-82. Factors Affecting the System Design and the Selection of Frequencies for Inter-Satellite Links of the Fixed-Satellite Service—Section 4B1—Systems Aspects. 13 pp.

—Spurious Radiation—Earth Stations

CCIR Report 713-1-82. Spurious Emissions from Earth Stations and Space Stations of the Fixed-Satellite Service—Section 4E—Frequency Sharing Between Networks of the Fixed-Satellite Service and Those of Other Space Radiocommunications Systems. 8 pp.

—Spurious Radiation—Space Stations

CCIR Report 713-1-82. Spurious Emissions from Earth Stations and Space Stations of the Fixed-Satellite Service—Section 4E—Frequency Sharing Between Networks of the Fixed-Satellite Service and Those of Other Space Radiocommunications Systems. 8 pp.

—Surveys—Interference Cancellation Systems

CCIR Report 875-1-90. Survey of Interference Cancellers for Application in the Fixed-Satellite Service—Section 4C—Earth Station and Baseband Characteristics—Earth Station Antennas—Maintenance of Earth Stations. 10 pp.

—Switching Discontinuities

CCIR Report 214-4-86. Effects of Doppler Frequency-Shifts and Switching Discontinuities in the Fixed-Satellite Service—Section 4B2—Performance and Availability. 8 pp.

CCIR QUESTION 7-2/4-86. Baseband Transmission Variability, Delay, Echoes and Switching Discontinuities in Systems in the Fixed-Satellite Service—Questions of Study Group 4 Fixed-Satellite Service. 1 p.

—Synchronous Digital Hierarchy

CCIR Report 1139-90. General System and Performance Aspects of Digital Transmission in the Fixed-Satellite Service—Section 4B1—Systems Aspects. 19 pp.

—Telecommunication Circuits—Television Transmission—Performance

CCIR Report 965-1-90. Transmission Performance of Television Circuits over Systems in the Fixed-Satellite Service—Section CMTT A—Television Transmission Standards and Performance Objectives. 5 pp.

—Telecommunication Traffic—Noise (Spurious Signals)—Telephony

CCIR RECMN 481-2-86. Measurement of Noise in Actual Traffic for Systems in the Fixed-Satellite Service for Telephony Using Frequency-Division Multiplex—Section 4C—Earth Station and Baseband Characteristics—Earth Station Antennas—Maintenance of Earth Stations. 2 pp.

—Telephony—Frequency Division Multiplexing—Performance Measurement

CCIR RECMN 482-2-86. Measurement of Performance by Means of a Signal of a Uniform Spectrum for Systems Using Frequency-Division Multiplex Telephony in the Fixed-Satellite Svc.—Section 4C—Earth Station and Baseband Characteristics—Earth Stn. Antennas—Maintenance of Earth Stns.. 3 pp.

—Telephony—Noise (Spurious Signals)

CCIR Report 208-7-90. Form of the Hypothetical Reference Circuit and Allowable Noise Standards for Frequency-Division Multiplex Telephony and Television in the Fixed-Satellite Service—Section 4B2—Performance and Availability. 6 pp.

—Television Systems—Energy Dispersal

CCIR Report 384-6-90. Energy Dispersal in the Fixed-Satellite Service—Section 4C—Earth Station and Baseband Characteristics—Earth Station Antennas—Maintenance of Earth Stations. 28 pp.

—Television Transmission—Earth Stations—Outside Broadcasts

CCIR Report 1090-1-90. Use of Transportable Transmitting Earth Stations for the Transmission of Television Outside Broadcasts over Satellites—Section CMTT A—Television Transmission Standards and Performance Objectives. 6 pp.

Fixed Satellite Services (Cont.)

—Television Transmission—Energy Dispersal Signals

CCIR Report 1094-1-90. Influence of an Energy Dispersal Signal on the Picture Quality of Satellite TV Transmissions and on the Automatic Measurement of Certain Parameters—Section CMTT B—Methods of Operation and Assessment of Performance of Television Transmissions. 2 pp.

—Television Transmission—Noise (Spurious Signals)

CCIR Report 208-7-90. Form of the Hypothetical Reference Circuit and Allowable Noise Standards for Frequency-Division Multiplex Telephony and Television in the Fixed-Satellite Service—Section 4B2—Performance and Availability. 6 pp.

—Terrestrial Services—Frequency Band Sharing

CCIR QUESTION 32-1/4-90. Frequency Sharing Between Systems in the Fixed-Satellite Service and Terrestrial Services—Questions of Study Group 4 Fixed-Satellite Service. 1 p.

—Time Signal Services—Frequency Band Sharing

CCIR Report 736-1-82. Frequency Sharing Between the Time-Signal Service and the Radiolocation Service, the Fixed-Satellite Service and the Fixed and Mobile Services Near 14, 21, 26 and 31 GHz—Section 7C—Systems for Dissemination and Comparison. 8 pp.

—Transmission Delay

CCIR QUESTION 7-2/4-86. Baseband Transmission Variability, Delay, Echoes and Switching Discontinuities in Systems in the Fixed-Satellite Service—Questions of Study Group 4 Fixed-Satellite Service. 1 p.

—Transmission Systems—Measurement

CCIR Report 553-3-86. Operation and Maintenance of Earth Stations in the Fixed-Satellite Service—Section 4C—Earth Station and Baseband Characteristics—Earth Station Antennas—Maintenance of Earth Stations. 7 pp.

—Transmission Variability

CCIR QUESTION 7-2/4-86. Baseband Transmission Variability, Delay, Echoes and Switching Discontinuities in Systems in the Fixed-Satellite Service—Questions of Study Group 4 Fixed-Satellite Service. 1 p.

—Voice Band Data—Switching Discontinuities—Compensation

CCIR RECMN 730-92. Compensation of the Effects of Switching Discontinuities for Voice Band Data and of Doppler Frequency-Shifts in the Fixed-Satellite Service—Section 4B2—Performance and Availability. 12 pp.

—White Noise Filters—Performance Measurement

CCIR Report 553-3-86. Operation and Maintenance of Earth Stations in the Fixed-Satellite Service—Section 4C—Earth Station and Baseband Characteristics—Earth Station Antennas—Maintenance of Earth Stations. 7 pp.

Fixed Services (Radio Communications)

See Also: Aeronautical Communications; Fixed Satellite Services; Radio Communications; Space Communications

CCIR Volume 3 Annex TOC-90. Table of Contents. 4 pp.

CCIR Volume 3 ANX Index-90. Numerical Index of Texts. 1 p.

CCIR Volume 3 ANX DELETED-90. Index of Texts Deleted. 1 p.

CCIR Volume XV-4 TOC-90. Table of Contents.

—Adjacent Channel Spacing

CCIR QUESTION 143/9-90. Signal-to-Noise Ratios and Protection Ratios; Bandwidth, Adjacent Channel Spacing and Frequency Stability—Questions of Study Group 9 Fixed Service. 3 pp.

—Antennas—Radiation Patterns

CCIR QUESTION 155/9-90. Performance over Real Ground of Antennas Operating at Frequencies Below About 30 MHz in the Fixed Service—Questions of Study Group 9 Fixed Service. 1 p.

—Bandwidth

CCIR QUESTION 143/9-90. Signal-to-Noise Ratios and Protection Ratios; Bandwidth, Adjacent Channel Spacing and Frequency Stability—Questions of Study Group 9 Fixed Service. 3 pp.

INDUSTRY STANDARDS

INTERNATIONAL AND NON-U.S. NATIONAL STANDARDS
SUBJECT INDEX

Fixed

Fixed Services (Radio Communications) (Cont.)

—**Broadcast Receivers—Sound Broadcasting—Digital**
CCIR Report 1203-90. Digital Sound Broadcasting to Mobile, Portable and Fixed Receivers Using Terrestrial Transmitters—Section 10B—Frequency-Modulation Sound Broadcasting in Bands 8 (VHF) and 9 (UHF). 14 pp.

—**Broadcasting—Frequency Band Sharing**
CCIR QUESTION 131/9-90. Criteria for Frequency Sharing Between the Fixed and Broadcasting Services—Questions of Study Group 9 Fixed Service. 1 p.

—**Broadcasting Satellite Services—Frequency Band Sharing**
CCIR Report 631-4-90. Frequency Sharing Between the Broadcasting-Satellite Service (Sound and Television) and Terrestrial Services—Section 10/11E—Sharing. 49 pp.
CCIR QUESTION 111/9-90. Protection Criteria Between the Broadcasting-Satellite Service and the Fixed Service—Questions of Study Group 9 Fixed Service. 1 p.
CCIR QUESTION 112/9-90. Frequency Sharing Between the Broadcasting-Satellite Service (Sound) and the Fixed Service in the Band 0.5 to 3 GHz—Questions of Study Group 9 Fixed Service. 1 p.

—**Cellular Mobile Radio Equipment—Rural Areas**
CCIR Report 1192-90. Application of Cellular Type Mobile Radiocommunication Systems for Use as Fixed Systems—Section 9E1—Line-of-Sight Radio-Relay Systems. 8 pp.

—**Cellular Mobile Radio Services**
CCIR RECMN 757-92. Basic System Requirements and Performance Objectives for Cellular Type Mobile Systems Used as Fixed Systems—Section 9E1—Line-of-Sight Radio Relay Systems. 7 pp.
CCIR QUESTION 140/9-90. Application of Cellular Type Mobile Radiocommunication Systems for Use as Fixed Systems—Questions of Study Group 9 Fixed Service. 1 p.

—**Directional Antennas**
CCIR RECMN 162-3-92. Use of Directional Transmitting Antennas in the Fixed Service Operating in Bands Below About 30 MHz—Section 3Ab—Antenna Characteristics. 12 pp.
CCIR Report 356-3-90. Use of Directional Antennas in the Bands 4 to 28 MHz—Section 3Ab—Antennas Characteristics. 11 pp.
CCIR QUESTION 150/9-90. Use of Directional Antennas in the Bands 4 to 27.5 MHz Limitation of Radiation Outside the Direction Necessary for the Service—Questions of Study Group 9 Fixed Service. 1 p.
CCIR QUESTION 151/9-90. Use of Directional Antennas in the Bands Below 30 MHz in the Fixed Service—Questions of Study Group 9 Fixed Service. 1 p.

—**EIRP—Feeder Links—Broadcasting Satellite Services**
CCIR Report 1006-86. Fixed Service e.i.r.p. Limits for the Protection of the Broadcasting-Satellite Feeder Links Around 18 GHz—Section 4/9A—Sharing Conditions. 2 pp.

—**Fixed Satellite Services—Emergency Relief Operations—Frequency Ba**
CCIR QUESTION 65/4-90. Frequency Sharing Between Systems in the Fixed-Satellite Service and the Fixed Service in the Case of Relief Operations and Other Temporary Applications—Questions of Study Group 4 Fixed-Satellite Service. 1 p.

—**Fixed Satellite Services—Frequency Band Sharing**
CCIR Report 876-82. Frequency Sharing Between Systems in the Fixed-Satellite Service and the Fixed Service in Frequency Bands Above 40 GHz—Section 4/9A—Sharing Conditions. 7 pp.
CCIR Report 1005-86. Frequency Sharing Between Systems of the Fixed Service and Systems of the Fixed-Satellite Service Comprising Forward Band Working (FBW) Networks and Reverse Band Working (RBW) Networks—Section 4/9A—Sharing Conditions. 7 pp.
CCIR Report 1143-90. Frequency Sharing Between the Fixed-Satellite Service and the Fixed Service Under Provisions of RR Article 14 in Region 2—Section 4/9A—Sharing Conditions. 3 pp.
CCIR QUESTION 61/4-90. Criteria for Frequency Sharing Between the Fixed Service and the Fixed-Satellite Service in Bidirectionally Allocated Frequency Bands—Questions of Study Group 4 Fixed-Satellite Service. 1 p.

Fixed Services (Radio Communications) (Cont.)

—**Fixed Satellite Services—Frequency Band Sharing (Cont.)**
CCIR QUESTION 62/4-90. Frequency Sharing of the Fixed-Satellite Service and the Inter-Satellite Service with the Fixed Service Under Provisions of RR Article 14—Questions of Study Group 4 Fixed-Satellite Service. 1 p.
CCIR QUESTION 115/9-90. Criteria for Frequency Sharing Between the Fixed Service and the Fixed-Satellite Service in Bidirectionally Allocated Frequency Bands—Questions of Study Group 9 Fixed Service. 1 p.

—**Fixed Satellite Services—Tropospheric Propagation—Sharing**
CCIR QUESTION 19-1/5-90. Propagation Factors Affecting Frequency-Sharing Between Fixed-Satellite Service and Fixed and Mobile Terrestrial Services—Questions Concerning Study Group 5—Radio Wave Propagation in Non-Ionized Media. 2 pp.

—**Fixed Satellites—Geostationary Satellite Orbit—Frequency Sharing**
CCIR Report 1142-90. Frequency Sharing Between the Fixed Service and the Fixed-Satellite Service Using Satellites in Slightly Inclined Geostationary Orbits—Section 4/9A—Sharing Conditions. 29 pp.
CCIR QUESTION 64/4-90. Preferred Technical Characteristics of Fixed-Satellite Service Networks Using Satellites in Slightly Inclined Geostationary Orbits to Facilitate Sharing with the Fixed Service—Questions of Study Group 4 Fixed-Satellite Service. 1 p.
CCIR QUESTION 117/9-90. Criteria for Frequency Sharing Between the Fixed Service and FSS Networks Using Satellites in Slightly Inclined Geostationary Orbits—Questions of Study Group 9 Fixed Service. 1 p.

—**Flux Density—Frequency Band Sharing**
CCIR RECMN 674-90. Power Flux-Density Values to Facilitate the Application of Article 14 for FSS in Relation to the Fixed-Satellite Service in the 11.7-12.2 GHz Band in Region 2—Section 4/9A—Sharing Conditions. 1 p.

—**Frequency Stability**
CCIR QUESTION 143/9-90. Signal-to-Noise Ratios and Protection Ratios; Bandwidth, Adjacent Channel Spacing and Frequency Stability—Questions of Study Group 9 Fixed Service. 3 pp.

—**HF—Frequency Band Sharing**
CCIR RECMN 831-92. Frequency Sharing Between Services in the Band 4-30 MHz. 3 pp.

—**HF—Frequency Stability—Automatic Frequency Control**
CCIR RECMN 349-4-86. Frequency Stability Required for Systems Operating in the HF Fixed Service to Make the Use of Automatic Frequency Control Superfluous—Section 3Aa—Technical Characteristics. 3 pp.

—**HF—Frequency Synthesizers**
CCIR Report 704-78. Characteristics of Frequency Synthesizers in the HF Fixed Service—Section 3Aa—Technical Characteristics. 5 pp.

—**HF—Mobile Radio Services—Frequency Band Sharing**
CCIR Report 911-82. Frequency Sharing Between Services in the Band 4 to 30 MHz—Section 8C—Maritime Mobile Service; Telephony and Related Subjects. 3 pp.

—**Intersatellite Services—Frequency Band Sharing**
CCIR QUESTION 62/4-90. Frequency Sharing of the Fixed-Satellite Service and the Inter-Satellite Service with the Fixed Service Under Provisions of RR Article 14—Questions of Study Group 4 Fixed-Satellite Service. 1 p.

—**Ionospheric Channel Sounding Systems—Frequencies**
CCIR RECMN 613-86. Use of Ionospheric Channel Sounding Systems Operating in the Fixed Service at Frequencies Below About 30 MHz—Section 3Ac—Influence of the Ionosphere. 1 p.

—**LF—Radio Frequency Interference**
CCIR Report 915-2-90. Interference Between Fixed, Maritime Mobile and Radionavigation Services in the Bands Between 70 kHz and 130 kHz—Section 8D—Radiodetermination, Global Maritime Distress and Safety System and Related Subjects. 17 pp.

Fixed Services (Radio Communications) (Cont.)

—**MF/HF—Broadcasting—Frequency Band Sharing**
CCIR Decision 97-89. Sharing Criteria Between Broadcasting and Fixed and Mobile Services in the Band 2-30 MHz—Annex to Volume III—Fixed Service at Frequencies Below About 30 MHz (See Annex to Volume X-1). 1 p.

—**Mobile Radio Services—Fixed Satellite—Frequency Band Sharing**
CCIR Report 540-1-82. Feasibility of Frequency Sharing Between an Earth Exploration-Satellite (EES) System and Fixed Satellite, Meteorological Satellite and Terrestrial Fixed and Mobile Services—Section 2F—Earth Exploration Satellites. 11 pp.

—**Mobile Radio Services—Frequency Band Sharing**
CCIR QUESTION 133/9-90. Sharing Criteria Between the Fixed and Mobile Services in the Frequency Bands Between About 0.5 and 3 GHz—Questions of Study Group 9 Fixed Service. 1 p.

—**Mobile Satellite Communications—Frequency Band Sharing**
CCIR QUESTION 118/9-90. Protection Criteria Between the Mobile Satellite Services and the Fixed Service in the Band 1 to 3 GHz—Questions of Study Group 9 Fixed Service. 1 p.

—**Packet Radio**
CCIR RECMN 764-92. Minimum Requirements for HF Radio Systems Using a Packet Transmission Protocol—Section 3Cb—Data Transmission. 7 pp.
CCIR QUESTION 158/9-90. Packet Data Transmission Protocols for Systems Operating Below About 30 MHz—Questions of Study Group 9 Fixed Service. 1 p.

—**Passive Detectors—Frequency Band Sharing**
CCIR Report 850-1-86. Frequency Sharing by Passive Sensors with the Fixed, Mobile Except Aeronautical Mobile, and Fixed-Satellite Services in the Band 18.6-18.8 GHz Minimum Restrictions to Other Services in Order to Ensure Satisfactory Operations of Passive Sensors—Section 2F—Earth. 16 pp.
CCIR RECMN 761-92. Frequency Sharing Between the Fixed Service and Passive Sensors in the Band 18.6-18.8 GHz—Section 9F—Frequency Sharing with Other Services. 1 p.
CCIR Report 942-1-86. Frequency Sharing Between the Fixed Service and Passive Sensors in the Band 18.6 to 18.8 GHz—Section 9F—Frequency Sharing with Other Services. 1 p.

—**Protection Ratio**
CCIR QUESTION 143/9-90. Signal-to-Noise Ratios and Protection Ratios; Bandwidth, Adjacent Channel Spacing and Frequency Stability—Questions of Study Group 9 Fixed Service. 3 pp.
CCIR QUESTION 154/9-90. Criteria to Be Used in Differentiating Between Classes of Operation—Questions of Study Group 9 Fixed Service. 1 p.

—**Quality of Performance**
CCIR Report 197-4-78. Factors Affecting the Quality of Performance of Complete Systems in the Fixed Service—Section 3Cd—Performance of Digital Transmission Systems. 6 pp.

—**Radio Equipment—Transportable—Emergency Relief Operations**
CCIR QUESTION 148/9-90. Transportable Fixed Service Radiocommunication Equipment for Relief Operations—Questions of Study Group 9 Fixed Service. 1 p.

—**Radio Frequency Interference**
CCIR QUESTION 141/9-90. Technical Criteria to Be Used in the Board's Examinations of the Probability of Harmful Interference Required by Provisions Nos. 1354, 1506 and 1509 of the Radio Regulations—Questions of Study Group 9 Fixed Service. 1 p.

—**Radio Receivers—Reciprocal Mixing—Measurement**
CCIR RECMN 612-86. Measurement of Reciprocal Mixing in HF Communication Receivers in the Fixed Service—Section 3Aa—Technical Characteristics. 4 pp.

—**Radio Relay Systems**
CCIR Volume IX TOC-90. Table of Contents. 3 pp.
CCIR Volume IX Index-90. Numerical Index of Texts. 1 p.
CCIR Volume IX INTRO-90. Terms of Reference of Study Group 9 and Introduction by the Chairman of Study Group 9. 12 pp.

INDEX and DIRECTORY of

INTERNATIONAL AND NON-U.S. NATIONAL STANDARDS
SUBJECT INDEX
Fixed

Fixed Services (Radio Communications) *(Cont.)*

—Radio Relay Systems *(Cont.)*

CCIR RECMN 697-90. Error Performance Objectives for the Local-Grade Portion at Each End of an ISDN Connection Utilizing Digital Radio-Relay Systems—Section 9A—Performance Objectives, Propagation and Interference Effects. 2 pp.
CCIR RECMN 697-1-91. Error Performance and Availability Objectives for the Local-Grade Portion at Each End of an ISDN Connection Utilizing Digital Radio-Relay Systems—Section 9A—Performance Objectives, Propagation and Interference Effects. 5 pp.
CCIR RECMN 756-92. TDMA Point-to-Multipoint Systems Used as Radio Concentrators—Section 9E1—Line-of-Sight Radio-Relay Systems. 12 pp.
CCIR RECMN 745-91. CCIR Recommendations for Analogue Radio-Relay Systems—Section 9T—General and Terminology. 3 pp.
CCIR Volume IX Annex TOC-90. Table of Contents. 5 pp.
CCIR Volume IX ANX Index-90. Numerical Index of Texts. 1 p.
CCIR Volume IX ANX DELETED-90. Index of Text Deleted. 1 p.
CCIR Report 287-4-90. Analogue Radio-Relay Systems of Capacity Greater Than 1800 Telephone Channels, or Their Equivalent (See Vol. IX-1, Dubrovnik 1986)—Section 9B1—Radio-Frequency Channel Arrangements. 1 p.
CCIR Report 378-6-90. Characteristics of Digital Radio-Relay Systems Below About 17 GHz—Section 9B2—System General Characteristics. 23 pp.
CCIR Report 610-1-90. Compatibility Between Digital and FDM-FM Radio-Relay Systems (See Vol. IX-1, Dubrovnik 1986)—Section 9B2—System General Characteristics. 1 p.
CCIR Report 783-3-90. Characteristics of Radio-Relay Systems in Frequency Bands Above About 17 GHz—Section 9B2—System General Characteristics. 10 pp.
CCIR Report 289-4-90. Preferred Characteristics for the Simultaneous Transmission of Television and a Maximum of Four Sound Channels on Analogue Radio-Relay Systems—Section 9C—Interconnection Characteristics (Baseband and Intermediate Frequency). 1 p.
CCIR Report 788-2-90. Choice of Intermediate Frequencies for Digital Radio-Relay Systems—Section 9C—Interconnection Characteristics (Baseband and Intermediate Frequency). 3 pp.
CCIR Report 940-2-90. Point-to-Multipoint Systems—Section 9E1—Line-of-Sight Radio-Relay Systems. 8 pp.
CCIR Report 1057-1-90. Point-to-Multipoint Systems Utilizing Time Division Multiple Access Techniques—Section 9E1—Line-of-Sight Radio-Relay Systems. 9 pp.
CCIR QUESTION 101/9-90. Analogue Radio-Relay Systems Using Amplitude or Frequency Modulation—Questions of Study Group 9 Fixed Service. 1 p.
CCIR QUESTION 107/9-90. Characteristics of Radio-Relay Systems Operating in Frequency Bands Above About 17 GHz —Questions of Study Group 9 Fixed Service. 1 p.
CCIR QUESTION 132/9-90. Use of Frequencies in the Band 0.5 to 3 GHz for Radio-Relay Systems—Questions of Study Group 9 Fixed Service. 1 p.
CCIR QUESTION 135/9-90. Characteristics of Digital Radio-Relay Systems Below About 17 GHz—Questions of Study Group 9 Fixed Service. 1 p.
CCIR QUESTION 137/9-90. Interconnection at Baseband and Intermediate Frequencies for Digital Radio-Relay Systems—Questions of Study Group 9 Fixed Service. 2 pp.
CCIR QUESTION 139/9-90. Measurements of Digital Radio-Relay Systems—Questions of Study Group 9 Fixed Service. 1 p.

—Radio Relay Systems—Antennas—Radiation Patterns

CCIR RECMN 699-90. Reference Radiation Patterns for Line-of-Sight Radio-Relay System Antennas for Use in Coordination Studies and Interference Assessment in the Frequency Range from 1 to About 40 GHz—Section 9B2—System General Characteristics. 2 pp.
CCIR RECMN 699-1-92. Reference Radiation Patterns for Line-of-Sight Radio-Relay System Antennas for Use in Coordination Studies and Interference Assessment in the Frequency Range from 1 to About 40 GHz—Section 9B2—System General Characteristics. 2 pp.
CCIR Report 614-3-90. Reference Radiation Patterns for Radio-Relay System Antennas—Section 9B2—System General Characteristics. 7 pp.
CCIR QUESTION 110/9-90. Antenna Radiation Diagrams of Radio-Relay Stations for Use in Sharing Studies—Questions of Study Group 9 Fixed Service. 1 p.

Fixed Services (Radio Communications) *(Cont.)*

—Radio Relay Systems—Availability

CCIR RECMN 700-1-92. Error Performance and Availability Measurement Algorithm for Digital Radio-Relay Links at the Bit Rate Interface—Section 9D1—Digital Systems. 4 pp.
CCIR Report 445-3-86. Availability and Reliability of Radio-Relay Systems—Section 9A—Performance Objectives, Propagation and Interference Effects. 5 pp.
CCIR QUESTION 102/9-90. Availability of Radio-Relay Systems—Questions of Study Group 9 Fixed Service. 1 p.

—Radio Relay Systems—Bit Error Rate

CCIR RECMN 700-90. Error Performance and Availability Measurement Algorithm for Digital Radio-Relay Links at the System Bit Rate Interface—Section 9D1—Digital Systems. 3 pp.
CCIR RECMN 700-1-92. Error Performance and Availability Measurement Algorithm for Digital Radio-Relay Links at the Bit Rate Interface—Section 9D1—Digital Systems. 4 pp.
CCIR Report 930-2-90. Performance Objectives for Digital Radio-Relay Systems—Section 9A—Performance Objectives, Propagation and Interference Effects. 15 pp.
CCIR Report 613-4-90. Performance Measurements for Digital Radio-Relay Systems—Section 9D—Maintenance. 9 pp.

—Radio Relay Systems—Broadcasting Satellite Services—Interference

CCIR RECMN 760-92. Protection of Terrestrial Line-of-Sight Radio-Relay Systems Against Interference from the Broadcasting-Satellite Service in the Band 22.5-23 GHz—Section 9F—Frequency Sharing with Other Services. 3 pp.
CCIR Report 789-1-82. Protection of Terrestrial Line-of-Sight Radio-Relay Systems Against Interference from the Broadcasting-Satellite Service in the Band 11.7 to 12.75 GHz—Section 9F—Frequency Sharing with Other Services. 2 pp.
CCIR Report 941-82. Protection of Terrestrial Line-of-Sight Radio-Relay Systems Against Interference from the Broadcasting-Satellite Service (Sound) in the Band 1427 to 1530 MHz—Section 9F—Frequency Sharing with Other Services. 5 pp.
CCIR Report 1189-90. Protection of Terrestrial Line-of-Sight Radio-Relay Systems Against Interference from the Broadcasting-Satellite Service in the Band 22.5—23 GHz—Section 9F—Frequency Sharing with Other Services. 4 pp.

—Radio Relay Systems—Communication Channels

CCIR RECMN 283-5-90. Radio-Frequency Channel Arrangements for Low and Medium Capacity Analogue or Digital Radio-Relay Systems Operating in the 2 GHz Band—Section 9B1—Radio-Frequency Channel Arrangements. 3 pp.
CCIR RECMN 382-5-90. Radio-Frequency Channel Arrangements for Medium and High Capacity Analogue Radio-Relay Systems Operating in the 2 and 4 GHz Bands, or for Medium and High Capacity Digital Radio-Relay Systems Operating in the 4 GHz Band—Section 9B1—Radio-Frequency Channel Argmts. 4 pp.
CCIR RECMN 382-6-91. Radio-Frequency Channel Arrangements for Radio-Relay Systems Operating in the 2 and 4 GHz Bands—Section 9B1—Radio-Frequency Channel Arrangements. 3 pp.
CCIR RECMN 383-4-90. Radio-Frequency Channel Arrangements, for High Capacity Analogue or Digital Radio-Relay Systems Operating in the Lower 6 GHz Band—Section 9B1—Radio-Frequency Channel Arrangements. 3 pp.
CCIR RECMN 383-5-92. Radio-Frequency Channel Arrangements for High Capacity Radio-Relay Systems Operating in the Lower 6 GHz Band—Section 9B1—Radio-Frequency Channel Arrangements. 4 pp.
CCIR RECMN 384-5-90. Radio-Frequency Channel Arrangements for Medium and High Capacity Analogue or High Capacity Digital Radio-Relay Systems Operating in the Upper 6 GHz Band—Section 9B1—Radio-Frequency Channel Arrangements. 3 pp.
CCIR RECMN 385-4-90. Radio-Frequency Channel Arrangements for Low Capacity Analogue Radio-Relay Systems Operating in the 7 GHz Band—Section 9B1—Radio-Frequency Channel Arrangements. 2 pp.
CCIR RECMN 385-5-92. Radio-Frequency Channel Arrangements for Radio-Relay Systems Operating in the 7 GHz Band—Section 9B1—Radio-Hz Frequency Channel Arrangements. 6 pp.
CCIR RECMN 386-3-86. Radio-Frequency Channel Arrangements for Systems with a Capacity of 960 Telephone Channels, or the Equivalent, Operating in the 8 GHz Band—Section 9B1—Radio-Frequency Channel Arrangements. 3 pp.

Fixed Services (Radio Communications) *(Cont.)*

—Radio Relay Systems—Communication Channels *(Cont.)*

CCIR RECMN 386-4-92. Radio-Frequency Channel Arrangements for Radio-Relay Systems Operating in the 8 GHz Band—Section 9B1—Radio-Frequency Channel Arrangements. 6 pp.
CCIR RECMN 387-5-90. Radio-Frequency Channel Arrangements for Medium and High Capacity Analogue or Digital Radio-Relay Systems Operating in the 11 GHz Band—Section 9B1—Radio-Frequency Channel Arrangements. 5 pp.
CCIR RECMN 387-6-92. Radio-Frequency Channel Arrangements for Radio-Relay Systems Operating in the 11 GHz Band—Section 9B1—Radio-Frequency Channel Arrangements. 6 pp.
CCIR RECMN 389-2-74. Preferred Characteristics of Auxiliary Radio-Relay Systems Operating in the 2, 4, 6 or 11 GHz Bands—Section 9B1—Radio-Frequency Channel Arrangements. 3 pp.
CCIR RECMN 497-3-90. Radio-Frequency Channel Arrangements for Low and Medium Capacity Analogue or Medium and High Capacity Digital Radio-Relay Systems Operating in the 13 GHz Band—Section 9B1—Radio-Frequency Channel Arrangements. 4 pp.
CCIR RECMN 497-4-92. Radio-Frequency Channel Arrangements for Radio-Relay Systems Operating in the 13 GHz Frequency Band—Section 9B1—Radio-Frequency Channel Arrangements. 5 pp.
CCIR RECMN 595-2-90. Radio-Frequency Channel Arrangements for Digital Radio-Relay Systems in the 17.7 to 19.7 GHz Frequency Band—Section 9B1—Radio-Frequency Channel Arrangements. 3 pp.
CCIR RECMN 595-3-92. Radio-Frequency Channel Arrangements for Radio-Relay Systems Operating in the 18 GHz Band—Section 9B1—Radio-Frequency Channel Arrangements. 8 pp.
CCIR RECMN 635-1-90. Radio-Frequency Channel Arrangements Based on a Homogeneous Pattern for High Capacity Digital Radio-Relay Systems Operating in the 4 GHz Band—Section 9B1—Radio-Frequency Channel Arrangements. 2 pp.
CCIR RECMN 635-2-92. Radio-Frequency Channel Arrangements Based on a Homogeneous Pattern for Radio-Relay Systems Operating in the 4 GHz Band—Section 9B1—Radio-Frequency Channel Arrangements. 5 pp.
CCIR RECMN 636-1-90. Radio-Frequency Channel Arrangements for Radio-Relay Systems Operating in the 15 GHz Band—Section 9B1—Radio-Frequency Channel Arrangements. 3 pp.
CCIR RECMN 636-2-92. Radio-Frequency Channel Arrangements for Radio-Relay Systems Operating in the 15 GHz Band—Section 9B1—Radio-Frequency Channel Arrangements. 4 pp.
CCIR RECMN 637-86. Radio-Frequency Channel Arrangements for Analogue and Digital Radio-Relay Systems in the 21.2 to 23.6 GHz Frequency Band—Section 9B1—Radio-Frequency Channel Arrangements. 1 p.
CCIR RECMN 637-1-92. Radio-Frequency Channel Arrangements for Radio-Relay Systems Operating in the 23 GHz Band—Section 9B1—Radio-Frequency Channel Arrangements. 6 pp.
CCIR RECMN 746-91. Radio-Frequency Channel Arrangements for Radio-Relay Systems—Section 9B1—Radio-Frequency Channel Arrangements. 14 pp.
CCIR RECMN 747-92. Radio-Frequency Channel Arrangements for Radio-Relay Systems Operating in the 10 GHz Band—Section 9B1—Radio-Frequency Channel Arrangements. 3 pp.
CCIR RECMN 748-92. Radio-Frequency Channel Arrangements for Radio-Relay Systems Operating in the 25.25 to 27.5 GHz and 27.5 to 29.5 GHz Bands—Section 9B1—Radio-Frequency Channel Arrangements. 1 p.
CCIR RECMN 749-92. Radio-Frequency Channel Arrangements for Radio-Relay Systems Operating in the 36.0 to 40.5 GHz Band—Section 9B1—Radio-Frequency Channel Arrangements. 4 pp.
CCIR RECMN 701-90. Radio-Frequency Channel Arrangements for Analogue and Digital Point-to-Multipoint Radio Systems Operating in Frequency Bands in the Range 1.427 to 2.690 GHz (1.5, 1.8, 2.0, 2.2, 2.4 and 2.6 GHz)—Section 9E1—Line-of-Sight Radio-Relay Systems. 3 pp.
CCIR RECMN 388-63. Radio-Frequency Channel Arrangements for Trans-Horizon Radio-Relay Systems—Section 9E2—Trans-Horizon Radio-Relay Systems. 1 p.
CCIR Report 607-4-90. Radio-Frequency Channel Arrangements for Radio-Relay Systems in the Ranges 10.5 to 10.68 GHz and 11.7 to 15.35 GHz—Section 9B1—Radio-Frequency Channel Arrangements. 7 pp.
CCIR Report 782-3-90. Radio-Frequency Channel Arrangements for High-Capacity Digital Radio-Relay Systems in the 11 GHz Frequency Band—Section 9B1—Radio-Frequency Channel Arrangements. 6 pp.

INDUSTRY STANDARDS
1099

INTERNATIONAL AND NON-U.S. NATIONAL STANDARDS
SUBJECT INDEX

Fixed

Fixed Services (Radio Communications) (Cont.)

—Radio Relay Systems—Communication Channels (Cont.)

CCIR Report 934-2-90. Radio-Frequency Channel Arrangements for High and Medium-High Capacity Digital Radio-Relay Systems Operating in the Frequency Bands Below About 10 GHz—Section 9B1—Radio-Frequency Channel Arrangements. 15 pp.

CCIR Report 936-2-90. Radio-Frequency Channel Arrangements in the Bands Above About 17 GHz—Section 9B1—Radio-Frequency Channel Arrangements. 11 pp.

CCIR Report 1055-1-90. Radio-Frequency Channel Arrangements for Analogue and Small and Medium Capacity Digital Radio-Relay Systems Operating in the Bands Below About 10 GHz—Section 9B1—Radio-Frequency Channel Arrangements. 8 pp.

CCIR QUESTION 103/9-90. Trans-Horizon Radio-Relay Systems—Questions of Study Group 9 Fixed Service. 1 p.

CCIR QUESTION 108/9-90. Channel Spacings and Arrangements for Radio-Relay Systems Operating in Frequency Bands Above About 17 GHz—Questions of Study Group 9 Fixed Service. 1 p.

CCIR QUESTION 136/9-90. Radio-Frequency Channel Arrangements for Digital Radio-Relay Systems Below About 17 GHz—Questions of Study Group 9 Fixed Service. 1 p.

—Radio Relay Systems—Continuity Pilots

CCIR RECMN 401-2-70. Frequencies and Deviations of Continuity Pilots for Frequency Modulation Radio-Relay Systems for Television and Telephony—Section 9D—Maintenance. 2 pp.

—Radio Relay Systems—Diversity

CCIR RECMN 752-92. Diversity Techniques for Radio-Relay Systems—Section 9B2—System General Characteristics. 9 pp.

CCIR Report 376-6-90. Diversity Techniques for Radio-Relay Systems—Section 9B2—System General Characteristics. 11 pp.

CCIR QUESTION 106/9-90. Diversity Techniques for Radio-Relay Systems—Questions of Study Group 9 Fixed Service. 1 p.

—Radio Relay Systems—Earth Exploration Satellites—Sharing

CCIR QUESTION 113/9-90. Frequency Sharing Between Radio-Relay Systems and Systems of the Earth Exploration-Satellite Service and the Space Research Service—Questions of Study Group 9 Fixed Service. 1 p.

—Radio Relay Systems—Fixed Satellite Systems—Frequency Sharing

CCIR QUESTION 109/9-90. Criteria for Frequency Sharing Between Radio-Relay Systems and Systems in the Fixed-Satellite Service—Questions of Study Group 9 Fixed Service. 1 p.

—Radio Relay Systems—Frequency Band Sharing

CCIR RECMN 759-92. Use of Frequencies in the Band 500 to 3000 MHz for Radio-Relay Systems—Section 9F—Frequency Sharing with Other Services. 4 pp.

CCIR Report 1196-90. Considerations in the Development of Criteria for Sharing Between the Terrestrial Fixed Service and Other Services—Section 9F—Frequency Sharing with Other Services. 5 pp.

CCIR Decision 89-89. Technical Characteristics of Systems in the Fixed Service Required for the Evaluation of Frequency Sharing—Annex to Volume IX-1—Fixed Service Using Radio-Relay Systems. 2 pp.

CCIR QUESTION 124/9-90. New Techniques for Spectrum Sharing and Band Utilization for Radio-Relay Systems—Questions of Study Group 9 Fixed Service. 1 p.

CCIR QUESTION 125/9-90. Point-to-Multipoint Systems for Telephony, Data or Video Applications Using New Techniques for Spectrum Sharing and Band Utilization—Questions of Study Group 9 Fixed Service. 1 p.

—Radio Relay Systems—Frequency Bands

CCIR RECMN 698-90. Preferred Frequency Bands for Trans-Horizon Radio-Relay Systems—Section 9E2—Trans-Horizon Radio-Relay Systems. 2 pp.

CCIR RECMN 698-1-92. Preferred Frequency Bands for Trans-Horizon Radio-Relay Systems—Section 9E2—Trans-Horizon Radio-Relay Systems. 8 pp.

CCIR Report 1191-90. Factors Affecting the Choice of Frequency Bands for Trans-Horizon Radio-Relay Systems—Section 9E2—Trans-Horizon Radio-Relay Systems. 8 pp.

CCIR QUESTION 103/9-90. Trans-Horizon Radio-Relay Systems—Questions of Study Group 9 Fixed Service. 1 p.

Fixed Services (Radio Communications) (Cont.)

—Radio Relay Systems—Frequency Deviation

CCIR RECMN 404-2-70. Frequency Deviation for Analogue Radio-Relay Systems for Telephony Using Frequency-Division Multiplex—Section 9C—Interconnection Characteristics (Baseband and Intermediate Frequency). 1 p.

—Radio Relay Systems—Frequency Tolerances

CCIR Report 785-2-90. Frequency Tolerances for Radio-Relay Systems—Section 9B2—System General Characteristics. 5 pp.

CCIR QUESTION 120/9-90. Frequency Tolerances of Radio-Relay Systems—Questions of Study Group 9 Fixed Service. 1 p.

—Radio Relay Systems—Glossaries

CCIR RECMN 592-2-90. Terminology Used for Radio-Relay Systems—Section 9T—Terminology. 3 pp.

—Radio Relay Systems—Hypothetical Reference Circuits

CCIR RECMN 390-4-82. Definitions of Terms and References Concerning Hypothetical Reference Circuits and Hypothetical Reference Digital Paths for Radio-Relay Systems—Section 9A—Performance Objectives, Propagation and Interference Effects. 2 pp.

CCIR RECMN 392-63. Hypothetical Reference Circuit for Radio-Relay Systems for Telephony Using Frequency-Division Multiplex with a Capacity of More Than 60 Telephone Channels—Section 9A—Performance Objectives, Propagation and Interference Effects. 2 pp.

CCIR RECMN 393-4-82. Allowable Noise Power in the Hypothetical Reference Circuit for Radio-Relay Systems for Telephony Using Frequency-Division Multiplex—Section 9A—Performance Objectives, Propagation and Interference Effects. 2 pp.

CCIR RECMN 396-1-66. Hypothetical Reference Circuit for Trans-Horizon Radio-Relay Systems for Telephony Using Frequency-Division Multiplex—Section 9A—Performance Objectives, Propagation and Interference Effects. 1 p.

CCIR RECMN 397-3-78. Allowable Noise Power in the Hypothetical Reference Circuit of Trans-Horizon Radio-Relay Systems for Telephony Using Frequency-Division Multiplex—Section 9A—Performance Objectives, Propagation and Interference Effects. 1 p.

CCIR RECMN 555-78. Permissible Noise in the Hypothetical Reference Circuit of Radio-Relay Systems for Television—Section 9A—Performance Objectives, Propagation and Interference Effects. 2 pp.

CCIR RECMN 556-1-86. Hypothetical Reference Digital Path for Radio-Relay Systems Which May Form Part of an Integrated Services Digital Network with a Capacity Above the Second Hierarchical Level—Section 9A—Performance Objectives, Propagation and Interference Effects. 2 pp.

CCIR RECMN 557-2-90. Availability Objective for a Hypothetical Reference Circuit and a Hypothetical Reference Digital Path—Section 9A—Performance Objectives, Propagation and Interference Effects. 2 pp.

CCIR RECMN 557-3-91. Availability Objective for Radio-Relay Systems over a Hypothetical Reference Circuit and a Hypothetical Reference Digital Path—Section 9A—Performance Objectives, Propagation and Interference Effects. 5 pp.

—Radio Relay Systems—Hypothetical Reference Digital Paths

CCIR RECMN 390-4-82. Definitions of Terms and References Concerning Hypothetical Reference Circuits and Hypothetical Reference Digital Paths for Radio-Relay Systems—Section 9A—Performance Objectives, Propagation and Interference Effects. 2 pp.

CCIR RECMN 557-2-90. Availability Objective for a Hypothetical Reference Circuit and a Hypothetical Reference Digital Path—Section 9A—Performance Objectives, Propagation and Interference Effects. 2 pp.

CCIR RECMN 557-3-91. Availability Objective for Radio-Relay Systems over a Hypothetical Reference Circuit and a Hypothetical Reference Digital Path—Section 9A—Performance Objectives, Propagation and Interference Effects. 5 pp.

CCIR RECMN 594-2-90. Allowable Bit Error Ratios at the Output of the Hypothetical Reference Digital Path for Radio-Relay Systems Which May Form Part of an Integrated Services Digital Network—Section 9A—Performance Objectives, Propagation and Interference Effects. 2 pp.

Fixed Services (Radio Communications) (Cont.)

—Radio Relay Systems—Hypothetical Reference Digital Paths (Cont.)

CCIR RECMN 594-3-91. Allowable Bit Error Ratios at the Output of the Hypothetical Reference Digital Path for Radio-Relay Systems Which May Form Part of an Integrated Services Digital Network—Section 9A—Performance Objectives, Propagation and Interference Effects. 4 pp.

CCIR RECMN 696-90. Error Performance and Availability Objectives for Hypothetical Reference Digital Sect. Utilizing Digital Radio-Relay Systems Forming Part or All of the Medium Grade Portion of an ISDN Connection—Section 9A—Performance Objectives, Propagation and Interference Effects. 3 pp.

CCIR RECMN 696-1-91. Error Performance and Availability Objectives for Hypothetical Reference Digital Sect. Utilizing Digital Radio-Relay Systems Forming Part or All of the Medium-Grade Portion of an ISDN Connection—Section 9A—Performance Objectives, Propagation and Interference Effects. 4 pp.

CCIR QUESTION 134/9-90. Hypothetical Reference Digital Paths and Performance Objectives for Digital Radio-Relay Systems—Questions of Group 9 Fixed Service. 1 p.

—Radio Relay Systems—Integrated Services Digital Networks

CCIR Report 1052-1-90. Error Performance and Availability Objectives for Digital Radio-Relay Systems Used in the "Medium Grade" Portion of an ISDN Connection—Section 9A—Performance Objectives, Propagation and Interference Effects. 6 pp.

CCIR Report 1053-1-90. Error Performance and Availability Objectives for Digital Radio-Relay Systems Used in the Local-Grade Portion of an ISDN Connection—Section 9A—Performance Objectives, Propagation and Interference Effects. 5 pp.

CCIR Report 1193-90. Requirements for Point-to-Multipoint Systems Used in the Local Grade Portion of an ISDN Connection—Section 9E1—Line-of-Sight Radio-Relay Systems. 12 pp.

CCIR QUESTION 126/9-90. Requirements for Point-to-Multipoint Systems Used in the Local Grade Portion of an ISDN Connection—Questions of Study Group 9 Fixed Service. 1 p.

—Radio Relay Systems—Interfaces

CCIR RECMN 403-3-78. Intermediate-Frequency Characteristics for the Interconnection of Analogue Radio-Relay Systems—Section 9C—Interconnection Characteristics (Baseband and Intermediate Frequency). 2 pp.

CCIR RECMN 596-82. Interconnection of Digital Radio-Relay Systems—Section 9C—Interconnection Characteristics (Baseband and Intermediate Frequency). 1 p.

CCIR Report 938-82. Baseband Interconnection of Digital Radio-Relay Systems—Section 9C—Interconnection Characteristics (Baseband and Intermediate Frequency). 4 pp.

CCIR QUESTION 101/9-90. Analogue Radio-Relay Systems Using Amplitude or Frequency Modulation—Questions of Study Group 9 Fixed Service. 1 p.

—Radio Relay Systems—Interference

CCIR RECMN 302-2-90. Limitation of Interference from Trans-Horizon Radio-Relay Systems—Section 9E2—Trans-Horizon Radio-Relay Systems. 2 pp.

CCIR Report 1187-90. Maximum Allowable Performance and Availability Degradations to Radio-Relay Systems Arising from Interference from Emissions and Radiations from Other Sources—Section 9A—Performance Objectives, Propagation and Interference Effects. 15 pp.

CCIR QUESTION 103/9-90. Trans-Horizon Radio-Relay Systems—Questions of Study Group 9 Fixed Service. 1 p.

CCIR QUESTION 127/9-90. Maximum Allowable Performance and Availability Degradations of Radio-Relay Systems Due to Various Sources of Interference—Questions of Study Group 9 Fixed Service. 2 pp.

CCIR QUESTION 128/9-90. Maximum Allowable Degradations of Radio-Relay Systems Due to Energy Spread from Services in the Adjacent Bands—Questions of Study Group 9 Fixed Service. 1 p.

CCIR QUESTION 129/9-90. Evaluation of Interference Among Line-of-Sight Radio-Relay Systems—Questions of Group 9 Fixed Service. 1 p.

—Radio Relay Systems—Interference—Terrain Scattering

CCIR Report 1054-1-90. Interference in Radio-Relay Systems Caused by Terrain Scattering—Section 9A—Performance Objectives, Propagation and Interference Effects. 7 pp.

INDEX and DIRECTORY of

INTERNATIONAL AND NON-U.S. NATIONAL STANDARDS
SUBJECT INDEX

Fixed

Fixed Services (Radio Communications) *(Cont.)*

—**Radio Relay Systems—Lightning Protection**

CCIR Report 932-82. Protection of Radio-Relay Stations Against Lightning Discharges—Section 9A—Performance Objectives, Propagation and Interference Effects. 10 pp.

—**Radio Relay Systems—Line of Sight Propagation**

CCIR Report 784-3-90. Effects of Propagation on the Design and Operation of Line-of-Sight Radio-Relay Systems—Section 9A—Performance Objectives, Propagation and Interference Effects. 62 pp.

—**Radio Relay Systems—Maintenance**

CCIR OPINION 50-74. Coordination of the Work of the CCIR and the IEC on Measurements for the Adjustment and Maintenance of Radio-Relay Systems—Volume IX-1—Fixed Service Using Radio-Relay Systems. 1 p.

—**Radio Relay Systems—Mobile Communications—Frequency Sharing**

CCIR Report 1195-90. Sharing Considerations Between the Fixed Service and the Aeronautical Mobile-Satellite Service—Section 9F—Frequency Sharing with Other Services. 2 pp.

—**Radio Relay Systems—Multiline Switching Systems**

CCIR RECMN 444-3-82. Preferred Characteristics for Multi-Line Switching Arrangements of Analogue Radio-Relay Systems—Section 9D—Maintenance. 2 pp.

CCIR Report 137-6-86. Preferred Characteristics for Multi-Line Switching Arrangements of Analogue Radio-Relay Systems—Section 9D—Maintenance. 5 pp.

—**Radio Relay Systems—Passive Detectors—Frequency Band Sharing**

CCIR RECMN 761-92. Frequency Sharing Between the Fixed Service and Passive Sensors in the Band 18.6-18.8 GHz—Section 9F—Frequency Sharing with Other Services. 1 p.

CCIR Report 942-1-86. Frequency Sharing Between the Fixed Service and Passive Sensors in the Band 18.6 to 18.8 GHz—Section 9F—Frequency Sharing with Other Services. 1 p.

—**Radio Relay Systems—Point to Multipoint Communications**

CCIR RECMN 755-92. Point-to-Multipoint Systems Used in the Fixed Service—Section 9E1—Line-of-Sight Radio-Relay Systems. 8 pp.

—**Radio Relay Systems—Protection**

CCIR RECMN 753-92. Preferred Methods and Characteristics for the Supervision and Protection of Digital Radio-Relay Systems Section 9D1—Digital Systems. 7 pp.

CCIR Report 787-3-90. Preferred Methods and Characteristics for the Supervision and Protection of Digital Radio-Relay Systems—Section 9D—Maintenance. 8 pp.

CCIR QUESTION 138/9-90. Preferred Methods and Characteristics for the Supervision and Protection of Digital Radio-Relay Systems—Questions of Study Group 9 Fixed Service. 1 p.

—**Radio Relay Systems—Radio Equipment—Noise Power**

CCIR Report 612-1-78. Measurements of Equipment Noise Performance of Analogue Radio-Relay Systems—Section 9D—Maintenance. 4 pp.

—**Radio Relay Systems—Radio Links**

CCIR RECMN 634-1-90. Error Performance Objectives for Real Digital Radio-Relay Links Forming Part of a High-Grade Circuit Within an Integrated Services Digital Network—Section 9A—Performance Objectives, Propagation and Interference Effects. 2 pp.

CCIR RECMN 634-2-91. Error Performance Objectives for Real Digital Radio-Relay Links Forming Part of a High-Grade Circuit Within an Integrated Services Digital Network—Section 9A—Performance Objectives, Propagation and Interference Effects. 12 pp.

CCIR RECMN 695-90. Availability Objectives for Real Digital Radio-Relay Links Forming Part of a High-Grade Circuit Within an Integrated Services Digital Network—Section 9A—Performance Objectives, Propagation and Interference Effects. 2 pp.

Fixed Services (Radio Communications) *(Cont.)*

—**Radio Relay Systems—Radio Links—Communication Channels**

CCIR OPINION 14-6-90. Preferred Radio-Frequency Channel Arrangements for Radio-Relay Links for International Connections—Volume IX-1—Fixed Service Using Radio-Relay Systems. 2 pp.

—**Radio Relay Systems—Radio Links—Emergency Relief Operations**

CCIR Report 615-1-82. Transportable Fixed Radiocommunications Equipment for Relief Operations—Section 9E1—Line-of-Sight Radio-Relay Systems. 4 pp.

—**Radio Relay Systems—Radio Spectra Utilization**

CCIR QUESTION 124/9-90. New Techniques for Spectrum Sharing and Band Utilization for Radio-Relay Systems—Questions of Study Group 9 Fixed Service. 1 p.

—**Radio Relay Systems—Radiowave Propagation**

CCIR QUESTION 122/9-90. Effects of Propagation on the Design and Operation of Radio-Relay Systems—Questions of Study Group 9 Fixed Service. 1 p.

—**Radio Relay Systems—Reliability**

CCIR Report 445-3-86. Availability and Reliability of Radio-Relay Systems—Section 9A—Performance Objectives, Propagation and Interference Effects. 5 pp.

—**Radio Relay Systems—Repeater Distance**

CCIR Report 1188-90. Transmitter Power and Repeater Distance of Radio-Relay Systems Operating in the Band 1 to About 10 GHz—Section 9B2—System General Characteristics. 2 pp.

CCIR QUESTION 130/9-90. Possible Limits for Reducing Transmitter Power and Repeater Distance of Radio-Relay Systems Operating in the Band 1 to About 10 GHz—Questions of Study Group 9 Fixed Service. 1 p.

—**Radio Relay Systems—Service Channels**

CCIR RECMN 400-2-70. Service Channels to Be Provided for the Operation and Maintenance of Radio-Relay Systems—Section 9D—Maintenance. 1 p.

—**Radio Relay Systems—Simultaneous Transmission**

CCIR Report 786-3-90. Analogue Radio-Relay Systems for Simultaneous Transmission of Analogue and Digital Signals—Section 9C—Interconnection Characteristics (Baseband and Intermediate Frequency). 7 pp.

CCIR QUESTION 123/9-90. Radio-Relay Systems for the Simultaneous Transmission of Analogue and Digital Signals—Questions of Study Group 9 Fixed Service. 1 p.

—**Radio Relay Systems—Sound Channels—Television Signals**

CCIR RECMN 402-2-78. Preferred Characteristics of a Single Sound Channel Simultaneously Transmitted with a Television Signal on an Analogue Radio-Relay System—Section 9C—Interconnection Characteristics (Baseband and Intermediate Frequency). 2 pp.

—**Radio Relay Systems—Sound-Program Circuits—Noise**

CCIR Report 375-3-82. Noise Objectives for Sound-Programme Circuits 2500 km Long Provided by Means of Analogue Radio-Relay Systems—Section 9A—Performance Objectives, Propagation and Interference Effects. 3 pp.

—**Radio Relay Systems—Space Research Systems—Frequency Band Sharing**

CCIR Report 1197-90. Sharing Considerations near 2 GHz Between the Fixed Service and near-Earth Space Research Systems—Section 9F—Frequency Sharing with Other Services. 13 pp.

CCIR QUESTION 113/9-90. Frequency Sharing Between Radio-Relay Systems and Systems of the Earth Exploration-Satellite Service and the Space Research Service—Questions of Study Group 9 Fixed Service. 1 p.

—**Radio Relay Systems—Spurious Radiation**

CCIR Report 937-2-90. Spurious Emissions of Radio-Relay Systems—Section 9B2—System General Characteristics. 18 pp.

CCIR QUESTION 119/9-90. Limitation of Unwanted Emissions from Radio-Relay Systems—Questions of Study Group 9 Fixed Service. 2 pp.

Fixed Services (Radio Communications) *(Cont.)*

—**Radio Relay Systems—Standby Arrangements**

CCIR RECMN 305-59. Stand-by Arrangements for Radio-Relay Systems for Television and Telephony—Section 9D—Maintenance. 1 p.

—**Radio Relay Systems—Subscriber Lines**

CCIR Report 380-3-90. Radio Systems Operating in Bands 8 and 9 for the Provision of Subscriber Telephone Connections in Rural Areas—Section 9E1—Line-of-Sight Radio-Relay Systems. 14 pp.

CCIR QUESTION 105/9-90. Radio Systems Operating in Bands 8 and 9 for the Provision of Subscriber Telephone Connections in Rural Areas—Questions of Study Group 9 Fixed Service. 1 p.

—**Radio Relay Systems—Supervision**

CCIR RECMN 753-92. Preferred Methods and Characteristics for the Supervision and Protection of Digital Radio-Relay Systems Section 9D1—Digital Systems. 7 pp.

CCIR Report 787-3-90. Preferred Methods and Characteristics for the Supervision and Protection of Digital Radio-Relay Systems—Section 9D—Maintenance. 8 pp.

CCIR QUESTION 138/9-90. Preferred Methods and Characteristics for the Supervision and Protection of Digital Radio-Relay Systems—Questions of Study Group 9 Fixed Service. 1 p.

—**Radio Relay Systems—Synchronous Digital Hierarchy**

CCIR RECMN 750-92. Architectures and Functional Aspects of Radio-Relay Systems for SDH-Based Networks—Section 9B2—System General Characteristics. 30 pp.

CCIR RECMN 751-92. Transmission Characteristics and Performance Requirements of Radio-Relay Systems for SDH-Based Networks—Section 9B2—System General Characteristics. 5 pp.

CCIR OPINION 89-90. Requirement of an Additional Interface Rate to the Synchronous Digital Hierarchy—Volume IX-1—Fixed Service Using Radio-Relay Systems. 6 pp.

CCIR Report 1190-90. Radio-Relay Systems in a Synchronous Digital Network—Section 9B2—System General Characteristics. 11 pp.

CCIR Decision 88-89. Radio-Relay Systems in a Synchronous Digital Network—Annex to Volume IX-1—Fixed Service Using Radio-Relay Systems. 2 pp.

CCIR QUESTION 160/9-91. Radio-Relay Systems in a Synchronous Digital Network—Questions of Study Group 9 Fixed Service. 2 pp.

—**Radio Relay Systems—Telephone Circuits**

CCIR RECMN 395-2-78. Noise in the Radio Portion of Circuits to Be Established Over Real Radio-Relay Links for FDM Telephony—Section 9A—Performance Objectives, Propagation and Interference Effects. 3 pp.

CCIR RECMN 593-82. Noise in Real Circuits of Multi-Channel Trans-Horizon FM Radio-Relay Systems of Less Than 2500 km—Section 9A—Performance Objectives, Propagation and Interference Effects. 1 p.

—**Radio Relay Systems—Telephony**

CCIR Report 781-2-86. Radio-Relay Systems for Telephony Using Single-Sideband Amplitude Modulation (SSB-AM)—Section 9B1—Radio-Frequency Channel Arrangements. 5 pp.

—**Radio Relay Systems—Telephony—Interfaces**

CCIR RECMN 268-1-70. Interconnection at Audio Frequencies of Radio-Relay Systems for Telephony—Section 9C—Interconnection Characteristics (Baseband and Intermediate Frequency). 1 p.

CCIR RECMN 380-4-86. Interconnection at Baseband Frequencies of Radio-Relay Systems for Telephony Using Frequency-Division Multiplex—Section 9C—Interconnection Characteristics (Baseband and Intermediate Frequency). 4 pp.

CCIR RECMN 381-2-70. Conditions Relating to Line Regulating and Other Pilots and to Limits for the Residues of Signals Outside the Baseband in the Interconnection of Radio-Relay and Line Systems for Telephony—Section 9C—Interconnection Characteristics. 3 pp.

—**Radio Relay Systems—Telephony—Maintenance**

CCIR RECMN 290-3-78. Maintenance Measurements on Radio-Relay Systems for Telephony Using Frequency-Division Multiplex—Section 9D—Maintenance. 2 pp.

INDUSTRY STANDARDS

Fixed Services (Radio Communications) (Cont.)

—Radio Relay Systems—Telephony—Noise
CCIR RECMN 398-3-74. Measurements of Noise in Actual Traffic over Radio-Relay Systems for Telephony Using Frequency-Division Multiplex—Section 9D—Maintenance. 2 pp.

—Radio Relay Systems—Telephony—Pre-Emphasis
CCIR RECMN 275-3-82. Pre-Emphasis Characteristic for Frequency Modulation Radio-Relay Systems for Telephony Using Frequency-Division Multiplex—Section 9C—Interconnection Characteristics (Baseband and Intermediate Frequency). 4 pp.

—Radio Relay Systems—Telephony—White Noise
CCIR RECMN 399-3-78. Measurement of Noise Using a Continuous Uniform Spectrum Signal on Frequency-Division Multiplex Telephony Radio-Relay Systems—Section 9D—Maintenance. 3 pp.

—Radio Relay Systems—Television Broadcasting—Frequency Sharing
CCIR Report 1194-90. Sharing Between the Fixed Service and the Broadcasting Service (Television) in the Band 790 to 862 MHz—Section 9F—Frequency Sharing with Other Services. 7 pp.

—Radio Relay Systems—Television Channels—Interference
CCIR Report 931-82. Study of the Effects of Co-Channel Interference in the TV Channels of FM Radio-Relay Systems—Section 9A—Performance Objectives, Propagation and Interference Effects. 4 pp.

—Radio Relay Systems—Television Systems—Frequency Deviation
CCIR RECMN 276-2-74. Frequency Deviation and the Sense of Modulation for Analogue Radio-Relay Systems for Television—Section 9C—Interconnection Characteristics (Baseband and Intermediate Frequency). 1 p.

—Radio Relay Systems—Television Systems—Interfaces
CCIR RECMN 270-2-78. Interconnection at Video Signal Frequencies of Radio-Relay Systems for Television—Section 9C—Interconnection Characteristics (Baseband and Intermediate Frequency). 1 p.
CCIR RECMN 463-1-78. Limits for the Residues of Signals Outside the Baseband of Radio-Relay Systems for Television—Section 9C—Interconnection Characteristics (Baseband and Intermediate Frequency). 1 p.

—Radio Relay Systems—Television Systems—Pre-Emphasis
CCIR RECMN 405-1-70. Pre-Emphasis Characteristics for Frequency Modulation Radio-Relay Systems for Television—Section 9C—Interconnection Characteristics (Baseband and Intermediate Frequency). 5 pp.

—Radio Relay Systems—Television Transmission
CCIR Report 1056-1-90. Radio-Relay Systems for Television Transmission Using Vestigial Sideband Amplitude Modulation (AM-VSB)—Section 9B1—Radio-Frequency Channel Arrangements. 11 pp.

—Radio Relay Systems—Transhorizon Propagation
CCIR Report 285-7-90. Propagation Effects on the Design and Operation of Trans-Horizon Radio-Relay Systems—Section 9E2—Trans-Horizon Radio-Relay Systems. 12 pp.

—Radio Relay Systems—Transmission Interruption
CCIR Report 443-1-78. Transmission Interruptions Due to the Switching of Analogue Radio-Relay Systems—Section 9D—Maintenance. 6 pp.

—Radio Relay Systems—Transmitter Power
CCIR Report 1188-90. Transmitter Power and Repeater Distance of Radio-Relay Systems Operating in the Band 1 to About 10 GHz—Section 9B2—System General Characteristics. 2 pp.
CCIR QUESTION 130/9-90. Possible Limits for Reducing Transmitter Power and Repeater Distance of Radio-Relay Systems Operating in the Band 1 to About 10 GHz—Questions of Study Group 9 Fixed Service. 1 p.

Fixed Services (Radio Communications) (Cont.)

—Radio Relay Systems—Transportable—Emergency Relief Operations
CCIR QUESTION 121/9-90. Transportable Fixed-Service Radiocommunication Equipment for Relief Operations—Questions of Study Group 9 Fixed Service. 1 p.

—Radio Relay Systems—Trunk Circuits
CCIR RECMN 754-92. Radio-Relay Systems in Bands 8 and 9 for the Provision of Telephone Trunk Connections in Rural Areas—Section 9E1—Line-of-Sight Radio-Relay Systems. 6 pp.
CCIR Report 379-5-86. Characteristics of Simple Radio-Relay Equipment Operating in Bands 8 and 9 for the Provision of Telephone Trunk Connections in Rural Areas—Section 9E1—Line-of-Sight Radio-Relay Systems. 12 pp.
CCIR QUESTION 104/9-90. Radio-Relay Systems Operating in Bands 8 and 9 for the Provision of Telephone Trunk Connections in Rural Areas—Questions of Group 9 Fixed Service. 1 p.

—Radio Systems—Automatic
CCIR QUESTION 147/9-90. Automatically Controlled Radio Systems in the HF Fixed Service—Questions of Study Group 9 Fixed Service. 1 p.

—Radio Transmitters—Frequency Tolerances
CCIR QUESTION 152/9-90. Frequency Tolerance of Transmitters for the Fixed Service at Frequencies Below About 30 MHz—Questions of Study Group 9 Fixed Service. 1 p.

—Radio Transmitters—Nonionizing Radiation
CCIR QUESTION 156/9-90. Non-Ionizing Radiation Harzards Due to Transmitting Systems Operating at Frequencies Below About 30 MHz—Questions of Study Group 9 Fixed Service. 1 p.

—Radiolocation Services—Radio Frequency Emissions
CCIR QUESTION 159/9-91. Effects of Unwanted Emissions from Radar Systems in the Radiodetermination Service on Systems in the Fixed Service—Questions of Study Group 9 Fixed Service; Addendum No. 1. 2 pp.

—Radiotelegraphy Systems—Quality of Performance
CCIR Report 197-4-78. Factors Affecting the Quality of Performance of Complete Systems in the Fixed Service—Section 3Cd—Performance of Digital Transmission Systems. 6 pp.

—Receiving Stations—Monitors
CCIR RECMN 762-92. Main Characteristics of Remote Control and Monitoring Systems for HF Receiving and Transmitting Stations—Section 3Ad—Operational Questions. 6 pp.

—Receiving Stations—Remote Control Equipment
CCIR RECMN 762-92. Main Characteristics of Remote Control and Monitoring Systems for HF Receiving and Transmitting Stations—Section 3Ad—Operational Questions. 6 pp.

—Satellite Communications—Frequency Band Sharing
CCIR Report 791-1-82. Inter-Satellite Service Sharing with the Fixed and Mobile Services—Section 4/9A—Sharing Conditions. 6 pp.

—Signal to Interference Ratio
CCIR RECMN 240-6-92. Signal-to-Interference Protection Ratios for Various Classes of Emission in the Fixed Service Below About 30 MHz—Section 3Aa—Technical Characteristics. 22 pp.

—Signal to Noise Ratio
CCIR QUESTION 143/9-90. Signal-to-Noise Ratios and Protection Ratios; Bandwidth, Adjacent Channel Spacing and Frequency Stability—Questions of Study Group 9 Fixed Service. 3 pp.

—Sound Broadcasting—Frequency Band Sharing
CCIR Decision 97-89. Sharing Criteria Between Broadcasting and Fixed and Mobile Services in the Band 2-30MHz—Volume X-1—Broadcasting Service (Sound). 3 pp.
CCIR QUESTION 63/10-90. Compatibility Between the MF Sound-Broadcasting Service and Other Services Using the Same Band—Questions Concerning Study Group 10—Broadcasting Service (Sound). 1 p.

Fixed Services (Radio Communications) (Cont.)

—Space Research Services—Frequency Band Sharing
CCIR Report 687-1-86. Feasibility of Frequency Sharing Between Space Research (Near-Earth) and Fixed and Mobile Services in the 7 to 8 GHz Spectral Region—Section 2E—Space Research. 5 pp.

—Space Stations—Fixed Satellite Services—Frequency Band Sharing
CCIR QUESTION 116/9-90. Sharing Criteria for Protecting the Fixed Service from Emissions of Space Stations in the Fixed-Satellite Service in Shared Frequency Bands—Questions of Study Group 9 Fixed Service. 1 p.

—Television Broadcasting—Frequency Band Sharing
CCIR Report 1087-1-90. Frequency Sharing Between the Broadcasting Service (Television) and the Fixed and Mobile Services Below 1 GHz—Section 11E—Planning of Television Networks, Protection Ratios, Television Receivers and Antennas. 7 pp.

—Terrestrial—Radio Relay Systems—Design—Line of Sight Propagation
CCIR RECMN 530-3-90. Propagation Data and Prediction Methods Required for the Design of Terrestrial Line-of-Sight Systems—Section 5E—Aspects Relative to the Terrestrial Fixed Service. 1 p.
CCIR Report 338-6-90. Propagation Data and Prediction Methods Required for Terrestrial Line-of-Sight Systems—Section 5E—Aspects Relative to the Terrestrial Fixed Service. 66 pp.
CCIR RECMN 530-4-92. Propagation Data and Prediction Methods Required for the Design of Terrestrial Line-of-Sight Systems. 23 pp.

—Terrestrial—Radio Relay Systems—Frequency Band Sharing
CCIR RECMN 758-92. Considerations in the Development of Criteria for Sharing Between the Terrestrial Fixed Service and Other Services—Section 9F—Frequency Sharing with Other Services. 14 pp.

—Time Signal Services—Frequency Band Sharing
CCIR Report 736-1-82. Frequency Sharing Between the Time-Signal Service and the Radiolocation Service, the Fixed-Satellite Service and the Fixed and Mobile Services Near 14, 21, 26 and 31 GHz—Section 7C—Systems for Dissemination and Comparison. 8 pp.

—Transmitting Stations—Lightning Protection—Netherlands
CCIR Report 861-82. Protection of Radio Stations Against Lightning and Other Electromagnetic Disturbances—Section 3Ad—Operational Questions. 17 pp.

—Transmitting Stations—Monitors
CCIR RECMN 762-92. Main Characteristics of Remote Control and Monitoring Systems for HF Receiving and Transmitting Stations—Section 3Ad—Operational Questions. 6 pp.

—Transmitting Stations—Remote Control Equipment
CCIR RECMN 762-92. Main Characteristics of Remote Control and Monitoring Systems for HF Receiving and Transmitting Stations—Section 3Ad—Operational Questions. 6 pp.

—UHF—Mobile Satellite Communications—Frequency Band Sharing
CCIR Report 1173-90. Technical and Operational Considerations for Aeronautical Mobile-Satellite Communications—Section 8F—Frequencies, Orbits and Systems. 40 pp.

—VHF/UHF—Broadcasting—Frequency Band Sharing
CCIR QUESTION 2/12-90. Sharing Between the Broadcasting Service and the Fixed and/or Mobile Services in VHF and UHF Bands—Questions Concerning Study Group 12—Inter-Service Sharing and Compatibility. 2 pp.
CCIR RECMN 851-92. Sharing Between the Broadcasting Service and the Fixed and/or Mobile Services in the VHF and UHF Bands. 26 pp.

Fixed Terminal Networks

Use For: Fixed Networks *See Also:* Integrated Services Digital Networks; Public Data Networks; Public Switched Telephone Networks

INTERNATIONAL AND NON-U.S. NATIONAL STANDARDS
SUBJECT INDEX

Fixed Terminal Networks (Cont.)
—Interfaces—Land Mobile Radio Services
CCITT RECMN E.202-92. Network Operational Principles for Future Public Mobile Systems and Services (Study Group II) 8 pp. 8 pp.

—Interfaces—Public Land Mobile Networks
CCITT RECMN E.202-92. Network Operational Principles for Future Public Mobile Systems and Services (Study Group II) 8 pp. 8 pp.

—Interfaces—Public Land Mobile Networks—Routing
CCITT RECMN E.173-91. Routing Plan for Interconnection Between Public Land Mobile Networks and Fixed Terminal Networks (Study Group II) 15 pp. 15 pp.

—Mobile Radio Equipment
CCITT RECMN F.111-91. Principles of Service for Mobile Systems (Study Group I) 5 pp. 5 pp.

Fixed Wing Aircraft
Use: Aircraft

Fixture Wire
Use: Cords (Electric)

Flags
See Also: Banners; Pennants; Signal Devices
MOD UK DSTAN 83-49: Part 1-01. Flags and Pennants: Signal Issue 2; Amendment 4. 65 pp.
MOD UK DSTAN 83-49: Part 1-91. Flags and Pennants Part 1: Signal Issue 3. 102 pp.

—Canadian National
CGSB CAN/CGSB-98.1-92. National Flag of Canada (Outdoor Use). 22 pp.
CGSB CAN/CGSB-98.2-92. National Flag of Canada (Indoor Use). 17 pp.

—Canadian National—Single Use
CGSB CAN/CGSB-98.3-M91. National Flag of Canada (One-Event-Only Use). 7 pp.

—MOD (UK)
MOD UK DSTAN 83-73-01. Flags and Pennants: Specification Issue 2; Amendment 1. 21 pp.
MOD UK DSTAN 83-77-87. Flags and Pennants: Distinguishing: UK Personal Ranks and Appointments Issue 1. 34 pp.

—Ships—Pulley Blocks
JIS F 3425-68. Ships' Steel Blocks for Signal Flags. 4 pp.

—Taffeta
CGSB CAN/CGSB-4.136-M90. Nylon Taffeta Cloth, 60 g/m2; (Corrigendum—Aug 1990). 10 pp.
CGSB CAN/CGSB-98.1-92. National Flag of Canada (Outdoor Use). 22 pp.

Flags (Stones)
Use: Flagstones

Flagstones
See Also: Rocks

—Concrete
BSI BS 368-71. (WITHDRAWN) 1971 Amd 3 Precast Concrete Flags (AMD 5174) August 29, 1986) (Superseded by BS 7263: Part 1: 1990). 25 pp.
CNS A2054-88. Sidewalk Concrete Flags (Nov)(4016). 2 pp.
CNS A3067-88. Method of Test for Sidewalk Concrete Flags (Nov)(4017). 2 pp.
DIN ENGL 485-87. Precast Concrete Paving Flags (Apr). 3 pp.
JIS A 5304-88. Concrete Flags. 10 pp.

—Precast Concrete
BSI BS 7263: Part 1-90. 1990 Precast Concrete Flags, Kerbs, Channels Edgings and Quadrants Part 1: Specification. 19 pp.

—Precast Concrete—Installation—Codes of Practice
BSI BS 7263: Part 2-90. 1990 Precast Concrete Flags, Kerbs, Channels Edgings and Quadrants Part 2: Code of Practice for Laying. 12 pp.

Flail Mowers
Use: Mowers

Flame Arresters
See Also: Fire Protection; Flashback Arresters
BSI BS 7244-90. 1990 Flame Arresters for General Use. 14 pp.

Flame Arresters (Cont.)
—Gas Cutting Equipment
ISO 5175-87. Equipment Used in Gas Welding, Cutting and Allied Processes—Safety Devices for Fuel Gases and Oxygen or Compressed Air—General Specifications, Requirements and Tests First Edition. 11 pp.

—Gas Welding Equipment
ISO 5175-87. Equipment Used in Gas Welding, Cutting and Allied Processes—Safety Devices for Fuel Gases and Oxygen or Compressed Air—General Specifications, Requirements and Tests First Edition. 11 pp.

—Safety
DIN ENGL 8521-81. Safety Devices Against Flashback and Backflow in Welding, Cutting and Allied Processes; Safety Requirements, Testing (Dec). 8 pp.

—Ships
CNS F3215-83. Flame Arresters for Ships (Mar)(10082).
JIS F 2416-82. Ships' Flame Arresters.

Flame Atomic Absorption Spectrometry
Use: Atomic Absorption Spectrometry

Flame Atomic Absorption Spectrophotometry
Use: Atomic Absorption Spectrophotometry

Flame Atomic Absorption Spectroscopy
Use: Atomic Absorption Spectroscopy

Flame Atomic Emission Spectrometry
Use: Atomic Emission Spectrometry

Flame Atomic Emission Spectrophotometry
Use: Atomic Emission Spectrophotometry

Flame Coating Equipment
Use: Flame Spraying Equipment

Flame Cuts
Use: Flame Cutting

Flame Cutting
See Also: Gas Cutting

—Defects—Glossaries
DIN ENGL 8518-74. Defects of Flame Cuts and Plasma Cuts; Classification, Designations, Definitions (Nov). 8 pp.

Flame Detectors
Use For: Flame Supervision Devices
See Also: Control Systems Equipment; Fire Alarm and Extinguishing Equipment; Heat Detectors; Smoke Detectors
BSI BS 5445: Part 8-84. 1984 Amd 1 Components of Automatic Fire Detection Systems Part 8: High Temperature Heat Detectors. 25 pp.
CEN PREN 54-10-91. Components of Automatic Fire Detection Systems—Part 10: Flame Detectors. 45 pp.
CEN PREN 109 (Part 10)-87. Components of Automatic Fire Detection and Alarm Systems Part 10: Specification for Flame Detectors. 18 pp.

—Appliances—Thermoelectric
BSI BS EN 125-91. 1991 Flame Supervision Devices for Gas-Burning Appliances—Thermo-Electric Types. 27 pp.
CEN PREN 125-90. Flame Supervision Devices for Gas Burning Appliances—Thermo-Electric Flame Supervision Devices. 38 pp.
CEN EN 125-91. Flame Supervision Devices for Gas Burning Appliances—Thermo-Electric Flame Supervision Devices. 38 pp.
DIN ENGL EN 125-91. Flame Supervision Devices for Gas Burning Appliances; Thermo-Electric Flame Supervision Devices (Sept). 38 pp.

—Comprehensive Testing
CEN PREN 109 (Part 10)-87. Components of Automatic Fire Detection and Alarm Systems Part 10: Specification for Flame Detectors. 18 pp.

—Oil Burners—Safety
DIN ENGL 4787 Pt 2-81. Atomizing Oil Burners; Flame Monitoring Devices; Flame Detectors and Automatic Firing Units; Safety Requirements, Testing, Marking (Sept). 11 pp.

Flame Emission Spectrometry
Use: Atomic Emission Spectrometry

Flame Emission Spectrophotometry
Use: Atomic Emission Spectrophotometry

Flame Emission Spectroscopy
Use: Emission Spectroscopy

Flame Hardening
Scope Note: See the subheading Hardening under the specific metal

Flame Monitors
Use: Flame Detectors

Flame Photometry
Use: Photometry

Flame Propagation
Use For: Fire Propagation *See Also:* Burning Rate; Combustion; Explosions; Fire Testing; Flammability Testing

—Electrical Insulating Liquids
CENELEC PREN 61197-92. Insulating Liquids—Test Method for the Linear Flame Propagation of Insulating Liquids Using a Glass-Fibre Tape. 13 pp.
IEC 1197-93. Insulating Liquids—Linear Flame Propagation—Test Method Using a Glass-Fibre Tape First Edition. 24 pp.

—Generators—Gaseous Mines
CSA C22.2 NO 145-M1986. Motors and Generators for Use in Hazardous Locations (R 1992); (Gen Instr 1 Thru 2). 47 pp.

—Generators—Hazardous Environments
CSA C22.2 NO 145-M1986. Motors and Generators for Use in Hazardous Locations (R 1992); (Gen Instr 1 Thru 2). 47 pp.

—Motors—Gaseous Mines
CSA C22.2 NO 145-M1986. Motors and Generators for Use in Hazardous Locations (R 1992); (Gen Instr 1 Thru 2). 47 pp.

—Motors—Hazardous Environments
CSA C22.2 NO 145-M1986. Motors and Generators for Use in Hazardous Locations (R 1992); (Gen Instr 1 Thru 2). 47 pp.

Flame Resistance
Use: Flammability Testing

Flame Retardant Clothing
Use For: Fire Resistant Clothing *See Also:* Fire Fighting Clothing; Fire Retardants; Heat Resistant Clothing; Protective Clothing
BSI BS 1547-59. 1959 Flameproof Industrial Clothing (Materials and Design). 8 pp.
BSI BS 2653-55. 1955 Amd 3 Protective Clothing for Welders. 12 pp.
BSI BS EN 366-93. 1993 Protective Clothing—Protection Against Heat and Fire Method of Test: Evaluation of Materials and Material Assemblies When Exposed to a Source of Radiant Heat (N). 17 pp.
BSI BS EN 367-92. 1992 Protective Clothing—Protection Against Heat and Fire—Method of Determining Heat Transmission on Exposure to Flame (Supersedes BS 3791: 1970). 19 pp.
BSI BS EN 367-01. 1992 Amd 1 Protective Clothing—Protection Against Heat and Flames—Test Method: Determination of the Transmission on Exposure to Flame (AMD 7667) May 15, 1993 (N). 20 pp.
BSI BS EN 373-93. 1993 Protective Clothing—Assessment of Resistance of Materials to Molten Metal Splash (N). 15 pp.
CEN PREN 366-90. Protective Clothing: Protection Against Heat and Fire: Method of Test: Evaluation of Materials and Material Assemblies when Exposed to a Source of Radiant Heat. 22 pp.
CEN EN 366-93. Protective Clothing—Protection Against Heat and Fire—Method of Test: Evaluation of Materials and Material Assemblies when Exposed to a Source of Radiant Heat. 16 pp.
CEN PREN 367-90. Clothing for Protection Against Heat and Flames: Method of Determining Heat Transmission on Exposure to Flames. 17 pp.
CEN EN 367-92. Protective Clothing—Protection Against Heat and Fire—Method of Determining Heat Transmission on Exposure to Flame. 13 pp.
CEN PREN 373-90. Protective Clothing—Assessment of Resistance of Materials to Molten Metal Splash. 16 pp.
CEN EN 373-93. Protective Clothing—Assessment of Resistance of Materials to Molten Metal Splash. 14 pp.
CEN PREN 533-91. Clothing for Protection Against Heat and Flame—Performance Specification for Limited Flame Spread of Materials. 9 pp.

Flame Retardant Clothing (Cont.)

DIN ENGL 32761-83. Clothing for Protection Against Flame for Short Periods; Safety Requirements, Testing (Apr) (Supersedes DIN 4845, July 1963 Edition). 6 pp.
DIN ENGL EN 367-92. Protective Clothing for Use Against Heat and Fire; Method of Determining Heat Transmission on Exposure to Flame (Nov). 12 pp.
ISO 2801-73. Clothing for Protection Against Heat and Fire—General Recommendations for Users and for Those in Charge of Such Users First Edition. 6 pp.
ISO TR9240-92. Textiles—Design of Apparel for Reduced Fire Hazard First Edition. 12 pp.

—Flammability Testing

BSI BS 6249: Part 1-82. 1982 Materials and Material Assemblies Used in Clothing for Protection Against Heat and Flame Part 1: Flammability Testing and Performance. 7 pp.
ISO 6942-93. Clothing for Protection Against Heat and Fire—Evaluation of Thermal Behaviour of Materials and Material Assemblies When Exposed to a Source of Radiant Heat Second Edition. 17 pp.

—Mining Equipment

DIN ENGL 23320 Pt 1-88. Flameproof Clothing for the Mining Industry; Safety Requirements and Testing (May). 8 pp.
DIN ENGL 23320 Pt 3-88. Flameproof Clothing for the Mining Industry; Two-Piece Outfits (May). 2 pp.

—Underwear—Mining Equipment

DIN ENGL 23320 Pt 4-88. Flameproof Clothing for the Mining Industry; Underwear (May). 2 pp.

—Welding

BSI BS 2653-55. 1955 Amd 3 Protective Clothing for Welders. 12 pp.

Flame Retardant Testing
Use: Flammability Testing

Flame Retardants
Use: Fire Retardants

Flame Spectrophotometry
Use: Spectrophotometry

Flame Spraying
See Also: Coatings; Detonation Flame Spraying Services; Plasma Spraying

MOD UK DTD-941-69. Surface Coating of Parts by Use of Detonation, Flame and Plasma Spraying Processes. 2 pp.

—Ceramic Coatings

BSI BS 4495-69. 1969 Recommendations for the Flame Spraying of Ceramic and Cement Coatings. 12 pp.

—Cermets

BSI BS 4495-69. 1969 Recommendations for the Flame Spraying of Ceramic and Cement Coatings. 12 pp.

—Metal Coatings

MOD UK DSTAN 03-6: Part 1-72. Guide to Flame Spraying Processes Part 1: Unfused Metal Coatings Issue 1. 19 pp.
MOD UK DSTAN 03-6: Part 2-72. Guide to Flame Spraying Processes Part 2: Fused Metal Coatings Issue 1. 16 pp.

—Ships

MOD UK NES 764-88. Flame Metal Spraying and Hot Dip Galvanizing of Ships Structure and Fittings Issue 3 (01.88). 23 pp.

—Submarines

MOD UK NES 764-88. Flame Metal Spraying and Hot Dip Galvanizing of Ships Structure and Fittings Issue 3 (01.88). 23 pp.

Flame Spraying Equipment
Use For: Flame Coating Equipment
See Also: Thermal Spraying; Thermal Spraying Equipment

—Filler Wire

DIN ENGL 8566 Pt 1-79. Metals for Thermal Spraying; Solid Wires for Flame Spraying (Mar). 9 pp.

Flame Supervision Devices
Use: Flame Detectors

Flame Testing
Use: Fire Testing

Flameproof Enclosures
Use: Explosive Atmospheres—Electrical Equipment

Flameproof Testing
Use: Fire Testing

Flammability Testing
Use For: Fire Retardant Testing; Flame Resistance; Ignitability Testing; Inflammability
See Also: Combustion; Fire Testing; Flame Propagation; Flammable Liquids; Flash Point; Materials Testing; Punking Behavior

CNS Z3006-72. Method of Test for Non-Combustibility of Elementary Material (Jun)(3406).
DIN ENGL 50051-77. Testing of Materials; Burning Behaviour of Materials; Burner (Feb). 3 pp.
DIN ENGL 53438 Pt 1-84. Testing of Combustible Materials; Response to Ignition by a Small Flame; General Data (June). 5 pp.
DIN ENGL 53438 Pt 2-84. Testing of Combustible Materials; Response to Ignition by a Small Flame; Edge Ignition (June). 3 pp.
DIN ENGL 53438 Pt 3-84. Testing of Combustible Materials; Response to Ignition by a Small Flame; Surface Ignition (June). 3 pp.
JIS Z 2150-66. Method of Flame Test for Thin Materials (45 Degree Meker Burner Method). 6 pp.

—Adhesive Tapes

SAA AS 1635.19.1-74. Methods of Testing Pressure Sensitive Adhesive Tape—Part 19.1: Flammability. 1 p.

—Aerosols

CNS Z6038-81. Test Methods for Inflammability of Aerosol Products (Mar) (7180). 4 pp.

—Aerospace

CEN PREN 2825-89. Burning Behaviour, Determination of Smoke Density and Gas Components in the Smoke of Materials Under the Influence of Radiating Heat and Flames—Determination of Smoke Density. 10 pp.
CEN PREN 2826-89. Burning Behaviour, Determination of Smoke Density and Gas Components in the Smoke of Materials Under the Influence of Radiating Heat and Flames—Determination of Gas Components in the Smoke. 9 pp.

—Aerospace—Testing Equipment

CEN PREN 2824-89. Burning Behaviour, Determination of Smoke Density and Gas Components in the Smoke of Materials Under the Influence of Radiating Heat and Flames Influence of Radiating Heat and Flames—Test Equipment, Apparatus and Media. 18 pp.

—Air Filters

BSI BS 7525-92. 1992 Flammability of Air Cleaner Elements for Internal Combustion Engines. 13 pp.

—Aircraft Cabins

CAA NOTICE #61 ISSUE 2. Improved Flammability Test Standards for Cabin Interior Materials (Airworthiness Notices). 4 pp.

—Automotive—Interior Trim Materials

BSI BS AU 169-78. (WITHDRAWN) 1978 Determination of Burning Behaviour of Interior Materials for Motor Vehicles (Superseded by BS AU 169a: 1992). 8 pp.
BSI BS AU 169A-92. 1992 Method for Determination of Burning Behaviour of Interior Materials for Road Vehicles, and Tractors and Machinery for Agriculture and Forestry (ISO 3795: 1989). 11 pp.
ISO 3795-89. Road Vehicles, and Tractors and Machinery for Agriculture and Forestry—Determination of Burning Behaviour of Interior Materials Second Edition. 10 pp.
JIS D 1201-77. Test Method for Flammability of Organic Interior Materials for Automobiles (R 1983). 38 pp.

—Bed Bases

BSI BS 6807-90. 1990 Methods of Test for Assessment of the Ignitability of Mattresses, Divans and Bed Bases with Primary and Secondary Sources of Ignition. 22 pp.
BSI BS 6807-01. 1990 Amd 1 Methods of Test for Assessment of the Ignitability of Mattresses, Divans and Bed Bases with Primary and Secondary Sources of Ignition (AMD 7551) January 15, 1993. 23 pp.
CEN PREN 597-1-91. Furniture—Assessment of the Ignitability of Mattresses and Bed Bases—Part 1: Ignition Source: Smouldering Cigarette. 15 pp.
CEN PREN 597-2-91. Furniture—Assessment of the Ignitability of Mattresses and Bed Bases—Part 2: Ignition Source: Match Flame Equivalent. 16 pp.

—Bedding

BSI BS 5815: Part 3-91. 1991 Sheets, Sheeting, Pillowslips, Towels, Napkins, Counterpanes and Continental Quilt Secondary Covers Suitable for Use in the Public Sector Part 3: Counterpanes and Continental Quilt Sec. Covers Including Flammability Performance. 12 pp.
BSI BS 7175-89. 1989 Methods of Test for the Ignitability of Bedcovers and Pillows by Smouldering and Flaming Ignition Sources. 20 pp.

—Beds

SNZ NZS 8720-92. Methods of Test for Assessment of the Ignitability of Mattresses, Divans and Bed Bases with Primary and Secondary Sources of Ignition Amend: A, 1992. 24 pp.

—Blankets

BSI BS 5866: Part 4-91. 1991 Blankets Suitable for Use in the Public Sector Part 4: Specification for Flammability Performance (L). 8 pp.
BSI BS 5866: Part 4-01. 1991 Amd 1 Blankets Suitable for Use in the Public Sector Part 4: Specification for Flammability Performance (AMD 7065) May 1, 1992. 9 pp.

—Cable Insulation—Communication Cables

JIS C 3521-86. Flame Test Method for Flame Retardant Sheath of Telecommunication Cables. 8 pp.

—Cables (Electric)—Aircraft

CEN PREN 3475 (Part 407)-92. Cables, Electrical, Aircraft Use Test Methods Part 407—Flammability. 3 pp.

—Cables (Electric)—Identification Systems

CSA CAN/CSA-C22. 2 NO 49-92. Flexible Cords and Cables; (Gen Instr 1). 121 pp.

—Cables (Electric)—Marine

MOD UK NES 641-91. Determination of the Vertical Flammability of Electric Cables Issue 2 (07.91). 13 pp.

—Carpets

CGSB CAN/CGSB-4.155-M88. Flammability of Soft Floor Coverings—Sampling Plans. 14 pp.

—Cellular Materials

BSI BS 5111: Part 1-74. 1974 Laboratory Methods of Test for Determination of Smoke Generation Characteristics of Cellular Plastics and Cellular Rubber Materials Part 1: Meth of Testing a 25 mm Cube Test Spec. of Low Density Mat. (up to 130 kg/Metres Cubed) to. 13 pp.
BSI BS 5111: Part 1-01. 1974 Amd 1 Laboratory Methods of Test for Determination of Smoke Generation Characteristics of Cellular Plastics and Cellular Rubber Materials Part 1: Method for Testing a 25 mm Cube Test Specimen of Low Density Material (up to 130 kg/m3) to. 14 pp.
ISO 3582-78. Cellular Plastic and Cellular Rubber Materials—Laboratory Assessment of Horizontal Burning Characteristics of Small Specimens Subjected to a Small Flame First Edition. 9 pp.

—Clothing

CNS L3197-83. Method of Test for Flammability of Clothes — Test for Burning Speed (May)(10286).
CNS L3198-83. Method of Test for Flammability of Clothes — Number Test of Ignition (May)(10287).

—Coated Fabrics

SAA AS 1441.13-73. Methods of Test for Coated Fabrics—Part 13: Method for Determination of Flammability.

—Color Filters

BSI BS 3944: Part 1-92. 1992 Colour and Diffusion Filter Material for Theatre, Television and Similar Entertainment Purposes Part 1: Specification for Flammability and Dimensional Stability. 8 pp.

—Connectors

CEN PREN 2591 (Part C17)-92. Elements of Electrical and Optical Connection Test Methods Part C17—Flammability. 6 pp.
CEN PREN 2591-FC17-93. Aerospace Series Elements of Electrical and Optical Connection Test Methods Part FC17—Optical Elements Flammability. 3 pp.

—Construction Materials

BSI BS 476: Part 4-70. 1970 Amd 2 Fire Tests on Building Materials and Structures Part 4: Non-Combustibility Test for Materials (AMD 4390) September 30, 1983. 13 pp.
BSI BS 476: Part 5-79. (WITHDRAWN) 1979 Amd 1 Fire Tests on Building Materials and Structures Part 5: Method of Test for Ignitability (Superseded by BS 476: Part 12 1991). 8 pp.
SAA AS 1530.1-84. Methods for Fire Tests on Building Materials, Components and Structures—Part 1: Combustibility Test for Materials Amdt 1 July 1993. 15 pp.
SAA AS 1530.2-93. Methods for Fire Tests on Building Materials, Components and Structures—Part 2: Test for Flammability of Materials (in Professional Package 63). 8 pp.

INTERNATIONAL AND NON-U.S. NATIONAL STANDARDS
SUBJECT INDEX
Flammability

Flammability Testing *(Cont.)*
—Construction Materials *(Cont.)*
SAA AS 1530.3-89. Methods for Fire Tests on Building Materials, Components and Struc-tures—Part 3: Simul-taneous Determination of Ignitability, Flame Propagation, Heat Release and Smoke Release Amdt 1 April 1992 (in Professional Packages 20, 21, 30, 44, 55, 62-69). 1 p.
SAA AS 1530.5-89. Methods for Fire Tests on Building Materials, Components and Structures—Part 5: Test for Piloted Ignitability. 29 pp.
SNZ NZS/BS 476: Part 4-70. Fire Tests on Building Materials and Structures Part 4: Non-Combustibility Test for Materials. 12 pp.
SNZ NZS/AS 1530. 1-84. Methods for Fire Tests on Building Materials, Components and Structures Part 1: Combustibility Test for Materials (This is a Joint Standard with SAA AS 1530.1). 20 pp.
SNZ NZS/AS 1530. 2-73. Methods for Fire Tests on Building Materials, Components and Structures Part 2: Test for Flammability of Materials (This is a Joint Standard with SAA AS 1530.2). 8 pp.
SNZ NZS/AS 1530. 3-89. Methods for Fire Tests on Building Materials, Components and Structures Part 3: Simultaneous Determination of Ignitability, Flame Propagation, Heat Release and Smoke Release (This is a Joint Standard with SAA AS 1530.3). 13 pp.

—Conveyor Belts
BSI BS 490: Part 3-91. 1991 Conveyor and Elevator Belting Part 3: Specification for Flammability and Anti-Static Properties of Rubber and of Plastics Conveyor Belting of Textile Construction for General Use. 7 pp.
BSI BS 490: Sec 11.1-90. (WITHDRAWN) 1990 Conveyor and Elevator Belting Part 11: Methods of Test for Safety Section 11.1: Laboratory Flame Tests (Superseded by BS EN 20340: 1973). 12 pp.
BSI BS EN 20340-93. 1993 Conveyor Belts—Flame Retardation—Specifications and Test Method (ISO 340: 1988) (Q). 15 pp.
CEN EN 20340-93. Conveyor Belts—Flame Retardation—Specifications and Test Method (ISO 340: 1988). 9 pp.
CNS K4078-83. Flame-Resistant Rubber Belts for Conveyors (Feb)(10026). 2 pp.
ISO 340-88. Conveyor Belts—Flame Retardation—Specifications and Test Method Second Edition; (CEN EN 20340: 1993). 10 pp.
JIS K 6324-77. Qualitative Standard for Flame Resistance of Conveyor Belts (R 1986). 6 pp.
SAA AS 1334.10-82. Methods of Testing Conveyor and Elevator Belting—Part 10: Determination of Ignitability and Flame Propagation Characteristics of Conveyor Belting. 3 pp.
SAA AS 1334.11-88. Methods of Testing Conveyor and Elevator Belting—Part 11: Determination of Ignitability and Maximum Surface Temperature of Belting Subjected to Friction Amdt 1 July 1989. 4 pp.
SAA AS 1334.12-86. Methods of Testing Conveyor and Elevator Belting—Part 12: Determination of Combustion Propagation Characteristics of Conveyor Belting Amdt 1 June 1988 Amdt 2 May 1993. 6 pp.
SNZ NZS/BS 490: Part 3-91. Conveyor and Elevator Belting Part 3: Specification for Flammability and Anti-Static Properties of Rubber and of Plastics Conveyor Belting of Textile Construction for General Use. 8 pp.
SNZ NZS/BS 490: Pt11:Sec11.1-90. Conveyor and Elevator Belting Part 11: Method of Test for Safety Section 11.1: Laboratory Flame Tests. 12 pp.

—Cords (Electric)—Identification Systems
CSA CAN/CSA-C22. 2 NO 49-92. Flexible Cords and Cables; (Gen Instr 1). 121 pp.

—Deck Coverings—Ships
MOD UK NES 750-81. Fire Tests for Deck Coverings with Under Deck Heat Source Issue 2 (04.81) (Withdrawn). 12 pp.

—Divans
BSI BS 6807-90. 1990 Methods of Test for Assessment of the Ignitability of Mattresses, Divans and Bed Bases with Primary and Secondary Sources of Ignition. 22 pp.
BSI BS 6807-01. 1990 Amd 1 Methods of Test for Assessment of the Ignitability of Mattresses, Divans and Bed Bases with Primary and Secondary Sources of Ignition (AMD 7551) January 15, 1993. 23 pp.
SNZ NZS 8720-92. Methods of Test for Assessment of the Ignitability of Mattresses, Divans and Bed Bases with Primary and Secondary Sources of Ignition Amend: A, 1992. 24 pp.

—Electric Blankets
CSA C22.2 NO 101-M1984. Electrically Heated Bedding Appliances for Household Use (R 1992); (Gen Instr 1 Thru 4). 61 pp.
CSA 1169 Bull. Electrical Bulletin 1169 June 27, 1978 to C22.2 NO 101. 2 pp.

Flammability Testing *(Cont.)*
—Electrical Components
BSI BS 2011: Part 2.1PZ-70. 1970 Amd 1 Basic Environmental Testing Procedures Part 2.1: Tests Part 2.1PZ: Flammability. 5 pp.

—Electrical Insulation
BSI BS 6334-83. 1983 Determination of the Flammability of Solid Electrical Insulating Materials When Exposed to an Igniting Source. 12 pp.
BSI BS 7447-91. 1991 Method of Test for Determination of the Ignitability of Soild Electrical Insulating Materials When Exposed to an Electrically Heated Wire Source. 10 pp.
CENELEC HD 441-83. Methods of Test for the Determination of the Flammability of Solid Electrical Insulating Materials When Exposed to an Igniting Source. 2 pp.
CENELEC HD 541 S1-91. Methods of Test for the Determination of the Ignitability of Solid Electrical Insulating Materials When Exposed to Electrically Heated Wire Sources. 7 pp.
DIN VDE 0304 Pt 3-85. Thermal Properties of Electrical Insulating Materials; Flammability Under the Action of Igniting Sources (National Foreword Only Corresponds to IEC 707—1981) (Sept). 6 pp.
IEC 707-81. Methods of Test for the Determination of the Flammability of Solid Electrical Insulating Materials When Exposed to an Igniting Source First Edition; (Amendment 1-1992). 33 pp.
IEC 829-88. Methods of Test for the Determination of the Ignitability of Solid Electrical Insulating Materials When Exposed to Electrically Heated Wire Sources First Edition. 26 pp.

—Electrical Insulation—Heat Shrinkable Tubing
CNS C3146-88. Method of Test for Combustion Resistance of Heat Shrinkable Tubing for Electrical Insulation (Sep)(8793).

—Electromechanical Components
BSI BS 5772: Part 9-93. 1993 Electromechanical Components for Electronic Equipment: Basic Testing Procedures and Measuring Methods Part 9: Miscellaneous Tests (IEC 512-9: 1992) (S). 20 pp.
BSI BS 5772: Part 9-79. 1979 Basic Testing Procedures and Measuring Methods for Electromechanical Components for Electronic Equipment Part 9: Cable Clamping Tests, Explosion Hazard Tests, r.f Resistance Test, Capacitance Tests Sheilding and Filtering. 12 pp.
IEC 512 Pt 9-92. Electromechanical Components for Electronic Equipment; Basic Testing Procedures and Measuring Methods Part 9: Miscellaneous Tests Second Edition. 42 pp.

—Electronic Components
CNS C6126-81. Flammability Tests for Electronic Components (Jul)(7650).

—Emergency Lighting
CSA C22.2 NO 141-M1985. Unit Equipment for Emergency Lighting (R 1992); (Gen Instr 1). 31 pp.
CSA 902B Bull. Electrical Bulletin 902B September 27, 1978 to C22.2 NO 141. 2 pp.

—Fabrics
BSI BS 4569-83. 1983 Ignitability (Surface Flash) of Pile Fabrics and Assemblies Having Pile on the Surface. 7 pp.
BSI BS 5438-89. 1989 Amd 1 Methods of Test for Flammability of Textile Fabrics When Subjected to a Small Igniting Flame Applied to the Face or Bottom Edge of Vertically Oriented Specimens (AMD 6509) June 29, 1990. 25 pp.
BSI BS 5438-76. 1976 Amd 1 Methods of Test for Flammability of Vertically Oriented Textile Fabrics and Fabric Assemblies Subjected to a Small Igniting Flame. 19 pp.
BSI BS 5651-89. 1989 Cleansing and Wetting Procedures for Use in the Assessment of the Effect of Cleansing and Wetting on the Flammability of Textile Fabrics and Fabric Assemblies. 12 pp.
BSI BS 5722-91. 1991 Flammability Performance of Fabrics and Fabric Combinations Used in Nightwear Garments. 12 pp.
BSI BS 5867: Part 2-80. 1980 Amd 1 Fabrics for Curtains and Drapes Part 2: Flammability Requirements. 8 pp.
BSI BS 7157-89. 1989 Method of Test for Ignitability of Fabrics Used in the Construction of Large Tented Structures. 8 pp.
BSI PD 2777-77. 1977 Fabric Flammability Burning Accidents and the Relevance of BS 5438. 14 pp.
CEN PREN 532-91. Clothing for Protection Against Heat and Flame—Method of Test for Limited Flame Spread. 14 pp.
CEN PREN 1101-93. Textiles and Textile Products—Burning Behaviour—Curtains and Drapes—Detailed Procedure to Determine the Ignitability of Vertically Oriented Specimens (Small Flame). 4 pp.

Flammability Testing *(Cont.)*
—Fabrics *(Cont.)*
CEN PREN 1102-93. Textiles and Textile Products—Burning Behaviour—Curtains and Drapes—Detailed Procedure to Determine the Flame Spread of Vertically Oriented Specimens. 6 pp.
CGSB CAN/CGSB-4.2 NO.27-M88. Textile Test Methods Burning Behaviour—Selection of Methods. 9 pp.
CGSB CAN/CGSB-4.2 NO.27.1-M87. Textile Test Methods Flame Resistance—Vertical Burning Test. 11 pp.
CGSB CAN/CGSB-4.2 NO.27.2-M87. Textile Test Methods Flame Resistance—Surface Burning Test. 10 pp.
CGSB CAN/CGSB-4.2 NO.27.3-M86. Textile Test Methods Textile Fabrics—Burning Behaviour—Measurement of Flame Spread Properties of Vertically Oriented Specimens (ISO 6941-1984). 27 pp.
CGSB CAN/CGSB-4.2 NO.27.4-M86. Textile Test Methods Textile Fabrics—Burning Behaviour—Determination of Ease of Ignition of Vertically Oriented Specimens (ISO 6940-1984). 29 pp.
CGSB CAN/CGSB-4.2 NO.27.5-M87. Textile Test Methods Flame-Resistance—45 Degree Angle Test—One Second Flame Impingement. 12 pp.
CNS L3196-83. Method of Test for Flammability of Clothes — Test for the Degree of Charring Extent (May)(10285).
CNS S2148-90. Method of Test for Tent—Flammability of Wall and Top Materials (Mar)(12693). 4 pp.
DIN ENGL 54336-86. Testing of Textiles; Determination of Burning Behaviour of Vertically Oriented Specimens with Ignition at Lower Edge (Nov). 6 pp.
ISO 6940-84. Textile Fabrics—Burning Behaviour—Determination of Ease of Ignition of Vertically Oriented Specimens First Edition; (Amendment 1-1993). 16 pp.
ISO 6941-84. Textile Fabrics—Burning Behaviour—Measurement of Flame Spread Properties of Vertically Oriented Specimens First Edition; (Amendment 1-1992). 16 pp.
ISO 10047-93. Textiles—Determination of Surface Burning Time of Fabrics First Edition. 10 pp.
JIS L 1091-92. Testing Methods for Flammability of Textiles. 23 pp.
SAA AS 1176.1-82. Textiles—Methods of Test for Combustion Properties—Part 1: The Determination of Ease of Ignition of Certain Textile Materials in a Horizontal Plane. 7 pp.
SAA AS 1176.1D-72. Textiles—Methods of Test for Combustion Properties—Part 1D: Drawings of Apparatus for Australian Standard Method for the Determination of Ease of Ignition of Textiles Corrig. Reconfirmed 1986.
SAA AS 1176.2D-72. Textiles—Methods of Test for Combustion Properties—Part 2D: Drawings of Apparatus for Australian Standard Method for the Determination of Rate of Burning and Heat Output of Textiles from Which Clothing May Be Made Reconfirmed 1986.
SAA AS 1176.2-82. Textiles—Methods of Test for Combustion Properties—Part 2: The Determination of Burning Time of Textile Materials. 7 pp.
SAA AS 1176.3-82. Textiles—Methods of Test for Combustion Properties—Part 3: Determination of Surface Burning Properties of Certain Textile Materials. 10 pp.
SNZ NZS 8703-77. Methods of Test for Combustion Characteristics of Textile Materials Amend: A, 1977. 24 pp.

—Fabrics—Gas Cutting/Welding
JIS A 1323-84. Flame Retardant Testing Method for Spark Droplets of Welding and Gas Cutting on Fabric Sheets in Construction Works. 9 pp.

—Fabrics—Glossaries
BSI BS 6373-85. 1985 Glossary of Terms Relating to Burning Behaviour of Textiles and Textile Products. 11 pp.
ISO 4880-84. Burning Behaviour of Textiles and Textile Products—Vocabulary First Edition; (Amendment 1-1992) (Amendment 2-1993). 21 pp.

—Fiber Optic Cables
CNS C6285-87. Method of Test for Fiber Optic Devices (FOTP-99 Gas Flame Test for Special Purpose Fiber Optic Cable) (Apr)(11907).

—Fiber Optic Connectors
CNS C6333-89. Method of Test for Fiber Optic Devices (FOTP-172 Flame Resistance of Firewall Connector) (Jul)(12564).

—Floor Coverings
BSI BS 4790-87. 1987 Amd 1 Determination of the Effects of a Small Source of Ignition on Textile Floor Coverings (Hot Metal Nut Method) (AMD 6444) May 31, 1990. 9 pp.

INDUSTRY STANDARDS
1105

INTERNATIONAL AND NON-U.S. NATIONAL STANDARDS SUBJECT INDEX

Flammability

Flammability Testing (Cont.)

—Floor Coverings (Cont.)

BSI BS 4790-02. 1987 Amd 2 Method for Determination of the Effects of a Small Source of Ignition on Textile Floor Coverings (Hot Metal Nut Method) (AMD 7376) January 15, 1993. 10 pp.

BSI BS 6307-82. 1982 Determination of the Effects of a Small Source of Ignition on Textile Floor Coverings (Methenamine Tablet Test). 7 pp.

CGSB CAN/CGSB-4.2 NO.27.6-M91. Textile Test Methods Flame Resistance—Methenamine Tablet Test for Textile Floor Coverings. 8 pp.

CGSB CAN/CGSB-4.2 NO.30.2-M90. Textile Test Methods Procedure for the Removal of Non-Permanent Flame-Retardant Treatments on Textile Floor Coverings. 8 pp.

CNS L3107-81. Flammability of Textile Floor Coverings Metal Nut Method (Mar) (7058).

CNS L3121-81. Method of Test for Flammability of Floor Coverings (Jun)(7496).

DIN ENGL 51960-75. Testing of Organic Floor Coverings (Excluding Textile Floor Coverings); Testing of Flammability (Aug). 1 p.

—Floor Tiles

CNS K6312-77. Method of Test for Polyvinyl Chloride Asbestos Tiles for Flooring (Dec)(3309). 4 pp.

—Furniture—Upholstered

EC SM 3. Fire Behaviour of Upholstered Furniture Etc.. 1 p.

SNZ NZS/AS 4088-93. Burning Behaviour of Upholstered Furniture (This is a Joint Standard with SAA AS 4088). 10 pp.

—Gases

CEN PREN 720-2-92. Classification of Gases and Gas Mixtures—Part 2: Gases and Gas Mixtures—Determination of Fire Potential and Oxidizing Ability. 26 pp.

—Heat Resistant Clothing

ISO 6942-93. Clothing for Protection Against Heat and Fire—Evaluation of Thermal Behaviour of Materials and Material Assemblies When Exposed to a Source of Radiant Heat Second Edition. 17 pp.

—Hoses

BSI BS 5173: Sec 103.6-90. 1990 Rubber and Plastics Hoses and Hose Assemblies Part 103: Physical Tests Section 103.6: Determination of Ignitability of Lining. 6 pp.

SAA AS 1180.10A-74. Methods of Test for Hose Made from Elastomeric Materials—Part 10A: Resistance of Hose Lining or Cover to Flame Reconfirmed 1988. 4 pp.

SAA AS 1180.10B-82. Methods of Test for Hose Made from Elastomeric Materials—Part 10B: Determination of Combustion Propagation Characteristics of a Horizontally Oriented Specimen of Hose Using Surface Ignition Reconfirmed 1988. 4 pp.

SAA AS 1180.12-74. Methods of Test for Hose Made from Elastomeric Materials—Part 12: Resistance to Ignition Reconfirmed 1989. 3 pp.

—Hoses—Coal Mining

BSI BS 5173: Sec 103.4-88. (WITHDRAWN) 1988 Rubber and Plastics Hoses and Hose Assemblies Part 103: Physical Test Section 103.4: Determination of Flammability of Mining Hoses. 6 pp.

BSI BS EN 28030-01. 1993 AMD 1 Rubber and Plastics Hoses for Underground Mining-Method of Test for Flammability (ISO 8030: 1987) (AMD 7522). 13 pp.

CEN EN 28030-93. Rubber and Plastics Hoses for Underground Mining—Methods of Test for Flammability (ISO 8030: 1987). 7 pp.

ISO 8030-87. Rubber and Plastics Hoses for Underground Mining—Method of Test for Flammability First Edition; (CEN EN 28030: 1993). 6 pp.

—Hydraulic Fluids

BSI DD 61-79. 1979 Flammability Spray Test for Hydraulic Fluids. 13 pp.

—Impregnated Papers

CNS P3063-84. Method of Test for Flame Resistance of Treated Paper and Paperboard (Jan)(10760). 4 pp.

—Lighting—Hazardous Locations

CSA C22.2 NO 137-M1981. Electric Luminaires for Use in Hazardous Locations (R 1993); (Gen Instr 1 Thru 3). 26 pp.

—Mattresses

BSI BS 6807-90. 1990 Methods of Test for Assessment of the Ignitability of Mattresses, Divans and Bed Bases with Primary and Secondary Sources of Ignition. 22 pp.

Flammability Testing (Cont.)

—Mattresses (Cont.)

BSI BS 6807-01. 1990 Amd 1 Methods of Test for Assessment of the Ignitability of Mattresses, Divans and Bed Bases with Primary and Secondary Sources of Ignition (AMD 7551) January 15, 1993. 23 pp.

BSI BS 7177-91. 1991 Resistance to Ignition of Mattresses, Divans and Bed Bases. 10 pp.

BSI BS 7177-01. 1991 Amd 1 Resistance to Ignition of Mattresses, Divans and Bed Bases (AMD 7345) October 15, 1992. 11 pp.

CEN PREN 597-1-91. Furniture—Assessment of the Ignitability of Mattresses and Bed Bases—Part 1: Ignition Source: Smouldering Cigarette. 15 pp.

CEN PREN 597-2-91. Furniture—Assessment of the Ignitability of Mattresses and Bed Bases—Part 2: Ignition Source: Match Flame Equivalent. 16 pp.

CGSB CAN/CGSB-4.2 NO.27.7-M89. Textile Test Methods Combustion Resistance of Mattresses-Cigarette Test. 13 pp.

SNZ NZS 8720-92. Methods of Test for Assessment of the Ignitability of Mattresses, Divans and Bed Bases with Primary and Secondary Sources of Ignition Amend: A, 1992. 24 pp.

—Medical Equipment—Fabrics

CGSB CAN/CGSB-4.162-M80. Hospital Textiles—Flammability Performance Requirements. 34 pp.

—Motion Picture Film

BSI BS 5550: SUBSEC 5.1.1-78. 1978 Cinematography Part 5: Common to More Than One Gauge Section 5.1: Raw Stock Subsection 5.1.1: Definition, Testing and Marking of Testing and Marking of Motion-Picture Safety Film. 11 pp.

ISO 543-90. Photography—Photographic Films—Specifications for Safety Film Second Edition; (ANSI IT9.6-1991). 9 pp.

—Nonmetallic Materials

AECMA PREN 2310-88. Test Methods for the Flame Resistance Rating of Non-Metallic Materials. 17 pp.

BSI BS EN 2310-91. 1991 Test Methods for the Flame Resistance Rating of Non-Metallic Materials. 22 pp.

CEN PREN2310-90. Test Methods for the Flame Resistance Rating of Non-Metallic Materials. 17 pp.

CEN EN 2310-91. Test Methods for the Flame Resistance Rating of Non-Metallic Materials. 18 pp.

SBAC TS 80 ISSUE 1. Recommendations for the Control of Flammability Testing of Non Metallic Materials and Textiles for Aeronautical Purposes.

—Paint Removers

CGSB 31-GP-0A METH 1.3-62. Methods of Testing Corrosion-Prevention Materials and Processes Flammability. 1 p.

—Paints

BSI BS 3900: Part A11-74. 1974 Methods of Test for Paints Group A: Tests on Liquid Paints (Excluding Chemical Tests) Part A11: Small Scale Test for Combustibility. 6 pp.

ISO TR9038-91. Paints and Varnishes—Determination of the Ability of Liquid Paints to Sustain Combustion First Edition. 9 pp.

—Paperboard

CNS P3063-84. Method of Test for Flame Resistance of Treated Paper and Paperboard (Jan)(10760). 4 pp.

—Plastic Sheets

CNS K6287-87. Method of Test for Polyvinyl Chloride Sheet (Jul)(3143). 7 pp.

—Plastics

BSI BS 2782:Pt1: METH 140A-92. 1992 Methods of Testing Plastics Part 1: Thermal Properties Method 140A: Determination of the Burning Behaviour of Horizontal and Vertical Specimens in Contact with a Small-Flame Ignition Source (ISO 1210: 1992). 15 pp.

BSI BS 2782:Pt1: METH 140B-93. 1993 Methods of Testing Plastics Part 1: Thermal Properties Method 140B: Determination of the Burning Behaviour of Flexible Vertical Specimens in Contact with a Small-Flame Ignition Source (ISO 9773: 1990) (V). 13 pp.

BSI BS 2782:Pt1: METH 140C-93. 1993 Methods of Testing Plastics Part 1: Thermal Properties Method 140C: Determination of the Combustibility of Specimens Using a 125 mm Flame Source (ISO 10351: 1992) (V). 16 pp.

BSI BS 2782:Pt1: METH 140D-80. 1980 Amd 1 Methods of Testing Plastics Part 1: Thermal Properties Method 140D: Flammability of a Test Piece 550 mm x 35 mm of Thin Polyvinyl Chloride Sheeting (Laboratory Method). 7 pp.

BSI BS 2782:Pt1: METH 140E-82. 1982 Amd 1 Methods of Testing Plastics Part 1: Thermal Properties Method 140E: Flammability of a Small, Inclined Test Piece Exposed to an Ethanol Flame (Laboratory Method). 5 pp.

Flammability Testing (Cont.)

—Plastics (Cont.)

BSI BS 2782:Pt1: METH 141-86. 1986 Methods of Testing Plastics Part 1: Thermal Properties Method 141: Determination of Flammability by Oxygen Index. 20 pp.

BSI BS 2782:Pt1: METH 143A&B-89. 1989 Methods of Testing Plastics Part 1: Thermal Properties Method 143A & B: Determination of Flammability Temperature of Materials. 14 pp.

BSI PD 6520-88. 1988 Guide to Fire test Methods for Building Materials and Elements of Construction. 14 pp.

ISO 181-81. Plastics—Determination of Flammability Characteristics of Rigid Plastics in the Form of Small Specimens in Contact with an Incandescent Rod First Edition. 6 pp.

ISO 1210-92. Plastics—Determination of the Burning Behaviour of Horizontal and Vertical Specimens in Contact with a Small-Flame Ignition Source Second Edition. 13 pp.

ISO R1326-70. Plastics Determination of Burning Rate of Plastics in the Form of Film First Edition. 6 pp.

ISO 4589-84. Plastics—Determination of Flammability by Oxygen Index First Edition. 17 pp.

ISO 9773-90. Plastics—Determination of Burning Behavior of Flexible Vertical Specimens in Contact with a Small-Flame Ignition Source First Edition. 11 pp.

ISO 10351-92. Plastics—Determination of the Combustibility of Specimens Using a 125 mm Flame Source First Edition. 14 pp.

—Plywood

CNS O2050-82. Method of Test for Flameproof of Flameproof Plywood (Apr)(8737). 2 pp.

—Polyester Resins

CNS K3062-82. Liquid Unsaturated Polyester Resin for Reinforced Plastics (Dec)(9715). 15 pp.

—Polymers—Electrical Equipment

CSA C22.2 NO 0.6-M1982. Flammability Testing of Polymeric Materials; (Gen Instr 1 Thru 4). 32 pp.

—Polymers—Oxygen Index Method

JIS K 7201-76. Testing Method for Flammability of Polymeric Materials Using the Oxygen Index Method (R 1979). 14 pp.

—Protective Clothing

BSI BS 6249: Part 1-82. 1982 Materials and Material Assemblies Used in Clothing for Protection Against Heat and Flame Part 1: Flammability Testing and Performance. 7 pp.

—Safety Film

BSI BS 6619-85. 1985 Flammability Performance of Safety Film for Still Photography. 7 pp.

—Steel Coatings

CGSB 1-GP-71 METH 118.5-79. Methods of Testing Paints and Pigments Flame Resistance Baked Films. 1 p.

CGSB 1-GP-71 METH 118.7-83. Methods of Testing Paints and Pigments Flame Resistance Two Foot Tunnel Method. 2 pp.

—Tent Floors

CNS S2147-90. Method of Test for Tent—Flammability of Flooring Material (Mar)(12692). 2 pp.

—Testing Equipment

DIN ENGL 50050 Pt 1-86. Testing of Materials; Burning Behaviour of Materials; Small Burning Cabinet (Apr) (Supersedes February 1977 Edition of DIN 50050). 3 pp.

DIN ENGL 50050 Pt 2-88. Testing the Burning Behaviour of Materials; Large Burning Cabinet (Jan). 3 pp.

—Thermal Insulation

BSI BS 5803: Part 4-85. 1985 Thermal Insulation for Pitched Roof Spaces in Dwellings Part 4: Methods for Determining Flammability and Resistance to Smouldering. 12 pp.

—Toys

BSI BS 5665: Part 2-89. 1989 Safety of Toys Part 2: Flammability Requirements (Supersedes BS 3443: 1968). 8 pp.

CEN EN 71-2-88. Safety of Toys—Part 2: Flammability. 6 pp.

CEN PREN 71-2-92. Safety of Toys—Part 2: Flammability. 10 pp.

CNS Z7066-86. Toy Safety (General Requirements) (Jun)(4797). 5 pp.

CNS Z7066-1-86. Toy Safety (Flammability Requirements) (Jun)(4797-1). 2 pp.

CNS Z8016-1-86. Method of Test for Toy Safety (Testing for Flammability of Textile Materials) (Jun)(4798-1). 3 pp.

INTERNATIONAL AND NON-U.S. NATIONAL STANDARDS
SUBJECT INDEX
Flammable

Flammability Testing (Cont.)
—Toys (Cont.)
  CNS Z8016-2-86. Method of Test for Toy Safety (Testing for Flammability of Rigid and Pliable Solids) (Jun)(4798-2). 3 pp.
  CNS Z8016-3-86. Method of Test for Toy Safety (Testing for Flammability of Self-Pressurized Container) (Jun)(4798-3). 2 pp.
  SAA AS 1647.4-80. Children's Toys (Safety Requirements)—Part 4: Flammability Requirements. 8 pp.

—Upholstery
  BSI BS 5852-90. 1990 Methods of Test for Assessment of the Ignitability of Upholstered Seating by Smouldering and Flaming Ignition Sources. 43 pp.
  BSI BS 5852-01. 1990 Amd 1 Methods of Test for Assessment of the Ignitability of Upholstered Seating by Smouldering and Flaming Ignition Sources (AMD 7349) October 15, 1992. 45 pp.
  BSI BS 5852: Part 1-79. 1979 Fire Tests for Furniture Part 1: Methods of Test for the Ignitability by Smokers' Materials of Upholstered Composites for Seating. 10 pp.
  BSI BS 5852: Part 2-82. 1982 Fire Tests for Furniture Part 2: Methods of Test for the Ignitability of Upholstered Composites for Seating by Flaming Sources. 18 pp.
  BSI BS 7176-91. 1991 Resistance to Ignition of Upholstered Furniture for Non-Domestic Seatings by Testing Composites (L). 13 pp.
  BSI BS 7176-89. 1989 Resistance to Ignition of Upholstered Furniture. 8 pp.
  ISO 8191 Pt 1-87. Furniture—Assessment of the Ignitability of Upholstered Furniture—Part 1: Ignition Source: Smouldering Cigarette First Edition. 12 pp.
  ISO 8191 Pt 2-88. Furniture—Assessment of Ignitability of Upholstered Furniture—Part 2: Ignition Source: Match-Flame Equivalent First Edition. 12 pp.
  SNZ NZS 8709: Part 1-84. Fire Tests for Furniture Part 1: Methods of Test for the Ignitability by Smokers' Materials of Upholstered Composites for Seating Amend: A, 1984. 12 pp.
  SNZ NZS 8709: Part 2-85. Fire Tests for Furniture Part 2: Methods of Test for the Ignitability of Upholstered Composites for Seating by Flaming Sources Amend: A, 1984. 20 pp.

—Wood
  CNS O2028-80. Method of Test for Flammability of Wood (Dec)(6718). 2 pp.
  JIS Z 2120-58. Method of Inflammability Test for Wood (R 1969). 4 pp.

Flammable Anesthetics
Use: Anesthetics—Flammable

Flammable Atmospheres
See Also: Fire Hazards

—Electromagnetic Radiation—Radiofrequency Interference
  BSI BS 6656-91. 1991 Prevention of Inadvertent Ignition of Flammable Atmospheres by Radio-Frequency Radiation. 55 pp.
  BSI BS 6656-86. 1986 Guide to Prevention of Inadvertent Ignition of Flammable Atmospheres by Radio-Frequency Radiation. 52 pp.

Flammable Gases
Use: Explosive Gases

Flammable Liquids
Scope Note: For additional listings, see also specific flammable liquids Use For: Combustible Liquids
See Also: Combustion; Flammability Testing; Flammable Materials; Liquids

—Fuel Dispensing Equipment—Electrical Equipment
  CSA C22.2 NO 22-M1986. Electrical Equipment for Flammable and Combustible Fuel Dispensers (R 1992); (Gen Instr 1). 19 pp.
  CSA 1169 Bull. Electrical Bulletin 1169 June 27, 1978 to C22.2 NO 22. 2 pp.
  SNZ NZS 6109: Part 1-88. Electrical Systems of Dispensing Equipment for Explosive Atmospheres Part 1: Flammable Liquids Dispensing Equipment. 16 pp.

—Handling
  SAA AS 1940-93. Storage and Handling of Flammable and Combustible Liquids (in Professional Package 17F, 19A, 21A, 30A, 32A, 44A, 47A, 55, 57, 66, 66). 103 pp.

Flammable Liquids (Cont.)
—Highway Transportation—Emergency Procedures
  SAA AS 1678. Emergency Procedure Guide—Transport Group Text EPGs for Class 3 Substances—Flammable Liquids Pads of 20 Forms.
  SAA AS 1678.3A1-87. Emergency Procedure Guide—Transport Group Text EPGs for Class 3 Substances—Flammable Liquids—Part 3A1: Flammable Liquids (Supersedes AS 1678.3.0.020—1986, AS 1678.3.1.004—1984, AS 1678.3.1.005—1984, AS 1678.3.1.007—1984, AS 1678.3.1.008—1984,.
  SAA AS 1678.3A2-87. Emergency Procedure Guide—Transport Group Text EPGs for Class 3 Substances—Flammable Liquids—Part 3A2: Flammable Liquid Poisonous.
  SAA AS 1678.3A3-87. Emergency Procedure Guide—Transport Group Text EPGs for Class 3 Substances—Flammable Liquids—Part 3A3: Flammable Liquid—Harmful (Supersedes AS 1678.3.1.006—1984).
  SAA AS 1678.3A4-87. Emergency Procedure Guide—Transport Group Text EPGs for Class 2 Substances—Flammable Liquids—Part 3A4: Flammable Liquid—Corrosive.
  SAA AS 1678.3A5-87. Emergency Procedure Guide—Transport Group Text EPGs for Class 3 Substances—Flammable Liquids—Part 3A5: Flammable Liquid, Poisonous, Corrosive.
  SAA AS 1678.3A6-87. Emergency Procedure Guide—Transport Group Text EPGs for Class 3 Substances—Flammable Liquids—Part 3A6: Flammable Liquid, Corrosive, Dangerous When Wet.
  SAA AS 1678.3B3-87. Emergency Procedure Guide—Transport Group Text EPGs for Class 3 Substances—Flammable Liquids—Part 3B3: Flammable Liquid—Some Harmful or Corrosive Properties.
  SAA AS 1678.3C1-87. Emergency Procedure Guide—Transport Group Text EPGs for Class 3 Substances—Flammable Liquids—Part 3C1: Flammable Liquid of Lesser Hazard.

—Ignition Temperature
  DIN ENGL 51794-78. Testing of Mineral Oil Hydrocarbons; Determination of Ignition Temperature (Jan). 5 pp.

—Seamless Pipes
  DIN ENGL 17172-78. Steel Pipes for Long-Distance Pipelines for Combustible Liquids and Gases; Technical Conditions of Delivery (May). 14 pp.

—Storage
  SAA AS 1940-93. Storage and Handling of Flammable and Combustible Liquids (in Professional Package 17F, 19A, 21A, 30A, 32A, 44A, 47A, 55, 57, 66, 66). 103 pp.

—Storage Tanks
  DIN ENGL 6608 Pt 1-89. Horizontal Single-Wall Steel Tanks for the Underground Storage of Flammable and Non-Flammable Water Polluting Liquids (Sept). 6 pp.
  DIN ENGL 6608 Pt 2-89. Horizontal Double-Wall Steel Tanks for the Underground Storage of Flammable and Non-Flammable Water Polluting Liquids (Sept). 3 pp.
  DIN ENGL 6616-89. Horizontal Single-Wall and Double-Wall Steel Tanks for the Above Ground Storage of Flammable and Non-Flammable Water Polluting Liquids (Sept). 10 pp.
  DIN ENGL 6618 Pt 1-89. Vertical Single-Wall Steel Tanks for the Above Ground Storage of Flammable and Non-Flammable Water Polluting Liquids (Sept). 10 pp.
  DIN ENGL 6618 Pt 2-89. Vertical Double-Wall Steel Tanks with Vacuum-Based Leak Detection System, for the Above Ground Storage of Flammable and Non-Flammable Water Polluting Liquids (Sept). 6 pp.
  DIN ENGL 6619 Pt 1-89. Vertical Single-Wall Steel Tanks for the Underground Storage of Flammable and Non-Flammable Water Polluting Liquids (Sept). 4 pp.
  DIN ENGL 6619 Pt 2-89. Vertical Double-Wall Steel Tanks for the Underground Storage of Flammable and Non-Flammable Water Polluting Liquids (Sept). 4 pp.
  DIN ENGL 6623 Pt 1-89. Vertical Single-Wall Steel Tanks with Less Than 1000 Litre Capacity, for the Above Ground Storage of Flammable and Non-Flammable Water Polluting Liquids (Sept). 5 pp.
  DIN ENGL 6623 Pt 2-89. Vertical Double-Wall Steel Tanks with Less Than 1000 Litre Capacity, for the Above Ground Storage of Flammable and Non-Flammable Water Polluting Liquids (Sept). 4 pp.
  DIN ENGL 6624 Pt 1-89. Horizontal Single-Wall Steel Tanks with Capacities Between 1000 and 5000 Litres, for the Above Ground Storage of Flammable and Non-Flammable Water Polluting Liquids (Sept). 5 pp.

Flammable Liquids (Cont.)
—Storage Tanks (Cont.)
  DIN ENGL 6624 Pt 2-89. Horizontal Double-Wall Steel Tanks with Capacities Between 1000 and 5000 Litres, for the Above Ground Storage of Flammable and Non-Flammable Water Polluting Liquids (Sept). 3 pp.
  DIN ENGL 6625 Pt 1-89. Steel Tanks Erected on Site, for the Above Ground Storage of Hazardous Flammable Water Polluting Liquids of Class A III and of Non-Flammable Water Polluting Liquids; Requirements and Testing (Sept). 3 pp.
  SAA AS 1692-89. Tanks for Flammable and Combustible Liquids. 13 pp.

—Storage Tanks—Design
  DIN ENGL 6625 Pt 2-89. Steel Tanks Erected on Site, for the Above Ground Storage of Hazardous Flammable Water Polluting Liquids of Class A III and of Non-Flammable Water Polluting Liquids; Design (Sept). 4 pp.

—Storage Tanks—Inspection
  DIN ENGL 6600-89. Steel Tanks for the Storage of Flammable and Non-Flammable Water Polluting Liquids; Concepts and Inspection (Sept). 4 pp.

—Storage Tanks—Leak Detectors
  DIN ENGL 6618 Pt 3-89. Vertical Double-Wall Steel Tanks with Liquid-Based Leak Detection System, for the Above Ground Storage of Flammable and Non-Flammable Water Polluting Liquids (Sept). 4 pp.

—Storage Tanks—Manholes
  DIN ENGL 6626-89. Steel Manhole Shafts for Underground Tanks Designed for the Storage of Flammable and Non-Flammable Water Polluting Liquids (Sept). 9 pp.

—Storage Tanks—Manholes—Collars
  DIN ENGL 6627-89. Collars for Masonry Manhole Shafts for Underground Tanks Designed for the Storage of Flammable and Non-Flammable Water Polluting Liquids (Sept). 2 pp.

—Welded Pipes
  DIN ENGL 17172-78. Steel Pipes for Long-Distance Pipelines for Combustible Liquids and Gases; Technical Conditions of Delivery (May). 14 pp.

Flammable Materials
Scope Note: For additional listings, see also specific flammable materials Use For: Combustible Materials
See Also: Combustion; Fire Hazards; Flammable Liquids; Flash Point; Hazardous Materials; Solvents

—Ash Content
  CPPA G.6U-77. Ashing by Oxygen Flask Combustion. 2 pp.

—Highway Transportation—Emergency Procedures
  SAA AS 1678.4.3. 002-91. Emergency Procedure Guide Transport —Part 4.3.002: Magnesium Phosphide (In Profressional Package 37).
  SAA AS 1678. Emergency Procedure Guide—Transport Group Text EPGs for Class 3 Substances—Flammable Liquids Pads of 20 Forms.
  SAA AS 1678.4A1-87. Emergency Procedure Guide—Transport Group Text EPGs for Class 4 Substances—Flammable Solids, Substances Liable to Spontaneous Combustion or Which Emit Flammable Gases When Wet—Part 4A1: Flammable Solids.
  SAA AS 1678.4A2-87. Emergency Procedure Guide—Transport Group Text EPGs for Class 4 Substances—Flammable Solids, Substances Liable to Spontaneous Combustion or Which Emit Flammable Gases When Wet—Part 4A2: Spontaneously Combustible.
  SAA AS 1678.4A3-87. Emergency Procedure Guide—Transport Group Text EPGs for Class 4 Substances—Flammable Solids, Substances Liable to Spontaneous Combustion or Which Emit Flammable Gases When Wet—Part 4A3: Dangerous When Wet.
  SAA AS 1678.4B1-87. Emergency Procedure Guide—Transport Group Text EPGs for Class 4 Substances—Flammable Solids, Liable to Spontaneous Combustion or Which Emit Flammable Gases When Wet—Part 4B1: Flammable Solids, Toxic.
  SAA AS 1678.4B2-87. Emergency Procedure Guide—Transport Group Text EPGs for Class 4 Substances—Flammable Solids, Substances Liable to Spontaneous Combustion or Which Emit Flammable Gases When Wet—Part 4B2: Spontaneously Combustible, Toxic.
  SAA AS 1678.4B3-87. Emergency Procedure Guide—Transport Group Text EPGs for Class 4 Substances—Flammable Solids, Substances Liable to Spontaneous Combustion or Which Emit Flammable Gases When Wet—Part 4B3: Dangerous When Wet, Toxic.

INDUSTRY STANDARDS

Flammable

INTERNATIONAL AND NON-U.S. NATIONAL STANDARDS
SUBJECT INDEX

Flammable Materials *(Cont.)*
—Highway Transportation—Emergency Procedures *(Cont.)*
SAA AS 1678.4C2-87. Emergency Procedure Guide—Transport Group Text EPGs for Class 4 Substances—Flammable Solids, Substances Liable to Spontaneous Combustion or Which Emit Flammable Gases When Wet—Part 4C2: Spontaneously Combustible, Corrosive.
SAA AS 1678.4C3-87. Emergency Procedure Guide—Transport Group Text EPGs for Class 4 Substances—Flammable Solids, Substances Liable to Spontaneous Combustion or Which Emit Flammable Gases When Wet—Part 4C3: Dangerous When Wet, Corrosive.
SAA AS 1678.4D2-87. Emergency Procedure Guide—Transport Group Text EPGs for Class 4 Substances—Flammable Solids, Substances Liable to Spontaneous Combustion or Which Emit Flammable Gases When Wet—Part 4D2: Spontaneously Combustible, Reacts Violently with Water.
SAA AS 1678.4E3-87. Emergency Procedure Guide—Transport Group Text EPGs for Class 4 Substances—Flammable Solids, Substances Liable to Spontaneous Combustion or Which Emit Flammable Gases When Wet—Part 4E3: Dangerous When Wet, Some Toxic Properties.
SAA AS 1678.4F1-87. Emergency Procedure Guide—Transport Group Text EPGs for Class 4 Substances—Flammable Solids, Substances Liable to Spontaneous Combustion or Which Emit Flammable Gases When Wet—Part 4F1: Flammable Solids, Explosive.
SAA AS 1678.4G1-87. Emergency Procedure Guide—Transport Group Text EPGs for Class 4 Substances—Flammable Solids, Substances Liable to Spontaneous Combustion or Which Emit Flammable Gases When Wet—Part 4G1: Flammable Solids, Harmful.
SAA AS 1678.4G2-87. Emergency Proc. Guide—Trans Group Text EPGs for Class 4 Substances—Flammable Solids, Substances Liable to Spontaneous Combustion or Which Emit Flammable Gases When Wet—Part 4G2: Spontaneously Combustible, Dangerous When Wet, May React Violently with Water.
SAA AS 1678.4H1-87. Emergency Procedure Guide—Transport Group Text EPGs for Class 4 Substances—Falmmable Solids, Substances Liable to Spontaneous Combustion or Which Emit Flammable Gases When Wet —Part 4H1: Flammable Solids, Some Toxic Properties.
SAA AS 1678.4K2-87. Emergency Procedure Guide—Transport Group Text EPGs for Class 4 Substances—Flammable Solids, Substances Liable to Spontaneous Combustion or Which Emit Flammable Gases When Wet—Part 4K2: Spontaneously Combustible, Some Toxic Properties.
SAA AS 1678.4M3-87. Emergency Proc. Guide—Trans. Group Text EPGs for Class 4 Substances—FlammableSolids, Substances Liable to Spontaneous Combustion or Which Emit Flammable Gases When Wet—Part 4M3: Spontaneously Combustible, Dangerous when Wet, Some Toxic Properties.
SAA AS 1678.4N3-87. Emergency Proc. Guide—Trans. Group Text EPGs for Class 4 Substances—FlammableSolids, Substances Liable to Spontaneous Combustion or Which Emit Flammable Gases When Wet—Part 4N3: Dangerous When Wet, Some Corrosive Properties, May Ignite Spontaneously.
SAA AS 1678.4O3-87. Emergency Procedure Guide—Transport Group Text EPGs for Class 4 Substances—Flammable Solids, Substances Liable to Spontaneous Combustion or Which Emit Flammable Gases When Wet—Part 4O3: Dangerous When Wet, Spontaneously Combustible.
SAA AS 1678.4P3-87. Emergency Procedure Guide—Transport Group Text EPGs for Class 4 Substances—Flammable Solids, Substances Liable to Spontaneous Combustion or Which Emit Flammable Gases When Wet—Part 4P3: Dangerous When Wet, Flammable Liquid, Some Toxic Properties.
SAA AS 1678.4Q3-87. Emergency Procedure Guide—Transport Group Text EPGs for Class 4 Substances—Flammable Solids, Substances Liable to Spontaneous Combustion or Which Emit Flammable Gases When Wet—Part 4Q3: Dangerous When Wet, Flammable Liquid, Corrosive.
SAA AS 1678.4R1-90. Emergency Procedur Guide—Transport Group Text EPGs for Class 4 Substances—Flammable Solids, Substances Liable to Spontaneous Combustion or Which Emit Flammable Gases When Wet—Part 4R1: Flammable Solids, Dangerous When Wet.
SAA AS 1678.4R3-87. Emergency Procedure Guide—Transport Group Text EPGs for Class 4 Substances—Flammable Solids, Substances Liable to Spontaneous Combustion or Which Emit Flammable Gases When Wet—Part 4R3: Dangerous When Wet, Harmful.
SAA AS 1678.8F1-87. Emergency Procedure Guide—Transport Group Text EPGs for Class 8 Substances—Corrosive Substances—Part 8F1: Corrosive—Flammable Solid.

Flammable Materials *(Cont.)*
—Highway Transportation—Emergency Procedures *(Cont.)*
SAA AS 1678.8F2-87. Emergency Procedure Guide—Transport Group Text EPGs for Class 8 Substances—Corrosive Substances—Part 8F2: Corrosive—Flammable—Solid—Violent Reaction with Water.
SAA AS 1678. Emergency Procedure Guide—Transport Group Text EPGs for Class 9 Substances Pads of 20 Forms.
SAA AS 1678.9A1-88. Emergency Procedure Guide—Transport Group Text EPGs for Class 9 Substances—Part 9A1: Flammable Solids.
SAA AS 1678.9A2-88. Emergency Procedure Guide—Transport Group Text EPGs for Class 9 Substances—Part 9A2: Spontaneously Combustible.
—Ovens
SAA AS 1681-74. Electrically-Heated Ovens in Which Flammable Volatiles Occur—Type 1 Ovens Corrig.. 29 pp.

Flanged Adapters
See Also: Adapters (Fittings)
—Aircraft
SBAC AS 6549 ISSUE 1. Flanged Adaptor.
SBAC AS 6557 ISSUE 1. Flanged Adaptor.

Flanged Nuts
See Also: Nuts (Fasteners)
—Aerospace—Hexagonal—Locking
CEN PREN 3536-91. Nuts, Hexagon, Self-Locking, in Heat Resisting Steel, MoS2 Lubricated Classification: 1100MPa (at Ambient Temperature)/315 Degrees C. 5 pp.
CEN PREN 3626-91. Nuts, Hexagon, Self-Locking, in Steel, Cadmium Plated, MoS2 Lubricated Classification: 1100 MPa (at Ambient Temperature)/235 Degrees C. 5 pp.
DIN ENGL LN 9161-77. Nuts Hexagon Flanged; Self-Locking for Temperatures up to 315 Degrees Celsius and up to 425 Degrees Celsius (May). 3 pp.
DIN ENGL LN 9338-85. Aerospace; Nuts Hexagon Flanged; Self-Locking for Temperatures up to 235 Degrees C (Feb). 3 pp.
—Aircraft—Hexagonal—Locking
SBAC AS 8600 ISSUE 4. Self-Locking Nuts, Hexagon, Flange Type, Steel, Cadmium Plated, 160,000LBS/SQ.IN. 250 Degrees Celsius.
SBAC AS 8623 ISSUE 2. Self-Locking Nuts, Hexagon, Flange Type, C.R. Steel, Silver Plated, 125,000 lbf/in2. 450 Degrees C.
SBAC AS 8650 ISSUE 4. Self-Locking Nuts, Hexagon, Flange Type, CR Steel, 125,000 lbf/in2 250 Degrees C.
—Hexagonal
DIN ENGL 6923-83. Hexagon Nuts with Flange (June). 4 pp.
JIS B 1190-87. Hexagon Nuts with Flange. 9 pp.
—Hexagonal—Prevailing Torque
ISO 7043-83. Prevailing Torque Type Hexagon Nuts with Flange (with Non-Metallic Insert) First Edition. 6 pp.
ISO 7044-83. Prevailing Torque Type All-Metal Hexagon Nuts with Flange First Edition. 6 pp.
—Hexagonal—Prevailing Torque—Nonmetallic Insert
DIN ENGL 6926-83. Prevailing Torque Type Hexagon Nuts with Flange and with Non-Metallic Insert (Nov). 4 pp.
—Prevailing Torque
DIN ENGL 6927-83. Prevailing Torque Type All-Metal Nuts with Flange (Nov). 3 pp.

Flanged Screws
Scope Note: See the subheading Flanged under specific types of screws

Flanged Valves
Scope Note: See the subheading Flanged under specific types of valves

Flanges
Scope Note: For additional listings, see the subheading Flanges under specific types of equipment
See Also: Locking Flanges; Pipe Flanges; Structural Flanges; Waveguide Flanges
ISO 7005 Pt 1-92. Metallic Flanges—Part 1: Steel Flanges First Edition. 86 pp.
—Aerospace—Clamped Fittings
NATO STANAG 3312 ED 4 AMD 5-75. Profile Dimensions of Flanges for V-Band Couplings for Piping and Ducting. 18 pp.

Flanges *(Cont.)*
—Aircraft
SBAC AS 2985 ISSUE 2. Locking Flange. 1 p.
—Aircraft—Accessories
SBAC RS 299 (V). Table of Accessory Flanges.
SBAC RS 338 (V). 2.175 P.C.DIA. Square Accessory Flange.
—Aircraft—Circular
SBAC RS 501-506, 514-521 (V). Circular Flanges and Drives.
—Aircraft—Circular—Accessories
SBAC RS 305 (V). Circular Flange and Drive 8.0 Inch P.C.D. X 1.125 Inches.
—Aircraft—Square
SBAC RS 327 (V). Square Flange and Drive 1.437 P.C.DIA..
—Aircraft—Square—Accessories
SBAC RS 538 (V). 2.175 Inch P.C.DIA. Square Accessory Flange—Unified Thread.
—Aircraft—Taper
SBAC RS 571-588 (V)(I). Accessory, Drive and Pad (Taper Flanges).
—Gaskets
DIN ENGL 2696-72. Lenticular Gaskets and Lenticular Seals for ND 64 to ND 400 Flanged Connections (Apr). 2 pp.
—Hydraulic Motors
BSI BS 6276-87. 1987 Dimensions and Identification Code for Mounting Flanges and Shaft Ends for Hydraulic Fluid Power Pumps and Motors. 18 pp.
ISO 3019 Pt 1-75. Hydraulic Fluid Power—Positive Displacement Pumps and Motors—Dimensions and Identification Code for Mounting Flanges and Shaft Ends—Part 1: Inch Series Shown in Metric Units First Edition; (Erratum—May 1975). 11 pp.
ISO 3019 Pt 2-86. Hydraulic Fluid Power—Positive Displacement Pumps and Motors—Dimensions and Identification Code for Mounting Flanges and Shaft Ends—Part 2: Two-and Four-Hole Flanges and Shaft Ends—Metric Series Second Edition. 15 pp.
ISO 3019 Pt 3-88. Hydraulic Fluid Power—Positive Displacement Pumps and Motors—Dimensions and Identification Code for Mounting Flanges and Shaft Ends—Part 3: Polygonal Flanges (Including Circular Flanges) Second Edition. 8 pp.
—Hydraulic Power Pumps
BSI BS 6276-87. 1987 Dimensions and Identification Code for Mounting Flanges and Shaft Ends for Hydraulic Fluid Power Pumps and Motors. 18 pp.
ISO 3019 Pt 1-75. Hydraulic Fluid Power—Positive Displacement Pumps and Motors—Dimensions and Identification Code for Mounting Flanges and Shaft Ends—Part 1: Inch Series Shown in Metric Units First Edition; (Erratum—May 1975). 11 pp.
ISO 3019 Pt 2-86. Hydraulic Fluid Power—Positive Displacement Pumps and Motors—Dimensions and Identification Code for Mounting Flanges and Shaft Ends—Part 2: Two-and Four-Hole Flanges and Shaft Ends—Metric Series Second Edition. 15 pp.
ISO 3019 Pt 3-88. Hydraulic Fluid Power—Positive Displacement Pumps and Motors—Dimensions and Identification Code for Mounting Flanges and Shaft Ends—Part 3: Polygonal Flanges (Including Circular Flanges) Second Edition. 8 pp.

Flanging Testing
—Metal Tubes
ISO 8494-86. Metallic Materials—Tube—Flanging Test First Edition. 4 pp.
—Steel Tubes
DIN ENGL 50139-65. Testing of Steel; Flanging Test on Tubes (Nov). 2 pp.

Flap Valves
Use For: Flapper Valves; Storm Valves
See Also: Check Valves; Drainage Systems; Hydraulic Valves; Swing Check Valves; Valves
—Flap Inserts—Ships
DIN ENGL 87104-75. Flap Inserts for Non-Return Flaps in Accordance with DIN 87 101 (June). 5 pp.
—Gaskets—Ships
DIN ENGL 87107-75. Gaskets for Non-Return Flaps in Accordance with DIN 87 101 (June). 1 p.
—Housings—Ships
DIN ENGL 87102-75. Casings for Non-Return Flaps in Accordance with DIN 87 101 (June). 3 pp.
—Locks (Hardware)—Ships
DIN ENGL 87106-75. Locking Devices for Non-Return Flaps in Accordance with DIN 87 101 (June). 3 pp.

INDEX and DIRECTORY of

INTERNATIONAL AND NON-U.S. NATIONAL STANDARDS
SUBJECT INDEX

Flap Valves *(Cont.)*
—Pipe Plugs—Ships
DIN ENGL 87105-75. Screw Plugs with Cylindrical Extension for Non-Return Flaps in Accordance with DIN 87 101 (June). 2 pp.

Flap Wheels
Use: Grinding Wheels

Flapper Valves
Use: Flap Valves

Flaps (Control Surfaces)
See Also: Control Surfaces
—Aircraft—Knobs
SBAC AS 1911 ISSUE 7. Knob (Flap Control). 1 p.
SBAC AS 1913 ISSUE 6. Knob (Undercarriage Control). 1 p.

Flare Nut Wrenches
Use For: Open Box End Flare Nut Wrenches
See Also: Box Wrenches; Open End Wrenches; Wrenches
CGSB 39-GP-11M-79. Wrenches, Box, Open End, Combination and Flare Nut, Standard for; (Amendment 1 Apr 1984). 29 pp.
—Torsional Strength
CNS B3348-80. Assembly Tools for Screws and Nuts Wrench Double End Flat, Test Torques Series C (Jul)(5659).

Flare Stacks
See Also: Chimneys
—Noise
DIN ENGL 45635 Pt 22-85. Measurement of Noise Emitted by Machines; Airborne Noise Emission; Enveloping Surface Method; Flares (June). 3 pp.

Flares (Pyrotechnic)
See Also: Battlefield Illumination; Distress Signal Devices; Parachute Flares; Pyrotechnics; Rocket Flares
—Acrylonitrile Copolymers
MOD UK TS 10251. Carboxy Terminated Acrylonitrile/Butadiene Reactive Liquid Polymer Type 1.
—Automotive
JIS D 5711-82. Red Fusee for Motor Vehicles.
JIS D 5711-72. Red Fusee for Motor Vehicles. 12 pp.
—Marine
CNS F1018-90. Method of Test for Hand Flares for Marine Use (Aug)(12772).
CNS F4022-90. Hand Flares for Marine Use (Aug)(12771).
—Papers
MOD UK DEF-86-A-01. Paper, Pure, Unbleached, Rolled; Amendment 1. 6 pp.
—Rayon Fabrics—Aircraft
MOD UK DTD-765A-53. Rayon Fabric for Slow Falling Flares. 2 pp.

Flash Bulbs
Use: Flashtakes

Flash Equipment
Use: Photographic Flash Equipment

Flash Guns
Use: Flashtakes

Flash Lamps
Use: Flashtubes

Flash Point
See Also: Abel Testers; Boiling Points; Fire Point; Flammability Testing; Flammable Materials; Pensky-Martens Closed Testers; Vapor Pressure
CNS K6365-87. Method of Test for Flash Point by Penky-Martens Closed Tester (Jul)(3574).
—Adhesives
CEN PREN 924-92. Adhesives—Solvent-Borne and Solvent-Free Adhesives—Determination of Flashpoint. 10 pp.
—Bitumens—Cutback
BSI BS 2000: Part 113-82. (WITHDRAWN) 1982 Petroleum and Its Products Part 113: Flash Point (Closed) of Cutback Bitumen. 4 pp.
—Crude Oil
JIS K 2265-89. Testing Methods for Flash Point of Crude Oil and Petroleum Products. 35 pp.

Flash Point *(Cont.)*
—Emulsifying Agents
MOD UK M 9536/68. Examination of Emulsifier for Solvent Cleaner (Superseded by Def Stan 68-188).
—Liquid Fuels
DIN ENGL 51755-74. Testing of Mineral Oils and Other Combustible Liquids; Determination of Flash Point by the Closed Tester According to Abel-Pensky (Mar). 5 pp.
—Mineral Oils
DIN ENGL 51755-74. Testing of Mineral Oils and Other Combustible Liquids; Determination of Flash Point by the Closed Tester According to Abel-Pensky (Mar). 5 pp.
—Mobile Liquids
CGSB 31-GP-0A METH 1.2-57. Methods of Testing Corrosion-Prevention Materials and Processes Flash Point by Tag Closed Cup; (Amended January 1974) (Re-Edited June 1982). 1 p.
—Paints
BSI BS 3900: Part A8-86. 1986 Amd 1 Methods of Test for Paints Group A: Tests on Liquid Paints (Excluding Chemical Tests) Part A8: Methods of Test for Flash/No Flash (Closed Cup Equilibrium Method). 8 pp.
BSI BS 3900: Part A9-86. 1986 Methods of Test for Paints Group A: Tests on Liquid Paints (Excluding Chemical Tests) Part A9: Determination of Flashpoint (Closed Cup Equilibrium Method). 9 pp.
BSI BS 3900: Part A13-86. 1986 Methods of Test for Paints: Group A: Tests Group A: Tests on Liquid Paints (Excluding Chemical Tests) Part A13: Test for Flash/No Flash (Rapid Equilibrium Method). 10 pp.
BSI BS 3900: Part A14-86. (WITHDRAWN) 1986 Methods of Test for Paints Group A: Tests on Liquid Paints (Excluding Chemical Tests)Part A14: Methods of Test for Determination of Flashpoint (Rapid Equilibrium Method) (Superseded by BS EN 456: 1991). 10 pp.
BSI BS 6664: Part 1-86. 1986 Flashpoint of Petroleum and Related Products Part 1: Methods of Test for Flash/No Flash (Closed Cup Equilibrium Method). 8 pp.
BSI BS 6664: Part 2-86. 1986 Flashpoint of Petroleum and Related Products Part 2: Methods for Determination of Flash-Point (Closed Cup Equillibrium Method). 9 pp.
BSI BS 6664: Part 3-86. 1986 Flashpoint of Petroleum and Related Products Part 3: Method of Test for Flash/No Flash (Rapid Equillibrium Method). 10 pp.
BSI BS 6664: Part 4-86. 1986 Flashpoint of Petroleum and Related Products Part 4: Method for Determination of Flashpoint (Rapid Equilibrium. 10 pp.
CEN EN 456-91. Paints, Varnishes and Related Products—Determination of Flashpoint—Rapid Equilibrium Method. 12 pp.
CGSB 1-GP-71 METH 3.1-75. Methods of Testing Paints and Pigments Flash Point Pigmented Materials by the Pensky-Martens Closed Cup Tester. 4 pp.
DIN ENGL EN 456-91. Paints, Varnishes and Related Products; Determination of Flashpoint by the Rapid Equilibrium Method (Modified Version of ISO 3679: 1983) (Sept). 8 pp.
ISO 1516-81. Paints, Varnishes, Petroleum and Related Products—Flash/No Flash Test—Closed Cup Equilibrium Method Second Edition. 7 pp.
ISO 1523-83. Paints, Varnishes, Petroleum and Related Products—Determination of Flashpoint—Closed Cup Equilibrium Method Second Edition. 8 pp.
ISO 3679-83. Paints, Varnishes, Petroleum and Related Products—Determination of Flashpoint—Rapid Equilibrium Method Second Edition. 9 pp.
ISO 3680-83. Paints, Varnishes, Petroleum and Related Products—Flash/No Flash Test—Rapid Equilibrium Method Third Edition. 8 pp.
—Petroleum Products
BSI BS 2000: Part 33-82. (WITHDRAWN) 1982 Petroleum and Its Products Part 33: Flash Point by the Able Apparatus (Statutory Method: Petroleum (Consolidation) Act 1928) (Supersedes BS 3442: Part 1: 1974). 10 pp.
BSI BS 2000: Part 34-90. (WITHDRAWN) 1990 Petroleum and Its Products Part 34: Flash Point by Pensky-Martens Closed Tester (W) (Superseded by BS 6664: Part 5: 1990). 15 pp.
BSI BS 2000: Part 35-93. 1993 Methods of Test for Petroleum and Its Products Part 35: Determination of Open Flash and Fire Point—Pensky-Martens Method (Supersedes BS 4688: 1991). 3 pp.
BSI BS 2000: Part 35-82. 1982 Petroleum and Its Products Part 35: Flash Point (Open) and Fire Point of Petroleum Products by the Pensky-Martens Apparatus. 4 pp.

Flash Point *(Cont.)*
—Petroleum Products *(Cont.)*
BSI BS 2000: Part 170-92. 1992 Methods of Test for Petroleum and Its Products Part 170: Determination of Flash Point—Abel Closed Cup Method (Identical with IP 170/90) (Supersedes BS 3442: Part 2: 1974). 10 pp.
BSI BS 2000: Part 170-82. 1982 Petroleum and Its Products Part 170: Flash Point by the Abel Apparatus (Non-Statutory Method). 8 pp.
BSI BS 3900: Part A8-86. 1986 Amd 1 Methods of Test for Paints Group A: Tests on Liquid Paints (Excluding Chemical Tests) Part A8: Methods of Test for Flash/No Flash (Closed Cup Equilibrium Method). 8 pp.
BSI BS 3900: Part A9-86. 1986 Methods of Test for Paints Group A: Tests on Liquid Paints (Excluding Chemical Tests) Part A9: Determination of Flashpoint (Closed Cup Equilibrium Method). 9 pp.
BSI BS 3900: Part A13-86. 1986 Methods of Test for Paints: Group A: Tests Group A: Tests on Liquid Paints (Excluding Chemical Tests) Part A13: Test for Flash/No Flash (Rapid Equilibrium Method). 10 pp.
BSI BS 3900: Part A14-86. (WITHDRAWN) 1986 Methods of Test for Paints Group A: Tests on Liquid Paints (Excluding Chemical Tests)Part A14: Methods of Test for Determination of Flashpoint (Rapid Equilibrium Method) (Superseded by BS EN 456: 1991). 10 pp.
BSI BS 4689-80. 1980 Amd 1 Determination of Flash and Fire Points of Petroleum Products: Cleveland Open Cup Method (AMD 6229) February 28, 1990 (W). 8 pp.
BSI BS 6664: Part 1-86. 1986 Flashpoint of Petroleum and Related Products Part 1: Methods of Test for Flash/No Flash (Closed Cup Equilibrium Method). 8 pp.
BSI BS 6664: Part 2-86. 1986 Flashpoint of Petroleum and Related Products Part 2: Methods for Determination of Flash-Point (Closed Cup Equillibrium Method). 9 pp.
BSI BS 6664: Part 3-86. 1986 Flashpoint of Petroleum and Related Products Part 3: Method of Test for Flash/No Flash (Rapid Equillibrium Method). 10 pp.
BSI BS 6664: Part 4-91. 1991 Flashpoint of Petroleum and Related Products Part 4: Method for Determination of Flashpoint (Rapid Equilibrium. 14 pp.
BSI BS 6664: Part 4-86. 1986 Flashpoint of Petroleum and Related Products Part 4: Method for Determination of Flashpoint (Rapid Equilibrium. 10 pp.
BSI BS 6664: Part 5-90. 1990 Flashpoint of Petroleum and Related Products Part 5: Method for Determination of Flashpoint by Pensky-Martens Closed Tester. 16 pp.
CEN EN 57-84. Petroleum Products: Determination of Flash Point; Abel-Pensky Closed Tester. 9 pp.
CNS K6377-90. Determination of Flash Fire Points for Petroleum Products with Cleveland Open Cup (Aug)(3775).
DIN ENGL 51755-74. Testing of Mineral Oils and Other Combustible Liquids; Determination of Flash Point by the Closed Tester According to Abel-Pensky (Mar). 5 pp.
DIN ENGL 51758-85. Testing of Liquid Petroleum Products and Other Combustible Liquids; Determination of Flash Point by Pensky-Martens Closed Tester (Aug). 4 pp.
ISO 1516-81. Paints, Varnishes, Petroleum and Related Products—Flash/No Flash Test—Closed Cup Equilibrium Method Second Edition. 7 pp.
ISO 1523-83. Paints, Varnishes, Petroleum and Related Products—Determination of Flashpoint—Closed Cup Equilibrium Method Second Edition. 8 pp.
ISO 2592-73. Petroleum Products—Determination of Flash and Fire Points—Cleveland Open Cup Method First Edition. 6 pp.
ISO 2719-88. Petroleum Products and Lubricants—Determination of Flash Point—Pensky-Martens Closed Cup Method Second Edition. 14 pp.
ISO 3679-83. Paints, Varnishes, Petroleum and Related Products—Determination of Flashpoint—Rapid Equilibrium Method Second Edition. 9 pp.
ISO 3680-83. Paints, Varnishes, Petroleum and Related Products—Flash/No Flash Test—Rapid Equilibrium Method Third Edition. 8 pp.
JIS K 2265-89. Testing Methods for Flash Point of Crude Oil and Petroleum Products. 35 pp.
—Polishes
CGSB 25-GP-1M METH 2-84. Methods of Sampling and Testing Waxes and Polishes Flash Point (Supersedes 2.1, 2.2, 2.3 of September 1974). 1 p.
—Solvents
BSI BS 3900: Part A8-86. 1986 Amd 1 Methods of Test for Paints Group A: Tests on Liquid Paints (Excluding Chemical Tests) Part A8: Methods of Test for Flash/No Flash (Closed Cup Equilibrium Method). 8 pp.

INDUSTRY STANDARDS

Flash Point (Cont.)

—Solvents (Cont.)

BSI BS 3900: Part A9-86. 1986 Methods of Test for Paints Group A: Tests on Liquid Paints (Excluding Chemical Tests) Part A9: Determination of Flashpoint (Closed Cup Equilibrium Method). 9 pp.

BSI BS 3900: Part A13-86. 1986 Methods of Test for Paints: Group A: Tests Group A: Tests on Liquid Paints (Excluding Chemical Tests) Part A13: Test for Flash/No Flash (Rapid Equilibrium Method). 10 pp.

BSI BS 3900: Part A14-86. (WITHDRAWN) 1986 Methods of Test for Paints Group A: Tests on Liquid Paints (Excluding Chemical Tests)Part A14: Methods of Test for Determination of Flashpoint (Rapid Equilibrium Method) (Superseded by BS EN 456: 1991). 10 pp.

BSI BS 6664: Part 1-86. 1986 Flashpoint of Petroleum and Related Products Part 1: Methods of Test for Flash/No Flash (Closed Cup Equilibrium Method). 8 pp.

BSI BS 6664: Part 2-86. 1986 Flashpoint of Petroleum and Related Products Part 2: Methods for Determination of Flash-Point (Closed Cup Equillibrium Method). 9 pp.

BSI BS 6664: Part 3-86. 1986 Flashpoint of Petroleum and Related Products Part 3: Method of Test for Flash/No Flash (Rapid Equillibrium Method). 10 pp.

BSI BS 6664: Part 4-86. 1986 Flashpoint of Petroleum and Related Products Part 4: Method for Determination of Flashpoint (Rapid Equilibrium. 10 pp.

BSI BS EN 456-91. 1991 Method for Determination of Flashpoint of Paints, Varnishes and Related Products by the Rapid Equilibrium Method (W). 16 pp.

CEN EN 456-91. Paints, Varnishes and Related Products—Determination of Flashpoint—Rapid Equilibrium Method. 12 pp.

DIN ENGL EN 456-91. Paints, Varnishes and Related Products; Determination of Flashpoint by the Rapid Equilibrium Method (Modified Version of ISO 3679: 1983) (Sept). 8 pp.

ISO 1516-81. Paints, Varnishes, Petroleum and Related Products—Flash/No Flash Test—Closed Cup Equilibrium Method Second Edition. 7 pp.

ISO 1523-83. Paints, Varnishes, Petroleum and Related Products—Determination of Flashpoint—Closed Cup Equilibrium Method Second Edition. 8 pp.

ISO 3679-83. Paints, Varnishes, Petroleum and Related Products—Determination of Flashpoint—Rapid Equilibrium Method Second Edition. 9 pp.

ISO 3680-83. Paints, Varnishes, Petroleum and Related Products—Flash/No Flash Test—Rapid Equilibrium Method Third Edition. 8 pp.

—Varnishes

BSI BS 3900: Part A8-86. 1986 Amd 1 Methods of Test for Paints Group A: Tests on Liquid Paints (Excluding Chemical Tests) Part A8: Methods of Test for Flash/No Flash (Closed Cup Equilibrium Method). 8 pp.

BSI BS 3900: Part A9-86. 1986 Methods of Test for Paints Group A: Tests on Liquid Paints (Excluding Chemical Tests) Part A9: Determination of Flashpoint (Closed Cup Equilibrium Method). 9 pp.

BSI BS 3900: Part A13-86. 1986 Methods of Test for Paints: Group A: Tests Group A: Tests on Liquid Paints (Excluding Chemical Tests) Part A13: Test for Flash/No Flash (Rapid Equilibrium Method). 10 pp.

BSI BS 3900: Part A14-86. (WITHDRAWN) 1986 Methods of Test for Paints Group A: Tests on Liquid Paints (Excluding Chemical Tests)Part A14: Methods of Test for Determination of Flashpoint (Rapid Equilibrium Method) (Superseded by BS EN 456: 1991). 10 pp.

BSI BS 6664: Part 1-86. 1986 Flashpoint of Petroleum and Related Products Part 1: Methods of Test for Flash/No Flash (Closed Cup Equilibrium Method). 8 pp.

BSI BS 6664: Part 2-86. 1986 Flashpoint of Petroleum and Related Products Part 2: Methods for Determination of Flash-Point (Closed Cup Equillibrium Method). 9 pp.

BSI BS 6664: Part 3-86. 1986 Flashpoint of Petroleum and Related Products Part 3: Method of Test for Flash/No Flash (Rapid Equillibrium Method). 10 pp.

BSI BS 6664: Part 4-86. 1986 Flashpoint of Petroleum and Related Products Part 4: Method for Determination of Flashpoint (Rapid Equilibrium. 10 pp.

CEN EN 456-91. Paints, Varnishes and Related Products—Determination of Flashpoint—Rapid Equilibrium Method. 12 pp.

DIN ENGL EN 456-91. Paints, Varnishes and Related Products; Determination of Flashpoint by the Rapid Equilibrium Method (Modified Version of ISO 3679: 1983) (Sept). 8 pp.

ISO 1516-81. Paints, Varnishes, Petroleum and Related Products—Flash/No Flash Test—Closed Cup Equilibrium Method Second Edition. 7 pp.

Flash Point (Cont.)

—Varnishes (Cont.)

ISO 1523-83. Paints, Varnishes, Petroleum and Related Products—Determination of Flashpoint—Closed Cup Equilibrium Method Second Edition. 8 pp.

ISO 3679-83. Paints, Varnishes, Petroleum and Related Products—Determination of Flashpoint—Rapid Equilibrium Method Second Edition. 9 pp.

ISO 3680-83. Paints, Varnishes, Petroleum and Related Products—Flash/No Flash Test—Rapid Equilibrium Method Third Edition. 8 pp.

—Waxes

CGSB 25-GP-1M METH 2-84. Methods of Sampling and Testing Waxes and Polishes Flash Point (Supersedes 2.1, 2.2, 2.3 of September 1974). 1 p.

Flash Tubes
Use: Flashtubes

Flash Welding
See Also: Welding

—Ferrous Metal

BSI BS 6944-88. 1988 Flash Welding of Butt Joints in Ferrous Metals (Excluding Pressure Piping Applications). 15 pp.

—Reinforcing Steels

DIN ENGL 4099-85. Welding of Reinforcing Steel; Execution of Welding Work and Testing (Nov). 16 pp.

—Steel Pipes

SAA AS CB15.5-68. Rules for Welding of Steel Pressure Piping (Known as the SAA Pipe Welding Code)—Part 5: Flash Butt Welding of Steel Pipes and Tubes (Incorporating Amdt 1). 23 pp.

—Steel Tubes

BSI BS 4204-89. 1989 Flash Welding of Steel Tubes for Pressure Applications. 20 pp.

Flashback Arresters
See Also: Fire Protection; Flame Arresters; Welding Equipment

—Safety

DIN ENGL 8521-81. Safety Devices Against Flashback and Backflow in Welding, Cutting and Allied Processes; Safety Requirements, Testing (Dec). 8 pp.

Flashcubes
Use: Flashtubes

Flasher Units
Use For: Blinker Units; Emergency Flasher Units; Flashers; Hazard Warning Signal Flashers
See Also: Anticollision Lights; Automotive Equipment; Lighting; Signal Lights; Turn Signals; Warning Lights

—Automotive

CNS D2034-86. Hazard Warning Signal Flasher for Automobiles (Nov)(5790).

JIS D 5707-86. Flasher Units for Automobiles. 15 pp.

—Automotive—Connectors—Mounting

BSI BS AU 201: Part 1-85. 1985 Relays and Flashers Part 1: Mounting and Positioning Dimensions of Male Tabs and Socket Apertures for Relays and Flashers for Road Vehicles. 5 pp.

ISO 7588-83. Road Vehicles—Relays and Flashers—Mounting and Positioning Dimensions of Male Tabs and Socket Apertures for Relays and Flashers First Edition. 5 pp.

—Automotive—Electric Outlets—Mounting

BSI BS AU 201: Part 1-85. 1985 Relays and Flashers Part 1: Mounting and Positioning Dimensions of Male Tabs and Socket Apertures for Relays and Flashers for Road Vehicles. 5 pp.

ISO 7588-83. Road Vehicles—Relays and Flashers—Mounting and Positioning Dimensions of Male Tabs and Socket Apertures for Relays and Flashers First Edition. 5 pp.

—Automotive—Electric Switches

JIS D 5811-74. Inspection of Hazard Warning Switches for Automobiles (R 1983). 10 pp.

—Electrical Properties

ISO 4082-81. Road Vehicles—Motor Vehicles—Flasher Units First Edition. 9 pp.

—Mopeds

ISO 7400-90. Mopeds—Alternating Current Flasher Units Second Edition. 10 pp.

ISO 8052-90. Mopeds—Direct Current Flasher Units Second Edition. 10 pp.

Flasher Units (Cont.)

—Motorcycles

ISO 7398-90. Motorcycles—Direct Current Flasher Units Second Edition. 10 pp.

ISO 7399-90. Motorcycles—Alternating Current Flasher Units Second Edition. 8 pp.

—Quality Assurance

ISO 6797-82. Road Vehicles—Motor Vehicles—Production Conformity Requirements for Flasher Units First Edition. 7 pp.

—Terminals

ISO TR8857-86. Road Vehicles—Flashers—Functional Allocation of Terminals First Edition. 3 pp.

Flashers
Use: Flasher Units

Flashing Christmas Tree Lights
Use: Lighting Chains—Flashing

Flashing Lights
See Also: Indicator Lights

—Color Coding

BSI BS 4099: Part 2-77. (WITHDRAWN) 1977 Amd 1 Colours of Indicator Lights, Push Buttons, Annunciators and Digital Readouts: Flashing Lights, Annunciators and Digital Readouts (Superseded by BS EN 60073: 1993). 8 pp.

Flashings
Use For: Vent Flashings *See Also:* Roofing

CSA B272-93. Prefabricated Self-Sealing Roof Vent Flashings; (Gen Instr 1). 18 pp.

—Construction Contracts

DIN ENGL 18339-88. Tendering and Performance Stipulations in Contracts for Construction Works (VOB); Part C: General Technical Specifications in Contracts for Construction Works (ATV); Sheet Metal Roof and Wall Covering Works (Sept) (This Standard, Together with DIN 18299,. 10 pp.

Flashlights
Use For: Hand-Portable Lamps; Handlamps; Torches (Handlamps) *See Also:* Lamps; Lighting; Trouble Lights

BSI BS 4533: Sec 102.8-90. 1990 Luminaires Part 102: Particular Requirements Section 102.8: Specification for Headlamps. 23 pp.

CNS C3019-82. Method of Test for Flash Light (Nov)(1030). 2 pp.

CNS C4022-90. Flash Lights (Dec)(1029).

CNS C4022-82. Flash Lights (Nov)(1029). 1 p.

MOD UK DSTAN 62-2: Part 2-01. Battery Operated Lights, Torches, and Lanterns, Electric Part 2: Torches, Hand, Electric Issue 1; Amendment 1. 39 pp.

—Bulbs

CNS C3013-84. Method of Test for Miniature Lamps (Jun)(720). 1 p.

CNS C4017-83. Miniature Lamp Bulb for Portable Flashlight (Feb)(719). 2 pp.

CNS C4137-85. Bulbs for Flashlight (Rated Voltage 1.1V to 7.2V) (Feb)(4612).

JIS C 7508-83. Bulbs for Flashlight. 10 pp.

—Ships—Explosive Atmospheres

CNS F5084-85. Explosion-Proof Portable Electric Lamps for Marine Use (Dry Cells and Batteries Type) (Sep)(10767).

JIS F 8425-62. Explosion-Proof Flash Lights for Marine Use (Dry Battery Type) (R 1968). 8 pp.

Flashtakes
Use For: Flash Bulbs; Flash Guns
See Also: Photographic Flash Equipment

CNS C3056-78. Method of Determining the Light Output of Photoflash Lamps (Oct)(4614).

CNS C4138-85. Photoflash Lamps (May)(4613). 4 pp.

ISO 1229-89. Photography—Expendable Photoflash Lamps—Determination of Light Output Second Edition. 7 pp.

ISO 2691-87. Photography—Expendable Photoflash Lamps (Without Integral Reflector)—Definitions and Requirements for Luminous Flux/Time Characteristics Second Edition. 4 pp.

ISO 2800-73. Photography—Expendable Photoflash Lamps—Definition and Evaluation of Flashability First Edition; (Erratum—June 1980). 6 pp.

JIS B 7114-75. Flash Guns for Expendable Photoflash Lamp (R 1983). 7 pp.

JIS C 7520-73. Photoflash Lamps (R 1984). 8 pp.

JIS C 7611-89. Method of Determining the Luminous Flux Versus Time Characteristics and the Light Output of Photoflash Lamps. 9 pp.

INTERNATIONAL AND NON-U.S. NATIONAL STANDARDS
SUBJECT INDEX

Flashtakes *(Cont.)*
—Guide Numbers
BSI BS 5841: Part 6-80. 1980 Photographic Flash Equipment Part 6: Determination of Flash Guide Numbers. 4 pp.
CNS Z9008-80. Determination of Flash Guide Numbers (Jan)(5149).
ISO 1230-92. Photography—Determination of Flash Guide Numbers Second Edition. 9 pp.

—Light Output
CNS C3056-78. Method of Determining the Light Output of Photoflash Lamps (Oct)(4614).
ISO 1229-89. Photography—Expendable Photoflash Lamps—Determination of Light Output Second Edition. 7 pp.
JIS C 7611-89. Method of Determining the Luminous Flux Versus Time Characteristics and the Light Output of Photoflash Lamps. 9 pp.

—Shoes
CNS Z9007-80. Camera Accessory Shoes, with and Without Electrical Contacts, for Photoflash Lamps and Electronic Photoflash Units (Jan)(5148).
ISO 518-77. Photography—Camera Accessory Shoes, with and Without Electrical Contacts, for Photoflash Lamps and Electronic Photoflash Units First Edition. 4 pp.

Flashtubes
Use For: Flash Lamps; Flash Tubes; Flashcubes; Photoflash Lamps *See Also:* Discharge Lamps; Electron Tubes; Gas Filled Tubes; Lamps; Light Sources; Photographic Flash Equipment; Photographic Lighting Equipment
MOD UK DSTAN 59-60: 90/087-80. Valve, Electronic, Flash Tube Issue 1. 20 pp.

—Luminance
ISO 10503-91. Photography—Expendable Reflectored Photoflash Lamp Arrays—Definitions and Requirements for Luminous Flux/Time Characteristics First Edition. 6 pp.

—Quality Assurance
BSI BS 9024-78. (WITHDRAWN) 1978 Rules for the Preparation of Detail Specifications for Flash Tubes of Assessed Quality. 7 pp.

—Spectral Energy Distribution
ISO 10503-91. Photography—Expendable Reflectored Photoflash Lamp Arrays—Definitions and Requirements for Luminous Flux/Time Characteristics First Edition. 6 pp.

—Topflash/Flipflash
BSI BS EN 60604-93. 1993 'Topflash/Flipflash' Photographic Flash Lamp Array (IEC 604: 1980) (H). 17 pp.
CENELEC HD 430-83. Topflash/Flipflash Photographic Flash Lamp Array. 2 pp.
CENELEC EN 60604-93. 'Topflash/Flipflash' Photographic Flash Lamp Array (IEC 604: 1980). 12 pp.
IEC 604-80. "Topflash/Flipflash" Photographic Flash Lamp Array First Edition (CENELEC EN 60604: 1993). 20 pp.

Flasks
See Also: Beakers; Bottles; Containers; Laboratory Ware
BSI BS 676-82. 1982 Flasks with Graduated Necks (Phenols Type). 8 pp.

—Bitumens—Thermal Stability
DIN ENGL 52016-88. Testing the Thermal Stability of Bitumen in a Rotating Flask (Dec) (Supersedes the December 1980 Edition of DIN 52017). 3 pp.

—Boiling
BSI BS 2734-84. 1984 Amd 1 Boiling Flasks (Narrow-Necked), Conical, Flat Bottom and Round Bottom. 7 pp.
BSI BS 6352-83. 1983 Flasks with Ground Glass Joints. 7 pp.

—Boiling—Conical
ISO 1773-76. Laboratory Glassware—Boiling Flasks (Narrow-Necked) First Edition. 5 pp.
ISO 4797-81. Laboratory Glassware—Flasks with Conical Ground Joints First Edition. 4 pp.

—Brewing Equipment
BSI BS 701-53. 1953 Amd 1 Brewers' Mash Flask. 10 pp.

—Density
BSI BS 2701-56. 1956 Amd 1 Rees-Hugill Powder Density Flask. 13 pp.

—Distillation—Splash Heads
BSI BS 6855-87. (WITHDRAWN) 1987 Design and Construction of Glass Splash Heads for Laboratory Use. 10 pp.

Flasks *(Cont.)*
—Filtering
BSI BS 1739-82. 1982 Filter Flasks. 10 pp.
ISO 6556-81. Laboratory Glassware—Filter Flasks First Edition. 7 pp.

—Food—Vacuum Insulated
BSI BS 6672: Part 1-86. 1986 Insulated Domestic Food Containers Part 1: Vacuum Ware and Insulated Flasks, Jars and Jugs. 12 pp.
BSI BS 6672: Part 1-01. 1986 Amd 1 Insulated Domestic Food Containers Part 1: Specification for Vacuum Ware and Insulated Flasks, Jars and Jugs (AMD 7199) June 15, 1992. 14 pp.
SNZ NZS 5847: Part 1-88. Insulated Domestic Food Containers Part 1: Specification for Vacuum Ware and Insulated Flasks, Jars and Jugs Amend: A, 1988. 12 pp.

—Glass—Chemical Analysis
CNS R2125-86. Glass Flasks for Chemical Analysis (Mar)(7301).

—Iodine
BSI BS 2735-56. 1956 Iodine Flasks. 9 pp.

—Sugar Analysis
BSI BS 675-53. 1953 Amd 2 Sugar Flasks. 13 pp.

—Volumetric
BSI BS 615-53. 1953 Amd 2 Kohlrausch Flasks. 13 pp.
BSI BS 1792-82. 1982 Amd 2 One-Mark Volumetric Flasks. 10 pp.
CNS R2152-82. Glass Volumetric Flask for Chemical Analysis (May)(8866).
ISO 1042-83. Laboratory Glassware—One-Mark Volumetric Flasks Third Edition. 5 pp.
JIS R 3505-83. Volumetric Glassware. 16 pp.

—Volumetric—Glass Joints
CNS R2144-86. Interchangeable Ground Glass Joints for Chemical Analysis (Mar) (7320).
JIS R 3649-81. Interchangeable Conical Ground Glass Joints for Volumetric Flask.

Flat Belts
See Also: Belts (Machinery)
BSI BS 351-76. (WITHDRAWN) 1976 Rubber, Balata or Plastics Flat Transmission Belting of Textile Construction for General Use. 13 pp.
CNS K4004-80. Flat Rubber Belts for Power Transmission (Dec)(747). 2 pp.
CNS K6053-80. Method of Test for Flat Rubber Belts (Dec)(748).
CNS K8019-82. Industrial Transmission Flat Leather Belt (Dec)(9722).
ISO 22-91. Belt Drives—Flat Transmission Belts and Corresponding Pulleys—Dimensions and Tolerances Second Edition; (Cancels and Replaces 63, 99, and 100). 8 pp.
JIS K 6321-77. Flat Rubber Belts for Power Transmission (R 1980). 7 pp.
JIS K 6501-76. Industrial Transmission Flat Leather Belt.

—Pulleys
ISO 22-91. Belt Drives—Flat Transmission Belts and Corresponding Pulleys—Dimensions and Tolerances Second Edition; (Cancels and Replaces 63, 99, and 100). 8 pp.
ISO 254-90. Belt Drives—Pulleys—Quality, Finish and Balance Second Edition. 6 pp.
JIS B 1852-80. Pulleys for Flat Transmission Belts. 8 pp.

Flat Cable Connectors
See Also: Connectors; Printed Circuit Connectors; Rectangular Connectors
CNS C7151-82. Electrical Flat Cable Type Connectors (Mar)(8562).

—Electric Terminals—Quick Connect—Safety
IEC 1210-93. Connecting Devices—Flat Quick-Connect Terminations for Electrical Copper Conductors—Safety Requirements First Edition. 50 pp.

Flat Cables
See Also: Cables (Electric); Ribbon Cables
—Coaxial—Construction
CSA CAN/CSA-C22. 2 NO 214-M90. Communications Cables; (Gen Instr 1 Thru 3). 60 pp.

—Coaxial—Multiconductor—Construction
CSA CAN/CSA-C22. 2 NO 214-M90. Communications Cables; (Gen Instr 1 Thru 3). 60 pp.

Flat Cables *(Cont.)*
—Coaxial—Multiconductor—Testing
CSA CAN/CSA-C22. 2 NO 214-M90. Communications Cables; (Gen Instr 1 Thru 3). 60 pp.

—Coaxial—Testing
CSA CAN/CSA-C22. 2 NO 214-M90. Communications Cables; (Gen Instr 1 Thru 3). 60 pp.

—Communication—Construction
CSA CAN/CSA-C22. 2 NO 214-M90. Communications Cables; (Gen Instr 1 Thru 3). 60 pp.

—Communication—Multiconductor—Construction
CSA CAN/CSA-C22. 2 NO 214-M90. Communications Cables; (Gen Instr 1 Thru 3). 60 pp.

—Communication—Multiconductor—Testing
CSA CAN/CSA-C22. 2 NO 214-M90. Communications Cables; (Gen Instr 1 Thru 3). 60 pp.

—Multiconductor—Portable—Construction
CSA 983B Bull. Electrical Bulletin 983B April 25, 1977 to C22.2 NO 38. 39 pp.
CSA 983B Bull. Electrical Bulletin 983B April 25, 1977 to C22.2 NO 96. 39 pp.

—Multiconductor—Portable—Testing
CSA 983B Bull. Electrical Bulletin 983B April 25, 1977 to C22.2 NO 38. 39 pp.
CSA 983B Bull. Electrical Bulletin 983B April 25, 1977 to C22.2 NO 96. 39 pp.

—Power
DIN VDE 0250 Pt 809-85. Cables, Wires and Flexible Cords for Power Installations; Flat Ordinary Tough-Polychloroprene-Sheathed Flexible Cable (May). 11 pp.
JIS C 3652-87. Installation Methods of Power Flat Conductor Cables. 22 pp.

—PVC Insulated—PVC Sheathed
DIN VDE 0281 Pt 403-89. PVC Cables, Wires and Flexible Cords for Power Installation; Flat PVC Sheathed Flexible Cable 05VVH6 (June) (Partially Supersedes DIN 57 281 Part 403/VDE 0281 Part 402/11.81). 11 pp.
DIN VDE 0281 Pt 404-89. PVC Cables, Wires and Flexible Cords for Power Installation; Flat PVC Sheathed Flexible Cable 07VVH6 (June) (Partially Supersedes DIN 57281 Part 404/VDE 0281 Part 404/11.81). 12 pp.

—Under Carpet—Safety
CSA C22.2 NO 222-M1986. Type FCC Under-Carpet Wiring System (R 1992); (Gen Instr 1). 25 pp.

Flat Cards
See Also: Cards (Textile Machinery); Textile Machinery
—Chains
CNS L2080-83. Flat Chains for Card (Apr)(10178).

—Comb Blades
CNS L2081-83. Comb Blades for Flat Cards (Apr)(10179).
JIS L 5143-81. Comb Blades for Flat Cards.

Flat Chisels
See Also: Chisels
CNS B3243-78. Flat Chisels (May)(3975). 1 p.

Flat Conductor Cables
Use: Flat Cables

Flat Files
Use: Files (Tools)—Flat

Flat Finishes
See Also: Coatings; Finishes; Matte Finishes; Semigloss
CGSB 1-GP-71 METH 150.2-74. Methods of Testing Paints and Pigments Standard Formulas for Test Materials Formula for Flat Black; (Re-Edited September 1982). 1 p.

—Enamel Paints
CGSB 1-GP-71 METH 134.1-78. Methods of Testing Paints and Pigments Applicability and Appearance Enamels on Sealed Boards. 1 p.

—Enamels
CGSB 1-GP-71 METH 134.3-78. Methods of Testing Paints and Pigments Applicability and Appearance Enamels on Primed Steel, Brush and Roller Application. 1 p.

Flat Finishes (Cont.)

—Enamels (Cont.)
CGSB 1-GP-71 METH 143.4-78. Methods of Testing Paints and Pigments Long-Term Outdoor Performance Test for One Coat Styrenated Flat Alkyd Enamel on Steel. 1 p.

—Latex Paints
CGSB 1-GP-71 METH 134.18-78. Methods of Testing Paints and Pigments Applicability and Appearance One Coat over Simulated Masonry Surface. 1 p.

Flat Glass
Use: Sheet Glass

Flat Head Bolts
Use For: Flush Bolts *See Also:* Bolts
JIS B 1179-76. Flat Head Bolts. 18 pp.

—Aircraft—Countersunk Head—Slotted—High Strength
SBAC AS 1242 ISSUE 18. Bolt, 90 Degree Countersunk Head H.T.S.. 1 p.

—Aircraft—Nut Assemblies—Chains
SBAC AS 168 ISSUE 4. Nut and Bolt Assembly for Chains. 1 p.

—Aircraft—Phillips—Countersunk
SBAC AS 2920 ISSUE 6. Bolt, 90 Degree Countersunk Head Stainless Steel. 1 p.
SBAC AS 2926 ISSUE 2(I). Bolt M.S.—"Phillip's" 90 Degree Countersunk Head. 1 p.

—Countersunk
DIN ENGL 604-81. Flat Countersunk Nib Bolts (Oct). 4 pp.

—Countersunk—Square Neck
DIN ENGL 605-81. Flat Countersunk Square Neck Bolts with Long Square (Oct). 4 pp.
DIN ENGL 608-81. Flat Countersunk Square Neck Bolts with Short Square (Oct). 4 pp.

Flat Head Rivets
See Also: Round Head Rivets

—Aircraft
SBAC AS 469 ISSUE 3. Rivet, Flat Head. 1 p.

—Countersunk
CNS B2576-81. Flat Countersunk Head Rivets Nominal Diameters 3 to 5mm (Jun)(7564).

—Round
CNS B2575-81. Flat Round Head Rivets Nominal Diameters 1.4 to 6mm (Jun)(7563).
DIN ENGL 674-77. Flat Round Head Rivets; Nominal Diameters 1.4 to 6 mm (July). 4 pp.

Flat Head Screws
Use For: Flush Head Screws *See Also:* Screws

—Aircraft—Phillips—Countersunk
SBAC AS 54600-699 ISSUE 2. Bolt, Close Tolerance—Double Hexagon Extended Washer Head, INCO 718, 0.2500—28 UNJF—3A MOD BS4084.
SBAC AS 54700-799 ISSUE 2. Bolt, Close Tolerance—Double Hexagon Extended Washer Head, INCO 718, 0.3125—24 UNJF—3A MOD BS4084.
SBAC AS 54800-899 ISSUE 2. Bolt, Close Tolerance—Double Hexagon Extended Washer Head, INCO 718, 03750—24 UNJF—3A MOD BS4084.
SBAC AS 54900-999 ISSUE 2. Bolt, Close Tolerance—Double Hexagon Extended Washer Head, INCO 718, 04375—20 UNJF—3A MOD BS4084.

—Aircraft—Slotted—Shoulder—Handles
SBAC AS 2314 ISSUE 2. Screw for Rotating Knob. 1 p.

—Phillips—Countersunk
CNS B2294-81. Cross Recessed Countersunk (Flat) Head Screws (Apr)(4560). 4 pp.
DIN ENGL 965-90. Cross Recessed Countersunk Flat Head Screws (Aug). 5 pp.
DIN ENGL 7987-72. Cross Recessed Countersunk (Flat) Head Screws; (Countersunk Heads of Type Hitherto Used) (May). 2 pp.
ISO 7046-83. Cross Recessed Countersunk Flat Head Screws (Common Head Style)—Product Grade A and Property Class 4.8 Only First Edition. 6 pp.
ISO 7046 Pt 2-90. Cross Recessed Countersunk Flat Head Screws (Common Head Style)—Grade A—Part 2: Steel of Property Class 8.8, Stainless Steel and Non-Ferrous Metals First Edition. 7 pp.
ISO 7721 Pt 2-90. Countersunk Flat Head Screws—Part 2: Penetration Depth of Cross Recesses First Edition. 7 pp.

Flat Head Screws (Cont.)

—Phillips—Countersunk—Machine
BSI BS 7433: Part 2-91. 1991 Cross Recessed Machine Screws Part 2: Specification for Countersunk Flat Head Screws of Product Grade A and Property Class 4.8 Only. 6 pp.

—Phillips—Countersunk—Tapping
CNS B2185-87. Cross Recessed Countersunk (Flat) Head Tapping Screws (Jul)(4306). 5 pp.
DIN ENGL 7982-90. Cross Recessed Countersunk Head Tapping Screws (Aug). 4 pp.
ISO 7050-83. Cross Recessed Countersunk (Flat) Head Tapping Screws (Common Head Style) First Edition. 6 pp.
MOD UK DSTAN 53-76-01. Screws, Tapping, Thread Forming (Self Tapping Screws) Issue 2; Amendment 1. 22 pp.

—Phillips—Countersunk—Wood
CNS B2297-81. Cross Recessed Countersunk (Flat) Head Wood Screws (Apr)(4563). 3 pp.

—Slotted—Countersunk
CNS B2231-81. Slotted Countersunk (Flat) Head Screws (Apr)(4411).
CNS B2232-81. Slotted Countersunk (Flat) Head Screws with Full Dog Point (Apr)(4412).
DIN ENGL 963-90. Slotted Countersunk Head Screws (Aug). 6 pp.

—Slotted—Countersunk—Fine Mechanics
CNS B2208-80. Slotted Countersunk (Flat) Head Screws for Fine Mechanics (MO.4 to M1.4) (Jul)(4363).
DIN ENGL 8245-72. Slotted Countersunk (Flat) Head Screws for Fine Mechanics M 0.4 to M 1.4 (Oct). 1 p.

—Slotted—Countersunk—Tapping
CNS B2182-87. Slotted Countersunk (Flat) Head Tapping Screws (Jul)(4303). 5 pp.
ISO 1482-83. Slotted Countersunk (Flat) Head Tapping Screws (Common Head Style) First Edition. 5 pp.
MOD UK DSTAN 53-76-01. Screws, Tapping, Thread Forming (Self Tapping Screws) Issue 2; Amendment 1. 22 pp.

—Slotted—Countersunk—Wood
DIN ENGL 97-86. Slotted Countersunk (Flat) Head Wood Screws (Dec). 2 pp.

—Slotted—Mushroom
CNS B2236-81. Slotted Flat Mushroom Head Screws (Apr)(4416).

—Slotted—Wood
CNS B3128-59. Flat Head Wood Screws (Sep)(1051)(R 1973). 2 pp.

—Tapping—Phillips—Countersunk
BSI BS 7496: Part 2-91. 1991 Cross-Recessed Tapping Screws Part 2: Specification for Countersunk Flat Head Screws. 8 pp.

Flat Rate Tariffs (Telecommunications)
See Also: Guarantor Services; Reverse Charging; Tariffs (Telecommunications); Telecommunication Services; Telephone Services

—Data Transmission—Telecommunication Administrations
CCITT RECMN D.160-89. Mode of Application of the Flat-Rate Price Procedure Set Forth in Recommendation D.67 and Recommendation D.150 for Remuneration of Facilities Made Available to the Administrations of Other Countries—General Tariff Principles—Charging and Accounting. 5 pp.
CCITT RECMN E.252-89. Mode of Application of the Flat-Rate Price Procedure Set Forthe in Recommendations D.67 and D.150 for Remuneration of Facilities Made Available to the Administrations of Other Countries—Telephone Network and ISDN—Operation, Numbering, Routing and Mobile. 1 p.

—Data Transmission—Telecommunication Administrations—Refunds
CCITT RECMN D.160-89. Mode of Application of the Flat-Rate Price Procedure Set Forth in Recommendation D.67 and Recommendation D.150 for Remuneration of Facilities Made Available to the Administrations of Other Countries—General Tariff Principles—Charging and Accounting. 5 pp.
CCITT RECMN E.252-89. Mode of Application of the Flat-Rate Price Procedure Set Forthe in Recommendations D.67 and D.150 for Remuneration of Facilities Made Available to the Administrations of Other Countries—Telephone Network and ISDN—Operation, Numbering, Routing and Mobile. 1 p.

Flat Rate Tariffs (Telecommunications) (Cont.)

—Groups—Telecommunication Administrations
CCITT RECMN D.160-89. Mode of Application of the Flat-Rate Price Procedure Set Forth in Recommendation D.67 and Recommendation D.150 for Remuneration of Facilities Made Available to the Administrations of Other Countries—General Tariff Principles—Charging and Accounting. 5 pp.
CCITT RECMN E.252-89. Mode of Application of the Flat-Rate Price Procedure Set Forthe in Recommendations D.67 and D.150 for Remuneration of Facilities Made Available to the Administrations of Other Countries—Telephone Network and ISDN—Operation, Numbering, Routing and Mobile. 1 p.

—Groups—Telecommunication Administrations—Refunds
CCITT RECMN D.160-89. Mode of Application of the Flat-Rate Price Procedure Set Forth in Recommendation D.67 and Recommendation D.150 for Remuneration of Facilities Made Available to the Administrations of Other Countries—General Tariff Principles—Charging and Accounting. 5 pp.
CCITT RECMN E.252-89. Mode of Application of the Flat-Rate Price Procedure Set Forthe in Recommendations D.67 and D.150 for Remuneration of Facilities Made Available to the Administrations of Other Countries—Telephone Network and ISDN—Operation, Numbering, Routing and Mobile. 1 p.

—Groups—Telephone Relations—Africa
CCITT RECMN D.600R-89. Determination of Accounting Rate Shares and Collection Charges in Telephone Relations Between Countries in Africa—General Tariff Principles—Charging and Accounting in International Telecommunications Services (Study Group III) 11 pp. 11 pp.

—Satellite Communications—Telegraph Circuits—Europe
CCITT RECMN D.301R-89. Determination of Accounting Rate Shares and Collection Charges in Telex Relations Between Countries in Europe and the Mediterranean Basin—General Tariff Principles—Charging and Accounting in Intl Telecom. Serv. (Study Group III) 12 pp. 12 pp.

—Satellite Communications—Telegraph Circuits—Mediterranean
CCITT RECMN D.301R-89. Determination of Accounting Rate Shares and Collection Charges in Telex Relations Between Countries in Europe and the Mediterranean Basin—General Tariff Principles—Charging and Accounting in Intl Telecom. Serv. (Study Group III) 12 pp. 12 pp.

—Satellite Communications—Telephone Circuits—Europe
CCITT RECMN D.300R-92. Determination of Accounting Rate Shares in Telephone Relations Between Countries in Europe and the Mediterranean Basin (Study Group III) 12 pp. 12 pp.
CCITT RECMN D.300R-91. Determination of Accounting Rate Shares in Telephone Relations Between Countries in Europe and the Mediterranean Basin (Study Group III) 16 pp. 16 pp.
CCITT RECMN D.300R-89. Determination of Accounting Rate Shares and Collection Charges in Telephone Relations Between Countries in Europe and the Mediterranean Basin—General Tariff Principles—Charging and Accounting in Intl Telecom. Serv. (Study Group III) 13 pp. 13 pp.

—Satellite Communications—Telephone Circuits—Mediterranean
CCITT RECMN D.300R-92. Determination of Accounting Rate Shares in Telephone Relations Between Countries in Europe and the Mediterranean Basin (Study Group III) 12 pp. 12 pp.
CCITT RECMN D.300R-91. Determination of Accounting Rate Shares in Telephone Relations Between Countries in Europe and the Mediterranean Basin (Study Group III) 16 pp. 16 pp.
CCITT RECMN D.300R-89. Determination of Accounting Rate Shares and Collection Charges in Telephone Relations Between Countries in Europe and the Mediterranean Basin—General Tariff Principles—Charging and Accounting in Intl Telecom. Serv. (Study Group III) 13 pp. 13 pp.

INTERNATIONAL AND NON-U.S. NATIONAL STANDARDS
SUBJECT INDEX

Flat

Flat Rate Tariffs (Telecommunications) *(Cont.)*

—Satellite Communications—Television Transmission—Europe

CCITT RECMN D.303R-89. Determination of Accounting Rate Shares and Collection Charges Applicable by Countries in Europe and the Mediterranean Basin to the Occasional Provision of Circuits for Sound-and Television-Programme Transmissions—General Tariff Principles—Charging and Accounting. 14 pp.

—Satellite Communications—Television Transmission—Mediterranean

CCITT RECMN D.303R-89. Determination of Accounting Rate Shares and Collection Charges Applicable by Countries in Europe and the Mediterranean Basin to the Occasional Provision of Circuits for Sound-and Television-Programme Transmissions—General Tariff Principles—Charging and Accounting. 14 pp.

—Sound-Program Circuits—Europe

CCITT RECMN D.303R-89. Determination of Accounting Rate Shares and Collection Charges Applicable by Countries in Europe and the Mediterranean Basin to the Occasional Provision of Circuits for Sound-and Television-Programme Transmissions—General Tariff Principles—Charging and Accounting. 14 pp.

—Sound-Program Circuits—Mediterranean

CCITT RECMN D.303R-89. Determination of Accounting Rate Shares and Collection Charges Applicable by Countries in Europe and the Mediterranean Basin to the Occasional Provision of Circuits for Sound-and Television-Programme Transmissions—General Tariff Principles—Charging and Accounting. 14 pp.

—Supergroups—Telecommunication Administrations

CCITT RECMN D.160-89. Mode of Application of the Flat-Rate Price Procedure Set Forth in Recommendation D.67 and Recommendation D.150 for Remuneration of Facilities Made Available to the Administrations of Other Countries—General Tariff Principles—Charging and Accounting. 5 pp.

CCITT RECMN E.252-89. Mode of Application of the Flat-Rate Price Procedure Set Forthe in Recommendations D.67 and D.150 for Remuneration of Facilities Made Available to the Administrations of Other Countries—Telephone Network and ISDN—Operation, Numbering, Routing and Mobile. 1 p.

—Supergroups—Telecommunication Administrations—Refunds

CCITT RECMN D.160-89. Mode of Application of the Flat-Rate Price Procedure Set Forth in Recommendation D.67 and Recommendation D.150 for Remuneration of Facilities Made Available to the Administrations of Other Countries—General Tariff Principles—Charging and Accounting. 5 pp.

CCITT RECMN E.252-89. Mode of Application of the Flat-Rate Price Procedure Set Forthe in Recommendations D.67 and D.150 for Remuneration of Facilities Made Available to the Administrations of Other Countries—Telephone Network and ISDN—Operation, Numbering, Routing and Mobile. 1 p.

—Supergroups—Telephone Relations—Africa

CCITT RECMN D.600R-89. Determination of Accounting Rate Shares and Collection Charges in Telephone Relations Between Countries in Africa—General Tariff Principles—Charging and Accounting in International Telecommunications Services (Study Group III) 11 pp. 11 pp.

—Telecommunication Circuits—Television Transmission—Europe

CCITT RECMN D.303R-89. Determination of Accounting Rate Shares and Collection Charges Applicable by Countries in Europe and the Mediterranean Basin to the Occasional Provision of Circuits for Sound-and Television-Programme Transmissions—General Tariff Principles—Charging and Accounting. 14 pp.

—Telecommunication Circuits—Television Transmission—Mediterranean

CCITT RECMN D.303R-89. Determination of Accounting Rate Shares and Collection Charges Applicable by Countries in Europe and the Mediterranean Basin to the Occasional Provision of Circuits for Sound-and Television-Programme Transmissions—General Tariff Principles—Charging and Accounting. 14 pp.

Flat Rate Tariffs (Telecommunications) *(Cont.)*

—Telecommunication Services—Digital Channels—Europe

CCITT RECMN D.307R-91. Remuneration of Digital Systems and Channels Used in Telecommunication Relations Between the Countries of Europe and the Mediterranean Basin (Study Group III) 7 pp. 7 pp.

CCITT RECMN D.307R-89. Remuneration of Digital Systems and Channels Used in Telecommunication Relations Between the Countries of Europe and the Mediterranean Basin—General Tariff Principles—Charging and Accounting in Intl Telecom. Serv. (Study Group III) 3 pp. 3 pp.

—Telecommunication Services—Digital Channels—Mediterranean

CCITT RECMN D.307R-91. Remuneration of Digital Systems and Channels Used in Telecommunication Relations Between the Countries of Europe and the Mediterranean Basin (Study Group III) 7 pp. 7 pp.

CCITT RECMN D.307R-89. Remuneration of Digital Systems and Channels Used in Telecommunication Relations Between the Countries of Europe and the Mediterranean Basin—General Tariff Principles—Charging and Accounting in Intl Telecom. Serv. (Study Group III) 3 pp. 3 pp.

—Telecommunication Services—ISDN—Europe

CCITT RECMN D.307R-91. Remuneration of Digital Systems and Channels Used in Telecommunication Relations Between the Countries of Europe and the Mediterranean Basin (Study Group III) 7 pp. 7 pp.

CCITT RECMN D.307R-89. Remuneration of Digital Systems and Channels Used in Telecommunication Relations Between the Countries of Europe and the Mediterranean Basin—General Tariff Principles—Charging and Accounting in Intl Telecom. Serv. (Study Group III) 3 pp. 3 pp.

—Telecommunication Services—ISDN—Mediterranean

CCITT RECMN D.307R-91. Remuneration of Digital Systems and Channels Used in Telecommunication Relations Between the Countries of Europe and the Mediterranean Basin (Study Group III) 7 pp. 7 pp.

CCITT RECMN D.307R-89. Remuneration of Digital Systems and Channels Used in Telecommunication Relations Between the Countries of Europe and the Mediterranean Basin—General Tariff Principles—Charging and Accounting in Intl Telecom. Serv. (Study Group III) 3 pp. 3 pp.

—Telecommunication Services—Telephone Relations—Europe

CCITT RECMN D.300R-92. Determination of Accounting Rate Shares in Telephone Relations Between Countries in Europe and the Mediterranean Basin (Study Group III) 12 pp. 12 pp.

CCITT RECMN D.300R-91. Determination of Accounting Rate Shares in Telephone Relations Between Countries in Europe and the Mediterranean Basin (Study Group III) 16 pp. 16 pp.

CCITT RECMN D.300R-89. Determination of Accounting Rate Shares and Collection Charges in Telephone Relations Between Countries in Europe and the Mediterranean Basin—General Tariff Principles—Charging and Accounting in Intl Telecom. Serv. (Study Group III) 13 pp. 13 pp.

—Telecommunication Services—Telephone Relations—Mediterranean

CCITT RECMN D.300R-92. Determination of Accounting Rate Shares in Telephone Relations Between Countries in Europe and the Mediterranean Basin (Study Group III) 12 pp. 12 pp.

CCITT RECMN D.300R-91. Determination of Accounting Rate Shares in Telephone Relations Between Countries in Europe and the Mediterranean Basin (Study Group III) 16 pp. 16 pp.

CCITT RECMN D.300R-89. Determination of Accounting Rate Shares and Collection Charges in Telephone Relations Between Countries in Europe and the Mediterranean Basin—General Tariff Principles—Charging and Accounting in Intl Telecom. Serv. (Study Group III) 13 pp. 13 pp.

—Telegraph Circuits—Telecommunication Administrations

CCITT RECMN D.160-89. Mode of Application of the Flat-Rate Price Procedure Set Forth in Recommendation D.67 and Recommendation D.150 for Remuneration of Facilities Made Available to the Administrations of Other Countries—General Tariff Principles—Charging and Accounting. 5 pp.

Flat Rate Tariffs (Telecommunications) *(Cont.)*

—Telegraph Circuits—Telecommunication Administrations *(Cont.)*

CCITT RECMN E.252-89. Mode of Application of the Flat-Rate Price Procedure Set Forthe in Recommendations D.67 and D.150 for Remuneration of Facilities Made Available to the Administrations of Other Countries—Telephone Network and ISDN—Operation, Numbering, Routing and Mobile. 1 p.

—Telegraph Circuits—Telecommunication Administrations—Refunds

CCITT RECMN D.160-89. Mode of Application of the Flat-Rate Price Procedure Set Forth in Recommendation D.67 and Recommendation D.150 for Remuneration of Facilities Made Available to the Administrations of Other Countries—General Tariff Principles—Charging and Accounting. 5 pp.

CCITT RECMN E.252-89. Mode of Application of the Flat-Rate Price Procedure Set Forthe in Recommendations D.67 and D.150 for Remuneration of Facilities Made Available to the Administrations of Other Countries—Telephone Network and ISDN—Operation, Numbering, Routing and Mobile. 1 p.

—Telegraphy—Telecommunication Administrations

CCITT RECMN D.160-89. Mode of Application of the Flat-Rate Price Procedure Set Forth in Recommendation D.67 and Recommendation D.150 for Remuneration of Facilities Made Available to the Administrations of Other Countries—General Tariff Principles—Charging and Accounting. 5 pp.

CCITT RECMN E.252-89. Mode of Application of the Flat-Rate Price Procedure Set Forthe in Recommendations D.67 and D.150 for Remuneration of Facilities Made Available to the Administrations of Other Countries—Telephone Network and ISDN—Operation, Numbering, Routing and Mobile. 1 p.

—Telegraphy—Telecommunication Administrations—Refunds

CCITT RECMN D.160-89. Mode of Application of the Flat-Rate Price Procedure Set Forth in Recommendation D.67 and Recommendation D.150 for Remuneration of Facilities Made Available to the Administrations of Other Countries—General Tariff Principles—Charging and Accounting. 5 pp.

CCITT RECMN E.252-89. Mode of Application of the Flat-Rate Price Procedure Set Forthe in Recommendations D.67 and D.150 for Remuneration of Facilities Made Available to the Administrations of Other Countries—Telephone Network and ISDN—Operation, Numbering, Routing and Mobile. 1 p.

—Telephone Circuits—Telecommunication Administrations

CCITT RECMN D.160-89. Mode of Application of the Flat-Rate Price Procedure Set Forth in Recommendation D.67 and Recommendation D.150 for Remuneration of Facilities Made Available to the Administrations of Other Countries—General Tariff Principles—Charging and Accounting. 5 pp.

CCITT RECMN E.252-89. Mode of Application of the Flat-Rate Price Procedure Set Forthe in Recommendations D.67 and D.150 for Remuneration of Facilities Made Available to the Administrations of Other Countries—Telephone Network and ISDN—Operation, Numbering, Routing and Mobile. 1 p.

—Telephone Circuits—Telecommunication Administrations—Refunds

CCITT RECMN D.160-89. Mode of Application of the Flat-Rate Price Procedure Set Forth in Recommendation D.67 and Recommendation D.150 for Remuneration of Facilities Made Available to the Administrations of Other Countries—General Tariff Principles—Charging and Accounting. 5 pp.

CCITT RECMN E.252-89. Mode of Application of the Flat-Rate Price Procedure Set Forthe in Recommendations D.67 and D.150 for Remuneration of Facilities Made Available to the Administrations of Other Countries—Telephone Network and ISDN—Operation, Numbering, Routing and Mobile. 1 p.

—Telephone Circuits—Telephone Relations—Africa

CCITT RECMN D.600R-89. Determination of Accounting Rate Shares and Collection Charges in Telephone Relations Between Countries in Africa—General Tariff Principles—Charging and Accounting in International Telecommunications Services (Study Group III) 11 pp. 11 pp.

INDUSTRY STANDARDS

Flat Rate Tariffs (Telecommunications) (Cont.)

—Telephone Services
CCITT RECMN D.150-92. New System for Accounting in International Telephony (Study Group III) 14 pp. 14 pp.

CCITT RECMN D.150-89. New System for Accounting in International Telephony—General Tariff Principles—Charging and Accounting in International Telecommunications Services (Study Group III) 12 pp. 12 pp.

CCITT RECMN E.250-89. New System for Accounting in International Telephony—Telephone Network and ISDN—Operation, Numbering, Routing and Mobile Serivce (Study Group II) 1 pp (Same as Recmn D.150).

—Telephone Services—Telecommunication Administrations
CCITT RECMN D.160-89. Mode of Application of the Flat-Rate Price Procedure Set Forth in Recommendation D.67 and Recommendation D.150 for Remuneration of Facilities Made Available to the Administrations of Other Countries—General Tariff Principles—Charging and Accounting. 5 pp.

CCITT RECMN E.252-89. Mode of Application of the Flat-Rate Price Procedure Set Forthe in Recommendations D.67 and D.150 for Remuneration of Facilities Made Available to the Administrations of Other Countries—Telephone Network and ISDN—Operation, Numbering, Routing and Mobile. 1 p.

—Telephone Services—Telecommunication Administrations—Refunds
CCITT RECMN D.160-89. Mode of Application of the Flat-Rate Price Procedure Set Forth in Recommendation D.67 and Recommendation D.150 for Remuneration of Facilities Made Available to the Administrations of Other Countries—General Tariff Principles—Charging and Accounting. 5 pp.

CCITT RECMN E.252-89. Mode of Application of the Flat-Rate Price Procedure Set Forthe in Recommendations D.67 and D.150 for Remuneration of Facilities Made Available to the Administrations of Other Countries—Telephone Network and ISDN—Operation, Numbering, Routing and Mobile. 1 p.

—Telephone Services—Telephone Relations—Africa
CCITT RECMN D.600R-89. Determination of Accounting Rate Shares and Collection Charges in Telephone Relations Between Countries in Africa—General Tariff Principles—Charging and Accounting in International Telecommunications Services (Study Group III) 11 pp. 11 pp.

—Telex Communications
CCITT RECMN D.67-91. Charging and Accounting in the International Telex Service (Study Group III) 7 pp. 7 pp.

CCITT RECMN D.67-89. Charging and Accounting in the International Telex Service—General Tariff Principles—Charging and Accounting in International Telecommunications Services (Study Group III) 4 pp. 4 pp.

—Telex Communications—Telecommunication Administrations
CCITT RECMN D.160-89. Mode of Application of the Flat-Rate Price Procedure Set Forth in Recommendation D.67 and Recommendation D.150 for Remuneration of Facilities Made Available to the Administrations of Other Countries—General Tariff Principles—Charging and Accounting. 5 pp.

CCITT RECMN E.252-89. Mode of Application of the Flat-Rate Price Procedure Set Forthe in Recommendations D.67 and D.150 for Remuneration of Facilities Made Available to the Administrations of Other Countries—Telephone Network and ISDN—Operation, Numbering, Routing and Mobile. 1 p.

—Telex Communications—Telecommunication Administrations—Refunds
CCITT RECMN D.160-89. Mode of Application of the Flat-Rate Price Procedure Set Forth in Recommendation D.67 and Recommendation D.150 for Remuneration of Facilities Made Available to the Administrations of Other Countries—General Tariff Principles—Charging and Accounting. 5 pp.

CCITT RECMN E.252-89. Mode of Application of the Flat-Rate Price Procedure Set Forthe in Recommendations D.67 and D.150 for Remuneration of Facilities Made Available to the Administrations of Other Countries—Telephone Network and ISDN—Operation, Numbering, Routing and Mobile. 1 p.

—Telex Communications—Telephone Relations—Africa
CCITT RECMN D.601R-89. Determination of Accounting Rate Shares and Collection Charges in Telex Relations Between Countries in Africa—General Tariff Principles—Charging and Accounting in International Telecommunications Services (Study Group III) 10 pp. 10 pp.

—Telex Communications—Telephone Relations—Europe
CCITT RECMN D.301R-89. Determination of Accounting Rate Shares and Collection Charges in Telex Relations Between Countries in Europe and the Mediterranean Basin—General Tariff Principles—Charging and Accounting in Intl Telecom. Serv. (Study Group III) 12 pp. 12 pp.

—Telex Communications—Telephone Relations—Mediterranean
CCITT RECMN D.301R-89. Determination of Accounting Rate Shares and Collection Charges in Telex Relations Between Countries in Europe and the Mediterranean Basin—General Tariff Principles—Charging and Accounting in Intl Telecom. Serv. (Study Group III) 12 pp. 12 pp.

Flat Ribbon Cables
Use: Ribbon Cables

Flat Springs
Use: Leaf Springs

Flat Warp Knitting Machines
Use: Warp Knitting Machines

Flat Washers
Use: Plain Washers

Flattening Testing
See Also: Compression Testing; Physical Testing

—Metal Tubes
AECMA PREN2002-10-87. Test Methods for Metallic Materials—Part 10—Tube Flattening Test. 5 pp.

ISO 8492-86. Metallic Materials—Tube—Flattening Test First Edition. 4 pp.

—Seamless Pipes
DIN ENGL 50136-79. Testing of Metallic Materials; Flattening Test on Tubes (June). 3 pp.

—Thermal Insulation
CEN PREN 825-92. Thermal Insulating Products for Building Applications—Determination of Flatness. 6 pp.

—Welded Pipes
DIN ENGL 50136-79. Testing of Metallic Materials; Flattening Test on Tubes (June). 3 pp.

Flatware
See Also: Cutlery; Tableware

—Cadmium Content
DIN ENGL 51031-86. Testing of Articles Intended for Use in Contact with Foodstuffs; Determination of Release of Lead and Cadmium from Silicate Surfaced Articles Intended for Use in Contact with Foodstuffs (Feb). 5 pp.

DIN ENGL 51032-86. Ceramics, Glass, Glass Ceramics, Vitreous Enamels; Permissable Limits for the Release of Lead and Cadmium from Articles Intended for Use in Contact with Foodstuffs (Feb). 4 pp.

—Glossaries
ISO 4481-77. Cutlery and Flatware—Nomenclature First Edition. 42 pp.

—Lead Content
DIN ENGL 51031-86. Testing of Articles Intended for Use in Contact with Foodstuffs; Determination of Release of Lead and Cadmium from Silicate Surfaced Articles Intended for Use in Contact with Foodstuffs (Feb). 5 pp.

DIN ENGL 51032-86. Ceramics, Glass, Glass Ceramics, Vitreous Enamels; Permissable Limits for the Release of Lead and Cadmium from Articles Intended for Use in Contact with Foodstuffs (Feb). 4 pp.

—Silver Coatings—Electroplated
CNS S1185-87. Electroplated Coatings of Silver for Cutlery, Flat-Ware and Hollow-Ware (Oct)(11316).

Flatwork Ironers
Use: Irons (Electric)—Flatwork

Flatwork Ironing Machines
Use: Irons (Electric)—Flatwork

Flavomycin Content Analysis
Use: Bambermycins Content Analysis

Flavor Testing
See Also: Sensory Analysis

—Canned Foods
CNS N6022-87. Method of Test for Canned Food: Test for Flavor and Odor (Feb)(978). 2 pp.

—Caseins
SAA AS 2300.10.2-91. Methods of Chemical and Physical Testing for the Dairying Industry—Part 10: Caseins, Caseinates and Coprecipitates—Part 10.2: Assessment of Odour and Flavour (Supersedes AS N60—1970 (in Part)). 1 p.

—Food
BSI BS 5929: Part 4-86. 1986 Methods for Sensory Analysis of Food Part 4: Flavour Profile Methods. 9 pp.

ISO 6564-85. Sensory Analysis—Methodology—Flavour Profile Methods First Edition. 8 pp.

—Vinegars
CNS N6025-83. Method of Test for Vinegar (Jul)(1071). 1 p.

Flavorings
See Also: Essential Oils; Food Additives

EC 88/388/EEC-88. Council Directive on the Approximation of the Laws of the Member States Relating to Flavourings for Use in Foodstuffs and to Source Materials for Their Production. 6 pp.

EC 88/389/EEC-88. Council Decision on the Establishment, by the Commission, of an Inventory of the Source Materials and Substances Used in the Preparation of Flavourings. 1 p.

Flaws
Use: Defects

Flax
See Also: Fibers; Linen; Linseed

BSI BS 7144-89. 1989 Flax and Jute Fabrics for Industrial Use. 16 pp.

CNS L3034-71. Method of Test for Flax Fiber (Jan)(2050).

ISO 2370-80. Textiles—Determination of Fineness of Flax Fibres—Permeametric Methods Second Edition. 13 pp.

MOD UK DSTAN 83-26-01. Cloths, Plain Woven, Flax Issue 2; Amendment 2. 20 pp.

—Carpets
CNS L4006-74. Ramie, Flax or Wool Rugs and Carpets (Mar)(1390). 2 pp.

—Linings
CNS L4135-83. Flax or Ramie Interlining Cloth (Jun)(7738).

—Solubility
BSI BS 6645-85. 1985 Method for Determination of Alkali Solubility of Flax Materials. 4 pp.

—Threads (Textiles)
CNS L3118-85. Method of Test for Flax and Ramie Sewing Thread (May)(7277).

CNS L4129-81. Flax and Ramie Sewing Thread (Apr)(7276).

JIS L 2403-78. Flax and Ramie Sewing Thread.

—Yarns—Tatami
JIS L 2404-78. Flax and Ramie Yarns for TATAMI.

Flax Fibers
Use: Flax

Fleece
See Also: Fabrics

—Polyester—Bitumen Sheeting
DIN ENGL 18192-85. Bonded Polyester Fleece Used as an Inlay for Bitumen and Polymer Bitumen Sheeting; Concept, Designation, Requirements, Testing (Aug). 4 pp.

FleetNET (BTN)
See Also: Message Handling Systems; SELCAL Systems

—Group Calls
CCIR Report 921-2-90. System Aspects of Digital Ship Earth Stations—Section 8F—Frequencies, Orbits and Systems. 46 pp.

Flesh Meal
Use: Bone Meal

INTERNATIONAL AND NON-U.S. NATIONAL STANDARDS
SUBJECT INDEX

Flexible

Flex Abrasion Resistance Testing
Use: Abrasion Testing

Flexibility
See Also: Bending Stress; Flexural Strength; Flexure; Plastic Properties; Rigidity; Softness

—**Aluminum Coatings (On Aluminum)**
CGSB 1-GP-71 METH 119.15-79. Methods of Testing Paints and Pigments Flexibility Finishes on Primed Aluminum. 1 p.

—**Belt Conveyors**
SAA AS 1334.4-82. Methods of Testing Conveyor and Elevator Belting—Part 4: Determination of Troughability of Conveyor Belting Reconfirmed 1989. 2 pp.

—**Bitumens**
CEN PREN 1109-93. Bitumen Sheets for Waterproofing—Determination of Flexibility at Low Temperature. 11 pp.

—**Coated Fabrics**
ISO 5979-82. Rubber or Plastics Coated Fabrics—Determination of Flexibility—Flat Loop Method First Edition. 5 pp.

—**Conveyor Belts**
ISO 703-88. Conveyor Belts—Troughability—Characteristics of Transverse Flexibility and Test Method Second Edition. 6 pp.

—**Electrical Insulating Varnishes**
MOD UK DEF-1053: METH NO. 40. Standard Methods of Testing Paint, Varnish, Lacquer and Related Products. Indices Method 40: The Effects of Heating on the Flexibility of Electrical Insulating Varnishes (Withdrawn).

—**Enamels**
CNS K6031-87. Method of Test for Lacquer Enamel (May) (628). 6 pp.

—**Fiberglass Reinforced Plastics**
CNS K6662-81. Method of Test for Fiber Glass Products (May)(7397). 9 pp.

—**Floor Coverings**
CEN PREN 435-90. Resilient Floorcoverings—Determination of Flexibility. 5 pp.

—**Gaskets**
CNS O2014-73. Method of Test for Cork Binder Discs (May)(2303). 1 p.

—**Lacquers**
CNS K6386-76. Method of Test for Lacquer in Surface of Leather Shoes (Mar) (3922). 2 pp.

—**Leathercloth**
CNS K6771-83. Method of Test for Polyurethane Leather (May)(10270). 3 pp.

—**Paints**
BSI BS AU 148: Part 3-69. 1969 Methods of Test for Motor Vehicle Paints Part 3: Flexibility and Adhesion. 3 pp.

—**Primers**
CGSB 1-GP-71 METH 119.12-79. Methods of Testing Paints and Pigments Flexibility Primers on Pretreated Aluminum. 1 p.
CGSB 1-GP-71 METH 119.14-79. Methods of Testing Paints and Pigments Flexibility Primers on Anodized Aluminum. 1 p.
CNS K6115-86. Method of Test for Oleoresinous Primer (Apr)(1249). 3 pp.

—**Sealants**
CGSB CAN2-19.0-M77METH11.1-78. Methods of Testing Putty, Caulking and Sealing Compounds Flexibility of Sealing Compounds. 1 p.

—**Tinplate Panels**
CGSB 1-GP-71 METH 119.1-79. Methods of Testing Paints and Pigments Flexibility Force-Dried Films. 1 p.
CGSB 1-GP-71 METH 119.4-79. Methods of Testing Paints and Pigments Flexibility Low-Temperature. 1 p.
CGSB 1-GP-71 METH 119.5-79. Methods of Testing Paints and Pigments Flexibility Air-Dried Films. 1 p.
CGSB 1-GP-71 METH 119.13-74. Methods of Testing Paints and Pigments Flexibility Baked Films. 1 p.

Flexible Cables
Use: Cables (Electric)

Flexible Circuits
Use: Flexible Printed Circuit Boards

Flexible Couplings
See Also: Shaft Couplings

Flexible Couplings (Cont.)
—**Shaft**
BSI BS 3170-72. 1972 Flexible Couplings for Power Transmission. 17 pp.
DIN ENGL 740 PT 1-86. Power Transmission Engineering; Flexible Shaft Couplings; Technical Delivery Conditions (Aug). 15 pp.
DIN ENGL 740 Pt 2-86. Power Transmission Engineering; Flexible Shaft Couplings; Parameters and Design Principles (Aug). 12 pp.

Flexible Disk Cartridges
Use: Disk Cartridges

Flexible Metal Conduit Fittings
See Also: Electric Conduit Fittings; Flexible Metal Conduits
CSA CAN/CSA-C22. 2 NO 18-92. Outlet Boxes, Conduit Boxes, and Fittings; (Gen Instr 1 Thru 2). 118 pp.
IEC 1035 Pt 2-3-93. Specification for Conduit Fittings for Electrical Installations Part 2: Particular Specifications Section 3: Fittings for Flexible Conduits of Metal, Insulating or Composite Materials and for Pliable Conduits of Metal or Composite Materials First Edition. 27 pp.
JIS C 8350-91. Fittings for Flexible Metal Conduits. 22 pp.

—**Connectors**
CSA CAN/CSA-C22. 2 NO 18-92. Outlet Boxes, Conduit Boxes, and Fittings; (Gen Instr 1 Thru 2). 118 pp.

—**Insulating Bushings**
CNS C4247-83. Insulated Bushings for Flexible Metal Conduit (Oct)(6103).

—**Steel**
BSI BS 31-40. 1940 Amd 5 Steel Conduit and Fittings for Electrical Wiring. 40 pp.
BSI BS 731: Part 1-52. 1952 Amd 3 Flexible Steel Conduit for Cable Protection and Flexible Steel Tubing to Enclose Flexible Drives Part 1: Flexible Steel Conduit and Adaptors for the Protection of Electric Cable. 16 pp.
BSI BS 4568: Part 1-70. 1970 Amd 2 Steel Conduit and Fittings with Metric Thread of ISO Form for Electrical Installations Part 1: Steel Conduit, Bends and Couplers (AMD 4568) January 30, 1987. 31 pp.
BSI BS 4568: Part 2-70. 1970 Amd 4 Steel Conduit and Fittings with Metric Thread of ISO Form for Electrical Installations Part 2: Fittings and Components. 39 pp.
CNS C4061-72. Steel Tubes for Electric Lines (with Insulating Varnish Coated) (Jun)(2607).
SNZ NZS 6221: Part 1-72. Specification for Steel Conduit and Fittings with Metric Threads of ISO Form for Electrical Installations Part 1: Steel Conduit, Bends and Couplers Amend: A, 1972; 1A, 1979. 32 pp.
SNZ NZS 6221: Part 2-72. Specification for Steel Conduit and Fittings with Metric Threads of ISO Form for Electrical Installations Part 2: Fittings and Components Amend: A, 1972; 1A, 1974; 2A 1978. 40 pp.

Flexible Metal Conduits
See Also: Electric Conduits; Flexible Metal Conduit Fittings; Liquidtight Flexible Conduits
BSI BS 6946-88. 1988 Amd 1 Metal Channel Cable Support Systems for Electrical Installations (AMD 6455) December 21, 1990. 13 pp.
CENELEC PREN 50086-2-2-92. Conduit Systems for Electrical Installations Part 2: General Requirements Section 2.2: Pliable Conduit Systems. 9 pp.
CENELEC PREN 50086-2-3-92. Conduit Systems for Electrical Installations Part 2: General Requirements Section 2.3: Flexible Conduit Systems. 8 pp.
CNS C4245-80. Flexible Metal Conduit (Aug)(6101).
IEC 614 Pt 2-1-82. Specification for Conduits for Electrical Installations Part 2: Particular Specifications for Conduits Section One—Metal Conduits First Edition. 19 pp.
IEC 614 Pt 2-5-92. Specification for Conduits for Electrical Installations Part 2: Particular Specifications for Conduits Section 5: Flexible Conduits First Edition. 37 pp.
IEC 614 Pt 2-6-92. Specification for Conduits for Electrical Installations Part 2: Particular Specifications for Conduits Section 6: Pliable Conduits of Metal or Composite Materials First Edition. 37 pp.
JIS C 8309-77. Flexible Metal Conduits (R 1983). 15 pp.

—**Cable Clamps**
CSA CAN/CSA-C22. 2 NO 18-92. Outlet Boxes, Conduit Boxes, and Fittings; (Gen Instr 1 Thru 2). 118 pp.

—**Electrical Codes**
CSA 851 Bull. Electrical Bulletin 851 November 17, 1971 to C22.2 NO 38. 2 pp.

Flexible Metal Conduits (Cont.)
—**Electrical Codes (Cont.)**
CSA 851 Bull. Electrical Bulletin 851 November 17, 1971 to C22.2 NO 56. 2 pp.

—**Electrical Insulation**
IEC 614 Pt 2-5-92. Specification for Conduits for Electrical Installations Part 2: Particular Specifications for Conduits Section 5: Flexible Conduits First Edition. 37 pp.

—**Security Current**
CNS C1118-88. Security Current of Conduit Distribution (Jul)(9829).

—**Staples**
CSA CAN/CSA-C22. 2 NO 18-92. Outlet Boxes, Conduit Boxes, and Fittings; (Gen Instr 1 Thru 2). 118 pp.

—**Steel**
BSI BS 31-40. 1940 Amd 5 Steel Conduit and Fittings for Electrical Wiring. 40 pp.
BSI BS 731: Part 1-52. 1952 Amd 3 Flexible Steel Conduit for Cable Protection and Flexible Steel Tubing to Enclose Flexible Drives Part 1: Flexible Steel Conduit and Adaptors for the Protection of Electric Cable. 16 pp.
BSI BS 4568: Part 1-70. 1970 Amd 2 Steel Conduit and Fittings with Metric Thread of ISO Form for Electrical Installations Part 1: Steel Conduit, Bends and Couplers (AMD 4568) January 30, 1987. 31 pp.
BSI BS 4568: Part 2-70. 1970 Amd 4 Steel Conduit and Fittings with Metric Thread of ISO Form for Electrical Installations Part 2: Fittings and Components. 39 pp.
CNS C4061-72. Steel Tubes for Electric Lines (with Insulating Varnish Coated) (Jun)(2607).
SNZ NZS 6221: Part 1-72. Specification for Steel Conduit and Fittings with Metric Threads of ISO Form for Electrical Installations Part 1: Steel Conduit, Bends and Couplers Amend: A, 1972; 1A, 1979. 32 pp.
SNZ NZS 6221: Part 2-72. Specification for Steel Conduit and Fittings with Metric Threads of ISO Form for Electrical Installations Part 2: Fittings and Components Amend: A, 1972; 1A, 1974; 2A 1978. 40 pp.

—**Steel—Screw Threads**
CNS B2283-78. Steel Conduit Thread (Dimensions) (Oct)(4549).
DIN ENGL 40430-71. Steel Conduit Thread; Dimensions (Feb). 1 p.
SNZ NZS 127-49. Steel Conduit and Fittings for Electrical Wiring Amend: 2A, 1967. 44 pp.

—**Straps (Fasteners)**
CSA CAN/CSA-C22. 2 NO 18-92. Outlet Boxes, Conduit Boxes, and Fittings; (Gen Instr 1 Thru 2). 118 pp.

—**Thermoplastic Covering**
CSA C22.2 NO 56-1977. Flexible Metal Conduit and Liquid-Tight Flexible Metal Conduit (R 1992); (Amd 1-2 February 1990) (Amd 3 August 1990) (Amd 4-6 January 1993). 28 pp.

Flexible Metal Hoses
Scope Note: See the subheading Flexible Metal under specific types of hoses

Flexible Nonmetallic Conduit Fittings
See Also: Electric Conduit Fittings; Flexible Nonmetallic Conduits
BSI BS 4607: Part 3-71. 1971 Amd 2 Non-Metallic Conduit Fittings for Electrical Installations Part 3: Pliable Corrugated, Plain and Reinforced Conduits of Self-Extinguishing Plastics Material. 27 pp.
CNS C4449-87. Fittings for Pliable Conduits (Nov)(12153).
IEC 1035 Pt 2-3-93. Specification for Conduit Fittings for Electrical Installations Part 2: Particular Specifications Section 3: Fittings for Flexible Conduits of Metal, Insulating or Composite Materials and for Pliable Conduits of Metal or Composite Materials First Edition. 27 pp.
JIS C 8412-92. Fittings for Pliable Plastics Conduits. 14 pp.

Flexible Nonmetallic Conduits
See Also: Electric Conduits; Flexible Nonmetallic Conduit Fittings; Liquidtight Flexible Conduits
CENELEC PREN 50086-2-2-92. Conduit Systems for Electrical Installations Part 2: General Requirements Section 2.2: Pliable Conduit Systems. 9 pp.
CENELEC PREN 50086-2-3-92. Conduit Systems for Electrical Installations Part 2: General Requirements Section 2.3: Flexible Conduit Systems. 8 pp.
CNS C4448-87. Pliable Plastic Conduits (Nov)(12152).
JIS C 8411-92. Pliable Plastics Conduits. 16 pp.

INDUSTRY STANDARDS

Flexible Nonmetallic Conduits (Cont.)

MOD UK DSTAN 59-77: Part 1-89. Conduit, Non-Metallic, Flexible Part 1: General Requirements and Test Methods Capability Approval of Flexible Polymeric Conduit Tubing Issue 2. 54 pp.

—PVC

CNS K3006-87. Unplasticized Polyvinyl Chloride Pipe for Electric Conduit Use (Mar)(1302). 3 pp.

CNS K6142-87. Method of Test for Unplasticized Polyvinyl Chloride Pipes for Electric Conduit Use (Mar)(1303).

JIS C 8430-77. Unplasticized Polyvinyl Chloride (UPVC) Conduit. 10 pp.

SNZ NZS 6207-82. Specification for Unplasticized PVC Conduit and Fittings for Electrical Wiring. 24 pp.

—PVC—Underground

CSA B196.3-M1983. PVC Underground Telecommunication Cable Ducting and Fittings (R 1992); (Gen Instr 1). 28 pp.

—Security Current

CNS C1117-88. Security Current of PVC Conduit Distribution (Jul)(9828).

Flexible Nonmetallic Tubing

Use: Electrical Flexible Tubing

Flexible Pavements

Use: Asphalt Pavements

Flexible Pipe Assemblies

Use: Hose Assemblies

Flexible Pipes

Use: Hoses

Flexible Printed Circuit Boards

See Also: Printed Circuit Boards

BSI BS 6221: Part 7-82. 1982 Amd 1 Printed Wiring Boards Part 7: Method for Specifying Single and Double Sided Flexible Printed Wiring Boards Without Through Connections (AMD 6700) April 30, 1991. 19 pp.

BSI BS 6221: Part 8-82. 1982 Amd 1 Printed Wiring Boards Part 8: Method for Specifying Single and Double Sided Flexible Printed Wiring Boards with Through Connections (AMD 6702) April 30, 1991. 21 pp.

BSI BS 6221: Part 10-91. 1991 Printed Wiring Boards Part 10: Specification for Flex-Rigid Double-Sided Printed Boards with Through Connections (IEC 326-10: 1991). 35 pp.

CECC CECC 23 400-001 ISSUE 1-89. Flexible Printed Boards Without Through Hole Connections (En). 29 pp.

IEC 326 Pt 7-81. Printed Boards Part 7: Specification for Single and Double Sided Flexible Printed Boards Without Through Connections First Edition; (Amendment 1-1989). 37 pp.

IEC 326 Pt 8-81. Printed Boards Part 8: Specification for Single and Double Sided Flexible Printed Boards with Through Connections First Edition; (Amendment 1-1989). 40 pp.

IEC 326 Pt 10-91. Printed Boards Part 10: Specification for Flex-Rigid Double-Sided Printed Boards with Through Connections First Edition. 56 pp.

IECQ PQC 92-90. Sectional Specification: Flexible Printed Boards Without Through Connections. 28 pp.

JIS C 5017-88. Flexible Printed Wiring Boards (Single-Sided, Double-Sided). 10 pp.

—Comprehensive Testing

JIS C 5016-88. Test Methods for Flexible Printed Wiring Boards. 24 pp.

—Multilayer

BSI BS 6221: Part 9-91. 1991 Printed Wiring Boards Part 9: Specification for Flexible Multilayer Printed Boards with Through Connections (IEC 326-9: 1991). 47 pp.

CENELEC PREN 123 800-91. Flexible Multilayer Printed Boards with Through-Connections. 105 pp.

IEC 326 Pt 9-91. Printed Boards Part 9: Specification for Flexible Multilayer Printed Boards with Through Connections First Edition. 68 pp.

—Quality Assurance

BSI BS 9764-88. (OBSOLESCENT) 1988 Sectional Specification for Capability Approval of Manufacturers of Single-Or Double-Sided Flexible Printed Wiring Boards of Assessed Quality Without Through Hole Connections and with or Without Rigidizing Component Materials. 37 pp.

Flexible Printed Circuit Boards (Cont.)

—Quality Assurance (Cont.)

BSI BS 9764-01. (OBSOLSCENT) 1988 Amd 1 Sectional Spec. for Capability Approval of Manufacturers of Single-Or Double-Sided Flexible Printed Wiring Boards of Assessed Quality Without Through Hole Connections and with or Without Rigidizing Component Materials (AMD 7356) December 15, 1992. 38 pp.

BSI BS 9765-88. (OBSOLESCENT) 1988 Sectional Specification for Capability Approval of Manufacturers of Double-Sided Flexible Printed Wiring Boards of Assessed Quality with Through Hole Connections and with or Without Rigidizng Component Materials. 37 pp.

BSI BS 9765-01. (OBSOLSCENT) 1988 Amd 1 Sectional Spec. for Capability Approval of Manufacturers of Double-Sided Flexible Printed Wiring Boards of Assessed Quality with Through Hole Connections and with or Without Rigidizng Component Materials (AMD 7357) December 15, 1992. 38 pp.

BSI BS CECC 23400-90. (WITHDRAWN) 1990 Flexible Printed Boards Without Through Connections (Renumbered as BS EN 123400: 1992). 29 pp.

BSI BS CECC 23500-90. 1990 Specification for Harmonized System of Quality Assessment for Electronic Components: Sectional Specification: Flexible Printed Boards with Through Connections (Renumbered as BS EN 123500: 1992). 31 pp.

BSI BS EN 123400-01. 1992 Amd 1 Sectional Specification: Flexible Printed Boards Without Through Connections (AMD 7367) November 15, 1992. 32 pp.

BSI BS EN 123500-01. 1992 Amd 1 Sectional Specification: Flexible Printed Boards with Through Connections (AMD 7369) November 15, 1992. 34 pp.

CENELEC EN 123400-92. Sectional Specification: Flexible Printed Boards Without Through Connections. 26 pp.

CENELEC EN 123400-800-92. Capability Detail Specification: Flexible Printed Boards Without Through Connections. 11 pp.

CENELEC EN 123500-92. Sectional Specification: Flexible Printed Boards with Through Connections. 28 pp.

CENELEC EN 123500-800-92. Capability Detail Specification: Flexible Printed Boards with Through Connections. 12 pp.

—Through Connections

IECQ PQC 93-90. Sectional Specification: Flexible Printed Boards With Through Connections. 30 pp.

Flexible Printed Circuits

Use: Flexible Printed Circuit Boards

Flexible Shafts

See Also: Shafts (Machine Elements)

—Automotive—Joints

DIN ENGL 75532 Pt 1-76. Transmission of Rotary Motions; Types of Connection to Gears, Intermediate Gears, Flexible Drive Shafts and Equipment (June). 8 pp.

—Automotive—Measuring Instruments

DIN ENGL 75532 Pt 2-79. Transmission of Rotary Motions; Flexible Drive Shafts (Apr). 2 pp.

—Automotive—Speedometers

CNS D2062-85. Flexible Shafts for Speedometers and Tachographs of Automobiles or Motorcycles (Jul)(6815).

JIS D 5602-91. Flexible Shafts for Speedometers and Tachographs of Automobiles. 16 pp.

—Automotive—Tachographs

CNS D2062-85. Flexible Shafts for Speedometers and Tachographs of Automobiles or Motorcycles (Jul)(6815).

JIS D 5602-91. Flexible Shafts for Speedometers and Tachographs of Automobiles. 16 pp.

—Construction Equipment—Meters

CNS A4009-88. Meter Flexible Shafts for Construction Machinery (Oct)(9457).

JIS A 8104-78. Meter Flexible Shafts for Construction Machinery.

Flexible Tubing (Electrical)

Use: Electric Conduits; Electrical Flexible Tubing

Flexible Tubing (Nonelectrical)

Use: Tubing (Flexible)

Flexible Waveguide Assemblies

See Also: Waveguides

IEC 636-79. Flexible Waveguide Assembly Performance First Edition. 39 pp.

Flexible Waveguide Assemblies (Cont.)

MOD UK DEF-5353-A-01. Specification for Waveguide Assemblies, Flexible (Non-Twistable & Twistable Type); Correction. 31 pp.

—Marine

MOD UK NES 1035-90. Requirements for Waveguide Assemblies—Flexible—(Non-Twistable & Twistable) Issue 1 (05.90). 56 pp.

Flexing

Use: Flexure

Flexometer Testing

Use: Bend Testing

Flexural Modulus

Use: Flexural Strength

Flexural Strength

Use For: Bending Strength; Flexural Strength Testing
See Also: Bend Testing; Flexibility; Flexure; Modulus of Rupture; Rigidity; Splitting Testing

—Adhesives

JIS K 6856-77. Testing Methods for Flexural Strength of Adhesives.

JIS K 6856-73. Testing Methods for Flexural Strength of Adhesives. 7 pp.

—Brackets—Masonry

CEN PREN 846-10-92. Methods of Test for Ancillary Components for Masonry—Part 10: Determination of Flexural Resistance and Stiffness of Brackets and Support Angles. 9 pp.

—Cellular Plastics

BSI BS 4370: Part 4-91. 1991 Methods of Test for Rigid Cellular Materials Part 4: Method 14. Determination of Flexural Properties. 8 pp.

DIN ENGL 53423-75. Testing of Rigid Cellular Materials; Flexural Test (Nov). 3 pp.

ISO 1209 Pt 2-90. Cellular Plastics, Rigid —Flexural Tests—Part 2: Determination of Flexural Properties First Edition. 6 pp.

—Cement Mortars

ISO 679-89. Methods of Testing Cements—Determination of Strength First Edition. 18 pp.

—Cements

CEN EN 196 (Part 1)-87. Methods of Testing Cement; Determination of Strength. 32 pp.

CEN EN 196 (Part 1)/A2-89. AMD 2 Methods of Testing Cement; Determination of Strength. 4 pp.

—Ceramic Matrix Composites

BSI DD ENV 658-3-93. 1993 Advanced Technical Ceramics—Mechanical Properties of Ceramic Composites at Room Temperature Part 3: Determination of Flexural Strength (V). 11 pp.

—Ceramics

CEN ENV 658-3-92. Advanced Technical Ceramics— Mechanical Properties of Ceramic Composites at Room Temperature—Part 3: Determination of Flexural Strength. 10 pp.

CEN PREN 843-1-92. Advanced Technical Ceramics—Mechanical Properties of Monolithic Ceramics at Room Temperature—Part 1: Determination of Flexural Strength. 19 pp.

JIS R 1601-81. Testing Method for Flexural Strength (Modulus of Rupture) of High Performance Ceramics. 8 pp.

JIS R 1604-87. Testing Method for Flexural Strength (Modulus of Rupture) of High Performance Ceramics at Elevated Temperature. 9 pp.

—Coated Fabrics

BSI BS 3424: Part 9-90. 1990 Testing Coated Fabrics Part 9: Methods 11A, 11B, 11C and 11D. Methods for Determination of Resistance to Damage by Flexing. 12 pp.

ISO 7854-84. Rubber-or Plastics-Coated Fabrics—Determination of Resistance to Damage by Flexing (Dynamic Method) First Edition. 6 pp.

—Concrete Blocks

CEN PREN 1052-2-93. Methods of Test for Masonry—Part 2: Determination of Flexural Strength. 13 pp.

—Concretes

BSI BS 1881: Part 118-83. 1983 Amd 1 Methods of Testing Concrete Part 118: Method for Determination of Flexural Strength (AMD 6095) July 31, 1989. 8 pp.

CNS A3046-84. Method of Test for Flexural Strength of Concrete (Using Simple Beam with Third-Point Loading) (Apr)(1233). 3 pp.

CNS A3047-84. Method of Test for Flexural Strength of Concrete (Using Simple Beam with Center-Point Loading) (Apr)(1234). 3 pp.

INTERNATIONAL AND NON-U.S. NATIONAL STANDARDS
SUBJECT INDEX

Flexural Strength *(Cont.)*

—Concretes *(Cont.)*
CNS A3051-84. Method of Test for Compressive and Flexural Strength of Drilled Cores and Sawed Beams of Concrete (Apr)(1238). 3 pp.
ISO 4013-78. Concrete—Determination of Flexural Strength of Test Specimens First Edition. 6 pp.
JIS A 1106-76. Method of Test for Flexural Strength of Concrete. 8 pp.
JIS A 1107-78. Method of Obtaining and Testing Drilled Cores and Sawed Beams of Concrete. 7 pp.
SAA AS 1012.11-85. Methods of Testing Concrete—Part 11: Method for the Determination of the Flexural Strength of Concrete Specimens. 5 pp.
SNZ NZS 3112: Part 2-86. Methods of Test for Concrete Part 2: Tests Relating to the Determination of Strength of Concrete Amend: 1. 22 pp.

—Concretes—Polyester Resin
JIS A 1184-78. Method of Test for Flexural Strength of Polyester Resin Concrete. 7 pp.

—Connectors
CSA C22.2 NO 188-M1983. Splicing Wire and Cable Connectors (R 1989); (Gen Instr 1 Thru 2). 38 pp.

—Fasteners—Masonry
CEN PREN 846-10-92. Methods of Test for Ancillary Components for Masonry—Part 10: Determination of Flexural Resistance and Stiffness of Brackets and Support Angles. 9 pp.

—Fiberglass Reinforced Plastics
AECMA PREN2746-90. Glass Fibre Reinforced Plastics Determination of Flexural Properties Three Point Bend Method. 9 pp.
CEN EN 63-77. Glass Reinforced Plastics: Determination of Flexural Properties; Three Point Method. 6 pp.

—Hydraulic Equipment—Tube Fittings
BSI M 55-84. 1984 Rotary Flexure Testing of Hydraulic Tubing Joints and Fittings for Aerospace Use. 12 pp.
ISO 7257-83. Aircraft—Hydraulic Tubing Joints and Fittings—Rotary Flexure Test First Edition. 11 pp.

—Hydraulic Equipment—Tubes
ISO 8574-90. Aerospace—Hydraulic System Tubing—Qualification Tests First Edition. 8 pp.

—Insulating Boards
CNS K6224-65. Method of Test for Foam Polystyrene Heat Insulating Material (Sep)(2536) (R 1971). 3 pp.

—Laminated Plastics
AECMA PREN2562-89. Unidirectional Laminates Carbon-Thermosetting Resin Flexural Test. 9 pp.

—Leather
CNS K6679-85. Method of Test for Flexing Endurance of Leather (Dec)(7705).
JIS K 6545-70. Testing Method for Flexing Endurance of Light Leathers and Their Surface Finishes.

—Leather Finishing Compounds
JIS K 6545-70. Testing Method for Flexing Endurance of Light Leathers and Their Surface Finishes.

—Lintels—Masonry
CEN PREN 846-9-92. Methods of Test for Ancillary Components for Masonry—Part 9: Determination of Flexural Resistance, Shear Load Resistance and Stiffness of Lintels. 14 pp.

—Masonry
CEN PREN 1052-2-93. Methods of Test for Masonry—Part 2: Determination of Flexural Strength. 13 pp.

—Metals
JIS Z 2203-56. Flexure Test Piece for Metals (R 1971). 3 pp.

—Mortars
CEN PREN 1015-11-93. Methods of Test for Mortar for Masonry—Part 11: Determination of Flexural and Compressive Strength of Hardened Mortar. 12 pp.
DIN ENGL 18555 Pt 3-82. Testing of Mortars Containing Mineral Binders; Hardened Mortars; Determination of Flexural Strength, Compressive Strength and Bulk Density (Sept). 4 pp.

—Plastics
BSI BS 2782:Pt3: METH 335A-93. 1993 Methods of Testing Plastics Part 3: Mechanical Properties Method 335A: Determination of Flexural Properties (ISO 178: 1993) (V). 13 pp.
BSI BS 2782:Pt3: METH 335A-78. 1978 Amd 1 Methods of Testing Plastics Part 3: Mechanical Properties Method 335A: Determination of Flexural Properties of Rigid Plastics. 10 pp.

Flexural Strength *(Cont.)*

—Plastics *(Cont.)*
BSI BS 2782:Pt10: METH 1005-77. 1977 Methods of Testing Plastics Part 10: Glass Reinforced Plastics Method 1005: Determination of Flexural Properties. Three Point Method. 8 pp.
BSI BS 4618: SUBSEC 1.1.2-76. 1976 Amd 1 Recommendations for the Presentation of Plastics Design Data Part 1: Mechanical Properties Section 1.1: Creep Subsection 1.1.2: Creep in Flexureat Low Strains. 5 pp.
CNS K6419-78. Method of Test for Flexural Properties of Rigid Plastics (May)(4392).
DIN ENGL 53452-77. Testing of Plastics; Flexural Test (Apr). 5 pp.
DIN ENGL 54852-86. Testing of Plastics; Determination of Flexural Creep of Plastics by Three-Point Loading and Four-Point Loading (Sept). 8 pp.
ISO 178-93. Plastics—Determination of Flexural Properties Third Edition. 11 pp.
ISO 6602-85. Plastics—Determination of Flexural Creep by Three-Point Loading First Edition. 8 pp.
JIS K 7055-87. Testing Method for Flexural Properties of Glass Fiber Reinforced Plastics. 12 pp.
JIS K 7074-88. Testing Methods for Flexural Properties of Carbon Fiber Reinforced Plastics. 11 pp.
JIS K 7203-82. Testing Method for Flexural Properties of Rigid Plastics. 13 pp.

—Plywood
CEN PREN 1072-93. Plywood—Testing of Mechanical Properties with Medium Sized Test Pieces—Bending Strength. 6 pp.

—Polyester Resins
CNS K3062-82. Liquid Unsaturated Polyester Resin for Reinforced Plastics (Dec)(9715). 15 pp.

—Resins
BSI BS 6319: Part 3-90. 1990 Testing of Resin Compositions for Use in Construction Part 3: Methods for Measurement of Modulus of Elasticity in Flexure and Flexural Strength. 9 pp.

—Rocks
DIN ENGL 52112-88. Testing the Flexural Strength of Natural Stone (Aug) (Together with DIN 52201, May 1985 Edition, Supersedes September 1942 Edition and DIN 52205, August 1933 Edition, Withdrawn in 1985). 3 pp.

—Roofing
SAA AS 3991.2-92. Methods of Testing Flat Cellulose-Cement Sheets—Part 2: Determination of Flexural Strength—Modulus of Rupture. 5 pp.

—Roofing Tiles
CEN PREN 538-91. Clay Roofing Tiles for Discontinuous Laying—Flexural Strength Test. 14 pp.

—Splicing
CSA C22.2 NO 188-M1983. Splicing Wire and Cable Connectors (R 1989); (Gen Instr 1 Thru 2). 38 pp.

—Thermal Insulation
CNS K6224-65. Method of Test for Foam Polystyrene Heat Insulating Material (Sep)(2536) (R 1971). 3 pp.

—Toys
CNS Z8016-29-87. Method of Test for Safety (Flexure Test for Toys with Stiffening Means for Retention of Form) (Dec)(4798-29). 2 pp.

—Waterproof Courses
BSI DD 86: Part 1-83. 1983 Damp-Proof Courses Part 1: Methods of Test for Flexural Bond Strength and Short Term Shear Strength. 9 pp.

Flexural Strength Testing
Use: Flexural Strength

Flexure
Use For: Flexing *See Also:* Bend Testing; Flexibility; Flexural Strength; Rigidity

—Elastomers
DIN ENGL 53522 Pt 1-79. Testing of Rubber and Elastomers; Flexing Endurance Test; Definitions, Apparatus, Preparation of Test Pieces (Jan). 2 pp.

—Elastomers—Cracking (Fracturing)
DIN ENGL 53522 Pt 2-79. Testing of Rubber and Elastomers; Flexing Endurance Test; Determination of Resistance to Flex Cracking (Jan). 2 pp.

—Elastomers—Cut Growth
DIN ENGL 53522 Pt 3-79. Testing of Rubber and Elastomers; Flexing Endurance Test; Determination of Resistance to Cut Growth (Jan). 2 pp.

Flexure *(Cont.)*

—Rubber
DIN ENGL 53522 Pt 1-79. Testing of Rubber and Elastomers; Flexing Endurance Test; Definitions, Apparatus, Preparation of Test Pieces (Jan). 2 pp.

—Rubber—Cracking (Fracturing)
DIN ENGL 53522 Pt 2-79. Testing of Rubber and Elastomers; Flexing Endurance Test; Determination of Resistance to Flex Cracking (Jan). 2 pp.

—Rubber—Cut Growth
DIN ENGL 53522 Pt 3-79. Testing of Rubber and Elastomers; Flexing Endurance Test; Determination of Resistance to Cut Growth (Jan). 2 pp.

Flickermeters
See Also: Voltage Fluctuations
BSI BS 6796-86. 1986 Flickermeters. 16 pp.
BSI BS 6796-01. (WITHDRAWN) 1986 Amd 1 Flickermeters (IEC 868: 1986) (Superseded by BS EN 60868: 1993). 42 pp.
BSI BS EN 60868-93. 1993 Flickermeter—Functional and Design Specifications (E). 32 pp.
BSI BS EN 60868-0-93. 1993 Flickermeter Part 0: Evaluation of Flicker Severity (IEC 868-0: 1991). 32 pp.
CENELEC HD 498 S1-88. Flickermeter Functional and Design Specifications. 3 pp.
CENELEC HD 498 S2-92. Flickermeter—Functional and Design Specifications. 5 pp.
CENELEC EN 60868-93. Flickermeter—Functional and Design Specifications (IEC 868: 1986 + A1: 1990). 21 pp.
CENELEC EN 60868-0-93. Flickermeter Part 0: Evaluation of Flicker Severity (IEC 868-0:1991). 5 pp.
CENELEC EN 60868-0-93. Flickermeter Part 0: Evaluation of Flicker Severity (IEC 868-0:1991). 27 pp.
IEC 868-86. Flickermeter—Functional and Design Specifications First Edition; (Amendment 1-1990) (CENELEC EN 60868: 1993). 43 pp.
IEC 868 Pt 0-91. Flickermeter Part 0: Evaluation of Flicker Severity First Edition; (CENELEC EN 60868-0: 1993). 45 pp.

Flight Control (Dynamics)
Use: Flight Dynamics

Flight Control Rods
Use: Control Rods

Flight Control Systems
Use For: Flight Operations; Primary Flight Controls; Secondary Flight Controls *See Also:* Air Navigation; Air Navigation Equipment; Air Transportation; Aircraft; Aircraft Equipment; Aircraft Landings; Instrument Flight; Trim Controls
CAA Chapter K2-1 09.66. Flight—General (Light Aeroplanes). 1 p.
CAA Chapter K2-1 App 09.66. Flight—General (Light Aeroplanes). 1 p.
CAA Chapter K2-2 App 04.72. Performance—General (Light Aeroplanes). 2 pp.
CAA Chapter K2-8 10.92. Handling—Controllability (Light Aeroplanes). 11 pp.
CAA Chapter K2-8 App 10.92. Handling—Controllability (Light Aeroplanes). 1 p.
CAA Chapter K7-1 10.72. Operating Limitations and Information (Light Aeroplanes). 1 p.
CAA Chapter P2-1. General (Provisional Airworthiness Requirements for Civil Powered Lift Aircraft). 4 pp.
CAA Chapter P2-2. Performance—General (Provisional Airworthiness Requirements for Civil Powered Lift Aircraft). 16 pp.
CAA Chapter P2-6. Handling—General (Provisional Airworthiness Requirements for Civil Powered Lift Aircraft). 5 pp.
CAA CAP 482 SUB-Part G 03.83. Operating Limitations and Information (Small Light Aeroplanes). 6 pp.
CAA PAPER FLIGHT 39 3.70. Flying Qualities—Requirements for Powered Lift Aircraft.
ICAO Annex 6 Part I. Operation of Aircraft Part I—International Commercial Air Transport—Aeroplanes Fifth Edition of Part I—July 1990; (Supplement 7/1/91). 114 pp.
ICAO Annex 6 Part II. Operation of Aircraft Part II—International General Aviation—Aeroplanes Fourth Edition of Part II—July 1990; (Supplement 7/1/91). 65 pp.
ICAO Annex 6 Part III. Operation of Aircraft Part III—International Operations—Helicopters Second Edition of Part III—July 1990; (Supplement 7/1/91). 110 pp.
ICAO Circular 207. Simultaneous Operations on Parallel or Near-Parallel Instrument Runways (Soir)—1988. 41 pp.
JIS W 0701-86. Flight Control Systems-Design, Installation and Test of Piloted Aircraft, General Specification for.

Flight
INTERNATIONAL AND NON-U.S. NATIONAL STANDARDS
SUBJECT INDEX

Flight Control Systems (Cont.)

—**Aerobatics**
CAA Chapter K2-12 10.92. Handling—Aerobatics (Light Aeroplanes). 3 pp.
CAA Chapter K2-12 App 10.92. Handling—Aerobatics (Light Aeroplanes). 2 pp.

—**Aerostats**
CAA 819 (Q). Section Q Non-Rigid Airships Performance Requirements (Blue Papers). 11 pp.
CAA Chapter Q2-2 12.79. Performance—General (Non-Rigid Airships). 4 pp.
CAA Chapter Q2-2 App 02.85. Performance—General (Non-Rigid Airships). 22 pp.
CAA Chapter Q7-1 12.79. Operating Limitations and Information—General (Non-Rigid Airships). 1 p.
CAA Chapter Q7-2 12.79. Operating Information (Non-Rigid Airships). 4 pp.

—**Aerostats—En Route Data**
CAA Chapter Q2-4 12.79. Performance—Climb and Level Flight (Non-Rigid Airships). 2 pp.
CAA Chapter Q2-4 App 02.85. Performance—Climb (Non-Rigid Airships). 1 p.

—**Aerostats—Landing Data**
CAA 819 (Q). Section Q Non-Rigid Airships Performance Requirements (Blue Papers). 11 pp.
CAA Chapter Q2-3 12.79. Performance—Take-Off and Landing (Non-Rigid Airships). 3 pp.
CAA Chapter Q2-3 App 02.85. Performance—Take-Off and Landing (Non-Rigid Airships). 1 p.

—**Aerostats—Takeoff Data**
CAA 819 (Q). Section Q Non-Rigid Airships Performance Requirements (Blue Papers). 11 pp.
CAA 843 (Q). Section Q Non-Rigid Airships Performance—Take off Distance Factors. 4 pp.
CAA Chapter Q2-3 12.79. Performance—Take-Off and Landing (Non-Rigid Airships). 3 pp.
CAA Chapter Q2-3 App 02.85. Performance—Take-Off and Landing (Non-Rigid Airships). 1 p.
CAA Chapter Q2-4 12.79. Performance—Climb and Level Flight (Non-Rigid Airships). 2 pp.
CAA Chapter Q2-4 App 02.85. Performance—Climb (Non-Rigid Airships). 1 p.

—**Altitude—Stability—Certification**
CAA. Stability and Stalling of Aircraft Above 5700 Kg AUW. 5 pp.

—**Altitude—Stalls—Certification**
CAA. Stability and Stalling of Aircraft Above 5700 Kg AUW. 5 pp.

—**Certification and Inspection**
ICAO 8335. Manual of Procedures for Operations Certification and Inspection Third Edition—1983. 136 pp.

—**Climbs**
CAA Chapter P2-4. Performance—Climb (Provisional Airworthiness Requirements for Civil Powered Lift Aircraft). 6 pp.

—**Controllability**
CAA Chapter P2-8. Handling Controllability and Manoeuverability (Provisional Airworthiness Requirements for Civil Powered Lift Aircraft). 34 pp.

—**Design—Instrument Panels**
NATO STANAG 3221 ED 6 AMD 6-77. Automatic Flight Control System (AFCS) in Aircraft—Design Standards and Location of Controls. 7 pp.
NATO STANAG 3221 ED 6 AMD 7-77. Automatic Flight Control System (AFCS) in Aircraft—Design Standards and Location of Controls. 8 pp.

—**Documentation**
CAA Chapter P2-8. Handling Controllability and Manoeuverability (Provisional Airworthiness Requirements for Civil Powered Lift Aircraft). 34 pp.

—**En Route Data**
CAA Chapter K2-4 10.92. Performance—Climb (Light Aeroplanes). 5 pp.

—**Glossaries**
JIS W 0131-91. Aircraft—Technical Operatios—Vocabulary.

—**Ground Effect Machines**
CAA 53.
CAA Chapter B3-1 01.91. Determination of Handling and Performance. 5 pp.
CAA Chapter B3-1 App 08.81. Determination of Handling and Performance. 2 pp.
CAA Chapter B3-2 08.81. Scheduling of Handling and Performance Information. 2 pp.

—**Ground Effect Machines—Limitations**
CAA Chapter B3-3 08.83. Operating Limitations and Approved Information. 2 pp.

Flight Control Systems (Cont.)

—**Ground Effect Machines—Limitations (Cont.)**
CAA Chapter B3-3 App 01.91. Operating Limitations and Approved Information. 2 pp.

—**Ground/Water Maneuvers**
CAA Chapter P2-7. Handling—Ground and Water Handling (Provisional Airworthiness Requirements for Civil Powered Lift Aircraft). 5 pp.

—**Inspection**
NATO STANAG 3374 ED 3 AMD 4-82. Flight Inspection of NATO Radio/Radar Navigation and Approach Aids, AEtP-1(B). 12 pp.

—**Landing Data**
CAA 789 (K). Section K Light Aeroplanes Landing Distance (Blue Papers). 6 pp.
CAA Chapter K2-5 10.92. Performance—Landing (Light Aeroplanes). 2 pp.
CAA Chapter K2-5 App 10.92. Performance—Landing (Light Aeroplanes). 1 p.

—**Microwave Landing Systems**
ICAO Circular 165. Microwave Landing System (MLS) Advisory Circular Issue No. 1—1981. 38 pp.

—**Procedures**
CAA Chapter G8-5 08.82. Flight Manuals (Rotocraft). 11 pp.
CAA Chapter G8-5 App #1 09.77. Flight Manuals—General (Rotocraft). 2 pp.
CAA Chapter G8-5 App #2 09.77. Flight Manuals—Limitations (Rotocraft). 1 p.
CAA Chapter G8-5 App #3 09.77. Flight Manuals—Procedures (Rotocraft). 1 p.
CAA Chapter G8-5 App #4 09.77. Flight Manuals—Performance (Rotocraft). 7 pp.
CAA Chapter G8-5 App #5 09.77. Flight Manuals—Supplements (Electrical). 2 pp.
CAA Chapter K2-2 04.72. Performance—General (Light Aeroplanes). 5 pp.
CAA Chapter K2-8 10.92. Handling—Controllability (Light Aeroplanes). 11 pp.
CAA Chapter K7-5 10.92. Flight Manuals (Light Aeroplanes). 14 pp.
CAA Chapter K7-5 App #1 10.92. Flight Manuals—General (Light Aeroplanes). 2 pp.
CAA Chapter K7-5 App #2 10.72. Flight Manuals—Limitations (Light Aeroplanes). 1 p.
CAA Chapter K7-5 App #3 10.72. Flight Manuals—Procedures (Light Aeroplanes). 1 p.
CAA Chapter K7-5 App #4 10.72. Flight Manuals—Performance (Light Aeroplanes). 17 pp.
CAA Chapter K7-5 App #5 10.92. Flight Manuals—Supplements (Light Aeroplanes). 3 pp.
CAA STANDARD NO. 9-3 & App. Aeroplane Flight Manual. 15 pp.
CAA JAR-22 SUBPART G. Operating Limitations and Information (Joint Airworthiness Requirements). 13 pp.
CAA JAR-22 Appendix H. Flight Manual for a Sailplane (Including a Powered Sailplane) (Joint Airworthiness Requirements). 45 pp.
ICAO 8168 Vol I. Procedures for Air Navigation Services Aircraft Operations Volume I Flight Procedures Third Edition—1986; (Amendment 6 11/15/90). 118 pp.
ICAO 8168 Vol II. Procedures for Air Navigation Services Aircraft Operations Volume II Construction of Visual and Instrument Flight Procedures Third Edition—1986; (Corrigendum 3 03/01/88) (Amendment 5 11/15/90). 513 pp.
ICAO 9368. Instrument Flight Procedures Construction Manual First Edition—1983. 251 pp.
ICAO 9371. Template Manual for Holding, Reversal and Racetrack Procedures Second Edition—1986. 195 pp.

—**Procedures—Flight Crews**
CAA Chapter P2-8. Handling Controllability and Manoeuverability (Provisional Airworthiness Requirements for Civil Powered Lift Aircraft). 34 pp.
CAA CAP 524 SUB-Part G 12.86. Operating Limitations and Information (Rotocraft). 13 pp.
CAA CAP524 SUB-Part G 12.86. (Rotocraft). 17 pp.

—**Quality Assurance**
MOD UK DSTAN 05-61: Part 12-87. Quality Assurance Procedural Requirements Part 12: Control of Wall Thickness in Design and Manufacture of Flying Control Systems Issue 1. 8 pp.

—**Rotary Wing Aircraft**
CAA Chapter G2-1 02.63. Flight—General (Rotocraft). 2 pp.
CAA Chapter G2-1 App 02.63. Flight—General (Rotocraft). 1 p.
CAA Chapter G2-2 11.75. Performance—General (Rotocraft). 5 pp.
CAA Chapter G7-4 App 06.66. Ground Endurance Test (Rotocraft). 1 p.

Flight Control Systems (Cont.)

—**Rotary Wing Aircraft (Cont.)**
CAA Chapter G8-1 09.77. Operating Limitations and Information—General (Rotocraft). 1 p.
CAA Chapter G8-2 09.77. Operating Information (Rotocraft). 5 pp.
CAA Chapter G8-2 App 09.77. Operating Limitations and Information (Rotocraft). 1 p.
CAA Chapter G8-3 09.77. Markings and Placards (Rotocraft). 4 pp.
CAA CAP 524 SUB-Part G 12.86. Operating Limitations and Information (Rotocraft). 13 pp.
CAA CAP524 SUB-Part G 12.86. (Rotocraft). 17 pp.

—**Rotary Wing Aircraft—En Route Data**
CAA Chapter G2-4 06.66. Performance—Climb (Rotocraft). 4 pp.
CAA Chapter G2-4 App 02.63. Performance—Climb (Rotocraft). 1 p.

—**Rotary Wing Aircraft—Landing Data**
CAA Chapter G2-5 11.75. Performance—Landing (Rotocraft). 3 pp.
CAA Chapter G2-5 App 02.63. Performance—Landing (Rotocraft). 1 p.

—**Rotary Wing Aircraft—Takeoff Data**
CAA Chapter G2-3 02.63. Performance—Take-Off (Rotocraft). 4 pp.
CAA Chapter G2-3 App 02.63. Performance—Take-Off (Rotocraft). 2 pp.

—**Rules**
ICAO Annex 2. Rules of the Air Ninth Edition—July 1990 (Incorporating Amendments 1-29); (Amendment 30 7/26/93). 55 pp.

—**Signs**
CAA Chapter K7-3 10.92. Markings and Placards (Light Aeroplanes). 4 pp.

—**Speed**
CAA Chapter K7-2 10.92. Operating Information (Light Aeroplanes). 4 pp.
CAA Chapter P2-11. Handling at Low Speeds (Provisional Airworthiness Requirements for Civil Powered Lift Aircraft). 8 pp.

—**Supersonic Transports—Information Systems**
CAA STANDARD NO. 7-5&AP 03.76. Basic Flight Information Systems. 14 pp.

—**Supersonic Transports—Limitations**
CAA STANDARD NO. 9-0 03.76. General. 2 pp.
CAA STANDARD NO. 9-1 12.69. Operating Limitations. 6 pp.

—**Surface Movement**
ICAO 9476. Manual of Surface Movement Guidance and Control Systems (SMGCS) First Edition—1986; (Amendment 1 05/06/87) (Amendment 2 12/01/87). 89 pp.

—**Takeoff Data**
CAA Chapter K2-3 10.92. Performance—Take-Off (Light Aeroplanes). 4 pp.

—**Technical Manuals**
ICAO 9376. Preparation of an Operations Manual First Edition—1990. 177 pp.

—**Weight Limits**
CAA Chapter K7-2 10.92. Operating Information (Light Aeroplanes). 4 pp.

Flight Crew Accommodations
Use: Aircraft Cabins—Flight Crew

Flight Crew Compartments
Use: Aircraft Cabins—Flight Crew

Flight Crews
Use For: Aircraft Personnel; Aircrews
See Also: Aircraft; Aviation Personnel; Pilots (Aircraft)

—**Breathing Apparatus**
BSI N5-92. 1992 Crew Portable Protective Breathing Equipment for Use During Aircraft Emergencies. 11 pp.

—**Equipment—Acceptance Testing**
MOD UK DTD-1292-81. Acceptance Tests for Aircrew Equipment Assemblies. 13 pp.

—**Ergonomics**
NATO STANAG 3994 ED 1 AMD 0-91. Application of Human Engineering to Advanced Aircrew Systems. 10 pp.
NATO STANAG 3994 ED 1 AMD 1-91. Application of Human Engineering to Advanced Aircrew Systems. 11 pp.

INTERNATIONAL AND NON-U.S. NATIONAL STANDARDS
SUBJECT INDEX

Flight

Flight Crews (Cont.)

—Flight Time Limitations
ICAO Circular 52. Flight Crew Fatigue and Flight Time Limitations Sixth Edition—1984. 314 pp.
NATO STANAG 3527 ED 1 AMD 6-69. Airborne Flying Time and Rest Periods. 7 pp.
NATO STANAG 3527 ED 1 AMD 7-69. Aircrew Flying Time and Rest Periods. 8 pp.

—Glossaries
BSI BS 185: Sec 16-73. 1973 Glossary of Aeronautical and Astronautical Terms Section 16: Personnel. 2 pp.

—Helmets—Visors
MOD UK DTD-1281-80. Optical Requirements for Aircrew Helmet Visors and Uncorrected Sunglasses. 21 pp.

—Laser Target Designators—Safety
NATO STANAG 3828 ED 1 AMD 5-81. Minimum Requirements for Aircrew Protection Against the Hazards of Laser Target Designators. 12 pp.
NATO STANAG 3828 ED 1 AMD 8-81. Minimum Requirements for Aircrew Protection Against the Hazards of Laser Target Designators. 12 pp.

—Medical Status—Interchangeability
NATO STANAG 3526 ED 4 AMD 4-82. Interchangeability of NATO Aircrew Medical Categories. 6 pp.
NATO STANAG 3526 ED 4 AMD 5-82. Interchangeability of NATO Aircrew Medical Categories. 7 pp.

—Oxygen Masks
BSI N 4-74. 1974 Full Face Mask Assemblies for Use in Contaminated Atmospheres. 6 pp.
ISO 5887-81. Aircraft—Joints for Connecting Crew Members' Regulator Masks to Oxygen Sources First Edition. 4 pp.

—Pressure Suits
NATO STANAG 3198 ED 3 AMD 4-73. Functional Requirements of Aircraft Oxygen Equipment and Pressure Suits. 7 pp.
NATO STANAG 3198 ED 4 AMD 0-91. Functional Requirements of Aircraft Oxygen Equipment and Pressure Suits. 11 pp.
NATO STANAG 3198 ED 4 AMD 1-91. Functional Requirements of Aircraft Oxygen Equipment and Pressure Suits. 15 pp.
NATO STANAG 3198 ED 4 AMD 2-91. Functional Requirements of Aircraft Oxygen Equipment and Pressure Suits. 14 pp.

—Protective Equipment—Flash Blindness—Nuclear Warfare
NATO STANAG 3830 ED 1 AMD 2-84. Aircrew Flash Blindness Protection. 14 pp.
NATO STANAG 3830 ED 1 AMD 5-84. Aircrew Nuclear Flash Blindness Protection. 9 pp.

—Respirators—NBC—Air Filters
NATO STANAG 3501 ED 2 AMD 1-90. Performance of Portable Filter—Blowers for Aircrew NBC Respirators. 9 pp.
NATO STANAG 3501 ED 2 AMD 2-90. Performance of Portable Filter—Blowers for Aircrew NBC Respirators. 12 pp.
NATO STANAG 3864 ED 1 AMD 5-81. The Measurement of Protection Provided to the Respiratory Tract and Eyes by Aircrew Equipment Assemblies Against NBC Agents in Particulate, Aerosol and Vapour Form. 16 pp.

—Safety—Restrictions
NATO STANAG 3474 ED 3 AMD 10-75. Temporary Flying Restrictions Due to Exogenous Factors Affecting Aircrew Efficiency. 8 pp.

—Tactical Air Transport Operations
NATO STANAG 3464 ED 3 AMD 2-79. Procedures for Aircrew and Airlifted Personnel in Tactical Air Transport Operations. 16 pp.
NATO STANAG 3464 ED 3 AMD 5-79. Procedures for Aircrew and Airlifted Personnel in Tactical Air Transport Operations. 13 pp.

—Training
ICAO Circular 217. Human Factors Digest No. 2 Flight Crew Training: Cockpit Resource Management (CRM) and Line-Oriented Flight Training (LOFT)—1989. 67 pp.
ICAO Circular 227. Human Factors Digest No. 3 Training of Operational Personnel in Human Factors—1991. 58 pp.

—Training—Aerospace Medicine
NATO STANAG 3114 ED 6 AMD 5-86. Aeromedical Training of Flight Personnel. 16 pp.
NATO STANAG 3114 ED 6 AMD 6-86. Aeromedical Training of Flight Personnel. 19 pp.
NATO STANAG 3114 ED 6 AMD 7-86. Aeromedical Training of Flight Personnel. 19 pp.

Flight Crews (Cont.)

—Training—Aerospace Medicine—NBC Equipment
NATO STANAG 3497 ED 1 AMD 2-88. Aeromedical Training of Aircrew NBC Equipment and Procedures. 12 pp.
NATO STANAG 3497 ED 1 AMD 3-88. Aeromedical Training of Aircrew in Aircrew NBC Equipment and Procedures. 13 pp.

—Training—High Sustained Gravity
NATO STANAG 3827 ED 2 AMD 2-89. Recommended Requirements for Training of Aircrew in High Sustained "G" Environment. 8 pp.
NATO STANAG 3827 ED 2 AMD 3-89. Recommended Requirements for Training of Aircrew in High Sustained "G" Environment. 9 pp.
NATO STANAG 3827 ED 2 AMD 4-89. Recommended Requirements for Training of Aircrew in High Sustained "G" Environment. 9 pp.

—Training Manuals
ICAO 7192 Part A-3. Training Manual Part A-3 Composite Ground Subject Curriculum First Edition—1975. 100 pp.
ICAO 7192 Part C-3. Training Manual Part C-3 Flight Engineer First Edition—1977 (OUT OF PRINT). 57 pp.
ICAO 7192 Part D-3. Training Manual Part D-3 Flight Operations Officer First Edition—1975 (OUT OF PRINT). 39 pp.
ICAO 7192 Part E-1. Training Manual Part E-1 Cabin Personnel First Edition—1976. 42 pp.

Flight Data Recorders
Use: Flight Recorders

Flight Decks
See Also: Aircraft; Aircraft Landing Areas; Aircraft Marshallers; Naval Ships
MOD UK NES 134-82. Requirements for Aviation Arrangements Issue 1 (11.82) (Withdrawn) (Refer to MoD). 32 pp.

—Automation—Ergonomics
ICAO Circular 234. Human Factors Digest No. 5 Operational Implications of Automation in Advanced Technology Flight Decks—1992. 46 pp.

—Coatings
MOD UK NES 754-84. Requirements for Preservation and Coating of Flight Decks Issue 3 (03.84). 20 pp.

—Communication Equipment
MOD UK NES 555-81. Guide to Internal Communication Equipment. Flight Deck Communications Issue 2 (01.81). 31 pp.

—Friction
NATO STANAG 1278 ED 1 AMD 2-82. Standard for the Required Level of Coefficient of Friction on Flight Decks. 6 pp.
NATO STANAG 1278 ED 1 AMD 3-82. Standard for the Required Level of Coefficient of Friction on Flight Decks. 7 pp.

—Protective Clothing—Color Coding
NATO STANAG 1050 ED 3 AMD 1-83. Distinctive Colour of Clothing for Flight Deck Personnel. 10 pp.
NATO STANAG 1050 ED 3 AMD 2-83. Distinctive Colour of Clothing for Flight Deck Personnel. 12 pp.

—Supersonic Transports—Visibility
CAA STANDARD NO. 5-1&AP 05.73. Crew Compartment Design and Flight Deck View. 29 pp.

Flight Dynamics
Scope Note: Study of the motion of an aircraft concerned with short-term effects relating to control and stability *Use For:* Stability (Aircraft)
See Also: Aerodynamics; Aeroelasticity; Aircraft; Automatic Control; Control Surfaces; Flight Testing; Helicopter Controllers (Personnel); Stability Testing
CAA Chapter K2-2 04.72. Performance—General (Light Aeroplanes). 5 pp.
CAA Chapter K2-6 04.72. Handling—General (Light Aeroplanes). 2 pp.
CAA Chapter K2-8 10.92. Handling—Controllability (Light Aeroplanes). 11 pp.
CAA Chapter K2-8 App 10.92. Handling—Controllability (Light Aeroplanes). 1 p.
CAA STANDARD NO. 3-0&AP 07.69. General and Definitions. 29 pp.
CAA STANDARD NO. 3-5&AP 07.69. Handling Qualities. 11 pp.
CAA CAP 482 SUB-Part B 03.83. Flight (Small Light Aeroplanes). 7 pp.
ISO 1151 Pt 1-88. Flight Dynamics—Concepts, Quantities and Symbols—Part 1: Aircraft Motion Relative to the Air Fourth Edition. 34 pp.

Flight Dynamics (Cont.)

ISO 1151 Pt 2-85. Flight Dynamics—Concepts, Quantities and Symbols—Part 2: Motions of the Aircraft and the Atmosphere Relative to the Earth Second Edition; (Addendum 1-1987). 11 pp.
ISO 1151 Pt 3-89. Flight Dynamics—Concepts, Quantities and Symbols—Part 3: Derivatives of Forces, Moments and Their Coefficients Second Edition. 15 pp.
ISO 1151 Pt 4-74. Terms and Symbols for Flight Dynamics—Part 4: Parameters Used in the Study of Aircraft Stability and Control First Edition; (Amendment Slip-1976) (Addendum 1-1986). 11 pp.
ISO 1151 Pt 5-87. Flight Dynamics—Concepts, Quantities and Symbols—Part 5: Quantities Used in Measurements Second Edition. 12 pp.
ISO 1151 Pt 6-82. Terms and Symbols for Flight Dynamics—Part 6: Aircraft Geometry Second Edition; (Amendment 1-1984). 31 pp.
ISO 1151 Pt 7-85. Flight Dynamics—Concepts, Quantities and Symbols—Part 7: Flight Points and Flight Envelopes First Edition. 10 pp.
JIS W 0111-91. Flight Dynamics—Concepts, Quantities and Symbols—Part 1: Aircraft Motion Relative to Air (ISO 1151/1:1988).
JIS W 0401-76. Helicopter Flying and Ground Handling Qualities; General Equipment for.
JIS W 0402-85. Flying Qualities of Piloted Airplanes.
MOD UK DSTAN 00-970: Pt 5:Vol 1-01. Aeroelasticity Issue 1; Amendment 10. 171 pp.
MOD UK DSTAN 00-970: Pt 5:Vol 1-02. Aeroelasticity Issue 1; Amendment 10. 143 pp.
MOD UK DSTAN 00-970: Pt 5:Vol 1-03. Aeroelasticity Issue 1; Amendment 12. 157 pp.
MOD UK DSTAN 00-970: Pt 6:Vol 1-01. Aerodynamics and Flying Qualities Issue 1; Amendment 10. 261 pp.
MOD UK DSTAN 00-970: Pt 6:Vol 1-02. Aerodynamics and Flying Qualities Issue 1; Amendment 10. 227 pp.
MOD UK DSTAN 00-970: Pt 6:Vol 1-03. Aerodynamics and Flying Qualities Issue 1; Amendment 12. 236 pp.
MOD UK DSTAN 00-970: Pt 6:Vol 2-01. Aerodynamics and Flying Qualities Issue 1; Amendment 7. 169 pp.
MOD UK DSTAN 00-970: Pt 6:Vol 2-02. Aerodynamics and Flying Qualities Issue 1; Amendment 7. 165 pp.
MOD UK DSTAN 00-970: Pt 6:Vol 2-03. Aerodynamics and Flying Qualities Issue 1; Amendment 7. 165 pp.

—Aerobatics
CAA Chapter K2-12 10.92. Handling—Aerobatics (Light Aeroplanes). 3 pp.
CAA Chapter K2-12 App 10.92. Handling—Aerobatics (Light Aeroplanes). 2 pp.

—Aerostats
CAA 819 (Q). Section Q Non-Rigid Airships Performance Requirements (Blue Papers). 11 pp.
CAA Chapter Q2-2 12.79. Performance—General (Non-Rigid Airships). 4 pp.
CAA Chapter Q2-8 12.79. Stability (Non-Rigid Airships). 1 p.

—Aerostats—Controllability
CAA Chapter Q2-5 12.79. Handling—General (Non-Rigid Airships). 2 pp.
CAA Chapter Q2-6 12.79. Controllability (Non-Rigid Airships). 3 pp.

—Aerostats—Trimming
CAA Chapter Q2-7 12.79. Handling—Ability to Trim (Non-Rigid Airships). 1 p.

—Aircraft
CAA 711 (K). Section K Light Aeroplanes Spinning. 6 pp.
JIS W 0112-92. Flight Dynamics—Concepts, Quantities and Symbols—Part 2: Motions of the Aircraft and the Atmosphere Relative to the Earth.

—Aircraft—Symbols
JIS W 0112-92. Flight Dynamics—Concepts, Quantities and Symbols—Part 2: Motions of the Aircraft and the Atmosphere Relative to the Earth.

—Certification
CAA. Stability and Stalling of Aircraft Above 5700 Kg AUW. 5 pp.

—Gliders
CAA JAR-22 SUBPART B. Flight (Joint Airworthiness Requirements). 17 pp.

—Glossaries
BSI BS 185: Sec 2-69. 1969 Amd 1 Glossary of Aeronautical and Astronautical Terms Section 2: Motion of Aircraft (In Flight and on the Earth's Surface). 6 pp.
CAA Chapter K2-2 04.72. Performance—General (Light Aeroplanes). 5 pp.

INDUSTRY STANDARDS

Flight Dynamics (Cont.)
—Glossaries (Cont.)
ISO 5843 Pt 4-90. Aerospace—List of Equivalent Terms—Part 4: Flight Dynamics Second Edition. 35 pp.

—Ground
CAA STANDARD NO. 3-7 03.76. Behaviour on Ground. 4 pp.

—Ground Testing
CAA Chapter K6-1 2 App2 10.92. Ground and Flight Tests. 1 p.

—Ground/Water Maneuvers
CAA Chapter K2-7 10.92. Handling—Ground Manoeuvres (Light Aeroplanes). 2 pp.
CAA Chapter K2-7 App 10.92. Handling—Ground Manoeuvres (Light Aeroplanes). 1 p.

—Lofting
SBAC TS 112 ISSUE 2. Manual of Recommended Lofting Procedures.

—Low Altitude
CAA STANDARD NO. 3-8 & AP. Low Speeds and Low Altitudes. 10 pp.

—Low Speed
CAA STANDARD NO. 3-8 & AP. Low Speeds and Low Altitudes. 10 pp.

—Rotary Wing Aircraft
CAA 235 (G). Section G Rotorcraft Miscellaneous Amendments Derived from Section D, "Aeroplanes" (Blue Papers). 6 pp.
CAA Chapter G2-6 02.63. Handling—General (Rotocraft). 3 pp.
CAA Chapter G2-7 02.63. Handling—Ground and Water Manoeuvres (Rotorcraft). 2 pp.
CAA Chapter G2-8 08.82. Handling—Controllability (Rotorcraft). 2 pp.
CAA Chapter G2-8 App 08.82. Handling—Controllability (Rotorcraft). 2 pp.
CAA Chapter G2-10 08.82. Handling—Stability (Rotorcraft). 1 p.
CAA Chapter G2-10 App 08.82. Handling—Stability (Rotorcraft). 2 pp.
CAA CAP 524 SUB-Part B 12.86. Flight (Rotorcraft). 12 pp.
CAA CAP524 SUB-Part B 12.86. (Rotorcraft). 11 pp.

—Rotary Wing Aircraft—Controllability
CAA Chapter G2-8 08.82. Handling—Controllability (Rotorcraft). 2 pp.

—Rotary Wing Aircraft—Ground/Water Maneuvers
CAA Chapter G2-7 02.63. Handling—Ground and Water Manoeuvres (Rotorcraft). 2 pp.

—Rotary Wing Aircraft—Trimming
CAA Chapter G2-9 08.82. Handling—Ability to Trim (Rotorcraft). 1 p.

—Stability Testing
CAA Chapter K2-10 10.92. Handling—Stability (Light Aeroplanes). 4 pp.

—Stalls
CAA Chapter K2-11 10.92. Handling—Stalling (Light Aeroplanes). 5 pp.
CAA Chapter K2-11 App 04.72. Handling—Stalling (Light Aeroplanes). 1 p.
CAA. Stability and Stalling of Aircraft Above 5700 Kg AUW. 5 pp.

—Strength/Rigidity
JIS W 0607-84. Airplane Strength and Rigidity, Flight Loads.
JIS W 0608-84. Airplane Strength and Rigidity, Landplane and Ground Handling Loads.
JIS W 0609-78. Airplane Strength and Rigidity, Water and Handling Loads for Seaplanes.
JIS W 0610-87. Airplane Strength and Rigidity, Miscellaneous Loads.
JIS W 0614-85. Airplane Strength and Rigidity, Vibration.
JIS W 0616-85. Airplane Strength and Rigidity, Flight and Ground Operations Tests.

—Strength/Rigidity—Ground Testing
JIS W 0612-83. Airplane Strength and Rigidity, Ground Tests.
JIS W 0616-85. Airplane Strength and Rigidity, Flight and Ground Operations Tests.

—Trimming
CAA Chapter K2-9 04.72. Handling—Ability to Trim (Light Aeroplanes). 2 pp.

Flight Dynamics (Cont.)
—Wind Velocity—Mathematical Models
ISO 1151 Pt 9-93. Flight Dynamics—Concepts, Quantities and Symbols—Part 9: Models of Atmospheric Motions Along the Trajectory of the Aircraft First Edition. 9 pp.

Flight Envelopes
See Also: Aircraft; Aircraft Safety
ISO 1151 Pt 7-85. Flight Dynamics—Concepts, Quantities and Symbols—Part 7: Flight Points and Flight Envelopes First Edition. 10 pp.

—Gliders
CAA JAR-22 SUBPART C. Structure (Joint Airworthiness Requirements). 17 pp.

—Supersonic Transports
CAA STANDARD NO. 3-1 07.69. Flight Envelopes and Associated Speeds. 5 pp.
CAA STANDARD NO. 3-2&AP 07.69. Limitations and Alarms. 6 pp.

—Supersonic Transports—Exceedance
CAA STANDARD NO. 3-6 03.76. Involuntary Exceeding of Absolute Limits. 3 pp.

Flight Information Publication
Use For: FLIP *See Also:* Documents
NATO STANAG 7005 ED 1 AMD 6-88. Exchange of Flight Information Publication (FLIP) Data. 18 pp.
NATO STANAG 7005 ED 2 AMD 0-92. Exchange of Flight Information Publication (Flip) Data. 12 pp.

Flight Levels
See Also: Air Navigation; Airspace; Atmospheric Pressure

—Vertical Separation
ICAO 9574. Manual on Implementation of a 300 m (1 000 ft) Vertical Separation Minimum Between FL 290 and FL 410 Inclusive First Edition—1992. 47 pp.

Flight Maneuvers
See Also: Aircraft Landings; Airspace; Airspeed; Flight Paths; Takeoff

—Supersonic Transports
CAA STANDARD NO. 3-4&AP 03.76. Maneouvrability. 10 pp.

Flight Operations
Use: Flight Control Systems

Flight Paths
See Also: Air Traffic Control; Airspace; Airspeed; Flight Maneuvers; Vertical Separation

—Documents—Clearance—Overflight
NATO STANAG 3961 ED 1 AMD 3-85. Application Form for Overflight and Landing Clearance. 10 pp.

—Supersonic Transports
CAA STANDARD NO. 2-3 03.76. On Route. 2 pp.

Flight Recorders
Use For: Black Boxes; Flight Data Recorders
See Also: Aircraft Equipment; Cockpit Voice Recorders; Recording Instruments
CAA Spec. NO. 10 ISSUE 1.
CAA Spec. NO. 10 A ISSUE 3.
CAA Spec. NO. 11 ISSUE 1.
EUROCAE ED-55 05.90. MOPS for Flight Data Recorder Systems. 311 pp.

—Aircraft
CAA LEAFLET 14-1 06.91. Midas Flight Data Recording System Type CMM/3RB/2. 3 pp.
CAA LEAFLET 14-2 06.91. Sadas Flight Data Recording Equipment. 3 pp.
CAA LEAFLET 14-3 06.91. Plessey/Davall Types PV710 and PV710A Data Recorders. 3 pp.
CAA LEAFLET 14-4 06.91. Sunstrand Flight Data Recorder Model FB542. 3 pp.
CAA LEAFLET 14-6 06.91. EFDAS Flight Data Recording Equipment AO and AO1 Systems. 3 pp.
CAA LEAFLET 14-7 06.91. Plessey/Davall Date Recorder Sustems. 3 pp.
CAA LEAFLET 14-10 06.91. Plessey Flight Data Recorder System Type PV1584 Series. 3 pp.
CAA LEAFLET 14-13 06.91. Universal/Sundstrand Flight Data Recorder Model 980-4100 Series. 3 pp.

—Helicopters
CAA Spec. NO. 18 ISSUE 1.

Flight Safety
Use: Aircraft Safety

Flight Stability
Use: Flight Dynamics

Flight Testing
See Also: Aerodynamics; Aircraft; Certificates; Flight Dynamics; Type Certificates
CAA NOTICE #15 ISSUE 3. U.K. Certification of Foreign Aircraft of MIWA Not Exceeding 5700 kg (Airworthiness Notices). 4 pp.
MOD UK DSTAN 00-970: Pt 9:Vol 1-01. Flight Tests—Handling Issue 1; Amendment 10. 252 pp.
MOD UK DSTAN 00-970: Pt 9:Vol 1-02. Flight Tests—Handling Issue 1; Amendment 10. 214 pp.
MOD UK DSTAN 00-970: Pt 9:Vol 1-03. Flight Tests—Handling Issue 1; Amendment 12. 217 pp.
MOD UK DSTAN 00-970: Pt10:Vol 1-01. Flight Tests—Installations and Structures Issue 1; Amendment 10. 116 pp.
MOD UK DSTAN 00-970: Pt10:Vol 1-02. Flight Tests—Installations and Structures Issue 1; Amendment 10. 73 pp.
MOD UK DSTAN 00-970: Pt10:Vol 1-03. Flight Tests—Installations and Structures Issue 1; Amendment 10. 74 pp.
MOD UK DSTAN 00-970: Pt 9:Vol 2-01. Flight Tests—Handling Issue 1; Amendment 8. 133 pp.
MOD UK DSTAN 00-970: Pt 9:Vol 2-02. Flight Tests—Handling Issue 1; Amendment 8. 125 pp.
MOD UK DSTAN 00-970: Pt 9:Vol 2-03. Flight Tests—Handling Issue 1; Amendment 8. 125 pp.
MOD UK DSTAN 00-970: Pt10:Vol 2-01. Flight Tests—Installations and Structures Issue 1; Amendment 7. 88 pp.
MOD UK DSTAN 00-970: Pt10:Vol 2-02. Flight Tests—Installations and Structures Issue 1; Amendment 8. 75 pp.
MOD UK DSTAN 00-970: Pt10:Vol 2-03. Flight Tests—Installations and Structures Issue 1; Amendment 9. 89 pp.

—Aerostats—Cooling Systems
CAA Chapter Q5-4 12.79. Cooling Systems (Non-Rigid Airships). 3 pp.

—Air Cushion Vehicles
CAA Chapter A5-2 App 08.81. Vehicle Trails. 2 pp.

—Aircraft
CAA 781 (K). Section K Light Aeroplanes Handling Requirements—Miscellaneous Amendments. 8 pp.
CAA Chapter B3-3 09.91. Flight Testing for Issue of Certifcate of Airworthiness or a Permit to Fly. 4 pp.
CAA Chapter B3-5 07.89. Flight Testing for Renewal of Certificates of Airworthiness or a Permit to Fly. 6 pp.
CAA Chapter B3-7 09.91. Issue and Renewal of Permits to Fly. 4 pp.
CAA Chapter B3-7 App 1 06.90. Evidenance to Substantiate Applications. 1 p.
CAA Chapter B3-7 App 2 07.89. Flight Release Certificate. 1 p.
CAA Chapter B6-8 07.89. Flight Testing After Modification or Repair. 1 p.
CAA LEAFLET 11-3 07.90. CAA Flight Testing Policy. 2 pp.
CAA JAR-25 Section 2. Acceptable Means of Compliance and Interpretations. 165 pp.
CAA JAR-25 Section 3. Advisory Material Joint (Joint Airworthiness Requirements). 138 pp.

—Aircraft—Certificate of Airworthiness
CAA Chapter A3-3 07.89. Flight Testing for Issue of Certificate of Airworthiness or a Permit to Fly. 4 pp.
CAA Chapter A3-5 06.90. Flight Testing for Renewal of Certificates of Airworthiness or Permits to Fly. 6 pp.

—Aircraft—Cooling Systems
CAA 600 (K). Section K Light Aeroplanes Powerplant Installations—Cooling Systems Chapter K5-4 (Blue Papers). 7 pp.
CAA Chapter K5-4 10.92. Cooling Systems (Light Aeroplanes). 3 pp.
CAA Chapter K5-4 App 10.92. Cooling Systems. 1 p.

—Aircraft—Electrical Installations
CAA 706 (K). Section K Light Aeroplanes Electrical Supply, Systems and Equipment (Blue Papers). 54 pp.
CAA 707 (G). Section G Rotorcraft Electrical Supply, Systems and Equipment (Blue Papers). 47 pp.
CAA Chap G6-13 App #2 11.85. Ground and Flight Tests (Rotocraft). 1 p.
CAA Chapter K6-1 2 App2 10.92. Ground and Flight Tests. 1 p.

—Aircraft Engines—Temperature
CAA Chapter P5-4. Temperature Suitability (Provisional Airworthiness Requirements for Civil Powered Lift Aircraft). 5 pp.

INTERNATIONAL AND NON-U.S. NATIONAL STANDARDS
SUBJECT INDEX

Floats

Flight Testing *(Cont.)*
—Aircraft—Noise
CAA 810 (N). Section N Amendments from ICAO Annex 16 (Not Including Helicopters) and to Include Microlight Aeroplanes (Blue Papers). 36 pp.
CAA Chapter N4-1 08.90. Flight Testing of Aircraft—General (Noise). 4 pp.
CAA Chapter N4-2 08.90. Flight Testing of Jet Aeroplanes and of Propeller-Driven Aeroplanes Whose Prototype Has a Maximum Take-Off Weight Exceeding 9000 kg (Noise). 4 pp.
CAA Chapter N4-3 08.90. Flight Testing of Propeller-Driven Aeroplanes Having a Maximum Take-Off Weight Not Exceeding 5700 kg Flight Testing of Propeller-Driven Aeroplanes Whose Prototype Has a Maximum Take-Off Weight Not Exceeding 9000 kg—Application for. 2 pp.
CAA Chapter N4-4 08.90. Flight Testing of Propeller-Driven Aeroplanes Whose Prototype Has a Maximum Take-Off Weight Not Exceeding 9000 kg Application for Certificate of Airworthiness for Prototype or Prototype (Modified) Accepted on or After 1 January 1988. 3 pp.
CAA Chapter N4-6 01.88. Instrumentation and Technique for Flight Path Measurement (Noise). 1 p.
CAA Chapter N4-7 08.90. Instrumentation and Technique for Measurement of Effective Perceived Noise Level in EPDdB (Noise). 8 pp.
CAA Chapter N4-9 08.90. Analysis and Adjustment of Flight Test Results for Jet and Large Propeller-Driven Aeroplanes (Noise). 3 pp.
CAA Chapter N4-9 App 08.90. Analysis and Adjustment of Flight Test Results for Jet and Large Propeller-Driven Aeroplanes (Noise). 14 pp.
CAA Chapter N4-10 08.90. Adjustment of Flight Test Results for Small Propeller-Driven Aeroplanes for Which an Application for Certificate of Airworthiness for Prototype Was Accepted Before 1 January 1988. 2 pp.
CAA Chapter N4-11 08.90. Adjustment of Flight Test Results for Small Propeller-Driven Aeroplanes for Which an Application for Certificate of Airworthiness for Prototype Was Accepted on or After 1 January 1988. 3 pp.
CAA Chapter N4-12 08.90. Adjustment of Flight Test Results for Microlight Aeroplanes. 1 p.

—Aircraft—Permit to Fly
CAA Chapter A3-3 07.89. Flight Testing for Issue of Certificate of Airworthiness or a Permit to Fly. 4 pp.
CAA Chapter A3-5 06.90. Flight Testing for Renewal of Certificates of Airworthiness or Permits to Fly. 6 pp.

—Aircraft—Type Certificates
CAA Chapter A2-3 09.91. Flight Testing for Type Certification. 4 pp.

—Airworthiness
CAA NOTICE #48 ISSUE 1. Airworthiness Flight Tests (Airworthiness Notices). 1 p.

—Airworthiness—Aircraft
CAA Chapter A6-8 09.91. Flight Testing After Modification or Repair. 2 pp.

—Confidential Information
CAA NOTICE #65 ISSUE 1. CAA Use of Confidential Information (Airworthiness Notices). 3 pp.

—Gliders
CAA JAR-22 SUBPART B. Flight (Joint Airworthiness Requirements). 17 pp.

—Ground Effect Machines—Certification
CAA Chapter A5-2 08.81. Vehicle Trails. 3 pp.

—Helicopters—Noise
CAA Chapter N4-5 08.90. Flight Testing of Helicopters (Noise). 7 pp.
CAA Chapter N4-13 08.90. Analysis and Adjustment of Flight Test Results for Helicopters. 3 pp.
CAA Chapter N4-13 App 08.90. Analysis and Adjustment of Flight Test Results for Helicopters. 4 pp.

FLIP
Use: Flight Information Publication

Flip Flop Relays
Use: Bistable Relays

Flip Flops
Use For: Bistable Multivibrators; Bistables
See Also: Bistable Latches; D-Type Flip Flops; Digital Circuits; Latches (Circuits); Oscillators

—Dual
CECC CECC 90 103-170 ISSUE 1-82. BS CECC 90 103-170; Dual Bistables with Preset, Common Clear and Common Clock (En) AMD 1 (En). 19 pp.

Flip Flops *(Cont.)*
—Dual—Preferred Products List
CECC CECC MUAHAG Vol 7 IS 8-92. Preferred Products List; Active Microcircuits (En, Fe, Ge). 89 pp.

—2-Bit—Dual—Preferred Products List
CECC CECC MUAHAG Vol 7 IS 8-92. Preferred Products List; Active Microcircuits (En, Fe, Ge). 89 pp.

Float and Sink Analysis
SAA AS 1661-79. Method for Float and Sink Testing and Presentation of Results. 48 pp.

—Anthracite
ISO 7936-92. Hard Coal—Determination and Presentation of Float and Sink Characteristics—General Directions for Apparatus and Procedures First Edition. 23 pp.

—Coal
CNS M3172-84. Method of Test for Float and Sink of Coal (Sep)(11025).

Float Gages
See Also: Gages; Level Indicators

—Aircraft—Fuel Tanks
EUROCAE ED-41 12.83. MPS for Airborne Fuel Quantity Gauging Systems. 30 pp.

—Aircraft—Fuel Tanks—Acceptance Testing
EUROCAE ED-41 12.83. MPS for Airborne Fuel Quantity Gauging Systems. 30 pp.

—Aircraft—Fuel Tanks—Accuracy Testing
EUROCAE ED-41 12.83. MPS for Airborne Fuel Quantity Gauging Systems. 30 pp.

—Aircraft—Fuel Tanks—Environmental Testing
EUROCAE ED-41 12.83. MPS for Airborne Fuel Quantity Gauging Systems. 30 pp.

—Ships
JIS F 7217-89. Marine Float Level Gauges.

—Ships—Tanks (Containers)—Light Hydrocarbons
ISO 10574-93. Refrigerated Light-Hydrocarbon Fluids—Measurement of Liquid Levels in Tanks Containing Liquefied Gases—Float-Type Level Gauges First Edition. 10 pp.

Float Glass
See Also: Glass; Plate Glass
CEN PREN 572-2-91. Glass in Building—Basic Products—Part 2: Float Glass. 9 pp.
CGSB CAN/CGSB-12.3-M91. Flat, Clear Float Glass. 10 pp.
CNS R2051-86. Float and Polished Plate Glasses (May)(2442).
CNS R3116-84. Method of Test for Float and Polished Plate Glasses (Apr)(10866).
JIS R 3202-85. Float and Polished Plate Glasses. 16 pp.

—Aircraft
MOD UK DTD-5625-80. Standard Annealed Float Glass for Aircraft Glazing. 8 pp.

—Heat Absorbing
CGSB CAN/CGSB-12.4-M91. Heat Absorbing Glass. 21 pp.
JIS R 3208-87. Heat Absorbing Glass. 8 pp.

—Reflecting
CGSB CAN/CGSB-12.10-M76. Glass, Light and Heat Reflecting. 17 pp.

Float Operated Valves
Use: Float Valves

Float Valves
Use For: Float Operated Valves *See Also:* Ball Valves; Control Valves; Valves
SAA AS 1910-76. Copper Alloy Float Control Valves for Use in Water Supply and Hot Water Services Corrig.. 20 pp.

—Aircraft—Ball
SBAC AS 6605 ISSUE 1. Float Operated Cut-Off Valve.

—Cisterns
BSI BS 1212: Part 4-91. 1991 Float Operated Valves Part 4: Specification for Compact Type Float Operated Valves for WC Flushing Cisterns (Including Floats). 20 pp.

Float Valves *(Cont.)*
—Diaphragm
BSI BS 1212: Part 2-90. 1990 Float Operated Valves (Excluding Floats) Part 2: Diaphragm Type Float Operated Valves (Copper Alloy Body) (Excluding Floats). 24 pp.
BSI BS 1212: Part 2-70. (WITHDRAWN) 1970 Amd 4 Ballvalves (Excluding Floats) Part 2: Diaphragm Type (Brass Body). 35 pp.

Floating Floors
See Also: Floors

—Acoustical Insulation
CNS A2166-83. Rock Wool Isolating Material for Floating Floors (Nov)(10637).
CNS A2167-83. Glass Wool Isolating Material for Floating Floors (Nov)(10638).
JIS A 6321-79. Rock Wool Isolating Material for Floating Floors.
JIS A 6322-79. Glass Wool Isolating Material for Floating Floors (R 1983). 12 pp.

—Acoustical Insulation—Dynamic Stiffness
BSI BS EN 29052-1-92. 1992 Acoustics—Method for the Determination of Dynamic Stiffness Part 1: Materials Used Under Floating Floors in Dwellings (ISO 9052-1: 1989). 11 pp.
CEN EN 29052-1-92. Acoustics—Determination of Dynamic Stiffness—Part 1: Materials Used Under Floating Floors in Dwellings. 6 pp.
DIN ENGL 52214 (S)-84. Testing of Acoustics in Buildings; Determination of the Dynamic Stiffness of Insulating Layers for Floating Screeds (Dec) (Superseded by DIN EN 29052 Part 1). 3 pp.
DIN ENGL EN 29052 Pt 1-92. Acoustics; Determination of Dynamic Stiffness; Materials Used Under Floating Screed in Dwellings (ISO 9052-1: 1989) (Aug) (Supersedes DIN 52214, December 1984 Edition). 8 pp.
ISO 9052 Pt 1-89. Acoustics—Determination of Dynamic Stiffness—Part 1: Materials Used Under Floating Floors in Dwellings First Edition. 7 pp.

—Aluminum Alloy
CNS A2171-91. Aluminum Alloy Floating Floors (Jan)(10678). 4 pp.

Floating Nuts
Use: Anchor Nuts

Floating Point Arithmetic
Use For: Binary Floating Point Arithmetic
See Also: Mathematics

—Microprocessors
BSI BS 7237-90. 1990 Binary Floating Point Arithmetic for Microprocessor Systems. 26 pp.
IEC 559-89. Binary Floating-Point Arithmetic for Microprocessor Systems Second Edition. 49 pp.
JTC1 559-89. Binary Floating-Point Arithmetic for Microprocessor Systems Second Edition. 49 pp.
OSI ISO/IEC DIS 10858-90. Information Technology—Radix-Independent Floating-Point Arithmetic. 18 pp.

Floating Point Processors
Use: Floating Point Units

Floating Point Units
Use For: Floating Point Processors
See Also: Microcomputers; Microprocessors

—CMOS—Microprocessors—Preferred Products List
CECC CECC MUAHAG Vol 7 IS 8-92. Preferred Products List; Active Microcircuits (En, Fe, Ge). 89 pp.

—32-Bit
CECC CECC 90 110-032 ISSUE 1-91. UTE C 86-241 ADD. 8/M 18; Microprocessor Integrated Circuits in Accordance with CECC 90 110 68881M-16 68881M-20 68882M-16 68882M-20 68882M-25 32 Bit Microprocessor 16,67 MHz 20 MHZ and 25 MHz ADD 8—11 (Fr) ADD 15—17 (Fr) (En). 34 pp.

Floating Screeds
Use: Screeds—Floating

Floating Structures
—Design
BSI BS 6349: Part 6-89. 1989 Code of Practice for Maritime Structures Part 6: Design of Inshore Moorings and Floating Structures. 54 pp.

Floats
Use For: Flotation Gear *See Also:* Landing Gear; Life Rafts

INDUSTRY STANDARDS

Floats (Cont.)

—Rotary Wing Aircraft
CAA Chapter G6-12 12.80. Emergency Flotation Gear (Rotocraft). 2 pp.
CAA CAP 524 SUB-Part F 12.86. Equipment (Rotorcraft). 17 pp.

Floats (Valves)

—Aircraft—Stop Valves
SBAC AS 4881 ISSUE 3. Float Assembly. 1 p.
SBAC AS 4882 ISSUE 1. Float for Cut off Valve Assembly. 1 p.

—Aircraft—Stop Valves—Brackets
SBAC AS 4887 ISSUE 1. Bracket for Float. 1 p.

—Aircraft—Stop Valves—Bumper Pads
SBAC AS 4886 ISSUE 1. Bumper Pad. 1 p.

—Aircraft—Stop Valves—Distance Bushings
SBAC AS 4884 ISSUE 1. Distance Bush for Float Assembly. 1 p.

—Aircraft—Stop Valves—Rivets
SBAC AS 4885 ISSUE 1. Rivet—Float. 1 p.
SBAC AS 4888 ISSUE 1. Rivet—Float Strap to Bracket. 1 p.

—Aircraft—Stop Valves—Straps
SBAC AS 4883 ISSUE 1. Float Support Strap. 1 p.

Flocked Fabrics
See Also: Fabrics
JIS L 1084-92. Testing Methods for Flocked Fabrics. 16 pp.

—Flammability Testing
ISO 10047-93. Textiles—Determination of Surface Burning Time of Fabrics First Edition. 10 pp.

FLON 11
Use: Trichlorofluoromethane

FLON 22
Use: Chlorodifluoromethane

FLON 113
Use: 1,1,2-Trichloro-1,2,2-Trifluoroethane

FLON 114
Use: Dichlorotetrafluoroethane

FLON 12
Use: Halon

Flood Lights
Use: Floodlights

Flood Protection
See Also: Fire Protection; Hydrology

—Pipelines—Drawings
DIN ENGL 2425 Pt 6-82. Plans for Public Utilities, Water Resources Management and for Long-Distance Pipelines; Maps and Plans for Water Engineering, Flood and Coastal Protection (Feb). 13 pp.

—Pipelines—Maps
DIN ENGL 2425 Pt 6-82. Plans for Public Utilities, Water Resources Management and for Long-Distance Pipelines; Maps and Plans for Water Engineering, Flood and Coastal Protection (Feb). 13 pp.

Floodlights
See Also: Lighting; Projectors
BSI BS 4533: Sec 102.5-90. 1990 Luminaires Part 102: Particular Requirements Section 102.5: Specification for Floodlights. 14 pp.
CENELEC EN 60 598-2-5-89. Luminaires. Part 2: Particular Requirements Section Five-Floodlights. 5 pp.
CNS C4169-80. Spot Light and Flood Light Lamps (Feb)(5201).
CSA C22.2 NO 9-M1989. Luminaires; (Gen Instr 1) (Supercedes C22.2 No 97-1969). 83 pp.
CSA 711 Bull. Electrical Bulletin 711 February 9, 1968 to C22.2 NO 9. 1 p.
CSA 1169 Bull. Electrical Bulletin 1169 June 27, 1978 to C22.2 NO 9. 2 pp.
CSA 1217A Bull. Electrical Bulletin 1217A November 5, 1980 to C22.2 NO 9. 2 pp.
CSA 1427 Bull. Electrical Bulletin 1427 November 28, 1984 to C22.2 NO 9. 2 pp.
IEC 598 Pt 2-5-79. Luminaires Part 2: Particular Requirements Section 5: Floodlights First Edition; (Amendment 1-1987) (Amendment 2-1993). 28 pp.

—Airport Lighting
NATO STANAG 3892 ED 2 AMD 2-86. Movement Area Floodlighting. 8 pp.
NATO STANAG 3892 ED 3 AMD 0-93. Movement Area Floodlighting. 7 pp.

—Electrical Codes
CSA 711 Bull. Electrical Bulletin 711 February 9, 1968 to C22.2 NO 9. 1 p.
CSA 1217A Bull. Electrical Bulletin 1217A November 5, 1980 to C22.2 NO 9. 2 pp.
CSA 1427 Bull. Electrical Bulletin 1427 November 28, 1984 to C22.2 NO 9. 2 pp.

—Halogen—Ships
CNS F5044-86. Marine Halogen Lamp Type Flood Lights (Jan)(9603).
JIS F 8445-83. Marine Halogen Lamp Type Flood Lights.

—Mercury Vapor—Ships
CNS F5043-86. Marine High Pressure Mercury Vapour Lamp Type Flood Lights (Jan)(9602).
JIS F 8444-83. Marine High Pressure Mercury Vapour Lamp Type Flood Lights.

—Projectors
CNS C4417-84. Projectors for Uncandescent Lamps (Flood Light Type) (Sep)(11007).
JIS C 8113-75. Projectors for Incandescent Lamps (Flood Light Type) (R 1983). 10 pp.

—Ships—Projectors
CNS F5040-86. Marine Floodlighting Projectors (Jan)(9599).
JIS F 8417-83. Marine Floodlighting Projectors.
JIS F 8417-65. Floodlighting Projectors for Marine Use (R 1968). 8 pp.

—Ships—Reflector Lamps
CNS F5042-86. Marine Reflector Lamp Type Flood Lights (Jan)(9601).
JIS F 8443-83. Marine Reflector Lamp Type Flood Lights.

—Sodium—Ships
CNS F5086-86. Marine High Pressure Sodium Lamp Type Flood Lights (Jan)(10769).
JIS F 8446-83. Marine High Pressure Sodium Lamp Type Flood Lights.

Floor Cleaning Equipment
Scope Note: For additional listings, use a more specific term See Also: Brooms; Carpet Cleaning Equipment; Floor Polishers; Floor Scrubbers; Floor Stripping Agents; Floors; Mops; Vacuum Cleaners; Water Suction Cleaning Equipment

—Commercial—Safety
BSI BS 5415: Sec 2.1-86. 1986 Amd 1 Safety of Electrical Motor-Operated Industrial Cleaning Appliances Part 2: Particular Requirements Section 2.1: Floor Polishing, Scrubbing, Grinding, Scarifying and Carpet Shampooing Appliances (AMD 6450) March 28,1991. 13 pp.
DIN VDE 0700 Pt 205-87. Safety of Household and Similar Electrical Appliances; Floor Treatment Appliances for Commercial Use (Feb). 42 pp.

—Motor Operated—Safety
CSA C22.2 NO 10-1965. Electric Floor Surfacing and Cleaning Machines (Second Edition) (R 1992). 24 pp.
CSA 1169 Bull. Electrical Bulletin 1169 June 27, 1978 to C22.2 NO 10. 2 pp.
IEC 335 Pt 2-67-92. Safety of Household and Similar Electrical Appliances Part 2: Particular Requirements for Floor Treatment and Floor Cleaning Machines, for Industrial and Commercial Use First Edition. 34 pp.

—Safety
BSI BS 3456: Sec 102.10-90. (WITHDRAWN) 1990 Safety of Household Electrical Appliances Part 102: Particular Requirements Section 102.10: Floor Treatment Machines and Wet Scrubbing Machines (Superseded by BS EN 60335-2-10: 1992). 18 pp.
BSI BS EN 60335-2-10-92. 1992 Safety of Household and Similar Electrical Appliances Part 2: Particular Requirements Section 2.10: Floor Treatment Machines and Wet Scrubbing Machines (Supersedes BS 3456: Section 102.10: 1992). 19 pp.
BSI BS EN 60335-2-10-01. 1992 Amd 1 Safety of Household and Similar Electrical Appliances Part 2: Particular Requirements Section 2.10: Floor Treatment Machines and Wet Scrubbing Machines (AMD 7792) July 15, 1993 (L). 22 pp.
CENELEC HD 281 S1-86. Safety of Household and Similar Electrical Appliances—Part 2: Particular Requirements for Floor Treatment Machines and Wet Scrubbing Machines. 10 pp.
CENELEC EN 60335-2-1 0-90. Safety of Household and Similar Electrical Appliances Part 2: Particular Requirements for Floor Treatment and Wet Scrubbing Machines. 10 pp.
CENELEC EN 60335-2-10-90. Safety of Household and Similar Electrical Appliances Part 2: Particular Requirements for Floor Treatment Machines and Wet Scrubbing Machines; (IEC 335-2-10: 1987, Modified). 12 pp.
CENELEC EN 60335-2-10/A1-92. AMD 1 Safety of Household and Similar Electrical Appliances Part 2: Particular Requirements for Floor Treatment Machines and Wet Scrubbing Machines (IEC 335-2-10:1987/A1:1991). 4 pp.
IEC 335 Pt 2-10-92. Safety of Household and Similar Electrical Appliances Part 2: Particular Requirements for Floor Treatment Machines and Wet Scrubbing Machines Fourth Edition; (CENELEC EN 60335-2-10: 1992). 28 pp.

—Spray Extraction
BSI BS 7460-91. 1991 Code of Practice for Use of Spray Extraction Machines on Carpets and Soft Floor Coverings in Hospitals. 9 pp.

—Spray Extraction—Safety
BSI BS 5415: Sec 2.3-86. 1986 Amd 1 Safety of Electrical Motor-Operated Industrial Cleaning Appliances Part 2: Particular Requirements Section 2.3: Spray Extraction Appliances (AMD 6454) December 21, 1990. 12 pp.
IEC 335 Pt 2-68-92. Safety of Household and Similar Electrical Appliances Part 2: Particular Requirements for Spray Extraction Appliances, for Industrial and Commercial Use First Edition. 42 pp.

Floor Coverings
Scope Note: For additional listings, use a more specific term See Also: Carpet Tiles; Carpets; Floor Mats; Floor Tiles; Linoleum; Rugs; Tatami; Textile Floor Coverings

—Accident Prevention
SNZ NZS 5841: Part 1-88. Code of Practice for the Reduction of Slip Hazards Part 1: Guidelines for the Selection, Installation, Care and Maintenance of Flooring and Other Surfaces. 12 pp.

—Adhesives
BSI BS 5350: Part F1-83. 1983 Amd 1 Methods of Test for Adhesives Group F: Performance of Flooring Adhesives Part F1: Performance Tests for Flooring Adhesives (AMD 6699) April 30, 1991. 7 pp.
BSI BS 5442: Part 1-89. 1989 Classification of Adhesives for Construction Part 1: Classification of Adhesives for Use with Flooring Materials. 4 pp.
BSI BS 5442: Part 1-77. (WITHDRAWN) 1977 Classification of Adhesives for Construction Part 1: Adhesives for Use with Flooring Materials. 4 pp.

—Colors
BSI BS 4902-76. (WITHDRAWN) 1976 Sheet and Tile Flooring Colours for Building Purposes. 11 pp.

—Cork
BSI BS 6263: Part 2-91. 1991 Care and Maintenance of Floor Surfaces Part 2: Code of Practice for Resilient Sheet and Tile Flooring. 14 pp.
BSI BS 6826-87. 1987 Linoleum and Cork Carpet Sheet and Tiles. 12 pp.

—Cork—Installation
BSI BS 8203-87. 1987 Code of Practice for Installation of Sheet and Tile Flooring. 31 pp.

—Cork—Mass
CEN PREN 672-92. Cork Floorcoverings—Determination of Mass Per Unit Volume. 4 pp.

—Elastomeric—Adhesives
DIN ENGL 16864-89. Dispersion, Solvent-Borne and Multi-Part Adhesives for Elastomer Floor, Wall and Ceiling Coverings; Requirements and Testing (July). 6 pp.

—Elastomeric—Electrical Insulating Mats
IEC 1111-92. Matting of Insulating Material for Electrical Purposes First Edition. 62 pp.

—Fire Testing
DIN ENGL 4102 Pt 14-90. Fire Behaviour of Building Materials and Elements; Determination of the Burning Behavior of Floor Covering Systems Using a Radiant Heat Source (May). 7 pp.

—Organic—Chemical Analysis
DIN ENGL 51958-79. Testing of Organic Floor Coverings (Except Textile Floor Coverings); Chemicophysical Effect of Test Agents up to 24 Hours (Mar). 3 pp.

INTERNATIONAL AND NON-U.S. NATIONAL STANDARDS
SUBJECT INDEX

Floor

Floor Coverings (Cont.)

—Organic—Dimensional Stability—Heat Treatment
DIN ENGL 51962-77. Testing of Organic Floor Coverings (Except Textile Floor Coverings); Determination of the Change in Dimensions After Heat Treatment (May). 2 pp.

—Organic—Electrostatic Protection—Hazardous Environments
DIN ENGL 51953-75. Testing of Organic Floor Coverings; Testing of the Ability of Floor Coverings in Explosion-Hazarded Rooms to Dissipate Electrostatic Charges (Aug). 2 pp.

—Organic—Flammability Testing
DIN ENGL 51960-75. Testing of Organic Floor Coverings (Excluding Textile Floor Coverings); Testing of Flammability (Aug). 1 p.

—Organic—Thickness Measurement—Wear Layer
DIN ENGL 51964-75. Testing of Organic Floorings (Except Textile Floorings); Determination of Thickness of the Wear Layer (Aug). 3 pp.

—Plastic
BSI BS 6263: Part 2-91. 1991 Care and Maintenance of Floor Surfaces Part 2: Code of Practice for Resilient Sheet and Tile Flooring. 14 pp.

—Plastic—Glowing Cigarette Test
DIN ENGL 51961-84. Testing of Plastics Surfaces; Behaviour on Exposure to Glowing Cigarettes (Aug). 2 pp.

—PVC
BSI BS 3261: Part 1-73. 1973 Unbacked Flexible PVC Flooring Part 1: Homogeneous Flooring. 20 pp.
BSI BS 5085: Part 1-74. 1974 Backed Flexible PVC Flooring Part 1: Needle-Loom Felt Backed Flooring. 15 pp.
BSI BS 5085: Part 2-76. 1976 Backed Flexible PVC Flooring Part 2: Cellular PVC Backing. 10 pp.
CNS K3023-82. PVC Floor Sheets (Feb)(3216). 3 pp.
CNS K6297-82. Method of Test for PVC Floor Sheets (Feb)(3217).
CSA CAN/CSA-A126.6-87. Unbacked Flexible Homogeneous PVC Floor Covering; (Gen Instr 1 Thru 2). 16 pp.
JIS A 5707-90. P.V.C. Floor Sheets. 17 pp.
SAA AS 2055.1-85. PVC Sheet Floor Covering—Part 1: Unbacked, Flexible. 16 pp.
SNZ NZS 2016: Part 1-81. Specification for Unbacked Flexible PVC Flooring Part 1: Homogeneous Flooring Amend: A, 1981. 20 pp.

—PVC—Installation
BSI BS 8203-87. 1987 Code of Practice for Installation of Sheet and Tile Flooring. 31 pp.

—PVC—Rubber Adhesives
DIN ENGL 16860-86. Adhesives for Use with Floor, Wall and Ceiling Coverings; Dispersion Adhesives and Adhesives Based on Synthetic Rubber Solutions for Use with Polyvinylchloride (PVC) Floor Coverings Without Backing; Requirements and Testing (Apr). 8 pp.

—Resilient—Chemical Analysis
BSI BS EN 423-93. 1993 Resilient Floor Coverings—Determination of the Effect of Stains (R). 9 pp.
CEN PREN 423-90. Resilient Floorcoverings—Determination of the Effect of Stains. 7 pp.

—Resilient—Classification
CEN PREN 685-92. Resilient Floorcoverings—Classification. 6 pp.

—Resilient—Dimensional Stability—Heat Treatment
CEN PREN 434-90. Resilient Floor Coverings—Determination of Dimensional Stability and Curling After Exposure to Heat. 6 pp.

—Resilient—Dimensional Stability—Water Treatment
CEN PREN 662-92. Resilient Floorcoverings—Determination of Curling on Exposure to Moisture. 5 pp.

—Resilient—Electrical Resistance
CEN PREN 1081-93. Resilient Floorcoverings—Determination of the Electrical Resistance. 9 pp.

—Resilient—Flexibility
CEN PREN 435-90. Resilient Floorcoverings—Determination of Flexibility. 5 pp.
CEN PREN 665-92. Resilient Floorcoverings—Determination of Exudation of Plasticizers. 4 pp.

—Resilient—Gel Time
CEN PREN 666-92. Resilient Floorcoverings—Determination of Gelling. 5 pp.

Floor Coverings (Cont.)

—Resilient—Installation
SAA AS 1884-85. Floor Coverings—Resilient Sheet and Tiles—Laying and Maintenance Practices (This is a Joint Standard with SANZ NZS 1884). 13 pp.
SNZ NZS/AS 1884-85. Floor Coverings—Resilient Sheet and Tiles—Laying and Maintenance Practices (This is a Joint Standard with SAA AS 1884). 13 pp.

—Resilient—Length Measurement
CEN PREN 426-90. Resilient Floorcoverings—Determination of Width, Length, Flatness and Straightness. 5 pp.

—Resilient—Loss of Mass
CEN PREN 664-92. Resilient Floorcoverings—Determination of Volatile Loss. 4 pp.

—Resilient—Maintenance
SAA AS 1884-85. Floor Coverings—Resilient Sheet and Tiles—Laying and Maintenance Practices (This is a Joint Standard with SANZ NZS 1884). 13 pp.
SNZ NZS/AS 1884-85. Floor Coverings—Resilient Sheet and Tiles—Laying and Maintenance Practices (This is a Joint Standard with SAA AS 1884). 13 pp.

—Resilient—Mass
CEN PREN 430-90. Resilient Floor Coverings—Determination of Mass Per Unit Area. 6 pp.
CEN PREN 436-90. Resilient Floorcoverings—Determination of Mass Per Unit Volume. 8 pp.
CEN PREN 718-92. Resilient Floorcoverings—Determination of Mass per Unit Area of a Reinforcement or a Backing. 4 pp.

—Resilient—Peel Testing
CEN PREN 431-90. Resilient Floorcoverings—Determination of Peel Resistance. 5 pp.

—Resilient—Shear Testing
CEN PREN 432-90. Resilient Floorcoverings—Determination of Shear Strength. 5 pp.

—Resilient—Tensile Testing
CEN PREN 684-92. Resilient Floorcoverings—Determination of Seam Strength. 6 pp.

—Resilient—Thickness Measurement
CEN PREN 428-90. Resilient Floorcoverings—Determination of Overall Thickness. 5 pp.
CEN PREN 429-90. Resilient Floorcoverings—Determination of the Thickness of Layers. 5 pp.
CEN PREN 663-92. Resilient Floorcoverings—Determination of Conventional Pattern Depth. 6 pp.

—Resilient—Water Absorption
CEN PREN 661-92. Resilient Floorcoverings—Determination of the Spreading of Water. 6 pp.

—Resilient—Wear Testing
BSI BS EN 424-93. 1993 Resilient Floor Coverings—Determination of the Effect of the Simulated Movement of a Furniture Leg (R). 9 pp.
CEN PREN 424-90. Resilient Floorcoverings—Determination of the Effect of the Simulated Movement of a Furniture Leg. 6 pp.
CEN PREN 425-90. Resilient Floorcoverings—Determination of the Effect of a Castor Chair. 6 pp.
CEN PREN 660-92. Resilient Floorcoverings—Determination of Wear Resistance. 7 pp.

—Resilient—Width Measurement
CEN PREN 426-90. Resilient Floorcoverings—Determination of Width, Length, Flatness and Straightness. 5 pp.

—Rubber
BSI BS 3187-78. 1978 Electrically Conducting Rubber Flooring. 8 pp.
BSI BS 6263: Part 2-91. 1991 Care and Maintenance of Floor Surfaces Part 2: Code of Practice for Resilient Sheet and Tile Flooring. 14 pp.

—Rubber—Abrasion Testing
CEN PREN 668-92. Rubber Floor Coverings—Determination of Abrasion Resistance Using a Rotating Cylindrical Drum Device and Non-Rotating Sample Holder. 14 pp.

—Rubber—Indentation Hardness Testing
CEN PREN 433-90. Resilient Floorcoverings—Determination of Static Indentation. 5 pp.
CEN PREN 667-92. Rubber Floor Coverings—Determination of Indentation Hardness by Means of a Durometer (Shore A Hardness). 7 pp.

—Rubber—Installation
BSI BS 8203-87. 1987 Code of Practice for Installation of Sheet and Tile Flooring. 31 pp.

Floor Coverings (Cont.)

—Thermoplastic—Installation
BSI BS 8203-87. 1987 Code of Practice for Installation of Sheet and Tile Flooring. 31 pp.

—Vinyl
CEN PREN 650-92. Resilient Floorcoverings—Specification for Jute or Polyester Backed Vinyl and Vinyl on Polyester Felt with Vinyl Backing. 13 pp.
CEN PREN 651-92. Resilient Floorcoverings—Specification for Flexible Vinyl with Vinyl Foam Layer. 12 pp.
CEN PREN 652-92. Resilient Floorcoverings—Specification for Flexible Vinyl with Cork Based Backing. 11 pp.
CEN PREN 653-92. Resilient Floorcoverings—Specification for Expanded (Cushioned) Vinyl. 12 pp.
CSA A126.3-M1984. Sheet Vinyl Flooring Products; (Gen Instr 1). 12 pp.

—Vinyl—Classification
CEN PREN 650-92. Resilient Floorcoverings—Specification for Jute or Polyester Backed Vinyl and Vinyl on Polyester Felt with Vinyl Backing. 13 pp.
CEN PREN 651-92. Resilient Floorcoverings—Specification for Flexible Vinyl with Vinyl Foam Layer. 12 pp.
CEN PREN 652-92. Resilient Floorcoverings—Specification for Flexible Vinyl with Cork Based Backing. 11 pp.
CEN PREN 653-92. Resilient Floorcoverings—Specification for Expanded (Cushioned) Vinyl. 12 pp.

Floor Drains

Use For: Trench Drains *See Also:* Drainpipes; Drains
CSA CAN3-B79-M79. Floor Drains and Trench Drains. 26 pp.
JIS A 4002-89. Floor Drain with Traps. 23 pp.

—Drainage Systems
CEN PREN 128-89. Performance Criteria and Requirements for Sanitary Appliances. 4 pp.

Floor Grinders

See Also: Grinders

—Safety
BSI BS 5415: Sec 2.1-86. 1986 Amd 1 Safety of Electrical Motor-Operated Industrial Cleaning Appliances Part 2: Particular Requirements Section 2.1: Floor Polishing, Scrubbing, Grinding, Scarifying and Carpet Shampooing Appliances (AMD 6450) March 28,1991. 13 pp.

Floor Mat Detectors

Use For: Pressure Mats *See Also:* Alarm Systems; Burglar Alarms; Burglar Detectors

—Burglar
BSI BS 4737: Sec 3.9-78. 1978 Amd 1 Intruder Alarm Systems in Buildings Part 3: Components Section 3.9: Pressure Mats. 5 pp.

Floor Mats

See Also: Mats; Mats (Gymnastic Equipment); Tatami

—Elastomeric—Electrical Protection
IEC 1111-92. Matting of Insulating Material for Electrical Purposes First Edition. 62 pp.

—Glass Fabrics
CNS K3059-82. Glass Fiber Mat for Floor-Covering Substrate (Jan)(8426). 2 pp.

—Plastic
CGSB 20-GP-32M-79. Matting, Floor, Rubber or Plastic, Standard for; (Amendment 1 May 1981). 11 pp.

—Pressure—Alarm Systems
BSI BS 4737: Sec 3.9-78. 1978 Amd 1 Intruder Alarm Systems in Buildings Part 3: Components Section 3.9: Pressure Mats. 5 pp.

—Rubber
CGSB 20-GP-32M-79. Matting, Floor, Rubber or Plastic, Standard for; (Amendment 1 May 1981). 11 pp.

—Rubber—Electrical Protection
BSI BS 921-76. 1976 Rubber Mats for Electrical Purposes. 7 pp.

—Straw
CNS Z7013-64. Straw Mat for One Man Use (Tentative) (Aug)(2399).
CNS Z8001-73. Testing Standard for Straw Mats (for Export) (May)(1162). 3 pp.

INTERNATIONAL AND NON-U.S. NATIONAL STANDARDS SUBJECT INDEX

Floor

Floor Matting
Use: Floor Mats

Floor Plates
See Also: Floors; Surface Plates; Tread Plates

—**Aluminum Alloys—Marine**
MOD UK NES 840-88. Requirements for Aluminium Alloy Patterned Floor and Tread Plate and Treadstrip Issue 1 (08.88). 20 pp.

Floor Polishers
See Also: Floor Cleaning Equipment
CNS B4009-68. Polishers (Nov)(2928). 2 pp.
IEC 369-71. Methods for Measuring Performance of Floor Polishers for Household and Similar Purposes First Edition. 33 pp.

—**Electric**
CGSB 152-GP-1M-83. Floor Machine, Electric, Polisher and Scrubber, Standard for; (Amendment 1 Mar 1985). 14 pp.
SAA AS 3157-86. Approval and Test Specifications—Electric Floor Polishers Amdt 1 May 1990 (This is a Joint Standard with SANZ NZS 3157) (Superseded by AS/NZS 3157:1993)
SNZ NZS/AS 3157-93. Approval and Test Specification—Electric Floor Polishers. 7 pp.

—**Household**
CSA C22.2 NO 67-M1985. Vacuum Cleaners and Floor Polishers; (Gen Instr 1 Thru 4). 57 pp.
CSA 1169 Bull. Electrical Bulletin 1169 June 27, 1978 to C22.2 NO 67. 2 pp.
CSA 1415A Bull. Electrical Bulletin 1415A February 12, 1986 to C22.2 NO 67. 4 pp.

—**Power Supply Cords**
CSA C22.2 NO 67-M1985. Vacuum Cleaners and Floor Polishers; (Gen Instr 1 Thru 4). 57 pp.

—**Safety**
BSI BS 5415: Sec 2.1-86. 1986 Amd 1 Safety of Electrical Motor-Operated Industrial Cleaning Appliances Part 2: Particular Requirements Section 2.1: Floor Polishing, Scrubbing, Grinding, Scarifying and Carpet Shampooing Appliances (AMD 6450) March 28,1991. 13 pp.

Floor Polishes
See Also: Polishes
CGSB 25-GP-1M METH 50.1C-78. Methods of Sampling and Testing Waxes and Polishes Floor Test (Supersedes 50.1b September 1974). 7 pp.
JIS K 3920-91. Test Methods for Floor Polishes. 29 pp.

—**Adhesion Testing**
CGSB 25-GP-1M METH 24.1-89. Methods of Sampling and Testing Waxes and Polishes Tackiness; (Amendment 2 Feb 1989). 2 pp.

—**Detergent Resistant**
CGSB CAN/CGSB-25.21-M89. Detergent-Resistant Floor Polish. 10 pp.
CGSB CAN/CGSB-25.24-M91. Burnishable, Detergent Resistant, Water Emulsion Floor Polish. 8 pp.
CGSB 25-GP-1M METH 50.1C-78. Methods of Sampling and Testing Waxes and Polishes Floor Test (Supersedes 50.1b September 1974). 7 pp.

—**Skid Resistance**
CGSB 25-GP-1M METH 28.2-86. Methods of Sampling and Testing Waxes and Polishes Resistance to Wet Traffic; (Amendment 1 Nov 1986). 2 pp.
CGSB 25-GP-1M METH 30.1-84. Methods of Sampling and Testing Waxes and Polishes Slip Resistance (Supersedes 30.1c of September 1974). 1 p.

Floor Scrubbers
See Also: Floor Cleaning Equipment

—**Safety**
BSI BS 3456: Sec 102.10-90. (WITHDRAWN) 1990 Safety of Household Electrical Appliances Part 102: Particular Requirements Section 102.10: Floor Treatment Machines and Wet Scrubbing Machines (Superseded by BS EN 60335-2-10: 1992). 18 pp.
BSI BS 5415: Sec 2.1-86. 1986 Amd 1 Safety of Electrical Motor-Operated Industrial Cleaning Appliances Part 2: Particular Requirements Section 2.1: Floor Polishing, Scrubbing, Grinding, Scarifying and Carpet Shampooing Appliances (AMD 6450) March 28,1991. 13 pp.
BSI BS EN 60335-2-10-92. 1992 Safety of Household and Similar Electrical Appliances Part 2: Particular Requirements Section 2.10: Floor Treatment Machines and Wet Scrubbing Machines (Supersedes BS 3456: Section 102.10: 1992). 19 pp.

Floor Scrubbers (Cont.)

—**Safety (Cont.)**
BSI BS EN 60335-2-10-01. 1992 Amd 1 Safety of Household and Similar Electrical Appliances Part 2: Particular Requirements Section 2.10: Floor Treatment Machines and Wet Scrubbing Machines (AMD 7792) July 15, 1993 (L). 22 pp.
CENELEC EN 60335-2-1 0-90. Safety of Household and Similar Electrical Appliances Part 2: Particular Requirements for Floor Treatment Machines and Wet Scrubbing Machines. 10 pp.
CENELEC EN 60335-2-10-90. Safety of Household and Similar Electrical Appliances Part 2: Particular Requirements for Floor Treatment Machines and Wet Scrubbing Machines; (IEC 335-2-10: 1987, Modified). 12 pp.
CENELEC EN 60335-2-10/A1-92. AMD 1 Safety of Household and Similar Electrical Appliances Part 2: Particular Requirements for Floor Treatment Machines and Wet Scrubbing Machines (IEC 335-2-10:1987/A1:1991). 4 pp.
IEC 335 Pt 2-10-92. Safety of Household and Similar Electrical Appliances Part 2: Particular Requirements for Floor Treatment Machines and Wet Scrubbing Machines Fourth Edition; (CENELEC EN 60335-2-10: 1992). 28 pp.

Floor Slabs
See Also: Floors

—**Composite—Profiled Steel—Design**
BSI BS 5950: Part 4-82. 1982 Structural Use of Steelwork in Building Part 4: Code of Practice for Design of Floors with Profiled Steel Sheeting. 22 pp.

—**Concrete—Prestressed—Reinforced**
DIN ENGL 51231 Pt 1-72. Testing Installations for Building Components; Floor Slabs or Beams (July). 4 pp.

—**Steel**
DIN ENGL 51231 Pt 1-72. Testing Installations for Building Components; Floor Slabs or Beams (July). 4 pp.

—**Terrazzo—Design**
BSI BS 5385: Part 5-90. 1990 Wall and Floor Tiling Part 5: Code of Practice of the Design and Installation of Terrazzo Tile and Slab, Natural Stone and Composition Block Floorings. 44 pp.
BSI BS 5385: Part 5-01. 1990 Amd 1 Wall and Floor Tiling Part 5: Code of Practice of the Design and Installation of Terrazzo Tile and Slab, Natural Stone and Composition Block Floorings (AMD 6666) November 30, 1990. 44 pp.
BSI BS 5385: Part 5-02. 1990 Amd 2 Wall and Floor Tiling Part 5: Code of Practice for the Design and Installation of Terrazzo Tile and Slab, Natural Stone and Composition Block Floorings (AMD 7060) February 28, 1992. 45 pp.

—**Terrazzo—Installation**
BSI BS 5385: Part 5-90. 1990 Wall and Floor Tiling Part 5: Code of Practice of the Design and Installation of Terrazzo Tile and Slab, Natural Stone and Composition Block Floorings. 44 pp.
BSI BS 5385: Part 5-01. 1990 Amd 1 Wall and Floor Tiling Part 5: Code of Practice of the Design and Installation of Terrazzo Tile and Slab, Natural Stone and Composition Block Floorings (AMD 6666) November 30, 1990. 44 pp.
BSI BS 5385: Part 5-02. 1990 Amd 2 Wall and Floor Tiling Part 5: Code of Practice for the Design and Installation of Terrazzo Tile and Slab, Natural Stone and Composition Block Floorings (AMD 7060) February 28, 1992. 45 pp.

Floor Stripping Agents
See Also: Cleaning Agents; Floor Cleaning Equipment

—**Polishes**
CGSB CAN/CGSB-2.60-92. Remover for Water-Emulsion Floor Polish and Wax. 7 pp.

—**Waxes**
CGSB CAN/CGSB-2.60-92. Remover for Water-Emulsion Floor Polish and Wax. 7 pp.
CGSB CAN/CGSB-2.112-92. Remover for Water-Emulsion Floor Polish. 6 pp.
CGSB 2-GP-11M METH 44.2-61. Methods of Testing and Analysis of Soaps and Detergents Deleterious Action on Floors. 1 p.

Floor Surfacing Equipment

—**Motor Operated—Safety**
CSA C22.2 NO 10-1965. Electric Floor Surfacing and Cleaning Machines (Second Edition) (R 1992). 24 pp.
CSA 1169 Bull. Electrical Bulletin 1169 June 27, 1978 to C22.2 NO 10. 2 pp.

Floor Sweeping Brushes
Use: Brooms—Push

Floor Tiles
See Also: Ceramic Tiles; Construction Materials; Floor Coverings; Floors; Tiles
BSI BS 5921-80. 1980 Amd 1 Determination of Size, Squareness and Straightness of Edge of Textile Floor Covering Tiles. 8 pp.

—**Adhesives**
CNS A2234-89. Adhesives for PVC Floor Tiles (Sep)(12596).
CNS A3306-89. Method of Test for Adhesives for PVC Floor Tiles (Sep)(12597).
JIS A 5536-91. Adhesives for P.V.C. Floor Tiles and P.V.C. Floor Sheets. 18 pp.

—**Ceramic**
BSI BS 6431: Part 10-84. 1984 Ceramic Floor and Wall Tiles Part 10: Method for Determination of Dimensions and Surface Quality. 11 pp.
BSI BS 6431: Part 10-01. 1984 Amd 1 Ceramic Floor and Wall Tiles Part 10: Method for Determination of Dimensions and Surface Quality (AMD 7099) July 15, 1992. 14 pp.
CEN EN 98-84. Ceramic Tiles: Determination of Dimensions and Surface Quality. 9 pp.
CEN EN 98-91. Ceramic Tiles: Determination of Dimensions and Surface Quality. 8 pp.

—**Ceramic—Acceptance Testing**
BSI BS 6431: Part 23-86. 1986 Ceramic Floor and Wall Tiles Part 23: Sampling and Basis for Acceptance. 9 pp.
BSI BS 6431: Part 23-01. 1986 Amd 1 Ceramic Floor and Wall Tiles Part 23: Specification for Sampling and Basis for Acceptance (AMD 7112) July 15, 1992. 12 pp.

—**Ceramic—Accessories**
DIN ENGL 18166 (S)-86. Ceramic Split Tiles and Split Tile Accessories (Oct) (Superseded by DIN EN 186 Parts 1 and 2, and DIN EN 188, December 1991 Editions). 5 pp.

—**Ceramic—Classification**
BSI BS 6431: Part 1-83. 1983 Ceramic Floor and Wall Tiles Part 1: Classification and Marking, Including Definitions and Characteristics. 9 pp.
BSI BS 6431: Part 1-01. 1983 Amd 1 Ceramic Floor and Wall Tiles Part 1: Specification for Classification and Marking, Including Definitions and Characteristics (AMD 7088) July 15, 1992. 12 pp.
CEN EN 87-82. Ceramic Floor and Wall Tiles Definitions, Classification, Characteristics and Marking. 7 pp.
CEN EN 87-91. Ceramic Floor and Wall Tiles—Definitions, Classification, Characteristics and Marking. 12 pp.
CEN EN 87-1-83. Ceramic Floor and Wall Tiles—Part 1: Specification for Classification and Marking, Including Definitions and Characteristics. 7 pp.

—**Ceramic—Design**
BSI BS 5385: Part 3-89. 1989 Wall and Floor Tiling Part 3: Code of Practice for the Design and Installation of Ceramic Floor Tiles and Mosaics. 49 pp.
BSI BS 5385: Part 3-01. 1989 Amd 1 Wall and Floor Tiling Part 3: Code of Practice for the Design and Installation of Ceramic Floor Tiles and Mosaics (AMD 7059) February 28, 1992. 50 pp.

—**Ceramic—Frost Resistance**
BSI BS 6431: Part 22-86. 1986 Ceramic Floor and Wall Tiles Part 22: Methods of Determination of Frost Resistance. 10 pp.
BSI BS 6431: Part 22-01. 1986 Amd 1 Ceramic Floor and Wall Tiles Part 22: Method for Determination of Frost Resistance (AMD 7111) July 15, 1992. 13 pp.

—**Ceramic—Glazed—Abrasion Testing**
BSI BS 6431: Part 20-84. 1984 Ceramic Floor and Wall Tiles Part 20: Method of Determination of Resistance to Surface Abrasion. Glazed Tiles. 10 pp.
BSI BS 6431: Part 20-01. 1984 Amd 1 Ceramic Floor and Wall Tiles Part 20: Method for Determination of Resistance to Surface Abrasion. Glazed Tiles (AMD 7109) July 15, 1992. 13 pp.
CEN EN 154-84. Ceramic Tiles: Determination of Resistance to Surface Abrasion; Glazed Tiles. 8 pp.
CEN EN 154-91. Ceramic Tiles—Determination of Resistance to Surface Abrasion—Glazed Tiles. 7 pp.

—**Ceramic—Glazed—Chemical Resistance**
BSI BS 6431: Part 19-01. 1984 Amd 1 Ceramic Floor and Wall Tiles Part 19: Method for Determination of Chemical Resistance. Glazed Tiles (AMD 7108) July 15, 1992 (R). 13 pp.
CEN EN 122-84. Ceramic Tiles: Determination of Chemical Resistance; Glazed Tiles. 8 pp.

INDEX and DIRECTORY of

INTERNATIONAL AND NON-U.S. NATIONAL STANDARDS
SUBJECT INDEX
Floor

Floor Tiles *(Cont.)*

—Ceramic—Glazed—Chemical Resistance *(Cont.)*

CEN EN 122-91. Ceramic Tiles—Determination of Chemical Resistance—Glazed Tiles. 14 pp.

—Ceramic—Glazed—Cracking (Fracturing)

BSI BS 6431: Part 17-83. 1983 Ceramic Floor and Wall Tiles Part 17: Method for Determination of Glazing Resistance. Glazed Tiles. 8 pp.

BSI BS 6431: Part 17-01. 1983 Amd 1 Ceramic Floor and Wall Tiles Part 17: Method for Determination of Glazing Resistance. Glazed Tiles (AMD 7106) July 15, 1992. 11 pp.

CEN EN 105-81. Ceramic Tiles: Determination of Crazing Resistance; Glazed Tiles. 6 pp.

CEN EN 105-91. Ceramic Tiles—Determination of Crazing Resistance—Glazed Tiles. 5 pp.

—Ceramic—Glossaries

BSI BS 6431: Part 1-83. 1983 Ceramic Floor and Wall Tiles Part 1: Classification and Marking, Including Definitions and Characteristics. 9 pp.

BSI BS 6431: Part 1-01. 1983 Amd 1 Ceramic Floor and Wall Tiles Part 1: Specification for Classification and Marking, Including Definitions and Characteristics (AMD 7088) July 15, 1992. 12 pp.

CEN EN 87-82. Ceramic Floor and Wall Tiles Definitions, Classification, Characteristics and Marking. 7 pp.

CEN EN 87-91. Ceramic Floor and Wall Tiles—Definitions, Classification, Characteristics and Marking. 12 pp.

CEN EN 87-1-83. Ceramic Floor and Wall Tiles—Part 1: Specification for Classification and Marking, Including Definitions and Characteristics. 7 pp.

—Ceramic—Identification Systems

BSI BS 6431: Part 1-83. 1983 Ceramic Floor and Wall Tiles Part 1: Classification and Marking, Including Definitions and Characteristics. 9 pp.

BSI BS 6431: Part 1-01. 1983 Amd 1 Ceramic Floor and Wall Tiles Part 1: Specification for Classification and Marking, Including Definitions and Characteristics (AMD 7088) July 15, 1992. 12 pp.

CEN EN 87-82. Ceramic Floor and Wall Tiles Definitions, Classification, Characteristics and Marking. 7 pp.

CEN EN 87-91. Ceramic Floor and Wall Tiles—Definitions, Classification, Characteristics and Marking. 12 pp.

CEN EN 87-1-83. Ceramic Floor and Wall Tiles—Part 1: Specification for Classification and Marking, Including Definitions and Characteristics. 7 pp.

—Ceramic—Installation

BSI BS 5385: Part 3-89. 1989 Wall and Floor Tiling Part 3: Code of Practice for the Design and Installation of Ceramic Floor Tiles and Mosaics. 49 pp.

BSI BS 5385: Part 3-01. 1989 Amd 1 Wall and Floor Tiling Part 3: Code of Practice for the Design and Installation of Ceramic Floor Tiles and Mosaics (AMD 7059) February 28, 1992. 50 pp.

BSI BS 5385: Part 4-92. 1992 Wall and Floor Tiling Part 4: Code of Practice for Tiling and Mosaics in Specific Conditions. 32 pp.

BSI BS 5385: Part 4-86. 1986 Wall and Floor Tiling Part 4: Code of Practice for Ceramic Tiling and Mosaics in Specific Conditions. 35 pp.

BSI BS 8000: Sec 11.1-89. 1989 Workmanship on Building Sites Part 11: Code of Practice for Wall and Floor Tiling Section 11.1: Ceramic Tiles, Terrazzo Tiles and Mosaics. 22 pp.

—Ceramic—Interior/Exterior

BSI BS 6431: Part 2-84. 1984 Ceramic Floor and Wall Tiles Part 2: Specification for Extruded Ceramic Tiles with a Low Water Absorption (EMV3%). Group A1. 11 pp.

BSI BS 6431: Part 2-01. 1984 Amd 1 Ceramic Floor and Wall Tiles Part 2: Specification for Extruded Ceramic Tiles with a Low Water Absorption (E less then or Equal to 3%) Group AI (AMD 7089) July 15, 1992 (R). 14 pp.

BSI BS 6431: Sec 3.1-86. 1986 Ceramic Floor and Wall Tiles Part 3: Extruded Ceramic Tiles with a Water Absorption of 3% Less Than EMV 6% Section 3.1: General Products. 11 pp.

BSI BS 6431: Sec 3.1-01. 1986 Amd 1 Ceramic Floor and Wall Tiles Part 3: Extruded Ceramic Tiles with a Water Absorption of 3% Less Than E Less Than or Equal to 6% Group AIIa. Section 3.1: Specification for General Products (AMD 7090) July 15, 1992. 14 pp.

BSI BS 6431: Sec 3.2-86. 1986 Ceramic Floor and Wall Tiles Part 3: Extruded Ceramic Tiles with a Water Absorption of 3% Less Than EMV 6% Section 3.2: Specific Products (Terre Cuite, Cotto, Baldosin Catalan). 11 pp.

Floor Tiles *(Cont.)*

—Ceramic—Interior/Exterior *(Cont.)*

BSI BS 6431: Sec 3.2-01. 1986 Amd 1 Ceramic Floor and Wall Tiles Part 3: Extruded Ceramic Tiles with a Water Absorption of 3% Less Than E Less Than or Equal to 6% Group AIIa. Section 3.2: Spec. for Specific Products (Terre Cuite, Cotto, Baldosin Catalan) (AMD 7091) July 15, 1992. 14 pp.

BSI BS 6431: Sec 4.1-86. 1986 Ceramic Floor and Wall Tiles Part 4: Extruded Ceramic Tiles with a Water Absorption of 6% Less Than EMV 10% Section 4.1: General Products. 11 pp.

BSI BS 6431: Sec 4.1-01. 1986 Amd 1 Ceramic Floor and Wall Tiles Part 4: Extruded Ceramic Tiles with a Water Absorption of 6% Less Than E Less Than or Equal to 10% Group AIIb. Sect. 4.1: Specification for General Products (AMD 7092) July 15, 1992. 14 pp.

BSI BS 6431: Sec 4.2-86. 1986 Ceramic Floor and Wall Tiles Part 4: Extruded Ceramic Tiles with a Water Absorption of 6% Less Than EMV 10% Section 4.2: Specific Products (Terre Cuite, Cotto, Baldosin Catalan). 11 pp.

BSI BS 6431: Sec 4.2-01. 1986 Amd 1 Ceramic Floor and Wall Tiles Part 4: Extruded Ceramic Tiles with a Water Absorption of 6% Less Than E Less Than or Equal to 10% Group AIIb. Sec 4.2: Spec. for Specific Products (Terre Cuite, Cotto, Baldosin Catalan) (AMD 7093) July 15, 1992. 14 pp.

BSI BS 6431: Part 5-86. 1986 Ceramic Floor and Wall Tiles Part 5: Extruded Ceramic Tiles with a Water of Absorption E Greater Than 10%. 11 pp.

BSI BS 6431: Part 5-01. 1986 Amd 1 Ceramic Floor and Wall Tiles Part 5: Specification for Extruded Ceramic Tiles with a Water Absorption of E Greater Than 10% Group AIII (AMD 7094) July 15, 1992. 14 pp.

BSI BS 6431: Part 6-84. 1984 Ceramic Floor and Wall Tiles Part 6: Specification for Dust-Pressed Ceramic Tiles with a Low Water Absorption (EMN3%). Group BI. 11 pp.

BSI BS 6431: Part 6-01. 1984 Amd 1 Ceramic Floor and Wall Tiles Part 6: Specification for Dust-Pressed Ceramic Tiles with a Low Water Absorption (E Less Than or Equal to 3%) Group BI (AMD 7095) July 15, 1992. 14 pp.

BSI BS 6431: Part 7-86. 1986 Ceramic Floor and Wall Tiles Part 7: Dust-Pressed Ceramic Tiles with a Water Absorption of 3% Less Than E Less Than 10%. 11 pp.

BSI BS 6431: Part 7-01. 1986 Amd 1 Ceramic Floor and Wall Tiles Part 7: Specification for Dust-Pressed Ceramic Tiles with a Water Absorption of 3% Less Than E Less Than or Equal to 6% Group BIIa (AMD 7096) July 15, 1992. 14 pp.

BSI BS 6431: Part 8-86. 1986 Ceramic Floor and Wall Tiles Part 8: Dust-Pressed Ceramic Tiles with a Water Absorption of 6% Less Than E Less Than 10%. 11 pp.

BSI BS 6431: Part 8-01. 1986 Amd 1 Ceramic Floor and Wall Tiles Part 8: Specification for Dust-Pressed Ceramic Tiles with a Water Absorption of 6% Less Than E Less Than or Equal to 10% Group BIIb (AMD 7097) July 15, 1992. 14 pp.

BSI BS 6431: Part 9-84. 1984 Ceramic Floor and Wall Tiles Part 9: Specification for Dust-Pressed Ceramic Tiles with a Water Absorption of (EMN10%). Group BIII. 11 pp.

BSI BS 6431: Part 9-01. 1984 Amd 1 Ceramic Floor and Wall Tiles Part 9: Specification for Dust-Pressed Ceramic Tiles with a Water Absorption of E Greater Than 10% Group BIII (AMD 7098) July 15, 1992. 14 pp.

CEN EN 121-84. Extruded Ceramic Tiles with Low Water Absorption (E Less Than or Equal to 3%); Group A1. 9 pp.

CEN EN 121-91. Extruded Ceramic Tiles with Low Water Absorption (E< =3%)—Group A 1. 8 pp.

CEN EN 159-84. Dust-Pressed Ceramic Tiles with Water Absorption E Greater Than 10%; Group B111. 9 pp.

CEN EN 159-91. Dust-Pressed Ceramic Tiles with Water Absorption E> 10%—Group BIII. 8 pp.

CEN EN 176-84. Dust-Pressed Ceramic Tiles with a Low Water Absorption (E Less Than or Equal to 3%) Group B1. 9 pp.

CEN EN 176-91. Dust-Pressed Ceramic Tiles with a Low Water Absorption (E< =3%)—Group BI. 8 pp.

CEN EN 177-84. Ceramic Tiles: Dust-Pressed Ceramic Tiles with a Water Absorption of 3% Less Than E Less Than or Equal to 6% (Group B11a). 9 pp.

CEN EN 177-91. Dust-Pressed Ceramic Tiles with a Water Absorption of 3%<E< =6% (Group BIIa). 8 pp.

CEN EN 178-84. Ceramic Tiles: Dust-Pressed Ceramic Tiles with a Water Absorption 6% Less than E Less Than or Equal to 10% (Group B11b). 9 pp.

CEN EN 178-91. Dust-Pressed Ceramic Tiles with a Water Absorption of 6%<E< =10% (Group BIIb). 8 pp.

Floor Tiles *(Cont.)*

—Ceramic—Interior/Exterior *(Cont.)*

CEN EN 186-1-85. Ceramic Tiles: Extruded Ceramic Tiles with a Water Absorption of 3% Less Than E Less Than or Equal to 6% (Group A11a) —Part 1. 9 pp.

CEN EN 186-1-91. Ceramic Tiles—Extruded Ceramic Tiles with a Water Absorption of 3%<E< =6% (Group AIIa) Part 1. 8 pp.

CEN EN 186-2-85. Ceramic Tiles: Extruded Ceramic Tiles with a Water Absorption of 3% Less Than E Less Than or Equal to 6% (Group A11a) —Part 2. 9 pp.

CEN EN 186-2-91. Ceramic Tiles—Extruded Ceramic Tiles with a Water Absorption of 3%<E< =6% (Group AIIa) Part 2. 8 pp.

CEN EN 187-1-85. Ceramic Tiles: Extruded Ceramic Tiles with a Water Absorption of 6% Less Than E Less Than or Equal to 10% (Group A11b)-Part 1. 9 pp.

CEN EN 187-1-91. Ceramic Tiles—Extruded Ceramic Tiles with a Water Absorption of 6%<E< =10% (Group AIIb) Part 1. 8 pp.

CEN EN 187-2-85. Ceramic Tiles: Extruded Ceramic Tiles with a Water Absorption of 6% Less Than E Less Than or Equal to 10% (Group A11b)-Part 2. 9 pp.

CEN EN 187-2-91. Ceramic Tiles—Extruded Ceramic Tiles with a Water Absorption of 6%<E< =10% (Group AIIb) Part 2. 8 pp.

CEN EN 188-85. Ceramic Tiles: Extruded Ceramic Tiles with a Water Absorption of E Greater Than 10% (Group A111). 9 pp.

CEN EN 188-91. Ceramic Tiles—Extruded Ceramic Tiles with a Water Absorption of E>10% (Group AIII). 8 pp.

DIN ENGL EN 121-91. Extruded Ceramic Tiles with a Low Water Absorption (E Less Than or Equal To 3%) (Group AI) (Dec)(This Standard, Together with DIN EN 186 Parts 1 and 2, DIN EN 187 Parts 1 and 2, and DIN EN 188, December 1991 Editions, Supersedes DIN 18 166 October 1986 Edition). 8 pp.

DIN ENGL EN 159-91. Dust-Pressed Ceramic Tiles with a High Water Absorption (E Greater Than 10%)—Group B III (Dec). 7 pp.

DIN ENGL EN 176-92. Dust-Pressed Ceramic Tiles with a Low Water Absorption (E Less Than or Equal To 3%) (Group B I) (Jan). 6 pp.

DIN ENGL EN 177-91. Dust-Pressed Ceramic Tiles with a Water Absorption of 3% Less Than E Less Than or Equal To 6% (Group B IIa) (Dec). 6 pp.

DIN ENGL EN 178-91. Dust-Pressed Ceramic Tiles with a Water Absorption of 6% Less Than E Less Than or Equal To 10% (Group B IIb) (Dec). 6 pp.

DIN ENGL EN 186 Pt 1-91. Extruded Ceramic Tiles with a Water Absorption of 3% Less Than E Less Than or Equal To 6% (Group AIIa)—Part 1 (Dec) (This Standard, Together with DIN EN 121, DIN EN 186 Part 2, DIN EN 187 Parts 1 and 2 and DIN EN 188 Dec 1991 Editions, Supersedes DIN 18166, Oct 1986 Edition). 7 pp.

DIN ENGL EN 186 Pt 2-91. Extruded Ceramic Tiles with a Water Absorption of 3% Less Than E Less Than or Equal To 6% (Group AIIa)—Part 2 (Dec) (This Standard, Together with DIN EN 121, DIN EN 186 Part 1, DIN EN 187 Parts 1 and 2 and DIN EN 188 Dec 1991 Editions, Supersedes DIN 18166, Oct 1986 Edition). 7 pp.

DIN ENGL EN 187 Pt 1-91. Extruded Ceramic Tiles with a Water Absorption of 6% Less Than E Less Than or Equal To 10% (Group AIIb)—Part 1 (Dec) (This Standard, Together with DIN EN 121, DIN EN 186 Parts 1 and 2, DIN EN 187 Part 2 and DIN EN 188 Dec 1991 Editions, Supersedes DIN 18166, Oct 1986 Edition). 7 pp.

DIN ENGL EN 187 Pt 2-91. Extruded Ceramic Tiles with a Water Absorption of 6% Less Than E Less Than or Equal To 10% (Group AIIb)—Part 2 (Dec) (This Standard, Together with DIN EN 121, DIN EN 186 Parts 1 and 2, DIN EN 187 Part 1 and DIN EN 188 Dec 1991 Editions, Supersedes DIN 18166, Oct 1986 Edition). 7 pp.

DIN ENGL EN 188-91. Extruded Ceramic Tiles with a Water Absorption of E Greater Than 10% (Group AIII) (Dec) (This Standard Together with DIN EN 121, DIN EN 186 Parts 1 and 2, December 1991 Editions, Supersedes DIN 18166, October 1986 Edition). 7 pp.

—Ceramic—Maintenance

BSI BS 5385: Part 3-89. 1989 Wall and Floor Tiling Part 3: Code of Practice for the Design and Installation of Ceramic Floor Tiles and Mosaics. 49 pp.

BSI BS 5385: Part 3-01. 1989 Amd 1 Wall and Floor Tiling Part 3: Code of Practice for the Design and Installation of Ceramic Floor Tiles and Mosaics (AMD 7059) February 28, 1992. 50 pp.

—Ceramic—Modulus of Rupture

BSI BS 6431: Part 12-83. 1983 Ceramic Floor and Wall Tiles Part 12: Method for Determination of Modulus of Rupture. 8 pp.

Floor Tiles (Cont.)

—Ceramic—Modulus of Rupture (Cont.)
BSI BS 6431: Part 12-01. 1983 Amd 1 Ceramic Floor and Wall Tiles Part 12: Method for Determination of Modulus of Rupture (AMD 7101) July 15, 1992. 11 pp.
CEN EN 100-82. Ceramic Tiles: Determination of Modulus of Rupture. 6 pp.
CEN EN 100-91. Ceramic Tiles—Determination of Modulus of Rupture. 5 pp.

—Ceramic—Sampling
BSI BS 6431: Part 23-86. 1986 Ceramic Floor and Wall Tiles Part 23: Sampling and Basis for Acceptance. 9 pp.
BSI BS 6431: Part 23-01. 1986 Amd 1 Ceramic Floor and Wall Tiles Part 23: Specification for Sampling and Basis for Acceptance (AMD 7112) July 15, 1992. 12 pp.

—Ceramic—Scratch Hardness Testing
BSI BS 6431: Part 13-86. 1986 Ceramic Floor and Wall Tiles Part 13: Method for Determination of Scratch Harness of Surface According to Mohs. 7 pp.
BSI BS 6431: Part 13-01. 1986 Amd 1 Ceramic Floor and Wall Tiles Part 13: Method for Determination of Scratch Hardness of Surface According to Mohs (AMD 7102) July 15, 1992. 10 pp.
CEN EN 101-84. Ceramic Tiles: Determination of Scratch Hardness of Surface According to Mohs. 5 pp.
CEN EN 101-91. Ceramic Tiles—Determination of Scratch Hardness of Surface According to Mohs. 4 pp.

—Ceramic—Thermal Expansion
BSI BS 6431: Part 15-83. 1983 Ceramic Floor and Wall Tiles Part 15: Method for Determination of Linear of Thermal Expansion. 7 pp.
BSI BS 6431: Part 15-01. 1983 Amd 1 Ceramic Floor and Wall Tiles Part 15: Method for Determination of Linear Thermal Expansion (AMD 7104) July 15, 1992. 10 pp.

—Ceramic—Thermal Shock
BSI BS 6431: Part 16-83. 1983 Ceramic Floor and Wall Tiles Part 16: Method for Determination of Resistance to Thermal Shock. 7 pp.
BSI BS 6431: Part 16-01. 1983 Amd 1 Ceramic Floor and Wall Tiles Part 16: Method for Determination of Resistance to Thermal Shock (AMD 7105) July 15, 1992. 10 pp.
CEN EN 104-82. Ceramic Tiles: Determination of Resistance to Thermal Shock. 5 pp.
CEN EN 104-91. Ceramic Tiles—Determination of Resistance to Thermal Shock. 4 pp.

—Ceramic—Unglazed—Abrasion Testing
BSI BS 6431: Part 14-83. 1983 Ceramic Floor and Wall Tiles Part 14: Method for Determination of Resistance to Deep Abrasion. Unglazed Tiles. 10 pp.
BSI BS 6431: Part 14-01. 1983 Amd 1 Ceramic Floor and Wall Tiles Part 14: Method for Determination of Resistance to Deep Abrasion. Unglazed Tiles (AMD 7103) July 15, 1992. 13 pp.
CEN EN 102-82. Ceramic Tiles: Determination of Resistance to Deep Abrasion; Unglazed Tiles. 8 pp.
CEN EN 102-91. Ceramic Tiles—Determination of Resistance to Deep Abrasion—Unglazed Tiles. 7 pp.

—Ceramic—Unglazed—Chemical Resistance
BSI BS 6431: Part 18-83. 1983 Ceramic Floor and Wall Tiles Part 18: Method for Determination of Chemical Resistance. Unglazed Tiles. 7 pp.
BSI BS 6431: Part 18-01. 1983 Amd 1 Ceramic Floor and Wall Tiles Part 18: Method for Determination of Chemical Resistance. Unglazed Tiles (AMD 7107) July 15, 1992. 10 pp.
BSI BS 6431: Part 19-84. 1984 Ceramic Floor and Wall Tiles Part 19: Method for Determination of Chemical Resistance. Glazed Tiles. 10 pp.
CEN EN 106-82. Ceramic Tiles: Determination of Chemical Resistance; Unglazed Tiles. 5 pp.
CEN EN 106-91. Ceramic Tiles—Determination of Chemical Resistance—Unglazed Tiles. 4 pp.

—Ceramic—Unglazed—Expansion
BSI BS 6431: Part 21-84. 1984 Ceramic Floor and Tiles Part 21: Method for Determination of Moisture Expansion Using Boiling Water. Unglazed Tiles. 7 pp.
BSI BS 6431: Part 21-01. 1984 Amd 1 Ceramic Floor and Tiles Part 21: Method for Determination of Moisture Expansion Using Boiling Water. Unglazed Tiles (AMD 7110) July 15, 1992. 10 pp.
CEN EN 155-84. Ceramic Tiles: Determination of Moisture Expansion Using Boiling Water; Unglazed Tiles. 5 pp.
CEN EN 155-91. Ceramic Tiles—Determination of Moisture Expansion Using Boiling Water—Unglazed Tiles. 4 pp.

Floor Tiles (Cont.)

—Ceramic—Water Absorption
BSI BS 6431: Part 11-83. 1983 Ceramic Floor and Wall Tiles Part 11: Method for Determination of Water Absorption. 7 pp.
BSI BS 6431: Part 11-01. 1983 Amd 1 Ceramic Floor and Wall Tiles Part 11: Method for Determination of Water Absorption (AMD 7100) July 15, 1992. 10 pp.
CEN EN 99-82. Ceramic Tiles: Determination of Water Absorption. 5 pp.
CEN EN 99-91. Ceramic Tiles—Determination of Water Absorption. 4 pp.

—Clinker
DIN ENGL 18158-86. Clinker Floor Tiles (Sept). 6 pp.

—Colors
BSI BS 4902-76. (WITHDRAWN) 1976 Sheet and Tile Flooring Colours for Building Purposes. 11 pp.

—Concrete
BSI BS 1197: Part 2-73. 1973 Concrete Flooring Tiles and Fittings Part 2: Metric Units. 12 pp.

—Cork
BSI BS 6263: Part 2-91. 1991 Care and Maintenance of Floor Surfaces Part 2: Code of Practice for Resilient Sheet and Tile Flooring. 14 pp.
BSI BS 8203-87. 1987 Code of Practice for Installation of Sheet and Tile Flooring. 31 pp.
ISO 3810-87. Floor Tiles of Agglomerated Cork—Methods of Test Second Edition; (Corrected and Reprinted -1992). 6 pp.
ISO 3813-87. Floor Tiles of Agglomerated Cork—Characteristics, Sampling and Packing Second Edition. 4 pp.

—Cork—Squareness
ISO 9366-90. Composition Cork Floor Tiles—Determination and Control of Squareness and Straightness of Edges First Edition. 8 pp.

—Cork—Straightness
ISO 9366-90. Composition Cork Floor Tiles—Determination and Control of Squareness and Straightness of Edges First Edition. 8 pp.

—Earthenware
CNS R2162-86. Ceramic Floor Tile-Earthenware (Aug)(9738). 6 pp.

—Installation
BSI BS 8000: Sec 11.2-90. 1990 Workmanship on Building Sites Part 11: Code of Practice for Wall and Floor Tiling Section 11.2: Natural Stone Tiles. 19 pp.

—Installation—Construction Contracts
DIN ENGL 18352-88. Tendering and Performance Stipulations in Contracts for Construction Works (VOB); Part C: General Technical Specifications in Contracts for Construction Works (ATV); Wall and Floor Tiling Works (Sept) (This Standard, Together with DIN 18299,. 6 pp.

—Linoleum
BSI BS 6263: Part 2-91. 1991 Care and Maintenance of Floor Surfaces Part 2: Code of Practice for Resilient Sheet and Tile Flooring. 14 pp.
BSI BS 6826-87. 1987 Linoleum and Cork Carpet Sheet and Tiles. 12 pp.
CEN PREN 548-91. Linoleum Floorcoverings—Specification for Plain and Decorative Linoleum. 11 pp.

—Linoleum—Installation
BSI BS 8203-87. 1987 Code of Practice for Installation of Sheet and Tile Flooring. 31 pp.

—Mosaic—Design
BSI BS 5385: Part 2-78. 1978 Amd 1 Wall and Floor Tiling Part 2: Code of Practice for External Ceramic Wall Tiling and Mosaics. 25 pp.
BSI BS 5385: Part 3-89. 1989 Wall and Floor Tiling Part 3: Code of Practice for the Design and Installation of Ceramic Floor Tiles and Mosaics. 49 pp.
BSI BS 5385: Part 3-01. 1989 Amd 1 Wall and Floor Tiling Part 3: Code of Practice for the Design and Installation of Ceramic Floor Tiles and Mosaics (AMD 7059) February 28, 1992. 50 pp.
BSI BS 5385: Part 5-90. 1990 Wall and Floor Tiling Part 5: Code of Practice of the Design and Installation of Terrazzo Tile and Slab, Natural Stone and Composition Block Floorings. 44 pp.

—Mosaic—Installation
BSI BS 5385: Part 1-90. 1990 Wall and Floor Tiling Part 1: Code of Practice for the Design and Installation of Internal Ceramic Wall Tiling and Mosaics in Normal Conditions. 36 pp.

Floor Tiles (Cont.)

—Mosaic—Installation (Cont.)
BSI BS 5385: Part 1-01. 1990 Amd 1 Wall and Floor Tiling Part 1: Code of Practice for the Design and Installation of Internal Ceramic Wall Tiling and Mosaics in Normal Conditions (AMD 7058) February 28, 1992. 37 pp.
BSI BS 5385: Part 2-91. 1991 Wall and Floor Tiling Part 2: Code of Practice for the Design and Installation of External Ceramic Wall Tiling and Mosaics (Including Terra Cotta and Faience Tiles). 32 pp.
BSI BS 5385: Part 2-78. 1978 Amd 1 Wall and Floor Tiling Part 2: Code of Practice for External Ceramic Wall Tiling and Mosaics. 25 pp.
BSI BS 5385: Part 3-89. 1989 Wall and Floor Tiling Part 3: Code of Practice for the Design and Installation of Ceramic Floor Tiles and Mosaics. 49 pp.
BSI BS 5385: Part 3-01. 1989 Amd 1 Wall and Floor Tiling Part 3: Code of Practice for the Design and Installation of Ceramic Floor Tiles and Mosaics (AMD 7059) February 28, 1992. 50 pp.
BSI BS 5385: Part 4-92. 1992 Wall and Floor Tiling Part 4: Code of Practice for Tiling and Mosaics in Specific Conditions. 32 pp.
BSI BS 5385: Part 4-86. 1986 Wall and Floor Tiling Part 4: Code of Practice for Ceramic Tiling and Mosaics in Specific Conditions. 35 pp.
BSI BS 5385: Part 5-90. 1990 Wall and Floor Tiling Part 5: Code of Practice of the Design and Installation of Terrazzo Tile and Slab, Natural Stone and Composition Block Floorings. 44 pp.
BSI BS 8000: Sec 11.1-89. 1989 Workmanship on Building Sites Part 11: Code of Practice for Wall and Floor Tiling Section 11.1: Ceramic Tiles, Terrazzo Tiles and Mosaics. 22 pp.

—Mosaic—Maintenance
BSI BS 5385: Part 3-89. 1989 Wall and Floor Tiling Part 3: Code of Practice for the Design and Installation of Ceramic Floor Tiles and Mosaics. 49 pp.
BSI BS 5385: Part 3-01. 1989 Amd 1 Wall and Floor Tiling Part 3: Code of Practice for the Design and Installation of Ceramic Floor Tiles and Mosaics (AMD 7059) February 28, 1992. 50 pp.

—Organic—Dimensional Stability—Heat Treatment
DIN ENGL 51962-77. Testing of Organic Floor Coverings (Except Textile Floor Coverings); Determination of the Change in Dimensions After Heat Treatment (May). 2 pp.

—Organic—Thickness Measurement—Wear Layer
DIN ENGL 51964-75. Testing of Organic Floorings (Except Textile Floorings); Determination of Thickness of the Wear Layer (Aug). 3 pp.

—Plastic
BSI BS 6263: Part 2-91. 1991 Care and Maintenance of Floor Surfaces Part 2: Code of Practice for Resilient Sheet and Tile Flooring. 14 pp.

—PVC
BSI BS 3260-69. 1969 Amd 2 Semi-Flexible PVC Floor Tiles. 27 pp.
BSI BS 3261: Part 1-73. 1973 Unbacked Flexible PVC Flooring Part 1: Homogeneous Flooring. 20 pp.
CNS A2138-91. Polyvinyl Chloride Floor Tiles (Jan)(8906). 3 pp.
CNS A3155-86. Method of Test for PVC Floor Tile (Jan)(8907).
JIS A 5705-92. Floorcovering-PVC. 23 pp.
SAA AS 1889.1-84. PVC Floor Tiles—Part 1: Semi-Rigid. 14 pp.
SAA AS 1889.2-85. PVC Floor Tiles—Part 2: Flexible. 14 pp.
SNZ NZS 2016: Part 1-81. Specification for Unbacked Flexible PVC Flooring Part 1: Homogeneous Flooring Amend: A, 1981. 20 pp.

—PVC Asbestos
CNS K3027-75. Polyvinyl Chloride Asbestos Tiles for Flooring (Aug)(3308). 2 pp.

—PVC Asbestos—Flammability Testing
CNS K6312-77. Method of Test for Polyvinyl Chloride Asbestos Tiles for Flooring (Dec)(3309). 4 pp.

—PVC Asbestos—Physical Properties
CNS K6312-77. Method of Test for Polyvinyl Chloride Asbestos Tiles for Flooring (Dec)(3309). 4 pp.

—PVC—Installation
BSI BS 8203-87. 1987 Code of Practice for Installation of Sheet and Tile Flooring. 31 pp.

INTERNATIONAL AND NON-U.S. NATIONAL STANDARDS
SUBJECT INDEX

Floors

Floor Tiles *(Cont.)*

—PVC—Ships
MOD UK NES 151: Part 1-81. Tiles and Sheet PVC Part 1: Technical Requirements for Production Issue 1 (8/81) (Withdrawn) (Refer to MoD). 15 pp.

—Resilient
CSA A126.10-M1984. Test Methods for Resilient Flooring; (Gen Instr 1 Thru 2). 26 pp.

—Resilient—Installation
SAA AS 1884-85. Floor Coverings—Resilient Sheet and Tiles—Laying and Maintenance Practices (This is a Joint Standard with SANZ NZS 1884). 13 pp.
SNZ NZS/AS 1884-85. Floor Coverings—Resilient Sheet and Tiles—Laying and Maintenance Practices (This is a Joint Standard with SAA AS 1884). 13 pp.

—Resilient—Length Measurement
CEN PREN 427-90. Resilient Floor Coverings—Determination of the Side Length and Squareness of Tiles. 5 pp.

—Resilient—Maintenance
SAA AS 1884-85. Floor Coverings—Resilient Sheet and Tiles—Laying and Maintenance Practices (This is a Joint Standard with SANZ NZS 1884). 13 pp.
SNZ NZS/AS 1884-85. Floor Coverings—Resilient Sheet and Tiles—Laying and Maintenance Practices (This is a Joint Standard with SAA AS 1884). 13 pp.

—Resilient—Squareness
CEN PREN 427-90. Resilient Floor Coverings—Determination of the Side Length and Squareness of Tiles. 5 pp.

—Rubber
BSI BS 1711-75. 1975 Solid Rubber Flooring. 10 pp.
BSI BS 6263: Part 2-91. 1991 Care and Maintenance of Floor Surfaces Part 2: Code of Practice for Resilient Sheet and Tile Flooring. 14 pp.
CSA A126.4-M1984. Rubber Floor Tile; (Gen Instr 1). 11 pp.

—Rubber—Installation
BSI BS 8203-87. 1987 Code of Practice for Installation of Sheet and Tile Flooring. 31 pp.

—Stone—Design
BSI BS 5385: Part 5-90. 1990 Wall and Floor Tiling Part 5: Code of Practice of the Design and Installation of Terrazzo Tile and Slab, Natural Stone and Composition Block Floorings. 44 pp.
BSI BS 5385: Part 5-01. 1990 Amd 1 Wall and Floor Tiling Part 5: Code of Practice of the Design and Installation of Terrazzo Tile and Slab, Natural Stone and Composition Block Floorings (AMD 6666) November 30, 1990. 44 pp.
BSI BS 5385: Part 5-02. 1990 Amd 2 Wall and Floor Tiling Part 5: Code of Practice for the Design and Installation of Terrazzo Tile and Slab, Natural Stone and Composition Block Floorings (AMD 7060) February 28, 1992. 45 pp.

—Stone—Installation
BSI BS 5385: Part 5-90. 1990 Wall and Floor Tiling Part 5: Code of Practice of the Design and Installation of Terrazzo Tile and Slab, Natural Stone and Composition Block Floorings. 44 pp.
BSI BS 5385: Part 5-01. 1990 Amd 1 Wall and Floor Tiling Part 5: Code of Practice of the Design and Installation of Terrazzo Tile and Slab, Natural Stone and Composition Block Floorings (AMD 6666) November 30, 1990. 44 pp.
BSI BS 5385: Part 5-02. 1990 Amd 2 Wall and Floor Tiling Part 5: Code of Practice for the Design and Installation of Terrazzo Tile and Slab, Natural Stone and Composition Block Floorings (AMD 7060) February 28, 1992. 45 pp.

—Stoneware
CNS R2163-86. Ceramic Floor Tile-Stoneware (Aug)(9739). 6 pp.

—Terrazzo
BSI BS 4131-73. 1973 Terrazzo Tiles. 16 pp.
BSI BS 8000: Sec 11.1-89. 1989 Workmanship on Building Sites Part 11: Code of Practice for Wall and Floor Tiling Section 11.1: Ceramic Tiles, Terrazzo Tiles and Mosaics. 22 pp.
CNS A2049-85. Terrazzo Tiles (Jan) (3803).
CNS A3062-85. Method of Test for Terrazzo Tiles (Jan)(3804). 2 pp.
JIS A 5415-77. Terrazzo Tiles (R 1982). 11 pp.

—Terrazzo—Design
BSI BS 5385: Part 3-89. 1989 Wall and Floor Tiling Part 3: Code of Practice for the Design and Installation of Ceramic Floor Tiles and Mosaics. 49 pp.

Floor Tiles *(Cont.)*

—Terrazzo—Design *(Cont.)*
BSI BS 5385: Part 3-01. 1989 Amd 1 Wall and Floor Tiling Part 3: Code of Practice for the Design and Installation of Ceramic Floor Tiles and Mosaics (AMD 7059) February 28, 1992. 50 pp.
BSI BS 5385: Part 5-90. 1990 Wall and Floor Tiling Part 5: Code of Practice of the Design and Installation of Terrazzo Tile and Slab, Natural Stone and Composition Block Floorings. 44 pp.
BSI BS 5385: Part 5-01. 1990 Amd 1 Wall and Floor Tiling Part 5: Code of Practice of the Design and Installation of Terrazzo Tile and Slab, Natural Stone and Composition Block Floorings (AMD 6666) November 30, 1990. 44 pp.
BSI BS 5385: Part 5-02. 1990 Amd 2 Wall and Floor Tiling Part 5: Code of Practice for the Design and Installation of Terrazzo Tile and Slab, Natural Stone and Composition Block Floorings (AMD 7060) February 28, 1992. 45 pp.

—Terrazzo—Installation
BSI BS 5385: Part 3-89. 1989 Wall and Floor Tiling Part 3: Code of Practice for the Design and Installation of Ceramic Floor Tiles and Mosaics. 49 pp.
BSI BS 5385: Part 3-01. 1989 Amd 1 Wall and Floor Tiling Part 3: Code of Practice for the Design and Installation of Ceramic Floor Tiles and Mosaics (AMD 7059) February 28, 1992. 50 pp.
BSI BS 5385: Part 5-90. 1990 Wall and Floor Tiling Part 5: Code of Practice of the Design and Installation of Terrazzo Tile and Slab, Natural Stone and Composition Block Floorings. 44 pp.
BSI BS 5385: Part 5-01. 1990 Amd 1 Wall and Floor Tiling Part 5: Code of Practice of the Design and Installation of Terrazzo Tile and Slab, Natural Stone and Composition Block Floorings (AMD 6666) November 30, 1990. 44 pp.
BSI BS 5385: Part 5-02. 1990 Amd 2 Wall and Floor Tiling Part 5: Code of Practice for the Design and Installation of Terrazzo Tile and Slab, Natural Stone and Composition Block Floorings (AMD 7060) February 28, 1992. 45 pp.

—Terrazzo—Maintenance
BSI BS 5385: Part 3-89. 1989 Wall and Floor Tiling Part 3: Code of Practice for the Design and Installation of Ceramic Floor Tiles and Mosaics. 49 pp.
BSI BS 5385: Part 3-01. 1989 Amd 1 Wall and Floor Tiling Part 3: Code of Practice for the Design and Installation of Ceramic Floor Tiles and Mosaics (AMD 7059) February 28, 1992. 50 pp.

—Thermoplastic
BSI BS 2592-73. 1973 Thermoplastic Flooring Tiles. 12 pp.

—Thermoplastic—Installation
BSI BS 8203-87. 1987 Code of Practice for Installation of Sheet and Tile Flooring. 31 pp.

—Vinyl
CEN PREN 650-92. Resilient Floorcoverings—Specification for Jute or Polyester Backed Vinyl and Vinyl on Polyester Felt with Vinyl Backing. 13 pp.
CEN PREN 651-92. Resilient Floorcoverings—Specification for Flexible Vinyl with Vinyl Foam Layer. 12 pp.
CEN PREN 652-92. Resilient Floorcoverings—Specification for Flexible Vinyl with Cork Based Backing. 11 pp.
CEN PREN 653-92. Resilient Floorcoverings—Specification for Expanded (Cushioned) Vinyl. 12 pp.
CEN PREN 654-92. Resilient Floorcoverings—Specification for Semi-Flexible Vinyl Tiles. 11 pp.
CEN PREN 655-92. Resilient Floorcoverings—Specification for Tiles on a Base of Agglomerated Composition Cork with a Vinyl Wear Layer. 11 pp.
CSA A126.2-M1984. High Vinyl Floor Tile; (Gen Instr 1). 11 pp.

—Vinyl Asbestos
CSA A126.1-M1984. Vinyl Asbestos and Vinyl Composition Floor Tile; (Gen Instr 1). 11 pp.

—Vinyl—Classification
CEN PREN 650-92. Resilient Floorcoverings—Specification for Jute or Polyester Backed Vinyl and Vinyl on Polyester Felt with Vinyl Backing. 13 pp.
CEN PREN 651-92. Resilient Floorcoverings—Specification for Flexible Vinyl with Vinyl Foam Layer. 12 pp.
CEN PREN 652-92. Resilient Floorcoverings—Specification for Flexible Vinyl with Cork Based Backing. 11 pp.
CEN PREN 653-92. Resilient Floorcoverings—Specification for Expanded (Cushioned) Vinyl. 12 pp.
CEN PREN 654-92. Resilient Floorcoverings—Specification for Semi-Flexible Vinyl Tiles. 11 pp.

Floor Tiles *(Cont.)*

—Vinyl—Classification *(Cont.)*
CEN PREN 655-92. Resilient Floorcoverings—Specification for Tiles on a Base of Agglomerated Composition Cork with a Vinyl Wear Layer. 11 pp.

—Vinyl Composition
CSA A126.1-M1984. Vinyl Asbestos and Vinyl Composition Floor Tile; (Gen Instr 1). 11 pp.

—Vitreous Ware
CNS R2164-86. Ceramic Floor Tile-Vitreous Ware (Aug) (9740). 6 pp.

Floor Waxes

See Also: Floors; Waxes
CGSB CAN/CGSB-25.3-92. Buffable, Water-Emulsion Floor Wax. 7 pp.

—Adhesion Testing
CGSB 25-GP-1M METH 24.1-89. Methods of Sampling and Testing Waxes and Polishes Tackiness; (Amendment 2 Feb 1989). 2 pp.

—Comprehensive Testing
MOD UK M 9503/59. Examination of Wax/Water Emulsion Floor Polish.

—Pastes
CGSB CAN/CGSB-25.2-92. Paste Floor Wax. 6 pp.
CGSB 25-GP-1M METH 27.1-84. Methods of Sampling and Testing Waxes and Polishes Polishing Properties of Paste Floor Wax (Supersedes 27.1a of September 1974). 1 p.

—Skid Resistance
CGSB 25-GP-1M METH 28.2-86. Methods of Sampling and Testing Waxes and Polishes Resistance to Wet Traffic; (Amendment 1 Nov 1986). 2 pp.
CGSB 25-GP-1M METH 30.1-84. Methods of Sampling and Testing Waxes and Polishes Slip Resistance (Supersedes 30.1c of September 1974). 1 p.

Floorings

Use: Floors

Floors

See Also: Concrete Floors; Decks; Floating Floors; Floor Cleaning Equipment; Floor Plates; Floor Slabs; Floor Tiles; Floor Waxes; Parquet Floors; Setts; Skid Resistance; Solums; Substructures; Waterproof Courses; Wooden Floors
JIS A 0030-73. Classification of Performance for Building Elements. 14 pp.

—Abrasion Testing—Rotating Disk
CNS A3160-82. Method of Abrasion Test for Building Material and Part of Building Construction (Method of Abrasion Test for Flooring Materials Method with Rotating Disk Fitted Friction and Impact) (Jun)(8912).
JIS A 1451-70. Method of Abrasion Test for Building Materials and Part of Building Construction (Method of Abrasion Test for Flooring Materials Method with Rotating Disk Fitted Friction and Impact).

—Acoustical Insulation
BSI BS 2750: Part 6-80. 1980 Measurement of Sound Insulation in Buildings and of Building Elements Part 6: Laboratory Measurements of Impact Sound Insulation of Floors. 10 pp.
BSI BS 2750: Part 7-80. 1980 Measurement of Sound Insulation in Buildings and of Building Elements Part 7: Field Measurements of Impact Sound Insulation of Floors. 8 pp.
BSI BS 2750: Part 8-80. 1980 Measurement of Sound Insulation in Buildings and of Building Elements Part 8: Laboratory Measurements of the Reduction of Transmitted Impact Noise by Floor Coverings on a Standard Floor. 11 pp.
CNS A3142-82. Method for Field Measurement of Floor Impact Sound Level (Feb)(8464).
ISO 140 Pt VI-78. Acoustics—Measurement of Sound Insulation in Buildings and of Building Elements—Part VI: Laboratory Measurements of Impact Sound Insulation of Floors First Edition. 7 pp.
ISO 140 Pt VII-78. Acoustics—Measurement of Sound Insulation in Buildings and of Building Elements—Part VII: Field Measurements of Impact Sound Insulation of Floors First Edition. 7 pp.
ISO 140 Pt VIII-78. Acoustics—Measurement of Sound Insulation in Buildings and of Building Elements—Part VIII: Laboratory Measurements of the Reduction of Transmitted Impact Noise by Floor Coverings on a Standard Floor First Edition. 9 pp.
JIS A 1418-78. Method for Field Measurement of Floor Impact Sound Level (R 1983). 11 pp.
SNZ NZS/ISO 140: Part 6-78. Acoustics Part 6: Laboratory Measurements of Impact Sound Insulation of Floors. 5 pp.

Floors (Cont.)

—Acoustical Insulation (Cont.)
SNZ NZS/ISO 140: Part 7-78. Acoustics Part 7: Field Measurements of Impact Sound Insulation of Floors. 5 pp.
SNZ NZS/ISO 140: Part 8-78. Acoustics Part 8: Laboratory Measurements of the Reduction of Transmitted Impact Noise by Floor Coverings on a Standard Floor. 7 pp.

—Agricultural Buildings
BSI BS 5502: Part 51-91. 1991 Buildings and Structures for Agriculture Part 51: Code of Practice for Design and Construction of Slatted, Perforated and Mesh Floors for Livestock. 14 pp.
BSI BS 5502: Part 51-01. 1991 Amd 1 Buildings and Structures for Agriculture Part 51: Code of Practice for Design and Construction of Slatted, Perforated and Mesh Floors for Livestock (AMD 7666) April 15, 1993 (R). 15 pp.

—Asphalt Mastic
BSI BS 988,1076, 1097,1451-73. 1973 Mastic Asphalt for Building (Limestone Aggregate). 15 pp.
BSI BS 6577-85. 1985 Mastic Asphalt for Building (Natural Rock Asphalt Aggregate). 13 pp.
BSI BS 6925-88. 1988 Mastic Asphalt for Building and Civil Engineering (Limestone Aggregate). 12 pp.
BSI BS 6925-01. 1988 Amd 1 Mastic Asphalt for Building and Civil Engineering (Limestone Aggregate) (AMD 7150) July 15, 1992 (R). 12 pp.
CGSB CAN/CGSB-37.26-M89. Mineral Colloid Type, Emulsified Asphalt for Mastic Flooring. 12 pp.
CGSB CAN/CGSB-37.27-M89. Chemical Emulsifier Type, Emulsified Asphalt for Mastic Flooring. 9 pp.
CGSB CAN/CGSB-37.38-M89. Application of Mineral-Colloid Type, Emulsified Asphalt Mixes for Cold Mastic Flooring. 10 pp.
CGSB CAN/CGSB-37.41-M89. Application of Chemical Type, Emulsified Asphalt Mixes for Cold Mastic Flooring. 11 pp.
CGSB CAN/CGSB-37.65-M88. Mastic Asphalt (Hot Process) for Flooring. 19 pp.
SNZ NZS/BS 6925-88. Specification for Mastic Asphalt for Building and Civil Engineering (Limestone Aggregate). 12 pp.

—Beech Wood
ISO 2457-76. Solid Wood Parquet—Classification of Beech Strips Second Edition. 5 pp.
ISO 3399-76. Broadleaved Wood Raw Parquet Blocks—Classification of Beech Parquet Blocks First Edition. 5 pp.

—Brick—Hollow Tile
DIN ENGL 4159-78. Structurally Cooperating Bricks for Floors and Wall Panels (Apr). 9 pp.

—Cleaning Agents
CGSB CAN/CGSB-2.4-92. Paste Soap for General Cleaning. 6 pp.
CGSB 2-GP-11M METH 44.2-61. Methods of Testing and Analysis of Soaps and Detergents Deleterious Action on Floors. 1 p.

—Conductive
CEPT T/TR 02-08-86. Conductivite Des Sols Dans Les Centres De Telecommunications. 4 pp.
CEPT T/TR 02-08 E-86. Conductivity of Floors in Telecommunication Centres. 4 pp.
CGSB 81-GP-1M-77. Standard for: Flooring, Conductive and Spark Resistant. 21 pp.

—Emissivity—Infrared Thermometers
CNS A3269-87. Simplified Test Method for Emissivity by Infrared Radio Meter (Sep)(12066).
JIS A 1423-83. Simplified Test Method for Emissivity by Infrared Radio Meter. 8 pp.

—Fire Testing
BSI BS 5268: Sec 4.2-90. 1990 Structural Use of Timber Part 4: Fire Resistance of Timber Structures Section 4.2: Recommendations for Calculating Fire Resistance of Timber Stud Walls and Joisted Floor Constructions. 28 pp.

—Glossaries
BSI BS 6100: SUBSEC 1.3.3-87. 1987 Glossary of Building and Civil Engineering Terms Part 1: General and Miscellaneous Section 1.3: Parts of Construction Works Subsection 1.3.3: Floors and Ceilings. 12 pp.
BSI BS 6100: SUBSEC 1.3.3-01. 1987 Amd 1 Building and Civil Engineering Terms Part 1: General and Miscellaneous Section 1.3: Parts of Construction Works Subsection 1.3.3: Floors and Ceilings (AMD 7233) August 15, 1992. 14 pp.

—Gymnasiums—Steel Furrings
CNS A2210-87. Steel Furring Components for Gymnasium Floors (Sep)(12053).
CNS A3263-87. Method of Test for Steel Furring Components for Gymnasium Floors (Sep)(12054).
JIS A 6519-89. Steel Furring Components for Gymnasium Floors. 24 pp.

Floors (Cont.)

—Heating—Electric
BSI CP 1018-71. 1971 Electric Floor-Warming Systems for Use with Off-Peak and Similar Supplies of Electricity. 34 pp.

—Hinges
CNS A2066-85. Floor Hinges (Feb)(4724). 3 pp.
CNS A3077-84. Method of Test for Checking Floor-Hinges and Door-Closers (Aug)(4725).
JIS A 1512-75. Testing Methods for Checking Floor-Hinges and Door-Closers.
JIS A 5543-75. Floor Hinges.

—Industrial Buildings—Loads (Forces)
ISO 2633-74. Determination of Imposed Floor Loads in Production Buildings and Warehouses First Edition. 4 pp.

—Industrial—Grilles
BSI BS 4592: Part 1-87. 1987 Amd 1 Industrial Type Metal Flooring, Walkways and Stair Treads Part 1: Opening Bar Gratings. 13 pp.
BSI BS 4592: Part 2-87. 1987 Industrial Type Metal Flooring, Walkways and Stair Treads Part 2: Expanded Metal Gratings Panels. 14 pp.
BSI BS 4592: Part 4-92. 1992 Industrial Type Flooring, Walkways and Stair Treads Part 4: Specification for Glass Reinforced Plastics Open Bar Gratings. 13 pp.

—Joists—Tables (Data)
BSI BS 5268: Sec 7.1-89. 1989 Structural Use of Timber Part 7: Recommendations for the Calculation Basis for Span Tables Section 7.1: Domestic Floor Joists. 15 pp.

—Magnesite
BSI BS 776: Part 2-72. 1972 Materials for Magnesium Oxychloride (Magnesite) Flooring Part 2: Metric Units. 20 pp.

—Metal Planks—Cold Formed
BSI BS 4592: Part 3-87. 1987 Industrial Type Metal Flooring, Walkways and Stair Treads Part 3: Cold Formed Planks. 12 pp.

—Panels—Precast Reinforced Concrete
CNS A2121-86. Building Components—Concrete Panel for Floors (Dec)(8076).
CNS A3131-81. Method of Test for Building Components (Concrete Panel for Floor) (Nov)(8077).
JIS A 6505-75. Building Components (Concrete Panel for Floor) (R 1983). 17 pp.

—Panels—Steel
CNS A2126-87. Building Components (Steel Panel for Floor) (Dec)(8186).
CNS A3137-87. Method of Test for Building Components (Steel Panel for Floor) (Dec)(8187).
JIS A 6507-75. Building Components (Steel Panel for Floor). 16 pp.

—Particle Board
BSI BS 5669: Part 5-89. 1989 Amd 1 Particleboard Part 5: Code of Practice for the Selection and Application of Particleboards for Specific Purposes (AMD 6612) April 30, 1991. 17 pp.
BSI BS 5669: Part 5-02. 1989 Amd 2 Particleboard Part 5: Code of Practice for the Selection and Application of Particleboards for Specific Purposes (AMD 6964) September 15, 1992. 17 pp.

—Particle Board—Installation
SAA AS 1860-91. Installation of Particleboard Flooring (in Professional Packages 20, 21, 30, 62-69). 9 pp.

—Pine Wood
SAA AS 1492-73. Flooring Milled from Radiata Pine Bound Together with Standards AS 1493-1498 Amdt 1 April 1977 Amdt 2 July 1977 Corrig..
SAA AS 1810-75. Flooring Milled from Cypress Pine. 16 pp.

—Plastic—Components
JIS A 5721-79. Plastic Floor Parts (R 1989). 12 pp.

—Polyurethane
CNS A2092-85. Polyurethane for Flooring (Feb)(6987). 2 pp.
CNS A3120-86. Method of Test for Polyurethane for Building Caulking and Flooring (May)(6988).

—Screeds
BSI BS 8000: Part 9-89. 1989 Workmanship on Building Sites Part 9: Code of Practice for Cement/Sand Floor Screeds and Concrete Floor Toppings. 17 pp.
BSI BS 8204: Part 1-87. 1987 In-Situ Floorings Part 1: Code of Practice for Concrete Bases and Screeds to Receive In-Situ Floorings. 19 pp.

Floors (Cont.)

—Screeds—Construction Contracts
DIN ENGL 18353-88. Tendering and Performance Stipulations in Contracts for Construction Works (VOB) Part C: General Technical Specifications in Contracts for Construction Works (ATV); Floor Screeding (Sept) (This Standard, Together with DIN 18299, September 1988 Edition,. 5 pp.

—Sealants
CGSB CAN/CGSB-25.20-M88. Surface Sealer for Floors. 10 pp.

—Seamless
CGSB 81-GP-10M-79. Application of Seamless Flooring, Standard for. 11 pp.

—Seamless—Epoxy Resin
CGSB 81-GP-4M-77. Flooring, Seamless, Decorative Epoxy, Troweled Finish, Standard for. 12 pp.
CGSB 81-GP-5M-78. Flooring, Seamless Epoxy, Broadcast Quartz, Standard for. 10 pp.

—Seamless—Epoxy Resin/Terrazzo
CGSB 81-GP-6M-77. Flooring, Seamless, Epoxy Terrazzo, Standard for. 11 pp.

—Seamless—Polyurethane
CGSB 81-GP-2M-78. Standard for: Flooring, Seamless, Polyurethane with Plastic Chips. 11 pp.

—Sheets—PVC—Ships
MOD UK NES 151: Part 1-81. Tiles and Sheet PVC Part 1: Technical Requirements for Production Issue 1 (8/81) (Withdrawn) (Refer to MoD). 15 pp.

—Sheets—Rubber
BSI BS 1711-75. 1975 Solid Rubber Flooring. 10 pp.

—Slipperiness
CNS A3159-82. Method of Test for Floor Slipperiness (Pendulum Type) (Jun)(8911).
JIS A 1407-63. Method of Test for Floor Slipperiness (Pendulum Type).

—Softwoods
SAA AS 1782-75. Flooring Milled from Australian-Grown Conifers (Softwoods) (Excluding Radiata Pine and Cypress Pine) Bound Together with AS 1782-1787.

—Spark Resistant
CGSB 81-GP-1M-77. Standard for: Flooring, Conductive and Spark Resistant. 21 pp.

—Tents—Flammability Testing
CNS S2147-90. Method of Test for Tent—Flammability of Flooring Material (Mar)(12692). 2 pp.

—Thermal Insulation—Heat Loss
DIN ENGL 52614-74. Testing of Thermal Insulations; Determination of Heat Dissipation of Floors (Dec). 2 pp.

—Varnishes (Polyurethane)
CNS K2165-87. Household Varnish for Floor (Oct)(12145). 2 pp.
CNS K6927-87. Method of Test for Household Varnish for Floor (Oct)(12146). 5 pp.
JIS K 5961-83. Household Varnish for Floor. 10 pp.

—Wood—Paints (Enamel)
CGSB 1-GP-71 METH 134.5-75. Methods of Testing Paints and Pigments Applicability and Appearance Enamels for Wooden Floors. 1 p.

Floppy Disk Controllers
See Also: Computer Equipment; Disk Drives (Computer); Peripheral Controllers

—Interfaces
JIS X 6052-90. Interface Between Flexible Disk Cartridge Drives and Their Host Controllers. 27 pp.

Flotation
See Also: Bulk Flotation; Froth Flotation Testing; Sewage Treatment; Water Treatment

—Combat Vehicles
NATO STANAG 2805 ED 4 AMD 0-90. Fording and Flotation Requirements for Combat and Support Ground Vehicles. 11 pp.

—Military Vehicles
NATO STANAG 2805 ED 4 AMD 0-90. Fording and Flotation Requirements for Combat and Support Ground Vehicles. 11 pp.

—Mobile Equipment
MOD UK DSTAN 00-6-01. Fording and Flotation Requirements for Mobile Equipment Issue 1; Amendment 1. 14 pp.

Flotation Agent Content Analysis

—Fluorite—Gravimetric Analysis
BSI BS 5659: Part 1-79. (WITHDRAWN) 1979 Acid-Grade Fluorspar Part 1: Determination of Flotation Agents. 4 pp.
ISO 3703-93. Acid-Grade and Ceramic-Grade Fluorspar—Determination of Flotation Agents Second Edition. 5 pp.

Flotation Gear
Use: Floats

Flours (Food)
See Also: Bread; Cereal Products; Food; Wheat

—Buckwheat
CGSB 32.6M-87. Flours, Specialty. 7 pp.

—Rye
CGSB 32.6M-87. Flours, Specialty. 7 pp.

—Soy
CGSB 32.6M-87. Flours, Specialty. 7 pp.
CNS N5030-84. Soybean Flour (Edible) (Apr)(1416). 2 pp.

—Soy—Ash Content
CNS N6028-73. Method of Test for Soybean Powder (Nov)(1418). 3 pp.

—Soy—Fat Content
CNS N6028-73. Method of Test for Soybean Powder (Nov)(1418). 3 pp.

—Soy—Fiber Content
CNS N6028-73. Method of Test for Soybean Powder (Nov)(1418). 3 pp.

—Soy—Moisture Content
CNS N6028-73. Method of Test for Soybean Powder (Nov)(1418). 3 pp.

—Soy—Protein Content
CNS N6028-73. Method of Test for Soybean Powder (Nov)(1418). 3 pp.

—Soy—Sampling
CNS N6028-73. Method of Test for Soybean Powder (Nov)(1418). 3 pp.

—Suspensions—Viscosity—Amylographs
BSI BS 4317: Part 28-93. 1993 Methods of Test for Cereals and Pulses Part 28: Determination of Viscosity of Flour Using an Amylograph (ISO 7973: 1992). 12 pp.
ISO 7973-92. Cereals and Milled Cereal Products—Determination of the Viscosity of Flour—Method Using an Amylograph First Edition. 11 pp.

—Wheat
CGSB 32.5M-88. Wheat Flour. 8 pp.
CNS N5007-80. Wheat Flour (May)(550). 2 pp.
CNS N6002-80. Methods of Test for Wheat Flour (May)(551). 3 pp.
ISO 6820-85. Wheat Flour and Rye Flour—General Guidance on the Drafting of Bread-Making Tests First Edition. 6 pp.

—Wheat—Doughs—Rheological Properties
BSI BS 4317: Part 20-89. 1989 Methods of Test for Cereals and Pulses Part 20: Determination of Water Absorption of Flour and Rheological Properties of Doughs Using a Farinograph. 12 pp.
BSI BS 4317: Part 21-89. 1989 Methods of Test for Cereals and Pulses Part 21: Determination of Rheological Properties of Doughs Using an Extensograph. 11 pp.
BSI BS 4317: Part 22-89. 1989 Methods of Test for Cereals and Pulses Part 22: Determination of Water Absorption of Flour and Rheological Properties of Doughs Using a Valorigraph. 10 pp.
BSI BS 4317: Part 27-92. 1992 Methods of Test for Cereals and Pulses Part 27: Determination of Rheological Properties of Doughs Using an Alveograph (ISO 5530-4: 1991) (W). 18 pp.
BSI BS 4317: Part 27-01. 1992 Amd 1 Methods of Test for Cereals and Pulses Part 27: Determination of Rheological Properties of Doughs Using an Alveograph (ISO 5530-4: 1991) (AMD 7607) May 15, 1993 (W). 19 pp.
ISO 5530 Pt 1-88. Wheat Flour—Physical Characteristics of Doughs—Part 1: Determination of Water Absorption and Rheological Properties Using a Farinograph First Edition. 11 pp.
ISO 5530 Pt 2-88. Wheat Flour—Physical Characteristics of Doughs—Part 2: Determination of Rheological Properties Using an Extensograph First Edition. 10 pp.
ISO 5530 Pt 3-88. Wheat Flour—Physical Characteristics of Doughs—Part 3: Determination of Water Absorption and Rheological Properties Using a Valorigraph First Edition. 8 pp.
ISO 5530 Pt 4-91. Wheat Flour—Physical Characteristics of Doughs—Part 4: Determination of Rheological Properties Using an Alveograph Second Edition; (Corrigendum 1-1992). 16 pp.

—Wheat—Fatty Acids Content
BSI BS 4317: Part 17-87. 1987 Cereals and Pulses Part 17: Determination of Fat Acidity. 6 pp.
ISO 7305-86. Milled Cereal Products—Determination of Fat Acidity First Edition. 5 pp.

—Wheat—Gluten Content
BSI BS 4317: Part 12-80. 1980 Methods of Test for Cereals and Pulses Part 12: Determination of Wet Gluten in Wheat Flour. 6 pp.
BSI BS 4317: Part 16-81. 1981 Methods of Test for Cereals and Pulses Part 16: Determination of Dry Gluten in Wheat Flour. 4 pp.
ISO 5531-78. Wheat Flour—Determination of Wet Gluten First Edition. 5 pp.
ISO 6645-81. Wheat Flour—Determination of Dry Gluten First Edition. 4 pp.
ISO 7495-90. Wheat Flour—Determination of Wet Gluten Content by Mechanical Means First Edition. 9 pp.

—Wheat—Impurities Content
BSI BS 4317: Part 29-93. 1993 Methods of Test for Cereals and Pulses Part 29: Determination of Impurities of Animal Origin in Wheat Flour and Durum Wheat Semolina (ISO 11050: 1993) (W). 16 pp.
ISO 11050-93. Wheat Flour and Durum Wheat Semolina—Determination of Impurities of Animal Origin First Edition. 14 pp.

—Wheat—Water Absorption
BSI BS 4317: Part 20-89. 1989 Methods of Test for Cereals and Pulses Part 20: Determination of Water Absorption of Flour and Rheological Properties of Doughs Using a Farinograph. 12 pp.
BSI BS 4317: Part 22-89. 1989 Methods of Test for Cereals and Pulses Part 22: Determination of Water Absorption of Flour and Rheological Properties of Doughs Using a Valorigraph. 10 pp.
ISO 5530 Pt 1-88. Wheat Flour—Physical Characteristics of Doughs—Part 1: Determination of Water Absorption and Rheological Properties Using a Farinograph First Edition. 11 pp.
ISO 5530 Pt 3-88. Wheat Flour—Physical Characteristics of Doughs—Part 3: Determination of Water Absorption and Rheological Properties Using a Valorigraph First Edition. 8 pp.

Flow Charts
Use For: Flow Diagrams; Flow Sheets
See Also: Charts; Computer Programming; Diagrams; Drawings; Process Charts; Symbols
ECMA ECMA 4-66. Flow Charts. 21 pp.

—Coal Preparation Plants
SAA AS 1414-90. Flowsheets and Symbols Relating to Coal Preparation Plant. 30 pp.

—Coal Preparation Plants—Symbols
BSI BS 3553-92. 1992 Graphical Symbols for Coal Preparation Plant Flowsheets (Q). 21 pp.
BSI BS 3553-86. 1986 Graphical Symbols for Coal Preparation Plant Flowsheets. 20 pp.
ISO 561-89. Coal Preparation Plant—Graphical Symbols Second Edition; (Corrected and Reprinted -1992). 17 pp.

—Coal Processing Plants
BSI BS 3567-85. 1985 Preparation of Flowsheets for the Design of Coal Preparation Plant. 12 pp.
BSI BS 3567-01. 1985 Amd 1 Preparation of Flowsheets for the Design of Coal Preparation Plant (AMD 6905) May 15, 1992 (Q). 13 pp.
ISO 924-89. Coal Preparation Plant—Principles and Conventions for Flowsheets Second Edition. 9 pp.
JIS M 4001-85. Methods of Presentation of Flowsheet for Mineral, Coal and Rock Processing.

—Data Processing
BSI BS 4058-87. 1987 Data Processing Flow Chart Symbols, Rules and Conventions. 31 pp.
CNS C5035-83. Flowchart Symbols for Information Processing (Feb)(5204).
CNS C5109-83. Information Processing — Conventions for Incorporating Flow Chart Symbols in Flow Charts (Feb)(7225).
ISO 5807-85. Information Processing—Documentation Symbols and Conventions for Data, Program and System Flowcharts, Program Network Charts and System Resources Charts First Edition. 29 pp.
JTC1 5807-85. Information Processing—Documentation Symbols and Conventions for Data, Program and System Flowcharts, Program Network Charts and System Resources Charts First Edition. 29 pp.
OSI ISO 5807-85. Information Processing—Documentation Symbols and Conventions for Data, Program and System Flowcharts, Programnet-work Charts and System Resources Charts. 29 pp.

—Data Processing—Symbols
JIS X 0121-86. Documentation Symbols and Conventions for Data Program and System Flowcharts, Program Networks Charts and System Resources Charts.

—Fluid Power Systems
BSI BS 5070: Part 3-88. 1988 Engineering Diagram Drawing Practice Part 3: Recommendations for Mechanical/Fluid Flow Diagrams. 44 pp.
SNZ BS 5070: Part 3-88. Engineering Diagram Drawing Practice Part 2: Recommendations for Mechanical/Fluid Flow Diagrams. 4 pp.

—Food Industry—Symbols
SAA AS 1109-81. Graphical Symbols for Process Flow Diagrams for the Food Industry. 16 pp.

—Industrial Plants
BSI BS 5070: Part 3-88. 1988 Engineering Diagram Drawing Practice Part 3: Recommendations for Mechanical/Fluid Flow Diagrams. 44 pp.
DIN ENGL 28004 Pt 1-88. Flow Sheets and Diagrams of Process Plants; Concepts, Types of Diagram, Information Content (May) (Supersedes DIN 28004 Part 10, August 1976 Edition). 16 pp.
DIN ENGL 28004 Pt 2-88. Flow Sheets and Diagrams of Process Plants; Drawing Instructions (May). 3 pp.
SNZ BS 5070: Part 3-88. Engineering Diagram Drawing Practice Part 2: Recommendations for Mechanical/Fluid Flow Diagrams. 4 pp.

—Industrial Plants—Symbols
DIN ENGL 28004 Pt 3-88. Flow Sheets and Diagrams of Process Plants; Graphical Symbols (May). 27 pp.
DIN ENGL 28004 Pt 4-77. Flow Sheets and Diagrams of Process Plants; Symbols (May). 2 pp.

—Microfilming
ISO 6829-83. Flowchart Symbols and Their Use in Micrographics First Edition. 17 pp.

—Microfilming—Symbols
CGSB CAN/CGSB-72.22-M87. Flowchart Symbols and Their Use in Micrographics (ISO 6829-1983). 36 pp.

—Mineral Processing
JIS M 4001-85. Methods of Presentation of Flowsheet for Mineral, Coal and Rock Processing.

—Pipelines
BSI BS 5070: Part 3-88. 1988 Engineering Diagram Drawing Practice Part 3: Recommendations for Mechanical/Fluid Flow Diagrams. 44 pp.
SNZ BS 5070: Part 3-88. Engineering Diagram Drawing Practice Part 2: Recommendations for Mechanical/Fluid Flow Diagrams. 4 pp.

—Rock Processing
JIS M 4001-85. Methods of Presentation of Flowsheet for Mineral, Coal and Rock Processing.

—Symbols
CNS Z7067-80. Graphical Symbols for Process Flow Diagrams (Feb)(5257).

—Telecommunication Networks—Planning Models
CCITT RECMN E.175-89. Models for International Network Planning—Telephone Network and ISDN—Operation, Numbering, Routing and Mobile Service (Study Group II) 3 pp. 3 pp.

Flow Control
Scope Note: Use a more specific term *See:* Control Systems Equipment; Control Valves; Flow Control Valves; Hydraulic Valves; Process Control Valves; Remote Control Equipment

Flow Control Protocols
See Also: Information Interchange; Protocols

—Integrated Services Digital Networks
CCITT RECMN I.320-89. ISDN Protocol Reference Model—Integrated Services Digital Network (ISDN)—Overall Network Aspects and Functions, ISDN User-Network Interfaces (Study Group XVIII) 10 pp. 10 pp.
CENELEC ETR 054-92. Integrated Services Digital Network (ISDN); the Protocol Reference Model and Its Relationship with the OSI Reference Model. 13 pp.

Flow Control Protocols (Cont.)
—Integrated Services Digital Networks (Cont.)
ETSI ETR 054-92. Integrated Services Digital Network (ISDN); the Protocol Reference Model and Its Relationship with the OSI Reference Model. 13 pp.

Flow Control Valves
See Also: Alarm Valves; Control Valves; Deceleration Valves; Proportional Valves; Throttle Valves; Valves

—Hydraulic
CNS B4028-80. Pressure Compensating Type Flow Control Valves for Hydraulic Use (Jul)(5707).

CNS B7149-80. Method of Test for Pressure Compensating Type Flow Control Valves for Hydraulic Use (Jul)(5708).

ISO 6403-88. Hydraulic Fluid Power—Valves Controlling Flow and Pressure—Test Methods First Edition; (Corrected and Reprinted -1989). 34 pp.

JIS B 8357-77. Pressure Compensating Type Flow Control Valves for Oil-Hydraulic Use.

JIS B 8357-67. Pressure Compensating Type Flow Control Valves for Oil-Hydraulic Use. 12 pp.

JIS B 8652-89. Test Methods for Electro-Hydraulic Proportional Pressure Reducing Valves and Electro-Hydraulic Proportional Pressure Reducing and Relieving Valves. 26 pp.

—Hydraulic—Proportional Control
JIS B 8654-89. Test Methods for Electro-Hydraulic Proportional Series Flow Control Valves. 23 pp.

JIS B 8655-89. Test Methods for Electro-Hydraulic Proportional Directional Series Flow Control Valves. 28 pp.

—Mounting Surfaces
ISO 6263-87. Hydraulic Fluid Power—Compensated Flow-Control Valves—Mounting Surfaces First Edition. 20 pp.

—Noise
BSI BS 5944: Part 4-84. 1984 Measurement of Airborne Noise From Hydraulic Fluid Power Systems and Components Part 4: Method of Determining Sound Power Levels from Valves Controlling Flow and Pressure. 12 pp.

Flow Diagrams
Use: Flow Charts

Flow Indicators, Sight
Use: Sight Glasses

Flow Measurement
Scope Note: For additional listings, use a more specific term *See Also:* Air Flow; Flow Rate; Flowmeters; Gaging Stations; Gas Flow; Gas Meters; Hydraulic Structures; Leak Detectors; Nozzles; Open Channel Flow; Orifice Flow; Pitot Tubes; Pressure Measurement; Rheological Properties; Steady State Performance; Tidal Channel Flow; Water Meters; Weirs

BSI BS 3680: Part 3M-90. 1990 Measurement of Liquid Flow in Open Channels Part 3: Stream Flow Measurement Part 3M: Guide to the Methods of Measuring High Discharges in Rivers. 14 pp.

BSI BS 5844-80. 1980 Measurement of Fluid Flow: Estimation of Uncertainty of a Flow-Rate Measurement. 28 pp.

BSI PP 7316-86. 1986 Streamflow Measurement. 30 pp.

DIN ENGL 19559 Pt 1-83. Measurement of Flow of Waste Water in Open Channels and Gravity Conduits; General Information (July). 10 pp.

ISO 5168-78. Measurement of Fluid Flow—Estimation of Uncertainty of a Flow-Rate Measurement First Edition. 29 pp.

ISO TR9824 Pt 1-90. Measurement of Free Surface Flow in Closed Conduits—Part 1: Methods First Edition. 11 pp.

ISO TR9824 Pt 2-90. Measurement of Free Surface Flow in Closed Conduits—Part 2: Equipment First Edition. 10 pp.

JIS Z 8766-89. Methods of Flow Measurement by Vortex Flowmeters. 20 pp.

—Adhesives
SAA AS 1937.1-91. Methods of Test for Sealers and Adhesives for Automotive Purposes—Part 1: Determination of Flow Properties in Absolute Units. 6 pp.

—Adhesives—Sagging Resistance Method
BSI BS 5350: Part B9-84. 1984 Methods of Test for Adhesives Group B: Adhesives Part B9: Determination of Resistance to Sagging (Flow After Application). 8 pp.

—Corrosion Inhibitors
CGSB 31-GP-0A METH 23.1-57. Methods of Testing Corrosion-Prevention Materials and Processes Flow Point. 1 p.

—Flowmeters
BSI BS 1042: Sec 2.4-89. 1989 Methods of Measurement of Fluid Flow in Closed Conduits Part 2: Pitot Tubes Section 2.4: Method of Measurement of Clean Water Flow Using Current Meters in Full Conduits and Under Regular Flow Conditions. 42 pp.

ISO 3354-88. Measurement of Clean Water Flow in Closed Conduits—Velocity-Area Method Using Current-Meters in Full Conduits and Under Regular Flow Conditions Second Edition. 42 pp.

—Flue Gases
CNS K9019-76. Method of Measuring Velocity and Flow Rate of Stack Gas (Dec)(4050).

—Fuel Injectors
ISO 7440 Pt 2-91. Road Vehicles—Fuel Injection Equipment Testing—Part 2: Orifice Plate Flow-Measurement Second Edition. 9 pp.

—Geotextiles
BSI BS 6906: Part 7-90. 1990 Methods of Test for Geotextiles Part 7: Determination of in-Plane Waterflow. 14 pp.

—Glazing Compounds—Sagging Resistance Method
CGSB CAN2-19.0-M77METH 7.4-78. Methods of Testing Putty, Caulking and Sealing Compounds Resistance to Sag or Flow of Glazing Compounds. 1 p.

—Glossaries
BSI BS 5875-91. 1991 Terms and Symbols for Measurement of Fluid Flow in Closed Conduits (ISO 4006: 1991). 59 pp.

BSI BS 5875-80. 1980 Glossary of Terms and Symbols for Measurement of Fluid Flow in Closed Conduits. 16 pp.

ISO 4006-91. Measurement of Fluid Flow in Closed Conduits—Vocabulary and Symbols Second Edition. 60 pp.

—Hydraulic Power Pumps
ISO 9632-92. Hydraulic Fluid Power—Fixed Displacement Pumps—Flow Degradation Due to Classified AC Fine Test Dust Contaminant—Test Method First Edition. 12 pp.

—Lubricating Oils
DIN ENGL 51568-74. Testing of Lubricating Oils; Determination of Ability to Flow; U-Tube Method (June). 3 pp.

—Membrane Filters
JIS K 3833-90. Testing Methods for Diffusive Flow Through Membrane Filters. 12 pp.

—Molding Materials
BSI BS 2782:Pt7: METH 722A-93. 1993 Methods of Testing Plastics Part 7: Rheological Properties Method 722A: Determination of Transfer Flow of Thermosetting Moulding Materials (ISO 7808: 1992). 5 pp.

ISO 7808-92. Plastics—Thermosetting Moulding Materials—Determination of Transfer Flow First Edition. 5 pp.

—Nozzles
BSI BS 1042: Sec 1.1-92. 1992 Measurement of Fluid Flow in Part 1: Pressure Differential Devices Section 1.1: Specification for Square-Edged Orifice Plates, Nozzles, and Venturi Tubes Inserted in Circular Cross-Section Conduits Running. 67 pp.

BSI BS 1042: Sec 1.1-81. 1981 Amd 2 Methods of Measurement of Fluid Flow in Closed Conduits Part 1: Pressure Differential Devices Section 1.1: Orifice Plates, Nozzles and Venturi Tubes Inserted in Circular Cross-Section Conduits Running Full. 69 pp.

BSI BS 1042: Sec 1.2-89. 1989 Methods of Measurement of Fluid Flow in Closed Conduits: Pt 1: Pressure Differential Devices: Sec 1.2: Specification for Squared-Edged-Orifice Plates and Nozzles (with Drain Holes, in Pipe Below. 19 pp.

BSI BS 1042: Sec 1.3-91. 1991 Measurement of Fluid Flow in Closed Conduits Part 1: Pressure Differential Devices Section 1.3: Method of Measurement of Gas Flow by Means of Critical Flow Venturi Nozzles (ISO 9300: 1990). 22 pp.

BSI BS 1042: Sec 1.4-92. 1992 Measurement of Fluid Flow in Closed Conduits Part 1: Pressure Differential Devices Section 1.4: Guide to the Use of Devices Specified in Sections 1.1 and 1.2. 40 pp.

BSI BS 1042: Sec 1.4-84. 1984 Amd 1 Methods of Measurement of Fluid Flow in Closed Conduits Part 1: Pressure Differential Devices Section 1.4: Guide to the Use of Devices Specified in Sections 1.1 and 1.2 (AMD 5456) November 30, 1987. 30 pp.

BSI BS 1042: Sec 1.5-87. 1987 Methods of Measurement of Fluid Flow in Closed Conduits Part 1: Pressure Differential Devices Section 1.5: Guide to the Effect of Departure from the Condition Specified in Section 1.1. 27 pp.

BSI BS 1042: Sec 1.6-93. 1993 Measurement of Fluid Flow in Closed Conduits Part 1: Pressure Differential Devices Section 1.6 Method of Measurement of Pulsating Fluid Flow in a Pipe, by Means of Orifice Plates, Nozzles or Venturi Tubes (ISO TR 3313: 1992) (Q). 27 pp.

CNS B1353-87. Measurement of Fluid Flow by Means of Orifice Plates and Nozzles (Mar)(11872).

ISO TR3313-92. Measurement of Pulsating Fluid Flow in a Pipe by Means of Orifice Plates, Nozzles or Venturi Tubes Second Edition. 25 pp.

ISO 5167 Pt 1-91. Measurement of Fluid Flow by Means of Pressure Differential Devices—Part 1: Orifice Plates, Nozzles and Venturi Tubes Inserted in Circular Cross-Section Conduits Running Full First Edition. 66 pp.

JIS Z 8762-88. Measurement of Fluid Flow by Means of Orifice Plates, Nozzles and Venturi Tubes. 93 pp.

—Orifice Plates
BSI BS 1042: Sec 1.1-92. 1992 Measurement of Fluid Flow in Part 1: Pressure Differential Devices Section 1.1: Specification for Square-Edged Orifice Plates, Nozzles, and Venturi Tubes Inserted in Circular Cross-Section Conduits Running. 67 pp.

BSI BS 1042: Sec 1.1-81. 1981 Amd 2 Methods of Measurement of Fluid Flow in Closed Conduits Part 1: Pressure Differential Devices Section 1.1: Orifice Plates, Nozzles and Venturi Tubes Inserted in Circular Cross-Section Conduits Running Full. 69 pp.

BSI BS 1042: Sec 1.2-89. 1989 Methods of Measurement of Fluid Flow in Closed Conduits: Pt 1: Pressure Differential Devices: Sec 1.2: Specification for Squared-Edged-Orifice Plates and Nozzles (with Drain Holes, in Pipe Below. 19 pp.

BSI BS 1042: Sec 1.3-91. 1991 Measurement of Fluid Flow in Closed Conduits Part 1: Pressure Differential Devices Section 1.3: Method of Measurement of Gas Flow by Means of Critical Flow Venturi Nozzles (ISO 9300: 1990). 22 pp.

BSI BS 1042: Sec 1.4-92. 1992 Measurement of Fluid Flow in Closed Conduits Part 1: Pressure Differential Devices Section 1.4: Guide to the Use of Devices Specified in Sections 1.1 and 1.2. 40 pp.

BSI BS 1042: Sec 1.4-84. 1984 Amd 1 Methods of Measurement of Fluid Flow in Closed Conduits Part 1: Pressure Differential Devices Section 1.4: Guide to the Use of Devices Specified in Sections 1.1 and 1.2 (AMD 5456) November 30, 1987. 30 pp.

BSI BS 1042: Sec 1.5-87. 1987 Methods of Measurement of Fluid Flow in Closed Conduits Part 1: Pressure Differential Devices Section 1.5: Guide to the Effect of Departure from the Condition Specified in Section 1.1. 27 pp.

BSI BS 1042: Sec 1.6-93. 1993 Measurement of Fluid Flow in Closed Conduits Part 1: Pressure Differential Devices Section 1.6 Method of Measurement of Pulsating Fluid Flow in a Pipe, by Means of Orifice Plates, Nozzles or Venturi Tubes (ISO TR 3313: 1992) (Q). 27 pp.

CNS B1353-87. Measurement of Fluid Flow by Means of Orifice Plates and Nozzles (Mar)(11872).

ISO TR3313-92. Measurement of Pulsating Fluid Flow in a Pipe by Means of Orifice Plates, Nozzles or Venturi Tubes Second Edition. 25 pp.

ISO 5167 Pt 1-91. Measurement of Fluid Flow by Means of Pressure Differential Devices—Part 1: Orifice Plates, Nozzles and Venturi Tubes Inserted in Circular Cross-Section Conduits Running Full First Edition. 66 pp.

ISO 9300-90. Measurement of Gas Flow by Means of Critical Flow Venturi Nozzles First Edition. 19 pp.

JIS Z 8762-88. Measurement of Fluid Flow by Means of Orifice Plates, Nozzles and Venturi Tubes. 93 pp.

—Orifice Valves
BSI BS 7350-90. 1990 Double Regulating Globe Valves and Flow Measurement Devices for Heating and Chilled Water Systems. 24 pp.

BSI BS 7350-01. 1990 Amd 1 Double Regulating Globe Valves and Flow Measurement Devices for Heating and Chilled Water Systems (AMD 6865) December 24, 1991. 25 pp.

—Petroleum Products
BSI BS 6439-83. 1983 Fidelity and Security of Dynamic Measurement of Petroleum Liquids and Gases in Cabled Transmission as Electric and/or Electronic Data. 15 pp.

INTERNATIONAL AND NON-U.S. NATIONAL STANDARDS SUBJECT INDEX

Flow

Flow Measurement (Cont.)
—Petroleum Products (Cont.)
ISO 6551-82. Petroleum Liquids and Gases—Fidelity and Security of Dynamic Measurement—Cabled Transmission of Electric and/or Electronic Pulsed Data First Edition. 14 pp.
SNZ NZS/ISO 6551-82. Petroleum Liquids and Gases. Fidelity and Security of Dynamic Measurement. Cabled Transmission of Electric and/or Electronic Pulsed Data. 12 pp.

—Phenolic Resins
ISO 8619-88. Plastics—Phenolic Resin Powder—Determination of Flow Distance on a Glass Plate First Edition. 6 pp.

—Powder Coatings—Spraying
BSI BS 3900: Part J9-93. 1993 Methods of Test for Paints Part J9: Determination of Flow Properties of a Coating Powder/Air Mixture (ISO 8130-5: 1992) (W). 11 pp.
ISO 8130 Pt 5-92. Coating Powders—Part 5: Determination of Flow Properties of a Powder/Air Mixture First Edition. 10 pp.

—Pumps
JIS B 8302-90. Measurement Methods of Pump Discharge. 39 pp.

—Sealing Materials
CEN PREN 3093-89. Sealants Test Method Determination of Flow. 4 pp.

—Sealing Materials—Sagging Resistance Method
CGSB CAN2-19.0-M77METH 7.1-78. Methods of Testing Putty, Caulking and Sealing Compounds Resistance to Sag or Flow in a Vertical Test Assembly. 1 p.
CGSB CAN2-19.0-M77METH 7.3-78. Methods of Testing Putty, Caulking and Sealing Compounds Resistance to Sag or Flow in an Inverted Horizontal Test Assembly. 1 p.

—Steam Traps
BSI BS EN 27842-91. 1991 Methods for Determination of Discharge Capacity of Automatic Steam Traps (ISO 7842: 1988). 18 pp.
CEN EN 27842-91. Automatic Steam Traps—Determination of Discharge Capacity—Test Methods. 3 pp.
DIN ENGL EN 27842-91. Automatic Steam Traps; Determination of Discharge Capacity Test Methods (ISO 7842: 1988) (Nov). 14 pp.
ISO 7842-88. Automatic Steam Traps—Determination of Discharge Capacity—Test Methods First Edition. 14 pp.

—Symbols
BSI BS 5875-91. 1991 Terms and Symbols for Measurement of Fluid Flow in Closed Conduits (ISO 4006: 1991). 59 pp.
BSI BS 5875-80. 1980 Glossary of Terms and Symbols for Measurement of Fluid Flow in Closed Conduits. 16 pp.
ISO 4006-91. Measurement of Fluid Flow in Closed Conduits—Vocabulary and Symbols Second Edition. 60 pp.

—Thermoplastic Resins—Capillary Rheometers
JIS K 7199-91. Testing Method for Flow Properties of Thermoplastics with a Capillary Rheometer. 13 pp.

—Tracer Methods
BSI BS 5857: Sec 1.1-80. 1980 Measurement of Fluid Flow in Closed Conduits, Using Tracers Part 1: Measurement of Water Flow Section 1.1: General. 15 pp.
BSI BS 5857: Sec 1.2-80. 1980 Measurement of Fluid Flow in Closed Conduits, Using Tracers Part 1: Measurement of Water Flow Section 1.1: Constant Rate Injection Method Using Non-Radioactive Tracers. 12 pp.
BSI BS 5857: Sec 1.3-80. 1980 Measurement of Fluid Flow in Closed Conduits, Using Tracers Part 1: Measurement of Water Flow Section 1.1: Constant Rate Injection Method Using Radioactive Tracers. 14 pp.
BSI BS 5857: Sec 1.4-80. 1980 Measurement of Fluid Flow in Closed Conduits, Using Tracers Part 1: Measurement of Water Flow Section 1.4: Transit Time Method Using Non-Radioactive Tracers. 16 pp.
BSI BS 5857: Sec 1.5-80. 1980 Measurement of Fluid Flow in Closed Conduits, Using Tracers Part 1: Measurement of Water Flow Section 1.5: Transit Time Method Using Radioactive Tracers. 12 pp.
BSI BS 5857: Sec 2.1-80. 1980 Measurement of Fluid Flow in Closed Conduits, Using Tracers Part 2: Measurement of Gas Flow Section 2.1: General. 13 pp.

Flow Measurement (Cont.)
—Tracer Methods (Cont.)
BSI BS 5857: Sec 2.4-80. 1980 Measurement of Fluid Flow in Closed Conduits, Using Tracers Part 2: Measurement of Gas Flow Section 2.4: Transit Time Method Using Radioactive Tracers. 12 pp.
ISO 2975 Pt I-74. Measurement of Water FLow in Closed Conduits—Tracer Methods—Part I: General First Edition. 14 pp.
ISO 2975 Pt II-75. Measurement of Water Flow in Closed Conduits—Tracer Methods—Part II: Constant Rate Injection Method Using Non-Radioactive Tracers First Edition. 11 pp.
ISO 2975 Pt III-76. Measurement of Water Flow in Closed Conduits—Tracer Methods—Part III: Constant Rate Injection Method Using Radioactive Tracers First Edition. 13 pp.
ISO 2975 Pt VI-77. Measurement of Water Flow in Closed Conduits—Tracer Methods—Part VI: Transit Time Method Using Non-Radioactive Tracers First Edition. 15 pp.
ISO 2975 Pt VII-77. Measurement of Water Flow in Closed Conduits—Tracer Methods—Part VII: Transit Time Method Using Radioactive Tracers First Edition. 11 pp.
ISO 4053 Pt I-77. Measurement of Gas Flow in Conduits—Tracer Methods—Part I: General First Edition. 12 pp.
ISO 4053 Pt IV-78. Measurement of Gas Flow in Conduits—Tracer Methods—Part IV: Transit Time Method Using Radioactive Tracers First Edition. 11 pp.

—Velocity Area Methods
BSI BS 1042: Sec 2.1-83. 1983 Methods of Measurement of Fluid Flow in Closed Conduits Part 2: Pitot Tubes Section 2.1: Method Using Pitot-Static Tubes. 42 pp.
BSI BS 1042: Sec 2.3-84. 1984 Methods of Measurement of Fluid Flow in Closed Conduits Part 2: Pitot Tubes Section 2.3: Methods of Flow Measurement in Swirling or Asymmetric Flow Conditions in Circular Ducts by Means of Current-Meters or Pitot Static Tubes. 27 pp.
BSI BS 1042: Sec 2.4-89. 1989 Methods of Measurement of Fluid Flow in Closed Conduits Part 2: Pitot Tubes Section 2.4: Method of Measurement of Clean Water Flow Using Current Meters in Full Conduits and Under Regular Flow Conditions. 42 pp.
BSI BS 3680: Part 3J-89. 1989 Measurement of Liquid Flow in Open Channels Part 3: Stream Flow Measurement Part 3J: Guide to the Measurement of Discharge Using Three Verticals. 12 pp.
ISO 3354-88. Measurement of Clean Water Flow in Closed Conduits—Velocity-Area Method Using Current-Meters in Full Conduits and Under Regular Flow Conditions Second Edition. 42 pp.
ISO 3966-77. Measurement of Fluid Flow in Closed Conduits—Velocity Area Method Using Pitot Static Tubes First Edition. 42 pp.
ISO 7194-83. Measurement of Fluid Flow in Closed Conduits—Velocity-Area Methods of Flow Measurement in Swirling or Asymmetric Flow Conditions in Circular Ducts by Means of Current-Meters or Pitot Static Tubes First Edition. 27 pp.

—Venturi Tubes
BSI BS 1042: Sec 1.1-92. 1992 Measurement of Fluid Flow in Part 1: Pressure Differential Devices Section 1.1: Specification for Square-Edged Orifice Plates, Nozzles, and Venturi Tubes Inserted in Circular Cross-Section Conduits Running. 67 pp.
BSI BS 1042: Sec 1.1-81. 1981 Amd 2 Methods of Measurement of Fluid Flow in Closed Conduits Part 1: Pressure Differential Devices Section 1.1: Orifice Plates, Nozzles and Venturi Tubes Inserted in Circular Cross-Section Conduits Running Full. 69 pp.
BSI BS 1042: Sec 1.3-91. 1991 Measurement of Fluid Flow in Closed Conduits Part 1: Pressure Differential Devices Section 1.3: Method of Measurement of Gas Flow by Means of Critical Flow Venturi Nozzles (ISO 9300: 1990). 22 pp.
BSI BS 1042: Sec 1.4-92. 1992 Measurement of Fluid Flow in Closed Conduits Part 1: Pressure Differential Devices Section 1.4: Guide to the Use of Devices Specified in Sections 1.1 and 1.2. 40 pp.
BSI BS 1042: Sec 1.4-84. 1984 Amd 1 Methods of Measurement of Fluid Flow in Closed Conduits Part 1: Pressure Differential Devices Section 1.4: Guide to the Use of Devices Specified in Sections 1.1 and 1.2 (AMD 5456) November 30, 1987. 30 pp.
BSI BS 1042: Sec 1.5-87. 1987 Methods of Measurement of Fluid Flow in Closed Conduits Part 1: Pressure Differential Devices Section 1.5: Guide to the Effect of Departure from the Condition Specified in Section 1.1. 27 pp.
BSI BS 1042: Sec 1.6-93. 1993 Measurement of Fluid Flow in Closed Conduits Part 1: Pressure Differential Devices Section 1.6 Method of Measurement of Pulsating Fluid Flow in a Pipe, by Means of Orifice Plates, Nozzles or Venturi Tubes (ISO TR 3313: 1992) (Q). 27 pp.

Flow Measurement (Cont.)
—Venturi Tubes (Cont.)
ISO TR3313-92. Measurement of Pulsating Fluid Flow in a Pipe by Means of Orifice Plates, Nozzles or Venturi Tubes Second Edition. 25 pp.
ISO 5167 Pt 1-91. Measurement of Fluid Flow by Means of Pressure Differential Devices—Part 1: Orifice Plates, Nozzles and Venturi Tubes Inserted in Circular Cross-Section Conduits Running Full First Edition. 66 pp.
ISO 9300-90. Measurement of Gas Flow by Means of Critical Flow Venturi Nozzles First Edition. 19 pp.
JIS Z 8762-88. Measurement of Fluid Flow by Means of Orifice Plates, Nozzles and Venturi Tubes. 93 pp.

—Volumetric Tank Method
BSI BS 6199: Part 1-81. 1981 Measurement of Liquid Flow in Closed Conduits Using Weighing and Volumetric Methods Part 1: Weighing Method. 24 pp.
BSI BS 6199: Part 2-88. 1988 Measurement of Liquid Flow in Closed Conduits Using Weighting and Volumetric Methods Part 2: Method for Measurement by Collection of the Liquid in a Volumetric Tank. 24 pp.
ISO 8316-87. Measurement of Liquid Flow in Closed Conduits—Method by Collection of the Liquid in a Volumetric Tank First Edition. 25 pp.

—Weighing Method
BSI BS 6199: Part 1-81. 1981 Measurement of Liquid Flow in Closed Conduits Using Weighing and Volumetric Methods Part 1: Weighing Method. 24 pp.
BSI BS 6199: Sec 1.2-91. 1991 Measurement of Liquid Flow in Closed Conduits Using Weighing and Volumetric Methods Part 1: Weighing Method Section 1.2: Procedures for Checking Static Weighing Systems (ISO 9368-1: 1990). 24 pp.
ISO 4185-80. Measurement of Liquid Flow in Closed Conduits—Weighing Method First Edition. 25 pp.
ISO 9368 Pt 1-90. Measurement of Liquid Flow in Closed Conduits by the Weighing Method—Procedures for Checking Installations—Part 1: Static Weighing Systems First Edition. 22 pp.

Flow Measurement Equipment
Use For: Liquid Flow Measurement Equipment
See Also: Diaphragms (Mechanics); Flow Rate; Flow Totalizers; Measuring Instruments; Orifice Plates

—Water
BSI BS 7350-01. 1990 Amd 1 Double Regulating Globe Valves and Flow Measurement Devices for Heating and Chilled Water Systems (AMD 6865) December 24, 1991. 25 pp.

Flow Measurement Systems
Use: Flow Measurement Equipment

Flow Rate
Scope Note: For additional listings, use a more specific term Use For: Flow Time; Pourability
See Also: Flow Measurement; Flow Measurement Equipment; Metal Powders; Weirs

—Air
ISO 5221-84. Air Distribution and Air Diffusion—Rules to Methods of Measuring Air Flow Rate in an Air Handling Duct First Edition. 36 pp.

—Air—Fan Coil Units
BSI BS 4856: Part 1-72. 1972 Methods of Testing and Rating Fan Coil Units; Unit Heaters and Unit Coolers Part 1: Thermal and Volumetric Performance for Heating Duties: Without Additional Ducting. 26 pp.
SNZ NZS/BS 4856: Part 1-72. Methods for Testing and Rating Fan Coil Units, Unit Heaters and Unit Coolers Part 1: Thermal and Volumetric Performance for Heating Duties: Without Additional Ducting. 24 pp.

—Air—Unit Coolers
BSI BS 4856: Part 2-75. 1975 Methods of Testing and Rating Fan Coil Units; Unit Heaters and Unit Coolers Part 2: Thermal and Volumetric Performance for Cooling Duties: Without Additional Ducting. 31 pp.
BSI BS 4856: Part 3-75. 1975 Methods of Testing and Rating Fan Coil Units; Unit Heaters and Unit Coolers Part 3: Thermal and Volumetric Performance for Heating and Cooling Duties: with Additional Ducting. 35 pp.

—Air—Unit Heaters
BSI BS 4856: Part 1-72. 1972 Methods of Testing and Rating Fan Coil Units; Unit Heaters and Unit Coolers Part 1: Thermal and Volumetric Performance for Heating Duties: Without Additional Ducting. 26 pp.

INDUSTRY STANDARDS

Flow Rate (Cont.)

—Air—Unit Heaters (Cont.)
BSI BS 4856: Part 3-75. 1975 Methods of Testing and Rating Fan Coil Units; Unit Heaters and Unit Coolers Part 3: Thermal and Volumetric Performance for Heating and Cooling Duties: with Additional Ducting. 35 pp.

SNZ NZS/BS 4856: Part 1-72. Methods for Testing and Rating Fan Coil Units, Unit Heaters and Unit Coolers Part 1: Thermal and Volumetric Performance for Heating Duties; Without Additional Ducting. 24 pp.

—Aircraft Fuels—Aircraft Instruments
NATO STANAG 3872 ED 1 AMD 5-79. Rate of Fuel Flow Indicators (Amd 5). 9 pp.

—Binders—Bituminous—Viscometers
DIN ENGL 52023 Pt 1-80. Testing of Bituminous Binders; Determination of Flow Time by the Tar Efflux Viscometer; Method of Measurement (Dec). 4 pp.

—Directional Control Valves
JIS B 8655-89. Test Methods for Electro-Hydraulic Proportional Directional Series Flow Control Valves. 28 pp.

—Electrically Conductive Liquids
BSI BS 5792: Part 1-93. 1993 Measurement of Conductive Liquid Flow in Closed Conduits Part 1: Method Using Eloectromagnetic Flowmeters (ISO 6817: 1992) (Supersedes BS 5792: 1980). 24 pp.

ISO 6817-92. Measurement of Conductive Liquid Flow in Closed Conduits—Method Using Electromagnetic Flowmeters First Edition. 22 pp.

—Fats
BSI BS 684: Sec 1.4-76. 1976 Methods of Analysis of Fats and Fatty Oils Part 1: Physical Methods Section 1.4: Determination of Flow and Drop Points. 8 pp.

—Faucets
BSI BS EN 200-92. 1992 Sanitary Tapware: General Technical Specifications for Single Taps and Mixer Taps (Nominal Size 1/2) PN 10: Minimum Flow Pressure of 0.05 MPa (0.5 Bar). 33 pp.

CEN EN 200-89. Sanitary Tapware; General Technical Specifications for Single Taps and Mixer Taps (Nominal Size 1/2) PN 10 Minimum Flow Pressure of 0.05 MPa (0.5 Bar). 46 pp.

—Flow Control Valves
JIS B 8654-89. Test Methods for Electro-Hydraulic Proportional Series Flow Control Valves. 23 pp.

—Flow Regulators—Faucets
BSI BS EN 246-92. 1992 Sanitary Tapware: General Specifications for Flow Rate Regulators. 18 pp.

CEN EN 246-89. Sanitary Tapware; General Specifications for Flow Rate Regulator. 17 pp.

DIN ENGL EN 246-90. Sanitary Taps; General Specifications for Flow Rate Regulators (Jan). 12 pp.

—Flue Gases
CNS K9019-76. Method of Measuring Velocity and Flow Rate of Stack Gas (Dec)(4050).

—Flue Gases—Particulates
BSI BS 6069: Sec 4.3-92. 1992 Characterization of Air Quality Part 4: Stationary Source Emissions Section 4.3: Method for the Manual Gravimetric Detn. of Concentration and Mass Flow Rate of Particulate Material in Gas-Carrying Ducts (ISO 9096: 1992). 37 pp.

ISO 9096-92. Stationary Source Emissions—Determination of Concentration and Mass Flow Rate of Particulate Material in Gas-Carrying Ducts—Manual Gravimetric Method First Edition. 36 pp.

—Fluids
BSI BS 1042: Sec 2.2-83. 1983 Methods of Measurement of Fluid Flow in Closed Conduits Part 2: Pitot Tubes Section 2.2: Method of Measurement of Velocity at One Point of a Conduit of Circular Cross Section. 14 pp.

BSI BS 7294-90. 1990 Method for Determination of Flow-Rate Characteristics of Pneumatic Fluid Power Components. 16 pp.

ISO 6358-89. Pneumatic Fluid Power—Components Using Compressible Fluids—Determination of Flow-Rate Characteristics First Edition. 19 pp.

ISO 7145-82. Determination of Flowrate of Fluids in Closed Conduits of Circular Cross-Section—Method of Velocity Measurement at One Point of the Cross-Section First Edition. 14 pp.

—Hexachloroethane
BSI BS 577-66. (WITHDRAWN) 1966 Amd 1 Hexachlorethane. 17 pp.

Flow Rate (Cont.)

—Membrane Filters
JIS K 3831-90. Testing Methods for Initial Flow Rate of Membrane Filters. 12 pp.

—Metal Powders
BSI BS 5600: Sec 2.6-85. 1985 Powder Metallurgical Materials and Products Part 2: Methods of Sampling and Testing Metallic Powders Section 2.6: Determination of Flowability by Means of a Calibrated Funnel (Hall Flowmeter). 4 pp.

CNS Z8038-82. Method for Determining Flow Rate of Metal Powders (Jul)(9202).

JIS Z 2502-79. Method for Determination of Flow Rate of Metal Powders. 6 pp.

—Paints and Varnishes
BSI BS 3900: Part A6-86. (WITHDRAWN) 1986 Methods of Test for Paints Group A: Tests on Liquid Paints (Excluding Chemical Tests) Part A6: Determination of Flow Time by Use of Flow Cups (Superseded by BS EN 535: 1991). 11 pp.

BSI BS EN 535-91. 1991 Method for Determination of Flow Time of Paints by Use of Flow Cups. 17 pp.

CEN EN 535-91. Paints and Varnishes—Determination of Flow Time by Use of Flow Cups. 12 pp.

DIN ENGL 53211-87. Paints, Varnishes and Similar Coating Materials; Determination of Flow Time Using the DIN Flow Cup (June). 5 pp.

ISO 2431-93. Paints and Varnishes—Determination of Flow Time by Use of Flow Cups Fourth Edition. 14 pp.

—Pipes—Ships
JIS F 7101-75. Standard Velocity of Flow in Pipes of Ship Machinery.

SAA AS 1322-87. Shipbuilding—Recommended Fluid Velocities in Ships' Pipes. 7 pp.

—Plastics
ISO 6186-80. Plastics—Determination of Pourability First Edition. 4 pp.

—Tables (Data)
SAA AS 1377.3-73. Conversion Tables—Part 3: Volume and Flow Rate (Incorporating Amdt 1). 17 pp.

SAA AS 1377.4-73. Conversion Tables—Part 4: Mass, Density, Concentration and Flow Rate (Incorporating Amdt 1). 28 pp.

—Thermoplastic Resins
CNS K6709-82. Method of Test for Flow Rates of Thermoplastics by Extrusion Plastometer (Feb)(8516).

—Vacuum Pumps—Positive Displacement
ISO 1607 Pt 1-80. Positive-Displacement Vacuum Pumps—Measurements of Performance Characteristics—Part 1: Measurement of Volume Rate of Flow (Pumping Speed) First Edition. 5 pp.

ISO 1608 Pt 1-80. Vapour Vacuum Pumps—Measurement of Performance Characteristics—Part 1: Measurement of Volume Rate of Flow (Pumping Speed) First Edition. 5 pp.

—Waste Water
CNS K9064-81. Method of Test for Mass Flow Rate of Industrial Waste Water (Apr)(7247).

—White Water
CPPA A.6-73. Flow Measurements of White-Waters and Wastes. 8 pp.

Flow Regulators
Use For: Jet Regulators *See Also:* Control Systems Equipment; Regulators; Valves

—Faucets
BSI BS EN 246-92. 1992 Sanitary Tapware: General Specifications for Flow Rate Regulators. 18 pp.

CEN EN 246-89. Sanitary Tapware; General Specifications for Flow Rate Regulator. 17 pp.

DIN ENGL EN 246-90. Sanitary Taps; General Specifications for Flow Rate Regulators (Jan). 12 pp.

Flow Resistance
Use: Viscosity

Flow Sheets
Use: Flow Charts

Flow Time
Use: Flow Rate

Flow Totalizers
See Also: Flow Measurement Equipment

—Pipelines—Tanker Ships
BSI BS MA 27-74. 1974 Sight Flow Indicators for Tank Overflow Systems. 6 pp.

Flowmeters
Use For: Current Meters (Fluids); Rotating Element Current Meters; Rotating Meters; Velocity Meters (Flow); Volumeters (Flowmeters); Volumetric Flowmeters *See Also:* Diaphragm Flowmeters; Differential Pressure Gages; Electrical Measurement; Electromagnetic Flowmeters; Flow Measurement; Flumes; Gas Meters; Machmeters; Magnetic Valves; Open Channel Flowmeters; Orifice Plates; Pitot Tubes; Positive Displacement Flowmeters; Pressure Measurement; Rheometers; Thermal Measurement; Turbine Flowmeters; Ultrasonic Flowmeters; Variable Area Flowmeters; Venturi Flumes; Vortex Flowmeters; Water Meters; Weirs

BSI BS 3680: Part 9A-71. 1971 Measurement of Liquid Flow in Open Channels: Water Level Instruments: Installation and Performance of Pressure Actuated Liquid Level Measuring Equipment. 17 pp.

BSI BS 3680: Part 9B-81. 1981 Measurement of Liquid Flow in Open Channels: Water Level Instruments: Float-Operated Water Level Recorders (Mechanical and Electromechanical). 14 pp.

BSI BS 7405-91. 1991 Selection and Application of Flowmeters for the Measurement of Fluid Flow in Closed Conduits. 233 pp.

ISO 4373-79. Measurement of Liquid Flow in Open Channels—Water Level Measuring Devices First Edition. 20 pp.

ISO 8368-85. Liquid Flow Measurement in Open Channels—Guidelines for the Selection of Flow Gauging Structures First Edition. 7 pp.

SAA AS 3778.2.5-91. Measurement of Water Flow in Open Channels—Part 2: General—Part 2.5: Guidelines for the Selection of Flow Gauging Structures (ISO 8368:1985). 5 pp.

SAA AS 3778.6.5-92. Measurement of Water Flow in Open Channels—Part 6: Measuring Devices, Instruments and Equipment—Part 6.5: Water Level Measuring Devices (ISO 4373:1979). 18 pp.

—Aircraft
MOD UK DSTAN 66-45-87. General Requirements for Aircraft Instruments and Displays. Circular Dial Rate of Fuel Flow Indicators Issue 1. 10 pp.

—Aircraft—Acceptance Testing
EUROCAE ED-42 10.83. MPS for a Fuel Flowmeter to Aircraft Standards. 24 pp.

—Aircraft—Accuracy Testing
EUROCAE ED-42 10.83. MPS for a Fuel Flowmeter to Aircraft Standards. 24 pp.

—Aircraft—Environmental Testing
EUROCAE ED-42 10.83. MPS for a Fuel Flowmeter to Aircraft Standards. 24 pp.

—Calibration
BSI BS 7118: Part 1-90. 1990 Measurement of Fluid Flow: Assessment of Uncertainty in the Calibration and Use of Flow Measurement Devices Part 1: Linear Calibration Relationships. 46 pp.

BSI BS 7118: Part 2-89. 1989 Measurement of Fluid Flow: Assessment of Uncertainty in the Calibration and Use of Flow Measurement Devices Part 2: Non-Linear Calibration Relationships. 32 pp.

ISO 7066 Pt 1-89. Assessment of Uncertainty in the Calibration and Use of Flow Measurement Devices—Part 1: Linear Calibration Relationships First Edition. 44 pp.

ISO 7066 Pt 2-88. Assessment of Uncertainty in the Calibration and Use of Flow Measurement Devices—Part 2: Non-Linear Calibration Relationships First Edition. 31 pp.

MOD UK DSTAN 66-32: Pt 2:Sec 1-84. Code of Practice for Flow Measurement in Closed Conduits Part 2: Calibration and Application of Flowmeters Covered in Part 1: Section 1 Issue 1. 73 pp.

—Condensate Wells
DIN ENGL 19211-80. Methods for Measurement of Fluid Flow; Rectifying Vessels for Throttling Appliances in Flow Measurement Devices (Sept). 3 pp.

—Differential Pressure
MOD UK DSTAN 66-32: Pt 1:Sec 1-84. Code of Practice for Flow Measurement in Closed Conduits Part 1: Specification of Flow Meters Section 1: Positive Displacement, Differential Pressure, Variable Area & Turbine Meters Issue 1. 37 pp.

—Float
JIS Z 8761-77. Method of Flow Measurement by Float Type Area Flowmeters (R 1982). 20 pp.

—Glass
CNS R2148-82. Glass Materials of Volumeter for Chemical Analysis (May)(8862).

CNS R2153-82. Glass Liquid Volumeters for Chemical Analysis (May)(8867).

INTERNATIONAL AND NON-U.S. NATIONAL STANDARDS
SUBJECT INDEX

Flowmeters (Cont.)
—**Liquids**
JIS B 7552-82. Proving Methods for Liquid Flowmeters. 14 pp.

—**Liquids—Hydrocarbons**
BSI BS 6866: Part 2-90. 1990 Proving Systems for Meters Used in Dynamic Measurement of Liquid Hydrocarbons Part 2: Methods for Design, Installation and Calibration of Pipe Provers. 24 pp.
BSI BS 6866: Part 3-87. 1987 Proving Systems for Meters Used in Dynamic Measurement of Liquid Hydrocarbons Part 3: Methods of Pulse Interpolation. 12 pp.
ISO 7278 Pt 1-87. Liquid Hydrocarbons—Dynamic Measurement—Proving Systems for Volumetric Meters—Part 1: General Principles First Edition. 8 pp.
ISO 7278 Pt 2-88. Liquid Hydrocarbons—Dynamic Measurement—Proving Systems for Volumetric Meters—Part 2: Pipe Provers First Edition. 22 pp.
ISO 7278 Pt 3-86. Liquid Hydrocarbons—Dynamic Measurement—Proving Systems for Volumetric Meters—Part 3: Pulse Interpolation Techniques First Edition. 10 pp.

—**Ships**
MOD UK NES 605: Part 4-88. Guide to the Selection of Sensors for the Measurement of System Parameters Part 4: Selection of Flow Sensors Issue 2 (12.88). 19 pp.

—**Submarines**
MOD UK NES 605: Part 4-88. Guide to the Selection of Sensors for the Measurement of System Parameters Part 4: Selection of Flow Sensors Issue 2 (12.88). 19 pp.

—**Valves**
BSI BS 6864: Part 3-87. 1987 Laboratory Tests on Noise Emission from Appliances and Equipment Intended for Use in Water Supply Installations Part 3: Method for Mounting and Operating in Line Valves and Appliances. 8 pp.
BSI BS 6866: Part 1-90. 1990 Proving Systems for Meters Used in Dynamic Measurement of Liquid Hydrocarbons Part 1: Introduction. 12 pp.

Flue Blocks
See Also: Chimneys
—**Clay**
BSI BS 1289: Part 2-89. 1989 Flue Blocks and Masonry Terminals for Gas Appliances Part 2: Clay Flue Blocks and Terminals. 20 pp.

Flue Dust
See Also: Dust; Fly Ash
—**Measurement**
JIS Z 8808-92. Methods of Measuring Dust Concentration in Flue Gas. 44 pp.

—**Particle Size Distribution**
JIS K 0302-89. Measuring Method for Particle-Size Distribution of Dusts in Flue Gas. 22 pp.

Flue Exhausters
Use For: Flue Gas Exhausters   See Also: Chimneys; Flue Gases; Gas Burners; Oil Burners
CSA CAN3-B255-M81. Mechanical Flue-Gas Exhausters; (Gen Instr 1 Thru 2). 17 pp.

Flue Gas Analyzers
See Also: Gas Analyzers
BSI BS 1756: Part 2-71. 1971 Methods for the Sampling and Analysis of Flue Gases Part 2: Analysis by the Orsat Apparatus. 17 pp.
BSI BS 1756: Part 3-71. 1971 Methods for the Sampling and Analysis of Flue Gases Part 3: Analysis by the Haldane Apparatus. 37 pp.
BSI BS 1756: Part 3-01. 1971 Amd 1 Methods for the Sampling and Analysis of Flue Gases Part 3: Analysis by the Haldane Apparatus (AMD 7754) August 15, 1993. 38 pp.
BSI BS 3048-58. 1958 Code for the Continuous Sampling and Automatic Analysis of Flue Gases. Indicators and Recorders. 77 pp.
BSI BS 3048-01. 1958 Amd 1 Code for the Continuous Sampling and Automatic Analysis of Flue Gases. Indicators and Recorders (AMD 7752) August 15, 1993 (Q). 78 pp.
JIS B 7981-84. Continuous Analyzers for Sulfur Dioxide in Flue Gas. 18 pp.
JIS B 7983-79. Continuous Analyzers for Oxygen in Flue Gas. 29 pp.

Flue Gas Exhausters
Use: Flue Exhausters

Flue Gases
Use For: Enclosed Gas Streams; Stack Gases

Flue Gases (Cont.)
See Also: Air Monitoring; Exhaust Gases; Flue Exhausters; Fumes; Gases; Ringelmann Charts; Vapors

—**Arsenic Content**
JIS K 0221-88. Methods for Determination of Arsenic in Flue Gas. 16 pp.

—**Beryllium Content**
JIS K 0224-83. Methods for Determination of Beryllium in Stack Gas (R 1988). 11 pp.

—**Cadmium Content**
JIS K 0097-79. Method for Determination of Cadmium and Lead in Stack Gas (R 1984). 22 pp.

—**Carbon Dioxide Content**
BSI BS 1756: Part 5-71. 1971 Methods for the Sampling and Analysis of Flue Gases Part 5: Semi-Routine Analysis. 25 pp.
BSI BS 1756: Part 5-01. 1971 Amd 1 Methods for the Sampling and Analysis of Flue Gases Part 5: Semi-Routine Analysis (AMD 7756) October 15, 1993 (Q). 26 pp.
CSA CAN/CSA-Z223.2-M86. Method for the Continuous Measurement of Oxygen, Carbon Dioxide, Carbon Monoxide, Sulphur Dioxide, and Oxides of Nitrogen in Enclosed Combustion Flue Gas Streams; (Gen Instr 1). 44 pp.

—**Carbon Monoxide Content**
BSI BS 1756: Part 4-77. 1977 Methods for the Sampling and Analysis of Flue Gases Part 4: Miscellaneous Analyses. 33 pp.
BSI BS 1756: Part 4-01. 1977 Amd 1 Methods for Sampling and Analysis of Flue Gases Part 4: Miscellaneous Analyses (AMD 7755) August 15, 1993. 35 pp.
BSI BS 1756: Part 5-71. 1971 Methods for the Sampling and Analysis of Flue Gases Part 5: Semi-Routine Analysis. 25 pp.
BSI BS 1756: Part 5-01. 1971 Amd 1 Methods for the Sampling and Analysis of Flue Gases Part 5: Semi-Routine Analysis (AMD 7756) October 15, 1993 (Q). 26 pp.
CSA CAN/CSA-Z223.2-M86. Method for the Continuous Measurement of Oxygen, Carbon Dioxide, Carbon Monoxide, Sulphur Dioxide, and Oxides of Nitrogen in Enclosed Combustion Flue Gas Streams; (Gen Instr 1). 44 pp.
CSA Z223.21-M1978. Method for the Measurement of Carbon Monoxide. 27 pp.

—**Chromium Content**
JIS K 0096-75. Methods for Determination of Chromium and Manganese in Stack Gas (R 1989). 16 pp.

—**Cleaning Equipment**
JIS B 9914-81. Method of Measuring Performance for Gas Treatment Equipment. 18 pp.

—**Dust Content**
CNS K9010-75. Determination for Location of Dust Content in Stack Gas (May)(3798).

—**Flow Measurement**
CNS K9019-76. Method of Measuring Velocity and Flow Rate of Stack Gas (Dec)(4050).

—**Flow Rate**
CNS K9019-76. Method of Measuring Velocity and Flow Rate of Stack Gas (Dec)(4050).

—**Flue Dust—Measurement**
JIS Z 8808-92. Methods of Measuring Dust Concentration in Flue Gas. 44 pp.

—**Flue Dust—Particle Size Distribution**
JIS K 0302-89. Measuring Method for Particle-Size Distribution of Dusts in Flue Gas. 22 pp.

—**Lead Content**
JIS K 0097-79. Method for Determination of Cadmium and Lead in Stack Gas (R 1984). 22 pp.

—**Manganese Content**
JIS K 0096-75. Methods for Determination of Chromium and Manganese in Stack Gas (R 1989). 16 pp.

—**Mercury Content**
JIS K 0222-81. Method for Determination of Mercury in Stack Gas. 12 pp.

—**Moisture Content**
BSI BS 1756: Part 4-77. 1977 Methods for the Sampling and Analysis of Flue Gases Part 4: Miscellaneous Analyses. 33 pp.
BSI BS 1756: Part 4-01. 1977 Amd 1 Methods for Sampling and Analysis of Flue Gases Part 4: Miscellaneous Analyses (AMD 7755) August 15, 1993. 35 pp.

Flue Gases (Cont.)
—**Nickel Content**
JIS K 0084-76. Method for Determination of Nickel in Stack Gas (R 1984). 11 pp.

—**Nitrogen Oxide Content**
BSI BS 1756: Part 4-77. 1977 Methods for the Sampling and Analysis of Flue Gases Part 4: Miscellaneous Analyses. 33 pp.
BSI BS 1756: Part 4-01. 1977 Amd 1 Methods for Sampling and Analysis of Flue Gases Part 4: Miscellaneous Analyses (AMD 7755) August 15, 1993. 35 pp.
CSA CAN/CSA-Z223.2-M86. Method for the Continuous Measurement of Oxygen, Carbon Dioxide, Carbon Monoxide, Sulphur Dioxide, and Oxides of Nitrogen in Enclosed Combustion Flue Gas Streams; (Gen Instr 1). 44 pp.
JIS B 7982-88. Continuous Analyzers for Oxides of Nitrogen in Flue Gas. 25 pp.
JIS K 0104-84. Methods for Determination of Oxides of Nitrogen in Flue Gases. 26 pp.

—**Oxygen Content**
CSA CAN/CSA-Z223.2-M86. Method for the Continuous Measurement of Oxygen, Carbon Dioxide, Carbon Monoxide, Sulphur Dioxide, and Oxides of Nitrogen in Enclosed Combustion Flue Gas Streams; (Gen Instr 1). 44 pp.
JIS B 7983-79. Continuous Analyzers for Oxygen in Flue Gas. 29 pp.
JIS K 0301-89. Methods for Determination of Oxygen in Flue Gas. 15 pp.

—**Particulates**
BSI BS 3405-83. 1983 Measurement of Particulate Emission Including Grit and Dust (Simplified Method). 20 pp.
CSA Z223.1-M1977. Method for the Determination of Particulate Mass Flows in Enclosed Gas Streams. 69 pp.

—**Particulates—Concentration (Composition)—Gravimetric Analysis**
BSI BS 6069: Sec 4.3-92. 1992 Characterization of Air Quality Part 4: Stationary Source Emissions Section 4.3: Method for the Manual Gravimetric Detn. of Concentration and Mass Flow Rate of Particulate Material in Gas-Carrying Ducts (ISO 9096: 1992). 37 pp.
ISO 9096-92. Stationary Source Emissions—Determination of Concentration and Mass Flow Rate of Particulate Material in Gas-Carrying Ducts—Manual Gravimetric Method First Edition. 36 pp.

—**Particulates—Flow Rate—Gravimetric Analysis**
BSI BS 6069: Sec 4.3-92. 1992 Characterization of Air Quality Part 4: Stationary Source Emissions Section 4.3: Method for the Manual Gravimetric Detn. of Concentration and Mass Flow Rate of Particulate Material in Gas-Carrying Ducts (ISO 9096: 1992). 37 pp.
ISO 9096-92. Stationary Source Emissions—Determination of Concentration and Mass Flow Rate of Particulate Material in Gas-Carrying Ducts—Manual Gravimetric Method First Edition. 36 pp.

—**Sampling**
BSI BS 1756: Part 1-71. 1971 Methods for the Sampling and Analysis of Flue Gases Part 1: Methods of Sampling. 32 pp.
BSI BS 1756: Part 1-01. 1971 Amd 1 Methods for the Sampling and Analysis of Flue Gases Part 1: Methods of Sampling (AMD 7798) August 15, 1993. 34 pp.
JIS K 0095-88. Method for Sampling of Flue.
JIS K 0095-73. Method for Sampling of Stack Gas. 27 pp.

—**Selenium Content**
JIS K 0223-83. Methods for Determination of Selenium in Stack Gas (R 1988). 15 pp.

—**Smoke Content—Charts**
BSI BS 2742-69. 1969 Notes on the Use of the Ringelmann and Miniature Smoke Charts. 9 pp.
BSI BS 2742: Add 1-72. 1972 The Calibration of Instruments in Ringelmann Number Addendum No. 1. 5 pp.
SNZ NZS 5201-73. Use of the Ringelmann and Miniature Smoke Charts Amend: A, 1973. 8 pp.
SNZ NZS 5201C-75. Ringelmann Chart. 1 p.
SNZ NZS 5201M-75. British Standard Miniature Smoke Chart. 1 p.

—**Solid Fuel Burning Equipment**
SAA AS 4013-92. Domestic Solid Fuel Burning Appliances—Method for Determination of Flue Gas Emission (NZS 7403:1992). 16 pp.

Flue Gases (Cont.)

—Solid Fuel Burning Equipment (Cont.)
SNZ NZS 7402-92. Domestic Solid Fuel Burning Appliances—Method for Determination of Power Output and Efficiency (NZS 7402:1992/AS 4012-1992). 17 pp.
SNZ NZS 7403-92. Domestic Solid Fuel Burning Appliances—Method for Determination of Flue Gas Emissions (NZS 7403:1992/AS 4013-1992). 16 pp.

—Sulfur Dioxide Content
CNS K9008-75. Method of Test for Total SO and SO2 Content in Flue Gas (May)(3796).
CSA CAN/CSA-Z223.2-M86. Method for the Continuous Measurement of Oxygen, Carbon Dioxide, Carbon Monoxide, Sulphur Dioxide, and Oxides of Nitrogen in Enclosed Combustion Flue Gas Streams; (Gen Instr 1). 44 pp.
JIS B 7981-84. Continuous Analyzers for Sulfur Dioxide in Flue Gas. 18 pp.

—Sulfur Dioxide Content—Automatic Test Equipment
BSI BS 6069: Sec 4.4-93. 1993 Characterization of Air Quality Part 4: Stationary Source Emissions Section 4.4: Determination of the Mass Concentration of Sulfur Dioxide—Performance Characteristics of Auto-mated Measuring Methods (ISO 7935: 1992). 16 pp.
ISO 7935-92. Stationary Source Emissions—Determination of the Mass Concentration of Sulphur Dioxide—Performance Characteristics of Automated Measuring Methods First Edition. 16 pp.

—Sulfur Oxide Content
BSI BS 1756: Part 4-77. 1977 Methods for the Sampling and Analysis of Flue Gases Part 4: Miscellaneous Analyses. 33 pp.
BSI BS 1756: Part 4-01. 1977 Amd 1 Methods for Sampling and Analysis of Flue Gases Part 4: Miscellaneous Analyses (AMD 7755) August 15, 1993. 35 pp.
BSI BS 1756: Part 5-71. 1971 Methods for the Sampling and Analysis of Flue Gases Part 5: Semi-Routine Analysis. 25 pp.
BSI BS 1756: Part 5-01. 1971 Amd 1 Methods for the Sampling and Analysis of Flue Gases Part 5: Semi-Routine Analysis (AMD 7756) October 15, 1993 (Q). 26 pp.
CNS K9008-75. Method of Test for Total SO and SO2 Content in Flue Gas (May)(3796).
JIS K 0103-88. Methods for Determination of Sulfur Oxides in Flue Gas. 17 pp.

—Sulfuric Acid Content
BSI BS 1756: Part 4-77. 1977 Methods for the Sampling and Analysis of Flue Gases Part 4: Miscellaneous Analyses. 33 pp.
BSI BS 1756: Part 4-01. 1977 Amd 1 Methods for Sampling and Analysis of Flue Gases Part 4: Miscellaneous Analyses (AMD 7755) August 15, 1993. 35 pp.

—Vanadium Content—Absorptiometric Analysis
JIS K 0083-76. Method for Determination of Vanadium in Stack Gas. 12 pp.

—Vanadium Content—Atomic Absorption Spectrophotometry
JIS K 0083-76. Method for Determination of Vanadium in Stack Gas. 12 pp.

—Water Content
CNS K9009-75. Method of Determination for Water Content in Stack Gas (May)(3797).

Flue Linings
Use: Flues

Flue Pipes
Use: Flues

Flue Terminals
Use For: Chimney Terminals *See Also:* Chimneys

—Clay
BSI BS 1289: Part 2-89. 1989 Flue Blocks and Masonry Terminals for Gas Appliances Part 2: Clay Flue Blocks and Terminals. 20 pp.

Flues
Use For: Flue Linings; Flue Pipes; Smoke Pipes
See Also: Chimneys; Ducts; Exhaust Systems; Ferrules; Firebacks; Vents
BSI BS 5854-80. 1980 Code of Practice for Flues and Flue Structures in Buildings. 55 pp.

—Air Preheaters
BSI BS 5440: Part 1-90. 1990 Code of Practice for Flues and Air Supply for Gas Appliances of Rated Input Not Exceeding 60 kW (1st and 2nd Family Gases) Part 1: Installation of Flues. 44 pp.

—Appliances
BSI BS 5440: Part 1-90. 1990 Code of Practice for Flues and Air Supply for Gas Appliances of Rated Input Not Exceeding 60 kW (1st and 2nd Family Gases) Part 1: Installation of Flues. 44 pp.
BSI BS 5440: Part 1-78. (WITHDRAWN) 1978 Amd 1 Code of Practice for Flues and Air Supply for Gas Appliances of Rated Input Not Exceeding 60 kW (1st and 2nd Family Gases) Part 1: Flues. 42 pp.

—Asbestos Cement
BSI BS 567-73. 1973 Amd 1 Asbestos-Cement Flue Pipes and Fittings, Light Quality (AMD 5963) April 28, 1989. 22 pp.
BSI BS 835-73. 1973 Amd 1 Asbestos-Cement Flue Pipes and Fittings, Heavy Quality. 26 pp.
BSI BS 4624-81. 1981 Amd 1 Asbestos-Cement Building Products. 13 pp.
BSI BS 4624-02. 1981 Amd 2 Methods of Test for Asbestos-Cement Building Products (AMD 7307) October 15, 1992. 14 pp.

—Asbestos Cement—Fittings
BSI BS 567-73. 1973 Amd 1 Asbestos-Cement Flue Pipes and Fittings, Light Quality (AMD 5963) April 28, 1989. 22 pp.
BSI BS 835-73. 1973 Amd 1 Asbestos-Cement Flue Pipes and Fittings, Heavy Quality. 26 pp.

—Boilers
BSI BS 5440: Part 1-90. 1990 Code of Practice for Flues and Air Supply for Gas Appliances of Rated Input Not Exceeding 60 kW (1st and 2nd Family Gases) Part 1: Installation of Flues. 44 pp.
BSI BS 5440: Part 1-78. (WITHDRAWN) 1978 Amd 1 Code of Practice for Flues and Air Supply for Gas Appliances of Rated Input Not Exceeding 60 kW (1st and 2nd Family Gases) Part 1: Flues. 42 pp.

—Boilers—Deposits—Sampling
BSI BS 2455: Part 2-83. 1983 Methods of Sampling and Examining Deposits from Boilers and Associated Industrial Plant Part 2: Methods for Sampling and Examining Free-Side Deposits. 41 pp.
BSI BS 2455: Part 2-01. 1983 Amd 1 Sampling and Examining Deposits from Boilers and Associated Industrial Plant Part 2: Methods for Sampling and Examining Fire-Side Deposits (AMD 7751) October 15, 1993 (Q). 42 pp.

—Cast Iron
BSI BS 41-73. 1973 Cast Iron Spigot and Socket Flue or Smoke Pipes and Fittings. 12 pp.

—Cast Iron—Fittings
BSI BS 41-73. 1973 Cast Iron Spigot and Socket Flue or Smoke Pipes and Fittings. 12 pp.

—Clay
CSA CAN/CSA-A324-M88. Clay Flue Liners; (Gen Instr 1). 35 pp.

—Clay—Appliances
BSI BS 1181-89. 1989 Clay Flue Linings and Flue Terminals. 16 pp.

—Concrete
BSI BS 1289: Part 1-86. 1986 Flue Blocks and Masonry Terminals for Gas Appliances Part 1: Precast Concrete Flue Blocks and Terminals. 20 pp.

—Fiber Cement
BSI BS 7435: Part 1-91. 1991 Fibre Cement Flue Pipes, Fittings and Terminals Part 1: Specification for Light Quality Fibre Cement Flue Pipes, Fittings and Terminals. 18 pp.
BSI BS 7435: Part 2-91. 1991 Fibre Cement Flue Pipes, Fittings and Terminals Part 2: Specification for Heavy Quality Fibre Cement Flue Pipes, Fittings and Terminals. 20 pp.

—Fiber Cement—Fittings
BSI BS 7435: Part 1-91. 1991 Fibre Cement Flue Pipes, Fittings and Terminals Part 1: Specification for Light Quality Fibre Cement Flue Pipes, Fittings and Terminals. 18 pp.
BSI BS 7435: Part 2-91. 1991 Fibre Cement Flue Pipes, Fittings and Terminals Part 2: Specification for Heavy Quality Fibre Cement Flue Pipes, Fittings and Terminals. 20 pp.

—Installation
BSI BS 6461: Part 1-84. 1984 Installation of Chimneys and Flues for Domestic Appliances Burning Solid-Fuel (Including Wood and Peat) Part 1: Code of Practice for Masonry Chimneys and Flue Pipes. 23 pp.

—Metal
BSI BS 715-89. 1989 Amd 2 Metal Flue Pipes, Fittings, Terminals and Accessories for Gas-Fired Appliances with a Rated Input Not Exceeding 60 kW (AMD 6335) July 31, 1991. 24 pp.
BSI BS 715-86. (WITHDRAWN) 1986 Metal Flue Pipes, Fittings, Terminals and Accessories for Gas-Fired Appliances with a Rated Input Not Exceeding 60 kW (E). 25 pp.

—Metal—Fittings
BSI BS 715-89. 1989 Amd 2 Metal Flue Pipes, Fittings, Terminals and Accessories for Gas-Fired Appliances with a Rated Input Not Exceeding 60 kW (AMD 6335) July 31, 1991. 24 pp.
BSI BS 715-86. (WITHDRAWN) 1986 Metal Flue Pipes, Fittings, Terminals and Accessories for Gas-Fired Appliances with a Rated Input Not Exceeding 60 kW (E). 25 pp.

—Refractory Materials—Installation
BSI BS 4207-89. 1989 Installation of Monolithic Linings for Steel Chimneys and Flues. 8 pp.

—Steel—Enameled—Solid Fuel Burning Equipment
BSI BS 6999-89. 1989 Vitreous-Enamelled Low-Carbon-Steel Fluepipes, Other Components and Accessories for Solid-Fuel-Burning Appliances with a Maximum Rated Output of 45 kW. 12 pp.

Fluid Coolers
Use For: Liquid Coolers *See Also:* Coolers

—Air Cooled
BSI DD ENV 1048-93. 1993 Heat Exchangers—Air Cooled Liquid Coolers 'Dry Coolers'—Test Procedure for Establishing the Performance. 17 pp.
CEN ENV 1048-93. Heat Exchangers—Air Cooled Liquid Coolers 'Dry Coolers'—Test Procedure for Establishing the Performance. 12 pp.

Fluid Couplings
Use: Hydraulic Couplings

Fluid Filled Transformers
Use: Liquid Filled Transformers

Fluid Flow Measurement
Use: Flow Measurement

Fluid Flow Rate
Use: Flow Rate

Fluid Immersion Testing
Use: Immersion Testing

Fluid Inlets
Use For: Inlets (Fluids)

—Discharge Stacks—Design
BSI BS 6367-83. 1983 Amd 1 Drainage of Roofs and Paved Areas. 58 pp.

—Gullies—Design
BSI BS 6367-83. 1983 Amd 1 Drainage of Roofs and Paved Areas. 58 pp.

Fluid Mechanics

—Aerospace—Glossaries
ISO 8625 Pt 2-91. Aerospace—Fluid Systems—Vocabulary—Part 2: General Terms and Definitions Relating to Flow First Edition. 7 pp.

Fluid Pipelines
Use: Pipelines

Fluid Power Cylinders
See Also: Fluid Power Equipment; Hydraulic Cylinders; Pneumatic Cylinders

—Accessories
JIS B 8369-91. Fluid Power Cylinders—Dimensions of Accessories. 26 pp.

—Barrels
BSI BS 5242: Part 1-87. 1987 Tubes for Fluid Power Cylinder Barrels Part 1: Steel Tubes with Specially Finished Surfaces. 16 pp.
ISO 4394 Pt 1-80. Fluid Power Systems and Components—Cylinder Barrels—Part 1: Requirements for Steel Tubes with Specially Finished Bores First Edition. 12 pp.

—Mounting
BSI BS 6331: Part 1-83. 1983 Mounting Dimensions of Single Rod Double-Acting Hydraulic Cylinders Part 1: Specification for 160 Bar Medium Series. 12 pp.

INTERNATIONAL AND NON-U.S. NATIONAL STANDARDS
SUBJECT INDEX

Fluid Power Cylinders (Cont.)

—Mounting (Cont.)

BSI BS 6331: Part 2-83. 1983 Mounting Dimensions of Single Rod Double-Acting Hydraulic Cylinders Part 2: Specification for 160 Bar Compact Series. 18 pp.

BSI BS 6331: Part 3-83. 1983 Mounting Dimensions of Single Rod Double-Acting Hydraulic Cylinders Part 3: Specification for 250 Bar Series. 10 pp.

BSI BS 6331: Part 4-85. 1985 Mounting Dimensions of Single Rod Double-Acting Hydraulic Cylinders Part 4: Specification for 63 Bar Series, Including Associated Components. 32 pp.

ISO 6020 Pt 2-91. Hydraulic Fluid Power—Mounting Dimensions for Single Rod Cylinders, 16 MPa (160 Bar) Series—Part 2: Compact Series Second Edition. 20 pp.

ISO 6022-81. Hydraulic Fluid Power—Single Rod Cylinders—Mounting Dimensions—250 Bar (25 000 kPa) Series First Edition. 7 pp.

JIS B 8366-90. Fluid Power Cylinders—General Rules for Requirement of Component and Identification Code of Mounting Dimensions and Mounting Types. 43 pp.

—Mounting—Identification Systems

ISO 6099-85. Fluid Power Systems and Components—Cylinders—Identification Code for Mounting Dimensions and Mounting Types Second Edition. 48 pp.

JIS B 8366-90. Fluid Power Cylinders—General Rules for Requirement of Component and Identification Code of Mounting Dimensions and Mounting Types. 43 pp.

—Piston Rods—Strokes

ISO 4393-78. Fluid Power Systems and Components—Cylinders—Basic Series of Piston Strokes First Edition. 4 pp.

—Piston Rods—Threads

ISO 4395-78. Fluid Power Systems and Components—Cylinders—Piston Rod Thread Dimensions and Types First Edition. 4 pp.

—Piston Rods—Wiper Ring Housings—Diameters

BSI BS 6852-87. 1987 Dimensions of Housings for Hydraulic Cylinder Piston Rod Wiper Rings in Reciprocating Applications. 8 pp.

ISO 6195-86. Fluid Power Systems and Components—Cylinders—Housings for Rod Wiper Rings in Reciprocating Applications—Dimensions and Tolerances First Edition. 8 pp.

—Wiper Ring

BSI BS 6852-87. 1987 Dimensions of Housings for Hydraulic Cylinder Piston Rod Wiper Rings in Reciprocating Applications. 8 pp.

ISO 6195-86. Fluid Power Systems and Components—Cylinders—Housings for Rod Wiper Rings in Reciprocating Applications—Dimensions and Tolerances First Edition. 8 pp.

Fluid Power Equipment

Use For: Fluid Power Systems; Fluid Power Systems and Equipment See Also: Fluid Power Cylinders; Hydraulic Equipment; Pneumatic Equipment; Pneumatic Systems

—Aerospace

BSI M 53-83. 1983 Thermal Shock Testing of Pipes and Fittings for Aerospace Fluid Systems. 4 pp.

BSI M 59-86. (WITHDRAWN) 1986 Interface of Metric Coupling for Aerospace Fluid Systems (Superseded by 2M59: 1993). 4 pp.

ISO 6773-82. Aerospace Fluid Systems—Thermal Shock Testing of Piping and Fittings First Edition. 4 pp.

—Aerospace—Flow—Glossaries

ISO 8625 Pt 2-91. Aerospace—Fluid Systems—Vocabulary—Part 2: General Terms and Definitions Relating to Flow First Edition. 7 pp.

—Aerospace—Hose Assemblies

ISO 7314-89. Aerospace—Fluid Systems—Hose Assembly, Metal First Edition. 15 pp.

—Aerospace—Pipe Couplings

BSI 2M 63-93. 1993 Dimensions of Aerospace Fluid System Port Connection, Seal and Fitting End (ISO 7320: 1992) (S). 11 pp.

BSI 2M 59-93. 1993 Interface of Metric Couplings, 24 Degree Cone, for Aerospace Fluid Systems (ISO 7319: 1992) (S). 8 pp.

ISO 7319-92. Aerospace—Fluid Systems—Interface of 24 Degree Cone Metric Couplings Second Edition. 7 pp.

ISO 7320-92. Aerospace—Couplings, Threaded and Sealed, for Fluid Systems—Dimensions Second Edition. 10 pp.

Fluid Power Equipment (Cont.)

—Aerospace—Pipe Fittings

AECMA PREN3635-90. Weld Lip Geometrical Configuration. 4 pp.

CEN PREN 3633-90. Installation Hole for Fluid Fittings, Flanged. 7 pp.

CEN PREN 3633-92. Installation Hole for Fluid Fittings, Flanged. 5 pp.

ISO 5855 Pt 3-88. Aerospace—MJ Threads—Part 3: Limit Dimensions for Fittings for Fluid Systems First Edition. 8 pp.

—Aerospace—Pipe Fittings—Hole Size

CEN PREN 3633-90. Installation Hole for Fluid Fittings, Flanged. 7 pp.

CEN PREN 3633-92. Installation Hole for Fluid Fittings, Flanged. 5 pp.

—Aerospace—Pipe Fittings—Sealing Rings

BSI M 63-87. (WITHDRAWN) 1987 Dimensions for Fluid System Port Connection, Seal and Fitting End (Superseded by 2M 63: 1993). 6 pp.

—Aerospace—Pressure

BSI M 51-82. (WITHDRAWN) 1982 Pressure and Temperature for Fluid Systems and Components (Superseded by BS 2M 51: 1989). 3 pp.

BSI 2M 51-89. 1989 Classification for Pressure and Temperature for Fluid Systems and Components. 2 pp.

—Aerospace—Pressure—Classification

ISO 6771-87. Aerospace—Fluid Systems and Components—Pressure and Temperature Classifications Second Edition. 4 pp.

—Aerospace—Pressure—Glossaries

ISO 8625 Pt 1-93. Aerospace—Fluid Systems—Vocabulary—Part 1: General Terms and Definitions Relating to Pressure First Edition. 17 pp.

—Aerospace—Temperature

BSI M 51-82. (WITHDRAWN) 1982 Pressure and Temperature for Fluid Systems and Components (Superseded by BS 2M 51: 1989). 3 pp.

BSI 2M 51-89. 1989 Classification for Pressure and Temperature for Fluid Systems and Components. 2 pp.

—Aerospace—Temperature—Classification

ISO 6771-87. Aerospace—Fluid Systems and Components—Pressure and Temperature Classifications Second Edition. 4 pp.

—Aerospace—Temperature—Glossaries

ISO 8625 Pt 3-91. Aerospace—Fluid Systems—Vocabulary—Part 3: General Terms and Definitions Relating to Temperature First Edition. 7 pp.

—Aerospace—Tube Assemblies

ISO 10583-93. Aerospace Fluid Systems—Test Methods for Tube/Fitting Assemblies First Edition. 8 pp.

—Aircraft—Bulkhead Fittings

AECMA PREN3246-89. Pipe coupling 8 Degrees 30' in Titanium Alloy Union, Bulkhead. 5 pp.

AECMA PREN3247-89. Pipe Coupling 8 Degrees 30' in Titanium Alloy Union, Bulkhead, Welded. 6 pp.

AECMA PREN3253-89. Pipe Coupling 8 Degrees 30' in Titanium Alloy Elbow 90 Degrees Swivel Nut, Bulkhead. 5 pp.

AECMA PREN3254-89. Pipe Coupling 8 Degrees 30' in Titanium Alloy Elbow 90 Degrees Swivel Nut, Bulkhead, Welded. 6 pp.

AECMA PREN3257-90. Pipe coupling 8 Degrees 30' in Titanium Alloy Elbow 45 Degrees, Bulk Head. 5 pp.

AECMA PREN3262-90. Pipe Coupling 8 Degrees 30' in Titanium Alloy Tee Bulkhead Branch. 5 pp.

AECMA PREN3263-90. Pipe Coupling 8 Degrees 30' in Titanium Alloy Bulkhead End. 5 pp.

AECMA PREN3266-90. Nut, Bulkhead in Titanium. 6 pp.

AECMA PREN3267-90. Washer, Bulkhead in Titanium Alloy. 6 pp.

CEN PREN3564-90. Pipe Coupling 8 Degrees 30 in Titanium Alloy Tee, Bulkhead Fitting on Limb. 5 pp.

—Aircraft—Hose Fittings

MOD UK DSTAN 47-22-01. End Fittings for Flexible Hose Assemblies for Aircraft—Metric Issue 1; Amendment 1. 15 pp.

—Aircraft—Pipe Fittings

AECMA PREN2656-89. Pipe Couplings up to 56000 kPa Fitting End, Welded, Geometrical Configuration. 5 pp.

AECMA PREN3242-89. Pipe Coupling 8 Degrees 30' in Titanium Alloy Union, Welded, Threaded. 6 pp.

AECMA PREN3243-89. Pipe Coupling 8 Degrees 30' in Titanium Alloy Ferrule, Welded, with Dynamic Beam Seal End. 5 pp.

Fluid Power Equipment (Cont.)

—Aircraft—Pipe Fittings (Cont.)

AECMA PREN3244-89. Pipe Coupling 8 Degrees 30' in Titanium Alloy Union Double Ended. 5 pp.

AECMA PREN3245-89. Pipe Coupling 8 Degrees 30' in Titanium Alloy Union Reducer. 5 pp.

AECMA PREN3249-89. Pipe Coupling 8 Degrees 30' in Titanium Alloy Elbow 90 Degress. 5 pp.

AECMA PREN3250-89. Pipe Coupling 8 Degrees in Titanium Alloy Elbow, 90 Degrees Swivel Nut. 5 pp.

AECMA PREN3251-89. Pipe Coupling 8 Degrees 30' in Titanium Alloy Elbow 90 Degrees, Welded. 6 pp.

AECMA PREN3252-89. Pipe Coupling 8 Degrees 30' in Titanium Alloy Elbow 90 Degrees Swivel Nut, Welded. 6 pp.

AECMA PREN3255-90. Pipe Coupling 8 Degrees 30' in Titanium Alloy Elbow 45 Degrees. 5 pp.

AECMA PREN3256-90. Pipe Coupling 8 Degrees 30' in Titanium Alloy Elbow 45 Degrees Swivel Nut, Welded. 6 pp.

AECMA PREN3258-90. Pipe Coupling 8 Degrees 30' in Titanium Alloy Tee. 5 pp.

AECMA PREN3260-90. Pipe Coupling 8 Degrees 30' in Titanium Alloy Tee, Branch with Swivel Nut. 5 pp.

AECMA PREN3261-90. Pipe Coupling 8 Degrees 30' in Titanium Alloy Tee with Swivel Nut. 5 pp.

AECMA PREN3264-90. Pipe Coupling 8 Degrees 30' in Titanium Alloy Nut, Swivel. 5 pp.

AECMA PREN3265-90. Pipe Coupling 8 Degrees 30' in Titanium Alloy Nut, Union. 5 pp.

AECMA PREN3268-90. Pipe Coupling 8 Degrees 30' in Titanium Alloy Plug, Pressure. 5 pp.

AECMA PREN3269-90. Pipe Coupling 8 Degrees 30' in Titanium Alloy Blind Ferrule with Dynamic Beam Seal End. 5 pp.

CEN PREN3248-90. Pipe Coupling 8 Degrees 30 in Titanium Alloy Adaptor-Reduced Pipe end with Locking Ring. 6 pp.

CEN PREN3270-90. Pipe coupling 8 Degrees 30 in Titanium Alloy Blanking Plug with Locking Ring. 5 pp.

CEN PREN3275-90. Pipe Coupling 8 Degrees 30 Dynamic Beam Seal up to 28000 KPa Metric Series Technical Specification. 32 pp.

CEN PREN3561-90. Pipe Coupling 8 Degrees 30 in Titanium Alloy Ferrule with Dynamic Beam Seal End, Welded and Reduced at Pipe End. 6 pp.

CEN PREN3562-90. Pipe Coupling in Titanium Alloy Elbow 90 Degrees Double Welded. 6 pp.

CEN PREN3563-90. Pipe Coupling in Titanium Alloy Elbow 45 Degrees Double Welded. 6 pp.

CEN PREN3566-90. Pipe Coupling 8 Degrees 30 in Titanium Alloy Adaptor with Locking Ring. 5 pp.

—Aircraft—Pipe Fittings—Oxygen Supply Equipment

ISO 1022-88. Aerospace—Gaseous Oxygen Replenishment Connection for Use in Fluid Systems (Old Type)—Dimensions (Inch Series) First Edition. 6 pp.

ISO 8775-88. Aerospace—Gaseous Oxygen Replenishment Connection for Use in Fluid Systems (New Type)—Dimensions (Inch Series) First Edition. 6 pp.

—Classification

BSI BS 1000: (621.1/22)-84. 1984 Universal Decimal Classificaton (UDC). English Full Edition (621.1/22): UDC 621.1 Heat Engines in General. Steam Power. Steam Engines. Boilers UDC 621.22 Water Power. Hydraulic Energy. 22 pp.

BSI BS 1000: (621.6)-73. 1973 Amd 1 Universal Decimal Classification (UDC). English Full Edition (621.6): Fluids Handling, Storage and Distribution Plant and Equipment. 24 pp.

SNZ NZS/BS 1000 (621.1/.22)-84. Universal Decimal Classification Heat Engines in General. Steam Power. Steam Engines. Boilers. Water Power. Hydraulic Energy. 24 pp.

SNZ NZS/BS 1000 (621.6)-73. Universal Decimal Classification Fluids Handling, Storage and Distribution Plant and Equipment Amend: 1. 24 pp.

—Connectors—Plugs

BSI BS 6361-88. 1988 Three-Pin Plug Connectors for Electrically Controlled Fluid Power Equipment. 8 pp.

ISO 4400-85. Fluid Power Systems and Components—Three-Pin Electrical Plug Connector—Characteristics and Requirements Second Edition. 6 pp.

ISO 6952-89. Fluid Power Systems and Components—Two-Pin Electrical Plug Connector with Earth Contact—Characteristics and Requirements First Edition. 7 pp.

—Diagrams—Symbols

CNS B1147-80. Graphical Symbols for Fluid Power Diagrams Line and Connection (Mar)(5270-1).

CNS B1148-80. Graphical Symbols for Fluid Power Diagrams Pump and Motor (Mar)(5270-2).

CNS B1149-80. Graphical Symbols for Fluid Power Diagrams Cylinder (Mar)(5270-3).

Fluid Power Equipment (Cont.)

—Diagrams—Symbols (Cont.)
CNS B1150-80. Graphical Symbols for Fluid Power Diagrams Control Method (Mar)(5270-4).
CNS B1151-80. Graphical Symbols for Fluid Power Diagrams Pressure Control Valve (Mar)(5270-5).
CNS B1152-80. Graphical Symbols for Fluid Power Diagrams Flow Control Valve (Mar)(5270-6).
CNS B1153-80. Graphical Symbols for Fluid Power Diagrams Directional Control Valve (Mar)(5270-7).
CNS B1154-80. Graphical Symbols for Fluid Power Diagrams Check Valve (Mar)(5270-8).
CNS B1155-80. Graphical Symbols for Fluid Power Diagrams Accessory Part (Mar)(5270-9).

—Earthmoving Equipment—Slope Limits—Static Testing
BSI BS 6911: Part 8-93. 1993 Testing Earth-Moving Machinery Part 8: Determination of Slope Limits for Machine Fluid Systems Operation (Static Test Method) (ISO 10266: 1992). 8 pp.
ISO 10266-92. Earth-Moving Machinery—Determination of Slope Limits for Machine Fluid Systems Operation—Static Test Method First Edition. 6 pp.

—Fatigue
BSI BS 7268-90. 1990 Method for Determination of Fatigue Pressure Rating of Metal Pressure Containing Envelopes in Hydraulic Fluid Power Systems. 12 pp.

—Flow Charts
BSI BS 5070: Part 3-88. 1988 Engineering Diagram Drawing Practice Part 3: Recommendations for Mechanical/Fluid Flow Diagrams. 44 pp.
SNZ BS 5070: Part 3-88. Engineering Diagram Drawing Practice Part 2: Recommendations for Mechanical/Fluid Flow Diagrams. 4 pp.

—Glossaries
ISO 5598-85. Fluid Power Systems and Components—Vocabulary First Edition. 99 pp.
JIS B 0133-79. Glossary of Terms for Fluid Control Device.
SAA AS B204-65. Glossary of Terms for Fluid Power Transmission and Control Systems (Superseded by AS 4061—1992).

—Hose Connectors—Diameters
ISO 4397-93. Fluid Power Systems and Components—Connectors and Associated Components—Nominal Outside Diameters of Tubes and Nominal Inside Diameters of Hoses Second Edition. 6 pp.

—O Rings
ISO 3601 Pt 1-88. Fluid Systems—Sealing Devices—O-Rings—Part 1: Inside Diameters, Cross-Sections, Tolerances and Size Identification Code Second Edition. 7 pp.
ISO 3601 Pt 3-87. Fluid Systems—Sealing Devices—O-Rings—Part 3: Quality Acceptance Criteria First Edition. 10 pp.

—Ports
ISO 6149-80. Fluid Power Systems and Components—Metric Ports—Dimensions and Design First Edition; (Corrected and Reprinted -1982). 4 pp.

—Ports—Earthmoving Equipment
ISO 8925-89. Earth-Moving Machinery—Diagnostic Ports First Edition. 4 pp.

—Pressure
ISO 2944-74. Fluid Power Systems and Components—Nominal Pressures First Edition; (Erratum—Aug 1974). 5 pp.

—Sealing Rings
ISO 3939-77. Fluid Power Systems and Components—Multiple Lip Packing Sets—Methods for Measuring Stack Heights First Edition. 6 pp.

—Ships—Cleaning
MOD UK NES 341: Part 1-89. Requirements for Cleaning of Items, Components and Equipment for Fluid Systems Part 1: Cleaning Issue 2 (05.89). 44 pp.

—Steady State Performance
ISO 9110 Pt 1-90. Hydraulic Fluid Power—Measurement Techniques—Part 1: General Measurement Principles First Edition. 7 pp.

—Submarines—Cleaning
MOD UK NES 341: Part 1-89. Requirements for Cleaning of Items, Components and Equipment for Fluid Systems Part 1: Cleaning Issue 2 (05.89). 44 pp.

—Submarines—Safety
MOD UK NES 341: Part 2-87. Requirements for Cleaning of Items Components and Equipment for Fluid Systems Part 2: Safety Precautions Issue 1 (03.87). 85 pp.

Fluid Power Equipment (Cont.)

—Symbols
BSI BS 2917-77. (WITHDRAWN) 1977 Graphical Symbols Used on Diagrams for Fluid Power Systems and Components (Superseded by BS 2917: Part1: 1993). 26 pp.
BSI BS 2917: Part 1-93. 1993 Graphic Symbols and Circuit Diagrams for Fluid Power Systems and Components Part 1: Specification for Graphic Symbols (ISO 1219-1: 1991) (E). 46 pp.
CNS B1146-80. Basic Graphical Symbols for Fluid Power Diagrams (Mar)(5270). 2 pp.
ISO 1219 Pt 1-91. Fluid Power Systems and Components—Graphic Symbols and Circuit Diagrams—Part 1: Graphic Symbols First Edition. 47 pp.
ISO 5784 Pt 1-88. Fluid Power Systems and Components—Fluid Logic Circuits—Part 1: Symbols for Binary Logic and Related Functions First Edition. 26 pp.
ISO 5784 Pt 2-89. Fluid Power Systems and Components—Fluid Logic Circuits—Part 2: Symbols for Supply and Exhausts as Related to Logic Symbols First Edition. 10 pp.
ISO 5784 Pt 3-89. Fluid Power Systems and Components—Fluid Logic Circuits—Part 3: Symbols for Logic Sequencers and Related Functions First Edition. 15 pp.
JIS B 0125-84. Graphic Symbols for Fluid Power Systems. 47 pp.

—Tube Connectors—Diameters
ISO 4397-93. Fluid Power Systems and Components—Connectors and Associated Components—Nominal Outside Diameters of Tubes and Nominal Inside Diameters of Hoses Second Edition. 6 pp.

Fluid Power Filters
Scope Note: Use a more specific term
See: Hydraulic Fluid Filters; Separators (Mechanical)

Fluid Power Systems
Use: Fluid Power Equipment

Fluid Power Systems and Equipment
Use: Fluid Power Equipment

Fluid Resistance

—Connectors
CEN PREN 2591 (Part C15)-92. Elements of Electrical and Optical Connection Test Methods Part C15—Fluid Resistance. 6 pp.
CEN PREN 2591-FC15-93. Aerospace Series Elements of Electrical and Optical Connection Test Methods Part FC15—Optical Elements Fluid Resistance. 3 pp.

Fluidity
Use: Viscosity

Fluids
See Also: Cryogenic Fluids; Damping Fluids; Deicer/Antiicing Fluids; Gases; Heat Transfer Fluids; Hydraulic Fluids; Lighter Fluid; Liquids; Newtonian Fluids; Water

—Synthetic—Emulsions
CNS K6340-73. Method of Test for Emulsion Characteristics of Petroleum Oils and Synthetic Fluids (Feb)(3518).

Flumes
See Also: Aqueducts; Flowmeters; Hydraulic Structures; Venturi Flumes; Weirs
BSI BS 3680: Part 4A-81. 1981 Amd 1 Measurement of Liquid Flow in Open Channels Part 4A: Weirs and Flumes: Thin-Plate Weirs. 33 pp.
BSI BS 3680: Part 4B-86. 1986 Measurement of Liquid Flow in Open Channels Part 4B: Weirs and Flumes: Triangular Profile Weirs. 16 pp.
BSI BS 3680: Part 4C-81. 1981 Measurement of Liquid Flow in Open Channels Part 4C: Weirs and Flumes: Flumes. 44 pp.
BSI BS 3680: Part 4D-89. 1989 Measurement of Liquid Flow in Open Channels Part 4: Weirs and Flumes Part 4D: Compound Gauging Structures. 27 pp.
BSI BS 3680: Part 4E-90. 1990 Measurement of Liquid Flow in Open Channels Part 4: Weirs and Flumes Part 4E: Rectangular Broad-Crested Weirs. 18 pp.
BSI BS 3680: Part 4F-90. 1990 Measurement of Liquid Flow in Open Channels Part 4F: Weirs and Flumes: Round-Nose Horizontal Broad-Crested Weirs. 24 pp.
BSI BS 3680: Part 4G-90. 1990 Measurement of Liquid Flow in Open Channels Part 4G: Flat-V Weirs. 24 pp.
BSI BS 3680: Part 4H-86. 1986 Measurement of Liquid Flow in Open Channels: Weirs an Flumes: Guide to the Selection of Flow Gauging Structures. 8 pp.

Flumes (Cont.)
BSI BS 3680: Part 4I-86. 1986 Measurement of Liquid Flow in Open Channels: Weirs and Flumes: V-Shaped Broad Crested Weirs. 19 pp.
ISO 3847-77. Liquid Flow Measurement in Open Channels by Weirs and Flumes—End-Depth Method for Estimation of Flow in Rectangular Channels with a Free Overfall First Edition. 9 pp.
ISO 4359-83. Liquid Flow Measurement in Open Channels—Rectangular, Trapezoidal and U-Shaped Flumes First Edition. 55 pp.
ISO 4371-84. Measurement of Liquid Flow in Open Channels by Weirs and Flumes—End Depth Method for Estimation of Flow in Non-Rectangular Channels with a Free Overfall (Approximate Method) First Edition. 13 pp.
ISO 8368-85. Liquid Flow Measurement in Open Channels—Guidelines for the Selection of Flow Gauging Structures First Edition. 7 pp.
ISO 9826-92. Measurement of Liquid Flow in Open Channels—Parshall and SANIIRI Flumes First Edition. 20 pp.
JIS B 7553-83. Parshall Flume Type Flowmeters.
SAA AS 3778.2.5-91. Measurement of Water Flow in Open Channels—Part 2: General—Part 2.5: Guidelines for the Selection of Flow Gauging Structures (ISO 8368:1985). 5 pp.
SAA AS 3778.4.7-91. Measurement of Water Flow in Open Channels—Part 4: Measurement Using Flow Gauging Structures—Part 4.7: Rectangular, Trapezoidal and U-Shaped Flumes (ISO 4359:1983). 51 pp.
SAA AS 3778.4.10-91. Measurement of Water Flow in Open Channels—Part 4: Measurement Using Flow Gauging Structures—Part 4.10: End-Depth Method for Estimation of Flow in Rectangular Channels with a Free Overfall (ISO 3847:1977). 7 pp.
SAA AS 3778.4.11-91. Measurement of Water Flow in Open Channels—Part 4: Measurement Using Flow Gauging Structures—Part 4.11: End-Depth Method for Estimation of Flow in Non-Rectangular Channels with a Free Overfall (Approximate Method) (ISO 4371:1984). 11 pp.

—Reinforced Concrete
CNS A2068-79. Reinforced Concrete Flumes (Apr)(4802). 6 pp.
CNS A3080-84. Method of Test for Reinforced Concrete Flumes (Sep)(4803).
JIS A 5318-90. Reinforced Concrete Flumes. 14 pp.
JIS A 5320-90. Reinforced Concrete Bench Flumes. 13 pp.

—Reinforced Concrete—Cradles
CNS A2069-87. Cradle for Reinforced Concrete Flumes (Aug)(4804). 2 pp.

Fluoboric Acid
See Also: Boric Acid
BSI BS 2657-74. 1974 Fluoroboric Acid and Metal Fluoroborates for Electroplating. 29 pp.

Fluor Chrome Arsenate Phenol
See Also: Phenol

—Wood Preservatives
CNS K1185-81. Wood Preservatives (Fluor Chrome Arsenate Phenol) (FCAP; Wolman Salt, Tanalith U) (May)(4133). 1 p.
CNS K6404-81. Method of Test for Fluor Chrome Arsenate Phenol (Wood Preservatives) (Jul)(4197).

Fluorescein
See Also: Eosin Yellowish-(YS)
JIS K 8829-91. Fluorescein.

Fluorescein Sodium
CNS K7232-66. Chemical Reagent (Fluorescein Sodium) (Jun)(1731).

Fluorescent Brighteners
Use For: Fabric Brighteners; Fluorescent Whitening Agents *See Also:* Dyes
CNS L1017-83. Scale for Assessing the Intensity of Fluorescent Whitening Effect (Jan)(9899).
CNS L1018-83. Scale for Assessing the Fastness of Fluorescent Whitened Textiles (Jan)(9900).
JIS L 0806-71. Scale for Assessing the Fastness of Fluorescent Whitened Textiles.
JIS L 0807-71. Scale for Assessing the Intensity of Fluorescent Whitening Effect.

—Colorfastness Testing—Light
CNS L3192-83. Method of Test for Colour Fastness to Light of Fluorescent Whitening Agents and Fluorescent Whitened Textiles (Jan)(9901).
JIS L 0887-75. Testing Method for Colour Fastness to Light of Fluorescent Whitening Agents and Fluorescent Whitened Textiles.

—Identification
JIS L 1064-85. Identification of Fluorescent Brightening Agents Classes on Textiles.

INTERNATIONAL AND NON-U.S. NATIONAL STANDARDS
SUBJECT INDEX

Fluorescent

Fluorescent Coatings
—Removable
CGSB CAN/CGSB-1.169-M91. High Visibility Coating System (Fluorescent) Removable. 11 pp.

Fluorescent Lamps
Use: Fluorescent Lighting

Fluorescent Lighting
Use For: Fluorescent Lights; Tubular Fluorescent Lamps *See Also:* Discharge Lamps; Lamps; Lighting; Starters

BSI BS 1853: Part 1-90. 1990 Tubular Fluorescent Lamps for General Lighting Service Part 1: Internationally Specified Lamps. 168 pp.
BSI BS 1853: Part 1-79. (WITHDRAWN) 1979 Tubular Fluorescent Lamps for General Lighting Service Part 1: Internationally Specified Lamps. 95 pp.
BSI BS 1853: Part 2-79. 1979 Tubular Fluorescent Lamps for General Lighting Service Part 2: Lamps Used in the United Kingdom Not Included in Part 1. 23 pp.
CENELEC EN 60 081-89. Tubular Fluorescent Lamps for General Lighting Service. 6 pp.
CNS C3157-82. Method of Test for Domestic Pendant Luminaires for Fluorescent Lamps (Jul)(9117).
CNS C3160-82. Method of Test for Table Lamps for Fluorescent Lamps (Jul)(9123).
CNS C4365-82. Domestic Pendant Luminaires for Fluorescent Lamps (Jul)(9116).
CNS C6028-89. Method of Test for Fluorescent Lamps (for General Lighting Service) (Jul)(3936).
CNS C7001-90. Fluorescent Lamps (for General Lighting Service) (Jun)(691).
CSA C22.2 NO 9-M1989. Luminaires; (Gen Instr 1) (Supercedes C22.2 No 97-1969). 83 pp.
CSA 719 Bull. Electrical Bulletin 719 March 20, 1968 to C22.2 NO 9. 3 pp.
CSA 917 Bull. Electrical Bulletin 917 July 23, 1973 to C22.2 NO 9. 1 p.
CSA 1169 Bull. Electrical Bulletin 1169 June 27, 1978 to C22.2 NO 9. 2 pp.
JIS C 7601-89. Fluorescent Lamps for General Lighting Service. 58 pp.
JIS C 8106-89. Fluorescent Lamp Luminaires for Commercial, Industrial and Public Lighting. 23 pp.
JIS C 8112-85. Table Study Lamps for Fluorescent Lamps. 14 pp.
JIS C 8115-89. Fluorescent Lamp Luminaires for Residential Lighting. 19 pp.
MOD UK DSTAN 62-6-01. Lamps, Electric Issue 2; Amendment 1. 109 pp.
SAA AS 1201-89. Tubular Fluorescent Lamps for General Lighting Service (IEC 81(1984)). 161 pp.
SNZ IEC 81-84. Tubular Fluorescent Lamps for General Lighting Service Amend: 1. 302 pp.

—Ballasts (Electric)
BSI BS 2818: Part 1-85. (WITHDRAWN) 1985 Amd 1 Ballasts for Tubular Fluorescent Lamps Part 1: Ballasts for Use Internationally (Superseded by BS EN 60920: 1991 and BS EN 60921: 1991). 73 pp.
BSI BS 2818: Part 2-85. (WITHDRAWN) 1985 Amd 1 Ballasts for Tubular Fluorescent Lamps Part 2: Ballasts for Use with Lamps Used in the UK Not Included in BS 1853 'Tubular Fluorescent Lamps for General Lighting Service 'Part 1' Specif for Internationally Specified Lamps'. 11 pp.
BSI BS 5717-84. (WITHDRAWN) 1984 Transistorized Ballasts for Tubular Fluorescent Lamps (Superseded by BS EN 60924: 1991). 30 pp.
BSI BS EN 60920-91. 1991 Ballasts for Tubular Fluorescent Lamps. General and Safety Requirements (Supersedes BS 2818: Part 1 & 2: 1985). 57 pp.
BSI BS EN 60921-91. 1991 Ballasts for Tubular Fluorescent Lamps. Performance Requirements. 34 pp.
BSI BS EN 60921-01. 1991 Amd 1 Ballasts for Tubular Fluorescent Lamps. Performance Requirements (AMD 7433) December 15, 1992 (Supersedes BS 2818: Parts 1 & 2: 1985). 45 pp.
BSI BS EN 60924-91. 1991 General and Safety Requirements for D.C. Supplied Electronic Ballasts for Tubular Fluorescent Lamps. 58 pp.
BSI BS EN 60925-91. 1991 Performance Requirements for D.C. Supplied Electronic Ballasts for Tubular Fluorescent Lamps. 26 pp.
BSI BS EN 60928-91. 1991 A.C. Supplied Electronic Ballasts for Tubular Fluorescent Lamps. General and Safety Requirements. 31 pp.
BSI BS EN 60929-92. 1992 A.C.-Supplied Electronic Ballasts for Tubular Fluorescent Lamps Performance Requirements. 46 pp.
CENELEC HD 302-75. Transistorized Ballast for Fluorescent Lamps. 4 pp.
CENELEC EN 60920-91. Ballasts for Tubular Fluorescent Lamps General and Safety Requirements. 7 pp.
CENELEC EN 60921-91. Ballasts for Tubular Fluorescent Lamps Performance Requirements. 6 pp.

—Ballasts (Electric) (Cont.)
CENELEC EN 60921/A1-92. AMD 1 Ballasts for Tubular Fluorescent Lamps Performance Requirements. 5 pp.
CENELEC EN 60924-91. D.C. Supplied Electronic Ballasts for Tubular Fluorescent Lamps General and Safety Requirements. 6 pp.
CENELEC EN 60925-91. D.C. Supplied Electronic Ballasts for Tubular Fluorescent Lamps—Performance Requirements. 5 pp.
CENELEC EN 60928-91. A.C. Supplied Electronic Ballasts for Tubular Fluorescent Lamps—General and Safety Requirements. 6 pp.
CENELEC EN 60929-92. A.C.-Supplied Electronic Ballasts for Tubular Fluorescent Lamps—Performance Requirements. 5 pp.
CNS C3041-85. Method of Test for Ballasts for Fluorescent Lamp (Oct)(3888). 5 pp.
CNS C4020-86. Ballasts for Fluorescent Lamp (Oct)(927).
CSA CAN/CSA-C22. 2 NO 74-92. Equipment for Use with Electric Discharge Lamps (Incorporating Electrical Bulletin Nos. 523F, 753, 846, 1124A, 1125A, 1325, and 1326); (Gen Instr 1). 82 pp.
CSA CAN/CSA-C654-M91. Fluorescent Lamp Ballast Efficacy Measurements; (Gen Instr 1 Thru 2). 54 pp.
IEC 82-84. Ballasts for Tubular Fluorescent Lamps Fifth Edition; (Amendment 1-1986). 131 pp.
IEC 920-90. Ballasts for Tubular Fluorescent Lamps General and Safety Requirements First Edition; (Amendment 1-1993). 121 pp.
IEC 921-88. Ballasts for Tubular Fluorescent Lamps Performance Requirements First Edition; (Corrigendum—April 1989) (Amendment 1-1990). 64 pp.
IEC 924-90. D.C. Supplied Electronic Ballasts for Tubular Fluorescent Lamps General and Safety Requirements First Edition; (Amendment 1-1993). 107 pp.
IEC 925-89. D.C. Supplied Electronic Ballasts for Tubular Fluorescent Lamps Performance Requirements First Edition. 40 pp.
IEC 928-90. A.C. Supplied Electronic Ballasts for Tubular Fluorescent Lamps General and Safety Requirements First Edition; (Amendment 1-1992) (Amendment 2-1993). 71 pp.
IEC 929-90. A.C.-Supplied Electronic Ballast for Tubular Fluorescent Lamps-Performance Requirements First Edition; (Corrigendum—June 1991). 77 pp.
JIS C 8108-91. Ballasts for Fluorescent Lamps. 43 pp.
JIS C 8117-92. AC Supplied Electronic Ballasts for Fluorescent Lamps. 45 pp.
SAA AS 3134-92. Approval and Test Specification—a.c. Supplied Electronic Ballasts for Tubular Fluorescent Lamps (in Professional Package 28) (Supersedes AS 3134(Int) —1992) (This is a Joint Standard with SNZ NZS 3134). 15 pp.
SAA AS 3963-91. a.c. Supplied Electronic Ballasts for Tubular Fluorescent Lamps—Performance Requirements. 18 pp.
SNZ NZS 6210-84. Ballasts for Tubular Fluorescent Lamps Amend: A, 1984. 104 pp.

—Ballasts (Electric)—Control Circuits
CSA C22.2 NO 184-M1988. Solid-State Lighting Controls; (Gen Instr 1 Thru 3). 33 pp.

—Ballasts (Electric)—Thermal Protectors
BSI BS EN 60730-2-3-92. 1992 Automatic Electrical Controls for Household and Similar Use Part 2: Particular Requirements Section 2.3: Thermal Protectors for Ballasts for Tubular Fluorescent Lamps. 19 pp.
CENELEC EN 60730-2-3-92. Automatic Electrical Controls for Household and Similar Use Part 2: Particular Requirements for Thermal Protectors for Ballasts for Tubular Fluorescent Lamps. 7 pp.
CSA CAN/CSA-C22. 2 NO 74-92. Equipment for Use with Electric Discharge Lamps (Incorporating Electrical Bulletin Nos. 523F, 753, 846, 1124A, 1125A, 1325, and 1326); (Gen Instr 1). 82 pp.
IEC 730 Pt 2-3-90. Automatic Electrical Controls for Household and Similar Use Part 2: Particular Requirements for Thermal Protectors for Ballasts for Tubular Fluorescent Lamps First Edition. 30 pp.

—Capacitors
BSI BS 4017-79. (WITHDRAWN) 1979 Capacitors for Use in Tubular Fluorescent, High Pressure Mercury, and Low Pressure Sodium Vapour Discharge Lamp Circuits (SUPERSEDED BY BS EN 61048 & BS EN 61049). 22 pp.
BSI BS EN 61048-93. 1993 Capacitors for Use in Tubular Fluorescent and Other Discharge Lamp Circuits—General and Safety Requirements (E). 31 pp.
BSI BS EN 61049-93. 1993 Capacitors for Use in Tubular Fluorescent and Other Discharge Lamp Circuits—Performance Requirements (E). 21 pp.

—Capacitors (Cont.)
CENELEC EN 61048-93. Capacitors for Use in Tubular Fluorescent and Other Discharge Lamp Circuits—General and Safety Requirements (IEC 1048:1991 + Corrigendum January 1992). 6 pp.
CENELEC EN 61048-93. Capacitors for Use in Tubular Fluorescent and Other Discharge Lamp Circuits—General and Safety Requirements (IEC 1048: 1991, Modified + Corrigendum: 1992). 25 pp.
CENELEC EN 61049-93. Capacitors for Use in Tubular Fluorescent and Other Discharge Lamp Circuits Performance Requirements (IEC 1049: 1991 + Corrigendum 1992, Modified). 14 pp.
DIN VDE 0560 Pt 6-86. Capacitors; Capacitors with Ratings up to 1.5 kvar for Installations with Discharge Lamps, and in Particular Fluorescent Lamps (Jan). 32 pp.
IEC 1048-91. Capacitors for Use in Tubular Fluorescent and Other Discharge Lamp Circuits General and Safety Requirements First Edition; (Replaces 566) (CENELEC EN 61048:1993). 54 pp.
IEC 1049-91. Capacitors for Use in Tubular Fluorescent and Other Discharge Lamp Circuits Performance Requirements First Edition; (Replaces 566) (CENELEC EN 61049: 1993). 30 pp.

—Capacitors—Safety
BSI BS EN 61048-93. 1993 Capacitors for Use in Tubular Fluorescent and Other Discharge Lamp Circuits—General and Safety Requirements (E). 31 pp.
CENELEC EN 61048-93. Capacitors for Use in Tubular Fluorescent and Other Discharge Lamp Circuits—General and Safety Requirements (IEC 1048:1991 + Corrigendum January 1992). 6 pp.
CENELEC EN 61048-93. Capacitors for Use in Tubular Fluorescent and Other Discharge Lamp Circuits—General and Safety Requirements (IEC 1048: 1991, Modified + Corrigendum: 1992). 25 pp.
IEC 1048-91. Capacitors for Use in Tubular Fluorescent and Other Discharge Lamp Circuits General and Safety Requirements First Edition; (Replaces 566) (CENELEC EN 61048:1993). 54 pp.

—Chromaticity
JIS Z 9112-90. Classification of Fluorescent Lamps by Chromaticity and Colour Rendering Property. 10 pp.

—Electric Switches
CSA C22.2 NO 111-M1986. General Use Switches (R 1992); (Gen Instr 1 Thru 2). 40 pp.

—Glass Tubes
CNS C3111-85. Method of Test for Glass Tubing for Fluorescent Lamps (Jul)(7007).
CNS C4296-85. Glass Tubing for Fluorescent Lamps (May)(7006).
JIS C 7708-84. Glass Tubing for Fluorescent Lamps. 12 pp.

—Insulated Wire
CNS C2072-89. 1000 V Grade Insulated Wires for Fluorescent Lamps (Jul)(6063). 6 pp.
JIS C 3309-91. 1000 V Grade Insulated Wires for Fluorescent Lamps. 10 pp.

—Lampholders
BSI BS 6702-91. (WITHDRAWN) 1991 Lampholders for Tubular Fluorescent Lamps and Starterholders (Superseded by BS EN 60400: 1992). 59 pp.
BSI BS EN 60400-92. 1992 Lampholders for Tubular Fluorescent Lamps and Starterholders. 82 pp.
CENELEC EN 60 400-85. Lampholders for Tubular Fluorescent Lamps and Starterholders. 42 pp.
CENELEC EN 60 400-89. Lampholders for Tubular Fluorescent Lamps and Starterholders. 8 pp.
CENELEC EN 60400-92. Lampholders for Tubular Fluorescent Lamps and Starterholders. 7 pp.
CNS C3094-80. Method of Test for Lampholders and Starterholders for Fluorescent Lamps (Aug)(6055).
CNS C4220-80. Lampholders and Starterholders for Fluorescent Lamps (Aug)(6054).
CSA C22.2 NO 43-M1984. Lampholders (R 1992); (Gen Instr 1 Thru 6). 54 pp.
CSA 1233A Bull. Electrical Bulletin 1233A June 17, 1981 to C22.2 NO 43. 2 pp.
IEC 400-91. Lampholders for Tubular Fluorescent Lamps and Starterholders Fourth Edition; (Amendment 1-1993, Including the Corrigendum of June 1992) (Corrigendum—Aug 1993). 130 pp.
JIS C 8324-79. Lampholders and Starterholders for Fluorescent Lamps. 21 pp.
SNZ IEC 400-87. Lampholders for Tubular Fluorescent Lamps and Startholders. 80 pp.
SNZ NZS 1123-53. Specification for Bi-Pin Lamp Caps and Lampholders for Tubular Fluorescent Lamps for Use in Circuits, the Declared Voltage of Which Does Not Exceed 250 Volts Amend: 1, 1963; 2, 1963; 3, 1981; 3A, 1981. 16 pp.

INDUSTRY STANDARDS

INTERNATIONAL AND NON-U.S. NATIONAL STANDARDS
SUBJECT INDEX — Fluorescent

Fluorescent Lighting (Cont.)

—**Light Sources—Colorimetry**
JIS Z 8716-91. Fluorescent Lamp as a Simulator of CIE Standard Illuminant D65 for a Visual Comparison of Surface Colours—Type and Characteristics. 8 pp.

—**Motor Vehicle**
CSA 1157 Bull. Electrical Bulletin 1157 February 9, 1978 to C22.2 NO 9. 4 pp.

—**Naval Ships**
NATO STANAG 1006 ED 4 AMD 3-86. Preferred Incandescent and Fluorescent Electric Lamps for Lighting Purposes on Naval Vessels of NATO Countries. 7 pp.

—**Photometry**
JIS C 7607-91. Total Luminous Flux Measurements on Discharge Lamps Used for Photometric Standards. 16 pp.

—**Portable**
CNS C4347-82. Portable Fluorescent Lamp (May)(8801).
CSA C22.2 NO 12-1982. Portable Luminaires; (Gen Instr 1 Thru 4). 52 pp.
CSA 1169 Bull. Electrical Bulletin 1169 June 27, 1978 to C22.2 NO 12. 2 pp.

—**Portable—Electric Plugs**
CSA 1276 Bull. Electrical Bulletin 1276 July 16, 1980 to C22.2 NO 12. 2 pp.

—**Power Reducers**
CSA CAN/CSA-C22. 2 NO 74-92. Equipment for Use with Electric Discharge Lamps (Incorporating Electrical Bulletin Nos. 523F, 753, 846, 1124A, 1125A, 1325, and 1326); (Gen Instr 1). 82 pp.

—**Preheating**
IEC 882-86. Pre-Heat Requirements for Starterless Tubular Fluorescent Lamps First Edition. 19 pp.

—**Preheating—Fittings**
CNS C3042-85. Method of Test for Lighting Fitting for Fluorescent Lamps (Pre-Heat Type) (May)(3889). 2 pp.
CNS C4065-84. Lighting Fitting for Fluorescent Lamps (Pre-Heat Type) (May)(2660). 4 pp.

—**Prewire Plates—Certification**
CSA 1330 Bull. Electrical Bulletin 1330 June 24, 1981 to C22.2 NO 9. 2 pp.

—**Radio Frequency Interference**
BSI BS 5394-01. 1988 Amd 1 Limits and Methods of Measurement of Radio Interference Characteristics of Fluorescent Lamps and Luminaires (AMD 6581) August 31, 1990. 44 pp.
BSI BS EN 55015-93. 1993 Limits and Methods of Measurement of Radio Disturbance Characteristics of Electrical Lighting and Similar Equipment (S). 48 pp.
BSI BS EN 55015-01. 1993 Amd 1 Limits and Methods of Measurement of Radio Disturbance Characteristics of Electrical Lighting and Similar Equipment (AMD 7878) July 15, 1993 (S). 49 pp.
CENELEC EN 55 015-87. Limits and Methods of Measurement of Radio Interference Characteristics of Fluorescent Lamps and Luminaires. 11 pp.
CENELEC EN 55015-93. Limits and Methods of Measurement of Radio Disturbance Characteristics of Electrical Lighting and Similar Equipment (IEC CISPR 15: 1992). 5 pp.
CENELEC EN 55 015/A1-90. AMD 1 Limits and Methods of Measurement of Radio Interference Characteristics of Fluorescent Lamps and Luminaires. 4 pp.
EC 76/890/EEC-76. Council Directive on the Approximation of the Laws of the Member States Relating to the Suppression of Radio Interference with Regard to Fluorescent Lighting Luminaires Fitted with Starters. 8 pp.
EC 83/447/EEC-83. Commission Directive Adopting the Measures Provided for in Article 3 (3) of Directive 76/889/EEC on the Approximation of the Laws of the Member States Relating to Radio Interference Caused by Electrical Household Appliances, Portable Tools and Similar. 1 p.
IEC CISPR 15-92. Limits and Methods of Measurement of Radio Disturbance Characteristics of Electrical Lighting and Similar Equipment Fourth Edition; (Corrigendum—Dec 1992) (CENELEC EN 55015:1993). 83 pp.
SAA AS/NZS 4051-92. Limits and Methods of Measurement of Radio Interference Characteristics of Fluorescent Lamps and Luminaires (IEC/CISPR 15:1985). 28 pp.
SNZ NZS/AS 4051-92. Limits and Methods of Measurement of Radio Interference Characteristics of Fluorescent Lamps and Luminaires (This is a Joint Standard with SAA AS 4051). 28 pp.

Fluorescent Lighting (Cont.)

—**Recreational Vehicles**
CSA 1157 Bull. Electrical Bulletin 1157 February 9, 1978 to C22.2 NO 9. 4 pp.

—**Self Ballasted**
BSI BS 7199-89. 1989 Self-Ballasted Lamps for General Lighting Services. Performance Requirements (Renumbered as BS EN 60969: 1993). 11 pp.
BSI BS 7199-01. 1989 Amd 1 Self-Ballasted Lamps for General Lighting Services Performance Requirements (IEC 969: 1988) (AMD 7013) July 15, 1992 (Renumbered as BS EN 60969: 1993) (E). 19 pp.
BSI BS EN 60969-93. 1993 Self-Ballasted Lamps for General Lighting Services. Performance Requirements (AMD 7717) May 15, 1993. 29 pp.
CENELEC HD 594 S1-91. Self-Ballasted Lamps for General Lighting Services—Performance Requirements. 6 pp.
CENELEC EN 60969-93. Self-Ballasted Lamps for General Lighting Services Performance Requirements (IEC 969:1988) (Supersedes HD 594 S1:1991). 5 pp.
CENELEC EN 60969-93. Self-Ballasted Lamps for General Lighting Services Performance Requirements (IEC 969:1988). 12 pp.
CENELEC EN 60969/A1-93. AMD 1 Self-Ballasted Lamps for General Lighting Services Performance Requirements (IEC 969:1988/A1: 1991). 4 pp.
IEC 969-88. Self-Ballasted Lamps for General Lighting Services Performance Requirements First Edition; (Amendment 1-1991) (CENELEC EN 60969/A1: 1993). 24 pp.

—**Self Ballasted—Safety**
BSI BS 7173-89. 1989 Amd 1 Self-Ballasted Lamps for General Lighting Services Safety Requirements (AMD 6707) September 30, 1991. 24 pp.
BSI BS 7173-02. 1989 Amd 2 Self-Ballasted Lamps for General Lighting Services Safety Requirements (AMD 7716) May 15, 1993. 30 pp.
CENELEC EN 60968-90. Self-Ballasted Lamps for General Lighting Services Safety Requirements. 5 pp.
CENELEC EN 60968/A1-93. AMD 1 Self-Ballasted Lamps for General Lighting Services Safety Requirements (IEC 968:1988/A1: 1991). 4 pp.
IEC 968-88. Self-Ballasted Lamps for General Lighting Services Safety Requirements First Edition; (Amendment 1-1991) (CENELEC EN 60968/A1: 1993). 34 pp.

—**Ships**
MOD UK NES 2003-89. Specification for Special Fluorescent Lamps by Thorn EMI Lighting Ltd for Use in Naval Design Lighting Fittings Issue 2 (12.89). 12 pp.

—**Ships—Ballasts (Electric)**
CNS F5034-87. Marine Ballasts for Fluorescent Lamp (Nov)(9390).
JIS F 8431-85. Marine Ballasts for Fluorescent Lam

—**Ships—Berths**
CNS F5039-82. Fluorescent Berth Lights with Spare Light for Marine Use (Sep)(9395).

—**Ships—Ceiling**
CNS F5037-85. Marine Fluorescent Ceiling Lights (Non-Watertight Type) (Dec)(9393).
CNS F5038-85. Marine Fluorescent Ceiling Lights (Watertight Type) (Dec)(9394).
JIS F 8436-83. Marine Non-Watertight Type Fluoroescent Ceiling Lights.
JIS F 8436-64. Fluorescent Ceiling Lights for Marine Use (Non-Watertight Type). 9 pp.
JIS F 8437-83. Marine Watertight Type Fluorescent Ceiling Lights.
JIS F 8437-64. Fluorescent Ceiling Lights for Marine Use (Watertight Type) (R 1970). 9 pp.
JIS F 8439-87. Non-Watertight Recessed Type Fluorescent Ceiling Lights for Small Ships.

—**Ships—Choke Coils**
MOD UK NES 2002-88. Specification for Chokes and Transformers Used in Fluorescent Fittings Issue 2 (12.88). 25 pp.
MOD UK NES 2002-92. Specification for Chokes and Transformers Used in Fluorescent Fittings Issue 3 (01.92). 25 pp.

—**Ships—Tables (Furniture)**
CNS F5035-85. Marine Fluorescent Table Lights (Dec)(9391).
JIS F 8434-83. Marine Fluorescent Table Lights.

—**Ships—Wall**
CNS F5036-85. Marine Fluorescent Wall Lights (Non-Watertight Type) (Dec)(9392).
JIS F 8435-83. Marine Non-Watertight Type Fluorescent Wall Lights.

Fluorescent Lighting (Cont.)

—**Signs**
CSA CAN/CSA-C22. 2 NO 207-M89. Portable and Stationary Electric Signs and Displays; (Gen Instr 1). 50 pp.

—**Single Ended—Safety**
BSI BS 6982-88. 1988 Amd 1 Single-Capped Fluorescent Lamps—Safety and Performance Requirements (IEC 901: 1987) (AMD 6572) February 28, 1991. 40 pp.
BSI BS 6982-02. 1988 Amd 2 Single-Capped Fluorscent Lamps—Safety and Performance Requirements (IEC 901: 1987) (AMD 6708) April 30, 1991. 48 pp.
CENELEC EN 60 901-90. Single-Capped Fluorescent Lamps—Safety and Performance Requirements. 6 pp.
CENELEC EN 60901/A1-90. Single-Capped Fluorescent Lamps Safety and Performance Requirements. 4 pp.
IEC 901-87. Single-Capped Fluorescent Lamps—Safety and Performance Requirements First Edition; (Amendment 1-1989) (Amendment 2-1992) (Corrigendum—June 1992). 211 pp.

—**Spectral Energy Distribution**
JIS Z 9112-90. Classification of Fluorescent Lamps by Chromaticity and Colour Rendering Property. 10 pp.
JIS Z 9301-82. Colour Rendering Classification of Fluorescent Lamps.

—**Starters**
BSI BS 3772-90. 1990 Starters for Fluorescent Lamps. 32 pp.
BSI BS 3772-81. (WITHDRAWN) 1981 Starters for Fluorescent Lamps. 32 pp.
BSI BS 6702-91. (WITHDRAWN) 1991 Lampholders for Tubular Fluorescent Lamps and Starterholders (Superseded by BS EN 60400: 1992). 59 pp.
BSI BS EN 60400-92. 1992 Lampholders for Tubular Fluorescent Lamps and Starterholders. 82 pp.
BSI BS EN 60926-91. 1991 General and Safety Requirements for Starting Devices (Other Than Glow Starters). 36 pp.
BSI BS EN 60927-91. 1991 Performance Requirements for Starting Devices (Other Than Glow Starters). 22 pp.
BSI BS EN 60927-01. 1991 Amd 1 Performance Requirements for Starting Devices (Other Than Glow Starters) (AMD 6882) May 1, 1992. 52 pp.
CENELEC EN 60 155-89. Starters for Tubular Fluorescent Lamps. 29 pp.
CENELEC EN 60 400-85. Lampholders for Tubular Fluorescent Lamps and Starterholders. 42 pp.
CENELEC EN 60 400-89. Lampholders for Tubular Fluorescent Lamps and Starterholders. 8 pp.
CENELEC EN 60400-92. Lampholders for Tubular Fluorescent Lamps and Starterholders. 7 pp.
CENELEC EN 60 926-90. Starting Devices (Other Than Glow Starters) General and Safety Requirements. 6 pp.
CENELEC EN 60927-90. Starting Devices (Other Than Glow Starters) Performance Requirements. 16 pp.
CENELEC EN 60927/A1-91. AMD 1 Starting Devices (Other Than Glow Starters) Performance Requirements. 5 pp.
CNS C3039-85. Method of Test for Glow Starters for Fluorescent Lamp (Jul)(3741).
CNS C3094-80. Method of Test for Lampholders and Starterholders for Fluorescent Lamps (Aug)(6055).
CNS C4025-87. Glow Starters for Fluorescent Lamps (Oct)(1092).
CNS C4220-80. Lampholders and Starterholders for Fluorescent Lamps (Aug)(6054).
IEC 155-83. Starters for Tubular Fluorescent Lamps Third Edition; (Amendment 1-1987) (Amendment 2-1991). 54 pp.
IEC 400-91. Lampholders for Tubular Fluorescent Lamps and Starterholders Fourth Edition; (Amendment 1-1993, Including the Corrigendum of June 1992) (Corrigendum—Aug 1993). 130 pp.
IEC 926-90. Starting Devices (Other Than Glow Starters) General and Safety Requirements First Edition; (Amendment 1-1992) (Amendment 2-1993). 85 pp.
IEC 927-88. Starting Devices (Other Than Glow Starters) Performance Requirements First Edition; (Corrigendum—Dec 1989) (Amendment 1-1990). 60 pp.
JIS C 7603-84. Glow Starters for Fluorescent Lamps. 15 pp.
JIS C 8324-79. Lampholders and Starterholders for Fluorescent Lamps. 21 pp.
SNZ IEC 155-83. Starters for Tubular Fluorescent Lamps. 39 pp.
SNZ IEC 400-87. Lampholders for Tubular Fluorescent Lamps and Startholders. 80 pp.

INTERNATIONAL AND NON-U.S. NATIONAL STANDARDS
SUBJECT INDEX

Fluorine

Fluorescent Lighting (Cont.)

—Starters—Radio Frequency Interference

EC 82/500/EEC-82. Commission Directive Adapting to Technical Progress Council Directive 76/890/EEC on the Approximation of the Laws of the Member States Relating to Suppression of Radio Interference with Regard to Fluorescent Lighting Luminaires Fitted with Starters. 11 pp.

EC 87/310/EEC-87. Commission Directive Adapting to Technical Progress Council Directive 76/890/EEC on the Approximation of the Laws of the Member States Relating to the Suppression of Radio Interference with Regard to Fluorescent Lighting Luminaires Fitted with Starters. 1 p.

—Submarines

MOD UK NES 2003-89. Specification for Special Fluorescent Lamps by Thorn EMI Lighting Ltd for Use in Naval Design Lighting Fittings Issue 2 (12.89). 12 pp.

—Submarines—Choke Coils

MOD UK NES 2002-88. Specification for Chokes and Transformers Used in Fluorescent Fittings Issue 2 (12.88). 25 pp.

MOD UK NES 2002-92. Specification for Chokes and Transformers Used in Fluorescent Fittings Issue 3 (01.92). 25 pp.

—Submersible—Swimming Pools—Accessories

CSA C22.2 NO 89-1976. Swimming-Pool Luminaires, Submersible Luminaires and Accessories (R 1992). 23 pp.

—Tables (Furniture)

CNS C4368-82. Table Lamps for Fluorescent Lamps (Jul)(9122).

—Thermal Cutoffs—Ballasts (Electric)

CSA C22.2 NO 209-M1985. Thermal Cut-Offs (R 1992); (Gen Instr 1 Thru 2). 28 pp.

—Transformers

CENELEC HD 388 S2-83. Transformers for Tubular Discharge Lamps Having No-Load Output Voltage Exceeding 1000V (Generally Called Neon Transformers). 2 pp.

—Transformers—Ships

MOD UK NES 2002-88. Specification for Chokes and Transformers Used in Fluorescent Fittings Issue 2 (12.88). 25 pp.

MOD UK NES 2002-92. Specification for Chokes and Transformers Used in Fluorescent Fittings Issue 3 (01.92). 25 pp.

—Transformers—Submarines

MOD UK NES 2002-88. Specification for Chokes and Transformers Used in Fluorescent Fittings Issue 2 (12.88). 25 pp.

MOD UK NES 2002-92. Specification for Chokes and Transformers Used in Fluorescent Fittings Issue 3 (01.92). 25 pp.

Fluorescent Lights

Use: Fluorescent Lighting

Fluorescent Materials

See Also: Phosphorescent Materials

—Colorimetry

JIS Z 8717-89. Methods of Measurement for Colour of Fluorescent Objects. 39 pp.

—Signs

CNS Z1034-87. Fluorescent Safety Signs Board (Mar)(9647).

CNS Z8045-87. Method of Test for Fluorescent Safety Signs (Mar)(9651).

JIS Z 9106-90. Fluorescent Safety Colours—General Rules for Application.

JIS Z 9108-90. Fluorescent Safety Signs. 13 pp.

Fluorescent Paints

Use: Luminous Paints

Fluorescent Penetration Testing

See Also: Liquid Penetrant Testing; Penetration Resistance

CNS Z8011-74. Method of Fluorescence Transmittance Test for Trouble Shooting (Mar)(3711).

JIS Z 2343-92. Method for Liquid Penetrant Testing and Classification of the Indication. 21 pp.

—Aircraft—Ultraviolet Lamps

CAA NOTICE #95 ISSUE 1. Use of High Intensity Ultra-Violet Lamps in Fluorescent Penetrant and Magnetic Particle Inspections (Airworthiness Notices). 3 pp.

Fluorescent Penetration Testing (Cont.)

—Paper Products

CNS P3082-87. Method of Test for Migratable Fluorescent Substances in Paper Products (Jan)(11830).

Fluorescent Screens

See Also: Screens (Projection)

—Photosensitive Materials—Sensitivity

JIS K 7607-76. Method for Determining Fluorographic Sensitivity of Photographic Sensitive Materials Used with Green Fluorescent Screen.

—Photosensitive Materials—Sensitivity—Light Source

JIS K 7606-76. Light Source for Determining Fluorographic Sensitivity of Photographic Sensitive Materials Used with Green Fluorescent Screen (R 1984). 5 pp.

—X-Rays

JIS Z 4911-89. Dimensions and Form for X-Ray Fluorescent Screen. 4 pp.

Fluorescent Whitening Agents

Use: Fluorescent Brighteners

Fluoride Content Analysis

Use For: Free Fluorides Content Analysis

—Cosmetics

CNS S2088-82. Methods of Hygienic Test for Cosmetics — Fluorides (Oct)(9542).

—Cryolite

CNS M3120-83. Method for Determining Free Fluorides Content in Natural and Artificial Cryolite (Conventional Test) (Mar)(10119).

ISO 4277-77. Cryolite, Natural and Artificial—Conventional Test for Evaluation of Free Fluorides Content First Edition. 5 pp.

—Distillation Methods

CNS J2042-82. Fluoride by Distillation-Spectrophotometric Method (Feb)(8512).

—Sludge (Sewage)—Electrochemistry

DIN ENGL 38405 Pt 4-85. German Standard Methods for the Examination of Water, Waste Water and Sludge; Anions (Group D); Determination of Fluoride (D 4) (July). 7 pp.

—Spectrophotometry

CNS J2042-82. Fluoride by Distillation-Spectrophotometric Method (Feb)(8512).

—Surface Water—Electrochemical Probes

BSI BS 6068: Sec 2.41-93. 1993 Water Quality Part 2: Physical, Chemical and Biochemical Methods Section 2.41: Determination of Fluoride: Electrochemical Probe Method for Potable and Lightly Polluted Water (ISO 10359-1: 1992). 11 pp.

ISO 10359 Pt 1-92. Water Quality—Determination of Fluoride—Part 1: Electrochemical Probe Method for Potable and Lightly Polluted Water First Edition. 10 pp.

—Waste Water

CNS K9104-83. Method of Test for Fluoride in Water and Wastewater (SPADNS Method) (Feb)(10010).

—Waste Water—Electrochemical Probes

BSI BS 6068: Sec 2.41-93. 1993 Water Quality Part 2: Physical, Chemical and Biochemical Methods Section 2.41: Determination of Fluoride: Electrochemical Probe Method for Potable and Lightly Polluted Water (ISO 10359-1: 1992). 11 pp.

ISO 10359 Pt 1-92. Water Quality—Determination of Fluoride—Part 1: Electrochemical Probe Method for Potable and Lightly Polluted Water First Edition. 10 pp.

—Waste Water—Electrochemistry

CNS K9103-83. Method of Test for Fluoride in Water and Wastewater (Electrode Method) (Feb)(10009).

DIN ENGL 38405 Pt 4-85. German Standard Methods for the Examination of Water, Waste Water and Sludge; Anions (Group D); Determination of Fluoride (D 4) (July). 7 pp.

—Water

CNS K9104-83. Method of Test for Fluoride in Water and Wastewater (SPADNS Method) (Feb)(10010).

—Water—Electrochemical Probes

BSI BS 6068: Sec 2.41-93. 1993 Water Quality Part 2: Physical, Chemical and Biochemical Methods Section 2.41: Determination of Fluoride: Electrochemical Probe Method for Potable and Lightly Polluted Water (ISO 10359-1: 1992). 11 pp.

Fluoride Content Analysis (Cont.)

—Water—Electrochemical Probes (Cont.)

ISO 10359 Pt 1-92. Water Quality—Determination of Fluoride—Part 1: Electrochemical Probe Method for Potable and Lightly Polluted Water First Edition. 10 pp.

—Water—Electrochemistry

CNS K9103-83. Method of Test for Fluoride in Water and Wastewater (Electrode Method) (Feb)(10009).

DIN ENGL 38405 Pt 4-85. German Standard Methods for the Examination of Water, Waste Water and Sludge; Anions (Group D); Determination of Fluoride (D 4) (July). 7 pp.

—Water—Spectrophotometry

BSI BS 2690: Part 108-79. 1979 Water Used in Industry Part 108: Fluoride: Spectrophotometric Method (Alizarin Fluorine Blue). 4 pp.

Fluoride Ion Content Analysis

—Water—Liquid Chromatography

DIN ENGL 38405 Pt 19-88. German Standard Methods for the Examination of Water, Waste Water and Sludge; Anions (Group D) Determination of Fluo-ride, Chloride, Nitrite, (Ortho)Phosphate, Brom-ide, Nitrate and Sulfate Anions in Water with a Low Pollution Level by Ion Exchange Chromatography (D 19) (Feb). 11 pp.

ISO 10304 Pt 1-92. Water Quality—Determination of Dissolved Fluoride, Chloride, Nitrite, Orthophosphate, Bromide, Nitrate and Sulfate Ions, Using Liquid Chromatography of Ions—Part 1: Method for Water with Low Contamination First Edition. 16 pp.

Fluorides

See Also: Aluminum Fluoride; Ammonium Fluoride; Hydrofluoric Acid; Potassium Bifluoride; Potassium Fluoride; Sodium Fluoride; Sulfur Hexafluoride; Uranium Hexafluoride

—Air—Sampling

SAA AS 3580.13.2-91. Methods for Sampling and Analysis of Ambient Air—Part 13: Determination of Fluorides—Gaseous and Acid-Soluble Particulate Fluorides—Part 13.2: Manual, Double Filter Paper Sampling (Supersedes AS 2618.2—1984). 8 pp.

Fluorinated Ethylene Propylene Resins

Scope Note: For additional listings, see also specific products made from fluorinated ethylene propylene resins *Use For:* FEP *See Also:* Thermoplastic Resins

—Sheets

MOD UK TS 10167A. FEP Fluorocarbon Sheet (Withdrawn).

—Sheets—Electrical Insulation

IEC 674 Pt 3-7-92. Specification for Plastic Films for Electrical Purposes Part 3: Specifications for Individual Materials Sheet 7: Requirements for Fluoroethylene-Propylene (FEP) Films Used for Electrical Insulation First Edition. 24 pp.

Fluorinated Hydrocarbons

Use: Fluorocarbons

Fluorine

See Also: Fluorine Content Analysis; Fluorspar

—Analyzers

JIS B 7958-80. Continuous Analyzers for Fluorine Compounds in Ambient Air.

Fluorine Compound Content Analysis

—Exhaust Gases

CNS K9069-81. Method of Test for Fluorine Compounds in Exhaust Gas (Aug)(7810).

JIS K 0105-82. Methods for Determination of Fluorine Compounds in Exhaust Gas. 25 pp.

Fluorine Content Analysis

See Also: Fluorine

—Aluminum Fluoride—Willard Winter Method

BSI BS 4993: Part 1-74. 1974 Methods of Test for Aluminium Fluoride for Industrial Use Part 1: Determination of Fluorine Content. 13 pp.

ISO 2362-72. Aluminium Fluoride for Industrial Use—Determination of Fluorine Content—Modified Willard-Winter Method First Edition. 7 pp.

—Aluminum Oxide—Spectrophotometry

BSI BS 4140: Part 17-86. 1986 Methods of Test for Aluminium Oxide Part 17: Determination of Fluorine Content. 5 pp.

INDUSTRY STANDARDS

Fluorine Content Analysis *(Cont.)*

—Aluminum Oxide—Spectrophotometry *(Cont.)*
ISO 2828-73. Aluminium Oxide Primarily Used for the Production of Aluminium—Determination of Fluorine Content—Alizarin Complexone and Lanthanum Chloride Spectrophotometric Method First Edition. 5 pp.

—Coal
SAA AS 1038.10.4-89. Methods for the Analysis and Testing of Coal and Coke—Part 10.4: Determinaton of Trace Elements—Coal, Coke and Fly-Ash—Determination of Fluorine Content—Phrohydrolysis Method Amdt 1 October 1990. 11 pp.

—Coke
SAA AS 1038.10.4-89. Methods for the Analysis and Testing of Coal and Coke—Part 10.4: Determinaton of Trace Elements—Coal, Coke and Fly-Ash—Determination of Fluorine Content—Phrohydrolysis Method Amdt 1 October 1990. 11 pp.

—Cryolite
BSI BS 5050-74. 1974 Methods of Test for Cryolite. 20 pp.
MOD UK M 874/68. Examination of Cryolite.

—Cryolite—Willard Winter Method
CNS M3114-83. Method for Determining Fluorine Content in Natural and Artificial Cryolite (Modified Willard Method) (Mar)(10113).
ISO 1693-76. Cryolite, Natural and Artificial—Determination of Fluorine Content—Modified Willard-Winter Method First Edition. 8 pp.

—Fluorite—Potentiometric Analysis
BSI BS 5659: Part 5-79. (WITHDRAWN) 1979 Acid-Grade Fluorspar Part 5: Determination of Available Fluorine Content. 8 pp.
ISO 5439-78. Acid-Grade Fluorspar—Determination of Available Fluorine Content—Potentiometric Method After Distillation First Edition. 8 pp.

—Fluorite—Willard-Winter Method
ISO 9503-91. Metallurgical-Grade Fluorspar—Determination of Available Fluorine Content—Modified Willard-Winter Method First Edition. 8 pp.

—Fluosilicic Acid—Potentiometric Analysis
BSI BS 6445: Part 1-83. (WITHDRAWN) 1983 Hexafluorosilicic Acid for Industrial Use Part 1: Method for Determination of Available Fluorine Content (Potentiometric Method After Distillation). 8 pp.
ISO 6677-83. Hexafluorosilicic Acid for Industrial Use—Determination of Available Fluorine Content—Potentiometric Method After Distillation First Edition. 7 pp.

—Fly Ash
SAA AS 1038.10.4-89. Methods for the Analysis and Testing of Coal and Coke—Part 10.4: Determinaton of Trace Elements—Coal, Coke and Fly-Ash—Determination of Fluorine Content—Phrohydrolysis Method Amdt 1 October 1990. 11 pp.

—Iron Ores—Electrolytic Analysis
BSI BS 7020: Part 19-88. 1988 Analysis of Iron Ores Part 19: Method for the Determination of Fluorine Content: Ion-Selective Electrode Method. 12 pp.
ISO 4694-87. Iron Ores—Determination of Fluorine Content—Ion-Selective Electrode Method First Edition. 11 pp.

—Phosphoric Acids—Photometry
BSI BS 4258: Part 4-75. 1975 Methods of Test for Phosphoric Acid (Orthophosphoric Acid) for Industrial Use Part 4: Determination of Fluorine Content. 8 pp.
ISO 3360-76. Phosphoric Acid and Sodium Phosphates for Industrial Use (Including Foodstuffs)—Determination of Fluorine Content—Alizarin Complexone and Lanthanum Nitrate Photometric Method First Edition. 7 pp.

—Potassium Cryolite
MOD UK M 828/68. Examination of Potassium Cryolite.

—Sodium Fluorides—Volumetric Analysis
ISO 2833-73. Sodium Fluoride for Industrial Use—Determination of Fluorine Content—Modified Willard-Winter Method First Edition. 7 pp.

—Sodium Fluorides—Willard Winter Method
BSI BS 5072: Part 3-75. 1975 Amd 1 Methods of Test for Sodium Fluoride for Industrial Use Part 3: Determination of Fluorine Content. 11 pp.

Fluorine Content Analysis *(Cont.)*

—Sodium Phosphates—Photometry
ISO 3360-76. Phosphoric Acid and Sodium Phosphates for Industrial Use (Including Foodstuffs)—Determination of Fluorine Content—Alizarin Complexone and Lanthanum Nitrate Photometric Method First Edition. 7 pp.

—Sodium Pyrophosphate—Photometry
BSI BS 4427: Part 9-75. (WITHDRAWN) 1975 Methods of Test for Sodium Tripolyphos-phate (Pentasodium Triphosphate) and Sodium Pyrophosphate (Tetrasodium Pyrophos-phate) for Industrial Use Part 9: Determination of Fluorine Content. 8 pp.

—Sodium Tripolyphosphate—Photometry
BSI BS 4427: Part 9-75. (WITHDRAWN) 1975 Methods of Test for Sodium Tripolyphos-phate (Pentasodium Triphosphate) and Sodium Pyrophosphate (Tetrasodium Pyrophos-phate) for Industrial Use Part 9: Determination of Fluorine Content. 8 pp.

—Tantalum
JIS H 1698-76. Method for Determination of Fluorine in Tantalum (R 1986). 8 pp.

—Toothpastes
EC 83/514/EEC-83. Third Commission Directive on the Approximation of the Laws of the Member States Relating to Methods of Analysis Necessary for Checking the Composition of Cosmetic Products. 38 pp.

—Uranium Dioxide—Ion Selective Electrodes
ISO 9892-92. Uranium Metal, Uranium Dioxide Powder and Pellets, and Uranyl Nitrate Solutions—Determination of Fluorine Content—Fluoride Ion Selective Electrode Method First Edition. 8 pp.

—Uranium—Ion Selective Electrodes
ISO 9892-92. Uranium Metal, Uranium Dioxide Powder and Pellets, and Uranyl Nitrate Solutions—Determination of Fluorine Content—Fluoride Ion Selective Electrode Method First Edition. 8 pp.

—Uranyl Nitrate—Ion Selective Electrodes
ISO 9892-92. Uranium Metal, Uranium Dioxide Powder and Pellets, and Uranyl Nitrate Solutions—Determination of Fluorine Content—Fluoride Ion Selective Electrode Method First Edition. 8 pp.

Fluorite
Use: Fluorspar

1-Fluoro-2,4-Dinitrobenzene
Use For: 2,4-Dinitrofluorobenzene
JIS K 8479-62. 2,4-Dinitrofluorobenzene.

Fluoroborates
BSI BS 2657-74. 1974 Fluoroboric Acid and Metal Fluoroborates for Electroplating. 29 pp.

Fluoroboric Acid
Use: Fluoboric Acid

Fluorocarbon Rubber
Scope Note: For additional listings, see also specific products made from fluorocarbon rubber
See Also: Rubber

—Aerospace
CEN PREN2795-89. Flurocarbon—Rubber (FPM) (Low Compression Set) Hardness 50 IRHD. 5 pp.
CEN PREN2796-89. Fluorocarbon—Rubber (FPM) (Low Compression Set) Hardness 60 IRHD. 5 pp.
CEN PREN2797-89. Fluorocarbon—Rubber (FPM) (Low Compression Set) Hardness 70 IRHD. 5 pp.
CEN PREN2798-89. Fluorocarbon—Rubber (FPM) (Low Compression Set) Hardness 80 IRHD. 5 pp.
CEN PREN2799-89. Fluorocarbon—Rubber (FPM) (Low Compression Set) Hardness 90 IRHD. 5 pp.
MOD UK DTD-5543B-01. Fluorocarbon Rubbers; Amendment 1. 6 pp.
MOD UK DTD-5612A-82. Fluoro-Carbon Rubber (Low Compression Set) (Incorporating Amendments 1 and 2). 6 pp.

—Aircraft
SBAC TS 77 ISSUE 2. Recommended Technical Requirements for a General Purpose Grade of Fluorocarbon Rubber.

Fluorocarbons
See Also: Chlorofluorocarbons; Hydrocarbons
CNS K1242-80. Flurocarbons (May)(5588). 1 p.
CNS K6485-80. Method of Test for Fluorocarbons (May)(5589).

Fluorocarbons *(Cont.)*

—Gas Cylinders
JIS B 8235-90. Refillable Welded Steel Gas Cylinders for Liquefied Fluoro Carbon. 26 pp.

—Machined Parts—Tolerances
SAA AS 1794-75. Commercial Tolerances on Finished Parts of Fluorocarbon Reconfirmed 1986. 8 pp.

—Residue Content—Evaporation
BSI BS 5598: Part 8-80. 1980 Methods of Sampling and Test for Halogenated Hydrocarbons Part 8: Determination of Non-Volatile Residue in Fluorinated Hydrocarbons. 4 pp.
ISO 5789-79. Fluorinated Hydrocarbons for Industrial Use—Determination of Non-Volatile Residue First Edition. 4 pp.

Fluorochlorinated Hydrocarbons
Use: Chlorofluorocarbons

Fluorometric Analysis
See Also: Chemical Analysis
JIS K 0120-86. General Rules for Fluorometric Analysis. 16 pp.

—Petroleum Products
BSI BS 2000: Part 156-93. 1993 Methods of Test for Petroleum and Its Products Part 156: Determination of Hydrocarbon Types—Fluorescent Indicator Adsorption Method (W). 6 pp.
BSI BS 2000: Part 156-85. 1985 Petroleum and Its Products Part 156: Hydrocarbon Types in Liquid Petroleum Products by Fluorescent Indicator Absorption. 8 pp.
CNS K6368-73. Method of Test for Hydrocarbon Types in Liquid Petroleum Products by Fluorescent Indicator Absorption (Nov)(3577).
JIS K 2536-80. Testing Method for Hydrocarbon Types in Petroleum Products by Fluorescent Indicator Adsorption (R 1991). 36 pp.

—Water
CNS J2019-81. Method of Test for Microquantities of Uranium in Water by Fluorometry (Jun)(7490).
CNS K9042-80. Method of Test for Aluminum in Water (Fluorometric Method) (Jul)(5861).

Fluoropolymer Paints
Use: Fluoropolymer Resin Paints

Fluoropolymer Resin Paints
See Also: Paints

—Exterior
JIS K 5658-92. Fluoro Resin Paint for Architecture. 15 pp.

—Steel
JIS K 5659-92. Fluoro Resin Paint for Steel Structures. 18 pp.

Fluorosilicic Acid Content Analysis

—Anhydrous Hydrogen Fluoride—Photometry
BSI BS 5365: Part 4-79. (WITHDRAWN) 1979 Methods of Sampling and Test for Anhydrous Hydrogen Fluoride for Industrial Use Part 4: Determination of Hexafluorosilicic Acid Content. 7 pp.

—Hydrofluoric Acid—Volumetric Analysis
BSI BS 5366-76. (WITHDRAWN) 1976 Amd 1 Methods of Sampling and Test for Aqueous Hydrofluoric Acid for Industrial Use (ISO 3139; 1976) (AMD 3784) October 30, 1981. 8 pp.

Fluorspar
Use For: Fluorite *See Also:* Fluorine; Hydrofluoric Acid; Minerals; Ores
CNS K6014-72. Method of Analysis for Fluorite (Jan)(189).
CNS M1012-72. Fluorspar (Oct)(2643).

—Antimony Content—Atomic Absorption Spectrometry
ISO 9504-93. Metallurgical-Grade Fluorspar—Determination of Antimony Content—Solvent Extraction Atomic Absorption Spectrometric Method Second Edition. 10 pp.

—Arsenic Content—Spectrometry
ISO 9505-92. All Grades of Fluorspar—Determination of Arsenic Content—Silver Diethyldithiocarbamate Spectrometric Method First Edition. 9 pp.

—Barium Sulfates Content—Gravimetric Analysis
BSI BS 5659: Part 8-89. (WITHDRAWN) 1989 Acid-Grade Fluorspar Part 8: Determination of Barium Sulphate Content. 6 pp.

Fluorspar (Cont.)

—Barium Sulfates Content—Gravimetric Analysis (Cont.)
ISO 5437-92. Acid-Grade and Ceramic-Grade Fluorspar—Determination of Barium Sulfate Content—Gravimetric Method Second Edition. 6 pp.

—Carbonate Content—Volumetric Analysis
BSI BS 5659: Part 4-79. (WITHDRAWN) 1979 Acid-Grade Fluorspar Part 4: Determination of Carbonate Content. 8 pp.
ISO 4283-93. Grades of Fluorspar—Determination of Carbonate Content—Titrimetric Method Third Edition. 8 pp.

—Chemical Analysis
JIS K 1468-78. Methods of Chemical Analysis for Fluorspar. 18 pp.
JIS M 8514-89. Methods for Chemical Analysis of Metallurgical Grade Fluorspar. 34 pp.

—Flotation Agent Content—Gravimetric Analysis
BSI BS 5659: Part 1-79. (WITHDRAWN) 1979 Acid-Grade Fluorspar Part 1: Determination of Flotation Agents. 4 pp.
ISO 3703-93. Acid-Grade and Ceramic-Grade Fluorspar—Determination of Flotation Agents Second Edition. 5 pp.

—Fluorine Content—Potentiometric Analysis
BSI BS 5659: Part 5-79. (WITHDRAWN) 1979 Acid-Grade Fluorspar Part 5: Determination of Available Fluorine Content. 8 pp.
ISO 5439-78. Acid-Grade Fluorspar—Determination of Available Fluorine Content—Potentiometric Method After Distillation First Edition. 8 pp.

—Fluorine Content—Willard-Winter Method
ISO 9503-91. Metallurgical-Grade Fluorspar—Determination of Available Fluorine Content—Modified Willard-Winter Method First Edition. 8 pp.

—Iron Content—Spectrometry
BSI BS 5659: Part 9-89. (WITHDRAWN) 1989 Acid-Grade Fluorspar Part 9: Determination of Iron Content. 7 pp.
ISO 9061-93. Acid-Grade and Ceramic-Grade Fluorspar—Determination of Iron Content—1,10-Phenanthroline Spectrometric Method Second Edition. 7 pp.

—Lead Content—Atomic Absorption Spectrometry
ISO 9779-93. Metallurgical-Grade Fluorspar—Determination of Lead Content—Solvent Extraction Atomic Absorption Spectrometric Method Second Edition. 9 pp.

—Loss of Mass
BSI BS 5659: Part 2-79. (WITHDRAWN) 1979 Acid-Grade Fluorspar Part 2: Determination of Loss in Mass at 105 Degrees C. 4 pp.
ISO 4282-92. Acid-Grade and Ceramic-Grade Fluorspar—Determination of Loss in Mass at 105 Degrees C Second Edition. 5 pp.

—Manganese Content—Spectrometry
ISO 9062-92. Acid-Grade and Ceramic-Grade Fluorspar—Determination of Manganese Content—Periodate Spectrometric Method Second Edition. 7 pp.

—Moisture Content
ISO 8875-92. Grades of Fluorspar—Determination of Moisture Content of a Lot Second Edition. 6 pp.

—Particle Size Distribution—Sieve Analysis
ISO 8876-89. Fluorspar—Determination of Particle Size Distribution by Sieving First Edition; (Corrigendum 1-1991). 15 pp.

—Phosphorus Content—Photometry
BSI BS 5659: Part 6-90. (WITHDRAWN) 1990 Acid-Grade Fluorspar Part 6: Determination of Total Phosphorus Content. 8 pp.
ISO 6676-93. Acid-Grade and Ceramic-Grade Fluorspar—Determination of Total Phosphorus Content—Reduced-Molybdophosphate Spectrometric Method Third Edition. 7 pp.

—Phosphorus Content—Spectrometry
ISO 9438-93. Metallurgical-Grade Fluorspar—Determination of Total Phosphorus Content—Reduced-Molybdophosphate Spectrometric Method Second Edition. 9 pp.

—Sampling
ISO 8868-89. Fluorspar—Sampling and Sample Preparation First Edition. 20 pp.
ISO 9497-93. Fluorspar—Experimental Methods for Evaluation of Quality Variation First Edition. 15 pp.

—Sampling—Bias
ISO 9498-93. Fluorspar—Experimental Methods for Checking the Bias of Sampling and Sample Preparation First Edition. 12 pp.

—Silica Content—Spectrometry
BSI BS 5659: Part 7-87. (WITHDRAWN) 1987 Acid-Grade Fluorspar Part 7: Determination of Silica Content. 8 pp.
ISO 5438-93. Acid-Grade and Ceramic-Grade Fluorspar—Determination of Silica Content—Reduced-Molybdosilicate Spectrometric Method Second Edition. 8 pp.
ISO 9502-93. Metallurgical-Grade Fluorspar—Determination of Silica Content—Reduced-Molybdosilicate Spectrometric Method Second Edition. 9 pp.

—Sulfide Content—Iodometry
BSI BS 5659: Part 3-89. (WITHDRAWN) 1989 Acid-Grade Fluorspar Part 3: Determination of Sulphide Content. 8 pp.
ISO 4284-93. Acid-Grade and Ceramic-Grade Fluorspar—Determination of Sulfide Content—Iodometric Method Third Edition. 7 pp.

—Sulfur Content—Iodometry
ISO 9501-91. Metallurgical-Grade Fluorspar—Determination of Total Sulfur Content—Iodometric Method After Combustion First Edition. 9 pp.

—Test Specimens
ISO 9498-93. Fluorspar—Experimental Methods for Checking the Bias of Sampling and Sample Preparation First Edition. 12 pp.

Fluosilicic Acid
Use For: Hexafluorosilicic Acid; Hydrofluosilicic Acid
See Also: Fluosilicic Acid Content Analysis
CNS K7248-67. Chemical Reagent (Hydrofluosilicic Acid) (Oct)(1747).

—Fluorine Content—Potentiometric Analysis
BSI BS 6445: Part 1-83. (WITHDRAWN) 1983 Hexafluorosilicic Acid for Industrial Use Part 1: Method for Determination of Available Fluorine Content (Potentiometric Method After Distillation). 8 pp.
ISO 6677-83. Hexafluorosilicic Acid for Industrial Use—Determination of Available Fluorine Content—Potentiometric Method After Distillation First Edition. 7 pp.

Fluosilicic Acid Content Analysis
Use For: Hexafluorosilicic Acid Content Analysis
See Also: Fluosilicic Acid

—Anhydrous Hydrogen Fluoride—Photometry
ISO 3701-76. Anhydrous Hydrogen Fluoride for Industrial Use—Determination of Hexafluorosilicic Acid Content—Reduced Molybdosilicate Photometric Method First Edition. 6 pp.

Flush Bolts
Use: Flat Head Bolts

Flush Device Cover Plates
Use: Wall Plates

Flush Head Screws
Use: Flat Head Screws

Flush Plates
Use: Wall Plates

Flush Tanks
See Also: Tanks (Containers)

—Toilets
BSI BS 1125-87. 1987 WC Flushing Cisterns (Including Dual Flush Cisterns and Flush Pipes). 15 pp.
BSI BS 5503: Part 3-90. 1990 Vitreous China Washdown WC Pans with Horizontal Outlet Part 3: WC Pans with Horizontal Outlet for Use with 7.5L Maximum Flush Capacity Cisterns. 16 pp.
BSI BS 5504: Part 4-90. 1990 Wall Hung WC Pan Part 4: Specificaton for Wall Hung WC Pans for Use with 7.5L Maximum Flush Capacity Cisterns. 13 pp.
BSI BS 7357-90. 1990 7.5 LWC Flushing Cisterns. 22 pp.
DIN ENGL EN 34-92. Wall-Hung W.C. Pans with Close-Coupled Cistern; Connecting Dimensions (July). 4 pp.
SAA AS 1218-90. Flushing Cisterns (Superseded by AS 1172.2—1993).
SNZ NZS 245-62. Specification for Water Closet Flushing Cisterns and Flush Pipes. 12 pp.

—Urinals
BSI BS 1876-90. 1990 Automatic Flushing Cisterns for Urinals. 8 pp.
BSI BS 1876-72. (WITHDRAWN) 1972 Automatic Flushing Cisterns for Urinals. 12 pp.

Flush Valves
See Also: Plumbing Equipment; Toilet Facilities; Valves

—Toilets
JIS A 5521-75. Flush Valves for Closet.
SNZ NZS 245-62. Specification for Water Closet Flushing Cisterns and Flush Pipes. 12 pp.

—Urinals
SNZ NZS 2038-66. Stainless Steel Urinals and Flushing Apparatus (Reconfirmed 1976). 7 pp.

Flushplates
Use: Wall Plates

Flutter
See Also: Vibration

—Aerostats
CAA Chapter Q3-9 12.79. Structrual Deformation, Flutter and Vibration (Non-Rigid Airships). 1 p.

—Aircraft
CAA 820 (K). Section K Light Aeroplanes Flutter Fail Safe Criteria (Blue Papers). 11 pp.
CAA Chapter K3-9 10.92. Flutter Prevention and Structural Stiffness (Light Aeroplanes). 4 pp.
CAA Chapter K3-9 App 10.69. Flutter Prevention and Structural Stiffness (Light Aeroplanes). 6 pp.
CAA Chapter P3-9. Flutter Preventions and Structural Stiffness of Aerodynamic Surfaces (Provisional Airworthiness Requirements for Civil Powered Lift Aircraft). 3 pp.

—Phonograph Test Records
DIN ENGL 45545-66. Wow and Flutter Test Records; 33 1/3 and 45 U/Min (R.P.M.) (Dec). 2 pp.

—Rotary Wing Aircraft
CAA Chapter G3-9 11.75. Flutter Prevention and Stiffness (Rotocraft). 1 p.
CAA Chapter G3-9 App 01.54. Flutter Prevention and Stiffness (Rotocraft). 1 p.
CAA Chapter G3-13 11.75. Additional Design Cases (Rotocraft). 1 p.

—Supersonic Transports
CAA STANDARD NO. 4-7 08.69. Flutter and Structural Distortion. 4 pp.

Flux Cored Arc Welding Equipment
See Also: Arc Welding Equipment

—Electrodes
JIS Z 3313-87. Arc Welding Flux Cored Wires for Mild Steel and High Strength Steel. 13 pp.
JIS Z 3318-91. MAG Welding Flux Cored Wires for Molybdenum Steel and Chromium Molybdenum Steel. 11 pp.
JIS Z 3319-91. Flux Cored Wires for Electrogas Arc Welding. 10 pp.
JIS Z 3320-87. Flux Cored Wires for CO2 Gas Shielded Arc Welding of Atmospheric Corrosion Resisting Steel. 12 pp.

—Electrodes—Hard Surfacing
JIS Z 3326-91. Arc Welding Flux Cored Wires for Hardfacing. 10 pp.

—Electrodes—Stainless Steel
JIS Z 3323-89. Stainless Steel Flux Cored Wires. 16 pp.

—Electrodes—Steel
CSA W48.5-M1990. Carbon Steel Electrodes for Flux- and Metal-Cored Arc Welding; (Gen Instr 1). 52 pp.

Flux Density
Scope Note: For additional references, use a more specific term **See Also:** Density; Dosimetry; Illuminance; Irradiation; Neutron Flux Density; Optical Density; Power Density; Sound Pressure

—Power—Earth Exploration Satellites
CCIR Report 692-2-86. Preferred Frequency Bands and Power Flux-Density Considerations for Earth Exploration Satellites—Section 2F—Earth Exploration Satellites. 6 pp.

Flux Density (Cont.)

—Power—Earth Surface—Fixed Satellite Services—Radio Relay Systems
CCIR RECMN 358-3-82. Maximum Permissible Values of Power Flux-Density at the Surface of the Earth Produced by Satellites in the Fixed-Satellite Service Using the Same Frequency Bands Above 1 GHz as Line-of-Sight Radio-Relay Systems—Section 4/9A—Sharing Conditions. 2 pp.

—Power—Fixed Satellite Services—Frequency Band Sharing
CCIR RECMN 674-90. Power Flux-Density Values to Facilitate the Application of Article 14 for FSS in Relation to the Fixed-Satellite Service in the 11.7-12.2 GHz Band in Region 2—Section 4/9A—Sharing Conditions. 1 p.

—Power—Fixed Services (Radio Communication)—Frequency Band Sharing
CCIR RECMN 674-90. Power Flux-Density Values to Facilitate the Application of Article 14 for FSS in Relation to the Fixed-Satellite Service in the 11.7-12.2 GHz Band in Region 2—Section 4/9A—Sharing Conditions. 1 p.

—Power—Meteorological Satellites—Small Earth Terminals
CCIR Report 851-1-90. Power Flux-Density Limitation in the Band 1 670–1 710 MHz for Dissemination of Meteorological Information to Small Earth Terminals—Section 2F—Earth Exploration Satellites. 4 pp.

—Power—Radio Navigation Equipment
CCIR RECMN 496-2-86. Limits of Power Flux-Density of Radionavigation Transmitters to Protect Space Station Receivers in the Fixed-Satellite Service in the 14 GHz Band—Section 8D—Radiodetermination, Global Maritime Distress and Safety System and Related Subjects. 1 p.

CCIR RECMN 496-3-92. Limits of Power Flux-Density of Radionavigation Transmitters to Protect Space Station Receivers in the Fixed-Satellite Service in the 14 GHz Band—Section 8D—Radiodetermination, Global Maritime Distress and Safety System and Related Subjects. 2 pp.

Flux Oil
See Also: Petroleum Products

—Loss of Mass—High Temperature Testing
BSI BS 2000: Part 45-93. 1993 Methods of Test for Petroleum and Its Products Part 45: Determination of Loss on Heating of Bitumen and Flux Oil. 4 pp.
BSI BS 2000: Part 45-82. 1982 Petroleum and Its Products Part 45: Loss on Heating of Bitumen and Flux Oil. 4 pp.

Fluxed Bitumeus
Use: Bitumens

Fluxes
Scope Note: Use a more specific term *See:* Brazing Fluxes; Flux Oil; Soldering Fluxes; Welding Fluxes

Fly Ash
Use For: Pulverized Fuel Ash
See Also: Admixtures; Air Pollution; Ashes; Flue Dust; Pozzolans

JIS A 6201-91. Fly Ash. 9 pp.

—Boron Content—Spectrophotometry
SAA AS 1038.10.3-88. Methods for the Analysis and Testing of Coal and Coke—Part 10.3: Determination of Trace Elements—Coal, Coke and Fly-Ash—Determination of Boron Content—Spectrophotometric Method. 5 pp.

—Calcium Oxides—Reference Method
CEN PREN 451 (Part 1)-91. Method of Testing Fly Ash—Part 1 Determination of Free Calcium Oxide. 5 pp.

—Concretes
SAA AS 1129-72. Fly Ash for Use in Concrete Withdrawn.

—Concretes—Physical Properties
CEN PREN 450-91. Fly Ash for Concrete—Definitions, Requirements and Quality Control. 7 pp.

—Concretes—Quality Assurance
CEN PREN 450-91. Fly Ash for Concrete—Definitions, Requirements and Quality Control. 7 pp.

Fly Ash (Cont.)

—Fluorine Content—Pyrohydrolysis
SAA AS 1038.10.4-89. Methods for the Analysis and Testing of Coal and Coke—Part 10.4: Determinaton of Trace Elements—Coal, Coke and Fly-Ash—Determination of Fluorine Content—Phrohydrolysis Method Amdt 1 October 1990. 11 pp.

—Portland Cements
BSI BS 3892: Part 1-93. 1993 Pulverized-Fuel Ash Part 1: Specification for Pulverized-Fuel Ash for Use with Portland Cement (R). 22 pp.
BSI BS 3892: Part 1-82. 1982 Amd 1 Pulverized-Fuel Ash Part 1: Pulverized-Fuel Ash for Use as a Cementitious Component in Structural Concrete. 14 pp.
BSI BS 3892: Part 2-84. 1984 Pulverized Fuel Ash Part 2: Pulverized Fuel-Ash for Use in Grouts and for Miscellaneous Uses in Concrete. 8 pp.
CNS A2040-84. Fly Ash and Raw or Calcined Natural Pozzolan for Use as a Mineral Admixture in Portland Cement Concrete (Jun)(3036). 5 pp.
CNS A3207-84. Method of Test for Fly Ash or Natural Pozzolans for Use as a Mineral Admixture in Portland Cement Concrete (Jun)(10896). 5 pp.
CNS R2181-85. Portland Fly Ash Cement (May)(11270).
CNS R2182-85. Fly Ash for Use in Portland Fly Ash Cement (May)(11271).
CSA CAN/CSA-A23.5-M86. Supplementary Cementing Materials (R 1992); (Gen Instr 1 Thru 2). 33 pp.
JIS R 5213-92. Portland Fly-Ash Cement. 13 pp.
SAA AS 3582.1-91. Supplementary Cementitious Materials for Use with Portland Cement—Part 1: Fly Ash (In Professional Packages 30,58). 5 pp.

—Pozzolanic Cements
BSI BS 6610-91. 1991 Pozzolanic Pulverized-Fuel Ash Cement. 18 pp.
BSI BS 6610-85. 1985 Pozzolanic Cement with Pulverized-Fuel Ash as Pozzolana. 8 pp.

—Trace Element Content—Atomic Absorption Spectrometry
SAA AS 1038.10.1-86. Methods for the Analysis and Testing of Coal and Coke—Part 10.1: Determination of Trace Elements—Determination of Eleven Trace Elements in Coal, Coke and Fly-Ash—Flame Atomic Absorption Spectrometric Method (R 1992). 10 pp.

Fly Ash Content Analysis

—Cements
CEN ENV 196 (Part 4)-89. Methods of Testing Cement: Quantitative Determination of Constituents. 31 pp.

—Portland Cements
CNS A3301-88. Method for Determination of Granulated (Water Quenched) Blast-Furnace Slag, Silica Material, Fly Ash and Limestone Content in Portland Cement (Nov)(12459).

Flywheels
See Also: Engines; Wheels

—Housings—Internal Combustion Engines
BSI BS 3529-62. 1962 Dimensions of Flywheel Housing Mounting Pads and Brackets for I.C. Engines. 12 pp.

—Housings—Internal Combustion Piston Engines
BSI BS AU 226-88. 1988 Nominal Dimensions and Tolerances of Flywheel Housings for Reciprocating Internal Combustion Engine. 11 pp.
ISO 7648-87. Flywheel Housings for Reciprocating Internal Combustion Engines—Nominal Dimensions and Tolerances First Edition. 11 pp.

FM Broadcast Transmitters
Use: Broadcast Transmitters—FM

FMEA
Use: Failure (Quality Control)

Foam Concentrates
Use: Blowing Agents

Foam Extinguishing Agents
Use: Fire Extinguishing Agents—Foam

Foam Insulation
See Also: Thermal Insulation

—Isocyanurate—Construction Sites
BSI BS 5241: Part 1-89. 1989 Rigid Polyurethane (PUR) and Polyisocyanurate (PIR) Foam When Dispensed or Sprayed on a Construction Site Part 1: Sprayed Foam Thermal Insulation Applied Externally. 8 pp.

Foam Insulation (Cont.)

—Isocyanurate—Roofs
BSI BS 7021-89. 1989 Code of Practice for Thermal Insulation of Roofs Externally by Means of Sprayed Rigid Polyurethane (PUR) or Polyisocyanurate (PIR) Foam. 18 pp.

—Polyethylene
CNS A2165-88. Polyethylene Foam for Heat Insulating (May)(10487). 12 pp.
JIS A 9515-89. Polyethylene Foam for Thermal Insulation. 13 pp.

—Polyisocyanurate
BSI BS 5241: Part 1-89. 1989 Rigid Polyurethane (PUR) and Polyisocyanurate (PIR) Foam When Dispensed or Sprayed on a Construction Site Part 1: Sprayed Foam Thermal Insulation Applied Externally. 8 pp.
BSI BS 5241: Part 2-91. 1991 Rigid Polyurethane (PUR) and Polyisocyanurate (PIR) Foam When Dispensed or Sprayed on a Construction Site Part 2: Specification for Dispensed Foam for Thermal Insulation or Buoyancy Applications. 10 pp.
BSI BS 5608-86. 1986 Preformed Rigid Polyurethane (PUR) and Polyisocyanurate (PIR) Foams for Thermal Insulation of Pipework and Equipment. 8 pp.

—Polystyrene
JIS A 9511-89. Thermal Insulation Material Made of Polystyrene Foam. 19 pp.

—Polyurethane
BSI BS 4840: Part 1-85. 1985 Rigid Polyurethane (PUR) Foam in Slab Form Part 1: PUR Foam for Use in Transport Containers and Insulated Vehicle Bodies. 10 pp.
BSI BS 4840: Part 2-85. 1985 Rigid Polyurethane (PUR) Foam in Slab Form Part 2: PUR Foam for Use in Refrigerator Cabinets, Cold Rooms and Stores. 11 pp.
BSI BS 5241: Part 1-89. 1989 Rigid Polyurethane (PUR) and Polyisocyanurate (PIR) Foam When Dispensed or Sprayed on a Construction Site Part 1: Sprayed Foam Thermal Insulation Applied Externally. 8 pp.
BSI BS 5241: Part 2-91. 1991 Rigid Polyurethane (PUR) and Polyisocyanurate (PIR) Foam When Dispensed or Sprayed on a Construction Site Part 2: Specification for Dispensed Foam for Thermal Insulation or Buoyancy Applications. 10 pp.
BSI BS 6586: Part 1-93. 1993 Rigid Polyurethane (PUR) Foam Produced by the Press Injection Method Part 1: Specification for PUR Foam for Insulated Panels for Transport Containers and Insulated Vehicle Bodies. 12 pp.
BSI BS 6586: Part 1-85. 1985 Rigid Polyurethane (PUR) Foam Produced by the Press Injection Method Part 1: PUR Foam for Insulated Panels for Transport Containers and Insulated Vehicle Bodies. 11 pp.
DIN ENGL 18159 Pt 1-78. Cellular Plastics as In-Situ Cellular Plastics in Building; In-Situ Polyurethane (PUR) Foam for Thermal and Cold Insulation; Application, Properties, Execution, Testing (June). 5 pp.
ISO 8873-87. Cellular Plastics, Rigid—Spray-Applied Polyurethane Foam for Thermal Insulation of Buildings—Specification First Edition. 8 pp.
JIS A 9526-89. Spray-Applied Rigid Polyurethane Foam for Thermal Insulation. 13 pp.

—Polyurethane—Walls
BSI BS 7456-91. 1991 Code of Practice for Stabilization and Thermal Insulation of Cavity Walls (with Masonry or Concrete Inner and Outer Leaves) by Filling with Polyurethane (PUR) Foam Systems. 17 pp.
BSI BS 7457-91. 1991 Polyurethane (PUR) Foam Systems Suitable for Stabilization and Thermal Insulation of Cavity Walls with Masonry or Concrete Inner and Outer Leaves. 12 pp.

—Urea Formaldehyde
DIN ENGL 18159 Pt 2-78. Cellular Plastics as In-Situ Foam in Building; In-Situ Foam Produced from Urea-Formaldehyde (UF) Resin for Thermal Insulation, Application, Properties, Execution, Testing (June). 5 pp.

—Urea Formaldehyde—Walls
BSI BS 5617-85. 1985 Urea-Formaldehyde (UF) Foam Systems Suitable for Thermal Insulation of Cavity Walls with Masonry or Concrete Inner and Outer Leaves. 16 pp.
BSI BS 5618-85. 1985 Amd 1 Thermal Insulation of Cavity Walls (with Masonry or Concrete Inner and Outer Leaves) by Filling with Urea-Formaldehyde (UF) Foam Systems (AMD 6262) March 30, 1990. 59 pp.

INTERNATIONAL AND NON-U.S. NATIONAL STANDARDS
SUBJECT INDEX

Foam Insulation (Cont.)
—Urea Formaldehyde—Walls (Cont.)
BSI BS 5618-02. 1985 Amd 2 Thermal Insulation of Cavity Walls (with Masonry or Concrete Inner and Outer Leaves) by Filling with Urea-Formaldehyde (UF) Foam Systems (AMD 7114) May 1, 1992. 62 pp.

SNZ NZS 4235-88. Specification for Urea-Formaldehyde (UF) Foam Systems Suitable for Thermal Insulation Within the Cavity in Walls of Buildings Amend: A, 1988. 16 pp.

—Urethane
BSI BS 5241-75. (WITHDRAWN) 1975 Rigid Urethane Foam When Dispensed or Sprayed on a Construction Site (Superseded by BS 5241: Part 1: 1989 + BS 5241: Part 2: 1991). 10 pp.

CNS A2108-81. Rigid Foam Urethane Heat Insulating Materials (Aug)(7774).

JIS A 9514-89. Thermal Insulation Material Made of Rigid Urethane Foam. 15 pp.

Foam Laminated Fabrics
See Also: Fabrics; Laminated Fabrics

—Polyurethane
BSI BS 4021-71. 1971 Flexible Polyurethane Foam Sheeting for Use in Laminates. 8 pp.

ISO 6915-91. Flexible Cellular Polymeric Materials—Polyurethane Foam for Laminate Use—Specification Second Edition. 9 pp.

JIS L 1066-63. Methods of Testing for Polyurethane Foam Laminated Fabrics (R 1985). 7 pp.

Foam Rubber
Scope Note: For additional listings, see also specific products made from foam rubber *Use For:* Cellular Rubber; Foamed Rubber; Sponge Rubber
See Also: Blowing Agents; Cellular Materials; Cellular Plastics; Polyurethane Resins; Rubber; Sponges (Materials); Vulcanized Rubber

ISO 1923-81. Cellular Plastics and Rubbers—Determination of Linear Dimensions Second Edition. 5 pp.

—Aging Testing
ISO 2440-83. Polymeric Materials, Cellular Flexible—Accelerated Ageing Tests Second Edition. 4 pp.

—Bulk Density
CNS K6667-81. Determination of Apparent Density of Cellular Rubbers and Plastics (May)(7407).

ISO 845-88. Cellular Plastics and Rubbers—Determination of Apparent (Bulk) Density Second Edition. 6 pp.

—Compression Testing
ISO 1856-80. Polymeric Materials, Cellular Flexible—Determination of Compression Set Second Edition; (Erratum—Nov 1981). 5 pp.

ISO 3386 Pt 1-86. Polymeric Materials, Cellular Flexible—Determination of Stress-Strain Characteristic in Compression—Part 1: Low-Density Materials Second Edition. 5 pp.

ISO 3386 Pt 2-84. Polymeric Materials, Cellular Flexible—Determination of Stress-Strain Characteristic in Compression—Part 2: High Density Materials First Edition. 4 pp.

—Fatigue Testing
BSI BS 4443: Part 5-80. 1980 Methods of Test for Flexible Cellular Materials Part 5: Method 13, Test for Dynamic Fatigue by Constant Load Pounding. 8 pp.

ISO 3385-89. Flexible Cellular Polymeric Materials—Determination of Fatigue by Constant-Load Pounding Third Edition. 7 pp.

—Fire Testing
BSI BS 4735-74. 1974 Laboratory Methods of Test for Assessment of the Horizontal Burning Characteristics of Specimens No Larger Than 150 mm X 50 mm X 13 mm (Nominal) of Cellular Plastics and Cellular Rubber Materials When Subjected to a Small Flame. 12 pp.

BSI BS 4735-01. 1974 Amd 1 Laboratory Method of Test for Assessment of the Horizontal Burning Char. of Specimens No Larger Than150 mm X 50 mm X 13 mm (Nominal) of Cellular Plastics and Cellular Rubber Materials When Subjected to a Small Flame (AMD 7687) July 15, 1993 (V). 13 pp.

—Flammability Testing
BSI BS 5111: Part 1-74. 1974 Laboratory Methods of Test for Determination of Smoke Generation Characteristics of Cellular Plastics and Cellular Rubber Materials Part 1: Meth of Testing a 25 mm Cube Test Spec. of Low Density Mat. (up to 130 kg/Metres Cubed) to. 13 pp.

Foam Rubber (Cont.)
—Flammability Testing (Cont.)
BSI BS 5111: Part 1-01. 1974 Amd 1 Laboratory Methods of Test for Determination of Smoke Generation Characteristics of Cellular Plastics and Cellular Rubber Materials Part 1: Method for Testing a 25 mm Cube Test Specimen of Low Density Material (up to 130 kg/m3) to. 14 pp.

ISO 3582-78. Cellular Plastic and Cellular Rubber Materials—Laboratory Assessment of Horizontal Burning Characteristics of Small Specimens Subjected to a Small Flame First Edition. 9 pp.

—Indentation Hardness Testing
ISO 2439-80. Polymeric Materials, Cellular Flexible—Determination of Hardness (Indentation Technique) Second Edition; (Erratum—Nov 1981). 6 pp.

Foamed Concretes
Use: Cellular Concretes

Foamed Glass
Use: Cellular Glass

Foamed Plastics
Use: Cellular Plastics

Foamed Rubber
Use: Foam Rubber

Foaming Power
See Also: Surface Properties

—Detergents
CGSB 2-GP-11M METH 27.1-89. Methods of Testing and Analysis of Soaps and Detergents Foaming Ability; (Amendment 3 August 1989). 2 pp.

—Engine Coolants
BSI BS 5117: Sec 1.4-85. 1985 Testing Corrosion Inhibiting, Engine Coolant Concentrate ('Antifreeze') Part 1: Methods of Test for Determination of Physical and Chemical Properties Section 1.4: Determination of Foaming Characteristics. 4 pp.

—Lubricating Oils
BSI BS 2000: Part 146-93. 1993 Methods of Test for Petroleum and Its Products Part 146: Determination of Foaming Characteristics of Lubricating Oils (W). 7 pp.

BSI BS 2000: Part 146-90. 1990 Petroleum and Its Products Part 146: Foaming Characteristics of Lubricating Oils. 11 pp.

CNS K6323-72. Method of Test for Foaming Characteristics of Lubricating Oils (Oct)(3384).

JIS K 2518-80. Petroleum Products—Lubricating Oils—Determination of Foaming Characteristics.

—Soaps
CGSB 2-GP-11M METH 27.1-89. Methods of Testing and Analysis of Soaps and Detergents Foaming Ability; (Amendment 3 August 1989). 2 pp.

—Surfactants
ISO 696-75. Surface Active Agents—Measurement of Foaming Power—Modified Ross-Miles Method First Edition. 7 pp.

Focal Length
See Also: Cameras

—Photographic Lenses
CNS Z9045-80. Focal Length Marking of Lenses (May)(5639).

CNS Z9051-80. General Rule of Measurement for Focal Length of Photographic Lenses (Aug)(6293).

CNS Z9052-80. Method of Measurement for Focal Length of Photographic Lenses Focal Length Comparing Method (Aug)(6294).

CNS Z9053-80. Method of Measurement for Focal Length of Photographic Lenses Distant Objective Aiming Method (Aug)(6295).

CNS Z9054-80. Method of Measurement for Focal Length of Photographic Lenses Nodal Slide Method (Aug)(6296).

JIS B 7094-78. Methods of Measurement for Focal Length of Photographic Lenses.

Focal Planes
—Photographic Lenses
JIS B 7096-78. Methods of Measurement for Vignetting Factor and Relative Focal Plane Illuminance (R 1983). 12 pp.

Focal Spots
—X-Ray Tubes—Industrial
BSI BS 6932-88. 1988 Method for Measurement of the Effective Focal Spot Size of Mini-Focus and Micro-Focus X-Ray Tubes Used for Industrial Radiography. 7 pp.

Focal Spots (Cont.)
—X-Ray Tubes—Medical
BSI BS 6530-84. 1984 Methods for Determining the Characteristics of Focal Spots in Diagnostic X-Ray Tube Assemblies for Medical Use. 36 pp.

BSI BS 6530-01. 1984 Amd 1 Methods for Determining the Characteristics of Focal Spots in Diagnostic X-Ray Tube Assemblies for Medical Use (IEC 336: 1982) (AMD 7002) February 28, 1992. 37 pp.

CENELEC HD 509 S1-88. Characteristics of Focal Spots in Diagnostic X-Ray Tube Assemblies for Medical Use. 3 pp.

IEC 336-93. X-Ray Tube Assemblies for Medical Diagnosis—Characteristics of Focal Spots Third Edition. 80 pp.

Fodders
Use: Forage

Foerderverein fuer Elektrotechnische Normung
Use: FEN

Fog Lamps
Use: Fog Lights

Fog Lights
See Also: Lighting; Motor Vehicle Lighting

—Automotive
EC 77/538/EEC-77. Council Directive on the Approximation of the Laws of the Member States Relating to Rear Fog Lamps for Motor Vehicles and Their Trailers. 11 pp.

EC 89/518/EEC-89. Commission Directive Adapting to Technical Progress Council Directive 77/538/EEC on the Approximation of the Laws of the Member States Relating to Rear Fog Lamps for Motor Vehicles and Their Trailers. 6 pp.

—Automotive—Incandescent
EC 76/762/EEC-76. Council Directive on the Approximation of the Laws of the Member States Relating to Front Fog Lamps for Motor Vehicles and Filament Lamps for Such Lamps. 13 pp.

—Trailers
EC 77/538/EEC-77. Council Directive on the Approximation of the Laws of the Member States Relating to Rear Fog Lamps for Motor Vehicles and Their Trailers. 11 pp.

EC 89/518/EEC-89. Commission Directive Adapting to Technical Progress Council Directive 77/538/EEC on the Approximation of the Laws of the Member States Relating to Rear Fog Lamps for Motor Vehicles and Their Trailers. 6 pp.

Foil Capacitors
Scope Note: See specific listings under Capacitors

Foil Laminates
Use: Foil Papers

Foil on Glass Intruder Systems
See Also: Alarm Systems; Burglar Alarms; Burglar Detectors

—Burglar
BSI BS 4737: Sec 3.2-77. 1977 Intruder Alarm Systems in Buildings Part 3: Components Section 3.2: Foil on Glass. 4 pp.

Foil Papers
Use For: Foil Laminates; Metal Foil/Paper Laminates; Paper/Metal Foil Laminates
See Also: Aluminum; Aluminum Foil; Foils; Laminated Papers; Laminates; Packaging Papers; Papers

—Aluminum
CNS H3024-81. Aluminium Foil, Laminated with Paper Lining (Aug) (2251).

CNS P2037-84. Aluminum Foil Laminating Paper (Oct)(2646). 2 pp.

CNS Z7016-84. Laminated Aluminum Foils (Aug)(2521).

JIS Z 1520-90. Laminated Aluminium Foils. 6 pp.

MOD UK DSTAN 81-75-87. Barrier Material, Aluminium Foil Laminate, Flexible, Heat Sealable, Water-Vapour Resistant Issue 1. 23 pp.

—Reflective
SAA AS 1903-76. Reflective Foil Laminate Amdt 1 November 1979 Bound with AS 1904.

SAA AS 1904-76. Code of Practice for Installation of Reflective Foil Laminate in Buildings Corrig. Bound with AS 1903.

Foil Resistors
Use: Metal Film Resistors

INDUSTRY STANDARDS

Foils
Scope Note: Use a more specific term
Use For: Metal Foils *See Also:* Aluminum Foil; Copper Foil; Foil Papers; Tantalum Foil
 CSA CAN3-G312.1-75. Preferred Metric Dimensions for Flat Metal Products (R 1992). 10 pp.

—**Detergents**
 MOD UK TS 10263. Metal-Foil/Plastics Laminate for Detergent Sachets.

Fokker F-27 Aircraft
Use: Fokker F27 Aircraft

Fokker F-28 Aircraft
Use: Fokker F28 Aircraft

Fokker F-28 2000 Aircraft
Use: Fokker F28 2000 Aircraft

Fokker F-28 4000 Aircraft
Use: Fokker F28 4000 Aircraft

Fokker F27 Aircraft
Use For: Fokker F-27 Aircraft *See Also:* Aircraft

—**Accidents**
 CAA. Fokker/Fairchild F-27 Series (World Airline Accident Summary). 2 pp.
 CAA. Fokker/Fairchild F-27 Series. 2 pp.

—**Antenna Positions**
 CAA. Fokker F.27 Series 200/500 (Approved Aerial Positions). 1 p.

—**Certification**
 CAA. Fokker F.27 (Friendship). 10 pp.

—**Foreign Airworthiness Directives**
 CAA. Fokker F27 Series Aircraft (Foreign Airworthiness Directives). 15 pp.

Fokker F28 Aircraft
Use For: Fokker F-28 Aircraft *See Also:* Aircraft

—**Accidents**
 CAA. Fokker F-28 Series (World Airline Accident Summary). 1 p.
 CAA. Fokker F-28 Series. 1 p.

—**Certification**
 CAA. Fokker F28 Mark 0100. 14 pp.

—**Foreign Airworthiness Directives**
 CAA. Fokker F28-0100 Series Aircraft. 15 pp.

Fokker F28 2000 Aircraft
Use For: Fokker F-28 2000 Aircraft
See Also: Aircraft

—**Antenna Positions**
 CAA. Fokker F.28 2000 and 4000 (Approved Aerial Positions). 1 p.

Fokker F28 4000 Aircraft
Use For: Fokker F-28 4000 Aircraft
See Also: Aircraft

—**Antenna Positions**
 CAA. Fokker F.28 2000 and 4000 (Approved Aerial Positions). 1 p.

—**Certification**
 CAA. Fokker F28 Mark 4000. 10 pp.

—**Data Sheets**
 CAA FA21 ISSUE 1. Fokker F28 Mk 4000. 4 pp.

Folding Endurance
See Also: Endurance Testing; Fatigue Testing

—**Footwear**
 BSI BS 5131: Sec 3.5-78. 1978 Methods of Test for Footwear and Footwear Materials Part 3: Uppers, Textiles and Threads Section 3.5: The Break/Pipiness Test. 2 pp.
 BSI BS 5131: Sec 4.2-90. 1990 Methods of Test for Footwear and Footwear Materials Part 4: Other Components Section 4.2: Flexing Index of Fibreboard. 6 pp.

—**Leather**
 CNS K6119-60. Method of Test for Leather (Folding Test) (May)(1275).

—**Leathercloth**
 DIN ENGL 53359-57. Testing of Artificial Leather; Repeated Flexure Test (Nov). 2 pp.

—**Paperboard**
 BSI BS 5131: Sec 4.2-90. 1990 Methods of Test for Footwear and Footwear Materials Part 4: Other Components Section 4.2: Flexing Index of Fibreboard. 6 pp.
 CNS P3057-83. Method of Test for Folding Endurance of Paper and Paperboard by MIT Tester (Jun)(10378). 2 pp.
 JIS P 8114-76. Testing Method for Folding Endurance of Paper and Paperboard by Schopper Tester (R 1984). 6 pp.
 JIS P 8115-76. Testing Method for Folding Endurance of Paper and Paperboard by MIT Tester.

—**Papers**
 BSI BS 4419-80. 1980 Determination of Folding Endurance of Paper. 12 pp.
 CNS P3008-91. Method of Test for Folding Endurance of Paper by Schopper Type Tester (Jun)(1358). 3 pp.
 CNS P3057-83. Method of Test for Folding Endurance of Paper and Paperboard by MIT Tester (Jun)(10378). 2 pp.
 CPPA D.17P-74. Folding Endurance of Paper (M.I.T. Method). 2 pp.
 ISO 5626-78. Paper—Determination of Folding Endurance First Edition. 10 pp.
 JIS P 8114-76. Testing Method for Folding Endurance of Paper and Paperboard by Schopper Tester (R 1984). 6 pp.
 JIS P 8115-76. Testing Method for Folding Endurance of Paper and Paperboard by MIT Tester.
 SAA AS 1301.423R P-89. Methods of Test for Pulp and Paper (Metric Units)—Part 423rp: Folding Endurance of Paper—Kohler-Molin Methods (This is a Joint Standard with SANZ NZS 1301). 3 pp.
 SNZ NZS/AS 1301. 423RP-89. Methods of Test for Pulp and Paper Folding Endurance of Paper—Kohler-Molin Methods (This is a Joint Standard with SAA AS 1301.423RP). 3 pp.

—**Photographic Film**
 ISO 8776-88. Photography—Photographic Film—Determination of Folding Endurance First Edition. 7 pp.

Folding Endurance Testing
Use: Folding Endurance

Follow-Up Milks
Use: Milk

Foncault Current Testing
Use: Eddy Current Testing

Fonts

—**Character Identification**
 OSI ISO DP 9541-3-87. Information Processing-Font and Character Information Interchange Part 3: Character Identification Method. 7 pp.

—**Character Models**
 OSI ISO DP 9541-4-87. Information Processing—Font and Character Information Interchange Part 4: Character Collections. 7 pp.
 OSI ISO DP 9541-5-87. Information Processing—Font and Character Information Interchange Part 5: Font Attributes and Character Model. 24 pp.

—**Glyphs**
 IEC DIS 9541 Pt 3-91. Information Technology—Font Information Interchange—Part 3: Glyph Shape Representation ***CD-ROM ONLY***. 73 pp.
 ISO DIS 9541 Pt 3-91. Information Technology—Font Information Interchange—Part 3: Glyph Shape Representation ***CD-ROM ONLY***. 73 pp.
 JTC1 DIS9541 Pt 3-91. Information Technology—Font Information Interchange—Part 3: Glyph Shape Representation ***CD-ROM ONLY***. 73 pp.

—**Glyphs—Registration Procedures**
 IEC 10036-93. Information Technology—Font Information Interchange—Procedure for Registration of Glyph and Glyph Collection Identifiers First Edition. 18 pp.
 ISO 10036-93. Information Technology—Font Information Interchange—Procedure for Registration of Glyph and Glyph Collection Identifiers First Edition. 18 pp.
 JTC1 10036-93. Information Technology—Font Information Interchange—Procedure for Registration of Glyph and Glyph Collection Identifiers First Edition. 18 pp.
 OSI ISO/IEC DIS 10036-90. Information Technology—Procedure for Registration at Glyph and Glyph Collection Identifiers. 21 pp.

—**Information Interchange**
 BSI BS ISO/IEC 9541-1-91. 1991 Information Technology—Font Information Interchange—Part 1: Architecture (S). 94 pp.
 BSI BS ISO/IEC 9541-1-01. 1991 Amd 1 Information Technology—Font Information Interchange—Part 1: Architecture (AMD 7571) February 15, 1993 (Technical Corr 1). 97 pp.
 BSI BS ISO/IEC 9541-2-91. 1991 Information Technology—Font Information Interchange—Part 2: Interchange Format. 35 pp.
 BSI BS ISO/IEC 9541-2-01. 1991 Amd 1 Information Technology—Font Information Interchange—Part 2: Interchange Format (AMD 7813) July 15, 1993 (Technical Corr 1) (S). 40 pp.
 IEC 9541 Pt 1-91. Information Technology—Font Information Interchange—Part 1: Architecture First Edition; (Corrigendum 1-1992). 94 pp.
 IEC 9541 Pt 2-91. Information Technology—Font Information Interchange—Part 2: Interchange Format First Edition; (Corrigendum 1-1993). 38 pp.
 ISO 9541 Pt 1-91. Information Technology—Font Information Interchange—Part 1: Architecture First Edition; (Corrigendum 1-1992). 94 pp.
 ISO 9541 Pt 2-91. Information Technology—Font Information Interchange—Part 2: Interchange Format First Edition; (Corrigendum 1-1993). 38 pp.
 JTC1 9541-91. Information Technology—Font Information Interchange—Part 1: Architecture First Edition; (Corrigendum 1-1992). 2 pp.
 JTC1 9541-91. Information Technology—Font Information Interchange—Part 2: Interchange Format First Edition; (Corrigendum 1-1993). 38 pp.
 OSI ISO/IEC DIS 9541-1.2-90. Information Technology—Font Information Interchange—Part 1: Architecture. 91 pp.
 OSI ISO/IEC DIS 9541-2.2-90. Information Technology—Font Information Interchange—Part 2: Interchange Format. 35 pp.

—**Registration Procedures**
 OSI ISO/IEC/DIS 9541-2-87. Information Processing—Font and Character Information Interchange—Part 2: Registration and Naming Procedures. 13 pp.
 OSI ISO DP 9541-2-87. Information Processing—Font and Character Information Interchange Part 2: Registration and Naming Procedures. 9 pp.

—**Subsets**
 OSI ISO DP 9541-6-87. Information Processing—Font and Character Information Interchange Part 6: Font and Character Attribute Subsets and Application. 14 pp.

Food
Scope Note: For additional listings, use a more specific term *Use For:* Foodstuffs *See Also:* Animal Feed; Apples; Apricots; Asparagus; Avocados; Baby Foods; Bacon; Bakery Products; Bananas; Beans; Beef; Beets; Beverages; Bilberries; Bitter Melons; Blueberries; Bread; Breakfast Cereals; Butter; Cabbages; Cake Mixes; Canned Foods; Carob; Carrots; Catsup; Cauliflowers; Cereal Products; Chayotes; Cherries; Chinese Cabbages; Citrus Fruits; Cocoa; Coconuts; Condiments; Cranberries; Cucumbers; Dairy Products; Dietary Foods; Dried Foods; Eggplants; Eggs; Emergency Rations; Fish; Flavorings; Flours (Food); Forage Crops; Frog Legs; Frozen Foods; Fruits; Garlic; Glutens; Grapefruits; Grapes; Ham; Honey; Horseradish; Jams (Food); Jellies (Food); Lamb; Leeks; Legumes; Lemons; Lettuce; Limes; Liver; Lobsters; Loquats; Mandarin Oranges; Mangos; Marmalades; Mayonnaise; Milk; Miso; Molasses; Mollusks; Mulberries; Mushrooms; Offals; Onions; Organ Meat; Pancake Mixes; Papayas; Pasta; Peaches; Peanut Butter; Pears; Peas; Peppers; Pickles; Pies; Pineapples; Plums; Pork; Potatoes; Poultry Meat; Rhubarb; Rice Vermicelli; Roe; Sausages *See Also:* Soups; Soybeans; Spices; Strawberries; Sugar Beets; Syrups; Tapioca; Tea; Tomatoes; Vanilla; Veal; Vegetable Fats; Vegetables; Waffle Mixes; Wakame; Watermelons
 EC 80/1276/EEC-80. Council Directive Amending, by Virtue of the Accession of Greece, Directives 76/893/EEC, 79/693/EEC and 80/777/EEC with Regard to the Majority Quorum of Votes Within the Standing Committee of Foodstuffs Procedure. 1 p.
 EC 85/7/EEC-84. Council Directive Amending a First Series of Directives on the Approximation of the Laws of the Member States in the Foodstuffs Sector, as Regards the Involvement of the Standing Committee for Foodstuffs. 2 pp.

—**Advertising**
 EC COM(86) 89-86. Proposal for a Council Directive Amending Directive 79/112/EEC on the Approximation of the Laws of the Member States Relating to the Labelling, Presentation and Advertising of Foodstuffs for Sale to the Ultimate Consumer. 18 pp.
 EC COM(87) 242-87. Amended proposal for a Council Directive Amending Directive 79/112/EEC on the Approximation of the Laws of the Member States Relating to the Labelling, Presentation and Advertising of Foodstuffs for Sale to the Ultimate Consumer (COM(86) 89 final of 22. 7 pp.

INTERNATIONAL AND NON-U.S. NATIONAL STANDARDS
SUBJECT INDEX

Food (Cont.)

—Antibiotics Content
CNS N6169-82. Method of Test for the Residue of Anti-Biotic in Foodstuffs (Babyfood Use) (May)(8860). 2 pp.

—Artificial Sweeteners Content—Gas Chromatography
CNS N6191-86. Method of Test for Artificial Sweeteners in Food (Sep)(10950). 4 pp.

—Artificial Sweeteners Content—Liquid Chromatography
CNS N6191-86. Method of Test for Artificial Sweeteners in Food (Sep)(10950). 4 pp.

—Artificial Sweeteners Content—Paper Chromatography
CNS N6191-86. Method of Test for Artificial Sweeteners in Food (Sep)(10950). 4 pp.

—Ash Content
CNS N6115-84. Method of Test for Ash in Food (Jan)(5034). 2 pp.

—Bacteria Content
BSI BS 5763: Part 2-91. 1991 Microbiological Examination of Food and Animal Feeding Stuffs Part 2: Enumeration of Coliforms—Colony Count Technique (ISO 4832: 1991). 11 pp.
BSI BS 5763: Part 3-91. 1991 Microbiological Examination of Food and Animal Feeding Stuffs Part 3: Enumeration of Coliforms—Most Probable Number Technique (ISO 4831: 1991). 19 pp.
BSI BS 5763: Part 4-90. 1990 Microbiological Examination of Food and Animal Feeding Stuffs Part 4: Detection of Salmonella. 22 pp.
BSI BS 5763: Part 7-83. 1983 Microbiological Examination of Food and Animal Feeding Stuffs Part 7: Enumeration of Staphylococcus Aureus by Colony Count Technique. 10 pp.
BSI BS 5763: Part 8-85. 1985 Microbiological Examination of Food and Animal Feeding Stuffs Part 8: Enumeration of Presumptive Escherichia Coli. 12 pp.
BSI BS 5763: Part 9-86. 1986 Microbiological Examination of Food and Animal Feeding Stuffs Part 9: Enumeration of Clostridium Perfringens. 10 pp.
BSI BS 5763: Part 10-86. 1986 Microbiological Examination of Food and Animal Feeding Stuffs Part 10: Enumeration of Enterobacteriaceae. 11 pp.
BSI BS 5763: Part 11-88. 1988 Microbiological Examination of Food and Animal Feeding Stuffs Part 11: Enumeration of Bacillus Cereus. 11 pp.
BSI BS 5763: Part 13-89. 1989 Microbiological Examination of Food and Animal Feeding Stuffs Part 13: Enumeration of Escherichia Coli: Colony Count Technique Using Membranes. 8 pp.
BSI BS 5763: Part 14-91. 1991 Methods for Microbiological Examination of Food and Animal Feeding Stuffs Part 14: Detection of Vibrio Parahaemolyticus. 17 pp.
BSI BS 5763: Part 15-91. 1991 Methods for Microbiological Examination of Food and Animal Feeding Stuffs Part 15: Detection of Enterobacteriaceae with Pre-Enrichment (W) (ISO 8523: 1991). 14 pp.
CNS N6192-88. Methods of Test for Food Microbiology-Test of Escherichia Coli (Nov)(10951). 7 pp.
CNS N6193-88. Methods of Test for Food Microbiology-Test of Salmonella (Apr)(10952). 11 pp.
CNS N6194-88. Methods of Test for Food Microbiology-Test of Coliform Bacteria (May)(10984). 3 pp.
CNS N6213-89. Method of Test for Shigella in Foods (Jun)(12541). 10 pp.
ISO 4831-91. Microbiology—General Guidance for the Enumeration of Coliforms —Most Probable Number Technique Second Edition. 16 pp.
ISO 4832-91. Microbiology—General Guidance for the Enumeration of Coliforms —Colony Count Technique Second Edition. 9 pp.
ISO 6579-93. Microbiology—General Guidance on Methods for the Detection of Salmonella Third Edition. 21 pp.
ISO 6888-83. Microbiology—General Guidance for Enumeration of Staphylococcus Aureus—Colony Count Technique First Edition. 9 pp.
ISO 7251-84. Microbiology—General Guidance for Enumeration of Presumptive Escherichia Coli—Most Probable Number Technique First Edition. 11 pp.
ISO 7402-93. Microbiology—General Guidance for the Enumeration of Enterobacteriaceae Without Resuscitation—MPN Technique and Colony-Count Technique Second Edition. 14 pp.
ISO 7932-87. Microbiology—General Guidance for Enumeration of Bacillus Cereus—Colony Count Technique at 30 Degrees Celsius First Edition. 11 pp.

Food (Cont.)

—Bacteria Content (Cont.)
ISO 7937-85. Microbiology—General Guidance for Enumeration of Clostridium Perfringens—Colony Count Technique First Edition. 9 pp.
ISO 8523-91. Microbiology—General Guidance for the Detection of Enterobacteriaceae with Pre-Enrichment First Edition. 11 pp.
ISO 8914-90. Microbiology—General Guidance for the Detection of Vibrio Parahaemolyticus. 15 pp.
SAA AS 1766.2.5-91. Methods for the Microbiological Examination of Food—Part 2: Examination for Specific Organisms—Part 2.5: Salmonellae Amdt 1 January/February 1993. 9 pp.
SAA AS 1766.2.6-91. Methods for the Microbiological Examination of Food—Part 2: Examination for Specific Organisms—Part 2.6: Bacillus Cereus (Supersedes AS 1766.2.1.6 Addendum 1—1976). 9 pp.
SAA AS 1766.2.8-91. Methods for the Microbiological Examination of Food—Part 2: Examination for Specific Organisms—Part 2.8: Clostridium Perfringens. 6 pp.
SAA AS 1766.2.9-91. Methods for the Microbiological Examination of Food—Part 2: Examination for Specific Organisms—Part 2.9: Vibrio Parahaemolyticus. 6 pp.
SAA AS 1766.2.13-91. Methods for the Microbiological Examination of Food—Part 2: Examination for Specific Organisms—Part 2.13: Campylobacter. 8 pp.

—Bacteria Count Methods
CNS N6186-91. Method of Test for Food Microbiology-Test of Standard Plate Count (Feb)(10890). 5 pp.
SAA AS 1766.2.3-92. Food Microbiology—Part 2: Examination for Specific Organisms—Part 2.3: Coliforms and Escherichia Coli (Supersedes AS 1095.3.1—1987). 7 pp.
SAA AS 1766.2.4-86. Methods for the Microbiological Examination of Food—Part 2: Examination for Specific Organisms—Part 2.4: Coagulase-Positive Staphylococci. 3 pp.
SAA AS 1766.2.5-91. Methods for the Microbiological Examination of Food—Part 2: Examination for Specific Organisms—Part 2.5: Salmonellae Amdt 1 January/February 1993. 9 pp.

—Benzoic Acid Content—Gas Chromatography
CNS N6190-84. Method of Test for Preservatives in Food (Jun)(10949). 8 pp.

—Benzoic Acid Content—Thin Line Chromatography
CNS N6190-84. Method of Test for Preservatives in Food (Jun)(10949). 8 pp.

—Calcium Content
CNS N5200-82. Method of Test for Calcium and Phosphorus in Food Stuffs (Nov)(9638). 6 pp.

—Chemical Analysis
EC 85/591/EEC-85. Council Directive Concerning the Introduction of Community Methods of Sampling and Analysis for the Monitoring of Foodstuffs Intended for Human Consumption. 2 pp.

—Classification
BSI BS 1000: (664)-85. 1985 Universal Decimal Classification (UDC). English Full Edition (664): Preparation and Preservation of Solid Foodstuffs. 33 pp.
SNZ NZS/BS 1000 (664)-85. Universal Decimal Classification Preparation and Preservation of Solid Foodstuffs. 36 pp.

—Clostridium Botulinum Content
SAA AS 1766.2.7-91. Methods for the Microbiological Examination of food—Part 2: Examination for Specific Organisms—Clostridium Botulinum and Clostridium Botulinum Toxin (Supersedes AS 1766.2.1.7 Addendum 1—1976). 5 pp.

—Contamination—Animal
CNS N6120-80. Methods of Analysis for Foodstuffs—Detection of Extraneous Matter (May)(5629). 3 pp.
EC EEC/315/93-93. Council Regulation Laying Down Community Procedures for Contaminants in Food. 3 pp.

—Contamination—Mineral
CNS N6120-80. Methods of Analysis for Foodstuffs—Detection of Extraneous Matter (May)(5629). 3 pp.
EC EEC/315/93-93. Council Regulation Laying Down Community Procedures for Contaminants in Food. 3 pp.

—Contamination—Plant
CNS N6120-80. Methods of Analysis for Foodstuffs—Detection of Extraneous Matter (May)(5629). 3 pp.
EC EEC/315/93-93. Council Regulation Laying Down Community Procedures for Contaminants in Food. 3 pp.

Food (Cont.)

—Defense Contracts
MOD UK DEFCON 112AD-85. Conditions of Contract for the Supply of Food 5/85. 2 pp.

—Dehydroacetic Acid Content—Gas Chromatography
CNS N6190-84. Method of Test for Preservatives in Food (Jun)(10949). 8 pp.

—Dehydroacetic Acid Content—Thin Line Chromatography
CNS N6190-84. Method of Test for Preservatives in Food (Jun)(10949). 8 pp.

—Dye Content
CNS N6185-87. Method of Test for Dyes in Foods (Jun)(10889). 8 pp.

—European Communities—Committees
EC 80/1073/EEC-80. Commission Decision Establishing a New Statute of the Advisory Committee on Foodstuffs. 4 pp.

—Fat Content
CNS N6117-84. Method of Test for Crude Fat in Food (Jan)(5036). 4 pp.

—Fiber Content
BSI BS 6215: Part 1-81. 1981 Amd 1 Agricultural Food Products Part 1: Determination of Crude Fibre Content (General Method). 12 pp.
BSI BS 6215: Part 1-02. 1981 Amd 2 Methods of Test for Agricultural Food Products Part 1: Determination of Crude Fibre Content (General Method) (ISO 5498: 1981) (AMD 7726) June 15, 1993 (W). 13 pp.
CNS N6118-79. Determination of Crude Fiber Content for Foodstuffs (Nov)(5037). 3 pp.
ISO 5498-81. Agricultural Food Products—Determination of Crude Fibre Content—General Method First Edition. 10 pp.

—Fiber Content—Scharrer Method
BSI BS 6215: Part 2-81. 1981 Amd 1 Agricultural Food Products Part 2: Determination of Crude Fibre Content (Modified Scharrer Method). 8 pp.
ISO 6541-81. Agricultural Food Products—Determination of Crude Fibre Content—Modified Scharrer Method First Edition. 6 pp.

—Fungus Content
BSI BS 5763: Part 12-88. 1988 Microbiological Examination of Food and Animal Feeding Stuffs Part 12: Enumeration of Yeasts and Moulds. 6 pp.
ISO 7954-87. Microbiology—General Guidance for Enumeration of Yeasts and Moulds—Colony Count Technique at 25 Degrees Celsius First Edition. 6 pp.
SAA AS 1766.2.2-80. Methods for the Microbiological Examination of Food—Part 2: Examination for Specific Organisms—Part 2.2: Colony Count of Yeast and Moulds. 2 pp.

—Heat Processed—pH—Potentiometric Analysis
ISO 11289-93. Heat-Processed Foods in Hermetically Sealed Containers—Determination of pH First Edition. 7 pp.

—Histamine Content
CNS N6159-81. Determination of Histamine for Food Stuff (Oct)(8052). 4 pp.

—Hydrogen Peroxide Content
CNS N6189-88. Method of Test for Bactericides in Food-Test of Hydrogen Peroxide (Oct)(10893). 1 p.

—Hydroxybenzoic Acid Content—Gas Chromatography
CNS N6190-84. Method of Test for Preservatives in Food (Jun)(10949). 8 pp.

—Hydroxybenzoic Acid Content—Thin Line Chromatography
CNS N6190-84. Method of Test for Preservatives in Food (Jun)(10949). 8 pp.

—Hydroxybenzoic Esters Content—Gas Chromatography
CNS N6190-84. Method of Test for Preservatives in Food (Jun)(10949). 8 pp.

—Hydroxybenzoic Esters Content—Thin Line Chromatography
CNS N6190-84. Method of Test for Preservatives in Food (Jun)(10949). 8 pp.

—Hygiene
EC 93/43/EEC-93. Council Directive on the Hygiene of Foodstuffs. 11 pp.

Food (Cont.)

—Identification Systems
EC COM(86) 89-86. Proposal for a Council Directive Amending Directive 79/112/EEC on the Approximation of the Laws of the Member States Relating to the Labelling, Presentation and Advertising of Foodstuffs for Sale to the Ultimate Consumer. 18 pp.
EC COM(87) 242-87. Amended proposal for a Council Directive Amending Directive 79/112/EEC on the Approximation of the Laws of the Member States Relating to the Labelling, Presentation and Advertising of Foodstuffs for Sale to the Ultimate Consumer (COM(86) 89 final of 22. 7 pp.
EC COM(88) 489-88. Proposals for Council Directives—on the Introduction of Compulsory Nutrition Labelling of Foodstuffs Intended for Sale to the Ultimate Consumer—on Nutrition Labelling Rules for Foodstuffs Intended for Sale to the Ultimate Consumer. 25 pp.
EC 83/463/EEC-83. Commission Directive Introducing Temporary Measures for the Designation of Certain Ingredients in the Labelling of Foodstuffs for Sale to the Ultimate Consumer. 6 pp.
EC 89/395/EEC-89. Council Directive Amending 79/112/EEC on the Approximation of the Laws of the Member States Relating to Labelling, Presentation and Advertising of Foodstuffs for Sale to the Ultimate Consumer. 4 pp.
EC 90/496/EEC-90. Council Directive on Nutrition Labelling for Foodstuffs. 5 pp.

—Inspection
EC COM(86) 747-86. Proposal for a Council Directive on the Official Inspection of Foodstuffs (Submitted to the Council by the Commission). 17 pp.
EC COM(88) 88-88. Amended Proposal for a Council Directive on the Official Inspection of Foodstuffs. 6 pp.
EC 89/397/EEC-89. Council Directive on the Official Control of Foodstuffs. 4 pp.

—Invitation for Bids
MOD UK DEFCON 47G-89. Invitation to Tender for the Supply of Food 11/89. 2 pp.

—Ionizing Radiation
EC COM(87) 242-87. Amended proposal for a Council Directive Amending Directive 79/112/EEC on the Approximation of the Laws of the Member States Relating to the Labelling, Presentation and Advertising of Foodstuffs for Sale to the Ultimate Consumer (COM(86) 89 final of 22. 7 pp.
EC COM(88) 654-88. Proposal for a Council Directive on the Approximation of the Laws of the Member States Concerning Foods and Food Ingredients Treated with Ionizing Radiation. 19 pp.

—Listeria Monocytogenes Content
SAA AS 1766.2.15 (INT)-91. Method for the Microbiological Examination of Food—Part 2: Examination for Specific Organisms—Part 2.15(Int): Listeria Monocytogenes in Dairy Products. 11 pp.

—Lots—Identification Systems
EC 89/396/EEC-89. Council Directive on Indications or Marks Identifying the Lot to Which a Foodstuff Belongs. 2 pp.
EC 91/238/EEC-91. Council Directive Amending Directive 89/396/EEC on Indications or Marks Identifying the Lot to Which a Foodstuff Belongs. 1 p.

—Materials Contact
EC 82/711/EEC-82. Council Directive Laying Down the Basic Rules Necessary for Testing Migration of the Constituents of Plastic Materials and Articles Intended to Come into Contact with Foodstuffs. 5 pp.
EC 83/229/EEC-83. Council Directive on the Approximation of the Laws of the Member States Relating to Materials and Articles Made of Regenerated Cellulose Film Intended to Come into Contact with Foodstuffs. 9 pp.
EC 84/500/EEC-84. Council Directive on the Approximation of the Laws of the Member States Relating to Ceramic Articles Intended to Come Into Contact with Foodstuffs. 5 pp.
EC 89/109/EEC-88. Council Directive on the Approximation of the Laws of the Member States Relating to Materials and Articles Intended to Come into Contact with Foodstuffs. 7 pp.
EC 90/128/EEC-90. Commission Directive Relating to Plastics Materials and Articles Intended to Come into Contact with Foodstuffs. 22 pp.
EC 93/8/EEC-93. Commission Directive Amending Council Directive 82/711/EEC Laying down the Basic Rules Necessary for Testing Migration of Constituents of Plastic Materials and Articles Intended to Come into Contact with Foodstuffs. 4 pp.
EC 93/9/EEC-93. Commission Directive Amending Directive 90/128/EEC Relating to Plastic Materials and Articles Intended to Come into Contact with Foodstuffs. 7 pp.
SAA AS 2070.1-92. Plastics Materials for Food Contact Use—Part 1: Polyethylene Amdt 1 August 1993. 7 pp.
SAA AS 2070.2-92. Plastics Materials for Food Contact Use—Part 2: Poly(Vinyl Chloride) (PVC) Compound Amdt 1 August 1992. 4 pp.
SAA AS 2070.3-92. Plastics Materials for Food Contact Use—Part 3: Styrene Plastics Materials Amdt 1 August 1993. 4 pp.
SAA AS 2070.4-92. Plastics Materials for Food Contact Use—Part 4: Acrylonitrile Plastics Materials Amdt 1 August 1993. 4 pp.
SAA AS 2070.5-92. Plastics Materials for Food Contact Use—Part 5: Polypropylene Amdt 1 August 1993. 8 pp.
SAA AS 2070.6-84. Plastics Materials for Food Contact Use—Part 6: Colourants. 12 pp.
SAA AS 2070.7-92. Plastics Materials for Food Contact Use—Part 7: Poly(vinylidene Chloride) (PVDC) Amdt 1 august 1993. 7 pp.
SAA AS 2070.8-92. Plastics Materials for Food Contact Use—Part 8: Miscellaneous Additives. 7 pp.
SNZ NZS 7608-87. Plastics Materials for Food Contact Use Amend: A. 20 pp.

—Materials Contact—Symbols
EC 80/590/EEC-80. Commission Directive Determining the Symbol That May Accompany Materials and Articles Intended to Come into Contact with Foodstuffs. 2 pp.

—Materials Contact—Vinyl Chloride Content—Gas Chromatography
EC 78/142/EEC-78. Council Directive on the Approximation of the Laws of the Member States Relating to Materials and Articles Which Contain Vinyl Chloride Monomer and Are Intended to Come into Contact with Foodstuffs. 3 pp.
EC 80/766/EEC-80. Commission Directive Laying Down the Community Method of Analysis for the Official Control of the Vinyl Chloride Monomer Level in Materials and Articles Which are Intended to Come into Contact with Foodstuffs. 5 pp.
EC 81/432/EEC-81. Commission Directive Laying Down the Community Method of Analysis for the Official Control of Vinyl Chloride Released by Materials and Articles into Foodstuffs. 6 pp.

—Microbiological Analysis
BSI BS 5763: Part 0-86. (OBSOLESCENT) 1986 Microbiological Examination of Food and Animal Feeding Stuffs Part 0: General Laboratory Practices. 16 pp.
BSI BS 5763: Part 5-81. 1981 Microbiological Examination of Food and Animal Feeding Stuffs Part 5: Enumeration of Micro-Organisms—Colony Count at 30 Degrees Celsius (Surface Plate Technique). 7 pp.
BSI BS 5763: Part 6-83. 1983 Microbiological Examination of Food and Animal Feeding Stuffs Part 6: Preparation of Dilutions. 7 pp.
SAA AS 1766. Methods for the Microbiological Examination of Food Complete Set in Binder.
SAA AS 1766.1-75. Methods for the Microbiological Examination of Food—Part 1: General Procedures and Techniques (Incorporating Amdt 1 and Corrig.) Reinstated December 1990 Withdrawn.
SAA AS 1766.1.1-91. Methods for the Microbiological Examination of Food—Part 1: General Procedures and Techniques—Part 1.1: Samples, Materials, Equipment, Laboratory Practice. 4 pp.
SAA AS 1766.1.2-91. Methods for the Microbiological Examination of Food—Part 1: General Procedures and Techniques—Part 1.2: Preparation of Dilutions. 2 pp.
SAA AS 1766.2-76. Methods for the Microbiological Examination of Food—Part 2: Examination for Specific Organisms Amdt 1 November 1980 Withdrawn.
SAA AS 1766.2.15 (INT)-91. Method for the Microbiological Examination of Food—Part 2: Examination for Specific Organisms—Part 2.15(Int): Listeria Monocytogenes in Dairy Products. 11 pp.
SAA AS 1766.3.7-86. Methods for the Microbiological Examination of Food—Part 3: Examination of Specific Products—Part 3.7: Heat-Processed Foods in Hermetically Sealed Containers. 11 pp.

—Microorganism Content
BSI BS 5763: Part 1-91. 1991 Microbiological Examination of Food and Animal Feeding Stuffs Part 1: Enumeration of Micro-Organisms—Colony Count Technique at 30 Degrees C (ISO 4833: 1991). 11 pp.
BSI BS 5763: Part 5-81. 1981 Microbiological Examination of Food and Animal Feeding Stuffs Part 5: Enumeration of Micro-Organisms—Colony Count at 30 Degrees Celsius (Surface Plate Technique). 7 pp.
ISO 4833-91. Microbiology—General Guidance for the Enumeration of Micro-Organisms—Colony Count Technique at 30 Degrees Celsius Second Edition. 9 pp.
SAA AS 1766.1.3-91. Methods for the Microbiological Examination of Food—Part 1: General Procedures and Techniques—Part 1.3: Colony Count—Pour Plate Method. 3 pp.
SAA AS 1766.1.4-91. Methods for the Microbiological Examination of Food—Part 1: General Procedures and Techniques—Part 1.4: Colony Count—Surface Spread Method. 2 pp.
SAA AS 1766.1.5-91. Methods for the Microbiological Examination of Food—Part 1: General Procedures and Techniques—Part 1.5: Colony Count—Membrane Filtration Method. 4 pp.
SAA AS 1766.1.6-91. Methods for the Microbiological Examination of Food—Part 1: General Procedures and Techniques—Part 1.6: Estimation of Most Probable Number of Micro-Organisms (MPN). 4 pp.
SAA AS 1766.2.1-91. Methods for the Microbiological Examination of Food—Part 2: Examination for Specific Organisms—Part 2.1: Standard Plate Count. 2 pp.

—Moisture Content
CNS N6114-84. Method of Test for Moisture in Food (Jan)(5033). 3 pp.
CNS N6119-87. Determination of Water Activity for Foodstuffs (Jul)(5255). 3 pp.

—Nitrogen Content
CNS N6116-86. Method of Test for Crude Protein in Food (Aug)(5035). 3 pp.

—Nitrogen Content—Kjeldahl Method
ISO 1871-75. Agricultural Food Products—General Directions for the Determination of Nitrogen by the Kjeldahl Method First Edition. 7 pp.

—Phosphorus Content
CNS N5200-82. Method of Test for Calcium and Phosphorus in Food Stuffs (Nov)(9638). 6 pp.

—Plastics—Migration Testing
EC 82/711/EEC-82. Council Directive Laying Down the Basic Rules Necessary for Testing Migration of the Constituents of Plastic Materials and Articles Intended to Come into Contact with Foodstuffs. 5 pp.
EC 85/572/EEC-85. Council Directive Laying Down the List of Simulants to be Used for Testing Migration of Constituents of Plastic Materials and Articles Intended to Come Into Contact with Foodstuffs. 2 pp.
EC 90/128/EEC-90. Commission Directive Relating to Plastics Materials and Articles Intended to Come into Contact with Foodstuffs. 22 pp.
EC 93/8/EEC-93. Commission Directive Amending Council Directive 82/711/EEC Laying down the Basic Rules Necessary for Testing Migration of Constituents of Plastic Materials and Articles Intended to Come into Contact with Foodstuffs. 4 pp.
EC 93/9/EEC-93. Commission Directive Amending Directive 90/128/EEC Relating to Plastic Materials and Articles Intended to Come into Contact with Foodstuffs. 7 pp.

—Prices—Identification Systems
EC 88/315/EEC-88. Council Directive Amending Directive 79/581/EEC on Consumer Protection in the Indication of the Prices of Foodstuffs. 5 pp.

—Protein Content
CNS N6116-86. Method of Test for Crude Protein in Food (Aug)(5035). 3 pp.

—Salicylic Acid Content—Gas Chromatography
CNS N6190-84. Method of Test for Preservatives in Food (Jun)(10949). 8 pp.

—Salicylic Acid Content—Thin Line Chromatography
CNS N6190-84. Method of Test for Preservatives in Food (Jun)(10949). 8 pp.

—Sampling
EC 85/591/EEC-85. Council Directive Concerning the Introduction of Community Methods of Sampling and Analysis for the Monitoring of Foodstuffs Intended for Human Consumption. 2 pp.
EC 89/397/EEC-89. Council Directive on the Official Control of Foodstuffs. 4 pp.
ISO 7002-86. Agricultural Food Products—Layout for a Standard Method of Sampling from a Lot First Edition. 19 pp.
SAA AS 1766.4-87. Methods for the Microbiological Examination of Food—Part 4: Sampling of Foods. 5 pp.

INTERNATIONAL AND NON-U.S. NATIONAL STANDARDS
SUBJECT INDEX

Food

Food (Cont.)

—**Sensory Analysis**
BSI BS 5929: Part 1-86. 1986 Methods for Sensory Analysis of Food Part 1: General Guide to Methodology. 17 pp.
BSI BS 5929: Part 4-86. 1986 Methods for Sensory Analysis of Food Part 4: Flavour Profile Methods. 9 pp.
BSI PP 999-83. (WITHDRAWN) 1983 Standardized Methods for the Sensory Analysis of Food. 25 pp.
ISO 4121-87. Sensory Analysis—Methodology—Evaluation of Food Products by Methods Using Scales First Edition. 9 pp.
ISO 5497-82. Sensory Analysis—Methodology—Guidelines for the Preparation of Samples for Which Direct Sensory Analysis Is Not Feasible First Edition. 4 pp.
ISO 6564-85. Sensory Analysis—Methodology—Flavour Profile Methods First Edition. 8 pp.
SAA AS 2542.2.5-91. Sensory Analysis of Foods—Part 2: Specific Methods—Part 2.5: 'A—Not A' Test. 8 pp.

—**Sensory Analysis—Comparative Testing**
BSI BS 5929: Part 2-82. 1982 Methods for Sensory Analysis of Food Part 2: Paired Comparison Test. 10 pp.
BSI BS 5929: Part 5-88. 1988 Methods for Sensory Analysis of Food Part 5: 'A'-'Not A' Test. 9 pp.

—**Sensory Analysis—Glossaries**
BSI BS 5098-75. 1975 Glossary of Terms Relating to Sensory Analysis of Food. 16 pp.

—**Sensory Analysis—Ranking**
BSI BS 5929: Part 6-89. 1989 Methods for Sensory Analysis of Food Part 6: Ranking. 12 pp.

—**Sensory Analysis—Triangular Testing**
BSI BS 5929: Part 3-84. 1984 Methods for Sensory Analysis of Food Part 3: Triangular Test. 11 pp.

—**Shipping—Customs and Excise Procedures**
MOD UK DEFCON 20-82. HM Customs and Excise Transhipment Procedure 2/82. 1 p.

—**Sodium Chloride**
BSI BS 998-69. (WITHDRAWN) 1969 Amd 1 Vacuum Salt for Butter and Cheese Making and Other Food Uses (Superseded by BS 998: 1990 & BS 7319: Parts 1 to 12: 1990). 22 pp.
BSI BS 998-90. 1990 Vacuum Salt for Food Use. 9 pp.

—**Solvents**
EC 88/344/EEC-88. Council Directive on the Approximation of the Laws of the Member States on Extraction Solvents Used in the Production of Foodstuffs and Food Ingredients. 6 pp.
EC 92/115/EEC-92. Council Directive Amending for the First Time Directive 88/344/EEC on the Approximation of the Laws of the Member States on Extraction Solvents Used in the Production of Foodstuffs and Food Ingredients. 2 pp.

—**Sorbic Acid Content—Gas Chromatography**
CNS N6190-84. Method of Test for Preservatives in Food (Jun)(10949). 8 pp.

—**Sorbic Acid Content—Thin Line Chromatography**
CNS N6190-84. Method of Test for Preservatives in Food (Jun)(10949). 8 pp.

—**Staphylococcus Content**
CNS N6187-88. Methods of Test for Food Microbiology-Test of Staphylococci (Apr)(10891). 5 pp.

—**Storage—Antimicrobial Agents**
CGSB CAN/CGSB-2.161-M91. Assessment of Efficacy of Antimicrobial Agents for Use on Environmental Surfaces and Medical Devices. 8 pp.

—**Vinyl Chloride Content**
EC 78/142/EEC-78. Council Directive on the Approximation of the Laws of the Member States Relating to Materials and Articles Which Contain Vinyl Chloride Monomer and Are Intended to Come into Contact with Foodstuffs. 3 pp.
EC 81/432/EEC-81. Commission Directive Laying Down the Community Method of Analysis for the Official Control of Vinyl Chloride Released by Materials and Articles into Foodstuffs. 6 pp.

—**Vitamin Content**
CNS N6168-90. Method of Test for Determination of Vitamin a Content in Foodstuff (May)(8859). 2 pp.

—**Vitamin E Content**
CNS N6229-90. Method of Test for Determining of Vitamin E Content for Foodstuff (May)(12724). 2 pp.

Food (Cont.)

—**Yeast Content**
SAA AS 1766.2.2-80. Methods for the Microbiological Examination of Food—Part 2: Examination for Specific Organisms—Part 2.2: Colony Count of Yeast and Moulds. 2 pp.

Food Additives

See Also: Additives; Flavorings; Food Colors
EC COM(93) 289-93. Amended Proposal for a Council Directive Amending Directive 89/107/EEC on the Approximation of the Laws of Member States Concerning Food Additives Intended for Human Consumption. 6 pp.
EC COM(93) 290-93. Amended Proposal for a Council Directive on Food Additives Other Than Colours and Sweeteners. 8 pp.
EC 89/107/EEC-88. Council Directive on the Approximation of the Laws of the Member States Concerning Food Additives Authorized for Use in Foodstuffs Intended for Human Consumption. 7 pp.

—**Antioxidants—Purity**
EC 78/664/EEC-78. Council Directive Laying Down Specific Criteria of Purity for Antioxidants Which May be Used in Foodstuffs Intended for Human Consumption. 18 pp.
EC 81/712/EEC-81. First Commission Directive Laying down Community Methods of Analysis for Verifying That Certain Additives Used in Foodstuffs Satisfy Criteria of Purity. 27 pp.
EC 82/712/EEC-82. Council Directive Amending Directive 78/664/EEC Laying down Specific Criteria of Purity for Antioxidants Which May be Used in Foodstuffs Intended for Human Consumption. 2 pp.

—**Emulsifying Agents**
EC 74/329/EEC-74. Council Directive on the Approximation of the Laws of the Member States Relating to Emulsifiers, Stabilizers, Thickeners and Gelling Agents for Use in Foodstuffs. 7 pp.
EC 78/612/EEC-78. Council Directive Amending for the First Time Directive 74/329/EEC on the Approximation of the Laws of the Member States Relating to Emulsifiers, Stabilizers, Thickeners and Gelling Agents for Use in Foodstuffs. 4 pp.
EC 80/597/EEC-80. Council Directive Amending for the Second Time Directive 74/329/EEC on the Approximation of the Laws of the Member States Relating to Emulsifiers, Stabilizers, Thickeners and Gelling Agents for Use in Foodstuffs. 3 pp.
EC 85/6/EEC-84. Council Directive Amending for the Third Time Directive 74/329/EEC on the Approximation of the Laws of the Member States Relating to Emulsifiers, Stabilizers, Thickeners and Gelling Agents for Use in Foodstuffs. 1 p.
EC 86/102/EEC-86. Council Directive Amending for the Fourth Time Directive 74/329/EEC on the Approximation of the Laws of the Member States Relating to Emulsifiers, Stabilizers, Thickeners and Gelling Agents for Use in Foodstuffs. 2 pp.
EC 89/393/EEC-89. Council Directive Amending for the Fifth Time Directive 74/329/EEC on the Approximation of the Laws of the Member States Relating to Emulsifiers, Stabilizers, Thickeners and Gelling Agents for Use in Foodstuffs. 1 p.
EC 90/612/EEC-90. Commission Directive Amending Council Directive 78/663/EEC Laying down Specific Purity Criteria for Emulsifiers, Stabilizers, Thickeners and Gelling Agents for Use in Foodstuffs. 2 pp.

—**Emulsifying Agents—Purity**
EC 78/663/EEC-78. Council Directive Laying Down Specific Criteria of Purity for Emulsifiers, Stabilizers, Thickeners and Gelling Agents for Use in Foodstuffs. 23 pp.
EC 82/504/EEC-82. Council Directive Amending Directive 78/663/EEC Laying down Specific Criteria of Purity for Emulsifiers, Stabilizers, Thickeners and Gelling Agents for Use in Foodstuffs. 3 pp.

—**Gelling Agents**
EC 74/329/EEC-74. Council Directive on the Approximation of the Laws of the Member States Relating to Emulsifiers, Stabilizers, Thickeners and Gelling Agents for Use in Foodstuffs. 7 pp.
EC 78/612/EEC-78. Council Directive Amending for the First Time Directive 74/329/EEC on the Approximation of the Laws of the Member States Relating to Emulsifiers, Stabilizers, Thickeners and Gelling Agents for Use in Foodstuffs. 4 pp.
EC 80/597/EEC-80. Council Directive Amending for the Second Time Directive 74/329/EEC on the Approximation of the Laws of the Member States Relating to Emulsifiers, Stabilizers, Thickeners and Gelling Agents for Use in Foodstuffs. 3 pp.
EC 85/6/EEC-84. Council Directive Amending for the Third Time Directive 74/329/EEC on the Approximation of the Laws of the Member States Relating to Emulsifiers, Stabilizers, Thickeners and Gelling Agents for Use in Foodstuffs. 1 p.

Food Additives (Cont.)

—**Gelling Agents (Cont.)**
EC 86/102/EEC-86. Council Directive Amending for the Fourth Time Directive 74/329/EEC on the Approximation of the Laws of the Member States Relating to Emulsifiers, Stabilizers, Thickeners and Gelling Agents for Use in Foodstuffs. 2 pp.
EC 89/393/EEC-89. Council Directive Amending for the Fifth Time Directive 74/329/EEC on the Approximation of the Laws of the Member States Relating to Emulsifiers, Stabilizers, Thickeners and Gelling Agents for Use in Foodstuffs. 1 p.
EC 90/612/EEC-90. Commission Directive Amending Council Directive 78/663/EEC Laying down Specific Purity Criteria for Emulsifiers, Stabilizers, Thickeners and Gelling Agents for Use in Foodstuffs. 2 pp.

—**Gelling Agents—Purity**
EC 78/663/EEC-78. Council Directive Laying Down Specific Criteria of Purity for Emulsifiers, Stabilizers, Thickeners and Gelling Agents for Use in Foodstuffs. 23 pp.
EC 82/504/EEC-82. Council Directive Amending Directive 78/663/EEC Laying down Specific Criteria of Purity for Emulsifiers, Stabilizers, Thickeners and Gelling Agents for Use in Foodstuffs. 3 pp.

—**Preservatives**
EC 85/585/EEC-85. Council Directive Amending Directive 64/54/EEC on the Approximation of the Laws of the Member States Concerning the Preservatives Authorized for Use in Foodstuffs Intended for Human Consumption. 1 p.

—**Preservatives—Purity**
EC 81/712/EEC-81. First Commission Directive Laying down Community Methods of Analysis for Verifying That Certain Additives Used in Foodstuffs Satisfy Criteria of Purity. 27 pp.

—**Purity**
EC 81/712/EEC-81. First Commission Directive Laying down Community Methods of Analysis for Verifying That Certain Additives Used in Foodstuffs Satisfy Criteria of Purity. 27 pp.

—**Safety**
EC 80/1089/EEC-80. Commission Recommendation to the Member States Concerning Tests Relating to the Safety Evaluation of Food Additives. 1 p.

—**Stabilizers**
EC 74/329/EEC-74. Council Directive on the Approximation of the Laws of the Member States Relating to Emulsifiers, Stabilizers, Thickeners and Gelling Agents for Use in Foodstuffs. 7 pp.
EC 78/612/EEC-78. Council Directive Amending for the First Time Directive 74/329/EEC on the Approximation of the Laws of the Member States Relating to Emulsifiers, Stabilizers, Thickeners and Gelling Agents for Use in Foodstuffs. 4 pp.
EC 80/597/EEC-80. Council Directive Amending for the Second Time Directive 74/329/EEC on the Approximation of the Laws of the Member States Relating to Emulsifiers, Stabilizers, Thickeners and Gelling Agents for Use in Foodstuffs. 3 pp.
EC 85/6/EEC-84. Council Directive Amending for the Third Time Directive 74/329/EEC on the Approximation of the Laws of the Member States Relating to Emulsifiers, Stabilizers, Thickeners and Gelling Agents for Use in Foodstuffs. 1 p.
EC 86/102/EEC-86. Council Directive Amending for the Fourth Time Directive 74/329/EEC on the Approximation of the Laws of the Member States Relating to Emulsifiers, Stabilizers, Thickeners and Gelling Agents for Use in Foodstuffs. 2 pp.
EC 89/393/EEC-89. Council Directive Amending for the Fifth Time Directive 74/329/EEC on the Approximation of the Laws of the Member States Relating to Emulsifiers, Stabilizers, Thickeners and Gelling Agents for Use in Foodstuffs. 1 p.
EC 90/612/EEC-90. Commission Directive Amending Council Directive 78/663/EEC Laying down Specific Purity Criteria for Emulsifiers, Stabilizers, Thickeners and Gelling Agents for Use in Foodstuffs. 2 pp.

—**Stabilizers—Purity**
EC 78/663/EEC-78. Council Directive Laying Down Specific Criteria of Purity for Emulsifiers, Stabilizers, Thickeners and Gelling Agents for Use in Foodstuffs. 23 pp.
EC 82/504/EEC-82. Council Directive Amending Directive 78/663/EEC Laying down Specific Criteria of Purity for Emulsifiers, Stabilizers, Thickeners and Gelling Agents for Use in Foodstuffs. 3 pp.

—**Thickeners**
EC 74/329/EEC-74. Council Directive on the Approximation of the Laws of the Member States Relating to Emulsifiers, Stabilizers, Thickeners and Gelling Agents for Use in Foodstuffs. 7 pp.

INDUSTRY STANDARDS

Food Additives (Cont.)
—Thickeners (Cont.)
EC 78/612/EEC-78. Council Directive Amending for the First Time Directive 74/329/EEC on the Approximation of the Laws of the Member States Relating to Emulsifiers, Stabilizers, Thickeners and Gelling Agents for Use in Foodstuffs. 4 pp.

EC 80/597/EEC-80. Council Directive Amending for the Second Time Directive 74/329/EEC on the Approximation of the Laws of the Member States Relating to Emulsifiers, Stabilizers, Thickeners and Gelling Agents for Use in Foodstuffs. 3 pp.

EC 85/6/EEC-84. Council Directive Amending for the Third Time Directive 74/329/EEC on the Approximation of the Laws of the Member States Relating to Emulsifiers, Stabilizers, Thickeners and Gelling Agents for Use in Foodstuffs. 1 p.

EC 86/102/EEC-86. Council Directive Amending for the Fourth Time Directive 74/329/EEC on the Approximation of the Laws of the Member States Relating to Emulsifiers, Stabilizers, Thickeners and Gelling Agents for Use in Foodstuffs. 2 pp.

EC 89/393/EEC-89. Council Directive Amending for the Fifth Time Directive 74/329/EEC on the Approximation of the Laws of the Member States Relating to Emulsifiers, Stabilizers, Thickeners and Gelling Agents for Use in Foodstuffs. 1 p.

EC 90/612/EEC-90. Commission Directive Amending Council Directive 78/663/EEC Laying down Specific Purity Criteria for Emulsifiers, Stabilizers, Thickeners and Gelling Agents for Use in Foodstuffs. 2 pp.

—Thickeners—Purity
EC 78/663/EEC-78. Council Directive Laying Down Specific Criteria of Purity for Emulsifiers, Stabilizers, Thickeners and Gelling Agents for Use in Foodstuffs. 23 pp.

EC 82/504/EEC-82. Council Directive Amending Directive 78/663/EEC Laying down Specific Criteria of Purity for Emulsifiers, Stabilizers, Thickeners and Gelling Agents for Use in Foodstuffs. 3 pp.

Food Colors
See Also: Colorants; Food Additives

EC COM(93) 153-93. Amended Proposal for a Council Directive on Colours for Use in Foodstuffs. 10 pp.

EC COM(93) 153-93. Amended Proposal for a Council Directive on Colours for Use in Foodstuffs. 22 pp.

SAA AS 2070.6-84. Plastics Materials for Food Contact Use—Part 6: Colourants. 12 pp.

—Medicinal Products
EC 78/25/EEC-78. Council Directive on the Approximation of the Laws of the Member States Relating to the Colouring Matters Which May Be Added to Medicinal Products. 3 pp.

EC 81/464/EEC-81. Council Directive Amending Council Directive 78/25/EEC on the Approximation of the Rules of the Member States Relating to Colouring Matters Which May Be Added to Medicinal Products. 1 p.

—Purity
EC 81/712/EEC-81. First Commission Directive Laying down Community Methods of Analysis for Verifying That Certain Additives Used in Foodstuffs Satisfy Criteria of Purity. 27 pp.

Food Containers
See Also: Containers; Food Packaging Equipment

BSI BS 4874-72. 1972 Catering Container Dimensions. 11 pp.

CEN PREN 631-1-92. Materials and Articles in Contact with Foodstuffs—Catering Containers—Part 1: Dimensions of Containers. 10 pp.

—Aerosol
CNS Z6056-82. Method of Test for Overrun on Food Aerosol Products (Nov)(9649).

—Aluminum
BSI BS 5313-76. 1976 Aluminium Catering Containers and Lids. 8 pp.

CEN PREN 602-91. Wrought Aluminium and Aluminium Alloys—Chemical Composition of the Metal Used for the Production of Materials and Articles Intended to Be in Contact with Food. 7 pp.

—Aluminum Alloys
CEN PREN 602-91. Wrought Aluminium and Aluminium Alloys—Chemical Composition of the Metal Used for the Production of Materials and Articles Intended to Be in Contact with Food. 7 pp.

—Aluminum Castings
CEN PREN 601-91. Cast Aluminium Alloys—Chemical Composition of the Metal Used for the Production of Materials and Articles Intended to Be in Contact with Food. 7 pp.

Food Containers (Cont.)
—Aluminum—Lids
BSI BS 5313-76. 1976 Aluminium Catering Containers and Lids. 8 pp.

—Glass
CNS Z6009-66. Method of Test for Off-Roundness of Glass for Food Storage (Mar)(2576)(R 1972).

CNS Z6010-72. Method of Test for Perpendicularity of Glass Containers for Food Storage (Oct)(2577).

CNS Z6011-72. Method of Test for Stability of Glass Containers for Food Storage (Oct)(2578).

—Glass—Capacity Measurement
BSI BS 5771-81. 1981 Packages for Certain Pre-Packed Foodstuffs; Capacities of Glass and Metal Containers. 9 pp.

—Glass—Hygiene
CNS Z6087-88. Method of Test for the Hygiene of Food Utensils, Containers and Packages Products (Jan)(12218).

—Insulated—Vacuum
BSI BS 6672: Part 1-86. 1986 Insulated Domestic Food Containers Part 1: Vacuum Ware and Insulated Flasks, Jars and Jugs. 12 pp.

BSI BS 6672: Part 1-01. 1986 Amd 1 Insulated Domestic Food Containers Part 1: Specification for Vacuum Ware and Insulated Flasks, Jars and Jugs (AMD 7199) June 15, 1992. 14 pp.

SNZ NZS 5847: Part 1-88. Insulated Domestic Food Containers Part 1: Specification for Vacuum Ware and Insulated Flasks, Jars and Jugs Amend: A, 1988. 12 pp.

—Metal—Hermetic Seals
CNS Z6014-72. Determination of Capacity for Hermetically Sealed Metal Food Containers (Jun)(3347). 2 pp.

—Metal—Hygiene
CNS Z6088-88. Method of Test for the Hygiene of Food Utensils, Containers and Packages Metallic (Jan)(12219).

—Plastic
BSI BS 5496-77. 1977 Plastics Catering Containers and Lids. 8 pp.

—Plastic—Hygiene
CNS Z6089-88. Method of Test for the Hygiene of Food Utensils, Containers and Packages Plastic Products (General Regulation) (Jan)(12220).

CNS Z6090-88. Method of Test for the Hygiene of Food Utensils, Containers and Packages Plastic Products (Classified Regulation) (Jan)(12221).

—Plastic—Lids
BSI BS 5496-77. 1977 Plastics Catering Containers and Lids. 8 pp.

—Plastic—Returnable
JIS Z 1655-84. Plastic Returnable Containers. 13 pp.

—Polypropylene—Bags
BSI BS 7303: Part 1-90. 1990 Sacks for the Transport of Polypropylene Fabric Part 1: Sacks Made of Polypropylene Fabric. 10 pp.

CEN EN 277-89. Sacks for the Transport of Food Aid; Sacks Made of Polypropylene Fabric. 9 pp.

CEN PREN 277-93. Sacks for the Transport of Food Aid—Sacks Made of Woven Polypropylene Fabric. 8 pp.

—Pressurized—Product Retention
CNS Z6051-82. Determination of Percent Product Retention in Pressurized Food Containers (Aug)(9333).

—Stainless Steel
BSI BS 5312-76. 1976 Stainless Steel Catering Containers and Lids. 8 pp.

—Stainless Steel—Lids
BSI BS 5312-76. 1976 Stainless Steel Catering Containers and Lids. 8 pp.

Food Cutters
See Also: Appliances; Cutting Tools; Food Processing Equipment; Meat Cutters

—Electric
CGSB CAN/CGSB-52.31-M87. Cutter, Electric, Food Service. 13 pp.

Food Industry
—Classification
BSI BS 1000: (664)-85. 1985 Universal Decimal Classification (UDC). English Full Edition (664): Preparation and Preservation of Solid Foodstuffs. 33 pp.

Food Industry (Cont.)
—Classification (Cont.)
SNZ NZS/BS 1000 (664)-85. Universal Decimal Classification Preparation and Preservation of Solid Foodstuffs. 36 pp.

—Flow Charts—Symbols
SAA AS 1109-81. Graphical Symbols for Process Flow Diagrams for the Food Industry. 16 pp.

—Pipe Fittings
SAA AS 1528.1-73. Tubes (Stainless Steel) and Tube Fittings for the Food Industry—Part 1: Tubes (R 1993). 4 pp.

SAA AS 1528.2-76. Tubes (Stainless Steel) and Tube Fittings for the Food Industry—Part 2: Screwed Couplings (R 1993). 52 pp.

SAA AS 1528.4-76. Tubes (Stainless Steel) and Tube Fittings for the Food Industry—Part 4: Clamp Liners with Gaskets (R 1993). 20 pp.

—Pipe Fittings—Steel
BSI BS 4825: Part 2-91. 1991 Stainless Steel Tubes and Fittings for the Food Industry and Other Hygienic Applications Part 2: Specification for Bends and Tees (Q). 10 pp.

BSI BS 4825: Part 2-73. 1973 Stainless Steel Pipes and Fittings for the Food Industry Part 2: Bends and Tees. 12 pp.

BSI BS 4825: Part 3-91. 1991 Stainless Steel Tubes and Fittings for the Food Industry and Other Hygienic Applications Part 3: Specification for Clamp Type Couplings (Q). 15 pp.

BSI BS 4825: Part 3-76. 1976 Stainless Steel Pipes and Fittings for the Food Industry Part 3: Clamp Type Couplings. 12 pp.

BSI BS 4825: Part 4-91. 1991 Stainless Steel Tubes and Fittings for the Food Industry and Other Hygienic Applications Part 4: Specification for Threaded (IDF Type) Couplings (Q). 25 pp.

BSI BS 4825: Part 4-77. 1977 Stainless Steel Pipes and Fittings for the Food Industry Part 4: Screwed Type Couplings. 12 pp.

BSI BS 4825: Part 5-91. 1991 Stainless Steel Tubes and Fittings for the Food Industry and Other Hygienic Applications Part 5: Specification for Recessed Ring Joint Type Couplings (Q). 22 pp.

ISO 2851-73. Metal Pipes and Fittings—Stainless Steel Bends and Tees for the Food Industry First Edition. 7 pp.

ISO 2852-93. Stainless Steel Clamp Pipe Couplings for the Food Industry Second Edition. 17 pp.

ISO 2853-93. Stainless Steel Threaded Couplings for the Food Industry Second Edition. 28 pp.

SAA AS 1528.3-75. Tubes (Stainless Steel) and Tube Fittings for the Food Industry—Part 3: Bends and Tees (R 1993). 14 pp.

—Pipes
SAA AS 1528.1-73. Tubes (Stainless Steel) and Tube Fittings for the Food Industry—Part 1: Tubes (R 1993). 4 pp.

—Pipes—Steel
BSI BS 4825: Part 1-91. 1991 Stainless Steel Tubes and Fittings for the Food Industry and Other Hygienic Applications Part 1: Specification for Tubes (Q). 10 pp.

BSI BS 4825: Part 1-72. 1972 Amd 1 Stainless Steel Pipes and Fittings for the Food Industry Part 1: Pipes. 14 pp.

CNS G3131-80. Stainless Steel Sanitary Tubing (Dec)(6668). 5 pp.

ISO 2037-92. Stainless Steel Tubes for the Food Industry Second Edition. 6 pp.

ISO 2851-73. Metal Pipes and Fittings—Stainless Steel Bends and Tees for the Food Industry First Edition. 7 pp.

SAA AS 1528.3-75. Tubes (Stainless Steel) and Tube Fittings for the Food Industry—Part 3: Bends and Tees (R 1993). 14 pp.

Food Mixers
See Also: Appliances; Food Processing Equipment; Food Service Equipment; Mixers

CGSB CAN/CGSB-52.33-M87. Mixing Machine, Electric, Food Service. 15 pp.

—Commercial
CSA C22.2 NO 195-M1987. Motor Operated Food Processing Appliances (Household and Commercial); (Gen Instr 1 Thru 2). 55 pp.

—Household
CSA C22.2 NO 195-M1987. Motor Operated Food Processing Appliances (Household and Commercial); (Gen Instr 1 Thru 2). 55 pp.

—Safety
CEN PREN 453-91. Food Processing Machinery—Dough Mixers—Safety and Hygiene Requirements. 12 pp.

INTERNATIONAL AND NON-U.S. NATIONAL STANDARDS
SUBJECT INDEX
Food

Food Mixers (Cont.)
—Safety (Cont.)
CEN PREN 454-91. Food Processing Machinery—Planetary Mixers—Safety and Hygiene Requirements. 14 pp.

Food Packaging
See Also: Packaging

—Greaseproof Papers
SAA AS 1814-75. High Wet-Strength Greaseproof Paper for Wrapping Dairy Products. 24 pp.

Food Packaging Equipment
Scope Note: For additional listings, see specific food items See Also: Food Containers; Packaging

—Aluminum Foil
BSI BS 5439-77. 1977 Aluminium Foil Catering Containers. 8 pp.
CGSB 43-GP-148M-84. Foil, Aluminum, Annealed. 9 pp.
SAA AS 1813-75. Aluminium Foil/Vegetable Parchment Laminates for Wrapping Dairy Products. 8 pp.

—Bags—Insulated
BSI BS 6672: Part 2-86. 1986 Insulated Domestic Food Containers Part 2: Insulated Bags and Boxes. 10 pp.
BSI BS 6672: Part 2-01. 1986 Amd 1 Insulated Domestic Food Containers Part 2: Specification for Insulated Bags and Boxes (AMD 7200) June 15, 1992. 11 pp.

—Boxes (Containers)—Insulated
BSI BS 6672: Part 2-86. 1986 Insulated Domestic Food Containers Part 2: Insulated Bags and Boxes. 10 pp.
BSI BS 6672: Part 2-01. 1986 Amd 1 Insulated Domestic Food Containers Part 2: Specification for Insulated Bags and Boxes (AMD 7200) June 15, 1992. 11 pp.

—Cans
CNS Z5018-85. Metal Cans of Irregular Shape for Foods (Dec)(2180). 4 pp.
CNS Z6019-85. Method of Test for Rounded Metal Cans for Foods (Nov)(4060). 6 pp.
ISO 1361-83. Light Gauge Metal Containers—Open-Top Cans—Round Cans—Internal Diameters Third Edition. 4 pp.

—Cans—Capacity Measurement
BSI BS 5771-81. 1981 Packages for Certain Pre-Packed Foodstuffs; Capacities of Glass and Metal Containers. 9 pp.
BSI BS 5774-84. 1984 Internal Diameters of Round Hermetically Sealed Metal Cans for Food and Non-Carbonated Drinks. 4 pp.
BSI BS 6992: Part 1-88. (OBSOLESCENT) 1988 Capacities and Related Cross Sections of Open-Top Cans for Food and Drinks Part 1: Recommendations for Cans for General Food. 7 pp.

—Cans—Hermetic Seals
JIS Z 1571-85. Hermetically Sealed Metal Cans for Food and Drink. 10 pp.

—Cans—Inspection
CNS N6012-82. Method of Test for Canned Food (Packaging Inspection) (May)(968). 1 p.
CNS N6017-78. Method of Test for Canned Food (Inspection for Interior of Can) (Jul)(973). 1 p.

—Cans—Inspection—Grading/Scoring
CNS N6024-78. Method of Test for Canned Food (Grading and Scoring) (Jul)(980). 1 p.

—Cans—Vacuum Testing
CNS N6015-78. Method of Test for Canned Food (Measurement for Degree of Vacuum) (Jul)(971). 2 pp.

—Cans—Visual Inspection
CNS N6013-83. Method of Test for Canned Food: Appearance of Can (Nov)(969). 2 pp.

—Greaseproof Papers
SAA AS 1764-75. Vegetable Parchment for Wrapping Dairy Products. 24 pp.
SAA AS 1813-75. Aluminium Foil/Vegetable Parchment Laminates for Wrapping Dairy Products. 8 pp.

—Identification Systems
CNS Z5031-82. Labeling of Packaged Foods (Jun)(3192). 2 pp.

—Materials—Identification Systems
DIN ENGL 7725-67. Articles Made of Rubber, Plastics, Other Polymers, Paper and Cardboard to Conform with the Foodstuffs Law; Marking (Jan). 3 pp.

Food Packaging Equipment (Cont.)
—Odors
BSI BS 3755-64. 1964 Amd 1 Methods of Test for the Assessment of Odour from Packaging Used for Foodstuffs. 17 pp.
BSI PD 6459-71. (WITHDRAWN) 1971 Guidance on Avoiding Odour from Packaging Materials Used for Foodstuffs. 48 pp.
DIN ENGL 10955-83. Sensory Analysis; Testing of Container Materials and Containers for Food Products (Apr). 6 pp.

—Plastic Sheets
CNS Z6075-86. Method of Test for Plastic Films for Food Packaging (Jan)(10591). 6 pp.
JIS Z 1707-75. Plastic Films for Food Packaging (R 1978). 18 pp.

—Pouches
CGSB CAN/CGSB-32.302-M87. Use of Flexible Laminated Pouches for Thermally Processed Foods. 18 pp.
CNS N5214-85. Retort Pouched Foods (General Rule) (Feb)(11210). 2 pp.

—Pouches—Heat Seal Strength
CNS N6198-85. Method of Test for Retort Pouch with Food (Apr)(11247). 2 pp.

—Pouches—Pressure Measurement
CNS N6198-85. Method of Test for Retort Pouch with Food (Apr)(11247). 2 pp.

—Pouches—Puncture Resistance
CNS N6198-85. Method of Test for Retort Pouch with Food (Apr)(11247). 2 pp.

Food Preparation Equipment
See Also: Appliances; Catering; Cooking Appliances; Food Processing Equipment; Rice Dryers
BSI BS EN 60619-93. 1993 Electrically Operated Food Preparation Appliances—Measuring Methods (L). 33 pp.
CEN PREN 1116-93. Kitchen Furniture—Co-Ordinating Sizes for Kitchen Furniture and Kitchen Appliances. 8 pp.
CENELEC EN 60619-93. Electrically Operated Food Preparation Appliances Measuring Methods (IEC 619: 1993). 28 pp.
CSA C22.2 NO 1335.2.14-93. Portable Electrical Motor-Operated and Heating Appliances: Particular Requirements for Electrical Motor-Operated Kitchen Appliances; (Gen Instr 1). 63 pp.
IEC 619-93. Electrically Operated Food Preparation Appliances—Measuring Methods Second Edition; (CENELEC EN 60619: 1993) (Corrigendum—July 1993). 56 pp.

—Antimicrobial Agents
CGSB CAN/CGSB-2.161-M91. Assessment of Efficacy of Antimicrobial Agents for Use on Environmental Surfaces and Medical Devices. 8 pp.

Food Processing Appliances
Use: Food Processing Equipment

Food Processing Equipment
Use For: Catering Equipment; Kitchen Fitments
See Also: Appliances; Blenders; Choppers; Coffee Grinders; Cooking Appliances; Cooking Utensils; Corn Grinders; Deep Fat Fryers; Food Cutters; Food Mixers; Food Preparation Equipment; Food Warmers; Juicers; Liquid Mixers; Measuring Cups; Measuring Spoons; Meat Cutters; Meat Grinders; Meat Slicers; Pasta Machines; Peelers; Potato Peelers
BSI BS 1195: Part 1-73. (WITHDRAWN) 1973 Amd 1 Kitchen Fitments and Equipment Part 1: Imperial Units with Metric Equivalents (Superseded by BS 6222: Part 1: 1982, Part 2: 1992 & Part 3: 1988). 41 pp.
BSI BS 1195: Part 2-72. (WITHDRAWN) 1972 Amd 1 Kitchen Fitments and Equipment Part 2: Metric Units. 36 pp.
BSI BS 3705-72. (OBSOLESCENT) 1972 Recommendations for Provision of Space for Domestic Kitchen Equipment. 12 pp.
BSI BS 3999: Part 3-67. (WITHDRAWN) 1967 Amd 2 Methods of Measuring the Performance of Household Electrical Appliances Part 3: Food Preparation Machines (Superseded by BS EN 60619: 1993). 36 pp.
BSI BS EN 60619-93. 1993 Electrically Operated Food Preparation Appliances—Measuring Methods (L). 33 pp.
CENELEC EN 60619-93. Electrically Operated Food Preparation Appliances Measuring Methods (IEC 619: 1993). 28 pp.
CNS S1120-81. Kitchen Equipment (Apr)(7323).
CNS S2051-81. Method of Test for Kitchen Equipments (May)(7444).

Food Processing Equipment (Cont.)
CSA C22.2 NO 1335.2.14-93. Portable Electrical Motor-Operated and Heating Appliances: Particular Requirements for Electrical Motor-Operated Kitchen Appliances; (Gen Instr 1). 63 pp.
IEC 619-93. Electrically Operated Food Preparation Appliances—Measuring Methods Second Edition; (CENELEC EN 60619: 1993) (Corrigendum—July 1993). 56 pp.
ISO 5731-78. Kitchen Equipment—Limit of Size First Edition. 4 pp.
ISO 5732-78. Kitchen Equipment—Sizes of Openings for Built-in Appliances First Edition. 5 pp.
JIS S 1005-84. Kitchen Equipments.

—Antimicrobial Agents
CGSB CAN/CGSB-2.161-M91. Assessment of Efficacy of Antimicrobial Agents for Use on Environmental Surfaces and Medical Devices. 8 pp.

—Combustion Control Equipment
BSI BS 6047: Part 1-81. 1981 Flame Supervision Devices for Domestic, Commercial and Catering Gas Appliances Part 1: Heat Sensitive Types. 11 pp.
BSI BS 6067-81. 1981 Amd 1 Multifunctional Gas Controls for Domestic, Commercial and Catering Appliances. 12 pp.

—Commercial
CSA C22.2 NO 195-M1987. Motor Operated Food Processing Appliances (Household and Commercial); (Gen Instr 1 Thru 2). 55 pp.

—Detergents
CNS S1085-75. Detergents for Vegetables and Kitchen Utensils (Jul)(3800).

—Energy Consumption
CEN PREN 203-2-92. Gas Heated Catering Equipment—Part 2: Rational Use of Energy. 10 pp.

—Ergonomics
ISO 3055-85. Kitchen Equipment—Coordinating Sizes Second Edition. 18 pp.

—Gas—Hose Fittings
BSI BS 669: Part 2-88. 1988 Flexible Hose, Covers, End Fittings and Sockets for Domestic Appliances Burning 1st and 2nd Family Gases Part 2: Corrugated Metallic Flexible Hoses, Covers, End Fittings and Sockets for Catering Appliances Burning 1st and 2nd Family Gases. 15 pp.

—Gas—Hoses
BSI BS 669: Part 2-88. 1988 Flexible Hose, Covers, End Fittings and Sockets for Domestic Appliances Burning 1st and 2nd Family Gases Part 2: Corrugated Metallic Flexible Hoses, Covers, End Fittings and Sockets for Catering Appliances Burning 1st and 2nd Family Gases. 15 pp.

—Household
BSI BS 6222: Part 1-82. 1982 Amd 1 Domestic Kitchen Equipment Part 1: Coordinating Dimensions. 8 pp.
BSI BS 6222: Part 2-92. 1992 Domestic Kitchen Equipment Part 2: Specification for Strength Requirements and Methods of Test for Fitted Kitchen Units (Supersedes BS 1195: Part 1: 1973 & Part 2: 1972) (L). 22 pp.
BSI BS 6222: Part 2-01. 1992 Amd 1 Domestic Kitchen Equipment Part 2: Specification for Strength Requirements and Methods of Test for Fitted Kitchen Units (AMD 7693) July 15, 1993 (L). 23 pp.
CSA C22.2 NO 195-M1987. Motor Operated Food Processing Appliances (Household and Commercial); (Gen Instr 1 Thru 2). 55 pp.
SAA AS 3162-90. Approval and Test Specification—Electric Kitchen Machines for Household Use (This is a Joint Standard with SANZ NZS 3162) Amdt 1 January 1991 Amdt 2 March 1992 Amdt 3 October 1992 Amdt 4 June 1993 (in Prof. Package 28). 1 p.
SNZ NZS/AS 3162-90. Approval and Test Specification—Electric Kitchen Machines for Household Use Amend: 1; 2; 3, 1992 (This is a Joint Standard with SAA AS 3162). 8 pp.

—Household—Safety
CSA C22.2 NO 1335.1-93. Portable Electrical Motor-Operated and Heating Appliances: General Requirements; (Gen Instr 1). 208 pp.

—Installation
BSI BS 6173-90. 1990 Installation of Gas Catering Appliances for Use in All Types of Catering Establishments (1st, 2nd and 3rd Family Gases). 24 pp.
BSI BS 6173-82. 1982 Installation of Gas Catering Appliances (2nd Family Gases). 18 pp.

—Lubricating Oils
MOD UK DSTAN 91-36-77. Lubricating Oil: White Joint Service Designation: OM-17 Issue 1. 7 pp.

INDUSTRY STANDARDS

Food Processing Equipment (Cont.)
—Safety
BSI BS 3456: Sec 202.14-90. 1990 Safety of Household Electrical Appliances Part 202: Particular Requirements Section 202.14: Electric Kitchen Machines. 31 pp.

BSI BS 3456: Sec 202.14-02. 1990 Amd 2 Safety of Household and Similar Electrical Appliances Part 202: Particular Requirements Section 202.14: Electric Kitchen Machines (AMD 7532) January, 15, 1993 (L). 50 pp.

BSI BS 3456: Sec 202.14-03. 1990 Amd 3 Safety of Household and Similar Electrical Appliances Part 202: Particular Requirements Section 202.14: Electric Kitchen Machines (AMD 7800) June 15, 1993 (SUPERSEDES BS 3456: SECTION 3. 12: 1979). 57 pp.

BSI BS EN 203-1-93. 1993 Gas Heated Catering Equipment Part 1: Safety Requirements (Supersedes BS 5314: Parts 1,2,3,4,5,6,7: 1976, 8,9,11,12: 1979, 10,13: 1982) (L). 40 pp.

CEN PREN 203-88. Gas-Heated Catering Equipment Part 1: Safety Requirements. 71 pp.

CEN EN 203-1-92. Gas Heated Catering Equipment—Part 1: Safety Requirements. 73 pp.

CEN EN 203-1-93. CORRIGENDUM Gas Heated Catering Equipment—Part 1: Safety Requirements. 2 pp.

CENELEC HD 261-80. Particular Specification for Kitchen Machines. 7 pp.

CENELEC HD 261 S1-84. Particular Specification for Kitchen Machines (Replaced by EN 60 335-2-14-1988). 3 pp.

CENELEC HD 261.2-77. Particular Specification for Kitchen Machines. 5 pp.

CENELEC HD 261.3-78. Particular Specification for Kitchen Machines. 5 pp.

CENELEC HD 261.4 S1-88. Particular Specification for Kitchen Machines. 5 pp.

CENELEC EN 60 335-2-14-88. Safety of Household and Similar Electrical Appliances Part 2: Particular Requirements for Electric Kitchen Machines (Replaces HD 261 S1-1984). 11 pp.

CENELEC EN 60335-2-14/A1-90. AMD 1 Safety of Household and Similar Electrical Appliances Part 2: Particular Requirements for Electric Kitchen Machines; (Replaces HD 261 S1:1984). 3 pp.

CENELEC EN 60335-2-14/A51-91. AMD 51 Safety of Household and Similar Electrical Appliances Part 2: Particular Requirements for Electric Kitchen Machines; (Replaces HD 261 S1:1984). 6 pp.

CENELEC EN 60335-2-14/A52-92. AMD 52 Safety of Household and Similar Electrical Appliances Part 2: Particular Requirements for Electric Kitchen Machines. 4 pp.

DIN VDE 0700 Pt 251-86. Safety of Household and Similar Electrical Appliances; Food Processors and Food Choppers (Oct). 21 pp.

IEC 335 Pt 2-14-84. Safety of Household and Similar Electrical Appliances Part 2: Particular Requirements for Electric Kitchen Machines Second Edition; (Amendment 1-1989) (Amendment 2-1990). 69 pp.

JIS B 9650-88. General Design Rules for Safety and Sanitation of Food Processing Machinery. 10 pp.

JIS B 9651-88. Design Rules for Safety and Sanitation of Baking Machinery. 29 pp.

JIS B 9652-88. Design Rules for Safety and Sanitation of Cake Making Machinery. 35 pp.

JIS B 9653-88. Design Rules for Safety and Sanitation of Meat Processing Machinery. 18 pp.

JIS B 9654-88. Design Rules for Safety and Sanitation of Marine Product Machinery. 17 pp.

JIS B 9655-90. Design Rules for Safety and Sanitation of Processing Machinery Used in Flouring Mill. 8 pp.

JIS B 9657-90. Design Rules for Safety and Sanitation of Drink Making Machinery. 18 pp.

—Sanitation
JIS B 9650-88. General Design Rules for Safety and Sanitation of Food Processing Machinery. 10 pp.

JIS B 9651-88. Design Rules for Safety and Sanitation of Baking Machinery. 29 pp.

JIS B 9652-88. Design Rules for Safety and Sanitation of Cake Making Machinery. 35 pp.

JIS B 9653-88. Design Rules for Safety and Sanitation of Meat Processing Machinery. 18 pp.

JIS B 9654-88. Design Rules for Safety and Sanitation of Marine Product Machinery. 17 pp.

JIS B 9655-90. Design Rules for Safety and Sanitation of Processing Machinery Used in Flouring Mill. 8 pp.

JIS B 9657-90. Design Rules for Safety and Sanitation of Drink Making Machinery. 18 pp.

—Schools
BSI HANDBOOK NO. 17-63. 1963 School Catering Equipment and Furniture. 59 pp.

—Ships
MOD UK NES 122-80. Catering Equipment Issue 1 (07.80). 81 pp.

MOD UK NES 122: Part 2-89. Catering Equipment Part 2: Proprietary Adapted Equipment Issue 2 (02.89). 109 pp.

—Ships—Design
MOD UK NES 122: Part 1-88. Catering Equipment Part 1: Equipment to MOD(PE) Design Standards Issue 3 (11.88). 112 pp.

—Submarines
MOD UK NES 122-80. Catering Equipment Issue 1 (07.80). 81 pp.

—Vitreous Enameled
BSI BS 3831-64. 1964 Vitreous Enamel Finishes for Domestic and Catering Appliances. 8 pp.

Food Service Equipment
Scope Note: For additional listings, use a more specific term *Use For:* Catering Equipment *See Also:* Cookware; Deep Fat Fryers; Food Mixers; Garbage Disposals; Hot Plates (Warmers); Kettles; Rotisseries; Steam Jacketed Kettles

—Containers
SAA AS 4027-92. Food-Service Container Dimensions. 6 pp.

—Microwave Ovens
CSA C22.2 NO 109-M1981. Commercial Cooking Appliances; (Gen Instr 1 Thru 3). 36 pp.

CSA CAN/CSA-C22. 2 NO 150-M89. Microwave Ovens; (Gen Instr 1). 50 pp.

CSA 1169 Bull. Electrical Bulletin 1169 June 27, 1978 to C22.2 NO 150. 2 pp.

CSA 1197 Bull. Electrical Bulletin 1197 November 6, 1978 to C22.2 NO 150. 4 pp.

—Refrigeration Equipment
CGSB 32.72M-91. Handling, Packaging and Labelling of Meat, Poultry and Fish for Food Services. 10 pp.

Food Starches
Use For: Starches (Food) *See Also:* Pasta; Tapioca
CNS N5012-84. Edible Starch (Apr)(1069). 1 p.

—Modified
EC COM(84) 726-84. Proposal for a Council Directive on the Approximation of the Laws of the Member States Relating to Modified Starches Intended for Human Consumption. 16 pp.

—Pasta
CNS N5033-87. Starch Vermicelli (Oct)(1484). 1 p.

—Potato—Paper Bags
JIS Z 1509-90. Kraft Paper Sacks for Potato Starch. 12 pp.

—Sweet Potato
CNS K1155-86. Sweet Potato Starch (for Industrial Use) (Jun)(3238).

CNS K6314-86. Method of Test for Sweet Potato Starch (for Industrial Use) (Jun)(3312).

—Sweet Potato—Paper Bags
JIS Z 1530-82. Sewn Kraft Paper Sacks for Sweet Potato Starch. 8 pp.

Food Storage
Scope Note: See the subheading Storage under specific types of food *See Also:* Food Containers

Food Warmers
Use For: Dish Warmers *See Also:* Appliances; Bottle Warmers; Broilers; Food Processing Equipment; Hot Plates (Warmers); Steam Tables
BSI BS 3547-62. 1962 Amd 2 Electrically-Heated Food Conveyors and Carriers. 21 pp.

BSI BS 4167: Part 11-70. 1970 Amd 1 Electrically-Heated Catering Equipment Part 11: Hot Cupboards. 43 pp.

BSI BS 5314: Part 11-79. (WITHDRAWN) 1979 Amd 2 Gas Heated Catering Equipment Part 11: Hot Cupboards (AMD 5872) May 31, 1989 (Superseded by BS EN 203-1: 1993). 25 pp.

CGSB CAN/CGSB-52.2-M88. Warm Food Tables, Steam or Dry. 13 pp.

CSA CAN/CSA-C22. 2 NO 64-M91. Household Cooking and Liquid-Heating Appliances; (Gen Instr 1 Thru 2). 91 pp.

—Buns
CSA CAN/CSA-C22. 2 NO 64-M91. Household Cooking and Liquid-Heating Appliances; (Gen Instr 1 Thru 2). 91 pp.

—Commercial—Electric
CSA C22.2 NO 109-M1981. Commercial Cooking Appliances; (Gen Instr 1 Thru 3). 36 pp.

CSA 1169 Bull. Electrical Bulletin 1169 June 27, 1978 to C22.2 NO 109. 2 pp.

—Commercial—Electric—Safety
BSI BS EN 60335-2-49-92. 1992 Safety of Household and Similar Electrical Appliances Part 2: Particular Requirements Section 2.49: Commercial Catering Electric Hot Cupboards (L). 22 pp.

BSI BS EN 60335-2-49-01. 1992 Amd 1 Safety of Household and Similar Electrical Appliances Part 2: Particular Requirements Section 2.49: Commercial Catering Electric Hot Cupboards (AMD 7516) January 15, 1993 (L). 32 pp.

CENELEC EN 60335-2-49-90. Safety of Household and Similar Electrical Appliances Part 2 Particular Requirements for Commercial Catering Electric Hot Cupboards; (IEC 335-2-49:1988, Modified). 9 pp.

CENELEC EN 60335-2-49-90. Safety of Household and Similar Electrical Appliances Part 2: Particular Requirements for Commercial Catering Electric Hot Cupboards; (IEC 335-2-49:1988, Modified). 17 pp.

CENELEC EN 60335-2-49/A1-92. AMD 1 Safety of Household and Similar Electrical Appliances Part 2: Particular Requirements for Commercial Catering Electric Hot Cupboards; (IEC 335-2-49:1988, Modified). 5 pp.

CENELEC EN 60335-2-49/A1-92. AMD 1 Safety of Household and Similar Electrical Appliances Part 2: Particular Requirements for Commercial Catering Electric Hot Cupboards; (IEC 335-2-49:1988, Modified). 7 pp.

CENELEC EN 60335-2-49/A1-92. CORRIGENDUM Safety of Household and Similar Electrical Appliances Part 2: Particular Requirements for Commercial Catering Electric Hot Cupboards (IEC 335-2-49:1988/A1: 1990). 1 p.

IEC 335 Pt 2-49-88. Safety of Household and Similar Electrical Appliances Part 2: Particular Requirements for Commercial Electric Hot Cupboards First Edition; (Amendment 1-1990) (CENELEC EN 60335-2-49: 1990). 44 pp.

—Electric—Safety
BSI BS EN 60335-2-12-91. 1991 Safety of Household and Similar Electrical Appliances Part 2: Particular Requirements Section 2.12: Warming Plates and Similar Appliances. 20 pp.

CENELEC EN 60 335-2-12-90. Safety of Household and Similar Electrical Appliances Part 2: Particular Requirements for Warming Plates and Similar Appliances. 7 pp.

IEC 335 Pt 2-12-92. Safety of Household and Similar Electrical Appliances Part 2: Particular Requirements for Warming Plates and Similar Appliances Fourth Edition. 30 pp.

—Household—Electric
BSI BS 3999: Part 16-93. 1993 Methods of Measuring the Performance of Household Electrical Appliances Part 16: Electric Warming Plates (IEC 496: 1975). 11 pp.

IEC 496-75. Methods for Measuring Performance of Electric Warming Plates for Household and Similar Purposes First Edition; (Amendment 1-1977) (Amendment 2-1992). 26 pp.

Food Waste Disposal Units
Use: Garbage Disposals

Foodstuffs
Use: Food

Foot and Mouth Disease
Use For: Hoof-and-Mouth Disease *See Also:* Animal Diseases

—Control Measures
EC COM(82) 529-82. Proposal for a Council Directive amending Directives 64/432/EEC 72/461/EEC as regards certain measures relating to foot-and-mouth disease, Aujeszky's disease and swine vesicular disease. 5 pp.

EC 85/511/EEC-85. Council Directive Introducing Measures for the Control of Foot and Mouth Disease. 8 pp.

EC 90/423/EEC-90. Council Directive Amending Directive 85/511/EEC Introducing Community Measures for the Control of Foot-and-Mouth Disease, Directive 64/432/EEC on Animal Health, Problems Affecting Intra-Community Trade in Bovine Animals and Swine. 9 pp.

—Control Measures—Financial Aid
EC 90/424/EEC-90. Council Decision on Expenditure in the Veterinary Field. 9 pp.

Foot Bridges
See Also: Bridges (Structures)

—Bearing Capacity
DIN ENGL 1072-85. Road and Foot Bridges; Design Loads (Dec). 14 pp.

INTERNATIONAL AND NON-U.S. NATIONAL STANDARDS
SUBJECT INDEX

Foot Bridges (Cont.)

—Steel—Construction
DIN ENGL 18809-87. Steel Road Bridges and Foot Bridges; Design and Construction (Sept) (Together with DIN 18800 Pt 1, Mar. 1981 Ed., and DIN 18800 Pt 7, May 1983 Ed., Supsds. DIN 1073, July 1974 Ed., Suppl. to DIN 1073, July 1974 Ed., DIN 1079, Sept. 1970 Ed., and DIN 4101, 7/74 Ed.). 13 pp.

—Steel—Design
DIN ENGL 18809-87. Steel Road Bridges and Foot Bridges; Design and Construction (Sept) (Together with DIN 18800 Pt 1, Mar. 1981 Ed., and DIN 18800 Pt 7, May 1983 Ed., Supsds. DIN 1073, July 1974 Ed., Suppl. to DIN 1073, July 1974 Ed., DIN 1079, Sept. 1970 Ed., and DIN 4101, 7/74 Ed.). 13 pp.

Foot Controls
See Also: Manual Controls

—Automotive
ISO 3409-75. Passenger Cars—Lateral Spacing of Foot Controls First Edition. 4 pp.
JIS D 0022-84. Lateral Spacing of Foot Controls for Automobiles.

Foot Pedals
Use: Pedals

Foot Powders

—Comprehensive Testing
MOD UK M 9507/65. Determination of Foot Powder (No Information).

Foot Prostheses
See Also: Joint Prostheses; Leg Prostheses
JIS T 9212-85. Artificial Feet and Ankle Joints.

Foot Valves
See Also: Backflow Preventers; Valves

—Ships
CNS F3063-80. Ships' Foot Valves (Sep)(6443).
JIS F 3056-90. Ships' Foot Valves.
JIS F 3056-68. Ships' Foot Valves (R 1971). 6 pp.

Foot Warmers (Electric)
See Also: Appliances
JIS C 9209-75. Electric KOTATSU.

Footings
See Also: Beams (Supports); Columns (Supports); Foundations; Structural Members; Walls

—Concrete—Asphalt Coatings
CGSB CAN/CGSB-37.3-M89. Application of Emulsified Asphalts for Dampproofing or Waterproofing. 9 pp.
CGSB 37-GP-36M-76. Application of Filled Cutback Asphalts for Dampproofing and Waterproofing, Standard for. 7 pp.
CGSB 37-GP-37M-77. Application of Hot Asphalt for Dampproofing or Waterproofing, Standard for. 8 pp.

—Masonry—Asphalt Coatings
CGSB CAN/CGSB-37.3-M89. Application of Emulsified Asphalts for Dampproofing or Waterproofing. 9 pp.
CGSB 37-GP-36M-76. Application of Filled Cutback Asphalts for Dampproofing and Waterproofing, Standard for. 7 pp.
CGSB 37-GP-37M-77. Application of Hot Asphalt for Dampproofing or Waterproofing, Standard for. 8 pp.

Foots Content Analysis
Use: Residue Content Analysis

Footways
Use: Pedestrian Roads

Footwear
See Also: Boots (Footwear); Clogs (Footwear); Clothing; Combat Uniforms; Footwear Linings; Heels (Footwear); Hosiery; Insoles (Footwear); Laces (Footwear); Protective Clothing; Protective Footwear; Sabots; Safety Shoes; Sandals; Shoe Machinery; Shoes (Footwear); Slippers (Footwear); Socks; Soles (Footwear); Uppers (Footwear)
BSI BS 5131: INTRODUCTION-75. (WITHDRAWN) 1975 Amd 2 Methods of Test for Footwear and Footwear Materials Introduction (Superseded by BS 5131: Part 0: 1990). 11 pp.
BSI BS 5131: Part 0-90. 1990 Methods of Test for Footwear and Footwear Materials Part 0: General Introduction. 11 pp.

Footwear (Cont.)

—Adhesives
BSI BS 5131: Sec 1.4-79. 1979 Methods of Test for Footwear and Footwear Materials Part 1: Adhesives Section 1.4: Heat Activation Life of Adhesives. 3 pp.
CEN PREN 522-91. Adhesives for Leather and Footwear Materials—Bond Strength—Minimum Requirements and Adhesive Classification. 17 pp.

—Adhesives—Creep Properties
BSI BS 5131: Sec 1.1-91. 1991 Methods of Test for Footwear and Footwear Materials Part 1: Adhesives Section 1.1: Resistance of Adhesive Joints to Heat (Creep Test). 9 pp.
BSI BS 5131: SUB SEC 1.1.1-76. (WITHDRAWN) 1976 Amd 1 Methods of Test for Footwear and Footwear Materials Part 1: Adhesives Section 1.1: Resistance of Adhesive Joints to Heat and to Peeling Subsec 1.1.1: Resist. to Heat (Creep Test) (Superseded by BS 5131: Section 1.1: 1991). 5 pp.
BSI BS 5131: Sec 1.3-91. 1991 Methods of Test for Footwear and Footwear Materials Part 1: Adhesives Section 1.3: Preparation of Test Assemblies Using Adhesives (Other Than Hot Melt Adhesives) for Heat Resistance (Creep) and Peel Tests. 15 pp.

—Adhesives—Green Strength
BSI BS 5131: Sec 1.9-85. 1985 Methods of Test for Footwear and Footwear Materials Part 1: Adhesives Section 1.9: Green Strength of Adhesive Joints. 5 pp.

—Adhesives—Peel Strength
BSI BS 5131: SUBSEC 1.1.2-76. (WITHDRAWN) 1976 Methods of Test for Footwear and Footwear Materials: Part 1: Adhesives: Section 1.1: Resistance of Adhesive Joints to Heat and to Peeling: Subsection 1.1.2: Resistance to Peeling (Superseded by BS 5131: Section 1.2: 1991). 2 pp.
BSI BS 5131: Sec 1.2-91. 1991 Methods of Test for Footwear and Footwear Materials Part 1: Adhesives Section 1.2: Resistance of Adhesive Joints to Peeling. 12 pp.
BSI BS 5131: Sec 1.3-91. 1991 Methods of Test for Footwear and Footwear Materials Part 1: Adhesives Section 1.3: Preparation of Test Assemblies Using Adhesives (Other Than Hot Melt Adhesives) for Heat Resistance (Creep) and Peel Tests. 15 pp.

—Adhesives—Shear Testing
BSI BS 5131: Sec 1.8-81. 1981 Methods of Test for Footwear and Footwear Materials Part 1: Adhesives Sectiion 1.8: Rate of Bond Strength Development in Shear of Hot Melt Adhesives for Lasting. 6 pp.

—Adhesives—Storage
BSI BS 5131: Sec 1.6-79. 1979 Methods of Test for Footwear and Footwear Materials Part 1: Adhesives Section 1.6: Recommended Environmental Storage Conditions for Adhesive Joints Prior to Heat Resistance or Peeling Tests. 2 pp.

—Adhesives—Test Specimens
BSI BS 5131: SUB SEC 1.1.3-76. (WITHDRAWN) 1976 Amd 1 Methods of Test for Footwear and Footwear Materials Part 1: Adhesives Section 1.1: Resistance of Adhesive Joints to Heat and to Peeling Subsection 1.1.3: Preparation of Test Assemblies for Adhesion Tests. 6 pp.
BSI BS 5131: Sec 1.7-91. 1991 Methods of Test for Footwear and Footwear Materials Part 1: Adhesives Section 1.7: Preparation of Test Assemblies Using Hot Melt Adhesives for Heat Resistance (Creep) and Peel Tests. 8 pp.

—Aging Testing
BSI BS 5131: Sec 5.3-90. 1990 Methods of Test for Footwear and Footwear Materials Part 5: Testing of Complete Footwear Section 5.3: Resistance of Complete Footwear to Heat. 5 pp.

—Antistatic
SNZ NZS 5808-80. Specification for Electrically Conducting and Antistatic Rubber Footwear Amend: A, 1980. 8 pp.

—Buckle Assemblies
BSI BS 5131: Sec 5.11-81. 1981 Footwear and Footwear Materials Part 5: Testing of Complete Footwear Section 5.11: Determination of the Strength of Buckle Fastening Assemblies. 12 pp.

—Conductive
ISO 2024-81. Rubber Footwear, Lined Conducting—Specification Second Edition. 9 pp.
SNZ NZS 5808-80. Specification for Electrically Conducting and Antistatic Rubber Footwear Amend: A, 1980. 8 pp.

—Glossaries
ISO 10335-90. Rubber and Plastics Footwear—Nomenclature First Edition; (Previously Published as Section 5100 of ISO 1382 -1982). 10 pp.

Footwear (Cont.)

—High Temperature Testing
JIS K 6543-77. Testing Method for Heat Resistance of Air Dry Leather of Footwear Construction.

—Identification Systems
BSI BS 4981-84. 1984 Mondopoint Footwear Sizing and Marking System. 4 pp.
BSI BS 5833-88. 1988 Scheme for Labelling of Footwear. 12 pp.

—Orthopedic—Measurement
BSI BS 5943-80. 1980 Amd 1 Measurement and Recording for Orthopaedic Footwear (AMD 5866) May 31, 1989. 11 pp.

—Paperboard
BSI BS 5131: Sec 4.4-90. 1990 Methods of Test for Footwear and Footwear Materials Part 4: Other Components Section 4.4: Heel Pin Strength of Fibreboard. 8 pp.
BSI BS 5131: Sec 4.4-75. (WITHDRAWN) 1975 Methods of Test for Footwear and Footwear Materials Part 4: Other Components Section 4.4: Heel Pin Holding Strength of Fibreboard. 2 pp.

—Paperboard—Abrasion Testing
BSI BS 5131: Sec 4.12-90. 1990 Methods of Test for Footwear and Footwear Materials Part 4: Other Components Section 4.12: Fastness of Fibreboard Finishes to Rubbing in the Presence of Water and Perspiration. 10 pp.

—Paperboard—Colorfastness Testing
BSI BS 5131: Sec 4.12-90. 1990 Methods of Test for Footwear and Footwear Materials Part 4: Other Components Section 4.12: Fastness of Fibreboard Finishes to Rubbing in the Presence of Water and Perspiration. 10 pp.

—Paperboard—Folding Endurance
BSI BS 5131: Sec 4.2-90. 1990 Methods of Test for Footwear and Footwear Materials Part 4: Other Components Section 4.2: Flexing Index of Fibreboard. 6 pp.

—Paperboard—Tear Strength
BSI BS 5131: Sec 4.3-90. 1990 Methods of Test for Footwear and Footwear Materials Part 4: Other Components Section 4.3: Resistance of Fibreboard to Stitch-Tear. 6 pp.

—Paperboard—Tensile Testing
BSI BS 5131: Sec 4.5-90. 1990 Methods of Test for Footwear and Footwear Materials Part 4: Other Components Section 4.5: Tensile Strength of Fibreboard. 7 pp.

—Sizes
BSI BS 4981-84. 1984 Mondopoint Footwear Sizing and Marking System. 4 pp.

—Sizes—Interchangeable—Military
NATO STANAG 2339 ED 1 AMD 4-75. Size Marking for Interchangeability of Operational Footwear. 9 pp.

—Steel Shanks
BSI BS 5131: Sec 4.18-85. 1985 Methods of Test for Footwear and Footwear Materials Part 4: Other Components Section 4.18: Longitudinal Stiffness of Steel Shanks. 6 pp.

—Water Resistance Testing
BSI BS 5131: Sec 5.5-78. 1978 Methods of Test for Footwear and Footwear Materials Part 5: Testing of Complete Footwear Section 5.5: Resistance of Finished Footwear to Water Penetration (Trough Test). 4 pp.

—Wear Testing
BSI BS 5131: Sec 6.1-81. 1981 Methods of Test for Footwear and Footwear Materials Part 6: Codes of Practice Section 6.1: Comparative Evaluation by Wear Trial of Materials, Components or Construction. 3 pp.
BSI BS 5131: Sec 6.2-79. 1979 Methods of Test for Footwear and Footwear Materials Part 6: Codes of Practice Sec. 6.2: Identification in Footwear Wear Trials of Major Weaknesses in Design or Construction and the Assessment of Fitness-For-Purpose. 4 pp.

Footwear Linings
See Also: Footwear; Linings
BSI BS 5131: Sec 5.13-80. 1980 Methods of Test for Footwear and Footwear Materials Part 5: Testing of Complete Footwear Section 5.13: Measurement of the Strength of Stitched Seams in Upper and Lining Materials. 4 pp.

—Leather
MOD UK DSTAN 83-9-01. Leathers, Lining, for Footwear Issue 2; Amendment 2. 14 pp.

INTERNATIONAL AND NON-U.S. NATIONAL STANDARDS SUBJECT INDEX

FOPS
Use: Falling Object Protective Structures

Forage
Use For: Fodders *See Also:* Animal Feed; Plants (Botany)

—Defense Contracts
MOD UK DEFCON 112BO-91. Conditions of Contract for Forage and Straw 9/91. 7 pp.

—Ducts—Conveyors—Agricultural Equipment
BSI BS 4286: Part 2-70. 1970 Steel Ducting for Grain and Fodder Conveying Part 2: Metric Units. 11 pp.

—Seeds
EC 88/380/EEC-88. Council Directive Amending Directives 66/400/EEC, 66/401/EEC, 66/ 402/EEC, 66/403/EEC, 69/ 208/EEC, 70/457/EEC AND 70/458/EEC on the Marketing of Beet Seed, Fodder Plant Seed, Cereal Seed, Seed Potatoes, Seed of Oil and Fibre Plants and Vegetable Seed. 18 pp.

Forage Crops
Use For: Forages *See Also:* Agricultural Products

—Harvesting Equipment—Compatibility
BSI BS 5137-74. 1974 Requirements for the Compatible Operation of Forage Harvesting Machinery. 12 pp.

—Harvesting Equipment—Safety
DIN ENGL 11001 Pt 3-80. Agricultural Machines and Tractors; Reaping Units, Hay Making Machines, Chopper Forage Harvesters; Special Technical Safety Requirements and Testing (Aug). 3 pp.

—Particle Size Distribution—Sieve Analysis
ISO TR10391-92. Forage Harvesters—Method of Determining by Screening and Expressing Particle Size of Chopped Forage Materials First Edition. 11 pp.

—Storage
BSI BS 5502: Part 75-93. 1993 Buildings and Structures for Agriculture Part 75: Code of Practice for the Design and Construction of Forage Stores (R). 13 pp.

Forages
Use: Forage Crops

Force
See Also: Contact Force; Loads (Forces); Pressure; Stresses; Torque; Traction

—Creep Testing Equipment
BSI BS 1610: Part 3-90. 1990 Materials Testing Machines and Force Verification Equipment: Part 3: Grading of the Forces Applied by Deadweight and Lever Creep Testing Machines. 13 pp.

—Force Proving Devices
BSI BS 1610: Part 2-85. 1985 Amd 1 Materials Testing Machines and Force Verification Equipment: Part 2: Grading of Equipment Used for the Verification of the Forces Applied By Materials Testing Machines (AMD 6174) October 31, 1989. 14 pp.
DIN ENGL 51301-86. Materials Testing Machines; Verification of Materials Testing Machines with the Aid of Static Force Measuring Devices (Feb). 11 pp.
SNZ NZS 6507: Part 2-86. Materials Testing Machines and Force Verification Equipment Part 2: Specification for the Grading of Equipment Used for the Verification of the Forces Applied by Materials Testing Machines Amend: A, 1986. 12 pp.

—Glossaries
DIN ENGL 1305-88. Mass, as Weighed Value, Force, Weight Force, Weight, Load; Concepts (Jan). 2 pp.

—Tables (Data)
SAA AS 1377.5-73. Conversion Tables—Part 5: Force, Pressure, Stress and Second Moment of Area (Incorporating Amdt 1). 40 pp.

Forceps
See Also: Medical Equipment; Medical Instruments; Surgical Instruments
CNS T1023-80. Medical Dressing Forceps (Aug)(6282).
CNS T1024-80. Medical Forceps (Aug)(6283).
CNS T1047-81. Eye Forceps (Apr)(7326).
JIS T 2302-53. Medical Forceps.

—Dental
BSI BS 4750-71. (WITHDRAWN) 1971 Dental Extracting Forceps (Performance Requirements) (Superseded by BS 7549: Part 1: 1992). 8 pp.
BSI BS 5211-75. (WITHDRAWN) 1973 Popular Patterns of Dental Extracting Forceps (Superseded by BS 7549: Part 1: 1992). 29 pp.
CNS T3030-81. Dental Pincettes (Mar)(7087).
CNS T3039-81. Dental Forceps (Nov)(8061).
JIS T 5401-84. Dental Pincettes. 7 pp.
JIS T 5410-88. Dental Forceps. 5 pp.

—Dissecting
BSI BS 5194: Part 3-85. 1985 Surgical Instruments Part 3: Specification for Dissecting Forceps. 8 pp.

—Dressing
JIS T 2301-85. Medical Dressing Forceps.

—Hemostatic
ISO 7151-88. Surgical Instruments—Non-Cutting, Articulated Instruments—General Requirements and Test Methods Second Edition. 6 pp.

—Ophthalmic
JIS T 2618-88. Eye Forceps. 10 pp.

Ford Cups
CNS K2143-86. Ford Cup for Determining Consistency of Coatings (Oct)(11722). 4 pp.

Forecasting
Use For: Predictions *See Also:* Maintainability; Market Research; Reliability; Statistical Analysis; Systems Engineering
IEC 863-86. Presentation of Reliability, Maintainability and Availability Predictions First Edition. 26 pp.

—Ionosphere
CCIR RECMN 434-5-92. CCIR Reference Ionospheric Characteristics and Methods of Basic MUF, Operational MUF and Ray-Path Prediction Section 6E—Ionospheric Propagation Prediction at Frequencies Between About 1.6 and 30 MHz. 13 pp.
CCIR Report 888-2-90. Short-Term Forecasting of Critical Frequencies, Operational Maximum Usable Frequencies and Total Electron Content—Section 6C—Ionospheric Propagation and Operational Forecasting. 9 pp.

—Ionosphere—Computer Programs
CCIR RESOLUTION 63-3-90. Computer Programs for the Prediction of Ionospheric Characteristics, Sky-Wave Transmission Loss and Noise—Volume VI—Propagation in Ionized Media. 7 pp.

—Ionosphere—Indices
CCIR RECMN 371-6-90. Choice of Indices for Long-Term Ionospheric Predictions—Section 6E—Ionospheric Propagation Prediction at Frequencies Between About 1.6 and 30 MHz. 6 pp.
CCIR QUESTION 34/6-90. Long-Term Predictions of Solar and Ionospheric Indices—Questions Concerning Study Group 6—Radio Wave Propagation in Ionized Media. 1 p.

—Ionosphere—Radio Communications
CCIR QUESTION 27-1/6-90. Short-Term Forecasting of Operational Parameters for Ionospheric and Trans-Ionospheric Radiocommunications—Questions Concerning Study Group 6—Radio Wave Propagation in Ionized Media. 1 p.

—Ionospheric Disturbances—Information Interchange
CCIR RECMN 313-6-90. Exchange of Information for Short-Term Forecasts and Transmission of Ionospheric Disturbance Warnings—Section 6C—Ionospheric Propagation and Operational Forecasting. 5 pp.
CCIR RECMN 313-7-92. Exchange of Information for Short-Term Forecasts and Transmission of Ionospheric Disturbance Warnings—Section 6C—Ionospheric Propagation and Operational Forecasting. 6 pp.

—Ionospheric Disturbances—Ionospheric Propagation
CCIR Report 763-3-90. Signal Level Variation Due to Multipath Effects and Blockage by Ship's Superstructure in Maritime Mobile-Satellite Service Links—Section 8I—Technical and Operating Characteristics of Mobile Satellite Services. 13 pp.

—Ionospheric Propagation
CCIR RECMN 434-5-92. CCIR Reference Ionospheric Characteristics and Methods of Basic MUF, Operational MUF and Ray-Path Prediction Section 6E—Ionospheric Propagation Prediction at Frequencies Between About 1.6 and 30 MHz. 13 pp.
CCIR RESOLUTION 112-90. CCIR Study Group 6 Report to the WARC HFBC(93)—Volume VI—Propagation in Ionized Media. 1 p.
CCIR OPINION 45-3-90. Evaluation of the CCIR HF Propagation Prediction Methods—Volume VI—Propagation in Ionized Media. 1 p.
CCIR Report 255-7-90. Long-Term Ionospheric Propagation Predictions—Section 6E—Ionospheric Propagation Prediction at Frequencies Between About 1.6 and 30 MHz. 9 pp.
CCIR Report 894-2-90. CCIR HF Propagation Prediction Method Third CCIR Computer-Based Method for Estimation of MUF, Sky-Wave Field Strength, Signal-to-Noise Ratio, LUF and Basic Circuit Reliability—Section 6E—Ionospheric Propagation Prediction at Frequencies Between. 12 pp.
CCIR Decision 85-89. Studies of the Propagation Prediction Method for HF Broadcasting—Annex to Volume VI—Propagation in Ionized Media. 1 p.
CCIR QUESTION 33/6-90. Ionospheric Propagation Predictions—Questions Concerning Study Group 6—Radio Wave Propagation in Ionized Media. 1 p.

—Ionospheric Propagation—Observation Stations
CCIR OPINION 67-82. Geophysical and Solar Observations Needed for Short-Term Forecasting of Ionospheric Propagation—Volume VI—Propagation in Ionized Media. 1 p.

—Magnetic Disturbances—Ionospheric Propagation
CCIR Report 763-3-90. Signal Level Variation Due to Multipath Effects and Blockage by Ship's Superstructure in Maritime Mobile-Satellite Service Links—Section 8I—Technical and Operating Characteristics of Mobile Satellite Services. 13 pp.

—Maximum Usable Frequencies
CCIR RECMN 434-5-92. CCIR Reference Ionospheric Characteristics and Methods of Basic MUF, Operational MUF and Ray-Path Prediction Section 6E—Ionospheric Propagation Prediction at Frequencies Between About 1.6 and 30 MHz. 13 pp.

—Radio Frequency Interference—Ionospheric Propagation
CCIR Report 763-3-90. Signal Level Variation Due to Multipath Effects and Blockage by Ship's Superstructure in Maritime Mobile-Satellite Service Links—Section 8I—Technical and Operating Characteristics of Mobile Satellite Services. 13 pp.

—Radio Wave Propagation—Earth Surface
CCIR QUESTION 9-1/5-90. Methods for Predicting Propagation over the Surface of the Earth—Questions Concerning Study Group 5—Radio Wave Propagation in Non-Ionized Media. 1 p.

—Ray Paths
CCIR RECMN 434-5-92. CCIR Reference Ionospheric Characteristics and Methods of Basic MUF, Operational MUF and Ray-Path Prediction Section 6E—Ionospheric Propagation Prediction at Frequencies Between About 1.6 and 30 MHz. 13 pp.

—Sky Wave Field Strength—Broadcasting
CCIR RECMN 435-7-92. Sky-Wave Field-Strength Prediction Method for the Broadcasting Service in the Frequency Range 150 to 1 600 kHz Section 6D—Ionospheric Propagation Prediction at Frequencies Below About 1.6 MHz. 25 pp.

—Space Research Services
CCIR Report 672-1-90. Forecast of Space Technology—Section 2A—Research in Space Technology. 6 pp.

—Sunspot Numbers—foF2
CCIR OPINION 82-86. Use of an Ionospherically Derived Solar Activity Index (IG) for the Prediction of foF2—Volume VI—Propagation in Ionized Media. 1 p.

—Telecommunication Services
CCITT RECMN E.508-89. Forecasting New International Services—Telephone Network and ISDN—Quality of Service, Network Management and Traffic Engineering (Study Group II) 9 pp. 9 pp.

—Telecommunication Services—Integrated Services Digital Networks
CCITT RECMN E.508-89. Forecasting New International Services—Telephone Network and ISDN—Quality of Service, Network Management and Traffic Engineering (Study Group II) 9 pp. 9 pp.

—Telecommunication Traffic
CCITT RECMN E.506 (REV 1)-92. Forecasting International Traffic (Study Group II) 22 pp. 22 pp.

INTERNATIONAL AND NON-U.S. NATIONAL STANDARDS
SUBJECT INDEX

Forecasting (Cont.)
—**Telecommunication Traffic** (Cont.)
CCITT RECMN E.506-89. Forecasting International Traffic—Telephone Network and ISDN—Quality of Service, Network Management and Traffic Engineering (Study Group II) 18 pp. 18 pp.

—**Telecommunication Traffic—Formulas**
CCITT RECMN E.506 (REV 1)-92. Forecasting International Traffic (Study Group II) 22 pp. 22 pp.
CCITT RECMN E.506-89. Forecasting International Traffic—Telephone Network and ISDN—Quality of Service, Network Management and Traffic Engineering (Study Group II) 18 pp. 18 pp.
CCITT RECMN E.507-89. Models for Forecasting International Traffic—Telephone Network and ISDN—Quality of Service, Network Management and Traffic Engineering (Study Group II) 20 pp. 20 pp.

—**Telecommunication Traffic—Models**
CCITT RECMN E.506 (REV 1)-92. Forecasting International Traffic (Study Group II) 22 pp. 22 pp.
CCITT RECMN E.506-89. Forecasting International Traffic—Telephone Network and ISDN—Quality of Service, Network Management and Traffic Engineering (Study Group II) 18 pp. 18 pp.
CCITT RECMN E.507-89. Models for Forecasting International Traffic—Telephone Network and ISDN—Quality of Service, Network Management and Traffic Engineering (Study Group II) 20 pp. 20 pp.

—**Telecommunication Traffic—Statistical Analysis**
CCITT RECMN E.507-89. Models for Forecasting International Traffic—Telephone Network and ISDN—Quality of Service, Network Management and Traffic Engineering (Study Group II) 20 pp. 20 pp.

—**Traffic Matrixes**
CCITT RECMN E.506 (REV 1)-92. Forecasting International Traffic (Study Group II) 22 pp. 22 pp.
CCITT RECMN E.506-89. Forecasting International Traffic—Telephone Network and ISDN—Quality of Service, Network Management and Traffic Engineering (Study Group II) 18 pp. 18 pp.

—**Traffic Units**
CCITT RECMN E.506 (REV 1)-92. Forecasting International Traffic (Study Group II) 22 pp. 22 pp.
CCITT RECMN E.506-89. Forecasting International Traffic—Telephone Network and ISDN—Quality of Service, Network Management and Traffic Engineering (Study Group II) 18 pp. 18 pp.

—**Transportation**
NATO STANAG 2165 ED 3 AMD 4-77. Forecast Movement/Transport Requirements—Rail, Road and Inland Waterways. 12 pp.
NATO STANAG 2165 ED 3 AMD 5-77. Forecast Movement/Transport Requirements—Rail, Road and Inland Waterways. 13 pp.

Foreign Acquisition
—**Weapons**
MOD UK DEFCON 112CT-89. Basic Set of Conditions of Contract for Overseas Contracts (Military Weapons) 4/89. 2 pp.

Foreign Airworthiness Directives
Scope Note: "Summary of FAA Airworthiness Directives for American constructed aircraft, engines, propellers and equipment" (Foreign Airworthiness Directives, Volume III, Foreword (Issue 5)). Latest hard copy revision available from the Civil Aviation Authority
CAA. Contents and Check List of Pages Issue 226 (Foreign Airworthiness Directives). 6 pp.
CAA. Foreword Issue 8 (Foreign Airworthiness Directives). 4 pp.
CAA. Contents and Check List of Pages Issue 235 (Foreign Airworthiness Directives). 11 pp.
CAA. Foreword (Foreign Airworthiness Directives). 4 pp.

—**Aircraft Engines**
CAA. Allison Series Engines. 2 pp.
CAA. Avco Lycoming Series Engines. 5 pp.
CAA. CFM International CFM56-3 Series Engines. 1 p.
CAA. Garrett Series Engines. 1 p.
CAA. General Electric Series Engines. 1 p.
CAA. Pratt and Whitney Engines. 10 pp.
CAA. Teledyne Continental Motors Series Engines. 1 p.
CAA. Textron Lycoming Series Engines. 4 pp.
CAA. Bombardier—Rotax Series Engines. 2 pp.
CAA. CFM International CFM56-5 Series Engines. 1 p.
CAA. Hirth F10 Engines (Foreign Airworthiness Directives). 1 p.

Foreign Airworthiness Directives (Cont.)
—**Aircraft Engines** (Cont.)
CAA. Limbach Engines (Foreign Airworthiness Directives). 1 p.
CAA. M462-RF Engines (Foreign Airworthiness Directives). 2 pp.
CAA. Microturbo APU Saphir I and II (Foreign Airworthiness Directives). 1 p.
CAA. Potez 4E20 Series Engines (Foreign Airworthiness Directives). 2 pp.
CAA. Pratt and Whitney Canada JT15D Series Engines (Foreign Airworthiness Directives). 3 pp.
CAA. Pratt and Whitney of Canada PT6 Series Engines (Foreign Airworthiness Directives). 5 pp.
CAA. Pratt and Whitney Canada PW100 Series Engines (Foreign Airworthiness Directives). 6 pp.
CAA. Pratt and Whitney of Canada ST6L Series Engines (Foreign Airworthiness Directives). 2 pp.
CAA. Renault Engines (Foreign Airworthiness Directives). 1 p.
CAA. Turbomeca Series Engines (Foreign Airworthiness Directives). 6 pp.
CAA. Walter M4-111 Engines (Foreign Airworthiness Directives). 1 p.
CAA. Walter Minor 6-111 Engines (Foreign Airworthiness Directives). 2 pp.

—**Aircraft Equipment**
CAA. Instruments (Foreign Airworthiness Directives). 2 pp.
CAA. Aircraft Radio (Foreign Airworthiness Directives). 7 pp.
CAA. Aircraft Instruments (Foreign Airworthiness Directives). 1 p.
CAA. Aircraft Equipment (Foreign Airworthiness Directives). 21 pp.
CAA. Aircraft—General (Foreign Airworthiness Directives). 1 p.

—**Aircraft Propellers**
CAA. Hartzell Propellers (Foreign Airworthiness Directives). 1 p.
CAA. McCauley Propellers (Foreign Airworthiness Directives). 1 p.
CAA. Avia Propellers (Foreign Airworthiness Directives). 2 pp.
CAA. Evra Propellers (Foreign Airworthiness Directives). 1 p.
CAA. Hoffman Propellers (Foreign Airworthiness Directives). 1 p.
CAA. Legere Propellers (Foreign Airworthiness Directives). 1 p.
CAA. MT Propellers. 1 p.
CAA. Ratier Propellers (Foreign Airworthiness Directives). 2 pp.

—**Aircraft Types**
CAA. Ayres S-2R (Foreign Airworthiness Directives). 1 p.
CAA. Beech Series Aircraft (Foreign Airworthiness Directives). 9 pp.
CAA. Bell 47 Series Helicopters. 3 pp.
CAA. Bell 206A and B Series Helicopters (Foreign Airworthiness Directives). 6 pp.
CAA. Bell 206 L Series Helicopters (Foreign Airworthiness Directives). 2 pp.
CAA. Bell 212 Series Helicopters. 8 pp.
CAA. Bell 214 Series Helicopters (Foreign Airworthiness Directives). 3 pp.
CAA. Bell 222 Series Helicopters (Foreign Airworthiness Directives). 2 pp.
CAA. Boeing 707/720 Series Aircraft (Foreign Airworthiness Directives). 10 pp.
CAA. Boeing 727 Series Aircraft. 9 pp.
CAA. Boeing 737 Series Aircraft (Foreign Airworthiness Directives). 28 pp.
CAA. Boeing 747 Series Aircraft (Foreign Airworthiness Directives). 13 pp.
CAA. Boeing 757 Series Aircraft (Foreign Airworthiness Directives). 7 pp.
CAA. Boeing 767 Series Aircraft (Foreign Airworthiness Directives). 8 pp.
CAA. Boeing Vertol Series Helicopters (Foreign Airworthiness Directives). 2 pp.
CAA. Cessna Series Aircraft (Foreign Airworthiness Directives). 18 pp.
CAA. Christen Industies (Pitts) S-1 and S-2 Series Aircraft. 1 p.
CAA. Enstrom Series Helicopters (Foreign Airworthiness Directives). 6 pp.
CAA. Fairchild Argus Series Aircraft. 1 p.
CAA. Fairchild SA227 Series Aircraft. 1 p.
CAA. Gulfstream Aerospace 112 and 114 Series Aircraft. 2 pp.
CAA. Gulfstream Aerospace G-159 Aircraft. 2 pp.
CAA. Gulfstream Aerospace 685, 690 and 695 Series Aircraft. 1 p.
CAA. Gulfstream Aerospace G1159 Series (Foreign Airworthiness Directives). 5 pp.
CAA. Gulfstream Aerospace GA-7 Aircraft. 1 p.
CAA. Gulfstream American AA-1 Series Aircraft and Gulfstream Aerospace AA-5 Series Aircraft. 3 pp.

Foreign Airworthiness Directives (Cont.)
—**Aircraft Types** (Cont.)
CAA. Hiller Aviation Series Helicopters (Foreign Airworthiness Directives). 2 pp.
CAA. Hughes 369 Series Helicopters (Foreign Airworthiness Directives). 7 pp.
CAA. Hynes (Brantly) Series Helicopters (Foreign Airworthiness Directives). 1 p.
CAA. Learjet Series Aircraft (Foreign Airworthiness Directives). 1 p.
CAA. Lockheed L-1011 Tristar (Foreign Airworthiness Directives). 26 pp.
CAA. McDonnell Douglas DC-9 and MD-88 Series Aircraft (Foreign Airworthiness Directives). 4 pp.
CAA. McDonnell Douglas DC-10 (Foreign Airworthiness Directives). 12 pp.
CAA. Maule Aerospace Technology Series Aircraft (Foreign Airworthiness Directives). 1 p.
CAA. Mooney Series Aircraft (Foreign Airworthiness Directives). 1 p.
CAA. North American Harvard AT-6 Series and T-6G Aircraft (Foreign Airworthiness Directives). 1 p.
CAA. Piper PA-60-601P (Foreign Airworthiness Directives). 3 pp.
CAA. Piper Series Aircraft (Foreign Airworthiness Directives). 22 pp.
CAA. Robinson Series Helicopters (Foreign Airworthiness Directives). 1 p.
CAA. Schweizer 269 Series Helcopters. 3 pp.
CAA. Sikorsky S58T Helcopters. 3 pp.
CAA. Sikorsky S61N Helicopters (Foreign Airworthiness Directives). 26 pp.
CAA. Sikorsky S76 Helicopters (Foreign Airworthiness Directives). 15 pp.
CAA. Thurston TSC-1A and TSC-1A1. 1 p.
CAA. Aerospatiale AS 332 Series Helcopters (Foreign Airworthiness Directives). 12 pp.
CAA. Aerospatiale ATR 42 Series Aircraft. 6 pp.
CAA. Aerospatiale SA 315B Lama Series Helcopters (Foreign Airworthiness Directives). 7 pp.
CAA. Aerospatiale SA 330 Puma Series Helicopters (Foreign Airworthiness Directives). 10 pp.
CAA. Aerospatiale SA 341 Series Helicopters (Foreign Airworthiness Directives). 6 pp.
CAA. Aerospatiale SA 365C Series Helicopters (Foreign Airworthiness Directives). 6 pp.
CAA. Aerospatiale SA 365N Series Helicopters (Foreign Airworthiness Directives). 5 pp.
CAA. Aerospatiale SE 316 and SA 319 Alouette III Series Helicopters. 8 pp.
CAA. Aerospatiale SE 3130 and SE 313B Alouette II and Aerospatiale SA 3180 and SA 318 B/C Alouette Astazou Series Helicopters. 10 pp.
CAA. AESL and Glos Air Airtourer Series (Foreign Airworthiness Directives). 14 pp.
CAA. Agusta A109 and A109A Helicopters (Foreign Airworthiness Directives). 11 pp.
CAA. Agusta Bell 47 Series (Foreign Airworthiness Directives). 17 pp.
CAA. Agusta Bell 206 Series Helicopters (Foreign Airworthiness Directives). 33 pp.
CAA. Airbus Industrie A300 and A310 Series Aircraft (Foreign Airworthiness Directives). 21 pp.
CAA. Airbus Industrie A320 Series Aircraft (Foreign Airworthiness Directives). 6 pp.
CAA. Anglo Polish Sailplanes SZD45A Ogar (Foreign Airworthiness Directives). 1 p.
CAA. Avions Mudry/CAARP CAP 10B and CAP 20 L/S 200 (Foreign Airworthiness Directives). 1 p.
CAA. Blanik Gliders (Foreign Airworthiness Directives). 3 pp.
CAA. Bolkow 107 and 207 Series (Foreign Airworthiness Directives). 2 pp.
CAA. Bolkow 208 Junior (Foreign Airworthiness Directives). 1 p.
CAA. Bolkow 209 Series (Foreign Airworthiness Directives). 3 pp.
CAA. Canadair CL-600 Series Aircraft. 8 pp.
CAA. Dassult Aviation Fan Jet Falcon and Mystere-Falcon 200 Series Aircraft. 5 pp.
CAA. Dassult Aviation Mystere-Falcon 900 Series. 1 p.
CAA. De Havilland DHC-2 (Foreign Airworthiness Directives). 5 pp.
CAA. De Havilland DHC-6 Twin Otter (Foreign Airworthiness Directives). 11 pp.
CAA. De Havilland DHC-7 Series Aircraft (Foreign Airworthiness Directives). 7 pp.
CAA. De Havilland DHC-8 Series Aircraft. 7 pp.
CAA. Dornier DO 27 Series. 2 pp.
CAA. Dornier DO 28 Series (Foreign Airworthiness Directives). 1 p.
CAA. Dornier 228 Series (Foreign Airworthiness Directives). 3 pp.
CAA. Embraer Bandeirante EMB-110 Series Aircraft. 21 pp.
CAA. Embraer EMB—120 Series Aircraft. 18 pp.
CAA. Embraer Bandeirante EMB-110K1, P1 and P2 (Foreign Airworthiness Directives). 30 pp.
CAA. Piel Emeraude Series Aircraft (Foreign Airworthiness Directives). 4 pp.
CAA. Extra EA-230 Series Aircraft. 2 pp.
CAA. Extra EA-300 Series Aircraft. 1 p.

Foreign Airworthiness Directives (Cont.)

—Aircraft Types (Cont.)

CAA. Falco F8L Series (Foreign Airworthiness Directives). 1 p.
CAA. FFA AS 202 Bravo Series Aircraft (Foreign Airworthiness Directives). 1 p.
CAA. Fokker F27 Series Aircraft (Foreign Airworthiness Directives). 15 pp.
CAA. Fokker F28-0100 Series Aircraft. 15 pp.
CAA. Fournier RF3 Series Motor Gliders and Fournier RF6 Series Aircraft (Foreign Airworthiness Directives). 1 p.
CAA. Fuji FA-200 Series Aircraft (Foreign Airworthiness Directives). 3 pp.
CAA. Gardan GY80 Series Aircraft (Foreign Airworthiness Directives). 1 p.
CAA. Glaser-Dirks DG-400 Series Motor Gliders (Foreign Airworthiness Directives). 2 pp.
CAA. Grob G109 Series Motor Gliders (Foreign Airworthiness Directives). 5 pp.
CAA. Grob G115 Series. 1 p.
CAA. Gyroflug SC01 Series Aircraft (Foreign Airworthiness Directives). 1 p.
CAA. Hoffman H36 Dimona Motor Glider (Foreign Airworthiness Directives). 3 pp.
CAA. ICA Brasov Motor Gliders (Foreign Airworthiness Directives). 2 pp.
CAA. Jodel Series (Foreign Airworthiness Directives). 11 pp.
CAA. L.40 Meta Sokol (Foreign Airworthiness Directives). 1 p.
CAA. M.100S Gliders (Foreign Airworthiness Directives). 1 p.
CAA. Messerschmitt Bolkow BLOHM BO 105 Series Helicopters. 11 pp.
CAA. Messerschmitt Bolkow Blohm BK 117 Series Helicopters. 2 pp.
CAA. Morane—Saulnier MS760 Series (Foreign Airworthiness Directives). 2 pp.
CAA. Morane—Saulnier 880 Series (Foreign Airworthiness Directives). 2 pp.
CAA. Morane—Saulnier 890 Series (Foreign Airworthiness Directives). 8 pp.
CAA. Morane—Saulnier Rallye 100, 110, 150, 180 and Variants (Foreign Airworthiness Directives). 3 pp.
CAA. Morane—Saulnier Rallye 235 Series (Foreign Airworthiness Directives). 2 pp.
CAA. Morava L200 Series (Foreign Airworthiness Directives). 5 pp.
CAA. Morava Zlin Z50L Series. 5 pp.
CAA. Noorduyn Aviation/Canadian Car and Foundry Harvard II and IV Series Aircraft. 2 pp.
CAA. Nord 1101 Series Aircraft. 1 p.
CAA. Piaggio FW P149D Series (Foreign Airworthiness Directives). 2 pp.
CAA. Piaggio P.166 Series. 10 pp.
CAA. Pierre Robin Series Aircraft (Foreign Airworthiness Directives). 13 pp.
CAA. PIK 16C and PIK-20 Series Sailplanes (Foreign Airworthiness Directives). 2 pp.
CAA. PIK 20E Motor Gliders. 1 p.
CAA. Procaer F-15 Series (Foreign Airworthiness Directives). 1 p.
CAA. PZL-104 Wilga Series (Foreign Airworthiness Directives). 5 pp.
CAA. Reims—Cessna Series (Foreign Airworthiness Directives). 22 pp.
CAA. SAAB-Fairchild 340 Series Aircraft (Foreign Airworthiness Directives). 8 pp.
CAA. SAAB Safir 91 Series Aircraft. 1 p.
CAA. Scheibe Motor Gliders (Foreign Airworthiness Directives). 6 pp.
CAA. Schempp-Hirth Motor Gliders (Foreign Airworthiness Directives). 1 p.
CAA. Schleicher Gliders and Motor Gliders (Foreign Airworthiness Directives). 8 pp.
CAA. SIAI—Marchetti F.260 Series (Foreign Airworthiness Directives). 10 pp.
CAA. SIAI—Marchetti S.205 Series (Foreign Airworthiness Directives). 14 pp.
CAA. Socata ST 10 Series (Foreign Airworthiness Directives). 3 pp.
CAA. Socata TB9, TB10 and TB20 Series Aircraft (Foreign Airworthiness Directives). 7 pp.
CAA. Sportavia—Putzer RF4 and RF5 Series Motor Gliders (Foreign Airworthiness Directives). 1 p.
CAA. Stampe SV4 Series (Foreign Airworthiness Directives). 13 pp.
CAA. Standard Austria Gliders (Foreign Airworthiness Directives). 3 pp.
CAA. Stemme S10 Series Motor Gliders. 1 p.
CAA. Super Aero 45 and Aero 145 Series. 3 pp.
CAA. Tipsy Nipper. 1 p.
CAA. Valentin Taifun 17E Motor Gliders (Foreign Airworthiness Directives). 1 p.
CAA. Victa Airtourer 100 and 115 Series (Foreign Airworthiness Directives). 15 pp.
CAA. Wassmer WA51, WA51A, WA52 and WA54 Series (Foreign Airworthiness Directives). 5 pp.
CAA. Zlin Z226, Z326 and Z526 Series (Foreign Airworthiness Directives). 8 pp.
CAA FR18 ISSUE 1. Robinson R22. 2 pp.

Foreign Matter Content Analysis
Use: Impurities Content Analysis

Forest Harvesting Machines
Use: Tree Harvesters

Forestry

—Classification

BSI BS 1000: (63/632)-81. 1981 Universal Decimal Classification (UDC). English Full Edition (63/632): Agriculture in General. Forestry. Plant Injuries, Diseases and Pests. Plant Protection. 58 pp.
SNZ NZS/BS 1000 (63/632)-81. Universal Decimal Classification Agriculture in General. Forestry. Plant Injuries, Diseases and Pests. Plant Protection. 60 pp.

Forestry Equipment
See Also: Brush Saws; Chain Saws; Construction Equipment; Earthmoving Equipment; Industrial Equipment; Lumbering Industry; Mowers; Off Road Equipment; Silage Cutters; Slashers (Forestry); Tree Harvesters

—Anchors (Fasteners)

CSA CAN/CSA-M6816-92. Machinery for Forestry—Winches—Classification and Nomenclature (ISO 6816-1984); (Gen Instr 1). 19 pp.

—Brakes

BSI BS 6384-83. 1983 Determination of Braking Performance of Agricultural and Forestry Vehicles. 18 pp.
BSI BS 7705-93. 1993 Brake Systems for Special Wheeled Forestry Machines (ISO 11169: 1993) (E). 12 pp.
ISO 5697-82. Agricultural and Forestry Vehicles—Determination of Braking Performance First Edition. 18 pp.
ISO 11169-93. Machinery for Forestry—Wheeled Special Machines—Vocabulary, Performance Test Methods and Criteria for Brake Systems First Edition. 10 pp.

—Brakes—Glossaries

BSI BS 7705-93. 1993 Brake Systems for Special Wheeled Forestry Machines (ISO 11169: 1993) (E). 12 pp.
ISO 11169-93. Machinery for Forestry—Wheeled Special Machines —Vocabulary, Performance Test Methods and Criteria for Brake Systems First Edition. 10 pp.

—Brakes—Hydraulic Couplings

BSI BS 4742: Part 4-85. 1985 Hydraulic Equipment for Agricultural Machinery Part 4: Hydraulic Couplings on Braking Systems for Trailers and Trailed Equipment. 4 pp.
ISO 5676-83. Tractors and Machinery for Agriculture and Forestry—Hydraulic Coupling—Braking Circuit First Edition. 4 pp.

—Cabs—Air Conditioners

ISO TR8953-87. Tractors and Self-Propelled Machines for Agriculture and Forestry—Test Method for Performance of Air-Conditioning System First Edition. 6 pp.

—Classification

ISO 3339 Pt 0-86. Tractors and Machinery for Agriculture and Forestry—Classification and Terminology—Part 0: Classification System and Classification Second Edition. 33 pp.

—Engines—Comprehensive Testing

BSI BS 4539-70. (WITHDRAWN) 1970 Methods of Testing Small Spark Ignition Engines for Agricultural Use. 33 pp.

—Ergonomics

CSA CAN/CSA-M6682-M91. Zones of Comfort and Reach for Controls—Machinery (EMM,FM) (Amd 1 Included); (ISO 6682-1986). 23 pp.

—Falling Object Protective Structures

ISO 8083-89. Machinery for Forestry—Falling-Object Protective Structures—Laboratory Tests and Performance Requirements First Edition. 8 pp.

—Glossaries

ISO 6814-83. Machinery for Forestry—Mobile and Self-Propelled Machinery—Identification Vocabulary First Edition. 5 pp.

—Hitches

ISO 6815-83. Machinery for Forestry—Hitches—Dimensions First Edition. 5 pp.

—Manual Controls

CSA CAN/CSA-M6682-M91. Zones of Comfort and Reach for Controls—Machinery (EMM,FM) (Amd 1 Included); (ISO 6682-1986). 23 pp.

Forestry Equipment (Cont.)
—Manual Controls (Cont.)

ISO 3789 Pt 1-82. Tractors, Machinery for Agriculture and Forestry, Powered Lawn and Garden Equipment—Location and Method of Operation of Operator Controls—Part 1: Common Controls First Edition. 6 pp.

—Manual Controls—Symbols

BSI BS 4964: Part 1-93. 1993 Symbols for Control Markings and Displays on Tractors and Machinery for Agricultural and Forestry, and on Powered Lawn and Garden Equipment Part 1: Specification for Common Symbols (ISO 3767-1: 1991) (E). 37 pp.
BSI BS 4964: Part 1-81. 1981 Symbols for Control Markings and Displays on Tractors and Machinery for Agricultural and Forestry, and on Powered Lawn and Garden Equipment Part 1: Common Symbols. 24 pp.
ISO 3767 Pt 1-91. Tractors, Machinery for Agriculture and Forestry, Powered Lawn and Garden Equipment—Symbols for Operator Controls and Other Displays—Part 1: Common Symbols Second Edition; (Incorporating Addendum 1). 33 pp.
ISO 3767 Pt 5-92. Tractors, Machinery for Agriculture and Forestry, Powered Lawn and Garden Equipment—Symbols for Operator Controls and Other Displays—Part 5: Symbols for Manual Portable Forestry Machinery First Edition. 7 pp.

—Noise

CSA CAN/CSA-Z107.32-M86. Test Procedure for the Measurement of Sound Emitted from Construction, Forestry, and Mining Machines to the Operator Station and Exterior of the Machine; (Gen Instr 1 Thru 2). 36 pp.
ISO 5131-82. Acoustics—Tractors and Machinery for Agriculture and Forestry—Measurement of Noise at the Operator's Position—Survey Method Amendment 1: Annex D—Forestry Forwarders and Skidders First Edition; (Amendment 1-1992). 15 pp.
ISO 7216-92. Acoustics—Agricultural and Forestry Wheeled Tractors and Self-Propelled Machines—Measurement of Noise Emitted When in Motion First Edition. 7 pp.

—Power Takeoffs

BSI BS 3417: Part 2-84. 1984 1974 Power Take-Off Shafts and Guards for Tractors and Machinery for Agriculture and Forestry Part 2: Methods for Testing Guards. 8 pp.
BSI BS 3417: Part 3-86. 1986 Power Take-Off Shaft and Guards for Tractors and Machinery for Agriculture and Forestry Part 3: Supplementary Requirements to Part 1 'Specification for Power Take-Off Drive Shafts' and to Part 2 'Methods for Testing Guards' (Sup's BS 3417: 1974). 4 pp.
CEN PREN 1152-93. Tractors and Machinery for Agriculture and Forestry—Guards for Power Take-Off (PTO) Drive Shafts—Wear and Strength Tests. 12 pp.
ISO 5674 Pt 1-92. Tractors and Machinery for Agriculture and Forestry—Guards for Power Take-off (PTO) Drive-Shafts—Part 1: Strength Test First Edition. 9 pp.

—Power Takeoffs—Wear Testing

CEN PREN 1152-93. Tractors and Machinery for Agriculture and Forestry—Guards for Power Take-Off (PTO) Drive Shafts—Wear and Strength Tests. 12 pp.
ISO 5674 Pt 2-92. Tractors and Machinery for Agriculture and Forestry—Guards for Power Take-off (PTO) Drive-Shafts—Part 2: Wear Test First Edition. 8 pp.

—Roll Over Protective Structures

CSA B352-M1980. Rollover Protective Structures (ROPS) for Agricultural, Construction, Earthmoving, Forestry, Industrial, and Mining Machines; (Gen Instr 1). 76 pp.

—Roll Over Protective Structures—Acceptance Testing

BSI BS 5947-92. 1992 Method of Test for Static Loading of Protective Cabs and Frames for Agricultural Wheeled Tractors, Including Acceptance Conditions (ISO 5700: 1989). 23 pp.
BSI BS 5947-85. 1985 Static Loading for Protective Cabs and Frames for Agricultural Wheeled Tractors, Including Acceptance Conditions. 20 pp.

—Safety

ISO 4254 Pt 1-89. Tractors and Machinery for Agriculture and Forestry—Technical Means for Ensuring Safety—Part 1: General Second Edition. 13 pp.

—Seat Index Point

DIN ENGL EN 25353-92. Earth-Moving Machinery, and Tractors and Machinery for Agriculture and Forestry; Seat Index Point (ISO 5353: 1978, A1: 1981, A2: 1984) (Jan). 10 pp.

INTERNATIONAL AND NON-U.S. NATIONAL STANDARDS
SUBJECT INDEX

Formaldehyde

Forestry Equipment *(Cont.)*

—**Seats**
ISO 3462-80. Tractors and Machinery for Agriculture and Forestry—Seat Reference Point—Method of Determination First Edition; (Amendment 1-1992). 8 pp.

—**Symbols**
SNZ NZS 5104: Part 1-86. Tractors, Machinery for Agriculture and Forestry, Powered Lawn and Garden Equipment Part 1: Common Symbols Addendum 1; Amend: A, 1986; 1A, 1986. 5 pp.

—**Technical Manuals**
BSI BS 5401-90. 1990 Guide to Information Content and Presentation of Operators' Manuals Provided for Tractors and Machinery for Agriculture and Forestry. 18 pp.
BSI BS 5401-82. (WITHDRAWN) 1982 Guide for Presentation of Operator Manuals and Technical Publications Dealing with Tractors and Machinery for Agriculture and Forestry. 18 pp.
ISO 3600-81. Tractors and Machinery for Agriculture and Forestry—Operator Manuals and Technical Publications—Presentation Second Edition. 5 pp.

—**Tires**
CGSB 20-GP-5D-80. Tires, Pneumatic, Low Speed, Off Highway, Standard for. 23 pp.

—**Trench Excavators**
ISO 6688-82. Machinery for Forestry—Disc Trenchers—Disc-to-Hub Flange Fixing Dimensions First Edition. 5 pp.

—**Winches**
BSI BS 6385-83. 1983 Method of Specifying Performance Requirements for Winches on Forestry Machinery. 4 pp.
CSA CAN/CSA-M6816-92. Machinery for Forestry—Winches—Classification and Nomenclature (ISO 6816-1984); (Gen Instr 1). 19 pp.
ISO 6816-84. Machinery for Forestry—Winches—Classification and Nomenclature First Edition; (Corrigendum 1-1990). 7 pp.

—**Winches—Safety**
CSA CAN/CSA-M4254.4-M92. Machinery for Forestry—Winches—Technical Means for Ensuring Safety (ISO 4254-4:1990); (Gen Instr 1). 19 pp.
ISO 4254 Pt 4-90. Tractors and Machinery for Agriculture and Forestry—Technical Means for Ensuring Safety—Part 4: Forestry Winches First Edition. 6 pp.

Forged Products
Scope Note: See the subheading Forged under specific types of metal

Forging
Scope Note: See the subheading Forged or Forging under specific metals *See Also:* Forgings; Hammer Forging; Metal Working; Press Forging

—**Glossaries**
JIS B 0112-81. Glossary of Terms for Forging.

—**Hot—Safety**
CSA CAN/CSA-Z615-87. Code for Hot Forging Producers, Health and Safety Requirements; (Gen Instr 1). 26 pp.

Forging Equipment
CNS B4068-87. Requirements of Construction System for Automatic Forging Machine (Sep)(12067).

Forging Hammers
See Also: Hammers
CNS B3197-86. Hammer for Forging (Apr)(3096).

—**Noise**
DIN ENGL 45635 Pt 1602-78. Measurement of Airborne Noise Emitted by Machines; Enveloping Surface Method; Metal Processing Machine Tools; Special Stipulations for Drop Forging Hammers (June). 2 pp.

Forging Tongs
Use: Tongs—Forging

Forgings
Scope Note: See the subheading Forgings under the specific metal *See Also:* Die Forgings; Forging

Fork Ends
Use: Shaft Ends—Forked

Fork Joints
See Also: Joints

—**Aerospace**
BSI 3SP 3-43. 1943 Amd 3 Fork Joints (Low Tensile Type). 8 pp.

Fork Joints *(Cont.)*

—**Aerospace** *(Cont.)*
BSI 2SP 7-43. 1943 Fork Joints (High Tensile Type). 9 pp.

—**Aircraft**
BSI SP 119-120-57. 1957 Amd 1 High Tensile Steel Fork Joints. 4 pp.

—**Turnbuckles—Aircraft**
BSI SP 8-39. 1939 Turnbuckles (Tension Rod Type). 7 pp.

Fork Trucks
Use: Forklifts

Forklift Trucks
Use: Forklifts

Forklifts
Use For: Fork Trucks; Forklift Trucks
See Also: Electric Trucks; Electric Vehicles; Ground Vehicles; Hand Trucks; Industrial Trucks; Lift Trucks; Materials Handling Equipment; Motor Vehicles; Trucks
CNS D1023-82. Standard Form of Specifications of Fork Lift Trucks (Mar)(8571). 9 pp.
CNS D2146-86. Fork Lift Trucks (Jun)(9250).
JIS D 6001-88. Fork Lift Trucks. 25 pp.
JIS D 6202-76. Standard Form of Specifications of Fork Lift Trucks.

—**Automatic Guided—Safety**
JIS D 6802-90. Safety Standards of Automatic Guided Vehicles. 11 pp.

—**Brakes**
JIS D 6023-85. Brake Performance and Brake Tests for Fork Lift Trucks.

—**Fork Arms**
BSI BS 5639: Part 3-78. 1978 Fork Arms for Fork Lift Trucks Part 3: Recommendations for Dimensions of Fork Arms. 4 pp.
BSI BS 5639: Part 4-78. 1978 Fork Arms for Fork Lift Trucks Part 4: Specification for Technical Characteristics and Testing. 4 pp.
BSI BS 5639: Part 5-78. 1978 Fork Arms for Fork Lift Trucks Part 5: Guide for Inspection and Repair of Fork Arms in Service. 4 pp.
DIN ENGL 15174-88. Industrial Trucks; Fork Arms for Fork Lift Trucks Equipped with ISO Fork Carriers; Principal Dimensions (Apr). 2 pp.
ISO 2330-74. Fork Lift Trucks—Fork Arms—Technical Characteristics and Testing First Edition. 4 pp.
JIS B 8926-90. Hand Lifters. 9 pp.
SNZ NZS/ISO 2330-74. Forklift Trucks. Fork Arms. Technical Characteristics and Testing. 2 pp.

—**Fork Arms—Glossaries**
BSI BS 5639: Part 1-78. 1978 Fork Arms for Fork Lift Trucks Part 1: Vocabulary for Hook-On Type Fork Arms. 7 pp.
ISO 2331-74. Fork Lift Trucks—Hook-on Type Fork Arms—Vocabulary First Edition. 7 pp.

—**Fork Arms—Hooks**
DIN ENGL 15178-86. Industrial Trucks; Fork Arms for High-Lift Trucks with ISO Fork Carrier; Fork Hooks and Fork Lock (Oct). 7 pp.

—**Fork Arms—Inspection and Repair**
ISO TR5057-77. Industrial Trucks—Inspection and Repair of Fork Arms in Service on Fork Lift Trucks First Edition. 2 pp.

—**Fork Arms—Locks**
DIN ENGL 15178-86. Industrial Trucks; Fork Arms for High-Lift Trucks with ISO Fork Carrier; Fork Hooks and Fork Lock (Oct). 7 pp.

—**Fork Arms—Mounting**
BSI BS 5639: Part 2-78. 1978 Fork Arms for Fork Lift Trucks Part 2: Mounting Dimensions of Hook-On Type Fork Arms and Fork Carriers. 7 pp.
ISO 2328-77. Fork Lift Trucks—Hook-on Type Fork Arms and Fork Carriers—Mounting Dimensions First Edition. 6 pp.
JIS D 6024-87. Fork Lift Trucks—Hook-on Type Fork Arms and Fork Carrier-Mounting Dimensions. 6 pp.

—**Fork Carriers—Mounting**
BSI BS 5639: Part 2-78. 1978 Fork Arms for Fork Lift Trucks Part 2: Mounting Dimensions of Hook-On Type Fork Arms and Fork Carriers. 7 pp.
ISO 2328-77. Fork Lift Trucks—Hook-on Type Fork Arms and Fork Carriers—Mounting Dimensions First Edition. 6 pp.
JIS D 6024-87. Fork Lift Trucks—Hook-on Type Fork Arms and Fork Carrier-Mounting Dimensions. 6 pp.

Forklifts *(Cont.)*

—**Glossaries**
JIS D 6201-89. Glossary of Terms Relating to Fork Lift Trucks.
JIS D 6201-82. Glossary of Terms Relating to Fork Lift Trucks. 30 pp.

—**Order Picking—Stability**
CNS D3172-86. Method of Test for Stability for Order Picking Truck (Apr)(11533).

—**Overhead Protectors**
JIS D 6021-84. Overhead Guards for Fork Lift Trucks.

—**Safety**
SNZ NZS/ASME/ANSI B56.6-87. Safety Standard for Rough Terrain Forklift Trucks. 3 pp.

—**Sideloader—Stability**
CNS D3171-86. Method of Test for Stability for Side Fork Lift Truck (Apr)(11532).

—**Stability Testing**
BSI BS 4436-78. 1978 Reach and Straddle Fork Lift Trucks. Stability Tests. 10 pp.
BSI BS 5767-79. 1979 Guide for Storage and Handling of Conveyor Belts. 4 pp.
BSI BS 5778-79. 1979 Methods of Test for Verification of Stability of Industrial Trucks Operating in Special Condition of Stacking with Mast Tilted Forward. 4 pp.
CNS D3169-86. Method of Test for Stability of Counterbalanced Fork Lift Truck (Apr)(11530).
CNS D3170-86. Method of Test for Stability for Reach Fork Lift Trucks and Straddle Fork Lift Trucks (Apr)(11531).
ISO 3184-74. Reach and Straddle Fork Lift Trucks—Stability Tests First Edition. 9 pp.
ISO 5767-92. Industrial Trucks Operating in Special Condition of Stacking with Mast Tilted Forward—Additional Stability Test Second Edition. 6 pp.
JIS D 6011-82. Counterbalanced Fork Lift Trucks—Stability and Stability Tests. 7 pp.

—**Tires**
CNS K4030-80. Dimensions of Tires for Industrial Vehicles and off the Road Service (Feb)(3673). 33 pp.

Form Interface Management System
Use: FIMS

Form Vibrators
Use: Concrete Vibrators—Form

Formaldehyde
Use For: Formalin *See Also:* Hazardous Materials; Paraformaldehyde
BSI BS 2942-57. (WITHDRAWN) 1957 Amd 2 Formaldehyde Solution. 12 pp.
CNS K1056-80. Formaldehyde (37 Percent Grade) (Jul)(1381). 1 p.
CNS K6149-74. Method of Test for Formaldehyde of Industrial Grade (Oct)(1382).
CNS K7233-66. Chemical Reagent (Formaldehyde, Neutral) (Jun)(1732).
JIS K 1502-69. Formalin. 11 pp.
JIS K 8872-78. Formalin, Formaldehyde Solution.

—**Acidity—Volumetric Analysis**
ISO 2225-72. Formaldehyde Solutions for Industrial Use—Determination of Acidity First Edition. 4 pp.

—**Ash Content**
ISO 2224-72. Formaldehyde Solutions for Industrial Use—Determination of Ash First Edition. 3 pp.

—**Carbinol Content**
ISO 2228-72. Formaldehyde Solutions for Industrial Use—Determination of Methanol Content First Edition. 5 pp.

—**Chloride Content—Visual Limit Testing**
ISO 2221-72. Formaldehyde Solutions for Industrial Use—Limit Test for Inorganic Chlorides First Edition. 4 pp.

—**Fabrics—Colorfastness Testing**
CNS L3172-82. Method of Test for Colour Fastness to Formaldehyde (Jun)(9015).
ISO 105 Pt X09-87. Textiles—Tests for Colour Fastness—Part X09: Colour Fastness to Formaldehyde Third Edition. 4 pp.
JIS L 0868-75. Testing Method for Colour Fastness to Formaldehyde. 5 pp.

—**Heavy Metals Content—Visual Limit Testing**
ISO 2223-72. Formaldehyde Solutions for Industrial Use—Limit Test for Heavy Metals (Excluding Iron) First Edition. 4 pp.

Formaldehyde (Cont.)

—Highway Transportation—Emergency Procedures
SAA AS 1678.3.0. 015-93. Emergency Procedure Guide—Transport—Part 3.0.015: Formaldehyde Solutions, Flammable (in Professional Package 37).
SAA AS 1678.6.0. 000-90. Emergency Procedure Guides—Transport—Part 6.0.000: Formaldehyde Solutions (Non-Flammable) (Superseded by AS 1678.8.0.005—1993) (Withdrawn).

—Iron Content—Photometry
ISO 2226-72. Formaldehyde Solutions for Industrial Use—Determination of Iron Content—2,2'-Bipyridyl Photometric Method First Edition. 5 pp.

—Plywood—Desiccation
CNS O2060-87. Method of Test for Emission of Formaldehyde from Plywood (Desiccator Method) (May)(11971). 3 pp.

—Sulfate Content—Visual Limit Testing
ISO 2222-72. Formaldehyde Solutions for Industrial Use—Limit Test for Inorganic Sulphates First Edition. 4 pp.

—Volumetric Analysis
ISO 2227-72. Formaldehyde Solutions for Industrial Use—Determination of Formaldehyde Content First Edition. 4 pp.

Formaldehyde Content Analysis
Use For: Free Formaldehyde Content Analysis

—Cosmetics
CNS S2084-82. Methods of Hygienic Test for Cosmetics — Formaldehyde (Oct)(9538).
EC 82/434/EEC-82. Second Commission Directive on the Approximation of the Laws of the Member States Relating to Methods of Analysis Necessary for Checking the Composition of Cosmetic Products. 28 pp.
EC 90/207/EEC-90. Commission Directive Amending the Second Directive 82/434/EEC on the Approximation of the Laws of the Member States Relating to Methods of Analysis Necessary for Checking the Composition of Cosmetic Products. 10 pp.

—Detergents
BSI BS 3762: Sec 3.11-89. 1989 Analysis of Formulated Detergents Part 3: Quantitative Test Methods Section 3.11: Method for Determination of Free Formaldehyde Content. 4 pp.

—Fabrics
BSI BS 6806: Part 1-87. 1987 Amd 1 Formaldehyde in Textiles Part 1: Method for Determination of Total Formaldehyde. 9 pp.
BSI BS 6806: Part 2-87. 1987 Amd 1 Formaldehyde in Textiles Part 2: Method for Determination of 'Free' Formaldehyde. 9 pp.
BSI BS 6806: Part 3-87. 1987 Amd 1 Formaldehyde in Textiles Part 3: Method for Determination of Released Formaldehyde. 9 pp.
CGSB CAN/CGSB-4.2 NO.63.2-M91. Textile Test Methods Formaldehyde in Textiles —Part 2: Method for Determination of "Free" Formaldehyde (BS 6806:Part 2:1987). 10 pp.
CGSB CAN/CGSB-4.2 NO.63.3-M91. Textile Test Methods Formaldehyde in Textiles — Part 3: Method for Determination of Released Formaldehyde (BS 6806:Part 3:1987); (Including Amendment 1 January 1988). 10 pp.

—Melamine Formaldehyde Resins
BSI BS 2782:Pt4: METH 462A-B-78. 1978 Methods of Testing Plastics Part 4: Chemical Properties Method 462A and 462B: Determination of Extractable Formaldehyde in Melamine Formaldehyde Mouldings. 6 pp.
ISO 4614-77. Plastics—Melamine-Formaldehyde Mouldings—Determination of Extractable Formaldehyde First Edition. 6 pp.

—Methenamine
MOD UK M 879/69. Examination of Hexamine.

—Paper Products—Acetylacetone Method
CNS P3085-87. Method of Test for Formaldehyde Content of Paper Products (Acetylacetone Method) (Sep)(12103).

—Phenol Formaldehyde Resins
DIN ENGL 53748-70. Chemical Analysis of Phenol-Formaldehyde Resins, Phenoplastic Moulding Materials and Moulded Materials (July). 3 pp.

Formaldehyde Content Analysis (Cont.)

—Phenolic Resins
BSI BS 2782:Pt4: METH 451F-J-78. 1978 Methods of Testing Plastics: Part 4: Chemical Properties: Method 451F; Determination of Formaldehyde in Phenolic Mouldings (Colorimetric Method) Method 451G; Deter of Formaldehyde in Phenolic Mouldings (Gravimetric Method). 4 pp.
DIN ENGL 53748-70. Chemical Analysis of Phenol-Formaldehyde Resins, Phenolplastic Moulding Materials and Moulded Materials (July). 3 pp.

—Phenolic Resins—Potentiometric Analysis
ISO 9397-89. Plastics—Phenolic Resins—Determination of Free Formaldehyde Content First Edition. 4 pp.

—Plywood—Gas Analysis
CEN PREN 1084-93. Plywood—Formaldehyde Release Classes. 5 pp.

—Thermosetting Resins
DIN ENGL 16746-86. Plastics, Paints and Varnishes; Thermosetting Resins; Determination of Free Formaldehyde in Condensation Resins (Apr). 4 pp.

—Water
CNS K3030-73. Phenol and Formaldehyde Content in Extract of Phenolic Resin or Other Thermosetting Resins Dinner Wares (Oct)(3707). 2 pp.
CNS K9068-81. Methods of Test for Formaldehyde in Water (May)(7440).

—Wood Panels—Gas Analysis
CEN PREN 717-2-92. Wood-Based Panels—Determination of Formaldehyde Emission—Part 2: Formaldehyde Release by the Gas Analysis Method. 11 pp.

—Wood Panels—Perforator Method
BSI BS EN 120-92. 1992 Wood Based Panels—Determination of Formaldehyde Content—Extraction Method Called the Perforator Method. 15 pp.
CEN EN 120-84. Particleboards: Determination of Formaldehyde Content Extraction Method Called Perforator Method. 4 pp.
CEN PREN 120-89. Wood-Based Panels—Determination of Formaldehyde Content—Extraction Method Called the Perforator Method. 11 pp.
CEN EN 120-91. Wood Based Panels—Determination of Formaldehyde Content—Extraction Method Called the Perforator Method. 10 pp.
CEN EN 120-92. Wood Based Panels—Determination of Formaldehyde Content—Extraction Method Called the Perforator Method. 13 pp.
DIN ENGL EN 120-92. Wood-Based Panel Products; Determination of Formaldehyde Content; Extraction Method (Known as Perforator Method) (Aug). 11 pp.

Formalin
Use: Formaldehyde

Formamides
CNS K7234-67. Chemical Reagent (Formamide) (Oct)(1733).
JIS K 8873-78. Formamide.

Formates
Scope Note: Use a more specific term
See: Ammonium Formate; Sodium Formate

Formation Gas
Use: Natural Gas

Formations (Military)
Use For: Military Units *See Also:* Logistics Operations; Military Operations; Military Personnel; Military Vehicles

—National Designations
NATO STANAG 2356 ED 1 AMD 5-75. Comparative Formation/Unit Designations. 22 pp.
NATO STANAG 2356 ED 1 AMD 6-75. Comparative Formation/Unit Designations. 23 pp.
NATO STANAG 2356 ED 2 AMD 0-91. Comparative Formation/Unit Designations. 21 pp.
NATO STANAG 2356 ED 2 AMD 1-91. Comparative Formation/Unit Designations. 22 pp.

Formboards
See Also: Formwork (Construction)

—Plywood—Concretes
CNS O1022-81. Plywood for Concrete-Form (Oct)(8057). 4 pp.

Formic Acid
See Also: Formic Acid Content Analysis
CNS K7235-67. Chemical Reagent (Formic Acid) (Oct)(1734).

Formic Acid (Cont.)
ISO 731 Pt I-77. Formic Acid for Industrial Use—Methods of Test—Part I: General First Edition. 3 pp.
JIS K 1356-60. Formic Acid.
JIS K 8264-76. Formic Acid.

—Acid Content—Potentiometric Analysis
ISO 731 Pt III-77. Formic Acid for Industrial Use—Methods of Test—Part III: Determination of Content of Other Acids—Potentiometric Method First Edition. 4 pp.

—Acid Content—Volumetric Analysis
BSI BS 4341-68. 1968 Methods of Test for Formic Acid. 13 pp.
ISO 731 Pt VII-77. Formic Acid for Industrial Use—Methods of Test—Part VII: Determination of Low Contents of Other Volatile Acids—Titrimetric Method After Distillation First Edition. 6 pp.

—Acidity—Volumetric Analysis
BSI BS 4341-68. 1968 Methods of Test for Formic Acid. 13 pp.
ISO 731 Pt II-77. Formic Acid for Industrial Use—Methods of Test—Part II: Determination of Total Acidity—Titrimetric Method First Edition. 4 pp.

—Chloride Content—Turbidity
BSI BS 4341-68. 1968 Methods of Test for Formic Acid. 13 pp.

—Chloride Content—Visual Limit Testing
ISO 731 Pt IV-77. Formic Acid for Industrial Use—Methods of Test—Part IV: Visual Limit Test for Inorganic Chlorides First Edition. 4 pp.

—Iron Content—Photometry
BSI BS 4341: Add 1-71. 1971 Methods of Test for Formic Acid Addendum No.1:. 6 pp.
ISO 731 Pt VI-77. Formic Acid for Industrial Use—Methods of Test—Part VI: Determination of Iron Content—2,2'-Bipyridyl Photometric Method First Edition. 5 pp.

—Sulfate Content—Turbidity
BSI BS 4341-68. 1968 Methods of Test for Formic Acid. 13 pp.

—Sulfate Content—Visual Limit Testing
ISO 731 Pt V-77. Formic Acid for Industrial Use—Methods of Test—Part V: Visual Limit Test for Inorganic Sulphates First Edition. 4 pp.

Formic Acid Content Analysis
See Also: Formic Acid

—Acetic Acid—Iodometry
ISO 753 Pt 3-83. Acetic Acid for Industrial Use—Methods of Test—Part 3: Determination of Formic Acid Content—Iodometric Method First Edition. 6 pp.

—Fruits—Gravimetric Analysis
ISO 6638 Pt 1-85. Fruit and Vegetable Products—Determination of Formic Acid Content—Part 1: Gravimetric Method First Edition. 5 pp.

—Fruits—Volumetric Analysis
ISO 6638 Pt 2-84. Fruit and Vegetable Products—Determination of Formic Acid Content—Part 2: Titrimetric Method First Edition; (Corrected and Reprinted -1985). 5 pp.

—Vegetables—Gravimetric Analysis
ISO 6638 Pt 1-85. Fruit and Vegetable Products—Determination of Formic Acid Content—Part 1: Gravimetric Method First Edition. 5 pp.

—Vegetables—Volumetric Analysis
ISO 6638 Pt 2-84. Fruit and Vegetable Products—Determination of Formic Acid Content—Part 2: Titrimetric Method First Edition; (Corrected and Reprinted -1985). 5 pp.

Forms (Paper)
Scope Note: For additional listings, use a more specific term *Use For:* Blanks (Forms)
See Also: Business Forms; Continuous Forms; Invoices; Papers; Printed Forms; Systems Engineering

—Air Force Operations—Air Movements
NATO STANAG 3345 ED 5 AMD 2-88. Data/Forms for Planning Air Movements. 13 pp.
NATO STANAG 3345 ED 5 AMD 3-88. Data/Forms for Planning Air Movements. 12 pp.

—Aircraft—Construction—Mass
DIN ENGL 9020 Pt 3-83. Aerospace; Mass Breakdown for Aircraft Heavier Than Air; Group Mass Statement (Oct). 16 pp.
DIN ENGL 9020 Pt 3 Suppl. 1-83. Aerospace; Mass Breakdown for Aircraft Heavier Than Air; Group Mass Statement; Condensed Version (Oct). 4 pp.

INTERNATIONAL AND NON-U.S. NATIONAL STANDARDS
SUBJECT INDEX

Forms (Paper) (Cont.)
—Aircraft—Construction—Mass (Cont.)
DIN ENGL 9020 Pt 4-83. Aerospace; Mass Breakdown for Aircraft Heavier Than Air; Detail Mass Statement (Oct). 77 pp.

—Defense Contracts—Acknowledgement of Receipt/Terms of Contract
MOD UK DEFCON 10-89. Acknowledgement of Receipt and Terms of Contract Form 9/89. 1 p.
MOD UK DEFCON 10A-90. Acknowledgement of Receipt and Terms of Contract Form 10/90. 1 p.

—Defense Contracts—Advice and Inspection Note Signature
MOD UK DEFCON 5-89. MOD Form 640—Advice and Inspection Note Signature by MOD QAR 12/89. 3 pp.
MOD UK DEFCON 5-92. MOD Forms 640—Advice and Inspection Notes 11/92. 1 p.

—Defense Contracts—Bank or Insurance Company Guarantee
MOD UK DEFCON 24A-90. Specimen—for Drafting Purposes Only Guarantee Given by a Bank or Insurance Company 1/90. 2 pp.

—Defense Contracts—Claims
MOD UK DEFCON 44-76. Contractor's Financial Claim 10/76. 4 pp.

—Defense Contracts—Government Supplied Property
MOD UK DEFCON 198-91. Clothing and Textile Contracts Issues of Government Property on Embodiment Loan Terms 2/91. 2 pp.
MOD UK DEFCON 198A-91. Clothing and Textile Contracts Issues of Government Property on Prepayment or Repayment Terms 2/91. 2 pp.
MOD UK DEFCON 198B-80. Materials to Be Issued to the Contractor 9/80. 1 p.

—Defense Contracts—Packaging Design Application and Authorization
MOD UK DEFCON 129A-85. Application for Packaging Designs and Authorisation for Package Design Work 5/85. 2 pp.
MOD UK DEFCON 129A CONT-78. Application for Packaging Specifications and Authorisation for Package Design Work 9/78. 2 pp.

—Defense Contracts—Parent Company Guarantee of a Subsidiary
MOD UK DEFCON 24-90. Specimen—for Drafting Purposes Only Guarantee Given by a Parent Company in Respect of a Subsidiary 1/90. 2 pp.

—Defense Contracts—Purchase Orders
MOD UK DEFCON 52-89. Local Purchase Order Form 11/89. 5 pp.
MOD UK DEFCON 52C-86. Local Purchase Order Form (Computerised Version) 12/86. 5 pp.
MOD UK DEFCON 52WS-89. Local Purchase Order Works Services—Schedule 12/89. 3 pp.

—Defense Contracts—Quarterly Financial Reports
MOD UK DEFCON 136-75. Quarterly Financial Report 11/75. 1 p.

—Defense Contracts—Repairs
MOD UK DEFCON 69-87. Repair Estimates (DGEME & General) 4/87. 2 pp.

—Design
BSI BS 5537-91. 1991 Forms Design Sheet and Layout Chart. 10 pp.

—Hazardous Materials Declaration
SNZ NZS 5433F-90. NZ Hazardous Substances Dangerous Goods Declaration. 50 pp.

—Lathes—Test Results
DIN ENGL 8613 Suppl. 1-70. Acceptance Conditions for Machine Tools; Multi-Spindle Automatic Lathes; Horizontal Type for Rotating Workpieces; Form for Test Results (Mar). 1 p.

—Materiel Defects
EURO CTI/88/18410-88. Basic Outlines for an Intercompany Defect Reporting and Investigation Procedure. 12 pp.

—Military Geographic Information—Data Transmission
NATO STANAG 3986 ED 1 AMD 1-87. Digital Data File Transmittal Form for Geographic Information. 14 pp.
NATO STANAG 3986 ED 1 AMD 2-87. Digital Data File Transmittal Form for Geographic Information. 15 pp.

Forms (Paper) (Cont.)
—Military Geographic Information—Data Transmission (Cont.)
NATO STANAG 3986 ED 1 AMD 3-87. Digital Data File Transmittal Form for Geographic Information. 16 pp.

—Occupational Safety and Health
SAA AS 1885.1 Supp 1-91. Measurement of Occupational Health and Safety Performance—Part 1: Describing and Reporting Occupational Injuries and Disease—Supplement 1: Recording Form (Supplement 1 to AS 1885.1—1990) (Supersedes AS 1885/A). 100 pp.
SAA AS 1885/B. Measurement of Occupational Health and Safety Performance—Part B: Register of Work Injuries Pads of 50 Forms.
SAA AS 1885/C. Measurement of Occupational Health and Safety Performance—Part C: Register of Work Injuries (Modified) Pads of 50 Forms.

—Pipelines—Axonometric Projections
DIN ENGL 2428-68. Drawings for Pipelines; Forms for Isometric Representation (Dec). 5 pp.

Formulas
Scope Note: See specific applications

Formwork (Construction)
Use For: Concrete Formwork; Shuttering (Formwork)
See Also: Concrete Construction; Formboards; Framing; Shoring
DIN ENGL 18217-81. Concrete Surfaces and Formwork Surface (Dec). 2 pp.

—Design
CSA CAN/CSA-S269.3-M92. Concrete Formwork; (Gen Instr 1). 43 pp.

—Fabrication
CSA CAN/CSA-S269.3-M92. Concrete Formwork; (Gen Instr 1). 43 pp.

—Glossaries
BSI BS 6100: Sec 6.5-87. 1987 Glossary of Building and Civil Engineering Terms Part 6: Concrete and Plaster Section 6.5: Formwork. 8 pp.

—Ties
DIN ENGL 18216-86. Formwork Ties; Requirements, Testing, Use (Dec). 11 pp.

Forney F1A Aircraft
See Also: Aircraft

—Antenna Positions
CAA. Forney F1A. 1 p.

FORTRAN
See Also: COBOL; PL/1; Programming Languages
BSI BS 6832-87. (WITHDRAWN) 1987 Method for Specifying Requirements for a FORTRAN Language Processor. 32 pp.
BSI BS 7146-90. (WITHDRAWN) 1990 Programming Language FORTRAN (Superseded by BS EN 21539: 1993). 4 pp.
BSI BS EN 21539-92. 1992 Information Technology—Programming Languages—FORTRAN (Supersedes BS 7146: 1990) (S). 391 pp.
CEN EN 21 539-89. Programming Languages FORTRAN. 434 pp.
CEN EN 21539-92. Information Technology—Programming Languages—FORTRAN (ISO/IEC 1539, 2nd Edition 1991). 388 pp.
IEC 1539-91. Information Technology—Programming Languages—Fortran Second Edition (CEN EN 21539: 1992). 387 pp.
ISO 1539-91. Information Technology—Programming Languages—Fortran Second Edition; (CEN EN 21539:1992). 387 pp.
JIS X 3001-82. Programming Language FORTRAN.
JTC1 1539-91. Information Technology—Programming Languages—Fortran Second Edition. 387 pp.
OSI ISO/IEC 1539-91. Information Technology—Programming Languages—Fortran. 415 pp.
SAA AS 1486-83. Programming Language FORTRAN. 424 pp.

—Language Binding—Graphical Kernel System
BSI BS 7040: Part 1-89. (WITHDRAWN) 1989 Computer Graphics: Graphical Kernel System (GKS) Language Bindings Part 1: GKS Language Binding for FORTRAN (Renumbered as BS EN 28651-1: 1992). 122 pp.
BSI BS EN 28651-1-92. 1992 Information Processing Systems—Computer Graphics—Graphical Kernel System (GKS) Language Bindings Part 1: FORTRAN (ISO 8651-1: 1988) (AMD 7537) December 15, 1992. 126 pp.

FORTRAN (Cont.)
—Language Binding—Graphical Kernel System (Cont.)
BSI DD 150: Part 1-87. (WITHDRAWN) 1987 Computer Graphics: Graphical Kernel System for Three Dimensions (GKS-3D) Language Bindings Part 1: Fortran (Superseded by ISO/DIS 8806-1). 48 pp.
CEN EN 28651-1-92. Information Processing Systems—Computer Graphics—Graphical Kernel System (GKS) Language Bindings Part 1 FORTRAN; (ISO 8651-1: 1988). 118 pp.
ISO 8651 Pt 1-88. Information Processing Systems—Computer Graphics—Graphical Kernel System (GKS) Language Bindings—Part 1: FORTRAN First Edition; (CEN EN 28651-1:1992). 119 pp.
JTC1 8651 Pt 1-88. Information Processing Systems—Computer Graphics—Graphical Kernel System (GKS) Language Bindings—Part 1: FORTRAN First Edition. 119 pp.
OSI ISO DP 8651-1-86. Computer Graphics-Graphical Kernel System (GKS) Language Bindings: Part 1: FORTRAN. 115 pp.
OSI ISO 8651-1-88. Information Processing Systems—Computer Graphics—Graphical Kernel Systems (GKS) Language Bindings—Part 1: FORTRAN. 119 pp.
OSI ISO/IEC DIS 8806-1-88. Information Processing Systems—Computer Graphics—Graphical Kernal System for Three Dimensions (GKS-3D) Language Bindings—Part 1: Fortran. 137 pp.
SNZ NZS/ISO 8651. 1-88. Computer Graphics—Graphical Kernel System (GKS) Language Bindings Part 1: FORTRAN. 116 pp.

—Language Binding—Programmers' Hierarchical Interactive Graphics
BSI BS ISO/IEC 9593/1-90. 1990 Information Technology—Computer Graphics—Programmers Hierarchical Interactive Graphics System (PHIGS) Language Bindings Part 1: FORTRAN. 223 pp.
IEC 9593 Pt 1-90. Information Processing Systems—Computer Graphics—Programmer's Hierarchical Interactive Graphics System (PHIGS) Language Bindings—Part 1: FORTRAN First Edition; (Technical Corrigendum 1 —1993). 221 pp.
IEC 9593 Pt 1 Draft AMD 1. Information Processing Systems—Computer Graphics—Programmer's Hierarchical Interactive Graphics System (PHIGS) Language Bindings—Part 1: FORTRAN; (1992) ***CD-ROM ONLY***. 162 pp.
ISO 9593 Pt 1-90. Information Processing Systems—Computer Graphics—Programmer's Hierarchical Interactive Graphics System (PHIGS) Language Bindings—Part 1: FORTRAN First Edition; (Technical Corrigendum 1 —1993). 221 pp.
ISO 9593 Pt 1 Draft AMD 1. Information Processing Systems—Computer Graphics—Programmer's Hierarchical Interactive Graphics System (PHIGS) Language Bindings—Part 1: FORTRAN; (1992) ***CD-ROM ONLY***. 162 pp.
JTC1 9593 Pt 1-90. Information Processing Systems—Computer Graphics—Programmer's Hierarchical Interactive Graphics System (PHIGS) Language Bindings—Part 1: FORTRAN First Edition; (Technical Corrigendum 1 —1993). 221 pp.
JTC1 9593 Pt1 Draft AMD 1. Information Processing Systems—Computer Graphics—Programmer's Hierarchical Interactive Graphics System (PHIGS) Language Bindings—Part 1: FORTRAN; (1992) ***CD-ROM ONLY***. 162 pp.
OSI ISO DIS 9593-1-88. Information Processing Systems—Computer Graphics—Programmers Hierarchical Interactive Graphics System (PHIGS) Language Bindings Part 1: Fortran Binding. 205 pp.
SAA AS 3794.1-91. Computer Graphics—Programmer's Hierarchical Interactive Graphics System (PHIGS) Language Bindings—Part 1: FORTRAN (ISO/IEC 9593-1) (In Professional Packages 26A, 26D). 213 pp.

—Process Control Systems
BSI BS 6831-87. 1987 Industrial Real-Time FORTRAN. 35 pp.
ISO 7846-85. Industrial Real-Time FORTRAN—Application for the Control of Industrial Processes First Edition. 38 pp.
JTC1 7846-85. Industrial Real-Time FORTRAN—Application for the Control of Industrial Processes First Edition. 38 pp.
OSI ISO 7846-85. Industrial Real-Time FORTRAN—Application for the Control of Industrial Processes. 38 pp.

—Source Programs
ECMA ECMA 53-78. Representation of Source Program for Program Interchange-APL, COBOL, FORTRAN, Minimal BASIC and PL/1. 16 pp.

Forward Address Information Signals
See Also: Signaling Systems

Forward Address Information Signals (Cont.)

CCITT RECMN Q.107-89. Standard Sending Sequence of Forward Address Information—General Recommendations on Telephone Switching and Signalling—Functions and Information Flows for Services in the ISDN—Supplements (Study Group XI) 9 pp. 9 pp.

CCITT RECMN Q.468-89. Signalling Procedures—Routing and Numbering for International Working. Termination of Interregister Signalling Specifications of Signalling Systems R1 and R2 (Study Group XI) 1 pp. 1 p.

—Routing

CCITT RECMN Q.107 BIS-89. Analysis of Forward Address Information for Routing—General Recommendations on Telephone Switching and Signalling—Functions and Information Flows for Services in the ISDN—Supplements (Study Group XI) 4 pp. 4 pp.

CCITT RECMN Q.126-89. Analysis and Transfer of Digital Information—Specifications of Signalling Systems Nos. 4 and 5 (Study Group XI) 1 pp. 1 p.

CCITT RECMN Q.155-89. Analysis of Digital Information for Routing—Specifications of Signalling Systems Nos. 4 and 5 (Study Group XI) 1 pp. 1 p.

CCITT RECMN Q.161-89. General Arrangements for Manual Testing—Specifications of Signalling Systems Nos. 4 and 5 (Study Group XI) 1 pp. 1 p.

CCITT RECMN Q.262-89. 4.2 Analysis of Digital Information for Routing—Specifications of Signalling System No. 6 (Study Group XI) 1 pp. 1 p.

CCITT RECMN Q.468-89. Signalling Procedures—Routing and Numbering for International Working. Termination of Interregister Signalling Specifications of Signalling Systems R1 and R2 (Study Group XI) 1 pp. 1 p.

Forward Air Controllers

See Also: Airborne Operations; Military Personnel; Tactical Warfare

—Qualifications

NATO STANAG 3797 ED 1 AMD 2-79. Minimum Qualifications for Forward Air Controllers. 7 pp.

NATO STANAG 3797 ED 1 AMD 3-79. Minimum Qualifications for Forward Air Controllers. 7 pp.

Forward and Reverse Audible Warning Alarms

Use: Backup Alarms

Forward Blocking Thyristors

See Also: Thyristors

CECC CECC 50 011-010 ISSUE 1-83. BS CECC 50 011-010; Type Numbers: BTW 63 Series (En). 18 pp.

Forward Error Correction

Use For: FEC *See Also:* Data Transmission; Error Correction

—Compensation—Multipath Fading

CCIR Report 921-2-90. System Aspects of Digital Ship Earth Stations—Section 8F—Frequencies, Orbits and Systems. 46 pp.

—Radiotelegraph Systems

CCIR Report 349-1-70. Single-Channel Radiotelegraph Systems Employing Forward Error Correction—Section 3Ca —Radiotelegraph Circuits. 14 pp.

—Radiotelegraph Systems—Maritime Mobile Services

CCIR RECMN 476-4-86. Direct-Printing Telegraph Equipment in the Maritime Mobile Service—Section 8B—Maritime Mobile Service; Telegraphy and Related Subjects. 10 pp.

—Telegraph Printers—Maritime Mobile Services

CCIR RECMN 492-4-90. Operational Procedures for the Use of Direct-Printing Telegraph Equipment in the Maritime Mobile Service—Section 8B—Maritime Mobile Service; Telegraphy and Related Subjects. 7 pp.

CCIR RECMN 492-5-92. Operational Procedures for the Use of Direct-Printing Telegraph Equipment in the Maritime Mobile Service—Section 8B—Maritime Mobile Service; Telegraphy and Related Subjects. 7 pp.

CCIR RECMN 625-1-90. Direct-Printing Telegraph Equipment Employing Automatic Identification in the Maritime Mobile Service—Section 8B—Maritime Mobile Service; Telegraphy and Related Subjects. 55 pp.

CCIR RECMN 625-2-92. Direct-Printing Telegraph Equipment Employing Automatic Identification in the Maritime Mobile Service—Section 8B—Maritime Mobile Service; Telegraphy and Related Subjects. 59 pp.

Forward Interworking Telephone Events

Use: FITES

Forward Signals

See Also: Signaling Systems

—Glossaries—CCITT R2 Signaling Systems

CCITT RECMN Q.400-89. Definitions and Functions of Signals—Forward Line Signals. Backward Line Signals. Forward Register Signals. Backward Register Signals—Specifications of Signalling Systems R1 and R2 (Study Group XI) 4 pp. 4 pp.

Forward Transfer Signals

See Also: Signaling Systems

—CCITT R2 Signaling Systems

CCITT FASCICLE VI.4. Specifications of Signalling Systems R1 and R2 Recommendations Q.310—Q.490 (Study Group XI). 181 pp.

CCITT RECMN Q.400-89. Definitions and Functions of Signals—Forward Line Signals. Backward Line Signals. Forward Register Signals. Backward Register Signals—Specifications of Signalling Systems R1 and R2 (Study Group XI) 4 pp. 4 pp.

CCITT RECMN Q.441-89. Interregister Signalling—Signalling Code—Specifications of Signalling Systems R1 and R2 (Study Group XI) 10 pp. 10 pp.

—Glossaries—CCITT R2 Signaling Systems

CCITT RECMN Q.400-89. Definitions and Functions of Signals—Forward Line Signals. Backward Line Signals. Forward Register Signals. Backward Register Signals—Specifications of Signalling Systems R1 and R2 (Study Group XI) 4 pp. 4 pp.

Forward Traveling Wave Amplifier Tubes

Use: Traveling Wave Tubes

Forward Travelling Wave Amplifier Tubes

Use: Traveling Wave Tubes

Fossil Fuels

Scope Note: For additional listings, use a more specific term *See Also:* Anthracite; Bituminous Coal; Coal; Crude Oil; Lignite; Liquefied Natural Gas; Liquefied Petroleum Gas; Minerals; Natural Gas; Ores; Peat; Sub-Bituminous Coal

—Classification

BSI BS 6843: Part 0-88. 1988 Classification of Petroleum Fuels Part 0: General Classification. 3 pp.

ISO 8216 Pt 0-86. Petroleum Products—Fuel (Class F)—Classification—Part 0: General First Edition. 3 pp.

Foundation Bolts

Use: Anchor Bolts

Foundation Garments

See Also: Clothing; Slips (Underwear); Underwear

—Sizes

JIS L 4006-87. Sizing Systems for Foundation Garments. 9 pp.

—Sizes—Girls'

ISO 4416-81. Size Designation of Clothes—Women's and Girls' Underwear, Nightwear, Foundation Garments and Shirts First Edition; (Corrigendum 1-1990). 9 pp.

—Sizes—Women's

ISO 4416-81. Size Designation of Clothes—Women's and Girls' Underwear, Nightwear, Foundation Garments and Shirts First Edition; (Corrigendum 1-1990). 9 pp.

Foundation Nuts

Use: Anchor Nuts

Foundation Studs

Use: Anchor Studs

Foundations

Use For: Shallow Foundations *See Also:* Bearing Capacity; Earth Fill; Footings; Grouting; Pile Foundations; Piles; Solums; Structural Members; Substructures

—Buildings

BSI CP 101-72. 1972 Foundations and Substructures for Non-Industrial Buildings of Not More Than Four Storeys. 29 pp.

Foundations (Cont.)

—Concrete—Asphalt Coatings

CGSB CAN/CGSB-37.1-M89. Chemical Emulsifer Type, Emulsified Asphalt for Damp-Proofing. 10 pp.

CGSB 37-GP-6MA-83. Asphalt, Cutback, Unfilled, for Dampproofing, Standard for. 9 pp.

CGSB 37-GP-12MA-84. Application of Unfilled Cutback Asphalt for Dampproofing, Standard for. 6 pp.

CGSB CAN/CGSB-37.16-M89. Filled, Cutback Asphalt for Dampproofing and Waterproofing. 10 pp.

—Concrete—Tar Coatings

CGSB CAN/CGSB-37.22-M89. Application of Unfilled, Cutback Tar Foundation Coating for Dampproofing. 7 pp.

—Construction

BSI BS 8004-86. 1986 Code of Practice for Foundations. 157 pp.

BSI CP 2004-72. 1972 Foundations. 161 pp.

—Design

BSI BS 8004-86. 1986 Code of Practice for Foundations. 157 pp.

BSI BS 8103: Part 1-86. 1986 Structural Design of Low-Rise Buildings Part 1: Code of Practice for Stability, Site Investigation, Foundations and Ground Floor Slabs for Housing. 27 pp.

BSI CP 2004-72. 1972 Foundations. 161 pp.

—Glossaries

BSI BS 6100: SUBSEC 2.2.2-90. 1990 Glossary of Building and Civil Engineering Terms Part 2: Civil Engineering Section 2.2: Substructures. Earthworks. Foundations. Tunnels Subsection 2.2.2: Substructures and Foundations. 12 pp.

BSI BS 6100: SUBSEC 2.2.2-01. 1990 Amd 1 Glossary of Building and Civil Engineering Terms Part 2: Civil Engineering Sec. 2.2: Substructures. Earthworks. Foundations. Tunnels Subsec 2.2.2: Substructures and Foundations (AMD 7250) August 15, 1992. 14 pp.

—Houses

BSI BS 8103: Part 1-86. 1986 Structural Design of Low-Rise Buildings Part 1: Code of Practice for Stability, Site Investigation, Foundations and Ground Floor Slabs for Housing. 27 pp.

—Machinery—Periodic Excitation

DIN ENGL 4024 Pt 2-91. Machine Foundations; Rigid Foundations for Machinery Subject to Periodic Vibration (Apr). 5 pp.

—Masonry—Asphalt Coatings

CGSB CAN/CGSB-37.1-M89. Chemical Emulsifer Type, Emulsified Asphalt for Damp-Proofing. 10 pp.

CGSB 37-GP-6MA-83. Asphalt, Cutback, Unfilled, for Dampproofing, Standard for. 9 pp.

CGSB 37-GP-12MA-84. Application of Unfilled Cutback Asphalt for Dampproofing, Standard for. 6 pp.

CGSB CAN/CGSB-37.16-M89. Filled, Cutback Asphalt for Dampproofing and Waterproofing. 10 pp.

—Masonry—Tar Coatings

CGSB 37-GP-18MA-85. Tar, Cutback, Unfilled, for Dampproofing,. 9 pp.

CGSB CAN/CGSB-37.22-M89. Application of Unfilled, Cutback Tar Foundation Coating for Dampproofing. 7 pp.

—Mobile Homes

CSA CAN3-Z240. 10.1-M86. Recommended Practice for the Site Preparation, Foundation, and Anchorage of Mobile Homes; (Gen Instr 1). 33 pp.

—Offshore Platforms

CSA CAN/CSA-S472-92. Foundations; (Gen Instr 1). 33 pp.

CSA S472.1-1992 SP. Commentary to CSA Standard CAN/CSA-S472-92, Foundations; (Gen Instr 1). 50 pp.

—Reciprocating Machines

BSI CP 2012: Part 1-74. 1974 Amd 2 Foundations for Machinery Part 1: Foundations for Reciprocating Machines (AMD 1938) April 30, 1976. 39 pp.

—Rotating Machines

DIN ENGL 4024 Pt 1-88. Machine Foundations; Flexible Structures That Support Machines with Rotating Elements (Apr). 10 pp.

—Subsoil—Bearing Pressure

DIN ENGL 4018-74. Subsoil; Calculation of the Bearing Pressure; Distribution Under Spread Foundations (Sept). 3 pp.

INTERNATIONAL AND NON-U.S. NATIONAL STANDARDS
SUBJECT INDEX

Foundations (Cont.)
—Subsoil—Shear Failure
DIN ENGL 4017 Pt 1-79. Subsoil; Shear Failure Calculations for Shallow Foundations with Vertical and Central Loading (Aug). 4 pp.
DIN ENGL 4017 Pt 1 Suppl. 1-79. Subsoil; Shear Failure Calculations for Shallow Foundations with Vertical and Central Loading; Explanations and Examples of Calculation (Aug). 12 pp.
DIN ENGL 4017 Pt 2-79. Subsoil; Shear Failure Calculations for Shallow Foundations with Oblique and Eccentric Loading (Aug). 4 pp.
DIN ENGL 4017 Pt 2 Suppl. 1-79. Subsoil; Shear Failure Calculations for Shallow Foundations with Oblique and Eccentric Loading; Explanations and Examples of Calculations (Aug). 13 pp.

—Symbols
DIN ENGL 1080 Pt 6-80. Terms, Symbols and Units Used in Civil Engineering; Soil Mechanics and Foundation Engineering (Mar). 8 pp.

—Wood—Construction
CSA CAN/CSA-S406-92. Construction of Preserved Wood Foundations; (Gen Instr 1). 103 pp.

—Wood—Preserved—Pressure Treated
CSA O322-1976. Procedure for Certification of Pressure-Treated Wood Materials for Use in Preserved Wood Foundations; (Amd 1-4 July 1986) (Erratum July 1986). 17 pp.

—Wood—Sills—Preservatives
JIS A 9108-92. Foundation Wood Sill Treated with Preservatives by Pressure Processes. 34 pp.

Foundry Equipment
Scope Note: Use a more specific term See: Foundry Patterns; Molding Boxes

Foundry Patterns
See Also: Molds (Casting); Patterns; Templates

—Classification
DIN ENGL 1511-78. Pattern Equipment for Foundries; Production and Quality (Apr). 10 pp.

—Color Coding
SAA AS B8-59. Methods for the Colouring and Marking of Foundry Patterns (Withdrawn).

—Identification Systems
SAA AS B8-59. Methods for the Colouring and Marking of Foundry Patterns (Withdrawn).

—Quality Assurance
DIN ENGL 1511-78. Pattern Equipment for Foundries; Production and Quality (Apr). 10 pp.

—Shrinkage
DIN ENGL 1511-78. Pattern Equipment for Foundries; Production and Quality (Apr). 10 pp.

Foundry Sands
Use For: Molding Sands See Also: Sands

—Clay Content
CNS Z8027-82. Method for Determining Clay Content of Foundry Sand (May)(8894).
JIS Z 2601-76. Method for Determining Clay Content of Foundry Sands.

—Compression Testing
CNS Z8030-82. Method for Determining Strength of Foundry Sand (May)(8897).
JIS Z 2604-76. Method for Determining Strength of Foundry Sands. 11 pp.

—Deterioration
CNS Z8033-82. Method for Determining Deterioration of Foundry Sand (May)(8900).

—Ignition Loss
CNS Z8032-82. Method for Determining Ignition Loss of Foundry Sand (May)(8899).
JIS Z 2606-63. Method for Determining Ignition Loss of Foundry Sands.

—Mixers
JIS B 6611-91. Green Sand Mixer. 13 pp.

—Moisture Content
CNS Z8031-82. Method for Determining Moisture for Foundry Sand (May)(8898).
JIS Z 2605-63. Method for Determining Moisture of Foundry Sands.

—Natural
JIS G 5902-74. Molding Natural Sand.

—Permeability
CNS Z8029-82. Method for Determining Permeability of Foundry Sands (May)(8896).
JIS Z 2603-76. Method for Determining Permeability of Foundry Sands. 10 pp.

Foundry Sands (Cont.)
—Sieve Analysis
CNS Z8028-82. Method for Determining Fireness of Foundry Sand (May)(8895).
JIS Z 2602-76. Method for Determining Fineness of Foundry Sands (R 1984). 5 pp.

—Silica
JIS G 5901-74. Molding Silica Sand.

Four Ball Testers
See Also: Greases; Lubricants; Testing Equipment
DIN ENGL 51350 Pt 4-84. Testing of Lubricants; Testing by the Shell Four-Ball Tester; Determination of Welding Load of Consistent Lubricants (Jan). 3 pp.
DIN ENGL 51350 Pt 5-84. Testing of Lubricants; Testing by the Shell Four-Ball Tester; Determination of Wear Characteristics for Consistent Lubricants (Jan). 3 pp.

Four Frequency Diplex Radiotelegraphy Systems
See Also: Radiotelegraphy; Telegraphy
CCIR RECMN 346-1-70. Four-Frequency Diplex Systems—Section 3Ca—Radiotelegraph Circuits. 2 pp.

Four Way Valves
Use: Directional Control Valves

Fourdrinier Machines
See Also: Headboxes; Paper Machines; Papermaking Equipment

—Wires—Wear Testing
CPPA A.4U-77. Measurement of Fourdrinier Wire Wear. 2 pp.

Fournier RF 3 Motor Gliders
See Also: Gliders

—Foreign Airworthiness Directives
CAA. Fournier RF3 Series Motor Gliders and Fournier RF6 Series Aircraft (Foreign Airworthiness Directives). 1 p.

Fournier RF 4 Aircraft
See Also: Aircraft

—Antenna Positions
CAA. Fournier RF.4 (Approved Aerial Positions). 1 p.

Fournier RF 6 Aircraft
See Also: Aircraft

—Foreign Airworthiness Directives
CAA. Fournier RF3 Series Motor Gliders and Fournier RF6 Series Aircraft (Foreign Airworthiness Directives). 1 p.

Fourrageres
Use For: Aguillettes; Braids (Uniform); Cords (Uniform) See Also: Cordage; Dress Uniforms; Uniforms

—Wool
CGSB 4-GP-112MA-90. Wool Braid Cloth; (Corrigendum—Aug 1990). 9 pp.

Fractional Distillation Methods
Use: Distillation Methods

Fractional Horsepower Motors
Use For: Subfractional Horsepower Motors
See Also: Motors
BSI BS 2048: Part 1-61. 1961 Amd 2 Dimensions of Fractional Horse-Power Motors Part 1: Motors for General Use. 15 pp.

—Packaging
MOD UK DSTAN 81-78-89. Packaging of Fractional Horsepower Motors Issue 1. 21 pp.

Fracture Strength
See Also: Cracking (Fracturing)

—Metals—Aerospace
AECMA PREN6019-90. Test Methods for Metallic Materials Recommended Practice for R-Curve and Kco Determination. 2 pp.

Fracture Testing
Use For: Fracture Toughness See Also: Macroscopic Examination

—Ceramics
JIS R 1607-90. Testing Methods for Fracture Toughness of High Performance Ceramics. 16 pp.

Fracture Testing (Cont.)
—Hoses
BSI BS 5173: Sec 103.7-91. 1991 Rubber and Plastics Hoses and Hose Assemblies Part 103: Physical Tests Section 103.7: Determination of Fracture Resistance of Rigid Polymer Helical Reinforcement in Thermoplastics Hoses. 7 pp.

—Metals
BSI BS 5447-77. (WITHDRAWN) 1977 Methods of Test for Plane Strain Fracture Toughness (K1) of Metallic Materials (Superseded by BS 7448: Part 1: 1991). 15 pp.
BSI BS 6729-87. 1987 Amd 1 Determination of the Dynamic Fracture Toughness of Metallic Materials. 34 pp.
BSI BS 7448: Part 1-91. 1991 Fracture Mechanics Toughness Tests Part 1: Method for Determination of KIc, Critical CTOD and Critical J Values of Metallic Materials. 39 pp.

—Mineral Aggregates
DIN ENGL 52116-88. Determination of the Degree of Surface Fracture of Particles of Mineral Aggregates (Aug). 2 pp.

Fracture Toughness
Use: Fracture Testing

Fragmentation
Use For: Shattering See Also: Spalling

—Coal
CNS M3161-84. Method of Drop Shatter Test for Coal (Jun) (10946).

—Coke
BSI BS 1016: Sec 108.1-92. 1992 Methods for Analysis and Testing of Coal and Coke Part 108: Tests Special to Coke Section 108.1: Determination of Shatter Indices (W). 11 pp.
CNS M3177-84. Method of Test for Shatter of Coke (Nov) (11119).
ISO 616-77. Coke—Determination of Shatter Indices First Edition. 8 pp.

Frame Alignment (Data Transmission)
See Also: Pulse Code Modulation

—Cyclic Redundancy Check—Frame Structure
CCITT RECMN G.706-91. Frame Alignment and Cyclic Redundancy Check (CRC) Procedures Relating to Basic Frame Structures Defined in Recommendation G.704 (Study Group XVIII) 19 pp. 19 pp.
CCITT RECMN G.706-89. Frame Alignment and Cyclic Redundancy Check (CRC) Procedures Relating to Basic Frame Structures Defined in Recommendation G.704—General Aspects of Digital Transmission Systems; Terminal Equipments (Study Groups XV and XVIII) 7 pp. 7 pp.

—Monitoring (Telecommunications)
CCITT RECMN O.163-89. Equipment to Perform In-Service Monitoring on 1544 kBit/s Signals—Specifications for Measuring Equipment (Study Group IV) 5 pp. 5 pp.

Frame Mode Bearer Services
Use: Bearer Services—Frame Mode

Frames
Scope Note: See the subheading Frames under specific items See: Chassis; Framing; Undercarriages

Framing
See Also: Construction Materials; Formwork (Construction); Shoring

—Safety
BSI BS 5531-88. 1988 Code of Practice for Safety in Erecting Structural Frames. 38 pp.

—Steel
CGSB CAN/CGSB-7.1-M86. Cold Formed Steel Framing Components. 16 pp.

—Structural Timber—Building Codes
SAA AS 1684-92. National Timber Framing Code (in Professional Packages 20, 21, 30, 62-69) Bound Together with Supplements 2, 3, 4, 5, 9, 15 Amdt 1—1993. 249 pp.
SAA AS 1684 Supp 0-92. National Timber Framing Code—Supplement 0: General Introduction and Index (Supplement to AS 1684—1992) (Supersedes AS 1684 Supplements 1 to 20—1975 (in Part) and Supplements 21 to 22—1978 (in Part)) (in Professional Packages 20, 21, 30,). 27 pp.

Framing (Cont.)
—Structural Timber—Tables (Data)—Building Codes

SAA AS 1684 Supp 1-92. National Timber Framing Code—Supplement 1: Timber Framing Span Tables—Unseasoned Timber—Stress Grade F4 (Supplement to AS 1684—1992) (Supersedes AS 1684 Supplement 1—1975 (in Part)) (in Professional Packages 20, 21, 30, 62-69). 27 pp.

SAA AS 1684 Supp 7-92. National Timber Framing Code—Supplement 7: Timber Framing Span Tables—Unseasoned Timber—Stress Grade F17 (Supplement to AS 1684—1992) (Supersedes AS 1684 Supplement 7—1975 (in Part)) (in Professional Packages 20, 21, 30, 62-69). 27 pp.

SAA AS 1684 Supp 8-92. National Timber Framing Code—Supplement 8: Timber Framing Span Tables—Unseasoned Timber—Stress Grade F22 (Supplement to AS 1684—1992) (Supersedes AS 1684 Supplement 8—1975 (in Part)) (in Professional Packages 20, 21, 30, 62-69). 27 pp.

SAA AS 1684 Supp 9-92. National Timber Framing Code—Supplement 9: Timber Framing Span Tables—Seasoned Softwood—Stress Grade F5 (Supplement to AS 1684—1992) (Supersedes AS 1684 Supplement 9—1975 (in Part)) (in Professional Packages 20, 21, 30, 62-69) Bound Together. 27 pp.

SAA AS 1684 Supp 10-75. Supplement to SA1684 SAA Timber Framing Code—Supplement 10: Light Timber Framing Span Tables—Seasoned Softwood—Stress Grade F7 (Superseded (in Part) by AS 1684 Supplement 0 and AS 1684 Supplement 10—1992). 15 pp.

SAA AS 1684 Supp 10-92. National Timber Framing Code—Supplement 10: Timber Framing Span Tables—Seasoned Softwood—Stress Grade F7 (Supplement to AS 1684—1992) (Supersedes AS 1684 Supplement 10—1975 (in Part)) (in Professional Packages 20, 21, 30, 62-69). 27 pp.

SAA AS 1684 Supp 11-75. Supplement to As 1684 SAA Timber Framing Code—Supplement 11: Light Timber Framing Span Tables—Seasoned Softwood—Stress Grade F8 Amdt 1 March 1991 (in Professional Package 30) (Superseded (in Part) by AS 1684 Supplement 0 and AS 1684 Supplement 11—1992). 15 pp.

SAA AS 1684 Supp 11-92. National Timber Framing Code—Supplement 11: Timber Framing Span Tables—Seasoned Softwood—Stress Grade F8 (Supplement to AS 1684—1992) (Supersedes AS 1684 Supplement 11—1975 (in Part)) (in Professional Packages 20, 21, 30, 62-69). 27 pp.

SAA AS 1684 Supp 12-75. Supplement to SA 1684 SAA Timber Framing Code—Supplement 12: Light Timber Framing Span Tables—Seasoned Softwood—Stress Grade F11 (Superseded (in Part) by AS 1684 Supplement 0 and AS 1684 Supplement 12—1992). 15 pp.

SAA AS 1684 Supp 12-92. National Timber Framing Code—Supplement 12: Timber Framing Span Tables—Seasoned Softwood—Stress Grade F11 (Supplement to AS 1684—1992) (Supersedes AS 1684 Supplement 12—1975 (in Part)) (in Professional Packages 20, 21, 30, 62-69). 27 pp.

SAA AS 1684 Supp 13-75. Supplement to AS 1684 SAA Timber Framing Code—Supplement 13: Light Timber Framing Span Tables—Seasoned Hardwood—Stress Grade F11 Amdt 1 March 1991 (in Professional Package 30) (Superseded (in Part) by AS 1684 Supplement 0 and AS 1684 Supplement 13—1992). 15 pp.

SAA AS 1684 Supp 13-92. National Timber Framing Code—Supplement 13: Timber Framing Span Tables—Seasoned Hardwood—Stress Grade F11 (Supplement to AS 1684—1992) (Supersedes AS 1684 Supplement 13—1975 (in Part)) (in Professional Packages 20, 21, 30, 62-69). 27 pp.

SAA AS 1684 Supp 14-75. Supplement to AS 1684 SAA Timber Framing Code—Supplement 14: Light Timber Framing Span Tables—Seasoned Hardwood—Stress Grade F14 Amdt 1 March 1991 (in Professional Package 30) (Superseded (in Part) by AS 1684 Supplement 0 and AS 1684 Supplement 14—1992). 15 pp.

SAA AS 1684 Supp 14-92. National Timber Framing Code—Supplement 14: Timber Framing Span Tables—Seasoned Hardwood—Stress Grade F14 (Supplement to AS 1684—1992) (Supersedes AS 1684 Supplement 14—1975 (in Part)) (in Professional Packages 20, 21, 30, 62-69). 27 pp.

SAA AS 1684 Supp 15-75. Supplement to AS 1684 SAA Timber Framing Code—Supplement 15: Light Timber Framing Span Tables—Seasoned Hardwood—Stress Code F17 Amdt 1 March 1991 (in Professional Package 30) (Superseded (in Part) by AS 1684 Supplement 0 and AS 1684 Supplement 15—1992). 15 pp.

SAA AS 1684 Supp 15-92. National Timber Framing Code—Supplement 15: Timber Framing Span Tables—Seasoned Hardwood—Stress Grade F17 (Supplement to AS 1684—1992) (Supersedes AS 1684 Supplement 15—1975 (in Part)) (in Professional Packages 20, 21, 30, 62-69) Bound Together. 27 pp.

SAA AS 1684 Supp 16-75. Supplement to AS 1684 SAA Timber Framing Code—Supplement 16: Light Timber Framing Span Tables—Seasoned Hardwood—Stress Grade F27 Amdt 1 March 1991 (in Professional Package 30) (Superseded (in Part) by AS 1684 Supplement 0 and AS 1684 Supplement 16—1992). 15 pp.

SAA AS 1684 Supp 16-92. National Timber Framing Code—Supplement 16: Timber Framing Span Tables—Seasoned Hardwood—Stress Grade F27 (Supplement to AS 1684—1992) (Supersedes AS 1684 Supplement 16—1975 (in Part)) (in Professional Packages 20, 21, 30, 62-69). 27 pp.

SAA AS 1684 Supp 17-75. Supplement to AS 1684 SAA Timber Framing Code—Supplement 17: Light Timber Framing Span Tables—Unseasoned Timber (Alternative Sizes)—Stress Grade F4 (Superseded (in Part) by AS 1684 Supplement 0 and AS 1684 Supplement 17—1992). 15 pp.

SAA AS 1684 Supp 17-92. National Timber Framing Code—Supplement 17: Timber Framing Span Tables—Unseasoned Timber (Alternative Sizes)—Stress Grade F4 (Supplement to AS 1684—1992) (Supersedes AS 1684 Supplement 17—1975 (in Part)) (in Professional Packages 20, 21, 30, 62-69). 27 pp.

SAA AS 1684 Supp 18-75. Supplement to AS 1684 SAA Timber Fraing Code—Supplement 18: Light Timber Framing Span Tables—Unseasoned Timber (Alternative Sizes)—Stress Grade F5 (Superseded (in Part) by AS 1684 Supplement 0 and AS 1684 Supplement 18—1992). 15 pp.

SAA AS 1684 Supp 18-92. National Timber Framing Code—Supplement 18: Timber Framing Span Tables—Unseasoned Timber (Alternative Sizes)—Stress Grade F5 (Supplement to AS 1684—1992) (Supersedes AS 1684 Supplement 18—1975 (in Part)) (in Professional Packages 20, 21, 30, 62-69). 27 pp.

SAA AS 1684 Supp 19-75. Supplement to AS 1684 SAA Timber Framing Code—Supplement 19: Light Timber Framing Span Tables—Unseasoned Timber (Alternative Sizes)—Stress Grade F8 (Superseded (in Part) by AS 1684 Supplement 0 and AS 1684 Supplement 19—1992). 15 pp.

SAA AS 1684 Supp 19-92. National Timber Framing Code—Supplement 19: Timber Framing Span Tables—Unseasoned Timber (Alternative Sizes)—Stress Grade F8 (Supplement to AS 1684—1992) (Supersedes AS 1684 Supplement 19—1975 (in Part)) (in Professional Packages 20, 21, 30, 62-69). 27 pp.

SAA AS 1684 Supp 20-92. National Timber Framing Code—Supplement 20: Timber Framing Span Tables—Unseasoned Timber (Alternative Sizes)—Stress Grade F11 (Supplement to AS 1684—1992) (Supersedes AS 1684 Supplement 20—1975 (in Part)) (in Professional Packages 20, 21, 30, 62-69). 27 pp.

SAA AS 1684 Supp 21-92. National Timber Framing Code—Supplement 21: Timber Framing Span Tables—Seasoned Softwood—Stress Grade F4 (Supplement to AS 1684—1992) (Supersedes AS 1684 Supplement 21—1978 (in Part)) (in Professional Packages 20, 21, 30, 62-69). 27 pp.

SAA AS 1684 Supp 22-92. National Timber Framing Code—Supplement 22: Timber Framing Span Tables—Seasoned Softwood—Stress Grade F14 (Supplement to AS 1684—1992) (Supersedes AS 1684 Supplement 22—1978 (in Part)) (in Professional Packages 20, 21, 30, 62-69). 27 pp.

Franking Machines
Use For: Postal Franking Machines *See Also:* Office Machines

BSI BS 6191: Part 4-84. 1984 Mail Processing Machines Part 4: Minimum Information to Be Included in Specification Sheets of Postal Franking Machines. 8 pp.

ISO 4232 Pt 3-84. Office Machines—Minimum Information to Be Included in Specification Sheets—Part 3: Postal Franking Machines First Edition. 7 pp.

JTC1 4232 Pt 3-84. Office Machines—Minimum Information to Be Included in Specification Sheets—Part 3: Postal Franking Machines First Edition. 7 pp.

OSI ISO 4232-3-84. Office Machines—Minimum Information to Be Included in Specification Sheets—Part 3: Postal Franking Machines. 7 pp.

—Glossaries

ISO 5138 Pt 7-86. Office Machines—Vocabulary—Part 07: Postal Franking Machines First Edition. 23 pp.

JTC1 5138 Pt 7-86. Office Machines—Vocabulary—Part 07: Postal Franking Machines First Edition. 23 pp.

OSI ISO 5138-7-86. Office Machines—Vocabulary—Section 07: Postal Franking Machines. 23 pp.

Free Acidity
Use: Acidity

Free Alcohols Content Analysis
Use: Alcohol Content Analysis

Free Alkali Content Analysis
Use: Alkali Content Analysis

Free Alkalinity
Use: Alkalinity

Free Ammonia Content Analysis
Use: Ammonia Content Analysis

Free Carbon Content Analysis
Use: Carbon Content Analysis

Free Cutting Steels
Use: Free Machining Steels

Free Fall Testing
Use: Impact Testing

Free Fat Content Analysis
Use: Fat Content Analysis

Free Fluorides Content Analysis
Use: Fluoride Content Analysis

Free Formaldehyde Content Analysis
Use: Formaldehyde Content Analysis

Free Machining Steels
Scope Note: For additional listings, see also products made from free machining steels *Use For:* Free Cutting Steels *See Also:* Alloy Steels; Austenitic Stainless Steels; Carbon Steels; Ferroalloys; Low Alloy Steels; Steels

ISO 683 Pt 9-88. Heat-Treatable Steels, Alloy Steels and Free-Cutting Steels—Part 9: Wrought Free-Cutting Steels First Edition. 16 pp.

JIS G 4804-83. Free Cutting Carbon Steels. 11 pp.

—Bars—Bright

DIN ENGL 1651-88. Free-Cutting Steels; Technical Delivery Conditions (Apr). 14 pp.

—Bars—Bright—Aerospace

BSI 2S 137-76. 1976 High Chromium-Nickel Corrosion-Resisting Steel Bright Bars (Free Machining) (880-1080 MPa; Limiting Ruling Section 70 mm). 4 pp.

—Hot Worked—Bars

DIN ENGL 1651-88. Free-Cutting Steels; Technical Delivery Conditions (Apr). 14 pp.

—Hot Worked—Rods

DIN ENGL 1651-88. Free-Cutting Steels; Technical Delivery Conditions (Apr). 14 pp.

—Rods—Bright

DIN ENGL 1651-88. Free-Cutting Steels; Technical Delivery Conditions (Apr). 14 pp.

Free Magnesium Content Analysis
Use: Magnesium Content Analysis

Free Silicon Content Analysis
Use: Silica Content Analysis; Silicon Content Analysis

Free Space
See Also: Antennas; Radiation

—Energy Transmission

CCIR Report 679-2-86. Characteristics and Effects of Radio Techniques for the Transmission of Energy from Space—Section 2B—Topics of General Interest. 7 pp.

Free Space Propagation
See Also: Radio Wave Propagation; Wave Propagation

—Attenuation

CCIR RECMN 525-1-82. Calculation of Free-Space Attenuation—Section 5A—Texts of General Interest. 5 pp.

Free Sulfur Trioxide Content Analysis
Use: Sulfur Trioxide Content Analysis

Freeness Testers
See Also: Pulps; Testing Equipment

CPPA M-15-75. Canadian Standard Freeness Tester. 1 p.

INTERNATIONAL AND NON-U.S. NATIONAL STANDARDS
SUBJECT INDEX

Freephone Service, European
Use: European Automatic Freephone Service

Freephone Services
Use For: IFS; International Freephone Services
See Also: European Automatic Freephone Service; Telephone Services

CCITT RECMN E.152-89. International Freephone Service (IFS)—Telephone Network and ISDN—Operation, Numbering, Routing and Mobile Service (Study Group II) 12 pp. 12 pp.

—Abusive Customers
CCITT RECMN E.152-89. International Freephone Service (IFS)—Telephone Network and ISDN—Operation, Numbering, Routing and Mobile Service (Study Group II) 12 pp. 12 pp.

—Accounting
CCITT RECMN D.115-89. Tariff Principles and Accounting for the International Freephone Service (IFS)—General Tariff Principles—Charging and Accounting in International Telecommunications Services (Study Group III) 2 pp. 2 pp.

—Call Diversion Services
CCITT RECMN E.152-89. International Freephone Service (IFS)—Telephone Network and ISDN—Operation, Numbering, Routing and Mobile Service (Study Group II) 12 pp. 12 pp.

—Directory Assistance
CCITT RECMN E.152-89. International Freephone Service (IFS)—Telephone Network and ISDN—Operation, Numbering, Routing and Mobile Service (Study Group II) 12 pp. 12 pp.

—Document Formats—Orders (Sales Documents)
CCITT RECMN E.152-89. International Freephone Service (IFS)—Telephone Network and ISDN—Operation, Numbering, Routing and Mobile Service (Study Group II) 12 pp. 12 pp.

—Network Management Systems
CCITT RECMN E.152-89. International Freephone Service (IFS)—Telephone Network and ISDN—Operation, Numbering, Routing and Mobile Service (Study Group II) 12 pp. 12 pp.

—Numbering Plans
CCITT RECMN E.152-89. International Freephone Service (IFS)—Telephone Network and ISDN—Operation, Numbering, Routing and Mobile Service (Study Group II) 12 pp. 12 pp.

—Quality of Service
CCITT RECMN E.152-89. International Freephone Service (IFS)—Telephone Network and ISDN—Operation, Numbering, Routing and Mobile Service (Study Group II) 12 pp. 12 pp.

—Recorded Information Services
CCITT RECMN E.152-89. International Freephone Service (IFS)—Telephone Network and ISDN—Operation, Numbering, Routing and Mobile Service (Study Group II) 12 pp. 12 pp.

—Routing
CCITT RECMN E.152-89. International Freephone Service (IFS)—Telephone Network and ISDN—Operation, Numbering, Routing and Mobile Service (Study Group II) 12 pp. 12 pp.

—Tariffs
CCITT RECMN D.115-89. Tariff Principles and Accounting for the International Freephone Service (IFS)—General Tariff Principles—Charging and Accounting in International Telecommunications Services (Study Group III) 2 pp. 2 pp.

—Telephone Directories
CCITT RECMN E.152-89. International Freephone Service (IFS)—Telephone Network and ISDN—Operation, Numbering, Routing and Mobile Service (Study Group II) 12 pp. 12 pp.

Freeze Thaw Testing
See Also: Environmental Testing; Temperature Change Testing; Temperature Testing

—Ceramic Tiles
CEN EN 202-85. Ceramic Tiles: Determination of Frost Resistance. 8 pp.

—Concretes
CNS A3032-84. Method of Test for Resistance of Concrete Specimens to Rapid Freezing and Thawing in Water (Apr)(1168).
CNS A3033-84. Method of Test for Resistance of Concrete Specimens to Rapid Freezing in Air and Rapid Thawing in Water (Apr)(1169).
CNS A3034-84. Method of Test for Resistance of Concrete Specimens to Slow Freezing and Thawing in Water (Apr)(1170).

Freeze Thaw Testing (Cont.)
—Facing Bricks
DIN ENGL 52252 Pt 1-86. Testing the Frost Resistance of Facing Bricks and Clinker Blocks; Freezing of Single Bricks on All Sides (Dec). 4 pp.
DIN ENGL 52252 Pt 2-86. Testing the Frost Resistance of Facing Bricks and Clinker Blocks; Freezing of Bricks Arranged in Test Blocks (Dec). 5 pp.
DIN ENGL 52252 Pt 3-86. Testing the Frost Resistance of Facing Bricks and Clinker Blocks; Freezing of Test Walls on One Side (Dec). 5 pp.

—Paints
CGSB 1-GP-71 METH 31.1-74. Methods of Testing Paints and Pigments Freeze-Thaw Test. 1 p.

—Polyvinyl Acetate Adhesives
BSI BS 3544-02. 1962 Amd 2 Methods of Test for Polyvinyl Acetate Adhesives for Wood (AMD 6873) January 31, 1992 (W). 18 pp.

—Rocks
DIN ENGL 52104 Pt 1-82. Testing of Natural Stone; Freeze-Thaw Cyclic Test; Methods A to Q (Nov). 6 pp.
DIN ENGL 52104 Pt 2-82. Testing of Natural Stone; Freeze-Thaw Cyclic Test; Method Z (Nov). 2 pp.

—Roofing
SAA AS 3991.7-92. Methods of Testing Flat Cellulose-Cement Sheets—Part 7: Determination of Resistance to Freeze/Thaw. 2 pp.

—Roofing Tiles
DIN ENGL 52253 Pt 1-88. Testing the Frost Resistance of Roofing Tiles; Freeze-Thaw Test with Upper Side Freezing After Sprinkling with Water (Dec) (Supersedes DIN 52251 Part 7, January 1981 Edition, Withdrawn in January 1986). 5 pp.
DIN ENGL 52253 Pt 2-88. Testing the Frost Resistance of Roofing Tiles; Freeze-Thaw Test with Freezing on All Sides After Impregnation with Water Under Vacuum (Dec). 4 pp.
SAA AS 4046.6-92. Methods of Testing Roof Tiles—Part 6: Determination of Resistance to Freeze/Thaw (in Professional Packages 20, 21, 40, 41, 58, 62, 63, 64, 65, 66, 67, 68, 69) (Supersedes AS 1757—1989 (in Part) and AS 2049—1989 (in Part)). 2 pp.

—Sealants
CGSB CAN2-19.0-M77METH 6.1-78. Methods of Testing Putty, Caulking and Sealing Compounds Freeze-Thaw Stability. 1 p.

—Walls
JIS A 1435-91. Test Methods for Frost Resistance of Exterior Wall Materials of Buildings (Freezing and Thawing Method). 15 pp.

Freezers
Use For: Frozen Food Storage Cabinets
See Also: Appliances; Coolers; Ice Makers; Refrigeration Equipment; Refrigerator/Freezers; Refrigerators

BSI BS EN 28187-92. 1992 Household Refrigerating Appliances Refrigerator-Freezers—Characteristics and Test Methods (ISO 8187: 1991) (L). 60 pp.
CEN EN 28187-91. Household Refrigerating Appliances —Refrigerator-Freezers—Characteristics and Test Methods. 3 pp.
CNS C3164-89. Method of Test for Electric Refrigerators and Freezers (Feb)(9577). 10 pp.
CNS C4048-89. Electric Refrigerators and Freezers (Feb)(2062). 33 pp.
ISO 5155-83. Household Frozen Food Storage Cabinets and Food Freezers—Essential Characteristics and Test Methods First Edition. 27 pp.
ISO 8187-91. Household Refrigerating Appliances—Refrigerator-Freezers—Characteristics and Test Methods First Edition. 56 pp.
JIS C 9607-86. Household Electric Refrigerators, Refrigerator-Freezers and Freezers. 61 pp.
SAA AS 1430-86. Household Refrigerators and Freezers. 33 pp.
SAA AS 1731-83. Frozen Food Retail Cabinets. 16 pp.
SNZ NZS 5236-90. Household Refrigerators and Freezers Amend: A, 1990. 33 pp.

—Capacity Measurement
CSA CAN/CSA-C300-M89. Capacity Measurement and Energy Consumption Test Methods for Refrigerators, Combination Refrigerator-Freezers, and Household Freezers; (Gen Instr 1). 90 pp.
CSA CAN/CSA-C300-M91. Capacity Measurement and Energy Consumption Test Methods for Refrigerators, Combination Refrigerator-Freezers, and Freezers; (Gen Instr 1). 95 pp.

—Electrical Codes
CSA 1311 Bull. Electrical Bulletin 1311 March 25, 1981 to C22.2 NO 63. 1 p.

Freezers (Cont.)
—Energy Consumption
BSI BS EN 153-90. 1990 Methods of Measuring the Energy Consumption of Electric Mains Operated Household Refrigerators, Frozen Food Storage Cabinets, Food Freezers and Their Combinations, Together with Associated Characteristics. 8 pp.
CEN EN 153-82. Methods of Measuring the Energy Consumption of Electric Mains Operated Household Refrigerators, Frozen Food Storage Cabinets, Food Freezers and Their Combinations, Together with Associated Characteristics. 7 pp.
CEN EN 153-90. Methods of Measuring the Energy Consumption of Electric Mains Operated Household Refrigerators, Frozen Food Storage Cabinets, Food Freezers and Their Combinations Together with Associated Characteristics. 7 pp.
CSA CAN/CSA-C300-M89. Capacity Measurement and Energy Consumption Test Methods for Refrigerators, Combination Refrigerator-Freezers, and Household Freezers; (Gen Instr 1). 90 pp.
CSA CAN/CSA-C300-M91. Capacity Measurement and Energy Consumption Test Methods for Refrigerators, Combination Refrigerator-Freezers, and Freezers; (Gen Instr 1). 95 pp.
SNZ NZS 6205: Part 1-89. Energy Labelling of Appliances Part 1: Refrigerators, Refrigerator/Freezers and Freezers—Specification for Appliance Energy Rating Label (NZS 6205.1:1989/AS 2575.1:1989). 12 pp.
SNZ NZS 6205: Part 2-89. Energy Labelling of Appliances Part 2: Refrigerators, Refrigerator/Freezers and Freezers—Determination of Energy Consumption and Efficiency Rating (NZS 6205.2:1989/AS 2575.2:1989). 36 pp.

—Household
BSI BS 922 & 1691-59. (WITHDRAWN) 1959 Amd 3 Electrical Refrigerators and Food Freezers for Household Use. 32 pp.
BSI BS 3999: Part 10-72. (WITHDRAWN) 1972 Amd 3 Methods of Measuring the Performance of Household Electrical Appliances Part 10: Food Freezers (Superseded by BS EN 153: 1990). 23 pp.
BSI BS 3999: Part 13-76. (WITHDRAWN) 1976 Methods of Measuring the Performance of Household Electrical Appliances Part 13: Frozen Food Storage Compartments in Refrigerator and Frozen Food Storage Cabinets (Superseded by BS EN 153: 1990). 12 pp.
CSA C22.2 NO 63-M1987. Household Refrigerators and Freezers; (Gen Instr 1 Thru 2). 53 pp.
CSA 708B Bull. Electrical Bulletin 708B December 12, 1969 to C22.2 NO 63. 5 pp.
CSA 1169 Bull. Electrical Bulletin 1169 June 27, 1978 to C22.2 NO 63. 2 pp.
CSA 1193 Bull. Electrical Bulletin 1193 October 30, 1978 to C22.2 NO 63. 1 p.
CSA 1436 Bull. Electrical Bulletin 1436 September 27, 1985 to C22.2 NO 63. 4 pp.

—Household—Electrical Grounding
CSA 649A Bull. Electrical Bulletin 649A January 10, 1967 to C22.2 NO 92. 1 p.

—Household—Noise
BSI BS EN 28960-93. 1993 Refrigerators, Frozen Food Storage Cabinets and Food Freezers for Household and Similar Use Measurement of Emission of Airborne Acoustical Noise (ISO 8960: 1991) (L). 12 pp.
CEN EN 28960-93. Refrigerators, Frozen-Food Storage Cabinets and Food Freezers for Household and Similar Use—Measurement of Emission of Airborne Acoustical Noise (ISO 8960: 1991). 7 pp.
ISO 8960-91. Refrigerators, Frozen-Food Storage Cabinets and Food Freezers for Household and Similar Use—Measurement of Emission of Airborne Acoustical Noise First Edition (CEN EN 28960: 1993). 9 pp.

—Household—Safety
BSI BS 3456: Sec 102.24-88. (WITHDRAWN) 1988 Safety of Household Electrical Appliances Part 102: Particular Requirements Section 102.24: Refrigerators and Food Freezers (Superseded by BS 3456: Section 202.24: 1990). 31 pp.
BSI BS 3456: Sec 202.24-90. 1990 Safety of Household Electrical Appliances Part 202: Particular Requirements Section 202.24: Refrigerators and Food Freezers (L). 42 pp.
BSI BS 5258: Part 6-88. 1988 Safety of Domestic Gas Appliances Part 6: Refrigerators and Food Freezers. 16 pp.
BSI BS 5258: Part 6-75. 1975 Safety of Domestic Gas Appliances Part 6: Refrigerators and Food Freezers. 15 pp.
CENELEC HD 269 S2-86. Safety of Household and Similar Electrical Appliances-Part: Particular Requirements for Refrigerators and Food Freezers. 7 pp.

Freezers (Cont.)
—Household—Safety (Cont.)
CENELEC HD 269 S2-86. Safety of Household and Similar Electrical Appliances Part: 2 Particular Requirements for Refrigerators and Food Freezers (IEC 335-2-24 (1976) Ed 1 Modified). 18 pp.

CENELEC HD 269 S2-85. Amd 1 Safety of Household Similar Electrical Appliances Particular Requirements for Refrigerators and Food Freezers. 4 pp.

CENELEC HD 270 S1-83. Safety of Household Similar Electrical Appliances Particular Requirements for Microwave Cooking Appliances. 17 pp.

CENELEC EN 60 335-2-24-89. Safety of Household and Similar Electrical Appliances-Part 2: Particular Requirements for Refrigerators and Food Freezers. 13 pp.

DIN VDE 0700 Pt 24-86. Safety of Household and Similar Electrical Appliances; Refrigerators and Food Freezers (Mar). 33 pp.

DIN VDE 0700 Pt 240-83. Safety of Household and Similar Electrical Appliances; Particular Requirements for Refrigerators and Freezers for Special Purposes and Ice Makers (Feb) (Supersedes VDE 0730 Part 2 ZK/10.78). 61 pp.

IEC 335 Pt 2-24-92. Safety of Household and Similar Electrical Appliances Part 2: Particular Requirements for Refrigerators, Food-Freezers and Ice-Makers Third Edition: (CENELEC HD 269 S2:1986). 78 pp.

—Safety
CNS Z1019-87. Safety Code of High Pressure for Freezing Devices (Oct)(3326).

SNZ NZS 6324-90. Approval and Test Specification—Particular Requirements for Refrigerators and Food Freezers Amend: 1, 1991; 2, 1992 (NZS 6324:1990/AS 3303:1990). 31 pp.

—Walk-In
CGSB CAN/CGSB-52.28-M87. Refrigerators and Freezers, Prefabricated, Mechanical, Commercial, Walk-in. 13 pp.

Freezing Points
See Also: Boiling Points; High Temperature Testing; Low Temperature Testing; Melting Points; Softening Points; Solidification Points; Temperature; Temperature Measurement; Thermodynamic Properties

—Aircraft Fuels
ISO 3013-74. Aviation Fuels—Determination of Freezing Point First Edition. 6 pp.

—Benzene
CNS K6261-74. Method of Test for Freezing Point of Benzene (Jun)(2762).

—Chemical Products
CNS K0017-82. Method for Determining Freezing Point of Chemical Products (May)(8836).

CNS K7007-82. Chemical Reagent (Freezing Point Determination) (Jun)(1507).

JIS K 0065-92. Test Method for Freezing Point of Chemical Products. 7 pp.

—Emulsified Asphalts
CNS K6786-83. Method of Test for Freezing of Emulsified Asphalts (Jun)(10369).

—Engine Coolants
BSI BS 5117: Sec 1.3-85. 1985 Testing Corrosion Inhibiting, Engine Coolant Concentrate ('Antifreeze') Part 1: Methods of Test for Determination of Physical and Chemical Properties Section 1.3: Determination of Freezing Point. 4 pp.

—Essential Oils
CNS K6604-80. Determination of Freezing Point of Essential Oils (Oct)(6577).

ISO 1041-73. Essential Oils—Determination of Freezing Point First Edition. 4 pp.

—Hydrocarbons
JIS K 0518-78. Testing Method for Freezing Points of High-Purity Hydrocarbons.

—Milk
BSI BS 3095: Sec 1.1-88. 1988 Determination of the Freezing-Point Depression of Milk Part 1: Method Sec 1.1: Thermistor Cryoscope Method (Supersedes BS 3095: part 1: 1980). 11 pp.

BSI BS 3095: Sec 1.2-88. 1988 Determination of the Freezing-Point Depression of Milk Part 1: Method Sec 1.2: Hortvet Method (Supersedes BS 3095: Part 1: 1980). 15 pp.

BSI BS 3095: Part 2-88. 1988 Determination of the Freezing-Point Depression of Milk Part 2: Recommendations for the Interpretation of the Freezing-Point Depression of Herd Milk. 4 pp.

BSI BS 3095: Part 3-81. 1981 Determination of the Freezing-Point Depression of Milk Part 3: Storage of Samples. 4 pp.

Freezing Points (Cont.)
—Milk (Cont.)
ISO 5764-87. Milk—Determination of Freezing Point—Thermistor Cryoscope Method First Edition. 9 pp.

SNZ NZS/ISO 5764-87. Milk. Determination of Freezing Point. Thermistor Cryoscope Method. 6 pp.

—Naphthalene
CNS K6183-82. Method of Test for Freezing Point of Naphthalene (Aug)(2332).

—Phenols
CNS K6215-65. Determination of Freezing Points of Phenols from Coal Distillation (Sep)(2497)(R 1971).

Freight Cars
Use For: Wagons (Railway) *See Also:* Ground Vehicles; Hopper Cars; Mine Cars; Railroad Cranes; Rolling Stock; Towed Vehicles

—Hazardous Materials
CGSB 43-GP-147P-90. Construction of Tank Car Tanks and Selection and Use of Tank Car Tanks, Portable Tanks and Rail Cars for the Transportation of Dangerous Goods by Rail. 241 pp.

—Inspection
JIS E 4045-73. General Rules for the Inspection of Freight Cars on Completion of Construction.

Freight Containers
Use: Shipping Containers

Freight Transport
Use: Shipping

French Chalk
See Also: Steatite
MOD UK CS 2164. French Chalk.

French Windows
Use: Windows

Freon 113
Use: 1,1,2-Trichloro-1,2,2-Trifluoroethane

Freon-114 (R)
Scope Note: Freon is a registered trademark
Use: Dichlorotetrafluoroethane

Freonte
Use: 1,1,2-Trichloro-1,2,2-Trifluoroethane

Frequencies
Scope Note: For additional listings, use a more specific term *See Also:* Audio Frequencies; Automatic Frequency Control; Bandwidth; Basebands; Carrier Frequencies; Doppler Shift; Frequency Assignment; Frequency Band Sharing; Frequency Bands; Frequency Control; Frequency Error; Frequency Meters; Frequency Offsets; Frequency Response; Frequency Reuse; Frequency Separation; Frequency Stability; Frequency Translation Equipment; Fundamental Frequencies; Harmonics; Intermediate Frequencies; Maximum Usable Frequencies; Optimum Working Frequencies; Phase Response; Pilot Channels; Radio Frequencies; Reference Frequencies; Resonance Frequency; Standard Frequencies; Ultrasonic Frequencies; Vibration; Video Frequencies; Virtual Carrier Frequencies; Wavelength

—0-30 MHz—Fixed Services—Ionospheric Channel Sounding Systems
CCIR RECMN 613-86. Use of Ionospheric Channel Sounding Systems Operating in the Fixed Service at Frequencies Below About 30 MHz—Section 3Ac—Influence of the Ionosphere. 1 p.

—0-30 MHz—Radiotelegraph Systems—Classification
CCIR RECMN 347-63. Classification of Multi-Channel Radiotelegraph Systems for Long-Range Circuits Operating at Frequencies Below About 30 MHz and the Designation of the Channels in These Systems—Section 3Aa—Technical Characteristics. 2 pp.

—Continuity Pilots—Radio Relay Systems
CCIR RECMN 401-2-70. Frequencies and Deviations of Continuity Pilots for Frequency Modulation Radio-Relay Systems for Television and Telephony—Section 9D—Maintenance. 1 p.

—Earth Exploration Satellites
CCIR RECMN 514-1-90. Telecommunication Links for Earth Exploration Satellites Frequencies, Bandwidths and Criteria for Protection from Interference—Section 2F—Earth Exploration Satellites. 2 pp.

Frequencies (Cont.)
—Emergency Position Indicating Radio Beacons—GMDSS
CCIR Report 1036-1-90. Frequencies for Homing and Locating in the Global Maritime Distress and Safety System (GMDSS)—Section 8D—Radiodetermination, Global Maritime Distress and Safety System and Related Subjects. 23 pp.

—Glossaries
CCIR RECMN 686-90. Glossary—Section 7A—Glossary. 7 pp.

CCIR RECMN 573-3-90. Radiocommunication Vocabulary—Section A—Terminology. 60 pp.

CCIR RECMN 662-1-90. Terms and Definitions—Section A—Terminology. 19 pp.

—Homing Systems—GMDSS
CCIR Report 1036-1-90. Frequencies for Homing and Locating in the Global Maritime Distress and Safety System (GMDSS)—Section 8D—Radiodetermination, Global Maritime Distress and Safety System and Related Subjects. 23 pp.

—Ionospheric Propagation
CCIR RECMN 373-6-90. Definitions of Maximum and Minimum Transmission Frequencies—Section 6C—Ionospheric Propagation and Operational Forecasting. 1 p.

—Line Amplifiers—Telephone Cables
CCITT RECMN G.323-89. Typical Transistorized System on Symmetric Cable Pairs—International Analogue Carrier Systems (Study Group XV) 3 pp. 3 pp.

—Meteorological Satellites
CCIR RECMN 362-2-82. Frequencies Technically Suitable for Meteorological Satellites—Section 2F—Earth Exploration Satellites. 1 p.

—NAVTEX Services
CCIR RECMN 540-2-90. Operational and Technical Characteristics for an Automated Direct-Printing Telegraph System for Promulgation of Navigational and Meteorological Warnings and Urgent Information to Ships—Section 8B—Maritime Mobile Service; Telegraphy and Related Subjects. 3 pp.

CCIR RECMN 688-90. Technical Characteristics for a High Frequency Direct-Printing Telegraph System for Promulgation of High Seas and Navtex-Type Maritime Safety Information—Section 8B —Maritime Mobile Service; Telegraphy and Related Subjects. 1 p.

—Power Systems
CNS C1004-88. Frequencies of Electrical Power Systems (Jan)(42).

—Radio Astronomy—Ionosphere
CCIR QUESTION 149/7-90. Frequency Utilization Above the Ionosphere and on the Far Side of the Moon—Questions Concerning Study Group 7—Science Services. 1 p.

—Radio Astronomy—Moon
CCIR QUESTION 149/7-90. Frequency Utilization Above the Ionosphere and on the Far Side of the Moon—Questions Concerning Study Group 7—Science Services. 1 p.

—Radio Astronomy—Protection
CCIR RECMN 314-7-90. Protection for Frequencies Used for Radioastronomical Measurements—Section 2G—Radioastronomy and Radar Astronomy. 4 pp.

—Radio Astronomy—Shielded Zones
CCIR RECMN 479-3-90. Protection of Frequencies for Radioastronomical Measurements in the Shielded Zone of the Moon—Section 2G—Radioastronomy and Radar Astronomy. 2 pp.

—Radio Determination Satellite Services
CCIR Report 1050-1-90. Technical and Operational Considerations for a Radiodetermination Satellite Service in Bands 9 and 10—Section 8J—Technical and Oper-ating Characteristics of Radiocommunications Using Satellite Distress and Safety Operation and of Radio Determination Satellite Services. 31 pp.

—Radiolocation—GMDSS
CCIR Report 1036-1-90. Frequencies for Homing and Locating in the Global Maritime Distress and Safety System (GMDSS)—Section 8D—Radiodetermination, Global Maritime Distress and Safety System and Related Subjects. 23 pp.

—Research Satellites
CCIR RECMN 364-4-86. Preferred Frequencies and Bandwidths for Manned and Unmanned Near-Earth Research Satellites—Section 2E—Space Research. 1 p.

Frequencies (Cont.)

—Research Satellites (Cont.)
CCIR RECMN 364-5-92. Preferred Frequencies and Bandwidths for Manned and Unmanned Near-Earth Research Satellites—Section 2E—Space Research. 11 pp.

—Satellite Communication Equipment
CCIR RECMN 363-4-90. Space Operation Systems Frequencies, Bandwidths and Protection Criteria—Section 2C—Space Operations. 2 pp.

—Satellites—Passive Detectors
CCIR RECMN 515-1-90. Frequency Bands and Performance Requirements for Satellite Passive Sensing—Section 2F—Earth Exploration Satellites. 2 pp.

—Scale (Ratio)
BSI BS 6397-83. 1983 Scales and Sizes for Plotting Frequency Characteristics and Polar Diagrams. 10 pp.
IEC 263-82. Scales and Sizes for Plotting Frequency Characteristics and Polar Diagrams Third Edition. 21 pp.

—SHF/EHF—Fixed Satellite Services—Electromagnetic Interference
CCIR Report 710-3-90. Interference Allocations in Systems Operating at Frequencies Greater Than 10 GHz in the Fixed-Satellite Service—Section 4D1—Permissible Levels of Interference. 11 pp.

—Sound Broadcasting
CCIR RECMN 702-1-92. Synchronization and Multiple Frequency Use per Programme in HF Broadcasting—Section 10A-1—Amplitude-Modulation Sound-Broadcasting in Bands 5 (LF), 6 (MF) and 7 (HF). 1 p.

—Sound Transmission
CCIR RECMN 702-90. Synchronization and Multiple Frequency Use per Programme in HF Broadcasting—Section 10A-1—Amplitude-Modulation Sound-Broadcasting in Bands 5 (LF), 6 (MF) and 7 (HF). 1 p.

—Space Communications
CCIR RECMN 610-1-90. Protection of Allocations for Deep-Space Research—Section 2E—Space Research. 1 p.

—Space Operation Systems
CCIR RECMN 363-4-90. Space Operation Systems Frequencies, Bandwidths and Protection Criteria—Section 2C—Space Operations. 2 pp.

—Space Research Services
CCIR RECMN 576-1-90. Preferred Frequencies and Bandwidths for Deep-Space Research—Section 2E—Space Research. 1 p.

—Spacecraft
CCIR RECMN 610-1-90. Protection of Allocations for Deep-Space Research—Section 2E—Space Research. 1 p.

—Telephone Systems
CCITT RECMN P.64-89. Determination of Sensitivity/Frequency Characteristics of Local Telephone Systems to Permit Calculation of Their Loudness Ratings—Telephone Transmission Quality (Study Group XII) 10 pp. 10 pp.

—Transponders—Satellite Communication Equipment
CCIR Report 923-1-86. Design of Frequency Plans for Satellite Transmission of SCPC Carriers Using Non-Linear Transponders—Section 8I—Technical and Operating Characteristics of Mobile Satellite Services. 7 pp.

—UHF/SHF—Earth Stations
CCIR RECMN 465-3-90. Reference Earth-Station Radiation Pattern for Use in Coordination and Interference Assessment in the Frequency Range from 2 to About 30 GHz—Section 4C—Earth Station and Baseband Characteristics—Earth Station Antennas—Maintenance of Earth Stations. 2 pp.
CCIR RECMN 465-4-92. Reference Earth-Station Radiation Pattern for Use in Coordination and Interference Assessment in the Frequency Range from 2 to About 30 GHz—Section 4C—Earth Station and Baseband Characteristics—Earth Station Antennas—Maintenance of Earth Stations. 2 pp.

—Units of Measurement
BSI BS 5775: Part 2-93. 1993 Quantities, Units and Symbols Part 2: Periodic and Related Phenomena (ISO 31-2: 1992) (G). 15 pp.
BSI BS 5775: Part 2-79. 1979 Amd 2 Specification for Quantities, Units and Symbols Part 2: Periodic and Related Phenomena (AMD 5847) July 31, 1989. 12 pp.

Frequencies (Cont.)

—Units of Measurement (Cont.)
CNS Z7196-2-85. Quantities and Units of Periodic and Related Phenomena (Sep)(11296-2).
ISO 31 Pt 2-92. Quantities and Units—Part 2: Periodic and Related Phenomena Second Edition; (Corrected and Reprinted —1993). 14 pp.

Frequency Allocation
Use: Frequency Assignment

Frequency Assignment
Use For: Communication Channel Assignment
See Also: Communication Channels; Frequencies; Frequency Bands

—Broadcasting Satellite Services
CCIR Decision 93-89. Preparatory Work for the WARC 1992—Annex to Volumes X and XI—Part 2—Broadcasting-Satellite Service (Sound and Television). 3 pp.

—Broadcasting Satellite Services—SHF
CCIR Report 811-2-86. Broadcasting-Satellite Service Planning Elements Including Those Used in the Establishment of Plans of Frequency Assignments and Orbital Positions for the Broadcasting-Satellite Service in the 12 GHz Band—Section 10/11D—Planning. 7 pp.

—Feeder Links—Broadcasting Satellite Services
CCIR QUESTION 86/11-90. Frequencies for the Feeder Links to a Broadcasting Satellite (Sound and Television)—Questions Concerning Study Group 11—Broadcasting Service (Television). 2 pp.

—Feeder Links—Broadcasting Satellite Services—SHF
CCIR Report 952-2-90. Technical Characteristics of Feeder Links to Broadcasting Satellites Elements Required for the Establishment of Plans of Frequency Assignments and Orbital Positions for the Broadcasting-Satellite Service and the Associated Feeder Links—Sharing in the Feeder-. 56 pp.
CCIR QUESTION 98/11-90. Technical Characteristics of and Planning Methods for Feeder Links to Broadcasting Satellites (Television) in the 12 GHz Band in Regions 1 and 3—Questions Concerning Study Group 11—Broadcasting Service (Television). 1 p.

—High Definition Television Systems
CCIR Decision 93-89. Preparatory Work for the WARC 1992—Annex to Volumes X and XI—Part 2—Broadcasting-Satellite Service (Sound and Television). 3 pp.

—Land Mobile Satellite Communications—UHF
CCIR Report 770-3-90. Technical and Operating Considerations for a Land Mobile-Satellite Service Operating in Band 9—Section 8F—Frequencies, Orbits and Systems. 27 pp.

—Land Mobile Services—HF/VHF/UHF
CCIR RESOLUTION 20-5-90. Characteristics of Equipment and Principles Governing the Allocation of Frequency Channels Between 25 and 3000 MHz in the Land Mobile Service—Volume VIII—Mobile, Radiodetermination, Amateur and Related Satellite Services. 1 p.
CCIR Report 319-7-90. Characteristics of Equipment and Principles Governing the Assignment of Frequency Channels Between 25 and 1000 MHz for Land Mobile Services—Section 8A—Land Mobile Service and Related Subjects. 24 pp.

—Land Mobile Services—UHF—Trunked
CCIR Report 901-2-90. Frequency Assignment Methods for Trunked Mobile Radio Systems—Section 8A—Land Mobile Service and Related Subjects. 3 pp.

—Mobile Satellite Communications
CCIR Report 1172-90. Intersystem Frequency Sharing and Reuse in the Mobile-Satellite Services Operating at Mid to High Portions of Band 9—Section 8H—Efficient Use of the Radio Spectrum Characteristics and Sharing of Frequency Resources. 25 pp.

—Mobile Satellite Communications—EHF
CCIR Report 771-3-90. Considerations for the Mobile-Satellite Service —Section 8F—Frequencies, Orbits and Systems. 7 pp.

—Mobile Satellite Communications—SHF
CCIR Report 771-3-90. Considerations for the Mobile-Satellite Service —Section 8F—Frequencies, Orbits and Systems. 7 pp.

Frequency Assignment (Cont.)

—Radio Spectra
CCIR QUESTION 66/1-90. Methods and Algorithms for Frequency Planning—Questions Concerning Study Group 1—Spectrum Management Techniques (Spectrum Engineering, Planning, Sharing, Monitoring and Utilization). 1 p.
CCIR QUESTION 69/1-90. Technical and Operational Methods Utilized for the Allocation and Improved Use of the Radio Frequency Spectrum—Questions Concerning Study Group 1—Spectrum Management Techniques (Spectrum Engineering, Planning, Sharing, Monitoring and Utilization). 1 p.
CCIR QUESTION 72/1-90. Optimum Network Planning and Frequency Assignment Techniques—Questions Concerning Study Group 1—Spectrum Management Techniques (Spectrum Engineering, Planning, Sharing, Monitoring and Utilization). 1 p.

—Radio Spectra—Algorithms
CCIR QUESTION 66/1-90. Methods and Algorithms for Frequency Planning—Questions Concerning Study Group 1—Spectrum Management Techniques (Spectrum Engineering, Planning, Sharing, Monitoring and Utilization). 1 p.

—Radio Spectra—World Administrative Radio Conference
CCIR RESOLUTION 100-90. CCIR Studies to Be Carried out for the Preparation of a Report to Be Submitted to the World Administrative Radio Conference (WARC-92) for Frequency Allocations in Certain Parts of the Spectrum—Volume XIV—Administrative Texts of the CCIR. 3 pp.

Frequency Band Sharing
Use For: Band Sharing; Frequency Sharing
See Also: Frequencies; Frequency Bands; Radio Communications

CCIR Decision 101-89. Establishment of Interim Working Party 2/2—Annex to Volume II—Space Research and Radioastronomy Services. 1 p.
CCIR Decision 97-89. Sharing Criteria Between Broadcasting and Fixed and Mobile Services in the Band 2-30 MHz—Annex to Volume III—Fixed Service at Frequencies Below About 30 MHz (See Annex to Volume X-1). 1 p.
CCIR Volume IV&IX TOC-90. Table of Contents. 1 p.
CCIR Volume IV&IX NUM INDX TXT-90. Numerical Index of Texts. 1 p.
CCIR Volume IV&IX TOC-90. Table of Contents. 2 pp.
CCIR Volume IV&IX NUM IND TXTS-90. Numerical Index of Texts. 1 p.

—0-1 GHz—Television Broadcasting—Fixed Services
CCIR Report 1087-1-90. Frequency Sharing Between the Broadcasting Service (Television) and the Fixed and Mobile Services Below 1 GHz—Section 11E—Planning of Television Networks, Protection Ratios, Television Receivers and Antennas. 7 pp.

—0-1 GHz—Television Broadcasting—Mobile Radio Services
CCIR Report 1087-1-90. Frequency Sharing Between the Broadcasting Service (Television) and the Fixed and Mobile Services Below 1 GHz—Section 11E—Planning of Television Networks, Protection Ratios, Television Receivers and Antennas. 7 pp.

—0-15 GHz—Radio Communications—Terrestrial—Earth Stations
CCIR Report 386-3-82. Determination of the Power in Any 4 kHz Band Radiated Towards the Horizon by Earth Stations of the Fixed-Satellite Service Sharing Frequency Bands Below 15 GHz with the Terrestrial Services—Section 4/9A—Sharing Conditions. 4 pp.

—0-30 MHz—Radio Communications
CCIR OPINION 66-82. Frequency Sharing Between Services Below 30 MHz—Volume III—Fixed Service at Frequencies Below About 30 MHz. 1 p.
CCIR Report 859-82. Frequency Sharing Between Services Below 30 MHz—Section 3Ad—Operational Questions. 2 pp.

—Band 9—Mobile Satellite Communications
CCIR Report 1172-90. Intersystem Frequency Sharing and Reuse in the Mobile-Satellite Services Operating at Mid to High Portions of Band 9—Section 8H—Efficient Use of the Radio Spectrum Characteristics and Sharing of Frequency Resources. 25 pp.

—Bands 9 and 10—Geostationary Satellites—Data Relay Satellites
CCIR Report 847-1-86. Data Relay Satellites Sharing with Other Services in Bands 9 and 10—Section 2D—Data Relay Satellites. 7 pp.

Frequency Band Sharing (Cont.)

—Broadcasting Satellite Services—Broadcasting

CCIR Report 631-4-90. Frequency Sharing Between the Broadcasting-Satellite Service (Sound and Television) and Terrestrial Services—Section 10/11E—Sharing. 49 pp.

—Broadcasting Satellite Services—Fixed Satellite Services

CCIR Report 631-4-90. Frequency Sharing Between the Broadcasting-Satellite Service (Sound and Television) and Terrestrial Services—Section 10/11E—Sharing. 49 pp.

CCIR QUESTION 66/4-90. Frequency Sharing Between the Fixed-Satellite Service and the Broadcasting-Satellite Service—Questions of Study Group 4 Fixed-Satellite Service. 1 p.

—Broadcasting Satellite Services—Fixed Services

CCIR Report 631-4-90. Frequency Sharing Between the Broadcasting-Satellite Service (Sound and Television) and Terrestrial Services—Section 10/11E—Sharing. 49 pp.

—Broadcasting Satellite Services—High Definition Television System

CCIR Decision 43-5-89. Satellite Sound Broadcasting for Portable and Vehicle Receivers and Sharing and Spectrum Aspects of Wide RF-Band HDTV Satellite Broadcasting—Annex to Volumes X and XI—Part 2—Broadcasting-Satellite Service (Sound and Television). 5 pp.

CCIR QUESTION 89/11-90. Sharing Studies Between High-Definition Television (HDTV) in the Broadcasting-Satellite Service and Other Services—Questions Concerning Study Group 11—Broadcasting Service (Television). 1 p.

—Broadcasting Satellite Services—Mobile Radio Services

CCIR Report 631-4-90. Frequency Sharing Between the Broadcasting-Satellite Service (Sound and Television) and Terrestrial Services—Section 10/11E—Sharing. 49 pp.

—Broadcasting Satellite Services—Space Stations

CCIR Report 1076-86. Considerations Affecting the Accommodation of Spacecraft Service Functions (TTC) Within the Broadcasting-Satellite and Feeder-Link Service Bands—Section 10/11E—Sharing. 14 pp.

—Communications Satellites

CCIR Decision 2-7-89. Frequency Sharing Between Radiocommunication Satellites Technical Considerations Affecting the Efficient Use of the Geostationary-Satellite Orbit—Annex to Volume IV-1—Fixed-Satellite Service. 4 pp.

—Communications Satellites—Fixed Satellite Services

CCIR QUESTION 48/4-90. Technical Factors Influencing the Efficiency of Use of the Geostationary-Satellite Orbit by Radiocommunication Sat. Networks Sharing Frequency Bands Allocated to the Fixed-Sat. Service—Questions of Study Group 4 Fixed-Satellite Service. 2 pp.

—Communications Satellites—Geostationary Satellite Orbit

CCIR Report 453-5-90. Technical Factors Influencing the Efficiency of Use of the Geostationary-Satellite Orbit by Radiocommunication Satellites Sharing the Same Frequency Bands General Summary—Section 4D2—Coordination Methods. 36 pp.

CCIR QUESTION 48/4-90. Technical Factors Influencing the Efficiency of Use of the Geostationary-Satellite Orbit by Radiocommunication Sat. Networks Sharing Frequency Bands Allocated to the Fixed-Sat. Service—Questions of Study Group 4 Fixed-Satellite Service. 2 pp.

—Data Relay Satellites—Longitude Separation Angle

CCIR Report 983-86. Minimum Longitude Separation Angle Necessary to Share Frequencies Between Two Data Relay Satellites—Section 2D—Data Relay Satellites. 6 pp.

—Data Relay Satellites—Radio Communications

CCIR QUESTION 118/7-90. Data Relay Satellite Systems and Factors Which Affect Frequency Sharing with Other Services—Questions Concerning Study Group 7—Science Services. 2 pp.

—Earth Exploration Satellites

CCIR Report 1122-90. Methodology for Deriving Sharing Criteria and Coordination Thresholds for the Earth Exploration-Satellite Service—Section 2F—Earth Exploration Satellites. 24 pp.

—Earth Exploration Satellites—Geostationary—Fixed

CCIR Report 540-1-82. Feasibility of Frequency Sharing Between an Earth Exploration-Satellite (EES) System and Fixed Satellite, Meteorological Satellite and Terrestrial Fixed and Mobile Services—Section 2F—Earth Exploration Satellites. 11 pp.

—Earth Exploration Satellites—Geostationary—Meteorological

CCIR Report 540-1-82. Feasibility of Frequency Sharing Between an Earth Exploration-Satellite (EES) System and Fixed Satellite, Meteorological Satellite and Terrestrial Fixed and Mobile Services—Section 2F—Earth Exploration Satellites. 11 pp.

—Earth Exploration Satellites—Meteorological Services

CCIR QUESTION 3/12-90. Sharing Between the Earth Exploration-Satellite Service or the Meteorological-Satellite Service on the One Hand and Other Space Services or the Metgl. Aids Service on the Other—Questions Concerning Study Group 12—Inter-Service Sharing and Compatibility. 1 p.

—Earth Exploration Satellites—Radiolocation Services—Radar

CCIR Report 695-2-86. Feasibility of Frequency Sharing Between Spaceborne Radars and Terrestrial Radars in the Radiolocation Service—Section 2F—Earth Exploration Satellites. 13 pp.

—Earth Exploration Satellites—Satellite Communications

CCIR QUESTION 3/12-90. Sharing Between the Earth Exploration-Satellite Service or the Meteorological-Satellite Service on the One Hand and Other Space Services or the Metgl. Aids Service on the Other—Questions Concerning Study Group 12—Inter-Service Sharing and Compatibility. 1 p.

—Earth Stations

CCIR RECMN 578-82. Protection Criteria and Sharing Considerations Relating to Deep-Space Research—Section 2E—Space Research. 1 p.

CCIR RECMN 848-92. Determination of the Coordination Area of a Transmitting Earth Station Using the Same Frequency Band as Receiving Earth Stations in Bidirectionally Allocated Frequency Bands. 9 pp.

—Earth Stations—Deep Space Research

CCIR QUESTION 114/7-90. Feasibility of Frequency Sharing Between Deep-Space Research Stations and Stations of Other Services—Questions Concerning Study Group 7—Science Services. 1 p.

—Earth Stations—Fixed Satellite Services—Coordinate Determination

CCIR RECMN 359-5-82. Determination of the Co-Ordination Area of Earth Stations in the Fixed-Satellite Service Using the Same Frequency Bands as the Systems in the Fixed Terrestrial Service—Section 4/9B—Co-Ordination and Interference Calculations. 1 p.

—Earth Stations—Fixed Satellite Services—Radio Navigation Services

CCIR QUESTION 5/12-90. Coordination Area of an Earth Station of the Fixed-Satellite Service Sharing the Same Frequency Band with the Radionavigation Service—Questions Concerning Study Group 12—Inter-Service Sharing and Compatibility. 1 p.

—Earth Stations—Fixed Satellite Services—Terrestrial Stations

CCIR QUESTION 57/4-90. Preferred Technical Characteristics and Selection of Sites for Earth Stations in the Fixed-Satellite Service to Facilitate Sharing with Terrestrial Services—Questions of Study Group 4 Fixed-Satellite Service. 1 p.

CCIR QUESTION 59/4-90. Preferred Technical Characteristics of Space Stations in the Fixed-Satellite Service to Facilitate Sharing with Terrestrial Services—Questions of Study Group 4 Fixed-Satellite Service. 1 p.

—Earth Stations—Terrestrial Stations

CCIR RECMN 847-92. Determination of the Coordination Area of an Earth Station Operating with a Geostationary Space Station and Using the Same Frequency Band as a System in a Terrestrial Service. 30 pp.

—Earth Stations—Terrestrial Stations (Cont.)

CCIR RECMN 849-92. Determination of Coordination Area for Earth Stations Operating with Non-Geostationary Spacecraft in Bands Shared with Terrestrial Services. 10 pp.

—EHF—Fixed Satellite Services—Fixed Services

CCIR Report 876-82. Frequency Sharing Between Systems in the Fixed-Satellite Service and the Fixed Service in Frequency Bands Above 40 GHz—Section 4/9A—Sharing Conditions. 7 pp.

—EHF—Geostationary Satellites—Radio Navigation Services—Links

CCIR Report 872-82. Sharing Criteria Between Inter-Satellite Links Connecting Geostationary Satellites in the Fixed-Satellite Service and the Radionavigation Service at 33 GHz—Section 4E—Frequency Sharing Between Networks of the Fixed-Satellite Service and Those of Other Space. 4 pp.

—EHF—Telecommunication Services

CCIR QUESTION 64/1-86. Spectrum Usage and Sharing Criteria Above 40 GHz—Questions Concerning Study Group 1—Spectrum Management Techniques (Spectrum Engineering, Planning, Sharing, Monitoring and Utilization). 1 p.

—Feeder Links—Space Stations

CCIR Report 1076-86. Considerations Affecting the Accommodation of Spacecraft Service Functions (TTC) Within the Broadcasting-Satellite and Feeder-Link Service Bands—Section 10/11E—Sharing. 14 pp.

—Fixed Satellite Services

CCIR Report 455-5-90. Frequency Sharing Between Networks of the Fixed-Satellite Service—Section 4D1—Permissible Levels of Interference. 43 pp.

—Fixed Satellite Services—Digital Radio Systems—Interference

CCIR Report 877-1-86. Interference Criteria for Digital Radio-Relay Systems Sharing Frequency Bands with the Fixed-Satellite Service—Section 4/9A—Sharing Conditions. 6 pp.

—Fixed Satellite Services—Earth Exploration Satellites

CCIR QUESTION 67/4-90. Frequency Sharing Between the Fixed-Satellite Service and the Earth Exploration-Satellite (Passive) and Space Research (Passive) Services—Questions of Study Group 4 Fixed-Satellite Service. 1 p.

—Fixed Satellite Services—Fixed Services (Radio Communications)

CCIR Report 1005-86. Frequency Sharing Between Systems of the Fixed Service and Systems of the Fixed-Satellite Service Comprising Forward Band Working (FBW) Networks and Reverse Band Working (RBW) Networks—Section 4/9A—Sharing Conditions. 7 pp.

CCIR Report 1143-90. Frequency Sharing Between the Fixed-Satellite Service and the Fixed Service Under Provisions of RR Article 14 in Region 2—Section 4/9A—Sharing Conditions. 3 pp.

CCIR QUESTION 61/4-90. Criteria for Frequency Sharing Between the Fixed Service and the Fixed-Satellite Service in Bidirectionally Allocated Frequency Bands—Questions of Study Group 4 Fixed-Satellite Service. 1 p.

CCIR QUESTION 62/4-90. Frequency Sharing of the Fixed-Satellite Service and the Inter-Satellite Service with the Fixed Service Under Provisions of RR Article 14—Questions of Study Group 4 Fixed-Satellite Service. 1 p.

CCIR QUESTION 109/9-90. Criteria for Frequency Sharing Between Radio-Relay Systems and Systems in the Fixed-Satellite Service—Questions of Study Group 9 Fixed Service. 1 p.

CCIR QUESTION 115/9-90. Criteria for Frequency Sharing Between the Fixed Service and the Fixed-Satellite Service in Bidirectionally Allocated Frequency Bands—Questions of Study Group 9 Fixed Service. 1 p.

—Fixed Satellite Services—Fixed Services—Emergency Relief Operatio

CCIR QUESTION 65/4-90. Frequency Sharing Between Systems in the Fixed-Satellite Service and the Fixed Service in the Case of Relief Operations and Other Temporary Applications—Questions of Study Group 4 Fixed-Satellite Service. 1 p.

INTERNATIONAL AND NON-U.S. NATIONAL STANDARDS
SUBJECT INDEX

Frequency

Frequency Band Sharing (Cont.)

—Fixed Satellite Services—Fixed Services—Propagation
CCIR QUESTION 19-1/5-90. Propagation Factors Affecting Frequency-Sharing Between Fixed-Satellite Service and Fixed and Mobile Terrestrial Services—Questions Concerning Study Group 5—Radio Wave Propagation in Non-Ionized Media. 2 pp.

—Fixed Satellite Services—Mobile Radio Communications—Propagation
CCIR QUESTION 19-1/5-90. Propagation Factors Affecting Frequency-Sharing Between Fixed-Satellite Service and Fixed and Mobile Terrestrial Services—Questions Concerning Study Group 5—Radio Wave Propagation in Non-Ionized Media. 2 pp.

—Fixed Satellite Services—Radio Communications
CCIR RECMN 355-3-82. Frequency Sharing Between Systems in the Fixed-Satellite Service and Terrestrial Radio Services in the Same Frequency Bands—Section 4/9A—Sharing Conditions. 2 pp.
CCIR Report 209-5-86. Frequency Sharing Between Systems in the Fixed-Satellite Service and Terrestrial Radio Services—Section 4/9A—Sharing Conditions. 8 pp.
CCIR QUESTION 63/4-90. Frequency Sharing of the Fixed-Satellite Service and the Inter-Satellite Service with Terrestrial Radio Services Other Than the Fixed Service Under Provisions of RR Article 14—Questions of Study Group 4 Fixed-Satellite Service. 1 p.

—Fixed Satellite Services—Radio Relay Systems
CCIR RECMN 355-3-82. Frequency Sharing Between Systems in the Fixed-Satellite Service and Terrestrial Radio Services in the Same Frequency Bands—Section 4/9A—Sharing Conditions. 2 pp.
CCIR RECMN 355-4-92. Frequency Sharing Between Systems in the Fixed-Satellite Service and Radio-Relay Systems in the Same Frequency Bands—Section 4/9A—Sharing Conditions. 6 pp.
CCIR Report 209-5-86. Frequency Sharing Between Systems in the Fixed-Satellite Service and Terrestrial Radio Services—Section 4/9A—Sharing Conditions. 8 pp.
CCIR QUESTION 109/9-90. Criteria for Frequency Sharing Between Radio-Relay Systems and Systems in the Fixed-Satellite Service—Questions of Study Group 9 Fixed Service. 1 p.

—Fixed Satellite Services—Satellite Communications
CCIR QUESTION 68/4-90. Frequency Sharing of the Fixed-Satellite Service and the Inter-Satellite Service with Other Space Radio Services Under Provisions of RR Article 14—Questions of Study Group 4 Fixed-Satellite Service. 1 p.

—Fixed Satellite Services—Space Research Services
CCIR QUESTION 67/4-90. Frequency Sharing Between the Fixed-Satellite Service and the Earth Exploration-Satellite (Passive) and Space Research (Passive) Services—Questions of Study Group 4 Fixed-Satellite Service. 1 p.

—Fixed Satellite Services—Terrestrial Services
CCIR QUESTION 32-1/4-90. Frequency Sharing Between Systems in the Fixed-Satellite Service and Terrestrial Services—Questions of Study Group 4 Fixed-Satellite Service. 1 p.

—Fixed Satellites—Geostationary Satellite Orbit—Fixed Services
CCIR Report 1142-90. Frequency Sharing Between the Fixed Service and the Fixed-Satellite Service Using Satellites in Slightly Inclined Geostationary Orbits—Section 4/9A—Sharing Conditions. 29 pp.
CCIR QUESTION 64/4-90. Preferred Technical Characteristics of Fixed-Satellite Service Networks Using Satellites in Slightly Inclined Geostationary Orbits to Facilitate Sharing with the Fixed Service—Questions of Study Group 4 Fixed-Satellite Service. 1 p.
CCIR QUESTION 117/9-90. Criteria for Frequency Sharing Between the Fixed Service and FSS Networks Using Satellites in Slightly Inclined Geostationary Orbits—Questions of Study Group 9 Fixed Service. 1 p.

—Fixed Satellites—Geostationary Satellite Orbit—Radio Astronomy
CCIR QUESTION 137/7-90. Effects on Study Group 2 Services of the Use of Inclined (near-Geostationary) Satellite Orbits by the Fixed-Satellite Service—Questions Concerning Study Group 7—Science Services. 1 p.

Frequency Band Sharing (Cont.)

—Fixed Satellites—Radio Relay Systems—Isotropic Radiated Power
CCIR RECMN 406-6-90. Maximum Equivalent Isotropically Radiated Power of Line-of-Sight Radio-Relay System Transmitters Operating in the Frequency Bands Shared with the Fixed-Satellite Service—Section 4/9A—Sharing Conditions. 3 pp.
CCIR RECMN 406-7-92. Maximum Equivalent Isotropically Radiated Power of Line-of-Sight Radio-Relay System Transmitters Operating in the Frequency Bands Shared with the Fixed-Satellite Service—Section 4/9A—Sharing Conditions. 2 pp.
CCIR QUESTION 114/9-90. Maximum e.i.r.p. for Line-of-Sight Radio-Relay Transmitters Operating in Frequency Bands Shared with the Fixed-Satellite Service—Questions of Study Group 9 Fixed Service. 1 p.

—Fixed Services—Broadcasting
CCIR QUESTION 131/9-90. Criteria for Frequency Sharing Between the Fixed and Broadcasting Services—Questions of Study Group 9 Fixed Service. 1 p.

—Fixed Services—Fixed Satellite Services—Space Stations
CCIR QUESTION 116/9-90. Sharing Criteria for Protecting the Fixed Service from Emissions of Space Stations in the Fixed-Satellite Service in Shared Frequency Bands—Questions of Study Group 9 Fixed Service. 1 p.

—Fixed Services—Mobile Radio—Fixed Satellite Services
CCIR Report 540-1-82. Feasibility of Frequency Sharing Between an Earth Exploration-Satellite (EES) System and Fixed Satellite, Meteorological Satellite and Terrestrial Fixed and Mobile Services—Section 2F—Earth Exploration Satellites. 11 pp.

—Fixed Services—Terrestrial—Radio Relay Systems
CCIR RECMN 758-92. Considerations in the Development of Criteria for Sharing Between the Terrestrial Fixed Service and Other Services—Section 9F—Frequency Sharing with Other Services. 14 pp.

—Geostationary Satellite Orbit—Inclined—Fixed Satellite Services
CCIR Report 1138-90. Intra-Service Implications of Using Slightly Inclined Geostationary Orbits for Fixed Satellite Service Networks Operational, Sharing and Coordination Considerations—Section 4D2—Coordination Methods. 22 pp.
CCIR QUESTION 51/4-90. Operation, Frequency Sharing and Coordination of Fixed Satellite Service Networks Using Satellites in Slightly Inclined Geostationary Orbits—Questions of Study Group 4 Fixed-Satellite Service. 1 p.

—Geostationary Satellites
CCIR Report 455-5-90. Frequency Sharing Between Networks of the Fixed-Satellite Service—Section 4D1—Permissible Levels of Interference. 43 pp.

—Geostationary Satellites—Coordination Calculations
CCIR RECMN 738-92. Procedure for Determining if Coordination is Required Between Geostationary-Satellite Networks Sharing the Same Frequency Bands—Section 4D2—Coordination Methods. 12 pp.
CCIR RECMN 739-92. Additional Methods for Determining if Detailed Coordination is Necessary Between Geostationary-Satellite Networks in the Fixed-Satellite Service Sharing the Same Frequency Bands—Section 4D2—Coordination Methods. 10 pp.
CCIR Report 454-5-90. Method of Calculation for Determining if Coordination is Required Between Geostationary-Satellite Networks Sharing the Same Frequency Bands—Section 4D2—Coordination Methods. 39 pp.

—Geostationary Satellites—Earth Stations—Coordination Calculations
CCIR Report 454-5-90. Method of Calculation for Determining if Coordination is Required Between Geostationary-Satellite Networks Sharing the Same Frequency Bands—Section 4D2—Coordination Methods. 39 pp.

—HF—Broadcasting
CCIR RECMN 831-92. Frequency Sharing Between Services in the Band 4-30 MHz. 3 pp.

Frequency Band Sharing (Cont.)

—HF—Earth Stations—Meteorological Satellites
CCIR Report 1124-90. Performance, Interference and Sharing Criteria for Receiving Earth Stations in the Meteorological-Satellite Service Operating in the 7 450—7 550 MHz Band with Satellites in Low Earth Orbit—Section 2F—Earth Exploration Satellites. 7 pp.

—HF—Fixed Satellites—Meteorological Satellites
CCIR Report 1125-90. Frequency Sharing Between Systems in the Fixed-Satellite Service and the Meteorological-Satellite Service Operating with Satellites in Low Earth Orbit in the Band 7 450—7 550 MHz—Section 2F—Earth Exploration Satellites. 10 pp.

—HF—Fixed Services
CCIR RECMN 831-92. Frequency Sharing Between Services in the Band 4-30 MHz. 3 pp.

—HF—Fixed Services—Mobile Radio Services
CCIR Report 911-82. Frequency Sharing Between Services in the Band 4 to 30 MHz—Section 8C—Maritime Mobile Service; Telephony and Related Subjects. 3 pp.

—HF—Maritime Mobile Services
CCIR QUESTION 64-2/8-90. HF Bands Allocated on an Exclusive or Shared Basis to the Maritime Mobile Service—Questions Concerning Study Group 8—Mobile, Radiodetermination, Amateur and Related Satellite Services. 1 p.

—HF—Microwave Landing Systems
CCIR Report 1181-90. Microwave Landing System (MLS) Spectrum Requirements and Signal Protection Criteria—Secton 8K—Aeronautical Mobile Service (Terrestrial). 2 pp.

—HF—Mobile Radio Services
CCIR QUESTION 56/8-82. Frequency Sharing Between Services in the Band 4-30 MHz—Questions Concerning Study Group 8—Mobile, Radiodetermination, Amateur and Related Satellite Services. 1 p.
CCIR RECMN 831-92. Frequency Sharing Between Services in the Band 4-30 MHz. 3 pp.

—Interference—Digital Satellite Systems—Radio Communications
CCIR Report 793-1-86. Derivation of Interference Criteria for Digital Systems in the Fixed-Satellite Service Sharing Bands with Terrestrial Systems—Section 4/9A—Sharing Conditions. 3 pp.

—Intersatellite Services—Fixed Satellite Services
CCIR QUESTION 56/4-90. Frequency Sharing Between the Inter-Satellite Service When Used for Links of the Fixed-Satellite Service and Terrestrial Radiocommunication Services—Questions of Study Group 4 Fixed-Satellite Service. 1 p.

—Intersatellite Services—Fixed Services (Radio Communications)
CCIR QUESTION 62/4-90. Frequency Sharing of the Fixed-Satellite Service and the Inter-Satellite Service with the Fixed Service Under Provisions of RR Article 14—Questions of Study Group 4 Fixed-Satellite Service. 1 p.

—Intersatellite Services—Radio Communications
CCIR QUESTION 56/4-90. Frequency Sharing Between the Inter-Satellite Service When Used for Links of the Fixed-Satellite Service and Terrestrial Radiocommunication Services—Questions of Study Group 4 Fixed-Satellite Service. 1 p.
CCIR QUESTION 63/4-90. Frequency Sharing of the Fixed-Satellite Service and the Inter-Satellite Service with Terrestrial Radio Services Other Than the Fixed Service Under Provisions of RR Article 14—Questions of Study Group 4 Fixed-Satellite Service. 1 p.

—Intersatellite Services—Satellite Communications
CCIR QUESTION 68/4-90. Frequency Sharing of the Fixed-Satellite Service and the Inter-Satellite Service with Other Space Radio Services Under Provisions of RR Article 14—Questions of Study Group 4 Fixed-Satellite Service. 1 p.

—Land Mobile Services
CCIR Decision 69-2-89. Future Public Land Mobile Telecommunication Systems (FPLMTS)—Annex 1 to Volume VIII—Land Mobile Service—Amateur Service—Amateur-Satellite Service. 4 pp.

INDUSTRY STANDARDS

INTERNATIONAL AND NON-U.S. NATIONAL STANDARDS
SUBJECT INDEX — Frequency

Frequency Band Sharing (Cont.)

—**Maritime Mobile Services**
CCIR QUESTION 53-3/8-90. Use of Frequencies by the Maritime Mobile Service in the Band 435-526.5 khz—Questions Concerning Study Group 8 —Mobile, Radiodetermination, Amateur and Related Satellite Services. 1 p.

—**Meteorological Satellites—Meteorological Services**
CCIR QUESTION 3/12-90. Sharing Between the Earth Exploration-Satellite Service or the Meteorological-Satellite Service on the One Hand and Other Space Services or the Metgl. Aids Service on the Other—Questions Concerning Study Group 12—Inter-Service Sharing and Compatibility. 1 p.

—**Meteorological Satellites—Satellite Communications**
CCIR QUESTION 3/12-90. Sharing Between the Earth Exploration-Satellite Service or the Meteorological-Satellite Service on the One Hand and Other Space Services or the Metgl. Aids Service on the Other—Questions Concerning Study Group 12—Inter-Service Sharing and Compatibility. 1 p.

—**MF—Earth Stations—Meteorological Satellites**
CCIR Report 1121-90. Performance, Interference and Sharing Criteria for Receiving Earth Stations in the Meteorological-Satellite Service Operating in the 1 670–1 710 MHz Band with Satellites in Low Earth Orbit—Section 2F—Earth Exploration Satellites. 12 pp.

—**MF/HF—Fixed Services—Broadcasting**
CCIR Decision 97-89. Sharing Criteria Between Broadcasting and Fixed and Mobile Services in the Band 2-30 MHz—Annex to Volume III—Fixed Service at Frequencies Below About 30 MHz (See Annex to Volume X-1). 1 p.

—**MF/HF—Mobile Radio Services—Broadcasting**
CCIR Decision 97-89. Sharing Criteria Between Broadcasting and Fixed and Mobile Services in the Band 2-30 MHz—Annex to Volume III—Fixed Service at Frequencies Below About 30 MHz (See Annex to Volume X-1). 1 p.

—**MF/HF—Sound Broadcasting—Fixed Services**
CCIR Decision 97-89. Sharing Criteria Between Broadcasting and Fixed and Mobile Services in the Band 2-30MHz—Volume X-1—Broadcasting Service (Sound). 3 pp.

—**MF/HF—Sound Broadcasting—Mobile Radio Services**
CCIR Decision 97-89. Sharing Criteria Between Broadcasting and Fixed and Mobile Services in the Band 2-30MHz—Volume X-1—Broadcasting Service (Sound). 3 pp.

—**MF—Sound Broadcasting—Fixed Services**
CCIR QUESTION 63/10-90. Compatibility Between the MF Sound-Broadcasting Service and Other Services Using the Same Band—Questions Concerning Study Group 10—Broadcasting Service (Sound). 1 p.

—**MF—Sound Broadcasting—Mobile Radio Services**
CCIR QUESTION 63/10-90. Compatibility Between the MF Sound-Broadcasting Service and Other Services Using the Same Band—Questions Concerning Study Group 10—Broadcasting Service (Sound). 1 p.

—**MF—Sound Broadcasting—Radio Navigation Services**
CCIR QUESTION 63/10-90. Compatibility Between the MF Sound-Broadcasting Service and Other Services Using the Same Band—Questions Concerning Study Group 10—Broadcasting Service (Sound). 1 p.

—**MF—Sound Broadcasting—Radiolocation Services**
CCIR QUESTION 63/10-90. Compatibility Between the MF Sound-Broadcasting Service and Other Services Using the Same Band—Questions Concerning Study Group 10—Broadcasting Service (Sound). 1 p.

—**Mobile Satellite Communications**
CCIR Report 1179-90. Methodology for the Derivation of Interference and Sharing Criteria for the Mobile-Satellite Services—Section 8H—Efficient Use of the Radio Spectrum Characteristics and Sharing of Frequency Resources. 13 pp.

Frequency Band Sharing (Cont.)

—**Mobile Satellite Communications (Cont.)**
CCIR QUESTION 83-1/8-90. Efficient Use of the Radio Spectrum and Sharing of Frequency Resources in the Mobile-Satellite Service—Questions Concerning Study Group 8 —Mobile, Radiodetermination, Amateur and Related Satellite Services. 1 p.

—**Radio Communications—Terrestrial—Ionospheric Propagation**
CCIR QUESTION 38/6-90. Propagation Factors Affecting the Sharing of the Radio-Frequency Spectrum Between Terrestrial Systems Involving Ionospheric Propagation—Questions Concerning Study Group 6—Radio Wave Propagation in Ionized Media. 1 p.

—**Radio Determination Satellite Services**
CCIR Report 1050-1-90. Technical and Operational Considerations for a Radiodetermination Satellite Service in Bands 9 and 10—Section 8J—Technical and Oper-ating Characteristics of Radiocommunications Using Satellite Distress and Safety Operation and of Radio Determination Satellite Services. 31 pp.
CCIR QUESTION 91-1/8-90. Technical and Operating Characteristics of the Radiodetermination-Satellite Service—Questions Concerning Study Group 8—Mobile, Radiodetermination, Amateur and Related Satellite Services. 1 p.

—**Radio Frequency Interference—Voltage Measurement**
CCIR Report 1102-90. Calculation of the Received Voltage Due to the Radiation from Multiple Co-Frequency Sources—Section 1B—Spectrum Sharing and Planning Principles and Techniques. 9 pp.

—**Radio Relay Systems—Earth Exploration Satellites**
CCIR Report 540-1-82. Feasibility of Frequency Sharing Between an Earth Exploration-Satellite (EES) System and Fixed Satellite, Meteorological Satellite and Terrestrial Fixed and Mobile Services—Section 2F—Earth Exploration Satellites. 11 pp.
CCIR QUESTION 113/9-90. Frequency Sharing Between Radio-Relay Systems and Systems of the Earth Exploration-Satellite Service and the Space Research Service—Questions of Study Group 9 Fixed Service. 1 p.

—**Radio Relay Systems—Fixed**
CCIR Report 1196-90. Considerations in the Development of Criteria for Sharing Between the Terrestrial Fixed Service and Other Services—Section 9F—Frequency Sharing with Other Services. 5 pp.
CCIR Decision 89-89. Technical Characteristics of Systems in the Fixed Service Required for the Evaluation of Frequency Sharing—Annex to Volume IX-1—Fixed Service Using Radio-Relay Systems. 2 pp.
CCIR QUESTION 124/9-90. New Techniques for Spectrum Sharing and Band Utilization for Radio-Relay Systems—Questions of Study Group 9 Fixed Service. 1 p.

—**Radio Relay Systems—Mobile Satellite Communications—Aeronautical**
CCIR Report 1195-90. Sharing Considerations Between the Fixed Service and the Aeronautical Mobile-Satellite Service—Section 9F—Frequency Sharing with Other Services. 2 pp.

—**Radio Relay Systems—Point to Multipoint Communications**
CCIR QUESTION 125/9-90. Point-to-Multipoint Systems for Telephony, Data or Video Applications Using New Techniques for Spectrum Sharing and Band Utilization—Questions of Study Group 9 Fixed Service. 1 p.

—**Radio Relay Systems—Space Research Systems**
CCIR QUESTION 113/9-90. Frequency Sharing Between Radio-Relay Systems and Systems of the Earth Exploration-Satellite Service and the Space Research Service—Questions of Study Group 9 Fixed Service. 1 p.

—**Radio Spectra**
CCIR QUESTION 45-2/1-86. Technical Criteria for Frequency Sharing—Questions Concerning Study Group 1—Spectrum Management Techniques (Spectrum Engineering, Planning, Sharing, Monitoring and Utilization). 2 pp.
CCIR QUESTION 71/1-90. Bandwidth Expansion Techniques and Spectrum Sharing—Questions Concerning Study Group 1 —Spectrum Management Techniques (Spectrum Engineering, Planning, Sharing, Monitoring and Utilization). 1 p.

Frequency Band Sharing (Cont.)

—**Radio Spectra—Protection Ratio**
CCIR RECMN 669-90. Protection Ratios for Spectrum Sharing Investigations—Section 1B—Spectrum Sharing and Planning Principles and Techniques. 1 p.

—**Satellite Communications—Fixed Services (Radio Communications)**
CCIR Report 791-1-82. Inter-Satellite Service Sharing with the Fixed and Mobile Services—Section 4/9A—Sharing Conditions. 6 pp.

—**Satellite Communications—Mobile Radio Services**
CCIR Report 791-1-82. Inter-Satellite Service Sharing with the Fixed and Mobile Services—Section 4/9A—Sharing Conditions. 6 pp.

—**Satellite Links**
CCIR Report 874-82. Frequency Sharing Between the Inter-Satellite Service When Used by the Fixed-Satellite Service and Other Space Services—Section 4E—Frequency Sharing Between Networks of the Fixed-Satellite Service and Those of Other Space Radiocommunications Systems. 2 pp.
CCIR QUESTION 135/7-90. Characteristics of Inter-Satellite Links—Questions Concerning Study Group 7—Science Services. 1 p.

—**Satellite Links—Fixed Satellite Services**
CCIR Report 874-82. Frequency Sharing Between the Inter-Satellite Service When Used by the Fixed-Satellite Service and Other Space Services—Section 4E—Frequency Sharing Between Networks of the Fixed-Satellite Service and Those of Other Space Radiocommunications Systems. 2 pp.

—**SHF—Broadcasting Satellite Services—Intersatellite Services**
CCIR Report 951-82. Sharing Between the Inter-Satellite Service and the Broadcasting-Satellite Service in the Vicinity of 23 GHz—Section 10/11E—Sharing. 7 pp.

—**SHF—Broadcasting Satellites—Fixed Satellites**
CCIR Report 809-3-90. Inter-Regional Sharing of the 11.7 to 12.75 GHz Frequency Band Between the Broadcasting-Satellite Service and the Fixed-Satellite Service—Section 10/11E—Sharing. 6 pp.

—**SHF—Digital Satellite Services—Radio Relay Systems**
CCIR Report 790-1-82. E.i.r.p. and Power Limits for Terrestrial Radio-Relay Transmitters Sharing with Digital Satellite Systems in Bands Between 11 to 14 GHz and Around 30 GHz—Section 4/9A—Sharing Conditions. 9 pp.

—**SHF/EHF—Fixed Satellite Services**
CCIR Report 552-4-90. Use of Frequency Bands Above 10 GHz in the Fixed-Satellite Service—Section 4B1—Systems Aspects. 34 pp.
CCIR QUESTION 47/4-90. Use of Frequency Bands Aboove 10 GHz in the Fixed-Satellite Service—Questions of Study Group 4 Fixed-Satellite Service. 1 p.

—**SHF—Fixed Services—Broadcasting Satellite Services**
CCIR QUESTION 111/9-90. Protection Criteria Between the Broadcasting-Satellite Service and the Fixed-Service—Questions of Study Group 9 Fixed Service. 1 p.

—**SHF—Flux Density—Fixed Satellite Services**
CCIR RECMN 674-90. Power Flux-Density Values to Facilitate the Application of Article 14 for FSS in Relation to the Fixed-Satellite Service in the 11.7-12.2 GHz Band in Region 2—Section 4/9A—Sharing Conditions. 1 p.

—**SHF—Flux Density—Fixed Services**
CCIR RECMN 674-90. Power Flux-Density Values to Facilitate the Application of Article 14 for FSS in Relation to the Fixed-Satellite Service in the 11.7-12.2 GHz Band in Region 2—Section 4/9A—Sharing Conditions. 1 p.

—**SHF—Passive Detectors—Fixed Satellite Services**
CCIR Report 850-1-86. Frequency Sharing by Passive Sensors with the Fixed, Mobile Except Aeronautical Mobile, and Fixed-Satellite Services in the Band 18.6-18.8 GHz Minimum Restrictions to Other Services in Order to Ensure Satisfactory Operations of Passive Sensors—Section 2F—Earth. 16 pp.

INTERNATIONAL AND NON-U.S. NATIONAL STANDARDS
SUBJECT INDEX

Frequency

Frequency Band Sharing (Cont.)

—SHF—Passive Detectors—Fixed Services
 CCIR Report 850-1-86. Frequency Sharing by Passive Sensors with the Fixed, Mobile Except Aeronautical Mobile, and Fixed-Satellite Services in the Band 18.6-18.8 GHz Minimum Restrictions to Other Services in Order to Ensure Satisfactory Operations of Passive Sensors—Section 2F—Earth. 16 pp.
 CCIR RECMN 761-92. Frequency Sharing Between the Fixed Service and Passive Sensors in the Band 18.6-18.8 GHz—Section 9F—Frequency Sharing with Other Services. 1 p.
 CCIR Report 942-1-86. Frequency Sharing Between the Fixed Service and Passive Sensors in the Band 18.6 to 18.8 GHz—Section 9F—Frequency Sharing with Other Services. 1 p.

—SHF—Passive Detectors—Radio Relay Systems
 CCIR RECMN 761-92. Frequency Sharing Between the Fixed Service and Passive Sensors in the Band 18.6-18.8 GHz—Section 9F—Frequency Sharing with Other Services. 1 p.
 CCIR Report 942-1-86. Frequency Sharing Between the Fixed Service and Passive Sensors in the Band 18.6 to 18.8 GHz—Section 9F—Frequency Sharing with Other Services. 1 p.

—SHF—Passive Detectors—Satellite Communications
 CCIR Report 850-1-86. Frequency Sharing by Passive Sensors with the Fixed, Mobile Except Aeronautical Mobile, and Fixed-Satellite Services in the Band 18.6-18.8 GHz Minimum Restrictions to Other Services in Order to Ensure Satisfactory Operations of Passive Sensors—Section 2F—Earth. 16 pp.

—SHF—Radio Navigation Services—Mobile Radio Services
 CCIR QUESTION 95/8-90. Sharing Between the Aeronautical Radionavigation Service and the Mobile Service in the Band 5000-5250 MHz—Questions Concerning Study Group 8 —Mobile, Radiodetermination, Amateur and Related Satellite Services. 1 p.

—SHF—Radio Relay Systems—Broadcasting Satellite Services
 CCIR RECMN 760-92. Protection of Terrestrial Line-of-Sight Radio-Relay Systems Against Interference from the Broadcasting-Satellite Service in the Band 22.5-23 GHz—Section 9F—Frequency Sharing with Other Services. 3 pp.
 CCIR Report 789-1-82. Protection of Terrestrial Line-of-Sight Radio-Relay Systems Against Interference from the Broadcasting-Satellite Service in the Band 11.7 to 12.75 GHz—Section 9F—Frequency Sharing with Other Services. 2 pp.
 CCIR Report 1189-90. Protection of Terrestrial Line-of-Sight Radio-Relay Systems Against Interference from the Broadcasting-Satellite Service in the Band 22.5—23 GHz—Section 9F—Frequency Sharing with Other Services. 4 pp.

—SHF—Space Research Services—Fixed Services
 CCIR Report 687-1-86. Feasibility of Frequency Sharing Between Space Research (Near-Earth) and Fixed and Mobile Services in the 7 to 8 GHz Spectral Region—Section 2E—Space Research. 5 pp.

—SHF—Space Research Services—Mobile Radio Services
 CCIR Report 687-1-86. Feasibility of Frequency Sharing Between Space Research (Near-Earth) and Fixed and Mobile Services in the 7 to 8 GHz Spectral Region—Section 2E—Space Research. 5 pp.

—SHF—Space Stations—Fixed Satellite Services
 CCIR Report 560-2-82. Sharing Criteria for the Protection of Space Stations in the Fixed-Satellite Service Receiving in the Band 14 to 14.4 GHz—Section 4E—Frequency Sharing Between Networks of the Fixed-Satellite Service and Those of Other Space Radiocommunications Systems. 3 pp.

—SHF—Space Stations—Radio Relay Systems
 CCIR Report 387-6-90. Protection of Terrestrial Line-of-Sight Radio-Relay Systems Against Interference Due to Emissions from Space Stations in the Fixed-Satellite Service in Shared Frequency Bands Between 1 and 23 GHz—Section 4/9A—Sharing Conditions. 19 pp.

—Sound Broadcasting—Interference—Tropical Regions
 CCIR Report 302-1-78. Interference to Sound Broadcasting in the Shared Bands in the Tropical Zone—Section 10A-2—Sound Broadcasting in the Tropical Zone. 11 pp.

Frequency Band Sharing (Cont.)

—Space Communications
 CCIR RESOLUTION 86-2-90. CCIR Recommendations to Be Drawn to the Attention of the Secretary-General in Accordance with the Provisions of Resolution No. 703 of the WARC-79—Volume XIV—Administrative Texts of the CCIR. 2 pp.

—Space Communications—Radar Astronomy
 CCIR Report 226-6-90. Factors Affecting the Possibility of Frequency Sharing Between Ground Based Radar Astronomy and Other Services—Section 2G—Radioastronomy and Radar Astronomy. 4 pp.

—Space Communications—Radio Astronomy
 CCIR Report 696-2-90. Feasibility of Frequency Sharing Between Radio Astronomy and Other Services—Section 2G—Radioastronomy and Radar Astronomy. 17 pp.

—Space Communications—Radio Communications—Terrestrial
 CCIR RESOLUTION 86-2-90. CCIR Recommendations to Be Drawn to the Attention of the Secretary-General in Accordance with the Provisions of Resolution No. 703 of the WARC-79—Volume XIV—Administrative Texts of the CCIR. 2 pp.

—Space Research Services
 CCIR RECMN 510-1-82. Feasibility of Frequency Sharing Between the Space Research Service and Other Services in Band 10 Potential Interference from Data Relay Satellite Systems—Section 2D—Data Relay Satellites. 1 p.
 CCIR RECMN 578-82. Protection Criteria and Sharing Considerations Relating to Deep-Space Research—Section 2E—Space Research. 1 p.

—Space Research Services—Radio Communications
 CCIR QUESTION 113/7-90. Frequency Sharing Between Space Research Links and Other Services—Questions Concerning Study Group 7—Science Services. 1 p.
 CCIR QUESTION 151/7-90. Feasibility of Frequency Sharing Between Space Research Satellites and Terrestrial Systems—Questions Concerning Study Group 7—Science Services. 2 pp.

—Space Research Services—Space Operation Services
 CCIR Report 678-78. Technical Feasibility of Frequency Sharing Between the Space Operation Service and the Space Research Service in the 1 to 10 GHz Band—Section 2C—Space Operations. 3 pp.

—Space Research Systems
 CCIR Report 847-1-86. Data Relay Satellites Sharing with Other Services in Bands 9 and 10—Section 2D—Data Relay Satellites. 7 pp.

—Space Research Systems—Near Earth—Deep Space
 CCIR QUESTION 119/7-90. Feasibility of Frequency Sharing Within and Among Space Research Systems—Questions Concerning Study Group 7—Science Services. 1 p.
 CCIR QUESTION 121/7-90. Frequency Sharing Between Deep-Space and Other Space Research Systems—Questions Concerning Study Group 7—Science Services. 1 p.

—Space Stations—Fixed Satellite Services—Radio Relay Transmitters
 CCIR QUESTION 60/4-90. Sharing Criteria for Protecting Receiving Space Stations in the Fixed-Satellite Service Against Interference from Line-of-Sight Radio-Relay Transmitters Operating in Shared Frequency Bands—Questions of Study Group 4 Fixed-Satellite Service. 1 p.

—Space Stations—Passive Microwave Detectors
 CCIR Report 694-3-90. Sharing Considerations and Protection Criteria Relating to Passive Microwave Sensors—Section 2F—Earth Exploration Satellites. 8 pp.

—Space Stations—Protection
 CCIR Report 685-3-90. Protection Criteria and Sharing Considerations Relating to Deep-Space Research—Section 2E—Space Research. 22 pp.

—Spread Spectrum Modulation
 CCIR Report 826-1-86. Examples of Band Sharing by Employing Spread-Spectrum Techniques—Section 1B—Spectrum Sharing and Planning Principles and Techniques. 18 pp.

Frequency Band Sharing (Cont.)

—Television Systems—Modulation—Interference—Fixed Satellite
 CCIR Report 449-1-74. Measured Interference into Frequency-Modulation Television Systems Using Frequencies Shared Within Systems in the Fixed-Satellite Service or Between These Systems and Terrestrial Systems—Section 4/9B—Co-Ordination and Inter-ference Calculations. 7 pp.

—Television Systems—Modulation—Interference—Radio Systems
 CCIR Report 449-1-74. Measured Interference into Frequency-Modulation Television Systems Using Frequencies Shared Within Systems in the Fixed-Satellite Service or Between These Systems and Terrestrial Systems—Section 4/9B—Co-Ordination and Inter-ference Calculations. 7 pp.

—Time Signal Services—Fixed Satellite Services
 CCIR Report 736-1-82. Frequency Sharing Between the Time-Signal Service and the Radiolocation Service, the Fixed-Satellite Service and the Fixed and Mobile Services Near 14, 21, 26 and 31 GHz—Section 7C—Systems for Dissemination and Comparison. 8 pp.

—Time Signal Services—Fixed Services (Radio Communications)
 CCIR Report 736-1-82. Frequency Sharing Between the Time-Signal Service and the Radiolocation Service, the Fixed-Satellite Service and the Fixed and Mobile Services Near 14, 21, 26 and 31 GHz—Section 7C—Systems for Dissemination and Comparison. 8 pp.

—Time Signal Services—Mobile Radio Equipment
 CCIR Report 736-1-82. Frequency Sharing Between the Time-Signal Service and the Radiolocation Service, the Fixed-Satellite Service and the Fixed and Mobile Services Near 14, 21, 26 and 31 GHz—Section 7C—Systems for Dissemination and Comparison. 8 pp.

—Time Signal Services—Radiolocation Services
 CCIR Report 736-1-82. Frequency Sharing Between the Time-Signal Service and the Radiolocation Service, the Fixed-Satellite Service and the Fixed and Mobile Services Near 14, 21, 26 and 31 GHz—Section 7C—Systems for Dissemination and Comparison. 8 pp.

—UHF—Broadcasting Satellite Services—Broadcasting
 CCIR QUESTION 87/11-90. Broadcasting-Satellite Service (Television) Criteria to Be Applied for Frequency Sharing Between the Broadcasting-Satellite Service and the Terrestrial Broadcasting Service in the Frequency Range 620 to 790 MHz—Questions Concerning Study Group 11-. 2 pp.

—UHF—Broadcasting Satellite Services—Fixed Satellite
 CCIR QUESTION 88/11-90. Broadcasting-Satellite Service (Television) Criteria to Be Applied for Frequency Sharing Between the Broadcasting-Satellite Service and the Terrestrial and Space Services in the Frequency Range 2500 MHz to 2690 MHz—Questions Concerning. 1 p.

—UHF—Fixed Services—Broadcasting Satellite Services
 CCIR QUESTION 112/9-90. Frequency Sharing Between the Broadcasting-Satellite Service (Sound) and the Fixed Service in the Band 0.5 to 3 GHz—Questions of Study Group 9 Fixed Service. 1 p.

—UHF—Fixed Services—Mobile Radio Services
 CCIR QUESTION 133/9-90. Sharing Criteria Between the Fixed and Mobile Services in the Frequency Bands Between About 0.5 and 3 GHz—Questions of Study Group 9 Fixed Service. 1 p.

—UHF—Geostationary Satellites—Meteorological Services
 CCIR Report 541-3-90. Feasibility of Frequency Sharing Between a Geostationary Meteorological Satellite System and the Meteorological Aids Service in the Region of 400 MHz and in the Upper Part of Band 9 (1 to 3 GHz)—Section 2F—Earth Exploration Satellites. 22 pp.

INDUSTRY STANDARDS

INTERNATIONAL AND NON-U.S. NATIONAL STANDARDS
SUBJECT INDEX
Frequency

Frequency Band Sharing (Cont.)

—UHF—Global Positioning Systems—Radio Navigation Services
CCIR Report 766-2-90. Feasibility of Frequency Sharing Between the GPS and Other Services—Section 8H—Efficient Use of the Radio Spectrum Characteristics and Sharing of Frequency Resources. 13 pp.

—UHF—Global Positioning Systems—Radiolocation Services
CCIR Report 766-2-90. Feasibility of Frequency Sharing Between the GPS and Other Services—Section 8H—Efficient Use of the Radio Spectrum Characteristics and Sharing of Frequency Resources. 13 pp.

—UHF—Land Mobile Services—Broadcasting Satellite Services
CCIR Report 770-3-90. Technical and Operating Considerations for a Land Mobile-Satellite Service Operating in Band 9—Section 8F—Frequencies, Orbits and Systems. 27 pp.

—UHF—Maritime Mobile Services—Radio Navigation Services
CCIR Report 910-1-86. Sharing Between the Maritime Mobile Service and the Aeronautical Radionavigation Service in the Band 415-526.5 kHz—Section 8C—Maritime Mobile Service; Telephony and Related Subjects. 12 pp.

—UHF—Mobile Radio Services—Broadcasting
CCIR Report 1098-90. Sharing Between the Land Mobile Service and the Broadcasting Service in the VHF and UHF Bands Where the Service Areas Are Geographically Seperated—Section 1B—Spectrum Sharing and Planning Principles and Techniques. 7 pp.

—UHF—Mobile Satellite Communications—Fixed Services
CCIR Report 1173-90. Technical and Operational Considerations for Aeronautical Mobile-Satellite Communications—Section 8F—Frequencies, Orbits and Systems. 40 pp.

—UHF—Mobile Satellite Communications—Radio Astronomy
CCIR Report 1182-90. Frequency Sharing in the 1 660–1 660.5 MHz Band Between the Mobile-Satellite Service and the Radio Astronomy Service—Section 8H—Efficient Use of the Radio Spectrum Characteristics and Sharing of Frequency Resources. 7 pp.
CCIR RECMN 829-92. Frequency Sharing in the 1 660-1 660.5 MHz Band Between the Mobile-Satellite Service and the Radioastronomy Service. 2 pp.

—UHF—Radio Relay—Sound Broadcasting—Satellite
CCIR Report 941-82. Protection of Terrestrial Line-of-Sight Radio-Relay Systems Against Interference from the Broadcasting-Satellite Service (Sound) in the Band 1427 to 1530 MHz—Section 9F—Frequency Sharing with Other Services. 5 pp.

—UHF—Radio Relay Systems—Fixed
CCIR RECMN 759-92. Use of Frequencies in the Band 500 to 3000 MHz for Radio-Relay Systems—Section 9F—Frequency Sharing with Other Services. 4 pp.

—UHF—Radio Relay Systems—Space Research Systems
CCIR Report 1197-90. Sharing Considerations near 2 GHz Between the Fixed Service and near-Earth Space Research Systems—Section 9F—Frequency Sharing with Other Services. 13 pp.

—UHF—Radio Relay Systems—Television Broadcasting
CCIR Report 1194-90. Sharing Between the Fixed Service and the Broadcasting Service (Television) in the Band 790 to 862 MHz—Section 9F—Frequency Sharing with Other Services. 7 pp.

—UHF—Satellite Links
CCIR Report 1179-90. Methodology for the Derivation of Interference and Sharing Criteria for the Mobile-Satellite Services—Section 8H—Efficient Use of the Radio Spectrum Characteristics and Sharing of Frequency Resources. 13 pp.

—UHF/SHF/EHF—Earth Stations—Space Communications—Coordinates
CCIR Report 382-6-90. Determination of Co-Ordination Area—Section 4/9B—Co-Ordination and Interference Calculations. 63 pp.

Frequency Band Sharing (Cont.)

—UHF/SHF/EHF—Fixed Satellite Services—Radio Relay Systems
CCIR RECMN 358-3-82. Maximum Permissible Values of Power Flux-Density at the Surface of the Earth Produced by Satellites in the Fixed-Satellite Service Using the Same Frequency Bands Above 1 GHz as Line-of-Sight Radio-Relay Systems—Section 4/9A—Sharing Conditions. 2 pp.

—UHF/SHF—Mobile Satellite Services—Fixed Services
CCIR QUESTION 118/9-90. Protection Criteria Between the Mobile Satellite Services and the Fixed Service in the Band 1 to 3 GHz—Questions of Study Group 9 Fixed Service. 1 p.

—UHF/SHF—Space Research Systems—Data Relay Satellites
CCIR Report 846-82. Data Relay Satellites Sharing with Other Space Research Systems Near 2 GHz—Section 2D—Data Relay Satellites. 5 pp.

—UHF/SHF—Space Stations—Radio Relay Systems
CCIR Report 387-6-90. Protection of Terrestrial Line-of-Sight Radio-Relay Systems Against Interference Due to Emissions from Space Stations in the Fixed-Satellite Service in Shared Frequency Bands Between 1 and 23 GHz—Section 4/9A—Sharing Conditions. 19 pp.

—UHF—Television Systems—Radio Navigation Transmitters
CCIR RECMN 565-78. Protection Ratios for 625-Line Television Against Radionavigation Transmitters Operating in the Shared Bands Between 582 and 606 MHz—Section 11E—Planning of Television Networks, Protection Ratios Television Receivers and Antennas. 2 pp.

—VHF—Ionospheric Propagation
CCIR RECMN 844-92. Ionospheric Factors Affecting Frequency Sharing in the VHF (30-300 MHz) Band. 6 pp.

—VHF—Mobile Radio Services—Broadcasting
CCIR Report 1098-90. Sharing Between the Land Mobile Service and the Broadcasting Service in the VHF and UHF Bands Where the Service Areas Are Geographically Seperated—Section 1B—Spectrum Sharing and Planning Principles and Techniques. 7 pp.
CCIR RECMN 851-92. Sharing Between the Broadcasting Service and the Fixed and/or Mobile Services in the VHF and UHF Bands. 26 pp.

—VHF/UHF—Fixed Services—Broadcasting
CCIR QUESTION 2/12-90. Sharing Between the Broadcasting Service and the Fixed and/or Mobile Services in VHF and UHF Bands—Questions Concerning Study Group 12—Inter-Service Sharing and Compatibility. 2 pp.
CCIR RECMN 851-92. Sharing Between the Broadcasting Service and the Fixed and/or Mobile Services in the VHF and UHF Bands. 26 pp.

—VHF/UHF—Land Mobile Services—Television Broadcasting
CCIR Report 1023-1-90. Frequency Sharing Between the Land Mobile Service and the Broadcasting Service (Television) Below 1 GHz—Section 8A—Land Mobile Service and Related Subjects. 6 pp.

—VHF/UHF—Mobile Radio Services—Broadcasting
CCIR QUESTION 2/12-90. Sharing Between the Broadcasting Service and the Fixed and/or Mobile Services in VHF and UHF Bands—Questions Concerning Study Group 12—Inter-Service Sharing and Compatibility. 2 pp.

Frequency Bands

See Also: Basebands; Frequencies; Frequency Assignment; Frequency Band Sharing; Frequency Management; Frequency Reuse; Frequency Separation; Intermediate Frequencies; Out of Band Emissions; Pilot Channels; Sidebands; Stability Loss (Transmission)

CCIR Report 1001-1-90. Off-Axis e.i.r.p. Density Limits for Fixed-Satellite Service Earth Stations—Section 4D1—Permissible Levels of Interference. 11 pp.
CCIR Report 1000-1-90. Spectrum Utilization Methodologies—Section 4D2—Coordination Methods. 8 pp.
CEPT T/R 72-01-81. Assignation De Frequences Dans Les Bandes De Frequences Comprises Entre 29,7 Et 960 MHz. 1 p.

Frequency Bands (Cont.)

CEPT T/R 72-01 E-88. Allocation of Frequencies in the Frequency Bands Between 29.7 and 960 MHz. 2 pp.

—

CCIR QUESTION 107/7-90. Standard-Frequency and Time-Signal Emissions in Additional Frequency Bands—Questions Concerning Study Group 7—Science Services. 1 p.

—

CCIR QUESTION 107/7-90. Standard-Frequency and Time-Signal Emissions in Additional Frequency Bands—Questions Concerning Study Group 7—Science Services. 1 p.

—30 MHz+—Standard Frequency Emissions
CCIR QUESTION 107/7-90. Standard-Frequency and Time-Signal Emissions in Additional Frequency Bands—Questions Concerning Study Group 7—Science Services. 1 p.

—30 MHz+—Standard Time Signal Emissions
CCIR QUESTION 107/7-90. Standard-Frequency and Time-Signal Emissions in Additional Frequency Bands—Questions Concerning Study Group 7—Science Services. 1 p.

—Active Detectors—Space Communications
CCIR RECMN 577-2-90. Preferred Frequency Bands for Active Sensing Measurements—Section 2F—Earth Exploration Satellites. 1 p.

—Aeronautical Passenger Communications
CEPT T/R 42-01 E-91. Designation of Frequency Bands for the Pan-European Terrestrial Flight Telephone System (TFTS). 2 pp.
EC COM(92) 314-92. Proposal for a Council Directive on Common Frequency Bands to Be Designated for the Coordinated Introduction of the Terrestrial Flight Telecommunications System (TFTS) in the Community. 2 pp.

—Amateur Radio Services
CCIR Report 1154-90. Techniques and Frequency Usage in the Amateur and Amateur-Satellite Services—Section 8L—Amateur Service; Amateur-Satellite Service. 24 pp.
CCIR QUESTION 48-2/8-90. Techniques and Frequency Usage in the Amateur Service and Amateur-Satellite Services—Questions Concerning Study Group 8—Mobile, Radiodetermination, Amateur and Related Satellite Services. 1 p.

—Amateur Radio Services—Satellite
CCIR QUESTION 48-2/8-90. Techniques and Frequency Usage in the Amateur Service and Amateur-Satellite Services—Questions Concerning Study Group 8—Mobile, Radiodetermination, Amateur and Related Satellite Services. 1 p.

—Amateur Radio Services—Satellite Communications
CCIR Report 1154-90. Techniques and Frequency Usage in the Amateur and Amateur-Satellite Services—Section 8L—Amateur Service; Amateur-Satellite Service. 24 pp.

—Assignment
CCIR Report 976-86. Systematic Planning Method for the Use of Unassigned Frequency Bands—Section 1B—Spectrum Sharing and Planning Principles and Techniques. 3 pp.

—Band 10—Television Broadcasting—Frequency Planning
CCIR QUESTION 43-1/11-90. Technical Bases Required for Planning the Broadcasting Service (Television) in Bands 8, 9 and 10—Questions Concerning Study Group 11—Broadcasting Service (Television). 1 p.

—Band 10—Television Signals
CCIR QUESTION 49/11-90. Characteristics of Television Signals Radiated in Band 10 (SHF) from Terrestrial Broadcasting Transmitters—Questions Concerning Study Group 11—Broadcasting Service (Television). 1 p.

—Band 5—Standard Frequency Emissions
CCIR Report 735-1-82. Importance of Standard-Frequency and Time-Signal Emissions in Band 5—Section 7C—Systems for Dissemination and Comparison. 1 p.

—Band 5—Time Signal Emissions
CCIR Report 735-1-82. Importance of Standard-Frequency and Time-Signal Emissions in Band 5—Section 7C—Systems for Dissemination and Comparison. 1 p.

Frequency Bands (Cont.)

—Band 6—Sound Broadcasting—AM
CCIR RECMN 598-1-90. Factors Influencing the Limits of Amplitude-Modulation Sound-Broadcasting Coverage in Band 6 (MF)—Section 10A-1—Amplitude-Modulation Sound Broadcasting in Bands 5 (LF), 6 (MF) and 7 (HF). 25 pp.

—Band 6—Television Broadcasting
CCIR Report 961-1-86. Terrestrial Television Broadcasting in the 12 GHz Band (Band VI)—Section 11E—Planning of Television Networks, Protection Ratios, Television Receivers and Antennas. 4 pp.

—Band 6—Time Signal Emissions—Frequency Emissions—Interference
CCIR Report 732-3-90. Proposed Reduction of Mutual Interference Between Standard-Frequency and Time-Signal Emissions in Bands 6 and 7—Section 7C—Systems for Dissemination and Comparison. 5 pp.

—Band 7—Broadcasting—Tropical Regions
CCIR QUESTION 65/10-90. Short-Distance Broadcasting in Band 7 (HF) in the Tropical Zone—Questions Concerning Study Group 10—Broadcasting Service (Sound). 1 p.

—Band 7—Sound Broadcasting—Channel Spacing
CCIR RECMN 597-1-86. Channel Spacing for Sound Broadcasting in Band 7 (HF)—Section 10A-1—Amplitude-Modulation Sound Broadcasting in Bands 5 (LF), 6 (MF) and 7 (HF). 1 p.

—Band 7—Sound Broadcasting—Signal to Noise Ratio
CCIR Report 1058-86. Minimum AF and RF Signal-to-Noise Ratio Required for Broadcasting in Band 7 (HF)—Section 10A-1—Amplitude-Modulation Sound Braodcasting in Bands 5 (LF), 6 (MF) and 7 (HF). 5 pp.

—Band 7—Time Signal Emissions—Frequency Emissions—Interference
CCIR Report 732-3-90. Proposed Reduction of Mutual Interference Between Standard-Frequency and Time-Signal Emissions in Bands 6 and 7—Section 7C—Systems for Dissemination and Comparison. 5 pp.

—Band 8—Radio Reception—Automotive
CCIR Report 1067-86. Improvement of the Reception Quality in Automobiles for Frequency Modulation Sound Broadcasts in Band 8 (VHF)—Section 10B—Frequency-Modulation Sound Broadcasting in Bands 8 (VHF) and 9 (UHF). 7 pp.

—Band 8—Radio Relay Systems—Subscriber Lines—Rural Areas
CCIR Report 380-3-90. Radio Systems Operating in Bands 8 and 9 for the Provision of Subscriber Telephone Connections in Rural Areas—Section 9E1—Line-of-Sight Radio-Relay Systems. 14 pp.
CCIR QUESTION 105/9-90. Radio Systems Operating in Bands 8 and 9 for the Provision of Subscriber Telephone Connections in Rural Areas—Questions of Study Group 9 Fixed Service. 1 p.

—Band 8—Radio Relay Systems—Trunk Circuits—Rural Areas
CCIR RECMN 754-92. Radio-Relay Systems in Bands 8 and 9 for the Provision of Telephone Trunk Connections in Rural Areas—Section 9E1—Line-of-Sight Radio-Relay Systems. 6 pp.
CCIR Report 379-5-86. Characteristics of Simple Radio-Relay Equipment Operating in Bands 8 and 9 for the Provision of Telephone Trunk Connections in Rural Areas—Section 9E1—Line-of-Sight Radio-Relay Systems. 12 pp.
CCIR QUESTION 104/9-90. Radio-Relay Systems Operating in Bands 8 and 9 for the Provision of Telephone Trunk Connections in Rural Areas—Questions of Group 9 Fixed Service. 1 p.

—Band 8—Sound Broadcasting
CCIR RECMN 412-5-90. Planning Standards for FM Sound Broadcasting at VHF—Section 10B—Frequency-Modulation Sound Broadcasting in Bands 8 (VHF) and 9 (UHF). 6 pp.
CCIR QUESTION 46-1/10-90. Frequency-Modulation Sound Broadcasting in Band 8 (VHF)—Questions Concerning Study Group 10—Broadcasting Service (Sound). 1 p.

—Band 8—Sound Broadcasting—Field Strength
CCIR QUESTION 101/10-90. Minimum Usable Field Strength for Sound Broadcasting in Band 8 (VHF)—Questions Concerning Study Group 10—Broadcasting Service (Sound). 1 p.

—Band 8—Sound Broadcasting—Frequency Planning
CCIR Report 946-1-90. Frequency-Planning Constraints on FM Sound Broadcasting in Band 8 (VHF)—Section 10B—Frequency-Modulation Sound Broadcasting in Bands 8 (VHF) and 9 (UHF). 4 pp.
CCIR QUESTION 74/10-90. Technical Bases for Planning of Frequency-Modulation Sound Broadcasting in Band 8 (VHF)—Questions Concerning Study Group 10—Broadcasting Service (Sound). 1 p.

—Band 8—Sound Broadcasting—Polarization
CCIR Report 464-5-90. Polarization of Emissions in Frequency-Modulation Broadcasting in Band 8 (VHF)—Section 10B—Frequency-Modulation Sound Broadcasting in Bands 8 (VHF) and 9 (UHF). 6 pp.
CCIR QUESTION 69/10-90. Polarization of Emissions in FM Sound Broadcasting—Questions Concerning Study Group 10—Broadcasting Service (Sound). 1 p.

—Band 8—Sound Broadcasting—Receiving Antennas
CCIR RECMN 599-82. Directivity of Antennas for the Reception of Sound Broadcasting in Band 8 (VHF)—Section 10B—Frequency-Modulation Sound Broadcasting in Bands 8 (VHF) and 9 (UHF). 1 p.

—Band 8—Sound Broadcasting—Transmission
CCIR RECMN 450-1-82. Transmission Standards for FM Sound Broadcasting at VHF—Section 10B—Frequency-Modulation Sound Broadcasting in Bands 8 (VHF) and 9 (UHF). 3 pp.

—Band 8—Sound Broadcasting—Tropical Regions
CCIR QUESTION 73/10-90. Sound Broadcasting in Band 8 (VHF) in the Tropical Zone—Questions Concerning Study Group 10—Broadcasting Service (Sound). 1 p.

—Band 8—Television Broadcasting—Frequency Planning
CCIR QUESTION 43-1/11-90. Technical Bases Required for Planning the Broadcasting Service (Television) in Bands 8, 9 and 10—Questions Concerning Study Group 11—Broadcasting Service (Television). 1 p.

—Band 8—Television Broadcasting—Polarization Discrimination
CCIR Report 122-4-90. Advantages to Be Gained by Using Orthogonal Wave Polarizations in the Planning of Television Broadcasting Services in Bands 8 (VHF) and 9 (UHF)—Section 11E—Planning of Television Networks, Protection Ratios, Television Receivers and Antennas. 2 pp.

—Band 9—Land Mobile Satellite Communications
CCIR Report 770-3-90. Technical and Operating Considerations for a Land Mobile-Satellite Service Operating in Band 9—Section 8F—Frequencies, Orbits and Systems. 27 pp.

—Band 9—Radio Relay Systems—Subscriber Lines—Rural Areas
CCIR Report 380-3-90. Radio Systems Operating in Bands 8 and 9 for the Provision of Subscriber Telephone Connections in Rural Areas—Section 9E1—Line-of-Sight Radio-Relay Systems. 14 pp.
CCIR QUESTION 105/9-90. Radio Systems Operating in Bands 8 and 9 for the Provision of Subscriber Telephone Connections in Rural Areas—Questions of Study Group 9 Fixed Service. 1 p.

—Band 9—Radio Relay Systems—Trunk Circuits—Rural Areas
CCIR RECMN 754-92. Radio-Relay Systems in Bands 8 and 9 for the Provision of Telephone Trunk Connections in Rural Areas—Section 9E1—Line-of-Sight Radio-Relay Systems. 6 pp.
CCIR Report 379-5-86. Characteristics of Simple Radio-Relay Equipment Operating in Bands 8 and 9 for the Provision of Telephone Trunk Connections in Rural Areas—Section 9E1—Line-of-Sight Radio-Relay Systems. 12 pp.
CCIR QUESTION 104/9-90. Radio-Relay Systems Operating in Bands 8 and 9 for the Provision of Telephone Trunk Connections in Rural Areas—Questions of Group 9 Fixed Service. 1 p.

—Band 9—Television Broadcasting—Frequency Planning
CCIR QUESTION 43-1/11-90. Technical Bases Required for Planning the Broadcasting Service (Television) in Bands 8, 9 and 10—Questions Concerning Study Group 11—Broadcasting Service (Television). 1 p.

—Band 9—Television Broadcasting—Polarization Discrimination
CCIR Report 122-4-90. Advantages to Be Gained by Using Orthogonal Wave Polarizations in the Planning of Television Broadcasting Services in Bands 8 (VHF) and 9 (UHF)—Section 11E—Planning of Television Networks, Protection Ratios, Television Receivers and Antennas. 2 pp.

—Busy Tones
CCITT RECMN E.180-89. Technical Characteristics of Tones for the Telephone Service—Telephone Network and ISDN—Operation, Numbering, Routing and Mobile Service (Study Group II) 9 pp. 9 pp.

—Call Waiting Tones
CCITT RECMN E.180-89. Technical Characteristics of Tones for the Telephone Service—Telephone Network and ISDN—Operation, Numbering, Routing and Mobile Service (Study Group II) 9 pp. 9 pp.

—Caller Waiting Tones
CCITT RECMN E.180-89. Technical Characteristics of Tones for the Telephone Service—Telephone Network and ISDN—Operation, Numbering, Routing and Mobile Service (Study Group II) 9 pp. 9 pp.

—Cellular Mobile Radio Communications
EC 87/372/EEC-87. Council Directive on the Frequency Bands to be Reserved for the Coordinated Introduction of Public Pan-European Cellular Digital Land-Based Mobile Communications in the Community. 2 pp.

—Congestion Tones
CCITT RECMN E.180-89. Technical Characteristics of Tones for the Telephone Service—Telephone Network and ISDN—Operation, Numbering, Routing and Mobile Service (Study Group II) 9 pp. 9 pp.

—Data Relay Satellites
CCIR Report 848-2-90. Characteristics of Data Relay Satellite Systems—Section 2D—Data Relay Satellites. 15 pp.

—Dial Tones
CCITT RECMN E.180-89. Technical Characteristics of Tones for the Telephone Service—Telephone Network and ISDN—Operation, Numbering, Routing and Mobile Service (Study Group II) 9 pp. 9 pp.

—Digital Radio Systems
CEPT T/R 22-05 E-93. Frequencies for Mobile Digital Trunked Radio Systems. 1 p.

—Downlinks—Geostationary Satellites
CCIR Report 557-2-86. Use of Frequency Bands Allocated to the Fixed-Satellite Service for Both the up Link and down Link Geostationary-Satellite Systems—Section 4D2—Coordination Methods. 9 pp.

—Earth Exploration Satellites
CCIR Report 692-2-86. Preferred Frequency Bands and Power Flux-Density Considerations for Earth Exploration Satellites—Section 2F—Earth Exploration Satellites. 6 pp.

—Earth Exploration Satellites—Active Detectors
CCIR RECMN 516-78. Frequency Bands for Active Sensors Used on Earth Exploration and Meteorological Satellites—Section 2F—Earth Exploration Satellites. 1 p.

—Earth Stations—Space Stations
CCIR Report 849-2-90. Frequency Bands in the 40 to 120 GHz Range That Are Preferred for Deep-Space Research—Section 2E—Space Research. 9 pp.

—EHF—Oscillators
CCIR Report 1152-90. General Sources of Highly Stable Signals in the UHF-EHF Bands Using Synchronized Oscillators—Section 7C—Systems for Dissemination and Comparison. 3 pp.

—European Digital Cordless Telecommunications
CEPT T/R 22-02 E-89. Frequency Band to Be Designated for the European Digital-Cordless Telecommunication System (DECT). 1 p.

Frequency

Frequency Bands (Cont.)

—European Digital Cordless Telecommunications (Cont.)

EC 91/287/EEC-91. Council Directive on the Frequency Band to be Designated for the Coordinated Introduction of Digital European Cordless Telecommunications (DECT) into the Community. 2 pp.

—Fixed Satellite Services—Geostationary Satellites—Station Keeping

CCIR RECMN 484-2-82. Station-Keeping in Longitude of Geostationary Satellites Using Frequency Bands Allocated to the Fixed-Satellite Service—Section 4D3—Spacecraft Station-Keeping—Satellite Antenna Radiation Pattern—Pointing Accuracy. 1 p.

CCIR RECMN 484-3-92. Station-Keeping in Longitude of Geostationary Satellites in the Fixed Satellite Service—Section 4D3—Spacecraft Station-Keeping—Satellite Antenna Radiation Pattern—Pointing Accuracy. 2 pp.

—Foreign Use

CEPT T/R 75-03-85. Utilisation De Frequences a L'Etranger Pour Des Applications Diverses. 5 pp.

CEPT T/R 75-03 E-85. Utilization of Frequencies Abroad for Various Applications. 6 pp.

—Glossaries

CCIR RECMN 573-3-90. Radiocommunication Vocabulary—Section A—Terminology. 60 pp.

CCIR RECMN 431-5-86. Nomenclature of the Frequency and Wavelength Bands Used in Telecommunications—Section C—Other Means of Expression. 3 pp.

—HF—Maritime Mobile Services

CCIR QUESTION 64-2/8-90. HF Bands Allocated on an Exclusive or Shared Basis to the Maritime Mobile Service—Questions Concerning Study Group 8—Mobile, Radiodetermination, Amateur and Related Satellite Services. 1 p.

—Land Mobile Services

CCIR RECMN 687-90. Future Public Land Mobile Telecommunication Systems (FPLMTS)—Section 8A—Land Mobile Service and Related Subjects. 11 pp.

CCIR RECMN 687-1-92. Future Public Land Mobile Telecommunication Systems (FPLMTS)—Section 8A—Land Mobile Service and Related Subjects. 21 pp.

—LF—Standard Frequencies

CCIR Report 271-8-90. Stability and Accuracy of Standard Frequency and Time Signals in VLF and LF Bands as Received—Section 7C—Systems for Dissemination and Comparison. 9 pp.

—LF—Standard Frequencies—Stability

CCIR Report 271-8-90. Stability and Accuracy of Standard Frequency and Time Signals in VLF and LF Bands as Received—Section 7C—Systems for Dissemination and Comparison. 9 pp.

—LF—Time Signals

CCIR Report 271-8-90. Stability and Accuracy of Standard Frequency and Time Signals in VLF and LF Bands as Received—Section 7C—Systems for Dissemination and Comparison. 9 pp.

—Low Power Devices

CEPT T/R 01-04 E-93. Use of Low Power Devices (LPD) Using Integral Antennas and Operating in Harmonized Frequency Bands. 2 pp.

CEPT T/R 01-04 E-92. Use of Low Power Devices (LPD) Using Integral Antennas and Operating in Harmonized Frequency. 2 pp.

—Maritime Mobile Services

CCIR Report 1029-86. Future Use of the Band 2170-2194 kHz—Section 8B—Maritime Mobile Service: Telegraphy and Related Subjects. 7 pp.

CCIR QUESTION 38-2/8-90. Use of Frequencies in the Bands Between About 1606 and 4000 khz Allocated to the Maritime Mobile Service—Questions Concerning Study Group 8—Mobile, Radiodetermination, Amateur and Related Satellite Services. 1 p.

—Meteorological Satellites—Active Detectors

CCIR RECMN 516-78. Frequency Bands for Active Sensors Used on Earth Exploration and Meteorological Satellites—Section 2F—Earth Exploration Satellites. 1 p.

—Mobile Radio Equipment

CEPT T/R 02-01-86. Planification Et Co-Ordination Des Services Mobiles Terrestres (Et Autres) Dans La Bande 47-68 MHz. 4 pp.

Frequency Bands (Cont.)

—Mobile Radio Equipment (Cont.)

CEPT T/R 02-01 E-86. Planning and Co-Ordination of Land Mobile and Other Services Operating in the Band 47-68 HMz. 5 pp.

CEPT T/R 25-01-74. Coordination Des Frequences Dans Le Service Mobile Terrestre Dans Les Bandes De 80, 160 Et 460 MHz. 8 pp.

CEPT T/R 25-01 E-88. Co-Ordination of Frequencies in the Land Mobile Service in the 80, 160 and 460 MHz Bands. 9 pp.

CEPT T/R 25-02-72. Donnees propres a Faciliter La Coordination Des Frequences Pour Le Service Mobile Terrestre Dans Les Bandes De 80, 160 Et 460 MHz. 1 p.

CEPT T/R 25-02 E-72. Information to Facilitate the Coordination Frequencies for the Land Mobile Service in the 80, 160 and 460 MHz Bands. 2 pp.

CEPT T/R 25-03-76. Coordination Des Frequences Du Service Mobile Terrestre Dans Les Bandes De 80, De 160 Et De 460 MHz, Et Aux Methodes a Utiliser Pour L'Evaluation Des Brouillages. 7 pp.

CEPT T/R 25-03 E-88. Co-Ordination of Frequencies for the Land Mobile Service in the 80, 160 and 460 MHz Bands and the Methods to Be Used for Assessing Interference. 7 pp.

CEPT T/R 25-03 E-76. Co-Ordination of Frequencies for the Land Mobile Service in the 80, 160 and 460 MHz Bands and the Methods to Be Used for Assessing Interference. 8 pp.

CEPT T/R 25-04-86. Coordination Dans Les Regions Frontalieres Des Frequences Pour Le Service Mobile Terrestre Dans Les Bandes Comprises Entre 862 Et 960 MHz. 2 pp.

CEPT T/R 25-04 E-88. Coordination in Frontier Regions of Frequencies for the Land Mobile Service in the Bands Between 862 and 960 MHz. 3 pp.

CEPT T/R 25-05-85. Plantification Et Coordination Des Services Mobiles Terrestres Dans La Bande 174-230 MHz (Bande III De La Television). 3 pp.

CEPT T/R 25-05 E-85. Planning and Coordination of the Land Mobile Services Operating in the Bands 174-230 MHz (Television Band III). 4 pp.

CEPT T/R 25-06-86. Parametres De Planification a Appliquer Pour L'Utilisation Efficace Des Bandes De Frequences Partagees Entre Le Service De Radiodiffusion (Television, Bandes I Et III Seulement) Et Le Service Mobile Terrestre. 5 pp.

CEPT T/R 25-06 E-86. Planning Parameters to Assist the Efficient and Effective Utilisation of Shared Frequency Bands Which Are Allocated to the Broadcasting Service (Television) and the Land Mobile Service, Using Assignments Which Overlap a Television Channel (Television Bands I and III Only). 6 pp.

CEPT T/R 25-08 E-89. Coordination of Frequencies in the Land Mobile Service in the Range 29.7-960 MHz. 27 pp.

CEPT T/R 32-01-70. Frequences a Utiliser Par Les Navires Etrangers Pour La Liaison Avec Les Stations Cotieres Europeennes Dans La Bande 1605 4000 kHz. 13 pp.

CEPT T/R 32-01 E-70. Frequencies to Be Used By Foreign Ships for Contact with European Coast Stations in the 1605-4000 kHz Band. 11 pp.

CEPT T/R 32-02 E-76. Frequencies to Be Used By on-Board Communications Stations. 2 pp.

CEPT T/R 75-02-82. Utilisation De Frequences De La Bande 862-960 MHz Par Les Services Mobiles Terrestre Et Maritime. 1 p.

CEPT T/R 75-02 E-91. Use of Frequencies in the Band 862-960 MHz by the Mobile Except Aeronautical Mobile Service. 3 pp.

—Mobile Radio Equipment—Maritime

CEPT T/R 75-02-82. Utilisation De Frequences De La Bande 862-960 MHz Par Les Services Mobiles Terrestre Et Maritime. 1 p.

CEPT T/R 75-02 E-91. Use of Frequencies in the Band 862-960 MHz by the Mobile Except Aeronautical Mobile Service. 3 pp.

—Mobile Radio Services—Railroads

CEPT T/R 22-01-75. Frequences Susceptibles D'etre Assignees Aux Chemins De Fer Internationaux. 1 p.

CEPT T/R 22-01 E-75. Frequencies Likely to Be Allocated to International Railways. 2 pp.

—Mobile Satellite Communications

CCIR Report 1180-90. Design of Mobile Satellite Systems Providing Aeronautical, Land and Maritime Services Using Shared Resources—Section 8F—Frequencies, Orbits and Systems. 38 pp.

—Mobile Stations—Sound Broadcasting—Tropical Regions

CCIR RECMN 48-2-86. Choice of Frequency for Sound Broadcasting in the Tropical Zone—Section 10A-2—Sound Broadcasting in the Tropical Zone. 1 p.

Frequency Bands (Cont.)

—Pay Station Identification Tones

CCITT RECMN E.180-89. Technical Characteristics of Tones for the Telephone Service—Telephone Network and ISDN—Operation, Numbering, Routing and Mobile Service (Study Group II) 9 pp. 9 pp.

—Radar Equipment—Radio Navigation

CCIR Report 1039-86. Present and Expected Use of the Band 9320-9500 MHz by Mobile Radars of the Radionavigation Service—Section 8D—Radiodetermination, Global Maritime Distress and Safety System and Related Subjects. 17 pp.

—Radio Astronomy

CCIR RECMN 314-8-92. Preferred Frequency Bands for Radioastronomical Measurements—Section 2G—Radioastronomy and Radar Astronomy. 3 pp.

CCIR Report 852-2-90. Characteristics of the Radio Astronomy Service and Preferred Frequency Bands—Section 2G—Radioastronomy and Radar Astronomy. 13 pp.

—Radio Beacons—Spacecraft

CCIR QUESTION 116/7-90. Preferred Frequency Bands for Spacecraft Transmitters Used as Beacons—Questions Concerning Study Group 7—Science Services. 1 p.

—Radio Determination Satellite Services

CCIR QUESTION 91-1/8-90. Technical and Operating Characteristics of the Radiodetermination-Satellite Service—Questions Concerning Study Group 8—Mobile, Radiodetermination, Amateur and Related Satellite Services. 1 p.

—Radio Local Area Networks

CEPT T/R 22-06 E-93. Harmonized Radio Frequency Bands for High Performance European Radio Local Area Networks (HIPERLANs) in the 5 GHz and 17 GHz Frequency Range. 2 pp.

—Radio Navigation Services

CCIR RECMN 629-86. Use for the Radionavigation Service of the Frequency Bands 2900-3100 MHz, 5470-5650 MHz, 9200-9300 MHz, 9300-9500 MHz and 9500-9800 MHz—Section 8D—Radiodetermination, Global Maritime Distress and Safety System and Related Subjects. 2 pp.

—Radio Paging Systems

CEPT T/R 25-07 E-91. Frequency Coordination for the European Radio Message System (ERMES) (Revised in Athens 1990). 8 pp.

CEPT T/R 25-07 E-89. Frequency Coordination for the European Radio Message System (ERMES). 8 pp.

EC 90/544/EEC-90. Council Directive on the Frequency Bands Designated for the Coordinated Introduction of Pan-European Land-Based Public Radio Paging in the Community. 2 pp.

—Radio Relay Systems

CCIR RECMN 698-90. Preferred Frequency Bands for Trans-Horizon Radio-Relay Systems—Section 9E2—Trans-Horizon Radio-Relay Systems. 2 pp.

CCIR RECMN 698-1-92. Preferred Frequency Bands for Trans-Horizon Radio-Relay Systems—Section 9E2—Trans-Horizon Radio-Relay Systems. 8 pp.

CCIR QUESTION 103/9-90. Trans-Horizon Radio-Relay Systems—Questions of Study Group 9 Fixed Service. 1 p.

—Radio Relay Systems—Interferences

CCIR Report 1191-90. Factors Affecting the Choice of Frequency Bands for Trans-Horizon Radio-Relay Systems—Section 9E2—Trans-Horizon Radio-Relay Systems. 8 pp.

—Radio Relay Systems—Transhorizon Propagation

CCIR Report 1191-90. Factors Affecting the Choice of Frequency Bands for Trans-Horizon Radio-Relay Systems—Section 9E2—Trans-Horizon Radio-Relay Systems. 8 pp.

—Radio Spectra Utilization

CCIR Report 1112-90. Measures of Aggregate Spectrum Use in a Frequency Band—Section 1B—Spectrum Sharing and Planning Principles and Techniques. 12 pp.

—Radiotelephones—Ships

CCIR RECMN 475-1-74. Improvements in the Performance of Radiotelephone Circuits in the MF and HF Maritime Mobile Bands—Section 8C—Maritime Mobile Service; Telephony and Related Subjects. 17 pp.

—Radiotelephony

CEPT T/R 33-01 E-72. Efficient Use of HF Single-Sideband Radiotelephone Channels. 1 p.

Frequency Bands (Cont.)

—Reassignment—Mobile Radio Services
CCIR Report 1113-90. Method of Frequency-Reassignment to Improve Channel Utilization—Section 1B—Spectrum Sharing and Planning Principles and Techniques. 8 pp.

—Ringing Tones
CCITT RECMN E.180-89. Technical Characteristics of Tones for the Telephone Service—Telephone Network and ISDN—Operation, Numbering, Routing and Mobile Service (Study Group II) 9 pp. 9 pp.

—Satellite Networks
CCIR Report 1140-90. Satellite Networks for More Than One Service in One or More Frequency Bands—Section 4D2—Coordination Methods. 7 pp.

—SHF/EHF—Atmospheric Attenuation
CCIR Report 1100-90. Impact of Atmospheric Attenuation on Spectrum Management Above 20 GHz—Section 1D—Spectrum Utilization and Applications. 6 pp.

—SHF/EHF—Atmospheric Attenuation—Radio Wave Propagation
CCIR QUESTION 79/1-90. Propagation Models for Spectrum Management and Planning Above 20 GHz—Questions Concerning Study Group 1—Spectrum Management Techniques (Spectrum Engineering, Planning, Sharing, Monitoring and Utilization). 1 p.

—SHF/EHF—Fixed Satellite Services
CCIR Report 552-4-90. Use of Frequency Bands Above 10 GHz in the Fixed-Satellite Service—Section 4B1—Systems Aspects. 34 pp.
CCIR QUESTION 47/4-90. Use of Frequency Bands Aboove 10 GHz in the Fixed-Satellite Service—Questions of Study Group 4 Fixed-Satellite Service. 1 p.

—SHF—Radio Altimeters
CCIR Report 1186-90. Use of Frequency Band 4 200 to 4 400 MHz by Radio Altimeters—Section 8K—Aeronautical Mobile Service (Terrestrial). 4 pp.

—Space Communications—Attenuation
CCIR Report 1119-90. Method of Calculating Attenuation, Noise Temperature, and Telecommunication Link Performance for the Selection of Preferred Frequency Bands—Section 2B—Topics of General Interest. 20 pp.

—Space Communications—Noise Temperature
CCIR Report 1119-90. Method of Calculating Attenuation, Noise Temperature, and Telecommunication Link Performance for the Selection of Preferred Frequency Bands—Section 2B—Topics of General Interest. 20 pp.

—Space Communications—Telecommunication Link Performance
CCIR Report 1119-90. Method of Calculating Attenuation, Noise Temperature, and Telecommunication Link Performance for the Selection of Preferred Frequency Bands—Section 2B—Topics of General Interest. 20 pp.

—Space Operation Services
CCIR Report 845-1-86. Space Operation Systems Frequencies, Bandwidths and Protection Criteria—Section 2C—Space Operations. 12 pp.

—Space Research Services
CCIR RECMN 576-1-90. Preferred Frequencies and Bandwidths for Deep-Space Research—Section 2E—Space Research. 1 p.

—Space Research Systems
CCIR Report 683-3-90. Frequency Bands in the 1 to 40 GHz Range That Are Preferred for Deep-Space Research—Section 2E—Space Research. 17 pp.
CCIR Report 984-86. Frequency Bands Preferred for Transmission to and from Manned and Unmanned Spacecraft for Near-Earth Space Research—Section 2E—Space Research. 13 pp.
CCIR QUESTION 132/7-90. Preferred Frequency Bands for Space Research—Questions Concerning Study Group 7—Science Services. 1 p.

—Spacecraft
CCIR Report 683-3-90. Frequency Bands in the 1 to 40 GHz Range That Are Preferred for Deep-Space Research—Section 2E—Space Research. 17 pp.
CCIR Report 984-86. Frequency Bands Preferred for Transmission to and from Manned and Unmanned Spacecraft for Near-Earth Space Research—Section 2E—Space Research. 13 pp.

Frequency Bands (Cont.)

—Spacecraft (Cont.)
CCIR QUESTION 133/7-90. Preferred Frequency Bands for Deep-Space Research Manned and Unmanned Spacecraft—Questions Concerning Study Group 7—Science Services. 1 p.
CCIR QUESTION 134/7-90. Preferred Frequency Bands for near-Earth Manned and Unmanned Spacecraft—Questions Concerning Study Group 7—Science Services. 1 p.

—Spacecraft—Beacons
CCIR Report 456-3-86. Preferred Frequency Bands for Spacecraft Transmitters Used as Beacons—Section 2E—Space Research. 5 pp.

—Spacecraft—Re-entry Communications
CCIR RECMN 367-63. Frequency Bands for Re-Entry Communications—Section 2B—Topics of General Interest. 1 p.

—Spacecraft—Transmitters
CCIR RECMN 513-1-86. Preferred Frequency Bands for Spacecraft Transmitters Used as Beacons—Section 2E—Space Research. 1 p.
CCIR Report 456-3-86. Preferred Frequency Bands for Spacecraft Transmitters Used as Beacons—Section 2E—Space Research. 5 pp.

—Special Information Tones
CCITT RECMN E.180-89. Technical Characteristics of Tones for the Telephone Service—Telephone Network and ISDN—Operation, Numbering, Routing and Mobile Service (Study Group II) 9 pp. 9 pp.

—Standard Frequency Emissions
CCIR Report 267-7-90. Standard Frequencies and Time Signals. Characteristics of Standard-Frequency and Time-Signal Emissions in Allocated Bands and Characteristics of Stations Emitting with Regular Schedules with Stabilized Frequencies, Outside of Allocated Bands—Section 7C-. 28 pp.

—Tables (Data)
CCITT RECMN B.15-89. Nomenclature of the Frequency and Wavelength Bands Used in Telecommunications—Terms and Definitions Abbreviations and Acronyms Recommendations on Means of Expression (Series B) General Telecommunications Statistics (Series C) 3 pp. 3 pp.

—Telecommunication Systems—Glossaries
CCITT RECMN B.15-89. Nomenclature of the Frequency and Wavelength Bands Used in Telecommunications—Terms and Definitions Abbreviations and Acronyms Recommendations on Means of Expression (Series B) General Telecommunications Statistics (Series C) 3 pp. 3 pp.

—Teleinformatic Services—Highway Transportation
CEPT T/R 22-04 E-91. Harmonisation of Frequency Bands for Road Transport Information Systems (RTI). 1 p.
EC COM(92) 341-92. Proposal for a Council Directive on the Frequency Bands to Be Designated for the Coordinated Introduction of Road Transport Telematic Systems in the Community, Including Road Information and Route Guidance Systems. 2 pp.

—Teleinformatic Services—Railroad Transportation
CEPT T/R 25-09 E-90. Designation of Frequencies in the 900 MHz Band for Railway Purposes. 1 p.

—Telephone Circuits—Carrier Systems
CCITT RECMN G.311-89. General Characteristics of Systems Providing 12 Carrier Telephone Circuits on an Open-Wire Pair—International Analogue Carrier Systems (Study Group XV) 6 pp. 6 pp.

—Telephone Circuits—Telegraphy
CCITT RECMN H.34-89. Subdivision of Frequency Band of a Telephone-Type Circuit Between Telegraphy and Other Services—Line Transmission of Non-Telephone Signals—Transmission of Sound-Programme and Television Signals (Study Group XV) 1 pp. 1 p.

—Television Broadcasting
CEPT T/R 25-06-86. Parametres De Planification a Appliquer Pour L'Utilisation Efficace Des Bandes De Frequences Partagees Entre Le Service De Radiodiffusion (Television, Bandes I et III Seulement) Et Le Service Mobile Terrestre. 5 pp.
CEPT T/R 25-06 E-86. Planning Parameters to Assist the Efficient and Effective Utilisation of Shared Frequency Bands Which Are Allocated to the Broadcasting Service (Television) and the Land Mobile Service, Using Assignments Which Overlap a Television Channel (Television Bands I and III Only). 6 pp.

Frequency Bands (Cont.)

—Time Signal Emissions
CCIR Report 267-7-90. Standard Frequencies and Time Signals. Characteristics of Standard-Frequency and Time-Signal Emissions in Allocated Bands and Characteristics of Stations Emitting with Regular Schedules with Stabilized Frequencies, Outside of Allocated Bands—Section 7C-. 28 pp.

—UHF—Mobile Satellite Communications—Monitoring
CCIR QUESTION 63-1/1-86. Monitoring of the Band 406-406.1 MHz—Questions Concerning Study Group 1—Spectrum Management Techniques (Spectrum Engineering, Planning, Sharing, Monitoring and Utilization). 1 p.

—UHF—Oscillators
CCIR Report 1152-90. General Sources of Highly Stable Signals in the UHF-EHF Bands Using Synchronized Oscillators—Section 7C—Systems for Dissemination and Comparison. 3 pp.

—Units of Measurement—Hertz
CCITT RECMN B.15-89. Nomenclature of the Frequency and Wavelength Bands Used in Telecommunications—Terms and Definitions Abbreviations and Acronyms Recommendations on Means of Expression (Series B) General Telecommunications Statistics (Series C) 3 pp. 3 pp.

—Uplinks—Geostationary Satellites
CCIR Report 557-2-86. Use of Frequency Bands Allocated to the Fixed-Satellite Service for Both the up Link and down Link Geostationary-Satellite Systems—Section 4D2—Coordination Methods. 9 pp.

—VHF/UHF—Attenuation
CCIR RECMN 833-92. Attenuation in Vegetation. 1 p.

—Videoconferencing Services
CEPT T/R 52-01 E-91. Designation of a Harmonized Frequency Band for Multipoint Video Distribution Systems in Europe. 1 p.

—VLF—Earth Stations—Fixed Satellite Services
CCIR Report 386-3-82. Determination of the Power in Any 4 kHz Band Radiated Towards the Horizon by Earth Stations of the Fixed-Satellite Service Sharing Frequency Bands Below 15 GHz with the Terrestrial Services—Section 4/9A—Sharing Conditions. 4 pp.

—VLF—Standard Frequencies
CCIR Report 271-8-90. Stability and Accuracy of Standard Frequency and Time Signals in VLF and LF Bands as Received—Section 7C—Systems for Dissemination and Comparison. 9 pp.

—VLF—Standard Frequencies—Stability
CCIR Report 271-8-90. Stability and Accuracy of Standard Frequency and Time Signals in VLF and LF Bands as Received—Section 7C—Systems for Dissemination and Comparison. 9 pp.

—VLF—Time Signals
CCIR Report 271-8-90. Stability and Accuracy of Standard Frequency and Time Signals in VLF and LF Bands as Received—Section 7C—Systems for Dissemination and Comparison. 9 pp.

—Warning Tones
CCITT RECMN E.180-89. Technical Characteristics of Tones for the Telephone Service—Telephone Network and ISDN—Operation, Numbering, Routing and Mobile Service (Study Group II) 9 pp. 9 pp.

—Wind Profiler Radar
CCIR QUESTION 102/8-90. Suitable Frequency Bands for the Operation of Wind Profiler Radars—Questions Concerning Study Group 8—Mobile, Radiodetermination, Amateur and Related Satellite Services. 1 p.

Frequency Control

See Also: Frequencies; Frequency Error

—Piezoelectric Devices—Glossaries
BSI BS 4727:Pt1: Group 08-92. 1992 Electrotechnical, Power, Telecommunication, Electronics, Lighting and Colour Terms Part 1: Terms Common to Power, Telecommunications and Electronics Group 08: Piezoelectric Devices for Frequency Control and Selection (IEC 50(561): 1991) (G). 56 pp.
BSI BS 4727:Pt1: Group 08-75. 1975 Electrotechnical, Power, Telecommunication, Electronics, Lighting and Colour Terms Part 1: Terms Common to Power, Telecommunications and Electronics Group 08: Piezo-Electric Device Terminology. 14 pp.

Frequency Control (Cont.)

—Piezoelectric Devices—Glossaries *(Cont.)*
IEC 50 Chap 561-91. International Electrotechnical Vocabulary Chapter 561: Piezoelectric Devices for Frequency Control and Selection First Edition. 60 pp.

—Quartz Crystal Units
BSI BS 5069: Part 1-80. 1980 Dimensions of Piezoelectric Devices Part 1: Standard Outlines and Pin Connections for Quartz Crystal Units. 61 pp.
IEC 122 Pt 1-76. Quartz Crystal Units for Frequency Control and Selection Part 1: Standard Values and Test Conditions Second Edition; (Amendment 1-1983). 46 pp.
IEC 122 Pt 2-83. Quartz Crystal Units for Frequency Control and Selection Part 2: Guide to the Use of Quartz Crystal Units for Frequency Control and Selection Second Edition. 91 pp.
IEC 122 Pt 3-77. Quartz Crystal Units for Frequency Control and Selection Pt 3: Standard Outlines and Lead Connections Second Ed.; (Supplement A-1979) (Supplement B-1980) (Supplement C-1981) (Supplement D-1989) (Amendment 2-1991, Incorporating Amd 1) (Amendment 3-1992). 103 pp.

—Quartz Crystal Units—Microprocessors
IEC 122 Pt 2-1-91. Quartz Crystal Units for Frequency Control and Selection Part 2: Guide to the Use of Quartz Crystal Units for Frequency Control and Selection—Section One: Quartz Crystal Units for Microprocessor Clock Supply First Edition. 44 pp.

Frequency Converters
Scope Note: Use a more specific term *See:* Down Converters; Up Converters; Voltage Converters

Frequency Coordination
Use: Frequency Management

Frequency Curves

—Field Strength—Ground Wave Propagation
CCIR RECMN 368-6-90. Ground-Wave Propagation Curves for Frequencies Between 10 kHz and 30 MHz—Section 5B—Effects of the Ground (Including Ground-Wave Propagation). 16 pp.
CCIR RECMN 368-7-92. Ground-Wave Propagation Curves for Frequencies Between 10 kHz and 30 MHz. 54 pp.

—Field Strength—Radio Wave Propagation—Broadcasting
CCIR RECMN 370-5-86. VHF and UHF Propagation Curves for the Frequency Range from 30 MHz to 1000 MHz. Broadcasting Services—Section 5D—Aspects Relative to the Terrestrial Broadcasting and Mobile Services. 28 pp.
CCIR Report 239-7-90. Propagation Statistics Required for Broadcasting Services Using the Frequency Range 30 to 1000 MHz—Section 5D—Aspects Relative to the Terrestrial Broadcasting and Mobile Services. 19 pp.

—Field Strength—Radio Wave Propagation—Mobile Radio Services
CCIR Report 567-4-90. Propagation Data and Prediction Methods for the Terrestrial Land Mobile Service Using the Frequency Range 30 MHz to 3 GHz—Section 5D—Aspects Relative to the Terrestrial Broadcasting and Mobile Services. 32 pp.

—Ground Wave Propagation
CCIR RESOLUTION 72-2-90. Handbook of Ground-Wave Propagation Curves—Volume V—Propagation in Non-Ionized Media. 1 p.
CCIR Report 717-3-90. World Atlas of Ground Conductivities (Published Separately)—Section 5B—Effects of the Ground (Including Ground-Wave Propagation). 9 pp.

—Transmission Loss—Radio Wave Propagation—Radio Communications
CCIR RECMN 528-2-86. Propagation Curves for Aeronautical Mobile and Radionavigation Services Using the VHF, UHF and SHF Bands—Section 5D—Aspects Relative to the Terrestrial Broadcasting and Mobile Services. 9 pp.

Frequency Demodulators
See Also: Demodulators; Frequency Modulators

—Earth Stations—Measurement
IEC 510 Pt 2-6-92. Methods of Measurement for Radio Equipment Used in Satellite Earth Stations Part 2: Measurements for Sub-Systems Section Six: Frequency Demodulators First Edition. 49 pp.

Frequency Demodulators (Cont.)

—Radio Relay Systems
IEC 487 Pt 2-5-84. Methods of Measurement for Equipment Used in Terrestrial Radio-Relay Systems Part 2: Measurements for Sub-Systems Section Five—Frequency Demodulators First Edition. 32 pp.

—Telegraph Channels
CCITT RECMN X.40-89. Standardization of Frequency-Shift Modulated Transmission Systems for the Provision of Telegraph and Data Channels by Frequency Division of a Group—Data Communication Networks—Transmission, Signalling and Switching, Network Aspects, Maintenance and. 3 pp.

Frequency Deviation
See Also: Carrier Frequencies

—Continuity Pilots—Radio Relay Systems
CCIR RECMN 401-2-70. Frequencies and Deviations of Continuity Pilots for Frequency Modulation Radio-Relay Systems for Television and Telephony—Section 9D—Maintenance. 2 pp.

—Radio Relay Systems—Telephony
CCIR RECMN 404-2-70. Frequency Deviation for Analogue Radio-Relay Systems for Telephony Using Frequency-Division Multiplex—Section 9C—Interconnection Characteristics (Baseband and Intermediate Frequency). 1 p.

—Radio Relay Systems—Television Systems
CCIR RECMN 276-2-74. Frequency Deviation and the Sense of Modulation for Analogue Radio-Relay Systems for Television—Section 9C—Interconnection Characteristics (Baseband and Intermediate Frequency). 1 p.

—Sound Broadcasting—Monitoring
CCIR QUESTION 67/1-90. Method of Measuring the Maximum Frequency Deviation of FM Broadcast Emissions at Monitoring Stations—Questions Concerning Study Group 1—Spectrum Management Techniques (Spectrum Engineering, Planning, Sharing, Monitoring and Utilization). 1 p.

Frequency Distance
Use For: FD *See Also:* Radio Frequency Interference

—Radio Equipment
CCIR RECMN 337-2-90. Frequency and Distance Separations—Section 1B—Spectrum Sharing and Planning Principles and Techniques. 1 p.
CCIR RECMN 337-3-92. Frequency and Distance Separations—Section 1B—Spectrum Sharing and Planning Principles and Techniques. 3 pp.

—Radio Transmitters
CCIR Report 842-2-90. Spectrum-Conserving Frequency Assignments for Given Frequency-Distance Separations—Section 1B—Spectrum Sharing and Planning Principles and Techniques. 13 pp.

Frequency Distortion
Use For: Waveform Amplitude Distortion

—Television Circuits
CCIR Report 636-3-86. Long-Time Waveform Distortion in Long-Distance Television Circuits—Section CMTT A—Television Transmission Standards and Performance Objectives. 3 pp.
CCIR QUESTION 27/CMTT-90. Long-Time Waveform Distortion in Analogue Television Circuits over Long Distances—Questions Concerning the CMTT CCIR/CCITT Joint Study Group for Television and Sound Transmission. 1 p.
CCIR QUESTION 45/CMTT-90. Long-Time Waveform Distortion in Analogue Television Circuits over Long Distances Methods of Measurement and Test Signals—Questions Concerning the CMTT CCIR/CCITT Joint Study Group for Television and Sound Transmission. 1 p.

Frequency Divider/Digital Timers
See Also: Digital Circuits; Timers

—Programmable
CECC CECC 90 109-838 ISSUE 1-89. Digital Integrated Circuits; Silicon Monolithic C MOS, Cavity or Non-Cavity Packages; Type(s) 54/74 HC 7292 Programmable Frequency Divider /Digital Timer Assessment Levels P, Y, L (En, Fr, Ge). 9 pp.
CECC CECC 90 109-839 ISSUE 1-89. Digital Integrated Circuits; Silicon Monolithic C MOS, Cavity or Non-Cavity Packages; Type(s) 54/74 HC 7294 Programmable Frequency Divider /Digital Timer Assessment Levels P, Y, L (En, Fr, Ge). 9 pp.

Frequency Divider/Digital Timers (Cont.)

—Programmable *(Cont.)*
CECC CECC 90 109-852 ISSUE 1-90. Digital Integrated Circuits; Silicon Monolithic C MOS, Cavity or Non-Cavity Packages; Type(s) 54/74 HC 292 Programmabble Frequency Divider/Digital Timer Assessment Levels P, Y, L (En, Fr, Ge). 9 pp.
CECC CECC 90 109-853 ISSUE 1-90. Digital Integrated Circuits; Silicon Monolithis C MOS, Cavity or Non-Cavity Packages; Type(s) 54/74 HC 294 Programmable Frequency Divider/Digital Timer Assessment Levels P, Y, L (En, Fr, Ge). 9 pp.

Frequency Dividers
See Also: Digital Circuits

—Digital
CECC CECC 90 104-074 ISSUE 2. NL CECC 90 104-074 Issue 2; Digital Integrated Circuits in Accordance with FS 90 104; HEC/HEF 4521B; 24-Stage Frequency Divider (En). 15 pp.

Frequency Division Multiplexers
Use For: FDM *See Also:* Frequency Division Multiplexing; Frequency Modulated FDM Carrier; Time Assignment Speech Interpolation

—Bearer Services—Digital Line Systems
CCITT RECMN G.941-89. Digital Line Systems Provided by FDM Transmission Bearers—Digital Networks, Digital Sections and Digital Line Systems (Study Groups XV and XVIII) 4 pp. 4 pp.

—Carrier Systems—Interfaces
CCITT RECMN G.371-89. FDM Carrier Systems for Submarine Cable—International Analogue Carrier Systems (Study Group XV) 3 pp. 3 pp.

—Earth Stations—Measurement
IEC 510 Pt 3-4-92. Methods of Measurement for Radio Equipment Used in Satellite Earth Stations Part 3: Methods of Measurement on Combinations of Sub-Systems Section Four: Measurements for Frequency Division Multiplex (f.d.m.) Transmission First Edition. 49 pp.

—Electrical Impedance—Carrier Systems
CCITT RECMN G.371-89. FDM Carrier Systems for Submarine Cable—International Analogue Carrier Systems (Study Group XV) 3 pp. 3 pp.

—Encoder/Decoders
CCITT RECMN G.795-89. Characteristics of Codecs for FDM Assemblies—General Aspects of Digital Transmission Systems; Terminal Equipments (Study Groups XV and XVIII) 4 pp. 4 pp.

—Encoder/Decoders—Transmultiplexers—(DIV) Systems
CCITT FASCICLE III.4-89. General Aspects of Digital Transmission Systems Terminal Equipments—Recommendations G.700-G.795. 621 pp.

—Encoder/Decoders—Transmultiplexers—(DOV) Systems
CCITT FASCICLE III.4-89. General Aspects of Digital Transmission Systems Terminal Equipments—Recommendations G.700-G.795. 621 pp.

—Noise
CCITT RECMN G.222-89. Noise Objectives for Design of Carrier-Transmission Systems of 2500 km—International Analogue Carrier Systems (Study Group XV) 4 pp. 4 pp.
IEC 487 Pt 3-4-82. Methods of Measurement for Equipment Used in Terrestrial Radio-Relay Systems Part 3: Simulated Systems Section Four—Measurements for F.D.M. Transmission First Edition. 33 pp.

—Power Levels—Carrier Systems
CCITT RECMN G.371-89. FDM Carrier Systems for Submarine Cable—International Analogue Carrier Systems (Study Group XV) 3 pp. 3 pp.

—Radio Relay Systems—Basebands
CCITT RECMN G.423-89. Interconnection at the Baseband Frequencies of Frequency-Division Multiplex Radio-Relay Systems—International Analogue Carrier Systems (Study Group XV) 8 pp. 8 pp.

—Radio Relay Systems—Circuit Noise
CCITT RECMN G.441-89. Permissable Circuit Noise on Frequency-Division Multiplex Radio-Relay Systems—International Analogue Carrier Systems (Study Group XV) 1 pp. 1 p.

INTERNATIONAL AND NON-U.S. NATIONAL STANDARDS
SUBJECT INDEX

Frequency

Frequency Division Multiplexers (Cont.)

—Radio Relay Systems—Hypothetical Reference Circuits

CCITT RECMN G.431-89. Hypothetical Reference Circuits for Frequency-Division Multiplex Radio-Relay Systems—International Analogue Carrier Systems (Study Group XV) 2 pp. 2 pp.

CCITT RECMN G.433-89. Hypothetical Reference Circuit for Trans-Horizon Radio-Relay Systems for Telephony Using Frequency-Division Multiplex—International Analogue Carrier Systems (Study Group XV) 1 pp. 1 p.

CCITT RECMN G.441-89. Permissable Circuit Noise on Frequency-Division Multiplex Radio-Relay Systems—International Analogue Carrier Systems (Study Group XV) 1 pp. 1 p.

CCITT RECMN G.444-89. Allowable Noise Power in the Hypothetical Reference Circuit of Trans-Horizon Radio-Relay Systems for Telephony Using Frequency-Division Multiplex—International Analogue Carrier Systems (Study Group XV) 1 pp. 1 p.

—Radio Relay Systems—Interfaces

CCITT RECMN G.423-89. Interconnection at the Baseband Frequencies of Frequency-Division Multiplex Radio-Relay Systems—International Analogue Carrier Systems (Study Group XV) 8 pp. 8 pp.

—Radio Relay Systems—Line Regulation Systems

CCITT RECMN G.423-89. Interconnection at the Baseband Frequencies of Frequency-Division Multiplex Radio-Relay Systems—International Analogue Carrier Systems (Study Group XV) 8 pp. 8 pp.

—Radio Relay Systems—Noise

CCITT RECMN G.444-89. Allowable Noise Power in the Hypothetical Reference Circuit of Trans-Horizon Radio-Relay Systems for Telephony Using Frequency-Division Multiplex—International Analogue Carrier Systems (Study Group XV) 1 pp. 1 p.

—Radio Relay Systems—Pilot Channels

CCITT RECMN G.423-89. Interconnection at the Baseband Frequencies of Frequency-Division Multiplex Radio-Relay Systems—International Analogue Carrier Systems (Study Group XV) 8 pp. 8 pp.

—Radio Relay Systems—Power Levels

CCITT RECMN G.423-89. Interconnection at the Baseband Frequencies of Frequency-Division Multiplex Radio-Relay Systems—International Analogue Carrier Systems (Study Group XV) 8 pp. 8 pp.

—Radio Relay Systems—Return Loss

CCITT RECMN G.423-89. Interconnection at the Baseband Frequencies of Frequency-Division Multiplex Radio-Relay Systems—International Analogue Carrier Systems (Study Group XV) 8 pp. 8 pp.

—Radiotelegraph Systems—Classification

CCIR RECMN 347-63. Classification of Multi-Channel Radiotelegraph Systems for Long-Range Circuits Operating at Frequencies Below About 30 MHz and the Designation of the Channels in These Systems—Section 3Aa—Technical Characteristics. 2 pp.

—Satellite Communications—Hypothetical Reference Circuits

CCITT RECMN G.445-89. Allowable Noise Power in the Hypothetical Reference Circuit for Frequency-Division Multiplex Telephony in the Fixed-Satellite Service—International Analogue Carrier Systems (Study Group XV) 1 pp. 1 p.

—Satellite Communications—Noise

CCITT RECMN G.445-89. Allowable Noise Power in the Hypothetical Reference Circuit for Frequency-Division Multiplex Telephony in the Fixed-Satellite Service—International Analogue Carrier Systems (Study Group XV) 1 pp. 1 p.

—Syllabic Compandors

CCITT RECMN G.166-89. Characteristics of Syllabic Compandors for Telephony on High Capacity Long Distance Systems—General Characteristics of International Telephone Connections and Circuits (Study Groups XII and XV) 7 pp. 7 pp.

—Telecommunication Systems

BSI BS 6328: Part 5-84. 1984 Amd 1 Apparatus for Connection to Private Circuits Run by Certain Public Telecommunication Operators Part 5: Apparatus for Connection to Wideband (FDM) Circuits. 33 pp.

Frequency Division Multiplexers (Cont.)

—Telephone Cables—Carrier Systems

CCITT RECMN G.371-89. FDM Carrier Systems for Submarine Cable—International Analogue Carrier Systems (Study Group XV) 3 pp. 3 pp.

—Telephone Circuits—Modems—Telephone Networks

CCITT RECMN V.22-89. 1200 Bits per Second Duplex Modem Standardized for Use in the General Switched Telephone Network and on Point-to-Point 2-Wire Leased Telephone-Type Circuits—Data Communication over the Telephone Network (Study Group XVII) 13 pp. 13 pp.

CCITT RECMN V.22 BIS-89. 2400 Bits per Second Duplex Modem Using the Frequency Division Technique Standardized for Use on the General Switched Telephone Network and on Point-to-Point 2-Wire Leased Telephone-Type Circuits—Data Communication over the Telephone Network. 16 pp.

—Test Signals—White Noise

CCITT RECMN G.228-89. Measurement of Circuit Noise in Cable Systems Using a Uniform-Spectrum Random Noise Loading—International Analogue Carrier Systems (Study Group XV) 11 pp. 11 pp.

Frequency Division Multiplexing

See Also: Frequency Division Multiplexers; Frequency Modulated FDM Carrier; Multiplexing

—Geostationary Satellites—Telephone Channels—Radio Interference

CCIR RECMN 466-5-90. Maximum Permissible Level of Interference in a Telephone Channel of a Geostationary-Satellite Network in the Fixed-Satellite Service Employing Frequency Modulation with Frequency-Division Multiplex, Caused by Other Networks of this Service—Section 4D1-. 3 pp.

CCIR RECMN 466-6-92. Maximum Permissible Level of Interference in a Telephone Channel of a Geostationary-Satellite Network in the Fixed-Satellite Service Employing Frequency Modulation with Frequency-Division Multiplex, Caused by Other Networks of this Service—Section 4D1-. 3 pp.

—Pre-Emphasis—Telephony—Frequency Modulation—Fixed Satellites

CCIR RECMN 464-1-82. Pre-Emphasis Characteristics for Frequency-Modulation Systems for Frequency-Division Multiplex Telephony in the Fixed-Satellite Service—Sec-tion 4C—Earth Station and Baseband Character-istics—Earth Station Antennas—Maintenance of Earth Stations. 4 pp.

CCIR RECMN 464-2-92. Pre-Emphasis Characteristics for Frequency-Modulation Systems for Frequency-Division Multiplex Telephony in the Fixed-Satellite Service—Sec 4C—Earth Station and Baseband Character-istics—Earth Station Antennas—Maintenance of Earth Stations. 5 pp.

CCIR Report 212-3-74. Use of Pre-Emphasis in Frequency-Modulation Systems for Frequency Division Multiplex Telephony and Television in the Fixed-Satellite Service—Section 4C—Earth Station and Baseband Characteristics —Earth Station Antennas —Maintenance of Earth Stations. 4 pp.

—Radio Relay Systems—Telephony—Maintenance

CCIR RECMN 290-3-78. Maintenance Measurements on Radio-Relay Systems for Telephony Using Frequency-Division Multiplex—Section 9D—Maintenance. 2 pp.

—Radio Relay Systems—Telephony—Noise

CCIR RECMN 398-3-74. Measurements of Noise in Actual Traffic over Radio-Relay Systems for Telephony Using Frequency-Division Multiplex—Section 9D—Maintenance. 2 pp.

—Radio Relay Systems—Telephony—White Noise

CCIR RECMN 399-3-78. Measurement of Noise Using a Continuous Uniform Spectrum Signal on Frequency-Division Multiplex Telephony Radio-Relay Systems—Section 9D—Maintenance. 3 pp.

—Telephony—Circuits—Radio Relay Systems

CCIR RECMN 395-2-78. Noise in the Radio Portion of Circuits to Be Established Over Real Radio-Relay Links for FDM Telephony—Section 9A—Performance Objectives, Propagation and Interference Effects. 3 pp.

Frequency Division Multiplexing (Cont.)

—Telephony—Fixed Satellite Services—Performance Measurement

CCIR RECMN 482-2-86. Measurement of Performance by Means of a Signal of a Uniform Spectrum for Systems Using Frequency-Division Multiplex Telephony in the Fixed-Satellite Svc.—Section 4C—Earth Station and Baseband Characteristics—Earth Stn. Antennas—Main-tenance of Earth Stns.. 3 pp.

—Telephony—Hypothetical Reference Circuits—Noise—Fixed Satellites

CCIR RECMN 353-6-90. Allowable Noise Power in the Hypothetical Reference Circuit for Frequency-Division Multiplex Telephony in the Fixed-Satellite Service—Section 4B2—Performance and Availability. 2 pp.

CCIR RECMN 353-7-92. Allowable Noise Power in the Hypothetical Reference Circuit for Frequency-Division Multiplex Telephony in the Fixed-Satellite Service—Section 4B2—Performance and Availability. 7 pp.

—Telephony—Hypothetical Reference Circuits—Radio Relay Systems

CCIR RECMN 391-63. Hypothetical Reference Circuit for Radio-Relay Systems for Telephony Using Frequency-Division Multiplex with a Capacity of 12 to 60 Telephone Channels—Section 9A—Performance Objectives, Propagation and Interference Effects. 1 p.

CCIR RECMN 393-4-82. Allowable Noise Power in the Hypothetical Reference Circuit for Radio-Relay Systems for Telephony Using Frequency-Division Multiplex—Section 9A—Performance Objectives, Propagation and Interference Effects. 2 pp.

CCIR RECMN 396-1-66. Hypothetical Reference Circuit for Trans-Horizon Radio-Relay Systems for Telephony Using Frequency-Division Multiplex—Section 9A—Performance Objectives, Propagation and Interference Effects. 1 p.

CCIR RECMN 397-3-78. Allowable Noise Power in the Hypothetical Reference Circuit of Trans-Horizon Radio-Relay Systems for Telephony Using Frequency-Division Multiplex—Section 9A—Performance Objectives, Propagation and Interference Effects. 1 p.

—Telephony—Noise (Spurious Signals)—Fixed Satellite Services

CCIR Report 208-7-90. Form of the Hypothetical Reference Circuit and Allowable Noise Standards for Frequency-Division Multiplex Telephony and Television in the Fixed-Satellite Service—Section 4B2—Performance and Availability. 6 pp.

—Telephony—Radio Relay Systems—Frequency Deviation

CCIR RECMN 404-2-70. Frequency Deviation for Analogue Radio-Relay Systems for Telephony Using Frequency-Division Multiplex—Section 9C—Interconnection Characteristics (Baseband and Intermediate Frequency). 1 p.

—Telephony—Radio Relay Systems—Interfaces

CCIR RECMN 380-4-86. Interconnection at Baseband Frequencies of Radio-Relay Systems for Telephony Using Frequency-Division Multiplex—Section 9C—Interconnection Characteristics (Baseband and Intermediate Frequency). 4 pp.

—Telephony—Radio Relay Systems—Pre-Emphasis

CCIR RECMN 275-3-82. Pre-Emphasis Characteristic for Frequency Modulation Radio-Relay Systems for Telephony Using Frequency-Division Multiplex—Section 9C—Interconnection Characteristics (Baseband and Intermediate Frequency). 4 pp.

—Television Signals—Sound/Picture

CCIR Report 488-5-90. Transmission of Sound and/or Vision Signals by Time-Division Multiplex or Frequency-Division Multiplex—Section CMTT E—Transmission of Signals with Multiplexing of Video, Sound and Data, and Signals of New Systems. 7 pp.

CCIR QUESTION 68/CMTT-90. Transmission of Sound and Picture Signals by Time-Division Multiplex or Frequency-Division Multiplex—Questions Concerning the CMTT CCIR/CCITT Joint Study Group for Television and Sound Transmission. 1 p.

Frequency Error

See Also: Connection Stability; Error Control; Frequencies; Frequency Control

INDUSTRY STANDARDS

Frequency Error (Cont.)

—Telephone Circuits—Voice Frequency Telegraphy
CCITT RECMN G.135-89. Error on the Reconstituted Frequency—General Characteristics of International Telephone Connections and Circuits (Study Groups XII and XV) 2 pp. 2 pp.

Frequency Generators
Use: Signal Generators

Frequency Hopping
See Also: Spread Spectrum Modulation

—Radio Equipment—Electromagnetic Compatibility
CCIR Report 1099-90. Procedure for Analyzing the Electromagnetic Compatibility of Frequency-Hopping Spread Spectrum Radios—Section 1A—Spectrum Engineering and Computer-Aided Principles and Techniques. 13 pp.

Frequency Instability
Use: Frequency Stability

Frequency Management
Use For: Frequency Coordination
See Also: Frequency Bands; Radio Spectra
CCIR Report 1105-90. Relationship Between Spectrum Management and Radio Frequency Monitoring—Section 1C—Spectrum Monitoring Techniques. 7 pp.

—Data Bases—Computer-Aided
CCIR Report 1110-90. Frequency Management Data Base Systems Using Small Computers—Section 1A—Spectrum Engineering and Computer-Aided Principles and Techniques. 11 pp.

—Earth Stations—Fixed Satellite Services
CCIR RECMN 731-92. Reference Earth-Station Cross-Polarized Radiation Pattern for Use in Frequency Coordination and Interference Assessment in the Frequency Range from 2 to About 30 GHz—Section 4C—Earth Station and Baseband Characteristics—Earth Station Antennas-. 15 pp.

—GSM System
CEPT T/R 20-08 E-89. Frequency Planning and Frequency Coordination for the GSM System. 11 pp.

—Radio Relay Systems
CCIR RECMN 699-90. Reference Radiation Patterns for Line-of-Sight Radio-Relay System Antennas for Use in Coordination Studies and Interference Assessment in the Frequency Range from 1 to About 40 GHz—Section 9B2—System General Characteristics. 2 pp.
CCIR RECMN 699-1-92. Reference Radiation Patterns for Line-of-Sight Radio-Relay System Antennas for Use in Coordination Studies and Interference Assessment in the Frequency Range from 1 to About 40 GHz—Section 9B2—System General Characteristics. 2 pp.

—Satellite Networks
CCIR RECMN 743-92. Coordination of Satellite Networks Using Slightly Inclined Geostationary-Satellite Orbits and Between Such Networks and Satellite Networks Using Non-Inclined GSO Satellites—Section 4D2—Coordination Methods. 14 pp.
CCIR RECMN 744-92. Orbit/Spectrum Improvement Measures for Satellite Networks Having More Than One Service in One or More Frequency Bands—Section 4D2—Coordination Methods. 4 pp.

Frequency Measurement

—Radio Spectra—Monitoring
CCIR RECMN 377-2-82. Accuracy of Frequency Measurements at Stations for International Monitoring—Section 1C—Spectrum Monitoring Techniques. 1 p.
CCIR QUESTION 22-1/1-82. Frequency Measurements at Monitoring Stations—Questions Concerning Study Group 1—Spectrum Management Techniques (Spectrum Engineering, Planning, Sharing, Monitoring and Utilization). 1 p.

—Sound-Program Connections
CCITT RECMN N.13-89. Measurements to Be Made by the Broadcasting Organizations During the Preparatory Period—Maintenance of International Sound-Programme and Television Transmission Circuits (Study Group IV) 3 pp. 3 pp.

—Sound-Program Links
CCITT RECMN N.12-89. Measurements to Be Made During the Line-up Period That Precedes a Sound-Programme Transmission—Maintenance of International Sound-Programme and Television Transmission Circuits (Study Group IV) 1 pp. 1 p.

Frequency Measurement (Cont.)

—Telephone Circuits
CCITT RECMN G.101-89. The Transmission Plan—General Characteristics of International Telephone Connections and Circuits (Study Groups XII and XV) 17 pp. 17 pp.

—Terrestrial Stations
CEPT T/R 22-03 E-91. Provisional Recommended Use of the Frequency Range 54.25-66 GHz by Terrestrial Fixed and Mobile Systems. 2 pp.

Frequency Meters
See Also: Electric Measuring Instruments; Electrical Measurement; Frequencies; Harmonic Analyzers; Tachometers; Vibration Meters

—Analog
BSI BS 89: Part 4-90. 1990 Direct Acting Indicating Analogue Electrical Measuring Instruments and Their Accessories Part 4: Special Requirements for Frequency Meters. 14 pp.
CENELEC EN 60 051-4-89. Direct Acting Indicating Analogue Electrical Measuring Instruments and Their Accessories Part 4: Special Requirements for Frequency Meters (Supersedes HD 233). 3 pp.
IEC 51 Pt 4-84. Direct Acting Indicating Analogue Electrical Measuring Instruments and Their Accessories Part 4: Special Requirements for Frequency Meters Fourth Edition. 20 pp.

Frequency Modulated FDM Carrier
See Also: Frequency Division Multiplexers; Frequency Division Multiplexing; Frequency Modulation Systems; Frequency Modulators

—Power Density
CCIR RECMN 675-90. Calculation of the Maximum Power Density (Averaged over 4 kHz) of a Frequency-Modulated FDM Carrier—Section 4/9B—Co-Ordination and Interference Calculations. 4 pp.
CCIR RECMN 675-1-92. Calculation of the Maximum Power Density (Averaged over 4 kHz) of an Angle-Modulated Carrier 4/9B—Co-Ordination and Interference Calculations. 6 pp.

Frequency Modulation Systems
See Also: Frequency Modulated FDM Carrier; Modulation; Narrowband Frequency Modulation

—Geostationary Satellites—Fixed—Television Channels—Interference
CCIR RECMN 483-1-78. Maximum Permissible Level of Interference in a Television Channel of a Geostationary-Satellite Network in the Fixed-Satellite Service Employing Frequency Modulation, Caused by Other Networks of This Service—Section 4D1—Permissible Levels of Interference. 1 p.
CCIR RECMN 483-2-92. Maximum Permissible Level of Interference in a Television Channel of a Geostationary-Satellite Network in the Fixed-Satellite Service Employing Frequency Modulation, Caused by Other Networks of This Service—Section 4D1—Permissible Levels of Interference. 2 pp.

—Mobile Satellite Communications
CCIR Report 509-5-90. Modulation and Coding Technique for Mobile Satellite Service—Section 8I—Technical and Operating Characteristics of Mobile Satellite Services. 23 pp.

—Pre-Emphasis—Radio Relay Systems—Telephony
CCIR RECMN 275-3-82. Pre-Emphasis Characteristic for Frequency Modulation Radio-Relay Systems for Telephony Using Frequency-Division Multiplex—Section 9C—Interconnection Characteristics (Baseband and Intermediate Frequency). 4 pp.

—Pre-Emphasis—Radio Relay Systems—Television Systems
CCIR RECMN 405-1-70. Pre-Emphasis Characteristics for Frequency Modulation Radio-Relay Systems for Television—Section 9C—Interconnection Characteristics (Baseband and Intermediate Frequency). 5 pp.

—Pre-Emphasis—Telephony—Fixed Satellite Services
CCIR RECMN 464-1-82. Pre-Emphasis Characteristics for Frequency-Modulation Systems for Frequency-Division Multiplex Telephony in the Fixed-Satellite Service—Sec-tion 4C—Earth Station and Baseband Character-istics—Earth Station Antennas—Maintenance of Earth Stations. 4 pp.
CCIR RECMN 464-2-92. Pre-Emphasis Characteristics for Frequency-Modulation Systems for Frequency-Division Multiplex Telephony in the Fixed-Satellite Service—Sec 4C—Earth Station and Baseband Character-istics—Earth Station Antennas—Maintenance of Earth Stations. 5 pp.

Frequency Modulation Systems (Cont.)

—Pre-Emphasis—Telephony—Fixed Satellite Services (Cont.)
CCIR Report 212-3-74. Use of Pre-Emphasis in Frequency-Modulation Systems for Frequency Division Multiplex Telephony and Television in the Fixed-Satellite Service—Section 4C—Earth Station and Baseband Characteristics —Earth Station Antennas —Maintenance of Earth Stations. 4 pp.

—Radio Relay Systems—Continuity Pilots
CCIR RECMN 401-2-70. Frequencies and Deviations of Continuity Pilots for Frequency Modulation Radio-Relay Systems for Television and Telephony—Section 9D—Maintenance. 2 pp.

—Television—Interference—Fixed Satellite—Frequency Band Sharing
CCIR Report 449-1-74. Measured Interference into Frequency-Modulation Television Systems Using Frequencies Shared Within Systems in the Fixed-Satellite Service or Between These Systems and Terrestrial Systems—Section 4/9B—Co-Ordination and Inter-ference Calculations. 7 pp.

—Television—Interference—Radio Systems—Terrestrial—Sharing
CCIR Report 449-1-74. Measured Interference into Frequency-Modulation Television Systems Using Frequencies Shared Within Systems in the Fixed-Satellite Service or Between These Systems and Terrestrial Systems—Section 4/9B—Co-Ordination and Inter-ference Calculations. 7 pp.

Frequency Modulators
See Also: Demodulators; Frequency Demodulators; Frequency Modulated FDM Carrier; Pulse Modulators

—Cellular Mobile Radio Equipment
CENELEC PRI-ETS 300 032-90. European Digital Cellular Telecommunications System (Phase 1); Modulation. 5 pp.
CENELEC PRI-ETS 300 032-91. European Digital Cellular Telcommunications System (Phase 1); Modulation. 7 pp.
CENELEC I-ETS 300 032-92. European Digital Cellular Telecommunications System (Phase 1); Modulation (GSM 05.04). 7 pp.
CENELEC GSM 05.04-92. See PRI-ETS 300 032. 4 pp.
ETSI PRI-ETS 300 032-90. European Digital Cellular Telecommunications System (Phase 1); Modulation (GSM 05.04). 5 pp.
ETSI I-ETS 300 032-92. European Digital Cellular Telecommunications System (Phase 1); Modulation (GSM 05.04). 7 pp.
ETSI PRI-ETS 300 032-91. European Digital Cellular Telcommunications System (Phase 1); Modulation. 7 pp.
ETSI GSM 05.04-92. See PRI-ETS 300 032. 4 pp.

—Earth Stations—Measurement
IEC 510 Pt 2-5-92. Methods of Measurement for Radio Equipment Used in Satellite Earth Stations Part 2: Measurements for Sub-Systems Section Five: Frequency Modulators First Edition. 35 pp.

—Radio Relay Systems
IEC 487 Pt 2-4-84. Methods of Measurement for Equipment Used in Terrestrial Radio-Relay Systems Part 2: Measurements for Sub-Systems Section Four—Frequency Modulators First Edition. 28 pp.

—Symbols
CNS C5132-82. Graphic Symbols for Frequencies, Bands, Modulation and Frequency Diagrams (Apr)(8664).

Frequency Offsets
See Also: Frequencies; Reference Frequencies

—Protection Ratio—Television Broadcasting
CCIR QUESTION 54/11-90. Protection Ratios for Television (Use of Frequency Offsets)—Questions Concerning Study Group 11—Broadcasting Service (Television). 1 p.

Frequency Response
See Also: Electrical Measurement; Frequencies; Line Regulation Systems (Telecommunications)

—Basebands—Fiber Optic Cables
CCITT RECMN G.651-89. Characteristics of a 50/125 Micrometer Multimode Graded Index Optical Fibre Cable—Transmission Media Characteristics (Study Group XV) 30 pp. 30 pp.

—Phonograph Test Records
DIN ENGL 45541-71. Frequency Test Record St 33 and M 33 (33 1/3 Rev/ Min; Stereo and Mono) (Mar). 2 pp.

INTERNATIONAL AND NON-U.S. NATIONAL STANDARDS
SUBJECT INDEX
Frequency

Frequency Reuse
See Also: Frequencies; Frequency Bands; Radio Transmission

—**Band 9—Mobile Satellite Communications**
CCIR Report 1172-90. Intersystem Frequency Sharing and Reuse in the Mobile-Satellite Services Operating at Mid to High Portions of Band 9—Section 8H—Efficient Use of the Radio Spectrum Characteristics and Sharing of Frequency Resources. 25 pp.

—**Mobile Satellite Communications—Spot Beams**
CCIR Report 1172-90. Intersystem Frequency Sharing and Reuse in the Mobile-Satellite Services Operating at Mid to High Portions of Band 9—Section 8H—Efficient Use of the Radio Spectrum Characteristics and Sharing of Frequency Resources. 25 pp.

Frequency Separation
Use For: Radio Frequency Separation
See Also: Communication Channels; Frequencies; Frequency Bands

—**Land Mobile Services**
CCIR QUESTION 72-1/8-90. Minimum Channel Separation and Optimum Systems of Modulation, Co-Channel and Adjacent-Channel Coordination Criteria for Simultaneous Use of Different Modulation Techniques in Systems of the Land Mobile Services Between 25 and 3000 MHz—Questions. 1 p.

—**Maritime Mobile Services—MF/HF—Radio Telephones**
CCIR Report 1035-86. Minimum Required Frequency Separation Between Receive and Transmit Frequencies Used for Duplex MF/HF Radiotelephony—Section 8C—Maritime Mobile Service; Telephony and Related Subjects. 2 pp.

—**Maritime Mobile Services—SELCAL Systems**
CCIR Report 1028-86. 3 kHz Duplex Separation for DSC Channels in the Band 435-526.5 kHz—Section 8B—Maritime Mobile Service: Telegraphy and Related Subjects. 4 pp.

—**Maritime Mobile Services—VHF—Radio Telephones**
CCIR RECMN 489-1-78. Technical Characteristics of VHF Radiotelephone Equipment Operating in the Maritime Mobile Service in Channels Spaced by 25 kHz—Section 8C—Maritime Mobile Service; Telephony and Related Subjects. 2 pp.

—**Protection Ratio—Radio Frequency Emissions**
CCIR Report 991-1-90. Measurement of Protection Ratios and Minimum Necessary Frequency Separation for Class of Emission J7B—Section 3Aa—Technical Characteristics. 5 pp.

Frequency Sharing
Use: Frequency Band Sharing

Frequency Shift Amplifiers
Use: Audio Amplifiers—Frequency Shift

Frequency Shift Keying
Use For: Frequency Shift Modulation; FSK
See Also: Frequency Standards; Minimum Shift Keying; Phase Shift Keying
CCIR RECMN 246-3-74. Frequency-Shift Keying—Section 3Ca—Radiotelegraph Circuits. 2 pp.

—**Data Transmission—Radio Beacons**
CCIR Report 1037-86. Choice Between the FSK and MSK Techniques for Data Transmission from Maritime Radiobeacons—Section 8D—Radiodetermination, Global Maritime Distress and Safety System and Related Subjects. 2 pp.

—**Telegraphy**
CCIR Report 702-78. Multi-Frequency-Shift-Keying Techniques for HF Telegraphy—Section 3Ca—Radiotelegraph Circuits. 8 pp.

—**Telemetering**
CCIR Report 702-78. Multi-Frequency-Shift-Keying Techniques for HF Telegraphy—Section 3Ca—Radiotelegraph Circuits. 8 pp.

—**Voice Frequency Telegraph Systems—Data Transmission**
CCIR RECMN 456-70. Data Transmission at 1200/600 Bit/s Over HF Circuits When Using Multi-Channel Voice-Frequency Telegraph Systems and Frequency-Shift Keying—Section 3Cb—Data Transmission. 3 pp.

Frequency Shift Measuring Instruments
Use: Measuring Instruments—Frequency Shift

Frequency Shift Modulation
Use: Frequency Shift Keying

Frequency Shift Standards
Use: Frequency Standards

Frequency Sources
Use: Frequency Synthesizers

Frequency Spectra
—**Mobile Satellite Communications**
CCIR Report 1173-90. Technical and Operational Considerations for Aeronautical Mobile-Satellite Communications—Section 8F—Frequencies, Orbits and Systems. 40 pp.

Frequency Stability
Use For: Frequency Instability
See Also: Frequencies; Phase Stability
CCIR RECMN 538-1-90. Frequency and Phase Stability Measures—Section 7D—Characterization of Sources and Time Scales Formation. 2 pp.
CCIR RECMN 538-2-92. Frequency and Time (Phase) Instability Measures Section 7D—Characterization of Sources and Time Scales Formation. 9 pp.
CCIR Report 580-3-90. Characterization of Frequency and Phase Noise—Section 7D—Characterization of Sources and Time Scales Formation. 12 pp.

—**Fixed Services (Radio Communications)**
CCIR QUESTION 143/9-90. Signal-to-Noise Ratios and Protection Ratios; Bandwidth, Adjacent Channel Spacing and Frequency Stability—Questions of Study Group 9 Fixed Service. 3 pp.

—**Fixed Services (Radio Communications)—Automatic Frequency Control**
CCIR RECMN 349-4-86. Frequency Stability Required for Systems Operating in the HF Fixed Service to Make the Use of Automatic Frequency Control Superfluous—Section 3Aa—Technical Characteristics. 3 pp.

—**Frequency Synthesizers**
CCIR Report 550-74. Short-Term Stability of Frequency Synthesizers—Section 3Aa—Technical Characteristics. 2 pp.

—**Standard Frequencies—Frequency Bands**
CCIR Report 271-8-90. Stability and Accuracy of Standard Frequency and Time Signals in VLF and LF Bands as Received—Section 7C—Systems for Dissemination and Comparison. 9 pp.

Frequency Standards
Use For: Frequency Shift Standards
See Also: Frequency Shift Keying; Standard Frequencies; Standard Frequency Emissions; Standards
CCIR QUESTION 101/7-90. Performance Characterization and Reliability of Frequency and Time Standards—Questions Concerning Study Group 7—Science Services. 1 p.

—**Aircraft—VOR Ground Stations**
EUROCAE ED-52 08.84. MPS for Ground Conventional and Doppler Very High Frequency Omni Range (CVOR and DVOR) Equipment. 61 pp.

—**Musical Instruments**
CNS S1171-83. Frequency Standards for Musical Instruments (Nov)(10663).
JIS Z 8702-76. Frequency Standard for Musical Instruments.

—**Radiotelegraphy**
MOD UK DSTAN 00-8-74. Frequency Shift Standards for HF/Ratt and VHF/Ratt Operations Issue 1. 3 pp.

—**Reliability**
CCIR QUESTION 101/7-90. Performance Characterization and Reliability of Frequency and Time Standards—Questions Concerning Study Group 7—Science Services. 1 p.

Frequency Synthesizers
Use For: Frequency Sources *See Also:* Linear Circuits; Oscillators; Signal Generators
CCIR Report 550-74. Short-Term Stability of Frequency Synthesizers—Section 3Aa—Technical Characteristics. 2 pp.

—**Fixed Services (Radio Communications)**
CCIR Report 704-78. Characteristics of Frequency Synthesizers in the HF Fixed Service—Section 3Aa—Technical Characteristics. 5 pp.

Frequency Synthesizers *(Cont.)*
—**Frequency Stability**
CCIR Report 550-74. Short-Term Stability of Frequency Synthesizers—Section 3Aa—Technical Characteristics. 2 pp.

Frequency Tolerances
See Also: Tolerances

—**Radio Relay Systems**
CCIR Report 785-2-90. Frequency Tolerances for Radio-Relay Systems—Section 9B2—System General Characteristics. 5 pp.
CCIR QUESTION 120/9-90. Frequency Tolerances of Radio-Relay Systems—Questions of Study Group 9 Fixed Service. 1 p.

—**Radio Transmitters**
CCIR Report 181-4-82. Frequency Tolerance of Transmitters—Section 1A—Spectrum Engineering and Computer-Aided Principles and Techniques. 3 pp.
CCIR QUESTION 54-1/1-86. Frequency Tolerance of Transmitters—Questions Concerning Study Group 1—Spectrum Management Techniques (Spectrum Engineering, Planning, Sharing, Monitoring and Utilization). 1 p.
CCIR QUESTION 152/9-90. Frequency Tolerance of Transmitters for the Fixed Service at Frequencies Below About 30 MHz—Questions of Study Group 9 Fixed Service. 1 p.

Frequency Translation Equipment
Use For: Translation Equipment, Frequency
See Also: Frequencies; Group Links; Mastergroup Links; Supergroup Links; Supermastergroup Links; Telecommunication Equipment; 15 Supergroup Assembly Links
CCITT RECMN G.233-89. Recommendations Concerning Translating Equipments—International Analogue Carrier Systems (Study Group XV) 10 pp. 10 pp.
CCITT RECMN G.602-89. Reliability and Availability of Analogue Cable Transmission Systems and Associated Equipments—Transmission Characteristics—(Study Group XV) 4 pp. 4 pp.

—**Carrier Frequencies**
CCITT RECMN G.233-89. Recommendations Concerning Translating Equipments—International Analogue Carrier Systems (Study Group XV) 10 pp. 10 pp.

—**Carrier Leaks**
CCITT RECMN G.233-89. Recommendations Concerning Translating Equipments—International Analogue Carrier Systems (Study Group XV) 10 pp. 10 pp.

—**Carrier Systems**
CCITT RECMN G.211-89. Make-up of a Carrier Link—International Analogue Carrier Systems (Study Group XV) 8 pp. 8 pp.

—**Carrier Systems—Noise—Measurement**
CCITT RECMN G.230-89. Measuring Methods for Noise Produced by Modulating Equipment and Through-Connection Filters—International Analogue Carrier Systems (Study Group XV) 4 pp. 4 pp.

—**Crosstalk**
CCITT RECMN G.233-89. Recommendations Concerning Translating Equipments—International Analogue Carrier Systems (Study Group XV) 10 pp. 10 pp.

—**Envelope Delay Distortion**
CCITT RECMN G.233-89. Recommendations Concerning Translating Equipments—International Analogue Carrier Systems (Study Group XV) 10 pp. 10 pp.

—**Modulation—Hypothetical Reference Circuits—Telephony**
CCITT RECMN G.229-89. Unwanted Modulation and Phase Jitter—International Analogue Carrier Systems (Study Group XV) 2 pp. 2 pp.

—**Noise**
CCITT RECMN G.233-89. Recommendations Concerning Translating Equipments—International Analogue Carrier Systems (Study Group XV) 10 pp. 10 pp.

—**Phase Jitter—Hypothetical Reference Circuits—Telephony**
CCITT RECMN G.229-89. Unwanted Modulation and Phase Jitter—International Analogue Carrier Systems (Study Group XV) 2 pp. 2 pp.

INDUSTRY STANDARDS

Frequency Translation Equipment (Cont.)

—Pilot Channels
CCITT RECMN G.233-89. Recommendations Concerning Translating Equipments—International Analogue Carrier Systems (Study Group XV) 10 pp. 10 pp.

—Return Loss
CCITT RECMN G.233-89. Recommendations Concerning Translating Equipments—International Analogue Carrier Systems (Study Group XV) 10 pp. 10 pp.

—Signal Levels
CCITT RECMN G.233-89. Recommendations Concerning Translating Equipments—International Analogue Carrier Systems (Study Group XV) 10 pp. 10 pp.

Fresh Concrete
Use: Concretes

Fretsaws
See Also: Coping Saws; Hand Tools; Saws

—Blades
MOD UK DSTAN 51-11: Part 9-75. Hand Tools, General Purpose Part 9: Frame, Hand Fretsaw and Blade, Hand Fretsaw Issue 1. 9 pp.

—Frames
MOD UK DSTAN 51-11: Part 9-75. Hand Tools, General Purpose Part 9: Frame, Hand Fretsaw and Blade, Hand Fretsaw Issue 1. 9 pp.

Friability
See Also: Mechanical Properties

—Drying Agents
BSI BS 3482: Part 11-91. 1991 Methods of Test for Desiccants Part 11: Determination of Friability and Dust Content. 5 pp.

—Polystyrene Foams
CSA CAN/CSA-Z217.2-M92. Test Method for Tumbling Friability of Degradable Polystyrene Foams; (Gen Instr 1). 25 pp.

Friction
Scope Note: See the subheading Friction under the specific material or product

Friction Clutches
See Also: Clutches; Disk Clutches; Slip Friction Clutches

—Internal Combustion Engines
BSI BS 3092-73. 1973 Main Friction Clutches, Main Power-Take-Off Assemblies and Associated Attachment for Internal Combustion Engines. 37 pp.

Friction Grip Bolts
See Also: Bolts

—Hexagonal Head—High Strength—Structural
BSI BS 4395: Part 1-69. 1969 Amd 2 High Strength Friction Grip Bolts and Associated Nuts and Washers for Structural Engineering Part 1: General Grade. 36 pp.
BSI BS 4395: Part 2-69. 1969 Amd 2 High Strength Friction Grip Bolts and Associated Nuts and Washers for Structural Engineering Part 2: Higher Grade Bolts and Nuts and General Grade Washers. 36 pp.
BSI BS 4604: Part 1-70. 1970 Amd 3 The Use of High Strength Friction Grip Bolts in Structural Steelwork. Metric Series Part 1: General Grade. 17 pp.
SNZ NZS/BS 4395: Part 1-69. Specification for High Strength Friction Grip Bolts and Associated Nuts and Washers for Structural Engineering Part 1: General Grade. 36 pp.
SNZ NZS/BS 4395: Part 2-69. Specification for High Strength Friction Grip Bolts and Associated Nuts and Washers for Structural Engineering Part 2: Higher Grade Bolts and Nuts and General Grade Washers. 36 pp.

—High Strength—Structural
BSI BS 3294: Part 1-60. (WITHDRAWN) 1960 Amd 6 The Use of High Strength Friction Grip Bolts in Structural Steelwork Part 1: General Grade Bolts. 15 pp.

—Parallel Shank—High Strength—Structural
BSI BS 4604: Part 2-70. 1970 Amd 2 The Use of High Strength Friction Grip Bolts in Structural Steelwork. Metric Series Part 2: Higher Grade (Parallel Shank). 17 pp.

Friction Grip Bolts (Cont.)

—Waisted Shank—High Strength—Structural
BSI BS 4604: Part 3-73. (WITHDRAWN) 1973 The Use of High Strength Friction Grip Bolts in Structural Steelwork. Metric Series Part 3: Higher Grade (Waisted Shank). 16 pp.
SNZ NZS/BS 4395: Part 3-73. Specification for High Strength Friction Grip Bolts and Associated Nuts and Washers for Structural Engineering Part 3: Higher Grade Bolts (Waisted Shank), Nuts and General Grade Washers. 24 pp.

Friction Pulleys
See Also: Pulleys

—Winders
SAA AS 3785.2-91. Underground Mining—Shaft Equipment—Part 2: Friction Winding Arresting Systems. 7 pp.

Friction Yielding Props
Use: Mechanical Yielding Props

Frictionless Bearings
Use: Pivots

Frigates
See Also: Naval Ships; Warships

—Seakeeping
NATO STANAG 4154 ED 2 AMD 2-86. General Criteria and Common Procedures for Seakeeping Performance Assessment. 49 pp.

Frog Legs
See Also: Food

—Frozen
CNS N5122-80. Frozen Frog Legs (May)(3925). 2 pp.

Front End Loaders
Use For: Wheel Loaders See Also: Bulk Transporters; Earthmoving Equipment; Excavating Equipment; Loaders; Materials Handling Equipment
CNS D1043-83. Standard Form of Specification of Shovel and Loaders (Front End) (Jan)(9850).
JIS D 6003-76. Shovel Loaders (Front End).

—Buckets (Machinery)—Capacity Measurement
BSI BS 6422-83. 1983 Volumetric Rating of Loader and Front Loading Excavator Buckets Used for Earth-Moving. 12 pp.
ISO 7546-83. Earth-Moving Machinery—Loader and Front Loading Excavator Buckets—Volumetric Ratings First Edition. 10 pp.
JIS A 8410-88. Calculation Method of Volumetric Ratings of Tractor Shovel and Loading Shovel Buckets. 11 pp.

—Noise—Limitation
EC COM(93) 154-93. Commission Proposal for a Council Directive to Amend Council Directive 86/662/EEC on the Limitation of Noise Emitted by Earth-Moving Machinery (93/C 157/11). 4 pp.
EC 86/662/EEC-86. Council Directive on the Limitation of Noise Emited by Hydraulic Excavators, Rope-Operated Excavators, Dozers, Loaders and Excavator-Loaders. 11 pp.

—Roll Over Protective Structures
BSI BS 5527: Part 1-87. 1987 Roll-Over Protective Structures on Earth-Moving Machinery: Laboratory Tests and Preformance Requirements Part 1: Crawler, Wheel Loaders and Tractors, Backhoe Loaders, Graders, Tractor Scrapers and Articulated Steel Dumpers. 19 pp.
ISO 3471 Pt 1-86. Earth-Moving Machinery—Roll-over Protective Structures—Laboratory Tests and Performance Requirements—Part 1: Crawler, Wheel Loaders and Tractors, Backhoe Loaders, Graders, Tractor Scrapers, Articulated Steer Dumpers First Edition. 18 pp.

—Tires
CNS K4030-80. Dimensions of Tires for Industrial Vehicles and off the Road Service (Feb)(3673). 33 pp.

Frontier Control
Use: Border Control

Frontplates
Use: Wall Plates

Frost Resistance
Scope Note: For frost resistance of specific products or materials, see the material or product
Use: Freezers

Froth Flotation Testing
See Also: Flotation

—Anthracite
ISO 8858 Pt 1-90. Hard Coal—Froth Flotation Testing—Part 1: Laboratory Procedure First Edition. 14 pp.

—Coal
BSI BS 7530: Part 1-92. 1992 Methods for Froth Flotation Testing of Hard Coal Part 1: Laboratory Procedure. 12 pp.

Frozen Food Storage Cabinets
Use: Freezers

Frozen Foods
See Also: Cold Storage; Food
CNS N5143-79. Frozen Prepared Foods (Feb)(4768). 2 pp.
EC 89/108/EEC-88. Council Directive on the Approximation of the Laws of the Member States Relating to Quick-Frozen Foodstuffs for Human Consumption. 4 pp.

—Eggs
CNS N5078-80. Frozen Eggs (Jul)(2812). 3 pp.

—Eggs—Chemical Analysis
CNS N6050-80. Method of Test for Frozen Eggs (Jul)(2923). 3 pp.

—Eggs—Inspection
CNS N6050-80. Method of Test for Frozen Eggs (Jul)(2923). 3 pp.

—Eggs—Microbiological Analysis
CNS N6050-80. Method of Test for Frozen Eggs (Jul)(2923). 3 pp.

—Eggs—Sampling
CNS N6050-80. Method of Test for Frozen Eggs (Jul)(2923). 3 pp.

—Fish
CGSB 32.141M-91. Fresh, Chilled or Frozen Fish and Fish Products. 13 pp.
CNS N5113-84. Frozen Fish (Dec)(3732). 2 pp.
CNS N5121-80. Frozen Roasted Eel (May)(3900). 1 p.
CNS N5140-82. Frozen Minced Fish Meat (Nov)(4640). 2 pp.

—Fish—Chemical Analysis
CNS N6029-84. Method of Test for Frozen Fishes (Aug)(1451). 4 pp.

—Fish—Inspection
CNS N6029-84. Method of Test for Frozen Fishes (Aug)(1451). 4 pp.

—Fish—Microbiological Analysis
CNS N6029-84. Method of Test for Frozen Fishes (Aug)(1451). 4 pp.

—Fish Paste—Chemical Analysis
CNS N6107-82. Method of Test for Frozen Minced Fish Meat (Nov)(4641). 2 pp.

—Fish Paste—Elastic Properties
CNS N6107-82. Method of Test for Frozen Minced Fish Meat (Nov)(4641). 2 pp.

—Fish Paste—Impurities Content
CNS N6107-82. Method of Test for Frozen Minced Fish Meat (Nov)(4641). 2 pp.

—Fish Paste—Temperature Measurement
CNS N6107-82. Method of Test for Frozen Minced Fish Meat (Nov)(4641). 2 pp.

—Fish—Physical Properties
CNS N6029-84. Method of Test for Frozen Fishes (Aug)(1451). 4 pp.

—Fish—Sampling
CNS N6029-84. Method of Test for Frozen Fishes (Aug)(1451). 4 pp.

—Fish—Temperature Measurement
CNS N6029-84. Method of Test for Frozen Fishes (Aug)(1451). 4 pp.

—Frog Legs
CNS N5122-80. Frozen Frog Legs (May)(3925). 2 pp.

—Fruits
CNS N5077-85. Frozen Fruits and Vegetables (Mar)(2772). 2 pp.

—Fruits—Microbiological Analysis
CNS N6048-85. Method of Test for Frozen Fruits and Vegetables (Mar)(2813). 2 pp.

—Fruits—Physical Properties
CNS N6048-85. Method of Test for Frozen Fruits and Vegetables (Mar)(2813). 2 pp.

INTERNATIONAL AND NON-U.S. NATIONAL STANDARDS
SUBJECT INDEX
Fruit

Frozen Foods (Cont.)

—**Fruits—Sampling**
CNS N6048-85. Method of Test for Frozen Fruits and Vegetables (Mar)(2813). 2 pp.

—**Fruits—Sensory Analysis**
CNS N6048-85. Method of Test for Frozen Fruits and Vegetables (Mar)(2813). 2 pp.

—**Fruits—Temperature Measurement**
CNS N6048-85. Method of Test for Frozen Fruits and Vegetables (Mar)(2813). 2 pp.

—**Meat**
CNS N5079-81. Frozen Rabbit Meat (May)(2814). 2 pp.

—**Meat—Inhibitors**
CNS N6121-87. Method of Test for the Inhibitory Substances in Fresh Meat (May)(5916). 2 pp.

—**Microbiological Analysis**
SAA AS 1766.3.4-91. Methods for the Microbiological Examination of Food—Part 3: Examination of Specific Products—Part 3.4: Frozen Foods. 3 pp.

—**Mollusks**
CNS N5114-80. Frozen Aquatic Mollusca (May)(3733). 2 pp.

—**Sausages**
CGSB 32.69M-90. Fresh or Cooked Sausages. 8 pp.

—**Shrimps**
CNS N5055-86. Frozen Shrimp (Mar)(2300). 2 pp.
CNS N5137-80. Freeze Dried Shrimps (May)(4453). 1 p.

—**Shrimps—Chemical Analysis**
CNS N6043-86. Method of Test for Frozen Shrimp (Mar)(2301). 5 pp.
CNS N6142-91. Method of Test for Frozen Shrimps Graded and Packaged on Board (Feb)(6507). 5 pp.

—**Shrimps—Impurities Content**
CNS N6043-86. Method of Test for Frozen Shrimp (Mar)(2301). 5 pp.

—**Shrimps—Microbiological Analysis**
CNS N6043-86. Method of Test for Frozen Shrimp (Mar)(2301). 5 pp.
CNS N6142-91. Method of Test for Frozen Shrimps Graded and Packaged on Board (Feb)(6507). 5 pp.

—**Shrimps—Physical Properties**
CNS N6043-86. Method of Test for Frozen Shrimp (Mar)(2301). 5 pp.
CNS N6142-91. Method of Test for Frozen Shrimps Graded and Packaged on Board (Feb)(6507). 5 pp.

—**Shrimps—Quality Assurance**
CNS N6043-86. Method of Test for Frozen Shrimp (Mar)(2301). 5 pp.
CNS N6142-91. Method of Test for Frozen Shrimps Graded and Packaged on Board (Feb)(6507). 5 pp.

—**Shrimps—Sampling**
CNS N6142-91. Method of Test for Frozen Shrimps Graded and Packaged on Board (Feb)(6507). 5 pp.

—**Shrimps—Temperature Measurement**
CNS N6043-86. Method of Test for Frozen Shrimp (Mar)(2301). 5 pp.
CNS N6142-91. Method of Test for Frozen Shrimps Graded and Packaged on Board (Feb)(6507). 5 pp.

—**Snails**
CNS N5136-79. Frozen Snail Meat (Jul)(4398). 2 pp.

—**Vegetables**
CGSB 32.254M-89. Frozen Fruits, Vegetables and Juices (Supersedes 32-GP-278 Jan 1970, and 32-GP-280C April 1971). 9 pp.
CNS N5067-85. Frozen Pea Pods (Mar)(2418). 2 pp.
CNS N5076-85. Frozen Mushrooms (Mar)(2771). 2 pp.
CNS N5077-85. Frozen Fruits and Vegetables (Mar)(2772). 2 pp.
CNS N5099-85. Frozen Green Asparagus (Mar)(3286). 4 pp.
CNS N5139-85. Frozen White Asparagus (Mar)(4528). 4 pp.

—**Vegetables—Microbiological Analysis**
CNS N6048-85. Method of Test for Frozen Fruits and Vegetables (Mar)(2813). 2 pp.

—**Vegetables—Physical Properties**
CNS N6048-85. Method of Test for Frozen Fruits and Vegetables (Mar)(2813). 2 pp.

—**Vegetables—Sampling**
CNS N6048-85. Method of Test for Frozen Fruits and Vegetables (Mar)(2813). 2 pp.

Frozen Foods (Cont.)

—**Vegetables—Sensory Analysis**
CNS N6048-85. Method of Test for Frozen Fruits and Vegetables (Mar)(2813). 2 pp.

—**Vegetables—Temperature Measurement**
CNS N6048-85. Method of Test for Frozen Fruits and Vegetables (Mar)(2813). 2 pp.

Fructose
See Also: Glucose; Lactose; Sugars
CNS K7280-66. Chemical Reagent (Levulose) (Fructose) (Dec)(1779).

Fructose Content Analysis
See Also: Sugar Content Analysis

—**Fruit Juices—Enzymatic Method**
CNS N6222-91. Method of Test for Fruit and Vegetable Juices and Drinks (Jun)(12633). 2 pp.

—**Vegetable Juices—Enzymatic Method**
CNS N6222-91. Method of Test for Fruit and Vegetable Juices and Drinks (Jun)(12633). 2 pp.

Fructose Syrups
See Also: Glucose Syrups; Maltose Syrups; Sugars
CNS N5215-85. High Fructose Syrup (Sep)(11369). 2 pp.

—**Ash Content**
CNS N6200-85. Method of Test for High Fructose Syrup (Sep)(11370). 3 pp.

—**Color Testing**
CNS N6200-85. Method of Test for High Fructose Syrup (Sep)(11370). 3 pp.

—**Moisture Content**
CNS N6200-85. Method of Test for High Fructose Syrup (Sep)(11370). 3 pp.

—**PH**
CNS N6200-85. Method of Test for High Fructose Syrup (Sep)(11370). 3 pp.

—**Sampling**
CNS N6200-85. Method of Test for High Fructose Syrup (Sep)(11370). 3 pp.

—**Sugar Content**
CNS N6200-85. Method of Test for High Fructose Syrup (Sep)(11370). 3 pp.

—**Turbidity**
CNS N6200-85. Method of Test for High Fructose Syrup (Sep)(11370). 3 pp.

Fruit Juices
Use For: Nectars See Also: Apple Juice; Beverages; Fruit Products; Fruits; Orange Juice
CNS N5065-91. Fruit and Vegetable Juice, Drink and Nectar (Aug)(2377). 13 pp.
EC COM(86) 688-86. Proposal for a Council Directive Amending for the Third Time Directive 75/726/EEC on the Approximation of the Laws of the Member States Concerning Fruit Juices and Certain Similar Products. 16 pp.
EC COM(88) 319-88. Amendment of the Proposal for a Council Directive Amending for the Third Time Directive 75/726/EEC on the Approximation of the Laws of the Member States Concerning Fruit Juices and Certain Similar Products. 3 pp.
EC 75/726/EEC-75. Council Directive on the Approximation of the Laws of the Member States Concerning Fruit Juices and Certain Similar Products. 10 pp.
EC 79/168/EEC-79. Council Directive Amending Directive 75/726/EEC on the Approximation of the Laws of the Member States Concerning Fruit Juices and Certain Similar Products. 2 pp.
EC 81/487/EEC-81. Council Directive Amending for the Second Time Directive 75/726/EEC on the Approximation of the Laws of the Member States Concerning Fruit Juices and Certain Similar Products. 3 pp.
EC 89/394/EEC-89. Council Directive Amending for the Third Time Directive 75/726/EEC on the Approximation of the Laws of the Member States Concerning Fruit Juices and Certain Similar Products. 3 pp.

—**Acidity**
CNS N6091-91. Method of Test for Fruit and Vegetable Juices and Drinks (General Rules) (Jun)(3736). 1 p.
CNS N6216-89. Method of Test for Fruit and Vegetable Juices and Drinks Determination of Acidity (Jul)(12570).

—**Amino Acid Content**
CNS N6221-91. Method of Test for Fruit and Vegetable Juices and Drinks-Determination of Free Amino Acids (Jun)(12632). 3 pp.

Fruit Juices (Cont.)

—**Artificial Sweeteners Content**
CNS N6091-91. Method of Test for Fruit and Vegetable Juices and Drinks (General Rules) (Jun)(3736). 1 p.

—**Ash Content**
CNS N6091-91. Method of Test for Fruit and Vegetable Juices and Drinks (General Rules) (Jun)(3736). 1 p.
CNS N6217-91. Method of Test for Fruit and Vegetable Juices and Drinks-Determination of Ash (Jun)(12571). 1 p.

—**Benzoic Acid Content—Molecular Absorption Spectrometry**
ISO 6560-83. Fruit and Vegetable Products—Determination of Benzoic Acid Content (Benzoic Acid Contents Greater Than 200 mg per Litre or per Kilogram)—Molecular Absorption Spectrometric Method First Edition. 5 pp.

—**Brix Testing**
CNS N6091-91. Method of Test for Fruit and Vegetable Juices and Drinks (General Rules) (Jun)(3736). 1 p.

—**Calcium Content**
CNS N6227-91. Method of Test for Fruit and Vegetable Juices and Drinks-Determination of Sodium, Potassium, Calcium and Magnesium (Jun)(12638). 4 pp.

—**Citric Acid Content—Enzymatic Method**
CNS N6225-91. Method of Test for Fruit and Vegetable Juices and Drinks-Determination of Citric Acid by Enzymatic Analysis (Jun)(12636). 2 pp.

—**Fructose Content—Enzymatic Method**
CNS N6222-91. Method of Test for Fruit and Vegetable Juices and Drinks (Jun)(12633). 2 pp.

—**Glucose Content—Enzymatic Method**
CNS N6222-91. Method of Test for Fruit and Vegetable Juices and Drinks (Jun)(12633). 2 pp.

—**Impurities Content**
CNS N6091-91. Method of Test for Fruit and Vegetable Juices and Drinks (General Rules) (Jun)(3736). 1 p.

—**Insoluble Matter Content**
CNS N6091-91. Method of Test for Fruit and Vegetable Juices and Drinks (General Rules) (Jun)(3736). 1 p.
CNS N6218-89. Method of Test for Fruit and Vegetable Juices and Drinks Determination of Water-Insoluble Solids (Nov) (12629).

—**Isocitric Acid Content—Enzymatic Method**
CNS N6226-91. Method of Test for Fruit and Vegetable Juices and Drinks-Determination of Isocitric Acid by Enzymatic Analysis (Jun)(12637). 3 pp.

—**Magnesium Content**
CNS N6227-91. Method of Test for Fruit and Vegetable Juices and Drinks-Determination of Sodium, Potassium, Calcium and Magnesium (Jun)(12638). 4 pp.

—**Nitrogen Content**
CNS N6091-91. Method of Test for Fruit and Vegetable Juices and Drinks (General Rules) (Jun)(3736). 1 p.
CNS N6219-91. Method of Test for Fruit and Vegetable Juices and Drinks-Determination of Amino Nitrogen (Jun)(12630). 2 pp.

—**Organic Acid Content—Liquid Chromatography**
CNS N6224-91. Method of Test for Fruit and Vegetable Juices and Drinks-Organic Acid Analysis by HPLC Method (Jun)(12635). 2 pp.

—**Phosphorus Content**
CNS N6220-91. Method of Test for Fruit and Vegetable Juices and Drinks-Determination of Total Phosphorus (Jun)(12631). 2 pp.

—**Potassium Content**
CNS N6227-91. Method of Test for Fruit and Vegetable Juices and Drinks-Determination of Sodium, Potassium, Calcium and Magnesium (Jun)(12638). 4 pp.

—**Powdered**
CGSB 32.283M-87. Beverage Powders, Fruit-Flavoured. 7 pp.

INDUSTRY STANDARDS

Fruit Juices (Cont.)

—Sodium Content
CNS N6227-91. Method of Test for Fruit and Vegetable Juices and Drinks-Determination of Sodium, Potassium, Calcium and Magnesium (Jun)(12638). 4 pp.

—Soluble Matter Content
CNS N6215-91. Method of Test for Fruit and Vegetable Juices and Drinks-Determination of Soluble Solids (Jun)(12569). 2 pp.

—Soluble Matter Content—Pycnometric Analysis
ISO 2172-83. Fruit Juice—Determination of Soluble Solids Content—Pyknometric Method First Edition. 5 pp.

—Soluble Matter Content—Refractive Index
ISO 2173-78. Fruit and Vegetable Products—Determination of Soluble Solids Content—Refractometric Method First Edition. 6 pp.

—Sucrose Content—Enzymatic Method
CNS N6222-91. Method of Test for Fruit and Vegetable Juices and Drinks (Jun)(12633). 2 pp.

—Sugar Content—Liquid Chromatography
CNS N6223-91. Method of Test for Fruit and Vegetable Juices and Drinks-Sugar Analysis by HPLC Method (Jun)(12634). 2 pp.

Fruit Products

See Also: Fruit Juices; Fruits

—Carotene Content
ISO 6558 Pt 2-92. Fruits, Vegetables and Derived Products—Determination of Carotene Content—Part 2: Routine Methods First Edition. 7 pp.

—Iron Content—Atomic Absorption Spectrometry
ISO 9526-90. Fruits, Vegetables and Derived Products—Determination of Iron Content by Flame Atomic Absorption Spectrometry First Edition. 6 pp.

—Pie Fillings—Canned
CGSB 32.234-92. Canned Pie Fillings. 7 pp.

Fruits

See Also: Apples; Apricots; Avocados; Bananas; Bilberries; Bitter Melons; Blueberries; Cherries; Citrus Fruits; Cranberries; Food; Fruit Juices; Fruit Products; Grapefruits; Grapes; Kiwifruit; Lemons; Limes; Litchis; Loquats; Mandarin Oranges; Mangos; Mulberries; Peaches; Pears; Pineapples; Plums; Raisins; Seeds; Strawberries; Tomatoes; Watermelons
EC EEC/220/92-92. Council Regulation Amending Regulation (EEC) No 3285/83 Laying down General Rules for the Extension of Certain Rules Issued by Producers' Organizations in the Fruit and Vegetable Sector. 2 pp.

—Acidity—Potentiometric Analysis
CNS N6167-91. Method of Test for Fruit and Vegetable Products-Determination of Titratable Acidity (Jun)(8626). 3 pp.
ISO 750-81. Fruit and Vegetable Products—Determination of Titratable Acidity First Edition. 5 pp.

—Ascorbic Acid Content
ISO 6557 Pt 1-86. Fruits, Vegetables and Derived Products—Determination of Ascorbic Acid—Part 1: Reference Method First Edition. 5 pp.
ISO 6557 Pt 2-84. Fruits, Vegetables and Derived Products—Determination of Ascorbic Acid Content—Part 2: Routine Methods First Edition. 6 pp.

—Ash Alkalinity
ISO 5520-81. Fruits, Vegetables and Derived Products—Determination of Alkalinity of Total Ash and of Water-Soluble Ash First Edition. 5 pp.

—Ash Content
CNS N6166-91. Method of Test for Fruit and Vegetable Products-Determination of Ash Insoluble in Hydrochloric Acid (Jun)(8625). 2 pp.
ISO 763-82. Fruit and Vegetable Products—Determination of Ash Insoluble in Hydrochloric Acid First Edition. 4 pp.

—Benzoic Acid Content—Molecular Absorption Spectrometry
ISO 6560-83. Fruit and Vegetable Products—Determination of Benzoic Acid Content (Benzoic Acid Contents Greater Than 200 mg per Litre or per Kilogram)—Molecular Absorption Spectrometric Method First Edition. 5 pp.

Fruits (Cont.)

—Benzoic Acid Content—Spectrophotometry
ISO 5518-78. Fruits, Vegetables and Derived Products—Determination of Benzoic Acid Content—Spectrophotometric Method First Edition; (Erratum—Aug 1978). 6 pp.

—Cadmium Content—Atomic Absorption Spectrometry
ISO 6561-83. Fruits, Vegetables and Derived Products—Determination of Cadmium Content—Flameless Atomic Absorption Spectrometric Method First Edition. 5 pp.

—Canned
CNS N5018-87. Canned Fruits (Sep)(1252). 8 pp.

—Carotene Content
ISO 6558 Pt 2-92. Fruits, Vegetables and Derived Products—Determination of Carotene Content—Part 2: Routine Methods First Edition. 7 pp.

—Containers
CNS Z5033-88. Corrugated Paperboard Containers for Fresh Fruits (Dec)(3247). 4 pp.

—Copper Content—Photometry
ISO 3094-74. Fruit and Vegetable Products—Determination of Copper Content—Photometric Method First Edition. 4 pp.

—Decomposition Methods
ISO 5515-79. Fruits, Vegetables and Derived Products—Decomposition of Organic Matter Prior to Analysis—Wet Method First Edition. 5 pp.
ISO 5516-78. Fruits, Vegetables and Derived Products—Decomposition of Organic Matter Prior to Analysis—Ashing Method First Edition. 4 pp.

—Dried
CGSB 32.276M-89. Dried Fruit. 9 pp.
CNS N5028-84. Preserved Fruits (May)(1346). 2 pp.

—Dried—Ash Content
CNS N6049-85. Method of Test for Dehydrated Fruits and Vegetables (Mar)(2896). 2 pp.

—Dried—Fumigant Residues
CNS N6049-85. Method of Test for Dehydrated Fruits and Vegetables (Mar)(2896). 2 pp.

—Dried—Glossaries
ISO 4125-91. Dry Fruits and Dried Fruits—Definitions and Nomenclature Second Edition. 9 pp.

—Dried—Impurities Content
CNS N6049-85. Method of Test for Dehydrated Fruits and Vegetables (Mar)(2896). 2 pp.

—Dried—Microbiological Analysis
CNS N6049-85. Method of Test for Dehydrated Fruits and Vegetables (Mar)(2896). 2 pp.

—Dried—Moisture Content
CNS N5027-85. Dehydrated Fruits and Vegetables (Mar)(1345). 3 pp.
CNS N6049-85. Method of Test for Dehydrated Fruits and Vegetables (Mar)(2896). 2 pp.

—Dried—Packaging
CNS N5027-85. Dehydrated Fruits and Vegetables (Mar)(1345). 3 pp.

—Dried—Sampling
CNS N6049-85. Method of Test for Dehydrated Fruits and Vegetables (Mar)(2896). 2 pp.

—Dried—Sensory Analysis
CNS N6049-85. Method of Test for Dehydrated Fruits and Vegetables (Mar)(2896). 2 pp.

—Dry Matter Content—Dehydration
ISO 1026-82. Fruit and Vegetable Products—Determination of Dry Matter Content by Drying Under Reduced Pressure and of Water Content by Azeotropic Distillation First Edition. 6 pp.

—Ethanol Content
ISO 2448-73. Fruit and Vegetable Products—Determination of Ethanol First Edition. 5 pp.

—Fatty Acid Content
ISO 6632-81. Fruits, Vegetables and Derived Products—Determination of Volatile Acidity First Edition. 11 pp.

—Formic Acid Content—Gravimetric Analysis
ISO 6638 Pt 1-85. Fruit and Vegetable Products—Determination of Formic Acid Content—Part 1: Gravimetric Method First Edition. 5 pp.

Fruits (Cont.)

—Formic Acid Content—Volumetric Analysis
ISO 6638 Pt 2-84. Fruit and Vegetable Products—Determination of Formic Acid Content—Part 2: Titrimetric Method First Edition; (Corrected and Reprinted -1985). 5 pp.

—Frozen
CNS N5077-85. Frozen Fruits and Vegetables (Mar)(2772). 2 pp.

—Frozen—Microbiological Analysis
CNS N6048-85. Method of Test for Frozen Fruits and Vegetables (Mar)(2813). 2 pp.

—Frozen—Sampling
CNS N6048-85. Method of Test for Frozen Fruits and Vegetables (Mar)(2813). 2 pp.

—Frozen—Sensory Analysis
CNS N6048-85. Method of Test for Frozen Fruits and Vegetables (Mar)(2813). 2 pp.

—Frozen—Temperature Measurement
CNS N6048-85. Method of Test for Frozen Fruits and Vegetables (Mar)(2813). 2 pp.

—Frozen—Weight Measurement
CNS N6048-85. Method of Test for Frozen Fruits and Vegetables (Mar)(2813). 2 pp.

—Glass Containers—Capacity Measurement
CEN EN 76-78. Packages for Certain Pre-Packed Foodstuffs: Capacities of Glass and Metal Containers. 6 pp.

—Glossaries
ISO 1956 Pt 1-82. Fruits and Vegetables—Morphological and Structural Terminology—Part 1 First Edition. 25 pp.
ISO 1956 Pt 2-89. Fruits and Vegetables—Morphological and Structural Terminology—Part 2 First Edition. 35 pp.
ISO 1990 Pt 1-82. Fruits—Nomenclature—First List First Edition. 14 pp.
ISO 1990 Pt 2-85. Fruits—Nomenclature—Second List First Edition. 7 pp.

—Hydroxymethylfurfural Content
ISO 7466-86. Fruit and Vegetable Products—Determination of 5-Hydroxymethylfurfural (5-HMF) Content First Edition. 5 pp.

—Insoluble Matter Content
CNS N6163-91. Method of Test for Fruit and Vegetable Products-Determination of Water-Insoluble Solids (Jun)(8622). 3 pp.
ISO 751-81. Fruit and Vegetable Products—Determination of Water-Insoluble Solids Content First Edition. 4 pp.

—Iron Content—Atomic Absorption Spectrometry
ISO 9526-90. Fruits, Vegetables and Derived Products—Determination of Iron Content by Flame Atomic Absorption Spectrometry First Edition. 6 pp.

—Iron Content—Photometry
ISO 5517-78. Fruits, Vegetables and Derived Products—Determination of Iron Content—1,10-Phenanthroline Photometric Method First Edition. 5 pp.

—Lead Content—Atomic Absorption Spectrometry
ISO 6633-84. Fruits, Vegetables and Derived Products—Determination of Lead Content—Flameless Atomic Absorption Spectrometric Method First Edition. 5 pp.

—Mercury Content—Atomic Absorption Spectrometry
ISO 6637-84. Fruits, Vegetables and Derived Products—Determination of Mercury Content—Flameless Atomic Absorption Method First Edition. 7 pp.

—Metal Containers—Capacity Measurement
CEN EN 76-78. Packages for Certain Pre-Packed Foodstuffs: Capacities of Glass and Metal Containers. 6 pp.

—Mineral Impurities Content
CNS N6165-87. Methods of Test for Fruit and Vegetable Products (Determination of Mineral Impurities) (Jun)(8624). 3 pp.
ISO 762-82. Fruit and Vegetable Products—Determination of Mineral Impurities Content First Edition. 4 pp.

INTERNATIONAL AND NON-U.S. NATIONAL STANDARDS
SUBJECT INDEX

Fruits (Cont.)
—Nitrate Content—Molecular Absorption Spectrometry
ISO 6635-84. Fruits, Vegetables and Derived Products—Determination of Nitrite and Nitrate Content—Molecular Absorption Spectrometric Method First Edition. 6 pp.

—Nitrite Content—Molecular Absorption Spectrometry
ISO 6635-84. Fruits, Vegetables and Derived Products—Determination of Nitrite and Nitrate Content—Molecular Absorption Spectrometric Method First Edition. 6 pp.

—Nursery Stock
BSI BS 3936: Part 3-90. 1990 Nursery Stock Part 3: Fruit Plants. 10 pp.

—Packaging
ISO 7558-88. Guide to the Prepacking of Fruits and Vegetables First Edition. 8 pp.

—Packaging—Arrangement—Cargo Vehicles
ISO 6661-83. Fresh Fruit and Vegetables—Arrangement of Parallelepipedic Packages in Land Transport Vehicles First Edition. 6 pp.

—Pesticide Residues
EC COM(82) 883-82. Proposal for a Council Directive Amending Annex 11 to Directive 76/895/EEC Relating to the Fixing of Maximum Levels for Pesticide Residues in and on Fruit and Vegetables. 4 pp.
EC COM(88) 798-89. Proposal for a Council Regulation (EEC) on the Fixing of Maximum Levels for Pesticide Residues in and on Certain Products of Plant Origin, Including Fruit and Vegetables, and Amending Directive 76/895/EEC as Regards Procedural Rules. 28 pp.
EC 76/895/EEC-76. Council Directive Relating to the Fixing of Maximum Levels for Pesticide Residues in and on Fruit and Vegetables. 6 pp.
EC 90/642/EEC-90. Council Directive on the Fixing of Maximum Levels for Pesticide Residues in and on Certain Products of Plant Origin, Including Fruit and Vegetables. 10 pp.
EC 93/58/EEC-93. Council Directive Amending Annex II to Directive 76/895/EEC Relating to the Fixing of Maximum Levels for Pesticide Residues in and on Fruit and Vegetables and the Annex to Directive 90/642/EEC Relating to the Fixing of Maximum Levels for Pesticide Residues in. 34 pp.

—PH—Potentiometric Analysis
ISO 1842-91. Fruit and Vegetable Products—Determination of pH Second Edition. 5 pp.

—Preserved—Sampling
CNS N6182-84. Method of Test for Preserved Fruits (May)(10886). 2 pp.

—Preserved—Sensory Analysis
CNS N6182-84. Method of Test for Preserved Fruits (May)(10886). 2 pp.

—Preserved—Visual Inspection
CNS N6182-84. Method of Test for Preserved Fruits (May)(10886). 2 pp.

—Ripening
ISO 3659-77. Fruit and Vegetables—Ripening After Cold Storage First Edition. 6 pp.

—Sampling
ISO 874-80. Fresh Fruits and Vegetables—Sampling First Edition. 5 pp.

—Arsenic Content—Spectrophotometry
ISO 6634-82. Fruits, Vegetables and Derived Products Determination of Arsenic Content—Silver Diethyldithiocarbamate Spectrophotometric Method First Edition. 7 pp.

—Silica Content
CNS N6166-91. Method of Test for Fruit and Vegetable Products-Determination of Ash Insoluble in Hydrochloric Acid (Jun)(8625). 2 pp.

—Sorbic Acid Content—Colorimetry
ISO 5519-78. Fruits, Vegetables and Derived Products—Determination of Sorbic Acid Content First Edition. 8 pp.

—Sorbic Acid Content—Spectrophotometry
ISO 5519-78. Fruits, Vegetables and Derived Products—Determination of Sorbic Acid Content First Edition. 8 pp.

—Storage
ISO 2169-81. Fruits and Vegetables—Physical Conditions in Cold Stores—Definitions and Measurement Second Edition. 7 pp.
ISO 3659-77. Fruit and Vegetables—Ripening After Cold Storage First Edition. 6 pp.

Fruits (Cont.)
—Storage (Cont.)
ISO 6949-88. Fruits and Vegetables—Principles and Techniques of the Controlled Atmosphere Method of Storage First Edition. 10 pp.

—Sulfur Dioxide Content
ISO 5521-81. Fruits, Vegetables and Derived Products—Qualitative Method for the Detection of Sulphur Dioxide First Edition. 4 pp.
ISO 5522-81. Fruits, Vegetables and Derived Products—Determination of Total Sulphur Dioxide Content First Edition. 9 pp.
ISO 5523-81. Liquid Fruit and Vegetable Products—Determination of Sulphur Dioxide Content (Routine Method) First Edition. 5 pp.

—Syrup Pack
CNS N5028-84. Preserved Fruits (May)(1346). 2 pp.

—Tin Content
CNS N6164-82. Methods of Test for Fruit and Vegetable Products (Determination of Tin) (Mar)(8623). 3 pp.
ISO 2447-74. Fruit and Vegetable Products—Determination of Tin First Edition. 4 pp.

—Water Content—Azeotropic Distillation
ISO 1026-82. Fruit and Vegetable Products—Determination of Dry Matter Content by Drying Under Reduced Pressure and of Water Content by Azeotropic Distillation First Edition. 6 pp.

—Zinc Content—Atomic Absorption Spectrometry
ISO 6636 Pt 2-81. Fruits, Vegetables and Derived Products—Determination of Zinc Content—Part 2: Atomic Absorption Spectrometric Method First Edition. 6 pp.

—Zinc Content—Polarographic Analysis
ISO 6636 Pt 1-86. Fruits, Vegetables and Derived Products—Determination of Zinc Content—Part 1: Polarographic Method First Edition. 5 pp.

—Zinc Content—Spectrometry
ISO 6636 Pt 3-83. Fruit and Vegetable Products—Determination of Zinc Content—Part 3: Dithizone Spectrometric Method First Edition. 5 pp.

Fryers (Food)
Use: Deep Fat Fryers

Frying Utensils
Use For: Electric Frying Pans See Also: Appliances; Cooking Appliances; Cooking Utensils; Deep Fat Fryers

—Electric
CGSB CAN/CGSB-52.25-M88. Frypans, Tilting, Electric or Gas. 10 pp.

—Electric—Safety
BSI BS 3456: Sec 102.13-91. (WITHDRAWN) 1991 Safety of Household and Similar Electrical Appliances Part 102: Particular Requirements Section 102.13: Frying Pans, Deep Fat Fryers and Similar Appliances (SUPERSEDED BY BS EN 60335-2-13: 1991). 24 pp.
BSI BS 3456: Sec 2.6-70. (WITHDRAWN) 1970 Amd 2 1972-1981 Edition. Specification for Safety of Household Electrical Appliances: Part 2: Particular Requirements: Section 2.6: Frying Pans, Grills, Plate Warmers, and Other Dry Cooking Appliances (Superseded by BS 3456: Sec 3.16 & 3.17). 18 pp.
BSI BS EN 60335-2-13-91. 1991 Safety of Household and Similar Electrical Appliances Part 2: Particular Requirements Section 2.13: Frying Pans, Deep Fat Fryers and Similar Appliances (L). 24 pp.
BSI BS EN 60335-2-13-01. 1991 Amd 1 Safety of Household and Similar Electrical Appliances Part 2: Particular Requirements Section 2.13: Frying Pans, Deep Fat Fryers and Similar Appliances (AMD 7585) March 15, 1993 (L). 30 pp.
CENELEC EN 60 335-2-13-90. Safety of Household and Similar Electrical Appliances Part 2: Particular Requirements for Frying Pans, Deep Fat Fryers and Similar Appliances. 7 pp.
CENELEC EN 60335-2-13/A1-92. AMD 1 Safety of Household and Similar Electrical Appliances Part 2: Particular Requirements for Frying Pans, Deep Fat Fryers and Similar Appliances. 4 pp.

—Gas
CGSB CAN/CGSB-52.25-M88. Frypans, Tilting, Electric or Gas. 10 pp.

—Tilting
CGSB CAN/CGSB-52.25-M88. Frypans, Tilting, Electric or Gas. 10 pp.

FSK
Use: Frequency Shift Keying

FTA
Use: Fault Tree Analysis

FTAM
Use: Application Layer (OSI)—File Transfer, Access, and Management

Fuchsin
Use For: Basic Fuchsin
CNS K7236-67. Chemical Reagent (Fuchsine) (Oct)(1735).
JIS K 8804-86. Fuchsine Basic.

Fuel Assemblies (Nuclear)
Use: Nuclear Fuel Assemblies

Fuel Burners
Use: Burners

Fuel Circuits
Use: Fuel Lines

Fuel Consumption
See Also: Consumption Rate; Energy Consumption

—Automotive Engines
CNS D3014-87. Method of Test for Fuel Consumption for Passenger Cars (Sep)(2733). 15 pp.
EC 80/1268/EEC-80. Council Directive on the Approximation of the Laws of the Member States Relating to the Fuel Consumption of Motor Vehicles. 10 pp.
ISO 1585-92. Road Vehicles—Engine Test Code—Net Power Third Edition. 29 pp.
ISO 2534-74. Road Vehicles—Engine Test Code—Gross Power First Edition; (Erratum—Aug 1974). 17 pp.
JIS D 1012-83. Fuel Consumption Test Methods of Automobiles. 26 pp.
SAA AS 2077-82. Methods of Test for Fuel Consumption of Passenger Cars, Their Derivatives and Multi-Purpose Passenger Cars. 32 pp.
SNZ NZS 5420-80. Methods of Test for Petrol Consumption of Passenger Cars. 19 pp.

—Automotive Engines—Identification Systems
SNZ NZS 5421-81. Specification for Fuel Consumption Labelling of Passenger Cars. 7 pp.

—Brush Saws
ISO 8893-89. Forestry Machinery—Portable Brush-Saws—Engine Performance and Fuel Consumption First Edition. 6 pp.

—Internal Combustion Engines
ISO 3046 Pt 1-86. Reciprocating Internal Combustion Engines—Performance—Part 1: Standard Reference Conditions and Declarations of Power, Fuel Consumption and Lubricating Oil Consumption Third Edition; (Amendment 1-1987). 19 pp.

—Military Operations—Land Forces
NATO STANAG 2115 ED 4 AMD 2-87. Fuel Consumption Unit. 7 pp.

—Mopeds
ISO 4164-78. Road Vehicles—Mopeds—Engine Test Code—Net Power First Edition. 13 pp.

—Motorcycles
ISO 7860-83. Road Vehicles—Motorcycles—Method of Measuring Fuel Consumption First Edition. 19 pp.
JIS D 1033-75. Method of Fuel Consumption Test for Motor Cycles.
JIS D 1033-75. Method of Fuel Consumption Test for Motor Cycles. 5 pp.

—Motorcycles—Dynamometers
CNS D3029-87. Method of Test for Fuel Consumption for Motorcycles (Sep)(3105). 13 pp.
ISO 11486-93. Two-Wheeled Motorcycles—Fuel Consumption Measurements—Chassis Dynamometer Setting by Coastdown Method First Edition. 17 pp.

Fuel Couplings
See Also: Couplings

—Aircraft—Blanking Caps
SBAC AS 6544 ISSUE 1. Blanking Cap for 1 1/2 Inch Fuel Coupling.
SBAC AS 6552 ISSUE 1. Blanking Cap for 2 1/2 Inch Fuel Coupling.

—Aircraft—Flanged—Half
SBAC AS 6543 ISSUE 1. G.A. Flanged Half for 1 1/2 Inch Fuelling Coupling (Non-Self Sealing Type).
SBAC AS 6550 ISSUE 1. G.A. Flanged Half for 1 1/2 Inch Re-Fuelling Coupling Self Sealing Type.
SBAC AS 6551 ISSUE 1. G.A. Flanged Half for 2 1/2 Inch Re-Fuelling Coupling (Open Ended Type).
SBAC AS 6558 ISSUE 1. G.A. Flanged Half for 2 1/2 Inch Re-Fuelling Coupling (Self Sealing Type).

INTERNATIONAL AND NON-U.S. NATIONAL STANDARDS
SUBJECT INDEX — Fuel

Fuel Couplings (Cont.)
—Aircraft—Pressure
NATO STANAG 3105 ED 4 AMD 0-91. Aircraft Pressure Fuelling Connections. 9 pp.
NATO STANAG 3105 ED 4 AMD 2-91. Aircraft Pressure Fuelling Connections. 11 pp.

—Replenishment at Sea
NATO STANAG 1222 ED 1 AMD 4-79. Replenishment at Sea—Single Probe Coupling. 10 pp.

Fuel Dispensing Equipment
See Also: Fuel Hoses; Gas Distribution; Gasoline
—Aircraft
JIS W 1904-87. Fitting, Flared Tube, Fluid Connection.

—Aircraft—Couplings
SBAC AS 2952 ISSUE 8. G.A. Flanged Half for 1 1/2 Inch Fuelling Coupling (Non-Self Sealing Type). 1 p.
SBAC AS 2953 ISSUE 8. G.A. Flanged Half for 1 1/2 Inch Re-Fuelling Coupling (Self Sealing Type). 1 p.
SBAC AS 2969 ISSUE 5. G.A. Flanged Half for 2 1/2 Inch Re-Fuelling Coupling (Open Ended Type). 1 p.

—Aircraft—Couplings—Blanking Caps
SBAC AS 2941 ISSUE 2. Blanking Cap for 1 1/2 Inch Fuel Coupling. 1 p.
SBAC AS 2946 ISSUE 2. Blanking Cap for 2 1/2 Inch Fuel Coupling. 1 p.

—Aircraft—Couplings—Springs (Elastic)
SBAC AS 2959 ISSUE 5. Spring. 1 p.
SBAC AS 2980 ISSUE 3. Spring. 1 p.

—Aircraft—Couplings—Valves
SBAC AS 2958 ISSUE 4. Valve. 1 p.
SBAC AS 2979 ISSUE 3. Valve. 1 p.

—Aircraft—Filling Connectors
ISO 45-90. Aircraft—Pressure Refuelling Connections Second Edition. 7 pp.

—Automotive
CSA B346-M1980. Power-Operated Dispensing Devices for Flammable Liquids; (Gen Instr 1 Thru 2). 21 pp.
ISO 9158-88. Road Vehicles—Nozzle Spouts for Unleaded Gasoline First Edition. 4 pp.
ISO 9159-88. Road Vehicles—Nozzle Spouts for Leaded Gasoline and Diesel Fuel First Edition. 4 pp.

—Compressed Natural Gas—Pipelines
CSA B51-M1991. Boiler, Pressure Vessel, and Pressure Piping Code; (Gen Instr 1). 96 pp.

—Compressed Natural Gas—Storage Tanks
CSA B51-M1991. Boiler, Pressure Vessel, and Pressure Piping Code; (Gen Instr 1). 96 pp.

—Earthmoving Equipment—Openings
JIS A 8912-88. Earth-Moving Machinery—Dimensions of Fuel Filler Opening (ISO 3541-1985).

—Flammable Liquids—Electrical Equipment
CSA C22.2 NO 22-M1986. Electrical Equipment for Flammable and Combustible Fuel Dispensers (R 1992); (Gen Instr 1). 19 pp.
CSA 1169 Bull. Electrical Bulletin 1169 June 27, 1978 to C22.2 NO 22. 2 pp.
SNZ NZS 6109: Part 1-88. Electrical Systems of Dispensing Equipment for Explosive Atmospheres Part 1: Flammable Liquids Dispensing Equipment. 16 pp.

—Gasoline—Vapor Control
CGSB CAN/CGSB-3.1000-M91. Standard for Vapour Control Systems in Gasoline Distribution Networks. 40 pp.

—Liquefied Petroleum Gas—Electrical Equipment
SNZ NZS 6109: Part 2-88. Electrical Systems of Dispensing Equipment for Explosive Atmospheres Part 2: Liquefied Petroleum Gas Dispensing Equipment. 16 pp.

—Liquid Fuels—Electrostatic Safety
NATO STANAG 3682 ED 4 AMD 0-90. Electrostatic Safety Connection Procedures for Liquid Fuel Loading/ Unloading Operations During Ground Transfer. 7 pp.
NATO STANAG 3682 ED 4 AMD 1-90. Electrostatic Safety Connection Procedures for Liquid Fuel Loading/ Unloading Operations During Ground Transfer. 7 pp.

Fuel Dispensing Equipment (Cont.)
—Liquid Fuels—Electrostatic Safety (Cont.)
NATO STANAG 3682 ED 4 AMD 2-90. Electrostatic Safety Connection Procedures for Liquid Fuel Loading/ Unloading Operations During Ground Transfer. 8 pp.

—Service Stations—Construction
BSI BS 7117: Part 1-91. 1991 Metering Pumps and Dispensers to Be Installed at Filling Stations and Used to Dispense Liquid Fuel Part 1: Specification for Construction. 29 pp.
BSI BS 7117: Part 1-89. 1989 Amd 1 Metering Pumps and Dispensers to Be Installed at Filling Stations and Used to Dispense Liquid Fuel Part 1: Specification for Construction (AMD 6287) April 28, 1989. 27 pp.

—Service Stations—Installation
BSI BS 7117: Part 2-91. 1991 Metering Pumps and Dispensers to Be Installed at Filling Stations and Used to Dispense Liquid Fuel Part 2: Guide to Installation. 13 pp.

—Service Stations—Maintenance
BSI BS 7117: Part 3-91. 1991 Metering Pumps and Dispensers to Be Installed at Filling Stations and Used to Dispense Liquid Fuel Part 3: Guide to Maintenance After Installation. 16 pp.

—Ships—Couplings
ISO 3926-80. Shipbuilding—Inland Navigation—Couplings for Oil and Fuel Reception—Mating Dimensions Second Edition. 5 pp.
NATO STANAG 1223 ED 2 AMD 2-83. Standardization of the Breakable Spool Coupling for Astern and Abeam Fuelling at Sea. 11 pp.

—Ships—Quick Disconnect Couplings
NATO STANAG 1221 ED 1 AMD 1-80. Standardization of the Use of the 6 Inch MK II Quick Release Coupling for Astern Fuelling. 10 pp.
NATO STANAG 1221 ED 1 AMD 2-80. Standardization of the Use of the 6 Inch MK II Quick Release Coupling for Astern Fuelling. 8 pp.

Fuel Filters
See Also: Oil Filters; Separators (Mechanical)
JIS B 9901-78. Method of Test for Performance of Gas-Removal Filters (R 1989). 15 pp.
MOD UK DSTAN 49-3-01. Performance Requirements and Test Methods for Filter Separators, Gasoline and Kerosine Fuels Issue 2; Amendment 1. 24 pp.
MOD UK DSTAN 49-4-01. Filter Elements, Fluid Pressure (Coalescer Elements), Gasoline and Kerosine Fuels Issue 1; Amendment 1. 13 pp.

—Aircraft
MOD UK DSTAN 49-5-01. Filter, Fluid, Pressure (Precoat Type for Aviation Fuels) Issue 1; Amendment 3. 12 pp.
SBAC AS 3220 ISSUE 1. Filter—Fuel Type 'A' Assembly 100 G.P.H. 1 p.
SBAC AS 3250 ISSUE 2. Filter Fuel Type 'B', Assembly 200 G.P.H.. 1 p.
SBAC AS 3310 ISSUE 2(I). 300 G.P.H. Fuel Filter Assembly. 1 p.

—Aircraft—Assembly
SBAC AS 4725 ISSUE 1. Assembly of 500 G.P.H. Filter. 1 p.
SBAC AS 5420 ISSUE 2. 300 GPH, Fuel Filter Assembly. 1 p.

—Aircraft—Bowls
SBAC AS 4728 ISSUE 2. Filter Bowl. 1 p.
SBAC AS 4731 ISSUE 2. Filter Bowl Sub Assembly. 1 p.

—Aircraft—Bowls—Assembly
SBAC AS 4726 ISSUE 2. Assembly of Filter Bowl. 1 p.

—Aircraft—Bowls—Seats
SBAC AS 4727 ISSUE 2. Filter Bowl, Seat. 1 p.

—Aircraft—Casings
SBAC AS 3221 ISSUE 1. Casing—Top Half Filter Fuel Type 'A'. 1 p.
SBAC AS 3223 ISSUE 1. Casing—Bottom Half—Filter Fuel Type 'A'. 1 p.
SBAC AS 3251 ISSUE 1. Casing—Top Half Filter Fuel Type B. 1 p.
SBAC AS 3253 ISSUE 1. Casing—Bottom Half Filter Fuel Type 'B'. 1 p.

—Aircraft—D Rings
SBAC AS 3273 ISSUE 1. 'D', Ring Filter Fuel Type 'B'. 1 p.

—Aircraft—Differential Pressure Gages
NATO STANAG 3583 ED 2 AMD 3-74. Standards of Accuracy for Differential Pressure Gauges for Aviation Fuel Filters and Filter Separators. 8 pp.

Fuel Filters (Cont.)
—Aircraft—Differential Pressure Gages (Cont.)
NATO STANAG 3583 ED 3 AMD 0-93. Standards of Accuracy for Differential Pressure Gauges for Aviation Fuel Filters and Filter Separator Vessels. 7 pp.
NATO STANAG 3583 ED 3 AMD 1-93. Standards of Accuracy for Differential Pressure Gauges for Aviation Fuel Filters and Filter Separator Vessels. 8 pp.

—Aircraft—Fin Centers
SBAC AS 4730 ISSUE 3. Fin Centre for Filter Pack Support. 1 p.

—Aircraft—Fins
SBAC AS 4729 ISSUE 2. Fin. 1 p.

—Aircraft—Heads
SBAC AS 5426 ISSUE 2. Filter Head. 1 p.

—Aircraft—Lids
SBAC AS 4751 ISSUE 1. Lid Assembly—Flush Filler Cap Mark II. 1 p.

—Aircraft—Packs—Assembly
SBAC AS 5421 ISSUE 1. Filter Pack Assembly. 1 p.
SBAC AS 5434 ISSUE 2. Filter Pack Assembly. 1 p.

—Aircraft—Packs—Rods
SBAC AS 5422 ISSUE 1. Rod for Filter Pack. 1 p.
SBAC AS 5435 ISSUE 1. Rod for Filter Pack. 1 p.

—Aircraft—Plugs
SBAC AS 5423 ISSUE 2. Flanged Lug. 1 p.

—Aircraft—Screens
SBAC AS 3222 ISSUE 1. Screen Filter Fuel Type 'A'. 1 p.
SBAC AS 3252 ISSUE 1. Screen-Filter Fuel Type B. 1 p.

—Aircraft—Springs (Elastic)
SBAC AS 3225 ISSUE 1. Spring Filter Fuel Type 'A'. 1 p.
SBAC AS 3255 ISSUE 1. Spring Filter Fuel Type B. 1 p.

—Aircraft—Stirrups
SBAC AS 3243 ISSUE 2. Stirrup—Filter Fuel Type 'A'. 1 p.
SBAC AS 3268 ISSUE 2. Stirrup—Filter Fuel Type 'B'. 1 p.
SBAC AS 3315 ISSUE 2. Stirrup Assembly. 1 p.
SBAC AS 3324 ISSUE 1. Stirrup—Filter Sump. 1 p.

—Aircraft—Stirrups—Clamping Screws
SBAC AS 3325 ISSUE 2. Clamping, Screw—Filter Stirrup. 1 p.

—Aircraft—Stirrups—Distance Pieces
SBAC AS 3245 ISSUE 1. Stirrup Distance, Piece—Filter Fuel Type 'A'. 1 p.
SBAC AS 3271 ISSUE 1. Stirrup Distance, Piece Filter Fuel Type 'B'. 1 p.

—Aircraft—Stirrups—Eyebolts
SBAC AS 3244 ISSUE 1. Stirrup Eyebolt—Filter Fuel Type 'A'. 1 p.
SBAC AS 3269 ISSUE 1. Stirrup Eyebolt—Filter Fuel Type 'B'. 1 p.

—Aircraft—Stirrups—Nuts (Fasteners)
SBAC AS 3242 ISSUE 1. Stirrup Nut Filter Fuel Type 'A'. 1 p.
SBAC AS 3270 ISSUE 1. Stirrup Nut—Filter Fuel Type 'B'. 1 p.

—Aircraft—Strainers
SBAC AS 3226 ISSUE 3. Strainer Assembly—Filter Fuel Type 'A'. 1 p.
SBAC AS 3256 ISSUE 2. Strainer Assembly—Filter Fuel Type 'B'. 1 p.
SBAC AS 3257 ISSUE 2. Bottom Cup—Strainer—Filter Fuel Type 'B'. 1 p.

—Aircraft—Strainers—Caps
SBAC AS 3235 ISSUE 2. Top Cap—Strainer—Filter Fuel Type 'A'. 1 p.
SBAC AS 3265 ISSUE 2. Top Cap—Strainer—Filter Fuel Type 'B'. 1 p.

—Aircraft—Strainers—Cups
SBAC AS 3227 ISSUE 2. Bottom Cup—Strainer—Filter Fuel Type 'A'. 1 p.
SBAC AS 3234 ISSUE 2. Spring Cup—Strainer—Filter Fuel Type 'A'. 1 p.
SBAC AS 3264 ISSUE 2. Spring Cup—Strainer Filter Fuel Type 'B'. 1 p.

—Aircraft—Strainers—Plates
SBAC AS 3233 ISSUE 2. Embossed, Top Plate Strainer—Filter Fuel Type 'A'. 1 p.

INTERNATIONAL AND NON-U.S. NATIONAL STANDARDS SUBJECT INDEX

Fuel Filters (Cont.)

—Aircraft—Strainers—Plates (Cont.)
SBAC AS 3263 ISSUE 2. Embossed, Top Plate—Strainer—Filter Fuel Type 'B'. 1 p.

—Aircraft—Strainers—Straps
SBAC AS 3236-3239 ISSUE 2. Connecting, Strap-Strainer—Filter Fuel Types A and B. 1 p.

—Aircraft—Strainers—Supports
SBAC AS 3229 ISSUE 2. Outer Wire, Support—Strainer—Filter Fuel Type 'A'. 1 p.
SBAC AS 3230 ISSUE 2. Inner Bottom Wire, Support—Strainer—Filter Fuel Type 'A'. 1 p.
SBAC AS 3231 ISSUE 2. Inner Top Wire, Support—Strainer Filter Fuel Type 'A'. 1 p.
SBAC AS 3232 ISSUE 2. Support for Embossed Plate—Strainer—Filter, Fuel Type 'A'. 1 p.
SBAC AS 3259 ISSUE 2. Outer Wire, Support—Strainer—Filter Fuel Type B. 1 p.
SBAC AS 3260 ISSUE 2. Inner Bottom Wire, Support—Strainer—Filter Fuel Type 'B'. 1 p.
SBAC AS 3261 ISSUE 2. Inner Top Wire, Support—Strainer Filter Fuel Type 'B'. 1 p.
SBAC AS 3262 ISSUE 2. Support for Embossed Plate Strainer Filter Fuel Type 'B'. 1 p.

—Aircraft—Strainers—U Rings
SBAC AS 3228 ISSUE 2. Outer 'U' Ring Strainer—Filter—Fuel Type 'A'. 1 p.
SBAC AS 3258 ISSUE 2. Outer 'U' Ring Strainer—Filter—Fuel Type 'B'. 1 p.

—Aircraft—Strainers—Wire Cloth
SBAC AS 3240 ISSUE 1. Inner, Gauze-Strainer—Filter Fuel Type 'A'. 1 p.
SBAC AS 3241 ISSUE 1. Outer, Gauze-Strainer—Filter Fuel Type 'A'. 1 p.
SBAC AS 3266 ISSUE 1. Inner, Gauze-Strainer—Filter Fuel Type 'B'. 1 p.
SBAC AS 3267 ISSUE 1. Outer, Gauze-Strainer—Filter Fuel Type 'B'. 1 p.

—Aircraft—Sumps
SBAC AS 3313 ISSUE 3. Sump. 1 p.

—Aircraft—Sumps—Sealing Rings
SBAC AS 3314 ISSUE 3. Filter Sump Joint Ring. 1 p.

—Aircraft—Tab Washers
SBAC AS 3274 ISSUE 2. Tab, Washer Filter Fuel Type B. 1 p.

—Aircraft—Union Locknuts
SBAC AS 3272 ISSUE 2. Union, Locknut—Filter Fuel Type 'B'. 1 p.

—Aircraft—Washers (Fasteners)
SBAC AS 3224 ISSUE 1. Jointing, Washer Filter Fuel Type 'A' Type 'A'. 1 p.
SBAC AS 3254 ISSUE 1. Jointing, Washer Filter Fuel Type 'B'. 1 p.

—Automotive
CNS D3054-89. Method of Test for Fuel Filters for Automobiles (Jul)(5651).

—Diesel Engines
BSI BS 4552: Part 1-79. 1979 Amd 1 Fuel Filters, Strainers and Sedimentors for Compression-Ignition Engines Part 1: Methods of Test. 43 pp.
BSI BS 4552: Part 2-80. 1980 Amd 1 Fuel Filters, Strainers and Sedimentors for Compression-Ignition Engines Part 2: Method of Classification. 5 pp.
BSI BS 4552: Part 3-92. 1992 Fuel Filters, Strainers and Sedimentors for Compression-Ignition Engines Part 3: Specification for Mounting and Connecting Dimensions of Spin-on Fuel Filters (ISO 7654: 1991). 12 pp.

—Diesel Engines—Automotive
CNS D3139-82. Method of Test of Fuel Filters for Automobile Diesel Engines (Sep)(9373).
ISO 4020 Pt 1-79. Road Vehicles—Fuel Filters for Automotive Compression Ignition Engines—Part 1: Test Methods First Edition; (Corrected and Reprinted -1980). 33 pp.
ISO 4020 Pt 2-79. Road Vehicles—Fuel Filters for Automotive Compression Ignition Engines—Part 2: Test Values and Classification First Edition. 6 pp.
ISO 7310-93. Diesel Engines—Heads for Spin-on Fuel Filters with Horizontal Flange—Mounting and Connecting Dimensions Second Edition. 7 pp.
ISO 7311-93. Diesel Engines—Heads for Fuel Filters with Vertical Flange—Mounting and Connecting Dimensions Second Edition. 9 pp.
ISO 7576-90. Road Vehicles—Two-Stage Fuel Filters for Compression-Ignition Engines—Mounting and Connecting Dimensions Second Edition. 6 pp.
ISO 7577-82. Road Vehicles—Fuel Filters for Compression-Ignition Engines—Mounting and Connecting Dimensions First Edition. 11 pp.
ISO 7654-91. Road Vehicles—Spin-on Fuel Filters for Compression-Ignition Engines—Mounting and Connecting Dimensions Second Edition. 9 pp.
ISO 7774-84. Road Vehicles—Compression Ignition Engines—Single Fuel Filters with Horizontal Flange and Centre Bolt Fixing—Mounting and Connecting Dimensions First Edition. 6 pp.
JIS D 1617-79. Test Method of Fuel Filters for Automobile Diesel Engines.

—Gasoline Engines—Automotive
CNS D3164-84. Fuel Filter Test Method for Automotive Gasoline Engines (Mar)(10800).
JIS D 1608-82. Fuel Filter Test Methods for Automotive Gasoline Engines. 11 pp.

Fuel Filters/Separators
Use: Fuel Separators/Filters

Fuel Gages
See Also: Level Indicators

—Aircraft
EUROCAE ED-41 12.83. MPS for Airborne Fuel Quantity Gauging Systems. 30 pp.
MOD UK DSTAN 66-35-86. Circular Dial Type Fuel Contents Gauges for Aircraft Issue 1. 5 pp.
NATO STANAG 3339 ED 4 AMD 2-73. Circular Dial Type Fuel Contents Gauges. 10 pp.
NATO STANAG 3339 ED 5 AMD 0-90. Fuel Contents Gauges. 6 pp.
NATO STANAG 3339 ED 5 AMD 1-90. Fuel Contents Gauges. 7 pp.

—Aircraft—Acceptance Testing
EUROCAE ED-41 12.83. MPS for Airborne Fuel Quantity Gauging Systems. 30 pp.

—Aircraft—Accuracy Testing
EUROCAE ED-41 12.83. MPS for Airborne Fuel Quantity Gauging Systems. 30 pp.

—Aircraft—Environmental Testing
EUROCAE ED-41 12.83. MPS for Airborne Fuel Quantity Gauging Systems. 30 pp.

—Automotive
CNS D2042-87. Fuel Level Gauges for Automobiles (Aug)(5798).
JIS D 5606-73. Fuel Level Gauges for Automobiles.

—Gliders
CAA JAR-22 SUBPART G. Operating Limitations and Information (Joint Airworthiness Requirements). 13 pp.

Fuel Gases
See Also: Gaseous Fuels

—Calorific Value
BSI BS 3156: Part 3-68. (WITHDRAWN) 1968 Methods for the Analysis of Fuel Gases Part 3: Combustion Characteristics. 24 pp.
BSI BS 3804: Part 1-64. (OBSOLESCENT) 1964 Amd 1 Methods for the Determination of Calorific Value of Fuel Gases Part 1: Non-Recording Methods. 61 pp.
JIS K 2301-92. Fuel Gases and Natural Gas—Methods for Chemical Analysis and Testing. 116 pp.

—Chemical Analysis
BSI BS 3156: Part 1-68. (WITHDRAWN) 1968 Methods for the Analysis of Fuel Gases Part 1: General Analysis. 27 pp.
JIS K 2301-92. Fuel Gases and Natural Gas—Methods for Chemical Analysis and Testing. 116 pp.

—Density
BSI BS 3156: Part 3-68. (WITHDRAWN) 1968 Methods for the Analysis of Fuel Gases Part 3: Combustion Characteristics. 24 pp.
JIS K 2301-92. Fuel Gases and Natural Gas—Methods for Chemical Analysis and Testing. 116 pp.

—Gas Chromatography
BSI BS 3156: Part 4-69. (OBSOLESCENT) 1969 Method for the Analysis of Fuel Gases Part 4: Gas Chromatographic Analysis. 24 pp.

—Pressure Measurement
BSI BS 3156: Part 5-69. (WITHDRAWN) 1969 Amd 1 Methods for the Analysis of Fuel Gases Part 5: Techniques for High Pressure Gases. 33 pp.

—Safety
BSI BS 6158-82. 1982 Safety Devices for Fuel Gases and Oxygen or Compressed Air for Welding, Cutting and Related Processes. 12 pp.

Fuel Gases (Cont.)

—Safety (Cont.)
CEN PREN 730-92. Equipment Used in Gas Welding, Cutting and Allied Processes—Safety Devices for Fuel Gases and Oxygen or Compressed Air—General Specifications, Requirements and Tests. 15 pp.

—Sampling
BSI BS 3156: Part 5-69. (WITHDRAWN) 1969 Amd 1 Methods for the Analysis of Fuel Gases Part 5: Techniques for High Pressure Gases. 33 pp.
DIN ENGL 51853-79. Testing of Fuel Gases, Protective Gases and Exhaust Gases; Sampling (Jan). 10 pp.

—Sulfur Compound Content
BSI BS 3156: Part 2-68. (WITHDRAWN) 1968 Amd 2 Methods for the Analysis of Fuel Gases Part 2: Special Determinations. 86 pp.

—Sulfur Dioxide Content
CPPA J.3H-65. Analysis of Sulphur-Burner Gas. 3 pp.
CPPA J.23P-66. Analysis of Sulphur-Burner Gas. 3 pp.

—Sulfur Trioxide Content
CPPA J.3H-65. Analysis of Sulphur-Burner Gas. 3 pp.
CPPA J.23P-66. Analysis of Sulphur-Burner Gas. 3 pp.

Fuel Gauges
Use: Fuel Gages

Fuel Hose Assemblies
See Also: Hose Assemblies

—Gases—Medical
BSI BS 5682-84. 1984 Amd 1 Terminal Units, Hose Assemblies and Their Connectors for Use with Medical Gas Pipeline Systems. 33 pp.
CEN PREN 739-92. Low Pressure Flexible Connecting Assemblies (Hose Assemblies) for Use with Medical Gas Supply Systems. 32 pp.
ISO 5359-89. Low-Pressure Flexible Connecting Assemblies (Hose Assemblies) for Use with Medical Gas Systems First Edition. 47 pp.

—Gases—Rubber
JIS S 2144-91. Reinforced Rubber Hose Assemblies for Gas. 26 pp.

—Liquefied Petroleum Gas
SAA AS 1869-91. Hose and Hose Assemblies for Liquefied Petroleum Gases (LPG), Natural Gas and Town Gas Amdt 1 May 1992 Amdt 2 October 1992. 2 pp.

—Liquefied Petroleum Gas—Rubber
BSI BS 4089-89. 1989 Hoses and Hose Assemblies for Liquefied Petroleum Gas. 21 pp.
ISO 2928-86. Rubber Hoses and Hose Assemblies for Liquefied Petroleum Gases (LPG)—Bulk Transfer Applications—Specification Second Edition. 4 pp.
JIS B 8261-80. Hose Assemblies for Liquefied Petroleum Gas (R 1985). 33 pp.

—Natural Gas
SAA AS 1869-91. Hose and Hose Assemblies for Liquefied Petroleum Gases (LPG), Natural Gas and Town Gas Amdt 1 May 1992 Amdt 2 October 1992. 2 pp.

—Plastic
BSI BS 3492-87. 1987 Road and Rail Tanker Hoses and Hose Assemblies for Petroleum Products, Including Aviation Fuels. 22 pp.
BSI BS 5842-80. 1980 Thermoplastic Hose Assemblies for Dock, Road and Rail Tanker Use. 4 pp.
SNZ NZS/BS 3492-87. Specification for Road and Rail Tanker Hoses and Hose Assemblies for Petroleum Products, Including Aviation Fuels. 12 pp.

—Plastic—Aircraft
BSI BS 3492-87. 1987 Road and Rail Tanker Hoses and Hose Assemblies for Petroleum Products, Including Aviation Fuels. 22 pp.
SNZ NZS/BS 3492-87. Specification for Road and Rail Tanker Hoses and Hose Assemblies for Petroleum Products, Including Aviation Fuels. 12 pp.

—Pressure Measurement
BSI BS 5173: Sec 102.3-88. 1988 Rubber and Plastics Hoses and Hose Assemblies Part 102: Hydraulic Pressure Tests Section 102.3: Determination of Volumetric Expansion of Fuel-Dispensing Pump Hoses. 4 pp.

—Rubber
BSI BS 3395-89. 1989 Electrically Bonded Hose and Hose Assemblies for Fuel Dispensing Petroleum Based Fuels. 13 pp.

Fuel Hose Assemblies (Cont.)

—Rubber (Cont.)

BSI BS 3395-01. 1989 Amd 1 Electrically Bonded Rubber Hoses and Hose Assemblies for Dispensing Petroleum Based Fuels (AMD 7229) September 15, 1992. 15 pp.

BSI BS 3492-87. 1987 Road and Rail Tanker Hoses and Hose Assemblies for Petroleum Products, Including Aviation Fuels. 22 pp.

SNZ NZS/BS 3395-89. Specification for Electrically Bonded Rubber Hoses and Hose Assemblies for Dispensing Petroleum Based Fuels. 8 pp.

SNZ NZS/BS 3492-87. Specification for Road and Rail Tanker Hoses and Hose Assemblies for Petroleum Products, Including Aviation Fuels. 12 pp.

—Rubber—Aircraft

BSI BS 3158-85. 1985 Rubber Hoses and Hose Assemblies for Aircraft Ground Fuelling and Defuelling. 19 pp.

BSI BS 3492-87. 1987 Road and Rail Tanker Hoses and Hose Assemblies for Petroleum Products, Including Aviation Fuels. 22 pp.

SNZ NZS/BS 3492-87. Specification for Road and Rail Tanker Hoses and Hose Assemblies for Petroleum Products, Including Aviation Fuels. 12 pp.

—Rubber—Oil Burners

BSI BS 6846-87. 1987 Rubber Hoses and Hose Assemblies for Use in Oil Burners. 12 pp.

ISO 6806-92. Rubber Hoses and Hose Assemblies for Use in Oil Burners—Specification Second Edition. 11 pp.

—Town Gas

SAA AS 1869-91. Hose and Hose Assemblies for Liquefied Petroleum Gases (LPG), Natural Gas and Town Gas Amdt 1 May 1992 Amdt 2 October 1992. 2 pp.

Fuel Hoses

Use For: Gas Hoses *See Also:* Fuel Dispensing Equipment; Fuel Lines; Fuels; Hoses

—Fire Resistant—Boats

BSI BS MA 102-86. 1986 Fire Resistant Fuel Hoses for Small Craft. 6 pp.

ISO 7840-85. Small Craft—Fire Resistant Fuel Hoses First Edition. 6 pp.

—Flexible Metal

JIS S 2145-91. Metallic Flexible Hoses for Gas. 27 pp.

—Flexible Metal—Appliances

BSI BS 669: Part 1-89. 1989 Flexible Hoses, End Fittings and Sockets for Gas Burning Appliances Part 1: Strip-Wound Metallic Flexible Hoses, Covers, End Fittings and Sockets for Domestic Appliances Burning 1st and 2nd Family Gases (L). 24 pp.

—Fuel Oil—Appliances

JIS S 3022-91. Oil Discharge Rubber Hoses for Oil Burning Appliances. 14 pp.

—Fuel Oil—Discharge—Rubber

CNS K4064-82. Oil Discharge Rubber Hose (Jun)(9011).

JIS K 6343-82. Rubber Hoses for Oil Discharge. 7 pp.

—Fuel Oil—Oil Burners

BSI BS 6846-87. 1987 Rubber Hoses and Hose Assemblies for Use in Oil Burners. 12 pp.

ISO 6806-92. Rubber Hoses and Hose Assemblies for Use in Oil Burners—Specification Second Edition. 11 pp.

—Fuel Oil—Suction/Discharge—Rubber

CNS K4065-82. Boundless Oil Suction and Discharge Rubber Hose (Jun)(9012).

ISO 1823-75. Rubber Hoses for Oil Suction and Discharge First Edition. 5 pp.

—Gases—Plastic

JIS S 2146-91. Rubber Tubes and Polyvinyl Chloride Hoses with Both End Convenient Joints for Gas. 21 pp.

—Gases—Rubber

JIS S 2146-91. Rubber Tubes and Polyvinyl Chloride Hoses with Both End Convenient Joints for Gas. 21 pp.

—Liquefied Petroleum Gas

SAA AS 1869-91. Hose and Hose Assemblies for Liquefied Petroleum Gases (LPG), Natural Gas and Town Gas Amdt 1 May 1992 Amdt 2 October 1992. 2 pp.

Fuel Hoses (Cont.)

—Liquefied Petroleum Gas—Rubber

BSI BS 4089-89. 1989 Hoses and Hose Assemblies for Liquefied Petroleum Gas. 21 pp.

ISO 2928-86. Rubber Hoses and Hose Assemblies for Liquefied Petroleum Gases (LPG)—Bulk Transfer Applications—Specification Second Edition. 4 pp.

JIS K 6347-80. Liquefied Petroleum Gas Hoses. 20 pp.

—Marine

JIS F 7150-91. Yachts Non-Fire Resistant Fuel Hoses.

SAA AS 2117-83. Hose and Hose Assemblies for Petroleum Products—Marine Suction and Discharge. 10 pp.

—Natural Gas

SAA AS 1869-91. Hose and Hose Assemblies for Liquefied Petroleum Gases (LPG), Natural Gas and Town Gas Amdt 1 May 1992 Amdt 2 October 1992. 2 pp.

—Nitrogen—Aircraft

MOD UK DSTAN 16-19-93. Nitrogen, Compressed, Oil-Free: for All Applications Issue 2. 9 pp.

NATO STANAG 3990 ED 1 AMD 1-90. Characteristics of High Quality Oil-Free Gaseous Nitrogen, Supply Pressure and Hoses, for Aircraft, Weapons and Electronic Equipment Servicing. 6 pp.

—Nitrogen—Electronic Equipment

MOD UK DSTAN 16-19-93. Nitrogen, Compressed, Oil-Free: for All Applications Issue 2. 9 pp.

NATO STANAG 3990 ED 1 AMD 1-90. Characteristics of High Quality Oil-Free Gaseous Nitrogen, Supply Pressure and Hoses, for Aircraft, Weapons and Electronic Equipment Servicing. 6 pp.

—Nitrogen—Weapons Systems

MOD UK DSTAN 16-19-93. Nitrogen, Compressed, Oil-Free: for All Applications Issue 2. 9 pp.

NATO STANAG 3990 ED 1 AMD 1-90. Characteristics of High Quality Oil-Free Gaseous Nitrogen, Supply Pressure and Hoses, for Aircraft, Weapons and Electronic Equipment Servicing. 6 pp.

—Oxygen—Aerospace—Construction—Rubber

BSI 3F 63-81. 1981 Amd 2 Low Pressure Oxygen Hose for Aeronautical Purposes. 12 pp.

—Oxygen—Aerospace—Identification Systems—Rubber

BSI 3F 63-81. 1981 Amd 2 Low Pressure Oxygen Hose for Aeronautical Purposes. 12 pp.

—Oxygen—Aircraft—Interdepartmental Procurement

NATO STANAG 3053 ED 4 AMD 2-70. Breathing Oxygen Characteristics, Supply Pressure and Hoses. 11 pp.

NATO STANAG 3053 ED 4 AMD 3-70. Breathing Oxygen Characteristics, Supply Pressure and Hoses. 7 pp.

NATO STANAG 3053 ED 5 AMD 0-92. Characteristics of Breathing Oxygen, Supply Pressure and Hoses. 6 pp.

—Plastic

BSI BS 3492-87. 1987 Road and Rail Tanker Hoses and Hose Assemblies for Petroleum Products, Including Aviation Fuels. 22 pp.

SNZ NZS/BS 3492-87. Specification for Road and Rail Tanker Hoses and Hose Assemblies for Petroleum Products, Including Aviation Fuels. 12 pp.

—Plastic—Aircraft

BSI BS 3492-87. 1987 Road and Rail Tanker Hoses and Hose Assemblies for Petroleum Products, Including Aviation Fuels. 22 pp.

SNZ NZS/BS 3492-87. Specification for Road and Rail Tanker Hoses and Hose Assemblies for Petroleum Products, Including Aviation Fuels. 12 pp.

—Rubber

BSI BS 3395-89. 1989 Electrically Bonded Hose and Hose Assemblies for Fuel Dispensing Petroleum Based Fuels. 13 pp.

BSI BS 3395-01. 1989 Amd 1 Electrically Bonded Rubber Hoses and Hose Assemblies for Dispensing Petroleum Based Fuels (AMD 7229) September 15, 1992. 15 pp.

BSI BS 3492-87. 1987 Road and Rail Tanker Hoses and Hose Assemblies for Petroleum Products, Including Aviation Fuels. 22 pp.

ISO 2929-91. Rubber Hoses for Bulk Fuel Truck Delivery—Specification Second Edition. 6 pp.

ISO 5772 Pt 1-86. Rubber Hoses—Measured Fuel Dispensing—Part 1: Conventional Petroleum Based Fuels—Specification First Edition. 6 pp.

Fuel Hoses (Cont.)

—Rubber (Cont.)

SNZ NZS/BS 3395-89. Specification for Electrically Bonded Rubber Hoses and Hose Assemblies for Dispensing Petroleum Based Fuels. 8 pp.

SNZ NZS/BS 3492-87. Specification for Road and Rail Tanker Hoses and Hose Assemblies for Petroleum Products, Including Aviation Fuels. 12 pp.

—Rubber—Aircraft

BSI BS 3158-85. 1985 Rubber Hoses and Hose Assemblies for Aircraft Ground Fuelling and Defuelling. 19 pp.

BSI BS 3492-87. 1987 Road and Rail Tanker Hoses and Hose Assemblies for Petroleum Products, Including Aviation Fuels. 22 pp.

BSI 2F 67-80. 1980 Hose for Aviation Fuel and Engine Lubricating Oil for Aeronautical Purposes. 7 pp.

ISO 1825-75. Rubber Hoses for Aircraft Ground Fuelling Without Static Conducting Wire First Edition. 4 pp.

MOD UK DTD-784-53. Rubber Materials for Fuel Hose. 4 pp.

SNZ NZS/BS 3492-87. Specification for Road and Rail Tanker Hoses and Hose Assemblies for Petroleum Products, Including Aviation Fuels. 12 pp.

—Rubber—Appliances

JIS K 6348-80. Gas Tubing (Extruded Rubber Tubing). 19 pp.

JIS K 6351-82. Wire Reinforced Rubber Hoses for Gaseous Fuels. 20 pp.

—Rubber—Automotive

BSI BS AU 108-65. 1965 Amd 1 Plain and Reinforced Hoses of Rubber. 8 pp.

—Suction/Discharge—Plastic

BSI BS 6847-87. 1987 Plastic Hoses for Suction and Low-Pressure Discharge of Petroleum Liquids. 10 pp.

ISO 6808-84. Plastics Hoses for Suction and Low-Pressure Discharge—Petroleum Liquids—Specification First Edition. 8 pp.

—Town Gas

SAA AS 1869-91. Hose and Hose Assemblies for Liquefied Petroleum Gases (LPG), Natural Gas and Town Gas Amdt 1 May 1992 Amdt 2 October 1992. 2 pp.

Fuel Injectors

See Also: Diesel Engines; Fuel Nozzles

—Automotive

ISO 8356-84. Road Vehicles—Diesel Engines—Screw-in Injector, Type 22 First Edition. 4 pp.

ISO 8984 Pt 2-87. Road Vehicles—Testing of Fuel Injectors for Compression-Ignition Engines—Part 2: Test Procedures First Edition. 4 pp.

—Automotive—Calibration Fluids

ISO 4113-88. Road Vehicles—Calibration Fluid for Diesel Injection Equipment Second Edition. 5 pp.

—Calibration Nozzles—Automotive

ISO 4010-77. Road Vehicles—Calibrating Nozzle, Delay Pintle Type First Edition; (Erratum—June 1977). 6 pp.

ISO 7440 Pt 1-91. Road Vehicles—Fuel Injection Equipment Testing—Part 1: Calibrating Nozzle and Holder Assemblies Second Edition. 17 pp.

—Flow Measurement—Automotive

ISO 7440 Pt 2-91. Road Vehicles—Fuel Injection Equipment Testing—Part 2: Orifice Plate Flow-Measurement Second Edition. 9 pp.

—Fuel Pumps

ISO 7612-83. Base Mounted in-Line Fuel Injection Pumps First Edition. 6 pp.

—Fuel Pumps—Automotive

ISO 4008 Pt 1-80. Road Vehicles—Fuel Injection Pump Testing—Part 1: Dynamic Conditions Second Edition; (Erratum—March 1982). 21 pp.

ISO 4008 Pt 2-83. Road Vehicles—Fuel Injection Pump Testing—Part 2: Static Conditions First Edition. 19 pp.

ISO 4008 Pt 3-87. Road Vehicles—Fuel Injection Pump Testing—Part 3: Application and Test Procedures First Edition. 20 pp.

ISO 7879-90. Road Vehicles—Cradle-Mounted in-Line Fuel Injection Pumps—Mounting Dimensions Second Edition. 7 pp.

—Fuel Pumps—Automotive—Dynamic Testing

JIS D 3633-90. Road Vehicles—Fuel Injection Pump Testing—Part 1: Dynamic Conditions; (ISO 4008/1-1980).

INTERNATIONAL AND NON-U.S. NATIONAL STANDARDS
SUBJECT INDEX

Fuel

Fuel Injectors (Cont.)
—Fuel Pumps—Automotive—Static Testing
JIS D 3634-91. Road Vehicles—Fuel Injection Pump Testing—Part 2: Static Conditions; (ISO 4008/2-1983).

—Fuel Pumps—Diesel
JIS B 8035-75. Testing Method for Fuel Injection Pump of Diesel Engine (R 1984). 12 pp.

—Fuel Pumps—Diesel—Automotive
CNS D2184-84. In-Line Injection Pumps for Automobile Diesel Engines (Oct)(11061).
JIS D 3603-86. In-Line Fuel Injection Pumps for Automobile Diesel Engines. 16 pp.

—Fuel Pumps—Flanges—Automotive
ISO 7299-84. Road Vehicles—End-Mounting Flanges for Fuel Injection Pumps First Edition. 11 pp.

—Fuel Pumps—Glossaries
ISO 7876 Pt 1-90. Fuel Injection Equipment—Vocabulary—Part 1: Fuel Injection Pumps Second Edition; (Incorporating Addendum 1). 22 pp.

—Fuel Pumps—Packaging
MOD UK DSTAN 81-63-89. Packaging of Fuel Injection Pumps and Fuel Injectors Issue 1. 24 pp.

—Fuel Pumps—Pressure Pipes
ISO 4093-86. Road Vehicles—Fuel Injection Pumps—High-Pressure Pipes for Testing Second Edition. 3 pp.

—Fuel Pumps—Speed Control
BSI BS AU 61-64. 1964 Amd 1 Hydraulic Governor for Size N Fuel Injection Pump. 4 pp.

—Fuel Pumps—Tapers
ISO 6519-93. Diesel Engines—Fuel Injection Pumps—Tapers for Shaft Ends and Hubs Second Edition. 7 pp.

—Glossaries
ISO 7876 Pt 2-91. Fuel Injection Equipment—Vocabulary—Part 2: Fuel Injectors First Edition. 19 pp.
ISO 7876 Pt 3-93. Fuel Injection Equipment—Vocabulary—Part 3: Unit Injectors First Edition. 11 pp.

—Nozzle Holders
ISO 9102-90. Road Vehicles—Compression-Ignition Engines—Screw-in Injection Nozzle Holders, Types 24, 25 and 26 First Edition. 7 pp.

—Nozzle Holders—Automotive
CNS D2186-84. Flange-Mounted and Screw-Mounted Injection Nozzle Holder Type S for Road Vehicles (Oct)(11063).
CNS D2187-84. Injection Nozzle Holder Type P for Road Vehicles (with Body and with Fixing Flats) (Oct)(11064).
ISO 2698-93. Diesel Engines—Clamp-Mounted Fuel Injectors, Types 7 and 28 Second Edition. 9 pp.
ISO 2699-83. Road Vehicles—Flange-Mounted Injection Nozzle Holders Size "S"—Types 2, 3, 4, 5 and 6 Second Edition. 7 pp.
ISO 3539-75. Road Vehicles—Injection Nozzle Holder with Body, Types 8 and 10, and Injection Nozzle Holder with Fixing Flats, Types 9 and 11 First Edition. 4 pp.
ISO 7026-90. Road Vehicles—Screw-in Injection Nozzle Holders, Types 20, 21 and 27 for Pintle Nozzle Size "S", Type "B" Second Edition. 7 pp.
ISO 7030-87. Road Vehicles—Screw-Mounted Injection Nozzle Holders, Types 12, 13, 14, 15, 16, 17, 18 and 19 Second Edition. 12 pp.
ISO 9103-87. Road Vehicles—Compression-Ignition Engines—Screw-in Injection Nozzle Holder, Type 23 First Edition. 5 pp.
JIS D 3631-85. Shapes and Dimensions of Flange-Mounted and Screw-Mounted Nozzle Holders Size S for Road Vehicles. 13 pp.
JIS D 3632-82. Injection Nozzle Holder Type P for Road Vehicles (with Body and with Fixing Flats). 6 pp.

—Nozzles—Automotive
CNS D2086-86. Shape and Dimension of Fuel Injection Nozzles for Automobiles Diesel Engines (Size "S") (Apr)(7134).
ISO 2697-74. Road Vehicles—Fuel Injection Nozzles—Size "S" First Edition. 7 pp.
JIS D 3604-84. Shape and Dimension of Fuel Injection Nozzles for Automobile Diesel Engines (Size "S"). 7 pp.

—Packaging
MOD UK DSTAN 81-63-89. Packaging of Fuel Injection Pumps and Fuel Injectors Issue 1. 24 pp.

—Pipe Fittings
BSI BS AU 68-64. 1964 Amd 1 Fuel Injection Low Pressure Pipe Fittings. 4 pp.

Fuel Injectors (Cont.)
—Pipe Fittings (Cont.)
CNS D2185-84. Shape and Dimension of Injection Pipe Ends and Fittings for Automobile Diesel Engines (Oct)(11062).

—Pressure Pipe Fittings
BSI BS AU 153-70. 1970 Fuel Injection High Pressure Pipe Fittings. 6 pp.
ISO 2974-90. Road Vehicles—High-Pressure Fuel Injection Pipe End-Connections with 60 Degree Female Cone Third Edition. 6 pp.

—Pressure Pipes—Steel Tubes
ISO 8535 Pt 1-90. Compression-Ignition Engines—Steel Tubes for High-Pressure Fuel Injection Pipes—Part 1: Requirements for Seamless Cold-Drawn Single-Wall Tubes Second Edition; (Amendment 1-1993). 16 pp.

—Testing Equipment—Automotive
ISO 8984 Pt 1-87. Road Vehicles—Testing of Fuel Injectors for Compression-Ignition Engines—Part 1: Hand-Lever-Operated Testing and Setting Apparatus First Edition. 9 pp.

Fuel Leakage
See Also: Leakage

—Automotive—Carburetors
CNS D3102-89. Fuel Leakage Test Code of Carburetors for Automobiles (Jan)(8259).

—Automotive—Collision Research
ISO 3437-75. Road Vehicles—Determination of Fuel Leakage in the Event of a Collision First Edition. 3 pp.
JIS D 1042-84. Determination of Fuel Leakage in the Event of a Collision for Passenger Cars. 5 pp.

Fuel Lines
Use For: Fuel Circuits *See Also:* Carburetors; Fuel Hoses

—Internal Combustion Engines
ISO 4639 Pt 1-87. Rubber Tubing and Hoses for Fuel Circuits for Internal Combustion Engines—Specification—Part 1: Conventional Liquid Fuels First Edition. 12 pp.

Fuel Nozzles
See Also: Fuel Injectors; Fuel Systems; Nozzles

—Aircraft
BSI 2C 13-88. 1988 Sizes of Aircraft Gravity Filling Orifices and Associated Replenishment Nozzles (Metric Series). 4 pp.

—Aircraft—Refueling Equipment
NATO STANAG 3447 ED 3 AMD 2-90. Aerial Refuelling Equipment Dimensional and Functional Characteristics. 10 pp.

—Aircraft—Tank Trucks—Electrical Grounding
BSI G 175-59. 1959 Amd 1 Aircraft Fuel Nozzle Grounding Plugs and Sockets. 2 pp.
ISO 46-73. Aircraft—Fuel Nozzle Grounding Plugs and Sockets First Edition. 3 pp.

Fuel Oil Tanks
Use: Fuel Tanks—Oil

Fuel Oils
See Also: Bunker Fuel Oil; Crude Oil; Diesel Fuels; Heating Oils; Kerosene; Liquid Fuels; Petroleum Products
CNS K5025-83. Fuel Oil (Dec)(1472). 2 pp.
JIS K 2205-91. Fuel Oil. 6 pp.

—Agricultural Equipment
BSI BS 2869: Part 2-88. 1988 Amd 1 Fuel Oils for Non-Marine Use Part 2: Specification for Fuel Oil for Agricultural and Industrial Engines and Burners (Classes A2,C1, C2,D,E,F,G and H) (AMD 6505) February 28, 1991 (Supersedes BS 2869: 1993). 16 pp.

—Aluminum Content—Atomic Absorption Spectrometry
DIN ENGL 51416 Pt 1-86. Testing of Liquid Fuels; Determination of the Aluminium Content of Fuel Oils; Determination by Atomic Absorption Spectrometry (AAS) After Ashing (Apr). 4 pp.

—Ash Content
CEN EN 7-74. Determination of Ash from Petroleum Products. 5 pp.

—Asphaltene Content
BSI BS 2000: Part 143-93. 1993 Methods of Test for Petroleum and Its Products Part 143: Determination of Asphaltenes (Heptane Insolubles) (W). 6 pp.

Fuel Oils (Cont.)
—Asphaltene Content (Cont.)
BSI BS 2000: Part 143-85. 1985 Petroleum and Its Products Part 143: Asphaltenes in Petroleum Products (Precipitation with Normal Heptane). 8 pp.

—Blending
JIS F 7010-85. Application of Fuel Oil Blending System.

—Boilers—Marine
CGSB 3-GP-12MA-88. Fuel Oil, Marine Boiler. 9 pp.

—Boilers—Ships
MOD UK DSTAN 91-5-01. Fuel, Residual: Light Viscosity, Boiler NATO Code No: F-77 Joint Service Designation: 50/50 FFO Fuel Residual: Medium Viscosity, Boiler NATO Code No: F-82 Joint Service Designation: 75/50 FFO Issue 3 Amendment 2. 28 pp.

—Carbon Residue Testing
DIN ENGL 51551-86. Testing of Lubricants and Liquid Fuels; Determination of Conradson Carbon Residue (Mar). 4 pp.

—Distillation Methods
BSI BS 7392-90. 1990 Method for Determination of Distillation Characteristics of Petroleum Products. 26 pp.
ISO 3405-88. Petroleum Products—Determination of Distillation Characteristics Second Edition. 24 pp.

—Flash Point
BSI BS 6664: Part 5-90. 1990 Flashpoint of Petroleum and Related Products Part 5: Method for Determination of Flashpoint by Pensky-Martens Closed Tester. 16 pp.
ISO 2719-88. Petroleum Products and Lubricants—Determination of Flash Point—Pensky-Martens Closed Cup Method Second Edition. 14 pp.

—Heavy
DIN ENGL 51603 Pt 3-86. Fuel Oils; S Fuel Oil; Minimum Requirements (July) (Supersedes Parts of October 1976 Edition of DIN 51603 Part 2). 3 pp.

—Hoses—Ships
MOD UK NES 2018-88. Specification for the Manufacture of 153mm (6 ins) Bore Hose for Abeam Replenishment at Sea Issue 1 (06.88). 24 pp.
MOD UK NES 2019-88. Specification for the Manufacture of 153mm (6 ins) Bore Hose for Astern Replenishment at Sea Issue 1 (06.88). 24 pp.

—Hoses—Submarines
MOD UK NES 2018-88. Specification for the Manufacture of 153mm (6 ins) Bore Hose for Abeam Replenishment at Sea Issue 1 (06.88). 24 pp.
MOD UK NES 2019-88. Specification for the Manufacture of 153mm (6 ins) Bore Hose for Astern Replenishment at Sea Issue 1 (06.88). 24 pp.

—Industrial Equipment
BSI BS 2869: Part 2-88. 1988 Amd 1 Fuel Oils for Non-Marine Use Part 2: Specification for Fuel Oil for Agricultural and Industrial Engines and Burners (Classes A2,C1, C2,D,E,F,G and H) (AMD 6505) February 28, 1991 (Supersedes BS 2869: 1993). 16 pp.

—Nickel Content—Atomic Absorption Spectroscopy
BSI BS 2000: Part 288-93. 1993 Methods of Test for Petroleum and Its Products Part 288: Determination of Nickel, Sodium and Vanadium—Atomic Absorption Spectroscopy Method (W). 5 pp.
BSI BS 2000: Part 288-83. 1983 Petroleum and Its Products Part 288: Sodium, Nickel and Vanadium in Fuel Oils and Crude Oils by Atomic Absorption Spectroscopy. 6 pp.

—Oil Burners
BSI BS 2869: Part 2-88. 1988 Amd 1 Fuel Oils for Non-Marine Use Part 2: Specification for Fuel Oil for Agricultural and Industrial Engines and Burners (Classes A2,C1, C2,D,E,F,G and H) (AMD 6505) February 28, 1991 (Supersedes BS 2869: 1993). 16 pp.
DIN ENGL 51603 Pt 5-90. Type SA Fuel Oils; Minimum Requirements (Feb). 3 pp.

—Pumpability
BSI BS 2000: Part 230-82. 1982 Petroleum and Its Products Part 230: Pumpability Test for Industrial Fuel Oils. 8 pp.
DIN ENGL 51427-82. Testing of Liquid Mineral Oil Hydrocarbons; Determination of Pumpability (Apr). 3 pp.

—Sediment Content—Centrifuging
BSI BS 2882-80. (WITHDRAWN) 1980 Determination of Water and Sediment in Crude Petroleum and Fuel Oils (Centrifuge Method). 8 pp.

INDUSTRY STANDARDS

Fuel Oils (Cont.)

—Sediment Content—Centrifuging (Cont.)
CNS K6577-80. Method of Test for Water and Sediment in Crude Oils and Fuel Oils by Centrifuge (Aug)(6358).
ISO 3734-76. Crude Petroleum and Fuel Oils—Determination of Water and Sediment—Centrifuge Method First Edition. 6 pp.

—Sediment Content—Extraction Analysis
BSI BS 4382-80. 1980 Determination of Sediment in Crude Petroleum and Fuel Oils (Extraction Method). 6 pp.
CNS K6576-80. Method of Test for Sediment in Crude and Fuel Oils by Extraction (Aug)(6357).
ISO 3735-75. Crude Petroleum and Fuel Oils—Determination of Sediment—Extraction Method First Edition. 5 pp.

—Sediment Content—Filtration
ISO 10307 Pt 1-93. Petroleum Products—Total Sediment in Residual Fuel Oils—Part 1: Determination by Hot Filtration First Edition. 9 pp.

—Sediments—Aging Testing
ISO 10307 Pt 2-93. Petroleum Products—Total Sediment in Residual Fuel Oils—Part 2: Determination Using Standard Procedures for Ageing First Edition. 9 pp.

—Sediments—Centrifuging
DIN ENGL 51793-71. Testing of Liquid Fuels; Determination of Water Content and Sediments in Fuel Oils and Crude Oils; Centrifuge Method (July). 3 pp.

—Separators (Mechanical)—Sewage Treatment Equipment
DIN ENGL 1999 Pt 2-89. Petrol Interceptors and Fuel Oil Interceptors; Design, Installation and Operation (Mar). 6 pp.

—Sodium Content—Atomic Absorption Spectroscopy
BSI BS 2000: Part 288-93. 1993 Methods of Test for Petroleum and Its Products Part 288: Determination of Nickel, Sodium and Vanadium—Atomic Absorption Spectroscopy Method (W). 5 pp.
BSI BS 2000: Part 288-83. 1983 Petroleum and Its Products Part 288: Sodium, Nickel and Vanadium in Fuel Oils and Crude Oils by Atomic Absorption Spectroscopy. 6 pp.

—Stability Testing
MOD UK DSTAN 05-50: Part 49-87. Methods for Testing Fuels, Lubricants and Associated Products Part 49: Storage Stability of Oils and Hydraulic Fluids Issue 1. 7 pp.

—Storage
BSI BS 2000: Part 230-93. 1993 Methods of Test for Petroleum and Its Products Part 230: Determination of Minimum Handling and Storage Temperatures of Fuel Oil—Ferranti Viscometer Method (W). 5 pp.

—Sulfur Content
DIN ENGL 51400 Pt 1-78. Testing of Mineral Oils and Fuels; Determination of the Sulfur Content (Total Sulfur); General Working Principles (Feb). 6 pp.
DIN ENGL 51400 Pt 3-78. Testing of Mineral Oils and Fuels; Determination of the Sulfur Content (Total Sulfur); Combustion According to Schoniger; Thorin-Sulfonazo-III Titration (Feb). 3 pp.
DIN ENGL 51400 Pt 4-90. Determination of Total Sulfur Content of Gaseous Petroleum Products by the Lingener Combustion Method (Oct). 8 pp.
DIN ENGL 51400 Pt 8-78. Testing of Mineral Oils and Fuels; Determination of Sulfur Content (Total Sulfur); Nickel Reduction Method; Dithizone Titration (Feb). 4 pp.

—Sulfur Content—Analyzers
JIS B 7995-84. Automatic Analyzers for Sulphur in Crude Oil and Petroleum Products. 24 pp.

—Traps (Drains)
DIN ENGL 4043-82. Traps for Light Liquids (Fuel Oil Traps); Principles of Construction, Installation and Operation, Testing (Oct). 2 pp.

—Vanadium Content
CNS K6623-80. Method of Test for Vanadium in Fuel Oil (Dec)(6838).

—Vanadium Content—Atomic Absorption Spectroscopy
BSI BS 2000: Part 288-93. 1993 Methods of Test for Petroleum and Its Products Part 288: Determination of Nickel, Sodium and Vanadium—Atomic Absorption Spectroscopy Method (W). 5 pp.

Fuel Oils (Cont.)

—Vanadium Content—Atomic Absorption Spectroscopy (Cont.)
BSI BS 2000: Part 288-83. 1983 Petroleum and Its Products Part 288: Sodium, Nickel and Vanadium in Fuel Oils and Crude Oils by Atomic Absorption Spectroscopy. 6 pp.

—Viscosity Temperature Chart
DIN ENGL 51563-76. Testing of Mineral Oils and Related Materials; Determination of Viscosity Temperature Relation; Slope m (Dec). 4 pp.

—Water Content—Centrifuging
BSI BS 2882-80. (WITHDRAWN) 1980 Determination of Water and Sediment in Crude Petroleum and Fuel Oils (Centrifuge Method). 8 pp.
CNS K6577-80. Method of Test for Water and Sediment in Crude Oils and Fuel Oils by Centrifuge (Aug)(6358).
DIN ENGL 51793-71. Testing of Liquid Fuels; Determination of Water Content and Sediments in Fuel Oils and Crude Oils; Centrifuge Method (July). 3 pp.
ISO 3734-76. Crude Petroleum and Fuel Oils—Determination of Water and Sediment—Centrifuge Method First Edition. 6 pp.

—Water Content—Crackling Testing
MOD UK DSTAN 05-50: Part 51-87. Methods for Testing Fuels, Lubricants and Associated Products Part 51: Detection of Free Water in Oil by the Crackle Test Issue 1. 7 pp.

Fuel Pumps
See Also: Feed Pumps; Pumps; Submersible Pumps

—Appliances—Oil Burning
JIS S 2037-92. Filler Pumps for Oil Burning Appliances. 26 pp.

—Automotive—Diaphragm
CNS D2119-90. Mechanical Fuel Pumps for Automobiles (Jan)(8386).
CNS D2120-82. Fuel Pump Diaphragms for Automobiles (Jan)(8387).
CNS D3187-90. Method of Test for Fuel Pumps Diaphragms for Automobiles (Jan)(12656).
JIS D 3601-88. Mechanical Fuel Pumps for Automobiles. 15 pp.

—Automotive—Electric Actuated
CNS D2089-90. Electric Fuel Pumps for Automobiles (Jan)(7140).
JIS D 3606-88. Electric Fuel Pumps for Automobiles. 11 pp.

—Electric Connectors—Automotive
ISO 9534-89. Road Vehicles—Fuel Pump Electric Connections First Edition. 7 pp.

—Fuel Injectors—Automotive
ISO 4008 Pt 1-80. Road Vehicles—Fuel Injection Pump Testing—Part 1: Dynamic Conditions Second Edition; (Erratum—March 1982). 21 pp.
ISO 4008 Pt 2-83. Road Vehicles—Fuel Injection Pump Testing—Part 2: Static Conditions First Edition. 19 pp.
ISO 4008 Pt 3-87. Road Vehicles—Fuel Injection Pump Testing—Part 3: Application and Test Procedures First Edition. 20 pp.
ISO 7879-90. Road Vehicles—Cradle-Mounted in-Line Fuel Injection Pumps—Mounting Dimensions Second Edition. 7 pp.

—Fuel Injectors—Automotive—Dynamic Testing
JIS D 3633-90. Road Vehicles—Fuel Injection Pump Testing—Part 1: Dynamic Conditions; (ISO 4008/1-1980).

—Fuel Injectors—Automotive—Static Testing
JIS D 3634-91. Road Vehicles—Fuel Injection Pump Testing—Part 2: Static Conditions; (ISO 4008/2-1983).

Fuel Reprocessing Plants, Nuclear
Use: Nuclear Fuel Reprocessing Plants

Fuel Separators/Filters
Use For: Fuel Filters/Separators *See Also:* Filters; Separators (Mechanical)

—Aircraft Fuels
NATO STANAG 3967 ED 1 AMD 0-93. Design and Performance Requirements for Aviation Fuel Filter Separator Vessels and Coalescer and Separator Elements. 33 pp.

Fuel Separators/Filters (Cont.)

—Aircraft Fuels—Design
NATO STANAG 3967 ED 1 AMD 0-93. Design and Performance Requirements for Aviation Fuel Filter Separator Vessels and Coalescer and Separator Elements. 33 pp.

—Aircraft Fuels—Differential Pressure Gages
NATO STANAG 3583 ED 2 AMD 3-74. Standards of Accuracy for Differential Pressure Gauges for Aviation Fuel Filters and Filter Separators. 8 pp.
NATO STANAG 3583 ED 3 AMD 0-93. Standards of Accuracy for Differential Pressure Gauges for Aviation Fuel Filters and Filter Separator Vessels. 7 pp.
NATO STANAG 3583 ED 3 AMD 1-93. Standards of Accuracy for Differential Pressure Gauges for Aviation Fuel Filters and Filter Separator Vessels. 8 pp.

Fuel Storage Tanks
See Also: Bulk Storage; Fuel Tanks; Oil Storage Tanks; Petroleum Storage Tanks; Storage Tanks; Tanks (Containers)

—Compressed Natural Gas—Dispensing Equipment
CSA B51-M1991. Boiler, Pressure Vessel, and Pressure Piping Code; (Gen Instr 1). 96 pp.

—Identification Systems
MOD UK DSTAN 05-52: Part 1-91. Markings for the Identification of Fuels, Lubricants and Associated Products: Containers Holding 210 Litres or Less Issue 2. 12 pp.
MOD UK DSTAN 05-52: Part 2-01. Markings for the Identification of Fuels, Lubricants, and Associated Products: Containers over 205 Litres and Pipe Lines Issue 1; Amendment 3. 16 pp.

—Liquefied Natural Gas
CSA CAN/CSA-Z276-M89. Liquefied Natural Gas (LNG)—Production, Storage, and Handling; (Gen Instr 1). 69 pp.
JIS B 8251-81. Construction of Welded Aluminium Alloy Liquefied Natural Gas Storage Tanks (R 1986). 41 pp.

—Maintenance
NATO STANAG 3609 ED 2 AMD 2-88. Standards for Maintenance of Fixed Aviation Fuel Receipt, Storage and Dispensing Systems. 34 pp.
NATO STANAG 3609 ED 2 AMD 3-88. Standards for Maintenance of Fixed Aviation Fuel Receipt, Storage and Dispensing Systems. 35 pp.

—Painting
CGSB 85-GP-12M-79. Painting Bulk Fuel Storage Tanks (Above Ground). 20 pp.

—Paints—Interior
MOD UK DSTAN 80-97-84. Paint System, for the Interior of Bulk Fuel Tanks and Fittings Issue 1. 19 pp.
MOD UK DSTAN 80-97-92. Paint System, Medium Build for the Interior of Bulk Fuel Tanks and Fitting Issue 2. 21 pp.
MOD UK DSTAN 80-155-92. Paint System, High Build for the Interior of Bulk Fuel Tanks, Airless Spray Paint, Primer, Two-Pack Paint, Finish, Two-Pack, Solvent Free Issue 1. 21 pp.
MOD UK DSTAN 80-156-92. Paint System, High Build Fibre Reinforced for the Interior of Bulk Fuel Tanks Paint, Primer, Two-Pack Paint, Finish, Two-Pack, Solvent Free, Fibre Reinforced Issue 1. 21 pp.

Fuel Systems
See Also: Aircraft Fuel Systems; Fuel Nozzles; Replenishment at Sea

—Automotive—Liquefied Petroleum Gas
SAA AS 1425-89. LP Gas Fuel Systems for Vehicle Engines (Known as the SAA Automotive LP Gas Code) Amdt 1 August 1990 Amdt 2 March 1991. 17 pp.

—Boats (Marine)
ISO 10088-92. Small Craft—Permanently Installed Fuel Systems and Fixed Fuel Tanks First Edition. 12 pp.

—Hoverplatforms
CAA Chapter C4 01.74. Systems and Machinery. 1 p.

—Liquid Fuels—Electrostatic Safety
NATO STANAG 3682 ED 4 AMD 0-90. Electrostatic Safety Connection Procedures for Liquid Fuel Loading/Unloading Operations During Ground Transfer. 7 pp.
NATO STANAG 3682 ED 4 AMD 1-90. Electrostatic Safety Connection Procedures for Liquid Fuel Loading/Unloading Operations During Ground Transfer. 7 pp.

INTERNATIONAL AND NON-U.S. NATIONAL STANDARDS
SUBJECT INDEX

Fuel Systems (Cont.)

—Liquid Fuels—Electrostatic Safety (Cont.)
NATO STANAG 3682 ED 4 AMD 2-90. Electrostatic Safety Connection Procedures for Liquid Fuel Loading/ Unloading Operations During Ground Transfer. 8 pp.

—Liquid Fuels—Installation
NATO STANAG 3756 ED 3 AMD 0-90. Facilities and Equipment for Receipt and Delivery of Liquid Fuels. 13 pp.
NATO STANAG 3756 ED 3 AMD 1-90. Facilities and Equipment for Receipt and Delivery of Liquid Fuels. 22 pp.
NATO STANAG 3756 ED 3 AMD 2-90. Facilities and Equipment for Receipt and Delivery of Liquid Fuels. 15 pp.

—Oil Burners—Safety
DIN ENGL 4755 Pt 2-84. Oil Burning Installations; Fuel Oil Supply, Fuel Oil Supply Systems; Safety Requirements, Testing (Feb). 11 pp.

—Oil—Ships—Automatic Control Equipment
JIS F 0803-79. Methods of Onboard Test on Automatic Control of Fuel Oil System for Smaller Ships.

—Ships—Diesel Engines
MOD UK NES 320-91. Requirements for Design and Installation of Fuel Systems for Gas Turbines and Diesel Engines in Surface Ships Issue 2 (07.91). 112 pp.

—Ships—Gas Turbine Engines
MOD UK NES 320-91. Requirements for Design and Installation of Fuel Systems for Gas Turbines and Diesel Engines in Surface Ships Issue 2 (07.91). 112 pp.

Fuel Tanks
See Also: Aircraft Fuel Systems; Containers; Fuel Storage Tanks; Storage Tanks; Tanks (Containers)

—Aircraft—Design
CAA Chapter K5-2 04.74. Fuel Systems (Light Aeroplanes). 8 pp.
CAA Chapter Q5-2 App 12.79. Fuel Systems (Non-Rigid Airships). 1 p.

—Aircraft—Explosion Suppressant—Polyamide
MOD UK DTD-5627-82. Fibrous Polyamide Material for Use as an Explosion Suppressant and as a Baffle Material in Aircraft Fuel Tanks. 8 pp.

—Aircraft—Identification Systems
MOD UK DTD-1101-50. External Marking of Aeroplane Fuel Tanks. 4 pp.

—Aircraft—Installation
CAA Chapter K5-2 04.74. Fuel Systems (Light Aeroplanes). 8 pp.

—Aircraft—Orifices—Diameters
NATO STANAG 3212 ED 5 AMD 2-88. Diameters for Gravity Filling Orifices. 7 pp.
NATO STANAG 3212 ED 5 AMD 3-88. Diameters for Gravity Filling Orifices. 8 pp.

—Aircraft—Seals—Rubber
MOD UK DTD-867-54. Cellular Vulcanised Rubber for Self Sealing Fuel Tanks (Fighter Type). 2 pp.

—Automotive
EC 79/490/EEC-79. Commission Directive Adapting to Technical Progress Council Directive 70/221/EEC on the Approximation of the Laws of the Member States Relating to the Liquid Fuel Tanks and Rear Underrun Protection of Motor Vehicles and Their Trailers. 7 pp.
EC 81/333/EEC-81. Commission Directive Amending Directive 79/490/EEC Adapting to Technical Progress Council Directive 70/221/EEC on the Approximation of the Laws of the Member Sts. Relating to Liquid Fuel Tanks and Rear Underrun Prot. of Motor Vehicles and Their Trailers. 2 pp.

—Automotive—Filler Caps
CNS D2043-89. Shape & Dimension of Fuel Filler Necks and Covers for Automobiles (Jul)(6159).
JIS D 2501-60. Shape and Dimension of Fuel Filler Necks and Covers for Automobiles.

—Automotive—Filler Necks
CNS D2043-89. Shape & Dimension of Fuel Filler Necks and Covers for Automobiles (Jul)(6159).
JIS D 2501-60. Shape and Dimension of Fuel Filler Necks and Covers for Automobiles.

Fuel Tanks (Cont.)

—Boats (Marine)
ISO 10088-92. Small Craft—Permanently Installed Fuel Systems and Fixed Fuel Tanks First Edition. 12 pp.

—Earthmoving Equipment—Filler Caps
BSI BS 5279-76. 1976 Dimensions of Fuel Filler Openings of Earth-Moving Machinery. 4 pp.
CNS A4019-88. Earth-Moving Machinery Dimensions of Fuel Filler Opening (Dec)(9950).
ISO 3541-85. Earth-Moving Machinery—Dimensions of Fuel Filler Opening Second Edition. 6 pp.

—Earthmoving Equipment—Filler Openings
BSI BS 5279-76. 1976 Dimensions of Fuel Filler Openings of Earth-Moving Machinery. 4 pp.
CNS A4019-88. Earth-Moving Machinery Dimensions of Fuel Filler Opening (Dec)(9950).
ISO 3541-85. Earth-Moving Machinery—Dimensions of Fuel Filler Opening Second Edition. 6 pp.

—Hydraulic Fluids—Painting—Ships
MOD UK NES 761-87. Preparation and Painting of Fuel and Hydraulic Oil Tanks Issue 3 (08.87). 28 pp.

—Marine Engines
CSA B306-M1977. Portable Fuel Tanks for Marine Use; (Amd 2-5 December 1979) (Amd 6-10 April 1988). 20 pp.

—Motorcycles—Cocks
CNS D2028-88. Dimensions of Cock Connectors of Fuel Tanks for Motor Cycles (Jul)(5784). 1 p.
JIS D 3903-66. Dimensions of Cock Connections of Fuel Tanks for Motorcycles (R 1983). 4 pp.

—Oil
JIS S 3020-76. Oil Tanks for Oil Burning Appliances.

—Oil—Automotive
CNS Z3012-86. Method of Test for Oil Tank Vehicles on Highway (Jan)(3591).

—Oil—Ships—Drain Valves
JIS F 7398-89. Marine Fuel Oil Tank Self-Closing Drain Valves.

—Oil—Ships—Stop Valves
CNS F3201-82. Marine Fuel Oil Tank Emergency Shut-Off Valves (Oct)(9484).
JIS F 7399-89. Marine Fuel Oil Tank Emergency Shut-off Valves.
JIS F 7456-76. Remote Shut-off Devices for Marine Fuel Oil Tank Emergency Shut-off Valves.
JIS F 7457-89. Pneumatically Operated Remote Shut-off Devices for Marine Fuel Oil Tank Emergency Shut-off Valves.

—Oil—Steel
SNZ NZS/ANSI/API 650-80. Welded Steel Tanks for Oil Storage. 84 pp.

—Ships—Painting
MOD UK NES 761-87. Preparation and Painting of Fuel and Hydraulic Oil Tanks Issue 3 (08.87). 28 pp.

—Tractors
EC 88/410/EEC-88. Commission Directive Adapting to Technical Progress Council Directive 74/151/EEC on the Approximation of the Laws of the Member States Relating to Certain Components or Characteristics of Wheeled Agricultural or Forestry Tractors. 3 pp.

—Water Detecting Pastes
MOD UK CS 2879A. Paste, Water Detecting (Withdrawn).
MOD UK M 9523/66. Examination of Paste, Water Detecting (Withdrawn).

Fuel Trucks
Use: Tank Trucks

Fuels
Scope Note: For additional listings, use a more specific term *See Also:* Aircraft Fuels; Aircraft Gasoline; Alcohol Fuels; Automotive Fuels; Biogas; Briquets; Bunker Fuel Oil; Coke; Diesel Fuels; Distillate Fuels; Energy Sources; Fossil Fuels; Fuel Dispensing Equipment; Fuel Hoses; Fuel Injectors; Fuel Oils; Fuel Storage Tanks; Gas Turbine Fuels; Gaseous Fuels; Gases; Gasoline; Heating Oils; Jet Engine Fuels; Lignite; Liquefied Gases; Liquefied Petroleum Gas; Liquid Fuels; Liquid Rocket Fuels; Marine Fuels; Natural Gas; Nuclear Fuels; Petroleum Products; Pulverized Fuels; Rocket Fuels; Solid Mineral Fuels; Town Gas
MOD UK DSTAN 01-5-89. Fuels, Lubricants and Associated Products Issue 8. 328 pp.
MOD UK DSTAN 01-5-91. Fuels, Lubricants and Associated Products Issue 9. 319 pp.

Fuels (Cont.)
MOD UK DSTAN 05-50: Part 0-86. Methods for Testing Fuels, Lubricants and Associated Products Part 0: Introduction and Index Issue 1. 14 pp.

—Civil/Military—Interchangeability
NATO STANAG 2945 ED 1 AMD 0-85. NATO Civil/Military Ground Fuels Interchangeability Catalogue (AFLP-1). 5 pp.
NATO STANAG 2945 ED 1 AMD 1-85. NATO Civil/Military Ground Fuels Interchangeability Catalogue (AFLP-1). 7 pp.
NATO STANAG 2945 ED 2 AMD 0-92. NATO Civil/Military Ground Fuels Interchangeability Catalogue—AFLP-1(A). 4 pp.

—Classification
BSI BS 1000: (662)-83. 1983 Universal Decimal Classification (UDC). English Full Edition (662): Explosives. Fuels. 38 pp.
SNZ NZS/BS 1000 (662)-83. Universal Decimal Classification Explosives. Fuels. 40 pp.

—Military—Interchangeability
NATO STANAG 1135 ED 3 AMD 6-83. Interchangeability of Fuels, Lubricants and Associated Products Used by the Armed Forces of the North Atlantic Treaty Nations. 82 pp.

—Military Operations
NATO STANAG 2845 ED 2 AMD 3-81. Guide Specifications for NATO Army Fuels, Lubricants and Associated Products. 27 pp.
NATO STANAG 2845 ED 3 AMD 1-90. Guide Specifications for NATO Army Fuels, Lubricants and Associated Products. 42 pp.
NATO STANAG 4362 ED 1 AMD 0-89. (DRAFT) Fuels for Future Ground Equipments Using Compression Ignition or Turbine Engines. 2 pp.

—Procurement
EC COM(88) 377-88. Proposal for a Council Directive on the Procurement Procedures of Entities Providing Water, Energy, and Transport Services. 127 pp.
NATO STANAG 2504 ED 1 AMD 0-90. NATO Standardization of Petroleum Specifications. 9 pp.

—Solids Content—Evaporation
MOD UK DSTAN 05-50: Part 3-01. Methods for Testing Fuels, Lubricants and Associated Products Part 3: Total Solids Content by Evaporation Issue 1; Amendment 1. 8 pp.

Fuji FA 200 Aircraft
See Also: Aircraft

—Antenna Positions
CAA. Fuji FA 200 (Approved Aerial Positions). 1 p.

—Foreign Airworthiness Directives
CAA. Fuji FA-200 Series Aircraft (Foreign Airworthiness Directives). 3 pp.

Full Adders
See Also: Carry Save Adders; Digital Circuits
CECC CECC 90 104-211 ISSUE 1-82. BS CECC 90 104-211; Natural Binary Coded Decimal Adder (En). 25 pp.
CECC CECC 90 104-211 ISSUE 1-86. CEI-CECC 90 104-211; Natural Binary Coded Decimal Adder (En). 2 pp.

—Preferred Products List
CECC CECC MUAHAG Vol 7 IS 8-92. Preferred Products List; Active Microcircuits (En, Fe, Ge). 89 pp.

—4-Bit
CECC CECC 90 102-040 ISSUE 1-82. BS CECC 90 102-040; 4-Bit Binary Full Adders with Fast Carry (En) AMD 1 (En). 19 pp.
CECC CECC 90 103-038 ISSUE 1-81. BS CECC 90 103-038; 4-Bit Binary Full Adders with Fast Carry (En) AMD 1 (En). 20 pp.
CECC CECC 90 103-106 ISSUE 1-81. BS CECC 90 103-106; 4-Bit Binary Full Adders with Fast Carry (En) AMD 1 (En). 21 pp.
CECC CECC 90 107-039 ISSUE 2-89. UTE C 86-218 ADD 3/FA 39; Digital Integrated Circuits in Accordance with FS 90 107; 54/74 F 283; 4-Bit Binary Full Adder with Fast Carry (En, Fr) ADD 3 (En, Fr). 9 pp.
CECC CECC 90 109-807 ISSUE 1-88. Digital Integrated Circuits in Accordance with FS 90 109; 54/74 HC 283; 4-Bit Binary Full Adder with Fast Carry (En, Fr). 6 pp.
CECC CECC 90 109-840 ISSUE 1-89. Digital Integrated Circuits; Silicon Monolithic C MOS, Cavity or Non-Cavity Packages; Type(s) 54/74 283 4-Bit Binary Full Adder with Fast Carry Assessment Levels P, Y, L (En, Fr, Ge). 9 pp.

Full Adders (Cont.)
—4-Bit—Parallel
CECC CECC 90 104-109 ISSUE 1-81. BS CECC 90 104-109; 4-Bit Full Adder with Carry Output (En). 24 pp.
CECC CECC 90 104-109 ISSUE 1-86. CEI CECC 90 104-109; 4-Bit Full Adder with Parallel Carry Output (En). 2 pp.

—4-Bit—Preferred Products List
CECC CECC MUAHAG Vol 7 IS 8-92. Preferred Products List; Active Microcircuits (En, Fe, Ge). 89 pp.

Fuller Chisels
See Also: Chisels
CGSB 39-GP-43A-74. Chisels, Hand, Cutting, and Swaging (For Metals), Standard for; (Amendment 1 Dec 1978). 15 pp.

Fuller's Earth
See Also: Absorbents
MOD UK DSTAN 68-142-91. Fuller's Earth, Special Issue 1. 11 pp.

Fume Cupboards
See Also: Cupboards; Ventilation Equipment
SNZ NZS 7203-92. Safety in Laboratories—Fume Cupboards Amend: A, 1992; 1, 1992. 28 pp.

—Installation
BSI BS 7258: Part 2-90. 1990 Laboratory Fume Cupboards Part 2: Recommendations for the Exchange of Information and Recommendations for Installation. 20 pp.

—Laboratory
BSI BS 7258: Part 3-90. 1990 Amd 1 Laboratory Fume Cupboards Part 3: Recommendations for Selection, Use and Maintenance (AMD 6746) September 30, 1991 (H). 16 pp.

—Laboratory—Safety
BSI BS 7258: Part 1-90. 1990 Laboratory Fume Cupboards Part 1: Safety and Performance. 16 pp.
BSI DD 191-90. 1990 Method for Determination of the Containment Value of a Laboratory Fume Cupboard. 9 pp.

—Safety
SAA AS 2243.9-91. Safety in Laboratories—Part 9: Recirculating Fume Cabinets (In Professional Packages 34B, 34C, 34D, 47). 17 pp.

Fumes
See Also: Aerosols; Air Pollution; Exhaust Gases; Flue Gases

—Thermal Cutting—Concentration
JIS Z 3952-90. Methods of Measurement for Gas Concentration in Welding Environment. 16 pp.

—Welding—Chemical Analysis
BSI BS 6691: Part 1-86. 1986 Fume from Welding and Allied Processes Part 1: Guide to Methods for the Sampling and Analysis of Particulate Matter. 12 pp.
BSI BS 6691: Part 2-86. 1986 Fume from Welding and Allied Processes Part 2: Guide to Methods for the Sampling and Analysis of Gases. 12 pp.
BSI BS 7383-90. 1990 Cold-Pour Resin-Based Compound for Use as a Filling Medium in Terminating Cables in Enclosures for Voltages Not Exceeding 11 kV for Use in Coal Mines. 14 pp.
JIS Z 3920-91. Methods for Chemical Analysis of Elements in Welding Fumes. 20 pp.

—Welding—Concentration
CNS Z8020-81. Method for Determining Average Concentration of Weld Fume in Welding Environment (Apr)(7250).
CNS Z8021-81. Method of Measuring Weld Fume Concentration (Apr)(7330).
JIS Z 3950-86. Methods of Measurement for Weld Fume Concentration.
JIS Z 3950-75. Methods of Measuring Weld Fume Concentration (R 1978). 6 pp.
JIS Z 3951-86. Methods for Determining Average Concentration of Weld Fume in Welding Environment.
JIS Z 3951-77. Methods for Determining Average Concentration of Weld Fume in Welding Environment (R 1980). 5 pp.
JIS Z 3952-90. Methods of Measurement for Gas Concentration in Welding Environment. 16 pp.

—Welding—Measurement
JIS Z 3930-79. Method of Measuring Total Amount of Weld Fumes Generated by Covered Electrode. 8 pp.

Fumes (Cont.)
—Welding—Sampling
BSI BS 6691: Part 1-86. 1986 Fume from Welding and Allied Processes Part 1: Guide to Methods for the Sampling and Analysis of Particulate Matter. 12 pp.
BSI BS 6691: Part 2-86. 1986 Fume from Welding and Allied Processes Part 2: Guide to Methods for the Sampling and Analysis of Gases. 12 pp.
BSI BS 7383-90. 1990 Cold-Pour Resin-Based Compound for Use as a Filling Medium in Terminating Cables in Enclosures for Voltages Not Exceeding 11 kV for Use in Coal Mines. 14 pp.
SAA AS 3853.1-91. Fume from Welding and Allied Processes—Part 1: Guide to Methods for the Sampling and Analysis of Particulate Matter. 8 pp.
SAA AS 3853.2-91. Fume from Welding and Allied Processes—Part 2: Guide to Methods for the Sampling and Analysis of Gases. 8 pp.

Fumigation
See Also: Pest Control
—Hair (Animal)
CNS N4025-85. Method of Test for Animal Hairs (May)(2826). 7 pp.

Fuming Nitric Acid
Use: Nitric Acid

Fuming Sulfuric Acid
Use: Sulfuric Acid

Function Cards
Use: Integrated Circuit Cards

Function Generators
See Also: Arithmetic Logic Unit/Function Generators
—Digital
CECC CECC 90 103-078 ISSUE 1-81. BS CECC 90 103-078; Arithmetic Logic Units/Function Generators (En) AMD 1 (En). 25 pp.

Function Generators/Arithmetic Logic Units
Use: Arithmetic Logic Unit/Function Generators

Function Keys
Use For: Control Keys *See Also:* Keyboards
BSI BS 5231-75. 1975 Principles Governing the Positioning of Control Keys on Keyboards of Office Machines and Data Processing Equipment. 9 pp.
CNS C5191-85. Office Machines and Data Processing Equipment—Principles Governing the Positioning of Control Keys on Keyboards (Mar)(11215).
IEC DIS 9995 Pt 6-91. Information Technology—Keyboard Layouts for Text and Office Systems—Part 6: Function Section ***CD-ROM ONLY***. 6 pp.
ISO 3244-84. Office Machines and Data Processing Equipment—Principles Governing the Positioning of Control Keys on Keyboards Second Edition. 6 pp.
ISO DIS 9995 Pt 6-91. Information Technology—Keyboard Layouts for Text and Office Systems—Part 6: Function Section ***CD-ROM ONLY***. 6 pp.
JTC1 3244-84. Office Machines and Data Processing Equipment—Principles Governing the Positioning of Control Keys on Keyboards Second Edition. 6 pp.
OSI ISO 3244-84. Office Machines and Data Processing Equipment—Principles Governing the Positioning of Control Keys on Keyboards. 6 pp.

—Symbols
IEC DIS 9995 Pt 7-91. Information Technology—Keyboard Layouts for Text and Office Systems—Part 7: Symbols Used to Represent Functions ***CD-ROM ONLY***. 20 pp.
ISO DIS 9995 Pt 7-91. Information Technology—Keyboard Layouts for Text and Office Systems—Part 7: Symbols Used to Represent Functions ***CD-ROM ONLY***. 20 pp.

Functional Testing
See Also: Testing
—Combines
BSI BS 7560-92. 1992 Method of Assessing, Testing and Reporting Characteristics and Performance of Combine Harvesters (ISO 8210: 1989). 12 pp.
ISO 8210-89. Equipment for Harvesting—Combine Harvesters—Test Procedure First Edition. 10 pp.

—Electrical Insulation
IEC 610-78. Principal Aspects of Functional Evaluation of Electrical Insulation Systems: Aging Mechanisms and Diagnostic Procedures First Edition. 23 pp.
IEC 791-84. Performance Evaluation of Insulation Systems Based on Service Experience and Functional Tests First Edition. 20 pp.

Functional Testing (Cont.)
—Electrical Insulation (Cont.)
IEC 792 Pt 1-85. Multi-Factor Functional Testing of Electrical Insulation Systems Part 1: Test Procedures First Edition. 34 pp.
IEC 941-88. Mechanical Endurance Functional Tests for Electrical Insulation Systems First Edition. 24 pp.

—Electrical Insulation—Bibliographies
IEC 792 Pt 2-93. Multi-Factor Functional Testing of Electrical Insulation Systems Part 2: Bibliography First Edition. 53 pp.

Fundamental Frequencies
See Also: Frequencies
—Concretes
CNS A3052-87. Method of Test for Fundamental Transverse, Longitudinal and Torsional Frequencies of Concrete Specimens (Dec)(1239). 6 pp.

Fundus Cameras
Use: Ophthalmic Cameras

Funerals
—Service Contracts
MOD UK DEFCON 112BJ-86. Conditions of Contract Funeral Services 8/86. 8 pp.

Fungi
Use For: Eumycophyta; Fungus; Molds (Organisms); Moulds (Organisms) *See Also:* Basidiomycetes; Mushrooms; Plants (Botany); Yeasts

—Cellular Plastics
ISO 846-78. Plastics—Determination of Behavior Under the Action of Fungi and Bacteria—Evaluation by Visual Examination or Measurement of Change in Mass or Physical Properties First Edition. 12 pp.

—Cereals—Count Methods
ISO 7698-90. Cereals, Pulses and Derived Products—Enumeration of Bacteria, Yeasts and Moulds First Edition. 10 pp.

—Classification
BSI BS 1000: (663)-85. 1985 Universal Decimal Classification (UDC). English Full Edition (663): Industrial Microbiology. Industrial Mycology. Fermentation Industry. Beverage Industry. Stimulant Industry. 30 pp.
SNZ NZS/BS 1000 (663)-85. Universal Decimal Classification Industrial Microbiology. Industrial Mycology. Fermentation Industry. Beverage Industry. Stimulant Industry.. 32 pp.

—Cork Stoppers—Count Methods
ISO 10718-93. Cork Stoppers—Enumeration of Colony-Forming Units of Yeasts, Moulds and Bacteria Capable of Growth in an Alcoholic Medium First Edition. 5 pp.

—Dairy Products
SAA AS 1095.3.3-82. Microbiological Methods for the Dairy Industry—Part 3: Methods of Examination for Specific Groups of Microorganisms—Part 3.3: Yeasts and Moulds. 2 pp.

—Dairy Products—Count Methods
BSI BS 4285: Sec 3.6-86. 1986 Microbiological Examination for Dairy Purposes Part 3: Methods Part 3: Methods for Detection and/or Enumeration of Specific Groups of Microorganisms Section 3.6: Enumeration of Yeasts and Moulds. 4 pp.

—Edible
CNS N5167-80. Edible Fungi and Fungus Products (Aug)(6394). 6 pp.

—Food
BSI BS 5763: Part 12-88. 1988 Microbiological Examination of Food and Animal Feeding Stuffs Part 12: Enumeration of Yeasts and Moulds. 6 pp.
ISO 7954-87. Microbiology—General Guidance for Enumeration of Yeasts and Moulds—Colony Count Technique at 25 Degrees Celsius First Edition. 6 pp.
SAA AS 1766.2.2-80. Methods for the Microbiological Examination of Food—Part 2: Examination for Specific Organisms—Part 2.2: Colony Count of Yeast and Moulds. 2 pp.

—Legumes—Count Methods
ISO 7698-90. Cereals, Pulses and Derived Products—Enumeration of Bacteria, Yeasts and Moulds First Edition. 10 pp.

Fungi (Cont.)

—Paints—Environmental Testing
SAA AS 1580.481. 1.13-92. Paints and Related Materials—Methods of Test—Pt. 481: Coatings —Pt. 481.1: Assmt. of Indiv. Defects of Exposed Films—Pt. 481. 1.13: Exposedto Weathering—Degree to Fungal or Algal Growth Amdt 1 January/February 1993 (in Professional Packages 30, 39). 5 pp.

—Plastics
DIN ENGL 53739-84. Testing of Plastics; Influence of Fungi and Bacteria; Visual Evaluation; Change in Mass or Physical Properties (Nov). 11 pp.

—Water Quality
ISO 6222-88. Water Quality—Enumeration of Viable Micro-Organisms—Colony Count by Inoculation in or on a Nutrient Agar Culture Medium First Edition. 4 pp.

—Wood Preservatives
BSI DD ENV 807-93. 1993 Wood Preservatives—Determination of the Toxic Effectiveness Against Soft Rotting Micro-Fungi and Other Soil Inhabiting Micro-Organisms (W). 30 pp.
CEN EN 113-80. Wood Preservatives: Determination of Toxic Values Against Wood Destroying Basidiomycetes Cultured on an Agar Medium. 17 pp.
CEN PREN 113-89. Wood Preservatives: Determination of the Toxic Valves Against Wood Destroying Basidiomycetes Cultured on an Agar Medium (Supersedes EN 113 Per Index). 18 pp.
CEN ENV 807-93. Wood Preservatives—Determination of the Toxic Effectiveness Against Soft Rotting Micro-Fungi and Other Soil Inhabiting Micro-Organisms. 35 pp.
DIN ENGL EN 118-91. Wood Preservatives; Determination of Preventive Effect on Reticulitermes Santonensis (De Feytaud) (Laboratory Method) (Feb). 10 pp.

Fungicides

Scope Note: Used for fungicides in general. For additional listings, see also specific chemicals used for fungal control *Use For:* Antifungal Agents
See Also: Pesticides

—Wood Preservatives
BSI BS 7066: Part 1-90. 1990 Wood Preservatives Laboratory Method for Determining the Protective Effectiveness of a Preservative Treatment Against Blue Stain in Service Part 1: Brushing Procedure. 20 pp.
BSI BS 7066: Part 2-90. 1990 Wood Preservatives Laboratory Method for Determining the Protective Effectiveness of a Preservative Treatment Against Blue Stain in Service Part 2: Application by Methods Other Than Brushing. 24 pp.
BSI BS 7282-90. 1990 Field Test Method for Determining the Relative Protective Effectiveness of a Wood Preservative in Ground Contact. 15 pp.
BSI BS EN 330-93. 1993 Wood Preservatives—Field Test Method for Determining the Relative Protective Effectiveness of a Wood Preservative for Use Under a Coating and Exposed Out-of-Ground Contact: L-Joint Method (W). 19 pp.
CEN EN 152 (Part 1)-88. Test Methods for Wood Preservatives; Laboratory Method for Determining the Preventive Effectiveness of a Preservative Treatment Against Blue Stain in Service Part 1: Brushing Procedure. 31 pp.
CEN EN 152 (Part 2)-88. Test Methods for Wood Preservatives; Laboratory Method for Determining the Protective Effectiveness of a Preservative Treatment Against Blue Stain in Service Part 2: Application by Methods Other Than Brushing. 20 pp.
CEN EN 252-89. Field Test Method for Determining the Relative Protection Effectiveness of a Wood Preservative in Ground Contact. 20 pp.
CEN EN 252 AC1-89. AMD 1 Field Test Method for Determining the Relative Protection Effectiveness of a Wood Preservative in Ground Contact. 3 pp.
CEN PREN 330-90. Wood Preservatives-L-Joint Field Test Method for Determining the Relative Protective Effectiveness of a Wood Perservative out of Ground Contact. 43 pp.
CEN EN 330-93. Wood Preservatives—Field Test Method for Determining the Relative Protective Effectiveness of a Wood Preservative for Use Under a Coating and Exposed out-of-Ground Contact: L-Joint Method. 14 pp.
DIN ENGL EN 252-90. Field Test Method for Determining the Relative Protective Effect of a Wood Preservative on Wood in Contact with Soil (Apr). 10 pp.

Fungus

Use: Fungi

Fungus Resistance Testing

Use For: Mildew Resistance Testing

Fungus Resistance Testing (Cont.)

See Also: Microbiological Resistance Testing
JIS Z 2911-92. Methods of Test for Fungus Resistance. 22 pp.
SAA AS 1157. Methods of Testing Materials for Resistance to Fungal Growth (Complete Set in Binder).
SAA AS 1157.1-72. Methods of Testing Materials for Resistance to Fungal Growth—Part 1: General Principles of Testing. 20 pp.

—Adhesives
SAA AS 1157.10-79. Methods of Testing Materials for Resistance to Fungal Growth—Part 10: Resistance of Adhesives and Glues to Fungal Growth. 6 pp.

—Canvas
MOD UK CS 2877A. Solution, Rot-Proofing and Water-Proofing (Withdrawn).

—Coated Fabrics
SAA AS 1157.4-78. Methods of Testing Materials for Resistance to Fungal Growth—Part 4: Resistance of Coated Fabrics to Fungal Growth. 10 pp.

—Connectors
CEN PREN 2591 (Part C6)-92. Elements of Electrical and Optical Connection Test Methods Part C6—Mould Growth. 11 pp.
CEN PREN 2591-FC6-93. Aerospace Series Elements of Electrical and Optical Connection Test Methods Part FC6—Optical Elements Mould Growth. 3 pp.

—Construction Materials
BSI BS 1982-68. (WITHDRAWN) 1968 Methods of Test for Fungal Resistance of Manufactured Building Materials Made of or Containing Materials of Organic Origin (R) (Superseded by BS 1982: Parts 0, 1, 2 and 3: 1990). 16 pp.
BSI BS 1982: Part 0-90. 1990 Fungal Resistance of Panel Products Made of or Containing Materials of Organic Origin Part 0: Guide to Methods for Determination. 10 pp.
BSI BS 1982: Part 1-90. 1990 Fungal Resistance of Panel Products Made of or Containing Materials of Organic Origin Part 1: Method for Determination of Resistance to Wood-Rotting Basidiomycetes. 19 pp.
BSI BS 1982: Part 1-01. 1990 Amd 1 Fungal Resistance of Panel Products Made of or Containing Materials of Organic Origin Part 1: Method for Determination of Resistance to Wood-Rotting Basidiomycetes (AMD 7780) June 15, 1993 (R). 20 pp.
BSI BS 1982: Part 2-90. 1990 Fungal Resistance of Panel Products Made of or Containing Materials of Organic Origin Part 2: Method for Determination of Resistance to Cellulose-Decomposing Microfungi. 16 pp.
BSI BS 1982: Part 3-90. 1990 Fungal Resistance of Panel Products Made of or Containing Materials of Organic Origin Part 3: Methods for Determination of Resistance to Mould or Mildew. 12 pp.

—Cordage
SAA AS 1157.3-78. Methods of Testing Materials for Resistance to Fungal Growth—Part 3: Resistance of Cordage and Yarns to Fungal Growth. 6 pp.

—Fabrics
CGSB CAN/CGSB-4.2 NO.28.1-M91. Textile Test Methods Resistance to Micro-Organisms—Fungus Damage Test—Pure Culture—Quantitative. 8 pp.
CGSB CAN/CGSB-4.2 NO.28.2-M91. Textile Test Methods Resistance to Micro-Organisms—Surface-Growing Fungus Test—Pure Culture. 7 pp.
CGSB CAN/CGSB-4.2 NO.28.4-M91. Textile Test Methods Resistance to Micro-Organisms—Fungus Damage Test—Pure Culture—Qualitative. 7 pp.
SAA AS 1157.2-72. Methods of Testing Materials for Resistance to Fungal Growth—Part 2: Resistance of Textiles to Fungal Growth. 9 pp.

—Fibers
CNS L3063-75. Method of Test for Mildew Resistance of Fiber Products (May) (2690).

—Leather
SAA AS 1157.6-71. Methods of Testing Materials for Resistance to Fungal Growth—Part 6: Resistance of Leather to Surface Fungal Growth. 3 pp.

—Materiel
MOD UK DSTAN 00-29-82. Fungal Contamination Affecting the Design of Military Materiel Issue 1. 13 pp.

—Optical Fibers
CNS C6281-87. Method of Test for Fiber Optic Devices (FOTP-56 Evaluating Fungus Resistance of Optical Fibers) (Apr)(11903).

Fungus Resistance Testing (Cont.)

—Paints
BSI BS 3900: Part G6-89. 1989 Methods of Test for Paints Group G: Environmental Tests on Paint Films (Including Tests for Resistance to Corrosion and Chemicals) Part G6: Asscssment of Resistance to Fungal Growth. 8 pp.
SNZ NZS/AS 1580. 481.1.13-92. Methods of Test for Paints and Related Materials Part 481.1.13: Coatings—Exposed to Weathering—Degree of Fungal or Algal Growth (This is a Joint Standard with SAA AS 1580.481.13). 3 pp.

—Paper Products
SAA AS 1157.7-78. Methods of Testing Materials for Resistance to Fungal Growth—Part 7: Resistance of Papers and Paper Products to Surface Fungal Growth. 7 pp.

—Papers
SAA AS 1157.7-78. Methods of Testing Materials for Resistance to Fungal Growth—Part 7: Resistance of Papers and Paper Products to Surface Fungal Growth. 7 pp.

—Plastics
SAA AS 1157.11-78. Methods of Testing Materials for Resistance to Fungal Growth—Part 11: Resistance of Rubbers and Plastics to Surface Fungal Growth. 10 pp.

—Rubber
SAA AS 1157.11-78. Methods of Testing Materials for Resistance to Fungal Growth—Part 11: Resistance of Rubbers and Plastics to Surface Fungal Growth. 10 pp.

—Sealants
CGSB CAN2-19.0-M77METH 9.5-78. Methods of Testing Putty, Caulking and Sealing Compounds Mildew Resistance of Sealing Compounds. 1 p.

—Timber
SAA AS 1157.5-71. Methods of Testing Materials for Resistance to Fungal Growth—Part 5: Resistance of Timber to Surface Fungal Growth. 3 pp.

—Yarns
SAA AS 1157.3-78. Methods of Testing Materials for Resistance to Fungal Growth—Part 3: Resistance of Cordage and Yarns to Fungal Growth. 6 pp.

Funnels

See Also: Laboratory Equipment; Laboratory Ware
CNS Z7065-79. 170 mm Phi Funnel (Apr)(4796).
ISO 7057-81. Plastics Laboratory Ware—Filter Funnels First Edition. 7 pp.

—Enamel
CNS R2159-85. Enamel Funnel for Laboratory (Jun)(9034).

—Filter—Plastic
BSI BS 5404: Part 3-77. 1977 Plastics Laboratory Ware Part 3: Filter Funnels. 4 pp.

—Filter—Porcelain
DIN ENGL 12905-80. Porcelain Laboratory Ware; Buechner Funnels and Hirsch Funnels (July). 2 pp.

—Glass
CNS R2127-86. Glass Funnels for Chemical Analysis (Mar) (7303).

—Glass—Dropping
BSI BS 2021-80. 1980 Separating and Dropping Funnels for Laboratory Use. 10 pp.
BSI BS 2021-01. 1980 Amd 1 Separating and Dropping Funnels for Laboratory Use (ISO 4800: 1977) (AMD 7865) August 15, 1993. 11 pp.
CNS R2102-80. Laboratory Glassware Separating Funnels and Dropping Funnels (Sep) (6513).
ISO 4800-77. Laboratory Glassware—Separating Funnels and Dropping Funnels First Edition. 9 pp.

—Glass—Long Stem
DIN ENGL 12446-70. Laboratory Instruments of Glass; Long-Stem Funnels (Bunsen Funnels) (Mar). 1 p.

—Glass—Separating
BSI BS 2021-80. 1980 Separating and Dropping Funnels for Laboratory Use. 10 pp.
BSI BS 2021-01. 1980 Amd 1 Separating and Dropping Funnels for Laboratory Use (ISO 4800: 1977) (AMD 7865) August 15, 1993. 11 pp.
CNS R2102-80. Laboratory Glassware Separating Funnels and Dropping Funnels (Sep) (6513).
ISO 4800-77. Laboratory Glassware—Separating Funnels and Dropping Funnels First Edition. 9 pp.

—Glass—Short Stem
DIN ENGL 12445-70. Laboratory Instruments of Glass; Short-Stem Funnels (Mar). 1 p.

Furan Coatings
Use: Polymer Coatings (Made From Polymers)

Furan Compound Content Analysis
See Also: Furfural Content Analysis

—Electrical Insulating Oils—Liquid Chromatography
CENELEC PREN 61198-92. Mineral Insulating Oils—Methods for the Determination of 2-Furfural and Related Compounds. 17 pp.

IEC 1198-93. Mineral Insulating Oils—Methods for the Determination of 2-Furfural and Related Compounds First Edition. 32 pp.

Furazolidone Content Analysis
—Animal Feed
CNS N4138-86. Determination of Furazolidone in Feeds (Aug)(11663).

Furfural
See Also: Furfural Content Analysis

CNS K7237-67. Chemical Reagent (Furfural) (Oct)(1736).

ISO 2511-74. Furfural for Industrial Use—List of Methods of Test First Edition. 4 pp.

JIS K 8833-78. Furfural.

—Acidity—Volumetric Analysis
ISO 2888-73. Furfural for Industrial Use—Determination of Acidity to Phenolphthalein—Volumetric Method First Edition. 4 pp.

—Carbonyl Compound Content—Volumetric Analysis
ISO 2512-74. Furfural for Industrial Use—Determination of Total Carbonyl Compounds—Volumetric Method First Edition. 4 pp.

Furfural Content Analysis
Use For: 2-Furfural Content Analysis
See Also: Furan Compound Content Analysis; Furfural

—Electrical Insulating Oils—Liquid Chromatography
CENELEC PREN 61198-92. Mineral Insulating Oils—Methods for the Determination of 2-Furfural and Related Compounds. 17 pp.

IEC 1198-93. Mineral Insulating Oils—Methods for the Determination of 2-Furfural and Related Compounds First Edition. 32 pp.

—Ethanols
ISO 1388 Pt 11-81. Ethanol for Industrial Use—Methods of Test—Part 11: Test for Detection of Furfural First Edition. 4 pp.

—Petroleum Products—Spectrophotometry
CEN PREN 214-86. Liquid Petroleum Products Determination of Furfural Content; Spectrophotometric Method. 9 pp.

2-Furfural Content Analysis
Use: Furfural Content Analysis

Furnace Cupolas
See Also: Furnaces

—Firebricks
CNS R1020-86. Shape and Dimension of Refractory Bricks for Cupola (Jul)(11637).

JIS R 2104-83. Shape and Dimension of Refractory Bricks for Cupola. 6 pp.

Furnaces
Scope Note: For additional listings, use a more specific term *Use For:* Boiler Furnaces; Central Furnaces; Combustion Furnaces *See Also:* Air Furnaces; Blast Furnaces; Boilers; Direct Arc Furnaces; Electric Arc Furnaces; Electroslag Remelting Furnaces; Fireplaces; Furnace Cupolas; Heat Treating Furnaces; Heating, Ventilating and Air Conditioning Equipment; Heating Equipment; Incinerators; Induction Furnaces; Kilns; Melting Furnaces; Oil Burners; Oil Furnaces; Open Hearth Furnaces; Ovens; Refractory Materials; Regenerative Furnaces; Resistance Furnaces; Separators (Mechanical); Solid Fuel Burning Equipment; Solid Fuel Furnaces; Steel Converters; Stokers; Submerged Arc Furnaces

BSI BS 1971-69. (WITHDRAWN) 1969 Corrugated Furnaces for Shell Boilers (Incorporated into BS 2790). 23 pp.

—Calcium Oxides—Heat Balance
CNS R3124-85. Method for Calculating Heat Balance of Kiln and Furnace for Lime (May) (11274).

JIS R 0305-91. Heat Balancing of Kiln and Furnace for Lime. 25 pp.

—Deposits—Sampling
BSI BS 2455: Part 2-83. 1983 Methods of Sampling and Examining Deposits from Boilers and Associated Industrial Plant Part 2: Methods for Sampling and Examining Free-Side Deposits. 41 pp.

BSI BS 2455: Part 2-01. 1983 Amd 1 Sampling and Examining Deposits from Boilers and Associated Industrial Plant Part 2: Methods for Sampling and Examining Fire-Side Deposits (AMD 7751) October 15, 1993 (Q). 42 pp.

—Electric—Forced Air—Residential
CSA C22.2 NO 23.1-M1986. Electric Furnaces in Combination with Solid Fuel Fired Furnaces (R 1992); (Gen Instr 1). 18 pp.

—Electrical Equipment
CENELEC PREN 50156-1-93. Electrical Equipment for Furnaces Part 1: Rules for Installation. 51 pp.

DIN VDE 0116-89. Electrical Equipment for Furnaces (Oct). 62 pp.

—Fire Testing—Building Components
DIN ENGL 4102 Pt 8-86. Fire Behaviour of Building Materials and Components; Small Scale Test Furnace (May). 4 pp.

—Fire Testing—Construction Materials
DIN ENGL 4102 Pt 8-86. Fire Behaviour of Building Materials and Components; Small Scale Test Furnace (May). 4 pp.

—Firebricks—Waterproof Coatings
MOD UK DSTAN 80-173-93. Waterproofing Compound for Refractory Furnace Linings Type: Brushing/Spraying Issue 1. 18 pp.

MOD UK TS 10192A. Waterproofing Compound for Refractory Furnace Linings. Type: Brushing/Spraying (Superseded by Def Stan 80-1730.

—Gas—Residential
CNS S1073-81. Gas Furnace for Household (Apr)(3101).

—Gas—Temperature Controllers
DIN ENGL EN 257-92. Mechanical Thermostats for Gas Burning Appliances (Mar). 45 pp.

—Glossaries
BSI BS 4642-70. 1970 Glossary of Industrial Furnace Terms. 33 pp.

—Heat Treatment—Steels
BSI DD 215-93. 1993 Validation of Metal Temperature in the Heat Treatment of Steels for Pressure Purposes (V). 16 pp.

—Industrial—Construction
CSA C22.2 NO 88-1958. Construction and Test of Industrial Heating Equipment (R 1992). 22 pp.

CSA 1169 Bull. Electrical Bulletin 1169 June 27, 1978 to C22.2 NO 88. 2 pp.

—Industrial—Testing
CSA C22.2 NO 88-1958. Construction and Test of Industrial Heating Equipment (R 1992). 22 pp.

CSA 1169 Bull. Electrical Bulletin 1169 June 27, 1978 to C22.2 NO 88. 2 pp.

—Linings—Ceramic Fibers—Design
BSI BS 6466-84. 1984 Amd 1 Design and Installation of Ceramic Fibre Furnace Linings (AMD 4870) June 28, 1985. 38 pp.

BSI BS 6466-02. 1984 Amd 2 Code of Practice for Design and Installation of Ceramic Fibre Furnace Linings (AMD 6748) October 31, 1991. 42 pp.

—Linings—Ceramic Fibers—Installation
BSI BS 6466-84. 1984 Amd 1 Design and Installation of Ceramic Fibre Furnace Linings (AMD 4870) June 28, 1985. 38 pp.

BSI BS 6466-02. 1984 Amd 2 Code of Practice for Design and Installation of Ceramic Fibre Furnace Linings (AMD 6748) October 31, 1991. 42 pp.

—Marine
BSI BS 1971-69. (WITHDRAWN) 1969 Corrugated Furnaces for Shell Boilers (Incorporated into BS 2790). 23 pp.

—Marine—Firebricks
CGSB CAN/CGSB-10.5-92. Insulating Refractory Brick. 7 pp.

—Oil
CSA B140.4-1974. Oil-Fired Warm Air Furnaces (R 1991); (Rev 1-3 September 1977) (Rev 4-8 June 1979) (Amd 9-10 April 1990) (Amd 11 July 1993) (Gen Instr 2). 72 pp.

—Oil Burners
BSI BS 5410: Part 3-76. 1976 Amd 1 Code of Practice for Oil Firing Part 3: Installations for Furnaces, Kilns, Ovens and Other Industrial Purposes. 40 pp.

—Refractory Mortars
CGSB CAN/CGSB-10.3-92. Air-Setting Refractory Mortar. 6 pp.

CGSB CAN/CGSB-10.6-92. Heat-Setting Refractory Mortar. 7 pp.

—Safety
DIN ENGL 3398 Pt 2-82. Pressure Cut-Off Switches for Air, Flue Gases and Exhaust Gases in Furnaces; Requirements, Testing (Nov). 8 pp.

JIS B 8415-82. General Safety Code for Industrial Combustion Furnaces. 15 pp.

—Stiffening Rings
BSI BS 2790: CASE 27-93. 1993 Design and Manufacture of Shell Boilers of Welded Construction Enquiry Cases: Case 27: Furnace Stiffening Rings. 1 p.

—Switches
DIN ENGL 3398 Pt 2-82. Pressure Cut-Off Switches for Air, Flue Gases and Exhaust Gases in Furnaces; Requirements, Testing (Nov). 8 pp.

Furnish
See Also: Additives; Caseins; Clays; Dyes; Fillers; Papermaking Equipment; Pigments; Pulps; Starches

—Paperboard
CPPA K.1-61. Standard Terms for the Furnish of Paper and Paperboard and for the Measurement of Fibre Losses. 2 pp.

—Papers
CPPA K.1-61. Standard Terms for the Furnish of Paper and Paperboard and for the Measurement of Fibre Losses. 2 pp.

Furnishing Fabrics
Use: Upholstery

Furniture
Scope Note: For additional listings, use a more specific term *Use For:* Storage Furniture
See Also: Beds; Bookcases; Bunk Beds; Cabinets (Furniture); Chairs (Seats); Conference Room Tables; Cots; Couches; Cradles (Furniture); Credenzas; Cribs (Furniture); Cushions; Desks; Drawers; Filing Cabinets; Futons; Laboratory Furniture; Office Furniture; Outdoor Furniture; Printer Stands; Recliners (Chairs); Rocking Chairs; School Furniture; Seats; Settees; Sofas; Step Stools; Stools; Tables (Stands); Upholstery; Waterbeds

BSI BS 5459: Part 3-83. 1983 Performance Requirements and Tests for Office Furniture Part 3: Storage Furniture. 11 pp.

CNS S2115-84. Method of Test for Furniture (General) (May)(10894).

JIS A 4415-84. Storage Furniture for Dwellings.

JIS S 1017-83. General Rule for Test Method of Furniture.

—Built In Appliances—Modular Construction
JIS A 0016-79. Modular Coordination-Coordinating Size of Opening for Built-in Appliances in Storage Furniture.

—Classification
BSI BS 1000: (684)-79. 1979 Universal Decimal Classification (UDC). English Full Edition (684): Furniture and Allied Industries. 8 pp.

SNZ NZS/BS 1000 (684)-79. Universal Decimal Classification Furniture and Allied Industries. 8 pp.

—Coatings—Adhesion Testing
CNS S2135-86. Method of Test for Adhesion of Furniture Coatings (Aug)(11684).

—Coatings—Corrosion Inhibitors
CNS S2136-86. Method of Test for Antirust Property of Furniture Coating (Aug)(11685).

—Electrical Installations
DIN VDE 0100 Pt 724-80. Erection of Power Installations with Rated Voltages up to 1000 V; Electrical Equipment in Furniture and Similar Fitments e.g. Curtain-Ledges, Decorative Coverings (June). 7 pp.

—Fatigue Testing
CNS S2133-86. Method of Test for Vertical Fatigue Load of Furniture (Aug)(11682).

CNS S2134-86. Method of Test for Horizontal Fatigue Load of Furniture (Aug)(11683).

INTERNATIONAL AND NON-U.S. NATIONAL STANDARDS
SUBJECT INDEX

Furniture (Cont.)
—Finishes—Gloss
BSI BS 3962: Part 1-80. 1980 Methods of Test for Clear Finishes for Wooden Furniture Part 1: Assessment of Specular Gloss at 85 Degrees. 7 pp.

—Foam Rubber Components
BSI BS 3129-59. 1959 Amd 1 Latex Foam Rubber Components for Furniture. 13 pp.

—Glossaries
CNS S1198-86. Glossary of Terms Used in Furniture—Common Terms (Aug)(11673).
CNS S1198-1-86. Glossary of Terms Used in Furniture—Parts (Aug)(11673-1).
CNS S1198-2-86. Glossary of Terms Used in Furniture—Style (Aug)(11673-2).

—Hardware—Glossaries
ISO 8555 Pt 1-87. Hardware for Furniture—Terms for Furniture Fittings—Part 1: Assembly Fittings First Edition. 15 pp.
ISO 8555 Pt 2-87. Hardware for Furniture—Terms for Furniture Fittings—Part 2: Hinges and Flap Hinges First Edition. 21 pp.
ISO 8555 Pt 3-87. Hardware for Furniture—Terms for Furniture Fittings—Part 3: Drawer Slides and Sliding Door Gears First Edition. 15 pp.
ISO 8555 Pt 4-87. Hardware for Furniture—Terms for Furniture Fittings—Part 4: Holding Devices, Flap Stays, Lid Stays First Edition. 10 pp.
ISO 8555 Pt 5-87. Hardware for Furniture—Terms for Furniture Fittings—Part 5: Height Adjusters, Furniture Legs, Underframes First Edition. 8 pp.
ISO 8555 Pt 6-87. Hardware for Furniture—Terms for Furniture Fittings—Part 6: Shelf Supports, Hanging Rails, Cabinet Suspension Brackets First Edition. 10 pp.
ISO 8555 Pt 7-87. Hardware for Furniture—Terms for Furniture Fittings—Part 7: Handles, Knobs, Escutcheons, Escutcheon Insets First Edition. 8 pp.
ISO 8555 Pt 8-87. Hardware for Furniture—Terms for Furniture Fittings—Part 8: Castors and Glides First Edition. 8 pp.

—Hazardous Materials
CNS S2128-86. Method of Test for Hazardous Substances of Furniture Surface (Aug)(11677).

—Identification Systems
BSI BS 6250: Part 4-92. 1992 Domestic and Contract Furniture Part 4: Guide to the Provision of Information. 9 pp.
BSI BS 6250: Part 4-85. 1985 Domestic and Contract Furniture Part 4: Guide to the Provision of Information. 8 pp.

—Impact Testing
ISO 4211 Pt 4-88. Furniture—Tests for Surfaces—Part 4: Assessment of Resistance to Impact First Edition. 7 pp.

—Kitchens—Sizes
CEN PREN 1116-93. Kitchen Furniture—Co-Ordinating Sizes for Kitchen Furniture and Kitchen Appliances. 8 pp.

—Lacquers—Print Resistance
CGSB 1-GP-71 METH 142.2-78. Methods of Testing Paints and Pigments Print Resistance Furniture Lacquers. 1 p.

—Leather
BSI BS 6608-85. 1985 Cattle Hide Leathers for Upholstered Furniture. 8 pp.

—Lighting
DIN VDE 0710 Pt 14-82. Luminaires with Operating Voltages Below 1000 V; Luminaires to Be Built into Furniture (Apr). 17 pp.

—Liquid Resistance
JIS A 1531-84. Test Method of Surface Resistance to Cold Liquids for Furniture.

—Loads (Forces)
CNS S2130-86. Method of Test for Vertical Load of Furniture (Aug)(11679).
CNS S2131-86. Method of Test for Horizontal Load of Furniture (Aug)(11680).
CNS S2132-86. Method of Test for Eccentric Load of Furniture (Aug)(11681).
CNS S2133-86. Method of Test for Vertical Fatigue Load of Furniture (Aug)(11682).
CNS S2134-86. Method of Test for Horizontal Fatigue Load of Furniture (Aug)(11683).

—Locks (Hardware)—Glossaries
ISO 8554 Pt 1-87. Hardware for Furniture—Terms for Furniture Locks—Part 1: Latch Lock, Dead Lock, Rod-Operating Lock, Central Locking System, Cylinder Lock, Combination Lock First Edition. 16 pp.

Furniture (Cont.)
—Locks (Hardware)—Glossaries (Cont.)
ISO 8554 Pt 2-87. Hardware for Furniture—Terms for Furniture Locks—Part 2: Surface-Mounted Lock, Inset-Type Lock, Mortise Lock First Edition. 6 pp.
ISO 8554 Pt 3-87. Hardware for Furniture—Terms for Furniture Locks—Part 3: Left Hand Lock, Right Hand Lock, down Lock, up Lock First Edition. 6 pp.
ISO 8554 Pt 4-87. Hardware for Furniture—Terms for Furniture Locks—Part 4: Key, Rotary Lock-Handle, Cylinder First Edition. 10 pp.

—Mechanical Testing
BSI BS 4875: Part 7-85. 1985 Strength and Stability of Furniture Part 7: Methods for Determination of Strength of Storage Furniture. 20 pp.

—Metal—Paints (Alkyd Enamel)
CGSB CAN/CGSB-1.88-92. Gloss Alkyd Enamel, Air Drying and Baking. 9 pp.

—Particle Board
BSI BS 5669: Part 5-89. 1989 Amd 1 Particleboard Part 5: Code of Practice for the Selection and Application of Particleboards for Specific Purposes (AMD 6612) April 30, 1991. 17 pp.
BSI BS 5669: Part 5-02. 1989 Amd 2 Particleboard Part 5: Code of Practice for the Selection and Application of Particleboards for Specific Purposes (AMD 6964) September 15, 1992. 17 pp.

—Removal—Service Contracts
MOD UK DEFCON 112BG-86. Conditions of Contract Furniture Removal 6/86. 8 pp.

—Stability Testing
BSI BS 4875: Part 8-85. 1985 Strength and Stability of Furniture Part 8: Methods for Determination of Stability of Storage Furniture. 10 pp.
ISO 7171-88. Furniture—Storage Units—Determination of Stability First Edition. 7 pp.

—Stretch Covers
BSI BS 4723-90. 1990 Amd 1 Stretch Covers for Upholstered Furniture. 10 pp.

—Surface Finishing
BSI BS 5910: Part 1-80. 1980 Methods of Test for Surface Finishes for Furniture Part 1: Assessment of Surface Resistance to Cold Liquids. 7 pp.
ISO 4211-79. Furniture—Assessment of Surface Resistance to Cold Liquids First Edition. 6 pp.

—Surface Finishing—Cold Resistance
BSI BS 3962: Part 4-80. 1980 Methods of Test for Clear Finishes for Wooden Furniture Part 4: Assessment of Surface Resistance to Cold Liquids. 7 pp.
BSI BS 3962: Part 5-80. 1980 Methods of Test for Clear Finishes for Wooden Furniture Part 5: Assessment of Surface Resistance to Cold Oils and Fats. 7 pp.

—Surface Finishing—Heat Resistance
BSI BS 3962: Part 2-83. 1983 Methods of Test for Clear Finishes for Wooden Furniture Part 2: Assessment of Surface Resistance to Wet Heat. 11 pp.
BSI BS 3962: Part 3-83. 1983 Methods of Test for Clear Finishes for Wooden Furniture Part 3: Assessment of Surface Resistance to Dry Heat. 10 pp.

—Surface Finishing—Mechanical Testing
BSI BS 3962: Part 6-80. 1980 Amd 1 Methods of Test for Clear Finishes for Wooden Furniture Part 6: Assessment of Resistance to Mechanical Damage (AMD 5937) April 28, 1989. 16 pp.

—Tumbling
JIS S 1018-85. Test Methods of Vibration and Earthquake Tumbling for Furniture.

—Upholstered—Flammability Testing
EC SM 3. Fire Behaviour of Upholstered Furniture Etc.. 1 p.
SNZ NZS/AS 4088-93. Burning Behaviour of Upholstered Furniture (This is a Joint Standard with SAA AS 4088). 10 pp.

—Upholstered—Labels
SAA AS 1957-87. Care Labelling of Clothing, Household Textiles, Furnishings, Upholstered Furniture, Bedding, Piece Goods and Yarns. 7 pp.

—Vibration
JIS S 1018-85. Test Methods of Vibration and Earthquake Tumbling for Furniture.

—Wood—Nitrocellulose Lacquers
CGSB CAN/CGSB-1.150-M91. Clear Lacquer for Wood Furniture. 9 pp.

Furniture Polishes
See Also: Polishes
CGSB CAN/CGSB-25.10-M88. Polish and Oil Finish for Furniture. 10 pp.

Furring
See Also: Construction Materials
CSA A82.30-M1980. Interior Furring, Lathing, and Gypsum Plastering (R 1992); (Gen Instr 1). 24 pp.

—Steel—Ceilings
CNS A2206-87. Steel Furrings for Wall and Ceiling in Buildings (Jun)(11984). 6 pp.
CNS A3259-87. Method of Test for Steel Furrings for Wall and Ceiling in Buildings (Jun)(11985).
JIS A 6517-89. Steel Furrings for Wall and Ceiling in Buildings. 16 pp.

—Steel—Floors—Gymnasium
CNS A2210-87. Steel Furring Components for Gymnasium Floors (Sep)(12053).
CNS A3263-87. Method of Test for Steel Furring Components for Gymnasium Floors (Sep)(12054).
JIS A 6519-89. Steel Furring Components for Gymnasium Floors. 24 pp.

—Steel—Walls
CNS A2206-87. Steel Furrings for Wall and Ceiling in Buildings (Jun)(11984). 6 pp.
CNS A3259-87. Method of Test for Steel Furrings for Wall and Ceiling in Buildings (Jun)(11985).
JIS A 6517-89. Steel Furrings for Wall and Ceiling in Buildings. 16 pp.

Furs
See Also: Hides

—Identification Systems
SNZ NZS 8721-88. Care Labelling of Clothing, Household Textiles, Furnishings, Upholstered Furniture, Bedding, Piece Goods and Yarns Amend: A, 1988. 7 pp.
SNZ NZS 8722-88. Care Labelling—Guide to the Selection of Correct Care Labelling Instructions from NZS 8721 Amend: A, 1988. 15 pp.

Fuse Bases
Use: Fuse Holders

Fuse Blocks
Use For: Fuseboards See Also: Fuse Holders; Fuses (Electric)
BSI BS 5486: Part 11-89. 1989 Low-Voltage Switchgear and Controlgear Assemblies Part 11: Particular Requirements of Fuseboards. 15 pp.
BSI BS 5486: Part 11-79. (WITHDRAWN) 1979 Factory-Built Assemblies of Switchgear and Controlgear for Voltages up to and Including 1000 V a.c. and 1200 V d.c. Part 11: Particular Requirements for Fuseboards. 14 pp.
MOD UK DSTAN 59-100: Part 1-83. Fuse Holders, Carriers and Bases Electrical Fuse (Block and Extractor Post Types) Part 1: Specifications Issue 2. 27 pp.
MOD UK DSTAN 59-100: Part 2-81. Fuseholders; Carriers, Cartridge Fuse Link and Wire Fuse Element; and Bases Electrical Fuse Part 2: Lists Issue 1. 44 pp.

—Ships
MOD UK NES 531-81. Requirements for Weapon Control and Communications Switchboards and Fuse Distribution Panels (Non Standard) Issue 2 (01.81) (Superseded by NES 530). 21 pp.

—Submarines
MOD UK NES 531-81. Requirements for Weapon Control and Communications Switchboards and Fuse Distribution Panels (Non Standard) Issue 2 (01.81) (Superseded by NES 530). 21 pp.

Fuse Boxes
See Also: Cutout Boxes; Electrical Enclosures; Fuses (Electric)
DIN VDE 0660 Pt 505-90. Switchgear and Controlgear; Low-Voltage Switchgear and Controlgear Assemblies; Specification for Domestic Connection Boxes and Fuseboxes (July). 38 pp.

—Automotive
CNS D2133-86. Fuse Box for Automobiles (Jun)(8947).
CNS D3126-86. Method of Test for Fuse Box for Automobiles (Jun)(8948).

Fuse Carriers
Use: Fuse Holders

Fuse Clips
Use: Fuse Holders

Fuse Cutouts
See Also: Fuse Holders; Fuse Links; Fuses (Electric)

INTERNATIONAL AND NON-U.S. NATIONAL STANDARDS
SUBJECT INDEX — Fuse

Fuse Cutouts (Cont.)
BSI BS 7657-93. 1993 Fuses (Cut-Outs), Ancillary Terminal Blocks and Interconnecting Units up to 100 A Rating, for Power Supplies to Buildidngs (E). 22 pp.

Fuse Holders
Use For: Fuse Bases; Fuse Carriers; Fuse Clips; Fuseholders *See Also:* Fuse Blocks; Fuse Cutouts; Fuses (Electric)

CSA 862 Bull. Electrical Bulletin 862 March 1, 1972 to C22.2 NO 39. 1 p.
DIN VDE 0636 Pt 1-83. Low Voltage Fuses; General Requirements (Dec). 39 pp.
MOD UK DSTAN 59-100: Part 1-83. Fuse Holders, Carriers and Bases Electrical Fuse (Block and Extractor Post Types) Part 1: Specifications Issue 2. 27 pp.
MOD UK DSTAN 59-100: Part 2-81. Fuseholders; Carriers, Cartridge Fuse Link and Wire Fuse Element; and Bases Electrical Fuse Part 2: Lists Issue 1. 44 pp.

— **500 V**
DIN VDE 0636 Pt 31-83. Low Voltage Fuses; D System; Protection of Cables and Wiring up to 100 A and 500 V and up to 63 A 660 V a.c. or 600 V d.c. (Dec). 59 pp.

— **500 V AC**
DIN VDE 0636 Pt 21-84. Low Voltage Fuses; NH System; Protection of Cables and Wiring up to 1250 A and 500 V a.c., 440 V d.c. Also 660 V a.c. (May). 75 pp.

— **600 V AC**
DIN VDE 0636 Pt 21-84. Low Voltage Fuses; NH System; Protection of Cables and Wiring up to 1250 A and 500 V a.c., 440 V d.c. Also 660 V a.c. (May). 75 pp.

— **600 V DC**
DIN VDE 0636 Pt 31-83. Low Voltage Fuses; D System; Protection of Cables and Wiring up to 100 A and 500 V and up to 63 A 660 V a.c. or 600 V d.c. (Dec). 59 pp.

— **660 V AC**
DIN VDE 0636 Pt 31-83. Low Voltage Fuses; D System; Protection of Cables and Wiring up to 100 A and 500 V and up to 63 A 660 V a.c. or 600 V d.c. (Dec). 59 pp.

— **1000 V AC**
DIN VDE 0636 Pt 22-84. Low Voltage Fuses; H.R.C. Fuses for Electrical Installations; Fuse Ratings up to 1250 A 1000 V a.c. Fuse Application Types aM, gTR, gB (May). 78 pp.

—**Cartridge Fuse Links**
CENELEC HD 119-74. Fuse-holders for Minature Cartridge Fuse-Links. 2 pp.
CENELEC EN 60 257-90. Fuse-Holders for Minature Cartridge Fuse-Links. 4 pp.
IEC 257-68. Fuse-Holders for Miniature Cartridge Fuse-Links First Edition; (Amendment 2-1989). 51 pp.

—**Cartridge Fuses**
CSA C22.2 NO 39-M1987. Fuseholder Assemblies (R 1992); (Gen Instr 1). 38 pp.
DIN VDE 0820 Pt 2-76. VDE Specification for Miniature Cartridge Fuses; Specification for Fuse-Holders (Feb). 7 pp.

—**HRC Fuses**
CSA C22.2 NO 39-M1987. Fuseholder Assemblies (R 1992); (Gen Instr 1). 38 pp.
CSA 1356 Bull. Electrical Bulletin 1356 January 13, 1982 to C22.2 NO 39. 2 pp.

—**Plug Fuses**
CSA C22.2 NO 39-M1987. Fuseholder Assemblies (R 1992); (Gen Instr 1). 38 pp.

—**Ships**
MOD UK NES 540-01. Fuselinks, Fuseholders and Fuse Panels for Power Distribution Issue 1 (10.88); Amendment 1. 106 pp.

—**Submarines**
MOD UK NES 540-01. Fuselinks, Fuseholders and Fuse Panels for Power Distribution Issue 1 (10.88); Amendment 1. 106 pp.

Fuse Links
Scope Note: For additional listings, use a more specific term *Use For:* Link Fuses; Thermal Links *See Also:* Cartridge Fuse Links; Fuse Cutouts; Fuses (Electric)

Fuse Links (Cont.)
BSI BS 88: Part 5-88. 1988 Amd 1 Cartridge Fuses for Voltages up to and Including 1000 V a.c. and 1500 V d.c. Part 5: Spec. of Supplementary Requirements for Fuse-Links for Use in a.c. Electricity Supply Networks (AMD 6695) April 30, 1991. 16 pp.
BSI BS 1362-73. 1973 Amd 2 General Purpose Fuse Links for Domestic and Similar Purposes (Primarily for Use in Plugs) (AMD 6691) April 30, 1991. 27 pp.
BSI BS 7283-90. 1990 Requirements and Application Guide for Thermal-Links. 24 pp.
BSI BS 7283-01. 1990 Amd 1 Requirements and Application Guide for Thermal-Links (AMD 7314) January 15, 1993. 25 pp.
CENELEC EN 60 691-87. Requirements and Application Guide for Thermal-Links. 23 pp.
CENELEC EN 60691-92. CORRIGENDUM Requirements and Application Guide for Thermal-Links. 2 pp.
CNS C2043-84. Low Voltage Link Fuses (Dec)(2226).
DIN VDE 0636 Pt 1-83. Low Voltage Fuses; General Requirements (Dec). 39 pp.
IEC 269 Pt 2-1-87. Low-Voltage Fuses Part 2: Supplementary Requirements for Fuses for Use by Authorized Persons (Fuses Mainly for Industrial Application) Sections I to III First Edition; (Amendment 1-1993). 81 pp.
IEC 691-93. Thermal-Links—Requirements and Application Guide Second Edition. 47 pp.
JIS C 8313-83. Particular Requirements for Link-Fuses. 8 pp.
MOD UK DSTAN 59-96: Part 2-01. Fuses, Electrical Part 2: Fuse Links, Electrical Lists Issue 1; Amendment 1. 74 pp.
SAA AS 1890-76. Thermally-Released Links Amdt 1 October 1977. 12 pp.
SNZ NZS 6208: Part 5-84. Specification for Cartridge Fuses for Voltages up to and Including 1000 V Alternating Current and 1500 V Direct Current Part 5: Supplementary Requirements for Fuse Links of Standardized Dimensions and Performance for Use in Alternating Current. 12 pp.
SNZ NZS 6208: Part 5 CONT-84. Electricity Supply Networks Amend: 1A, 1984.

— **440 V DC**
DIN VDE 0636 Pt 21-84. Low Voltage Fuses; NH System; Protection of Cables and Wiring up to 1250 A and 500 V a.c., 440 V d.c. Also 660 V a.c. (May). 75 pp.

— **500 V**
DIN VDE 0635-84. Low Voltage Fuses; D-Type Fuses E 16 up to 25 A, 500 V; D-Type Fuses up to 100 A, 750 V; D-Type Fuses up to 100 A, 500 V (Feb) (Partially Supersedes VDE 0635/03.63 and VDE 0635d/08.75). 35 pp.
DIN VDE 0636 Pt 31-83. Low Voltage Fuses; D System; Protection of Cables and Wiring up to 100 A and 500 V and up to 63 A 660 V a.c. or 600 V d.c. (Dec). 59 pp.

— **500 V AC**
DIN VDE 0636 Pt 21-84. Low Voltage Fuses; NH System; Protection of Cables and Wiring up to 1250 A and 500 V a.c., 440 V d.c. Also 660 V a.c. (May). 75 pp.

— **600 V DC**
DIN VDE 0636 Pt 31-83. Low Voltage Fuses; D System; Protection of Cables and Wiring up to 100 A and 500 V and up to 63 A 660 V a.c. or 600 V d.c. (Dec). 59 pp.

— **660 V AC**
DIN VDE 0636 Pt 21-84. Low Voltage Fuses; NH System; Protection of Cables and Wiring up to 1250 A and 500 V a.c., 440 V d.c. Also 660 V a.c. (May). 75 pp.
DIN VDE 0636 Pt 31-83. Low Voltage Fuses; D System; Protection of Cables and Wiring up to 100 A and 500 V and up to 63 A 660 V a.c. or 600 V d.c. (Dec). 59 pp.

— **750 V**
DIN VDE 0635-84. Low Voltage Fuses; D-Type Fuses E 16 up to 25 A, 500 V; D-Type Fuses up to 100 A, 750 V; D-Type Fuses up to 100 A, 500 V (Feb) (Partially Supersedes VDE 0635/03.63 and VDE 0635d/08.75). 35 pp.

— **1000 V AC**
DIN VDE 0636 Pt 22-84. Low Voltage Fuses; H.R.C. Fuses for Electrical Installations; Fuse Ratings up to 1250 A 1000 V a.c. Fuse Application Types aM, gTR, gB (May). 78 pp.

—**Aircraft**
ISO 1547-75. Aircraft—Precision Fuse-Links—General Requirements First Edition. 6 pp.
ISO 1548-76. Aircraft—Precision Fuse-Links—Type A First Edition. 22 pp.

Fuse Links (Cont.)
—**Aircraft** (Cont.)
ISO 1549-76. Aircraft—Precision Fuse-Links—Type B First Edition. 14 pp.

—**Automotive**
CNS D1050-83. Selection of Fuses and Fusible-Links for Automobile Wiring (Apr)(10157).

—**Miniature**
BSI BS EN 60127-1-91. 1991 Miniature Fuses Part 1: Definitions for Miniature Fuses and General Requirements for Miniature Fuse-Links. 31 pp.
CENELEC HD 109.1 S1-89. Miniature Fuses Part 1: Definitions for Miniature Fuses and General Requirements for Miniature Fuse-Links. 4 pp.
CENELEC HD 109.1 S1-90. Miniature Fuses Part 1: Definitions for Miniature Fuses and General Requirements for Miniature Fuse-Links. 2 pp.
CENELEC HD 109.3 S1-89. Miniature Fuses Part 3: Sub-Miniature Fuse-Links. 3 pp.
CENELEC EN 60 127-1-91. Miniature Fuses Part 1: Definitions for Miniature Fuses and General Requirements for Miniature Fuse-Links. 7 pp.
IEC 127 Pt 1-88. Miniature Fuses Part 1: Definitions for Miniature Fuses and General Requirements for Miniature Fuse-Links First Edition; (Corrigendum—March 1990). 49 pp.
SNZ IEC 127: Part 1-88. Miniature Fuses Part 1: Definitions for Miniature Fuses and General Requirements for Miniature Fuse-Links. 44 pp.

—**Miniature—Quality Assurance**
BSI BS EN 60127-5-91. 1991 Miniature Fuses Part 5: Guide for the Quality Assessment of Miniature Fuse-Links. 13 pp.
CENELEC HD 109.5 S1-89. Miniature Fuses Part 5: Guidelines for Quality Assessment of Miniature Fuse-Links. 3 pp.
CENELEC EN 60127-5-91. Miniature Fuses Part 5: Guidelines for Quality Assessment of Miniature Fuse-Links. 4 pp.
IEC 127 Pt 5-88. Miniature Fuses Part 5: Guidelines for Quality Assessment of Miniature Fuse-Links First Edition. 16 pp.

—**Modular**
BSI DD 183-89. 1989 Universal Modular Fuses. 7 pp.
IEC 127 Pt 4 TTD-89. Universal Modular Fuses (UMF). 11 pp.

—**Motors**
BSI BS 5907-80. 1980 Amd 1 High Voltage Fuse-Links for Motor Circuit Applications (IEC 644: 1979) (AMD 6687) April 30, 1991. 13 pp.
CENELEC HD 424-83. Specification for High-Voltage Fuse-Links for Motor Circuit Applications. 2 pp.
CENELEC EN 60644-93. Specification for High-Voltage Fuse-Links for Motor Circuit Applications (Supersedes HD 424 S1:1983); (IEC 644:1979). 5 pp.
IEC 644-79. Specification for High-Voltage Fuse-Links for Motor Circuit Applications First Edition; (CENELEC EN 60644:1993). 21 pp.

—**Semiconductor Devices**
BSI BS 88: Part 4-88. 1988 Amd 2 Cartridge Fuses for Voltages up to and Including 1000 V a.c. and 1500 V d.c. Part 4: Spec. of Supplementary Requirements for Fuse-Links for the Protection of Semiconductor Devices (IEC 269-4: 1986) (AMD 6696) April 30, 1991. 33 pp.
IEC 269 Pt 4-86. Low-Voltage Fuses Part 4: Supplementary Requirements for Fuse-Links for the Protection of Semiconductor Devices Third Edition. 57 pp.
SAA AS 2005.40-89. Low Voltage Fuses—Fuses with Enclosed Fuse-Links—Part 40: Supplementary Requirements for Fuse-Links for the Protection of Semiconductor Devices (IEC 269-4:1986). 28 pp.
SNZ NZS 6208: Part 4-80. Specification for Cartridge Fuses for Voltages up to and Including 1000 V Alternating Current and 1500 V Direct Current Part 4: Supplementary Requirements for Fuse Links for the Protection of Semiconductor Devices Amend: 2A, 1980; 3, 1984; 3A, 1980. 20 pp.

—**Ships**
MOD UK NES 540-01. Fuselinks, Fuseholders and Fuse Panels for Power Distribution Issue 1 (10.88); Amendment 1. 106 pp.

—**Signal Lines—Railroad**
BSI BS 714-50. 1950 Amd 1 Cartridge Fuse-Links for Use in Railway Signalling Circuits. 12 pp.

—**Submarines**
MOD UK NES 540-01. Fuselinks, Fuseholders and Fuse Panels for Power Distribution Issue 1 (10.88); Amendment 1. 106 pp.

—**Transformers**
BSI BS 6553-84. 1984 Amd 2 Selection of Fuse Links of High-Voltage Fuses for Transformer Circuit Applications (IEC 787: 1983) (AMD 6685) April 30, 1991. 10 pp.

Fuse Links (Cont.)
—Transformers (Cont.)
IEC 787-83. Application Guide for the Selection of Fuse-Links of High-Voltage Fuses for Transformer Circuit Applications First Edition; (Amendment 1-1985). 16 pp.

Fuse Setters
MOD UK DSTAN 12-1-70. Fuze Setters, Hand, for Time Fuzed Ammunition Issue 1. 5 pp.
MOD UK DSTAN 12-1: Addendum. Fuze Setters, Hand for Time Fuzed Ammunition Issue 1. 2 pp.

Fuse Switch Combinations
Use: Fusible Switches

Fuse Switches
Use: Fusible Switches

Fuseboards
Use: Fuse Blocks

Fused Fabrics
See Also: Bonded Fabrics; Fabrics; Laminated Fabrics

—Bonding Strength
CGSB CAN/CGSB-4.2 NO.65-M91. Textile Test Methods Determination of Strength of Bonds of Bonded, Laminated and Fused Fabrics. 9 pp.

—Dry Cleaning—Dimensional Stability
CGSB CAN/CGSB-4.2 NO.66-M91. Textile Test Methods Dimensional Change and Appearance After Dry Cleaning of Coated, Bonded, Laminated and Fused Fabrics. 9 pp.

Fused Switches
Use: Fusible Switches

Fuseholders
Use: Fuse Holders

Fusel Oil
See Also: Oils
CNS K5002-65. Refined Fusel Oil (Apr)(77). 1 p.
CNS K6007-65. Method of Test for Refined Fusel Oil (Apr)(78). 1 p.

Fuselages
See Also: Aircraft; Hulls

—Break-In Points—Emergency Evacuation
CAA Spec. NO. 7 ISSUE 2.

Fuses (Electric)
See Also: Cartridge Fuses; Circuit Protectors; Current Limiting Fuses; Cutout Boxes; Disconnecting Switches; Expulsion Fuses; Fuse Blocks; Fuse Boxes; Fuse Cutouts; Fuse Holders; Fuse Links; Fusible Switches; Plug Fuses; Switchgear; Thermal Cutoffs; Time Delay Relays
BSI BS EN 60127-1-91. 1991 Miniature Fuses Part 1: Definitions for Miniature Fuses and General Requirements for Miniature Fuse-Links. 31 pp.
CNS C4218-84. Low Voltage Screw Fuses and D-Type Fuses (Dec)(6051).
CNS C4452-88. Fuses for Instrument and Power Style F20 (Jul)(12353).
CNS C7100-84. Fuses for Instrument, Power and Telephone (Nonindicating), Style F01 (May)(4978).
CNS C7101-84. Fuses for Instrument, Power and Telephone (Nonindicating), Style F02 (May)(4979).
CNS C7102-84. Fuses for Instrument, Power and Telephone (Nonindicating), Style F05 (May)(4980).
CNS C7103-84. Fuses for Instrument, Power and Telephone (Nonindicating), Style F06 (May)(4981).
CNS C7148-84. Fuses for Instrument, Power and Telephone (Nonindicating), Style F03 (May)(8231).
CNS C7149-84. Fuses for Instrument, Power and Telephone (Nonindicating), Style 107 (May)(8232).
CNS C7164-83. Fuses for Instrument, Power and Telephone (Nonindicating), Style F15 (Nov)(10650).
CNS C7165-83. Fuses for Instrument, Power and Telephone (Nonindicating), Style F16 (Nov)(10651).
CNS C7178-84. Fuses for Instrument and Power (Nonindicating), Style F19, Knife Type (Nov)(11095).
JIS C 8319-83. Particular Requirements for Screw Fuses and D-Type Fuses. 11 pp.
SNZ NZS 1951-65. Specification for Semi-Enclosed Electric Fuses, Ratings up to 200 Amperes and 250 Volts to Earth Amend: 1, 1966; 2, 1968. 46 pp.

— 600 V—Overcurrent Protection Equipment
CSA C22.2 NO 59.2-M1986. Supplemental Fuses; (Gen Instr 1 Thru 3). 22 pp.

— 600 V—Renewable
CSA CAN/CSA-C22. 2 NO59.3-M90. DC Rated Fuses; (Gen Instr 1). 25 pp.

—Automotive
CNS D1050-83. Selection of Fuses and Fusible-Links for Automobile Wiring (Apr)(10157).
CNS D2134-86. Fuses for Automobiles (Jun)(8949).
CNS D3127-86. Method of Test for Fuses for Automobiles (Jun)(8950).

—Cables (Electric)—Low Voltage
JIS C 8352-83. General Requirements for Fuses for the Protection of Low-Voltage Cables and Lines. 25 pp.

—Capacitors—Internal
BSI BS 7631-93. 1993 Internal Fuses and Internal Overpressure Disconnectors for Shunt Capacitors (IEC 593: 1977) (E). 16 pp.
BSI BS 7632-93. 1993 Internal Fuses and Internal Overpressure Disconnectors for Capacitors for Inductive Heat Generating Plants (IEC 594: 1977) (E). 16 pp.
BSI BS 7633-93. 1993 Internal Fuses for Series Capacitors (IEC 595: 1977) (E). 12 pp.
IEC 593-77. Internal Fuses and Internal Overpressure Disconnectors for Shunt Capacitors First Edition; (Amendment 1-1980) (Amendment 2-1986). 36 pp.
IEC 594-77. Internal Fuses and Internal Overpressure Disconnectors for Capacitors for Inductive Heat Generating Plants First Edition; (Amendment 2-1987). 36 pp.
IEC 595-77. Internal Fuses for Series Capacitors First Edition; (Amendment 2-1987). 30 pp.

—Glossaries
BSI BS 4727:Pt2: Group 06-85. 1985 Electrotechnical, Power, Telecommunication, Electronics, Lighting and Colour Terms Part 2: Terms Particular to Power Engineering Group 06: Switchgear and Controlgear Terminology (Including Fuse Terminology). 27 pp.
CNS C1128-84. Glossary of Terms for Fuses (Jan)(10740).
IEC 50 Chap 441-84. International Electrotechnical Vocabulary Chapter 441: Switchgear, Controlgear and Fuses Second Edition. 102 pp.
IEC 291-69. Fuse Definitions First Edition; (Supplement A-1975). 21 pp.
JIS C 0201-82. Glossary of Terms for Fuses. 18 pp.
SAA AS 1852.441-85. International Electrotechnical Vocabulary—Part 441: Switchgear, Controlgear and Fuses. 61 pp.
SNZ IEC 50: 50(441)-84. International Electrotechnical Vocabulary 50(441): Switchgear, Controlgear and Fuses. 90 pp.

—High Voltage
BSI BS 7426-91. 1991 High-Voltage Alternating Current Switch-Fuse Combinations. 55 pp.
IEC 282 Pt 2-70. High-Voltage Fuses Part 2: Expulsion and Similar Fuses First Edition; (Amendment 1-1978). 62 pp.
IEC 420-90. High-Voltage Alternating Current Switch-Fuse Combinations Second Edition. 101 pp.

—High Voltage—Capacitors—External
BSI BS 5564-78. 1978 Amd 1 High-Voltage Fuses for the External Protection of Shunt Power Capacitors (AMD 6686) April 30, 1991. 16 pp.
IEC 549-76. High-Voltage Fuses for the External Protection of Shunt Power Capacitors First Edition. 26 pp.

—High Voltage—Short Circuit Power Factor
BSI BS 2692: Part 3-90. 1990 Fuses for Voltages Exceeding 1000 V a.c. Part 3: Guide to the Determination of Short Circuit Power Factor. 7 pp.
IEC 282 Pt 3-76. High-Voltage Fuses Part 3: Determination of Short-Circuit Power Factor for Testing Current-Limiting Fuses and Expulsion and Similar Fuses First Edition. 10 pp.

—Highway Lighting
BSI BS 7654-93. 1993 Single Phase Street Lighting Fuses (Cut-Outs) for Low-Voltage Public Electricity Distribution Systems—25 A Rating for Highway Power Supplies and Street Furniture (F). 20 pp.

—Low Voltage
BSI BS 3036-58. 1958 Amd 5 Semi-Enclosed Electric Fuses (Ratings up to 100 Amperes and 240 Volts to Earth) (AMD 6689) January 31, 1991. 51 pp.
BSI BS 7656-93. 1993 Low-Voltage Pole-Mounting Fuses (Cut-Outs)—315 A Rating (E). 17 pp.
CENELEC EN 60 269-1-89. Low-Voltage Fuses—Part 1: General Requirements. 6 pp.
CNS C1072-84. General Requirements for Fuses for the Protection of Low-Voltage Cables and Lines (Dec)(6056).
CNS C3095-84. Method of Test for Fuses for the Protection of Low-Voltage Cables and Lines (Dec)(6057).

—Low Voltage (Cont.)
DIN VDE 0635-84. Low Voltage Fuses; D-Type Fuses E 16 up to 25 A, 500 V; D-Type Fuses up to 100 A, 750 V; D-Type Fuses up to 100 A, 500 V (Feb) (Partially Supersedes VDE 0635/03.63 and VDE 0635d/08.75). 35 pp.
DIN VDE 0636 Pt 1-83. Low Voltage Fuses; General Requirements (Dec). 39 pp.
DIN VDE 0636 Pt 21-84. Low Voltage Fuses; NH System; Protection of Cables and Wiring up to 1250 A and 500 V a.c., 440 V d.c. Also 660 V a.c. (May). 75 pp.
DIN VDE 0636 Pt 22-84. Low Voltage Fuses; H.R.C. Fuses for Electrical Installations; Fuse Ratings up to 1250 A 1000 V a.c. Fuse Application Types aM, gTR, gB (May). 78 pp.
DIN VDE 0636 Pt 31-83. Low Voltage Fuses; D System; Protection of Cables and Wiring up to 100 A and 500 V and up to 63 A 660 V a.c. or 600 V d.c. (Dec). 59 pp.
IEC 241-68. Fuses for Domestic and Similar Purposes First Edition. 75 pp.
IEC 269 Pt 1-86. Low-Voltage Fuses Part 1: General Requirements Second Edition; (Corrigendum—June 1993). 127 pp.
IEC 269 Pt 2-86. Low-Voltage Fuses Part 2: Supplementary Requirements for Fuses for Use by Authorized Persons (Fuses Mainly for Industrial Application) Second Edition. 17 pp.
IEC 269 Pt 2-1-87. Low-Voltage Fuses Part 2: Supplementary Requirements for Fuses for Use by Authorized Persons (Fuses Mainly for Industrial Application) Sections I to III First Edition; (Amendment 1-1993). 81 pp.
IEC 269 Pt 3-87. Low-Voltage Fuses Part 3: Supplementary Requirements for Fuses for Use by Unskilled Persons (Fuses Mainly for Household and Similar Applications) Second Edition. 24 pp.
JIS C 0502-87. Standard Conditions for Operation in Service of Low-Voltage Switchgear and Fuses. 5 pp.
SAA AS 2005.1-81. Low Voltage Fuses—Fuses with Enclosed Fuse-Links—Part 1: General Requirements. 34 pp.
SAA AS 2005.10-88. Low Voltage Fuses—Fuses with Enclosed Fuse-Links—Part 10: General Requirements Amdt 1 April 1991.
SAA AS 2005.20-90. Low Voltage Fuses—Fuses with Enclosed Fuse-Links—Part 20: Supplementary Requirements for Fuses for Use by Authorized Persons (Fuses Mainly for Industrial Application)—Common Requirements. 6 pp.
SAA AS 2005.21.1-90. Low Voltage Fuses—Fuses with Enclosed Fuse-Links—Part 21.1: Supplementary Requirements for Fuses for Use by Authorized Persons (Fuses Mainly for Industrial Applica-tion)—Standardized Fuse Systems—Fuses with Fuse-Links with Blade Contacts. 20 pp.
SAA AS 2005.21.2-90. Low Voltage Fuses—Fuses with Enclosed Fuse-Links—Part 21.2: Supplementary Requirements for Fuses for Use by Authorized Persons (Fuses Mainly for Industrial Applica-tion)—Standardized Fuse Systems—Fuses with Fuse-Links for Bolted Connections. 12 pp.
SAA AS 2005.29-90. Low Voltage Fuses—Fuses with Enclosed Fuse-Links—Part 29: Supplementary Requirements for Fuses for Use by Authorized Persons (Fuses Mainly for Industrial Applica-tion)—Standardized Fuses with Compact Dimensions. 10 pp.
SAA AS 2005.30-91. Low Voltage Fuses—Fuses with Enclosed Fuse-Links—Part 30: Supplementary Requirements for Fuses for Use by Unskilled Persons (Fuses Mainly for Household and Similar Applications) (In Professional Package 57). 14 pp.

—Miniature
CENELEC EN 60 127-1-91. Miniature Fuses Part 1: Definitions for Miniature Fuses and General Requirements for Miniature Fuse-Links. 7 pp.
IEC 127 Pt 1-88. Miniature Fuses Part 1: Definitions for Miniature Fuses and General Requirements for Miniature Fuse-Links First Edition; (Corrigendum—March 1990). 49 pp.
SNZ IEC 127: Part 1-88. Miniature Fuses Part 1: Definitions for Miniature Fuses and General Requirements for Miniature Fuse-Links. 44 pp.

—Miniature—Glossaries
CENELEC HD 109.1 S1-89. Miniature Fuses Part 1: Definitions for Miniature Fuses and General Requirements for Miniature Fuse-Links. 4 pp.
CENELEC HD 109.1 S1-90. Miniature Fuses Part 1: Definitions for Miniature Fuses and General Requirements for Miniature Fuse-Links. 2 pp.

—Overcurrent Relays—Short Circuit Current Testing
CSA CAN/CSA-C22. 2 NO 14-M91. Industrial Control Equipment; (Gen Instr 1 Thru 4). 142 pp.
CSA 933 Bull. Electrical Bulletin 933 January 21, 1974 to C22.2 NO 14. 2 pp.

Fuses (Electric) (Cont.)

—Power Converters—Semiconductor
IEC 146 Pt 6-92. Semiconductor Convertors Part 6: Application Guide for the Protection of Semiconductor Convertors Against Overcurrent by Fuses First Edition. 37 pp.

—Power Lines—Low Voltage
CNS C2042-86. Low Voltage Fuses (May)(2225). 2 pp.

—Rolling Stock
JIS E 4901-91. DC Fuses for Railway Rolling Stock.

—Spacers (Electric)
CSA CAN/CSA-C22. 2 NO 0-M91. General Requirements—Canadian Electrical Code, Part II; (Gen Instr 1). 48 pp.

—Telephone
CNS C7100-84. Fuses for Instrument, Power and Telephone (Nonindicating), Style F01 (May)(4978).
CNS C7101-84. Fuses for Instrument, Power and Telephone (Nonindicating), Style F02 (May)(4979).
CNS C7102-84. Fuses for Instrument, Power and Telephone (Nonindicating), Style F05 (May)(4980).
CNS C7103-84. Fuses for Instrument, Power and Telephone (Nonindicating), Style F06 (May)(4981).
CNS C7148-84. Fuses for Instrument, Power and Telephone (Nonindicating), Style F03 (May)(8231).
CNS C7149-84. Fuses for Instrument, Power and Telephone (Nonindicating), Style 107 (May)(8232).
CNS C7164-83. Fuses for Instrument, Power and Telephone (Nonindicating), Style F15 (Nov)(10650).
CNS C7165-83. Fuses for Instrument, Power and Telephone (Nonindicating), Style F16 (Nov)(10651).

—0-3000 V—Power Converters—Semiconductor
DIN VDE 0636 Pt 23-84. Low Voltage Fuses; LV HRC Fuses for Protection of Semiconductors up to 1600 A and up to 3000 V (Dec). 31 pp.

Fuses (Ordnance)

Use For: Fuzes (Ordnance) *See Also:* Detonating Fuses; Initiators (Explosives); Munitions; Safety Fuses
MOD UK DSTAN 07-14-78. Fuzeheads, Wirebridge Issue 2. 16 pp.

—Airborne—Weapons Systems—Design
NATO STANAG 3525 ED 5 AMD 0-87. Design Safety Principles and General Design Criteria for Airborne Weapon Fuzing Systems. 15 pp.
NATO STANAG 3525 ED 5 AMD 1-87. Design Safety Principles and General Design Criteria for Airborne Weapon Fuzing Systems. 16 pp.
NATO STANAG 3525 ED 5 AMD 2-87. Design Safety Principles and General Design Criteria for Airborne Weapon Fuzing Systems. 16 pp.

—Airborne—Weapons Systems—Safety
NATO STANAG 3525 ED 5 AMD 0-87. Design Safety Principles and General Design Criteria for Airborne Weapon Fuzing Systems. 15 pp.
NATO STANAG 3525 ED 5 AMD 1-87. Design Safety Principles and General Design Criteria for Airborne Weapon Fuzing Systems. 16 pp.
NATO STANAG 3525 ED 5 AMD 2-87. Design Safety Principles and General Design Criteria for Airborne Weapon Fuzing Systems. 16 pp.
NATO STANAG 4326 ED 1 AMD 0-90. NATO Fuze Characteristics Data—AOP-8. 5 pp.
NATO STANAG 4326 ED 1 AMD 1-90. NATO Fuze Characteristics Data—AOP-8. 8 pp.

—Aircraft—Stores Management Systems
NATO STANAG 3605 ED 2 AMD 1-86. Compatibility of Mechanical Fuzing Systems and Arming Devices for Expendable Aircraft Stores for Fixed Wing Aircraft. 12 pp.

—Artillery—Interchangeable
NATO STANAG 2916 ED 1 AMD 0-89. Nose Fuze Contours and Matching Projectile Cavities for Artillery and Mortar Projectiles. 31 pp.

—Design—Safety
NATO STANAG 4187 ED 1 AMD 2-89. Fuzing System—Safety Design Requirements. 19 pp.
NATO STANAG 4187 ED 2 AMD 0-00. Fuzing Systems—Safety Design Requirements. 21 pp.
NATO STANAG 4326 ED 1 AMD 0-90. NATO Fuze Characteristics Data—AOP-8. 5 pp.
NATO STANAG 4326 ED 1 AMD 1-90. NATO Fuze Characteristics Data—AOP-8. 8 pp.

—Gaskets—Rubber
MOD UK TS 10230. Polychloroprene Rubber Mixes, Special Purpose, Type QX.

—Jute
JIS L 2401-92. Jute Yarns. 6 pp.

Fuses (Ordnance) (Cont.)

—Lacquers (Polyvinyl Alcohol)
MOD UK TS 10235A. Polyvinyl Alcohol for Lacquers Type QX.
MOD UK TS 10236A. Lacquer Polyvinyl Alcohol for Fuzeheads, Type QX.

—Missiles—Safety
NATO STANAG 4326 ED 1 AMD 0-90. NATO Fuze Characteristics Data—AOP-8. 5 pp.
NATO STANAG 4326 ED 1 AMD 1-90. NATO Fuze Characteristics Data—AOP-8. 8 pp.

—Mortars (Weapons)—Interchangeable
NATO STANAG 2916 ED 1 AMD 0-89. Nose Fuze Contours and Matching Projectile Cavities for Artillery and Mortar Projectiles. 31 pp.

—Projectiles—Inductive Settings
NATO STANAG 4369 ED 1 AMD 0-00. (Draft) Design Requirements for Inductive Setting of Electronic Projectile Fuzes. 20 pp.

—Projectiles—Safety
NATO STANAG 4157 ED 1 AMD 0-91. Development of Safety Test Methods and Procedures for Fuzes for Unguided Tube Launched Projectiles. 182 pp.

—Seals—Rubber
MOD UK TS 10230. Polychloroprene Rubber Mixes, Special Purpose, Type QX.

—Tissue Papers
MOD UK DEF-50-A-67. Paper, Tissue (for Fuses). 4 pp.

Fusibility

See Also: Thermodynamic Properties; Welding

—Ashes
BSI BS 1016: Part 15-70. 1970 Methods for the Analysis and Testing of Coal and Coke Part 15: Fusibility of Coal Ash and Coke Ash. 16 pp.
ISO 540-81. Solid Mineral Fuels—Determination of Fusibility of Ash—High-Temperature Tube Method Second Edition. 5 pp.

—Coal
CNS M3163-84. Method of Test for Fusibility of Coal and Coke Ash (Jun)(10948).
JIS M 8801-79. Methods for Testing of Coal. 71 pp.

—Coke
CNS M3163-84. Method of Test for Fusibility of Coal and Coke Ash (Jun)(10948).

Fusible Link PROM

Use: PROMs—Fusible Link

Fusible Plugs

Use: Safety Plugs

Fusible Switches

Use For: Fuse-Switch Combinations; Fuse/Switch Combinations; Fuse Switches; Switch/Fuse Combinations *See Also:* Circuit Protectors; Electric Switches; Fuses (Electric); Switches
BSI BS 7426-91. 1991 High-Voltage Alternating Current Switch-Fuse Combinations. 55 pp.
BSI BS EN 60947-3-92. 1992 Low-Voltage Switchgear and Controlgear Part 3: Switches, Disconnectors, Switch-Disconnectors and Fuse-Combination Units. 62 pp.
BSI BS EN 60947-3-01. 1992 Amd 1 Low-Voltage Switchgear and Controlgear Part 3: Switches, Disconnectors, Switch-Disconnectors and Fuse-Combination Units (AMD 7854) July 15, 1993 (E). 63 pp.
CENELEC EN 60947-3-92. Low-Voltage Switchgear and Controlgear Part 3: Switches, Disconnectors, Switch-Disconnectors and Fuse-Combination Units. 8 pp.
CENELEC EN 60947-3-92. CORRIGENDUM Low-Voltage Switchgear and Controlgear Part 3: Switches, Disconnectors, Switch Diconnectors and Fuse-Combination Units. 1 p.
DIN VDE 0660 Pt 107 A1 (D)-91. Switchgear and Controlgear; Low Voltage Switchgear and Controlgear; Part 3: Switches, Disconnectors, Switch-Disconnectors and Fuse-Combination Units (Nov). 5 pp.
DIN VDE 0660 Pt 107 A2 (D)-92. Switchgear and Controlgear; Low-Voltage Switchgear and Controlgear; Part 3: Switches, Disconnectors and Fuse Combination Units (Feb). 5 pp.
IEC 420-90. High-Voltage Alternating Current Switch-Fuse Combinations Second Edition. 101 pp.
IEC 947 Pt 3-90. Low-Voltage Switchgear and Controlgear Part 3: Switches, Disconnectors, Switch-Disconnectors and Fuse-Combination Units First Edition; (Corrigendum—Dec 1991). 115 pp.

Fusible Switches (Cont.)

SAA AS 1775-84. Low Voltage Switchgear and Controlgear—Air-Break Switches, Isolators and Fuse-Combination Units (up to and Including 1 000 V a.c. and 1 200 V d.c.) Amdt 1 January 1985. 38 pp.
SAA AS 2024-91. High Voltage a.c. Switchgear and Controlgear—Switch-Fuse Combinations. 28 pp.

—High Voltage
BSI BS 7426-91. 1991 High-Voltage Alternating Current Switch-Fuse Combinations. 55 pp.
IEC 420-90. High-Voltage Alternating Current Switch-Fuse Combinations Second Edition. 101 pp.

—Motors
BSI BS EN 60947-3-92. 1992 Low-Voltage Switchgear and Controlgear Part 3: Switches, Disconnectors, Switch-Disconnectors and Fuse-Combination Units. 62 pp.
BSI BS EN 60947-3-01. 1992 Amd 1 Low-Voltage Switchgear and Controlgear Part 3: Switches, Disconnectors, Switch-Disconnectors and Fuse-Combination Units (AMD 7854) July 15, 1993 (E). 63 pp.
CENELEC EN 60947-3-92. Low-Voltage Switchgear and Controlgear Part 3: Switches, Disconnectors, Switch-Disconnectors and Fuse-Combination Units. 8 pp.
CENELEC EN 60947-3-92. CORRIGENDUM Low-Voltage Switchgear and Controlgear Part 3: Switches, Disconnectors, Switch Diconnectors and Fuse-Combination Units. 1 p.
DIN VDE 0660 Pt 107 A1 (D)-91. Switchgear and Controlgear; Low Voltage Switchgear and Controlgear; Part 3: Switches, Disconnectors, Switch-Disconnectors and Fuse-Combination Units (Nov). 5 pp.
DIN VDE 0660 Pt 107 A2 (D)-92. Switchgear and Controlgear; Low-Voltage Switchgear and Controlgear; Part 3: Switches, Disconnectors and Fuse Combination Units (Feb). 5 pp.
IEC 947 Pt 3-90. Low-Voltage Switchgear and Controlgear Part 3: Switches, Disconnectors, Switch-Disconnectors and Fuse-Combination Units First Edition; (Corrigendum—Dec 1991). 115 pp.

—1000 V AC
BSI BS 5419-77. (WITHDRAWN) 1977 Amd 1 Air-Break Switches, Air-Break Disconnectors, Air-Break Switch Disconnectors and Fuse Combination Units for Voltages up to and Including 1000 V a.c. and 1200 V d.c. (Superseded by BS EN 60947-3: 1992). 51 pp.

—1200 V DC
BSI BS 5419-77. (WITHDRAWN) 1977 Amd 1 Air-Break Switches, Air-Break Disconnectors, Air-Break Switch Disconnectors and Fuse Combination Units for Voltages up to and Including 1000 V a.c. and 1200 V d.c. (Superseded by BS EN 60947-3: 1992). 51 pp.

Fusion (Melting)

Use For: Melting *See Also:* Adhesion Testing; Heat of Fusion; Induction Melting

—Fabrics—PVC Coatings
BSI BS 3424: Part 22-83. 1983 Testing Coated Fabrics Part 22: Methods for Determination of Fusion of PVC Coatings and the State of Cure of Vulcanized Rubber Coatings. 4 pp.
ISO 6451-82. Plastics Coated Fabrics—Polyvinyl Chloride Coatings—Rapid Method for Checking Fusion First Edition. 3 pp.

—Silica—Water Analysis
BSI BS 2690: Part 104-83. 1983 Water Used in Industry Part 104: Silica: Reactive, Total and Suspended. 6 pp.

Fusion Welding

Scope Note: For additional listings, use a more specific term *See Also:* Arc Welding; Braze Welding; Electric Welding; Electron Beam Welding; Flash Welding; Gas Metal Arc Welding; Gas Shielded Arc Welding; Gas Tungsten Arc Welding; Gas Welding; Oxyacetylene Welding; Pressure Gas Welding; Resistance Welding; Shielded Metal Arc Welding; Spot Welds; Stud Welding; Welding
BSI BS 4870: Part 4-88. 1988 Approval Testing of Welding Procedures Part 4: Automatic Fusion Welding of Metallic Materials, Including Welding Operator Approval. 8 pp.

—Aluminum Alloys—Aerospace
JIS W 2016-85. Fusion Welding Process of Aerospace Aluminium Alloys.

—Aluminum Alloys—Filler Metal
DIN ENGL 1732 Pt 1-88. Filler Metals for Welding Aluminium and Aluminium Alloys; Composition, Application and Technical Delivery Conditions (June). 8 pp.

INTERNATIONAL AND NON-U.S. NATIONAL STANDARDS SUBJECT INDEX

Fusion Welding (Cont.)

—Aluminum Alloys—Filler Metal—Test Specimens
DIN ENGL 1732 Pt 3-88. Filler Metals for Welding Aluminium and Aluminium Alloys; Sample and Specimen Preparation, Mechanical Properties of All-Weld Metal (June). 2 pp.

—Aluminum Alloys—Quality Assurance
DIN ENGL 8563 Pt 30-85. Quality Assurance of Welding Operations; Fusion Welded Joints in Aluminium and Aluminium Alloys (Except Beam Welded Joints); Requirements, Classification (Oct). 13 pp.

—Aluminum—Company Certification
CSA W47.2-M1987. Certification of Companies for Fusion Welding of Aluminum. 75 pp.

—Aluminum—Filler Metal
DIN ENGL 1732 Pt 1-88. Filler Metals for Welding Aluminium and Aluminium Alloys; Composition, Application and Technical Delivery Conditions (June). 8 pp.

DIN ENGL 1732 Pt 2-77. Welding Filler Metals for Aluminium; Testing by Welded Joints (Dec). 4 pp.

—Aluminum—Filler Metal—Test Specimens
DIN ENGL 1732 Pt 3-88. Filler Metals for Welding Aluminium and Aluminium Alloys; Sample and Specimen Preparation, Mechanical Properties of All-Weld Metal (June). 2 pp.

—Aluminum—Quality Assurance
DIN ENGL 8563 Pt 30-85. Quality Assurance of Welding Operations; Fusion Welded Joints in Aluminium and Aluminium Alloys (Except Beam Welded Joints); Requirements, Classification (Oct). 13 pp.

—Copper Alloys—Filler Metal
DIN ENGL 1733 Pt 1-88. Filler Metals for Welding Copper and Copper Alloys; Composition, Application and Technical Delivery Conditions (June). 8 pp.

—Copper—Filler Metal
DIN ENGL 1733 Pt 1-88. Filler Metals for Welding Copper and Copper Alloys; Composition, Application and Technical Delivery Conditions (June). 8 pp.

—Defects—Glossaries
SAA AS Z5.2-68. Glossary of Metal Welding Terms and Definitions—Part 2: Terminology of and Abbreviations for Fusion Weld Imperfections as Revealed by Radiography Being BS 499:Part 3:1965 Endorsed Subject to Australian Amendment. 130 pp.

—Metal Containers
DIN ENGL 8562-75. Welding of Vessels; Vessels of Metallic Materials; Principles for Welding (Jan). 5 pp.

—Metals
BSI BS EN 288: Part 1-92. 1992 Specification and Approval of Welding Procedures for Metallic Materials Part 1: General Rules for Fusion Welding. 14 pp.

CEN PREN 288-1-91. Specification and Qualification of Welding Procedures for Metallic Materials—Part 1: General Rules for Fusion Welding. 15 pp.

CEN EN 288-1-92. Specification and Qualification of Welding Procedures for Metallic Materials—Part 1: General Rules for Fusion Welding. 11 pp.

CEN PREN 288-5-92. Specification and Approval of Welding Procedures for Metallic Materials—Part 5: Approval by Using Approved Welding Consumables for Arc Welding. 9 pp.

DIN ENGL 1910 Pt 2-77. Welding; Welding of Metals, Processes (Aug). 14 pp.

DIN ENGL EN 288 Pt 1-92. Specification and Approval of Procedures for Welding Metallic Materials; General Rules for Fusion Welding (Apr). 11 pp.

—Metals—Quality Assurance
CEN PREN 288-6-92. Specification and Approval of Welding Procedures for Metallic Materials—Part 6: Approval Related to Previous Experience. 6 pp.

CEN PREN 288-8-93. Specification and Approval of Welding Procedures for Metallic Materials—Part 8: Approval by a Pre-Production Welding Test. 7 pp.

CEN PREN 729-2-92. Quality Systems for Welding—Fusion Welding of Metallic Materials—Part 2: Complete Quality System. 15 pp.

CEN PREN 729-3-92. Quality Systems for Welding—Fusion Welding of Metallic Materials—Part 3: Standard Quality System. 14 pp.

CEN PREN 729-4-92. Quality Systems for Welding—Fusion Welding of Metallic Materials—Part 4: Elementary Quality System. 7 pp.

Fusion Welding (Cont.)

—Nickel Alloys—Filler Metal
DIN ENGL 1736 Pt 1-85. Welding Filler Metals for Nickel and Nickel Alloys; Composition, Application and Technical Delivery Conditions (Aug). 9 pp.

—Nickel—Filler Metal
DIN ENGL 1736 Pt 1-85. Welding Filler Metals for Nickel and Nickel Alloys; Composition, Application and Technical Delivery Conditions (Aug). 9 pp.

—Steel Castings
BSI BS 4570-85. 1985 Fusion Welding of Steel Castings. 32 pp.

—Steels
BSI BS 4870: Part 1-81. (WITHDRAWN) 1981 Amd 4 Approval Testing of Welding Procedures Part 1: Fusion Welding of Steel (AMD 6293) February 28, 1990 (Superseded by BS EN 288: Part 3: 1992). 23 pp.

—Steels—Company Certification
CSA W47.1-92. Certification of Companies for Fusion Welding of Steel Structures; (Gen Instr 1). 113 pp.

—Steels—Filler Metal
CEN PREN 759-92. Technical Delivery Conditions for Steel Welding Filler Metals—Including Type of Product, Dimensions, Tolerances and Marking. 11 pp.

—Steels—Quality Assurance
DIN ENGL 8563 Pt 3 (S)-85. Quality Assurance of Welding Operations; Fusion Welded Joints in Steel (Except Beam Welded Joints); Requirements, Classification (Oct) (Superseded by DIN EN 25817). 15 pp.

—Structural Steels—Weldability
DIN ENGL 8528 Pt 2-75. Weldability; Welding Suitability of General Structural Steels for Fusion Welding (Mar). 3 pp.

—Titanium Alloys—Filler Metal—Test Specimens
DIN ENGL 1737 Pt 2-88. Filler Metals for Welding Titanium and Titanium-Palladium Alloys; Sample and Specimen Preparation, Mechanical Properties of All-Weld Metal (June). 2 pp.

—Titanium—Filler Metal—Test Specimens
DIN ENGL 1737 Pt 2-88. Filler Metals for Welding Titanium and Titanium-Palladium Alloys; Sample and Specimen Preparation, Mechanical Properties of All-Weld Metal (June). 2 pp.

—Waterproofing Membranes—Bituminous
DIN ENGL 52131-85. Bitumen Sheeting for Fusion Welding; Terms and Definitions, Designation, Requirements (Aug). 2 pp.

DIN ENGL 52133-85. Polymer Bitumen Sheeting for Fusion Welding; Terms and Definitions, Designation, Requirements (Aug). 2 pp.

—Welders (Personnel)—Acceptance Testing
BSI BS 4871: Part 1-82. (WITHDRAWN) 1982 Amd 2 Approval Testing of Welders Working to Approved Welding Procedures Part 1: Fusion Welding of Steel (Superseded by BS EN 287: Part 1: 1992). 15 pp.

BSI BS 4871: Part 2-82. (WITHDRAWN) 1982 Approval Testing of Welders Working to Approved Welding Procedures Part 2: TIG or MIG Welding of Aluminium and Its Alloys (Superseded by BS EN 287: Part 2: 1992). 16 pp.

BSI BS 4872: Part 1-82. 1982 Amd 1 Approval Testing of Welders When Welding Procedure Approval is Not Required Part 1: Fusion Welding of Steel. 19 pp.

BSI BS 4872: Part 2-76. 1976 Approval Testing of Welders When Welding Procedure Approval is Not Required Part 2: TIG or MIG Welding of Aluminium and its Alloys. 21 pp.

BSI BS EN 287: Part 1-92. 1992 Approval Testing of Welders for Fusion Welding Part 1: Steels. 32 pp.

BSI BS EN 287: Part 2-92. 1992 Approval Testing of Welders for Fusion Welding Part 2: Aluminium and Aluminium Alloys. 30 pp.

CEN PREN 287 (Part 1)-91. Approval Testing of Welders—Fusion Welding—Part 1: Steels. 28 pp.

CEN EN 287-1-92. Approval Testing of Welders—Fusion Welding—Part 1: Steels. 31 pp.

CEN PREN 287 (Part 2)-91. Approval Testing of Welders—Fusion Welding—Part 2: Aluminium and Aluminium Alloys. 26 pp.

CEN EN 287-2-92. Approval Testing of Welders—Fusion Welding—Part 2: Aluminium and Aluminium Alloys. 29 pp.

DIN ENGL EN 287 Pt 1-92. Approval Testing of Welders; Fusion Welding of Steel; English Version of DIN EN 287 Part 1 (Apr) (Supersedes DIN 8560, May 1982 Edition). 30 pp.

Fusion Welding (Cont.)

—Welders (Personnel)—Acceptance Testing (Cont.)
DIN ENGL EN 287 Pt 2-92. Approval Testing of Welders; Fusion Welding of Aluminium and Aluminium Alloys; English Version of DIN EN 287 Part 2 (Apr) (Supersedes DIN 8561, February 1974 Edition). 28 pp.

—Welders—Certification—Aerospace
JIS W 0901-85. Aerospace Fusion Welders Qualification.

Fusion Welding Equipment

Scope Note: Use a more specific term *See:* Arc Welding Equipment; Braze Welding Equipment; Electron Beam Welding Machines; Electroslag Welding Equipment; Gas Metal Arc Welding Equipment; Gas Shielded Arc Welding Equipment; Gas Welding; Gas Welding Equipment; Oxyacetylene Welding Equipment; Plasma Arc Welding Equipment; Resistance Welding Equipment; Seam Welding Machines; Spot Welding Equipment; Submerged Arc Welding Equipment; Welding Equipment

Fusion Welds

See Also: Welded Joints
BSI BS 3451-73. 1973 Amd 1 Methods of Testing Fusion Welds in Aluminium and Aluminium Alloys. 22 pp.

—Acceptance Testing
BSI PD 6493-91. 1991 Guidance on Methods for Assessing the Acceptability of Flaws in Fusion Welded Structures. 122 pp.

—Aerospace—Nondestructive Testing
BSI M 42-72. 1972 Methods for Non-Destructive Testing of Fusion Welds in Thin Gauge Materials. 10 pp.

—Aerospace—Radiography
BSI M 42-72. 1972 Methods for Non-Destructive Testing of Fusion Welds in Thin Gauge Materials. 10 pp.

—Copper
BSI BS 4206-67. 1967 Amd 1 Methods of Testing Fusion Welds in Copper and Copper Alloys. 26 pp.

—Copper Alloys
BSI BS 4206-67. 1967 Amd 1 Methods of Testing Fusion Welds in Copper and Copper Alloys. 26 pp.

—Destructive Testing
BSI BS 709-83. 1983 Methods of Destructive Testing Fusion Welded Joints and Weld Metal in Steel. 24 pp.

—Metals—Defects—Classification
BSI BS EN 26520-92. 1992 Imperfections in Metallic Fusion Welds, with Explanations (ISO 6520: 1982). 19 pp.

CEN EN 26520-91. Classification of Imperfections in Metallic Fusion Welds, with Explanations. 3 pp.

DIN ENGL EN 26520-91. Imperfections in Metallic Fusion Welds; Classification and Terminlogy (ISO 6520: 1982) (English Version of DIN EN 26 520) (Dec) (Supersedes DIN 8524 Part 1, July 1986 Edition). 16 pp.

ISO 6520-82. Classification of Imperfections in Metallic Fusion Welds, with Explanations First Edition. 15 pp.

—Radiography
BSI BS 2600: Part 1-83. 1983 Radiographic Examination of Fusion Welded Butt Joints in Steel Part 1: Methods for Steel 2 mm up to and Including 50 mm Thick. 15 pp.

BSI BS 2600: Part 2-73. 1973 Amd 2 Radiographic Examination of Fusion Welded Butt Joints in Steel Part 2: Over 50 mm up to and Including 200 mm Thick. 15 pp.

BSI BS 7257-89. 1989 Amd 1 Methods for Radiographic Examination of Fusion Welded Branch and Nozzle Joints in Steel (AMD 6641) December 21, 1990. 34 pp.

—Visual Inspection
CEN PREN 970-92. Welding—Visual Examination of Fusion Welded Joints. 12 pp.

Futons

See Also: Bedding; Furniture
JIS L 4403-74. Futon (R 1988). 8 pp.

Future Air Navigation Systems (ICAO)

Use: ICAO Special Committee on Future Air Navigation Systems

Fuze Setters
Use: Fuse Setters

Fuzeheads
Use: Fuses (Ordnance)

Fuzes (Ordnance)
Use: Fuses (Ordnance)

G.A.L. Cygnet Aircraft
See Also: Aircraft

—Antenna Positions
CAA. GAL Cygnet (Approved Aerial Positions). 1 p.

G Acid
Use: 2-Naphthol-6,8-disulfonic Acid

G Suits
Use: Antigravity Suits

GaAs FETs
Use For: Gallium Arsenide Field Effect Transistors
See Also: FETs; Transistors

—Microwave—Preferred Products List
CECC CECC MUAHAG Vol 9 IS 3-90. Preferred Products List; Semiconductors (En, Fr, Ge) AMD 1 (En, Fr, Ge). 51 pp.

Gabions
See Also: Construction Equipment; Mining Equipment

—Galvanized Steel—Sheathing
JIS A 5513-84. Zinc Coated Low Carbon Steel Wire Gabions. 11 pp.

Gadolinium Oxide

—Powder—Nuclear Grade
CNS J3006-81. Nuclear-Grade Gadolinium Oxide (Gd2O3) Powder (Aug)(7802).

Gage Blocks
BSI BS 4311: Part 1-93. 1993 Gauge Blocks and Accessories Part 1: New Gauge Blocks (H). 16 pp.
BSI BS 4311: Part 1-01. 1993 Amd 1 Gauge Blocks and Accessories Part 1: New Gauge Blocks (AMD 7883) July 15, 1993 (H). 17 pp.
BSI BS 4311: Part 1-68. 1968 Amd 2 Metric Gauge Blocks Part 1: Gauge Blocks. 20 pp.
BSI BS 4311: Part 3-93. 1993 Gauge Blocks and Accessories Part 3: Gauge Blocks in Use. 12 pp.
CNS B6058-81. Gauge Blocks (Nov)(8092).
DIN ENGL 861 Pt 1-80. Gauge Blocks; Concepts, Requirements, Testing (Jan). 9 pp.
ISO 3650-78. Gauge Blocks First Edition; (Erratum—March 1979). 13 pp.
JIS B 7506-89. Gauge Blocks. 23 pp.
MOD UK DSTAN 52-1-73. Gauge Blocks—Metric Issue 2 (Withdrawn). 14 pp.
SAA AS 1457-89. Gauge Blocks and Rectangular Length Bars and Their Accessories. 23 pp.

—Accessories
BSI BS 4311: Part 2-77. 1977 Amd 1 Metric Gauge Blocks Part 2: Accessories. 9 pp.

Gage Glasses

—Boilers
CNS B2743-83. Gage Glasses for Boilers (Feb)(9970).
JIS B 8211-75. Gauge Glasses for Boilers.

—Pressure Vessels
SNZ NZS 5301-77. Specification for Observation and Gauge Glasses for Pressure Vessels Amend: A, 1977. 24 pp.

Gage Pressure Transmitters
Use For: Gauge Pressure Transmitters
See Also: Absolute Pressure Gages; Differential Pressure Transmitters; Pressure Transmitters; Transmitters
BSI BS 6447-84. 1984 Amd 1 Absolute and Gauge Pressure Transmitters with Electrical Outputs. 13 pp.

Gage Ties
See Also: Rolling Stock

—Railroad Signals
JIS E 3351-80. Gauge Ties for Railway Signaling.

Gages
Scope Note: For additional listings, use a more specific term *See Also:* Air Gages; Angle Gages; Bourdon Tubes; Capacitance Level Indicators; Center Gages; Check Gages; Crusher Gages; Cylinder Gages; Deposit Gages; Depth Gages; Dial Gages; Diaphragm Gages; Differential Pressure Gages; Draft Gages; Feeler Gages; Float Gages; Fuel Gages; Gage

Gages (Cont.)
See Also: (Cont.)
Glasses; Gaging Stations; Gap Gages; Go Gages; Height Gages; Inspection Gages; Ionization Gages; Level Indicators; Limit Gages; Master Gages; Measuring Instruments; Newton Gages; No Go Gages; Plug Gages; Pressure Gages; Radionuclide Gages; Rain Gages; Ring Gages; Setting Gages; Sight Glasses; Surgical Instruments; Taper Gages; Taper Plug Gages; Taper Ring Gages; Temperature Gages; Thread Gages; Thread Pitch Gages; Thread Plug Gages; Thread Ring Gages; Thread Snap Gages; Tire Gages; Track Gages; Vernier Gages; Water Gages; Wire Gages; Workshop Gages
DIN ENGL 2239-89. Gauges in Dimensional Metrology; Requirements and Testing (May). 3 pp.

—Coaxial Connectors
CENELEC HD 489 S1-87. Recommended Dimensions for Hexagonal and Square Crimping-Die Cavities, Indentors, Gauges, Outer Conductor Crimp Sleeves and Centre Contact Crimp Barrels for r.f. Cables and Connectors. 4 pp.
IEC 803-84. Recommended Dimensions for Hexagonal and Square Crimping-Die Cavities, Indentors, Gauges, Outer Conductor Crimp Sleeves and Centre Contact Crimp Barrels for R.F. Cables and Connectors First Edition. 40 pp.

—Construction Equipment—Mechanical Shock
CNS A3164-82. Testing Method of Vibration and Shock for Construction Machinery Gauges (Jul)(9054).
JIS A 8101-78. Vibration and Shock Testing Method for Construction Machinery Gauges.

—Construction Equipment—Vibration
CNS A3164-82. Testing Method of Vibration and Shock for Construction Machinery Gauges (Jul)(9054).
JIS A 8101-78. Vibration and Shock Testing Method for Construction Machinery Gauges.

—Defense Contracts
MOD UK DEFCON 23-76. Special Jigs, Tools, Etc 6/76. 1 p.

—Diameter—Glass Joints
CNS R3090-80. Suitable Gauging System for Diameter and Length of Conical Joints (Apr) (5480).

—Electric Contacts—Contact Force
CEN PREN 2591 (Part D18)-92. Elements of Electrical and Optical Connection Test Methods Part D18—Gauge Insertion and Extraction Forces in a Female Contact. 4 pp.

—Electric Contacts—Insertion Force
CEN PREN 2591 (Part D18)-92. Elements of Electrical and Optical Connection Test Methods Part D18—Gauge Insertion and Extraction Forces in a Female Contact. 4 pp.

—Handles
DIN ENGL 2240 Pt 1-89. Gauge Handles for Gauging Members up to 40 mm Nominal Diameter with 1: 50 Taper Shank (Nov). 2 pp.
DIN ENGL 2240 Pt 2-89. Gauge Handles for Gauging Members over 40 mm Nominal Diameter; Handles, Fixing Screws and Locking Prongs (Nov). 4 pp.
DIN ENGL 2240 Pt 3-89. Gauge Handles for Gauging Members for Use in Precision Engineering (Nov). 3 pp.

—Internal Combustion Piston Engines—Symbols
BSI BS 7712-93. 1993 Graphical Symbols for Reciprocating Internal Combustion Engines (ISO 8999: 1993) (G). 23 pp.
ISO 8999-93. Reciprocating Internal Combustion Engines—Graphical Symbols First Edition. 20 pp.

—Invitation for Bids
MOD UK DEFCON 47V-86. Invitation to Tender Supplementary Conditions—Jigs, Tools, Etc 10/86. 1 p.

—Lampholders
CENELEC HD 65.3-87. Lamp Caps and Holders Together with Gauges for the Control of Interchangeability and Safety Part 3: Gauges. 3 pp.

—Lampholders—Safety
CENELEC HD 65.3-87. Lamp Caps and Holders Together with Gauges for the Control of Interchangeability and Safety Part 3: Gauges. 3 pp.

—Length—Glass Joints
CNS R3090-80. Suitable Gauging System for Diameter and Length of Conical Joints (Apr) (5480).

Gages (Cont.)

—Penetration—Cross Recesses
CNS B6024-78. Cross Recesses Penetration Gauge (May)(3541).

Gaging
Use For: Gauging *See Also:* Radiation
MOD UK DSTAN 05-38-75. General Principles of Gauging Issue 1. 4 pp.
MOD UK DSTAN 05-38-92. General Principles of Gauging Issue 2. 6 pp.

Gaging Members
Use: Gages

Gaging Stations
Use For: Gauging Station *See Also:* Flow Measurement; Water Level Recorders
BSI BS 3680: Part 3B-83. 1983 Measurement of Liquid Flow in Open Channels Part 3B: Stream Flow Measurement Guide for Establishment and Operation of a Gauging Station. 24 pp.
BSI BS 3680: Part 3Q-93. 1993 Measurement of Liquid Flow in Open Channels Part 3: Stream Flow Measurement Part 3Q: Code of Practice for Safe Practice in Stream Gauging (Q). 18 pp.
BSI BS 3680: Part 8D-80. 1980 Measurement of Liquid Flow in Open Channels: Measuring Instruments and Equipment: Cableway System for Stream Gauging. 11 pp.
CGSB CAN/CGSB-157.3-M91. Liquid Flow Measurement in Open Channels—Establishment and Operation of a Gauging Station. 32 pp.
ISO 1100 Pt 1-81. Liquid Flow Measurement in Open Channels—Part 1: Establishment and Operation of a Gauging Station Second Edition. 23 pp.
ISO 4375-79. Measurement of Liquid Flow in Open Channels—Cableway System for Stream Gauging First Edition. 10 pp.
SAA AS 3778.6.6-92. Measurement of Water Flow in Open Channels—Part 6: Measuring Devices, Instruments and Equipment—Part 6.6: Cableway System for Stream Gauging (ISO 4375:1979). 8 pp.

—Electromagnetic
BSI BS 3680: Part 3H-93. 1993 Measurement of Liquid Flow in Open Channels Part 3: Stream Flow Measurement Part 3H: Electromagnetic Method Using a Full-Channel-Width Coil (ISO 9213: 1992) (Q). 20 pp.
ISO 9213-92. Measurement of Total Discharge in Open Channels—Electromagnetic Method Using a Full-Channel-Width Coil First Edition. 17 pp.

—Stage-Discharge Relations
CGSB CAN/CGSB-157.4-M91. Liquid Flow Measurement in Open Channels—Part 2: Determination of the Stage-Discharge Relation (ISO 1100-2:1982). 38 pp.
ISO 1100 Pt 2-82. Liquid Flow Measurement in Open Channels—Part 2: Determination of the Stage-Discharge Relation First Edition. 35 pp.

Gain (Amplification)
Use For: Amplification (Gain) *See Also:* Electrical Properties; Gain Hits; Voltage

—Telephone Repeaters—Carrier Systems
CCITT RECMN G.312-89. Intermediate Repeaters for Open-Wire Carrier Systems Conforming to Recommendation G.311—International Analogue Carrier Systems (Study Group XV) 2 pp. 2 pp.

Gain Hits
See Also: Gain (Amplification); Surges

—Telephone Circuits
CCITT FASCICLE III.1. General Characteristics of International Telephone Connections and Circuits Recommendations G.101-G.181. 332 pp.

—Voice Band Data Transmission
CCITT RECMN G.113-89. Transmission Impairments—General Characteristics of International Telephone Connections and Circuits (Study Groups XII and XV) 22 pp. 22 pp.

Gaiters
Use For: Leggings *See Also:* Clothing
BSI BS 4676-83. 1983 Gaiters and Footwear for Protection Against Burns and Impact Risks in Foundries. 12 pp.
CEN PREN 381 (Part 5)-90. Protective Clothing for Users of Hand Held Chain Saws—Part 5: Requirements for Leg Protection. 8 pp.
CEN PREN 381-8-91. Protective Clothing for Users of Hand-Held Chainsaws: Part 8—Test Method for Chainsaw Protective Gaiters. 15 pp.
CEN PREN 381-9-91. Protective Clothing for Users of Hand-Held Chainsaws: Part 9—Requirements for Chainsaw Protective Gaiters. 9 pp.

Galactose
JIS K 8250-80. Galactose.

Galilean Binoculars
Use: Binoculars—Galilean

Gallates Content Analysis
See Also: Antioxidant Content Analysis

—**Fats—Molecular Absorption Spectrophotometry**
BSI BS 684: Sec 2.44-84. 1984 Methods of Analysis of Fats and Fatty Oils Part 2: Other Methods Section 2.44: Determination of Gallates. 4 pp.
ISO 6464-83. Animal and Vegetable Fats and Oils—Determination of Gallates Content—Molecular Absorption Spectrometric Method First Edition. 5 pp.

—**Oils—Molecular Absorption Spectrophotometry**
BSI BS 684: Sec 2.44-84. 1984 Methods of Analysis of Fats and Fatty Oils Part 2: Other Methods Section 2.44: Determination of Gallates. 4 pp.
ISO 6464-83. Animal and Vegetable Fats and Oils—Determination of Gallates Content—Molecular Absorption Spectrometric Method First Edition. 5 pp.

Galleys
See Also: Ships
MOD UK NES 121-81. Requirements for Galleys and Associated Spaces in Surface Ships Issue 1 (10.81) (Superseded in Part by NES 121: Part 1-4). 82 pp.
MOD UK NES 135-83. Galleys & Associated Spaces—Submarines Issue 1 (05.83) (Withdrawn). 54 pp.

—**Aircraft**
CAA 706 (K). Section K Light Aeroplanes Electrical Supply, Systems and Equipment (Blue Papers). 54 pp.
CAA NOTICE #99 ISSUE 1. Galley Equipment (Airworthiness Notices). 4 pp.

—**Aircraft—Design**
CAA Chapter K6-1 3 App4 10.92. Design and Installation of Galleys and Domestic Equipment. 1 p.

—**Aircraft—Installation**
CAA Chapter K6-1 3 App4 10.92. Design and Installation of Galleys and Domestic Equipment. 1 p.

—**Naval Ships**
MOD UK NES 121: Part 1-92. Requirements for Galleys and Associated Spaces Part 1: Common Requirements Issue 2 (10.92). 42 pp.
MOD UK NES 121: Part 1-01. Requirements for Galleys and Associated Spaces Part 1: Common Requirements Issue 2 (10.92); Amendment 1. 59 pp.
MOD UK NES 121: Part 2-92. Requirements for Galleys and Associated Spaces Part 2: Specific Requirements—Surface Ships Issue 2 (10.92). 45 pp.

—**Nuclear Submarines**
MOD UK NES 121: Part 4-92. Requirements for Galleys and Associated Spaces Part 4: Specific Requirements—Nuclear Submarines Issue 2 (10.92). 17 pp.

—**Rotary Wing Aircraft**
CAA 707 (G). Section G Rotorcraft Electrical Supply, Systems and Equipment (Blue Papers). 47 pp.
CAA Chap G6-14 App #4 11.85. Design and Installation of Galleys and Domestic Equipment (Rotorcraft). 2 pp.

—**Ships—Ventilation**
ISO 9943-91. Shipbuilding—Ventilation and Air-Treatment of Galleys and Pantries with Cooking Appliances First Edition. 8 pp.

—**Submarines**
MOD UK NES 121: Part 1-92. Requirements for Galleys and Associated Spaces Part 1: Common Requirements Issue 2 (10.92). 42 pp.
MOD UK NES 121: Part 1-01. Requirements for Galleys and Associated Spaces Part 1: Common Requirements Issue 2 (10.92); Amendment 1. 59 pp.
MOD UK NES 121: Part 3-92. Requirements for Galleys and Associated Spaces Part 3: Specific Requirements—Conventional Submarines Issue 2 (10.92). 12 pp.

Gallic Acid
See Also: Benzoic Acid; Phenol
CNS K7238-67. Chemical Reagent (Gallic Acid) (Oct)(1737).
JIS K 8898-61. Gallic Acid.

Gallium Arsenide
—**Boron Content**
JIS H 1191-91. Methods for Chemical Analysis of Gallium Arsenide. 57 pp.

—**Chromium Content**
JIS H 1191-91. Methods for Chemical Analysis of Gallium Arsenide. 57 pp.

—**Copper Content**
JIS H 1191-91. Methods for Chemical Analysis of Gallium Arsenide. 57 pp.

—**Indium Content**
JIS H 1191-91. Methods for Chemical Analysis of Gallium Arsenide. 57 pp.

—**Iron Content**
JIS H 1191-91. Methods for Chemical Analysis of Gallium Arsenide. 57 pp.

—**Manganese Content**
JIS H 1191-91. Methods for Chemical Analysis of Gallium Arsenide. 57 pp.

—**Silicon Content**
JIS H 1191-91. Methods for Chemical Analysis of Gallium Arsenide. 57 pp.

—**Tellurium Content**
JIS H 1191-91. Methods for Chemical Analysis of Gallium Arsenide. 57 pp.

—**Zinc Content**
JIS H 1191-91. Methods for Chemical Analysis of Gallium Arsenide. 57 pp.

Gallium Arsenide Field Effect Transistors
Use: GaAs FETs

Gallium Inorganic Compounds
Scope Note: Use a more specific term *See:* Gallium Arsenide

Galvanic Anodes
—**Magnesium**
JIS H 6125-61. Magnesium Galvanic Anodes (R 1968). 4 pp.

Galvanized Materials
Scope Note: For additional listings, see also specific products made from galvanized materials
See Also: Cadmium Coatings (Made From Cadmium); Galvanized Steel Coatings; Galvanized Steels; Zinc Coatings (Made From Zinc)

—**Sheets**
CNS G2209-84. Method of Test for Precoated Galvanized Sheets (Mar)(10805).
CNS G3217-84. Precoated Galvanized Sheets (Mar)(10804).

—**Sheets—Corrugated**
DIN ENGL 59231-53. Corrugated Sheet; Tile Sheet; Galvanized (Apr). 3 pp.

Galvanized Steel Coatings
See Also: Coatings; Galvanized Materials; Metal Coatings (On Metal); Steel Coatings

—**Threaded Fasteners**
SAA AS 1214-83. Hot-Dip Galvanized Coatings on Threaded Fasteners (ISO Metric Coarse Thread Series). 8 pp.

—**Windows—Organic—Powder**
BSI BS 6497-84. 1984 Powder Organic Coatings for Application & Stov-ing to Hot-Dip Galvaniz-ed Hot-Rolled Steel Sec-tions & Steel Sheet for Windows & Associated Ex-ternal Architectural Purposes, & for the Fin-ish on Galvanized Steel Sections & Sheet Coated w/Powder Org. Coatings. 11 pp.
BSI BS 6497-01. 1984 Amd 1 Powder Organic Coatings for Application and Stoving to Hot-Dip Galvanized Hot-Rolled Steel Sections and Preformed Steel Sheet for Windows and Associated External Architectural Purposes, and for the Finish on Galvanized Steel. 12 pp.

Galvanized Steels
Scope Note: For additional listings, see also specific products made from galvanized steel
See Also: Galvanized Materials; Steels; Zinc Coatings (Made From Zinc)

—**Coils—Hot Dip**
CNS G3027-83. Hot-Dip Galvanized Sheets (Oct)(1244). 11 pp.
JIS G 3302-87. Hot-Dip Zinc-Coated Steel Sheets and Coils. 34 pp.
JIS G 3312-87. Prepainted Hot-Dip Zinc-Coated Steel Sheets and Coils. 25 pp.

—**Electrodeposited**
CNS H3079-87. Electroplated Coatings of Zinc on Iron and Steel (Sep)(4827).

—**Electroplated**
DIN ENGL 50961-87. Electroplated Coatings; Zinc and Cadmium Coatings on Iron and Steel; Chromate Treatment of Zinc and Cadmium Coatings (June) (Supersedes DIN 50941, May 1978 Edition, and DIN 50962, April 1976 Edition). 6 pp.
ISO 2081-86. Metallic Coatings—Electroplated Coatings of Zinc on Iron or Steel Second Edition. 7 pp.
JIS H 8610-91. Electroplated Coatings of Zinc on Iron or Steel. 8 pp.

—**Hot Dip**
BSI BS 729-71. 1971 Hot Dip Galvanized Coatings on Iron and Steel Articles. 15 pp.
CNS H3102-82. Recommended Practice for Zinc Coatings (Hot-Dipped) (Feb)(8503).
CNS H3116-83. Zinc Coating (Hot Dipped) on Iron and Steel (Feb)(10007).
DIN ENGL 50976-89. Corrosion Protection; Hot-Dip Batch Galvanizing; Requirements and Testing (May). 7 pp.
ISO 1459-73. Metallic Coatings—Protection Against Corrosion by Hot Dip Galvanizing—Guiding Principles First Edition. 4 pp.
JIS H 8641-82. Zinc Hot Dip Galvanizings. 7 pp.
JIS H 9124-87. Recommended Practice for Zinc Coating (Hot-Dipped). 9 pp.
SNZ NZS 3441-78. Specification for Hot-Dipped Zinc-Coated Steel Coil and Cut Lengths Amend: 1, 1988, 2. 20 pp.

—**Plates—Hot Dip**
BSI BS 2858-57. 1957 Amd 1 Steel Plates for Use in the Manufacture of Galvanizing Pots. 10 pp.
BSI BS 2989-92. (WITHDRAWN) 1992 Continuously Hot-Dip Zinc Coated and Iron-Zinc Alloy Coated Steel Flat Products: Tolerances on Dimensions and Shape (Superseded by BS EN 10143: 1993) (V). 10 pp.
BSI BS 2989-91. 1991 Continuously Hot-Dip Zinc Coated and Iron-Zinc Alloy Coated Steel of Structural Qualities: Wide Strip, Sheet/Plate and Slit Wide Strip (V). 16 pp.
BSI BS 2989-82. 1982 Continuously Hot-Dip Zinc Coated and Iron-Zinc Alloy Coated Steel: Wide Strip, Sheet/Plate and Slit Wide Strip. 16 pp.

—**Sheets—Corrugated**
DIN ENGL 25512-80. Corrugated Sheet Sections for Rail Vehicles; Dimensions; Weights; Static Values (Oct). 2 pp.

—**Sheets—Corrugated—Hot Dip**
BSI BS 3083-88. 1988 Amd 1 Hot-Dip Zinc Coated and Hot-Dip Aluminium/Zinc Coated Corrugated Steel Sheets for General Purposes (AMD 6054) May 31, 1990. 13 pp.
BSI BS 3083-80. 1980 Hot-Dip Zinc Coated Corrugated Steel Sheets for General Purposes. 8 pp.
SNZ NZS 3403-78. Specification for Hot-Dip Galvanized Corrugated Steel Sheet for Building Purposes. 7 pp.

—**Sheets—Electrodeposited**
CNS G2191-83. Method of Test for Electrolytic Zinc-Coated Steel Sheets and Strips (Sep)(10569).
CNS G3211-83. Electrolytic Zinc-Coated Steel Sheets and Strips (Sep)(10568).
DIN ENGL 17163-88. Steel Flat Products; Electrolytically Zinc Coated Cold Rolled Strip and Sheet; Technical Delivery Conditions (Mar). 6 pp.
ISO 5001-80. Cold-Reduced Carbon Steel Sheet for Vitreous Enamelling First Edition; DoD Adopted. 15 pp.
ISO 5002-82. Hot-Rolled and Cold-Reduced Electrolytic Zinc-Coated Carbon Steel Sheet of Commercial and Drawing Qualities First Edition; DoD Adopted. 14 pp.
JIS G 3313-90. Electrolytic Zinc-Coated Steel Sheets and Coils. 77 pp.

—**Sheets—Hot Dip**
BSI BS 2989-92. (WITHDRAWN) 1992 Continuously Hot-Dip Zinc Coated and Iron-Zinc Alloy Coated Steel Flat Products: Tolerances on Dimensions and Shape (Superseded by BS EN 10143: 1993) (V). 10 pp.
BSI BS 2989-91. 1991 Continuously Hot-Dip Zinc Coated and Iron-Zinc Alloy Coated Steel of Structural Qualities: Wide Strip, Sheet/Plate and Slit Wide Strip (V). 16 pp.
BSI BS 2989-82. 1982 Continuously Hot-Dip Zinc Coated and Iron-Zinc Alloy Coated Steel: Wide Strip, Sheet/Plate and Slit Wide Strip. 16 pp.

INTERNATIONAL AND NON-U.S. NATIONAL STANDARDS
SUBJECT INDEX

Galvanized

Galvanized Steels (Cont.)
—Sheets—Hot Dip (Cont.)
BSI BS EN 10142-91. 1991 Continuously Hot-Dip Zinc Coated Low Carbon Steel Sheet and Strip for Cold Forming: Technical Delivery Conditions. 15 pp.
BSI BS EN 10142-01. 1991 Amd 1 Continuously Hot-Dip Zinc Coated Low Carbon Steel Sheet and Strip for Cold Forming: Technical Delivery Conditions (AMD 6932) November 15, 1992. 16 pp.
BSI BS EN 10143-93. 1993 Continuously Hot-Dip Metal Coated Steel Sheet and Strip—Tolerances on Dimensions and Shape (Supersedes BS 2989: 1992). 13 pp.
BSI BS EN 10147-92. 1992 Continuously Hot-Dip Zinc Coated Structural Steel Sheet and Strip—Technical Delivery Conditions (V). 16 pp.
CEN PREN 10 142-88. Continuously Hot-Dip Zinc Coated Mild Steel Sheet and Strip for Cold Forming; Technical Delivery Conditions. 24 pp.
CEN EN 10 142-90. Continuously Hot-Dip Zinc Coated Low Carbon Steel Sheet and Strip for Cold Forming—Technical Delivery Conditions. 21 pp.
CEN PREN 10143-91. Continuously Hot-Dip Metal Coated Steel Sheet and Strip: Tolerances on Dimensions and Shape. 16 pp.
CEN EN 10143-93. Continuously Hot-Dip Metal Coated Steel Sheet and Strip—Tolerances on Dimensions and Shape. 7 pp.
CEN PREN 10 147-90. Continuously Hot-Dip Zinc Coated Unalloyed Structural Steel Sheet and Strip; Technical Delivery Conditions. 21 pp.
CEN EN 10147-91. Continuously Hot-Dip Zinc Coated Structural Steel Sheet and Strip—Technical Delivery Conditions. 22 pp.
CNS G2012-83. Method of Test for Hot-Dip Galvanized Sheet (Oct)(1361).
CNS G3027-83. Hot-Dip Galvanized Sheets (Oct)(1244). 11 pp.
DIN ENGL 17162 Pt 2-80. Flat Steel Products; Hot-Dip Zinc-Coated Strip and Sheet; Technical Conditions of Delivery; Structural Steels for General Use (Sept). 7 pp.
DIN ENGL 59232 (S)-78. Flat Steel Products; Hot Galvanized Wide Strip and Sheet of Mild Unalloyed Steels and General Structural Steels; Dimensions, Permissible Deviations on Dimension and Form (July) (Superseded by DIN EN 10143). 5 pp.
DIN ENGL EN 10142-91. Continuously Hot-Dip Galvanized Mild Steel Sheet and Strip for Cold Forming; Technical Delivery Conditions; English Version of DIN EN 10142 (Mar) (Supersedes DIN 17162 Part 1, September 1977 Edition). 10 pp.
DIN ENGL EN 10143-93. Continuously Hot-Dip Metal Coated Steel Sheet and Strip; Tolerances on Dimensions and Shape (Mar) (Supersedes DIN 59232, July 1978 Edition). 8 pp.
DIN ENGL EN 10147-92. Continuously Hot-Dip Galvanized Structural Steel Sheet and Strip; Technical Delivery Conditions; English Version of DIN EN 10 147: 1991 (Jan) (Supersedes DIN 17 162 Part 2, September 1980 Edition). 10 pp.
ISO 3575-76. Continuous Hot-Dip Zinc-Coated Carbon Steel Sheet of Commercial, Lock-Forming and Drawing Qualities First Edition; DoD Adopted. 17 pp.
ISO 4998-91. Continuous Hot-Dip Zinc-Coated Carbon Steel Sheet of Structural Quality Second Edition. 14 pp.
JIS G 3302-87. Hot-Dip Zinc-Coated Steel Sheets and Coils. 34 pp.
JIS G 3312-87. Prepainted Hot-Dip Zinc-Coated Steel Sheets and Coils. 25 pp.

—Sheets—Prefinished
CGSB CAN/CGSB-93.3-M91. Prefinished Galvanized and Aluminum-Zinc Alloy Steel Sheet for Residential Use. 8 pp.

—Sprayed
CNS H2053-81. Methods of Test for Zinc Spray Products (Oct)(8010). 6 pp.
CNS H3103-82. Recommended Practice for Zinc Spray Coatings on Iron and Steel (Feb)(8504).
ISO 2063-91. Metallic and Other Inorganic Coatings—Thermal Spraying—Zinc, Aluminium and Their Alloys Second Edition. 14 pp.
JIS H 8300-90. Zinc Spraying on Iron and Steel. 6 pp.
JIS H 8305-82. Sprayed Zinc-Aluminum Alloy Coatings on Iron or Steel. 9 pp.
JIS H 8661-61. Testing Methods for Zinc Spray Products. 10 pp.
JIS H 9300-91. Recommended Practice for Zinc Spraying on Iron and Steel. 11 pp.

—Strips—Electrodeposited
DIN ENGL 17163-88. Steel Flat Products; Electrolytically Zinc Coated Cold Rolled Strip and Sheet; Technical Delivery Conditions (Mar). 6 pp.

Galvanized Steels (Cont.)
—Strips—Hot Dip
BSI BS 2989-92. (WITHDRAWN) 1992 Continuously Hot-Dip Zinc Coated and Iron-Zinc Alloy Coated Steel Flat Products: Tolerances on Dimensions and Shape (Superseded by BS EN 10143: 1993) (V). 10 pp.
BSI BS 2989-91. 1991 Continuously Hot-Dip Zinc Coated and Iron-Zinc Alloy Coated Steel of Structural Qualities: Wide Strip, Sheet/Plate and Slit Wide Strip (V). 16 pp.
BSI BS 2989-82. 1982 Continuously Hot-Dip Zinc Coated and Iron-Zinc Alloy Coated Steel: Wide Strip, Sheet/Plate and Slit Wide Strip. 16 pp.
BSI BS EN 10142-91. 1991 Continuously Hot-Dip Zinc Coated Low Carbon Steel Sheet and Strip for Cold Forming: Technical Delivery Conditions. 15 pp.
BSI BS EN 10142-01. 1991 Amd 1 Continuously Hot-Dip Zinc Coated Low Carbon Steel Sheet and Strip for Cold Forming: Technical Delivery Conditions (AMD 6932) November 15, 1992. 16 pp.
BSI BS EN 10143-93. 1993 Continuously Hot-Dip Metal Coated Steel Sheet and Strip—Tolerances on Dimensions and Shape (Supersedes BS 2989: 1992). 13 pp.
BSI BS EN 10147-92. 1992 Continuously Hot-Dip Zinc Coated Structural Steel Sheet and Strip—Technical Delivery Conditions (V). 16 pp.
CEN PREN 10 142-88. Continuously Hot-Dip Zinc Coated Mild Steel Sheet and Strip for Cold Forming; Technical Delivery Conditions. 24 pp.
CEN EN 10 142-90. Continuously Hot-Dip Zinc Coated Low Carbon Steel Sheet and Strip for Cold Forming—Technical Delivery Conditions. 21 pp.
CEN PREN 10143-91. Continuously Hot-Dip Metal Coated Steel Sheet and Strip: Tolerances on Dimensions and Shape. 16 pp.
CEN EN 10143-93. Continuously Hot-Dip Metal Coated Steel Sheet and Strip—Tolerances on Dimensions and Shape. 7 pp.
CEN PREN 10 147-90. Continuously Hot-Dip Zinc Coated Unalloyed Structural Steel Sheet and Strip; Technical Delivery Conditions. 21 pp.
CEN EN 10147-91. Continuously Hot-Dip Zinc Coated Structural Steel Sheet and Strip—Technical Delivery Conditions. 22 pp.
DIN ENGL 17162 Pt 2-80. Flat Steel Products; Hot-Dip Zinc-Coated Strip and Sheet; Technical Conditions of Delivery; Structural Steels for General Use (Sept). 7 pp.
DIN ENGL 59232 (S)-78. Flat Steel Products; Hot Galvanized Wide Strip and Sheet of Mild Unalloyed Steels and General Structural Steels; Dimensions, Permissible Deviations on Dimension and Form (July) (Superseded by DIN EN 10143). 5 pp.
DIN ENGL EN 10142-91. Continuously Hot-Dip Galvanized Mild Steel Sheet and Strip for Cold Forming; Technical Delivery Conditions; English Version of DIN EN 10142 (Mar) (Supersedes DIN 17162 Part 1, September 1977 Edition). 10 pp.
DIN ENGL EN 10143-93. Continuously Hot-Dip Metal Coated Steel Sheet and Strip; Tolerances on Dimensions and Shape (Mar) (Supersedes DIN 59232, July 1978 Edition). 8 pp.
DIN ENGL EN 10147-92. Continuously Hot-Dip Galvanized Structural Steel Sheet and Strip; Technical Delivery Conditions; English Version of DIN EN 10 147: 1991 (Jan) (Supersedes DIN 17 162 Part 2, September 1980 Edition). 10 pp.

—Tubes—Hot Dip
DIN ENGL 2444-84. Zinc Coatings on Steel Tubes; Quality Standard for Hot Dip Galvanizing of Steel Tubes for Installation Purposes (Jan). 4 pp.

—Wash Primers
CGSB CAN/CGSB-31.116-M90. Pretreatment Solution for Galvanized Steel. 8 pp.
MOD UK TS 10229A. Paint, Priming, for Use in Galvinized Steel; Types: Brushing, Dipping, Spraying.

Game (Meat)
See Also: Meat
—Health Requirements
EC 92/45/EEC-92. Council Directive on Public Health and Animal Health Problems Relating to the Killing of Wild Game and the Placing on the Market of Wild-Game Meat. 19 pp.

Gamma Acid
JIS K 4136-91. 6-Amino-4-Hydroxy-2-Naphthalenesulfonic Acid (Gamma-Acid).

Gamma Beam Equipment
Use: Gamma Ray Equipment

Gamma Cameras
BSI BS 6609-92. 1992 Methods of Test for Anger Type Gamma Cameras (IEC 789: 1992). 23 pp.
BSI BS 6609-85. 1985 Methods of Test for Radionuclide Imaging Devices. 19 pp.

Gamma Cameras (Cont.)
IEC 789-92. Characteristics and Test Conditions of Radionuclide Imaging Devices; Anger Type Gamma Cameras Second Edition. 38 pp.

Gamma Contamination Meters
Use: Radiation Meters

Gamma Decay
Use: Gamma Emission

Gamma Dosimeters
Use: Dosimeters—Gamma Radiation

Gamma Emission
Use For: Gamma Decay See Also: Gamma Rays
—Uranium Hexafluoride
CNS J2090-83. Method for Determining Gamma-Energy Emission Rate from Fission Products in UF6 (Jan)(9889). 3 pp.

Gamma Irradiation
See Also: Gamma Rays; Irradiation
—Sterilization—Medical
CEN PREN 552-91. Sterilization of Medical Devices—Validation and Routine Control of Sterilization by Irradiation—Requirements. 21 pp.
CEN PREN 553-91. Sterilization of Medical Devices—Validation and Routine Control of Sterilization by Irradiation—Guidance. 22 pp.

Gamma Radiation
Use: Gamma Rays

Gamma Radiation Monitors
See Also: Gamma Ray Equipment; Radiation Measuring Instruments; Radiation Monitors
—Airborne—Environmental Conditions
IEC 1134-92. Airborne Instrumentation for Measurement of Terrestrial Gamma Radiation First Edition. 57 pp.
—Airborne—Geological Mapping
IEC 1134-92. Airborne Instrumentation for Measurement of Terrestrial Gamma Radiation First Edition. 57 pp.
—Airborne—Prospecting
IEC 1134-92. Airborne Instrumentation for Measurement of Terrestrial Gamma Radiation First Edition. 57 pp.
—Effluents
IEC 861-87. Equipment for Continuously Monitoring for Beta and Gamma Emitting Radionuclides in Liquid Effluents First Edition. 71 pp.
IEC 951 Pt 3-89. Radiation Monitoring Equipment for Accident and Post-Accident Conditions in Nuclear Power Plants Part 3: High Range Area Gamma Radiation Dose Rate Monitoring Equipment First Edition. 29 pp.
IEC 1031-90. Design, Location and Application Criteria for Installed Area Gamma Radiation Dose Rate Monitoring Equipment for Use in Nuclear Power Plants During Normal Operation and Anticipated Operational Occurrences First Edition. 26 pp.
—Personnel—Surface Contamination
IEC 1137-92. Radiation Protection Instrumentation—Installed Personnel Surface Contamination Monitoring Assemblies—Low Energy X and Gamma Emitters First Edition. 63 pp.

Gamma Radiation Shielding
Use: Radiation Shields (Protective)

Gamma Radiography Apparatus
Use: Gamma Ray Equipment

Gamma Ray Detectors
Scope Note: Use a more specific term See: Gamma Ray Equipment; Germanium Detectors; Ionization Chambers; Scintillation Counters; Semiconductor Detectors

Gamma Ray Equipment
Use For: Gamma Beam Equipment; Gamma Radiography Apparatus See Also: Gamma Radiation Monitors; Radiography
—Airborne—Environmental Conditions
IEC 1134-92. Airborne Instrumentation for Measurement of Terrestrial Gamma Radiation First Edition. 57 pp.
—Airborne—Geological Mapping
IEC 1134-92. Airborne Instrumentation for Measurement of Terrestrial Gamma Radiation First Edition. 57 pp.

Gamma Ray Equipment (Cont.)

—Airborne—Prospecting
IEC 1134-92. Airborne Instrumentation for Measurement of Terrestrial Gamma Radiation First Edition. 57 pp.

—Medical—Safety
BSI BS 5724: Sec 2.11-89. 1989 Medical Electrical Equipment Part 2: Particular Requirements for Safety Section 2.11: Specification for Gamma Beam Therapy Equipment (IEC 601-2-11: 1987). 35 pp.

BSI BS 5724: Sec 2.11-01. 1989 Amd 1 Medical Electrical Equipment Part 2: Particular Requirements for Safety Section 2.11: Specification for Gamma Beam Therapy Equipment (IEC 601-2-11: 1987) (AMD 6804) November 29, 1991. 36 pp.

BSI BS 5724: Sec 2.11: Supp 1-89. 1989 Medical Electrical Equipment Part 2: Particular Requirements for Safety Section 2.11: Specification for Gamma Beam Therapy Equipment Supplement 1: Revised and Additional Text (IEC 601-2-11: 1988). 15 pp.

BSI BS 5724: Sec 2.11: Supp 1-01. 1989 Amd 1 Medical Electrical Equipment Part 2: Particular Requirements Safety Section 2.11: Specification for Gamma Beam Therapy Equipment Supplement 1: Revised and Additional Text (IEC 601-2-11: 1988) (AMD 6805) November 29, 1991. 16 pp.

BSI BS 5724: Sec 2.11: Supp 2-93. 1993 Medical Electrical Equipment Part 2: Particular Requirements Safety Section 2.11: Specification for Gamma Beam Therapy Equipment Supplement 2: Methods of Test for Radiation Safety (IEC 601-2-11: Amendment 2: 1993) (N). 28 pp.

BSI BS 5724: Sec 2.17-90. 1990 Medical Electrical Equipment Part 2: Particular Requirements for Safety Section 2.17: Remote Controlled Automatically Driven Gamma Ray After-loading Equipment. 31 pp.

BSI BS 5724: Sec 2.17-01. 1990 Amd 1 Medical Electrical Equipment Part 2: Particular Requirements for Safety Section 2.17: Specification for Remote-Controlled Automatically-Driven Gamma-Ray Afterloading Equipment (IEC 601-2-17: 1989) (AMD 7210) September 15, 1992. 33 pp.

CENELEC HD 395.2.11 S1-89. Medical Electrical Equipment-Part 2: Particular Requirements for the Safety of Gamma Beam Therapy Equipment. 3 pp.

CENELEC HD 395.2.11 S2-90. Medical Electrical Equipment Part 2: Particular Requirements for the Safety of Gamma Beam Therapy Equipment. 3 pp.

CENELEC HD 395.2.11 S2-92. Medical Electrical Equipment Part 2: Particular Requirements for the Safety of Gamma Beam Therapy Equipment; (IEC 601-2-11:1987 + A1:1988). 3 pp.

CENELEC HD 395.2.17 S1-92. Medical Electrical Equipment Part 2: Particular Requirements for the Safety of Remote-Controlled Automatically-Driven Gamma-Ray Afterloading Equipment. 6 pp.

CSA CAN/CSA-C22. 2601.2.11-92. Medical Electrical Equipment Part 2: Particular Requirements for the Safety of Gamma Beam Therapy Equipment (IEC 601-2-11:1988); (Gen Instr 1). 60 pp.

IEC 601 Pt 2-11-87. Medical Electrical Equipment Part 2: Particular Requirements for the Safety of Gamma Beam Therapy Equipment First Edition; (Amendment 1-1988) (Amendment 2-1993) (CENELEC HD 395.2.11 S2:1990). 133 pp.

IEC 601 Pt 2-17-89. Medical Electrical Equipment Part 2: Particular Requirements for the Safety of Remote-Controlled Automatically-Driven Gamma-Ray Afterloading Equipment First Edition. 55 pp.

Gamma Ray Spectrometry
See Also: Spectrometry

—Drinking Water
DIN ENGL 38404 Pt 16-89. German Standard Methods for the Examination of Water, Waste Water and Sludge; Physical and Physicochemical Para-meters (Group C); Deter-mination of Radio-nuclides in Drinking Water, Ground Water, Surface Water and Waste Water by Y-Ray Spectro-metry (C 16) (Apr). 7 pp.

—Ground Water
DIN ENGL 38404 Pt 16-89. German Standard Methods for the Examination of Water, Waste Water and Sludge; Physical and Physicochemical Para-meters (Group C); Deter-mination of Radio-nuclides in Drinking Water, Ground Water, Surface Water and Waste Water by Y-Ray Spectro-metry (C 16) (Apr). 7 pp.

—Plutonium Nitrate
CNS J2047-82. Method of Test for Gamma-Emitting Fission Products, Uranium and Thorium in Nuclear-Grade Plutonium Nitrate Solutions by Gamma Ray Spectroscope (Mar)(8594).

Gamma Ray Spectrometry (Cont.)

—Surface Water
DIN ENGL 38404 Pt 16-89. German Standard Methods for the Examination of Water, Waste Water and Sludge; Physical and Physicochemical Para-meters (Group C); Deter-mination of Radio-nuclides in Drinking Water, Ground Water, Surface Water and Waste Water by Y-Ray Spectro-metry (C 16) (Apr). 7 pp.

—Waste Water
DIN ENGL 38404 Pt 16-89. German Standard Methods for the Examination of Water, Waste Water and Sludge; Physical and Physicochemical Para-meters (Group C); Deter-mination of Radio-nuclides in Drinking Water, Ground Water, Surface Water and Waste Water by Y-Ray Spectro-metry (C 16) (Apr). 7 pp.

—Water
CNS J2022-81. Method of Test for Gamma Spectrometry in Water (Jun)(7493).

Gamma Rays
Use For: Gamma Radiation
See Also: Electromagnetic Radiation; Gamma Emission; Gamma Irradiation; Ionizing Radiation; X-Rays

—Calculation
CNS J2027-81. Calculation of Absorbed Dose from Gamma Radiation (Jul)(7700).

—Dosimetry
CNS J2024-81. Method of Test for Absorbed Gamma and Electron Radiation Dose with the Ferrous Sulfate-Cupric Sulfate Dosimeter (Jul)(7697).

CNS J2025-81. Method of Test for Absorbed Gamma and Electron Radiation Dose with the Ceric Sulfate Dosimeter (Jul)(7698).

CNS J2026-81. Method of Test for Absorbed Gamma Radiation Dose in the Fricke Dosimeter (Jul)(7699).

ISO 8963-88. Dosimetry of X and Gamma Reference Radiations for Radiation Protection over the Energy Range from 8 keV to 1,3 MeV First Edition. 11 pp.

—Impurities
CNS J2013-81. Method of Analysis of Radioisotopes Determination of Gamma Ray Impurities with NaI (TL) Detectors (May)(7390).

—Impurities—Germanium Detectors
CNS J2014-81. Method for Analysis of Radioisotopes Determination of Gamma Ray Impurities with Germanium Detector (May)(7391).

—Photographic Dosimeters
JIS K 7559-77. Badge Films for Gamma and Hard X-Rays (R 1985). 9 pp.

—Radiation Detection and Measurement
IEC 1017 Pt 1-91. Portable, Transportable or Installed X or Gamma Radiation Ratemeters for Environmental Monitoring Part 1: Ratemeters First Edition. 63 pp.

—Water
CNS J2006-81. Method of Test for Gamma—Radioactivity of Water (Mar)(7151). 4 pp.

Gang Channel Nuts
Use: Channel Nuts

Gangplanks
Use: Gangways

Gangways
Use For: Gangplanks; Swing Derricks
See Also: Guardrails; Stanchions; Walkways

—Ships
CEN PREN 526-91. Inland Navigation Vessels—Gangways with a Length Not Exceeding 8 m—Requirements, Types. 11 pp.

ISO 4085-79. Shipbuilding—Inland Navigation—Swing Derricks First Edition. 4 pp.

—Ships—Aluminum
BSI BS MA 78-78. 1978 Aluminium Shore Gangways. 12 pp.

ISO 7061-93. Shipbuilding—Aluminium Shore Gangways for Seagoing Vessels Second Edition. 10 pp.

Gantry Cranes
See Also: Cranes; Ground Support Equipment
SAA AS 1418.3-90. Cranes (Including Hoists and Winches) (Known as the SAA Crane Code)—Part 3: Bridge, Gantry and Portal Cranes (Including Container Cranes). 22 pp.

Gantry Cranes (Cont.)

—Air Cargo—Self-Propelled
ISO 6965-82. Aircraft—Self-Propelled Gantry for Lifting Air Cargo Containers and Outside Cargoes—Functional Requirements First Edition. 7 pp.

—Ships
MOD UK NES 113: Part 2-90. Requirements for Mechanical Handling Part 2: Fixed Cranes and Gantries Design Parameters Issue 2 (10.90). 17 pp.

Gap Brazed Joints
Use: Brazed Joints

Gap Gages
See Also: Feeler Gages; Gages
DIN ENGL 7163-66. Workshop Gap Gauges and Check Gauges for ISO Fit Sizes from 1 to 500 mm Nominal Dimension; Gauge Dimensions and Manufacturing Tolerances (Aug). 15 pp.

SAA AS 1655-74. Feeler Gauges (Metric Units) Reconfirmed 1987. 5 pp.

—No Go—Serrated
DIN ENGL 2266-55. Measuring Instruments; Serrated "Not Go" Gap Gauges for Serrations According to DIN 5481 Part 1 (Oct). 1 p.

—No Go—Serrated—Setting Gages
DIN ENGL 2267-55. Measuring Instruments; Serrated Setting Gauges for Serrated "Not Go" Gap Gauges for Serrations According to DIN 5481 Part 1 (Oct). 2 pp.

Gap Gauges
Use: Gap Gages

Garbage Bags
Use For: Refuse Sacks; Trash Bags   See Also: Bags; Waste Disposal

—Plastic
CGSB CAN/CGSB-156.1-M87. Plastic Garbage Bags. 9 pp.

SNZ NZS 7603-79. Specification for Refuse Bags for Local Authority Collection (Low Density Polyethylene) (Reconfirmed 1985). 8 pp.

—Polyethylene
BSI BS 6642-85. 1985 Disposable Plastics Refuse Sacks Made from Polyethylene. 9 pp.

SAA AS 1251.1-82. Polyethylene (Polythene) Garbage Bags—Part 1: Low Density. 12 pp.

Garbage Cans
Use: Waste Containers

Garbage Chutes
Use: Refuse Chutes

Garbage Disposals
See Also: Appliances; Food Service Equipment
CSA CAN/CSA-C22. 2 NO 68-92. Motor-Operated Appliances (Household and Commercial); (Gen Instr 1 Thru 2). 115 pp.

CSA 1169 Bull. Electrical Bulletin 1169 June 27, 1978 to C22.2 NO 68. 2 pp.

—Safety
BSI BS 3456: Sec 202.16-90. 1990 Safety of Household Electrical Appliances Part 202: Particular Requirements Section 202.16: Food Waste Disposers. 24 pp.

BSI BS 3456: Sec 2.30-71. 1971 Testing and Approval of Household Electrical Appliances Part 2: Particular Requirements Section 2.30: Food Waste Disposal Units. 16 pp.

BSI BS 3456: Sec 3.8-79. (WITHDRAWN) 1979 1972-1981 Edition. Specification for Safety of Household Electrical Appliances Part 3: Complete Particular Specifications Section 3.8: Food Waste Disposal Units. 46 pp.

CENELEC HD 259-76. Particular Specification for Food Waste Disposal Units. 6 pp.

CENELEC HD 259.2-77. Particular Specification for Food Waste Disposal Units. 5 pp.

CENELEC EN 60 335-2-16-89. Safety of Household and Similar Electrical Appliances Part 2: Particular Requirements for Food Waste Disposers. 9 pp.

IEC 335 Pt 2-16-86. Safety of Household and Similar Electrical Appliances Part 2: Particular Requirements for Food Waste Disposers Third Edition. 37 pp.

—Ships
MOD UK NES 721: Part 1-91. Garbage Disposal in Surface Ships Part 1: General Requirements Issue 2 (09.91). 33 pp.

MOD UK NES 721: Part 2-91. Garbage Disposal in Surface Ships Part 2: Requirements for Equipment Issue 2 (09.91). 20 pp.

Garbage

Garbage Hoppers
Use: Refuse Hoppers

Garbage Trucks
Use: Waste Disposal Vehicles

Gardan GY 80 Aircraft
See Also: Aircraft

—**Antenna Positions**
CAA. Gardan GY.80 (Approved Aerial Positions). 1 p.

—**Foreign Airworthiness Directives**
CAA. Gardan GY80 Series Aircraft (Foreign Airworthiness Directives). 1 p.

Garden Equipment
Use: Lawn and Garden Equipment

Garden Forks
Use: Pitchforks

Garden Hoses
Use: Lawn and Garden Hoses

Garden Shears
Use: Shears (Hand Tools)—Garden

Garlic
Use For: Green Garlic *See Also:* Food; Spices; Vegetables

—**Defects**
CNS.N4037-72. Method of Test for Garlic (Jun)(3358). 2 pp.

—**Dried**
BSI BS 6194-81. 1981 Dehydrated Garlic. 10 pp.
ISO 5560-83. Dehydrated Garlic—Specification Second Edition. 8 pp.

—**Dried—Sulfur Compound Content**
BSI BS 4585: Part 11-83. 1983 Methods of Test for Spices and Condiments Part 11: Determination of Volatile Organic Sulphur Compounds in Dehydrated Garlic. 10 pp.
ISO 5567-82. Dehydrated Garlic—Determination of Volatile Organic Sulphur Compounds First Edition. 5 pp.

—**Grading**
CNS N1012-82. Grading and Packaging of Garlic (Jan)(1119). 3 pp.

—**Inspection**
CNS N4037-72. Method of Test for Garlic (Jun)(3358). 2 pp.

—**Packaging**
CNS N1012-82. Grading and Packaging of Garlic (Jan)(1119). 3 pp.

—**Sampling**
CNS N4037-72. Method of Test for Garlic (Jun)(3358). 2 pp.

—**Storage**
ISO 6663-83. Garlic—Guide to Cold Storage First Edition. 4 pp.

Garment Racks
Use: Clothing Racks

Garment Steamers
Use For: Steamers (Garment) *See Also:* Steamers

—**Acceptance Testing**
SAA AS 3312-89. Approval and Test Specification—Particular Requirements for Garment or Fabric Steamers Amdt 1 January 1991 Amdt 2 March 1992 (in Professional Package 28).

Garments
Scope Note: See listings under Clothing

Garnett Wire
See Also: Textile Machinery; Wire

—**Spinning (Textiles)**
CNS L2077-83. Garnett Wires for Cotton Spinning System (Apr)(10175).
JIS L 5134-81. Garnett Wires for Cotton Spinning.
JIS L 5150-81. Garnett Wires for Worsted Spinning.
JIS L 5151-81. Garnett Wires for Woollen Spinning.

Garnierite

—**Sampling—Moisture Content**
JIS M 8109-74. Method for Sampling and Method of Determination of Moisture Content of Garnierite Nickel Ores.

Gas Adsorption
See Also: Adsorption

—**Powders**
JIS Z 8830-90. Determination of the Specific Surface Area of Powders by Gas Adsorption Methods. 13 pp.

—**Solids**
BSI BS 7591: Part 2-92. 1992 Porosity and Pore Size Distribution of Materials Part 2: Method of Evaluation by Gas Adsorption (H). 21 pp.

Gas Alarms
See Also: Alarm Systems; Gas Detectors; Mining Equipment; Warning Systems

—**Hot Wire**
CNS M2055-80. Hot Wire Type Gas Alarms (Jul)(5902).
JIS M 7626-65. Hot Wire Type Gas Alarms.

—**Hydrogen Sulfide**
JIS T 8205-86. Hydrogen Sulfide Indicator, Hydrogen Sulfide Alarm and Hydrogen Sulfide Indicator/Alarm. 15 pp.

—**Interferometer**
CNS M2054-80. Interferometer Type Gas Alarms (Jul)(5901).
JIS M 7625-77. Interferometer Type Gas Alarms.

—**Liquefied Petroleum Gas**
CNS Z2039-81. Alarm Equipments of Liquefied Petroleum Gas Leakage (Jul)(7760).

Gas Analysis
See Also: Calibration Gases; Chemical Analysis; Gas Analyzers; Gas Chromatography; Gas Chromatography-Mass Spectrometry; Infrared Analysis; Manometry; Permeation Method; Quantitative Analysis; Volumetric Analysis

—**Aircraft—Technical Manuals**
NATO STANAG 3977 ED 1 AMD 0-91. Manual of Techniques of Sampling and Analysis of Gases and Liquefied Gases for Aircraft Servicing—AEP-6. 5 pp.
NATO STANAG 3977 ED 1 AMD 1-91. Manual of Techniques of Sampling and Analysis of Gases and Liquefied Gases for Aircraft Servicing—AEP-6. 6 pp.

—**Calibration—Gas Mixtures**
ISO 6711-81. Gas Analysis—Checking of Calibration Gas Mixtures by a Comparison Method First Edition. 4 pp.

—**Calorific Value**
JIS K 2301-92. Fuel Gases and Natural Gas—Methods for Chemical Analysis and Testing. 116 pp.

—**Carbon Dioxide Content**
ISO TR6567 Pt 1-81. Gas Analysis—Determination of Carbon Dioxide—Part 1: General Guidance for the Choice of Methods First Edition. 9 pp.

—**Concentration—Glossaries**
DIN ENGL 1310-84. Composition of (Gaseous, Liquid and Solid) Mixtures; Concepts, Symbols (Feb). 3 pp.

—**Concentration—Symbols**
DIN ENGL 1310-84. Composition of (Gaseous, Liquid and Solid) Mixtures; Concepts, Symbols (Feb). 3 pp.

—**Density**
JIS K 2301-92. Fuel Gases and Natural Gas—Methods for Chemical Analysis and Testing. 116 pp.

—**Electrical Insulating Oils**
BSI BS 5574-78. (WITHDRAWN) 1978 Guide for the Sampling of Gases and of Oil from Oil-Filled Electrical Equipment and for the Analysis of Free and Dissolved Gases (Superseded by BS EN 60567: 1993). 32 pp.
BSI BS 5800-79. 1979 Guide for the Interpretation of the Analysis of Gases in Transformers and Other Oil-Filled Electrical Equipment in Service. 15 pp.
BSI BS EN 60567-93. 1993 Guide for the Sampling of Gases and of Oil from Oil-Filled Electrical Equipment and for the Analysis of Free and Dissolved Gases (IEC 567: 1992) (Supersedes BS 5574: 1978). 54 pp.
BSI BS EN 61181-93. 1993 Impregnated Insulating Materials—Application of Dissolved Gas Analysis (DGA) to Factory Tests on Electrical Equipment (IEC 1181: 1993) (Q). 15 pp.
CENELEC HD 397-79. Interpretation of the Analysis of Gases in Transformers and Other Oil-Filled Electrical Equipment in Service. 1 p.
CENELEC EN 60567-92. Guide for the Sampling of Gases and of Oil From Oil-Filled Electrical Equipment and for the Analysis of Free and Dissolved Gases. 5 pp.

Gas Analysis (Cont.)

—**Electrical Insulating Oils** *(Cont.)*
CENELEC PREN 61181-92. Impregnated Insulating Materials—Application of Dissolved Gas Analysis (DGA) to Factory Tests on Electrical Equipment. 13 pp.
IEC 567-92. Guide for the Sampling of Gases and of Oil from Oil-Filled Electrical Equipment and for the Analysis of Free and Dissolved Gases Second Edition. 85 pp.
IEC 599-78. Interpretation of the Analysis of Gases in Transformers and Other Oil-Filled Electrical Equipment in Service First Edition. 28 pp.
IEC 1181-93. Impregnated Insulating Materials—Application of Dissolved Gas Analysis (DGA) to Factory Tests on Electrical Equipment First Edition. 21 pp.

—**Fire Effluents**
ISO TR9122 Pt 3-93. Toxicity Testing of Fire Effluents—Part 3: Methods for the Analysis of Gases and Vapours in Fire Effluents First Edition. 41 pp.

—**Glossaries**
ISO 7504-84. Gas Analysis—Vocabulary First Edition. 32 pp.

—**Plastics—Combustion**
JIS K 7217-83. Analytical Method for Determining Gases Evolved from Burning Plastics. 15 pp.

—**Plywood**
CEN PREN 1084-93. Plywood—Formaldehyde Release Classes. 5 pp.

—**Sulfur Dioxide Content**
ISO TR6566 Pt 1-80. Gas Analysis—Determination of Sulphur Dioxide—Part 1: General Guidance for the Choice of Methods First Edition; (Amendment Slip-1981). 12 pp.

—**Wood Panels**
CEN PREN 717-2-92. Wood-Based Panels—Determination of Formaldehyde Emission—Part 2: Formaldehyde Release by the Gas Analysis Method. 11 pp.

Gas Analyzers
See Also: Carbon Dioxide Analyzers; Carbon Monoxide Analyzers; Flue Gas Analyzers; Gas Analysis; Infrared Analyzers; Nitrogen Oxide Analyzers; Nitrometers; Orsat Analyzers; Oxygen Analyzers; Sulfur Dioxide Analyzers
ISO 8158-85. Evaluation of the Performance Characteristics of Gas Analysers First Edition. 17 pp.

—**Exhaust Gases**
BSI BS 1756: Part 3-71. 1971 Methods for the Sampling and Analysis of Flue Gases Part 3: Analysis by the Haldane Apparatus. 37 pp.
BSI BS 1756: Part 3-01. 1971 Amd 1 Methods for the Sampling and Analysis of Flue Gases Part 3: Analysis by the Haldane Apparatus (AMD 7754) August 15, 1993. 38 pp.
BSI BS 3048-58. 1958 Code for the Continuous Sampling and Automatic Analysis of Flue Gases. Indicators and Recorders. 77 pp.
BSI BS 3048-01. 1958 Amd 1 Code for the Continuous Sampling and Automatic Analysis of Flue Gases. Indicators and Recorders (AMD 7752) August 15, 1993 (Q). 78 pp.
CNS D4009-83. Reflection Type Smokemeters for Measuring Smoke Concentration of Exhaust Gas for Diesel Automobiles (Jan)(9845).
CNS D4010-83. Light Extinction Type Smokemeters for Measuring Smoke Concentration of Exhaust Gas for Diesel Automobiles (Jan)(9846).
ISO 3930-76. Road Vehicles—Carbon Monoxide Analyser Equipment—Technical Specifications First Edition. 4 pp.
JIS D 1101-85. Diesel Engine Smoke Measurement. 5 pp.
JIS D 8004-86. Reflection Type Smokemeters for Automobile Diesel Engine.
JIS D 8004-71. Reflection Type Smokemeters for Measuring Carbon Concentration of Exhaust Smoke for Diesel Automobiles. 12 pp.
JIS D 8005-73. Light Extinction Type Smokemeters for Measuring Automotive Diesel Exhaust.
JIS D 8006-83. Simplified Carbonmonoxide Testers for Automobiles. 9 pp.
JIS K 0055-86. General Rules for Calibration Method of Gas Analyzer. 10 pp.

—**Exhaust Gases—Air Monitoring**
JIS K 0055-86. General Rules for Calibration Method of Gas Analyzer. 10 pp.

—**Fluorine**
JIS B 7958-80. Continuous Analyzers for Fluorine Compounds in Ambient Air.

INTERNATIONAL AND NON-U.S. NATIONAL STANDARDS
SUBJECT INDEX
Gas

Gas Analyzers *(Cont.)*

—Hydrocarbons
- JIS B 7956-79. Continuous Analyzers for Hydrocarbons in Ambient Air.

—Nitrogen
- JIS B 7953-81. Continuous Analysers for Oxides of Nitrogen in Ambient Air.
- JIS B 7953-79. Continuous Analyzers for Oxides of Nitrogen in Ambient Air. 14 pp.
- JIS B 7982-88. Continuous Analyzers for Oxides of Nitrogen in Flue Gas. 25 pp.

—Oxidizers
- JIS B 7957-92. Continuous Analyzers for Oxidants in Ambient Air. 29 pp.

—Sampling/Transfer Equipment
- ISO 6712-82. Gas Analysis—Sampling and Transfer Equipment for Gases Supplying an Analytical Unit First Edition. 21 pp.

—Ventilation Equipment
- IEC 79 Pt 16-90. Electrical Apparatus for Explosive Gas Atmospheres Part 16: Artificial Ventilation for the Protection of Analyzer(s) Houses First Edition. 44 pp.

Gas Appliances
Use: Appliances

Gas Boilers
Use: Boilers

Gas Burners
See Also: Flue Exhausters
- DIN ENGL 4788 Pt 1-77. Gas Burners; Gas Burners Without Blowers (June). 15 pp.

—Atmospheric—Boilers
- BSI BS 5978: Part 2-89. 1989 Amd 1 Safety and Performance of Gas-Fired Hot Water Boilers (60kW to 2MW Input) Part 2: Additional Requirements for Boilers with Atmospheric Burners. 16 pp.
- CEN PREN 297-89. Gas-Fired Central Heating Boilers Fitted with Atmospheric Burners; Type B11 Boilers of Nominal Heat Input Not Exceeding 70kw. 105 pp.
- CEN PREN 483-91. Gas-Fired Central Heating Boilers Fitted with Atmospheric Burners—Type C Boilers of Nominal Heat Input Not Exceeding 70kw. 158 pp.
- CEN PREN 625-92. Gas-Fired Central Heating Boilers Fitted with Atmospheric Burners—Specific Requirements for Combination Boilers of a Nominal Heat Input Not Exceeding 70 kW. 25 pp.
- CEN PREN 677-92. Gas-Fired Central Heating Boilers—Specific Requirements for Condensing Boilers with a Nominal Heat Input Not Exceeding 70 kW. 16 pp.

—Atmospheric—Boilers—Safety
- BSI BS 5978: Part 2-89. 1989 Amd 1 Safety and Performance of Gas-Fired Hot Water Boilers (60kW to 2MW Input) Part 2: Additional Requirements for Boilers with Atmospheric Burners. 16 pp.
- CEN PREN 297-89. Gas-Fired Central Heating Boilers Fitted with Atmospheric Burners; Type B11 Boilers of Nominal Heat Input Not Exceeding 70kw. 105 pp.
- CEN PREN 483-91. Gas-Fired Central Heating Boilers Fitted with Atmospheric Burners—Type C Boilers of Nominal Heat Input Not Exceeding 70kw. 158 pp.

—Automatic
- BSI BS 5885: Part 1-88. 1988 Amd 1 Automatic Gas Burners Part 1: Specification for Burners with Input Rating 60kW and Above (AMD 6631) December 21, 1990. 25 pp.
- BSI BS 5885: Part 2-87. 1987 Automatic Gas Burners Part 2: Packaged Burners with Input Rating 7.5kW up to but Excluding 60kW. 14 pp.

—Automatic Control Equipment
- BSI BS EN 60730-2-5-92. 1992 Automatic Electrical Controls for Household and Similar Use Part 2: Particular Requirements Section 2.5: Automatic Electrical Burner Control Systems. 30 pp.
- CEN PREN 298-89. Automatic Burner Control Systems for Gas Burners and Gas Burning Appliances with or Without Fans. 39 pp.
- CENELEC EN 60730-2-5-91. Automatic Electrical Controls for Household and Similar Use Part 2: Particular Requirements for Automatic Electrical Burner Control Systems. 7 pp.
- IEC 730 Pt 2-5-90. Automatic Electrical Controls for Household and Similar Use Part 2: Particular Requirements for Automatic Electrical Burner Control Systems First Edition. 48 pp.

Gas Burners *(Cont.)*

—Automatic—Forced Draft—Design
- DIN ENGL 4788 Pt 2-90. Induced and Forced Draught Gas Burners; Concepts, Safety Requirements, Testing and Marking (Feb). 14 pp.

—Automatic—Forced Draft—Safety
- DIN ENGL 4788 Pt 2-90. Induced and Forced Draught Gas Burners; Concepts, Safety Requirements, Testing and Marking (Feb). 14 pp.

—Automatic—Induced Draft—Design
- DIN ENGL 4788 Pt 2-90. Induced and Forced Draught Gas Burners; Concepts, Safety Requirements, Testing and Marking (Feb). 14 pp.

—Automatic—Induced Draft—Safety
- DIN ENGL 4788 Pt 2-90. Induced and Forced Draught Gas Burners; Concepts, Safety Requirements, Testing and Marking (Feb). 14 pp.

—Codes
- CNS B1356-89. Boiler Code (Gas Burning Equipment) (Jul)(12554).

—Combustion Control Equipment
- BSI BS 6505-84. 1984 Amd 2 Rubber-Type Materials Used for Controls Components for Use with 1st, 2nd and 3rd Family Gases. 13 pp.
- BSI BS 6505-03. 1984 Amd 3 Rubber-Type Controls Components for Use with 1st, 2nd and 3rd Family Gases (AMD 7178) July 15, 1992. 15 pp.
- CSA CAN/CSA-C22. 2 NO 199-M89. Combustion Safety Controls and Solid-State Igniters for Gas-and Oil-Burning Equipment; (Gen Instr 1). 94 pp.

—Cooking Appliances
- BSI BS 5314: Part 2-76. (WITHDRAWN) 1976 Amd 2 Gas Heated Catering Equipment Part 2: Boiling Burners (AMD 5397) July 31, 1987 (Superseded by BS EN 203-1: 1993). 27 pp.

—Electrical Components
- CSA C22.2 NO 3-M1988. Electrical Features of Fuel-Burning Equipment (R 1993); (Gen Instr 1 Thru 2). 64 pp.

—Flammability Testing
- DIN ENGL 50051-77. Testing of Materials; Burning Behaviour of Materials; Burner (Feb). 3 pp.

—Forced Draft
- CEN PREN 676-92. Forced Draught Burners for Gaseous Fuels. 49 pp.

—Forced Draft—Boilers
- BSI BS 5978: Part 3-89. 1989 Safety and Performance of Gas-Fired Hot Water Boilers (60 kW to 2MW Input) Part 3: Additional Requirements for Boilers with Forced or Induced Draught Burners. 4 pp.

—Forced Draft—Boilers—Safety
- BSI BS 5978: Part 3-89. 1989 Safety and Performance of Gas-Fired Hot Water Boilers (60 kW to 2MW Input) Part 3: Additional Requirements for Boilers with Forced or Induced Draught Burners. 4 pp.

—Igniters
- BSI BS 3929-65. 1965 Domestic Solid Fuel Ignition Pokers and Portable Undergrate Ignition Burners for Use with Commerical Butane. 9 pp.

—Igniters—Solid State
- CSA CAN/CSA-C22. 2 NO 199-M89. Combustion Safety Controls and Solid-State Igniters for Gas-and Oil-Burning Equipment; (Gen Instr 1). 94 pp.

—Ignition Cables
- CSA C22.2 NO 17-1973. Cable for Luminous-Tube Signs and for Oil-and Gas-Burner Ignition Equipment (R 1992); (Gen Instr 1) (Amd 1 February 1981) (Amd 2-8 March 1988). 28 pp.

—Ignition Transformers
- CSA C22.2 NO 13-1962. Transformers for Luminous-Tube Signs, Oil-or Gas-Burner Ignition Equipment, Cold-Cathode Interior Lighting (R 1992). 46 pp.
- CSA 1169 Bull. Electrical Bulletin 1169 June 27, 1978 to C22.2 NO 13. 2 pp.
- CSA 1238 Bull. Electrical Bulletin 1238 September 17, 1979 to C22.2 NO 13. 2 pp.

—Induced Draft—Boilers
- BSI BS 5978: Part 3-89. 1989 Safety and Performance of Gas-Fired Hot Water Boilers (60 kW to 2MW Input) Part 3: Additional Requirements for Boilers with Forced or Induced Draught Burners. 4 pp.

—Induced Draft—Boilers—Safety
- BSI BS 5978: Part 3-89. 1989 Safety and Performance of Gas-Fired Hot Water Boilers (60 kW to 2MW Input) Part 3: Additional Requirements for Boilers with Forced or Induced Draught Burners. 4 pp.

Gas Burners *(Cont.)*

—Installation
- BSI CP 335: Part 1-73. 1973 Selection and Installation of Miscellaneous Town Gas Appliances Part 1: Domestic Laundering and Miscellaneous Appliances. 17 pp.

—Liquefied Petroleum
- CNS B7050-81. Method of Test for Liquescent Petroleum Gas Burner (May)(3662).

—Liquefied Petroleum—Outdoor
- CEN PREN 497-91. Boiling Burners Burning Liquefied Petroleum Gases for Outdoor Use. 41 pp.

—Mechanical Draft
- SAA AS 1853-83. Automatic Oil and Gas Burners—Mechanical Draught. 24 pp.

—Safety
- DIN ENGL 4756-86. Gas Burning Equipment in Heating Plant; Safety Requirements (Feb). 8 pp.

—Stop Valves
- BSI BS EN 161-91. 1991 Automatic Shut-Off Valves for Gas Burners and Gas Appliances. 27 pp.
- CEN EN 161-91. Automatic Shut-Off Valves for Gas Burners and Gas Appliances. 38 pp.

—Taps (Valves)
- CEN PREN 1106-93. Manually Operated Taps for Gas Burning Appliances. 32 pp.

Gas Chromatography
Use For: Charcoal Tube/Gas Chromatographic Method; Gas Chromatography Analysis; Gas Liquid Chromatography *See Also:* Gas Analysis; Gas Chromatography-Mass Spectrometry; Infrared Analysis
- BSI BS 5443-77. 1977 Recommendations for a Standard Layout for Methods of Chemical Analysis by Gas Chromatography. 16 pp.
- BSI BS 5443-01. 1977 Amd 1 Recommendations for a Standard Layout for Methods of Chemical Analysis by Gas Chromatography (AMD 7605) May 15, 1993 (H). 18 pp.
- CNS K0021-82. General Rules for Analytical Method in Gas Chromatography (Jul)(9178).
- ISO 2718-74. Standard Layout for a Method of Chemical Analysis by Gas Chromatography First Edition. 9 pp.
- JIS K 0114-82. General Rules for Gas Chromatographic Analysis. 24 pp.

—Acrylate Esters
- CNS K6610-80. Purity of Acrylate Esters by Gas Chromatography (Dec)(6672).

—Air
- BSI BS 6069: Sec 3.1-89. 1989 Characterization of Air Quality Part 3: Workplace Atmospheres Section 3.1: Method for the Determination of Vinyl Chloride Using a Charcoal Tube and a Gas Chromatograph. 12 pp.
- BSI BS 6069: Sec 3.3-91. 1991 Characterization of Air Quality Part 3: Workplace Atmospheres Section 3.3: Method for the Determination of Vaporous Chlorinated Hydrocarbons by Charcoal Tube/Solvent Desorption/ Gas Chromatography (ISO 9486: 1991). 18 pp.
- BSI BS 6069: Sec 3.4-91. 1991 Characterization of Air Quality Part 3: Workplace Atmospheres Section 3.4: Method for the Determination of Vaporous Aromatic Hydrocarbons by Charcoal Tube/Solvent Desorption/ Gas Chromatography (ISO 9487: 1991). 18 pp.
- CSA CAN3-Z223.25-M86. Method for the Measurement of Vinyl Chloride in Air; (Gen Instr 1). 27 pp.
- ISO 8186-89. Ambient Air—Determination of the Mass Concentration of Carbon Monoxide—Gas Chromatographic Method First Edition. 12 pp.
- ISO 8762-88. Workplace Air—Determination of Vinyl Chloride—Charcoal Tube/Gas Chromatographic Method First Edition. 12 pp.
- ISO 9486-91. Workplace Air—Determination of Vaporous Chlorinated Hydrocarbons—Charcoal Tube/Solvent Desorption / Gas Chromatographic Method First Edition. 15 pp.
- ISO 9487-91. Workplace Air—Determination of Vaporous Aromatic Hydrocarbons—Charcoal Tube/ Solvent Desorption/ Gas Chromatographic Method First Edition. 15 pp.

—Alcohols
- BSI BS 5711: Part 11-79. 1979 Methods of Sampling and Test for Glycerol Part 11: Determination of Propane, 1,3-Diol Content: Gas Chromatographic Method. 4 pp.

—Ammonia
- ISO 7104-85. Liquefied Anhydrous Ammonia for Industrial Use—Determination of Water Content—Gas Chromatographic Method First Edition. 6 pp.

INDUSTRY STANDARDS

Gas Chromatography (Cont.)

—Animal Fats
BSI BS 684: Sec 2.38-92. 1992 Methods of Analysis of Fats and Fatty Oils Part 2: Other Methods Section 2.38: Determination of the Proportions of Individual Sterols in the Sterol Fraction (ISO 6799: 1991). 12 pp.
BSI BS 684: Sec 2.38-83. 1983 Methods of Analysis of Fats and Fatty Oils Part 2: Other Methods Section 2.38: Determination of the Proportions of Individual Sterols in the Sterol Fraction. 8 pp.
ISO 6799-83. Animal and Vegetable Fats and Oils—Determination of Composition of the Sterol Fraction—Method by Gas-Liquid Chromatography First Edition. 7 pp.

—Animal Oils
BSI BS 684: Sec 2.38-92. 1992 Methods of Analysis of Fats and Fatty Oils Part 2: Other Methods Section 2.38: Determination of the Proportions of Individual Sterols in the Sterol Fraction (ISO 6799: 1991). 12 pp.
BSI BS 684: Sec 2.38-83. 1983 Methods of Analysis of Fats and Fatty Oils Part 2: Other Methods Section 2.38: Determination of the Proportions of Individual Sterols in the Sterol Fraction. 8 pp.
ISO 6463-82. Animal and Vegetable Fats and Oils—Determination of Butylhydroxyanisole (BHA) and Butylhydroxytoluene (BHT)—Gas-Liquid Chromatography Method First Edition. 5 pp.
ISO 6799-83. Animal and Vegetable Fats and Oils—Determination of Composition of the Sterol Fraction—Method by Gas-Liquid Chromatography First Edition. 7 pp.

—Aromatic Hydrocarbons
CNS K6497-80. Nonaromatic Hydrocarbon in Monocyclic Aromatic Hydrocarbon by Gas Chromatography (May)(5603).

—Beverages
CNS N6180-83. Method of Test for Beverage: Determination of Methanol Content (May)(10291). 1 p.
CNS N6181-82. Method of Test for Beverage—Determination of Ethanol Content (May)(10292). 1 p.

—Butadienes
ISO 6378-81. Butadiene for Industrial Use—Determination of Hydrocarbon Impurities—Gas Chromatographic Method First Edition. 9 pp.
ISO 6792-82. Butadiene for Industrial Use—Determination of Oxygen and Argon in the Gaseous Phase Above Liquid Butadiene—Gas Chromatographic Method First Edition. 4 pp.
ISO 7381-86. Butadiene for Industrial Use—Determination of Oligomers—Gas Chromatographic Method First Edition. 10 pp.

—Butanes
BSI BS 7276-90. 1990 Method for Gas Chromatographic Analysis of Liquified Petroleum Gas. 16 pp.
CEN PREN 242-85. Commercial Propane and Butane (LPG); Determination of Hydrocarbons Content by Gas Chromatographic Method. 24 pp.
ISO 7941-88. Commercial Propane and Butane—Analysis by Gas Chromatography First Edition. 15 pp.

—Calculation Methods
CNS K6390-76. Calibration and Calculation of Gas Chromatography (Jun)(3949).

—Calibration
CNS K6390-76. Calibration and Calculation of Gas Chromatography (Jun)(3949).

—Chlorofluorocarbons
BSI BS 5598: Part 11-83. 1983 Methods of Sampling and Test for Halogenated Hydrocarbons Part 11: Determination of 'Inert' Gas Content of Chlorofluorohydrocarbons Gas Chromatographic Method, General Purposes. 7 pp.
ISO 5918-82. Chlorofluorohydrocarbons for Industrial Use—Determination of Inert Gas Content—Gas Chromatographic Method—General Principles First Edition. 6 pp.
ISO 5921-82. Chlorofluorohydrocarbons for Industrial Use—Analysis by Gas Chromatography—General Principles First Edition. 9 pp.

—Chloromethanes
BSI BS 5598: Part 10-83. 1983 Methods of Sampling and Test for Halogenated Hydrocarbons Part 10: Determination of Impurities in Methyl Chloride, Gas Chromatographic Method. 9 pp.
ISO 5916-82. Methyl Chloride for Industrial Use—Determination of Impurities Chromatographic Methods First Edition. 8 pp.

Gas Chromatography (Cont.)

—Cigarette Smoke
BSI BS 5202: Part 12-91. 1991 Methods for Chemical Analysis of Tobacco and Tobacco Products Part 12: Determination of Nicotine in Smoke Condensate of Cigarettes (Gas-Chromatographic Method) (ISO 10315: 1991). 11 pp.
BSI BS 5202: Part 12-01. 1991 Amd 1 Methods for Chemical Analysis of Tobacco and Tobacco Products Part 12: Determination of Nicotine in Smoke Condensate of Cigarettes (Gas-Chromatographic Method) (ISO 10315: 1991) (AMD 7155) July 15, 1992. 12 pp.
BSI BS 5202: Part 13-91. 1991 Methods for Chemical Analysis of Tobacco and Tobacco Products Part 13: Determination of Water in Smoke Condensate of Cigarettes (Gas-Chromatographic Method) (W) (ISO 10362-1: 1991). 11 pp.
BSI BS 5202: Part 13-01. 1991 Amd 1 Methods for Chemical Analysis of Tobacco and Tobacco Products Part 13: Determination of Water in Smoke Condensate of Cigarettes (Gas-Chromatographic Method) (ISO 10362-1: 1991) (AMD 7156) July 15, 1992. 12 pp.
ISO 10315-91. Cigarettes—Determination of Nicotine in Smoke Condensates—Gas-Chromatographic Method First Edition. 9 pp.
ISO 10362 Pt 1-91. Cigarettes—Determination of Water in Smoke Condensates—Part 1: Gas-Chromatographic Method First Edition. 8 pp.

—Cyclohexane
CNS K6651-81. Method of Test for Purity and Benzene Content of Cyclohexane by Gas Chromatography (Mar)(7170).

—Dairy Products
CNS N6203-86. Method of Detection for Vegetable Fat in Milk Fat (Oct)(11737). 3 pp.

—Detergents
BSI BS 3762: Sec 3.24-88. 1988 Analysis of Formulated Detergents Part 3: Quantitative Test Methods Section 3.24: Methods for Determination of Low Molecular Mass Alcohols Content. 6 pp.

—Diesel Motor Oils
DIN ENGL 51380-90. Determination of Readily Volatile Components in Used Automotive Engine Oils by Gas Chromatography (Nov). 2 pp.

—Electrical Insulating Oils
BSI PD 6528-90. 1990 Method for Verifying Accuracy of Tan Delta Measurements Applicable to Capacitors. 9 pp.
IEC 996-89. Method for Verifying Accuracy of Tan Delta Measurements Applicable to Capacitors First Edition. 11 pp.

—Essential Oils
ISO 7355-85. Oils of Sassafras and Nutmeg—Determination of Safrole and cis-and trans-isosafrole Content—Gas Chromatographic Method on Packed Columns First Edition. 6 pp.
ISO 7356-85. Oil of Thujone-Containing Artemisia and Oil of Sage (Salvia Officinalis Linnaeus)—Determination of Alpha Particle and Beta Ray-Thujone Content—Gas Chromatographic Method on Packed Columns First Edition. 6 pp.
ISO 7357-85. Oil of Calamus—Determination of cis-Beta Ray-Asarone Content—Gas Chromatographic Method on Packed Columns First Edition. 5 pp.
ISO 7359-85. Essential Oils—Analysis by Gas Chromatography on Packed Columns—General Method First Edition. 9 pp.
ISO 7609-85. Essential Oils—Analysis by Gas Chromatography on Capillary Columns—General Method First Edition. 9 pp.

—Ethylene
ISO 6379-81. Ethylene for Industrial Use—Determination of Hydrocarbon Impurities—Gas Chromatographic Method First Edition. 6 pp.
ISO 6381-81. Ethylene and Propylene for Industrial Use—Determination of Traces of Carbon Monoxide and Carbon Dioxide—Gas Chromatographic Method First Edition. 8 pp.
ISO 8174-86. Ethylene and Propylene for Industrial Use—Determination of Acetone, Acetonitrile, Propan-2-ol and Methanol—Gas Chromatographic Method First Edition. 8 pp.

—Exhaust Gases
JIS K 0109-82. Methods for Determination of Hydrogen Cyanide in Exhaust Gas. 23 pp.

—Fats
BSI BS 684: Sec 2.36-83. 1983 Methods of Analysis of Fats and Fatty Oils Part 2: Other Methods Section 2.36: Determination of Butylhydroxyanisole (BHA) and Butylhydroxytoluene (BHT). 4 pp.
ISO 6463-82. Animal and Vegetable Fats and Oils—Determination of Butylhydroxyanisole (BHA) and Butylhydroxytoluene (BHT)—Gas-Liquid Chromatographic Method First Edition. 5 pp.

Gas Chromatography (Cont.)

—Fatty Acids
DIN ENGL 55957-83. Binders for Paints and Varnishes; Determination of Fatty Acids by Gas Chromatography (Oct). 4 pp.

—Fatty Oils
BSI BS 684: Sec 2.36-83. 1983 Methods of Analysis of Fats and Fatty Oils Part 2: Other Methods Section 2.36: Determination of Butylhydroxyanisole (BHA) and Butylhydroxytoluene (BHT). 4 pp.

—Food
CNS N6190-84. Method of Test for Preservatives in Food (Jun)(10949). 8 pp.
CNS N6191-86. Method of Test for Artificial Sweeteners in Food (Sep)(10950). 4 pp.
EC 78/142/EEC-78. Council Directive on the Approximation of the Laws of the Member States Relating to Materials and Articles Which Contain Vinyl Chloride Monomer and Are Intended to Come into Contact with Foodstuffs. 3 pp.
EC 81/432/EEC-81. Commission Directive Laying Down the Community Method of Analysis for the Official Control of Vinyl Chloride Released by Materials and Articles into Foodstuffs. 6 pp.

—Food Contacting Materials
EC 78/142/EEC-78. Council Directive on the Approximation of the Laws of the Member States Relating to Materials and Articles Which Contain Vinyl Chloride Monomer and Are Intended to Come into Contact with Foodstuffs. 3 pp.
EC 80/766/EEC-80. Commission Directive Laying Down the Community Method of Analysis for the Official Control of the Vinyl Chloride Monomer Level in Materials and Articles Which are Intended to Come into Contact with Foodstuffs. 5 pp.

—Gaseous Fuels
BSI BS 3156: Part 4-69. (OBSOLESCENT) 1969 Method for the Analysis of Fuel Gases Part 4: Gas Chromatographic Analysis. 24 pp.

—Gasoline
CGSB CAN/CGSB-3.0 NO.14.3-M91. Methods of Testing Petroleum and Associated Products Standard Test Method for the Identification of Hydrocarbon Components in Automotive Gasoline Using Gas Chromatography. 32 pp.
CNS K6326-72. Method of Test for C2 Through C5 Hydrocarbons in Gasolines by Gas Chromatography Determination (Oct)(3387).
DIN ENGL 51380-90. Determination of Readily Volatile Components in Used Automotive Engine Oils by Gas Chromatography (Nov). 2 pp.
DIN ENGL 51413 Pt 1-84. Analysis of Liquid Petroleum Products by Gas Chromatography; Determination of Alcohol Content (Nov). 3 pp.
DIN ENGL 51413 Pt 2-84. Analysis of Liquid Petroleum Products by Gas Chromatography; Determination of Benzene Content (Nov). 2 pp.

—Glossaries
BSI BS 3282-69. 1969 Glossary of Terms Relating to Gas Chromatography. 20 pp.
JIS K 0214-83. Technical Terms for Analytical Chemistry (Chromatography Part).

—Lemon Oil
ISO 7611-85. Oils of Lemon and Petitgrain Citronnier, and Oil of Lime Obtained by a Mechanical Process—Determination of Citral (Neral + Geranial) Content—Gas Chromatographic Method on Capillary Columns First Edition. 7 pp.

—Lime Oil
ISO 7611-85. Oils of Lemon and Petitgrain Citronnier, and Oil of Lime Obtained by a Mechanical Process—Determination of Citral (Neral + Geranial) Content—Gas Chromatographic Method on Capillary Columns First Edition. 7 pp.

—Liquefied Petroleum Gas
DIN ENGL 51619-85. Testing of Petroleum Products; Determination of the Composition of Liquefied Petroleum Gas; Analysis by Gas Chromatography (Aug). 2 pp.

—Methyl Esters
BSI BS 684: Sec 2.35-90. 1990 Methods of Analysis of Fats and Fatty Oils Part 2: Other Methods Section 2.35: Analysis by Gas Chromatography of Methyl Esters of Fatty Acids. 12 pp.
ISO 5508-90. Animal and Vegetable Fats and Oils—Analysis by Gas Chromatography of Methyl Esters of Fatty Acids Second Edition. 10 pp.

—Methyl Ethyl Ketone
CNS K6608-80. Method of Test for Methyl Ethyl Ketone by Gas Chromatography (Dec)(6670).

—Methyl Isobutyl Ketone
CNS K6496-80. Analysis of Methyl Isobutyl Ketones by Gas Chromatography (May)(5602).

INTERNATIONAL AND NON-U.S. NATIONAL STANDARDS
SUBJECT INDEX

Gas Chromatography (Cont.)

—Milk—Fats
ISO 3594-76. Milk Fat—Detection of Vegetable Fat by Gas-Liquid Chromatography of Sterols (Reference Method) First Edition. 7 pp.

—Natural Gas
BSI BS 3156: SUB SEC 11.1.1-86. 1986 Methods for the Analysis of Fuel Gases Part 11: Methods for Non-Manufactured Gases Section 11.1: Hydrocarbons and Inert Gases Subsection 11.1.1: Determination of Hydro-carbons to C<of12>8 and Inert Gases by Gas Chromatography. 19 pp.

BSI BS 3156: SUB SEC 11.1.2-87. 1987 Methods for Analysis of Fuel Gases Part 11:Methods for Non-Manufactured Gases Sec 11.1: Hydrocarbons and Inert Gases Sub 11.1.2:Determination of Hydrocarbons from Butane (C<of12>4) to Hexadecane (C<of12>16) by Gas Chromatography. 13 pp.

ISO 6326 Pt 2-81. Gas Analysis—Determination of Sulphur Compounds in Natural Gas—Part 2: Gas Chromatographic Method Using an Electrochemical Detector for the Determination of Odoriferous Sulphur Compounds First Edition. 7 pp.

ISO 6568-81. Natural Gas—Simple Analysis by Gas Chromatography First Edition. 6 pp.

ISO 6974-84. Natural Gas—Determination of Hydrogen, Inert Gases and Hydrocarbons up to C8—Gas Chromatographic Method First Edition. 16 pp.

ISO 6975-86. Natural Gas—Determination of Hydrocarbons from Butane (C4) to Hexadecane (C16)—Gas Chromatographic Method First Edition. 11 pp.

—Oilseeds
BSI BS 4325: Part 8-93. 1993 Methods for Analysis of Oilseed Residues Part 8: Determination of Total Isothiocyanate Content and Vinylthiooxazolidone Content (ISO 5504: 1992). 16 pp.

BSI BS 4325: Part 8-84. 1984 Analysis of Oilseed Residues Part 8: Determination of Isothiocyanates and Vinyl Thiooxazolidone. 10 pp.

ISO 5504-92. Oilseed Residues—Determination of Total Isothiocyanate Content and Vinylthiooxazolidone Content Second Edition. 13 pp.

—Olefins
ISO 6377-81. Light Olefins for Industrial Use—Determination of Hydrocarbon Impurities by Gas Chromatography—General Considerations First Edition. 6 pp.

—Paints
CGSB 1-GP-71 METH 28.5-79. Methods of Testing Paints and Pigments Analysis of Solvents by Gas-Liquid Chromatography. 4 pp.

—Petitgrain Oil
ISO 7611-85. Oils of Lemon and Petitgrain Citronnier, and Oil of Lime Obtained by a Mechanical Process—Determination of Citral (Neral + Geranial) Content—Gas Chromatographic Method on Capillary Columns First Edition. 7 pp.

—Petroleum Products
DIN ENGL 51413 Pt 3-88. Testing of Liquid Petroleum Products; Determination of Aromatics Content by Gas Chromatography (Oct). 3 pp.

DIN ENGL 51527 Pt 1-87. Testing of Petroleum Products; Determination of Polychlorinated Biphenyls (PCB); Preseparation by Liquid Chromatography and Determination of Six Selected PCB Compounds by Gas Chromatography Using an Electron Capture Detector (May). 6 pp.

ISO 3924-77. Petroleum Products—Determination of Boiling Range Distribution—Gas Chromatography Method First Edition. 11 pp.

—Phenolic Resins
ISO 8974-88. Plastics—Phenolic Resins—Determination of Residual Phenol Content by Gas Chromatography First Edition. 5 pp.

—Polystyrene
BSI BS 2782:Pt4: METH 453A-78. 1978 Methods of Testing Plastics Part 4: Chemical Properties Method 453A: Determination of Residual Styrene Monomer in Polystyrene by Gas Chromatography. 5 pp.

DIN ENGL 53741-71. Testing of Plastics; Determination of Volatile Aromatic Hydrocarbons in Polystyrene; Gas Chromatographic Method (Sept). 4 pp.

ISO 2561-74. Plastics—Determination of Residual Styrene Monomer in Polystyrene by Gas Chromatography First Edition. 5 pp.

—Propane
BSI BS 7276-90. 1990 Method for Gas Chromatography Analysis of Liquified Petroleum Gas. 16 pp.

CEN PREN 242-85. Commercial Propane and Butane (LPG); Determination of Hydrocarbons Content by Gas Chromatographic Method. 24 pp.

Gas Chromatography (Cont.)

—Propane (Cont.)
ISO 7941-88. Commercial Propane and Butane—Analysis by Gas Chromatography First Edition. 15 pp.

—Propylene
ISO 6380-81. Propylene for Industrial Use—Determination of Hydrocarbon Impurities—Gas Chromatographic Method First Edition. 7 pp.

ISO 6381-81. Ethylene and Propylene for Industrial Use—Determination of Traces of Carbon Monoxide and Carbon Dioxide—Gas Chromatographic Method First Edition. 8 pp.

ISO 8174-86. Ethylene and Propylene for Industrial Use—Determination of Acetone, Acetonitrile, Propan-2-ol and Methanol—Gas Chromatographic Method First Edition. 8 pp.

ISO 8175-86. Propylene for Industrial Use—Determination of Oligomers—Gas Chromatographic Method First Edition. 6 pp.

—PVC
ISO 6401-85. Plastics—Homopolymer and Copolymer Resins of Vinyl Chloride—Determination of Residual Vinyl Chloride Monomer—Gas Chromatographic Method First Edition. 6 pp.

—Reinforced Plastics
BSI BS 2782:Pt4: METH 432A-91. 1991 Methods of Testing Plastics Part 4: Chemical Properties Method 432A: Determin. of Residual Styrene MonomerContent in Reinforced Plastics Based on Unsaturated Polyester Resins (ISO 4901: 1985). 9 pp.

ISO 4901-85. Reinforced Plastics Based on Unsaturated Polyester Resins—Determination of Residual Styrene Monomer Content First Edition. 7 pp.

—Rosewood Oil
ISO 7353-85. Oil of Rosewood—Determination of Alpha-Terpineol Content—Gas Chromatographic Method on Packed Columns First Edition. 5 pp.

—Rosins
CNS K6462-80. Method of Test for Resin Acids in Rosin by Gas Chromatography (Mar)(5347).

—Rubber
ISO 6225 Pt 2-90. Rubber, Raw, Natural—Determination of Castor Oil Content—Part 2: Determination of Total Ricinoleic Acid Content by Gas Chromatography Second Edition. 7 pp.

ISO 7270-87. Rubber—Identification of Polymers (Single Polymers and Blends)—Pyrolytic Gas Chromatographic Method First Edition; (Supersedes 5475). 7 pp.

—Sandalwood Oil
ISO 7610-85. Oil of Sandalwood—Determination of Santalols Content (in the Form of Their Trimethylsilyl Derivative)—Gas Chromatographic Method on Capillary Columns First Edition. 6 pp.

—Sludge (Sewage)
DIN ENGL 38407 Pt 9-91. German Standard Methods for the Examination of Water, Waste Water and Sludge; Substance Group Analysis (Group F); Determination of Benzene and Some of Its Derivatives by Gas Chromatography (F9) (May). 16 pp.

—Sodium Alkylbenzene Sulfonates
BSI BS 6829: Sec 3.1-89. 1989 Analysis of Surface Active Agents (Raw Materials) Part 3: Sodium Alkylbenzenesulphonates Section 3.1: Method for Determination of Mean Relative Molecular Mass. 8 pp.

ISO 6841-88. Surface Active Agents—Technical Straight-Chain Sodium Alkylbenzenesulfonates—Determination of Mean Relative Molecular Mass by Gas-Liquid Chromatography Second Edition. 9 pp.

—Solvents
BSI BS 6455-84. 1984 Monitoring the Levels of Residual Solvents in Flexible Packaging Materials. 8 pp.

—Styrene Acrylonitrile Resins
ISO 4581-87. Plastics—Styrene/Acrylonitrile Copolymers—Determination of Residual Acrylonitrile Monomer Content—Gas Chromatography Method First Edition. 8 pp.

—Timber
BSI BS 5666: Part 6-83. 1983 Amd 1 Wood Preservatives and Treated Timber: Part 6: Quantitative Analysis or Preservative Solution and Treated Timber Containing Pentachloro-phenol, Phenachloro-phenyl Laurate, GS Hexachlorocyclohexane and Dieldrin (AMD 6224). 11 pp.

Gas Chromatography (Cont.)

—Tobacco
BSI BS 5202: Part 10-82. 1982 Methods for Chemical Analysis of Tobacco and Tobacco Products Part 10: Determination of Organochlorine Pesticide Residues. 8 pp.

ISO 4389-81. Tobacco and Tobacco Products—Determination of Organochlorine Pesticide Residues (Reference Method) First Edition. 8 pp.

—Toluene
ISO 5279-80. Toluene—Determination of Hydrocarbon Impurities—Gas Chromatographic Method First Edition; (Erratum—Dec 1981). 7 pp.

—Vegetable Fats
BSI BS 684: Sec 2.38-92. 1992 Methods of Analysis of Fats and Fatty Oils Part 2: Other Methods Section 2.38: Determination of the Proportions of Individual Sterols in the Sterol Fraction (ISO 6799: 1991). 12 pp.

BSI BS 684: Sec 2.38-83. 1983 Methods of Analysis of Fats and Fatty Oils Part 2: Other Methods Section 2.38: Determination of the Proportions of Individual Sterols in the Sterol Fraction. 8 pp.

ISO 6799-83. Animal and Vegetable Fats and Oils—Determination of Composition of the Sterol Fraction—Method by Gas-Liquid Chromatography First Edition. 7 pp.

—Vegetable Oils
BSI BS 684: Sec 2.38-92. 1992 Methods of Analysis of Fats and Fatty Oils Part 2: Other Methods Section 2.38: Determination of the Proportions of Individual Sterols in the Sterol Fraction (ISO 6799: 1991). 12 pp.

BSI BS 684: Sec 2.38-83. 1983 Methods of Analysis of Fats and Fatty Oils Part 2: Other Methods Section 2.38: Determination of the Proportions of Individual Sterols in the Sterol Fraction. 8 pp.

ISO 6463-82. Animal and Vegetable Fats and Oils—Determination of Butylhydroxyanisole (BHA) and Butylhydroxytoluene (BHT)—Gas-Liquid Chromatographic Method First Edition. 5 pp.

ISO 6799-83. Animal and Vegetable Fats and Oils—Determination of Composition of the Sterol Fraction—Method by Gas-Liquid Chromatography First Edition. 7 pp.

—Waste Water
DIN ENGL 38407 Pt 9-91. German Standard Methods for the Examination of Water, Waste Water and Sludge; Substance Group Analysis (Group F); Determination of Benzene and Some of Its Derivatives by Gas Chromatography (F9) (May). 16 pp.

JIS K 0093-74. Method for Determination of Polychlorinated Biphenyl in Industrial Waste Water. 19 pp.

—Water
DIN ENGL 38407 Pt 4-88. German Standard Methods for the Examination of Water, Waste Water and Sludge; Substance Group Analysis (Group F); Determination of Readily Volatile Halogenated Hydrocarbons (F4) (May). 16 pp.

DIN ENGL 38407 Pt 9-91. German Standard Methods for the Examination of Water, Waste Water and Sludge; Substance Group Analysis (Group F); Determination of Benzene and Some of Its Derivatives by Gas Chromatography (F9) (May). 16 pp.

DIN ENGL 38413 Pt 2-88. German Standard Methods for the Examination of Water, Waste Water and Sludge; Individual Constituents (Group P); Determination of Vinyl Chloride by Headspace; Gas Chromatography (P 2) (May). 4 pp.

ISO 8165 Pt 1-92. Water Quality—Determination of Selected Monovalent Phenols—Part 1: Gas-Chromatographic Method After Enrichment by Extraction First Edition. 11 pp.

—Wood Preservatives
BSI BS 5666: Part 6-83. 1983 Amd 1 Wood Preservatives and Treated Timber: Part 6: Quantitative Analysis or Preservative Solution and Treated Timber Containing Pentachloro-phenol, Phenachloro-phenyl Laurate, GS Hexachlorocyclohexane and Dieldrin (AMD 6224). 11 pp.

Gas Chromatography Analysis
Use: Gas Chromatography

Gas Chromatography-Mass Spectrometry
See Also: Gas Analysis; Gas Chromatography; Mass Spectrometry

CNS K0042-89. General Rules for Analytical Methods in Gas Chromatography/Mass Spectrometry (Apr)(12509).

JIS K 0123-82. General Rules for Analytical Methods in Gas Chromatography Mass Spectrometry. 32 pp.

INTERNATIONAL AND NON-U.S. NATIONAL STANDARDS
SUBJECT INDEX

Gas

Gas Cleaning Equipment
See Also: Air Cleaners
JIS B 9914-81. Method of Measuring Performance for Gas Treatment Equipment. 18 pp.

—**Glossaries**
BSI BS 6202-82. 1982 Cleaning Equipment for Air or Other Gases. 10 pp.
ISO 3649-80. Cleaning Equipment for Air or Other Gases—Vocabulary First Edition. 10 pp.
Scope Note: Use a more specific term

Gas Cutting
See Also: Cutting; Flame Cutting

—**Fabrics—Flammability Testing**
JIS A 1323-84. Flame Retardant Testing Method for Spark Droplets of Welding and Gas Cutting on Fabric Sheets in Construction Works. 9 pp.

—**Safety**
CSA CAN/CSA-W117.2-M87. Safety in Welding, Cutting, and Allied Processes; (Gen Instr 2 Thru 3). 63 pp.

Gas Cutting Equipment
See Also: Cutting Tools; Oxygen Cutting Equipment

—**Blowpipes**
BSI BS 6503-84. 1984 Handheld Blowpipes and Nozzles, Using Fuel Gas and Oxygen, for Gas Welding, Cutting and Related Processes. 20 pp.
CEN PREN 874-92. Welding—Oxygen/Fuel Gas Blowpipes (Cutting Machine Type) of Cylindrical Barrel—Type of Construction, General Specifications, Test Methods. 25 pp.

—**Bourdon Tubes**
CEN PREN 562-91. Pressure Gauges Used in Welding, Cutting and Allied Processes. 14 pp.
DIN ENGL 8549-86. Bourdon-Tube Pressure Gauges Used in Welding, Cutting and Related Processes with 50 mm or 63 mm Diameter Casing (Dec). 8 pp.

—**Construction Materials**
BSI BS 7329-90. (WITHDRAWN) 1990 Materials for Equipment Used in Gas Welding, Cutting and Allied Processes (Superseded by BS EN 29539: 1992). 8 pp.
BSI BS EN 29539-92. 1992 Materials for Equipment Used in Gas Welding, Cutting and Allied Processes (ISO 9539: 1988). 9 pp.
CEN EN 29539-92. Materials for Equipment Used in Gas Welding, Cutting and Allied Processes. 4 pp.
ISO 9539-88. Materials for Equipment Used in Gas Welding, Cutting and Allied Processes First Edition. 4 pp.

—**Flame Arresters**
DIN ENGL 8521-81. Safety Devices Against Flashback and Backflow in Welding, Cutting and Allied Processes; Safety Requirements, Testing (Dec). 8 pp.
ISO 5175-87. Equipment Used in Gas Welding, Cutting and Allied Processes—Safety Devices for Fuel Gases and Oxygen or Compressed Air—General Specifications, Requirements and Tests First Edition. 11 pp.

—**Flashback Arresters**
DIN ENGL 8521-81. Safety Devices Against Flashback and Backflow in Welding, Cutting and Allied Processes; Safety Requirements, Testing (Dec). 8 pp.

—**Gases**
CEN PREN 439-90. Gases for Gas Shielded Arc Welding and Cutting. 9 pp.

—**Hose Assemblies**
BSI BS 1389-86. 1986 Hose Connections and Hose Assemblies for Equipment for Gas Welding, Cutting and Related Processes. 14 pp.

—**Hose Fittings**
BSI BS 1389-86. 1986 Hose Connections and Hose Assemblies for Equipment for Gas Welding, Cutting and Related Processes. 14 pp.
CEN PREN 560-91. Hose Connections for Equipment for Welding, Cutting and Allied Processes. 7 pp.
CNS B2533-80. Rubber Hose Coupling for Gas Welding and Cutting Torches (Dec)(6741).
DIN ENGL 8542-83. Hose Connections and Hose Couplers for Equipment for Welding, Cutting and Allied Processes (Sept). 9 pp.
JIS B 6805-86. Rubber Hose Connections for Welding and Cutting Equipment.
JIS B 6805-68. Rubber Hose Coupling for Gas Welding and Cutting Torches. 5 pp.

—**Hoses**
CEN PREN 559-91. Rubber Hoses for Welding, Cutting and Allied Processes. 12 pp.
CNS K4008-86. Rubber Hoses for Oxygen (Oct)(809).

Gas Cutting Equipment (Cont.)
—**Hoses** (Cont.)
CNS K4009-86. Rubber Hoses for Acetylene (Oct)(810).
DIN ENGL 8541 Pt 1-86. Hoses for Welding, Cutting and Allied Processes; Hoses Without Protective Cover for Use with Fuel Gas, Oxygen and Other Non-Combustible Gases (Aug). 5 pp.
DIN ENGL 8541 Pt 2-87. Hoses for Welding, Cutting and Allied Processes; Hoses with Protective Cover for Use with Fuel Gas, Oxygen and Other Non-Combustible Gases (Dec). 4 pp.
ISO 3821-92. Welding—Rubber Hoses for Welding, Cutting and Allied Processes Second Edition. 12 pp.
JIS K 6333-84. Rubber Hoses for Oxygen.
JIS K 6333-78. Oxygen Hose. 6 pp.
JIS K 6334-84. Rubber Hoses for Acetylene. 7 pp.

—**Hoses—Flammability Testing**
BSI BS 5173: Sec 103.6-90. 1990 Rubber and Plastics Hoses and Hose Assemblies Part 103: Physical Tests Section 103.6: Determination of Ignitability of Lining. 6 pp.

—**Hoses—Safety**
DIN ENGL 8541 Pt 3-79. Hoses for Welding, Cutting and Allied Processes; Oxygen Hoses Without Protective Cover, Subject to Special Requirements; Safety Requirements and Testing (Dec). 3 pp.
DIN ENGL 8541 Pt 4-83. Hoses for Welding, Cutting and Allied Processes; Oxygen Hoses with Outer Protective Cover for Special Requirements; Safety Requirements and Testing (Mar). 4 pp.

—**Leakage**
BSI BS 7278-90. (WITHDRAWN) 1990 Gas Tightness of Equipment for Gas Welding and Allied Processes (Superseded by BS EN 29090: 1992). 8 pp.
BSI BS EN 29090-92. 1992 Gas Tightness of Equipment for Gas Welding and Allied Processes (ISO 9090: 1989) (F). 11 pp.
CEN EN 29090-92. Gas Tightness of Equipment for Gas Welding and Allied Processes. 6 pp.
ISO 9090-89. Gas Tightness of Equipment for Gas Welding and Allied Processes First Edition. 7 pp.

—**Nozzles**
BSI BS 6503-84. 1984 Handheld Blowpipes and Nozzles, Using Fuel Gas and Oxygen, for Gas Welding, Cutting and Related Processes. 20 pp.

—**Pressure Regulators**
BSI BS 5741-79. 1979 Pressure Regulators Used in Welding, Cutting and Related Processes. 11 pp.
DIN ENGL 8545-81. Manifold Regulators for Welding, Cutting and Related Processes; Concepts, Requirements and Testing (Sept). 8 pp.
ISO 7291-90. Welding, Cutting and Allied Processes—Manifold Regulators First Edition. 11 pp.
JIS B 6803-87. Pressure Regulators for Welding, Cutting and Allied Processes. 19 pp.

—**Pressure Regulators—Gas Cylinders**
CEN PREN 585-91. Pressure Regulators for Gas Cylinders Used in Welding, Cutting and Allied Processes. 21 pp.
CEN PREN 961-92. Welding—Cutting and Allied Processes—Manifold Regulators. 19 pp.
DIN ENGL 8546-88. Pressure Regulators for Gas Cylinders Used in Welding, Cutting and Related Processes; Terminology, Requirements and Testing (Aug). 8 pp.
ISO 2503-83. Welding—Regulators for Gas Cylinders Used in Welding, Cutting and Related Processes Second Edition; (Addendum 1-1984). 15 pp.

—**Quick Disconnect Couplings—Stop Valves**
CEN PREN 561-91. Quick-Action Couplings with Shut-Off Valve for Welding, Cutting and Allied Processes. 10 pp.

—**Safety Valves**
DIN ENGL 8521-81. Safety Devices Against Flashback and Backflow in Welding, Cutting and Allied Processes; Safety Requirements, Testing (Dec). 8 pp.
DIN ENGL 32509-85. Hand-Operated Shut-Off Valves for Welding, Cutting and Allied Processes; Designs, Safety Requirements, Testing (Aug). 5 pp.
ISO 5175-87. Equipment Used in Gas Welding, Cutting and Allied Processes—Safety Devices for Fuel Gases and Oxygen or Compressed Air—General Specifications, Requirements and Tests First Edition. 11 pp.

—**Torches**
JIS B 6802-91. Manual Gas Cutting Torches. 16 pp.

Gas Cylinders
See Also: Containers; Cylinder Valves; Liquefied Petroleum Gas Cylinders; Oxygen Supply Equipment; Pressure Vessels

Gas Cylinders (Cont.)
MOD UK DSTAN 81-13-01. Cylinders, Compressed Gas, and Valves; Cylinder, Gas, for Ground, Marine, Airborne, and Medical Uses Issue 1; Amendment 2. 20 pp.
MOD UK DSTAN 81-13-92. Containers, Compressed Gas, and Valves Issue 2. 24 pp.
SAA AS B239-70. Welded Steel Cylinders for Compressed Gases of Capacity Over 10 Litres (Approx. 22 lb of Water) up to and Including 130 Litres (Approx. 286 lb of Water) Corrig. Amdt 1 September 1975 Amdt 2 June 1979 Obsolescent 1987 See AS 2470—1981. 22 pp.
SAA AS 2030.1-89. Approval, Filing, Inspection, Testing and Maintenance of Cylinders for the Storage and Transport of Compressed Gases (Known as the SAA Gas Cylinders Code)—Part 1: Cylinders for Compressed Gases Other Than Acetylene. 23 pp.
SAA AS 2030.3-82. Approval, Filing, Inspection, Testing and Maintenance of Cylinders for the Storage and Transport of Compressed Gases (Known as the SAA Gas Cylinders Code)—Part 3: Non-Refillable Cylinders for Compressed Gases. 4 pp.
SAA AS 2030.4-85. Approval, Filing, Inspection, Testing and Maintenance of Cylinders for the Storage and Transport of Compressed Gases (Known as the SAA Gas Cylinders Code)—Part 4: Welded Cylinders—Insulated. 3 pp.
SAA AS 2030 Supp 1-86. Approval, Filing, Inspection, Testing and Maintenance of Cylinders for the Storage and Transport of Compressed Gases—Supplement 1: Foreign Gas Cylinder Specifications. 40 pp.

—**Acetylene**
SAA AS CB4. Gas Cylinders Code—Interpretations to SAA Gas Cylinders Code No 6 (April 1973) Welded Three-Piece Type Cylinders for Acetylene Amdt 1 January 1976 See AS 2527—1982 Annex 2. 21 pp.
SAA AS 2030.2-85. Approval, Filing, Inspection, Testing and Maintenance of Cylinders for the Storage and Transport of Compressed Gases (Known as the SAA Gas Cylinders Code)—Part 2: Cylinders for Dissolved Acetylene. 5 pp.

—**Acetylene—Steel**
CNS B5053-89. Steel Cylinder for Acetylene (Jul)(2724).
CNS Z1008-63. Safety Code for Steel Cylinder of Highly Compressed Acetylene Gas (Nov)(2224)(R 1970).

—**Acetylene—Valves**
CNS B5077-90. Valve for Dissolved Acetylene Cylinder (May)(4152).
CNS B5077-77. Valves of Dissolved Acetylene Cylinder (Jun)(4152). 3 pp.

—**Aircraft—Identification Systems**
ISO R443-65. Marking of Aircraft Gas Cylinders First Edition; (Erratum—Jan 1966). 5 pp.

—**Aircraft—Identification Systems—Interdepartmental Procurement**
NATO STANAG 3056 ED 5 AMD 3-83. Marking of Airborne and Ground Gas and Cryogenic Fluid Containers. 7 pp.
NATO STANAG 3056 ED 5 AMD 4-83. Marking of Airborne and Ground Gas and Cryogenic Fluid Containers. 9 pp.

—**Alloy Steel—High Pressure—Installation—Marine**
MOD UK NES 317: Part 2-91. Requirements for Non—Transportable Pressure Vessels for the Storage of High Pressure Gases Part 2: Manufacture, Installation and Testing Issue 2 (06.91). 32 pp.
MOD UK NES 317: Part 2-01. Requirements for Non—Transportable Pressure Vessels for the Storage of High Pressure Gases Part 2: Manufacture, Installation and Testing Issue 2 (06.91); Amendment 1. 37 pp.

—**Alloy Steel—High Pressure—Storage—Marine**
MOD UK NES 317: Part 3-91. Requirements for Non—Transportable Pressure Vessels for the Storage of High Pressure Gases Part 3: Storage Requirements Issue 2 (06.91). 18 pp.
MOD UK NES 317: Part 3-01. Requirements for Non—Transportable Pressure Vessels for the Storage of High Pressure Gases Part 3: Storage Requirements Issue 2 (06.91); Amendment 1. 21 pp.

—**Aluminum**
EC 84/526/EEC-84. Council Directive on the Approximation of the Laws of the Member States Relating to Seamless, Unalloyed Aluminium and Aluminium Alloy Gas Cylinders. 28 pp.

INTERNATIONAL AND NON-U.S. NATIONAL STANDARDS
SUBJECT INDEX

Gas Cylinders (Cont.)

—Aluminum (Cont.)
SAA AS 1777-89. Aluminium Cylinders for Compressed Gases—Seamless 0.1 kg to 130 kg. 14 pp.

—Aluminum Alloy
BSI BS 5045: Part 3-84. 1984 Amd 4 Transportable Gas Containers Part 3: Spec. for Seamless Aluminium Alloy Gas Containers Above 0.5 Litre Water Capacity and up to 300 Bar Charged Pressure at 15 Degrees C (AMD 6320) January 31, 1991. 34 pp.

BSI BS 5045: Part 5-86. 1986 Amd 1 Transportable Gas Containers Part 5: Specification for Aluminium Alloy Gas Containers Above 0.5 Litre up to 130 Litres Water Capacity with Welded Seams (AMD 6321) January 31, 1991. 38 pp.

BSI BS 5045: Part 6-87. 1987 Transportable Gas Containers Part 6: Seamless Containers of Less Than 0.5 Litre Water Capacity. 16 pp.

BSI BS 5430: Part 3-90. 1990 Amd 1 Periodic Inspection, Testing and Maintenance of Transportable Gas Containers (Excluding Dissolved Acetylene Containers) Part 3: Specification for Seamless Aluminium Alloy Containers of Water Capacity 0.5 Litres and Above. 28 pp.

EC 84/526/EEC-84. Council Directive on the Approximation of the Laws of the Member States Relating to Seamless, Unalloyed Aluminium and Aluminium Alloy Gas Cylinders. 28 pp.

MOD UK DSTAN 81-96-90. Periodic Inspection, Testing and Maintenance of Transportable Gas Containers and Fittings Issue 1. 11 pp.

—Aluminum Alloy—Inspection
ISO 10461-93. Seamless Aluminium-Alloy Gas Cylinders—Periodic Inspection and Testing First Edition. 31 pp.

MOD UK DSTAN 81-96-90. Periodic Inspection, Testing and Maintenance of Transportable Gas Containers and Fittings Issue 1. 11 pp.

—Aluminum Alloy—Maintenance
MOD UK DSTAN 81-96-90. Periodic Inspection, Testing and Maintenance of Transportable Gas Containers and Fittings Issue 1. 11 pp.

—Breathing Apparatus—Valve Fittings
BSI BS 341: Part 2-63. 1963 Amd 1 Valve Fittings for Compressed Gas Cylinders Part 2: Valves with Taper Stems for Use with Breathing Apparatus (Excluding Medical Gas Cylinders to BS 1319). 29 pp.

BSI BS EN 144: Part 1-91. 1991 Respiratory Protective Devices—Gas Cylinder Valves Part 1: Specification for Thread Connection for Insert Connector (N). 12 pp.

CEN EN 144-01-91. Respiratory Protective Devices—Gas Cylinder Valves—Thread Connection for Insert Connector. 9 pp.

—Caps (Lids)—Screw Threads
DIN ENGL 4668-68. Gas Cylinders; Screw Threads for Valve Sockets, Neck Rings and Protecting Caps (Sept). 2 pp.

—Carbon Steel—Inspection
ISO 10460-93. Welded Carbon Steel Gas Cylinders—Periodic Inspection and Testing First Edition. 29 pp.

—Color Coding
CEN PREN 1089-3-93. Cylinder Identification—Part 3: Colour Coding System for Gas Cylinders for Use in Europe. 8 pp.

—Compressors—Refueling
SNZ NZS 5425: Part 2-82. Code of Practice for CNG Compressor and Refuelling Stations Part 2: Compressor Equipment. 16 pp.

—Cooking Appliances—Portable
JIS S 2148-91. Gas Cylinders for Portable Gas Cooker. 19 pp.

—Diving—Gas Purity
MOD UK DSTAN 68-75-01. Breathing Gas Purity for Diving Issue 1; Amendment 1. 9 pp.

MOD UK DSTAN 68-75-93. Breathing Gas Purity for Diving Issue 2. 11 pp.

—Dollies
BSI BS 2718-79. 1979 Gas Cylinder Trolleys. 7 pp.

—Fittings
MOD UK DSTAN 81-96-90. Periodic Inspection, Testing and Maintenance of Transportable Gas Containers and Fittings Issue 1. 11 pp.

—Glossaries
ISO 10286-92. Gas Cylinders—Terminology First Edition. 24 pp.

Gas Cylinders (Cont.)

—Highway Transportation—Emergency Procedures
SAA AS 1678.2M1-87. Emergency Procedure Guide—Transport Group Text EPGs for Class 2 Substances—Compressed and Liquefied Gases—Part 2M1: Mixed Load of Gases in Cylinders.

—Identification Systems
BSI BS 349-73. 1973 Amd 1 Identification of Contents of Industrial Gas Containers (AMD 6132) May 31, 1989. 13 pp.

CEN PREN 1089-1-93. Cylinder Identification—Part 1: Stampmarking for Refillable Transportable Gas Cylinders. 14 pp.

CEN PREN 1089-2-93. Cylinder Identification—Part 2: Precautionary Labels for Gas Cylinders. 9 pp.

CNS Z7027-87. Marking for Content of Movable High Pressure Compressed Gas Containers (Mar)(2994).

ISO 448-81. Gas Cylinders for Industrial Use—Marking for Identification of Content Second Edition. 4 pp.

MOD UK DSTAN 81-24-90. Identification Marking of Containers, Compressed Gas Issue 3. 49 pp.

—Industrial—Identification Systems
SAA AS 1943-87. Industrial Gas Cylinder Identification. 5 pp.

—Medical
BSI BS 1319-76. 1976 Amd 2 Medical Gas Cylinders, Valves and Yoke Connections (AMD 4603) June 29, 1984. 23 pp.

SNZ NZS 7101-81. Specification for Medical Gas Cylinders, Valves and Yoke Connections Amend: 1, 1981; 1A, 1981; 2; 2A. 20 pp.

—Medical Electrical—Connections
CSA CAN/CSA-C22. 2NO601.1-M90. Medical Electrical Equipment, Part 1: General Requirements for Safety; (Gen Instr 1). 240 pp.

—Medical Electrical—Identification Systems
CSA CAN/CSA-C22. 2NO601.1-M90. Medical Electrical Equipment, Part 1: General Requirements for Safety; (Gen Instr 1). 240 pp.

—Medical—Identification Systems
ISO 32-77. Gas Cylinders for Medical Use—Marking for Identification of Content First Edition; (Erratum—Aug 1979). 5 pp.

SAA AS 1944-87. Medical Gas Cylinder Identification. 2 pp.

—Medical—Interdepartmental Procurement
NATO STANAG 2121 ED 2 AMD 5-79. Cross Servicing of Medical Gas Cylinders. 13 pp.

—Medical—Valve Yokes
BSI BS 1319-76. 1976 Amd 2 Medical Gas Cylinders, Valves and Yoke Connections (AMD 4603) June 29, 1984. 23 pp.

ISO 407-91. Small Medical Gas Cylinders—Pin-Index Yoke-Type Valve Connections Second Edition; (Incorporating Amendment 1). 23 pp.

SNZ NZS 7101-81. Specification for Medical Gas Cylinders, Valves and Yoke Connections Amend: 1, 1981; 1A, 1981; 2; 2A. 20 pp.

—Metal
BSI BS 5188-90. 1990 Non-Refillable Metallic Containers up to 1.4 Litres Capacity and 85 MM Diameter for Liquefied or Compressed Non-Flammable Gases. 12 pp.

BSI BS 5188-75. (WITHDRAWN) 1975 Non-Refillable Metallic Containers up to 1.4 Litres Capacity and 85 mm Diameter for Liquefied or Compressed Non-Flammable Gases. 10 pp.

—Meters
SNZ NZS 5425: DIV 3.1P-85. Code of Practice for CNG Compressor and Refuelling Stations Part 3: Metering Devices. Division 3.1P: Metering Method Using Tables and Calculator Programmes. 16 pp.

SNZ NZS 5425: DIV 3.2-84. Code of Practice for CNG Compressor and Refuelling Stations Part 3: Metering Devices. Division 3.2: Installation and Operation of On-Line Metering Devices. 12 pp.

SNZ NZS 5425: DIV 3.3-84. Code of Practice for CNG Compressor and Refuelling Stations Part 3: Metering Devices. Division 3.3: Requirements for Type Approval of On-Line Metering Devices Amend: 1, 1993. 25 pp.

—Neck Rings—Screw Threads
DIN ENGL 4668-68. Gas Cylinders; Screw Threads for Valve Sockets, Neck Rings and Protecting Caps (Sept). 2 pp.

—Nozzles—Screw Threads
DIN ENGL 4668-68. Gas Cylinders; Screw Threads for Valve Sockets, Neck Rings and Protecting Caps (Sept). 2 pp.

Gas Cylinders (Cont.)

—Pressure Regulators
CEN PREN 585-91. Pressure Regulators for Gas Cylinders Used in Welding, Cutting and Allied Processes. 21 pp.

CEN PREN 961-92. Welding—Cutting and Allied Processes—Manifold Regulators. 19 pp.

DIN ENGL 8546-88. Pressure Regulators for Gas Cylinders Used in Welding, Cutting and Related Processes; Terminology, Requirements and Testing (Aug). 8 pp.

ISO 2503-83. Welding—Regulators for Gas Cylinders Used in Welding, Cutting and Related Processes Second Edition; (Addendum 1-1984). 15 pp.

—Refrigerants—Identification Systems
SAA AS 1942-87. Refrigerant Gas Cylinder Identification. 3 pp.

—Riot Control Agents—Identification Systems
NATO STANAG 4201 ED 1 AMD 1-85. Marking of Riot Control and Training Canisters. 7 pp.

—Screw Threads
CNS B5105-84. Screw Threads of Valves for High Pressure Gas Cylinder (Apr)(10849).

—Scuba Diving
JIS S 7302-88. SCUBA Diving Goods—Cylinders. 10 pp.

—Scuba Diving—Pressure Regulators
JIS S 7303-88. SCUBA Diving Goods—Cylinders Valves. 9 pp.

—Steel
BSI BS 5045: Part 1-82. 1982 Amd 2 Transportable Gas Containers Part 1: Seamless Steel Gas Containers Above 0.5 Litre Water Capacity (AMD 6318) January 31, 1991. 34 pp.

BSI BS 5045: Part 2-89. 1989 Amd 1 Transportable Gas Containers Part 2: Specification for Steel Containers of 0.5 L up to 450 L Water Capacity with Welded Seams (AMD 6319) January 31, 1991. 41 pp.

BSI BS 5045: Part 2-02. 1989 Amd 2 Transportable Gas Containers Part 2: Specification for Steel Containers of 0.5 L up to 450 L Water Capacity with Welded Seams (AMD 7911) October 15, 1993 (E). 42 pp.

BSI BS 5045: Part 2-78. (WITHDRAWN) 1978 Amd 1 Transportable Gas Containers Part 2: Steel Containers up to 130 Litres Water Capacity with Welded Seams. 30 pp.

BSI BS 5045: Part 6-87. 1987 Transportable Gas Containers Part 6: Seamless Containers of Less Than 0.5 Litre Water Capacity. 16 pp.

BSI BS 5430: Part 1-90. 1990 Periodic Inspection, Testing and Maintenance of Transportable Gas Containers (Excluding Dissolved Acetylene Containers) Part 1: Seamless Steel Containers of Water Capacity 0.5 Litres and Above. 24 pp.

BSI BS 5430: Part 1-77. (WITHDRAWN) 1977 Periodic Inspection, Testing and Maintenance of Transportable Gas Containers (Excluding Dissolved Acetylene Containers) Part 1: Seamless Steel Containers. 20 pp.

BSI BS 5430: Part 2-90. 1990 Periodic Inspection, Testing and Maintenance of Transportable Gas Containers (Excluding Dissolved Acetylene Containers) Part 2: Welded Steel Containers of Water Capacity 0.5l up to 150L. 22 pp.

CNS B5046-67. Steel Cylinder for Compressed Natural Gas (Mar)(1337).

CNS B5107-88. Seamless Steel Gas Cylinders (Apr)(12242).

CNS G2044-78. Method of Test for Steel Sheets, Plates and Strip for Gas Cylinders (Mar)(4274).

CNS G3101-78. Steel Sheets, Plates and Strip for Gas Cylinders (Mar)(4273). 3 pp.

EC 84/525/EEC-84. Council Directive on the Approximation of the Laws of the Member States Relating to Seamless, Steel Gas Cylinders. 19 pp.

EC 84/527/EEC-84. Council Directive on the Approximation of the Laws of the Member States Relating to Welded Unalloyed Steel Gas Cylinders. 24 pp.

ISO 4705-83. Refillable Seamless Steel Gas Cylinders First Edition. 24 pp.

ISO 4706-89. Refillable Welded Steel Gas Cylinders First Edition. 21 pp.

ISO 4978-83. Flat Rolled Steel Products for Welded Gas Cylinders First Edition. 6 pp.

ISO 6406-92. Periodic Inspection and Testing of Seamless Steel Gas Cylinders First Edition. 26 pp.

JIS B 8230-89. Small Type Seamless Steel Gas Cylinders. 20 pp.

JIS B 8241-89. Seamless Steel Gas Cylinders. 29 pp.

JIS G 3116-90. Steel Sheets, Plates and Strip for Gas Cylinders. 10 pp.

MOD UK DSTAN 81-96-90. Periodic Inspection, Testing and Maintenance of Transportable Gas Containers and Fittings Issue 1. 11 pp.

INDUSTRY STANDARDS

Gas Cylinders (Cont.)

—Steel (Cont.)
MOD UK DEF-168-01. Cylinders, Compressed Gas, High Pressure, MK12A; Amendment 1. 8 pp.
MOD UK DEF-176-68. Cylinders, Compressed Gas, High Pressure, MK6. 7 pp.
SNZ NZS/BS 5045: Part 1-82. Transportable Gas Containers Part 1: Specification for Seamless Steel Gas Containers Above 0.5 Litre Water Capacity. 32 pp.
SNZ NZS 5454-89. Standard Requirements for Lightweight Steel Automotive Compressed Natural Gas Cylinders for Use in New Zealand. 20 pp.

—Steel—Fire Extinguishers—Ships
BSI BS 5396-76. 1976 Seamless Steel Co2 Containers for Fixed Fire-Fighting Installations on Ships. 4 pp.
ISO 3500-90. Seamless Steel CO2 Cylinders for Fixed Fire-Fighting Installations on Ships Second Edition. 4 pp.

—Steel—Fluorocarbons
JIS B 8235-90. Refillable Welded Steel Gas Cylinders for Liquefied Fluoro Carbon. 26 pp.

—Steel—Inspection
ISO 6406-92. Periodic Inspection and Testing of Seamless Steel Gas Cylinders First Edition. 26 pp.
MOD UK DSTAN 81-96-90. Periodic Inspection, Testing and Maintenance of Transportable Gas Containers and Fittings Issue 1. 11 pp.

—Steel—Maintenance
MOD UK DSTAN 81-96-90. Periodic Inspection, Testing and Maintenance of Transportable Gas Containers and Fittings Issue 1. 11 pp.

—Steel Pipes
CNS G2080-81. Method of Test for Seamless Steel Tubes for High Pressure Gas Cylinder (May)(7378).
CNS G3139-81. Seamless Steel Tubes for High Pressure Gas Cylinder (May)(7377). 2 pp.
JIS G 3429-88. Seamless Steel Tubes for High Pressure Gas Cylinder. 9 pp.

—Stoppers
MOD UK DSTAN 81-2-77. Plugs, Screwed, Gas Cylinder, (Non-Ferrous) Issue 2. 9 pp.
MOD UK DSTAN 81-2-92. Plugs, Screwed, Gas Container (Non Ferrous) Issue 3. 11 pp.

—Storage
SNZ NZS 5425: Part 1-80. Code of Practice for CNG Compressor and Refuelling Stations Part 1: On Site Storage and Location of Equipment Amend: 1, 1986. 27 pp.

—Threads
CNS B2285-78. Gas Cylinders; Thread for Nozzles, Neck Rings and Protection Caps (Oct)(4551).

—Transportable—Inspection
ISO 10463-93. Cylinders for Permanent Gases—Inspection at Time of Filling First Edition. 7 pp.

—Valve Fittings
BSI BS 341: Part 1-91. 1991 Transportable Gas Container Valves Part 1: Specification for Industrial Valves for Working Pressures up to and Including 300 Bar (E). 70 pp.
BSI BS 341: Part 1-01. 1991 Amd 1 Transportable Gas Container Valves Part 1: Specification for Industrial Valves for Working Pressures up to and Including 300 Bar (AMD 7641) May 15, 1993. 73 pp.
CEN PREN 144 (Part 1)-88. Respiratory Protective Devices; Gas Cylinder Valves; Thread Connection for Insert Connector. 5 pp.
CEN PREN 144-2-90. Respiratory Protective Devices; Gas Cylinder Valves; Thread Connection for Side Connector. 4 pp.

—Valve Outlets
BSI PD 6488-79. 1979 Gas Cylinder Valve Outlets in Use Within the Countries of ISO Member Bodies. 12 pp.
ISO 5145-90. Cylinder Valve Outlets for Gases and Gas Mixtures—Selection and Dimensioning First Edition. 26 pp.
ISO TR7470-88. Valve Outlets for Gas Cylinders—List of Provisions Which Are Either Standardized or in Use Second Edition. 13 pp.
ISO 10156-90. Gases and Gas Mixtures—Determination of Fire Potential and Oxidizing Ability for the Selection of Cylinder Valve Outlets First Edition. 17 pp.

—Valves
BSI BS 341: Part 1-91. 1991 Transportable Gas Container Valves Part 1: Specification for Industrial Valves for Working Pressures up to and Including 300 Bar (E). 70 pp.
BSI BS 341: Part 1-01. 1991 Amd 1 Transportable Gas Container Valves Part 1: Specification for Industrial Valves for Working Pressures up to and Including 300 Bar (AMD 7641) May 15, 1993. 73 pp.
CEN PREN 144 (Part 1)-88. Respiratory Protective Devices; Gas Cylinder Valves; Thread Connection for Insert Connector. 5 pp.
CEN PREN 144-2-90. Respiratory Protective Devices; Gas Cylinder Valves; Thread Connection for Side Connector. 4 pp.
CNS B5104-84. Valves for High Pressure Pressure Cas Cylinder (Apr)(10848).
DIN ENGL 477 Pt 1-90. Gas Cylinder Valves Rated for Test Pressures up to 300 Bar; Types, Sizes and Outlets (May). 28 pp.
DIN ENGL 477 Pt 5-90. Gas Cylinder Valves Rated for Test Pressures up to 450 Bar; Types, Sizes and Outlets (Feb). 5 pp.
JIS B 8246-89. Valves for High Pressure Gas Cylinder. 28 pp.
MOD UK DSTAN 81-4-77. Valves, Cylinder, Gas Issue 2. 11 pp.
MOD UK DSTAN 81-4-92. Valves, Container, Gas Issue 3. 17 pp.
MOD UK DSTAN 81-13-01. Cylinders, Compressed Gas, and Valves, Cylinder, Gas, for Ground, Marine, Airborne, and Medical Uses Issue 1; Amendment 2. 20 pp.
MOD UK DSTAN 81-13-92. Containers, Compressed Gas, and Valves Issue 2. 24 pp.
MOD UK DSTAN 81-44-01. Adaptors for Gas Cylinder Valves and Regulating Equipment Issue 1; Amendment 1 (Withdrawn). 43 pp.

—Valves—Caps (Lids)
CEN PREN 962-92. Gas Cylinders—Valve Protection Caps and Valve Guards for Industrial and Medical Gas Cylinders—Design, Construction and Tests. 12 pp.

—Valves—Gages
CEN PREN 628-92. Gas Cylinders—Requirements for Gauge Inspection of Tapered Threads for Connection of Valves to Gas Cylinders of 25.8 mm Nominal Diameter. 17 pp.

—Valves—Guards (Protective)
CEN PREN 962-92. Gas Cylinders—Valve Protection Caps and Valve Guards for Industrial and Medical Gas Cylinders—Design, Construction and Tests. 12 pp.

—Valves—Taper Threads
CEN PREN 628-92. Gas Cylinders—Requirements for Gauge Inspection of Tapered Threads for Connection of Valves to Gas Cylinders of 25.8 mm Nominal Diameter. 17 pp.

—Valves—Type Testing
CEN PREN 849-92. Gas Cylinder Valves—Specification and Type Testing. 23 pp.

Gas Detectors

See Also: Accident Prevention; Combustible Gas Detectors; Fire Protection; Gas Alarms; Leak Detectors; Safety Lamps

BSI BS 7348-90. 1990 Electrical Apparatus for the Detection of Combustible Gases in Domestic Premises (Q). 11 pp.
BSI BS 7348-01. 1990 Amd 1 Electrical Apparatus for the Detection of Combustible Gases in Domestic Premises (AMD 6869) October 31, 1991. 12 pp.
CENELEC EN 50054-91. Electrical Apparatus for the Detection and Measurement of Combustible Gases General Requirements and Test Methods. 55 pp.
CENELEC EN 50057-91. Electrical Apparatus for the Detection and Measurement of Combustible Gases Performance Requirements for Group II Apparatus Indicating up to 100 % Lower Explosive Limit. 13 pp.
CENELEC EN 50058-91. Electrical Apparatus for the Detection and Measurement of Combustible Gases Performance Requirements for Group II Apparatus Indicating up to 100 % (v/v) Gases. 13 pp.
JIS K 0804-85. Detector Tube Type Gas Measuring Instruments.

—Aspiration Pumps
BSI BS 5343: Part 1-86. 1986 Gas Detector Tubes Part 1: Short Term Gas Detector Tubes. 8 pp.

—Carbon Monoxide
CNS M2024-80. Carbon Monoxide Detector with Color-Intensity Type Tubes (Feb)(5252).
CNS M2025-80. Carbon Monoxide Detector with Length-of-Stain Type Detector Tubes (Feb)(5253).
JIS M 7605-74. Carbon Monoxide Detector with Colour-Intensity Type Tubes.
JIS M 7650-74. Carbon Monoxide Detector with Length-of-Stain Type Detector Tubes.

—Hydrogen Sulfide
JIS T 8204-86. Detector Tube Type Hydrogen Sulfide Measuring Instruments (Length-of-Stain Type). 13 pp.

—Interferometer—Inflammable Gas
CNS M2014-80. Interferometer Type Instrumental Gas Detector for Inflammable Gas (Feb)(5242).
JIS M 7602-77. Interferometer Type Instrumental Gas Detector for Inflammable Gas.

—Methane
CENELEC EN 50055-91. Electrical Apparatus for the Detection and Measurement of Combustible Gases Performance Requirements for Group I Apparatus Indicating up to 5 % (v/v) Methane in Air. 13 pp.
CENELEC EN 50056-91. Electrical Apparatus for the Detection and Measurement of Combustible Gases Performance Requirements for Group I Apparatus Indicating up to 100 % (v/v) Methane. 13 pp.
JIS M 7653-87. Portable Type Methane Detector. 9 pp.

—Portable
CSA CAN/CSA-Z433-M92. Portable Gas Monitors; (Gen Instr 1). 27 pp.

—Tubes
BSI BS 5343: Part 1-86. 1986 Gas Detector Tubes Part 1: Short Term Gas Detector Tubes. 8 pp.
BSI BS 5343: Part 2-91. 1991 Gas Detector Tubes Part 2: Specification for Long Term Gas Detector Tubes. 11 pp.

Gas Discharge Tubes

Use For: Discharge Tubes; Gas Tubes
See Also: Electron Tubes; Ignitrons; Luminous Tube Signs; Neon Tubes; Phototubes; Thyratrons; Voltage Regulator Tubes

CNS K4071-82. Household Gas Tube (Nov)(9620).

—Connectors
CSA C22.2 NO 34-M1987. Electrode Receptacles, Fittings, and Connectors for Gas Tubes (R 1993); (Gen Instr 1 Thru 2). 23 pp.

—Electrode Fittings
CSA C22.2 NO 34-M1987. Electrode Receptacles, Fittings, and Connectors for Gas Tubes (R 1993); (Gen Instr 1 Thru 2). 23 pp.

—Lighting—Cables (Electric)
CSA C22.2 NO 17-1973. Cable for Luminous-Tube Signs and for Oil-and Gas-Burner Ignition Equipment (R 1992); (Gen Instr 1) (Amd 1 February 1981) (Amd 2-8 March 1988). 28 pp.

—Stage Lighting
CSA C22.2 NO 166-M1983. Stage and Studio Luminaires; (Gen Instr 1) (Erratum May 1983). 26 pp.

—Studio Lighting
CSA C22.2 NO 166-M1983. Stage and Studio Luminaires; (Gen Instr 1) (Erratum May 1983). 26 pp.

—Surge Arresters
IECQ PQC 78-88. Sectional Specification for Surge Protective Gas Discharge Tubes. 16 pp.
IECQ PQC 79-88. Blank Detail Specification for Surge Protective Gas Discharge Tubes. 11 pp.

—Telecommunication Equipment
CCITT RECMN K.12-89. Characteristics of Gas Discharge Tubes for the Protection of Telecommunications Installations—Protection Against Interference (Study Group V) 13 pp. 13 pp.

Gas Distribution

Use For: Gas Supply *See Also:* Distribution Systems; Fuel Dispensing Equipment; Gas Pipelines; Gas Supply Installations

SAA AS 1697-81. Gas Transmission and Distribution Systems (Known as the SAA Gas Pipeline Code). 126 pp.
SNZ NZS 5258-89. Code of Practice for Gas Distribution. 70 pp.

—Engineering Drawings
SAA AS 1100.401 Supp 1-84. Technical Drawing—Part 401: Engineering Survey and Engineering Survey Design Drawing—Supplement 1: Aviation Facilities (Supplement to AS 1100. 401—1984). 7 pp.
SAA AS 1100.401 Supp 2-84. Technical Drawing—Part 401: Engineering Survey and Engineering Survey Design Drawing—Supplement 2: Gas Distribution (Supplement to AS 1100. 401—1984). 3 pp.

INTERNATIONAL AND NON-U.S. NATIONAL STANDARDS
SUBJECT INDEX

Gas

Gas Distribution (Cont.)
—Engineering Drawings (Cont.)
SNZ NZS/AS 1100: Pt401: Supp1-84. Technical Drawing Part 401: Engineering Survey and Engineering Survey Design Drawing Supplement 1: Aviation Facilities. 7 pp.
SNZ NZS/AS 1100: Pt401: Supp2-84. Technical Drawing Part 401: Engineering Survey and Engineering Survey Design Drawing Supplement 2: Gas Distribution. 3 pp.

—Vapor Control
CGSB CAN/CGSB-3.1000-M91. Standard for Vapour Control Systems in Gasoline Distribution Networks. 40 pp.

Gas Expansion Turbines
Use: Gas Turbines

Gas Filled Cable Fittings
DIN VDE 0257-67. Specifications for Gas-Filled Cables in Steel Conduit and Associated Fittings for a.c. and d.c. Installations up to 275 kV Rated Voltage (Nov). 23 pp.

—Internal Cables
DIN VDE 0258A-72. Specifications for Internal Gas-Pressure Cables and Their Accessories for Single and 3-Phase a.c. Systems with Nominal Voltages up to 275 kV: Amendment A (Sept). 9 pp.
DIN VDE 0258 A2-81. Internal Gas-Pressure Cables and Accessories for Alternating and Rotary Current at Voltages up to 275 kV: Amendment 2 (Oct). 1 p.

Gas Filled Cables
See Also: Cables (Electric); Oil Filled Cables

—External
DIN VDE 0257-67. Specifications for Gas-Filled Cables in Steel Conduit and Associated Fittings for a.c. and d.c. Installations up to 275 kV Rated Voltage (Nov). 23 pp.
IEC 141 Pt 3-63. Tests on Oil-Filled and Gas-Pressure Cables and Their Accessories Part 3: External Gas-Pressure (Gas Compression) Cables and Accessories for Alternating Voltages up to 275 kV First Edition; (Amendment 1-1967). 30 pp.

—Internal
DIN VDE 0258A-72. Specifications for Internal Gas-Pressure Cables and Their Accessories for Single and 3-Phase a.c. Systems with Nominal Voltages up to 275 kV: Amendment A (Sept). 9 pp.
DIN VDE 0258 A2-81. Internal Gas-Pressure Cables and Accessories for Alternating and Rotary Current at Voltages up to 275 kV: Amendment 2 (Oct). 1 p.
IEC 141 Pt 2-63. Tests on Oil-Filled and Gas-Pressure Cables and Their Accessories Part 2: Internal Gas-Pressure Cables and Accessories for Alternating Voltages up to 275 kV First Edition; (Amendment 1-1967). 28 pp.

Gas Filled Lamps
See Also: Incandescent Lighting
JIS C 7526-90. Standard Incandescent Lamps of Luminous Intensity. 10 pp.

Gas Filled Tubes
Use For: Gas Filled Valves See Also: Antitransmit Receive Tubes; Corona Stabilizer Tubes; Electron Tubes; Flashtubes; Transmit Receive Limiters; Transmit Receive Tubes
BSI BS 9043-77. 1977 Amd 1 Rules for the Preparation of Detail Specifications for Gas-Filled Microwave Switching Tubes of Assessed Quality: Pre-TR Tubes. Full Assessment Level. 14 pp.

—Electrical Measurement
IEC 151 Pt 17-69. Measurements of the Electrical Properties of Electronic Tubes and Valves Part 17: Methods of Measurement of Gasfilled Tubes and Valves Second Edition. 51 pp.

—Microwave—Switching
MOD UK DSTAN 59-60: Pt 8:Sec E-01. Valves, Electronic (Electronic Tubes) (Listed on EPIC Database) Part 8: Gas-Filled Microwave Switching Tubes Section E: Detail Specifications Issue 1; Amendment 2. 5 pp.
MOD UK DSTAN 59-60: 08/001-71. Valve, Electronic, Gas-Filled Microwave Switching Tube Issue 1. 9 pp.
MOD UK DSTAN 59-60: 08/006-72. Valve Electronic, Gas-Filled Microwave Switching Tube Issue 1. 8 pp.
MOD UK DSTAN 59-60: 08/013-72. Electronic Valve, Gas-Filled Microwave Switching Tube Issue 1. 2 pp.
MOD UK DSTAN 59-60: 08/014-01. Valve Electronic, Gas-Filled Microwave Switching Tube Issue 1; Amendment 1. 6 pp.

Gas Filled Tubes (Cont.)
—Microwave—Switching (Cont.)
MOD UK DSTAN 59-60: 08/032-81. Valve Electronic, Gas-Filled Microwave Switching Tube Issue 1. 10 pp.
MOD UK DSTAN 59-60: 08/033-81. Valve Electronic, Gas-Filled Microwave Switching Tube Issue 1. 2 pp.
MOD UK DSTAN 59-60: 90/006-70. Valve, Electronic, Gas Filled Microwave Switching Tube Issue 1. 6 pp.
MOD UK DSTAN 59-60: 90/027-70. ELectronic Valve, Gas-Filled Microwave Switching Tube Issue 1. 7 pp.
MOD UK DSTAN 59-60: 90/053-70. Valve, Electronic, Gas Filled Microwave Switching Tube Issue 1. 4 pp.

—Microwave—Switching—Pre-TR—Quality Assurance
BSI BS 9043-77. 1977 Amd 1 Rules for the Preparation of Detail Specifications for Gas-Filled Microwave Switching Tubes of Assessed Quality: Pre-TR Tubes. Full Assessment Level. 14 pp.

—Microwave—Switching—Quality Assurance
BSI BS 9040-78. 1978 Gas-Filled Microwave Switching Tubes of Assessed Quality: Generic Data and Methods of Test. 41 pp.

Gas Filled Valves
Use: Gas Filled Tubes

Gas Fires
Use: Heaters

Gas Flow
See Also: Air Flow; Flow Measurement

—Gas Pipelines—Tracer Method
BSI BS 5857: Sec 2.1-80. 1980 Measurement of Fluid Flow in Closed Conduits, Using Tracers Part 2: Measurement of Gas Flow Section 2.1: General. 13 pp.
BSI BS 5857: Sec 2.4-80. 1980 Measurement of Fluid Flow in Closed Conduits, Using Tracers Part 2: Measurement of Gas Flow Section 2.4: Transit Time Method Using Radioactive Tracers. 12 pp.
ISO 4053 Pt I-77. Measurement of Gas Flow in Conduits—Tracer Methods—Part I: General First Edition. 12 pp.
ISO 4053 Pt IV-78. Measurement of Gas Flow in Conduits—Tracer Methods—Part IV: Transit Time Method Using Radioactive Tracers First Edition. 11 pp.

—Gas Pipelines—Venturi Tubes
BSI BS 1042: Sec 1.3-91. 1991 Measurement of Fluid Flow in Closed Conduits Part 1: Pressure Differential Devices Section 1.3: Method of Measurement of Gas Flow by Means of Critical Flow Venturi Nozzles (ISO 9300: 1990). 22 pp.
ISO 9300-90. Measurement of Gas Flow by Means of Critical Flow Venturi Nozzles First Edition. 19 pp.

Gas Generators
Use For: Kipp Gas Generators See Also: Gases; Pneumatic Equipment

—Chemical Analysis
CNS R2139-86. Kipp's Gas-Generator for Chemical Analysis (Mar) (7315).

Gas Grills
Use: Grills (Appliances)—Gas

Gas Hoses
Use: Fuel Hoses

Gas Industry
—Glossaries
BSI BS 1179: Part 6-80. 1980 Terms Used in the Gas Industry Part 6: Combustion and Utilization Including Installation at Consumers' Premises. 44 pp.
BSI BS 1179-67. 1967 Amd 2 Glossary of Terms Used in the Gas Industry. 141 pp.

—Pipe Flanges
CSA CAN/CSA-Z245.12-M91. Steel Flanges; (Gen Instr 1). 58 pp.

—Valves
CSA CAN/CSA-Z245.15-M91. Steel Valves; (Gen Instr 1). 81 pp.

Gas Installations
Use: Gas Supply Installations

Gas Insulation
—Glossaries
BSI BS 4727:Pt1: Group 10-91. 1991 Electrotechnical, Power, Telecommunication, Electronics, Lighting and Colour Terms Part 1: Terms Common to Power Telecommunications and Electronics Group 10: Insulating Solids, Liquids and Gases. 101 pp.
IEC 50 Chap 212-90. International Electrotechnical Vocabulary Chapter 212: Insulating Solids, Liquids and Gases First Edition. 105 pp.

Gas Lifts
Use For: Air Lifts (Materials Handling)
See Also: Lifting Equipment

—Automobiles
CNS D4004-77. Lifts for Automobiles (Apr)(4077). 9 pp.
CNS D4007-85. Airlifts for Automobiles (Jul)(8389).
JIS D 8107-77. Lifts for Automobiles (R 1983). 12 pp.

Gas Lines
Use: Gas Pipelines

Gas Liquid Chromatography
Use: Gas Chromatography

Gas Loaded Accumulators
Use: Hydropneumatic Accumulators

Gas Logs
—Installation
BSI BS 6714-86. (WITHDRAWN) 1986 Installation of Decorative Gas Log and Other Fuel Effect Appliances (1st, 2nd, and 3rd Family Gases) (Superseded by BS 5871: Part 3: 1991). 18 pp.

—Safety
BSI BS 5258: Part 12-90. 1990 Safety of Domestic Gas Appliances Part 12: Decorative Fuel Effect Gas Appliances (2nd and 3rd Family Gases). 34 pp.
BSI BS 5258: Part 12-80. 1980 Amd 1 Safety of Domestic Gas Appliances Part 12: Decorative Gas Log and Other Fuel Effect Appliances (2nd and 3rd Family Gases) (AMD 5434) May 29, 1987. 20 pp.

Gas Mains
Use: Gas Pipelines

Gas Masks
See Also: Protective Masks; Respirators
CNS Z2023-87. Gas Masks (Jul) (6636).
JIS T 8152-81. Gas Masks. 29 pp.

Gas Metal Arc Welding
Use For: Electrogas Welding; Metal Active Gas Welding; Metal Inert Gas Welding See Also: Arc Welding; Electric Welding; Fusion Welding; Gas Metal Arc Welding Equipment; Gas Shielded Arc Welding; Gas Tungsten Arc Welding; Shielded Arc Welding; Shielded Metal Arc Welding; Welding

—Aluminum
BSI BS 3571: Part 1-85. 1985 General Recommendations for Manual Inert-Gas Metal-Arc Welding Part 1: MIG Welding of Aluminium and Aluminium Alloys. 30 pp.
BSI BS 4870: Part 2-82. (WITHDRAWN) 1982 Approval Testing of Welding Procedures Part 2: TIG or MIG of Aluminium and Its Alloys (Superseded by BS EN 288: Part 4: 1992). 16 pp.
JIS Z 3604-85. Recommended Practice for Inert Gas Shielded Arc Welding of Aluminium and Aluminium Alloy. 22 pp.
JIS Z 3811-76. Standard Qualification Procedure for Welding Technique of Aluminium and Aluminium Alloy (R 1980). 18 pp.

—Aluminum Alloys
BSI BS 3571: Part 1-85. 1985 General Recommendations for Manual Inert-Gas Metal-Arc Welding Part 1: MIG Welding of Aluminium and Aluminium Alloys. 30 pp.
BSI BS 4870: Part 2-82. (WITHDRAWN) 1982 Approval Testing of Welding Procedures Part 2: TIG or MIG of Aluminium and Its Alloys (Superseded by BS EN 288: Part 4: 1992). 16 pp.
JIS Z 3604-85. Recommended Practice for Inert Gas Shielded Arc Welding of Aluminium and Aluminium Alloy. 22 pp.
JIS Z 3811-76. Standard Qualification Procedure for Welding Technique of Aluminium and Aluminium Alloy (R 1980). 18 pp.

—Reinforcing Steels
DIN ENGL 4099-85. Welding of Reinforcing Steel; Execution of Welding Work and Testing (Nov). 16 pp.

INDUSTRY STANDARDS

Gas Metal Arc Welding (Cont.)
—Steels
ISO 9692-92. Metal-Arc Welding with Covered Electrode, Gas-Shielded Metal-Arc Welding and Gas Welding—Joint Preparations for Steel First Edition. 17 pp.

Gas Metal Arc Welding Equipment
See Also: Arc Welding Equipment; Gas Metal Arc Welding; Welding Equipment

—Electrodes
CNS G3180-87. Solid Wires for MAG Welding of Mild Steel and High Strength Steel (Oct)(8967).
JIS Z 3312-89. MAG Welding Solid Wires for Mild Steel and High Strength Steel. 13 pp.
JIS Z 3317-91. MAG Welding Solid Wires for Molybdenum Steel and Chromium Molybdenum Steel. 12 pp.
JIS Z 3331-88. Titanium and Titanium Alloy Rods and Wires for Inert Gas Shielded Arc Welding. 9 pp.
JIS Z 3334-88. Nickel and Nickel Alloy Filler Rods and Wires for Arc Welding. 8 pp.
SAA AS 2717.3-92. Welding—Electrodes—Gas Metal Arc—Part 3: Corrosion-Resisting Chromium and Chromium-Nickel Steel Electrodes (in Professional Package 36). 16 pp.

—Electrodes—Classification
CEN PREN 440-90. Classification of Wire Electrodes and Deposits for Gas Metal Arc Welding of Carbon Steels, Carbon-Manganese Steels and Micro Alloyed Steels. 8 pp.

—Electrodes—Flux Cored
JIS Z 3318-91. MAG Welding Flux Cored Wires for Molybdenum Steel and Chromium Molybdenum Steel. 11 pp.
JIS Z 3319-91. Flux Cored Wires for Electrogas Arc Welding. 10 pp.

—Torches
BSI BS 638: Part 8-84. 1984 Amd 1 Arc Welding Power Sources, Equipment and Accessories Part 8: Holders and Handheld Torches and Guns for MIG, MAG and TIG Welding. 24 pp.

—Welding Rods
JIS Z 3334-88. Nickel and Nickel Alloy Filler Rods and Wires for Arc Welding. 8 pp.

Gas Meters
See Also: Flow Measurement; Flowmeters

—Acceptance Testing
SNZ NZS 5259-91. Gas Metering. Gas Meter Acceptance Testing, Meter Installation and Calculation of Energy from Measured Volume Amend: 1, 1993. 61 pp.

—Coin Operated
BSI BS 4161: Part 3-89. 1989 Amd 1 Gas Meters Part 3: Diaphragm Meters of 6 Cubic Metres (or 212 Cubic Feet) per Hour Rating for Working Pressures up to 50 mbar (AMD 6648) October 31, 1990. 39 pp.
BSI BS 4161: Part 3-77. (WITHDRAWN) 1977 Amd 2 Gas Meters Part 3: Unit Construction Meters of 6 Cubic Metres (or 212 Cubic Feet) per Hour Rating. 30 pp.

—Diaphragms
BSI BS 2797-76. 1976 Amd 1 Leathers for Gas Meter Diaphrams. 9 pp.

—Energy Consumption
SNZ NZS 5259-91. Gas Metering. Gas Meter Acceptance Testing, Meter Installation and Calculation of Energy from Measured Volume Amend: 1, 1993. 61 pp.

—Fittings
BSI BS 746-87. 1987 Amd 1 Gas Meter Unions and Adaptors. 16 pp.
BSI BS 746-02. 1987 Amd 2 Gas Meter Unions and Adaptors (E) (AMD 6811) December 24, 1991. 20 pp.
BSI BS 746-03. 1987 Amd 3 Gas Meter Unions and Adaptors (AMD 7159) July 15, 1992. 21 pp.

—Gas Cylinders
SNZ NZS 5425: DIV 3.1P-85. Code of Practice for CNG Compressor and Refuelling Stations Part 3: Metering Devices. Division 3.1P: Metering Method Using Tables and Calculator Programmes. 16 pp.
SNZ NZS 5425: DIV 3.2-84. Code of Practice for CNG Compressor and Refuelling Stations Part 3: Metering Devices. Division 3.2: Installation and Operation of On-Line Metering Devices. 12 pp.

Gas Meters (Cont.)
—Gas Cylinders (Cont.)
SNZ NZS 5425: DIV 3.3-84. Code of Practice for CNG Compressor and Refuelling Stations Part 3: Metering Devices. Division 3.3: Requirements for Type Approval of On-Line Metering Devices Amend: 1, 1993. 25 pp.

—Installation
BSI BS 6400-85. 1985 Amd 1 Installation of Domestic Gas Meters (2nd Family Gases). 20 pp.
SNZ NZS 5259-91. Gas Metering. Gas Meter Acceptance Testing, Meter Installation and Calculation of Energy from Measured Volume Amend: 1, 1993. 61 pp.

—Medical
CEN PREN 738-92. Pressure Regulators and Pressure Regulators with Flow Metering Devices for Use with Medical Gas Supply Systems. 32 pp.
CSA CAN/CSA-Z305.3-M87. Pressure Regulators, Gauges, and Flow-Metering Devices for Medical Gases; (Gen Instr 1). 25 pp.

—Volume Correctors
BSI BS 4161: Part 8-87. 1987 Gas Meters Part 8: Electronic Correctors Volume. 11 pp.

Gas/Oil Separators
Use For: Gas Separators; Oil Separators
See Also: Separators (Mechanical)

—Design
CEN PREN 858-1-92. Installations for Separation of Light Liquids (e.g. Oil and Petrol)—Part 1: Principles of Design—Performance and Testing—Marking and Quality Control. 47 pp.

—Identification Systems
CEN PREN 858-1-92. Installations for Separation of Light Liquids (e.g. Oil and Petrol)—Part 1: Principles of Design—Performance and Testing—Marking and Quality Control. 47 pp.

—Quality Control
CEN PREN 858-1-92. Installations for Separation of Light Liquids (e.g. Oil and Petrol)—Part 1: Principles of Design—Performance and Testing—Marking and Quality Control. 47 pp.

—Refrigeration Equipment
CSA C22.2 NO 140.3-M1987. Refrigerant-Containing Components for Use in Electrical Equipment (R 1993). 19 pp.

Gas Oils
See Also: Petroleum Products
JIS K 2204-92. Gas Oil. 7 pp.

—Asphaltene Content
BSI BS 2000: Part 143-93. 1993 Methods of Test for Petroleum and Its Products Part 143: Determination of Asphaltenes (Heptane Insolubles) (W). 6 pp.
BSI BS 2000: Part 143-85. 1985 Petroleum and Its Products Part 143: Asphaltenes in Petroleum Products (Precipitation with Normal Heptane). 8 pp.

—Distillation Methods
BSI BS 7392-90. 1990 Method for Determination of Distillation Characteristics of Petroleum Products. 26 pp.
ISO 3405-88. Petroleum Products—Determination of Distillation Characteristics Second Edition. 24 pp.

—Low Temperature Testing
BSI BS 6380-83. (WITHDRAWN) 1983 Amd 1 Low Temperature Properties and Cold Weather Use of Diesel Fuels and Gas Oils (Classes A1, A2 and D of BS 2869) (AMD 5452) September 30, 1986. 16 pp.

—Plugging Point
JIS K 2288-87. Testing Method for Cold Filter Plugging Point of Gas Oil. 13 pp.

—Sulfur Content
DIN ENGL 51400 Pt 1-78. Testing of Mineral Oils and Fuels; Determination of the Sulfur Content (Total Sulfur); General Working Principles (Feb). 6 pp.
DIN ENGL 51400 Pt 3-78. Testing of Mineral Oils and Fuels; Determination of the Sulfur Content (Total Sulfur); Combustion According to Schoniger; Thorin-Sulfonazo-III Titration (Feb). 3 pp.
DIN ENGL 51400 Pt 4-90. Determination of Total Sulfur Content of Gaseous Petroleum Products by the Lingener Combustion Method (Oct). 8 pp.
DIN ENGL 51400 Pt 8-78. Testing of Mineral Oils and Fuels; Determination of Sulfur Content (Total Sulfur); Nickel Reduction Method; Dithizone Titration (Feb). 4 pp.

—Sulfur Content—Analyzers
JIS B 7995-84. Automatic Analyzers for Sulphur in Crude Oil and Petroleum Products. 24 pp.

Gas Permeability
Use For: Air Permeability See Also: Leakage

—Cellular Plastics
ISO 4638-84. Polymeric Materials, Cellular Flexible—Determination of Air Flow Permeability First Edition. 9 pp.

—Cements
BSI BS EN 196-6-92. 1992 Methods of Testing Cement Part 6: Determination of Fineness. 23 pp.
CEN EN 196 (Part 6)-89. Methods of Testing Cement; Determination of Fineness. 17 pp.

—Cigarette Papers
BSI BS 5381: Part 4-80. (WITHDRAWN) 1980 Determination of Physical Properties of Tobacco and Tobacco Products Part 4: Determination of Air Permeability of Material Used as Cigarette Papers. 6 pp.
ISO 2965-79. Material Used as Cigarette Papers—Determination of Air Permeability First Edition. 5 pp.

—Containment Cabinets
SAA AS 1807.25-90. Cleanrooms, Workstations and Safety Cabinets—Methods of Test—Part 25: Determination of Gas Tightness of Outer Shell of Biological Safety Cabinets. 1 p.

—Cotton Fibers
BSI BS 3181: Part 1-87. 1987 Method for the Determination of Cotton Fiber Properties the Airflow Method Part 1: Determination of Micronaire Value by the Single Compression Airflow Method. 8 pp.
ISO 2403-72. Textiles—Cotton Fibres—Determination of Micronaire Value First Edition. 8 pp.

—Doors
CEN PREN 1026-93. Windows and Doors—Air Permeability—Test Method. 8 pp.
ISO 8272-85. Doorsets—Air Permeability Test First Edition. 5 pp.

—Elastomers
DIN ENGL 53536-85. Testing of Rubber and Elastomers; Determination of Permeability to Gases (Feb). 5 pp.

—Fabrics
BSI BS 5636-90. 1990 Amd 1 Determination of Permeability of Fabrics to Air (AMD 6341) June 28, 1991. 6 pp.
CGSB CAN/CGSB-4.2 NO.36-M89. Textile Test Methods Air Permeability. 8 pp.
CNS L3081-80. Method of Test for Air Permeability of Fabrics (May)(5612).
DIN ENGL 53887-86. Testing of Textiles; Determination of Air Permeability of Textile Fabrics (Aug). 4 pp.

—Firebricks
JIS R 2115-87. Testing Method for Permeability to Gases of Refractory Bricks. 8 pp.

—Foundry Sands
CNS Z8029-82. Method for Determining Permeability of Foundry Sands (May)(8896).
JIS Z 2603-76. Method for Determining Permeability of Foundry Sands. 10 pp.

—Glass Fabrics
CNS K6706-82. Method of Test of Glass Fiber Mat (Jan)(8428). 2 pp.

—Hoses
BSI BS 5173: Sec 103.11-92. 1992 Methods of Test for Rubber and Plastics Hoses and Hose Assemblies Part 103: Physical Tests Section 103.11: Determination of Gas Permeance (ISO 4080: 1991). 9 pp.
BSI BS 5173: Sec 103.11-88. 1988 Rubber and Plastics Hoses and Hose Assemblies Part 103: Physical Tests Section 103.11: Determination of Gas Permeance. 6 pp.
ISO 4080-91. Rubber and Plastics Hoses and Hose Assemblies—Determination of Permeability to Gas Third Edition. 8 pp.

—Joints
BSI BS 6181-81. 1981 Air Permeability of Joints in Building. 8 pp.
ISO 2445-72. Joints in Building—Fundamental Principles for Design First Edition. 3 pp.
ISO 3447-75. Joints in Building—General Check-List of Joint Functions First Edition. 4 pp.
ISO 6589-83. Joints in Building—Laboratory Method of Test for Air Permeability of Joints Second Edition. 5 pp.
ISO 7727-84. Joints in Building—Principles for Jointing of Building Components—Accommodation of Dimensional Deviations During Construction First Edition. 7 pp.

INTERNATIONAL AND NON-U.S. NATIONAL STANDARDS
SUBJECT INDEX

Gas Permeability (Cont.)

—Joints (Cont.)

JIS B 1186-79. Sets of High Strength Hexagon Bolt, Hexagon Nut and Plain Washers for Friction Grip Joints. 21 pp.

—Metal Powders

ISO 10070-91. Metallic Powders—Determination of Envelope-Specific Surface Area from Measurements of the Permeability to Air of a Powder Bed Under Steady-State Flow Conditions First Edition. 16 pp.

—Paperboard

BSI BS 6538: Part 1-85. 1985 Air Permeance of Paper and Board Part 1: Method for Determination of Air Permeance: General Aspects of Testing (Supersedes BS 2925: 1958). 8 pp.

BSI BS 6538: Part 2-92. 1992 Air Permeance of Paper and Board Part 2: Method for Determination of Air Permeance Using the Bendtsen Apparatus (ISO 5636-3: 1992) (Supersedes BS 2925: 1958) (H). 15 pp.

BSI BS 6538: Part 2-85. 1985 Air Permeance of Paper and Board Part 2: Method for Determination of Air Permeance Using the Bendtsen Apparatus. 11 pp.

BSI BS 6538: Part 3-90. 1990 Amd 1 Air Permeance of Paper and Board Part 3: Method for Determination of Our Permeance Using the Gurley Apparatus (AMD 6516) November 30, 1990 (Supersedes BS 2925: 1958). 10 pp.

CNS P3007-82. Method of Test for Air Permeability of Paper and Paper Board (Nov)(1357). 3 pp.

DIN ENGL 53120 Pt 1-79. Testing of Paper and Board; Determination of Air Permeability; Method for Medium Rates of Air Permeability According to Bendtsen (Aug). 5 pp.

DIN ENGL 53120 Pt 2-79. Testing of Paper and Board; Determination of Air Permeability; Method for Medium Rates of Air Permeability According to Schopper (Aug). 5 pp.

ISO 5636 Pt 1-84. Paper and Board—Determination of Air Permeance (Medium Range)—Part 1: General Method First Edition. 6 pp.

ISO 5636 Pt 2-84. Paper and Board—Determination of Air Permeance (Medium Range)—Part 2: Schopper Method First Edition. 5 pp.

ISO 5636 Pt 3-92. Paper and Board—Determination of Air Permeance (Medium Range)—Part 3: Bendtsen Method Second Edition. 12 pp.

ISO 5636 Pt 4-86. Paper and Board—Determination of Air Permeance (Medium Range)—Part 4: Sheffield Method First Edition. 9 pp.

ISO 5636 Pt 5-86. Paper and Board—Determination of Air Permeance (Medium Range)—Part 5: Gurley Method First Edition; (Corrigendum 1-1990). 8 pp.

ISO 11004-92. Paper and Board—Determination of Air Permeance—Low Range First Edition. 10 pp.

JIS P 8117-80. Testing Method for Air Permeability of Paper and Paperboard. 6 pp.

SAA AS 1301.440S-91. Methods of Test for Pulp and Paper —Part 440s: Bendtsen Air Permeance of Paper and Board (This is a Joint Standard with SANZ NZS 1301). 7 pp.

SNZ NZS/AS 1301. 440S-91. Methods of Test for Pulp and Paper Bendtsen Air Permeance of Paper and Board (This is a Joint Standard with SAA AS 1301.440S). 7 pp.

—Papers

BSI BS 6538: Part 1-85. 1985 Air Permeance of Paper and Board Part 1: Method for Determination of Air Permeance: General Aspects of Testing (Supersedes BS 2925: 1958). 8 pp.

BSI BS 6538: Part 2-92. 1992 Air Permeance of Paper and Board Part 2: Method for Determination of Air Permeance Using the Bendtsen Apparatus (ISO 5636-3: 1992) (Supersedes BS 2925: 1958) (H). 15 pp.

BSI BS 6538: Part 2-85. 1985 Air Permeance of Paper and Board Part 2: Method for Determination of Air Permeance Using the Bendtsen Apparatus. 11 pp.

BSI BS 6538: Part 3-90. 1990 Amd 1 Air Permeance of Paper and Board Part 3: Method for Determination of Our Permeance Using the Gurley Apparatus (AMD 6516) November 30, 1990 (Supersedes BS 2925: 1958). 10 pp.

CNS P3007-82. Method of Test for Air Permeability of Paper and Paper Board (Nov)(1357). 3 pp.

DIN ENGL 53120 Pt 1-79. Testing of Paper and Board; Determination of Air Permeability; Method for Medium Rates of Air Permeability According to Bendtsen (Aug). 5 pp.

DIN ENGL 53120 Pt 2-79. Testing of Paper and Board; Determination of Air Permeability; Method for Medium Rates of Air Permeability According to Schopper (Aug). 5 pp.

ISO 5636 Pt 1-84. Paper and Board—Determination of Air Permeance (Medium Range)—Part 1: General Method First Edition. 6 pp.

Gas Permeability (Cont.)

—Papers (Cont.)

ISO 5636 Pt 2-84. Paper and Board—Determination of Air Permeance (Medium Range)—Part 2: Schopper Method First Edition. 5 pp.

ISO 5636 Pt 3-92. Paper and Board—Determination of Air Permeance (Medium Range)—Part 3: Bendtsen Method Second Edition. 12 pp.

ISO 5636 Pt 4-86. Paper and Board—Determination of Air Permeance (Medium Range)—Part 4: Sheffield Method First Edition. 9 pp.

ISO 5636 Pt 5-86. Paper and Board—Determination of Air Permeance (Medium Range)—Part 5: Gurley Method First Edition; (Corrigendum 1-1990). 8 pp.

ISO 11004-92. Paper and Board—Determination of Air Permeance—Low Range First Edition. 10 pp.

JIS P 8117-80. Testing Method for Air Permeability of Paper and Paperboard. 6 pp.

SAA AS 1301.420S-89. Methods of Test for Pulp and Paper (Metric Units)—Part 420s: Gurley Air Permeance of Paper (This is a Joint Standard with SANZ NZS 1301). 4 pp.

SAA AS 1301.440S-91. Methods of Test for Pulp and Paper —Part 440s: Bendtsen Air Permeance of Paper and Board (This is a Joint Standard with SANZ NZS 1301). 7 pp.

SAA AS 1301.447S-91. Methods of Test for Pulp and Paper—Part 447S: Sheffield Air Permeance of Paper (This is a Joint Standard with SANZ NZS 1301.447S). 6 pp.

SNZ NZS/AS 1301. 420S-89. Methods of Test for Pulp and Paper Gurley Air Permeance of Paper (This is a Joint Standard with SAA AS 1301.420S). 4 pp.

SNZ NZS/AS 1301. 440S-91. Methods of Test for Pulp and Paper Bendtsen Air Permeance of Paper and Board (This is a Joint Standard with SAA AS 1301.440S). 7 pp.

SNZ NZS/AS 1301. 447S-91. Methods of Test for Pulp and Paper Sheffield Air Permeance of Paper (This is a Joint Standard with SAA AS 1301.447S). 6 pp.

—Pillows

BSI BS 4578-70. 1970 Amd 1 Methods of Test for Hardness of, and Air Flow Through, Infants' Pillows. 10 pp.

—Plastic Sheets

DIN ENGL 53380-69. Testing of Plastics Films; Determination of the Gas Transmission Rate (June). 4 pp.

—Plastics

BSI BS 2782:Pt8: METH 821A-79. 1979 Methods of Testing Plastics Part 8: Other Properties Method 821A: Determination of the Gas Transmission Rate of Films and Thin Sheets Under Atmospheric Pressure Manometric Method). 8 pp.

ISO 2556-74. Plastics—Determination of the Gas Transmission Rate of Films and Thin Sheets Under Atmospheric Pressure—Manometric Method First Edition. 8 pp.

JIS K 7126-87. Testing Method for Gas Transmission Rate Through Plastic Film and Sheeting. 12 pp.

—Portland Cements

CNS R3059-84. Method of Test for Fineness of Portland Cement (by Air Permeability Apparatus) (Jan)(2924).

—Powders (Particles)

BSI BS 4359: Part 2-82. 1982 Amd 1 Determination of Specific Surface of Powders Part 2: Recommended Air Permeability Methods. 23 pp.

—Refractory Materials

BSI BS 1902: Sec 3.9-81. 1981 Amd 2 Methods for Testing Refractory Materials Part 3: General and Textural Properties Section 3.9: Determination of Permability to Gases (Method 1902-309). 11 pp.

CEN PREN 993-4-93. Methods of Test for Dense Shaped Refractory Products—Part 4: Determination of Permeability to Gases. 8 pp.

SAA AS 1774.7-91. Refractories and Refractory Materials—Physical Test Methods—Part 7: Determination of Permeability to Gases (Revision of AS R31.7—1966). 6 pp.

—Refractory Shapes

ISO 8841-91. Dense, Shaped Refractory Products—Determination of Permeability to Gases First Edition. 9 pp.

—Surface Properties

MOD UK M 825/62. Specific Surface Area and Surface Mean Diameter Determination by Air Permeability.

—Vulcanized Rubber

BSI BS 903: Part A17-73. 1973 Methods of Testing Vulcanized Rubber Part A17: Determination of the Permeability of Rubber to Gases (Constant Volume Method). 11 pp.

Gas Permeability (Cont.)

—Vulcanized Rubber (Cont.)

BSI BS 903: Part A30-75. 1975 Methods of Testing Vulcanized Rubber Part A30: Determination of the Permeability of Rubber to Gases (Constant Pressure Method). 10 pp.

ISO 1399-82. Rubber, Vulcanized—Determination of Permeability to Gases—Constant Volume Method Second Edition. 7 pp.

ISO 2782-77. Rubber, Vulcanized—Determination of Permeability to Gases—Constant Pressure Method First Edition. 7 pp.

—Wall Panels

BSI BS 4315: Part 2-70. 1970 Amd 1 Methods of Test for Resistance to Air and Water Penetration Part 2: Permeable Walling Constructions (Water Penetration). 17 pp.

—Windows

BSI BS 4315: Part 1-68. 1968 Amd 1 Methods of Test for Resistance to Air and Water Penetration Part 1: Windows and Structural Gasket-Glazing Systems. 18 pp.

BSI BS 5368: Part 1-76. 1976 Method of Testing Windows Part 1: Air Permeability Test. 9 pp.

BSI BS 6375: Part 1-89. 1989 Performance of Windows Part 1: Classification for Weathertightness (Including Guidance on Selection and Specification). 12 pp.

BSI BS 6375: Part 1-83. (WITHDRAWN) 1983 Amd 1 Performance of Windows Part 1: Classification for Weathertightness (Including Guidance on Selection and Specification). 12 pp.

CEN EN 42-75. Methods of Festing Windows—Part 1: Air Permeability Test. 7 pp.

CEN PREN 1026-93. Windows and Doors—Air Permeability—Test Method. 8 pp.

ISO 6613-80. Windows and Door Height Windows—Air Permeability Test First Edition. 5 pp.

Gas Pipe Fittings

See Also: Gas Pipelines; Gas Pipes; Pipe Fittings

—Collars—Ductile Iron

DIN ENGL 28624-90. Ductile Iron Collars for Use with Gas and Water Pipes; Dimensions and Mass (Jan). 3 pp.

—Iron

DIN ENGL 28600-83. Ductile Iron Pressure Pipes and Fittings for Gas and Water Pipelines; Technical Delivery Conditions (Jan). 12 pp.

—Medical

CSA CAN/CSA-Z305.2-M88. Low-Pressure Connecting Assemblies for Medical Gas Systems. 34 pp.

—Polyethylene

SAA AS 1667.2-84. Plastic Pipes and Fittings for Gas Reticulation—Polyethylene—Nominal Size Series—Part 2: Fittings Amdt 1 December 1985.

—Pressure—Ductile Iron

CEN PREN 969-92. Ductile Iron Pipes, Fittings Accessories and Their Joints for Gas Pipelines—Requirements and Test Methods. 57 pp.

DIN ENGL 28600-83. Ductile Iron Pressure Pipes and Fittings for Gas and Water Pipelines; Technical Delivery Conditions (Jan). 12 pp.

DIN ENGL 28624-90. Ductile Iron Collars for Use with Gas and Water Pipes; Dimensions and Mass (Jan). 3 pp.

DIN ENGL 28625-90. Ductile Iron Double Socket 90 Degree Bends for Use with Gas and Water Pipes; Dimensions and Mass (Jan). 3 pp.

DIN ENGL 28626-90. Ductile Iron Double Socket 45 Degree Bends for Use with Gas and Water Pipes; Dimensions and Mass (Jan). 3 pp.

DIN ENGL 28627-90. Ductile Iron Double Socket 30 Degree Bends for Use with Gas and Water Pipes; Dimensions and Mass (Jan). 3 pp.

DIN ENGL 28628-90. Ductile Iron Double Socket 22 1/2 Degree Bends for Use with Gas and Water Pipes; Dimensions and Mass (Jan). 3 pp.

DIN ENGL 28629-90. Ductile Iron Double Socket 11 1/4 Degrees Bends for Use with Gas and Water Pipes; Dimensions and Mass (Jan). 3 pp.

DIN ENGL 28632-90. Ductile Iron Double Socket Tees with Socket Branch for Use with Gas and Water Pipes; Dimensions and Mass (Jan). 3 pp.

DIN ENGL 28634-90. Ductile Iron Double Socket Tapers for Use with Gas and Water Pipes; Dimensions and Mass (Jan). 4 pp.

—Pressure—Flanged—Ductile Iron

DIN ENGL 28622-90. Ductile Iron Flanged Socket Pieces for Use with Gas and Water Pipes; Dimensions and Mass (Jan). 3 pp.

DIN ENGL 28645-90. Ductile Iron Double-Flanged Tapers for Use with Gas and Water Pipes; Dimensions and Mass (Jan). 3 pp.

Gas Pipe Fittings (Cont.)

—PVC—Unplasticized
SAA AS 1464.2-84. Plastic Pipes and Fittings for Gas Reticulation—Unplasticized PVC (UPVC)—Part 2: Fittings Amdt 1 May 1986.

—Socket—Iron
BSI BS 78: Part 2-65. (OBSOLESCENT) 1965 Amd 1 Cast Iron Spigot and Socket Pipes (Vertically Cast) and Spigot and Socket Fittings Part 2: Fittings (Partially Superseded by BS 4772). 73 pp.

—Steel
DIN ENGL 2470 Pt 1-87. Steel Gas Pipelines for Permissible Working Pressures up to 16 Bar; Pipes and Fittings (Dec). 6 pp.
DIN ENGL 2470 Pt 2-83. Steel Gas Pipelines for Permissible Working Pressures Exceeding 16 Bar; Requirements for Pipeline Components (May). 8 pp.

—Welded—Polyethylene
JIS K 6775-89. Polyethylene Pipe-Fittings for the Supply of Gaseous Fuels. 52 pp.

—Welded—Socket—Polyethylene
CSA CAN/CSA-B137.4.1-M89. Electrofusion-Type Polyethylene Fittings for Gas Services; (Gen Instr 1). 21 pp.
JIS K 6775-89. Polyethylene Pipe-Fittings for the Supply of Gaseous Fuels. 52 pp.

Gas Pipe Joints
See Also: Pipe Joints

—Joint Sealants
BSI BS 5292-80. 1980 Amd 2 Jointing Materials and Compounds for Installations Using Water, Low-Pressure Steam or 1st, 2nd and 3rd Family Gases (AMD 5960) November 30, 1988. 22 pp.
BSI BS 5292-03. 1980 Amd 3 Jointing Materials and Compounds for Installations Using Water, Low-Pressure Steam or 1st, 2nd and 3rd Family Gases (AMD 6918) January 31, 1992 (Q). 24 pp.
BSI BS 5292-04. 1980 Amd 4 Jointing Materials and Compounds for Installations Using Water, Low-Pressure Steam or 1st, 2nd and 3rd Family Gases (AMD 7047) April 1, 1992. 26 pp.

—Pressure—Ductile Iron
CEN PREN 969-92. Ductile Iron Pipes, Fittings Accessories and Their Joints for Gas Pipelines—Requirements and Test Methods. 57 pp.

—Sealing Rings—Vulcanized Rubber
CEN PREN 682-92. Elastomeric Seals—Materials Requirements for Joint Seals Used in Pipes and Fittings Carrying Gas and Hydrocarbon Fluids. 14 pp.

Gas Pipelines
Use For: Gas Lines; Gas Mains *See Also:* Curb Boxes; Gas Distribution; Gas Pipe Fittings; Natural Gas; Oil Pipelines; Pipelines
CSA CAN/CSA-Z184-M92. Gas Pipeline Systems; (Gen Instr 1). 326 pp.
CSA CAN/CSA-Z187-M87. Offshore Pipelines (R 1992). 74 pp.
SAA AS 1697-81. Gas Transmission and Distribution Systems (Known as the SAA Gas Pipeline Code). 126 pp.

—Compressed Natural Gas—Fuel Dispensing Equipment
CSA B51-M1991. Boiler, Pressure Vessel, and Pressure Piping Code; (Gen Instr 1). 96 pp.

—Ductile Iron—Pressure
CEN PREN 969-92. Ductile Iron Pipes, Fittings Accessories and Their Joints for Gas Pipelines—Requirements and Test Methods. 57 pp.
DIN ENGL 28614-90. Ductile Iron Pipes with Cast-On Flanges for Use with Gas and Water Pipelines; Dimensions and Mass (Jan). 3 pp.
DIN ENGL 28615 Pt 1-90. Ductile Iron Pipes with Welded-On Flanges for Use with Gas and Water Pipelines; Dimensions and Mass (Jan) (Together with the January 1990 Edition of DIN 28615 Part 2, Supersedes the March 1976 Edition of DIN 28615). 3 pp.
DIN ENGL 28615 Pt 2-90. Ductile Iron Pipes with Screwed-On Flanges for Use with Gas and Water Pipelines; Dimensions and Mass (Jan) (Together with the January 1990 Edition of DIN 28615 Part 1, Supersedes the March 1976 Edition of DIN 28615). 3 pp.

—Engineering Drawings—DVGW Codes
DIN ENGL 2425 Pt 3-80. Plans for Public Utilities, Water Resources and Long-Distance Lines; Plans for Long-Distance Pipelines; Technical Regulation of the DVGW (German Gas and Water Engineers Association) (May). 14 pp.

Gas Pipelines (Cont.)

—Gas Flow—Tracer Method
BSI BS 5857: Sec 2.1-80. 1980 Measurement of Fluid Flow in Closed Conduits, Using Tracers Part 2: Measurement of Gas Flow Section 2.1: General. 13 pp.
BSI BS 5857: Sec 2.4-80. 1980 Measurement of Fluid Flow in Closed Conduits, Using Tracers Part 2: Measurement of Gas Flow Section 2.4: Transit Time Method Using Radioactive Tracers. 12 pp.
ISO 4053 Pt I-77. Measurement of Gas Flow in Conduits—Tracer Methods—Part I: General First Edition. 12 pp.
ISO 4053 Pt IV-78. Measurement of Gas Flow in Conduits—Tracer Methods—Part IV: Transit Time Method Using Radioactive Tracers First Edition. 11 pp.

—Gas Flow—Venturi Tubes
BSI BS 1042: Sec 1.3-91. 1991 Measurement of Fluid Flow in Closed Conduits Part 1: Pressure Differential Devices Section 1.3: Method of Measurement of Gas Flow by Means of Critical Flow Venturi Nozzles (ISO 9300: 1990). 22 pp.
ISO 9300-90. Measurement of Gas Flow by Means of Critical Flow Venturi Nozzles First Edition. 19 pp.

—Marine—Polyethylene Coatings
CSA CAN/CSA-Z245.21-M92. External Polyethylene Coating for Pipe; (Gen Instr 1 Thru 2). 87 pp.

—Medical
CSA CAN/CSA-Z305.1-92. Nonflammable Medical Gas Piping Systems; (Gen Instr 1). 104 pp.
ISO 7396-87. Non-Flammable Medical Gas Pipeline Systems First Edition. 44 pp.

—Medical—Hose Assemblies
BSI BS 5682-84. 1984 Amd 1 Terminal Units, Hose Assemblies and Their Connectors for Use with Medical Gas Pipeline Systems. 33 pp.

—Medical—Identification Systems
CGSB CAN/CGSB-24.2-M86. Identification of Medical Gas Containers, Pipelines and Valves. 11 pp.

—Medical—Oxygen Concentrators
BSI BS 7634-93. 1993 Oxygen Concentrators for Use with Medical Gas Pipeline Systems (ISO 10083: 1992) (N). 15 pp.
CSA CAN/CSA-Z305.6-92. Medical Oxygen Concentrator Central Supply System: with Nonflammable Medical Gas Piping Systems; (Gen Instr 1). 47 pp.
ISO 10083-92. Oxygen Concentrators for Use with Medical Gas Pipeline Systems First Edition. 14 pp.

—Medical—Terminal Units
BSI BS 5682-84. 1984 Amd 1 Terminal Units, Hose Assemblies and Their Connectors for Use with Medical Gas Pipeline Systems. 33 pp.
CEN PREN 737-1-92. Medical Gas Pipeline Systems—Part 1: Terminal Units for Compressed Medical Gases and Vacuum. 18 pp.
CSA CAN/CSA-Z305.5-M86. Medical Gas Terminal Units. 21 pp.
ISO 9170-90. Terminal Units for Use in Medical Gas Pipeline Systems First Edition. 21 pp.

—Medical—Testing—Company Certification
CSA CAN3-Z305.4-M85. Qualification Requirements for Agencies Testing Non-Flammable Medical Gas Piping Systems; (Gen Instr 1). 15 pp.

—Polyethylene
CSA CAN/CSA-B137.4-92. Polyethylene Piping Systems for Gas Services; (Gen Instr 1). 40 pp.
SNZ NZS 7646-78. Specification for Polyethylene Pipes and Fittings for Gas Reticulation. 23 pp.

—Polyethylene—Installation
SNZ NZS 7647-79. Code of Practice for the Installation of Polyethylene Gas Pipes. 15 pp.

—Pressure
SNZ NZS 5223: Part 1-85. Code of Practice for High Pressure Gas and Petroleum Liquids Pipelines Part 1: High Pressure Gas Pipelines. 77 pp.

—Pressure Measurement
SAA AS 1978-87. Pipelines—Gas and Liquid Petroleum—Field Pressure Testing (Known as the SAA Code for Fied Pressure Testing of Pipelines). 43 pp.

—Sealing Rings
BSI BS 6956: Part 1-88. 1988 Jointing Materials and Compounds Part 1: Corrugated Metal Joint Rings. 8 pp.

—Submarine
SAA AS 1958-81. Gas and Liquid Petroleum Submarine Pipelines (Known as the SAA Submarine Pipeline Code). 111 pp.

Gas Pipelines (Cont.)

—Underground—Polyethylene Coatings
CSA CAN/CSA-Z245.21-M92. External Polyethylene Coating for Pipe; (Gen Instr 1 Thru 2). 87 pp.

Gas Pipes
See Also: Gas Pipe Fittings; Pipes

—Aluminum—Line—Coils
CSA CAN/CSA-Z245.6-92. Coiled Aluminum Line Pipe and Accessories; (Gen Instr 1). 28 pp.

—Asbestos Cement—Pressure
CNS A2077-81. Asbestos Cement Pressure Pipes (Apr)(5179). 6 pp.

—Cast Iron—Pressure
BSI BS 1211-58. (OBSOLESCENT) 1958 Centrifugally Cast (Spun) Iron Pressure Pipes for Water, Gas and Sewage (Partially Superseded by BS 4772). 26 pp.

—Construction Contracts
DIN ENGL 18307-88. Tendering and Performance Stipulations in Contracts for Construction Works (VOB); Part C: General Technical Specifications in Contracts for Construction Works (ATV); Laying of Gas and Water Pipes Below Ground (Sept) (This Standard, Together with DIN 18299,. 4 pp.

—Copper
BSI BS 2871: Part 1-71. 1971 Amd 2 Copper and Copper Alloys. Tubes Part 1: Copper Tubes for Water, Gas and Sanitation. 18 pp.
BSI BS 6891-88. 1988 Installation of Low Pressure Gas Pipework of up to 28mm (R1) in Domestic Premises (2nd Family Gas). 18 pp.
SAA AS 1432-90. Copper Tubes for Plumbing, Gasfitting and Drainage Applications. 9 pp.
SNZ NZS 3501-76. Specification for Copper Tubes for Water, Gas, and Sanitation Amend: 1, 1983; 2; 3. 13 pp.

—Ductile Iron—Pressure
DIN ENGL 28600-83. Ductile Iron Pressure Pipes and Fittings for Gas and Water Pipelines; Technical Delivery Conditions (Jan). 12 pp.
DIN ENGL 28610 Pt 2-83. Ductile Iron Pressure Pipes with Socket for Gas and Water Pipelines Rated for Pressures over 4 Bar up to and Including 16 Bar; Dimensions and Masses (Jan). 6 pp.
DIN ENGL 28614-90. Ductile Iron Pipes with Cast-On Flanges for Use with Gas and Water Pipelines; Dimensions and Mass (Jan). 3 pp.
DIN ENGL 28615 Pt 1-90. Ductile Iron Pipes with Welded-On Flanges for Use with Gas and Water Pipelines; Dimensions and Mass (Jan) (Together with the January 1990 Edition of DIN 28615 Part 2, Supersedes the March 1976 Edition of DIN 28615). 3 pp.
DIN ENGL 28615 Pt 2-90. Ductile Iron Pipes with Screwed-On Flanges for Use with Gas and Water Pipelines; Dimensions and Mass (Jan) (Together with the January 1990 Edition of DIN 28615 Part 1, Supersedes the March 1976 Edition of DIN 28615). 3 pp.

—Installation
SNZ NZS 5261-90. Code of Practice for the Installation of Gas Burning Appliances and Equipment Amend: 1, 1993. 164 pp.

—Iron—Pressure
BSI BS 1211-58. (OBSOLESCENT) 1958 Centrifugally Cast (Spun) Iron Pressure Pipes for Water, Gas and Sewage (Partially Superseded by BS 4772). 26 pp.

—Iron—Pressure—Cement Mortar Lining
DIN ENGL 28610 Pt 1-83. Ductile Iron Pressure Pipes with Socket with Cement-Mortar Lining for Gas and Water Pipelines; Dimensions and Masses (Jan). 8 pp.

—Plastic
BSI BS 7281-90. 1990 Polyethylene Pipes for the Supply of Gaseous Fuels. 37 pp.
SAA AS 1685-84. Plastics Pipes and Fittings for Gas Reticulation—Polyethylene Compound for Manufacturing. 14 pp.
SNZ NZS/AS 1685-84. Plastics Pipes and Fittings for Gas Reticulation—Polyethylene Compound for Manufacturing (This is a Joint Standard with SAA AS 1685). 14 pp.

—Plastic—Automobiles
CNS D3155-86. Method of Test for Nylon for Air Piping of Automobiles (Oct)(10333).

—Plastic—Thermal Stability
ISO TR10837-91. Determination of the Thermal Stability of Polyethylene (PE) for Use in Gas Pipes and Fittings First Edition. 10 pp.

—Polyethylene
CNS K3097-91. Polyethylene Pipes for Natural Gas Supply (Jan)(12835). 4 pp.

INTERNATIONAL AND NON-U.S. NATIONAL STANDARDS
SUBJECT INDEX

Gas Pipes (Cont.)

—Polyethylene (Cont.)
JIS K 6774-89. Polyethylene Pipes for the Supply of Gaseous Fuels. 33 pp.
SAA AS 1667. Plastic Pipes and Fittings for Gas Reticulation—Polyethylene—Nominal Size Series Bound Together.
SAA AS 1667.1-84. Plastic Pipes and Fittings for Gas Reticulation—Polyethylene—Nominal Size Series—Part 1: Pipes.

—Polyethylene—Underground
ISO 4437-88. Buried Polyethylene (PE) Pipes for the Supply of Gaseous Fuels—Metric Series—Specification First Edition. 11 pp.

—PVC
ISO 2703-73. Buried Unplasticized Polyvinyl Chloride (PVC) Pipes for the Supply of Gaseous Fuels—Metric Series—Specification First Edition. 5 pp.

—PVC—Underground
ISO 6993-90. Buried, High-Impact Poly(Vinyl Chloride) (PVC-HI) Pipes for the Supply of Gaseous Fuels—Specification First Edition. 8 pp.

—PVC—Unplasticized
SAA AS 1464. Plastics Pipes and Fittings for Gas Reticulation—Unplasticized PVC (UPVC) Bound Together. 15 pp.
SAA AS 1464.1-84. Plastics Pipes and Fittings for Gas Reticulation—Unplasticized PVC (UPVC)—Part 1: Pipes Amdt 1 May 1986.

—Spigots—Flanged—Ductile Iron
DIN ENGL 28623-90. Ductile Iron Flanged Spigot Pieces for Use with Gas and Water Pipes; Dimensions and Mass (Jan). 3 pp.

—Steel
BSI BS 6891-88. 1988 Installation of Low Pressure Gas Pipework of up to 28mm (R1) in Domestic Premises (2nd Family Gas). 18 pp.
DIN ENGL 17172-78. Steel Pipes for Long-Distance Pipelines for Combustible Liquids and Gases; Technical Conditions of Delivery (May). 14 pp.
ISO 559-91. Steel Tubes for Water and Sewage Second Edition. 19 pp.
SAA AS 1579-93. Arc Welded Steel Pipes and Fittings for Water and Waste Waste Water (in Professional Packages 61A, 61B). 24 pp.

—Steel—Pressure
SNZ NZS 4442-88. Welded Steel Pipes and Fittings for Water, Sewage and Medium Pressure Gas.

—Stop Valves
BSI BS 1552-89. 1989 Manual Shut-off Valves for Use with 1st, 2nd and 3rd Family Gases. 12 pp.

Gas Plants
Use: Gas Processing Plants

Gas Pokers
Use: Ignition Pokers

Gas Pressure Cables
Use: Gas Filled Cables

Gas Processing Plants
Use For: Gas Plants See Also: Industrial Plants

—Liquefied Petroleum Gas—Electrical Equipment
CNS Z2073-88. Electrical Device for Liquefied Petroleum Gas Plant Use (Dec)(12478).

—Liquefied Petroleum Gas—Measuring Instruments
CNS Z2074-88. Measuring Instruments for Liquefied Petroleum Gas Plant Use (Dec)(12479).

—Liquefied Petroleum Gas—Valves
CNS Z2072-88. Valve for Liquefied Petroleum Gas Plant Use (Dec)(12477).

Gas Ranges
Use: Ranges—Gas

Gas Reagent Persuaders
Use: Governors

Gas Sampling
See Also: Chemical Analysis; Sampling

—Electrical Equipment
BSI BS 5574-78. (WITHDRAWN) 1978 Guide for the Sampling of Gases and of Oil from Oil-Filled Electrical Equipment and for the Analysis of Free and Dissolved Gases (Superseded by BS EN 60567: 1993). 32 pp.

Gas Sampling (Cont.)

—Electrical Equipment (Cont.)
BSI BS EN 60567-93. 1993 Guide for the Sampling of Gases and of Oil from Oil-Filled Electrical Equipment and for the Analysis of Free and Dissolved Gases (IEC 567: 1992) (Supersedes BS 5574: 1978). 54 pp.
CENELEC EN 60567-92. Guide for the Sampling of Gases and of Oil From Oil-Filled Electrical Equipment and for the Analysis of Free and Dissolved Gases. 5 pp.
IEC 567-92. Guide for the Sampling of Gases and of Oil from Oil-Filled Electrical Equipment and for the Analysis of Free and Dissolved Gases Second Edition. 85 pp.

—Tubes
BSI BS 2069-54. 1954 Gas Sampling Tubes. 11 pp.

Gas Scavenging Systems, Anesthetic
Use: Anesthetic Gas Scavenging Systems

Gas Separators
Use: Gas/Oil Separators

Gas Shielded Arc Welding
See Also: Fusion Welding; Gas Metal Arc Welding; Gas Shielded Arc Welding Equipment; Shielded Arc Welding; Welding

—Aluminum—Edge Preparation
DIN ENGL 8552 Pt 1-81. Weld Preparation; Groove Forms for Aluminium and Aluminium Alloys; Gas Welding and Gas-Shielded Arc Welding (May). 5 pp.

—Copper—Edge Preparation
DIN ENGL 8552 Pt 3-82. Edge Preparation for Welding; Groove Forms on Copper and Copper Alloys; Gas Welding and Gas-Shielded Arc Welding (July). 5 pp.

—Magnesium—Aerospace
BSI L 515-73. 1973 Sheet and Strip of Magnesium-1 1/4 Per Cent Zinc-Zirconium Alloy. 2 pp.

—Steels—Edge Preparation
DIN ENGL 2559 Pt 1-73. Edge Preparation for Welding; Directions Regarding Edge Forms; Fusion Welding of Butt Joints in Steel Tubes (May). 3 pp.
DIN ENGL 2559 Pt 2-84. Edge Preparation for Welding; Matching of Inside Diameter for Circumferential Welds on Seamless Pipes (Feb). 4 pp.
DIN ENGL 8551 Pt 1-76. Edge Preparation for Welding; Edge Forms on Steel; Gas Welding, Manual Arc Welding and Gas-Shielded Arc Welding (June). 5 pp.

Gas Shielded Arc Welding Equipment
See Also: Arc Welding Equipment; Gas Shielded Arc Welding; Shielded Metal Arc Welding; Welding Equipment

—Electrodes
BSI BS 2901: Part 1-83. 1983 Filler Rods and Wires for Gas-Shielded Arc Welding Part 1: Ferritic Steels. 11 pp.
BSI BS 2901: Part 2-90. 1990 Filler Rods and Wires for Gas-Shielded Arc Welding Part 2: Stainless Steels. 14 pp.
BSI BS 2901: Part 3-90. 1990 Filler Rods and Wires for Gas-Shielded Arc Welding Part 3: Specification for Copper and Copper Alloys. 11 pp.
BSI BS 2901: Part 4-90. 1990 Filler Rods and Wires for Gas-Shielded Arc Welding Part 4: Specification for Aluminium and Aluminium Alloys and Magnesium Alloys. 13 pp.
BSI BS 2901: Part 5-90. 1990 Filler Rods and Wires for Gas-Shielded Arc Welding Part 5: Specification for Nickel and Nickel Alloys. 12 pp.
BSI BS 6678-86. (WITHDRAWN) 1986 Tungsten Electrodes for Inert Gas Shielded Arc Welding and for Plasma Cutting and Welding (Superseded by BS EN 26848: 1991). 4 pp.
BSI BS EN 26848-91. 1991 Tungsten Electrodes for Inert Gas Shielded Arc Welding and for Plasma Cutting and Welding (ISO 6848: 1984). 12 pp.
CEN EN 26848-91. Tungsten Electrodes for Inert Gas Shielded Arc Welding and for Plasma Cutting and Welding—Codification. 6 pp.
DIN ENGL EN 26848-91. Tungsten Electrodes for Inert Gas Shielded Arc Welding and for Plasma Cutting and Welding; Codification (ISO 6848: 1984) (Oct). 7 pp.
ISO 6848-84. Tungsten Electrodes for Inert Gas Shielded Arc Welding, and for Plasma Cutting and Welding—Codification First Edition. 6 pp.
JIS Z 3315-87. Solid Wires for CO2 Gas Shielded Arc Welding of Atmospheric Corrosion Resisting Steel. 11 pp.

Gas Shielded Arc Welding Equipment (Cont.)

—Electrodes—Flux Cored
JIS Z 3313-87. Arc Welding Flux Cored Wires for Mild Steel and High Strength Steel. 13 pp.
JIS Z 3320-87. Flux Cored Wires for CO2 Gas Shielded Arc Welding of Atmospheric Corrosion Resisting Steel. 12 pp.
JIS Z 3323-89. Stainless Steel Flux Cored Wires. 16 pp.

—Electrodes—Flux Cored—Hard Surfacing
JIS Z 3326-91. Arc Welding Flux Cored Wires for Hardfacing. 10 pp.

—Filler Metal
CSA W48.4-M1980. Solid Mild Steel Filler Metals for Gas Shielded Arc Welding; (Gen Instr 1 Thru 2). 27 pp.
DIN ENGL 1737 Pt 1-84. Filler Metals for Welding Titanium and Titanium-Palladium Alloys; Chemical Composition, Technical Delivery Conditions (June). 3 pp.
DIN ENGL 8559 Pt 1-84. Filler Metals for Gas-Shielded Arc Welding; Wire Electrodes, Filler Wires, Solid Rods and Solid Wires for Gas-Shielded Arc Welding of Unalloyed and Alloyed Steels (July). 8 pp.
DIN ENGL 8573 Pt 1-83. Filler Metals for Welding Unalloyed and Low Alloy Cast Iron Materials; Designation; Technical Delivery Conditions (Jan). 8 pp.
DIN ENGL 8575 Pt 1-84. Filler Metals for Arc Welding of Creep-Resisting Steels; Classification, Designation, Technical Delivery Conditions (Apr). 11 pp.

—Gases
CEN PREN 439-90. Gases for Gas Shielded Arc Welding and Cutting. 9 pp.
DIN ENGL 32526-78. Shielding Gases for Welding (Aug). 3 pp.

Gas Supply
Use: Gas Distribution

Gas Supply Installations
See Also: Gas Distribution

—Plumbing—Construction Contracts
DIN ENGL 18381-90. Tendering and Performance Stipulations in Contracts for Construction Works (VOB); Part C: General Technical Specifications in Contracts for Construction Works (ATV); Installation of Pipework for Gas, Water and Drainage Services (July). 7 pp.

—Pressure Regulators
CEN PREN 334-90. Gas Pressure Governors for Inlet Pressures up to 100 Bar. 50 pp.
DIN ENGL 3381-84. Safety Devices for Gas Supply Installations Operating at Working Pressures up to 100 Bar; Pressure Relief Governors and Safety Shut-Off Devices (June). 16 pp.

—Safety Valves
DIN ENGL 3381-84. Safety Devices for Gas Supply Installations Operating at Working Pressures up to 100 Bar; Pressure Relief Governors and Safety Shut-Off Devices (June). 16 pp.

—Stop Valves
CEN PREN 331-90. General Requirements for Manually Operated Metallic Shut-Off Valves for Domestic Gas Installations. 34 pp.
CEN PREN 332-90. Manually Operated Metallic Taper Plug Shut-Off Valves with Closed Bottom for Domestic Gas Installations for Pressures up to PN 0.2 and with DN 15 to DN 50. 12 pp.
CEN PREN 333-90. Manually Operated Metallic Ball Shut-Off Valves for Domestic Gas Installations up to PN 4 and with DN 15 to DN 50. 13 pp.

—Valves
DIN ENGL 3230 Pt 5-84. Technical Delivery Conditions; Valves for Gas Installations and Gas Pipelines; Requirements and Testing (Aug). 6 pp.

Gas Supply Meters
Use: Gas Meters

Gas Taps
Use: Gas Valves

Gas Torches
Use: Torches

Gas Treatment Equipment
Use: Gas Cleaning Equipment

Gas Tubes
Use: Gas Discharge Tubes

Gas Tungsten Arc Welding

Use For: GTAW; TIG Welding; Tungsten Inert Gas Welding *See Also:* Arc Welding; Electric Welding; Fusion Welding; Gas Metal Arc Welding; Gas Tungsten Arc Welding Equipment; Shielded Arc Welding; Welding

—**Aluminum**

BSI BS 3019: Part 1-84. 1984 TIG Welding Part 1: TIG Welding of Aluminium, Magnesium and Their Alloys. 42 pp.

BSI BS 4870: Part 2-82. (WITHDRAWN) 1982 Approval Testing of Welding Procedures Part 2: TIG or MIG of Aluminium and Its Alloys (Superseded by BS EN 288: Part 4: 1992). 16 pp.

JIS Z 3604-85. Recommended Practice for Inert Gas Shielded Arc Welding of Aluminium and Aluminium Alloy. 22 pp.

JIS Z 3811-76. Standard Qualification Procedure for Welding Technique of Aluminium and Aluminium Alloy (R 1980). 18 pp.

—**Aluminum Alloys**

BSI BS 3019: Part 1-84. 1984 TIG Welding Part 1: TIG Welding of Aluminium, Magnesium and Their Alloys. 42 pp.

BSI BS 4870: Part 2-82. (WITHDRAWN) 1982 Approval Testing of Welding Procedures Part 2: TIG or MIG of Aluminium and Its Alloys (Superseded by BS EN 288: Part 4: 1992). 16 pp.

JIS Z 3604-85. Recommended Practice for Inert Gas Shielded Arc Welding of Aluminium and Aluminium Alloy. 22 pp.

JIS Z 3811-76. Standard Qualification Procedure for Welding Technique of Aluminium and Aluminium Alloy (R 1980). 18 pp.

—**Magnesium**

BSI BS 3019: Part 1-84. 1984 TIG Welding Part 1: TIG Welding of Aluminium, Magnesium and Their Alloys. 42 pp.

—**Magnesium Alloys**

BSI BS 3019: Part 1-84. 1984 TIG Welding Part 1: TIG Welding of Aluminium, Magnesium and Their Alloys. 42 pp.

—**Power Supplies**

BSI BS 638: Part 1-79. 1979 Arc Welding Power Sources, Equipment and Accessories Part 1: Oil Cooled Power Sources for Manual, Semi-Automatic and Automatic Metal-Arc Welding for TIG Welding. 18 pp.

BSI BS 638: Part 2-79. 1979 Arc Welding Power Sources, Equipment and Accessories Part 2: Air Cooled Power Sources for Manual Metal-Arc Welding with Covered Electrodes and for TIG Welding. 20 pp.

ISO 700-82. Power Sources for Manual Metal Arc Welding with Covered Electrodes and for the TIG Process First Edition. 14 pp.

—**Steel**

BSI BS 3019: Part 2-60. (WITHDRAWN) 1960 Amd 1 TIG Welding Part 2: Austenitic Stainless and Heat-Resisting Steels (Superseded by BS 7475: 1991). 21 pp.

Gas Tungsten Arc Welding Equipment

See Also: Gas Tungsten Arc Welding; Inert Gas Welding Equipment; Welding Equipment

—**Electrodes**

JIS Z 3316-89. TIG Welding Rods and Wires for Mild Steel and Low Alloy Steel. 15 pp.

JIS Z 3332-90. Filler Rods and Wires for TIG Welding of 9% Nickel Steel. 10 pp.

JIS Z 3334-88. Nickel and Nickel Alloy Filler Rods and Wires for Arc Welding. 8 pp.

—**Torches**

BSI BS 638: Part 8-84. 1984 Amd 1 Arc Welding Power Sources, Equipment and Accessories Part 8: Holders and Handheld Torches and Guns for MIG, MAG and TIG Welding. 24 pp.

—**Welding Rods**

JIS Z 3316-89. TIG Welding Rods and Wires for Mild Steel and Low Alloy Steel. 15 pp.

JIS Z 3332-90. Filler Rods and Wires for TIG Welding of 9% Nickel Steel. 10 pp.

JIS Z 3334-88. Nickel and Nickel Alloy Filler Rods and Wires for Arc Welding. 8 pp.

—**Welding Rods—Symbols**

ISO 636-89. Bare Solid Filler Rods for Oxy-Acetylene and Tungsten Inert Gas Arc (TIG) Welding, Depositing an Unalloyed or Low Alloyed Steel—Codification Second Edition. 4 pp.

Gas Turbine Engines

See Also: Aircraft Engines; Gasoline Engines; Internal Combustion Engines; Turbine Engines; Turbofan Engines; Turbojet Engines; Turboprop Engines

MOD UK DSTAN 28-6-01. Engines Gas Turbine for General Purpose Applications Issue 2; Amendment 2. 10 pp.

—**Acceptance Testing**

NATO STANAG 4195 ED 1 AMD 1-85. NATO Standard Engine Laboratory Test for Diesel and Gasoline Engines and Gas Turbine Engines. 6 pp.

—**Aircraft**

MOD UK DSTAN 00-971-87. General Specification for Aircraft Gas Turbine Engines Issue 1. 199 pp.

MOD UK DSTAN 28-8-86. General Requirements for the Overhaul of Gas Turbine Aero Engines Engine Change Units, Engine Modules and Jet Pipes Issue 1. 30 pp.

MOD UK DSTAN 28-8-93. General Requirements for the Overhaul of Aero Engines, Engine Modules, Jet Pipes and Engine Accessories Including Propellers Issue 2. 31 pp.

MOD UK DSTAN 28-8-01. General Requirements for the Overhaul of Aero Engines, Engine Modules, Jet Pipes and Engine Accessories Including Propellers Issue 2; Amendment 1. 33 pp.

—**Aircraft—Air Intakes**

CAA Chapter K5-5 04.74. Engine Air Intake and Ice Protection Systems (Light Aeroplanes). 3 pp.

—**Aircraft—Decarbonizing**

MOD UK TS 10127. 2-Butoxyethanol, Technical.

—**Aircraft—Design**

CAA Chapter K5-1 10.92. Power Plant Installations—General (Light Aeroplanes). 6 pp.

—**Aircraft—Fatigue (Materials)**

CAA NOTICE #44 ISSUE 5. Gas Turbine Engine Parts Subject to Retirement or Ultimate (Scrap) Lives (Airworthiness Notices). 3 pp.

—**Aircraft—Flight Testing**

JIS W 4012-87. Tests, Ground and Flight, Aircraft Gas Turbine Propulsion System Installation.

—**Aircraft—Fuel**

NATO STANAG 4270 ED 1 AMD 1-84. Aviation Fuel Design Parameter for Future NATO Land Based Turbine Powered Military Aircraft. 5 pp.

—**Aircraft—Fuel Contamination**

MOD UK DERD 2153-78. Fuel and Control Systems for Gas Turbine and Ramjet Engines: Rig Tests for Assessing Sensitivity to Abrasion and Blockage by Fuel Borne Solids Issue 4. 13 pp.

—**Aircraft—Glossaries**

BSI BS 185: Sec 8-70. 1970 Glossary of Aeronautical and Astronautical Terms Section 8: Power Plant (Piston Engines, Gas Turbines and Jet Propulsion). 20 pp.

—**Aircraft—Ice Formation**

CAA Chapter K5-5 04.74. Engine Air Intake and Ice Protection Systems (Light Aeroplanes). 3 pp.

—**Aircraft—Life (Durability)**

CAA NOTICE #44 ISSUE 5. Gas Turbine Engine Parts Subject to Retirement or Ultimate (Scrap) Lives (Airworthiness Notices). 3 pp.

—**Aircraft—Lubricating Oils**

MOD UK DERD 2490-01. Lubricating Oil, Aircraft Turbine Engine, Petroleum Oil OM-11 NATO Code No. 0-135 Issue 2; Amendment 1. 23 pp.

—**Aircraft—Motor Oils—Pressure Replenishment Connections**

NATO STANAG 3595 ED 3 AMD 5-83. Aircraft Fitting For Pressure Replenishment Of Gas Turbine Engines with Oil. 16 pp.

—**Aircraft—Preservation**

MOD UK DERD 2028-73. Preservation of Gas Turbine Aero-Engines and Modules During Storage and Transit Issue 5. 5 pp.

—**Aircraft—Vibration**

CAA Chapter K5-1 App 03.67. Power-Plant Installations—General (Light Aeroplanes). 1 p.

—**Compressors—Cleaning Agents**

MOD UK TS 10268A. Cleaning Fluid for Compressors of Gas Turbine Engines.

—**Compressors—Corrosion Inhibitors**

MOD UK DSTAN 68-10-89. Corrosion Preventive Compound: Water Displacing NATO Code: C-634 Joint Service Designation: PX-24 Issue 3. 28 pp.

Gas Turbine Engines (Cont.)

—**Compressors—Corrosion Inhibitors** (Cont.)

MOD UK DSTAN 68-10-01. Corrosion Preventive Compound: Water Displacing NATO Code: C-634 Joint Service Designation: PX-24 Issue 3; Amendment 1. 29 pp.

—**Ground Effect Machines—Comprehensive Testing**

CAA Chapter B7-3 App 04.79. Engine Testing—Type Tests for Gas Turbine Shaft Power Engines. 7 pp.

—**Lubricating Oils**

MOD UK DERD 2499-01. Lubricating Oil, Gas Turbine Engine, Synthetic Issue 1; Amendment 1. 12 pp.

—**Rotary Wing Aircraft**

CAA 774 (G). Section G Rotorcraft Installational Assumptions Involved in Engine Certification (Blue Papers). 1 p.

CAA 775 (K). Section K Light Aeroplanes Installational Assumptions Involved in Engine Certification (Blue Papers). 1 p.

CAA Chapter G5-1 06.76. Power-Plant Installations—General (Rotocraft). 6 pp.

CAA Chapter G5-1 App 06.76. Power-Plant Installations—General (Rotocraft). 2 pp.

CAA Chapter G5-8 06.76. Fire Precautions (Rotocraft). 8 pp.

CAA Chapter G7-2 06.66. Test Conditions—Piston and Turbine Engine Installations (Rotocraft). 3 pp.

CAA Chapter G7-3 06.66. Ground and Flight Tests—Single Turbine Engine Installations (Rotocraft). 6 pp.

CAA Chapter G7-4 06.66. Ground Flight Tests—Twin Turbine Engine Installations (Rotocraft). 8 pp.

—**Ships**

MOD UK NES 312-88. Requirements for Gas Turbine Intakes and Uptakes Issue 3 (10.88). 129 pp.

—**Ships—Fuel Systems**

MOD UK NES 320-91. Requirements for Design and Installation of Fuel Systems for Gas Turbines and Diesel Engines in Surface Ships Issue 2 (07.91). 112 pp.

—**Type Testing**

MOD UK DSTAN 28-5-89. Type Testing of Engines for General Purpose Applications Using Diesel and AVTUR Fuels Issue 2. 26 pp.

MOD UK DSTAN 28-7-01. Type Testing of Gas Turbine Engines for General Purpose Applications Issue 1; Amendment 1. 15 pp.

Gas Turbine Fuels

See Also: Gas Turbines; Jet Engine Fuels; Liquid Fuels

—**Classification**

BSI BS 6843: Part 2-88. 1988 Classification of Petroleum Fuels Part 2: Gas Turbine Fuels for Industrial and Marine Applications. 3 pp.

ISO 8216 Pt 2-86. Petroleum Products—Fuels (Class F)—Classification—Part 2: Categories of Gas Turbine Fuels for Industrial and Marine Applications First Edition. 3 pp.

—**Vanadium Content—Atomic Absorption Spectroscopy**

DIN ENGL 51790 Pt 3-78. Testing of Liquid Fuels; Determination of Vanadium Content; Vanadium Content in the Range of 0.4 to 4.0 mg/kg; Determination by Flameless Atomic Absorption Spectroscopy After Incineration (Apr). 6 pp.

Gas Turbines

See Also: Gas Turbine Fuels; Turbines

JIS B 8041-89. Test Method for Gas Turbines. 44 pp.

JIS B 8042-81. General Specification of Gas Turbines. 31 pp.

—**Acceptance Testing**

BSI BS 3135-89. 1989 Gas Turbines: Acceptance Tests. 24 pp.

ISO 2314-89. Gas Turbines—Acceptance Tests Second Edition. 23 pp.

—**Aircraft—Auxiliary Power Units—Certificate of Airworthiness**

CAA. Contents (Joint Airworthiness Requirements). 1 p.

CAA. Foreword. 1 p.

CAA. Check List of Pages (Joint Airworthiness Requirements). 1 p.

CAA JAR-APU Section 1. Regulations (Joint Airworthiness Requirements). 12 pp.

CAA JAR-APU Section 2. Acceptable Means of Compliance and Interpretations (Joint Airworthiness Requirements). 4 pp.

INTERNATIONAL AND NON-U.S. NATIONAL STANDARDS
SUBJECT INDEX
Gas

Gas Turbines (Cont.)
—Aircraft—Auxiliary Power Units—Certificate of Airworthiness (Cont.)
CAA JAR-APU Section 3. Advisory Material Joint—AMJ (Joint Airworthiness Requirements). 5 pp.
CAA JAR-APU Section 4. Approval of Gas Turbine Auxiliary Power Units and Associated Equipment. 5 pp.

—Construction
DIN ENGL 4312-83. Industrial-Type Turbines; Steam Turbines and Gas Expansion Turbines; Construction Principles for Industrial-Type Turbines (Dec). 74 pp.

—Designations
DIN ENGL 4340-76. Gas Turbines; Definitions; Designations (Aug). 12 pp.

—Glossaries
DIN ENGL 4340-76. Gas Turbines; Definitions; Designations (Aug). 12 pp.
JIS B 0128-83. Glossary of Terms for Thermal Power Plant (Gas Turbines and Auxiliary Equipment).

—Mineral Oils
ISO 8068-87. Petroleum Products and Lubricants—Petroleum Lubricating Oils for Turbines (Categories ISO-L-TSA and ISO-L-TGA)—Specifications First Edition. 7 pp.

—Procurement
BSI BS 3863-92. 1992 Gas Turbines Procurement (ISO 3977: 1991) (Q). 43 pp.
BSI BS 3863-79. 1979 Gas Turbines Procurement. 36 pp.
ISO 3977-91. Gas Turbines—Procurement Second Edition. 40 pp.

—Ships—Acoustical Insulation
MOD UK NES 802: Part 6-89. Requirements for Acoustic and Thermal Insulation Material Part 6: Mineral Wool Products Rock Wool Mattresses for Gas Turbine Uptakes Issue 1 (08.89). 13 pp.

—Ships—Auxiliary Power Units
MOD UK NES 309-01. Requirements for Gas Turbines Issue 3 (01.89); Amendment 1. 58 pp.

—Ships—Propulsion
MOD UK NES 309-01. Requirements for Gas Turbines Issue 3 (01.89); Amendment 1. 58 pp.

—Ships—Thermal Insulation
MOD UK NES 802: Part 6-89. Requirements for Acoustic and Thermal Insulation Material Part 6: Mineral Wool Products Rock Wool Mattresses for Gas Turbine Uptakes Issue 1 (08.89). 13 pp.

—Sound Pressure
ISO 6190-88. Acoustics—Measurement of Sound Pressure Levels of Gas Turbine Installations for Evaluating Environmental Noise—Survey Method First Edition. 8 pp.
ISO 10494-93. Gas Turbines and Gas Turbine Sets—Measurement of Emitted Airborne Noise—Engineering/Survey Method First Edition. 27 pp.

Gas Valves
See Also: Ball Valves; Butterfly Valves; Cocks; Expansion Valves; Gate Valves; Globe Valves; Hydraulic Valves; Pneumatic Valves; Thermostatic Valves; Valves
BSI BS 341: Part 1-91. 1991 Transportable Gas Container Valves Part 1: Specification for Industrial Valves for Working Pressures up to and Including 300 Bar (E). 70 pp.
BSI BS 341: Part 1-01. 1991 Amd 1 Transportable Gas Container Valves Part 1: Specification for Industrial Valves for Working Pressures up to and Including 300 Bar (AMD 7641) May 15, 1993. 73 pp.
BSI BS 5494-78. 1978 Amd 2 Gas Taps for Domestic and Catering Appliances. 14 pp.
CEN PREN 144 (Part 1)-88. Respiratory Protective Devices; Gas Cylinder Valves; Thread Connection for Insert Connector. 5 pp.
CEN PREN 144-2-90. Respiratory Protective Devices; Gas Cylinder Valves; Thread Connection for Side Connector. 4 pp.
CNS B5104-84. Valves for High Pressure Pressure Gas Cylinder (Apr)(10848).
DIN ENGL 3230 Pt 5-84. Technical Delivery Conditions; Valves for Gas Installations and Gas Pipelines; Requirements and Testing (Aug). 6 pp.
JIS B 8246-89. Valves for High Pressure Gas Cylinder. 28 pp.

—Acceptance Testing
DIN ENGL 3537 Pt 1-90. Gas Stopvalves Rated for Pressures up to 4 Bar; Requirements and Acceptance Testing (June). 10 pp.

Gas Valves (Cont.)
—Acetylene—Gas Cylinders
CNS B5077-90. Valve for Dissolved Acetylene Cylinder (May)(4152).

—Adapters
MOD UK DSTAN 81-44-01. Adaptors for Gas Cylinder Valves and Regulating Equipment Issue 1; Amendment 1 (Withdrawn). 43 pp.

—Aerostats
CAA Chapter Q6-2 12.79. Gas and Air Supply Systems (Non-Rigid Airships). 2 pp.

—Appliances
BSI BS 1963-90. 1990 Pressure Operated Relay Valves for Domestic, Commercial and Catering Gas Appliances. 14 pp.
BSI BS 1963-69. (WITHDRAWN) 1969 Amd 1 Pressure Operated Relay Valves for Domestic, Commercial and Catering Gas Appliances. 21 pp.
BSI BS 7673-93. 1993 Domestic Gas Cooker Oven Control Valves with or Without Manual Override (L). 14 pp.

—Ball
CEN PREN 331-93. Manually Operated Ball Valves and Closed Bottom Taper Plug Valves for Gas Installations for Buildings. 31 pp.

—Control—Plug
SAA AS A72-54. Control Plug Cocks for Low Pressure Gas Amdt 1 February 1959. 10 pp.

—Cylinders
MOD UK DSTAN 81-4-77. Valves, Cylinder, Gas Issue 2. 11 pp.
MOD UK DSTAN 81-4-92. Valves, Container, Gas Issue 3. 17 pp.
MOD UK DSTAN 81-13-01. Cylinders, Compressed Gas, and Valves, Cylinder, Gas, for Ground, Marine, Airborne, and Medical Uses Issue 1; Amendment 2. 20 pp.
MOD UK DSTAN 81-13-92. Containers, Compressed Gas, and Valves Issue 2. 24 pp.

—Laboratory
DIN ENGL 3537 Pt 3-92. Gas Stopvalves Rated for Pressures up to 4 Bar; Requirements and Acceptance Testing for Laboratory Valves (Jan). 2 pp.

—Lubricants
DIN ENGL 3536-82. Lubricants for Gas Valves and Controls; Requirements, Testing (Nov). 5 pp.

—Plug
CEN PREN 331-93. Manually Operated Ball Valves and Closed Bottom Taper Plug Valves for Gas Installations for Buildings. 31 pp.

—Rubber—Cork Sealants—Safety
DIN ENGL 3535 Pt 5-83. Sealants for Gas Supply; Rubber-Cork and Rubber-Cork-Asbestos Sealants for Gas Valves and Gas Appliances, Safety Requirements, Testing (Apr). 6 pp.

—Safety
BSI BS 6759: Part 2-84. 1984 Amd 1 Safety Valves Part 2: Specification for Safety Valves for Compressed Air or Inert Gases (AMD 5494) March 31, 1987. 25 pp.
DIN ENGL 3381-84. Safety Devices for Gas Supply Installations Operating at Working Pressures up to 100 Bar; Pressure Relief Governors and Safety Shut-Off Devices (June). 16 pp.
DIN ENGL 8521-81. Safety Devices Against Flashback and Backflow in Welding, Cutting and Allied Processes; Safety Requirements, Testing (Dec). 8 pp.
DIN ENGL 32509-85. Hand-Operated Shut-Off Valves for Welding, Cutting and Allied Processes; Designs, Safety Requirements, Testing (Aug). 5 pp.
ISO 5175-87. Equipment Used in Gas Welding, Cutting and Allied Processes—Safety Devices for Fuel Gases and Oxygen or Compressed Air—General Specifications, Requirements and Tests First Edition. 11 pp.

—Stop
CEN PREN 331-90. General Requirements for Manually Operated Metallic Shut-Off Valves for Domestic Gas Installations. 34 pp.

—Stop—Acceptance Testing
DIN ENGL 3437-90. Gas Stopvalves Rated for Pressures over 16 Bar; Requirements and Acceptance Testing (June). 7 pp.
DIN ENGL 3547 Pt 1-90. PN 4 to PN 16 Gas and Water Stopvalves; Requirements and Acceptance Testing (June). 9 pp.

Gas Valves (Cont.)
—Stop—Automatic Shutoff
BSI BS 5963-81. (WITHDRAWN) 1981 Amd 4 Electrically Operated Automatic Gas Shut-off Valves (E) (AMD 6266) December 22, 1989 (Superseded by BS 7461: 1991 and BS EN 161: 1991). 28 pp.
BSI BS 7461-91. 1991 Electrically Operated Automatic Gas Shut-Off Valves Fitted with Throughput Adjusters, Proof of Closure Switches, Closed Position Indicator Switches or Gas Flow Control. 10 pp.

—Stop—Ball
CEN PREN 333-90. Manually Operated Metallic Ball Shut-Off Valves for Domestic Gas Installations up to PN 4 and with DN 15 to DN 50. 13 pp.

—Stop—Plug
CEN PREN 332-90. Manually Operated Metallic Taper Plug Shut-Off Valves with Closed Bottom for Domestic Gas Installations for Pressures up to PN 0.2 and with DN 15 to DN 50. 12 pp.

—Stop—Safety
DIN ENGL 3399-86. Gas Low-Pressure Cut-Off Valves; Safety Requirements, Testing (Aug). 5 pp.

Gas Welding
Use For: Torch Welding See Also: Braze Welding; Brazing; Gas Welding Equipment; Oxyacetylene Welding; Pressure Gas Welding; Welding
JIS Z 3801-79. Standard Qualification Procedure for Welding Technique. 19 pp.

—Aluminum—Edge Preparation
DIN ENGL 8552 Pt 1-81. Weld Preparation; Groove Forms for Aluminium and Aluminium Alloys; Gas Welding and Gas-Shielded Arc Welding (May). 5 pp.

—Bronze
BSI BS 1724-90. 1990 Bronze Welding by Gas. 23 pp.

—Copper—Edge Preparation
DIN ENGL 8552 Pt 3-82. Edge Preparation for Welding; Groove Forms on Copper and Copper Alloys; Gas Welding and Gas-Shielded Arc Welding (July). 5 pp.

—Steels—Edge Preparation
DIN ENGL 2559 Pt 1-73. Edge Preparation for Welding; Directions Regarding Edge Forms; Fusion Welding of Butt Joints in Steel Tubes (May). 3 pp.
DIN ENGL 2559 Pt 2-84. Edge Preparation for Welding; Matching of Inside Diameter for Circumferential Welds on Seamless Pipes (Feb). 4 pp.
DIN ENGL 8551 Pt 1-76. Edge Preparation for Welding; Edge Forms on Steel; Gas Welding, Manual Arc Welding and Gas-Shielded Arc Welding (June). 5 pp.

Gas Welding Equipment
See Also: Brazing Equipment; Gas Welding; Oxyacetylene Welding Equipment; Welding Equipment

—Blowpipes
BSI BS 6503-84. 1984 Handheld Blowpipes and Nozzles, Using Fuel Gas and Oxygen, for Gas Welding, Cutting and Related Processes. 20 pp.
CEN PREN 874-92. Welding—Oxygen/Fuel Gas Blowpipes (Cutting Machine Type) of Cylindrical Barrel—Type of Construction, General Specifications, Test Methods. 25 pp.

—Construction Materials
BSI BS 7329-90. (WITHDRAWN) 1990 Materials for Equipment Used in Gas Welding, Cutting and Allied Processes (Superseded by BS EN 29539: 1992). 9 pp.
BSI BS EN 29539-92. 1992 Materials for Equipment Used in Gas Welding, Cutting and Allied Processes (ISO 9539: 1988). 9 pp.
CEN EN 29539-92. Materials for Equipment Used in Gas Welding, Cutting and Allied Processes. 4 pp.
ISO 9539-88. Materials for Equipment Used in Gas Welding, Cutting and Allied Processes First Edition. 4 pp.

—Electrodes
JIS Z 3232-86. Aluminium and Aluminium Alloy Welding Rods and Wires. 17 pp.
JIS Z 3233-90. Tungsten Electrodes for Inert Gas Shielded Arc Welding. 9 pp.

—Filler Metal
BSI BS 1453-72. 1972 Amd 1 Filler Materials for Gas Welding. 25 pp.
BSI BS EN 20544-91. 1991 Sizes for Filler Metals for Manual Welding (F) (ISO 544: 1989). 9 pp.
CEN PREN 20 544-90. Filler Materials for Manual Welding—Size Requirements. 3 pp.

INDUSTRY STANDARDS

Gas Welding Equipment *(Cont.)*

—Filler Metal *(Cont.)*
CEN EN 20544-91. Filler Materials for Manual Welding—Size Requirements. 3 pp.
DIN ENGL 8573 Pt 1-83. Filler Metals for Welding Unalloyed and Low Alloy Cast Iron Materials; Designation; Technical Delivery Conditions (Jan). 8 pp.
DIN ENGL EN 20544-91. Filler Materials for Manual Welding; Size Requirements (ISO 544: 1989) (Dec). 4 pp.
ISO 544-89. Filler Materials for Manual Welding—Size Requirements Second Edition; (Supersedes 545, 546 and 547). 4 pp.

—Flame Arresters
ISO 5175-87. Equipment Used in Gas Welding, Cutting and Allied Processes—Safety Devices for Fuel Gases and Oxygen or Compressed Air—General Specifications, Requirements and Tests First Edition. 11 pp.

—Hose Assemblies
BSI BS 1389-86. 1986 Hose Connections and Hose Assemblies for Equipment for Gas Welding, Cutting and Related Processes. 14 pp.
SAA AS 1335-74. Hose and Hose Assemblies for General Purpose Gas Welding and Allied Processes Amdt 1 June 1979 Reconfirmed 1988. 11 pp.

—Hose Connectors
BSI BS 1389-86. 1986 Hose Connections and Hose Assemblies for Equipment for Gas Welding, Cutting and Related Processes. 14 pp.
DIN ENGL 8542-83. Hose Connections and Hose Couplers for Equipment for Welding, Cutting and Allied Processes (Sept). 9 pp.

—Hose Couplings
CNS B2533-80. Rubber Hose Coupling for Gas Welding and Cutting Torches (Dec)(6741).
JIS B 6805-86. Rubber Hose Connections for Welding and Cutting Equipment.
JIS B 6805-68. Rubber Hose Coupling for Gas Welding and Cutting Torches. 5 pp.

—Hoses
BSI BS 5120-87. 1987 Rubber Hose for Gas Welding and Allied Processes. 13 pp.
CEN PREN 559-91. Rubber Hoses for Welding, Cutting and Allied Processes. 12 pp.
CGSB 20-GP-13MA-89. Pneumatic Hose (Acetylene, Hydrogen, Oxygen, Air). 9 pp.
CNS K4008-86. Rubber Hoses for Oxygen (Oct)(809).
CNS K4009-86. Rubber Hoses for Acetylene (Oct)(810).
DIN ENGL 8541 Pt 1-86. Hoses for Welding, Cutting and Allied Processes; Hoses Without Protective Cover for Use with Fuel Gas, Oxygen and Other Non-Combustible Gases (Aug). 5 pp.
DIN ENGL 8541 Pt 2-87. Hoses for Welding, Cutting and Allied Processes; Hoses with Protective Cover for Use with Fuel Gas, Oxygen and Other Non-Combustible Gases (Dec). 4 pp.
ISO 3821-92. Welding—Rubber Hoses for Welding, Cutting and Allied Processes Second Edition. 12 pp.
JIS K 6333-84. Rubber Hoses for Oxygen.
JIS K 6333-78. Oxygen Hose. 6 pp.
JIS K 6334-84. Rubber Hoses for Acetylene. 7 pp.
SNZ NZS/BS 5120-87. Specification for Rubber Hoses for Gas Welding and Allied Processes. 16 pp.

—Hoses—Flammability Testing
BSI BS 5173: Sec 103.6-90. 1990 Rubber and Plastics Hoses and Hose Assemblies Part 103: Physical Tests Section 103.6: Determination of Ignitability of Lining. 6 pp.

—Kits
BSI BS 6942: Part 1-88. 1988 Design and Construction of Small Kits for Oxy-Fuel Gas Welding and Allied Processes Part 1: Kits Using One or More Non-Refillable Gas Containers for Oxygen and Fuel Gas. 8 pp.
BSI BS 6942: Part 2-89. 1989 Design and Construction of Small Kits for Oxy-Fuel Gas Welding and Allied Processes Part 2: Kits Using Refillable Gas Containers for Oxygen and Fuel Gas. 7 pp.

—Leakage
BSI BS 7278-90. (WITHDRAWN) 1990 Gas Tightness of Equipment for Gas Welding and Allied Processes (Superseded by BS EN 29090: 1992). 8 pp.
BSI BS EN 29090-92. 1992 Gas Tightness of Equipment for Gas Welding and Allied Processes (ISO 9090: 1989) (F). 11 pp.
CEN EN 29090-92. Gas Tightness of Equipment for Gas Welding and Allied Processes. 6 pp.
ISO 9090-89. Gas Tightness of Equipment for Gas Welding and Allied Processes First Edition. 7 pp.

—Nozzles
BSI BS 6503-84. 1984 Handheld Blowpipes and Nozzles, Using Fuel Gas and Oxygen, for Gas Welding, Cutting and Related Processes. 20 pp.

—Pressure Regulators
BSI BS 5741-79. 1979 Pressure Regulators Used in Welding, Cutting and Related Processes. 11 pp.
CEN PREN 585-91. Pressure Regulators for Gas Cylinders Used in Welding, Cutting and Allied Processes. 21 pp.
CEN PREN 961-92. Welding—Cutting and Allied Processes—Manifold Regulators. 19 pp.
DIN ENGL 8545-81. Manifold Regulators for Welding, Cutting and Related Processes; Concepts, Requirements and Testing (Sept). 8 pp.
ISO 2503-83. Welding—Regulators for Gas Cylinders Used in Welding, Cutting and Related Processes Second Edition; (Addendum 1-1984). 15 pp.
ISO 7291-90. Welding, Cutting and Allied Processes—Manifold Regulators First Edition. 11 pp.
JIS B 6803-87. Pressure Regulators for Welding, Cutting and Allied Processes. 19 pp.

—Pressure Regulators—Gas Cylinders
CEN PREN 585-91. Pressure Regulators for Gas Cylinders Used in Welding, Cutting and Allied Processes. 21 pp.
CEN PREN 961-92. Welding—Cutting and Allied Processes—Manifold Regulators. 19 pp.
DIN ENGL 8546-88. Pressure Regulators for Gas Cylinders Used in Welding, Cutting and Related Processes; Terminology, Requirements and Testing (Aug). 8 pp.
ISO 2503-83. Welding—Regulators for Gas Cylinders Used in Welding, Cutting and Related Processes Second Edition; (Addendum 1-1984). 15 pp.

—Safety Valves
DIN ENGL 8521-81. Safety Devices Against Flashback and Backflow in Welding, Cutting and Allied Processes; Safety Requirements, Testing (Dec). 8 pp.
DIN ENGL 32509-85. Hand-Operated Shut-Off Valves for Welding, Cutting and Allied Processes; Designs, Safety Requirements, Testing (Aug). 5 pp.
ISO 5175-87. Equipment Used in Gas Welding, Cutting and Allied Processes—Safety Devices for Fuel Gases and Oxygen or Compressed Air—General Specifications, Requirements and Tests First Edition. 11 pp.

—Welding Rods
CNS Z7026-83. Gas Welding Rods for Mild Steel (Aug)(2957).
DIN ENGL 8554 Pt 1-86. Unalloyed and Low Alloy Filler Rods for Gas Welding; Designation, Technical Delivery Conditions (May). 5 pp.
JIS Z 3201-90. Gas Welding Rods for Mild Steel. 10 pp.

—Welding Rods—Aluminum
JIS Z 3232-86. Aluminium and Aluminium Alloy Welding Rods and Wires. 17 pp.

—Welding Rods—Copper
JIS Z 3202-83. Copper and Copper Alloy Gas Welding Rods. 6 pp.

Gaseous Fuels
See Also: Fuel Gases; Gases; Liquefied Petroleum Gas; Natural Gas

—Calorific Value
BSI BS 7420-91. 1991 Determination of Calorific Values of Solid, Liquid and Gaseous Fuels (Including Definitions). 11 pp.
DIN ENGL 51858-82. Gaseous Fuels and Other Gases; Calculation of Gross and Nett Calorific Values and of Relative Density of Gas Mixtures (Nov). 7 pp.

—Density
CNS K6245-71. Method for Determination of Specific Gravity of Gas Fuels (Jul)(2705).
DIN ENGL 51858-82. Gaseous Fuels and Other Gases; Calculation of Gross and Nett Calorific Values and of Relative Density of Gas Mixtures (Nov). 7 pp.

—Dispensing Equipment—Pipelines
CSA B51-M1991. Boiler, Pressure Vessel, and Pressure Piping Code; (Gen Instr 1). 96 pp.

—Dispensing Equipment—Storage Tanks
CSA B51-M1991. Boiler, Pressure Vessel, and Pressure Piping Code; (Gen Instr 1). 96 pp.

—Oxygen Content
DIN ENGL 51856-79. Testing of Gaseous Fuels and Other Gases; Determination of Oxygen Content (June). 6 pp.

Gaseous Fuels *(Cont.)*

—Sulfur Compound Content
DIN ENGL 51855 Pt 4-79. Testing of Gaseous Fuels and Other Gases; Determination of Sulphur Compound Content; Hydrogen Sulphite Content, Cadmium Acetate Method (Jan). 6 pp.

—Sulfur Content
DIN ENGL 51400 Pt 1-78. Testing of Mineral Oils and Fuels; Determination of the Sulfur Content (Total Sulfur); General Working Principles (Feb). 6 pp.
DIN ENGL 51400 Pt 3-78. Testing of Mineral Oils and Fuels; Determination of the Sulfur Content (Total Sulfur); Combustion According to Schoniger; Thorin-Sulfonazo-III Titration (Feb). 3 pp.
DIN ENGL 51400 Pt 4-90. Determination of Total Sulfur Content of Gaseous Petroleum Products by the Lingener Combustion Method (Oct). 8 pp.
DIN ENGL 51400 Pt 8-78. Testing of Mineral Oils and Fuels; Determination of Sulfur Content (Total Sulfur); Nickel Reduction Method; Dithizone Titration (Feb). 4 pp.
DIN ENGL 51855 Pt 1-79. Testing of Gaseous Fuels and Other Gases; Determination of Sulphur Compound Content; Range of Application, Purpose, Terms (Jan). 2 pp.

—Water Vapor Content
CNS K6379-75. Determination of Water Vapor Content in Gaseous Fuel (Dew-Point Method) (May)(3777).

Gaseous Tritium Lighting
See Also: Lamps; Lighting
MOD UK DSTAN 62-4-01. Lamps, Nuclear (Gaseous Tritium Light Sources) Issue 3; Amendment 2. 43 pp.
MOD UK DSTAN 62-4-92. Lamps, Nuclear (Gaseous Tritium Light Sources) Issue 4. 55 pp.

Gases
Scope Note: For additional listings, use a more specific term *Use For:* Poisonous Gases
See Also: Air; Argon; Butanes; Calibration Gases; Coal Gas; Diluent Gases; Exhaust Gases; Explosive Gases; Flue Gases; Fuel Gases; Fuels; Gas Generators; Gaseous Fuels; Krypton; Liquefied Gases; Natural Gas; Neon; Nitrogen; Noble Gases; Phosphine; Protective Gases; Shielding Gases; 1,1,2-Trichloro-1,2,2-Trifluoroethane; Trichlorofluoromethane; Vapors; Zero Gases
BSI BS 4314: Part 1-68. 1968 Apparatus for Physical Methods of Gas Analysis Part 1: Infra-Red Gas Analysers for Industrial Use. 25 pp.

—Aircraft—Sampling—Technical Manuals
NATO STANAG 3977 ED 1 AMD 0-91. Manual of Techniques of Sampling and Analysis of Gases and Liquefied Gases for Aircraft Servicing—AEP-6. 5 pp.
NATO STANAG 3977 ED 1 AMD 1-91. Manual of Techniques of Sampling and Analysis of Gases and Liquefied Gases for Aircraft Servicing—AEP-6. 6 pp.

—Carbon Dioxide Content
CNS K6230-67. Method of Analysis for Carbon Dioxide, Oxygen, Carbon Monoxide Hydrogen and Methane in Gases (Mar)(2592)(R 1971).

—Carbon Monoxide Content
CNS K6230-67. Method of Analysis for Carbon Dioxide, Oxygen, Carbon Monoxide Hydrogen and Methane in Gases (Mar)(2592)(R 1971).

—Density—Glossaries
CNS Z7183-84. Definitions of Terms Relating to Density and Specific Gravity of Solids Liquids and Gases (Oct)(11027).

—Flammability Testing
CEN PREN 720-2-92. Classification of Gases and Gas Mixtures—Part 2: Gases and Gas Mixtures—Determination of Fire Potential and Oxidizing Ability. 26 pp.

—Humidity
JIS Z 8806-81. Methods of Humidity Measurement. 42 pp.

—Hydrogen Content
CNS K6230-67. Method of Analysis for Carbon Dioxide, Oxygen, Carbon Monoxide Hydrogen and Methane in Gases (Mar)(2592)(R 1971).

—Ignition Temperature
SAA AS 1896-76. Method of Test for Ignition Temperature of Gases and Vapours Corrig.. 20 pp.

—Lifting—Aerostats
CAA Chapter Q6-2 12.79. Gas and Air Supply Systems (Non-Rigid Airships). 2 pp.

INTERNATIONAL AND NON-U.S. NATIONAL STANDARDS
SUBJECT INDEX

Gaskets

Gases (Cont.)

—**Methane Content**
CNS K6230-67. Method of Analysis for Carbon Dioxide, Oxygen, Carbon Monoxide Hydrogen and Methane in Gases (Mar)(2592)(R 1971).

—**Oxygen Content**
CNS K6230-67. Method of Analysis for Carbon Dioxide, Oxygen, Carbon Monoxide Hydrogen and Methane in Gases (Mar)(2592)(R 1971).
DIN ENGL 51856-79. Testing of Gaseous Fuels and Other Gases; Determination of Oxygen Content (June). 6 pp.

—**Pressure Reduction**
CNS Z6058-82. Method of Test for Pressure Drop Rate of Compressed Gas-Propelled Products (Dec)(9762).

—**Sampling**
BSI BS 5309: Part 2-76. 1976 Methods for Sampling Chemical Products Part 2: Sampling of Gases. 21 pp.

—**Sulfur Compound Content**
DIN ENGL 51855 Pt 1-79. Testing of Gaseous Fuels and Other Gases; Determination of Sulphur Compound Content; Range of Application, Purpose, Terms (Jan). 2 pp.
DIN ENGL 51855 Pt 4-79. Testing of Gaseous Fuels and Other Gases; Determination of Sulphur Compound Content; Hydrogen Sulphite Content, Cadmium Acetate Method (Jan). 6 pp.

Gasification Plants
See Also: Industrial Plants
BSI BS 4445-69. 1969 Schedule of Tests for Gasification and Reforming Plants Using Hydrocarbon Feedstocks. 97 pp.

Gasified Wines
Use: Wines

Gaskets
Use For: Binder Discs *See Also:* Hydraulic Seals; O Rings; Oil Seals; Sealing Rings; Seals

—**Aircraft**
SBAC AS 4650 ISSUE 4. Accessory Gasket. 1 p.
SBAC AS 4652-4657 ISSUE 3. Accessory Gaskets. 1 p.
SBAC AS 4658-4666 ISSUE 1. Accessory Gaskets. 1 p.
SBAC AS 42700-748 ISSUE 4. Gasket.
SBAC AS 42749-799 ISSUE 5. Gasket.
SBAC AGS 3098-99 ISSUE 2. Gasket 'O' Ring.
SBAC AGS 3870-89 ISSUE 2. Gasket (Corrujoint).
SBAC AGS 3890-909 ISSUE 1. Gasket (Corrujoint).

—**Aircraft—Copper—Thermal Relief Valves**
SBAC AS 3357 ISSUE 2. Gasket. 1 p.

—**Aircraft—Cork**
BSI 2F 66-82. 1982 Rubber Bonded Cork Sheets. 7 pp.

—**Aircraft Engines**
JIS W 1102-56. Gasket Sheet for Aircraft Engine.

—**Aircraft—Grooves**
SBAC RS 734 (V). Groove Dimensions for Gaskets (Metallic Seals).

—**Aircraft—Metal**
SBAC TS 104 ISSUE 2. SBAC Manufacturing Specification for Gaskets (Metallic Seals).

—**Aircraft—Pipe Adapters**
SBAC AS 6691 (V). Pipe, Coupling (Aluminium Alloy) Cone Union Adaptor/Unified Attachment End 3/16 Inch to 1/2 Inch for Use with Parallel Gaskets (e.g. Bonded Seals).
SBAC AS 6692 (V). Pipe Coupling (Aluminium Alloy) Union Adaptor/Unified Attachment End 3/16 Inch to 1 Inch for Use with Parallel Gaskets (e.g. Bonded Seal).
SBAC AS 6693 (V). Pipe, Coupling Aluminium Alloy, Cone Union Adaptor/Unified Attachment End, 3/16 Inch to 1/2 Inch for Use with 'O' Ring Gaskets (In American)(And10050) or Equivalent to AGS3018 Tapped Bosses.
SBAC AS 6694 (V). Pipe, Coupling (Aluminium Alloy) Union Adaptor/Unified Attachment End 3/16 Inch to 1 Inch for Use with 'O' Ring Gaskets in American (and 10050) or Equivalent A.G.S.3018 Tapped Bosses.
SBAC AGS 3012 (V). Pipe Couplings (Steel), Cone, Union Adaptor/Unified Attachment End 3/16" to 1/2" for Use with Parallel Gaskets (e.g. Bonded Seals).
SBAC AGS 3013 (V). Pipe Couplings (Steel) Union Adaptor/Unified Attachement End, 3/16" to 1/2" for Use with Parallel Gaskets. (e.g. Bonded Seals).

Gaskets (Cont.)

—**Aircraft—Pipe Adapters** (Cont.)
SBAC AGS 3016 (V)(I). Pipe, Couplings (Steel) Cone Union Adaptor Unified Attachment End 3/16 Inch to 1/2 Inch for Use with 'O' Ring Gaskets (AND 10050) or Equivalent AGS 3018 Tapped Bosses.

—**Aircraft—Tachometers**
SBAC AS 4651 ISSUE 3. Percentage R.P.M. Tachometer Gasket. 1 p.

—**Asbestos**
CNS R2161-86. Compressed Asbestos Sheet Packing (May)(9326). 2 pp.
CNS R2178-85. Asbestos Mill Boards (May)(11267).
DIN ENGL 3754 Pt 1-84. Gasket Materials; Compressed Asbestos Fibre Sheets; Dimensions, Requirements, Testing (May). 5 pp.
JIS R 3453-79. Compressed Asbestos Sheets. 11 pp.
JIS R 3454-79. Asbestos Mill Boards. 6 pp.
MOD UK DSTAN 53-99-85. Ropes, Asbestos and Packing (Braids), Asbestos: Dust Suppressed and Fibres, Asbestos Issue 1. 21 pp.

—**Automotive—Cylinder Heads**
CNS D2090-81. Cylinder Head Gaskets for Automobile Engines (Jul)(7667).
JIS D 3105-92. Cylinder Head Gaskets for Automobile Engines. 26 pp.

—**Automotive—Master Cylinders—Hydraulic Brakes**
CNS D2177-83. Diaphragm Gaskets of Hydraulic Brake Master Cylinder Reservoirs for Automobiles (Dec)(10712).
ISO 4929-78. Road Vehicles—Diaphragm Gaskets for Hydraulic Brake Master Cylinder Reservoirs Using a Non-Petroleum Base Hydraulic Brake Fluid First Edition. 6 pp.
JIS D 2610-82. Diaphragm Gaskets of Hydraulic Brake Master Cylinder Reservoirs for Automobiles.

—**Cork**
BSI BS 4243-67. (WITHDRAWN) 1967 Amd 1 Cork/Paper Jointing (Superseded by BS 4249: 1989). 24 pp.
BSI BS 4332-89. 1989 Phenol-Formaldehyde Resin-Bonded Cork Jointing. 12 pp.
CNS K6099-59. Method of Test for Cork Sheet for Gasket (Sep)(1068).
CNS K8001-59. Cork Sheet for Gasket (Sep)(1067)(R 1971).
CNS O1013-64. Cork Binder Discs (May)(2302). 2 pp.
CNS O2014-73. Method of Test for Cork Binder Discs (May)(2303). 1 p.
ISO 4708-85. Cork—Composition Cork Gasket Material—Test Methods First Edition. 4 pp.
ISO 4709-85. Cork—Composition Cork Gasket Material—Specifications First Edition. 4 pp.

—**Cork—Acid Resistance Testing**
CNS O2014-73. Method of Test for Cork Binder Discs (May)(2303). 1 p.

—**Cork—Alcohol Resistance**
CNS O2014-73. Method of Test for Cork Binder Discs (May)(2303). 1 p.

—**Cork—Boiling Water Testing**
CNS O2014-73. Method of Test for Cork Binder Discs (May)(2303). 1 p.

—**Cork—Compression Testing**
CNS O2014-73. Method of Test for Cork Binder Discs (May)(2303). 1 p.

—**Cork—Flexibility**
CNS O2014-73. Method of Test for Cork Binder Discs (May)(2303). 1 p.

—**Cork—Preservation**
MOD UK DEF-1234-A: LEAFLET C32. Preservation of Paper, Cork and Felt Gaskets with Pentachlorophenyl Laurate Solution, DEF-117 Issue 2 (Withdrawn).

—**Doors**
CNS A2154-88. Gaskets for Windows, Doors and Joints of Panel in Buildings (Dec)(10209).
CNS A3186-88. Method of Test for Gaskets for Windows, Doors and Joints of Panel in Buildings (Dec)(10210).
JIS A 5756-89. Gaskets for Windows, Doors and Joints of Panel in Buildings. 26 pp.

—**EMI/RFI Shielding**
MOD UK DSTAN 59-103: Part 1-93. EMI/EMP Gasket Components Part 1: General Requirements Issue 1. 14 pp.
MOD UK DSTAN 59-103: Part 3-93. EMI/EMP Gasket Components Part 3: Gasket Test Methods Issue 1. 47 pp.

Gaskets (Cont.)

—**EMI/RFI Shielding—Components—Quality Assurance**
MOD UK DSTAN 59-103: Part 2-93. EMI/EMP Gasket Components Part 2: Gasket Capability Approval Procedures Issue 1. 20 pp.

—**EMI/RFI Shielding—Flanges**
MOD UK DSTAN 59-103: Part 4-93. EMI/EMP Gasket Components Part 4: Gasket/Flange Design Considerations Issue 1. 25 pp.

—**Felt—Preservation**
MOD UK DEF-1234-A: LEAFLET C32. Preservation of Paper, Cork and Felt Gaskets with Pentachlorophenyl Laurate Solution, DEF-117 Issue 2 (Withdrawn).

—**Flanges**
BSI BS 3381-89. 1989 Spiral Wound Gaskets for Steel Flanges to BS 1560 (Q). 10 pp.
DIN ENGL 2696-72. Lenticular Gaskets and Lenticular Seals for ND 64 to ND 400 Flanged Connections (Apr). 2 pp.

—**Glossaries**
JIS B 0116-78. Glossary of Terms for Packings and Gaskets. 35 pp.

—**Joints**
CNS A2154-88. Gaskets for Windows, Doors and Joints of Panel in Buildings (Dec)(10209).
CNS A3186-88. Method of Test for Gaskets for Windows, Doors and Joints of Panel in Buildings (Dec)(10210).
JIS A 5756-89. Gaskets for Windows, Doors and Joints of Panel in Buildings. 26 pp.

—**Joints—Flanged**
DIN ENGL 28040-89. Gaskets for Use with Flanged Joints (Feb). 3 pp.

—**Marine—Pipe Flanges**
BSI BS MA 9-70. 1970 Amd 3 Flanges, Bolting and Gaskets for Exhaust Gas Piping for Diesel Engines and Boiler Uptakes. 10 pp.

—**Paper**
BSI BS 4243-67. (WITHDRAWN) 1967 Amd 1 Cork/Paper Jointing (Superseded by BS 4249: 1989). 24 pp.
BSI BS 4249-89. 1989 Paper and Cork/Paper Jointing. 14 pp.

—**Paper—Preservation**
MOD UK DEF-1234-A: LEAFLET C32. Preservation of Paper, Cork and Felt Gaskets with Pentachlorophenyl Laurate Solution, DEF-117 Issue 2 (Withdrawn).

—**Pipe Fittings—Refrigeration Equipment**
DIN ENGL 8914-90. 90 Degree Tapered Seal Rings and Caps for Use with PN 40 Flared Flange Solderless Compression Couplings in Refrigerating Systems (June). 2 pp.

—**Pipe Flanges**
BSI BS 3063-65. 1965 Amd 1 Dimensions of Gaskets for Pipe Flanges. 35 pp.
BSI BS 4865: Part 1-89. 1989 Dimensions of Gaskets for Pipe Flanges to BS 4504 Part 1: Non-Metallic Flat Gaskets (Including Gaskets for Flanges to BS 4772). 18 pp.
BSI BS 4865: Part 1-01. 1989 Amd 1 Dimensions of Gaskets for Flanges to BS 4504 Part 1: Specification for Non-Metallic Flat Gaskets (Including Gaskets for Flanges to BS 4772) (AMD 7561) March 15, 1993. 19 pp.
BSI BS 4865: Part 2-89. 1989 Dimensions of Gaskets for Pipe Flanges to BS 4504 Part 2: Spiral Wound Gaskets for Use with Steel Flanges. 8 pp.
BSI BS 4865: Part 3-89. 1989 Dimensions of Gaskets for Pipe Flanges to BS 4504 Part 3: Non-Metallic Envelope Gaskets. 6 pp.
BSI BS 4865: Part 4-89. 1989 Dimensions of Gaskets for Pipe Flanges to BS 4504 Part 4: Corrugated, Flat or Grooved Metallic and Filled Metallic Gaskets for Use with Steel Flanges. 8 pp.
BSI BS 7076: Part 1-89. 1989 Dimensions of Gaskets for Flanges to BS 1560 Part 1: Non-Metallic Flat Gaskets. 12 pp.
BSI BS 7076: Part 2-89. 1989 Dimensions of Gaskets for Flanges to BS 1560 Part 2: Metallic Ring-Joint Gaskets for Use with Steel Flanges. 12 pp.
BSI BS 7076: Part 3-89. 1989 Dimensions of Gaskets for Flanges to BS 1560 Part 3: Specification for Non-Metallic Envelope Gaskets. 8 pp.
BSI BS 7076: Part 4-89. 1989 Dimensions of Gaskets for Flanges to BS 1560 Part 4: Corrugated, Flat or Grooved Metallic and Filled Metallic Gaskets. 8 pp.
ISO 7483-91. Dimensions of Gaskets for Use with Flanges to ISO 7005 First Edition. 39 pp.

INDUSTRY STANDARDS

Gaskets

Gaskets (Cont.)

—Polyester Resin—Pipe Joints
DIN ENGL 16966 Pt 6-82. Glass Fibre Reinforced Polyester Resin (UP-GF) Pipe Fittings and Joint Assemblies; Collars, Flanges, Joint Rings, Dimensions (July). 11 pp.

—Rubber
ISO 3934-78. Rubber Building Gaskets—Materials in Preformed Solid Vulcanizates Used for Sealing Glazing and Panels—Specification First Edition. 4 pp.
ISO 5892-81. Rubber Building Gaskets—Materials for Preformed Solid Vulcanized Structural Gaskets—Specification First Edition. 7 pp.
MOD UK DSTAN 93-21-91. Chloroprene Rubber, Cellular, Closed Cell, Minimum Tarnishing Issue 3. 16 pp.

—Rubber—Buildings
BSI BS 4255: Part 1-86. 1986 Rubber Used in Preformed Gaskets for Weather Exclusion from Buildings Part 1: Non-Cellular Gaskets. 7 pp.
SNZ NZS/BS 4255: Part 1-86. Rubber Used in Preformed Gaskets for Weather Exclusion from Buildings Part 1: Specification for Non-Cellular Gaskets. 8 pp.

—Rubber—Fuses (Ordnance)
MOD UK TS 10230. Polychloroprene Rubber Mixes, Special Purpose, Type QX.

—Ships—Flap Valves
DIN ENGL 87107-75. Gaskets for Non-Return Flaps in Accordance with DIN 87 101 (June). 1 p.

—Ships—Pipelines
JIS F 7102-89. Application for Gaskets and Packings to Piping System in Machinery Space.

—Ships—Windows
ISO 3902-90. Shipbuilding and Marine Structures—Gaskets for Rectangular Windows and Side Scuttles Second Edition. 4 pp.

—Spiral Wound—Pipe Flanges
CNS B2797-89. Spiral-Wound Gaskets for Pipe Flanges (Sep) (12606).
JIS B 2404-79. Spiral-Wound Gaskets for Pipe Flanges. 24 pp.

—Sponge—Doors
CNS A2226-88. Sponge Gaskets for Windows, Doors and Joints of Panel in Buildings (Jul)(12351).
CNS A3279-88. Method of Test for Sponge Gaskets for Windows, Doors and Joints of Panel in Building (Jul)(12352).
JIS A 5750-87. Sponge Gaskets for Windows, Doors and Joints of Panel in Buildings. 21 pp.

—Sponge—Joints
CNS A2226-88. Sponge Gaskets for Windows, Doors and Joints of Panel in Buildings (Jul)(12351).
CNS A3279-88. Method of Test for Sponge Gaskets for Windows, Doors and Joints of Panel in Building (Jul)(12352).
JIS A 5750-87. Sponge Gaskets for Windows, Doors and Joints of Panel in Buildings. 21 pp.

—Sponge—Windows
CNS A2226-88. Sponge Gaskets for Windows, Doors and Joints of Panel in Buildings (Jul)(12351).
CNS A3279-88. Method of Test for Sponge Gaskets for Windows, Doors and Joints of Panel in Building (Jul)(12352).
JIS A 5750-87. Sponge Gaskets for Windows, Doors and Joints of Panel in Buildings. 21 pp.

—Steel—Corrugated
DIN ENGL 2698-72. Corrugated Steel Sheet Gaskets Incorporating Asbestos Ropes for ND 25 to ND 250 Flanged Connections (Jan). 2 pp.

—Vulcanized Fibers
MOD UK DEF-143-62. Vulcanized Fibre Sheet Jointing, Compressible Type. 5 pp.

—Windows
BSI BS 4315: Part 1-68. 1968 Amd 1 Methods of Test for Resistance to Air and Water Penetration Part 1: Windows and Structural Gasket-Glazing Systems. 18 pp.
CNS A2154-88. Gaskets for Windows, Doors and Joints of Panel in Buildings (Dec)(10209).
CNS A3186-88. Method of Test for Gaskets for Windows, Doors and Joints of Panel in Buildings (Dec)(10210).
JIS A 5756-89. Gaskets for Windows, Doors and Joints of Panel in Buildings. 26 pp.

Gasohol
See Also: Alcohol Fuels; Diesel Fuels; Exhaust Gases; Gasoline

Gasohol (Cont.)

—Octane Number
DIN ENGL 51756 Pt 7-86. Testing of Gasolines; Determination of Knock Characteristics (Octane Number) of Alcohols and Alcohol/Fuel Mixtures Using the CFR Engine (Feb). 4 pp.

—Unleaded
CGSB CAN/CGSB-3.511-M90. Oxygenated Unleaded Automotive Gasoline Containing Ethanol. 17 pp.
CGSB CAN/CGSB-3.515-M90. Oxygenated Unleaded Automotive Gasoline Containing Methanol and Cosolvent. 17 pp.

Gasoline
Use For: Petrol See Also: Aircraft Gasoline; Automotive Fuels; Fuel Dispensing Equipment; Gasohol; Jet Engine Fuels; Kerosene; Liquid Fuels; Octane Number; Petroleum Products
CAA NOTICE #98 ISSUE 11. Use of Motor Gasoline (Mogas) in Certain Light Aircraft (Airworthiness Notices). 11 pp.
CAA NOTICE #98A ISSUE 1. Use of Filling Station Forecourt Motor Gasoline (Mogas) in Certain Light Aircraft. 4 pp.
CGSB CAN/CGSB-3.1-92. Leaded Gasoline. 13 pp.
CNS K5022-89. Gasoline for Automobile (Oct)(1469).
CNS K6003-47. Method of Test for Gasoline (Mar)(46).
JIS K 2201-91. Gasoline for Industrial Purpose. 5 pp.
JIS K 2202-91. Motor Gasoline. 6 pp.
MOD UK DSTAN 91-13-01. Gasoline, Automotive: Military (91 RON) NATO Code No: F-46 Joint Service Designation: COMBATGAS Gasoline, Automotive: (91 RON) NATO Code No: F-50 Joint Service Designation: CIVGAS Gasoline, Automotive: Combat, Sub-Zero (95 RON) Joint Service. 12 pp.
SAA AS 1876-90. Petrol (Gasoline) for Motor Vehicles. 5 pp.

—Alcohol Content—Gas Chromatography
DIN ENGL 51413 Pt 1-84. Analysis of Liquid Petroleum Products by Gas Chromatography; Determination of Alcohol Content (Nov). 3 pp.

—Alkane Content—Gas Chromatography
CGSB CAN/CGSB-3.0 NO.14.3-M91. Methods of Testing Petroleum and Associated Products Standard Test Method for the Identification of Hydrocarbon Components in Automotive Gasoline Using Gas Chromatography. 32 pp.

—Aromatic Hydrocarbon Content—Gas Chromatography
CGSB CAN/CGSB-3.0 NO.14.3-M91. Methods of Testing Petroleum and Associated Products Standard Test Method for the Identification of Hydrocarbon Components in Automotive Gasoline Using Gas Chromatography. 32 pp.

—Benzene Content—Gas Chromatography
DIN ENGL 51413 Pt 2-84. Analysis of Liquid Petroleum Products by Gas Chromatography; Determination of Benzene Content (Nov). 2 pp.

—Benzene Content—Infrared Spectroscopy
DIN ENGL 51414-85. Testing of Petroleum Products; Determination of the Benzene Content of Gasolines by Infrared Spectroscopy (June). 3 pp.

—Distillation Methods
BSI BS 2000: Part 191-83. 1983 Petroleum and Its Products Part 191: Distillation of Natural Gasoline. 7 pp.
BSI BS 7392-90. 1990 Method for Determination of Distillation Characteristics of Petroleum Products. 26 pp.
ISO 3405-88. Petroleum Products—Determination of Distillation Characteristics Second Edition. 24 pp.

—Fuel Dilution—Gas Chromatography
DIN ENGL 51380-90. Determination of Readily Volatile Components in Used Automotive Engine Oils by Gas Chromatography (Nov). 2 pp.

—Gum Testing
JIS K 2261-92. Petroleum Products—Motor Gasoline and Aviation Fuels—Determination of Existent Gum—Jet Evaporation Method. 20 pp.

—Gum Testing—Jet Evaporation Method
BSI BS 4348-76. 1976 Determination of Existent Gum in Fuels by Jet Evaporation. 12 pp.
CEN EN 5-74. Determination of Existent Gum in Fuels by Jet Evaporation. 10 pp.
CNS K6321-74. Method of Test for Existing Gum Content in Jet and Motor Fuels (Oct)(3382).
ISO 6246-81. Petroleum Products—Motor Gasoline and Aviation Fuels—Determination of Existent Gum—Jet Evaporation Method First Edition. 10 pp.

Gasoline (Cont.)

—Highway Transportation—Emergency Procedures
SAA AS 1678.3.1. 001-86. Emergency Procedure Guide—Transport—Part 3.1.001: Petrol—as Cargo.

—Hoses
CNS K4064-82. Oil Discharge Rubber Hose (Jun)(9011).
JIS K 6343-82. Rubber Hoses for Oil Discharge. 7 pp.

—Hydrocarbon Content—Gas Chromatography
CGSB CAN/CGSB-3.0 NO.14.3-M91. Methods of Testing Petroleum and Associated Products Standard Test Method for the Identification of Hydrocarbon Components in Automotive Gasoline Using Gas Chromatography. 32 pp.
CNS K6326-72. Method of Test for C2 Through C5 Hydrocarbons in Gasolines by Gas Chromatography Determination (Oct)(3387).

—Lead Content
EC 85/210/EEC-85. Council Directive on the Approximation of the Laws of the Member States Concerning the Lead Content of Petrol. 5 pp.
EC 87/416/EEC-87. Council Directive Amending Directive 85/210/EEC on the Approximation of the Laws of the Member States Concerning the Lead Content of Petrol. 2 pp.

—Lead Content—Atomic Absorption Spectrometry
CNS K6912-87. Method of Test for Lead in Gasoline by Atomic Absorption Spectrometry (Jul)(12013).

—Lead Content—Complexometric Titrations
DIN ENGL 51769 Pt 5-84. Testing of Petroleum Products; Determination of Lead Content (Total Lead) of Gasolines; Complexometric Method (June). 7 pp.

—Lead Content—Iodine Monochloride Method
BSI BS 5657-88. 1988 Amd 1 Method for Determination of Lead Content of Gasoline by the Iodine Monochloride Method (AMD 6512) February 28, 1991. 11 pp.
CEN EN 23 830-89. Petroleum Products, Gasoline, Chemical Analysis, Determination of Content, Lead, Analysis Method. 2 pp.
ISO 3830-81. Petroleum Products—Gasoline—Determination of Lead Content—Iodine Monochloride Method Second Edition. 5 pp.
JIS K 2255-87. Testing Methods for Lead in Gasoline. 22 pp.

—Lead Content—Volumetric Analysis
BSI BS 5290-76. 1976 Determination of Lead Content of Gasoline Volumetric Chromate Method (Supersedes BS 2878: 1976). 11 pp.
CEN EN 13-74. Determination of Lead Content of Gasoline: Volumetric Chromate Method. 9 pp.

—Lead Content—X-Ray Spectrometry
CNS K6973-90. Method of Test for Lead in Gasoline by X-Ray Spectrometry (Aug)(12762).
DIN ENGL 51769 Pt 6-90. Determination of Total Lead Content of Petrol with a Lead Content Exceeding 25mg/l by Wavelength-Dispersive X-Ray Spectrometry (XRS) (Nov). 3 pp.

—Leaded
BSI BS 4040-88. 1988 Amd 1 Leaded Petrol (Gasoline) for Motor Vehicles (AMD 6524) June 29, 1990. 17 pp.
CGSB CAN/CGSB-3.1-92. Leaded Gasoline. 13 pp.

—Leaded—Spark Ignition Engines
MOD UK DSTAN 91-76-93. Gasoline Automotive, Leaded: CIVGAS (93 RON) Issue 1. 11 pp.

—Naphthene Content—Gas Chromatography
CGSB CAN/CGSB-3.0 NO.14.3-M91. Methods of Testing Petroleum and Associated Products Standard Test Method for the Identification of Hydrocarbon Components in Automotive Gasoline Using Gas Chromatography. 32 pp.

—Octane Number
DIN ENGL 51756 Pt 1-90. Determination of Knock Characteristics (Octane Number) of Petrol; General (June) (Supersedes July 1987 Edition). 4 pp.

—Olefin Content—Gas Chromatography
CGSB CAN/CGSB-3.0 NO.14.3-M91. Methods of Testing Petroleum and Associated Products Standard Test Method for the Identification of Hydrocarbon Components in Automotive Gasoline Using Gas Chromatography. 32 pp.

INTERNATIONAL AND NON-U.S. NATIONAL STANDARDS
SUBJECT INDEX

Gasoline (Cont.)

—Oxidation Resistance
BSI BS 2000: Part 40-92. 1992 Methods of Test for Petroleum and Its Products Part 40: Determination of Oxidation Stability of Gasoline—Induction Period Method (Identical with IP 40/92). 7 pp.
BSI BS 2000: Part 40-91. 1991 Petroleum and Its Products Part 40: Oxidation Stability of Gasoline (Induction Period Method) (Identical with IP 40/89). 8 pp.
CNS K6913-87. Method of Test for Oxidation Stability of Gasoline (Induction Period Method) (Jul)(12014).
JIS K 2287-86. Testing Method for Oxidation Stability of Gasoline (Induction Period Method). 14 pp.

—Phosphorus Content—X-Ray Spectrometry
DIN ENGL 51440 Pt 1-88. Determination of Phosphorus Content of Petrol by Wavelength-Dispersive X-Ray Spectrometry (XRS) (Oct). 3 pp.

—Separators (Mechanical)—Sewage Treatment Equipment
DIN ENGL 1999 Pt 2-89. Petrol Interceptors and Fuel Oil Interceptors; Design, Installation and Operation (Mar). 6 pp.

—Ships
MOD UK NES 775-01. Gasoline Systems and Stowages Issue 1 (02.88); Amendment 1. 19 pp.

—Ships—Stowage
MOD UK NES 775-01. Gasoline Systems and Stowages Issue 1 (02.88); Amendment 1. 19 pp.

—Tetraethyl Lead Content—Separation
CNS K6974-90. Method for Separation of Tetraethyl Lead and Tetamethyl Lead in Gasoline (Aug)(12763).

—Tetramethyl Lead Content—Separation
CNS K6974-90. Method for Separation of Tetraethyl Lead and Tetamethyl Lead in Gasoline (Aug)(12763).

—Thiol Content—Potentiometric Analysis
CNS K6547-80. Method of Test for Mercaptan Sulfur in Gasoline, Kerosine, Aviation Turbine and Distillate Fuels (Potentiometric Method) (Aug)(6192).

—Unleaded
BSI BS 7070-88. (WITHDRAWN) 1988 Amd 1 Unleaded Petrol (Gasoline) for Motor Vehicles (AMD 6525) June 29, 1990 (Superseded by BS EN 228: 1993). 17 pp.
BSI BS 7800-92. 1992 High Octane (Super) Unleaded Petrol (Gasoline) for Motor Vehicles. 18 pp.
BSI BS EN 228-93. 1993 Unleaded Petrol (Gasoline) for Motor Vehicles (W). 16 pp.
CEN EN 228-87. Liquid Petroleum Products: Unleaded Petrol; Specification. 5 pp.
CEN PREN 228-91. Automotive Fuels—Unleaded Petrol—Requirements and Methods of Test. 20 pp.
CEN EN 228-93. Automotive Fuels—Unleaded Petrol—Requirements and Methods of Test. 6 pp.
CGSB CAN/CGSB-3.5-92. Unleaded Automotive Gasoline. 12 pp.
CNS K5140-89. Unleaded Gasoline for Automobile (Oct)(12614).
EC 89/491/EEC-89. Commission Directive Adapting to Technical Progress Council Directives 70/157/EEC, 70/220/EEC, 72/245/EEC, 72/306/EEC, 80/1268/EEC and 80/1269/EEC Relating to Motor Vehicles. 2 pp.

—Vapor Pressure
DIN ENGL 51754-83. Testing of Liquid Fuels; Determination of Vapour Pressure; Reid Method (Sept). 4 pp.

Gasoline Content Analysis

—Lubricating Oils
DIN ENGL 51565-85. Testing of Lubricants; Determination of Gasoline Content of Used Engine Lubricating Oils (Aug). 3 pp.

Gasoline Engines
See Also: Gas Turbine Engines; Internal Combustion Engines; Piston Engines
MOD UK DSTAN 28-3-87. Engines, Gasoline for General Purpose Applications Issue 4. 10 pp.

—Acceptance Testing
NATO STANAG 4195 ED 1 AMD 1-85. NATO Standard Engine Laboratory Test for Diesel and Gasoline Engines and Gas Turbine Engines. 6 pp.

—Agricultural Equipment
CNS B4012-69. Air-Cooled Gasoline Engines for Land Use (Small Size) (Jul)(3003). 6 pp.

Gasoline Engines (Cont.)

—Air Cooled
JIS B 8017-87. Performance Test Method of Small Size Air Cooled Gasoline Engines for Land Use. 26 pp.

—Cylinder Blocks—Preservation
MOD UK DSTAN 81-72-90. Preservation of Cylinder Blocks (with or Without Pistons) and Cylinder Heads Issue 1. 9 pp.

—Cylinder Heads—Preservation
MOD UK DSTAN 81-72-90. Preservation of Cylinder Blocks (with or Without Pistons) and Cylinder Heads Issue 1. 9 pp.

—Exhaust Gases
CNS D3077-87. Method of Test for Exhaust Pollution for Gasoline Engine Automobiles (Sep)(7895). 16 pp.

—Preservation
MOD UK DSTAN 81-70-01. Preservation of Engines (Gasoline and Diesel) Issue 1; Amendment 1. 22 pp.

—Pressure Measurement
CNS B6074-82. Test Device for Cylinder Pressure of Gasoline Engine (Oct)(9466).
CNS B7209-82. Method of Test for Test Device for Cylinder Pressure of Gasoline Engine (Oct)(9467).

—Starting Fluids
MOD UK DSTAN 29-6-72. Cold Starting Aids (Fluid Type) for Gasoline and Diesel Engines Issue 1 (Withdrawn). 11 pp.

—Type Testing
MOD UK DSTAN 28-4-01. Type Testing of Diesel and Gasoline Engines for General Purpose Applications Issue 3; Corrigendum. 23 pp.

Gasoline Stations
Use: Service Stations

Gassing

—Electrical Insulating Liquids
BSI BS 5797-86. 1986 Amd 1 Methods for Measurement of Gassing of Insulating Liquids Under Electrical Stress and Ionization. 24 pp.
CENELEC HD 488 S1-88. Gassing of Insulating Liquids Under Electrical Stress and Ionization. 3 pp.
IEC 628-85. Gassing of Insulating Liquids Under Electrical Stress and Ionization Second Edition; (Corrigendum—Oct 1986). 38 pp.

—Sealing Materials
CGSB CAN2-19.0-M77METH18.1-78. Methods of Testing Putty, Caulking and Sealing Compounds Gassing of Sealing Compounds. 1 p.

Gastrofiberscopes
CNS T1049-81. Gastrofiberscopes (Apr)(7328).

Gastroscopes
See Also: Endoscopes; Medical Electrical Equipment; Medical Equipment
JIS T 4408-79. Gastroscopes.

GAT
Use: Group Audio Terminals

Gate Circuits
See Also: AND Gates; AND NAND Gates; AND OR Gates; AND OR INVERT Gates; AND OR SELECT Gates; Digital Circuits; Exclusive NOR Gates; Exclusive OR Gates; Exclusive OR NOR Gates; Extenders (Logic Circuits); Integrated Circuits; Inverting Buffer Gates; Inverting Gates; Logic Circuits; NAND Gates; Non-Inverting Buffer Gates; Non-Inverting Gates; NOR INVERT Gates; OR Gates; Shift Registers; Switching Circuits
MOD UK DSTAN 59-62: Pt 99:Sec 14. Integrated Circuits Part 99: Data Sheets for CECC/BS 9000 Detail Specification (Listed on EPIC Database) Section 14: TTL Low Power Circuits Interconnected Gates. 3 pp.

—Bipolar—Monolithic
IECQ QC 790132-91. Semiconductor Devices Integrated Circuits Part 2: Digital Integrated Circuits Section One—Blank Detail Specification for Bipolar Monolithic Digital Integrated Circuit Gates (Excluding Uncommitted Logic Arrays) (IEC 748-2-1 ED 1). 36 pp.

—Bipolar—Monolithic—Quality Assurance
IEC 748 Pt 2-1-91. Semiconductor Devices Integrated Circuits Part 2: Digital Integrated Circuits Section One—Blank Detail Specification for Bipolar Monolithic Digital Integrated Circuit Gates (Excluding Uncommitted Logic Arrays) First Edition (IECQ QC 790132). 36 pp.

Gate Circuits (Cont.)

—Noise Margin
CNS C6102-80. Standard Test Procedure for Noise Margin Measurements for Semiconductor Logic Gating Microcircuits (Dec)(6802).

—Quality Assurance
BSI BS 9401-70. (WITHDRAWN) 1970 Amd 2 Rules for the Preparation of Detail Specifications for Integrated Circuits of Assessed Quality: TTL Digital Gate Circuits. General Application Category (AMD 1821) December 31, 1989. 12 pp.
BSI BS 9402-72. (WITHDRAWN) 1972 Amd 1 Rules for the Preparation of Detail Specifications for Integrated Circuits of Assessed Quality: DTL Digital Gate Circuits. General Application Category. 16 pp.

—Reliability Assured
JIS C 7312-82. Reliability Assured Complementary MOS Digital Semiconductor Integrated Circuits (Gates) (R 1987). 18 pp.

Gate Openers
Use: Gate Operators

Gate Operators
Use For: Gate Openers See Also: Door Openers; Gates (Barriers)
CSA CAN/CSA-C22. 2 NO 247-92. Operators and Systems of Doors, Gates, Draperies, and Louvres; (Gen Instr 1). 98 pp.

Gate Turn Off Switches
Use: Gate Turn Off Thyristors

Gate Turn Off Thyristors
Use For: Gate Turn Off Switches; GTOs; Turn Off Thyristors See Also: Silicon Controlled Rectifiers; Thyristors
CECC CECC 50 011-012 ISSUE 1-86. CEI CECC 50 011-012; Case-Rated Thyristor Type Number: 50 RIA (En). 13 pp.

Gate Valves
See Also: Faucets; Gas Valves; Globe Valves; Hydraulic Valves; Lock Gates; Pneumatic Valves; Stop Valves; Valves
BSI BS 1414-75. 1975 Amd 2 Steel Wedge Gate Valves (Flanged and Butt-Welding Ends) for the Petroleum, Petrochemical and Allied Industries (AMD 6562) August 31, 1990. 32 pp.
BSI BS 5150-90. 1990 Cast Iron Gate Valves. 21 pp.
BSI BS 5150-74. (WITHDRAWN) 1974 Amd 2 Specification for Cast Iron Wedge and Double Disk Gate Valves for General Purposes. 18 pp.
BSI BS 5151-74. 1974 Amd 2 Cast Iron Gate (Parallel Slide) Valves for General Purposes. 18 pp.
BSI BS 5154-91. 1991 Copper Alloy Globe, Globe Stop and Check, Check and Gate Valves. 22 pp.
BSI BS 5157-89. 1989 Steel Gate (Parallel Slide) Valves. 23 pp.
BSI BS 5163-86. 1986 Amd 2 Predominantly Key-Operated Cast Iron Gate Valves for Waterworks Purposes (AMD 6057) July 31, 1989. 24 pp.
BSI BS 5352-81. 1981 Amd 2 Steel Wedge Gate, Globe and Check Valves 50 mm and Smaller for the Petroleum, Petrochemical and Allied Industries (AMD 6560) August 31, 1990 (Supersedes BS 2995: 1966). 25 pp.
CNS B2107-87. Cast Iron Flanged Gate Valves (10kgf/cm2) (Inside Screw Non-Rising Stem Type) (Mar)(713). 6 pp.
CNS B2108-81. Cast Iron Flanged Gate Valves (5 kgf/cm2) (Outside Screw Type) (Mar)(714).
CNS B2109-87. Cast Iron Flanged Gate Valves (10 kgf/cm2) (Outside Screw Type) (Mar)(715). 7 pp.
CNS B2110-62. Gate Valve (Flange Type) (Nominal Pressure 16 kgf/cm2) (Sep)(716)(R 1973).
CNS B2111-62. Gate Valve (Flange Type) (Nominal Pressure 25 kgf/cm2) (Jul)(717)(R 1973).
CNS B2112-62. Gate Valve (Flange Type) (Nominal Pressure 40 kgf/cm2) (Jul)(718)(R 1973).
CNS B2494-80. Face-to-Face and End-to-End Dimensions of Gate Valves (Jul)(5710).
CNS B2503-88. Bronze Screwed Gate Valves (Jul)(5964).
CNS B2505-88. Bronze Screwed Gate Valves (Jul)(5966).
CNS B2510-85. Bronze Flanged Gate Valves (Mar)(5971).
CNS B2537-82. Cast Steel Flanged Gate Valves (10 kgf/cm2) (Outside Screw Type) (Sep)(6884).
CNS B2551-82. Cast Steel Flanged Gate Valves (20 kgf/cm2) (Outside Screw Type) (Sep)(7114).
CNS B2740-87. Brass Screwed Gate Valves (8.5 kgf/cm2) (Mar)(9805). 3 pp.
CNS B2764-87. Bronze Screwed Gate Valves (15kgf/cm2) (Mar)(11089).
DIN ENGL 3352 Pt 1-79. Gate Valves; General Information (May). 7 pp.
DIN ENGL 3352 Pt 2-88. Cast Iron Gate Valves with Metallic Seat and Inside Screw Stem (Aug). 4 pp.

INDUSTRY STANDARDS

Gate Valves (Cont.)

DIN ENGL 3352 Pt 3-88. Cast Iron Gate Valves with Metallic Seat and Outside Screw Stem (Aug). 4 pp.
DIN ENGL 3352 Pt 4-86. Cast Iron Gate Valves with Elastomeric Obturator Seatings and Inside Screw Stem (Jan). 4 pp.
DIN ENGL 3352 Pt 6-79. Gate Valves of Unalloyed and Low-Alloyed Steel with Internal Stem Thread (May). 5 pp.
DIN ENGL 3352 Pt 7-79. Gate Valves of Unalloyed and Low-Alloyed Steel with External Stem Thread (May). 4 pp.
DIN ENGL 3352 Pt 10-79. Gate Valves of Stainless Steel (May). 4 pp.
DIN ENGL 3352 Pt 11-81. Flanged Copper Alloy Gate Valves (Dec). 3 pp.
DIN ENGL 3352 Pt 12-81. Socket End Copper Alloy Gate Valves (Dec). 3 pp.
DIN ENGL 3352 Pt 13-87. Double-Socket Cast Iron Gate Valves with Elastomeric Obturator Seat and Inside Screw Stem (Oct). 3 pp.
DIN ENGL 3441 Pt 6-88. Unplasticized Polyvinyl Chloride (UPVC) Valves; Gate Valves with Inside Screw Stem Dimensions (Mar). 2 pp.
DIN ENGL 86720-68. Bronze Wedge-Type Flat-Sided Gate Valves with Screwed Bonnet and Flanges; ND 16; NW 20 to 100 (Sept). 2 pp.
ISO 5752-82. Metal Valves for Use in Flanged Pipe Systems—Face-to-Face and Centre-to-Face Dimensions Second Edition. 13 pp.
ISO 5996-84. Cast Iron Gate Valves First Edition. 11 pp.
ISO 6002-92. Bolted Bonnet Steel Gate Valves First Edition. 17 pp.
ISO 7259-88. Predominantly Key-Operated Cast Iron Gate Valves for Underground Use First Edition. 12 pp.
JIS B 2002-87. Face-to-Face and End-to-End Dimensions of Valves.
JIS B 2002-68. Face-to-Face and End-to-End Dimensions of Valves. 14 pp.
JIS B 2011-88. Bronze Gate, Globe, Angle and Check Valves. 41 pp.
SAA AS 1628-77. Copper Alloy Gate Valves and Non-Return Valves for Use in Water Supply and Hot Water Services Amdt 1 November 1977 Amdt 2 October 1983 Amdt 3 February 1988 Amdt 4 September 1989. 34 pp.
SNZ NZS/BS 5150-90. Specification for Cast Iron Gate Valves. 18 pp.
SNZ NZS/BS 5151-74. Specification for Cast Iron Gate (Parallel Slide) Valves for General Purposes Amend: 1; 2. 16 pp.
SNZ NZS/BS 5154-89. Specification for Copper Alloy Globe, Globe Stop and Check, Check and Gate Valves. 20 pp.
SNZ NZS/BS 5154-91. Specification for Copper Alloy Globe, Globe Stop and Check, Check and Gate Valves. 24 pp.
SNZ NZS/BS 5163-86. Specification for Predominantly Key-Operated Cast Iron Gate Valves for Waterworks Purposes Amend: 1; 2. 20 pp.
SNZ NZS/BS 5352-81. Specification for Steel Wedge Gate, Globe and Check Valves 50 mm and Smaller for the Petroleum, Petrochemical and Allied Industries Amend: 1. 24 pp.

—**Fire Hydrants**
BSI BS 5041: Part 2-87. 1987 Amd 1 Fire Hydrant Systems Equipment Part 2: Landing Valves for Dry Risers (AMD 5776) July 29, 1988. 12 pp.
BSI BS 5041: Part 2-76. 1976 Fire Hydrant Systems Equipment Part 2: Landing Valves for Dry Risers. 7 pp.

—**Gas Industry**
CSA CAN/CSA-Z245.15-M91. Steel Valves; (Gen Instr 1). 81 pp.

—**Heat Resistant**
DIN ENGL 3352 Pt 9-79. Gate Valves of Heat-Resistant Steel (May). 4 pp.

—**Petroleum Industry**
CSA CAN/CSA-Z245.15-M91. Steel Valves; (Gen Instr 1). 81 pp.

—**Pipelines—Flanged**
CEN PREN 558-1-91. Metal Valves for Use in Flanged Pipe Systems—Face-to-Face and Centre-to-Face Dimensions—Part 1: General. 5 pp.
CEN PREN 558-2-91. Metal Valves for Use in Flanged Pipe Systems—Face-to-Face and Centre-to-Face Dimensions—Part 2: PN Designated Valves. 13 pp.
CEN PREN 558-3-91. Metal Valves for Use in Flanged Pipe Systems—Face-to-Face and Centre-to-Face Dimensions—Part 3: Class-Designation Valves. 17 pp.

—**Ships**
CNS F3006-86. Cast Iron Gate Valves for Marine Use (5kgf/cm2) (Mar)(3811).

Gate Valves (Cont.)

—**Ships (Cont.)**
CNS F3007-86. Cast Iron Gate Valves for Marine Use (10kgf/cm2) (Mar)(3812).
CNS F3175-82. Marine Hull Cast Steel Gate Valves (Apr)(8691).
CNS F3177-82. Marine Cast Steel 10 kgf/cm2 Gate Valves (May)(8818).
CNS F3178-82. Marine Bronze 5 kgf/cm2 Rising Stem Type Gate Valves (May)(8819).
CNS F3179-82. Marine Bronze 10 kgf/cm2 Rising Stem Type Gate Valves (May)(8820).
CNS F3180-82. Marine Cast Iron 16 kgf/cm2 Gate Valves (May)(8821).
CNS F3197-82. Marine Cast Iron 3 kgf/cm2 Gate Valves (Aug)(9259).
DIN ENGL 86500-68. Valves and Gate Valves with Screwed Connections; Survey of Types for Shipbuilding (Aug). 8 pp.
JIS F 7360-86. Marine Hull Cast Steel Gate Valves.
JIS F 7363-89. Cast Iron 5 K Gate Valves for Marine Use.
JIS F 7364-89. Cast Iron 10 K Gate Valves for Marine Use.
JIS F 7366-89. Cast Steel 10 K Gate Valves for Marine Use.
JIS F 7367-88. Bronze 5 K Rising Stem Type Gate Valves for Marine Use.
JIS F 7368-88. Bronze 10 K Rising Stem Type Gate Valves for Marine Use.
JIS F 7369-88. Cast Iron 16 K Gate Valves for Marine Use.

—**Sounding Pipes—Ships**
CNS F3059-80. Short Sounding Pipe Heads of Self-Closing Gate Valve Type (Sep)(6439).
JIS F 3019-90. Self-Closing Gate Valve Heads for Short Sounding Pipe.
JIS F 3019-68. Self-Closing Gate Valve Heads for Short Sounding Pipe. 6 pp.

—**Water**
CNS B2803-90. Resilient-Seated Gate Valves for Water Works (Oct)(12795).
DIN ENGL 3500-90. PN 10 Piston Type Gate Valves for Use in Drinking Water Supply Systems (Feb). 3 pp.

Gates (Barriers)

See Also: Access Control Systems; Doors; Gate Operators; Lock Gates; Roller Gates; Stiles

—**Aluminum Alloy**
JIS A 6513-83. Fences and Gates with Metals. 21 pp.

—**Equestrian**
BSI BS 5709-79. 1979 Amd 1 Stiles, Bridle Gates and Kissing Gates. 14 pp.

—**Fences—Agricultural**
BSI BS 3470-75. 1975 Field Gates and Posts. 16 pp.

—**Fences—Chain Link**
CGSB CAN/CGSB-138.4-M82. Fence, Chain Link, Gates. 12 pp.

—**Hardware**
BSI BS 3827: Part 4-67. (WITHDRAWN) 1967 Glossary of Terms Relating to Builders' Hardware Part 4: Door Drawer, Cupboard and Gate Furniture (Superseded by BS 6100: Subsection 1.3.6:1991). 24 pp.

—**Iron**
BSI BS 4092: Part 1-66. 1966 Domestic Front Entrance Gates Part 1: Metal Gates. 10 pp.

—**Pedestrian**
BSI BS 5709-79. 1979 Amd 1 Stiles, Bridle Gates and Kissing Gates. 14 pp.

—**Power Drives—Safety**
DIN VDE 0700 Pt 238-83. Safety of Household and Similar Electrical Appliances; Particular Requirements for Power Drive for Gates, Doors, Windows and Similar Equipment (Oct). 15 pp.

—**Ships**
BSI BS MA 40: Part 2-75. 1975 Marine Guardrails, Stanchions, Etc Part 2: Gates and Portable Guardrail Sections for Merchant Ships (Excluding Passenger Ships). 8 pp.

—**Steel**
BSI BS 4092: Part 1-66. 1966 Domestic Front Entrance Gates Part 1: Metal Gates. 10 pp.
JIS A 6513-83. Fences and Gates with Metals. 21 pp.

—**Swimming Pools**
SAA AS 1926-86. Fences and Gates for Private Swimming Pools Amdt 1 March 1987 (Superseded by AS 1926.1 —1993 but will Remain Avalible Superseded). 10 pp.

Gates (Barriers) (Cont.)

—**Wood**
BSI BS 4092: Part 2-66. 1966 Domestic Front Entrance Gates: Wooden Gates. 8 pp.

Gates (Circuits)
Use: Gate Circuits

Gates Learjets
Scope Note: See listings under Learjet

Gateways
See Also: Communication Equipment; Controllers; Protocols; Telecommunication Equipment

—**Military Communications**
NATO STANAG 4206 ED 1 AMD 2-83. The NATO Multi-Channel Tactical Digital Gateway—System Standards. 20 pp.
NATO STANAG 4206 ED 2 AMD 0-89. (DRAFT) NATO Multi-Channel Tactical Digital Gateway—System Standards. 18 pp.

—**Military Communications—Analog to Digital Converters**
NATO STANAG 4209 ED 1 AMD 2-83. The NATO Multi-Channel Tactical Digital Gateway—Standards for Analogue to Digital Conversion of Speech Signals. 38 pp.

—**Military Communications—Cables (Electric)**
NATO STANAG 4210 ED 1 AMD 2-83. The NATO Multi-Channel Tactical Digital Gateway—Cable Link Standards. 18 pp.
NATO STANAG 4210 ED 2 AMD 0-89. (DRAFT) The NATO Multi-Channel Tactical Digital Gateway—Cable Link Standards. 17 pp.

—**Military Communications—Data Transmission**
NATO STANAG 4213 ED 1 AMD 0-88. The NATO Multi-Channel Tactical Digital Gateway—Data Transmission Standards. 24 pp.
NATO STANAG 4213 ED 2 AMD 0-89. (DRAFT) The NATO Multi-Channel Tactical Digital Gateway—Data Transmission Standards. 27 pp.

—**Military Communications—Digital Multiplexers**
NATO STANAG 4207 ED 1 AMD 2-83. The NATO Multi-Channel Tactical Digital Gateway—Multiplex Group Framing Standards. 28 pp.
NATO STANAG 4207 ED 2 AMD 0-89. (DRAFT) NATO Multi-Channel Tactical Digital Gateway—Multiplex Group Framing Standards. 26 pp.

—**Military Communications—Digital Multiplexers—Signaling Format**
NATO STANAG 4208 ED 1 AMD 2-83. The NATO Multi-Channel Tactical Digital Gateway—Signalling Standards. 51 pp.
NATO STANAG 4208 ED 2 AMD 0-89. (DRAFT) The NATO Multi-Channel Tactical Digital Gateway—Signalling Standards. 51 pp.

—**Military Communications—Packet Switched Networks**
NATO STANAG 4249 ED 1 AMD 0-88. The NATO Multi-Channel Tactical Digital Gateway—Data Transmission Standards (Packet Switching Service). 91 pp.

—**Military Communications—Radio Relay Systems**
NATO STANAG 4212 ED 1 AMD 3-83. The NATO Multi-Channel Tactical Digital Gateway-Radio Relay Link Standards. 18 pp.
NATO STANAG 4212 ED 2 AMD 0-89. (DRAFT) NATO Multi-Channel Tactical Digital Gateway—Radio Relay Link Standards. 19 pp.

—**Military Communications—System Management**
NATO STANAG 4211 ED 1 AMD 2-83. The NATO Multi-Channel Tactical Digital Gateway-System Control Standards. 38 pp.
NATO STANAG 4211 ED 2 AMD 0-89. (DRAFT) The NATO Multi-Channel Tactical Digital Gateway—System Control Standards. 39 pp.

—**Public Data Networks—Interworking**
CCITT RECMN D.10-91. General Tariff Principles for International Public Data Communication Services (Study Group III) 6 pp. 6 pp.
CCITT RECMN D.10-89. General Tariff Principles for International Public Data Communication Services—General Tariff Principles—Charging and Accounting in International Telecommunications Services (Study Group III) 3 pp. 3 pp.

INTERNATIONAL AND NON-U.S. NATIONAL STANDARDS
SUBJECT INDEX
Gear

Gateways (Cont.)
—Videotex Communications
CCITT RECMN F.300-89. Videotex Service—Telematic, Data Transmission and Teleconference Services—Operations and Quality of Service (Study Group I) 25 pp. 25 pp.

—Videotex Equipment—Interworking
CCITT RECMN T.564-89. Gateway Characteristics for Videotex Interworking—Terminal Equipment and Protocols for Telematic Services (Study Group VIII) 30 pp. 30 pp.

Gauge Blocks
Use: Gage Blocks

Gauge Glasses
Use: Gage Glasses

Gauge Pressure Transmitters
Use: Gage Pressure Transmitters

Gauge Rings
Use: Ring Gages

Gauge Ties
Use: Gage Ties

Gauges
Use: Gages

Gauging
Use: Gaging

Gauging Station
Use: Gaging Stations

Gauze
See Also: Wire Gauze
—Absorbent
CNS L4070-78. Gauze, Absorbent (May) (4397).

Gazetteer
—Streets—Data Management
BSI BS 7666: Part 1-93. 1993 Spatial Data-Sets for Geographic Referencing Part 1: Specification for a Street Gazetteer (G). 24 pp.

Gear Boxes
See Also: Bearings; Boxes (Containers); Gears; Transmissions (Power Sources)

—Aircraft—Quick Release Fasteners
BSI M 32-69. 1969 Amd 3 Dimensions for Aircraft Accessory Drives and Mounting Pads. 20 pp.
ISO 1971-75. Aircraft—Accessory Drives and Mounting Pads First Edition. 19 pp.

—Automotive—Flanges
BSI BS AU 214A: Part 1-87. 1987 Amd 1 Gearbox Flanges Part 1: Dimensions of Cross-Tooth Gearbox Flanges, Type T, for Commercial Vehicles and Buses. 5 pp.
BSI BS AU 214: Part 2-88. 1988 Gearbox Flanges Part 2: Dimensions for Type A Gearbox Flanges for Commercial Vehicles and Buses. 5 pp.
BSI BS AU 214: Part 3-88. 1988 Gearbox Flanges Part 3: Dimensions for Type S Gearbox Flanges for Commercial Vehicles and Buses. 5 pp.
ISO 7646-86. Commercial Vehicles and Buses—Gearbox Flanges—Type A First Edition. 5 pp.
ISO 7647-86. Commercial Vehicles and Buses—Gearbox Flanges—Type S First Edition. 5 pp.
ISO 8667-92. Commercial Vehicles and Buses—Cross-Tooth Gearbox Flanges, Type T Second Edition. 7 pp.

—Buses (Vehicles)—Flanges
BSI BS AU 214A: Part 1-87. 1987 Amd 1 Gearbox Flanges Part 1: Dimensions of Cross-Tooth Gearbox Flanges, Type T, for Commercial Vehicles and Buses. 5 pp.
BSI BS AU 214: Part 2-88. 1988 Gearbox Flanges Part 2: Dimensions for Type A Gearbox Flanges for Commercial Vehicles and Buses. 5 pp.
BSI BS AU 214: Part 3-88. 1988 Gearbox Flanges Part 3: Dimensions for Type S Gearbox Flanges for Commercial Vehicles and Buses. 5 pp.
ISO 7646-86. Commercial Vehicles and Buses—Gearbox Flanges—Type A First Edition. 5 pp.
ISO 7647-86. Commercial Vehicles and Buses—Gearbox Flanges—Type S First Edition. 5 pp.
ISO 8667-92. Commercial Vehicles and Buses—Cross-Tooth Gearbox Flanges, Type T Second Edition. 7 pp.

—Ground Effect Machines
CAA Chapter B7-4 04.79. Transmission Items. 1 p.

Gear Boxes (Cont.)
—Lubricating Oils
MOD UK DSTAN 91-21-89. Lubricating Oil, Compounded NATO Code: 0-254 Joint Service Designation: OC-160 Issue 3. 12 pp.

—Paints (Cellulose Nitrate)—Oil Resistant
MOD UK DSTAN 80-37-73. Paint, Finishing, Cellulose Nitrate, Oil-Resisting Types: Brushing Spraying Issue 1. 7 pp.

—Preservation
MOD UK DSTAN 81-62-91. Preservation of Mechanical Power Transmission Assemblies Issue 2. 10 pp.

—Ships
MOD UK NES 305-91. Requirements for Gearing—Main Propulsion Issue 4 (09.91). 103 pp.

—Ships—Shaft Bearings
JIS F 7455-78. Marine Transmission Shaft Bearings.

—Ships—Shaft Couplings
JIS F 7453-78. Marine Transmission Shaft Joints.
JIS F 7454-78. Marine Transmission Shaft Loose Joints.

—Worm Gears—Modules
DIN ENGL 780 Pt 2-77. Series of Modules for Gears; Modules for Cylindrical Worm Gear Transmissions (May). 1 p.

Gear Cutters
Use: Hobbing Machines

Gear Drives
See Also: Gears
—Helical Gears—Gear Teeth
CNS B2418-84. Helical Gear Drives for Fine Mechanics, Tooth Distance Table (Mar)(5275).

—Spur Gears
DIN ENGL 58405 Pt 1-72. Spur Gear Drives for Fine Mechanics; Scope, Definitions, Principal Design Data, Classification (May). 12 pp.
DIN ENGL 58405 Pt 2-72. Spur Gear Drives for Fine Mechanics; Gear Fit Selection, Tolerances, Allowances (May). 16 pp.
DIN ENGL 58405 Pt 4-72. Spur Gear Drives for Fine Mechanics; Tables (May). 30 pp.

—Spur Gears—Engineering Drawings
DIN ENGL 58405 Pt 3-72. Spur Gear Drives for Fine Mechanics; Indication in Drawings, Examples for Calculation (May). 19 pp.

Gear Grinders
See Also: Grinders
—Glossaries
CNS B1232-80. Glossary of Terms Relating to Parts of Gear Grinding Machine (Dec)(6756).

Gear Hobbing Machines
Use: Hobbing Machines

Gear Hobs
See Also: Hobbing Machines; Milling Cutters
BSI BS 1498-54. (OBSOLESCENT) 1954 Amd 3 Gear Hobbing Machines for Turbines and Similar Drives. 28 pp.
BSI BS 2062: Part 2-60. 1960 Gear Hobs Part 2: Hobs for Gears for Turbine Reduction and Similar Drives. 13 pp.
BSI BS 4582: Part 2-78. 1978 Fine Pitch Gears (Metric Module) Part 2: Hobs and Cutters. 31 pp.
JIS B 6216-79. Test Code for Performance and Accuracy of Gear Hobbing Machines. 21 pp.

—Cylindrical Gears
CNS B3410-81. Metal-Cutting Tools; Hobs for Cylindrical Gears with Clutch Drive Slot or Keyway (Apr)(7194).

—Helical Gears
BSI BS 978: Part 5-65. 1965 Fine Pitch Gears Part 5: Hobs and Cutters. 27 pp.
BSI BS 2062: Part 1-59. 1959 Amd 1 Gear Hobs Part 1: Hobs for General Purposes, 1 d.p. tp 20 d.p. Inclusive. 18 pp.
BSI BS 4656: Part 19-76. 1976 Accuracy of Machine Tools and Methods of Test Part 19: Gear Hobbing Machines. 12 pp.
BSI BS 5221-87. 1987 General Purpose, Metric Module Year Hobs. 22 pp.

—Involute Splines
CNS B1336-83. Involute Splines Hobs, Pinion-Type Cutters, Broaches (Sep)(10538).
DIN ENGL 5480 Pt 16-86. Involute Splines with 30 Degree Pressure Angle; Hobs; Pinion Type Cutters; Broaches (Mar). 11 pp.

Gear Hobs (Cont.)
—Single Start
ISO 2490-75. Single-Start Solid (Monobloc) Gear Hobs with Axial Keyway, 1 to 20 Module and 1 to 20 Diametral Pitch—Nominal Dimensions First Edition. 5 pp.
ISO 4468-82. Gear Hobs—Single Start—Accuracy Requirements First Edition. 18 pp.

—Single Start—Spur Gears
DIN ENGL 3968-60. Tolerances for Single-Start Hobs for Involute Spur Gears (Sept). 5 pp.

—Single Thread
JIS B 4354-88. Single Thread Gear Hobs. 22 pp.
JIS B 4355-88. Single Thread Fine Pitch Gear Hobs. 17 pp.

—Splines
BSI BS 4656: Part 19-76. 1976 Accuracy of Machine Tools and Methods of Test Part 19: Gear Hobbing Machines. 12 pp.

—Spur Gears
BSI BS 978: Part 5-65. 1965 Fine Pitch Gears Part 5: Hobs and Cutters. 27 pp.
BSI BS 2062: Part 1-59. 1959 Amd 1 Gear Hobs Part 1: Hobs for General Purposes, 1 d.p. tp 20 d.p. Inclusive. 18 pp.
BSI BS 4656: Part 19-76. 1976 Accuracy of Machine Tools and Methods of Test Part 19: Gear Hobbing Machines. 12 pp.
BSI BS 5221-87. 1987 General Purpose, Metric Module Year Hobs. 22 pp.
DIN ENGL 8002-55. Metal-Cutting Tools; Hobs for Spur Gears with Clutch Drive Slot or Keyway; Modules 1 to 20 (Jan). 2 pp.

—Spur Gears—Glossaries
DIN ENGL 8000-62. Design Dimensions and Errors of Hobs for Involute Spur Gears; Fundamental Terms (Oct). 11 pp.

—Worm Gears
BSI BS 4656: Part 19-76. 1976 Accuracy of Machine Tools and Methods of Test Part 19: Gear Hobbing Machines. 12 pp.

Gear Lubricants
See Also: Bearing Lubricants; Gear Oils; Gears; Lubricants
BSI BS 6413: Part 6-91. 1991 Lubricants, Industrial Oils and Related Products (Class L) Part 6: Classification for Family C (Gears) (ISO 6743-6: 1990). 9 pp.
ISO 6743 Pt 6-90. Lubricants, Industrial Oils and Related Products (Class L)—Classification—Part 6: Family C (Gears) First Edition. 7 pp.

—Axles
CNS K6940-88. Method of Test for Performance of Gear Lubricants in Axles at High Speed, Low Torque, Followed by Low Speed, High Torque (Jan)(12213).
CNS K6941-88. Method of Test for Performance of Gear Lubricants in Axles Under High Speed and Shock Loading (Jan)(12214).

—Thermal Stability
CNS K6939-88. Method of Test for Thermal Oxidation Stability of Gear Lubricants (Jan)(12212).

Gear Motors
See Also: Gears; Motors
—Helical
BSI BS 4517-69. 1969 Amd 1 Dimensions of Spur and Helical Geared Motor Units (Metric Series). 11 pp.

—Spur
BSI BS 4517-69. 1969 Amd 1 Dimensions of Spur and Helical Geared Motor Units (Metric Series). 11 pp.

Gear Oils
See Also: Crankcase Oil; Gear Lubricants; Lubricating Oils
CNS K5053-88. Gear Oil (Ordinary Use) (Jan)(2978). 2 pp.
CNS K5137-83. Industrial Gear Oil (Dec)(10722).
JIS K 2219-83. Gear Oils. 10 pp.
MOD UK DSTAN 91-65-86. Lubricating Oil, Gear, Compounded Joint Service Designation: OC-300 Lubricating Oil, Gear, Compounded NATO Code Number: 0-208 Joint Service Designation: OC-600 Issue 1. 13 pp.
MOD UK DSTAN 91-71-89. Lubricating Fluid, Gear, Synthetic Joint Service Designation: OX-165 Issue 1. 13 pp.
MOD UK CS 3000B. Lubricating Oil, Gear, Extreme Pressure, SAE Grades 75 and 90.
MOD UK TS 10033E. Lubricating Oil, Engine: OMD-30, OMD-80 and OMD-85.

INDUSTRY STANDARDS

Gear Oils (Cont.)
MOD UK TS 10134A. Lubricating Fluid, Gear, Synthetic OX-165 (Superseded by Def Stan 91-71).

—Automotive
CNS K5052-88. Automotive Multi-Purpose Gear Oil (Jan)(2977). 3 pp.

—Channelling Characteristics
MOD UK DSTAN 05-50: Part 6-89. Methods for Testing Fuels, Lubricants and Associated Products Part 6: Channelling Characteristics of Gear Oils Issue 1. 8 pp.

—Marine
MOD UK DSTAN 91-21-89. Lubricating Oil, Compounded NATO Code: 0-254 Joint Service Designation: OC-160 Issue 3. 12 pp.
MOD UK DSTAN 91-32-78. Lubricating Oil, Steam Turbine and Gear: Extreme Pressure NATO Code No. O-249 Joint Service Designation: 0EP-69 Issue 1. 15 pp.

—Ships
MOD UK DSTAN 91-32-78. Lubricating Oil, Steam Turbine and Gear: Extreme Pressure NATO Code No. O-249 Joint Service Designation: 0EP-69 Issue 1. 15 pp.

Gear Pairs
Use: Gears

Gear Pumps
See Also: Centrifugal Pumps; Hydraulic Power Pumps; Pumps; Rotary Pumps; Vane Pumps
JIS B 8312-91. Gear Pumps and Screw Pumps—Testing Methods. 14 pp.

—Hydraulic Equipment
CNS B4025-80. Gear Pumps for Hydraulic Use —Discharge Diameter 10-50mm (Jul)(5701).
CNS B7146-80. Method of Test for Gear Pumps for Hydraulic Use Diameter 10-50mm (Jul)(5702).
JIS B 8352-76. Gear Pumps for Hydraulic Use (R 1979). 24 pp.

—Hydraulic Equipment—Ships
JIS F 6721-90. Stop Test Method for Oil Pressure Pumps at Hydraulic Steering Gears for Ships.

—Oil Burners
JIS B 8408-85. Gear Pumps for Gun Type Oil Burners.

Gear Ratios
—Worms (Mechanical Drives)
CNS B2422-84. Dimensions of Cylindrical Worm; Coordinations of Shaft Centre Distances and Transmission Ratios of Worm Drives (Aug)(5279).
DIN ENGL 3976-80. Cylindrical Worms; Dimensions; Correlation of Shaft Centre Distances and Gear Ratios of Worm Gear Drives (Nov). 8 pp.

Gear Reducers
Use For: Reduction Gears *See Also:* Axles; Bevel Gears; Gears; Pinions; Spur Gears; Worm Gears; Worms (Mechanical Drives)

—Cranes
CNS B1243-81. Crans; Reduction Gears; Connecting Dimensions Contour Gauges Output Torques (Apr)(7191).

Gear Teeth
See Also: Gears; Hobbing Machines; Involute Gear Teeth

—Bevel Gears
JIS B 1741-77. Tooth Contact Marking of Gears. 10 pp.

—Bevel Gears—Engineering Drawings
DIN ENGL 3966 Pt 2-78. Information on Gear Teeth in Drawings; Information on Straight Bevel Gear Teeth (Aug). 6 pp.

—Bevel Gears—Loads (Forces)
CNS B1204-84. Calculation of Load Capacity on Tooth Root for Bevel Gears (Aug)(5984).
CNS B1205-84. Calculation of Load Capacity on Tooth Flank for Bevel Gears (Aug)(5985).
CNS B1206-84. Calculation of Load Capacity for Cylindrical and Bevel Gears-Tooth Form Factor YF (Aug)(5986).

—Change Gears—Machine Tools
CNS B2025-84. Number of Teeth of Change Gears for Geared Lathe, Milling Machine and Gear Generating Equipment (Aug)(186).

—Cylindrical Gears
CNS B2415-84. Accuracy of Cylindrical Gears Teeth; Tolerance for Errors of Cumulative Circular Pitch (Aug)(5267).
CNS B2416-84. Accuracy of Cylindrical Gears Teeth; Tolerance for Errors of Composite (Aug)(5268).
CNS B2426-84. Accuracy of Cylindrical Gears Teeth; Tolerance for Errors of Tooth (Aug)(5283).
CNS B2428-84. Accuracy of Cylindrical Gear Teeth; Tolerance for Errors of Individual Parameters (Aug)(5285).
DIN ENGL 3961-78. Tolerances for Cylindrical Gear Teeth; Bases (Aug). 12 pp.
DIN ENGL 3962 Pt 1-78. Tolerances for Cylindrical Gear Teeth; Tolerances for Deviations of Individual Parameters (Aug). 18 pp.
DIN ENGL 3962 Pt 2-78. Tolerances for Cylindrical Gear Teeth; Tolerances for Tooth Trace Deviations (Aug). 2 pp.
DIN ENGL 3962 Pt 3-78. Tolerances for Cylindrical Gear Teeth; Tolerances for Pitch-Span Deviations (Aug). 2 pp.
DIN ENGL 3963-78. Tolerances for Cylindrical Gear Teeth; Tolerances for Working Deviations (Aug). 18 pp.
ISO TR4467-82. Addendum Modification of the Teeth of Cylindrical Gears for Speed-Reducing and Speed-Increasing Gear Pairs First Edition. 12 pp.
JIS B 1741-77. Tooth Contact Marking of Gears. 10 pp.

—Cylindrical Gears—Backlash—Glossaries
CNS B1189-84. Definitions and Denominations for Glossary of Gear Terms — Errors for the Axial Positions and Backlash for a Cylindrical Gear Teeth (May)(5731).

—Cylindrical Gears—Glossaries
CNS B1185-84. Definitions and Denominations for Glossary of Gear Terms — Cylindrical Gear Teeth (May)(5727).
CNS B1187-84. Definitions and Denominations for Glossary of Gear Terms — Deviation, Tolerances and Change Factor for Cylindrical Gear Teeth (May)(5729).
CNS B1188-84. Definitions and Denominations for Glossary of Gear Terms — Errors for Cylindrical Gear Teeth (May)(5730).

—Cylindrical Gears—Loads (Forces)
CNS B1202-84. Calculation of Load Capacity on Tooth Root for Cylindrical Gears (Aug)(5982).
CNS B1203-84. Calculation of Load Capacity on Tooth Flank for Cylindrical Gears (Aug)(5983).
CNS B1206-84. Calculation of Load Capacity for Cylindrical and Bevel Gears-Tooth Form Factor YF (Aug)(5986).

—Gear Trains—Damage
DIN ENGL 3979-79. Tooth Damage on Gear Trains; Designation, Characteristics, Causes (July). 19 pp.

—Glossaries
CNS B1184-84. Definitions and Denomination for Glossary of Gear Terms — Crest, Root, Surface, Tooth Flank, Tooth Profiles, Backlash, and Clearance (May)(5726).

—Helical Gears
CNS B2418-84. Helical Gear Drives for Fine Mechanics, Tooth Distance Table (Mar)(5275).

—Micrometers
CNS B6051-81. Gear Tooth Micrometers (Sep)(7872).
JIS B 7530-80. Gear Tooth Micrometer Callipers (R 1986). 14 pp.

—Profiles—Cutters
JIS B 4350-91. Gear Cutter—Tooth Profiles and Dimensions. 16 pp.

—Spur Gears
CNS B1255-84. Normal Chordal Tooth Thickness and Chordal Height of Spur Gears in the 0.5 System (Aug)(7578).
CNS B1257-84. Sliding Speed at Tooth Tip of Spur Gears in the 0.5 System (Aug)(7580).

—Thickness Measurement
DIN ENGL 3967-78. System of Gear Fits; Backlash; Tooth Thickness Allowances; Tooth Thickness Tolerances; Principles (Aug). 24 pp.

—Vernier Calipers
CNS B6052-81. Gear Tooth Vernier Calipers (Sep)(7873).
JIS B 7531-82. Gear Tooth Vernier Callipers. 16 pp.

—Worm Gears
JIS B 1741-77. Tooth Contact Marking of Gears. 10 pp.

Gear Tooth Micrometers
Use: Micrometers—Gear Tooth

Gear Trains
See Also: Gears

—Gear Teeth—Damage
DIN ENGL 3979-79. Tooth Damage on Gear Trains; Designation, Characteristics, Causes (July). 19 pp.

—Glossaries
DIN ENGL 868-76. General Definitions and Specification Factors for Gears, Gear Pairs and Gear Trains (Dec). 20 pp.

Gearing
Use: Gears

Gearmotors
Use: Gear Motors

Gears
Scope Note: For additional listings, use a more specific term *Use For:* Gear Pairs; Transmission Gears *See Also:* Backlash; Bevel Gears; Change Gears; Crown Gears; Cylindrical Gears; Gear Boxes; Gear Drives; Gear Lubricants; Gear Motors; Gear Reducers; Gear Teeth; Gear Trains; Gearshifts; Helical Gears; Hobbing Machines; Hypoid Gears; Idlers; Intermediate Gears; Internal Gears; Master Gears; Pinions; Propulsion Gears; Racks (Gears); Ring Gears; Spindle Drives; Spur Gears; Transmissions (Power Sources); Wheels; Worm Gears; Worms (Mechanical Drives)
CNS B1198-84. Symbols, Denominations and Units for Gears Calculation (Aug)(5978).
CNS B1199-84. General Theory of Calculation for Gears (Aug)(5979).
CNS B1201-84. Basic Formulas for Gears Calculation (Aug)(5981).

—Assembly—Pilot Seat—Aircraft
SBAC AS 6230 ISSUE 2(I). Assembly of Pilot's Seat and Raising Gear—25G. 1 p.
SBAC AS 6235 ISSUE 2(I). Quadrant Assembly Pilots Seat Raising Gear. 1 p.

—Automotive
DIN ENGL 75532 Pt 1-76. Transmission of Rotary Motions; Types of Connection to Gears, Intermediate Gears, Flexible Drive Shafts and Equipment (June). 8 pp.

—Bolts—Pilot Seat—Aircraft
SBAC AS 6234 ISSUE 3(I). Bearing Bolt—Pilot's Seat Raising Gear. 1 p.

—Classification
BSI BS 1000: (621.8)-76. 1976 Amd 1 Universal Decimal Classification (UDC). English Full Edition (621.8): Mechanical Power Transmission. Machine Elements. Gearing. Materials Handling. Mechanical Attachments, Fixing, Fasteners. Lubrication. 30 pp.
SNZ NZS/BS 1000 (621.8)-76. Universal Decimal Classification Mechanical Power Transmission. Machine Elements. Gearing. Materials Handling. Mechanical Attachments, Fixing, Fasteners. Lubrication Amend: 1. 28 pp.

—Drafting
DIN ENGL 37-61. Conventional and Simplified Representation of Gears and Gear Pairs (Dec). 4 pp.
JIS B 0003-89. Drawing Office Practice for Gears. 20 pp.

—Drafting—Rolling Stock
CNS E1021-82. Rubber Draft Gears for Railway Rolling Stock (Dec)(9702).
CNS E1022-82. Rubber Material for Draft Gears of Railway Rolling Stock (Dec)(9703).
JIS E 4204-90. Rubber Draft Gears for Railway Rolling Stock.

—Engineering Drawings—Symbols
ISO 2203-73. Technical Drawings—Conventional Representation of Gears First Edition. 8 pp.

—Glossaries
BSI BS 2519: Part 1-76. 1976 Glossary for Gears Part 1: Geometrical Definitions. 40 pp.
CNS B1175-84. Glossary of Gears Geometrical Definitions (General) (Feb)(5717).
CNS B1179-84. Glossary of Gear Terms (Aug)(5721).
CNS B1180-84. Definitions and Denominations for Glossary of Gear Terms — General Terms (May)(5722).
CNS B1181-84. Definitions and Denominations for Glossary of Gear Terms — General Terms on Gear Pairs (May)(5723).
CNS B1183-84. Definitions and Denomination for Glossary of Gear Terms — Types of Gears and Gear Pairs (May)(5725).
CNS B1184-84. Definitions and Denomination for Glossary of Gear Terms — Crest, Root, Surface, Tooth Flank, Tooth Profiles, Backlash, and Clearance (May)(5726).

INTERNATIONAL AND NON-U.S. NATIONAL STANDARDS
SUBJECT INDEX

Gears (Cont.)
—Glossaries (Cont.)
CNS B1193-84. Definition and Denominations for Glossary of Gear Terms —Errors and Allowances (May)(5735).
DIN ENGL 3998 Suppl. 1-76. Denominations on Gears and Gear Pairs; Alphabetical Index of Equivalent Terms (Sept). 12 pp.
DIN ENGL 3998 Pt 1-76. Denominations on Gears and Gear Pairs; General Definitions (Sept). 17 pp.
ISO 1122 Pt 1-83. Glossary of Gear Terms—Part 1: Geometrical Definitions First Edition. 39 pp.
JIS B 0102-88. Glossary of Gear Terms. 126 pp.
SAA AS 2075-91. Glossary of Terms and Notations for Gears. 47 pp.

—Glossaries—Kinematics
CNS B1182-84. Definitions and Denominations for Glossary of Gear Terms — Kinematics (May)(5724).

—Glossaries/Symbols
DIN ENGL 868-76. General Definitions and Specification Factors for Gears, Gear Pairs and Gear Trains (Dec). 20 pp.

—Greases
CGSB 3-GP-691MB-90. General Purpose Grease. 9 pp.
DIN ENGL 51826-84. Lubricants; Type G Lubricating Greases (Nov). 3 pp.

—Greases—Wear Testing
MOD UK DSTAN 05-50: Part 28-01. Methods for Testing Fuels, Lubricants and Associated Products Part 28: Gear Wear Test for Grease Issue 1; Amendment 1. 11 pp.

—Harness Release—Assembly—Aircraft
SBAC AS 2830 ISSUE 5. 25 Gram Harness Release Gear (Locking In). 1 p.

—Harness Release—Casings—Aircraft
SBAC AS 2819 ISSUE 4. Half Casing for Harness Release Gear. 1 p.

—Harness Release—Casings—Aircraft—Assembly
SBAC AS 2820 ISSUE 4. Assembly of Front Casing. 1 p.
SBAC AS 2821 ISSUE 5. Assembly of Back Casing. 1 p.

—Harness Release—Latches—Aircraft
SBAC AS 2829 ISSUE 4. Catch for Harness Release Gear. 1 p.

—Harness Release—Springs (Elastic)—Aircraft
SBAC AS 2814 ISSUE 4. Cable Spring for Harness Release Gear. 1 p.
SBAC AS 2815 ISSUE 1. Cable Spring for Harness Release Gear. 1 p.
SBAC AS 2817 ISSUE 4. Catch Spring for Harness Release Gear. 1 p.

—Loads (Forces)
CNS B1200-84. Load Capacity for Gears (Aug)(5980).

—Loads (Forces)—Capacity Measurement
CNS B2525-80. Calculation of Load Capacity of Gears; Examples (Sep)(6413).

—Lubricants
BSI BS 6413: Part 6-91. 1991 Lubricants, Industrial Oils and Related Products (Class L) Part 6: Classification for Family C (Gears) (ISO 6743-6: 1990). 9 pp.
ISO 6743 Pt 6-90. Lubricants, Industrial Oils and Related Products (Class L)—Classification—Part 6: Family C (Gears) First Edition. 7 pp.

—Lubricants—Axles
CNS K6940-88. Method of Test for Performance of Gear Lubricants in Axles at High Speed, Low Torque, Followed by Low Speed, High Torque (Jan)(12213).
CNS K6941-88. Method of Test for Performance of Gear Lubricants in Axles Under High Speed and Shock Loading (Jan)(12214).

—Lubricants—Thermal Stability
CNS K6939-88. Method of Test for Thermal Oxidation Stability of Gear Lubricants (Jan)(12212).

—Lubricating Oils
MOD UK DSTAN 91-74-93. Lubricating Oil, Steam Turbine and Gear, Extreme Pressure Joint Service Designation: OEP-80 Issue 1. 16 pp.

—Modular
CNS B2022-84. Series of Modules for Cylindrical Gears (Aug)(183).
CNS B2414-84. Series of Modules for Worm Gear Pairs (Aug)(5266).
CNS B2527-84. Series of Modules for Straight Bevel Gears (Aug)(6415).
DIN ENGL 780 Pt 2-77. Series of Modules for Gears; Modules for Cylindrical Worm Gear Transmissions (May). 1 p.

—Multispindle Heads
DIN ENGL 69001 Pt 52-81. Machine Tools; Multi-Spindle Heads; Gears; Type A (Oct). 1 p.

—Noise
CNS B7125-80. Measuring Method of Noise of Gears (Feb)(5192).
JIS B 1753-76. Measuring Method of Noise of Gears (R 1985). 9 pp.

—Pilot Seats—Stop Collars—Aircraft
SBAC AS 2112 ISSUE 1. Stop Collar, Pilots Seat-Raising Gear. 1 p.

—Planers
BSI BS 4656: Part 25-80. 1980 Accuracy of Machine Tools and Methods of Test Part 25: Gear Planing Machines. 12 pp.

—Plug Ends—Pilot Seat—Aircraft
SBAC AS 6236 ISSUE 2(I). Plug End—Pilot's Seat Raising Gear. 1 p.

—Pushbuttons—Pilot Seat—Aircraft
SBAC AS 6223 ISSUE 2(I). Press Button Pilot's Seat Raising Gear. 1 p.

—Remote Control—Installation
CNS B1215-80. Remote Control Gear for Manual Operation; Installation Principles (Sep)(6424).

—Remote Control—Transmission Tubing
CNS B1213-80. Remote Control Gear with Transmission Tubing for Manual Operation; Fork Links with Square Hole for Setting-Up Connection (Sep)(6422).
CNS B1214-80. Remote Control Gear with Transmission Tubing for Manual Operation; Nominal Size Transmission Tubes, Survey of Component Parts (Sep)(6423).
CNS B2647-82. Remote Control Gear with Transmission Tubing for Manual Operation (Mar)(8557).

—Servomechanisms
MOD UK DSTAN 61-14: Part 10-89. Precisiion Instrument, Rotating, Servo—Components Part 10: Gearheads Issue 1. 20 pp.

—Shaft Ends
CNS B2008-83. Cylindrical Shaft End (for Assembling Coupling, Pulley and Gear) (May)(147).
CNS B2009-83. Taper Shaft End (for Assembling Coupling and Gear) (May)(148).

—Shaping Machines
BSI BS 4656: Part 26-80. 1980 Accuracy of Machine Tools and Methods of Test Part 26: Gear Shaping Machines (Supersedes BS 3329: 1961). 12 pp.

—Sound Power
BSI BS 7676: Part 1-93. 1993 Acceptance Code for Gears Part 1: Determination of Airborne Sound Power Levels Emitted by Gear Units (ISO 8579-1: 1993) (E). 23 pp.
ISO 8579 Pt 1-93. Acceptance Code for Gears—Part 1: Determination of Airborne Sound Power Levels Emitted by Gear Units First Edition. 21 pp.

—Stops—Pilot Seat—Aircraft
SBAC AS 6238 ISSUE 2(I). Quadrant, Stop Pilot's Seat Raising Gear. 1 p.

—Supersonic Transports—Ground Loads
CAA STANDARD NO. 4-4 08.60. Ground Loads. 10 pp.

—Symbols
BSI BS 2519: Part 2-76. 1976 Glossary for Gears Part 2: Notation. 8 pp.
CNS B1198-84. Symbols, Denominations and Units for Gears Calculation (Aug)(5978).
ISO 701-76. International Gear Notation—Symbols for Geometrical Data First Edition. 7 pp.
JIS B 0121-88. Gear Notation-Symbols for Geometrical Data. 10 pp.

—Trim Controls—Aircraft
SBAC AS 897 ISSUE 1. Gear—Trim Control. 1 p.
SBAC AS 1233 ISSUE 2. Gear—Trim Control. 1 p.

—Trunnions—Aircraft
SBAC AS 2264 ISSUE 4. Trunnion for Harness Release Gear. 1 p.

—Valves—Ships
MOD UK NES 361-81. Valve Rod Gearing Issue 2 (06.81). 50 pp.

—Vibration
BSI BS 7676: Part 2-93. 1993 Acceptance Code for Gears Part 2: Determination of Mechanical Vibrations of Gear Units During Acceptance Testing (ISO 8579-2: 1993) (E). 20 pp.
ISO 8579 Pt 2-93. Acceptance Code for Gears—Part 2: Determination of Mechanical Vibrations of Gear Units During Acceptance Testing First Edition. 18 pp.

Gearshifts
See Also: Derailleurs; Gears

—Automotive—Drive Positions
JIS D 0012-87. Transmission Shift Positions for Passenger Cars. 6 pp.

—Automotive—Reverse
EC 75/443/EEC-75. Council Directive on the Approximation of the Laws of the Member States Relating to the Reverse and Speedometer Equipment of Motor Vehicles. 5 pp.

—Buses (Vehicles)—Drive Positions
CNS D1014-81. Transmission Shift Positions for Trucks and Buses (Sep)(7888).
JIS D 0011-79. Transmission Shift Positions for Trucks and Buses (R 1983). 4 pp.

—Motorcycles
CNS D2150-88. Gear Shift Assembly for Motorcycles (Jul)(9376). 3 pp.

—Trucks—Drive Positions
CNS D1014-81. Transmission Shift Positions for Trucks and Buses (Sep)(7888).
JIS D 0011-79. Transmission Shift Positions for Trucks and Buses (R 1983). 4 pp.

Geiger Mueller Counter Tubes
Use: Geiger Mueller Tubes

Geiger Mueller Probes
See Also: Probes (Sensors); Radiation Measuring Instruments; Radiation Probes
IEC 256-67. External Diameters of Cylindrical Radiation Probes Containing Geiger-Muller or Proportional Counter Tubes or Scintillation Detectors First Edition. 5 pp.

Geiger Mueller Tubes
Use For: Geiger Mueller Counter Tubes
See Also: Electron Tubes
MOD UK DSTAN 59-60: 90/045-01. Valve, Electronic Geiger Muller Tube Issue 1; Amendment 1. 8 pp.
MOD UK DSTAN 59-60: 90/079-74. Valve, Electronic, Geiger Muller Tube Issue 1. 15 pp.

—Electrical Measurement
IEC 151 Pt 25-71. Measurements of the Electrical Properties of Electronic Tubes Part 25: Methods of Measurement of Geiger-Muller Counter Tubes First Edition. 34 pp.

—Probes
IEC 256-67. External Diameters of Cylindrical Radiation Probes Containing Geiger-Muller or Proportional Counter Tubes or Scintillation Detectors First Edition. 5 pp.

—Radiation Meters
IEC 421-73. Portable Prospecting Radiation Meters with Geiger-Muller Counter Tube (Linear Scale Instruments) First Edition. 33 pp.
JIS Z 4202-88. Geiger-Muller Counter Tubes. 13 pp.

Gelatin (Explosives)
See Also: Explosives
CNS M2002-88. Gelatinous Permissible Explosives for Coal Mines (for Taiwan Area) (Jun)(2764).
CNS M2004-88. Gelatinous Explosives (Jun)(2766).

Gelatins
BSI BS 757-75. (WITHDRAWN) 1975 Methods for Sampling and Testing Gelatine (Physical and Chemical Methods). 35 pp.
JIS K 6503-77. Animal Glues and Gelatins.
JIS K 6503-70. Animal Glues and Gelatins. 19 pp.

—Sampling
BSI BS 757-75. (WITHDRAWN) 1975 Methods for Sampling and Testing Gelatine (Physical and Chemical Methods). 35 pp.

Gelation
See Also: Rheological Properties

Gelation (Cont.)

—Powder Coatings

BSI BS 3900: Part J3-93. 1993 Methods of Test for Paints Part J3: Determination of the Gelation Time of Coating Powders (ISO 8130-6: 1992). 6 pp.

BSI BS 3900: Part J3-87. 1987 Methods of Test for Paints Group J: Testing of Coating Powders Part J3: Determination of the Gelation Time of Coating Powders. 4 pp.

ISO 8130 Pt 6-92. Coating Powders—Part 6: Determination of Gel Time of Thermosetting Coating Powders at a Given Temperature First Edition. 7 pp.

Gelation Time

Use: Gelation

Gelidium

Use: Algae

Gelling Agents

See Also: Additives

—Food

EC 74/329/EEC-74. Council Directive on the Approximation of the Laws of the Member States Relating to Emulsifiers, Stabilizers, Thickeners and Gelling Agents for Use in Foodstuffs. 7 pp.

EC 78/612/EEC-78. Council Directive Amending for the First Time Directive 74/329/EEC on the Approximation of the Laws of the Member States Relating to Emulsifiers, Stabilizers, Thickeners and Gelling Agents for Use in Foodstuffs. 4 pp.

EC 80/597/EEC-80. Council Directive Amending for the Second Time Directive 74/329/EEC on the Approximation of the Laws of the Member States Relating to Emulsifiers, Stabilizers, Thickeners and Gelling Agents for Use in Foodstuffs. 3 pp.

EC 85/6/EEC-84. Council Directive Amending for the Third Time Directive 74/329/EEC on the Approximation of the Laws of the Member States Relating to Emulsifiers, Stabilizers, Thickeners and Gelling Agents for Use in Foodstuffs. 1 p.

EC 86/102/EEC-86. Council Directive Amending for the Fourth Time Directive 74/329/EEC on the Approximation of the Laws of the Member States Relating to Emulsifiers, Stabilizers, Thickeners and Gelling Agents for Use in Foodstuffs. 2 pp.

EC 89/393/EEC-89. Council Directive Amending for the Fifth Time Directive 74/329/EEC on the Approximation of the Laws of the Member States Relating to Emulsifiers, Stabilizers, Thickeners and Gelling Agents for Use in Foodstuffs. 1 p.

EC 90/612/EEC-90. Commission Directive Amending Council Directive 78/663/EEC Laying down Specific Purity Criteria for Emulsifiers, Stabilizers, Thickeners and Gelling Agents for Use in Foodstuffs. 2 pp.

—Food—Purity

EC 78/663/EEC-78. Council Directive Laying Down Specific Criteria of Purity for Emulsifiers, Stabilizers, Thickeners and Gelling Agents for Use in Foodstuffs. 23 pp.

EC 82/504/EEC-82. Council Directive Amending Directive 78/663/EEC Laying down Specific Criteria of Purity for Emulsifiers, Stabilizers, Thickeners and Gelling Agents for Use in Foodstuffs. 3 pp.

Gemini Aircraft

See Also: Aircraft

—Antenna Positions

CAA. Gemini (Approved Aerial Positions). 1 p.

General Practice

Use: Family Practice

General Switched Telephone Network

Use: Telephone Networks

Generating Sets

Use: Generator Sets

Generating Stations

Use For: Electric Generating Stations; Electric Power Generation; Electric Power Generation Equipment; Power Generation Equipment; Power Stations
See Also: Electrical Equipment; Power Plants; Power Systems; Power Transmission; Substations (Electric)

—Electrical Grounding

CSA C22.3 NO 2-1975. General Grounding Requirements and Grounding Requirements for Electrical Supply Stations (R 1980). 12 pp.

—Glossaries

BSI BS 4727:Pt2: Group 11-86. 1986 Amd 1 Electrotechnical, Power, Telecommunication, Electronics, Lighting and Colour Terms Part 2: Terms Particular to Power Engin. Group 11: Generation, Trans. and Distribution of Electricity Terminology (AMD 5524) February 28, 1989. 72 pp.

IEC 50 Chap 601-85. International Electrotechnical Vocabulary Chapter 601: Generation, Transmission and Distribution of Electricity—General. 42 pp.

IEC 50 Chap 602-83. International Electrotechnical Vocabulary Chapter 602: Generation, Transmission and Distribution of Electricity—Generation. 55 pp.

IEC 50 Chap 604-87. International Electrotechnical Vocabulary Chapter 604: Generation, Transmission and Distribution of Electricity—Operation First Edition; (Corrigendum—July 1987). 72 pp.

SAA AS 1852.601-88. International Electrotechnical Vocabulary—Part 601: Generation, Transmission and Distribution of Electricity—General. 19 pp.

SAA AS 1852.602-88. International Electrotechnical Vocabulary—Part 602: Generation, Transmission and Distribution of Electricity—Generation. 26 pp.

SAA AS 1852.604-88. International Electrotechnical Vocabulary—Part 604: Generation, Transmission and Distribution of Electricity—Operation. 39 pp.

SNZ IEC 50: 50(602)-83. International Electrotechnical Vocabulary 50(602): Generation, Transmission and Distribution of Electricity. Generation. 45 pp.

SNZ IEC 50: 50(604)-87. International Electrotechnical Vocabulary 50(604): Generation, Transmission and Distribution of Electricity. Operation. 60 pp.

—Pulverized Coal—Pneumatic Conveyors—Sampling

ISO 9931-91. Coal—Sampling of Pulverized Coal Conveyed by Gases in Direct Fired Coal Systems First Edition. 22 pp.

Generator Sets

Use For: Generating Sets *See Also:* Generators; Power Supplies

MOD UK DSTAN 61-10-84. Generator Sets (Ground, Engine Driven) Issue 4. 9 pp.

MOD UK DEF-44-65. Ground Generating Sets. 9 pp.

—AC Generators—Internal Combustion Piston Engines

ISO 8528 Pt 3-93. Reciprocating Internal Combustion Engine Driven Alternating Current Generating Sets—Part 3: Alternating Current Generators for Generating Sets First Edition. 16 pp.

—Design

MOD UK DSTAN 61-5: Pt 2: Sec 2-92. Electrical Power Supply Systems Below 650 Volts Part 2: Ground Generating Set Characteristics Section 2: General Design Requirements Issue 1. 21 pp.

—Electrical Properties

MOD UK DSTAN 61-5: Pt 2: Sec 1-78. Electrical Power Supply Systems Below 650 Volts Part 2: Ground Generating Set Characteristics Section 1: Electrical Performance Characteristics Issue 1. 18 pp.

MOD UK DSTAN 61-5: Pt 2: Sec 1-92. Electrical Power Supply Systems Below 650 Volts Part 2: Ground Generating Set Characteristics Section 1: Electrical Performance Characteristics and Methods of Measurement Issue 2. 24 pp.

MOD UK DSTAN 61-5: Pt 2: Sec 1-01. Electrical Power Supply Systems Below 650 Volts Part 2: Ground Generating Set Characteristics Section 1: Electrical Performance Characteristics and Methods of Measurement Issue 2; Amendment 1. 27 pp.

NATO STANAG 4135 ED 2 AMD 0-92. Electrical Characteristics of Rotating Alternating Current Generating Sets. 15 pp.

—Electromagnetic Compatibility

MOD UK DSTAN 61-5: Pt 2: Sec 5-92. Electrical Power Supply Systems Below 650 Volts Part 2: Ground Generating Set Characteristics Section 5: Electromagnetic Compatibility Requirements Issue 1. 10 pp.

—Environmental Engineering

MOD UK DSTAN 61-5: Pt 2: Sec 4-92. Electrical Power Supply Systems Below 650 Volts Part 2: Ground Generating Set Characteristics Section 4: Environmental Requirements Issue 1. 18 pp.

—Glossaries

JIS B 0149-90. Terms and Definitions of Engine-Driven Generating Sets.

—Internal Combustion Piston Engines

BSI BS 7698: Part 6-93. 1993 Reciprocating Internal Combustion Engine Driven Alternating Current Generating Sets Part 6: Test Methods (ISO 8528-6: 1993) (E). 12 pp.

DIN ENGL 6281-78. Generator Sets with Reciprocating Internal Combustion Engines; Connection Dimensions for Generators and Reciprocating Internal Combustion Engines (Apr). 3 pp.

ISO 8528 Pt 1-93. Reciprocating Internal Combustion Engine Driven Alternating Current Generating Sets—Part 1: Application, Ratings and Performance First Edition. 14 pp.

ISO 8528 Pt 2-93. Reciprocating Internal Combustion Engine Driven Alternating Current Generating Sets—Part 2: Engines First Edition. 14 pp.

ISO 8528 Pt 5-93. Reciprocating Internal Combustion Engine Driven Alternating Current Generating Sets—Part 5: Generating Sets First Edition. 32 pp.

ISO 8528 Pt 6-93. Reciprocating Internal Combustion Engine Driven Alternating Current Generating Sets—Part 6: Test Methods First Edition. 11 pp.

JIS B 8022-90. Standard Form of Specifications of Engine-Driven Generating Sets. 7 pp.

—Internal Combustion Piston Engines—Controlgear

ISO 8528 Pt 4-93. Reciprocating Internal Combustion Engine Driven Alternating Current Generating Sets—Part 4: Controlgear and Switchgear First Edition. 16 pp.

—Internal Combustion Piston Engines—Switchgear

ISO 8528 Pt 4-93. Reciprocating Internal Combustion Engine Driven Alternating Current Generating Sets—Part 4: Controlgear and Switchgear First Edition. 16 pp.

—Noise

MOD UK DSTAN 61-5: Pt 2: Sec 6-92. Electrical Power Supply Systems Below 650 Volts Part 2: Ground Generating Set Characteristics Section 6: Audio Noise, Methods of Measurement Issue 2. 15 pp.

Generators

Scope Note: For additional listings, use a more specific term *Use For:* Dynamas; Electric Generators; Power Generators *See Also:* AC Generators; Alternators; Arithmetic Logic Unit/Function Generators; Auxiliary Power Units; Boilers; DC Generators; Generator Sets; Look Ahead Carry Generators; Noise Generators; Oxygen Generators; Parity Generator/Checkers; Parity Generators; Power Supplies; Rotating Machines; Signal Generators; Square Wave Generators; Tachometer Generators; Three Phase Turbogenerators; Turbine Generators; Wind Turbines; X-Ray Generators

BSI BS 5000: Part 11-73. 1973 Amd 3 Rotating Electrical Machines of Particular Types or for Particular Applications Part 11: Small-Power Electric Motors and Generators (AMD 5634) July 29, 1988. 26 pp.

—Aircraft

CAA 706 (K). Section K Light Aeroplanes Electrical Supply, Systems and Equipment (Blue Papers). 54 pp.

CAA Chapter J2-1 09.66. Supply and Distribution. 5 pp.

CAA LEAFLET 11-10 06.92. Electrical Generation Systems—Bus-Bar Low Voltage Warning Single Engined Aircraft with a UK Certificate of Airworthiness. 5 pp.

—Aircraft—Switchgear

CAA Chapter J2-1 App #2 09.66. Generator Switchgear and Battery Installation. 3 pp.

—Automobive—Power Ratings

CNS D3052-87. Rated Output and Its Test Methods for Automobile Generators (Oct)(5564).

—Construction—Hazardous Locations

CSA C22.2 NO 145-M1986. Motors and Generators for Use in Hazardous Locations (R 1992); (Gen Instr 1 Thru 2). 47 pp.

—DC Servomotors—Preferred Products List

CECC CECC MUAHAG Vol 13 IS 1-91. Preferred Products List; Servo Components (En, Fr, Ge). 27 pp.

—Flame Propagation—Hazardous Locations

CSA C22.2 NO 145-M1986. Motors and Generators for Use in Hazardous Locations (R 1992); (Gen Instr 1 Thru 2). 47 pp.

—Ground Effect Machines

CAA Chapter B6-7 01.91. Electrical Systems. 8 pp.

INTERNATIONAL AND NON-U.S. NATIONAL STANDARDS
SUBJECT INDEX

Generators *(Cont.)*

—**Hazardous Locations**
CSA 1169 Bull. Electrical Bulletin 1169 June 27, 1978 to C22.2 NO 145. 2 pp.

—**Internal Combustion Engines—Military Vehicles**
MOD UK DSTAN 29-3-71. Generators, Engine Accessory for Tactical and Logistical Vehicles and IC Engine Powered Equipments Issue 1. 5 pp.
MOD UK DSTAN 29-3: Addendum. Generators, Engine Accessory, for Motor Transport Vehicles and IC Engine Powered Equipments Issue 1. 7 pp.

—**Internal Combustion Engines—Recreational Vehicles**
CSA 946 Bull. Electrical Bulletin 946 April 19, 1974 to C22.2 NO 0. 8 pp.
CSA 946 Bull. Electrical Bulletin 946 April 19, 1974 to C22.2 NO 100. 8 pp.

—**Internal Combustion Piston Engines**
BSI BS 5000: Part 3-80. 1980 Amd 1 Rotating Electrical Machines of Particular Types or for Particular Applications Part 3: Generators to Be Driven by Reciprocating Internal Combustion Engines. 9 pp.
MOD UK DSTAN 29-3: Addendum. Generators, Engine Accessory, for Motor Transport Vehicles and IC Engine Powered Equipments Issue 1. 7 pp.

—**Naval Ships**
MOD UK NES 634-89. Requirements for Motor Generators and Associated Control Gear Issue 4 (02.89). 25 pp.
MOD UK NES 634-93. Requirements for Motor Generators and Associated Control Gear Issue 5 (01.93). 30 pp.

—**Recreational Vehicles**
CSA CAN/CSA-C22. 2 NO 100-92. Motors and Generators; (Gen Instr 1). 74 pp.
CSA 1169 Bull. Electrical Bulletin 1169 June 27, 1978 to C22.2 NO 100. 2 pp.

—**Rotary Wing Aircraft**
CAA 707 (G). Section G Rotorcraft Electrical Supply, Systems and Equipment (Blue Papers). 47 pp.
CAA Chapter G6-13 11.85. Electrical Generation, Supply and Distribution (Rotocraft). 11 pp.

—**Safety**
CSA CAN/CSA-C22. 2 NO 100-92. Motors and Generators; (Gen Instr 1). 74 pp.
CSA 1169 Bull. Electrical Bulletin 1169 June 27, 1978 to C22.2 NO 100. 2 pp.

—**Ships**
IEC 92 Pt 301-80. Electrical Installations in Ships Part 301: Equipment—Generators and Motors Third Edition. 38 pp.
JIS F 8064-86. Electrical Installations in Ships Part 301 Equipment-Generators and Motors.

—**Ships—Diesel Engines**
JIS F 4306-87. Water-Cooled Four Cycle Marine Diesel Engines for Electric Generator.

—**Ships—Glossaries**
JIS F 0022-85. Glossary of Terms for Shipbuilding (Machinery Part—Propulsion Machinery, Boilers, Generator Engines and Prime Mover for Auxiliary Machinery).

—**Sound Power**
EC 84/535/EEC-84. Council Directive on the Approximation of the Laws of the Member States Relating to the Permissible Sound Power Level of Welding Generators. 7 pp.
EC 84/536/EEC-84. Council Directive on the Approximation of the Laws of the Member States Relating to the Permissible Sound Power Level of Power Generators. 7 pp.

—**Static Electricity—Hazardous Locations**
CSA C22.2 NO 145-M1986. Motors and Generators for Use in Hazardous Locations (R 1992); (Gen Instr 1 Thru 2). 47 pp.

—**Submarines**
MOD UK NES 634-89. Requirements for Motor Generators and Associated Control Gear Issue 4 (02.89). 25 pp.
MOD UK NES 634-93. Requirements for Motor Generators and Associated Control Gear Issue 5 (01.93). 30 pp.

—**Switchgear**
CSA CAN/CSA-C22. 2 NO 31-M89. Switchgear Assemblies; (Gen Instr 1 Thru 3). 54 pp.
CSA 1169 Bull. Electrical Bulletin 1169 June 27, 1978 to C22.2 NO 31. 2 pp.

Generators *(Cont.)*

—**Temperature Testing—Hazardous Locations**
CSA C22.2 NO 145-M1986. Motors and Generators for Use in Hazardous Locations (R 1992); (Gen Instr 1 Thru 2). 47 pp.

Generators (Gas)
Use: Gas Generators

Genotoxicity
Use: Toxicity

Gentex Call Numbers
See Also: Gentex Networks; Telex Call Numbers
CCITT RECMN F.20-89. International Gentex Service—Telegraph and Mobile Services Operations and Quality of Service (Study Group I) 5 pp. 5 pp.

Gentex Code
See Also: Codes; Gentex Networks; Telegraph Networks; Telegraphy
CCITT RECMN F.1-89. Operational Provisions for the International Public Telegram Service—Telegraph and Mobile Services Operations and Quality of Service (Study Group I) 54 pp. 54 pp.

Gentex Networks
Use For: Switched Telegraph Networks
See Also: Gentex Call Numbers; Gentex Code; Switched Networks; Telecommunication Networks; Telegraph Networks; Telegraphy; Telex Call Numbers; Telex Destination Codes; Telex Network Identification Codes
CCITT RECMN F.1-92. Operational Provisions for the International Public Telegram Service (Study Group I) 57 pp. 57 pp.
CCITT RECMN F.20-89. International Gentex Service—Telegraph and Mobile Services Operations and Quality of Service (Study Group I) 5 pp. 5 pp.

—**Answerback**
CCITT RECMN F.1-89. Operational Provisions for the International Public Telegram Service—Telegraph and Mobile Services Operations and Quality of Service (Study Group I) 54 pp. 54 pp.
CCITT RECMN F.20-89. International Gentex Service—Telegraph and Mobile Services Operations and Quality of Service (Study Group I) 5 pp. 5 pp.
CCITT RECMN F.21-89. Composition of Answer-Back Codes for the International Gentex Service—Telegraph and Mobile Services Operations and Quality of Service (Study Group I) 3 pp. 3 pp.
CCITT RECMN S.6-89. Characteristics of Answerback Units (ITA2)—Telegraph Services Terminal Equipment (Study Group IX) 2 pp. 2 pp.

—**Call Set Up**
CCITT RECMN F.1-89. Operational Provisions for the International Public Telegram Service—Telegraph and Mobile Services Operations and Quality of Service (Study Group I) 54 pp. 54 pp.

—**Fault Finders**
CCITT RECMN R.90-89. Organization for Locating and Clearing Faults in International Telegraph Switched Networks—Telegraph Transmission (Study Group IX) 5 pp. 5 pp.

—**Faulty Station Line**
CCITT RECMN U.31-89. Prevention of Connection to Faulty Stations and/or Station Lines in the Gentex Service—Telegraph Switching (Study Group IX) 1 pp. 1 p.

—**Glossaries**
CCITT RECMN F.68-89. Establishment of the Automatic Intercontinental Telex Network—Telegraph and Mobile Services Operations and Quality of Service (Study Group I) 6 pp. 6 pp.

—**Hypothetical Reference Connections**
CCITT RECMN U.8-89. Hypothetical Reference Connections for Telex and Gentex Networks—Telegraph Switching (Study Group IX) 5 pp. 5 pp.

—**Message Formats**
CCITT RECMN F.1-89. Operational Provisions for the International Public Telegram Service—Telegraph and Mobile Services Operations and Quality of Service (Study Group I) 54 pp. 54 pp.

—**Muldexes—Time Division Multiplexers**
CCITT RECMN R.105-89. Duplex Muldex Concentrator, Connecting a Group of Gentex and Telex Subscribers to a Telegraph Exchange by Assigning Virtual Channels to Time Slots of a Bit-Interleaved TDM System—Telegraph Transmission (Study Group IX) 2 pp. 2 pp.

Gentex Networks *(Cont.)*

—**Quality Assurance**
CCITT RECMN R.57-89. Standard Limits of Transmission Quality for Planning Code-Independent Intl Point-to-Point Telegraph Communications and Switched Networks Using 50-Baud Start-Stop Equipment—Telegraph Transmission (Study Group IX) 2 pp. 2 pp.
CCITT RECMN R.58-89. Standard Limits of Transmission Quality for the Gentex and Telex Networks—Telegraph Transmission (Study Group IX) 4 pp. 4 pp.

—**Routing**
CCITT RECMN F.20-89. International Gentex Service—Telegraph and Mobile Services Operations and Quality of Service (Study Group I) 5 pp. 5 pp.

—**Routing—Documentation—ITU**
CCITT RECMN F.20-89. International Gentex Service—Telegraph and Mobile Services Operations and Quality of Service (Study Group I) 5 pp. 5 pp.
CCITT RECMN F.93-89. Routing Table for Offices Connected to the Gentex Service—Telegraph and Mobile Services Operations and Quality of Service (Study Group I) 1 pp. 1 p.

—**Routing—Overflow Traffic**
CCITT RECMN F.20-89. International Gentex Service—Telegraph and Mobile Services Operations and Quality of Service (Study Group I) 5 pp. 5 pp.

—**Service Messages—Answerback**
CCITT RECMN F.21-89. Composition of Answer-Back Codes for the International Gentex Service—Telegraph and Mobile Services Operations and Quality of Service (Study Group I) 3 pp. 3 pp.

—**Signal Transfer Delay**
CCITT RECMN R.58 BIS-89. Limits on Signal Transfer Delay for Telegraph, Telex and Gentex Networks—Telegraph Transmission (Study Group IX) 2 pp. 2 pp.

—**Signaling Systems**
CCITT RECMN U.30-89. Signalling Conditions for Use in the International Gentex Network—Telegraph Switching (Study Group IX) 1 pp. 1 p.

—**Telegram Identification Groups**
CCITT RECMN F.31-89. Telegram Retransmission System—Telegraph and Mobile Services Operations and Quality of Service (Study Group I) 12 pp. 12 pp.

—**Telegraph Circuits—Grade of Service**
CCITT RECMN F.23-89. Grade of Service for Long-Distance International Gentex Circuits—Telegraph and Mobile Services Operations and Quality of Service (Study Group I) 1 pp. 1 p.
CCITT RECMN F.24-89. Average Grade of Service from Country to Country in the Gentex Service—Telegraph and Mobile Services Operations and Quality of Service (Study Group I) 1 pp. 1 p.

—**Telegraph Networks—Interworking**
CCITT RECMN F.1-89. Operational Provisions for the International Public Telegram Service—Telegraph and Mobile Services Operations and Quality of Service (Study Group I) 54 pp. 54 pp.

—**Teleprinters**
CCITT RECMN F.1-89. Operational Provisions for the International Public Telegram Service—Telegraph and Mobile Services Operations and Quality of Service (Study Group I) 54 pp. 54 pp.

—**Teleprinters—Answerback**
CCITT RECMN F.21-89. Composition of Answer-Back Codes for the International Gentex Service—Telegraph and Mobile Services Operations and Quality of Service (Study Group I) 3 pp. 3 pp.

—**Teleprinters—International Telegraph Alphabet No. 2**
CCITT RECMN F.20-89. International Gentex Service—Telegraph and Mobile Services Operations and Quality of Service (Study Group I) 5 pp. 5 pp.

—**Telex Communications—Prohibitions**
CCITT RECMN F.20-89. International Gentex Service—Telegraph and Mobile Services Operations and Quality of Service (Study Group I) 5 pp. 5 pp.

—**Time Division Multiplexers**
CCITT RECMN U.24-89. Requirements for Telex and Gentex Operation to be Met by Synchronous Multiplex Equipment Described in Recommendation R.44—Telegraph Switching (Study Group IX) 5 pp. 5 pp.
CCITT RECMN U.25-89. Requirements for Telex and Gentex Operation to be Met by Code-and Speed-Dependent TDM Systems Conforming to Recommendation R.101—Telegraph Switching (Study Group IX) 4 pp. 4 pp.

Gentex Networks (Cont.)
—Transit Traffic
CCITT RECMN U.11-89. Telex and Gentex Signalling on Intercontinental Circuits Used for Intercontinental Automatic Transit Traffic (Type C Signalling)—Telegraph Switching (Study Group IX) 15 pp. 15 pp.

Gentex Services
Use: Gentex Networks

Gentian Violet
Use: Methyl Violet

Geodesy
See Also: Earth Sciences

—Classification
BSI BS 1000: (52)-77. 1977 Universal Decimal Classification (UDC). English Full Edition (52): Astronomy. Astrophysics. Space Research. Geodesy. 47 pp.
SNZ NZS/BS 1000 (52)-77. Universal Decimal Classification Astronomy. Astrophysics. Space Research. Geodesy. 48 pp.

—Satellites
CCIR Report 988-1-90. Satellite Systems for Geodesy and Geodynamics—Section 2F—Earth Exploration Satellites. 12 pp.

—Satellites—Radio Communications
CCIR QUESTION 143/7-90. Radiocommunications for Satellite Systems for Geodesy and Geodynamics—Questions Concerning Study Group 7—Science Services. 1 p.

Geodetic Instruments
See Also: Distance Measuring Equipment; Levels (Instruments); Measuring Instruments; Surveying Instruments; Surveys; Theodolites; Topography

—Glossaries
ISO 9849-91. Optics and Optical Instruments—Geodetic Instruments—Vocabulary First Edition. 29 pp.

Geodynamics
See Also: Earth Sciences

—Satellites
CCIR Report 988-1-90. Satellite Systems for Geodesy and Geodynamics—Section 2F—Earth Exploration Satellites. 12 pp.

—Satellites—Radio Communications
CCIR QUESTION 143/7-90. Radiocommunications for Satellite Systems for Geodesy and Geodynamics—Questions Concerning Study Group 7—Science Services. 1 p.

Geographical Coordinates
Scope Note: Use a more specific term *See:* Altitude; Latitude; Longitude

Geographical Location Systems
Use: GEOLOC Systems

Geography
See Also: Earth Sciences; Place Names; Polar Regions; Tropical Regions

—Classification
BSI BS 1000: (9)-72. 1972 Amd 1 Universal Decimal Classification (UDC). English Full Edition (9): Geography. Biography. History. 55 pp.
SNZ NZS/BS 1000 (9)-72. Universal Decimal Classification Geography. Biography. History Amend: 1. 52 pp.

GEOLOC Systems
Use For: Geographical Location Systems
See Also: Radiolocation; Radiolocation Equipment

—MF—Spread Spectrum Modulation
CCIR Report 1041-1-90. Maritime Radiolocation Operating in the Medium Frequency Band and Using Spread-Spectrum Techniques—Section 8D—Radiodetermination, Global Maritime Distress and Safety System and Related Subjects. 11 pp.

Geological Ages
Use: Stratigraphic Columns

Geological Maps
—Symbols
CNS M4001-80. Graphical Symbols for Use on Detailed Maps Plans and Geological Cross-Sections: General Rule of Representation (Oct) (6589).
CNS M4002-80. Graphical Symbols for Use on Detailed Maps Plans and Geological Cross-Sections Part II: Representation of Sedimentary Rocks (Oct)(6590).
CNS M4003-80. Graphical Symbols for Use on Detailed Maps Plans and Geological Cross-Sections III: Representation of Magmatic Rocks (Oct) (6591).
ISO 710 Pt I-74. Graphical Symbols for Use on Detailed Maps, Plans and Geological Cross-Sections—Part I: General Rules of Representation First Edition. 3 pp.
ISO 710 Pt II-74. Graphical Symbols for Use on Detailed Maps, Plans and Geological Cross-Sections—Part II: Representation of Sedimentary Rocks First Edition. 9 pp.
ISO 710 Pt III-74. Graphical Symbols for Use on Detailed Maps, Plans and Geological Cross-Sections—Part III: Representation of Magmatic Rocks First Edition. 8 pp.
ISO 710 Pt 4-82. Graphical Symbols for Use on Detailed Maps, Plans and Geological Cross-Sections—Part 4: Representation of Metamorphic Rocks First Edition. 6 pp.
ISO 710 Pt 5-89. Graphical Symbols for Use on Detailed Maps, Plans and Geological Cross-Sections—Part 5: Representation of Minerals Second Edition. 6 pp.
ISO 710 Pt 6-84. Graphical Symbols for Use on Detailed Maps, Plans and Geological Cross-Sections—Part 6: Re-presentation of Contact Rocks and Rocks Which Have Undergone Metasom-atic, Pneumatolytic or Hydrothermal Transform-ation or Transformation by Weathering First Edition. 6 pp.
ISO 710 Pt 7-84. Graphical Symbols for Use on Detailed Maps, Plans and Geological Cross-Sections—Part 7: Tectonic Symbols First Edition. 12 pp.

Geology
Scope Note: Use a more specific term
See: Petrography

Geomembranes
See Also: Geotextiles

—Glossaries
CGSB CAN/CGSB-148.2-M89. Geotextiles and Geomembranes—Definitions; (Amendment 1 January 1993). 12 pp.

—Mass
CGSB CAN/CGSB-148. 1 NO.2-M85. Methods of Testing Geotextiles and Geomembranes Mass per Unit Area. 9 pp.

—Polyethylene—High Density—Landfills
CNS K3093-89. High Density Polyethylene Geomembrane for Sanitary Landfill (Feb)(12493). 3 pp.
CNS K6961-89. Method of Test for High Density Polyethylene Geomembrane for Sanitary Landfill (Feb)(12494). 5 pp.

—Sampling
CGSB CAN/CGSB-148. 1 NO.1-M85. Methods of Testing Geotextiles and Geomembranes Sampling and Preparation of Specimens. 8 pp.

—Test Specimens
CGSB CAN/CGSB-148. 1 NO.1-M85. Methods of Testing Geotextiles and Geomembranes Sampling and Preparation of Specimens. 8 pp.

—Thickness Measurement
CGSB CAN/CGSB-148. 1 NO.3-M85. Methods of Testing Geotextiles and Geomembranes Thickness of Geotextiles. 9 pp.

Geometric Quantities, Cutting Tools
Use: Cutting Tools—Geometric and Kinematic Quantities

Geometrical Orientation
Use: Orientation

Geostationary Orbits
Use For: Clarke Orbits *See Also:* Altitude; Geostationary Satellites; Orbits; Position (Location); Satellites

CCIR RESOLUTION 99-90. Technical and Operational Methods Utilized for the Allocations and Improved Use of the Radio Frequency Spectrum—Volume XIV—Administrative Texts of the CCIR. 2 pp.

Geostationary Orbits (Cont.)
—Communications Satellites—Efficiency—Frequency Band Sharing
CCIR Report 453-5-90. Technical Factors Influencing the Efficiency of Use of the Geostationary-Satellite Orbit by Radiocommunication Satellites Sharing the Same Frequency Bands General Summary—Section 4D2—Coordination Methods. 36 pp.
CCIR QUESTION 48/4-90. Technical Factors Influencing the Efficiency of Use of the Geostationary-Satellite Orbit by Radiocommunication Sat. Networks Sharing Frequency Bands Allocated to the Fixed-Sat. Service—Questions of Study Group 4 Fixed-Satellite Service. 2 pp.

—Fixed Satellite Services—Earth Stations—Interference
CCIR QUESTION 70/4-90. Protection of the Geostationary-Satellite Orbit Against Unacceptable Interference from Transmitting Earth Stations in the Fixed-Satellite Service at Frequencies Above 10 GHz—Questions of Study Group 4 Fixed-Satellite Service. 1 p.

—Fixed Satellite Services—Frequency Management
CCIR RECMN 743-92. Coordination of Satellite Networks Using Slightly Inclined Geostationary-Satellite Orbits and Between Such Networks and Satellite Networks Using Non-Inclined GSO Satellites—Section 4D2—Coordination Methods. 14 pp.

—Fixed Satellite Services—Physical Interference
CCIR QUESTION 34/4-86. Physical Interference in the Geostationary-Satellite Orbit—Questions of Study Group 4 Fixed-Satellite Service. 1 p.

—Fixed Satellite Services—Space Stations—Spot Beams
CCIR QUESTION 36/4-87. Use of Steerable Spot Beams by Space Stations in the Geostationary-Satellite Orbit—Questions of Study Group 4 Fixed-Satellite Service. 1 p.

—Fixed Satellites—Fixed Services—Frequency Band Sharing
CCIR Report 1142-90. Frequency Sharing Between the Fixed Service and the Fixed-Satellite Service Using Satellites in Slightly Inclined Geostationary Orbits—Section 4/9A—Sharing Conditions. 29 pp.
CCIR QUESTION 64/4-90. Preferred Technical Characteristics of Fixed-Satellite Service Networks Using Satellites in Slightly Inclined Geostationary Orbits to Facilitate Sharing with the Fixed Service—Questions of Study Group 4 Fixed-Satellite Service. 1 p.
CCIR QUESTION 117/9-90. Criteria for Frequency Sharing Between the Fixed Service and FSS Networks Using Satellites in Slightly Inclined Geostationary Orbits—Questions of Study Group 9 Fixed Service. 1 p.

—Inclined—Fixed Satellite Services
CCIR Report 1138-90. Intra-Service Implications of Using Slightly Inclined Geostationary Orbits for Fixed Satellite Service Networks Operational, Sharing and Coordination Considerations—Section 4D2—Coordination Methods. 22 pp.
CCIR QUESTION 51/4-90. Operation, Frequency Sharing and Coordination of Fixed Satellite Service Networks Using Satellites in Slightly Inclined Geostationary Orbits—Questions of Study Group 4 Fixed-Satellite Service. 1 p.

—Inclined—Fixed Satellite Services—Frequency Band Sharing
CCIR Report 1138-90. Intra-Service Implications of Using Slightly Inclined Geostationary Orbits for Fixed Satellite Service Networks Operational, Sharing and Coordination Considerations—Section 4D2—Coordination Methods. 22 pp.
CCIR QUESTION 51/4-90. Operation, Frequency Sharing and Coordination of Fixed Satellite Service Networks Using Satellites in Slightly Inclined Geostationary Orbits—Questions of Study Group 4 Fixed-Satellite Service. 1 p.

—Inclined—Fixed Satellite Services—Frequency Management
CCIR RECMN 743-92. Coordination of Satellite Networks Using Slightly Inclined Geostationary-Satellite Orbits and Between Such Networks and Satellite Networks Using Non-Inclined GSO Satellites—Section 4D2—Coordination Methods. 14 pp.

INTERNATIONAL AND NON-U.S. NATIONAL STANDARDS
SUBJECT INDEX

Geostationary Orbits *(Cont.)*

—Inclined—Mobile Satellite Communications
CCIR Report 1170-90. Mobile-Satellite Communication Systems Using Highly Inclined Elliptical Orbits—Section 8F—Frequencies, Orbits and Systems. 16 pp.

—Mobile Satellite Communications
CCIR Report 1170-90. Mobile-Satellite Communication Systems Using Highly Inclined Elliptical Orbits—Section 8F—Frequencies, Orbits and Systems. 16 pp.

—Physical Interference
CCIR Report 1004-1-90. Physical Interference in the Geostationary-Satellite Orbit—Section 4D2—Coordination Methods. 5 pp.

—Spacing—Mobile Satellite Communications
CCIR Report 772-2-86. Orbital Spacing Considerations for a Mobile-Satellite Service —Section 8H—Efficient Use of the Radio Spectrum Characteristics and Sharing of Frequency Resources. 13 pp.

—Utilization
CCIR OPINION 88-90. CCIR Terms of Reference—Volume XIV—Administrative Texts of the CCIR. 1 p.

Geostationary Satellite Orbits
Use: Geostationary Orbits

Geostationary Satellites
Use For: Geosynchronous Satellites; Synchronous Satellites *See Also:* Fixed Satellite Services; Geostationary Orbits; Satellites

—Angular Distance—Satellite Networks
CCIR Report 772-2-86. Orbital Spacing Considerations for a Mobile-Satellite Service —Section 8H—Efficient Use of the Radio Spectrum Characteristics and Sharing of Frequency Resources. 13 pp.

—Antennas—Accurate Beam Pointing
CCIR Report 1136-90. Geostationary Satellite Antenna Pointing Accuracy—Section 4D3—Spacecraft Station Keeping—Satellite Antenna Radiation Pattern—Pointing Accuracy. 13 pp.

—Attitude Control Errors
CCIR Report 546-4-90. Space Systems Technology: Attitude Control Technology—Section 2A—Research in Space Technology. 22 pp.

—Communications
CCIR Report 771-3-90. Considerations for the Mobile-Satellite Service —Section 8F—Frequencies, Orbits and Systems. 7 pp.

—Data Relay—Space Communications—Frequency Band Sharing
CCIR Report 847-1-86. Data Relay Satellites Sharing with Other Services in Bands 9 and 10—Section 2D—Data Relay Satellites. 7 pp.

—Deep Space—Electromagnetic Interference
CCIR Report 844-1-90. Potential Interference Between Deep-Space Telecommunications and Some Other Services in Harmonically Related Bands—Section 2B—Topics of General Interest. 12 pp.

—Downlinks—Frequency Bands
CCIR Report 557-2-86. Use of Frequency Bands Allocated to the Fixed-Satellite Service for Both the up Link and down Link Geostationary-Satellite Systems—Section 4D2—Coordination Methods. 9 pp.

—Earth Stations—Antennas—Design
CCIR RECMN 580-2-90. Radiation Diagrams for Use as Design Objectives for Antennas of Earth Stations Operating with Geostationary Satellites —Section 4C—Earth Station and Baseband Characteristics—Earth Station Antennas—Maintenance of Earth Stations. 3 pp.

CCIR RECMN 580-3-92. Radiation Diagrams for Use as Design Objectives for Antennas of Earth Stations Operating with Geostationary Satellites —Section 4C—Earth Station and Baseband Characteristics—Earth Station Antennas—Maintenance of Earth Stations. 3 pp.

—Earth Stations—Coordination Calculations—Frequency Band Sharing
CCIR Report 454-5-90. Method of Calculation for Determining if Coordination is Required Between Geostationary-Satellite Networks Sharing the Same Frequency Bands—Section 4D2—Coordination Methods. 39 pp.

Geostationary Satellites *(Cont.)*

—EPIRB
CCIR RECMN 632-1-90. Transmission Characteristics of a Satellite Emergency Position-Indicating Radiobeacon (Satellite EPIRB) System Operating Through Geostationary Satellites in the 1.6 GHz Band—Section 8J—Technical and Operating Characteristics of Radiocommunications. 7 pp.

—EPIRB—INMARSAT
CCIR Report 1184-90. Pre-Operational Demonstrations of the 1.6 GHz Satellite EPIRB System Using the Inmarsat Geostationary Space Segment—Section 8J—Technical and Operating Characteristics of Radiocommunications Using Satellite Distress and Safety Operation and. 8 pp.

—Fixed—Coordination Calculation—Frequency Band Sharing
CCIR RECMN 738-92. Procedure for Determining if Coordination is Required Between Geostationary-Satellite Networks Sharing the Same Frequency Bands—Section 4D2—Coordination Methods. 12 pp.

CCIR RECMN 739-92. Additional Methods for Determining if Detailed Coordination is Necessary Between Geostationary-Satellite Networks in the Fixed-Satellite Service Sharing the Same Frequency Bands—Section 4D2—Coordination Methods. 10 pp.

CCIR Report 454-5-90. Method of Calculation for Determining if Coordination is Required Between Geostationary-Satellite Networks Sharing the Same Frequency Bands—Section 4D2—Coordination Methods. 39 pp.

—Fixed—Earth Exploration—Frequency Band Sharing
CCIR Report 540-1-82. Feasibility of Frequency Sharing Between an Earth Exploration-Satellite (EES) System and Fixed Satellite, Meteorological Satellite and Terrestrial Fixed and Mobile Services—Section 2F—Earth Exploration Satellites. 11 pp.

—Fixed—Frequency Modulation—Television Channels—Interference
CCIR RECMN 483-1-78. Maximum Permissible Level of Interference in a Television Channel of a Geostationary-Satellite Network in the Fixed-Satellite Service Employing Frequency Modulation, Caused by Other Networks of This Service—Section 4D1—Permissible Levels of Interference. 1 p.

CCIR RECMN 483-2-92. Maximum Permissible Level of Interference in a Television Channel of a Geostationary-Satellite Network in the Fixed-Satellite Service Employing Frequency Modulation, Caused by Other Networks of This Service—Section 4D1—Permissible Levels of Interference. 2 pp.

—Fixed—ISDN—Hypothetical Reference Digital Paths—Interference
CCIR RECMN 735-92. Maximum Permissible Levels of Interference in a Geostationary-Satellite Network for an HRDP When Forming Part of the ISDN in the Fixed-Satellite Service Caused by Other Networks of This Service Below 15 GHz—Section 4D1—Permissible Levels of Interference. 5 pp.

—Fixed—Multiplexing—Telephone Channels—Radio Interference
CCIR RECMN 466-5-90. Maximum Permissible Level of Interference in a Telephone Channel of a Geostationary-Satellite Network in the Fixed-Satellite Service Employing Frequency Modulation with Frequency-Division Multiplex, Caused by Other Networks of this Service—Section 4D1-. 3 pp.

CCIR RECMN 466-6-92. Maximum Permissible Level of Interference in a Telephone Channel of a Geostationary-Satellite Network in the Fixed-Satellite Service Employing Frequency Modulation with Frequency-Division Multiplex, Caused by Other Networks of this Service—Section 4D1-. 3 pp.

—Fixed—Pulse Code Modulation—Telephony—Radio Interference
CCIR RECMN 523-3-90. Maximum Permissible Levels of Interference in a Geostationary-Satellite Network in the Fixed-Satellite Service Using 8-Bit PCM Encoded Telephony, Caused by Other Networks of this Service—Section 4D1—Permissible Levels of Interference. 3 pp.

CCIR RECMN 523-4-92. Maximum Permissible Levels of Interference in a Geostationary-Satellite Network in the Fixed-Satellite Service Using 8-Bit PCM Encoded Telephony, Caused by Other Networks of this Service—Section 4D1—Permissible Levels of Interference. 2 pp.

—Fixed—Radio Spectra Utilization
CCIR RECMN 742-92. Spectrum Utilization Methodologies—Section 4D2—Coordination Methods. 5 pp.

Geostationary Satellites *(Cont.)*

—Fixed Satellite Services—Frequency Bands—Station Keeping
CCIR RECMN 484-2-82. Station-Keeping in Longitude of Geostationary Satellites Using Frequency Bands Allocated to the Fixed-Satellite Service—Section 4D3—Spacecraft Station-Keeping—Satellite Antenna Radiation Pattern—Pointing Accuracy. 1 p.

CCIR RECMN 484-3-92. Station-Keeping in Longitude of Geostationary Satellites in the Fixed Satellite Service—Section 4D3—Spacecraft Station-Keeping—Satellite Antenna Radiation Pattern—Pointing Accuracy. 2 pp.

—Fixed Satellite Services—Polarization Discrimination
CCIR RECMN 736-92. Estimation of Polarization Discrimination in the Interference Calculations Between Geostationary-Satellite Networks in the Fixed-Satellite Service—Section 4D1—Permissible Levels of Interference. 7 pp.

—Fixed Satellite Services—Radio Frequency Interference
CCIR QUESTION 39/4-89. Technical Criteria to Be Used in the Board's Examinations of the Probability of Harmful Interference Required by Provisions Nos. 1354, 1506 and 1509 of the Radio Regulations—Questions of Study Group 4 Fixed-Satellite Service. 1 p.

—Fixed Satellite Services—Station Keeping
CCIR Report 556-4-90. Factors Affecting Station-Keeping of Geostationary Satellites of the Fixed-Satellite Service—Section 4D3—Spacecraft Station Keeping—Satellite Antenna Radiation Pattern—Pointing Accuracy. 4 pp.

—Frequency Band Sharing
CCIR Report 455-5-90. Frequency Sharing Between Networks of the Fixed-Satellite Service—Section 4D1—Permissible Levels of Interference. 43 pp.

—Meteorological—Earth Exploration—Frequency Band Sharing
CCIR Report 540-1-82. Feasibility of Frequency Sharing Between an Earth Exploration-Satellite (EES) System and Fixed Satellite, Meteorological Satellite and Terrestrial Fixed and Mobile Services—Section 2F—Earth Exploration Satellites. 11 pp.

—Meteorological—Meteorological Services—Frequency Band Sharing
CCIR Report 541-3-90. Feasibility of Frequency Sharing Between a Geostationary Meteorological Satellite System and the Meteorological Aids Service in the Region of 400 MHz and in the Upper Part of Band 9 (1 to 3 GHz)—Section 2F—Earth Exploration Satellites. 22 pp.

—Radio Navigation Services—Satellite Links—Frequency Band Sharing
CCIR Report 872-82. Sharing Criteria Between Inter-Satellite Links Connecting Geostationary Satellites in the Fixed-Satellite Service and the Radionavigation Service at 33 GHz—Section 4E—Frequency Sharing Between Networks of the Fixed-Satellite Service and Those of Other Space. 4 pp.

—Spot Beams
CCIR Report 1171-90. Coordination Considerations of Geostationary Satellites Using Steerable Spot Beams with Other Systems —Section 8H—Efficient Use of the Radio Spectrum Characteristics and Sharing of Frequency Resources. 3 pp.

—Station Keeping
CCIR Report 843-2-90. Station-Keeping Techniques for Geostationary Satellites —Section 2A—Research in Space Technology. 9 pp.

—UHF—Distress Alerting
CCIR Report 1175-90. 406 MHz Geostationary Satellite Distress Alerting Experiment—Section 8J—Technical and Operating Characteristics of Radiocommunications Using Satellite Distress and Safety Operation and of Radio Determination Satellite Services. 18 pp.

—Uplinks—Frequency Bands
CCIR Report 557-2-86. Use of Frequency Bands Allocated to the Fixed-Satellite Service for Both the up Link and down Link Geostationary-Satellite Systems—Section 4D2—Coordination Methods. 9 pp.

Geosynchronous Satellites
Use: Geostationary Satellites

Geotextiles
See Also: Fabrics; Geomembranes

INDUSTRY STANDARDS

Geotextiles (Cont.)

—Breaking Load
CGSB CAN/CGSB-148. 1 NO.7.3-92. Methods of Testing Geotextiles and Geomembranes Grab Tensile Test for Geotextiles. 9 pp.

—Creep Properties
BSI BS 6906: Part 5-91. 1991 Geotextiles Part 5: Determination of Creep. 10 pp.

—Elongation
CGSB CAN/CGSB-148. 1 NO.7.3-92. Methods of Testing Geotextiles and Geomembranes Grab Tensile Test for Geotextiles. 9 pp.

—Flow Measurement
BSI BS 6906: Part 7-90. 1990 Methods of Test for Geotextiles Part 7: Determination of in-Plane Waterflow. 14 pp.

—Glossaries
CGSB CAN/CGSB-148.2-M89. Geotextiles and Geomembranes—Definitions; (Amendment 1 January 1993). 12 pp.
ISO 10318-90. Geotextiles—Vocabulary First Edition. 7 pp.

—Identification Systems
BSI BS EN 30320-93. 1993 Geotextiles—Identification on Site (ISO 10320: 1991) (L). 9 pp.
CEN EN 30320-93. Geotextiles—Identification on Site (ISO 10320: 1991). 4 pp.
ISO 10320-91. Geotextiles—Identification on Site First Edition; (CEN EN 30320: 1993). 5 pp.

—Life (Durability)
BSI PD 6533-93. 1993 Methods for Assessing the Durability of Geotextiles—an Interim Document (L). 18 pp.

—Mass
CEN PREN 965-92. Geotextiles and Geotextile-Related Products—Determination of Mass per Unit Area. 4 pp.
CGSB CAN/CGSB-148. 1 NO.2-M85. Methods of Testing Geotextiles and Geomembranes Mass per Unit Area. 9 pp.
ISO 9864-90. Geotextiles—Determination of Mass per Unit Area First Edition. 4 pp.

—Perforation Resistance
BSI BS 6906: Part 6-90. 1990 Methods of Test for Geotextiles Part 6: Determination of Resistance to Perforation (Cone Drop Test). 12 pp.
CEN PREN 918-92. Dynamic Perforation Test for Geotextiles and Geotextile-Related Products (Cone Drop Test). 11 pp.

—Porosity
BSI BS 6906: Part 2-89. 1989 Methods of Test for Geotextiles Part 2: Determination of the Apparent Pore Size Distribution by Dry Sieving. 8 pp.

—Pressure
CEN PREN 964-92. Geotextiles and Geotextile-Related Products—Determination of Thickness at Specified Pressures. 6 pp.
ISO 9863-90. Geotextiles—Determination of Thickness at Specified Pressures First Edition. 6 pp.

—Puncture Resistance
BSI BS 6906: Part 4-89. 1989 Methods of Test for Geotextiles Part 4: Determination of the Puncture Resistance (CBR Puncture Test). 12 pp.
CEN PREN 776-92. Static Puncture Test for Geotextiles and Geotextile-Related Products (CBR-Test). 10 pp.

—Sampling
CEN PREN 963-92. Geotextiles and Geotextile-Related Products—Sampling and Preparation of Test Specimens. 5 pp.
CGSB CAN/CGSB-148. 1 NO.1-M85. Methods of Testing Geotextiles and Geomembranes Sampling and Preparation of Specimens. 8 pp.
ISO 9862-90. Geotextiles—Sampling and Preparation of Test Specimens First Edition. 6 pp.

—Seams—Tensile Testing
ISO 10321-92. Geotextiles—Tensile Test for Joints/Seams by Wide-Width Method First Edition. 10 pp.

—Shear Strength
BSI BS 6906: Part 8-91. 1991 Geotextiles Part 8: Determination of Sand-Geotextile Frictional Behaviour by Direct Shear. 12 pp.

—Tensile Testing
BSI BS 6906: Part 1-87. 1987 Methods of Test for Geotextiles Part 1: Determination of the Tensile Properties Using a Wide Width Strip. 12 pp.
ISO 10319-93. Geotextiles—Wide-Width Tensile Test First Edition. 11 pp.

—Test Specimens
CEN PREN 963-92. Geotextiles and Geotextile-Related Products—Sampling and Preparation of Test Specimens. 5 pp.
CGSB CAN/CGSB-148. 1 NO.1-M85. Methods of Testing Geotextiles and Geomembranes Sampling and Preparation of Specimens. 8 pp.
ISO 9862-90. Geotextiles—Sampling and Preparation of Test Specimens First Edition. 6 pp.

—Thickness Measurement
CEN PREN 964-92. Geotextiles and Geotextile-Related Products—Determination of Thickness at Specified Pressures. 6 pp.
CGSB CAN/CGSB-148. 1 NO.3-M85. Methods of Testing Geotextiles and Geomembranes Thickness of Geotextiles. 9 pp.
ISO 9863-90. Geotextiles—Determination of Thickness at Specified Pressures First Edition. 6 pp.

—Transmissivity
SAA AS 3706.10.1-91. Geotextiles—Methods of Test—Part 10: Determination of Transmissivity—Part 10.1: Radial Method. 3 pp.

—Water Permeability
BSI BS 6906: Part 3-89. 1989 Methods of Test for Geotextiles Part 3: Determination of Water Flow Normal to the Plane of the Geotextile Under a Constant Head. 9 pp.

Geothermal Resources

—Heating Equipment
SNZ NZS 2402P-87. Code of Practice for Geothermal Heating Equipment in Rotorua. 60 pp.

Geraniol
CNS K1137-69. Geraniol (Jul)(3046)(R 1973).

Geranium Oil
See Also: Essential Oils
ISO 4731-78. Oil of Geranium First Edition. 5 pp.

—Ester Value
CNS K6616-80. Determination of Ester Value After Hot Formulation of Essential Oils of Geranium and Rose (Dec)(6682).

Geranium Rose Oil
Use: Geranium Oil

German Characters
DIN ENGL 1455-74. Handwritings (Feb). 2 pp.

German Standards Institute
Use: DIN

Germanium
Scope Note: For additional listings, see also specific products made from germanium

—Conductivity Type—Thermoelectric Method
JIS H 0607-78. Determination of Conductivity Type in Germanium by Thermoelectromotive Method (R 1983). 5 pp.

—Crystals—Electrical Resistivity—Four Point Probe DC Method
DIN ENGL 50431-88. Testing of Semiconductor Materials; Measurement of the Resistivity of Silicon or Germanium Single Crystals by Means of the Four Probe/Direct Current Method with Collinear Array (May) (ASTM F43-88). 5 pp.

—Crystals—Etch Pit Density
JIS H 0610-66. Method of Measurement of Etch Pit Density of Germanium Crystal (R 1983). 7 pp.

—Electrical Resistivity
JIS H 0601-62. Testing Methods of Resistivity for Germanium (R 1983). 6 pp.

—Minority Carrier Lifetime—Photoconductive Decay Method
JIS H 0603-78. Measurement of Minority Carrier Life Time in Germanium by Photoconductive Decay Method (R 1983). 7 pp.

Germanium Detectors
See Also: Radiation Measuring Instruments
CENELEC HD 303-75. Test Procedures for Germanium Gamma-Ray Detectors. 2 pp.
IEC 937-88. Cryostat End-Cap Dimensions for Germanium Semiconductor Detectors for Gamma-Ray Spectrometers First Edition. 11 pp.
IEC 973-89. Test Procedures for Germanium Gamma-Ray Detectors First Edition. 83 pp.
JIS Z 4520-79. Test Procedures for Germanium Gamma-Ray Detectors.

Germanium Detectors (Cont.)

—Cryostat—End Caps
IEC 937-88. Cryostat End-Cap Dimensions for Germanium Semiconductor Detectors for Gamma-Ray Spectrometers First Edition. 11 pp.

—Glossaries
CENELEC HD 396-79. Definitions of Test Method Terms for Semiconductor Radiation Detectors and Scintillation Counting. 1 p.
IEC 596-78. Definitions of Test Method Terms for Semiconductor Radiation Detectors and Scintillation Counting First Edition. 24 pp.

—Impurities
CNS J2014-81. Method for Analysis of Radioisotopes Determination of Gamma Ray Impurities with Germanium Detector (May)(7391).

Germanium Diodes

—Quality Assurance
BSI BS 9300 C776-777-71. (OBSOLESCENT) 1971 Detail Requirements for Germanium Coaxial Mixer Diodes. 12 pp.
BSI BS 9300 C778-71. (OBSOLESCENT) 1971 Detail Requirements for a Matched Pair of Germanium Coaxial Mixer Diodes. 3 pp.

Germanium Semiconductor Detectors
Use: Germanium Detectors

Germanium Slices
Use: Germanium Wafers

Germanium Transistors
See Also: Bipolar Transistors; Transistors

—PNP
MOD UK DSTAN 59-61: 90/103-72. Semiconductor Device, Transistor Issue 1. 3 pp.

Germanium Wafers
See Also: Semiconductor Devices; Semiconductors

—Electrical Resistivity
DIN ENGL 50435-88. Testing of Semiconductor Materials; Determination of the Radial Resistivity Variation of Silicon or Germanium Slices by Means of the Four-Probe/Direct Current Method (May) (ASTM F81-89). 3 pp.

Germicide Content Analysis
Use: Bactericide Content Analysis

Ghee
Use: Vegetable Fats

Gibbs Reagent
Use: 2,6-Dichloroquinone Chlorimide

Gilding Metals
See Also: Ammunition; Copper Alloys

—Bullet Envelopes
MOD UK DSTAN 95-11-80. Bullet Envelope Materials Issue 1. 14 pp.

Gin Blocks
Use: Gin Wheels

Gin Wheels
Use For: Gin Blocks *See Also:* Pulley Blocks; Scaffolds
BSI BS 1692-71. 1971 Gin Blocks. 13 pp.

Ginger
See Also: Herbs; Spices
BSI BS 7087: Part 8-90. 1990 Herbs and Spices Ready for Food Use Part 8: Dried Ginger (Whole or Ground). 9 pp.
ISO 1003-80. Spices and Condiments—Ginger, Whole, in Pieces, or Ground—Specification First Edition. 8 pp.

—Canned
CNS N5021-82. Canned Ginger (Jan)(1255). 3 pp.

—Defects
CNS N4039-72. Method of Test for Gingers (Jun)(3360). 2 pp.

—Grading
CNS N1013-82. Grades and Packaging of Gingers (Jan)(1120). 2 pp.

—Inspection
CNS N4039-72. Method of Test for Gingers (Jun)(3360). 2 pp.

—Packaging
CNS N1013-82. Grades and Packaging of Gingers (Jan)(1120). 2 pp.

INTERNATIONAL AND NON-U.S. NATIONAL STANDARDS SUBJECT INDEX

Ginger (Cont.)

—Salted
CNS N5081-84. Salted Gingers (Jul)(2816). 3 pp.

—Salted—Chemical Analysis
CNS N6108-78. Method of Test for Salted Gingers (Nov)(4680). 2 pp.

—Salted—Inspection
CNS N6108-78. Method of Test for Salted Gingers (Nov)(4680). 2 pp.

—Salted—Sampling
CNS N6108-78. Method of Test for Salted Gingers (Nov)(4680). 2 pp.

—Sampling
CNS N4039-72. Method of Test for Gingers (Jun)(3360). 2 pp.

Girders

See Also: Plates (Supports); Splice Plates; Structural Forms; Structural Members

—Steel—Glossaries
DIN ENGL 1080 Pt 4-80. Terms, Symbols and Units Used in Civil Engineering; Concrete Construction, Composite Steel Construction and Steel Girders in Concrete (Mar). 10 pp.

—Steel—Symbols
DIN ENGL 1080 Pt 4-80. Terms, Symbols and Units Used in Civil Engineering; Concrete Construction, Composite Steel Construction and Steel Girders in Concrete (Mar). 10 pp.

Glacial Acetic Acid
Use: Acetic Acid

Glacial Phosphoric Acid
Use: Phosphoric Acid, Meta

Gland Nuts
See Also: Nuts (Fasteners)

—Aircraft
SBAC AGS 1658 ISSUE 14. Nut Gland.

—Aircraft—Blanking Discs
SBAC AGS 1725 ISSUE 7(I). Blanking Disc (for Use Under Gland Nuts Where Cable Sleeves Are Not Fitted).

Glands (Cables)
Use: Cable Glands

Glands (Seals)
See Also: Cable Glands; Hydraulic Seals; O Rings; Oil Seals; Packings (Seals); Seals

DIN ENGL 86005-66. Screwed Glands for Fittings up to Nominal Pressure 100; Summary, Base Rings, Glands, Union Nuts (Nov). 4 pp.

—Aircraft Hydraulic Systems
JIS W 2006-79. Gland Design; Packings, Hydraulic, General Requirements for.

—Cocks
JIS B 2191-84. Screwed Bronze Plug Cocks and Gland Cocks. 14 pp.

—Cocks—Ships
DIN ENGL 87020-67. Gland Cocks in Red Brass with Unions with 25 Deg. Seat Angle; NW 6 to 25; ND 16: up to 80 Deg. C; ND 25: up to 40 Deg. C (Feb). 4 pp.

—O Rings—Design
JIS B 2406-91. O-Rings Housings—Design Criteria. 49 pp.

—Pipe Joints
CNS B5048-71. Cast-Iron Pipes and Screw for Assembly, Screw-Gland flexible Joint (Jan)(2581).
CNS B5049-66. Cast-Iron Pipes and Sealing Rings for Assembly, Screw-Gland Flexible Joint (Mar)(2582).
CNS B5050-71. Cast-Iron Pipes, Screw-Gland Flexiable Joint, Class LA (Jan)(2583).
CNS B5051-71. Cast-Iron Pipes, Screw-Gland Flexiable Joint, Class A (Jan)(2584).
CNS B5052-71. Cast-Iron Pipes, Screw-Gland Flexiable Joint, Class B (Jan)(2585).
CNS B5054-67. Nodular Graphite Cast-Iron Pipes, Bolt-Gland Flexible Joint, Class 70 (Aug)(2781)(R 1973).
CNS B5055-67. Nodular Graphite Cast-Iron Pipes, Bolt-Gland Flexible Joint, Class 80 (Aug)(2782)(R 1973).
CNS B5056-67. Nodular Graphite Cast-Iron Pipes, Screw-Gland Flexible Joint, Class 70 (Aug)(2783).
CNS B5057-67. Nodular Graphite Cast-Iron Pipes, Screw-Gland Flexible Joint, Class 80 (Aug)(2784)(R 1973).

Glands (Seals) (Cont.)

—Pipe Joints (Cont.)
CNS B5058-67. Cast-Iron Pipes and Sealing Rings for Assembly, Bolt-Gland Flexible Joint (Aug)(2794).
CNS B5059-72. Bolts and Nuts for Bolt-Gland Flexible Joint Cast-Iron Pipes and Fittings (Jun)(2795).
CNS B5060-72. Cast-Iron Pipes, Socket and Sealing Washers for Assembly, Bolt-Gland Flexible Joint (Jun)(2796).
CNS B5064-68. Cast-Iron Pipes, Bolt-Gland Flexible Joint, Class A (Jan)(2858)(R 1973).
CNS B5065-68. Cast-Iron Pipes, Bolt-Gland Flexible Joint, Class A (Jan)(2859)(R 1973).
CNS B5066-68. Cast-Iron Pipes, Bolt-Gland Flexible Joint, Class B (Jan)(2860)(R 1973).

Glaser-Dirks DG-400 Motor Gliders
See Also: Gliders

—Foreign Airworthiness Directives
CAA. Glaser-Dirks DG-400 Series Motor Gliders (Foreign Airworthiness Directives). 2 pp.

Glass

Scope Note: For additional listings, see also specific products and materials made from glass
See Also: Alkali Glass; Annealing Point; Borosilicate Glass; Cellular Glass; Ceramics; Coated Glass; Crown Glass; Figured Glass; Float Glass; Glass Beads; Glass Blocks; Glass Ceramics; Glass Fabrics; Glassy Alloys; Glazing Materials; Heat Treated Glass; Insulating Glass; Laminated Glass; Lead Glass; Optical Glass; Optical Properties; Plate Glass; Reinforced Glass; Safety Glass; Sheet Glass; Spandrel Glass; Window Glass; Wire Glass

CNS R2052-84. Multiple Glasses (Aug) (2541).

—Acid Resistance Testing—Atomic Absorption Spectrometry
BSI BS 3473: Part 5-87. 1987 Chemical Resistance of Glass Used in the Production of Laboratory Glassware Part 5: Method for Determination of Resistance of Glass to Attack by 6 mol/L Hydrochloric Acid at 100 Degrees Celcius. 10 pp.
ISO 1776-85. Glass—Resistance to Attack by Hydrochloric Acid at 100 Degrees Celsius—Flame Emission or Flame Atomic Absorption Spectrometric Method First Edition. 7 pp.

—Alkali Resistance—Boiling Temperature Testing
BSI BS 3473: Part 1-91. 1991 Chemical Resistance of Glass Used in the Production of Laboratory Glassware Part 1: Method for Determination of Resistance of Glass to Attack by a Boiling Aqueous Solution of Mixed Alkali. 11 pp.
ISO 695-91. Glass—Resistance to Attack by a Boiling Aqueous Solution of Mixed Alkali—Method of Test and Classification Third Edition. 8 pp.

—Aluminum Oxide Content—Molecular Absorption Spectrophotometry
BSI BS 7709: Part 4-93. 1993 Analysis of Extract Solutions of Glass Part 4: Method for Determination of Aluminium Oxide by Molecular Absorption Spectrometry (ISO 10136-4: 1993) (V). 11 pp.
ISO 10136 Pt 4-93. Glass and Glassware—Analysis of Extract Solutions—Part 4: Determination of Aluminium Oxide by Molecular Absorption Spectrometry First Edition. 9 pp.

—Boron Oxide Content—Molecular Absorption Spectrophotometry
BSI BS 7709: Part 6-93. 1993 Analysis of Extract Solutions of Glass Part 6: Method for Determination of Boron(III) Oxide by Molecular Absorption Spectrometry (ISO 10136-6: 1993) (V). 11 pp.
ISO 10136 Pt 6-93. Glass and Glassware—Analysis of Extract Solutions—Part 6: Determination of Boron(III) Oxide by Molecular Absorption Spectrometry First Edition. 9 pp.

—Buildings—Installation
SAA AS 1288-89. Glass in Buildings—Selection and Installation. 82 pp.

—Buildings—Selection
SAA AS 1288-89. Glass in Buildings—Selection and Installation. 82 pp.

—Calcium Oxide Content—Atomic Absorption Spectrometry
ISO 10136 Pt 3-93. Glass and Glassware—Analysis of Extract Solutions—Part 3: Determination of Calcium Oxide and Magnesium Oxide by Flame Atomic Absorption Spectrometry First Edition. 10 pp.

Glass (Cont.)

—Cast
CEN PREN 572-5-91. Glass in Building—Basic Products—Part 5: Cast Glass. 8 pp.
CEN PREN 572-6-91. Glass in Building—Basic Products—Part 6: Wired Cast Glass. 9 pp.
CEN PREN 572-7-91. Glass in Building—Basic Products—Part 7: Wired or Unwired Channel Shaped Glass. 6 pp.

—Chemical Analysis
BSI BS 2649: Part 1-88. 1988 Methods for the Analysis of Glass Part 1: Glasses of the Soda-Lime-Magnesia-Silica-Type. 16 pp.
CNS R3035-86. Method of Chemical Analysis of Soda-Lime-Magnesia-Silica Glass (Jan)(1046).
JIS R 3101-91. Methods for Chemical Analysis of Soda-Lime-Magnesia-Silica Glasses. 32 pp.

—Chemical Plants
CNS R2194-88. Glass Components for Chemical Plants (Oct)(12445).

—Chemical Properties
DIN ENGL 1249 Pt 10-90. Glass for Use in Building Construction; Chemical and Physical Properties (Aug). 4 pp.

—Classification
BSI BS 1000: (666)-84. 1984 Universal Decimal Classification (UDC). English Full Edition (666): Glass Industry. Ceramics. Cement and Concrete. 60 pp.
SNZ NZS/BS 1000 (666)-84. Universal Decimal Classification Glass Industry. Ceramics. Cement and Concrete. 60 pp.

—Cleaning Cloths
CNS L4062-74. Glass Cleaning Cloth (Tentative) (Mar)(3713).

—Fire Testing
BSI PD 6512: Part 3-87. 1987 Use of Elements of Structural Protection with Particular Reference to the Recommendations Given in BS 5588 Fire Precautions in the Design and Construction of Buildings Part 3: To the Fire Performance of Glass. 10 pp.
CEN PREN 357-90. Glass in Building—Glazed Assemblies Containing Fire Resistant Transparent or Translucent Glass, for Use in Buildings. 7 pp.

—Glossaries
BSI BS 3446: Part 2-90. 1990 Glossary of Terms Associated with Refractory Materials Part 2: Applications in the Coke, Glass, Cement and Other Non-Metallurgical Industries (Supersedes BS 3446: 1962). 49 pp.
BSI BS 3447-62. 1962 Amd 2 Glossary of Terms Used in the Glass Industry. 57 pp.
CEN PREN 572-1-91. Glass in Building—Basic Products—Part 1: Definitions and General Physical and Mechanical Properties. 8 pp.

—Hydrolytic Resistance—High Temperature Testing
BSI BS 3473: Part 2-87. 1987 Chemical Resistance of Glass Used in the Production of Laboratory Glassware Part 2: Method for Determination of Hydrolytic Resistance of Glass Grains at 98 Degrees Centigrade. 10 pp.
BSI BS 3473: Part 3-87. 1987 Chemical Resistance of Glass Used in the Production of Laboratory Glassware Part 3: Method for Determination of Hydrolytic Resistance of Glass Grains at 121 Degrees Centigrade. 10 pp.
ISO 719-85. Glass—Hydrolytic Resistance of Glass Grains at 98 Degrees C—Method of Test and Classification Second Edition. 7 pp.
ISO 720-85. Glass—Hydrolytic Resistance of Glass Grains at 121 Degrees C—Method of Test and Classification Second Edition. 7 pp.

—Iron Oxide Content—Atomic Absorption Spectrometry
BSI BS 7709: Part 5-93. 1993 Analysis of Extract Solutions of Glass Part 5: Method for Determination of Iron(III) Oxide by Molecular Absorption Spectrometry and Flame Atomic Absorption Spectrometry (ISO 10136-5: 1993) (V). 13 pp.
ISO 10136 Pt 5-93. Glass and Glassware—Analysis of Extract Solutions—Part 5: Determination of Iron(III) Oxide by Molecular Absorption Spectrometry and Flame Atomic Absorption Spectrometry First Edition. 10 pp.

INDUSTRY STANDARDS

Glass (Cont.)

—Iron Oxide Content—Molecular Absorption Spectrophotometry
BSI BS 7709: Part 5-93. 1993 Analysis of Extract Solutions of Glass Part 5: Method for Determination of Iron(III) Oxide by Molecular Absorption Spectrometry and Flame Atomic Absorption Spectrometry (ISO 10136-5: 1993) (V). 13 pp.
ISO 10136 Pt 5-93. Glass and Glassware—Analysis of Extract Solutions—Part 5: Determination of Iron(III) Oxide by Molecular Absorption Spectrometry and Flame Atomic Absorption Spectrometry First Edition. 10 pp.

—Isotropic—Bend Testing
BSI BS 7604: Part 2-92. 1992 Method for Determination of Stress-Optical Coefficient of Glass Part 2: Bending Test (ISO 10345-2: 1992). 11 pp.
ISO 10345 Pt 2-92. Glass—Determination of Stress-Optical Coefficient—Part 2: Bending Test First Edition. 8 pp.

—Isotropic—Tensile Testing
BSI BS 7604: Part 1-92. 1992 Method for Determination of Stress-Optical Coefficient of Glass Part 1: Tensile Test (ISO 10345-1: 1992). 10 pp.
ISO 10345 Pt 1-92. Glass—Determination of Stress-Optical Coefficient—Part 1: Tensile Test First Edition. 7 pp.

—Knoop Hardness Testing
CNS Z9068-86. Method of Test for Knoop Hardness of Optical Glasses (Dec)(9939).
ISO 9385-90. Glass and Glass-Ceramics —Knoop Hardness Test First Ediiton. 7 pp.

—Limestone
BSI BS 3108-80. 1980 Amd 1 Limestone for Making Colourless Glasses. 10 pp.

—Magnesium Oxide Content—Atomic Absorption Spectrometry
ISO 10136 Pt 3-93. Glass and Glassware—Analysis of Extract Solutions—Part 3: Determination of Calcium Oxide and Magnesium Oxide by Flame Atomic Absorption Spectrometry First Edition. 10 pp.

—Physical Properties
DIN ENGL 1249 Pt 10-90. Glass for Use in Building Construction; Chemical and Physical Properties (Aug). 4 pp.

—Potassium Oxide Content—Atomic Emission Spectrometry
ISO 10136 Pt 2-93. Glass and Glassware—Analysis of Extract Solutions—Part 2: Determination of Sodium Oxide and Potassium Oxide by Flame Spectrometric Methods First Edition. 10 pp.

—Powder—Blasting Caps
MOD UK DSTAN 13-76-91. Glass Powder Issue 2. 7 pp.

—Powder—Sieve Analysis
MOD UK M 809/91. Examination of Glass Powder.

—Powder—Visual Inspection
MOD UK M 809/91. Examination of Glass Powder.

—Powder—Water Extract—Alkalinity
MOD UK M 809/91. Examination of Glass Powder.

—Pressure Vessels
BSI BS 3463-75. 1975 Observation and Gauge Glasses for Pressure Vessels. 21 pp.

—Reflective
JIS R 3221-90. Solar Reflective Glass. 9 pp.

—Rods
CNS R2145-86. Glass Rods (Mar)(7321).
JIS R 3645-79. Glass Rods.

—Silica Content—Molecular Absorption Spectrophotometry
ISO 10136 Pt 1-93. Glass and Glassware—Analysis of Extract Solutions—Part 1: Determination of Silicon Dioxide by Molecular Absorption Spectrometry First Edition. 9 pp.

—Sodium Oxide Content—Atomic Emission Spectrometry
ISO 10136 Pt 2-93. Glass and Glassware—Analysis of Extract Solutions—Part 2: Determination of Sodium Oxide and Potassium Oxide by Flame Spectrometric Methods First Edition. 10 pp.

—Softening Points
BSI BS 7034: Part 6-88. 1988 Viscosity and Viscometric Fixed Points of Glass Part 6: Method for the Determination of Softening Point. 16 pp.
CNS R3126-85. Method of Test for Softening Point of Glass (Jun)(11293).
ISO 7884 Pt 6-87. Glass—Viscosity and Viscometric Fixed Points—Part 6: Determination of Softening Point First Edition. 8 pp.
JIS R 3104-70. Method of Test for Softening Point of Glass.

—Strain Points
BSI BS 7034: Part 7-88. 1988 Viscosity and Viscometric Fixed Points of Glass Part 7: Method for the Determination of Annealing Point and Strain Point by Beam Bending. 12 pp.
CNS R3125-85. Method of Test for Strain Point of Glass (Jun)(11292).
ISO 7884 Pt 7-87. Glass—Viscosity and Viscometric Fixed Points—Part 7: Determination of Annealing Point and Strain Point by Beam Bending First Edition. 12 pp.
JIS R 3103-78. Method of Test for Strain Point of Glass. 9 pp.

—Thermal Expansion
BSI BS 7030-88. 1988 Method for Determination of the Coefficient of Mean Linear Thermal Expansion of Glass. 11 pp.
BSI BS 7034: Part 8-88. 1988 Viscosity and Viscometric Fixed Points of Glass Part 8: Method for the Determination of (Dilatometric) Transformation Temperature. 8 pp.
CNS R3093-84. Method of Test for Average Linear Thermal Expansion of Glass (Aug)(6614).
DIN ENGL 52328-85. Testing of Glass; Determination of Mean Linear Thermal Expansion Coefficient (Mar). 4 pp.
ISO 7884 Pt 8-87. Glass—Viscosity and Viscometric Fixed Points—Part 8: Determination of (Dilatometric) Transformation Temperature First Edition. 7 pp.
ISO 7991-87. Glass—Determination of Coefficient of Mean Linear Thermal Expansion First Edition. 10 pp.
JIS R 3102-78. Testing Method for Average Linear Thermal Expansion of Glass.

—Thermal Expansion—Interferometry
JIS R 3251-90. Measuring Method of the Linear Thermal Expansion Coefficient for Low Expansion Glass by Laser Interferometry. 14 pp.

—Viscometric Analysis
BSI BS 7034: Part 1-88. 1988 Viscosity and Viscometric Fixed Points of Glass Part 1: Recommendations for Use in the Determination of Viscosity and Viscometric Fixed Points. 12 pp.
BSI BS 7034: Part 2-88. 1988 Viscosity and Viscometric Fixed Points of Glass Part 2: Method for the Determination of Viscosity by Rotation Viscometers. 14 pp.
BSI BS 7034: Part 4-88. 1988 Viscosity and Viscometric Fixed Points of Glass Part 4: Method for the Determination of Viscosity by Beam Bending. 16 pp.
BSI BS 7034: Part 5-88. 1988 Viscosity and Viscometric Fixed Points of Glass Part 5: Method for the Determination of Working Point by Sinking Bar Viscometer. 8 pp.
ISO 7884 Pt 1-87. Glass—Viscosity and Viscometric Fixed Points—Part 1: Principles for Determining Viscosity and Viscometric Fixed Points First Edition. 11 pp.
ISO 7884 Pt 2-87. Glass—Viscosity and Viscometric Fixed Points—Part 2: Determination of Viscosity by Rotation Viscometers First Edition. 12 pp.
ISO 7884 Pt 4-87. Glass—Viscosity and Viscometric Fixed Points—Part 4: Determination of Viscosity by Beam Bending First Edition. 15 pp.
ISO 7884 Pt 5-87. Glass—Viscosity and Viscometric Fixed Points—Part 5: Determination of Working Point by Sinking Bar Viscometer First Edition. 8 pp.

—Viscometric Analysis—Annealing Points
BSI BS 7034: Part 7-88. 1988 Viscosity and Viscometric Fixed Points of Glass Part 7: Method for the Determination of Annealing Point and Strain Point by Beam Bending. 12 pp.
ISO 7884 Pt 7-87. Glass—Viscosity and Viscometric Fixed Points—Part 7: Determination of Annealing Point and Strain Point by Beam Bending First Edition. 12 pp.

—Viscometric Analysis—Elongation
BSI BS 7034: Part 3-88. 1988 Viscosity and Viscometric Fixed Points of Glass Part 3: Method for the Determination of Viscosity by Fibre Elongation Viscometer (ISO 7884/3: 1987). 11 pp.
ISO 7884 Pt 3-87. Glass—Viscosity and Viscometric Fixed Points—Part 3: Determination of Viscosity by Fibre Elongation Viscometer First Edition. 11 pp.

—Viscometric Analysis—Softening Points
BSI BS 7034: Part 6-88. 1988 Viscosity and Viscometric Fixed Points of Glass Part 6: Method for the Determination of Softening Point. 16 pp.
ISO 7884 Pt 6-87. Glass—Viscosity and Viscometric Fixed Points—Part 6: Determination of Softening Point First Edition. 8 pp.

—Viscometric Analysis—Strain Points
BSI BS 7034: Part 7-88. 1988 Viscosity and Viscometric Fixed Points of Glass Part 7: Method for the Determination of Annealing Point and Strain Point by Beam Bending. 12 pp.
ISO 7884 Pt 7-87. Glass—Viscosity and Viscometric Fixed Points—Part 7: Determination of Annealing Point and Strain Point by Beam Bending First Edition. 12 pp.

—Viscometric Analysis—Transformation Temperature—Thermal Expansion
BSI BS 7034: Part 8-88. 1988 Viscosity and Viscometric Fixed Points of Glass Part 8: Method for the Determination of (Dilatometric) Transformation Temperature. 8 pp.
ISO 7884 Pt 8-87. Glass—Viscosity and Viscometric Fixed Points—Part 8: Determination of (Dilatometric) Transformation Temperature First Edition. 7 pp.

—Viscosity
BSI BS 7034: Part 6-88. 1988 Viscosity and Viscometric Fixed Points of Glass Part 6: Method for the Determination of Softening Point. 16 pp.
BSI BS 7034: Part 8-88. 1988 Viscosity and Viscometric Fixed Points of Glass Part 8: Method for the Determination of (Dilatometric) Transformation Temperature. 8 pp.

Glass Bead Content Analysis
See Also: Glass Beads; Glass Material Content Analysis

—Paints
CNS K6143-87. Method of Test for Traffic Paints (May)(1334). 19 pp.

Glass Beads
Use For: Beads, Glass See Also: Glass; Glass Bead Content Analysis
BSI BS 6088-81. 1981 Amd 3 Solid Glass Beads for Use with Road Marking Compounds and for Other Industrial Uses (AMD 5600) September 30, 1987. 19 pp.

—Reflective Materials
CGSB 1-GP-71 METH 149.1-78. Methods of Testing Paints and Pigments Imperfections in Glass Beads. 1 p.

—Roads—Marking
SAA AS 2009-91. Glass Beads for Road-Marking Materials Amdt 1 March 1992.

—Traffic Paints
CNS R2095-89. Glass Beads for Reflective Traffic Paint (Dec)(4342). 2 pp.
CNS R3080-89. Method of Test for Glass Beads for Reflective Traffic Paint (Dec) (4343). 3 pp.
JIS R 3301-87. Glass Beads for Traffic Paint. 9 pp.

Glass Blocks
See Also: Building Blocks; Construction Materials; Glass
CEN PREN 1051-93. Glass Blocks and Paver Units. 16 pp.
JIS A 5212-79. Hollow Glass Blocks.

—Quality Assurance
CEN PREN 1051-93. Glass Blocks and Paver Units. 16 pp.

Glass Capacitors
See Also: Ceramic Capacitors; Fixed Capacitors; Mica Capacitors; Variable Capacitors
MOD UK DEF-5138-41-01. Specification for Capacitors, Fixed, Glass or Porcelain Dielectric; Amendment 1. 32 pp.

—Preferred Products List
CECC CECC MUAHAG Vol 1 IS 4-90. Preferred Products List; Capacitors (En, Fr, Ge). 64 pp.

Glass Ceramics
Use For: Devitrified Glass See Also: Ceramics; Glass; Glassware

—Knoop Hardness Testing
ISO 9385-90. Glass and Glass-Ceramics —Knoop Hardness Test First Ediiton. 7 pp.

INTERNATIONAL AND NON-U.S. NATIONAL STANDARDS
SUBJECT INDEX

Glass

Glass Ceramics *(Cont.)*
—Sheets—Bend Testing
  DIN ENGL 52292 Pt 1-84. Testing of Glass and Glass Ceramics; Determination of Bending Strength; Coaxial Double Ring Bending Test on Flat Specimens with Small Test Area (Apr). 6 pp.

Glass Cloth
Use: Glass Fabrics

Glass Coatings (On Glass)
See Also: Coatings
  CGSB 1-GP-71 METH 98.2-74. Methods of Testing Paints and Pigments Coating of Metal and Glass Panels for Test Purposes Panels with One Coat of Material. 2 pp.
  CGSB 1-GP-71 METH 98.3-82. Methods of Testing Paints and Pigments Coating of Metal and Glass Panels for Test Purposes Panels with Two Coats of the Same Material. 2 pp.
  CGSB 1-GP-71 METH 98.4-74. Methods of Testing Paints and Pigments Coating of Metal and Glass Panels for Test Purposes Primed Panels. 1 p.

—Air Dried—Alkali Resistance Testing
  CGSB 1-GP-71 METH 106.1-79. Methods of Testing Paints and Pigments Alkali Resistance Air-Dried Films. 1 p.

—Antireflection—Display Devices—Aircraft Cabins
  NATO STANAG 3643 ED 2 AMD 7-73. Coating, Reflection Reducing for Glass Elements Used in Aircrew Station Displays. 6 pp.
  NATO STANAG 3643 ED 2 AMD 9-73. Coating, Reflection Reducing for Glass Elements Used in Aircrew Station Displays. 6 pp.

—Baked—Alkali Resistance Testing
  CGSB 1-GP-71 METH 106.2-79. Methods of Testing Paints and Pigments Alkali Resistance Baked Films. 1 p.

—Enamels—Acid Resistance Testing
  CGSB 1-GP-71 METH 105.2-79. Methods of Testing Paints and Pigments Acid Resistance Enamel Films on Glass Panels. 1 p.

—Paints—Test Panels
  CGSB 1-GP-71 METH 98.1-82. Methods of Testing Paints and Pigments Coating of Metal and Glass Panels for Test Purposes General Procedures for Applying Paints for the Preparation of Test Panels and Films. 2 pp.

—Paints—Washability
  CGSB 1-GP-71 METH 114.1-79. Methods of Testing Paints and Pigments Cleansability Paint Films on Unprimed Panels. 1 p.
  CGSB 1-GP-71 METH 114.2-74. Methods of Testing Paints and Pigments Cleansability Paint Films on Primed Panels. 1 p.

—Primers—Scrub Resistance
  CGSB 1-GP-71 METH 125.3-79. Methods of Testing Paints and Pigments Scrubbability Films Applied over a Primer. 1 p.

—Scrub Resistance
  CGSB 1-GP-71 METH 125.1-82. Methods of Testing Paints and Pigments Scrubbability Films Applied Directly over Glass (General Procedure). 1 p.

Glass Containers
See Also: Bottles; Containers; Glass Packaging; Glass Tubes; Test Tubes
  BSI BS 1133: Sec 18-91. 1991 Packaging Code Section 18: Packaging in Glass (H). 14 pp.
  BSI BS 4602-77. 1977 Amd 1 The Use of Metric Units in Specifications for Glass Containers and Finishes. 5 pp.
  ISO 8164-90. Glass Containers—520 ml Euro-Form Bottles—Dimensions First Edition. 4 pp.
  ISO 9058-92. Glass Containers—Tolerances First Edition. 5 pp.

—Capacity Measurement—Gravimetric Analysis
  ISO 8106-85. Glass Containers—Determination of Capacity by Gravimetric Method—Test Method First Edition. 4 pp.

—Caps (Lids)—Aluminum Foil—Dairy Products
  BSI BS 3313: Part 1-68. 1968 Amd 1 Aluminium Capping Foil and Strip for Dairy Product Containers Part 1: Aluminium Capping Foil for Glass Container. 10 pp.

—Compression Testing
  DIN ENGL 52346-82. Testing of Glass; Thrust Load Test on Containers; Testing by Attributes and by Variables (June). 2 pp.

Glass Containers *(Cont.)*
—Compression Testing *(Cont.)*
  ISO 8113-85. Glass Containers—Resistance to Vertical Load—Test Method First Edition. 4 pp.

—Continuous Threads
  BSI BS 1918: Part 1-78. 1978 Amd 2 Glass Container Finishes Part 1: Continuous Thread Finish. 14 pp.

—Corrosion Testing
  CNS R3031-73. Method of Test for Chemical Corrosion Resistance of Glass Containers (Nov)(996).

—Finishes—Heights
  ISO 9009-91. Glass Containers—Height and Non-Parallelism of Finish with Reference to Container Base—Test Methods First Edition. 7 pp.

—Food—Capacity Measurement
  BSI BS 5771-81. 1981 Packages for Certain Pre-Packed Foodstuffs; Capacities of Glass and Metal Containers. 9 pp.
  CEN EN 76-78. Packages for Certain Pre-Packed Foodstuffs: Capacities of Glass and Metal Containers. 6 pp.

—Food—Hygiene
  CNS Z6087-88. Method of Test for the Hygiene of Food Utensils, Containers and Packages Products (Jan)(12218).

—Glossaries
  BSI BS 3130: Part 3-74. 1974 Glossary of Packaging Terms Part 3: Glass Containers and Closures. 21 pp.
  ISO 7348-92. Glass Containers—Manufacture—Vocabulary First Edition. 245 pp.

—Hydrolytic Resistance—Atomic Emission Spectrometry
  BSI BS 3473: Sec 4.2-89. 1989 Chemical Resistance of Glass Used in the Pro-duction of Laboratory Glassware: Part 4: Methods for determination of Hydrolytic Resistance of the Interior Surfaces of Glass Containers: Section 4.2:. 13 pp.
  ISO 4802 Pt 2-88. Glassware—Hydrolytic Resistance of the Interior Surfaces of Glass Containers—Part 2: Determination by Flame Spectrometry and Classification First Edition. 11 pp.

—Hydrolytic Resistance—High Temperature Testing
  BSI BS 3473: Sec 4.1-89. 1989 Chemical Resistance of Glass Used in the Pro-duction of Laboratory Glassware: Part 4: Methods for Determination of hydrolytic Resistance of the Interior Surfaces of Glass Containers: Section 4.1:. 11 pp.

—Hydrolytic Resistance—Volumetric Analysis
  ISO 4802 Pt 1-88. Glassware—Hydrolytic Resistance of the Interior Surfaces of Glass Containers—Part 1: Determination by Titration Method and Classification First Edition. 11 pp.

—Impact Testing
  DIN ENGL 52295-83. Testing of Glass; Pendulum Impact Test on Containers; Inspection by Attributes and by Variables (Sept). 4 pp.

—Liquids—Pressurized—Tamper Evident
  ISO 9057-91. Glass Containers—28 mm Tamper-Evident Finish for Pressurized Liquids—Dimensions First Edition. 5 pp.

—Manufacturing—Glossaries
  ISO 7348-92. Glass Containers—Manufacture—Vocabulary First Edition. 245 pp.

—Mouths—Flatness
  ISO 9885-91. Wide-Mouth Glass Containers—Deviation from Flatness of Top Sealing Surface—Test Methods First Edition. 5 pp.

—Packaging—Vegetables
  CEN EN 76-78. Packages for Certain Pre-Packed Foodstuffs: Capacities of Glass and Metal Containers. 6 pp.

—Pressure Measurement
  CNS R3029-73. Method of Static Pressure Test for Glass Containers (Nov)(994).
  DIN ENGL 52320 Pt 1-78. Testing of Glass; Internal Pressure Test on Hollow Glassware, Especially Glass Containers; Inspection by Attributes (Nov). 2 pp.
  DIN ENGL 52320 Pt 2-81. Testing of Glass; Internal Pressure Test on Hollow Glassware, Especially Glass Containers; Inspection by Variables (Oct). 2 pp.
  ISO 7458-84. Glass Containers—Internal Pressure Resistance—Test Methods First Edition. 4 pp.

Glass Containers *(Cont.)*
—Sampling
  CNS R3028-73. Method of Sampling for Glass Containers (Nov) (993).

—Screw Threads
  CNS B2164-78. Finishes with External Screw Thread for Glass Containers (Mar)(4228).

—Screw Threads—Nominal Sizes
  DIN ENGL 168 Pt 1-79. Round Screw Threads; Especially for Glass Containers Screw Thread Sizes (Dec). 2 pp.

—Shape
  CNS Z6009-66. Method of Test for Off-Roundness of Glass for Food Storage (Mar)(2576)(R 1972).
  CNS Z6010-72. Method of Test for Perpendicularity of Glass Containers for Food Storage (Oct)(2577).

—Sodium Oxide Content
  CNS Z6008-73. Method of Test Extracted Sodium Oxide from Glass Containers for Food Storage (Jan)(2575).

—Stability Testing
  CNS Z6011-72. Method of Test for Stability of Glass Containers for Food Storage (Oct)(2578).

—Tamper Proof
  ISO 9056-90. Glass Containers—Series of Pilferproof Finish—Dimensions First Edition. 8 pp.

—Thermal Shock
  CNS R3030-73. Method of Thermal Shock Test for Glass Containers (Nov)(995).
  ISO 7459-84. Glass Containers—Thermal Shock Resistance and Thermal Shock Endurance—Test Methods First Edition. 4 pp.

—Vacuum Lug Finishes
  ISO 9100-92. Wide-Mouth Glass Containers—Vacuum Lug Finishes—Dimensions First Edition. 8 pp.

Glass Crystal
Use: Lead Glass

Glass Fabrics
Use For: Glass Cloth; Glass Textiles; Textile Glass Fabrics *See Also:* Fabrics; Glass; Rovings
  BSI BS 3749-91. 1991 E Glass Fibre Woven Roving Fabrics for the Reinforcement of Polyester and Epoxy Resin Systems. 12 pp.
  BSI BS 3749-74. 1974 Woven Roving Fabrics of E Glass Fibre for the Reinforcement of Polyester Resin Systems. 8 pp.
  BSI BS 4045-66. 1966 Epoxide Resin Pre-Impregnated Glass Fibre Fabrics. 12 pp.
  CNS K3063-82. Glass Clothes (Dec)(9717). 7 pp.
  CNS K3064-82. Finished Glass Clothes (Dec)(9718). 4 pp.
  ISO 2113-81. Textile Glass—Woven Fabrics—Basis for Specification Second Edition. 7 pp.
  ISO 5025-78. Textile Glass—Woven Fabric—Determination of Width and Length First Edition. 3 pp.
  JIS R 3414-92. Textile Glass Fabrics. 11 pp.
  JIS R 3416-92. Finished Textile Glass Fabrics. 8 pp.
  JIS R 3420-89. Testing Methods for Textile Glass Products. 38 pp.

—Aerospace—Filament Yarns—Sampling
  CEN PREN 3675-90. Sampling Plan for Acceptance Testing of Aramid, Carbon Fibre and Textile Glass Filament Yarns. 8 pp.
  CEN PREN 3675-92. Sampling Plan for Acceptance Testing of Aramid, Carbon Fibre and Textile Glass Filament Yarns. 7 pp.

—Aerospace—Mass
  AECMA PREN2329-83. Test Method for the Determination of Mass per Unit Area of Woven Textile Glass Fibre Fabric Preimpregnate. 6 pp.
  BSI BS EN 2329-93. 1993 Textile Glass Fibre Preimpregnates—Test Method for the Determination of Mass per Unit Area (S). 12 pp.
  CEN PREN 2329-91. Textile Glass Fibre Preimpregnates Test Method for the Determination of Mass Per Unit Area. 8 pp.
  CEN EN 2329-93. Aerospace Series—Textile Glass Fibre Preimpregnates—Test Method for the Determination of Mass per Unit Area (S). 7 pp.

—Aerospace—Resin Content
  AECMA PREN2331-83. Test Methods for the Determination of the Resin Content of Woven Textile Glass Fibre Fabric Preimpregnate. 9 pp.
  BSI BS EN 2331-93. 1993 Textile Glass Fibre Preimpregnates—Test Method for the Determination of the Resin and Fibre Content and Mass of Fibre Per Unit Area (S). 18 pp.

INDUSTRY STANDARDS

Glass Fabrics (Cont.)

—Aerospace—Resin Content (Cont.)
CEN PREN 2331-91. Textile Glass Fibre Preimpregnates Test Method for the Determination of the Resin and Fibre Content and Mass of Fibre Per Unit Area. 14 pp.
CEN EN 2331-93. Aerospace Series—Textile Glass Fibre Preimpregnates—Test Method for the Determination of the Resin and Fibre Content and Mass of Fibre per Unit Area. 13 pp.

—Aerospace—Resin Flow
AECMA PREN2332-83. Test Method for the Determination of the Resin Flow of Woven Textile Glass Fibre Preimpregnate. 7 pp.
BSI BS EN 2332-93. 1993 Textile Glass Fibre Preimpregnates—Test Method for the Determination of the Resin Flow (S). 13 pp.
CEN PREN 2332-91. Textile Glass Fibre Preimpregnates Test Method for the Determination of the Resin Flow. 9 pp.
CEN EN 2332-93. Aerospace Series—Textile Glass Fibre Preimpregnates—Test Method for the Determination of the Resin Flow. 8 pp.

—Aerospace—Volatile Matter Content
AECMA PREN2330-83. Test Method for the Determination of the Percentage of Volatile Matter in Woven Textile Glass Fibre Fabric Preimpregnate. 7 pp.
BSI BS EN 2330-93. 1993 Textile Glass Fibre Preimpregnates—Test Method for the Determination of the Content of Volatile Matter (S). 13 pp.
CEN PREN 2330-91. Textile Glass Fibre Preimpregnates Test Method for the Determination of the Percentage of Volatile Matter. 9 pp.
CEN EN 2330-93. Aerospace Series—Textile Glass Fibre Preimpregnates—Test Method for the Determination of the Content of Volatile Matter. 8 pp.

—Aircraft
MOD UK DTD-5559A-64. Contour Woven Glass Fibre Fabric. 2 pp.

—Bag Filters
JIS R 3421-91. Textile Finished Glass Fabrics for Bag Filter. 7 pp.

—Breaking Load
BSI BS 3396: Part 2-87. 1987 Woven Glass Fibre Fabrics for Plastics Reinforcement Part 2: Desized Fabrics. 4 pp.
BSI BS 3396: Part 3-87. 1987 Woven Glass Fibre Fabrics for Plastics Reinforcement Part 3: Finished Fabrics for Use with Polyester Resin Systems. 8 pp.
ISO 4606-79. Textile Glass—Woven Fabric—Determination of Tensile Breaking Force and Breaking Elongation by the Strip Method First Edition. 7 pp.

—Combustible Matter Content
ISO 1887-80. Textile Glass—Determination of Combustible Matter Content Second Edition. 5 pp.

—Density
ISO 4602-78. Textile Glass—Woven Fabrics—Determination of Number of Yarns per Unit Length of Warp and Weft First Edition. 4 pp.
ISO 4605-78. Textile Glass—Woven Fabrics—Determination of Mass per Unit Area First Edition. 4 pp.

—Draperies
JIS R 3418-91. Textile Glass Fabric Curtains for Casement and Drapery. 7 pp.

—Elongation
ISO 4606-79. Textile Glass—Woven Fabric—Determination of Tensile Breaking Force and Breaking Elongation by the Strip Method First Edition. 7 pp.

—Fiberglass Reinforced Plastics
DIN ENGL 61854 Pt 2-87. Textile Glass; Woven Glass Fabrics for Plastics Reinforcement; Woven Glass Filament Fabric and Woven Roving; Types (Apr). 4 pp.

—Filters
MOD UK TS 10066B. Paper, Filter, Glass Fibre (Water Repellant Treated).

—Glossaries
JIS R 3410-87. Glossary of Terms Relating to Textile Glass. 33 pp.

—Mats
BSI BS 3496-89. 1989 E Glass Fibre Chopped Strand Mat for Reinforcement of Polyester and Other Liquid Laminating Systems. 12 pp.
CNS K3053-81. Fiber Glass Chopped Strand Mats (May)(7398). 3 pp.

Glass Fabrics (Cont.)

—Mats (Cont.)
DIN ENGL 61853 Pt 1-87. Textile Glass; Textile Glass Mats for Plastics Reinforcement; Technical Delivery Conditions (Apr) (Together with DIN 61853 Part 2, April 1987 Edition, Supersedes DIN 61853, June 1975 Edition). 7 pp.
ISO 2559-91. Textile Glass—Mats (Made from Chopped or Continuous Strands)—Basis for a Specification Third Edition. 10 pp.
JIS R 3411-91. Textile Glass Chopped Strand Mats. 8 pp.

—Mats—Air Permeability
CNS K6706-82. Method of Test of Glass Fiber Mat (Jan)(8428). 2 pp.

—Mats—Binder Dissolution
ISO 2558-74. Textile Glass Chopped-Strand Mats for Reinforcement of Plastics—Determination of Time of Dissolution of the Binder in Styrene First Edition. 5 pp.

—Mats—Bursting Strength
CNS K6706-82. Method of Test of Glass Fiber Mat (Jan)(8428). 2 pp.

—Mats—Classification
DIN ENGL 61853 Pt 2-87. Textile Glass; Textile Glass Mats for Plastics Reinforcement; Classification and Application (Apr) (Together with DIN 61853 Part 1, April 1987 Edition, Supersedes DIN 61853, June 1975 Edition). 10 pp.

—Mats—Density
ISO 3374-90. Textile Glass Mats—Determination of Mass per Unit Area Second Edition. 6 pp.

—Mats—Floor Coverings
CNS K3059-82. Glass Fiber Mat for Floor-Covering Substrate (Jan)(8426). 2 pp.

—Mats—Moldability
ISO 4900-90. Textile Glass—Mats and Fabrics—Determination of Contact Mouldability First Edition. 6 pp.

—Mats—Pipe Insulation
CNS K3058-82. Glass Fiber Mat for Steel Pipe Over Wrapping (Jan)(8425). 1 p.

—Mats—Printed Circuit Base Materials
CNS K3060-82. Glass Fiber Mat for Printed Circuit Board (Jan)(8427). 1 p.

—Mats—Tensile Testing
CNS K6706-82. Method of Test of Glass Fiber Mat (Jan)(8428). 2 pp.
ISO 3342-87. Textile Glass—Mats—Determination of Tensile Breaking Force Second Edition. 5 pp.

—Mats—Thickness Measurement
CNS K6706-82. Method of Test of Glass Fiber Mat (Jan)(8428). 2 pp.
ISO 3616-77. Textile Glass—Mats—Determination of Average Thickness, Thickness Under Load and Recovery After Compression First Edition. 6 pp.

—Mats—Weight Measurement
CNS K6706-82. Method of Test of Glass Fiber Mat (Jan)(8428). 2 pp.

—Membranes
JIS R 3424-90. Glass Fabrics for Membrane Structure. 9 pp.

—Moisture Content
ISO 3344-77. Textile Glass Products—Determination of Moisture Content First Edition. 5 pp.

—Moldability
ISO 4900-90. Textile Glass—Mats and Fabrics—Determination of Contact Mouldability First Edition. 6 pp.

—Prepreg—Aerospace—Fiber Content
CEN PREN 3783-90. Procedure for the Normalisation of Test Results of Fibre Dominated Composite Mechanical Properties. 6 pp.

—Printed Circuit Base Materials
BSI BS 4584: Part 13-77. 1977 Metal-Clad Base Materials for Printed Circuits Part 13: Silicone Woven Glass Fabric Copper-Clad Laminated Sheet Si-GC-Cu-13. 7 pp.
BSI BS 4584: Sec 102.4-90. 1990 Metal-Clad Base Materials for Printed Circuits Part 102: Copper-Clad Base Materials Section 102.4: Epoxide Woven Glass Fabric Copper-Clad Laminated Sheet, General Purpose Grade. 12 pp.
BSI BS 4584: Sec 102.5-90. 1990 Metal-Clad Base Materials for Printed Circuits Part 102: Copper-Clad Base Materials Section 102.5: Epoxide Woven Glass Fabric Copper-Clad Laminated Sheet of Defined Flammability (Vertical Burning Test). 14 pp.

Glass Fabrics (Cont.)

—Printed Circuit Base Materials (Cont.)
BSI BS 4584: Sec 102.11-90. 1990 Metal-Clad Base Materials for Printed Circuits Part 102: Copper-Clad Base Materials Section 102. 11: Thin Epoxide Woven Glass Fabric Copper-Clad Laminated Sheet, General Purpose Grade, for Use in the Fabrication of Multilayer Prtd Boards. 10 pp.
BSI BS 4584:Sec 102.11-01. 1990 Amd 1 Metal-Clad Base Materials for Printed Wiring Boards Part 102: Copper-Clad Base Materials Section 102.11: Specification for Thin Epoxide Woven Glass Fabric Copper-Clad Laminated Sheet, General Purpose Grade, for Use. 13 pp.
BSI BS 4584: Sec 102.12-90. 1990 Metal-Clad Base Materials for Printed Circuits Part 102: Copper-Clad Base Materials Section 102. 12: Thin Epoxide Woven Glass Fabric Copper-Clad Lamin. Sheet of Defined Flammability, for Use in the Fabrication of Multilayer Prtd Boards. 10 pp.
BSI BS 4584:Sec 102.12-01. 1990 Amd 1 Metal-Clad Base Materials for Printed Wiring Boards Part 102: Copper-Clad Base Materials Section 102.12: Specification for Thin Epoxide Woven Glass Fabric Copper-Clad Laminated Sheet of Defined Flammability,. 12 pp.
BSI BS 4584: Sec 102.16-92. 1992 Metal-Clad Base Materials for Printed Wiring Boards Part 102: Copper-Clad Base Materials Sec 102.16: Spec for Polyimide Woven Glass Fabric Copper-Clad Laminated Sheet of Defined Flammability (Vertical Burning Test) (IEC 249-2-16: 1992). 19 pp.
BSI BS 4584: Sec 102.17-92. 1992 Metal-Clad Base Materials for Printed Wiring Boards Part 102: Copper-Clad Base Materials Section 102.17: Specification for Thin Polyimide Woven Glass Fabric Copper-Clad Laminated Sheet of Defined Flammability for Use in. 16 pp.
BSI BS 4584: Sec 102.18-92. 1992 Metal-Clad Base Materials for Printed Wiring Boards Part 102: Copper-Clad Base Materials Section 102.18: Specification for Bismaleimide/Triazine Modified Epoxide Woven Glass Fabric Copper-Clad Laminated Sheet of. 19 pp.
BSI BS 4584: Sec 102.19-92. 1992 Metal-Clad Base Materials for Printed Wiring Boards Part 102: Copper-Clad Base Materials Section 102. 19: Spec. for Thin Bismaleimide/Triazine Modified Epoxide Glass Fabric Copper-Clad Laminated Sheets of Defined Flammability for. 16 pp.
BSI BS 4584: Sec 103.1-90. 1990 Metal-Clad Base Materials for Printed Circuits Part 103: Materials Used in Connection with Printed Circuits Section 103.1: Prepreg for Use as a Bonding Sheet Material in the Fabrication of Multilayer Prtd Boards. 12 pp.
CENELEC HD 313.2.3 S2-90. Base Materials for Printed Circuits Part 2: Specifications Specification No.3: Epoxide Cellulose Paper Copper-Clad Laminated Sheet of Defined Flammability (Vertical Burning Test). 3 pp.
CENELEC HD 313.2.4 S2-90. Base Materials for Printed Circuits Part 2: Specifications Specification No.4: Epoxide Woven Glass Fabric Copper-Clad Laminated Sheet, General Purpose Grade. 3 pp.
CENELEC HD 313.2.5 S2-90. Base Materials for Printed Circuits Part 2: Specifications Specification No.5: Epoxide Woven Glass Fabric Copper-Clad Laminated Sheet of Defined Flammability (Vertical Burning Test). 3 pp.
CENELEC HD 313.2.11 S1-89. Base Materials for Printed Circuits-Part 2: Specifications Specification No 11: Thin Epoxide Woven Glass Fabric Copper-Clad Laminated Sheet, General Purpose Grade for Use in the Fabrication of Multilayer Printed Boards. 3 pp.
CENELEC HD 313.2.12 S1-89. Base Materials for Printed Circuits-Part 2: Specifications Specification No 12: Thin Epoxide Woven Glass Fabric Copper-Clad Laminated Sheet of Defined Flammability for Use in the Fabrication of Multilayer Printed Boards. 3 pp.
CENELEC HD 313.2.12 S2-90. Base Materials for Printed Circuits Part 2: Specifications Specification No. 12: Thin Epoxide Woven Glass Fabric Copper-Clad Laminated Sheet of Defined Flammability, for Use in the Fabrication of Multilayer Printed Boards. 3 pp.
IEC 249 Pt 2-4-87. Base Materials for Printed Circuits Part 2: Specifications Specification No. 4: Epoxide Woven Glass Fabric Copper-Clad Laminated Sheet, General Purpose Grade Second Edition; (Amendment 3-1993, Incorporating Amendment 1 and 2). 39 pp.
IEC 249 Pt 2-5-87. Base Materials for Printed Circuits Part 2: Specifications Specification No. 5: Epoxide Woven Glass Fabric Copper-Clad Laminated Sheet of Defined Flammability (Vertical Burning Test) Second Edition; (Amd. 3-1993, Incorporating Amendment 1 and 2). 43 pp.

INTERNATIONAL AND NON-U.S. NATIONAL STANDARDS
SUBJECT INDEX
Glass

Glass Fabrics (Cont.)
—Printed Circuit Base Materials (Cont.)
IEC 249 Pt 2-11-87. Base Materials for Printed Circuits Part 2: Specifications Spec. No. 11: Thin Epoxide Woven Glass Fabric Copper-Clad Laminated Sheet, General Purpose Grade, for Use in the Fabrication of Multilayer Printed Boards First Edition (Amendment 2-1993, Incorporating Amd 1). 33 pp.

IEC 249 Pt 2-12-87. Base Materials for Printed Circuits Part 2: Specifications Spec. No. 12: Thin Epoxide Woven Glass Fabric Copper-Clad Laminated Sheet of Defined Flammability, for Use in the Fabrication of Multi-layer Printed Boards Second Ed.; (Amd 2-1993, Incorporating Amd 1). 35 pp.

IEC 249 Pt 2-16-92. Base Materials for Printed Circuits Part 2: Specifications Specification No. 16: Polyimide Woven Glass Fabric Copper-Clad Laminated Sheet of Defined Flammability (Vertical Burning Test) First Edition; (Amendment 1-1993). 41 pp.

IEC 249 Pt 2-17-92. Base Materials for Printed Circuits Part 2: Specifications Spec. No. 17: Thin Polyimide Woven Glass Fabric Copper-Clad Laminated Sheet of Defined Flammability for Use in the Fabrication of Multilayer Printed Board First Ed.; (Cor—June 1992) (Amd 1-1993). 40 pp.

IEC 249 Pt 2-18-92. Base Materials for Printed Circuits Part 2: Specifications Specification No. 18: Bismaleimide/Triazine Modified Epoxide Woven Glass Fabric Copper-Clad Laminated Sheet of Defined Flammability (Vertical Burning Test) First Edition; (Amendment 1-1993). 41 pp.

IEC 249 Pt 2-19-92. Base Materials for Printed Circuits Part 2: Specifications Specification No. 19: Thin Bismaleimide/Triazine Modified Epoxide Woven Glass Fabric Copper-Clad Laminated Sheet of Defined Flammability for Use in the Fabrication of Multilayer Printed. 37 pp.

IEC 249 Pt 3-1-81. Base Materials for Printed Circuits Part 3: Special Materials Used in Connection with Printed Circuits Specification No. 1: Prepreg for Use as Bonding Sheet Material in the Fabrication of Multilayer Printed Boards Second Edition. 23 pp.

JIS C 6522-90. Prepreg for Multilayer Printed Wiring Boards (Epoxy Resin-Impregnated Glass Cloth). 10 pp.

JIS R 3423-87. Finished Glass Fabrics Used for Copper Clad Laminates and Other Electric Purposes.

MOD UK DSTAN 59-50-01. Requirements for Plastics Sheet, Laminated Copper Clad, Epoxide Resin Bonded, Woven Glass Fabric Base-Fire-Retardant (Metal Clad Base Materials for Printed Circuits) Issue 1; Amendment 1. 18 pp.

—Resin Content
BSI BS 2782:Pt10: METH 1006-78. 1978 Methods of Testing Plastics Part 10: Glass Reinforced Plastics Method 1006: Determination of Volatile Matter and Resin Content of Synthetic Resin-Impregnated Textile Glass Fibre. 2 pp.

—Resins—Viscosity
BSI BS 2782:Pt7: METH 721A-88. 1988 Methods of Testing Plastics Part 7: Rheological Properties Method 721A: Determination of Resin Flow from Resin Impregnated Glass Fabric. 4 pp.

—Rovings
ISO 2797-86. Textile Glass—Rovings—Basis for a Specification Second Edition. 5 pp.
JIS R 3412-91. Textile Glass Rovings. 7 pp.
JIS R 3417-84. Woven Roving Glass Fabrics. 7 pp.

—Sampling
ISO 1886-90. Reinforcement Fibres—Sampling Plans Applicable to Received Batches Third Edition. 13 pp.

—Sheets—Thickness Measurement
BSI BS 3953-90. 1990 Synthetic Resin Bonded Woven Glass Fabric Laminated Sheet (V). 19 pp.
BSI BS 3953-01. 1990 Amd 1 Synthetic Resin Bonded Woven Glass Fabric Laminated Sheet (AMD 6967) February 28, 1992 (V). 20 pp.

—Stiffness
ISO 4604-78. Textile Glass—Woven Fabrics—Determination of Conventional Flexural Stiffness—Fixed-Angle Flexometer Method First Edition. 5 pp.

—Thermal Insulation—Marine
MOD UK NES 801: Part 2-91. Requirements for Insulation Material Part 2: Glass-Fibre Products Double-Woven Glass Cloth Issue 2 (03.91). 13 pp.
MOD UK NES 801: Part 4-89. Requirements for Insulation Material Part 4: Glass Fibre Products Rewettable Adhesive Impregnated Glass Cloth Issue 2 (04.89). 12 pp.

Glass Fabrics (Cont.)
—Thickness Measurement
ISO 4603-78. Textile Glass—Woven Fabrics—Determination of Thickness First Edition. 4 pp.

—Volatile Matter Content
BSI BS 2782:Pt10: METH 1006-78. 1978 Methods of Testing Plastics Part 10: Glass Reinforced Plastics Method 1006: Determination of Volatile Matter and Resin Content of Synthetic Resin-Impregnated Textile Glass Fibre. 2 pp.

Glass Fiber Content Analysis
See Also: Fiber Content Analysis

—Aerospace
CEN PREN 3783-90. Procedure for the Normalisation of Test Results of Fibre Dominated Composite Mechanical Properties. 6 pp.

—Fiberglass Reinforced Plastics
CNS K6976-90. Method of Test for Fiber Content of Glass Fiber Reinforced Plastics (Sep)(12777).
JIS K 7052-87. Measuring Method for Fiber Content of Glass Fiber Reinforced Plastics. 8 pp.

Glass Fiber Reinforced Plastics
Use: Fiberglass Reinforced Plastics

Glass Fibre Reinforced Cements
Use: Fiberglass Reinforced Cements

Glass Joints
See Also: Joints; Laboratory Ware

—Conical
BSI BS 572-85. 1985 Amd 1 Interchangeable Conical Ground Glass Joints. 11 pp.
BSI BS 6352-83. 1983 Flasks with Ground Glass Joints. 7 pp.
CNS R2144-86. Interchangeable Ground Glass Joints for Chemical Analysis (Mar) (7320).
CNS R3090-80. Suitable Gauging System for Diameter and Length of Conical Joints (Apr) (5480).
CNS R4002-80. Interchangeable Conical Ground Joints for Laboratory Glassware (Apr)(5479).
ISO 383-76. Laboratory Glassware—Interchangeable Conical Ground Joints First Edition. 7 pp.
ISO 4797-81. Laboratory Glassware—Flasks with Conical Ground Joints First Edition. 4 pp.
JIS R 3646-81. Interchangeable Conical Ground Glass Joints (Full Length). 9 pp.
JIS R 3647-81. Interchangeable Conical Ground Glass Joints (Medium Length) (R 1987). 9 pp.
JIS R 3648-81. Interchangeable Conical Ground Glass Joints (Short Length).
JIS R 3649-81. Interchangeable Conical Ground Glass Joints for Volumetric Flask.
JIS R 3650-81. Interchangeable Conical Ground Glass Joints for Narrow Mouth Bottle.

—Conical—Leakage
CNS R3091-80. Leakage Test for Conical Joints (Apr) (5481).

—Spherical
BSI BS 2761-83. 1983 Spherical Ground Glass Joints. 8 pp.
CNS Z7096-81. Clamps for Spherical-Ground Joints (Nov)(8173).
CNS Z7097-81. Laboratory Glassware; Connections with Spherical Joint (Nov)(8174).
ISO 641-75. Laboratory Glassware—Interchangeable Spherical Ground Joints First Edition. 7 pp.
JIS R 3651-81. Interchangeable Spherical Glass Joints.

Glass Laminates
Use: Laminated Glass

Glass Material Content Analysis
See Also: Glass Bead Content Analysis

—Slags
CNS A3300-88. Method of Determination of Glass Material Content in Granulate (Water Quenched) Blast-Furnace Slag (Nov)(12458).

Glass Packaging
See Also: Glass Containers; Glassware; Packaging

—Continuous Thread Finishes
BSI BS 1918: Part 1-78. 1978 Amd 2 Glass Container Finishes Part 1: Continuous Thread Finish. 14 pp.
DIN ENGL 6094 Pt 12-83. Packaging; Finishes; 7,5 R Continuous Thread Finishes for Bottles Subjected to Internal Pressure (Jan). 2 pp.

—Crown Finishes
BSI BS 1918: Part 2-81. 1981 Glass Container Finishes Part 2: Crown Finish. 4 pp.
DIN ENGL 6094 Pt 1-82. Packaging; Finishes; Crown Finishes (Oct). 2 pp.

Glass Packaging (Cont.)
—Finishes
BSI BS 4602-77. 1977 Amd 1 The Use of Metric Units in Specifications for Glass Containers and Finishes. 5 pp.

—Glossaries
ISO 7348-92. Glass Containers—Manufacture—Vocabulary First Edition. 245 pp.

—Vacuum Lug Finishes
DIN ENGL 6094 Pt 11-82. Packaging; Finishes; Vacuum Lug Type Closure Finish (Nov). 5 pp.

Glass Paper
See Also: Abrasive Sheets; Abrasives; Papers
SAA AS B292-69. Emery Cloth and Glass Paper. 12 pp.

Glass Plants
See Also: Industrial Plants

—Fittings
BSI BS 2598: Part 2-80. 1980 Glass Plant, Pipeline and Fittings Part 2: Testing, Handling and Use. 9 pp.
ISO 3586-76. Glass Plant, Pipeline and Fittings—General Rules for Testing, Handling and Use First Edition. 6 pp.

—Fittings—Borosilicate
BSI BS 2598: Part 1-80. 1980 Glass Plant, Pipeline and Fittings Part 1: Properties of Borosilicate Glass 3.3. 4 pp.
BSI BS 2598: Part 3-80. 1980 Glass Plant, Pipeline and Fittings Part 3: Pipeline and Fittings of Nominal Bore 15 to 150 mm: Compatibility and Interchangeability. 21 pp.
BSI BS 2598: Part 4-80. 1980 Glass Plant, Pipeline and Fittings Part 4: Glass Plant Components. 18 pp.
ISO 3587-76. Glass Plant, Pipeline and Fittings—Pipeline and Fittings of Nominal Bore 15 to 150 mm—Compatibility and Interchangeability First Edition; (Erratum—Aug 1979). 20 pp.
ISO 4704-77. Glass Plant, Pipeline and Fittings—Glass Plant Components First Edition; (Erratum—Aug 1979). 18 pp.

—Pipelines
BSI BS 2598: Part 1-91. 1991 Glass Plant, Pipeline and Fittings Part 1: Specification for Properties of Borosilicate Glass 3.3 (ISO 3585: 1991). 8 pp.
BSI BS 2598: Part 1-80. 1980 Glass Plant, Pipeline and Fittings Part 1: Properties of Borosilicate Glass 3.3. 4 pp.
BSI BS 2598: Part 2-80. 1980 Glass Plant, Pipeline and Fittings Part 2: Testing, Handling and Use. 9 pp.
BSI BS 2598: Part 3-80. 1980 Glass Plant, Pipeline and Fittings Part 3: Pipeline and Fittings of Nominal Bore 15 to 150 mm: Compatibility and Interchangeability. 21 pp.
BSI BS 2598: Part 4-80. 1980 Glass Plant, Pipeline and Fittings Part 4: Glass Plant Components. 18 pp.
ISO 3585-91. Borosilicate Glass 3.3—Properties Second Edition. 7 pp.
ISO 3586-76. Glass Plant, Pipeline and Fittings—General Rules for Testing, Handling and Use First Edition. 6 pp.
ISO 3587-76. Glass Plant, Pipeline and Fittings—Pipeline and Fittings of Nominal Bore 15 to 150 mm—Compatibility and Interchangeability First Edition; (Erratum—Aug 1979). 20 pp.
ISO 4704-77. Glass Plant, Pipeline and Fittings—Glass Plant Components First Edition; (Erratum—Aug 1979). 18 pp.

—Pipelines—Borosilicate
BSI BS 2598: Part 1-91. 1991 Glass Plant, Pipeline and Fittings Part 1: Specification for Properties of Borosilicate Glass 3.3 (ISO 3585: 1991). 8 pp.
BSI BS 2598: Part 1-80. 1980 Glass Plant, Pipeline and Fittings Part 1: Properties of Borosilicate Glass 3.3. 4 pp.
BSI BS 2598: Part 3-80. 1980 Glass Plant, Pipeline and Fittings Part 3: Pipeline and Fittings of Nominal Bore 15 to 150 mm: Compatibility and Interchangeability. 21 pp.
BSI BS 2598: Part 4-80. 1980 Glass Plant, Pipeline and Fittings Part 4: Glass Plant Components. 18 pp.
ISO 3587-76. Glass Plant, Pipeline and Fittings—Pipeline and Fittings of Nominal Bore 15 to 150 mm—Compatibility and Interchangeability First Edition; (Erratum—Aug 1979). 20 pp.
ISO 4704-77. Glass Plant, Pipeline and Fittings—Glass Plant Components First Edition; (Erratum—Aug 1979). 18 pp.

Glass Sands
See Also: Sands

Glass Sands (Cont.)

—Physical Properties
BSI BS 2975-88. 1988 Methods for Sampling and Analysis of Glass-Making Sands. 16 pp.

—Sampling
BSI BS 2975-88. 1988 Methods for Sampling and Analysis of Glass-Making Sands. 16 pp.

Glass Seals
See Also: Oil Seals; Seals

—Glass to Glass—Stresses
BSI BS 7603-92. 1992 Method for Determination of Stresses in Glass-to-Glass Sealings (ISO 4790: 1992) (H). 11 pp.
ISO 4790-92. Glass-to-Glass Sealings-Determination of Stresses First Edition. 8 pp.

Glass Textiles
Use: Glass Fabrics

Glass Transition Temperature
See Also: Temperature

—Electrical Insulation
BSI BS EN 61006-93. 1993 Methods of Test for Determination of the Glass Transition Temperature of Electrical Insulating Materials (IEC 1006: 1991) (F). 29 pp.
CENELEC EN 61006-93. Methods of Test for the Determination of the Glass Transition Temperature of Electrical Insulating Materials (IEC 1006: 1991). 24 pp.
IEC 1006-91. Methods of Test for the Determination of the Glass Transition Temperature of Electrical Insulating Materials First Edition; (CENELEC EN 61006:1993). 50 pp.

Glass Tubes
See Also: Combustion Tubes; Glass Containers; Test Tubes; U Tubes
CNS R2146-86. Glass Tubes (Mar)(7322).
JIS R 3644-79. Glass Tubes.

—Borosilicate—Laboratory
BSI BS 5895-80. 1980 Amd 1 Borosilicate Glass Tubing for Laboratory Apparatus. 5 pp.
BSI BS EN 28362-1-93. 1993 Injection Containers for Inectables and Accessories Part 1: Injection Vials Made of Glass Tubing (ISO 8362-1: 1989) (N). 12 pp.
CEN EN 28362-1-93. Injection Containers for Injectables and Accessories—Part 1. Injectable Vials Made of Glass Tubing (ISO 8362-1: 1989). 7 pp.
ISO 4803-78. Laboratory Glassware—Borosilicate Glass Tubing First Edition. 4 pp.
ISO 8362 Pt 1-89. Injection Containers for Injectables and Accessories—Part 1: Injection Vials Made of Glass Tubing First Edition; (CEN EN 28362-1: 1993). 8 pp.

—Fluorescent Lighting
CNS C3111-85. Method of Test for Glass Tubing for Fluorescent Lamps (Jul)(7007).
CNS C4296-85. Glass Tubing for Fluorescent Lamps (May)(7006).
JIS C 7708-84. Glass Tubing for Fluorescent Lamps. 12 pp.

Glass Wool
See Also: Fibers; Rock Wool
CNS K7239-67. Chemical Reagent (Glass Wool) (Oct)(1738).
JIS K 8251-61. Glass Wool.

—Acoustic Insulation
CNS A2140-82. Glass Wool Acoustic Materials (Jul)(9057).
CNS A3166-82. Method of Test for Glass Wool Acoustic Materials (Jul)(9058).
JIS A 6306-91. Glass Wool Acoustic Materials. 20 pp.

—Acoustic Insulation—Floating Floors
CNS A2167-83. Glass Wool Isolating Material for Floating Floors (Nov)(10638).
JIS A 6322-79. Glass Wool Isolating Material for Floating Floors (R 1983). 12 pp.

—Belts—Thermal Insulation
CNS R2059-85. Glass Fiber Heat Insulation Materials (May)(3065). 10 pp.
JIS A 9505-89. Thermal Insulation Material Made of Glass Wool. 13 pp.

—Blanket Insulation—Thermal
CNS R2059-85. Glass Fiber Heat Insulation Materials (May)(3065). 10 pp.
JIS A 9505-89. Thermal Insulation Material Made of Glass Wool. 13 pp.

—Insulating Board—Thermal
CNS R2059-85. Glass Fiber Heat Insulation Materials (May)(3065). 10 pp.

Glass Wool (Cont.)

—Insulating Board—Thermal (Cont.)
JIS A 9505-89. Thermal Insulation Material Made of Glass Wool. 13 pp.

—Molds (Castings)—Thermal
JIS A 9505-89. Thermal Insulation Material Made of Glass Wool. 13 pp.

—Pipe Insulation—Thermal
CNS R2059-85. Glass Fiber Heat Insulation Materials (May)(3065). 10 pp.

—Thermal Insulation—Buildings
CNS A2211-87. Glass Wool Thermal Insulation for Dwellings (Sep)(12055).
CNS A3264-87. Method of Test for Glass Wool Thermal Insulation for Dwellings (Sep)(12056).
JIS A 9522-85. Glass Wool Thermal Insulation Material for Dwellings.
JIS A 9522-81. Glass Wool Thermal Insulating Material for Dwellings. 10 pp.

—Thermal Insulation—Ceilings
CNS A2212-87. Glass Wool Loose Fill Thermal Insulation (Sep)(12057).
CNS A3265-87. Method of Test for Glass Wool Loose Fill Thermal Insulation (Sep)(12058).
JIS A 9523-90. Glass Wool Loose Fill Thermal Insulation. 12 pp.

—Thermal Insulation—Thermal Resistance
JIS A 1427-86. Testing Method for Thermal Resistance of Glass Wool Heat Insulating Material. 13 pp.

Glass Yarns
Use For: Textile Glass Yarns *See Also:* Rovings; Yarns
ISO 1888-79. Textile Glass—Determination of the Average Diameter of Staple Fibres or Continuous Filaments Constituting a Textile Glass Yarn—Cross-Section Method First Edition. 5 pp.
ISO 3598-86. Textile Glass—Yarns—Basis for a Specification Second Edition. 5 pp.
JIS R 3413-90. Textile Glass Yarns. 10 pp.
JIS R 3420-89. Testing Methods for Textile Glass Products. 38 pp.

—Breaking Load
ISO 3341-84. Textile Glass—Yarns—Determination of Breaking Force and Breaking Elongation Second Edition. 7 pp.

—Designations
ISO 2078-93. Textile Glass—Yarns—Designation Fifth Edition. 7 pp.

—Elongation
ISO 3341-84. Textile Glass—Yarns—Determination of Breaking Force and Breaking Elongation Second Edition. 7 pp.

—Linear Density
ISO 1889-87. Textile Glass—Continuous Filament Yarns, Staple Fibre Yarns, Textured Yarns and Rovings (Packages)—Determination of Linear Density Second Edition. 6 pp.

—Textured
ISO 8516-87. Textile Glass—Textured Yarns—Basis for a Specification First Edition. 4 pp.

—Twist Balance Index
ISO 3343-84. Textile Glass—Yarns—Determination of Twist Balance Index Second Edition. 3 pp.

—Twist Determination
ISO 1890-86. Textile Glass—Continuous Filament Yarns and Staple Fibre Yarns—Determination of Twist Second Edition. 5 pp.

Glasses, Eye
Use: Eyeglasses

Glassine Paper
Use: Greaseproof Papers

Glassware
See Also: Alkali Glass; Borosilicate Glass; Crown Glass; Glass Ceramics; Glass Packaging; Lead Glass; Optical Glass; Tasting Glasses

—Aluminum Oxide Content—Molecular Absorption Spectrophotometry
BSI BS 7709: Part 4-93. 1993 Analysis of Extract Solutions of Glass Part 4: Method for Determination of Aluminium Oxide by Molecular Absorption Spectrometry (ISO 10136-4: 1993) (V). 11 pp.
ISO 10136 Pt 4-93. Glass and Glassware—Analysis of Extract Solutions—Part 4: Determination of Aluminium Oxide by Molecular Absorption Spectrometry First Edition. 9 pp.

Glassware (Cont.)

—Boron Oxide Content—Molecular Absorption Spectrophotometry
BSI BS 7709: Part 6-93. 1993 Analysis of Extract Solutions of Glass Part 6: Method for Determination of Boron(III) Oxide by Molecular Absorption Spectrometry (ISO 10136-6: 1993) (V). 11 pp.
ISO 10136 Pt 6-93. Glass and Glassware—Analysis of Extract Solutions—Part 6: Determination of Boron(III) Oxide by Molecular Absorption Spectrometry First Edition. 9 pp.

—Borosilicate
BSI BS 2598: Part 1-91. 1991 Glass Plant, Pipeline and Fittings Part 1: Specification for Properties of Borosilicate Glass 3.3 (ISO 3585: 1991). 8 pp.
BSI BS 2598: Part 1-80. 1980 Glass Plant, Pipeline and Fittings Part 1: Properties of Borosilicate Glass 3.3. 4 pp.
ISO 3585-91. Borosilicate Glass 3.3—Properties Second Edition. 7 pp.

—Cadmium Content
BSI BS 6748-86. 1986 Limits of Metal Release from Ceramic Ware, Glassware, Glass Ceramic Ware and Vitreous Enamel Ware. 8 pp.

—Calcium Oxide Content—Atomic Absorption Spectrometry
ISO 10136 Pt 3-93. Glass and Glassware—Analysis of Extract Solutions—Part 3: Determination of Calcium Oxide and Magnesium Oxide by Flame Atomic Absorption Spectrometry First Edition. 10 pp.

—Engineering Drawings
ISO 6414-82. Technical Drawings for Glassware First Edition. 8 pp.

—Heat Resistant
CNS R2189-87. Heat Resistant Glass Wares (Aug)(12050).
CNS R3155-87. Method of Test for Heat Resistant Glass Wares (Aug)(12051).
JIS S 2030-79. Heat Resistance Glass Wares.

—Iron Oxide Content—Atomic Absorption Spectrometry
BSI BS 7709: Part 5-93. 1993 Analysis of Extract Solutions of Glass Part 5: Method for Determination of Iron(III) Oxide by Molecular Absorption Spectrometry and Flame Atomic Absorption Spectrometry (ISO 10136-5: 1993) (V). 13 pp.
ISO 10136 Pt 5-93. Glass and Glassware—Analysis of Extract Solutions—Part 5: Determination of Iron(III) Oxide by Molecular Absorption Spectrometry and Flame Atomic Absorption Spectrometry First Edition. 10 pp.

—Iron Oxide Content—Molecular Absorption Spectrophotometry
BSI BS 7709: Part 5-93. 1993 Analysis of Extract Solutions of Glass Part 5: Method for Determination of Iron(III) Oxide by Molecular Absorption Spectrometry and Flame Atomic Absorption Spectrometry (ISO 10136-5: 1993) (V). 13 pp.
ISO 10136 Pt 5-93. Glass and Glassware—Analysis of Extract Solutions—Part 5: Determination of Iron(III) Oxide by Molecular Absorption Spectrometry and Flame Atomic Absorption Spectrometry First Edition. 10 pp.

—Lampholders
CNS C1104-82. Glass Ware for Lighting Fittings (Dimensions for Fitted Portions) (Jul)(9115).

—Lead Content
BSI BS 6748-86. 1986 Limits of Metal Release from Ceramic Ware, Glassware, Glass Ceramic Ware and Vitreous Enamel Ware. 8 pp.

—Magnesium Oxide Content—Atomic Absorption Spectrometry
ISO 10136 Pt 3-93. Glass and Glassware—Analysis of Extract Solutions—Part 3: Determination of Calcium Oxide and Magnesium Oxide by Flame Atomic Absorption Spectrometry First Edition. 10 pp.

—Petroleum Products—Testing Equipment
JIS K 2839-90. Glassware for Testing Apparatus of Petroleum Products.
JIS K 2839-83. Glassware for Testing Apparatus of Petroleum Products. 139 pp.

—Potassium Oxide Content—Atomic Emission Spectrometry
ISO 10136 Pt 2-93. Glass and Glassware—Analysis of Extract Solutions—Part 2: Determination of Sodium Oxide and Potassium Oxide by Flame Spectrometric Methods First Edition. 10 pp.

INTERNATIONAL AND NON-U.S. NATIONAL STANDARDS
SUBJECT INDEX
Glazing

Glassware *(Cont.)*

—Pressure Measurement
DIN ENGL 52320 Pt 1-78. Testing of Glass; Internal Pressure Test on Hollow Glassware, Especially Glass Containers; Inspection by Attributes (Nov). 2 pp.
DIN ENGL 52320 Pt 2-81. Testing of Glass; Internal Pressure Test on Hollow Glassware, Especially Glass Containers; Inspection by Variables (Oct). 2 pp.

—Silica Content—Molecular Absorption Spectrophotometry
ISO 10136 Pt 1-93. Glass and Glassware—Analysis of Extract Solutions—Part 1: Determination of Silicon Dioxide by Molecular Absorption Spectrometry First Edition. 9 pp.

—Sodium Oxide Content—Atomic Emission Spectrometry
ISO 10136 Pt 2-93. Glass and Glassware—Analysis of Extract Solutions—Part 2: Determination of Sodium Oxide and Potassium Oxide by Flame Spectrometric Methods First Edition. 10 pp.

Glassy Alloys
Use For: Metallic Glass *See Also:* Alloys; Glass; Metals; Plastics; Polymers

—Crystallization—Calorimetry
JIS H 7151-91. Method of Determining the Crystallization Temperatures of Amorphous Metals. 7 pp.

—Crystallization—Thermal Analysis
JIS H 7151-91. Method of Determining the Crystallization Temperatures of Amorphous Metals. 7 pp.

—Glossaries
JIS H 7004-90. Glossary of Terms Used in Amorphous Metals. 14 pp.

—Strips—Magnetic Measurement
JIS H 7152-91. Methods of Test for Magnetic Properties of Amorphous Metals Using Single Sheet Specimen. 23 pp.

Glauber's Salt
Use: Sodium Sulfate

Glazed Assemblies
Use: Glazing Systems

Glazed Coatings
Use: Glazes

Glazes
Use For: Glazed Coatings *See Also:* Ceramic Coatings (Made From Ceramics); Coatings; Nonmetallic Coatings; Organic Coatings; Paints; Shellacs; Varnishes; Vitreous Enamels

—Interior
CGSB CAN/CGSB-1.186-M89. High Performance Glazed Coating System, Interior. 14 pp.

—Laboratory Equipment—Acid Resistance Testing
CNS R3089-80. Method of Test for Resistance of Glaze to Acid and Alkali for Porcelain Laboratory Apparatus (Apr)(5478).

—Laboratory Equipment—Alkali Resistance Testing
CNS R3089-80. Method of Test for Resistance of Glaze to Acid and Alkali for Porcelain Laboratory Apparatus (Apr)(5478).

—Laboratory Equipment—Defects
CNS R3085-80. Method of Test for Porosity of Body and Imperfections of Glaze for Porcelain Laboratory Apparatus (Apr)(5474).

—Laboratory Equipment—High Temperature Testing
CNS R3087-80. Method of Test for Resistance of Glaze to High Temperature for Porcelain Laboratory Apparatus (Apr)(5476).

—Laboratory Equipment—Porosity
CNS R3085-80. Method of Test for Porosity of Body and Imperfections of Glaze for Porcelain Laboratory Apparatus (Apr)(5474).

—Plastic—Automotive—Safety
BSI BS AU 182-82. 1982 Plastics Safety Glazing Materials for Use in Road Vehicles. 19 pp.

Glazing
Use For: Transparent Glazing; Window Glazing
See Also: Safety Glazing

Glazing *(Cont.)*

—Automotive
EC 92/22/EEC-92. Council Directive on Safety Glazing and Glazing Materials on Motor Vehicles and Their Trailers. 84 pp.

—Thermal Transmittance
BSI BS 6993: Part 1-89. 1989 Thermal and Radiometric Properties of Glazing Part 1: Method for Calculation of the Steady State U-Value (Thermal Transmittance). 14 pp.
CEN PREN 673-92. Thermal Insulation of Glazing—Calculation Rules for Determining the Steady State "U" Value (Thermal Transmittance) of Glazing. 11 pp.
CEN PREN 674-92. Measuring Procedures for the Determination of the Thermal Transmittance (U Value) of Multiple Glazing (Guarded Hot Plate Method). 8 pp.
CEN PREN 675-92. Measuring Procedures for the Determination of the Thermal Transmittance (U Value) of Multiple Glazing (Heat Flow Meter Method). 9 pp.
CEN PREN 1098-93. Measuring Method for the Determination of the Thermal Transmittance of Multiple Glazing (U Value)—Calibrated and Guarded Hot Box Method. 15 pp.

—Trailers
EC 92/22/EEC-92. Council Directive on Safety Glazing and Glazing Materials on Motor Vehicles and Their Trailers. 84 pp.

Glazing Compounds
See Also: Caulking Compounds; Glazing Materials; Putty; Sealants
JIS A 5758-92. Sealing Compounds for Sealing and Glazing in Building. 36 pp.

—Adhesion Testing—Metals
CGSB CAN2-19.0-M77METH 9.3-78. Methods of Testing Putty, Caulking and Sealing Compounds Oil Bleeding and Spotting of Face Glazing Compounds. 1 p.

—Aircraft
MOD UK DTD-5548-58. Glazing Compound. 3 pp.

—Bleeding
CGSB CAN2-19.0-M77METH 9.3-78. Methods of Testing Putty, Caulking and Sealing Compounds Oil Bleeding and Spotting of Face Glazing Compounds. 1 p.

—Cracking (Fracturing)
CGSB CAN2-19.0-M77METH 9.3-78. Methods of Testing Putty, Caulking and Sealing Compounds Oil Bleeding and Spotting of Face Glazing Compounds. 1 p.
CGSB CAN2-19.0-M77METH19.1-78. Methods of Testing Putty, Caulking and Sealing Compounds Susceptibility of Glazing Compounds to Cracking. 1 p.

—Flow Measurement—Sagging Resistance
CGSB CAN2-19.0-M77METH 7.4-78. Methods of Testing Putty, Caulking and Sealing Compounds Resistance to Sag or Flow of Glazing Compounds. 1 p.

—Sealants
DIN ENGL 18545 Pt 2-92. Glazing with Sealants; Sealants (Feb). 3 pp.
DIN ENGL 18545 Pt 3-92. Glazing with Sealants; Glazing Systems (Feb). 3 pp.

—Sealants—Adhesion Testing
DIN ENGL 52455 Pt 3-74. Testing of Materials for Joint and Glazing Seals in Building Construction; Adhesion and Extension Test; Exposure to Light (Sept). 2 pp.

—Sealants—Rebates (Grooves)
DIN ENGL 18545 Pt 1-92. Glazing with Sealants; Rebates (Feb). 3 pp.
DIN ENGL 18545 Pt 3-92. Glazing with Sealants; Glazing Systems (Feb). 3 pp.

—Volatile Matter Content
CGSB CAN2-19.0-M77METH 5.2-78. Methods of Testing Putty, Caulking and Sealing Compounds Volatile Content of Putty and Glazing Compounds. 1 p.

—Window Frames—Metal—Set
BSI DD 178-89. (WITHDRAWN) 1989 Methods of Test for Setting Glazing Compounds for Use in Softwood and Steel Frames. 13 pp.

—Window Frames—Wood—Set
BSI DD 178-89. (WITHDRAWN) 1989 Methods of Test for Setting Glazing Compounds for Use in Softwood and Steel Frames. 13 pp.

Glazing Compounds *(Cont.)*

—Wrinkling
CGSB CAN2-19.0-M77METH 9.3-78. Methods of Testing Putty, Caulking and Sealing Compounds Oil Bleeding and Spotting of Face Glazing Compounds. 1 p.

Glazing Materials
Use For: Fenestration Materials; Glazing Panels; Transparent Glazing *See Also:* Aircraft Glazing Materials; Bullet Resistant Glazing; Glass; Glazing Compounds; Glazing Systems; Patent Glazing; Plastic Sheets; Plate Glass; Security Glazing

—Acrylic—Aircraft—Heat Resistance
AECMA PREN2342-82. Heat and Crazing Resistant Acrylic Sheets for Aircraft Glazing—Technical Specification. 12 pp.

—Gaskets
BSI BS 4315: Part 1-68. 1968 Amd 1 Methods of Test for Resistance to Air and Water Penetration Part 1: Windows and Structural Gasket-Glazing Systems. 18 pp.

—Glass—Classification
BSI BS 952: Part 1-78. 1978 Glass for Glazing Part 1: Classification. 12 pp.

—Glass—Glossaries
BSI BS 952: Part 2-80. 1980 Glass for Glazing Part 2: Terminology for Work on Glass. 14 pp.

—Glossaries
BSI BS 6100: SUBSEC 1.4.1-90. 1990 Glossary of Building and Civil Engineering Terms Part 1: General and Miscellaneous Section 1.4: Materials Subsection 1.4.1: Glazing. 10 pp.
BSI BS 6100: SUBSEC 1.4.1-01. 1990 Amd 1 Building and Civil Engineering Terms Part 1: General and Miscellaneous Section 1.4: Materials Subsection 1.4.1: Glazing (AMD 7239) August 15, 1992. 11 pp.
CGSB CAN/CGSB-19.28-91. Glossary of Terms Related to Sealants. 24 pp.

—Plastic—Buildings
BSI BS 6262-82. 1982 Amd 2 Code of Practice for Glazing for Buildings (AMD 4582) June 29, 1982. 81 pp.

Glazing Panels
Use: Glazing Materials

Glazing Sealants
Use: Glazing Compounds—Sealants

Glazing Seals
Use: Glazing Compounds—Sealants

Glazing Systems
Use For: Glazed Assemblies *See Also:* Construction Materials; Glazing Materials
BSI BS 6262-82. 1982 Amd 2 Code of Practice for Glazing for Buildings (AMD 4582) June 29, 1982. 81 pp.
BSI BS 8000: Part 7-90. 1990 Workmanship on Building Sites Part 7: Code of Practice for Glazing. 32 pp.
SNZ NZS 4223-85. Code of Practice for Glazing in Buildings Amend: 1, 1992. 69 pp.

—Construction Contracts
DIN ENGL 18361-88. Tendering and Performance Stipulations in Contracts for Construction Works (VOB); Part C: General Technical Specifications in Contracts for Construction Works (ATV); Glazing (Sept) (This Standard, Together with DIN 18299, September 1988 Edition,. 6 pp.

—Fire Resistant—Classification
ISO 9051-90. Glass in Building—Glazed Assemblies Containing Fire-Resistant Transparent or Translucent Glass, for Use in Building First Edition. 4 pp.

—Fire Resistant—Fire Testing
DIN ENGL 4102 Pt 5-77. Fire Behaviour of Building Materials and Building Components; Fire Barriers, Barriers in Lift Wells and Glazings Resistant Against Fire; Definitions, Requirements and Tests (Sept) (Superseded in Part by DIN 4102 Part 13). 8 pp.
DIN ENGL 4102 Pt 13-90. Fire Behaviour of Building Materials and Elements; Fire Resistant Glazing; Concepts, Requirements and Testing (May) (Supersedes Parts of DIN 4102 Part 5, September 1977 Edition). 4 pp.
ISO 3009-76. Fire-Resistance Tests—Glazed Elements First Edition; (Amendment Slip-1977) (Amendment 1-1984). 13 pp.

Glazing Systems *(Cont.)*

—Fire Resistant—Identification Systems
ISO 9051-90. Glass in Building—Glazed Assemblies Containing Fire-Resistant Transparent or Translucent Glass, for Use in Building First Edition. 4 pp.

Glide Path Receiving Instruments
See Also: Instrument Landing Systems

—Aircraft
CAA Part 3 CAP 208. ILS Glide Path Receiving Apparatus (Civil Air Publications: Airborne Radio Apparatus). 11 pp.

—Aircraft—Accuracy Testing
EUROCAE ED-47A 01.88. MPS for Airborne ILS Receiving Equipment (Glide Path). 223 pp.

—Aircraft—Electrical Properties
EUROCAE ED-47A 01.88. MPS for Airborne ILS Receiving Equipment (Glide Path). 223 pp.

—Aircraft—Environmental Testing
EUROCAE ED-47A 01.88. MPS for Airborne ILS Receiving Equipment (Glide Path). 223 pp.

—Aircraft—Failure (Quality Control)
EUROCAE ED-47A 01.88. MPS for Airborne ILS Receiving Equipment (Glide Path). 223 pp.

—Design
EUROCAE ED-47A 01.88. MPS for Airborne ILS Receiving Equipment (Glide Path). 223 pp.

Glide Slope Guidance Systems
See Also: Aircraft Equipment; Navigation; Radio Guidance Systems

—Helicopters—Shipborne
NATO STANAG 1236 ED 2 AMD 2-88. Glide Slope Indicators for Helicopter Operations from NATO Ships. 10 pp.
NATO STANAG 1236 ED 2 AMD 3-92. Glide Slope Indicators for Helicopter Operations from NATO Ships. 11 pp.

—Landing Lights
NATO STANAG 1236 ED 2 AMD 2-88. Glide Slope Indicators for Helicopter Operations from NATO Ships. 10 pp.
NATO STANAG 1236 ED 2 AMD 3-92. Glide Slope Indicators for Helicopter Operations from NATO Ships. 11 pp.

Gliders
Use For: Sailplanes *See Also:* Aircraft; Blanik Gliders; Fournier RF 3 Motor Gliders; Glaser-Dirks DG-400 Motor Gliders; Grob G 109 Motor Gliders; Hoffman H36 Dimona Motor Gliders; M 100S Gliders; PIK 20E Motor Gliders; Stemme S10 Motor Gliders
CAA. Contents (Joint Airworthiness Requirements). 15 pp.
CAA. Foreword (Joint Airworthiness Requirements). 2 pp.
CAA. Check List of Pages (Joint Airworthiness Requirements). 3 pp.
CAA. Preambles (Joint Airworthiness Requirements). 5 pp.

—Aerobatics
CAA JAR-22 Appendix F. Glossary of Aerobatic Manoeuvres (Joint Airworthiness Requirements). 2 pp.

—Airworthiness
CAA JAR-22 SUBPART A. General (Joint Airworthiness Requirements). 3 pp.
CAA JAR-22 Appendix 1. Self-Sustaining Powered Sailplanes. 1 pp.

—Control Knobs
SBAC AS 6393 ISSUE 2. Knob (Glider Release Control).

—Control Surfaces
CAA JAR-22 SUBPART C. Structure (Joint Airworthiness Requirements). 17 pp.
CAA JAR-22 SUBPART D. Design and Construction (Joint Airworthiness Requirements). 15 pp.

—Design
CAA JAR-22 SUBPART D. Design and Construction (Joint Airworthiness Requirements). 15 pp.

—Design Speeds
CAA JAR-22 SUBPART C. Structure (Joint Airworthiness Requirements). 17 pp.

—Engines
CAA JAR-22 SUBPART E. Power Plant (Joint Airworthiness Requirements). 10 pp.
CAA JAR-22 SUBPART H. Engines (Joint Airworthiness Requirements). 3 pp.

Gliders *(Cont.)*

—Equipment
CAA JAR-22 Appendix F. Glossary of Aerobatic Manoeuvres (Joint Airworthiness Requirements). 2 pp.

—Flight Dynamics
CAA JAR-22 SUBPART B. Flight (Joint Airworthiness Requirements). 17 pp.

—Flight Envelopes
CAA JAR-22 SUBPART C. Structure (Joint Airworthiness Requirements). 17 pp.

—Flight Operations
CAA JAR-22 SUBPART D. Design and Construction (Joint Airworthiness Requirements). 15 pp.
CAA JAR-22 Appendix H. Flight Manual for a Sailplane (Including a Powered Sailplane) (Joint Airworthiness Requirements). 45 pp.

—Flight Testing
CAA JAR-22 SUBPART B. Flight (Joint Airworthiness Requirements). 17 pp.

—Instruments
CAA JAR-22 Appendix F. Glossary of Aerobatic Manoeuvres (Joint Airworthiness Requirements). 2 pp.

—Landing Gear
CAA JAR-22 SUBPART C. Structure (Joint Airworthiness Requirements). 17 pp.

—Loads (Forces)
CAA JAR-22 SUBPART C. Structure (Joint Airworthiness Requirements). 17 pp.

—Manual Controls
SBAC AS 3194 ISSUE 3. Knob (Glider Release Control). 1 p.

—Ropes—Rudders
JIS W 9001-77. Rubber Rope for Glider.

—Structural Design
CAA JAR-22 SUBPART C. Structure (Joint Airworthiness Requirements). 17 pp.

Global Land Distress and Safety Systems
Use For: Distress and Safety Systems (Land); Land Distress and Safety Systems *See Also:* Global Maritime Distress and Safety Systems; Mobile Radio Services; Mobile Satellite Communications; Telecommunication Services; Telecommunication Systems
CCIR QUESTION 45-4/8-90. Technical and Operating Considerations for a Global Land and Maritime Distress and Safety System—Questions Concerning Study Group 8 —Mobile, Radiodetermination, Amateur and Related Satellite Services. 1 p.

Global Maritime Distress and Safety Systems
Use For: GMDSS; Maritime Distress and Safety Systems *See Also:* Global Land Distress and Safety Systems; Telecommunication Services; Telecommunication Systems
CCIR Report 1167-90. Study on General Questions Relating to the Global Maritime Distress and Safety System—Section 8D—Radiodetermination, Global Maritime Distress and Safety System and Related Subjects. 4 pp.
CCIR QUESTION 45-4/8-90. Technical and Operating Considerations for a Global Land and Maritime Distress and Safety System—Questions Concerning Study Group 8 —Mobile, Radiodetermination, Amateur and Related Satellite Services. 1 p.
CCIR QUESTION 92/8-88. Study on General Questions Relating to the Global Maritime Distress and Safety System (GMDSS)—Questions Concerning Study Group 8—Mobile, Radiodetermination, Amateur and Related Satellite Services. 1 p.

—Call Attempts
CCIR RECMN 541-3-90. Operational Procedures for the Use of Digital Selective-Calling (DSC) Equipment in the Maritime Mobile Service—Section 8B—Maritime Mobile Service; Telegraphy and Related Subjects. 13 pp.
CCIR RECMN 541-4-92. Operational Procedures for the Use of Digital Selective-Calling (DSC) Equipment in the Maritime Mobile Service—Section 8B—Maritime Mobile Service; Telegraphy and Related Subjects. 14 pp.

—Data Communication
CCIR Report 1043-86. Characteristics of a Data Exchange System for Use with Maritime Navigation and Radiocommunication Equipment—Section 8D—Radiodetermination, Global Maritime Distress and Safety System and Related Subjects. 7 pp.

Global Maritime Distress and Safety Systems *(Cont.)*

—Emergency Position Indicating Radio Beacons
CCIR Report 761-3-90. Technical and Operating Characteristics of Distress Systems in the Mobile-Satellite Service —Section 8J—Technical and Operating Characteristics of Radiocommunications Using Satellite Distress and Safety Operation and of Radio Determination Satellite Services. 19 pp.
CCIR QUESTION 90/8-88. Technical and Operating Characteristics of Systems Providing Radiocommunication Using Satellite Techniques for Distress and Safety Operations—Questions Concerning Study Group 8 —Mobile, Radiodetermination, Amateur and Related Satellite Services. 1 p.

—Frequencies—Emergency Position Indicating Radio Beacons
CCIR Report 1036-1-90. Frequencies for Homing and Locating in the Global Maritime Distress and Safety System (GMDSS)—Section 8D—Radiodetermination, Global Maritime Distress and Safety System and Related Subjects. 23 pp.

—Frequencies—Homing Systems
CCIR Report 1036-1-90. Frequencies for Homing and Locating in the Global Maritime Distress and Safety System (GMDSS)—Section 8D—Radiodetermination, Global Maritime Distress and Safety System and Related Subjects. 23 pp.

—Frequencies—Radiolocation
CCIR Report 1036-1-90. Frequencies for Homing and Locating in the Global Maritime Distress and Safety System (GMDSS)—Section 8D—Radiodetermination, Global Maritime Distress and Safety System and Related Subjects. 23 pp.

—Mobile Satellite Communications
CCIR Report 761-3-90. Technical and Operating Characteristics of Distress Systems in the Mobile-Satellite Service —Section 8J—Technical and Operating Characteristics of Radiocommunications Using Satellite Distress and Safety Operation and of Radio Determination Satellite Services. 19 pp.
CCIR QUESTION 90/8-88. Technical and Operating Characteristics of Systems Providing Radiocommunication Using Satellite Techniques for Distress and Safety Operations—Questions Concerning Study Group 8 —Mobile, Radiodetermination, Amateur and Related Satellite Services. 1 p.
CCIR RECMN 830-92. Operational Procedures for Mobile-Satellite Networks or Systems in the Bands 1 530-1 544 MHz and 1 626.5-1 645.5 MHz Which Are Used for Distress and Safety Purposes as Specified for GMDSS. 2 pp.

—Radar Transponders
BSI BS EN 61097-1-93. 1993 Global Maritime Distress and Safety System (GMDSS)—Part 1: Radar Transponder—Marine Search and Rescue (SART) —Operational and Performance Requirements, Methods of Testing and Required Test Results (IEC 1097-1: 1992) (Q). 30 pp.
CCIR RECMN 628-1-90. Technical Characteristics for Search and Rescue Radar Transponders—Section 8D—Radiodetermination, Global Maritime Distress and Safety System and Related Subjects. 2 pp.
CCIR RECMN 628-2-92. Technical Characteristics for Search and Rescue Radar Transponders—Section 8D—Radiodetermination, Global Maritime Distress and Safety System and Related Subjects. 2 pp.
CCIR RECMN 630-86. Main Characteristics of Two Frequency Shipborne Interrogator Transponders (SIT)—Section 8D—Radiodetermination, Global Maritime Distress and Safety System and Related Subjects. 3 pp.
CCIR Report 774-3-90. Technical Parameters of Radar Beacons (RACONS)—Section 8D—Radiodetermination, Global Maritime Distress and Safety System and Related Subjects. 17 pp.
CCIR Report 775-3-90. Frequency Requirements for Shipborne Transponders Operating in Frequency Bands Allocated to the Maritime Radionavigation Service—Section 8D—Radiodetermination, Global Maritime Distress and Safety System and Related Subjects. 31 pp.
CCIR Report 1036-1-90. Frequencies for Homing and Locating in the Global Maritime Distress and Safety System (GMDSS)—Section 8D—Radiodetermination, Global Maritime Distress and Safety System and Related Subjects. 23 pp.
CENELEC EN 61097-1-93. Global Maritime Distress and Safety System (GMDSS) Part 1: Radar Transponder—Marine Search and Rescue (SART) —Operational and Performance Requirements, Methods of Testing and Required Test Results (IEC 1097-1: 1992). 21 pp.

INTERNATIONAL AND NON-U.S. NATIONAL STANDARDS
SUBJECT INDEX

Global Maritime Distress and Safety Systems (Cont.)

—Radar Transponders (Cont.)
IEC 1097 Pt 1-92. Global Maritime Distress and Safety System (GMDSS)—Part 1: Radar Transponder—Marine Search and Rescue (SART)—Operational and Performance Requirements, Methods of Testing and Required Test Results First Edition. 39 pp.

—Radio Frequency Emissions—Direction Finders
CCIR Report 744-2-86. Use of Class J3E Emissions for Distress and Safety Purposes—Section 8C—Maritime Mobile Service; Telephony and Related Subjects. 7 pp.

—Radio Frequency Emissions—Homing Systems
CCIR Report 744-2-86. Use of Class J3E Emissions for Distress and Safety Purposes—Section 8C—Maritime Mobile Service; Telephony and Related Subjects. 7 pp.

—Radiolocation
CCIR Report 1167-90. Study on General Questions Relating to the Global Maritime Distress and Safety System—Section 8D—Radiodetermination, Global Maritime Distress and Safety System and Related Subjects. 4 pp.

—SELCAL Systems
CCIR RECMN 541-3-90. Operational Procedures for the Use of Digital Selective-Calling (DSC) Equipment in the Maritime Mobile Service—Section 8B—Maritime Mobile Service; Telegraphy and Related Subjects. 13 pp.
CCIR RECMN 541-4-92. Operational Procedures for the Use of Digital Selective-Calling (DSC) Equipment in the Maritime Mobile Service—Section 8B—Maritime Mobile Service; Telegraphy and Related Subjects. 14 pp.

—SELCAL Systems—Frequencies
CCIR RECMN 541-3-90. Operational Procedures for the Use of Digital Selective-Calling (DSC) Equipment in the Maritime Mobile Service—Section 8B—Maritime Mobile Service; Telegraphy and Related Subjects. 13 pp.
CCIR RECMN 541-4-92. Operational Procedures for the Use of Digital Selective-Calling (DSC) Equipment in the Maritime Mobile Service—Section 8B—Maritime Mobile Service; Telegraphy and Related Subjects. 14 pp.

Global Navigation Satellite Systems
Use For: GLONASS See Also: Global Positioning Systems; Radio Navigation; Radio Navigation Equipment; Satellite Communication Equipment; Satellite Communications

—Radio Beacons—Differential Modulation
CCIR RECMN 823-92. Technical Characteristics of Differential Transmissions for Global Navigation Satellite Systems (GNSS) from Maritime Radio Beacons in the Frequency Band 285-325 kHz (283.5-315 kHz in Region 1). 7 pp.

—Time Transfer
CCIR RECMN 767-92. Use of the Global Positioning System (GPS) and the Global Navigation Satellite System (GLONASS) for High-Accuracy Time Transfer. 2 pp.

Global Positioning Systems
Use For: GPS; NAVSTAR; Position Determining Systems; Position Location Systems See Also: Air Navigation; Air Navigation Equipment; Computer Systems; Global Navigation Satellite Systems; Navigation Computers; Navigational Aids; Radio Navigation; Radio Navigation Equipment; Radiolocation; Radiolocation Equipment; Satellite Communication Equipment

CCIR Report 766-2-90. Feasibility of Frequency Sharing Between the GPS and Other Services—Section 8H—Efficient Use of the Radio Spectrum Characteristics and Sharing of Frequency Resources. 13 pp.
NATO STANAG 4294 ED 1 AMD 0-93. Navstar Global Positionning System (GPS) System Characteristics. 173 pp.

—Earth Exploration Satellites
CCIR QUESTION 142/7-90. Radiocommunications for Earth Exploration Satellites Data Collection and Position Location Systems—Questions Concerning Study Group 7—Science Services. 1 p.

Global Positioning Systems (Cont.)

—Radio Beacons—Differential Modulation
CCIR Report 1166-90. Technical Characteristics of GPS Differential Transmissions from Maritime Radiobeacons—Section 8D—Radiodetermination, Global Maritime Distress and Safety System and Related Subjects. 5 pp.

—Receivers
CCIR Report 766-2-90. Feasibility of Frequency Sharing Between the GPS and Other Services—Section 8H—Efficient Use of the Radio Spectrum Characteristics and Sharing of Frequency Resources. 13 pp.

—Time Transfer
CCIR RECMN 767-92. Use of the Global Positioning System (GPS) and the Global Navigation Satellite System (GLONASS) for High-Accuracy Time Transfer. 2 pp.

—UHF—Radio Navigation Services—Frequency Band Sharing
CCIR Report 766-2-90. Feasibility of Frequency Sharing Between the GPS and Other Services—Section 8H—Efficient Use of the Radio Spectrum Characteristics and Sharing of Frequency Resources. 13 pp.

—UHF—Radiolocation Services—Frequency Band Sharing
CCIR Report 766-2-90. Feasibility of Frequency Sharing Between the GPS and Other Services—Section 8H—Efficient Use of the Radio Spectrum Characteristics and Sharing of Frequency Resources. 13 pp.

Globe Check Valves
See Also: Check Valves; Globe Valves
BSI BS 5152-74. 1974 Amd 4 Cast Iron Globe and Globe Stop and Check Valves for General Purposes (AMD 6042) July 31, 1989. 25 pp.
BSI BS 5160-89. 1989 Steel Globe Valves, Globe Stop and Check Valves and Lift Type Check Valves. 24 pp.
BSI BS 5352-81. 1981 Amd 2 Steel Wedge Gate, Globe and Check Valves 50 mm and Smaller for the Petroleum, Petrochemical and Allied Industries (AMD 6560) August 31, 1990 (Supersedes BS 2995:1966). 25 pp.
SNZ NZS/BS 5152-74. Specification for Cast Iron Globe and Globe Stop and Check Valves for General Purposes Amend: 1; 2; 3; 4. 20 pp.
SNZ NZS/BS 5352-81. Specification for Steel Wedge Gate, Globe and Check Valves 50 mm and Smaller for the Petroleum, Petrochemical and Allied Industries Amend: 1. 24 pp.

—Bronze—Ships
CNS F3162-88. Bronze Screwdown Check Globe Valves for Marine Use (5 kgf/cm2) (Apr)(8492).
CNS F3206-88. Bronze Screw-Down Check Globe Valves for Marine Use (16 kgf/cm2) (Union Bonnet Type) (May)(9489).
CNS F3208-88. Bronze Lift Check Valves for Marine Use (5 kgf/cm2) (Union Bonnet Type) (May)(9491).
JIS F 7351-87. Marine Bronze 5 K Screw-Down Check Globe Valves.
JIS F 7409-87. Marine Bronze 16 K Screw-Down Check Globe Valves.
JIS F 7411-88. Bronze 5 K Screw-Down Check Globe Valves for Marine Use (Union Bonnet Type).
JIS F 7413-88. Bronze 16 K Screw-Down Check Globe Valves for Marine Use (Union Bonnet Type).
JIS F 7415-88. Bronze 5 K Lift Check Globe Valves for Marine Use (Union Bonnet Type).
JIS F 7417-88. Bronze 16 K Lift Check Globe Valves for Marine Use (Union Bonnet Type).

—Cast Iron
BSI BS 5152-74. 1974 Amd 4 Cast Iron Globe and Globe Stop and Check Valves for General Purposes (AMD 6042) July 31, 1989. 25 pp.
SNZ NZS/BS 5152-74. Specification for Cast Iron Globe and Globe Stop and Check Valves for General Purposes Amend: 1; 2; 3; 4. 20 pp.

—Cast Iron—Ships
CNS F3005-88. Cast Iron Screw-Down Check Globe Valves for Marine Use (5 kgf/cm2) (May)(3810).
CNS F3164-88. Cast Iron Screw-Down Check Globe Valves for Marine Use (5 kgf/cm2) (Apr)(8494).
CNS F3184-88. Cast Iron Screw-Down Check Globe Valves for Marine Use (10 kgf/cm2) (May)(8825).
CNS F3185-88. Cast Iron Screw-Down Check Globe Valves for Marine Use (16 kgf/cm2) (May)(8961).
JIS F 7353-89. Cast Iron 5 K Screw-Down Check Globe Valves for Marine Use.
JIS F 7375-89. Cast Iron 10 K Screw-Down Check Globe Valves for Marine Use.
JIS F 7377-89. Cast Iron 16 K Screw-Down Check Globe Valves for Marine Use.

Globe Check Valves (Cont.)

—Cast Steel—Ships
JIS F 7471-88. Cast Steel 10 K Screw-Down Check Globe Valves for Marine Use.
JIS F 7473-88. Cast Steel 20 K Screw-Down Check Globe Valves for Marine Use.

—Flanged
BSI BS 1873-75. 1975 Amd 1 Steel Globe and Globe Stop and Check Valves (Flanged and Butt-Welding Ends) for the Petroleum, Petrochemical and Allied Industries (AMD 6564) July 31, 1990. 34 pp.

—Ships
CNS F3005-88. Cast Iron Screw-Down Check Globe Valves for Marine Use (5 kgf/cm2) (May)(3810).
CNS F3162-88. Bronze Screwdown Check Globe Valves for Marine Use (5 kgf/cm2) (Apr)(8492).
CNS F3164-88. Cast Iron Screw-Down Check Globe Valves for Marine Use (5 kgf/cm2) (Apr)(8494).
CNS F3184-88. Cast Iron Screw-Down Check Globe Valves for Marine Use (10 kgf/cm2) (May)(8825).
CNS F3185-88. Cast Iron Screw-Down Check Globe Valves for Marine Use (16 kgf/cm2) (May)(8961).
CNS F3204-88. Marine Bronze 5 kgf/cm2 Screw-Down Check Globe Valves (Union Bonnet Type) (May)(9487).
CNS F3206-88. Bronze Screw-Down Check Globe Valves for Marine Use (16 kgf/cm2) (Union Bonnet Type) (May)(9489).
CNS F3208-88. Bronze Lift Check Valves for Marine Use (5 kgf/cm2) (Union Bonnet Type) (May)(9491).
JIS F 7351-87. Marine Bronze 5 K Screw-Down Check Globe Valves.
JIS F 7353-89. Cast Iron 5 K Screw-Down Check Globe Valves for Marine Use.
JIS F 7375-89. Cast Iron 10 K Screw-Down Check Globe Valves for Marine Use.
JIS F 7377-89. Cast Iron 16 K Screw-Down Check Globe Valves for Marine Use.
JIS F 7409-87. Marine Bronze 16 K Screw-Down Check Globe Valves.
JIS F 7411-88. Bronze 5 K Screw-Down Check Globe Valves for Marine Use (Union Bonnet Type).
JIS F 7413-88. Bronze 16 K Screw-Down Check Globe Valves for Marine Use (Union Bonnet Type).
JIS F 7415-88. Bronze 5 K Lift Check Globe Valves for Marine Use (Union Bonnet Type).
JIS F 7417-88. Bronze 16 K Lift Check Globe Valves for Marine Use (Union Bonnet Type).
JIS F 7471-88. Cast Steel 10 K Screw-Down Check Globe Valves for Marine Use.
JIS F 7473-88. Cast Steel 20 K Screw-Down Check Globe Valves for Marine Use.

—Steel
BSI BS 1873-75. 1975 Amd 1 Steel Globe and Globe Stop and Check Valves (Flanged and Butt-Welding Ends) for the Petroleum, Petrochemical and Allied Industries (AMD 6564) July 31, 1990. 34 pp.
BSI BS 5160-89. 1989 Steel Globe Valves, Globe Stop and Check Valves and Lift Type Check Valves. 24 pp.
BSI BS 5352-81. 1981 Amd 2 Steel Wedge Gate, Globe and Check Valves 50 mm and Smaller for the Petroleum, Petrochemical and Allied Industries (AMD 6560) August 31, 1990 (Supersedes BS 2995:1966). 25 pp.
SNZ NZS/BS 5352-81. Specification for Steel Wedge Gate, Globe and Check Valves 50 mm and Smaller for the Petroleum, Petrochemical and Allied Industries Amend: 1. 24 pp.

Globe Valves
See Also: Angle Globe Valves; Ball Valves; Faucets; Gas Valves; Gate Valves; Globe Check Valves; Hydraulic Valves; Pneumatic Valves; Stop Valves; Valves; Y Valves
BSI BS 5152-74. 1974 Amd 4 Cast Iron Globe and Globe Stop and Check Valves for General Purposes (AMD 6042) July 31, 1989. 25 pp.
BSI BS 5160-89. 1989 Steel Globe Valves, Globe Stop and Check Valves and Lift Type Check Valves. 24 pp.
BSI BS 5352-81. 1981 Amd 2 Steel Wedge Gate, Globe and Check Valves 50 mm and Smaller for the Petroleum, Petrochemical and Allied Industries (AMD 6560) August 31, 1990 (Supersedes BS 2995:1966). 25 pp.
CNS B2495-80. Face-to-Face and End-to-End Dimensions of Glove Valves (Jul)(5711).
DIN ENGL 3356 Pt 1-82. Globe Valves; General Data (May). 7 pp.
DIN ENGL 3356 Pt 2-82. Globe Valves; Cast Iron Stop Valves (May). 6 pp.
DIN ENGL 3356 Pt 3-82. Globe Valves; Unalloyed Steel Stop Valves (May). 8 pp.
DIN ENGL 3356 Pt 5-82. Globe Valves; Stainless Steel Stop Valves (May). 7 pp.
ISO 5752-82. Metal Valves for Use in Flanged Pipe Systems—Face-to-Face and Centre-to-Face Dimensions Second Edition. 13 pp.
JIS B 2002-87. Face-to-Face and End-to-End Dimensions of Valves.

INDUSTRY STANDARDS

Globe Valves (Cont.)

JIS B 2002-68. Face-to-Face and End-to-End Dimensions of Valves. 14 pp.
JIS B 2011-88. Bronze Gate, Globe, Angle and Check Valves. 41 pp.
SNZ NZS/BS 5152-74. Specification for Cast Iron Globe and Globe Stop and Check Valves for General Purposes Amend: 1; 2; 3; 4. 20 pp.
SNZ NZS/BS 5352-81. Specification for Steel Wedge Gate, Globe and Check Valves 50 mm and Smaller for the Petroleum, Petrochemical and Allied Industries Amend: 1. 24 pp.

—Compressed Air—Marine

CNS F3146-87. Marine Forged Steel Globe Valves for Compressed Air (Dec)(8110).
CNS F3148-87. Marine Cast Steel Globe Valves for Compressed Air (Dec)(8112).
JIS F 7336-88. Forged Steel Globe Air Valves for Marine Use.
JIS F 7340-88. Cast Steel Globe Air Valves for Marine Use.

—Copper Alloy

BSI BS 5154-91. 1991 Copper Alloy Globe, Globe Stop and Check, Check and Gate Valves. 22 pp.
SNZ NZS/BS 5154-89. Specification for Copper Alloy Globe, Globe Stop and Check, Check and Gate Valves. 20 pp.
SNZ NZS/BS 5154-91. Specification for Copper Alloy Globe, Globe Stop and Check, Check and Gate Valves. 24 pp.

—Copper Alloy—Check

BSI BS 5154-91. 1991 Copper Alloy Globe, Globe Stop and Check, Check and Gate Valves. 22 pp.
SNZ NZS/BS 5154-89. Specification for Copper Alloy Globe, Globe Stop and Check, Check and Gate Valves. 20 pp.
SNZ NZS/BS 5154-91. Specification for Copper Alloy Globe, Globe Stop and Check, Check and Gate Valves. 24 pp.

—Fire Hydrants

BSI BS 5041: Part 1-87. 1987 Amd 1 Fire Hydrant System Equipment Part 1: Landing Valves for Wet Risers (AMD 5912) September 30, 1988. 15 pp.

—Flanged

BSI BS 1873-75. 1975 Amd 1 Steel Globe and Globe Stop and Check Valves (Flanged and Butt-Welding Ends) for the Petroleum, Petrochemical and Allied Industries (AMD 6564) July 31, 1990. 34 pp.
CNS B2508-85. Bronze Flanged Globe Valves (Mar)(5969).
CNS B2511-87. Cast Iron Flanged Globe Valves (10 kgf/cm2) (Mar)(5972). 6 pp.
CNS B2535-82. Cast Steel Flanged Globe Valves (10 kgf/cm2) (Sep)(6882).
CNS B2539-82. Cast Steel Flanged Globe Valves (20 kgf/cm2) (Sep)(6886).
JIS B 2071-87. Cast Steel Flanged Valves. 23 pp.

—Heat Resistant

DIN ENGL 3356 Pt 4-82. Globe Valves; High Temperature Steel Stop Valves (May). 7 pp.

—Pipelines—Flanged

CEN PREN 558-1-91. Metal Valves for Use in Flanged Pipe Systems—Face-to-Face and Centre-to-Face Dimensions—Part 1: General. 5 pp.
CEN PREN 558-2-91. Metal Valves for Use in Flanged Pipe Systems—Face-to-Face and Centre-to-Face Dimensions—Part 2: PN Designated Valves. 13 pp.
CEN PREN 558-3-91. Metal Valves for Use in Flanged Pipe Systems—Face-to-Face and Centre-to-Face Dimensions—Part 3: Class-Designation Valves. 17 pp.

—Process Control

IEC 534 Pt 3-76. Industrial-Process Control Valves Part 3: Dimensions Section One—Face-to-Face Dimensions for Flanged, Two-Way, Globe-Type Control Valves First Edition. 13 pp.

—Screwed

CNS B2106-87. Brass Screwed Globe Valves (10 kgf/cm2) (Mar)(712). 3 pp.
CNS B2501-85. Bronze Screwed Globe Valves (5kgf/cm2) (Mar)(5962).
CNS B2502-88. Bronze Screwed Globe Valves (Jul)(5963).

—Ships

CNS F3003-88. Cast Iron Globe Valves for Marine Use (5 kgf/cm2) (May)(3808).
CNS F3096-89. Bronze Globe Valves for Marine Use (5 kgf/cm2) (Apr)(7239).
CNS F3098-88. Bronze Globe Valves for Marine Use (16 kgf/cm2) (Apr)(7241).
CNS F3100-87. Marine Cast Iron Globe Valve (Dec)(7243).
CNS F3102-87. Marine Cast Iron Globe Valves (16 kgf/cm2) (Dec)(7245).
CNS F3110-87. Marine Cast Steel Globe Valves (5 kgf/cm2) (Dec)(7682).
CNS F3112-87. Marine Cast Steel Globe Valves (20 kgf/cm2) (Dec)(7684).
CNS F3114-87. Marine Cast Steel Globe Valves (30 kgf/cm2) (Dec)(7686).
CNS F3116-87. Marine Cast Steel Globe Valves (40 kgf/cm2) (Dec)(7688).
CNS F3131-87. Marine Cast Steel Globe Valves (10 kgf/cm2) (Dec)(7898).
CNS F3133-87. Marine Malleable Iron Globe Valves (5 kgf/cm2) (Dec)(7900).
CNS F3135-87. Marine Malleable Iron Globe Valves (16 kgf/cm2) (Dec)(7902).
CNS F3137-87. Marine Forged Steel Globe Valves (40 kgf/cm2) (Dec)(7904).
CNS F3146-87. Marine Forged Steel Globe Valves for Compressed Air (Dec)(8110).
CNS F3148-87. Marine Cast Steel Globe Valves for Compressed Air (Dec)(8112).
CNS F3157-88. Bronze Globe Valves for Marine Use (Union Bonnet Type) (5 kgf/cm2) (Apr)(8272).
CNS F3159-88. Bronze Globe Valves for Marine Use (Union Bonnet Type) (16 kgf/cm—S) (Apr)(8274).
CNS F3176-87. Marine Hull Cast Steel Globe Valves (Dec)(8692).
CNS F3190-88. Bronze Globe Valves for Marine Use (20 kgf/cm2) (May)(8966).
CNS F3193-87. Marine Cast Iron Globe Valves (3 kgf/cm2) (Dec)(9255).
CNS F3195-88. Bronze Globe Valves for Marine Use (3 kgf/cm2) (May)(9257).
JIS F 7301-87. Marine Bronze 5 K Globe Valves.
JIS F 7303-87. Marine Bronze 16 K Globe Valves.
JIS F 7305-89. Cast Iron 5 K Globe Valves for Marine Use.
JIS F 7307-89. Cast Iron 10 K Globe Valves for Marine Use.
JIS F 7309-88. Cast Iron 16 K Globe Valves for Marine Use.
JIS F 7311-88. Cast Steel 5 K Globe Valves for Marine Use.
JIS F 7313-89. Cast Steel 20 K Globe Valves for Marine Use.
JIS F 7315-88. Cast Steel 30 K Globe Valves for Marine Use.
JIS F 7317-88. Cast Steel 40 K Globe Valves for Marine Use.
JIS F 7319-88. Cast Steel 10 K Globe Valves for Marine Use.
JIS F 7321-88. Malleable Iron 5 K Globe Valves for Marine Use.
JIS F 7323-88. Malleable Iron 16 K Globe Valves for Marine Use.
JIS F 7329-88. Forged Steel 40 K Globe Valves for Marine Use.
JIS F 7336-88. Forged Steel Globe Air Valves for Marine Use.
JIS F 7340-88. Cast Steel Globe Air Valves for Marine Use.
JIS F 7346-88. Bronze 5 K Globe Valves (Union Bonnet Type) for Marine Use.
JIS F 7348-88. Bronze 16 K Globe Valves (Union Bonnet Type) for Marine Use.
JIS F 7365-86. Marine Hull Cast Steel Globe Valves.
JIS F 7388-88. Bronze 20 K Globe Valves for Marine Use.
JIS F 7403-87. Marine Hull Bronze Globe Valves.
JIS F 7421-89. Forged Steel 20 K Globe Valves for Marine Use.

—Water

BSI BS 7350-90. 1990 Double Regulating Globe Valves and Flow Measurement Devices for Heating and Chilled Water Systems. 24 pp.
BSI BS 7350-01. 1990 Amd 1 Double Regulating Globe Valves and Flow Measurement Devices for Heating and Chilled Water Systems (AMD 6865) December 24, 1991. 25 pp.
DIN ENGL 3502-85. Stopvalves for Drinking Water Supplies on and in Private Property; Straight Pattern Globe Valves with Oblique Bonnet, Rated for Nominal Pressure PN 10 (Nov). 4 pp.
DIN ENGL 3512-85. Stopvalves for Drinking Water Supplies on and in Private Property; Straight Pattern Globe Valves with Vertical Bonnet, Rated for Nominal Pressure PN 10 (Nov). 4 pp.

—Water—Heating Chilling Equipment

BSI BS 7350-01. 1990 Amd 1 Double Regulating Globe Valves and Flow Measurement Devices for Heating and Chilled Water Systems (AMD 6865) December 24, 1991. 25 pp.

Globulins

Use For: Ig; Immunoglobulins **See Also:** Monoclonal Antibodies

—IgG—Monoclonal—Quantitative Analysis

JIS K 0604-92. Methods for Quantitative Analysis of Monoclonal IgG. 10 pp.

—Quantitative Analysis

JIS K 0603-92. Methods for Quantitative Analysis of Immuno Globulin. 12 pp.

Glockenspiels

See Also: Musical Instruments; Xylophones
JIS S 8506-81. Glockenspiels.

GLONASS

Use: Global Navigation Satellite Systems

Glos Air Airtourer Aircraft

See Also: Aircraft

—Foreign Airworthiness Directives

CAA. AESL and Glos Air Airtourer Series (Foreign Airworthiness Directives). 14 pp.

Glos Airtourer Super 150 Aircraft

See Also: Aircraft

—Data Sheets

CAA BA2 ISSUE 27. Glos-Airtourer T3, 115, 150, Super 150. 4 pp.

Glos Airtourer T 3 Aircraft

See Also: Aircraft

—Data Sheets

CAA BA2 ISSUE 27. Glos-Airtourer T3, 115, 150, Super 150. 4 pp.

Glos Airtourer 115 Aircraft

See Also: Aircraft

—Data Sheets

CAA BA2 ISSUE 27. Glos-Airtourer T3, 115, 150, Super 150. 4 pp.

Glos Airtourer 150 Aircraft

See Also: Aircraft

—Data Sheets

CAA BA2 ISSUE 27. Glos-Airtourer T3, 115, 150, Super 150. 4 pp.

Gloss

Use For: Specular Gloss **See Also:** Optical Properties; Reflectance; Reflection; Semigloss
CNS Z8023-81. Method of Measurement Glossiness (Jul)(7773).

—Anodic Coatings

BSI BS 6161: Part 12-87. 1987 Methods of Test for Anodic Oxidation Coatings on Aluminium and Its Alloys Part 12: Measurement of Specular Reflectance and Specular Gloss at Angles of 20 Degrees, 45 Degrees, 60 Degrees or 85 Degrees. 14 pp.
ISO 7668-86. Anodized Aluminium and Aluminium Alloys—Measurement of Specular Reflectance and Specular Gloss at Angles of 20 Degrees, 45 Degrees, 60 Degrees or 85 Degrees First Edition. 12 pp.

—Coatings

CGSB 1-GP-71 METH 13.1-75. Methods of Testing Paints and Pigments Gloss After 48 Hours. 1 p.
CGSB 1-GP-71 METH 13.2-75. Methods of Testing Paints and Pigments Gloss After 7 Days. 1 p.
CGSB 1-GP-71 METH 13.3-75. Methods of Testing Paints and Pigments Gloss Gloss After One Hour Baking. 1 p.

—Enamel Paints

CGSB 1-GP-71 METH 134.1-78. Methods of Testing Paints and Pigments Applicability and Appearance Enamels on Sealed Boards. 1 p.

—Enamels

CGSB 1-GP-71 METH 134.3-78. Methods of Testing Paints and Pigments Applicability and Appearance Enamels on Primed Steel, Brush and Roller Application. 1 p.

—Finishes—Furniture

BSI BS 3962: Part 1-80. 1980 Methods of Test for Clear Finishes for Wooden Furniture Part 1: Assessment of Low Angle Glare by Measurements of Specular Gloss at 85 Degrees. 7 pp.

—Lacquers

CNS K6386-76. Method of Test for Lacquer in Surface of Leather Shoes (Mar) (3922). 2 pp.
SAA AS 1580.602. 1-75. Paints and Related Materials—Methods of Test—Part 602.1: Visual Assessment of Gloss Reconfirmed 1988. 2 pp.
SAA AS 1580.602. 2-74. Paints and Related Materials—Methods of Test—Part 602.2: Specular Gloss Corrig. Reconfirmed 1988. 4 pp.

INTERNATIONAL AND NON-U.S. NATIONAL STANDARDS
SUBJECT INDEX

Gloss *(Cont.)*
—Lacquers *(Cont.)*
SNZ NZS/AS 1580. 602.1-75. Methods of Test for Paints and Related Materials Part 602.1: Visual Assessment of Gloss (Reconfirmed 1988) (This is a Joint Standard with SAA AS 1580.602.1). 2 pp.
SNZ NZS/AS 1580. 602.2-74. Methods of Test for Paints and Related Materials Part 602.2: Specular Gloss (Reconfirmed 1988) (This is a Joint Standard with SAA AS 1580.602.2). 4 pp.

—Manufacturing Products
JIS Z 8741-83. Method of Measurement for Specular Glossiness. 11 pp.

—Mining Products
JIS Z 8741-83. Method of Measurement for Specular Glossiness. 11 pp.

—Paints
BSI BS 3900: Part D5-80. 1980 Methods of Test for Paints Group D: Optical Tests on Paint Films Part D5: Measurement of Specular Gloss of Non-Metallic Paint Films 20 Degrees, 60 Degrees and 85 Degrees. 8 pp.
BSI BS AU 148: Part 5-69. 1969 Methods of Test for Motor Vehicle Paints Part 5: Gloss Measurement. 5 pp.
ISO 2813-78. Paints and Varnishes—Measurement of Specular Gloss of Non-Metallic Paint Films at 20 Degrees, 60 Degrees and 85 Degrees Second Edition. 9 pp.
SAA AS 1580.481. 1.5-91. Paints and Related Materials—Methods of Test—Part 481: Coatings—Part 481.1: Assessment of Individual Defects of Exposed Films —Part 481.1.5: Exposed to Weathering—Change in Gloss (Supersedes AS 1580.481.1—1975 (in Part)) (in Professional Packages 30, 39). 3 pp.
SAA AS 1580.602. 1-75. Paints and Related Materials—Methods of Test—Part 602.1: Visual Assessment of Gloss Reconfirmed 1988. 2 pp.
SAA AS 1580.602. 2-74. Paints and Related Materials—Methods of Test—Part 602.2: Specular Gloss Corrig. Reconfirmed 1988. 4 pp.
SNZ NZS/AS 1580. 481.1.5-91. Methods of Test for Paints and Related Materials Part 481.1.5: Coatings—Exposed to Weathering—Change in Gloss (This is a Joint Standard with SAA AS 1580.481.1.5). 3 pp.
SNZ NZS/AS 1580. 602.1-75. Methods of Test for Paints and Related Materials Part 602.1: Visual Assessment of Gloss (Reconfirmed 1988) (This is a Joint Standard with SAA AS 1580.602.1). 2 pp.
SNZ NZS/AS 1580. 602.2-74. Methods of Test for Paints and Related Materials Part 602.2: Specular Gloss (Reconfirmed 1988) (This is a Joint Standard with SAA AS 1580.602.2). 4 pp.

—Paper Products
CNS P3049-83. Method of Test for 75 Degrees Gloss of Paper and Paperboard (Nov)(7299). 2 pp.
JIS P 8142-65. Testing Method for 75 Degree Specular Gloss of Paper and Paperboard (R 1989). 7 pp.

—Paperboard
CNS P3049-83. Method of Test for 75 Degrees Gloss of Paper and Paperboard (Nov)(7299). 2 pp.
CPPA E.3P-81. Seventy-Five Degree Specular Gloss of Paper and Paperboard. 4 pp.
JIS P 8142-65. Testing Method for 75 Degree Specular Gloss of Paper and Paperboard (R 1989). 7 pp.

—Papers
CNS P3049-83. Method of Test for 75 Degrees Gloss of Paper and Paperboard (Nov)(7299). 2 pp.
CPPA E.3P-81. Seventy-Five Degree Specular Gloss of Paper and Paperboard. 4 pp.
JIS P 8142-65. Testing Method for 75 Degree Specular Gloss of Paper and Paperboard (R 1989). 7 pp.

—Pigments
BSI BS 3483: Part E3-91. 1991 Testing Pigments for Paints Part E3: Assessment of Dispersion Characteristics from the Change in Gloss (ISO 8781-3: 1990). 9 pp.
ISO 8781 Pt 3-90. Pigments and Extenders—Methods of Assessment of Dispersion Characteristics—Part 3: Assessment from the Change in Gloss First Edition. 7 pp.

—Plastics
BSI BS 2782:Pt5: METH 520A-92. 1992 Methods of Testing Plastics Part 5: Optical and Colour Properties, Weathering Method 520A: Determination of Specular Gloss (V). 9 pp.

—Varnishes
SAA AS 1580.602. 1-75. Paints and Related Materials—Methods of Test—Part 602.1: Visual Assessment of Gloss Reconfirmed 1988. 2 pp.
SAA AS 1580.602. 2-74. Paints and Related Materials—Methods of Test—Part 602.2: Specular Gloss Corrig. Reconfirmed 1988. 4 pp.
SNZ NZS/AS 1580. 602.1-75. Methods of Test for Paints and Related Materials Part 602.1: Visual Assessment of Gloss (Reconfirmed 1988) (This is a Joint Standard with SAA AS 1580.602.1). 2 pp.
SNZ NZS/AS 1580. 602.2-74. Methods of Test for Paints and Related Materials Part 602.2: Specular Gloss (Reconfirmed 1988) (This is a Joint Standard with SAA AS 1580.602.2). 4 pp.

—Waxes
CGSB 25-GP-1M METH 10.1-84. Methods of Sampling and Testing Waxes and Polishes Gloss and Drying Time (Supersedes 10.1e of September 1974). 2 pp.

Glossaries
Scope Note: For glossaries about specific subjects, products, or materials, see the subheading Glossaries under specific subjects, products or materials
BSI BS 3669-63. 1963 Recommendations for the Selection, Formation and Definition of Technical Terms. 18 pp.
CSA CAN/CSA-Z780-92. Principles and Methods of Terminology (ISO 704-1987); (Gen Instr 1 Thru 2). 31 pp.
CSA CAN/CSA-Z781-92. Terminology—Vocabulary (ISO 1087:1990); (Gen Instr 1 Thru 2). 36 pp.
ISO 704-87. Principles and Methods of Terminology First Edition; (CAN/CSA-Z780-92). 20 pp.
ISO 1087-90. Terminology—Vocabulary First Edition. 23 pp.

—Layouts
ISO 10241-92. International Terminology Standards—Preparation and Layout First Edition. 27 pp.

Gloster Meteor NFII Aircraft
See Also: Aircraft

—Antenna Positions
CAA. Gloster Meteor NFII (Approved Aerial Positions). 1 p.

Glove Boxes
See Also: Laboratory Equipment

—Radioactive Materials
JIS Z 4808-75. Glove Box for Handling Radioactive Substance.

Gloves
Use For: Mitts *See Also:* Clothing; Finger Cots; Medical Gloves; Oven Mitts; Protective Clothing; Surgical Gloves
BSI BS 1651-86. 1986 Industrial Gloves. 28 pp.

—Chemical Resistant—Safety
DIN ENGL 4841 Pt 5-87. Protective Gloves; Grade 2 Gloves Affording Protection Against Chemicals; Safety Requirements, Testing (Apr). 2 pp.

—Cotton
CNS L4047-67. Cotton Gloves (for General Use) (Tentative) (Aug)(2768)(R 1973).

—Cotton—Moisture Content
CNS L3065-67. Method of Test for Cotton Gloves (for General Use) (Aug)(2769)(R 1973). 2 pp.

—Cotton—Sampling
CNS L3065-67. Method of Test for Cotton Gloves (for General Use) (Aug)(2769)(R 1973). 2 pp.

—Cotton—Shrinkage
CNS L3065-67. Method of Test for Cotton Gloves (for General Use) (Aug)(2769)(R 1973). 2 pp.

—Cotton—Tensile Testing
CNS L3065-67. Method of Test for Cotton Gloves (for General Use) (Aug)(2769)(R 1973). 2 pp.

—Cotton—Visual Inspection
CNS L3065-67. Method of Test for Cotton Gloves (for General Use) (Aug)(2769)(R 1973). 2 pp.

—Hand Prostheses
JIS T 9223-90. Cosmetic Gloves for Artificial Hands. 9 pp.

—Linemen's Equipment
BSI BS EN 60903-93. 1993 Gloves and Mitts of Insulating Material for Live Working (N). 54 pp.
CENELEC PREN 60903-92. Specification for Gloves and Mitts of Insulating Material for Live Working. 40 pp.
CENELEC EN 60903-92. Specification for Gloves and Mitts of Insulating Material for Live Working; (IEC 903: 1988). 12 pp.
IEC 903-88. Specification for Gloves and Mitts of Insulating Material for Live Working First Edition; (CENELEC EN 60903:1992). 73 pp.

—Linemen's Equipment—Rubber
BSI BS 697-86. 1986 Rubber Gloves for Electrical Purposes. 11 pp.

—Ovenmitts
BSI BS 6526-84. 1984 Amd 1 Domestic Oven Gloves. 9 pp.

—Protective
CEN PREN 340-90. General Requirements for Protective Clothing Including Hand and Arm Protection and Lifejackets. 20 pp.
CEN PREN 381-4-92. Protective Clothing for Users of Hand-Held Chainsaws—Part 4: Test Method for Chainsaw Protective Gloves. 13 pp.
CEN PREN 381-7-92. Protective Clothing for Users of Hand-Held Chainsaws—Part 7: Requirements for Chainsaw Protective Gloves. 7 pp.
CEN PREN 420-90. General Requirements for Gloves. 28 pp.
CEN PREN 659-92. Firefighters' Gloves—Protection Against Heat and Flames. 6 pp.
CEN PREN 1082-93. Protective Chain Mail Gloves and Protective Armguards for Use with Hand Knives. 48 pp.
SNZ NZS 5812-82. Industrial Protective Gloves (Reconfirmed 1989). 19 pp.

—Protective—Mechanical Testing
CEN PREN 388-90. Protective Gloves-Mechanical Test Methods and Specifications. 21 pp.

—PVC—Household
CNS S1101-89. Household Polyvinyl Chloride Gloves (Jun)(6632). 3 pp.
CNS S2047-89. Method of Test for Household Polyvinyl Chloride Gloves (Jun)(6633). 5 pp.
JIS S 2045-77. Household Polyvinyl Chloride Gloves.

—Rubber—Household
CNS S1169-83. Household Rubber Gloves (Aug)(10518).
CNS S2110-89. Method of Test for Household Rubber Gloves (Jun)(10519). 4 pp.
JIS S 2042-77. Household Rubber Gloves.

—Safety
JIS T 8116-79. Protective Gloves for Occupational Health.

—Sizes
ISO 4418-78. Size Designation of Clothes—Gloves First Edition. 4 pp.

—Thermal Protection
CEN PREN 407-90. Protective Gloves Against Thermal Hazards. 58 pp.
CEN PREN 511-91. Protective Gloves Against Cold. 13 pp.

—Vibration Isolation
JIS T 8114-87. Vibration Isolation Gloves. 9 pp.

—Welding
CNS Z2035-81. Protective Leather Gloves for Welders (Mar)(7178).
JIS T 8113-76. Protective Leather Gloves for Welders (R 1979). 9 pp.

—Welding—Safety
DIN ENGL 4841 Pt 4-87. Protective Gloves; Leather Protective Gloves for Welders; Safety Requirements and Testing (Jan). 3 pp.

—Work
JIS L 4131-86. Working Gloves. 9 pp.

Glow Lamps
See Also: Discharge Lamps; Lamps
JIS C 7606-85. Negative-Glow Lamps. 11 pp.

Glow Modulator Tubes
—Quality Assurance
BSI BS 9047-78. (OBSOLESCENT) 1978 Glow Modulator Tubes of Assessed Quality. Generic Data and Methods of Test. 12 pp.
BSI BS 9048-78. (OBSOLESCENT) 1978 Rules for the Preparation of Detail Specifications for Glow Modulator Tubes of Assessed Quality. 6 pp.

Glow Plugs
Use For: Heat Plugs *See Also:* Diesel Engines; Engine Heaters; Heaters

—Automotive
CNS D2115-86. Glow Plugs for Automobiles (Jun)(7979).

INDUSTRY STANDARDS

INTERNATIONAL AND NON-U.S. NATIONAL STANDARDS
SUBJECT INDEX

Glow

Glow Plugs (Cont.)
—Automotive (Cont.)
ISO 6550-89. Road Vehicles—M12 X 1,25 and M14 X 1,25 Sheath-Type Glow-Plugs with Conical Seating and Their Cylinder Head Housing Second Edition. 6 pp.
JIS D 5103-92. Glow Plugs for Automobiles. 11 pp.

—Automotive—Comprehensive Testing
ISO 7578-86. Road Vehicles—Sheath-Type Glow Plugs—General Requirements and Test Methods Second Edition. 5 pp.

GLP
Use: Good Laboratory Practice

Gluconic Acid
MOD UK DSTAN 68-125-90. Gluconic Acid, 30% Solution, Technical Issue 1. 18 pp.
MOD UK TS 10104. Gluconic Acid, 30% Solution, Technical (Superseded by Def Stan 68-125.

Glucono-Delta-Lactone Content Analysis
—Meat
BSI BS 4401: Part 13-79. 1979 Methods of Test for Meat and Meat Products Part 13: Determination of Gluconodelta Lactone Content (Reference Method). 7 pp.
ISO 4133-79. Meat and Meat Products—Determination of Glucono-Delta-Lactone Content (Reference Method) First Edition. 6 pp.

Glucose
Use For: Dextrose See Also: Fructose; Sugars
CNS K7180-65. Chemical Reagent (Dextrose, Anhydrous) (Jan)(1680).
CNS N5101-79. Solid Glucose (Jan)(3348). 1 p.
CNS N5153-79. Crystalline Dextrose (May)(4867). 2 pp.
JIS K 8824-61. Dextrose, Anhydrous.

—Ash Content
CNS N6054-72. Method of Test for Solid Glucose (Jun)(3349). 5 pp.

—Dextrin Content
CNS N6054-72. Method of Test for Solid Glucose (Jun)(3349). 5 pp.

—Dried—Arsenic Content
CNS N6053-73. Method of Test for Dried Glucose and Glucose Syrup (Jan)(3218). 9 pp.

—Dried—Ash Content
CNS N6053-73. Method of Test for Dried Glucose and Glucose Syrup (Jan)(3218). 9 pp.

—Dried—Color Testing
CNS N6053-73. Method of Test for Dried Glucose and Glucose Syrup (Jan)(3218). 9 pp.

—Dried—Lead Content
CNS N6053-73. Method of Test for Dried Glucose and Glucose Syrup (Jan)(3218). 9 pp.

—Dried—Moisture Content
CNS N6053-73. Method of Test for Dried Glucose and Glucose Syrup (Jan)(3218). 9 pp.

—Dried—Solids Content
CNS N6053-73. Method of Test for Dried Glucose and Glucose Syrup (Jan)(3218). 9 pp.

—Dried—Sulfur Dioxide Content
CNS N6053-73. Method of Test for Dried Glucose and Glucose Syrup (Jan)(3218). 9 pp.

—Dry Matter Content—Oven Drying
EC 79/786/EEC-79. First Commission Directive Laying down Community Methods of Analysis for Testing Certain Sugars Intended for Human Consumption. 29 pp.

—Iron Content
CNS N6054-72. Method of Test for Solid Glucose (Jun)(3349). 5 pp.

—Loss of Mass
ISO 1741-80. Dextrose—Determination of Loss in Mass on Drying—Vacuum Oven Method First Edition. 4 pp.

—Moisture Content
CNS N6054-72. Method of Test for Solid Glucose (Jun)(3349). 5 pp.

—Powdered
CNS N5102-79. Powdered Glucose (Jan)(3350). 1 p.

—Powdered—Arsenic Content
CNS N6055-72. Method of Test for Powdered Glucose (Jun)(3351). 3 pp.

Glucose (Cont.)
—Powdered—Ash Content
CNS N6055-72. Method of Test for Powdered Glucose (Jun)(3351). 3 pp.

—Powdered—Chloride Content
CNS N6055-72. Method of Test for Powdered Glucose (Jun)(3351). 3 pp.

—Powdered—Dextrin Content
CNS N6055-72. Method of Test for Powdered Glucose (Jun)(3351). 3 pp.

—Powdered—Heavy Metals Content
CNS N6055-72. Method of Test for Powdered Glucose (Jun)(3351). 3 pp.

—Powdered—Hydroxy Methyl Furfurol Content
CNS N6055-72. Method of Test for Powdered Glucose (Jun)(3351). 3 pp.

—Powdered—Moisture Content
CNS N6055-72. Method of Test for Powdered Glucose (Jun)(3351). 3 pp.

—Reducing Sugar Content
EC 79/786/EEC-79. First Commission Directive Laying down Community Methods of Analysis for Testing Certain Sugars Intended for Human Consumption. 29 pp.

—Sulfated Ash Content
EC 79/786/EEC-79. First Commission Directive Laying down Community Methods of Analysis for Testing Certain Sugars Intended for Human Consumption. 29 pp.

Glucose Content Analysis
Use For: D-Glucose Content Analysis
See Also: Glucoside Content Analysis; Sugar Content Analysis

—Analyzers—Enzyme Electrode Method
JIS K 0701-88. Glucose Analyzer. 11 pp.

—Cane Sugar
CNS N6026-79. Method of Test for Cane Sugar (Jan)(1338). 7 pp.

—Dairy Products—Enzyme Assay
DIN ENGL 10326-86. Determination of Sucrose and Glucose Content in Milk Products and Ice Cream; Enzymatic Method (Feb). 5 pp.

—Fruit Juices—Enzymatic Method
CNS N6222-91. Method of Test for Fruit and Vegetable Juices and Drinks (Jun)(12633). 2 pp.

—Honey
CNS N6027-74. Method of Test for Honey (Oct)(1344). 8 pp.

—Maltose Syrups
CNS N6073-82. Method of Test for Malt Sugar Syrup (May)(3584). 5 pp.

—Vegetable Juices—Enzymatic Method
CNS N6222-91. Method of Test for Fruit and Vegetable Juices and Drinks (Jun)(12633). 2 pp.

D-Glucose Content Analysis
Use: Glucose Content Analysis

Glucose Isomerase
JIS K 7002-88. Glucose Isomerase for Industrial Use. 11 pp.

Glucose Syrups
See Also: Fructose Syrups; Maltose Syrups
CNS N5084-79. Glucose Syrup (Jan)(2875). 1 p.

—Arsenic Content
CNS N6053-73. Method of Test for Dried Glucose and Glucose Syrup (Jan)(3218). 9 pp.

—Ash Content
CNS N6053-73. Method of Test for Dried Glucose and Glucose Syrup (Jan)(3218). 9 pp.

—Color Testing
CNS N6053-73. Method of Test for Dried Glucose and Glucose Syrup (Jan)(3218). 9 pp.

—Dried
CNS N5083-79. Dried Glucose Syrup (Tentative) (Jan)(2874). 1 p.

—Dry Matter Content—Oven Drying
EC 79/786/EEC-79. First Commission Directive Laying down Community Methods of Analysis for Testing Certain Sugars Intended for Human Consumption. 29 pp.

Glucose Syrups (Cont.)
—Lead Content
CNS N6053-73. Method of Test for Dried Glucose and Glucose Syrup (Jan)(3218). 9 pp.

—Loss of Mass
ISO 1742-80. Glucose Syrups—Determination of Dry Matter—Vacuum Oven Method First Edition. 4 pp.

—Moisture Content
CNS N6053-73. Method of Test for Dried Glucose and Glucose Syrup (Jan)(3218). 9 pp.

—Reducing Sugar Content
EC 79/786/EEC-79. First Commission Directive Laying down Community Methods of Analysis for Testing Certain Sugars Intended for Human Consumption. 29 pp.

—Refractive Index
ISO 1743-82. Glucose Syrup—Determination of Dry Matter Content—Refractive Index Method Second Edition. 13 pp.

—Solids Content
CNS N6053-73. Method of Test for Dried Glucose and Glucose Syrup (Jan)(3218). 9 pp.

—Sulfated Ash Content
EC 79/786/EEC-79. First Commission Directive Laying down Community Methods of Analysis for Testing Certain Sugars Intended for Human Consumption. 29 pp.

—Sulfur Dioxide Content
CNS N6053-73. Method of Test for Dried Glucose and Glucose Syrup (Jan)(3218). 9 pp.

Glucoside Content Analysis
Use For: Glucosinolates Content Analysis
See Also: Glucose Content Analysis

—Rapeseeds—Liquid Chromatography
BSI BS 4289: Part 9-93. 1993 Methods for the Analysis of Oilseeds Part 9: Determination of Glucosinolates Content of Rapeseed by High Performance Liquid Chromatography (ISO 9167-1: 1992) (W). 15 pp.
ISO 9167 Pt 1-92. Rapeseed—Determination of Glucosinolates Content—Part 1: Method Using High-Performance Liquid Chromatography First Edition. 12 pp.

Glucosinolates Content Analysis
Use: Glucoside Content Analysis

Glue
See Also: Adhesives; Glue Guns
JIS S 6023-92. Glues. 9 pp.

Glue Guns
See Also: Glue

—Portable
CSA C22.2 NO 122-M1989. Hand-Held Electrically Heated Tools; (Gen Instr 1 Thru 2). 50 pp.
CSA 1169 Bull. Electrical Bulletin 1169 June 27, 1978 to C22.2 NO 122. 2 pp.

Glue Spreaders
See Also: Woodworking Equipment
CNS B1307-84. Nominal Dimension of Glue Spreaders (Jun)(10059).
CNS B7247-86. Test Code for Accuracy of Glue Spreaders (Oct)(10227).
CNS B7265-86. Test Code for Performance of Glue Spreaders (Oct)(10788).
JIS B 6549-91. Glue Spreaders—Test and Inspection Methods. 11 pp.

Glued Laminated Wood
Use: Laminated Wood

Glulam
—Structural Timber
CNS O1031-84. Glued-Laminated Timber (Nonstructure Members) (Sep)(11029). 3 pp.
CNS O1034-84. Decorative Glued-Laminated Timber (Structure Members) (Sep)(11032). 4 pp.
CNS O2056-84. Method of Test for Glued-Laminated Timber (Sep)(11033). 5 pp.
CSA CAN/CSA-O122-M89. Structural Glued-Laminated Timber; (Gen Instr 1). 31 pp.
CSA CAN/CSA-O177-M89. Qualification Code for Manufacturers of Structural Glued-Laminated Timber; (Gen Instr 1). 31 pp.

Glutamic Acid Content Analysis
—Meat
BSI BS 4401: Part 14-79. 1979 Methods of Test for Meat and Meat Products Part 14: Determination of L(+)-Glutamic Acid Content (Reference Method). 7 pp.
CNS N6148-80. Method of Test for Meat and Meat Products—Determination of L—(+) -Glutamic Acid Content (Oct)(6609). 4 pp.
ISO 4134-78. Meat and Meat Products—Determination of L-(+)-Glutamic Acid Content (Reference Method) First Edition; (Erratum—June 1979). 7 pp.

L-Glutamic Acid
JIS K 9047-72. L-Glutamic Acid.

L-Glutamine
JIS K 9103-76. L-Glutamine.

Glutathione
JIS K 9509-78. Glutathione, Reduced Form.

Gluten Content Analysis
—Wheat Flour
BSI BS 4317: Part 12-80. 1980 Methods of Test for Cereals and Pulses Part 12: Determination of Wet Gluten in Wheat Flour. 6 pp.
BSI BS 4317: Part 16-81. 1981 Methods of Test for Cereals and Pulses Part 16: Determination of Dry Gluten in Wheat Flour. 4 pp.
ISO 5531-78. Wheat Flour—Determination of Wet Gluten First Edition. 5 pp.
ISO 6645-81. Wheat Flour—Determination of Dry Gluten First Edition. 4 pp.
ISO 7495-90. Wheat Flour—Determination of Wet Gluten Content by Mechanical Means First Edition. 9 pp.

Glutens
See Also: Food
CNS N5070-72. Gluten (Jun)(2434). 1 p.

Glyceride Content Analysis
—Fats
BSI BS 684: Sec 2.42-89. 1989 Methods of Analysis of Fats and Fatty Oils Section 2.42: Determination of 1-Monoglycerides and Free Glycerol Contents. 6 pp.
ISO 7366-87. Animal and Vegetable Fats and Oils—Determination of 1-Monoglycerides and Free Glycerol Contents First Edition. 6 pp.

—Oils
BSI BS 684: Sec 2.42-89. 1989 Methods of Analysis of Fats and Fatty Oils Section 2.42: Determination of 1-Monoglycerides and Free Glycerol Contents. 6 pp.
ISO 7366-87. Animal and Vegetable Fats and Oils—Determination of 1-Monoglycerides and Free Glycerol Contents First Edition. 6 pp.

Glycerine
Use: Glycerol

Glycerol
Use For: Glycerine; Glycyl Alcohol
See Also: Alcohols; Glycerol Content Analysis
BSI BS 2621-5-79. 1979 Amd 1 Glycerol (Glycerine). 11 pp.
CNS K1039-87. Glycerines for Industrial Use (Jun)(633).
CNS K5065-80. High-Gravity Glycerin (Jul)(5834).
CNS K6010-87. Method of Test for Glycerines for Industrial Use (Jun)(117). 16 pp.
CNS K7240-67. Chemical Reagent (Glycerol) (Oct)(1739).
JIS K 8295-78. Glycerol.

—Acidity—Volumetric Analysis
BSI BS 5711: Part 5-79. 1979 Methods of Sampling and Test for Glycerol. 2 pp.
ISO 1615-76. Glycerines for Industrial Use—Determination of Alkalinity or Acidity—Titrimetric Method First Edition. 4 pp.

—Alkalinity—Volumetric Analysis
BSI BS 5711: Part 5-79. 1979 Methods of Sampling and Test for Glycerol. 2 pp.
ISO 1615-76. Glycerines for Industrial Use—Determination of Alkalinity or Acidity—Titrimetric Method First Edition. 4 pp.

—Arsenic Content—Photometry
BSI BS 5711: Part 10-79. 1979 Methods of Sampling and Test for Glycerol Part 10: Determination of Arsenic Content: Silver Diethyldithiocarbamate Method. 4 pp.
ISO 2465-74. Glycerols for Industrial Use—Determination of Arsenic Content—Silver Diethyldithiocarbamate Photometric Method First Edition. 5 pp.

Glycerol (Cont.)
—Ash Content—Gravimetric Analysis
BSI BS 5711: Part 6-79. 1979 Methods of Sampling and Test for Glycerol Part 6: Determination of Ash: Gravimetric Method. 2 pp.
ISO 2098-72. Glycerols for Industrial Use—Determinaton of Ash—Gravimetric Method First Edition. 3 pp.

—Chemical Analysis
BSI BS 5711: Part 9-79. 1979 Methods of Sampling and Test for Glycerol Part 9: Calculation of Matter (Organic) Non-Glycerol (MONG). 2 pp.
ISO 2464-73. Crude Glycerine for Industrial Use—Calculation of Matter (Organic) Non-Glycerol (MONG) First Edition. 3 pp.

—Chloride Content—Turbidimetry
BSI BS 5711: Part 13-79. 1979 Methods of Sampling and Test for Glycerol Part 13: Limit Test for Chloride. 2 pp.
BSI BS 5711: Part 14-79. 1979 Methods of Sampling and Test for Glycerol Part 14: Limit Test for Organic Chloride. 2 pp.

—Chloride Content—Volumetric Analysis
BSI BS 5711: Part 12-79. 1979 Methods of Sampling and Test for Glycerol Part 12: Determination of Chloride Content. 2 pp.

—Density
BSI BS 5711: Part 4-79. 1979 Methods of Sampling and Test for Glycerol Part 4: Determination of Density and Relative Density. 6 pp.
ISO 2099-72. Purified Glycerol for Industrial Use—Determination of Density at 20 Degrees Celsius First Edition. 6 pp.

—Dynamite
JIS K 3351-56. Dynamite Glycerine. 3 pp.

—Glycerol Content
BSI BS 5711: Part 3-79. 1979 Methods of Sampling and Test for Glycerol Part 3: Determination of Glycerol Content. 6 pp.

—Heavy Metals Content
BSI BS 5711: Part 15-79. 1979 Amd 1 Methods of Sampling and Test for Glycerol Part 15: Determination of Heavy Metals Content. 3 pp.

—Industrial
JIS K 3351-84. Glycerines for Industrial Use.
JIS K 3351-56. Dynamite Glycerine. 3 pp.

—Iron Content—Photometry
BSI BS 5711: Part 16-79. 1979 Methods of Sampling and Test for Glycerol Part 16: Determination of Iron Content. 2 pp.

—Lead Content
BSI BS 5711: Part 17-83. 1983 Methods of Sampling and test for Glycerol Part 17: Limit Test for Lead. 2 pp.

—Nitroglycerin
MOD UK TS 648B. Glycerine, Special.

—Odor Detection
BSI BS 5711: Part 19-79. 1979 Methods of Sampling and Test for Glycerol Part 19: Assessment of Odour. 2 pp.

—Propane Content—Gas Chromatography
BSI BS 5711: Part 11-79. 1979 Methods of Sampling and Test for Glycerol Part 11: Determination of Propane, 1,3-Diol Content: Gas Chromatographic Method. 4 pp.

—Reduction
BSI BS 5711: Part 20-79. 1979 Methods of Sampling and Test for Glycerol Part 20: Test for Reducing Substances. 2 pp.

—Sampling
BSI BS 2621-5-79. 1979 Amd 1 Glycerol (Glycerine). 11 pp.
BSI BS 5711: Part 0-79. 1979 Methods of Sampling and Test for Glycerol Part 0: General Introduction. 4 pp.
BSI BS 5711: Part 1-79. 1979 Methods of Sampling and Test for Glycerol Part 1: Samples and Test Methods: General. 4 pp.
BSI BS 5711: Part 2-79. 1979 Methods of Sampling and Test for Glycerol Part 2: Methods of Sampling (ISO 2096-1972). 8 pp.
BSI BS 5711: Part 22-79. 1979 Amd 1 Methods of Sampling and Test for Glycerol Part 22: Additional Methods of Test (Non-Specification). 9 pp.
ISO 1614-76. Glycerines for Industrial Use—Samples and Test Methods—General First Edition. 4 pp.
ISO 2096-72. Glycerols for Industrial Use—Methods of Sampling First Edition. 8 pp.

Glycerol (Cont.)
—Saponification Number—Volumetric Analysis
BSI BS 5711: Part 21-79. 1979 Methods of Sampling and Test for Glycerol Part 21: Determination of Saponification Equivalent. 2 pp.

—Standard Atmospheres
DIN ENGL 50008 Pt 1-81. Atmospheres and Their Technical Application; Standard Atmospheres over Aqueous Solutions; Saturated Salt Solutions, Glycerol Solutions (Feb). 4 pp.

—Sugar Content—Colorimetric Analysis
BSI BS 5711: Part 18-79. 1979 Methods of Sampling and Test for Glycerol Part 18: Detection of Sugars. 2 pp.

—Sulfated Ash Content—Gravimetric Analysis
BSI BS 5711: Part 7-79. 1979 Methods of Sampling and Test for Glycerol Part 7: Determination of Sulphated Ash: Gravimetric Method. 2 pp.
ISO 1616-76. Glycerines for Industrial Use—Determination of Sulphated Ash—Gravimetric Method First Edition. 4 pp.

—Volumetric Analysis
ISO 2879-75. Glycerine for Industrial Use—Determination of Glycerol Content—Titrimetric Method First Edition. 4 pp.

—Water Content—Karl Fischer Method
BSI BS 5711: Part 8-79. 1979 Methods of Sampling and Test for Glycerol Part 8: Determination of Water Content: Karl Fischer Method. 2 pp.
ISO 2097-72. Glycerols for Industrial Use—Determination of Water Content—Karl Fischer Method First Edition. 4 pp.

Glycerol 1-(4-Aminobenzoate) Content Analysis
Use: Glyceryl p-Aminobenzoate Content Analysis

Glycerol Content Analysis
See Also: Glycerol
—Fats
BSI BS 684: Sec 2.42-89. 1989 Methods of Analysis of Fats and Fatty Oils Section 2.42: Determination of 1-Monoglycerides and Free Glycerol Contents. 6 pp.
ISO 7366-87. Animal and Vegetable Fats and Oils—Determination of 1-Monoglycerides and Free Glycerol Contents First Edition. 6 pp.

—Glycerol
BSI BS 5711: Part 3-79. 1979 Methods of Sampling and Test for Glycerol Part 3: Determination of Glycerol Content. 6 pp.

—Oils
BSI BS 684: Sec 2.42-89. 1989 Methods of Analysis of Fats and Fatty Oils Section 2.42: Determination of 1-Monoglycerides and Free Glycerol Contents. 6 pp.
ISO 7366-87. Animal and Vegetable Fats and Oils—Determination of 1-Monoglycerides and Free Glycerol Contents First Edition. 6 pp.

—Soaps—Spectrophotometry
ISO 2272-89. Surface Active Agents—Soaps—Determination of Low Contents of Free Glycerol by Molecular Absorption Spectrometry Second Edition. 6 pp.

—Soaps—Volumetric Analysis
BSI BS 1715: Sec 2.8-89. 1989 Analysis of Soaps Part 2: Quantitative Test Methods Section 2.8: Method for Determination of Glycerol Content (ISO 1066-1975). 7 pp.
ISO 1066-75. Analysis of Soaps—Determination of Glycerol Content—Titrimetric Method First Edition. 5 pp.
MOD UK M 9532/67. Examination of Soap, Saddle, Glycerine (No Information) (Withdrawn).

Glyceryl p-Aminobenzoate Content Analysis
Use For: Glycerol 1-(4-Aminobenzoate) Content Analysis
—Cosmetics
EC 85/490/EEC-85. Fourth Commission Directive on the Approximation of the Laws of the Member States Relating to Methods of Analysis Necessary for Checking the Composition of Cosmetic Products. 16 pp.

Glyceryl Diacetate
Use: Diacetin

Glyceryl Triacetate
Use: Triacetin

gamma-Glycidoxypropyltrimethoxysilane
Use For: 3-Glycidoxypropyltrimethoxysilane
See Also: Ethers
    MOD UK DSTAN 80-133-89. 3-Glycidoxypropyltrimethoxy Silane Issue 1. 10 pp.

3-Glycidoxypropyltrimethoxysilane
Use: gamma-Glycidoxypropyltrimethoxysilane

Glycidyl Esters
See Also: Esters

—Chlorine Content
    BSI BS 2782:Pt4: METH 433B-79. 1979 Methods of Testing Plastics Part 4: Chemical Properties Method 433B: Determination of Easily Saponifiable Chlorine in Epoxide Resins and Related Materials. 4 pp.
    ISO 4583-78. Plastics—Epoxide Resins and Related Materials—Determination of Easily Saponifiable Chlorine First Edition. 5 pp.

—Chlorine Content—Potentiometric Analysis
    BSI BS 2782:Pt4: METH 433A-79. 1979 Methods of Testing Plastics Part 4: Chemical Properties Method 433A: Determination of Inorganic Chlorine in Epoxide Resins and Glycidyl Esters. 4 pp.
    ISO 4573-78. Plastics—Epoxide Resins and Glycidyl Esters—Determination of Inorganic Chlorine First Edition. 4 pp.

Glycine
Use For: Aminoacetic Acid
    CNS K7040-61. Chemical Reagent (Aminoacetic Acid) (Glycocoll; Glycine) (Dec)(1540).
    JIS K 8291-91. Glycine (Aminoacetic Acid).

Glycols
Scope Note: Use a more specific term
See: Alcohols; Diethylene Glycol; Dipropylene Glycols; Ethylene Glycol; Hexylene Glycol; Propylene Glycols; Triethylene Glycol

Glycyl Alcohol
Use: Glycerol

Glyoxal Content Analysis
—Paperboard
    DIN ENGL 54603-81. Testing of Paper, Boxboard and Paper Board; Determination of Glyoxal Content (Sept). 3 pp.

—Papers
    DIN ENGL 54603-81. Testing of Paper, Boxboard and Paper Board; Determination of Glyoxal Content (Sept). 3 pp.

GMDSS
Use: Global Maritime Distress and Safety Systems

Go Gages
See Also: Limit Gages
    DIN ENGL 2248 Pt 1-89. GO Gauging Members from 1 mm to 40 mm Nominal Diameter (Nov). 2 pp.
    DIN ENGL 2248 Pt 2-89. GO Gauging Members over 40 mm up to 120 mm Nominal Diameter (Nov). 2 pp.
    DIN ENGL 2248 Pt 3-89. GO Segmental Cylindrical Bar Gauging Members over 120 mm up to 200 mm Nominal Diameter (Nov). 2 pp.
    DIN ENGL 2248 Pt 4-89. GO Gauging Members over 40 mm up to 250 mm Nominal Diameter for Use in Precision Engineering (Nov). 2 pp.

—Plug
    DIN ENGL 2245 Pt 1-89. GO and NOT GO Plug Gauges from 1 mm to 40 mm Nominal Diameter (Nov). 2 pp.
    DIN ENGL 2245 Pt 2-89. GO and NOT GO Plug Gauges over 40 mm up to 65 mm Nominal Diameter (Nov). 2 pp.
    DIN ENGL 2246 Pt 1-89. GO Plug Gauges from 1 mm to 40 mm Nominal Diameter (Nov). 2 pp.
    DIN ENGL 2246 Pt 2-89. GO Segmental Cylindrical Bar Gauges over 120 mm up to 200 mm Nominal Diameter (Nov). 2 pp.
    DIN ENGL 2246 Pt 3-89. GO Plug Gauges over 40 mm up to 120 mm Nominal Diameter (Nov). 2 pp.
    DIN ENGL 2246 Pt 4-89. GO Plug Gauges over 40 mm up to 250 mm Nominal Diameter for Use in Precision Engineering (Nov). 2 pp.
    DIN ENGL 2261-55. Measuring Instruments; Serrated Limit Plug Gauges for Serrations 7 x 8 to 26 x 30 According to DIN 5481 Part 1 (Oct). 1 p.

Go Gages (Cont.)
—Plug (Cont.)
    DIN ENGL 2262 Pt 1-55. Measuring Instruments; Serrated "Go" Plug Gauges for Serrations 7 x 8 to 26 x 30 According to DIN 5481 Part 1 (Oct). 2 pp.
    DIN ENGL 2262 Pt 2-55. Measuring Instruments; Serrated "Go" Plug Gauges for Serrations 30 x 34 to 120 x 125 According to DIN 5481 Part 1 (Oct). 2 pp.

—Ring
    DIN ENGL 2250 Pt 1-89. General-Purpose GO Ring Gauges and Setting Ring Gauges and Setting Ring Gauges for Pneumatic Length Measuring Instruments, from 1 mm to 315 mm Nominal Diameter (Nov). 3 pp.
    DIN ENGL 2250 Pt 2-89. GO Ring Gauges and Setting Ring Gauges from 1 mm to 315 mm Nominal Diameter for Use in Precision Engineering (Nov). 3 pp.
    DIN ENGL 2264-55. Measuring Instruments; Serrated "Go" Ring Gauges for Serrations According to DIN 5481 Part 1 (Oct). 2 pp.

—Thread
    CNS B6016-55. Inspection Gauge for Metric Thread (Go End Wear Limit, Fine Medium and Coarse Fits) (Sep)(536)(R 1970).

—Thread Plug
    CNS B6007-55. Plug Gauge for Whitworth Thread (Go End Limit for Fine, Medium and Coarse Fits) (Sep)(523)(R 1970).
    CNS B6011-55. Inspection Plug Gauge for Whitworth Thread Ring Gauge (Go End Limit for Fine, Medium and Coarse Fits) (Sep)(527)(R 1970).
    CNS B6012-55. Adjustable Plug Gauge for Whitworth Thread Ring Gauge (for Testing of Go and No Go Ring Gauge, Fine, Medium and Coarse Fits) (Sep)(528)(R 1970).
    CNS B6018-55. Inspection Plug Gauge for Metric Thread Ring Gauge (Go End Limit, Fine, Medium and Coarse Fits) (Sep)(538)(R 1970).
    CNS B6019-55. Adjustable Plug Gauge for Metric Thread (for Testing of Go and No Go Ring Gauge, Fine, Medium and Coarse Fits) (Sep)(539)(R 1970).
    DIN ENGL 2280-89. GO and NOT GO Screw Plug Gauges for ISO Metric Screw Threads from 1 mm to 40 mm Nominal Diameter (Nov). 2 pp.
    DIN ENGL 2281 Pt 1-89. GO Workshop Screw Plug Gauges, Check Plugs and Setting Plugs for ISO Metric Screw Threads from 1 mm to 40 mm Nominal Diameter (Nov). 2 pp.
    DIN ENGL 2281 Pt 2-89. GO Workshop Screw Plug Gauges, Check Plugs and Setting Plugs for ISO Metric Screw Threads over 40 mm up to 200 mm Nominal Diameter (Nov). 2 pp.
    DIN ENGL 2281 Pt 3-89. GO Workshop Screw Plug Gauges and Check Plugs for ISO Metric Screw Threads over 40 mm up to 250 mm Nominal Diameter for Use in Precision Engineering (Nov). 2 pp.
    DIN ENGL 2282 Pt 1-89. GO Screw Gauging Members for Workshop Plug Gauges, Check Plugs and Setting Plugs for ISO Metric Screw Threads from 1 mm to 40 mm Nominal Diameter (Nov). 4 pp.
    DIN ENGL 2282 Pt 2-89. GO Screw Gauging Members for Workshop Plug Gauges, Check Plugs and Setting Plugs for ISO Metric Screw Threads over 40 mm up to 200 mm Nominal Diameter (Nov). 4 pp.
    DIN ENGL 2282 Pt 3-89. GO Screw Gauging Members for Workshop Plug Gauges and Check Plugs for ISO Metric Screw Threads over 40 mm up to 250 mm Nominal Diameter for Use in Precision Engineering (Nov). 3 pp.
    DIN ENGL 2285 Pt 1-89. General-Purpose GO Screw Ring Gauges for ISO Metric Screw Threads from 1 mm to 200 mm Nominal Diameter (Nov). 4 pp.
    DIN ENGL 40431 Pt 2-72. Steel Conduit Thread; Thread Gauges; "Go" and "Not Go" Thread Plug Gauges (Nov). 3 pp.

—Thread Ring
    CNS B6010-55. Ring Gauge for Whitworth Thread (Go End Limit for Fine, Medium and Coarse Fits) (Sep)(526)(R 1970).
    CNS B6011-55. Inspection Plug Gauge for Whitworth Thread Ring Gauge (Go End Limit for Fine, Medium and Coarse Fits) (Sep)(527)(R 1970).
    CNS B6012-55. Adjustable Plug Gauge for Whitworth Thread Ring Gauge (for Testing of Go and No Go Ring Gauge, Fine, Medium and Coarse Fits) (Sep)(528)(R 1970).
    CNS B6017-55. Ring Gauge for Metric Thread (Go End Limit, Fine Medium and Coarse Fits) (Sep)(537)(R 1970).
    CNS B6018-55. Inspection Plug Gauge for Metric Thread Ring Gauge (Go End Limit, Fine, Medium and Coarse Fits) (Sep)(538)(R 1970).

Go Gages (Cont.)
—Thread Ring (Cont.)
    CNS B6019-55. Adjustable Plug Gauge for Metric Thread (for Testing of Go and No Go Ring Gauge, Fine, Medium and Coarse Fits) (Sep)(539)(R 1970).
    DIN ENGL 2285 Pt 2-89. GO Screw Ring Gauges for ISO Metric Screw Threads over 50 mm up to 200 mm Nominal Diameter for Use in Precision Engineering (Nov). 2 pp.
    DIN ENGL 40431 Pt 1-70. Steel Conduit Thread; Thread Gauges; "Go" and "Not Go" Thread Ring Gauges (Nov). 2 pp.

—Thread Snap
    CNS B6009-55. Adjustable Snap Gauge for Whitworth Thread (for Go End Limit and Wear Test; Fine, Medium and Coarse Fits) (Sep)(525) (R 1970).

Go Gauges
Use: Go Gages

Goals (Sports Equipment)
See Also: Sports Equipment

—Football
    CEN PREN 748-92. Football Goals. 10 pp.

—Handball
    CEN PREN 749-92. Handball Goals. 10 pp.

—Hockey
    CEN PREN 750-92. Hockey Goals. 8 pp.

Goats
Use For: Caprines *See Also:* Livestock

—Breeding Stock
    EC COM(87) 591-87. Proposal for a Council Directive Concerning Pure-Bred Breeding Sheep and Goats. 7 pp.
    EC COM(88) 742-88. Proposal for a Council Regulation (EEC) on Animal Health Conditions Governing Intra-Community Trade in Ovine and Caprine Animals: Proposal for a Council Directive Amending Directive 72/462/EEC. 42 pp.
    EC 89/361/EEC-89. Council Directive Concerning Pure-Bred Breeding Sheep and Goats. 2 pp.

—Hair
    CNS N1078-85. Goat Hair (May)(2824).

—Health and Veterinary Inspection
    EC COM(88) 742-88. Proposal for a Council Regulation (EEC) on Animal Health Conditions Governing Intra-Community Trade in Ovine and Caprine Animals: Proposal for a Council Directive Amending Directive 72/462/EEC. 42 pp.
    EC 90/425/EEC-90. Council Directive Concerning Veterinary and Zootechnical Checks Applicable in Intra-Community Trade in Certain Live Animals and Products with a View to the Completion of the Internal Market. 13 pp.
    EC 91/68/EEC-91. Council Directive on Animal Health Conditions Governing Intra-Community Trade in Ovine and Caprine Animals. 19 pp.
    EC 91/628/EEC-91. Council Directive on the Protection of Animals During Transport and Amending Directives 90/425/EEC and 91/496/EEC. 12 pp.

—Health and Veterinary Inspection—Importation
    EC 91/69/EEC-91. Council Directive Amending Directive 72/462/EEC on Health and Veterinary Inspection Problems upon Importation of Bovine Animals and Swine, Fresh Meat or Meat Products from Third World Countries, in Order to Include Ovine and Caprine Animals. 4 pp.
    EC 93/451/EEC-93. Commission Decision Concerning the Animal Health Conditions and Veterinary Certification of Imports of Fresh Meat from Austria. 6 pp.

Goats Milk
See Also: Animal Products; Milk
    CNS N5092-87. Raw Milk (Oct)(3055). 2 pp.

—Powdered
    CNS N5217-86. Full Fat Goat Milk Powder (Feb)(11509). 2 pp.

Goggles
See Also: Eye Protectors; Night Vision Goggles

—Swimming
    BSI BS 5883-80. 1980 Surface Swimming Goggles. 7 pp.
    JIS S 7301-92. Swimming Goggles. 8 pp.

Goggles, Safety
Use: Eye Protectors

INTERNATIONAL AND NON-U.S. NATIONAL STANDARDS
SUBJECT INDEX

Gold

Gold
See Also: Gold Castings; Metals; Precious Metals

—Atomic Absorption Spectroscopy
SAA AS 3895.1-91. Methods for the Analysis of Copper, Lead, Zinc, Gold and Silver Ores—Part 1: Determination of Gold (Fire Assay—Flame AAS Method). 9 pp.

—Composition
SAA AS 2140-78. Composition and Marking Requirements of Gold Articles Amdt 1 March 1980 (This is a Joint Standard with SANZ NZS 2140). 8 pp.

SNZ NZS/AS 2140-78. Composition and Marking Requirements of Gold Articles Amend: 1 (This is a Joint Standard with SAA AS 2140). 8 pp.

—Dry Matter Content
JIS M 8101-88. Methods for Sampling, Preparation and Determination of Moisture Content of Non-Ferrous Metal Bearing Ores. 25 pp.

—Identification Systems
SAA AS 2140-78. Composition and Marking Requirements of Gold Articles Amdt 1 March 1980 (This is a Joint Standard with SANZ NZS 2140). 8 pp.

SNZ NZS/AS 2140-78. Composition and Marking Requirements of Gold Articles Amend: 1 (This is a Joint Standard with SAA AS 2140). 8 pp.

—Moisture Content
JIS M 8101-88. Methods for Sampling, Preparation and Determination of Moisture Content of Non-Ferrous Metal Bearing Ores. 25 pp.

—Sampling
CNS M3032-81. Methods for Sampling of Metal Bearing Ores of Copper, Lead, Zinc, Tin, Gold, Silver and Others (Apr)(7280).

SAA AS 3988-91. Copper, Lead, Zinc, Gold and Silver Ores—Guide to Sample Preparation for the Determination of Gold. 16 pp.

Gold (Dental)
Use: Dental Materials

Gold Alloy Castings
Use: Gold Castings

Gold Alloys
Scope Note: For additional listings, see also specific products made from gold alloys *See Also:* Alloys; Gold Castings

—Colors
BSI BS 7031-88. (WITHDRAWN) 1988 Colours of Gold Alloys (Superseded by BS EN 28654: 1993). 4 pp.

BSI BS EN 28654-93. 1993 Colours of Gold Alloys—Definition, Range of Colours and Designation (ISO 8654: 1987) (Supersedes BS 7031: 1988) (H). 10 pp.

CEN EN 28654-92. Colours of Gold Alloys—Definition, Range of Colours and Designation (ISO 8654: 1987). 5 pp.

ISO 8654-87. Colours of Gold Alloys—Definition, Range of Colours and Designation First Edition; (CEN EN 28654:1992). 6 pp.

—Gold Content—Cupellation
ISO 11426-93. Determination of Gold in Gold Jewellery Alloys—Cupellation Method (Fire Assay) First Edition. 7 pp.

Gold Castings
Use For: Gold Alloy Castings *See Also:* Castings; Gold; Gold Alloys

—Dental
BSI BS 4425-88. 1988 Amd 1 Dental Casting Gold Alloys (N) (ISO 1562-1984) (AMD 6469) April 30, 1991. 9 pp.

BSI BS 6042-01. 1990 Amd 1 Dental Semi-Precious Metal Casting Alloys (ISO 8891: 1990) (AMD 7019) May 1, 1992. 16 pp.

BSI BS EN 27490-92. 1992 Dental Gypsum-Bonded Casting Investments for Gold Alloys (ISO 7490: 1990). 24 pp.

CEN EN 21 562-89. Dentistry Dental Casting Gold Alloys. 8 pp.

CEN PREN 28 891-90. Dental Casting Alloys with Noble Metal Content of 25 % up to but Not Including 75 %. 3 pp.

CEN EN 28891-91. Dental Casting Alloys with Noble Metal Content of 25 % up to but Not Including 75 %. 3 pp.

CNS T3006-80. Dental Casting Gold-Silver-Palladium Alloy (Oct)(6616).

CNS T3013-80. Dental Casting 14K Gold Alloy (Oct)(6623).

CNS T3014-80. Plus Metals for Dental Casting 14K Gold Alloy (Oct)(6624).

Gold Castings (Cont.)
—Dental (Cont.)
CSA CAN/CSA-Z349.44-93. Dental Casting Alloys with Noble Metal Content of 25% up to but Not Including 75% (ISO 8891-1990); (Gen Instr 1). 32 pp.

ISO 1562-84. Dental Casting Gold Alloys Second Edition. 6 pp.

ISO 8891-90. Dental Casting Alloys with Noble Metal Content of 25 % up to but Not Including 75 % First Edition. 9 pp.

JIS T 6106-91. Dental Casting Gold-Silver-Palladium Alloys. 9 pp.

JIS T 6113-87. Dental Casting 14K Gold Alloy. 6 pp.

JIS T 6114-87. Plus Metals for Dental Casting 14K Gold Alloys. 5 pp.

JIS T 6116-90. Dental Casting Gold Alloys. 6 pp.

—Jewelry
ISO 10713-92. Jewellery—Gold Alloy Coatings First Edition. 6 pp.

Gold Chloride
Use: Gold Trichloride

Gold Coatings (Made From Gold)
Scope Note: Includes coatings made from gold alloys *See Also:* Coatings; Metal Coatings (Made From Metal)

—Electrodeposited
ISO 4523-85. Metallic Coatings—Electrodeposited Gold and Gold Alloy Coatings for Engineering Purposes First Edition. 8 pp.

MOD UK DSTAN 03-17-79. Electro-Deposition of Gold Issue 1. 21 pp.

MOD UK DSTAN 03-17-92. Electro—Deposition of Gold Issue 2. 22 pp.

—Electrodeposited—Adhesion Testing
BSI BS 6670: Part 5-86. 1986 Methods of Test for Electroplated Gold and Gold Alloy Coatings Part 5: Adhesion Tests. 4 pp.

ISO 4524 Pt 5-85. Metallic Coatings—Test Methods for Electrodeposited Gold and Gold Alloy Coatings—Part 5: Adhesion Tests First Edition. 3 pp.

—Electrodeposited—Environmental Testing
BSI BS 6670: Part 2-86. 1986 Methods of Test for Electroplated Gold and Gold Alloy Coatings Part 2: Environmental Tests. 7 pp.

ISO 4524 Pt 2-85. Metallic Coatings—Test Methods for Electrodeposited Gold and Gold Alloy Coatings—Part 2: Environmental Tests First Edition. 5 pp.

—Electrodeposited—Gold Content
BSI BS 6670: Part 4-86. 1986 Methods of Test for Electroplated Gold and Gold Alloy Coatings Part 4: Determination of Gold Content. 4 pp.

ISO 4524 Pt 4-85. Metallic Coatings—Test Methods for Electrodeposited Gold and Gold Alloy Coatings—Part 4: Determination of Gold Content First Edition. 4 pp.

—Electrodeposited—Porosity
BSI BS 6670: Part 3-86. 1986 Methods of Test for Electroplated Gold and Gold Alloy Coatings Part 3: Electrographic Tests for Porosity. 7 pp.

ISO 4524 Pt 3-85. Metallic Coatings—Test Methods for Electrodeposited Gold and Gold Alloy Coatings—Part 3: Electrographic Tests for Porosity First Edition; (Corrected and Reprinted -1986). 5 pp.

—Electrodeposited—Residual Salt Content
BSI BS 6670: Part 6-89. 1989 Methods of Test for Electroplated Gold and Gold Alloy Coatings Part 6: Detection of Residual Salts. 4 pp.

ISO 4524 Pt 6-88. Metallic Coatings—Test Methods for Electrodeposited Gold and Gold Alloy Coatings—Part 6: Determination of the Presence of Residual Salts First Edition. 4 pp.

—Electrodeposited—Thickness Measurement
BSI BS 6670: Part 1-86. 1986 Methods of Test for Electroplated Gold and Gold Alloy Coatings Part 1: Determination of Coating Thickness. 8 pp.

ISO 4524 Pt 1-85. Metallic Coatings—Test Methods for Electrodeposited Gold and Gold Alloy Coatings—Part 1: Determination of Coating Thickness First Edition. 7 pp.

—Electroplated
BSI BS 4292: Part 1-89. 1989 Method for Specifying Electroplated Coatings of Gold and Gold Alloys Part 1: Gold and Gold Alloys for Engineering Purposes. 12 pp.

BSI BS 4292: Part 2-89. 1989 Method for Specifying Electroplated Coatings of Gold and Gold Alloys Part 2: Gold and Gold Alloys for Decorative and Protective Purposes. 10 pp.

Gold Coatings (Made From Gold) (Cont.)
—Electroplated (Cont.)
BSI BS 5658-79. 1979 Gold Potassium Cyanide for Electro-Plating. 12 pp.

CNS H3076-87. Electroplated Coatings of Gold for Engineering Use (Aug)(4824).

CNS H3144-87. Electroplated Coatings of Gold for Decorative Use (Aug)(12048).

JIS H 8620-93. Electroplated Coatings of Gold and Gold Alloy for Engineering Purposes. 19 pp.

JIS H 8622-93. Electroplated Coatings of Gold and Gold Alloy for Decorative Purposes. 14 pp.

SAA AS 1901-93. Electroplated Coatings—Gold and Gold Alloys (Supersedes AS 1902—1976). 9 pp.

SAA AS 1902-76. Electroplated Coatings of Gold and Gold Alloys for Decorative Applications (Superseded by AS 1901—1993).

—Jewelry
SAA AS 2335-80. Rolled Gold, Gold-Plated and Silver-Plated Jewellery—Composition and Marking (This is a Joint Standard with SANZ NZS 2335). 8 pp.

SNZ NZS/AS 2335-80. Rolled Gold, Gold-Plated and Silver-Plated Jewellery—Composition and Marking (This is a Joint Standard with SAA AS 2335). 8 pp.

—Watch Cases
ISO 3160 Pt 1-82. Watch Cases and Accessories—Gold Alloy Coverings—Part 1: General Requirements First Edition. 4 pp.

—Watch Cases—Abrasion Testing
ISO 3160 Pt 3-93. Watch Cases and Accessories—Gold Alloy Coverings—Part 3: Abrasion Resistance Tests of a Type of Coating on Standard Guages First Edition. 7 pp.

—Watch Cases—Adhesion Testing
ISO 3160 Pt 2-92. Watch Cases and Accessories—Gold Alloy Coverings—Part 2: Determination of Fineness, Thickness, Corrosion Resistance and Adhesion Second Edition. 12 pp.

—Watch Cases—Corrosion Testing
ISO 3160 Pt 2-92. Watch Cases and Accessories—Gold Alloy Coverings—Part 2: Determination of Fineness, Thickness, Corrosion Resistance and Adhesion Second Edition. 12 pp.

—Watch Cases—Fineness
ISO 3160 Pt 2-92. Watch Cases and Accessories—Gold Alloy Coverings—Part 2: Determination of Fineness, Thickness, Corrosion Resistance and Adhesion Second Edition. 12 pp.

—Watch Cases—Thickness Measurement
ISO 3160 Pt 2-92. Watch Cases and Accessories—Gold Alloy Coverings—Part 2: Determination of Fineness, Thickness, Corrosion Resistance and Adhesion Second Edition. 12 pp.

Gold Content Analysis
—Blister Copper
CNS M3041-81. Methods for Determination of Gold and Silver in Blister Coppers (Apr) (7289).

JIS M 8114-50. Determination of Gold and Silver in Blister Copper.

—Bullion
CNS M3042-81. Methods for Determination of Gold and Silver in Crude Bullions (Apr) (7290).

JIS M 8115-50. Determination of Gold and Silver in Crude Bullion.

—Cyanide Precipitates
CNS M3039-81. Methods for Determination of Gold and Silver in Cyanidation Precipitates (Apr)(7287).

—Gold Alloys—Cupellation
ISO 11426-93. Determination of Gold in Gold Jewellery Alloys—Cupellation Method (Fire Assay) First Edition. 7 pp.

—Gold Coatings
BSI BS 6670: Part 4-86. 1986 Methods of Test for Electroplated Gold and Gold Alloy Coatings Part 4: Determination of Gold Content. 4 pp.

ISO 4524 Pt 4-85. Metallic Coatings—Test Methods for Electrodeposited Gold and Gold Alloy Coatings—Part 4: Determination of Gold Content First Edition. 4 pp.

—Mattes
CNS M3040-81. Methods for Determination of Gold and Silver in Mattes and Speises (Apr) (7288).

—Ores
CNS M3033-81. General Rules for Determination of Gold and Silver in Ores (Apr) (7282).

JIS M 8111-63. Methods for Determination of Gold and Silver in Ores (R 1972). 19 pp.

INDUSTRY STANDARDS

INTERNATIONAL AND NON-U.S. NATIONAL STANDARDS SUBJECT INDEX

Gold Content Analysis (Cont.)

—Ores—Bismuth
CNS M3038-81. Methods for Determination of Gold and Silver in Ores with Bismuth (Apr) (7286).

—Ores—Tellurium
CNS M3037-81. Methods for Determination of Gold and Silver in Ores with Tellurium (Apr)(7285).

—Oxide Ores
CNS M3035-81. Methods for Determination of Gold and Silver in Oxide Ores (Apr)(7283).

—Siliceous Ores
CNS M3034-81. Methods for Determination of Gold and Silver in Silicate Ores (Apr) (7282).

—Speisses
CNS M3040-81. Methods for Determination of Gold and Silver in Mattes and Speises (Apr) (7288).

—Sulfide Ores
CNS M3036-81. Methods for Determination of Gold and Silver in Sulphide Ores (Apr) (7284).

Gold Trichloride
Use For: Chloroauric Acid; Gold Chloride
CNS K7241-67. Chemical Reagent (Gold Chloride) (Oct)(1740).
CNS K7552-81. Chemical Reagent (Chloroauric Acid) (Jul)(7710).
JIS K 8127-76. Chloroauric Acid.

Golden Mushrooms
Use: Mushrooms

Golf Balls
See Also: Balls (Sports Equipment); Golf Equipment
CNS S1153-82. Golf-Balls (Dec)(9750).
CNS S2093-82. Methods of Test for Golf-Balls (Dec)(9751).
JIS S 7005-88. Golf Balls. 6 pp.

Golf Equipment
See Also: Golf Balls; Sports Equipment

—Putting Machines
CSA CAN/CSA-C22. 2 NO 68-92. Motor-Operated Appliances (Household and Commercial); (Gen Instr 1 Thru 2). 115 pp.

Goligth Cranes
Use: Bridge Cranes

Gongs
See Also: Signal Devices

—Ships
MOD UK NES 739-82. Sound Signal Appliances (Whistles, Gongs, Bells Etc) Issue 1 (10.82). 14 pp.

—Submarines
MOD UK NES 739-82. Sound Signal Appliances (Whistles, Gongs, Bells Etc) Issue 1 (10.82). 14 pp.

Goniophotometers
Use: Photometry

Good Laboratory Practice
Use For: GLP
EC 88/320/EEC-88. Council Directive on the Inspection and Verification of Good Laboratory Practice (GLP). 3 pp.

Goods Vehicles
Use: Trucks

Goosenecks
See Also: Pipes

—Ventilation Equipment—Ships
CNS F3041-80. Gooseneck Ventilators (May)(5574).
JIS F 2408-91. Gooseneck Ventilators.
JIS F 2408-74. Gooseneck Ventilators. 7 pp.
JIS F 3012-90. Goose Neck Air Pipe Heads with Ball Float.
JIS F 3012-68. Goose Neck Air Pipe Heads (Ball Float Type). 11 pp.

GOS
Use: Grade of Service

GOSIP
Use: Government Open System Interconnection Profile

Gossypol Content Analysis

—Animal Feed
BSI BS 5766: Part 12-86. 1986 Methods for Analysis of Animal Feeding Stuffs Part 12: Determination of Gossypol. 6 pp.

Gossypol Content Analysis (Cont.)

—Animal Feed (Cont.)
CNS N4071-81. Determination of Free Gossypol in Cottonseed Meal of Cake (Feed Grade)(Feb)(7067).
ISO 6866-93. Animal Feeding Stuffs—Determination of Free and Total Gossypol First Edition. 5 pp.

Gothic Characters
DIN ENGL 1455-74. Handwritings (Feb). 2 pp.

Gouges
Use: Chisels

Gourds
Use For: Calabash; Sponge Gourds
See Also: Cucumbers; Pumpkins; Watermelons; Winter Melons

—Grading
CNS N1049-63. Grades of Calabash (for Export) (Mar)(2134) (R 1973). 2 pp.
CNS N1101-82. Grades of Sponge Gourd (Nov)(9629). 2 pp.

Government Open System Interconnection Profile
Use For: GOSIP
SNZ NZMP 6600-91. Essential Guide to GOSIP. 24 pp.
SNZ NZS 6623-91. GOSIP V2.0 Volume 1: The Australian/New Zealand Government Open Systems Interconnection Profile Volume 2: The Australian/New Zealand GOSIP User Manual.

Government Supplied Property
Scope Note: Government property which is made available to a contractor *See Also:* Defense Contracts
MOD UK DEFCON Guide no. 7-01. Issue of Ministry of Defence (MOD) Owned Equipment 1/84; Amendment 1. 19 pp.

—Contractor Liability
MOD UK DEFCON 147-86. Negative Acquital of Issue Transaction Summaries-Consignee Receipt 10/86. 1 p.

—Fabrics/Clothing
MOD UK DEFCON 198B-80. Materials to Be Issued to the Contractor 9/80. 1 p.

—Fabrics/Clothing—Order Forms
MOD UK DEFCON 198-91. Clothing and Textile Contracts Issues of Government Property on Embodiment Loan Terms 2/91. 2 pp.
MOD UK DEFCON 198A-91. Clothing and Textile Contracts Issues of Government Property on Prepayment or Repayment Terms 2/91. 2 pp.

Governors
Use For: Meter Governors *See Also:* Speed Controls

—Automotive
BSI BS AU 217: Part 1A-87. 1987 Amd 1 Maximum Road Speed Limiters for Motor Vehicles Part 1: Installed Requirements (AMD 5969) June 30, 1988. 8 pp.
BSI BS AU 217: Part 2-89. 1989 Maximum Road Speed Limiters for Motor Vehicles Part 2: System and Component Requirements. 9 pp.

Gowns (Protective)
See Also: Clothing; Protective Clothing; Surgical Gowns

—Hospital Patient
CGSB CAN/CGSB-38.23-93. Multipurpose Hospital Patient Gowns. 8 pp.

—Isolation
CGSB CAN/CGSB-38.39-M79. Gown, Isolation, Polyester/Cotton. 16 pp.

GPS
Use: Global Positioning Systems

Gracilaria
Use: Algae

Grade of Service
Use For: GOS *See Also:* Quality of Service

—Automatic Services
CCITT RECMN E.541-89. Overall Grade of Service for International Connections (Subscriber-to-Subscriber)—Telephone Network and ISDN—Quality of Service, Network Management and Traffic Engineering (Group Study II) 3 pp. 3 pp.

Grade of Service (Cont.)

—Automatic Services—Traffic Units
CCITT RECMN E.541-89. Overall Grade of Service for International Connections (Subscriber-to-Subscriber)—Telephone Network and ISDN—Quality of Service, Network Management and Traffic Engineering (Group Study II) 3 pp. 3 pp.

—Call Release Time—ISDN
CCITT RECMN E.721-91. Network Grade of Service Parameters and Target Values for Circuit-Switched Services in the Evolving ISDN (Study Group II) 8 pp. 8 pp.
CCITT RECMN E.721-89. Network Grade of Service Parameters in ISDN—Telephone Network and ISDN—Quality of Service, Network Management and Traffic Engineering (Study Group II) 2 pp. 2 pp.

—CCITT No. 7 Signaling Systems—ISDN
CCITT RECMN E.723-92. Grade-of-Service Parameters for Signalling System No. 7 Networks (Study Group II) 8 pp. 8 pp.

—Circuit Groups
CCITT RECMN E.540-89. Overall Grade of Service of the International Part of an International Connection —Telephone Network and ISDN—Quality of Service, Network Management and Traffic Engineering (Study Group II) 1 pp. 1 p.

—Circuit Groups—Busy Hour
CCITT RECMN E.521-89. Calculation of the Number of Circuits in a Group Carrying Overflow Traffic—Telephone Network and ISDN—Quality of Service, Network Management and Traffic Engineering (Study Group II) 11 pp. 11 pp.

—Circuit Switched Data Transmission Services—ISDN
CCITT RECMN E.701-92. Reference Connections for Traffic Engineering—Telephone Network and ISDN—Quality of Service, Network Management and Traffic Engineering (Study Group II) 4 pp. 4 pp.
CCITT RECMN E.721-91. Network Grade of Service Parameters and Target Values for Circuit-Switched Services in the Evolving ISDN (Study Group II) 8 pp. 8 pp.
CCITT RECMN E.721-89. Network Grade of Service Parameters in ISDN—Telephone Network and ISDN—Quality of Service, Network Management and Traffic Engineering (Study Group II) 2 pp. 2 pp.
CCITT RECMN E.723-92. Grade-of-Service Parameters for Signalling System No. 7 Networks (Study Group II) 8 pp. 8 pp.

—Digital Exchanges
CCITT RECMN E.543-89. Grades of Service in Digital International Telephone Exchanges—Telephone Network and ISDN—Quality of Service, Network Management and Traffic Engineering (Study Group II) 4 pp. 4 pp.

—Digital Exchanges—Exchange Call Set Up Delay
CCITT RECMN E.543-89. Grades of Service in Digital International Telephone Exchanges—Telephone Network and ISDN—Quality of Service, Network Management and Traffic Engineering (Study Group II) 4 pp. 4 pp.

—Digital Exchanges—Incoming Response Delay
CCITT RECMN E.543-89. Grades of Service in Digital International Telephone Exchanges—Telephone Network and ISDN—Quality of Service, Network Management and Traffic Engineering (Study Group II) 4 pp. 4 pp.

—Digital Exchanges—Internal Loss Probability
CCITT RECMN E.543-89. Grades of Service in Digital International Telephone Exchanges—Telephone Network and ISDN—Quality of Service, Network Management and Traffic Engineering (Study Group II) 4 pp. 4 pp.

—Digital Exchanges—Measurement
CCITT RECMN E.543-89. Grades of Service in Digital International Telephone Exchanges—Telephone Network and ISDN—Quality of Service, Network Management and Traffic Engineering (Study Group II) 4 pp. 4 pp.

—Digital Exchanges—Through Connection Delay
CCITT RECMN E.543-89. Grades of Service in Digital International Telephone Exchanges—Telephone Network and ISDN—Quality of Service, Network Management and Traffic Engineering (Study Group II) 4 pp. 4 pp.

Grade of Service (Cont.)
—Digital Satellite Dedicated Networks
CCIR Report 1134-90. Digital Satellite Dedicated Networks—Section 4B1—Systems Aspects. 20 pp.

—Integrated Services Digital Networks—Design
CCITT RECMN E.525 (REV 1)-92. Designing Networks to Control Grade of Service (Study Group II) 10 pp. 10 pp.

—International Exchanges—Failure (Quality Control)
CCITT RECMN E.550-89. Grade-of-Service and New Performance Criteria Under Failure Conditions in International Telephone Exchanges—Telephone Network and ISDN—Quality of Service, Network Management and Traffic Engineering (Study Group II) 8 pp. 8 pp.

—Packet Switched Data Transmission Services—ISDN
CCITT RECMN E.701-92. Reference Connections for Traffic Engineering—Telephone Network and ISDN—Quality of Service, Network Management and Traffic Engineering (Study Group II) 4 pp. 4 pp.
CCITT RECMN E.721-91. Network Grade of Service Parameters and Target Values for Circuit-Switched Services in the Evolving ISDN (Study Group II) 8 pp. 8 pp.
CCITT RECMN E.721-89. Network Grade of Service Parameters in ISDN—Telephone Network and ISDN—Quality of Service, Network Management and Traffic Engineering (Study Group II) 2 pp. 2 pp.

—Postselection Delay—ISDN
CCITT RECMN E.721-91. Network Grade of Service Parameters and Target Values for Circuit-Switched Services in the Evolving ISDN (Study Group II) 8 pp. 8 pp.
CCITT RECMN E.721-89. Network Grade of Service Parameters in ISDN—Telephone Network and ISDN—Quality of Service, Network Management and Traffic Engineering (Study Group II) 2 pp. 2 pp.

—Preselection Delay—ISDN
CCITT RECMN E.721-91. Network Grade of Service Parameters and Target Values for Circuit-Switched Services in the Evolving ISDN (Study Group II) 8 pp. 8 pp.
CCITT RECMN E.721-89. Network Grade of Service Parameters in ISDN—Telephone Network and ISDN—Quality of Service, Network Management and Traffic Engineering (Study Group II) 2 pp. 2 pp.

—Private Branch Exchanges—Private Switched Networks
CCITT RECMN G.171-89. Transmission Plan Aspects of Privately Operated Networks—General Characteristics of International Telephone Connections and Circuits (Study Groups XII and XV) 19 pp. 19 pp.

—Telecommunication Circuits
CCITT RECMN E.540-89. Overall Grade of Service of the International Part of an International Connection—Telephone Network and ISDN—Quality of Service, Network Management and Traffic Engineering (Study Group II) 1 pp. 1 p.

—Telecommunication Circuits—Transit Traffic
CCITT RECMN E.540-89. Overall Grade of Service of the International Part of an International Connection—Telephone Network and ISDN—Quality of Service, Network Management and Traffic Engineering (Study Group II) 1 pp. 1 p.

—Telecommunication Services—Alternative Routing
CCITT RECMN E.525-89. Service Protection Methods—Telephone Network and ISDN—Quality of Service, Network Management and Traffic Engineering (Study Group II) 5 pp. 5 pp.

—Telecommunication Services—ISDN
CCITT RECMN E.700-92. Framework of the E.700-Series Recommendations—Telephone Network and ISDN Service, Network Management and Traffic Engineering (Study Group II) 3 pp. 3 pp.
CCITT RECMN E.720-89. ISDN Grade of Service Concept—Telephone Network and ISDN—Quality of Service, Network Management and Traffic Engineering (Study Group II) 3 pp. 3 pp.

—Telecommunication Traffic—ISDN—Dimensioning
CCITT RECMN E.730-92. ISDN Dimensioning Methods Overview (Study Group II) 4 pp. 4 pp.

Grade of Service (Cont.)
—Telecommunication Traffic—Measurement
CCITT RECMN E.502 (REV 1)-92. Traffic Measurement Requirements for Digital Telecommunication Exchanges (Study Group II) 23 pp. 23 pp.
CCITT RECMN E.502-89. Traffic Measurement Requirements for SPC (Especially Digital) Telecommunication Exchanges—Telephone Network and ISDN—Quality of Service, Network Management and Traffic Engineering (Study Group II) 15 pp. 15 pp.

—Telegraph Circuits—Gentex Networks
CCITT RECMN F.23-89. Grade of Service for Long-Distance International Gentex Circuits—Telegraph and Mobile Services Operations and Quality of Service (Study Group I) 1 pp. 1 p.
CCITT RECMN F.24-89. Average Grade of Service from Country to Country in the Gentex Service—Telegraph and Mobile Services Operations and Quality of Service (Study Group I) 1 pp. 1 p.

—Telephone Networks—Design
CCITT RECMN E.525 (REV 1)-92. Designing Networks to Control Grade of Service (Study Group II) 10 pp. 10 pp.

Graders
See Also: Earthmoving Equipment; Excavating Equipment; Motor Graders

—Cutting Edges—Shapes
JIS D 6103-83. Shapes and Dimensions of Cutting Edges for Graders.
JIS D 6103-63. Cutting Edges for Motor Graders (R 1970). 7 pp.

—Glossaries
ISO 7134-85. Earth-Moving Machinery—Graders—Terminology and Commercial Specifications First Edition. 21 pp.

—Roll Over Protective Structures
BSI BS 5527: Part 1-87. 1987 Roll-Over Protective Structures on Earth-Moving Machinery: Laboratory Tests and Preformance Requirements Part 1: Crawler, Wheel Loaders and Tractors, Backhoe Loaders,Graders, Tractor Scrapers and Articulated Steel Dumpers. 19 pp.
ISO 3471 Pt 1-86. Earth-Moving Machinery—Rollover Protective Structures—Laboratory Tests and Performance Requirements—Part 1: Crawler, Wheel Loaders and Tractors, Backhoe Loaders, Graders, Tractor Scrapers, Articulated Steer Dumpers First Edition. 18 pp.

—Roll Over Protective Structures—Static Loads
ISO 3471 Pt 1-86. Earth-Moving Machinery—Rollover Protective Structures—Laboratory Tests and Performance Requirements—Part 1: Crawler, Wheel Loaders and Tractors, Backhoe Loaders, Graders, Tractor Scrapers, Articulated Steer Dumpers First Edition. 18 pp.

—Tires
CGSB 20-GP-5D-80. Tires, Pneumatic, Low Speed, Off Highway, Standard for. 23 pp.
CNS K4030-80. Dimensions of Tires for Industrial Vehicles and off the Road Service (Feb)(3673). 33 pp.

Graduated Cylinders
Use: Measuring Cylinders—Graduated

Grain Content Analysis
—Grinding Wheels
CNS R3077-87. Method of Test for Grinding Wheels (Nov) (3786).
JIS R 6240-86. Testing Methods of Grinding Wheels. 18 pp.

Grain Dryers
Use: Crop Dryers

Grain Elevators
Use For: Elevators (Grain) *See Also:* Silos

—Safety—Dust
CNS Z1020-88. Safety Code of Preventing Dust Explosion for Grain Elevator (Oct)(3364).

Grain Size Analysis
See Also: Particle Size Analysis; Particle Size Distribution; Physical Testing; Sieve Analysis

—Abrasive Grains
BSI BS 5851-80. 1980 Grain Sizes of Diamond or Cubic Boron Nitride. 5 pp.
CNS R2082-84. Abrasive Grain Sizes (Nov)(3787).
CNS R3098-86. Method of Tests for Abrasive Grain Size (Jan)(7530). 10 pp.

Grain Size Analysis (Cont.)
—Abrasive Grains (Cont.)
ISO 6106-79. Abrasive Products—Grain Sizes of Diamond or Cubic Boron Nitride First Edition. 8 pp.
ISO 8486-86. Bonded Abrasives—Grain Size Analysis—Designation and Determination of Grain Size Distribution of Macrogrits F4 to F220 First Edition. 5 pp.
JIS B 4130-82. Grain Sizes of Diamond or Cubic Boron Nitride.
JIS R 6001-87. Abrasive Grain Sizes. 8 pp.
JIS R 6002-87. Testing Methods for Abrasive Grain Size. 17 pp.
JIS R 6011-91. Testing Method for Grain Size of Coated Abrasive Macrogrits (P12—P220). 9 pp.

—Abrasives
SAA AS 1487-75. Abrasive Grain Size. 30 pp.

—Ceramics
BSI DD ENV 623-3-93. 1993 Advanced Technical Ceramics Monolithic Ceramics—General and Textural Properties Part 3: Determination of Grain Size (V). 15 pp.
CEN ENV 623-3-93. Advanced Technical Ceramics—Monolithic Ceramics—General and Textural Properties—Part 3: Determination of Grain Size. 15 pp.
CEN ENV 623-3-93. Advanced Technical Ceramics—Monolithic Ceramics—General and Textural Properties—Part 3: Determination of Grain Size. 10 pp.

—Coated Abrasives
JIS R 6010-91. Coated Abrasive Grain Sizes. 7 pp.
JIS R 6012-91. Testing Method for Grain Size of Coated Abrasive Microgrits (P240—P1200). 14 pp.

—Copper
BSI BS 7428-91. 1991 Estimating the Average Grain Size of Copper and Copper Alloys (ISO 2624: 1990). 12 pp.
CNS H2069-82. Methods for Estimating the Average Grain Size Test of Wrought Copper and Copper Base Alloys (Oct)(9504).
ISO 2624-90. Copper and Copper Alloys—Estimation of Average Grain Size Second Edition. 11 pp.
JIS H 0501-86. Methods for Estimating Average Grain Size of Wrought Copper and Copper-Alloys. 8 pp.

—Copper Alloys
BSI BS 7428-91. 1991 Estimating the Average Grain Size of Copper and Copper Alloys (ISO 2624: 1990). 12 pp.
CNS H2069-82. Methods for Estimating the Average Grain Size Test of Wrought Copper and Copper Base Alloys (Oct)(9504).
ISO 2624-90. Copper and Copper Alloys—Estimation of Average Grain Size Second Edition. 11 pp.
JIS H 0501-86. Methods for Estimating Average Grain Size of Wrought Copper and Copper-Alloys. 8 pp.

—Grit
CNS G2130-82. Method of Test for Cast Shot and Grit Grain Size (Aug)(9271).
JIS G 5904-66. Testing Method of Cast Shot and Grit Grain Size (R 1969). 3 pp.

—Metals
AECMA PREN2002-08-86. Test Methods for Metallic Materials—Part 8—Micrographic Determination of Grain Size. 1 p.
SAA AS 1733-76. Methods for the Determination of Grain Size in Metals Reconfirmed 1986. 24 pp.

—Refractory Materials
CNS R3129-85. Method of Test for Grain Size of Refractory Mortar (Nov) (11423).
CNS R3147-87. Method of Test for Grain Size of Castable Refractories (May) (11973).
CNS R3150-87. Method of Test for Grain Size of Light Weight Castable Refractories (May) (11976).
JIS R 2502-76. Testing Method for Grain Size of Refractory Mortar.
JIS R 2552-77. Testing Method for Grain Size of Castable Refractories.
JIS R 2652-91. Testing Method for Grain Size of Light Weight Castable Refractories. 6 pp.
SAA AS 1774.25-92. Refractories and Refractory Materials—Physical Test Methods—Part 25: Determination of Density by the Rees-Hugill Method. 3 pp.

—Shot
CNS G2130-82. Method of Test for Cast Shot and Grit Grain Size (Aug)(9271).
JIS G 5904-66. Testing Method of Cast Shot and Grit Grain Size (R 1969). 3 pp.

—Soil Analysis
CNS A3085-86. Method of Test for Dry Preparation of Soil Samples for Grain—Size Analysis and Determination of Soil Constancy (Dec)(5086).

Grain

Grain Size Analysis (Cont.)
—Soil Analysis (Cont.)

DIN ENGL 19683 Pt 2-73. Methods of Soil Analysis for Water Management for Agricultural Purposes; Physical Laboratory Tests; Determination of Grain Size Distribution After Pretreatment Using Sodium Pyrophosphate (Apr). 3 pp.

DIN ENGL 19683 Pt 3-73. Methods of Soil Analysis for Water Management for Agricultural Purposes; Physical Laboratory Tests; Determination of Grain Size Distribution After Treatment Using Water (Apr). 2 pp.

JIS A 1201-90. Practice for Preparing Disturbed Soil Samples for Soil Testing. 8 pp.

JIS A 1204-90. Test Method for Particle Size Distribution of Soils. 14 pp.

—Soil Analysis—Dry Sieving

SNZ NZS 4402: Pt 2:TEST 2.8.2-86. Methods of Testing Soils for Civil Engineering Purposes Part 2: Soil Classification Tests. Section 2.8: Determination of the Particle-Size Distribution. Test 2.8.2: Subsidiary Method by Dry Sieving. 5 pp.

—Soil Analysis—Hydrometer Method

SNZ NZS 4402: Pt 2:TEST 2.8.4-86. Methods of Testing Soils for Civil Engineering Purposes Part 2: Soil Classification Tests. Section 2.8: Determination of the Particle-Size Distribution. Test 2.8.4: Subsidiary Method for Fine Soils (Hydrometer Method). 11 pp.

—Soil Analysis—Pipette Method

SNZ NZS 4402: Pt 2:TEST 2.8.3-86. Methods of Testing Soils for Civil Engineering Purposes Part 2: Soil Classification Tests. Section 2.8: Determination of the Particle-Size Distribution. Test 2.8.3: Standard Method for Fine Soils (Pipette Method). 9 pp.

—Soil Analysis—Wet Sieving

SNZ NZS 4402: Pt 2:TEST 2.8.1-86. Methods of Testing Soils for Civil Engineering Purposes Part 2: Soil Classification Tests. Section 2.8: Determination of the Particle-Size Distribution. Test 2.8.1: Standard Method by Wet Sieving. 7 pp.

—Steels

BSI BS 4490-89. 1989 Amd 1 Methods for Micrographic Determination of the Grain Size of Steel (AMD 6597) January 31, 1991. 36 pp.

CNS G2177-83. Method of Austenite Grain Size Test for Steel (Jul)(10436).

CNS G2178-83. Method of Ferrite Grain Size Test for Steel (Jul)(10437).

DIN ENGL 50601-85. Metallographic Examination; Determination of the Ferritic or Austenitic Grain Size of Steel and Ferrous Materials (Aug). 13 pp.

ISO 643-83. Steels—Micrographic Determination of the Ferritic or Austenitic Grain Size First Edition. 31 pp.

JIS G 0551-77. Method of Austenite Grain Size Test for Steel. 16 pp.

JIS G 0552-77. Methods of Ferrite Grain Size Test for Steel. 15 pp.

Grains (Food)
Use: Cereals

Grammage
Use: Weight Measurement

Granular Materials
—Asphalt Shingles

CSA A123.1-M1979. Asphalt Shingles Surfaced with Mineral Granules (R 1992); (Amd 1 May 1985). 14 pp.

CSA CAN/CSA-A123.5-M90. Asphalt Shingles Made from Glass Felt and Surfaced with Mineral Granules; (Gen Instr 1). 18 pp.

—Polymethyl Methacrylate

MOD UK TS 626. Acrylic Granules.

Granules
Use: Granular Materials

Grapefruit Oil
See Also: Essential Oils

CNS K5075-80. Oil of Grapefruit (Obtain by Expression) (Aug)(6354).

ISO 3053-75. Oil of Grapefruit (Obtained by Expression) First Edition. 4 pp.

Grapefruits
Use For: Citrus Grandis; Pomelos; Shaddocks
See Also: Citrus Fruits; Food; Fruits

—Defects

CNS N4032-72. Method of Test for Pomelo (Shaddocks) (Jun)(3353). 2 pp.

Grapefruits (Cont.)
—Grading

CNS N1053-63. Grades and Packaging of Pomelo (for Export) (Jun)(2212)(R 1973). 3 pp.

—Inspection

CNS N4032-72. Method of Test for Pomelo (Shaddocks) (Jun)(3353). 2 pp.

—Packaging

CNS N1053-63. Grades and Packaging of Pomelo (for Export) (Jun)(2212)(R 1973). 3 pp.

—Sampling

CNS N4032-72. Method of Test for Pomelo (Shaddocks) (Jun)(3353). 2 pp.

—Storage

ISO 3631-78. Citrus Fruits—Guide to Storage First Edition. 11 pp.

Grapes
See Also: Food; Fruits; Raisins

EC EEC/291/92-92. Commission Regulation Amending Regulation (EEC) No 1730/87 Laying down Quality Standards for Table Grapes. 1 p.

—Grading

CNS N1116-83. Grades of Grape (Sep)(10572). 3 pp.

—Storage

ISO 2168-74. Table Grapes—Guide to Cold Storage First Edition. 7 pp.

Graphic Arts
See Also: Diagrams; Drawings; Electronic Prepress Systems; Photography; Printing

—Classification

BSI BS 1000: (655)-86. 1986 Universal Decimal Classification (UDC). English Full Edition (655): Graphic Industries. Printing. Publishing. Book Trade. 27 pp.

BSI BS 1000: (681.6)-89. 1989 Universal Decimal Classification (UDC). English Full Edition (681.6): Graphic Reproduction Machines and Equipment. 28 pp.

SNZ NZS/BS 1000 (655)-86. Universal Decimal Classification Graphic Industries. Printing. Publishing. Book Trade. 28 pp.

SNZ NZS/BS 1000 (681.6)-71. Universal Decimal Classification Graphic Reproduction Machines and Equipment. 24 pp.

—Glossaries

CGSB 94-GP-1-72. Glossary of Photographic Terms in General Use in the Graphic Arts. 51 pp.

ISO 5127 Pt 3-88. Documentation and Information—Vocabulary—Part 3: Iconic Documents First Edition. 22 pp.

—Photographic Film—Sizes

SAA AS 1933-81. Sizes of Photographic Film for Use in Graphic Arts and Reprography. 4 pp.

Graphic Characters
Scope Note: For additional listings, use a more specific term *See Also:* African Characters; Arabic Characters; Character Sets; Chinese Characters; Cyrillic Characters; German Characters; Gothic Characters; Greek Characters; Hebraic Characters; Japanese Characters; Keyboards; Korean Characters; Roman Characters; Russian Characters; Symbols

—Coded Character Sets

BSI BS ISO/IEC 7350-91. 1991 Information Technology—Registration of Repertoires of Graphic Characters from ISO/IEC 10367. 19 pp.

BSI BS ISO/IEC 10367-91. 1991 Information Technology—Standardized Coded Graphic Character Sets for Use in 8-Bit Codes. 58 pp.

BSI DD ENV 41501-91. 1991 Information Systems Interconnection—Graphic Character Repertoire for Information Received from or Transmitted to CEPT Videotex Services or Private Videotex Systems. 14 pp.

BSI DD ENV 41502-91. 1991 Information Systems Interconnection—Character Repertoires for Information Received from or Transmitted to the Teletex Service or Private Information Processing Systems Using Teletex Technology. 12 pp.

BSI DD ENV 41503-91. 1991 Information Systems Interconnection—European Character Repertoires and Their Coding (Incorporating Corrigendum June 1991). 15 pp.

BSI DD ENV 41504-91. 1991 Information Systems Interconnection—Character Repertoire and Coding for Interworking with Telex Services (Incorporating Corrigendum June 1991). 14 pp.

BSI DD ENV 41505-91. 1991 Information Systems Interconnection—Graphic Character Repertoire and Coding for Line Drawing. 9 pp.

Graphic Characters (Cont.)
—Coded Character Sets (Cont.)

BSI DD ENV 41507-91. 1991 Information Systems Interconnection—Data String Formats for Information Received from or Transmitted to CEPT Videotex Services or Private Videotex Systems. 11 pp.

BSI DD ENV 41508-91. 1991 Information Systems Interconnection—East European Graphic Character Repertoires and Their Coding (Incorporating Corrigendum June 1991). 13 pp.

CEN ENV 41 501-87. Information Systems Interconnection: Graphic Chartacter Repertoire and Coding for Interworking with CEPT Videotex Services. 7 pp.

CEN ENV 41501-91. Information Systems Interconnection—Graphic Character Repertoire for Information Received from or Transmitted to CEPT Videotex Services or Private Videotex Systems. 12 pp.

CEN ENV 41 502-87. Information Systems Interconnection: Graphic Character Repertoire and Coding for Interworking with CEPT Teletex Services. 7 pp.

CEN ENV 41502-91. Information Systems Interconnection Character Repertoires for Information Received from or Transmitted to the Teletex Service or Private Information Processing Systems Using Teletex Technology. 10 pp.

CEN ENV 41 502/A1-92. AMD 1 Information Systems Interconnection: Character Repertoires for Information Received from or Transmitted to the Teletex Service or Private Information Processing Systems Using Teletex Technology. 5 pp.

CEN ENV 41 503-87. Information Systems Interconnection: European Graphic Character Repertoires and Their Coding. 9 pp.

CEN ENV 41503-90. Information Systems Interconnection—European Character Repertoires and Their Coding. 14 pp.

CEN ENV 41504-90. Information Systems Interconnection—Character Repertoire and Coding for Interworking with Telex Services. 12 pp.

CEN PRENV 41505-91. Information Systems Interconnection—Graphic Character Repertoire and Coding for Line Drawing. 7 pp.

CEN ENV 41505-91. Information Systems Interconnection—Graphic Character Repertoire and Coding for Line Drawing. 7 pp.

CEN ENV 41507-91. Information Systems Interconnection—Data String Formats for Information Received from or Transmitted to CEPT Videotex Services or Private Videotex Systems. 9 pp.

CEN ENV 41508-90. Information Systems Interconnection—East European Graphic Character Repertoires and Their Coding. 11 pp.

IEC 7350-91. Information Technology—Registration of Repertoires of Graphic Characters from ISO/IEC 10367 Second Edition. 17 pp.

IEC 10367-91. Information Technology—Standardized Coded Graphic Character Sets for Use in 8-Bit Codes First Edition. 55 pp.

ISO 7350-91. Information Technology—Registration of Repertoires of Graphic Characters from ISO/IEC 10367 Second Edition. 17 pp.

ISO 10367-91. Information Technology—Standardized Coded Graphic Character Sets for Use in 8-Bit Codes First Edition. 55 pp.

JTC1 7350-91. Information Technology—Registration of Repertoires of Graphic Characters from ISO/IEC 10367 Second Edition. 17 pp.

JTC1 10367-91. Information Technology—Standardized Coded Graphic Character Sets for Use in 8-Bit Codes First Edition. 55 pp.

OSI ISO DIS 6862.2-90. Documentation—Mathematical Coded Character Set for Bibliographic Information Interchange. 18 pp.

OSI ISO DIS 6937-3-84. Information Processing—Coded Character Sets for Text Communication. 54 pp.

OSI ISO 7350-84. Text Communication—Registration of Graphic Character Subrepertoires. 13 pp.

OSI ISO/IEC DIS 7350.2. Information Technology—Registration of Subrepertoires of the Graphic Character Repertoire of ISO 10367. 23 pp.

OSI ISO IEC DIS 10367-90. Information Processing—Repertoire of Standardized Coded Graphic Character Sets for Use in 8-Bit Codes. 55 pp.

OSI ISO/IEC DIS 10646-90. Information Technology—Universal Coded Character Set (UCS). 162 pp.

—Display Terminals

CCITT RECMN Z.322-89. Capabilities of Visual Display Terminals—Man-Machine Language (MML) (Study Group X) 7 pp. 7 pp.

Graphic Methods
Use For: Graphic Statistics; Graphical Analysis; Graphics *See Also:* Charts; Drawings; Statistical Analysis

INTERNATIONAL AND NON-U.S. NATIONAL STANDARDS SUBJECT INDEX

Graphic Methods (Cont.)
—Vibration Damping Compounds—Complex Modulus
BSI BS 7544-91. 1991 Method for Graphical Presentation of the Complex Modulus of Damping Materials (ISO 10112: 1991) (V). 15 pp.
ISO 10112-91. Damping Materials—Graphical Presentation of the Complex Modulus First Edition. 13 pp.

Graphic Statistics
Use: Graphic Methods

Graphic Symbols
Use: Symbols

Graphical Analysis
Use: Graphic Methods

Graphical Kernel System
Use: Computer Graphics Software—Graphical Kernel System

Graphical Symbols
Use: Symbols

Graphics
Use: Graphic Methods

Graphics Software
Use: Computer Graphics Software

Graphite
Scope Note: For additional listings, see also specific products made from graphite *See Also:* Carbon; Graphite Powders; Minerals; Ores
CNS K6006-72. Method of Analysis for Graphite (Oct)(60).
JIS M 8601-60. Natural Graphite.
JIS R 7221-79. High Purity Graphite Material.
MOD UK DEF-173-69. Graphite, Sizes 120, 170 and 200. 5 pp.

—Ash Content
CNS K6806-84. Method of Test for Ash in Graphite (Nov)(11114).
JIS R 7223-79. Chemical Analysis of High Purity Graphite Material (R 1985). 32 pp.

—Boron Content
JIS R 7223-79. Chemical Analysis of High Purity Graphite Material (R 1985). 32 pp.

—Chemical Analysis
JIS M 8511-76. Methods for Industrial Analysis and Testing of Natural Graphite (R 1984). 14 pp.

—Microstructure—Cast Iron
ISO 945-75. Cast Iron—Designation of Microstructure of Graphite First Edition. 10 pp.

—Modulus of Elasticity—Resonance Method
DIN ENGL 51915-86. Testing of Carbonaceous Materials; Determination of Dynamic Modulus of Elasticity by the Resonance Method; Solids (Apr). 3 pp.

—Moisture Content
CNS K6805-84. Method of Test for Moisture in Graphite (Nov)(11113).

—Nuclear—Lattice Parameters
CNS J2028-81. Method for Measurement of Lattice Spacing of Nuclear Graphite (Nov)(8128).

—Physical Testing
JIS R 7222-79. Physical Testing Methods for High Purity Graphite Material (R 1985). 10 pp.

—Pigments—Red Lead Content
CGSB 1-GP-71 METH 50.3-80. Methods of Testing Paints and Pigments Pigment Analysis Red Lead with Graphite or Iron Oxide. 1 p.

Graphite Content Analysis
—Greases
MOD UK DSTAN 05-50: Part 39-89. Methods for Testing Fuels, Lubricants and Associated Products Part 39: Graphite Content of Greases Issue 1. 8 pp.

Graphite Electrodes
Use: Carbon Electrodes

Graphite Fibers
Use: Carbon Fibers

Graphite Greases
See Also: Greases

Graphite Greases (Cont.)
MOD UK DSTAN 91-18-77. Grease, Graphite: Medium NATO Code No: G-412 Joint Service Designation: XG-264 Issue 2. 8 pp.

—Aircraft
MOD UK DSTAN 91-54-01. Grease, Aircraft: Graphite NATO Code No: G-355 Joint Service Designation: XG-285 Issue 1; Amendment 1. 10 pp.

Graphite Oils
See Also: Lubricating Oils

—Artillery
MOD UK DSTAN 91-30-77. Lubricating Oil, Colloidal Graphite NATO Code No: O-218 Joint Service Designation: OX-320 Issue 1 (Withdrawn). 14 pp.

—Guided Missiles
MOD UK DSTAN 91-30-77. Lubricating Oil, Colloidal Graphite NATO Code No: O-218 Joint Service Designation: OX-320 Issue 1 (Withdrawn). 14 pp.

Graphite-Petrolatum
Use For: Graphite-Petroleum Jelly
MOD UK DSTAN 80-80-93. Anti-Seize Compound, Graphite NATO Code: S-720 Joint Service Designation: ZX-13 Issue 2. 9 pp.

—Aircraft
MOD UK DSTAN 80-80-01. Anti-Seize Compound: Aircraft, Graphite-Petrolatum NATO Code No: S-720 Joint Service Designation: ZX-13 Issue 1; Amendment 1. 7 pp.

Graphite-Petroleum Jelly
Use: Graphite-Petrolatum

Graphite Powders
Use For: Powdered Graphite *See Also:* Graphite
CNS K1164-72. Graphite Powder (Oct)(3480).
MOD UK DSTAN 68-49-01. Graphite Powder Issue 1; Amendment 1. 7 pp.
MOD UK DSTAN 96-1-01. Graphite Powder, Lubricating ZX-20, NATO S-732 Issue 1; Amendment 3. 15 pp.

Graphs (Charts)
See Also: Charts
BSI BS 7581-92. 1992 Presentation of Tables and Graphs (Supersedes DD 52: 1977) (G). 28 pp.
BSI DD 52-77. (WITHDRAWN) 1977 Recommendations for the Presentation of Tables, Graphs and Charts (Superseded by BS 7581: 1992). 16 pp.

—Coordinates
DIN ENGL 461-73. Graphical Representation in Systems of Coordinates (Mar). 6 pp.

—Optical Fibers
CCITT FASCICLE III.3. Transmission Media Characteristics Recommendations G.601–G.654. 134 pp.

—Optical Fibers—Cutoff Wavelength
CCITT FASCICLE III.3. Transmission Media Characteristics Recommendations G.601–G.654. 134 pp.

—Project Management
BSI BS 6046: Part 2-92. 1992 Use of Network Techniques in Project Management Part 2: Guide to the Use of Graphical and Estimating Techniques (G). 25 pp.
BSI BS 6046: Part 2-81. 1981 Use of Network Techniques in Project Management Part 2: Guide to the Use of Graphical and Estimating Techniques. 20 pp.

Grasping Devices
Use: Gripping Devices

Grass Jelly
Use: Jellies (Food)—Grass

Grass Shears
Use: Shears (Hand Tools)—Grass

Graticules
See Also: Optical Equipment

—Microscopes
BSI BS 7012: Part 8-90. 1990 Light Microscopes Part 8: Surface Marked Graticules for Eyepieces. 4 pp.

—Microscopes—Particle Size Distribution
BSI BS 3625-63. 1963 Eyepiece and Screen Graticules for the Determination of Particle Size of Powders. 7 pp.

Gratings (Metal)
Use: Grilles

Gravel
See Also: Aggregates; Sands; Sediments

—Bulk Density
DIN ENGL 52110-85. Testing of Natural Stone; Determination of Bulk Density of Stone Gradings (Aug). 3 pp.

—Impact Testing
DIN ENGL 52115 Pt 3-88. Determination of Impact Resistance of Mineral Aggregates; Testing of Gravel (Aug). 2 pp.

—Water Filters
DIN ENGL 19623-78. Filter Sands and Filter Gravels for Water Purification Filters; Technical Conditions of Delivery (Jan). 5 pp.

—Well Filters
DIN ENGL 4924-72. Filter Sands and Filter Gravels for Well Filters (Feb). 1 p.

Gravimetric Analysis
Scope Note: For gravimetric analysis of specific products or materials, see the subheading Gravimetric Analysis under the specific material or product. For additional references, consult the following list
See: Electrogravimetric Analysis; Evaporation Residue Analysis; Materials Testing; Quantitative Analysis; Thermogravimetric Analysis; Volumetric Analysis; Water Analysis

Gravity Conveyors
Use: Roller Conveyors

Gravity Flow
Use: Open Channel Flow

Gravity Roller Conveyors
Use: Roller Conveyors

Gravity Wheel Conveyors
Use: Roller Conveyors

Gravure Printing
—Papers
DIN ENGL 19306-77. Uncoated Printing Papers; Letterpress Paper, Offset Paper, Rotogravure Paper; Technical Conditions of Delivery (Sept). 2 pp.

Gray Cast Iron
Use: Gray Iron

Gray Iron
Scope Note: For additional listings, see also specific products made from gray iron *See Also:* Cast Iron; Ferroalloys; Pig Iron; White Iron
DIN ENGL 1686 Pt 1-80. Rough Castings of Grey Iron with Flake Graphite; General Tolerances; Machining Allowances (Oct). 3 pp.

—Brinell Hardness Testing
DIN ENGL 1691-85. Flake Graphite Cast Iron (Grey Cast Iron); Properties (May). 7 pp.

—Impact Testing
ISO 946-75. Grey Cast Iron—Beam Unnotched Impact Test First Edition. 5 pp.

—Mechanical Properties
DIN ENGL 1691 Suppl. 1-85. Flake Graphite Cast Iron (Grey Cast Iron); General Information on the Selection of Material and Design; Guide Values of Mechanical and Physical Properties (May). 4 pp.

—Physical Properties
DIN ENGL 1691 Suppl. 1-85. Flake Graphite Cast Iron (Grey Cast Iron); General Information on the Selection of Material and Design; Guide Values of Mechanical and Physical Properties (May). 4 pp.

—Tensile Testing
DIN ENGL 1691-85. Flake Graphite Cast Iron (Grey Cast Iron); Properties (May). 7 pp.
DIN ENGL 50109-89. Tensile Testing of Lamellar Graphite Cast Iron (Apr). 3 pp.
ISO 185-88. Grey Cast Iron—Classification First Edition. 15 pp.

Gray Iron Castings
See Also: Iron Castings
CNS G3038-84. Gray Iron Castings (Jan)(2472).
JIS G 5501-89. Grey Iron Castings. 8 pp.
SAA AS 1830-86. Iron Castings—Grey Cast Iron. 17 pp.
SNZ NZS/AS 1830-76. Iron Castings—Grey Cast Iron (This is a Joint Standard with SAA As 1830). 17 pp.

INDUSTRY STANDARDS

INTERNATIONAL AND NON-U.S. NATIONAL STANDARDS
SUBJECT INDEX

Gray Iron Pipes
See Also: Iron Pipes; Pipes
BSI BS 4622-70. (OBSOLESCENT) 1970 Amd 3 Grey Iron Pipes and Fittings. 64 pp.

—**Pressure**
ISO 13-78. Grey Iron Pipes, Special Castings and Grey Iron Parts for Pressure Main Lines First Edition. 37 pp.

Gray Scale
See Also: Colorimetry

—**Color Testing**
CNS L1005-82. Grey Scale for Assessing Change in Color (Feb)(3839).
JIS L 0804-83. Grey Scale for Assessing Change in Colour. 6 pp.

—**Fabrics—Fastness**
CNS L1018-83. Scale for Assessing the Fastness of Fluorescent Whitened Textiles (Jan)(9900).
JIS L 0806-71. Scale for Assessing the Fastness of Fluorescent Whitened Textiles.

—**Fabrics—Intensity**
CNS L1017-83. Scale for Assessing the Intensity of Fluorescent Whitening Effect (Jan)(9899).
JIS L 0807-71. Scale for Assessing the Intensity of Fluorescent Whitening Effect.

—**Fabrics—Staining**
CNS L1006-82. Grey Scale for Assessing Staining (Feb)(3840).
JIS L 0805-83. Grey Scale for Assessing Staining. 7 pp.

—**Upholstery—Staining**
BSI BS 4948-73. 1973 Amd 1 Assessment of the Visible Soiling by Body Contact of Upholstery Fabrics. 13 pp.

Grease Content Analysis
—**Waste Water**
CNS K9004-75. Method of Test for Oil and Grease in Waste Water (May)(3792).

Grease Guns
See Also: Lubrication; Lubricators
BSI BS 5077-74. 1974 Grease Guns (Hand-Operated Lever Type) for Use with Cartridges Complying with BS 4507. 7 pp.
CNS B3447-84. Grease Guns Hand Guns (Dec)(11154).
CNS B3448-84. Grease Guns Hand Guns (Dec)(11155).
JIS B 9808-91. Grease Guns. 13 pp.
MOD UK DSTAN 49-2-01. Lubricating Guns, Hand, and Guns, Fluid, Direct Delivery Issue 2; Amendment 3. 21 pp.

—**Cartridges**
BSI BS 4507-69. 1969 Grease Gun Cartridges (Composite Type). 8 pp.
CNS B2766-84. Grease Guns Dimensions of Grease Cartridges for Lever Hand Guns (Dec)(11156).

Grease Nipples
Use For: Lubricating Nipples *See Also:* Lubrication; Lubricators; Oil Rings
CNS B2445-83. Grease Nipple with R1/8" (Mar)(5400).
CNS B2584-81. Lubricating Nipples, Ball Type (Jun)(7581).
CNS B2585-81. Lubricating Nipples, Flat Type (Jun)(7582).
CNS B2586-81. Lubricating Nipples, Funnel Type (Jun)(7583).
JIS B 1575-79. Grease Nipples. 12 pp.
MOD UK DSTAN 47-7-01. Lubricating Nipples, and Adaptors, Lubricating Nipple Issue 2; Amendment 1. 11 pp.
MOD UK DSTAN 47-7: Addendum. Lubricating Nipples, and Adaptors, Lubricating Nipple Issue 2. 10 pp.

—**Aerospace**
CEN PREN 2646-91. Nipples, Lubricating, Axial Type. 10 pp.

—**Aircraft**
BSI SP 21-50. 1950 Lubricating Nipples for Aircraft Purposes. 2 pp.
ISO 413-74. Aircraft—Heads of Lubricating Nipples First Edition. 3 pp.
NATO STANAG 3766 ED 2 AMD 3-88. Grease Nipples. 8 pp.
NATO STANAG 3766 ED 2 AMD 4-88. Grease Nipples. 9 pp.
SBAC AS 44401-403 (V). Lubricating Nipples Hydraulic Surface—Check Type Metric.
SBAC TS 85 ISSUE 1. Lubricating Nipples, Hydraulic, Surface—Check Type, Metric.

—**Aircraft—Hydraulic**
SBAC RS 720 ISSUE 1. Minimum Clearance Around Hydraulic Lubricating, Nipples.

—**Aircraft—Interdepartmental Procurement**
NATO STANAG 3766 ED 2 AMD 3-88. Grease Nipples. 8 pp.
NATO STANAG 3766 ED 2 AMD 4-88. Grease Nipples. 9 pp.

—**Earthmoving Equipment**
BSI BS 6060-81. 1981 Nipple-Type Lubrication Fittings for Earth-Moving Machinery. 7 pp.
ISO 6392-80. Earth Moving Machinery—Lubrication Fittings—Nipple Type First Edition. 6 pp.

—**Hydraulic**
CNS B2528-80. Lubricating Nipples, Hydraulic Type with Metric Thread (Sep)(6420).
CNS B2529-80. Lubricating Nipples, Hydraulic Type with Whitworth-Pipe-Thread (Sep)(6421).

—**Hydraulic—Aircraft**
BSI SP 115-117-56. 1956 Amd 4 Hydraulic Surface-Check Type Lubricating Nipples for Aircraft. 11 pp.

—**Hydraulic—Machinery**
BSI BS 1486: Part 4-81. (OBSOLESCENT) 1981 Lubricating Nipples Part 4: Hydraulic Grease Nipples. 8 pp.

—**Hydraulic—Vehicles**
BSI BS 1486: Part 4-81. (OBSOLESCENT) 1981 Lubricating Nipples Part 4: Hydraulic Grease Nipples. 8 pp.

—**Locomotives**
BSI BS 1486: Part 2-61. (OBSOLESCENT) 1961 Lubricating Nipples Part 2: Heavy Duty Lubricating Nipples. 19 pp.

—**Machinery**
BSI BS 1486: Part 1-59. (OBSOLESCENT) 1959 Amd 1 Lubricating Nipples Part 1: Lubricating Nipples and Adaptors for Use on Machinery and Vehicles. 34 pp.
BSI BS 1486: Part 2-61. (OBSOLESCENT) 1961 Lubricating Nipples Part 2: Heavy Duty Lubricating Nipples. 19 pp.

—**Ships**
ISO 7824-86. Shipbuilding and Marine Structures—Lubrication Nipples—Cone and Flat Types First Edition. 6 pp.

—**Textile Machinery**
ISO 3799-76. Textile Machinery and Accessories—Hydraulic Lubricating Fittings for Textile Machinery First Edition. 4 pp.

—**Vehicles**
BSI BS 1486: Part 1-59. (OBSOLESCENT) 1959 Amd 1 Lubricating Nipples Part 1: Lubricating Nipples and Adaptors for Use on Machinery and Vehicles. 34 pp.

Grease Permeability
Use: Grease Resistance

Grease Resistance
—**Paperboard**
BSI BS 6890-87. 1987 Determination of Grease Resistance of Paper and Board. 8 pp.
BSI BS 6890-01. 1987 Amd 1 Method for Determination of Grease Resistance of Paper and Board (ISO 5634: 1986) (AMD 7214) September 15, 1992. 9 pp.
ISO 5634-86. Paper and Board—Determination of Grease Resistance First Edition; (Corrigendum 1-1992). 7 pp.

—**Papers**
BSI BS 6890-87. 1987 Determination of Grease Resistance of Paper and Board. 8 pp.
BSI BS 6890-01. 1987 Amd 1 Method for Determination of Grease Resistance of Paper and Board (ISO 5634: 1986) (AMD 7214) September 15, 1992. 9 pp.
CNS P3046-90. Method of Test for Grease Resistance of Paper (Jun)(7296). 2 pp.
CPPA F.6P-88. Turpentine Test for Grease-Resistant Papers. 1 p.
ISO 5634-86. Paper and Board—Determination of Grease Resistance First Edition; (Corrigendum 1-1992). 7 pp.
JIS P 8146-76. Testing Method for Grease Resistance of Paper (R 1984). 6 pp.

Grease Seals
See Also: Hydraulic Seals; Oil Seals; Seals

—**Rolling Stock**
CNS E1040-85. Oil Seals for Railway Rolling Stock (Feb)(11205).
JIS E 4704-92. Grease Seals for Railway Rolling Stock.

Grease Traps
Use: Traps (Drains)

Greaseproof Papers
Use For: Buttler Paper; Glassine Paper; Vegetable Parchment *See Also:* Packaging Materials; Packaging Papers; Papers
CNS P2060-79. Glassine Paper (Nov)(5040). 1 p.
MOD UK DSTAN 81-93-89. Paper, Wrapping, Grease-Resisting Issue 1. 15 pp.

—**Grease Resistance**
BSI BS 6890-87. 1987 Determination of Grease Resistance of Paper and Board. 8 pp.
BSI BS 6890-01. 1987 Amd 1 Method for Determination of Grease Resistance of Paper and Board (ISO 5634: 1986) (AMD 7214) September 15, 1992. 9 pp.
ISO 5634-86. Paper and Board—Determination of Grease Resistance First Edition; (Corrigendum 1-1992). 7 pp.

—**Packaging—Dairy Products**
SAA AS 1764-75. Vegetable Parchment for Wrapping Dairy Products. 24 pp.
SAA AS 1813-75. Aluminium Foil/Vegetable Parchment Laminates for Wrapping Dairy Products. 8 pp.
SAA AS 1814-75. High Wet-Strength Greaseproof Paper for Wrapping Dairy Products. 24 pp.

—**Punched Tapes**
MOD UK TS 288A. Paper, Vegetable Parchment (for Telegraph Machines).

Greaseproofing
Use For: Oil Proofing

—**Barrier Materials**
CNS Z5127-83. Corrosion Preventive Greaseproofed Barried Materials (Jun)(10396).
CNS Z6065-83. Method of Test for Corrosion Preventive Greaseproofed Barried Materials (Jun)(10397).
JIS Z 1705-76. Corrosion Preventive Greaseproofed Barrier Materials. 13 pp.

Greases
See Also: Bearing Greases; Extreme Pressure Lubricants; Fats; Four Ball Testers; Graphite Greases; Lubricants; Petrolatums; Silicone Greases
CNS K6433-79. Method of Analysis for Lubricating Grease (Apr)(4831).
DIN ENGL 51825-90. Type K Lubricating Greases (Aug) (This Standard Supersedes the June 1981 Edition of DIN 51825 Part 1, the December 1979 Edition of DIN 51825 Part 2, and the June 1981 Edition of DIN 51825 Part 3 Withdrawn in January 1986). 5 pp.
JIS K 2220-84. Lubricating Grease.
JIS K 2220-80. Lubricating Grease. 96 pp.
MOD UK DEF-2221-A-54. Grease LG-380. 3 pp.

—**Aircraft**
MOD UK DSTAN 91-12-72. Grease, Aircraft: General Purpose XG-271, NATO G-382 Issue 1. 9 pp.
MOD UK DSTAN 91-51-01. Grease, Aircraft: Helicopter Oscillating Bearing NATO Code No: G-366 Joint Service Designation: XG-284 Issue 1; Amendment 1. 13 pp.
MOD UK DSTAN 91-52-01. Grease, Aircraft: Multi-Purpose NATO Code No: G-395 Joint Service Designation: XG-293 Issue 1; Amendment 1. 17 pp.
MOD UK DSTAN 91-53-01. Grease, Aircraft: Synthetic, Extreme Pressure NATO Code No: G-354 Joint Service Designation: XG-287 Issue 1; Amendment 1. 14 pp.
MOD UK DSTAN 91-55-01. Grease, Aircraft: Synthetic, High Temperature NATO Code No: G-372 Joint Service Designation: XG-300 Issue 1; Amendment 1. 12 pp.
MOD UK DSTAN 91-57-01. Grease, Aircraft: Synthetic, Molybdendum Disulphide NATO Code No: G-353 Joint Service Designation: XG-276 Issue 1; Amendment 1. 12 pp.

—**Artillery**
MOD UK DSTAN 91-17-90. Grease, Calcium Base Joint Service Designation LG-280 Issue 2. 12 pp.
MOD UK DSTAN 91-27-76. Grease, Automotive and Artillery NATO Code No: G-403 Joint Service Designation: XG-279 Issue 1. 13 pp.

INTERNATIONAL AND NON-U.S. NATIONAL STANDARDS
SUBJECT INDEX

Greases (Cont.)

—Automotive
MOD UK DSTAN 91-17-90. Grease, Calcium Base Joint Service Designation LG-280 Issue 2. 12 pp.

MOD UK DSTAN 91-27-76. Grease, Automotive and Artillery NATO Code No: G-403 Joint Service Designation: XG-279 Issue 1. 13 pp.

—Classification
BSI BS 6413: Part 9-88. 1988 Lubricants, Industrial Oils and Related Products (Class L) Part 9: Classification for Family X (Greases). 7 pp.

ISO 6743 Pt 9-87. Lubricants, Industrial Oils and Related Products (Class L)—Classification—Part 9: Family X (Greases) First Edition. 7 pp.

—Cone Penetration
BSI BS 2000: Part 50-93. 1993 Methods of Test for Petroleum and Its Products Part 50: Determination of Cone Penetration of Lubricating Grease. 9 pp.

BSI BS 2000: Part 50-85. 1985 Petroleum and Its Products Part 50: Cone Penetration of Lubricating Grease. 12 pp.

ISO 2137-85. Petroleum Products—Lubricating Grease and Petrolatum—Determination of Cone Penetration Second Edition. 17 pp.

—Containers
BSI BS 2469-73. (WITHDRAWN) 1973 Round Tins for Lubricating Greases. 8 pp.

—Copper Strip Corrosion
BSI BS 2000: Part 112-93. 1993 Methods of Test for Petroleum and Its Products Part 112: Determination of Corrosiveness to Copper of Lubricating Grease—Copper Strip Method (Supersedes BS 4455: 1969). 3 pp.

BSI BS 2000: Part 112-82. 1982 Petroleum and Its Products Part 112: Corrosive Substances in Grease (Copper Strip Test). 4 pp.

—Corrosion Inhibitive—Sea Waterproof
MOD UK DSTAN 91-34-01. Grease, Sea Water Resisting NATO Code No. G-460 Joint Service Designation: XG-286 Issue 1; Amendment 2. 12 pp.

—Corrosion Inhibitors
MOD UK CS 3120. Protective PX-19.

—Corrosion Prevention
BSI BS 2000: Part 220-93. 1993 Methods of Test for Petroleum and Its Products Part 220: Determination of Rust Prevention Characteristics of Lubricating Greases (W). 6 pp.

BSI BS 2000: Part 220-82. 1982 Petroleum and Its Products Part 220: Dynamic Anti-Rust Test for Lubricating Greases. 7 pp.

—Defense Contracts
MOD UK DEFCON 112CV-90. Conditions of Contract for the Supply of Oils and Greases 4/90. 8 pp.

—Dropping Point
BSI BS 2000: Part 132-93. 1993 Methods of Test for Petroleum and Its Products Part 132: Determination of Dropping Point of Lubricating Grease. 4 pp.

BSI BS 2000: Part 132-90. 1990 Petroleum and Its Products Part 132: Dropping Point of Lubricating Grease (Supersedes BS 2877: 1976). 7 pp.

ISO 2176-72. Petroleum Products—Lubricating Grease—Determination of Dropping Point First Edition. 4 pp.

—Evaporation Loss
CNS K6769-83. Method of Test for Evaporation Loss of Lubricating Greases and Oils (May)(10267).

CNS K6770-83. Method of Test for Evaporation Loss of Lubricating Greases Over Wide-Temperature Range (May)(10268).

—Fabrics
CNS L4010-62. Grease for Textile Products (May)(1442) (R 1973). 1 p.

—Gears
DIN ENGL 51826-84. Lubricants; Type G Lubricating Greases (Nov). 3 pp.

—Gears—Naval
CGSB 3-GP-691MB-90. General Purpose Grease. 9 pp.

—Graphite Content
MOD UK DSTAN 05-50: Part 39-89. Methods for Testing Fuels, Lubricants and Associated Products Part 39: Graphite Content of Greases Issue 1. 8 pp.

—Immersion Testing
DIN ENGL 51807 Pt 1-79. Testing of Lubricants; Test of the Behaviour of Lubricating Greases in the Presence of Water; Static Test (Apr). 5 pp.

Greases (Cont.)

—Interchangeability
NATO STANAG 1135 ED 3 AMD 6-83. Interchangeability of Fuels, Lubricants and Associated Products Used by the Armed Forces of the North Atlantic Treaty Nations. 82 pp.

—Invitation for Bids
MOD UK DEFCON 47Q-87. Invitation to Tender for Supply of Oils and Greases 10/87. 2 pp.

—Lead Content
CNS K6759-83. Method of Test for Lead in New and Used Greases (Mar)(10094).

—Lead Content—Atomic Absorption Spectrometry
DIN ENGL 51827-89. Determination of Lead Content of Lubricating Greases by Atomic Absorption Spectrometry After Incineration (Jan). 3 pp.

—Letter Contracts—Small Value Orders
MOD UK DEFCON 53U (TU)-87. Small Value Order and Schedule for the Supply of Oils and Greases 10/87. 2 pp.

—Machine Tools
BSI BS 5063-92. 1992 Rationalized Range of Lubricants for Machine Tool Application. 10 pp.

BSI BS 5063-82. 1982 Rationalized Range of Lubricants for Machine Tool Applications. 7 pp.

—Marine
MOD UK DSTAN 91-17-90. Grease, Calcium Base Joint Service Designation LG-280 Issue 2. 12 pp.

MOD UK DSTAN 91-28-76. Grease, Naval: General Purpose, Ball and Roller Bearing NATO Code No: G-450 Joint Service Designation: XG-274 Issue 1. 11 pp.

—Molybdenum Disulfide
MOD UK DSTAN 91-64-91. Grease Molybdenum Disulphide Joint Service Designation XG—305 Issue 2. 12 pp.

—Naval
CGSB 3-GP-691MB-90. General Purpose Grease. 9 pp.

—Oil Separation
BSI BS 2000: Part 121-93. 1993 Methods of Test for Petroleum and Its Products Part 121: Determination of Oil Separation from Lubricating Grease—Pressure Filtration Method (W). 4 pp.

BSI BS 2000: Part 121-82. 1982 Petroleum and Its Products: Part 121: Oil Separation on Storage of Greases. 4 pp.

—Oxidation Resistance
BSI BS 2000: Part 142-93. 1993 Methods of Test for Petroleum and Its Products Part 142: Determination of Oxidation Stability of Lubricating Grease—Oxygen Bomb Method (W). 5 pp.

BSI BS 2000: Part 142-82. 1982 Petroleum and Its Products Part 142: Oxidation Stability of Lubricating Greases by the Oxygen Bomb Method. 7 pp.

—Plug Valves—Hydrocarbon Resistant
MOD UK DSTAN 91-6-91. Grease, Plug Valve, Hydrocarbon Resistant NATO Code G—363 Joint Service Designation XG—235 Issue 3. 18 pp.

—Pneumatic Systems—Aircraft
MOD UK DSTAN 91-56-80. Grease, Aircraft: Silicone, Pneumatic System NATO Code No: G-394 Joint Service Designation: XG-315 Issue 1. 8 pp.

—Pressure Measurement
DIN ENGL 51816 Pt 2-78. Testing of Lubricants; Conveying Characteristics of Lubricating Greases; Determination of Decompression Characteristics (Feb). 5 pp.

—Separators (Mechanical)—Sewage Treatment Equipment
DIN ENGL 4040 Pt 1-89. Grease Interceptor Systems; Concepts, Nominal Sizes, Requirements and Testing (Mar). 6 pp.

DIN ENGL 4040 Pt 2-89. Grease Interceptor Systems; Design, Installation and Operation (Mar). 4 pp.

—Shafts (Machine Elements)
CNS K5041-73. Lubricating Grease for Journal (Tentative) (May)(2590).

CNS K6229-73. Method of Test for Lubricating Grease for Journal (Aug)(2591).

—Solids Content
DIN ENGL 51813-89. Determination of Solid Matter Content of Lubricating Greases; (Particle Sizes Above 25 Micrometer) (Sept) (Supersedes DIN 51813 Part 1, May 1979 Edition). 4 pp.

Greases (Cont.)

—Solvents
MOD UK DSTAN 68-189-93. Solvent for Grease, Emulsifiable Issue 1. 12 pp.

MOD UK TS 414. Solvent for Grease, Emulsifiable.

—Stability Testing
MOD UK DSTAN 05-50: Part 24-87. Methods for Testing Fuels, Lubricants and Associated Products Part 24: Working Stability of Grease Issue 1. 8 pp.

MOD UK DSTAN 05-50: Part 50-87. Methods for Testing Fuels, Lubricants and Associated Products Part 50: Stability of Grease in the Presence of Water Issue 1. 8 pp.

—Steel Wire Rope
CGSB 3-GP-659M-77. Grease, Wire Rope, Standard for. 7 pp.

—Storage Stability
MOD UK DSTAN 05-50: Part 33-90. Methods for Testing Fuels, Lubricants and Associated Products Part 33: Storage Stability of Grease Issue 2. 8 pp.

—Sulfated Ash Content
DIN ENGL 51803-82. Testing of Lubricants; Determination of Ash in Lubricating Greases (Mar). 3 pp.

—Viscosity
CNS K6768-83. Method of Test for Apparent Viscosity of Lubricating Greases (May)(10266).

Greasing
Use: Lubrication

Greek Characters

—Coded Character Sets
BSI DD ENV 41503-91. 1991 Information Systems Interconnection—European Character Repertoires and Their Coding (Incorporating Corrigendum June 1991). 15 pp.

BSI DD ENV 41505-91. 1991 Information Systems Interconnection—Graphic Character Repertoire and Coding for Line Drawing. 9 pp.

BSI DD ENV 41506-91. 1991 Information Systems Interconnection—Data Stream Formats for Information Received from or Transmitted to the Teletex Service or Private Information Processing Systems Using Teletex Technology. 11 pp.

CEN ENV 41 503-87. Information Systems Interconnection: European Graphic Character Repertoires and Their Coding. 9 pp.

CEN ENV 41503-90. Information Systems Interconnection—European Character Repertoires and Their Coding. 14 pp.

CEN ENV 41508-90. Information Systems Interconnection—East European Graphic Character Repertoires and Their Coding. 11 pp.

ECMA ECMA 118-86. 8-Bit Single-Byte Coded Graphic Character Sets, Latin/Greek Alphabet. 15 pp.

ISO 8859 Pt 7-87. Information Processing—8-Bit Single-Byte Coded Graphic Character Sets—Part 7: Latin/Greek Alphabet First Edition. 9 pp.

JTC1 8859 Pt 7-87. Information Processing—8-Bit Single-Byte Coded Graphic Character Sets—Part 7: Latin/Greek Alphabet First Edition. 9 pp.

OSI ISO DIS 8859-7-87. Information Processing—8-Bit Single-Byte Coded Graphic Character Sets—Part 7: Latin/Greek Alphabet. 9 pp.

OSI ISO 8859 7-87. Information Processing–8-Bit Single-Byte Coded Graphic Character Sets-Part 7: Latin/Greek Alphabet.

—Coded Character Sets—Bibliographic References
BSI BS 6474: Part 2-85. 1985 Coded Character Sets for Bibliographic Information Interchange Part 2: Greek Alphabet Coded Character Set. 8 pp.

ISO 5428-84. Greek Alphabet Coded Character Set for Bibliographic Information Interchange Second Edition. 7 pp.

OSI ISO 5428-84. Greek Alphabet Coded Character Set for Bibliographic Information Interchange. 7 pp.

—Engineering Drawings
ISO 3098 Pt 2-84. Technical Drawings—Lettering—Part 2: Greek Characters First Edition. 7 pp.

—Romanization
ISO R843-68. International System for the Transliteration of Greek Characters into Latin Characters First Edition. 6 pp.

Green Algae
Use: Algae

Green Beans
Use For: Snap Beans; String Beans See Also: Beans; Kidney Beans; Vegetables

Green Beans (Cont.)

—Cold Storage

ISO 9930-93. Green Beans—Storage and Refrigerated Transport First Edition. 5 pp.

—Refrigerated Transportation

ISO 9930-93. Green Beans—Storage and Refrigerated Transport First Edition. 5 pp.

Green Coffee
Use: Coffee

Green Garlic
Use: Garlic

Green Index
Use: Chlorophyll Content Analysis

Green Liquors
See Also: Black Liquors; Calcium Oxide; Kraft Liquors; Papermaking Equipment; Spent Liquors; White Liquors

—Alkalinity

CPPA G.11U-77. Alkalinity of Sulphate Green and White Liquor. 2 pp.

—Kraft—Chemical Analysis

CPPA J.12-84. Analysis of Sulphate Green and White Liquors. 6 pp.

Green Onions
Use: Scallions

Green Peppers
Use: Peppers

Green Strength
See Also: Compatibility; Compression Testing; Molds (Casting); Pellets; Porosity; Powder Metallurgy

—Metal Powders

BSI BS 5600: Sec 2.5-88. (WITHDRAWN) 1988 Powder Metallurgical Materials and Products Part 2: Methods of Sampling and Testing Metallic Powders Section 2.5: Determination of Green Strength by Transverse Rupture of Rectangular Compacts. 7 pp.

BSI BS EN 23995-93. 1993 Metallic Powders—Determination of Green Strength by Transverse Rupture of Rectangular Compacts (ISO 3995: 1985) (V). 12 pp.

CEN EN 23995-93. Metallic Powders—Determination of Green Strength by Transverse Rupture of Rectangular Compacts (ISO 3995: 1985). 7 pp.

ISO 3995-85. Metallic Powders—Determination of Green Strength by Transverse Rupture of Rectangular Compacts Second Edition (CEN EN 23995: 1993). 7 pp.

—Rubber

BSI BS 903: Part A62-93. 1993 Physical Testing of Rubber Part A62: Method for Determination of Green Strength of Raw Rubber of Unvulcanized Compounds (ISO 9026: 1991) (V). 11 pp.

ISO 9026-91. Raw Rubber or Unvulcanized Compounds—Determination of Green Strength First Edition. 9 pp.

Greenhouses
See Also: Agricultural Buildings; Shelters

—Design

BSI BS 5502: Sec 2.4-81. (WITHDRAWN) 1981 Design of Buildings and Structures for Agriculture Part 2: General Considerations Section 2.4: Crop Production Buildings (Superseded by BS 5502: Part 66: 1992). 6 pp.

—Electrical Installations

CENELEC HD 384.7.705 S1-91. Electrical Installations of Buildings Part 7: Requirements for Special Installations or Locations Section 705—Electrical Installations of Agricultural and Horticultural Premises. 5 pp.

IEC 364 Pt 7-705-84. Electrical Installations of Buildings Part 7: Requirements for Special Installations or Locations Section 705—Electrical Installations of Agricultural and Horticultural Premises First Edition. 13 pp.

Grenades
See Also: Ammunition; Armor Piercing Ammunition; Pyrotechnics; Small Arms Ammunition; Weapons

—Baking Paints

MOD UK DSTAN 80-116-91. Paint, System, Stoving, Textured Issue 1. 11 pp.

MOD UK TS 10183. Paint System, Stoving Textured for Grenades, LF Quality.

Grenades (Cont.)

—Cases—Identification Systems—Inks

MOD UK DSTAN 80-170-93. Ink, Marking, for Rubber Grenade Cases Type: Silk Screen Stencilling Issue 1. 14 pp.

MOD UK TS 10156B. Ink, Marking for Rubber Grenade Cases Type: Silk, Screen Stencilling (Superseded by Def Stan 80-170).

Grey Cast Iron
Use: Gray Iron

Grey Scale
Use: Gray Scale

Griddle Grills
Use: Griddles

Griddles
See Also: Appliances; Cooking Appliances

BSI BS 5314: Part 8-79. (WITHDRAWN) 1979 Amd 1 Gas Heated Catering Equipment Part 8: Griddle Plates (AMD 5404) July 31, 1987 (Superseded by BS EN 203-1: 1993). 23 pp.

CSA CAN/CSA-C22. 2 NO 64-M91. Household Cooking and Liquid-Heating Appliances; (Gen Instr 1 Thru 2). 91 pp.

JIS C 9215-88. Electric Griddles. 29 pp.

—Commercial

CSA C22.2 NO 109-M1981. Commercial Cooking Appliances; (Gen Instr 1 Thru 3). 36 pp.

CSA 1169 Bull. Electrical Bulletin 1169 June 27, 1978 to C22.2 NO 109. 2 pp.

—Identification Systems

CEN EN 30-79. Domestic Cooking Appliances Burning Gas. 50 pp.

CEN PREN 30 (Part 1)-90. Domestic Cooking Appliances Burning Gas—Part 1: Safety. 122 pp.

—Safety

BSI BS 3456: Sec 102.38-89. (WITHDRAWN) 1989 Safety of Household Electrical Appliances Part 102: Particular Requirements Section 102.38: Commercial Electric Griddles and Griddle Grills (Superseded by BS 3456: Section 202.38: 1989). 18 pp.

BSI BS 3456: Sec 202.38-90. (WITHDRAWN) 1990 Safety of Household Electrical Appliances Part 202: Particular Requirements Section 202.38: Commercial Electric Griddles and Griddle Grills (Supersedes BS 3456: Section 102.38: 1989). 22 pp.

BSI BS 3456: Sec 202.38-01. 1990 Amd 1 Safety of Household and Similar Electrical Appliances Part 202: Particular Requirements Section 202.38: Commercial Electric Griddles and Griddle Grills (AMD 7821) July 15, 1993 (L). 37 pp.

BSI BS 3456: Sec 202.38-02. 1990 Amd 2 Safety of Household and Similar Electrical Appliances Part 202: Particular Requirements Section 202.38: Commercial Electric Griddles and Griddle Grills (AMD 7822) JULY 15, 1993 SUPERSEDES BS 3456: SECTION 102. 38: 1989 (L). 45 pp.

BSI BS 5784: Part 3-84. 1984 Amd 1 Safety of Electrical Commercial Catering Equipment Part 3: Griddles and Griddle Grills. 59 pp.

CEN EN 30-79. Domestic Cooking Appliances Burning Gas. 50 pp.

CEN PREN 30 (Part 1)-90. Domestic Cooking Appliances Burning Gas—Part 1: Safety. 122 pp.

CENELEC HD 286 S1-87. Safety of Household and Similar Electrical Appliances—Part 2: Particular Requirements for Commercial Electric Griddles and Griddle Grills (Replaced by EN 60 335-2-38-1989). 10 pp.

CENELEC EN 60 335-2-38-89. Safety of Household and Similar Electrical Appliances Part 2: Particular Requirements for Commercial Electric Griddles and Griddle Grills (Replaces HD 286 S1:1987). 11 pp.

CENELEC EN 60335-2-38/A1-92. AMD 1 Safety of Household and Similar Electrical Appliances Part 2: Particular Requirements for Commercial Electric Griddles and Griddle Grills; (Replaces HD 286 S1:1987). 5 pp.

CENELEC EN 60335-2-38/A51-92. AMD 51 Safety of Household and Similar Electrical Appliances Part 2: Particular Requirements for Commercial Electric Griddles and Griddle Grills; (Replaces HD 286 S1:1987). 6 pp.

DIN VDE 0700 Pt 6 A8 (D)-89. Safety of Household and Similar Electrical Appliances; Cooking Ranges, Cooking Tables, Ovens and Similar Appliances for Household Use; Amendment 8 to Draft DIN 57700 Part 6/VDE 0700 Part 6/02.84 (Apr). 15 pp.

IEC 335 Pt 2-38-86. Safety of Household and Similar Electrical Appliances Part 2: Particular Requirements for Commercial Electric Griddles and Griddle Grills Second Edition; (Amendment 1-1990). 46 pp.

Grids
Use: Grilles

Grids (Coordinates)
See Also: Latitude; Longitude; Maps

—Military Operations

NATO STANAG 2211 ED 5 AMD 0-91. Geodetic Datums, Ellipsoids, Grids and Grid References. 97 pp.

NATO STANAG 2211 ED 5 AMD 1-91. Geodetic Datums, Ellipsoids, Grids and Grid References. 75 pp.

NATO STANAG 2211 ED 5 AMD 2-91. Geodetic Datums, Ellipsoids, Grids and Grid References. 89 pp.

Grill Charcoal
Use: Charcoal

Grilles
Use For: Gratings (Metal); Grids *See Also:* Gullies

—Gullies

CNS A2173-84. Gully Grating (Apr)(10841).

—Industrial

BSI BS 4592: Part 1-87. 1987 Amd 1 Industrial Type Metal Flooring, Walkways and Stair Treads Part 1: Opening Bar Gratings. 13 pp.

BSI BS 4592: Part 2-87. 1987 Industrial Type Metal Flooring, Walkways and Stair Treads Part 2: Expanded Metal Gratings Panels. 14 pp.

BSI BS 4592: Part 4-92. 1992 Industrial Type Flooring, Walkways and Stair Treads Part 4: Specification for Glass Reinforced Plastics Open Bar Gratings. 13 pp.

—Roads

SAA AS 3996-92. Metal Access Covers, Road Grates and Frames. 20 pp.

—Roller—Construction Contracts

DIN ENGL 18358-88. Tendering and Performance Stipulations in Contracts for Construction Works (VOB); Part C: General Technical Specifications in Contracts for Construction Works (ATV); Installation of Roller Shutters and Similar Equipment (Sept). 3 pp.

—Roller—Installation—Construction Contracts

DIN ENGL 18358-88. Tendering and Performance Stipulations in Contracts for Construction Works (VOB); Part C: General Technical Specifications in Contracts for Construction Works (ATV); Installation of Roller Shutters and Similar Equipment (Sept). 3 pp.

—Ships—Scuppers

JIS F 3015-77. Gratings of Ships' Scupper Pipes (R 1984). 11 pp.

—Ships—Water Inlets

BSI BS MA 62-74. 1974 Weed Grids for Sea Inlets. 13 pp.

—Storage Heaters

CENELEC HD 283 S1-92. Safety of Household and Similar Electrical Appliances Particular Requirement for the Maximum Temperature Allowed for the Surfaces of Air-Outlet Grilles of Thermal Storage Room Heating Appliances. 9 pp.

Grills (Appliances)
See Also: Appliances; Barbecues; Cooking Appliances

CSA CAN/CSA-C22. 2 NO 64-M91. Household Cooking and Liquid-Heating Appliances; (Gen Instr 1 Thru 2). 91 pp.

—Electric

BSI BS 4167: Part 3-69. 1969 Amd 1 Electrically-Heated Catering Equipment Part 3: Grillers, Grillers on Ranges, Toasters. 49 pp.

—Electric—Household—Safety

BSI BS EN 60335-2-9-91. 1991 Safety of Household and Similar Electrical Appliances Part 2: Particular Requirements Section 2.9: Toasters, Grills, Roasters and Similar Appliances. 26 pp.

BSI BS EN 60335-2-9-01. 1991 Amd 1 Safety of Household and Similar Electrical Appliances Part 2: Particular Requirements Section 2.9: Toasters, Grills, Roasters and Similar Appliances (AMD 6897) April 1, 1992. 32 pp.

BSI BS EN 60335-2-9-02. 1991 Amd 2 Safety of Household and Similar Electrical Appliances Part 2: Particular Requirements Section 2.9: Toasters, Grills, Roasters and Similar Appliances (AMD 7584) March 15, 1993 (L). 42 pp.

CENELEC EN 60 335-2-9-90. Safety of Household and Similar Electrical Appliances Part 2: Particular Requirements for Toasters, Grills, Roasters and Similar Appliances. 8 pp.

CENELEC EN 60335-2-9 /A2-92. AMD 2 Safety of Household and Similar Electrical Appliances Part 2: Particular Requirments for Toasters, Grills, Roasters and Similar Appliances. 5 pp.

Grills (Appliances) (Cont.)

—Electric—Household—Safety (Cont.)

CSA C22.2 NO 1335.2.9-93. Portable Electrical Motor-Operated and Heating Appliances: Particular Requirements for Portable Electric Cooking Appliances; (Gen Instr 1). 40 pp.

IEC 335 Pt 2-9-93. Safety of Household and Similar Electrical Appliances Part 2: Particular Requirements for Toasters, Grills, Roasters and Similar Appliances Fourth Edition. 46 pp.

—Electric—Safety

BSI BS 3456: Sec 102.9-91. (WITHDRAWN) 1991 Safety of Household and Similar Electrical Appliances Part 102: Particular Requirements Section 102.9: Toasters, Grills, Roasters and Similar Appliances (Superseded by BS EN 60335-2-9: 1991). 26 pp.

CENELEC HD 265 S2-84. Safety of Household and Similar Electrical Appliances—Particular Requirements for Appliances for Toasters, Grills, Roasters and Similar Appliances. 8 pp.

CENELEC HD 265 S2/A1-88. AMD 1 Safety of Household and Similar Electrical Appliances—Particular Requirements for Appliances for Toasters, Grills, Roasters and Similar Appliances. 4 pp.

CENELEC HD 275 S1-86. Safety of Household and Similar Electrical Appliances—Part 2: Particular Requirements for Cooking Ranges, Cooking Tables, Ovens and Similar Appliances for Household Use. 30 pp.

CENELEC HD 275 S1/A1-89. AMD 1 Safety of Household and Similar Electrical Appliances—Part 2: Particular Requirements for Cooking Ranges, Cooking Tables, Ovens and Similar Appliances for Household Use. 9 pp.

—Gas

BSI BS 5386: Part 4-91. 1991 Gas Burning Appliances Part 4: Built-In Domestic Cooking Appliances (L). 18 pp.

BSI BS 5386: Part 4-01. 1991 Amd 1 Gas Burning Appliances Part 4: Built-In Domestic Cooking Appliances (AMD 6940) July 15, 1992. 19 pp.

BSI BS 5386: Part 4-83. 1983 Gas Burning Appliances Part 4: Built-In Domestic Cooking Appliances. 12 pp.

JIS S 2124-81. Hot Plates Grill Used with Town Gas.

JIS S 2126-81. Grills Used with Town Gas.

—Gas—Household

BSI BS 5386: Part 3-80. 1980 Amd 5 Gas Burning Appliances Part 3: Domestic Cooking Appliances Burning Gas (L) (AMD 6642) March 28, 1991. 107 pp.

BSI BS 5386: Part 3-06. 1980 Amd 6 Gas Burning Appliances Part 3: Domestic Cooking Appliances Burning Gas (AMD 6883) July 15, 1992. 118 pp.

—Gas—Identification Systems

CEN EN 30-79. Domestic Cooking Appliances Burning Gas. 50 pp.

CEN PREN 30 (Part 1)-90. Domestic Cooking Appliances Burning Gas—Part 1: Safety. 122 pp.

—Gas—Safety

CEN EN 30-79. Domestic Cooking Appliances Burning Gas. 50 pp.

CEN PREN 30 (Part 1)-90. Domestic Cooking Appliances Burning Gas—Part 1: Safety. 122 pp.

—Safety

BSI BS 3456: Sec 2.6-70. (WITHDRAWN) 1970 Amd 2 1972-1981 Edition. Specification for Safety of Household Electrical Appliances: Part 2: Particular Requirements: Section 2.6: Frying Pans, Grills, Plate Warmers, and Other Dry Cooking Appliances (Superseded by BS 3456: Sec 3.16 & 3.17). 18 pp.

BSI BS 3456: Sec 3.17-80. (WITHDRAWN) 1980 Amd 2 1972-1981 Edition. Specification for Safety of Household Electrical Appliances: Part 3: Complete Particular Specif-ications: Section 3.17: Toasters, Grills, Waffle Irons, Roasters & Other Dry Cooking Appliances (Superseded by BS 3456: Section 102.9: 1991). 48 pp.

BSI BS EN 60335-2-6-91. 1991 Safety of Household and Similar Electrical Appliances Part 2: Particular Requirements Section 2.6: Cooking Ranges, Cooking Tables, Ovens and Similar Appliances for Household Use. 52 pp.

BSI BS EN 60335-2-6-01. 1991 Amd 1 Safety of Household and Similar Electrical Appliances Part 2: Particular Requirements Section 2.6: Cooking Ranges, Cooking Tables, Ovens and Similar Appliances for Household Use (AMD 7677) July 15, 1993 (L). 70 pp.

BSI BS EN 60335-2-6-02. 1991 Amd 2 Safety of Household and Similar Electrical Appliances Part 2: Particular Requirements Section 2.6: Cooking Ranges, Cooking Tables, Ovens and Similar Appliances for Household Use (AMD 7915) September 15, 1993 (L). 81 pp.

BSI BS EN 60335-2-6-03. 1991 Amd 3 Safety of Household and Similar Electrical Appliances Part 2: Particular Requirements Section 2.6: Cooking Ranges, Cooking Tables, Ovens and Similar Appliances for Household Use (AMD 7916) October 15, 1993 (L). 84 pp.

BSI BS EN 60335-2-48-92. 1992 Safety of Household and Similar Electrical Appliances Part 2: Particular Requirements Section 2.48: Commercial Electric Grillers and Toasters. 22 pp.

BSI BS EN 60335-2-48-01. 1992 Amd 1 Safety of Household and Similar Electrical Appliances Part 2: Particular Requirements Section 2.48: Commercial Electric Grillers and Toasters (AMD 7832) August 15, 1993 (L). 33 pp.

CENELEC EN 60335-2-6-90. Safety of Household and Similar Electrical Appliances Part 2: Particular Requirements for Cooking Ranges, Cooking Tables, Ovens and Similar Appliances for Household Use (Supercedes HD 275 S1: 1986 and Its Amendment 1: 1988). 38 pp.

CENELEC EN 60335-2-6-91. CORRIGENDUM Safety of Household and Similar Electrical Appliances Part 2: Particular Requirements for Cooking Ranges, Cooking Tables, Ovens and Similar Appliances for Household Use. 2 pp.

CENELEC EN 60335-2-6 /A2-92. AMD 2 Safety of Household and Similar Electrical Appliances Part 2: Particular Requirements for Cooking Ranges, Cooking Tables, Ovens and Similar Appliances for Household Use; (IEC 335-2-6:1986/A2: 1990). 6 pp.

CENELEC EN 60335-2-48-90. Safety of Household and Similar Electrical Appliances Part 2: Particular Requirements for Commercial Electric Grillers and Toasters. 10 pp.

CENELEC EN 60335-2-48/A1-92. AMD 1 Safety of Household and Similar Electrical Appliances Part 2: Particular Requirements for Commercial Electric Grillers and Toasters. 5 pp.

DIN VDE 0700 Pt 6 A8 (D)-89. Safety of Household and Similar Electrical Appliances; Cooking Ranges, Cooking Tables, Ovens and Similar Appliances for Household Use; Amendment 8 to Draft DIN 57700 Part 6/VDE 0700 Part 6/02.84 (Apr). 15 pp.

IEC 335 Pt 2-6-86. Safety of Household and Similar Electrical Appliances Part 2: Part-icular Requirements for Cooking Ranges, Cooking Tables, Ovens and Sim-ilar Appliances for Household Use Third Edition; (Amendment 2-1990, Incorporating Amendment 1) (Amendment 3-1992). 106 pp.

IEC 335 Pt 2-48-88. Safety of Household and Similar Electrical Appliances Part 2: Particular Requirements for Commercial Electric Grillers and Toasters First Edition; (Amendment 1-1990). 44 pp.

Grindability

See Also: Grindability Index; Grinding (Material Removal)

—Cellular Plastics

BSI BS 4370: Part 3-88. 1988 Methods of Test for Rigid Cellular Materials Part 3: Methods 12 and 13. 12 pp.

—Coal

JIS M 8801-79. Methods for Testing of Coal. 71 pp.

Grindability Index

See Also: Grindability

JIS M 4002-76. Testing Method of Grinding Work Index (R 1984). 7 pp.

Grinders

Scope Note: For additional listings, use a more specific term *See Also:* Angle Grinders; Bench Grinders; Broaching Machines; Coffee Grinders; Corn Grinders; Cutting Tools; Cylindrical Grinders; Die Grinders; Disintegrators; Disk Grinders; Floor Grinders; Gear Grinders; Grinding (Material Removal); Grinding Wheels; Honing Machinery; Internal Grinders; Knife Grinders; Lathes; Machine Tools; Meat Grinders; Milling Machines; Planers (Tools); Pneumatic Grinders; Sanders (Tools); Surface Grinders; Tool Holders; Tools; Universal Grinders

CNS B1125-79. Specified Items of Machine Tools (Universal Tool Grinder) (Apr)(4670-28). 1 p.

CNS B7198-82. Test Code for Performance of Universal Tool Grinder (Aug)(9213).

CNS B7199-82. Accuracy Inspection of Universal Tool Grinder (Aug)(9214). 7 pp.

CNS C3182-85. Method of Test for Portable Electric Grinders (Jan)(10609).

CNS C4104-83. Portable Electric Grinders (Oct)(3265). 9 pp.

JIS B 6229-83. Test Code for Performance and Accuracy of Universal Tool-Grinding Machines. 30 pp.

—Electric—Portable

CSA CAN/CSA-C22. 2 NO71.1-M89. Portable Electric Tools; (Gen Instr 1 Thru 4). 89 pp.

—Electric—Portable—Safety

BSI BS 2769: Sec 2.3-84. 1984 Hand-Held Electric-Motor-Operated Tools Part 2: Particular Requirements Section 2.3: Grinders, Polishers and Disc-Type Sanders. 16 pp.

BSI BS 2769: Sec 2.3-01. 1984 Amd 1 Hand-Held Electric Motor-Operated Tools Part 2: Particular Requirements Section 2.3: Specification for Grinders, Polishers and Disc-Type Sanders (AMD 6924) May 1, 1992. 17 pp.

CENELEC HD 400.2C/A1-91. AMD 1 Hand-Held Motor Operated Tools Part II: Particular Specifications (Section C). 5 pp.

CENELEC HD 400.2C S1 PRA2-93. AMD 2 Hand-Held Motor Operated Tools Part II: Particular Specifications Section C: Grinders, Polishers and Disc Type Sanders. 2 pp.

—Electric—Valve Seats

CSA CAN/CSA-C22. 2 NO71.1-M89. Portable Electric Tools; (Gen Instr 1 Thru 4). 89 pp.

—Energy

ISO 3002 Pt 4-84. Basic Quantities in Cutting and Grinding—Part 4: Forces, Energy, Power First Edition. 14 pp.

—Food Processing

CSA C22.2 NO 195-M1987. Motor Operated Food Processing Appliances (Household and Commercial); (Gen Instr 1 Thru 2). 55 pp.

—Forces

ISO 3002 Pt 4-84. Basic Quantities in Cutting and Grinding—Part 4: Forces, Energy, Power First Edition. 14 pp.

—Geometric Quantities

ISO 3002 Pt 3-84. Basic Quantities in Cutting and Grinding—Part 3: Geometric and Kinematic Quantities in Cutting First Edition. 12 pp.

—Glossaries

CNS B1229-80. Glossary of Terms Relating to Parts of Grinding Machine (Dec)(6753).

ISO 3002 Pt 3-84. Basic Quantities in Cutting and Grinding—Part 3: Geometric and Kinematic Quantities in Cutting First Edition. 12 pp.

—Identification Systems

CEN EN 68-77. Hand Held (Portable) Power Driven Grinding Machines: Mechanical Safety. 29 pp.

CEN EN 68/A1-78. AMD 1 Hand Held (Portable) Power Driven Grinding Machines: Mechanical Safety. 2 pp.

CEN EN 68/2-78. AMD 2 Hand Held (Portable) Power Driven Grinding Machines: Mechanical Safety. 4 pp.

—Kinematic Quantities

ISO 3002 Pt 3-84. Basic Quantities in Cutting and Grinding—Part 3: Geometric and Kinematic Quantities in Cutting First Edition. 12 pp.

—Portable

CGSB 45-GP-4C-72. Grinders and Disk Sanders, Electric, Portable, Heavy Duty, Standard for. 9 pp.

JIS C 9610-76. Portable Electric Grinders (R 1984). 25 pp.

—Portable—Safety

CEN EN 68-77. Hand Held (Portable) Power Driven Grinding Machines: Mechanical Safety. 29 pp.

CEN EN 68/A1-78. AMD 1 Hand Held (Portable) Power Driven Grinding Machines: Mechanical Safety. 2 pp.

CEN EN 68/2-78. AMD 2 Hand Held (Portable) Power Driven Grinding Machines: Mechanical Safety. 4 pp.

IEC 745 Pt 2-3-84. Safety of Hand-Held Motor-Operated Electric Tools Part 2: Particular Requirements for Grinders, Polishers and Disk-Type Sanders First Edition. 21 pp.

—Power

ISO 3002 Pt 4-84. Basic Quantities in Cutting and Grinding—Part 4: Forces, Energy, Power First Edition. 14 pp.

—Safety

DIN VDE 0741-63. Specifications for Grinding and Polishing Machines (Aug). 26 pp.

IEC 745 Pt 2-3-84. Safety of Hand-Held Motor-Operated Electric Tools Part 2: Particular Requirements for Grinders, Polishers and Disk-Type Sanders First Edition. 21 pp.

—Spindle Noses

BSI BS 1089-73. 1973 Work Head Spindle Noses for Grinding Machines: Cylindrical External, Internal and Universal Types. 11 pp.

INTERNATIONAL AND NON-U.S. NATIONAL STANDARDS
SUBJECT INDEX

Grinding

Grinding (Material Removal)
See Also: Grindability; Grinders; Grinding Wheels; Sanding; Tumbling

CNS M3026-80. Methods of Test for Grinding Work Index (Dec)(6698).

JIS B 0711-76. Removal in Grindings (R 1984). 11 pp.

—Glossaries
ISO 3002 Pt 5-89. Basic Quantities in Cutting and Grinding—Part 5: Basic Terminology for Grinding Processes Using Grinding Wheels First Edition. 26 pp.

—Steel Products—Quality Assurance
MOD UK QTR 14/AQD. Inspection of Ground Steel Parts for Grinding Damage Issue 1.

Grinding Discs
Use: Grinding Wheels

Grinding Disks
Use: Grinding Wheels

Grinding Machines
Use: Grinders

Grinding Spindles
See Also: Spindles

—Acceptance Testing
DIN ENGL 8637-90. Machine Tools; Grinding Spindles with Cartridge Diameters up to 200 mm; Acceptance Conditions (May). 6 pp.

Grinding Wheels
Use For: Abrasive Flap Wheels; Abrasive Wheels; Cup Wheels; Flap Wheels; Grinding Discs; Grinding Disks *See Also:* Abrasives; Carbide Tools; Cutting Tools; Diamond Wheels; Diamonds (Industrial); Grinders; Grinding (Material Removal); Machine Tools; Magnesite Wheels; Sanders (Tools); Tools; Wheels

BSI BS 4481: Part 1-89. 1989 Bonded Abrasive Products Part 1: General Features of Abrasive Wheels, Segments, Bricks and Sticks. 16 pp.

BSI BS 4481: Part 2-83. 1983 Bonded Abrasive Products Part 2: Dimensions of Grinding Wheels and Segments. 38 pp.

CNS R2086-87. Shape and Dimensions of Grinding Wheels (Nov) (3965).

CNS R2174-87. Segment Grinding Wheels (Oct)(11141).

DIN ENGL 69104-81. Grinding Wheels; Outside Diameters; Thicknesses; Hole Diameters (Feb). 3 pp.

ISO R603 Pt 1-67. Bonded Abrasive Products Grinding-Wheel Dimensions (Part 1) First Edition; (Erratum—Oct 1968). 13 pp.

ISO 1117-75. Bonded Abrasive Products—Grinding-Wheel Dimensions (Part 2) First Edition. 24 pp.

ISO 2933-74. Bonded Abrasive Products—Grinding Wheel Dimensions (Part 3) First Edition. 8 pp.

JIS B 4051-88. Recommendation of Grinding Wheels. 17 pp.

JIS B 4052-88. Recommendation of Wheels for Free Hand Grinding. 9 pp.

JIS R 6211-86. Shapes and Dimensions of Grinding Wheels. 74 pp.

JIS R 6218-86. Segment Grinding Wheels. 14 pp.

SAA AS B172.2-68. Bonded Abrasive Products —Part 2: Shapes and Dimensions of Grinding Wheels. 32 pp.

SAA AS B172.3-68. Bonded Abrasive Products —Part 3: Shapes and Dimensions of Segments of Grinding Wheels. 6 pp.

SAA AS B172.4-69. Bonded Abrasive Products —Part 4: Shapes, Sizes and Identification of Mounted Wheels (Inch Series). 15 pp.

SAA AS 1788.1-87. Abrasive Wheels—Part 1: Design, Construction and Safeguarding. 46 pp.

SAA AS 1788.2-87. Abrasive Wheels—Part 2: Selection, Care, and Use. 24 pp.

SNZ NZS/BS 4481: Part 2-83. Bonded Abrasive Products Part 2: Specification for Dimensions of Grinding Wheels and Segments. 40 pp.

SNZ NZS 5246: Part 1-90. Specification for Abrasive Wheels Part 1: Design, Construction and Safeguarding Amend: A, 1990. 46 pp.

SNZ NZS 5246: Part 2-90. Specification for Abrasive Wheels Part 2: Selection, Care and Use Amend: A, 1990. 24 pp.

—Aluminum Oxide—Depressed Center—Fabric Reinforced
CNS R2084-87. Depressed Center Wheels with Fabric Reinforcement (Oct) (3789).

JIS R 6213-86. Depressed Center Wheels with Fabric Reinforcement. 8 pp.

—Aluminum Oxide—Inserted Nut—Ring
CNS R2093-87. Inserted Nut Abrasive Discs and Inserted Nut Ring Wheels (Oct)(4208).

Grinding Wheels (Cont.)

—Aluminum Oxide—Inserted Nut—Ring (Cont.)
JIS R 6216-86. Inserted Nut Abrasive Discs and Inserted Nut Ring Wheels. 12 pp.

—Aluminum Oxide—Magnesite
CNS R2175-87. Magnesia Grinding Wheels (Oct)(11142).

JIS R 6219-86. Magnesia Grinding Wheels. 13 pp.

—Aluminum Oxide—Mounted
CNS R2089-87. Mounted Wheels (Oct) (4204).

JIS R 6217-86. Mounted Wheels. 13 pp.

—Aluminum Oxide—Resinoid
CNS R2088-87. Resinoid Grinding Wheels (Oct)(3967).

JIS R 6212-86. Resinoid Grinding Wheels. 12 pp.

—Aluminum Oxide—Resinoid Cutoff
CNS R2085-87. Resinoid Cutting-Off Wheels (Oct)(3790).

JIS R 6214-86. Resinoid Cutting-off Wheels. 9 pp.

—Aluminum Oxide—Rubber Cutoff
CNS R2090-87. Rubber Cutting-off Wheels (Oct)(4205).

JIS R 6215-86. Rubber Cutting-off Wheels. 8 pp.

—Aluminum Oxide—Segmental
CNS R2174-87. Segment Grinding Wheels (Oct)(11141).

JIS R 6218-86. Segment Grinding Wheels. 14 pp.

—Aluminum Oxide—Vitrified
CNS R2034-87. Vitrified Grinding Wheels (Oct)(991). 7 pp.

JIS R 6210-86. Vitrified Grinding Wheels. 12 pp.

—Balancing
BSI BS 4481: Part 3-88. 1988 Bonded Abrasive Products Part 3: Methods for Assessment of Static Balancing of Abrasive Wheels. 8 pp.

CNS R3077-87. Method of Test for Grinding Wheels (Nov) (3786).

ISO 6103-86. Bonded Abrasive Products—Static Balancing of Grinding Wheels—Testing First Edition. 7 pp.

JIS R 6240-86. Testing Methods of Grinding Wheels. 18 pp.

SNZ NZS/ISO 6103-86. Bonded Abrasive Products —Static Balancing of Grinding Wheels—Testing. 5 pp.

—Bonding Strength
CNS R3077-87. Method of Test for Grinding Wheels (Nov) (3786).

JIS R 6240-86. Testing Methods of Grinding Wheels. 18 pp.

—Boron Nitride
ISO 6168-80. Abrasive Products—Diamond or Cubic Boron Nitride Grinding Wheels—Dimensions First Edition. 17 pp.

JIS B 4131-82. Diamond or Cubic Boron Nitride Grinding Wheels.

—Boron Nitride—Glossaries
BSI BS 5831-79. 1979 Designation and Multilingual Nomenclature for Diamond or Cubic Boron Nitride Grinding Wheels and Saws. 20 pp.

ISO 6104-79. Abrasive Products—Diamond or Cubic Baron Nitride Grinding Wheels and Saws—General Survey, Designation and Multilingual Nomenclature First Edition. 16 pp.

—Cutting Off
CNS R2085-87. Resinoid Cutting-Off Wheels (Oct)(3790).

CNS R2090-87. Rubber Cutting-off Wheels (Oct)(4205).

ISO 603 Pt 2-81. Bonded Abrasive Products—Grinding Wheel Dimensions—Part 2: Cutting-off Wheels, Type 1 First Edition. 3 pp.

JIS R 6214-86. Resinoid Cutting-off Wheels. 9 pp.

JIS R 6215-86. Rubber Cutting-off Wheels. 8 pp.

—Designations
BSI BS 5587-78. 1978 Amd 1 Coated Abrasives Flap Wheels with Incorporated Flanges or Separate Flanges. Designation and Dimensions. 5 pp.

BSI BS 7007: Part 2-88. 1988 Abrasive Products Part 2: Designation and Dimensions of Coated Abrasive Flap Wheels with Shafts. 3 pp.

CNS R2191-87. Coated Abrasives-Flap Wheels with Shafts-Designation and Dimensions (Nov)(12168).

CNS R2192-87. Coated Abrasives-Flap Wheels with Incorporated Flanges or Separate Flange-Designation and Dimensions (Nov)(12169).

ISO 3919-93. Coated Abrasives—Flap Wheels with Shafts—Dimensions and Designation Third Edition. 5 pp.

ISO 5429-77. Coated Abrasives—Flap Wheels with Incorporated Flanges or Separate Flanges— Designation and Dimensions First Edition. 3 pp.

Grinding Wheels (Cont.)

—Designations (Cont.)
JIS R 6258-81. Coated Abrasives—Flap Wheels with Shafts—Designation and Dimensions. 5 pp.

JIS R 6259-81. Coated Abrasives—Flap Wheels with Incorporated Flanges or Separate Flanges— Designation and Dimensions. 5 pp.

—Flanges
BSI BS 4581-70. 1970 Amd 4 Dimensions of Flanges for the Mounting of Plain Grinding Wheels. 18 pp.

BSI BS 4581: Part 2-84. 1984 Flanges for Mounting Grinding Wheels on Grinding Machine Tools Part 2: Specification for Dimensions and Materials of Flanges for Grinding Wheels of Less Than 76.2mm Bore. 12 pp.

SNZ NZS/BS 4581: Part 1-70. Flanges for Mounting Grinding Wheels or Grinding Machine Tools Part 1: Specification for Dimensions of Flanges for Grinding Wheels of 76.2 mm Bore and Greater Amend: 1; 2; 3; 4. 12 pp.

SNZ NZS/BS 4581: Part 2-84. Flanges for Mounting Grinding Wheels or Grinding Machine Tools Part 2: Specification for Dimensions and Materials of Flanges for Grinding Wheels of Less Than 76.2 mm Bore. 12 pp.

—Glossaries
ISO 3002 Pt 5-89. Basic Quantities in Cutting and Grinding—Part 5: Basic Terminology for Grinding Processes Using Grinding Wheels First Edition. 26 pp.

JIS R 6004-90. Glossary of Terms and Marks Used in Abrasives, Grinding Wheels and Coated Abrasives. 20 pp.

—Grain Content
CNS R3077-87. Method of Test for Grinding Wheels (Nov) (3786).

JIS R 6240-86. Testing Methods of Grinding Wheels. 18 pp.

—Hub Flanges
ISO 666-75. Machine Tools—Mounting of Plain Grinding Wheels by Means of Hub Flanges First Edition. 5 pp.

—Phenolic Resins—Comprehensive Testing
JIS K 6909-77. Testing Methods of Phenolic Resins for Grinding Wheels (R 1980). 15 pp.

—Plastics—Abrasion Resistance
BSI BS 2782:Pt3: METH 370-90. 1990 Methods of Testing Plastics Part 3: Mechanical Properties Method 370: Determination of Resistance to Wear by Abrasive Wheels. 11 pp.

ISO 9352-89. Plastics—Determination of Resistance to Wear by Abrasive Wheels First Edition. 8 pp.

JIS K 7204-77. Testing Method for Abrasion Resistance of Plastics by Abrasive Wheels (R 1980). 11 pp.

—Safety
CNS R3077-87. Method of Test for Grinding Wheels (Nov) (3786).

CNS Z1007-87. Safety Code for Grinding Wheels (Oct)(2223).

JIS R 6240-86. Testing Methods of Grinding Wheels. 18 pp.

SNZ NZS 5246: Part 2-90. Specification for Abrasive Wheels Part 2: Selection, Care and Use Amend: A, 1990. 24 pp.

—Silicon Carbide—Depressed Center—Fabric Reinforced
CNS R2084-87. Depressed Center Wheels with Fabric Reinforcement (Oct) (3789).

JIS R 6213-86. Depressed Center Wheels with Fabric Reinforcement. 8 pp.

—Silicon Carbide—Inserted Nut—Ring
CNS R2093-87. Inserted Nut Abrasive Discs and Inserted Nut Ring Wheels (Oct)(4208).

JIS R 6216-86. Inserted Nut Abrasive Discs and Inserted Nut Ring Wheels. 12 pp.

—Silicon Carbide—Magnesia
CNS R2175-87. Magnesia Grinding Wheels (Oct)(11142).

JIS R 6219-86. Magnesia Grinding Wheels. 13 pp.

—Silicon Carbide—Mounted
CNS R2089-87. Mounted Wheels (Oct) (4204).

JIS R 6217-86. Mounted Wheels. 13 pp.

—Silicon Carbide—Resinoid
CNS R2085-87. Resinoid Cutting-Off Wheels (Oct)(3790).

CNS R2088-87. Resinoid Grinding Wheels (Oct)(3967).

JIS R 6212-86. Resinoid Grinding Wheels. 12 pp.

JIS R 6214-86. Resinoid Cutting-off Wheels. 9 pp.

—Silicon Carbide—Rubber
CNS R2090-87. Rubber Cutting-off Wheels (Oct)(4205).

INTERNATIONAL AND NON-U.S. NATIONAL STANDARDS
SUBJECT INDEX

Grinding Wheels (Cont.)
—Silicon Carbide—Rubber (Cont.)
JIS R 6215-86. Rubber Cutting-off Wheels. 8 pp.
—Silicon Carbide—Vitrified
CNS R2034-87. Vitrified Grinding Wheels (Oct)(991). 7 pp.
JIS R 6210-86. Vitrified Grinding Wheels. 12 pp.
—Speed
CNS R2087-87. Maximum Operating Speed of Grinding Wheels (Nov)(3966).
JIS R 6241-86. Maximum Operating Speed of Grinding Wheels. 7 pp.
—Symbols
JIS R 6004-90. Glossary of Terms and Marks Used in Abrasives, Grinding Wheels and Coated Abrasives. 20 pp.

Gripping Devices
Use For: Grasping Devices *See Also:* Grips; Handles; Tongs
—Beams (Supports)—Ships
JIS F 7604-89. Beam Grabs for Machine Use.
—Conveyors—Railroad
JIS E 9302-76. Fixed Grips for Lifts.
—Wire Rope
BSI BS 462-83. (WITHDRAWN) 1983 Amd 1 Wire Rope Grips (AMD 4957) September 30, 1985. 11 pp.
SAA AS 2076-77. Wire Rope Grips Reconfirmed 1983. 12 pp.

Grips
See Also: Gripping Devices; Handles
CNS B3307-79. Star Grips (Nov)(5012).
CNS B3308-79. Palm Grips (Nov)(5013).
JIS B 2603-82. Grips. 11 pp.
JIS B 2807-78. Grip Rings. 7 pp.
—Bicycles
CNS B2041-75. Handle Grips for Bicycles (Dec)(351). 2 pp.
CNS B7057-90. Method of Test for the Handle Grips of Bicycles (Nov)(3870).
CNS B7057-75. Method of Test for the Handle Grips of Bicycles (Dec)(3870). 2 pp.
CNS K4055-81. Grip Rubber for Bicycles (Jan)(6921). 2 pp.
JIS D 9413-90. Handle Grips for Bicycles. 10 pp.
—Motorcycles
CNS D2139-88. Grip for Motorcycle (Jul)(8956).

Grit
See Also: Abrasive Grains; Abrasives; Sands
BSI BS 2451-63. (OBSOLESCENT) 1963 Chilled Iron Shot and Grit. 8 pp.
CNS G3191-82. Cast Shot and Grit (Aug) (9270).
JIS G 5903-75. Cast Shot and Grit. 8 pp.
—Grain Size Analysis
CNS G2130-82. Method of Test for Cast Shot and Grit Grain Size (Aug)(9271).
JIS G 5904-66. Testing Method of Cast Shot and Grit Grain Size (R 1969). 3 pp.
—Suction Boxes
CPPA A.3U-77. Grit in Suction Boxes. 1 p.

Grit Blasting
Use: Sandblasting

Grit Content Analysis
—Beeswax
MOD UK M 2012/71. Examination of Beeswax GS and Beeswax, Lead Free.
—Calcium Hydroxides
MOD UK M 9514/65. Examination of Calcium Hydroxide.
—Camphor
MOD UK M 9555/72. Examination of Camphor for Celluloid (Superseded by CS 2914).
—Dammar
MOD UK M 9535/68. Examination of Gum, Dammar, and Gum, Dammar, Lead Free.
—Dyes
MOD UK M 818/63. Examination of Dyestuffs for Use in Pyrotechnic Compositions and HE Substitutes.
—Lactose Monohydrate
MOD UK M 821/71. Examination of Lactose.

Grit Content Analysis (Cont.)
—Magnesium Carbonate
MOD UK M 870/67. Examination of Magnesium Carbonate, Heavy.
—Sodium Nitrate
MOD UK M 869/91. Examination of Sodium Nitrate, Grade 1.
—Soldering Fluxes
MOD UK M 9524/66. Examination of Flux Soldering Solution (Withdrawn).
—Strontium Peroxide
MOD UK M 884/70. Examination of Strontium Peroxide (Anhydrous).
—Tetranitro Oxanilide
MOD UK M 887/72. Examination of Tetranitro-Oxanilide.
—Zinc Oxides
MOD UK M 806/91. Examination of Zinc Oxide.

Grob G 109 Motor Gliders
See Also: Gliders
—Foreign Airworthiness Directives
CAA. Grob G109 Series Motor Gliders (Foreign Airworthiness Directives). 5 pp.

Grob G 109B Aircraft
See Also: Aircraft
—Antenna Positions
CAA. Grob G109B (Approved Aerial Positions). 1 p.

Grob G 115 Aircraft
See Also: Aircraft
—Antenna Positions
CAA. Grob G115. 1 p.
—Foreign Airworthiness Directives
CAA. Grob G115 Series. 1 p.

Grommet Nuts
See Also: Blind Nuts; Fasteners; Nuts (Fasteners)
—Aerospace
CEN PREN 3660-003-91. Part 003—Detail Specification Sheet Grommet Nut, Style A. 11 pp.
CEN PREN 3660-003-93. Aerospace Series Cable Outlet Accessories for Circular and Rectangular Electrical and Optical Connectors Part 003—Grommet Nut, Type A, 003 Product Standard. 7 pp.

Grommet Washers
Use: Grommets

Grommets
See Also: Bushings; Fasteners; Joints; Washers (Fasteners)
CNS Z7022-67. Grommet (with Washer) (Oct)(2827)(R 1972).
CNS Z7060-79. 10mm Phi Grommet Washer (Apr)(4791).
CNS Z7061-79. 14mm Phi Grommet Washer (Apr)(4792).
MOD UK DSTAN 53-13-78. Grommets, Rubber Issue 2. 56 pp.
MOD UK DSTAN 53-22-01. Eyelets, Metallic, and Grommets, Metallic Issue 2; Amendment 1. 11 pp.
—Aerospace
BSI 2SP 93-74. 1974 Amd 1 Elastomeric Grommets (Acrylonitrile-Butadiene) for Aerospace Use. 5 pp.
BSI 2SP 95-74. 1974 Amd 1 Elastomeric Grommets (Polychloroprene) for Aerospace Use (AMD 3515) Februray 27, 1981. 5 pp.
—Aircraft
SBAC AS 44404-406 (V). Elastomeric Grommets (Various) for Aeronautical Purposes.
SBAC AS 44413-417 (V). Elastomeric Blind Grommets (Various).
—Automobiles—Tire Valves
CNS D2112-84. Tire Valve Washers and Grommets for Automobiles (Feb)(7975).
—Ships—Cables (Electric)
MOD UK NES 512: Part 11-01. Guide to Cables, Electrical and Associated Items Part 11: Glands, Grommets and Deck Tubes Issue 2 (03.91); Amendment 1. 81 pp.
—Ships—Electric Wiring
MOD UK NES 512: Part 11-01. Guide to Cables, Electrical and Associated Items Part 11: Glands, Grommets and Deck Tubes Issue 2 (03.91); Amendment 1. 81 pp.

Grommets (Cont.)
—Submarines—Cables (Electric)
MOD UK NES 512: Part 11-01. Guide to Cables, Electrical and Associated Items Part 11: Glands, Grommets and Deck Tubes Issue 2 (03.91); Amendment 1. 81 pp.
—Submarines—Electric Wiring
MOD UK NES 512: Part 11-01. Guide to Cables, Electrical and Associated Items Part 11: Glands, Grommets and Deck Tubes Issue 2 (03.91); Amendment 1. 81 pp.

Grooming Appliances
Scope Note: Use a more specific term
See: Appliances; Hair Care Appliances; Hair Clippers; Shavers; Skin Care Appliances

Groove Cutters
See Also: Cutting Tools; Woodworking Equipment
—Woodworking Equipment
CNS B3295-85. Groove Cutters for Woodworking Machines (Oct)(4967).
JIS B 4710-72. Groove Cutters for Wood Working Machines.

Grooved Pins
See Also: Pins
CNS B2668-82. Grooved Pins, Half Length Grooved with Gorge (Jun)(8925).
DIN ENGL 1469-78. Grooved Pins, Half Length Grooved with Gorge (Nov). 4 pp.
—Aircraft—Headless
SBAC AS 20740-759 ISSUE 3. Pin, Grooved, Headless.
—Center
BSI BS 7048-89. (WITHDRAWN) 1989 Third-Length Centre Grooved Pins (Renumbered as BS EN 28742: 1992). 7 pp.
BSI BS 7049-89. (WITHDRAWN) 1989 Half-Length Centre Grooved Pins (Renumbered as BS EN 28743: 1992). 7 pp.
BSI BS EN 28742-92. 1992 Amd 1 Grooved Pins—Thrid-Length Centre Grooved (AMD 7489) November 15, 1992. 13 pp.
BSI BS EN 28743-92. 1992 Amd 1 Grooved Pins—Half-Length Centre Grooved (ISO 8743: 1986) (AMD 7490) November 15, 1992. 13 pp.
CEN EN 28742-92. Grooved Pins—Third-Length Centre Grooved. 5 pp.
CEN EN 28743-92. Grooved Pins—Half-Length Centre Grooved. 5 pp.
CNS B2690-82. Grooved Pins, Third Length Center Grooved (Aug)(9217).
DIN ENGL 1475 (S)-78. Grooved Pins, Third Length Centre-Grooved (Nov) (Superseded by DIN EN 28742). 4 pp.
DIN ENGL EN 28742-92. Third-Length Centre Grooved Pins; (ISO 8742: 1986) (Oct) Supersedes DIN 1475, November 1978 Edition). 7 pp.
DIN ENGL EN 28743-92. Half-Length Centre Grooved Pins; (ISO 8743: 1986) (Oct). 7 pp.
ISO 8742-86. Grooved Pins—Third-Length Centre Grooved First Edition. 5 pp.
ISO 8743-86. Grooved Pins—Half-Length Centre Grooved First Edition. 5 pp.
—Countersunk Head
BSI BS 7053-89. (WITHDRAWN) 1989 Countersunk Head Grooved Pins (Renumbered as BS EN 28747: 1992). 7 pp.
BSI BS EN 28747-92. 1992 Amd 1 Grooved Pins with Countersunk Head (ISO 8747: 1986) (AMD 7494) November 15, 1992. 13 pp.
CEN EN 28747-92. Grooved Pins with Countersunk Head. 5 pp.
CNS B2692-82. Countersunk Head Grooved Pins (Aug)(9219).
DIN ENGL 1477 (S)-78. Countersunk Head Grooved Pins (Nov) (Superseded by DIN EN 28747). 4 pp.
DIN ENGL EN 28747-92. Grooved Pins with Countersunk Head; (ISO 8747: 1986) (Oct) (Supersedes DIN 1477, November 1978 Edition). 7 pp.
ISO 8747-86. Grooved Pins with Countersunk Head First Edition. 5 pp.
—Headless
MOD UK DSTAN 53-83-01. Pins, Grooved, Headless, Steel Issue 1; Amendment 1. 50 pp.
—Parallel
BSI BS 7045-89. (WITHDRAWN) 1989 Full-Length Parallel Grooved Pins, with Pilot (Renumbered as BS EN 28739: 1992). 6 pp.
BSI BS 7046-89. (WITHDRAWN) 1989 Full-Length Parallel Grooved Pins, with Chamfer (Renumbered as BS EN 28740: 1992). 7 pp.
BSI BS EN 28739-92. 1992 Amd 1 Grooved Pins—Full-Length Parallel Grooved, with Pilot (ISO 8739: 1986) (AMD 7486) November 15, 1992. 11 pp.

INDUSTRY STANDARDS

Grooved Pins (Cont.)
—Parallel (Cont.)
BSI BS EN 28740-92. 1992 Amd 1 Grooved Pins—Full-Length Parallel Grooved, with Chamfer (ISO 8740: 1986) (AMD 7487) November 15, 1992. 13 pp.

CEN EN 28739-92. Grooved Pins—Full-Length Parallel Grooved, with Pilot. 5 pp.

CEN EN 28740-92. Grooved Pins—Full-Length Parallel Grooved, with Chamfer. 5 pp.

CNS B2669-82. Grooved Pins, Full Length Parallel Grooved with Pilot (Jun)(8926).

CNS B2688-82. Grooved Pins, Full Length Parallel Grooved with Chamfer (Aug)(9215).

CNS B2689-82. Grooved Pins, Half Length Reverse Grooved (Aug)(9216).

DIN ENGL 1470 (S)-78. Grooved Pins, Full Length Parallel-Grooved with Pilot (Nov) (Superseded by DIN EN 28739). 4 pp.

DIN ENGL 1473 (S)-78. Grooved Pins, Full Length Parallel-Grooved with Chamfer (Nov) (Superseded by DIN EN 28740). 4 pp.

DIN ENGL EN 28739-92. Full-Length Parallel Grooved Pins, with Pilot; (ISO 8739: 1986) (Oct) (Supersedes DIN 1470, November 1978 Edition). 7 pp.

DIN ENGL EN 28740-92. Full-Length Parallel Grooved Pins, with Chamfer (ISO 8740: 1986) (Oct) (Supersedes DIN 1473, November 1978 Edition). 7 pp.

ISO 8739-86. Grooved Pins—Full-Length Parallel Grooved, with Pilot First Edition. 5 pp.

ISO 8740-86. Grooved Pins—Full-Length Parallel Grooved, with Chamfer First Edition. 5 pp.

—Reverse Taper
BSI BS 7047-89. (WITHDRAWN) 1989 Half-Length Reverse Taper Grooved Pins (Renumbered as BS EN 28741: 1992). 7 pp.

BSI BS EN 28741-92. 1992 Grooved Pins—Half-Length Reverse Taper Grooved (ISO 8741: 1986) (AMD 7488) November 15, 1992. 13 pp.

CEN EN 28741-92. Grooved Pins—Half-Length Reverse Taper Grooved. 5 pp.

DIN ENGL 1474 (S)-78. Grooved Pins, Half Length Reverse-Grooved (Nov) (Superseded by DIN EN 28741). 4 pp.

DIN ENGL EN 28741-92. Half-Length Reverse Taper Grooved Pins; (ISO 8741: 1986) (Oct) (Supersedes DIN 1474, November 1978 Edition). 7 pp.

ISO 8741-86. Grooved Pins—Half-Length Reverse Taper Grooved First Edition. 5 pp.

—Round Head
BSI BS 7052-89. (WITHDRAWN) 1989 Round Head Grooved Pins (Renumbered as BS EN 28746: 1992). 7 pp.

BSI BS EN 28746-92. 1992 Amd 1 Grooved Pins with Round Head (ISO 8746: 1986) (AMD 7493) November 15, 1992. 13 pp.

CEN EN 28746-92. Grooved Pins with Round Head. 5 pp.

CNS B2691-82. Round Head Grooved Pins (Aug)(9218).

DIN ENGL 1476 (S)-78. Round Head Grooved Pins (Nov) (Superseded by DIN EN 28746). 4 pp.

DIN ENGL EN 28746-92. Grooved Pins with Round Head; (ISO 8746: 1986) (Oct) (Supersedes DIN 1476, November 1978 Edition). 7 pp.

ISO 8746-86. Grooved Pins with Round Head First Edition. 5 pp.

—Shear Testing
BSI BS 7054-89. (WITHDRAWN) 1989 Method for Shear Testing of Pins and Grooved Pins (Superseded by BS EN 28749: 1992). 3 pp.

BSI BS EN 28749-92. 1992 Pins and Grooved Pins—Shear Test (ISO 8749: 1986) (E). 8 pp.

CEN EN 28749-92. Pins and Grooved Pins—Shear Test. 3 pp.

—Taper
BSI BS 7050-89. (WITHDRAWN) 1989 Full-Length Taper Grooved Pins (Renumbered as BS EN 28744: 1992). 7 pp.

BSI BS 7051-89. (WITHDRAWN) 1989 Half-Length Taper Grooved Pins (Renumbered as BS EN 28745: 1992). 7 pp.

BSI BS EN 28744-92. 1992 Amd 1 Grooved Pins—Full-Length Taper Grooved (ISO 8744: 1986) (AMD 7491) November 15, 1992. 13 pp.

BSI BS EN 28745-92. 1992 Amd 1 Grooved Pins—Half-Length Taper Grooved (ISO 8745: 1986) (AMD 7492) November 15, 1992. 13 pp.

CEN EN 28744-92. Grooved Pins—Full-Length Taper Grooved. 5 pp.

CEN EN 28745-92. Grooved Pins—Half-Length Taper Grooved. 5 pp.

CNS B2670-82. Grooved Pins, Full Length Taper Grooved (Jun)(8927).

CNS B2671-82. Grooved Pins, Half Length Taper Grooved (Jun)(8928).

Grooved Pins (Cont.)
—Taper (Cont.)
DIN ENGL 1471 (S)-78. Grooved Pins, Full Length Taper-Grooved (Nov) (Superseded by DIN EN 28744). 4 pp.

DIN ENGL 1472 (S)-78. Grooved Pins, Half Length Taper-Grooved (Nov) (Superseded by DIN EN 28745). 4 pp.

DIN ENGL EN 28744-92. Full-Length Centre Grooved Pins; (ISO 8744: 1986) (Oct) (Supersedes DIN 1471, November 1978 Edition). 7 pp.

DIN ENGL EN 28745-92. Half-Length Taper Grooved Pins; (ISO 8745: 1986) (Oct) (Supersedes DIN 1472, November 1978 Edition). 7 pp.

ISO 8744-86. Grooved Pins—Full-Length Taper Grooved First Edition. 5 pp.

ISO 8745-86. Grooved Pins—Half-Length Taper Grooved First Edition. 5 pp.

Grooved Pulleys
See Also: Pulleys

ISO 254-90. Belt Drives—Pulleys—Quality, Finish and Balance Second Edition. 6 pp.

ISO 255-90. Belt Drives—Pulleys for V-Belts (System Based on Datum Width)—Geometrical Inspection of Grooves Second Edition. 11 pp.

ISO 1604-89. Belt Drives—Endless Wide V-Belts for Industrial Speed-Changers and Groove Profiles for Corresponding Pulleys Second Edition. 12 pp.

—Glossaries
ISO 1081-80. Drives Using V-Belts and Grooved Pulleys—Terminology First Edition. 16 pp.

—Groove Sections
ISO 5290-93. Belt Drives—Grooved Pulleys for Joined Narrow V-Belts—Groove Sections 9J, 15J, 20J and 25J (Effective System) Third Edition. 11 pp.

—Groove Sections—Agricultural Equipment
ISO 3410-89. Agricultural Machinery—Endless Variable-Speed V-Belts and Groove Sections of Corresponding Pulleys Second Edition. 8 pp.

ISO 5289-92. Agricultural Machinery—Endless Hexagonal Belts and Groove Sections of Corresponding Pulleys Second Edition. 8 pp.

—Hexagonal Belts—Agricultural Equipment
ISO 5289-92. Agricultural Machinery—Endless Hexagonal Belts and Groove Sections of Corresponding Pulleys Second Edition. 8 pp.

—Motors
DIN ENGL 2211 Pt 3-86. Power Transmission Elements; Grooved Pulleys for Narrow V-Belts; Assignment of Pulleys to Electric Motors (Jan). 4 pp.

—V Belts
DIN ENGL 2211 Pt 1-84. Power Transmission Elements; Grooved Pulleys for Narrow V-Belts; Dimensions, Materials (Mar). 10 pp.

DIN ENGL 2217 Pt 1-73. Driving Elements; Grooved Pulleys for V-Belts; Dimensions; Material (Feb). 10 pp.

DIN ENGL 2217 Pt 2-73. Driving Elements; Grooved Pulleys for V-Belts; Testing of Grooves (Feb). 1 p.

DIN ENGL 7753 Pt 3-86. Endless Narrow V-Belts for the Automotive Industry; Dimensions of Belts and Pulley Groove Profiles (Feb). 6 pp.

ISO 4183-89. Belt Drives—Classical and Narrow V-Belts—Grooved Pulleys (System Based on Datum Width) Second Edition. 7 pp.

ISO 5290-93. Belt Drives—Grooved Pulleys for Joined Narrow V-Belts—Groove Sections 9J, 15J, 20J and 25J (Effective System) Third Edition. 11 pp.

ISO 5291-93. Belt Drives—Grooved Pulleys for Joined Classical V-Belts—Groove Sections AJ, BJ, CJ and DJ (Effective System) Second Edition. 8 pp.

ISO 9980-90. Belt Drives—Grooved Pulleys for V-Belts (System Based on Effective Width)—Geometrical Inspection of Grooves First Edition. 11 pp.

JIS B 1854-87. Grooved Pulleys for Classical V-Belts. 23 pp.

JIS B 1855-91. Grooved Pulleys for Narrow V-Belts. 20 pp.

—V Belts—Agricultural Equipment
BSI BS 3733-74. 1974 Endless V-Belt Drivers for Agricultural Purposes. 37 pp.

ISO 3410-89. Agricultural Machinery—Endless Variable-Speed V-Belts and Groove Sections of Corresponding Pulleys Second Edition. 8 pp.

—V Belts—Automotive
BSI BS AU 150B-84. 1984 Automotive V-Belts and Pulleys. 18 pp.

BSI BS AU 248-93. 1993 Automotive V-Ribbed Belt Drives—Dimensions of Pulleys and Belts of PK Profile (ISO 9981: 1990) (E). 12 pp.

ISO 2790-89. Belt Drives—Narrow V-Belts for the Automotive Industry and Corresponding Pulleys—Dimensions Third Edition. 6 pp.

Grooved Pulleys (Cont.)
—V Belts—Automotive (Cont.)
ISO 9981-90. Belt Drives—Pulleys and V-Ribbed Belts for the Automotive Industry—Dimensions—PK Profile First Edition. 10 pp.

—V Belts—Industrial
BSI BS 7620-93. 1993 Industrial Belt Drives—Dimensions of Pulleys and V-Ribbed Belts of PH, PJ, PK, PL, and PM Profiles (ISO 9982: 1991) (F). 12 pp.

ISO 2790-89. Belt Drives—Narrow V-Belts for the Automotive Industry and Corresponding Pulleys—Dimensions Third Edition. 6 pp.

ISO 9982-91. Belt Drives—Pulleys and V-Ribbed Belts for Industrial Applications—Dimensions—PH, PJ, PK, PL and PM Profiles First Edition. 11 pp.

—V Belts—Inspection
DIN ENGL 2211 Pt 2-84. Power Transmission Elements; Grooved Pulleys for Narrow V-Belts; Inspection of Grooves (Mar). 2 pp.

Grooves
—Aircraft—Snap Rings
SBAC AGS 2030 (V)(I). Internal Circlips and Grooves.

SBAC AGS 2031 (V)(I). External Circlips and Grooves.

Ground Anchors
See Also: Anchors (Fasteners); Structural Members

BSI BS 8081-89. 1989 Code of Practice for Ground Anchorages. 180 pp.

BSI BS 8081-01. 1989 Amd 1 Ground Anchorages. 182 pp.

—Prestressing—Construction
DIN ENGL 4125-90. Ground Anchorages; Design, Construction and Testing (Nov). 25 pp.

—Prestressing—Design
DIN ENGL 4125-90. Ground Anchorages; Design, Construction and Testing (Nov). 25 pp.

Ground Clamps
Use: Grounding Clamps

Ground Conductivity
CCIR RESOLUTION 73-1-90. World Atlas of Ground Conductivities—Volume V—Propagation in Non-Ionized Media. 1 p.

—Atlases
CCIR Report 717-3-90. World Atlas of Ground Conductivities (Published Separately)—Section 5B—Effects of the Ground (Including Ground-Wave Propagation). 9 pp.

CCIR RECMN 832-92. World Atlas of Ground Conductivities. 48 pp.

Ground Crews
See Also: Ground Handling Equipment

—Aircraft Cross Servicing—Training
NATO STANAG 3812 ED 5 AMD 1-90. Responsibilities for Aircraft Cross-Servicing Ground Crew Training. 14 pp.

NATO STANAG 3812 ED 5 AMD 2-90. Responsibilities for Aircraft Cross-Servicing Ground Crew Training. 14 pp.

NATO STANAG 3812 ED 5 AMD 3-90. Responsibilities for Aircraft Cross-Servicing Ground Crew Training. 16 pp.

NATO STANAG 3812 ED 6 AMD 0-93. Responsibilities for Aircraft Cross-Servicing Ground Crew Training. 18 pp.

Ground Effect Machines
Use: Air Cushion Vehicles

Ground Electrodes
Use For: Earth Electrodes; Grounding Electrodes
See Also: Electrical Grounding; Electrodes

CSA C22.2 NO 41-M1987. Grounding and Bonding Equipment (R 1993); (Gen Instr 1). 30 pp.

—Corrosion
DIN VDE 0151-86. Material & Minimum Dimensions of Earth Electrodes with Respect to Corrosion (June). 26 pp.

—Corrosion Prevention
CENELEC PREN 50114-93. Materials and Size Requirements for Earth Electrodes from the Corrosion Point of View. 23 pp.

—Plates
CSA C22.2 NO 41-M1987. Grounding and Bonding Equipment (R 1993); (Gen Instr 1). 30 pp.

INTERNATIONAL AND NON-U.S. NATIONAL STANDARDS
SUBJECT INDEX
Ground

Ground Fault Circuit Interrupters
See Also: Circuit Breakers; Electrical Grounding; Ground Fault Protection; Ground Fault Protection Equipment

CENELEC HD 408-79. Alternating Current Disconnectors (Isolators) and Earthing Switches. 3 pp.
CENELEC HD 408 S2-90. Alternating Current Disconnectors and Earthing Switches. 4 pp.
CSA 1164 Bull. Electrical Bulletin 1164 April 13, 1978 to C22.2 NO 68. 2 pp.
CSA CAN/CSA-C22. 2 NO 144-M91. Ground Fault Circuit Interrupters; (Gen Instr 1 Thru 2). 69 pp.
CSA 1164 Bull. Electrical Bulletin 1164 April 13, 1978 to C22.2 NO 144. 2 pp.

—**Discharge Lamps**
DIN VDE 0713 Pt 2-85. Equipment for Tubular Discharge Installations over 1000 V; Earth Leakage Protection (Sept). 8 pp.

Ground Fault Protection
Use For: Earth Fault Protection *See Also:* Electrical Grounding; Electrical Protection Equipment; Ground Fault Circuit Interrupters

—**Electrical Installations—Telephone Cables**
CCITT RECMN K.8-89. Separation in the Soil Between Telecommunication Cables and Earthing System of Power Facilities—Protection Against Interference (Study Group V) 5 pp. 5 pp.

—**Electrical Installations—Telephone Cables—Railroad Tracks**
CCITT RECMN K.9-89. Protection of Telecommunication Staff and Plant Against a Large Earth Potential Due to a Neighbouring Electric Traction Line (Study Group V) 2 pp. 2 pp.

Ground Fault Protection Equipment
Scope Note: For additional listings, use a more specific term *See Also:* Electrical Protection Equipment; Ground Fault Circuit Interrupters; Ground Resistance Meters; Ground Rods; Grounding Devices

—**Current Limiters—Mining**
SAA AS 2081.3-88. Electrical Equipment for Coal and Shale Mines—Electrical Protection Devices—Part 3: Earth-Leakage Protection Devices for Use on Systems Incorporating Earth-Fault Current Limiters. 5 pp.
SAA AS 2081.5-88. Electrical Equipment for Coal and Shale Mines—Electrical Protection Devices—Part 5: Earth-Fault Current Limiters. 2 pp.

—**Mining**
SAA AS 2081.4-88. Electrical Equipment for Coal and Shale Mines—Electrical Protection Devices—Part 4: Lockout Earth-Fault Protection Devices. 3 pp.

Ground Handling Equipment
See Also: Ground Crews; Materials Handling Equipment; Shipboard Handling Equipment

—**Air Cargo—Unit Loads**
ISO 4116-86. Air Cargo Equipment—Ground Equipment Requirements for Compatibility with Aircraft Unit Load Devices Second Edition. 6 pp.
ISO 7715-85. Air Cargo Equipment—Ground Handling and Transport Systems for Unit Load Devices—Minimum Requirements First Edition. 10 pp.

—**Aircraft**
CAA LEAFLET 10-1 07.90. Aircraft Handling. 15 pp.

—**Packages**
BSI BS 1133: Sec 4-91. 1991 Packaging Code Section 4: Mechanical Aids in Package Handling. 84 pp.
BSI BS 1133: Sec 4-65. 1965 Packaging Code Section 4: Mechanical Aids in Package Handling. 53 pp.

Ground Loads
See Also: Aerodynamic Loads; Loads (Forces)

—**Aerostats**
CAA Chapter Q3-5 12.79. Ground Loads (Non-Rigid Airships). 3 pp.

—**Aerostats—Moorings**
CAA Chapter Q3-13 12.79. Mooring and Groundhandling. 2 pp.

—**Aircraft**
CAA Chapter K3-13 10.69. Ground Handling. 2 pp.
CAA STANDARD NO. 4-4 08.60. Ground Loads. 10 pp.
CAA CAP 482 SUB-Part C 03.83. Structure (Small Light Aeroplanes). 12 pp.

Ground Loads *(Cont.)*
—**Aircraft—Landing Gear**
CAA Chapter P3-5. Ground Loads (Provisional Airworthiness Requirements for Civil Powered Lift Aircraft). 18 pp.

—**Gliders**
CAA JAR-22 SUBPART C. Structure (Joint Airworthiness Requirements). 17 pp.

—**Rotary Wing Aircraft**
CAA Chapter G3-5 12.80. Ground and Water Loads (Rotocraft). 4 pp.
CAA CAP 524 SUB-Part C 12.86. Structure (Rotorcraft). 11 pp.

Ground Proximity Warning Systems
See Also: Aircraft Instruments; Warning Systems
CAA Spec. NO. 14 ISSUE 1.

Ground Relays
See Also: Electrical Grounding; Protective Relays; Relays

—**50-599 V—Sets**
JIS C 4601-91. Ground Relay Set for 6 600 V or 3 300 V Consumer. 18 pp.

—**Directional—Sets**
JIS C 4609-90. Directional Ground Relay Set for 6.6 kV Consumer. 24 pp.

Ground Resistance Meters
Use For: Earth Resistance Meters *See Also:* Ground Fault Protection Equipment; Insulation Resistance Meters

DIN VDE 0413 Pt 5-77. VDE Specification for Equipment for Testing the Protective Measures in Electrical Installations; Earth Resistance Meters Based on the Compensation Measuring Method (July). 22 pp.
DIN VDE 0413 Pt 7-82. Measurement and Control; Equipment for Testing the Protective Measures in Electrical Installations; Earth Resistance Meters Using the Ammeter/Voltmeter Method (July). 24 pp.

Ground Rods
Use For: Grounding Rods *See Also:* Electrical Grounding; Electrical Protection Equipment; Ground Fault Protection Equipment

MOD UK DSTAN 59-28-91. Rods, Earthing Issue 3. 6 pp.

Ground Stations
See Also: Satellite Communication Equipment

—**Aircraft—Frequency Standards**
EUROCAE ED-52 08.84. MPS for Ground Conventional and Doppler Very High Frequency Omni Range (CVOR and DVOR) Equipment. 61 pp.

—**Aircraft—Maintainability**
EUROCAE ED-52 08.84. MPS for Ground Conventional and Doppler Very High Frequency Omni Range (CVOR and DVOR) Equipment. 61 pp.

—**Aircraft—Reliability**
EUROCAE ED-52 08.84. MPS for Ground Conventional and Doppler Very High Frequency Omni Range (CVOR and DVOR) Equipment. 61 pp.

—**Aircraft—Siting Criteria**
EUROCAE ED-52 08.84. MPS for Ground Conventional and Doppler Very High Frequency Omni Range (CVOR and DVOR) Equipment. 61 pp.

—**Aircraft—VOR**
EUROCAE ED-52 08.84. MPS for Ground Conventional and Doppler Very High Frequency Omni Range (CVOR and DVOR) Equipment. 61 pp.

Ground Support Equipment
Use For: Aerospace Ground Equipment *See Also:* Aircraft Refueling Equipment; Command Guidance; Equipment; Gantry Cranes; Materials Handling Equipment; Military Facilities; Oxygen Supply Equipment; Rocket Launchers

MOD UK DSTAN 16-17-83. Air Supply and Test Connections for Aircraft Issue 2. 8 pp.
MOD UK DSTAN 17-5-86. Power Driven Hydraulic Ground Servicing Equipment for Aircraft Issue 2. 7 pp.

—**Air Conditioners—Connectors**
NATO STANAG 3917 ED 2 AMD 2-84. Air Conditioning Connections, Ground Half Connectors. 9 pp.

Ground Support Equipment *(Cont.)*
—**Air Conditioners—Connectors** *(Cont.)*
NATO STANAG 3917 ED 2 AMD 3-84. Air Conditioning Connections, Ground Half Connectors. 10 pp.

—**Air Conditioners—Fittings**
BSI C 10-58. 1958 Amd 1 Coupling Dimensions for Aircraft Ground Air-Conditioning Connections. 4 pp.
ISO 1034-73. Aircraft—Ground Air-Conditioning Connections First Edition; (Erratum—March 1974). 5 pp.

—**Auxiliary Power Units**
MOD UK DERD 2080-90. General Procedures for Design and Development of Ground Support Equipment for Aircraft Propulsion and Auxiliary Power Unit Systems Issue 6. 14 pp.
MOD UK DERD 2081-90. Design, Development and Production of Tools, Ground Support Equipment (GSE) for Aircraft Propulsion and Auxiliary Power Unit Systems Issue 5. 8 pp.

—**Connectors**
BSI 4G 173: Part 1-85. 1985 Connectors for Ground Electrical Supplies for Aircraft Part 1: Design, Performance and Test Requirements. 6 pp.
BSI 4G 173: Part 2-85. 1985 Connectors for Ground Electrical Supplies for Aircraft Part 2: Dimensions. 9 pp.
ISO 461 Pt 1-85. Aircraft—Connectors for Ground Electrical Supplies—Part 1: Design, Performance and Test Requirements First Edition. 6 pp.
ISO 461 Pt 2-85. Aircraft—Connectors for Ground Electrical Supplies—Part 2: Dimensions First Edition. 9 pp.

—**Connectors—Electrical Grounding**
NATO STANAG 3632 ED 3 AMD 2-86. Electrical Safety Connections for Aircraft and Ground Support Equipment. 12 pp.
NATO STANAG 3632 ED 4 AMD 0-91. Aircraft and Ground Support Equipment Electrical Connections for Static Grounding. 9 pp.
NATO STANAG 3632 ED 4 AMD 1-91. Aircraft and Ground Support Equipment Electrical Connections for Static Grounding. 10 pp.

—**Environmental Testing**
MOD UK DTD-1086C-60. Climatic Testing of Ground Instruments and Equipment (Reprinted April 1962). 3 pp.

—**Filling Connectors**
ISO 45-90. Aircraft—Pressure Refuelling Connections Second Edition. 7 pp.

—**Glossaries**
BSI BS 185: Sec 13-72. 1972 Glossary of Aeronautical and Astronautical Terms Section 13: Air-Traffic and Ground Services (General, Air-Traffic Control, Ground Services, Naval Terms). 12 pp.

—**Ground Vehicles—Drinking Water**
ISO 9678-91. Aircraft—Self-Propelled Potable-Water Vehicle First Edition. 7 pp.

—**Hydraulic Couplings**
ISO 3174-81. Aircraft—Connections for Checking Hydraulic Systems by Ground Appliances First Edition. 6 pp.

—**Immersion Testing**
MOD UK DSTAN 00-7-01. Immersion Requirements for Ground Role Service Equipment Issue 2; Amendment 1. 12 pp.

—**Mobile—Packaging**
MOD UK DSTAN 81-67-01. Packaging of RAF Mobile Ground Support Equipment Issue 1; Amendment 1. 16 pp.

—**Power Supplies**
BSI 2G 219-83. 1983 General Requirements for Ground Support Electrical Supplies for Aircraft. 12 pp.
ISO 6858-82. Aircraft—Ground Support Electrical Supplies—General Requirements First Edition. 9 pp.
JIS W 7007-91. Aircraft—Ground Support Electrical Supplies—General Requirements (ISO 6858:1982).
NATO STANAG 3457 ED 5 AMD 3-86. Ground Electrical Power Supplies for Aircraft. 11 pp.

—**Power Supplies—Safety**
ISO 6858-82. Aircraft—Ground Support Electrical Supplies—General Requirements First Edition. 9 pp.
JIS W 7007-91. Aircraft—Ground Support Electrical Supplies—General Requirements (ISO 6858:1982).

—**Propulsion Systems**
MOD UK DERD 2080-90. General Procedures for Design and Development of Ground Support Equipment for Aircraft Propulsion and Auxiliary Power Unit Systems Issue 6. 14 pp.

INDUSTRY STANDARDS

Ground Support Equipment *(Cont.)*
—Propulsion Systems *(Cont.)*
MOD UK DERD 2081-90. Design, Development and Production of Tools, Ground Support Equipment (GSE) for Aircraft Propulsion and Auxiliary Power Unit Systems Issue 5. 8 pp.

—Reference Books
EURO EC/EEPSG/73/2248-73. Product Support Reference Book. 104 pp.

Ground Terminals
Use For: Earth Terminals *See Also:* Electric Terminals; Electrical Grounding

—Hospital Grade
CNS C4268-80. Hospital Grade Earth Centers and Terminals (Dec)(6767).

Ground Vehicles
Scope Note: For additional listings, use a more specific term *See Also:* Ambulances; Armored Vehicles; Automobiles; Bicycles; Bulk Transporters; Buses (Vehicles); Caterpillars (Vehicles); Combat Vehicles; Dollies; Dump Trucks; Electric Trucks; Electric Vehicles; Forklifts; Freight Cars; Hopper Cars; Industrial Trucks; Military Vehicles; Mine Cars; Motor Vehicles; Motorcycles; Prime Movers (Vehicles); Rolling Stock; Tank Trucks; Towed Vehicles; Tracked Vehicles; Trailers; Trucks; Vehicles (Transportation)

—Air Cargo—Unit Loads
ISO 7715-85. Air Cargo Equipment—Ground Handling and Transport Systems for Unit Load Devices—Minimum Requirements First Edition. 10 pp.

ISO 7716-85. Air Cargo Equipment—Unit Load Devices Transport Vehicle (UTV)—Functional Requirements First Edition. 9 pp.

—Decals
CGSB 62-GP-6M-82. Prefabricated Transparent-Base Markings, General-Purpose, for Use on Exterior and Interior Surfaces, Standard for. 12 pp.

CGSB 62-GP-7M-82. Prefabricated "Mirrorized" Markings, General Purpose, for Use on Exterior and Interior Surfaces, Standard for. 12 pp.

CGSB 62-GP-8M-80. Prefabricated Opaque Markings, General Purpose, for Use on Exterior and Interior Surfaces, Standard for. 12 pp.

CGSB 62-GP-9M-80. Prefabricated Markings, Positionable, Exterior, for Aircraft, Ground Equipment and Facilities, Standard for. 17 pp.

—Inner Tubes
CGSB 20-GP-6E-81. Tubes, Inner, Vehicle and Mobile Ground Equipment, Standard for. 26 pp.

—Installed Equipment—Environmental Conditions
BSI BS 7527: Sec 3.5-91. 1991 Classification of Environmental Conditions Part 3: Classification of Groups of Environmental Parameters and Their Severities Section 3.5: Ground Vehicle Installations (IEC 721-3-5: 1985) (Renumbered as BS EN 60721-3-5: 1993) (Q). 28 pp.

BSI BS EN 60721-3-5-93. 1993 Amd 1 Classification of Environmental Conditions Part 3: Classification of Groups of Environmental Parameters and Their Severities Section 3.5: Ground Vehicle Installations (AMD September 15, 1993 (G). 43 pp.

CENELEC EN 60721-3-5-93. Classification of Environmental Conditions Part 3: Classification of Groups of Environmental Parameters and Their Severities Ground Vehicle Installations (IEC 721-3-5: 1985 + A1: 1991). 25 pp.

CENELEC PRETS 300 019-1-5-91. Equipment Engineering (EE); Environmental Conditions and Environmental Tests for Telecommunications Equipment Part 1-5: Classification of Environmental Conditions Ground Vehicle Installations. 16 pp.

CENELEC ETS 300 019-1-5-92. Equipment Engineering (EE); Environmental Conditions and Environmental Tests for Telecommunications Equipment Part 1-5: Classification of Environmental Conditions Ground Vehicle Installations. 16 pp.

CENELEC PRETS 300 019-2-5-92. Equipment Engineering (EE); Environmental Conditions and Environmental Tests for Telecommunications Equipment Part 2-5: Specification of Environmental Tests T 5.1 and T 5.2 Ground Vehicle Installations. 15 pp.

ETSI PRETS 300 019-1-5-91. Equipment Engineering (EE); Environmental Conditions and Environmental Tests for Telecommunications Equipment Part 1-5: Classification of Environmental Conditions Ground Vehicle Installations. 15 pp.

ETSI ETS 300 019-1-5-92. Equipment Engineering (EE); Environmental Conditions and Environmental Tests for Telecommunications Equipment Part 1-5: Classification of Environmental Conditions Ground Vehicle Installations. 16 pp.

Ground Vehicles *(Cont.)*
—Installed Equipment—Environmental Conditions *(Cont.)*
ETSI PRETS 300 019-1-5-91. Equipment Engineering (EE); Environmental Conditions and Environmental Tests for Telecommunications Equipment Part 1-5: Classification of Environmental Conditions Ground Vehicle Installations. 16 pp.

ETSI PRETS 300 019-2-5-92. Equipment Engineering (EE); Environmental Conditions and Environmental Tests for Telecommunications Equipment Part 2-5: Specification of Environmental Tests T 5.1 and T 5.2 Ground Vehicle Installations. 15 pp.

IEC 721 Pt 3-5-85. Classification of Environmental Conditions Part 3: Classification of Groups of Environmental Parameters and Their Severities Ground Vehicle Installations First Edition; (Amendment 1-1991) (CENELEC EN 60721-3-5: 1993). 57 pp.

—Unit Loads—Environmental Testing
ISO 10531-92. Packaging—Complete, Filled Transport Packages—Stability Testing of Unit Loads First Edition. 10 pp.

—Unit Loads—Stability Testing
ISO 10531-92. Packaging—Complete, Filled Transport Packages—Stability Testing of Unit Loads First Edition. 10 pp.

—Vibration
BSI BS 6794-86. 1986 Reporting Measured Vibration Data for Land Vehicles. 14 pp.

ISO 8002-86. Mechanical Vibrations—Land Vehicles—Method for Reporting Measured Data First Edition. 12 pp.

Ground Water
See Also: Aquifers; Hydrology; Precipitation (Meteorology); Spring Water; Water

—Arsenic Content—Atomic Absorption Spectrometry
DIN ENGL 38405 Pt 18-85. German Standard Methods for the Examination of Water, Waste Water and Sludge; Anions (Group D); Determination of Arsenic by Atomic Absorption Spectrometry (D 18) (Sept). 5 pp.

—Arsenic Content—Photometry
DIN ENGL 38405 Pt 12 (S)-81. German Standard Methods for the Analysis of Water, Waste Water and Sludge; Anions (Group D); Determination of Arsenic (D 12) (June) (Superseded by DIN EN 26595). 5 pp.

—Borehole Logs
DIN ENGL 4022 Pt 1-87. Subsoil and Groundwater; Classification and Description of Soil and Rock; Borehole Logging of Soil and Rock Not Involving Continuous Core Sample Recovery (Sept). 21 pp.

DIN ENGL 4022 Pt 3-82. Subsoil and Groundwater; Designation and Description of Soil Types and Rock; Borehole Log for Boring in Soil (Loose Rock) by Continuous Extraction of Cores (May). 16 pp.

DIN ENGL 4023-84. Borehole Logging; Graphical Representation of the Results (Mar). 11 pp.

—Buildings—Protection
BSI BS 8102-90. 1990 Code of Practice for Protection of Structures Against Water from the Ground. 40 pp.

BSI CP 102-73. 1973 Amd 3 Protection of Buildings Against Water from the Ground. 30 pp.

—Cadmium Content—Voltametry
DIN ENGL 38406 Pt 16-90. German Standard Methods for the Examination of Water, Waste Water and Sludge; Cations (Group E); Determination of Zinc, Cadmium, Lead, Copper, Thallium, Nickel, Cobalt by Voltammetry (E16) (Mar). 8 pp.

—Cobalt Content—Voltametry
DIN ENGL 38406 Pt 16-90. German Standard Methods for the Examination of Water, Waste Water and Sludge; Cations (Group E); Determination of Zinc, Cadmium, Lead, Copper, Thallium, Nickel, Cobalt by Voltammetry (E16) (Mar). 8 pp.

—Copper Content—Voltametry
DIN ENGL 38406 Pt 16-90. German Standard Methods for the Examination of Water, Waste Water and Sludge; Cations (Group E); Determination of Zinc, Cadmium, Lead, Copper, Thallium, Nickel, Cobalt by Voltammetry (E16) (Mar). 8 pp.

Ground Water *(Cont.)*
—Cyanide Content—Photometry
DIN ENGL 38405 Pt 14-88. German Standard Methods for the Examination of Water, Waste Water and Sludge; Anions (Group D); Determination of Cyanides in Drinking Water, and in Groundwater and Surface Water with Low Pollution Levels (D 14) (Dec). 7 pp.

—Halogenated Hydrocarbon Content—Gas Chromatography
DIN ENGL 38407 Pt 4-88. German Standard Methods for the Examination of Water, Waste Water and Sludge; Substance Group Analysis (Group F); Determination of Readily Volatile Halogenated Hydrocarbons (F4) (May). 16 pp.

—Lead Content—Voltametry
DIN ENGL 38406 Pt 16-90. German Standard Methods for the Examination of Water, Waste Water and Sludge; Cations (Group E); Determination of Zinc, Cadmium, Lead, Copper, Thallium, Nickel, Cobalt by Voltammetry (E16) (Mar). 8 pp.

—Lowering—Construction Contracts
DIN ENGL 18305-88. Tendering and Performance Stipulations in Contracts for Construction Works (VOB) Part C: General Technical Specifications in Contracts for Construction Works (ATV); Groundwater Lowering (Sept) (This Standard, Together with DIN 18299, September. 4 pp.

—Microbiological Analysis
DIN ENGL 38411 Pt 1-83. German Standard Methods for the Examination of Water, Waste Water and Sludge; Microbiological Methods (Group K); Preparation for the Microbiological Examination of Water Samples (K 1) (Feb). 4 pp.

—Nickel Content—Voltametry
DIN ENGL 38406 Pt 16-90. German Standard Methods for the Examination of Water, Waste Water and Sludge; Cations (Group E); Determination of Zinc, Cadmium, Lead, Copper, Thallium, Nickel, Cobalt by Voltammetry (E16) (Mar). 8 pp.

—Radioisotopes
DIN ENGL 38404 Pt 15-87. German Standard Methods for the Examination of Water, Waste Water and Sludge; Physical and Physico-Chemical Para-meters (Group C); Determination of Beta Activity per Unit Volume in Drinking Water, Ground Water, Surface Water and Waste Water (C 15) (Sept). 4 pp.

—Radioisotopes—Gamma Ray Spectrometry
DIN ENGL 38404 Pt 16-89. German Standard Methods for the Examination of Water, Waste Water and Sludge; Physical and Physicochemical Para-meters (Group C); Deter-mination of Radio-nuclides in Drinking Water, Ground Water, Surface Water and Waste Water by γ-Ray Spectro-metry (C 16) (Apr). 7 pp.

—Sampling
BSI BS 6068: Sec 6.11-93. 1993 Water Quality Part 6: Sampling Section 6.11: Guidance on Sampling of Groundwaters (ISO 5667-11: 1993). 15 pp.

ISO 5667 Pt 11-93. Water Quality—Sampling —Part 11: Guidance on Sampling of Groundwaters First Edition. 15 pp.

—Selenium Content—Atomic Absorption Spectrometry
ISO 9965-93. Water Quality—Determination of Selenium—Atomic Absorption Spectrometric Method (Hydride Technique) First Edition. 8 pp.

—Silver Content—Atomic Absorption Spectrometry
DIN ENGL 38406 Pt 18-90. German Standard Methods for the Examination of Water, Waste Water and Sludge; Cations (Group E); Determination of Dissolved Silver by Atomic Absorption Spectrometry Using Electrothermal Atomization (E 18) (May). 5 pp.

—Stratigraphic Columns
DIN ENGL 4022 Pt 2-81. Subsoil and Groundwater; Designation and Description of Soil Types and Rock; Stratigraphic Representation for Borings in Rock (Mar). 15 pp.

—Sulfate Content
BSI BS 1377: Part 3-90. 1990 Methods of Test for Soils for Civil Engineering Purposes Part 3: Chemical and Electro-Chemical Tests. 45 pp.

SAA AS 1289.D2.1-77. Methods of Testing Soil for Engineering Purposes—Part D2.1: Soil Chemical Tests—Determination of the Sulphate Content of an Undisturbed Soil and the Sulphate Content of the Ground Water Corrig.. 3 pp.

INTERNATIONAL AND NON-U.S. NATIONAL STANDARDS
SUBJECT INDEX

Ground Water *(Cont.)*
—Suspended Solids—Filtration
CEN PREN 870-92. Water Quality—Determination of Suspended Solids—Determination by 0.4—0.45 um Pore Width Filters. 7 pp.

—Thallium Content—Voltametry
DIN ENGL 38406 Pt 16-90. German Standard Methods for the Examination of Water, Waste Water and Sludge; Cations (Group E); Determination of Zinc, Cadmium, Lead, Copper, Thallium, Nickel, Cobalt by Voltammetry (E16) (Mar). 8 pp.

—Tritium Content
DIN ENGL 38404 Pt 13-88. German Standard Methods for the Examination of Water, Waste Water and Sludge; Physical and Physicochemical Parameters (Group C); Determination of Tritium (C13) (May). 6 pp.

—Zinc Content—Voltametry
DIN ENGL 38406 Pt 16-90. German Standard Methods for the Examination of Water, Waste Water and Sludge; Cations (Group E); Determination of Zinc, Cadmium, Lead, Copper, Thallium, Nickel, Cobalt by Voltammetry (E16) (Mar). 8 pp.

Ground Wave Propagation
See Also: Ground Waves; Radio Wave Propagation; Wave Propagation

—Backscattering
CCIR Report 726-2-90. Ground and Ionospheric Side- and Back-Scatter—Section 6C—Ionospheric Propagation and Operational Forecasting. 7 pp.

—Electrical Properties—Earth Surface
CCIR RECMN 527-2-90. Electrical Characteristics of the Surface of the Earth—Section 5B—Effects of the Ground (Including Ground-Wave Propagation). 3 pp.
CCIR RECMN 527-3-92. Electrical Characteristics of the Surface of the Earth. 5 pp.

—Field Strength
CCIR RECMN 527-2-90. Electrical Characteristics of the Surface of the Earth—Section 5B—Effects of the Ground (Including Ground-Wave Propagation). 3 pp.
CCIR RECMN 527-3-92. Electrical Characteristics of the Surface of the Earth. 5 pp.
CCIR RECMN 832-92. World Atlas of Ground Conductivities. 48 pp.

—Field Strength—Diffraction
CCIR RECMN 526-1-82. Propagation by Diffraction—Section 5B—Effects of the Ground (Including Ground-Wave Propagation). 1 p.
CCIR RECMN 526-2-92. Propagation by Diffraction. 16 pp.

—Frequency Curves
CCIR RECMN 368-6-90. Ground-Wave Propagation Curves for Frequencies Between 10 kHz and 30 MHz—Section 5B—Effects of the Ground (Including Ground-Wave Propagation). 16 pp.
CCIR RESOLUTION 72-2-90. Handbook of Ground-Wave Propagation Curves—Volume V—Propagation in Non-Ionized Media. 1 p.
CCIR Report 717-3-90. World Atlas of Ground Conductivities (Published Separately)—Section 5B—Effects of the Ground (Including Ground-Wave Propagation). 9 pp.
CCIR RECMN 368-7-92. Ground-Wave Propagation Curves for Frequencies Between 10 kHz and 30 MHz. 54 pp.

—Isothermal Atmospheres
CCIR Report 714-2-90. Ground-Wave Propagation in an Exponential Atmosphere—Section 5B—Effects of the Ground (Including Ground-Wave Propagation). 3 pp.

—Sidescattering
CCIR Report 762-2-82. Effects of Multipath on Digital Transmission over Links in the Maritime Mobile-Satellite Service—Section 8I—Technical and Operating Characteristics of Mobile Satellite Services. 16 pp.

Ground Waves
See Also: Ground Wave Propagation

—Wave Phases
CCIR Report 716-3-90. Phase of the Ground Wave—Section 5B—Effects of the Ground (Including Ground-Wave Propagation). 4 pp.

Grounding (Electrical)
Use: Electrical Grounding

Grounding Clamps
Use For: Earth Clamps *See Also:* Clamps; Electrical Grounding
BSI BS 951-86. 1986 Amd 1 Clamps for Earthing and Bonding Purposes (AMD 6635) April 30, 1991. 11 pp.
CSA C22.2 NO 41-M1987. Grounding and Bonding Equipment (R 1993); (Gen Instr 1). 30 pp.
SAA AS 1882-76. Earth and Bonding Clamps Reconfirmed 1985. 12 pp.
SNZ NZS 2265-69. Specification for Earthing Clamps. 14 pp.

—Direct Burial
CSA C22.2 NO 41-M1987. Grounding and Bonding Equipment (R 1993); (Gen Instr 1). 30 pp.

Grounding Conductors
Use For: Earthing Conductors *See Also:* Bonding Conductors; Electric Conductors; Electrical Grounding
MOD UK DSTAN 61-12: Pt 16:Sec 2-80. Wires, Cords, and Cables, Electrical-Metric Units Part 16 Section 2: Wires, Electrical (for HF Aerial (Antenna) Feeder, or Earthing Systems) Issue 1. 11 pp.

—Electrical Installations
CSA C22.2 NO 41-M1987. Grounding and Bonding Equipment (R 1993); (Gen Instr 1). 30 pp.

—Medical Electrical Equipment—Color Coding
CSA CAN/CSA-C22. 2NO601.1-M90. Medical Electrical Equipment, Part 1: General Requirements for Safety; (Gen Instr 1). 240 pp.

—Power Cables
CSA 1369 Bull. Electrical Bulletin 1369 May 4, 1982 to C22.2 NO 96. 3 pp.

Grounding Connectors
Use For: Earthing Connectors *See Also:* Bonding Conductors; Connectors; Electrical Grounding
BSI BS 196-61. 1961 Amd 4 Protected-Type Non-Reversible Plugs, Socket-Outlets, Cable-Couplers and Appliance Couplers, with Earthing Contacts for Single Phase a.c. Circuits up to 250 Volts. 64 pp.

—Aerospace—Contact Force
CEN PREN 2591 (Part D13)-92. Elements of Electrical and Optical Connection Test Methods Part D13—Holding Force of Grounding Spring System. 4 pp.

Grounding Devices
Scope Note: For additional listings, use a more specific term *Use For:* Earthing Devices
See Also: Electrical Grounding; Ground Electrodes; Ground Fault Protection Equipment; Ground Rods; Ground Terminals; Grounding Clamps; Grounding Conductors; Grounding Connectors; Grounding Reactors; Grounding Switches; Grounding Transformers; Short Circuiting/Grounding Devices

—Portable—Live Wire Working
IEC 1230-93. Live Working—Portable Equipment for Earthing or Earthing and Short-Circuiting First Edition. 77 pp.

Grounding Electrodes
Use: Ground Electrodes

Grounding Reactors
Use For: Earthing Reactors *See Also:* Electrical Grounding
BSI BS 4944-73. 1973 Reactors, Arc-Suppression Coils and Earthing Transformers for Electric Power Systems. 19 pp.
SAA AS 1028-92. Power Reactors and Earthing Transformers (IEC 289:1988). 45 pp.

Grounding Rods
Use: Ground Rods

Grounding/Short Circuiting Devices
Use: Short Circuiting/Grounding Devices

Grounding Switches
Use For: Earthing Switches *See Also:* Electric Switches; Electrical Grounding
BSI BS 5253-90. 1990 Alternation Current Disconnectors and Earthing Switches. 47 pp.
IEC 129-84. Alternating Current Disconnectors and Earthing Switches Third Edition; (Amendment 1-1993). 92 pp.
SAA AS 1306-85. High Voltage a.c. Switchgear and Controlgear—Disconnectors (Isolators) and Earthing Switches. 28 pp.

Grounding Switches *(Cont.)*
— 52 kV
IEC 1129-92. Alternating Current Earthing Switches Induced Current Switching First Edition. 31 pp.

Grounding Transformers
Use For: Earthing Transformers *See Also:* Electrical Grounding; Transformers
BSI BS 4944-73. 1973 Reactors, Arc-Suppression Coils and Earthing Transformers for Electric Power Systems. 19 pp.
SAA AS 1028-92. Power Reactors and Earthing Transformers (IEC 289:1988). 45 pp.

Groundnut Oils
Use: Peanut Oil

Group Audio Terminals
Use For: GAT

—Transmission—Quality of Service
CCITT Volume V. Telephone Transmission Quality Series P Recommendations (Study Group XII). 438 pp.
CCITT RECMN P.30-89. Transmission Performance of Group Audio Terminals (GATs)—Telephone Transmission Quality (Study Group XII) 4 pp. 4 pp.

Group B Signals
See Also: Signaling Systems

—Interregister Signaling—Termination—CCITT R2 Signaling Systems
CCITT RECMN Q.474-89. Signalling Procedures—Use of Group B Signals—Specifications of Signalling Systems R1 and R2 (Study Group XI) 3 pp. 3 pp.

Group Calling Services
Use: Group Calls

Group Calls
See Also: Message Handling Systems; Telephone Services

—FleetNET (BTN)
CCIR Report 921-2-90. System Aspects of Digital Ship Earth Stations—Section 8F—Frequencies, Orbits and Systems. 46 pp.

—Maritime Mobile Satellite Communications—VHF/UHF
CCITT RECMN F.120-89. Ship Station Identification for VHF/UHF and Maritime Mobile-Satellite Services—Telegraph and Mobile Services Operations and Quality of Service (Study Group I) 6 pp (Same as Recmn E.210). 6 pp.

—Mobile Satellite Communications—INMARSAT
CCITT RECMN F.122-89. Operational Procedures for the Maritime Satellite Data Transmission Service—Telegraph and Mobile Services Operations and Quality of Service (Study Group I) 8 pp. 8 pp.

—Numbering Plans—INMARSAT
CCITT RECMN E.215-89. Telephone/ISDN Numbering Plan for the Mobile-Satellite Services of INMARSAT—Telephone Network and ISDN—Operation, Numbering, Routing and Mobile Service (Study Group II)14 pp. 14 pp.

—SafetyNET (BTN)
CCIR Report 921-2-90. System Aspects of Digital Ship Earth Stations—Section 8F—Frequencies, Orbits and Systems. 46 pp.

—Telex Communications—INMARSAT Mobile Numbers
CCITT RECMN F.125-89. Telex Numbering Plan for the Mobile-Satellite Services of INMARSAT—Telegraph and Mobile Services Operations and Quality of Service (Study Group I) 9 pp. 9 pp.

Group Delay
See Also: Delay; Envelope Delay Distortion; Propagation Time; Time

—Communication Terminal Equipment—Groups
CCITT RECMN G.235-89. 16-Channel Terminal Equipments—International Analogue Carrier Systems (Study Group XV) 4 pp. 4 pp.

—Data Circuits
CCITT RECMN O.82-89. Group-Delay Measuring Equipment for the Range 5 to 600 kHz—Specifications for Measuring Equipment (Study Group IV) 6 pp. 6 pp.

Group Delay (Cont.)

—Echo Cancellers
CCITT RECMN G.165-89. Echo Cancellers—General Characteristics of International Telephone Connections and Circuits (Study Groups XII and XV) 23 pp. 23 pp.

—Echo Suppressors
CCITT RECMN G.164-89. Echo Suppressors—General Characteristics of International Telephone Connections and Circuits (Study Groups XII and XV) 36 pp. 36 pp.

—Telephone Circuits
CCITT RECMN G.151-89. General Performance Objectives Applicable to All Modern International Circuits and National Extension Circuits—General Characteristics of International Telephone Connections and Circuits (Study Groups XII and XV) 6 pp. 6 pp.

CCITT RECMN O.81-89. Group-Delay Measuring Equipment for Telephone-Type Circuits—Specifications for Measuring Equipment (Study Group IV) 6 pp. 6 pp.

Group Delay Distortion
Use: Envelope Delay Distortion

Group Links
Use For: International Leased Group Links
See Also: Frequency Translation Equipment; Groups (Communications); Line Links; Telecommunication Equipment; Through Group Connection Points

CCITT RECMN M.460-89. Bringing International Group, Supergroup, Etc., Links into Service—General Maintenance Principles—Maintenance of International Transmission Systems and Telephone Circuits (Study Group IV) 18 pp. 18 pp.

—Carrier Systems
CCITT RECMN G.211-89. Make-up of a Carrier Link—International Analogue Carrier Systems (Study Group XV) 8 pp. 8 pp.

—Data Transmission
CCITT RECMN H.14-89. Characteristics of Group Links for the Transmission of Wide-Spectrum Signals—Line Transmission of Non-Telephone Signals—Transmission of Sound-Programme and Television Signals (Study Group XV) 1 pp. 1 p.

CCITT RECMN H.52-89. Transmission of Wide-Spectrum Signals (Data, Facsimile, Etc.) on Wideband Group Links—Line Transmission of Non-Telephone Signals—Transmission of Sound-Programme and Television Signals (Study Group XV) 1 pp. 1 p.

CCITT RECMN M.900-89. Use of Leased Group and Supergroup Links for Wide-Spectrum Signal Transmission (Data, Facsimile, Etc.)—Maintenance of International Telegraph, Phototelegraph and Leased Circuits—Maintenance of the International Public Telephone Network-. 4 pp.

—Data Transmission—Line Up
CCITT RECMN M.910-89. Setting up and Lining up an International Leased Group Link for Wide-Spectrum Signal Transmission—Maintenance of International Telegraph, Phototelegraph and Leased Circuits—Maintenance of the International Public Telephone Network-. 4 pp.

—Data Transmission—Set Up
CCITT RECMN M.910-89. Setting up and Lining up an International Leased Group Link for Wide-Spectrum Signal Transmission—Maintenance of International Telegraph, Phototelegraph and Leased Circuits—Maintenance of the International Public Telephone Network-. 4 pp.

—Designations
CCITT RECMN M.140-89. Designations of International Circuits, Groups, Group and Line Links, Digital Blocks, Digital Paths, Data Transmission Systems and Related Information—General Maintenance Principles—Maintenance of International Transmission Systems and Telephone Circuits. 53 pp.

—Distributing Frames—Power Levels
CCITT RECMN G.233-89. Recommendations Concerning Translating Equipments—International Analogue Carrier Systems (Study Group XV) 10 pp. 10 pp.

—Hypothetical Reference Circuits—Intermodulation Noise
CCITT RECMN G.223-89. Assumptions for the Calculation of Noise on Hypothetical Reference Circuits for Telephony—International Analogue Carrier Systems (Study Group XV) 9 pp. 9 pp.

Group Links (Cont.)

—Line Regulation Systems—Maintenance
CCITT RECMN M.530-89. Readjustment to the Nominal Value of an International Group, Supergroup, Etc., Link—General Maintenance Principles—Maintenance of International Transmission Systems and Telephone Circuits (Study Group IV) 3 pp. 3 pp.

—Maintenance
CCITT RECMN M.520-89. Routine Maintenance on International Group, Supergroup, Etc., Links—General Maintenance Principles—Maintenance of International Transmission Systems and Telephone Circuits (Study Group IV) 1 pp. 1 p.

CCITT RECMN M.525-89. Automatic Maintenance Procedures for International Group, Supergroup, Etc., Links—General Maintenance Principles—Maintenance of International Transmission Systems and Telephone Circuits (Study Group IV) 2 pp. 2 pp.

CCITT RECMN M.530-89. Readjustment to the Nominal Value of an International Group, Supergroup, Etc., Link—General Maintenance Principles—Maintenance of International Transmission Systems and Telephone Circuits (Study Group IV) 3 pp. 3 pp.

CCITT RECMN M.535-89. Special Maintenance Procedures for Multiple Destination, Undirectional (MU) Group and Supergroup Links—General Maintenance Principles—Maintenance of International Transmission Systems and Telephone Circuits (Study Group IV) 1 pp. 1 p.

—Pilot Channels
CCITT RECMN M.241-89. Pilots on Groups, Supergroups, Etc.—International Analogue Carrier Systems (Study Group XV) 6 pp. 6 pp.

—Pilot Channels—Harmonics
CCITT RECMN M.241-89. Pilots on Groups, Supergroups, Etc.—International Analogue Carrier Systems (Study Group XV) 6 pp. 6 pp.

—Readjustments
CCITT RECMN M.525-89. Automatic Maintenance Procedures for International Group, Supergroup, Etc., Links—General Maintenance Principles—Maintenance of International Transmission Systems and Telephone Circuits (Study Group IV) 2 pp. 2 pp.

Groups (Communications)
Use For: 12 Channel Units (Telephony)
See Also: Carrier Systems; Circuit Groups (Telecommunications); Communication Channels; Group Links; Mastergroups (Communications); Supergroups (Communications); Supermastergroups (Communications); Telecommunication Circuits; Telephone Channel Units; Telephone Circuits; Through Group Connection Points; Trunk Groups

—Accounting—Telecommunication Administrations
CCITT RECMN E.252-89. Mode of Application of the Flat-Rate Price Procedure Set Forthe in Recommendations D.67 and D.150 for Remuneration of Facilities Made Available to the Administrations of Other Countries—Telephone Network and ISDN—Operation, Numbering, Routing and Mobile. 1 p.

—Carrier Frequencies
CCITT RECMN G.225-89. Recommendations Relating to the Accuracy of Carrier Frequencies—International Analogue Carrier Systems (Study Group XV) 3 pp. 3 pp.

—Carrier Systems—Attenuation
CCITT RECMN G.313-89. Open-Wire Lines for Use with 12-Channel Carrier Systems—International Analogue Carrier Systems (Study Group XV) 3 pp. 3 pp.

—Carrier Systems—Circuit Noise
CCITT RECMN G.311-89. General Characteristics of Systems Providing 12 Carrier Telephone Circuits on an Open-Wire Line—International Analogue Carrier Systems (Study Group XV) 6 pp. 6 pp.

CCITT RECMN G.322-89. General Characteristics Recommended for Systems on Symmetric Pair Cables—International Analogue Carrier Systems (Study Group XV) 8 pp. 8 pp.

CCITT RECMN G.325-89. General Characteristics Recommended for Systems Providing 12 Telephone Carrier Circuits on a Symmetric Cable Pair ((12 + 12) Systems)—International Analogue Carrier Systems (Study Group XV) 5 pp. 5 pp.

—Carrier Systems—Crosstalk
CCITT RECMN G.313-89. Open-Wire Lines for Use with 12-Channel Carrier Systems—International Analogue Carrier Systems (Study Group XV) 3 pp. 3 pp.

Groups (Communications) (Cont.)

—Carrier Systems—Hypothetical Reference Circuits
CCITT RECMN G.325-89. General Characteristics Recommended for Systems Providing 12 Telephone Carrier Circuits on a Symmetric Cable Pair ((12 + 12) Systems)—International Analogue Carrier Systems (Study Group XV) 5 pp. 5 pp.

—Carrier Systems—Pilot Channels
CCITT RECMN G.322-89. General Characteristics Recommended for Systems on Symmetric Pair Cables—International Analogue Carrier Systems (Study Group XV) 8 pp. 8 pp.

CCITT RECMN G.325-89. General Characteristics Recommended for Systems Providing 12 Telephone Carrier Circuits on a Symmetric Cable Pair ((12 + 12) Systems)—International Analogue Carrier Systems (Study Group XV) 5 pp. 5 pp.

—Carrier Systems—Surge Protectors
CCITT RECMN G.313-89. Open-Wire Lines for Use with 12-Channel Carrier Systems—International Analogue Carrier Systems (Study Group XV) 3 pp. 3 pp.

—Carrier Systems—Telephone Cables
CCITT RECMN G.322-89. General Characteristics Recommended for Systems on Symmetric Pair Cables—International Analogue Carrier Systems (Study Group XV) 8 pp. 8 pp.

CCITT RECMN G.325-89. General Characteristics Recommended for Systems Providing 12 Telephone Carrier Circuits on a Symmetric Cable Pair ((12 + 12) Systems)—International Analogue Carrier Systems (Study Group XV) 5 pp. 5 pp.

CCITT RECMN G.327-89. Valve-Type Systems Offering 12 Carrier Telephone Circuits on a Symmetric Cable Pair ((12 + 12) Systems)—International Analogue Carrier Systems (Study Group XV) 2 pp. 2 pp.

—Communication Channels—Numbering
CCITT RECMN M.320-89. Numbering of the Channels in a Group—General Maintenance Principles—Maintenance of International Transmission Systems and Telephone Circuits (Study Group IV) 2 pp. 2 pp.

—Communication Terminal Equipment
CCITT RECMN G.232-89. 12-Channel Terminal Equipments—International Analogue Carrier Systems (Study Group XV) 13 pp. 13 pp.

CCITT RECMN G.235-89. 16-Channel Terminal Equipments—International Analogue Carrier Systems (Study Group XV) 4 pp. 4 pp.

—Communication Terminal Equipment—Attenuation Distortion
CCITT RECMN G.232-89. 12-Channel Terminal Equipments—International Analogue Carrier Systems (Study Group XV) 13 pp. 13 pp.

CCITT RECMN G.235-89. 16-Channel Terminal Equipments—International Analogue Carrier Systems (Study Group XV) 4 pp. 4 pp.

—Communication Terminal Equipment—Carrier Leaks
CCITT RECMN G.232-89. 12-Channel Terminal Equipments—International Analogue Carrier Systems (Study Group XV) 13 pp. 13 pp.

CCITT RECMN G.235-89. 16-Channel Terminal Equipments—International Analogue Carrier Systems (Study Group XV) 4 pp. 4 pp.

—Communication Terminal Equipment—Crosstalk
CCITT RECMN G.232-89. 12-Channel Terminal Equipments—International Analogue Carrier Systems (Study Group XV) 13 pp. 13 pp.

CCITT RECMN G.235-89. 16-Channel Terminal Equipments—International Analogue Carrier Systems (Study Group XV) 4 pp. 4 pp.

—Communication Terminal Equipment—Electrical Impedance
CCITT RECMN G.232-89. 12-Channel Terminal Equipments—International Analogue Carrier Systems (Study Group XV) 13 pp. 13 pp.

—Communication Terminal Equipment—Envelope Delay Distortion
CCITT RECMN G.232-89. 12-Channel Terminal Equipments—International Analogue Carrier Systems (Study Group XV) 13 pp. 13 pp.

—Communication Terminal Equipment—Group Delay
CCITT RECMN G.235-89. 16-Channel Terminal Equipments—International Analogue Carrier Systems (Study Group XV) 4 pp. 4 pp.

INTERNATIONAL AND NON-U.S. NATIONAL STANDARDS
SUBJECT INDEX

Groups (Communications) (Cont.)
—Communication Terminal Equipment—Linearity
- CCITT RECMN G.232-89. 12-Channel Terminal Equipments—International Analogue Carrier Systems (Study Group XV) 13 pp. 13 pp.
- CCITT RECMN G.235-89. 16-Channel Terminal Equipments—International Analogue Carrier Systems (Study Group XV) 4 pp. 4 pp.

—Communication Terminal Equipment—Pilot Channels
- CCITT RECMN G.232-89. 12-Channel Terminal Equipments—International Analogue Carrier Systems (Study Group XV) 13 pp. 13 pp.
- CCITT RECMN G.235-89. 16-Channel Terminal Equipments—International Analogue Carrier Systems (Study Group XV) 4 pp. 4 pp.

—Communication Terminal Equipment—Psophometric Power
- CCITT RECMN G.235-89. 16-Channel Terminal Equipments—International Analogue Carrier Systems (Study Group XV) 4 pp. 4 pp.

—Communication Terminal Equipment—Return Loss
- CCITT RECMN G.232-89. 12-Channel Terminal Equipments—International Analogue Carrier Systems (Study Group XV) 13 pp. 13 pp.

—Communication Terminal Equipment—Signal Levels
- CCITT RECMN G.232-89. 12-Channel Terminal Equipments—International Analogue Carrier Systems (Study Group XV) 13 pp. 13 pp.

—Communication Terminal Equipment—Signal Limiters
- CCITT RECMN G.232-89. 12-Channel Terminal Equipments—International Analogue Carrier Systems (Study Group XV) 13 pp. 13 pp.
- CCITT RECMN G.235-89. 16-Channel Terminal Equipments—International Analogue Carrier Systems (Study Group XV) 4 pp. 4 pp.

—Communication Terminal Equipment—Virtual Carrier Frequencies
- CCITT RECMN G.235-89. 16-Channel Terminal Equipments—International Analogue Carrier Systems (Study Group XV) 4 pp. 4 pp.

—Designations
- CCITT RECMN M.140-89. Designations of International Circuits, Groups, Group and Line Links, Digital Blocks, Digital Paths, Data Transmission Systems and Related Information—General Maintenance Principles—Maintenance of International Transmission Systems and Telephone Circuits. 53 pp.

—Numbering—Supergroups
- CCITT RECMN M.330-89. Numbering of Groups Within a Supergroup—General Maintenance Principles—Maintenance of International Transmission Systems and Telephone Circuits (Study Group IV) 1 pp. 1 p.

—Numbering—Transmission Systems
- CCITT RECMN M.380-89. Numbering in Coaxial Systems—General Maintenance Principles—Maintenance of International Transmission Systems and Telephone Circuits (Study Group IV) 6 pp. 6 pp.
- CCITT RECMN M.390-89. Numbering in Systems on Symmetric Pair Cable—General Maintenance Principles—Maintenance of International Transmission Systems and Telephone Circuits (Study Group IV) 4 pp. 4 pp.
- CCITT RECMN M.400-89. Numbering in Radio-Relay Links or Open-Wire Line Systems—General Maintenance Principles—Maintenance of International Transmission Systems and Telephone Circuits (Study Group IV) 1 pp. 1 p.

—Open Wire Lines—Carrier Systems
- CCITT RECMN G.311-89. General Characteristics of Systems Providing 12 Carrier Telephone Circuits on an Open-Wire Pair—International Analogue Carrier Systems (Study Group XV) 6 pp. 6 pp.
- CCITT RECMN G.313-89. Open-Wire Lines for Use with 12-Channel Carrier Systems—International Analogue Carrier Systems (Study Group XV) 3 pp. 3 pp.

Groups (Communications) (Cont.)
—Tariffs—Telecommunication Administrations
- CCITT RECMN D.160-89. Mode of Application of the Flat-Rate Price Procedure Set Forth in Recommendation D.67 and Recommendation D.150 for Remuneration of Facilities Made Available to the Administrations of Other Countries—General Tariff Principles—Charging and Accounting. 5 pp.
- CCITT RECMN E.252-89. Mode of Application of the Flat-Rate Price Procedure Set Forthe in Recommendations D.67 and D.150 for Remuneration of Facilities Made Available to the Administrations of Other Countries—Telephone Network and ISDN—Operation, Numbering, Routing and Mobile. 1 p.

—Telephone Cables
- CCITT RECMN G.326-89. Typical Systems on Symmetric Cable Pairs ((12 + 12) Systems)—International Analogue Carrier Systems (Study Group XV) 3 pp. 3 pp.

—Telephone Cables—Signal Levels
- CCITT RECMN G.326-89. Typical Systems on Symmetric Cable Pairs ((12 + 12) Systems)—International Analogue Carrier Systems (Study Group XV) 3 pp. 3 pp.

—Telephone Relations—Accounting—Africa
- CCITT RECMN D.600R-89. Determination of Accounting Rate Shares and Collection Charges in Telephone Relations Between Countries in Africa—General Tariff Principles—Charging and Accounting in International Telecommunications Services (Study Group III) 11 pp. 11 pp.

—Telephone Relations—Tariffs—Africa
- CCITT RECMN D.600R-89. Determination of Accounting Rate Shares and Collection Charges in Telephone Relations Between Countries in Africa—General Tariff Principles—Charging and Accounting in International Telecommunications Services (Study Group III) 11 pp. 11 pp.

—Virtual Carrier Frequencies
- CCITT RECMN G.225-89. Recommendations Relating to the Accuracy of Carrier Frequencies—International Analogue Carrier Systems (Study Group XV) 3 pp. 3 pp.

—Virtual Carrier Frequencies—Connection Integrity
- CCITT RECMN G.232-89. 12-Channel Terminal Equipments—International Analogue Carrier Systems (Study Group XV) 13 pp. 13 pp.
- CCITT RECMN G.235-89. 16-Channel Terminal Equipments—International Analogue Carrier Systems (Study Group XV) 4 pp. 4 pp.

Grout
Use For: Masonry Grout *See Also:* Cements; Clays; Concretes; Mortars
- CSA A179-1975. Mortar and Grout for Unit Masonry. 20 pp.
- CSA A179-M1976. Mortar and Grout for Unit Masonry; (Erratum September 1977). 20 pp.

—Admixtures
- CEN PREN 104.301 (Part 1)-91. Admixtures for Concrete, Mortar and Grout Test Methods Reference Concrete and Reference Mortar for Testing. 3 pp.
- SAA AS 2073-77. Methods for the Testing of Expanding Admixtures for Concrete, Mortar and Grout. 28 pp.
- SAA MP20.3-77. Part 3: Expanding Admixtures for Use in Concrete, Mortar and Grout. 12 pp.

—Admixtures—Sampling
- SAA AS 2072-77. Methods for the Sampling of Expanding Admixtures for Concrete, Mortar and Grout. 4 pp.

—Bleeding
- SNZ NZS 3112: Part 4-86. Methods of Test for Concrete Part 4: Tests Relating to Grout. 10 pp.

—Compressive Testing
- SNZ NZS 3112: Part 4-86. Methods of Test for Concrete Part 4: Tests Relating to Grout. 10 pp.

—Flow Time
- SNZ NZS 3112: Part 4-86. Methods of Test for Concrete Part 4: Tests Relating to Grout. 10 pp.

—Life (Durability)
- BSI DD 88-83. 1983 Method for the Assessment of Pot Life of Non-Flowing Resin Compositions for Use in Civil Engineering. 6 pp.

Grout (Cont.)
—Setting Time
- SAA AS 1012.18-75. Methods of Testing Concrete—Part 18: Determination of Setting Time of Fresh Concrete, Mortar and Grout by Penetration Resistance. 10 pp.

—Tendons
- CEN PREN 445-90. Grout for Prestressing Tendons—Test Methods. 12 pp.
- CEN PREN 447-90. Grout for Prestressing Tendons—Specification for Common Grout. 6 pp.

—Volume Change
- SNZ NZS 3112: Part 4-86. Methods of Test for Concrete Part 4: Tests Relating to Grout. 10 pp.

Grouting
See Also: Foundations; Mortars; Sealing

—Aggregates
- CNS A2036-70. Aggregate of Masonry Concrete for Grouting (Jan)(2466). 3 pp.

—Construction Contracts
- DIN ENGL 18309-92. Construction Contract Procedures (VOB); Part C: General Technical Specifications in Construction Contracts (ATV); Ground Treatment by Grouting (Dec). 4 pp.

—Prestressed Concrete
- CEN PREN 446-90. Grout for Prestressing Tendons—Grouting Procedures. 8 pp.

—Subsoils
- DIN ENGL 4093-87. Ground Treatment by Grouting; Planning, Grouting Procedure and Testing (Sept). 15 pp.

Growth Hormones
See Also: Medicinal Products

—Livestock—Control Measures
- EC 85/358/EEC-85. Council Directive Supplementing Directive 81/602/EEC Concerning the Prohibition of Certain Substances Having a Hormonal Action and of any Substances Having a Thyrostatic Action. 4 pp.
- EC 88/146/EEC-88. Council Directive Prohibiting the Use in Livestock Farming of Certain Substances Having a Hormonal Action. 3 pp.

Growth Regulators
—Plants—Glossaries
- ISO 257-88. Pesticides and Other Agrochemicals—Principles for the Selection of Common Names Second Edition. 11 pp.
- ISO 765-76. Pesticides Considered Not to Require Common Names First Edition. 17 pp.
- ISO 1750-81. Pesticides and Other Agrochemicals—Common Names First Edition; (Amendment 1-1982) (Addendum 1-1983) (Addendum 2-1983). 132 pp.

Growth Rings
Use: Annual Rings

Grub Screws
See Also: Screws
- DIN ENGL 6324-71. Operating Elements for Clamping Devices; Survey (Jan). 2 pp.
- MOD UK DSTAN 07-63: Part 6-77. Standardization Status of General Purpose Threaded Fasteners, Steel, Held in COSA Section G1 Part 6: Screws, Socket Head and Screws, Grub, Steel, BSF and BSW to Be Read in Conjunction with COSA Section G1—Part 5 Issue 1. 28 pp.
- MOD UK DSTAN 53-19-01. Screws, Grub Metric—Steel Issue 1; Amendment 1. 9 pp.
- MOD UK DSTAN 53-56-73. Screws, Grub, Unified Threads Issue 1. 10 pp.

—Aircraft
- SBAC AS 3374 ISSUE 5. Grub Screw. 1 p.

—Aircraft—Filler Caps
- SBAC AS 6750 ISSUE 2. Grub Screw.

—Aircraft—Plungers
- SBAC AS 3346 ISSUE 1. Grub Screw. 1 p.
- SBAC AS 6383 ISSUE 1. Grub Screw.

—Slotted
- BSI BS 768-58. 1958 Amd 1 Slotted Grub Screws. 15 pp.

—Thrust Point
- CNS B2335-82. Grub Screws with Thrust Point (Dec)(4657).
- DIN ENGL 6332-81. Grub Screws with Thrust Point (Jan). 3 pp.

Grub Screws (Cont.)

—Whitworth Thread
MOD UK DSTAN 07-63: Part 6-77. Standardization Status of General Purpose Threaded Fasteners, Steel, Held in COSA Section G1 Part 6: Screws, Socket Head and Screws, Grub, Steel, BSF and BSW to Be Read in Conjunction with COSA Section G1—Part 5 Issue 1. 28 pp.

Grumman American AA 1 Aircraft
See Also: Aircraft

—Antenna Positions
CAA. Grumman American AA1 Series (Approved Aerial Positions). 1 p.

—Foreign Airworthiness Directives
CAA. Gulfstream American AA-1 Series Aircraft and Gulfstream Aerospace AA-5 Series Aircraft. 3 pp.

Grumman American AA 5 Aircraft
See Also: Aircraft

—Antenna Positions
CAA. Grumman American AA5 Series/American General AG-5B (Approved Aerial Positions). 1 p.

Grumman American GA 7 Aircraft
See Also: Aircraft

—Antenna Positions
CAA. Grumman American GA-7 (Approved Aerial Positions). 1 p.

—Certification
CAA. Grumman Gulfstream. 12 pp.

Grumman American General AG-5B Aircraft
See Also: Aircraft

—Antenna Positions
CAA. Grumman American AA5 Series/American General AG-5B (Approved Aerial Positions). 1 p.

Grumman G 159 Gulfstream 1 Aircraft
See Also: Aircraft

—Antenna Positions
CAA. Grumman G159 Gulfstream 1 (Approved Aerial Positions). 1 p.

Grumman G 164A Aircraft
See Also: Aircraft

—Antenna Positions
CAA. Grumman G164A (Approved Aerial Positions). 1 p.

Grumman G 1159 Gulfstream 2 Aircraft
See Also: Aircraft

—Antenna Positions
CAA. Grumman G1159 Gulfstream 2 (Approved Aerial Positions). 1 p.

—Data Sheets
CAA FA7 ISSUE 1. Grumman G-1159 Gulfstream II & III. 4 pp.

Grumman G 1159 Gulfstream 3 Aircraft
See Also: Aircraft

—Data Sheets
CAA FA7 ISSUE 1. Grumman G-1159 Gulfstream II & III. 4 pp.

Grumman G 1159A Gulfstream 3 Aircraft
See Also: Aircraft

—Antenna Positions
CAA. Grumman G1159A Gulfstream 3. 1 p.

Grumman Gulfstream II Aircraft
See Also: Aircraft

—Certification
CAA. Grumman Gulfstream II. 14 pp.

GTAW
Use: Gas Tungsten Arc Welding

GTOs
Use: Gate Turn Off Thyristors

Guanylurea Sulfate
See Also: Urea

—Fertilizers
CNS N3071-88. Guanylurea Sulfate Fertilizer (Jul)(11958).

Guarantor Services
See Also: Flat Rate Tariffs (Telecommunications); Tariffs (Telecommunications); Telephone Services

—Public Data Communication—Reverse Charging
CCITT RECMN D.30-89. Implementation of Reverse Charging on International Public Data Communication Services—General Tariff Principles—Charging and Accounting in International Telecommunications Services (Study Group III) 5 pp. 5 pp.

Guard Rails
Use: Guardrails

Guardrails
Use For: Safety Fences; Slide Rails
See Also: Barriers; Gangways; Guide Rails; Handrails

—Boats (Marine)
CEN PREN 711-92. Inland Navigation Vessels—Railings for Decks—Requirements, Types. 9 pp.

—Corrugated Beam
BSI BS 6579: Part 7-89. 1989 Safety Fences and Barriers for Highways Part 7: Components for Untensioned Corrugated Beam Safety Fence. 18 pp.

—Corrugated Beam—Tensioned
BSI BS 6579: Part 1-88. 1988 Amd 1 Safety Fences and Barriers for Highways Part 1: Components for Tensioned Corrugated Beam Safety Fence on Z Posts. 29 pp.

—Open Box Beam
BSI BS 6579: Part 5-86. 1986 Safety Fences and Barriers for Highways Part 5: Open Box Beam Safety Fence (Single Height). 27 pp.
BSI BS 6579: Part 6-88. 1988 Safety Fences and Barriers for Highways Components for Open Box Beam Safety Fence (Double Height). 18 pp.

—Pedestrian
BSI BS 3049-76. 1976 Pedestrian Guard Rails (Metal). 7 pp.

—Rectangular Hollow Beam—Tensioned
BSI BS 6579: Part 3-88. 1988 Safety Fences and Barriers for Highways Part 3: Components for Tensioned Rectangular Hollow Section Beam (100mm x100mm) Safety Fence. 18 pp.
BSI BS 6579: Part 4-90. 1990 Safety Fences and Barriers for Highways Part 4: Specification for Components for Tensioned Rectangular Hollow Section Beam (200 mm x 100 mm) Safety Fence. 28 pp.

—Ships
BSI BS MA 40: Part 1-75. 1975 Marine Guardrails, Stanchions, Etc Part 1: Guardrails for Merchant Ships (Excluding Passenger Ships). 11 pp.
ISO 3674-76. Shipbuilding—Inland Vessels—Deck Rail First Edition. 8 pp.
ISO 5480-79. Shipbuilding—Guardrails for Cargo Ships First Edition. 8 pp.
MOD UK NES 162: Part 1-01. Guard and Awning Stanchions, Rails and Associated Fittings Part 1: Guard Stanchions, Rails and Associated Fittings—Weatherdecks Issue 2 (07.89); Amendment 1. 40 pp.
SAA AS 1986-87. Shipbuilding—Guardrails for Cargo Ships. 6 pp.

—Ships—Portable
BSI BS MA 40: Part 2-75. 1975 Marine Guardrails, Stanchions, Etc Part 2: Gates and Portable Guardrail Sections for Merchant Ships (Excluding Passenger Ships). 8 pp.

—Ships—Steel
BSI BS MA 40: Part 3-75. 1975 Marine Guardrails, Stanchions, Etc Part 3: Stormrails (Exterior). 4 pp.

Guards (Covers)
Use: Guards (Protective)

Guards (Machines)
Use: Guards (Protective)

Guards (Personnel)
Use For: Security Guards See Also: Personnel
CGSB CAN/CGSB-133.1-87. Security Guards, Uniformed. 29 pp.

—Supervisors
CGSB CAN/CGSB-133.2-92. Uniformed Security Guard Supervisors. 17 pp.

Guards (Protective)
Use For: Guards (Covers) See Also: Chain Cases; Dust Covers; Occupational Safety and Health
CEN PREN 1082-93. Protective Chain Mail Gloves and Protective Armguards for Use with Hand Knives. 48 pp.
SNZ NZS 5801-74. Specification for the Construction and Fitting of Machinery Guards Amend: 1, 1977 (Reconfirmed 1983). 11 pp.

—Brush Saws
ISO 7918-85. Forestry Machinery—Portable Brush-Saws—Circular Saw-Blade Guard—Dimensions First Edition. 5 pp.

—Brush Saws—Impact Testing
ISO 8380-85. Forestry Machinery—Portable Brush-Saws—Circular Saw-Blade Guard—Strength First Edition. 4 pp.

—Chain Saws
ISO 6533-93. Forestry Machinery—Portable Chain-Saw Front Hand-Guard—Dimensions Second Edition. 6 pp.

—Chain Saws—Impact Testing
BSI BS 6916: Part 3-88. 1988 Chain Saws Part 3: Specification for Dimensions of Front Hand Guards (ISO 6533: 1983). 4 pp.
BSI BS 6916: Part 4-88. 1988 Chain Saws Part 4: Specification for Strength of Front Hand Guards (ISO 6534: 1985). 4 pp.
ISO 6534-92. Portable Chain-Saws—Hand-Guards—Mechanical Strength Second Edition. 6 pp.

—Check Valves—Aircraft—Pipe Threads
SBAC AGS 3415 ISSUE 1. Non Return Valve, Safety Guard for Use with 1/4 Inch B.S.P.F. Thread.

—Drive Shafts
CEN PREN 1152-93. Tractors and Machinery for Agriculture and Forestry—Guards for Power Take-Off (PTO) Drive Shafts—Wear and Strength Tests. 12 pp.
ISO 5674 Pt 1-92. Tractors and Machinery for Agriculture and Forestry—Guards for Power Take-off (PTO) Drive-Shafts—Part 1: Strength Test First Edition. 9 pp.

—Drive Shafts—Wear Testing
CEN PREN 1152-93. Tractors and Machinery for Agriculture and Forestry—Guards for Power Take-Off (PTO) Drive Shafts—Wear and Strength Tests. 12 pp.
ISO 5674 Pt 2-92. Tractors and Machinery for Agriculture and Forestry—Guards for Power Take-off (PTO) Drive-Shafts—Part 2: Wear Test First Edition. 8 pp.

—Earthmoving Equipment—Glossaries
BSI BS 5945-87. 1987 Guards and Shields for Earth-Moving Machinery. 8 pp.
ISO 3457-86. Earth-Moving Machinery—Guards and Shields—Definitions and Specifications Third Edition. 7 pp.
JIS A 8307-91. Earth-Moving Machinery—Guards and Shields—Definitions and Specifications (ISO 3457-1986).

—Machinery
CEN PREN 953-92. Safety of Machinery—General Requirements for the Design and Construction of Guards (Fixed, Movable). 30 pp.
CEN PREN 1088-93. Safety of Machinery—Interlocking Devices with and Without Guard Locking—General Principles and Provisions for Design. 38 pp.

—Milling Machines
SAA AS CZ14-71. Code of Recommended Practice for Guarding and Safe Use of Milling Machines (Incorporating Amdt 1). 62 pp.

—Punch Presses
CSA CAN/CSA-Z142-M90. Code for Punch Press and Brake Press Operation: Health, Safety, and Guarding Requirements; (Gen Instr 1). 44 pp.

—Screw Conveyors
ISO TR9172-87. Continuous Mechanical Handling Equipment—Safety Code for Screw Conveyors—Examples of Guards for Trapping and Shearing Points First Edition. 16 pp.

—Valves—Gas Cylinders
CEN PREN 962-92. Gas Cylinders—Valve Protection Caps and Valve Guards for Industrial and Medical Gas Cylinders—Design, Construction and Tests. 12 pp.

Gudgeon Pins
Use: Piston Pins

Guide Bushes
Use: Guide Bushings

INTERNATIONAL AND NON-U.S. NATIONAL STANDARDS
SUBJECT INDEX

Guide Bushings
See Also: Bushings

—Ball Cage—Press Tools
BSI BS 7568: Part 18-92. 1992 Press Tools Part 18: Guide Bushes—Gliding Bushes, Plain, Type 1 (ISO 9448-2: 1991). 7 pp.
BSI BS 7568: Part 19-92. 1992 Press Tools Part 19: Guide Bushes—Form B, Ball Cage Bushes, Plain, Type 1 (ISO 9448-3: 1991). 9 pp.
BSI BS 7568: Part 21-92. 1992 Press Tools Part 21: Guide Bushes—Form D, Ball Cage Bushes, Headed, Type 1 (ISO 9448-5: 1991). 9 pp.
BSI BS 7568: Part 23-92. 1992 Press Tools Part 23: Guide Bushes—Form F, Ball Cage Bushes, Flanged, Type 1 (ISO 9448-7: 1991). 9 pp.
BSI BS 7568: Part 31-93. 1993 Press Tools Part 31: Specification for Guide Bushes: Form E, Gliding Bushes, Flanged, Type 2 (ISO 9448-10: 1992) (F). 9 pp.
ISO 9448 Pt 3-91. Tools for Pressing—Guide Bushes—Part 3: Form B, Ball Cage Bushes, Plain, Type 1 First Edition. 8 pp.
ISO 9448 Pt 5-91. Tools for Pressing—Guide Bushes—Part 5: Form D, Ball Cage Bushes, Headed, Type 1 First Edition. 8 pp.
ISO 9448 Pt 7-91. Tools for Pressing—Guide Bushes—Part 7: Form F, Ball Cage Bushes, Flanged, Type 1 First Edition. 8 pp.
ISO 9448 Pt 9-92. Tools for Pressing—Guide Bushes—Part 9: Form B, Ball Cage Bushes, Plain, Type 2 First Edition. 7 pp.
ISO 9448 Pt 10-92. Tools for Pressing—Guide Bushes—Part 10: Form E, Gliding Bushes, Flanged, Type 2 First Edition. 8 pp.
ISO 9448 Pt 11-92. Tools for Pressing—Guide Bushes—Part 11: Form F, Ball Cage Bushes, Flanged, Type 2 First Edition. 8 pp.

—Dies
CNS B3373-80. Guide Pin Bushings of Dies for Die Casting (Aug)(5999).
JIS B 5105-89. Guide Pin Bushes of Dies for Die Casting. 8 pp.

—Gliding—Press Tools
BSI BS 7568: Part 20-92. 1992 Press Tools Part 20: Guide Bushes—Form C, Gliding Bushes, Headed, Type 1 (ISO 9448-4: 1991). 9 pp.
BSI BS 7568: Part 22-92. 1992 Press Tools Part 22: Guide Bushes—Form E, Gliding Bushes, Flanged, Type 1 (ISO 9448-6: 1991). 9 pp.
BSI BS 7568: Part 24-92. 1992 Press Tools Part 24: Guide Bushes—Form G, Gliding Bushes, Stepped, Type 1 (ISO 9448-8: 1991). 9 pp.
ISO 9448 Pt 4-91. Tools for Pressing—Guide Bushes—Part 4: Form C, Gliding Bushes, Headed, Type 1 First Edition. 8 pp.
ISO 9448 Pt 6-91. Tools for Pressing—Guide Bushes—Part 6: Form E, Gliding Bushes, Flanged, Type 1 First Edition. 8 pp.
ISO 9448 Pt 8-91. Tools for Pressing—Guide Bushes—Part 8: Form G, Gliding Bushes, Stepped, Type 1 First Edition. 8 pp.

—Molds (Casting)
BSI BS 7564: Part 4-92. 1992 Tools for Moulding Part 4: Mould Guide Bushes, Headed, and Locating Guide Bushes, Headed (ISO 8018: 1985). 10 pp.
CNS B3388-80. Guide Pin Bushings of Molds for Plastics (Oct)(6549).
ISO 8018-85. Mould Guide Bushes, Headed, and Locating Guide Bushes, Headed First Edition. 6 pp.
JIS B 5110-89. Guide Pin Bushings of Moulds for Plastics. 7 pp.

—Press Dies
JIS B 5007-86. Guide Posts and Guide Bushes for Stamping Dies. 13 pp.

—Press Tools
BSI BS 7568: Part 17-92. 1992 Press Tools Part 17: Guide Bushes—Forms (ISO 9448-1: 1991). 6 pp.
ISO 9448 Pt 1-91. Tools for Pressing—Guide Bushes—Part 1: Forms First Edition. 6 pp.
ISO 9448 Pt 2-91. Tools for Pressing—Guide Bushes—Part 2: Form A, Gliding Bushes, Plain, Type 1 First Edition. 7 pp.

—Punches
DIN ENGL 9845-79. Piercing Die Bushes and Punch Guide Bushes (Feb). 5 pp.

Guide Cards
Use For: File Guides
CGSB CAN/CGSB-53.72-M89. Guides, File, Pressboard. 9 pp.
CGSB CAN/CGSB-53.93-92. Guide Card Sets for Filing. 9 pp.

Guide Pins
See Also: Pins

—Dies
CNS B2783-87. Ejector Plate Guide Pin of Die Casting Molds (Sep)(12070).
CNS B3370-80. Guide Pins of Dies for Die Casting (Aug)(5996).
JIS B 5102-89. Guide Pins of Dies for Die Casting. 9 pp.

—Molds (Casting)
CNS B3385-80. Guide Pins of Molds for Plastics (Oct)(6546).
JIS B 5107-89. Guide Pins of Moulds for Plastics. 9 pp.

Guide Rails
Use For: Rails (Guide) *See Also:* Guardrails; Handrails; Mounting Rails

—Elevators (Lifts)
BSI BS 5655: Part 9-85. 1985 Amd 2 Lifts and Service Lifts Part 9: Guide Rails. 22 pp.
ISO 7465-83. Passenger Lifts and Service Lifts—Guide Rails for Lifts and Counterweights—T-Type First Edition. 11 pp.

—Sliding Doors
CNS A2086-83. Sliding Door Rails (Jul)(6537). 3 pp.
JIS A 5509-71. Sliding Door Rails.

Guided Missiles
See Also: Spacecraft; Weapons

—Defense Contracts
MOD UK DEFCON 112CO-86. Basic Set of Conditions of Contract for Surface Weapons—Navy 10/86. 3 pp.
MOD UK DEFCON 112U-86. Basic Set of Conditions of Contract—Guided Weapons 8/86. 2 pp.

—Designations—Warsaw Pact—Codes
NATO STANAG 3236 ED 7 AMD 0-88. Designation System for Warsaw Pact Aircraft and Guided Missiles and Designation of Role Codes of Aircraft. 10 pp.

—Glossaries
BSI BS 185: Sec 6-70. 1970 Amd 1 Glossary of Aeronautical and Astronautical Terms Section 6: Ballistic and Guided Missiles (Propulsion, Launching, Control and Guidance). 8 pp.

—Graphite Oils
MOD UK DSTAN 91-30-77. Lubricating Oil, Colloidal Graphite NATO Code No: 0-218 Joint Service Designation: OX-320 Issue 1 (Withdrawn). 14 pp.

Guided Vehicles, Automatic
Use: Automatic Guided Vehicles

Guides (Buyers)
Use: Buyers Guides

Guides (Mechanical)
Use: Mechanical Guides

Guillotine Shears
See Also: Cutting Tools

—Acceptance Testing
DIN ENGL 55804-79. Machine Tools; Guillotine Shears with Parallel Guided Knife Beam; Acceptance Conditions (Dec). 5 pp.

—Safety
SAA AS 1893-77. Code of Practice for the Guarding and Safe Use of Metal and Paper Cutting Guillotines. 32 pp.

Guillotines (Paper)
Use: Paper Cutters

Guitars
CNS S1087-75. Guitars (Jul)(3863).

—Acoustic
JIS S 8510-84. Acoustic Guitars.

Gulfstream Aerospace G-159 Aircraft
See Also: Aircraft

—Foreign Airworthiness Directives
CAA. Gulfstream Aerospace G-159 Aircraft. 2 pp.

Gulfstream Aerospace GA-7 Aircraft
See Also: Aircraft

—Foreign Airworthiness Directives
CAA. Gulfstream Aerospace GA-7 Aircraft. 1 p.

Gulfstream Aerospace Gulfstream III Aircraft
See Also: Aircraft

—Certification
CAA. Gulfstream—Aerospace Gulfstream III. 14 pp.

Gulfstream Aerospace Gulfstream IV Aircraft
See Also: Aircraft

—Certification
CAA. Grumman Gulfstream IV. 14 pp.

Gulfstream Aerospace 112 Aircraft
See Also: Aircraft

—Foreign Airworthiness Directives
CAA. Gulfstream Aerospace 112 and 114 Series Aircraft. 2 pp.

Gulfstream Aerospace 114 Aircraft
See Also: Aircraft

—Foreign Airworthiness Directives
CAA. Gulfstream Aerospace 112 and 114 Series Aircraft. 2 pp.

Gulfstream Aerospace 685 Aircraft
See Also: Aircraft

—Foreign Airworthiness Directives
CAA. Gulfstream Aerospace 685, 690 and 695 Series Aircraft. 1 p.

Gulfstream Aerospace 690 Aircraft
See Also: Aircraft

—Foreign Airworthiness Directives
CAA. Gulfstream Aerospace 685, 690 and 695 Series Aircraft. 1 p.

Gulfstream Aerospace 695 Aircraft
See Also: Aircraft

—Foreign Airworthiness Directives
CAA. Gulfstream Aerospace 685, 690 and 695 Series Aircraft. 1 p.

Gulfstream Aircraft
See Also: Aircraft

—Data Sheets
CAA FA29 ISSUE 1. Gulfstream. 2 pp.

Gulfstream American AA 1 Aircraft
See Also: Aircraft

—Foreign Airworthiness Directives
CAA. Gulfstream American AA-1 Series Aircraft and Gulfstream Aerospace AA-5 Series Aircraft. 3 pp.

Gulfstream American G 159 Aircraft
See Also: Aircraft

—Foreign Airworthiness Directives
CAA. Gulfstream Aerospace G1159 Series (Foreign Airworthiness Directives). 5 pp.

Gulfstream American GA 7 Aircraft
See Also: Aircraft

—Certification
CAA. Grumman Gulfstream. 12 pp.

Gullies
See Also: Grilles; Soils; Streams

—Buckets—Design
DIN ENGL 1236 Pt 1-81. Class A and Class B Concrete Elements and Buckets for Gullies; Design, Installation and Assemblies (Nov). 3 pp.
DIN ENGL 1236 Pt 3-81. Classes A and B Concrete Elements and Buckets for Gullies; Buckets (Nov). 4 pp.

—Buckets—Installation
DIN ENGL 1236 Pt 1-81. Class A and Class B Concrete Elements and Buckets for Gullies; Design, Installation and Assemblies (Nov). 3 pp.
DIN ENGL 1236 Pt 3-81. Classes A and B Concrete Elements and Buckets for Gullies; Buckets (Nov). 4 pp.

—Concrete Elements
BSI BS 5911: Part 2-82. 1982 Amd 1 Precast Concrete Pipes and Fittings for Drainage and Sewerage Part 2: Inspection Chambers and Street Gullies. 12 pp.

Gullies (Cont.)

—Concrete Elements—Design
DIN ENGL 1236 Pt 1-81. Class A and Class B Concrete Elements and Buckets for Gullies; Design, Installation and Assemblies (Nov). 3 pp.
DIN ENGL 1236 Pt 2-81. Class A and Class B Concrete Elements and Buckets for Gullies; Concrete Elements (Nov). 3 pp.

—Concrete Elements—Installation
DIN ENGL 1236 Pt 1-81. Class A and Class B Concrete Elements and Buckets for Gullies; Design, Installation and Assemblies (Nov). 3 pp.
DIN ENGL 1236 Pt 2-81. Class A and Class B Concrete Elements and Buckets for Gullies; Concrete Elements (Nov). 3 pp.

—Fluid Inlets—Design
BSI BS 6367-83. 1983 Amd 1 Drainage of Roofs and Paved Areas. 58 pp.

—Grilles
CNS A2173-84. Gully Grating (Apr)(10841).

—Grilles—Cast Iron
BSI BS 497: Part 1-76. 1976 Amd 2 Manhole Covers, Road Gully Gratings and Frames for Drainage Purposes Part 1: Cast Iron and Cast Steel (AMD 6643) December 21, 1990. 17 pp.

—Grilles—Cast Steel
BSI BS 497: Part 1-76. 1976 Amd 2 Manhole Covers, Road Gully Gratings and Frames for Drainage Purposes Part 1: Cast Iron and Cast Steel (AMD 6643) December 21, 1990. 17 pp.

—Manhole Covers
CEN EN 124-86. Gully Tops and Manhole Tops for Vehicular and Pedestrian Areas; Design Requirements, Type Testing, Marking. 11 pp.
DIN ENGL 1229-86. Gully Tops and Manhole Tops for Vehicular and Pedestrian Areas; Classification, Design Requirements, Testing, Inspection and Marking (Mar). 4 pp.
DIN ENGL 19599-90. Gullies and Manhole Tops for Use in Buildings Classification, Design and Construction, Testing and Inspection (Nov). 15 pp.

—Manhole Covers—Cast Iron
BSI BS 497: Part 1-76. 1976 Amd 2 Manhole Covers, Road Gully Gratings and Frames for Drainage Purposes Part 1: Cast Iron and Cast Steel (AMD 6643) December 21, 1990. 17 pp.

—Manhole Covers—Design
CEN EN 124-86. Gully Tops and Manhole Tops for Vehicular and Pedestrian Areas; Design Requirements, Type Testing, Marking. 11 pp.
CEN PREN 124-92. Gully Tops and Manhole Tops for Vehicular and Pedestrian Areas—Design Requirements, Type Testing, Marking, Quality Control. 35 pp.
DIN ENGL 1229-86. Gully Tops and Manhole Tops for Vehicular and Pedestrian Areas; Classification, Design Requirements, Testing, Inspection and Marking (Mar). 4 pp.

—Manhole Covers—Identification Systems
CEN EN 124-86. Gully Tops and Manhole Tops for Vehicular and Pedestrian Areas; Design Requirements, Type Testing, Marking. 11 pp.
CEN PREN 124-92. Gully Tops and Manhole Tops for Vehicular and Pedestrian Areas—Design Requirements, Type Testing, Marking, Quality Control. 35 pp.
DIN ENGL 1229-86. Gully Tops and Manhole Tops for Vehicular and Pedestrian Areas; Classification, Design Requirements, Testing, Inspection and Marking (Mar). 4 pp.
DIN ENGL 19599-90. Gullies and Manhole Tops for Use in Buildings Classification, Design and Construction, Testing and Inspection (Nov). 15 pp.

Gum Dammar
Use: Dammar

Gum Testing

—Aircraft Fuels
JIS K 2261-92. Petroleum Products—Motor Gasoline and Aviation Fuels—Determination of Existent Gum—Jet Evaporation Method. 20 pp.

—Aircraft Fuels—Jet Evaporation Method
BSI BS 4348-76. 1976 Determination of Existent Gum in Fuels by Jet Evaporation. 12 pp.
CEN EN 5-74. Determination of Existent Gum in Fuels by Jet Evaporation. 10 pp.
CNS K6321-72. Method of Test for Existing Gum Content in Jet and Motor Fuels (Oct)(3382).
ISO 6246-81. Petroleum Products—Motor Gasoline and Aviation Fuels—Determination of Existent Gum—Jet Evaporation Method First Edition. 10 pp.

Gum Testing (Cont.)

—Gasoline
JIS K 2261-92. Petroleum Products—Motor Gasoline and Aviation Fuels—Determination of Existent Gum—Jet Evaporation Method. 20 pp.

—Gasoline—Jet Evaporation Method
BSI BS 4348-76. 1976 Determination of Existent Gum in Fuels by Jet Evaporation. 12 pp.
CEN EN 5-74. Determination of Existent Gum in Fuels by Jet Evaporation. 10 pp.
CNS K6321-72. Method of Test for Existing Gum Content in Jet and Motor Fuels (Oct)(3382).
ISO 6246-81. Petroleum Products—Motor Gasoline and Aviation Fuels—Determination of Existent Gum—Jet Evaporation Method First Edition. 10 pp.

—Jet Engine Fuels—Jet Evaporation Method
BSI BS 4348-76. 1976 Determination of Existent Gum in Fuels by Jet Evaporation. 12 pp.
CEN EN 5-74. Determination of Existent Gum in Fuels by Jet Evaporation. 10 pp.
CNS K6321-72. Method of Test for Existing Gum Content in Jet and Motor Fuels (Oct)(3382).
ISO 6246-81. Petroleum Products—Motor Gasoline and Aviation Fuels—Determination of Existent Gum—Jet Evaporation Method First Edition. 10 pp.

—Volatile Fluids—Jet Evaporation Method
CEN EN 5-74. Determination of Existent Gum in Fuels by Jet Evaporation. 10 pp.

Gummed Concrete
Use: Shotcrete

Gummed Papers

—Kraft—Tapes
MOD UK TS 447. Tape, Kraft Paper, Gummed, Non-Corrosive (Withdrawn).

Gums
See Also: Arabic Gum

—Stamps
CNS S1104-80. Stamp Gums (Dec)(6854).

Gun Barrels
See Also: Guns

—Aircraft—Chambers—30mm
NATO STANAG 3231 ED 2 AMD 6-66. Aden and Defa 30 mm Gun Barrel Chambers for Aircraft. 10 pp.
NATO STANAG 3232 ED 2 AMD 6-66. Aden and Defa 30 mm Gun Barrel Chambers for Aircraft. 10 pp.

—Blanket Insulation—Coatings
MOD UK TS 10258. Coating for Thermal Insulating Blankets.

—Chambers
MOD UK DSTAN 07-18-01. Barrel Chamber, 30 mm ADEN Gun Issue 1; Amendment 1. 6 pp.

Gun Components
Scope Note: Use a more specific term *See:* Breech Mechanisms; Gun Barrels

Gun Propellants
See Also: Guns; Propellants

—Ballistics
NATO STANAG 4115 ED 1 AMD 0-70. Definition and Determination of Ballistic Properties of Gun Propellants. 16 pp.

Gunite
Scope Note: Gunite is a formerly registered trademark *Use:* Shotcrete

Gunmetal
See Also: Alloys; Copper Tin Alloys; Metals

—Ingots—Marine
MOD UK NES 830: Part 1-01. Requirements for Gunmetal Ingots and Castings Part 1: Gunmetal Sand Castings and Ingots Class III Castings Only Issue 1 (11.88); Amendment 1. 21 pp.

Gunmetal Castings
See Also: Bronze Castings; Castings; Tin Bronze Castings
DIN ENGL 1705-81. Copper-Tin and Copper-Tin-Zinc Casting Alloys; (Cast Tin Bronze and Gunmetal); Castings (Nov). 8 pp.

—Chemical Analysis
DIN ENGL 1705-81. Copper-Tin and Copper-Tin-Zinc Casting Alloys; (Cast Tin Bronze and Gunmetal); Castings (Nov). 8 pp.

Gunmetal Castings (Cont.)

—Marine
MOD UK NES 830: Part 2-87. Requirements for Gunmetal Ingots and Castings Part 2: Continuously Cast Issue 1 (11.87). 13 pp.

—Mechanical Properties
DIN ENGL 1705-81. Copper-Tin and Copper-Tin-Zinc Casting Alloys; (Cast Tin Bronze and Gunmetal); Castings (Nov). 8 pp.
DIN ENGL 1705 Suppl. 1-81. Copper-Tin and Copper-Tin-Zinc Casting Alloys; (Cast Tin Bronze and Gunmetal); Castings; Reference Data on Mechanical and Physical Properties (Nov). 3 pp.

—Physical Properties
DIN ENGL 1705 Suppl. 1-81. Copper-Tin and Copper-Tin-Zinc Casting Alloys; (Cast Tin Bronze and Gunmetal); Castings; Reference Data on Mechanical and Physical Properties (Nov). 3 pp.

—Sand—Marine
MOD UK NES 830: Part 1-01. Requirements for Gunmetal Ingots and Castings Part 1: Gunmetal Sand Castings and Ingots Class III Castings Only Issue 1 (11.88); Amendment 1. 21 pp.

Gunn Diodes
See Also: Diodes; Microwave Diodes; Semiconductor Devices
MOD UK DSTAN 59-61: 90/050-72. Semiconductor Device, Gunn Effect Issue 2. 9 pp.
MOD UK DSTAN 59-61: 90/130-74. Semiconductor Device R.F. Source (Gunn Effect) Issue 1. 17 pp.
MOD UK DSTAN 59-61: 90/184-77. Semiconductor Device, Diode Issue 1. 22 pp.
MOD UK DSTAN 59-61: 90/196-82. Semiconductor Devices Issue 1. 23 pp.
MOD UK DSTAN 59-61: 90/206-82. Semiconductor Devices Issue 1. 22 pp.
MOD UK DSTAN 59-61: 90/222-01. Semiconductor Device Microwave Source Issue 1; Amendment 1. 15 pp.

—Microwave—Preferred Products List
CECC CECC MUAHAG Vol 9 IS 3-90. Preferred Products List; Semiconductors (En, Fr, Ge) AMD 1 (En, Fr, Ge). 51 pp.

Gunn Oscillators
See Also: Diodes; Microwave Oscillators; Semiconductor Devices
BSI BS 6493: Sec 1.4-92. 1992 Semiconductor Devices Part 1: Discrete Devices Section 1.4: Recommendations for Microwave Diodes and Transistors (IEC 747-4: 1991). 83 pp.
IEC 747 Pt 4-91. Semiconductor Devices Discrete Devices Part 4: Microwave Diodes and Transistors First Edition. 163 pp.

—Quality Assurance
BSI BS 9325-77. 1977 Rules for the Preparation of Detail Specifications for Semiconductor Devices of Assessed Quality: Microwave Gunn Oscillators (C.W. Operation). 8 pp.
BSI BS 9326-77. (WITHDRAWN) 1977 Rules for the Preparation of Detail Specifications for Semiconductor Devices of Assessed Quality: Microwave Gunn Oscillators (Pulse Operation). 9 pp.

Gunny Cloth
Use: Burlaps

Gunny Sacks
See Also: Bags; Burlaps
JIS Z 1513-76. Gunny Bags.

Gunpowder
See Also: Explosives
MOD UK DSTAN 13-134-93. Composition SR 239 Issue 1. 12 pp.
MOD UK M 602/92. Examination of Gunpowder and Gunpowder, Sulfurless.

—Shotguns
JIS K 4817-86. Shot Gun Powders.

Guns
Use For: Cannons *See Also:* Armor Piercing Ammunition; Artillery; Cartridges (Explosives); Explosives; Gun Barrels; Gun Propellants; Gunpowder; Muzzles (Ordnance); Obturators (Ordnance); Rifles; Shotguns; Weapons

—Aircraft—Ammunition
NATO STANAG 3820 ED 1 AMD 2-88. 27 mm X 145 Ammunition and Links for Aircraft Guns. 16 pp.

—Cabinets (Furniture)
BSI BS 7558-92. 1992 Gun Cabinets. 11 pp.

Guns (Cont.)
—Classification
BSI BS 1000: (682/683)-73. 1973 Universal Decimal Classification (UDC). English Full Edition (682/683): Smithery. Farriery. Hand-Forged Ironwork. Ironmongery. Hardware. Locksmithing. Gunsmithing. Bottling. Lamps. Stoves. 27 pp.
SNZ NZS/BS 1000 (682/683)-73. Universal Decimal Classification Smithery. Farriery. Hand-Forged Ironwork. Ironmongery. Hardware. Locksmithing. Gunsmithing. Bottling. Lamps. Stoves. 28 pp.

—Design—Ballistics
NATO STANAG 4110 ED 2 AMD 0-91. Definition of Pressure Terms and Their Inter-Relationship for Use in the Design and Proof of Cannons and Ammunition. 18 pp.

—Firing Table Format
MOD UK DSTAN 12-2-82. Standard Cannon Artillery Firing Table Format Issue 1. 4 pp.
NATO STANAG 4119 ED 1 AMD 2-72. Adoption of a Standard Cannon Artillery Firing Table Format. 41 pp.

—Firing Table Format—Ballistics
NATO STANAG 4144 ED 1 AMD 1-84. Dynamic Firing Techniques to Determine Ballistic Data for Cannon Artillery Firing Tables and Associated Fire Control Equipment. 13 pp.

—Spring Steels
MOD UK DEF-103-60. Steel for Hardened and Tempered High Quality Coil Springs for Guns and Armoured Fighting Vehicles (Reprinted December 1971). 4 pp.
MOD UK DEF-104-60. Hardened and Tempered Coil Springs of High Quality for Guns and Armoured Fighting Vehicles (Reprinted May 1973, Incorporating Amendment No. 1). 4 pp.

Guns (Welding)
Use: Torches

Gurley Stiffness Testers
Use: Stiffness Testers—Gurley

Gurley Testers
Use For: Densometers (Gurley) *See Also:* Testing Equipment
CPPA M-1-70. Gurley Densometer—Model 4110. 1 p.
CPPA M-4-80. Gurley S-P-S Tester. 1 p.

Gurneys
Use For: Patient Trolleys *See Also:* Medical Equipment; Stretchers

—Safety
BSI BS 5402-76. 1976 Amd 1 Patient Trolleys. 14 pp.

Gust Loads
See Also: Aerodynamic Loads

—Aerostats
CAA Chapter Q3-1 12.79. Structures—General (Non-Rigid Airships). 3 pp.
CAA Chapter Q3-3 12.79. Gust Loads (Non-Rigid Airships). 2 pp.

—Aircraft
CAA Chapter K3-3 10.69. Gust Loads (Light Aeroplanes). 4 pp.
CAA Chapter P3-3. Gust Loads (Provisional Airworthiness Requirements for Civil Powered Lift Aircraft). 11 pp.
CAA STANDARD NO. 4-3 08.69. Design Airspeeds, Gust Cases and Flight Manoeuvring Loads. 13 pp.

—Rotary Wing Aircraft
CAA Chapter G3-2 01.75. Flight Loads (Rotocraft). 3 pp.
CAA CAP 524 SUB-Part C 12.86. Structure (Rotorcraft). 11 pp.

Gutta Percha
—Weapons
MOD UK CS 3052. Gutta Percha (Reinstated).

Gutter Fittings
Use For: Rain Gutter Fittings

—Aluminum
BSI BS 2997-58. 1958 Amd 1 Aluminium Rainwater Goods. 44 pp.

—Asbestos Cement
BSI BS 569-73. 1973 Amd 1 Asbestos-Cement Rainwater Goods (AMD 5755) October 31, 1988. 35 pp.

—Copper
BSI BS 1431-60. 1960 Wrought Copper and Wrought Zinc Rainwater Goods. 25 pp.

Gutter Fittings (Cont.)
—PVC
BSI BS 4576: Part 1-89. 1989 Amd 1 Specification for Unplasticized Polyvinyl Chloride (PVC-U) Rainwater Goods and Accessories Part 1: Half-Round Gutters and Pipes of Circular Cross Section (AMD 6350) June 28, 1991. 17 pp.
BSI BS 4576: Part 1-70. (WITHDRAWN) 1970 Amd 3 Unplasticized PVC Rainwater Goods Part 1: Half-Round Gutters and Circular Pipe. 17 pp.
CEN PREN 607-91. Eaves Gutters from PVC-U—Definitions, Requirements and Testing. 11 pp.

—Steel
BSI BS 1091-63. 1963 Pressed Steel Gutters, Rainwater Pipes, Fittings and Accessories. 20 pp.

—Zinc
BSI BS 1431-60. 1960 Wrought Copper and Wrought Zinc Rainwater Goods. 25 pp.

Gutters
Use For: Rain Gutters *See Also:* Conduits (Channels); Downspouts; Drains
DIN ENGL 19580-88. Surface Water Drainage Channels for Traffic Areas; Classification, Design, Marking, Testing and Inspection (Dec). 6 pp.

—Aluminum
BSI BS 2997-58. 1958 Amd 1 Aluminium Rainwater Goods. 44 pp.

—Asbestos Cement
BSI BS 569-73. 1973 Amd 1 Asbestos-Cement Rainwater Goods (AMD 5755) October 31, 1988. 35 pp.

—Cast Iron
BSI BS 460-64. 1964 Cast Iron Rainwater Goods. 29 pp.
SNZ NZS 773-67. Specification for Cast Iron Rainwater Goods. 32 pp.

—Concrete
SAA AS A175-70. Concrete Kerbs and Channels (Gutters) Corrig.. 15 pp.

—Covers—Reinforced Concrete
CNS A2056-86. Reinforced Concrete Covers for U-Type Ditches (Mar)(4063). 3 pp.
JIS A 5334-87. Covers for Reinforced Concrete Gutters. 9 pp.

—Covers—Reinforced Concrete—Bend Testing
CNS A3069-86. Method of Test for Reinforced Concrete Covers for Ditches (U Type) (Mar)(4064). 2 pp.

—Covers—Reinforced Concrete—Water Absorption
CNS A3068-86. Method of Test for Reinforced Concrete Ditches (U Type) (Mar)(4062). 2 pp.
CNS A3096-87. Method of Test for Internal Vibrators for Concrete (Mar)(5647). 1 p.

—Covers—Roadside—Reinforced Concrete
CNS A2185-85. Reinforced Resin Concrete Gutter Covers for Roadside (Aug)(11323).
CNS A3230-85. Method of Test for Reinforced Resin Concrete Gutter Covers for Roadside (Aug)(11324).

—Curbs—Concrete
CNS A2057-84. L-Type Concrete and Reinforced Concrete Curb-Gutters (Sep)(4065). 4 pp.
CNS A3070-84. Method of Test for Concrete and Reinforced Concrete Curb-Gutters (L Type) (Sep)(4066).
JIS A 5306-88. Concrete Curb-Gutters and Reinforced Concrete Curb-Gutters. 10 pp.

—Curbs—Reinforced Concrete
CNS A2057-84. L-Type Concrete and Reinforced Concrete Curb-Gutters (Sep)(4065). 4 pp.
CNS A3070-84. Method of Test for Concrete and Reinforced Concrete Curb-Gutters (L Type) (Sep)(4066).
JIS A 5306-88. Concrete Curb-Gutters and Reinforced Concrete Curb-Gutters. 10 pp.

—Design
BSI BS 6367-83. 1983 Amd 1 Drainage of Roofs and Paved Areas. 58 pp.

—Eaves Troughs—Copper
BSI BS 1431-60. 1960 Wrought Copper and Wrought Zinc Rainwater Goods. 25 pp.

—Eaves Troughs—Design
DIN ENGL 18460-89. External Rainwater Pipes and Eaves Gutters; Concepts and Design Principles (May). 3 pp.

Gutters (Cont.)
—Eaves Troughs—Metal
CEN PREN 612-91. Eaves Gutters and Rainwater Down-Pipes from Metal Sheet—Definitions, Classifications, Requirements and Testing. 13 pp.

—Eaves Troughs—PVC
CEN PREN 607-91. Eaves Gutters from PVC-U—Definitions, Requirements and Testing. 11 pp.
JIS A 5706-90. Unplasticized Polyvinyl Chloride Eaves Gutters and Downspouts. 11 pp.

—Eaves Troughs—PVC—Unplasticized
BSI BS 4576: Part 1-89. 1989 Amd 1 Specification for Unplasticized Polyvinyl Chloride (PVC-U) Rainwater Goods and Accessories Part 1: Half-Round Gutters and Pipes of Circular Cross Section (AMD 6350) June 28, 1991. 17 pp.
BSI BS 4576: Part 1-70. (WITHDRAWN) 1970 Amd 3 Unplasticized PVC Rainwater Goods Part 1: Half-Round Gutters and Circular Pipe. 17 pp.
DIN ENGL 18469-88. Unplasticized Polyvinyl Chloride Eaves Gutters; Requirements, Testing (May). 3 pp.

—Eaves Troughs—Zinc
BSI BS 1431-60. 1960 Wrought Copper and Wrought Zinc Rainwater Goods. 25 pp.

—PVC—Elongation
BSI BS 2782:Pt11: METH 1110-89. 1989 Methods of Testing Plastics Part 11: Thermoplastics Pipes, Fittings and Valves Method 1110: Tensile Properties of Dumb-Bell Specimens from PVC Gutter Profiles or Pipes for Non-Pressure Applications. 5 pp.

—PVC—Tensile Testing
BSI BS 2782:Pt11: METH 1110-89. 1989 Methods of Testing Plastics Part 11: Thermoplastics Pipes, Fittings and Valves Method 1110: Tensile Properties of Dumb-Bell Specimens from PVC Gutter Profiles or Pipes for Non-Pressure Applications. 5 pp.

—Reinforced Concrete
CNS A2055-86. U-Type Reinforced Concrete Ditches (Mar)(4061). 3 pp.
JIS A 5305-87. Reinforced Concrete Gutters. 10 pp.

—Reinforced Concrete—Bend Testing
CNS A3068-86. Method of Test for Reinforced Concrete Ditches (U Type) (Mar)(4062). 2 pp.

—Roadside—Reinforced Concrete
JIS A 5345-92. Reinforced Concrete Gutters for Roadside. 16 pp.

—Steel
BSI BS 1091-63. 1963 Pressed Steel Gutters, Rainwater Pipes, Fittings and Accessories. 20 pp.

Gymnasiums
See Also: Sports Facilities

—Equipment
SAA AS Z12-62. Gymnasium Equipment. 36 pp.

—Floors—Steel Furrings
CNS A2210-87. Steel Furring Components for Gymnasium Floors (Sep)(12053).
CNS A3263-87. Method of Test for Steel Furring Components for Gymnasium Floors (Sep)(12054).
JIS A 6519-89. Steel Furring Components for Gymnasium Floors. 24 pp.

Gymnastic Equipment
See Also: Climbing Ropes; Playground Equipment
BSI BS 1892: Part 1-86. 1986 Gymnasium Equipment Part 1: General Requirements. 8 pp.

—Balancing Beams
CNS S1193-85. Balancing Beams (Dec)(11469).
CNS S2125-85. Method of Test for Balancing Beams (Dec)(11470).
JIS S 7015-83. Balancing Beams.

—Horizontal Bars
CNS S1190-85. Horizontal Bars (Dec)(11464).
CNS S2123-85. Method of Test for Horizontal Bars (Dec)(11465).
ISO 379-80. Gymnastic Equipment—Horizontal Bar First Edition. 4 pp.
JIS S 7011-83. Horizontal Bars.

—Mats
BSI BS 1892: Sec 2.3-86. 1986 Gymnasium Equipment Part 2: Particular Requirements Section 2.3: Boxes, Bucks and Horses. 16 pp.
CNS S1223-90. Gymnastic Equipment-Landing Mats (Dec)(12829). 2 pp.
ISO 5903-81. Gymnastic Equipment—Landing Mats and Surfaces for Floor Exercises—Determination of Hardness and Impact Damping First Edition. 4 pp.
ISO 5904-81. Gymnastic Equipment—Landing Mats and Surfaces for Floor Exercises—Determination of Resistance to Slipping First Edition. 5 pp.

Gymnastic — INTERNATIONAL AND NON-U.S. NATIONAL STANDARDS SUBJECT INDEX

Gymnastic Equipment (Cont.)
—Mats (Cont.)
ISO 5905-80. Gymnastic Equipment—Landing Mats (2 000 mm X 1 250 mm X 60 mm) First Edition. 4 pp.
ISO 5906-80. Gymnastic Equipment—Surfaces for Floor Exercises—Mats First Edition. 4 pp.
ISO 5907-80. Gymnastic Equipment—Surfaces for Floor Exercises—Boards First Edition. 4 pp.
JIS S 7021-84. Gymnastic Equipment—Landing Mats.
JIS S 7022-84. Gymnastic Equipment—Surfaces for Floor Exercise-Boards.

—Mats—Safety
BSI BS 1892: Sec 2.10-90. 1990 Gymnasium Equipment Part 2: Particular Requirements Section 2.10: Safety Requirements for Mats, Mattresses and Landing Areas. 27 pp.

—Parallel Bars
CNS S1191-85. Parallel Bars (Dec)(11466).
CNS S2124-85. Method of Test for Parallel Bars (Dec)(11467).
ISO 378-80. Gymnastic Equipment—Parallel Bars First Edition. 5 pp.

—Parallel Bars—Safety
CEN PREN 914-92. Gymnastic Equipment—Parallel Bars and Combination Asymmetric/Parallel Bars. 11 pp.

—Parallel Bars—Uneven
CNS S1194-85. Uneven Parallel Bars (Dec)(11471).
CNS S2126-85. Method of Test for Uneven Parallel Bars (Dec)(11472).

—Schools—Safety
CEN PREN 913-92. Gymnastic Equipment—General Safety Requirements and Test Methods. 11 pp.

—Trampolines
BSI BS 1892: Sec 2.8-86. 1986 Gymnasium Equipment Part 2: Particular Requirements Section 2.8: Trampolines, Mini-Trampolines and Safety Harnesses (Trampoline Spotting Rig). 10 pp.

—Uneven Bars—Safety
CEN PREN 914-92. Gymnastic Equipment—Parallel Bars and Combination Asymmetric/Parallel Bars. 11 pp.
CEN PREN 915-92. Gymnastic Equipment—Asymmetric Bars. 6 pp.

—Vaulting Boxes
BSI BS 1892: Sec 2.3-86. 1986 Gymnasium Equipment Part 2: Particular Requirements Section 2.3: Boxes, Bucks and Horses. 16 pp.

—Vaulting Boxes—Safety
CEN PREN 916-92. Gymnastic Equipment—Vaulting Boxes. 8 pp.

—Vaulting Horses
BSI BS 1892: Sec 2.3-86. 1986 Gymnasium Equipment Part 2: Particular Requirements Section 2.3: Boxes, Bucks and Horses. 16 pp.
CNS S1192-85. Vaulting Horses (Dec)(11468).

—Vaulting Horses—Pommelled
CNS S1195-85. Pommeled Horses (Dec)(11473).

Gymnastic Rings
See Also: Rings
CNS S2108-85. Method of Test for Hanging Rings (Dec)(10514).

Gypsum
See Also: Alabaster; Gypsum Plaster; Minerals; Ores
CNS K1029-70. Gypsum (Jan)(379).
CNS K6284-70. Method of Test for Gypsum (Jan)(3081).
CSA CAN/CSA-A82.21-M91. Gypsum and Terms Relating to Gypsum Products; (Gen Instr 1). 17 pp.

—Binders (Materials)
ISO 1587-75. Gypsum Rock for the Manufacture of Binders—Specifications First Edition. 5 pp.

—Chemical Analysis
CNS R3083-79. Method of Chemical Analysis for Gypsum and Gypsum Products (Dec) (5080).
CSA CAN/CSA-A82. 20Series-M91. Methods of Testing Gypsum and Gypsum Products; (Includes a82.20.1, A82.20.2, A82.20.3) (Gen Instr 1). 54 pp.
JIS R 9101-86. Chemical Analysis of Gypsum.
JIS R 9101-65. Chemical Analysis of Gypsum (R 1971). 18 pp.

—Dental Materials
BSI BS 7013-89. 1989 Dental Gypsum Products (ISO 6873: 1983). 14 pp.
BSI BS 7013-01. 1989 Amd 1 Dental Gypsum Products (ISO 6873: 1983) (AMD 7038) May 1, 1992. 17 pp.

Gypsum (Cont.)
—Dental Materials (Cont.)
CEN EN 26873-91. Dentistry—Dental Gypsum Products. 3 pp.
ISO 6873-83. Dental Gypsum Products First Edition. 10 pp.

—Glossaries
JIS R 9200-86. Glossary of Terms Related to Gypsum, Lime and Magnesia Cement. 32 pp.

—Portland Cements—Retardants
JIS R 9151-79. Gypsum for Portland Cement Retarder.

Gypsum Cements
Use: Gypsum Plaster

Gypsum Laths
See Also: Lathing
CNS A2063-90. Gypsum Lath Boards (Oct)(4643).
CNS A3075-90. Method of Test for Gypsum Lath Boards (Oct)(4644).
JIS A 6906-83. Gypsum Lath Boards. 9 pp.

Gypsum Plaster
Use For: Building Plaster; Gypsum Cements; Plaster of Paris See Also: Gypsum; Plaster
BSI BS 1191: Part 1-73. 1973 Gypsum Building Plasters Part 1: Excluding Premixed Lightweight Plasters. 17 pp.
CNS A2169-83. Gypsum Plaster (Nov)(10640).
CSA A82.22-M1977. Gypsum Plasters (R 1992). 12 pp.
CSA A82.30-M1980. Interior Furring, Lathing, and Gypsum Plastering (R 1992); (Gen Instr 1). 24 pp.
DIN ENGL 1168 Pt 1-86. Building Plasters; Term and Definition, Types and Application, Delivery and Marking (Jan). 3 pp.
ISO 3048-74. Gypsum Plasters—General Test Conditions First Edition. 3 pp.
JIS A 6904-76. Gypsum Plaster.
SNZ NZS/AS 2592-83. Gypsum Plaster for Building Purposes (Reconfirmed 1991) (This is a Joint Standard with SAA AS 2592:1983). 10 pp.

—Comprehensive Testing
DIN ENGL 1168 Pt 2-75. Building Plasters; Requirements, Testing, Control (July). 9 pp.

—Content Analysis
MOD UK M 9519/66. Examination of Plaster of Paris (Withdrawn).

—Ignition Loss
MOD UK M 9519/66. Examination of Plaster of Paris (Withdrawn).

—Mechanical Properties
ISO 3051-74. Gypsum Plasters—Determination of Mechanical Properties First Edition. 6 pp.

—Molds (Casting)
CNS R2099-79. Gypsum Casting and Molding Plaster (Dec) (5079).

—Molds (Casting)—Pottery
CNS R2100-85. Plaster of Paris for Pottery Mold (May)(5081).
CNS R3084-85. Method of Physical Test for Plaster of Paris for Pottery Mold (May) (5082).
JIS R 9111-77. Plaster of Paris Mold for Pottery (R 1985). 6 pp.
JIS R 9112-78. Method of Physical Test for Plaster of Paris Mold for Pottery (R 1986). 6 pp.

—Physical Properties
ISO 3049-74. Gypsum Plasters—Determination of Physical Properties of Powder First Edition. 5 pp.

—Physical Testing
CNS R3084-85. Method of Physical Test for Plaster of Paris for Pottery Mold (May) (5082).
JIS R 9112-78. Method of Physical Test for Plaster of Paris Mold for Pottery (R 1986). 6 pp.

—Polyvinyl Acetate Adhesives
BSI BS 5270: Part 1-89. 1989 Bonding Agents for Use with Gypsum Plasters and Cement Part 1: Polyvinyl Acetate (PVAC) Emulsion Bonding Agents for Indoor Use with Gypsum Building Plasters. 12 pp.

—Premixed
BSI BS 1191: Part 2-73. 1973 Gypsum Building Plasters Part 2: Premixed Lightweight Plasters. 11 pp.

—Setting Time
MOD UK M 9519/66. Examination of Plaster of Paris (Withdrawn).

Gypsum Plaster (Cont.)
—Sieve Analysis
MOD UK M 9519/66. Examination of Plaster of Paris (Withdrawn).

—Thermal Insulation
BSI BS 3958: Part 6-72. 1972 Amd 1 Thermal Insulating Materials Part 6: Finishing Materials; Hard Setting Composition, Self-Setting Cement and Gypsum Plaster. 19 pp.

—Water of Crystallization Content—Gravimetric Analysis
ISO 3052-74. Gypsum Plasters—Determination of Water of Crystallization Content First Edition. 4 pp.

Gypsum Plasterboard
Use: Gypsum Wallboard

Gypsum Wallboard
Use For: Drywall (Gypsum); Gypsum Plasterboard; Sheetrock See Also: Construction Materials; Plasterboard; Wallboard
BSI BS 1230: Part 1-85. 1985 Gypsum Plasterboard Part 1: Plasterboard Excluding Materials Submitted to Secondary Operations. 14 pp.
CNS A2061-90. Gypsum Boards (Oct)(4458). 4 pp.
CNS A3073-90. Method of Test for Gypsum Boards (Oct)(4459). 4 pp.
CSA CAN/CSA-A82.27-M91. Gypsum Board; (Gen Instr 1). 21 pp.
CSA CAN/CSA-A82.31-M91. Gypsum Board Application; (Gen Instr 1). 47 pp.
DIN ENGL 18180-89. Gypsum Plasterboard; Types, Requirements and Testing (Sept). 6 pp.
DIN ENGL 18181-90. Dry Lining and Partitioning Using Gypsum Plasterboard (Sept). 6 pp.
ISO 6308-80. Gypsum Plasterboard—Specification First Edition. 10 pp.
JIS A 6901-83. Gypsum Boards. 12 pp.
SNZ ISO 6308-80. Gypsum Plasterboard. 8 pp.

—Acoustic Insulation
CNS A2070-90. Perforated Gypsum Boards for Acoustic Use (Oct)(4965).
CNS A3081-90. Method of Test for Perforated Gypsum Boards for Acoustic Use (Oct)(4966).
JIS A 6301-84. Perforated Gypsum Boards for Acoustic Use. 17 pp.

—Decorative
CNS A2062-85. Decorated Gypsum Board (May)(4460). 5 pp.
CNS A3074-90. Method of Test for Decorated Gypsum Boards (Oct)(4461).
JIS A 6911-83. Decorated Gypsum Boards. 13 pp.

—Fiber Reinforced
CNS A2178-90. Inorganic-Fiber Reinforced Gypsum Boards (Oct)(10996).
CNS A3214-90. Method of Test for Inorganic-Fiber Reinforced Gypsum Boards (Oct)(10997).
SNZ NZS 4221-72. Specification for Fibrous Plaster Sheet (Reconfirmed 1985). 16 pp.

—Joint Treatment Materials
CNS A2209-87. Joint Treatment Materials for Gypsum Boards (Jun)(11990).
JIS A 6914-85. Joint Treatment Materials for Gypsum Boards. 9 pp.

—Linings
BSI BS 8212-88. 1988 Amd 1 Code of Practice for Dry Lining and Partitioning Using Gypsum Plasterboard (AMD 6245) June 30, 1989. 46 pp.

—Partitions
BSI BS 8212-88. 1988 Amd 1 Code of Practice for Dry Lining and Partitioning Using Gypsum Plasterboard (AMD 6245) June 30, 1989. 46 pp.
DIN ENGL 4103 Pt 2-85. Internal Non-Loadbearing Partitions; Gypsum Wallboard Partitions (Dec). 4 pp.
DIN ENGL 18183-88. Prefabricated Gypsum Plasterboard Metal Stud Partitions (Nov). 10 pp.

—Prefabricated
BSI BS 4022-70. 1970 Amd 1 Prefabricated Gypsum Wallboard Panels. 13 pp.
DIN ENGL 18183-88. Prefabricated Gypsum Plasterboard Metal Stud Partitions (Nov). 10 pp.

—Sandwich Panels
DIN ENGL 18184-91. Gypsum Plasterboard Composites with Polystyrene or Polyurethane Rigid Foam Insulation (June). 3 pp.

—Steel Siding
CNS A2224-88. Steel Sidings-Backing with Gypsum Boards (Jun)(12332).
CNS A3277-88. Method of Test for Steel Sidings Backing with Gypsum Boards (Jun)(12333).
JIS A 6711-79. Steel Sidings-Backing by Gypsum Boards.

INTERNATIONAL AND NON-U.S. NATIONAL STANDARDS
SUBJECT INDEX

Gyrocompasses
Use For: Directional Gyroscopes *See Also:* Air Navigation Equipment; Aircraft Instruments; Compasses (Indicating Instruments); Navigational Aids

—**Aircraft**
BSI 2G 112-56. 1956 Air-Driven Directional Gyroscopes for Aircraft. 5 pp.

—**Aircraft—Remote Reading**
BSI G 122-49. 1949 Aircraft Magnetic Compass, Remote Reading (Gyro-Stablized). 3 pp.

—**Ships**
BSI BS 7366-90. 1990 Marine Gyro-Compasses. 12 pp.
ISO 8728-87. Shipbuilding—Marine Gyro-Compasses First Edition. 8 pp.
JIS F 9602-92. Marine Gyro—Compasses.

Gyroflug SC 01 Aircraft
See Also: Aircraft

—**Foreign Airworthiness Directives**
CAA. Gyroflug SC01 Series Aircraft (Foreign Airworthiness Directives). 1 p.

Gyroflug SC 01B 160 Aircraft
See Also: Aircraft

—**Antenna Positions**
CAA. Gyroflug SCO1B 160. 1 p.

Gyrohorizons
Use For: Gyroscopic Bank and Pitch Indicators
See Also: Aircraft Instruments; Attitude Indicators; Indicating Instruments
CAA 731 (K). Section K Light Aeroplanes Gyroscopic Rate-of-Turn Indicators (Blue Papers). 1 p.

—**Emergency and Standby Power Supplies**
CAA NOTICE #81 ISSUE 1. Emergency Power Supply for Electrically Operated Gyroscopic Bank and Pitch Indicators (Artificial Horizons) (Airworthiness Notices). 5 pp.

Gyromagnetic Materials
See Also: Ferrites; Magnetic Materials

—**Microwave Equipment**
IEC 556-82. Measuring Methods for Properties of Gyromagnetic Materials Intended for Application at Microwave Frequencies First Edition. 106 pp.

Gyroplanes
See Also: Aircraft

—**Mandatory Aircraft Modifications and Inspections Summaries**
CAA. Gyroplanes. 1 p.

Gyroscopic Bank and Pitch Indicators
Use: Gyrohorizons

Gyroscopic Compasses
Use: Gyrocompasses

Gyrotrons
See Also: Electron Tubes; Microwave Tubes; Oscillators

—**Preferred Products List**
CECC CECC MUAHAG Vol 12 IS 1-90. Preferred Products List; Microwave Components (En, Fr, Ge). 76 pp.

H Acid

—**Monosodium Salt**
CNS K2228-88. H-Acid (Monosodium Salt) (Jun)(12345). 3 pp.
JIS K 4134-90. 4-Amino-5-Hydroxy-2,7-Naphthalenedisulfonic Acid Monosodium Salt (H-Acid Monosodium Salt).

H Beams
See Also: Beams (Supports); Structural Members

—**Steel—Light Gauge**
CNS G2064-87. Method of Test for Welded Light Gauge Steels H Sections for General Structure (Sep)(6186). 3 pp.
JIS G 3353-90. Welded Light Gauge H Steels for General Structures. 15 pp.

H.I.D. Lights
Use: High Intensity Discharge Lamps

H Point

—**Seats—Automotive**
JIS D 0024-85. Road Vehicles-Procedure for H-Point Determination.

—**Seats—Automotive—Testing Equipment**
ISO 6549-80. Road Vehicles—Procedure for H-Point Determination First Edition. 8 pp.

Hacksaw Blades
See Also: Hacksaw Frames; Hacksaws; Saw Blades
BSI BS 1919-83. 1983 Amd 2 Hacksaw Blades. 22 pp.
BSI BS 6271-82. 1982 Miniature Hacksaw Blades. 7 pp.
CGSB 46-GP-1A-84. Blades, Hacksaw, Standard for. 21 pp.
CNS B3186-86. Saw Blades for Hand Hack Saws (Sep)(1433). 5 pp.
CNS B3422-86. Flexable Saw Blades for Hand Hack Saw (Sep)(8102).
CNS B3435-86. Saw Blades for Power Hack Saws (Sep)(10236).
ISO 2336-80. Hand and Machine Hacksaw Blades—Dimensions for Lengths up to 450 mm and Pitches up to 6,3 mm Second Edition. 5 pp.
JIS B 4751-72. Hand Hacksaw Blades.
JIS B 4752-72. Machine Hacksaw Blade.
MOD UK DSTAN 51-11: Part 7-75. Hand Tools, General Purpose Part 7: Frames, Hand Hacksaw and Blades, Hand Hacksaw Issue 1. 27 pp.
SAA AS 1912-76. Hacksaw Blades (Reconfirmed 1987). 16 pp.

Hacksaw Frames
See Also: Hacksaw Blades; Hacksaws; Saws
BSI BS 7398-91. 1991 Hand Hacksaw Frames. 13 pp.
CNS B3241-73. Hack Saw Frames (for Hand Saw, Jewelry and Bow Form Saw) (Nov)(3659).

Hacksaws
See Also: Hacksaw Blades; Hacksaw Frames; Hand Tools; Saws; Tools

—**Accuracy Testing**
CNS B7115-82. Accuracy Inspection of Hack Sawing Machines (Dec)(4895). 3 pp.

Haematite
Use: Ferric Oxide

Haemoconcentrators
Use: Hemoconcentrators

Haemocytometers
Use: Hemocytometers

Haemodialysers
Use: Hemodialyzers

Haemodialysis Equipment
Use: Hemodialyzers

Haemofilters
Use: Hemofiltration Units

Haemoglobin Cyanide
Use: Cyanmethemoglobin

Haemolysis

—**Medical Equipment—Biological Hazards**
BSI BS 5736: Part 11-90. 1990 Evaluation of Medical Devices for Biological Hazards Part 11: Method of Test for Haemolysis. 10 pp.

Hafnium Content Analysis

—**Uranium Hexafluoride—Spectrography**
CNS J2072-82. Method for Determining Hafnium, Molybdenum, Niobium (Columbium), Tantalum, Titanium, Tungsten and Zirconium After Seperation from UF6 with BPHA Spectrographic Method (Nov)(9615).
CNS J2073-82. Method for Determining Hafnium, Molybdenum, Niobium (Columbium), Tantalum, Titanium, Tungsten, Vanadium and Zirconium After Seperation from UF6 as Cupferrides Spectrographic Method (Nov)(9616).

—**Zirconium**
JIS H 1667-86. Method for Determination of Hafnium in Zirconium and Zirconium Alloys. 6 pp.

—**Zirconium Alloys**
JIS H 1667-86. Method for Determination of Hafnium in Zirconium and Zirconium Alloys. 6 pp.

Hair (Animal)
See Also: Fibers; Wool Fibers
CNS N1072-85. Animal Hair, Fine (May)(2817).
CNS N1079-85. Hog and Other Animal Hairs, Not Carded or Combed (May)(2825).

Hair (Animal) (Cont.)

—**Bleaching**
CNS N4025-85. Method of Test for Animal Hairs (May)(2826). 7 pp.

—**Cattle**
CNS N1075-85. Cow Tail Hair (May)(2820).
CNS N1077-85. Cow Tail (May)(2822).

—**Dogs**
CNS N1074-85. Dog Hair (May)(2819).

—**Fumigation**
CNS N4025-85. Method of Test for Animal Hairs (May)(2826). 7 pp.

—**Goats**
CNS N1078-85. Goat Hair (May)(2824).

—**Horses**
CNS N1073-85. Horse Hair, Not Carded or Combed (May)(2818).
CNS N1076-85. Horse Bristle and Tail Hair (May)(2821).

—**Impurities Content**
CNS N4025-85. Method of Test for Animal Hairs (May)(2826). 7 pp.

—**Macroscopic Examination**
CNS N4025-85. Method of Test for Animal Hairs (May)(2826). 7 pp.

—**Moisture Content**
CNS N4025-85. Method of Test for Animal Hairs (May)(2826). 7 pp.

—**Rabbits**
CNS N1084-85. Rabbit Hair (May)(2881).

—**Sampling**
CNS N4025-85. Method of Test for Animal Hairs (May)(2826). 7 pp.

—**Sanitation**
CNS N4025-85. Method of Test for Animal Hairs (May)(2826). 7 pp.

—**Swine**
CNS N1079-85. Hog and Other Animal Hairs, Not Carded or Combed (May)(2825).

Hair Care Appliances
See Also: Appliances; Curling Irons; Hair Care Products; Hair Clippers; Hair Dryers
CNS C3181-87. Method of Test for Hair Curling Appliances (Aug)(10608).
DIN VDE 0720 Pt 2J-80. Electric Cooking and Heating Appliances for Domestic and Similar Purposes; Appliances for Skin and Hair Treatment (VDE Specifications) (Sept). 17 pp.
DIN VDE 0720 Pt 2J A1 (D)-80. Electric Cooking and Heating Appliances for Domestic and Similar Purposes; Appliances for Skin and Hair Treatment: (Amendment 1) (VDE Specification) (Sept). 5 pp.
DIN VDE 0720 Pt 2J A2-83. Electric Cooking and Heating Appliances for Domestic and Similar Purposes; Appliances for Skin and Hair Treatment: Amendment 2 (VDE Specification) (Oct). 5 pp.
DIN VDE 0720 Pt 2J A3 (D)-82. Electric Cooking and Heating Appliances for Domestic and Similar Purposes; Appliances for Skin and Hair Care: Amendment 3 (VDE Specification) (Nov). 5 pp.
SAA AS 3304-91. Approval and Test Specification—Particular Requirements for Skin or Hair Care Appliances (NZS 6323:1991) Amdt 1 March 1992 Amdt 2 October 1992 Amdt 3 March 1993 Amdt 4 August 1993 (in Professional Package 28). 2 pp.

—**Acceptance Testing**
SNZ NZS 6323-91. Approval and Test Specification—Particular Requirements for Skin or Hair Care Appliances Amend: 1, 1992 (NZS 6323:1991/AS 3304-1991). 14 pp.

—**Hand Held**
CSA C22.2 NO 36-M1989. Hairdressing Equipment; (Gen Instr 1 Thru 4). 79 pp.
CSA 1169 Bull. Electrical Bulletin 1169 June 27, 1978 to C22.2 NO 36. 2 pp.

—**Pedestal**
CSA C22.2 NO 36-M1989. Hairdressing Equipment; (Gen Instr 1 Thru 4). 79 pp.
CSA 1169 Bull. Electrical Bulletin 1169 June 27, 1978 to C22.2 NO 36. 2 pp.

—**Safety**
BSI BS 3456: Sec 3.13-79. (WITHDRAWN) 1979 Amd 2 1972-1981 Edition. Specification for Safety of Household Electrical Appliances Part 3: Complete Particular Specifications Section 3.13: Appliances for Skin or Hair Treatment (Superseded by BS EN 60335-2-23: 1991). 60 pp.

INDUSTRY STANDARDS

Hair Care Appliances (Cont.)
—Safety (Cont.)

BSI BS EN 60335-2-23-91. 1991 Safety of Household and Similar Electrical Appliances Part 2: Particular Requirements Section 2.23: Appliances for Skin or Hair Care (Supersedes BS 3456: Section 3.13: 1979). 32 pp.

BSI BS EN 60335-2-23-01. 1991 Amd 1 Safety of Household and Similar Electrical Appliances Part 2: Particular Requirements Section 2.23: Appliances for Skin or Hair Care (AMD 7793) July 15, 1993 (L). 44 pp.

BSI BS EN 60335-2-23-02. 1991 Amd 2 Safety of Household and Similar Electrical Appliances Part 2: Particular Requirements Section 2.23: Appliances for Skin or Hair Care (AMD 7901) August 15, 1993 (L). 50 pp.

CENELEC HD 266-75. Particular Specification for Appliances for Skin or Hair Treatment. 6 pp.

CENELEC HD 266.2-77. Particular Specification for Appliances for Skin and Hair Treatment. 9 pp.

CENELEC HD 266.3-83. Particular Specification for Appliances for Skin or Hair Treatment. 7 pp.

CENELEC HD 266.4-83. Particular Specification for Appliances for Skin or Hair Treatment. 6 pp.

CENELEC EN 60 335-2-23-90. Safety of Household and Similar Electrical Appliances Part 2: Particular Requirements for Appliances for Skin or Hair Care. 11 pp.

CENELEC EN 60335-2-23/A1-92. AMD 1 Safety of Household and Similar Electrical Appliances Part 2: Particular Requirements for Appliances for Skin or Hair Care. 5 pp.

CENELEC EN 60335-2-23/A51-92. AMD 51 Safety of Household and Similar Electrical Appliances Part 2: Particular Requirements for Appliances for Skin or Hair Care. 5 pp.

IEC 335 Pt 2-23-86. Safety of Household and Similar Electrical Appliances Part 2: Particular Requirements for Appliances for Skin or Hair Care Third Edition; (Amendment 1-1990). 50 pp.

—Table Mounted

CSA C22.2 NO 36-M1989. Hairdressing Equipment; (Gen Instr 1 Thru 4). 79 pp.

CSA 1169 Bull. Electrical Bulletin 1169 June 27, 1978 to C22.2 NO 36. 2 pp.

Hair Care Products
See Also: Hair Care Appliances; Hair Dyes; Shampoos

—Alkalinity

EC 80/1335/EEC-80. First Commission Directive on the Approximation of the Laws of the Member States Relating to Methods of Analysis Necessary for Checking the Composition of Cosmetic Products. 20 pp.

—Hydrogen Peroxide Content

EC 82/434/EEC-82. Second Commission Directive on the Approximation of the Laws of the Member States Relating to Methods of Analysis Necessary for Checking the Composition of Cosmetic Products. 28 pp.

—Oxalic Acid Content

EC 80/1335/EEC-80. First Commission Directive on the Approximation of the Laws of the Member States Relating to Methods of Analysis Neccessary for Checking the Composition of Cosmetic Products. 20 pp.

—Oxidizer Content

EC 82/434/EEC-82. Second Commission Directive on the Approximation of the Laws of the Member States Relating to Methods of Analysis Necessary for Checking the Composition of Cosmetic Products. 28 pp.

—Resorcinol Content

EC 82/434/EEC-82. Second Commission Directive on the Approximation of the Laws of the Member States Relating to Methods of Analysis Necessary for Checking the Composition of Cosmetic Products. 28 pp.

—Thioglycolic Acid Content

EC 83/514/EEC-83. Third Commission Directive on the Approximation of the Laws of the Member States Relating to Methods of Analysis Necessary for Checking the Composition of Cosmetic Products. 38 pp.

Hair Clippers
See Also: Appliances; Hair Care Appliances

CNS S2041-80. Method of Test for Electric Hair-Clipper (Aug)(6271).

CSA CAN/CSA-C22. 2 NO 68-92. Motor-Operated Appliances (Household and Commercial); (Gen Instr 1 Thru 2). 115 pp.

CSA 1169 Bull. Electrical Bulletin 1169 June 27, 1978 to C22.2 NO 68. 2 pp.

Hair Clippers (Cont.)

SAA AS 3125-90. Approval and Test Specification—Electric Dry Shavers and Hair Clippers Amdt 1 June 1993. 4 pp.

SNZ NZS/AS 3125-90. Approval and Test Specification—Electric Dry Shavers and Hair Clippers (This is a Joint Standard with SAA AS 3125). 4 pp.

—Safety

BSI BS 3456: Sec 102.8-88. (WITHDRAWN) 1988 Safety of Household Electrical Appliances Part 102: Particular Requirements Section 102.8: Electric Shavers, Hair Clippers and Similar Appliances (Superseded by BS EN 60335-2-8: 1991). 19 pp.

BSI BS 3456: Sec 202.19-90. 1990 Safety of Household Electrical Appliances Part 202: Particular Requirements Section 202.19: Battery-Powered Shavers, Hair Clippers and Similar Appliances and Their Charging and Battery Assemblies. 28 pp.

BSI BS EN 60335-2-8-91. 1991 Safety of Household and Similar Electrical Appliances Part 2: Particular Requirements Section 2.8: Electric Shavers, Hair Clippers and Similar Appliances (Supersedes BS 3456: Section 102.8: 1988) (L). 19 pp.

CENELEC HD 254 S2-84. Safety of Household and Similar Electric Appliances Particular Requirements for Electric Shavers, Hair Clippers and Similar Appliances (Superseded by EN 60335-2-8). 10 pp.

CENELEC EN 60335-2-8-90. Safety of Household and Similar Electrical Appliances Part 2: Particular Requirements for Electric Shavers, Hair Clippers and Similar Appliances (Supersedes HD 254 S2). 7 pp.

CENELEC EN 60 335-2-19-90. Safety of Household and Similar Electrical Appliances Part 2: Particular Requirements for Battery—Powered Shavers, Hair Clippers and Similar Appliances and Their Charging and Battery Assemblies. 7 pp.

IEC 335 Pt 2-8-92. Safety of Household and Similar Electrical Appliances Part 2: Particular Requirements for Shavers, Hair Clippers and Similar Appliances Fourth Edition. 29 pp.

IEC 335 Pt 2-19-84. Safety of Household and Similar Electrical Appliances Part 2: Particular Requirements for Battery-Powered Shavers, Hair Clippers and Similar Appliances and Their Charging and Battery Assemblies Second Edition. 45 pp.

Hair Curling Appliances
Use: Curling Irons

Hair Dryers
See Also: Appliances; Curling Irons; Dryers; Hair Care Appliances

CNS C4122-90. Hand-Supported Hair Dryers (Jun)(3714).

CSA C22.2 NO 36-M1989. Hairdressing Equipment; (Gen Instr 1 Thru 4). 79 pp.

CSA 1169 Bull. Electrical Bulletin 1169 June 27, 1978 to C22.2 NO 36. 2 pp.

JIS C 9613-76. Hand-Hold Hair Dryers.

Hair Dyes
See Also: Dyes; Hair Care Products

EC 83/341/EEC-83. Third Commission Directive Adapting to Technical Progress Annexes II, III and V of Council Directive 76/768/EEC on the Approximation of the Laws of the Member States Relating to Cosmetic Products. 1 p.

—Oxidation Dye Content

EC 82/434/EEC-82. Second Commission Directive on the Approximation of the Laws of the Member States Relating to Methods of Analysis Necessary for Checking the Composition of Cosmetic Products. 28 pp.

Hair Felts
See Also: Felts

—Packaging

CNS L3061-80. Method of Test for Cattle Hair Felt (Packing) (Aug)(2687).

CNS L4043-80. Cattle Hair Felt (for Packing) (Aug) (2686).

—Thermal Insulation

CNS A2157-89. Hair Felt (Feb)(10408).

CNS A3189-89. Method of Test for Hair Felts (Feb)(10409).

Hair Hygrometers

DIN ENGL 50012 Pt 3-86. Climates and Their Technical Application; Methods of Measuring Humidity; Hair Hygrometer (Jan). 2 pp.

Haircloth
See Also: Fabrics

—Interlinings

CNS L3199-83. Method of Test for Interlining Hair Cloth (Jun)(10375).

CNS L4148-83. Interlining Hair Cloth (Jun)(10374).

Haircloth (Cont.)
—Interlinings (Cont.)

JIS L 3202-76. Interlining Hair Cloth.

Hairdressing Appliances
Use: Hair Care Appliances

Hairdressing Equipment
Use: Hair Care Appliances

Half Adders
—Serial

CECC CECC 90 104-126 ISSUE 1-81. BS CECC 90 104-125; Silicon Complementary MOS with (B) Buffered Outputs and Cavity Packaging (En). 25 pp.

CECC CECC 90 104-126 ISSUE 1-86. CEI CECC 90 104-126; Silicon Complementary MOS with (B) Buffered Outputs Cavity and Non Cavity Packaging (En). 2 pp.

CECC CECC 90 104-129 ISSUE 1-81. BS CECC 90 104-129; Silicon Complementary MOS with (B) Buffered Outputs and Cavity Packaging (En). 25 pp.

CECC CECC 90 104-129 ISSUE 1-86. CEI CECC 90 104-129; Silicon Complementary MOS with (B) Buffered Outputs Cavity and Non Cavity Packaging (En). 2 pp.

Half Duplex Transmission
See Also: Communication Circuits; Data Transmission

—Link Access Procedures—Public Switched Telephone Networks

CCITT RECMN T.71-89. Link Access Protocol Balanced (LAPB) Extended for Half-Duplex Physical Level Facility—Terminal Equipment and Protocols for Telematic Services (Study Group VIII) 7 pp. 7 pp.

—Link Access Procedures—Teletex Equipment

CCITT RECMN T.71-89. Link Access Protocol Balanced (LAPB) Extended for Half-Duplex Physical Level Facility—Terminal Equipment and Protocols for Telematic Services (Study Group VIII) 7 pp. 7 pp.

Half Round Files
Use: Files (Tools)—Half Round

Halide Content Analysis
—Photographic Chemicals

ISO 10349 Pt 6-92. Photography—Photographic-Grade Chemicals—Test Methods—Part 6: Determination of Halide Content First Edition. 6 pp.

Halides
Scope Note: Use a more specific term
See: Aluminum Alkyl Halides; Aluminum Chloride; Cadmium Chloride; Cadmium Iodide; Calcium Chloride; Ferric Chloride; Halide Content Analysis; Hydrobromic Acid; Hydrochloric Acid; Hydrofluoric Acid; Hydrogen Chloride; Manganese Chloride Tetrahydrate; Nickel Chloride Hexahydrate; Potassium Bromide; Potassium Chloride; Silver Halides; Sodium Bromide; Sodium Chloride; Sodium Fluoride; Sodium Iodide; Stannic Chloride; Strontium Chloride; Sulfur Chloride; Sulfur Hexafluoride; Uranium Hexafluoride; Zirconium Oxychloride

Halite
Use: Rock Salt

Halocarbon 11
Use: Trichlorofluoromethane

Halocarbon 113
Use: 1,1,2-Trichloro-1,2,2-Trifluoroethane

Halofuginone Content Analysis
—Animal Feed

CNS N4119-84. Method of Test for Feed Additives: Determination of Halofuginone (Nov)(11125).

Halogen Content Analysis
—Camphor

CNS K6064-57. Method of Test for Natural Camphor (Jul)(818)(R 1973). 2 pp.

Halogen Lamps
See Also: Incandescent Lighting; Lamps

—Ships—Floodlights

CNS F5044-86. Marine Halogen Lamp Type Flood Lights (Jan)(9603).

JIS F 8445-83. Marine Halogen Lamp Type Flood Lights.

INTERNATIONAL AND NON-U.S. NATIONAL STANDARDS
SUBJECT INDEX

Halogen Minerals
Use: Minerals

Halogen Organic Compound Content Analysis
See Also: Halogenated Hydrocarbons

—Sediments—Coulometric Analysis
DIN ENGL 38414 Pt 17-89. German Standard Methods for the Examination of Water, Waste Water and Sludge; Sludge and Sediments (Group S); Determination of Strippable and Extractable Organically Bound Halogens (S 17) (Nov). 5 pp.

DIN ENGL 38414 Pt 18-89. German Standard Methods for the Examination of Water, Waste Water and Sludge; Sludge and Sediments (Group S); Determination of Adsorbed Organically Bound Halogens (AOX) (S18) (Nov). 4 pp.

—Sediments—Ion Exchange Chromatography
DIN ENGL 38414 Pt 17-89. German Standard Methods for the Examination of Water, Waste Water and Sludge; Sludge and Sediments (Group S); Determination of Strippable and Extractable Organically Bound Halogens (S 17) (Nov). 5 pp.

DIN ENGL 38414 Pt 18-89. German Standard Methods for the Examination of Water, Waste Water and Sludge; Sludge and Sediments (Group S); Determination of Adsorbed Organically Bound Halogens (AOX) (S18) (Nov). 4 pp.

—Sediments—Photometry
DIN ENGL 38414 Pt 17-89. German Standard Methods for the Examination of Water, Waste Water and Sludge; Sludge and Sediments (Group S); Determination of Strippable and Extractable Organically Bound Halogens (S 17) (Nov). 5 pp.

—Sludge (Sewage)—Coulometric Analysis
DIN ENGL 38414 Pt 17-89. German Standard Methods for the Examination of Water, Waste Water and Sludge; Sludge and Sediments (Group S); Determination of Strippable and Extractable Organically Bound Halogens (S 17) (Nov). 5 pp.

DIN ENGL 38414 Pt 18-89. German Standard Methods for the Examination of Water, Waste Water and Sludge; Sludge and Sediments (Group S); Determination of Adsorbed Organically Bound Halogens (AOX) (S18) (Nov). 4 pp.

—Sludge (Sewage)—Ion Exchange Chromatography
DIN ENGL 38414 Pt 17-89. German Standard Methods for the Examination of Water, Waste Water and Sludge; Sludge and Sediments (Group S); Determination of Strippable and Extractable Organically Bound Halogens (S 17) (Nov). 5 pp.

DIN ENGL 38414 Pt 18-89. German Standard Methods for the Examination of Water, Waste Water and Sludge; Sludge and Sediments (Group S); Determination of Adsorbed Organically Bound Halogens (AOX) (S18) (Nov). 4 pp.

—Sludge (Sewage)—Photometry
DIN ENGL 38414 Pt 17-89. German Standard Methods for the Examination of Water, Waste Water and Sludge; Sludge and Sediments (Group S); Determination of Strippable and Extractable Organically Bound Halogens (S 17) (Nov). 5 pp.

—Waste Water—Adsorption
CPPA H.6P-91. Determination of Adsorbable Organic Halogens (AOX) in Waters and Wastewaters. 7 pp.

DIN ENGL 38409 Pt 14-85. German Standard Methods for the Examination of Water, Waste Water and Sludge; Summary Indices of Actions and Substances (Group H); Determination of Adsorbable Organically Bonded Halogens (AOX) (H 14) (Mar). 6 pp.

—Waste Water—Extraction Analysis
DIN ENGL 38409 Pt 8-84. German Standard Methods for the Examination of Water, Waste Water and Sludge; Summary Indices of Actions and Substances (Group H); Determination of Extractable Organically Bonded Halogens (EOX) (H 8) (Sept). 3 pp.

—Water—Adsorption
CPPA H.6P-91. Determination of Adsorbable Organic Halogens (AOX) in Waters and Wastewaters. 7 pp.

DIN ENGL 38409 Pt 14-85. German Standard Methods for the Examination of Water, Waste Water and Sludge; Summary Indices of Actions and Substances (Group H); Determination of Adsorbable Organically Bonded Halogens (AOX) (H 14) (Mar). 6 pp.

ISO 9562-89. Water Quality—Determination of Adsorbable Organic Halogens (AOX) First Edition. 11 pp.

Halogen Organic Compound Content Analysis (Cont.)

—Water—Extraction Analysis
DIN ENGL 38409 Pt 8-84. German Standard Methods for the Examination of Water, Waste Water and Sludge; Summary Indices of Actions and Substances (Group H); Determination of Extractable Organically Bonded Halogens (EOX) (H 8) (Sept). 3 pp.

Halogenated Hydrocarbon Content Analysis
Use For: Chlorinated Hydrocarbon Content Analysis
See Also: Halogenated Hydrocarbons; Hydrocarbon Content Analysis

—Air—Gas Chromatography
BSI BS 6069: Sec 3.3-91. 1991 Characterization of Air Quality Part 3: Workplace Atmospheres Section 3.3: Method for the Determination of Vaporous Chlorinated Hydrocarbons by Charcoal Tube/Solvent Desorption/ Gas Chromatography (ISO 9486: 1991). 18 pp.

ISO 9486-91. Workplace Air—Determination of Vaporous Chlorinated Hydrocarbons—Charcoal Tube/Solvent Desorption/ Gas Chromatographic Method First Edition. 15 pp.

—Gas Chromatography
DIN ENGL 38407 Pt 4-88. German Standard Methods for the Examination of Water, Waste Water and Sludge; Substance Group Analysis (Group F); Determination of Readily Volatile Halogenated Hydrocarbons (F4) (May). 16 pp.

—Ground Water—Gas Chromatography
DIN ENGL 38407 Pt 4-88. German Standard Methods for the Examination of Water, Waste Water and Sludge; Substance Group Analysis (Group F); Determination of Readily Volatile Halogenated Hydrocarbons (F4) (May). 16 pp.

—Industrial Water
JIS K 0125-90. Testing Methods for Determination of Low Molecular Weight Halogenated Hydrocarbons in Industrial Water and Waste Water. 21 pp.

—Sludge (Sewage)—Gas Chromatography
DIN ENGL 38407 Pt 4-88. German Standard Methods for the Examination of Water, Waste Water and Sludge; Substance Group Analysis (Group F); Determination of Readily Volatile Halogenated Hydrocarbons (F4) (May). 16 pp.

—Uranium Hexafluoride
CNS J2066-82. Method for Determining Hydrocarbons, Chlorocarbons and Partially Substituted Halohydrocarbons in UF6 (Sept)(9399).

—Waste Water
JIS K 0125-90. Testing Methods for Determination of Low Molecular Weight Halogenated Hydrocarbons in Industrial Water and Waste Water. 21 pp.

—Waste Water—Gas Chromatography
DIN ENGL 38407 Pt 4-88. German Standard Methods for the Examination of Water, Waste Water and Sludge; Substance Group Analysis (Group F); Determination of Readily Volatile Halogenated Hydrocarbons (F4) (May). 16 pp.

Halogenated Hydrocarbons
Use For: Halohydrocarbons See Also: Benzyl Chloride; Carbon Tetrachloride; Chlorobenzene; Chlorofluorocarbons; Chloroform; Halogen Organic Compound Content Analysis; Halogenated Hydrocarbon Content Analysis; Hexachlorobenzene; Methyl Chloride; Perchloroethylene; 1,2,4-Trichlorobenzene; 1,1,1-Trichloroethane; Trichloroethylene

—Acidity—Volumetric Analysis
BSI BS 5598: Part 4-79. 1979 Methods of Sampling and Test for Halogenated Hydrocarbons Part 4: Determination of Acidity, Titrimetric Method. 4 pp.

CNS K6561-80. Method of Test for Acidity of Liquid Halogenated Hydrocarbons (Titrimetric Method) (Aug)(6206).

ISO 1393-77. Liquid Halogenated Hydrocarbons for Industrial Use—Determination of Acidity—Titrimetric Method First Edition. 4 pp.

—Cloud Point Determination
BSI BS 5598: Part 5-79. 1979 Methods of Sampling and Test for Halogenated Hydrocarbons Part 5: Determination of Cloud Point. 4 pp.

ISO 1394-77. Liquid Halogenated Hydrocarbons for Industrial Use—Determination of Cloud Point First Edition. 4 pp.

Halogenated Hydrocarbons (Cont.)

—Fire Extinguishers
SAA AS 3689.2-91. Automatic Fire Extinguishing Systems Using Halogenated Hydrocarbons—Part 2: Mechanical Components Amdt 1 June 1991 (in Professional Package 44). 30 pp.

SNZ NZS 4551-74. Specification for Portable Fire Extinguishers of the Halogenated Hydrocarbon Type. 20 pp.

—Fire Extinguishing Agents
BSI BS 6535: Part 2-84. (WITHDRAWN) 1984 Fire Extinguishing Media Part 2: Specification for Halogenated Hydrocarbons (ISO 7201: 1982) (Superseded by BS 6535: Section 2.1: 1990). 8 pp.

BSI BS 6535: Sec 2.1-90. 1990 Fire Extinguishing Media Part 2: Halons Section 2.1: Halon 1211 and Halon 1301. 8 pp.

ISO 7201 Pt 1-89. Fire Protection—Fire Extinguishing Media—Halogenated Hydrocarbons—Part 1: Specifications for Halon 1211 and Halon 1301 Second Edition. 7 pp.

—Fire Extinguishing Agents—Safety
BSI BS 6535: Sec 2.2-89. 1989 Fire Extinguishing Media Part 2: Halons Section 2.2: Code of Practice for Safe Handling and Transfer Procedures. 8 pp.

ISO 7201 Pt 2-91. Fire Extinguishing Media —Halogenated Hydrocarbons—Part 2: Code of Practice for Safe Handling and Transfer Procedures of Halon 1211 and Halon 1301 First Edition. 8 pp.

—Liquefied Gases—Sampling
BSI BS 5598: Part 2-79. 1979 Methods of Sampling and Test for Halogenated Hydrocarbons Part 2: Sampling of Liquefied Gases. 7 pp.

ISO 3427-76. Gaseous Halogenated Hydrocarbons (Liquefied Gases)—Taking of a Sample First Edition. 6 pp.

—Refrigerants—Heat Pumps—Safety
DIN ENGL 8901-83. Heat Pumps; Heat Pumps Using Halogenated Hydrocarbons; Protection of Soil, Ground Water and Surface Water; Requirements and Teesting (Jan). 4 pp.

—Residue Content—Evaporation
BSI BS 5598: Part 3-79. 1979 Medthods of Sampling and Test for Halogenated Hydrocarbons Part 3: Determination of Residue on Evaporation. 4 pp.

CNS K6562-80. Method of Test for Evaporation Residue of Liquid Halogenated Hydrocarbons (Aug)(6207).

ISO 2210-72. Liquid Halogenated Hydrocarbons for Industrial Use—Determination of Residue on Evaporation First Edition. 4 pp.

—Sampling
BSI BS 5598: Part 1-78. 1978 Methods of Sampling and Test for Halogenated Hydrocarbons Part 1: Sampling of Liquid Products. 8 pp.

ISO 2209-73. Liquid Halogenated Hydrocarbons for Industrial Use—Sampling First Edition. 7 pp.

Halogenation
Scope Note: Use a more specific term
See: Bromination; Chlorination

Halohydrocarbons
Use: Halogenated Hydrocarbons

Halon
Use For: Dichlorodifluoromethane; Difluorodichloromethane; FLON 12
See Also: Chlorofluorocarbons

—Fire Extinguishers
BSI BS 5306: Sec 5.2-84. 1984 Fire Extinguishing Installations and Equipment on Premises Part 5: Halon Systems Section 5.2: Halon 1211 Total Flooding Systems. 40 pp.

SAA AS 1851.11-91. Maintenance of Fire Protection Equipment—Part 11: Halon 1301 Total Flooding Systems (in Professional Packages 21, 30, 44). 7 pp.

—Fire Extinguishing Agents
BSI BS 5306: Sec 5.1-92. 1992 Fire Extinguishing Installations and Equipment on Premises Part 5: Halon Systems Section 5.1: Specification for Halon 1301 Total Flooding Systems. 66 pp.

BSI BS 5306: Sec 5.1-82. 1982 Amd 1 Fire Extinguishing Installations and Equipment on Premises Part 5: Halon Systems Section 5.1: Halon 1301 Total Flooding Systems. 48 pp.

—Refrigerants
CNS K1079-63. Difluorodichloromethane for Refrigerant (Nov)(2235)(R 1973).

Ham
See Also: Food; Pork
CNS N5038-87. Chinese Ham (Jun)(2104). 3 pp.
CNS N5128-81. Western Ham (May)(4144). 2 pp.

INTERNATIONAL AND NON-U.S. NATIONAL STANDARDS
SUBJECT INDEX

Ham

Ham (Cont.)
—**Canned**
CGSB 32.65M-90. Canned or Boil-in-Bag Ham. 7 pp.

—**Chemical Analysis**
CNS N6100-81. Method of Test for Western Ham (May)(4145). 1 p.

—**Identification Systems**
CNS N6033-87. Method of Test for Chinese Ham (Jun)(2105). 2 pp.

—**Inspection**
CNS N6033-87. Method of Test for Chinese Ham (Jun)(2105). 2 pp.
CNS N6100-81. Method of Test for Western Ham (May)(4145). 1 p.

—**Nitrite Content**
CNS N6184-88. Method of Test for Nitrates in Food (Jan)(10888). 2 pp.

—**Sampling**
CNS N6033-87. Method of Test for Chinese Ham (Jun)(2105). 2 pp.
CNS N6100-81. Method of Test for Western Ham (May)(4145). 1 p.

—**Smoked**
CGSB 32.63M-91. Smoked Ham. 7 pp.

Ham Radio
Use: Amateur Radio Equipment

Ham Radio Services
Use: Amateur Radio Services

Hammer Drills
See Also: Rock Drilling Equipment; Rock Drills
MOD UK DSTAN 38-1-68. Drills, Pneumatic, Hand Hammer (Rock Drills) Issue 1 (Withdrawn). 7 pp.

—**Electric—Portable**
MOD UK DSTAN 51-19: Part 2-92. Hand Tools, Powered Part 2: Electric Issue 1. 31 pp.

—**Safety**
CEN PREN 792-5-92. Handheld Non-Electric Power Tools—Safety Requirements—Part 5: Rotary Percussive Drills. 8 pp.

Hammer Forging
See Also: Forging; Metal Working

—**Die Blocks**
JIS B 6470-83. Die Blocks for Closed-Die Hammer Forging (R 1987). 9 pp.

Hammers
See Also: Anvils; Chipping Hammers; Forging Hammers; Hand Tools; Hatchets; Presses; Riveting Hammers; Sledgehammers; Soft Face Hammers; Tools
BSI BS 876-81. 1981 Amd 1 Hand Hammers. 29 pp.
CGSB 39-GP-1C-75. Hammer, Hand, (Forged Steel Head), Standard for; (Amendment 1 Sept 1980). 23 pp.
CNS B3126-86. Hand Hammers (Dec)(1026).
MOD UK DSTAN 51-11: Part 31-89. Hand Tools, General Purpose Part 31: Hammers, Hand: and Replacement Piece Parts Issue 2. 23 pp.
MOD UK DSTAN 51-11: Part 31-01. Hand Tools, General Purpose Part 31: Hammers, Hand: and Replacement Piece Parts Issue 2; Amendment 1. 32 pp.
SAA AS 3797.1-91. Hand Hammers—Part 1: General Requirements Amdt 1 August 1991.
SAA AS 3797.2-91. Hand Hammers—Part 2: Specific Requirements for Heavy Hammers. 5 pp.
SAA AS 3797.3-91. Hand Hammers—Part 3: Specific Requirements for Light Hammers. 6 pp.

—**Copper Beryllium Alloy—Nonsparking**
CNS M2058-86. Non-Sparking Beryllium Copper Alloy Tools Hand Hammers (Nov) (5906).

—**Dovetail**
CNS B3125-83. Hand Hammer, Dove-Tail Shaped (Mar)(1025).

—**Electric**
BSI BS 2769: Sec 2.6-84. 1984 Hand-Held Electric-Motor-Operated Tools Part 2: Particular Requirements Section 2.6: Hammers. 6 pp.
CENELEC PREN 50XXX-2-6-92. Safety of Hand-Held Electric Motor Operated Tools Part 2: Particular Requirements for Hammers. 15 pp.
CSA CAN/CSA-C22. 2 NO71.1-M89. Portable Electric Tools; (Gen Instr 1 Thru 4). 89 pp.
MOD UK DSTAN 51-19: Part 2-92. Hand Tools, Powered Part 2: Electric Issue 1. 31 pp.

Hammers (Cont.)
—**Electric—Safety**
BSI BS 2769: Sec 2.6-84. 1984 Hand-Held Electric-Motor-Operated Tools Part 2: Particular Requirements Section 2.6: Hammers. 6 pp.
CENELEC HD 400.2F S1 PRA1-93. AMD 1 Hand-Held Motor Operated Tools Part II: Particular Specifications Section F: Hammers. 4 pp.
CENELEC PREN 50XXX-2-6-92. Safety of Hand-Held Electric Motor Operated Tools Part 2: Particular Requirements for Hammers. 15 pp.
IEC 745 Pt 2-6-89. Safety of Hand-Held Motor-Operated Electric Tools Part 2: Particular Requirements for Hammers First Edition; (Corrigendum—Dec 1989) (Amendment 1-1992). 36 pp.

—**Handles**
CGSB 39-GP-10M-79. Handle, Hickory, Striking Tool (Metric), Standard for; (Amendment 2 Jan 1984). 15 pp.

—**Replacement Parts**
MOD UK DSTAN 51-11: Part 31-89. Hand Tools, General Purpose Part 31: Hammers, Hand: and Replacement Piece Parts Issue 2. 23 pp.
MOD UK DSTAN 51-11: Part 31-01. Hand Tools, General Purpose Part 31: Hammers, Hand: and Replacement Piece Parts Issue 2; Amendment 1. 32 pp.

—**Spike—Rail Track**
JIS E 1501-78. Spike Hammers.

Hand Braces
See Also: Drills; Hand Tools; Tools
BSI BS 1978-65. 1965 Amd 1 Bit Braces (AMD 5973) July 31, 1989. 21 pp.

Hand Cleaners
See Also: Cleaning Agents; Soaps

—**Industrial**
SAA AS 1223-91. Industrial Hand Cleaners (Petroleum Solvent Type) Amdt 1 September 1991 (in Professional Packages 32, 47). 8 pp.

—**Liquid**
CGSB CAN/CGSB-2.1-M89. Skin Cleaning Lotion; (Amendment 1 May 1992). 11 pp.

—**Waterless**
CGSB CAN/CGSB-2.16-M90. Waterless Hand Cleaner. 8 pp.

Hand Controls
Use: Manual Controls

Hand Drills
Use: Drills

Hand Finishing Sticks
See Also: Hand Tools; Oilstones
ISO 2220-72. Hand Finishing Sticks and Oil Stones—Dimensions First Edition. 3 pp.

Hand Luggage
Use: Luggage

Hand Nuts
Use: Wing Nuts

Hand Off
Use For: Handoff; Handover *See Also:* Cellular Mobile Radio Services; Switching; Switching Systems

—**Public Land Mobile Networks**
CCIR RECMN 624-86. Public Land Mobile Communication Systems Location Registration—Section 8A—Land Mobile Service and Related Subjects. 6 pp.
CCITT RECMN Q.1005-89. Handover Procedures—Public Land Mobile Network Interworking with ISDN and PSTN (Study Group XI) 32 pp. 32 pp.
CENELEC GSM 03.09-92. European Digital Cellular Telecommunication System (Phase 1); Handover Procedures. 38 pp.
ETSI GSM 03.09-92. European Digital Cellular Telecommunication System (Phase 1); Handover Procedures. 38 pp.

Hand Over
Use: Hand Off

Hand-Portable Lamps
Use: Flashlights

Hand Prostheses
See Also: Arm Prostheses; Prosthetic Devices
JIS T 9224-90. Cosmetic Hands for Arm Prostheses. 10 pp.

Hand Prostheses (Cont.)
—**Gloves**
JIS T 9223-90. Cosmetic Gloves for Artificial Hands. 9 pp.

Hand Rails
Use: Handrails

Hand Reamers
See Also: Hand Tools; Reamers; Tools
BSI BS 3088-59. 1959 Adjustable Hand Reamers with Inserted Blades. 8 pp.
CNS B3066-80. Hand Reamer (Aug)(240).
CNS B3067-52. Adjustable Hand Reamer (Sep)(241)(R 1970).
CNS B3085-52. Hand Reamer (Sep)(259) (R 1970).
ISO 236 Pt I-76. Hand Reamers First Edition. 6 pp.
JIS B 4405-90. Hand Reamers. 12 pp.
MOD UK DSTAN 51-9. Reamers, Hand. 8 pp.
MOD UK DSTAN 51-9-93. Reamers, Hand Issue 2. 17 pp.
MOD UK DSTAN 51-9: Addendum. Reamers, Hand Issue 1. 10 pp.

—**Taper Pin**
ISO 3465-75. Hand Taper Pin Reamers First Edition. 4 pp.
JIS B 4411-88. Hand Taper Pin Reamers. 10 pp.

Hand Set Telephones
Use: Telephones

Hand Sheets
Use: Handsheets

Hand Shovels
Use: Shovels

Hand Signals

—**Agricultural Equipment**
BSI BS 6736-86. 1986 Code of Practice for Hand Signalling for Use in Agricultural Operations. 7 pp.

—**Aircraft Marshallers**
NATO STANAG 3117 ED 5 AMD 0-85. Aircraft Marshalling Signals. 35 pp.
NATO STANAG 3117 ED 5 AMD 3-85. Aircraft Marshalling Signals. 36 pp.

—**Cranes**
SNZ NZS 5818-82. Specification for Hand Signals for the Direction of Cranes and Similar Lifting Devices Including Helicopters (Other Than for Cargo Handling in Wharf Areas) (Reconfirmed 1989). 12 pp.

—**Earthmoving Equipment**
DIN ENGL 24081-78. Safety in the Organization of Work; Earth-Moving Machinery; Hand Signals (Nov). 4 pp.

—**Helicopters**
SNZ NZS 5818-82. Specification for Hand Signals for the Direction of Cranes and Similar Lifting Devices Including Helicopters (Other Than for Cargo Handling in Wharf Areas) (Reconfirmed 1989). 12 pp.

—**Lifting Equipment**
SNZ NZS 5818-82. Specification for Hand Signals for the Direction of Cranes and Similar Lifting Devices Including Helicopters (Other Than for Cargo Handling in Wharf Areas) (Reconfirmed 1989). 12 pp.

Hand Tools
Scope Note: For additional listings, use a more specific term *See Also:* Adjustable Wrenches; Augers; Axes; Bevels; Bits (Tools); Box Spanners; Breast Drills; Center Punches; Chipping Hammers; Chisels; Cold Chisels; Compass Saws; Coping Saws; Crimping Tools; Crosscut Saws; Cutting Tools; Dies; Drilling Hammers; Drills; Entrenching Tools; Files (Tools); Fretsaws; Hacksaws; Hammers; Hand Braces; Hand Finishing Sticks; Hand Reamers; Hoes; Impact Wrenches; Insertion Tools; Keyhole Saws; Knives; Machine Tools; Mallets; Mattocks; Monkey Wrenches; Nail Pullers; Nested Saws; Nippers; Nonsparking Tools; Nutdrivers; Oilstones; Picks; Piercing Saws; Pincers; Pinch Bars; Pipe Cutters; Pitchforks; Planes (Tools); Pliers; Polishers; Pruners; Rammers (Tools); Rasps (Tools); Ratchet Braces; Ratchet Screwdrivers; Reamer Wrenches; Reamers; Riveting Hammers; Rock Drills; Sanders (Tools); Saw Files; Saws; Screwdrivers; Sealing Irons; Setting Punches; Shears (Hand Tools); Shovels; Snow Shovels; Soldering Guns; Soldering Irons; Spades; Spuds; Tackers; Tap Wrenches; Taps (Threading Tools); Tenon Saws; Tire Regroovers; Tools; Trimmers; Turning Saws; Tweezers; Woodburning Tools; Wrenches

INTERNATIONAL AND NON-U.S. NATIONAL STANDARDS
SUBJECT INDEX

Hand

Hand Tools (Cont.)
MOD UK DSTAN 51-11: Index-01. Hand Tools-General Purpose Index of Item Names and Related Parts of Defence Standard 51-11 Issue 1; Amendment 1. 25 pp.
MOD UK DL-9-A. Common User Hand Tools.
MOD UK DL-9-A: Sup Common User Hand Tools.

—**Agricultural—Handles**
CGSB 39-GP-14M-79. Handles, Tool, Agricultural, Standard for. 9 pp.

—**Construction Equipment**
JIS A 8905-86. Construction Machinery—Service Tools—Hand Tools for Maintenance and Adjustment Works. 11 pp.

—**Electric**
BSI BS 2769: Part 1-84. 1984 Hand-Held Electric-Motor-Operated Tools Part 1: General Requirements. 33 pp.
BSI BS 2769: Part 1-01. 1984 Amd 1 Hand-Held Electric Motor-Operated Tools Part 1: Specification for General Requirements (AMD 7158) August 15, 1992. 35 pp.
BSI BS 2769: Sec 2.0-84. 1984 Hand-Held Electric-Motor-Operated Tools Part 2: Particular Requirements Section 2.0: General Introduction and List of Sections. 4 pp.
CSA CAN/CSA-C22. 2 NO71.1-M89. Portable Electric Tools; (Gen Instr 1 Thru 4). 89 pp.
DIN VDE 0701 Pt 260-86. Repair, Modification & Inspection of Electrical Appliances; Particular Requirements for Hand-Held Motor-Operated Tools (June). 7 pp.
DIN VDE 0740 Pt 1-81. Hand-Held Motor Operated Tools; General Requirements (Apr). 97 pp.
DIN VDE 0740 Pt 1 A1-82. Hand-Held Motor Operated Electric Tools; General Specifications; Amendment 1; (VDE Specification) (July). 8 pp.
DIN VDE 0740 Pt 22-91. Hand-Held Motor-Operated Tools; Additional Specified Requirements (Apr). 41 pp.
SAA AS 3160-89. Approval and Test Specification—Hand-Held Portable Electric Tools Amdt 1 May 1990 Amdt 2 October 1990 Amdt 3 March 1992 Amdt 4 October 1992 (Superseded by AS/NZS 3160:1993) (in Professional Package 28).
SNZ NZS/AS 3160-93. Approval and Test Specification—Hand-Held Portable Electric Tools. 21 pp.

—**Electric—Glossaries**
DIN VDE 0740 Pt 1000-85. Hand-Held Motor-Operated Electric Tools; Definitions (Jan) (Supersedes DIN 44720 Part 1/06.72). 15 pp.

—**Electric—Radio Frequency Interference**
CENELEC HD 400.1-79. Hand-Held Motor Operated Tools-Part 1: General Specifications. 59 pp.
CENELEC HD 400.1-80. Hand-Held Motor Operated Tools Part I: General Specifications. 60 pp.
CENELEC HD 400.1/A1-81. AMD 1 Hand-Held Motor Operated Tools Part I: General Specifications. 8 pp.
CENELEC HD 400.1 S1/A1-91. AMD 1 Hand-Held Motor Operated Tools Part I: General Specifications. 6 pp.

—**Electric—Safety**
CENELEC HD 400.1-79. Hand-Held Motor Operated Tools-Part 1: General Specifications. 59 pp.
CENELEC HD 400.1-80. Hand-Held Motor Operated Tools Part I: General Specifications. 60 pp.
CENELEC HD 400.1/A1-81. AMD 1 Hand-Held Motor Operated Tools Part I: General Specifications. 8 pp.
CENELEC HD 400.1 S1/A1-91. AMD 1 Hand-Held Motor Operated Tools Part I: General Specifications. 6 pp.
CENELEC HD 400.2 S2-79. Hand-Held Motor Operated Tools Part II: Particular Specifications Sections A to G. 34 pp.
CENELEC HD 400.2 S2-89. Hand-Held Motor Operated Tools—Part II: Particular Specifications Section E: Circular Saws and Circular Knives. 29 pp.
CENELEC HD 400.2E/A1-91. AMD 1 Hand-Held Motor Operated Tools Part II: Particular Specifications Section E: Circular Saws and Circular Knives. 3 pp.
CENELEC HD 400.2E S2 /A1-91. AMD 1 Hand-Held Motor Operated Tools Part II: Particular Specifications Section E: Circular Saws and Circular Knives. 4 pp.

—**Electrical Codes**
CSA 803A Bull. Electrical Bulletin 803A September 3, 1971 to C22.2 NO 37. 1 p.
CSA 649B Bull. Electrical Bulletin 649B October 11, 1967 to C22.2 NO 122. 1 p.
CSA 649E Bull. Electrical Bulletin 649E December 31, 1969 to C22.2 NO 122. 1 p.
CSA 649F Bull. Electrical Bulletin 649F December 15, 1970 to C22.2 NO 122. 1 p.

Hand Tools (Cont.)
—**Electrical Codes (Cont.)**
CSA 803 Bull. Electrical Bulletin 803 December 15, 1970 to C2.2 NO 122. 7 pp.
CSA 803B Bull. Electrical Bulletin 803B February 9, 1973 to C22.2 NO 122. 1 p.

—**Handles**
BSI BS 3823-90. 1990 Grading of Ash and Hickory Wood Handles for Hand Tools. 16 pp.
BSI BS 3823: Part 1-65. (WITHDRAWN) 1965 Grading of Wood Handles for Hand Tools Part 1: Ash and Hickory Handles (Superseded by BS 3823: 1990). 22 pp.

—**Heated**
CSA C22.2 NO 122-M1989. Hand-Held Electrically Heated Tools; (Gen Instr 1 Thru 2). 50 pp.
CSA 1169 Bull. Electrical Bulletin 1169 June 27, 1978 to C22.2 NO 122. 2 pp.

—**Linemen's**
CENELEC PREN 60900-92. Hand Tools for Live Working up to 1000 V A.C. and 1500 V D.C. 39 pp.
DIN VDE 0680 Pt 201 (D)-83. Hand Tools for Live Working up to 1000 V a.c. and 1500 V d.c. (July). 42 pp.
IEC 900-87. Hand Tools for Live Working up to 1000 V A.C. and 1500 V D.C. First Edition. 54 pp.

—**Power—Safety**
CEN PREN 792-1-92. Handheld Non-Electric Power Tools—Safety Requirements—Part 1: General Safety Requirements for all Types of Handheld Non-Electric Power Tools. 17 pp.
CEN PREN 792-2-92. Handheld Non-Electric Power Tools—Safety Requirements—Part 2: Safety Requirements Related to the Energy Supply of Power Tools. 9 pp.
CEN PREN 792-10-92. Handheld Non-Electric Power Tools—Safety Requirements—Part 10: Compression Power Tools, Squeeze Riveters, Presses, Punches. 6 pp.
IEC 745 Pt 1-82. Safety of Hand-Held Motor-Operated Electric Tools Part 1: General Requirements First Edition. 155 pp.

—**Power—Vibration**
BSI BS 6842-87. 1987 Measurement and Evaluation of Human Exposure to Vibration Transmitted to the Hand. 18 pp.
BSI BS 7482: Part 2-91. 1991 Instrumentation for the Measurement of Vibration Exposure of Human Beings Part 2: Specification for Instrumentation for Measuring Vibration Transmitted to the Hand. 11 pp.
BSI BS EN 28662-1-93. 1993 Hand-Held Portable Power Tools—Measurement of Vibrations at the Handle Part 1: General (ISO 8662-1: 1988) (F). 11 pp.
BSI DD ENV 25349-93. 1993 Mechanical Vibration—Guidelines for the Measurement and the Assessment of Human Esposure to Hand-Transmitted Vibration (ISO 5349: 1986) (F). 22 pp.
CEN ENV 25349-92. Mechanical Vibration—Guidelines for the Measurement and the Assessment of Human Exposure to Hand-Transmitted Vibration; (ISO 5349: 1986). 15 pp.
CEN EN 28662-1-92. Hand-Held Portable Power Tools—Measurement of Vibrations at the Handle—Part 1: General; (ISO 8662-1: 1988). 6 pp.
DIN ENGL 45675 Pt 1-87. Exposure to Mechanical Vibration Transmitted to the Hand-Arm System; Principles of Measurement (Sept) (Superseded in Part by DIN EN 28662 Part 1). 8 pp.
DIN ENGL EN 28662 Pt 1-93. Hand-Held Portable Power Tools; Measurement of Vibration at the Handle; General (ISO 8662-1: 1988) (Jan) (Supersedes Parts of DIN 45675 Part 1, September 1987 Edition). 8 pp.
ISO 5349-86. Mechanical Vibration—Guidelines for the Measurement and the Assessment of Human Exposure to Hand-Transmitted Vibration First Edition; (CEN ENV 25349:1992). 14 pp.
ISO 8662 Pt 1-88. Hand-Held Portable Power Tools—Measurement of Vibrations at the Handle—Part 1: General First Edition; (Corrected and Reprinted -1988) (CEN EN 28662-1:1992). 7 pp.
JIS B 4900-86. Method of Measurement and Description of Hand-Transmitted Vibration Level. 9 pp.
JIS C 1511-79. Vibration Level Meters for Hand Tools.

—**Safety**
BSI BS EN 50 059-91. 1991 Electrostatic Hand-Held Spraying Equipment for Non-Flammable Material for Painting and Finishing. 15 pp.
MOD UK DSTAN 51-11: Part 28-01. Hand Tools, General Purpose Part 28 (A): Miscellaneous Items Part 28 (B): Hand Tools for Use in Hazardous or Critical Conditions Issue 1; Amendment 7. 166 pp.

Hand Tools (Cont.)
—**Safety (Cont.)**
MOD UK DSTAN 51-11: Part 28-93. Hand Tools, General Purpose Part 28 (A): Miscellaneous Items (B): Hand Tools for Use in Hazardous or Critical Conditions Issue 2. 146 pp.

—**Ships**
MOD UK NES 43-88. Ancillary Support Equipment, Test Equipment and Tools Issue 2 (09.88). 45 pp.

—**Studs**
MOD UK DSTAN 51-11: Part 30-01. Hand Tools, General Purpose Part 30: Stud Removers and Setters (Including Extractors and Extractor Sets) Issue 1; Amendment 1. 18 pp.

—**Thread Inserts**
MOD UK DSTAN 51-7-76. Hand Tools for Free Running Wire Thread Inserts—Inch Issue 2. 31 pp.
MOD UK DSTAN 51-17-78. Hand Tools for Metric Coiled Wire Screw Thread Inserts Issue 1. 21 pp.

Hand Trucks
See Also: Dollies; Forklifts; Industrial Trucks; Materials Handling Equipment; Pallet Trucks; Platform Trucks; Trucks
JIS B 8920-87. Hand Trucks. 11 pp.
JIS B 8925-90. Hand Trucks with Lifting Table. 10 pp.

—**Axles—Springs**
CNS B2606-81. Industrial Trucks up to 20 km/h; Rubberboned-Springs for Axels of Handcarts and Trailers (Sep)(7867).

—**Glossaries**
BSI BS 3810: Part 1-64. 1964 Glossary of Terms Used in Materials Handling Part 1: Terms Used in Connection with Pallets, Stillages, Hand and Powered Trucks. 37 pp.

—**Wheels**
JIS B 8922-84. Industrial Wheels; (Erratum). 18 pp.

Hand Wheels
Use For: Handwheels **See Also:** Control Wheels; Wheels
DIN ENGL 6324-71. Operating Elements for Clamping Devices; Survey (Jan). 2 pp.
JIS B 2601-82. Handwheels. 16 pp.

—**Aircraft**
SBAC AS 152 ISSUE 6. Handwheel 3 1/2 Inch Diameter. 1 p.
SBAC AS 3205 ISSUE 1. Handwheel Assembly Showing Switch Fixing. 1 p.
SBAC AS 6604 ISSUE 1. Handwheel Assembly Showing Switch Fixing.

—**Aircraft—Bosses—Trim Controls**
SBAC AS 887 ISSUE 8. Boss for Handwheel—Trim Control. 1 p.

—**Curved Arm**
CNS B2028-54. Hand Wheels with Crooked Rims and Curved Arms (Jul)(332)(R 1970).

—**Ergonomics**
DIN ENGL 33411 Pt 3-86. Human Physical Strength; Maximum Static Action Moments Applied by Male Operators When Actuating Hand-Wheels (Dec) 5 pp.

—**Handles**
DIN ENGL 39-62. Fixed Ball Handles (Oct). 2 pp.

—**Handles—Rotatable**
DIN ENGL 98-62. Rotatable Ball Handles (Oct). 2 pp.

—**Heat Resistant**
CNS B2026-54. Hand Wheels of Low Heat Insulation Materials (Jul)(330)(R 1973).

—**Offset Arm**
CNS B2029-80. Hand Wheels, Offset-Arm Type, Round Hub Hole (Jan)(333).
CNS B2030-80. Hand Wheels, Offset-Arm Type, Square Hub Hole (Jan)(334).
CNS B2031-80. Hand Wheels, Offset-Arm Type, Tapered Square Hub Hole (Jan)(335).
DIN ENGL 950-70. Handwheels; Offset-Arm Type; Round Hub Hole (June). 3 pp.
DIN ENGL 951-72. Handwheels; Offset Arm Type; Square Hub Hole (Apr). 2 pp.

—**Plastic—Offset Arm**
CNS B2396-80. Hand Wheels of Plastics, Offset-Arm Type, Hole of Hub Round or Parallel Square (Jan)(5111).

—**Square Head**
CNS B3002-72. Square Heads and Holes (for Shaft, Hand Wheel and Handles) (Jun)(66).

INDUSTRY STANDARDS

INTERNATIONAL AND NON-U.S. NATIONAL STANDARDS
SUBJECT INDEX

Hand

Hand Wheels (Cont.)
—**Straight Arm**
CNS B2027-54. Hand Wheels with Crooked Rims and Straight Arms (Jul)(331)(R 1973).

Handbooks
See Also: CCIR Handbooks; Information Services; Publications; Technical Manuals; Training Manuals

—**CECC—Management**
CECC CECC 00 700 ISSUE 1-84. Handbook of Administration (En) AMD 1 & ADDENDA (En) ADDENDA 2 (En) AMD 2 (En) AMD 3 (En) AMD 4 (En) AMD 5 (En) AMD 6 (En) AMD 7 (En) AMD 8 (En). 80 pp.

—**CECC—Specifications Preparation**
CECC CECC 00 400 ISSUE 3-86. Handbook for the Production of CECC Documents (En) ERRATUM (En, Fr, Ge) AMD 1 (En, Fr, Ge) AMD 2 (En, Fr, Ge) AMD 4 (En, Fr, Ge) AMD 5 (En, Fr, Ge) AMD 6 (En, Fr, Ge) AMD 7 (En, Fr, Ge) AMD 8 (En, Fr, Ge). 103 pp.

—**CECC—Standards Preparation**
CECC CECC 00 400 ISSUE 3-86. Handbook for the Production of CECC Documents (En) ERRATUM (En, Fr, Ge) AMD 1 (En, Fr, Ge) AMD 2 (En, Fr, Ge) AMD 4 (En, Fr, Ge) AMD 5 (En, Fr, Ge) AMD 6 (En, Fr, Ge) AMD 7 (En, Fr, Ge) AMD 8 (En, Fr, Ge). 103 pp.

—**Inductive Coordination**
CSA C22.3 NO 3.1-1974SP. Inductive Coordination Handbook for Use with CSA Standard C22.3 No. 3 of the Canadian Electrical Code, Part III. 54 pp.

—**Military—Oceanography**
NATO STANAG 1171 ED 6 AMD 0-86. NATO Handbook of Military Oceanographic Information and Services—ATP-32(A)(Navy/Air). 4 pp.

—**Telephone Services—Cost Analysis**
CCITT FASCICLE II.1-88. General Tariff Principles—Charging and Accounting in International Telecommunications Services Series D Recommendations. 371 pp.

—**Telephone Services—Expenses**
CCITT FASCICLE II.1-88. General Tariff Principles—Charging and Accounting in International Telecommunications Services Series D Recommendations. 371 pp.

—**Telephone Services—Tariffs**
CCITT FASCICLE II.1-88. General Tariff Principles—Charging and Accounting in International Telecommunications Services Series D Recommendations. 371 pp.

—**Telex Communications—Cost Analysis**
CCITT FASCICLE II.1-88. General Tariff Principles—Charging and Accounting in International Telecommunications Services Series D Recommendations. 371 pp.

—**Telex Communications—Expenses**
CCITT FASCICLE II.1-88. General Tariff Principles—Charging and Accounting in International Telecommunications Services Series D Recommendations. 371 pp.

—**Telex Communications—Tariffs**
CCITT FASCICLE II.1-88. General Tariff Principles—Charging and Accounting in International Telecommunications Services Series D Recommendations. 371 pp.

Handholes
See Also: Inspection

—**Ships**
ISO 3876-86. Shipbuilding—Inland Vessels—Hand-Holes Second Edition. 5 pp.

Handicap Architectural Design
Use: Barrier Free Design

Handicapped Accessible Equipment
Scope Note: Includes equipment which has been modified to provide access to disabled persons. For assistive equipment manufactured for use by disabled persons, see the listings under Disability Equipment
Use For: Wheelchair Accessible Equipment
See Also: Disability Equipment; Handicapped Persons

—**Children**
SAA AS 1428.3-92. Design for Access and Mobility—Part 3: Requirements for Children and Adolescents with Physical Disabilities (in Professional Packages 17F, 20, 21, 55, 62-69). 16 pp.

Handicapped Accessible Equipment (Cont.)
—**Elevators (Lifts)**
SAA AS 1735.12-86. Lifts, Escalators, and Moving Walks (Known as the SAA Lift Code)—Part 12: Facilities for Persons with Disabilities Amdt 1 December 1991. 29 pp.

—**Escalators**
SAA AS 1735.12-86. Lifts, Escalators, and Moving Walks (Known as the SAA Lift Code)—Part 12: Facilities for Persons with Disabilities Amdt 1 December 1991. 29 pp.

—**Glossaries**
JIS T 0102-91. Glossary of Terms Used in Technical Systems and Aids for Disabled or Handicapped Persons.

—**Passenger Conveyors**
SAA AS 1735.12-86. Lifts, Escalators, and Moving Walks (Known as the SAA Lift Code)—Part 12: Facilities for Persons with Disabilities Amdt 1 December 1991. 29 pp.

Handicapped Equipment
Use: Disability Equipment

Handicapped Persons
Use For: Disabled Persons *See Also:* Aging Persons; Disability Equipment; Handicapped Accessible Equipment; Hearing Impaired Persons; Visually Impaired Persons

—**Automobiles—Adaptive Controls**
CSA CAN3-Z323 .1.2-M85. Adaptive Automotive Control Systems for Physically Disabled Persons; (Gen Instr 1). 16 pp.
SAA AS 3954.1-91. Motor Vehicle Controls—Adaptive Systems for People with Disabilities—Part 1: General Requirements. 8 pp.

—**Automobiles—Manual Controls**
SAA AS 3954.2-91. Motor Vehicle Controls—Adaptive Systems for People with Disabilities—Part 2: Hand Controls—Product Requirements. 5 pp.
SNZ NZS 5832: Part 1-88. Driving Controls for People with Disabilities Part 1: Hand Controls. 24 pp.

—**Buildings—Design**
BSI BS 5619-78. 1978 Amd 1 Design of Housing for the Convenience of Disabled People. 9 pp.
BSI BS 5810-79. 1979 Code of Practice for Access for the Disabled to Buildings. 15 pp.
BSI PD 6523-89. 1989 Information on Access to and Movement Within and Around Buildings and on Certain Facilities for Disabled People. 49 pp.
SAA AS 1428.2-92. Design for Access and Mobility—Part 2: Enhanced and Additional Requirements—Buildings and Facilities (Supersedes AS 1428 Supplement 1—1988) (in Professional Packages 17F, 20, 21, 55, 62-69). 49 pp.
SNZ NZS 4121-85. Code of Practice for Design for Access and Use of Buildings and Facilities by Disabled Persons. 37 pp.
SNZ NZMP 4122-89. Guide to Approachability, Accessibility and Usability of Buildings. 60 pp.

—**Buildings—Fire Protection**
BSI BS 5588: Part 8-88. 1988 Fire Precautions in the Design and Construction of Buildings Part 8: Code of Practice for Means of Escape for Disabled People. 24 pp.

—**Data Terminal Equipment**
CEPT T/SF 3-77. Services Et Facilities Pour Les Personnes Handicapees. 7 pp.

—**Electrical Aids—Safety**
CSA Z323.3.1-1982. Electrical Aids for Physically Disabled Persons. 17 pp.

—**Elevators (Lifts)**
SAA AS 1735.13-86. Lifts, Escalators, and Moving Walks (Known as the SAA Lift Code)—Part 13: Lifts for Persons with Limited Mobility—Manually Powered Amdt 1 July 1989. 6 pp.
SAA AS 1735.14-90. Lifts, Escalators, and Moving Walks (Known as the SAA Lift Code)—Part 14: Lifts for People with Limited Mobility—Restricted Use—Low-Rise Platforms. 7 pp.
SAA AS 1735.15-90. Lifts, Escalators, and Moving Walks (Known as the SAA Lift Code)—Lifts for People with Limited Mobility—Restricted Use—Non-Automatically Controlled. 17 pp.

—**Elevators (Lifts)—Control Systems Equipment**
CSA CAN/CSA-B355-M86. Elevating Devices for the Handicapped. 35 pp.

Handicapped Persons (Cont.)
—**Elevators (Lifts)—Control Systems Equipment (Cont.)**
CSA CAN/CSA-B613-M87. Elevating Devices for the Handicapped in Private Residences; (Gen Instr 1). 30 pp.

—**Stairlifts**
BSI BS 5776-79. 1979 Amd 2 Powered Stairlifts (AMD 6523) December 21, 1990. 19 pp.

—**Stairlifts—Control Systems Equipment**
CSA CAN/CSA-B355-M86. Elevating Devices for the Handicapped. 35 pp.
CSA CAN/CSA-B613-M87. Elevating Devices for the Handicapped in Private Residences; (Gen Instr 1). 30 pp.

—**Technical Aids—Glossaries**
JIS T 0102-91. Glossary of Terms Used in Technical Systems and Aids for Disabled or Handicapped Persons.

—**Telecommunication Systems**
CEPT T/SF 3-77. Services Et Facilities Pour Les Personnes Handicapees. 7 pp.

Handicrafts
Use For: Crafts *See Also:* Arts

—**Classification**
BSI BS 1000: (688/689)-80. 1980 Universal Decimal Classification (UDC). English Full Edition (688/689): Fancy Goods. Toys. Hobbies. 15 pp.
BSI BS 1000: (73/76)-77. 1977 Universal Decimal Classification (UDC). English Full Edition (73/76): Various Arts and Crafts. 23 pp.
SNZ NZS/BS 1000 (688/689)-80. Universal Decimal Classification Fancy Goods. Toys. Hobbies. 16 pp.
SNZ NZS/BS 1000 (73/76)-77. Universal Decimal Classification Various Arts and Crafts. 24 pp.

Handlamps
Use: Flashlights

Handles
Scope Note: For additional listings, see the subheading Handles under the specific item
Use For: Knobs (Handles) *See Also:* Gripping Devices; Grips; Hardware; Knobs; Knurls; Levers
CNS A2015-89. Level Handles, Two-Sided Operation (Jul)(865). 2 pp.
CNS A2016-89. Handles, One-Sided Operation (Jul)(866). 2 pp.

—**Aircraft**
SBAC AS 2519 ISSUE 6. Inner Handle. 1 p.
SBAC AS 2520 ISSUE 5. Outer Handle. 1 p.

—**Aircraft—Assembly**
SBAC AS 2549 ISSUE 5. Outer Handle Assembly Double Handle Lock. 1 p.
SBAC AS 2552 ISSUE 5. Outer Handle Assembly Single Handle Lock. 1 p.
SBAC AS 6221 ISSUE 2(I). Assembly of Handle. 1 p.

—**Aircraft—Escape Hatch Control**
SBAC RS 433 ISSUE 1. Handle for Escape Hatch Control.

—**Aircraft—Hood Jettison Release**
SBAC AS 4556 ISSUE 2. Hood Jettison Release Handle. 1 p.

—**Aircraft—Lock—Assembly**
SBAC AS 6523 ISSUE 1. Outer Handle Assembly—Double Handle Lock.
SBAC AS 6536 ISSUE 1. Outer Handle Assembly—Single Handle Lock.

—**Aircraft—Sealing Rings**
SBAC AS 2559 ISSUE 3. Seal—Handle. 1 p.

—**Ball**
CNS B3329-83. Ball Handles (Mar)(5110).
DIN ENGL 39-62. Fixed Ball Handles (Oct). 2 pp.
DIN ENGL 6337-72. Ball Handles (Apr). 2 pp.

—**Ball—Rotatable**
DIN ENGL 98-62. Rotatable Ball Handles (Oct). 2 pp.

—**Bow**
CNS A2018-89. Handles, Bow Form (Jul)(868). 2 pp.

—**Crank**
DIN ENGL 468-62. Crank Handles; Offset Type (Oct). 2 pp.
DIN ENGL 469-62. Crank Handles; Straight Pattern (Oct). 2 pp.

—**Machine**
CNS B3101-80. Fixed Machine Handle (Mar)(319).

INTERNATIONAL AND NON-U.S. NATIONAL STANDARDS
SUBJECT INDEX

Handles (Cont.)
—Machine (Cont.)
CNS B3102-80. Revolving Machine Handles (Mar)(320).
DIN ENGL 830-62. Machine Handles; Long Pattern (Aug). 1 p.

—Mops
CGSB 22.41-92. Mop Stick Assemblies, Screw-Clamp Type. 7 pp.

—Palm
DIN ENGL 6335-68. Star Handles (May). 2 pp.

—Rolling Stock—Bow
CNS E1010-81. Bow-Type Handles for Rail Vehicles (Oct)(7981).

—Square Head
CNS B3002-72. Square Heads and Holes (for Shaft, Hand Wheel and Handles) (Jun)(66).

—Star
DIN ENGL 6336-68. Machine Knobs (May). 2 pp.

—Taper
CNS B3106-54. Fixed Handle (Taper Form) (Jul)(324)(R 1973).
DIN ENGL 99-72. Clamping Levers (June). 2 pp.

—Taper—Rotatable
CNS B3107-54. Rotatable Handle (Taper Form) (Jul)(325) (R 1973).

—Tools
SAA AS 1729-75. Timber Handles for Tools Amdt 1 March 1976. 28 pp.

Handley Page Halifax Aircraft
See Also: Aircraft
—Accidents
CAA. Handley Page Halifax/Halton (World Airline Accident Summary). 1 p.

Handley Page Halton Aircraft
See Also: Aircraft
—Accidents
CAA. Handley Page Halifax/Halton (World Airline Accident Summary). 1 p.

Handley Page Herald Aircraft
See Also: Aircraft
—Accidents
CAA. Handley Page Herald (World Airline Accident Summary). 1 p.
CAA. Handley Page Herald. 1 p.
—Mandatory Aircraft Modifications and Inspections Summaries
CAA. Handley Page Herald. 19 pp.

Handley Page Hermes Aircraft
See Also: Aircraft
—Accidents
CAA. Handley Page Hermes (World Airline Accident Summary). 1 p.

Handley Page Marathon Aircraft
See Also: Aircraft
—Accidents
CAA. Handley Page Marathon (World Airline Accident Summary). 1 p.

Handoff
Use: Hand Off

Handover
Use: Hand Off

Handrails
See Also: Guardrails; Guide Rails
—PVC
CNS K3047-81. Flexible PVC Handrail (Sep)(6484). 2 pp.
CNS K6592-80. Method of Test for Nonrigid PVC Handrail (Sep)(6485).

—Residential Buildings
JIS A 6601-92. Metal Components for Balcony and Handrails of Dwellings. 32 pp.

—Ships—Stanchions
JIS F 2607-75. Ships' Handrail Stanchion. 7 pp.

—Ships—Steel
BSI BS MA 40: Part 3-75. 1975 Marine Guardrails, Stanchions, Etc Part 3: Stormrails (Exterior). 4 pp.

Handrails (Cont.)
—Ships—Steel (Cont.)
JIS F 7502-89. Steel Ladders and Steel Handrails for Marine Use.

—Ships—Wood
CNS F3022-80. Ships' Wooden Handrails (Mar)(5330).
JIS F 2606-58. Ships' Wooden Handrails (R 1964). 3 pp.

Handsaws
Use: Saws

Handsets
See Also: Audio Equipment; Headsets
—Aircraft
EUROCAE ED-18 07.85. Audio Systems Characteristics and MPS Aircraft Microphones (Except Carbon), Aircraft Headsets, Handsets and Loudspeakers, Aircraft Audio Selector Panels and Amplifiers. 108 pp.

—Aircraft—Environmental Testing
EUROCAE ED-18 07.85. Audio Systems Characteristics and MPS Aircraft Microphones (Except Carbon), Aircraft Headsets, Handsets and Loudspeakers, Aircraft Audio Selector Panels and Amplifiers. 108 pp.

—Packaging
MOD UK DSTAN 81-87-91. Packaging of Handsets, Headsets, Microphones, Ancillary Parts and Accessories Issue 1. 24 pp.

—Sidetone—Maritime Mobile Satellite Communications
CCIR RECMN 549-1-82. Side Tone Reference Equivalent of Handset Used on Board a Ship in the Maritime Mobile-Satellite Service and in Automated VHF/UHF Mari-time Mobile Radiotele-phone Systems—Section 8G—Availability, Performance Objectives and Interworking with Terrestrial Networks. 1 p.

Handsheets
See Also: Papers
SAA AS 1301.214S-89. Methods of Test for Pulp and Paper (Metric Units)—Part 214s: Equipment for Preparation of Handsheets (This is a Joint Standard with SANZ NZS 1301) (Superseded by AS/NZS 1301.214s:1993).
SAA AS 1301.215S-89. Methods of Test for Pulp and Paper (Metric Units)—Part 215s: Removal of Latency (This is a Joint Standard with SANZ NZS 1301). 2 pp.
SNZ NZS/AS 1301. 214S-93. Methods of Test for Pulp and Paper Equipment for Preparation of Handsheets (This is a Joint Standard with SAA AS 1301.214S). 8 pp.
SNZ NZS/AS 1301. 215S-89. Methods of Test for Pulp and Paper Removal of Latency (This is a Joint Standard with SAA AS 1301.215S). 2 pp.

—Dyed
CNS P3075-85. Method of Test for Forming Dyed Handsheet (Sep)(11375).

—Physical Testing
CNS P3072-85. Method of Test for Physical Properties of Handsheet (Jun)(11291). 4 pp.
SAA AS/NZS 1301. 203S-93. Methods of Test for Pulp and Paper (Metric Units)—Part P203s: Forming Handsheets for Physical Testing of Pulp NZS 1301). 8 pp.
SAA AS 1301.208S-89. Methods of Test for Pulp and Paper (Metric Units) —Part 208s: Physical Testing of Handsheets (This is a Joint Standard with SANZ NZS 1301). 6 pp.
SNZ NZS/AS 1301. P203S-80. Methods of Test for Pulp and Paper Forming Handsheets for Physical Testing of Pulp. Plus Amendments 1 and 2 (This is a Joint Standard with SAA AS 1301.P203S). 4 pp.
SNZ NZS/AS 1301. 208S-89. Methods of Test for Pulp and Paper Physical Testing of Handsheets (This is a Joint Standard with SAA AS 1301.208S). 6 pp.

—Pulps—Optical Properties
CPPA C.5-88. Forming Handsheets for Optical Tests of Pulp (British Sheet Machine Method). 3 pp.

—Pulps—Physical Properties
CPPA C.4-50. Forming Handsheets for Physical Tests of Pulp. 9 pp.
CPPA D.12-91. Physical Testing of Pulp Handsheets. 2 pp.

—Pulps—Tear Strength
CPPA D.9-77. Internal Tearing Resistance of Paper, Paperboard and Pulp Handsheets Determined with the Appita-Elmendorf Instrument. 6 pp.

Handtools
Use: Hand Tools

Handwheels
Use: Hand Wheels

Hangers (Clothing)
CNS S1110-77. Clothes Pegs (Single Arm Light Duty) (Mar)(7071). 1 p.
CNS S1111-81. Clothes-Pegs (Single Arm, Heavy Duty) (Mar)(7072). 1 p.
CNS S1112-81. Hanger for Clothes-Pegs (Mar)(7073). 1 p.
CNS S1115-81. Clothes-Pegs (Double Arm) (Mar)(7076). 1 p.

—Collar
CNS Z7014-66. Collar Hooks (Large Size) (for Clothing) (Sep)(2515).
CNS Z7015-66. Collar Hooks (Small Size) (for Clothing) (Sep)(2516).

Hangers (Fasteners)
See Also: Brackets; Hooks; Pipe Supports
—Electrical Metallic Tubing
CSA CAN/CSA-C22. 2 NO 18-92. Outlet Boxes, Conduit Boxes, and Fittings; (Gen Instr 1 Thru 2). 118 pp.

—Masonry
CEN PREN 845-1-92. Specification for Ancillary Components for Masonry—Part 1: Ties, Straps, Hangers, Brackets and Support Angles. 26 pp.

—Rigid Metal Conduits
CSA CAN/CSA-C22. 2 NO 18-92. Outlet Boxes, Conduit Boxes, and Fittings; (Gen Instr 1 Thru 2). 118 pp.

Hanging Hooks
Use: Hooks

Harbin Y-12 II Aircraft
See Also: Aircraft
—Certification
CAA. Harbin Y12 II. 8 pp.
—Data Sheets
CAA FA49. Harbin Y12 (11). 2 pp.

Harbors
Use For: Ports See Also: Marinas
—Military Geographic Documentation
NATO STANAG 2255 ED 4 AMD 8-76. MGD-Ports. 55 pp.
NATO STANAG 2255 ED 4 AMD 9-76. MGD-Ports. 58 pp.

—Military Geographic Information
NATO STANAG 2255 ED 4 AMD 9-76. MGD-Ports. 58 pp.

—Naval Ships—Radhaz Control
NATO STANAG 1233 ED 1 AMD 1-84. Procedures for Radhaz Control in Ports and the Territorial Sea. 11 pp.

Hard Coal
Use: Anthracite

Hard Contact Lenses
Use: Contact Lenses—Hard

Hard Facing
Use: Hard Surfacing

Hard Lead
See Also: Lead Alloys
CNS H2006-52. Method of Test for Hard Lead (Mar)(203) (R 1973).
—Plates
JIS H 4302-56. Hard Lead Plates (R 1968). 5 pp.

Hard Metals
See Also: Alloys; Carbides; Metal Powders; Sintered Materials
—Atomic Absorption Spectrometry
BSI BS 5600: SUB SEC 4.17.1-86. (WITHDRAWN) 1986 Powder Metallurgical Materials and Products Part 4: Methods of Testing and Chemical Analysis of Hardmetals Section 4.17: Chemical Analysis by Flame Atomic Absorption Spectrometry Subsection 4.17.1: General Requirements. 4 pp.
BSI BS EN 27627-1-93. 1993 Hardmetals—Chemical Analysis by Flame Atomic Absorption Spectrometry Part 1: General Requirements (ISO 7627-1: 1983) (V). 9 pp.

INDUSTRY STANDARDS

Hard Metals *(Cont.)*

—Atomic Absorption Spectrometry *(Cont.)*

CEN EN 27627-1-93. Hardmetals—Chemical Analysis by Flame Atomic Absorption Spectrometry—Part 1: General Requirements (ISO 7627-1: 1983). 4 pp.

ISO 7627 Pt 1-83. Hardmetals—Chemical Analysis by Flame Atomic Absorption Spectrometry—Part 1: General Requirements First Edition (CEN EN 27627-1: 1993). 4 pp.

—Calcium Content—Atomic Absorption Spectrometry

BSI BS 5600: SUB SEC 4.17.2-86. (WITHDRAWN) 1986 Part 4: Methods of Testing and Chemical Analysis of Hardmetals Section 4.17: Chemical Analysis by Flame Atomic Absorption Spectrometry Subsection 4.17.2: De-termination of Calcium, Potassium, Magnesium and Sodium in Contents from 0.001% to 0.02% (m/m). 4 pp.

BSI BS EN 27627-2-93. 1993 Hardmetals—Chemical Analysis by Flame Atomic Absorption Spectrometry Part 2: Determination of Calcium, Potassium, Magnesium and Sodium, in Contents from 0.001 to 0.02 % (m/m) (ISO 7627-2: 1983) (V). 10 pp.

CEN EN 27627-2-93. Hardmetals—Chemical Analysis by Flame Atomic Absorption Spectrometry—Part 2: Determination of Calcium, Potassium, Magnesium and Sodium in Contents from 0,001 to 0,02 % (m/m) (ISO 7627-2: 1983). 5 pp.

ISO 7627 Pt 2-83. Hardmetals—Chemical Analysis by Flame Atomic Absorption Spectrometry—Part 2: Determination of Calcium, Potassium, Magnesium and Sodium in Contents from 0,001 to 0,02 % (m/m) First Edition (CEN EN 27627-2: 1993). 4 pp.

—Carbon Content—Gravimetric Analysis

BSI BS 5600: Sec 4.1-87. (WITHDRAWN) 1987 Powder Metallurgical Materials and Products Part 4: Methods of Testing and Chemical Analysis of Hardmetals Section 4.1: Determination of Total Carbon: Gravimetric Method (Superseded by BS EN 23907: 1993). 5 pp.

BSI BS 5600: Sec 4.2-87. 1987 Powder Metallurgical Materials and Products Part 4: Methods of Testing and Chemical Analysis of Hardmetals Section 4.2: Determination of Insoluble (Free) Carbon: Gravimetric Method. 4 pp.

BSI BS EN 23907-93. 1993 Hardmetals—Determination of Total Carbon Content—Gravimetric Method (ISO 3907: 1985) (V). 10 pp.

BSI BS EN 23908-93. 1993 Hardmetals—Determination of Insoluble (Free) Carbon Content—Gravimetric Method (ISO 3908: 1985) (V). 10 pp.

CEN EN 23908-93. Hardmetals—Determination of Insoluble (Free) Carbon Content—Gravimetric Method (ISO 3908: 1985). 5 pp.

ISO 3907-85. Hardmetals—Determination of Total Carbon Content—Gravimetric Method Second Edition. 5 pp.

ISO 3908-85. Hardmetals—Determination of Insoluble (Free) Carbon Content—Gravimetric Method Second Edition (CEN EN 23908: 1993). 4 pp.

—Carbon Content—Metallography

BSI BS 5600: Sec 4.14-80. (WITHDRAWN) 1980 Powder Metallurgical Materials and Products Part 4: Methods of Testing and Chemical Analysis of Hardmetals Section 4.14: Metallographic Determination of Porosity and Uncombined Carbon (Superseded by BS EN 24505: 1993). 7 pp.

BSI BS EN 24505-93. 1993 Hardmetals—Metallographic Determination of Porosity and Uncombined Carbon (ISO 4505: 1978) (V). 12 pp.

CEN EN 24505-93. Hardmetals—Metallographic Determination of Porosity and Uncombined Carbon (ISO 4505: 1978). 7 pp.

ISO 4505-78. Hardmetals—Metallographic Determination of Porosity and Uncombined Carbon First Edition (CEN EN 24505: 1993). 7 pp.

—Chromium Content—Atomic Absorption Spectrometry

BSI BS 5600: SUB SEC 4.17.6-86. (WITHDRAWN) 1986 Part 4: Methods of Testing and Chemical Analysis of Hardmetals Section 4.17: Chemical Analysis by Flame Atomic Absorption Spectrometry Subsection 4.17.6: Determination of Chromium in Contents from 0.01% to 2% (m/m). 4 pp.

BSI BS EN 27627-6-93. 1993 Hardmetals—Chemical Analysis by Flame Atomic Absorption Spectrometry Part 6: Determination of Chromium in Contents from 0.01 to 2 % (m/m) (ISO 7627-6: 1985) (V). 10 pp.

CEN EN 27627-6-93. Hardmetals—Chemical Analysis by Flame Atomic Absorption Spectrometry—Part 6: Determination of Chromium in Contents from 0,01 to 2 % (m/m) (ISO 7627-6: 1985). 5 pp.

ISO 7627 Pt 6-85. Hardmetals—Chemical Analysis by Flame Atomic Absorption Spectrometry—Part 6: Determination of Chromium in Contents from 0,01 to 2 % (m/m) First Edition (CEN EN 27627-6: 1993). 4 pp.

Hard Metals *(Cont.)*

—Cobalt Content—Atomic Absorption Spectrometry

BSI BS 5600: SUB SEC 4.17.3-86. (WITHDRAWN) 1986 Part 4: Methods of Testing and Chemical Analysis of Hardmetals Section 4.17: Chemical Analysis by Flame Atomic Absorption Spectrometry Subsection 4.17.3: De-termination of Cobalt, Iron, Manganese and Nickel in Contents from 0.01% to 0.05% (m/m). 4 pp.

BSI BS 5600: SUB SEC 4.17.5-86. (WITHDRAWN) 1986 Part 4: Methods of Testing and Chemical Analysis of Hardmetals Section 4.17: Chemical Analysis by Flame Atomic Absorption Spectrometry Sub 4.17.5:Determination of Cobalt, Iron, Manganese, Molybdenum, Nickel Titanium and Vanadium in Cont from 0.5% to 2% m/m. 4 pp.

BSI BS EN 27627-3-93. 1993 Hardmetals—Chemical Analysis by Flame Atomic Absorption Spectrometry Part 3: Determination of Cobalt, Iron, Manganese and Nickel in Contents from 0.01 to 0.5 % (m/m) (ISO 7627-3: 1983) (V). 10 pp.

BSI BS EN 27627-5-93. 1993 Hardmetals—Chemical Analysis by Flame Atomic Absorption Spectrometry Part 5: Determination of Cobalt, Iron, Manganese, Molybdenum, Nickel, Titanium and Vanadium in Contents from 0.5 to 2 % (m/m) (ISO 7627-5: 1985) (V). 10 pp.

CEN EN 27627-3-93. Hardmetals—Chemical Analysis by Flame Atomic Absorption Spectrometry—Part 3: Determination of Cobalt, Iron, Manganese and Nickel in Contents from 0,01 to 0,5 % (m/m) (ISO 7627-3: 1983). 5 pp.

CEN EN 27627-5-93. Hardmetals—Chemical Analysis by Flame Atomic Absorption Spectrometry—Part 5: Determination of Cobalt, Iron, Manganese, Molybdenum, Nickel, Titanium and Vanadium in Contents from 0,5 to 2 % (m/m) (ISO 7627-5: 1983). 5 pp.

ISO 7627 Pt 3-83. Hardmetals—Chemical Analysis by Flame Atomic Absorption Spectrometry—Part 3: Determination of Cobalt, Iron, Manganese and Nickel in Contents from 0,01 to 0,5 % (m/m) First Edition (CEN EN 27627-3: 1993). 4 pp.

ISO 7627 Pt 5-83. Hardmetals—Chemical Analysis by Flame Atomic Absorption Spectrometry—Part 5: Determination of Cobalt, Iron, Manganese, Molybdenum, Nickel, Titanium and Vanadium in Contents from 0,5 to 2 % (m/m) First Edition (CEN EN 27627-5: 1993). 4 pp.

—Cobalt Content—Potentiometric Analysis

BSI BS 5600: Sec 4.3-79. 1979 Powder Metallurgical Materials and Products Part 4: Methods of Testing and Chemical Analysis of Hardmetals Section 4.3: Determination of Cobalt: Potentiometric Method. 4 pp.

ISO 3909-76. Hardmetals—Determination of Cobalt—Potentiometric Method First Edition. 5 pp.

—Coercive Force

BSI BS 5600: Sec 4.9-79. (WITHDRAWN) 1979 Powder Metallurgical Materials and Products Part 4: Methods of Testing and Chemical Analysis of Hardmetals Section 4.9: Determination of (the Magnetization) Coercivity (Superseded by BS EN 23326: 1993). 4 pp.

BSI BS EN 23326-93. 1993 Hardmetals—Determination of (the Magnetization) Coercivity (ISO 3326: 1975) (V). 9 pp.

CEN EN 23326-93. Hardmetals—Determination of (the Magnetization) Coercivity (ISO 3326: 1975). 4 pp.

ISO 3326-75. Hardmetals—Determination of (the Magnetization) Coercivity First Edition (CEN EN 23326: 1993). 4 pp.

—Compression Testing

BSI BS 5600: Sec 4.16-81. (WITHDRAWN) 1981 Powder Metallurgical Materials and Products Part 4: Methods of Testing and Chemical Analysis of Hardmetals Section 4.16: Compression Test (Superseded by BS EN 24506: 1993). 6 pp.

BSI BS EN 24506-93. 1993 Hardmetals—Compression Test (ISO 4506: 1979) (V). 11 pp.

CEN EN 24506-93. Hardmetals—Compression Test (ISO 4506: 1979). 6 pp.

ISO 4506-79. Hardmetals—Compression Test First Edition (CEN EN 24506: 1993). 6 pp.

—Density

BSI BS 5600: Sec 4.4-79. (WITHDRAWN) 1979 Powder Metallurgical Materials and Products Part 4: Methods of Testing and Chemical Analysis of Hardmetals Section 4.4: Determination of Density (Superseded by BS EN 23369: 1993). 6 pp.

—Drawing Dies

BSI BS 4276-68. 1968 Hard Metal for Wire, Bar and Tube Drawing Dies. 12 pp.

BSI BS 4276-01. 1968 Amd 1 Hard Metal for Wire, Bar and Tube Drawing Dies (AMD 7746) August 15, 1993 (F). 13 pp.

Hard Metals *(Cont.)*

—Iron Content—Atomic Absorption Spectrometry

BSI BS 5600: SUB SEC 4.17.3-86. (WITHDRAWN) 1986 Part 4: Methods of Testing and Chemical Analysis of Hardmetals Section 4.17: Chemical Analysis by Flame Atomic Absorption Spectrometry Subsection 4.17.3: De-termination of Cobalt, Iron, Manganese and Nickel in Contents from 0.01% to 0.05% (m/m). 4 pp.

BSI BS 5600: SUB SEC 4.17.5-86. (WITHDRAWN) 1986 Part 4: Methods of Testing and Chemical Analysis of Hardmetals Section 4.17: Chemical Analysis by Flame Atomic Absorption Spectrometry Sub 4.17.5:Determination of Cobalt, Iron, Manganese, Molybdenum, Nickel Titanium and Vanadium in Cont from 0.5% to 2% m/m. 4 pp.

BSI BS EN 27627-3-93. 1993 Hardmetals—Chemical Analysis by Flame Atomic Absorption Spectrometry Part 3: Determination of Cobalt, Iron, Manganese and Nickel in Contents from 0.01 to 0.5 % (m/m) (ISO 7627-3: 1983) (V). 10 pp.

BSI BS EN 27627-5-93. 1993 Hardmetals—Chemical Analysis by Flame Atomic Absorption Spectrometry Part 5: Determination of Cobalt, Iron, Manganese, Molybdenum, Nickel, Titanium and Vanadium in Contents from 0.5 to 2 % (m/m) (ISO 7627-5: 1985) (V). 10 pp.

CEN EN 27627-3-93. Hardmetals—Chemical Analysis by Flame Atomic Absorption Spectrometry—Part 3: Determination of Cobalt, Iron, Manganese and Nickel in Contents from 0,01 to 0,5 % (m/m) (ISO 7627-3: 1983). 5 pp.

CEN EN 27627-5-93. Hardmetals—Chemical Analysis by Flame Atomic Absorption Spectrometry—Part 5: Determination of Cobalt, Iron, Manganese, Molybdenum, Nickel, Titanium and Vanadium in Contents from 0,5 to 2 % (m/m) (ISO 7627-5: 1983). 5 pp.

ISO 7627 Pt 3-83. Hardmetals—Chemical Analysis by Flame Atomic Absorption Spectrometry—Part 3: Determination of Cobalt, Iron, Manganese and Nickel in Contents from 0,01 to 0,5 % (m/m) First Edition (CEN EN 27627-3: 1993). 4 pp.

ISO 7627 Pt 5-83. Hardmetals—Chemical Analysis by Flame Atomic Absorption Spectrometry—Part 5: Determination of Cobalt, Iron, Manganese, Molybdenum, Nickel, Titanium and Vanadium in Contents from 0,5 to 2 % (m/m) First Edition (CEN EN 27627-5: 1993). 4 pp.

—Magnesium Content—Atomic Absorption Spectrometry

BSI BS 5600: SUB SEC 4.17.2-86. (WITHDRAWN) 1986 Part 4: Methods of Testing and Chemical Analysis of Hardmetals Section 4.17: Chemical Analysis by Flame Atomic Absorption Spectrometry Subsection 4.17.2: De-termination of Calcium, Potassium, Magnesium and Sodium in Contents from 0.001% to 0.02% (m/m). 4 pp.

BSI BS EN 27627-2-93. 1993 Hardmetals—Chemical Analysis by Flame Atomic Absorption Spectrometry Part 2: Determination of Calcium, Potassium, Magnesium and Sodium, in Contents from 0.001 to 0.02 % (m/m) (ISO 7627-2: 1983) (V). 10 pp.

CEN EN 27627-2-93. Hardmetals—Chemical Analysis by Flame Atomic Absorption Spectrometry—Part 2: Determination of Calcium, Potassium, Magnesium and Sodium in Contents from 0,001 to 0,02 % (m/m) (ISO 7627-2: 1983). 5 pp.

ISO 7627 Pt 2-83. Hardmetals—Chemical Analysis by Flame Atomic Absorption Spectrometry—Part 2: Determination of Calcium, Potassium, Magnesium and Sodium in Contents from 0,001 to 0,02 % (m/m) First Edition (CEN EN 27627-2: 1993). 4 pp.

—Manganese Content—Atomic Absorption Spectrometry

BSI BS 5600: SUB SEC 4.17.3-86. (WITHDRAWN) 1986 Part 4: Methods of Testing and Chemical Analysis of Hardmetals Section 4.17: Chemical Analysis by Flame Atomic Absorption Spectrometry Subsection 4.17.3: De-termination of Cobalt, Iron, Manganese and Nickel in Contents from 0.01% to 0.05% (m/m). 4 pp.

BSI BS 5600: SUB SEC 4.17.5-86. (WITHDRAWN) 1986 Part 4: Methods of Testing and Chemical Analysis of Hardmetals Section 4.17: Chemical Analysis by Flame Atomic Absorption Spectrometry Sub 4.17.5:Determination of Cobalt, Iron, Manganese, Molybdenum, Nickel Titanium and Vanadium in Cont from 0.5% to 2% m/m. 4 pp.

BSI BS EN 27627-3-93. 1993 Hardmetals—Chemical Analysis by Flame Atomic Absorption Spectrometry Part 3: Determination of Cobalt, Iron, Manganese and Nickel in Contents from 0.01 to 0.5 % (m/m) (ISO 7627-3: 1983) (V). 10 pp.

BSI BS EN 27627-5-93. 1993 Hardmetals—Chemical Analysis by Flame Atomic Absorption Spectrometry Part 5: Determination of Cobalt, Iron, Manganese, Molybdenum, Nickel, Titanium and Vanadium in Contents from 0.5 to 2 % (m/m) (ISO 7627-5: 1985) (V). 10 pp.

INTERNATIONAL AND NON-U.S. NATIONAL STANDARDS
SUBJECT INDEX

Hard

Hard Metals (Cont.)
—Manganese Content—Atomic Absorption Spectrometry (Cont.)

CEN EN 27627-3-93. Hardmetals—Chemical Analysis by Flame Atomic Absorption Spectrometry—Part 3: Determination of Cobalt, Iron, Manganese and Nickel in Contents from 0,01 to 0,5 % (m/m) (ISO 7627-3: 1983). 5 pp.

CEN EN 27627-5-93. Hardmetals—Chemical Analysis by Flame Atomic Absorption Spectrometry—Part 5: Determination of Cobalt, Iron, Manganese, Molybdenum, Nickel, Titanium and Vanadium in Contents from 0,5 to 2 % (m/m) (ISO 7627-5: 1983). 5 pp.

ISO 7627 Pt 3-83. Hardmetals—Chemical Analysis by Flame Atomic Absorption Spectrometry—Part 3: Determination of Cobalt, Iron, Manganese and Nickel in Contents from 0,01 to 0,5 % (m/m) First Edition (CEN EN 27627-3: 1993). 4 pp.

ISO 7627 Pt 5-83. Hardmetals—Chemical Analysis by Flame Atomic Absorption Spectrometry—Part 5: Determination of Cobalt, Iron, Manganese, Molybdenum, Nickel, Titanium and Vanadium in Contents from 0,5 to 2 % (m/m) First Edition (CEN EN 27627-5: 1993). 4 pp.

—Metallic Element Content—X-Ray Fluorescence Spectrometry

BSI BS 5600: Sec 4.10-80. (WITHDRAWN) 1980 Powder Metallurgical Materials and Products Part 4: Methods of Testing and Chemical Analysis of Hardmetals Section 4.10: Determination of Contents of Metallic Elements by X-ray Fluorescence: Solution Method. 4 pp.

BSI BS 5600: Sec 4.15-80. (WITHDRAWN) 1980 Powder Metallurgical Materials and Products Part 4: Methods of Testing and Chemical Analysis of Hardmetals Section 4.15: Determination of Contents of Metallic Elements by X-ray Fluorescence: Fusion Method. 4 pp.

BSI BS EN 24503-93. 1993 Hardmetals—Determination of Contents of Metallic Elements by X-Ray Fluorescence—Fusion Method (ISO 4503: 1978) (V). 10 pp.

BSI BS EN 24883-93. 1993 Hardmetals—Determination of Contents of Metallic Elements by X-Ray Fluorescence—Solution Method (ISO 4883: 1978) (V). 9 pp.

CEN EN 24503-93. Hardmetals—Determination of Contents of Metallic Elements by X-Ray Fluorescence—Fusion Method (ISO 4503: 1978). 5 pp.

CEN EN 24883-93. Hardmetals—Determination of Contents of Metallic Elements by X-Ray Fluorescence—Solution Method (ISO 4883: 1978). 4 pp.

ISO 4503-78. Hardmetals—Determination of Contents of Metallic Elements by X-Ray Fluorescence—Fusion Method First Edition (CEN EN 24503: 1993). 5 pp.

ISO 4883-78. Hardmetals—Determination of Contents of Metallic Elements by X-Ray Fluorescence—Solution Method First Edition (CEN EN 24883: 1993). 4 pp.

—Microstructure—Metallography

BSI BS 5600: Sec 4.13-80. (WITHDRAWN) 1980 Powder Metallurgical Materials and Products Part 4: Methods of Testing and Chemical Analysis of Hardmetals Section 4.13: Metallographic Determination of Microstructure (Superseded by BS EN 24499: 1993). 6 pp.

BSI BS EN 24499-93. 1993 Hardmetals—Metallographic Determination of Microstructure (ISO 4499: 1978) (V). 11 pp.

CEN EN 24499-93. Hardmetals—Metallographic Determination of Microstructure (ISO 4499: 1978). 6 pp.

ISO 4499-78. Hardmetals—Metallographic Determination of Microstructure First Edition (CEN EN 24499: 1993). 6 pp.

—Modulus of Elasticity

BSI BS 5600: Sec 4.7-79. (WITHDRAWN) 1979 Powder Metallurgical Materials and Products Part 4: Methods of Testing and Chemical Analysis of Hardmetals Section 4.7: Determination of Young's Modulus (Superseded by BS EN 23312: 1993). 4 pp.

BSI BS EN 23312-93. 1993 Sintered Metal Materials and Hardmetals—Determination of Young Modulus (ISO 3312: 1987) (V). 10 pp.

CEN EN 23312-93. Sintered Metal Materials and Hardmetals—Determination of Young Modulus (ISO 3312: 1987). 5 pp.

ISO 3312-87. Sintered Metal Materials and Hardmetals—Determination of Young Modulus Second Edition (CEN EN 23312: 1993). 4 pp.

Hard Metals (Cont.)
—Molybdenum Content—Atomic Absorption Spectrometry

BSI BS 5600: SUB SEC 4.17.4-86. (WITHDRAWN) 1986 Part 4: Methods of Testing and Chemical Analysis of Hardmetals Section 4.17: Chemical Analysis by Flame Atomic Absorption Spectrometry Subsection 4.17.4: De-termination of Malybdenum, Titanium and Vanadium in Contents From 0.01% to 0.05% m/m. 4 pp.

BSI BS 5600: SUB SEC 4.17.5-86. (WITHDRAWN) 1986 Part 4: Methods of Testing and Chemical Analysis of Hardmetals Section 4.17: Chemical Analysis by Flame Atomic Absorption Spectrometry Sub 4.17.5:Determination of Cobalt, Iron, Manganese, Molybdenum, Nickel Titanium and Vanadium in Cont from 0.5% to 2% m/m. 4 pp.

BSI BS EN 27627-4-93. 1993 Hardmetals—Chemical Analysis by Flame Atomic Absorption Spectrometry Part 4: Determination of Molybdenum, Titanium and Vanadium in Contents from 0.01 to 0.5 % (m/m) (ISO 7627-4: 1983) (V). 10 pp.

BSI BS EN 27627-5-93. 1993 Hardmetals—Chemical Analysis by Flame Atomic Absorption Spectrometry Part 5: Determination of Cobalt, Iron, Manganese, Molybdenum, Nickel, Titanium and Vanadium in Contents from 0.5 to 2 % (m/m) (ISO 7627-5: 1985) (V). 10 pp.

CEN EN 27627-4-93. Hardmetals—Chemical Analysis by Flame Atomic Absorption Spectrometry—Part 4: Determination of Molybdenum, Titanium and Vanadium in Contents from 0,01 to 0,5 % (m/m) (ISO 7627-4: 1983). 5 pp.

CEN EN 27627-5-93. Hardmetals—Chemical Analysis by Flame Atomic Absorption Spectrometry—Part 5: Determination of Cobalt, Iron, Manganese, Molybdenum, Nickel, Titanium and Vanadium in Contents from 0,5 to 2 % (m/m) (ISO 7627-5: 1983). 5 pp.

ISO 7627 Pt 4-83. Hardmetals—Chemical Analysis by Flame Atomic Absorption Spectrometry—Part 4: Determination of Molybdenum, Titanium and Vanadium in Contents from 0,01 to 0,5 % (m/m) First Edition (CEN EN 27627-4: 1993). 4 pp.

ISO 7627 Pt 5-83. Hardmetals—Chemical Analysis by Flame Atomic Absorption Spectrometry—Part 5: Determination of Cobalt, Iron, Manganese, Molybdenum, Nickel, Titanium and Vanadium in Contents from 0,5 to 2 % (m/m) First Edition (CEN EN 27627-5: 1993). 4 pp.

—Nickel Content—Atomic Absorption Spectrometry

BSI BS 5600: SUB SEC 4.17.3-86. (WITHDRAWN) 1986 Part 4: Methods of Testing and Chemical Analysis of Hardmetals Section 4.17: Chemical Analysis by Flame Atomic Absorption Spectrometry Subsection 4.17.3: De-termination of Cobalt, Iron, Manganese and Nickel in Contents from 0.01% to 0.05% (m/m). 4 pp.

BSI BS 5600: SUB SEC 4.17.5-86. (WITHDRAWN) 1986 Part 4: Methods of Testing and Chemical Analysis of Hardmetals Section 4.17: Chemical Analysis by Flame Atomic Absorption Spectrometry Sub 4.17.5:Determination of Cobalt, Iron, Manganese, Molybdenum, Nickel Titanium and Vanadium in Cont from 0.5% to 2% m/m. 4 pp.

BSI BS EN 27627-3-93. 1993 Hardmetals—Chemical Analysis by Flame Atomic Absorption Spectrometry Part 3: Determination of Cobalt, Iron, Manganese and Nickel in Contents from 0.01 to 0.5 % (m/m) (ISO 7627-3: 1983) (V). 10 pp.

BSI BS EN 27627-5-93. 1993 Hardmetals—Chemical Analysis by Flame Atomic Absorption Spectrometry Part 5: Determination of Cobalt, Iron, Manganese, Molybdenum, Nickel, Titanium and Vanadium in Contents from 0.5 to 2 % (m/m) (ISO 7627-5: 1985) (V). 10 pp.

CEN EN 27627-3-93. Hardmetals—Chemical Analysis by Flame Atomic Absorption Spectrometry—Part 3: Determination of Cobalt, Iron, Manganese and Nickel in Contents from 0,01 to 0,5 % (m/m) (ISO 7627-3: 1983). 5 pp.

CEN EN 27627-5-93. Hardmetals—Chemical Analysis by Flame Atomic Absorption Spectrometry—Part 5: Determination of Cobalt, Iron, Manganese, Molybdenum, Nickel, Titanium and Vanadium in Contents from 0,5 to 2 % (m/m) (ISO 7627-5: 1983). 5 pp.

ISO 7627 Pt 3-83. Hardmetals—Chemical Analysis by Flame Atomic Absorption Spectrometry—Part 3: Determination of Cobalt, Iron, Manganese and Nickel in Contents from 0,01 to 0,5 % (m/m) First Edition (CEN EN 27627-3: 1993). 4 pp.

ISO 7627 Pt 5-83. Hardmetals—Chemical Analysis by Flame Atomic Absorption Spectrometry—Part 5: Determination of Cobalt, Iron, Manganese, Molybdenum, Nickel, Titanium and Vanadium in Contents from 0,5 to 2 % (m/m) First Edition (CEN EN 27627-5: 1993). 4 pp.

Hard Metals (Cont.)
—Porosity—Metallography

BSI BS 5600: Sec 4.14-80. (WITHDRAWN) 1980 Powder Metallurgical Materials and Products Part 4: Methods of Testing and Chemical Analysis of Hardmetals Section 4.14: Metallographic Determination of Porosity and Uncombined Carbon (Superseded by BS EN 24505: 1993). 7 pp.

BSI BS EN 24505-93. 1993 Hardmetals—Metallographic Determination of Porosity and Uncombined Carbon (ISO 4505: 1978) (V). 12 pp.

CEN EN 24505-93. Hardmetals—Metallographic Determination of Porosity and Uncombined Carbon (ISO 4505: 1978). 7 pp.

ISO 4505-78. Hardmetals—Metallographic Determination of Porosity and Uncombined Carbon First Edition (CEN EN 24505: 1993). 7 pp.

—Potassium Content—Atomic Absorption Spectrometry

BSI BS 5600: SUB SEC 4.17.2-86. (WITHDRAWN) 1986 Part 4: Methods of Testing and Chemical Analysis of Hardmetals Section 4.17: Chemical Analysis by Flame Atomic Absorption Spectrometry Subsection 4.17.2: De-termination of Calcium, Potassium, Magnesium and Sodium in Contents from 0.001% to 0.02% (m/m). 4 pp.

BSI BS EN 27627-2-93. 1993 Hardmetals—Chemical Analysis by Flame Atomic Absorption Spectrometry Part 2: Determination of Calcium, Potassium, Magnesium and Sodium, in Contents from 0.001 to 0.02 % (m/m) (ISO 7627-2: 1983) (V). 10 pp.

CEN EN 27627-2-93. Hardmetals—Chemical Analysis by Flame Atomic Absorption Spectrometry—Part 2: Determination of Calcium, Potassium, Magnesium and Sodium in Contents from 0,001 to 0,02 % (m/m) (ISO 7627-2: 1983). 5 pp.

ISO 7627 Pt 2-83. Hardmetals—Chemical Analysis by Flame Atomic Absorption Spectrometry—Part 2: Determination of Calcium, Potassium, Magnesium and Sodium in Contents from 0,001 to 0,02 % (m/m) First Edition (CEN EN 27627-2: 1993). 4 pp.

—Rockwell Hardness Testing

BSI BS 5600: Sec 4.5-79. 1979 Powder Metullurgical Materials and Products Part 4: Methods of Testing and Chemical Analysis of Hardmetals Section 4.5: Rockwell Hardness Test (Scale A). 4 pp.

ISO 3738 Pt 1-82. Hardmetals—Rockwell Hardness Test (Scale A)—Part 1: Test Method First Edition. 5 pp.

—Sintered—Density

BSI BS EN 23369-93. 1993 Impermeable Sintered Metal Materials and Hardmetals—Determination of Density (ISO 3369: 1975) (V). 10 pp.

CEN EN 23369-93. Impermeable Sintered Metal Materials and Hardmetals—Determination of Density (ISO 3369: 1975). 5 pp.

ISO 3369-75. Impermeable Sintered Metal Materials and Hardmetals—Determination of Density First Edition (CEN EN 23369: 1993). 5 pp.

—Sintered—Pellets

BSI BS 3821: Part 2-74. 1974 Hardmetal Dies and Associated Hardmetal Tools Part 2: As-Sintered Pellets and Finished Dies for Drawing Round Wire. 15 pp.

BSI BS 3821: Part 3-74. 1974 Hardmetal Dies and Associated Hardmetal Tools Part 3: As-Sintered Pellets and Finished Dies for Drawing Round Bar. 15 pp.

BSI BS 3821: Part 4-82. 1982 Hardmetal Dies and Associated Hardmetal Tools Part 4: Dimensions and Tolerances of As-Sintered Hard-Metal Pellets Used in Heading Dies. 7 pp.

ISO 5407-81. As-Sintered Hardmetal Pellets Used in Heading Dies—Dimensions and Tolerances First Edition. 6 pp.

—Sintered—Pellets—Identification Systems

BSI BS 3821: Part 1-74. 1974 Hardmetal Dies and Associated Hardmetal Tools Part 1: Designation Marking of As-Sintered Pellets and Finished Dies. 12 pp.

—Sintered—Sampling

BSI BS 5600: Sec 4.11-80. (WITHDRAWN) 1980 Powder Metallurgical Materials and Products Part 4: Methods of Testing and Chemical Analysis of Hardmetals Section 4.11: Sampling and Testing of Sintered Hardmetals (Superseded by BS EN 24489: 1993). 4 pp.

BSI BS EN 24489-93. 1993 Sintered Hardmetals—Sampling and Testing (ISO 4489: 1978) (V). 9 pp.

CEN EN 24489-93. Sintered Hardmetals—Sampling and Testing (ISO 4489: 1978). 4 pp.

ISO 4489-78. Sintered Hardmetals—Sampling and Testing First Edition (CEN EN 24489: 1993). 3 pp.

INDUSTRY STANDARDS

Hard Metals (Cont.)

—Sodium Content—Atomic Absorption Spectrometry

BSI BS 5600: SUB SEC 4.17.2-86. (WITHDRAWN) 1986 Part 4: Methods of Testing and Chemical Analysis of Hardmetals Section 4.17: Chemical Analysis by Flame Atomic Absorption Spectrometry Subsection 4.17.2: De-termination of Calcium, Potassium, Magnesium and Sodium in Contents from 0.001% to 0.02% (m/m).

BSI BS EN 27627-2-93. 1993 Hardmetals—Chemical Analysis by Flame Atomic Absorption Spectrometry Part 2: Determination of Calcium, Potassium, Magnesium and Sodium in Contents from 0.001 to 0.02 % (m/m) (ISO 7627-2: 1983) (V). 10 pp.

CEN EN 27627-2-93. Hardmetals—Chemical Analysis by Flame Atomic Absorption Spectrometry—Part 2: Determination of Calcium, Potassium, Magnesium and Sodium in Contents from 0,001 to 0,02 % (m/m) (ISO 7627-2: 1983). 5 pp.

ISO 7627 Pt 2-83. Hardmetals—Chemical Analysis by Flame Atomic Absorption Spectrometry—Part 2: Determination of Calcium, Potassium, Magnesium and Sodium in Contents from 0,001 to 0,02 % (m/m) First Edition (CEN EN 27627-2: 1993). 4 pp.

—Test Specimens—Rockwell Hardness Testing

ISO 3738 Pt 2-88. Hardmetals—Rockwell Hardness Test (Scale A)—Part 2: Preparation and Calibration of Standard Test Blocks First Edition. 7 pp.

—Titanium Content—Atomic Absorption Spectrometry

BSI BS 5600: SUB SEC 4.17.4-86. (WITHDRAWN) 1986 Part 4: Methods of Testing and Chemical Analysis of Hardmetals Section 4.17: Chemical Analysis by Flame Atomic Absorption Spectrometry Subsection 4.17.4: De-termination of Malybdenum, Titanium and Vanadium in Contents From 0.01% to 0.05% m/m. 4 pp.

BSI BS 5600: SUB SEC 4.17.5-86. (WITHDRAWN) 1986 Part 4: Methods of Testing and Chemical Analysis of Hardmetals Section 4.17: Chemical Analysis by Flame Atomic Absorption Spectrometry Sub 4.17.5:Determination of. Cobalt, Iron, Manganese, Molybdenum, Nickel Titanium and Vanadium in Cont from 0.5% to 2% m/m. 4 pp.

BSI BS EN 27627-4-93. 1993 Hardmetals—Chemical Analysis by Flame Atomic Absorption Spectrometry Part 4: Determination of Molybdenum, Titanium and Vanadium in Contents from 0.01 to 0.5 % (m/m) (ISO 7627-4: 1983) (V). 10 pp.

BSI BS EN 27627-5-93. 1993 Hardmetals—Chemical Analysis by Flame Atomic Absorption Spectrometry Part 5: Determination of Cobalt, Iron, Manganese, Molybdenum, Nickel, Titanium and Vanadium in Contents from 0.5 to 2 % (m/m) (ISO 7627-5: 1985) (V). 10 pp.

CEN EN 27627-4-93. Hardmetals—Chemical Analysis by Flame Atomic Absorption Spectrometry—Part 4: Determination of Molybdenum, Titanium and Vanadium in Contents from 0,01 to 0,5 % (m/m) (ISO 7627-4: 1983). 5 pp.

CEN EN 27627-5-93. Hardmetals—Chemical Analysis by Flame Atomic Absorption Spectrometry—Part 5: Determination of Cobalt, Iron, Manganese, Molybdenum, Nickel, Titanium and Vanadium in Contents from 0,5 to 2 % (m/m) (ISO 7627-5: 1983). 5 pp.

ISO 7627 Pt 4-83. Hardmetals—Chemical Analysis by Flame Atomic Absorption Spectrometry—Part 4: Determination of Molybdenum, Titanium and Vanadium in Contents from 0,01 to 0,5 % (m/m) First Edition (CEN EN 27627-4: 1993). 4 pp.

ISO 7627 Pt 5-83. Hardmetals—Chemical Analysis by Flame Atomic Absorption Spectrometry—Part 5: Determination of Cobalt, Iron, Manganese, Molybdenum, Nickel, Titanium and Vanadium in Contents from 0,5 to 2 % (m/m) First Edition (CEN EN 27627-5: 1993). 4 pp.

—Titanium Content—Photometry

BSI BS 5600: Sec 4.12-80. 1980 Powder Metallurgical Materials and Products Part 4: Methods of Testing and Chemical Analysis of Hardmetals Section 4.12: Determination of Titanium Content; Photometric Peroxide Method. 4 pp.

ISO 4501-78. Hardmetals—Determination of Titanium—Photometric Peroxide Method First Edition. 5 pp.

—Transverse—Strain Testing

BSI BS 5600: Sec 4.8-87. (WITHDRAWN) 1987 Powder Metallurgical Materials and Products Part 4: Methods of Testing and Chemical Analysis of Hardmetals Section 4.8: Determination of Transverse Rupture Strength (Superseded by BS EN 23326: 1993). 4 pp.

BSI BS EN 23327-93. 1993 Hardmetals—Determination of Transverse Rupture Strength (ISO 3327: 1982) (V). 9 pp.

Hard Metals (Cont.)

—Transverse—Strain Testing (Cont.)

CEN EN 23327-93. Hardmetals—Determination of Transverse Rupture Strength (ISO 3327: 1982). 4 pp.

ISO 3327-82. Hardmetals—Determination of Transverse Rupture Strength Second Edition (CEN EN 23327: 1993). 4 pp.

—Vanadium Content—Atomic Absorption Spectrometry

BSI BS 5600: SUB SEC 4.17.4-86. (WITHDRAWN) 1986 Part 4: Methods of Testing and Chemical Analysis of Hardmetals Section 4.17: Chemical Analysis by Flame Atomic Absorption Spectrometry Subsection 4.17.4: De-termination of Malybdenum, Titanium and Vanadium in Contents From 0.01% to 0.05% m/m. 4 pp.

BSI BS 5600: SUB SEC 4.17.5-86. (WITHDRAWN) 1986 Part 4: Methods of Testing and Chemical Analysis of Hardmetals Section 4.17: Chemical Analysis by Flame Atomic Absorption Spectrometry Sub 4.17.5:Determination of Cobalt, Iron, Manganese, Molybdenum, Nickel Titanium and Vanadium in Cont from 0.5% to 2% m/m. 4 pp.

BSI BS EN 27627-4-93. 1993 Hardmetals—Chemical Analysis by Flame Atomic Absorption Spectrometry Part 4: Determination of Molybdenum, Titanium and Vanadium in Contents from 0.01 to 0.5 % (m/m) (ISO 7627-4: 1983) (V). 10 pp.

BSI BS EN 27627-5-93. 1993 Hardmetals—Chemical Analysis by Flame Atomic Absorption Spectrometry Part 5: Determination of Cobalt, Iron, Manganese, Molybdenum, Nickel, Titanium and Vanadium in Contents from 0.5 to 2 % (m/m) (ISO 7627-5: 1985) (V). 10 pp.

CEN EN 27627-4-93. Hardmetals—Chemical Analysis by Flame Atomic Absorption Spectrometry—Part 4: Determination of Molybdenum, Titanium and Vanadium in Contents from 0,01 to 0,5 % (m/m) (ISO 7627-4: 1983). 5 pp.

CEN EN 27627-5-93. Hardmetals—Chemical Analysis by Flame Atomic Absorption Spectrometry—Part 5: Determination of Cobalt, Iron, Manganese, Molybdenum, Nickel, Titanium and Vanadium in Contents from 0,5 to 2 % (m/m) (ISO 7627-5: 1983). 5 pp.

ISO 7627 Pt 4-83. Hardmetals—Chemical Analysis by Flame Atomic Absorption Spectrometry—Part 4: Determination of Molybdenum, Titanium and Vanadium in Contents from 0,01 to 0,5 % (m/m) First Edition (CEN EN 27627-4: 1993). 4 pp.

ISO 7627 Pt 5-83. Hardmetals—Chemical Analysis by Flame Atomic Absorption Spectrometry—Part 5: Determination of Cobalt, Iron, Manganese, Molybdenum, Nickel, Titanium and Vanadium in Contents from 0,5 to 2 % (m/m) First Edition (CEN EN 27627-5: 1993). 4 pp.

—Vickers Hardness Testing

BSI BS 5600: Sec 4.6-87. (WITHDRAWN) 1987 Powder Metallurgical Materials and Products Part 4: Methods of Testing and Chemical Analysis of Hardmetals Section 4.6: Vickers Hardness Test (Superseded by BS EN 23878: 1993). 6 pp.

BSI BS EN 23878-93. 1993 Hardmetals—Vickers Hardness Test (ISO 3878: 1983) (V). 12 pp.

CEN EN 23878-93. Hardmetals—Vickers Hardness Test (ISO 3878: 1983). 5 pp.

ISO 3878-83. Hardmetals—Vickers Hardness Test Second Edition (CEN EN 23878: 1993). 4 pp.

Hard Radiation
Use: Ionizing Radiation

Hard Rubber
Use: Ebonite

Hard Soldering Equipment
Use: Brazing Equipment

Hard Surfacing
Use For: Hard Facing; Hardfacing
See Also: Coatings; Conversion Coatings; Metallizing; Plating; Welding

—Coated Electrodes

JIS Z 3251-91. Covered Electrodes for Hardfacing. 14 pp.

—Flux Cored Arc Welding—Electrodes

JIS Z 3326-91. Arc Welding Flux Cored Wires for Hardfacing. 10 pp.

Hard Water
See Also: Water; Water Analysis; Water Content Analysis

—Appliances

BSI BS EN 60734-93. 1993 Hard Water to be Used for Testing the Performance of Some Household Electrical Appliances (IEC 734: 1993) (L). 15 pp.

Hard Water (Cont.)

—Appliances (Cont.)

CENELEC EN 60734-93. Hard Water to Be Used for Testing the Performance of Some Household Electrical Appliances (IEC 734: 1993). 10 pp.

IEC 734-93. Hard Water to be Used for Testing the Performance of Some Household Electrical Appliances Second Edition (CENELEC EN 60734: 1993). 23 pp.

—Detergents—Stability Testing

CGSB 2-GP-11M METH 19.1-83. Methods of Testing and Analysis of Soaps and Detergents Stability to Hard Water. 1 p.

—Lubricants—Stability Testing

DIN ENGL 51367-78. Testing of Cooling Lubricants; Determination of the Stability of Emulsifiable Cooling Lubricants in Mixture with Hard Water (May). 4 pp.

—Steam Cleaning Compounds—Stability Testing

CGSB 31-GP-0A METH 46.1-57. Methods of Testing Corrosion-Prevention Materials and Processes Stability to Water Hardness. 1 p.

—Surfactants—Stability Testing

ISO 1063-74. Surface Active Agents—Determination of Stability in Hard Water First Edition. 5 pp.

Hardboard
See Also: Fiberboard

CEN PREN 622-2-93. Fibreboards—Specifications—Part 2: Requirements for General Purpose Boards for Use in Dry Conditions. 10 pp.

CEN PREN 622-4-93. Fibreboards—Specifications—Part 4: Requirements for General Purpose Boards for Use in Humid Conditions. 10 pp.

CGSB CAN/CGSB-11.3-M87. Hardboard. 11 pp.

CGSB CAN/CGSB-11.5-M87. Hardboard, Precoated, Factory Finished, for Exterior Cladding. 10 pp.

JIS A 5907-77. Hard Fibreboards (R 1980). 10 pp.

—Dimensional Stability

ISO TR7469-81. Dimensional Stability of Hardboard First Edition. 5 pp.

—Dressed

JIS A 5910-75. Dressed Hard Fibreboard for Exterior Use. 9 pp.

—Installation

CGSB CAN/CGSB-11.6-M87. Installation of Exterior Hardboard Cladding. 11 pp.

—Loads (Forces)

CEN PREN 622-3-93. Fibreboards—Specifications—Part 3: Requirements for Loadbearing Boards for Use in Dry Conditions. 10 pp.

CEN PREN 622-5-93. Fibreboards—Specifications—Part 5: Requirements for Loadbearing Boards for Use in Humid Conditions. 10 pp.

—Quality Assurance

DIN ENGL 68750-58. Wood Fibre Building Boards; Porous and Hard Wood Fibre Building Boards; Quality Conditions (Apr). 2 pp.

—Surface Absorption

CEN PREN 382-2-92. Fibreboards—Determination of Surface Absorbtion—Part 2: Test Methods for Hardboards. 6 pp.

Hardenability
See Also: Jominy Testing

—Steels

CNS G2021-86. Method of Hardenability Test for Steel (Jun)(2911).

—Steels—End Quenching

DIN ENGL 50191-87. Hardenability Testing of Steel by End Quenching (Sept). 7 pp.

—Steels—Selection

DIN ENGL 17021 Pt 1-76. Heat Treatment of Ferrous Metals; Material Selection; Steel Selection According to Hardenability (Feb). 11 pp.

—Structural Steels

CNS G3230-87. Structural Steels with Specific Hardenability Bands (Jun)(11999).

JIS G 4052-79. Structural Steels with Specified Hardenability Bands. 39 pp.

Hardened Castor Oils
Use: Castor Oil

Hardened Concrete
Use: Concretes

INTERNATIONAL AND NON-U.S. NATIONAL STANDARDS
SUBJECT INDEX
Hardness

Hardeners
DIN ENGL 16945-89. Testing of Resins, Hardeners and Accelerators, and Catalyzed Resins (Mar). 16 pp.

—**Epoxy Resins**
ISO 4597 Pt 1-83. Plastics—Hardeners and Accelerators for Epoxide Resins—Part 1: Designation First Edition. 4 pp.
JIS K 7231-86. General Rules for Testing Methods of Epoxide Resins and Hardeners. 6 pp.

—**Epoxy Resins—Acidity**
JIS K 7239-91. Determination of Free Acid in Acid Anhydride-Based Hardeners for Epoxy Resins. 8 pp.

—**Epoxy Resins—Amine Values**
JIS K 7237-86. Testing Methods for Total Amine Values of Amine-Based Hardeners of Epoxide Resins. 10 pp.

—**Epoxy Resins—Density**
JIS K 7232-86. Testing Methods for Specific Gravity of Epoxide Resins and Hardeners. 16 pp.

—**Epoxy Resins—Viscosity**
JIS K 7233-86. Testing Methods for Viscosity of Epoxide Resins and Hardeners. 13 pp.

Hardening (Materials)
Scope Note: See the subheading Hardening or Hardened under specific metals or products
See Also: Strain Hardening

Hardfacing
Use: Hard Surfacing

Hardgrove Grindability Index
—**Coal**
BSI BS 1016: Part 20-81. 1981 Methods for the Analysis and Testing of Coal and Coke Part 20: Determination of Hardgrove Grindability Index of Hard Coal. 12 pp.
CNS M3164-84. Method of Test for Grindability of Coal by the Hardgrove-Machine Method (Sep)(11017).
ISO 5074-80. Hard Coal—Determination of Hardgrove Grindability Index First Edition. 10 pp.
SAA AS 1038.20-92. Coal and Coke—Analysis and Testing—Part 20: Higher Rank Coal—Hardgrove Grindability Index (in Professional Package 32). 6 pp.
SAA CRM 011E-91. Certified Reference Materials—Certified Reference Coal (Hardgrove Grindability Index)—Part E: Hardgrove Grindability Index: 36 (Superseded by CRM 011-3A—1992).
SAA CRM 011F-91. Certified Reference Materials—Certified Reference Coal (Hardgrove Grindability Index)—Part F: Hardgrove Grindability Index: 48 (Superseded by CRM 011-3B—1992).
SAA CRM 011G-91. Certified Reference Materials—Certified Reference Coal (Hardgrove Grindability Index)—Part G: Hardgrove Grindability Index: 68 (Superseded by CRM 011-3C—1992).
SAA CRM 011H-91. Certified Reference Materials—Certified Reference Coal (Hardgrove Grindability Index)—Part H: Hardgrove Grindability Index: 96 (Superseded by CRM 011-3D—1992).
SAA TR2.11-92. Certified Reference Materials—Part 11: Coal—Preparation and Certification of ASCRM-011-3 (Hardgrove Grindability Index) (in Professional Package 32). 2 pp.

Hardgrove Number
Use: Hardgrove Grindability Index

Hardness
Use: Hardness Testing

Hardness Testers
See Also: Brinell Hardness Testers; Hardness Testing; Knoop Hardness Testers; Microhardness Testers; Rockwell Hardness Testers; Rockwell Superficial Hardness Testers; Testing Equipment; Vickers Hardness Testers

BSI BS 5411: Part 6-81. 1981 Methods of Test for Metallic and Related Coatings Part 6: Vickers and Knoop Microhardness Tests. 11 pp.
CNS B6067-89. Shore Hardness Testing Machines (Jan)(8766).
CNS B7191-89. Method of Test for Shore Hardness Testing Machines (Jan)(8767).
DIN ENGL 51224-85. Materials Testing Machines; Hardness Testing Machines with a Depth-Measuring Device (Jan). 7 pp.
DIN ENGL 51304-83. Material Testing Machines; Verification of Hardness Testing Machines Fitted with Indentation Depth Measuring Device (Sept). 6 pp.

Hardness Testers *(Cont.)*
DIN ENGL 51305-83. Material Testing Machines; Verification of Hardness Testing Machines Fitted with an Optical Indentation Depth Measuring Device (Sept). 7 pp.
ISO 726-82. Metallic Materials—Hardness Test—Calibration of Standardized Blocks to Be Used for Brinell Hardness Testing Machines First Edition. 5 pp.
JIS B 7724-86. Brinell Hardness Testing Machines. 10 pp.
JIS B 7727-88. Shore Hardness Testing Machines. 9 pp.

—**Blocks**
DIN ENGL 51303-83. Material Testing Machines; Standardized Blocks for the Verification of Static Hardness Testing Machines (Nov). 6 pp.
ISO 1355-89. Metallic Materials—Hardness Test—Calibration of Standardized Blocks to Be Used for Rockwell Superficial Hardness Testing Machines (Scales 15N, 30N, 45N, 15T, 30T and 45T) First Edition. 7 pp.

—**Holders—Test Specimens**
DIN ENGL 51200-85. Materials Testing Machines; Design and Application of Test Piece Holding Devices in Hardness Testing Machines (Oct). 6 pp.

Hardness Testing
See Also: Brinell Hardness Testing; Brittleness Testing; Compression Testing; Ductility; Durometer Hardness Testing; Dynamic Testing; Hardness Testers; High Temperature Testing; Impact Testing; Indentation Hardness Testing; Jominy Testing; Knoop Hardness Testing; Materials Testing; Microhardness Testing; Penetration Resistance; Plastic Properties; Rocker Hardness Testing; Rockwell Hardness Testing; Rockwell Superficial Hardness Testing; Scratch Hardness Testing; Shore Hardness Testing; Softness; Static Hardness Testing; Static Testing; Strain Hardening; Vickers Hardness Testing
CNS Z8004-83. Method of Vickers Hardness Test (Jul)(2115).

—**Bearing Alloys**
ISO 4384 Pt 1-82. Plain Bearings—Hardness Testing of Bearing Metals—Part 1: Compound Materials First Edition. 4 pp.
ISO 4384 Pt 2-82. Plain Bearings—Hardness Testing of Bearing Metals—Part 2: Solid Materials First Edition. 3 pp.

—**Brake Pads**
JIS D 4421-87. Method of Hardness Test for Brake Linings, Pads and Clutch Facings of Automobiles. 11 pp.

—**Butter**
DIN ENGL 10331-85. Determination of the Hardness of Butter (Aug). 3 pp.

—**Cellular Materials**
DIN ENGL 53579 Pt 2-85. Testing of Flexible Cellular Materials; Hardness Test on Finished Parts; Compressibility of Profiles (Apr). 3 pp.

—**Clutch Facings**
JIS D 4421-87. Method of Hardness Test for Brake Linings, Pads and Clutch Facings of Automobiles. 11 pp.

—**Concretes**
BSI BS 1881: Part 202-86. 1986 Methods of Testing Concrete Part 202: Recommendations for Surface Hardness Testing by Rebound Hammer. 8 pp.

—**Corrugated Board**
SAA AS 1301.445S-89. Methods of Test for Pulp and Paper (Metric Units)—Part 445s: Hardness of Corrugated Board Amdt 1 June 1990 (This is a Joint Standard with SANZ NZS 1301). 2 pp.
SNZ NZS/AS 1301. 445S-89. Methods of Test for Pulp and Paper Hardness of Corrugated Board (This is a Joint Standard with SAA AS 1301.445S). 2 pp.

—**Deposited Metal**
JIS Z 3114-90. Method of Hardness Test for Deposited Metal. 6 pp.

—**Fiberglass Reinforced Plastics—Barcol Impressors**
CEN EN 59-77. Glass Reinforced Plastics: Measurement of Hardness by Means of a Barcol Impressor. 5 pp.
CNS K6984-90. Method of Test for Barcol Hardness of Glass Fiber Reinforced Plastics (Sep)(12785).

—**Lacquers**
SAA AS 1580.401. 5-93. Paints and Related Materials—Methods of Test—Part 401.5: Hard Dry Condition—Sanding Test (in Professional Package 39A). 2 pp.

Hardness Testing *(Cont.)*
—**Lacquers** *(Cont.)*
SAA AS 1580.401. 6-92. Paints and Related Materials—Methods of Test—Part 401.6: Hard Dry Condition (Mechanical Thumb Test) (in Professional Packages 30, 39). 6 pp.
SNZ NZS/AS 1580. 401.5-80. Methods of Test for Paints and Related Materials Part 401.5: Hard Dry Condition—Sanding Test (This is a Joint Standard with SAA AS 1580.401.5). 2 pp.
SNZ NZS/AS 1580. 401.6-92. Methods of Test for Paints and Related Materials Part 401.6: Hard Dry Condition (Mechanical Thumb Test) (This is a Joint Standard with SAA AS 1580.401.6). 6 pp.

—**Mats (Gymnastic Equipment)**
ISO 5903-81. Gymnastic Equipment—Landing Mats and Surfaces for Floor Exercises—Determination of Hardness and Impact Damping First Edition. 4 pp.

—**Metals**
AECMA PREN2002-07-87. Test Methods for Metallic Materials—Part 7—Hardness Test. 4 pp.

—**Metals—Aircraft**
BSI 4A 4-66. 1966 Amd 1 Test Pieces and Test Methods for Metallic Materials for Aircraft. 13 pp.

—**Paints**
BSI BS AU 148: Part 6-69. 1969 Methods of Test for Motor Vehicle Paints Part 6: Hardness. 5 pp.
SAA AS 1580.401. 5-93. Paints and Related Materials—Methods of Test—Part 401.5: Hard Dry Condition—Sanding Test (in Professional Package 39A). 2 pp.
SAA AS 1580.401. 6-92. Paints and Related Materials—Methods of Test—Part 401.6: Hard Dry Condition (Mechanical Thumb Test) (in Professional Packages 30, 39). 6 pp.
SAA AS 1580.405. 1-78. Paints and Related Materials—Methods of Test—Part 405.1: Determination of Pencil Hardness of Paint Film. 2 pp.
SNZ NZS/AS 1580. 401.5-80. Methods of Test for Paints and Related Materials Part 401.5: Hard Dry Condition—Sanding Test (This is a Joint Standard with SAA AS 1580.401.5). 2 pp.
SNZ NZS/AS 1580. 401.6-92. Methods of Test for Paints and Related Materials Part 401.6: Hard Dry Condition (Mechanical Thumb Test) (This is a Joint Standard with SAA AS 1580.401.6). 6 pp.
SNZ NZS/AS 1580. 405.1-78. Methods of Test for Paints and Related Materials Part 405.1: Determination of Pencil Hardness of Paint Film (This is a Joint Standard with SAA AS 1580.405.1). 2 pp.

—**Pillows**
BSI BS 4578-70. 1970 Amd 1 Methods of Test for Hardness of, and Air Flow Through, Infants' Pillows. 10 pp.

—**Plastics**
BSI BS 2782:Pt10: METH 1001-77. 1977 Methods of Testing Plastics Part 10: Glass Reinforced Plastics Method 1001: Measurement of Hardness by Means of a Barcol Impressor. 7 pp.
JIS K 7060-87. Testing Method for Barcol Hardness of Glass Fiber Reinforced Plastics. 8 pp.

—**Resins—Athletic Fields**
CNS K6591-86. Method of Test for Polyurethane (PU) Athletic Installation Material (May)(6483). 6 pp.

—**Rollers**
BSI BS 7442: Sec 3.1-91. 1991 Rubber or Plastics Covered Rollers Part 3: Methods of Test Section 3.1: Determination of Apparent Hardness (IRHD Method) (ISO 7267/1: 1986). 7 pp.
BSI BS 7442: Sec 4.1-92. 1992 Rubber or Plastics Covered Rollers Part 4: Requirements Section 4.1: Specification for Apparent Hardness. 8 pp.
ISO 6123 Pt 1-82. Rubber or Plastics Covered Rollers—Specifications—Part 1: Requirements for Hardness First Edition. 5 pp.
ISO 7267 Pt 1-86. Rubber-Covered Rollers—Determination of Apparent Hardness—Part 1: IRHD Method First Edition. 4 pp.

—**Screeds**
DIN ENGL 272-86. Testing of Magnesium Oxychloride Screeds (Feb). 8 pp.

—**Sintered Materials**
BSI BS 5600: SUB SEC 3.11.1-91. (WITHDRAWN) 1991 Powder Metallurgical Materials and Products Part 3: Methods of Test-ing Sintered Metal Materials and Products, Excluding Hardmetals Section 3.11: Determin. of Apparent Hardness: Subsec 3.11.1: Materials of Essentially Uniform Section Hardness. 6 pp.

INDUSTRY STANDARDS

INTERNATIONAL AND NON-U.S. NATIONAL STANDARDS SUBJECT INDEX

Hardness

Hardness Testing (Cont.)
—Sintered Materials (Cont.)
BSI BS 5600: SUB SEC 3.11.2-84. 1984 Methods of Testing Sintered Metal Materials and Products, Excluding Hardmetals Section 3.11: Determination of Apparent Hardness Subsection 3.11.2: Case Hardened Ferrous Materials, Surface Enriched by Carbon or Carbon and Nitrogen. 4 pp.
BSI BS EN 24498-1-93. 1993 Sintered Metal Materials, Excluding Hardmetals—Determination of Apparent Hardness Part 1: Materials of Essentially Uniform Section Hardness (ISO 4498-1: 1990) (V). 11 pp.
CEN EN 24498-1-93. Sintered Metal Materials, Excluding Hardmetals—Determination of Apparent Hardness—Part 1: Materials of Essentially Uniform Section Hardness (ISO 4498-1: 1990). 6 pp.
ISO 4498 Pt 1-90. Sintered Metal Materials, Excluding Hardmetals—Determination of Apparent Hardness—Part 1: Materials of Essentially Uniform Section Hardness Second Edition (CEN EN 24498-1: 1993). 7 pp.
ISO 4498 Pt 2-81. Sintered Metal Materials, Excluding Hardmetals—Determination of Apparent Hardness—Part 2: Case-Hardened Ferrous Materials, Surface Enriched by Carbon or Carbon and Nitrogen First Edition. 4 pp.
ISO 4507-78. Sintered Ferrous Materials, Carburized or Carbonitrided—Determination and Verification of Effective Case Depth by the Vickers Microhardness Testing Method First Edition. 5 pp.

—Steels
BSI BS 6481-84. 1984 Determination of Effective Depth of Hardening of Steel After Flame or Induction Hardening. 4 pp.
ISO 3754-76. Steel—Determination of Effective Depth of Hardening After Flame or Induction Hardening First Edition. 4 pp.
ISO 4964-84. Steel—Hardness Conversions First Edition. 9 pp.
JIS Z 3101-90. Testing Method of Maximum Hardness in Weld Heat-Affected Zone. 6 pp.
JIS Z 3115-73. Method of Taper Hardness Test in Weld Heat-Affected Zone. 5 pp.
SAA AS B161-60. Charts for Approximate Comparison of Hardness Scales for Steels. 8 pp.

—Tables (Data)
BSI BS 860-67. 1967 Tables for Comparison of Hardness Scales. 13 pp.

—Varnishes
CNS K6103-59. Method of Test for Low Temperature Baking Varnish (Apr)(1113). 3 pp.
SAA AS 1580.401. 5-93. Paints and Related Materials—Methods of Test—Part 401.5: Hard Dry Condition—Sanding Test (in Professional Package 39A). 2 pp.
SAA AS 1580.401. 6-92. Paints and Related Materials—Methods of Test—Part 401.6: Hard Dry Condition (Mechanical Thumb Test) (in Professional Packages 30, 39). 6 pp.
SNZ NZS/AS 1580. 401.5-80. Methods of Test for Paints and Related Materials Part 401.5: Hard Dry Condition—Sanding Test (This is a Joint Standard with SAA AS 1580.401.5). 2 pp.
SNZ NZS/AS 1580. 401.6-92. Methods of Test for Paints and Related Materials Part 401.6: Hard Dry Condition (Mechanical Thumb Test) (This is a Joint Standard with SAA AS 1580.401.6). 6 pp.

—Vulcanized Rubber
BSI BS 903: Part A26-69. 1969 Amd 1 Methods of Testing Vulcanized Rubber Part A26: Determination of Hardness. 43 pp.
CNS K6346-85. Method of Test for Hardness of Vulcanized Rubber (Nov)(3555).
ISO 48-79. Vulcanized Rubbers—Determination of Hardness (Hardness Between 30 and 85 IRHD) Second Edition. 7 pp.
ISO 1400-75. Vulcanized Rubbers of High Hardness (85 to 100 IRHD)—Determination of Hardness First Edition. 6 pp.
ISO 1818-75. Vulcanized Rubbers of Low Hardness (10 to 35 IRHD)—Determination of Hardness First Edition. 6 pp.
SAA AS 1180.9A-73. Methods of Test for Hose Made from Elastomeric Materials—Part 9A: Hardness of Vulcanized Rubbers of Standard Hardness (35 to 85 IRHD) Reconfirmed 1988.
SAA AS 1180.9B-73. Methods of Test for Hose Made from Elastomeric Materials—Part 9B: Hardness of Vulcanized Rubbers of Low Hardness (10 to 35 IRHD) Reconfirmed 1988.
SAA AS 1180.9C-73. Methods of Test for Hose Made from Elastomeric Materials—Part 9C: Hardness of Vulcanized Rubbers of High Hardness (85 to 100 IRHD) (Bound Together) Reconfirmed 1988. 26 pp.
SAA AS 1683.15.1-90. Methods of Test for Elastomers—Part 15.1: International Rubber Hardness. 4 pp.
SAA AS 1683.15.2-90. Methods of Test for Elastomers—Part 15.2: Durometer Hardness. 5 pp.

Hardness Testing (Cont.)
—Welded Joints
CNS Z8008-72. Method of Test for Hardness of Weld Deposit of Metals (Oct)(3508).
DIN ENGL 50163 Pt 1-82. Testing of Metallic Materials; Hardness Test on Welds; Transverse Sections of Joint Welds (Apr). 5 pp.

—Wood
CNS O2011-81. Method of Test for Hardness of Wood (Mar)(460). 2 pp.
JIS Z 2117-77. Method of Test for Hardness of Wood (R 1982). 5 pp.

Hardware
See Also: Appliances; Construction Materials; Doors; Fasteners; Handles; Hinges; Locks (Hardware); Mounting Hardware; Mounting Rails; Power Line Hardware; Printed Circuit Board Hardware; Telephone Line Hardware; Tools
BSI BS 3827: Part 4-67. (WITHDRAWN) 1967 Glossary of Terms Relating to Builders' Hardware Part 4: Door Drawer, Cupboard and Gate Furniture (Superseded by BS 6100: Subsection 1.3.6:1991). 24 pp.

—Aerospace—Engineering Drawings
SBAC AGS 1 (V). Procedure for the Establishment of Standards for Proprietary Items.

—Aerospace—Specifications Preparation—SBAC
SBAC AGS 1 (V). Procedure for the Establishment of Standards for Proprietary Items.

—Classification
BSI BS 1000: (682/683)-73. 1973 Universal Decimal Classification (UDC). English Full Edition (682/683): Smithery. Farriery. Hand-Forged Ironwork. Ironmongery. Hardware. Locksmithing. Gunsmithing. Bottling. Lamps. Stoves. 27 pp.
SNZ NZS/BS 1000 (682/683)-73. Universal Decimal Classification Smithery. Farriery. Hand-Forged Ironwork. Ironmongery. Hardware. Locksmithing. Gunsmithing. Bottling. Lamps. Stoves. 28 pp.

—Doors
CGSB CAN/CGSB-69.30-M90. Sliding and Folding Door Hardware (ANSI/BHMA A156.14-1985) (Supersedes 69-GP-6M). 36 pp.

—Doors—Construction Contracts
DIN ENGL 18357-88. Tendering and Performance Stipulations in Contracts for Construction Works (VOB); Part C: General Technical Specifications in Contracts for Construction Works (ATV); Mounting of Door and Window Hardware (Sept) (This Standard, Together with DIN 18299,. 6 pp.

—Finishes
CGSB CAN/CGSB-69.34-M90. Materials and Finishes (ANSI/BHMA A156.18-1984). 15 pp.

—Packaging
MOD UK DSTAN 81-94-89. Guide on the Packaging of Consumable Stores Where No Military Packaging Level is Stipulated Issue 1. 10 pp.

Hardwoods
Scope Note: For additional listings, see the subheading Hardwoods under specific products made from hardwoods Use For: Broadleaved Woods; Deciduous Woods See Also: Ash Wood; Beech Wood; Lumber; Oak Wood; Pulps; Pulpwood; Softwoods; Wood

—Crossarms
SAA AS O61-55. Cross-Arms from Eastern and South-Eastern Australian Hardwoods. 7 pp.

—Grades
BSI BS 5756-80. 1980 Tropical Hardwoods Graded for Structural Use. 10 pp.

—Grades—Structural Timber
SAA AS 2082-79. Visually Stress-Graded Hardwood for Structural Purposes. 32 pp.

—Joinery
BSI BS 1186: Part 1-91. 1991 Timber for and Workmanship in Joinery Part 1: Specification for Timber. 30 pp.

—Moisture Content Analysis
BSI BS 5450-77. 1977 Sizes of Hardwoods and Methods of Measurement. 8 pp.

—Poles (Supports)
SNZ NZS 168-69. Specification for New South Wales Desapped, Dressed Desapped, and Shaped Desapped Hardwood Poles (Reconfirmed 1984). 18 pp.

Hardwoods (Cont.)
—Sizes
BSI BS 5450-77. 1977 Sizes of Hardwoods and Methods of Measurement. 8 pp.

—Solid Fuel Burning Equipment
SAA AS 4014.1-92. Domestic Solid Fuel Burning Appliances—Test Fuels—Part 1: Hardwood (NZS 7404.1:1992). 2 pp.
SNZ NZS 7404.1-92. Domestic Solid Fuel Burning Appliances Part 1: Hardwood (NZS 7404.1:1992/AS 4014.1-1992). 2 pp.

—Standard Names
SNZ NZS 3621-87. Standard Names of Commercial Timbers in New Zealand. 44 pp.

—Structural Timber
SAA AS O98-66. Seasoned Size—Matched Framing Timber (Including Finger-Jointed Pieces) From South-Eastern Australian Hardwoods (Incorporating Amdt 1). 10 pp.
SNZ NZS 485-69. Specification for New South Wales Hardwoods. 38 pp.

—Structural Timber—Decks
SAA AS O80-63. Decking Timbers from Eastern and South-Eastern Australian Hardwoods. 8 pp.

—Structural Timber—Framing—Tables (Data)—Building Codes
SAA AS 1684 Supp 13-75. Supplement to AS 1684 SAA Timber Framing Code—Supplement 13: Light Timber Framing Span Tables—Seasoned Hardwood—Stress Grade F11 Amdt 1 March 1991 (in Professional Package 30) (Superseded (in Part) by AS 1684 Supplement 0 and AS 1684 Supplement 13—1992). 15 pp.
SAA AS 1684 Supp 13-92. National Timber Framing Code—Supplement 13: Timber Framing Span Tables—Seasoned Hardwood—Stress Grade F11 (Supplement to AS 1684—1992) (Supersedes AS 1684 Supplement 13—1975 (in Part)) (in Professional Packages 20, 21, 30, 62-69). 27 pp.
SAA AS 1684 Supp 14-75. Supplement to AS 1684 SAA Timber Framing Code—Supplement 14: Light Timber Framing Span Tables—Seasoned Hardwood—Stress Grade F14 Amdt 1 March 1991 (in Professional Package 30) (Superseded (in Part) by AS 1684 Supplement 0 and AS 1684 Supplement 14—1992). 15 pp.
SAA AS 1684 Supp 14-92. National Timber Framing Code—Supplement 14: Timber Framing Span Tables—Seasoned Hardwood—Stress Grade F14 (Supplement to AS 1684—1992) (Supersedes AS 1684 Supplement 14—1975 (in Part)) (in Professional Packages 20, 21, 30, 62-69). 27 pp.
SAA AS 1684 Supp 15-75. Supplement to AS 1684 SAA Timber Framing Code—Supplement 15: Light Timber Framing Span Tables—Seasoned Hardwood—Stress Code F17 Amdt 1 March 1991 (in Professional Package 30) (Superseded (in Part) by AS 1684 Supplement 0 and AS 1684 Supplement 15—1992). 15 pp.
SAA AS 1684 Supp 15-92. National Timber Framing Code—Supplement 15: Timber Framing Span Tables—Seasoned Hardwood—Stress Grade F17 (Supplement to AS 1684—1992) (Supersedes AS 1684 Supplement 15—1975 (in Part)) (in Professional Packages 20, 21, 30, 62-69) Bound Together. 27 pp.
SAA AS 1684 Supp 16-75. Supplement to AS 1684 SAA Timber Framing Code—Supplement 16: Light Timber Framing Span Tables—Seasoned Hardwood—Stress Grade F27 Amdt 1 March 1991 (in Professional Package 30) (Superseded (in Part) by AS 1684 Supplement 0 and AS 1684 Supplement 16—1992). 15 pp.
SAA AS 1684 Supp 16-92. National Timber Framing Code—Supplement 16: Timber Framing Span Tables—Seasoned Hardwood—Stress Grade F27 (Supplement to AS 1684—1992) (Supersedes AS 1684 Supplement 16—1975 (in Part)) (in Professional Packages 20, 21, 30, 62-69). 27 pp.

Harmful Organisms
Use: Noxious Organisms

Harmonic Analysis
See Also: Electromagnetic Compatibility; Harmonic Analyzers; Harmonic Distortion

—Electrical Installations—Safety
CSA CAN/CSA-C22. 2 NO0.16-M92. Measurement of Harmonic Currents; (Gen Instr 1 Thru 2). 26 pp.

—Power Systems
BSI BS EN 61000-4-7-93. 1993 Electromagnetic Compatibility (EMC) Part 4: Testing and Measurement Techniques Section 7: General Guide on Harmonics and Interharmonics Msts and Instrumentation, for Power Supply Systems and Equipment Connected Thereto (E). 40 pp.

INTERNATIONAL AND NON-U.S. NATIONAL STANDARDS
SUBJECT INDEX

Harmonic Analysis (Cont.)
—Power Systems (Cont.)
CENELEC EN 61000-4-7-93. Electromagnetic Compatibility (EMC) Part 4: Testing and Measurement Techniques Section 7: General Guide on Harmonics and Interharmonics Measurements and Instrumentation, for Power Supply Systems and Equipment Connected Thereto(IEC 1000-4-7:91). 5 pp.

CENELEC EN 61000-4-7-93. Electromagnetic Compatibility (EMC) Part 4: Testing and Measurement Techniques Section 7: General Guide on Harmonics and Interharmonics Measurements and Instrumentation, for Power Supply Systems and Equipment Connected Thereto(IEC 1000-4-7:91). 34 pp.

IEC 1000 Pt 4-7-91. Electromagnetic Compatibility (EMC) Part 4: Testing and Measurement Techniques Section 7: General Guide on Harmonics and Interharmonics Measurements and Instrumentation, for Power Supply Systems and Equipment Connected Thereto Fitst Edition (CENE EN 61000-4-7:1993). 70 pp.

Harmonic Analyzers
Use For: Harmonic Distortion Analyzers
See Also: Analyzers; Distortion Analyzers; Electric Measuring Instruments; Frequency Meters; Harmonic Analysis; Harmonic Distortion

—Power Systems
BSI BS EN 61000-4-7-93. 1993 Electromagnetic Compatibility (EMC) Part 4: Testing and Measurement Techniques Section 7: General Guide on Harmonics and Interharmonics Msts and Instrumentation, for Power Supply Systems and Equipment Connected Thereto (E). 40 pp.

CENELEC EN 61000-4-7-93. Electromagnetic Compatibility (EMC) Part 4: Testing and Measurement Techniques Section 7: General Guide on Harmonics and Interharmonics Measurements and Instrumentation, for Power Supply Systems and Equipment Connected Thereto(IEC 1000-4-7:91). 5 pp.

CENELEC EN 61000-4-7-93. Electromagnetic Compatibility (EMC) Part 4: Testing and Measurement Techniques Section 7: General Guide on Harmonics and Interharmonics Measurements and Instrumentation, for Power Supply Systems and Equipment Connected Thereto(IEC 1000-4-7:91). 34 pp.

IEC 1000 Pt 4-7-91. Electromagnetic Compatibility (EMC) Part 4: Testing and Measurement Techniques Section 7: General Guide on Harmonics and Interharmonics Measurements and Instrumentation, for Power Supply Systems and Equipment Connected Thereto Fitst Edition (CENE EN 61000-4-7:1993). 70 pp.

Harmonic Distortion
See Also: Distortion (Electrical); Harmonic Analysis; Harmonic Analyzers; Nonlinear Distortion

—Compandors—Telephone Circuits
CCITT RECMN G.162-89. Characteristics of Compandors for Telephony—General Characteristics of International Telephone Connections and Circuits (Study Groups XII and XV) 8 pp. 8 pp.

CCITT RECMN G.166-89. Characteristics of Syllabic Compandors for Telephony on High Capacity Long Distance Systems—General Characteristics of International Telephone Connections and Circuits (Study Groups XII and XV) 7 pp. 7 pp.

—Echo Suppressors
CCITT RECMN G.164-89. Echo Suppressors—General Characteristics of International Telephone Connections and Circuits (Study Groups XII and XV) 36 pp. 36 pp.

—Radio Receivers—Mobile
CNS C6210-83. Method of Test for Land Mobile Communication FM or PM Receiver, 25-947 MHz (Audio Frequency, Harmonic Distortion Factor, Hum and Noise) (Dec)(10695).

—Telephone Repeaters—Carrier Systems
CCITT RECMN G.322-89. General Characteristics Recommended for Systems on Symmetric Pair Cables —International Analogue Carrier Systems (Study Group XV) 8 pp. 8 pp.

—Voice Band Data Transmission
CCITT RECMN G.113-89. Transmission Impairments—General Characteristics of International Telephone Connections and Circuits (Study Groups XII and XV) 22 pp. 22 pp.

Harmonic Distortion Analyzers
Use: Harmonic Analyzers

Harmonic Measurements
Use: Harmonic Analysis

Harmonic Oscillators
Use: Sinusoidal Oscillators

Harmonicas
JIS S 8501-82. Harmonicas.

Harmonics
See Also: Acoustics; Frequencies; Resonance Frequency; Vibration; Wavelength

—Appliances
BSI BS 5406: Part 2-88. 1988 Disturbances in Supply Systems Caused by Household Appliances and Similar Electrical Equipment Part 2: Specification of Harmonics. 19 pp.

BSI BS 5406: Part 2-01. 1988 Amd 1 Disturbances in Supply Systems Caused by Household Appliances and Similar Electrical Equipment Part 2: Specification of Harmonics (AMD 7420) October 15, 1992 (F). 21 pp.

CENELEC EN 60 555 (Part 2)-87. Disturbances in Supply Systems Caused by Household Appliance and Similar Electrical Equipment—Part 2: Harmonics. 16 pp.

IEC 555 Pt 2-82. Disturbances in Supply Systems Caused by Household Appliances and Similar Electrical Equipment Part 2: Harmonics First Edition; (Second Impression, Incorporating Amendments 1 Thru 3). 44 pp.

SAA AS 2279.1-91. Disturbances in Mains Supply Networks—Part 1: Limitation of Harmonics Caused by Household and Similar Electrical Appliances. 8 pp.

—Industrial Equipment
SAA AS 2279.2-91. Disturbances in Mains Supply Networks—Part 2: Limitation of Harmonics Caused by Industrial Equipment. 12 pp.

—Pilot Channels—Group Links
CCITT RECMN G.241-89. Pilots on Groups, Supergroups, Etc.—International Analogue Carrier Systems (Study Group XV) 6 pp. 6 pp.

—Pilot Channels—Supergroup Links
CCITT RECMN G.241-89. Pilots on Groups, Supergroups, Etc.—International Analogue Carrier Systems (Study Group XV) 6 pp. 6 pp.

—Power Lines
DIN VDE 0846 Pt 1-85. Measuring Apparatus for Judgement of Electromagnetic Compatibility; Measuring Harmonics of the Mains Voltages and Currents up to 2500 Hz (Aug). 18 pp.

Harnesses (Electric)
See Also: Cables (Electric); Electric Wire; Wiring Harnesses
MOD UK DSTAN 59-82-86. Wiring Harness and Connector Cable Assembles Generic Data and Methods of Test for Capability Approval Issue 1. 50 pp.

—Automotive
CNS D1053-83. Tolerance in Length for Wiring Harness for Automobiles (Apr)(10160).

—Automotive—Color Coding
CNS D1052-83. Colour Code for Wiring Harness for Automobiles (Apr)(10159).

Harnesses (Safety)
See Also: Safety Belts
BSI BS EN 361-93. 1993 Personal Protective Equipment Against Falls from a Height—Full Body Harnesses (Supersedes BS 1397: 1979) (N). 11 pp.

CEN PREN 361-90. Personal Fall Arresting Systems: Harnesses. 6 pp.

CEN PREN 813-92. Personal Protective Equipment for Prevention of Falls from a Height—Sit Harnesses and Associated Equipment. 25 pp.

CSA CAN/CSA-Z259.10-M90. Full Body Harnesses; (Gen Instr 1). 29 pp.

—Aerostats
CAA Chapter Q4-4 12.79. Seats, Safety Belts and Harnesses (Non-Rigid Airships). 3 pp.

—Aircraft
CAA 647 (K). Section K Light Aeroplanes Chapter K4-4, Seats, Safety Belts and Harnesses (Blue Papers). 11 pp.

CAA Chapter K4-4 10.92. Seats, Safety Belts and Harnesses (Light Aeroplanes). 4 pp.

CAA Chapter P4-4. Seats and Berths: Safety Belts and Harnesses (Provisional Airworthiness Requirements for Civil Powered Lift Aircraft). 6 pp.

CAA LEAFLET 14-15 06.92. Maintenance and Inspection of Crew Harnesses and Passenger Seat Belts (Metal to Metal Attachment). 1 p.

CAA Spec. NO. 4 ISSUE 2.

Harnesses (Safety) (Cont.)
—Aircraft—Gears—Assembly
SBAC AS 2830 ISSUE 5. 25 Gram Harness Release Gear (Locking In). 1 p.

—Aircraft—Gears—Casings
SBAC AS 2819 ISSUE 4. Half Casing for Harness Release Gear. 1 p.

—Aircraft—Gears—Casings—Assembly
SBAC AS 2820 ISSUE 4. Assembly of Front Casing. 1 p.

SBAC AS 2821 ISSUE 5. Assembly of Back Casing. 1 p.

—Aircraft—Gears—Latches
SBAC AS 2829 ISSUE 4. Catch for Harness Release Gear. 1 p.

—Aircraft—Gears—Springs (Elastic)
SBAC AS 2814 ISSUE 4. Cable Spring for Harness Release Gear. 1 p.

SBAC AS 2815 ISSUE 1. Cable Spring for Harness Release Gear. 1 p.

SBAC AS 2817 ISSUE 4. Catch Spring for Harness Release Gear. 1 p.

—Aircraft—Lugs (Fasteners)
SBAC AS 2111 ISSUE 1. Attachment Lug for 'Z' Type Harness. 1 p.

—Aircraft—Parachutes—Quality Assurance
MOD UK DSTAN 05-61: Part 18-92. Quality Assurance Procedural Requirements Part 18: Aircraft and Munitions Parachutes and Parachute Assemblies, Harnesses and Personnel Restraint Harnesses for Use in Aircraft Issue 1. 12 pp.

MOD UK DSTAN 16-27-87. Additional Quality Requirements for Aircraft and Munitions Parachutes and Parachute Assemblies, Harnesses and Personnel Restraint Harnesses for Use in Aircraft Issue 1 (Superseded by Def Stan 05-61: Part 18). 11 pp.

—Children's
BSI BS 3785-64. 1964 Amd 3 Webbing Safety Harness for Baby Carriages and Chairs and Walking Reins. 13 pp.

BSI BS 6684-89. 1989 Amd 1 Safety Harnesses (Including Detachable Walking Reins) for Restraining Children When in Perambulators (Baby Carriages), Pushchairs and High Chairs and When Walking (AMD 6531) June 29, 1990. 16 pp.

SAA AS 3747-89. Harnesses for Use in Prams, Strollers, and High Chairs (Including a Detachable Walking Rein) (This is a Joint Standard with SANZ NZS 3747). 8 pp.

SNZ NZS/AS 3747-89. Harnesses for Use in Prams, Strollers and High Chairs, (Including a Detachable Walking Rein) (This is a Joint Standard with SAA AS 3747). 8 pp.

—Children's—Yachts
BSI BS 4474-79. 1979 Safety Harnesses and Safety Lines for Use by Children in Yachts. 4 pp.

—Energy Absorbers
DIN ENGL 32766-81. Protective Equipment Against Fall; Energy Absorbers; Safety Requirements, Testing (Dec). 2 pp.

—Ground Effect Machines
CAA Chapter B5-5 11.72. Seats, Safety Belts and Harnesses. 2 pp.

—Industrial
SAA AS 1891-83. Industrial Safety Belts and Harnesses. 18 pp.

—Industrial—Anchors (Fasteners)
BSI BS 6858-87. 1987 Manually Operated Positioning Devices and Associated Anchorage Lines for Use with Industrial Safety Belts and Harnesses. 11 pp.

—Industrial Equipment
BSI BS 1397-79. (WITHDRAWN) 1979 Amd 3 Industrial Safety Belts, Harnesses and Safety Lanyards (AMD 4905) October 31,. 1985 (Superseded by BS EN 354, 355, 358, 361, 362, 363, 364, 365: 1993). 21 pp.

—Mountaineering Equipment
BSI BS 7323-90. 1990 Sit Harnesses for Rock Climbing. 14 pp.

—Positioning Devices
BSI BS 6858-87. 1987 Manually Operated Positioning Devices and Associated Anchorage Lines for Use with Industrial Safety Belts and Harnesses. 11 pp.

—Rotary Wing Aircraft
CAA Chapter G4-4 11.75. Seats, Safety Belts and Harnesses (Rotocraft). 4 pp.

INDUSTRY STANDARDS

Harnesses (Safety) (Cont.)
—Yachts
BSI BS 4224-82. 1982 Yachtsmen's Safety Harnesses and Safety Lines. 10 pp.
CEN PREN 1095-93. Deck Safety Harness and Safety Line for Use on Recreational Craft. 11 pp.

Harnesses (Weaving)
See Also: Looms
—Numbering
BSI BS 6876-87. (WITHDRAWN) 1987 Numbering of Machine Parts for Drawing-in on Jacquard Machines (Superseded by ISO 7506). 12 pp.
ISO 7506-84. Textile Machinery and Accessories—Numbering of Harnesses for Drawing-in on Jacquard Machines First Edition; (Corrected and Reprinted -1992). 11 pp.

Harpoon Missiles
See Also: Missiles
—Interchangeability
NATO STANAG 1341 ED 1 AMD 1-88. Harpoon Missile Interchangeable Within NATO Naval Forces. 17 pp.
NATO STANAG 1341 ED 1 AMD 2-88. Harpoon Missile Interchangeable Within NATO Naval Forces. 30 pp.

Harrow Tines
Use: Tines—Harrow

Harvard AT 6D Aircraft
See Also: Aircraft
—Antenna Positions
CAA. Harvard AT6D (Approved Aerial Positions). 1 p.

Harvesting Equipment
Scope Note: For additional listings, use a more specific term See Also: Agricultural Equipment; Balers; Binders (Agricultural); Combines; Haymakers; Mowers; Reapers; Sickles; Tree Harvesters
—Beets—Safety
DIN ENGL 11001 Pt 7-80. Agricultural Machines and Tractors; Beet-Harvesters; Special Safety Requirements and Testing (Dec). 3 pp.
—Carrots—Rod Links
BSI BS 4026: Part 4-69. (OBSOLESCENT) 1969 Rod Links for Root Machinery Part 4: Carrot and Sugar Beet Harvesters: Metric Units. 8 pp.
—Components
BSI BS 1562: Part 2-73. (OBSOLESCENT) 1973 Agricultural Mower and Combine-Harvester Parts Part 2: Metric Units. 15 pp.
—Forage Crops—Compatibility
BSI BS 5137-74. 1974 Requirements for the Compatible Operation of Forage Harvesting Machinery. 12 pp.
ISO 5715-83. Equipment for Harvesting—Dimensional Compatibility of Forage Harvesting Machinery First Edition. 7 pp.
—Forage Crops—Safety
DIN ENGL 11001 Pt 3-80. Agricultural Machines and Tractors; Reaping Units, Hay Making Machines, Chopper Forage Harvesters; Special Technical Safety Requirements and Testing (Aug). 3 pp.
—Potatoes—Rod Links
BSI BS 4026: Part 3-69. (OBSOLESCENT) 1969 Rod Links for Root Machinery Part 3: Potato Elevator Diggers and Main Elevators on Complete Potato Harvesters: Metric Units. 9 pp.
—Potatoes—Safety
DIN ENGL 11001 Pt 8-80. Agricultural Machines and Tractors; Equipment for Harvesting Potatoes; Special Technical Safety Requirements and Testing (Dec). 3 pp.
—Sugar Beets—Rod Links
BSI BS 4026: Part 4-69. (OBSOLESCENT) 1969 Rod Links for Root Machinery Part 4: Carrot and Sugar Beet Harvesters: Metric Units. 8 pp.
—Sugar Beets—Topping Knives
BSI BS 4394-69. 1969 Topping Knives for Sugar Beet Harvesters. 9 pp.

Hash Functions
See Also: Information Security; Message Authentication Codes
IEC DIS 10118 Pt 1-93. Information Technology—Security Techniques—Hash-Functions—Part 1: General ***CD-ROM ONLY***. 9 pp.
ISO DIS 10118 Pt 1-93. Information Technology—Security Techniques—Hash-Functions—Part 1: General ***CD-ROM ONLY***. 9 pp.
JTC1 DIS10118 Pt 1-93. Information Technology—Security Techniques—Hash-Functions—Part 1: General ***CD-ROM ONLY***. 9 pp.
—Algorithms—n-bit Block Cipher
IEC DIS 10118 Pt 2-93. Information Technology—Security Techniques—Hash-Functions—Part 2: Hash-Functions Using an n-Bit Cipher Algorithm ***CD-ROM ONLY***. 10 pp.
ISO DIS 10118 Pt 2-93. Information Technology—Security Techniques—Hash-Functions—Part 2: Hash-Functions Using an n-Bit Cipher Algorithm ***CD-ROM ONLY***. 10 pp.
JTC1 DIS10118 Pt 2-93. Information Technology—Security Techniques—Hash-Functions—Part 2: Hash-Functions Using an n-Bit Cipher Algorithm ***CD-ROM ONLY***. 10 pp.

Hasps
See Also: Fasteners
CGSB CAN/CGSB-69.36-M90. Strap and Tee Hinges and Hasps (ANSI/BHMA A156.20-1984); (Amendment 1 August 1993). 26 pp.
—Ships—Hatch Covers
CNS F3170-82. Fittings for Ships' Small Size Steel Hatch Covers (Apr)(8684).
JIS F 2322-87. Fittings of Ships' Steel Small Hatch Covers.
JIS F 2322-61. Fittings of Ships' Steel Small Hatch Covers (R 1970). 9 pp.

Hatch Covers
See Also: Hatches; Ships
—Air
CNS F3216-83. Air Hatch Covers for Ships (Mar)(10083).
—Cleaning Holes—Tanks
CNS F3212-83. Cover for Tank Cleaning Holes (Mar)(10073).
JIS F 2331-91. Covers for Tank Cleaning Holes.
JIS F 2331-75. Covers for Tank Cleaning Holes. 5 pp.
—Oil Tight
CNS F3168-82. Ships' Oiltight Hatch Covers (Apr)(8682).
JIS F 2320-78. Ships' Oiltight Hatch Covers.
JIS F 2320-69. Oiltight Hatch Covers. 19 pp.
—Rubber Seals
ISO 4089-79. Shipbuilding—Inland Navigation—Sealing Rubber for Covers of Cargo Hatches First Edition. 4 pp.
—Spanners
CNS F3171-82. Ships' Ratchet Spanners (Apr)(8685).
JIS F 2323-82. Ships' Ratchet Spanners.
JIS F 2323-76. Ships' Ratchet Spanners. 7 pp.
—Steel
CNS F3169-82. Ships' Small Size Steel Hatch Covers (Apr)(8683).
JIS F 2321-84. Ships' Steel Small Hatch Covers.
JIS F 2321-70. Ships' Steel Small Hatch Covers. 8 pp.
—Steel—Fittings
CNS F3170-82. Fittings for Ships' Small Size Steel Hatch Covers (Apr)(8684).
JIS F 2322-87. Fittings of Ships' Steel Small Hatch Covers.
JIS F 2322-61. Fittings of Ships' Steel Small Hatch Covers (R 1970). 9 pp.
—Ullage Holes
CNS F3049-80. Ships' Ullage Holes (Aug)(6181). 3 pp.
JIS F 2317-81. Ships' Ullage Holes.
JIS F 2317-75. Ships' Ullage Holes. 8 pp.
—Wood—Identification Systems
BSI BS 4268-67. 1967 Marking of Wooden Hatchway Covers. 8 pp.

Hatches
See Also: Hatch Covers; Scuttles; Ships; Side Scuttles
—Battens
CNS F3019-83. Hatch Battens (Mar)(5327).
CNS F3167-82. Hatch Board (Apr)(8681).
—Battens—Identification Systems
CNS F4009-82. Marking of Hatch Board (Apr)(8687).
—Beams—Hoisting Slings
CNS F3082-81. Ships' Hatch Beam Slings (Jan)(6912).
—Beams—Identification Systems
BSI BS 4263-67. 1967 Marking of Hatchway Beams. 8 pp.
CNS F4010-82. Marking of Hatchway Beam (Apr)(8688).
—Cleats
CNS F3017-80. Hatch Cleats (Mar)(5325).
CNS F3018-80. Hatch Cleats (Mar)(5326).
CNS F3172-82. Hatch Cleats (Simple Type) (Apr)(8686).
—Steel
BSI BS MA 8-83. 1983 Small Weathertight Hatches. 12 pp.
ISO 5778-79. Shipbuilding—Small Weathertight Steel Hatches First Edition. 8 pp.
—Weatherproof
SAA AS 1987-77. Small Square Weathertight Hatches Amdt 1 February 1984 Reconfirmed 1984. 12 pp.
—Wedges
CNS F3020-80. Hatch Wedges (Mar)(5328).

Hatchets
See Also: Axes; Hammers
BSI BS 2945-58. 1958 Amd 1 Axes and Hatchets. 20 pp.

Hats
Use For: Caps (Headgear) See Also: Clothing; Headgear; Helmets; Uniforms
CNS Z7034-71. Plastic Cloth for Anti-Perspiration Hats and Caps (Tentative) (Apr)(3251).
—Leather
MOD UK DSTAN 83-4-92. Leather, Sheep, Head Leathers Issue 2. 12 pp.
—Park Wardens'—Felts
CGSB CAN/CGSB-38.21-93. Park Warden Felt Hat. 12 pp.
—Plastic Fabrics
CNS Z7034-71. Plastic Cloth for Anti-Perspiration Hats and Caps (Tentative) (Apr)(3251).

Hawker Siddeley Argosy Aircraft
See Also: Aircraft
—Accidents
CAA. Hawker Siddeley Argosy (World Airline Accident Summary). 1 p.
CAA. Hawker Siddely Argosy. 1 p.

Hawker Siddeley Comet 1 Aircraft
See Also: Aircraft
—Accidents
CAA. Hawker Siddeley Comet 1 (World Airline Accident Summary). 1 p.

Hawker Siddeley/De Havilland 89A Rapide Aircraft
See Also: Aircraft
—Mandatory Aircraft Modifications and Inspections Summaries
CAA. Hawker Siddeley/De Havilland 89a Rapide. 1 p.

Hawker Siddeley/De Havilland 104 Dove Aircraft
See Also: Aircraft
—Mandatory Aircraft Modifications and Inspections Summaries
CAA. Hawker Siddeley/De Havilland 104 Dove. 7 pp.

Hawker Siddeley/De Havilland 114 Heron Aircraft
See Also: Aircraft
—Mandatory Aircraft Modifications and Inspections Summaries
CAA. Hawker Siddeley/De Havilland 114 Heron. 10 pp.

Hawker Siddeley/De Havilland 121 Trident Aircraft
See Also: Aircraft
—Mandatory Aircraft Modifications and Inspections Summaries
CAA. Hawker Siddeley/De Havilland 121 Trident. 24 pp.

INTERNATIONAL AND NON-U.S. NATIONAL STANDARDS
SUBJECT INDEX

Hazardous

Hawker Siddeley DH.106 Comet 4 Aircraft
See Also: Aircraft
—Accidents
CAA. Hawker Siddeley DH.106 Comet 4 Series (World Airline Accident Summary). 1 p.

Hawker Siddeley 114 Heron Aircraft
See Also: Aircraft
—Accidents
CAA. Hawker Siddeley 114 Heron/Riley Heron (World Airline Accident Summary). 1 p.
CAA. Hawker Siddely 114 Heron/Riley Heron. 1 p.

Hawker Siddeley 121 Trident Aircraft
See Also: Aircraft
—Accidents
CAA. Hawker Siddeley 121 Trident (World Airline Accident Summary). 1 p.
CAA. Hawker Siddely 121 Trident. 1 p.

Hawker Siddeley 125 Aircraft
See Also: Aircraft
—Accidents
CAA. Hawker Siddeley/125 Series (World Airline Accident Summary). 1 p.
CAA. Hawker Siddely/BAe 125 Series. 1 p.

Hawker Siddeley 748 Aircraft
See Also: Aircraft
—Accidents
CAA. Hawker Siddeley/748 Series (World Airline Accident Summary). 1 p.
CAA. Hawker Siddely/BAe 748 Series. 1 p.
—Mandatory Aircraft Modifications and Inspections Summaries
CAA. Hawker Siddeley 748. 26 pp.

Hayforks
Use: Pitchforks

Haymakers
See Also: Agricultural Equipment; Harvesting Equipment
—Safety
DIN ENGL 11001 Pt 3-80. Agricultural Machines and Tractors; Reaping Units, Hay Making Machines, Chopper Forage Harvesters; Special Technical Safety Requirements and Testing (Aug). 3 pp.

Hazard Analysis
—Software
MOD UK DSTAN 00-55. Requirements for the Procurement of Safety Critical Software in Defence Equipment (Draft) (Superseded by 00-55: Parts 1 & 2).

Hazard Warning Lights
Use: Warning Lights

Hazard Warning Signal Flashers
Use: Flasher Units

Hazardous Areas
Use: Hazardous Locations

Hazardous Electrical Locations
Use: Hazardous Locations

Hazardous Environments
See Also: Explosive Atmospheres; Hazardous Locations
—Hand Tools
MOD UK DSTAN 51-11: Part 28-01. Hand Tools, General Purpose Part 28 (A): Miscellaneous Items Part 28 (B): Hand Tools for Use in Hazardous or Critical Conditions Issue 1; Amendment 7. 166 pp.
MOD UK DSTAN 51-11: Part 28-93. Hand Tools, General Purpose Part 28 (A): Miscellaneous Items (B): Hand Tools for Use in Hazardous or Critical Conditions Issue 2. 146 pp.
—Warning Signs
NATO STANAG 2002 ED 7 AMD 2-80. Warning Signs for the Marking of Contaminated or Dangerous Land Areas, Complete Equipments, Supplies and Stores. 9 pp.
NATO STANAG 2002 ED 7 AMD 3-80. Warning Signs for the Marking of Contaminated or Dangerous Land Areas, Complete Equipments, Supplies and Stores. 9 pp.

Hazardous Locations
Use For: Hazardous Areas; Hazardous Electrical Locations *See Also:* Explosive Atmospheres; Hazardous Environments; Hazardous Materials
CEN PREN 1127-1-93. Safety of Machinery—Fire and Explosions—Part 1: Explosion Prevention and Protection. 50 pp.
—Aircraft Refueling Equipment
NATO STANAG 7012 ED 2 AMD 0-91. Minimum Radiotelephony (R/T) Aerodrome Departure Procedures. 6 pp.
NATO STANAG 7013 ED 1 AMD 0-90. Aircraft Fuelling Hazard Zones. 7 pp.
NATO STANAG 7013 ED 1 AMD 1-90. Aircraft Fuelling Hazard Zones. 8 pp.
NATO STANAG 7013 ED 1 AMD 2-92. Aircraft Fuelling Hazard Zones. 9 pp.
NATO STANAG 7013 ED 1 AMD 3-90. Aircraft Fuelling Hazard Zones. 10 pp.
—Anesthetics—Flammable
CSA CAN/CSA-C22. 2NO601.1-M90. Medical Electrical Equipment, Part 1: General Requirements for Safety; (Gen Instr 1). 240 pp.
—Cable Glands
CSA C22.2 NO 174-M1984. Cables and Cable Glands for Use in Hazardous Locations (R 1992); (Gen Instr 1 Thru 5). 31 pp.
—Cables (Electric)
CSA C22.2 NO 174-M1984. Cables and Cable Glands for Use in Hazardous Locations (R 1992); (Gen Instr 1 Thru 5). 31 pp.
—Combustible Gas Detectors
CSA C22.2 NO 152-M1984. Combustible Gas Detection Instruments (R 1989); (Gen Instr 1 Thru 3). 37 pp.
—Combustible Gas Detectors—Portable
CSA C22.2 NO 152-M1984. Combustible Gas Detection Instruments (R 1989); (Gen Instr 1 Thru 3). 37 pp.
—Electric Conduits—Overpressure Testing
CSA C22.2 NO 30-M1986. Explosion-Proof Enclosures for Use in Class I Hazardous Locations (R 1992); (Gen Instr 1 Thru 2). 41 pp.
—Electric Plug Receptacles
CSA C22.2 NO 159-M1987. Attachment Plugs, Receptacles, and Similar Wiring Devices for Use in Hazardous Locations: Class I, Groups A, B, C, and D; Class II, Group G, in Coal or Coke Dust, and in Gaseous Mines (R 1993). 39 pp.
—Electric Plugs
CSA C22.2 NO 159-M1987. Attachment Plugs, Receptacles, and Similar Wiring Devices for Use in Hazardous Locations: Class I, Groups A, B, C, and D; Class II, Group G, in Coal or Coke Dust, and in Gaseous Mines (R 1993). 39 pp.
—Electrical Enclosures
CSA C22.2 NO 25-1966. Enclosures for Use in Class II Groups E, F, and G Hazardous Locations (R 1992); (Gen Instr 1). 18 pp.
CSA 929 Bull. Electrical Bulletin 929 November 16, 1973 to C22.2 NO 25. 2 pp.
CSA 1273 Bull. Electrical Bulletin 1273 July 2, 1980 to C22.2 NO 25. 2 pp.
CSA 1310 Bull. Electrical Bulletin 1310 March 24, 1981 to C22.2 NO 25. 2 pp.
CSA 1273A Bull. Electrical Bulletin 1273A February 8, 1983 to C22.2 NO 25. 2 pp.
CSA C22.2 NO 30-M1986. Explosion-Proof Enclosures for Use in Class I Hazardous Locations (R 1992); (Gen Instr 1 Thru 2). 41 pp.
CSA 1273 Bull. Electrical Bulletin 1273 July 2, 1980 to C22.2 NO 30. 2 pp.
CSA 1273A Bull. Electrical Bulletin 1273A February 8, 1983 to C22.2 NO 30. 2 pp.
—Electrical Enclosures—Bonding
CSA C22.2 NO 25-1966. Enclosures for Use in Class II Groups E, F, and G Hazardous Locations (R 1992); (Gen Instr 1). 18 pp.
CSA 1273 Bull. Electrical Bulletin 1273 July 2, 1980 to C22.2 NO 25. 2 pp.
CSA 1273A Bull. Electrical Bulletin 1273A February 8, 1983 to C22.2 NO 25. 2 pp.
CSA C22.2 NO 30-M1986. Explosion-Proof Enclosures for Use in Class I Hazardous Locations (R 1992); (Gen Instr 1 Thru 2). 41 pp.
—Electrical Enclosures—Dust
IEC 1241 Pt 1-1-93. Electrical Apparatus for Use in the Presence of Combustible Dust—Part 1: Electrical Apparatus Protected by Enclosures—Section 1: Specification for Apparatus First Edition. 47 pp.

Hazardous Locations (Cont.)
—Electrical Enclosures—Dust (Cont.)
IEC 1241 Pt 1-2-93. Electrical Apparatus for Use in the Presence of Combustible Dust—Part 1: Electrical Apparatus Protected by Enclosures—Section 2: Selection, Installation, and Maintenance First Edition. 39 pp.
—Electrical Enclosures—Grounding
CSA C22.2 NO 25-1966. Enclosures for Use in Class II Groups E, F, and G Hazardous Locations (R 1992); (Gen Instr 1). 18 pp.
CSA 1273 Bull. Electrical Bulletin 1273 July 2, 1980 to C22.2 NO 25. 2 pp.
CSA 1273A Bull. Electrical Bulletin 1273A February 8, 1983 to C22.2 NO 25. 2 pp.
CSA C22.2 NO 30-M1986. Explosion-Proof Enclosures for Use in Class I Hazardous Locations (R 1992); (Gen Instr 1 Thru 2). 41 pp.
—Electrical Equipment—Intrinsically Safe
CSA CAN/CSA-C22. 2 NO 157-92. Intrinsically Safe and Non-Incendive Equipment for Use in Hazardous Locations; (Gen Instr 1). 61 pp.
—Electrical Installations
DIN VDE 0100 Pt 720-83. Erection of Power Installations with Rated Voltages up to 1000 V; Locations Exposed to Fire Hazards (Mar). 15 pp.
—Electrical Protection Equipment
CSA C22.2 NO 213-M1987. Non-Incendive Electrical Equipment for Use in Class I, Division 2 Hazardous Locations (R 1992). 30 pp.
—Floor Coverings—Electrostatic Charge
DIN ENGL 51953-75. Testing of Organic Floor Coverings; Testing of the Ability of Floor Coverings in Explosion-Hazarded Rooms to Dissipate Electrostatic Charges (Aug). 2 pp.
—Gaseous Mines—Electrical Equipment—Intrinsically Safe
CSA CAN/CSA-C22. 2 NO 157-92. Intrinsically Safe and Non-Incendive Equipment for Use in Hazardous Locations; (Gen Instr 1). 61 pp.
—Generators
CSA C22.2 NO 145-M1986. Motors and Generators for Use in Hazardous Locations (R 1992); (Gen Instr 1 Thru 2). 47 pp.
CSA 1169 Bull. Electrical Bulletin 1169 June 27, 1978 to C22.2 NO 145. 2 pp.
—Heat Tracing Systems—Heating Cables—Gaseous Mines
CSA C22.2 NO 138-M1989. Heat Tracing Cable and Cable Sets for Use in Hazardous Locations; (Gen Instr 1). 25 pp.
—Heat Tracing Systems—Heating Cables—Pipes
CSA C22.2 NO 138-M1989. Heat Tracing Cable and Cable Sets for Use in Hazardous Locations; (Gen Instr 1). 25 pp.
—Heat Tracing Systems—Heating Cables—Tanks (Containers)
CSA C22.2 NO 138-M1989. Heat Tracing Cable and Cable Sets for Use in Hazardous Locations; (Gen Instr 1). 25 pp.
—Internal Combustion Engines—Ignition Systems
CSA C22.2 NO 30-M1986. Explosion-Proof Enclosures for Use in Class I Hazardous Locations (R 1992); (Gen Instr 1 Thru 2). 41 pp.
CSA CAN/CSA-C22. 2 NO 157-92. Intrinsically Safe and Non-Incendive Equipment for Use in Hazardous Locations; (Gen Instr 1). 61 pp.
—Lighting—Exterior
CSA C22.2 NO 137-M1981. Electric Luminaires for Use in Hazardous Locations (R 1993); (Gen Instr 1 Thru 3). 26 pp.
—Lighting—Flammability Testing
CSA C22.2 NO 137-M1981. Electric Luminaires for Use in Hazardous Locations (R 1993); (Gen Instr 1 Thru 3). 26 pp.
—Lighting—Interior
CSA C22.2 NO 137-M1981. Electric Luminaires for Use in Hazardous Locations (R 1993); (Gen Instr 1 Thru 3). 26 pp.
—Lighting—Temperature Testing
CSA C22.2 NO 137-M1981. Electric Luminaires for Use in Hazardous Locations (R 1993); (Gen Instr 1 Thru 3). 26 pp.

INDUSTRY STANDARDS

INTERNATIONAL AND NON-U.S. NATIONAL STANDARDS SUBJECT INDEX

Hazardous

Hazardous Locations *(Cont.)*

—Lighting—Underground
CSA C22.2 NO 137-M1981. Electric Luminaires for Use in Hazardous Locations (R 1993); (Gen Instr 1 Thru 3). 26 pp.

—Motors
CSA C22.2 NO 145-M1986. Motors and Generators for Use in Hazardous Locations (R 1992); (Gen Instr 1 Thru 2). 47 pp.
CSA 1169 Bull. Electrical Bulletin 1169 June 27, 1978 to C22.2 NO 145. 2 pp.

—Occupancies
SAA AS 2430.3-91. Classification of Hazardous Areas—Part 3: Specific Occupancies (in Professional Packages 21, 40, 44, 67). 36 pp.

—Panel Boards (Electrical)
CSA C22.2 NO 30-M1986. Explosion-Proof Enclosures for Use in Class I Hazardous Locations (R 1992); (Gen Instr 1 Thru 2). 41 pp.

—Process Control Equipment
CSA C22.2 NO 213-M1987. Non-Incendive Electrical Equipment for Use in Class I, Division 2 Hazardous Locations (R 1992). 30 pp.

—Sprayers—Electrostatic
BSI BS 6742: Part 3-90. 1990 Electrostatic Painting and Finishing Equipment Using Flammable Materials Part 3: Selection, Installation and Use of Hand Held Powder Spray Guns (Energy Limit of 5 mJ) and Their Associated Apparatus. 12 pp.
BSI BS 6742: Part 4-90. 1990 Electrostatic Painting and Finishing Equipment Using Flammable Materials Part 4: Selection, Installation and Use of Hand Held Flock Spray Guns (Energy Limit of 0.24 mJ or 5 mJ) and Their Associated Apparatus. 12 pp.

—Thread Gages
CSA C22.2 NO 30-M1986. Explosion-Proof Enclosures for Use in Class I Hazardous Locations (R 1992); (Gen Instr 1 Thru 2). 41 pp.

—Warning Systems—Route Marking
NATO STANAG 2889 ED 3 AMD 2-84. Marking of Hazardous Areas and Routes Through Them. 16 pp.
NATO STANAG 2889 ED 3 AMD 3-84. Marking of Hazardous Areas and Routes Through Them. 16 pp.

—Wiring Devices
CSA C22.2 NO 159-M1987. Attachment Plugs, Receptacles, and Similar Wiring Devices for Use in Hazardous Locations: Class I, Groups A, B, C, and D; Class II, Group G, in Coal or Coke Dust, and in Gaseous Mines (R 1993). 39 pp.

Hazardous Materials

Use For: Dangerous Goods *See Also:* Acetic Acid; Acetone; Acetylene; Aluminum Chloride; Ammonia; Ammonium Fluoride; Ammonium Hydroxide; Ammonium Nitrate; Ammonium Thioglycolate; Amyl Acetate; Aniline; Argon; Arsenic; Arsenic Acid; Benzene; Beryllium; Biological Hazards; 1,3-Butadiene; n-Butyl Acetate; sec-Butyl Acetate; tert-Butyl Acetate; n-Butyl Acrylate; Cadmium; Cadmium Compounds; Calcium Carbide; Calcium Hypochlorite; Carbon Disulfide; Carbon Tetrachloride; Chlorine; Chlorobenzene; Chloropicrin; Copper-Chrome-Arsenic; Cresol; Cyanides; Cyclohexane; Cyclohexanone; Diacetone Alcohol; Diethylamine; Diphenylmethane-4, 4'-Diisocyanate; Epichlorohydrin; Ethanol; Ethyl Acetate; Ethyl Acrylate; Ethylbenzene; Ethylene Dichloride; Ethylene Oxide; Explosives; Flammable Materials; Formaldehyde; Hazardous Locations; Helium; Hydrochloric Acid; Hydrogen; Hydrogen Peroxide; Hypochlorites; Isobutyl Acetate; Isopropyl Acetate; Isopropylamine; Krypton; Lead Acetate; Lead Alkyls; Lead Arsenate; Lead Arsenite; Liquefied Petroleum Gas; Liquid Oxygen; Maleic Anhydride; Mercury; Methacrylates; Methacrylic Acid; Methanols; Methyl Ethyl Ketone; Methyl Ethyl Ketone Peroxide; Methyl Isobutyl Ketone; Methylamyl Alcohol; Methylene Chloride; Neon; Nitric Acid; Nitrogen; Organic Peroxides; Oxygen; Pentachlorophenol; Perchloroethylene; Pesticides; Phenol; Phenylmercuric Compounds; Phosphoric Acid; Potassium Cyanide; Potassium Fluoride; n-Propyl Acetate; Radioactive Effluents; Radioactive Materials; Radioactive Ores; Sodium Cyanide; Sodium Hydroxide; Sodium Hypochlorite; Sodium Pentachlorophenate; Styrene; Sulfuric Acid; Tetrahydrofuran *See Also:* Toluene-2,4-Diisocyanate; Toxicity; Tributyltin Oxide; 1,1,1-Trichloroethane; Triethylamine; Vinyl Acetate; Vinyl Chloride; Xylene
EC EEC/2455/92-92. Council Regulation Concerning the Export and Import of Certain Dangerous Chemicals. 10 pp.

Hazardous Materials *(Cont.)*

SAA AS 1216.2-81. Classification, Hazard Identification and Information Systems for Dangerous Goods—Part 2: HAZCHEM Emergency Action Code Bound Together with 1216.3 and 1216.4. 15 pp.

—Accumulators—Disposal
EC 91/157/EEC-91. Council Directive on Batteries and Accumulators Containing Certain Dangerous Substances. 4 pp.

—Air Cargo Transportation
NATO STANAG 3854 ED 2 AMD 2-88. Policies and Procedures Governing the Air Transportation of Dangerous Cargo. 29 pp.
NATO STANAG 3854 ED 2 AMD 3-88. Policies and Procedures Governing the Air Transportation of Dangerous Cargo. 31 pp.
NATO STANAG 3854 ED 2 AMD 4-88. Policies and Procedures Governing the Air Transportation of Dangerous Cargo. 32 pp.

—Air Transportation
CAA Section 9 CAP 393. Air Navigation (Dangerous Goods) Regulations 1985. 6 pp.
ICAO Annex 18. Safe Transport of Dangerous Goods by Air Second Edition—July 1989; (Supplement 8/1/90). 40 pp.
ICAO 9284 Supplement. Technical Instructions for the Safe Transport of Dangerous Goods by Air—1986 Edition. 117 pp.
ICAO 9375 Book 1. Dangerous Goods Training Programme Book 1 Shippers and Packers Second Edition—October 1985. 102 pp.
ICAO 9375 Book 2. Dangerous Goods Training Programme Book 2 Cargo Agents Second Edition—October 1985. 91 pp.
ICAO 9375 Book 3. Dangerous Goods Training Programme Book 3 Operator's Cargo Acceptance Staff Second Edition—October 1985. 89 pp.
ICAO 9375 Book 4. Dangerous Goods Training Programme Book 4 Load Planners and Flight Crew Second Edition—January 1986. 36 pp.
ICAO 9375 Book 5. Dangerous Goods Training Programme Book 5 Passenger Handling Staff and Flight Attendants Second Edition—January 1986. 19 pp.
ICAO 9375 Book 6. Dangerous Goods Training Programme Book 6 Loading and Warehouse Personnel Second Edition—January 1986. 35 pp.

—Aircraft Accidents—Emergency Planning
ICAO 9481. Emergency Response Guidance for Aircraft Incidents Involving Dangerous Goods 1993-1994 Edition. 78 pp.

—Batteries—Disposal
EC 91/157/EEC-91. Council Directive on Batteries and Accumulators Containing Certain Dangerous Substances. 4 pp.

—Classification
EC 67/548/EEC-67. Council Directive on the Approximation of Laws, Regulations and Administrative Provisions Relating to the Classification, Packaging and Labelling of Dangerous Substances. 23 pp.
EC 79/831/EEC-79. Council Directive Amending for the Sixth Time Directive 67/548/EEC on the Approximation of the Laws, Regulations and Administrative Provisions Relating to the Classification, Packaging and Labelling of Dangerous Substances. 19 pp.
EC 88/379/EEC-88. Council Directive on the Approximation of the Laws, Regulations and Administrative Provisions of the Member States Relating to the Classification, Packaging and Labelling of Dangerous Preparations. 13 pp.
EC 93/18/EEC-93. Commission Directive Adapting for the Third Time to Technical Progress Council Directive 88/379/EEC on the Approximation of the Laws, Regulations and Administrative Provisions of the Member States Relating to the Classification, Packaging and Labelling. 11 pp.
EC 93/21/EEC-93. Commission Directive Adapting to Technical Progress for the 18th Time Council Directive 67/548/EEC on the Approximation of the Laws, Regulations and Administrative Provisions Relating to the Classification, Packaging and Labelling of Dangerous Substances. 2 pp.
EC 93/67/EEC-93. Commission Directive Laying Down the Principles for Assessment of Risks to Man and the Environment of Substances Notified in Accordance with Council Directive 67/548/EEC. 10 pp.
EC EEC/2455/92-92. Council Regulation Concerning the Export and Import of Certain Dangerous Chemicals. 10 pp.
EC 85/71/EEC-84. Commission Decision Concerning the List of Chemical Substances Notified Pursuant to Council Directive 67/548/EEC on the Approximation of Laws, Regs. and Administrative Provisions Relating to the Classification, Packaging and Labeling of Dangerous Substances. 2 pp.

Hazardous Materials *(Cont.)*

—Classification *(Cont.)*
SAA AS 1216.1-84. Classification, Hazard Identification and Information Systems for Dangerous Goods—Part 1: Classification and Class Labels for Dangerous Goods. 11 pp.
SNZ NZS 5433-88. Code of Practice for Transportation of Hazardous Substances on Land Amend: 1, 1990. 172 pp.
SNZ NZS 5433B-88. Code of Practice for Transportation of Hazardous Substances on Land. 1 p.

—Defense Contracts
MOD UK DEFCON 68-90. Control of Dangerous Articles and Substances (Navy Contracts) 2/90. 2 pp.

—Disposal—Service Contracts
MOD UK DEFCON 112BD-87. Conditions of Contract Disposal of Poisonous Waste 1/87. 3 pp.

—Freight Cars
CGSB 43-GP-147P-90. Construction of Tank Car Tanks and Selection and Use of Tank Car Tanks, Portable Tanks and Rail Cars for the Transportation of Dangerous Goods by Rail. 241 pp.

—Furniture
CNS S2128-86. Method of Test for Hazardous Substances of Furniture Surface (Aug)(11677).

—Highway Transportation
SNZ NZS 5433B-88. Code of Practice for Transportation of Hazardous Substances on Land. 1 p.
SNZ NZS 5433D-89. Code of Practice for Transportation of Hazardous Substances on Land. 1 p.

—Highway Transportation—Emergency Planning
SAA AS 1678. Emergency Procedure Guide—Transport Group Text EPGs for Class 5 Substances—Oxidizing Substances and Organic Peroxides (in Professional Package 37).
SAA AS 1678.2.1. 005-85. Emergency Procedure Guide—Transport—Part 2.1.005: Ethylene Oxide.
SAA AS 1678.2.1. 008-87. Emergency Procedure Guide—Transport—Part 2.1.008: Vinyl Chloride (Inhibited).
SAA AS 1678.2.1. 010-86. Emergency Procedure Guide—Transport—Part 2.1.010: Butadiene (Inhibited).
SAA AS 1678.2.2. 000-86. Emergency Procedure Guides—Transport—Part 2.2.000: Oxygen (Refrigerated Liquid).
SAA AS 1678.2.2. 002-91. Emergency Procedure Guide—Transport—Part 2.2.002: Chlorine (in Professional Package 37).
SAA AS 1678.2.3. 001-84. Emergency Procedure Guides—Transport—Part 2.3.001: Sulphur Dioxide (Liquefied).
SAA AS 1678.3.0. 015-93. Emergency Procedure Guide—Transport—Part 3.0.015: Formaldehyde Solutions, Flammable (in Professional Package 37).
SAA AS 1678.3.0. 021-86. Emergency Procedure Guide—Transport—Part 3.0.021: Methyl Ethyl Ketone (MEK).
SAA AS 1678.3.0. 022-89. Emergency Procedure Guide—Transport—Part 3.0.022: Acrylonitrile (Inhibited) (Vinyl Cyanide).
SAA AS 1678.3.0. 023-80. Emergency Procedure Guide—Transport—Part 3.0.023: Acrylates (Methyl, Ethyl, and Butyl).
SAA AS 1678.3.1. 001-86. Emergency Procedure Guide—Transport—Part 3.1.001: Petrol—as Cargo.
SAA AS 1678.3.1. 003-89. Emergency Procedure Guide—Transport—Part 3.1.003: Carbon Disulfide (Carbon Bisulfide).
SAA AS 1678.3.2. 001-86. Emergency Procedure Guides—Transport—Part 3.2.001: Toluene (Toluol).
SAA AS 1678.3.2. 002-86. Emergency Procedure Guides—Transport—Part 3.2.002: Xylenes (Xylol).
SAA AS 1678.4.1. 001-84. Emergency Procedure Guides—Transport—Part 4.1.001: Nitrocelluose (Spirit Wet).
SAA AS 1678.4.2. 001-93. Emergency Procedure Guides—Transport—Part 4.2.001: Aluminium Alkyls and Aluminium Alkyl Halides (in Professional Package 37A).
SAA AS 1678.4.3. 002-91. Emergency Procedure Guide Transport—Part 4.3.002: Magnesium Phosphide (In Profressional Package 37).
SAA AS 1678.5.1. 002-91. Emergency Procedure Guides—Transport—Part 5.1.002: Ammonium Nitrate. 2 pp.
SAA AS 1678.5.1. 004-93. Emergency Procedure Guides—Transport—Part 5.1.004: Calcium Hypochlorite (in Professional Package 37A).
SAA AS 1678.5.1. 005-93. Emergency Procedure Guides—Transport—Part 5.1.005: Hydrogen Peroxide (20% or More) (in Professional Package 37A).

INTERNATIONAL AND NON-U.S. NATIONAL STANDARDS
SUBJECT INDEX

Hazardous

Hazardous Materials *(Cont.)*
—**Highway Transportation—Emergency Planning** *(Cont.)*

SAA AS 1678.5.2. 001-83. Emergency Procedure Guides—Transport—Part 5.2.001: Organic Peroxides (Withdrawn).

SAA AS 1678.6.0. 000-90. Emergency Procedure Guides—Transport—Part 6.0.000: Formaldehyde Solutions (Non-Flammable) (Superseded by AS 1678.8.0.005—1993) (Withdrawn).

SAA AS 1678.6.0. 001-83. Emergency Procedure Guides—Transport—Part 6.0.001—Aniline (Amino benzene).

SAA AS 1678.6.0. 002-83. Emergency Procedure Guides—Transport—Part 6.0.002: Cyanides (Sodium and Potassium).

SAA AS 1678.6.0. 004-93. Emergency Procedure Guides—Transport—Part 6.0.004: Phenol (Carbolic Acid), Cresols and Cresylic Acid (in Professional Package 37A).

SAA AS 1678.6.0. 006-84. Emergency Procedure Guides—Transport—Part 6.0.006: Toluene Diisocyanate (TDI).

SAA AS 1678.6.0. 007-83. Emergency Procedure Guides—Transport—Part 6.0.007: Epichlorhydrin (ECH or 1-Chloro-2, 3-Epoxypropane).

SAA AS 1678.6.0. 009-84. Emergency Procedure Guides—Transport—Part 6.0.009: Cyanides, Inorganic.

SAA AS 1678.6.0. 010-84. Emergency Procedure Guides—Transport—Part 6.0.010: Phenylmercuric Compounds.

SAA AS 1678.6.0. 011-88. Emergency Procedure Guides—Transport—Part 6.0.011: Copper Chrome Arsenic (C.C.A. Wood Preservative).

SAA AS 1678.6.0. 012-84. Emergency Procedure Guides—Transport—Part 6.0.012: Sodium Pentachlorophenate.

SAA AS 1678.6.0. 013-84. Emergency Procedure Guides—Transport—Part 6.0.013: Chlorolpicrin (Trichloronitromethane).

SAA AS 1678.6.0. 014-84. Emergency Procedure Guides—Transport—Part 6.0.014: Lead Arsenate and Lead Arsenite.

SAA AS 1678.6.0. 015-84. Emergency Procedure Guides—Transport—Part 6.0.015: Beryllium and Compounds.

SAA AS 1678.6.0. 016-84. Emergency Procedure Guides—Transport—Part 6.0.016: Selenious Acid, Selenates, Selenites.

SAA AS 1678.6.0. 017-93. Emergency Procedure Guides—Transport—Part 6.0.017: Motor Fuel Anti-Knock Mixtures (Lead Aklyls, N.O.S.) (in Professional Package 37A).

SAA AS 1678.6.1. 010-83. Emergency Procedure Guides—Transport—Part 6.1.010: Arsenic Acid—Ortho-Arsenic Acid (Liquid) Meta-Arsenic Acid (Solid).

SAA AS 1678.6.1. 012-84. Emergency Procedure Guides—Transport—Part 6.1.012: Pentachlorophenol (PCP, PENTA).

SAA AS 1678.6.1. 014-84. Emergency Procedure Guides—Transport—Part 6.1.014: Ammonium Thioglycolate.

SAA AS 1678.6.1. 015-84. Emergency Procedure Guides—Transport—Part 6.1.015: Fluorides (Ammonium, Potassium or Sodium Fluoride).

SAA AS 1678.6.1. 016-84. Emergency Procedure Guides—Transport—Part 6.1.016: Diphenylmethane DI-ISO-Cyanate (Methylene Bis-(Phenylene) Diisocyanate) (MDI).

SAA AS 1678.6.1. 017-84. Emergency Procedure Guides—Transport—Part 6.1.017: Cadmium Compounds (Other Than Selenide and Sulphide).

SAA AS 1678.8.0. 001-83. Emergency Procedure Guides—Transport—Part 8.0.001: Hydrochloric Acid (Muriatic Acid).

SAA AS 1678.8.0. 004-93. Emergency Procedure Guides—Transport—Part 8.0.004: Hypochlorite Solutions (with more than 5% Avalible Chlorine) (in Professional Package 37A).

SAA AS 1678.8.0. 005-93. Emergency Procedure Guides—Transport—Part 8.0.005: Acetic Anhydride (in Professional Package 37A).

SAA AS 1678.8.0. 008-87. Emergency Procedure Guides—Transport—Part 8.0.008: Hydrofluoric Acid.

SAA AS 1678.2A1-87. Emergency Procedure Guide—Transport Group Text EPGs for Class 2 Substances—Compressed and Liquefied Gases—Part 2A1: Flammable, Compressed Gas (Supersedes AS 1678.2.1.007—1983 and AS 1678.2.1.009—1987).

SAA AS 1678.2A2-87. Emergency Procedure Guide—Transport Group Test EPGs for Class 2 Substances—Compressed and Liquefied Gases—Part 2A2: Flammable, Liquefied Gas (Supersedes AS 1678.2.1.002—1984 and AS 1678.2.1.006—1984).

SAA AS 1678.2A3-87. Emergency Precedure Guide—Transport Group Text EPGs for Class 2 Substances—Compressed and Liquefied Gases—Part 2A3: Falmmable, Low-Temperature, Liquefied Gas.

SAA AS 1678.2A4-87. Emergency Precedure Guide—Transport Group Text EPGs for Class 2 Substances—Compressed and Liquefied Gases—Part 2A4: Flammable, Poisonous Gas.

SAA AS 1678.2A5-87. Emergency Precedure Guide—Transport Group Text EPGs for Class 2 Substances—Compressed and Liquefied Gases—Part 2A5: Flammable, Poisonous—Special Low-Temperature Control Gas.

SAA AS 1678.2A6-87. Emergency Precedure Guide—Transport Group Text EPGs for Class 2 Substances—Compressed and Liquefied Gases—Part 2A6: Liquefied Gas—Corrosive—Dangerous When Wet.

SAA AS 1678.2B1-87. Emergency Precedure Guide—Transport Group Text EPGs for Class 2 Substances—Compressed and Liquefied Gases—Part 2B1: Poisonous (Non-Flammable) Gas.

SAA AS 1678.2B2-87. Emergency Precedure Guide—Transport Group Text EPGs for Class 2 Substances—Compressed and Liquefied Gases—Part 2B2: Poisonous, Flammable Gas.

SAA AS 1678.2B3-87. Emergency Procedure Guide—Transport Group Text EPGs for Class 2 Substances—Compressed and Liquefied Gases—Part 2B3: Poisonous Gas (Will Burn, Corrode) (Supersedes AS 1678.8.0.010—1984 (in Part)).

SAA AS 1678.2B4-87. Emergency Precedure Guide—Transport Group Text EPGs for Class 2 Substances—Compressed and Liquefied Gases—Part 2B4: Poisonous, Flammable (Spontaneously Combustible) Gas.

SAA AS 1678.2B5-87. Emergency Precedure Guide—Transport Group Text EPGs for Class 2 Substances—Compressed and Liquefied Gases—Part 2B5: Poisonous, Oxidizing Gas.

SAA AS 1678.2B6-87. Emergency Precedure Guide—Transport Group Text EPGs for Class 2 Substances—Compressed and Liquefied Gases—Part 2B6: Poisonous, Oxidizing Gas (Flourine-Type).

SAA AS 1678.2B7-87. Emergency Precedure Guide—Transport Group Text EPGs for Class 2 Substances—Compressed and Liquefied Gases—Part 2B7: Poisonous, Oxiding Gas, Corrosive Gas.

SAA AS 1678.2B8-87. Emergency Procedure Guide—Transport Group Text EPGs for Class 2 Substances—Compressed and Liquefied Gases—Part 2B8: Poisonous and Corrosive Gas.

SAA AS 1678.2C2-87. Emergency Procedure Guide—Transport Group Text EPGs for Class 2 Substances—Compressed and Liquefied Gases—Part 2C2: Non-Flammable, Liquefied Gas (Supersedes AS 1678.2.2.003—1984).

SAA AS 1678.2C3-87. Emergency Precedure Guide—Transport Group Text EPGs for Class 2 Substances—Compressed and Liquefied Gases—Part 2C3: Non-Flammable, Low-Temperature, Liquefied Gas.

SAA AS 1678.2C4-87. Emergency Procedure Guide—Transport Group Text EPGs for Class 2 Substances—Compressed and Liquefied Gases—Part 2C4: Non-Flammable, Corrosive, Liquefied Gas (Fluorine-Type).

SAA AS 1678.2C5-87. Emergency Procedure Guide—Transport Group Text EPGs for Class 2 Substances—Compressed and Liquefied Gases—Part 2C5: Oxidizing, Compressed Gas—Breathable, Non-Flammable.

SAA AS 1678.2C6-87. Emergency Procedure Guide—Transport Group Text EPGs for Class 2 Substances—Compressed and Liquefied Gases—Part 2C6: Non-Flammable, Oxidizing, Compressed Gas—Breathable.

SAA AS 1678.2C7-87. Emergency Procedure Guide—Transport Group Text EPGs for Class 2 Substances—Compressed and Liquefied Gases—Part 2C7: Non-Flammable, Oxidizing, Low-Temperature, Liquefied Gas—Breathable.

SAA AS 1678.2C8-87. Emergency Procedure Guide—Transport Group Text EPGs for Class 2 Substances—Compressed and Liquefied Gases—Part 2C8: Non-Flammable, Oxidizing, Compressed Gas—Asphyxiating.

SAA AS 1678.2C9-87. Emergency Procedure Guide—Transport Group Text EPGs for Class 2 Substances—Compressed and Liquefied Gases—Part 2C9: Non-Flammable, Oxidizing, Low-Temperature, Liquefied Gas—Asphyxiating.

SAA AS 1678.2D1-87. Emergency Procedure Guide—Transport Group Text EPGs for Class 2 Substances—Compressed and Liquefied Gases—Part 2D1: Aerosol Dispensers.

SAA AS 1678.2M1-87. Emergency Procedure Guide—Transport Group Text EPGs for Class 2 Substances—Compressed and Liquefied Gases—Part 2M1: Mixed Load of Gases in Cylinders.

SAA AS 1678. Emergency Procedure Guide—Transport Group Text EPGs for Class 3 Substances—Flammable Liquids Pads of 20 Forms.

SAA AS 1678.3A1-87. Emergency Procedure Guide—Transport Group Text EPGs for Class 3 Substances—Flammable Liquids—Part 3A1: Flammable Liquids (Supersedes AS 1678.3.0.020—1986, AS 1678.3.1.004—1984, AS 1678.3.1.005—1984, AS 1678.3.1.007—1984, AS 1678.3.1.008—1984,.

SAA AS 1678.3A2-87. Emergency Procedure Guide—Transport Group Text EPGs for Class 3 Substances—Flammable Liquids—Part 3A2: Flammable Liquid Poisonous.

SAA AS 1678.3A3-87. Emergency Procedure Guide—Transport Group Text EPGs for Class 3 Substances—Flammable Liquids—Part 3A3: Flammable Liquid—Harmful (Supersedes AS 1678.3.1.006—1984).

SAA AS 1678.3A4-87. Emergency Procedure Guide—Transport Group Text EPGs for Class 3 Substances—Flammable Liquids—Part 3A4: Flammable Liquid—Corrosive.

SAA AS 1678.3A5-87. Emergency Procedure Guide—Transport Group Text EPGs for Class 3 Substances—Flammable Liquids—Part 3A5: Flammable Liquid, Poisonous, Corrosive.

SAA AS 1678.3A6-87. Emergency Procedure Guide—Transport Group Text EPGs for Class 3 Substances—Flammable Liquids—Part 3A6: Flammable Liquid, Corrosive, Dangerous When Wet.

SAA AS 1678.3B3-87. Emergency Procedure Guide—Transport Group Text EPGs for Class 3 Substances—Flammable Liquids—Part 3B3: Flammable Liquid—Some Harmful or Corrosive Properties.

SAA AS 1678.3C1-87. Emergency Procedure Guide—Transport Group Text EPGs for Class 3 Substances—Flammable Liquids—Part 3C1: Flammable Liquid of Lesser Hazard.

SAA AS 1678.3T1-87. Emergency Procedure Guide—Transport Group Text EPGs for Class 3 Substances—Flammable Liquids—Part 3T1: Bitumen Products.

SAA AS 1678.4A1-87. Emergency Procedure Guide—Transport Group Text EPGs for Class 4 Substances—Flammable Solids, Substances Liable to Spontaneous Combustion or Which Emit Flammable Gases When Wet—Part 4A1: Flammable Solids.

SAA AS 1678.4A2-87. Emergency Procedure Guide—Transport Group Text EPGs for Class 4 Substances—Flammable Solids, Substances Liable to Spontaneous Combustion or Which Emit Flammable Gases When Wet—Part 4A2: Spontaneously Combustible.

SAA AS 1678.4A3-87. Emergency Procedure Guide—Transport Group Text EPGs for Class 4 Substances—Flammable Solids, Substances Liable to Spontaneous Combustion or Which Emit Flammable Gases When Wet—Part 4A3: Dangerous When Wet.

SAA AS 1678.4B1-87. Emergency Procedure Guide—Transport Group Text EPGs for Class 4 Substances—Flammable Solids, Liable to Spontaneous Combustion or Which Emit Flammable Gases When Wet—Part 4B1: Flammable Solids, Toxic.

SAA AS 1678.4B2-87. Emergency Procedure Guide—Transport Group Text EPGs for Class 4 Substances—Flammable Solids, Substances Liable to Spontaneous Combustion or Which Emit Flammable Gases When Wet—Part 4B2: Spontaneously Combustible, Toxic.

SAA AS 1678.4B3-87. Emergency Procedure Guide—Transport Group Text EPGs for Class 4 Substances—Flammable Solids, Substances Liable to Spontaneous Combustion or Which Emit Flammable Gases When Wet—Part 4B3: Dangerous When Wet, Toxic.

SAA AS 1678.4C2-87. Emergency Procedure Guide—Transport Group Text EPGs for Class 4 Substances—Flammable Solids, Substances Liable to Spontaneous Combustion or Which Emit Flammable Gases When Wet—Part 4C2: Spontaneously Combustible, Corrosive.

SAA AS 1678.4C3-87. Emergency Procedure Guide—Transport Group Text EPGs for Class 4 Substances—Flammable Solids, Substances Liable to Spontaneous Combustion or Which Emit Flammable Gases When Wet—Part 4C3: Dangerous When Wet, Corrosive.

SAA AS 1678.4D2-87. Emergency Procedure Guide—Transport Group Text EPGs for Class 4 Substances—Flammable Solids, Substances Liable to Spontaneous Combustion or Which Emit Flammable Gases When Wet—Part 4D2: Spontaneously Combustible, Reacts Violently with Water.

SAA AS 1678.4E3-87. Emergency Procedure Guide—Transport Group Text EPGs for Class 4 Substances—Flammable Solids, Substances Liable to Spontaneous Combustion or Which Emit Flammable Gases When Wet—Part 4E3: Dangerous When Wet, Some Toxic Properties.

INDUSTRY STANDARDS

Hazardous
INTERNATIONAL AND NON-U.S. NATIONAL STANDARDS
SUBJECT INDEX

Hazardous Materials *(Cont.)*

—Highway Transportation—Emergency Planning *(Cont.)*

SAA AS 1678.4F1-87. Emergency Procedure Guide —Transport Group Text EPGs for Class 4 Substances—Flammable Solids, Substances Liable to Spontaneous Combustion or Which Emit Flammable Gases When Wet—Part 4F1: Flammable Solids, Explosive.

SAA AS 1678.4G1-87. Emergency Procedure Guide —Transport Group Text EPGs for Class 4 Substances—Flammable Solids, Substances Liable to Spontaneous Combustion or Which Emit Flammable Gases When Wet—Part 4G1: Flammable Solids, Harmful.

SAA AS 1678.4G2-87. Emergency Proc. Guide—Trans Group Text EPGs for Class 4 Substances—Flammable Solids, Substances Liable to Spontaneous Combustion or Which Emit Flammable Gases When Wet—Part 4G2: Spontaneously Combustible, Dangerous When Wet, May React Violently with Water.

SAA AS 1678.4H1-87. Emergency Procedure Guide —Transport Group Text EPGs for Class 4 Substances—Falmmable Solids, Substances Liable to Spontaneous Combustion or Which Emit Flammable Gases When Wet —Part 4H1: Flammable Solids, Some Toxic Properties.

SAA AS 1678.4K2-87. Emergency Procedure Guide —Transport Group Text EPGs for Class 4 Substances—Flammable Solids, Substances Liable to Spontaneous Combustion or Which Emit Flammable Gases When Wet—Part 4K2: Spontaneously Combustible, Some Toxic Properties.

SAA AS 1678.4M3-87. Emergency Proc. Guide—Trans. Group Text EPGs for Class 4 Substances—FlammableSolids, Substances Liable to Spontaneous Combustion or Which Emit Flammable Gases When Wet—Part 4M3: Spontaneously Combustible, Dangerous when Wet, Some Toxic Properties.

SAA AS 1678.4N3-87. Emergency Proc. Guide—Trans. Group Text EPGs for Class 4 Substances—FlammableSolids, Substances Liable to Spontaneous Combustion or Which Emit Flammable Gases When Wet—Part 4N3: Dangerous When Wet, Some Corrosive Properties, May Ignite Spontaneously.

SAA AS 1678.4O3-87. Emergency Procedure Guide —Transport Group Text EPGs for Class 4 Substances—Flammable Solids, Substances Liable to Spontaneous Combustion or Which Emit Flammable Gases When Wet—Part 4O3: Dangerous When Wet, Spontaneously Combustible.

SAA AS 1678.4P3-87. Emergency Procedure Guide —Transport Group Text EPGs for Class 4 Substances—Flammable Solids, Substances Liable to Spontaneous Combustion or Which Emit Flammable Gases When Wet—Part 4P3: Dangerous When Wet, Flammable Liquid, Some Toxic Properties.

SAA AS 1678.4Q3-87. Emergency Procedure Guide —Transport Group Text EPGs for Class 4 Substances—Flammable Solids, Substances Liable to Spontaneous Combustion or Which Emit Flammable Gases When Wet—Part 4Q3: Dangerous When Wet, Flammable Liquid, Corrosive.

SAA AS 1678.4R1-90. Emergency Procedur Guide —Transport Group Text EPGs for Class 4 Substances—Flammable Solids, Substances Liable to Spontaneous Combustion or Which Emit Flammable Gases When Wet—Part 4R1: Flammable Solids, Dangerous When Wet.

SAA AS 1678.4R3-87. Emergency Procedure Guide —Transport Group Text EPGs for Class 4 Substances—Flammable Solids, Substances Liable to Spontaneous Combustion or Which Emit Flammable Gases When Wet—Part 4R3: Dangerous When Wet, Harmful.

SAA. Emergency Procedure Guide —Transport Group Text EPGs for Class 5 Substances—Oxidizing Substances and Organic Peroxides Pads of 20 Forms.

SAA AS 1678.5A1-87. Emergency Procedure Guide—Transport Group Text EPGs for Class 5 Substances—Oxidizing Substances and Organic Peroxides—Part 5A1: Oxidizing Agents—Harmful and Corrosive.

SAA AS 1678.5B1-87. Emergency Procedure Guide—Transport Group Text EPGs for Class 5 Substances—Oxidizing Substances and Organic Peroxides—Part 5B1: Oxidizing Agents—Poisonous.

SAA AS 1678.5B2-87. Emergency Procedure Guide—Transport Group Text EPGs for Class 5 Substances—Oxidizing Substances and Organic Peroxides—Part 5B2: Oxidizing Agents—Harmful.

SAA AS 1678.5C1-87. Emergency Procedure Guide—Transport Group Text EPGs for Class 5 Substances—Oxidizing Substances and Organic Peroxides—Part 5C1: Oxidizing Agents—Poisonous, Corrosive, Reactive.

SAA AS 1678.5C2-87. Emergency Procedure Guide—Transport Group Text EPGs for Class 5 Substances—Oxidizing Substances and Organic Peroxides—Part 5C2: Oxidizing Agents—Corrosive, Reactive.

Hazardous Materials *(Cont.)*

—Highway Transportation—Emergency Planning *(Cont.)*

SAA AS 1678.5D1-87. Emergency Procedure Guide—Transport Group Text EPGs for Class 5 Substances—Oxidizing Substances and Organic Peroxides—Part 5D1: Oxidizing Agents—Containing Ammonium Nitrate.

SAA AS 1678.5E1-87. Emergency Procedure Guide—Transport Group Text EPGs for Class 5 Substances—Osidizing Substances and Organic Peroxides—Part 5E1: Osidizing Agents—Corrosive, Unstable.

SAA AS 1678.5E2-87. Emergency Procedure Guide—Transport Group Text EPGs for Class 5 Substances—Oxidizing Substances and Organic Peroxides—Part 5E2: Oxidizing Agents—Corrosive, Unstable, Flammable.

SAA AS 1678.5F1-87. Emergency Procedure Guide—Transport Group Text EPGs for Class 5 Substances—Oxidizing Substances and Organic Peroxides—Part 5F1: Oxidizing Agents—Thermally Unstable.

SAA AS 1678.5K1-93. Emergency Procedure Guide—Transport Group Text EPGs for Class 5 Substances—Oxidizing Substances and Organic Peroxides—Part 5K1: Organic Peroxides.

SAA AS 1678.5K2-93. Emergency Procedure Guide—Transport Group Text EPGs for Class 5 Substances—Oxidizing Substances and Organic Peroxides—Part 5K2: Organic Peroxides—Unstable.

SAA AS 1678.5L1-87. Emergency Procedure Guide—Transport Group Text EOGs for Class 5 Substances—Oxidizing Substances and Organic Peroxides—Part 5L1: Organic Peroxides, Highly Flammable, Unstable (Withdrawn).

SAA AS 1678.5M1-93. Emergency Procedure Guide—Transport Group Text EPGs for Class 5 Substances—Oxidizing Substances and Organic Peroxides—Part 5M1: Organic Peroxides—Corrosive.

SAA AS 1678.5N2-87. Emergency Procedure Guide—Transport Group Text EPGs for Class 5 Substances—Oxidizing Substances and Organic Peroxides—Part 5N2: Organic Peroxides, Highly Unstable (Withdrawn).

SAA AS 1678.5P1-93. Emergency Procedure Guide—Transport Group Text EPGs for Class 5 Substances—Oxidizing Substances and Organic Peroxides—Part 5P1: Organic Peroxides—Unstable, Keep Refrigerated.

SAA AS 1678.5Q1-93. Emergency Procedure Guide—Transport Group Text EPGs for Class 5 Substances—Oxidizing Substances and Organic Peroxides—Part 5Q1: Organic Peroxides—Keep Refrigerated.

SAA. Emergency Procedure Guide —Transport Group Text EPGs for Class 6 Substances—Poison Substances Pads of 20 Forms.

SAA AS 1678.6A1-87. Emergency Procedure Guide —Transport Group Text EPGs for Class 6 Substances—Poison Substances—Part 6A1: Poisonous Substance, Will Burn, Do Not Use Water on Fires (Induce Vomiting).

SAA AS 1678.6A2-87. Emergency Procedure Guide —Transport Group Text EPGs for Class 6 Substances—Poison Substances—Part 6A2: Poisonous Substance, Will Not Burn, Do Not Use Water on Fires (Induce Vomiting).

SAA AS 1678.6A3-87. Emergency Procedure Guide —Transport Group Text EPGs for Class 6 Substances—Poison Substances—Part 6A3: Poisonous Substance, Will Burn, Water May Be Used on Fires (Induce Vomiting).

SAA AS 1678.6A4-87. Emergency Procedure Guide —Transport Group Text EPGs for Class 6 Substances—Poison Substances—Part 6A4: Poisonous Substance, Will Burn, Do Not Use Water on Fires (Do Not Induce Vomiting).

SAA AS 1678.6A5-87. Emergency Procedure Guide —Transport Group Text EPGs for Class 6 Substances—Poison Substances—Part 6A5: Poisonous Substance, Will Not Burn, Water May Be Used on Fires (Induce Vomiting) (Supersedes AS 1678.6.0.003—1982).

SAA AS 1678.6A6-87. Emergency Procedure Guide —Transport Group Text EPGs for Class 6 Substances—Poison Substances—Part 6A6: Poisonous Substance, Will Burn, Water May Be Used on Fires (Do Not Induce Vomiting).

SAA AS 1678.6A7-87. Emergency Procedure Guide —Transport Group Text EPGs for Class 6 Substances—Poison Substances—Part 6A7: Poisonous Substance, Will Not Burn, Water May Be Used on Fires (Do Not Induce Vomiting).

SAA AS 1678.6B1-87. Emergency Procedure Guide —Transport Group Text EPGs for Class 6 Substances—Poison Substances—Part 6B1: Harmful Substance, Will Burn, Do Not Use Water on Fires (Induce Vomiting).

Hazardous Materials *(Cont.)*

—Highway Transportation—Emergency Planning *(Cont.)*

SAA AS 1678.6B3-87. Emergency Procedure Guide —Transport Group Text EPGs for Class 6 Substances—Poison Substances—Part 6B3: Harmful Substance, Will Burn, Water May Be Used on Fires (Induce Vomiting) (Supersedes AS 1678.6.1.011—1984).

SAA AS 1678.6B5-87. Emergency Procedure Guide —Transport Group Text EPGs for Class 6 Substances—Poison Substances—Part 6B5: Harmful Substance, Will Not Burn, Water May Be Used on Fires (Induce Vomiting).

SAA AS 1678.6B6-87. Emergency Procedure Guide —Transport Group Text EPGs for Class 6 Substances—Poison Substances—Part 6B6: Harmful Substance, Will Burn, Water May Be Used on Fires (Do Not Induce Vomiting).

SAA AS 1678.6B7-87. Emergency Procedure Guide —Transport Group Text EPGs for Class 6 Substances—Poison Substances—Part 6B7: Harmful Substance, Will Not Burn, Water May Be Used on Fires (Do Not Induce Vomiting) (Supersedes AS 1678.6.1.007—1987 and AS 1678.6.1.008—1982).

SAA AS 1678.6C1-87. Emergenc Procedure —Transport Group Text EPGs for Class 6 Substances—Poison Substances—Part 6C1: Poisonous Substance, Flammable, Do Not Use Water on Fires (Induce Vomiting).

SAA AS 1678.6C3-87. Emergency Procedure Guide —Transport Group Text EPGs for Class 6 Substances—Poison Substances—Part 6C3: Poisonous Substance, Flammable, Water May Be Used on Fires (Induce Vomiting).

SAA AS 1678.6C6-87. Emergency Procedure Guide —Transport Group Text EPGs for Class 6 Substances—Poison Substances—Part 6C6: Poisonous Substance, Flammable, Do Not Use Water on Fires (Do Not Induce Vomiting).

SAA AS 1678.6D1-87. Emergency Procedure Guide —Transport Group Text EPGs for Class 6 Substances—Poison Substances—Part 6D1: Harmful Substance, Flammable, Do Not Use Water on Fires (Induce Vomiting).

SAA AS 1678.6D3-87. Emergency Procedure Guide —Transport Group Text EPGs for Class 6 Substances—Poison Substances—Part 6D3: Harmful Substance, Flammable, Water May Be Used on Fires (Induce Vomiting).

SAA AS 1678.6E3-87. Emergency Procedure Guide —Transport Group Text EPGs for Class 6 Substances—Poison Substances—Part 6E3: Poisonous Substance, Flammable Solid, Water May Be Used on Fires (Induce Vomiting).

SAA AS 1678.6J4-87. Emergency Procedure Guide —Transport Group Text EPGs for Class 6 Substances—Poison Substances—Part 6J4: Poisonous Substance, Corrosive, Will Burn, Reacts with Water (Give Milk or Water, Do Not Induce Vomiting).

SAA AS 1678.6J6-87. Emergency —Transport Group Text EPGs for Class 6 Substances—Poison Substances—Part 6J6: Poisonous Substance, Corrosive, Will Burn, Reacts with Water But Water May Be Used on Fires (Give Milk or Water, Do Not Induce Vomiting).

SAA AS 1678.6L4-87. Emergency Procedure Guide —Transport Group Text EPGs for Class 6 Substances—Poison Substances—Part 6L4: Poisonous Substance, Corrosive, Flammable, Reacts with Water (Give Milk or Water, Do Not Induce Vomiting).

SAA AS 1678.6L6-87. Emergency Procedure Guide —Transport Group Text EPGs for Class 6 Substances—Poison Substances—Part 6L6: Poisonous Substance, Corrosive, Flammable, Reacts with Water but Water May Be Used on Fires (Give Milk or Water, Do Not Induce Vomiting).

SAA AS 1678. Emergency Procedure Guide —Transport Group Text EPGs for Class 8 Substances—Corrosive Substances Pads of 20 Forms.

SAA AS 1678.8A1-87. Emergency Procedure Guide—Transport Group Text EPGs for Class 8 Substances—Corrosive Substances—Part 8A1: Corrosive (Supersedes AS 1678.8.0.007—1986 and AS 1678.8.0.010—1984 (in Part)).

SAA AS 1678.8A2-87. Emergency Procedure Guide —Transport Group Text EPGs for Class 8 Substances—Corrosive Substances—Part 8A2: Corrosive—Violent Reaction with Water.

SAA AS 1678.8A3-87. Emergency Procedure Guide —Transport Group Text EPGs for Class 8 Substances—Corrosive Substances—Part 8A3: Corrosive—Oxidizing Organic Materials, Violent Reaction with Water.

SAA AS 1678.8A4-87. Emergency Procedure Guide —Transport Group Text EPGs for Class 8 Substances—Corrosive Substances—Part 8A4: Corrosive—Can Release Hydrogen Fluoride.

SAA AS 1678.8B1-87. Emergency Procedure Guide —Transport Group Text EPGs for Class 8 Substances—Corrosive Substances—Part 8B1: Corrosive—Flammable.

INTERNATIONAL AND NON-U.S. NATIONAL STANDARDS
SUBJECT INDEX
Hazardous

Hazardous Materials (Cont.)
—Highway Transportation—Emergency Planning (Cont.)
SAA AS 1678.8B2-87. Emergency Procedure Guide —Transport Group Text EPGs for Class 8 Substances—Corrosive Substances—Part 8B2: Corrosive—Flammable—Violent Reaction with Water.
SAA AS 1678.8B3-87. Emergency Procedure Guide —Transport Group Text EPGs for Class 8 Substances—Corrosive Substances—Part 8B3: Corrosive—Flammable—Self Reactive.
SAA AS 1678.8C1-87. Emergency Procedure Guide —Transport Group Text EPGs for Class 8 Substances—Corrosive Substances—Part 8C1: Corrosive—Poison.
SAA AS 1678.8C2-87. Emergency Procedure Guide —Transport Group Text EPGs for Class 8 Substances—Corrosive Substances—Part 8C2: Corrosive—Poison—Can Relaese Hydrogen Fluoride.
SAA AS 1678.8C3-87. Emergency Procedure Guide —Transport Group Text EPGs for Class 8 Substances—Corrosive Substances—Part 8C3: Corrosive—Poison—Violent Reaction with Water.
SAA AS 1678.8C4-87. Emergency Procedure Guide —Transport Group Text EPGs for Class 8 Substances—Corrosive Substances—Part 8C4: Corrosive—Poison—Violent Reaction with Water, Glass and Sand.
SAA AS 1678.8C5-87. Emergency Procedure Guide —Transport Group Text EPGs for Class 8 Substances—Corrosive Substances—Part 8C5: Corrosive—Poison—Oxidizing Organic Materials.
SAA AS 1678.8C6-87. Emergency Procedure Guide —Transport Group Text EPGs for Class 8 Substances—Corrosive Substances—Part 8C6: Corrosive—Poisonous Substance—Polymerizes Readily.
SAA AS 1678.8C8-87. Emergency Procedure Guide —Transport Group Text EPGs for Class 8 Substances—Corrosive Substnaces—Part 8C8: Corrosive—Poison—Flammable.
SAA AS 1678.8C9-87. Emergency Procedure Guide —Transport Group Text EPGs for Class 8 Substances—Corrosive Substances—Corrosive—Harmful.
SAA AS 1678.8D1-87. Emergency Procedure Guide —Transport Group Text EPGs for Class 8 Substances—Corrosive Substances—Part 8D1: Corrosive—Oxidizing Agent (Supersedes AS 1678.8.0.003—1984 (in Part)).
SAA AS 1678.8D2-87. Emergency Procedure Guide —Transport Group Text EPGs for Class 8 Substances—Corrosive Substances—Part 8D2: corrosive—Oxidizing Substance—Reacts Violently with Water.
SAA AS 1678.8E1-87. Emergency Procedure Guide —Transport Group Text EPGs for Class 8 Substances—Corrosive Substances—Part 8E1: Corrosive—Oxidizing Agent—Poison (Supersedes AS 1678.8.0.003—1984 (in Part)).
SAA AS 1678.8F1-87. Emergency Procedure Guide —Transport Group Text EPGs for Class 8 Substances—Corrosive Substances—Part 8F1: Corrosive—Flammable Solid.
SAA AS 1678.8F2-87. Emergency Procedure Guide —Transport Group Text EPGs for Class 8 Substances—Corrosive Substances—Part 8F2: Corrosive—Flammable—Solid—Violent Reaction with Water.
SAA AS 1678. Emergency Procedure Guide —Transport Group Text EPGs for Class 9 Substances Pads of 20 Forms.
SAA AS 1678.9A1-88. Emergency Procedure Guide —Transport Group Text EPGs for Class 9 Substances—Part 9A1: Flammable Solids.
SAA AS 1678.9A2-88. Emergency Procedure Guide —Transport Group Text EPGs for Class 9 Substances—Part 9A2: Spontaneously Combustible.
SAA AS 1678.9A3-88. Emergency Procedure Guide —Transport Group Text EPGs for Class 9 Substances—Part 9A3: Dangerous When Wet.
SAA AS 1678.9A5-88. Emergency Procedure Guide —Transport Group Text EPGs for Class 9 Substances—Part 9A5: Poisonous Substance, Will Not Burn, Water May Be Used on Fires (Induce Vomiting).
SAA AS 1678.9B3-88. Emergency Procedure Guide —Transport Group Text EPGs for Class 9 Substances—Part 9B3: Dangerous When Wet, Toxic.
SAA AS 1678.9B7-88. Emergency Procedure Guide —Transport Group Text EPGs for Class 9 Substances—Part 9B7: Harmful Substance, Will Not Burn, Water May Be Used on Fires (Do Not Induce Vomiting).
SAA AS 1678.9E1-88. Emergency Procedure Guide —Transport Group Text EPGs for Class 9 Substances—Part 9E1: Oxidizing Agents—Corrosive, Unstable.
SAA AS 1678. Emergency Procedure Guide —Transport Group Text EPGs for Mixed, Packaged Dangerous Goods of Various Classes Pads of 20 Forms.
SAA AS 1678.X1-93. Emergency Procedure Guide —Transport Group Text EPGs for Mixed, Packaged Dangerous Goods of Various Classes—Part X1: Mixed Dangerous Goods (Packaged Goods Only)—Substances Can be Liquids, Solids and/or Gases—Do Not React Violently with Water (in Prof. Package 37A).
SAA AS 1678.X2-87. Emergency Procedure Guide —Transport Group Text EPGs for Mixed, Packaged Dangerous Goods of Various Classes—Part X2: Mixed Dangerous Goods (Compatible Packaged Goods Only)—Substances Can be Liquids, Solids and/or Gases—React Violently with Water.

—Identification Systems
EC 67/548/EEC-67. Council Directive on the Approximation of Laws, Regulations and Administrative Provisions Relating to the Classification, Packaging and Labelling of Dangerous Substances. 23 pp.
EC 79/831/EEC-79. Council Directive Amending for the Sixth Time Directive 67/548/EEC on the Approximation of the Laws, Regulations and Administrative Provisions Relating to the Classification, Packaging and Labelling of Dangerous Substances. 19 pp.
EC 88/379/EEC-88. Council Directive on the Approximation of the Laws, Regulations and Administrative Provisions of the Member States Relating to the Classification, Packaging and Labelling of Dangerous Preparations. 13 pp.
EC 93/18/EEC-93. Commission Directive Adapting for the Third Time to Technical Progress Council Directive 88/379/EEC on the Approximation of the Laws, Regulations and Administrative Provisions of the Member States Relating to the Classification, Packaging and Labelling. 11 pp.
EC 93/21/EEC-93. Commission Directive Adapting to Technical Progress for the 18th Time Council Directive 67/548/EEC on the Approximation of the Laws, Regulations and Administrative Provisions Relating to the Classification, Packaging and Labelling of Dangerous Substances. 2 pp.
EC 93/67/EEC-93. Commission Directive Laying Down the Principles for Assessment of Risks to Man and the Environment of Substances Notified in Accordance with Council Directive 67/548/EEC. 10 pp.
EC EEC/2455/92-92. Council Regulation Concerning the Export and Import of Certain Dangerous Chemicals. 10 pp.
EC 85/71/EEC-84. Commission Decision Concerning the List of Chemical Substances Notified Pursuant to Council Directive 67/548/EEC on the Approximation of Laws, Regs. and Administrative Provisions Relating to the Classification, Packaging and Labeling of Dangerous Substances. 2 pp.
SAA AS 1216.1-84. Classification, Hazard Identification and Information Systems for Dangerous Goods—Part 1: Classification and Class Labels for Dangerous Goods. 11 pp.
SAA AS 1216.3-81. Classification, Hazard Identification and Information Systems for Dangerous Goods—Part 3: NFPA Hazard Identification System.
SAA AS 1216.4-81. Classification, Hazard Identification and Information Systems for Dangerous Goods—Part 4: UN Substance Identification Numbers.
SNZ NZS 5433F-90. NZ Hazardous Substances Dangerous Goods Declaration. 50 pp.

—Machinery
CEN PREN 626-92. Safety of Machinery—Principles for Machinery Manufacturers on the Reduction of Risk to Health from Hazardous Substances Emitted by Machinery. 15 pp.

—Occupational Safety and Health—Military
NATO STANAG 2908 ED 2 AMD 3-81. Preventive Measures for an Occupational Health Programme. 6 pp.

—Packaging
EC 67/548/EEC-67. Council Directive on the Approximation of Laws, Regulations and Administrative Provisions Relating to the Classification, Packaging and Labelling of Dangerous Substances. 23 pp.
EC 79/831/EEC-79. Council Directive Amending for the Sixth Time Directive 67/548/EEC on the Approximation of the Laws, Regulations and Administrative Provisions Relating to the Classification, Packaging and Labelling of Dangerous Substances. 19 pp.
EC 88/379/EEC-88. Council Directive on the Approximation of the Laws, Regulations and Administrative Provisions of the Member States Relating to the Classification, Packaging and Labelling of Dangerous Preparations. 13 pp.
EC 93/18/EEC-93. Commission Directive Adapting for the Third Time to Technical Progress Council Directive 88/379/EEC on the Approximation of the Laws, Regulations and Administrative Provisions of the Member States Relating to the Classification, Packaging and Labelling. 11 pp.
EC 93/21/EEC-93. Commission Directive Adapting to Technical Progress for the 18th Time Council Directive 67/548/EEC on the Approximation of the Laws, Regulations and Administrative Provisions Relating to the Classification, Packaging and Labelling of Dangerous Substances. 2 pp.
EC 93/67/EEC-93. Commission Directive Laying Down the Principles for Assessment of Risks to Man and the Environment of Substances Notified in Accordance with Council Directive 67/548/EEC. 10 pp.
EC EEC/2455/92-92. Council Regulation Concerning the Export and Import of Certain Dangerous Chemicals. 10 pp.

—Packaging—Identification Systems
BSI BS 7280-90. 1990 Requirements for Tactile Danger Warnings for Packaging. 8 pp.
CNS Z5071-88. Labels and Placards for Dangerous Materials (Jan)(6864).
SNZ NZS 5433-88. Code of Practice for Transportation of Hazardous Substances on Land Amend: 1, 1990. 172 pp.

—Packaging—Identification Systems—Visually Impaired Persons
CEN EN 272-89. Packaging: Tactile Danger Warnings: Requirements. 6 pp.
ISO 11683-93. Packaging—Tactile Danger Warnings—Requirements First Edition. 7 pp.

—Restricted Use
EC 79/663/EEC-79. Council Directive Supplementing the Annex to Coun. Dir. 76/769/EEC on the Approximation of the Laws, Regs. and Ad. Provisions of the Member States Relating to Restrictions on the Marketing and Use of Certain Dangerous Substances and Preparations. 2 pp.
EC 83/264/EEC-83. Council Directive Amending for the Fourth Time Directive 76/769/EEC on the Approximation of the Laws, Regs. and Ad. Provs. of the Member States Relating to Restriction on the Mktg. and Use of Certain Dangerous Substances and Preparations. 2 pp.
EC 89/677/EEC-89. Council Directive Amending for the Eighth Time Directive 76/769/EEC on the Approximation of the Laws, Regs. and Ad. Provs. of the Member States Relating to Restrictions on the Mktg. and Use of Certain Dangerous Substances and Preparations. 5 pp.

—Safety Data Sheets
DIN ENGL 52900-83. DIN Safety Data Sheet for Chemical Substances and Preparations; Form and Instructions on How to Fill in the Form (Feb). 10 pp.

—Shipping Containers
BSI BS 6939: Part 1-87. 1987 Amd 2 Intermediate Bulk Containers (IBCs) for Dangerous Goods Part 1: General Provisions Applicable to All Types of IBC (H) (AMD 6379) May 31, 1991. 11 pp.
CGSB 43-GP-57M-81. Polyethylene, Inside Containers, (TC-2E, TC-2S, TC-2SL, TC-2T, TC-2TL, TC-2U), Standard for. 12 pp.
CGSB CAN/CGSB-43.124-M89. Inside Metal Containers and Liners (TC-2F and TC-2N). 9 pp.
CGSB 43-GP-147P-90. Construction of Tank Car Tanks and Selection and Use of Tank Car Tanks, Portable Tanks and Rail Cars for the Transportation of Dangerous Goods by Rail. 241 pp.
CSA CAN/CSA-B339-88. Cylinders, Spheres, and Tubes for the Transportation of Dangerous Goods; (Gen Instr 1 Thru 3). 112 pp.
CSA CAN/CSA-B340-M88. Selection and Use of Cylinders, Spheres, Tubes, and Other Containers for the Transportation of Dangerous Goods, Class 2; (Gen Instr 1 Thru 3). 50 pp.
CSA B616-M1989P. Rigid Polyethylene Intermediate Bulk Containers for the Transportation of Dangerous Goods; (Gen Instr 1). 26 pp.
SNZ NZS 5418: Part 2-83. Transportation Containers for Hazardous Substances Part 2: Specification for Small Packages for Transportation by Land, Sea and Air. 15 pp.

—Shipping Containers—Aerosol
CGSB CAN/CGSB-43.123-M86. Containers, Metal, Aerosol, (TC-2P, TC-2Q). 7 pp.

—Shipping Containers—Aerosol—Emergency Procedures
SAA AS 1678.2D1-87. Emergency Procedure Guide —Transport Group Text EPGs for Class 2 Substances—Compressed and Liquefied Gases—Part 2D1: Aerosol Dispensers.

Hazardous Materials (Cont.)

—Shipping Containers—Bags
CGSB CAN/CGSB-43.67-93. Lined Cloth and Paper Bags (TC-45B). 7 pp.
CGSB CAN/CGSB-43.110-93. Multiwall Paper Bags (TC-44B). 10 pp.
CGSB CAN/CGSB-43.111-93. Multiwall Paper Bags (TC-44C). 10 pp.
CGSB CAN/CGSB-43.112-93. Multiwall Paper Bags (TC-44D). 7 pp.
CGSB CAN/CGSB-43.120-M89. Single-Trip Steel Drums, Removable Head (TC-17H). 12 pp.

—Shipping Containers—Barrels
CGSB CAN/CGSB-43.96-M89. Steel Drums, Removable, or Non-Removable Head (TC-5B). 12 pp.
CGSB CAN/CGSB-43.97-M90. Stainless Steel Drums, Non-Removable Head (TC-5C). 12 pp.
CGSB CAN/CGSB-43.99-M88. Barrels or Drums, Steel (TC-6B). 11 pp.

—Shipping Containers—Bottles
CGSB 43-GP-57M-81. Polyethylene, Inside Containers, (TC-2E, TC-2S, TC-2SL, TC-2T, TC-2TL, TC-2U), Standard for. 12 pp.

—Shipping Containers—Boxes
CGSB CAN/CGSB-43.80-M90. Plywood Boxes (TC-19A and TC-19B). 15 pp.
CGSB CAN/CGSB-43.84-M89. Fibreboard Boxes for Inside Plastic Containers (TC-12P). 14 pp.
CGSB CAN/CGSB-43.90-M89. Nonreusable Fibreboard Boxes (TC-12A). 14 pp.
CGSB CAN/CGSB-43.91-M89. Fiberboard Boxes (TC-12B). 21 pp.
CGSB CAN/CGSB-43.93-M89. Fibreboard Boxes (TC-12D). 16 pp.
CGSB CAN/CGSB-43.95-M89. Fibreboard Boxes (TC-12H). 19 pp.

—Shipping Containers—Buckets
CGSB CAN/CGSB-43.117-M89. Single-Trip Steel Drums, Removable or Non-Removable Head (TC-17C). 12 pp.
CGSB CAN/CGSB-43.118-M89. Single-Trip Steel Drums, Non-Removable Head (TC-17E). 12 pp.
CGSB CAN/CGSB-43.119-M89. Single-Trip Steel Drums, Non-Removable Head (TC-17F). 12 pp.
CGSB CAN/CGSB-43.121-M90. Single-Trip Steel Drums, Removable Head (TC-37A). 11 pp.
CGSB CAN/CGSB-43.122-M89. Single-Trip Steel Drums, Non-Removable Head (TC-37B). 11 pp.

—Shipping Containers—Drums
CGSB CAN/CGSB-43.55-M87. Drums, Polyethylene, Reusable, Non-Removable Head, Without Overpack (TC-34). 11 pp.
CGSB CAN/CGSB-43.60-92. Non-Reusable Polyethylene Drum, Removable Head (TC-35). 9 pp.
CGSB CAN/CGSB-43.69-M90. Cylindrical Steel Overpack for Inside Plastic Container (TC-6D). 10 pp.
CGSB CAN/CGSB-43.86-M90. Fibre Drums (TC-21C). 12 pp.
CGSB CAN/CGSB-43.87-M90. Fibre Drum for an Inside Plastic Container (TC-21P). 11 pp.
CGSB CAN/CGSB-43.96-M89. Steel Drums, Removable, or Non-Removable Head (TC-5B). 12 pp.
CGSB CAN/CGSB-43.97-M90. Stainless Steel Drums, Non-Removable Head (TC-5C). 12 pp.
CGSB CAN/CGSB-43.98-M90. Steel Drums, Removable, or Non-Removable Head (TC-6J). 11 pp.
CGSB CAN/CGSB-43.99-M88. Barrels or Drums, Steel (TC-6B). 11 pp.
CGSB CAN/CGSB-43.114-M90. Non-Reusable, Steel Drums, Removable Head (TC-37C). 10 pp.
CGSB CAN/CGSB-43.115-M90. Non-Reusable, Cylindrical, Steel Overpack for Inside Plastic Containers (TC-37M). 10 pp.
CGSB CAN/CGSB-43.116-M90. Non-Reusable Steel Drums with Polyethylene Liners (TC-37P). 11 pp.
CGSB CAN/CGSB-43.117-M89. Single-Trip Steel Drums, Removable or Non-Removable Head (TC-17C). 12 pp.
CGSB CAN/CGSB-43.118-M89. Single-Trip Steel Drums, Non-Removable Head (TC-17E). 12 pp.
CGSB CAN/CGSB-43.119-M89. Single-Trip Steel Drums, Non-Removable Head (TC-17F). 12 pp.
CGSB CAN/CGSB-43.121-M90. Single-Trip Steel Drums, Removable Head (TC-37A). 11 pp.
CGSB CAN/CGSB-43.122-M89. Single-Trip Steel Drums, Non-Removable Head (TC-37B). 11 pp.

—Shipping Containers—Fiberboard
BSI BS 6939: Part 6-87. 1987 Amd 2 Intermediate Bulk Containers (IBCs) for Dangerous Goods Part 6: Specific Requirements for Fibreboard IBCs (AMD 6384) May 31, 1991. 9 pp.

Hazardous Materials (Cont.)

—Shipping Containers—Highway Transportation
CSA B621-1987P. Selection and Use of Highway Tanks, Portable Tanks, Cargo Compartments and Containers for the Transportation of Dangerous Goods, Classes 3, 4, 5, 6, and 8, in Bulk by Road. 26 pp.
CSA B622-1987P. Selection and Use of Highway Tanks, Multi-Unit Tank Car Tanks, and Portable Tanks for the Transportation of Dangerous Goods, Class 2, by Road; (Gen Instr 1 Thru 2). 37 pp.
SNZ NZS 5433-88. Code of Practice for Transportation of Hazardous Substances on Land Amend: 1, 1990. 172 pp.

—Shipping Containers—Liners
CGSB CAN/CGSB-43.124-M89. Inside Metal Containers and Liners (TC-2F and TC-2N). 9 pp.

—Shipping Containers—Metal
BSI BS 6939: Part 2-90. 1990 Amd 1 Intermediate Bulk Containers (IBCs) for Dangerous Goods Part 2: Specific Requirements for Metal IBCs (AMD 6380) May 31, 1991. 11 pp.

—Shipping Containers—Packaging
BSI BS 7569-92. 1992 Recommendations for the Packaging of Dangerous Goods for Transport (H). 28 pp.
CGSB CAN/CGSB-43.150-M89. Performance Packagings for Transportation of Dangerous Goods; (Amendment 2 Incorporates Amendment 1 November 1990). 53 pp.
CGSB 43-GP-152MP-85. Packing for Transportation of Dangerous Goods in Prescribed Packagings; (Corrigendum—June 1989). 429 pp.
CGSB 43-GP-153MP-85. Packing for Transportation of Dangerous Goods in United Nations Performance Packagings; (Amendment 1 March 1987). 37 pp.

—Shipping Containers—Paper
BSI BS 6939: Part 3-87. 1987 Amd 2 Intermediate Bulk Containers (IBCs) for Dangerous Goods Part 3: Specific Requirements for Flexible IBCs (H) (AMD 6381) May 31, 1991. 11 pp.
CGSB CAN/CGSB-43.113-93. Multiwall Paper Bags (TC-44E). 10 pp.

—Shipping Containers—Plastic
BSI BS 6939: Part 3-87. 1987 Amd 2 Intermediate Bulk Containers (IBCs) for Dangerous Goods Part 3: Specific Requirements for Flexible IBCs (H) (AMD 6381) May 31, 1991. 11 pp.
BSI BS 6939: Part 4-87. 1987 Amd 2 Intermediate Bulk Containers (IBCs) for Dangerous Goods Part 4: Specific Requirements for Rigid Plastics IBCs (AMD 6382) May 31, 1991. 10 pp.
BSI BS 6939: Part 5-87. 1987 Amd 2 Intermediate Bulk Containers (IBCs) for Dangerous Goods Part 5: Specific Requirements for Composite IBCs with Plastics Inner Receptacles (AMD 6383) May 31, 1991 (H). 12 pp.

—Shipping Containers—Textile
BSI BS 6939: Part 3-87. 1987 Amd 2 Intermediate Bulk Containers (IBCs) for Dangerous Goods Part 3: Specific Requirements for Flexible IBCs (H) (AMD 6381) May 31, 1991. 11 pp.

—Shipping Containers—Wood
BSI BS 6939: Part 7-87. 1987 Amd 2 Intermediate Bulk Containers (IBCs) for Dangerous Goods Part 7: Specific Requirements for Wooden IBCs (H) (AMD 6385) May 31, 1991. 9 pp.

—Shipping—Documentation
CSA CAN/CSA-B629-M90. Recommended Practice on Shipping Documentation for the Transportation of Dangerous Goods; (Gen Instr 1). 47 pp.

—Storage Tanks
DIN ENGL 6625 Pt 1-89. Steel Tanks Erected on Site, for the Above Ground Storage of Hazardous Flammable Water Polluting Liquids of Class A III and of Non-Flammable Water Polluting Liquids; Requirements and Testing (Sept). 3 pp.

—Storage Tanks—Design
DIN ENGL 6625 Pt 2-89. Steel Tanks Erected on Site, for the Above Ground Storage of Hazardous Flammable Water Polluting Liquids of Class A III and of Non-Flammable Water Polluting Liquids; Design (Sept). 4 pp.

—Submarines—Maximum Allowable Limits
NATO STANAG 1184 ED 1 AMD 4-77. Maximum Concentration of Toxic Substances in Conventional Submarines and Submersibles at Normal Operational Pressure About a Bar for Emergency Conditions (Withdrawn). 6 pp.

Hazardous Materials (Cont.)

—Tank Cars
CGSB 43-GP-147P-90. Construction of Tank Car Tanks and Selection and Use of Tank Car Tanks, Portable Tanks and Rail Cars for the Transportation of Dangerous Goods by Rail. 241 pp.

—Tanks (Containers)
SNZ NZS 5418: Part 1-83. Transportation Containers for Hazardous Substances Part 1: Specification for Tanks for the Mulit-Modal Transportation of Hazardous Liquids. 79 pp.

—Tanks (Containers)—Highway Transportation
CSA B620-1987P. Highway Tanks and Portable Tanks for the Transportation of Dangerous Goods (Formerly Designated B338); (Gen Instr 1 Thru 2). 120 pp.
CSA B621-1987P. Selection and Use of Highway Tanks, Portable Tanks, Cargo Compartments and Containers for the Transportation of Dangerous Goods, Classes 3, 4, 5, 6, and 8, in Bulk by Road. 26 pp.
CSA B622-1987P. Selection and Use of Highway Tanks, Multi-Unit Tank Car Tanks, and Portable Tanks for the Transportation of Dangerous Goods, Class 2, by Road; (Gen Instr 1 Thru 2). 37 pp.

—Tanks (Containers)—Portable—Shipping
CGSB 43-GP-147P-90. Construction of Tank Car Tanks and Selection and Use of Tank Car Tanks, Portable Tanks and Rail Cars for the Transportation of Dangerous Goods by Rail. 241 pp.
CSA B623-1987P. Selection, Handling, and Use of Intermodal Portable Tanks for the Transportation of Dangerous Goods, Other Than by Air; (Gen Instr 1). 69 pp.

—Transportation
SNZ NZMP 200-88. Transport of Hazardous Substances on Land Amend: 1, 1990. 32 pp.
SNZ NZMP 201-90. Driver Training Packs.

—Transportation—Labels
SNZ NZS 5417-86. Specification for Transportation Labels for Hazardous Substances Amend: 1, 1988. 36 pp.

Hazards
Scope Note: For hazards of specific products or materials, see the subheading Hazards under the specific product or material. For additional references, consult the following list *See:* Accident Prevention; Alarm Systems; Biological Hazards; Electric Shock; Explosions; Explosive Atmospheres; Explosive Gases; Fire Hazards; Fire Protection; Fires; Hazard Analysis; Hazardous Materials; Inhalation Hazards; Protective Clothing; Space Hazards; Warning Systems

Haze
See Also: Turbidity

—Glazes—Aircraft
AECMA PREN2155-09-81. Test Methods for Transparent Materials for Aircraft Glazing—Part 9—Determination of Haze (C7/SC2/01). 6 pp.
BSI BS EN 2155: Part 9-89. 1989 Determination of Haze. 10 pp.
CEN EN 2155 (Part 9)-89. Test Methods for Transparent Materials for Aircraft Glazing Part 9: Determination of Haze. 7 pp.

HDDR
Use: High Density Digital Recording

HDLC
Use: High Level Data Link Control

HDPE
Use: Polyethylene—High Density

HDTV
Use: High Definition Television Equipment

Head Down Displays
Use: Multifunction Displays

Head Lamps
Use: Headlights

Head Lights
Use: Headlights

Head Restraints
Use: Headrests

Head Sets
Use: Headsets

INTERNATIONAL AND NON-U.S. NATIONAL STANDARDS
SUBJECT INDEX
Health

Head Up Display Systems
Use: Head Up Displays

Head Up Displays
Use For: Head Up Display Systems; HUD
See Also: Aircraft Instruments; Cathode Ray Tube Displays; Display Devices; Multifunction Displays

—Cathode Ray Tubes
CECC CECC 11 001-021 ISSUE 1-83. BS CECC 11 001-021; Head-Up Display Cathode Ray Tube (En). 27 pp.

CECC CECC 11 001-027 ISSUE 1-84. BS CECC 11 001-027; Radar Display Cathode Ray Tube (En). 18 pp.

Headboxes
See Also: Fourdrinier Machines; Paper Machines; Papermaking Equipment; Papers

—Drainage
CPPA D.1U-77. Drainage Test for Paper-Machine Headbox Stock. 1 p.

Headers (Supports)
—Steel Pipes
DIN ENGL 2917-82. Seamless Steel Tubes for Superheated Steam Pipelines and Headers (May). 2 pp.

Headframes
See Also: Mining Equipment
DIN ENGL 4118-81. Head Frames and Winding Towers for Mines; Design Loads, Calculation Principles and Design Principles (June). 4 pp.

SAA AS 3785.5-91. Underground Mining—Shaft Equipment—Part 5: Headframes Amdt 1 May 1993 (in Professional Package 32). 10 pp.

Headgear
Scope Note: For additional listings, use a more specific term *Use For:* Bump Caps
See Also: Clothing; Hats; Helmets; Hoods (Headgear); Scalp Protectors

—Industrial
CEN PREN 812-92. Industrial Bump Caps. 11 pp.

—Sizes
BSI BS 5592-78. 1978 Size Designation of Clothes. Headwear. 4 pp.

ISO 4417-77. Size Designation of Clothes—Headwear First Edition; (Amendment Slip-1977). 5 pp.

Heading Dies
See Also: Dies

—Glossaries
ISO 5396-77. Hardmetal Heading Dies—Terminology First Edition. 10 pp.

Headlamps
Use: Headlights

Headless Screws (Slotted)
Use: Slotted Headless Screws

Headlights
Use For: Head Lamps *See Also:* Motor Vehicle Lighting; Running Lights
BSI BS AU 40: Part 4A-66. 1966 Amd 1 Motor Vehicle Lighting and Signalling Equipment Part 4a: Sealed Beam Headlamps. 20 pp.

CNS D2101-82. Sealed Beam Head Lamp Units for Motor Vehicles (May)(7887). 18 pp.

CNS D4013-84. Head Lamp Testing Machines (Jan)(10742).

EC 76/761/EEC-76. Council Directive on the Approximation of the Laws of the Member States Relating to Motor-Vehicle Headlamps Which Function as Main-Beam and/or Dipped-Beam Headlamps and to Incandescent Electric Filament Lamps for Such Headlamps. 26 pp.

EC 89/517/EEC-89. Commission Directive Adapting to Technical Progress Council Directive 76/761/EEC on the Approximation of the Laws of the Mem. Sts. Rel. to Motor-Vehicle Headlamps Which Function as Main-Beam and/or Dipped-Beam Headlamps Incandescent Elec. Fil. Lamps for Such Headlamps. 2 pp.

ISO 4182-86. Motor Vehicles—Measurement of Variations in Dipped-Beam Headlamp Angle as a Function of Load Second Edition. 10 pp.

ISO 9987-90. Motorcycles—Measurement of Variation of Dipped Beam Inclination as a Function of Load First Edition. 10 pp.

JIS D 5504-81. Sealed Beam Headlamp Units for Motor Vehicles. 22 pp.

Headlights *(Cont.)*
—Aiming
BSI BS AU 156-71. 1971 Recommendations for Aiming of Headlamps and Auxiliary Lamps on Motor Vehicles. 13 pp.

CNS D2178-84. Headlamp Cleaners for Automobiles (Mar)(10801).

CNS D3159-83. Method of Test for Front Light Aiming of Automobiles (Dec)(10704).

CNS D4014-84. Head Lamp Aiming Device for Mechanically Aimable Sealed Beam Head Lamp Units (Jan)(10743).

ISO 3267-91. Road Vehicles—Headlamp Cleaners Second Edition. 7 pp.

JIS D 5715-81. Headlamp Cleaners for Automobiles.

—Beams—Orientation
ISO 10604-93. Road Vehicles—Measurement Equipment for Orientation of Headlamp Luminous Beams First Edition. 11 pp.

—Beams—Visual Inspection
ISO 10604-93. Road Vehicles—Measurement Equipment for Orientation of Headlamp Luminous Beams First Edition. 11 pp.

—Cleaners
ISO 3267-91. Road Vehicles—Headlamp Cleaners Second Edition. 7 pp.

—Collimators
BSI BS AU 162A-76. 1976 Amd 1 Headlamp Optical Aiming Devices. 8 pp.

—Dimmer Switches
CNS D2033-87. Dimmer Switches for Automobiles (Aug)(5789).

—Incandescent
EC 76/761/EEC-76. Council Directive on the Approximation of the Laws of the Member States Relating to Motor-Vehicle Headlamps Which Function as Main-Beam and/or Dipped-Beam Headlamps and to Incandescent Electric Filament Lamps for Such Headlamps. 26 pp.

EC 89/517/EEC-89. Commission Directive Adapting to Technical Progress Council Directive 76/761/EEC on the Approximation of the Laws of the Mem. Sts. Rel. to Motor-Vehicle Headlamps Which Function as Main-Beam and/or Dipped-Beam Headlamps Incandescent Elec. Fil. Lamps for Such Headlamps. 2 pp.

—Incandescent—Inspection
SAA AS 3118-86. Approval and Test Specification—Electric Inspection Handlamps Amdt 1 July 1989 (This is a Joint Standard with SANZ NZS 3118) (Superseded by AS/NZS 3118:1993).

SNZ NZS/AS 3118-93. Approval and Test Specification Electric Inspection Handlamps. 6 pp.

—Levelling Devices
ISO 8218-89. Road Vehicles—Levelling Devices for Headlamp Dipped Beam First Edition. 6 pp.

Headphones
See Also: Audio Equipment; Electroacoustic Transducers; Headsets; Monitor Headphones; Speakers

CNS C5186-86. General Rules for Headphones (Jan)(11096).

CNS C6234-84. Method of Test for Headphones (Nov)(11097).

IEC 581 Pt 10-86. High Fidelity Audio Equipment and Systems; Minimum Performance Requirements Part 10: Headphones First Edition. 15 pp.

—Conference Systems
BSI BS 7154-89. 1989 Amd 1 Conference Systems: Electrical and Audio Requirements (AMD 6408) September 30, 1991. 47 pp.

IEC 914-88. Conference Systems—Electrical and Audio Requirements First Edition. 91 pp.

—Infrared Detectors—Sound Transmission
BSI BS 6418-83. 1983 Cordless Audio Transmission Devices Using Infra-Red Radiation. 4 pp.

CENELEC HD 455-85. Sound Transmission Using Infra-Red Radiation. 2 pp.

IEC 764-83. Sound Transmission Using Infra-Red Radiation First Edition. 11 pp.

—Printed Circuit Boards—Audio Connectors
IEC 603 Pt 11-92. Connectors for Frequencies Below 3 MHz for Use with Printed Boards Part 11: Detail Specification for Concentric Connectors (Dimensions for Free Connectors and Fixed Connectors) First Edition. 50 pp.

Headrests
Use For: Head Restraints *See Also:* Cushions; Safety Belts

Headrests *(Cont.)*
—Automotive
CNS D2181-84. Head Restraints for Automobile Occupants (Integral Type) (Oct)(11058).

CNS D2183-84. Head Restraints for Automobile Occupants (Separable Type) (Oct)(11060).

EC 78/932/EEC-78. Council Directive on the Approximation of the Laws of the Member States Relating to Head Restraints of Seats of Motor Vehicles. 15 pp.

JIS D 4606-84. Head Restraints for Automobiles Occupants.

JIS D 4606-70. Head Restraints for Automobile Occupants. 12 pp.

Headsets
Use For: Head Sets *See Also:* Earphones; Handsets; Headphones; Operator Telephone Systems; Telephones

—Aircraft
EUROCAE ED-18 07.85. Audio Systems Characteristics and MPS Aircraft Microphones (Except Carbon), Aircraft Headsets, Handsets and Loudspeakers, Aircraft Audio Selector Panels and Amplifiers. 108 pp.

—Aircraft—Environmental Testing
EUROCAE ED-18 07.85. Audio Systems Characteristics and MPS Aircraft Microphones (Except Carbon), Aircraft Headsets, Handsets and Loudspeakers, Aircraft Audio Selector Panels and Amplifiers. 108 pp.

—Audio Equipment
IEC 268 Pt 7-84. Sound System Equipment Part 7: Headphones and Headsets First Edition. 39 pp.

—Connectors—Preferred Products List
CECC CECC MUAHAG Vol 3A IS 4-91. Preferred Products List; Connectors; L. F. (En, Fr, Ge). 46 pp.

—Earphones
MOD UK DSTAN 59-39-01. Standardization Status for Earphone Elements (Insets) for Headsets, Electrical and Headsets, Microphone Issue 2; Amendment 1. 9 pp.

—Packaging
MOD UK DSTAN 81-87-91. Packaging of Handsets, Headsets, Microphones, Ancillary Parts and Accessories Issue 1. 24 pp.

—Ships
MOD UK NES 556-01. Guide to Internal Communication Equipment Headsets and Associated Items Including Microphone and Receiver Insets Issue 3 (08.89); Amendment 2. 159 pp.

—Submarines
MOD UK NES 556-01. Guide to Internal Communication Equipment Headsets and Associated Items Including Microphone and Receiver Insets Issue 3 (08.89); Amendment 2. 159 pp.

Headstocks
See Also: Machining Centers
ISO 2727-73. Modular Units for Machine Tool Construction—Headstocks First Edition. 5 pp.

SNZ NZS 5216-74. Modular Units for Machine Tool Construction. Headstocks Amend: A, 1974. 3 pp.

Healds
Use: Heddles

Health Care Facilities
See Also: Hospitals; Hyperbaric Chambers; Medical Electrical Equipment; Medical Equipment; Medical Records

—Antimicrobial Agents
CGSB CAN/CGSB-2.161-M91. Assessment of Efficacy of Antimicrobial Agents for Use on Environmental Surfaces and Medical Devices. 8 pp.

—Area
CSA CAN3-Z317.11-M82. Area Measurement for Health Care Facilities; (Gen Instr 1). 27 pp.

—Bedding
SAA AS 3789.1-91. Textiles for Health Care Facilities and Institutions—Part 1: General Ward Linen Amdt 1 January/February 1992 (in Professional Package 17G).

SAA AS 3789.2-91. Textiles for Health Care Facilities and Institutions—Part 2: Theatre Linen and Pre-Packs Amdt 1 January/February 1992 (in Professional Package 17G). 71 pp.

INDUSTRY STANDARDS

Health

Health Care Facilities (Cont.)
—Heating, Ventilating and Air Conditioning Equipment
CSA CAN/CSA-Z317.2-M91. Special Requirements for Heating, Ventilation, and Air Conditioning (HVAC) Systems in Health Care Facilities; (Gen Instr 1). 53 pp.

—Lighting
CSA CAN/CSA-Z317.5-M89. Illumination Systems in Health Care Facilities; (Gen Instr 1). 39 pp.
SAA AS 1765-75. Artificial Lighting for Clinical Observation Amdt 1 September 1979. 24 pp.

—Plumbing
CSA CAN/CSA-Z317.1-88. Special Requirements for Plumbing Installations in Health Care Facilities; (Gen Instr 1 Thru 2). 30 pp.

—Prepacks
SAA AS 3789.2-91. Textiles for Health Care Facilities and Institutions—Part 2: Theatre Linen and Pre-Packs Amdt 1 January/February 1992 (in Professional Package 17G). 71 pp.

—Waste Disposal
CSA CAN/CSA-Z317.10-88. Handling of Waste Materials Within Health Care Facilities; (Gen Instr 1). 31 pp.
SNZ NZS 4304-90. Health Care Waste Management. 16 pp.

Health Care Institutions
Use: Health Care Facilities

Health Insurance
See Also: Insurance

—Medicinal Products—Pricing
EC 89/105/EEC-88. Council Directive Relating to the Transparency of Measures Regulating the Pricing of Medicinal Products for Human Use and Their Inclusion in the Scope of National Health Insurance Systems. 4 pp.

Hearing Aids
See Also: Audio Amplifiers; Disability Equipment; Earphones; Sound Recording and Reproduction Equipment; Speakers

BSI BS 6083: Part 7-85. (WITHDRAWN) 1985 AMD 1 Hearing Aids Part 7: Methods for Measurement of the Performance Characteristics of Hearing Aids for Quality Inspection on Delivery (AMD 6184) October 31, 1990 (Renumbered as BS EN 60118-7: 1993). 17 pp.
BSI BS 6083: Part 8-85. 1985 Hearing Aids Part 8: Methods for Measurement of the Performance Characteristics of Hearing Aids Under Simulated In Situ Working Conditions. 19 pp.
BSI BS 6083: Part 10-88. 1988 Hearing Aids Part 10: Guide to Hearing Aid Standards. 26 pp.
BSI BS EN 60118-7-93. 1993 Amd 2 Hearing Aids Part 7: Methods for Measurement of Performance Characteristics of Hearing Aids for Quality Inspection on Delivery (IEC 118-7: 1983) (AMD 7837) August 15, 1993 (N). 21 pp.
CENELEC HD 450.7-87. Hearing Aids Part 7: Measurement of Performance Characterist ics of Hearing Aids for Quality Inspection for Delivery Purposes. 3 pp.
CENELEC EN 60118-7-93. Hearing Aids Part 7: Measurement of Performance Characteristics of Hearing Aids for Quality Inspection for Delivery Purposes. 5 pp.
CENELEC EN 60118-7-93. Hearing Aids Part 7: Measurement of Performance Characteristics of Hearing Aids for Quality Inspection for Delivery Purposes (IEC 118-7: 1983) (Supersedes HD 450.7 S1: 1985). 15 pp.
CNS C7125-88. Hearing Aids (Apr)(6135).
IEC 118 Pt 7-83. Hearing Aids Part 7: Measurement of the Performance Characteristics of Hearing Aids for Quality Inspection for Delivery Purposes First Edition (CENELEC EN 60118-7: 1993). 31 pp.
IEC 118 Pt 8-83. Hearing Aids Part 8: Methods of Measurement of Performance Characteristics of Hearing Aids Under Simulated In Situ Working Conditions First Edition. 43 pp.
IEC 118 Pt 10-86. Hearing Aids Part 10: Guide to Hearing Aid Standards First Edition. 56 pp.
JIS C 5512-86. Hearing Aids. 28 pp.
SAA AS 1088.7-87. Hearing Aids—Part 7: Measurement of the Performance Characteristics of Hearing Aids for Quality Inspection for Delivery Purposes. 12 pp.
SAA AS 1088.8-87. Hearing Aids—Part 8: Method of Measurement of Performance Characteristic of Hearing Aids Under Simulated In Situ Working Conditions. 16 pp.

Hearing Aids (Cont.)
—Acoustic Testing
IEC 959-90. Provisional Head and Torso Simulator for Acoustic Measurements on Air Conduction Hearing Aids First Edition. 60 pp.

—Automatic Gain Control
BSI BS 6083: Part 2-84. (WITHDRAWN) 1984 Hearing Aids Part 2: Methods for Measurement of Electroacoustical Characteristics of Hearing Aids with Automatic Gain Control Circuits. 11 pp.
BSI BS 6083: Part 2-01. 1984 Amd 1 Hearing Aids Part 2: Methods for Measurement of Electroacoustical Characteristics of Hearing Aids with Automatic Gain Control Circuits (AMD 6181) October 31, 1990. 12 pp.
CENELEC HD 450.2-84. Hearing Aids Part 2: Hearing Aids with Automatic Gain Control-Circuits. 2 pp.
IEC 118 Pt 2-83. Hearing Aids Part 2: Hearing Aids with Automatic Gain Control Circuits Second Edition; (Amendment 1-1993). 36 pp.
SAA AS 1088.2-87. Hearing Aids—Part 2: Hearing Aids with Automatic Gain Control Circuits. 7 pp.

—Bone Vibrators
BSI BS 6083: Part 9-86. 1986 Amd 1 Hearing Aids Part 9: Methods for Measurement of Characteristics of Hearing Aids with Bone Vibrator Output (AMD 6186) October 31, 1990. 16 pp.
CENELEC HD 450.9-87. Hearing Aids Part 9: Methods of Measurement of Characteristics of Hearing Aids with Bone Vibrator Output. 3 pp.
IEC 118 Pt 9-85. Hearing Aids Part 9: Methods of Measurement of Characteristics of Hearing Aids with Bone Vibrator Output First Edition. 27 pp.

—Earphones
SAA AS 1089-71. Reference Coupler for the Measurement of the Electro-Acoustic Characteristics of Hearing Aid Earphones. 8 pp.

—Earphones—Couplers
BSI BS 6111-81. 1981 Amd 1 Reference Coupler for the Measurement of Hearing Aids Using Earphones Coupled to the Ear by Means of Ear Inserts (AMD 6232) April 30, 1990. 10 pp.
CENELEC HD 305-75. IEC Reference Coupler for the Measurement of Hearing Aids Using Earphones Coupled to the Ear by Means of Ear Inserts. 2 pp.
IEC 126-73. IEC Reference Coupler for the Measurement of Hearing Aids Using Earphones Coupled to the Ear by Means of Ear Inserts Second Edition. 20 pp.

—Earphones—Nipples
BSI BS 6083: Part 5-84. 1984 Amd 1 Hearing Aids Part 5: Specifications for Dimensions of the Nipple and Sealing Device for Insert Earphones (AMD 6182) October 31, 1990. 8 pp.
CENELEC HD 450.5-85. Hearing Aids Part 5: Nipples for Insert Earphones. 2 pp.
IEC 118 Pt 5-83. Hearing Aids Part 5: Nipples for Insert Earphones First Edition. 11 pp.
SAA AS 1088.5-87. Hearing Aids—Part 5: Nipples for Insert Earphones. 3 pp.

—Earphones—Pressure Knobs
DIN ENGL 45602-57. Electric Hearing Aids; Earphone Pressure Knob and Earphone Pressure Ring; Mating Dimensions (July). 1 p.

—Earphones—Pressure Rings
DIN ENGL 45602-57. Electric Hearing Aids; Earphone Pressure Knob and Earphone Pressure Ring; Mating Dimensions (July). 1 p.

—Electrical Input Circuits
BSI BS 6083: Part 6-85. 1985 Amd 1 Hearing Aids Part 6: Characteristics of Electrical Input Circuits for Hearing Aids (AMD 6183) October 31, 1990. 8 pp.
CENELEC HD 450.6 S1-86. Hearing Aids Part 6: Characteristics of Electrical Input Circuits for Hearing Aids. 2 pp.
IEC 118 Pt 6-84. Hearing Aids Part 6: Characteristics of Electrical Input Circuits for Hearing Aids First Edition. 13 pp.
SAA AS 1088.6-87. Hearing Aids—Part 6: Characteristics of Electrical Input Circuits for Hearing Aids. 3 pp.

—Electroacoustics
BSI BS 6083: Part 0-84. (WITHDRAWN) 1984 AMD 1 Hearing Aids Part 0: Methods for Measurement of Electroacoustical Characteristics (AMD 6179) October 31, 1990 (Renumbered as BS EN 60118-0: 1993). 27 pp.
BSI BS 6083: Part 3-91. 1991 Amd 1 Hearing Aids Part 3: Methods for Measurement of Electroacoustical Characteristics of Hearing Aid Equipment Not Entirely Worn on the Listener (AMD 6398) May 31 1991. 11 pp.

Hearing Aids (Cont.)
—Electroacoustics (Cont.)
BSI BS EN 60118-0-93. 1993 Amd 2 Hearing Aids Part 0: Methods for Measurement of Electroacoustical Characteristics (IEC 118-0: 1983) (AMD 7836) August 15, 1993 (N). 32 pp.
CENELEC EN 60118-0-93. Hearing Aids Part 0: Measurement of Electroacoustical Characteristics (IEC 118-0:1983) (Supersedes HD 450.0 S1:1984). 5 pp.
CENELEC EN 60118-0-93. Hearing Aids Part 0: Measurement of Electroacoustical Characteristics (IEC 118-0: 1983) (Supersedes HD 450.0 S1: 1984). 25 pp.
IEC 118 Pt 0-83. Hearing Aids Part 0: Measurement of Electroacoustical Characteristics Second Edition; (CENELEC EN 60118-0: 1993). 54 pp.
SAA AS 1088.0-87. Hearing Aids—Part 0: Measurement of Electroacoustical Characteristics. 21 pp.

—Equipment
CENELEC HD 450.3-84. Hearing Aids Part 3: Hearing Aids Equipment Not Entirely Worn on the Listener. 2 pp.
IEC 118 Pt 3-83. Hearing Aids Part 3: Hearing Aid Equipment Not Entirely Worn on the Listener Second Edition. 17 pp.
SAA AS 1088.3-87. Hearing Aids—Part 3: Hearing Aid Equipment Not Entirely Worn on the Listener. 5 pp.

—Glossaries
BSI BS 6083: Part 10-88. 1988 Hearing Aids Part 10: Guide to Hearing Aid Standards. 26 pp.

—Induction Loops—Magnetic Field Strength
BSI BS 6083: Part 4-81. 1981 Amd 1 Hearing Aids Part 4: Magnetic Field Strength in Audio-Frequency Induction-Loops for Hearing Aid Purposes (AMD 6323) May 31, 1991. 7 pp.
CENELEC HD 450.4-84. Hearing Aids Part 4: Magnetic Field Strength in Audio-Frequency Induction Loops for Hearing Aid Purposes. 2 pp.
IEC 118 Pt 4-81. Methods of Measurement of Electro-Acoustical Characteristics of Hearing Aids Part 4: Magnetic Field Strength in Audio-Frequency Induction Loops for Hearing Aid Purposes First Edition. 13 pp.
SAA AS 1088.4-87. Hearing Aids—Part 4: Magnetic Field Strength in Audio-Frequency Induction Loops for Hearing Aid Purposes. 3 pp.

—Induction Pickup Coil Inputs
BSI BS 6083: Part 1-84. 1984 Amd 1 Hearing Aids Part 1: Method for Measurement of Characteristics of Hearing Aids with Induction Pick-up Coil Input (AMD 6180) October 31, 1990. 9 pp.
CENELEC HD 450.1-84. Hearing Aids Part 1: Hearing Aids with Induction Pick-up Coil Input. 2 pp.
IEC 118 Pt 1-83. Hearing Aids Part 1: Hearing Aids with Induction Pick-up Coil Input Second Edition. 15 pp.
SAA AS 1088.1-87. Hearing Aids—Part 1: Hearing Aids with Induction Pick-up Coil Input. 4 pp.

—Plugs
BSI BS 2813-81. 1981 Dimensions of Plugs for Hearing Aids. 8 pp.
CENELEC HD 304-77. Dimensions of Plugs for Hearing Aids. 2 pp.
IEC 90-73. Dimensions of Plugs for Hearing Aids Second Edition. 11 pp.

—Radio Communications
CCIR QUESTION 49/8-78. Radiocommunication for Short-Range Hearing Aids —Questions Concerning Study Group 8—Mobile, Radiodetermination, Amateur and Related Satellite Services. 1 p.

—Symbols—Identification Systems
BSI BS 6083: Part 11-84. 1984 Amd 1 Hearing Aids Part 11: Specification for Symbols and Other Markings on Hearing Aids and Related Equipment (AMD 6187) October 31, 1990. 16 pp.
CENELEC HD 450.11-87. Hearing Aids Part 11: Symbols and Other Markings on Hearing Aids and Related Equipment. 3 pp.
IEC 118 Pt 11-83. Hearing Aids Part 11: Symbols and Other Markings on Hearing Aids and Related Equipment First Edition. 29 pp.

—Telephone Sets
CSA CAN3-T515-M85. Requirements for Handset Telephones Intended for Use by the Hard of Hearing (R 1992); (Gen Instr 1). 21 pp.

INTERNATIONAL AND NON-U.S. NATIONAL STANDARDS
SUBJECT INDEX

Hearing Aids (Cont.)

—Telephones—Quality of Service
CCITT RECMN P.37-89. Magnetic Field Strength Around the Earcap of Telephone Handsets Which Provide for Coupling to Hearing Aids—Telephone Transmission Quality (Study Group XII) 4 pp. 4 pp.

Hearing Impaired Persons
See Also: Handicapped Persons; Visually Impaired Persons

—Sound Transmission
CCIR QUESTION 84/10-90. Sound Systems for the Hearing Impaired—Questions Concerning Study Group 10—Broadcasting Service (Sound). 1 p.

—Speech Transmission Systems
CCIR Report 778-2-90. Wireless Communication Systems for Persons with Impaired Hearing—Section 8A—Land Mobile Service and Related Subjects. 6 pp.

Hearing Protectors
See Also: Ear Muffs; Ear Plugs (Hearing Protection); Helmets; Noise Reduction
CEN PREN 458-91. Hearing Protectors—Recommendations for Selection, Use, Care and Maintenance. 38 pp.
CNS T2012-82. Ear Protectors (Jan)(8454).
CSA Z94.2-M1984. Hearing Protectors; (Gen Instr 1). 20 pp.
DIN ENGL 32760-85. Hearing Protectors; Terminology, Safety Requirements, Testing (June). 11 pp.
JIS T 8161-83. Ear Protectors. 11 pp.
SAA AS 1270-88. Acoustics—Hearing Protectors. 12 pp.

—Glossaries
DIN ENGL 32760-85. Hearing Protectors; Terminology, Safety Requirements, Testing (June). 11 pp.

—Noise
BSI BS 5108-83. (WITHDRAWN) 1983 Measurement of Sound Attenuation of Hearing Protectors (Superseded by BS 5108: Part 1: 1991). 7 pp.
BSI BS 5108: Part 1-91. (WITHDRAWN) 1991 Sound Attenuation of Hearing Protectors Part 1: Subjective Method of Measurement (ISO 4869-1: 1990) (Renumbered as BS EN 24869-1: 1993). 15 pp.
BSI BS EN 24869-1-93. 1993 Amd 1 Sound Attenuation of Hearing Protectors Part 1: Subjective Method of Measurement (ISO 4869-1: 1990) (AMD 7552) February 15, 1993 (N). 20 pp.
ISO 4869 Pt 1-90. Acoustics—Hearing Protectors—Part 1: Subjective Method for the Measurement of Sound Attenuation First Edition. 11 pp.

—Safety
DIN ENGL 32760-85. Hearing Protectors; Terminology, Safety Requirements, Testing (June). 11 pp.

Heart Valve Prostheses
Use: Cardiac Valves

Hearths
Use: Fireplaces

Heat
See Also: Decay Heat; Heat Exchangers; Heat Loss; Heat of Combustion; Heat of Hydration; Heat Release; Heat Transfer Coefficient; Heat Transfer Equipment; Specific Heat; Temperature Coefficient; Thermodynamic Properties

—Units of Measurement
BSI BS 5775: Part 4-93. 1993 Quantities, Units and Symbols Part 4: Heat (ISO 31-4: 1992) (G). 23 pp.
BSI BS 5775: Part 4-79. 1979 Amd 4 Specification for Quantities, Units and Symbols Part 4: Heat (AMD 5849) July 31, 1989. 23 pp.
CNS Z7172-84. British Thermal Unit-Kilocalorie Conversion Tables (Sep)(11041).
CNS Z7196-4-85. Quantities and Units of Heat (Oct)(11296-4).
ISO 31 Pt 4-92. Quantities and Units—Part 4: Heat Second Edition. 21 pp.
JIS Z 8418-71. British Thermal Unit-Kilocalorie Conversion Tables.

Heat Affected Zones
See Also: Weld Metal; Welding

—Steels—Hardness Testing
JIS Z 3101-90. Testing Method of Maximum Hardness in Weld Heat-Affected Zone. 6 pp.
JIS Z 3115-73. Method of Taper Hardness Test in Weld Heat-Affected Zone. 6 pp.

Heat Balance
See Also: Boilers
JIS Z 9202-91. General Rules for Heat Balance. 27 pp.

—Boilers
JIS B 8222-86. Heat Balancing of Boilers for Land Use.

—Dryers
CNS R3121-85. Method for Calculating Heat Balance of Continuous Dryer for Mineral Materials and Their Related Products (Apr)(11250).

—Furnaces
CNS R3124-85. Method for Calculating Heat Balance of Kiln and Furnace for Lime (May) (11274).
JIS G 0702-77. Method of Heat Balance for Continuous Furnace for Steel.
JIS G 0703-77. Heat Balance of Arc Furnace.
JIS R 0305-91. Heat Balancing of Kiln and Furnace for Lime. 25 pp.

—Kilns
CNS R3118-84. Method for Calculating Heat Balancing of Cement Rotary Kiln (Oct)(11081).
CNS R3119-85. Method for Calculating Heat Balancing of Pottery and Refractory Firing Tunnel Kiln (Jan)(11191).
CNS R3120-85. Method for Calculating Heat Balance of Firing Periodic Kiln for Pottery and Refractory (Apr)(11249).
CNS R3124-85. Method for Calculating Heat Balance of Kiln and Furnace for Lime (May) (11274).
JIS R 0301-91. Heat Balancing of Pottery and Refractory Firing Tunnel Kiln. 47 pp.
JIS R 0302-91. Heat Balancing of Pottery and Refractory Firing Periodic Kiln. 44 pp.
JIS R 0303-91. Heat Balancing of Cement Rotary Kiln. 46 pp.
JIS R 0305-91. Heat Balancing of Kiln and Furnace for Lime. 25 pp.

—Symbols—Diagrams
JIS Z 9201-58. Symbol of Heat Balance Diagram.

Heat Blankets
Use: Heating Blankets

Heat Blowers

—Portable
CSA C22.2 NO 122-M1989. Hand-Held Electrically Heated Tools; (Gen Instr 1 Thru 2). 50 pp.
CSA 1169 Bull. Electrical Bulletin 1169 June 27, 1978 to C22.2 NO 122. 2 pp.

Heat Capacity
Use: Specific Heat

Heat Coils
BSI BS 5141: Part 2-77. 1977 Air Heating and Cooling Coils Part 2: Method of Testing for Rating of Heating Coils. 35 pp.

Heat Detectors
Use For: Heat Sensitive Detectors See Also: Fire Alarm and Extinguishing Equipment; Flame Detectors; Smoke Detectors
BSI BS 5445: Part 5-77. 1977 Amd 1 Components of Automatic Fire Detection Systems Part 5: Heat Sensitive Detectors Containing a Static Element (Supersedes BS 3116: Part 1: 1970). 22 pp.
CEN EN 54-6-88. AMD 1 Components of Automatic Fire Detection Systems—Part 6: Heat-Sensitive Detectors; Rate-of-Rise Point Detectors Without a Static Element. 4 pp.
SAA AS 1603.1-90. Automatic Fire Detection and Alarm Systems—Part 1: Heat Detectors. 11 pp.
SNZ NZS 2139-67. Specification for Heat Actuated Fire Detectors Amend: 1, 1968. 16 pp.

—Appliances
BSI BS 5445: Part 1-77. 1977 Components of Automatic Fire Detection Systems Part 1: Introduction. 9 pp.
BSI BS 6047: Part 1-81. 1981 Flame Supervision Devices for Domestic, Commercial and Catering Gas Appliances Part 1: Heat Sensitive Types. 11 pp.

—Comprehensive Testing
CEN EN 54-5-76. Components of Automatic Fire Detection Systems Part 5 Heat Sensitive Detectors—Point Detectors Containing a Static Element. 20 pp.
CEN EN 54 (Part 5)-88. AMD 1 Components of Automatic Fire Detection Systems Part 5 Heat Sensitive Detectors—Point Detectors Containing a Static Element. 21 pp.
CEN EN 54-8-88. Components of Automatic Fire Detection Systems—Part 8: Hig Temperature Heat Detectors. 23 pp.

Heat Detectors (Cont.)

—Identification Systems
CEN EN 54-5-76. Components of Automatic Fire Detection Systems Part 5 Heat Sensitive Detectors—Point Detectors Containing a Static Element. 20 pp.
CEN EN 54 (Part 5)-88. AMD 1 Components of Automatic Fire Detection Systems Part 5 Heat Sensitive Detectors—Point Detectors Containing a Static Element. 21 pp.
CEN EN 54-8-88. Components of Automatic Fire Detection Systems—Part 8: Hig Temperature Heat Detectors. 23 pp.

Heat Dissipation
Use: Heat Loss

Heat Exchanger Tubes
See Also: Heat Exchangers
CNS G3208-83. Seamless Nickel-Chromium-Iron Alloy Heat Exchanger Tubes (Feb) (10001).

—Alloy Steel
CNS G2082-87. Method of Test for Alloy Steel Tube for Boiler and Heat Exchanger (Aug)(7382).
CNS G3141-87. Alloy Steel Tube for Boiler and Heat Exchanger (Aug)(7381).
JIS G 3462-88. Alloy Steel Boiler and Heat Exchanger Tubes. 21 pp.

—Aluminum Alloy
CNS H2032-81. Method of Test for Aluminium-Alloy Drawn Seamless Tubes for Condenser and Heat Exchangers Use (Jun) (2967).
CNS H3037-81. Aluminium Alloy Drawn Seamless Tubes for Condensers and Heat Exchangers (Jun)(2966).

—Carbon Steel
CNS G2081-87. Method of Test for Carbon Steel Tubes for Boiler and Heat Exchanger (Aug)(7380).
CNS G3140-87. Carbon Steel Tubes for Boiler and Heat Exchanger (Aug)(7379). 12 pp.
JIS G 3461-88. Carbon Steel Boiler and Heat Exchanger Tubes. 24 pp.

—Copper and Copper Alloy
BSI BS 2871: Part 3-72. 1972 Amd 2 Copper and Copper Alloys. Tubes Part 3: Tubes for Heat Exchangers. 15 pp.
DIN ENGL 1785-83. Wrought Copper and Copper Alloy Tubes for Condensers and Heat Exchangers (Oct). 7 pp.
DIN ENGL 17679-83. Wrought Copper and Copper Alloy Tubes with Rolled Fins for Use in Heat Exchangers (Oct). 8 pp.
ISO 1635 Pt 2-87. Seamless Wrought Copper and Copper Alloy Tube—Part 2: Technical Conditions of Delivery for Condenser and Heat-Exchanger Tubes First Edition. 7 pp.

—Nickel
ISO 6207-92. Seamless Nickel and Nickel Alloy Tube First Edition. 17 pp.

—Nickel Alloy
CNS G2165-83. Method of Test for Seamless Nickel-Chromium -Iron Alloy Heat Exchanger Tubes (Feb)(10002).
ISO 6207-92. Seamless Nickel and Nickel Alloy Tube First Edition. 17 pp.
JIS G 4904-91. Seamless Nickel-Chromium-Iron Alloy Heat Exchanger Tubes. 16 pp.

—Stainless Steel
CNS G2083-88. Method of Test for Stainless Steel Boiler and Heat Exchanger Tubes (Mar)(7384).
CNS G3142-88. Stainless Steel Boiler and Heat Exchanger Tubes (Mar)(7383). 23 pp.
JIS G 3463-88. Stainless Steel Boiler and Heat Exchanger Tubes. 33 pp.

—Stainless Steel—Seamless
BSI BS 3606-01. 1992 Amd 1 Steel Tubes for Heat Exchangers (AMD 7442) February 15, 1993. 24 pp.

—Stainless Steel—Welded
BSI BS 3606-01. 1992 Amd 1 Steel Tubes for Heat Exchangers (AMD 7442) February 15, 1993. 24 pp.

—Steel
BSI BS 3606-92. 1992 Steel Tubes for Heat Exchangers (Q). 24 pp.
BSI BS 3606-78. 1978 Amd 3 Steel Tubes for Heat Exchangers. 19 pp.
CNS G2084-88. Method of Test for Steel Heat Exchanger Tubes for Low Temperature Service (Mar)(7386).
CNS G3143-88. Steel Heat Exchanger Tubes for Low Temperature Service (Mar)(7385). 12 pp.
DIN ENGL 28181-85. Welded Steel Tubes for Tubular Heat Exchangers; Dimensions, Dimensional Deviations and Materials (Aug). 4 pp.

Heat Exchanger Tubes *(Cont.)*
—Steel *(Cont.)*
ISO 1129-80. Steel Tubes for Boilers, Superheaters and Heat Exchangers—Dimensions, Tolerances and Conventional Masses per Unit Length Second Edition. 6 pp.
ISO 6758-80. Welded Steel Tubes for Heat Exchangers First Edition. 5 pp.
JIS G 3464-88. Steel Heat Exchanger Tubes for Low Temperature Service. 25 pp.
JIS G 3467-88. Steel Tubes for Fired Heater. 24 pp.

—Steel—Seamless
DIN ENGL 28180-85. Seamless Steel Tubes for Tubular Heat Exchangers; Dimensions, Dimensional Deviations and Materials (Aug). 4 pp.
ISO 6759-80. Seamless Steel Tubes for Heat Exchangers First Edition. 5 pp.

—Titanium
JIS H 4631-88. Titanium Tubes for Heat Exchangers. 13 pp.
MOD UK NES 310-90. Requirements for the Manufacture of Commercially Pure Titanium Seamless Tubes Type A—Heat Exchanger, Type B—General Purpose Issue 3 (08.90). 21 pp.

—Titanium Alloy
JIS H 4636-86. Titanium-Palladium Alloy Pipes and Tubes for Heat Exchanger. 13 pp.

—U Tubes—Steel
DIN ENGL 28179-89. Steel U-Tubes for Tubular Heat Exchangers; Technical Delivery Conditions (Nov). 12 pp.

Heat Exchangers
See Also: Calorifiers; Condensing Units; Cooling Systems; Fan Coil Units; Heat; Heat Exchanger Tubes; Heat/Moisture Exchangers; Heat Transfer Equipment; Heating Equipment; Steam Condensers; Unit Coolers; Unit Heaters; Water Heaters
BSI BS 3274-60. 1960 Amd 4 Tubular Heat Exchangers for General Purposes. 48 pp.
CEN ENV 305-90. Heat Exchangers—Definitions of Performance of Heat Exchangers and the General Test Procedure for Establishing Performance of All Heat Exchangers. 27 pp.
SAA AS 1361-73. Automatic Electric Heat Exchange Water Heaters (Metric Units) Amdt 1 November 1974 Amdt 2 May 1976. 23 pp.

—Air to Air
BSI DD ENV 308-91. 1991 Heat Exchangers—Test Procedures for Establishing Performance of Air to Air and Flue Gases Heat Recovery Devices. 23 pp.
CEN PRENV 308-89. Heat Exchangers; Test Procedures for Establishing Performance of Heat Recovery Devices. 21 pp.
CEN ENV 308-91. Heat Exchangers—Test Procedures for Establishing Performance of Air to Air and Flue Gases Heat Recovery Devices. 20 pp.

—Aircraft
SBAC AGS 519 ISSUE 3. Protective, Rubber Caps for Heat Exchange Equipment.
SBAC TS 92 ISSUE 1. Code of Practice for Airborne Heat Exchangers.

—Aircraft—Caps (Lids)
SBAC AGS 519 ISSUE 3. Protective, Rubber Caps for Heat Exchange Equipment.

—Exhaust—Aircraft
CAA Chapter G5-6 06.76. Exhaust Systems (Rotorcraft). 2 pp.

—Indirect—Drinking Water—Identification Systems
DIN ENGL 4753 Pt 11-90. Water Heaters and Hot Water Systems for Drinking and Service Water; Indirect Heat Exchangers; Requirements, Testing and Marking (Feb). 3 pp.

—Indirect—Drinking Water—Safety
DIN ENGL 4753 Pt 11-90. Water Heaters and Hot Water Systems for Drinking and Service Water; Indirect Heat Exchangers; Requirements, Testing and Marking (Feb). 3 pp.

—Marine
JIS F 0502-78. Sea Water Temperature for Designing Marine Heat Exchangers.

—Motor Vehicles—Packaging
MOD UK DSTAN 81-61-89. Packaging of Vehicle Type Radiators and Heat Exchangers Issue 1. 11 pp.

—Noise
DIN ENGL 45635 Pt 14-80. Noise Measurement on Machines; Airborne Noise Measurement, Enveloping Surface Method, Air-Cooled Heat Exchangers (Air Coolers) (July). 4 pp.

Heat Exchangers *(Cont.)*
—Performance Measurement
CEN ENV 306-90. Heat Exchangers—Methods of Measuring the Parameters Necessary for Establishing the Performance. 47 pp.

—Plates—Copper and Copper Alloys
DIN ENGL 17675 Pt 1-80. Plates of Copper and Wrought Copper Alloys for Condensers and Heat Exchangers; Strength Properties (July). 4 pp.
DIN ENGL 17675 Pt 2-80. Plates of Copper and Wrought Copper Alloys for Condensers and Heat Exchangers; Technical Conditions of Delivery (July). 3 pp.
DIN ENGL 17675 Pt 3-80. Plates of Copper and Wrought Copper Alloys for Condensers and Heat Exchangers; Dimensions (July). 3 pp.
ISO 1634 Pt 2-87. Wrought Copper and Copper Alloy Plate, Sheet and Strip Part 2: Technical Conditions of Delivery for Plate and Sheet for Boilers, Pressure Vessels and Heat-Exchangers First Edition. 6 pp.

—Refrigeration Equipment
CSA C22.2 NO 140.3-M1987. Refrigerant-Containing Components for Use in Electrical Equipment (R 1993). 19 pp.

—Sheets—Copper and Copper Alloys
ISO 1634 Pt 2-87. Wrought Copper and Copper Alloy Plate, Sheet and Strip Part 2: Technical Conditions of Delivery for Plate and Sheet for Boilers, Pressure Vessels and Heat-Exchangers First Edition. 6 pp.

—Shell/Tube
CNS B5108-90. Shell and Tube Heat Exchangers (Jan)(12652).
JIS B 8249-78. Shell and Tube Heat Exchangers. 103 pp.

—Thermal Measurement
ISO 3147-75. Heat Exchangers—Verification of Thermal Balance of Water-Fed or Steam-Fed Primary Circuits—Principles and Test Requirements First Edition. 8 pp.

—Transformers
DIN ENGL 42556 Pt 1-75. Oil-to-Water Heat Exchanger for Transformers; Vertical Arrangement (Jan). 5 pp.
DIN ENGL 42557-69. Oil-to-Air Heat Exchangers for Transformers (Jan). 2 pp.

—Water Formed Deposits—Sampling
BSI BS 2455: Part 1-73. 1973 Methods of Sampling and Examining Deposits from Boilers and Associated Industrial Plant Part 1: Water-Side Deposits. 31 pp.
BSI BS 2455: Part 1-01. 1973 Amd 1 Methods of Sampling and Examining Deposits from Boilers and Associated Industrial Plant Part 1: Water-Side Deposits (AMD 7753) October 15, 1993 (Q). 32 pp.

Heat Flow Meters
—Thermal Insulation
ISO 8301-91. Thermal Insulation—Determination of Steady-State Thermal Resistance and Related Properties—Heat Flow Meter Apparatus First Edition. 46 pp.

Heat Insulation
Use: Thermal Insulation

Heat Loss
Use For: Heat Dissipation *See Also:* Heat Release; Thermodynamic Properties

—Thermal Insulation—Floors
DIN ENGL 52614-74. Testing of Thermal Insulations; Determination of Heat Dissipation of Floors (Dec). 2 pp.

Heat/Moisture Exchangers
See Also: Breathing Apparatus; Heat Exchangers; Heat Transfer Equipment; Humidifiers

—Anesthesia Equipment
ISO 9360-92. Anaesthetic and Respiratory Equipment—Heat and Moisture Exchangers for Use in Humidifying Respired Gases in Humans First Edition. 14 pp.

—Anesthesia Equipment—Safety
ISO 9360-92. Anaesthetic and Respiratory Equipment—Heat and Moisture Exchangers for Use in Humidifying Respired Gases in Humans First Edition. 14 pp.

—Breathing Apparatus
ISO 9360-92. Anaesthetic and Respiratory Equipment—Heat and Moisture Exchangers for Use in Humidifying Respired Gases in Humans First Edition. 14 pp.

Heat/Moisture Exchangers *(Cont.)*
—Breathing Apparatus—Safety
ISO 9360-92. Anaesthetic and Respiratory Equipment—Heat and Moisture Exchangers for Use in Humidifying Respired Gases in Humans First Edition. 14 pp.

Heat of Combustion
See Also: Adsorption; Calorific Value; Combustion; Heat; Heat Release; Punking Behavior; Thermodynamic Properties

—Aircraft Fuels
ISO 3648-76. Aviation Fuels—Estimation of Net Heat of Combustion First Edition. 8 pp.

—Bunker Fuel Oils
CPPA J-24(B)P-75. Thermal Value of Bunker "C" Oil. 3 pp.

—Hydrocarbons
BSI BS 2000: Part 12-82. 1982 Petroleum and Its Products Part 12: Heat of Combustion of Liquid Hydrocarbon Fuels. 16 pp.
CNS K6578-80. Method of Test for Heat of Combustion of Liquid Hydrocarbon Fuels by Bomb Calorimeter (Aug)(6359).
JIS K 2279-85. Testing Method for Heat of Combustion of Liquid Hydrocarbon Fuel by Bomb Calorimeter. 27 pp.

—Liquid Fuels
BSI BS 2000: Part 12-93. 1993 Methods of Test for Petroleum and Its Products Part 12: Determination of Specific Energy. 11 pp.

Heat of Crystallization
See Also: Crystallization; Thermodynamic Properties

—Electrical Insulation
BSI BS EN 61074-93. 1993 Method of Test for Determination of Heats and Temperatures of Melting and Crystallization of Electrical Insulating Materials by Differential Scanning Calorimetry (IEC 1074: 1991) (F).
CENELEC EN 61074-93. Determination of Heats and Temperatures of Melting and Crystallization of Electrical Insulating Materials by Differential Scanning Calorimetry (IEC 1074: 1991). 13 pp.
IEC 1074-91. Determination of Heats and Temperatures of Melting and Crystallization of Electrical Insulating Materials by Differential Scanning Calorimetry First Edition; (CENELEC EN 61074:1993). 29 pp.

—Plastics
JIS K 7122-87. Testing Methods for Heat of Transitions of Plastics. 10 pp.

Heat of Fusion
See Also: Fusion (Melting); Low Temperature Testing; Melting Points; Temperature; Thermodynamic Properties

—Electrical Insulation
BSI BS EN 61074-93. 1993 Method of Test for Determination of Heats and Temperatures of Melting and Crystallization of Electrical Insulating Materials by Differential Scanning Calorimetry (IEC 1074: 1991) (F). 18 pp.
CENELEC EN 61074-93. Determination of Heats and Temperatures of Melting and Crystallization of Electrical Insulating Materials by Differential Scanning Calorimetry (IEC 1074: 1991). 13 pp.
IEC 1074-91. Determination of Heats and Temperatures of Melting and Crystallization of Electrical Insulating Materials by Differential Scanning Calorimetry First Edition; (CENELEC EN 61074:1993). 29 pp.

—Plastics
JIS K 7122-87. Testing Methods for Heat of Transitions of Plastics. 10 pp.

Heat of Hydration
See Also: Heat; Thermodynamic Properties

—Cements
BSI BS 4550: Sec 3.8-78. 1978 Methods of Testing Cement Part 3: Physical Tests Section 3.8: Test for Heat Hydration. 8 pp.

—Portland Cements
CNS R3047-89. Standard Test Method for Heat of Hydration of Hydraulic Cement (Jul)(2248). 9 pp.
DIN ENGL 1164 Pt 8-78. Portland-, Iron Portland-, Blast-Furnace-, and Trass Cement; Determination of the Heat of Hydration with the Solution Calorimeter (Nov). 6 pp.
JIS R 5203-87. Testing Method for Heat of Hydration of Cement. 9 pp.

INTERNATIONAL AND NON-U.S. NATIONAL STANDARDS
SUBJECT INDEX

Heat of Hydration (Cont.)
—Slag Cements
DIN ENGL 1164 Pt 8-78. Portland-, Iron Portland-, Blast-Furnace-, and Trass Cement; Determination of the Heat of Hydration with the Solution Calorimeter (Nov). 6 pp.

—Trass Cements
DIN ENGL 1164 Pt 8-78. Portland-, Iron Portland-, Blast-Furnace-, and Trass Cement; Determination of the Heat of Hydration with the Solution Calorimeter (Nov). 6 pp.

Heat Plugs
Use: Glow Plugs

Heat Pumps
Use For: Unitary Heat Pumps *See Also:* Air Conditioners; Cooling Systems; Heat Sinks; Heaters; Pumps; Refrigeration Equipment

CSA CAN/CSA-C22. 2 NO 236-M90. Heating and Cooling Equipment; (Gen Instr 1) (ANSI/UL 1995). 136 pp.

CSA CAN/CSA-C445-M92. Design and Installation of Earth Energy Heat Pump Systems for Residential and Other Small Buildings; (Gen Instr 1). 32 pp.

CSA CAN/CSA-C446-M90. Performance of Ground and Water Source Heat Pumps; (Gen Instr 1). 30 pp.

CSA CAN/CSA-C746-93. Performance Standard for Rating Large Air Conditioners and Heat Pumps; (Gen Instr 1). 31 pp.

—Electric
BSI BS 6901-87. 1987 Rating and Performance of Air Source Heat Pumps with Electrically Driven Compressors. 12 pp.

CEN PREN 255-88. Heat Pumps; Heat Pump Units with Electrically Driven Compressors for Heating or for Heating and Cooling, Part 4: Air/Air Pump Units, Testing and Requirements for Marking. 10 pp.

CEN PREN 255 (Part 2)-88. Heat Pumps; Heat Pump Units with Electrically Driven Compressors for Heating or for Heating and Cooling; Part 2: Air/Water Heat Pump Units, Testing and Requirements for Marking. 15 pp.

CEN PREN 255 (Part 3)-88. Heat Pumps, Heat Pump Units with Electrically Driven Compressors for Heating or for Heating and Cooling, Part 3: Water/Water and Brine/Water Heat Pump Units, Testing and Requirements for Marking. 11 pp.

CSA CAN/CSA-C655-M91. Performance Standard for Internal Water-Loop Heat Pumps; (Gen Instr 1). 23 pp.

DIN ENGL 8900 Pt 2-80. Heat Pumps; Heat Pump Units with Electrically Driven Compressors; Test Conditions; Extent of Testing; Marking (Oct). 7 pp.

DIN ENGL 8900 Pt 3-82. Heat Pumps; Heat Pump Units with Electrically Driven Compressors; Testing of Water/Water and Brine/Water Heat Pumps (Sept). 11 pp.

DIN ENGL 8900 Pt 4-82. Heat Pumps; Heat Pump Units with Electrically Driven Compressors; Testing of Air/Water Heat Pumps (June). 14 pp.

—Electric—Air to Air—Installation
CSA C273.5-1980. Installation Requirements for Air-to-Air Heat Pumps (R 1991); (Amd 1-2 September 1980). 16 pp.

—Electric—Glossaries
BSI BS 7326: Part 1-90. 1990 Rating and Performance of Heat Pumps with Electrically Driven Compressors Part 1: Terms, Definitions and Designations. 8 pp.

CEN EN 255-1-88. Heat Pumps; Heat Pump Units with Electrically Driven Compressors for Heating or for Heating and Cooling Part 1: Terms Definitions and Designations. 5 pp.

—Electric—Sound Power
CEN PREN 255-7-92. Heat Pumps—Heat Pump Units with Electrically Driven Compressors—Part 7: Heat Pump Units and Heat Pumps for Heating Sanitary Water—Measurement of Airborne Noise—Determination of the Sound Power Level. 19 pp.

—Electric—Water Heaters
CEN PREN 255-6-92. Heat Pumps—Heat Pump Units with Electrically Driven Compressors Used for Heating or for Heating and Cooling—Part 6: Heat Pump Units for Heating Sanitary Water—Testing, Definitions and Requirements for Marking. 23 pp.

CEN PREN 255-8-92. Heat Pumps—Heat Pump Units with Electrically Driven Compressors—Part 8: Heat Pump Units and Heat Pumps for Heating Sanitary Water—Requirements. 14 pp.

—Electrical Codes
CSA CAN/CSA-C273.3-M91. Performance Standard for Split-System Central Air-Conditioners and Heat Pumps; (Gen Instr 1 Thru 3). 54 pp.

Heat Pumps (Cont.)
—Electrical Codes (Cont.)
CSA C273.5-1980. Installation Requirements for Air-to-Air Heat Pumps (R 1991); (Amd 1-2 September 1980). 16 pp.

—Glossaries
CEN PREN 378-2-92. Refrigerating Systems and Heat Pumps—Safety and Environmental Requirements—Part 2: General Definitions. 11 pp.

—Noise
DIN ENGL 45635 Pt 35-86. Measurement of Noise Emitted by Machines; Airborne Noise Emission; Enveloping Surface Method; Heat Pump Units with Electrically Driven Compressors (Apr). 5 pp.

—Packaged Units
CSA CAN/CSA-C656-M92. Performance Standard for Single Package Central Air-Conditioners and Heat Pumps; (Gen Instr 1). 51 pp.

—Refrigerants—Halogenated Hydrocarbon—Soil/Water Protection
DIN ENGL 8901-83. Heat Pumps; Heat Pumps Using Halogenated Hydrocarbons; Protection of Soil, Ground Water and Surface Water; Requirements and Teesting (Jan). 4 pp.

—Safety
CEN PREN 378 (Part 1)-90. Refrigerating Systems and Heat Pumps—Safety and Environmental Requirements—Part 1: Basic Requirements and Definitions. 10 pp.

CENELEC PREN 60 335-2-402-89. Safety of Household and Similar Electrical Appliances Part 2: Particular Requirements for Electric Heat Pumps, Air Conditioners and Dehumidifiers for Commercial and Light Industrial Applications. 37 pp.

DIN VDE 0700 Pt 222-84. Safety of Household and Similar Electrical Appliances; Heat Pumps for Room Heating Purposes (VDE Specification) (Sept). 40 pp.

IEC 335 Pt 2-40-92. Safety of Household and Similar Electrical Appliances Part 2: Particular Requirements for Electrical Heat Pumps, Air-Conditioners and Dehumidifiers Second Edition; (Replaces 378). 74 pp.

—Water Heaters
CSA CAN/CSA-C22. 2 NO 236-M90. Heating and Cooling Equipment; (Gen Instr 1) (ANSI/UL 1995). 136 pp.

Heat Radiators
Use: Radiators (Heating)

Heat Release
See Also: Heat; Heat Loss; Heat of Combustion; Heat Transfer Coefficient; Thermodynamic Properties

—Construction Materials—Fire Testing
BSI BS 476: Part 15-93. 1993 Fire Tests on Building Materials and Structures Part 15: Method of Measuring the Rate of Heat Release of Products (ISO 5660-1: 1993) (R). 38 pp.

ISO 5660 Pt 1-93. Fire Tests—Reaction to Fire—Part 1: Rate of Heat Release from Building Products (Cone Calorimeter Method) First Edition. 36 pp.

Heat Release Rate
Use: Heat Release

Heat Resistant Alloy Castings
Use: Heat Resistant Castings

Heat Resistant Alloys
Scope Note: For additional listings, see also specific products made from heat resistant alloys
Use For: High Temperature Alloys; Superalloys
See Also: Alloys; Cobalt Alloys; Ferromolybdenum; Ferrotungsten; Heat Resistant Castings; Heat Resistant Steels; Nickel Alloys; Shape Memory Alloys

ISO 4955-83. Heat-Resisting Steels and Alloys First Edition. 17 pp.

—Aerospace
CEN PREN 2500-3-91. Instructions for the Drafting and Use of Metallic Material Standards Part 3—Specific Requirements for Heat Resisting Alloys. 14 pp.

CEN PREN 2500-3-92. Instructions for the Drafting and Use of Metallic Material Standards Part 3—Specific Requirements for Heat Resisting Alloys. 13 pp.

CEN PREN 3225 (Part 1)-90. Heat Resisting Alloys Produced from Atomized Powder Powder, Compacted Forging Stock and Parts—Technical Specification—Part 1—General Requirements. 10 pp.

CEN PREN 3225 (Part 4)-90. Heat Resisting Alloys Produced from Atomized Powder Powder, Compacted Forging Stock and Parts—Technical Specification—Part 4—Parts. 9 pp.

Heat Resistant Alloys (Cont.)
—Aerospace—Electropolishing
CEN PREN 3769-90. Electrolytic Polishing of Steels and Heat Resisting Alloys. 6 pp.

—Atomized Power—Aerospace
CEN PREN 3225 (Part 2)-90. Heat Resisting Alloys Produced from Atomized Powder Powder, Compacted Forging Stock and Parts—Technical Specification—Part 2—Atomized Powder. 9 pp.

—Bars
CNS G2144-87. Method of Test for Corrosion-Resisting and Heat-Resisting Superalloy Bars (Apr)(9605).

—Bars—Aerospace
BSI 2HR 1-73. 1973 Amd 2 Nickel-Chromium-Titanium-Aluminium Heat-Resisting Alloy Billets, Bars, Forgings and Parts (Nickel Base, Cr 19.5, Ti 2.2, Al 1.4). 6 pp.

BSI 2HR 2-73. 1973 Amd 1 Nickel-Chromium-Cobalt-Titanium-Aluminium Heat-Resisting Alloy Billets, Bars, Forgings and Parts (Nickel Base, Cr 19.5, Co 18.0, Ti 2.5, Al 1.5). 4 pp.

BSI HR 3-72. 1972 Nickel-Cobalt-Chromium-Molybdenum-Aluminium-Titanium Heat-Resisting Alloy Billets, Bars, Forgings and Parts (Nickel Base Cr 14.8, Mo 5, Al 4.7, Ti 1.2). 3 pp.

BSI HR 4-72. 1972 Amd 2 Nickel-Chromium-Cobalt-Aluminium-Molybdenum-Titanium Heat-Resisting Alloy Billets, Bars, Forgings and Parts (Nickel Base, Cr 15, Co 14.2, Al 5, Mo 4, Ti 4). 5 pp.

BSI HR 5-72. 1972 Amd 1 Nickel-Chromium-Titanium Heat-Resisting Alloy Billets, Bars Forgings and Parts (Nickel Base Cr 19.5, TI 0.4). 3 pp.

BSI HR 6-72. 1972 Amd 1 Nickel-Chromium-Iron-Molybdenum-Cobalt-Tungsten Heat-Resisting Alloy Billets, Bars, Forgings and Parts (Nickel Base, Cr 21.7, Fe 18.5, Mo 9, Co 1.5, W 0.6). 3 pp.

BSI HR 10-72. 1972 Amd 1 Nickel-Cobalt-Chromium-Molybdenum-Titanium-Aluminium Heat-Resisting Alloy Billets, Bars, Forgings and Parts (Nickel Base, Co 20, Cr 20, Mo 5.9, Ti 2.1, Al 0.5). 4 pp.

BSI HR 40-72. 1972 Amd 1 Cobalt-Chromium-Tungsten-Nickel-Manganese Heat-Resisting Alloy Billets, Bars and Forgings(Cobalt Base Cr 20, W 15, Ni 10, Mn 1.5). 3 pp.

BSI HR 53-73. 1973 Amd 2 Nickel-Iron-Chromium-Molybdenum-Titanium Heat-Resisting Alloy Billets, Bars, Forgings and Parts (Ni/Co 42.5, Cr 12.5, Mo 5.8, Ti 3.0, Fe Remainder). 5 pp.

BSI HR 55-80. 1980 Nickel-Iron-Chromium-Molybdenum-Aluminium-Titanium Heat-Resisting Alloy Billets, Bars, Forgings and Parts (Ni/Co 43.5, Cr 16.5, Mo 3.3, Al 1.2, Ti 1.2, Fe Remainder). 3 pp.

BSI 2HR 504-73. 1973 Amd 2 Nickel-Chromium-Titanium Heat-Resisting Alloy Bar, and Wire for Rivets, and Rivets (Nickel Base, Cr 19.5, Ti 0.4). 4 pp.

BSI 2HR 601-75. (WITHDRAWN) 1975 Nickel-Chromium-Titanium Aluminium Heat-Resisting Alloy Bar and Wire for the Manufacture of Fasteners (Maximum Diameter or Minor Sectional Dimensions 25 mm) (Nickel Base Cr 19.5, Ti 2.2, Al 1.4) (Superseded by 3HR 601: 1989). 3 pp.

CEN PREN 3235 (Part 3)-89. Heat Resisting Alloy Wrought Products Technical Specification Part 3-Bar and Section. 7 pp.

CEN PREN 3666-91. Heat Resisting Alloy N1-P100 Solution Treated and Cold Worked Rm Equal to or Greater Than 1550 MPa Bar for Fasteners 3 Equal to or Less Than D Equal to or Less Than 30 mm. 5 pp.

—Bars—Aerospace—Comprehensive Testing
AECMA PREN2098-03-81. Inspection and Testing Requirements for Titanium and Heat Resisting Alloy Wrought Products—Part 3—Inspection and Testing Requirements for Bars and Sections (C5/27). 6 pp.

AECMA PREN2098-06-81. Inspection and Testing Requirements for Titanium and Heat Resisting Alloy Wrought Products—Part 6—Inspection and Testing Requirements for Bars and Wires for Fasteners (C5/27). 6 pp.

—Bars—Aerospace—Inspection
AECMA PREN2098-03-81. Inspection and Testing Requirements for Titanium and Heat Resisting Alloy Wrought Products—Part 3—Inspection and Testing Requirements for Bars and Sections (C5/27). 6 pp.

—Bars—Round—Aerospace
AECMA PREN2368-83. Round Bars, Heat Resisting Alloys—Machined Diameter 12 mm Less Than or Equal to D Less Than or Equal to 180 mm—Dimensions. 5 pp.

CEN PREN 2368-91. Round Bars, Heat Resisting Alloys Machined Diameter 10 mm Equal to or Less Than D Equal to or Less Than 180 mm Dimensions. 6 pp.

INTERNATIONAL AND NON-U.S. NATIONAL STANDARDS
SUBJECT INDEX — Heat

Heat Resistant Alloys (Cont.)
—Bars—Round—Aerospace (Cont.)

CEN PREN 2368-93. Aerospace Series Round Bars, Heat Resisting Alloys Machined Diameter 10 mm Less Than or Equal to D Less Than or Equal to 180 mm Dimensions. 5 pp.

—Billets—Aerospace

BSI 2HR 1-73. 1973 Amd 2 Nickel-Chromium-Titanium-Aluminium Heat-Resisting Alloy Billets, Bars, Forgings and Parts (Nickel Base, Cr 19.5, Ti 2.2, Al 1.4). 6 pp.

BSI 2HR 2-73. 1973 Amd 1 Nickel-Chromium-Cobalt-Titanium-Aluminium Heat-Resisting Alloy Billets, Bars, Forgings and Parts (Nickel Base, Cr 19.5, Co 18.0, Ti 2.5, Al 1.5). 4 pp.

BSI HR 3-72. 1972 Nickel-Cobalt-Chromium-Molybdenum-Aluminium-Titanium Heat-Resisting Alloy Billets, Bars, Forgings and Parts (Nickel Base Cr 14.8, Mo 5, Al 4.7, Ti 1.2). 3 pp.

BSI HR 4-72. 1972 Amd 2 Nickel-Chromium-Cobalt-Aluminium-Molybdenum-Titanium Heat-Resisting Alloy Billets, Bars, Forgings and Parts (Nickel Base, Cr 15, Co 14.2, Al 5, Mo 4, Ti 4). 5 pp.

BSI HR 5-72. 1972 Amd 1 Nickel-Chromium-Titanium Heat-Resisting Alloy Billets, Bars Forgings and Parts (Nickel Base Cr 19.5, TI 0.4). 3 pp.

BSI HR 6-72. 1972 Amd 1 Nickel-Chromium-Iron-Molybdenum-Cobalt-Tungsten Heat-Resisting Alloy Billets, Bars, Forgings and Parts (Nickel Base, Cr 21.7, Fe 18.5, Mo 9, Co 1.5, W 0.6). 3 pp.

BSI HR 10-72. 1972 Amd 1 Nickel-Cobalt-Chromium-Molybdenum-Titanium-Aluminium Heat-Resisting Alloy Billets, Bars, Forgings and Parts (Nickel Base, Co 20, Cr 20, Mo 5.9, Ti 2.1, Al 0.5). 4 pp.

BSI HR 40-72. 1972 Amd 1 Cobalt-Chromium-Tungsten-Nickel-Manganese Heat-Resisting Alloy Billets, Bars and Forgings(Cobalt Base Cr 20, W 15, Ni 10, Mn 1.5). 3 pp.

BSI HR 53-73. 1973 Amd 2 Nickel-Iron-Chromium-Molybdenum-Titanium Heat-Resisting Alloy Billets, Bars, Forgings and Parts (Ni/Co 42.5, Cr 12.5, Mo 5.8, Ti 3.0, Fe Remainder). 5 pp.

BSI HR 55-80. 1980 Nickel-Iron-Chromium-Molybdenum-Aluminium-Titanium Heat-Resisting Alloy Billets, Bars, Forgings and Parts (Ni/Co 43.5, Cr 16.5, Mo 3.3, Al 1.2, Ti 1.2, Fe Remainder). 3 pp.

—Cold Worked—Bars—Aerospace

CEN PREN 2952-93. Aerospace Series Heat Resisting Alloy NI-PH 2601 Solution Treated and Cold Worked Bars for Forged Fasteners D Less Than or Equal to 50 mm to 1270 MPa Less Than or Equal to Rm Less Less Than or Equal to 1470 MPa. 4 pp.

—Cold Worked—Heat Treated—Tubes—Seamless—Aerospace

BSI HR 404-73. 1973 Amd 2 Nickel-Cobalt-Chromium-Molybdenum-Titanium-Aluminium Heat-Resisting Alloy Cold Worked Solution Treated Seamless Tubes (Nickel Base, Co 20, Cr 20, Mo 5.9, Ti 2.1, Al 0.5). 5 pp.

—Cold Worked—Softened—Bars—Aerospace

CEN PREN 2398-93. Aerospace Series Heat Resisting Alloy FE-PA 2601 Softened, Cold Worked, Solution Treated, Precipitation Treated Bars for Machined Fasteners D Less Than or Equal to 50 mm 900 MPa Less Than or Equal to Rm Less Than or Equal to 1100 MPa. 4 pp.

CEN PREN 2399-93. Aerospace Series Heat Resisting Alloy FE-PA 2601 Softened and Cold Worked Bars for Forged Fasteners D Less Than or Equal to 50 mm 900 MPa Less Than or Equal to Rm Less Than or Equal to 1100 MPa. 4 pp.

CEN PREN 3761-93. Aerospace Series Heat Resisting Alloy FE-PA 2601 Softened and Cold Worked Bars for Forged Fasteners D Less Than or Equal to 50 mm 1100 MPa Less Than or Equal to Rm Less Than or Equal to 1300 MPa. 4 pp.

—Cold Worked—Softened—Tubes—Seamless—Aerospace

BSI HR 401-73. 1973 Amd 1 Nickel-Chromium-Titanium-Aluminium Heat-Resisting Alloy Cold Worked and Softened Seamless Tubes (Nickel Base, Cr 19.5, Ti 2.2, Al 1.4). 4 pp.

BSI HR 402-73. 1973 Nickel-Chromium-Cobalt-Titanium-Aluminium Heat-Resisting Alloy Cold Worked and Softened Seamless Tubes (Nickel Base Ti 2.5, Al 1.5). 3 pp.

BSI HR 403-73. 1973 Amd 1 Nickel-Chromium-Titanium Heat-Resisting Alloy Cold Worked and Softened Seamless Tubes (Nickel Base Ti 0.4). 3 pp.

Heat Resistant Alloys (Cont.)
—Cold Worked—Softened—Wire—Aerospace

CEN PREN 3639-93. Aerospace Series Heat Resisting Alloy FE-PA 2601 Softened and Cold Worked Wires for Forged Fasteners D Less Than or Equal to 15 mm 900 MPa Less Than or Equal to Rm Less Than or Equal to 1100 MPa. 4 pp.

CEN PREN 3762-93. Aerospace Series Heat Resisting Alloy FE-PA 2601 Softened and Cold Worked Wires for Forged Fasteners D Less Than or Equal to 15 mm 1100 MPa Less Than or Equal to Rm Less Than or Equal to 1300 MPa. 4 pp.

—Compacted—Aerospace—Porosity

CEN PREN 3641-88. Test Methods Heat Resisting Alloys Compacted Material Procedure for Quantitative Thermally Induced Porosity Test (T.I.P.). 4 pp.

CEN PREN 3641-92. Test Methods Heat Resisting Alloys Compacted Material Procedure for Quantitative Determination of Thermally Induced Porosity (T.I.P.). 4 pp.

—Forgings—Aerospace

BSI 2HR 1-73. 1973 Amd 2 Nickel-Chromium-Titanium-Aluminium Heat-Resisting Alloy Billets, Bars, Forgings and Parts (Nickel Base, Cr 19.5, Ti 2.2, Al 1.4). 6 pp.

BSI 2HR 2-73. 1973 Amd 1 Nickel-Chromium-Cobalt-Titanium-Aluminium Heat-Resisting Alloy Billets, Bars, Forgings and Parts (Nickel Base, Cr 19.5, Co 18.0, Ti 2.5, Al 1.5). 4 pp.

BSI HR 3-72. 1972 Nickel-Cobalt-Chromium-Molybdenum-Aluminium-Titanium Heat-Resisting Alloy Billets, Bars, Forgings and Parts (Nickel Base Cr 14.8, Mo 5, Al 4.7, Ti 1.2). 3 pp.

BSI HR 4-72. 1972 Amd 2 Nickel-Chromium-Cobalt-Aluminium-Molybdenum-Titanium Heat-Resisting Alloy Billets, Bars, Forgings and Parts (Nickel Base, Cr 15, Co 14.2, Al 5, Mo 4, Ti 4). 5 pp.

BSI HR 5-72. 1972 Amd 1 Nickel-Chromium-Titanium Heat-Resisting Alloy Billets, Bars Forgings and Parts (Nickel Base Cr 19.5, TI 0.4). 3 pp.

BSI HR 6-72. 1972 Amd 1 Nickel-Chromium-Iron-Molybdenum-Cobalt-Tungsten Heat-Resisting Alloy Billets, Bars, Forgings and Parts (Nickel Base, Cr 21.7, Fe 18.5, Mo 9, Co 1.5, W 0.6). 3 pp.

BSI HR 10-72. 1972 Amd 1 Nickel-Cobalt-Chromium-Molybdenum-Titanium-Aluminium Heat-Resisting Alloy Billets, Bars, Forgings and Parts (Nickel Base, Co 20, Cr 20, Mo 5.9, Ti 2.1, Al 0.5). 4 pp.

BSI HR 40-72. 1972 Amd 1 Cobalt-Chromium-Tungsten-Nickel-Manganese Heat-Resisting Alloy Billets, Bars and Forgings(Cobalt Base Cr 20, W 15, Ni 10, Mn 1.5). 3 pp.

BSI HR 53-73. 1973 Amd 2 Nickel-Iron-Chromium-Molybdenum-Titanium Heat-Resisting Alloy Billets, Bars, Forgings and Parts (Ni/Co 42.5, Cr 12.5, Mo 5.8, Ti 3.0, Fe Remainder). 5 pp.

BSI HR 55-80. 1980 Nickel-Iron-Chromium-Molybdenum-Aluminium-Titanium Heat-Resisting Alloy Billets, Bars, Forgings and Parts (Ni/Co 43.5, Cr 16.5, Mo 3.3, Al 1.2, Ti 1.2, Fe Remainder). 3 pp.

CEN PREN 2860 (Part 1)-90. Heat Resisting Alloys Forging Stock and Forgings—Technical Specification—Part 1 General Requirements. 10 pp.

CEN PREN 2860 (Part 2)-90. Heat Resisting Alloys Forging Stock and Forgings—Technical Specification—Part 2—Forging Stock. 7 pp.

CEN PREN 2860 (Part 3)-90. Heat Resisting Alloys Forging Stock and Forgings—Technical Specification—Part 3—Pre-Production and Production Forgings. 11 pp.

CEN PREN 3225 (Part 3)-90. Heat Resisting Alloys Produced from Atomized Powder Powder, Compacted Forging Stock and Parts—Technical Specification Part 3—Hot Compacted Forging Stock. 10 pp.

—Heat Treated—Bars—Aerospace

BSI HR 51-73. 1973 Amd 2 Heat-Resisting Alloy Billets, Bars, Forgings and Parts (Solution Treated at 980 Degrees C) (Ni 25.5, Cr 14.7, Ti 2, Mn 1.5, Mo 1.2, Si 0.7, V 0.3, Fe Remainder). 5 pp.

BSI HR 52-73. 1973 Amd 2 Heat-Resisting Alloy Billets, Bars, Forgings and Parts (Ni 25.5, Cr 14.7, Ti 1.8, Mn 1.5, Mo 1.2, V 0.3, Fe Remainder). 5 pp.

CEN PREN 2961-93. Aerospace Series Heat Resisting Alloy NI-PH 2601 Solution Treated Bars for Machined Fasteners D Less Than or Equal to 50 mm Rm Greater Than or Equal to 1270 MPa. 4 pp.

—Heat Treated—Billets—Aerospace

BSI HR 51-73. 1973 Amd 2 Heat-Resisting Alloy Billets, Bars, Forgings and Parts (Solution Treated at 980 Degrees C) (Ni 25.5, Cr 14.7, Ti 2, Mn 1.5, Mo 1.2, Si 0.7, V 0.3, Fe Remainder). 5 pp.

Heat Resistant Alloys (Cont.)
—Heat Treated—Billets—Aerospace (Cont.)

BSI HR 52-73. 1973 Amd 2 Heat-Resisting Alloy Billets, Bars, Forgings and Parts (Ni 25.5, Cr 14.7, Ti 1.8, Mn 1.5, Mo 1.2, V 0.3, Fe Remainder). 5 pp.

—Heat Treated—Forgings—Aerospace

BSI HR 51-73. 1973 Amd 2 Heat-Resisting Alloy Billets, Bars, Forgings and Parts (Solution Treated at 980 Degrees C) (Ni 25.5, Cr 14.7, Ti 2, Mn 1.5, Mo 1.2, Si 0.7, V 0.3, Fe Remainder). 5 pp.

BSI HR 52-73. 1973 Amd 2 Heat-Resisting Alloy Billets, Bars, Forgings and Parts (Ni 25.5, Cr 14.7, Ti 1.8, Mn 1.5, Mo 1.2, V 0.3, Fe Remainder). 5 pp.

—Plates

CNS G2145-87. Method of Test for Corrosion-Resisting and Heat-Resisting Superalloy Sheets and Plates (Apr)(9607).

ISO 9328 Pt 2-91. Steel Plates and Strips for Pressure Purposes—Technical Delivery Conditions—Part 2: Unalloyed and Low-Alloyed Steels with Specified Room Temperature and Elevated Temperature Properties First Edition. 16 pp.

—Plates—Aerospace

BSI 2HR 201-74. 1974 Amd 1 Nickel-Chromium-Titanium-Aluminium Heat-Resisting Alloy Plate, Sheet, and Strip (Nickel Base, Cr 19.5, Ti 2.2, Al 1.4) (AMD 5996) August 31, 1988. 5 pp.

BSI HR 203-72. 1972 Amd 2 Nickel-Chromium-Titanium Heat-Resisting Alloy Plate, Sheet and Strip (Nickel Base, Cr 19.5, Ti 0.4). 4 pp.

BSI HR 204-72. 1972 Amd 1 Nickel-Chromium-Iron-Molybdenum-Cobalt-Tungsten Heat-Resisting Alloy Plate, Sheet and Strip (Nickel Base, Cr 22, Fe 18.5, Mo 9, Co 1.5, W 0.6) (AMD 1809) September 30, 1975. 4 pp.

BSI HR 206-73. 1973 Amd 2 Nickel-Cobalt-Chromium-Molybdenum-Titanium-Aluminium Heat-Resisting Alloy Plate, Sheet and Strip (Nickel Base, Co 20, Cr 20, Mo 5.9, Ti 2.1, Al 0.5). 6 pp.

BSI HR 207-73. 1973 Amd 1 Nickel-Iron-Chromium-Molybdenum-Aluminium-Titanium Heat-Resisting Alloy Plate, Sheet and Strip (Ni/Co 43.5, Cr 16.5, Mo 3.3, Al 1.2, Ti 1.2, Fe Remainder). 5 pp.

BSI HR 240-72. 1972 Amd 1 Cobalt-Chromium-Tungsten Nickel-Manganese Heat-Resisting Alloy Plate, Sheet and Strip (Cobalt Base W 15, Ni 10, Mn 1.5). 3 pp.

CEN PREN 3235 (Part 2)-89. Heat Resisting Alloy Wrought Products Technical Specification Part 2—Plate, Sheet and Strip. 6 pp.

CEN PREN 3506-91. Hot Finished Flat Products Heat Resisting Alloys. 9 pp.

—Plates—Aerospace—Comprehensive Testing

AECMA PREN2098-02-81. Inspection and Testing Requirements for Titanium and Heat Resisting Alloy Wrought Products—Part 2—Inspection and Testing Requirements for Sheets Strips and Plates (C5/27). 5 pp.

—Plates—Aerospace—Inspection

AECMA PREN2098-02-81. Inspection and Testing Requirements for Titanium and Heat Resisting Alloy Wrought Products—Part 2—Inspection and Testing Requirements for Sheets Strips and Plates (C5/27). 5 pp.

—Sections—Aerospace

CEN PREN 3235 (Part 3)-89. Heat Resisting Alloy Wrought Products Technical Specification Part 3-Bar and Section. 7 pp.

—Sections—Aerospace—Comprehensive Testing

AECMA PREN2098-03-81. Inspection and Testing Requirements for Titanium and Heat Resisting Alloy Wrought Products—Part 3—Inspection and Testing Requirements for Bars and Sections (C5/27). 6 pp.

—Sections—Aerospace—Inspection

AECMA PREN2098-03-81. Inspection and Testing Requirements for Titanium and Heat Resisting Alloy Wrought Products—Part 3—Inspection and Testing Requirements for Bars and Sections (C5/27). 6 pp.

—Sheets

CNS G2145-87. Method of Test for Corrosion-Resisting and Heat-Resisting Superalloy Sheets and Plates (Apr)(9607).

—Sheets—Aerospace

BSI 2HR 201-74. 1974 Amd 1 Nickel-Chromium-Titanium-Aluminium Heat-Resisting Alloy Plate, Sheet, and Strip (Nickel Base, Cr 19.5, Ti 2.2, Al 1.4) (AMD 5996) August 31, 1988. 5 pp.

INTERNATIONAL AND NON-U.S. NATIONAL STANDARDS SUBJECT INDEX

Heat

Heat Resistant Alloys (Cont.)

—Sheets—Aerospace (Cont.)

BSI 2HR 202-73. 1973 Amd 1 Nickel-Chromium-Cobalt-Titanium-Aluminium Heat-Resisting Alloy Sheet and Strip (Nickel Base, Cr 19.5, Co 18.0, Ti 2.5, Al 1.5) (AMD 6143) June 29, 1990. 4 pp.

BSI HR 203-72. 1972 Amd 2 Nickel-Chromium-Titanium Heat-Resisting Alloy Plate, Sheet and Strip (Nickel Base, Cr 19.5, Ti 0.4). 4 pp.

BSI HR 204-72. 1972 Amd 1 Nickel-Chromium-Iron-Molybdenum-Cobalt-Tungsten Heat-Resisting Alloy Plate, Sheet and Strip (Nickel Base, Cr 22, Fe 18.5, Mo 9, Co 1.5, W 0.6) (AMD 1809) September 30, 1975. 4 pp.

BSI HR 206-73. 1973 Amd 2 Nickel-Cobalt-Chromium-Molybdenum-Titanium-Aluminium Heat-Resisting Alloy Plate, Sheet and Strip (Nickel Base, Co 20, Cr 20, Mo 5.9, Ti 2.1, Al 0.5). 6 pp.

BSI HR 207-73. 1973 Amd 1 Nickel-Iron-Chromium-Molybdenum-Aluminium-Titanium Heat-Resisting Alloy Plate, Sheet and Strip (Ni/Co 43.5, Cr 16.5, Mo 3.3, Al 1.2, Ti 1.2, Fe Remainder). 5 pp.

BSI HR 240-72. 1972 Amd 1 Cobalt-Chromium-Tungsten Nickel-Manganese Heat-Resisting Alloy Plate, Sheet and Strip (Cobalt Base W 15, Ni 10, Mn 1.5). 3 pp.

CEN PREN 3235 (Part 2)-89. Heat Resisting Alloy Wrought Products Technical Specification Part 2—Plate, Sheet and Strip. 6 pp.

—Sheets—Aerospace—Comprehensive Testing

AECMA PREN2098-02-81. Inspection and Testing Requirements for Titanium and Heat Resisting Alloy Wrought Products—Part 2—Inspection and Testing Requirements for Sheets Strips and Plates (C5/27). 5 pp.

—Sheets—Aerospace—Inspection

AECMA PREN2098-02-81. Inspection and Testing Requirements for Titanium and Heat Resisting Alloy Wrought Products—Part 2—Inspection and Testing Requirements for Sheets Strips and Plates (C5/27). 5 pp.

—Sheets—Cold Rolled—Aerospace

AECMA PREN2366-83. Sheets and Strips—Heat Resisting Alloys—Cold Rolled—Thickness a Less Than or Equal to 3 mm Dimensions. 6 pp.

—Strips

ISO 9328 Pt 2-91. Steel Plates and Strips for Pressure Purposes—Technical Delivery Conditions—Part 2: Unalloyed and Low-Alloyed Steels with Specified Room Temperature and Elevated Temperature Properties First Edition. 16 pp.

—Strips—Aerospace

BSI 2HR 201-74. 1974 Amd 1 Nickel-Chromium-Titanium-Aluminium Heat-Resisting Alloy Plate, Sheet, and Strip (Nickel Base, Cr 19.5, Ti 2.2, Al 1.4) (AMD 5996) August 31, 1988. 5 pp.

BSI 2HR 202-73. 1973 Amd 1 Nickel-Chromium-Cobalt-Titanium-Aluminium Heat-Resisting Alloy Sheet and Strip (Nickel Base, Cr 19.5, Co 18.0, Ti 2.5, Al 1.5) (AMD 6143) June 29, 1990. 4 pp.

BSI HR 203-72. 1972 Amd 2 Nickel-Chromium-Titanium Heat-Resisting Alloy Plate, Sheet and Strip (Nickel Base, Cr 19.5, Ti 0.4). 4 pp.

BSI HR 204-72. 1972 Amd 1 Nickel-Chromium-Iron-Molybdenum-Cobalt-Tungsten Heat-Resisting Alloy Plate, Sheet and Strip (Nickel Base, Cr 22, Fe 18.5, Mo 9, Co 1.5, W 0.6) (AMD 1809) September 30, 1975. 4 pp.

BSI HR 206-73. 1973 Amd 2 Nickel-Cobalt-Chromium-Molybdenum-Titanium-Aluminium Heat-Resisting Alloy Plate, Sheet and Strip (Nickel Base, Co 20, Cr 20, Mo 5.9, Ti 2.1, Al 0.5). 6 pp.

BSI HR 207-73. 1973 Amd 1 Nickel-Iron-Chromium-Molybdenum-Aluminium-Titanium Heat-Resisting Alloy Plate, Sheet and Strip (Ni/Co 43.5, Cr 16.5, Mo 3.3, Al 1.2, Ti 1.2, Fe Remainder). 5 pp.

BSI HR 240-72. 1972 Amd 1 Cobalt-Chromium-Tungsten Nickel-Manganese Heat-Resisting Alloy Plate, Sheet and Strip (Cobalt Base W 15, Ni 10, Mn 1.5). 3 pp.

CEN PREN 3235 (Part 2)-89. Heat Resisting Alloy Wrought Products Technical Specification Part 2—Plate, Sheet and Strip. 6 pp.

—Strips—Aerospace—Comprehensive Testing

AECMA PREN2098-02-81. Inspection and Testing Requirements for Titanium and Heat Resisting Alloy Wrought Products—Part 2—Inspection and Testing Requirements for Sheets Strips and Plates (C5/27). 5 pp.

—Strips—Aerospace—Inspection

AECMA PREN2098-02-81. Inspection and Testing Requirements for Titanium and Heat Resisting Alloy Wrought Products—Part 2—Inspection and Testing Requirements for Sheets Strips and Plates (C5/27). 5 pp.

Heat Resistant Alloys (Cont.)

—Strips—Cold Rolled—Aerospace

AECMA PREN2366-83. Sheets and Strips—Heat Resisting Alloys—Cold Rolled—Thickness a Less Than or Equal to 3 mm Dimensions. 6 pp.

—Tubes—Aerospace

CEN PREN 3235 (Part 4)-89. Heat Resisting Alloy Wrought Products Technical Specification Part 4—Seamless Tube. 7 pp.

—Tubes—Aerospace—Comprehensive Testing

AECMA PREN2098-04-81. Inspection and Testing Requirements for Titanium and Heat Resisting Alloy Wrought Products—Part 4—Inspection and Testing Requirements for Tubes (C5/27). 5 pp.

—Tubes—Aerospace—Inspection

AECMA PREN2098-04-81. Inspection and Testing Requirements for Titanium and Heat Resisting Alloy Wrought Products—Part 4—Inspection and Testing Requirements for Tubes (C5/27). 5 pp.

—Wire—Aerospace

AECMA PREN2369-83. Wires, Heat Resisting Alloys—Diameter 0.2 mm Less Than or Equal to D Less Than or Equal to 8 mm—Dimensions. 5 pp.

CEN PREN 3235 (Part 5)-89. Heat Resisting Alloy Wrought Products Technical Specification Part 5—Wire. 6 pp.

—Wire—Aerospace—Comprehensive Testing

AECMA PREN2098-05-81. Inspection and Testing Requirements for Titanium and Heat Resisting Alloy Wrought Products—Part 5—Inspection and Testing Requirements for Wires (C5/27). 5 pp.

—Wire—Aerospace—Inspection

AECMA PREN2098-05-81. Inspection and Testing Requirements for Titanium and Heat Resisting Alloy Wrought Products—Part 5—Inspection and Testing Requirements for Wires (C5/27). 5 pp.

AECMA PREN2098-06-81. Inspection and Testing Requirements for Titanium and Heat Resisting Alloy Wrought Products—Part 6—Inspection and Testing Requirements for Bars and Wires for Fasteners (C5/27). 6 pp.

—Wrought—Aerospace

BSI 3HR 100-78. 1978 Amd 3 Inspection and Testing Procedure for Aerospace Material. Wrought Heat Resisting Alloys. 41 pp.

BSI 3HR 100-04. 1978 Amd 4 Inspection and Testing of Wrought Heat-Resisting Alloys (AMD 7514) January 15, 1993. 42 pp.

—Wrought—Aerospace—Inspection

AECMA PREN2098-01-81. Inspection and Testing Requirements for Titanium and Heat Resisting Alloy Wrought Products—Part 1—General Requirements (C5/27). 12 pp.

CEN PREN 3235 (Part 1)-89. Heat Resisting Alloy Wrought Products Technical Specification Part 1—General Requirements. 9 pp.

Heat Resistant Castings

See Also: Castings; Heat Resistant Alloys; Heat Resistant Steel Castings

BSI 2HC 204-89. 1989 Nickel-Cobalt-Chromium-Aluminium-Titanium-Molybdenum Alloy Castings (Nickel Base Co 15.0, Cr 10.0, Al 5.5, Ti 4.8, Mo 3.0). 3 pp.

BSI 2HC 207-89. 1989 Nickel-Cobalt-Tungsten-Chromium-Aluminium-Tantalum-Hafnium-Titanium Alloy Castings, (Nickel Base Co 10.0, W 10.0, Cr 9.0, Al 5.5, Ta 2.5, Hf 1.5, Ti 1.5). 3 pp.

BSI 2HC 208-89. 1989 Nickel-Cobalt-Tungsten-Chromium-Aluminium-Molybdenum Titanium-Tantalum Alloy Castings (Nickel Base Co 10.0, W 10.0, Cr 9.0, Al 5.5, Mo 2.5, Ti 1.5, Ta 1.5). 3 pp.

BSI 2HC 210-89. 1989 Nickel-Tungsten-Aluminium-Chromium-Molybdenum Alloy Castings. 3 pp.

—Aerospace

BSI HC 301-73. 1973 Cobalt Base Chromium-Nickel-Tungsten Alloy Castings (Cr 25.5, Ni 105, W 7.5). 2 pp.

—Centrifugal—Aerospace

BSI HC 100-72. 1972 Amd 3 Inspection and Testing Procedure for Iron, Nickel, Copper, Cobalt and Refractory Metal Base Alloy Castings (AMD 6574) May 28, 1991. 17 pp.

—Investment—Aerospace

BSI HC 100-72. 1972 Amd 3 Inspection and Testing Procedure for Iron, Nickel, Copper, Cobalt and Refractory Metal Base Alloy Castings (AMD 6574) May 28, 1991. 17 pp.

Heat Resistant Castings (Cont.)

—Investment—Precipitation Hardening—Aerospace

BSI HC 206-75. 1975 Amd 1 Precipitation Hardening Nickel Base Chromium-Cobalt-Molybdenum-Titanium Alloy Precision Castings (Cr 20, Co 20, Mo 6, Ti 2). 3 pp.

—Precipitation Hardening—Aerospace

BSI HC 202-73. 1973 Amd 1 Precipitation Hardening Nickel Base Chromium-Niobium-Molybdenum-Iron-Tungsten Alloy Castings (Cr 20.0, Nb 6.6, Mo 6.0, Fe 3.0, W 2.5) (AMD 3884) April 30, 1982. 3 pp.

BSI HC 205-74. 1974 Amd 1 Precipitation Hardening Nickel Base Chromium-Cobalt-Molybdenum-Titanium Alloy Castings (Cr 20, Co 20, Mo 6, Ti 2). 3 pp.

—Sand—Aerospace

BSI HC 100-72. 1972 Amd 3 Inspection and Testing Procedure for Iron, Nickel, Copper, Cobalt and Refractory Metal Base Alloy Castings (AMD 6574) May 28, 1991. 17 pp.

Heat Resistant Clothing

See Also: Exposure Suits; Fire Fighting Clothing; Flame Retardant Clothing; Protective Clothing

BSI BS 3791-70. (WITHDRAWN) 1970 Clothing for Protection Against Intense Heat for Short Periods (Superseded by BS EN 366: 93 & BS EN 367: 92). 29 pp.

BSI BS EN 366-93. 1993 Protective Clothing—Protection Against Heat and Fire Method of Test: Evaluation of Materials and Material Assemblies When Exposed to a Source of Radiant Heat (N). 17 pp.

BSI BS EN 367-92. 1992 Protective Clothing—Protection Against Heat and Fire—Method of Determining Heat Transmission on Exposure to Flame (Supersedes BS 3791: 1970). 19 pp.

BSI BS EN 367-01. 1992 Amd 1 Protective Clothing—Protection Against Heat and Flames—Test Method: Determination of the Transmission on Exposure to Flame (AMD 7667) May 15, 1993 (N). 20 pp.

BSI BS EN 373-93. 1993 Protective Clothing—Assessment of Resistance of Materials to Molten Metal Splash (N). 15 pp.

CEN PREN 366-90. Protective Clothing: Protection Against Heat and Fire: Method of Test: Evaluation of Materials and Material Assemblies when Exposed to a Source of Radiant Heat. 22 pp.

CEN EN 366-93. Protective Clothing—Protection Against Heat and Fire—Method of Test: Evaluation of Materials and Material Assemblies when Exposed to a Source of Radiant Heat. 16 pp.

CEN PREN 367-90. Clothing for Protection Against Heat and Flames: Method of Determining Heat Transmission on Exposure to Flames. 17 pp.

CEN EN 367-92. Protective Clothing—Protection Against Heat and Fire—Method of Determining Heat Transmission on Exposure to Flame. 13 pp.

CEN PREN 373-90. Protective Clothing—Assessment of Resistance of Materials to Molten Metal Splash. 16 pp.

CEN EN 373-93. Protective Clothing—Assessment of Resistance of Materials to Molten Metal Splash. 14 pp.

CEN PREN 531-91. Protective Clothing for Industrial Workers Exposed to Heat (Excluding Firefighters' and Welders' Clothing). 11 pp.

CEN PREN 533-91. Clothing for Protection Against Heat and Flame—Performance Specification for Limited Flame Spread of Materials. 9 pp.

DIN ENGL 32764 Pt 1-87. Clothing for Protection Against Radiant Heat; Type WS (Heavy Duty) Protective Clothing; Safety Requirements, Testing (Feb). 4 pp.

DIN ENGL EN 367-92. Protective Clothing for Use Against Heat and Fire; Method of Determining Heat Transmission on Exposure to Flame (Nov). 12 pp.

ISO 2801-73. Clothing for Protection Against Heat and Fire—General Recommendations for Users and for Those in Charge of Such Users First Edition. 6 pp.

ISO 9150-88. Protective Clothing—Determination of Behaviour of Materials on Impact of Small Splashes of Molten Metal First Edition. 11 pp.

—Flammability Testing

BSI BS 6249: Part 1-82. 1982 Materials and Material Assemblies Used in Clothing for Protection Against Heat and Flame Part 1: Flammability Testing and Performance. 7 pp.

ISO 6942-93. Clothing for Protection Against Heat and Fire—Evaluation of Thermal Behaviour of Materials and Material Assemblies When Exposed to a Source of Radiant Heat Second Edition. 17 pp.

Heat Resistant Steel Castings

Use For: High Temperature Steel Castings
See Also: Castings; Heat Resistant Castings; Heat Resistant Steels; Steel Castings

Heat Resistant Steel Castings (Cont.)

CNS G3093-87. Heat Resisting Steel Castings (Nov)(4002). 4 pp.
DIN ENGL 17465-93. Heat Resisting Steel Castings; Technical Delivery Conditions (Aug). 8 pp.
JIS G 5122-91. Heat Resisting Steel Castings. 10 pp.

—Aerospace

BSI HC 3-73. 1973 1% Chromium-Molybdenum Low Alloy Steel Castings (700 N/Square mm). 2 pp.
BSI HC 4-73. 1973 3% Chromium-Molybdenum Steel Castings (620-770 N/Square mm). 3 pp.
BSI HC 7-74. 1974 Amd 1 3% Chromium-Molybdenum Steel Castings (880-1080 Mpa). 3 pp.
BSI HC 8-74. 1974 Amd 1 3% Chromium-Molybdenum Steel Castings (1150-1300 MPa). 3 pp.
BSI HC 9-74. 1974 Amd 1 Nickel-Chromium-Molybdenum Steel Castings (880-1080 MPa). 3 pp.
BSI HC 10-74. 1974 Amd 1 Nickel-Chromium-Molybdenum Steel Castings (1150-1300 MPa). 3 pp.
BSI HC 103-74. 1974 Amd 1 23% Chromium-Nickel-Tungsten Corrosion-Resisting Steel Castings (Cr 23, Ni 11, W 3). 3 pp.
BSI HC 104-74. 1974 Amd 1 19% Chromium-10% Nickel Niobium-Stabilized Corrosion-Resisting Steel Castings (460 MPa). 3 pp.
BSI HC 105-74. 1974 Amd 1 18% Chromium-11% Nickel-2.5% Molybdenum Niobium-Stabilized Corrosion-Resisting Steel Castings. 3 pp.

—Case Hardening—Aerospace

BSI HC 5-73. 1973 3% Nickel Case-Hardening Steel Castings (700 N/Square mm). 2 pp.

—Investment

BSI BS 3146: Part 2-75. 1975 Amd 1 Investment Castings in Metal Part 2: Corrosion and Heat Resisting Steels, Nickel and Cobalt Base Alloys. 28 pp.

—Investment—Aerospace

BSI HC 401-74. 1974 18% Nickel Maraging Steel Precision Castings (1600-1850 MPa). 3 pp.
MOD UK DTD-5279-68. Chromium-Nickel-2.5 Per Cent Molybdenum Heat-Resisting and Corrosion-Resisting Steel Investment Castings (50 Hbar) (High Temperature Properties Not Verified). 2 pp.
MOD UK DTD-5289-68. Chromium-Nickel-3.5 Per Cent Molybdenum Heat-Resisting and Corrosion-Resisting Steel Investment Castings (50 Hbar) (High Temperature Properties Not Verified). 2 pp.

—Nitriding—Aerospace

BSI HC 6-73. 1973 3% Chromium-Molybdenum Nitriding Steel Castings (850-1000 N/Square mm). 2 pp.

—Precipitation Hardening—Aerospace

BSI HC 101-73. (WITHDRAWN) 1973 Amd 1 Precipitation Hardening Chromium-Nickel-Copper-Molybdenum Steel Castings (950 N/Square mm) (Cr 14.0, Ni 4.5, Cu 2.5, Mo 1.5) (Superseded by 2HC 101: 1989). 3 pp.
BSI 2HC 101-89. 1989 Amd 2 Precipitation Hardening Chromium-Nickel-Copper-Molybdenum Steel Castings (950 MPa) (Cr 14.0, Ni 4.5, Cu 2.5, Mo 1.5) (AMD 6704) April 30, 1991. 6 pp.

Heat Resistant Steels

Scope Note: For additional listings, see also specific products made from heat resistant steels
Use For: High Temperature Steels *See Also:* Alloy Steels; Austenitic Stainless Steels; Austenitic Steels; Chromium Molybdenum Steels; Chromium Molybdenum Vanadium Steels; Chromium Steels; Corrosion Resistant Steels; Ferroalloys; Heat Resistant Alloys; Heat Resistant Steel Castings; Maraging Steels; Nickel Chromium Steels; Nickel Steels; Stainless Steels; Steels; Tool Steels; Tungsten Steels

ISO 4954-93. Steels for Cold Heading and Cold Extruding Second Edition. 42 pp.
ISO 4955-83. Heat-Resisting Steels and Alloys First Edition. 17 pp.

—Bars

CNS G2146-88. Method of Test for Heat-Resisting Steel Bars (Sep)(9609).
CNS G3197-87. Corrosion-Resisting and Heat-Resisting Superalloy Bars (Apr) (9604).
CNS G3199-88. Heat-Resisting Steel Bars (Sep)(9608).
JIS G 4311-91. Heat-Resisting Steel Bars. 23 pp.
JIS G 4901-91. Corrosion-Resisting and Heat-Resisting Superalloy Bars. 15 pp.

—Bars—Aerospace

AECMA PREN2171-79. Heat Resisting Steel FE-PA92-HT—Rm Less Than or Equal to 900 MPa —Bars (C5/48). 3 pp.
AECMA PREN2303-80. Heat Resisting Steel FE-PA92-HT—Rm Greater Than or Equal to 960 MPa—Bars (C5/48). 3 pp.
AECMA PREN2493-81. Heat Resisting Steel FE-PM38—Rm 1000-1140 MPa—Bars (C5/48). 3 pp.
BSI HR 650-72. 1972 Amd 2 High Expansion Heat-Resisting Steel Bar and Wire for the Manufacture of Bolts, Studs, Set Screws and Nuts (Ni 25.5, Cr 15, Ti 2, Mn 1.5, Mo 1.25, Si 0.7, V 0.3) (Limiting Ruling Section 20 mm). 5 pp.
BSI S 150-75. 1975 Amd 1 Chromium-Molybdenum-Vanadium-Niobium Heat-Resisting Steel Billets, Bars, Forgings and Parts (930-1080 MPa) (Cr 10.5, Mo 0.6, V 0.2, Nb 0.3). 4 pp.
BSI S 151-75. 1975 Amd 1 Chromium-Nickel-Molybdenum-Vanadium Heat-Resisting Steel Billets, Bars, Forgings and Parts (930-1130 MPa: Limiting Ruling Section 150 mm) (Cr 12, Ni 2.5, Mo 1.7, V 0.3). 3 pp.
BSI S 152-75. 1975 Amd 1 Chromium-Cobalt-Molybdenum-Vanadium-Niobium Heat-Resisting Steel Billets, Bars, Forgings and Parts (Consumable Electrode Remelted)(1000-1140 MPa) (Cr 10.5, Co 6, Mo 0.8, V 0.2, Nb 0.3). 4 pp.
BSI S 159-81. 1981 Amd 1 12% Chromium-Nickel-Molybdenum-Vanadium Heat-Resisting Steel Bars for the Manufacture of Fasteners (1100-1300 MPa: Limiting Ruling Section 50 mm). 4 pp.

—Bars—Bolts—Aerospace

AECMA PREN2398-80. Heat Resisting Steel FE-PA92-HT—Rm Greater Than or Equal to 900 MPa—Bars for Machined Bolts—D Less Than or Equal to 25 mm (C5/48). 3 pp.
AECMA PREN2399-80. Heat Resisting Steel FE-PA92-HT—Rm Greater Than or Equal to 900 MPa—Bars for Forged Bolts D Less Than or Equal to 25 mm (C5/48). 3 pp.

—Billets—Aerospace

BSI S 150-75. 1975 Amd 1 Chromium-Molybdenum-Vanadium-Niobium Heat-Resisting Steel Billets, Bars, Forgings and Parts (930-1080 MPa) (Cr 10.5, Mo 0.6, V 0.2, Nb 0.3). 4 pp.
BSI S 151-75. 1975 Amd 1 Chromium-Nickel-Molybdenum-Vanadium Heat-Resisting Steel Billets, Bars, Forgings and Parts (930-1130 MPa: Limiting Ruling Section 150 mm) (Cr 12, Ni 2.5, Mo 1.7, V 0.3). 3 pp.
BSI S 152-75. 1975 Amd 1 Chromium-Cobalt-Molybdenum-Vanadium-Niobium Heat-Resisting Steel Billets, Bars, Forgings and Parts (Consumable Electrode Remelted)(1000-1140 MPa) (Cr 10.5, Co 6, Mo 0.8, V 0.2, Nb 0.3). 4 pp.

—Filler Metal—Welding

DIN ENGL 8556 Pt 1-86. Filler Metals for Welding Stainless and Heat-Resisting Steels; Designation, Technical Delivery Conditions (May). 10 pp.

—Forgings—Aerospace

AECMA PREN2172-79. Heat Resisting Steel FE-PA92-HT—Rm Less Than or Equal to 900 MPa —Forgings (C5/48). 3 pp.
AECMA PREN2304-80. Heat Resisting Steel FE-PA92-HT—Rm Greater Than or Equal to 960 MPa—Forgings (C5/48). 3 pp.
AECMA PREN2494-81. Heat Resisting Steel FE-PM38—Rm 1000-1140 MPa—Forgings (C5/48). 3 pp.
BSI S 150-75. 1975 Amd 1 Chromium-Molybdenum-Vanadium-Niobium Heat-Resisting Steel Billets, Bars, Forgings and Parts (930-1080 MPa) (Cr 10.5, Mo 0.6, V 0.2, Nb 0.3). 4 pp.
BSI S 151-75. 1975 Amd 1 Chromium-Nickel-Molybdenum-Vanadium Heat-Resisting Steel Billets, Bars, Forgings and Parts (930-1130 MPa: Limiting Ruling Section 150 mm) (Cr 12, Ni 2.5, Mo 1.7, V 0.3). 3 pp.
BSI S 152-75. 1975 Amd 1 Chromium-Cobalt-Molybdenum-Vanadium-Niobium Heat-Resisting Steel Billets, Bars, Forgings and Parts (Consumable Electrode Remelted)(1000-1140 MPa) (Cr 10.5, Co 6, Mo 0.8, V 0.2, Nb 0.3). 4 pp.

—Forgings—Weldable

DIN ENGL 17243-87. Weldable Heat Resisting Steel Forgings and Rolled or Forged Steel Bars; Technical Delivery Conditions (Jan). 18 pp.

—Heat Treated—Bars—Aerospace

AECMA PREN2167-79. Heat Resisting Steel FE-PA91-HT-Solution Treated —Bars (C5/48). 3 pp.
AECMA PREN2173-79. Heat Resisting Steel FE-PA93-HT—Solution Treated and Precipitation Treated—Bars (C5/48). 3 pp.
AECMA PREN2238-79. Heat Resisting Steel FE-PA91-HT—Solution Treated and Precipitation Treated—Bars (C5/48). 3 pp.

—Heat Treated—Bars—Rings—Aerospace

AECMA PREN2169-79. Heat Resisting Steel FE-PA91-HT—Solution Treated—Bars and Sections for Welded Rings (C5/48). 3 pp.
AECMA PREN2237-79. Heat Resisting Steel FE-PA91-HT—Solution Treated and Precipitation Treated—Bars and Sections for Welded Rings (C5/48). 3 pp.

—Heat Treated—Forgings—Aerospace

AECMA PREN2168-79. Heat Resisting Steel FE-PA91-HT—Solution Treated—Forgings (C5/48). 3 pp.
AECMA PREN2174-79. Heat Resisting Steel FE-PA93-HT—Solution Treated and Precipitation Treated—Forgings (C5/48). 3 pp.
AECMA PREN2239-79. Heat Resisting Steel FE-PA91-HT—Solution Treated and Precipitation Treated—Forgings (C5/48). 3 pp.

—Heat Treated—Plates—Aerospace

AECMA PREN2416-80. Heat Resisting Steel FE-PA91-HT—Solution Treated—Plates a Greater than 3 mm (C5/48). 3 pp.
AECMA PREN2417-80. Heat Resisting Steel FE-PA93-HT—Solution Treated and Precipitation Treated—Plates a Greater Than 3 mm (C5/48). 3 pp.

—Heat Treated—Sections—Rings—Aerospace

AECMA PREN2169-79. Heat Resisting Steel FE-PA91-HT—Solution Treated—Bars and Sections for Welded Rings (C5/48). 3 pp.
AECMA PREN2237-79. Heat Resisting Steel FE-PA91-HT—Solution Treated and Precipitation Treated—Bars and Sections for Welded Rings (C5/48). 3 pp.

—Heat Treated—Sheets—Aerospace

AECMA PREN2170-80. Heat Resisting Steel FE-PA91-HT—Solution Treated—Sheets and Strips a Less Than or Equal to 3 mm (C5/48). 3 pp.
AECMA PREN2175-80. Heat Resisting Steel FE-PA93-HT—Solution Treated and Precipitation Treated—Sheets and Strips a Less Than or Equal to 3 mm (C5/48). 3 pp.

—Heat Treated—Strips—Aerospace

AECMA PREN2175-80. Heat Resisting Steel FE-PA93-HT—Solution Treated and Precipitation Treated—Sheets and Strips a Less Than or Equal to 3 mm (C5/48). 3 pp.

—Plates

BSI BS 1501: Part 3-90. 1990 Amd 1 Steels for Pressure Purposes: Plates, Sheet and Strip Part 3: Specification for Corrosion-and Heat-Resisting Steels (AMD 6744) July 31, 1991. 24 pp.
BSI BS 1501: Part 3-02. 1990 Amd 2 Steels for Pressure Purposes: Plates, Sheet and Strip Part 3: Specification for Corrosion—and Heat-Resisting Steels (AMD 6868) October 31, 1991. 28 pp.
BSI BS 1501: Part 3-03. 1990 Amd 3 Steels for Pressure Purposes Part 3: Specification for Corrosion—and Heat-Resisting Steels: Plates, Sheet and Strip (AMD 7026) June 15, 1992 (Q). 29 pp.
CNS G2147-88. Method of Test for Heat-Resisting Steel Sheets and Plates (Sep)(9611).
CNS G3198-87. Corrosion-Resisting and Heat-Resisting Superalloy Sheets and Plates (Apr)(9606).
CNS G3200-88. Heat-Resisting Steel Sheets and Plates (Sep)(9610).
ISO 9328 Pt 2-91. Steel Plates and Strips for Pressure Purposes—Technical Delivery Conditions—Part 2: Unalloyed and Low-Alloyed Steels with Specified Room Temperature and Elevated Temperature Properties First Edition. 16 pp.
JIS G 4312-91. Heat-Resisting Steel Plates and Sheets. 26 pp.
JIS G 4902-91. Corrosion-Resisting and Heat-Resisting Superalloy Plates and Sheets. 18 pp.
SAA AS 1449-80. Wrought Alloy Steels—Stainless and Heat-Resisting Steel Plate, Sheet and Strip. 19 pp.
SNZ NZS/BS 1501: Part 3-90. Steels for Pressure Purposes: Plates Part 3: Specification for Corrosion-and Heat-Resisting Steels Amend: 1, 1991; 2, 1991. 20 pp.

—Plates—Cold Rolled

BSI BS 1449: Part 2-83. 1983 Amd 2 Steel Plate, Sheet and Strip Part 2: Specification for Stainless and Heat-Resisting Steel Plate, Sheet and Strip (AMD 6646) April 30, 1991. 18 pp.
BSI BS 1449: Part 2-75. 1975 Steel Plate, Sheet and Strip Part 2: Stainless and Heat Resisting Steel Plate, Sheet and Strip. 31 pp.
SNZ NZS/BS 1449: Part 2-83. Steel Plate, Sheet and Strip Part 2: Specification for Stainless and Heat-Resisting Steel Plate, Sheet and Strip. 16 pp.

—Plates—Hot Rolled

BSI BS 1449: Part 2-83. 1983 Amd 2 Steel Plate, Sheet and Strip Part 2: Specification for Stainless and Heat-Resisting Steel Plate, Sheet and Strip (AMD 6646) April 30, 1991. 18 pp.
BSI BS 1449: Part 2-75. 1975 Steel Plate, Sheet and Strip Part 2: Stainless and Heat Resisting Steel Plate, Sheet and Strip. 31 pp.
SNZ NZS/BS 1449: Part 2-83. Steel Plate, Sheet and Strip Part 2: Specification for Stainless and Heat-Resisting Steel Plate, Sheet and Strip. 16 pp.

INTERNATIONAL AND NON-U.S. NATIONAL STANDARDS
SUBJECT INDEX
Heat

Heat Resistant Steels (Cont.)
—Plates—Mass
- JIS G 4310-91. Method of Mass Calculation for Stainless Steel Plates and Sheets, and Heat-Resisting Steel Plates and Sheets. 7 pp.

—Sections—Bolts—Aerospace
- MOD UK DTD-5066-01. 12 Per Cent Chromium Heat Resisting Steel Suitable for Bolts, Studs, Set Screws and Nuts (Limiting Ruling Section 50 mm; 110 Hbar); Amendment 1. 4 pp.
- MOD UK DTD-5076-69. High Expansion Heat Resisting Steel for the Manufacture of Bolts, Studs, Set Screws and Nuts (97 Hbar) (Vacuum Remelted: Limiting Ruling Section 20 mm). 3 pp.

—Sections—Nuts—Aerospace
- MOD UK DTD-5066-01. 12 Per Cent Chromium Heat Resisting Steel Suitable for Bolts, Studs, Set Screws and Nuts (Limiting Ruling Section 50 mm; 110 Hbar); Amendment 1. 4 pp.
- MOD UK DTD-5076-69. High Expansion Heat Resisting Steel for the Manufacture of Bolts, Studs, Set Screws and Nuts (97 Hbar) (Vacuum Remelted: Limiting Ruling Section 20 mm). 3 pp.

—Sections—Screws—Aerospace
- MOD UK DTD-5066-01. 12 Per Cent Chromium Heat Resisting Steel Suitable for Bolts, Studs, Set Screws and Nuts (Limiting Ruling Section 50 mm; 110 Hbar); Amendment 1. 4 pp.
- MOD UK DTD-5076-69. High Expansion Heat Resisting Steel for the Manufacture of Bolts, Studs, Set Screws and Nuts (97 Hbar) (Vacuum Remelted: Limiting Ruling Section 20 mm). 3 pp.

—Sections—Studs—Aerospace
- MOD UK DTD-5066-01. 12 Per Cent Chromium Heat Resisting Steel Suitable for Bolts, Studs, Set Screws and Nuts (Limiting Ruling Section 50 mm; 110 Hbar); Amendment 1. 4 pp.
- MOD UK DTD-5076-69. High Expansion Heat Resisting Steel for the Manufacture of Bolts, Studs, Set Screws and Nuts (97 Hbar) (Vacuum Remelted: Limiting Ruling Section 20 mm). 3 pp.

—Sheets
- CNS G2147-88. Method of Test for Heat-Resisting Steel Sheets and Plates (Sep)(9611).
- CNS G3198-87. Corrosion-Resisting and Heat-Resisting Superalloy Sheets and Plates (Apr)(9606).
- CNS G3200-88. Heat-Resisting Steel Sheets and Plates (Sep)(9610).
- JIS G 4312-91. Heat-Resisting Steel Plates and Sheets. 26 pp.
- JIS G 4902-91. Corrosion-Resisting and Heat-Resisting Superalloy Plates and Sheets. 18 pp.
- SAA AS 1449-80. Wrought Alloy Steels—Stainless and Heat-Resisting Steel Plate, Sheet and Strip. 19 pp.

—Sheets—Aerospace
- BSI S 530-70. 1970 24/17 Chromium-Nickel Heat-Resisting Steel Sheet and Strip (Titanium Stabilized: 54 Hbar) (Elevated Temperature Properties Not Verified). 2 pp.
- BSI S 538-75. 1975 Chromium-Cobalt-Molybdenum-Vanadium Heat-Resisting Steel Sheet and Strip (930-1130 MPa: Maximum Thickness 6 mm) (Cr 12, Co 2.5, Mo 1.7, V 0.3). 3 pp.

—Sheets—Cold Rolled
- BSI BS 1449: Part 2-83. 1983 Amd 2 Steel Plate, Sheet and Strip Part 2: Specification for Stainless and Heat-Resisting Steel Plate, Sheet and Strip (AMD 6646) April 30, 1991. 18 pp.
- BSI BS 1449: Part 2-75. 1975 Steel Plate, Sheet and Strip Part 2: Stainless and Heat Resisting Steel Plate, Sheet and Strip. 31 pp.
- SNZ NZS/BS 1449: Part 2-83. Steel Plate, Sheet and Strip Part 2: Specification for Stainless and Heat-Resisting Steel Plate, Sheet and Strip. 16 pp.

—Sheets—Hot Rolled
- BSI BS 1449: Part 2-83. 1983 Amd 2 Steel Plate, Sheet and Strip Part 2: Specification for Stainless and Heat-Resisting Steel Plate, Sheet and Strip (AMD 6646) April 30, 1991. 18 pp.
- BSI BS 1449: Part 2-75. 1975 Steel Plate, Sheet and Strip Part 2: Stainless and Heat Resisting Steel Plate, Sheet and Strip. 31 pp.
- SNZ NZS/BS 1449: Part 2-83. Steel Plate, Sheet and Strip Part 2: Specification for Stainless and Heat-Resisting Steel Plate, Sheet and Strip. 16 pp.

—Sheets—Mass
- JIS G 4310-91. Method of Mass Calculation for Stainless Steel Plates and Sheets, and Heat-Resisting Steel Plates and Sheets. 7 pp.

—Strips
- ISO 9328 Pt 2-91. Steel Plates and Strips for Pressure Purposes—Technical Delivery Conditions—Part 2: Unalloyed and Low-Alloyed Steels with Specified Room Temperature and Elevated Temperature Properties First Edition. 16 pp.

Heat Resistant Steels (Cont.)
—Strips (Cont.)
- SAA AS 1449-80. Wrought Alloy Steels—Stainless and Heat-Resisting Steel Plate, Sheet and Strip. 19 pp.

—Strips—Aerospace
- BSI S 530-70. 1970 24/17 Chromium-Nickel Heat-Resisting Steel Sheet and Strip (Titanium Stabilized: 54 Hbar) (Elevated Temperature Properties Not Verified). 2 pp.
- BSI S 538-75. 1975 Chromium-Cobalt-Molybdenum-Vanadium Heat-Resisting Steel Sheet and Strip (930-1130 MPa: Maximum Thickness 6 mm) (Cr 12, Co 2.5, Mo 1.7, V 0.3). 3 pp.

—Strips—Cold Rolled
- BSI BS 1449: Part 2-83. 1983 Amd 2 Steel Plate, Sheet and Strip Part 2: Specification for Stainless and Heat-Resisting Steel Plate, Sheet and Strip (AMD 6646) April 30, 1991. 18 pp.
- BSI BS 1449: Part 2-75. 1975 Steel Plate, Sheet and Strip Part 2: Stainless and Heat Resisting Steel Plate, Sheet and Strip. 31 pp.
- DIN ENGL 59381-80. Flat Products of Steel; Cold Rolled Strip of Stainless Steel and of Heat Resisting Steels; Dimensions, Permissible Dimensional Deviations, Deviations of Form and Weight (Aug). 4 pp.
- SNZ NZS/BS 1449: Part 2-83. Steel Plate, Sheet and Strip Part 2: Specification for Stainless and Heat-Resisting Steel Plate, Sheet and Strip. 16 pp.

—Strips—Hot Rolled
- BSI BS 1449: Part 2-83. 1983 Amd 2 Steel Plate, Sheet and Strip Part 2: Specification for Stainless and Heat-Resisting Steel Plate, Sheet and Strip (AMD 6646) April 30, 1991. 18 pp.
- BSI BS 1449: Part 2-75. 1975 Steel Plate, Sheet and Strip Part 2: Stainless and Heat Resisting Steel Plate, Sheet and Strip. 31 pp.
- SNZ NZS/BS 1449: Part 2-83. Steel Plate, Sheet and Strip Part 2: Specification for Stainless and Heat-Resisting Steel Plate, Sheet and Strip. 16 pp.

—Wrought
- JIS G 0321-66. Product Analysis and Its Tolerance for Wrought Steel. 8 pp.

Heat Sensitive Detectors
Use: Heat Detectors

Heat Shrink Tubing
Use For: Electrical Heat Shrink Tubing
See Also: Cables (Electric); Electrical Insulation; Tubing (Flexible)
- CNS C3134-88. General Rules for Testing of Heat Shrinkable Tubing for Electrical Insulation (Sep)(8781).
- CNS C3135-88. Method for Determining Dimension of Heat Shrinkable Tubing for Electrical Insulation (Sep)(8782).
- CNS C3136-88. Method of Test for Contraction Percentage of Heat Shrinkable Tubing for Electrical Insulation (Sep)(8736).

—Corrosion Testing—Copper
- CNS C3148-88. Method of Test for Copper Corrosion of Heat Shrinkable Tubing for Electrical Insulation (Sep)(8795).

—Dielectric Properties
- CNS C3138-88. Method of Test for Electric Strength of Heat Shrinkable Tubing for Electrical Insulation (Sep)(8785).
- CNS C3140-88. Method for Determining Dielectric Constant and Dielectric Dissipation Factor of Heat Shrinkable Tubing for Electrical Insulation (Sep)(8787).

—Electrical Resistivity
- CNS C3139-88. Method for Determining Resistivity of Heat Shrinkable Tubing for Electrical Insulation (Sep)(8786).

—Extruded
- CSA C22.2 NO 198.1-M1986. Extruded Insulating Tubing (R 1992); (Gen Instr 1 Thru 4). 89 pp.

—Flammable Testing
- CNS C3146-88. Method of Test for Combustion Resistance of Heat Shrinkable Tubing for Electrical Insulation (Sep)(8793).

—Low Temperature Testing
- CNS C3144-88. Method of Test for Low-Temperature Characteristics of Heat Shrinkable Tubing for Electrical Insulation (Sep)(8791).

—Preservation
- CNS C3147-88. Method of Test for Preservation of Heat Shrinkable Tubing for Electrical Insulation (Sep)(8794).

Heat Shrink Tubing (Cont.)
—PVC
- MOD UK TS 10279. PVC Tubing, Heat Shrinkable, Natural.

—Solvent Resistance Testing
- CNS C3142-88. Method of Test for Solvent Resistance of Heat Shrinkable Tubing for Electrical Insulation (Sep)(8789).

—Tensile Testing
- CNS C3137-88. Method of Test for Tensile of Heat Shrinkable Tubing for Electrical Insulation (Sep)(8784).

—Thermal Resistance
- CNS C3143-88. Method of Test for Thermal Resistance of Heat Shrinkable Tubing for Electrical Insulation (Sep)(8790).
- CNS C3145-88. Method of Test for Thermal Impact Resistance of Heat Shrinkable Tubing for Electrical Insulation (Sep)(8792).

—Water Absorption
- CNS C3141-88. Method of Test for Water Absorption of Heat Shrinkable Tubing for Electrical Insulation (Sep)(8788).

Heat Sinks
Use For: Dissipators See Also: Cooling Systems; Heat Pumps

—Resistors
- CECC CECC 40 203 ISSUE 1-81. Blank Detail Specification: Fixed Power Resistors; (Assessment Level H) (En, Fr, Ge) AMD 1 (En, Fr, Ge). 48 pp.
- CECC CECC 40 203-004. CEI CECC 40 203-004 Issue 1; Fixed Power Resistors Wirewound Insulated Heat Sink Resistors with Rigid Terminations (En). 10 pp.
- CECC CECC 40 203-006. BS CECC 40 203-006 Issue 1; Fixed Power Resistors Wirewound Insulated Heat Sink Resistors with Rigid Terminations (En). 11 pp.

—Semiconductor Devices
- MOD UK DEF-5254-66. General Requirements for Retainers, Mounting Pads and Heat Dissipators for Semi-Conductor Devices. 10 pp.

Heat Stability
Use: Thermal Stability

Heat Strengthened Glass
Use: Heat Treated Glass

Heat Stress
—Ergonomics—Sweat Rate
- ISO 7933-89. Hot Environments—Analytical Determination and Interpretation of Thermal Stress Using Calculation of Required Sweat Rate First Edition. 24 pp.

—Ergonomics—Wet Bulb Temperature
- ISO 7243-89. Hot Environments—Estimation of the Heat Stress on Working Man, Based on the WBGT-Index (Wet Bulb Globe Temperature) Second Edition. 13 pp.
- SNZ NZS/ISO 7243-89. Hot Environments. Estimation of the Heat Stress on Working Man, Based on the WBGT-Index (Wet Bulb Globe Temperature). 9 pp.

Heat Test Source Lamps
Use For: HTS Lamps See Also: Lamps; Lighting
- BSI BS 6012-80. 1980 Amd 1 Heat Test Source (H.T.S.) Lamps for Carrying out Heating Tests on Luminaires. 9 pp.
- CENELEC HD 392 S2-85. Heat Test Source (H.T.S.) Lamps for Carrying Out Heating Tests on Luminaires. 2 pp.
- IEC 634-93. Heat Test Source (H.T.S.) Lamps for Carrying out Heating Tests on Luminaires Second Edition. 24 pp.

Heat Tracing Systems
See Also: Electric Tracing Equipment

—Aerospace
- CEN PREN 3703-91. Heat Release Rate for Materials and Products Under the Influence of Radiating Heat and Flames Test Method. 26 pp.

—Heating Cables—Gaseous Mines
- CSA C22.2 NO 138-M1989. Heat Tracing Cable and Cable Sets for Use in Hazardous Locations; (Gen Instr 1). 25 pp.

—Heating Cables—Pipes—Hazardous Locations
- CSA C22.2 NO 138-M1989. Heat Tracing Cable and Cable Sets for Use in Hazardous Locations; (Gen Instr 1). 25 pp.

INDUSTRY STANDARDS

Heat Tracing Systems (Cont.)
—Heating Cables—Tanks (Containers)—Hazardous Locations
CSA C22.2 NO 138-M1989. Heat Tracing Cable and Cable Sets for Use in Hazardous Locations; (Gen Instr 1). 25 pp.

Heat Transfer Coefficient
See Also: Heat Release; Thermodynamic Properties
—Glass Doors
DIN ENGL 52619 Pt 1-82. Testing of Thermal Insulation; Determination of Thermal Resistance and Overall Heat Transfer Coefficient of Windows; Measurement of the Whole Construction (Nov). 6 pp.
—Windows
DIN ENGL 52619 Pt 1-82. Testing of Thermal Insulation; Determination of Thermal Resistance and Overall Heat Transfer Coefficient of Windows; Measurement of the Whole Construction (Nov). 6 pp.

Heat Transfer Equipment
See Also: Dimensionless Numbers; Heat; Heat Exchangers; Heat/Moisture Exchangers; Heat Transfer Fluids; Mass Transfer; Reflective Coatings; Thermal Insulation; Thermal Transmittance
—Organic Media—Safety
DIN ENGL 4754-90. Heat Transfer Systems Operating with Organic Media; Safety Requirements and Testing (Dec). 22 pp.
—Pipelines—Thermally Insulated
ISO 9489-91. Thermally Insulated Fibre-Cement Piping Systems First Edition. 12 pp.
—Thermal Insulation—Glossaries
ISO 9251-87. Thermal Insulation—Heat Transfer Conditions and Properties of Materials—Vocabulary First Edition. 6 pp.
ISO 9288-89. Thermal Insulation—Heat Transfer by Radiation—Physical Quantities and Definitions First Edition. 22 pp.

Heat Transfer Fluids
See Also: Heat Transfer Equipment; Heat Transfer Oils
—Dimethyl Silicone
MOD UK DSTAN 91-46-79. Damping Fluids: Dimethyl Silicone Issue 1. 11 pp.
—Ethylene Glycol
CGSB 3-GP-855M-79. Ethylene Glycol, Uninhibited, Standard for. 7 pp.
—Fire Resistant—Oxidation Resistance
DIN ENGL 51373-84. Testing of Fire Resistant Heat Transfer Fluids; Determination of Resistance to Oxidation Including an Assessment of the Catalyst Plates (Mar). 5 pp.
—Lubricants
ISO 6743 Pt 12-89. Lubricants, Industrial Oils and Related Products (Class L)—Classification—Part 12: Family Q (Heat Transfer Fluids) First Edition. 4 pp.

Heat Transfer Oils
See Also: Heat Transfer Fluids
DIN ENGL 51522-82. Mineral Oils and Related Hydrocarbons; Q Heat Transfer Oils; Liquid, Unused Heat Transfer Media; Requirements, Testing (Nov). 3 pp.
—Pressure Switches
DIN ENGL 3398 Pt 4-86. Pressure Cut-Off Switches for Liquid Fuels and Heat Transfer Oils (Oct). 8 pp.

Heat Treated Glass
Use For: Tempered Glass See Also: Glass; Safety Glass
CGSB CAN/CGSB-12.1-M90. Tempered or Laminated Safety Glass. 15 pp.
CNS R2044-85. Tempered Glasses (Nov)(2217). 5 pp.
CNS R3046-85. Method of Test for Tempered Glasses (Nov) (2218).
JIS R 3206-89. Tempered Glasses. 15 pp.
—Aircraft
CEN PREN 3001-91. Tempered Float Glass Plies for Aircraft Applications Technical Specification. 8 pp.
CEN PREN 3001-93. Aerospace Series Tempered Float Glass Plies for Aircraft Applications Technical Specification. 7 pp.

Heat Treated Glass (Cont.)
—Buildings
DIN ENGL 1249 Pt 12-90. Glass for Use in Building Construction; Toughened Glass; Dimensions, Working and Requirements (Sept). 4 pp.
JIS R 3222-90. Heat-Strengthened Glass. 10 pp.
—Patterned
CGSB CAN/CGSB-12.13-M91. Patterned Glass. 7 pp.
—Side Scuttles
JIS F 2410-90. Toughened Safety Glass Panes for Ships' Side Scuttles.
JIS F 2410-55. Tempered Glasses for Ship's Side Scuttles (R 1976). 8 pp.
—Spandrel
CGSB CAN/CGSB-12.9-M91. Spandrel Glass. 9 pp.

Heat Treating Furnaces
See Also: Furnaces; Metals; Steels
—Marine
MOD UK NES 746-01. Approval and Control of Heat Treatment Furnaces Issue 3 (12.88); Amendment 1. 24 pp.
—Temperature
DIN ENGL 17052-85. Temperature Uniformity in Heat Treatment Furnaces; Requirements (Aug). 4 pp.

Heat Treating Oils
JIS K 2242-91. Heat Treating Oils. 16 pp.

Heat Treatment
Scope Note: See the subheadings Heat Treatment or Heat Treated under the specific material or product

Heater Cord Sets
Use: Heater Cords

Heater Cords
See Also: Cords (Electric)
—Electric Plugs—Appliances
CSA C22.2 NO 57-M1985. Appliance Plugs for Heater Cord Sets (R 1993); (Gen Instr 1 Thru 3). 26 pp.

Heater Elements
Use: Heating Elements

Heaters
Scope Note: For additional listings, use a more specific term Use For: Combustion Heaters; Gas Fires; Heating Appliances See Also: Appliances; Baseboard Heaters; Ceiling Heaters; Conduction Heating; Convectors; Duct Heaters; Engine Heaters; Glow Plugs; Heat Pumps; Heating Equipment; Immersion Heaters; Immersion Testing; Oil Heaters; Radiant Heaters; Radiators (Heating); Room Heaters; Space Heaters; Storage Heaters; Storage Water Heaters; Unit Heaters; Water Heaters
JIS B 8411-89. Testing Methods for Air Preheaters. 21 pp.
—Aerostats
CAA Chapter Q6-5 12.79. Combustion Heater Systems (Non-Rigid Airships). 1 p.
—Aircraft
CAA Chapter K6-5 10.69. Combustion Heater Systems (Light Aeroplanes). 1 p.
CAA Chapter P6-5. Combustion Heaters (Provisional Airworthiness Requirements for Civil Powered Lift Aircraft). 5 pp.
—Aircraft—Inspection
CAA NOTICE #41 ISSUE 8. Maintenance of Cockpit and Cabin Combustion Heaters and Their Associated Exhaust Systems (Airworthiness Notices). 2 pp.
—Aircraft—Repair/Maintenance
CAA NOTICE #41 ISSUE 8. Maintenance of Cockpit and Cabin Combustion Heaters and Their Associated Exhaust Systems (Airworthiness Notices). 2 pp.
—Automotive
EC 78/548/EEC-78. Council Directive on the Approximation of the Laws of the Member States Relating to Heating Systems for the Passenger Compartment of Motor Vehicles. 2 pp.
—Bitumens
BSI BS 1676-70. 1970 Heaters for Tar and Bitumen (Mobile and Transportable). 12 pp.
—Electric
CSA C22.2 NO 46-M1988. Electric Air-Heaters. 76 pp.

Heaters (Cont.)
—Electric (Cont.)
CSA 1169 Bull. Electrical Bulletin 1169 June 27, 1978 to C22.2 NO 46. 2 pp.
CSA 1219 Bull. Electrical Bulletin 1219 April 2, 1979 to C22.2 NO 46. 2 pp.
CSA CAN/CSA-C22. 2 NO 64-M91. Household Cooking and Liquid-Heating Appliances; (Gen Instr 1 Thru 2). 91 pp.
SAA AS 1996-77. Special Requirements for Electrically Operated Cooking and Heating Appliances for Marine Use Amdt 1 June 1984 Reconfirmed 1984. 16 pp.
—Electric—Animal Husbandry—Safety
IEC 335 Pt 2-71-93. Safety of Household and Similar Electrical Appliances Part 2: Particular Requirements for Electrical Heating Appliances for Breeding and Rearing Animals First Edition. 40 pp.
—Electric—Commercial
CSA 1036 Bull. Electrical Bulletin 1036 December 16, 1975 to C22.2 NO 46. 2 pp.
—Electric—Floors
BSI CP 1018-71. 1971 Electric Floor-Warming Systems for Use with Off-Peak and Similar Supplies of Electricity. 34 pp.
—Electric—Foot
CSA C22.2 NO 200-M1985. Electric Waterbed Heater Systems (R 1992); (Gen Instr 1 Thru 2). 31 pp.
—Electric—Forced Air
CSA C22.2 NO 46-M1988. Electric Air-Heaters. 76 pp.
CSA 1063A Bull. Electrical Bulletin 1063A June 7, 1977 to C22.2 NO 46. 4 pp.
CSA 1169 Bull. Electrical Bulletin 1169 June 27, 1978 to C22.2 NO 46. 2 pp.
—Electric—Laboratory Equipment—Safety
BSI BS 7687: Sec 2.10-93. 1993 Safety Requirements for Electrical Equipment for Measurement, Control and Laboratory Use Part 2: Particular Requirements Section 2.10: Specification for Laboratory Equipment for the Heating of Materials (IEC 1010-2-010: 1992) (S). 21 pp.
IEC 1010 Pt 2-010-92. Safety Requirements for Electrical Equipment for Measurement, Control, and Laboratory Use Part 2-010: Particular Requirements for Laboratory Equipment for the Heating of Materials First Edition. 38 pp.
—Electric—Residential
CSA 1036 Bull. Electrical Bulletin 1036 December 16, 1975 to C22.2 NO 46. 2 pp.
DIN VDE 0720 Pt 1-72. Specification for Electrical Heating Appliances for Household and Similar Applications Part 1: General Provisions (Feb) (This Document is Not a Complete Translation and Must Only Be Used in Conjunction with the Following Standard CEE 11 (1964)). 10 pp.
DIN VDE 0720 Pt 1 AE-80. Electric Heating Appliances for Domestic and Similar Purposes; General Regulations: Amendment E (Mar). 10 pp.
—Electric—Residential—Safety
CNS C4125-90. Safety Testing Code for Household Electric Heaters (Jan)(3765).
CSA C22.2 NO 1335.1-93. Portable Electrical Motor-Operated and Heating Appliances: General Requirements; (Gen Instr 1). 208 pp.
—Electric—Safety
BSI BS 3456: Sec 2.22-72. 1972 Amd 4 Safety of Household Electrical Appliances Part 2: Particular Requirements Section 2.22: Safety of Household Electrical Appliances (AMD 6226) February 28, 1990. 38 pp.
CNS C4125-86. Safety Testing Code for Household Electric Heaters (Nov)(3765). 3 pp.
—Electric—Saunas
CSA C22.2 NO 46-M1988. Electric Air-Heaters. 76 pp.
CSA 1169 Bull. Electrical Bulletin 1169 June 27, 1978 to C22.2 NO 46. 2 pp.
CSA CAN/CSA-C22. 2 NO 164-M91. Electric Sauna Heating Equipment; (Gen Instr 1). 67 pp.
—Electric—Saunas—Acceptance Testing
SAA AS 3316-92. Approval and Test Specification—Particular Requirements for Electric Sauna Heating Appliances (NZS 6353:1992) (in Professional Package 28). 20 pp.
SNZ NZS 6353-92. Approval and Test Specification—Particular Requirements for Electric Sauna Heating Appliances (NZS 6353:1992/AS 3316-1992). 20 pp.
—Electric—Saunas—Safety
BSI BS EN 60335-2-53-91. 1991 Safety of Household and Similar Electrical Appliances Part 2: Particular Requirements Section 2.53: Electric Sauna Heating Appliances. 30 pp.

INTERNATIONAL AND NON-U.S. NATIONAL STANDARDS
SUBJECT INDEX

Heating

Heaters (Cont.)
—**Electric—Saunas—Safety** (Cont.)
CENELEC EN 60 335-2-53-91. Safety of Household and Similar Electrical Appliances Part 2: Particular Requirements for Electric Sauna Heating Appliances. 7 pp.
IEC 335 Pt 2-53-88. Safety of Household and Similar Electical Appliances Part 2: Particular Requirements for Electric Sauna Heating Appliances First Edition. 44 pp.

—**Electric—Thermal Cutoffs**
CSA C22.2 NO 209-M1985. Thermal Cut-Offs (R 1992); (Gen Instr 1 Thru 2). 28 pp.

—**Electric—Waterbeds**
CSA C22.2 NO 200-M1985. Electric Waterbed Heater Systems (R 1992); (Gen Instr 1 Thru 2). 31 pp.

—**Electric—Waterbeds—Safety**
IEC 335 Pt 2-66-93. Safety of Household and Similar Electrical Appliances Part 2: Particular Requirements for Water-Bed Heaters First Edition. 35 pp.

—**Electrical Codes**
CSA 1084 Bull. Electrical Bulletin 1084 December 3, 1976 to C22.2 NO 46. 2 pp.

—**Fans**
CSA CAN/CSA-C22. 2 NO 236-M90. Heating and Cooling Equipment; (Gen Instr 1) (ANSI/UL 1995). 136 pp.

—**Fire Guards**
BSI BS 6778-86. 1986 Fireguards for Use with Portable Free-Standing or Wall-Mounted Heating Appliances. 11 pp.
BSI PD 6516-87. 1987 Guarding Fires and Heating Appliances. 8 pp.

—**Fire Hazards**
BSI BS 6458: Sec 2.3-85. 1985 Fire Hazard Testing for Electrotechnical Products Part 2: Methods of Test Section 2.3: Bad-Connection Test with Heaters. 20 pp.
IEC 695 Pt 2-1-91. Fire Hazard Testing Part 2: Test Methods Section 1: Glow-Wire Test and Guidance Second Edition. 28 pp.
IEC 695 Pt 2-2-91. Fire Hazard Testing Part 2: Test Methods Section 2—Needle-Flame Test Second Edition; (CENELEC HD 444.2.2 S2:1992). 21 pp.
IEC 695 Pt 2-3-84. Fire Hazard Testing Part 2: Test Methods Bad-Connection Test with Heaters First Edition. 33 pp.
JIS C 0062-87. Fire Hazard Testing Part 2: Test Methods Bad-Connection Test with Heaters. 17 pp.

—**Fired—Pipe Fittings—Steel**
JIS B 2312-91. Steel Butt-Welding Pipe Fittings. 36 pp.

—**Gas—Residential—Installation**
BSI BS 5864-89. 1989 Installation in Domestic Premises of Gas-Fired Ducted-Air Heaters of Rated Input Not Exceeding 60 kW. 16 pp.
BSI BS 5864-80. (WITHDRAWN) 1980 Amd 1 Code of Practice for the Installation of Gas-Fired Ducted-Air Heaters of Rated Input not Exceeding 60kW (2nd Family Gases) (Formerly CP 332: Part 4). 12 pp.

—**Gas—Safety**
BSI BS 5258: Part 4-87. 1987 Safety of Domestic Gas Appliances Part 4: Fanned-Circulation Ducted-Air Heaters. 28 pp.
BSI BS 5258: Part 5-89. 1989 Safety of Domestic Gas Appliances Part 5: Gas Fires. 27 pp.
BSI BS 5258: Part 5-75. 1975 Amd 2 Safety of Domestic Gas Appliances Part 5: Gas Fires (AMD 4745) March 29, 1985. 37 pp.
BSI BS 5258: Part 8-80. 1980 Safety of Domestic Gas Appliances Part 8: Combined Appliances: Gas Fire/Back Boiler. 24 pp.
BSI BS 5258: Part 9-89. 1989 Safety of Domestic Gas Appliances Part 9: Combined Appliances: Fanned-Circulation Ducted-Air Heaters/Circulators. 32 pp.

—**Gas—Thermal Efficiency**
BSI BS 6332: Part 2-83. 1983 Thermal Performance of Domestic Gas Appliances Part 2: Thermal Performance of Gas Fires. 14 pp.
BSI BS 6332: Part 3-84. 1984 Thermal Performance of Domestic Gas Appliances Part 3: Specification for Thermal Performance of Combined Appliances: Gas Fire/Back Boilers. 15 pp.
BSI BS 6332: Part 5-86. 1986 Thermal Performance of Domestic Gas Appliances Part 5: Thermal performance of Fanned-Circulation Ducted-Air Heaters. 12 pp.
CEN PREN 442-91. Radiators, Convectors and Similar Appliances—Testing and Rating Standard. 83 pp.

Heaters (Cont.)
—**Identification Systems**
DIN ENGL 2404-42. Identification Colour Code for Heating System Pipelines (Dec). 1 p.

—**Industrial Buildings—Installation**
BSI BS 6230-91. 1991 Installation of Gas-Fired Forced Convection Air Heaters for Commercial and Industrial Space Heating (2nd Family Gases). 27 pp.

—**Installation**
BSI BS 5871-80. (WITHDRAWN) 1980 Amd 2 Installation of Gas Fires, Convectors and Fire/Back Boilers (2nd Family Gas) (AMD 4638) March 29, 1985 (Superseded by BS 5871: Parts 1&2: 1991). 22 pp.
BSI BS 5871: Part 1-91. 1991 Installation of Gas Fires, Convector Heaters, Fire/Back Boilers and Decorative Fuel Effect Gas Appliances Part 1: Gas Fires, Convector Heaters and Fire/Back Boilers (1st, 2nd and 3rd Family Gases). 32 pp.
BSI BS 5871: Part 2-91. 1991 Installation of Gas Fires, Convector Heaters, Fire/Back Boilers and Decorative Fuel Effect Gas Appliances Part 2: Inset Live Fuel Effect Gas Fires of Heat Input Not Exceeding 15 kW (2nd and 3rd Family Gases). 22 pp.
BSI BS 8303-86. 1986 Amd 1 Code of Practice for Installation of Domestic Heating and Cooking Appliances Burning Solid Mineral Fuels (AMD 5723) October 30, 1987. 27 pp.

—**Installation—Construction Contracts**
DIN ENGL 18380-90. Tendering and Performance Stipulations in Contracts for Construction Works (VOB); Part C: General Technical Specifications in Contracts for Construction Works (ATV); Installation of Central Heating Systems and Hot Water Supply. 10 pp.

—**Oil**
BSI BS 4256: Part 2-72. 1972 Oil-Burning Air Heaters Part 2: Fixed, Flued, Fan-Assisted Heaters. 13 pp.

—**Oil—Mobile Homes**
CSA B140.10-1974. Oil-Fired Warm Air Heating Appliances for Mobile Housing and Recreational Vehicles (R 1991); (Rev 1-2 September 1977) (Amd 3 June 1988). 47 pp.

—**Oil—Recreational Vehicles**
CSA B140.10-1974. Oil-Fired Warm Air Heating Appliances for Mobile Housing and Recreational Vehicles (R 1991); (Rev 1-2 September 1977) (Amd 3 June 1988). 47 pp.

—**Recreational Vehicles**
BSI BS 6762: Part 3-89. 1989 Amd 1 Services for Leisure Accommodation Vehicles and Transportable Accommodation Units Part 3: Specification for Oil-Fired Heating Systems (ISO 7420: 1987) (AMD 6667) April 30, 1991. 5 pp.

—**Safety**
BSI BS 5990-90. 1990 Amd 1 Direct Gas-Fired Forced Convection Air Heaters with Rated Heat Inputs up to 2 MW for Ind. and Commercial Space Heating Safety and Performance Requirements (Excl. Electrical Reqmts.) (2nd Family Gases) (AMD 6649) October 31, 1990. 26 pp.
BSI BS 5991-89. 1989 Indirect Gas Fired Forced Convection Air Heaters with Rated Heat Inputs up to 2MW for Industrial and Commercial Space Heating: Safety and Performance Requirements (Excluding Electrical Requirements) (2nd Family Gases). 28 pp.
BSI BS 5991-83. (WITHDRAWN) 1983 Amd 1 Indirect Gas Fired Forced Convection Air Heaters for Space Heating (60 kW up to 2MW Input Safety and Performance Requirements (Excluding Electrical Requirements) (2nd Family Gases). 32 pp.

—**Self Propelled Machinery—Cabs**
ISO 6097-89. Tractors and Self-Propelled Machines for Agriculture—Performance of Heating and Ventilation Systems in Closed Cabs—Test Method Second Edition. 10 pp.

—**Solid Fuel**
BSI BS 4834-72. (WITHDRAWN) 1972 Amd 1 Inset Open Fires Without Convection. 39 pp.
CSA CAN/CSA-B366.1-M91. Solid-Fuel-Fired Central Heating Appliances; (Gen Instr 1). 70 pp.

—**Solid Fuel—Fire Guards**
BSI BS 6539-84. 1984 Amd 1 Fireguards for Use with Solid Fuel Appliances. 13 pp.
SNZ NZS/BS 6539-84. Specification for Fireguards for Use with Solid Fuel Appliances Amend: 1. 12 pp.

—**Solid Fuel—Residential—Clearances**
ISO 10774-92. Solid Fuelled Heaters—Test Method for Determining Allowable Clearances from Combustible Surfaces First Editon. 26 pp.

Heaters (Cont.)
—**Symbols—Engineering Drawings**
ISO 4067 Pt 1-84. Technical Drawings—Installations—Part 1: Graphical Symbols for Plumbing, Heating, Ventilation and Ducting First Edition. 12 pp.

—**Tars**
BSI BS 1676-70. 1970 Heaters for Tar and Bitumen (Mobile and Transportable). 12 pp.

—**Tractors—Cabs**
ISO 6097-89. Tractors and Self-Propelled Machines for Agriculture—Performance of Heating and Ventilation Systems in Closed Cabs—Test Method Second Edition. 10 pp.

Heating, Ventilating and Air Conditioning Equipment
Use For: Environmental Systems (HVAC); HVAC
See Also: Air Cleaners; Air Conditioners; Air Handling Equipment; Boilers; Cooling Systems; Fan Coil Units; Furnaces; Heating Equipment; Heating Plants; Humidifiers; Refrigerant Compressors; Ventilation Equipment

—**Ducts—Construction**
SNZ NZS/SMACNA 15D-85. HVAC Duct Construction Standards-Metal and Flexible First Edition. 216 pp.

—**Health Care Facilities**
CSA CAN/CSA-Z317.2-M91. Special Requirements for Heating, Ventilation, and Air Conditioning (HVAC) Systems in Health Care Facilities; (Gen Instr 1). 53 pp.

—**Hospitals**
DIN ENGL 1946 Pt 4-89. Heating, Ventilation and Air Conditioning; HVAC Systems in Hospitals; (VDI Code of Practice) (Dec). 24 pp.

—**Operating Rooms**
DIN ENGL 4799-90. Heating, Ventilation and Air-Conditioning; Testing of Air Distributions Systems Serving Operating Theatres (June). 14 pp.

—**Residential Buildings**
CSA CAN/CSA-F326-M91. Residential Mechanical Ventilation Systems; (Gen Instr 1). 97 pp.

—**Temperature Controllers**
IEC 730 Pt 2-9-92. Automatic Electrical Controls for Household and Similar Use Part 2: Particular Requirements for Temperature Sensing Controls First Edition. 52 pp.

—**Temperature Controllers—Safety**
IEC 730 Pt 2-9-92. Automatic Electrical Controls for Household and Similar Use Part 2: Particular Requirements for Temperature Sensing Controls First Edition. 52 pp.

Heating Appliances
Use: Heaters

Heating Blankets
Use For: Heat Blankets *See Also:* Electric Blankets; Heating Cables; Heating Equipment; Heating Pads

—**Safety**
BSI BS EN 60967-91. 1991 Safety of Electrically Heated Blankets, Pads and Similar Flexible Heating Appliances for Household Use. 123 pp.
CENELEC EN 60 967-90. Safety of Electrically Heated Blankets, Pads and Similar Flexible Heating Appliances for Household Use. 13 pp.
CENELEC EN 60967-92. Safety of Electrically Heated Blankets, Pads and Similar Flexible Heating Appliances for Household Use. 3 pp.
IEC 967-88. Safety of Electrically Heated Blankets, Pads and Similar Flexible Heating Appliances for Household Use First Edition; (Amendment 1-1991). 260 pp.

Heating Boilers
Use: Boilers

Heating Cable Sets
Use: Heating Cables

Heating Cables
See Also: Cables (Electric); Heating Blankets; Resistance Wire
IEC 800-92. Heating Cables with a Rated Voltage of 300/500 V for Comfort Heating and Prevention of Ice Formation Second Edition. 91 pp.

—**Commercial**
CSA C22.2 NO 130.2-93. Heat Cable Systems for Use in Other Than Industrial Establishments; (Gen Instr 1) (Supercedes C22.2 NO 130-M1985). 43 pp.

INDUSTRY STANDARDS

Heating Cables (Cont.)

—Electric Tracing Equipment—Pipes—Industrial—0-600 V
CSA CAN/CSA-C22. 2NO130.1-M90. Heat-Tracing Cable Systems for Use in Industrial Locations; (Gen Instr 1). 32 pp.

—Electric Tracing Equipment—Tanks (Containers)—Industrial—0-600 V
CSA CAN/CSA-C22. 2NO130.1-M90. Heat-Tracing Cable Systems for Use in Industrial Locations; (Gen Instr 1). 32 pp.

—Heat Tracing Systems—Gaseous Mines
CSA C22.2 NO 138-M1989. Heat Tracing Cable and Cable Sets for Use in Hazardous Locations; (Gen Instr 1). 25 pp.

—Heat Tracing Systems—Pipes—Hazardous Locations
CSA C22.2 NO 138-M1989. Heat Tracing Cable and Cable Sets for Use in Hazardous Locations; (Gen Instr 1). 25 pp.

—Heat Tracing Systems—Tanks (Containers)—Hazardous Locations
CSA C22.2 NO 138-M1989. Heat Tracing Cable and Cable Sets for Use in Hazardous Locations; (Gen Instr 1). 25 pp.

—Ice Prevention
IEC 800-92. Heating Cables with a Rated Voltage of 300/500 V for Comfort Heating and Prevention of Ice Formation Second Edition. 91 pp.

—Residential
CSA C22.2 NO 130.2-93. Heat Cable Systems for Use in Other Than Industrial Establishments; (Gen Instr 1) (Supercedes C22.2 NO 130-M1985). 43 pp.

Heating Circuits
See Also: Circuits

—Dry Type Transformers
CSA CAN/CSA-C22. 2 NO 47-M90. Air-Cooled Transformers (Dry Type); (Gen Instr 1 Thru 2). 32 pp.
CSA 1169 Bull. Electrical Bulletin 1169 June 27, 1978 to C22.2 NO 47. 2 pp.

Heating Elements
Use For: Electric Heating Elements; Heater Elements
See Also: Heating Equipment

CNS C4091-84. Electric Heating Wire and Ribbon (May)(2962).
JIS C 2520-86. Wires and Rolled Wires for Electrical Heating. 23 pp.

—Cartridge
CSA C22.2 NO 72-M1984. Heater Elements (R 1992); (Gen Instr 1 Thru 4). 30 pp.

—Evaporative Coolers—Motor Operated
CSA C22.2 NO 104-M1983. Humidifiers and Evaporative Coolers; (Gen Instr 1 Thru 2). 37 pp.

—Humidifiers—Motor Operated
CSA C22.2 NO 104-M1983. Humidifiers and Evaporative Coolers; (Gen Instr 1 Thru 2). 37 pp.

—Installation
JIS C 3651-87. Installation Methods of Heating Facilities. 35 pp.

—Life (Durability)
CNS C3030-75. Method of Life for Electric Heating Wires and Bands (Feb)(2963).
JIS C 2524-79. Accelerated Life Test of Electric Heating Wires and Rolled Wires.
JIS C 2524-65. Methods of Life Test for Electric Heating Wires and Bands (R 1971). 6 pp.

—Mat
CSA C22.2 NO 72-M1984. Heater Elements (R 1992); (Gen Instr 1 Thru 4). 30 pp.

—Metal—Sheathed
BSI BS 7351-90. 1990 Metal-Sheathed Heating Elements for Industrial Use. 11 pp.
CSA C22.2 NO 72-M1984. Heater Elements (R 1992); (Gen Instr 1 Thru 4). 30 pp.

—Open Coil—Air Preheaters—Portable
CSA C22.2 NO 46-M1988. Electric Air-Heaters. 76 pp.

—Quartz Tube
CSA C22.2 NO 72-M1984. Heater Elements (R 1992); (Gen Instr 1 Thru 4). 30 pp.

—Ribbon
CSA C22.2 NO 72-M1984. Heater Elements (R 1992); (Gen Instr 1 Thru 4). 30 pp.

Heating Elements (Cont.)

—Rope
CSA C22.2 NO 72-M1984. Heater Elements (R 1992); (Gen Instr 1 Thru 4). 30 pp.

—Silicon Carbide
CNS R2183-85. Silicon Carbide Electric Heating Element (Nov) (11417).
JIS R 7501-78. Silicon Carbide Electric Heating Element.

—Strip
CSA C22.2 NO 72-M1984. Heater Elements (R 1992); (Gen Instr 1 Thru 4). 30 pp.

—Thermoplastic Insulated—Heating Pads
CSA CAN/CSA-C22. 2 NO 15-M91. Electrically Heated Warming Pads; (Gen Instr 1 Thru 3). 45 pp.
CSA 1169 Bull. Electrical Bulletin 1169 June 27, 1978 to C22.2 NO 15. 2 pp.

—Variable Resistance
CSA C22.2 NO 72-M1984. Heater Elements (R 1992); (Gen Instr 1 Thru 4). 30 pp.

—Waterbeds
CSA C22.2 NO 200-M1985. Electric Waterbed Heater Systems (R 1992); (Gen Instr 1 Thru 2). 31 pp.

—Wire
CSA C22.2 NO 72-M1984. Heater Elements (R 1992); (Gen Instr 1 Thru 4). 30 pp.

Heating Equipment
Scope Note: For additional listings, use a more specific term *See Also:* Air Handling Equipment; Air Preheaters; Boilers; Conduction Heating; Convectors; Dampers; Dielectric Heating; Electric Blankets; Environmental Engineering; Fireplaces; Furnaces; Heat Exchangers; Heaters; Heating, Ventilating and Air Conditioning Equipment; Heating Blankets; Heating Elements; Heating Pads; Induction Heating Equipment; Infrared Heating Equipment; Microwave Heating Equipment; Oil Heaters; Ovens; Plasma Arc Heating Equipment; Radiators (Heating); Resistance Heaters; Solar Heating and Cooling Equipment; Solid Fuel Burning Equipment; Space Heaters; Stoves; Thermal Insulation; Water Heaters

BSI BS 3456: Sec 2.26-73. 1973 Amd 4 1972-1981 Edition. Specification for Safety of Household Electrical Appliances Part 2: Section 2.26: Thermal Storage Electrical Room Heaters. 21 pp.

—Construction
CSA CAN/CSA-C22. 2 NO 236-M90. Heating and Cooling Equipment; (Gen Instr 1) (ANSI/UL 1995). 136 pp.

—Electric
BSI BS 6351: Part 1-83. 1983 Electric Surface Heating Part 1: Specification for Electric Surface Heating Devices. 19 pp.

—Electric—Design
BSI BS 6351: Part 2-83. 1983 Electric Surface Heating Part 2: Guide to the Design of Electric Surface Heating Systems. 24 pp.

—Electric—Installation
BSI BS 6351: Part 3-83. 1983 Electric Surface Heating Part 3: Code of Practice of the Installation, Testing and Maintenance of Electric Surface Heating Systems. 18 pp.

—Electrical Insulation—Mica
CENELEC HD 352.3.3 S1-89. Specification for Insulating Materials Based on Mica-Part 3: Specifications for Individual Materials. Sheet Three: Specifications for Rigid Mica Materials for Heating Equipment. 3 pp.
IEC 371 Pt 3-3-83. Specification for Insulating Materials Based on Mica Part 3: Specifications for Individual Materials Sheet 3: Specification for Rigid Mica Materials for Heating Equipment First Edition. 13 pp.

—Evaporative Coolers—Motor Operated
CSA C22.2 NO 104-M1983. Humidifiers and Evaporative Coolers; (Gen Instr 1 Thru 2). 37 pp.

—Geothermal
SNZ NZS 2402P-87. Code of Practice for Geothermal Heating Equipment in Rotorua. 60 pp.

—Glossaries
DIN ENGL 1946 Pt 1-88. Heating, Ventilation and Air Conditioning; Terminology and Graphical Symbols (VDI Code of Practice) (Oct). 38 pp.

—Heat Treatment—Effective Working Zone
JIS B 6901-87. Test Methods for Effective Working Zone of Heating Equipment for Metals Heat Treatment Use. 14 pp.

Heating Equipment (Cont.)

—Humidifiers—Motor Operated
CSA C22.2 NO 104-M1983. Humidifiers and Evaporative Coolers; (Gen Instr 1 Thru 2). 37 pp.

—Industrial—Electric
CENELEC HD 353-76. General Test Conditions for Industrial Electro-Heating Equipment. 4 pp.
DIN VDE 0721 Pt 1 & 1B (S)-75. Regulations for Industrial Heating Equipment (Apparatus and Accessories); General Regulations (Nov) (Superseded by VDE 0721 Part 911). 35 pp.
DIN VDE 0721 Pt 2-75. Regulations for Industrial Heating Equipment (Apparatus and Accessories) (Nov). 8 pp.
IEC 398-72. General Test Conditions for Industrial Electro-Heating Equipment First Edition. 14 pp.

—Industrial—Electric—Construction
CSA C22.2 NO 88-1958. Construction and Test of Industrial Heating Equipment (R 1992). 22 pp.
CSA 1169 Bull. Electrical Bulletin 1169 June 27, 1978 to C22.2 NO 88. 2 pp.

—Industrial—Electric—Portable—Construction
CSA C22.2 NO 88-1958. Construction and Test of Industrial Heating Equipment (R 1992). 22 pp.
CSA 1169 Bull. Electrical Bulletin 1169 June 27, 1978 to C22.2 NO 88. 2 pp.

—Industrial—Electric—Portable—Testing
CSA C22.2 NO 88-1958. Construction and Test of Industrial Heating Equipment (R 1992). 22 pp.
CSA 1169 Bull. Electrical Bulletin 1169 June 27, 1978 to C22.2 NO 88. 2 pp.

—Industrial—Electric—Testing
CSA C22.2 NO 88-1958. Construction and Test of Industrial Heating Equipment (R 1992). 22 pp.
CSA 1169 Bull. Electrical Bulletin 1169 June 27, 1978 to C22.2 NO 88. 2 pp.

—Industrial—Glossaries
BSI BS 4727:Pt2: Group 10-85. 1985 Electrotechnical, Power, Telecommunication, Electronics, Lighting and Colour Terms Part 2: Terms Particular to Power Engineering Group 10: Industrial Electroheating Terminology. 36 pp.
IEC 50 Chap 841-83. International Electrotechnical Vocabulary Chapter 841: Industrial Electroheating. 166 pp.
JIS B 0113-89. Glossary of Terms-Relating to Industrial Combustion Equipments.
SNZ IEC 50: 50(841)-83. International Electrotechnical Vocabulary 50(841): Industrial Electroheating. 154 pp.

—Industrial—Safety
CENELEC HD 491.1 S1-88. Safety in Electroheat Installations—Part 1: General Requirements. 3 pp.
DIN ENGL 4756-86. Gas Burning Equipment in Heating Plant; Safety Requirements (Feb). 8 pp.
DIN VDE 0721 Pt 911-88. Industrial Electroheat Installations; General Safety Regulations (June) (Supersedes VDE 0721 Part 1/11.75 and VDE 0721 Part 1b/03.78). 36 pp.
IEC 519 Pt 1-84. Safety in Electroheat Installations Part I: General Requirements Second Edition. 42 pp.

—Industrial—Triodes
CECC CECC 45 002 ISSUE 1-75. Blank Detail Specification: Industrial Heating Triodes (En, Fr, Ge). 21 pp.

—Recreational Vehicles—Installation
BSI BS 5601: Part 2-78. (WITHDRAWN) 1978 Code of Practice for Ventilation and Heating of Caravans Part 2: Installation of Solid Fuel Fired Heating Appliances (Superseded by BS 6762: Part 2: 1991). 8 pp.
BSI BS 5601: Part 3-79. (WITHDRAWN) 1979 Code of Practice for Ventilation and Heating of Caravans Part 3: Installation of Oil Fired Heating Appliances (Superseded by BS 6762: Part 3: 1989). 12 pp.
BSI BS 5601: Part 4-80. 1980 Code of Practice for Ventilation and Heating of Caravans Part 4: Installation of Electrical Heating Appliances. 4 pp.
BSI BS 6762: Part 2-91. 1991 Services for Leisure Accommodation Vehicles and Transportable Accommodation Units Part 2: Code of Practice for the Installation of Solid Fuel Fired Heating in Park Homes and Transportable Accommodation Units. 17 pp.

—RFI
BSI BS 4809-72. (WITHDRAWN) 1972 Radio Interference Limits and Measurements for Radio Frequency Heating Equipment (Superseded by BS EN 55011: 1991). 17 pp.

INTERNATIONAL AND NON-U.S. NATIONAL STANDARDS
SUBJECT INDEX

Heating Equipment (Cont.)
—Rolling Stock
JIS E 4015-89. Measuring Methods for Air Conditioning and Herting Temperature of Railway Rolling Stock.

—Safety
BSI BS 3456: Sec 202.43-90. 1990 Safety of Household Electrical Appliances Part 202: Particular Requirements Section 202.43: Clothes Dryers and Towel Rails. 20 pp.
BSI BS 3456: Sec 202.43-01. 1990 Amd 1 Safety of Household and Similar Electrical Appliances Part 202: Particular Requirements Section 202.43: Clothes Dryers and Towel Rails (AMD 6901) April 1, 1992 (SUPERSEDES BS 3456 SECTION 2.9: 1970 & 3. 6: 1979). 26 pp.
BSI BS 3456: Sec 202.43-02. 1990 Amd 2 Safety of Household and Similar Electrical Appliances Part 202: Particular Requirements Section 202.43: Clothes Dryers and Towel Rails (AMD 7550) JAN 15, 1993 SUPERSEDES BS 3456 SECTION 2. 9: 1970 & 3. 6: 1979. 31 pp.
BSI BS 3456: Sec A1-66. (WITHDRAWN) 1966 Amd 4 Safety of Household Electrical Appliances Part A: Heating and Cooking Appliances Section A1: General Requirements. 57 pp.
BSI BS 3456: Sec A4-71. (WITHDRAWN) 1971 Amd 9 Safety of Household Electrical Appliances Part A: Heating and Cooking Appliances Section A4: Electrically Heated Blankets (Withdrawn, superseded by BS En 60967: 1991). 67 pp.
BSI BS EN 60335-2-45-91. 1991 Safety of Household and Similar Electrical Appliances Part 2: Particular Requirements Section 2.45: Portable Electric Heating Tools and Similar Appliances. 23 pp.
BSI BS EN 60335-2-45-01. 1991 Amd 1 Safety of Household and Similar Electrical Appliances Part 2: Particular Requirements Section 2.45: Portable Electric Heating Tools And Similar Appliances. 44 pp.
CENELEC HD 251-75. General Specification for Electric Cooking and Heating Appliances for Household and Similar Purposes. 9 pp.
CENELEC HD 251.2-78. General Specification for Electric Cooking and Heating Appliances for Household and Similar Purposes. 3 pp.
CENELEC EN 60 335-2-43-89. Safety of Household and Similar Electrical Appliances—Part 2: Particular Requirements for Clothes Dryers and Towel Rails. 7 pp.
CENELEC EN 60 335-2-43/A1-90. AMD 1 Safety of Household and Similar Electrical Appliances—Part 2: Particular Requirements for Clothes Dryers and Towel Rails. 6 pp.
CENELEC EN 60335-2-43/A51-92. AMD 51 Safety of Household and Similar Electrical Appliances—Part 2: Particular Requirements for Clothes Dryers and Towel Rails. 4 pp.
CENELEC EN 60 335-2-45-90. Safety of Household and Similar Electrical Appliances—Part 2: Particular Requirements for Portable Electric Heating Tools and Similar Appliances. 11 pp.
CENELEC EN 60335-2-45/PRAB-93. AMD AB Safety of Household and Similar Electric Appliances. Part 2: Particular Requirements for Portable Electric Heating Tools and Similar Appliances—Sub-Clauses 8.1 and 22.107. 2 pp.
CENELEC EN 60335-2-45/A1-92. AMD 1 Safety of Household and Similar Electrical Appliances Part 2: Particular Requirements for Portable Electric Heating Tools and Similar Appliances (IEC 335-2-45:1986/A1:1990). 5 pp.
DIN VDE 0700 Pt 30-83. Test Certificate, Safety Test of Household & Similar Electrical Appliances. Portable Electric Heating Tools (Jan). 26 pp.
DIN VDE 0700 Pt 30 A1 (D)-84. Safety of Household and Similar Electrical Appliances; Particular Specifications for Portable Electric Heating Tools (VDE Specification): Amendment 1 (Nov). 19 pp.
IEC 335 Pt 2-43-84. Safety of Household and Similar Electrical Appliances Part 2: Particular Requirements for Clothes Dryers and Towel Rails First Edition; (Amendment 1-1988). 35 pp.
IEC 335 Pt 2-45-86. Safety of Household and Similar Electrical Appliances Part 2: Particular Requirements for Portable Electric Heating Tools and Similar Appliances First Edition; (Amendment 1-1990) (CENELEC EN 60335-2-45/A1:1992). 48 pp.

—Ships
IEC 92 Pt 307-80. Electrical Installations in Ships Part 307: Equipment—Heating and Cooking Appliances Third Edition. 18 pp.
JIS F 8070-86. Electrical Installations in Ships Part 307 Equipment—Heating and Cooking Appliances (IEC 92-307-1980).
MOD UK NES 596-89. Preferred Range of Heating and Thermostatic Components Issue 3 (06.89). 133 pp.

Heating Equipment (Cont.)
—Submarines
MOD UK NES 596-89. Preferred Range of Heating and Thermostatic Components Issue 3 (06.89). 133 pp.

—Symbols
DIN ENGL 1946 Pt 1-88. Heating, Ventilation and Air Conditioning; Terminology and Graphical Symbols (VDI Code of Practice) (Oct). 38 pp.

—Temperature
BSI BS 4086-66. 1966 Amd 1 Recommendations for Maximum Surface Temperatures of Heated Domestic Equipment. 13 pp.

—Temperature Controllers
CSA C22.2 NO 24-93. Temperature-Indicating and-Regulating Equipment; (Gen Instr 1). 88 pp.
DIN ENGL 3440-84. Temperature Control and Limiting Devices for Heat Generating Systems; Safety Requirements and Testing (July). 15 pp.

—Temperature Indicators
CSA C22.2 NO 24-93. Temperature-Indicating and-Regulating Equipment; (Gen Instr 1). 88 pp.

—Thermostats
CSA C22.2 NO 24-93. Temperature-Indicating and-Regulating Equipment; (Gen Instr 1). 88 pp.

Heating Fuels
Use: Heating Oils

Heating Oils
See Also: Fuel Oils; Oils; Petroleum Products
CGSB CAN/CGSB-3.2-M89. Fuel Oil, Heating. 12 pp.
DIN ENGL 51603 Pt 1-88. Liquid Fuels; EL Domestic Fuel Oils; Minimum Requirements (Mar). 3 pp.

—Cold Filter Plugging Point
CEN EN 116-81. Determination of Cold Filter Plugging Point of Diesel and Domestic Heating Fuels. 13 pp.

—Plugging Point
BSI BS 6188-82. 1982 Determination of Cold Filter Plugging Point of Diesel and Domestic Heating Fuels. 14 pp.

Heating Pads
See Also: Heating Blankets; Heating Equipment
CNS C3065-87. Method of Test for Electric Warm Pad (Aug)(4874).
CNS C4149-87. Electric Warm Pad (Aug)(4873). 12 pp.

—Electric—Household
CSA CAN/CSA-C22. 2 NO 15-M91. Electrically Heated Warming Pads; (Gen Instr 1 Thru 3). 45 pp.
CSA 1169 Bull. Electrical Bulletin 1169 June 27, 1978 to C22.2 NO 15. 2 pp.

—Electric—Medical
CSA CAN/CSA-C22. 2 NO 15-M91. Electrically Heated Warming Pads; (Gen Instr 1 Thru 3). 45 pp.
CSA 1169 Bull. Electrical Bulletin 1169 June 27, 1978 to C22.2 NO 15. 2 pp.

—Heating Elements
CSA CAN/CSA-C22. 2 NO 15-M91. Electrically Heated Warming Pads; (Gen Instr 1 Thru 3). 45 pp.
CSA 1169 Bull. Electrical Bulletin 1169 June 27, 1978 to C22.2 NO 15. 2 pp.

Heating Panels
Use: Radiators (Heating)

Heating Plants
Use For: Electroheating Installations
See Also: Heating, Ventilating and Air Conditioning Equipment

—Electron Guns
BSI BS 7689-93. 1993 Test Methods for Electroheating Installations with Electron Guns (IEC 703: 1981) (F). 11 pp.

—Glossaries
SAA AS 1852.841-88. International Electrotechnical Vocabulary—Part 841: Industrial Electroheating. 154 pp.

Heating Tools
Use: Heating Equipment

Heating Value
Use: Calorific Value

Heavier-Than-Air Craft
Use: Aircraft

Heavy Metals
See Also: Lead; Metals

—Soldering Fluxes
DIN ENGL 8511 Pt 2-88. Fluxes for Brazing and Soldering; Fluxes for Soldering (May). 5 pp.
DIN ENGL 8527 Pt 1-70. Fluxes for Soft Soldering Heavy Metals; Testing (June). 4 pp.

Heavy Metals Content Analysis
—Acetic Acid—Visual Limit Testing
ISO 753 Pt 10-81. Acetic Acid for Industrial Use—Methods of Test—Part 10: Visual Limit Test for Heavy Metals (Including Iron) First Edition. 4 pp.

—Aluminum Sulfate
CNS K6163-74. Method of Test for Aluminium Sulfate (for Water Works) (Jun)(2075). 4 pp.

—Formaldehyde—Visual Limit Testing
ISO 2223-72. Formaldehyde Solutions for Industrial Use—Limit Test for Heavy Metals (Excluding Iron) First Edition. 4 pp.

—Glucose
CNS N6055-72. Method of Test for Powdered Glucose (Jun)(3351). 3 pp.

—Glycerol
BSI BS 5711: Part 15-79. 1979 Amd 1 Methods of Sampling and Test for Glycerol Part 15: Determination of Heavy Metals Content. 3 pp.

—Monosodium Glutamate
CNS N6007-82. Method of Test for Monosodium L-Glutamate (May)(765). 4 pp.

—Phosphoric Acids
CNS K6235-72. Method of Test for Phosphoric Acid (Industrial Use) (Jun)(2620). 2 pp.

—Photographic Chemicals
ISO 10349 Pt 5-92. Photography—Photographic-Grade Chemicals—Test Methods—Part 5: Determination of Heavy Metals and Iron Content First Edition. 7 pp.

—Shortening
CNS N6111-86. Method of Test for Shortening (Nov)(4990). 2 pp.

—Sodium Hydroxide
MOD UK M 808/73. Examination of Sodium Hydroxide Pure (Withdrawn).

—Toys
CNS Z8016-10-86. Method of Test for Toy Safety (Testing for Lead and Other Hazardous Heavy Metal Contents) (Nov)(4798-10). 4 pp.

Heavy Water
Use: Deuterium Oxide

Heavyweight Aggregates
Use: Aggregates

Hebraic Characters
Use For: Hebrew Characters *See Also:* Graphic Characters

—Coded Character Sets
ECMA ECMA 121-87. 8-Bit Single-Byte Coded Graphic Character Sets. 14 pp.
ISO 8859 Pt 8-88. Information Processing—8-Bit Single-Byte Coded Graphic Character Sets—Part 8: Latin/Hebrew Alphabet First Edition. 10 pp.
JTC1 8859 Pt 8-88. Information Processing—8-Bit Single-Byte Coded Graphic Character Sets—Part 8: Latin/Hebrew Alphabet First Edition. 10 pp.
OSI ISO 8859-8-88. Information Processing—8-Bit Single-Byte Coded Graphic Character Sets—Part 8: Latin/Hebrew Alphabet. 10 pp.
OSI ISO DIS 8859-8-87. Information Processing—8-Bit Single-Byte Coded Graphic Character Sets—Part 8: Latin/Hebrew Alphabet. 16 pp.
OSI ISO DIS 8859-8-87. Information Processing-8-Bit Single-Byte Coded Graphic Character Sets-Part 8: Latin/Hebrew Alphabet.

—Romanization
ISO 259-84. Documentation—Transliteration of Hebrew Characters into Latin Characters First Edition. 10 pp.

Hebrew Characters
Use: Hebraic Characters

Heddles
Use For: Healds *See Also:* Looms; Textile Machinery

INDUSTRY STANDARDS

Heddles (Cont.)

—Carrying Rods
BSI BS 3182: Part 6-84. 1984 Textile Machinery and Accessories: Heals, Heald Frames and Reeds Part 6: Specification for Dimensions for Heald-Carrying Rods and for C Shaped End Loops of Heals for Weaving Machines with Heald Frames. 4 pp.
ISO 6457-82. Textile Machinery and Accessories—Heald Carrying Rod for "C" Shaped End Loop of Flat Steel Heals—Dimensions First Edition. 3 pp.
JIS L 6313-83. Heald Carrying Rods.

—Flat Steel
BSI BS 3182: Part 1-78. (WITHDRAWN) 1978 Textile Machinery and Accessories: Heals, Heald Frames and Reeds Part 1: Flat Steel Heals for General Use: Dimensions (Superseded by BS ISO 363: 1992). 4 pp.
BSI BS ISO 363-92. 1992 Textile Machinery and Accessories—Flat Steel Heals with Closed End Loops—Dimensions (L). 7 pp.
ISO 363-92. Textile Machinery and Accessories—Flat Steel Heals with Closed End Loops—Dimensions Second Edition. 5 pp.

—Frames
BSI BS 3182: Part 3-78. 1978 Textile Machinery and Accessories: Heals, Heald Frames and Reeds Part 3: Heald Frames for Single or Double Row of Heals: Designation of Dimensions. 3 pp.
ISO 568-76. Textile Machinery and Accessories—Heald Frames for Single or Double Row of Heals—Designation of Dimensions First Edition. 3 pp.
ISO 569-82. Textile Machinery and Accessories—Heald Frames—Co-Ordinated Dimensions in Relation to the Pitch of the Harness First Edition. 4 pp.
JIS L 6508-84. Heald Frames.

—Frames—Numbering
BSI BS 4802: Part 3-78. 1978 Specification for Terminology and Classification of Weaving Machinery Part 3: Textile Machinery and Accessories. Numbering of Heald Frames in a Loom. 3 pp.
ISO 5243-77. Textile Machinery and Accessories—Numbering of Heald Frames in a Loom First Edition. 3 pp.

—Hooks
CNS L2044-82. Heald Hook (for Woollen and Worsted Loom) (Apr)(8719).

—Jacquard Weaving
BSI BS 3182: Part 2-83. 1983 Textile Machinery and Accessories: Heals, Heald Frames and Reeds Part 2: Twin Wire Heals with Inset Mail for Jacquard Weaving. 4 pp.
ISO 365-82. Textile Machinery and Accessories—Twin Wire Heals with Inset Mail for Jacquard Weaving Second Edition. 5 pp.
JIS L 6312-83. Wire Heals for Jacquard Weaving.

—Looms
CNS L2047-82. Flat Heals for Looms (Apr)(8722).
CNS L2061-84. Wire Heals for Looms (May)(9528).
JIS L 6405-78. Wire Heals for Frame Weaving of Looms.
JIS L 6405-61. Wire Heald. 6 pp.
JIS L 6507-79. Flat Heals for Looms.

—Wire
BSI BS 3182: Part 8-84. 1984 Textile Machinery and Accessories: Heals, Heald Frames and Reeds Part 8: Types and Dimensions of Twin Wire Heals for Weaving Machines with Heald Frames for Closed O Shaped End Loops. 4 pp.
ISO 364-83. Textile Machinery and Accessories—Twin Wire Heals for Weaving Machines with Heald Frames First Edition. 5 pp.

Heels (Footwear)
See Also: Footwear
BSI BS 5131: Sec 4.20-85. 1985 Methods of Test for Footwear Materials Part 4: Other Components Section 4.20: Force Required to Pull Heel Pins out of Shoe Heels. 5 pp.
BSI BS 5131: Sec 5.9-79. 1979 Methods of Test for Footwear and Footwear Materials Part 5: Testing of Complete Footwear Section 5.9: Strength of Top Piece Attachment to Shoe Heels. 4 pp.

—Fatigue Testing
BSI BS 5131: Sec 4.9-91. 1991 Methods of Test for Footwear and Footwear Materials Part 4: Other Components Section 4.9: Fatigue Resistance of Heels of Ladies' Shoes. 7 pp.

—Impact Testing
BSI BS 5131: Sec 4.8-90. 1990 Methods of Test for Footwear and Footwear Materials Part 4: Other Components Section 4.8: Resistance of Heels of Ladies' Shoes to Lateral Impact. 7 pp.

Height
See Also: Altitude; Dimensions

—Motor Vehicles
EC 85/3/EEC-84. Council Directive on the Weights, Dimensions and Certain Other Technical Characteristics of Certain Road Vehicles. 5 pp.

Height Gages
JIS B 7517-82. Height Gauges. 16 pp.

—Personal
BSI BS 1887-66. 1966 Person Weighing Machines and Height-Measuring Equipment for Hospitals, Welfare and Health Services. 12 pp.

—Vernier
CGSB 39-GP-19M-79. Calipers and Gages, Vernier and Dial, Standard for. 22 pp.

Helical Coil Thread Inserts
Use: Helical Thread Inserts

Helical Gear Pair
Use: Helical Gears

Helical Gears
Use For: Helical Gear Pair See Also: Gears; Hypoid Gears; Internal Gears; Ring Gears; Spur Gears; Worm Gears
BSI BS 436: Part 1-67. 1967 Amd 1 Spur and Helical Gears Part 1: Basic Rack Form, Pitches and Accuracy (Diametral Pitch Series). 31 pp.
BSI BS 436: Part 2-70. 1970 Amd 1 Spur and Helical Gears Part 2: Basic Rack Form, Modules and Accuracy (1 to 50 Metric Module). 33 pp.
BSI BS 978: Part 1-68. 1968 Amd 1 Fine Pitch Gears Part 1: Involute Spur and Helical Gears. 28 pp.
BSI BS 978: Pt 2: Add 1-59. 1959 Fine Pitch Gears Part 2: Cycloidal Type Gears Addendum 1: Double Circular Arc Type Gears. 7 pp.
BSI BS 4582: Part 1-70. 1970 Amd 2 Fine Pitch Gears (Metric Module) Part 1: Involute Spur and Helical Gears. 28 pp.
CNS B2423-84. Helix Angles for Helical Gears (Aug)(5280).
ISO 54-77. Cylindrical Gears for General Engineering and for Heavy Engineering—Modules and Diametral Pitches First Edition. 3 pp.
JIS B 1702-76. Accuracy for Spur and Helical Gears. 36 pp.
JIS B 1722-74. Shapes and Dimensions of Helical Gears for General Use. 26 pp.
JIS B 1752-89. Methods of Measurement for Spur and Helical Gears. 37 pp.
SNZ NZS 177-69. Specification for Spur and Helical Gears. Part 1 Basic Rack Form, Pitches and Accuracy (Diametral Pitch Series). 32 pp.

—Accuracy Testing
BSI BS 4185: Part 11-83. 1983 Machine Tool Components Part 11: Recommendations for Accuracy Grades of Gears. 4 pp.
CNS B1241-84. Accuracy of Spur Gear and Helical Gear (Aug)(7187).
JIS B 1702-76. Accuracy for Spur and Helical Gears. 36 pp.

—Backlash
JIS B 1703-76. Backlash for Spur and Helical Gears (R 1979). 22 pp.

—Gear Cutters
BSI BS 978: Part 5-65. 1965 Fine Pitch Gears Part 5: Hobs and Cutters. 27 pp.

—Gear Drives—Gear Teeth
CNS B2418-84. Helical Gear Drives for Fine Mechanics, Tooth Distance Table (Mar)(5275).

—Gear Teeth
BSI PD 6457-70. 1970 Guide to the Application of Addendum Modification to Involute Spur and Helical Gears. 20 pp.
JIS B 1701-73. Involute Gear Tooth Profile and Dimensions (R 1979). 7 pp.

—Gearmotors
BSI BS 4517-69. 1969 Amd 1 Dimensions of Spur and Helical Geared Motor Units (Metric Series). 11 pp.

—Hobs
BSI BS 978: Part 5-65. 1965 Fine Pitch Gears Part 5: Hobs and Cutters. 27 pp.
BSI BS 2062: Part 1-59. 1959 Amd 1 Gear Hobs Part 1: Hobs for General Purposes, 1 d.p. tp 20 d.p. Inclusive. 18 pp.
BSI BS 4656: Part 19-76. 1976 Accuracy of Machine Tools and Methods of Test Part 19: Gear Hobbing Machines. 12 pp.
BSI BS 5221-87. 1987 General Purpose, Metric Module Year Hobs. 22 pp.

Helical Gears (Cont.)

—Loads (Forces)—Capacity Measurement
CNS B2520-84. Calculation of Load Capacity of Spur, Helical and Bevel Gears; Zone Factor ZH (Aug)(6408).
CNS B2521-84. Calculation of Load Capacity of Spur, Helical and Bevel Gears; Material Factor ZM (Aug)(6409).

—Master Gears
BSI BS 3696: Part 1-77. 1977 Amd 1 Master Gears: Spur and Helical Gears (Metric Module). 12 pp.

—Rolling Stock
BSI BS 235-87. 1987 Gears for Electric Traction. 14 pp.

—Stresses
BSI BS 436: Part 3-86. 1986 Amd 1 Spur and Helical Gears Part 3: Method for Calculation of Contact and Root Bending Stress Limitations for Metallic Involute Gears. 46 pp.

Helical Scan Data Cartridges
Use: Data Cartridges—Helical Scan

Helical Scan Video Cassette Recorders
Use: Video Cassette Recorders

Helical Scan Video Tape Cassettes
Use: Video Cassettes

Helical Scan Video Tape Recorders
Use: Video Tape Recorders

Helical Springs
See Also: Coil Springs; Leaf Springs; Spiral Springs; Springs (Elastic); Torsion Springs
JIS B 2702-87. Hot Coiled Helical Springs. 10 pp.

—Aircraft
SBAC AS 140 ISSUE 5. Spring. 1 p.
SBAC AS 2534 ISSUE 3. Spring for Catch. 1 p.
SBAC AS 2541 ISSUE 4. Main Spring. 1 p.
SBAC AS 3117 ISSUE 4(I). Spring. 1 p.
SBAC AS 5380 ISSUE 2. Spring. 1 p.

—Aircraft—Caps (Lids)
SBAC AS 6276 ISSUE 1. Cap, Spring. 1 p.
SBAC AS 6366 ISSUE 3. Spring—Cap—Oil Drain Valve.

—Aircraft—Compression—Harness Release Gears
SBAC AS 2817 ISSUE 4. Catch Spring for Harness Release Gear. 1 p.

—Aircraft—Control Wheels
SBAC AS 3054 ISSUE 2. Brake Lock Spring for Pilot's Control Wheel. 1 p.
SBAC AS 3071 ISSUE 3. Switch Spring for Pilot's Control Wheel. 1 p.
SBAC AS 3145 ISSUE 2. Switch Spring for Pilot's Control Wheel. 1 p.

—Aircraft—Control Wheels—Covers
SBAC AS 3133 ISSUE 2. Spring Cover Pilot's Control Wheel Switch. 1 p.

—Aircraft—Cradles
SBAC AS 1899 ISSUE 3. Spring for Oxygen Cradle. 1 p.

—Aircraft—Drain Valves
SBAC AS 6366 ISSUE 3. Spring—Cap—Oil Drain Valve.

—Aircraft—Fuel Couplings
SBAC AS 2959 ISSUE 5. Spring. 1 p.

—Aircraft—Harness Release Gears
SBAC AS 2814 ISSUE 4. Cable Spring for Harness Release Gear. 1 p.
SBAC AS 2815 ISSUE 1. Cable Spring for Harness Release Gear. 1 p.

—Aircraft—Levers
SBAC AS 2270 ISSUE 3. Spring—Control Lever. 1 p.

—Aircraft—Pins
SBAC AS 2460-2466 ISSUE 2. Springs for Quick-Release Pins. 1 p.

—Aircraft—Pushbuttons
SBAC AS 3144 ISSUE 3. Spring for Auto Pilot Switch. 1 p.

—Aircraft—Pushbuttons—Covers
SBAC AS 3134 ISSUE 2. Spring for Switch Cover. 1 p.

INTERNATIONAL AND NON-U.S. NATIONAL STANDARDS
SUBJECT INDEX
Helicopters

Helical Springs (Cont.)

—Aircraft—Torsion—Parachutes
SBAC AS 481 ISSUE 2(I). Spring Anti-Spin Parachute Jettison Slip. 1 p.

—Aircraft—Valves
SBAC AS 6365 ISSUE 3. Spring Valve.

—Compression
CNS B1258-81. Helical Springs Made of Round Wire; Information to be Given in Order to Obtain the Tension Springs Required (Jun)(7590).
CNS B1259-81. Helical Springs Made of Round Wire and Rod; Information to be Given in Order to Obtain the Compression Springs Required (Jun)(7591).
CNS B1260-81. Helical Compression Springs Made of Flat Barsteel Calculation (Jun)(7592).
CNS B1269-81. Calculation and Design for Helical Compression Spring Circular Wire or Bar (Sep)(7857).
CNS B2593-81. Helical Springs Made of Round Wire; Dimensions for Cold Coiled Compression Springs of Less Than 9.5mm Wire Diameter (Jun)(7593).
CNS B2594-81. Helical Springs Made of Round Wire; Dimensions for Cold Coiled Compression Springs with Wire Diameter from 0.5 Upwards (Jun)(7594).
CNS B2596-81. Quality Specifications for Cold Coiled Compression Springs Made of Round Wire (Jun)(7596). 9 pp.
CNS B2598-82. Helical Compression Spring Made of Round Rod; Quality Requirements for Mass Productions (Nov)(7855).
DIN ENGL 2089 Pt 1-84. Helical Compression Springs Made from Round Wire or Rod; Calculation and Design (Dec). 28 pp.
DIN ENGL 2090-71. Helical Compression Springs Made of Flat Bar Steel; Calculation (Jan). 2 pp.
DIN ENGL 2098 Pt 1-68. Helical Springs Made of Round Wire; Dimensions for Cold Coiled Compression Springs with Wire Diameter from 0.5 mm Upwards (Oct). 4 pp.
DIN ENGL 2098 Pt 2-70. Helical Springs Made of Round Wire; Dimensions for Cold-Coiled Compression Springs of Less Than 0.5 mm Wire Diameter (Aug). 4 pp.
DIN ENGL 2099 Pt 1-73. Helical Springs Made of Round Wire and Rod; Data for Compression Springs; Printed Form (Nov). 3 pp.
JIS B 2704-87. Design of Helical Compression and Extension Springs.
JIS B 2704-78. Design of Helical Compression and Extension Springs. 30 pp.
JIS B 2707-87. Cold Coiled Helical Compression Springs. 11 pp.

—Compression—Design
BSI BS 1726: Part 1-87. 1987 Amd 2 Coil Springs Part 1: Guide for the Design of Helical Compression Springs (AMD 6218) April 30, 1990. 32 pp.

—Compression—Quality Assurance
DIN ENGL 2095-73. Helical Springs Made of Round Wire; Quality Specifications for Cold Coiled Compression Springs (May). 8 pp.
DIN ENGL 2096 Pt 1-81. Helical Compression Springs Made of Round Wire and Rod; Quality Requirements for Hot Formed Compression Springs (Nov). 7 pp.

—Extension
CNS B1268-81. Calculation and Design for Helical Tension Springs Round Wire or Round Bar (Sep)(7856).
CNS B1270-81. Calculation and Design for Helical Tensional Springs Wire or Rod (Sep)(7858).
CNS B2595-81. Quality Specification for Cold Coiled Helical Tension Springs Made of Round Metal (Jun)(7595). 15 pp.
DIN ENGL 2099 Pt 2-73. Helical Springs Made of Round Wire; Data for Tension Springs; Printed Form (Nov). 3 pp.
JIS B 2704-87. Design of Helical Compression and Extension Springs.
JIS B 2704-78. Design of Helical Compression and Extension Springs. 30 pp.
JIS B 2708-87. Cold Coiled Helical Tension Springs. 8 pp.

—Extension—Design
BSI BS 1726: Part 2-88. 1988 Amd 1 Coil Springs Part 2: Guide for the Design of Helical Extension Springs (AMD 6110) October 31, 1989. 18 pp.

—Extension—Quality Assurance
DIN ENGL 2097-73. Helical Springs Made of Round Wire; Quality Specifications for Cold Coiled Tension Springs (May). 10 pp.

—Marine
JIS F 0503-87. Coiled Springs for Marine Machinery.
JIS F 0503-60. Coil Springs for Ship Machinery (R 1976). 19 pp.

Helical Springs (Cont.)

—Pipes
BSI BS 5431-76. 1976 Bending Springs for Use with Copper Tubes for Water, Gas and Sanitation. 7 pp.

—Press Dies
CNS B3367-87. Coil Springs for Press Dies (Apr)(5993).
JIS B 5012-86. Coiled Helical Springs for Stamping Dies. 38 pp.

—Ships
JIS F 0503-87. Coiled Springs for Marine Machinery.
JIS F 0503-60. Coil Springs for Ship Machinery (R 1976). 19 pp.

—Steel Wire
BSI BS 4637-70. 1970 Amd 1 Carbon Steel Wire for Coiled Springs (Bedding and Seating). 12 pp.
BSI BS 4637-03. 1970 Amd 3 Carbon Steel Wire for Coiled Springs (Bedding and Seating) (AMD 7609) June 15, 1993. 14 pp.

—Testing Equipment
JIS B 7738-84. Testing Machines for Helical Compression and Extension Springs. 11 pp.

—Torsion
BSI BS 1726: Part 3-88. 1988 Amd 1 Coil Springs Part 3: Guide for the Design of Helical Torsion Springs (AMD 6668) February 28, 1991. 17 pp.
DIN ENGL 2088-69. Helical Springs Made of Round Wire and Rod; Calculation and Design of Torsional Springs; (Leg Springs) (July). 11 pp.
JIS B 2709-87. Design of Helical Torsion Springs. 23 pp.

—Torsion—Design
BSI BS 1726: Part 3-88. 1988 Amd 1 Coil Springs Part 3: Guide for the Design of Helical Torsion Springs (AMD 6668) February 28, 1991. 17 pp.
JIS B 2709-87. Design of Helical Torsion Springs. 23 pp.

Helical Thread Inserts
Use For: Helical Coil Thread Inserts; Wire Thread Inserts (Helical Thread) *See Also:* Thread Inserts
DIN ENGL 8140 Pt 1-88. Wire Thread Inserts for ISO Metric Screw Threads; Dimensions and Technical Delivery Conditions (Oct). 10 pp.
MOD UK DSTAN 53-14-01. Inserts Screw Thread (Inch) Coiled Wire Type Issue 2; Amendment 3. 60 pp.

—Aerospace
DIN ENGL LN 9039 Pt 1-89. Aerospace; Wire Thread Inserts, Class: 1100 MPa/235 Degrees C/425 Degrees C (July). 10 pp.
DIN ENGL LN 9039 Pt 2-89. Aerospace; Wire Thread Inserts, Assembly and Special Tools (July). 3 pp.

—Aircraft
SBAC AS 6316 ISSUE 6. Wire Thread, Insert UNF-Unplated. 1 p.
SBAC AS 6317 ISSUE 6. Wire Thread, Insert UNC-Unplated. 1 p.
SBAC AS 6695 ISSUE 4(I). Wire Thread Inserts (U.N.F.)—Unplated.
SBAC AS 6696 ISSUE 4(I). Wire Thread Inserts (U.N.C.)—Unplated.
SBAC AS 6733 ISSUE 5. Wire Thread, Inserts (UNF)—Unplated.
SBAC AS 6734 ISSUE 6. Wire Thread, Insert (U.N.C.) Unplated.
SBAC AS 8451 ISSUE 5. Wire Thread, Inserts U.N.F. Cadmium Plated.
SBAC AS 8452 ISSUE 5. Wire Thread, Inserts (UN.C), Cadmium Plated.
SBAC AS 8453 ISSUE 4(I). Wire Thread, Inserts (UNF) Cadmium Plated.
SBAC AS 8454 ISSUE 4(I). Wire Thread, Inserts (U.N.C.) Cadmium Plated.
SBAC AS 8455 ISSUE 5. Wire Thread, Inserts (U.N.F.) Cadmium Plated.
SBAC AS 8456 ISSUE 6. Wire Thread, Inserts (UNC) Cadmium Plated.
SBAC AGS 3600-99 ISSUE 2. Insert, Screw Thread, Unified, Screw Locking, Helical Coil, C.R. Steel, DTD 734, Cadmium Plated.
SBAC AGS 3700-99 (V). Insert-Screw Thread, Unified, Screw Locking, Helical Coil, H.R. Alloy BS HR503.
SBAC AGS 4677 (V). Insert, Screw, Thread, Metric, Screw Locking, Helical Coil, Corrosion Resistant Steel DTD734, Cadmium Plated.
SBAC RS 428 (V). Tapped Holes for, Wire Thread, Inserts, (BA, BSF,BSW Threads)—Unplated.
SBAC RS 429 (V). Tapped Holes for, Wire Thread Inserts, (BSPF Threads), Unplated.
SBAC RS 430 (V). Tapped Holes for, Wire Thread Inserts, (Unified Threads), Unplated.
SBAC RS 679 ISSUE 1. Insert, Screw Thread, Unified, Helical Coil, Tang and Notch Details.

Helical Thread Inserts (Cont.)

—Aircraft (Cont.)
SBAC RS 729 (V). Metric Thread, Insert—Wire Section.
SBAC TS 17 ISSUE 1. Insert, Screw Thread, Metric, Screw Locking Helical Coil, AMS 7246 (Inconel X750) Silver Coated.
SBAC TS 40 ISSUE 4. Inserts, Screw Thread, Unified, Screw Locking, Helical Coil, Corrosion Resistant Steel, DTD 734 Cadmium Coated.
SBAC TS 41 ISSUE 4. Insert, Screw Thread, Unified, Screw Locking, Helical Coil, BS HR 503 (Nimonic 90), Silver Coated.
SBAC TS 55 ISSUE 2. Installation of Inserts, Screw Thread, Unified Screw Locking, Helical Coil.
SBAC TS 70 ISSUE 3. Inserts, Screw Thread, Unified, Screw Locking Helical Coil, Corrosion Resistant Steel DTD 734 Cadmium Plated.
SBAC TS 81 ISSUE 1. Insert, Screw Thread, Metric, Screw Locking, Helical Coil, Corrosion Resistant Steel DTD 734, Cadmium Plated.
SBAC TS 82 ISSUE 3. Insert, Screw Thread, Metric, Screw Locking, Helical Coil, Corrosion Resistant Steel DTD 734, Cadmium Plated.
SBAC TS 141 ISSUE 1. Inserts, Screw Thread, Unified, Screw Locking, Helical Coil, BS HR503.

—Aircraft—Screw Threads
SBAC RS 728 ISSUE 2. Insert, Screw Thread, Metric, Helical Coil, Tang and Notch Details.

—Aircraft—Sizes
SBAC RS 727 (V). Tapped Hole and Fitted Size Data for Helical Coil Wire Thread (Metric).

—Aircraft—Tapped Hole Geometries
SBAC RS 727 (V). Tapped Hole and Fitted Size Data for Helical Coil Wire Thread (Metric).

—Tapped Holes—Nominal Size
DIN ENGL 8140 Pt 2-88. Wire Thread Inserts for ISO Metric Screw Threads; Tapped Holes for Thread Inserts, Thread Tolerances (Oct). 5 pp.

Helicopter Controllers (Personnel)
See Also: Flight Dynamics; Helicopters; Personnel

—Qualifications
NATO STANAG 1154 ED 7 AMD 4-90. NATO Qualifications for Helicopter Controllers at Sea. 11 pp.

Helicopter Engines
See Also: Aircraft Engines; Engines; Helicopters

—Gas Turbine
CAA Chapter G7-2 06.66. Test Conditions—Piston and Turbine Engine Installations (Rotocraft). 3 pp.

—Gas Turbine—Comprehensive Testing
CAA Chapter G7-3 06.66. Ground and Flight Tests—Single Turbine Engine Installations (Rotocraft). 6 pp.
CAA Chapter G7-4 06.66. Ground Flight Tests—Twin Turbine Engine Installations (Rotocraft). 8 pp.

—Gas Turbine—Design
CAA E-3(C). Section C Engines and Propellers Engine Identification (Blue Papers). 2 pp.

—Piston
CAA Chapter G7-2 06.66. Test Conditions—Piston and Turbine Engine Installations (Rotocraft). 3 pp.

Helicopter Rotors
Use: Rotary Wings

Helicopters
Scope Note: See also specific types of helicopters, e.g., Bell, Sikorsky *See Also:* Aerospatiale AS 332C Helicopters; Aerospatiale AS 332L Helicopters; Aerospatiale AS 350B Helicopters; Aerospatiale AS 350B1 Helicopters; Aerospatiale AS 350B2 Helicopters; Aerospatiale AS 355F II Helicopters; Aerospatiale AS 365N2 Helicopters; Aerospatiale SA 330 Puma Helicopters; Aerospatiale SA 330G Puma Helicopters; Aerospatiale SA 330J Puma Helicopters; Aerospatiale SA 332 Puma Helicopters; Aerospatiale SA 365 Helicopters; Aerospatiale SA 365C Helicopters; Aerospatiale SA 365C1 Helicopters; Aerospatiale SA 365C2 Helicopters; Aerospatiale SA 365C3 Helicopters; Aerospatiale SA 365N Helicopters; Aerospatiale SA 365N1 Helicopters; Agusta A 1 Helicopters; Agusta A 109 Helicopters; Agusta A 109A Helicopters; Agusta A 109C Helicopters; Agusta Bell 47 Helicopters; Agusta Bell 206 Helicopters; Agusta Bell 212 Helicopters; Air Medical Transport Units; Airborne Operations; Aircraft; AS 332L2 Super Puma Helicopters; Bell 205 Helicopters; Bell 206A Helicopters; Bell 206B Helicopters; Bell 206L Helicopters; Bell 206L1 Helicopters; Bell 206L3 Helicopters; Bell 212

INTERNATIONAL AND NON-U.S. NATIONAL STANDARDS
SUBJECT INDEX

Helicopters

Helicopters (Cont.)
See Also: (Cont.)
Helicopters; Bell 214 Helicopters; Bell 214B Helicopters; Bell 214ST Helicopters; Bell 222 Helicopters; Bell 412 Helicopters; Bell 47 Helicopters; Bell 47D Helicopters; Bell 47G Helicopters; Bell 47G5 Helicopters; Bell 47J Helicopters; Boeing Vertol Helicopters; Enstrom Helicopters; Helicopter Controllers (Personnel); Helicopter Engines; Hiller Helicopters; Hughes 269A Helicopters; Hughes 369 Helicopters; Hughes 369D Helicopters; Hughes 369E Helicopters; Hyne (Brantly) Helicopters; Messerschmitt-Bolkow-Blohm BK 117 Helicopters; Messerschmitt-Bolkow-Blohm BO 105 Helicopters; Robinson Helicopters; Rotary Wing Aircraft; Rotors (Machine Elements); Russian-Built Helicopters; Schweizer 269 Helicopters; Sikorsky S-58 ET Helicopters; Sikorsky S-58 Helicopters; Sikorsky S-58T Helicopters; Sikorsky S-61 Helicopters; Sikorsky S-61N Helicopters; Sikorsky S-61N2 Helicopters; Sikorsky S-64 Helicopters; Sikorsky S-67 Helicopters; Sikorsky S-70 Helicopters; Sikorsky S-76 Helicopters; Sikorsky S-76A Helicopters; Sikorsky S-76B Helicopters; Sikorsky S-76C Helicopters; Vertical Takeoff Aircraft

—**Aerial Recovery**
NATO STANAG 2970 ED 1 AMD 0-90. Aerial Recovery Equipment and Techniques for Helicopters. 8 pp.
NATO STANAG 2970 ED 2 AMD 0-92. Aerial Recovery Equipment and Techniques for Helicopters. 9 pp.

—**Air Navigation—Military Communication**
NATO STANAG 2863 ED 5 AMD 2-88. Navigational and Communicational Capabilities for Helicopters in Multinational Land Operations. 26 pp.
NATO STANAG 2863 ED 5 AMD 4-88. Navigational and Communicational Capabilities for Helicopters in Multinational Land Operations. 28 pp.

—**Airmobile Operations**
NATO STANAG 2904 ED 1 AMD 3-79. Airmobile Operations (ATP-41). 4 pp.

—**Bearings—Greases**
MOD UK DSTAN 91-51-01. Grease, Aircraft: Helicopter Oscillating Bearing NATO Code No: G-366 Joint Service Designation: XG-284 Issue 1; Amendment 1. 13 pp.

—**Cargo Handling Equipment**
MOD UK DSTAN 16-9-69. Nets and Slings, Cargo, Helicopter (Section on Slings is Superseded) Issue 1. 5 pp.
MOD UK DSTAN 16-9: Addendum. Nets and Slings, Helicopter Issue 1. 3 pp.
MOD UK DSTAN 16-9: Part 2-01. Nets and Slings, Cargo, Helicopter Part 2: Slings, Cargo, Helicopter Issue 1; Amendment 2. 12 pp.
NATO STANAG 2403 ED 1 AMD 1-90. Technical Criteria for External Cargo Carrying Strops/Pendants. 7 pp.
NATO STANAG 2949 ED 2 AMD 3-88. Technical Criteria for External Cargo Carrying Sling. 11 pp.
NATO STANAG 3542 ED 2 AMD 0-81. Technical Criteria for the Transport of Cargo by Helicopter. 16 pp.
NATO STANAG 3542 ED 3 AMD 1-90. Technical Criteria for the Transport of Cargo by Helicopter. 14 pp.
NATO STANAG 3542 ED 4 AMD 0-93. Technical Criteria for the Transport of Cargo by Helicopter. 15 pp.

—**Cargo Handling Equipment—Shipborne**
NATO STANAG 1277 ED 1 AMD 0-85. Vertrep Equipment and Procedures. 13 pp.
NATO STANAG 1277 ED 1 AMD 1-92. Vertrep Equipment and Procedures. 14 pp.

—**Cargo Nets**
NATO STANAG 2950 ED 1 AMD 3-83. Technical Criteria for External Cargo Carrying Nets. 8 pp.
NATO STANAG 2950 ED 2 AMD 0-91. Technical Criteria for External Cargo Carrying Nets. 7 pp.

—**Crash Locator Beacons**
CAA Spec. NO. 16 ISSUE 1.

—**Evacuation Equipment**
CAA NOTICE #27 ISSUE 2. Helicopter Emergency Escape Facilities. 4 pp.

—**Exposure Suits**
CAA Spec. NO. 19.
CGSB CAN/CGSB-65.17-M88. Helicopter Passenger Transportation Suit Systems. 35 pp.

Helicopters (Cont.)
—**Flight Dynamics**
JIS W 0401-76. Helicopter Flying and Ground Handling Qualities; General Equipment for.

—**Flight Recorders**
CAA Spec. NO. 18 ISSUE 1.

—**Glide Slope Guidance Systems—Shipborne**
NATO STANAG 1236 ED 2 AMD 2-88. Glide Slope Indicators for Helicopter Operations from NATO Ships. 10 pp.
NATO STANAG 1236 ED 2 AMD 3-92. Glide Slope Indicators for Helicopter Operations from NATO Ships. 11 pp.

—**Hand Signals**
SNZ NZS 5818-82. Specification for Hand Signals for the Direction of Cranes and Similar Lifting Devices Including Helicopters (Other Than for Cargo Handling in Wharf Areas) (Reconfirmed 1989). 12 pp.

—**Hauldown Cables—Shipborne**
NATO STANAG 1274 ED 1 AMD 3-81. Shipboard Hauldown Cable End Fitting for Helicopter Wire Recovery Assist Systems. 7 pp.

—**Hoisting Slings**
NATO STANAG 3295 ED 6 AMD 4-81. Horse Collar/Rescue Strop Type Helicopter Hoisting Gear. 9 pp.

—**In Flight Refueling**
NATO STANAG 1280 ED 2 AMD 2-85. Helicopter in-Flight Refuelling (HIFR) Operating Requirements. 12 pp.
NATO STANAG 1280 ED 2 AMD 3-85. Helicopter in-Flight Refuelling (HIFR) Operating Requirements. 6 pp.
NATO STANAG 1354 ED 1 AMD 0-86. Helicopter in-Flight Refuelling (HIFR) Procedures (Withdrawn). 8 pp.

—**In Flight Refueling—Lighting**
NATO STANAG 1251 ED 2 AMD 3-83. Shipboard Helicopter in-Flight Refuelling (HIFR) Operating Area Marking, Clearances and Lighting. 19 pp.

—**In Flight Refueling—Markings**
NATO STANAG 1251 ED 2 AMD 3-83. Shipboard Helicopter in-Flight Refuelling (HIFR) Operating Area Marking, Clearances and Lighting. 19 pp.

—**Interdepartmental Procurement**
NATO STANAG 3430 ED 8 AMD 0-91. Responsibilities for Aircraft Cross-Servicing. 24 pp.

—**Microwave Landing Systems—Shipborne**
NATO STANAG 1375 ED 1 AMD 0-89. Microwave Landing Systems for Small Ships. 6 pp.

—**Military Operations—Land Forces**
NATO STANAG 2999 ED 1 AMD 1-87. Use of Helicopters in Land Operations—ATP-49. 5 pp.
NATO STANAG 2999 ED 2 AMD 0-92. Use of Helicopters in Land Operations—ATP-49. 5 pp.

—**Mine Countermeasures—Technical Manuals**
NATO STANAG 1137 ED 3 AMD 1-84. Helicopter Mine Countermeasures Manual—AMP-7(B)(Navy/Air). 9 pp.

—**Naval Operations**
NATO STANAG 1288 ED 2 AMD 0-83. Helicopter Operations from Ships Other Than Aircraft Carriers Supplement—APP-2 Supp 1. 4 pp.

—**Naval Ships—Inspection**
NATO STANAG 1237 ED 1 AMD 5-79. Stanag 1237—Inspection of Aviation Facilities for Helicopter Operations from Ships Other Than Aircraft Carriers. 24 pp.

—**Noise**
CAA Chapter N3-6 08.90. Maximum Permissible Noise Levels for Microlight Aeroplanes (Noise). 1 p.
CAA Chapter N3-7 08.90. Maximum Permissible Noise Levels for Helicopters (Noise). 2 pp.
CAA Chapter N3-7 App 08.90. Maximum Permissible Noise Levels for Helicopters (Noise). 1 p.

—**Noise—Flight Testing**
CAA Chapter N4-5 08.90. Flight Testing of Helicopters (Noise). 7 pp.
CAA Chapter N4-13 08.90. Analysis and Adjustment of Flight Test Results for Helicopters. 3 pp.
CAA Chapter N4-13 App 08.90. Analysis and Adjustment of Flight Test Results for Helicopters. 4 pp.

Helicopters (Cont.)
—**Noise—Sonar Detection**
NATO STANAG 1136 ED 2 AMD 1-84. Standards for Use When Measuring and Reporting Radiated Noise Characteristics of Surface Ships, Submarines, Helicopters, Etc in Relation to Sonar Detention and Torpedo Aquisition Risk. 36 pp.

—**Noise—Torpedo Acquisition Risk**
NATO STANAG 1136 ED 2 AMD 1-84. Standards for Use When Measuring and Reporting Radiated Noise Characteristics of Surface Ships, Submarines, Helicopters, Etc in Relation to Sonar Detention and Torpedo Aquisition Risk. 36 pp.

—**Radiation Protection and Safety—Shipborne**
NATO STANAG 1305 ED 1 AMD 0-89. Radhaz Procedures for Receiving Helicopters (and VSTOL Aircraft) of Other Nations on Ships Other Than Aircraft Carriers. 16 pp.
NATO STANAG 1305 ED 1 AMD 2-89. Radhaz Procedures for Receiving Helicopters (and VSTOL Aircraft) of Other Nations on Ships Other Than Aircraft Carriers. 16 pp.
NATO STANAG 1305 ED 1 AMD 3-89. Radhaz Procedures for Receiving Helicopters (and VSTOL Aircraft) of Other Nations on Ships Other Than Aircraft Carriers. 18 pp.
NATO STANAG 1308 ED 1 AMD 0-88. Radhaz to Ships Personnel During Helicopter (and VSTOL Aircraft) Operations on Ships Other Than Aircraft Carriers. 11 pp.
NATO STANAG 1308 ED 2 AMD 0-89. Radhaz to Ships Personnel During Helicopter (and VSTOL Aircraft) Operations on Ships Other Than Aircraft Carriers. 12 pp.
NATO STANAG 1308 ED 2 AMD 1-89. Radhaz to Ships Personnel During Helicopter (and VSTOL Aircraft) Operations on Ships Other Than Aircraft Carriers. 13 pp.

—**Refueling Equipment—In Flight**
NATO STANAG 3847 ED 1 AMD 3-84. Helicopter in-Flight Refuelling (HIFR) Equipment. 7 pp.

—**Refueling—Shipborne**
NATO STANAG 1281 ED 1 AMD 2-82. Pressure Refuelling of Shipborne Helicopters with Engines and/or Rotors Running (Withdrawn). 12 pp.

—**Shipborne**
MOD UK NES 134-82. Requirements for Aviation Arrangements Issue 1 (11.82) (Withdrawn) (Refer to MoD). 32 pp.
NATO STANAG 1194 ED 5 AMD 1-91. Helicopter Operations from Ships Other Than Aircraft Carriers—APP-2(D). 6 pp.

—**Structural Design**
JIS W 0605-76. Structural Design Requirement, Helicopters.

—**Tactical Formation**
NATO STANAG 3627 ED 1 AMD 6-74. Helicopter Day and Night Tactical Formation Flying. 30 pp.
NATO STANAG 3627 ED 3 AMD 0-91. Helicopter Day and Night Tactical Formation Flying. 9 pp.

—**Tasking**
NATO STANAG 2956 ED 1 AMD 1-88. Helicopter Tasking Messages. 10 pp.

—**Tie Downs—Shipborne**
NATO STANAG 1276 ED 1 AMD 5-81. Shipborne Helicopter Harpoon/Grid Rapid Securing System. 9 pp.
NATO STANAG 1276 ED 1 AMD 6-81. Shipborne Helicopter Harpoon/Grid Rapid Securing System. 11 pp.

—**Vertical Operating Areas—Shipborne**
NATO STANAG 1211 ED 2 AMD 2-83. Shipboard Helicopter Facility Designations (Withdrawn). 14 pp.

Helio H-295 Aircraft
See Also: Aircraft

—**Antenna Positions**
CAA. Helio H-295 and H-395 (Approved Aerial Positions). 1 p.

Helio H-395 Aircraft
See Also: Aircraft

—**Antenna Positions**
CAA. Helio H-295 and H-395 (Approved Aerial Positions). 1 p.

Heliports
See Also: Airports
ICAO Annex 14 Vol II. Aerodromes Volume II Heliports First Edition—July 1990; (Corrigendum 1 05/31/91) (Supplement 1/2/91). 82 pp.

INTERNATIONAL AND NON-U.S. NATIONAL STANDARDS
SUBJECT INDEX

Heliports *(Cont.)*
ICAO 9261. Heliport Manual Second Edition—1985; (Amendment 1 04/30/86). 86 pp.

—**Fire Fighting—Identification Systems**
NATO STANAG 3861 ED 2 AMD 3-83. Heliport Rescue and Fire-Fighting Services—Identification Categories. 6 pp.

—**Helipads—Identification Systems**
NATO STANAG 3619 ED 2 AMD 4-80. Helipad Marking. 8 pp.

—**Helipads—Lighting**
NATO STANAG 3652 ED 1 AMD 6-74. Helipad Lighting (VMC). 15 pp.
NATO STANAG 3652 ED 2 AMD 0-92. Helipad Lighting (VMC). 14 pp.
NATO STANAG 3652 ED 2 AMD 1-92. Helipad Lighting (VMC). 15 pp.

—**Lighting**
NATO STANAG 1162 ED 4 AMD 0-91. Vertical Replenishment (Vertrep) Operating Area Marking, Clearances, and Lighting. 20 pp.

—**Lighting—Marking**
NATO STANAG 3535 ED 1 AMD 3-81. Heliport Marking and Lighting. 18 pp.

—**Lighting—Shipborne**
NATO STANAG 1275 ED 2 AMD 0-91. Minimum Standard Requirements for the Marking and Lighting of Helicopter Deck Landing Areas. 13 pp.
NATO STANAG 1275 ED 2 AMD 1-92. Minimum Standard Requirements for the Marking and Lighting of Helicopter Deck Landing Areas. 14 pp.

—**Marking—Shipborne**
NATO STANAG 1275 ED 2 AMD 0-91. Minimum Standard Requirements for the Marking and Lighting of Helicopter Deck Landing Areas. 13 pp.
NATO STANAG 1275 ED 2 AMD 1-92. Minimum Standard Requirements for the Marking and Lighting of Helicopter Deck Landing Areas. 14 pp.

—**Vertical Operating Areas**
NATO STANAG 1162 ED 4 AMD 0-91. Vertical Replenishment (Vertrep) Operating Area Marking, Clearances, and Lighting. 20 pp.

Helium
See Also: Hazardous Materials; Noble Gases; Nonmetals

—**Leakage**
JIS Z 2330-92. Standard Recommended Guide for the Selection of Helium Leak Testing. 14 pp.
JIS Z 2331-92. Method for Helium Leak Testing. 53 pp.

—**Purity—Gas Cylinders—Diving**
MOD UK DSTAN 68-75-01. Breathing Gas Purity for Diving Issue 1; Amendment 1. 9 pp.
MOD UK DSTAN 68-75-93. Breathing Gas Purity for Diving Issue 2. 11 pp.

—**Storage and Handling—Information Cards**
SAA AS 2508.2.01 2-85. Safe Storage and Handling Information Cards for Hazardous Materials—Part 2.012: Non-Flammable, Inert, Compressed Gases (Includes: Argon, Helium, Krypton, Neon, Nitrogen and Xenon) Double Sided Card.

Helix Angles
—**Gear Teeth**
DIN ENGL 3978-76. Helix Angles for Cylindrical Gear Teeth (Aug). 6 pp.

—**Gears**
CNS B2423-84. Helix Angles for Helical Gears (Aug)(5280).

Helmets
Use For: Safety Helmets *See Also:* Breathing Apparatus; Cap Lamps; Clothing; Combat Uniforms; Face Shields; Hats; Headgear; Hearing Protectors; Hoods (Headgear); Occupational Safety and Health; Protective Clothing; Protective Masks; Safety; Scalp Protectors; Uniforms; Visors

BSI BS 5240: Part 1-87. 1987 Industrial Saftey Helmets Part 1: Construction and Performance. 12 pp.
CEN PREN 397-90. Industrial Safety Helmets. 21 pp.
CNS Z3001-88. Testing Standard for Safety Helmet (for Work Site) (Apr)(1336).
ISO 3873-77. Industrial Safety Helmets First Edition. 10 pp.
JIS T 8131-90. Industrial Safety Helmets. 18 pp.
SAA AS 1800-81. Selection, Care and Use of Industrial Safety Helmets. 5 pp.
SAA AS 1801-81. Industrial Safety Helmets (Incorporating Amdt 1). 20 pp.

Helmets *(Cont.)*
SAA AS 2063.1-86. Lightweight Protective Helmets (for Use in Pedal Cycling, Horseriding and Other Activities Requiring Similar Protection)—Part 1: Basic Performance Requirements Amdt 1 November 1987 Amdt 2 May 1988. 4 pp.
SNZ NZS 2264-70. Specification for Industrial Safety Helmets (Maximum Protection) Amend: 2, 1981 (Reconfirmed 1990). 21 pp.

—**Airborne Sports**
CEN PREN 966-92. Airborne Sports Helmets. 22 pp.

—**Automotive**
BSI BS 2495-77. 1977 Amd 5 Protective Helmets for Vehicle Users (High Protection). 29 pp.
BSI BS 5361-76. 1976 Amd 5 Protective Helmets for Vehicle Users. 28 pp.
BSI BS 6658-85. 1985 Amd 1 Protective Helmets for Vehicle Users. 30 pp.
CEN PREN 819-92. Helmets for Landbased Motor Vehicle Competition Use (Motor Cycles and Automobiles). 40 pp.
CSA CAN3-D230-M85. Protective Headgear in Motor Vehicle Applications. 26 pp.
ISO R1511-70. Protective Helmets for Road Users First Edition. 30 pp.
JIS T 8133-82. Protective Helmets for Vehicular Users. 28 pp.
SAA AS 1698-88. Protective Helmets for Vehicle Users. 4 pp.
SNZ NZS 1884-69. Specification for Protective Helmets and Peaks for Racing Car Drivers (Reconfirmed 1984) Amend: 1, 1971; 2, 1973. 23 pp.
SNZ NZS 5430-92. Protective Helmets for Vehicle Users Amend: A, 1992. 4 pp.

—**Ballistic Limit Protection**
NATO STANAG 2920 ED 1 AMD 0-00. Ballistic Test Method for Personal Armours. 9 pp.

—**Bicycling**
BSI BS 6863-89. 1989 Pedal Cyclists' Helmets. 19 pp.
BSI BS 6863-01. 1989 Amd 1 Pedal Cyclists' Helmets (AMD 6954) November 29, 1991. 18 pp.
CEN PREN 1078-93. Helmets for Pedal Cyclists. 19 pp.
CSA CAN/CSA-D113.2-M89. Cycling Helmets; (Gen Instr 1 Thru 2). 28 pp.
JIS T 8133-82. Protective Helmets for Vehicular Users. 28 pp.
JIS T 8134-82. Protective Helmets for Bicycle Users. 28 pp.
SAA AS 2063.1-86. Lightweight Protective Helmets (for Use in Pedal Cycling, Horseriding and Other Activities Requiring Similar Protection)—Part 1: Basic Performance Requirements Amdt 1 November 1987 Amdt 2 May 1988. 4 pp.
SAA AS 2063.2-90. Lightweight Protective Helmets (for Use in Pedal Cycling, Horse Riding and Other Activities Requiring Similar Protection)—Part 2: Helmets for Pedal Cyclists Amdt 1 September 1990 Amdt 2 April 1991.
SNZ NZS 5439-86. Specification for Lightweight Protective Helmets for Pedal Cyclists Amend: 1, 1989. 11 pp.

—**Children's**
CEN PREN 1080-93. Protective Helmets for Young Children—Play Activities. 18 pp.

—**Climbers'**
BSI BS 4423-69. 1969 Amd 1 Climbers' Helmets. 21 pp.

—**Combat**
NATO STANAG 2902 ED 1 AMD 2-79. Criteria for a NATO Combat Helmet. 9 pp.
NATO STANAG 2902 ED 1 AMD 3-79. Criteria for a NATO Combat Helmet. 10 pp.

—**Ear Muffs—Safety**
CEN PREN 352-3-92. Hearing Protectors—Safety Requirements and Testing—Part 3: Ear Muffs Attached to an Industrial Safety Helmet. 41 pp.

—**Electrical Workers**
CNS Z2022-88. Safety Helmets for Electrical Workers (Apr)(4598).
CNS Z3015-88. Method of Test for Safety Helmets for Electrical Workers (Apr)(4599).

—**Firefighters'**
BSI BS 3864-89. 1989 Protective Helmets for Firefighters. 26 pp.
CEN PREN 443-90. Protective Helmets for Firefighters. 26 pp.
SAA AS 4067-92. Firefighters' Helmets—Specifications (in Professional Package 47). 31 pp.

—**Flight Crews'—Sunglasses**
MOD UK DTD-1281-80. Optical Requirements for Aircrew Helmet Visors and Uncorrected Sunglasses. 21 pp.

Helmets *(Cont.)*
—**Flight Crews'—Visors**
MOD UK DTD-1281-80. Optical Requirements for Aircrew Helmet Visors and Uncorrected Sunglasses. 21 pp.

—**Headforms**
BSI BS 6489-84. 1984 Amd 1 Headforms for Use in the Testing of Protective Helmets. 22 pp.
CEN PREN 960-93. Headforms for Use in the Testing of Protective Helmets. 17 pp.

—**Hockey**
CEN PREN 967-92. Head Protectors for Ice Hockey Players. 38 pp.
CSA CAN/CSA-Z262.1-M90. Ice Hockey Helmets; (Gen Instr 1). 30 pp.

—**Horseback Riding**
BSI BS 4472-88. 1988 Protective Skull Caps for Jockeys. 16 pp.
BSI BS 6473-84. 1984 Amd 2 Protective Hats for Horse and Pony Riders. 14 pp.
SAA AS 2063.1-86. Lightweight Protective Helmets (for Use in Pedal Cycling, Horseriding and Other Activities Requiring Similar Protection)—Part 1: Basic Performance Requirements Amdt 1 November 1987 Amdt 2 May 1988. 4 pp.
SAA AS 2063.3-88. Lightweight Protective Helmets (for use in Pedal Cycling, Horseriding and Other Activities Requiring Similar Protection—Part 3: Helmets for Horse Riders. 2 pp.
SNZ NZS 8602-89. Protective Helmets for Horse and Pony Riders. 16 pp.

—**Industrial**
CSA CAN/CSA-Z94.1-92. Industrial Protective Headwear; (Gen Instr 1). 37 pp.
SAA AS 1800-81. Selection, Care and Use of Industrial Safety Helmets. 5 pp.
SAA AS 1801-81. Industrial Safety Helmets (Incorporating Amdt 1). 20 pp.
SNZ NZS 5806-80. Specification for Industrial Safety Helmets (Medium Protection) (Reconfirmed 1986). 12 pp.

—**Mopeds**
CEN PREN 398-90. Protective Helmets for Drivers and Passengers of Motor Cycles and Mopeds. 42 pp.
CEN PREN 968-92. Protective Helmets for Users of Mopeds in Hot Climates. 29 pp.

—**Motorcycle**
CEN PREN 398-90. Protective Helmets for Drivers and Passengers of Motor Cycles and Mopeds. 42 pp.
CNS Z2009-86. Protective Helmets for Motor Cyclists (Dec)(2396). 10 pp.
CNS Z3014-85. Method of Test for Protective Helmets of Motor Cyclists (Sep)(3902). 10 pp.
SNZ NZS 1215-69. Specification for Protective Helmets for Motor Cyclists Amend: 1, 1971; 2, 1973 (Reconfirmed 1984). 19 pp.
SNZ NZS 5430-92. Protective Helmets for Vehicle Users Amend: A, 1992. 4 pp.

—**Motorcycle—Visors**
CEN PREN 173-91. Personal Eye-Protection—Visors for Motorcycle Helmets. 12 pp.

—**Police**
CSA CAN/CSA-Z611-M86. Police Riot Helmets and Faceshield Protection; (Gen Instr 1). 28 pp.

—**Rollerskating**
CEN PREN 1079-93. Helmets for Users of Skateboards and Roller Skates. 19 pp.

—**Skateboarding**
CEN PREN 1079-93. Helmets for Users of Skateboards and Roller Skates. 19 pp.

—**Skiing**
CEN PREN 1077-93. Helmets for Alpine Skiers. 28 pp.

—**Welders**
BSI BS 1542-82. 1982 Equipment for Eye, Face and Neck Protection Against Non-Ionizing Radiation Arising During Welding and Similar Operations. 8 pp.
CNS Z2032-87. Shields and Helmets for Welders (Jul)(7175).
JIS T 8142-89. Helmet Type and Handshield Type Protectors for Welders. 12 pp.

Hemacytometers
Use: Hemocytometers

Hematite
Use: Ferric Oxide

Hemoconcentrators

Hemoconcentrators
Use For: Haemoconcentrators
See Also: Hemodialysis; Medical Electrical Equipment; Medical Equipment
BSI BS 7297: Part 1-90. 1990 Haemodialysers and Related Equipment Part 1: Haemodialysers, Haemofilters and Haemoconcentrators. 22 pp.
CSA CAN3-Z364.1. 1-M84. Hemodialysers, Hemofilters, and Hemoconcentrators; (Gen Instr 1). 31 pp.
ISO 8637-89. Haemodialysers, Haemofilters and Haemoconcentrators First Edition. 23 pp.

—**Extracorporeal Blood Systems**
BSI BS 7297: Part 2-90. 1990 Haemodialysers and Related Equipment Part 2: Extracorporeal Circuits for Use with Haemodialysers, Haemofilters and Haemoconcentrators. 12 pp.
CSA CAN3-Z364.1. 2-M84. Extracorporeal Circuits for Use with a Hemodialyser, Hemofilter, or Hemoconcentrator; (Gen Instr 1). 21 pp.
ISO 8638-89. Extracorporeal Blood Circuit for Haemodialysers, Haemofilters and Haemoconcentrators First Edition. 14 pp.

Hemocytometers
See Also: Medical Electrical Equipment; Medical Equipment
BSI BS 748-82. 1982 Haemacytometer and Particle Counting Chambers. 11 pp.
CNS T1056-82. Haemocytometer (Jan)(8333).
JIS T 4204-74. Haemocytometer.

Hemodialysis
See Also: Hemoconcentrators; Hemodialyzers; Hemofiltration Units

—**Dialysing Fluids**
CSA Z364.2.1-M1986P. Fluid Supply and Monitoring Systems for Hemodialysis; (Gen Instr 1 Thru 3). 48 pp.

—**Monitors**
CSA Z364.2.1-M1986P. Fluid Supply and Monitoring Systems for Hemodialysis; (Gen Instr 1 Thru 3). 48 pp.
CSA CAN3-Z364.2. 2-M86. Water Treatment Equipment and Water Quality Requirements for Hemodialysis; (Gen Instr 1 Thru 2). 25 pp.

—**Water Treatment**
CSA CAN3-Z364.2. 2-M86. Water Treatment Equipment and Water Quality Requirements for Hemodialysis; (Gen Instr 1 Thru 2). 25 pp.

Hemodialysis Equipment
Use: Hemodialyzers

Hemodialyzers
Use For: Haemodialysers; Haemodialysis Equipment; Hemodialysis Equipment *See Also:* Hemodialysis; Medical Electrical Equipment; Medical Equipment
BSI BS 7297: Part 1-90. 1990 Haemodialysers and Related Equipment Part 1: Haemodialysers, Haemofilters and Haemoconcentrators. 22 pp.
CSA CAN3-Z364.1. 1-M84. Hemodialysers, Hemofilters, and Hemoconcentrators; (Gen Instr 1). 31 pp.
ISO 8637-89. Haemodialysers, Haemofilters and Haemoconcentrators First Edition. 23 pp.

—**Extracorporeal Blood Systems**
BSI BS 7297: Part 2-90. 1990 Haemodialysers and Related Equipment Part 2: Extracorporeal Circuits for Use with Haemodialysers, Haemofilters and Haemoconcentrators. 12 pp.
CSA CAN3-Z364.1. 2-M84. Extracorporeal Circuits for Use with a Hemodialyser, Hemofilter, or Hemoconcentrator; (Gen Instr 1). 21 pp.
ISO 8638-89. Extracorporeal Blood Circuit for Haemodialysers, Haemofilters and Haemoconcentrators First Edition. 14 pp.

—**Safety**
BSI BS 5724: Sec 2.16-89. 1989 Medical Electrical Equipment Part 2: Particular Requirements for Safety Section 2.16: Specification for Haemodialysis Equipment (IEC 601-2-16: 1989). 31 pp.
BSI BS 5724: Sec 2.16-01. 1989 Amd 1 Medical Electrical Equipment Part 2: Particular Requirements for Safety Section 2.16: Specification for Haemodialysis Equipment (IEC 601-2-16: 1989) (AMD 6504) December 21, 1990. 32 pp.
BSI BS 5724: Sec 2.16-02. 1989 Amd 2 Medical Electrical Equipment Part 2: Particular Requirements for Safety Section 2.16: Specification for Haemodialysis Equipment (IEC 601-2-16: 1989) (AMD 6783) November 29, 1991. 33 pp.
CENELEC HD 395.2.16 S1-89. Medical Electrical Equipment Part 2: Particular Requirements for the Safety of Haemodialysis Equipment. 3 pp.

Hemodialyzers (Cont.)
—**Safety (Cont.)**
CSA CAN/CSA-C22. 2601.2.16-92. Medical Electrical Equipment Part 2: Particular Requirements for the Safety of Haemodialysis Equipment (IEC 601-2-16:1989); (Gen Instr 1). 49 pp.
DIN VDE 0750 Pt 206-84. Medical Electrical Equipment; Haemodialysis Equipment; Particular Requirements for Safety (Sept). 32 pp.
IEC 601 Pt 2-16-89. Medical Electrical Equipment Part 2: Particular Requirements for Safety of Haemodialysis Equipment First Edition. 61 pp.
SAA AS 3200.2.16-92. Approval and Test Specification—Medical Electrical Equipment—Part 2: Particular Requirements for Safety—Part 2.16: Haemodialysis Equipment (Supersedes AS 3207—1981) (in Professional Packages 17C, 28) (This is a Joint Standard with SNZ NZS 3200.2.16). 27 pp.

Hemofilters
Use: Hemofiltration Units

Hemofiltration Units
Use For: Haemofilters; Hemofilters
See Also: Hemodialysis; Medical Electrical Equipment; Medical Equipment
BSI BS 7297: Part 1-90. 1990 Haemodialysers and Related Equipment Part 1: Haemodialysers, Haemofilters and Haemoconcentrators. 22 pp.
CSA CAN3-Z364.1. 1-M84. Hemodialysers, Hemofilters, and Hemoconcentrators; (Gen Instr 1). 31 pp.
ISO 8637-89. Haemodialysers, Haemofilters and Haemoconcentrators First Edition. 23 pp.

—**Extracorporeal Blood Systems**
BSI BS 7297: Part 2-90. 1990 Haemodialysers and Related Equipment Part 2: Extracorporeal Circuits for Use with Haemodialysers, Haemofilters and Haemoconcentrators. 12 pp.
CSA CAN3-Z364.1. 2-M84. Extracorporeal Circuits for Use with a Hemodialyser, Hemofilter, or Hemoconcentrator; (Gen Instr 1). 21 pp.
ISO 8638-89. Extracorporeal Blood Circuit for Haemodialysers, Haemofilters and Haemoconcentrators First Edition. 14 pp.

Hemoglobinometers
Use For: Hemometers *See Also:* Medical Electrical Equipment; Medical Equipment
CNS T1057-82. Haemometers (Jan)(8334).
JIS T 4205-79. Haemometers.

Hemometers
Use: Hemoglobinometers

Hemp
Scope Note: For additional listings, see also specific products made from hemp *See Also:* Fibers; Jute; Ramie; Sisal

—**Ambria**
CNS L3020-62. Method of Test for Jute and Ambria Hemp Fibers (Jul)(1407)(R 1973).
CNS L4009-62. Ambria Hemp Fiber (Jul) (1406)(R 1973).

Hempseed Oil
See Also: Oils
CNS K5019-90. Hemp Twisting Oil (Aug)(1378). 2 pp.
CNS K5102-81. Crude Hempseed Oil (Sep)(7413).

Heng Shan Pears
Use: Pears

Heptane
See Also: n-Heptane; Hydrocarbons
JIS K 0505-83. High Purity Heptane.
JIS K 9701-83. Heptane.

n-Heptane
See Also: Hydrocarbons
CNS K0002-88. High Purity Heptance (Sep)(6217).

Herald (Dart) 100 Aircraft
See Also: Aircraft

—**Antenna Positions**
CAA. Herald (Dart) 100 Series (Approved Aerial Positions). 1 p.

Herald (Dart) 200 Aircraft
See Also: Aircraft

—**Antenna Positions**
CAA. Herald (Dart) 200 Series (Approved Aerial Positions). 1 p.

Herbicide Distribution Equipment
BSI BS 7293-90. 1990 Methods of Test for Equipment for Distributing Granulated Pesticides or Herbicides. 22 pp.
ISO 8524-86. Equipment for Distributing Granulated Pesticides or Herbicides—Test Method First Edition. 19 pp.

Herbicides
Use For: Weed Killers
SAA AS 1175-76. Herbicides of the Phenoxyacetic Acid Type. 52 pp.

—**Calcium Chloride Content—Volumetric Analysis**
MOD UK M 9529/66. Examination of Solution Weedkiller Chlorate.

—**Density**
MOD UK M 9529/66. Examination of Solution Weedkiller Chlorate.

—**Sampling**
MOD UK M 9544/69. Examination of Powder Weed Killer (Withdrawn).

—**Sieve Analysis**
MOD UK M 9544/69. Examination of Powder Weed Killer (Withdrawn).

—**Sodium Chlorate Content—Volumetric Analysis**
MOD UK M 9529/66. Examination of Solution Weedkiller Chlorate.
MOD UK M 9544/69. Examination of Powder Weed Killer (Withdrawn).

—**Sodium Chloride Content—Volumetric Analysis**
MOD UK M 9544/69. Examination of Powder Weed Killer (Withdrawn).

—**Sodium Sulfate Content—Volumetric Analysis**
MOD UK M 9544/69. Examination of Powder Weed Killer (Withdrawn).

—**Trisodium Phosphate Content—Volumetric Analysis**
MOD UK M 9544/69. Examination of Powder Weed Killer (Withdrawn).

—**Visual Inspection**
MOD UK M 9529/66. Examination of Solution Weedkiller Chlorate.
MOD UK M 9544/69. Examination of Powder Weed Killer (Withdrawn).

Herbs
Scope Note: For additional listings, use a more specific term *See Also:* Aniseed; Caraway; Coriander; Cumin; Ginger; Marjoram; Mint; Oregano; Peppermint; Savory; Spearmint; Spices; Tarragon; Thyme

—**Nursery Stock**
BSI BS 3936: Part 11-84. 1984 Amd 1 Nursery Stock Part 11: Container-Grown Culinary Herbs (AMD 6256) April 30, 1990. 5 pp.

—**Volatile Oil Content**
BSI BS 4585: Part 15-85. 1985 Methods of Test for Spices and Condiments Part 15: Determination of Volatile Oil Content. 7 pp.
ISO 6571-84. Spices, Condiments and Herbs—Determination of Volatile Oil Content First Edition. 6 pp.

Hermaphroditic Connectors
Use For: Sexless Connectors *See Also:* Connectors

—**Circular—Preferred Products List**
CECC CECC MUAHAG Vol 3A IS 4-91. Preferred Products List; Connectors; L. F. (En, Fr, Ge). 46 pp.

—**Precision—Coaxial**
CENELEC HD 351.1-76. Rigid Precision Coaxial Lines and Their Associated Precision Connectors Part 1: General Requirements and Measuring Methods. 2 pp.
CENELEC HD 351.2-76. Rigid Precision Coaxial Lines and Their Associated Precision Connectors Part 2: 50ohm 7mm Rigid Precision Coaxial Line and Associated Hermaphroditic Precision Coaxial Connector. 2 pp.
CENELEC HD 351.3 S2-81. Rigid Precision Coaxial Lines and Their Associated Precision Connectors—Part 3: 14 mm Rigid Precision Coaxial Line and Associated Hermaphroditic Precision Coaxial Connector—Characteristic Impedances 50 Ohms and 75 Ohms. 2 pp.

INTERNATIONAL AND NON-U.S. NATIONAL STANDARDS
SUBJECT INDEX

Hermaphroditic Connectors (Cont.)
—Precision—Coaxial (Cont.)
CENELEC HD 351.4 S2-84. Rigid Precision Coaxial Lines and Their Associated Precision Connectors—Part 4: 21 mm Rigid Precision Coaxial line and Associated Hermaphroditic Precision Coaxial Connector Characteristic Impedance 50 OHMs (Type 9/21) Characteristic Impedance. 2 pp.
IEC 457 Pt 2-74. Rigid Precision Coaxial Lines and Their Associated Precision Connectors Part 2: 50 Ohm 7 mm Rigid Precision Coaxial Line and Associated Hermaphroditic Precision Coaxial Connector First Edition. 15 pp.
IEC 457 Pt 3-80. Rigid Precision Coaxial Lines and Their Associated Precision Connectors Part 3: 14 mm Rigid Precision Coaxial Line and Associated Hermaphroditic Precision Coaxial Connector—Characteristic Impedances 50 Ohms and 75 Ohms Second Edition. 17 pp.
IEC 457 Pt 4-78. Rigid Precision Coaxial Lines and Their Associ-ated Precision Connec-tors Part 4: 21 mm Rigid Precision Coaxial Line and Associated Hermaphr-oditic Precision Coaxial Connector Characteristic Impedance 50 Ohms (Type 9/21)—Characteristic Impedance 75 Ohms (Type 6/21) Second Edition. 16 pp.

—Rigid Cables
CENELEC HD 351.1-76. Rigid Precision Coaxial Lines and Their Associated Precision Connectors Part 1: General Requirements and Measuring Methods. 2 pp.
CENELEC HD 351.2-76. Rigid Precision Coaxial Lines and Their Associated Precision Connectors Part 2: 50ohm 7mm Rigid Precision Coaxial Line and Associated Hermaphroditic Precision Coaxial Connector. 2 pp.
CENELEC HD 351.3 S2-81. Rigid Precision Coaxial Lines and Their Associated Precision Connectors—Part 3: 14 mm Rigid Precision Coaxial Line and Associated Hermaphroditic Precision Coaxial Connector—Characteristic Impedances 50 Ohms and 75 Ohms. 2 pp.
CENELEC HD 351.4 S2-84. Rigid Precision Coaxial Lines and Their Associated Precision Connectors—Part 4: 21 mm Rigid Precision Coaxial line and Associated Hermaphroditic Precision Coaxial Connector Characteristic Impedance 50 OHMs (Type 9/21) Characteristic Impedance. 2 pp.
IEC 457 Pt 2-74. Rigid Precision Coaxial Lines and Their Associated Precision Connectors Part 2: 50 Ohm 7 mm Rigid Precision Coaxial Line and Associated Hermaphroditic Precision Coaxial Connector First Edition. 15 pp.
IEC 457 Pt 3-80. Rigid Precision Coaxial Lines and Their Associated Precision Connectors Part 3: 14 mm Rigid Precision Coaxial Line and Associated Hermaphroditic Precision Coaxial Connector—Characteristic Impedances 50 Ohms and 75 Ohms Second Edition. 17 pp.
IEC 457 Pt 4-78. Rigid Precision Coaxial Lines and Their Associ-ated Precision Connec-tors Part 4: 21 mm Rigid Precision Coaxial Line and Associated Hermaphr-oditic Precision Coaxial Connector Characteristic Impedance 50 Ohms (Type 9/21)—Characteristic Impedance 75 Ohms (Type 6/21) Second Edition. 16 pp.

Hermetic Seals
See Also: Leakage; Seals
—Connectors
SBAC TS 347 ISSUE 1. Test for Electrical Connectors Hermeticity.
—Connectors—Aerospace
CEN PREN 2591 (Part C22)-92. Elements of Electrical and Optical Connection Test Methods Part C22—Hermeticity. 4 pp.
—Connectors—Aerospace—Thermal Shock
CEN PREN 2591 (Part C23)-92. Elements of Electrical and Optical Connection Test Methods Part C23—Thermal Shock. 5 pp.
—Insulating Glass
BSI BS 5713-79. 1979 Specification for Hermetically Sealed Flat Double Glazing Units. 15 pp.
—Metal Containers—Food
CNS Z6014-72. Determination of Capacity for Hermetically Sealed Metal Food Containers (Jun)(3347). 2 pp.
JIS Z 1571-85. Hermetically Sealed Metal Cans for Food and Drink. 10 pp.

Heron Aircraft
See Also: Aircraft
—Antenna Positions
CAA. Heron (Approved Aerial Positions). 1 p.

Herrings
See Also: Anchovies; Fish; Sardines

Herrings (Cont.)
—Dried
CNS N5073-81. Dried Herring and Horse Mackerel (Aug)(2538). 3 pp.

Hessian
Use: Burlaps

Heterocyclic Compounds
Scope Note: Use a more specific term
See: Carbazoles; Diiodofluorescein; p-Dimethylaminobenzalrhodanine; 5-(4-Dimethylaminobenzilidene)-2-Thioxo-4-Thiazolidinone; Dioxane; Epichlorohydrin; Ethylene Oxide; Fluorescein; Hexamethylenetetramine; Hydroxyproline; Maleic Anhydride; Phthalic Anhydride; Phthalic Anhydride-13-8; Propylene Oxide; Pyridine; Quinoline; L-Tryptophan; Uric Acid

Heterogeneity
—Aluminum Ores—Sampling
ISO 6138-91. Aluminium Ores—Experimental Determination of the Heterogeneity of Constitution First Edition. 6 pp.
ISO 6139-93. Aluminium Ores—Experimental Determination of the Heterogeneity of Distribution of a Lot First Edition. 10 pp.

Hex Inverters
Use: NOT Circuits—Hex

Hexachlorobenzene
See Also: Halogenated Hydrocarbons
CNS K1170-74. Hexachlorobenzene (Jun)(3716).
CNS K6371-74. Method of Test for Hexachlorobenzene (Jun)(3717).
ISO 2756-73. Hexachlorobenzene for Industrial Use—List of Methods of Test First Edition. 3 pp.

Hexachlorocyclohexane
Use: Benzene Hexachloride

1,2,3,4,5,6-Hexachlorocyclohexane
Use: Benzene Hexachloride

Hexachloroethane
BSI BS 577-66. (WITHDRAWN) 1966 Amd 1 Hexachlorethane. 17 pp.

Hexachlorophene Content Analysis
EC 83/514/EEC-83. Third Commission Directive on the Approximation of the Laws of the Member States Relating to Methods of Analysis Necessary for Checking the Composition of Cosmetic Products. 38 pp.
—Cosmetics
EC 83/514/EEC-83. Third Commission Directive on the Approximation of the Laws of the Member States Relating to Methods of Analysis Necessary for Checking the Composition of Cosmetic Products. 38 pp.

Hexacyanoferrate Content Analysis
—Effluents—Photographic Processing—Spectrometry
ISO 7766 Pt 1-93. Photography—Processing Wastes—Analysis of Cyanides—Part 1: Determination of Hexacyanoferrate(II) and Hexacyanoferrate(III) by Spectrometry First Edition. 9 pp.
—Sodium Chloride
BSI BS 7319: Part 12-90. 1990 Analysis of Sodium Chloride for Industrial Use Part 12: Methods of Determination of Anti-Caking Additives Content of Salt for Food Use. 6 pp.
BSI BS 7319: Part 12-01. 1990 Amd 1 Analysis of Sodium Chloride for Industrial Use Part 12: Methods for Determination of Anti-Caking Additives Content of Salt for Food Use (AMD 7786) June 15, 1993 (W). 7 pp.

Hexafluorosilicate Content Analysis
—Sodium Hexafluorosilicate—Volumetric Analysis
BSI BS 5705: Part 1-79. (WITHDRAWN) 1979 Sodium Hexafluorosilicate for Industrial Use Part 1: Determination of Free Acidity and Total Hexafluorosilicate Content. 4 pp.
ISO 4281-77. Sodium Hexafluorosilicate for Industrial Use—Determination of Free Acidity and Total Hexafluorosilicate Content—Titrimetric Method First Edition. 4 pp.

Hexafluorosilicic Acid
Use: Fluosilicic Acid

Hexafluorosilicic Acid Content Analysis
Use: Fluosilicic Acid Content Analysis

Hexagonal Belts
See Also: Belts (Machinery)
—Agricultural Equipment
ISO 5289-92. Agricultural Machinery—Endless Hexagonal Belts and Groove Sections of Corresponding Pulleys Second Edition. 8 pp.

Hexagonal Dienuts
Use: Threading Dies

Hexagonal Head Bolts
See Also: Bolts; Double Hexagonal Bolts; Structural Bolts
BSI BS 1083-65. (OBSOLESCENT) 1965 Amd 2 Precision Hexagon Bolts, Screws and Nuts (B.S.W. and B.S.F. Threads). 31 pp.
BSI BS 3692-67. 1967 Amd 1 ISO Metric Precision Hexagon Bolts, Screws and Nuts. 60 pp.
BSI BS EN 24014-92. 1992 Hexagon Head Bolts—Product Grades A and B (ISO 4014: 1988). 15 pp.
BSI BS EN 24016-92. 1992 Hexagon Head Bolts—Product Grade C (ISO 4016: 1988). 15 pp.
CEN PREN 781-92. Hexagon Bolts for High-Strength Bolting with Large Width Across Flats (Thread Lengths According to ISO 888)—Product Grade C—Property Classes 8.8 and 10.9. 11 pp.
CEN PREN 782-92. Hexagon Bolts for High-Strength Structural Bolting with Large Width Across Flats (Short Thread Length)—Product Grade C—Property Class 10.9. 11 pp.
CEN EN 24014-91. Hexagon Head Bolts—Product Grades A and B. 3 pp.
CEN EN 24016-91. Hexagon Head Bolts—Product Grade C. 3 pp.
CNS B1006-47. Minimum Distance Across Hexagonal Bolts (Mar)(79)(R 1973).
CNS B2120-88. Hexagon Head Bolts (Finished and Semi-Finished, Metric Coarse Thread) (Aug)(3121). 9 pp.
CNS B2121-87. Hexagon Bolts with Thread Approximately to Head, Finished and Semi-Finished, Metric Coarse Thread (Oct)(3122). 5 pp.
CNS B2122-87. Hexagon Head Bolts, Regular, Metric Thread (Oct)(3123). 4 pp.
CNS B2196-87. Hexagon Head Bolts, Finished and Semi-Finished Metric Fine Thread (Oct)(4320).
CNS B2197-87. Hexagon Bolts with Thread Approximately to Head, Finished and Semi-Finished, Metric Fine Thread (Oct)(4321).
CNS B2198-82. Hexagon Bolts with Thread Approximately to Head, Regular, Metric Thread (Dec)(4322).
CNS B2209-81. Hexagon Head Fitting Bolts with Long Threaded Portion (Apr)(4364).
CNS B2210-81. Hexagon Head Fitting Bolts with Short Threaded Portion (Apr)(4365).
CNS B2359-82. Hexagon Head Bolts with Reduced Body (Oct)(4702).
DIN ENGL 601 (S)-87. M 5 to M 52 Hexagon Head Bolts; Product Grade C (Sept) (Superseded by DIN EN 24016). 8 pp.
DIN ENGL 609-84. Hexagon Fit Bolts with Long Threaded Dog Point (July). 8 pp.
DIN ENGL 610-84. Hexagon Fit Bolts with Short Threaded Dog Point (July). 8 pp.
DIN ENGL 931 Pt 1 (S)-87. M 1,6 to M 39 Hexagon Head Bolts; Product Grades A and B (Sept) (Superseded by DIN EN 24014). 7 pp.
DIN ENGL 931 Pt 2-87. M 42 to M 160 x 6 Hexagon Head Bolts; Product Grade B (Sept). 5 pp.
DIN ENGL 960 (S)-90. M8 x 1 to M100 x 4 Hexagon Head Bolts with Fine Pitch Thread; Product Grades A and B (Jan) (Superseded by DIN EN 28765). 8 pp.
DIN ENGL 961 (S)-90. M8 x 1 to M52 x 3 Hexagon Head Bolts with Fine Pitch Thread; Product Grades A and B (Jan) (Superseded by DIN EN 28676). 6 pp.
DIN ENGL EN 24014-92. Hexagon Head Bolts; Product Grades A and B; (ISO 4014: 1988) (Oct) (Supersedes DIN 931 Part 1, September 1987 Edition, and DIN ISO 4014, September 1989 Edition). 12 pp.
DIN ENGL EN 24016-92. Hexagon Head Bolts Product Grade C (ISO 4016: 1988) (Feb) (Supersedes DIN 601, September 1987 Edition, and DIN ISO 4016, October 1989 Edition). 12 pp.
ISO 4014-88. Hexagon Head Bolts—Product Grades A and B Second Edition. 11 pp.
ISO 4016-88. Hexagon Head Bolts—Product Grade C Second Edition. 11 pp.
JIS B 1180-85. Hexagon Head Bolts and Hexagon Head Screws. 52 pp.

INTERNATIONAL AND NON-U.S. NATIONAL STANDARDS
SUBJECT INDEX

Hexagonal

Hexagonal Head Bolts (Cont.)

SAA AS B148-56. Unified Black Hexagon Bolts, Screws and Nuts (UNC and UNF Threads) and Plain Washers—Heavy Series Being BS 1769:1951 (Including Amendments of March 1955, May 1957, and January 1960), Endorsed Without Australian Amendment Corrig.. 20 pp.

SAA AS 1110-84. ISO Metric Hexagon Precision Bolts and Screws. 22 pp.

SAA AS 1111-80. ISO Metric Hexagon Commercial Bolts and Screws. 22 pp.

SNZ NZS 1069-66. Specification for Precision Hexagon Bolts, Screws and Nuts (B.S.W. and B.S.F. Threads). 32 pp.

SNZ NZS/AS 1110-84. ISO Metric Hexagon Precision Bolts and Screws (This is a Joint Standard with SAA AS 1110). 22 pp.

SNZ NZS/AS 1111-80. ISO Metric Hexagon Commercial Bolts and Screws (This is a Joint Standard with SAA AS 1111). 22 pp.

SNZ NZS/AS 2465-81. Unified Hexagon Bolts, Screws and Nuts (UNC and UNF Threads) (This is a Joint Standard with SAA AS 2465). 30 pp.

—Aerospace

BSI A 241-73. 1973 Amd 3 General Requirements for Steel Protruding-Head Bolts of Tensile Strength 1250 MPa (180 000 Lbf/Square Inches) or Greater. 44 pp.

BSI A 254-A 259-76. 1976 Amd 2 Hexagonal Head Titanium Alloy Bolts 160 000 Lbf/ Square Inches (1100 MPa). 12 pp.

DIN ENGL LN 29930-84. Aerospace; Bolts, Hexagon, Close Tolerance with Reduced Head and Short Thread Length (Apr). 4 pp.

DIN ENGL LN 29943-81. Aerospace; Bolts, Hexagon, Close Tolerance, Short Thread, Titanium Alloy (Dec). 4 pp.

—Aerospace—Close Shank

AECMA PREN2413-87. Bolts, Shouldered, Thin Hexagonal Head, Close Tolerance Shank, Short Thread, in Steel Cadmium Plated Classification 1100 MPa/235 Degrees Celsius. 7 pp.

AECMA PREN2549-87. Bolts, Hexagonal Normal Head, Close Tolerance Shank, Short Thread in Titanium, Anodised Classification 1100 MPa/315 Degrees Celsius. 7 pp.

AECMA PREN2859-89. Bolts, Hexagonal Normal Head, Close Tolerance Shank, Short Thread in Steel Cadmium Plated Classification 1100 MPa/235 Degrees Celsius. 7 pp.

AECMA PREN3052-87. Bolts, Hexagonal Normal Head, Close Tolerance Shank, Short Thread in Corrosion Resisting Steel, Passivated Classification 1100 MPa/425 Degrees Celsius. 7 pp.

BSI 3A 212-93. 1993 Cadmium-Plated Steel Bolts, 55/65 Tonf/Square Inches (Unified Hexagons, 3A Unified and Close Tolerance Shanks) (Supersedes 2A 212: 1973) (S). 9 pp.

CEN PREN2859-87. Bolts, Hexagonal Normal Head, Close Tolerance Shank, Short Thread in Steel Cadmium Plated Classification: 1100 MPa/235 Degrees C. 9 pp.

CEN PREN 3740-92. Bolts, Shouldered, Thin Hexagonal Head, Close Tolerance Shank, Short Thread, in Titanium Alloy, MoS2 Lubricated Classificaton: 1 100 MPa (at Ambient Temperature) /315 Degrees C. 5 pp.

—Aerospace—Coarse Shank

AECMA PREN 2888-88. Bolts, Hexagonal Normal Head, Coarse Tolerance Shank, Short Thread, in Corrosion Resisting Steel, Passivated Classification: 600 MPa/425 Degrees Celsius. 7 pp.

AECMA PREN2889-89. Bolts, Hexagonal Normal Head, Coarse Tolerance Shank, Short Thread, in Steel, Cadmium Plated Classification 900 MPa/235 Degrees Celsius. 2 pp.

CEN PREN2889-87. Bolts, Hexagonal Normal Head, Coarse Tolerance Shank, Short Thread, in Steel, Cadmium Plated Classification: 900 MPa/235 Degrees C. 9 pp.

—Aerospace—Relieved Shank

AECMA PREN3006-87. Bolts, Hexagon Head, Relieved Shank, Long Thread, in Heat Resisting Steel FE-PA92HT (A286) Classification 900 MPa/ 650 Degrees Celsius Unplated. 7 pp.

AECMA PREN3007-87. Bolts, Hexagon Head, Relieved Shank, Long Thread, in Heat Resisting Steel FE-PA92HT (A286) Classification 900 MPa/ 650 Degrees Celsius Silver Plated. 7 pp.

AECMA PREN3008-87. Bolts, Hexagon Head, Relieved Shank, Long Thread, in Heat Resisting Nickel Base Alloy NI-P100HT (INCO 718) Classification 1275 MPa/650 Degrees Celsius Unplated. 7 pp.

AECMA PREN3009-87. Bolts, Hexagon Head, Relieved Shank, Long Thread, in Heat Resisting Nickel Base Alloy NI-P100HT (INCO 718) Classification 1275 MPa/650 Degrees Celsius Silver Plated. 7 pp.

Hexagonal Head Bolts (Cont.)

—Aerospace—Relieved Shank (Cont.)

AECMA PREN3010-87. Bolts, Hexagon Head, Relieved Shank, Long Thread, in Heat Resisting Nickel Base Alloy N1-P101HT (Waspaloy) Classification 1210 MPa/730 Degrees Celsius Unplated. 7 pp.

AECMA PREN3011-87. Bolts, Hexagon Head, Relieved Shank, Long Thread, in Heat Resisting Nickel Base Alloy N1-P101HT (Waspaloy) Classification 1210 MPa/730 Degrees Celsius Silver Plated. 7 pp.

AECMA PREN3613. Bolts, Normal Hexagon Head, Relieved Shank, Long Thread, in Heat Resisting Nickel Base Alloy NI-P100HT (Inconel 718), Silver Plated Classification: 1275 MPa/650 Degrees. 8 pp.

AECMA PREN3614-90. Bolts, Normal Hexagon Head, Relieved Shank, Long Thread, in Heat Resisting Steel FE-PA92HT (A286), Silver Plated Classification: 900 MPa/650 Degrees. 7 pp.

CEN PREN 3324-91. Bolts, Hexagon Head, Relieved Shank, Long Thread, in Heat Resisting Steel FR-PM 38 (FV 535) Classification: 1000 MPa/550 Degrees C Unplated. 7 pp.

CEN PREN 3687-91. Bolts, Normal Hexagon Head, Relieved Shank, Long Thread, in Heat Resisting Steel, FE-PA92HT (A286), Silver. 7 pp.

CEN PREN 3724-91. Bolts, Double Hexagon Head Relieved Shank, Long Thread, in Titanium Alloy Tl-P63, MoS2 Coated Classification: 1100 MPa (at Ambient Temperature)/350 Degrees C. 7 pp.

CENELEC PREN 3687-89. Bolts, Normal Hexagon Head,Relieved Shank, Long Thread, in Heat Resisting Steel FE PA92HT (A286) Silver Plated, Classification 1100 MPa/650 Degrees C. 6 pp.

—Aerospace—Shank

ISO 3193-91. Aerospace—Bolts, Normal Hexagonal Head, Normal Shank, Short or Medium Length MJ Threads, Metallic Material, Coated or Uncoated, Strength Classes Less Than or Equal to 1 100 MPa—Dimensions First Edition; (Corrigendum 1-1992). 7 pp.

—Aerospace—Shear

BSI 3A 109-85. 1985 Amd 1 Steel Bolts (Unified Hexagons and Unified Threads) (AMD 5563) July 29, 1988. 11 pp.

BSI 4A 112-92. 1992 Cadmium-Plated Steel Shear Bolts (Unified Hexagons and Unified Threads) (Supersedes 3A 112: 1985). 12 pp.

BSI 3A 112-85. (WITHDRAWN) 1985 Amd 1 Cadmium-Plated Steel Shear Bolts (Unified Hexagons and Unified Threads) (AMD 5564) September 30, 1988 (Superseded by 4A 112: 1992). 11 pp.

—Aerospace—Socket—Cap

CEN PREN 3303-92. Bolts, Cap Head, Hexagon Socket, Coarse Tolerance Normal Shank, Medium Length Thread, in Alloy Steel, Cadmium Plated Classification: 1 100 MPa (at Ambient Temperature) /235 Degrees C. 6 pp.

—Aircraft

BSI 2A 60-05. 1962 Amd 5 Cadmium-Plated Steel Hexagonal-Headed Shear Bolts (B.S.F. Threads) for Aircraft (AMD 7660) April 15, 1993 (S). 10 pp.

BSI 3A 102-62. 1962 Amd 4 Steel Bolts (Unified Hexagons and Unified Threads) for Aircraft (AMD 5388) August 31, 1988. 14 pp.

BSI 3A 111-62. (WITHDRAWN) 1962 Amd 4 Cadmium-Plated Steel Bolts (Unified Hexagons and Unified Threads) with Close Tolerance Shanks for Aircraft. 11 pp.

BSI 4A 111-93. 1993 Cadmium-Plated Steel Bolts (Unified Hexagons and Unified Threads) with Close Tolerance Shanks (S). 9 pp.

BSI 3A 169-62. 1962 Amd 1 Aluminium Alloy Bolts (Unified Hexagons and Unified Threads) for Aircraft. 9 pp.

BSI A 226-65. 1965 Amd 2 Steel Hexagon Head Bolts (Short Thread: Class 3A). 7 pp.

BSI 2A 229-72. 1972 Amd 3 Hexagonal Head Steel Bolts 160 000 Lbf/Square Inches (1100 MPa) (AMD 6142) December 21, 1990. 16 pp.

SBAC AS 2504 ISSUE 10. Bolts, Hexagon, Close Tolerance. 1 p.

SBAC AS 4569 ISSUE 7. Bolts, 2 BA Hexagon, Close Tolerance. 1 p.

SBAC AS 52000-599 ISSUE 1. Bolt—Hexagon—Self Retaining—Impedance Type—Metric Series.

SBAC AGS 3817 (V). 'Jo-Bolt' Fastners Hexagon Head.

—Aircraft—Blanking Caps

SBAC AS 2942 ISSUE 3. Bolt. 1 p.

—Aircraft—Control Columns

SBAC AS 2292 ISSUE 6. Bolt 3/8 Inch Dia-Control Column. 1 p.

SBAC AS 2293 ISSUE 6. Bolt 5/16 Inch Dia-Control Column. 1 p.

Hexagonal Head Bolts (Cont.)

—Aircraft—Corrosion Resistant

BSI 3A 104-62. 1962 Amd 1 Corrosion-Resisting Steel Bolts (Unified Hexagons and Unified Threads) for Aircraft. 12 pp.

—Aircraft—Machine

SBAC AS 44702 ISSUE 2. Bolt, Machine—Hexagon Head—.190-32—(Balance Weight).

—Aircraft—Rings

SBAC AS 2006-2020 ISSUE 9. Ring, Bolt. 1 p.
SBAC AS 2021-2025 ISSUE 5. Bolt for Inner Ring. 1 p.

—Aircraft—Rudder Bars

SBAC AS 2196 ISSUE 6. Special Bolt (Rudder Bar). 1 p.
SBAC AS 6332 ISSUE 1. Special Bolt (Rudder Bar).

—Aircraft—Self Retaining

SBAC AS 53761-543 60 ISSUE 1. Bolt—Hexagon—Self Retaining—Impedance Type—Inch Series.

—Aircraft—Self Retaining—Hole Size

SBAC RS 747 ISSUE 1. Hole Sizes for Self Retaining "Impedance Type" Bolts. Inch Series.
SBAC RS 748 ISSUE 1. Hole Sizes for Self Retaining "Impedance Type" Bolts. Metric Series.

—Aircraft—Shank

SBAC AS 49800-899 ISSUE 1. Bolt, Hexagon Head, P.D. Shank BS. HR650. 190-32 UNJF-3A. BS4084.
SBAC AS 49900-999 ISSUE 1. Bolt, Hexagon Head, P.D. Shank BS.HR650. 250-28 UNJF-3A. BS4084.
SBAC AS 50000-099 ISSUE 1. Bolt, Hexagon Head, P.D. Shank BS.HR650. 3125-24 UNJF-3A. BS4084.
SBAC AS 50100-199 ISSUE 1. Bolt, Hexagon Head, P.D. Shank BS.HR650 UNJF-3A. BS4084.
SBAC AS 50200-299 ISSUE 2. Bolt, Hexagon Head, P.D. Shank INCO 718 0.1900-32 UNJF-3A. BS4084.
SBAC AS 50300-399 ISSUE 2. Bolt, Hexagon Head, P.D. Shank INCO 718 0.2500-28 UNJF-3A. BS4084.
SBAC AS 50400-499 ISSUE 2. Bolt, Hexagon Head, P.D. Shank INCO 718 0.3125-24 UNJF-3A. BS4084.
SBAC AS 50500-599 ISSUE 2. Bolt, Hexagon Head, P.D. Shank INCO 718 0.3750-24 UNJF-3A. BS4084.
SBAC AS 56300-399 ISSUE 1. Bolt, Hexagon Head, PD Shank, INCO 718, 0.4375—20 UNJF—3A. BS4084.
SBAC AS 56400-99 ISSUE 1. Bolt, Hexagon Head, PD Shank, INCO 718, 0.5000-20 UNJF-3A. BS4084.
SBAC AS 56600-699 ISSUE 1. Bolt, Hexagon Head, P.D. Shank, Waspaloy, 0.2500—28 UNJF—3A. BS4084.
SBAC AS 56700-799 ISSUE 1. Bolt, Hexagon Head, P.D. Shank, Waspaloy, 0.3125—24 UNJF—3A. BS4084.
SBAC AS 56800-899 ISSUE 1. Bolt, Hexagon Head, P.D. Shank, Waspaloy, 0.3750—24 UNJF—3A. BS4084.
SBAC AS 56900-999 ISSUE 1. Bolt, Hexagon Head, P.D. Shank, Waspaloy, 0.4375—20 UNJF—3A. BS4084.
SBAC AS 57000-099 ISSUE 1. Bolt, Hexagon Head, P.D. Shank, Waspaloy, 0.5000—20 UNJF—3A. BS 4084.
SBAC AS 58900-999 ISSUE 1. Bolt, Hexagon Head, P.D. Shank, FV535, 0.1900-32 UNJF-3A. BS4084.
SBAC AS 59000-099 ISSUE 1. Bolt, Hexagon Head, P.D. Shank, FV535, 0.2500—28 UNJF—3A. BS4084.
SBAC AS 59100-199 ISSUE 1. Bolt, Hexagon Head, P.D. Shank, FV535, 0.3125-24 UNJF-3A. BS4084.
SBAC AS 59200-299 ISSUE 1. Bolt, Hexagon Head, P.D. Shank, FV535, 0.3750-24 UNJF-3A. BS4084.
SBAC AS 59300-399 ISSUE 1. Bolt, Hexagon Head, P.D. Shank, FV535, 0.4375-20 UNJF-3A. BS4084.
SBAC AS 59400-499 ISSUE 1. Bolt, Hexagon Head, P.D. Shank, FV535, 0.5000-20 UNJF-3A. BS4084.

—Aircraft—Washers (Fasteners)

BSI SP 107-112-54. 1954 Amd 2 Tab Washers for Unified Hexagons for Aircraft. 4 pp.
BSI SP 122-125-58. 1958 Washers for Unified Hexagons for Aircraft (Primarily for Packing Purposes). 5 pp.
BSI SP 126-127-59. 1959 Washers for Unified Hexagons for Aircraft (Primarily for Facing Purposes). 4 pp.

—Automotive—Wheels

MOD UK DSTAN 53-7-68. Bolts and Studs for Vehicle Wheels Issue 1. 12 pp.

—Black

BSI BS 4190-67. 1967 Amd 2 ISO Metric Black Hexagon Bolts, Screws and Nuts. 41 pp.

INTERNATIONAL AND NON-U.S. NATIONAL STANDARDS
SUBJECT INDEX
Hexagonal

Hexagonal Head Bolts (Cont.)

—Counterbores
DIN ENGL 74 Pt 3-91. Counterbores (Holes) for Hexagon Bolts and Nuts (May). 6 pp.
DIN ENGL 974 Pt 2-91. Diameters of Counterbores (Holes) for Hexagon Bolts and Nuts (May). 4 pp.

—Fine Pitch Thread
BSI BS EN 28765-92. 1992 Hexagon Head Bolts with Metric Fine Pitch Thread—Product Grades A and B (ISO 8765: 1988) (E). 15 pp.
BSI BS EN 28765-01. 1992 Amd 1 Hexagon Head Bolts with Metric Fine Pitch Thread—Product Grades A and B (ISO 8765: 1988) (AMD 7385) November 15, 1992. 20 pp.
CEN EN 28765-91. Hexagon Head Bolts with Metric Fine Pitch Thread—Product Grades A and B. 3 pp.
DIN ENGL EN 28765-92. Hexagon Head Bolts with Metric Fine Pitch Thread; Product Grades A and B (ISO 8765:1988) (Feb) (Supersedes DIN 960, January 1990 Edition, and DIN ISO 8765, January 1990 Edition). 12 pp.
ISO 8765-88. Hexagon Head Bolts with Metric Fine Pitch Thread—Product Grades A and B First Edition. 11 pp.

—Flanged
DIN ENGL 6921-83. Hexagon Flange Bolts (June). 6 pp.
DIN ENGL 6922-83. Hexagon Flange Bolts with Reduced Shank (June). 6 pp.
ISO 4162-90. Hexagon Flange Bolts—Small Series First Edition. 11 pp.
JIS B 1189-87. Hexagon Bolts with Flange. 16 pp.

—Flat Widths
CNS B2119-87. Hexagon Bolts with Small Widths Across Flats (Oct)(3120).
CNS B2357-87. Hexagon Head Bolts with Thread Approximately to Head and with Small Widths Across Flats (Oct)(4700).
ISO 272-82. Fasteners—Hexagon Products—Widths Across Flats Second Edition. 5 pp.

—Friction Grip—High Strength—Structural
BSI BS 4395: Part 1-69. 1969 Amd 2 High Strength Friction Grip Bolts and Associated Nuts and Washers for Structural Engineering Part 1: General Grade. 36 pp.
BSI BS 4395: Part 2-69. 1969 Amd 2 High Strength Friction Grip Bolts and Associated Nuts and Washers for Structural Engineering Part 2: Higher Grade Bolts and Nuts and General Grade Washers. 36 pp.
BSI BS 4604: Part 1-70. 1970 Amd 3 The Use of High Strength Friction Grip Bolts in Structural Steelwork. Metric Series Part 1: General Grade. 17 pp.
SNZ NZS/BS 4395: Part 1-69. Specification for High Strength Friction Grip Bolts and Associated Nuts and Washers for Structural Engineering Part 1: General Grade. 36 pp.
SNZ NZS/BS 4395: Part 2-69. Specification for High Strength Friction Grip Bolts and Associated Nuts and Washers for Structural Engineering Part 2: Higher Grade Bolts and Nuts and General Grade Washers. 36 pp.

—High Strength—Structural
CNS B2124-80. Hexagon Head Fitting Bolts for Steel Structures (May)(3125).
DIN ENGL 6914-89. High-Strength Hexagon Head Bolts with Large Widths Across Flats for Structural Steel Bolting (Oct). 6 pp.
DIN ENGL 7999-83. High Strength Hexagon Fit Bolts with Large Widths Across Flats for Structural Steel Bolting (Dec). 5 pp.
ISO 7411-84. Hexagon Bolts for High-Strength Structural Bolting with Large Width Across Flats (Thread Lengths According to ISO 888)—Product Grade C—Property Classes 8.8 and 10.9 First Edition. 9 pp.
ISO 7412-84. Hexagon Bolts for High-Strength Structural Bolting with Large Width Across Flats (Short Thread Length)—Product Grade C—Property Classes 8.8 and 10.9 First Edition. 9 pp.

—Machine
MOD UK DSTAN 53-18-01. Bolts, Machine, Steel, Hexagon Head, Metric (Coarse Pitch Precision Bolts) Issue 3; Amendment 1. 19 pp.
MOD UK DSTAN 53-29-70. Bolts, Machine, Metric, Hexagon Head, Brass (Coarse Pitch) Issue 1. 7 pp.
MOD UK DSTAN 53-39-83. Bolts, Machine, Hexagon Head, Corrosion-Resistant Steel, Metric (Coarse Pitch) Issue 2. 5 pp.
MOD UK DSTAN 53-51-72. Bolts, Machine, Steel, Hexagon Head, UNC Issue 1. 9 pp.
MOD UK DSTAN 53-62-73. Bolts, Machine, Steel Hexagon Head, UNF Issue 1. 33 pp.
MOD UK DSTAN 53-62: Addendum. Bolts, Machine, Steel Hexagon Head, UNF Issue 1. 66 pp.

Hexagonal Head Bolts (Cont.)

—Mass
DIN ENGL 931 Suppl. 1-86. Hexagon Head Bolts with Shank; Masses (Sept). 3 pp.

—Reduced Shank
BSI BS EN 24015-92. 1992 Hexagon Head Bolts—Product Grades B—Reduced Shank (Shank Diameter Approximately Equal to Pitch Diameter) (ISO 4015: 1979). 12 pp.
CEN EN 24015-91. Hexagon Head Bolts—Product Grade B—Reduced Shank (Shank Diameter = Pitch Diameter). 3 pp.
DIN ENGL EN 24015-91. Hexagon Head Bolts with Reduced Shank; (Shank Diameter = Pitch Diameter); Product Grade B (ISO 4015: 1979) (Dec). 9 pp.
ISO 4015-79. Hexagon Head Bolts—Product Grade B—Reduced Shank (Shank Diameter is Approximately Equal to Pitch Diameter) First Edition. 7 pp.

—Sets—High Strength
CNS B2768-85. Sets of High Strength Hexagon Bolt, Hexagon Nut and Plain Washers for Friction Grip Joints (Aug)(11328).
CNS B7272-85. Method of Test for the Sets of High Strength Hexagon Bolt, Hexagon Nut and Plain Washers for Friction Grip Joints (Aug)(11329).
JIS B 1186-79. Sets of High Strength Hexagon Bolt, Hexagon Nut and Plain Washers for Friction Grip Joints. 21 pp.

—Socket—Counterbores
JIS B 4236-89. Counterbores for Hexagon Socket Head Bolts. 9 pp.

—Socket Wrenches
DIN ENGL 3124-80. Square Drive Socket Wrenches for Hexagon Bolts; Hand-Operated (Nov). 4 pp.

—Structural
BSI BS 7419-91. 1991 Holding Down Bolts. 13 pp.
CNS B2123-87. Hexagon Head Bolts for Steel Structures (Oct)(3124).
DIN ENGL 7968-89. Hexagon Fit Bolts for Structural Steel Bolting for Supply with or Without Nut (Oct). 6 pp.
DIN ENGL 7990-89. Hexagon Head Bolts for Structural Steel Bolting for Supply with Nut (Oct). 6 pp.

—Structural—Flat Widths
CNS B2211-87. Hexagon Bolts with Large Widths Across Flats for Steel Structures (Joints with High-Tensile, Prestressed Bolts) (Oct)(4366).

—Towers
CSA B33.4-1973. Galvanized Steel Tower Bolts and Nuts. 14 pp.

—Washer
JIS B 1187-87. Hexagon Head Screws with Washer. 30 pp.

—Washers (Fasteners)
BSI BS 5814-79. (WITHDRAWN) 1979 Metal Tab Washers for General Engineering Purposes. Metric Series. 8 pp.
CNS B2011-47. Semi-Finished Washer for Hexagonal Bolts and Nuts (M1—M20) (Mar)(151). 2 pp.
CNS B2013-78. Washers Coarse Type Primarily for Hexagon Bolts and Nuts (Mar)(153). 2 pp.
DIN ENGL 125 Pt 1-90. Product Grade A Washers with a Hardness up to 250 HV Designed for Use with Hexagon Head Bolts and Nuts (Mar). 4 pp.
DIN ENGL 125 Pt 2-90. Product Grade A Washers with a Hardness from 300 HV Designed for Use with Hexagon Head Bolts and Nuts (Mar). 4 pp.
DIN ENGL 126-90. Product Grade C Washers; Designed for Use with Hexagon Head Bolts and Nuts (Mar). 2 pp.

Hexagonal Head Screws

See Also: Allen Wrenches; Screws
BSI BS 1083-65. (OBSOLESCENT) 1965 Amd 2 Precision Hexagon Bolts, Screws and Nuts (B.S.W. and B.S.F. Threads). 31 pp.
BSI BS 3692-67. 1967 Amd 1 ISO Metric Precision Hexagon Bolts, Screws and Nuts. 60 pp.
BSI BS EN 24017-92. 1992 Hexagon Head Screws—Product Grades A and B (ISO 4017: 1988). 15 pp.
BSI BS EN 24018-92. 1992 Hexagon Head Screws—Product Grade C (ISO 4018: 1988). 15 pp.
BSI BS EN 28676-92. 1992 Hexagon Head Screws with Metric Fine Pitch Thread—Product Grades A and B (ISO 8676: 1988). 15 pp.
CEN EN 24017-91. Hexagon Head Screws—Product Grades A and B. 3 pp.
CEN EN 24018-91. Hexagon Head Screws—Product Grade C. 3 pp.
CEN EN 28676-91. Hexagon Head Screws with Metric Fine Pitch Thread—Product Grades A and B. 3 pp.

Hexagonal Head Screws (Cont.)

CNS B1059-87. Hexagon Screws and Nuts and Thicknesses of Nuts (Oct)(4244).
DIN ENGL 558 (S)-87. M 5 to M 36 Hexagon Head Screws Threaded up to the Head; Product Grade C (Sept) (Superceded by DIN EN 24017). 4 pp.
DIN ENGL 933 (S)-87. M 1,6 to M 52 Hexagon Head Screws Threaded up to the Head; Product Grades A and B (Sept) (Superseded by DIN EN 24017). 9 pp.
DIN ENGL EN 24017-92. Hexagon Head Screws; Product Grades A and B (ISO 4017: 1988) (Feb) (Supersedes DIN 933, September 1987 Edition, and DIN ISO 4017, October 1989 Edition). 12 pp.
DIN ENGL EN 24018-92. Hexagon Head Screws; Product Grade C (ISO 4018: 1988) (Feb) (Supersedes DIN 558, September 1987 Edition, and DIN ISO 4018, October 1989 Edition). 12 pp.
DIN ENGL EN 28676-92. Hexagon Head Screws with Metric Fine Pitch Thread; Product Grades A and B (ISO 8676:1988) (Feb) (Supersedes DIN 961, January 1990 Edition, and DIN ISO 8676, January 1990 Edition). 12 pp.
ISO 4017-88. Hexagon Head Screws—Product Grades A and B Second Edition. 11 pp.
ISO 4018-88. Hexagon Head Screws—Product Grade C Second Edition. 11 pp.
ISO 8676-88. Hexagon Head Screws with Metric Fine Pitch Thread—Product Grades A and B First Edition. 11 pp.
JIS B 1180-85. Hexagon Head Bolts and Hexagon Head Screws. 52 pp.
SAA AS B148-56. Unified Black Hexagon Bolts, Screws and Nuts (UNC and UNF Threads) and Plain Washers—Heavy Series Being BS 1769:1951 (Including Amendments of March 1955, May 1957, and January 1960), Endorsed Without Australian Amendment Corrig.. 20 pp.
SAA AS 1110-84. ISO Metric Hexagon Precision Bolts and Screws. 22 pp.
SAA AS 1111-80. ISO Metric Hexagon Commercial Bolts and Screws. 22 pp.
SNZ NZS 1069-66. Specification for Precision Hexagon Bolts, Screws and Nuts (B.S.W. and B.S.F. Threads). 32 pp.
SNZ NZS/AS 1110-84. ISO Metric Hexagon Precision Bolts and Screws (This is a Joint Standard with SAA AS 1110). 22 pp.
SNZ NZS/AS 1111-80. ISO Metric Hexagon Commercial Bolts and Screws (This is a Joint Standard with SAA AS 1111). 22 pp.
SNZ NZS/AS 2465-81. Unified Hexagon Bolts, Screws and Nuts (UNC and UNF Threads) (This is a Joint Standard with SAA AS 2465). 30 pp.

—Aerospace
AECMA PREN2887-88. Screws, Hexagonal Normal Head, Fully Threaded, in Corrosion Resisting Steel, Passivated Classification: 600 MPa/425 Degrees Celsius. 8 pp.
AECMA PREN2937-86. Screws, Hexagon Head, Thread to Head, in Heat Resisting Steel FE-PA92HT (A286) Unplated Classification 900 MPa/650 Degrees Celsius. 5 pp.
AECMA PREN2938-86. Screws, Hexagon Head, Thread to Head, in Heat Resisting Steel FE-PA92HT (A286) Silver Plated Classification 900 MPa/650 Degrees Celsius. 5 pp.
AECMA PREN3112-87. Screws, Hexagonal Normal Head, Fully Threaded in Steel, Cadmium Plated Classification 900 MPa/235 Degrees Celsius. 7 pp.
AECMA PREN3113-87. Screws, Hexagonal Normal Head, Fully Threaded in Titanium Alloy Anodized Classification 900 MPa/315 Degrees Celsius. 7 pp.
CEN PREN 3308-91. Screws, Normal Hexagonal Head, Threaded to Head, in Titanium Alloy, MoS2 Lubricated Classification: 1100MPa (at Ambient Temperature)/315 Degrees C. 7 pp.
DIN ENGL LN 9038-81. Aerospace; Screws, Hexagon Head, Fully Threaded (Dec). 4 pp.

—Aerospace—Socket—Cap
BSI A 287-292-82. 1982 Hexagon Socket Head Cap Screws. Metric Series. 10 pp.

—Aircraft—Machine
SBAC AS 36500-599 (V). Screw, Machine, Hexagon Head, Washer Face, HR650 4(.112)-40 UNJC-3A BS4084.
SBAC AS 36600-699 (V). Screw, Machine, Hexagon Head, Washer Face, HR650 6(.138)-32 UNJF-3A BS4084.
SBAC AS 36700-799 (V). Screw, Machine, Hexagon Head, Washer Face, HR650 8(.164)-32 UNJC-3A BS4084.

—Black
BSI BS 4190-67. 1967 Amd 2 ISO Metric Black Hexagon Bolts, Screws and Nuts. 41 pp.

—Coach
SAA AS 1393-73. Coach Screws (Metric Series) (with ISO Hexagon Heads). 15 pp.

INDUSTRY STANDARDS

Hexagonal — INTERNATIONAL AND NON-U.S. NATIONAL STANDARDS SUBJECT INDEX

Hexagonal Head Screws (Cont.)

—Counterbores
CNS B1130-83. Counterbores for Hexagon Socket Head Screws and Slotted Cheese Head Screws (Mar)(4807).
CNS B1131-79. Counterbores for Hexagon Screws and Nuts (Apr)(4808).

—Flat Widths
ISO 272-82. Fasteners—Hexagon Products—Widths Across Flats Second Edition. 5 pp.

—Machine
MOD UK DSTAN 53-21-01. Screws, Machine, Steel, Hexagon Head, Metric (Coarse Pitch Precision Screws) Issue 3; Amendment 1. 19 pp.
MOD UK DSTAN 53-28-77. Screw, Machine, Metric, Hexagon Head, Brass (Coarse Pitch Precision Screws) Issue 2. 4 pp.
MOD UK DSTAN 53-40-83. Screws, Machine, Hexagon Head, Corrosion-Resistant Steel, Metric (Coarse Pitch) Issue 2. 4 pp.
MOD UK DSTAN 53-41-81. Screws, Machine, Hexagon Head, UNC Issue 2. 67 pp.
MOD UK DSTAN 53-57-01. Screws, Machine, Steel, Hexagon Head, UNF Issue 1; Amendment 1. 12 pp.

—Machine Tools
BSI BS 4185: Part 9-76. 1976 Amd 1 Machine Tool Components Part 9: Levelling Screws, Nuts and Seating Plates. 8 pp.

—Machine—Whitworth Thread
MOD UK DSTAN 07-63: Part 2-76. Standardization Status of General Purpose Threaded Fasteners, Steel, Held in COSA Section G1 Part 2: Screws, Machine, Steel, Hexagon Head, BSF and BSW to Be Read in Conjunction with COSA Section G1-Part 5 Issue 1. 27 pp.

—Reduced Shank
DIN ENGL 7964-90. Reduced Shank Bolts and Screws with Coarse Thread (July). 6 pp.

—Set
DIN ENGL 561-85. Hexagon Head Set Screws with Small Hexagon and Full Dog Point (Feb). 4 pp.
DIN ENGL 564-85. Hexagon Head Set Screws with Small Hexagon and Coned Half Dog Point (Feb). 4 pp.

—Socket—Bits
CNS B3305-79. Assembly Tools for Screws and Nuts Insert Bits for Hexagon (Nov)(5010).
ISO 3109-84. Assembly Tools for Screws and Nuts—Hexagon Insert Bits for Hexagon Socket Head Screws Second Edition. 4 pp.

—Socket—Button Head
BSI BS 4168: Part 6-82. 1982 Amd 1 Hexagon Socket Screws and Wrench Keys; Metric Series Part 6: Specification for Hexagon Socket Button Head Screws. 8 pp.
SNZ NZS/BS 4168: Part 6-82. Hexagon Socket Screws and Wrench Keys: Metric Series Part 6: Specification for Hexagon Socket Button Head Screws Amend: 1. 8 pp.

—Socket—Cap
BSI BS 4168: Part 1-81. 1981 Amd 2 Hexagon Socket Screws and Wrench Keys; Metric Series Part 1: Hexagon Socket Head Cap Screws (AMD 5652) July 29, 1987. 11 pp.
CNS B2142-88. Hexagon Socket Head Cap Screws (M3 to M48) (Oct)(3932).
CNS B2289-81. Hexagon Socket Head Cap Screw (M1.4 to M2.5) (Apr)(4555). 3 pp.
CNS B2290-81. Hexagon Socket Head Cap Screws with Reduced Height of Head (Apr)(4556). 4 pp.
CNS B7075-76. Method of Test for Hexagon Socket Head Cap Screws (Metric) (Jun)(3933). 2 pp.
DIN ENGL 912-83. Hexagon Socket Head Cap Screws; (Modified Version of ISO 4762) (Dec). 12 pp.
DIN ENGL 3871-82. Solderless and Soldered Pipe Unions; Male Fittings (Nov). 2 pp.
ISO 4762-89. Hexagon Socket Head Cap Screws—Product Grade A Second Edition. 8 pp.
JIS B 1176-88. Hexagon Socket Head Cap Screws. 19 pp.
MOD UK DSTAN 53-25-01. Screws, Socket Head, Steel—Metric (Course Pitch, Cap Head) Issue 1; Amendment 2. 13 pp.
MOD UK DSTAN 53-67-01. Screws, Socket Head, Unified Coarse Thread (Cap Head) Issue 1; Amendment 2. 22 pp.
MOD UK DSTAN 53-79-01. Screw, Socket Head Unified Fine Thread (Cap Head) Issue 1; Amendment 1. 33 pp.
SAA AS 1420-73. ISO Metric Hexagon Socket Head Cap Screws. 29 pp.
SNZ NZS/BS 4168: Part 1-81. Hexagon Socket Screws and Wrench Keys: Metric Series Part 1: Specification for Hexagon Socket Head Cap Screws Amend: 2. 8 pp.

Hexagonal Head Screws (Cont.)

—Socket—Cap—Counterbores
DIN ENGL 74 Pt 2-91. Counterbores (Holes) for Cheese Head, Pan Head and Hexagon Socket Head Cap Screws (May). 5 pp.
DIN ENGL 974 Pt 1-91. Diameters of Counterbores (Holes) for Cheese Head, Pan Head and Hexagon Socket Head Cap Screws (May). 4 pp.

—Socket—Cap—Countersunk
DIN ENGL 7991-86. Hexagon Socket Countersunk Head Cap Screws (Jan). 6 pp.

—Socket—Cap—Holes
BSI BS 4185: Part 13-85. 1985 Machine Tool Components Part 13: Dimensions of Counterbored Holes. 4 pp.

—Socket—Cap—Thin
CNS B2291-81. Hexagon Socket Head Cap Screws (Shallow Head with Pilot Recess for Wrench Key) (Apr)(4557). 4 pp.
DIN ENGL 6912-85. Hexagon Socket Thin Head Cap Screws with Pilot Recess for Wrench Key (May). 7 pp.
DIN ENGL 7984-85. Hexagon Socket Thin Head Cap Screws (May). 6 pp.

—Socket—Counterbores
CNS B1130-83. Counterbores for Hexagon Socket Head Screws and Slotted Cheese Head Screws (Mar)(4807).

—Socket—Countersunk
BSI BS 4168: Part 8-82. 1982 Amd 1 Hexagon Socket Screws and Wrench Keys; Metric Series Part 8: Specification for Hexagon Socket Countersunk Head Screws (AMD 5569) August 28, 1987. 9 pp.
CNS B2292-81. Hexagon Socket Countersunk Head Screws (Apr)(4558). 4 pp.
SNZ NZS/BS 4168: Part 8-82. Hexagon Socket Screws and Wrench Keys: Metric Series Part 8: Specification for Hexagon Socket Countersunk Head Screws. 8 pp.

—Socket—Keys—Surgical Implants
BSI BS 3531: Sec 5.1-90. 1990 Surgical Implants Part 5: Bone Screws and Auxiliary Equipment Section 5.1: Keys for Use with Screws Having Heads with Hexagon Socket. 8 pp.
ISO 8319 Pt 1-86. Orthopaedic Instruments—Drive Connections—Part 1: Keys for Use with Screws with Hexagon Socket Heads First Edition. 6 pp.

—Socket—Set
BSI BS 4168: Part 2-81. 1981 Amd 2 Hexagon Socket Screws and Wrench Keys; Metric Series Part 2: Hexagon Socket Set Screws with Flat Point. 9 pp.
BSI BS 4168: Part 3-81. 1981 Amd 2 Hexagon Socket Screws and Wrench Keys; Metric Series Part 3: Hexagon Socket Set Screws with Cone Point. 9 pp.
BSI BS 4168: Part 4-81. 1981 Amd 2 Hexagon Socket Screws and Wrench Keys; Metric Series Part 4: Hexagon Socket Set Screws with Dog Point. 9 pp.
BSI BS 4168: Part 5-81. 1981 Amd 2 Hexagon Socket Screws and Wrench Keys; Metric Series Part 5: Hexagon Socket Set Screws with Cup Point. 9 pp.
CNS B2272-81. Hexagon Socket Set Screws with Full Dog Point (Apr)(4479). 4 pp.
CNS B2273-81. Hexagon Socket Set Screws with Cup Point (Apr)(4480). 4 pp.
CNS B2274-81. Hexagon Socket Set Screws with Cone End (Apr)(4481). 4 pp.
CNS B2275-81. Hexagon Socket Set Screws with Flat Point (Apr)(4482). 4 pp.
DIN ENGL 913-80. Hexagon Socket Set Screws with Flat Point; ISO 4026 Modified (Dec). 4 pp.
DIN ENGL 914-80. Hexagon Socket Set Screws with Cone Point; ISO 4027 Modified (Dec). 5 pp.
DIN ENGL 915-80. Hexagon Socket Set Screws with Dog Point; ISO 4028 Modified (Dec). 4 pp.
DIN ENGL 916-80. Hexagon Socket Set Screws with Cup Point; ISO 4029 Modified (Dec). 4 pp.
ISO 4026-77. Hexagon Socket Set Screws with Flat Point First Edition; (Erratum—Nov 1979). 6 pp.
ISO 4027-77. Hexagon Socket Set Screws with Cone Point First Edition. 5 pp.
ISO 4028-77. Hexagon Socket Set Screws with Dog Point First Edition; (Erratum—Nov 1979). 6 pp.
ISO 4029-77. Hexagon Socket Set Screws with Cup Point First Edition; (Erratum—Nov 1979). 6 pp.
JIS B 1177-88. Hexagon Socket Set Screws. 17 pp.
SAA AS 1421-73. ISO Metric Hexagon Socket Set Screws. 22 pp.
SNZ NZS/BS 4168: Part 2-81. Hexagon Socket Screws and Wrench Keys: Metric Series Part 2: Specification for Hexagon Socket Set Screws with Flat Point Amend: 2. 8 pp.
SNZ NZS/BS 4168: Part 3-81. Hexagon Socket Screws and Wrench Keys: Metric Series Part 3: Specification for Hexagon Socket Set Screws with Cone Point Amend: 2. 8 pp.

Hexagonal Head Screws (Cont.)

—Socket—Set (Cont.)
SNZ NZS/BS 4168: Part 4-81. Hexagon Socket Screws and Wrench Keys: Metric Series Part 4: Specification for Hexagon Socket Set Screws with Dog Point Amend: 2. 8 pp.
SNZ NZS/BS 4168: Part 5-81. Hexagon Socket Screws and Wrench Keys: Metric Series Part 5: Specification for Hexagon Socket Set Screws with Cup Point Amend: 2. 8 pp.

—Socket—Shoulder
BSI BS 4168: Part 7-82. 1982 Amd 1 Hexagon Socket Screws and Wrench Keys; Metric Series Part 7: Specification for Hexagon Socket Shoulder Screws. 11 pp.
ISO 7379-83. Hexagon Socket Head Shoulder Screws First Edition. 7 pp.
JIS B 1175-88. Hexagon Socket Head Shoulder Screws. 13 pp.
SNZ NZS/BS 4168: Part 7-82. Hexagon Socket Screws and Wrench Keys: Metric Series Part 7: Specification for Hexagon Socket Shoulder Screws Amend: 1. 8 pp.

—Tapping
CNS B2187-87. Hexagon Head Tapping Screws (Jul)(4308). 5 pp.
DIN ENGL 7513-86. Hexagon Head and Slotted Head Thread Cutting Screws; Dimensions, Requirements, Testing (Nov). 5 pp.
DIN ENGL 7976-90. Hexagon Head Tapping Screws (Aug). 3 pp.
ISO 1479-83. Hexagon Head Tapping Screws First Edition. 5 pp.
JIS B 1123-88. Hexagon Head Tapping Screws. 20 pp.

—Tapping—Flanged
BSI BS 7600-92. 1992 Hexagon Flange Head Tapping Screws (ISO 10509: 1992). 11 pp.
ISO 10509-92. Hexagon Flange Head Tapping Screws First Edition. 8 pp.

—Tapping—Washer
BSI BS 7599-92. 1992 Hexagon Washer Head Tapping Screws (ISO 7053: 1992). 9 pp.
DIN ENGL 6928-90. Hexagon Washer Head Tapping Screws (Nov). 2 pp.
ISO 7053-92. Hexagon Washer Head Tapping Screws First Edition. 7 pp.

—Wood
CNS B3131-87. Hexagon Head Wood Screws (Oct)(1054). 2 pp.
DIN ENGL 571-86. Hexagon Head Wood Screws (Dec). 2 pp.

Hexagonal Nuts

Use For: Plain Nuts, Hexagonal *See Also:* Cap Nuts; Castle Nuts; Double Hexagonal Nuts; Locknuts; Nuts (Fasteners);
BSI BS 1083-65. (OBSOLESCENT) 1965 Amd 2 Precision Hexagon Bolts, Screws and Nuts (B.S.W. and B.S.F. Threads). 31 pp.
BSI BS 3692-67. 1967 Amd 1 ISO Metric Precision Hexagon Bolts, Screws and Nuts. 60 pp.
BSI BS EN 24032-92. 1992 Hexagon Nuts, Style 1—Product Grades A and B (ISO 4032: 1986). 10 pp.
BSI BS EN 24033-92. 1992 Hexagon Nuts, Style 2—Product Grades A and B (ISO 4033: 1979). 11 pp.
BSI BS EN 24034-92. 1992 Hexagon Nuts—Product Grade C (ISO 4034: 1986). 10 pp.
BSI BS EN 28673-92. 1992 Hexagon Nuts, Style 1, with Metric Fine Pitch Thread—Product Grades A and B (ISO 8673: 1988). 11 pp.
BSI BS EN 28674-92. 1992 Hexagon Nuts, Style 2, with Metric Fine Pitch Thread—Product Grades A and B (ISO 8674: 1988). 10 pp.
BSI BS EN 28675-92. 1992 Hexagon Thin Nuts with Metric Fine Pitch Thread—Product Grades A and B (ISO 8675: 1988). 11 pp.
CEN PREN 780-92. Hexagon Nuts for High-Strength Structural Bolting with Large Width Across Flats—Product Grade B—Property Classes 8 and 10. 11 pp.
CEN PREN 783-92. Hexagon Nuts for Structural Bolting with Large Width Across Flats, Style 1—Product Grade B—Property Class 10. 12 pp.
CEN EN 24032-91. Hexagon Nuts, Style 1—Product Grades A and B. 3 pp.
CEN EN 24033-91. Hexagon Nuts, Style 2—Product Grades A and B. 3 pp.
CEN EN 24034-91. Hexagon Nuts—Product Grade C. 3 pp.
CEN EN 28673-91. Hexagon Nuts, Style 1, with Metric Fine Pitch Thread—Product Grades A and B. 3 pp.
CEN EN 28674-91. Hexagon Nuts, Style 2, with Metric Fine Pitch Thread—Product Grades A and B. 3 pp.
CEN EN 28675-91. Hexagon Thin Nuts with Metric Fine Pitch Thread—Product Grades A and B. 3 pp.
CNS B1059-87. Hexagon Screws and Nuts and Thicknesses of Nuts (Oct)(4244).

INTERNATIONAL AND NON-U.S. NATIONAL STANDARDS
SUBJECT INDEX
Hexagonal

Hexagonal Nuts *(Cont.)*
- CNS B2126-83. Hexagon Nuts, Finished and Semi-Finished (Feb)(3128). 4 pp.
- CNS B2128-83. Hexagon Nuts (Regular) (Feb)(3130). 2 pp.
- CNS B2386-79. Hexagon Nuts 1.5d High Metric Thread (Nov)(5014).
- CNS B2393-79. Hexagon Nuts 1.5d High with Collar Metric Thread (Dec)(5055).
- CNS B7274-88. Method of Test for Sets of High Strength Torque Control Bolt, Hexagon Nut and Plain Washers (Jan)(12210).
- DIN ENGL 555-87. M 5 to M 100 X 6 Hexagon Nuts; Product Grade C (Oct). 4 pp.
- DIN ENGL 971 Pt 1 (S)-87. Style 1 Hexagon Nuts with Metric Fine Pitch Thread; Property Classes 6 and 8 (Oct) (Superseded by DIN EN 28673). 4 pp.
- DIN ENGL 971 Pt 2 (S)-87. Style 2 Hexagon Nuts with Metric Fine Pitch Thread; Property Classes 10 and 12 (Oct) (Superseded by DIN EN 28674). 4 pp.
- DIN ENGL 6330-91. Hexagon Nuts with a Height of 1,5 d (Aug). 2 pp.
- DIN ENGL EN 24032-92. Hexagon Nuts, Style 1; Product Grades A and B; (ISO 4032: 1986) (Feb) (This Standard Together with DIN EN 28673, February 1992 Edition, Supersedes DIN 934, October 1987 Edition, and DIN ISO 4032, October 1987 Edition). 6 pp.
- DIN ENGL EN 24033-91. Style 2 Hexagon Nuts; Product Grades A and B; (ISO 4033: 1979) (Dec). 8 pp.
- DIN ENGL EN 24034-92. Hexagon Nuts; Product Grade C; (ISO 4034: 1986) (Feb). 6 pp.
- DIN ENGL EN 28673-92. Hexagon Nuts, Style 1, with Metric Fine Pitch Thread; Product Grades A and B (ISO 8673:1988) (Feb) (Together with DIN EN 24032, February 1992 Edition, Supersedes DIN 934, October 1987 Edition, DIN 971 Part 1, October 1987 Edition, and DIN ISO 8673, January 1990 Edition). 7 pp.
- DIN ENGL EN 28674-92. Hexagon Nuts, Style 2, with Metric Fine Pitch Thread; Product Grades A and B (ISO 8674:1988) (Feb) (Supersedes DIN 971 Part 2, October 1987 Edition, and DIN ISO 8674, January 1990 Edition). 6 pp.
- ISO 4032-86. Hexagon Nuts, Style 1—Product Grades A and B Second Edition. 5 pp.
- ISO 4033-79. Hexagon Nuts, Style 2—Product Grades A and B First Edition. 6 pp.
- ISO 4034-86. Hexagon Nuts—Product Grade C Second Edition. 5 pp.
- ISO 4166-79. Hexagon Nuts for Fine Mechanics—Product Grade F First Edition. 4 pp.
- ISO 8673-88. Hexagon Nuts, Style 1, with Metric Fine Pitch Thread—Product Grades A and B First Edition. 7 pp.
- ISO 8674-88. Hexagon Nuts, Style 2, with Metric Fine Pitch Thread—Product Grades A and B First Edition. 6 pp.
- ISO 8675-88. Hexagon Thin Nuts with Metric Fine Pitch Thread—Product Grades A and B First Edition. 7 pp.
- JIS B 1181-93. Hexagon Nuts and Hexagon Thin Nuts. 43 pp.
- MOD UK DSTAN 53-27-77. Nuts, Plain, Hexagon, Steel, Metric (Coarse Pitch Precision Nuts) Issue 3. 10 pp.
- MOD UK DSTAN 53-37-01. Nut, Plain, Hexagon, Brass, Metric (Coarse Pitch) Issue 1; Amendment 1. 6 pp.
- MOD UK DSTAN 53-45-72. Nuts, Plain, Hexagon, Steel, UNF Issue 1. 9 pp.
- MOD UK DSTAN 53-48-72. Nuts, Plain, Hexagon, Steel, UNC Issue 1. 8 pp.
- MOD UK DSTAN 53-60-83. Nuts, Plain, Hexagon, Corrosion-Resistant Steel, Metric (Coarse Pitch) Issue 2. 4 pp.
- MOD UK DSTAN 53-66-74. Nuts, Plain, Hexagon, Brass, Unified Issue 1. 21 pp.
- SAA AS B148-56. Unified Black Hexagon Bolts, Screws and Nuts (UNC and UNF Threads) and Plain Washers—Heavy Series Being BS 1769:1951 (Including Amendments of March 1955, May 1957, and January 1960), Endorsed Without Australian Amendment Corrig.. 20 pp.
- SAA AS 1112-82. ISO Metric Hexagon Nuts, Including Thin Nuts, Slotted Nuts and Castle Nuts. 28 pp.
- SAA AS 1474-74. Hexagon and Square Pressed Nuts—Metric Series. 7 pp.
- SNZ NZS 1069-66. Specification for Precision Hexagon Bolts, Screws and Nuts (B.S.W. and B.S.F. Threads). 32 pp.
- SNZ NZS/AS 1112-80. ISO Metric Hexagon Nuts, Including Thin Nuts, Slotted Nuts and Castle Nuts (This is a Joint Standard with SAA AS 1112). 28 pp.
- SNZ NZS/AS 2465-81. Unified Hexagon Bolts, Screws and Nuts (UNC and UNF Threads) (This is a Joint Standard with SAA AS 2465). 30 pp.

—**Aerospace**
- BSI A 237-A 240-71. 1971 Amd 1 Plain Hexagonal Nuts (of Class 3B UNJ Thread). 5 pp.
- BSI A 237-A 240-02. 1971 Amd 2 Plain Hexagonal Nut (of Class 3B UNJ Thread) (AMD 7770) August 15, 1993 (S). 6 pp.
- CEN PREN 2895-91. Nuts, Hexagonal, Plain, Normal Height, Normal Across Flats, in Corrosion Resisting Steel, Passivated Classification: 900 MPa (at Ambient Temperature)/425 Degrees C. 6 pp.
- CEN PREN 2895-93. Aerospace Series Nuts, Hexagonal, Plain, Normal Height, Normal Across Flats, in Corrosion Resisting Steel, Passivated Classification: 900 MPa (at Ambient Temperature) /425 Degrees C. 4 pp.
- CEN PREN 2922-92. Nuts, Hexagon, Plain, Reduced Height, Reduced Across Flats, in Heat Resisting Steel, Passivated, Left Hand Thread Classification: 600 MPa (at Ambient Temperature) /650 Degrees C. 5 pp.
- CEN PREN 2923-92. Nuts, Hexagon, Plain, Reduced Height, Reduced Across Flats, in Heat Resisting Steel, Silver Plated Classification: 600 MPa (at Ambient Temperature) /425 Degrees C. 5 pp.
- CEN PREN 2924-92. Nuts, Hexagon, Plain, Reduced Height, Reduced Across Flats, in Heat Resisting Steel, Silver Plated, Left Hand Thread Classification: 600 MPa (at Ambient Temperature) /425 Degrees C. 5 pp.
- CEN PREN 3226-90. Nuts, Hexagon, Plain, Normal Height, Normal Across Flats, in Steel, Cadmium Plated Classification: 1100 MPa (at Ambient Temperature) 235 Degrees Celsius. 7 pp.
- CEN PREN 3226-92. Nuts, Hexagon, Plain, Normal Height, Normal Across Flats, in Steel, Cadmium Plated Classification: 1 100 MPa (at Ambient Temperature) /235 Degrees C. 5 pp.
- CEN PREN 3227-90. Nuts, Hexagon, Plain, Normal Heights, Normal Across Flats, in Steel, Cadmium Plated, Left Hand Thread, Classification: 1100 MPa (at Ambient Temperature) 1235 Degrees Celsius. 7 pp.
- CEN PREN 3227-92. Nuts, Hexagon, Plain, Normal Height, Normal Across Flats, in Steel, Cadmium Plated, Left Hand Thread Classification: 1 100 MPa (at Ambient Temperature) /235 Degrees C. 5 pp.
- CEN PREN 3228-90. Nuts, Hexagon, Plain Reduced Height, Normal Across Flats, in Steel, Cadmium Plated Classification: 900 MPa (at Ambient Temperature) 1235 Degrees Celsius. 7 pp.
- CEN PREN 3228-92. Nuts, Hexagon, Plain, Reduced Height, Normal Across Flats, in Steel, Cadmium Plated Classification: 900 MPa (at Ambient Temperature) /235 Degrees C. 5 pp.
- CEN PREN 3229-90. Nuts, Hexagon, Plain, Reduced Height, Normal Across Flats in Steel, Cadmium Plated, Left Hand Thread, Classification: 900 MPa (at Ambient Temperature) /235 Degrees Celsius. 7 pp.
- CEN PREN 3229-92. Nuts, Hexagon, Plain, Reduced Height, Normal Across Flats, in Steel, Cadmium Plated, Left Hand Thread, Classification: 900 MPa (at Ambient Temperature) /235 Degrees C. 5 pp.
- CEN PREN 3742-90. Nuts, Hexagonal, Slotted/Catellated, Thin, Reduced Across Flats, in Heat Resisting Steel, Passivated Classification: 600 MPa (at Ambient Temperature) /650 Degrees Celsius. 5 pp.
- CEN PREN 3742-92. Nuts, Hexagonal, Slotted/Castellated, Reduced Height, Reduced Across Flats, in Heat Resisting Steel, Passivated Classification: 600 MPa (at Ambient Temperature)/650 Degrees C. 5 pp.
- DIN ENGL LN 9343-85. Aerospace; Nuts, Hexagon (Feb). 3 pp.
- DIN ENGL LN 29669-80. Aerospace; Hexagon Nuts for Screws and Bolts with a Nominal Tensile Strength of 1100 N/mm2 (Mar). 2 pp.

—**Aerospace—Locking**
- AECMA PREN3034-89. Nuts, Self-Locking, Hexagonal with Captive Washer, in Heat Resisting Steel FE-PA92HT (A286), Silver Coated Classification: 1100 MPA/425 Degrees Celsius. 5 pp.
- AECMA PREN3196-87. Nuts, Self-Locking, Hexagonal, in Heat Resisting Steel FE-PA92HT (A286), Silver Coated Classification 1100 MPa/425 Degrees Celsius. 4 pp.
- AECMA PREN3377-89. Nuts, Self-Locking, Hexagonal in Heat Resisting Steel FE-PA92HT (A286), Uncoated Classification: 1100 MPa/425 Degrees Celsius. 5 pp.
- BSI A 275-A 280-81. 1981 Amd 1 Hexagon Self-Locking Nuts with Non-Metallic Locking Inserts. Metric Series. 10 pp.
- CEN PREN 3723-91. Nuts, Hexagonal,Self-Locking, in Heat Resisting Steel FE-PA92HT (A286), MoS2 Coated Classification: 1100 MPa (at Ambient Temperature)/425 Degrees C. 5 pp.
- CEN PREN 3763-90. Nuts, Hexagonal, Self-Locking, Ball Seat, in Heat Resisting Steel, Passivated, MOS2 Lubricated Classification; 900MPa (at Ambient Temperature)/315 Degrees C. 7 pp.
- CEN PREN 3763-93. Aerospace Series Nuts, Hexagonal, Self-Locking, Ball Seat, in Heat Resisting Steel, MoS2 Lubricated Classification: 900 MPa (at Ambient Temperature) /315 Degrees C. 6 pp.
- CEN PREN 3763-93. CORRIGENDUM Aerospace Series Nuts, Hexagonal, Self-Locking, Ball Seat, in Heat Resisting Steel, MoS2 Lubricated Classification: 900 MPa (at Ambient Temperature) /315 Degrees C. 1 p.

—**Aerospace—Locking—Captive Washers**
- AECMA PREN3033-89. Nuts, Self-Locking, Hexagonal with Captive Washer, in Heat Resisting Steel FE-PA92HT (A286), Uncoated Classification: 1100 MPa/425 Degrees Celsius. 5 pp.
- DIN ENGL LN 29790-76. Nuts Hexagon with Captive Washer; Self-Locking for Temperatures up to 235 Degrees C (Oct). 3 pp.

—**Aerospace—Locking—Counterbore**
- SBAC TS 135 ISSUE 1. All Metal Self-Locking Nuts (MJ Threads) Reduced Hexagon—Deep Counterbored Tensile Strength:—1100 MPa at Ambient Temperature.

—**Aerospace—Locking—Counterbore—Captive Washers**
- CEN PREN2882-90. Nuts, Hexagonal, Self-Locking, with Counterbore and Captive Washer, in Steel, Cadmium Plated, MoS2 Lubricated Classification: 1100MPa (at Ambient Temperature) /235 Degrees Celsius. 9 pp.

—**Aerospace—Locking—Flanged**
- CEN PREN 3536-91. Nuts, Hexagon, Self-Locking, in Heat Resisting Steel, MoS2 Lubricated Classification: 1100MPa (at Ambient Temperature)/315 Degrees C. 5 pp.
- CEN PREN 3626-91. Nuts, Hexagon, Self-Locking, in Steel, Cadmium Plated, MoS2 Lubricated Classification: 1100 MPa (at Ambient Temperature)/235 Degrees C. 5 pp.
- DIN ENGL LN 9161-77. Nuts Hexagon Flanged; Self-Locking for Temperatures up to 315 Degrees Celsius and up to 425 Degrees Celsius (May). 3 pp.
- DIN ENGL LN 9338-85. Aerospace; Nuts Hexagon Flanged; Self-Locking for Temperatures up to 235 Degrees C (Feb). 3 pp.

—**Aerospace—Locking—Self Aligning**
- DIN ENGL LN 29795-80. Aerospace; Hexagon Nuts; Self-Aligning; Self-Locking for Temperatures up to 235 Degrees C (Sept). 3 pp.

—**Aerospace—Locking—Thin**
- BSI A 281-A 286-81. 1981 Hexagon (Thin) Self-Locking Nuts with Non-Metallic Locking Inserts. Metric Series. 8 pp.

—**Aerospace—Thin**
- BSI A 233-A 236-71. 1971 Amd 1 Plain Hexagonal Thin Nuts (of Class 3B UNJ Thread). 5 pp.
- BSI A 233-A 236-02. 1971 Amd 2 Plain Hexagonal Thin Nut (of Class 3B UNJ Thread) (AMD 7769) August 15, 1993 (S). 6 pp.
- CEN PREN 2876-91. Nuts, Hexagonal, Plain, Thin Normal Across Flats, in Aluminium Alloy, Anodized Classification: 450 MPa (at Ambient Temperature) /120 Degrees C. 8 pp.
- DIN ENGL LN 9342-84. Aerospace; Nuts Hexagon, Thin, Steel (Feb). 3 pp.

—**Aircraft**
- AECMA PREN2370-83. Nuts, Hexagon Steel, Cadmium Plated Classification 1100 MPa/235 Degrees Celsius. 6 pp.
- AECMA PREN2372-83. Nuts, Hexagon, Thin Steel, Cadmium Plated Classification 1100 MPa/235 Degrees Celsius. 6 pp.
- AECMA PREN2852-87. Nuts, Hexagonal, Plain, Normal Height, Normal Across Flats, Heat Resisting Steel Passivated Classification 1100 MPa/650 Degrees Celsius. 5 pp.
- AECMA PREN2921-87. Nuts, Hexagon, Thin, Reduced Across Flats, Heat Resisting Steel, Passivated Classification 900 MPa/650 Degrees Celsius. 5 pp.
- BSI 3A 107-62. 1962 Amd 1 Aluminium Alloy Nuts (Unified Hexagons and Unified Threads) (Ordinary and Slotted) for Aircraft. 5 pp.
- BSI 2A 110-62. 1962 Amd 4 Steel Nuts (Unified Hexagons and Unified Threads) for Shear Bolts for Aircraft. 8 pp.
- BSI 2A 210-62. 1962 Brass Nuts (Unified Hexagons and Unified Threads) for Aircraft. 4 pp.
- ISO 8279-85. Aerospace—Plain Hexagon Nuts with Strength Classification 1 100 MPa and Maximum Operating Temperature 235 Degrees Celsius First Edition. 5 pp.

INDUSTRY STANDARDS

INTERNATIONAL AND NON-U.S. NATIONAL STANDARDS
SUBJECT INDEX

Hexagonal

Hexagonal Nuts (Cont.)

—Aircraft—Bolt Assemblies—Chains
SBAC AS 168 ISSUE 4. Nut and Bolt Assembly for Chains. 1 p.

—Aircraft—Corrosion Resistant
BSI 2A 24-62. 1962 Amd 2 Corrosion-Resisting Steel Hexagon Nuts (B.A. and B.S.F. Threads) (Ordinary, Thin, Slotted and Castle) for Aircraft (AMD 6066) May 31, 1990. 8 pp.
BSI 3A 105-62. 1962 Corrosion-Resisting Steel Nuts (Unified Hexagons and Unified Threads) (Ordinary, Thin, Slotted and Castle) for Aircraft. 6 pp.

—Aircraft—Left Hand Thread
AECMA PREN2371-83. Nuts, Hexagon Steel, Cadmium Plated Left Hand Thread Classification 1100 MPa/235 Degrees Celsius. 6 pp.
AECMA PREN2373-83. Nuts, Hexagon, Thin Steel, Cadmium Plated Left Hand Thread Classification 1100 MPa/235 Degrees Celsius. 6 pp.

—Aircraft—Locking
AECMA PREN2647-87. Nuts, Hexagon, Self Locking Ball Seat Classification 900 MPa/235 Degrees Celsius. 6 pp.
ISO 7995-88. Aerospace—Nuts, Hexagonal, Self-Locking, with MJ Threads, Coated or Uncoated, Classification 1 100 MPa/235 Degrees Celsius, 1 100 MPa/315 Degrees Celsius or 1 100 MPa/425 Degrees Celsius—Dimensions Second Edition. 5 pp.

—Aircraft—Locking—Captive Washers
SBAC AS 27830-832 (V). Nut, Self-Locking, Six-Point, Captive Washer.
SBAC AS 28000 ISSUE 2. Nut, Self-Locking, Non CR Steel, Captive Washer, Cadmium Coated.

—Aircraft—Locking—Counterbore
ISO 8538-86. Aerospace—Self-Locking Hexagon Nuts with Counterbore and Captive Washer, Classification 1 100 MPa/235 Degrees Celsius First Edition. 5 pp.
SBAC AS 52600-699 ISSUE 1. Nut—Reduced-Hexagon—Self Locking—Deep Counterbored—Metric Series.
SBAC AS 54361-54460 (V). Nut—Reduced Hexagon—Self Locking—Deep Counterbored—Inch Series.

—Aircraft—Locking—Countersunk
SBAC AGS 2003 ISSUE 14(I). Stiffnuts—Hexagon—Countersunk.

—Aircraft—Locking—Flanged
SBAC AS 8600 ISSUE 4. Self-Locking Nuts, Hexagon, Flange Type, Steel, Cadmium Plated, 160,000LBS/SQ.IN. 250 Degrees Celsius.
SBAC AS 8623 ISSUE 2. Self-Locking Nuts, Hexagon, Flange Type, C.R. Steel, Silver Plated, 125,000 lbf/in2. 450 Degrees C.
SBAC AS 8650 ISSUE 4. Self-Locking Nuts, Hexagon, Flange Type, CR Steel, 125,000 lbf/in.2 250 Degrees C.

—Aircraft—Locking—Thick
SBAC AGS 2001 ISSUE 14(I). Stiffnuts—Hexagon—Thick.

—Aircraft—Locking—Thin
SBAC AGS 2002 ISSUE 14(I). Stiffnuts—Hexagon—Thin.

—Aircraft—Pipe Fittings—Thin
SBAC AS 43099 ISSUE 1. Nut—Thin (Aluminium Alloy).

—Aircraft—Pipe Unions
SBAC AS 43042 ISSUE 1. Elbow Assembly—90 Degree, Nipple End, Female Nut Union (Steel).
SBAC AS 43046 ISSUE 1. Tee Assembly—Nipple Branch, Female Nut Union (Steel).
SBAC AS 43048 ISSUE 1. Tee Assembly—Nipple End, Female Nut Union (Steel).
SBAC AS 43064 ISSUE 1. Elbow Assembly—135 Degree, Nipple End, Female Nut Union (Steel).
SBAC AS 43191 ISSUE 1. Reducer-Adaptor Assembly—Nipple End, Male Nut Union (Steel).
SBAC AS 43192 ISSUE 1. Reducer-Adaptor, Nipple End, Male Nut Union (Steel).
SBAC AS 43194 ISSUE 1. Nut Union, Cone Blank (Steel).
SBAC AS 46733 ISSUE 1. Union Adaptor Assembly—Nipple End Male Nut Union (Steel).
SBAC AS 46734 ISSUE 1. Union—Adaptor, Nipple End, Male Nut Union (Steel).
SBAC AS 46735 ISSUE 1. Tee Assembly—Nipple Branch, Male Nut Union (Steel).
SBAC AS 46737 ISSUE 1. Tee Assembly—Nipple End, Male Nut Union (Steel).
SBAC AS 46739 ISSUE 1. Elbow Assembly, 90 Degree—Nipple End Male Nut Union (Steel).
SBAC AS 46741 ISSUE 1. Elbow Assembly, 135 Degree Nipple End Male Nut Union (Steel).

Hexagonal Nuts (Cont.)

—Aircraft—Rudder Bars
SBAC AS 2197 ISSUE 5. Special Nut Rudder Bar. 1 p.
SBAC AS 6333 ISSUE 1. Special, Nut (Rudder Bar).
SBAC AS 6342 ISSUE 1. Cap, Nut (Rudder Bar).

—Aircraft—Slotted
AECMA PREN3230-89. Nuts, Hexagon, Slotted/Castellated, Thin Normal Across Flats, in Steel, Cadmium Plated Classification: 900 MPa/235 Degrees Celsius. 6 pp.

—Aircraft—Stiffnuts
SBAC AS 8624 ISSUE 2(I). Stiffnuts—Hexagon, (Standard Hexagon).
SBAC AS 8651 ISSUE 2(I). Stiffnuts—Hexagon, (Standard Hexagon).

—Aircraft—Thin
BSI 3A 103-62. 1962 Amd 3 Steel Nuts (Unified Hexagons and Unified Threads) (Ordinary, Thin, Slotted and Castle) for Aircraft. 9 pp.
BSI 2A 210-62. 1962 Brass Nuts (Unified Hexagons and Unified Threads) for Aircraft. 4 pp.

—Aircraft—Thin—Corrosion Resistant
BSI 2A 24-62. 1962 Amd 2 Corrosion-Resisting Steel Hexagon Nuts (B.A. and B.S.F. Threads) (Ordinary, Thin, Slotted and Castle) for Aircraft (AMD 6066) May 31, 1990. 8 pp.
BSI 3A 105-62. 1962 Corrosion-Resisting Steel Nuts (Unified Hexagons and Unified Threads) (Ordinary, Thin, Slotted and Castle) for Aircraft. 6 pp.

—Aircraft—Union
SBAC AS 15700-711 (V). Nut, Union—Tube and Hose Coupling.
SBAC AS 43021 ISSUE 1. Nut Union (Steel).
SBAC AS 43025 ISSUE 1. Nut Union—Male (Steel).
SBAC AS 43042 ISSUE 1. Elbow Assembly—90 Degree, Nipple End, Female Nut Union (Steel).
SBAC AS 43044 ISSUE 1. Nut Union—THRUST WIRE—FEMALE (Steel).
SBAC AS 43046 ISSUE 1. Tee Assembly—Nipple Branch, Female Nut Union (Steel).
SBAC AS 43048 ISSUE 1. Tee Assembly—Nipple End, Female Nut Union (Steel).
SBAC AS 43064 ISSUE 1. Elbow Assembly—135 Degree, Nipple End, Female Nut Union (Steel).
SBAC AS 43091 ISSUE 1. Nut Union—Female (Aluminium Alloy).
SBAC AS 43095 ISSUE 1. Nut Union—Male (Aluminium Alloy).
SBAC AS 43191 ISSUE 1. Reducer-Adaptor Assembly—Nipple End, Male Nut Union (Steel).
SBAC AS 43194 ISSUE 1. Nut Union, Cone Blank (Steel).
SBAC AS 44441 ISSUE 2. Union Nut 3/8 Inch O/D Pipe.
SBAC AS 44444 ISSUE 1. Assembly Union Nut—3/8 Inch Diameter Pipe.
SBAC AS 46733 ISSUE 1. Union Adaptor Assembly—Nipple End Male Nut Union (Steel).
SBAC AS 46734 ISSUE 1. Union—Adaptor, Nipple End, Male Nut Union (Steel).
SBAC AS 46735 ISSUE 1. Tee Assembly—Nipple Branch, Male Nut Union (Steel).
SBAC AS 46737 ISSUE 1. Tee Assembly—Nipple End, Male Nut Union (Steel).
SBAC AS 46739 ISSUE 1. Elbow Assembly, 90 Degree—Nipple End Male Nut Union (Steel).
SBAC AS 46741 ISSUE 1. Elbow Assembly, 135 Degree Nipple End Male Nut Union (Steel).

—Aircraft—Union—Thrust Wires
SBAC AS 46730 ISSUE 1. Nut Union—Thrust Wire—Male (Steel).
SBAC AS 46731 ISSUE 1. Nut Union—Thrust Wire, Male 5,0 mm Size (Steel).

—Black
BSI BS 4190-67. 1967 Amd 2 ISO Metric Black Hexagon Bolts, Screws and Nuts. 41 pp.

—Butt Joints
DIN ENGL 3872-85. Solderless Pipe Couplings with Cutting Ring; Coupling Nuts for Butt Joints (June). 2 pp.

—Cap—Prevailing Torque
DIN ENGL 986-87. Prevailing Torque Type Hexagon Domed Cap Nuts with Nonmetallic Insert (June). 4 pp.

—Coarse Thread
DIN ENGL 934 (S)-87. Hexagon Nuts with Metric Coarse and Fine Pitch Thread; Product Classes A and B (Oct) (Superseded by DIN EN 24032). 7 pp.

—Collar Type
CNS B2366-84. Hexagon Nuts with Collar (Jul)(4733).
CNS B2394-83. Thrust Pads (Mar)(5108).
DIN ENGL 6331-91. Hexagon Collar Nuts with a Height of 1,5 d (Aug). 2 pp.

Hexagonal Nuts (Cont.)

—Counterbores
CNS B1131-79. Counterbores for Hexagon Screws and Nuts (Apr)(4808).
DIN ENGL 74 Pt 3-91. Counterbores (Holes) for Hexagon Bolts and Nuts (May). 6 pp.
DIN ENGL 974 Pt 2-91. Diameters of Counterbores (Holes) for Hexagon Bolts and Nuts (May). 4 pp.

—Flanged
DIN ENGL 6923-83. Hexagon Nuts with Flange (June). 4 pp.
ISO 4161-83. Hexagon Nuts with Flange —Product Grade A First Edition. 5 pp.
JIS B 1190-87. Hexagon Nuts with Flange. 9 pp.

—Flanged—Prevailing Torque
DIN ENGL 6926-83. Prevailing Torque Type Hexagon Nuts with Flange and with Non-Metallic Insert (Nov). 4 pp.
ISO 7043-83. Prevailing Torque Type Hexagon Nuts with Flange (with Non-Metallic Insert) First Edition. 6 pp.
ISO 7044-83. Prevailing Torque Type All-Metal Hexagon Nuts with Flange First Edition. 6 pp.

—Flat Widths
DIN ENGL 70615-90. Hexagon Nuts with Small Widths Across Flats for Use in Automotive Engineering (Not Intended for New Designs) (Mar) (This Standard, Together with DIN 70616, March 1990 Edition, Supersedes DIN 70615/DIN 70616, February 1964 Edition). 3 pp.
DIN ENGL 70616-90. Hexagon Thin Nuts with Small Widths Across Flats for Use in Automotive Engineering (Not Intended for New Designs) (Mar) (This Standard, Together with DIN 70615, March 1990 Edition, Supersedes DIN 70615/DIN 70616, February 1964 Edition). 3 pp.
ISO 272-82. Fasteners—Hexagon Products—Widths Across Flats Second Edition. 5 pp.

—Flat Widths—Motorcycles
CNS B2130-83. Hexagon Nuts with Small Widths Across Flats (for Motorcycle)(Feb)(3132). 2 pp.

—High Strength—Structural
BSI BS 4395: Part 1-69. 1969 Amd 2 High Strength Friction Grip Bolts and Associated Nuts and Washers for Structural Engineering Part 1: General Grade. 36 pp.
BSI BS 4395: Part 2-69. 1969 Amd 2 High Strength Friction Grip Bolts and Associated Nuts and Washers for Structural Engineering Part 2: Higher Grade Bolts and Nuts and General Grade Washers. 36 pp.
CNS B2170-83. Hexagon Nuts with Large Widths Across Flats for Steel Structures (Feb)(4236). 2 pp.
DIN ENGL 6915-89. High-Strength Hexagon Nuts with Large Widths Across Flats for Structural Steel Bolting (Oct). 3 pp.
ISO 4775-84. Hexagon Nuts for High-Strength Structural Bolting with Large Width Across Flats—Product Grade B—Property Classes 8 and 10 First Edition. 8 pp.
SNZ NZS/BS 4395: Part 1-69. Specification for High Strength Friction Grip Bolts and Associated Nuts and Washers for Structural Engineering Part 1: General Grade. 36 pp.
SNZ NZS/BS 4395: Part 2-69. Specification for High Strength Friction Grip Bolts and Associated Nuts and Washers for Structural Engineering Part 2: Higher Grade Bolts and Nuts and General Grade Washers. 36 pp.

—High Torque
CNS B2791-88. Sets of High Strength Torque Control Bolt, Hexagon Nut and Plain Washers (Jan)(12209). 12 pp.

—Jigs
JIS B 5226-89. Hexagon Nuts for Jigs and Fixtures. 6 pp.

—Locking—Heavy
CNS B2192-83. Self-Locking Hexagon Nuts Heavy Type (Feb)(4313).

—Locking—Thin
CNS B2193-83. Self-Locking Hexagon Nuts Thin Type (Feb)(4314).

—Pipe
DIN ENGL 431-82. Pipe Nuts with Thread in Accordance with DIN ISO 228 Part 1 (Nov). 4 pp.

—Pipe Couplings
DIN ENGL 3870-85. Soldered and Solderless Pipe Couplings; Coupling Nuts (May). 4 pp.

—Prevailing Torque
BSI BS 4929: Part 1-73. 1973 Amd 1 Steel Hexagon Prevailing-Torque Type Nuts Part 1: Metric Series. 16 pp.

INDEX and DIRECTORY of

INTERNATIONAL AND NON-U.S. NATIONAL STANDARDS
SUBJECT INDEX

Hexagonal Nuts *(Cont.)*
—Prevailing Torque *(Cont.)*
BSI BS 4929: Part 2-73. 1973 Steel Hexagon Prevailing-Torque Type Nuts Part 2: Unified (Inch) Series. 16 pp.
DIN ENGL 980-87. All-Metal Prevailing Torque Type Hexagon Nuts (May). 4 pp.
DIN ENGL 6925-87. Prevailing Torque Type All-Metal Hexagon Nuts (July). 4 pp.
ISO 2320-83. Prevailing Torque Type Steel Hexagon Nuts—Mechanical and Performance Properties Second Edition. 11 pp.
ISO 7040-83. Prevailing Torque Type Hexagon Nuts (with Non-Metallic Insert), Style 1—Property Classes 5, 8 and 10 First Edition. 5 pp.
ISO 7041-83. Prevailing Torque Type Hexagon Nuts (with Non-Metallic Insert), Style 2—Property Classes 9 and 12 First Edition. 5 pp.
ISO 7042-83. Prevailing Torque Type All-Metal Hexagon Nuts, Style 2—Property Classes 5, 8, 10 and 12 First Edition. 5 pp.
ISO 7719-83. Prevailing Torque Type All-Metal Hexagon Nuts, Style 1—Property Classes 5, 8 and 10 First Edition. 5 pp.
ISO 7720-83. Prevailing Torque Type All-Metal Hexagon Nuts, Style 2—Property Class 9 First Edition. 5 pp.
JIS B 1056-87. Mechanical and Performance Properties for Prevailing Torque Type Steel Nuts. 21 pp.
SAA AS 1285-73. Prevailing Torque Steel Hexagon Locknuts (ISO Metric Series). 34 pp.
SNZ NZS/BS 4929: Part 1-73. Specification for Steel Hexagon Prevailing-Torque Type Nuts Part 1: Metric Series Amend: 1. 16 pp.
SNZ NZS/BS 4929: Part 2-73. Specification for Steel Hexagon Prevailing-Torque Type Nuts Part 2: Unified (Inch) Series. 16 pp.

—Prevailing Torque—Nonmetallic Inserts
DIN ENGL 982-87. Prevailing Torque Type Hexagon Nuts with Nonmetallic Insert (May). 4 pp.
DIN ENGL 985-87. Prevailing Torque Type Hexagon Thin Nuts with Nonmetallic Insert (May). 4 pp.
DIN ENGL 6924-87. Prevailing Torque Type Hexagon Nuts with Nonmetallic Insert (July). 4 pp.

—Sets—High Strength
CNS B2768-85. Sets of High Strength Hexagon Bolt, Hexagon Nut and Plain Washers for Friction Grip Joints (Aug)(11328).
CNS B7272-85. Method of Test for the Sets of High Strength Hexagon Bolt, Hexagon Nut and Plain Washers for Friction Grip Joints (Aug)(11329).
JIS B 1186-79. Sets of High Strength Hexagon Bolt, Hexagon Nut and Plain Washers for Friction Grip Joints. 21 pp.

—Structural
DIN ENGL 7968-89. Hexagon Fit Bolts for Structural Steel Bolting for Supply with or Without Nut (Oct). 6 pp.
DIN ENGL 7990-89. Hexagon Head Bolts for Structural Steel Bolting for Supply with Nut (Oct). 6 pp.
ISO 7413-84. Hexagon Nuts for Structural Bolting, Style 1, Hot-Dip Galvanized (Oversize Tapped)—Product Grades A and B—Property Classes 5, 6 and 8 First Edition. 6 pp.
ISO 7414-84. Hexagon Nuts for Structural Bolting with Large Width Across Flats, Style 1—Product Grade B—Property Class 10 First Edition. 8 pp.
ISO 7417-84. Hexagon Nuts for Structural Bolting—Style 2, Hot-Dip Galvanized (Oversize Tapped)—Product Grade A—Property Class 9 First Edition. 6 pp.

—Studs (Fasteners)
CNS B2311-78. Bolted Connections with Reduced Body (Hexagon Nuts) (Oct)(4581).
DIN ENGL 2510 Pt 5-71. Bolted Connections with Reduced Shank; Hexagon Nuts (Aug). 2 pp.

—Thin
BSI BS EN 24035-92. 1992 Hexagon Thin Nuts (Chamfered)—Product Grades A and B (ISO 4035: 1986). 10 pp.
BSI BS EN 24036-92. 1992 Hexagon Thin Nuts—Product Grade B (Unchamfered) (ISO 4036: 1979). 10 pp.
CEN EN 24035-91. Hexagon Thin Nuts (Chamfered)—Product Grades A and B. 3 pp.
CEN EN 24036-91. Hexagon Thin Nuts—Product Grade B (Unchamfered). 3 pp.
CNS B2127-83. Hexagon Nuts, Thin Type (Feb)(3129). 3 pp.
DIN ENGL 439 Pt 1 (S)-87. Unchamfered Hexagon Thin Nuts; Product Grade B (Oct) (Superseded by DIN EN 24036). 4 pp.
DIN ENGL 439 Pt 2 (S)-87. Chamfered Hexagon Thin Nuts; Product Grades A and B (Oct) (Superseded by DIN EN 28675 and DIN EN 24035). 6 pp.

Hexagonal Nuts *(Cont.)*
—Thin *(Cont.)*
DIN ENGL 936-85. M 8 to M 52 and M 8 x 1 to M 52 x 3 Hexagon Thin Nuts; Product Grades A and B (Feb). 4 pp.
DIN ENGL 985-87. Prevailing Torque Type Hexagon Thin Nuts with Nonmetallic Insert (May). 4 pp.
DIN ENGL EN 24035-92. Hexagon Thin Nuts (Chamfered); Product Grades A and B (ISO 4035: 1986) Supersedes DIN 439 Part 2, October 1987 Edition. 6 pp.
DIN ENGL EN 24036-92. Hexagon Thin Nuts; Product Grade B (Unchamfered): (ISO 4036: 1979) (Feb) (Supersedes DIN 439 Part 1, October 1987 Edition, and DIN ISO 4036, October 1987 Edition). 6 pp.
DIN ENGL EN 28675-92. Hexagon Thin Nuts with Metric Fine Pitch Thread; Product Grades A and B (ISO 8675:1988) (Feb) (This Standard, Together with DIN EN 24035, February 1992 Edition, Supersedes DIN 439 Part 2, October 1987 Edition and DIN ISO 8675, January 1990 Edition). 7 pp.
ISO 4035-86. Hexagon Thin Nuts (Chamfered)—Product Grades A and B Second Edition. 5 pp.
ISO 4036-79. Hexagon Thin Nuts—Product Grade B (Unchamfered) First Edition; (Erratum—June 1979). 6 pp.
JIS B 1181-93. Hexagon Nuts and Hexagon Thin Nuts. 43 pp.

—Towers
CSA B33.4-1973. Galvanized Steel Tower Bolts and Nuts. 14 pp.

—Trapezoidal Thread
CNS B2248-83. Hexagon Nuts, Rounded Trapezoidal Thread (Feb)(4428).

—Turnbuckles
CNS B2368-79. Hexagon and Castle Nuts with Round Threads (for Motive Power Unit Turnbuckles and Drawbars) (Apr)(4772).
DIN ENGL 1479-75. Turnbuckle Nuts Made from Hexagon Bar (Sept). 2 pp.

—Union
DIN ENGL 3870-85. Soldered and Solderless Pipe Couplings; Coupling Nuts (May). 4 pp.

—Union—Aerospace
SBAC AS 62350 ISSUE 1. Nut Union (Cres). 2 pp.
SBAC AS 62351 ISSUE 1. Nut Union-Male (Cres). 2 pp.

—Washers
CNS B2011-47. Semi-Finished Washer for Hexagonal Bolts and Nuts (M1—M20) (Mar)(151). 2 pp.
CNS B2013-78. Washers Coarse Type Primarily for Hexagon Bolts and Nuts (Mar)(153). 2 pp.
DIN ENGL 125 Pt 1-90. Product Grade A Washers with a Hardness up to 250 HV Designed for Use with Hexagon Head Bolts and Nuts (Mar). 4 pp.
DIN ENGL 125 Pt 2-90. Product Grade A Washers with a Hardness from 300 HV Designed for Use with Hexagon Head Bolts and Nuts (Mar). 4 pp.
DIN ENGL 126-90. Product Grade C Washers; Designed for Use with Hexagon Head Bolts and Nuts (Mar). 2 pp.

—Weld
CNS B2250-83. Hexagon Weld Nuts (Feb)(4430).
DIN ENGL 929-87. Hexagon Weld Nuts (Sept). 4 pp.

—Wheels—Automotive
ISO 7575-93. Commercial Road Vehicles—Flat Attachment Wheel Fixing Nuts Second Edition. 5 pp.
MOD UK DSTAN 53-7-68. Bolts and Studs for Vehicle Wheels Issue 1. 12 pp.

—Whitworth Thread
MOD UK DSTAN 07-63: Part 5-76. Standardization Status of General Purpose Threaded Fasteners, Steel, Held in COSA Section G1 Part 5: Nuts, Various, Steel, Hexagon, BSF and BSW to Be Read in Conjunction with COSA Section G1-Part 5 Issue 1. 22 pp.

Hexahydrophthalic Anhydride
MOD UK TS 434. Hexahydrophthalic Anhydride.

Hexamethyldisilazane
JIS K 9556-78. 1,1,1,3,3,3-Hexamethyldisilazane.

Hexamethylenetetramine
Use For: Hexamine; Methenamine
CNS K1236-80. Hexamethylene Tetramine (Apr)(5451).
CNS K6475-80. Method of Test for Hexamethylenetetramine (Apr)(5456).
CNS K7242-67. Chemical Reagent (Hexamethylenetramine) (Oct)(1741).
CNS K7315-71. Chemical Reagent (Methenamine) (Jan)(1814).
JIS K 1532-78. Hexamethylene-Tetramine (Hexamine).

Hexamethylenetetramine *(Cont.)*
JIS K 8847-86. Hexamethylenetetramine.

—Angle of Slide
MOD UK M 879/69. Examination of Hexamine.

—Sieve Analysis
MOD UK M 879/69. Examination of Hexamine.

—Visual Inspection
MOD UK M 879/69. Examination of Hexamine.

Hexamethylenetetramine Content Analysis
Use For: Methenamine Content Analysis
MOD UK M 879/69. Examination of Hexamine.

—Phenolic Resins—Kjeldahl Method
ISO 8988-89. Plastics—Phenolic Resins—Determination of Hexamethylenetetramine Content First Edition. 6 pp.

—Phenolic Resins—Perchloric Acid Method
ISO 8988-89. Plastics—Phenolic Resins—Determination of Hexamethylenetetramine Content First Edition. 6 pp.

Hexamine
Use: Hexamethylenetetramine

Hexane
Use: n-Hexane

Hexane Content Analysis
—Animal Fats
BSI BS 684: Sec 2.37-93. 1993 Methods of Analysis of Fats and Fatty Oils Part 2: Other Methods Section 2.37: Determination of Residual Technical Hexane Content (ISO 9832: 1992). 9 pp.
ISO 9832-92. Animal and Vegetable Fats and Oils—Determination of Residual Technical Hexane Content First Edition. 8 pp.

—Animal Oils
BSI BS 684: Sec 2.37-93. 1993 Methods of Analysis of Fats and Fatty Oils Part 2: Other Methods Section 2.37: Determination of Residual Technical Hexane Content (ISO 9832: 1992). 9 pp.
ISO 9832-92. Animal and Vegetable Fats and Oils—Determination of Residual Technical Hexane Content First Edition. 8 pp.

—Oilseeds
BSI BS 4325: Part 10-88. 1988 Analysis of Oilseed Residues Part 10: Determination of Total Residual Hexane. 6 pp.
BSI BS 4325: Part 11-91. 1991 Analysis of Oilseed Residues Part 11: Determination of Free Residual Hexane Content (ISO 9289: 1991). 9 pp.
ISO 8892-87. Oilseed Residues—Determination of Total Residual Hexane First Edition. 6 pp.
ISO 9289-91. Oilseed Residues—Determination of Free Residual Hexane First Edition. 8 pp.

—Vegetable Fats
BSI BS 684: Sec 2.37-93. 1993 Methods of Analysis of Fats and Fatty Oils Part 2: Other Methods Section 2.37: Determination of Residual Technical Hexane Content (ISO 9832: 1992). 9 pp.
ISO 9832-92. Animal and Vegetable Fats and Oils—Determination of Residual Technical Hexane Content First Edition. 8 pp.

—Vegetable Oils
BSI BS 684: Sec 2.37-93. 1993 Methods of Analysis of Fats and Fatty Oils Part 2: Other Methods Section 2.37: Determination of Residual Technical Hexane Content (ISO 9832: 1992). 9 pp.
ISO 9832-92. Animal and Vegetable Fats and Oils—Determination of Residual Technical Hexane Content First Edition. 8 pp.

n-Hexane Extract Content Analysis
—Waste Water
CNS K9074-82. Method of Test for n-Hexane Extract in Waste Water Method (Jan)(8303).

—Waste Water—Extraction Analysis
CNS K9073-82. Method of Test for n-Hexane Extract in Waste Water Extractor Method (Jan)(8302).
CNS K9075-82. Method of Test for n-Hexane Extract in Waste Water Extraction Method (Jan)(8304).

n-Hexane
Use For: Hexane *See Also:* Cyclohexane; Hydrocarbons
CNS K0001-88. High Purity N-Hexane (Sep)(6216).
CNS K1061-62. N-Hexane, for Industrial Use (Dec)(1414). 1 p.
CNS K6153-62. Method of Test for Normal Hexane (for Industrial Use) (Dec)(1415)(R 1973).
CNS K7667-83. Chemical Reagent (Hexane) (Mar)(10106).

INTERNATIONAL AND NON-U.S. NATIONAL STANDARDS
SUBJECT INDEX

n-Hexane (Cont.)
JIS K 0504-83. High Purity N-Hexane.
JIS K 8848-76. Hexane.

Hexanitrostilbene
Use For: HNS *See Also:* Benzene; Ethylene; Explosives
NATO STANAG 4230 ED 1 AMD 0-88. (DRAFT) Specification for HNS (Hexanitrostilbene) Explosive for Deliveries from One NATO Nation to Another. 21 pp.

Hexogenes
—Delivery Systems
NATO STANAG 4022 ED 3 AMD 0-83. Specification for RDX (Hexogene) for Deliveries from One NATO Nation to Another. 9 pp.
NATO STANAG 4022 ED 3 AMD 1-83. Specification for RDX (Hexogene) for Deliveries from One NATO Nation to Another. 14 pp.

Hexone
Use: Methyl Isobutyl Ketone

Hexylene Glycol
CNS K1231-80. Hexylene Glycol (Feb)(5237).
ISO 2519-74. Hexylene Glycol for Industrial Use—List of Methods of Test First Edition. 4 pp.

—Acidity—Volumetric Analysis
ISO 2887-73. secButyl Alcohol, Methyl Ethyl Ketone, isoButyl Methyl Ketone, isoAmyl Ethyl Ketone, Diacetone Alcohol, and Hexylene Glycol for Industrial Use—Determination of Acidity to Phenolphthalein—Volumetric Method First Edition. 4 pp.

—Miscibility
ISO 2518-74. Diacetone Alcohol and Hexylene Glycol for Industrial Use—Test for Miscibility with Water First Edition. 4 pp.

HID Lights
Use: High Intensity Discharge Lamps

Hides
See Also: Furs; Leather
CNS N1068-72. Animal Hides (Jun)(2510). 5 pp.
CNS N1085-68. Dry Hides of Small Animals (Aug)(2882). 3 pp.
ISO 2821-74. Leather—Raw Hides of Cattle and Horses—Preservation by Stack Salting First Edition. 4 pp.

—Classification
BSI BS 3935-91. 1991 Classification and Marking of Cattle Hides and Calfskins. 12 pp.
SNZ NZS 8401-88. Treatment, Grading and Classification of Hides and Calfskins. 12 pp.
SNZ NZS 8404-75. Specification for the Treatment, Grading and Classification of Freezer and Abattoir Quality Green Hides Amend: 1, 1986. 7 pp.

—Grading
CNS N4021-65. Method of Test for Animal Hides (Sep) (2511). 4 pp.
SNZ NZS 8401-88. Treatment, Grading and Classification of Hides and Calfskins. 12 pp.
SNZ NZS 8404-75. Specification for the Treatment, Grading and Classification of Freezer and Abattoir Quality Green Hides Amend: 1, 1986. 7 pp.

—Identification Systems
BSI BS 3935-91. 1991 Classification and Marking of Cattle Hides and Calfskins. 12 pp.

—Macroscopic Inspection
CNS N4021-65. Method of Test for Animal Hides (Sep) (2511). 4 pp.

—Powder—Reagent
CNS K7645-82. Chemical Reagent (Hide Powder) (Sep)(9407).

—Sampling
CNS N4021-65. Method of Test for Animal Hides (Sep) (2511). 4 pp.

—Sanitation
CNS N4021-65. Method of Test for Animal Hides (Sep) (2511). 4 pp.

Hiding Power
Scope Note: See the subheading Hiding Power under specific coating

High Alloy Steels
See Also: Alloy Steels; Ferroalloys; Low Alloy Steels; Steels

—Bars—Valves
DIN ENGL 17480-92. Valve Materials; Technical Delivery Conditions (May) (Supersedes September 1984 Edition). 13 pp.

—Forgings—Valves
DIN ENGL 17480-92. Valve Materials; Technical Delivery Conditions (May) (Supersedes September 1984 Edition). 13 pp.

—Rods—Valves
DIN ENGL 17480-92. Valve Materials; Technical Delivery Conditions (May) (Supersedes September 1984 Edition). 13 pp.

High Alumina Bricks
Use: Firebricks

High Alumina Refractories
Use: Aluminous Refractories

High Chairs
See Also: Chairs (Seats)
ISO 9221 Pt 2-92. Furniture—Children's High Chairs—Part 2: Test Methods First Edition. 12 pp.

—Safety
BSI BS 5799-86. 1986 Amd 1 Safety Requirements for Children's High Chairs and Multi-Purpose High Chairs for Domestic Use (AMD 6681) April 30, 1991 April 30, 1991. 20 pp.
ISO 9221 Pt 1-92. Furniture—Children's High Chairs—Part 1: Safety Requirements First Edition. 6 pp.
SNZ NZS/BS 5799-86. Specification for Safety Requirements for Children's High Chairs and Multi-Purpose High Chairs for Domestic Use Amend: 1. 20 pp.

—Safety Belts
BSI BS 6684-89. 1989 Amd 1 Safety Harnesses (Including Detachable Walking Reins) for Restraining Children When in Perambulators (Baby Carriages), Pushchairs and High Chairs and When Walking (AMD 6531) June 29, 1990. 16 pp.
SAA AS 3747-89. Harnesses for Use in Prams, Strollers, and High Chairs (Including a Detachable Walking Rein) (This is a Joint Standard with SANZ NZS 3747). 8 pp.
SNZ NZS/AS 3747-89. Harnesses for Use in Prams, Strollers and High Chairs, (Including a Detachable Walking Rein) (This is a Joint Standard with SAA AS 3747). 8 pp.

High Conductivity Materials
Scope Note: See the subheading High Conductivity under the specific type of material

High Definition Television Equipment
Use For: HDTV *See Also:* Enhanced Quality Television Systems; Television Equipment; Television Systems
CCIR Report 801-4-90. Present State of High-Definition Television—Section 11A—Characteristics of Systems for Monochrome and Colour Television. 126 pp.
CCIR Report 1217-90. Future Development of HDTV—Section 11A—Characteristics of Systems for Monochrome and Colour Television. 15 pp.
CCIR Report 1224-90. HDTV Standards Harmonization—Section 11A—Characteristics of Systems for Monchrome and Color Television. 6 pp.
CCIR Report 1218-90. Measurement in HDTV—Section 11C—Control, Measurement and International Exchange of Television Programmes. 8 pp.
CCIR Decision 58-4-89. High-Definition Television Standard—Annex to Volume XI-1—Broadcasting Service (Television). 4 pp.
CCIR Decision 60-3-89. Digital Television Standards—Annex to Volume XI-1—Broadcasting Service (Television). 4 pp.
CCIR QUESTION 27-2/11-90. High-Definition Television—Questions Concerning Study Group 11—Broadcasting Service (Television). 1 p.
CCIR QUESTION 71/11-90. Objective Measurement in an HDTV Environment—Questions Concerning Study Group 11—Broadcasting Service (Television). 1 p.

—Audio Systems
CCIR Report 1072-1-90. Suitable Sound Systems to Accompany High-Definition and Enhanced Television Systems—Section 10C—Audio-Frequency Characteristics of Sound-Broadcasting Signals. 9 pp.
CCIR QUESTION 79/10-90. Suitable Sound Systems to Accompany High-Definition Television and Enhanced Television Systems—Questions Concerning Study Group 10—Broadcasting Service (Sound). 1 p.

—Audio Systems—Multichannel
CCIR Decision 94-89. Multi-Channel Sound Systems (Especially Suited to Accompany High-Definition and Enhanced Television Systems)—Volume X-1—Broadcasting Service (Sound). 3 pp.

—Broadcasting
CCIR Decision 91-89. Harmonization of HDTV Standards Between Broadcast and Non-Broadcast Applications—Annex to Volume XI-1—Broadcasting Service (Television). 3 pp.

—Broadcasting Satellite Services
CCIR Report 1075-1-90. High Definition Television Broadcasting by Satellite—Section 10/11B—Systems. 98 pp.
CCIR Decision 51-4-89. Satellite Broadcasting of High Definition Television (HDTV) Signals and Accomodation of Several Audio and/or Data Signals Either Associated with Television Signals or for Sound/Data Broadcasting in Terrestrial and Satellite Broadcasting. 6 pp.
CCIR QUESTION 100/11-90. Satellite Broadcasting of High Definition Television (HDTV)—Questions Concerning Study Group 11—Broadcasting Service (Television). 1 p.

—Broadcasting Satellite Services—Bit Rate Reduction
CCIR RECMN 788-92. Coding Rate for the Wide RF-Band HDTV Broadcasting-Satellite Service. 1 p.

—Broadcasting Satellite Services—Frequency Assignment
CCIR Decision 93-89. Preparatory Work for the WARC 1992—Annex to Volumes X and XI—Part 2—Broadcasting-Satellite Service (Sound and Television). 3 pp.

—Broadcasting Satellite Services—Frequency Band Sharing
CCIR Decision 43-5-89. Satellite Sound Broadcasting for Portable and Vehicle Receivers and Sharing and Spectrum Aspects of Wide RF-Band HDTV Satellite Broadcasting—Annex to Volumes X and XI—Part 2—Broadcasting-Satellite Service (Sound and Television). 5 pp.
CCIR QUESTION 89/11-90. Sharing Studies Between High-Definition Television (HDTV) in the Broadcasting-Satellite Service and Other Services—Questions Concerning Study Group 11—Broadcasting Service (Television). 1 p.

—Broadcasting Satellite Services—MAC/Packet
CCIR RECMN 787-92. MAC/Packet Based System for HDTV Broadcasting-Satellite Services. 18 pp.

—Broadcasting Satellite Services—MUSE
CCIR RECMN 786-92. MUSE System for HDTV Broadcasting-Satellite Services. 19 pp.

—Data Broadcasting
CCIR Report 1225-90. Data Broadcasting Systems and Services in and HDTV Environment—Section 11B—Ancillary Television Services. 8 pp.

—Digital
CCIR QUESTION 47/11-90. Standards for Digital High Definition Television—Questions Concerning Study Group 11—Broadcasting Service (Television). 1 p.

—Display Devices
CCIR QUESTION 70/11-90. Effect of Display Technology on the HDTV Standard—Questions Concerning Study Group 11—Broadcasting Service (Television). 1 p.

—Earth Stations—Outside Broadcasts
CCIR QUESTION 74/CMTT-90. Use of Portable and Transportable Satellite Earth Stations for the Transmission of High Definition Television Outside Broadcasts over Satellites—Questions Concerning the CMTT CCIR/CCITT Joint Study Group for Television and Sound Transmission. 1 p.

—Earth Stations—Satellite News Gathering
CCIR QUESTION 75/CMTT-90. Use of Portable and Transportable Satellite Earth Stations for the Transmission of High Definition Television for News Gathering Via Satellite—Questions Concerning the CMTT CCIR/CCITT Joint Study Group for Television and Sound Transmission. 1 p.

—Integrated Services Digital Networks
CCIR RESOLUTION 103-90. Liasion and Joint Studies with the CCITT in Mobile and Fixed-Satellite Services and High-Definition Television in ISDNs—Volume XIV—Administrative Texts of the CCIR. 2 pp.

INTERNATIONAL AND NON-U.S. NATIONAL STANDARDS
SUBJECT INDEX

High

High Definition Television Equipment (Cont.)

—Motion Picture Film—Image Area
CCIR RECMN 716-90. Scanned Area of 35 mm Motion Picture Film in HDTV Telecines (Non-Anamorphic Pictures)—Section 10/11H—Use of Film in Television. 2 pp.
CCIR RECMN 716-1-92. Area of 35 mm Motion Picture Film Used by HDTV Telecines Section 10/11H—Use of Film in Television. 6 pp.
JIS B 7230-92. Image Area Produced by High-Definition Television (HDTV) on 35 mm Motion-Picture Film—Position and Dimensions. 4 pp.
JIS B 7231-92. HDTV Monitor—Picture Area for 35 mm Motion Picture Filming. 5 pp.

—Nonbroadcasting
CCIR Decision 91-89. Harmonization of HDTV Standards Between Broadcast and Non-Broadcast Applications—Annex to Volume XI-1—Broadcasting Service (Television). 3 pp.

—Standards—Compatibility
CCIR QUESTION 69/11-90. Compatibility of the HDTV Standard with Existing and Future Standards—Questions Concerning Study Group 11—Broadcasting Service (Television). 1 p.
EC 89/630/EEC-89. Council Decision on the Common Action to be Taken by the Member States with Respect to the Adoption of a Single World-Wide High-Definition Television Production Standard by the Plenary Assembly of the International Radio Consultative Committee (CCIR) in 1990. 1 p.

—Studios
CCIR Decision 90-89. Special Meeting Concerning the Draft Recommendation for the HDTV Studio Standard—Annex to Volume XI-1—Broadcasting Service (Television). 2 pp.

—Telecines—Color Balance
CCIR RECMN 781-92. HDTV Telecine Colour Balance for Film Programmes. 1 p.

—Telecommunication Circuits—Television Transmission
CCIR QUESTION 73/CMTT-90. Characteristics and Design Objectives for Analogue HDTV Transmission Circuits—Questions Concerning the CMTT CCIR/CCITT Joint Study Group for Television and Sound Transmission. 1 p.

—Television Pictures—Quality Assurance—Subjective Assessment
CCIR RECMN 710-90. Subjective Assessment Methods for Image Quality in High-Definition Television—Section 11D—Picture Quality and the Parameters Affecting It. 2 pp.
CCIR RECMN 710-1-92. Subjective Assessment Methods for Image Quality in High-Definition Television—Section 11D—Picture Quality and the Parameters Affecting It. 15 pp.
CCIR Report 1216-90. Subjective Assessment of HDTV Pictures—Section 11D—Picture Quality and the Parameters Affecting It. 20 pp.
CCIR QUESTION 52/11-90. Subjective Assessment Procedures for Pictures Originating in a High Definition Television Studio—Questions Concerning Study Group 11—Broadcasting Service (Television). 1 p.

—Television Program Exchange
CCIR RECMN 714-90. International Exchange of Programmes Electronically Produced by Means of High-Definition Television—Section 10/1G—Exchange of Recorded Television Programmes. 1 p.
CCIR Report 1231-90. International Exchange of Programmes Produced Electronically by Means of High-Definition Television—Section 10/11G—Exchange of Recorded Television Programmes. 2 pp.
CCIR RECMN 709-90. Basic Parameter Values for the HDTV Standard for the Studio and for International Programme Exchange—Section 11A—Characteristics of Systems for Monochrome and Colour Television. 9 pp.

—Television Recording
CCIR Report 1230-90. Recording of High-Definition Television on Videotape and Disk—Section 10/11G—Exchange of Recorded Television Programmes. 5 pp.
CCIR QUESTION 108/11-90. Recording of High Definition Television Programmes—Questions Concerning Study Group 11—Broadcasting Service (Television). 1 p.

—Television Recording—Digital—Magnetic Tape
CCIR Decision 59-3-89. Television Programmes on Digital Tape and on Film —Annex to Volumes X and XI—Part 3—Sound and Television Recording. 3 pp.

High Definition Television Equipment (Cont.)

—Television Recording—Motion Picture Film
CCIR RECMN 713-90. Recording of HDTV Images on Film—Section 10/11H —Use of Film in Television. 3 pp.
CCIR Report 1229-90. Recording of High-Definition Television Programmes on Cinematographic Film—Section 10/11H—Use of Film in Television. 6 pp.
CCIR Decision 59-3-89. Television Programmes on Digital Tape and on Film —Annex to Volumes X and XI—Part 3—Sound and Television Recording. 3 pp.
CCIR QUESTION 109/11-90. Recording of High Definition Television Programmes on Cinematographic Film—Questions Concerning Study Group 11—Broadcasting Service (Television). 1 p.

—Television Signals—Digital Transmission
CCIR Decision 67-2-89. Digital Transmission of Component-Coded Television and High Definition Television Signals—Annex to Volume XII—Television and Sound Transmission (CMTT). 3 pp.

—Television Studios
CCIR RECMN 709-90. Basic Parameter Values for the HDTV Standard for the Studio and for International Programme Exchange—Section 11A—Characteristics of Systems for Monochrome and Colour Television. 9 pp.

—Television Transmission
CCIR Report 1092-1-90. Transmission of High-Definition Television Signals—Section CMTT A —Television Transmission Standards and Performance Objectives. 9 pp.

—Television Transmission—Long Distances
CCIR QUESTION 22-1/CMTT-86. Standards, Operational Characteristics, Test Signals and Methods of Measurement Relevant to Systems for the Transmission of High Definition Television Signals over Long Distances—Questions Concerning the CMTT CCIR/CCITT Joint Study Group for Television. 1 p.

—Television Transmission—Test Signals
CCIR QUESTION 22-1/CMTT-86. Standards, Operational Characteristics, Test Signals and Methods of Measurement Relevant to Systems for the Transmission of High Definition Television Signals over Long Distances—Questions Concerning the CMTT CCIR/CCITT Joint Study Group for Television. 1 p.

—Video Cassette Recorders
CCIR Report 1233-90. HDTV Recording/Reproduction Equipment for Consumer and Industrial Uses—Section 10/11G—Exchange of Recorded Television Programmes. 5 pp.

—Video Cassettes
CCIR QUESTION 110/11-90. Transfer of High Definition Television Programmes to Non-Broadcast Media for Domestic Use—Questions Concerning Study Group 11—Broadcasting Service (Television). 1 p.

—Video Disk Recorders
CCIR Report 1233-90. HDTV Recording/Reproduction Equipment for Consumer and Industrial Uses—Section 10/11G—Exchange of Recorded Television Programmes. 5 pp.

—Video Disk Recording
CCIR Report 1230-90. Recording of High-Definition Television on Videotape and Disk—Section 10/11G—Exchange of Recorded Television Programmes. 5 pp.

—Video Disks
CCIR QUESTION 110/11-90. Transfer of High Definition Television Programmes to Non-Broadcast Media for Domestic Use—Questions Concerning Study Group 11—Broadcasting Service (Television). 1 p.

—Video Tape Recording
CCIR Report 1230-90. Recording of High-Definition Television on Videotape and Disk—Section 10/11G—Exchange of Recorded Television Programmes. 5 pp.

High Density Digital Recording
Use For: HDDR *See Also:* Digital Recording; Recording Instruments
OSI ISO IEC DIS 8441-2-89. Information Technology—High Density Digital Recording (HDDR)—Part 2: Guide for Interchange Practice. 59 pp.

High Density Digital Recording (Cont.)

—Magnetic Tapes
BSI BS ISO/IEC 8441-1-91. 1991 Information Technology—High Density Digital Recording (HDDR)—Part 1: Unrecorded Magnetic Tape for (HDDR) Applications. 38 pp.
IEC 8441 Pt 1-91. Information Technology—High Density Digital Recording (HDDR)—Part 1: Unrecorded Magnetic Tape for (HDDR) Applications First Edition. 35 pp.
ISO 8441 Pt 1-91. Information Technology—High Density Digital Recording (HDDR)—Part 1: Unrecorded Magnetic Tape for (HDDR) Applications First Edition. 35 pp.
JTC1 8441 Pt 1-91. Information Technology—High Density Digital Recording (HDDR)—Part 1: Unrecorded Magnetic Tape for (HDDR) Applications First Edition. 35 pp.
OSI ISO IEC DIS 8441-1-89. Information Technology—High Density Digital Recording (HDDR)—Part 1: Unrecorded Magnetic Tape for (HDDR) Applications. 53 pp.
SAA AS 3977.1-91. Informaiton Processing Systems—High Density Digital Recording (HDDR) —Part 1: Unrecorded Magnetic Tape for HDDR Applications (ISO/IEC 8441-1:1991) (in Professional Package 26A). 27 pp.

—Magnetic Tapes—Information Interchange
BSI BS ISO/IEC 8441-2-91. 1991 Information Technology—High Density Digital Recording (HDDR)—Part 2: Guide for Interchange Practice. 41 pp.
IEC 8441 Pt 2-91. Information Technology—High Density Digital Recording (HDDR)—Part 2: Guide for Interchange Practice First Edition. 38 pp.
ISO 8441 Pt 2-91. Information Technology—High Density Digital Recording (HDDR)—Part 2: Guide for Interchange Practice First Edition. 38 pp.
JTC1 8441 Pt 2-91. Information Technology—High Density Digital Recording (HDDR)—Part 2: Guide for Interchange Practice First Edition. 38 pp.
SAA AS 3977.2-92. Information Processing Systems—High Density Digital Recording (HDDR) —Part 2: Guide for Interchange Practice (ISO/IEC 8441-2:1991) (in Professional Package 26A). 31 pp.

High Density Polyethylene
Use: Polyethylene—High Density

High Expansion Steels
See Also: Steels

—Bars—Aerospace
BSI HR 650-72. 1972 Amd 2 High Expansion Heat-Resisting Steel Bar and Wire for the Manufacture of Bolts, Studs, Set Screws and Nuts (Ni 25.5, Cr 15, Ti 2, Mn 1.5, Mo 1.25, Si 0.7, V 0.3) (Limiting Ruling Section 20 mm). 5 pp.
BSI 2S 131-77. 1977 Amd 1 High Thermal Expansion Steel Billets, Bars, Forgings and Parts (620 MPa: Limiting Ruling Section 150 mm). 5 pp.

—Billets—Aerospace
BSI 2S 131-77. 1977 Amd 1 High Thermal Expansion Steel Billets, Bars, Forgings and Parts (620 MPa: Limiting Ruling Section 150 mm). 5 pp.

—Forgings—Aerospace
BSI 2S 131-77. 1977 Amd 1 High Thermal Expansion Steel Billets, Bars, Forgings and Parts (620 MPa: Limiting Ruling Section 150 mm). 5 pp.

High Explosives
See Also: Explosives

—Ammonium Perchlorate
NATO STANAG 4299 ED 1 AMD 0-00. Specification Ammonium Perchlorate (NH4C104) for Deliveries from One NATO Nation to Another. 24 pp.

—Density
MOD UK M 216/91. Determination of Density as Applied to High Explosives in Biscuit and Pellet Form.

—Qualification Testing
NATO STANAG 4170 ED 1 AMD 1-85. Principles and Methodology for the Qualification of Explosive Materials for Military Use. 19 pp.

High Fidelity Audio Equipment
Use: Audio Equipment

High Intensity Discharge Lamps
Use For: H.I.D. Lights *See Also:* Discharge Lamps; Lighting; Mercury Vapor Lamps; Metal Halide Lamps; Quartz Tungsten Halogen Lamps
CSA C22.2 NO 9-M1989. Luminaires; (Gen Instr 1) (Supercedes C22.2 No 97-1969). 83 pp.

INDUSTRY STANDARDS

High Intensity Discharge Lamps (Cont.)
CSA 922A Bull. Electrical Bulletin 922A April 12, 1978 to C22.2 NO 9. 1 p.
CSA 1169 Bull. Electrical Bulletin 1169 June 27, 1978 to C22.2 NO 9. 2 pp.

—Ballasts (Electric)—Branch Circuits
CSA CAN/CSA-C22. 2 NO 74-92. Equipment for Use with Electric Discharge Lamps (Incorporating Electrical Bulletin Nos. 523F, 753, 846, 1124A, 1125A, 1325, and 1326); (Gen Instr 1). 82 pp.

—Ballasts (Electric)—Control Circuits
CSA C22.2 NO 184-M1988. Solid-State Lighting Controls; (Gen Instr 1 Thru 3). 33 pp.

High Intensity Lights
Use: High Intensity Discharge Lamps

High Level Data Link Control
Use For: HDLC *See Also:* Data Links; Open Systems Interconnection; Protocols

BSI BS 5397: Part 6-90. 1990 High-Level Data Link Control Procedures for Data Communication Part 6: Multilink Procedures. 15 pp.
BSI BS 5397: Part 7-90. 1990 High-Level Data Link Control Procedures for Data Communication Part 7: X.25 LAPB-Compatible DTE Data Link Procedures. 28 pp.
CEPT T/A 07-01 E-86. Harmonization of Teletex Terminal Equipment. 133 pp.
CSA CAN/CSA-Z243.53-88. Information Processing Systems—Data Communication—High-Level Data Link Procedures—Description of the X.25 LAPB-Compatible DTE Data Link Procedures (ISO 7776-1986). 35 pp.
ECMA ECMA 71-81. HDLC Selected Procedures. 41 pp.
SNZ NZS/ISO 7776-86. Information Processing Systems—Data Communication—High-Level Data Link Control Procedures—Description of the X.25 LAPB-Compatible DTE Data Link Procedures. 23 pp.

—Classes
BSI BS 5397: Part 5-90. (WITHDRAWN) 1990 High-Level Data Link Control Procedures for Data Communication Part 5: Consolidation of Classes of Procedures (Superseded by BS ISO/IEC 7809: 1991). 27 pp.
BSI BS ISO/IEC 7809-91. 1991 Amd 7 Information Technology—Telecommunications and Information Exchange Between Systems—High-Level Data Link Control (HDLC) Procedures—Classes of Procedures. 52 pp.
CNS C5208-86. Information Processing Systems-Data Communications-High-Level Data Link Control Procedures-Consolidation of Classes of Procedures (Jul)(11628).
CSA CAN/CSA-Z243.54-88. Information Processing Systems—Data Communication—High-Level Data Link Control Procedures—Consolidation of Classes of Procedures (ISO 7809-1984). 33 pp.
CSA CAN/CSA-Z243.55-88. Information Processing Systems—Data Communication—High-Level Data Link Control Balanced Classes of Procedures—Data-Link Layer Address Resolution/Negotiation in Switched Environments (ISO 8471-1987). 15 pp.
IEC 7809-91. Information Technology—Telecommunications and Information Exchange Between Systems—High-Level Data Link Control (HDLC) Procedures—Classes of Procedures—Amendment 6: Extended Transparency Options for Start/Stop Transmission Second Edition; (Amd 7-1991)(Amd 5-6:1992). 46 pp.
IEC 7809 Draft AMD 9. Information Technology—Telecommunications and Information Exchange Between Systems—High-Level Data Link Control (HDLC) Procedures—Classes of Procedures Amendment 9: Seven-Bit Data Path Transparency Option for Start/Stop Transmission; (1992) ***CD-ROM ONLY***. 4 pp.
ISO 7809-91. Information Technology—Telecommunications and Information Exchange Between Systems—High-Level Data Link Control (HDLC) Procedures—Classes of Procedures—Amendment 6: Extended Transparency Options for Start/Stop Transmission Second Edition; (Amd 7-1991)(Amd 5-6:1992). 46 pp.
ISO 7809 Draft AMD 9. Information Technology—Telecommunications and Information Exchange Between Systems—High-Level Data Link Control (HDLC) Procedures—Classes of Procedures Amendment 9: Seven-Bit Data Path Transparency Option for Start/Stop Transmission; (1992) ***CD-ROM ONLY***. 4 pp.
ISO 8471-87. Information Processing Systems—Data Communication—High-Level Data Link Control Balanced Classes of Procedures—Data-Link Layer Address Resolution/Negotiation in Switched Environments First Edition. 6 pp.
JIS X 5106-78. High Level Data Link Control Procedures—Classes of Procedures.

High Level Data Link Control (Cont.)
—Classes (Cont.)
JTC1 7809-91. Information Technology—Telecommunications and Information Exchange Between Systems—High-Level Data Link Control (HDLC) Procedures—Classes of Procedures—Amendment 6: Extended Transparency Options for Start/Stop Transmission Second Edition; (Amd 7-1991)(Amd 5-6:1992). 11 pp.
JTC1 7809 Draft AMD 9. Information Technology—Telecommunications and Information Exchange Between Systems—High-Level Data Link Control (HDLC) Procedures—Classes of Procedures Amendment 9: Seven-Bit Data Path Transparency Option for Start/Stop Transmission; (1992) ***CD-ROM ONLY***. 4 pp.
JTC1 8471-87. Information Processing Systems—Data Communication—High-Level Data Link Control Balanced Classes of Procedures—Data-Link Layer Address Resolution/Negotiation in Switched Environments First Edition. 6 pp.
OSI ISO 7809 DAM 5-91. Information Processing Systems—Data Communication—High-Level Data Link Control Procedures—Classes of Procedures—Amendment 5: Connectionless Classes of Procedures.
OSI ISO 7809 DAM7-84. Information Processing Systems—Data Communication—High-Level Data Link Control Procedures—Consolidation of Classes of Procedures: Amendment 7: Multiselective Reject Option. 5 pp.
OSI ISO 7809 PDAD 4-88. Telecommunications and Information Exchange Between Systems. Info. Proc. Systems—Data Comm.—High Level Data Link Control Proc.-Consol. of Classes of Procedures Addendum 4: List of Std. Data Link Layer Protocols that Utilize HDLC Classes of Procedures.
OSI ISO 7809 PDAD 3-88. Telecommunications and Information Exchange Between Systems. Information Processing Systems—Data Communication—High Level Data Link Control Procedures—Consolidation of Procedures Addendum 3: Start/Stop Transmission.
OSI ISO 7809-84. Information Processing Systems—Data Communications—High Level Data Link Control Procedures—Consolidation of Classes of Procedures. 17 pp.
OSI ISO 7809 Add.1-87. Information Processing Systems—Data Communication—High-Level Data Link Control Procedures—Consolidation of Classes of Procedures ADDENDUM 1. 2 pp.
OSI ISO 7809 Add.2-87. Information Processing Systems—Data Communication—High-Level Data Link Control Procedures—Consolidation of Classes of Procedures Addendum 2: Description of Optional Functions. 4 pp.
OSI ISO 7809 DAD 3-89. Information Processing Systems—Data Communication—High-Level Data Link Control Procedures—Consolidation of Classes of Procedures Addendum 3: Start/Stop Transmission. 5 pp.
OSI ISO/IEC DIS 7809-2-90. Information Technology—Data Communication—High-Level Data Link Control (HDLC) Procedures—Classes of Procedures. 20 pp.
OSI ISO 8471-87. Information Processing Systems—Data Communication—High-Level Data Link Control Balanced Classes of Procedures—Data-Link Layer Address Resolution/Negotiation in Switched Environments.
SNZ NZS/ISO 7809-84. Information Processing Systems—High-Level Data Link Control Procedures—Consolidation of Classes of Procedures. 15 pp.
SNZ NZS/ISO 7809: Addendum 1-87. Information Processing Systems—Data Communication—High-Level Data Link Control Procedures—Consolidation of Classes. 2 pp.
SNZ NZS/ISO 7809: Addendum 2-87. Information Processing Systems—Data Communication—High-Level Data Link Control Procedures—Consolidation of Classes of Procedures. Desciption of Optional Functions. 4 pp.
SNZ NZS/ISO 8471-87. Information Processing Systems—Data Communication—High-Level Data Link Control Balanced Classes of Procedures—Data-Link Layer Address Resolution/Negotiation in Switched Environments. 3 pp.

—Conformance
IEC 7776 AMD 1. Amendment 1—Information Processing Systems—Data Communication—High-Level Data Link Control Procedures—Description of the X.25 LAPB-Compatible DTE Data Link Procedures Amendment 1: Conformance Requirements; (1992). 20 pp.

—Data Terminal Equipment—Conformance
BSI BS ISO/IEC 8882-2-92. 1992 Information Technology—Telecommunications and Information Exchange Between Systems—X.25 DTE Conformance Testing—Part 2: Data Link Layer Conformance Test Suite. 241 pp.

High Level Data Link Control (Cont.)
—Data Terminal Equipment—Conformance (Cont.)
IEC 8882 Pt 2-92. Information Technology—Telecommunications and Information Exchange Between Systems—X.25 DTE Conformance Testing—Part 2: Data Link Layer Conformance Test Suite First Edition. 241 pp.
ISO 8882 Pt 2-92. Information Technology—Telecommunications and Information Exchange Between Systems—X.25 DTE Conformance Testing—Part 2: Data Link Layer Conformance Test Suite First Edition. 241 pp.
JTC1 8882-92. Information Technology—Telecommunications and Information Exchange Between Systems—X.25 DTE Conformance Testing—Part 2: Data Link Layer Conformance Test Suite First Edition. 241 pp.

—Elements
BSI BS 5397: Part 2-90. (WITHDRAWN) 1990 High-Level Data Link Control Procedures for Data Communication Part 2: Elements of Procedures (Superseded by BS ISO/IEC 4335: 1991). 48 pp.
BSI BS ISO/IEC 4335-91. 1991 Amd 4 Information Technology—Telecommunications and Information Exchange Between Systems—High-Level Data Link Control (HDLC) Procedures—Elements of Procedures. 61 pp.
CNS C5207-86. Data Communication-High-Level Data Link Control Procedures-Consolidation of Elements of Procedures (Jul)(11627).
CSA CAN/CSA-Z243.51-88. Information Processing Systems—Data Communication—High-Level Data Link Control Elements of Procedures (ISO 4335-1987). 55 pp.
IEC 4335-91. Information Technology—Telecommunications and Information Exchange Between Systems—High-Level Data Link Control (HDLC) Procedures—Elements of Procedures Amendment 4: Multi-Selective Reject Option Fourth Edition; (Amendment 4-1991). 58 pp.
ISO 4335-91. Information Technology—Telecommunications and Information Exchange Between Systems—High-Level Data Link Control (HDLC) Procedures—Elements of Procedures Amendment 4: Multi-Selective Reject Option Fourth Edition; (Amendment 4-1991). 58 pp.
JIS X 5105-91. Information Processing Systems—Data Communication—High-Level Data Link Control.
JTC1 4335-91. Information Technology—Telecommunications and Information Exchange Between Systems—High-Level Data Link Control (HDLC) Procedures—Elements of Procedures Amendment 4: Multi-Selective Reject Option Fourth Edition; (Amendment 4-1991). 58 pp.
OSI ISO 4335-91. Information Technology—Telecommunications and Information Exchange Between Systems—High-Level Data Link Control (HDLC) Procedures—Elements of Procedures—Amendment 4: Multi-Selective Reject Option. 60 pp.
OSI ISO 4335-2-88. High-Level Data Link Control (HDLC) Procedures—Part 2: Specification for Consolidation of Elements of Procedures. 4 pp.
OSI ISO/IEC DIS 4335-2-90. Information Technology—Data Communication—High-Level Data Link Control (HDLC) Procedures—Elements of Procedures. 46 pp.
SNZ NZS/ISO 4335-87. Information Processing Systems—Data Communications—High-Level Data Link Control Elements of Procedures. 43 pp.

—Formats
BSI BS 5397: Part 9-90. (WITHDRAWN) 1990 High-Level Data Link Control Procedures for Data Communication Part 9: General Purpose XID Frame Information Field Content and Format (Including Parameters for Para. Negot. and Multilink Subfields) (Superseded by BS ISO/IEC 8885: 1991). 15 pp.
BSI BS 7404: Sec 5.1-91. 1991 Telecontrol Equipment and Systems Part 5: Transmission Protocols Section 5.1: Specification for Transmission Frame Formats (IEC 870-5-1: 1990). 42 pp.
BSI BS ISO/IEC 8885-91. 1991 Amd 5 Information Technology—Telecommunications and Information Exchange Between Systems—High-Level Data Link Control (HDLC) Procedures—General Purpose XID Frame Information Field Content and Format. 32 pp.
CSA CAN/CSA-Z243.56-88. Information Processing Systems—Data Communication—High-Level Data Link Control Procedures—General Purpose XID Frame Information Field Content and Format (ISO 8885-1987). 20 pp.
IEC 870 Pt 5-1-90. Telecontrol Equipment and Systems Part 5: Transmission Protocols Section One—Transmission Frame Formats First Edition. 82 pp.

INTERNATIONAL AND NON-U.S. NATIONAL STANDARDS
SUBJECT INDEX High

High Level Data Link Control *(Cont.)*

—Formats *(Cont.)*

IEC 8885-91. Information Technology—Telecommunications and Information Exchange Between Systems—High-Level Data Link Control (HDLC) Procedures—General Purpose XID Frame Information Field Content and Format Amendment 5: Multi-Selective Reject Option Second Edition;. 37 pp.

ISO 8885-91. Information Technology—Telecommunications and Information Exchange Between Systems—High-Level Data Link Control (HDLC) Procedures—General Purpose XID Frame Information Field Content and Format Amendment 5: Multi-Selective Reject Option Second Edition;. 37 pp.

JTC1 8885-91. Information Technology—Telecommunications and Information Exchange Between Systems—High-Level Data Link Control (HDLC) Procedures—General Purpose XID Frame Information Field Content and Format Amendment 5: Multi-Selective Reject Option Second Edition;. 5 pp.

OSI ISO 8885 DAM 4-87. Information Processing Systems—Data Communication—High-Level Data Link Control Procedures—General Pupose XID Frame Information Field Content and Format Amendment 4: Extended Transparency Options for Start/Stop Transmission. 4 pp.

OSI ISO 8885 DAM 5-90. Information Processing Systems—Data Communications—High Level Data Link Control Procedures—General Purpose XID Frame Information Field Content and Format—Amendment 5: Multi-Selective Reject Option. 5 pp.

OSI ISO 8885/DAM 3-90. Information Processing Systems—Data Communication—High-Level Data Link Control Procedures—General Purpose XID Frame Information Field Content and Format—Amendment 3: Definition of a Private Parameter Data Link Layer Subfield. 5 pp.

OSI ISO 8885 AD 1-89. Information Processing Systems—Data Communication—High-Level Data Link Control Procedures—General Purpose XID Frame Information Field Content and Format—Addendum 1: Additional Operational Parameters. 5 pp.

OSI ISO 8885 DAD 1-87. General Purpose XID Frame Information Field Content and Format—Addendum 1: Additional Operational Parameters for the Parameter Negotiation Data Link Layer Subfield and Definition of a Multilink Parameter Negotiation Data Link Subfield. 6 pp.

OSI ISO 8885 DAD 2-89. Information Processing Systems—Data Communication—High Level Data Link Control Procedures—General Purpose XID Frame Information Field Content and Format ADDENDUM 2: Start/Stop Transmission. 4 pp.

OSI ISO 8885-87. Information Processing Systems—Data Communication—High-Level Data Link Control Procedures—General Purpose XID Frame Information Field Content and Format. 11 pp.

OSI ISO IEC 8885-91. Information Technology—Telecommunication and Information Exchange Between Systems—High-Level Data Link Control (HDLC) Procedures—General Purpose XID Frame Information Field Content and Format (Incorporating Amendment 5).

SNZ NZS/ISO 8885-87. Information Processing Systems—Data Communications—High Level Data Link Control Procedures—General Purpose XID Frame Information Field Control Format. 8 pp.

—Frame Structure

BSI BS 5397: Part 1-85. (WITHDRAWN) 1985 High-Level Data Link Control Procedures for Data Communication Part 1: Frame Structure (Superseded by BS ISO/IEC 3309: 1991). 9 pp.

BSI BS 5397: Part 8-90. 1990 High-Level Data Link Control Procedures for Data Communication Part 8: Frame Level Address Assignment. 7 pp.

BSI BS 5397: Part 9-90. (WITHDRAWN) 1990 High-Level Data Link Control Procedures for Data Communication Part 9: General Purpose XID Frame Information Field Content and Format (Including Parameters for Para. Negot. and Multilink Subfields) (Superseded by BS ISO/IEC 8885: 1991). 15 pp.

BSI BS ISO/IEC 3309-91. 1991 Information Technology—Telecommunications and Information Exchange Between Systems—High-Level Data Link Control (HDLC) Procedures—Frame Structure. 15 pp.

BSI BS ISO/IEC 3309-01. 1991 Amd 1 Information Technology—Telecommunications and Information Exchange Between Systems—High-Level Data Link Control (HDLC) Procedures—Frame Structure (AMD 7171) July 15, 1992. 21 pp.

BSI BS ISO/IEC 8885-91. 1991 Amd 5 Information Technology—Telecommunications and Information Exchange Between Systems—High-Level Data Link Control (HDLC) Procedures—General Purpose XID Frame Information Field Content and Format. 32 pp.

High Level Data Link Control *(Cont.)*

—Frame Structure *(Cont.)*

CNS C5206-86. Information Processing System-Data Communication-High-Level Data Link Control Procedures-Frame Structure (Jul)(11626).

CSA CAN/CSA-Z243.50-88. Information Processing Systems—Data Communication—High-Level Data Link Control Procedures—Frame Structure (ISO 3309-1984). 18 pp.

CSA CAN/CSA-Z243.56-88. Information Processing Systems—Data Communication—High-Level Data Link Control Procedures—General Purpose XID Frame Information Field Content and Format (ISO 8885-1987). 20 pp.

IEC 3309-91. Information Technology—Telecommunications and Information Exchange Between Systems—High-Level Data Link Control (HDLC) Procedures—Frame Structure Amendment 2: Extended Transparency Options for Start/Stop Transmission Fourth Edition; (Amendment 2-1992). 17 pp.

IEC 3309 Draft AMD 3. Information Technology—Telecommunications and Information Exchange Between Systems—High-Level Data Link Control (HDLC) Procedures—Frame Structure Amendment 3: Seven-Bit Data Path Transparency Option for Start/Stop Transmission; (1992) ***CD-ROM ONLY***. 5 pp.

IEC 8885-91. Information Technology—Telecommunications and Information Exchange Between Systems—High-Level Data Link Control (HDLC) Procedures—General Purpose XID Frame Information Field Content and Format Amendment 5: Multi-Selective Reject Option Second Edition;. 37 pp.

IEC 8885 Draft AMD 7. Information Technology—Telecommunications and Information Exchange Between Systems—High-Level Data Link Control (HDLC) Procedures—General Purpose XID Frame Information Field Content and Format Amendment 7: Frame Check Sequence (FCS) Negotiation Using the. 6 pp.

ISO 3309-91. Information Technology—Telecommunications and Information Exchange Between Systems—High-Level Data Link Control (HDLC) Procedures—Frame Structure Amendment 2: Extended Transparency Options for Start/Stop Transmission Fourth Edition; (Amendment 2-1992). 17 pp.

ISO 3309 Draft AMD 3. Information Technology—Telecommunications and Information Exchange Between Systems—High-Level Data Link Control (HDLC) Procedures—Frame Structure Amendment 3: Seven-Bit Data Path Transparency Option for Start/Stop Transmission; (1992) ***CD-ROM ONLY***. 5 pp.

ISO 8885-91. Information Technology—Telecommunications and Information Exchange Between Systems—High-Level Data Link Control (HDLC) Procedures—General Purpose XID Frame Information Field Content and Format Amendment 5: Multi-Selective Reject Option Second Edition;. 37 pp.

ISO 8885 Draft AMD 7. Information Technology—Telecommunications and Information Exchange Between Systems—High-Level Data Link Control (HDLC) Procedures—General Purpose XID Frame Information Field Content and Format Amendment 7: Frame Check Sequence (FCS) Negotiation Using the. 6 pp.

JIS X 5104-91. High Level Data Link Control Procedures—Frame Structure.

JTC1 3309-91. Information Technology—Telecommunications and Information Exchange Between Systems—High-Level Data Link Control (HDLC) Procedures—Frame Structure Amendment 2: Extended Transparency Options for Start/Stop Transmission Fourth Edition; (Amendment 2-1992). 17 pp.

JTC1 3309 Draft AMD 3. Information Technology—Telecommunications and Information Exchange Between Systems—High-Level Data Link Control (HDLC) Procedures—Frame Structure Amendment 3: Seven-Bit Data Path Transparency Option for Start/Stop Transmission; (1992) ***CD-ROM ONLY***. 5 pp.

JTC1 8885-91. Information Technology—Telecommunications and Information Exchange Between Systems—High-Level Data Link Control (HDLC) Procedures—General Purpose XID Frame Information Field Content and Format Amendment 5: Multi-Selective Reject Option Second Edition;. 5 pp.

JTC1 8885 Draft AMD 7-92. Information Technology—Telecommunications and Information Exchange Between Systems—High-Level Data Link Control (HDLC) Procedures—General Purpose XID Frame Information Field Content and Format Amendment 7: Frame Check Sequence (FCS) Negotiation Using the. 6 pp.

High Level Data Link Control *(Cont.)*

—Frame Structure *(Cont.)*

OSI ISO/IEC 3309-91. Information Technology—Telecommunications and Information Exchange Between Systems—High-Level Data Link Control (HDLC) Procedures—Frame Structure. 17 pp.

OSI ISO 3309 DAM 2-91. Information Processing Systems—Data Communication—High-Level Data Link Control Procedures—Frame Structure Amendment 2: Extended Transparency Options for Start/Stop Transmission. 4 pp.

OSI ISO/IEC DIS 3309-2-90. Information Technology—Data Communication—High-Level Data Link Control (HDLC) Procedures—Frame Structure. 18 pp.

OSI ISO/IEC 3309 DAM 3-92. Information Technology—Telecommunications and Information Exchange Between Systems—High-Level Data Link Control (HDLC) Procedures—Frame Structure—Amendment 3: Seven-Bit Data Path Transparency Option for Start/Stop Transmission. 5 pp.

OSI ISO 8885 DAM 4-87. Information Processing Systems—Data Communication—High-Level Data Link Control Procedures—General Pupose XID Frame Information Field Content and Format Amendment 4: Extended Transparency Options for Start/Stop Transmission. 4 pp.

OSI ISO 8885 DAM 5-90. Information Processing Systems—Data Communications—High Level Data Link Control Procedures—General Purpose XID Frame Information Field Content and Format—Amendment 5: Multi-Selective Reject Option. 5 pp.

OSI ISO 8885/DAM 3-90. Information Processing Systems—Data Communication—High-Level Data Link Control Procedures—General Purpose XID Frame Information Field Content and Format—Amendment 3: Definition of a Private Parameter Data Link Layer Subfield. 5 pp.

OSI ISO 8885 AD 1-89. Information Processing Systems—Data Communication—High-Level Data Link Control Procedures—General Purpose XID Frame Information Field Content and Format—Addendum 1: Additional Operational Parameters. 5 pp.

OSI ISO 8885 DAD 1-87. General Purpose XID Frame Information Field Content and Format—Addendum 1: Additional Operational Parameters for the Parameter Negotiation Data Link Layer Subfield and Definition of a Multilink Parameter Negotiation Data Link Subfield. 6 pp.

OSI ISO 8885 DAD 2-89. Information Processing Systems—Data Communication—High Level Data Link Control Procedures—General Purpose XID Frame Information Field Content and Format ADDENDUM 2: Start/Stop Transmission. 4 pp.

OSI ISO 8885-87. Information Processing Systems—Data Communication—High-Level Data Link Control Procedures—General Purpose XID Frame Information Field Content and Format. 11 pp.

OSI ISO IEC 8885-91. Information Technology—Telecommunication and Information Exchange Between Systems—High-Level Data Link Control (HDLC) Procedures—General Purpose XID Frame Information Field Content and Format (Incorporating Amendment 5).

SNZ NZS/ISO 3309-84. Information Processing—Data Communications—High-Level Data Link Control Procedures—Frame Structure. 6 pp.

SNZ NZS/ISO 8885-87. Information Processing Systems—Data Communications—High Level Data Link Control Procedures—General Purpose XID Frame Information Field Control Format. 8 pp.

—Packet Switched Networks—Data Terminal Equipment

IEC DIS 7776-93. Information Technology—Telecommunications and Information Exchange Between Systems—High-Level Data Link Control Procedures—Description of the X.25 LAPB-Compatible DTE Data Link Procedures ***CD-ROM ONLY***. 44 pp.

ISO 7776-86. Information Processing Systems—Data Communi-cation—High-Level Data Link Control Procedures—Description of the X.25 LAPB-Compatible DTE Data Link Procedures Amendment 1: Conformance Requirements First Ed.; (Corrigendum 1-2:1989) (Corrigendum 3-1991) (Amendment 1-1992). 47 pp.

ISO DIS 7776-93. Information Technology—Telecommunications and Information Exchange Between Systems—High-Level Data Link Control Procedures—Description of the X.25 LAPB-Compatible DTE Data Link Procedures ***CD-ROM ONLY***. 44 pp.

JTC1 7776-86. Information Processing Systems—Data Communi-cation—High-Level Data Link Control Procedures—Description of the X.25 LAPB-Compatible DTE Data Link Procedures Amendment 1: Conformance Requirements First Ed.; (Corrigendum 1-2:1989) (Corrigendum 3-1991) (Amendment 1-1992). 31 pp.

INDUSTRY STANDARDS 1309

High Level Data Link Control (Cont.)

—Packet Switched Networks—Data Terminal Equipment (Cont.)

JTC1 DIS7776-93. Information Technology—Telecommunications and Information Exchange Between Systems—High-Level Data Link Control Procedures—Description of the X.25 LAPB-Compatible DTE Data Link Procedures ***CD-ROM ONLY***. 44 pp.

OSI ISO 7776-86. Information Processing Systems—Data Communication—High-Level Data Link Control Procedures—Description of the X.25 LAPB—Compatible DTE Data Link Procedures (with Technical Corrigendums 1-3). 33 pp.

High-Lift Trucks
Use: Lift Trucks

High Performance Liquid Chromatography
Use: Liquid Chromatography

High Pressure Laminates
Use: Sheet Laminates—High Pressure

High Pressure Water Cleaning Equipment
See Also: Cleaning Equipment and Supplies

BSI BS 5415: Sec 2.4-86. 1986 Amd 1 Safety of Electrical Motor-Operated Industrial Cleaning Appliances Part 2: Particular Requirements Section 2.4: High Pressure Water/Steam Cleaning Appliances (AMD 6453) January 31, 1991. 12 pp.

High Purity Water
Use For: Biomedical Water; Laboratory Water
See Also: Water

BSI BS 3978-87. 1987 Water for Laboratory Use. 8 pp.

CNS K1165-74. Purified Water for Chemical Analysis (Mar)(3699).

ISO 3696-87. Water for Analytical Laboratory Use—Specification and Test Methods First Edition. 8 pp.

High Rupturing Capacity Fuses
Use: HRC Fuses

High Speed Buses
Use: Data Buses

High Speed Steel Tools
Scope Note: For additional listings, see the subheading High Speed Steels under specific types of tools *See Also:* Broaches; Burrs (Tools); Cutting Tools; Drills; Hobbing Machines; Milling Cutters; Reamers; Saws; Taps (Threading Tools); Tool Steels; Tools

CNS B3279-78. High Speed Steel Tipped Tools (Mar)(4266). 19 pp.

JIS B 4152-88. High Speed Steel Tipped Tools. 27 pp.

High Speed Tool Steels
Use: Tool Steels

High Strength Bolts
See Also: Bolts; Structural Bolts

—Aircraft

MOD UK DTD-5162-61. High Tensile Steel Bolts (55-65 Ton) of High Metallurgical Quality (Reprinted March 1967). 3 pp.

—Aircraft—Countersunk—Flat Head—Slotted

SBAC AS 1242 ISSUE 18. Bolt, 90 Degree Countersunk Head H.T.S.. 1 p.

—Aircraft—Oval Head—Slotted

SBAC AS 1248 ISSUE 16. Bolt Mushroom Head (H.T.S.). 1 p.

—Aircraft—Round Head—Slotted

SBAC AS 1246 ISSUE 18. Bolt, Round Head (H.T.S.). 1 p.

—Friction Grip—Hexagonal Head—Structural

BSI BS 4395: Part 1-69. 1969 Amd 2 High Strength Friction Grip Bolts and Associated Nuts and Washers for Structural Engineering Part 1: General Grade. 36 pp.

BSI BS 4395: Part 2-69. 1969 Amd 2 High Strength Friction Grip Bolts and Associated Nuts and Washers for Structural Engineering Part 2: Higher Grade Bolts and Nuts and General Grade Washers. 36 pp.

High Strength Bolts (Cont.)

—Friction Grip—Hexagonal Head—Structural (Cont.)

BSI BS 4604: Part 1-70. 1970 Amd 3 The Use of High Strength Friction Grip Bolts in Structural Steelwork. Metric Series Part 1: General Grade. 17 pp.

SNZ NZS/BS 4395: Part 1-69. Specification for High Strength Friction Grip Bolts and Associated Nuts and Washers for Structural Engineering Part 1: General Grade. 36 pp.

SNZ NZS/BS 4395: Part 2-69. Specification for High Strength Friction Grip Bolts and Associated Nuts and Washers for Structural Engineering Part 2: Higher Grade Bolts and Nuts and General Grade Washers. 36 pp.

—Friction Grip—Parallel Shank—Structural

BSI BS 4604: Part 2-70. 1970 Amd 2 The Use of High Strength Friction Grip Bolts in Structural Steelwork. Metric Series Part 2: Higher Grade (Parallel Shank). 17 pp.

—Friction Grip—Structural

BSI BS 3294: Part 1-60. (WITHDRAWN) 1960 Amd 6 The Use of High Strength Friction Grip Bolts in Structural Steelwork Part 1: General Grade Bolts. 15 pp.

SNZ NZS/BS 4395: Part 3-73. Specification for High Strength Friction Grip Bolts and Associated Nuts and Washers for Structural Engineering Part 3: Higher Grade Bolts (Waisted Shank), Nuts and General Grade Washers. 24 pp.

—Friction Grip—Waisted Shank—Structural

BSI BS 4604: Part 3-73. (WITHDRAWN) 1973 The Use of High Strength Friction Grip Bolts in Structural Steelwork. Metric Series Part 3: Higher Grade (Waisted Shank). 16 pp.

—Hexagonal Head—Structural

CNS B2124-80. Hexagon Head Fitting Bolts for Steel Structures (May)(3125).

DIN ENGL 6914-89. High-Strength Hexagon Head Bolts with Large Widths Across Flats for Structural Steel Bolting (Oct). 6 pp.

DIN ENGL 7999-83. High Strength Hexagon Fit Bolts with Large Widths Across Flats for Structural Steel Bolting (Dec). 5 pp.

ISO 7411-84. Hexagon Bolts for High-Strength Structural Bolting with Large Width Across Flats (Thread Lengths According to ISO 888)—Product Grade C—Property Classes 8.8 and 10.9 First Edition. 9 pp.

ISO 7412-84. Hexagon Bolts for High-Strength Structural Bolting with Large Width Across Flats (Short Thread Length)—Product Grade C—Property Classes 8.8 and 10.9 First Edition. 9 pp.

—Sets

CNS B2768-85. Sets of High Strength Hexagon Bolt, Hexagon Nut and Plain Washers for Friction Grip Joints (Aug)(11328).

CNS B7272-85. Method of Test for the Sets of High Strength Hexagon Bolt, Hexagon Nut and Plain Washers for Friction Grip Joints (Aug)(11329).

CNS B7274-88. Method of Test for Sets of High Strength Torque Control Bolt, Hexagon Nut and Plain Washers (Jan)(12210).

JIS B 1186-79. Sets of High Strength Hexagon Bolt, Hexagon Nut and Plain Washers for Friction Grip Joints. 21 pp.

—Steel Structures

SAA AS 1252-83. High Strength Steel Bolts with Associated Nuts and Washers for Structural Engineering (Incorporating Amdt 1). 30 pp.

SNZ NZS/AS 1252-83. High-Strength Steel Bolts with Associated Nuts and Washers for Structural Engineering (This is a Joint Standard with SAA AS 1252). 30 pp.

High Strength Steel Castings
See Also: Alloy Steel Castings; Carbon Steel Castings; Steel Castings

—Heat Treated

ISO 9477-92. High Strength Cast Steels for General Engineering and Structural Purposes First Edition. 6 pp.

High Strength Steels
Scope Note: For additional listings, see also specific products made from high strength steels
Use For: High Tensile Steels; High Yield Strength Steels *See Also:* Alloy Steels; Carbon Steels; Ferroalloys; Low Alloy Steels; Low Carbon Steels; Maraging Steels; Reinforcing Steels; Steels; Structural Steels; Tungsten Steels

CEN PREN 10138-1-91. Prestressing Steel—Part 1: General Requirements. 29 pp.

High Strength Steels (Cont.)

—Bars—Hot Rolled

BSI BS 4486-80. 1980 Amd 1 Hot Rolled and Hot Rolled and Processed High Tensile Alloy Steel Bars for the Prestressing of Concrete. 11 pp.

ISO 4951-79. High Yield Strength Steel Bars and Sections First Edition. 9 pp.

—Bearing Capacity

DIN ENGL 50969-90. Testing of High-Strength Steel Building Elements for Resistance to Hydrogen-Induced Brittle Fracture and Advice on the Prevention of Such Fracture (Dec). 5 pp.

—Normalized—Plates

ISO 9328 Pt 4-91. Steel Plates and Strips for Pressure Purposes—Technical Delivery Conditions—Part 4: Weldable Fine Grain Steels with High Proof Stress Supplied in the Normalized or Quenched and Tempered Condition First Edition. 15 pp.

—Normalized—Plates—Hot Rolled—Flat

ISO 4950 Pt 2-81. High Yield Strength Flat Steel Products—Part 2: Products Supplied in the Normalized or Controlled Rolled Condition First Edition. 5 pp.

—Normalized—Strips

ISO 9328 Pt 4-91. Steel Plates and Strips for Pressure Purposes—Technical Delivery Conditions—Part 4: Weldable Fine Grain Steels with High Proof Stress Supplied in the Normalized or Quenched and Tempered Condition First Edition. 15 pp.

—Plates

CNS G2233-84. Method of Test for 70fg/mm2 High Yield Strength Steel Plates for Welded Structure (Nov)(11110).

CNS G3224-84. 70kgf/mm2 High Yield Strength Steel Plates for Welded Structure (Nov)(11109).

ISO 9328 Pt 1-91. Steel Plates and Strips for Pressure Purposes—Technical Delivery Conditions—Part 1: General Requirements First Edition. 24 pp.

ISO 9328 Pt 2-91. Steel Plates and Strips for Pressure Purposes—Technical Delivery Conditions—Part 2: Unalloyed and Low-Alloyed Steels with Specified Room Temperature and Elevated Temperature Properties First Edition. 16 pp.

JIS G 3128-87. High Yield Strength Steel Plates for Welded Structure. 16 pp.

—Plates—Heat Resistant

ISO 9328 Pt 2-91. Steel Plates and Strips for Pressure Purposes—Technical Delivery Conditions—Part 2: Unalloyed and Low-Alloyed Steels with Specified Room Temperature and Elevated Temperature Properties First Edition. 16 pp.

—Plates—Hot Rolled—Flat

ISO 4950 Pt 1-81. High Yield Strength Flat Steel Products—Part 1: General Requirements First Edition. 7 pp.

ISO 6930-83. High Yield Strength Flat Steel Products for Cold Forming First Edition. 7 pp.

—Quenched—Plates

ISO 9328 Pt 4-91. Steel Plates and Strips for Pressure Purposes—Technical Delivery Conditions—Part 4: Weldable Fine Grain Steels with High Proof Stress Supplied in the Normalized or Quenched and Tempered Condition First Edition. 15 pp.

—Quenched—Plates—Hot Rolled—Flat

ISO 4950 Pt 3-81. High Yield Strength Flat Steel Products—Part 3: Products Supplied in the Heat-Treated (Quenched + Tempered) Condition First Edition. 5 pp.

—Quenched—Strips

ISO 9328 Pt 4-91. Steel Plates and Strips for Pressure Purposes—Technical Delivery Conditions—Part 4: Weldable Fine Grain Steels with High Proof Stress Supplied in the Normalized or Quenched and Tempered Condition First Edition. 15 pp.

—Quenched—Welding

SAA AS 1554.4-89. Structural Steel Welding (Known as the SAA Structural Steel Welding Code)—Part 4: Welding of High Strength Quenched and Tempered Steels. 47 pp.

—Sheets—Cold Rolled

BSI BS EN 10131-91. 1991 Cold-Rolled Uncoated Low Carbon and High Yield Strength Steel Flat Products for Cold Forming—Tolerances on Dimensions and Shape (V). 7 pp.

BSI BS EN 10131-01. 1991 Amd 1 Cold-Rolled Uncoated Low Carbon and High Yield Strength Steel Flat Products for Cold Forming—Tolerances on Dimensions and Shape (AMD 7327) Februray 15, 1993 (Supersedes BS 1449: Section 1.7: 1991). 8 pp.

INTERNATIONAL AND NON-U.S. NATIONAL STANDARDS
SUBJECT INDEX

High Strength Steels (Cont.)
—Sheets—Cold Rolled (Cont.)
CEN PREN 10 131-89. Cold Rolled Uncoated Low Carbon and High Yield Strength Steel Flat Products for Cold Forming: Tolerances on Dimensions and Shape. 13 pp.

CEN EN 10131-91. Cold Rolled Uncoated Low Carbon and High Yield Strength Steel Flat Products for Cold Forming—Tolerances on Dimensions and Shape. 15 pp.

DIN ENGL EN 10131-92. Cold Rolled Uncoated Low Carbon and High Yield Strength Steel Flats for Cold Forming; Tolerances on Size and Geometrical Tolerances; English Version of DIN EN 10 131 (Jan) (Supersedes DIN 1541, August 1975 Edition). 8 pp.

—Sheets—Hot Rolled
ISO 4996-91. Hot-Rolled Steel Sheet of High Yield Stress Structural Quality Second Edition. 12 pp.

ISO 5951-80. Hot-Rolled Steel Sheet of Higher Yield Strength with Improved Formability First Edition; (Erratum—Nov 1981); DoD Adopted. 15 pp.

—Strips
ISO 9328 Pt 1-91. Steel Plates and Strips for Pressure Purposes—Technical Delivery Conditions—Part 1: General Requirements First Edition. 24 pp.

ISO 9328 Pt 2-91. Steel Plates and Strips for Pressure Purposes—Technical Delivery Conditions—Part 2: Unalloyed and Low-Alloyed Steels with Specified Room Temperature and Elevated Temperature Properties First Edition. 16 pp.

—Strips—Heat Resistant
ISO 9328 Pt 2-91. Steel Plates and Strips for Pressure Purposes—Technical Delivery Conditions—Part 2: Unalloyed and Low-Alloyed Steels with Specified Room Temperature and Elevated Temperature Properties First Edition. 16 pp.

—Tempered—Plates
ISO 9328 Pt 4-91. Steel Plates and Strips for Pressure Purposes—Technical Delivery Conditions—Part 4: Weldable Fine Grain Steels with High Proof Stress Supplied in the Normalized or Quenched and Tempered Condition First Edition. 15 pp.

—Tempered—Plates—Hot Rolled—Flat
ISO 4950 Pt 3-81. High Yield Strength Flat Steel Products—Part 3: Products Supplied in the Heat-Treated (Quenched + Tempered) Condition First Edition. 5 pp.

—Tempered—Strips
ISO 9328 Pt 4-91. Steel Plates and Strips for Pressure Purposes—Technical Delivery Conditions—Part 4: Weldable Fine Grain Steels with High Proof Stress Supplied in the Normalized or Quenched and Tempered Condition First Edition. 15 pp.

—Tempered—Welding
SAA AS 1554.4-89. Structural Steel Welding (Known as the SAA Structural Steel Welding Code)—Part 4: Welding of High Strength Quenched and Tempered Steels. 47 pp.

High Temperature Alloys
Use: Heat Resistant Alloys

High Temperature Steel Castings
Use: Heat Resistant Steel Castings

High Temperature Steels
Use: Heat Resistant Steels

High Temperature Testing
Scope Note: For additional listings, use a more specific term *Use For:* Oven Testing
See Also: Bend Testing; Boiling Points; Brittleness Testing; Calorific Value; Calorimeters; Compression Testing; Creep Properties; Damp Heat Testing; Dry Heat Testing; Ductility; Dynamic Testing; Environmental Testing; Fatigue Testing; Freezing Points; Hardness Testing; Hydrostatic Testing; Impact Testing; Life (Durability); Low Temperature Testing; Materials Testing; Melting Points; Physical Testing; Shear Testing; Static Testing; Stiffness Testing; Stress (Testing); Temperature; Temperature Controllers; Temperature Testing; Tensile Testing; Thermal Expansion; Thermal Shock; Thermal Stability; Thermal Stresses; Thermodynamic Properties; Torsion; Wear Testing

—Automobiles
CNS D3047-89. Method of High and Low Temperature Test for Automobile Parts (Jul)(5434).

JIS D 0204-67. Method of High and Low Temperature Test for Automobile Parts (R 1979). 6 pp.

High Temperature Testing (Cont.)
—Bitumens
BSI BS 2000: Part 45-93. 1993 Methods of Test for Petroleum and Its Products Part 45: Determination of Loss on Heating of Bitumen and Flux Oil. 4 pp.

BSI BS 2000: Part 45-82. 1982 Petroleum and Its Products Part 45: Loss on Heating of Bitumen and Flux Oil. 4 pp.

—Brake Pads
BSI BS AU 180: Part 4-82. 1982 Brake Linings Part 4: Method for Determining Effects of Heat on Dimensions and Form of Disc Brake Pads. 6 pp.

ISO 6313-80. Road Vehicles—Brake Linings—Effects of Heat on Dimensions and Form of Disc Brake Pads—Test Procedure First Edition. 6 pp.

—Cable Insulation
DIN VDE 0472 Pt 632-87. Testing of Cables and Insulated Cords; Behaviour at High Temperatures (Oct) (Partially Supersedes VDE 0271/03.69 Which Was Withdrawn in December 1988). 4 pp.

—Carbon Black
BSI BS 5293: Part 5-92. 1992 Sampling and Testing Carbon Black for Use in the Rubber Industry Part 5: Method for Determination of Loss of Mass on Heating (ISO 1126: 1992). 7 pp.

BSI BS 5293: Part 5-88. 1988 Sampling and Testing Carbon Black for Use in the Rubber Industry Part 5: Method for Determination of Loss of Mass on Heating. 3 pp.

CNS K6510-80. Methods of Test for Heating Loss of Carbon Black (Jul)(5814).

ISO 1126-92. Rubber Compounding Ingredients—Carbon Black—Determination of Loss on Heating Third Edition. 5 pp.

—Cellular Materials
DIN ENGL 53424-78. Testing of Rigid Cellular Materials; Determination of Dimensional Stability at Elevated Temperatures with Flexural Load and with Compressive Load (Dec). 3 pp.

—Ceramics
BSI BS 7134: Sec 3.5-89. 1989 Testing of Engineering Ceramics Part 3: Thermo-Mechanical Properties Section 3.5: Methods of Determination of Pyroplastic Deformation (Sagging). 10 pp.

BSI BS 7134: Sec 4.2-90. 1990 Testing of Engineering Ceramics Part 4: Thermo-Physical Properties Section 4.2: Method for the Determination of Thermal Diffusivity, by the Laser Flash (or Heat Pulse) Method. 14 pp.

CEN ENV 820-2-92. Advanced Technical Ceramics—Methods of Test for Monolithic Ceramics—Thermo-Mechanical Properties—Part 2: Determination of Selfloaded Deformation. 10 pp.

JIS R 1609-90. Testing Methods for Oxidation Resistance of Non-Oxide High Performance Ceramics. 8 pp.

—Coffee
BSI BS 5752: Part 7-84. 1984 Coffee and Coffee Products Part 7: Determination of Loss in Mass at 105 Degrees Celsius. 4 pp.

ISO 6673-83. Green Coffee—Determination of Loss in Mass at 105 Degrees C First Edition. 4 pp.

—Connectors
BSI BS EN 2591-C1-92. 1992 Elements of Electrical and Optical Connection—Test Methods Part C1: Endurance at Temperature. 10 pp.

CEN PREN 2591 (Part C1)-91. Elements of Electrical and Optical Connection Test Methods Part C1—Endurance at Temperature. 6 pp.

CEN EN 2591-C1-92. Aerospace Series—Elements of Electrical and Optical Connection—Test Methods Part C1: Endurance at Temperature. 5 pp.

CEN PREN 2591-FC1-93. Aerospace Series Elements of Electrical and Optical Connection Test Methods Part FC1—Optical Elements Endurance at Temperature. 3 pp.

SBAC TS 332 ISSUE 2. Test for Electrical Connectors High Temperature Cyclic Endurance.

SBAC TS 377 ISSUE 1. High Temperature Endurance and Fluid Resistance.

—Drinking Water
BSI BS 6920: Part 3-90. 1990 Suitability of Non-Metallic Products for Use in Contact with Water Intended for Human Consumption with Regard to Their Effect on the Quality of the Water Part 3: High Temperature Tests. 6 pp.

SNZ NZS/BS 6920: Part 3-90. Suitability of Non-Metallic Products for Use in Contact with Water Intended for Human Consumption with Regard to Their Effect on the Quality of Water Part 3: High Temperature Tests. 8 pp.

—Elastomeric Sheets
CEN PREN 495-1-91. Thermoplastic and Elastomeric Roofing and Sealing Sheets—Determination of High Temperature Dimensional Stability. 6 pp.

High Temperature Testing (Cont.)
—Electric Switches
CNS C6091-88. Method of Test for Electromechanical Switches (High/Low Temperature Operation) (Dec)(6154).

—Extenders
BSI BS 3483: Part B6-74. 1974 Methods for Testing Pigments for Paints Part B6: Determination of Matter Volatile at 105 Degrees Centigrade. 4 pp.

ISO 787 Pt 2-81. General Methods of Test for Pigments and Extenders—Part 2: Determination of Matter Volatile at 105 Degrees Celsius First Edition. 5 pp.

—Fiber Optic Cables
CNS C6276-86. Method of Test for Fiber Optic Devices (FOTP-37 Cable Bend Test Low and High Temperature) (Dec)(11788).

—Fire Alarms
CEN EN 54-5-76. Components of Automatic Fire Detection Systems Part 5 Heat Sensitive Detectors—Point Detectors Containing a Static Element. 20 pp.

CEN EN 54 (Part 5)-88. AMD 1 Components of Automatic Fire Detection Systems Part 5 Heat Sensitive Detectors—Point Detectors Containing a Static Element. 21 pp.

CEN EN 54-7-88. Components of Automatic Fire Detection Systems—Part 7: Point-Type Smoke Detectors; Detectors Using Scattered Light, Transmitted Light or Ionization. 30 pp.

CEN EN 54-8-88. Components of Automatic Fire Detection Systems—Part 8: Hig Temperature Heat Detectors. 23 pp.

CEN EN 54-9-82. Components of Automatic Fire Fetection Systems Part 9 Fire Sensitivity Test. 10 pp.

—Flux Oil
BSI BS 2000: Part 45-93. 1993 Methods of Test for Petroleum and Its Products Part 45: Determination of Loss on Heating of Bitumen and Flux Oil. 4 pp.

BSI BS 2000: Part 45-82. 1982 Petroleum and Its Products Part 45: Loss on Heating of Bitumen and Flux Oil. 4 pp.

—Footwear
BSI BS 5131: Sec 2.10-91. 1991 Methods of Test for Footwear and Footwear Materials Part 2: Solings Section 2.10: Heat Shrinkage of Cellular Solings. 7 pp.

—Glass
BSI BS 3473: Part 2-87. 1987 Chemical Resistance of Glass Used in the Production of Laboratory Glassware Part 2: Method for Determination of Hydrolytic Resistance of Glass Grains at 98 Degrees Centigrade. 10 pp.

BSI BS 3473: Part 3-87. 1987 Chemical Resistance of Glass Used in the Production of Laboratory Glassware Part 3: Method for Determination of Hydrolytic Resistance of Glass Grains at 121 Degrees Centigrade. 10 pp.

BSI BS 3473: Sec 4.1-89. 1989 Chemical Resistance of Glass Used in the Pro-duction of Laboratory Glassware: Part 4: Methods for Determination of hydrolytic Resistance of the Interior Surfaces of Glass Containers: Section 4.1:. 11 pp.

ISO 719-85. Glass—Hydrolytic Resistance of Glass Grains at 98 Degrees C—Method of Test and Classification Second Edition. 7 pp.

ISO 720-85. Glass—Hydrolytic Resistance of Glass Grains at 121 Degrees C—Method of Test and Classification Second Edition. 7 pp.

—Glazes
CNS R3087-80. Method of Test for Resistance of Glaze to High Temperature for Porcelain Laboratory Apparatus (Apr)(5476).

—Laboratory Equipment—Porcelain
CNS R3088-80. Method of Test for Constancy of Mass on Ignition for Porcelain Laboratory Apparatus (Apr)(5477).

—Leather—Footwear
JIS K 6543-77. Testing Method for Heat Resistance of Air Dry Leather of Footwear Construction.

—Masking Tapes
SAA AS 1635.13.1-74. Methods of Testing Pressure Sensitive Adhesive Tape—Part 13.1: Oven Performance of Masking Tape. 1 p.

—Metals
BSI 4A 4:Pt1: Sec 2-67. 1967 Amd 1 Test Pieces and Test Methods for Metallic Materials for Aircraft Part 1: Tensile Tests Section 2: Elevated Temperature. 10 pp.

JIS Z 2252-91. Test Methods for Vickers Hardness at Elevated Temperatures. 8 pp.

JIS Z 2279-92. Method of High Temperature Low Cycle Fatigue Testing for Metallic Materials. 12 pp.

INDUSTRY STANDARDS

INTERNATIONAL AND NON-U.S. NATIONAL STANDARDS SUBJECT INDEX

High Temperature Testing (Cont.)

—Metals (Cont.)
JIS Z 2280-93. Test Method for Young's Modulus of Metallic Materials at Elevated Temperature. 13 pp.

—Mineral Fibers
DIN ENGL 52271-81. Testing of Mineral Fibre Insulating Materials; Behaviour at Elevated Temperatures (June). 3 pp.

—Mineral Wool Insulation—Melting Points
DIN ENGL 4102 Pt 17-90. Fire Behaviour of Building Materials and Elements; Determination of Melting Point of Mineral Fibre Insulating Materials; Concepts, Requirements and Testing (Dec). 2 pp.

—Munitions
NATO STANAG 4382 ED 1 AMD 0-00. Slow Heating Tests for Munitions. 11 pp.

—Oil Filters
ISO 4548 Pt 3-82. Methods of Test for Full-Flow Lubricating Oil Filters for Internal Combustion Engines—Part 3: Resistance to High Pressure Drop and to Elevated Temperature First Edition; (Corrigendum 1-1990). 7 pp.

—Phenolic Resins
ISO 8619-88. Plastics—Phenolic Resin Powder—Determination of Flow Distance on a Glass Plate First Edition. 6 pp.

—Pigments
BSI BS 3483: Part B6-74. 1974 Methods for Testing Pigments for Paints Part B6: Determination of Matter Volatile at 105 Degrees Centigrade. 4 pp.
ISO 787 Pt 2-81. General Methods of Test for Pigments and Extenders—Part 2: Determination of Matter Volatile at 105 Degrees Celsius First Edition. 5 pp.

—Pipe Fittings
CNS K6530-80. Method of Oven Test for Moulded fittings in Unplasticized Polyvinyl Chloride (PVC) for Use Under Pressure (Jul)(5852).
CNS K6531-80. Method of Oven Test for Moulded Fittings in Unplasticized Polyvinyl Chloride (PVC) with Elastic Sealing Ring Type Joints for Use Under Pressure (Jul)(5853).
ISO 580-90. Injection-Moulded Unplasticized Poly (Vinyl Chloride) (PVC-U) Fittings—Oven Test—Test Method and Basic Specifications Second Edition; (Supersedes 2043). 8 pp.

—Pipe Fittings—PVC—Unplasticized
SAA AS 1462.11-90. Methods of Test for Unplasticized PVC (UPVC) Pipes and Fittings—Part 11: Method for High Temperature Stress-Relief Testing of UPVC Fittings. 2 pp.

—Pipes
BSI BS 2782:Pt11: METH 1102B-81. 1981 Methods of Testing Plastics Part 11: Thermoplastic Pipes, Fittings and Valves Method 1102B: Longitudinal Reversion of Pipes: Oven Method. 2 pp.
CNS K6593-80. Polyethylene (PE) Pipes — Determination of Longitudinal Reversion — Liquid Bath Immersion Method (Sep)(6486).
CNS K6594-80. Polyethylene (PP) Pipes Determination of Longitudinal Reversion (Sep)(6487).
CNS K6595-80. Unplasticized Polyvinyl Chloride (PVC) Pipes Determination of Longitudinal Reversion — Liquid Bath Immersion Method (Sep)(6488).
ISO 2505-81. Unplasticized Polyvinyl Chloride (PVC) Pipes—Longitudinal Reversion—Test Methods and Specification Second Edition. 5 pp.
ISO 2506-81. Polyethylene Pipes (PE)—Longitudinal Reversion—Test Methods and Specification Second Edition. 5 pp.

—Plastic Pipes—PVC—Unplasticized
SAA AS 1462.16-86. Methods of Test for Unplasticized PVC (UPVC) Pipes and Fittings—Part 16: Method for High Temperature Testing of UPVC Pipe. 1 p.

—Plastics
BSI BS 2782:Pt1: METH 131B-83. 1983 Methods of Testing Plastics Part 1: Thermal Properties Method 131B: Determination of Extensibility After Heat Ageing of Flexible Polyvinyl Chloride Sheet. 2 pp.
BSI BS 2782:Pt4: METH 461A-78. 1978 Methods of Testing Plastics Part 4: Chemical Properties Method 461A: Determination of Volatile Matter in Aminoplastic Moulding Materials. 2 pp.
CNS K6618-80. Aminoplastic Moulding Materials Determination of Volatile Matter (Dec)(6684).
DIN ENGL 53383 Pt 1-75. Testing of Plastics; Testing of Oxidation Stability by Means of Ageing in an Oven; Polypropylene (Sept). 2 pp.
ISO 871-80. Plastics—Determination of Temperature of Evolution of Flammable Gases (Decomposition Temperature) from a Small Sample of Pulverized Material First Edition. 5 pp.

—Plastics (Cont.)
ISO 1625-77. Plastics—Aqueous Dispersions of Polymers and Copolymers—Determination of Residue at 105 Degrees Celsius First Edition. 5 pp.
ISO 3671-76. Plastics—Aminoplastic Moulding Materials—Determination of Volatile Matter First Edition. 3 pp.
ISO 4577-83. Plastics—Polypropylene and Propylene-Copolymers—Determination of Thermal Oxidative Stability in Air—Oven Method First Edition. 6 pp.

—Potassium Sulfate
CNS K6529-80. Determination of Loss of Mass at 105 Degrees Celsius of Potassium Sulfate for Industrial Use (Jul)(5849).
ISO 2850-73. Potassium Sulphate for Industrial Use—Determination of Loss of Mass at 105 Degrees Celsius First Edition. 3 pp.

—PVC—Unplasticized
CEN PREN 478-91. Unplasticized Polyvinylchloride (PVC-U) Profiles for the Construction of Windows—Appearance After Heating at 150 Degrees C—Test Method. 6 pp.

—Railroad Signals
JIS E 3019-79. High and Low Temperature Testing Methods for Parts of Railway Signaling.

—Refractory Materials
ISO 5013-85. Refractory Products—Determination of Modulus of Rupture at Elevated Temperatures First Edition. 6 pp.

—Safety Glass—Automotive
ISO 3917-92. Road Vehicles—Safety Glazing Materials—Test Methods for Resistance to Radiation, High Temperature, Humidity, Fire and Simulated Weathering Second Edition. 8 pp.

—Semiconductor Devices
CNS C6056-88. Environmental Testing Methods and Endurance Testing Methods for Discrete Semiconductor Devices (High Temperature for Reverse Bias Test of Variable Capacitance Diodes) (Sep)(5540).
CNS C6061-88. Environmental Testing Methods and Endurance Testing Methods for Discrete Semiconductor Devices (Reverse Bias Test of Transistor Under High Temperature) (Sep)(5545).
CNS C6062-88. Environmental Testing Methods and Endurance Testing Methods of Discrete Semiconductor Devices (Reverse Bias Test of Field-Effect Transistor Under High Temperature) (Sep)(5546).
CNS C6063-88. Environmental Testing Methods and Endurance Testing Methods for Discrete Semiconductor Devices (Test of Storage Under High Temperature) (Sep)(5547).
CNS C6080-88. Environmental Testing Methods and Endurance Testing Methods for Discrete Semiconductor Devices (High Temperature for Applying Voltage Test of Rectifier Diodes) (Nov)(6125).
CNS C6081-88. Environmental Testing Methods and Endurance Testing Methods for Discrete Semiconductor Devices (High Temperature for Applying Voltage Test of Thyristors) (Nov)(6126).

—Soaps
BSI BS 1715: Sec 2.6-89. 1989 Analysis of Soaps Part 2: Quantitative Test Methods Section 2.6: Method for Determination of Moisture and Volatile Matter Content (ISO 672-1978). 4 pp.
CGSB 2-GP-11M METH 13.1-83. Methods of Testing and Analysis of Soaps and Detergents Moisture and Volatile Matter (Oven Method). 1 p.
ISO 672-78. Soaps—Determination of Moisture and Volatile Matter Content—Oven Method First Edition. 4 pp.

—Sodium Borates
BSI BS 5688: Part 9-79. 1979 Orthoboric Acid (Boric Acid), Diboron Trioxide (Boric Oxide), Disodium Tetraborates, Sodium Perborates and Crude Sodium Borages for Ind. Use Part 9: Detn. of Loss in Mass of Crude Sodium Borates After Heating ataTemperature of 900 Degrees C. 3 pp.
CNS K6563-80. Determination of Lost Weight After Heating at 900 Degrees Celsius of Crude Sodium Borates for Industrial Use (Aug)(6208).
ISO 2218-72. Crude Sodium Borates for Industrial Use—Determination of Loss in Mass After Heating at 900 Degrees Celsius First Edition. 3 pp.

—Sodium Carbonates
BSI BS 6070: Part 3-81. 1981 Methods of Sampling and Test for Sodium Carbonate for Industrial Use Part 3: Determination of Non-Volatile Matter. 4 pp.
ISO 745-76. Sodium Carbonate for Industrial Use—Determination of Loss of Mass and of Non-Volatile Matter at 250 Degrees Celsius First Edition. 4 pp.

—Sodium Chloride
BSI BS 7319: Part 2-90. 1990 Analysis of Sodium Chloride for Industrial Use Part 2: Method for Determination of Moisture Content. 6 pp.
ISO 2483-73. Sodium Chloride for Industrial Use—Determination of the Loss of Mass at 110 Degrees Celsius First Edition. 4 pp.

—Sodium Hexafluorosilicate
BSI BS 5705: Part 4-79. (WITHDRAWN) 1979 Sodium Hexafluorosilicate for Industrial Use Part 4: Determination of Loss in Mass at 105 Degrees Celsius. 4 pp.
ISO 5444-78. Sodium Hexafluorosilicate for Industrial Use—Determination of Loss in Mass at 105 Degrees Celsius First Edition. 4 pp.

—Sodium Sulfates
ISO 3234-75. Sodium Sulphate for Industrial Use—Determination of Loss in Mass at 110 Degrees Celsius First Edition. 5 pp.

—Soles (Footwear)
BSI BS 5131: Sec 2.11-83. 1983 Methods of Test for Footwear and Footwear Materials Part 2: Solings Section 2.11: Resistance of Soilings to Short-Term Contact with a Hot Surface. 4 pp.

—Steels
BSI BS 3228: Part 2-60. (WITHDRAWN) 1960 Procedures for Obtaining Properties of Steel at Elevated Temperatures Part 2: Rupture Strength (Superseded by ISO 6303). 8 pp.
BSI BS 3228: Part 3-60. (WITHDRAWN) 1960 Procedures for Obtaining Properties of Steel at Elevated Temperatures Part 3: Creep Strength (Superseded by ISO 6303). 8 pp.
BSI PD 6525: Part 1-90. 1990 Elevated Temperature Properties for Steels for Pressure Purposes Part 1: Stress Rupture Properties. 96 pp.
CNS G2058-80. Method of High Temperature Tensile Test for Steels and Heat-Resisting Alloys (Mar)(5336).
ISO R206-61. Creep Stress Rupture Testing of Steel at Elevated Temperatures First Edition. 8 pp.
JIS G 0567-78. Method of High Temperature Tension Test for Steels and Heat-Resisting Alloys. 8 pp.

—Sulfur
ISO 3425-75. Sulphur for Industrial Use—Determination of Ash at 850-900 Degrees Celsius and of Residue at 200 Degrees Celsius First Edition. 4 pp.
ISO 3426-75. Sulphur for Industrial Use-Determination of Loss in Mass at 80 Degree Celsius First Edition. 4 pp.

—Thermoplastic Resins
DIN ENGL 53497-69. Testing of Plastics; Hot Storage Test on Mouldings Made of Thermoplastic Moulding Materials Without External Mechanical Stressing (Oct). 3 pp.

—Thermoplastic Sheets
CEN PREN 495-1-91. Thermoplastic and Elastomeric Roofing and Sealing Sheets—Determination of High Temperature Dimensional Stability. 6 pp.

—Thermosetting Resins
DIN ENGL 53498-67. Testing of Plastics; Heat Storage of Compression Moulding Made of Thermosetting Compression Moulding Materials (Feb). 2 pp.

—Viscous Hydrocarbons
ISO 9200-93. Crude Petroleum and Liquid Petroleum Products—Volumetric Metering of Viscous Hydrocarbons First Edition. 13 pp.

—Vulcanized Rubber
BSI BS 903: Part A5-93. 1993 Physical Testing of Rubber Part A5: Method for Determination of Tension Set at Normal and High Temperatures (ISO 2285: 1988) (V). 10 pp.
BSI BS 903: Part A5-74. 1974 Amd 1 Methods of Testing Vulcanized Rubber Part A5: Determination of Tension Set. 11 pp.
BSI BS 903: Part A6-92. 1992 Physical Testing of Rubber Part A6: Method for Determination of Compression Set at Ambient, Elevated or Low Temperatures (ISO 815: 1991) (V). 16 pp.
ISO 815-91. Rubber, Vulcanized or Thermoplastic—Determination of Compression Set at Ambient, Elevated or Low Temperatures Second Edition; (Corrigendum 1-1993). 15 pp.
ISO 2285-88. Rubber, Vulcanized or Thermoplastic—Determination of Tension Set at Normal and High Temperatures Third Edition. 7 pp.
SAA AS 1683.9-92. Methods of Test for Elastomers—Part 9: Rubber, Vulcanized or Thermoplastic—Determination of Tension Set at Normal and High Temperatures (ISO 2285:1988). 4 pp.

INTERNATIONAL AND NON-U.S. NATIONAL STANDARDS
SUBJECT INDEX

High Tensile Steels
Use: High Strength Steels

High Test Peroxide
Use: Hydrogen Peroxide

High Voltage Arresters
Use: Surge Arresters

High Voltage Direct Current Power Transmission
Use: Power Transmission—DC

High Voltage Insect Killers
Use: Insect Electrocution Devices

High Voltage Installations
See Also: Electrical Installations

—Circuit Breakers
CNS C4356-82. Cubicle Type High Voltage Power Receiving Unit (Jul)(9090).
JIS C 4620-92. Cubicle Type High Voltage Power Receiving Units. 43 pp.

—Electrical Faults
IEC 919 Pt 2-91. Performance of High-Voltage d.c. (HVDC) Systems Part 2: Faults and Switching First Edition. 161 pp.

—Electrical Faults—Low Voltage Installations
IEC 364 Pt 4-442-93. Electrical Installations of Buildings Part 4: Protection for Safety Chapter 44: Protection Against Overvoltages Section 442—Protection of Low-Voltage Installations Against Faults Between High-Voltage Systems and Earth First Edition. 45 pp.

—Electrical Protection Equipment
IEC 919 Pt 2-91. Performance of High-Voltage d.c. (HVDC) Systems Part 2: Faults and Switching First Edition. 161 pp.

—Surges
IEC 919 Pt 2-91. Performance of High-Voltage d.c. (HVDC) Systems Part 2: Faults and Switching First Edition. 161 pp.

—Switching—Surges
IEC 919 Pt 2-91. Performance of High-Voltage d.c. (HVDC) Systems Part 2: Faults and Switching First Edition. 161 pp.

High Voltage Luminous Tube Signs
Use: Luminous Tube Signs

High Voltage Testing
BSI BS 923: Part 1-90. 1990 High-Voltage Testing Techniques Part 1: General (IEC 60-1: 1989). 77 pp.
BSI BS 923: Part 1-01. 1990 Amd 1 High-Voltage Testing Techniques Part 1: General (IEC 60-1: 1989) (AMD 6880) February 28, 1992 (E). 78 pp.
BSI BS 923: Part 2-80. (WITHDRAWN) 1980 Guide on High-Voltage Testing Techniques Part 2: Test Procedures (Superseded by BS 923: Part 1: 1990). 31 pp.
CENELEC HD 588 S1-91. High-Voltage Test Techniques Part 1: General Definitions and Test Requirements. 5 pp.
DIN VDE 0432 Pt 1-78. High-Voltage Test Techniques; General Definitions and Test Requirements (Oct). 23 pp.
DIN VDE 0432 Pt 2-78. High-Voltage Test Techniques; Test Procedures (Oct). 48 pp.
IEC 60 Pt 1-89. High-Voltage Test Techniques Part 1: General Definitions and Test Requirements Second Edition; (Corrigendum—March 1990) (Corrigendum—March 1992). 136 pp.
SAA AS 1931.1-76. High Voltage Testing Techniques—Part 1: General Definitions, Test Requirements, Test Procedures and Measuring Devices Reconfirmed 1988. 72 pp.
SAA AS 1931.2-77. High Voltage Testing Techniques—Part 2: Application Guide for Measuring Devices Reconfirmed 1988. 44 pp.

—Electric Wiring—Aircraft
AECMA PREN2283-88. Testing of Aircraft Wiring. 9 pp.

—Electrical Equipment—Aircraft
BSI 3G 100:Pt 4: SUBSEC 1.1-82. 1982 General Requirements for Equipment for Use in Aircraft Part 4: Electrical Equipment Section 1: Construction and General Subsection 1.1: Electrical Insulation Tests. 4 pp.
ISO 2678-85. Environmental Tests for Aircraft Equipment—Insulation Resistance and High Voltage Tests for Electrical Equipment Second Edition. 5 pp.

High Voltage Testing (Cont.)
—Enamels
BSI BS 1344: Part 11-75. 1975 Methods of Testing Vitreous Enamel Finishes Part 11: High Voltage Test for Articles for Use Under Highly Corrosive Conditions. 7 pp.
ISO 2746-73. Vitreous and Porcelain Enamels—Enamelled Articles for Service Under Highly Corrosive Conditions—High Voltage Test First Edition. 4 pp.

—Low Voltage Installations
BSI BS 7640: Part 1-93. 1993 High-Voltage Test Techniques for Low-Voltage Equipment Part 1: Definitions, Test and Procedure Requirements (IEC 1180-1: 1992) (E). 31 pp.
IEC 1180 Pt 1-92. High-Voltage Test Techniques for Low-Voltage Equipment Part 1: Definitions, Test and Procedure Requirements First Edition. 60 pp.

—Measuring Instruments
BSI BS 923: Part 3-80. 1980 Guide on High-Voltage Testing Techniques Part 3: Measuring Devices. 12 pp.
BSI BS 923: Part 4-80. 1980 Guide on High-Voltage Testing Techniques Part 4: Application Guide for Measuring Devices. 43 pp.
IEC 60 Pt 3-76. High-Voltage Test Techniques Part 3: Measuring Devices First Edition. 23 pp.
IEC 60 Pt 4-77. High-Voltage Test Techniques Part 4: Application Guide for Measuring Devices First Edition. 80 pp.
SAA AS 1931.1-76. High Voltage Testing Techniques—Part 1: General Definitions, Test Requirements, Test Procedures and Measuring Devices Reconfirmed 1988. 72 pp.
SAA AS 1931.2-77. High Voltage Testing Techniques—Part 2: Application Guide for Measuring Devices Reconfirmed 1988. 44 pp.

—Partial Discharge
DIN VDE 0434-83. High Voltage Testing Techniques; Partial Discharge Measurements (May). 48 pp.

High Wet Strength Papers
Use: Wet Strength Papers

High Yield Strength Steels
Use: High Strength Steels

Higher Alcohol Content Analysis
Use: Alcohol Content Analysis

Higher Alcohols
Use: Alcohols

Higher Education
—Diplomas—European Communities
EC 89/48/EEC-88. Council Directive on a General System for the Recognition of Higher-Education Diplomas Awarded on Completion of Professional Education and Training of at Least Three Years' Duration. 8 pp.
EURO 1989 Oct. European Community and Recognition of Diplomas for Professional Purposes. 11 pp.

—European Communities
EURO 1990-5. Education and Training-the Approach to 1992. 12 pp.
EURO 1991 JUN. Education and Training. 9 pp.

Highway Bridges
See Also: Bridge Bearings; Bridges (Structures)
CGSB 48-GP-3M-79. Radiographic Inspection of Steel Castings, Standard for. 7 pp.

—Bearing Capacity
DIN ENGL 1072-85. Road and Foot Bridges; Design Loads (Dec). 14 pp.

—Steel—Construction
DIN ENGL 18809-87. Steel Road Bridges and Foot Bridges; Design and Construction (Together with DIN 18800 Pt 1, Mar. 1981 Ed., and DIN 18800 Pt 7, May 1983 Ed., Supsds. DIN 1073, July 1974 Ed., Suppl. to DIN 1073, July 1974 Ed., DIN 1079, Sept. 1970 Ed., and DIN 4101, 7/74 Ed.). 13 pp.

—Steel—Design
DIN ENGL 18809-87. Steel Road Bridges and Foot Bridges; Design and Construction (Sept) (Together with DIN 18800 Pt 1, Mar. 1981 Ed., and DIN 18800 Pt 7, May 1983 Ed., Supsds. DIN 1073, July 1974 Ed., Suppl. to DIN 1073, July 1974 Ed., DIN 1079, Sept. 1970 Ed., and DIN 4101, 7/74 Ed.). 13 pp.

—Structural Design
CSA CAN/CSA-S6-88. Design of Highway Bridges; (Gen Instr 1 Thru 2); (Supplement 1 1990, Existing Bridge Evaluation). 357 pp.

Highway Bridges (Cont.)
—Structural Members
DIN ENGL 4141 Pt 2-84. Structural Bearings; Bearing Systems for Civil Engineering Structures Forming Part of Traffic Routes (Bridges) (Sept). 5 pp.

Highway Engineering
See Also: Highways

—Classification
BSI BS 1000: (625)-76. 1976 Universal Decimal Classification (UDC). English Full Edition (625): Civil Engineering of Land Transport. Railway Engineering. Highway Engineering. 25 pp.
SNZ NZS/BS 1000 (625)-76. Universal Decimal Classification Civil Engineering of Land Transport. Railway Engineering. Highway Engineering. 28 pp.

—Glossaries
BSI BS 6100: SUBSEC 2.4.1-90. 1990 Glossary of Building and Civil Engineering Terms Part 2: Civil Engineering Section 2.4: Highway and Railway Engineering Subsection 2.4.1: Highway Engineering. 40 pp.
BSI BS 6100: SUBSEC 2.4.1-92. 1992 Glossary of Building and Civil Engineering Terms Part 2: Civil Engineering Section 2.4: Highway, Railway and Airport Engineering Subsection 2.4.1: Highway Engineering. 44 pp.
BSI BS 6100: SUBSEC 2.4.1-86. (WITHDRAWN) 1986 Glossary of Building and Civil Engineering Terms Part 2: Civil Engineering Section 2.4: Highway and Railway Engineering Subsection 2.4.1: Highway Engineering. 40 pp.

—Structures—Inspection
DIN ENGL 1076-83. Engineering Structures Connected with Roads and Tracks; Observation and Inspection (Mar). 32 pp.

Highway Lighting
Use For: Road Lighting; Roadway Lighting; Street Lighting *See Also:* Highways; Lighting
BSI BS 4533: Sec 102.3-90. 1990 Luminaires Part 102: Particular Requirements Section 102.3: Specification for Luminaires for Road and Street Lighting. 14 pp.
BSI BS 5489: Part 1-92. 1992 Road Lighting Part 1: Guide to the General Principles. 16 pp.
BSI BS 5489: Part 1-87. 1987 Road Lighting Part 1: Guide to the General Principles. 14 pp.
BSI BS 5489: Part 2-92. 1992 Road Lighting Part 2: Code of Practice for Lighting for Traffic Routes. 33 pp.
BSI BS 5489: Part 2-87. 1987 Road Lighting Part 2: Code of Practice for Lighting for Traffic Routes. 35 pp.
BSI BS 5489: Part 3-89. 1989 Road Lighting Part 3: Code of Practice for Lighting for Subsidiary Roads and Associated Pedestrian Areas. 19 pp.
BSI BS 5489: Part 3-92. 1992 Road Lighting Part 3: Code of Practice for Lighting for Subsidiary Roads and Associated Pedestrian Areas. 20 pp.
BSI BS 5489: Part 3-77. (WITHDRAWN) 1977 Amd 1 Code of Practice for Road Lighting Part 3: Lighting for Subsidiary Roads (Group B) (Formerly CP 1004: Part 3). 7 pp.
BSI BS 5489: Part 8-92. 1992 Road lighting Part 8: Code of Practice for Lighting for Roads Near Aerodromes, Railways, Docks and Navigable Waterways. 7 pp.
BSI BS 5489: Part 8-87. 1987 Road lighting Part 8: Code of Practice for Lighting for Roads Near Aerodromes, Railways, Docks and Navigable Waterways. 4 pp.
BSI BS 5489: Part 10-92. 1992 Road Lighting Part 10: Code of Practice for Lighting for Motorways. 19 pp.
BSI BS 5489: Part 10-90. 1990 Road Lighting Part 10: Code of Practice for Lighting for Motorways. 19 pp.
CENELEC EN 60 598-2-3-89. Luminaires Part 2: Particular Requirements Section Three-Luminaires for Road and Street Lighting. 5 pp.
CNS C3158-82. Method of Test for Roadway Lighting Fittings for Traffic Route (Jul)(9119).
CNS C4366-82. Roadway Lighting Fittings for Traffic Route (Jul)(9118).
CNS Z1039-86. Lighting for Traffic Route (Sep)(10779).
CNS Z3025-86. Method of Test for Luminance for Traffic Route (Sep)(10780).
IEC 598 Pt 2-3-93. Luminaires Part 2: Particular Requirements Section Three: Luminaires for Road and Street Lighting Second Edition. 23 pp.
JIS C 8131-91. Luminaires for Road Lighting. 12 pp.
JIS Z 9111-88. Lighting for Roads. 19 pp.
SAA AS 1158.1-86. Code of Practice for Public Lighting (Known as the SAA Public Lighting Code)—Part 1: Performance and Installation Design Requirements Amdt 1 May 1987. 25 pp.
SNZ NZS 6701-83. Code of Practice for Road Lighting. 44 pp.

Highway Lighting (Cont.)

SNZ NZS 6705: Pt 2: Sec 3-86. Luminaires Part 2: Particular Requirements. Section 3: Luminaires for Road and Street Lighting Amend: A, 1986; 1, 1986; 1A, 1986. 11 pp.

—Bridges (Structures)

BSI BS 5489: Part 6-92. 1992 Road Lighting Part 6: Code of Practice for Lighting for Bridges and Elevated Roads. 11 pp.

BSI BS 5489: Part 6-90. 1990 Road Lighting Part 6: Code of Practice for Bridges and Elevated Roads (Group D). 11 pp.

BSI BS 5489: Part 6-67. (WITHDRAWN) 1967 Amd 1 Road Lighting Part 6: Lighting for Bridges and Elevated Roads. 17 pp.

—Elevated Roads

BSI BS 5489: Part 6-92. 1992 Road Lighting Part 6: Code of Practice for Lighting for Bridges and Elevated Roads. 11 pp.

BSI BS 5489: Part 6-90. 1990 Road Lighting Part 6: Code of Practice for Bridges and Elevated Roads (Group D). 11 pp.

BSI BS 5489: Part 6-67. (WITHDRAWN) 1967 Amd 1 Road Lighting Part 6: Lighting for Bridges and Elevated Roads. 17 pp.

—Fuses (Electric)

BSI BS 7654-93. 1993 Single Phase Street Lighting Fuses (Cut-Outs) for Low-Voltage Public Electricity Distribution Systems—25 A Rating for Highway Power Supplies and Street Furniture (F). 20 pp.

—High Pressure Sodium

CSA CAN/CSA-C653-92. Performance Standard for Roadway Lighting Luminaires; (Gen Instr 1). 22 pp.

—Intersections

BSI BS 5489: Part 4-92. 1992 Road Lighting Part 4: Code of Practice for Lighting for Single-Level Road Junctions Including Roundabouts. 24 pp.

BSI BS 5489: Part 4-87. 1987 Road Lighting Part 4: Code of Practice for Lighting for Single-Level Road Junctions Including Roundabouts. 22 pp.

BSI BS 5489: Part 5-92. 1992 Road Lighting Part 5: Code of Practice for Lighting for Grade-Separated Junctions. 15 pp.

BSI BS 5489: Part 5-89. 1989 Amd 1 Road Lighting Part 5: Code of Practice for Lighting for Grade-Separated Junctions (AMD 6559) June 29, 1990. 14 pp.

BSI BS 5489: Part 5-73. (WITHDRAWN) 1973 Amd 1 Code of Practice for Road Lighting Part 5: Lighting for Grade-Separated Junctions. 11 pp.

—Light Distribution

BSI BS 4533: Sec 103.1-81. 1981 Amd 1 Luminaires Part 103: Performance Requirements Section 103.1: Light Distribution from Road-Lighting Lanterns. 10 pp.

—Lighting Poles

CNS C4134-87. Lamp Pole for Illumination of Road (Jul)(4117).

CNS C4134-77. Lamp Pole for Illumination of Road (Jul)(4117). 6 pp.

SAA AS 4065-92. Concrete Poles for Overhead Lines and Street Lighting. 20 pp.

—Pedestrian Crossings

BSI BS 5489: Part 3-92. 1992 Road Lighting Part 3: Code of Practice for Lighting for Subsidiary Roads and Associated Pedestrian Areas. 20 pp.

BSI BS 5489: Part 3-77. (WITHDRAWN) 1977 Amd 1 Code of Practice for Road Lighting Part 3: Lighting for Subsidiary Roads (Group B) (Formerly CP 1004: Part 3). 7 pp.

CNS Z1040-84. Lighting for Pedestrian Crossing (Feb)(10781).

SAA AS 1158.4-87. Code of Practice for Public Lighting (Known as the SAA Public Lighting Code)—Part 4: Supplementary Lighting at Pedestrian Crossings. 6 pp.

—Photoelectric Controls

BSI BS 5972-80. 1980 Amd 2 Photoelectric Control Units for Road Lighting. 14 pp.

—Software

SAA AS 1158.2-86. Code of Practice for Public Lighting (Known as the SAA Public Lighting Code)—Part 2: Computer Procedures for the Calculation of Light Technical Parameters for Category A Lighting Amdt 1 May 1987. 32 pp.

—Switches

CNS C3093-80. Method of Test for Small Switches for Street Lamps (Aug)(6053).

CNS C4219-80. Small Switches for Street Lamps (Aug)(6052).

Highway Lighting (Cont.)

—Traffic Signs

BSI BS 873: Part 5-83. 1983 Road Traffic Signs and Internally Illuminated Bollards Part 5: Internally Illuminated Signs and External Lighting Luminaires. 9 pp.

—Tunnels

BSI BS 5489: Part 7-92. 1992 Road Lighting Part 7: Code of Practice for the Lighting of Tunnels and Underpasses. 40 pp.

BSI BS 5489: Part 7-90. 1990 Road Lighting Part 7: Code of Practice for the Lighting of Tunnels and Underpasses. 40 pp.

CNS Z1045-88. Recommendation for Lighting of Traffic Tunnels (Apr)(12281).

JIS Z 9116-90. Lighting of Tunnels for Motorized Traffic. 12 pp.

—Underpasses

BSI BS 5489: Part 7-92. 1992 Road Lighting Part 7: Code of Practice for the Lighting of Tunnels and Underpasses. 40 pp.

BSI BS 5489: Part 7-90. 1990 Road Lighting Part 7: Code of Practice for the Lighting of Tunnels and Underpasses. 40 pp.

—Urban Areas

BSI BS 5489: Part 9-92. 1992 Road Lighting Part 9: Code of Practice for Lighting for Urban Centres and Public Amenity Areas. 25 pp.

BSI BS 5489: Part 9-90. 1990 Road Lighting Part 9: Code of Practice for Lighting for Urban Centres and Public Amenity Areas. 26 pp.

—Warning Lights

BSI BS 3143: Part 1-85. 1985 Amd 1 Road Danger Lamps Part 1: Kerosine Burning Lamps (AMD 4846) February 28, 1985. 11 pp.

BSI BS 3143: Part 2-90. 1990 Road Danger Lamps Part 2: Low Intensity Battery Operated Lamps. 13 pp.

BSI BS 3143: Part 2-73. (WITHDRAWN) 1973 Road Danger Lamps Part 2: Low Intensity Battery Operated Lamps. 11 pp.

BSI BS 3143: Part 4-85. 1985 Amd 1 Road Danger Lamps Part 4: High Intensity Battery Operated Beacons. 13 pp.

SAA AS 1165-82. Traffic Hazard Warning Lamps. 22 pp.

SNZ NZS 5415-88. Traffic Hazard Warning Lamps Amend: A. 22 pp.

Highway Signs
Use: Traffic Signs

Highway Transportation
Use For: Road Transport *See Also:* Highways; Materials Handling Equipment; Roads; Transportation; Trucks

—Acetic Anhydride—Emergency Procedures

SAA AS 1678.8.0. 005-93. Emergency Procedure Guides—Transport—Part 8.0.005: Acetic Anhydride (in Professional Package 37A).

—Acrylates—Emergency Procedures

SAA AS 1678.3.0. 023-80. Emergency Procedure Guide—Transport—Part 3.0.023: Acrylates (Methyl, Ethyl, and Butyl).

—Acrylonitriles—Emergency Procedures

SAA AS 1678.3.0. 022-89. Emergency Procedure Guide—Transport—Part 3.0.022: Acrylonitrile (Inhibited) (Vinyl Cyanide).

—Aluminum Alkyl Halides—Emergency Procedures

SAA AS 1678.4.2. 001-93. Emergency Procedure Guides—Transport—Part 4.2.001: Aluminium Alkyls and Aluminium Alkyl Halides (in Professional Package 37A).

—Aluminum Alkyls—Emergency Procedures

SAA AS 1678.4.2. 001-93. Emergency Procedure Guides—Transport—Part 4.2.001: Aluminium Alkyls and Aluminium Alkyl Halides (in Professional Package 37A).

—Ammonium Fluoride—Emergency Procedures

SAA AS 1678.6.1. 015-84. Emergency Procedure Guides—Transport—Part 6.1.015: Fluorides (Ammonium, Potassium or Sodium Fluoride).

—Ammonium Nitrate—Emergency Procedures

SAA AS 1678.5.1. 002-91. Emergency Procedure Guides—Transport—Part 5.1.002: Ammonium Nitrate. 2 pp.

Highway Transportation (Cont.)

—Ammonium Thioglycolate—Emergency Procedures

SAA AS 1678.6.1. 014-84. Emergency Procedure Guides—Transport—Part 6.1.014: Ammonium Thioglycolate.

—Anilines—Emergency Procedures

SAA AS 1678.6.0. 001-83. Emergency Procedure Guides—Transport—Part 6.0.001 (Amino benzene).

—Arsenic Acid—Emergency Procedures

SAA AS 1678.6.1. 010-83. Emergency Procedure Guides—Transport—Part 6.1.010: Arsenic Acid—Ortho-Arsenic Acid (Liquid) Meta-Arsenic Acid (Solid).

—Beryllium—Emergency Procedures

SAA AS 1678.6.0. 015-84. Emergency Procedure Guides—Transport—Part 6.0.015: Beryllium and Compounds.

—Bitumess—Emergency Procedures

SAA AS 1678.3T1-87. Emergency Procedure Guide—Transport Group Text EPGs for Class 3 Substances—Flammable Liquids—Part 3T1: Bitumen Products.

—Butadienes—Emergency Procedures

SAA AS 1678.2.1. 010-86. Emergency Procedure Guide—Transport—Part 2.1.010: Butadiene (Inhibited).

—Cadmium Compounds—Emergency Procedures

SAA AS 1678.6.1. 017-84. Emergency Procedure Guides—Transport—Part 6.1.017: Cadmium Compounds (Other Than Selenide and Sulphide).

—Calcium Hyprochlorite—Emergency Procedures

SAA AS 1678.5.1. 004-93. Emergency Procedure Guides—Transport—Part 5.1.004: Calcium Hypochlorite (in Professional Package 37A).

—Carbon Disulfide—Emergency Procedures

SAA AS 1678.3.1. 003-89. Emergency Procedure Guide—Transport—Part 3.1.003: Carbon Disulfide (Carbon Bisulfide).

—Cargo—Quotas

EC EEC/1841/88-88. Council Regulation Amending Regulation (EEC) No 3164/76 on the Community Quota for the Carriage of Goods by Road Between Member States. 2 pp.

—Chlorine—Emergency Procedures

SAA AS 1678.2.2. 002-91. Emergency Procedure Guide—Transport—Part 2.2.002: Chlorine (in Professional Package 37).

—Chloropicrin—Emergency Procedures

SAA AS 1678.6.0. 013-84. Emergency Procedure Guides—Transport—Part 6.0.013: Chlorolpicrin (Trichloronitromethane).

—Compressed Gases—Emergency Procedures

SAA AS 1678.2A1-87. Emergency Procedure Guide—Transport Group Text EPGs for Class 2 Substances—Compressed and Liquefied Gases—Part 2A1: Flammable, Compressed Gas (Supersedes AS 1678.2.1.007—1983 and AS 1678.2.1.009—1987).

SAA AS 1678.2A4-87. Emergency Procedure Guide—Transport Group Text EPGs for Class 2 Substances—Compressed and Liquefied Gases—Part 2A4: Flammable, Poisonous Gas.

SAA AS 1678.2A5-87. Emergency Precedure Guide—Transport Group Text EPGs for Class 2 Substances—Compressed and Liquefied Gases—Part 2A5: Flammable, Poisonous—Special Low-Temperature Control Gas.

SAA AS 1678.2B1-87. Emergency Precedure Guide—Transport Group Text EPGs for Class 2 Substances—Compressed and Liquefied Gases—Part 2B1: Poisonous (Non-Flammable) Gas.

SAA AS 1678.2B2-87. Emergency Procedure Guide—Transport Group Text EPGs for Class 2 Substances—Compressed and Liquefied Gases—Part 2B2: Poisonous, Flammable Gas.

SAA AS 1678.2B3-87. Emergency Procedure Guide—Transport Group Text EPGs for Class 2 Substances—Compressed and Liquefied Gases—Part 2B3: Poisonous Gas (Will Burn, Corrode) (Supersedes AS 1678.8.0.010—1984 (in Part)).

SAA AS 1678.2B4-87. Emergency Procedure Guide—Transport Group Text EPGs for Class 2 Substances—Compressed and Liquefied Gases—Part 2B4: Poisonous, Flammable (Spontaneously Combustible) Gas.

INTERNATIONAL AND NON-U.S. NATIONAL STANDARDS
SUBJECT INDEX
Highway

Highway Transportation (Cont.)
—**Compressed Gases—Emergency Procedures** (Cont.)

SAA AS 1678.2B5-87. Emergency Procedure Guide—Transport Group Text EPGs for Class 2 Substances—Compressed and Liquefied Gases—Part 2B5: Poisonous, Oxidizing Gas.

SAA AS 1678.2B6-87. Emergency Precedure Guide—Transport Group Text EPGs for Class 2 Substances—Compressed and Liquefied Gases—Part 2B6: Poisonous, Oxidizing Gas (Flourine-Type).

SAA AS 1678.2B7-87. Emergency Procedure Guide—Transport Group Text EPGs for Class 2 Substances—Compressed and Liquefied Gases—Part 2B7: Poisonous, Oxiding Gas, Corrosive Gas.

SAA AS 1678.2B8-87. Emergency Procedure Guide—Transport Group Text EPGs for Class 2 Substances—Compressed and Liquefied Gases—Part 2B8: Poisonous and Corrosive Gas.

SAA AS 1678.2C5-87. Emergency Procedure Guide—Transport Group Text EPGs for Class 2 Substances—Compressed and Liquefied Gases—Part 2C5: Oxidizing, Compressed Gas—Breathable, Non-Flammable.

SAA AS 1678.2C6-87. Emergency Procedure Guide—Transport Group Text EPGs for Class 2 Substances—Compressed and Liquefied Gases—Part 2C6: Non-Flammable, Oxidizing, Compressed Gas—Breathable.

SAA AS 1678.2C8-87. Emergency Procedure Guide—Transport Group Text EPGs for Class 2 Substances—Compressed and Liquefied Gases—Part 2C8: Non-Flammable, Oxidizing, Compressed Gas—Asphyxiating.

SAA AS 1678.2D1-87. Emergency Procedure Guide—Transport Group Text EPGs for Class 2 Substances—Compressed and Liquefied Gases—Part 2D1: Aerosol Dispensers.

SAA AS 1678. Emergency Procedure Guide—Transport Group Text EPGs for Class 3 Substances—Flammable Liquids Pads of 20 Forms.

—**Copper Chrome Arsenic—Emergency Procedures**

SAA AS 1678.6.0. 011-88. Emergency Procedure Guides—Transport—Part 6.0.011: Copper Chrome Arsenic (C.C.A. Wood Preservative).

—**Cyanides—Emergency Procedures**

SAA AS 1678.6.0. 009-84. Emergency Procedure Guides—Transport—Part 6.0.009: Cyanides, Inorganic.

—**Diphenylmethane Diisocyanate—Emergency Procedures**

SAA AS 1678.6.1. 016-84. Emergency Procedure Guides—Transport—Part 6.1.016: Diphenylmethane DI-ISO-Cyanate (Methylene Bis-(Phenylene) Diisocyanate) (MDI).

—**Epichlorohydrin—Emergency Procedures**

SAA AS 1678.6.0. 007-83. Emergency Procedure Guides—Transport—Part 6.0.007: Epichlorhydrin (ECH or 1-Chloro-2, 3-Epoxypropane).

—**Ethylene Oxide—Emergency Procedures**

SAA AS 1678.2.1. 005-85. Emergency Procedure Guide—Transport—Part 2.1.005: Ethylene Oxide.

—**Flammable Liquids—Emergency Procedures**

SAA AS 1678. Emergency Procedure Guide—Transport Group Text EPGs for Class 5 Substances—Oxidizing Substances and Organic Peroxides (in Professional Package 37).

SAA AS 1678.3A1-87. Emergency Procedure Guide—Transport Group Text EPGs for Class 3 Substances—Flammable Liquids—Part 3A1: Flammable Liquids (Supersedes AS 1678.3.0.020—1986, AS 1678.3.1.004—1984, AS 1678.3.1.005—1984, AS 1678.3.1.007—1984, AS 1678.3.1.008—1984,..

SAA AS 1678.3A2-87. Emergency Procedure Guide—Transport Group Text EPGs for Class 3 Substances—Flammable Liquids—Part 3A2: Flammable Liquid Poisonous.

SAA AS 1678.3A3-87. Emergency Procedure Guide—Transport Group Text EPGs for Class 3 Substances—Flammable Liquids—Part 3A3: Flammable Liquid—Harmful (Supersedes AS 1678.3.1.006—1984).

SAA AS 1678.3A4-87. Emergency Procedure Guide—Transport Group Text EPGs for Class 2 Substances—Flammable Liquids—Part 3A4: Flammable Liquid—Corrosive.

SAA AS 1678.3A5-87. Emergency Procedure Guide—Transport Group Text EPGs for Class 3 Substances—Flammable Liquids—Part 3A5: Flammable Liquid, Poisonous, Corrosive.

SAA AS 1678.3A6-87. Emergency Procedure Guide—Transport Group Text EPGs for Class 3 Substances—Flammable Liquids—Part 3A6: Flammable Liquid, Corrosive, Dangerous When Wet.

Highway Transportation (Cont.)
—**Flammable Liquids—Emergency Procedures** (Cont.)

SAA AS 1678.3B3-87. Emergency Procedure Guide—Transport Group Text EPGs for Class 3 Substances—Flammable Liquids—Part 3B3: Flammable Liquid—Some Harmful or Corrosive Properties.

SAA AS 1678.3C1-87. Emergency Procedure Guide—Transport Group Text EPGs for Class 3 Substances—Flammable Liquids—Part 3C1: Flammable Liquid of Lesser Hazard.

—**Flammable Materials—Emergency Procedures**

SAA AS 1678.4.3. 002-91. Emergency Procedure Guide Transport —Part 4.3.002: Magnesium Phosphide (In Profressional Package 37).

SAA. Emergency Procedure Guide —Transport Group Text EPGs for Class 4 Substances—Flammable Solids, Substances Liable to Spontaneous Combustion or Which Emit Flammable Gases When Wet Pads of 20 Forms.

SAA AS 1678.4A1-87. Emergency Procedure Guide—Transport Group Text EPGs for Class 4 Substances—Flammable Solids, Substances Liable to Spontaneous Combustion or Which Emit Flammable Gases When Wet—Part 4A1: Flammable Solids.

SAA AS 1678.4A2-87. Emergency Procedure Guide—Transport Group Text EPGs for Class 4 Substances—Flammable Solids, Substances Liable to Spontaneous Combustion or Which Emit Flammable Gases When Wet—Part 4A2: Spontaneously Combustible.

SAA AS 1678.4A3-87. Emergency Procedure Guide—Transport Group Text EPGs for Class 4 Substances—Flammable Solids, Substances Liable to Spontaneous Combustion or Which Emit Flammable Gases When Wet—Part 4A3: Dangerous When Wet.

SAA AS 1678.4B1-87. Emergency Procedure Guide—Transport Group Text EPGs for Class 4 Substances—Flammable Solids, Liable to Spontaneous Combustion or Which Emit Flammable Gases When Wet—Part 4B1: Flammable Solids, Toxic.

SAA AS 1678.4B2-87. Emergency Procedure Guide—Transport Group Text EPGs for Class 4 Substances—Flammable Solids, Substances Liable to Spontaneous Combustion or Which Emit Flammable Gases When Wet—Part 4B2: Spontaneously Combustible, Toxic.

SAA AS 1678.4B3-87. Emergency Procedure Guide—Transport Group Text EPGs for Class 4 Substances—Flammable Solids, Substances Liable to Spontaneous Combustion or Which Emit Flammable Gases When Wet—Part 4B3: Dangerous When Wet, Toxic.

SAA AS 1678.4C2-87. Emergency Procedure Guide—Transport Group Text EPGs for Class 4 Substances—Flammable Solids, Substances Liable to Spontaneous Combustion or Which Emit Flammable Gases When Wet—Part 4C2: Spontaneously Combustible, Corrosive.

SAA AS 1678.4C3-87. Emergency Procedure Guide—Transport Group Text EPGs for Class 4 Substances—Flammable Solids, Substances Liable to Spontaneous Combustion or Which Emit Flammable Gases When Wet—Part 4C3: Dangerous When Wet, Corrosive.

SAA AS 1678.4D2-87. Emergency Procedure Guide—Transport Group Text EPGs for Class 4 Substances—Flammable Solids, Substances Liable to Spontaneous Combustion or Which Emit Flammable Gases When Wet—Part 4D2: Spontaneously Combustible, Reacts Violently with Water.

SAA AS 1678.4E3-87. Emergency Procedure Guide—Transport Group Text EPGs for Class 4 Substances—Flammable Solids, Substances Liable to Spontaneous Combustion or Which Emit Flammable Gases When Wet—Part 4E3: Dangerous When Wet, Some Toxic Properties.

SAA AS 1678.4F1-87. Emergency Procedure Guide—Transport Group Text EPGs for Class 4 Substances—Flammable Solids, Substances Liable to Spontaneous Combustion or Which Emit Flammable Gases When Wet—Part 4F1: Flammable Solids, Explosive.

SAA AS 1678.4G1-87. Emergency Procedure Guide—Transport Group Text EPGs for Class 4 Substances—Flammable Solids, Substances Liable to Spontaneous Combustion or Which Emit Flammable Gases When Wet—Part 4G1: Flammable Solids, Harmful.

SAA AS 1678.4G2-87. Emergency Proc. Guide—Trans Group Text EPGs for Class 4 Substances—Flammable Solids, Substances Liable to Spontaneous Combustion or Which Emit Flammable Gases When Wet—Part 4G2: Spontaneously Combustible, Dangerous When Wet, May React Violently with Water.

Highway Transportation (Cont.)
—**Flammable Materials—Emergency Procedures** (Cont.)

SAA AS 1678.4H1-87. Emergency Procedure Guide—Transport Group Text EPGs for Class 4 Substances—Falmmable Solids, Substances Liable to Spontaneous Combustion or Which Emit Flammable Gases When Wet —Part 4H1: Flammable Solids, Some Toxic Properties.

SAA AS 1678.4K2-87. Emergency Procedure Guide—Transport Group Text EPGs for Class 4 Substances—Flammable Solids, Substances Liable to Spontaneous Combustion or Which Emit Flammable Gases When Wet—Part 4K2: Spontaneously Combustible, Some Toxic Properties.

SAA AS 1678.4M3-87. Emergency Proc. Guide—Trans. Group Text EPGs for Class 4 Substances—FlammableSolids, Substances Liable to Spontaneous Combustion or Which Emit Flammable Gases When Wet—Part 4M3: Spontaneously Combustible, Dangerous when Wet, Some Toxic Properties.

SAA AS 1678.4N3-87. Emergency Proc. Guide—Trans. Group Text EPGs for Class 4 Substances—FlammableSolids, Substances Liable to Spontaneous Combustion or Which Emit Flammable Gases When Wet—Part 4N3: Dangerous When Wet, Some Corrosive Properties, May Ignite Spontaneously.

SAA AS 1678.4O3-87. Emergency Procedure Guide—Transport Group Text EPGs for Class 4 Substances—Flammable Solids, Substances Liable to Spontaneous Combustion or Which Emit Flammable Gases When Wet—Part 4O3: Dangerous When Wet, Spontaneously Combustible.

SAA AS 1678.4P3-87. Emergency Procedure Guide—Transport Group Text EPGs for Class 4 Substances—Flammable Solids, Substances Liable to Spontaneous Combustion or Which Emit Flammable Gases When Wet—Part 4P3: Dangerous When Wet, Flammable Liquid, Some Toxic Properties.

SAA AS 1678.4Q3-87. Emergency Procedure Guide—Transport Group Text EPGs for Class 4 Substances—Flammable Solids, Substances Liable to Spontaneous Combustion or Which Emit Flammable Gases When Wet—Part 4Q3: Dangerous When Wet, Flammable Liquid, Corrosive.

SAA AS 1678.4R1-90. Emergency Procedur Guide—Transport Group Text EPGs for Class 4 Substances—Flammable Solids, Substances Liable to Spontaneous Combustion or Which Emit Flammable Gases When Wet—Part 4R1: Flammable Solids, Dangerous When Wet.

SAA AS 1678.4R3-87. Emergency Procedure Guide—Transport Group Text EPGs for Class 4 Substances—Flammable Solids, Substances Liable to Spontaneous Combustion or Which Emit Flammable Gases When Wet—Part 4R3: Dangerous When Wet, Harmful.

SAA AS 1678.8F1-87. Emergency Procedure Guide—Transport Group Text EPGs for Class 8 Substances—Corrosive Substances—Part 8F1: Corrosive—Flammable Solid.

SAA AS 1678.8F2-87. Emergency Procedure Guide—Transport Group Text EPGs for Class 8 Substances—Corrosive Substances—Part 8F2: Corrosive—Flammable—Solid—Violent Reaction with Water.

SAA AS 1678. Emergency Procedure Guide—Transport Group Text EPGs for Class 9 Substances Pads of 20 Forms.

SAA AS 1678.9A1-88. Emergency Procedure Guide—Transport Group Text EPGs for Class 9 Substances—Part 9A1: Flammable Solids.

SAA AS 1678.9A2-88. Emergency Procedure Guide—Transport Group Text EPGs for Class 9 Substances—Part 9A2: Spontaneously Combustible.

—**Formaldehyde—Emergency Procedures**

SAA AS 1678.3.0. 015-93. Emergency Procedure Guide—Transport—Part 3.0.015: Formaldehyde Solutions, Flammable (in Professional Package 37).

SAA AS 1678.6.0. 000-90. Emergency Procedure Guides—Transport—Part 6.0.000: Formaldehyde Solutions (Non-Flammable) (Superseded by AS 1678.8.0.005—1993) (Withdrawn).

—**Gas Cylinders—Emergency Procedures**

SAA AS 1678.2M1-87. Emergency Procedure Guide—Transport Group Text EPGs for Class 2 Substances—Compressed and Liquefied Gases—Part 2M1: Mixed Load of Gases in Cylinders.

—**Gasoline—Emergency Procedures**

SAA AS 1678.3.1. 001-86. Emergency Procedure Guide—Transport—Part 3.1.001: Petrol—as Cargo.

—**Hazardous Materials**

CSA B620-1987P. Highway Tanks and Portable Tanks for the Transportation of Dangerous Goods (Formerly Designated B338); (Gen Instr 1 Thru 2). 120 pp.

INDUSTRY STANDARDS

INTERNATIONAL AND NON-U.S. NATIONAL STANDARDS
SUBJECT INDEX

Highway Transportation (Cont.)
—Hazardous Materials (Cont.)

CSA B621-1987P. Selection and Use of Highway Tanks, Portable Tanks, Cargo Compartments and Containers for the Transportation of Dangerous Goods, Classes 3, 4, 5, 6, and 8, in Bulk by Road. 26 pp.

CSA B622-1987P. Selection and Use of Highway Tanks, Multi-Unit Tank Car Tanks, and Portable Tanks for the Transportation of Dangerous Goods, Class 2, by Road; (Gen Instr 1 Thru 2). 37 pp.

CSA B623-1987P. Selection, Handling, and Use of Intermodal Portable Tanks for the Transportation of Dangerous Goods, Other Than by Air; (Gen Instr 1). 69 pp.

SNZ NZS 5433-88. Code of Practice for Transportation of Hazardous Substances on Land Amend: 1, 1990. 172 pp.

SNZ NZS 5433B-88. Code of Practice for Transportation of Hazardous Substances on Land. 1 p.

SNZ NZS 5433D-89. Code of Practice for Transportation of Hazardous Substances on Land. 1 p.

—Hazardous Materials—Emergency Planning

SAA AS 1678. Emergency Procedure Guide—Transport Group Text EPGs for Class 5 Substances—Oxidizing Substances and Organic Peroxides (in Professional Package 37).

SAA AS 1678.6A1-87. Emergency Procedure Guide—Transport Group Text EPGs for Class 6 Substances—Poison Substances—Part 6A1: Poisonous Substance, Will Burn, Do Not Use Water on Fires (Induce Vomiting).

SAA AS 1678.6A2-87. Emergency Procedure Guide—Transport Group Text EPGs for Class 6 Substances—Poison Substances—Part 6A2: Poisonous Substance, Will Not Burn, Do Not Use Water on Fires (Induce Vomiting).

SAA AS 1678.6A3-87. Emergency Procedure Guide—Transport Group Text EPGs for Class 6 Substances—Poison Substances—Part 6A3: Poisonous Substance, Will Burn, Water May Be Used on Fires (Induce Vomiting).

SAA AS 1678.6A4-87. Emergency Procedure Guide—Transport Group Text EPGs for Class 6 Substances—Poison Substances—Part 6A4: Poisonous Substance, Will Burn, Do Not Use Water on Fires (Do Not Induce Vomiting).

SAA AS 1678.6A5-87. Emergency Procedure Guide—Transport Group Text EPGs for Class 6 Substances—Poison Substances—Part 6A5: Poisonous Substance, Will Not Burn, Water May Be Used on Fires (Induce Vomiting) (Supersedes AS 1678.6.0.003—1982).

SAA AS 1678.6A6-87. Emergency Procedure Guide—Transport Group Text EPGs for Class 6 Substances—Poison Substances—Part 6A6: Poisonous Substance, Will Burn, Water May Be Used on Fires (Do Not Induce Vomiting).

SAA AS 1678.6A7-87. Emergency Procedure Guide—Transport Group Text EPGs for Class 6 Substances—Poison Substances—Part 6A7: Poisonous Substance, Will Not Burn, Water May Be Used on Fires (Do Not Induce Vomiting).

SAA AS 1678.6B1-87. Emergency Procedure Guide—Transport Group Text EPGs for Class 6 Substances—Poison Substances—Part 6B1: Harmful Substance, Will Burn, Do Not Use Water on Fires (Induce Vomiting).

SAA AS 1678.6B3-87. Emergency Procedure Guide—Transport Group Text EPGs for Class 6 Substances—Poison Substances—Part 6B3: Harmful Substance, Will Burn, Water May Be Used on Fires (Supersedes AS 1678.6.1.011—1984).

SAA AS 1678.6B5-87. Emergency Procedure Guide—Transport Group Text EPGs for Class 6 Substances—Poison Substances—Part 6B5: Harmful Substance, Will Not Burn, Water May Be Used on Fires (Induce Vomiting).

SAA AS 1678.6B6-87. Emergency Procedure Guide—Transport Group Text EPGs for Class 6 Substances—Poison Substances—Part 6B6: Harmful Substance, Will Burn, Water May Be Used on Fires (Do Not Induce Vomiting).

SAA AS 1678.6B7-87. Emergency Procedure Guide—Transport Group Text EPGs for Class 6 Substances—Poison Substances—Part 6B7: Harmful Substance, Will Not Burn, Water May Be Used on Fires (Do Not Induce Vomiting) (Supersedes AS 1678.6.1.007—1987 and AS 1678.6.1.008—1982).

SAA AS 1678.6C1-87. Emergenc Procedure—Transport Group Text EPGs for Class 6 Substances—Poison Substances—Part 6C1: Poisonous Substance, Flammable, Do Not Use Water on Fires (Induce Vomiting).

SAA AS 1678.6C3-87. Emergency Procedure Guide—Transport Group Text EPGs for Class 6 Substances—Poison Substances—Part 6C3: Poisonous Substance, Flammable, Water May Be Used on Fires (Induce Vomiting).

SAA AS 1678.6C6-87. Emergency Procedure Guide—Transport Group Text EPGs for Class 6 Substances—Poison Substance, Flammable, Do Not Use Water on Fires (Do Not Induce Vomiting).

SAA AS 1678.6D1-87. Emergency Procedure Guide—Transport Group Text EPGs for Class 6 Substances—Poison Substances—Part 6D1: Harmful Substance, Flammable, Do Not Use Water on Fires (Induce Vomiting).

SAA AS 1678.6D3-87. Emergency Procedure Guide—Transport Group Text EPGs for Class 6 Substances—Poison Substances—Part 6D3: Harmful Substance, Flammable, Water May Be Used on Fires (Induce Vomiting).

SAA AS 1678.6E3-87. Emergency Procedure Guide—Transport Group Text EPGs for Class 6 Substances—Poison Substances—Part 6E3: Poisonous Substance, Flammable Solid, Water May Be Used on Fires (Induce Vomiting).

SAA AS 1678.6J4-87. Emergency Procedure Guide—Transport Group Text EPGs for Class 6 Substances—Poison Substances—Part 6J4: Poisonous Substance, Corrosive, Will Burn, Reacts with Water (Give Milk or Water, Do Not Induce Vomiting).

SAA AS 1678.6J6-87. Emergency —Transport Group Text EPGs for Class 6 Substances—Poison Substances—Part 6J6: Poisonous Substance, Corrosive, Will Burn, Reacts with Water But Water May Be Used on Fires (Give Milk or Water, Do Not Induce Vomiting).

SAA AS 1678.6L4-87. Emergency Procedure Guide—Transport Group Text EPGs for Class 6 Substances—Poison Substances—Part 6L4: Poisonous Substance, Corrosive, Flammable, Reacts with Water (Give Milk or Water, Do Not Induce Vomiting).

SAA AS 1678.6L6-87. Emergency Procedure Guide—Transport Group Text EPGs for Class 6 Substances—Poison Substances—Part 6L6: Poisonous Substance, Corrosive, Flammable, Reacts with Water but Water May Be Used on Fires (Give Milk or Water, Do Not Induce Vomiting).

SAA AS 1678. Emergency Procedure Guide—Transport Group Text EPGs for Class 8 Substances—Corrosive Substances Pads of 20 Forms.

SAA AS 1678.8A1-87. Emergency Procedure Guide—Transport Group Text EPGs for Class 8 Substances—Corrosive Substances—Part 8A1: Corrosive (Supersedes AS 1678.8.0.007—1986 and AS 1678.8.0.010—1984 (in Part)).

SAA AS 1678.8A2-87. Emergency Procedure Guide—Transport Group Text EPGs for Class 8 Substances—Corrosive Substances—Part 8A2: Corrosive—Violent Reaction with Water.

SAA AS 1678.8A3-87. Emergency Procedure Guide—Transport Group Text EPGs for Class 8 Substances—Corrosive Substances—Part 8A3: Corrosive—Oxidizing Organic Materials, Violent Reaction with Water.

SAA AS 1678.8A4-87. Emergency Procedure Guide—Transport Group Text EPGs for Class 8 Substances—Corrosive Substances—Part 8A4: Corrosive—Can Release Hydrogen Fluoride.

SAA AS 1678.8B1-87. Emergency Procedure Guide—Transport Group Text EPGs for Class 8 Substances—Corrosive Substances—Part 8B1: Corrosive—Flammable.

SAA AS 1678.8B2-87. Emergency Procedure Guide—Transport Group Test EPGs for Class 8 Substances—Corrosive Substances—Part 8B2: Corrosive—Flammable—Violent Reaction with Water.

SAA AS 1678.8B3-87. Emergency Procedure Guide—Transport Group Text EPGs for Class 8 Substances—Corrosive Substances—Part 8B3: Corrosive—Flammable—Self Reactive.

SAA AS 1678.8C1-87. Emergency Procedure Guide—Transport Group Text EPGs for Class 8 Substances—Corrosive Substances—Part 8C1: Corrosive—Poison.

SAA AS 1678.8C2-87. Emergency Procedure Guide—Transport Group Text EPGs for Class 8 Substances—Corrosive Substances—Part 8C2: Corrosive—Poison—Can Relaese Hydrogen Fluoride.

SAA AS 1678.8C3-87. Emergency Procedure Guide—Transport Group Text EPGs for Class 8 Substances—Corrosive Substances—Part 8C3: Corrosive—Poison—Violent Reaction with Water.

SAA AS 1678.8C4-87. Emergency Procedure Guide—Transport Group Text EPGs for Class 8 Substances—Corrosive Substances—Part 8C4: Corrosive—Poison—Violent Reaction with Water, Glass and Sand.

SAA AS 1678.8C5-87. Emergency Procedure Guide—Transport Group Text EPGs for Class 8 Substances—Corrosive Substances—Part 8C5: Corrosive—Poison—Oxidizing Organic Materials.

SAA AS 1678.8C6-87. Emergency Procedure Guide—Transport Group Text EPGs for Class 8 Substances—Corrosive Substnaces—Part 8C6: Corrosive—Poisonous Substance—Polymerizes Readily.

SAA AS 1678.8C8-87. Emergency Procedure Guide—Transport Group Text EPGs for Class 8 Substances—Corrosive Substnaces—Part 8C8: Corrosive—Poison—Flammable.

SAA AS 1678.8C9-87. Emergency Procedure Guide—Transport Group Text EPGs for Class 8 Substances—Corrosive Substances—Corrosive—Harmful.

SAA AS 1678.9A3-88. Emergency Procedure Guide—Transport Group Text EPGs for Class 9 Substances—Part 9A3: Dangerous When Wet.

SAA AS 1678.9A5-88. Emergency Procedure Guide—Transport Group Text EPGs for Class 9 Substances—Part 9A5: Poisonous Substance, Will Not Burn, Water May Be Used on Fires (Induce Vomiting).

SAA AS 1678.9B3-88. Emergency Procedure Guide—Transport Group Text EPGs for Class 9 Substances—Part 9B3: Dangerous When Wet, Toxic.

SAA AS 1678.9B7-88. Emergency Procedure Guide—Transport Group Text EPGs for Class 9 Substances—Part 9B7: Harmful Substance, Will Not Burn, Water May Be Used on Fires (Do Not Induce Vomiting).

—Hydrochloric Acid—Emergency Procedures

SAA AS 1678.8.0. 001-83. Emergency Procedure Guides—Transport—Part 8.0.001: Hydrochloric Acid (Muriatic Acid).

SAA AS 1678.8.0. 008-87. Emergency Procedure Guides—Transport—Part 8.0.008: Hydrofluoric Acid.

—Hydrogen Peroxide—Emergency Procedures

SAA AS 1678.5.1. 005-93. Emergency Procedure Guides—Transport—Part 5.1.005: Hydrogen Peroxide (20% or More) (in Professional Package 37A).

—Invitation for Bids

MOD UK DEFCON 47T-88. Invitation to Tender for the Hire of Motor Transport with or Without Drivers and Conditions of Contract 3/88. 2 pp.

—Lead Alkyls—Emergency Procedures

SAA AS 1678.6.0. 017-93. Emergency Procedure Guides—Transport—Part 6.0.017: Motor Fuel Anti-Knock Mixtures (Lead Aklyls, N.O.S.) (in Professional Package 37A).

—Lead Arsenate—Emergency Procedures

SAA AS 1678.6.0. 014-84. Emergency Procedure Guides—Transport—Part 6.0.014: Lead Arsenate and Lead Arsenite.

—Lead Arsenite—Emergency Procedures

SAA AS 1678.6.0. 014-84. Emergency Procedure Guides—Transport—Part 6.0.014: Lead Arsenate and Lead Arsenite.

—Liquefied Gases

BSI BS 7122-89. 1989 Welded Steel Tanks for the Road Transport of Liquefiable Gases. 17 pp.

—Liquefied Gases—Emergency Procedures

SAA AS 1678.2A2-87. Emergency Procedure Guide—Transport Group Test EPGs for Class 2 Substances—Compressed and Liquefied Gases—Part 2A2: Flammable, Liquefied Gas (Supersedes AS 1678.2.1.002—1984 and AS 1678.2.1.006—1984).

SAA AS 1678.2A3-87. Emergency Precedure Guide—Transport Group Text EPGs for Class 2 Substances—Compressed and Liquefied Gases—Part 2A3: Falmmable, Low -Temperature, Liquefied Gas.

SAA AS 1678.2A4-87. Emergency Procedure Guide—Transport Group Text EPGs for Class 2 Substances—Compressed and Liquefied Gases—Part 2A4: Flammable, Poisonous Gas.

SAA AS 1678.2A5-87. Emergency Precedure Guide—Transport Group Text EPGs for Class 2 Substances—Compressed and Liquefied Gases—Part 2A5: Flammable, Poisonous—Special Low-Temperature Control Gas.

SAA AS 1678.2A6-87. Emergency Precedure Guide—Transport Group Text EPGs for Class 2 Substances—Compressed and Liquefied Gases—Part 2A6: Liquefied Gas—Corrosive—Dangerous When Wet.

SAA AS 1678.2B1-87. Emergency Precedure Guide—Transport Group Text EPGs for Class 2 Substances—Compressed and Liquefied Gases—Part 2B1: Poisonous (Non-Flammable) Gas.

SAA AS 1678.2B2-87. Emergency Procedure Guide—Transport Group Text EPGs for Class 2 Substances—Compressed and Liquefied Gases—Part 2B2: Poisonous, Flammable Gas.

INTERNATIONAL AND NON-U.S. NATIONAL STANDARDS
SUBJECT INDEX
Highway

Highway Transportation *(Cont.)*

—**Liquefied Gases—Emergency Procedures** *(Cont.)*

SAA AS 1678.2B3-87. Emergency Procedure Guide—Transport Group Text EPGs for Class 2 Substances—Compressed and Liquefied Gases—Part 2B3: Poisonous Gas (Will Burn, Corrode) (Supersedes AS 1678.8.0.010—1984 (in Part)).

SAA AS 1678.2B4-87. Emergency Procedure Guide—Transport Group Text EPGs for Class 2 Substances—Compressed and Liquefied Gases—Part 2B4: Poisonous, Flammable (Spontaneously Combustible) Gas.

SAA AS 1678.2B5-87. Emergency Procedure Guide—Transport Group Text EPGs for Class 2 Substances—Compressed and Liquefied Gases—Part 2B5: Poisonous, Oxidizing Gas.

SAA AS 1678.2B6-87. Emergency Precedure Guide—Transport Group Text EPGs for Class 2 Substances—Compressed and Liquefied Gases—Part 2B6: Poisonous, Oxidizing Gas (Flourine-Type).

SAA AS 1678.2B7-87. Emergency Procedure Guide—Transport Group Text EPGs for Class 2 Substances—Compressed and Liquefied Gases—Part 2B7: Poisonous, Oxiding Gas, Corrosive Gas.

SAA AS 1678.2B8-87. Emergency Procedure Guide—Transport Group Text EPGs for Class 2 Substances—Compressed and Liquefied Gases—Part 2B8: Poisonous and Corrosive Gas.

SAA AS 1678.2C2-87. Emergency Procedure Guide—Transport Group Text EPGs for Class 2 Substances—Compressed and Liquefied Gases—Part 2C2: Non-Flammable, Liquefied Gas (Supersedes AS 1678.2.2.003—1984).

SAA AS 1678.2C3-87. Emergency Precedure Guide—Transport Group Text EPGs for Class 2 Substances—Compressed and Liquefied Gases—Part 2C3: Non-Flammable, Low-Temperature, Liquefied Gas.

SAA AS 1678.2C4-87. Emergency Procedure Guide—Transport Group Text EPGs for Class 2 Substances—Compressed and Liquefied Gases—Part 2C4: Non-Flammable, Corrosive, Liquefied Gas (Fluorine-Type).

SAA AS 1678.2C7-87. Emergency Procedure Guide—Transport Group Text EPGs for Class 2 Substances—Compressed and Liquefied Gases—Part 2C7: Non-Flammable, Oxidizing, Low-Temperature, Liquefied Gas—Breathable.

SAA AS 1678.2C9-87. Emergency Procedure Guide—Transport Group Text EPGs for Class 2 Substances—Compressed and Liquefied Gases—Part 2C9: Non-Flammable, Oxidizing, Low-Temperature, Liquefied Gas—Asphyxiating.

SAA AS 1678. Emergency Procedure Guide—Transport Group Text EPGs for Class 3 Substances—Flammable Liquids Pads of 20 Forms.

—**Liquefied Petroleum Gas—Emergency Procedures**

SAA AS 1678.2.1. 001-86. Emergency Procedure Guide—Transport—Part 2.1.001: Liquefied Petroleum Gas (LPG).

—**Liquid Oxygen—Emergency Procedures**

SAA AS 1678.2.2. 000-86. Emergency Procedure Guides—Transport—Part 2.2.000: Oxygen (Refrigerated Liquid).

—**Livestock Crates**

SNZ NZS 5413-77. Specification for Stock Crates Amend: 1 (Reconfirmed 1984). 12 pp.

—**Load Restraints**

BSI BS 6451-84. 1984 Netting and Fibre Rope Load Restraint Systems in Surface Transport. 8 pp.

—**Meat—Truck Doors**

SNZ NZS 5412-82. Specification for Loading Door Aperture Dimensions of Road Vehicles Used for the Bulk Transportation of Meat. 4 pp.

—**Methyl Ethyl Ketone—Emergency Procedures**

SAA AS 1678.3.0. 021-86. Emergency Procedure Guide—Transport—Part 3.0.021: Methyl Ethyl Ketone (MEK).

—**Nitrocellulose—Emergency Procedures**

SAA AS 1678.4.1. 001-84. Emergency Procedure Guides—Transport—Part 4.1.001: Nitrelluose (Spirit Wet).

—**Nonresident Carriers**

EC COM(85) 611-85. Proposal for a Council Regulation (EEC) Laying Down the Conditions Under Which Non-Resident Carriers May Operate National Road Haulage Services Within a Member State. 15 pp.

EC COM(88) 596-88. Amendment to the Proposal for a Council Regulation (EEC) Laying Down the Conditions under Which Non-Resident Carriers may Operate National Road Passenger Transport Services Within a Member State. 4 pp.

Highway Transportation *(Cont.)*

—**Nonresident Carriers** *(Cont.)*

EC COM(87) 31-87. Proposal for a Council Regulation (EEC) Laying Down the Conditions Under Which Non-Resident Carriers May Operate National Road Passenger Transport Services Within a Member State. 12 pp.

EC EEC/4059/89-89. Council Regulation Laying Down the Conditions Under Which Non-resident Carriers May Operate National Road Haulage Services Within a Member State. 15 pp.

—**Organic Peroxides—Emergency Procedures**

SAA AS 1678.5.2. 001-83. Emergency Procedure Guides—Transport—Part 5.2.001: Organic Peroxides (Withdrawn).

SAA. Emergency Procedure Guide —Transport Group Text EPGs for Class 5 Substances—Oxidizing Substances and Organic Peroxides Pads of 20 Forms.

SAA AS 1678.5K1-93. Emergency Procedure Guide—Transport Group Text EPGs for Class 5 Substances—Oxidizing Substances and Organic Peroxides—Part 5K1: Organic Peroxides.

SAA AS 1678.5K2-93. Emergency Procedure Guide—Transport Group Text EPGs for Class 5 Substances—Oxidizing Substances and Organic Peroxides—Part 5K2: Organic Peroxides—Unstable.

SAA AS 1678.5L1-87. Emergency Procedure Guide—Transport Group Text EOGs for Class 5 Substances—Oxidizing Substances and Organic Peroxides—Part 5L1: Organic Peroxides, Highly Flammable, Unstable (Withdrawn).

SAA AS 1678.5M1-93. Emergency Procedure Guide—Transport Group Text EPGs for Class 5 Substances—Oxidizing Substances and Organic Peroxides—Part 5M1: Organic Peroxides—Corrosive.

SAA AS 1678.5N2-87. Emergency Procedure Guide—Transport Group Text EPGs for Class 5 Substances—Oxidizing Substances and Organic Peroxides—Part 5N2: Organic Peroxides, Highly Unstable (Withdrawn).

SAA AS 1678.5P1-93. Emergency Procedure Guide—Transport Group Text EPGs for Class 5 Substances—Oxidizing Substances and Organic Peroxides—Part 5P1: Organic Peroxides—Unstable, Keep Refrigerated.

SAA AS 1678.5Q1-93. Emergency Procedure Guide—Transport Group Text EPGs for Class 5 Substances—Oxidizing Substances and Organic Peroxides—Part 5Q1: Organic Peroxides—Keep Refrigerated.

—**Oxidizers—Emergency Procedures**

SAA. Emergency Procedure Guide —Transport Group Text EPGs for Class 5 Substances—Oxidizing Substances and Organic Peroxides Pads of 20 Forms.

SAA AS 1678.5A1-87. Emergency Procedure Guide—Transport Group Text EPGs for Class 5 Substances—Oxidizing Substances and Organic Peroxides—Part 5A1: Oxidizing Agents—Harmful and Corrosive.

SAA AS 1678.5B1-87. Emergency Procedure Guide—Transport Group Text EPGs for Class 5 Substances—Oxidizing Substances and Organic Peroxides—Part 5B1: Oxidizing Agents—Poisonous.

SAA AS 1678.5B2-87. Emergency Procedure Guide—Transport Group Text EPGs for Class 5 Substances—Oxidizing Substances and Organic Peroxides—Part 5B2: Oxidizing Agents—Harmful.

SAA AS 1678.5C1-87. Emergency Procedure Guide—Transport Group Text EPGs for Class 5 Substances—Oxidizing Substances and Organic Peroxides—Part 5C1: Oxidizing Agents—Poisonous, Corrosive, Reactive.

SAA AS 1678.5C2-87. Emergency Procedure Guide—Transport Group Text EPGs for Class 5 Substances—Oxidizing Substances and Organic Peroxides—Part 5C2: Oxidizing Agents—Corrosive, Reactive.

SAA AS 1678.5D1-87. Emergency Procedure Guide—Transport Group Text EPGs for Class 5 Substances—Oxidizing Substances and Organic Peroxides—Part 5D1: Oxidizing Agents—Containing Ammonium Nitrate.

SAA AS 1678.5E1-87. Emergency Procedure Guide—Transport Group Text EPGs for Class 5 Substances—Osidizing Substances and Organic Peroxides—Part 5E1: Osidizing Agents—Corrosive, Unstable.

SAA AS 1678.5E2-87. Emergency Procedure Guide—Transport Group Text EPGs for Class 5 Substances—Oxidizing Substances and Organic Peroxides—Part 5E2: Oxidizing Agents—Corrosive, Unstable, Flammable.

SAA AS 1678.5F1-87. Emergency Procedure Guide—Transport Group Text EPGs for Class 5 Substances—Oxidizing Substances and Organic Peroxides—Part 5F1: Oxidizing Agents—Thermally Unstable.

Highway Transportation *(Cont.)*

—**Oxidizers—Emergency Procedures** *(Cont.)*

SAA AS 1678.8D1-87. Emergency Procedure Guide—Transport Group Text EPGs for Class 8 Substances—Corrosive Substances—Part 8D1: Corrosive—Oxidizing Agent (Supersedes AS 1678.8.0.003—1984 (in Part)).

SAA AS 1678.8D2-87. Emergency Procedure Guide—Transport Group Text EPGs for Class 8 Substances—Corrosive Substances—Part 8D2: corrosive—Oxidizing Substance—Reacts Violently with Water.

SAA AS 1678.8E1-87. Emergency Procedure Guide—Transport Group Text EPGs for Class 8 Substances—Corrosive Substances—Part 8E1: Corrosive—Oxidizing Agent—Poison (Supersedes AS 1678.8.0.003—1984 (in Part)).

SAA AS 1678.9E1-88. Emergency Procedure Guide—Transport Group Text EPGs for Class 9 Substances—Part 9E1: Oxidizing Agents—Corrosive, Unstable.

SAA AS 1678. Emergency Procedure Guide—Transport Group Text EPGs for Mixed, Packaged Dangerous Goods of Various Classes Pads of 20 Forms.

SAA AS 1678.X1-93. Emergency Procedure Guide—Transport Group Text EPGs for Mixed, Packaged Dangerous Goods of Various Classes—Part X1: Mixed Dangerous Goods (Packaged Goods Only)—Substances Can be Liquids, Solids and/or Gases—Do Not React Violently with Water (in Prof. Package 37A).

SAA AS 1678.X2-87. Emergency Procedure Guide—Transport Group Text EPGs for Mixed, Packaged Dangerous Goods of Various Classes—Part X2: Mixed Dangerous Goods (Compatible Packaged Goods Only)—Substances Can be Liquids, Solids and/or Gases—React Violently with Water.

—**Passenger**

EC COM(87) 79-87. Proposal for a Council Regulation (EEC) on Common Rules for the International Carriage of Passengers by Coach and Bus. 33 pp.

EC COM(87) 595-88. Amendments to the Proposal for a Council Regulation (EEC) on Common Rules for the International Carriage of Passengers by Coach and Bus. 11 pp.

EC COM(88) 596-88. Amendment to the Proposal for a Council Regulation (EEC) Laying Down the Conditions under Which Non-Resident Carriers may Operate National Road Passenger Transport Services Within a Member State. 4 pp.

EC COM(88) 770-88. Amendment to the Proposal for a Council Regulation (EEC) on Common Rules for the International Carriage of Passengers by Coach and Bus. 4 pp.

EC COM(87) 31-87. Proposal for a Council Regulation (EEC) Laying Down the Conditions Under Which Non-Resident Carriers May Operate National Road Passenger Transport Services Within a Member State. 12 pp.

EC EEC/684/92-92. Council Regulation on Common Rules for the International Carriage of Passengers by Coach and Bus. 9 pp.

—**Pentachlorophenol—Emergency Procedures**

SAA AS 1678.6.1. 012-84. Emergency Procedure Guides—Transport—Part 6.1.012: Pentachlorophenol (PCP, PENTA).

—**Pesticides—Emergency Procedures**

SAA AS 1678.10.0 01-85. Emergency Procedures Guides—Transport-Part 10.001: Pesticides.

—**Phenylmercuric Compounds—Emergency Procedures**

SAA AS 1678.6.0. 010-84. Emergency Procedure Guides—Transport—Part 6.0.010: Phenylmercuric Compounds.

—**Potassium Cyanide—Emergency Procedures**

SAA AS 1678.6.0. 002-83. Emergency Procedure Guides—Transport—Part 6.0.002: Cyanides (Sodium and Potassium).

—**Potassium Fluoride—Emergency Procedures**

SAA AS 1678.6.1. 015-84. Emergency Procedure Guides—Transport—Part 6.1.015: Fluorides (Ammonium, Potassium or Sodium Fluoride).

—**Selenates—Emergency Procedures**

SAA AS 1678.6.0. 016-84. Emergency Procedure Guides—Transport—Part 6.0.016: Selenious Acid, Selenates, Selenites.

—**Selenious Acid—Emergency Procedures**

SAA AS 1678.6.0. 016-84. Emergency Procedure Guides—Transport—Part 6.0.016: Selenious Acid, Selenates, Selenites.

INDUSTRY STANDARDS

Highway Transportation (Cont.)

—Selenites—Emergency Procedures
SAA AS 1678.6.0. 016-84. Emergency Procedure Guides—Transport—Part 6.0.016: Selenious Acid, Selenates, Selenites.

—Sodium Cyanide—Emergency Procedures
SAA AS 1678.6.0. 002-83. Emergency Procedure Guides—Transport—Part 6.0.002: Cyanides (Sodium and Potassium).

—Sodium Fluorides—Emergency Procedures
SAA AS 1678.6.1. 015-84. Emergency Procedure Guides—Transport—Part 6.1.015: Fluorides (Ammonium, Potassium or Sodium Fluoride).

—Sodium Hypochlorite—Emergency Procedures
SAA AS 1678.8.0. 004-93. Emergency Procedure Guides—Transport—Part 8.0.004: Hypochlorite Solutions (with more than 5% Avalible Chlorine) (in Professional Package 37A).

—Sodium Pentachlorophenate—Emergency Procedures
SAA AS 1678.6.0. 012-84. Emergency Procedure Guides—Transport—Part 6.0.012: Sodium Pentachlorophenate.

—Sulfur Dioxide—Emergency Procedures
SAA AS 1678.2.3. 001-84. Emergency Procedure Guides—Transport—Part 2.3.001: Sulphur Dioxide (Liquefied).

—Surface Transport Request/Reply
NATO STANAG 2156 ED 4 AMD 0-88. Surface Transport Request and Surface Transport Reply. 25 pp.

—Teleinformatic Services—Frequency Bands
CEPT T/R 22-04 E-91. Harmonisation of Frequency Bands for Road Transport Information Systems (RTI). 1 p.
EC COM(92) 341-92. Proposal for a Council Directive on the Frequency Bands to Be Designated for the Coordinated Introduction of Road Transport Telematic Systems in the Community, Including Road Information and Route Guidance Systems. 2 pp.

—Toluene—Emergency Procedures
SAA AS 1678.3.2. 001-86. Emergency Procedure Guides—Transport—Part 3.2.001: Toluene (Toluol).

—Toluene 2,4-Diisocyanate—Emergency Procedures
SAA AS 1678.6.0. 006-84. Emergency Procedure Guides—Transport—Part 6.0.006: Toluene Diisocyanate (TDI).

—Toluene 2,4-Diisocyanite—Emergency Procedures
SAA AS 1678.6.0. 004-93. Emergency Procedure Guides—Transport—Part 6.0.004: Phenol (Carbolic Acid), Cresols and Cresylic Acid (in Professional Package 37A).

—Vinyl Chloride—Emergency Procedures
SAA AS 1678.2.1. 008-87. Emergency Procedure Guide—Transport—Part 2.1.008: Vinyl Chloride (Inhibited).

—Xylenes—Emergency Procedures
SAA AS 1678.3.2. 002-86. Emergency Procedure Guides—Transport—Part 3.2.002: Xylenes (Xylol).

Highways
See Also: Civil Engineering; Culverts (Pipes); Highway Engineering; Highway Lighting; Highway Transportation; Landscaping; Pavements; Roads

—Barriers—Concrete
BSI BS 6579: Part 8-87. 1987 Safety Fences and Barriers for Highways Part 8: Concrete Safety Barriers. 31 pp.

—Flexible Pavements—Static Loads
CNS A3290-88. Method of Test for Nonrepetitive Static Plate Load of Soils and Flexible Pavement Components, for Use in Evaluation and Design of Airport and Highway Pavements (Aug)(12392).
CNS A3291-88. Method of Test for Repetitive Static Plate Load of Soils and Flexible Pavement Components, for Use in Evaluation and Design of Airport and Highway Pavements (Aug)(12393).

—Safety Fences
BSI BS 6579: Part 1-88. 1988 Amd 1 Safety Fences and Barriers for Highways Part 1: Components for Tensioned Corrugated Beam Safety Fence on Z Posts. 29 pp.
BSI BS 6579: Part 3-88. 1988 Safety Fences and Barriers for Highways Part 3: Components for Tensioned Rectangular Hollow Section Beam (100mm x100mm) Safety Fence. 18 pp.
BSI BS 6579: Part 4-90. 1990 Safety Fences and Barriers for Highways Part 4: Specification for Components for Tensioned Rectangular Hollow Section Beam (200 mm x 100 mm) Safety Fence. 28 pp.
BSI BS 6579: Part 5-86. 1986 Safety Fences and Barriers for Highways Part 5: Open Box Beam Safety Fence (Single Height). 27 pp.
BSI BS 6579: Part 6-88. 1988 Safety Fences and Barriers for Highways Components for Open Box Beam Safety Fence (Double Height). 18 pp.
BSI BS 6579: Part 7-89. 1989 Safety Fences and Barriers for Highways Part 7: Components for Untensioned Corrugated Beam Safety Fence. 18 pp.

—Salt—Maintenance
BSI BS 3247-91. 1991 Salt for Spreading on Highways for Winter Maintenance. 13 pp.

—Traffic Signs
SAA AS 1742.8-90. Manual of Uniform Traffic Control Devices—Part 8: Freeways. 51 pp.

—Vehicles—Natural Gas
CSA 1169 Bull. Electrical Bulletin 1169 June 27, 1978 to C22.2 NO 22. 2 pp.

Hijacking
See Also: Aircraft Accidents; Aircraft Safety

—Aircraft
ICAO Annex 17. Security Safeguarding International Civil Aviation Against Acts of Unlawful Interference Fifth Edition—December 1992. 31 pp.
ICAO 8364. Convention on Offences and Certain Other Acts Committed on Board Aircraft Signed at Tokyo on 14 September 1963. 15 pp.
ICAO 8920. Convention for the Suppression of Unlawful Seizure of Aircraft Signed at the Hague on 16 December 1970. 20 pp.
ICAO 8966. Convention for the Suppression of Unlawful Acts Against the Safety of Civil Aviation Signed at Montreal on 23 September 1971. 30 pp.
NATO STANAG 3771 ED 3 AMD 0-90. Ground Security Measures Against Aircraft Sabotage/Hijacking. 7 pp.
NATO STANAG 3771 ED 3 AMD 1-90. Ground Security Measures Against Aircraft Sabotage/Hijacking. 8 pp.

Hiller FH 1100 Aircraft
See Also: Aircraft

—Antenna Positions
CAA. Hiller UH12A/UH12C/UH12E and FH 1100 (Approved Aerial Positions). 1 p.

Hiller Helicopters
See Also: Helicopters

—Foreign Airworthiness Directives
CAA. Hiller Aviation Series Helicopters (Foreign Airworthiness Directives). 2 pp.

Hiller UH 12A Aircraft
See Also: Aircraft

—Antenna Positions
CAA. Hiller UH12A/UH12C/UH12E and FH 1100 (Approved Aerial Positions). 1 p.

Hiller UH 12C Aircraft
See Also: Aircraft

—Antenna Positions
CAA. Hiller UH12A/UH12C/UH12E and FH 1100 (Approved Aerial Positions). 1 p.

Hiller UH 12E Aircraft
See Also: Aircraft

—Antenna Positions
CAA. Hiller UH12A/UH12C/UH12E and FH 1100 (Approved Aerial Positions). 1 p.

Hiller UH 12E4 Aircraft
See Also: Aircraft

—Antenna Positions
CAA. Hiller UH12E4 (Approved Aerial Positions). 1 p.

Hinge Pins
Use: Kingpins

Hinges
See Also: Butt Hinges; Fasteners; Hardware; Kingpins; Spring Hinges
BSI BS 1227: Part 1A-67. (WITHDRAWN) 1967 Amd 2 Hinges Part 1A: Hinges for General Building Purposes (Superseded by BS 7352: 1990). 98 pp.
BSI BS 7352-90. 1990 Strength and Durability Performance of Metal Hinges for Side Hanging Applications and Dimensional Requirements for Template Drilled Hinges. 24 pp.
CGSB CAN/CGSB-69.36-M90. Strap and Tee Hinges and Hasps (ANSI/BHMA A156.20-1984); (Amendment 1 August 1993). 26 pp.
CNS A2008-89. Hinges, Butterfly Form (Jul)(858). 2 pp.
CNS A3078-84. Performance Test of Hinges (Aug)(4726).
JIS A 1511-79. Performance Test of Hinges.
JIS T 9220-92. Outside Locking Elbow Hinge Units. 11 pp.

—Aerospace
DIN ENGL LN 9372 Pt 1-84. Aerospace; Hinges; Extruded; Half Hinges and Hinges (Feb). 3 pp.
DIN ENGL LN 9372 Pt 3-84. Aerospace; Hinges; Extruded; Blanks (Feb). 2 pp.

—Aircraft
SBAC AS 1876-1879 ISSUE 10. Hinges Aluminium Alloy Single Flange Plates. 1 p.
SBAC AS 1880-1881 ISSUE 13. Hinges Aluminium Alloy Double Flange Plates. 1 p.

—Brackets—Aircraft
SBAC AS 3290 ISSUE 2. Hinge Bracket. 1 p.

—Doors
CGSB CAN/CGSB-69.36-M90. Strap and Tee Hinges and Hasps (ANSI/BHMA A156.20-1984); (Amendment 1 August 1993). 26 pp.
CNS A2087-86. Door Hinges (with Bushings or Washers) (Dec)(6538). 5 pp.
CNS A2102-81. Flag-Form Hinges, Buffer and Flush Bolts for Steel Doors (Apr)(7185). 8 pp.
JIS A 5511-72. Door Hinges (with Bushing or Washers).
JIS A 5516-72. Door Hinges (with Ball Bearings).
JIS A 5518-92. Fittings for Doors and Doorsets. 16 pp.
SNZ NZS 844-51. Specification for Butt Hinges Amend: 1, 1965. 10 pp.

—Doors—Automobiles
CNS D2098-89. Door Hinges for Automobiles (Jul)(7681).

—Doors—Automobiles—Static Loads
JIS D 1621-88. Test Method of Side Door Hinge Systems for Automobiles. 7 pp.

—Doors—Ships
CNS F3050-80. Ships' Steel Watertight Doors (Aug)(6182). 9 pp.
CNS F3211-83. Fitting for Weathertight Steel Doors of Small Ships (Mar)(10072).
JIS F 2316-87. Fittings for Weather-Tight Steel Doors.
JIS F 2316-70. Fittings for Ships' Weathertight Steel Doors. 9 pp.
JIS F 2330-80. Fittings for Small Ships' Weather-Tight Steel Doors.
JIS F 2330-75. Fittings for Small Ships' Weathertight Steel Doors. 13 pp.

—Endurance Testing
BSI BS 7352-90. 1990 Strength and Durability Performance of Metal Hinges for Side Hanging Applications and Dimensional Requirements for Template Drilled Hinges. 24 pp.

—Finger—Handhole Covers—Aircraft
SBAC AS 3174 ISSUE 2. Finger, Hinge Handhole Cover. 1 p.

—Floors
CNS A2066-85. Floor Hinges (Feb)(4724). 3 pp.
CNS A3077-84. Method of Test for Checking Floor-Hinges and Door-Closers (Aug)(4725).
JIS A 1512-75. Testing Methods for Checking Floor-Hinges and Door-Closers.
JIS A 5543-75. Floor Hinges.

—Furniture
ISO 8555 Pt 2-87. Hardware for Furniture—Terms for Furniture Fittings—Part 2: Hinges and Flap Hinges First Edition. 21 pp.

—Hatch Covers—Ships
CNS F3170-82. Fittings for Ships' Small Size Steel Hatch Covers (Apr)(8684).
JIS F 2322-87. Fittings of Ships' Steel Small Hatch Covers.
JIS F 2322-61. Fittings of Ships' Steel Small Hatch Covers (R 1970). 9 pp.

INTERNATIONAL AND NON-U.S. NATIONAL STANDARDS
SUBJECT INDEX

Hinges (Cont.)
—Self Closing
CGSB CAN/CGSB-69.33-M90. Self-Closing Hinges and Pivots (ANSI/BHMA A156.17-1987). 35 pp.

—Sewing Machine Tables—Household
CNS B2514-89. Hinges and Locks for Household Sewing Machine Table (Sep)(6000).

JIS B 9037-78. Hinges and Locks for Household Sewing Machine Table (R 1983). 13 pp.

—Sewing Machines—Industrial
CNS B2460-89. Rubber Hinges for Industrial Sewing Machines (Feb)(5489).

JIS B 9067-76. Rubber Hinges for Industrial Sewing Machines.

—Straps—Aircraft
SBAC AS 2378 ISSUE 2. Tank Strap End (1/4 Inch B.S.F.). 1 p.

SBAC AS 2379 ISSUE 2. Tank Strap End (5/16 Inch B.S.F.). 1 p.

SBAC AS 2380 ISSUE 2. Tank Strap End (3/8 Inch B.S.F.). 1 p.

—Template—Doors
CGSB CAN/CGSB-69.23-M90. Template Hinge Dimensions (ANSI/BHMA A156.7-1981). 17 pp.

Hip Prostheses
See Also: Joint Prostheses; Leg Prostheses; Prosthetic Devices

—Classification
BSI BS 7251: Part 3-90. 1990 Orthopaedic Joint Prostheses Part 3: Classification, Designation of Dimensions and Requirements for Hip Joint Prostheses. 10 pp.

ISO 7206 Pt 1-85. Implants for Surgery—Partial and Total Hip Joint Prostheses—Part 1: Classification, Designation of Dimensions and Requirements First Edition. 7 pp.

—Designations
BSI BS 7251: Part 3-90. 1990 Orthopaedic Joint Prostheses Part 3: Classification, Designation of Dimensions and Requirements for Hip Joint Prostheses. 10 pp.

ISO 7206 Pt 1-85. Implants for Surgery—Partial and Total Hip Joint Prostheses—Part 1: Classification, Designation of Dimensions and Requirements First Edition. 7 pp.

—Endurance Testing
BSI BS 7251: Part 5-90. 1990 Orthopaedic Joint Prostheses Part 5: Method for Determination of Endurance Properties of Stemmed Femoral Components of Hip Joint Prostheses with Application of Torsion. 11 pp.

BSI BS 7251: Part 6-90. 1990 Orthopaedic Joint Prostheses Part 6: Method for Determination of Endurance Properties of Stemmed Femoral Components of Hip Joint Prostheses Without Application of Torsion. 12 pp.

BSI BS 7251: Part 10-92. 1992 Orthopaedic Joint Prostheses Part 10: Method of Determination of Endurance Properties of the Head and Neck Region of Stemmed Femoral Components of Hip Joint Prostheses (ISO 7206-6: 1992). 11 pp.

BSI BS 7251: Part 11-93. 1993 Orthopaedic Joint Prostheses Part 11: Specification for Endurance of Stemmed Femoral Components Without Application of Torsion (ISO 7206-7: 1993) (N). 7 pp.

BSI DD 91-86. (WITHDRAWN) 1986 Method for Determination of Endurance Properties of Stemmed Femoral Components of Hip Joint Prostheses (Superseded by BS 7251: Part 5: 1990). 11 pp.

ISO 7206 Pt 3-88. Implants for Surgery—Partial and Total Hip Joint Prostheses—Part 3: Determination of Endurance Properties of Stemmed Femoral Components Without Application of Torsion First Edition. 14 pp.

ISO 7206 Pt 4-89. Implants for Surgery—Partial and Total Hip Joint Prostheses—Part 4: Determination of Endurance Properties of Stemmed Femoral Components with Application of Torsion First Edition. 12 pp.

ISO 7206 Pt 6-92. Implants for Surgery—Partial and Total Hip Joint Prostheses—Part 6: Determination of Endurance Properties of Head and Neck Region of Stemmed Femoral Components First Edition. 10 pp.

ISO 7206 Pt 7-93. Implants for Surgery—Partial and Total Hip Joint Prostheses—Part 7: Endurance Performance of Stemmed Femoral Components Without Application of Torsion First Edition. 6 pp.

—Simulators
BSI BS 7251: Part 7-90. 1990 Orthopaedic Joint Prostheses Part 7: Guide to Hipjoint Simulators. 14 pp.

Hip Prostheses (Cont.)
—Simulators (Cont.)
ISO TR9325-89. Implants for Surgery—Partial and Total Hip Joint Prostheses—Recommendations for Simulators for Evaluation of Hip Joint Prostheses First Edition. 12 pp.

—Static Loads
BSI BS 7251: Part 9-92. 1992 Orthopaedic Joint Prostheses Part 9: Method of Determination of Resistance to Static Load of the Head and Neck Region of Stemmed Femoral Components of Hip Joint Prostheses (ISO 7206-5: 1992). 10 pp.

ISO 7206 Pt 5-92. Implants for Surgery—Partial and Total Hip Joint Prostheses—Part 5: Determination of Resistance to Static Load of Head and Neck Region of Stemmed Femoral Components First Edition. 8 pp.

—Surface Finishing
BSI BS 7251: Part 4-90. 1990 Orthopaedic Joint Prostheses Part 4: Bearing Surfaces of Hip Joint Prostheses. 5 pp.

ISO 7206 Pt 2-87. Implants for Surgery—Partial and Total Hip Joint Prostheses—Part 2: Bearing Surfaces Made of Metallic and Plastics Materials First Edition. 4 pp.

—Wear Testing
BSI BS 7251: Part 8-90. 1990 Orthopaedic Joint Prostheses Part 8: Guide to Laboratory Evaluation of Change of Form of Bearing Surfaces of Hip Joint Prostheses (ISO/TR 9326: 1989). 11 pp.

ISO TR9326-89. Implants for Surgery—Partial and Total Hip Joint Prostheses—Guidance for Laboratory Evaluation of Change of Form of Bearing Surfaces First Edition. 11 pp.

HIRAGANA Character Sets
Use: Optical Character Recognition—Japanese Character Sets—HIRAGANA

HIRAGANA Characters
Use: Japanese Characters

Hirsch Funnels
Use: Funnels—Filter

Histamine Content Analysis
See Also: Drugs

—Food
CNS N6159-81. Determination of Histamine for Food Stuff (Oct)(8052). 4 pp.

L-Histidine Monohydrochloride Monohydrate
JIS K 9050-72. L-Histidine Monohydrochloride, Monohydrate.

Histograms
—Particle Size Distribution
ISO 9276 Pt 1-90. Representation of Results of Particle Size Analysis—Part 1: Graphical Representation First Edition. 11 pp.

History
See Also: Political Science

—Classification
BSI BS 1000: (9)-72. 1972 Amd 1 Universal Decimal Classification (UDC). English Full Edition (9): Geography. Biography. History. 55 pp.

SNZ NZS/BS 1000 (9)-72. Universal Decimal Classification Geography. Biography. History Amend: 1. 52 pp.

Hitches
See Also: Fifth Wheels; Towing Attachments; Yokes

—Agricultural Equipment
CNS D2082-86. Dimensions of Hitches for Agricultural Tractors (Jul)(7031).

JIS D 6704-84. Dimensions of Hitchs for Agricultural Tractors. 4 pp.

—Agricultural Equipment—Clevis
ISO 6489 Pt 2-80. Agricultural Vehicles—Mechanical Connections on Towing Vehicles—Part 2: Clevis Type—Dimensions First Edition. 4 pp.

—Agricultural Equipment—Four Point
ISO 11374-93. Agricultural Tractors and Implements—Four-Point Rigid Hitch—Specifications First Edition. 8 pp.

—Agricultural Equipment—Linchpins
ISO 7072-93. Tractors and Machinery for Agriculture and Forestry—Linch Pins and Spring Pins—Dimensions Second Edition. 6 pp.

Hitches (Cont.)
—Agricultural Equipment—Ring Type
BSI BS 5891-80. 1980 Ring-Type Hitches on Agricultural Trailers and Towed Implements. 4 pp.

ISO 5692-79. Agricultural Vehicles—Mechanical Connections on Towed Vehicles—Hitch-Rings—Specifications First Edition. 5 pp.

—Agricultural Equipment—Three Point
BSI BS 1841: Part 2-73. (WITHDRAWN) 1973 Attachment of Mounted Implements to Agricultural Wheeled Tractors Part 2: Metric Units (Superseded by BS 1841: Part 1: 1991). 18 pp.

BSI BS 6509-84. 1984 Clearance Zone on Implements for the Three-Point Linkage to Agricultural Tractors. 6 pp.

BSI BS 6818: Part 2-87. 1987 Front-Mounted Linkage and Power Take-off for Agricultural Wheeled Tractors Part 2: Front Linkage. 7 pp.

ISO 730 Pt 1-90. Agricultural Wheeled Tractors—Rear-Mounted Three-Point Linkage—Part 1: Categories 1, 2 and 3 Second Edition. 8 pp.

ISO 730 Pt II-79. Agricultural Wheeled Tractors—Three-Point Linkage—Part II: Category 1 N (Narrow Hitch) First Edition. 6 pp.

ISO 730 Pt 3-82. Agricultural Wheeled Tractors—Three-Point Linkage—Part 3: Category 4 First Edition. 6 pp.

ISO 2332-83. Agricultural Tractors and Machinery—Connections—Clearance Zone for the Three-Point Linkage of Implements First Edition. 5 pp.

ISO 8759 Pt 2-85. Agricultural Wheeled Tractors—Front-Mounted Linkage and Power Take-off—Part 2: Front Linkage First Edition; (ASAE S513). 6 pp.

JIS D 6703-78. Dimensions of Three-Point Linkages for Agricultural Wheeled Tractors.

JIS D 6704-84. Dimensions of Hitchs for Agricultural Tractors. 4 pp.

—Agricultural Equipment—Towing Hooks
BSI BS 6108-81. (WITHDRAWN) 1981 Dimensions of Hook-Type Couplings on Agricultural Towing Vehicles (Superseded by BS 6108: Part 1: 1993). 3 pp.

BSI BS 6108: Part 1-93. 1993 Couplings on Agricultural Towing Vehicles Part 1: Specification for Hook Type (ISO 6489-1: 1991) (E). 7 pp.

ISO 6489 Pt 1-91. Agricultural Vehicles—Mechanical Connections on Towing Vehicles—Part 1: Hook Type Second Edition. 5 pp.

—Forestry Equipment
ISO 6815-83. Machinery for Forestry—Hitches—Dimensions First Edition. 5 pp.

—Lawn and Garden Equipment—One Point
ISO 9192-91. Lawn and Garden Ride-on (Riding) Tractors—One-Point Tubular Sleeve Hitch First Edition. 8 pp.

—Lawn and Garden Equipment—Three Point
ISO 9191-91. Lawn and Garden Ride-on (Riding) Tractors—Three-Point Hitch First Edition. 8 pp.

—Tillers
CNS B4031-80. Dimensions of Hitch for Power Tillers (Aug)(6006).

JIS B 9209-76. Dimensions of Hitch for Power Tillers. 4 pp.

—Tractors—Four Point
ISO 11374-93. Agricultural Tractors and Implements—Four-Point Rigid Hitch—Specifications First Edition. 8 pp.

—Trailers
CNS D1019-82. Interchange Coupling Dimensions of Truck Tractors and Semitrailers (Jan)(8238).

CSA D264-1976. Trailer Hitches; (Amd 1 June 1986). 15 pp.

JIS D 6603-82. Interchange Coupling Dimensions of Truck Tractors and Semitrailers (R 1987). 9 pp.

SNZ NZS 5232-71. Specification for Ball-and-Socket Type Trailer Couplings and Their Associated Safety Connections. 17 pp.

—Trucks
JIS D 4010-82. Mounting of Mechanical Coupling Devices on Rear Cross Members of Trucks. 6 pp.

HME
Use: Heat/Moisture Exchangers

5-HMF Content Analysis
Use: 5-Hydroxymethylfurfural Content Analysis

HMX
Use: Octogene

HNS
Use: Hexanitrostilbene

INTERNATIONAL AND NON-U.S. NATIONAL STANDARDS
SUBJECT INDEX

Ho Leaf Oil
See Also: Oils
CNS K5042-66. Ho-Leaf Oil (Sep)(2682) (R 1973). 1 p.

Ho Oil
CNS K5043-66. Ho-Oil (Sep)(2683) (R 1973). 1 p.
CNS K6147-74. Method of Test for Ho-Oil (Jun)(1376).

Hob Arbors
Use: Hobbing Arbors

Hob Elements (Ranges)
Use: Ranges

Hobbing Arbors
See Also: Arbors
DIN ENGL 2086 Pt 2-64. Hob Arbors for Hobs with Longitudinal Keyway (July). 2 pp.
DIN ENGL 2086 Pt 3-64. Milling Machine Arbors for Hobs with Clutch Drive (July). 2 pp.

—**Bearing Sleeves**
DIN ENGL 2083 Pt 2-64. Bearing Sleeves for Hob Arbors; Parallel Type (July). 2 pp.

—**Clutch Drive Rings**
DIN ENGL 6366 Pt 2-64. Clutch Drive Rings for Hob Arbors (July). 1 p.

—**Nuts**
DIN ENGL 2082 Pt 2-64. Nuts for Hob Arbors (July). 2 pp.

—**Spacers**
DIN ENGL 2084 Pt 2-64. Spacer Collars for Hob Arbors for Hobs with Longitudinal Keyway (July). 2 pp.
DIN ENGL 2084 Pt 3-64. Spacer Collars for Hob Arbors for Hobs with Transverse Slot (July). 1 p.

Hobbing Machines
Use For: Gear Cutters *See Also:* Cutting Tools; Gear Hobs; Gear Teeth; Gears; High Speed Steel Tools; Milling Cutters
BSI BS 4582: Part 2-78. 1978 Fine Pitch Gears (Metric Module) Part 2: Hobs and Cutters. 31 pp.
CNS B1119-79. Specified Items of Machine Tools (Hobbing Machine) (Apr)(4670-22).
CNS B7099-82. Accuracy Inspection of Hobbing Machines (Dec)(4351).
CNS B7182-82. Performance Inspection of Hobbing Machine (Jan)(8353).

—**Accuracy Testing**
CNS B7099-82. Accuracy Inspection of Hobbing Machines (Dec)(4351).
ISO 6545-92. Acceptance Conditions for Gear Hobbing Machines—Testing of the Accuracy First Edition. 33 pp.

—**Bevel Gears**
JIS B 4351-85. Straight Bevel Gear Generating Cutters (Type G). 15 pp.

—**Change Gears**
CNS B2025-84. Number of Teeth of Change Gears for Geared Lathe, Milling Machine and Gear Generating Equipment (Aug)(186).

—**Disk—Spur Gears**
CNS B3381-80. Pinion-Type Cutters for Cylindrical Gears; Disc-Gear Cutters for Spur Gears (Sep)(6417).

—**External Gears**
BSI BS 2007-75. 1975 Circular Gear Shaving Cutters, 1 to 8 Metric Module, Accuracy Requirements. 8 pp.

—**Glossaries**
CNS B1231-80. Glossary of Terms Relating to Parts of Gear Cutting Machine (Dec)(6755).
JIS B 0174-91. Gear Cutters—Vocabulary. 68 pp.
JIS B 0174-78. Glossary of Terms for Gear Cutters. 68 pp.

—**Helical**
BSI BS 978: Part 5-65. 1965 Fine Pitch Gears Part 5: Hobs and Cutters. 27 pp.
BSI BS 2518: Part 2-77. 1977 Rotary Form Relieved Gear Cutters Part 2: Metric Module. 29 pp.
BSI BS 2697-76. 1976 Specification for Rack Type Gear Cutters, Metric Module. 12 pp.
JIS B 4358-91. Rack Type Cutters. 19 pp.

—**Helical—Pitch**
BSI BS 2518: Part 1-54. 1954 Amd 2 Rotary Form Relieved Gear Cutters Part 1: Diametral Pitch. 29 pp.

Hobbing Machines (Cont.)
—**Internal Gears**
BSI BS 2007-75. 1975 Circular Gear Shaving Cutters, 1 to 8 Metric Module, Accuracy Requirements. 8 pp.

—**Involute Gears**
BSI BS 5246-76. 1976 Amd 1 Pinion Type Cutters for Spur Gears; 1 to 8 Metric Module. 17 pp.
CNS B2425-84. Basic Rack of Gear Cutting Tools for Involute Tooth System (Aug)(5282).
DIN ENGL 3972-52. Reference Profiles of Gear-Cutting Tools for Involute Tooth Systems According to DIN 867 (Feb). 2 pp.
JIS B 4232-85. Involute Gear Milling Cutters. 10 pp.
JIS B 4358-91. Rack Type Cutters. 19 pp.

—**Pinion Type—Cylindrical Gears**
CNS B1212-80. Pinion-Type Cutters for Cylindrical Gears; Tolerances (Sep)(6419).

—**Pinion Type—Spur Gears**
CNS B3380-80. Pinion-Type Cutters for Cylindrical Gear Cutters for Spur Gears (Sep)(6416).

—**Profiles**
JIS B 4350-91. Gear Cutter—Tooth Profiles and Dimensions. 16 pp.

—**Rotary Gears**
JIS B 4357-88. Rotary Gear Shaving Cutters. 14 pp.

—**Shank—Spur Gears**
CNS B3382-80. Pinion-Type Cutters for Cylindrical Gears; Inter Shank-Gear Cutters for Spur Gears (Sep)(6418).

—**Spur Gears**
BSI BS 978: Part 5-65. 1965 Fine Pitch Gears Part 5: Hobs and Cutters. 27 pp.
BSI BS 2518: Part 2-77. 1977 Rotary Form Relieved Gear Cutters Part 2: Metric Module. 29 pp.
BSI BS 5246-76. 1976 Amd 1 Pinion Type Cutters for Spur Gears; 1 to 8 Metric Module. 17 pp.
JIS B 4358-91. Rack Type Cutters. 19 pp.

—**Spur Gears—Pitch**
BSI BS 2518: Part 1-54. 1954 Amd 2 Rotary Form Relieved Gear Cutters Part 1: Diametral Pitch. 29 pp.

Hobs
Scope Note: Use a more specific term *See:* Gear Hobs

Hobs (Appliances)
Use: Hot Plates (Warmers)

Hockey Fields
Use: Athletic Fields

Hoes
See Also: Agricultural Equipment; Cultivators; Hand Tools; Tools
CGSB 39-GP-5-53. Hoes, Mattocks and Picks, Spec. for; (Amendment 1 Oct 1976). 7 pp.

—**Triangular**
CNS B3414-81. Hand Hoe Triangular Shaped (May)(7345).

Hoffman H36 Dimona Motor Gliders
See Also: Gliders

—**Foreign Airworthiness Directives**
CAA. Hoffman H36 Dimona Motor Glider (Foreign Airworthiness Directives). 3 pp.

Hog Cholera
Use For: African Swine Fever; Swine Fever

—**Belgium—Control Measures**
EC 89/224/EEC-89. Commission Decision Amending Commission Decision 89/224/EEC Recognizing Certain Parts of Belgium as Being Officially Swine Fever Free. 1 p.
EC 89/446/EEC-89. Commission Decision Amending Commission Decision 89/224/EEC Recognizing Certain Parts of Belgium as Being Officially Swine Fever Free (Only the French and Dutch Texts are Authentic). 2 pp.

—**Control Measures**
EC 85/320/EEC-85. Council Directive Amending Directive 64/432/EEC as Regards Certain Measures Relating to Classical Swine Fever and African Swine Fever. 3 pp.
EC 85/321/EEC-85. Council Directive Amending Directive 80/215/EEC as Regards Certain Measures Relating to African Swine Fever. 2 pp.
EC 85/322/EEC-85. Council Directive Amending Directive 72/461/EEC as Regards Certain Measures Relating to Classical Swine Fever and African Swine Fever. 2 pp.

Hog Cholera (Cont.)
—**Control Measures (Cont.)**
EC 87/486/EEC-87. Council Directive Amending Directive 80/217/EEC Introducing Community Measures for the Control of Classical Swine Fever. 3 pp.
EC 87/487/EEC-87. Council Directive Amending Directive 80/1095/EEC Laying Down Conditions Designed to Render and Keep the Territory of the Community Free from Classical Swine Fever. 2 pp.
EC 87/489/EEC-87. Council Directive Amending Directives 64/432/EEC and 72/461/EEC as Regards Certain Measures Relating to Swine Fever. 2 pp.
EC 87/230/EEC-87. Council Decision Amending Directive 80/1095/EEC and 82/18/EEC with Regard to the Duration and the Financial Means of Measures for the Eradication of Classical Swine Fever. 3 pp.
EC 87/231/EEC-87. Council Decision Amending Directives 64/432/EEC and 72/461/EEC as Regards Certain Measures Relating to Swine Fever. 2 pp.
EC 87/488/EEC-87. Council Decision Supplementing and Amending Decision 80/1096/EEC Introducing Community Financial Measures for the Eradication of Classical Swine Fever. 3 pp.

—**Italy—Control Measures**
EC 90/217/EEC-90. Council Directive on Financial Aid from the Community for the Eradication of African Swine Fever in Sardinia. 4 pp.

—**Portugal—Control Measures**
EC 86/649/EEC-86. Council Decision Introducing a Community Financial Measure for the Eradication of African Swine Fever in Portugal. 4 pp.

—**Spain—Control Measures**
EC 86/650/EEC-86. Council Decision Introducing a Community Financial Measure for the Eradication of African Swine Fever in Spain. 4 pp.

Hog Rings
Use: Retaining Rings

Hogs
Use: Swine

Hoisting Drums
Use For: Drums (Hoisting) *See Also:* Hoisting Slings; Lifting Equipment

—**Ships—Wire Rope**
CNS F4005-80. Fastening Method of Wire Rope to Drum for Marine Use (Dec)(6823). 1 p.
JIS F 3439-69. Fastening Method of Wire Ropes to Drum for Ship Use. 4 pp.

Hoisting Slings
Use For: Chain Slings; Lifting Slings; Rope Slings; Slings (Hoists) *See Also:* Hoisting Drums; Lifting Equipment; Lifting Tackle
BSI BS 6304-82. 1982 Chain Slings of Welded Construction: Grades M(4), S(6) and T(8). 19 pp.
DIN ENGL 695-86. Grade 2 Chain Slings with Hook or Ring Type Terminal Fittings (July). 8 pp.
DIN ENGL 5688 Pt 1-86. Grade 5 Chain Slings with Hook or Ring Type Terminal Fittings (July). 8 pp.
DIN ENGL 5688 Pt 3-86. Grade 8 Chain Slings with Hook or Ring Type Terminal Fittings and Endless Slings (July). 8 pp.
ISO 4778-81. Chain Slings of Welded Construction—Grades M (4), S (6) and T (8) First Edition. 17 pp.
ISO 7593-86. Chain Slings Assembled by Methods Other Than Welding—Grade T(8) First Edition. 17 pp.
JIS B 8816-89. Chain Slings for Lifting Purposes. 28 pp.
MOD UK DSTAN 39-4-01. Materials Handling Equipment Bridge Plates Lifting Slings Vehicle Loading Ramps Issue 1; Amendment 2. 10 pp.

—**Air Cargo—Helicopters**
MOD UK DSTAN 16-9-69. Nets and Slings, Cargo, Helicopter (Section on Slings is Superseded) Issue 1. 5 pp.
MOD UK DSTAN 16-9: Addendum. Nets and Slings, Helicopter Issue 1. 3 pp.

—**Aircraft—Link**
NATO STANAG 3237 ED 3 AMD 3-82. Aperture of Terminal Ring or Link of Aircraft Lifting Slings. 7 pp.
NATO STANAG 3237 ED 3 AMD 4-82. Aperture of Terminal Ring or Link of Aircraft Lifting Slings. 8 pp.

—**Aircraft—Terminal Fittings**
MOD UK DSTAN 17-1-84. Aircraft Lifting Slings Aperture of Terminal Ring or Link Issue 2. 5 pp.

—**Alloy Steel—Link**
BSI BS 3458-62. 1962 Alloy Steel Chain Slings. 49 pp.

INTERNATIONAL AND NON-U.S. NATIONAL STANDARDS
SUBJECT INDEX

Hoisting Slings (Cont.)

—Disposable
- BSI BS 3481: Part 3-74. 1974 Flat Lifting Slings Part 3: Disposable Flat Lifting Slings. 16 pp.
- SNZ NZS/BS 3481: Part 3-74. Specification for Flat Lifting Slings Part 3: Disposable Flat Lifting Slings. 16 pp.

—Helicopters
- NATO STANAG 3295 ED 6 AMD 4-81. Horse Collar/Rescue Strop Type Helicopter Hoisting Gear. 9 pp.

—Identification Systems
- BSI BS 6166: Part 2-86. 1986 Lifting Slings Part 2: Marking. 4 pp.

—Lifting Chains
- ISO 8539-86. Forged Steel Lifting Components for Use with Grade T(8) Chain First Edition. 8 pp.

—Lifting Chains—Link
- BSI BS 4942: Part 2-81. 1981 Short Link Chain for Lifting Purposes Part 2: Grade M (4) Non-Calibrated Chain. 11 pp.
- BSI BS 4942: Part 4-81. 1981 Short Link Chain for Lifting Purposes Part 4: Grade S (6) Non-Calibrated Chain. 11 pp.
- BSI BS 4942: Part 5-81. 1981 Amd 1 Short Link Chain for Lifting Purposes Part 5: Grade T (8) Non-Calibrated Chain (AMD 4934) March 31, 1986. 13 pp.
- BSI BS 6968-88. 1988 Guide for Use and Maintenance of Non-Calibrated Round Steel Lifting Chain and Chain Slings. 13 pp.
- CEN PREN 818-1-92. Short Link Chain for Lifting Purposes—Safety—Part 1: General Conditions of Acceptance. 20 pp.
- CEN PREN 818-2-92. Short Link Chain for Lifting Purposes—Safety—Part 2: Medium Tolerance Chain for Chain Slings—Grade 8. 18 pp.
- DIN ENGL 5687 Pt 3-80. Round Steel Chains; Quality Grade 8; Not-True-to-Gauge-Size; Tested (June). 6 pp.
- ISO 1835-80. Short Link Chain for Lifting Purposes—Grade M (4), Non-Calibrated, for Chain Slings, Etc. First Edition. 10 pp.
- ISO 3056-86. Non-Calibrated Round Steel Link Lifting Chain and Chain Slings—Use and Maintenance Second Edition. 12 pp.
- ISO 3075-80. Short Link Chain for Lifting Purposes—Grade S (6), Non-Calibrated, for Chain Slings, Etc. First Edition. 10 pp.
- ISO 3076-84. Short Link Chain for Lifting Purposes—Grade T (8), Non-Calibrated, for Chain Slings, Etc. Second Edition. 11 pp.
- SNZ NZS/ISO 1835-80. Short Link Chain for Lifting Purposes—Grade M (4), Non-Calibrated, for Chain Slings Etc.. 8 pp.
- SNZ NZS/ISO 3075-80. Short Link Chain for Lifting Purposes-Grade S(6) Non-Calibrated, for Chain Slings Etc.. 8 pp.
- SNZ NZS/ISO 3076-84. Short Link Chain for Lifting Purposes-Grade T(8), Non-Calibrated, for Chain Slings Etc.. 9 pp.

—Lifting Chains—Link—Repair—Boats
- MOD UK NES 744-90. Requirements for the Manufacture and Repair of Rings and Links Issue 2 (10.90). 11 pp.

—Lifting Chains—Ships
- CNS F3068-80. Chain Slings (Oct)(6565).
- CNS F3070-80. Small Size Chain Slings (Oct)(6567).

—Lifting Hooks
- BSI BS 2903-80. 1980 Amd 1 Higher Tensile Steel Hooks for Chains, Slings, Blocks and General Engineering Purposes. 28 pp.
- BSI BS 3458-62. 1962 Alloy Steel Chain Slings. 49 pp.
- SNZ NZS/BS 2903-80. Specification for Higher Tensile Steel Hooks for Chains, Slings, Blocks and General Engineering Purposes Amend: 1. 28 pp.

—Military Vehicles
- NATO STANAG 4062 Pt1 ED1 AMD2-70. Attachments for Lifting Vehicles and Heavy Military Equipment by Land and Sea. 12 pp.
- NATO STANAG 4062 ED 2 AMD 0-89. Slinging and Tie-Down Facilities for Lifting and Tying Down Military Equipment for Movement by Land and Sea. 14 pp.

—Mountaineering Equipment—Safety
- BSI BS EN 566-93. 1993 Mountaineering Equipment Slings Safety Requirements and Test Method (N). 8 pp.
- CEN PREN 566-91. Mountaineering Equipment—Slings—Safety Requirements, Testing, Marking. 5 pp.
- CEN EN 566-92. Mountaineering Equipment—Slings—Safety Requirements and Test Method. 3 pp.

Hoisting Slings (Cont.)

—Natural Fibers
- BSI BS 6668: Part 1-86. 1986 Textile Lifting Slings Part 1: Lifting Slings for General Service Made from Certain Natural and Man-Made Fibre Ropes. 31 pp.
- SAA AS 1380-72. Fibre-Rope Slings (of Natural or Synthetic Rope) Corrig.. 40 pp.

—Plastic Tubes
- MOD UK DSTAN 40-13-91. Roundslings, Seamless Tube, Polyester Issue 2. 10 pp.

—Ratings
- BSI BS 6166: Part 1-86. 1986 Lifting Slings Part 1: Methods of Rating. 12 pp.

—Safety
- BSI BS 6166: Part 3-88. 1988 Lifting Slings Part 3: Guide to Selection and Safe use of Lifting Slings for Multi-Purpose. 28 pp.
- CEN PREN 818-8-92. Short Link Chain for Lifting Purposes—Safety—Part 8: Chain Slings—Grade 8. 23 pp.
- SAA AS 1353.2-90. Flat Synthetic-Webbing Slings—Part 2: Care and Use. 10 pp.

—Ships
- CNS F3082-81. Ships' Hatch Beam Slings (Jan)(6912).

—Steel—Link
- BSI BS 2902-57. 1957 Amd 2 Higher Tensile Steel Chain Slings and Rings, Links Alternative to Rings, Egg Links and Intermediate Links. 52 pp.

—Steel Wire Rope
- DIN ENGL 3078-90. Steel Wire for Hoisting Ropes (May). 4 pp.

—Steel Wire Rope—Assembly
- DIN ENGL 3088-89. Steel Wire Rope Slings for Lifting Purposes; Safety Requirements, Marking and Assembly (May). 13 pp.

—Steel Wire Rope—Identification Systems
- DIN ENGL 3088-89. Steel Wire Rope Slings for Lifting Purposes; Safety Requirements, Marking and Assembly (May). 13 pp.

—Steel Wire Rope—Safety
- DIN ENGL 3088-89. Steel Wire Rope Slings for Lifting Purposes; Safety Requirements, Marking and Assembly (May). 13 pp.

—Synthetic Fibers
- BSI BS 6668: Part 1-86. 1986 Textile Lifting Slings Part 1: Lifting Slings for General Service Made from Certain Natural and Man-Made Fibre Ropes. 31 pp.
- BSI BS 6668: Part 2-87. 1987 Textile Lifting Slings Part 2: Roundslings Made of Man-Made Fibre for General Service. 12 pp.
- DIN ENGL 61360 Pt 1-86. Man-Made Fibre Lifting Slings; Concepts, Dimensions, Modes of Assembly (Mar). 4 pp.
- SAA AS 1380-72. Fibre-Rope Slings (of Natural or Synthetic Rope) Corrig.. 40 pp.

—Synthetic Fibers—Safety
- DIN ENGL 61360 Pt 2-86. Synthetic Fibre Lifting Slings; Safety Requirements and Testing (Mar). 7 pp.

—Webbing
- BSI BS 3481: Part 2-83. 1983 Amd 1 Flat Lifting Slings Part 2: Flat Woven Webbing Slings Made of Man-Made Fibre for General Service. 13 pp.
- ISO 4878-81. Flat Woven Webbing Slings Made of Man-Made Fibre First Edition. 15 pp.
- JIS B 8818-89. Webbing Slings for Lifting Purposes. 26 pp.
- SAA AS 1353.1-90. Flat Synthetic-Webbing Slings—Part 1: Product Specification. 11 pp.
- SNZ NZS 5227-86. Specification for Flat Lifting Slings. Part 2 Flat Woven Webbing Lifting Slings Made of Man-Made Fibre for General Service Amend: A, 1986; 1, 1987; 1A, 1987. 12 pp.

—Webbing—Sampling
- BSI BS 5053-85. 1985 Cordage and Webbing Slings and for Fibre Cores for Wire Ropes. 22 pp.
- SNZ NZS/BS 5053-85. Methods of Test for Cordage and Webbing Slings and for Fibre Cores for Wire Ropes. 24 pp.

—Wire Coil
- SAA AS 1438-74. Wire-Coil Flat Slings. 14 pp.

—Wire Rope
- BSI BS 1290-83. 1983 Wire Rope Slings and Sling Legs for General Lifting Purposes. 16 pp.
- ISO 7531-87. Wire Rope Slings for General Purposes—Characteristics and Specifications First Edition. 8 pp.

Hoisting Slings (Cont.)

—Wire Rope (Cont.)
- JIS B 8817-91. Wire Rope Slings for Lifting Purposes. 31 pp.
- SAA AS 1666-76. Wire-Rope Slings. 52 pp.
- SNZ NZS/ISO 7531-87. Wire Rope Slings for General Purposes—Characteristics and Specifications. 5 pp.

—Wire Rope—Eye Terminations
- ISO 8794-86. Steel Wire Ropes—Spliced Eye Terminations for Slings First Edition. 7 pp.

—Wire Rope—Safety
- BSI BS 6210-83. 1983 Safe Use of Wire Rope Slings for General Lifting Purposes. 16 pp.
- ISO 8792-86. Wire Ropes Slings—Safety Criteria and Inspection Procedures for Use First Edition. 13 pp.

Hoists

Use For: Chain Hoists; Material Hoists; Wire Rope Hoists. *See Also:* Booms (Equipment); Construction Equipment; Cranes; Derricks; Drilling Rigs; Elevators (Lifts); Escalators; Lifting Equipment; Mast Hoists; Materials Handling Equipment; Mine Hoists; Personnel Hoists; Platforms; Pulleys; Winches

- BSI BS 4898-73. 1973 Chain Lever Hoists. 12 pp.
- CNS C3057-84. Method of Test for Electric Hoists (Sep)(4617).
- CNS C4140-84. Electric Hoists (Sep)(4616).
- JIS B 8802-89. Manually Operated Chain Hoists. 22 pp.
- JIS B 8810-89. Testing Methods for Electric Chain Hoists. 12 pp.
- JIS B 8815-87. Electric Chain Hoists; (Erratum). 34 pp.
- JIS B 8819-90. Manually Operated Chain Lever Hoists. 20 pp.
- JIS C 9620-88. Electric Wire Rope Hoists. 22 pp.
- SAA AS 1418.2-90. Cranes (Including Hoists and Winches) (Known as the SAA Crane Code)—Part 2: Serial Hoists and Winches. 18 pp.
- SAA AS 1418.8-89. Cranes (Including Hoists and Winches) (Known as the SAA Crane Code)—Part 8: Special Purpose Appliances. 17 pp.
- SAA AS 1418.12-91. Cranes (Including Hoists and Winches) (Known as the SAA Crane Code)—Part 12: Crane Collector Systems. 27 pp.

—Agricultural Equipment
- ISO 5699-79. Agricultural Machines, Implements and Equipment—Dimensions for Mechanical Loading with Bulk Goods First Edition. 3 pp.

—Construction Equipment
- SAA AS 1418.7-88. Cranes (Including Hoists and Winches) (Known as the SAA Crane Code)—Part 7: Builders' Hoists and Equipment Amdt 1 September 1989. 24 pp.

—Construction Equipment—Safety
- BSI BS 7212-89. 1989 Code of Practice for Safe Use of Construction Hoists. 27 pp.

—Design
- BSI BS 4465-89. 1989 Design and Construction of Electric Hoists for Both Passengers and Materials. 57 pp.

—Electric
- CSA C22.2 NO 33-M1984. Construction and Test of Electric Cranes and Hoists (R 1992); (Gen Instr 1 Thru 2). 30 pp.
- CSA 1169 Bull. Electrical Bulletin 1169 June 27, 1978 to C22.2 NO 33. 2 pp.

—Electric—Construction
- CSA C22.2 NO 33-M1984. Construction and Test of Electric Cranes and Hoists (R 1992); (Gen Instr 1 Thru 2). 30 pp.
- CSA 1169 Bull. Electrical Bulletin 1169 June 27, 1978 to C22.2 NO 33. 2 pp.

—Electric—Sheave Blocks
- DIN ENGL 15410-82. Serial Lifting Appliances; Single and Twin Sheave Bottom Blocks for Electric Hoists; Assembly (July). 4 pp.

—Electric—Slip Friction Clutches
- CSA C22.2 NO 33-M1984. Construction and Test of Electric Cranes and Hoists (R 1992); (Gen Instr 1 Thru 2). 30 pp.
- CSA 891 Bull. Electrical Bulletin 891 August 24, 1972 to C22.2 NO 33. 1 pp.
- CSA 1169 Bull. Electrical Bulletin 1169 June 27, 1978 to C22.2 NO 33. 2 pp.

—Electrical Codes
- BSI BS 4465-89. 1989 Design and Construction of Electric Hoists for Both Passengers and Materials. 57 pp.

INDUSTRY STANDARDS

INTERNATIONAL AND NON-U.S. NATIONAL STANDARDS
SUBJECT INDEX

Hoists

Hoists (Cont.)

—Electrical Installations
DIN VDE 0100 Pt 726-90. Erection of Power Installations with Nominal Voltages up to 1000 V; Lifting and Hoisting Devices (Mar). 35 pp.

—Glossaries
JIS B 0148-84. Glossary of Terms for Hoists. 47 pp.

—Induction Motors
CNS C1037-70. Standard Dimensions of Electric Motor for Hoist (Dec)(3202). 4 pp.

—Link Chains
BSI BS 4942: Part 1-81. 1981 Short Link Chain for Lifting Purposes Part 1: General Conditions of Acceptance. 11 pp.
BSI BS 4942: Part 3-81. 1981 Short Link Chain for Lifting Purposes Part 3: Grade M (4) Calibrated Chain. 12 pp.
BSI BS 4942: Part 6-81. 1981 Amd 1 Short Link Chain for Lifting Purposes Part 6: Grade T (8) Calibrated Chain (AMD 4953) March 31, 1986. 13 pp.
CNS B2713-82. Round-Steel Link Chains for General Purposes and for Haisting Equipments, Chain in Fixed Length, Short Linked (Nov)(9557).
CNS B2714-82. Round-Steel Link Chains Haisting Equipments Calibrated, Tested, Short Linked (Nov)(9558).
ISO 1836-80. Short Link Chain for Lifting Purposes—Grade M (4), Calibrated, for Chain Hoists and Other Lifting Appliances First Edition. 11 pp.
ISO 3077-84. Short Link Chain for Lifting Purposes—Grade T (8), Calibrated, for Chain Hoists and Other Lifting Appliances Second Edition. 11 pp.
JIS B 8812-89. Link Chains for Manually Operated Chain Hoists. 17 pp.
SNZ NZS/ISO 1836-80. Short Link Chain for Lifting Purposes—Grade M (4), Calibrated, for Chain Hoists and Other Lifting Appliances. 9 pp.
SNZ NZS/ISO 3077-84. Short Link Chain for Lifting Purposes-Grade T(8), Calibrated, for Chain Hoists and Other Lifting Appliances. 9 pp.

—Link Chains—Ships
MOD UK NES 113: Part 7-89. Requirements for Mechanical Handling Part 7: Lever Hoists Issue 2 (11.89). 13 pp.

—Link Chains—Submarines
MOD UK NES 113: Part 7-89. Requirements for Mechanical Handling Part 7: Lever Hoists Issue 2 (11.89). 13 pp.

—Mobile Cranes
ISO 8087-85. Mobile Cranes—Drum and Sheave Sizes First Edition. 3 pp.

—Rotary Wing Aircraft
CAA 839 (G). Section G Rotorcraft Helicopter Hoists. 3 pp.

—Safety
CEN PREN 109-86. Safety Rules for the Construction and Installation of Building Hoists for Passengers and Materials (Category 1). 45 pp.
CSA CAN/CSA-Z256-M87. Safety Code for Material Hoists; (Gen Instr 1). 44 pp.

—Vehicles
SAA AS 1418.9-87. Cranes (Including Hoists and Winches) (Known as the SAA Crane Code)—Part 9: Vehicle Hoists. 22 pp.

—Wire Rope
DIN ENGL 21251 Pt 1-84. Ropes for Hoisting Purposes; 6 x 19 Standard Round Strand Ropes (June). 3 pp.
DIN ENGL 21251 Pt 2-84. Ropes for Hoisting Purposes; 6 x 33 Warrington Compound Round Strand Ropes (June). 3 pp.
DIN ENGL 21252-84. Ropes for Hoisting Purposes; Flat Ropes (June). 3 pp.

—Wire Rope—Ships
MOD UK NES 113: Part 7-89. Requirements for Mechanical Handling Part 7: Lever Hoists Issue 2 (11.89). 13 pp.

—Wire Rope—Submarines
MOD UK NES 113: Part 7-89. Requirements for Mechanical Handling Part 7: Lever Hoists Issue 2 (11.89). 13 pp.

Hold (Telephone Services)
See Also: Holding Time; Telephone Services
CEPT T/CS 20-19-82. Mise En Garde. 2 pp.
CEPT T/CS 20-19 E-82. Hold. 2 pp.

Holders
Scope Note: For additional listings, use a more specific term *Use For:* Bridge Pieces; Holding Devices; Work Holding Devices

Holders (Cont.)
See Also: Anchors (Fasteners); Bands; Bolts; Brackets; Chucks; Clamps; Clevis; Clips; Cradles; Electrode Holders; Electron Tube Holders; Fasteners; Jigs (Positioners); Latches (Fasteners); Lugs (Fasteners); Nails (Fasteners); Nuts (Fasteners); Pins; Quick Release Fasteners; Retaining Links; Retaining Rings; Rivets; Screws; Spikes (Fasteners); Splines; Spools; Staples; Straps (Fasteners); Studs (Fasteners); Tool Holders

BSI BS 5043-01. (OBSOLESCENT) 1973 Amd 1 Book Holders, and Prismatic Spectacles for Use as Reading Aids in Hospitals and the Home (AMD 6895) February 28, 1992. 14 pp.
CNS C4217-84. Catch-Holder (May)(6050).

—Aircraft—Cables (Electric)
SBAC AS 2507 ISSUE 1. Bridge Piece. 1 p.

—Aircraft—Fasteners
SBAC AS 12980-984 ISSUE 3. Retainer, Nut and Bolt, with Rivet Holes.
SBAC AS 12985-989 ISSUE 1. Retainer, Nut and Bolt, Without Rivet Holes.
SBAC AS 26070-074 ISSUE 2. Retainer, Nut and Bolt, with Rivet Holes.

—Aircraft—Maps
SBAC AS 432 ISSUE 3(I). Stowage—Map. 1 p.

—Aircraft—Maps—Springs
SBAC AS 431 ISSUE 2(I). Spring:—Stowage—Map. 1 p.

—Aircraft—Parachutes—Assembly
SBAC AS 423 ISSUE 4(I). Assembly of Stowage for Parachute Pack. 1 p.

—Aircraft—Seals
SBAC AS 2956 ISSUE 4. Seal Retainer. 1 p.
SBAC AS 2981 ISSUE 3. Seal Retainer. 1 p.

—Aircraft—Self Locking Nuts
SBAC RS 698 ISSUE 1. Application of Self Locking Nut Retainers.

—Aircraft—Signal Lights—Storage
SBAC AS 3166 ISSUE 1. Elastic, Retainer. 1 p.

—Automotive—Nozzles—Injection
ISO 3539-75. Road Vehicles—Injection Nozzle Holder with Body, Types 8 and 10, and Injection Nozzle Holder with Fixing Flats, Types 9 and 11 First Edition. 4 pp.

—Cable—Ships
MOD UK NES 113: Part 3-89. Requirements for Mechanical Handling Part 3: Anchor Capstans, Capstans and Cable Holders Issue 2 (11.89). 14 pp.

—Capacitors
MOD UK DSTAN 59-67-01. Retainers, Capacitor Issue 1; Corrigenda. 23 pp.

—Card Labels
MOD UK DSTAN 99-2-81. Holders, Card Label, Tinplated Issue 2. 6 pp.

—Marine—Thermometers
JIS F 7210-89. Marine Thermometer Pockets.

—Semiconductor Devices
MOD UK DEF-5254-66. General Requirements for Retainers, Mounting Pads and Heat Dissipators for Semi-Conductor Devices. 10 pp.
MOD UK DEF-5254-1-67. Retainers, Semiconductor Device, Insulated, Pattern RSC 1. 13 pp.

—Test Specimens—Hardness Testers
DIN ENGL 51200-85. Materials Testing Machines; Design and Application of Test Piece Holding Devices in Hardness Testing Machines (Oct). 6 pp.

—Transmission Tubing
CNS B2590-81. Remote Control Gear with Transmission Tubing for Manual Operation; Holder for Transmission Tubing (Jun)(7587).

Holding Devices
Use: Holders

Holding Fixtures
Use: Mandrels

Holding Force
Use: Contact Force

Holding Time
See Also: Hold (Telephone Services); Telephone Services; Time

Holding Time (Cont.)

—Telecommunication Traffic—Integrated Services Digital Networks
CCITT RECMN E.711-92. User Demand Modelling Study Group II) 11 pp. 11 pp.

Holds (Cargo)
Use: Cargo Holds

Hole Size
See Also: Diameters; Dimensions; Holes; Lightening Holes

AECMA PREN3149-87. Shank Nuts, Metric Installation Holes. 5 pp.
BSI BS 4500: Sec 1.2-90. (WITHDRAWN) 1990 ISO Limits and Fits Part 1: General, Tolerances and Deviations Section 1.2: Tables of Commonly Used Tolerance Grades and Limit Deviations for Holes and Shafts (Renumbered as BS EN 20286-2: 1993). 48 pp.
BSI BS EN 20286-2-93. 1993 Amd 1 ISO System of Limits and Fits Part 2: Tables of Standard Tolerance Grades and Limit Deviations for Holes and Shafts (ISO 286-2: 1988) (AMD 7631) August 15, 1993 (H). 51 pp.
CEN EN 20286-2-93. ISO System of Limits and Fits—Part 2: Tables of Standard Tolerance Grades and Limit Deviations for Holes and Shafts (ISO 286-2: 1988). 46 pp.
CNS B1013-79. 60 Degree Center Holes; Forms R, A, B and C (Jan)(300).
CNS B1126-79. 60 Degree Center Holes for Big Workpieces (Forms A, B and C) (Jan)(4685).
CNS B2384-79. Tapping Screws, Applications and Core Holes (Aug)(4949).
DIN ENGL 7500 Pt 2-84. Thread Rolling Screws for ISO Metric Thread; Guideline Values for Hole Diameters (Dec). 2 pp.
DIN ENGL 7975-89. Tapping Screw Connections; Guideline Values for Core Hole Diameters and Use (Aug). 7 pp.
ISO 286 Pt 2-88. ISO System of Limits and Fits—Part 2: Tables of Standard Tolerance Grades and Limit Deviations for Holes and Shafts First Edition (CEN EN 20286-2: 1993). 47 pp.
ISO 1829-75. Selection of Tolerance Zones for General Purposes First Edition. 4 pp.
ISO R1938-71. ISO System of Limits and Fits Part II: Inspection of Plain Workpieces First Edition. 28 pp.
MOD UK DSTAN 53-63-01. Metric Tapped Holes for Desiccating Unions Issue 1; Amendment 1. 6 pp.
SNZ NZS/ISO 286: Part 2-88. ISO System of Limits and Fits Part 2: Tables of Standard Tolerance Grades and Limit Deviations for Holes and Shafts. 43 pp.
SNZ NZS/ISO R1938-71. ISO System of Limits and Fits. Part 2 Inspection of Plain Workpieces. 28 pp.

—Aerospace—Clip Nuts
CEN PREN 3741-91. Clip Nuts, Metric Installation Holes and Assembly. 6 pp.
CEN PREN 3741-93. Aerospace Series Nuts, Clip, Metric Installation Holes and Assembly. 5 pp.

—Aerospace—Countersunk Head Screws
CEN PREN 3782-92. Holes for 100 Degrees Countersunk Head Screws Design Standard. 5 pp.

—Aerospace—Locating Pins
AECMA PREN3368-90. Aerospace Design Standard Holes for Locating Pins. 8 pp.

—Aerospace—Pipe Fittings
CEN PREN 3633-91. Installation Hole for Fluid Fittings, Flanged. 7 pp.
CEN PREN 3633-92. Installation Hole for Fluid Fittings, Flanged. 5 pp.

—Aerospace Rivets
BSI BS EN 2309-90. 1990 Hole Sizes for Solid Rivets. 6 pp.
CEN EN 2309-89. Hole Sizes for Solid Rivets. 3 pp.

—Aerospace—Screw Plugs
CEN PREN 3707-91. Headless Threaded Plugs Installation Holes. 5 pp.

—Aerospace—Threaded Fasteners
AECMA PREN3201-90. Design Standard Holes for Metric Threaded Fasteners. 9 pp.

—Aircraft—Hexagonal Head Bolts
SBAC RS 747 ISSUE 1. Hole Sizes for Self Retaining "Impedance Type" Bolts. Inch Series.
SBAC RS 748 ISSUE 1. Hole Sizes for Self Retaining "Impedance Type" Bolts. Metric Series.

—Aircraft—Pipe Couplings
SBAC AGS 1219 (V). Dimensions of Tapped Holes in Components for Connecting A.G.S. Pipe Couplings Direct.

INTERNATIONAL AND NON-U.S. NATIONAL STANDARDS
SUBJECT INDEX

Hole Size (Cont.)
—Aircraft—Split Cotter Pins
SBAC RS 722 ISSUE 1. Selection of Cotter Pins —Metric.

—Aircraft—Standard Diameters
SBAC RS 178 ISSUE 1. Standard Diameters.

—Bone Plates
BSI BS 3531: Sec 23.2-93. 1993 Implants for Osteosynthesis Part 23: Bone Plates Section 23.2: Specification for Holes and Slots for Use with Screws of 4.5 mm, 4.2 mm, 4.0 mm, 3.9 mm, 3.5 mm and 2.9 mm Nominal Sizes (ISO 9269: 1988) (SUP'S BS 3531: PART 5: 1982). 12 pp.
BSI BS 3531: Sec 23.2-90. 1990 Surgical Implants Part 23: Bone Plates Section 23.2: Holes and Slots for Use with Screws of 4mm, 3.5mm and 3mm Nominal Sizes. 8 pp.
BSI BS 3531: Sec 23.3-91. 1991 Implants for Osteosynthesis Part 23: Bone Plates Section 23.3: Specification for Holes Corresponding to Screws with Asymmetrical Thread and Spherical Undersurfaces (ISO 5836: 1988). 12 pp.
ISO 5836-88. Implants for Surgery—Metal Bone Plates—Holes Corresponding to Screws with Asymmetrical Thread and Spherical Under-Surface First Edition. 8 pp.
ISO 9269-88. Implants for Surgery—Metal Bone Plates—Holes and Slots Corresponding to Screws with Conical Under-Surface First Edition; (Supersedes 5836 Pt 4). 8 pp.

—Counterbores
BSI BS 4185: Part 13-85. 1985 Machine Tool Components Part 13: Dimensions of Counterbored Holes. 4 pp.

Holes
See Also: Hole Size

—Aircraft Instruments
SBAC AGS 3090 ISSUE 3. 4375-20 UNJF-3B Tapped Hole for Instruments (Pressure).
SBAC AGS 3091 ISSUE 3. 500-20 UNJF-3B Tapped Hole for Instruments (Static).

—Binders (Files)
SAA AS P5-69. Punching Patterns for Round Holes Used in Files and Loose Leaf Binders Reconfirmed 1986. 4 pp.

—Files
SAA AS P5-69. Punching Patterns for Round Holes Used in Files and Loose Leaf Binders Reconfirmed 1986. 4 pp.

Holiday Caravans
Use: Recreational Vehicles

Hollow Bricks
See Also: Bricks

—Clay
CNS R2003-53. Hollow Bricks (Jul)(383) (R 1973). 1 p.
CSA CAN3-A82.8-M78. Hollow Clay Brick (R 1992). 19 pp.

Hollow Walls
Use: Walls

Hollow Ware
See Also: Tableware

—Plastic—Medical Equipment
BSI BS 5452-77. 1977 Hospital Hollow-Ware Made of Plastics Material. 6 pp.

—Silver Coatings—Electroplated
CNS S1185-87. Electroplated Coatings of Silver for Cutlery, Flat-Ware and Hollow-Ware (Oct)(11316).

—Stainless Steel—Medical Equipment
BSI BS 1823-73. 1973 Amd 1 Stainless Steel Hollow-Ware for Use in Hospital Operating-Theatres and Wards (AMD 6006) July 31, 1989. 15 pp.

Holocellulose Content Analysis
See Also: Cellulose Content Analysis

—Unbleached Pulps
CPPA G.9U-77. Determination of Holocellulose. 1 p.

—Wood
CPPA G.9U-77. Determination of Holocellulose. 1 p.

Home Computers
Use: Personal Computers

Home Economics
—Classification
BSI BS 1000: (64)-84. 1984 Universal Decimal Classification (UDC). English Full Edition (64): Home Economics. 50 pp.
SNZ NZS/BS 1000 (64)-84. Universal Decimal Classification Home Economics. 52 pp.

Home Electronic Systems
Use: Building Automation Systems—Residential

Homelifts
Use: Wheelchair Lifts

Hometaxial Transistors
See Also: Transistors

—Silicon—NPN—LF
CECC CECC 50 003-022. BS CECC 50 003-022 Issue 1; Case Rated Low Frequency Bipolar Transistor (En). 8 pp.

Homing Systems
See Also: Direction Finders; Navigational Aids

—Distress Frequency—Ships
CCIR RECMN 428-3-90. Direction-Finding and/or Homing in the 2 MHz Band on Board Ships—Section 8C—Maritime Mobile Service; Telephony and Related Subjects. 3 pp.

—Global Maritime Distress and Safety Systems
CCIR Report 744-2-86. Use of Class J3E Emissions for Distress and Safety Purposes—Section 8C—Maritime Mobile Service; Telephony and Related Subjects. 7 pp.

—Monitoring—Radio Spectra
CCIR Report 834-82. Recent Developments in Low-Cost Mobile Homing Systems—Section 1C—Spectrum Monitoring Techniques. 5 pp.

—SHF—Global Maritime Distress and Safety Systems
CCIR Report 1036-1-90. Frequencies for Homing and Locating in the Global Maritime Distress and Safety System (GMDSS)—Section 8D—Radiodetermination, Global Maritime Distress and Safety System and Related Subjects. 23 pp.

—UHF—Global Maritime Distress and Safety Systems
CCIR Report 1036-1-90. Frequencies for Homing and Locating in the Global Maritime Distress and Safety System (GMDSS)—Section 8D—Radiodetermination, Global Maritime Distress and Safety System and Related Subjects. 23 pp.

—VHF—Global Maritime Distress and Safety Systems
CCIR Report 1036-1-90. Frequencies for Homing and Locating in the Global Maritime Distress and Safety System (GMDSS)—Section 8D—Radiodetermination, Global Maritime Distress and Safety System and Related Subjects. 23 pp.

Homopolymers
Use: Polymers

Homosulfamine Content Analysis
Use: Mafenide Content Analysis

Hones
See Also: Honing Machinery; Oilstones
ISO 3920-76. Honing Stones of Square Section—Designation and Dimensions First Edition. 3 pp.
ISO 3921-76. Honing Stones of Rectangular Section—Designation and Dimensions First Edition. 3 pp.
SAA AS B209-69. Diamond Abrasive Wheels and Hones. 39 pp.

Honey
See Also: Food
EC 74/409/EEC-74. Council Directive on the Harmonization of the Laws of the Member States Relating to Honey. 5 pp.

—Acidity
CNS N6027-74. Method of Test for Honey (Oct)(1344). 8 pp.

—Ash Content
CNS N6027-74. Method of Test for Honey (Oct)(1344). 8 pp.

—Glucose Content
CNS N6027-74. Method of Test for Honey (Oct)(1344). 8 pp.

—Grading
CNS N5024-84. Honey (Apr)(1305). 2 pp.

Honey (Cont.)
—Impurities Content
CNS N6027-74. Method of Test for Honey (Oct)(1344). 8 pp.

—Inspection
CNS N6027-74. Method of Test for Honey (Oct)(1344). 8 pp.

—Jars
BSI BS 1777-81. 1981 Glass Honey Jars. 4 pp.

—Moisture Content
CNS N6027-74. Method of Test for Honey (Oct)(1344). 8 pp.

—Sampling
CNS N6027-74. Method of Test for Honey (Oct)(1344). 8 pp.

—Sensory Analysis
CNS N6027-74. Method of Test for Honey (Oct)(1344). 8 pp.

—Sucrose Content
CNS N6027-74. Method of Test for Honey (Oct)(1344). 8 pp.

Honeycomb Cores
See Also: Structures

—Paper—Aerial Delivery
NATO STANAG 3778 ED 1 AMD 3-80. Performance Criteria for Honeycomb Paper Used as Energy Dissipating Material. 6 pp.
NATO STANAG 3778 ED 1 AMD 4-80. Performance Criteria for Honeycomb Paper Used as Energy Dissipating Material. 6 pp.

Honeycomb Structures
Scope Note: Use a more specific term
See: Honeycomb Cores; Insulation

Honing Machinery
See Also: Finishing Machinery; Grinders; Hones; Lapping Machines; Machine Tools

—Vertical—Acceptance Testing
DIN ENGL 8635-71. Acceptance Conditions for Machine Tools; Vertical Honing Machines up to 500 mm Length of Stroke (Jan). 3 pp.

Honing Stones
Use: Hones

Hoods (Headgear)
See Also: Exposure Suits; Helmets; Protective Clothing; Respirators

—Dust
BSI BS 4771-71. 1971 Positive Pressure, Powered Dust Hoods and Blouses. 10 pp.
BSI BS DD 97: Part 13-87. (WITHDRAWN) 1987 Respiratory Protective Equipment Part 13: Powered Particle Filtering Devices Incorporating Helmets and Hoods (Superseded by BS EN 146: 1992). 18 pp.

—Fire Resistant—Mining Equipment
DIN ENGL 23320 Pt 5-88. Flameproof Clothing for the Mining Industry; Protective Hoods (May). 3 pp.

Hoof-and-Mouth Disease
Use: Foot and Mouth Disease

Hook and Loop Fastener Tapes
Use: Velcro

Hook and Loop Fasteners
Use: Velcro

Hook and Loop Touch Fasteners
Use: Velcro

Hook Spanners
See Also: Spanners; Wrenches
BSI BS 2090-54. 1954 Hook Spanners, Peg Spanners, Coupling Wrenches and the Related Slots, Holes and Horns. 27 pp.

—Round Nuts
DIN ENGL 1804-71. Slotted Round Nut for Hook Spanner; ISO Metric Fine Thread (Mar). 2 pp.

—Round Nuts—Tab Washers
CNS B2180-78. Machine Tools Internal Tab Washers for Slotted Round Nuts for Hook Spanner (Mar)(4254).
DIN ENGL 462-73. Machine Tools; Internal Tab Washers for Slotted Round Nuts for Hook Spanner According to DIN 1804 (Sept). 2 pp.

Hook Sticks
Use: Linemen's Sticks

INDUSTRY STANDARDS

Hook Wrenches
See Also: Wrenches
CNS B3324-80. Assembly Tools for Screws and Nuts and Pin Wrench (Jan)(5103).

Hooks
Use For: Hanging Hooks See Also: Eyebolts; Fasteners; Hangers (Fasteners); Lifting Hooks; Pelican Hooks; Ramshorn Hooks; Spring Hooks; Swivel Hooks; Swivels; Towing Hooks
CNS A2009-89. Hooks (Jul)(859). 2 pp.
CNS A2025-89. Hanging Hooks (Plain) (Jun)(875). 2 pp.
CNS A2026-89. Hanging Hooks (with Flanges) (Jun)(876). 2 pp.
CNS B2441-80. Hooks (Apr)(5394).
CNS Z7068-80. 25mm Copper Hook (Mar)(5384).
JIS B 2803-87. Hooks. 19 pp.

—Cargo—Ships
CNS F3140-81. Ships' Cargo Hooks (Oct)(7985).
DIN ENGL 82017-71. Cargo Hooks (Nov). 4 pp.
JIS F 2105-90. Ships' Cargo Hooks.

—Chimneys
BSI BS 3678-86. 1986 Access Hooks for Chimneys and Other High Structures. 8 pp.

—Containers
CNS S1117-81. Vessel-Hooks (Mar)(7078). 2 pp.

—Draperies
CNS S1113-81. Curtain Hooks (Mar)(7074). 1 p.

—Hand
SNZ NZS 2251-69. Specification for Industrial Hand Hooks (Reconfirmed 1983). 17 pp.

—Keys
CNS S1116-81. Key-Hooks (Mar)(7077). 1 p.

—Laboratory Equipment—Connectors
CNS Z7103-81. Metal Laboratory Ware; Hook Connector (Nov)(8180).

—Medical
CNS T1027-80. Medical Hooks (Aug)(6286).
JIS T 2603-53. Medical Hooks.

—Plenum Cables
CSA CAN/CSA-C22. 2 NO 18-92. Outlet Boxes, Conduit Boxes, and Fittings; (Gen Instr 1 Thru 2). 118 pp.

—Sewing Machines—Industrial
CNS B2452-89. Rotating Hooks on Horizontal Axis of Sewing Machines for Industrial Use (Jan)(5409).
JIS B 9072-72. Rotating Hooks on Horizontal Axis of Sewing Machines for Industrial Use.

—Structures
BSI BS 3678-86. 1986 Access Hooks for Chimneys and Other High Structures. 8 pp.

—Towels
CNS S1114-81. Towel Hooks (Mar)(7075). 1 p.

Hookstick Switches
Use: Disconnecting Switches

Hookup Wire
See Also: Copper Wire; Electric Wire

—Appliances
CSA CAN/CSA-C22. 2NO210.2-M90. Appliance Wiring Material Products (R 1992); (Gen Instr 1 Thru 4). 73 pp.

—Appliances—Multiconductor
CSA CAN/CSA-C22. 2NO210.2-M90. Appliance Wiring Material Products (R 1992); (Gen Instr 1 Thru 4). 73 pp.

—Communication Equipment
DIN VDE 0881-86. Jumper Wires and Stranded Hook-Up Wires with Extended Temperature Range for Telecommunications Systems and Data Processing Systems (Mar). 34 pp.

—Data Processing Equipment
DIN VDE 0881-86. Jumper Wires and Stranded Hook-Up Wires with Extended Temperature Range for Telecommunications Systems and Data Processing Systems (Mar). 34 pp.
DIN VDE 0891 Pt 2-90. Use of Cables and Insulated Cords for Telecommunications and Information Processing Systems; Special Provisions for Hook-Up Wires with Solid and Stranded Conductors in Accordance with DIN VDE 0812 (May). 6 pp.

—Dielectric Strength
DIN VDE 0472 Pt 509-86. Testing of Cables, Wires and Flexible Cords; Dielectric Strength on Cables, Wires and Cords for Telecommunications and Information Processing Systems (Oct). 11 pp.

Hookup Wire (Cont.)
—Telecommunication Systems
DIN VDE 0881-86. Jumper Wires and Stranded Hook-Up Wires with Extended Temperature Range for Telecommunications Systems and Data Processing Systems (Mar). 34 pp.
DIN VDE 0891 Pt 2-90. Use of Cables and Insulated Cords for Telecommunications and Information Processing Systems; Special Provisions for Hook-Up Wires with Solid and Stranded Conductors in Accordance with DIN VDE 0812 (May). 6 pp.

Hopper Cars
See Also: Freight Cars; Ground Vehicles; Rolling Stock

—Hazardous Materials
CGSB 43-GP-147P-90. Construction of Tank Car Tanks and Selection and Use of Tank Car Tanks, Portable Tanks and Rail Cars for the Transportation of Dangerous Goods by Rail. 241 pp.

Hoppers
See Also: Bulk Storage

—Agricultural
ISO 5699-79. Agricultural Machines, Implements and Equipment—Dimensions for Mechanical Loading with Bulk Goods First Edition. 3 pp.

—Agricultural—Aircraft
CAA Chapter K4-9 04.72. Agricultural Equipment (Light Aeroplanes). 3 pp.

—Agricultural—Capacity Measurement
BSI BS 6621-85. 1985 Determination of Combine Harvester Capacity and Unloading Rate. 4 pp.
ISO 5687-81. Equipment for Harvesting—Combine Harvester—Determination and Designation of Grain Tank Capacity and Unloading Device Performance First Edition. 3 pp.

—Agricultural—Loading Heights
ISO 5698-79. Agricultural Machinery—Hoppers—Manual Loading Height First Edition. 3 pp.

—Agricultural—Rotary Wing Aircraft
CAA Chapter G4-11 01.75. Agricultural Equipment (Rotorcraft). 3 pp.

—Agricultural—Unloading Rate
BSI BS 6621-85. 1985 Determination of Combine Harvester Capacity and Unloading Rate. 4 pp.
ISO 5687-81. Equipment for Harvesting—Combine Harvester—Determination and Designation of Grain Tank Capacity and Unloading Device Performance First Edition. 3 pp.

—Marine—Steel Plate
JIS F 7219-78. Marine Steel Plate Hoppers.

Horizontal Bars (Gymnastic Equipment)
See Also: Parallel Bars (Gymnastic Equipment); Uneven Bars (Gymnastic Equipment)
CNS S1190-85. Horizontal Bars (Dec)(11464).
CNS S2123-85. Method of Test for Horizontal Bars (Dec)(11465).
ISO 379-80. Gymnastic Equipment—Horizontal Bar First Edition. 4 pp.
JIS S 7011-83. Horizontal Bars.

Horizontal Bars (Playground Equipment)
See Also: Playground Equipment
JIS S 7203-76. Horizontal Bars Use Only for Children.

Horizontal Boring Machines
Use: Boring Machines—Horizontal

Horizontal Impact Testing
Use: Impact Testing

Horizontal Indicators
See Also: Aircraft Instruments; Attitude Indicators
MOD UK DSTAN 66-40-86. General Requirements for Aircraft Instruments and Displays Horizontal Situation Indicators (HSI) Issue 1. 12 pp.
NATO STANAG 3741 ED 4 AMD 3-84. Horizontal Situation Indicators (HSI). 10 pp.
NATO STANAG 3741 ED 4 AMD 4-84. Horizontal Situation Indicators (HSI). 11 pp.

Horizontal Lathes
See Also: Lathes

—Acceptance Testing
DIN ENGL 8611 Pt 1-76. Machine Tools; Horizontal Single-Spindle Automatic Lathes; Acceptance Conditions (Jan). 6 pp.

Horizontal Lathes (Cont.)
—Acceptance Testing (Cont.)
DIN ENGL 8611 Pt 2-76. Machine Tools; Horizontal Automatic Lathes; Front-Loaded; Acceptance Conditions (Jan). 7 pp.
DIN ENGL 8611 Pt 3-76. Machine Tools; Horizontal Automatic Lathes; Swiss-Type Automatic Lathes; Acceptance Conditions (Jan). 5 pp.
DIN ENGL 8613-70. Acceptance Conditions for Machine Tools; Multi-Spindle Automatic Lathes; Horizontal Type for Rotating Workpieces (Mar). 6 pp.

—Acceptance Testing—Forms (Paper)
DIN ENGL 8613 Suppl. 1-70. Acceptance Conditions for Machine Tools; Multi-Spindle Automatic Lathes; Horizontal Type for Rotating Workpieces; Form for Test Results (Mar). 1 p.

Hormones
See: Drugs; Growth Hormones; Medicinal Products

Hormones, Growth
Use: Growth Hormones

Horns
See Also: Signal Devices; Sirens; Speakers; Warning Systems
CSA C22.2 NO 205-M1983. Signal Equipment (R 1992); (Gen Instr 1). 32 pp.

—Automotive
CNS D2060-86. Electric Horns for Automobiles (Jun)(6813). 8 pp.
CNS D2114-86. Horn, Air-Actuated Electrically Controlled (Jun)(7978).
CNS D3056-86. Method of Test for Electric Horns for Automobiles (Jun)(5776).
JIS D 5701-82. Electric Horns for Automobiles.
JIS D 5701-74. Electric Horns for Automobiles. 9 pp.

—Automotive—Relays
CNS D2031-86. Horn Relays for Automobiles (Jun)(5787). 5 pp.
JIS D 5706-74. Horn Relays for Automobiles. 9 pp.

—Marine
CNS F5080-87. Marine Electronic Horns (Mar)(10627).
JIS F 8504-84. Marine Electronic Horns.

—Motorcycles—Acoustic Testing
CNS D3036-88. Method of Sound Level Test of the Horn for Motorcycles (May)(3113). 2 pp.

Horology
Use: Time Measuring Instruments

Horse Mackerel
Use: Mackerel

Horseradish
See Also: Condiments; Food; Vegetables

—Storage
ISO 4187-80. Horse-Radish—Guide to Storage First Edition. 4 pp.

Horses
—Air Cargo Handling Equipment—Unit Loads
BSI M 68-92. 1992 Air Cargo Unit Load Devices for Transportation of Horses (ISO 9469: 1991). 10 pp.
ISO 9469-91. Air Cargo Equipment—Unit Load Devices for Transportation of Horses First Edition. 9 pp.

—Glossaries
ISO 3975-77. Definitions of Living Animals for Slaughter—Horses First Edition. 8 pp.

—Hair
CNS N1073-85. Horse Hair, Not Carded or Combed (May)(2818).
CNS N1076-85. Horse Bristle and Tail Hair (May)(2821).

Hose Adapters
Use For: Hydraulic Hose Adapters See Also: Hose Fittings
JIS B 8363-88. End Fittings and Adapters for Hydraulic Hose Assemblies. 25 pp.

—Aircraft
SBAC AS 6374 ISSUE 2. Adaptor—Hose.

Hose Assemblies
Use For: Flexible Pipe Assemblies; Hydraulic Hose Assemblies See Also: Fire Hose Assemblies; Fuel Hose Assemblies; Hose Fittings; Hoses; Petroleum Hose Assemblies; Tube Assemblies; Water Hose Assemblies

INTERNATIONAL AND NON-U.S. NATIONAL STANDARDS
SUBJECT INDEX
Hose

Hose Assemblies *(Cont.)*
ISO 6605-86. Hydraulic Fluid Power—Hose Assemblies—Method of Test First Edition. 8 pp.

JIS B 8363-88. End Fittings and Adapters for Hydraulic Hose Assemblies. 25 pp.

—Aerospace—Plastic

CEN PREN 3571-91. PTFE Flexible Hose Assembly with Convoluted Inner Tube of a Nominal Pressure up to 6800 kPa and 8 Degrees 30' Fitting in Corrosion-Resistant Steel Product Standard. 7 pp.

CEN PREN3572-89. PTFE Flexible Hose Assembly with Convoluted Inner Tube and 8 Degrees 30 Fitting in Titanium Product Standard. 6 pp.

CEN PREN 3572-91. PTFE Flexible Hose Assembly with Convoluted Inner Tube of a Nominal Pressure up to 6800 kPa and 8 Degree 30' Fitting in Titanium Product Standard. 7 pp.

CEN PREN 3573-91. PTFE Flexible Hose Assembly with Convoluted Inner Tube of a Nominal Pressure up to 6800 kPa and 24 Degrees Fitting in Corrosion-Resistant Steel Product Standard. 7 pp.

CEN PREN 3574-91. PTFE Flexible Hose Assembly with Convoluted Inner Tube of a Nominal Pressure up to 6800 kPa and 24 Degrees Fitting in Titanium Product Standard. 7 pp.

CEN PREN 3575-91. PTFE Flexible Hose Assembly of a Nominal Pressure Equal to 10500 kPa and 8 Degrees 30' Fitting in Corrosion-Resistant Steel Product Standard. 7 pp.

CEN PREN3576-89. PTFE Flexible Hose Assembly of a Nominal Pressure Equal to 10500 KPa with 8 Degrees 30 Fitting in Titanium Product Standard. 6 pp.

CEN PREN 3576-91. PTFE Flexible Hose Assembly of a Nominal Pressure Equal to 10500 kPa with 8 Degree 30' Fitting in Titanium Product Standard. 7 pp.

CEN PREN 3577-91. PTFE Flexible Hose Assembly of a Nominal Pressure Equal to 10500 kPa with 24 Degrees Fitting in Corrosion Resistant Steel Product Standard. 7 pp.

CEN PREN 3578-91. PTFE Flexible Hose Assembly of a Nominal Pressure Equal to 10500 kPa with 24 Degrees Fitting in Titanium Product Standard. 7 pp.

CEN PREN 3579-91. Standard-Weight PTFE Flexible Hose Assembly of a Nominal Pressure Equal to 21000 kPa with 8 Degrees 30' Fitting in Corrosion-Resistant Steel Product Standard. 7 pp.

CEN PREN3580-89. Aerospace Series PTFE Flexible Hose Assembly of a Nominal Presser Equal to 21000 kPa with 8 Degree 30 Fitting in Titanium Product Standard. 7 pp.

CEN PREN 3580-91. Standard Weight PTFE Flexible Hose Assembly of a Nominal Pressure Equal to 21000 kPa with 8 Degree 30' Fitting in Titanium Product Standard. 7 pp.

CEN PREN 3581-91. Standard-Weight PTFE Flexible Hose Assembly of a Nominal Pressure Equal to 21000 kPa with 24 Degrees Fitting in Corrosion Resistant Steel Product Standard. 7 pp.

CEN PREN 3582-91. Standard-Weight PTFE Flexible Hose Assembly of a Nominal Pressure Equal to 21000 kPa with 24 Degrees Fitting in Titanium Product Standard. 7 pp.

CEN PREN 3583-91. Lightweight PTFE Flexible Hose Assembly of a Nominal Pressure Equal to 21000 kPa and 8 Degrees 30' Fitting in Corrosion Resistant Steel Product Standard. 7 pp.

CEN PREN3584-89. Light Weight PTFE Flexible Hose Assembly of Nominal Pressure Equal to 21000 KPa with 8 Degrees 30 Fitting in Titanium Product Standard. 6 pp.

CEN PREN 3584-91. Lightweight PTFE Flexible Hose Assembly of a Nominal Pressure Equal to 21000 kPa with 8 Degree 30' Fitting in Titanium Product Standard. 7 pp.

CEN PREN 3585-91. Lightweight PTFE Flexible Hose Assembly of a Nominal Pressure Equal to 21000 kPa with 24 Degrees Fitting in Corrosion-Resistant Steel Product Standard. 7 pp.

CEN PREN 3586-91. Lightweight PTFE Flexible Hose Assembly of a Nominal Pressure Equal to 21000 kPa with 24 Degrees Fitting in Titanium Product Standard. 7 pp.

CEN PREN 3587-91. PTFE Flexible Hose Assembly of a Nominal Pressure Equal to 28000 kPa with 8 Degrees 30' Fitting in Corrosion-Resistant Steel Product Standard. 7 pp.

CEN PREN3588-89. PTFE Flexible Hose Assembly of Nominal Pressure Equal to 21000 kPa with 8 Degrees 30 Fitting in Titanium Product Standard. 6 pp.

CEN PREN 3588-91. Lightweight PTFE Flexible Hose Assembly of a Nominal Pressure Equal to 28000 kPa with 8 Degree 30' Fitting in Titanium Product Standard. 7 pp.

CEN PREN 3589-91. PTFE Flexible Hose Assembly of a Nominal Pressure Equal to 28000 kPa with 24 Degrees Fitting in Corrosion-Resistant Steel Product Standard. 7 pp.

CEN PREN 3590-91. PTFE Flexible Hose Assembly of a Nominal Pressure Equal to 28000 kPa with 24 Degrees Fitting in Titanium Product Standard. 7 pp.

Hose Assemblies *(Cont.)*
—Aerospace—Pressure Measurement

BSI 2M 52-89. 1989 Impulse Testing of Hydraulic Hose, Tubing and Fitting Assemblies for Aerospace Fluid Systems. 6 pp.

ISO 6772-88. Aerospace—Fluid Systems—Impulse Testing of Hydraulic Hose, Tubing and Fitting Assemblies Second Edition. 6 pp.

—Aircraft

MOD UK DSTAN 47-21-79. Flexible Hose Assemblies for Pitot and Static Systems in Aircraft (Metric) Issue 1 (Withdrawn). 4 pp.

—Aircraft—Elastomeric

AECMA PREN2253-84. Flexible Hose Assemblies in Elastomers—Type 1. 21 pp.

—Aircraft—Nominal Pressures—Nonmetallic

AECMA PREN2245-84. Pipelines for Liquids and Gases for Aircraft—Definitions (C4/01). 4 pp.

—Aircraft—Nominal Sizes—Nonmetallic

AECMA PREN2245-84. Pipelines for Liquids and Gases for Aircraft—Definitions (C4/01). 4 pp.

—Aircraft—Plastic

BSI M 57-85. 1985 High Temperature Convoluted Hose Assemblies in Polytetrafluorethylene (PTFE) for Aerospace Applications. 10 pp.

ISO 7313-84. Aircraft—High Temperature Convoluted Hose Assemblies in Polytetrafluoroethylene (PTFE) First Edition. 10 pp.

ISO 8829-90. Aerospace—Polytetrafluoroethylene (PTFE) Hose Assemblies—Test Methods First Edition. 15 pp.

MOD UK DSTAN 47-12-87. Polytetrafluoroethylene (PTFE) Hose Assemblies (Wire Reinforced) for Medium and High Pressure Fluid Systems in Aircraft Issue 4. 58 pp.

—Aircraft—Plastic—Procurement

ISO 8913-89. Aerospace—Lightweight Polytetrafluoroethylene (PTFE) Hose Assemblies, Classification 204 Degrees Celsius/21 000 kPa—Procurement Specification First Edition; (Corrigendum 1-1990). 17 pp.

ISO 9528-89. Aerospace—Standard-Weight Polytetrafluoroethylene (PTFE) Hose Assemblies, Classification 204 Degrees Celsius /21 000 kPa—Procurement Specification First Edition; (Corrigendum 1-1990). 17 pp.

ISO 9938-90. Aerospace—Polytetrafluoroethylene (PTFE) Hose Assemblies, Classification 204 Degrees Celsius/28 000 kPa—Procurement Specification First Edition. 16 pp.

ISO 10502-92. Aerospace—Hose Assemblies in Polytetrafluoroethylene (PTFE) for Use up to 232 Degrees C and 10 500 kPa—Procurement Specification First Edition. 17 pp.

—Aircraft—Rubber

CAA LEAFLET 5-5 07.90. Hose and Hose Assemblies. 17 pp.

JIS W 1822-79. Hose Assemblies, Rubber, Hydraulic, High Pressure 210 kgf/cm2 (20.6 MPa) (3000 psi).

—Aircraft—Synthetic

CAA LEAFLET 5-5 07.90. Hose and Hose Assemblies. 17 pp.

—Automotive—Brakes

MOD UK DSTAN 25-1-93. Vehicle Braking Systems, Hose Assemblies and Air Brake Coupling Adaptors, Brake, Intervehicular Issue 2. 10 pp.

SNZ NZS/SAEJ1401-85. Road Vehicle-Hydraulic Brake Hose Assemblies for Use with Non-Petroleum Base Hydraulic Fluid. 6 pp.

—Automotive—Brakes—Rubber

ISO 3996-86. Road Vehicles—Brake Hose Assemblies for Hydraulic Braking Systems Used with a Non-Petroleum-Base Hydraulic Fluid Second Edition. 13 pp.

ISO 6120-86. Road Vechicles—Brake Hose Assemblies for Hydraulic Braking Systems Used with a Petroleum-Base Hydraulic Fluid First Edition. 13 pp.

—Flexible—Metal

BSI BS 6130-81. 1981 Amd 1 Hose and Hose Assemblies for Asphalt and Bitumen. 5 pp.

BSI BS 6501: Part 1-91. 1991 Flexible Metallic Hose Assemblies Part 1: Specification for Corrugated Hose Assemblies. 35 pp.

BSI BS 6501: Part 2-91. 1991 Flexible Metallic Hose Assemblies Part 2: Specification for Strip Wound Hoses and Hose Assemblies. 16 pp.

BSI BS 6501: Part 2-88. 1988 Flexible Metallic Hose Assemblies Part 2: Strip Wound Hoses and Assemblies. 12 pp.

Hose Assemblies *(Cont.)*
—Fluid Power Equipment—Aerospace—Metal

ISO 7314-89. Aerospace—Fluid Systems—Hose Assembly, Metal First Edition. 15 pp.

—Hydrogen Peroxide—Plastic

MOD UK DEF-1214-60. PVC Hose and Hose Assemblies for Use with High Test Peroxide in Temperate and Tropical Conditions. 6 pp.

—Pitot Tubes—Aircraft—Rubber

SBAC AGS 3429 ISSUE 2. Flexible Hose Assembly—Pitot.

—Plastic

BSI BS 4983-92. 1992 Textile Reinforced Thermoplastics Hydraulic Hoses and Hose Assemblies (ISO 3949: 1991). 10 pp.

BSI BS 4983-84. 1984 Textile Reinforced Thermoplastics Hydraulic Hoses and Hose Assemblies. 11 pp.

BSI BS 5173: Part 100-92. 1992 Methods of Test for Rubber and Plastics Hoses and Hose Assemblies Part 100: General Introduction. 6 pp.

BSI BS 5173: Part 100-86. 1986 Rubber and Plastics Hoses and Hose Assemblies Part 100: General Introduction. 5 pp.

BSI BS 5173: Sec 101.1-85. (WITHDRAWN) 1985 Rubber and Plastics Hoses and Hose Assemblies Part 101: Determination of Dimensions Section 101.1: Measurement of Dimensions (Excluding Length) (Renumbered as BS EN 24671: 1993). 7 pp.

BSI BS 5173: Sec 101.2-87. 1987 Rubber and Plastics Hoses and Hose Assemblies Part 101: Determination of Dimensions Section 101.2: Measurement of Length of Hoses and Hose Assemblies. 6 pp.

BSI BS EN 24671-93. 1993 Amd 1 Rubber and Plastics Hose and Hose Assemblies—Methods of Measurement of Dimensions (ISO 4671: 1984) (AMD 7519) April 15, 1993. 14 pp.

CEN PREN 855-92. Plastics Hoses and Hose Assemblies—Thermoplastics Textile Reinforced Hydraulic Type—Specification. 9 pp.

CEN EN 24671-93. Rubber and Plastic Hose and Hose Assemblies—Methods of Measurement of Dimensions (ISO 4671: 1984). 8 pp.

ISO 3949-91. Plastics Hoses and Hose Assemblies—Thermoplastics, Textile-Reinforced, Hydraulic Type—Specification Second Edition. 7 pp.

ISO 4671-84. Rubber and Plastics Hose and Hose Assemblies—Methods of Measurement of Dimensions First Edition (CEN EN 24671: 1993). 7 pp.

ISO 8331-91. Rubber and Plastics Hoses and Hose Assemblies—Guide to Selection, Storage, Use and Maintenance First Edition. 17 pp.

JIS B 8362-89. Textile Reinforced Thermoplastic Hose Assemblies for Hydraulic Use. 13 pp.

—Plastic—Chemical Resistance

BSI BS 5173: Part 5-77. 1977 Rubber and Plastics Hoses and Hose Assemblies Part 5: Chemical Resistance Tests. 8 pp.

—Plastic—Electrical Resistance

BSI BS 5173: Part 4-77. (WITHDRAWN) 1977 Rubber and Plastics Hoses and Hose Assemblies Part 4: Electrical Tests (Superseded by BS 5173: Sections 104.1 & 104.2: 1991). 8 pp.

BSI BS 5173: Sec 104.1-91. (WITHDRAWN) 1991 Rubber and Plastics Hoses and Hose Assemblies Part 104: Electrical Tests Section 104.1: Measurement of Electrical Resistance of Hoses and Hose Assemblies (Superseded BS BS EN 28031: 1993). 9 pp.

BSI BS 5173: Sec 104.2-91. (WITHDRAWN) 1991 Rubber and Plastics Hoses and Hose Assemblies Part 104: Electrical Test Section 104.2: Measurement of Electrical Continuity and Discontinuity of Hoses and Hose Assemblies (Superseded by BS EN 28031: 1993). 6 pp.

BSI BS EN 28031-93. 1993 Ruber and Plastics Hoses and Hose Assemblies—Determination of Electrical Resistance (ISO 8031: 1987) (Supersedes BS 5173: Sections 104.1: 1991 & 104.2: 1990) (E). 11 pp.

CEN EN 28031-93. Rubber and Plasitics Hoses and Hose Assemblies—Determination of Electrical Resistance (ISO 8031: 1987). 6 pp.

ISO 8031-87. Rubber and Plastics Hoses and Hose Assemblies—Determination of Electrical Resistance First Edition; (CEN EN 28031:1993). 7 pp.

—Plastic—Maintenance

ISO 8331-91. Rubber and Plastics Hoses and Hose Assemblies—Guide to Selection, Storage, Use and Maintenance First Edition. 17 pp.

—Plastic—Packaging

MOD UK DSTAN 81-39-89. Packaging of Rubber Hoses, Plastics Hoses, and Hose Assemblies Issue 2. 11 pp.

INDUSTRY STANDARDS

Hose Assemblies (Cont.)

—Plastic—Paint Sprayers

BSI BS EN 28028-93. 1993 Rubber and/or Plastics Hose Assemblies for Airless Paint Spraying—Specification (ISO 8028: 1987) (E). 10 pp.

CEN EN 28028-93. Rubber and/or Plastics Hose Assemblies for Airless Paint Spraying—Specification (ISO 8028: 1987). 5 pp.

ISO 8028-87. Rubber and/or Plastics Hose Assemblies for Airless Paint Spraying—Specification First Edition; (CEN EN 28028:1993). 6 pp.

—Plastic—Pressure Measurement

BSI BS 5173: Part 2-76. (WITHDRAWN) 1976 Amd 4 Rubber and Plastics Hoses and Hose Assemblies Part 2: Hydraulic Pressure Tests (AMD 5620) May 29, 1987 (Superseded by Sections of BS 5173: Part 102). 21 pp.

BSI BS 5173: Sec 102.4-90. 1990 Rubber and Plastics Hoses and Hose Assemblies Part 102: Hydraulic Pressure Tests Section 102.4: Determination of Volumetric Expansion of Hydraulic Hoses. 7 pp.

BSI BS 5173: Sec 102.5-85. 1985 Rubber and Plastics Hoses and Hose Assemblies Part 102: Hydraulic Pressure Tests Section 102.5: Pressure Impulse Test for High Pressure Hydraulic Hoses. 9 pp.

BSI BS 5173: Sec 102.6-91. (WITHDRAWN) 1991 Methods of Test for Rubber and Plastics Hoses and Hose Assemblies Part 102: Hydraulic Pressure Tests Section 102.6: Pressure Impulse Test with Flexing for High Pressure Hydraulic Hoses (Inverted 'U' Config.) (ISO 6802: 1991). 5 pp.

BSI BS 5173: Sec 102.7-88. 1988 Rubber and Plastics Hoses and Hose Assemblies Part 102: Hydraulic Pressure Tests Section 102.7: Pressure Impulse Test with Flexing for High Pressure Hydraulic Hoses (Half Omega Configuration). 5 pp.

BSI BS 6596-92. 1992 Ratios of Proof and Burst Pressure to Design Working Pressure for Rubber and Plastics Hoses and Hose Assemblies (ISO 7751: 1991). 6 pp.

BSI BS 6596-85. 1985 Amd 1 Ratios of Proof and Burst Pressure to Design Working Pressure for Rubber and Plastic Hoses and Hose Assemblies. 5 pp.

BSI BS EN 26802-01. 1993 Amd 1 Rubber and Plastic Hose and Hose Assemblies—Wire Reinforced—Hydraulic Impulse Test with Flexing (ISO 6802: 1991) (AMD 7521) April 15, 1993 (E). 12 pp.

CEN EN 26802-93. Rubber and Plastic Hose and Hose Assemblies—Wire Reinforced—Hydraulic Impulse Test with Flexing (ISO 6802: 1991). 5 pp.

ISO 6802-91. Rubber and Plastics Hose and Hose Assemblies with Wire Reinforcements—Hydraulic Impulse Test with Flexing Second Edition; (CEN EN 26802: 1993). 6 pp.

ISO 6803-84. Rubber or Plastics Hoses and Hose Assemblies—Hydraulic Pressure Impulse Test Without Flexing First Edition. 8 pp.

ISO 7751-91. Rubber and Plastics Hoses and Hose Assemblies—Ratios of Proof and Burst Pressure to Design Working Pressure Second Edition. 4 pp.

ISO 8032-87. Rubber and Plastics Hose Assemblies—Flexing Combined with Hydraulic Impulse Test (Half Omega Test) First Edition. 6 pp.

—Power Steering Systems—Automotive—Rubber

BSI BS 6784-86. 1986 Amd 1 Rubber Hoses and Hose Assemblies for Automobile Power Steering Systems. 12 pp.

—Refrigeration Equipment—Rubber

ISO 5771-81. Rubber Hose and Hose Assemblies for Transferring Anhydrous Ammonia First Edition. 7 pp.

—Rubber

BSI BS 3832-91. 1991 Wire Reinforced Rubber Hoses and Hose Assemblies for Hydraulic Installations (ISO 1436: 1991). 10 pp.

BSI BS 3832-81. 1981 Wire Reinforced Rubber Hoses and Hose Assemblies for Hydraulic Installations. 12 pp.

BSI BS 4586-92. 1992 Spiral Wire Reinforced Rubber Covered Hydraulic Hoses and Hose Assemblies (ISO 3862: 1991). 12 pp.

BSI BS 4586-84. 1984 Spiral Wire Reinforced Rubber Covered Hydraulic Hoses and Hose Assemblies. 14 pp.

BSI BS 4749-91. 1991 Textile Reinforced Rubber Hydraulic Hoses and Hose Assemblies (ISO 4079: 1991). 10 pp.

BSI BS 4749-84. 1984 Textile Reinforced Rubber Hydraulic Hoses and Hose Assemblies. 7 pp.

BSI BS 5173: Part 100-92. 1992 Methods of Test for Rubber and Plastics Hoses and Hose Assemblies Part 100: General Introduction. 6 pp.

BSI BS 5173: Part 100-86. 1986 Rubber and Plastics Hoses and Hose Assemblies Part 100: General Introduction. 5 pp.

BSI BS 5173: Sec 101.1-85. (WITHDRAWN) 1985 Rubber and Plastics Hoses and Hose Assemblies Part 101: Determination of Dimensions Section 101.1: Measurement of Dimensions (Excluding Length) (Renumbered as BS EN 24671: 1993). 7 pp.

BSI BS 5173: Sec 101.2-87. 1987 Rubber and Plastics Hoses and Hose Assemblies Part 101: Determination of Dimensions Section 101.2: Measurement of Length of Hoses and Hose Assemblies. 6 pp.

BSI BS 5244-86. 1986 Recommendations for Application, Storage and Life Expiry of Hydraulic Rubber Hose and Assemblies. 11 pp.

BSI BS 6130-81. 1981 Amd 1 Hose and Hose Assemblies for Asphalt and Bitumen. 5 pp.

BSI BS EN 24671-93. 1993 Amd 1 Rubber and Plastics Hose and Hose Assemblies—Methods of Measurement of Dimensions (ISO 4671: 1984) (AMD 7519) April 15, 1993. 14 pp.

CEN PREN 853-92. Rubber Hoses and Hose Assemblies—Wire Braid Reinforced Hydraulic Type—Specification. 11 pp.

CEN PREN 854-92. Rubber Hoses and Hose Assemblies—Textile Reinforced Hydraulic Type—Specification. 10 pp.

CEN PREN 856-92. Rubber Hoses and Hose Assemblies—Spiral Wire Reinforced Hydraulic Type—Specification. 10 pp.

CEN PREN 857-92. Rubber Hoses and Hose Assemblies—Wire Braid Reinforced Compact Type for Hydraulic Applications—Specification. 9 pp.

CEN EN 24671-93. Rubber and Plastic Hose and Hose Assemblies—Methods of Measurement of Dimensions (ISO 4671: 1984). 8 pp.

ISO 1436-91. Rubber Hoses and Hose Assemblies—Wire-Reinforced Hydraulic Type—Specification Third Edition. 7 pp.

ISO 3862-91. Rubber Hoses and Hose Assemblies—Rubber-Covered, Spiral Wire Reinforced, Hydraulic Type—Specification Second Edition. 9 pp.

ISO 4079-91. Rubber Hoses and Hose Assemblies—Textile-Reinforced Hydraulic Type—Specification Second Edition. 7 pp.

ISO 4671-84. Rubber and Plastic Hose and Hose Assemblies—Methods of Measurement of Dimensions First Edition (CEN EN 24671: 1993). 7 pp.

ISO 8331-91. Rubber and Plastics Hoses and Hose Assemblies—Guide to Selection, Storage, Use and Maintenance First Edition. 17 pp.

JIS B 8360-88. Wire Reinforced Rubber Hose Assemblies for Hydraulic Use. 16 pp.

JIS B 8364-88. Textile Reinforced Rubber Hose Assemblies for Hydraulic Use. 17 pp.

MOD UK DSTAN 47-2-01. Rubber Hose Assemblies for Hydraulic Systems Issue 2; Amendment 3. 24 pp.

MOD UK DSTAN 47-2-91. Rubber Hose Assemblies for Hydraulic Systems Issue 3. 39 pp.

SNZ NZS/ISO 1436-78. Rubber Products—Hose and Hose Assemblies—Wire Reinforced Hydraulic Type. 7 pp.

SNZ NZS/ISO 3862-80. Rubber Hoses and Hose Assemblies—Rubber-Covered, Spiral Wire Reinforced, Hydraulic Type. 9 pp.

SNZ NZS/ISO 4079-78. Rubber Products—Hoses and Hose Assemblies, Textile Reinforced, for Hydraulic Purposes. 4 pp.

—Rubber—Chemical Resistance

BSI BS 5173: Part 5-77. 1977 Rubber and Plastics Hoses and Hose Assemblies Part 5: Chemical Resistance Tests. 8 pp.

—Rubber—Electrical Resistance

BSI BS 5173: Part 4-77. (WITHDRAWN) 1977 Rubber and Plastics Hoses and Hose Assemblies Part 4: Electrical Tests (Superseded by BS 5173: Sections 104.1 & 104.2: 1991). 8 pp.

BSI BS 5173: Sec 104.1-91. (WITHDRAWN) 1991 Rubber and Plastics Hoses and Hose Assemblies Part 104: Electrical Tests Section 104.1: Measurement of Electrical Resistance of Hoses and Hose Assemblies (Superseded BS BS EN 28031: 1993). 9 pp.

BSI BS 5173: Sec 104.2-91. (WITHDRAWN) 1991 Rubber and Plastics Hoses and Hose Assemblies Part 104: Electrical Test Section 104.2: Measurement of Electrical Continuity and Discontinuity of Hoses and Hose Assemblies (Superseded by BS EN 28031: 1993). 6 pp.

BSI BS EN 28031-93. 1993 Rubber and Plastics Hoses and Hose Assemblies—Determination of Electrical Resistance (ISO 8031: 1987) (Supersedes BS 5173: Sections 104.1: 1991 & 104.2: 1990) (E). 11 pp.

CEN EN 28031-93. Rubber and Plasitics Hoses and Hose Assemblies—Determination of Electrical Resistance (ISO 8031: 1987). 6 pp.

ISO 8031-87. Rubber and Plastics Hoses and Hose Assemblies—Determination of Electrical Resistance First Edition; (CEN EN 28031:1993). 7 pp.

—Rubber—Identification Systems

MOD UK DEFCON 67-90. Specifications Covering Identification, Marking and Age on Delivery of Rubber Materiel, Assemblies and Rubber Containing Composites 3/90. 2 pp.

MOD UK DEFCON 67-92. Specifications Covering Identification, Marking and Age on Delivery of Rubber Materiel, Assemblies and Rubber Containing Composites (Navy Contracts Only) 2/92. 2 pp.

—Rubber—Liquefied Petroleum Gas—Safety

BSI BS 3212-91. 1991 Flexible Rubber Tubing, Rubber Hose and Rubber Hose Assemblies for Use in LPG Vapour Phase and LPG/Air Installations. 13 pp.

—Rubber—Maintenance

ISO 8331-91. Rubber and Plastics Hoses and Hose Assemblies—Guide to Selection, Storage, Use and Maintenance First Edition. 17 pp.

—Rubber—Marine

MOD UK NES 324: Part 5-83. Requirements for Hydraulic Systems Part 5: Flexible Hose, Multiple Spiral Wire Reinforced Rubber Hose Assemblies Issue 1 (03.83). 12 pp.

MOD UK NES 324: Part 6-83. Requirements for Hydraulic Systems Part 6: Flexible Hose 2 Layer Wire Braid Reinforced Rubber Hose Assemblies Issue 1 (03.83). 12 pp.

MOD UK NES 345-92. Requirements for Flexible Rubber Pipe Assemblies and Bellows for Use in Systems from Vacuum to 10 Bar Issue 1 (10.92). 73 pp.

—Rubber—Packaging

MOD UK DSTAN 81-39-89. Packaging of Rubber Hoses, Plastics Hoses, and Hose Assemblies Issue 2. 11 pp.

—Rubber—Paint Sprayers

BSI BS EN 28028-93. 1993 Rubber and/or Plastics Hose Assemblies for Airless Paint Spraying—Specification (ISO 8028: 1987) (E). 10 pp.

CEN EN 28028-93. Rubber and/or Plastics Hose Assemblies for Airless Paint Spraying—Specification (ISO 8028: 1987). 5 pp.

ISO 8028-87. Rubber and/or Plastics Hose Assemblies for Airless Paint Spraying—Specification First Edition; (CEN EN 28028:1993). 6 pp.

—Rubber—Pressure Measurement

BSI BS 5173: Part 2-76. (WITHDRAWN) 1976 Amd 4 Rubber and Plastics Hoses and Hose Assemblies Part 2: Hydraulic Pressure Tests (AMD 5620) May 29, 1987 (Superseded by Sections of BS 5173: Part 102). 21 pp.

BSI BS 5173: Sec 102.4-90. 1990 Rubber and Plastics Hoses and Hose Assemblies Part 102: Hydraulic Pressure Tests Section 102.4: Determination of Volumetric Expansion of Hydraulic Hoses. 7 pp.

BSI BS 5173: Sec 102.5-85. 1985 Rubber and Plastics Hoses and Hose Assemblies Part 102: Hydraulic Pressure Tests Section 102.5: Pressure Impulse Test for High Pressure Hydraulic Hoses. 9 pp.

BSI BS 5173: Sec 102.6-91. (WITHDRAWN) 1991 Methods of Test for Rubber and Plastics Hoses and Hose Assemblies Part 102: Hydraulic Pressure Tests Section 102.6: Pressure Impulse Test with Flexing for High Pressure Hydraulic Hoses (Inverted 'U' Config.) (ISO 6802: 1991). 5 pp.

BSI BS 5173: Sec 102.7-88. 1988 Rubber and Plastics Hoses and Hose Assemblies Part 102: Hydraulic Pressure Tests Section 102.7: Pressure Impulse Test with Flexing for High Pressure Hydraulic Hoses (Half Omega Configuration). 5 pp.

BSI BS 6596-92. 1992 Ratios of Proof and Burst Pressure to Design Working Pressure for Rubber and Plastics Hoses and Hose Assemblies (ISO 7751: 1991). 6 pp.

BSI BS 6596-85. 1985 Amd 1 Ratios of Proof and Burst Pressure to Design Working Pressure for Rubber and Plastic Hoses and Hose Assemblies. 5 pp.

BSI BS EN 26802-01. 1993 Amd 1 Rubber and Plastic Hose and Hose Assemblies—Wire Reinforced—Hydraulic Impulse Test with Flexing (ISO 6802: 1991) (AMD 7521) April 15, 1993 (E). 12 pp.

CEN EN 26802-93. Rubber and Plastic Hose and Hose Assemblies—Wire Reinforced—Hydraulic Impulse Test with Flexing (ISO 6802: 1991). 5 pp.

ISO 6802-91. Rubber and Plastics Hose and Hose Assemblies with Wire Reinforcements—Hydraulic Impulse Test with Flexing Second Edition; (CEN EN 26802: 1993). 6 pp.

ISO 6803-84. Rubber or Plastics Hoses and Hose Assemblies—Hydraulic Pressure Impulse Test Without Flexing First Edition. 8 pp.

ISO 7751-91. Rubber and Plastics Hoses and Hose Assemblies—Ratios of Proof and Burst Pressure to Design Working Pressure Second Edition. 4 pp.

INTERNATIONAL AND NON-U.S. NATIONAL STANDARDS SUBJECT INDEX

Hose

Hose Assemblies (Cont.)

—Rubber—Pressure Measurement (Cont.)
ISO 8032-87. Rubber and Plastics Hose Assemblies—Flexing Combined with Hydraulic Impulse Test (Half Omega Test) First Edition. 6 pp.

—Static Tubes—Aircraft
SBAC AGS 3430 ISSUE 2. Flexible Hose Assembly—Static.

—Static Tubes—Aircraft—Rubber
SBAC AGS 3915 (V). Hose Assembly Flexible—Static (Metric).

Hose Clamps

Use For: Hose Clips   See Also: Clamps; Hose Fittings
CNS B2440-84. Hose Clamps for the Lower Pressure (Jul)(5298).

—Aerospace
AECMA PREN2903-89. Clamps Worm Drive Dimensions-Masses. 5 pp.

—Aircraft
SBAC AGS 605 (V)(I). Clip Hose (Type J).
SBAC AGS 1000 (V)(I). Clip-Hose, Type 'S' General Arrangement.

—Aircraft—Assembly
SBAC AGS 606 (V)(I). Assembly of Hose Cli

—Aircraft—Worms (Mechanical Drives)
SBAC AGS 3924 ISSUE 2. Clamp, Hose, Worm Drive, "Microflex", Extra Light Duty.
SBAC AGS 3925 ISSUE 2. Clamp, Hose, Worm Drive, "Minox" Light Duty.

—Automotive
CNS D2164-83. Hose Clamps for Automobiles (Feb)(9983).
JIS D 3621-82. Hose Clamps for Automobiles.

—Worms (Mechanical Drives)
BSI BS 5315-91. 1991 Hose Clamps (Worm Drive Type) for General Purpose Use (Metric Series). 12 pp.
BSI BS 5315-76. 1976 Amd 2 Hose Clamps (Worm Drive Type) for General Purpose Use (Metric Series). 9 pp.
SNZ NZS/BS 5315-91. Specification for Hose Clamps (Worm Drive Type) for General Purpose Use (Metric Series). 12 pp.

—Worms (Mechanical Drives)—Aerospace
AECMA PREN2903-89. Clamps Worm Drive Dimensions-Masses. 5 pp.

—Worms (Mechanical Drives)—Aircraft
BSI SP 144-147-71. 1971 Amd 2 Corrosion Resisting Worm Drive Clamps for Aircraft. 16 pp.

Hose Clips
Use: Hose Clamps

Hose Connections
Use: Hose Connectors; Hose Fittings

Hose Connectors
Use For: Hose Connections   See Also: Hose Fittings
ISO 8434-86. Hydraulic Fluid Power—Connection for Tubes and Hoses—Dimensions and Designs for 37 Degree Flare and 24 Degree Flareless Fittings First Edition. 10 pp.

—Dental Instruments
CEN EN 29 168-90. Dentistry—Dental Handpieces—Hose Connectors. 4 pp.
ISO 9168-91. Dental Handpieces—Hose Connectors Second Edition; (Incorporating Draft Addendum 1-1989). 9 pp.

—Fluid Power Equipment—Diameters
ISO 4397-93. Fluid Power Systems and Components—Connectors and Associated Components—Nominal Outside Diameters of Tubes and Nominal Inside Diameters of Hoses Second Edition. 6 pp.

—Welding Equipment
DIN ENGL 8542-83. Hose Connections and Hose Couplers for Equipment for Welding, Cutting and Allied Processes (Sept). 9 pp.

Hose Couplers
Use: Hose Couplings

Hose Couplings
See Also: Couplings; Hose Fittings; Quick Disconnect Couplings
BSI BS 1782-51. (OBSOLESCENT) 1951 Amd 5 Hose Couplings (1 and a Half Inch to 8 Inch Nominal Sizes) Other Than Fire Hose Couplings. 48 pp.

Hose Couplings (Cont.)

—Air
BSI BS 1906-52. (OBSOLESCENT) 1952 Amd 6 Hose Couplings (Air and Water) One Eighth of an Inch to 1 and a Quarter Inch in Nominal Sizes. 60 pp.

—Air—Automobiles—Brakes
JIS D 6605-84. Air Brake Line Couplings Between Truck Tractors and Trailers. 9 pp.

—Air—Rolling Stock
JIS E 4301-89. Air Hose Couplings.

—Aircraft—Oxygen Supply Equipment
BSI 2C 5-76. 1976 Mating Dimensions of Liquid Oxygen Replenishment Couplings. 5 pp.
ISO 1465-89. Aircraft—Liquid Oxygen Replenishment Couplings—Mating Dimensions Second Edition. 5 pp.

—Aircraft—Thrust Wires
SBAC AGS 1231 (V). Union Nuts Re-Usable End Fittings for Non Fire Resistant Hose D.T.D. (R.D.I.) 3951.
SBAC AGS 1239 ISSUE 2. Union Nut (Steel) 1/4" B.S.P.F. Re-Usable End Fittings for 1/4" I/D Non Fire Resistant Hose to Spec. D.T.D. (R.D.I.) 3951.
SBAC AGS 4715 ISSUE 1. Union Nut Re-Usable End Fittings for Non Fire Resistant Hose D.T.D. (R.D.1.) 3951.

—Fire
BSI BS 336-89. 1989 Fire Hose Couplings and Ancillary Equipment. 31 pp.
JIS B 9911-68. Dimensions of Insert Type Couplings for Fire-Hose.

—Fire—Quick Disconnect
CNS Z2051-86. Quick Connector for Fire Hose (Jun)(10206).

—Fire—Threaded
JIS B 9912-70. Classification and Dimensions of Screwed Type Couplings for Fire Hose.

—Fuel—Diesel Locomotives
BSI BS 3818-64. (OBSOLESCENT) 1964 Self-Sealing Fuelling Couplings for Diesel Locomotives and Diesel Railcars. 21 pp.

—Fuel—Hydrogen Peroxide
MOD UK DEF-1403-61. Couplings, Pressure Fuelling, for High Test Peroxide. 12 pp.

—Fuel—Pressure
BSI BS 2464: Part 2-69. (OBSOLESCENT) 1969 Hose Couplings for Petrol, Oil and Lubricants Part 2: High Pressure Couplings. 21 pp.

—Fuel—Quick Disconnect
BSI BS 2464: Part 3-68. (OBSOLESCENT) 1968 Hose Couplings for Petrol, Oil and Lubricants Part 3: Quick-Acting Couplings. 39 pp.
BSI BS 2464: Part 4-80. (OBSOLESCENT) 1980 Hose Couplings for Petrol, Oil and Lubricants Part 4: 100 mm Quick-Acting Couplings: European Type. 13 pp.

—Fuel—Replenishment at Sea
NATO STANAG 1357 ED 1 AMD 0-89. NATO Standard F-44 Hose Coupling. 8 pp.
NATO STANAG 1357 ED 1 AMD 1-92. NATO Standard F-44 Hose Coupling. 9 pp.

—Fuel—Threaded
MOD UK DEF-1056-52. Screw Threads for Petroleum Hose Couplings. 4 pp.

—Hydrogen Peroxide
MOD UK DEF-1215-60. Open-Ended Hose Couplings for High Test Peroxide. 16 pp.

—Prime Movers—Brakes
SAA AS D8-71. Hose Couplings for Use with Vacuum and Air-Pressure Braking Systems on Prime Movers, Trailers and Semitrailers. 22 pp.

—Rolling Stock—Brakes
BSI BS 3710-64. (OBSOLESCENT) 1964 Brake Hose Couplings for Locomotives and Rolling Stock. 29 pp.

—Rolling Stock—Packings (Seals)
JIS E 4303-91. Packing Rubbers for Hose Couplings.

—Rolling Stock—Test Specimens
JIS E 4302-91. Dummy Hose Couplings.

—Rubber—Welding Equipment
JIS B 6805-86. Rubber Hose Connections for Welding and Cutting Equipment.
JIS B 6805-68. Rubber Hose Coupling for Gas Welding and Cutting Torches. 5 pp.

Hose Couplings (Cont.)

—Semitrailers—Brakes
SAA AS D8-71. Hose Couplings for Use with Vacuum and Air-Pressure Braking Systems on Prime Movers, Trailers and Semitrailers. 22 pp.

—Ships
ISO TR5987-84. Inland Navigation—Water Fire-Fighting System—Couplings of Fire Hoses—General Technical Requirements First Edition. 4 pp.

—Spraying Equipment
CNS B2534-81. Hose Couplings and Hose Joints for Sprayer (Jan)(6875).

—Trailers—Brakes
SAA AS D8-71. Hose Couplings for Use with Vacuum and Air-Pressure Braking Systems on Prime Movers, Trailers and Semitrailers. 22 pp.

—Water
BSI BS 1906-52. (OBSOLESCENT) 1952 Amd 6 Hose Couplings (Air and Water) One Eighth of an Inch to 1 and a Quarter Inch in Nominal Sizes. 60 pp.

—Welding Equipment
CNS B2533-80. Rubber Hose Coupling for Gas Welding and Cutting Torches (Dec)(6741).
DIN ENGL 8544-81. Probe Couplings for Equipment and Gas Hoses for Welding, Cutting and Related Processes; Couplings for Hoses from 4 to 10 mm Internal Diameter (Nov). 7 pp.

Hose Ends
See Also: Hose Fittings

—Aircraft
SBAC AGS 1232 (V). Tailpiece Re-Usable End Fittings for Non Resistant Hose D.T.D. (RDI) 3951.
SBAC AGS 1237 (V). End Fitting 90 Degree G.A. Re-Usable, for 1/4 Inch I/D Non Fire Resistant Hose DTD(RD. 1.) 3951.
SBAC AGS 1238 (V). End Fitting 45 Degree G.A. Re-Usable, for 1/4 Inch I/D Non Fire Resistant Hose DTD (RD.1.) 5951.

—Aircraft—High Pressure
JIS W 1821-79. Fitting End, Attachable, Hydraulic, High-Pressure Hose.

—Automotive—Brakes
ISO 4038-84. Road Vehicles—Hydraulic Braking Systems—Pipes, Tapped Holes, Male Fittings and Hose End Fittings Second Edition. 10 pp.

—Hydraulic
JIS B 8363-88. End Fittings and Adapters for Hydraulic Hose Assemblies. 25 pp.

Hose Fittings
See Also: Couplings; Hose Adapters; Hose Assemblies; Hose Clamps; Hose Connectors; Hose Couplings; Hose Ends; Hose Joints; Hose Unions; Union Nuts
BSI BS 1906-52. (OBSOLESCENT) 1952 Amd 6 Hose Couplings (Air and Water) One Eighth of an Inch to 1 and a Quarter Inch in Nominal Sizes. 60 pp.
ISO 8434-86. Hydraulic Fluid Power—Connection for Tubes and Hoses—Dimensions and Designs for 37 Degree Flare and 24 Degree Flareless Fittings First Edition. 10 pp.
SAA AS 1180.11-82. Methods of Test for Hose Made from Elastomeric Materials—Part 11: Hose and Coupling Compatability—Tensile Method. 1 p.

—Aerospace—Locking Rings
AECMA PREN2603-87. Straight Metric-Size Unions with Locking Ring Port End Dimensions. 4 pp.
AECMA PREN2604-87. Straight Metric-Size Unions with Locking Ring 8 Degrees 30' Union Interface Dimensions. 5 pp.
AECMA PREN2605-87. Straight Metric-Size Unions with Locking Ring 24 Degrees Union Interface Dimensions. 6 pp.
AECMA PREN2607-87. Straight Metric-Size Unions with Locking Ring Configuration O-Ring Dimensions. 5 pp.
AECMA PREN2645-87. Straight Metric-Size Unions with Locking Ring Locking Ring Dimensions. 5 pp.

—Aerospace—Ports (Openings)—Locking Rings
AECMA PREN2602-87. Parts for Installation of Straight Metric-Size Unions with Locking Ring Dimensions. 6 pp.

—Aerospace—Titanium
CEN PREN3580-89. Aerospace Series PTFE Flexible Hose Assembly of a Nominal Presser Equal to 21000 kPa with 8 Degree 30 Fitting in Titanium Product Standard. 7 pp.

INDUSTRY STANDARDS

INTERNATIONAL AND NON-U.S. NATIONAL STANDARDS
SUBJECT INDEX

Hose Fittings (Cont.)
—Aircraft
BSI 2C 2-73. 1973 Coupling Dimensions for Aircraft Connections for Water of Drinkable Quality. 4 pp.
BSI 2C 4-73. 1973 Amd 2 Coupling Dimensions for Aero-Engine Refrigerant Pressure Replenishment Connections. 9 pp.
BSI C 17-73. 1973 Coupling Dimensions for Low Pressure Air Connections for Engine Starting. 4 pp.
ISO 8829-90. Aerospace—Polytetrafluoroethylene (PTFE) Hose Assemblies—Test Methods First Edition. 15 pp.
JIS W 1904-87. Fitting, Flared Tube, Fluid Connection.
SBAC AS 28950-959 ISSUE 1. 'P.' Section Rubber.
SBAC AGS 1230 ISSUE 2. G.A. Re-Usable End Fitting for Non-Fire Resistant Hose D.T.D. (R.D.1.) 3951.
SBAC AGS 1234 (V). G.A. Re-Usable 90 Degree End Fittings for Non Fire Resistant Hoe D.T.D. (R.D.I.) 3951.
SBAC AGS 1235 (V). G.A. Re-Usable 45 Degree End Fittings for Non Fire Resistant Hose D.T.D. (R.D.I.) 3951.

—Aircraft Cabins
BSI 2C 11-88. 1988 Ground Pressure Test Connections for Aircraft Pressure Cabins. 4 pp.
ISO 11-87. Aircraft—Ground Pressure Testing Connections for Pressure Cabins Second Edition. 5 pp.
JIS W 3104-91. Aircraft—Ground Pressure Testing Connections for Pressure Cabins (ISO 11:1976).

—Aircraft—Fluid Power Equipment
MOD UK DSTAN 47-22-01. End Fittings for Flexible Hose Assemblies for Aircraft—Metric Issue 1; Amendment 1. 15 pp.

—Aircraft—Thrust Wires
SBAC AS 15776-787 ISSUE 5. Wire, Thrust-Tube and Hose Coupling Nut.

—Aircraft—Titanium
SBAC AS 56500-599 ISSUE 1. Bolt, Hexagon Head P.D. Shank Waspaloy, 0.1900—32 BS4084.

—Aircraft—Valve Lifters—Assembly
SBAC AS 2583 ISSUE 5. Assembly of Valve Lifter and Adaptor. 1 p.
SBAC AS 2587 ISSUE 5. Assembly of Valve Lifter and Adaptor. 1 p.

—Automotive—Brakes
CNS D2099-86. Air Brake Line Couplings Between Truck Tractors and Trailers (Jan)(7882).

—Automotive—Hydraulic Brakes
ISO 4038-84. Road Vehicles—Hydraulic Braking Systems—Pipes, Tapped Holes, Male Fittings and Hose End Fittings Second Edition. 10 pp.

—Cutting Equipment
BSI BS 1389-86. 1986 Hose Connections and Hose Assemblies for Equipment for Gas Welding, Cutting and Related Processes. 14 pp.
DIN ENGL 8542-83. Hose Connections and Hose Couplers for Equipment for Welding, Cutting and Allied Processes (Sept). 9 pp.
DIN ENGL 8544-81. Probe Couplings for Equipment and Gas Hoses for Welding, Cutting and Related Processes; Couplings for Hoses from 4 to 10 mm Internal Diameter (Nov). 7 pp.
ISO 3253-75. Hose Connections for Equipment for Welding, Cutting and Related Processes First Edition. 6 pp.

—Dental Instruments
BSI BS 7077: Sec 2.1-89. 1989 Dental Handpieces and Accessories Part 2: Accessories Section 2.1: Specification for Hose Connectors. 7 pp.
BSI BS 7077: Sec 2.1-01. 1989 Amd 1 Dental Handpieces and Accessories Part 2: Accessories Section 2.1: Specification for Hose Connectors (AMD 6679) May 31, 1991. 10 pp.

—Diffusers
MOD UK DEF-1229-60. Standard Diffuser Branch. 3 pp.

—Fire
BSI BS 336-89. 1989 Fire Hose Couplings and Ancillary Equipment. 31 pp.
MOD UK DSTAN 42-27-01. Wrench, Suction Hose Coupling, Fire Appliance Issue 1; Amendment 1. 10 pp.
SNZ NZS 4505-77. Specification for Fire-Fighting Waterway Equipment Amend: 1, 1982; 2, 1985. 20 pp.

—Fire—Nozzles
JIS B 9913-72. Classification and Dimensions of Connections for Fire Department Nozzle-Tips (R 1984). 24 pp.

Hose Fittings (Cont.)
—Fuel
JIS S 2146-91. Rubber Tubes and Polyvinyl Chloride Hoses with Both End Convenient Joints for Gas. 21 pp.

—Gas
JIS S 2146-91. Rubber Tubes and Polyvinyl Chloride Hoses with Both End Convenient Joints for Gas. 21 pp.

—Gas—Appliances
BSI BS 669: Part 1-89. 1989 Flexible Hoses, End Fittings and Sockets for Gas Burning Appliances Part 1: Strip-Wound Metallic Flexible Hoses, Covers, End Fittings and Sockets for Domestic Appliances Burning 1st and 2nd Family Gases (L). 24 pp.

—Gas Cutting Equipment
CEN PREN 560-91. Hose Connections for Equipment for Welding, Cutting and Allied Processes. 7 pp.

—Marine
CNS F3145-81. Marine Hose Connections and Fittings (Nov)(8109).
JIS F 7335-89. Marine Hose Connections and Fittings.

—Medical Gas Equipment
BSI BS 5682-84. 1984 Amd 1 Terminal Units, Hose Assemblies and Their Connectors for Use with Medical Gas Pipeline Systems. 33 pp.

—Replenishment at Sea—Thread Design
NATO STANAG 1234 ED 1 AMD 5-80. Standardization of Thread Design for 65 mm (Nominal 2 1/2 Inch) Bore Replenishment Fitting. 10 pp.

—Rock Drills
CNS M2053-80. Shape and Dimension of Air and Water Hose Connections for Rock Drill (May)(5626).
JIS M 3915-64. Shape and Dimension of Air and Water Hose Connections for Rock Drill.

—Sewage—Ships
NATO STANAG 4167 ED 1 AMD 3-82. NATO Pollutant Discharge Connection for Sewage and for Oily Water. 28 pp.
NATO STANAG 4167 ED 1 AMD 4-82. NATO Pollutant Discharge Connection for Sewage and for Oily Water. 26 pp.

—Spraying Equipment
JIS B 9119-89. Hose Couplings and Hose Joints for Sprayer. 16 pp.

—Threaded
SAA AS 1610-74. Threaded Brass Hose Couplings for Use with Taps and Watering Devices. 8 pp.

—Welding Equipment
BSI BS 1389-86. 1986 Hose Connections and Hose Assemblies for Equipment for Gas Welding, Cutting and Related Processes. 14 pp.
CEN PREN 560-91. Hose Connections for Equipment for Welding, Cutting and Allied Processes. 7 pp.
DIN ENGL 8542-83. Hose Connections and Hose Couplers for Equipment for Welding, Cutting and Allied Processes (Sept). 9 pp.
ISO 3253-75. Hose Connections for Equipment for Welding, Cutting and Related Processes First Edition. 6 pp.

Hose Joints
See Also: Hose Fittings

—Spraying Equipment
CNS B2534-81. Hose Couplings and Hose Joints for Sprayer (Jan)(6875).
JIS B 9119-89. Hose Couplings and Hose Joints for Sprayer. 16 pp.

Hose Pipes
Use: Hoses

Hose Reels
See Also: Cable Reels; Reels

—Fire Hoses
BSI BS 5306: Part 1-76. 1976 Amd 2 Fire Extinguishing Installations and Equipment on Premises Part 1: Hydrant Systems, Hose Reels and Foam Inlets (Formerly CP 402.101) (AMD 5756) February 29, 1988. 24 pp.
CEN PREN 671-1-92. Fixed Firefighting Systems—Hose Systems—Part 1: Hose Reels with Semi-Rigid Hose. 21 pp.

—Tanker Ships
NATO STANAG 1199 ED 2 AMD 2-83. Reelable Astern Refuelling Rig for the Conversion of Merchant Tankers. 21 pp.

Hose Reels (Cont.)
—Tanker Ships (Cont.)
NATO STANAG 1199 ED 2 AMD 3-83. Reelable Astern Refuelling Rig for the Conversion of Merchant Tankers. 20 pp.
NATO STANAG 1199 ED 2 AMD 4-83. Reelable Astern Refuelling Rig for the Conversion of Merchant Tankers. 21 pp.

Hose Unions
See Also: Hose Fittings; Pipe Fittings

—Aerospace—Locking Rings
AECMA PREN2603-87. Straight Metric-Size Unions with Locking Ring Port End Dimensions. 4 pp.
AECMA PREN2604-87. Straight Metric-Size Unions with Locking Ring 8 Degrees 30' Union Interface Dimensions. 5 pp.
AECMA PREN2605-87. Straight Metric-Size Unions with Locking Ring 24 Degrees Union Interface Dimensions. 6 pp.
AECMA PREN2607-87. Straight Metric-Size Unions with Locking Ring Configuration O-Ring Dimensions. 5 pp.
AECMA PREN2645-87. Straight Metric-Size Unions with Locking Ring Locking Ring Dimensions. 5 pp.

—Aerospace—Ports (Openings)—Locking Rings
AECMA PREN2602-87. Parts for Installation of Straight Metric-Size Unions with Locking Ring Dimensions. 6 pp.

—Screwed
DIN ENGL 7591-81. Water Fittings for Domestic Installation; Screwed Unions with Flat Gasket for Hose Connection on Either Side, Designed for Nominal Pressure 10 (Jan). 2 pp.

Hoses
Use For: Hydraulic Hoses *See Also:* Air Hoses; Breathing Apparatus; Chemical Hoses; Fire Hoses; Fuel Hoses; Hose Assemblies; Lawn and Garden Hoses; Pipes; Pneumatic Hoses; Sandblast Hoses; Steam Hoses; Suction/Discharge Hoses; Suction Hoses; Water Hoses; Welding Hoses

SAA AS 3791-91. Hydraulic Hose Amdt 1 June 1991 (in Professional Package 55).

—Aircraft
MOD UK DTD-503A-54. Steel Tubes for High-Pressure Hydraulic Systems (Reprinted February 1965). 2 pp.

—Aircraft—Oxygen
MOD UK DSTAN 16-1-01. Breathing Oxygen Characteristics, Supply Pressure, and Hoses for Aircraft Systems Issue 2; Amendment 1. 9 pp.
MOD UK DSTAN 16-1-92. Oxygen; Liquid and Gaseous (Breathing) for Aircraft Systems Issue 3. 13 pp.

—Aircraft—Rubber
CAA LEAFLET 5-5 07.90. Hose and Hose Assemblies. 17 pp.

—Aircraft—Synthetic
CAA LEAFLET 5-5 07.90. Hose and Hose Assemblies. 17 pp.

—Automotive
BSI BS 5409: Part 2-78. (OBSOLESCENT) 1978 Specification for Nylon Tubing Part 2: Plasticized and Unplasticized Nylon Tubing Types 11 and 12 for Use Primarily in the Automobile Industry. 16 pp.

—Brakes—Automotive
SNZ NZS/SAEJ1403-85. Vacuum Brake Hose. 2 pp.

—Brakes—Rolling Stock
BSI BS 3682-69. 1969 Amd 2 Railway Brake Hose. 14 pp.

—Brewing Equipment
CNS K4068-82. Brewing Hoses (Nov)(9617).

—Carbon Steel Wire
BSI BS 3592: Part 2-92. 1992 Steel Wire for Hose Reinforcement Part 2: Specification for Metallic Coated Steel Wire for the Bonded Reinforcement of Hydraulic Hoses (V). 12 pp.
BSI BS 3592: Part 2-86. 1986 Steel Wire for Rubber Hose Reinforcement Part 2: Metallic Coated Steel Wire for the Bonded Reinforcement of Hydraulic Hoses. 8 pp.

—Chemical Resistant
CNS K4070-82. Chemical Resistant Hoses (Nov)(9619).

INTERNATIONAL AND NON-U.S. NATIONAL STANDARDS
SUBJECT INDEX

Hoses

Hoses (Cont.)

—Cooling Systems—Automotive—Rubber

BSI BS 7038-88. 1988 Rubber Coolant Hoses and Tubing for Use on Private Cars and Light Commercial Vehicles. 12 pp.

BSI BS AU 108-65. 1965 Amd 1 Plain and Reinforced Hoses of Rubber. 8 pp.

CNS D2046-86. Coolant System Hose for Automobiles (Aug)(6162).

ISO 4081-87. Rubber—Coolant Hoses and Tubing for Use on Private Cars and Light Commercial Vehicles—Specification First Edition. 10 pp.

—Cooling Systems—Combat Vehicles

NATO STANAG 4043 ED 1 AMD 4-58. Cooling System Hoses for Wheeled Transport Tactical Vehicles. 7 pp.

—Cooling Systems—Internal Combustion Engines—Rubber

BSI BS 2952-58. 1958 Amd 2 Rubber Hose for I.C. Engine Cooling Systems. 13 pp.

—Corrugated—Low Pressure—Aircraft

SBAC RS 718 ISSUE 1. Connection, Threaded, Geometry for Low Pressure Corrugated Hose.

—Cotton Fabrics—Rubber

BSI BS 1103-89. 1989 Fabrics for the Reinforcement of Rubber Hoses. 8 pp.

BSI BS 1103-57. (WITHDRAWN) 1957 Cotton Fabrics for the Reinforcement of Rubber Hose. 8 pp.

CNS L3134-81. Method of Test for Cotton Cloth for Rubber Hose (Jul)(7743).

CNS L4137-81. Cotton Cloth for Rubber Hose (Jul)(7742).

—Current Carrying—Vacuum Cleaners

CSA C22.2 NO 67-M1985. Vacuum Cleaners and Floor Polishers; (Gen Instr 1 Thru 4). 57 pp.

CSA 1169 Bull. Electrical Bulletin 1169 June 27, 1978 to C22.2 NO 67. 2 pp.

CSA 1415A Bull. Electrical Bulletin 1415A February 12, 1986 to C22.2 NO 67. 4 pp.

—Elastomeric

SAA AS 1180. Methods of Test for Hose Made from Elastomeric Materials (Complete Set in Binder).

SAA AS 1257-73. Bore Sizes, Test Pressures and Tolerances on Lengths of Elastomeric Hose Amdt 1 June 1979 Reconfirmed 1988. 4 pp.

—Elastomeric—Adhesion Testing

SAA AS 1180.4A-72. Methods of Test for Hose Made from Elastomeric Materials—Part 4A: Ply Adhesion—Dead Weight Method Reconfirmed 1988.

SAA AS 1180.4B-72. Methods of Test for Hose Made from Elastomeric Materials—Part 4B: Ply Adhesion—Autographic Method Corrig. Reconfirmed 1988.

—Elastomeric—Aging Testing

SAA AS 1180.3-72. Methods of Test for Hose Made from Elastomeric Materials—Part 3: Accelerated Ageing Reconfirmed 1988.

—Elastomeric—Bend Testing

SAA AS 1180.8A-72. Methods of Test for Hose Made from Elastomeric Materials—Part 8A: Resistance to Cold Flexing of Hose Assembly Corrig. Reconfirmed 1988.

SAA AS 1180.8B-72. Methods of Test for Hose Made from Elastomeric Materials—Part 8B: Resistance to Cold Flexing of Hose Lining and Cover (Bound Together) Reconfirmed 1988. 42 pp.

SAA AS 1180.14-91. Methods of Test for Hose Made from Elastomeric Materials—Part 14: Determination of Reeling Properties of Non-Collapsible Hose. 3 pp.

—Elastomeric—Corrosion Testing

SAA AS 1180.7G-72. Methods of Test for Hose Made from Elastomeric Materials—Part 7G: Corrosion Resistance Reconfirmed 1988.

—Elastomeric—Detergent Resistance Testing

SAA AS 1180.7D-72. Methods of Test for Hose Made from Elastomeric Materials—Part 7D: Resistance to Detergent Reconfirmed 1988.

—Elastomeric—Dimensions

SAA AS 1180.1-72. Methods of Test for Hose Made from Elastomeric Materials—Part 1: Dimensions Reconfirmed 1988.

—Elastomeric—Electrical Continuity

SAA AS 1180.13C-83. Methods of Test for Hose Made from Elastomeric Materials—Part 13C: Determination of Electrical Continuity of a Hose Assembly with Reinforcing Wire(s). 1 p.

Hoses (Cont.)

—Elastomeric—Electrical Resistance

SAA AS 1180.13A-83. Methods of Test for Hose Made from Elastomeric Materials—Part 13A: Determination of Electrical Resistance of Hose and Hose Components. 1 p.

SAA AS 1180.13B-87. Methods of Test for Hose Made from Elastomeric Materials—Part 13B: Determination of Electrical Resistance of Hose Assembly. 1 p.

—Elastomeric—Flammability Testing

SAA AS 1180.10A-74. Methods of Test for Hose Made from Elastomeric Materials—Part 10A: Resistance of Hose Lining or Cover to Flame Reconfirmed 1988. 4 pp.

SAA AS 1180.10B-82. Methods of Test for Hose Made from Elastomeric Materials—Part 10B: Determination of Combustion Propagation Characteristics of a Horizontally Oriented Specimen of Hose Using Surface Ignition Reconfirmed 1988. 4 pp.

SAA AS 1180.12-74. Methods of Test for Hose Made from Elastomeric Materials—Part 12: Resistance to Ignition Reconfirmed 1989. 3 pp.

—Elastomeric—Hydrostatic Testing

SAA AS 1180.5A-72. Methods of Test for Hose Made from Elastomeric Materials—Part 5A: Hydrostatic Pressure—Burst Test (Superseded by AS 1180.5—1993).

SAA AS 1180.5B-72. Methods of Test for Hose Made from Elastomeric Materials—Part 5B: Hydrostatic Pressure—Proof Test (Superseded by AS 1180—1993).

SAA AS 1180.5C-72. Methods of Test for Hose Made from Elastomeric Materials—Part 5C: Hydrostatic Pressure—Change-in-Length Test (Supresed by AS 1180—1993).

SAA AS 1180.5D-72. Methods of Test for Hose Made from Elastomeric Materials—Part 5D: Hydrostatic Pressure—Leakage Test (Superseded by AS 1180—1993).

SAA AS 1180.5E-72. Methods of Test for Hose Made from Elastomeric Materials—Part 5E: Hydrostatic Pressure—Expansion and Distortion Test (Superseded by AS 1180—1993).

—Elastomeric—Liquid Resistance

SAA AS 1180.7A-72. Methods of Test for Hose Made from Elastomeric Materials—Part 7A: Resistance of Lining and Cover to Liquids Reconfirmed 1988.

SAA AS 1180.7B-72. Methods of Test for Hose Made from Elastomeric Materials—Part 7B: Resistance to Liquids—Physical Reconfirmed 1988.

—Elastomeric—Oil Resistance Testing

SAA AS 1180.7E-72. Methods of Test for Hose Made from Elastomeric Materials—Part 7E: Resistance to Oil Reconfirmed 1988.

—Elastomeric—Ozone Resistance Testing

SAA AS 1180.7F-72. Methods of Test for Hose Made from Elastomeric Materials—Part 7F: Resistance of Lining and Cover to Ozone Reconfirmed 1988.

—Elastomeric—Pressure Measurement

SAA AS 1180.6-72. Methods of Test for Hose Made from Elastomeric Materials—Part 6: Impulse Test (Bound Together) Reconfirmed 1988. 38 pp.

—Elastomeric—Steam Resistance

SAA AS 1180.7C-72. Methods of Test for Hose Made from Elastomeric Materials—Part 7C: Resistance to Steam Reconfirmed 1988.

—Elastomeric—Tensile Testing

SAA AS 1180.2-72. Methods of Test for Hose Made from Elastomeric Materials—Part 2: Tensile Strength and Elongation Reconfirmed 1988.

—Elastomeric—Vacuum Resistance

SAA AS 1180.7J-72. Methods of Test for Hose Made from Elastomeric Materials—Part 7J: Resistance to Vacuum Corrig. Reconfirmed 1988.

—Hydrogen Peroxide—Plastic

MOD UK DEF-1214-60. PVC Hose and Hose Assemblies for Use with High Test Peroxide in Temperate and Tropical Conditions. 6 pp.

—Laboratory Equipment

BSI BS 2775-87. 1987 Rubber Stoppers and Tubing for General Laboratory Use. 11 pp.

—Metal

BSI BS 6130-81. 1981 Amd 1 Hose and Hose Assemblies for Asphalt and Bitumen. 5 pp.

BSI BS 6501: Part 2-91. 1991 Flexible Metallic Hose Assemblies Part 2: Specification for Strip Wound Hoses and Hose Assemblies. 16 pp.

BSI BS 6501: Part 2-88. 1988 Flexible Metallic Hoses and Hose Assemblies Part 2: Strip Wound Hoses and Hose Assemblies. 12 pp.

Hoses (Cont.)

—Metal (Cont.)

ISO 7657-84. Pipework—Stripwound Flexible Metal Hoses—Specifications and Temperature-Related Requirements for Use First Edition. 5 pp.

ISO 7658-84. Pipework—Stripwound Flexible Metal Hoses—Testing and Verification of Characteristics First Edition. 6 pp.

ISO 8446-85. Pipework—Double Overlap Flexible Metal Hoses (Asbestos Packing, Leakproof, Circular Section, in Austenitic Stainless Steel) First Edition. 4 pp.

ISO 8449-86. Pipework—Single Overlap Flexible Metal Hoses (Asbestos Packing, Limited Tightness, Circular or Polygonal Section, in Protected Carbon Steel) First Edition. 4 pp.

JIS S 2145-91. Metallic Flexible Hoses for Gas. 27 pp.

—Metal—Carbon Steel

ISO 8444-85. Pipework—Double Overlap Flexible Metal Hoses (Copper Packing, Limited Tightness, Circular Section, in Protected Carbon Steel) First Edition. 4 pp.

ISO 8445-85. Pipework—Double Overlap Flexible Metal Hoses (Asbestos Packing, Leakproof, Circular Section, in Protected Carbon Steel) First Edition. 4 pp.

ISO 8447-86. Pipework—Single Overlap Flexible Metal Hoses (Rubber Packing, Limited Tightness, Circular or Polygonal Section, in Protected Carbon Steel) First Edition. 4 pp.

ISO 8448-86. Pipework—Double Overlap Flexible Metal Hoses (Unpacked, Limited Tightness, Circular or Polygonal Section, in Protected Carbon Steel) First Edition. 4 pp.

ISO 8449-86. Pipework—Single Overlap Flexible Metal Hoses (Asbestos Packing, Limited Tightness, Circular or Polygonal Section, in Protected Carbon Steel) First Edition. 4 pp.

ISO 8450-86. Pipework—Single Overlap Flexible Metal Hoses (Unpacked, No Tightness, Circular or Polygonal Section, in Protected Carbon Steel) First Edition. 4 pp.

—Metal—Glossaries

ISO 7369-83. Pipework—Flexible Metallic Hoses—Vocabulary of General Terms First Edition. 9 pp.

—Plastic

BSI BS 4983-92. 1992 Textile Reinforced Thermoplastics Hydraulic Hoses and Hose Assemblies (ISO 3949: 1991). 10 pp.

BSI BS 4983-84. 1984 Textile Reinforced Thermoplastics Hydraulic Hoses and Hose Assemblies. 11 pp.

BSI BS 5173: Part 3-77. (WITHDRAWN) 1977 Amd 4 Rubber and Plastics Hoses and Hose Assemblies Part 3: General Physical Tests (AMD 6471) November 30, 1990 (Superseded by Various Sections of Part 103 of the New edition). 32 pp.

BSI BS 5173: Part 100-92. 1992 Methods of Test for Rubber and Plastics Hoses and Hose Assemblies Part 100: General Introduction. 6 pp.

BSI BS 5173: Part 100-86. 1986 Rubber and Plastics Hoses and Hose Assemblies Part 100: General Introduction. 5 pp.

BSI BS 5173: Sec 101.1-85. (WITHDRAWN) 1985 Rubber and Plastics Hoses and Hose Assemblies Part 101: Determination of Dimensions Section 101.1: Measurement of Dimensions (Excluding Length) (Renumbered as BS EN 24671: 1993). 7 pp.

BSI BS 5173: Sec 101.2-87. 1987 Rubber and Plastics Hoses and Hose Assemblies Part 101: Determination of Dimensions Section 101.2: Measurement of Length of Hoses and Hose Assemblies. 6 pp.

BSI BS 5173: Sec 102.8-87. 1987 Rubber and Plastics Hoses and Hose Assemblies Part 102: Hydraulic Pressure Tests Section 102.8: Pressure Impulse Test for Rigid Helix Reinforced Thermoplastic Hoses. 6 pp.

BSI BS 5409: Part 1-76. 1976 Amd 1 Specification for Nylon Tubing Part 1: Fully Placticized Nylon Tubing Types 11 and 12 for Use Primarily in Pneumatic Installations. 11 pp.

BSI BS 6597-85. 1985 Rubber and Plastics Hoses: Internal Diameters and Tolerances on Length. 3 pp.

BSI BS EN 23994-93. 1993 Plastics Products—Hose of Polymer Reinforced Thermoplastics for Suction and Discharge (ISO 3994: 1977) (Supersedes BS 5780: 1979) (E). 13 pp.

BSI BS EN 24671-93. 1993 Amd 1 Rubber and Plastics Hose and Hose Assemblies—Methods of Measurement of Dimensions (ISO 4671: 1984) (AMD 7519) April 15, 1993. 14 pp.

CEN PREN 855-92. Plastics Hoses and Hose Assemblies—Thermoplastics Textile Reinforced Hydraulic Type—Specification. 9 pp.

CEN EN 23994-93. Plastics Products—Hose of Polymer Reinforced Thermoplastics for Suction and Discharge (ISO 3994: 1977). 8 pp.

CEN EN 24671-93. Rubber and Plastic Hose and Hose Assemblies—Methods of Measurement of Dimensions (ISO 4671: 1984). 8 pp.

INDUSTRY STANDARDS

INTERNATIONAL AND NON-U.S. NATIONAL STANDARDS
SUBJECT INDEX

Hoses

Hoses (Cont.)

—Plastic (Cont.)

ISO 1307-92. Rubber and Plastics Hoses for General-Purpose Industrial Applications—Bore Diameters and Tolerances, and Tolerances on Length Third Edition. 5 pp.

ISO 3949-91. Plastics Hoses and Hose Assemblies—Thermoplastics, Textile-Reinforced, Hydraulic Type—Specification Second Edition. 7 pp.

ISO 3994-77. Plastics Products—Hose of Polymer Reinforced Thermoplastics for Suction and Discharge First Edition; (CEN EN 23994:1993). 8 pp.

ISO 4671-84. Rubber and Plastics Hose and Hose Assemblies—Methods of Measurement of Dimensions First Edition (CEN EN 24671: 1993). 7 pp.

ISO 8331-91. Rubber and Plastics Hoses and Hose Assemblies—Guide to Selection, Storage, Use and Maintenance First Edition. 17 pp.

JIS K 6375-89. Textile Reinforced Thermoplastic Hoses for Hydraulic Use. 14 pp.

JIS K 6771-77. Flexible Vinyl Tube. 13 pp.

—Plastic—Abrasion Testing

BSI BS 5173: Sec 103.9-86. 1986 Rubber and Plastics Hoses and Hose Assemblies Part 103: Physical Tests Section 103.9: Determination of Abrasion Resistance of the Outer Cover. 6 pp.

ISO 7662-88. Rubber and Plastics Hoses—Determination of Abrasion of Lining First Edition. 6 pp.

—Plastic—Adhesion Testing

BSI BS 5173: Sec 103.1-92. (WITHDRAWN) 1992 Rubber and Plastics Hoses and Hose Assemblies Part 103: Physical Tests Section 103.1: Determination of Adhesion Between Components (ISO 8033: 1991) (NOW KNOWN AS BS EN 28033: 1993). 13 pp.

BSI BS 5173: Sec 103.1-86. 1986 Rubber and Plastics Hoses and Hose Assemblies Part 103: Physical Tests Section 103.1: Determination of Adhesion Between Components (Renumbered as BS EN 28033: 1993). 11 pp.

BSI BS EN 28033-93. 1993 Amd 1 Rubber and Plastics Hose —Determination of Adhesion Between Components (ISO 8033: 1991) (AMD 7563) August 15, 1993 (E). 16 pp.

CEN EN 28033-93. Rubber and Plastics Hose—Determination of Adhesion Between Components (ISO 8033: 1991). 12 pp.

ISO 8033-91. Rubber and Plastics Hose—Determination of Adhesion Between Components Second Edition (CEN EN 28033: 1993). 12 pp.

—Plastic—Agricultural Spraying Equipment

JIS K 6339-82. Spray Hoses for Agricultural Use. 7 pp.

—Plastic—Aircraft

ISO 8829-90. Aerospace—Polytetrafluoroethylene (PTFE) Hose Assemblies—Test Methods First Edition. 15 pp.

—Plastic—Automotive

BSI BS 5409: Part 2-78. (OBSOLESCENT) 1978 Specification for Nylon Tubing Part 2: Plasticized and Unplasticized Nylon Tubing Types 11 and 12 for Use Primarily in the Automobile Industry. 16 pp.

—Plastic—Bend Testing

BSI BS 5173: Sec 103.5-85. (WITHDRAWN) 1985 Rubber and Plastics Hoses and Hose Assemblies Part 103: Physical Tests Section 103.5: Bending Tests (Renumbered as BS EN 21746: 1993). 7 pp.

BSI BS EN 21746-93. 1993 Amd 1 Rubber and Plastics Hoses and Tubing—Bending Tests (ISO 1746: 1983) (AMD 7523) April 15, 1993. 14 pp.

CEN EN 21746-93. Rubber and Plastics Hoses and Tubing—Bending Tests (ISO 1746: 1983). 8 pp.

ISO 1746-83. Rubber or Plastics Hoses and Tubing—Bending Tests Second Edition (CEN EN 21746:1993). 6 pp.

—Plastic—Chemical Resistance

BSI BS 5173: Part 5-77. 1977 Rubber and Plastics Hoses and Hose Assemblies Part 5: Chemical Resistance Tests. 8 pp.

—Plastic—Compression Testing

BSI BS 5173: Sec 103.2-90. 1990 Rubber and Plastics Hoses and Hose Assemblies Part 103: Physical Tests Section 103.2: Determination of Crush Resistance of Hoses. 8 pp.

—Plastic—Creep Rupture Strength

CNS K6640-81. Method of Test for Short-Time Rupture Strength of Plastic Rope, Tubing and Fittings (Mar)(7051).

—Plastic—Electrical Insulation

CNS C4370-82. Poly Insulating Line Hose for High Voltage (Nov)(9576).

Hoses (Cont.)

—Plastic—Electrical Resistance

BSI BS 5173: Part 4-77. (WITHDRAWN) 1977 Rubber and Plastics Hoses and Hose Assemblies Part 4: Electrical Tests (Superseded by BS 5173: Sections 104.1 & 104.2: 1991). 8 pp.

BSI BS 5173: Sec 104.1-91. (WITHDRAWN) 1991 Rubber and Plastics Hoses and Hose Assemblies Part 104: Electrical Tests Section 104.1: Measurement of Electrical Resistance of Hoses and Hose Assemblies (Superseded BS BS EN 28031: 1993). 9 pp.

BSI BS EN 28031-93. 1993 Ruber and Plastics Hoses and Hose Assemblies—Determination of Electrical Resistance (ISO 8031: 1987) (Supersedes BS 5173: Sections 104.1: 1991 & 104.2: 1990) (E). 11 pp.

CEN EN 28031-93. Rubber and Plasitics Hoses and Hose Assemblies—Determination of Electrical Resistance (ISO 8031: 1987). 6 pp.

ISO 8031-87. Rubber and Plastics Hoses and Hose Assemblies—Determination of Electrical Resistance First Edition; (CEN EN 28031:1993). 7 pp.

—Plastic—Environmental Testing

BSI BS 5173: Part 6-77. (WITHDRAWN) 1977 Amd 1 Rubber and Plastics Hoses and Hose Assemblies Part 6: Environmental Tests (AMD 5188) July 31, 1986 (Superseded by BS 5173: Sections 106.1: 1989, 106.2: 1990, 106.3: 1986, 106.4: 1989). 9 pp.

—Plastic—Expansion

BSI BS 5173: Sec 102.2-85. (WITHDRAWN) 1985 Amd 1 Rubber and Plastics Hoses and Hose Assemblies Part 102: Hydraulic Pressure Tests Section 102.2: Determination of Volumetric Expansion (AMD 5965) May 31, 1989 (Renumbered as BS EN 26801: 1993). 7 pp.

BSI BS EN 26801-93. 1993 Amd 2 Rubber and Plastics Hoses—Determination of Volumetric Expansion (AMD 7520) April 15, 1993. 14 pp.

CEN EN 26801-93. Rubber and Plastics Hoses—Determination of Volumetric Expansion (ISO 6801: 1983). 6 pp.

ISO 6801-83. Rubber or Plastics Hoses—Determination of Volumetric Expansion First Edition (CEN EN 26801: 1993). 5 pp.

—Plastic—Glossaries

ISO TR8330-86. Rubber and Plastics—Glossary of Terms Used by the Hose Industry First Edition. 7 pp.

—Plastic—Hydrostatic Testing

BSI BS 5173: Sec 102.1-91. 1991 Rubber and Plastics Hoses and Hose Assemblies Part 102: Hydraulic Pressure Tests Section 102.1: Hydrostatic Tests. 10 pp.

ISO 1402-84. Rubber and Plastic Hoses and Hose Assemblies—Hydrostatic Testing Second Edition. 6 pp.

—Plastic—Linings—Flammability Testing

BSI BS 5173: Sec 103.6-90. 1990 Rubber and Plastics Hoses and Hose Assemblies Part 103: Physical Tests Section 103.6: Determination of Ignitability of Lining. 6 pp.

—Plastic—Low Temperature Testing

BSI BS 5173: Sec 106.1-89. (WITHDRAWN) 1989 Rubber and Plastics Hoses and Hose Assemblies Part 106: Environmental Tests Section 106.1: Determination of Low Temperature Flexibility (Renumbered as BS EN 24672: 1993). 6 pp.

BSI BS EN 24672-93. 1993 Amd 1 Rubber and Plastics Hoses—Sub-Ambient Temperature Flexibility Tests (ISO 4672: 1988) (AMD 7524) April 15, 1993. 13 pp.

CEN EN 24672-93. Rubber and Plastics Hoses—Sub-Ambient Temperature Flexibility Tests (ISO 4672: 1988). 6 pp.

ISO 4672-88. Rubber and Plastics Hoses—Sub-Ambient Temperature Flexibility Tests Second Edition (CEN EN 24672: 1993). 6 pp.

—Plastic—Maintenance

ISO 8331-91. Rubber and Plastics Hoses and Hose Assemblies—Guide to Selection, Storage, Use and Maintenance First Edition. 17 pp.

—Plastic—Ozone Resistance

BSI BS 5173: Sec 106.3-91. (WITHDRAWN) 1991 Methods of Test for Rubber and Plastics Hoses and Hose Assemblies Part 106: Environmental Tests Section 106.3: Setermination of Ozone Resistance (ISO 7326: 1991) (Renumbered as BS EN 27326: 1993). 7 pp.

BSI BS EN 27326-93. 1993 Amd 1 Rubber and Plastics Hoses—Assessment of Ozone Resistance Under Static Conditions (ISO 7326: 1991) (AMD 7526) August 15, 1993 (E). 11 pp.

CEN EN 27326-93. Rubber and Plastics Hoses—Assessment of Ozone Resistance Under Static Conditions (ISO 7326: 1991). 6 pp.

Hoses (Cont.)

—Plastic—Ozone Resistance (Cont.)

ISO 7326-91. Rubber and Plastics Hoses—Assessment of Ozone Resistance Under Static Conditions Second Edition (CEN EN 27326: 1993). 7 pp.

—Plastic—Packaging

MOD UK DSTAN 81-39-89. Packaging of Rubber Hoses, Plastics Hoses, and Hose Assemblies Issue 2. 11 pp.

—Plastic—Permeability

BSI BS 5173: Sec 103.11-92. 1992 Methods of Test for Rubber and Plastics Hoses and Hose Assemblies Part 103: Physical Tests Section 103.11: Determination of Gas Permeance (ISO 4080: 1991). 9 pp.

BSI BS 5173: Sec 103.11-88. 1988 Rubber and Plastics Hoses and Hose Assemblies Part 103: Physical Tests Section 103.11: Determination of Gas Permeance. 6 pp.

BSI BS 5173: Sec 103.12-88. 1988 Rubber and Plastics Hoses and Hose Assemblies Part 103: Physical Tests Section 103.12: Determination of Vapour Transmission of Liquids Through Walls. 6 pp.

ISO 4080-91. Rubber and Plastics Hoses and Hose Assemblies—Determination of Permeability to Gas Third Edition. 8 pp.

ISO 8308-93. Rubber and Plastics Hoses and Tubing—Determination of Transmission of Liquids Through Hose and Tubing Walls Second Edition. 8 pp.

—Plastic—Pressure Measurement

BSI BS 5173: Part 2-76. (WITHDRAWN) 1976 Amd 4 Rubber and Plastics Hoses and Hose Assemblies Part 2: Hydraulic Pressure Tests (AMD 5620) May 29, 1987 (Superseded by Sections of BS 5173: Part 102). 21 pp.

BSI BS 5173: Sec 102.4-90. 1990 Rubber and Plastics Hoses and Hose Assemblies Part 102: Hydraulic Pressure Tests Section 102.4: Determination of Volumetric Expansion of Hydraulic Hoses. 7 pp.

BSI BS 5173: Sec 102.5-85. 1985 Rubber and Plastics Hoses and Hose Assemblies Part 102: Hydraulic Pressure Tests Section 102.5: Pressure Impulse Test for High Pressure Hydraulic Hoses. 9 pp.

BSI BS 5173: Sec 102.6-91. (WITHDRAWN) 1991 Methods of Test for Rubber and Plastics Hoses and Hose Assemblies Part 102: Hydraulic Pressure Tests Section 102.6: Pressure Impulse Test with Flexing for High Pressure Hydraulic Hoses (Inverted 'U' Config.) (ISO 6802: 1991). 5 pp.

BSI BS 6596-92. 1992 Ratios of Proof and Burst Pressure to Design Working Pressure for Rubber and Plastics Hoses and Hose Assemblies (ISO 7751: 1991). 6 pp.

BSI BS 6596-85. 1985 Amd 1 Ratios of Proof and Burst Pressure to Design Working Pressure for Rubber and Plastic Hoses and Hose Assemblies. 5 pp.

BSI BS EN 26802-01. 1993 Amd 1 Rubber and Plastic Hose and Hose Assemblies—Wire Reinforced—Hydraulic Impulse Test with Flexing (ISO 6802: 1991) (AMD 7521) April 15, 1993 (E). 12 pp.

CEN EN 26802-93. Rubber and Plastic Hose and Hose Assemblies—Wire Reinforced—Hydraulic Impulse Test with Flexing (ISO 6802: 1991). 5 pp.

ISO 6802-91. Rubber and Plastics Hose and Hose Assemblies with Wire Reinforcements—Hydraulic Impulse Test with Flexing Second Edition; (CEN EN 26802: 1993). 6 pp.

ISO 6803-84. Rubber or Plastics Hoses and Hose Assemblies—Hydraulic Pressure Impulse Test Without Flexing First Edition. 8 pp.

ISO 7751-91. Rubber and Plastics Hoses and Hose Assemblies—Ratios of Proof and Burst Pressure to Design Working Pressure Second Edition. 4 pp.

—Plastic—Radiation Effects

BSI BS 5173: Sec 106.4-89. 1989 Rubber and Plastics Hoses and Hose Assemblies Part 103: Environmental Tests Section 106.4: Determination of Resistance to Ultraviolet Light. 8 pp.

ISO 8580-87. Rubber and Plastics Hoses—Determination of Ultra-Violet Resistance Under Static Conditions First Edition. 7 pp.

—Plastic—Reinforcement—Fracture Testing

BSI BS 5173: Sec 103.7-91. 1991 Rubber and Plastics Hoses and Hose Assemblies Part 103: Physical Tests Section 103.7: Determination of Fracture Resistance of Rigid Polymer Helical Reinforcement in Thermoplastics Hoses. 7 pp.

—Plastic—Vacuum Resistance

BSI BS 5173: Sec 102.9-92. 1992 Methods of Test for Rubber and Plastics Hoses and Hose Assemblies Part 102: Hydraulic Pressure Tests Section 102.9: Determination of Suction Resistance (ISO 7233: 1991). 6 pp.

INTERNATIONAL AND NON-U.S. NATIONAL STANDARDS
SUBJECT INDEX

Hoses

Hoses (Cont.)

—Plastic—Vacuum Resistance (Cont.)

BSI BS 5173: Sec 102.9-85. 1985 Rubber and Plastics Hoses and Hose Assemblies Part 102: Hydraulic Pressure Tests Section 102.9: Determination of Resistance to Vacuum. 3 pp.

ISO 7233-91. Rubber and Plastics Hoses and Hose Assemblies—Determination of Suction Resistance Second Edition. 6 pp.

—PVC

DIN ENGL 16940-64. Extruded Hoses of Non-Rigid PVC (Non-Rigid Polyvinyl Chloride); Permissible Variations for Dimensions for Which Tolerances are Not Indicated (July). 1 p.

—Refrigerants—Refrigeration Systems—Plants—Safety

DIN ENGL 8975 Pt 9-83. Refrigeration Plants; Safety Principles for Design, Equipment and Installation; Flexible Pipe Elements in the Refrigerant Circuit (Oct). 4 pp.

—Rubber

BSI BS 3832-91. 1991 Wire Reinforced Rubber Hoses and Hose Assemblies for Hydraulic Installations (ISO 1436: 1991). 10 pp.

BSI BS 3832-81. 1981 Wire Reinforced Rubber Hoses and Hose Assemblies for Hydraulic Installations. 12 pp.

BSI BS 4586-92. 1992 Spiral Wire Reinforced Rubber Covered Hydraulic Hoses and Hose Assemblies (ISO 3862: 1991). 12 pp.

BSI BS 4586-84. 1984 Spiral Wire Reinforced Rubber Covered Hydraulic Hoses and Hose Assemblies. 14 pp.

BSI BS 4749-91. 1991 Textile Reinforced Rubber Hydraulic Hoses and Hose Assemblies (ISO 4079: 1991). 10 pp.

BSI BS 4749-84. 1984 Textile Reinforced Rubber Hydraulic Hoses and Hose Assemblies. 7 pp.

BSI BS 5173: Part 3-77. (WITHDRAWN) 1977 Amd 4 Rubber and Plastics Hoses and Hose Assemblies Part 3: General Physical Tests (AMD 6471) November 30, 1990 (Superseded by Various Sections of Part 103 of the New edition). 32 pp.

BSI BS 5173: Part 100-92. 1992 Methods of Test for Rubber and Plastics Hoses and Hose Assemblies Part 100: General Introduction. 6 pp.

BSI BS 5173: Part 100-86. 1986 Rubber and Plastics Hoses and Hose Assemblies Part 100: General Introduction. 5 pp.

BSI BS 5173: Sec 101.1-85. (WITHDRAWN) 1985 Rubber and Plastics Hoses and Hose Assemblies Part 101: Determination of Dimensions Section 101.1: Measurement of Dimensions (Excluding Length) (Renumbered as BS EN 24671: 1993). 7 pp.

BSI BS 5173: Sec 101.2-87. 1987 Rubber and Plastics Hoses and Hose Assemblies Part 101: Determination of Dimensions Section 101.2: Measurement of Length of Hoses and Hose Assemblies. 6 pp.

BSI BS 5244-86. 1986 Recommendations for Application, Storage and Life Expiry of Hydraulic Rubber Hose and Assemblies. 11 pp.

BSI BS 6130-81. 1981 Amd 1 Hose and Hose Assemblies for Asphalt and Bitumen. 5 pp.

BSI BS 6597-85. 1985 Rubber and Plastics Hoses: Internal Diameters and Tolerances on Length. 3 pp.

BSI BS EN 24671-93. 1993 Amd 1 Rubber and Plastics Hose and Hose Assemblies—Methods of Measurement of Dimensions (ISO 4671: 1984) (AMD 7519) April 15, 1993. 14 pp.

CEN PREN 853-92. Rubber Hoses and Hose Assemblies—Wire Braid Reinforced Hydraulic Type—Specification. 11 pp.

CEN PREN 854-92. Rubber Hoses and Hose Assemblies—Textile Reinforced Hydraulic Type—Specification. 10 pp.

CEN PREN 856-92. Rubber Hoses and Hose Assemblies—Spiral Wire Reinforced Hydraulic Type—Specification. 10 pp.

CEN PREN 857-92. Rubber Hoses and Hose Assemblies—Wire Braid Reinforced Compact Type for Hydraulic Applications—Specification. 9 pp.

CEN EN 24671-93. Rubber and Plastic Hose and Hose Assemblies—Methods of Measurement of Dimensions (ISO 4671: 1984). 8 pp.

CGSB 20-GP-12MA-89. Braided Water Hose, Knitted or Spiral Wound Reinforcement. 8 pp.

CNS K4005-57. Rubber Hoses, General Rules (Dec)(806). 3 pp.

CNS K6619-86. Method of Test for Rubber Hoses (Sep)(6834).

ISO 1307-92. Rubber and Plastics Hoses for General-Purpose Industrial Applications—Bore Diameters and Tolerances, and Tolerances on Length Third Edition. 5 pp.

ISO 1436-91. Rubber Hoses and Hose Assemblies—Wire-Reinforced Hydraulic Type—Specification Third Edition. 7 pp.

ISO 3862-91. Rubber Hoses and Hose Assemblies—Rubber-Covered, Spiral Wire Reinforced, Hydraulic Type—Specification Second Edition. 9 pp.

Hoses (Cont.)

—Rubber (Cont.)

ISO 4079-91. Rubber Hoses and Hose Assemblies—Textile-Reinforced Hydraulic Type—Specification Second Edition. 7 pp.

ISO 4671-84. Rubber and Plastics Hose and Hose Assemblies—Methods of Measurement of Dimensions First Edition (CEN EN 24671: 1993). 7 pp.

ISO 8331-91. Rubber and Plastics Hoses and Hose Assemblies—Guide to Selection, Storage, Use and Maintenance First Edition. 17 pp.

JIS K 6330-82. Testing Methods for Rubber Hoses. 14 pp.

JIS K 6349-88. Wire Reinforced Rubber Hoses for Hydraulic Use. 13 pp.

JIS K 6379-88. Textile Reinforced Rubber Hoses for Hydraulic Use. 13 pp.

JIS S 2190-83. Bands for Rubber Tube.

MOD UK DSTAN 47-2-01. Rubber Hose Assemblies for Hydraulic Systems Issue 2; Amendment 3. 24 pp.

MOD UK DSTAN 93-71-93. Vulcanized Butyl Rubber for Tubing of Baby Viper Hose Type QX Issue 1. 13 pp.

MOD UK CS 3109. Vulcanized Butyl Rubber for Tubing of Baby Viper Hose (Superseded by Def Stan 93-71).

SNZ NZS/ISO 1436-78. Rubber Products—Hose and Hose Assemblies—Wire Reinforced Hydraulic Type. 7 pp.

SNZ NZS/ISO 3862-80. Rubber Hoses and Hose Assemblies—Rubber-Covered, Spiral Wire Reinforced, Hydraulic Type. 9 pp.

SNZ NZS/ISO 4079-78. Rubber Products—Hoses and Hose Assemblies, Textile Reinforced, for Hydraulic Purposes. 4 pp.

—Rubber—Abrasion Testing

BSI BS 5173: Sec 103.9-86. 1986 Rubber and Plastics Hoses and Hose Assemblies Part 103: Physical Tests Section 103.9: Determination of Abrasion Resistance of the Outer Cover. 6 pp.

ISO 6945-91. Rubber Hoses—Determination of Abrasion Resistance of the Outer Cover Second Edition. 7 pp.

ISO 7662-88. Rubber and Plastics Hoses—Determination of Abrasion of Lining First Edition. 6 pp.

—Rubber—Adhesion Testing

BSI BS 5173: Sec 103.1-92. (WITHDRAWN) 1992 Rubber and Plastics Hoses and Hose Assemblies Part 103: Physical Tests Section 103.1: Determination of Adhesion Between Components (ISO 8033: 1991) (NOW KNOWN AS BS EN 28033: 1993). 13 pp.

BSI BS 5173: Sec 103.1-86. 1986 Rubber and Plastics Hoses and Hose Assemblies Part 103: Physical Tests Section 103.1: Determination of Adhesion Between Components (Renumbered as BS EN 28033: 1993). 11 pp.

BSI BS EN 28033-93. 1993 Amd 1 Rubber and Plastics Hose —Determination of Adhesion Between Components (ISO 8033: 1991) (AMD 7563) August 15, 1993 (E). 16 pp.

CEN EN 28033-93. Rubber and Plastics Hose—Determination of Adhesion Between Components (ISO 8033: 1991). 12 pp.

ISO 8033-91. Rubber and Plastics Hose—Determination of Adhesion Between Components Second Edition (CEN EN 28033: 1993). 12 pp.

—Rubber—Agricultural Spraying Equipment

CNS K4054-80. Sprayer Rubber Hoses for Agricultural Purpose (Sep)(6491).

ISO 1401-87. Rubber Hoses for Agricultural Spraying First Edition. 4 pp.

JIS K 6339-82. Spray Hoses for Agricultural Use. 7 pp.

—Rubber—Air Pumps

CNS K4011-86. Rubber Hoses for Air Pump (Oct)(812).

—Rubber—Appliances

ISO 6804-91. Rubber Hoses and Hose Assemblies for Washing-Machines and Dishwashers —Specification for Inlet Hoses Second Edition. 8 pp.

—Rubber—Automotive—Power Steering Systems

BSI BS 6784-86. 1986 Amd 1 Rubber Hoses and Hose Assemblies for Automobile Power Steering Systems. 12 pp.

—Rubber—Bend Testing

BSI BS 5173: Sec 103.5-85. (WITHDRAWN) 1985 Rubber and Plastics Hoses and Hose Assemblies Part 103: Physical Tests Section 103.5: Bending Tests (Renumbered as BS EN 21746: 1993). 6 pp.

BSI BS EN 21746-93. 1993 Amd 1 Rubber and Plastics Hoses and Tubing—Bending Tests (ISO 1746: 1983) (AMD 7523) April 15, 1993. 14 pp.

Hoses (Cont.)

—Rubber—Bend Testing (Cont.)

CEN EN 21746-93. Rubber and Plastics Hoses and Tubing—Bending Tests (ISO 1746: 1983). 8 pp.

ISO 1746-83. Rubber or Plastics Hoses and Tubing—Bending Tests Second Edition (CEN EN 21746:1993). 6 pp.

—Rubber—Brazing Equipment

CEN PREN 559-91. Rubber Hoses for Welding, Cutting and Allied Processes. 12 pp.

—Rubber—Brewing Equipment

JIS K 6344-82. Rubber Hoses for Brewing.

—Rubber—Chemical Resistance

BSI BS 5173: Part 5-77. 1977 Rubber and Plastics Hoses and Hose Assemblies Part 5: Chemical Resistance Tests. 8 pp.

—Rubber—Compression Testing

BSI BS 5173: Sec 103.2-90. 1990 Rubber and Plastics Hoses and Hose Assemblies Part 103: Physical Tests Section 103.2: Determination of Crush Resistance of Hoses. 8 pp.

—Rubber—Electrical Resistance

BSI BS 5173: Part 4-77. (WITHDRAWN) 1977 Rubber and Plastics Hoses and Hose Assemblies Part 4: Electrical Tests (Superseded by BS 5173: Sections 104.1 & 104.2: 1991). 8 pp.

BSI BS 5173: Sec 104.1-91. (WITHDRAWN) 1991 Rubber and Plastics Hoses and Hose Assemblies Part 104: Electrical Tests Section 104.1: Measurement of Electrical Resistance of Hoses and Hose Assemblies (Superseded BS BS EN 28031: 1993). 9 pp.

BSI BS EN 28031-93. 1993 Ruber and Plastics Hoses and Hose Assemblies—Determination of Electrical Resistance (ISO 8031: 1987) (Supersedes BS 5173: Sections 104.1: 1991 & 104.2: 1990) (E). 11 pp.

CEN EN 28031-93. Rubber and Plasitics Hoses and Hose Assemblies—Determination of Electrical Resistance (ISO 8031: 1987). 6 pp.

ISO 8031-87. Rubber and Plastics Hoses and Hose Assemblies—Determination of Electrical Resistance First Edition; (CEN EN 28031:1993). 7 pp.

—Rubber—Environmental Testing

BSI BS 5173: Part 6-77. (WITHDRAWN) 1977 Amd 1 Rubber and Plastics Hoses and Hose Assemblies Part 6: Environmental Tests (AMD 5188) July 31, 1986 (Superseded by BS 5173: Sections 106.1: 1989, 106.2: 1990, 106.3: 1986, 106.4: 1989). 9 pp.

—Rubber—Expansion

BSI BS 5173: Sec 102.2-85. (WITHDRAWN) 1985 Amd 1 Rubber and Plastics Hoses and Hose Assemblies Part 102: Hydraulic Pressure Tests Section 102.2: Determination of Volumetric Expansion (AMD 5965) May 31, 1989 (Renumbered as BS EN 26801: 1993). 6 pp.

BSI BS EN 26801-93. 1993 Amd 2 Rubber and Plastics Hoses—Determination of Volumetric Expansion (AMD 7520) April 15, 1993. 14 pp.

CEN EN 26801-93. Rubber and Plastics Hoses—Determination of Volumetric Expansion (ISO 6801: 1983). 6 pp.

ISO 6801-83. Rubber or Plastics Hoses—Determination of Volumetric Expansion First Edition (CEN EN 26801: 1993). 5 pp.

—Rubber—Fuel Oils—Ships—Replenishment at Sea

MOD UK NES 2018-88. Specification for the Manufacture of 153mm (6 ins) Bore Hose for Abeam Replenishment at Sea Issue 1 (06.88). 24 pp.

MOD UK NES 2019-88. Specification for the Manufacture of 153mm (6 ins) Bore Hose for Astern Replenishment at Sea Issue 1 (06.88). 24 pp.

—Rubber—Fuel Oils—Submarines—Replenishment at Sea

MOD UK NES 2018-88. Specification for the Manufacture of 153mm (6 ins) Bore Hose for Abeam Replenishment at Sea Issue 1 (06.88). 24 pp.

MOD UK NES 2019-88. Specification for the Manufacture of 153mm (6 ins) Bore Hose for Astern Replenishment at Sea Issue 1 (06.88). 24 pp.

—Rubber—Gas Cutting Equipment

CEN PREN 559-91. Rubber Hoses for Welding, Cutting and Allied Processes. 12 pp.

—Rubber—Gas Welding

CEN PREN 559-91. Rubber Hoses for Welding, Cutting and Allied Processes. 12 pp.

—Rubber—Glossaries

ISO TR8330-86. Rubber and Plastics—Glossary of Terms Used by the Hose Industry First Edition. 7 pp.

SAA AS 1179-72. Glossary of Terms for Rubber Hose Reconfirmed 1988. 22 pp.

INDUSTRY STANDARDS

Hoses (Cont.)

—Rubber—High Pressure
CNS K4066-82. Hydraulic High Pressure Rubber Hose (Class A) (Jun)(9013).
CNS K4067-82. Hydraulic High Pressure Rubber Hose (Class B) (Jun)(9014).

—Rubber—Hydrostatic Testing
BSI BS 5173: Sec 102.1-91. 1991 Rubber and Plastics Hoses and Hose Assemblies Part 102: Hydraulic Pressure Tests Section 102.1: Hydrostatic Tests. 10 pp.
ISO 1402-84. Rubber and Plastic Hoses and Hose Assemblies—Hydrostatic Testing Second Edition. 6 pp.

—Rubber—Identification Systems
MOD UK DEFCON 67-90. Specifications Covering Identification, Marking and Age on Delivery of Rubber Materiel, Assemblies and Rubber Containing Composites 3/90. 2 pp.
MOD UK DEFCON 67-92. Specifications Covering Identification, Marking and Age on Delivery of Rubber Materiel, Assemblies and Rubber Containing Composites (Navy Contracts Only) 2/92. 2 pp.

—Rubber—Linings—Flammability Testing
BSI BS 5173: Sec 103.6-90. 1990 Rubber and Plastics Hoses and Hose Assemblies Part 103: Physical Tests Section 103.6: Determination of Ignitability of Lining. 6 pp.

—Rubber—Low Temperature Testing
BSI BS 5173: Sec 106.1-89. (WITHDRAWN) 1989 Rubber and Plastics Hoses and Hose Assemblies Part 106: Environmental Tests Section 106.1: Determination of Low Temperature Flexibility (Renumbered as BS EN 24672: 1993). 6 pp.
BSI BS EN 24672-93. 1993 Amd 1 Rubber and Plastics Hoses—Sub-Ambient Temperature Flexibility Tests (ISO 4672: 1988) (AMD 7524) April 15, 1993. 13 pp.
CEN EN 24672-93. Rubber and Plastics Hoses—Sub-Ambient Temperature Flexibility Tests (ISO 4672: 1988). 6 pp.
ISO 4672-88. Rubber and Plastics Hoses—Sub-Ambient Temperature Flexibility Tests Second Edition (CEN EN 24672: 1993). 6 pp.

—Rubber—Lubricating Oils—Ships—Replenishment at Sea
MOD UK NES 2016-88. Specification for the Manufacture of 64mm (2 1/2 ins) Bore Hose for Replenishment at Sea Issue 1 (06.88). 21 pp.
MOD UK NES 2017-88. Specification for the Manufacture of 76mm (3 ins) Bore Hose for Replenishment at Sea Issue 1 (06.88). 21 pp.

—Rubber—Lubricating Oils—Submarines—Replenishment at Sea
MOD UK NES 2016-88. Specification for the Manufacture of 64mm (2 1/2 ins) Bore Hose for Replenishment at Sea Issue 1 (06.88). 21 pp.
MOD UK NES 2017-88. Specification for the Manufacture of 76mm (3 ins) Bore Hose for Replenishment at Sea Issue 1 (06.88). 21 pp.

—Rubber—Maintenance
ISO 8331-91. Rubber and Plastics Hoses and Hose Assemblies—Guide to Selection, Storage, Use and Maintenance First Edition. 17 pp.

—Rubber—Metallizing
CEN PREN 559-91. Rubber Hoses for Welding, Cutting and Allied Processes. 12 pp.

—Rubber—Ozone Resistance
BSI BS 5173: Sec 106.3-91. (WITHDRAWN) 1991 Methods of Test for Rubber and Plastics Hoses and Hose Assemblies Part 106: Environmental Tests Section 106.3: Setermination of Ozone Resistance (ISO 7326: 1991) (Renumbered as BS EN 27326: 1993). 7 pp.
BSI BS EN 27326-93. 1993 Amd 1 Rubber and Plastics Hoses—Assessment of Ozone Resistance Under Static Conditions (ISO 7326: 1991) (AMD 7526) August 15, 1993 (E). 11 pp.
CEN EN 27326-93. Rubber and Plastics Hoses—Assessment of Ozone Resistance Under Static Conditions (ISO 7326: 1991). 6 pp.
ISO 7326-91. Rubber and Plastics Hoses—Assessment of Ozone Resistance Under Static Conditions Second Edition (CEN EN 27326: 1993). 7 pp.

—Rubber—Packaging
MOD UK DSTAN 81-39-89. Packaging of Rubber Hoses, Plastics Hoses, and Hose Assemblies Issue 2. 11 pp.

Hoses (Cont.)

—Rubber—Permeability
BSI BS 5173: Sec 103.11-92. 1992 Methods of Test for Rubber and Plastics Hoses and Hose Assemblies Part 103: Physical Tests Section 103.11: Determination of Gas Permeance (ISO 4080: 1991). 9 pp.
BSI BS 5173: Sec 103.11-88. 1988 Rubber and Plastics Hoses and Hose Assemblies Part 103: Physical Tests Section 103.11: Determination of Gas Permeance. 6 pp.
BSI BS 5173: Sec 103.12-88. 1988 Rubber and Plastics Hoses and Hose Assemblies Part 103: Physical Tests Section 103.12: Determination of Vapour Transmission of Liquids Through Walls. 6 pp.
ISO 4080-91. Rubber and Plastics Hoses and Hose Assemblies—Determination of Permeability to Gas Third Edition. 8 pp.
ISO 8308-93. Rubber and Plastics Hoses and Tubing—Determination of Transmission of Liquids Through Hose and Tubing Walls Second Edition. 8 pp.

—Rubber—Pneumatic Tools—Ships
MOD UK NES 741-89. Hose—Rubber—Armoured for Pneumatic Tool Air Supply Issue 2 (06.89). 13 pp.

—Rubber—Pneumatic Tools—Submarines
MOD UK NES 741-89. Hose—Rubber—Armoured for Pneumatic Tool Air Supply Issue 2 (06.89). 13 pp.

—Rubber—Pressure Measurement
BSI BS 5173: Part 2-76. (WITHDRAWN) 1976 Amd 4 Rubber and Plastics Hoses and Hose Assemblies Part 2: Hydraulic Pressure Tests (AMD 5620) May 29, 1987 (Superseded by Sections of BS 5173: Part 102). 21 pp.
BSI BS 5173: Sec 102.4-90. 1990 Rubber and Plastics Hoses and Hose Assemblies Part 102: Hydraulic Pressure Tests Section 102.4: Determination of Volumetric Expansion of Hydraulic Hoses. 7 pp.
BSI BS 5173: Sec 102.5-85. 1985 Rubber and Plastics Hoses and Hose Assemblies Part 102: Hydraulic Pressure Tests Section 102.5: Pressure Impulse Test for High Pressure Hydraulic Hoses. 9 pp.
BSI BS 5173: Sec 102.6-91. (WITHDRAWN) 1991 Methods of Test for Rubber and Plastics Hoses and Hose Assemblies Part 102: Hydraulic Pressure Tests Section 102.6: Pressure Impulse Test with Flexing for High Pressure Hydraulic Hoses (Inverted 'U' Config.) (ISO 6802: 1991). 5 pp.
BSI BS 6596-92. 1992 Ratios of Proof and Burst Pressure to Design Working Pressure for Rubber and Plastics Hoses and Hose Assemblies (ISO 7751: 1991). 6 pp.
BSI BS 6596-85. 1985 Amd 1 Ratios of Proof and Burst Pressure to Design Working Pressure for Rubber and Plastic Hoses and Hose Assemblies. 5 pp.
BSI BS EN 26802-01. 1993 Amd 1 Rubber and Plastic Hose and Hose Assemblies—Wire Reinforced—Hydraulic Impulse Test with Flexing (ISO 6802: 1991) (AMD 7521) April 15, 1993 (E). 12 pp.
CEN EN 26802-93. Rubber and Plastic Hose and Hose Assemblies—Wire Reinforced—Hydraulic Impulse Test with Flexing (ISO 6802: 1991). 5 pp.
ISO 6802-91. Rubber and Plastics Hose and Hose Assemblies with Wire Reinforcements—Hydraulic Impulse Test with Flexing Second Edition; (CEN EN 26802: 1993). 6 pp.
ISO 6803-84. Rubber or Plastics Hoses and Hose Assemblies—Hydraulic Pressure Impulse Test Without Flexing First Edition. 8 pp.
ISO 7751-91. Rubber and Plastics Hoses and Hose Assemblies—Ratios of Proof and Burst Pressure to Design Working Pressure Second Edition. 4 pp.

—Rubber—Radiation Resistance—Ultraviolet
BSI BS 5173: Sec 106.4-89. 1989 Rubber and Plastics Hoses and Hose Assemblies Part 103: Environmental Tests Section 106.4: Determination of Resistance to Ultraviolet Light. 8 pp.
ISO 8308-87. Rubber and Plastics Hoses—Determination of Ultra-Violet Resistance Under Static Conditions First Edition. 7 pp.

—Rubber—Refrigeration Equipment
ISO 5771-81. Rubber Hose and Hose Assemblies for Transferring Anhydrous Ammonia First Edition. 7 pp.

—Rubber—Safety
BSI BS 3212-91. 1991 Flexible Rubber Tubing, Rubber Hose and Rubber Hose Assemblies for Use in LPG Vapour Phase and LPG/Air Installations. 13 pp.

—Rubber—Ships
MOD UK NES 324: Part 5-83. Requirements for Hydraulic Systems Part 5: Flexible Hose, Multiple Spiral Wire Reinforced Rubber Hose Assemblies Issue 1 (03.83). 12 pp.

Hoses (Cont.)

—Rubber—Ships (Cont.)
MOD UK NES 324: Part 6-83. Requirements for Hydraulic Systems Part 6: Flexible Hose 2 Layer Wire Braid Reinforced Rubber Hose Assemblies Issue 1 (03.83). 12 pp.

—Rubber—Steel Wire
BSI BS 3592: Part 1-86. 1986 Steel Wire for Rubber Hose Reinforcement Part 1: Coated Round and Flat Steel Wire for Rubber Hose Reinforcement. 8 pp.
BSI BS 3592: Part 1-01. 1986 Amd 1 Steel Wire for Hose Reinforcement Part 1: Specification for Coated Round and Flat Steel Wire for Rubber Hose Reinforcement (AMD 7283) October 15, 1992. 9 pp.

—Rubber—Submarines
MOD UK NES 374-82. Rubber Flexible Pipes and Bellows for Use in Submarines Issue 1 (02.82). 28 pp.

—Rubber—Vacuum Resistance
BSI BS 5173: Sec 102.9-92. 1992 Methods of Test for Rubber and Plastics Hoses and Hose Assemblies Part 102: Hydraulic Pressure Tests Section 102.9: Determination of Suction Resistance (ISO 7233: 1991). 6 pp.
BSI BS 5173: Sec 102.9-85. 1985 Rubber and Plastics Hoses and Hose Assemblies Part 102: Hydraulic Pressure Tests Section 102.9: Determination of Resistance to Vacuum. 3 pp.
ISO 7233-91. Rubber and Plastics Hoses and Hose Assemblies—Determination of Suction Resistance Second Edition. 6 pp.

—Showers
CEN PREN 1113-93. Showers Hoses for Sanitary Valves (Valve Pressure Range PN 10). 13 pp.

—Sprayers
MOD UK DSTAN 47-28-89. Hose, for Low Pressure Spraying Applications Issue 1. 16 pp.

Hosiery
Use For: Pantyhose; Stockings *See Also:* Clothing; Footwear; Socks

—Colorfastness Testing
CNS L3045-64. Inspection Standard of Knitted Hoses and Socks (for Export) (Jul)(2280) (R 1970). 10 pp.

—Grading
CNS L1003-65. Gradation Standard of Ladies Full-Fashioned Stockings (for Export) (Aug)(2537)(R 1973). 2 pp.

—Identification Systems
BSI BS 7672-93. 1993 Compression, Stiffness and Labelling of Anti-Embolism Hosiery (L). 16 pp.

—Inspection
CNS L3045-64. Inspection Standard of Knitted Hoses and Socks (for Export) (Jul)(2280) (R 1970). 10 pp.

—Sampling
CNS L3045-64. Inspection Standard of Knitted Hoses and Socks (for Export) (Jul)(2280) (R 1970). 10 pp.

—Sizes
ISO 5971-81. Size Designation of Clothes—Pantyhose First Edition. 4 pp.
ISO 7070-82. Size Designation of Clothes—Hosiery First Edition. 4 pp.
JIS L 4006-87. Sizing Systems for Foundation Garments. 9 pp.
JIS L 4007-87. Sizing Systems for Hosiery. 7 pp.
SAA AS 1923-76. Size Coding Scheme for Hosiery for Infants, Children, Women and Men. 12 pp.

—Support
BSI BS 6612-85. 1985 Graduated Compression Hoisery. 14 pp.

—Support—Compression Testing
BSI BS 6612-01. 1985 Amd 1 Graduated Compression Hoisery (AMD 7728) June 15, 1993. 15 pp.
BSI BS 7563-92. 1992 Method for Determination of Compression Values and Stiffness of Non-Prescriptive Graduated Support Hosiery (N). 13 pp.
BSI BS 7672-93. 1993 Compression, Stiffness and Labelling of Anti-Embolism Hosiery (L). 16 pp.

Hospital Beds
See Also: Beds; Medical Electrical Equipment; Medical Equipment
CNS T2029-85. Hospital Beds (Fence Type) (Apr)(11252).
CNS T4014-85. Method of Test for Hospital Beds (Fence Type) (Apr)(11253).
JIS T 9205-83. Hospital Beds.

—Cushions
CNS Z7191-84. Coconut Fibre Cushion for Sick Bed (Dec)(11179).

INTERNATIONAL AND NON-U.S. NATIONAL STANDARDS
SUBJECT INDEX

Hospital Carts
See Also: Carts; Medical Equipment
BSI BS 4068-77. 1977 Amd 1 Hospital Trolleys for Instruments, Dressings and for Anaesthetists' Use Including Angular Trolleys (Supersedes BS 3236: 1960). 9 pp.
BSI BS 5853-80. 1980 Medicine Trolleys. 8 pp.

—Bumpers
BSI BS 4322-68. 1968 Recommendations for Buffering on Hospital Vehicles Such as Trolleys. 8 pp.

—Medicine
CNS T2024-83. Cart, Medicine (Sep)(10582).

Hospital Equipment
Use: Medical Equipment

Hospital Gowns
Use: Gowns (Protective)—Hospital Patient

Hospital Trolleys
Use: Hospital Carts

Hospital Wastes
Use: Medical Wastes

Hospitals
Scope Note: For additional listings, see the subheading Hospitals under specific types of equipment and materials See Also: Casualties; Field Hospitals; Health Care Facilities; Medical Electrical Equipment; Medical Equipment; Medical Services; Military Facilities; Outpatient Clinics

—Design—Firesafety
SNZ NZS 4208-73. Code of Practice for the Design of Hospitals with Respect to Fire Amend: 1, 1976. 20 pp.

—Electrical Installations
DIN VDE 0107-89. Electrical Installations in Hospitals and Locations for Medical Use Outside Hospitals (Nov) (Supersedes DIN 57107 A1/VDE 0107 A1/11.82 with DIN VDE 0108 Part 1 to Part 8 (Issues of 10.89) Supersedes DIN 57108/VDE 0108/12.79). 55 pp.
DIN VDE 0107 Suppl 1-89. Electrical Installations in Hospitals and Locations for Medical Use Outside Hospitals; Extracts from Building and Industrial Safety Regulations (Nov). 64 pp.

—Heating, Ventilating and Air Conditioning Equipment
DIN ENGL 1946 Pt 4-89. Heating, Ventilation and Air Conditioning; HVAC Systems in Hospitals; (VDI Code of Practice) (Dec). 24 pp.

—Lighting
BSI BS 4533: Sec 102.55-86. 1986 Luminaires Part 102: Particular Requirements Section 102.55: Luminaires for Hospitals and Health Care Buildings. 18 pp.
BSI BS 4533: Sec 103.2-86. 1986 Luminaires Part 103: Performance Requirements Section 103.2: Photometric Characteristics of Luminaires for Hospitals and Health Care Buildings. 15 pp.
DIN ENGL 5035 Pt 3-88. Artificial Lighting of Interiors; Hospital Lighting (Sept). 7 pp.
SAA AS 1765-75. Artificial Lighting for Clinical Observation Amdt 1 September 1979. 24 pp.

—Medical Records—Military
NATO STANAG 2348 ED 2 AMD 2-84. Basic Military Hospital (Clinical) Records. 10 pp.

Host Adapters
Use: Small Computer System Interfaces

Hostile Environments
Scope Note: See the subheading Severe Environments under specific types of equipment or materials

Hot Air Balloons
Use: Balloons

Hot Carrier Diodes
Use: Schottky Barrier Diodes

Hot Cathode Tubes
Use: Thermionic Tubes

Hot Cupboards
Use: Food Warmers

Hot Dip Coatings
See Also: Coatings

—Aluminum
BSI BS 6536-85. 1985 Continuously Hot-Dip Aluminium/Silicon Coated Cold Reduced Carbon Steel Sheet and Strip. 16 pp.

Hot Dip Coatings (Cont.)
—Aluminum (Cont.)
CNS G2164-83. Method of Test for Hot-Dip Aluminium Coated Steel Sheets and Strips (Feb)(9999).
CNS G3206-83. Hot-Dip Aluminium Coated Steel Sheets and Strips (Feb)(9998).
CNS H2058-82. Methods of Test for Aluminium Coating (Hot-Dipped) on Iron and Steel (Jan)(8296). 5 pp.
CNS H3099-82. Aluminum Coatings (Hot-Dipped) on Iron and Steel (Jan)(8295).
CNS H3100-82. Recommended Practice for Aluminum Coatings (Hot-Dipped) (Jan)(8297).
ISO 5000-80. Continuous Hot-Dip Aluminium-Silicon-Coated Cold-Reduced Carbon Steel Sheet of Commercial and Drawing Qualities First Edition; DoD Adopted. 15 pp.
JIS G 3314-77. Hot-Dip Aluminium Coated Steel Sheets (R 1980). 15 pp.
JIS H 8642-72. Aluminium Coating (Hot-Dipped) on Iron or Steel. 5 pp.
JIS H 8672-72. Methods of Test for Aluminium Coating (Hot-Dipped) on Iron or Steel. 12 pp.
JIS H 9126-68. Recommended Practice for Aluminium Coating (Hot-Dipped). 9 pp.
SAA AS 1397-93. Steel Sheet and Strip—Hot-Dipped Zinc-Coated or Aluminium/Zinc Coated. 18 pp.
SAA AS 1445-86. Hot-Dipped Zinc-Coated or Aluminium/Zinc-Coated Steel Sheet—76mm Pitch Corrugated. 4 pp.

—Ferroalloys
SAA AS 1650-89. Hot-Dipped Galvanized Coatings on Ferrous Articles (This is a Joint Standard with SANZ NZS 1650). 20 pp.
SNZ NZS/AS 1650-89. Hot-Dipped Galvanized Coatings on Ferrous Articles (This is a Joint Standard with SAA AS 1650). 20 pp.

—Lead
BSI BS 6582-85. 1985 Continuously Hot-Dip Alloy (Terne) Coated Cold Reduced Carbon Steel Flat Rolled Products. 12 pp.
ISO 4999-91. Continuous Hot-Dip Terne (Lead Alloy) Coated Cold-Reduced Carbon Steel Sheet of Commercial and Drawing Qualities Second Edition. 17 pp.

—Silicon
BSI BS 6536-85. 1985 Continuously Hot-Dip Aluminium/Silicon Coated Cold Reduced Carbon Steel Sheet and Strip. 16 pp.
ISO 5000-80. Continuous Hot-Dip Aluminium-Silicon-Coated Cold-Reduced Carbon Steel Sheet of Commercial and Drawing Qualities First Edition; DoD Adopted. 15 pp.

—Steels
MOD UK NES 764-88. Flame Metal Spraying and Hot Dip Galvanizing of Ships Structure and Fittings Issue 3 (01.88). 23 pp.
SAA AS 1214-83. Hot-Dip Galvanized Coatings on Threaded Fasteners (ISO Metric Coarse Thread Series). 8 pp.

—Tin
BSI BS 4479: Part 6-90. 1990 Design of Articles That Are to Be Coated Part 6: Recommendations for Hot-Dip Metal Coatings. 18 pp.
ISO 1111 Pt 1-83. Single Cold-Reduced Tinplate and Single Cold-Reduced Blackplate—Part 1: Electrolytic and Hot-Dipped Tinplate Sheet and Blackplate Sheet First Edition. 16 pp.

—Zinc
BSI BS 729-71. 1971 Hot Dip Coatings on Iron and Steel Articles. 15 pp.
BSI BS 4479: Part 6-90. 1990 Design of Articles That Are to Be Coated Part 6: Recommendations for Hot-Dip Metal Coatings. 18 pp.
CEN PREN 1029-93. Hot Dip Galvanized Coatings on Fabricated Ferrous Products—Specification. 17 pp.
CNS H3102-82. Recommended Practice for Zinc Coatings (Hot-Dipped) (Feb)(8503).
CNS H3116-83. Zinc Coating (Hot Dipped) on Iron and Steel (Feb)(10007).
CSA CAN/CSA-G164-M92. Hot Dip Galvanizing of Irregularly Shaped Articles; (Gen Instr 1 Thru 2). 35 pp.
DIN ENGL 267 Pt 10-88. Fasteners; Technical Delivery Conditions; Hot-Dip Galvanized Parts (Jan). 5 pp.
DIN ENGL 2444-84. Zinc Coatings on Steel Tubes; Quality Standard for Hot Dip Galvanizing of Steel Tubes for Installation Purposes (Jan). 4 pp.
DIN ENGL 50976-89. Corrosion Protection; Hot-Dip Batch Galvanizing; Requirements and Testing (May). 7 pp.
DIN ENGL 50978-85. Testing of Metallic Coatings; Adherence of Hot-Dip Zinc Coatings (Oct). 3 pp.
ISO 1459-73. Metallic Coatings—Protection Against Corrosion by Hot Dip Galvanizing—Guiding Principles First Edition. 4 pp.

Hot Dip Coatings (Cont.)
—Zinc (Cont.)
ISO 1460-92. Metallic Coatings—Hot Dip Galvanized Coatings on Ferrous Materials—Gravimetric Determination of the Mass per Unit Area Second Edition. 5 pp.
ISO 1461-73. Metallic Coatings—Hot Dip Galvanized Coatings on Fabricated Ferrous Products—Requirements First Edition. 4 pp.
ISO 3575-76. Continuous Hot-Dip Zinc-Coated Carbon Steel Sheet of Commercial, Lock-Forming and Drawing Qualities First Edition; DoD Adopted. 17 pp.
ISO 4998-91. Continuous Hot-Dip Zinc-Coated Carbon Steel Sheet of Structural Quality Second Edition. 14 pp.
JIS G 3302-87. Hot-Dip Zinc-Coated Steel Sheets and Coils. 34 pp.
JIS G 3312-87. Prepainted Hot-Dip Zinc-Coated Steel Sheets and Coils. 25 pp.
JIS G 3317-90. Hot-Dip Zinc—5 Percent Aluminium Alloy-Coated Steel Sheets and Coils. 33 pp.
JIS G 3318-90. Prepainted Hot-Dip Zinc—5 Percent Aluminium Alloy-Coated Steel Sheets and Coils. 28 pp.
JIS H 0401-83. Methods of Test for Hot Dip Galvanized Coatings. 16 pp.
JIS H 8641-82. Zinc Hot Dip Galvanizings. 7 pp.
JIS H 9124-87. Recommended Practice for Zinc Coating (Hot-Dipped). 9 pp.
SAA AS 1397-93. Steel Sheet and Strip—Hot-Dipped Zinc-Coated or Aluminium/Zinc Coated. 18 pp.
SAA AS 1445-86. Hot-Dipped Zinc-Coated or Aluminium/Zinc-Coated Steel Sheet—76mm Pitch Corrugated. 4 pp.
SAA AS 1650-89. Hot-Dipped Galvanized Coatings on Ferrous Articles (This is a Joint Standard with SANZ NZS 1650). 20 pp.
SNZ NZS 3403-78. Specification for Hot-Dip Galvanized Corrugated Steel Sheet for Building Purposes. 7 pp.
SNZ NZS 3441-78. Specification for Hot-Dipped Zinc-Coated Steel Coil and Cut Lengths Amend: 1, 1988, 2. 20 pp.

Hot Filtration
Use: Filtration

Hot Finishing
Scope Note: See the subheading Hot Worked under the specific metal or alloy

Hot Forged Products
Scope Note: See the subheading Hot Forged under the specific type of metal

Hot Forging
Use: Forging—Hot

Hot Line Maintenance
Use: Live Wire Working

Hot Peppers
Use: Peppers

Hot Plates (Warmers)
Use For: Hobs (Appliances); Stockpot Stands
See Also: Appliances; Cooking Appliances; Food Service Equipment; Food Warmers
BSI BS 5386: Part 4-91. 1991 Gas Burning Appliances Part 4: Built-In Domestic Cooking Appliances (L). 18 pp.
BSI BS 5386: Part 4-01. 1991 Amd 1 Gas Burning Appliances Part 4: Built-In Domestic Cooking Appliances (AMD 6940) July 15, 1992. 19 pp.
BSI BS 5386: Part 4-83. 1983 Gas Burning Appliances Part 4: Built-In Domestic Cooking Appliances. 12 pp.
CEN PREN 484-91. Dedicated Liquefied Petroleum Gas Appliances—Independent Hotplates (Including Those Incorporating a Grill) for Outdoor Use. 48 pp.
CNS C3061-87. Method of Test for Electric Hotplates (Jan)(4775).
CNS C4143-87. Electric Hotplates (Oct)(4774).
CSA CAN/CSA-C22. 2 NO 64-M91. Household Cooking and Liquid-Heating Appliances; (Gen Instr 1 Thru 2). 91 pp.
JIS C 9201-75. Electric Hotplates.
JIS S 2103-91. Gas Burning Cooking Appliances for Domestic Use. 93 pp.
JIS S 2123-80. Hotplates Used with Town Gas.
SNZ NZS 1910-65. Electric Hot Plates for Domestic Purposes. 12 pp.

—Gas—Identification Systems
CEN EN 30-79. Domestic Cooking Appliances Burning Gas. 50 pp.
CEN PREN 30 (Part 1)-90. Domestic Cooking Appliances Burning Gas—Part 1: Safety. 122 pp.

INDUSTRY STANDARDS

Hot Plates (Warmers) *(Cont.)*
—Gas—Safety
CEN EN 30-79. Domestic Cooking Appliances Burning Gas. 50 pp.
CEN PREN 30 (Part 1)-90. Domestic Cooking Appliances Burning Gas—Part 1: Safety. 122 pp.
—Household—Electric—Safety
CSA C22.2 NO 1335.2.9-93. Portable Electrical Motor-Operated and Heating Appliances: Particular Requirements for Portable Electric Cooking Appliances; (Gen Instr 1). 40 pp.
—Household—Gas
BSI BS 5386: Part 3-80. 1980 Amd 5 Gas Burning Appliances Part 3: Domestic Cooking Appliances Burning Gas (L) (AMD 6642) March 28, 1991. 107 pp.
BSI BS 5386: Part 3-06. 1980 Amd 6 Gas Burning Appliances Part 3: Domestic Cooking Appliances Burning Gas (AMD 6883) July 15, 1992. 118 pp.
—Induction Heating
CSA CAN/CSA-C22. 2 NO 64-M91. Household Cooking and Liquid-Heating Appliances; (Gen Instr 1 Thru 2). 91 pp.
—Ranges
SAA AS 1550-83. Performance of Household Electrical Appliances—Hotplates for Use in Ranges and Built-in Cooking Tops. 16 pp.

Hot Presses
See Also: Presses
CNS B1306-83. Nominal Dimension of Hot Presses (Mar)(10058).
CNS B7239-83. Test Code for Accuracy of Hot Presses (May)(10219).
CNS B7256-83. Test Code for Performance of Hot Presses (Sep)(10539).
JIS B 6548-91. Hot Presses—Test and Inspection Methods. 16 pp.
—Safety
JIS B 6609-83. Safety Standards for Construction of Hot Presses. 8 pp.

Hot Pressing (Ironing)
Use: Ironing

Hot Rolled Products
Scope Note: See the subheading Hot Rolled under specific types of metal

Hot Soak Testing
Use: Soaking Testing

Hot Stamping
Use: Stamping (Forming)

Hot Tubs
See Also: Hydrobaths; Spas; Swimming Pools
—Indoor/Outdoor
CSA CAN/CSA-C22. 2 NO 108-M89. Liquid Pumps; (Gen Instr 1 Thru 2). 68 pp.
CSA 1282B Bull. Electrical Bulletin 1282B April 8, 1985 to C22.2 NO 108. 16 pp.
CSA 1283B Bull. Electrical Bulletin 1283B April 8, 1985 to C22.2 NO 108. 14 pp.
—Lighting
CSA CAN/CSA-C22. 2NO218.1-M89. Spas, Hot Tubs, and Associated Equipment; (Gen Instr 1). 74 pp.
—Remote Control Equipment
CSA CAN/CSA-C22. 2NO218.1-M89. Spas, Hot Tubs, and Associated Equipment; (Gen Instr 1). 74 pp.

Hot Water Bags
Use: Hot Water Bottles

Hot Water Boilers
Use: Boilers

Hot Water Bottles
Use For: Hot Water Bags *See Also:* Bags; Medical Equipment
—Polyethylene
BSI BS 6728: Part 1-86. 1986 Thermoplastics Hot Water Bottles Part 1: Hot Water Bottles Manufactured from PVC Compounds. 8 pp.
BSI BS 6728: Part 2-93. 1993 Thermoplastics Hot Water Bottles Part 2: Specification for Hot Water Bottles Manufactured from Polyolefin Materials. 8 pp.
—Polypropylene
BSI BS 6728: Part 2-93. 1993 Thermoplastics Hot Water Bottles Part 2: Specification for Hot Water Bottles Manufactured from Polyolefin Materials. 8 pp.

Hot Water Bottles *(Cont.)*
—Rubber
BSI BS 1970-84. 1984 Amd 1 Rubber Hot Water Bottles. 8 pp.
CNS T2002-80. Rubber Hot Water Bags (Jul)(5943).
JIS T 9102-75. Rubber Hot Water Bags.
—Steel
CNS S1142-88. Steel Hot Water Bag (Jan)(8743).
CNS S2072-88. Method of Test for Steel Hot Water Bag (Jan)(8744).
JIS S 3013-77. Steel Hot Water Bag.

Hot Water Central Heating Systems
Use: Water Heaters

Hot Water Heaters
Use: Water Heaters

Hot Water Heating Boilers
Use: Boilers

Hot Water Heating Equipment
Use: Water Heaters

Hot Water Space Heating Boilers
Use: Boilers

Hot Water Supply Boilers
Use: Boilers

Hot Work Tool Steels
Use: Tool Steels

Hot Working
Scope Note: See the subheading Hot Worked under the specific metal or alloy

Hotplates (Warmers)
Use: Hot Plates (Warmers)

Hotsticks
Use: Linemen's Sticks

Household Appliances
Use: Appliances

Household Automation Systems
Use: Building Automation Systems—Residential

Houses
Scope Note: See the subheading Houses under the specific type of equipment and products
See Also: Residential Buildings

Housings
Scope Note: For additional listings, see the subheading Housings under the specific types of equipment or products *See Also:* Bearing Housings; Bell Housings; Connector Housings; Instrument Housings; Nacelles; Pillow Blocks; Plunger Housings
—Aircraft—Assembly
SBAC AS 6323 ISSUE 2. Assembly of Housing and Bonding Socket.

Hovercraft
Use: Air Cushion Vehicles

Hoverplatforms
See Also: Air Cushion Vehicles
CAA. Foreword 01.91 (British Hovercraft Safety Requirements). 3 pp.
CAA. Index 01.91 (British Hovercraft Safety Requirements). 15 pp.
CAA Chapter C1 01.74. General Requirements. 2 pp.
—Civil Aviation Authority—Approved Authorities
CAA Chapter A2-1 08.81. General. 4 pp.
CAA. Introductory Note 4.79.
—Construction—Certification
CAA Chapter A2-5 01.74. Appoval of Organisations for Design and Construction of Hoverplatforms. 3 pp.
—Cooling Systems
CAA Chapter C4 01.74. Systems and Machinery. 1 p.
—Design
CAA Chapter C3 01.74. Craft Design. 2 pp.
—Design—Certification
CAA Chapter A2-5 01.74. Appoval of Organisations for Design and Construction of Hoverplatforms. 3 pp.
—Fuel Systems
CAA Chapter C4 01.74. Systems and Machinery. 1 p.

Hoverplatforms *(Cont.)*
—Glossaries
CAA 45.
CAA Chapter A1-2 09.91. Definitions. 8 pp.
—Hydraulic Systems
CAA Chapter C4 01.74. Systems and Machinery. 1 p.
—Loads (Forces)
CAA Chapter C2 01.74. Structures. 1 p.
—Machinery
CAA Chapter C4 01.74. Systems and Machinery. 1 p.
—Safety
CAA Chapter C3 01.74. Craft Design. 2 pp.
CAA Chapter C5 01.74. Safety Equipment. 1 p.
—Structural Design
CAA Chapter C2 01.74. Structures. 1 p.

HP 137 Jetstream Mark 1 Aircraft
See Also: Aircraft
—Antenna Positions
CAA. HP.137 Jetstream and BAe Jetstream 31 (Approved Aerial Positions). 1 p.
—Data Sheets
CAA BA4 ISSUE 4. HP 137 Jetstream, Mark 1 and SAL Jetstream Series 200. 5 pp.

HPLC
Use: Liquid Chromatography

HRC
Use: Hypothetical Reference Circuits

HRC Fuses
Use For: High Rupturing Capacity Fuses
See Also: Current Limiting Fuses
—Cartridge—600 V—Nonrenewable
CSA CAN/CSA-C22. 2 NO 106-M92. HRC Fuses; (Gen Instr 1 Thru 2). 63 pp.
CSA 1437 Bull. Electrical Bulletin 1437 October 23, 1985 to C22.2 NO 106. 2 pp.
—Fuse Holders
CSA C22.2 NO 39-M1987. Fuseholder Assemblies (R 1992); (Gen Instr 1). 38 pp.

HRDP
Use: Hypothetical Reference Digital Paths

HS 125-1000 Aircraft
See Also: Aircraft
—Antenna Positions
CAA. HS 125-1000. 1 p.

HS 125-200 Aircraft
See Also: Aircraft
—Antenna Positions
CAA. HS125-200 and 400 Series (Approved Aerial Positions). 1 p.

HS 125-400 Aircraft
See Also: Aircraft
—Antenna Positions
CAA. HS125-200 and 400 Series (Approved Aerial Positions). 1 p.

HS 125-600 Aircraft
See Also: Aircraft
—Antenna Positions
CAA. HS125-600 and 700 Series (Approved Aerial Positions). 1 p.

HS 125-700 Aircraft
See Also: Aircraft
—Antenna Positions
CAA. HS125-600 and 700 Series (Approved Aerial Positions). 1 p.

HS 125-800 Aircraft
See Also: Aircraft
—Antenna Positions
CAA. HS125-800 (Approved Aerial Positions). 1 p.

HS 748 Aircraft
See Also: Aircraft
—Antenna Positions
CAA. Avro/HS 748 (Approved Aerial Positions). 1 p.

HTP
Use: Hydrogen Peroxide

HTS Lamps
Use: Heat Test Source Lamps

Hub and Bushings Sets
Use: Bushings

Hub Bolts
See Also: Bolts; Wheels

—**Automotive**
CNS D2157-82. Hub Bolts for Automobiles (Oct)(9482).

Hubs
—**Alternator**
BSI BS AU 227-88. 1988 Dimensions of Cylindrical Shaft Ends and Hubs for Alterations for Commercial Vehicles and Buses. 4 pp.
ISO 7467-87. Commercial Vehicles and Buses—Cylindrical Shaft Ends and Hubs for Alternators First Edition; (Corrigendum 1-1990). 7 pp.

—**Bicycles**
BSI BS 6102: Part 13-91. 1991 Cycles Part 13: Specification for Mating Dimensions for Splined Hub and Sprocket (ISO 10230: 1990). 9 pp.
CNS B2049-75. Hub Cogs for Bicycles (Dec)(359). 2 pp.
CNS B2050-75. Front Hubs for Bicycles (Dec)(360). 3 pp.
CNS B2051-75. Rear Hubs for Bicycles (Dec)(361). 13 pp.
CNS B7054-75. Method of Test for the Front Hubs of Bicycles (Dec)(3867). 2 pp.
CNS B7060-75. Method of Test for the Rear Hubs of Bicycles (Dec)(3873). 2 pp.
CNS B7064-75. Method of Test for the Hub Cogs of Bicycles (Dec)(3877). 1 p.
ISO 10230-90. Cycles—Splined Hub and Sprocket—Mating Dimensions First Edition. 4 pp.
JIS D 9418-87. Free Wheels and Hub Cogs for Bicycles.

—**Keys**
CNS B3033-83. Flat Fitting Key (for Thin Hub) (May)(170).

—**Magnetic Tapes—Computer**
JTC1 1858-77. Information Processing—General Purpose Hubs and Reels, with 76 mm (3 in) Centrehole, for Magnetic Tape Used in Interchange Instrumentation Applications First Edition. 6 pp.

—**Magnetic Tapes—Data Storage**
CNS C5083-88. Information Processing — General Purpose Hubs and Reels, with 76mm (3in) Centre Hole for Magnetic Tape Used in Interchange Instrumentation Applications (Dec)(6892).
CNS C7111-80. Type A Hubs and Reels for Magnetic Tape (Requirements for Interchange) (Jul)(5768).
ISO 1858-77. Information Processing—General Purpose Hubs and Reels, with 76 mm (3 in) Centrehole, for Magnetic Tape Used in Interchange Instrumentation Applications First Edition. 6 pp.
OSI ISO 1858-77. Information Processing—General Purpose Hubs and Reels, with 76mm (3in) Centrehole, for Magnetic Tape Used in Interchange Instrumentation Applications. 6 pp.
SAA AS 1073.3-79. Dimensions of Hubs and Reels for Magnetic Tape for Instrumentation Interchange—Part 3: General Purpose Reels with 8 mm (5/16 in) Centre Hole (Bound with Part 1). 15 pp.

—**Magnetic Tapes—Instrumentation**
SAA AS 1073.1-79. Dimensions of Hubs and Reels for Magnetic Tape for Instrumentation Interchange—Part 1: General Purpose Hubs and Reels, with 76 mm (3 in) Centre Hole (Bound with Part 3).

—**Pulleys**
DIN ENGL 15062 Pt 2-82. Cranes; Rope Pulleys; Dimensions of Hubs and Bearings (July). 4 pp.

HUD
Use: Head Up Displays

Hue
Use: Color

Hughes 269A Helicopters
See Also: Helicopters

—**Antenna Positions**
CAA. Hughes 269A (Approved Aerial Positions). 1 p.

Hughes 369 Helicopters
See Also: Helicopters

—**Antenna Positions**
CAA. Hughes 369 (Approved Aerial Positions). 1 p.

Hughes 369 Helicopters *(Cont.)*
—**Certification**
CAA. Hughes 369-369D and E. 3 pp.

Hughes 369D Helicopters
See Also: Helicopters

—**Certification**
CAA. Hughes 369-369D and E. 3 pp.

—**Foreign Airworthiness Directives**
CAA. Hughes 369 Series Helicopters (Foreign Airworthiness Directives). 7 pp.

Hughes 369E Helicopters
See Also: Helicopters

—**Certification**
CAA. Hughes 369-369D and E. 3 pp.

Hull Shell Fittings
Scope Note: Use a more specific term *See:* Sea Tubes

Hulls
See Also: Boats (Marine); Bulkheads; Fuselages; Ships; Structural Forms
JIS F 0201-83. General Rules for Automatic Drawing of Basic Hull Construction Plans.

—**Distance Measuring Equipment**
JIS F 7131-87. Distance Pieces for Ships' Hull.

—**Flanges—Outboard**
DIN ENGL 86041 Pt 2-77. Welding Flanges for Outboard Connections; Nominal Pressure 10, 16, 40, 160 (Dec). 4 pp.

—**Glossaries**
JIS F 0014-79. Glossary of Terms of Shipbuilding (Hull Part—Piping).
JIS F 0015-79. Glossary of Terms for Shipbuilding (Hull Part—Accomodation and Hold).
JIS F 0016-82. Glossary of Terms for Shipbuilding (Hull Part—Testing and Working Practice).
JIS F 0045-92. Pleasure Boat—Vocabulary—Hull.

—**Ground Effect Machines—Design**
CAA Chapter B5-1 App 08.83. Hull Hydrodynamic Design (Amphibious Craft). 1 p.

—**Identification Systems**
BSI BS 7490-91. 1991 Coding System for Hull Identification of Small Craft (ISO 10087: 1990). 7 pp.
ISO 10087-90. Small Craft—Hull Identification—Coding System First Edition. 4 pp.

—**Inventories**
JIS F 0903-74. Small Ships' Supply Standard for Hull Inventory Articles.

—**Painting**
MOD UK NES 760-89. Preparation and Painting of Outer Bottoms of Steel Hulled Surface Ships Issue 2 (11.89). 68 pp.

—**Preservation**
MOD UK NES 766-85. Requirements for the Presentation of the Class Hull Preservation Schedule for Surface Ships Issue 2 (06.85). 29 pp.

—**Structural Members**
ISO 9203 Pt 1-89. Shipbuilding—Topology of Ship Hull Structure Elements—Part 1: Location of Elements First Edition. 4 pp.
ISO 9203 Pt 2-89. Shipbuilding—Topology of Ship Hull Structure Elements—Part 2: Description of Elements First Edition. 4 pp.
ISO 9203 Pt 3-89. Shipbuilding—Topology of Ship Hull Structure Elements—Part 3: Relations of Elements First Edition. 4 pp.

Hum (Electrical)
See Also: Electromagnetic Noise

—**Radio Receivers—Mobile**
CNS C6210-83. Method of Test for Land Mobile Communication FM or PM Receiver, 25-947 MHz (Audio Frequency, Harmonic Distortion Factor, Hum and Noise) (Dec)(10695).

Human Effects (Biomechanical)
Use: Biomechanical Effects (Human Body)

Human Engineering
Use: Ergonomics

Human Factors Engineering
Use: Ergonomics

Human Resources
Use: Personnel Management

Humane Treatment (Laboratory Animals)
Use: Laboratory Animals—Humane Treatment

Humic Acid Content Analysis
—**Lignite**
ISO 5073-85. Brown Coals and Lignites—Determination of Humic Acids First Edition. 5 pp.

Humidifiers
See Also: Dehumidifiers; Heat/Moisture Exchangers; Heating, Ventilating and Air Conditioning Equipment; Humidity

—**Medical—Safety**
BSI BS 5724: Sec 2.24-89. 1989 Medical Electrical Equipment Part 2: Particular Requirements for Safety Section 2.24: Specification for Humidifiers (ISO 8185: 1988). 24 pp.
ISO 8185-88. Humidifiers for Medical Use—Safety Requirements First Edition. 23 pp.

—**Motor Operated**
CSA C22.2 NO 104-M1983. Humidifiers and Evaporative Coolers; (Gen Instr 1 Thru 2). 37 pp.

—**Motor Operated—Cooling Systems**
CSA C22.2 NO 104-M1983. Humidifiers and Evaporative Coolers; (Gen Instr 1 Thru 2). 37 pp.

—**Motor Operated—Heating Equipment**
CSA C22.2 NO 104-M1983. Humidifiers and Evaporative Coolers; (Gen Instr 1 Thru 2). 37 pp.

—**Motor Operated—In Duct**
CSA C22.2 NO 104-M1983. Humidifiers and Evaporative Coolers; (Gen Instr 1 Thru 2). 37 pp.

—**Motor Operated—Residential**
CSA C22.2 NO 104-M1983. Humidifiers and Evaporative Coolers; (Gen Instr 1 Thru 2). 37 pp.

—**Safety**
BSI BS 3456: Sec 202.139-91. 1991 Safety of Household and Similar Electrical Appliances Part 202: Particular Requirements Section 202.139: Room Humidifiers (SUPERSEDES BS 3456 SECTION 2. 39: 1973). 9 pp.
BSI BS 3456: Sec 2.39-73. (WITHDRAWN) 1973 Amd 1 1972-1981 Edition. Specification for Safety of Household Electrical Appliances Part 2: Section 2.39: Room Humidifiers (Superseded by BS 3456: Section 202.139: 1991). 12 pp.

Humidity
See Also: Absolute Humidity; Atmospheric Corrosion Testing; Climatology; Controlled Atmospheres; Corrosion; Damp Heat Testing; Dehumidifiers; Environmental Conditions; Humidifiers; Humidity Test Chambers; Moisture Content Analysis; Perspiration; Precipitation (Meteorology); Psychrometers; Temperature; Vapor Transmission; Water; Wet Bulb Temperature
BSI BS 1339-65. 1965 Definitions, Formulae and Constants Relating to the Humidity of the Air. 22 pp.
CNS C6335-89. Basic Environmental Testing Procedures, Part 2: Tests, Test Z/AD: Composite Temperature /Humidity Cyclic Test (Jul)(12566).
DIN ENGL 50012 Pt 1-86. Climates and Their Technical Application; Methods of Measuring Humidity; General (Jan). 3 pp.
JIS C 0028-88. Basic Environmental Testing Procedures Part 2: Tests, Test Z/AD: Composite Temperature/Humidity Cyclic Test. 13 pp.
MOD UK DEF-1151-66. Determination of Atmospheric Conditions (Relative Humidity) in Magazines, Filling and Storage Rooms. 10 pp.

—**Aircraft**
BSI 3G 100:Pt 2: SUBSEC 3.7-72. 1972 General Requirements for Equipment for Use in Aircraft Part 2: All Equipment Section 3: Environmental Conditions Subsection 3.7: Tropical Exposure. 2 pp.

—**Bituminous Coatings**
MOD UK M 3004/67. Protective PX-2 and Composition Rust Preventative Type B.

—**Clean Rooms**
SAA AS 1807.13-89. Cleanrooms, Workstations and Safety Cabinets—Methods of Test—Part 13: Determination of Relative Humidity in Cleanrooms. 2 pp.

—**Coal Tar Coatings**
MOD UK M 3004/67. Protective PX-2 and Composition Rust Preventative Type B.

—**Coatings**
CGSB 1-GP-71 METH 113.5-74. Methods of Testing Paints and Pigments Corrosion Humidity Cabinet Test. 5 pp.

INTERNATIONAL AND NON-U.S. NATIONAL STANDARDS
SUBJECT INDEX — Humidity

Humidity (Cont.)
—**Connectors**
CEN PREN 2591 (Part C4)-92. Elements of Electrical and Optical Connection Test Methods Part C4—Damp Heat Steady State. 6 pp.
CNS C6142-86. Method of Test for Low Frequency (Below 3 MHz) Electrical Connectors (TP-31 Humidity Test) (Sep)(8220).

—**Corrosion Inhibitors**
CGSB 31-GP-0A METH 13.1-57. Methods of Testing Corrosion-Prevention Materials and Processes Humidity Test. 9 pp.

—**Doors**
BSI BS 5369-87. 1987 Methods of Testing Doors; Behaviour Under Humidity Variations of Door Leaves Placed in Successive Uniform Climates. 7 pp.
CEN EN 43-85. Methods of Testing Doors: Behaviour Under Humidity Variations of Door Leaves Placed in Successive Uniform Climates. 5 pp.
DIN ENGL EN 43-90. Methods of Testing Doors; Behaviour Under Humidity Variations of Door Leaves Placed in Successive Uniform Climates (Nov). 3 pp.
ISO 6444-80. Door Leaves—Test of Behaviour Under Humidity Variations (Successive Uniform Climates) First Edition. 3 pp.

—**Electrotechnical Equipment—Environmental Conditions**
BSI BS 7527: Sec 2.1-91. 1991 Classification of Environmental Conditions Part 2: Environmental Conditions Appearing in Nature Section 2.1: Temperature and Humidity (IEC 721-2-1: 1982). 31 pp.
IEC 721 Pt 2-1-82. Classification of Environmental Conditions Part 2: Environmental Conditions Appearing in Nature Temperature and Humidity First Edition; (Amendment 1-1987). 46 pp.

—**Fiberboard**
BSI BS EN 318-93. 1993 Fibreboards—Determination of Dimensional Changes Associated with Changes in Relative Humidity (R). 11 pp.
CEN PREN 318-90. Fibreboards—Determination of Dimensional Changes Associated with Changes in Relative Humidity. 7 pp.
CEN EN 318-93. Fibreboards—Determination of Dimensional Changes Associated with Changes in Relative Humidity. 8 pp.

—**Floor Coverings—Textile**
BSI BS 4682: Part 2-88. 1988 Amd 1 Methods of Test for the Dimensional Stability of Textile Floor Coverings Part 2: Determination of Dimensional Changes Due to Changes in Ambient Humidity. 4 pp.

—**Gases—Manufacturing**
JIS Z 8806-81. Methods of Humidity Measurement. 42 pp.

—**Gases—Mining**
JIS Z 8806-81. Methods of Humidity Measurement. 42 pp.

—**Lacquers**
SAA AS 1580.101. 5-86. Paints and Related Materials—Methods of Test—Part 101.5: Conditions of Test, Temperature and Humidity Controlled (R 1992). 1 p.
SAA AS 1580.452. 1-92. Paints and Related Materials—Methods of Test—Part 452.1: Resistance to Humidity Under Condensation Conditions (in Professional Packages 30, 39). 3 pp.
SNZ NZS/AS 1580. 101.5-86. Methods of Test for Paints and Related Materials Part 101.5: Conditions of Test, Temperature and Humidity Controlled (This is a Joint Standard with SAA AS 1580.101.5). 1 p.
SNZ NZS/AS 1580. 452.1-92. Methods of Test for Paints and Related Materials Part 452.1: Resistance to Humidity (This is a Joint Standard with SAA AS 1580.452.1). 3 pp.

—**Medical Electrical Equipment**
CSA CAN/CSA-C22. 2NO601.1-M90. Medical Electrical Equipment, Part 1: General Requirements for Safety; (Gen Instr 1). 240 pp.

—**Optical Waveguide Connectors**
CNS C6117-81. Method of Test for Fiber Optic Devices (Humidity Test Procedure for Fiber Optic Connecting Devices) (Jul)(7639).

—**Paints**
SAA AS 1580.101. 5-86. Paints and Related Materials—Methods of Test—Part 101.5: Conditions of Test, Temperature and Humidity Controlled (R 1992). 1 p.
SAA AS 1580.452. 1-92. Paints and Related Materials—Methods of Test—Part 452.1: Resistance to Humidity Under Condensation Conditions (in Professional Packages 30, 39). 3 pp.

Humidity (Cont.)
—**Paints (Cont.)**
SNZ NZS/AS 1580. 101.5-86. Methods of Test for Paints and Related Materials Part 101.5: Conditions of Test, Temperature and Humidity Controlled (This is a Joint Standard with SAA AS 1580.101.5). 1 p.
SNZ NZS/AS 1580. 452.1-92. Methods of Test for Paints and Related Materials Part 452.1: Resistance to Humidity (This is a Joint Standard with SAA AS 1580.452.1). 3 pp.

—**Papers**
SAA AS 1301.P416 S-86. Methods of Test for Pulp and Paper (Metric Units)—Part P416s: Determination of Relative Humidity for Paper and Paperboard Testing Amdt 1 August 1987 (This is a Joint Standard with SANZ NZS 1301).
SNZ NZS/AS 1301. P416S-86. Methods of Test for Pulp and Paper Determination of Relative Humidty for Paper and Paperboard Testing. Plus Amendment 1 Amend: 1, 1987 (This is a Joint Standard with SAA AS 1301.P416S). 6 pp.

—**Radio Receivers—Mobile**
CNS C6214-83. Method of Test for Land Mobile Communication FM or PM Receiver, 25-947 MHz (Audio Sensitivity, Temperature and High Humidity) (Dec)(10699).

—**Reinforced Plastics**
CEN PREN 2823-91. Fibre Reinforced Plastics Test Method for the Determination of the Effect of Exposure to Humid Atmosphere on Physical and Mechanical Characteristics. 9 pp.

—**Safety Glass—Automotive**
ISO 3917-92. Road Vehicles—Safety Glazing Materials—Test Methods for Resistance to Radiation, High Temperature, Humidity, Fire and Simulated Weathering Second Edition. 8 pp.

—**Semiconductor Devices**
CNS C6046-88. Environmental Testing Methods and Endurance Testing Methods for Discrete Semiconductor Devices (Cycle Test for Temperature and Humidity) (Jun)(5071).

—**Steel Coatings**
CGSB 1-GP-71 METH 113.6-79. Methods of Testing Paints and Pigments Corrosion Humidity Cabinet Test, One Coat of Material Applied to Steel Panels. 1 p.
CGSB 1-GP-71 METH 113.8-79. Methods of Testing Paints and Pigments Corrosion Humidity Cabinet Test, Two Coats of Material Applied to Steel Panels. 1 p.

—**Varnishes**
SAA AS 1580.101. 5-86. Paints and Related Materials—Methods of Test—Part 101.5: Conditions of Test, Temperature and Humidity Controlled (R 1992). 1 p.
SAA AS 1580.452. 1-92. Paints and Related Materials—Methods of Test—Part 452.1: Resistance to Humidity Under Condensation Conditions (in Professional Packages 30, 39). 3 pp.
SNZ NZS/AS 1580. 101.5-86. Methods of Test for Paints and Related Materials Part 101.5: Conditions of Test, Temperature and Humidity Controlled (This is a Joint Standard with SAA AS 1580.101.5). 1 p.
SNZ NZS/AS 1580. 452.1-92. Methods of Test for Paints and Related Materials Part 452.1: Resistance to Humidity (This is a Joint Standard with SAA AS 1580.452.1). 3 pp.

Humidity Cabinets
See Also: Containers; Humidity Test Chambers
CENELEC HD 98 S1-88. Test Enclosures of Non-Injection Type Constant Relative Humidity. 2 pp.
IEC 260-68. Test Enclosures of Non-Injection Type for Constant Relative Humidity First Edition; (Errata—Dec 1969). 30 pp.
ISO 483-88. Plastics—Small Enclosures for Conditioning and Testing Using Aqueous Solutions to Maintain Relative Humidity at Constant Value First Edition. 8 pp.

—**Coatings**
CGSB 1-GP-71 METH 113.5-74. Methods of Testing Paints and Pigments Corrosion Humidity Cabinet Test. 5 pp.

—**Hydraulic Cements**
CNS A2041-86. Moist Tank Cabinet and Room Used in the Testing of Hydraulic Cements (Dec)(3037). 2 pp.

—**Steel Coatings**
CGSB 1-GP-71 METH 113.6-79. Methods of Testing Paints and Pigments Corrosion Humidity Cabinet Test, One Coat of Material Applied to Steel Panels. 1 p.

Humidity Cabinets (Cont.)
—**Steel Coatings (Cont.)**
CGSB 1-GP-71 METH 113.8-79. Methods of Testing Paints and Pigments Corrosion Humidity Cabinet Test, Two Coats of Material Applied to Steel Panels. 1 p.

Humidity Indicators
See Also: Dehumidifiers; Indicating Instruments; Moisture Content Analysis
MOD UK DSTAN 44-2: Part 1-83. Desiccant Containers, Dehumidifier and Indicators, Humidity, Plug Part 1: General Requirements Issue 3. 37 pp.
MOD UK DSTAN 44-2: Part 2-01. Desiccant Containers, Dehumidifier and Indicators, Humidity, Plug Part 2: Desiccant Containers, Dehumidifier (Tubular Type) Issue 3; Amendment 1. 27 pp.
MOD UK DSTAN 44-2: Part 3-01. Desiccant Containers, Dehumidifier and Indicators, Humidity, Plug Part 3: Desiccant Containers, Dehumidifier (Breather Tube Type) Issue 3; Amendment 1. 43 pp.
MOD UK DSTAN 44-2: Part 3-93. Desiccant Containers, Dehumidifier Part 3: Desiccator and Indicator Types Issue 4. 20 pp.
MOD UK DSTAN 44-2: Part 4-83. Desiccant Containers, Dehumidifier and Indicators, Humidity, Plug Part 4: Desiccant Containers, Dehumidifier (Sachet Type, with Sight Window) (Sachet Type, Without Sight Window) Issue 2. 15 pp.
MOD UK DSTAN 66-14-01. Desiccant Containers, Dehumidifier and Indicators, Humidity, Plug Indicator, Humidity, Plug Issue 3; Amendment 1. 19 pp.
MOD UK DSTAN 66-14-93. Indicators, Humidity, Plug Issue 4. 9 pp.

—**Paper**
MOD UK DSTAN 81-50-88. Paper, Humidity Indicator (Colour Intensified) Issue 1. 10 pp.
MOD UK DEF-1296-58. Paper, Humidity Indicator, Cobalt Chloride (Reprinted April 1970). 3 pp.
MOD UK TS 10232. Base Material for Paper Humidity Indicator, Colour Intensified.

Humidity Ovens
See Also: Ovens

—**Laboratory Equipment**
BSI BS 3718-64. (WITHDRAWN) 1964 Laboratory Humidity Ovens (Non-Injection Type). 22 pp.
BSI BS 3898-65. (WITHDRAWN) 1965 Amd 1 Laboratory Humidity Ovens (Injection Type). 27 pp.
SNZ NZS 2112-66. Specification for Laboratory Humidity Ovens (Injection Type) Amend: 1, 1979, 1A, 1979. 28 pp.

Humidity Rooms
Use: Humidity Test Chambers

Humidity Tables
Use: Tables (Data)—Humidity

Humidity Test Chambers
Use For: Humidity Rooms; Moist Rooms
See Also: Controlled Atmospheres; Humidity; Test Chambers

—**Hydraulic Cements**
CNS A2041-86. Moist Tank Cabinet and Room Used in the Testing of Hydraulic Cements (Dec)(3037). 2 pp.

—**Plastics**
ISO 483-88. Plastics—Small Enclosures for Conditioning and Testing Using Aqueous Solutions to Maintain Relative Humidity at Constant Value First Edition. 8 pp.

—**Tables (Data)**
BSI BS 4833-86. 1986 Hygrometric Tables for Use in the Testing and Operation of Environmental Enclosures. 58 pp.
BSI BS 4833-01. 1986 Amd 1 Hygrometric Tables for Use in the Testing and Operation of Environmental Enclosures (AMD 6957) February 28, 1992 (G). 59 pp.

Humidity Testing
Use: Humidity

Humus Content Analysis
—**Soil**
DIN ENGL 19684 Pt 2-77. Methods of Soil Analysis for Water Management for Agricultural Purposes; Chemical Laboratory Tests; Determination of the Humus Content of the Soil (Feb). 2 pp.

INTERNATIONAL AND NON-U.S. NATIONAL STANDARDS
SUBJECT INDEX
Hydraulic

Humus Content Analysis (Cont.)
—Soil Analysis—Ignition
SNZ NZS 4402: Pt 3:TEST 3.1.2-86. Methods of Testing Soils for Civil Engineering Purposes Part 3: Soil Chemical Tests. Section 3.1: Determination of the Organic Matter Content. Test 3.1.2: Subsidiary Method by Ignition. 2 pp.

—Soil Analysis—Volumetric Analysis
SNZ NZS 4402: Pt 3:TEST 3.1.1-86. Methods of Testing Soils for Civil Engineering Purposes Part 3: Soil Chemical Tests. Section 3.1: Determination of the Organic Matter Content. Test 3.1.1: Standard Method by Titration. 4 pp.

Hunting
See Also: Arrows

—Classification
BSI BS 1000: (636/639)-81. 1981 Universal Decimal Classification (UDC). English Full Edition (636/639): Animal Breeding. Animal Produce. Hunting. Fishing. 56 pp.
SNZ NZS/BS 1000 (636/639)-81. Universal Decimal Classification Animal Breeding. Animal Produce. Hunting. Fishing. 56 pp.

Hunting President 2 Aircraft
See Also: Aircraft

—Mandatory Aircraft Modifications and Inspections Summaries
CAA. Hunting President 2 and 2A. 2 pp.

Hunting President 2A Aircraft
See Also: Aircraft

—Mandatory Aircraft Modifications and Inspections Summaries
CAA. Hunting President 2 and 2A. 2 pp.

HVAC
Use: Heating, Ventilating and Air Conditioning Equipment

HVIK
Use: Insect Electrocution Devices

HVOC Power Transmission
Use: Power Transmission—DC

Hybrid Circuits
Use For: Hybrid Integrated Circuits
See Also: Integrated Circuits
CECC CECC 63 000 ISSUE 1-84. Generic Specification: Film and Hybrid Integrated Circuits (En, Fr, Ge) AMD 1 (En, Fr, Ge) AMD 2 (En, Fr, Ge) AMD 3 (En, Fr, Ge) AMD 4 (En, Fr, Ge) AMD 5 (En, Fr, Ge). 230 pp.
CECC CECC 63 100 ISSUE 1-84. Sectional Specification: Film and Hybrid Integrated Circuits (En, Fr, Ge) AMD 1 (En, Fr, Ge) AMD 2 (En, Fr, Ge) AMD 3 (En, Fr, Ge). 67 pp.
CECC CECC 63 200 ISSUE 1-84. Sectional Specification: Film and Hybrid Integrated Circuits; (Capability Approval) (En, Fr, Ge) AMD 1 (En, Fr, Ge) AMD 2 (En, Fr, Ge) AMD 3 (En, Fr, Ge) AMD 4 (En, Fr, Ge). 275 pp.
CECC CECC 63 201 ISSUE 1-84. Blank Detail Specification: Film and Hybrid Integrated Circuits; (Capability Approval) (En, Fr, Ge) AMD 1 (En, Fr, Ge) AMD 2 (En, Fr, Ge). 36 pp.
IECQ QC 760200-92. Semiconductor Devices Integrated Circuits Part 22: Sectional Specification for Film Integrated Circuits and Hybrid Film Integrated Circuits on the Basis of the Capability Approval Procedures (IEC 748-22 ED 1). 116 pp.

—Film
BSI BS QC 763000-90. 1990 Film and Hybrid Film Integrated Circuits. Generic Specification (Document Number Changed to BS QC 760000: 1990 by Amendment 1). 45 pp.
IEC 748 Pt 20-88. Semiconductor Devices Integrated Circuits Part 20: Generic Specification for Film Integrated Circuits and Hybrid Film Integrated Circuits First Edition (IECQ QC 760000). 87 pp.
IEC 748 Pt 21-91. Semiconductor Devices Integrated Circuits Part 21: Sectional Specification for Film Integrated Circuits and Hybrid Film Intergrated Circuits on the Basis of Qualification Approval Procedure First Edition (IECQ QC 760100). 36 pp.
IECQ QC 760000-88. Semiconductor Devices Integrated Circuits Part 20: Generic Specification for Film Integrated Circuits and Hybrid Film Integrated Circuits (IEC 748-20 ED 1). 87 pp.
IECQ QC 760100-91. Semiconductor Devices Integrated Circuits Part 21: Sectional Specification for Film Integrated Circuits and Hybrid Film Integrated Circuits on the Basis of Qualification Approval Procedure (IEC 748-21 ED 1). 36 pp.

Hybrid Circuits (Cont.)
—Film (Cont.)
IECQ QC 760101-91. Semiconductor Devices—Integrated Circuits Part 21: Section One: Blank Detail Specification for Film Integrated Circuits and Hybrid Film Integrated Circuits on the Basis of Qualification Approval Procedure (IEC 748-21-1 ED 1). 24 pp.
IECQ QC 760201-91. Semiconductor Devices Integrated Circuits Part 22: Section One: Blank Detail Specification for Film Integrated Circuits and Hybrid Film Integrated Circuits on the Basis of the Capability Approval Procedures (IEC 748-22-1 ED 1). 30 pp.

—Film—Quality Assurance
BSI BS QC 760000-90. 1990 Amd 1 Film and Hybrid Film Integrated Circuits. General Specification (IEC 748-20: 1988) (AMD 6754) September 1991. 46 pp.
BSI BS QC 760200-92. 1992 Semiconductor Devices Integrated Circuits. Sectional Specification for Film Integrated Circuits and Hybrid Film Integrated Circuits: Capability Approval (IEC 748-22: 1992). 59 pp.
BSI BS QC 760201-92. 1992 Semiconductor Devices. Integrated Circuits Blank Detail Specification Film and Hybrid Integrated Circuits: Capability Approval (IEC 748-22-1: 1991). 12 pp.
IEC 748 Pt 21-1-91. Semiconductor Devices—Integrated Circuits Part 21: Section One: Blank Detail Specification for Film Integrated Circuits and Hybrid Film Integrated Circuits on the Basis of Qualification Approval Procedure First Edition (IECQ QC 760101). 24 pp.
IEC 748 Pt 22-92. Semiconductor Devices Integrated Circuits Part 22: Sectional Specification for Film Integrated Circuits and Hybrid Integrated Circuits on the Basis of the Capability Approval Procedures First Edition (IECQ QC 760200). 116 pp.
IEC 748 Pt 22-1-91. Semiconductor Devices Integrated Circuits Part 22: Section One: Blank Detail Specification for Film Integrated Circuits and Hybrid Film Integrated Circuits on the Basis of the Capability Approval Procedures First Edition (IECQ QC 760201). 30 pp.

—Preferred Products List
CECC CECC MUAHAG Vol 2 IS 4-90. Preferred Products List; Resistors and Potentiometers (En, Fr, Ge). 59 pp.

—Quality Assurance
BSI BS CECC 63000-90. 1990 Film and Hybrid Integrated Circuits: Generic Specification. 76 pp.
BSI BS CECC 63100-85. 1985 Amd 3 Harmonized System of Quality Assessment for Electronic Components: Sectional Specification: Film and Hybrid Integrated Circuits (T) (AMD 6621) April 30, 1991. 36 pp.
BSI BS CECC 63101-85. 1985 Amd 1 Film and Hybrid Integrated Circuits: Blank Detail Specification (AMD 6595) July 31, 1990. 13 pp.
BSI BS CECC 63200-85. 1985 Amd 4 Harmonized System of Quality Assessment for Electronic Components: Sectional Specification: Film and Hybrid Integrated Circuits (Capability Approval) (AMD 6623) April 30, 1991. 90 pp.
BSI BS CECC 63201-85. 1985 Amd 2 Film and Hybrid Integrated Circuits: Blank Detail Specification (Capability Approval) (AMD 6583) July 31, 1990. 18 pp.

—Thin Film—Cable Television
CECC CECC 63 101-007 ISSUE 1-88. NL-CECC 63 101-007 Issue 1; CATV Hybrid Modules Manufacturer's Type Numbers BGY 84 H, BGY 85 H (En). 12 pp.
CECC CECC 63 101-009 ISSUE 2-91. NL CECC 63 101-009 Issue 1; CATV Hybrid Module Manufacterer's Type Number BGD 502 (En). 13 pp.
CECC CECC 63 101-010 ISSUE 2-91. NL CECC 63 101-010 Issue 2; CATV Hybrid Module Manufacturer's Type Number BGY80 (En). 12 pp.
CECC CECC 63 101-011 ISSUE 2-91. NL CECC 63 101-011 Issue 2; CATV Hybrid Module Manufacturer's Type Number BGY81 (En). 12 pp.
CECC CECC 63 101-012 ISSUE 2-91. NL CECC 63 101-012 Issue 2; CATV Hybrid Module Manufacturer's Type Number BGY81/01 (En). 12 pp.
CECC CECC 63 101-013 ISSUE 2-91. NL CECC 63 101-013 Issue 2; CATV Hybrid Module Manufacturer's Type Number BGY85/04 (En). 12 pp.
CECC CECC 63 101-014 ISSUE 2-91. NL CECC 63 101-014 Issue 2; CATV Hybrid Module Manufacturer's Type Number BGY85A/04 (En). 12 pp.
CECC CECC 63 101-015 ISSUE 2-91. NL CECC 63 101-015 Issue 2; CATV Hybrid Module Manufacturer's Type Number BGY85H/01 (En). 11 pp.

Hybrid Circuits (Cont.)
—Thin Film—Cable Television (Cont.)
CECC CECC 63 101-016 ISSUE 2-91. NL CECC 63 101-016 Issue 2; CATV Hybrid Module Manufacturer's Type Number BGY 82 (En). 11 pp.
CECC CECC 63 101-017 ISSUE 2-91. NL CECC 63 101-017 Issue 2; CATV Hybrid Module Manufacturer's Type Number BGY83 (En). 11 pp.
CECC CECC 63 101-018 ISSUE 2-91. NL CECC 63 101-018 Issue 2; CATV Hybrid Module Manufacturer's Type Number BGD504 (En). 11 pp.
CECC CECC 63 101-019 ISSUE 2-91. NL CECC 63 101-019 Issue 2; CATV Hybrid Module Manufacturer's Type Number BGY580 (En). 11 pp.
CECC CECC 63 101-020 ISSUE 2-91. NL CECC 63 101-020 Issue 2; CATV Hybrid Module Manufacturer's Type Number BGY581 (En). 11 pp.
CECC CECC 63 101-021 ISSUE 2-91. NL CECC 63 101-021 Issue 2; CATV Hybrid Module Manufacturer's Type Number BGY582 (En). 11 pp.
CECC CECC 63 101-022 ISSUE 2-91. NL CECC 63 101-022 Issue 2; CATV Hybrid Module Manufacturer's Type Number BGY583 (En). 11 pp.
CECC CECC 63 101-023 ISSUE 2-91. NL CECC 63 101-023 Issue 2; CATV Hybrid Module Manufacturer's Type Number BGY584 (En). 11 pp.
CECC CECC 63 101-024 ISSUE 2-91. NL CECC 63 101-024 Issue 2; CATV Hybrid Module Manufacturer's Type Number BGY584A (En). 11 pp.
CECC CECC 63 101-025 ISSUE 2-91. NL CECC 63 101-025 Issue 2; CATV Hybrid Module Manufacturer's Type Number BGY585 (En). 11 pp.

—Thin Film—Preferred Products List
CECC CECC MUAHAG Vol 2 IS 4-90. Preferred Products List; Resistors and Potentiometers (En, Fr, Ge). 59 pp.

Hybrid Integrated Circuits
Use: Hybrid Circuits

Hybrid Junctions
Use For: Microwave Hybrids See Also: Coaxial Connectors; Couplers; Microwave Couplers; Microwave Isolators; Power Dividers

—Preferred Products List
CECC CECC MUAHAG Vol 12 IS 1-90. Preferred Products List; Microwave Components (En, Fr, Ge). 76 pp.

Hybrid Rings
See Also: Local Area Networks

—Fiber Distributed Data Interface
IEC DIS 9314 Pt 5-91. Information Processing Systems—Fibre Distributed Data Interface (FDDI)—Part 5: Hybrid Ring Control (HRC) ***CD-ROM ONLY***. 129 pp.
ISO DIS 9314 Pt 5-91. Information Processing Systems—Fibre Distributed Data Interface (FDDI)—Part 5: Hybrid Ring Control (HRC) ***CD-ROM ONLY***. 129 pp.
JTC1 DIS9314 Pt 5-91. Information Processing Systems—Fibre Distributed Data Interface (FDDI)—Part 5: Hybrid Ring Control (HRC) ***CD-ROM ONLY***. 129 pp.

Hydrated Lime
Use: Calcium Hydroxides

Hydrated Silica
Use: Silicic Acid

Hydraulic Accumulators
See Also: Accumulators; Hydraulic Cylinders; Hydraulic Equipment; Hydraulic Presses; Hydropneumatic Accumulators
CNS B2615-81. Bladder Type Accumulators for Oil Hydraulic Use (Oct)(7953).
CNS B7172-81. Method of Test for Bladder Type Accumulators (Oil Hydraulic Use) (Oct)(7954).
JIS B 8358-76. Bladder Type Accumulators for Oil Hydraulic Use (R 1984). 21 pp.

Hydraulic Brakes
See Also: Automotive Equipment; Brakes; Disk Brakes; Drum Brakes; Hydraulic Equipment; Wheel Cylinders

—Agricultural Equipment—Couplings
BSI BS 4742: Part 7-89. 1989 Hydraulic Equipment for Agricultural Machinery Part 7: Port Connections for Agricultural Trailed Vehicle Hydraulic Brake Couplings. 4 pp.

—Automotive
ISO 7635-91. Road Vehicles—Air or Air over Hydraulic Braking Systems—Measurement of Braking Performance First Edition. 22 pp.

INDUSTRY STANDARDS

Hydraulic

Hydraulic Brakes (Cont.)

—Automotive—Boots—Rubber
JIS D 2611-84. Rubber Boots of Hydraulic Disc Brakes for Automobiles.

—Automotive—Cylinders
CNS D2011-85. Hydraulic Brake Master Cylinders for Automobiles (Jun)(3474). 14 pp.
JIS D 2603-88. Hydraulic Brake Master Cylinders for Automotive Hydraulic Brake Systems Using a Non-Petroleum Base Hydraulic Brake Fluid. 21 pp.

—Automotive—Cylinders—Bleed Screws
CNS D2179-84. Bleed Screw of Hydraulic Brake Wheel Cylinders for Automobiles (Jun)(10934).

—Automotive—Cylinders—Boots—Rubber
CNS D2009-85. Rubber Cups of Hydraulic Brake Cylinders for Automobiles (Jun)(3472).
CNS D2175-83. Rubber Boots of Hydraulic Brake Wheel Cylinders for Automobiles (Dec)(10710).
JIS D 2608-89. Rubber Boots for Automotive Hydraulic Brake Wheel Cylinders Using a Non-Petroleum Base Hydraulic Brake Fluid.

—Automotive—Cylinders—Cups—Rubber
JIS D 2605-88. Rubber Cups for Hydraulic Cylinders for Automotive Hydraulic Brake Systems Using a Non-Petroleum Base Hydraulic Brake Fluid. 35 pp.

—Automotive—Cylinders—Gaskets
ISO 4929-78. Road Vehicles—Diaphragm Gaskets for Hydraulic Brake Master Cylinder Reservoirs Using a Non-Petroleum Base Hydraulic Brake Fluid First Edition. 6 pp.
JIS D 2610-82. Diaphragm Gaskets of Hydraulic Brake Master Cylinder Reservoirs for Automobiles.

—Automotive—Cylinders—Seals
ISO 6118-80. Road Vehicles—Elastomeric Cups and Seals for Hydraulic Brake Actuating Cylinders Using a Non-Petroleum Base Hydraulic Brake Fluid (Service Temperature 70 Degrees Celsius Maximum) First Edition. 14 pp.

—Automotive—Cylinders—Seals—Elastomer
CNS D2176-83. Elastomeric Seals of Hydraulic Disc Brake Cylinders for Automobiles (Dec)(10711).
JIS D 2609-82. Elastomeric Seals of Hydraulic Disc Brake Cylinders for Automobiles.

—Automotive—Cylinders—Seals—Rubber
ISO 4928-80. Road Vehicles—Elastomeric Cups and Seals for Cylinders for Hydraulic Braking Systems Using a Non-Petroleum Base Hydraulic Brake Fluid (Service Temperature 120 Degrees Celsius Maximum) Second Edition. 14 pp.
ISO 7631-85. Road Vehicles—Elastomeric Cups and Seals for Cylinders for Hydraulic Braking Systems Using a Petroleum Base Hydraulic Brake Fluid (Service Temperature 120 Degrees Celsius Max.) First Edition. 13 pp.

—Automotive—Fittings
ISO 4038-84. Road Vehicles—Hydraulic Braking Systems—Pipes, Tapped Holes, Male Fittings and Hose End Fittings Second Edition. 10 pp.

—Automotive—Hose Assemblies
ISO 3996-86. Road Vehicles—Brake Hose Assemblies for Hydraulic Braking Systems Used with a Non-Petroleum-Base Hydraulic Fluid Second Edition. 13 pp.
ISO 6120-86. Road Vechicles—Brake Hose Assemblies for Hydraulic Braking Systems Used with a Petroleum-Base Hydraulic Fluid First Edition. 13 pp.
SNZ NZS/SAEJ1401-85. Road Vehicle-Hydraulic Brake Hose Assemblies for Use with Non-Petroleum Base Hydraulic Fluid. 6 pp.

—Automotive—Hoses
CNS D2045-86. Hydraulic Brake Rubber Hoses for Automobiles (Oct)(6161).
JIS D 2601-77. Hydraulic Brake Hose for Automobiles. 12 pp.

—Automotive—Pipes
ISO 4038-84. Road Vehicles—Hydraulic Braking Systems—Pipes, Tapped Holes, Male Fittings and Hose End Fittings Second Edition. 10 pp.

—Automotive—Pressure Measurement
ISO 3803-84. Road Vehicles—Hydraulic Pressure Test Connection for Braking Equipment Second Edition. 4 pp.

—Automotive—Tapped Holes
ISO 4038-84. Road Vehicles—Hydraulic Braking Systems—Pipes, Tapped Holes, Male Fittings and Hose End Fittings Second Edition. 10 pp.

Hydraulic Brakes (Cont.)

—Master Cylinders—Gaskets
CNS D2177-83. Diaphragm Gaskets of Hydraulic Brake Master Cylinder Reservoirs for Automobiles (Dec)(10712).

Hydraulic Cement Mortars
See Also: Cement Mortars; Mortars

—Air Content
CNS R3023-83. Method of Test for Air Content in Hydraulic Cement Mortar (Jul)(787).

—Compression Testing
CNS R3032-83. Method of Test for Compressive Strength of Hydraulic Cement Mortar (Using 50mm Cube Specimens) (Jul)(1010).

—Mixing
CNS R2079-73. Mechanical Mixing Method for Hydraulic Cement Paste and Mortars of Plastic Consistency (Nov)(3655).

—Tensile Testing
CNS R3033-83. Method of Test for Tensile Strength of Hydraulic Cement Mortar (Jul)(1011).

Hydraulic Cements
See Also: Blended Cements; Clinkers; Gypsum Plaster; Portland Cements; Portland Oil Shale Cements; Portland Slag Cements; Slag Cements; Supersulfated Cements; Trass Cements

CSA CAN/CSA-A362-93. Blended Hydraulic Cement; (Gen Instr 1). 29 pp.

—Chemical Analysis
CNS R3039-83. Method of Test for Chemical Analysis of Hydraulic Cement (Aug) (1078).

—Consistency
CNS R3075-88. Method of Test for Normal Consistency of Hydraulic Cement (Jun)((3590).

—Density
CNS R3122-85. Method of Test for Density of Hydraulic Cement (May)(11272).

—Float Tables
CNS R3034-88. Float Table for Test of Hydraulic Cement (Jun)(1012).

—Humidity Test Chambers
CNS A2041-86. Moist Tank Cabinet and Room Used in the Testing of Hydraulic Cements (Dec)(3037). 2 pp.

—Pastes—Mixing
CNS R2079-73. Mechanical Mixing Method for Hydraulic Cement Paste and Mortars of Plastic Consistency (Nov)(3655).

—Processing
CNS R2073-85. Processing Additions for Use in the Manufacture of Hydraulic Cements (May)(3459).

—Sampling
CNS R3020-83. Method of Sampling for Hydraulic Cement (Jul) (784).

—Setting Time
CNS R3021-83. Method of Test for Time of Setting of Hydraulic Cement (by Gilmore Needle) (Jul)(785).
CNS R3022-83. Method of Test for Time of Setting of Hydraulic Cement (by Vicat Needle) (Jul)(786).

—Sieve Analysis
CNS R3115-83. Method of Test for Fineness of Hydraulic Cement by 0.150mm and 0.075mm Sieves (Jul) (10473).
CNS R3123-85. Method of Test for Fineness of Hydraulic Cement by the Test Sieve 0.045mm CNS Z386 (May) (11273).

Hydraulic Clutches
See Also: Clutches; Hydraulic Equipment

—Cylinders—Automotive
CNS D2123-82. Hydraulic Clutch Operating Cylinders for Automobiles (Feb)(8481). 7 pp.

—Master Cylinders—Automotive
CNS D2122-82. Hydraulic Clutch Master Cylinder for Automobiles (Feb)(8480). 16 pp.

Hydraulic Conveyors
See Also: Conveyors; Hydraulic Equipment

—Glossaries
BSI BS 3810: Part 3-67. 1967 Glossary of Terms Used in Materials Handling Part 3: Terms Used in Connection with Pneumatic and Hydraulic Handling. 30 pp.

Hydraulic Couplings
Use For: Fluid Couplings *See Also:* Couplings; Hydraulic Equipment; Hydraulic Seals; Hydraulic Turbines

BSI BS 5380-84. 1984 Hydraulic Port and Stud Couplings Using O'-Ring Sealing and G' Series Fastening Threads. 8 pp.
BSI BS 6537-84. 1984 Staple Type Connectors for Hydraulic Fluid Power Applications. 15 pp.
ISO 4399-77. Fluid Power Systems and Components—Connectors and Associated Components—Nominal Pressures First Edition. 3 pp.

—Agricultural Equipment
BSI BS 4742: Part 2-93. 1993 Hydraulic Equipment for Agricultural Machinery Part 2: Specification for Quick-Action Hydraulic Couplers for General Purposes (ISO 5675: 1992) (E). 8 pp.
BSI BS 4742: Part 2-83. 1983 Hydraulic Equipment for Agricultural Machinery Part 2: Specification for Hydraulic Couplers for General Purposes. 6 pp.
BSI BS 4742: Part 4-85. 1985 Hydraulic Equipment for Agricultural Machinery Part 4: Hydraulic Couplings on Braking Systems for Trailers and Trailed Equipment. 4 pp.
CSA CAN/CSA-M5675-92. Agricultural Tractors and Machinery—General Purpose Quick-Action Hydraulic Couplers (ISO 5675:1992); (Gen Instr 1). 19 pp.
ISO 5675-92. Agricultural Tractors and Machinery—General Purpose Quick-Action Hydraulic Couplers Second Edition. 7 pp.
ISO 5676-83. Tractors and Machinery for Agriculture and Forestry—Hydraulic Coupling—Braking Circuit First Edition. 4 pp.

—Aircraft—Ground Support Equipment
ISO 3174-81. Aircraft—Connections for Checking Hydraulic Systems by Ground Appliances First Edition. 6 pp.

—Brakes—Agricultural Equipment
BSI BS 4742: Part 7-89. 1989 Hydraulic Equipment for Agricultural Machinery Part 7: Port Connections for Agricultural Trailed Vehicle Hydraulic Brake Couplings. 4 pp.

—Fluid Power Systems
BSI DD 195: Part 1-91. 1991 Hydraulic Couplings Part 1: Method of Test for Coupling Port Threads and Their Seals Intended for Use in Fluid Power Systems. 12 pp.
BSI DD 195: Part 2-91. 1991 Hydraulic Couplings Part 2: Method of Test for Couplings Intended for Use in Fluid Power Systems. 12 pp.

—Forestry Equipment
BSI BS 4742: Part 4-85. 1985 Hydraulic Equipment for Agricultural Machinery Part 4: Hydraulic Couplings on Braking Systems for Trailers and Trailed Equipment. 4 pp.
ISO 5676-83. Tractors and Machinery for Agriculture and Forestry—Hydraulic Coupling—Braking Circuit First Edition. 4 pp.

—Quick Disconnect
BSI BS 7198: Part 1-89. 1989 Hydraulic Fluid Power Quick-Action Couplings: Part 1: Dimensions and Performance. 8 pp.
BSI BS 7198: Part 2-89. 1989 Hydraulic Fluid Power Quick-Action Couplings: Part 2: Methods of Test. 19 pp.
ISO 7241 Pt 1-87. Hydraulic Fluid Power—Quick-Action Couplings—Part 1: Dimensions and Requirements First Edition. 8 pp.
ISO 7241 Pt 2-86. Hydraulic Fluid Power—Quick-Action Couplings—Part 2: Test Methods First Edition. 18 pp.

—Seals
BSI BS 7417-91. 1991 Interface Dimensions of Flat Face 'O'-Ring Seal Hydraulic Couplings. 10 pp.
CNS B2149-83. Flat Seal for Hydraulic Couplings (Tentative) (Mar)(4107).

Hydraulic Cylinders
See Also: Brakes; Fluid Power Cylinders; Hydraulic Accumulators; Hydraulic Equipment; Hydraulic Motors; Hydraulic Presses; Wheel Cylinders

BSI BS 5755-86. 1986 Dimensions of Basic Features of Fluid Power Cylinders. 10 pp.
CNS B1049-77. Hydraulic Cylinders Internal Diameters and Piston Rod Diameters (Jun)(4106).
CNS B2614-81. Hydraulic Cylinder (Oct)(7951).
CNS B7171-81. Method of Test for Hydraulic Cylinder (Oct)(7952).
ISO 8131-92. Hydraulic Fluid Power—Single Rod Cylinders, 16 MPa (160 Bar) Compact Series—Tolerances Second Edition. 8 pp.
ISO 8135-86. Hydraulic Fluid Power—Single Rod Cylinders, 160 Bar (16 MPa) Medium and 250 Bar (25 MPa) Series—Tolerances First Edition. 4 pp.
JIS B 8662-89. Test Method for Precise Hydraulic Cylinder with Positioning Sensor. 22 pp.

INTERNATIONAL AND NON-U.S. NATIONAL STANDARDS
SUBJECT INDEX

Hydraulic

Hydraulic Cylinders (Cont.)
SAA AS 2019-86. Fluid Power—Hydraulic and Pneumatic Cylinders—Bore and Rod Dimensions. 4 pp.

—**Acceptance Testing**
ISO 10100-90. Hydraulic Fluid Power—Cylinders—Acceptance Test First Edition. 6 pp.

—**Accessories—Mounting Dimensions**
ISO 8133-91. Hydraulic Fluid Power—Single Rod Cylinders, 16 MPa (160 Bar) Compact Series—Accessory Mounting Dimensions First Edition. 16 pp.

—**Agricultural Equipment**
BSI BS 4742: Part 5-85. 1985 Hydraulic Equipment for Agricultural Machinery Part 5: Dimensions of Braking Cylinders for Trailers and Trailed Equipment. 8 pp.
ISO 5669-82. Agricultural Trailers and Trailed Equipment—Braking Cylinders—Specifications First Edition; (Corrected and Reprinted -1984). 7 pp.

—**Agricultural Equipment—Remote Control**
ISO 2057-81. Agricultural Tractors—Remote Control Hydraulic Cylinders for Trailed Implements Second Edition. 7 pp.

—**Agricultural Equipment—Tippers**
BSI BS 4742: Part 3-85. 1985 Hydraulic Equipment for Agricultural Machinery Part 3: Dimensions and Pressure Rating of Single-Acting Telescopic Tipping Cylinders for Trailers. 8 pp.
ISO 5670-84. Agricultural Trailers—Single-Acting Telescopic Tipping Cylinders—25 MPa (250 Bar) Series—Types 1, 2, and 3—Interchangeability Dimensions First Edition. 7 pp.

—**Barrels—Carbon Steel**
JIS G 3473-88. Carbon Steel Tubes for Cylinder Barrels. 16 pp.

—**Bore and Rod Area Ratios**
ISO 7181-91. Hydraulic Fluid Power—Cylinders—Bore and Rod Area Ratios Second Edition. 7 pp.

—**Bores**
ISO 3320-87. Fluid Power Systems and Components—Cylinder Bores and Piston Rod Diameters—Metric Series Second Edition. 4 pp.
ISO 3321-75. Fluid Power Systems and Components—Cylinder Bores and Piston Rod Diameters—Inch Series First Edition. 4 pp.
ISO 7181-91. Hydraulic Fluid Power—Cylinders—Bore and Rod Area Ratios Second Edition. 7 pp.

—**Disk Brakes—Seals—Elastomer—Automotive**
CNS D2176-83. Elastomeric Seals of Hydraulic Disc Brake Cylinders for Automobiles (Dec)(10711).
JIS D 2609-82. Elastomeric Seals of Hydraulic Disc Brake Cylinders for Automobiles.

—**Disk Brakes—Seals—Rubber**
ISO 4930-78. Road Vehicles—Elastomeric Seals for Hydraulic Disc Brake Cylinders Using a Non-Petroleum Base Hydraulic Brake Fluid (Service Temperature 150 Degrees Celsius Maximum) First Edition; (Erratum—June 1979). 9 pp.
ISO 6119-80. Road Vehicles—Elastomeric Seals for Hydraulic Disc Brake Cylinders Using a Non-Petroleum Base Hydraulic Brake Fluid (Service Temperature 120 Degrees Celcius Maximum) First Edition. 8 pp.

—**Double Acting**
JIS B 8354-92. Double Acting Hydraulic Cylinders. 51 pp.

—**Double Acting—Mounting Dimensions**
BSI BS 6331: Part 1-83. 1983 Mounting Dimensions of Single Rod Double-Acting Hydraulic Cylinders Part 1: Specification for 160 Bar Medium Series. 12 pp.
BSI BS 6331: Part 2-83. 1983 Mounting Dimensions of Single Rod Double-Acting Hydraulic Cylinders Part 2: Specification for 160 Bar Compact Series. 18 pp.
BSI BS 6331: Part 3-83. 1983 Mounting Dimensions of Single Rod Double-Acting Hydraulic Cylinders Part 3: Specification for 250 Bar Series. 10 pp.
BSI BS 6331: Part 4-85. 1985 Mounting Dimensions of Single Rod Double-Acting Hydraulic Cylinders Part 4:Specification for 63 Bar Series, Including Associated Components. 32 pp.

—**Drum Brakes—Boots—Rubber**
ISO 4927-78. Road Vehicles—Elastomeric Boots for Drum Type Hydraulic Brake Wheel Cylinders Using a Non-Petroleum Base Hydraulic Brake Fluid (Service Temperature 120 Degrees Celsius Maximum) First Edition. 6 pp.

Hydraulic Cylinders (Cont.)
—**Drum Brakes—Boots—Rubber** (Cont.)
ISO 6117-80. Road Vehicles—Elastomeric Boots for Drum Type Hydraulic Brake Wheel Cylinders Using A Non-Petroleum Base Hydraulic Brake Fluid (Service Temperature 100 Degrees Celsius Maximum) First Edition. 6 pp.
ISO 7633-85. Road Vehicles—Elastomeric Boots for Drum Type Hydraulic Brake Wheel Cylinders Using a Petroleum Base Hydraulic Brake Fluid (Service Temperature 120 Degrees Celsius Max.) First Edition. 6 pp.

—**Drum Brakes—O Rings—Rubber**
ISO 7630-85. Road Vehicles—Elastomeric O-Rings for Hydraulic Drum Brake Wheel Cylinders Using a Petroleum Base Hydraulic Brake Fluid (Service Temperature 120 Degrees Celsius Max.) First Edition. 7 pp.

—**Hydraulic Brakes—Automotive**
CNS D2011-85. Hydraulic Brake Master Cylinders for Automobiles (Jun)(3474). 14 pp.
JIS D 2603-88. Hydraulic Brake Master Cylinder for Automotive Hydraulic Brake Systems Using a Non-Petroleum Base Hydraulic Brake Fluid. 21 pp.

—**Hydraulic Brakes—Bleed Screws—Automotive**
CNS D2179-84. Bleed Screw of Hydraulic Brake Wheel Cylinders for Automobiles (Jun)(10934).

—**Hydraulic Brakes—Cups—Rubber**
ISO 4928-80. Road Vehicles—Elastomeric Cups and Seals for Cylinders for Hydraulic Braking Systems Using a Non-Petroleum Base Hydraulic Brake Fluid (Service Temperature 120 Degrees Celsius Maximum) Second Edition. 14 pp.
ISO 7631-85. Road Vehicles—Elastomeric Cups and Seals for Cylinders for Hydraulic Braking Systems Using a Petroleum Base Hydraulic Brake Fluid (Service Temperature 120 Degrees Celsius Max.) First Edition. 13 pp.

—**Hydraulic Brakes—Cups—Rubber—Automotive**
JIS D 2605-88. Rubber Cups for Hydraulic Cylinders for Automotive Hydraulic Brake Systems Using a Non-Petroleum Base Hydraulic Brake Fluid. 35 pp.

—**Hydraulic Brakes—Gaskets—Automotive**
CNS D2177-83. Diaphragm Gaskets of Hydraulic Brake Master Cylinder Reservoirs for Automobiles (Dec)(10712).
JIS D 2610-82. Diaphragm Gaskets of Hydraulic Brake Master Cylinder Reservoirs for Automobiles.

—**Hydraulic Brakes—Seals—Rubber**
ISO 4928-80. Road Vehicles—Elastomeric Cups and Seals for Cylinders for Hydraulic Braking Systems Using a Non-Petroleum Base Hydraulic Brake Fluid (Service Temperature 120 Degrees Celsius Maximum) Second Edition. 14 pp.
ISO 7631-85. Road Vehicles—Elastomeric Cups and Seals for Cylinders for Hydraulic Braking Systems Using a Petroleum Base Hydraulic Brake Fluid (Service Temperature 120 Degrees Celsius Max.) First Edition. 13 pp.
ISO 7632-85. Road Vehicles—Elastomeric Seals for Hydraulic Disc Brake Cylinders Using a Petroleum Base Hydraulic Brake Fluid (Service Temperature 120 Degrees Celsius Max.) First Edition. 8 pp.

—**Hydraulic Clutches—Automotive**
CNS D2123-82. Hydraulic Clutch Operating Cylinders for Automobiles (Feb)(8481). 7 pp.

—**Mounting Dimensions**
ISO 6020 Pt 1-81. Hydraulic Fluid Power—Single Rod Cylinders—Mounting Dimensions—160 Bar (16 000 kPa) Series—Part 1: Medium Series First Edition. 10 pp.
ISO 6020 Pt 2-91. Hydraulic Fluid Power—Mounting Dimensions for Single Rod Cylinders, 16 MPa (160 Bar) Series—Part 2: Compact Series Second Edition. 20 pp.
ISO 6022-81. Hydraulic Fluid Power—Single Rod Cylinders—Mounting Dimensions—250 Bar (25 000 kPa) Series First Edition. 7 pp.
ISO 8132-86. Hydraulic Fluid Power—Single Rod Cylinders, 160 Bar (16 MPa) Medium and 250 Bar (25 MPa) Series—Mounting Dimensions for Accessories First Edition. 10 pp.
JIS B 8367-90. Hydraulic Cylinder—Mounting Dimensions. 33 pp.

—**Piston Rods—Mounting Dimensions**
ISO 6981-92. Hydraulic Fluid Power—Cylinders—Rod End Plain Eyes—Mounting Dimensions Second Edition. 8 pp.
ISO 6982-92. Hydraulic Fluid Power—Cylinders—Rod End Spherical Eyes—Mounting Dimensions Second Edition. 8 pp.

Hydraulic Cylinders (Cont.)
—**Ports**
ISO 8136-86. Hydraulic Fluid Power—Single Rod Cylinders, 160 Bar (16 MPa) Medium Series—Port Dimensions First Edition. 4 pp.
ISO 8137-86. Hydraulic Fluid Power—Single Rod Cylinders, 250 Bar (25 MPa) Series—Port Dimensions First Edition. 4 pp.
ISO 8138-86. Hydraulic Fluid Power—Single Rod Cylinders, 160 Bar (16 MPa) Compact Series—Port Dimensions First Edition. 4 pp.

—**Pressure Measurement**
ISO 3322-85. Fluid Power Systems and Components—Cylinders—Nominal Pressures Second Edition. 3 pp.

—**Sealing Housings**
BSI BS 5751-87. 1987 Dimensions of Housings for Hydraulic Seals for Reciprocating Applications. 11 pp.
BSI BS 6241-82. 1982 Dimensions of Housings for Hydraulic Seals Incorporating Bearing Rings for Hydraulic Applications. 4 pp.
BSI BS 6984-88. 1988 Dimensions of Housings for Elastomer-Energized Plastic-Faced Hydraulic Seals in Reciprocating Applications. 11 pp.
ISO 5597-87. Hydraulic Fluid Power—Cylinders—Housings for Piston and Rod Seals in Reciprocating Applications—Dimensions and Tolerances First Edition. 12 pp.
ISO 6547-81. Hydraulic Fluid Power—Cylinders—Piston Seal Housings Incorporating Bearing Rings—Dimensions and Tolerances First Edition. 5 pp.
ISO 7425 Pt 1-88. Hydraulic Fluid Power—Housings for Elastomer-Energized, Plastic-Faced Seals—Dimensions and Tolerances—Part 1: Piston Seal Housings First Edition. 9 pp.
ISO 7425 Pt 2-89. Hydraulic Fluid Power—Housings for Elastomer-Energized, Plastic-Faced Seals—Dimensions and Tolerances—Part 2: Rod Seal Housings First Edition. 7 pp.

Hydraulic Equipment
Scope Note: For additional listings, use a more specific term *See Also:* Fluid Power Equipment; Hydraulic Accumulators; Hydraulic Brakes; Hydraulic Clutches; Hydraulic Conveyors; Hydraulic Couplings; Hydraulic Cylinders; Hydraulic Excavating Equipment; Hydraulic Fluid Filters; Hydraulic Fluid Reservoirs; Hydraulic Fluids; Hydraulic Jacks; Hydraulic Motors; Hydraulic Power Pumps; Hydraulic Presses; Hydraulic Pressure Pumps; Hydraulic Seals; Hydraulic Structures; Hydraulic Turbines; Hydraulic Valves; Hydropneumatic Accumulators; Pressure Switches

BSI BS 4575: Part 3-88. 1988 Guide for Application of Fluid Power Equipment to Transmission and Control Systems Part 3: Code of Practice for the Technical Specification, Design, Construction, Commissioning and Safe Application of Hydraulic and Pneumatic Fluid Power Systems. 22 pp.
DIN ENGL 24346-84. Hydraulic Fluid Power; Hydraulic Systems; General Rules for Application (Dec). 15 pp.
JIS B 8361-82. General Rules for Hydraulic Systems. 29 pp.
JIS B 8654-89. Test Methods for Electro-Hydraulic Proportional Series Flow Control Valves. 23 pp.
JIS B 8655-89. Test Methods for Electro-Hydraulic Proportional Directional Series Flow Control Valves. 28 pp.

—**Adapters**
BSI BS 5200-86. 1986 Dimensions of Hydraulic Connectors and Adaptors. 19 pp.

—**Aerospace—Pressure Measurement**
AECMA PREN2624-90. Pressure Impulse Testing of Hydraulic System Components. 7 pp.

—**Aerospace—Symbols**
ISO 5859-91. Aerospace—Graphic Symbols for Schematic Drawings of Hydraulic and Pneumatic Systems and Components First Edition. 5 pp.

—**Aerospace—Tubes**
BSI T 72-73-77. 1977 18/10 Chromium-Nickel Corrosion-Resisting Steel Tube for Hydraulic Purposes (Niobium/Titanium Stabilized: 550 MPa). 7 pp.
BSI T 72-73-01. 1977 Amd 1 18/10 Chromium-Nickel Corrosion-Resisting Steel Tube for Hydraulic Purposes (Niobium/Titanium Stabilized: 550 MPa) (AMD 6852) December 24, 1991. 8 pp.

—**Agricultural Equipment**
BSI BS 6142-81. 1981 Presentation of Technical Data on Hydraulic Systems of Agricultural Tractors and Machinery. 2 pp.
BSI BS 6347: Part 2-83. 1983 Performance Assessment of Agricultural Tractors Part 2: Methods for Determination of Hydraulic Power and Lifting Capacity. 8 pp.

INDUSTRY STANDARDS

Hydraulic Equipment (Cont.)

—Aircraft

CAA Chapter K6-2 04.72. Hydraulic Systems (Light Aeroplanes). 4 pp.
CAA Chapter P6-2. Hydraulic Systems (Provisional Airworthiness Requirements for Civil Powered Lift Aircraft). 5 pp.
JIS W 0105-84. Glossary of Terms for Aircraft Hydraulic and Pneumatic Systems.
JIS W 2909-82. Test for Aircraft Hydraulic and Emergency Pneumatic Systems.
JIS W 2913-82. Hydraulic System Components, Aircraft, General Specification for.
JIS W 2914-82. Hydraulic Systems, Aircraft, Type I and II, Design and Installation Requirements for.
SBAC TS 103 ISSUE 1. Recommendations for Design Considerations for Aircraft Hydraulic Systems.

—Aircraft—Contamination Control

SBAC TS 74 ISSUE 1. Code of Practice for Controlling Contamination Introduced into Aircraft Hydraulic Components and Systems During Manufacture and Assembly.

—Aircraft—Cross Servicing

NATO STANAG 3510 ED 2 AMD 9-73. The Provision of Hydraulic Power for Servicing Aircraft Hydraulic Systems. 5 pp.
NATO STANAG 3510 ED 3 AMD 0-92. The Provision of Hydraulic Power for Servicing Aircraft Hydraulic Systems. 5 pp.
NATO STANAG 3510 ED 3 AMD 1-92. Provision of Hydraulic Power for Servicing Aircraft Hydraulic Systems. 7 pp.

—Aircraft—Glossaries

BSI BS 185: Sec 10-70. 1970 Glossary of Aeronautical and Astronautical Terms Section 10: Auxiliary Services (Hydraulic, Pneumatic, Electrical, Air Conditioning and Refuelling). 7 pp.

—Aircraft—Identification Systems

BSI M 49-77. 1977 Marking of Hydraulic Components. 4 pp.
ISO 3323-87. Aircraft—Hydraulic Components—Marking to Indicate Fluid for Which Component Is Approved Second Edition. 4 pp.
MOD UK DSTAN 16-21-01. Marking of Aircraft Hydraulic Components Issue 2; Amendment 1. 7 pp.

—Aircraft—Pipe Fittings—Flexural Strength

BSI M 55-84. 1984 Rotary Flexure Testing of Hydraulic Tubing Joints and Fittings for Aerospace Use. 12 pp.
ISO 7257-83. Aircraft—Hydraulic Tubing Joints and Fittings—Rotary Flexure Test First Edition. 11 pp.

—Aircraft—Pipe Joints—Flexural Strength

BSI M 55-84. 1984 Rotary Flexure Testing of Hydraulic Tubing Joints and Fittings for Aerospace Use. 12 pp.
ISO 7257-83. Aircraft—Hydraulic Tubing Joints and Fittings—Rotary Flexure Test First Edition. 11 pp.

—Aircraft—Ratings

SBAC RS 183 ISSUE 2. Standard Capacity, Ratings for Hydraulic Components.

—Aircraft—Rubber

MOD UK DTD-458A-45. Rubber Parts for Use with Mineral Base Hydraulic Fluid (Reprinted March 1961, Incorporating Amendments Nos. 1 & 2). 4 pp.

—Aircraft—Symbols

MOD UK DSTAN 05-56-80. Graphical Symbols for Aircraft Hydraulic and Pneumatic Systems Issue 1. 6 pp.

—Aircraft Valves—Charging

MOD UK DSTAN 16-26-83. Charging Valves for Aircraft High Pressure Nitrogen and Hydraulic Fluid Systems Issue 1. 10 pp.

—Ball Bearings—Life Testing

BSI PD 6487-79. 1979 Calculating the Life of Ball Bearings When Used in Contact with Fire Resistant Hydraulic Fluids. 7 pp.

—Cleanness

SAA AS 4002.1-92. Fluid Power Systems and Components—Cleanness Levels in Hydraulic Systems—Part 1: Classification and Determination of Cleanness. 9 pp.

—Combat Vehicles—Pipe Fittings

MOD UK DSTAN 47-20-78. Brazed Nipple Pipe Fittings-Metric-High and Low Pressure for Fighting Vehicle Applications Issue 1. 20 pp.

—Connectors

BSI BS 5200-86. 1986 Dimensions of Hydraulic Connectors and Adaptors. 19 pp.
BSI BS 5630-78. 1978 Interface Dimensions of Snap-On Connections for Use with Electrically Controlled Hydraulic Equipment. 4 pp.

—Contamination

BSI BS 7265-90. 1990 Code of Practice for Assessing the Cleanliness of Hydraulic Fluid Power Components. 11 pp.

—Corrosion Inhibitors—Crude Oils—Aerospace

MOD UK DTD-5540B-75. Corrosion Preventive Oil: Hydraulic System NATO Code Number: C-635 Joint Service Designation: PX-26. 8 pp.

—Design

BSI BS 7388-91. 1991 Guide for Prevention of Leaks from Hydraulic Fluid Power Systems. 33 pp.

—Earthmoving Equipment—Movement Time

BSI BS 6198-81. (WITHDRAWN) 1981 Measurement of Tool Movement Time of Earth-Moving Machinery. 7 pp.
JIS A 8306-90. Earth-Moving Machinery—Test Method for Measurement of Tool Movement Time (ISO 5004-1987).

—Engineering Drawings—Symbols

CNS B1001-9-81. Engineering Drawing Symbols for Hydraulic and Pneumatic Power System (Jul)(3-9).
SAA AS 1101.1-93. Graphical Symbols for General Engineering—Part 1: Hydraulic and Pneumatic Systems (in Profesional Package 56A). 35 pp.

—Fit Assemblies—Naval Ships

MOD UK NES 1023-01. Interference Fit—Oil Injection Method Issue 1 (10.85); Amendment 1. 71 pp.
MOD UK NES 1023-93. Interference Fit Oil Injection Method Issue 2 (04.93). 82 pp.

—Fit Assemblies—Submarines

MOD UK NES 1023-01. Interference Fit—Oil Injection Method Issue 1 (10.85); Amendment 1. 71 pp.
MOD UK NES 1023-93. Interference Fit Oil Injection Method Issue 2 (04.93). 82 pp.

—Forms (Paper)—Noise

DIN ENGL 45635 Pt 41 Suppl. 1-86. Measurement of Noise Emitted by Machines; Airborne Noise Emission; Enveloping Surface Method; Hydraulic Assemblies (Test Report Specimen Form) (Apr). 4 pp.

—Glossaries

CNS B8007-86. Glossary of Basic Terms for Oil Hydraulics and Pneumatics (Aug)(11642).
CNS B8007-1-86. Glossary of Energy Conversion Terms for Oil Hydraulics and Pneumatics (Aug)(11642-1).
CNS B8007-2-86. Glossary of Energy Controlling Terms for Oil Hydraulics and Pneumatics (Aug)(11642-2).
CNS B8007-3-86. Glossary of Controlling Device and Circuit Terms for Oil Hydraulics and Pneumatics (Aug)(11642-3).
CNS B8007-4-86. Glossary of Accessory Device and Part Terms for Oil Hydraulics and Pneumatics (Aug)(11642-4).
JIS B 0142-84. Glossary of Terms for Oil Hydraulics and Pneumatics. 60 pp.

—Ground Effect Machines

CAA 56.
CAA Chapter B6-4 01.91. Hydraulic Systems. 3 pp.

—Hose Assemblies

MOD UK DSTAN 47-2-01. Rubber Hose Assemblies for Hydraulic Systems Issue 2; Amendment 3. 24 pp.

—Hose Assemblies—Ships

MOD UK NES 324: Part 5-83. Requirements for Hydraulic Systems Part 5: Flexible Hose, Multiple Spiral Wire Reinforced Rubber Hose Assemblies Issue 1 (03.83). 12 pp.
MOD UK NES 324: Part 6-83. Requirements for Hydraulic Systems Part 6: Flexible Hose 2 Layer Wire Braid Reinforced Rubber Hose Assemblies Issue 1 (03.83). 12 pp.

—Hoses

MOD UK DSTAN 47-2-01. Rubber Hose Assemblies for Hydraulic Systems Issue 2; Amendment 3. 24 pp.

—Hoses—Ships

MOD UK NES 324: Part 5-83. Requirements for Hydraulic Systems Part 5: Flexible Hose, Multiple Spiral Wire Reinforced Rubber Hose Assemblies Issue 1 (03.83). 12 pp.
MOD UK NES 324: Part 6-83. Requirements for Hydraulic Systems Part 6: Flexible Hose 2 Layer Wire Braid Reinforced Rubber Hose Assemblies Issue 1 (03.83). 12 pp.

—Hoverplatforms

CAA Chapter C4 01.74. Systems and Machinery. 1 p.

—Liquid Automatic Particle Counters—Calibration

BSI BS 5540: Part 2-92. 1992 Evaluating Particulate Contamination of Hydraulic Fluids Part 2: Method of Calibrating Liquid Automatic Particle-Count Instruments (Using AC Fine Test Dust Contaminant) (ISO 4402: 1991). 13 pp.
BSI BS 5540: Part 2-78. 1978 Evaluating Particulate Contamination of Hydraulic Fluids Part 2: Method of Calibrating Liquid Automatic Particle Count Instruments (Using AC Fine Test Dust Contaminant). 8 pp.
BSI BS 5540: Part 6-90. 1990 Evaluating Particulate Contamination of Hydraulic Fluids Part 6: Method of Calibrating Liquid Automatic Particle-Count Instruments (Using Mono-Sized Latex Spheres) (W). 12 pp.
ISO 4402-91. Hydraulic Fluid Power—Calibration of Automatic -Count Instruments for Particles Suspended in Liquids—Method Using Classified AC Fine Test Dust Contaminant Second Edition. 10 pp.

—Lubricants

BSI BS 6413: Part 4-83. 1983 Lubricants, Industrial Oils and Related Products (Class L) Part 4: Classification for Family H (Hydraulic Systems). 4 pp.
ISO 6743 Pt 4-82. Lubricants, Industrial Oils and Related Products (Class L)—Classification—Part 4: Family H (Hydraulic Systems) First Edition. 4 pp.

—Mineral Oil Content—Gravimetric Analysis

JIS B 9931-90. Fluid Contamination-Determination of Contaminants by the Gravimetric Method. 10 pp.

—Noise

DIN ENGL 45635 Pt 41-86. Measurement of Noise Emitted by Machines; Airborne Noise Emission; Enveloping Surface Method; Hydraulic Assemblies (Apr). 4 pp.

—Pipe Flanges

CNS B2573-81. Socket Welding Pipe-Flanges for Hydraulic Use (210/kgf/cm2) (Jun)(7561).
JIS B 2291-76. 210 kgf/cm2 Slip-on Welding Pipe Flanges for Hydraulic Use. 9 pp.

—Rotary Wing Aircraft

CAA Chapter G6-2 11.75. Hydraulic Systems (Rotorcraft). 4 pp.
CAA CAP 524 SUB-Part F 12.86. Equipment (Rotorcraft). 17 pp.

—Safety

BSI BS 4575: Part 1-79. 1979 Amd 1 Guide for Application of Fluid Power Equipment to Transmission and Control Systems Part 1: Hydraulic Equipment. 23 pp.
CEN PREN 982-92. Safety Requirements for Fluid Power Systems and Components—Hydraulics. 36 pp.
ISO 4413-79. Hydraulic Fluid Power—General Rules for the Application of Equipment to Transmission and Control Systems First Edition. 20 pp.

—Ships

MOD UK NES 324: Part 1-85. Requirements for Hydraulic Systems Part 1—General Issue 1 (01.85). 80 pp.

—Silencers—Marine

MOD UK NES 859-92. Guide to the Selection of Reactive Type Hydraulic Silencers in Design and Installation of Liquid Systems Issue 1 (12.92). 47 pp.

—Solenoids

CNS B3424-82. Solenoid Assembly for Oil-Hydraulic Use Type (Jan)(8196).

—Steering Gear—Ships

JIS F 6720-90. Shop Test Method for Ships' Hydraulic Steering Gears.

—Submarines

MOD UK NES 324: Part 1-85. Requirements for Hydraulic Systems Part 1—General Issue 1 (01.85). 80 pp.

—Supersonic Transports—Reliability

CAA STANDARD NO. 7-2 03.76. Hydraulic System. 6 pp.

INTERNATIONAL AND NON-U.S. NATIONAL STANDARDS SUBJECT INDEX

Hydraulic

Hydraulic Equipment (Cont.)

—Tractors—Identification Systems
BSI BS 4742: Part 8-90. 1990 Hydraulic Equipment for Agricultural Machinery Part 8: Method for Colour Coding of Remote Hydraulic Power Services and Controls on Agricultural Tractors. 7 pp.

—Tractors—Sump Port Connection
BSI BS 4742: Part 6-88. 1988 Hydraulic Equipment for Agricultural Machinery Part 6: Dimensions and Location of Agricultural Tractor Hydraulic Return to Sump Porting. 8 pp.

—Tubes—Aerospace
CEN PREN 3504-91. Tube for Hydraulic Systems Titanium and Titanium Alloys. 5 pp.
ISO 8575-90. Aerospace—Fluid Systems—Hydraulic System Tubing First Edition. 6 pp.

—Tubes—Aerospace—Fatigue Testing
ISO 8574-90. Aerospace—Hydraulic System Tubing—Qualification Tests First Edition. 8 pp.

—Tubes—Aerospace—Pressure Measurement
ISO 8574-90. Aerospace—Hydraulic System Tubing—Qualification Tests First Edition. 8 pp.

—Vibrators (Machinery)
BSI BS 7285-90. 1990 Method for Describing Characteristics of Servo-Hydraulic Test Equipment for Generating Vibration. 43 pp.
ISO 8626-89. Servo-Hydraulic Test Equipment for Generating Vibration—Method of Describing Characteristics First Edition. 42 pp.

—Welding Equipment—Pipe Flanges
JIS B 2291-76. 210 kgf/cm2 Slip-on Welding Pipe Flanges for Hydraulic Use. 9 pp.

Hydraulic Excavating Equipment

See Also: Earthmoving Equipment; Excavating Equipment; Hydraulic Equipment

—Booms—Control Systems and Equipment
BSI BS 6912: Part 1-88. 1988 Safety of Earth-Moving Machinery Part 1: Hydraulic Excavator and Backhoe Loader Boom Lowering Control Device. 6 pp.
ISO 8643-88. Earth-Moving Machinery—Hydraulic Excavator and Backhoe Loader Boom Lowering Control Device—Requirements and Tests First Edition. 6 pp.

—Buckets (Machinery)—Capacity Measurement
ISO 7451-83. Earth-Moving Machinery—Hydraulic Excavators—Hoe Type Buckets—Volumetric Ratings First Edition. 7 pp.

—Capacity Measurement
BSI BS 6421-83. 1983 Volumetric Rating of Hoe Type Buckets of Hydraulic Excavators Used for Earth-Moving. 8 pp.

—Glossaries
DIN ENGL 24080-79. Earth-Moving Machinery; Hydraulic Excavators; Cable-Operated, Excavators; Terms (Mar). 5 pp.
DIN ENGL 24086-78. Earth-Moving Machinery; Hydraulic Excavators, Digging Forces; Terms and Definitions; Nominal Values (Oct). 3 pp.
ISO 7135-93. Earth-Moving Machinery—Hydraulic Excavators—Terminology and Commercial Specifications First Edition. 53 pp.

—Lift Capacity
BSI BS 6912: Part 9-92. 1992 Safety of Earth-Moving Machinery Part 9: Methods for the Calculation and Verification of the Lift Capacity of Hydraulic Excavators (ISO 10567: 1992). 16 pp.
ISO 10567-92. Earth-Moving Machinery—Hydraulic Excavators—Lift Capacity First Edition; (Corrected and Reprinted -1993). 13 pp.

—Loads
DIN ENGL 24083-78. Earth-Moving Machinery; Hydraulic Excavators; Indication of Load-Lifting Capacity (Nov). 2 pp.

—Movement Time
BSI BS 6198-81. (WITHDRAWN) 1981 Measurement of Tool Movement Time of Earth-Moving Machinery. 7 pp.
JIS A 8306-90. Earth-Moving Machinery—Test Method for Measurement of Tool Movement Time (ISO 5004-1987).

—Noise—Limitation
EC COM(93) 154-93. Commission Proposal for a Council Directive to Amend Council Directive 86/662/EEC on the Limitation of Noise Emitted by Earth-Moving Machinery (93/C 157/11). 4 pp.

Hydraulic Excavating Equipment (Cont.)

—Noise—Limitation (Cont.)
EC 86/662/EEC-86. Council Directive on the Limitation of Noise Emitted by Hydraulic Excavators, Rope-Operated Excavators, Dozers, Loaders and Excavator-Loaders. 11 pp.

—Stability Testing
DIN ENGL 24087-79. Earth-Moving Machinery: Determination of Stability of Hydraulic Excavators; Safety Requirements (Mar). 4 pp.

—Tool Forces
BSI BS 6825: Part 1-87. (WITHDRAWN) 1987 Hydraulic Excavators for Earth-Moving Part 1: Methods for Measurement of Tool Forces (Superseded by BS 6911: Part 3: 1990). 11 pp.
BSI BS 6911: Part 3-90. 1990 Testing Earth-Moving Machinery Part 3: Measurement of Tool Forces of Hydraulic Excavators. 12 pp.
ISO 6015-89. Earth-Moving Machinery—Hydraulic Excavators—Methods of Measuring Tool Forces First Edition. 10 pp.
JIS A 8406-91. Earth-Moving Machinery—Hydraulic Excavators—Methods of Measuring Tool Forces.

Hydraulic Fluid Filters

See Also: Hydraulic Equipment; Oil Filters; Separators (Mechanical)

BSI BS 6275: Part 1-82. 1982 Hydraulic Fluid Power Filter Elements Part 1: Method of Evaluating Filtration Performance (Multi-Pass Method). 32 pp.
BSI BS 6851-87. 1987 Method for Preparing a Statement of Requirements for Hydraulic Fluid Power Filters. 8 pp.
CNS B2442-80. Filters for Hydraulic Use (Apr)(5396).
CNS B7127-80. Method of Test of Filters for Hydraulic Use (Apr)(5397).
ISO 2942-85. Hydraulic Fluid Power—Filter Elements—Verification of Fabrication Integrity Second Edition. 4 pp.
ISO 4572-81. Hydraulic Fluid Power—Filters—Multi-Pass Method for Evaluating Filtration Performance First Edition. 31 pp.
ISO 7744-86. Hydraulic Fluid Power—Filters—Statement of Requirements First Edition. 7 pp.
JIS B 8356-81. Filters for Oil Hydraulic Use. 31 pp.

—Bursting Strength
BSI BS 6275: Part 2-84. 1984 Hydraulic Fluid Power Filter Elements Part 2: Method of Test to Verify Structural Integrity. 8 pp.
ISO 2941-74. Hydraulic Fluid Power—Filter Elements—Verification of Collapse/Burst Resistance First Edition. 4 pp.

—End Load Testing
BSI BS 6275: Part 2-84. 1984 Hydraulic Fluid Power Filter Elements Part 2: Method of Test to Verify Structural Integrity. 8 pp.
ISO 3723-76. Hydraulic Fluid Power—Filter Elements—Method for End Load Test First Edition. 3 pp.

—Flow Fatigue Testing
BSI BS 6275: Part 2-84. 1984 Hydraulic Fluid Power Filter Elements Part 2: Method of Test to Verify Structural Integrity. 8 pp.
ISO 3724-76. Hydraulic Fluid Power—Filter Elements—Verification of Flow Fatigue Characteristics First Edition. 4 pp.

—Hydraulic Fluids—Compatibility
BSI BS 6275: Part 2-84. 1984 Hydraulic Fluid Power Filter Elements Part 2: Method of Test to Verify Structural Integrity. 8 pp.
ISO 2943-74. Hydraulic Fluid Power—Filter Elements—Verification of Material Compatibility with Fluids First Edition. 4 pp.

—Identification Systems
BSI BS 6275: Part 3-88. 1988 Hydraulic Fluid Power Filter Elements Part 3: Marking. 2 pp.

—Packaging—Marine
MOD UK NES 324: Part 7-85. Requirements for Hydraulic Systems Part 7: Packaging of Elastomeric Separators Issue 1 (07.85). 12 pp.
MOD UK NES 324: Part 7-93. Requirements for Hydraulic Systems Part 7: Packaging of Elastomeric Separators Issue 2 (01.93). 13 pp.

—Pressure Measurement
BSI BS 6277-82. 1982 Determination of Pressure Drop/Flow Characteristics of Hydraulic Fluid Power Filters. 10 pp.
ISO 3968-81. Hydraulic Fluid Power—Filters—Evaluation of Pressure Drop Versus Flow Characteristics First Edition. 9 pp.

Hydraulic Fluid Reservoirs

See Also: Hydraulic Equipment

Hydraulic Fluid Reservoirs (Cont.)
BSI BS 6525-84. 1984 Amd 1 Design and Construction of Reservoirs Used in Hydraulic Fluid Power Systems (AMD 6075) February 28, 1990. 14 pp.

Hydraulic Fluids

See Also: Brake Fluids; Hydraulic Equipment; Hydraulic Oils; Hydraulic Seals

CGSB 3-GP-36MB-90. Petroleum Hydraulic Fluid, Inhibited. 9 pp.
CNS K5048-88. Hydraulic Fluid (Jan)(2973). 2 pp.
ISO 6073-80. Hydraulic Fluid Power—Petroleum Fluids—Prediction of Bulk Moduli First Edition. 10 pp.
MOD UK DSTAN 91-39-01. Hydraulic Fluid, Petroleum: Anti-Wear NATO Code No: H-576 Joint Service Designation: OM-33 Issue 1; Amendment 2. 22 pp.
MOD UK DSTAN 91-48-01. Hydraulic Fluid, Petroleum: Superclean NATO Code No: H-515 Joint Service Designation: OM-15 Hydraulic Fluid, Petroleum: Normal NATO Code No: H-520 Joint Service Designation: OM-18 Issue 1; Amendment 1. 28 pp.

—Aerospace—Cleanliness—Classification
ISO 11218-93. Aerospace—Cleanliness Classfication for Hydraulic Fluids First Edition. 4 pp.

—Aerospace—Particulate Contamination
ISO 5884-87. Aerospace—Fluid Systems and Components—Methods for System Sampling and Measuring the Solid Particle Contamination of Hydraulic Fluids First Edition. 19 pp.
NATO STANAG 3713 ED 3 AMD 1-90. Determination of Particulate Matter in Aerospace Hydraulic Fluids Using a Particle Size Analyser. 15 pp.
NATO STANAG 3713 ED 3 AMD 2-90. Determination of Particulate Matter in Aerospace Hydraulic Fluids Using a Particle Size Analyser. 16 pp.

—Aerospace—Particulate Contamination—Sampling
ISO 11217-93. Aerospace—Hydraulic System Fluid Contamination—Location of Sampling Points and Criteria for Sampling First Edition. 33 pp.

—Aircraft—Particulate Contamination
SBAC TS 76 ISSUE 1. Standard Method for Sampling and Processing Aerospace Hydraulic Fluids for Subsequent Evaluation of Particulate Contamination.

—Aircraft—Quality Assurance
NATO STANAG 3748 ED 2 AMD 3-85. Hydraulic Fluids, Petroleum (H-515 and H-520). 13 pp.
NATO STANAG 3748 ED 2 AMD 5-85. Hydraulic Fluids, Petroleum (H-515 and H-520). 11 pp.

—Aircraft—Silcodyne H
SBAC TS 14 ISSUE 1. Recommendations for the Handling of High Temperature Low-Inflammability Hydraulic Fluid—Silcodyne H.

—Compatibility—Hydraulic Fluid Filters
BSI BS 6275: Part 2-84. 1984 Hydraulic Fluid Power Filter Elements Part 2: Method of Test to Verify Structural Integrity. 8 pp.
ISO 2943-74. Hydraulic Fluid Power—Filter Elements—Verification of Material Compatibility with Fluids First Edition. 4 pp.

—Compatibility—Seals
BSI BS 4832-87. 1987 Compatibility Between Elastomeric Materials and Hydraulic Fluids. 18 pp.
ISO 6072-86. Hydraulic Fluid Power—Compatibilty Between Elastomeric Materials and Fluids First Edition. 16 pp.

—Corrosion Testing
MOD UK DSTAN 05-50: Part 29-01. Methods for Testing Fuels, Lubricants and Associated Products Part 29: Corrosiveness and Oxidation Stability of Oils and Hydraulic Fluids Issue 1; Amendment 1. 13 pp.

—Emulsification
MOD UK DSTAN 91-35-01. Hydraulic Fluid Petroleum: Emulsifying Joint Service Designation: OX-30 Issue 1; Amendment 2. 15 pp.

—Fire Resistance
BSI BS 7287-90. 1990 Guide for the Use of Fire-Resistant Hydraulic Fluids. 15 pp.

—Fire Resistant
ISO 7745-89. Hydraulic Fluid Power—Fire-Resistant (FR) Fluids—Guidelines for Use First Edition. 14 pp.

INDUSTRY STANDARDS

INTERNATIONAL AND NON-U.S. NATIONAL STANDARDS
SUBJECT INDEX

Hydraulic

Hydraulic Fluids *(Cont.)*

—Fire Resistant—Classification
SAA AS 3997.1-92. Fluid Power—Fire-Resistant Hydraulic Fluids—Part 1: Classification Amdt 1 October 1992. 3 pp.

—Fire Resistant—Compatibility—Metals
DIN ENGL 51345-86. Testing the Compatibility of Fire-Resistant Hydraulic Fluids with Metals (Aug). 4 pp.

—Fire Resistant—Stability
DIN ENGL 51346-86. Testing the Stability of Fire-Resistant Hydraulic Fluids (Aug). 2 pp.

—Flammability Testing
BSI DD 61-79. 1979 Flammability Spray Test for Hydraulic Fluids. 13 pp.

—Interchangeability
NATO STANAG 1135 ED 3 AMD 6-83. Interchangeability of Fuels, Lubricants and Associated Products Used by the Armed Forces of the North Atlantic Treaty Nations. 82 pp.

—Low Temperature Testing
MOD UK DSTAN 05-50: Part 7-01. Methods for Testing Fuels, Lubricants and Associated Products Part 7: Low Temperature Stability of Lubricating Oil and Hydraulic Fluids Issue 1; Amendment 1. 9 pp.

—Membrane Filters—Aircraft
SBAC TS 64 ISSUE 1. Standard Method for Microscopic Evaluation of Hydraulic Fluid Samples for Particle Contamination.

—Mining Equipment—Fire Resistant
CSA CAN/CSA-M423-M87. Fire Resistant Hydraulic Fluids. 23 pp.

—Particulate Contamination
BSI BS 5540: Part 1-78. 1978 Evaluating Particulate Contamination of Hydraulic Fluids Part 1: Qualifying and Controlling of Cleaning Methods for Sample Containers. 4 pp.
BSI BS 5540: Part 2-92. 1992 Evaluating Particulate Contamination of Hydraulic Fluids Part 2: Method of Calibrating Liquid Automatic Particle-Count Instruments (Using AC Fine Test Dust Contaminant) (ISO 4402: 1991). 13 pp.
BSI BS 5540: Part 2-78. 1978 Evaluating Particulate Contamination of Hydraulic Fluids Part 2: Method of Calibrating Liquid Automatic Particle Count Instruments (Using AC Fine Test Dust Contaminant). 8 pp.
BSI BS 5540: Part 4-88. 1988 Evaluating Particulate Contamination of Hydraulic Fluids Part 4: Method of Defining Levels of Contamination (Solid Contaminant Code). 7 pp.
BSI BS 5540: Part 5-87. 1987 Evaluating Particulate Contamination of Hydraulic Fluids Part 5: Method of Reporting Contamination Analysis Data. 10 pp.
BSI BS 5540: Part 6-90. 1990 Evaluating Particulate Contamination of Hydraulic Fluids Part 6: Method of Calibrating Liquid Automatic Particle-Count Instruments (Using Mono-Sized Latex Spheres) (W). 12 pp.
ISO 3722-76. Hydraulic Fluid Power—Fluid Sample Containers—Qualifying and Controlling Cleaning Methods First Edition. 4 pp.
ISO 3938-86. Hydraulic Fluid Power—Contamination Analysis—Method for Reporting Analysis Data First Edition. 9 pp.
ISO 4402-91. Hydraulic Fluid Power—Calibration of Automatic -Count Instruments for Particles Suspended in Liquids—Method Using Classified AC Fine Test Dust Contaminant Second Edition. 10 pp.
ISO 4406-87. Hydraulic Fluid Power—Fluids—Method for Coding Level of Contamination by Solid Particles First Edition. 7 pp.
JIS B 9930-77. Determination of Particulate Contamination in Hydraulic Fluids by the Particle Count Method (R 1986). 22 pp.
JIS B 9931-90. Fluid Contamination-Determination of Contaminants by the Gravimetric Method. 10 pp.
MOD UK DSTAN 05-42-82. Particulate Contamination Classes for Fluids in Hydraulic Systems Issue 2. 12 pp.
MOD UK DSTAN 05-44-77. Comparison Slide Method for Assessing the Particulate Contamination Class of Hydraulic System Fluid Issue 1. 10 pp.
MOD UK DSTAN 05-45-01. Sizing and Counting Particulate Contamination in Hydraulic System Fluid Using a Back Projection Microscope Issue 1; Corrigendum. 21 pp.
MOD UK DSTAN 05-46-01. Determination of Particulate Matter in Hydraulic Fluids Using an Automatic Particle Size Analyser Employing the Light Interruption Principle Issue 2; Corrigendum. 13 pp.

Hydraulic Fluids *(Cont.)*

—Particulate Contamination—Gravimetric Analysis
BSI BS 5540: Part 7-91. 1991 Evaluating Particulate Contamination of Hydraulic Fluids Part 7: Method of Determining the Level of Contamination (by the Gravimetric Method) (ISO 4405: 1991). 11 pp.
ISO 4405-91. Hydraulic Fluid Power—Fluid Contamination—Determination of Particulate Contamination by the Gravimetric Method First Edition. 8 pp.

—Particulate Contamination—Microscopy
BSI BS 5540: Part 8-91. 1991 Evaluating Particulate Contamination of Hydraulic Fluids Part 8: Method of Determining the Level of Contamination (by the Counting Method Using a Microscope) (W) (ISO 4407: 1991). 13 pp.
ISO 4407-91. Hydraulic Fluid Power—Fluid Contamination—Determination of Particulate Contamination by the Counting Method Using a Microscope First Edition. 10 pp.

—Particulate Contamination—Sampling
BSI BS 5540: Part 3-78. 1978 Evaluating Particulate Contamination of Hydraulic Fluids Part 3: Method of Bottling Fluid Samples. 4 pp.
ISO 4021-92. Hydraulic Fluid Power—Particulate Contamination Analysis—Extraction of Fluid Samples from Lines of an Operating System Second Edition. 10 pp.
ISO 5884-87. Aerospace—Fluid Systems and Components—Methods for System Sampling and Measuring the Solid Particle Contamination of Hydraulic Fluids First Edition. 19 pp.

—Rubber—Swelling
MOD UK TS 338A. Vulcanized Rubber Sheet, Standard L (Superseded by Def Stan 93-52: Part 5).

—Triaryl Phosphate Esters
IEC 978-89. Maintenance and Use Guide for Triaryl Phosphate Ester Turbine Control Fluids First Edition. 47 pp.

—Wear Testing—Vane Pumps
BSI BS 2000: Part 281-93. 1993 Methods of Test for Petroleum and Its Products Part 281: Determination of Anti-Wear Properties of Hydraulic Fluids—Vane Pump Method (W). 5 pp.
BSI BS 2000: Part 281-83. 1983 Petroleum and Its Products Part 281: Anti-Wear Properties of Hydraulic Fluids by the Vane Pump Method. 8 pp.

Hydraulic Hand Pumps

—Aircraft
SBAC RS 201 ISSUE 1. Hydraulic Hand Pump (Double Acting Type).

Hydraulic Hose Adapters
Use: Hose Adapters

Hydraulic Hose Assemblies
Use: Hose Assemblies

Hydraulic Hose Fittings
Use: Hose Fittings

Hydraulic Hoses
Use: Hoses

Hydraulic Impulse Testing
Use: Pressure Measurement

Hydraulic Jacks
See Also: Hydraulic Equipment; Jacks (Lifts)
MOD UK DSTAN 49-1-68. Jacks, Dolly Type, Hydraulic (Trolley Jacks) Issue 1. 7 pp.
MOD UK DEF-1009-51. Jacks, Lifting and Traversing, Hydraulic 30-Ton. 6 pp.
MOD UK DEF-1012-51. Jacks, Lever, Hydraulic 3 1/2-Ton. 5 pp.

—Automotive
CNS D3046-85. Method of Test for Hydraulic Trolley Jacks (Jul)(5433).
CNS D4001-85. Portable Hydraulic Jacks for Automobiles (Nov)(4074). 8 pp.
CNS D4002-85. Hydraulic Garage Jacks for Automobiles (Jul)(4075). 4 pp.
ISO 11530-93. Road Vehicles—Hydraulic Jacks—Specifications First Edition. 6 pp.
JIS D 8101-77. Portable Hydraulic Jacks for Automobiles.
JIS D 8101-66. Portable Hydraulic Jacks for Automobiles (R 1971). 10 pp.
JIS D 8102-77. Hydraulic Garage Jacks for Automobiles (R 1983). 8 pp.

Hydraulic Lifts
Use: Elevators (Lifts)

Hydraulic Motors
See Also: Hydraulic Cylinders; Hydraulic Equipment; Hydraulic Turbines; Motors; Vane Motors
BSI BS 4617-83. 1983 Methods for Determining the Performance of Pumps and Motors for Hydraulic Fluid Power Transmission. 20 pp.
BSI BS 7275: Part 1-90. 1990 Determination of Hydraulic Fluid Power Motor Characteristics Part 1: Method at Constant Low Speed and at Constant Pressure. 11 pp.
BSI BS 7275: Part 2-90. 1990 Determination of Hydraulic Fluid Power Motor Characteristics Part 2: Method for Startability. 12 pp.
ISO 4392 Pt 1-89. Hydraulic Fluid Power—Determination of Characteristics of Motors—Part 1: At Constant Low Speed and at Constant Pressure Second Edition. 11 pp.
ISO 4392 Pt 2-89. Hydraulic Fluid Power—Determination of Characteristics of Motors—Part 2: Startability Second Edition. 14 pp.
ISO 4409-86. Hydraulic Fluid Power—Positive Displacement Pumps, Motors and Integral Transmissions—Determination of Steady-State Performance First Edition. 18 pp.
SNZ NZS/BS 4617-83. Methods for Determining the Performance of Pumps and Motors for Hydraulic Fluid Power Transmission. 20 pp.

—Aircraft
ISO 9206-90. Aerospace—Constant Displacement Hydraulic Motors—General Specifications First Edition. 22 pp.

—Capacity Measurement
BSI BS 7250-89. 1989 Methods of Determining the Derived Capacity of Hydraulic Fluid Power Positive Displacement Pumps and Motors. 12 pp.
ISO 8426-88. Hydraulic Fluid Power—Positive Displacement Pumps and Motors—Determination of Derived Capacity First Edition. 12 pp.

—Flanges—Identification Systems
BSI BS 6276-87. 1987 Dimensions and Identification Code for Mounting Flanges and Shaft Ends for Hydraulic Fluid Power Pumps and Motors. 18 pp.
ISO 3019 Pt 1-75. Hydraulic Fluid Power—Positive Displacement Pumps and Motors—Dimensions and Identification Code for Mounting Flanges and Shaft Ends—Part 1: Inch Series Shown in Metric Units First Edition; (Erratum—May 1975). 11 pp.
ISO 3019 Pt 2-86. Hydraulic Fluid Power—Positive Displacement Pumps and Motors—Dimensions and Identification Code for Mounting Flanges and Shaft Ends—Part 2: Two-and Four-Hole Flanges and Shaft Ends—Metric Series Second Edition. 15 pp.
ISO 3019 Pt 3-88. Hydraulic Fluid Power—Positive Displacement Pumps and Motors—Dimensions and Identification Code for Mounting Flanges and Shaft Ends—Part 3: Polygonal Flanges (Including Circular Flanges) Second Edition. 8 pp.

—Geometric Displacements
ISO 3662-76. Hydraulic Fluid Power—Pumps and Motors—Geometric Displacements First Edition. 4 pp.

—Glossaries
ISO 4391-83. Hydraulic Fluid Power—Pumps, Motors and Integral Transmissions—Parameter Definitions and Letter Symbols Second Edition. 13 pp.

—Noise
BSI BS 5944: Part 2-92. 1992 Measurement of Airborne Noise from Hydraulic Fluid Power Systems and Components Part 2: Method of Test for Motors (ISO 4412-2: 1991) (E). 28 pp.
BSI BS 5944: Part 2-80. (WITHDRAWN) 1980 Measurement of Airborne Noise from Hydraulic Fluid Power Systems and Components Part 2: Method of Test for Motors (Superseded by BS 5944: Part 1 and 2: 1992). 8 pp.
BSI BS 5944: Part 3-80. (WITHDRAWN) 1980 Measurement of Airborne Noise from Hydraulic Fluid Power Systems and Components Part 3: Guide to the Application of Part 1 and Part 2 (Superseded by BS 5944: Part 1 and 2: 1992). 15 pp.
ISO 4412 Pt 2-91. Hydraulic Fluid Power—Test Code for Determination of Airborne Noise Levels—Part 2: Motors Second Edition. 25 pp.
JIS B 8350-89. Methods of Noise Level Measurement for Oil Hydraulic Pumps and Motors. 41 pp.

—Shaft Ends—Identification Systems
BSI BS 6276-87. 1987 Dimensions and Identification Code for Mounting Flanges and Shaft Ends for Hydraulic Fluid Power Pumps and Motors. 18 pp.
ISO 3019 Pt 1-75. Hydraulic Fluid Power—Positive Displacement Pumps and Motors—Dimensions and Identification Code for Mounting Flanges and Shaft Ends—Part 1: Inch Series Shown in Metric Units First Edition; (Erratum—May 1975). 11 pp.

INTERNATIONAL AND NON-U.S. NATIONAL STANDARDS
SUBJECT INDEX

Hydraulic Motors (Cont.)
—Shaft Ends—Identification Systems (Cont.)
ISO 3019 Pt 2-86. Hydraulic Fluid Power—Positive Displacement Pumps and Motors—Dimensions and Identification Code for Mounting Flanges and Shaft Ends—Part 2: Two-and Four-Hole Flanges and Shaft Ends—Metric Series Second Edition. 15 pp.

Hydraulic Oils
See Also: Hydraulic Fluids; Hydraulic Seals; Lubricating Oils; Oils
DIN ENGL 51524 Pt 1-85. Pressure Fluids; Hydraulic Oils; HL Hydraulic Oils; Minimum Requirements (June). 4 pp.
DIN ENGL 51524 Pt 2-85. Pressure Fluids; Hydraulic Oils; HLP Hydraulic Oils; Minimum Requirements (June). 4 pp.
DIN ENGL 51524 Pt 3-90. HVLP Hydraulic Oils; Minimum Requirements (Aug). 4 pp.
MOD UK DSTAN 91-44-01. Lubricating Oil, General Purpose: Petroleum, Light NATO Code No: O-134 Joint Service Designation: OM-13 Issue 1; Amendment 1. 16 pp.
MOD UK TS 10033E. Lubricating Oil, Engine: OMD-30, OMD-80 and OMD-85.

—Aging Testing
DIN ENGL 51554 Pt 3-78. Testing of Mineral Oils; Test of Susceptibility to Ageing According to Baader; Testing at 95 Deg. C (Sept). 2 pp.
DIN ENGL 51587-74. Testing of Lubricants; Determination of the Ageing Behaviour of Steam Turbine Oils and Hydraulic Oils Containing Additives (Aug). 6 pp.

—Air Release Value
BSI BS 2000: Part 313-93. 1993 Methods of Test for Petroleum and Its Products Part 313: Determination of Air Release Value of Hydraulic, Turbine and Lubricating Oils (W). 5 pp.
BSI BS 2000: Part 313-82. 1982 Petroleum and Its Products Part 313: Air Relase Value of Industrial Oils. 6 pp.

—Corrosion Testing
DIN ENGL 51585-71. Testing of Lubricants; Testing of Corrosion-Protection Properties of Steam-Turbine Oils and Hydraulic Oils Containing Additives (Dec). 4 pp.
MOD UK DSTAN 05-50: Part 29-01. Methods for Testing Fuels, Lubricants and Associated Products Part 29: Corrosiveness and Oxidation Stability of Oils and Hydraulic Fluids Issue 1; Amendment 1. 13 pp.

—Demulsification
DIN ENGL 51599-75. Testing of Lubricating Oils; Determination of Demulsification Capacity According to the Stirring Method (Oct). 3 pp.

—Low Temperature Testing
MOD UK DSTAN 05-50: Part 7-01. Methods for Testing Fuels, Lubricants and Associated Products Part 7: Low Temperature Stability of Lubricating Oil and Hydraulic Fluids Issue 1; Amendment 1. 9 pp.

Hydraulic Power Pumps
See Also: Gear Pumps; Hydraulic Equipment; Hydraulic Presses; Hydraulic Pressure Pumps; Positive Displacement Pumps; Pumps
BSI BS 4617-83. 1983 Methods for Determining the Performance of Pumps and Motors for Hydraulic Fluid Power Transmission. 20 pp.
ISO 4409-86. Hydraulic Fluid Power—Positive Displacement Pumps, Motors and Integral Transmissions—Determination of Steady-State Performance First Edition. 18 pp.
SNZ NZS/BS 4617-83. Methods for Determining the Performance of Pumps and Motors for Hydraulic Fluid Power Transmission. 20 pp.

—Capacity Measurement
BSI BS 7250-89. 1989 Methods of Determining the Derived Capacity of Hydraulic Fluid Power Positive Displacement Pumps and Motors. 12 pp.
ISO 8426-88. Hydraulic Fluid Power—Positive Displacement Pumps and Motors—Determination of Derived Capacity First Edition. 12 pp.

—Contamination—Flow Measurement
ISO 9632-92. Hydraulic Fluid Power—Fixed Displacement Pumps—Flow Degradation Due to Classified AC Fine Test Dust Contaminant—Test Method First Edition. 12 pp.

—Flanges
BSI BS 6276-87. 1987 Dimensions and Identification Code for Mounting Flanges and Shaft Ends for Hydraulic Fluid Power Pumps and Motors. 18 pp.

Hydraulic Power Pumps (Cont.)
—Flanges (Cont.)
ISO 3019 Pt 1-75. Hydraulic Fluid Power—Positive Displacement Pumps and Motors—Dimensions and Identification Code for Mounting Flanges and Shaft Ends—Part 1: Inch Series Shown in Metric Units First Edition; (Erratum—May 1975). 11 pp.
ISO 3019 Pt 2-86. Hydraulic Fluid Power—Positive Displacement Pumps and Motors—Dimensions and Identification Code for Mounting Flanges and Shaft Ends—Part 2: Two-and Four-Hole Flanges and Shaft Ends—Metric Series Second Edition. 15 pp.
ISO 3019 Pt 3-88. Hydraulic Fluid Power—Positive Displacement Pumps and Motors—Dimensions and Identification Code for Mounting Flanges and Shaft Ends—Part 3: Polygonal Flanges (Including Circular Flanges) Second Edition. 8 pp.

—Flanges—Identification Systems
BSI BS 6276-87. 1987 Dimensions and Identification Code for Mounting Flanges and Shaft Ends for Hydraulic Fluid Power Pumps and Motors. 18 pp.
ISO 3019 Pt 1-75. Hydraulic Fluid Power—Positive Displacement Pumps and Motors—Dimensions and Identification Code for Mounting Flanges and Shaft Ends—Part 1: Inch Series Shown in Metric Units First Edition; (Erratum—May 1975). 11 pp.
ISO 3019 Pt 2-86. Hydraulic Fluid Power—Positive Displacement Pumps and Motors—Dimensions and Identification Code for Mounting Flanges and Shaft Ends—Part 2: Two-and Four-Hole Flanges and Shaft Ends—Metric Series Second Edition. 15 pp.
ISO 3019 Pt 3-88. Hydraulic Fluid Power—Positive Displacement Pumps and Motors—Dimensions and Identification Code for Mounting Flanges and Shaft Ends—Part 3: Polygonal Flanges (Including Circular Flanges) Second Edition. 8 pp.

—Geometric Displacements
ISO 3662-76. Hydraulic Fluid Power—Pumps and Motors—Geometric Displacements First Edition. 4 pp.

—Glossaries
ISO 4391-83. Hydraulic Fluid Power—Pumps, Motors and Integral Transmissions—Parameter Definitions and Letter Symbols Second Edition. 13 pp.

—Noise
BSI BS 5944: Part 1-92. 1992 Measurement of Airborne Noise from Hydraulic Fluid Power Systems and Components Part 1: Method of Test for Pumps (ISO 4412-1: 1991) (Supersedes BS 5944: Parts 1-3: 1980). 28 pp.
BSI BS 5944: Part 1-80. (WITHDRAWN) 1980 Measurement of Airborne Noise from Hydraulic Fluid Power Systems and Components Part 1: Method of Test for Pumps (Superseded by BS 5944: Part 1 and 2: 1992). 10 pp.
BSI BS 5944: Part 3-80. (WITHDRAWN) 1980 Measurement of Airborne Noise from Hydraulic Fluid Power Systems and Components Part 3: Guide to the Application of Part 1 and Part 2 (Superseded by BS 5944: Part 1 and 2: 1992). 15 pp.
BSI BS 5944: Part 5-85. 1985 Measurement of Airborne Noise from Hydraulic Fluid Power Systems and Components Part 5: Simplified Method of Determining Sound Power Levels from Pumps Using an Anechoic Chamber. 12 pp.
BSI BS 5944: Part 6-92. 1992 Measurement of Airborne Noise from Hydraulic Fluid Power Systems and Components Part 6: Method of Test for Pumps Using a Parallelepiped Microphone Array (ISO 4412-3: 1991) (E). 15 pp.
ISO 4412 Pt 1-91. Hydraulic Fluid Power—Test Code for Determination of Airborne Noise Levels—Part 1: Pumps Second Edition. 25 pp.
ISO 4412 Pt 3-91. Hydraulic Fluid Power—Test Code for Determination of Airborne Noise Levels—Part 3: Pumps—Method Using a Parallelepiped Microphone Array First Edition. 14 pp.
JIS B 8350-89. Methods of Noise Level Measurement for Oil Hydraulic Pumps and Motors. 41 pp.

—Pressure Measurement
BSI BS 6335: Part 1-90. 1990 Methods for Determining Pressure Ripple Levels Generated in Hydraulic Fluid Power Systems and Components Part 1: High Impedance Method for Pumps. 18 pp.

—Shaft Ends
BSI BS 6276-87. 1987 Dimensions and Identification Code for Mounting Flanges and Shaft Ends for Hydraulic Fluid Power Pumps and Motors. 18 pp.
ISO 3019 Pt 1-75. Hydraulic Fluid Power—Positive Displacement Pumps and Motors—Dimensions and Identification Code for Mounting Flanges and Shaft Ends—Part 1: Inch Series Shown in Metric Units First Edition; (Erratum—May 1975). 11 pp.
ISO 3019 Pt 2-86. Hydraulic Fluid Power—Positive Displacement Pumps and Motors—Dimensions and Identification Code for Mounting Flanges and Shaft Ends—Part 2: Two-and Four-Hole Flanges and Shaft Ends—Metric Series Second Edition. 15 pp.

Hydraulic Power Pumps (Cont.)
—Shaft Ends (Cont.)
ISO 3019 Pt 3-88. Hydraulic Fluid Power—Positive Displacement Pumps and Motors—Dimensions and Identification Code for Mounting Flanges and Shaft Ends—Part 3: Polygonal Flanges (Including Circular Flanges) Second Edition. 8 pp.

—Shaft Ends—Identification Systems
BSI BS 6276-87. 1987 Dimensions and Identification Code for Mounting Flanges and Shaft Ends for Hydraulic Fluid Power Pumps and Motors. 18 pp.
ISO 3019 Pt 1-75. Hydraulic Fluid Power—Positive Displacement Pumps and Motors—Dimensions and Identification Code for Mounting Flanges and Shaft Ends—Part 1: Inch Series Shown in Metric Units First Edition; (Erratum—May 1975). 11 pp.
ISO 3019 Pt 2-86. Hydraulic Fluid Power—Positive Displacement Pumps and Motors—Dimensions and Identification Code for Mounting Flanges and Shaft Ends—Part 2: Two-and Four-Hole Flanges and Shaft Ends—Metric Series Second Edition. 15 pp.

Hydraulic Presses
See Also: Hydraulic Accumulators; Hydraulic Cylinders; Hydraulic Equipment; Hydraulic Power Pumps; Presses
CNS B7118-90. Accuracy Inspection of Hydraulic Presses (Nov)(4961).
CNS B7126-80. Test Code for Performance of Hydraulic Presses (Apr)(5395).
JIS B 6403-77. Test Code for Performance and Accuracy of Hydraulic Presses.
JIS B 6450-85. Tooling Dimensions for the Straight-Side Hydraulic Presses. 32 pp.

—Buttress Threads
CNS B2261-78. Buttress Thread 45 degree (for Hydraulic Presses) (Aug)(4468).

—Noise
DIN ENGL 45635 Pt 1603-78. Measurements of Airborne Noise Emitted by Machines; Enveloping Surface Method; Metal Processing Machine Tools; Special Stipulations for Universal Presses (June). 2 pp.

—Safety
CEN PREN 693-92. Hydraulic Presses—Safety. 41 pp.

Hydraulic Pressure Pumps
See Also: Hydraulic Equipment; Hydraulic Power Pumps; Pumps

—Aircraft
BSI M 65-87. 1987 Pressure Compensated Variable Delivery Hydraulic Pumps. 25 pp.
ISO 8278-86. Aerospace—Hydraulic, Pressure Compensated, Variable Delivery Pumps—General Requirements First Edition. 25 pp.

—Dynamic Testing
JIS B 8661-89. Test Methods for Electronically Controlled Oil Hydraulic Pumps. 21 pp.

—Static Testing
JIS B 8661-89. Test Methods for Electronically Controlled Oil Hydraulic Pumps. 21 pp.

Hydraulic Pressure Switches
Use: Pressure Switches—Hydraulic

Hydraulic Seals
See Also: Gaskets; Glands (Seals); Grease Seals; Hydraulic Couplings; Hydraulic Equipment; Hydraulic Fluids; Hydraulic Oils; O Rings; Oil Seals

—Aircraft—Design
JIS W 2006-79. Gland Design; Packings, Hydraulic, General Requirements for.

—Compatibility—Hydraulic Fluids
BSI BS 4832-87. 1987 Compatibility Between Elastomeric Materials and Hydraulic Fluids. 18 pp.
ISO 6072-86. Hydraulic Fluid Power—Compatibility Between Elastomeric Materials and Fluids First Edition. 16 pp.

—Couplings
CNS B2149-83. Flat Seal for Hydraulic Couplings (Tentative) (Mar)(4107).

—Housings
BSI BS 1658-70. 1970 Amd 1 Housings for Hydraulic Seals for Reciprocating Applications. 16 pp.

Hydraulic Structures
See Also: Dams; Flow Measurement; Flumes; Hydraulic Equipment; Hydroelectric Power Plants; Locks (Waterways); Reservoirs (Lakes); Water Storage Tanks; Weirs

INDUSTRY STANDARDS

Hydraulic Structures (Cont.)
—**Classification**
BSI BS 1000: (626/627)-77. 1977 Universal Decimal Classification (UDC). English Full Edition (626/627): Hydraulic (Water) Construction Works. 29 pp.
SNZ NZS/BS 1000 (626/627)-77. Universal Decimal Classification Hydraulic (Water) Construction Works. 32 pp.

Hydraulic Systems
Use: Hydraulic Equipment

Hydraulic Tools
See Also: Insertion Tools

—**Hand Held—Safety**
CEN PREN 792-1-92. Handheld Non-Electric Power Tools—Safety Requirements—Part 1: General Safety Requirements for all Types of Handheld Non-Electric Power Tools. 17 pp.

—**Safety**
CEN PREN 792-1-92. Handheld Non-Electric Power Tools—Safety Requirements—Part 1: General Safety Requirements for all Types of Handheld Non-Electric Power Tools. 17 pp.
CEN PREN 792-2-92. Handheld Non-Electric Power Tools—Safety Requirements—Part 2: Safety Requirements Related to the Energy Supply of Power Tools. 9 pp.

Hydraulic Torque Converters
See Also: Torque
JIS D 1007-76. Test Code of Hydraulic Torque Converters for Construction Machineries and Industrial Vehicles (R 1984). 22 pp.

—**Lubricating Oils**
MOD UK TS 10033E. Lubricating Oil, Engine: OMD-30, OMD-80 and OMD-85.

Hydraulic Transmission Systems
Use For: Integral Transmissions
ISO 4409-86. Hydraulic Fluid Power—Positive Displacement Pumps, Motors and Integral Transmissions—Determination of Steady-State Performance First Edition. 18 pp.

—**Glossaries**
ISO 4391-83. Hydraulic Fluid Power—Pumps, Motors and Integral Transmissions—Parameter Definitions and Letter Symbols Second Edition. 13 pp.

Hydraulic Tube Couplings
Use: Tube Couplings

Hydraulic Tube Fittings
Use: Tube Fittings

Hydraulic Tube Joints
Use: Tube Joints

Hydraulic Tubes
Use: Tubes

Hydraulic Turbines
Use For: Water Turbines *See Also:* Draft Tubes; Hydraulic Couplings; Hydraulic Equipment; Hydraulic Motors; Turbine Generators; Turbines

—**Acceptance Testing**
IEC 41-91. Field Acceptance Tests to Determine the Hydraulic Performance of Hydraulic Turbines, Storage Pumps and Pump-Turbines Third Edition. 422 pp.
IEC 193-65. International Code for Model Acceptance Tests of Hydraulic Turbines First Edition; (Supplement A-1972) (Amendment 1-1977). 147 pp.

—**Cavitation Pitting**
IEC 609-78. Cavitation Pitting Evaluation in Hydraulic Turbines, Storage Pumps and Pump-Turbines First Edition. 33 pp.

—**Efficiency Measurement—Thermodynamic Method**
BSI BS 5860-80. 1980 Measuring the Efficiency of Hydraulic Turbines, Storage Pumps and Pump-Turbines (Thermodynamic Method). 25 pp.

—**Glossaries**
JIS B 0119-92. Hydraulic Turbine and Reversible Pump Turbines —Vocabulary.

—**Hydroelectric Power Plants—Scale Effects**
IEC 995-91. Determination of the Prototype Performance from Model Acceptance Tests of Hydraulic Machines with Consideration of Scale Effects First Edition. 68 pp.

Hydraulic Turbines (Cont.)
—**Lubricating Oils**
CNS K5047-88. Turbine Oil (Jan)(2972).
ISO 8068-87. Petroleum Products and Lubricants—Petroleum Lubricating Oils for Turbines (Categories ISO-L-TSA and ISO-L-TGA)—Specifications First Edition. 7 pp.
JIS K 2213-83. Turbine Oils. 7 pp.

—**Maintenance**
BSI BS 5671-79. 1979 Commissioning, Operation and Maintenance of Hydraulic Turbines. 21 pp.
IEC 545-76. Guide for Commissioning, Operation and Maintenance of Hydraulic Turbines First Edition. 43 pp.

—**Model Testing**
JIS B 8103-89. Methods for Model Tests of Hydraulic Turbine and Reversible Pump-Turbine. 83 pp.

—**Pressure Measurement**
BSI BS EN 60994-93. 1993 Guide for Field Measurement of Vibrations and Pulsations in Hydraulic Machines (Turbines, Storage Pumps and Pump-Turbines) (IEC 994: 1991) (Q). 73 pp.
CENELEC EN 60994-92. Guide for Field Measurement of Vibrations and Pulsations in Hydraulic Machines (Turbines, Storage Pumps and Pump-Turbines); (IEC 994: 1991). 5 pp.
IEC 994-91. Guide for Field Measurement of Vibrations and Pulsations in Hydraulic Machines (Turbines, Storage Pumps and Pump-Turbines) First Edition; (CENELEC EN 60994:1992). 136 pp.

—**Speed Control**
IEC 308-70. International Code for Testing of Speed Governing Systems for Hydraulic Turbines First Edition. 95 pp.
SAA AS 1156-71. Code for Testing of Speed Governing Systems for Hydraulic Turbines (Being IEC Publication 308(1970) Endorsed as the Australian Standard Without Australian Amendment. 95 pp.

—**Vibration**
BSI BS EN 60994-93. 1993 Guide for Field Measurement of Vibrations and Pulsations in Hydraulic Machines (Turbines, Storage Pumps and Pump-Turbines) (IEC 994: 1991) (Q). 73 pp.
CENELEC EN 60994-92. Guide for Field Measurement of Vibrations and Pulsations in Hydraulic Machines (Turbines, Storage Pumps and Pump-Turbines); (IEC 994: 1991). 5 pp.
IEC 994-91. Guide for Field Measurement of Vibrations and Pulsations in Hydraulic Machines (Turbines, Storage Pumps and Pump-Turbines) First Edition; (CENELEC EN 60994:1992). 136 pp.

Hydraulic Valves
Scope Note: For additional listings, see the subheading Hydraulic under specific types of Valves
See Also: Ball Valves; Butterfly Valves; Bypass Valves; Check Valves; Cocks; Control Valves; Damper Valves; Directional Control Valves; Faucets; Flap Valves; Gas Valves; Gate Valves; Globe Valves; Hydraulic Equipment; Plug Valves; Safety Valves; Servovalves; Swing Check Valves; Thermostatic Valves; Valves
BSI BS 4062: Part 2-90. 1990 Valves for Hydraulic Fluid Power Systems Part 2: Methods for Determining Performance. 32 pp.
BSI BS 4062: Part 2-82. (WITHDRAWN) 1982 Valves for Hydraulic Fluid Power Systems Part 2: Methods for Determining Performance. 23 pp.

—**Cartridge—Cavities**
ISO 7368-89. Hydraulic Fluid Power—Two-Port Slip-in Cartridge Valves—Cavities First Edition. 21 pp.

—**Controllers—Identification Systems**
ISO 9461-92. Hydraulic Fluid Power—Identification of Valve Ports, Subplates, Control Devices and Solenoids First Edition. 7 pp.

—**Irrigation Equipment**
ISO 9635-90. Irrigation Equipment—Hydraulically Operated Irrigation Valves First Edition. 8 pp.

—**Modular Stack—Clamping**
BSI BS 6494: Part 4-89. 1989 Hydraulic Fluid Power Valve Mounting Surfaces Part 4: Clamping Dimensions of Four-Port Sizes 03 and 05, Modular Stack Valves and Directional Control Valves. 6 pp.
ISO 7790-86. Hydraulic Fluid Power—Four-Port Modular Stack Valves and Four-Port Directional Control Valves, Sizes 03 and 05—Clamping Dimensions First Edition. 4 pp.

—**Mounting Surfaces**
BSI BS 6494: Part 6-89. 1989 Hydraulic Fluid Power Valve Mounting Surfaces Part 6: Pressure-Control Valves (Excluding Pressure-Relief Valves), SequenceValves, Unloading Valves, Throttle Valves and Check Valves. 22 pp.

Hydraulic Valves (Cont.)
—**Mounting Surfaces—Identification Systems**
ISO 5783-81. Hydraulic Fluid Power—Code for Identification of Valve Mounting Surfaces First Edition. 4 pp.

—**Ports—Identification Systems**
ISO 9461-92. Hydraulic Fluid Power—Identification of Valve Ports, Subplates, Control Devices and Solenoids First Edition. 7 pp.

—**Pressure Measurement**
BSI BS 4062: Part 1-82. 1982 Valves for Hydraulic Fluid Power Systems Part 1: Methods for Determining Pressure Differential Pressure/Flow Characteristics. 12 pp.
ISO 4411-86. Hydraulic Fluid Power—Valves—Determination of Pressure Differential/Flow Characteristics First Edition. 14 pp.

—**Proportional Control**
JIS B 8653-89. Test Methods for Electro-Hydraulic Proportional Metering Valves. 21 pp.

—**Subplates—Identification Systems**
ISO 9461-92. Hydraulic Fluid Power—Identification of Valve Ports, Subplates, Control Devices and Solenoids First Edition. 7 pp.

Hydrazine Content Analysis
—**Water—Photometry**
DIN ENGL 38413 Pt 1-82. German Standard Methods for the Analysis of Water, Waste Water and Sludge; Individual Components (Group P); Determination of Hydrazine (P 1) (Mar). 3 pp.

—**Water—Spectrophotometry**
BSI BS 2690: Part 102-79. 1979 Water Used in Industry Part 102: Hydrazine: Spectrophotometric Method (4-Dimethylamino-benzaldehyde). 2 pp.

Hydrazine Sulfate
Use For: Hydrazinium Sulfate
CNS K7243-67. Chemical Reagent (Hydrazine Sulfate) (Oct)(1742).
JIS K 8992-80. Hydrazinium (2+) Sulfate.

Hydrazinium Dichloride
CNS K7590-81. Chemical Reagent (Hydrazinium (II) Dichloride) (Oct)(8028).
JIS K 8200-80. Hydrazinium (2+) Dichloride.

Hydrazinium Sulfate
Use: Hydrazine Sulfate

Hydrazinobenzene
Use: Phenylhydrazine

Hydrides
Use For: Hydrogen Absorbing Alloys

—**Glossaries**
JIS H 7003-89. Glossary of Terms Used in Hydrogen Absorbing Alloys. 30 pp.

—**Pressure Concentration Temperature**
JIS H 7201-91. Method of Determining the PCT Relations of Hydrogen—Absorbing Alloys. 9 pp.

Hydriodic Acid
CNS K7244-67. Chemical Reagent (Hydriodic Acid) (Oct)(1743).

Hydro Extractors
Use: Water Extractors

Hydrobaths
Use For: Hydromassage Bathtub Units; Immersion Hydrobaths *See Also:* Bathtubs; Hot Tubs; Medical Electrical Equipment; Medical Equipment; Spas

—**Air Pumps—Household—Portable**
CSA 1268 Bull. Electrical Bulletin 1268 May 16, 1980 to C22.2 NO 108. 5 pp.

—**Indoor/Outdoor**
CSA CAN/CSA-C22. 2 NO 108-M89. Liquid Pumps; (Gen Instr 1 Thru 2). 68 pp.
CSA 1282B Bull. Electrical Bulletin 1282B April 8, 1985 to C22.2 NO 108. 16 pp.
CSA 1283B Bull. Electrical Bulletin 1283B April 8, 1985 to C22.2 NO 108. 14 pp.

—**Integral Electric Pumps**
CSA CAN/CSA-C22. 2 NO 108-M89. Liquid Pumps; (Gen Instr 1 Thru 2). 68 pp.
CSA 1260 Bull. Electrical Bulletin 1260 February 26, 1980 to C22.2 NO 108. 5 pp.

INTERNATIONAL AND NON-U.S. NATIONAL STANDARDS
SUBJECT INDEX

Hydrocarbons

Hydrobaths (Cont.)
—Water Pumps—Household—Portable
CSA 1270 Bull. Electrical Bulletin 1270 May 16, 1980 to C22.2 NO 108. 5 pp.

Hydrobromic Acid
CNS K7245-67. Chemical Reagent (Hydrobromic Acid) (Oct)(1744).
JIS K 8509-75. Hydrobromic Acid.

Hydrocarbon Analyzers
See Also: Analyzers; Hydrocarbon Content Analysis; Hydrocarbons
JIS B 7956-79. Continuous Analyzers for Hydrocarbons in Ambient Air.

Hydrocarbon Content Analysis
See Also: Halogenated Hydrocarbon Content Analysis; Hydrocarbon Analyzers; Hydrocarbons

—Aromatic Hydrocarbons—Gas Chromatography
CNS K6497-80. Nonaromatic Hydrocarbon in Monocyclic Aromatic Hydrocarbon by Gas Chromatography (May)(5603).

—Butadienes—Gas Chromatography
ISO 6378-81. Butadiene for Industrial Use—Determination of Hydrocarbon Impurities—Gas Chromatographic Method First Edition. 9 pp.

—Butanes—Gas Chromatography
BSI BS 7276-90. 1990 Method for Gas Chromatographic Analysis of Liquified Petroleum Gas. 16 pp.
CEN PREN 242-85. Commercial Propane and Butane (LPG); Determination of Hydrocarbons Content by Gas Chromatographic Method. 24 pp.
ISO 7941-88. Commercial Propane and Butane—Analysis by Gas Chromatography First Edition. 15 pp.

—Corrosion Inhibitors
CGSB 31-GP-0A METH 49.1-57. Methods of Testing Corrosion-Prevention Materials and Processes Hydrocarbons Volatile with Steam; (Re-Edited April 1982). 1 p.
CGSB 31-GP-0A METH 50.1-57. Methods of Testing Corrosion-Prevention Materials and Processes Hydrocarbons Absorbed in Sulfuric Acid. 1 p.

—Crude Oils—Volumetric Analysis
BSI BS 7340-91. 1991 Schedule for Compressibility Factors for Petroleum Hydrocarbons in the Density Range 638kg/m3 to 1074 kg/m3. 14 pp.
ISO 9770-89. Crude Petroleum and Petroleum Products—Compressibility Factors for Hydrocarbons in the Range 638 kg/m3 to 1 074 kg/m3 First Edition; (Corrected and Reprinted -1989). 11 pp.

—Dodecylbenzene
CNS K6266-68. Method of Test for Unsulfonatable Hydrocarbons in Detergent Alkylate (Oct)(2913)(R 1973).

—Ethanols—Distillation Methods
ISO 1388 Pt 10-81. Ethanol for Industrial Use—Methods of Test—Part 10: Estimation of Hydrocarbons Content—Distillation Method First Edition. 7 pp.

—Ethylene—Gas Chromatography
ISO 6379-81. Ethylene for Industrial Use—Determination of Hydrocarbon Impurities—Gas Chromatographic Method First Edition. 6 pp.

—Exhaust Gases—Automotive
CNS D3173-87. Method of Test for Evaporative Emission Control System for Gasoline Engine Automobiles (Sep)(11534). 10 pp.
JIS D 1030-76. Analytical Procedure for Continuous Measurement of Carbon Monoxide, Carbon Dioxide and Hydrocarbon in Automobile Exhaust Gas.
SNZ NZS 5429-82. Code of Practice for In-Field Testing of Road Vehicles for Exhaust Carbon Monoxide and Hydrocarbons Concentration at Idle Speed. 8 pp.

—Gasoline—Gas Chromatography
CGSB CAN/CGSB-3.0 NO.14.3-M91. Methods of Testing Petroleum and Associated Products Standard Test Method for the Identification of Hydrocarbon Components in Automotive Gasoline Using Gas Chromatography. 32 pp.
CNS K6326-72. Method of Test for C2 Through C5 Hydrocarbons in Gasolines by Gas Chromatography Determination (Oct)(3387).

—Lacquer Thinners
CGSB 1-GP-71 METH 66.2-82. Methods of Testing Paints and Pigments Hydrocarbons and Esters Hydrocarbons in Lacquer Thinners. 1 p.

Hydrocarbon Content Analysis (Cont.)

—Natural Gas
BSI BS 3156: SUB SEC 11.2.1-86. 1986 Methods for the Analysis of Fuel Gases Part 11: Methods for Non-Manufactured Gases Section 11.2: Determination of Potential Hydrocarbon Liquid Content Subsection 11.2.1: General Introduction. 12 pp.
ISO 6570 Pt 1-83. Natural Gas—Determination of Potential Hydrocarbon Liquid Content—Part 1: Principles and General Requirements First Edition. 11 pp.

—Natural Gas—Gas Chromatography
BSI BS 3156: SUB SEC 11.1.1-86. 1986 Methods for the Analysis of Fuel Gases Part 11: Methods for Non-Manufactured Gases Section 11.1: Hydrocarbons and Inert Gases Subsection 11.1.1: Determination of Hydro-carbons to C<of12>8 and Inert Gases by Gas Chromatography. 19 pp.
ISO 6568-81. Natural Gas—Simple Analysis by Gas Chromatography First Edition. 6 pp.
ISO 6974-84. Natural Gas—Determination of Hydrogen, Inert Gases and Hydrocarbons up to C8—Gas Chromatographic Method First Edition. 16 pp.

—Natural Gas—Volumetric Analysis
BSI BS 3156: SUB SEC 11.2.3-86. 1986 Methods for the Analysis of Fuel Gases Part 11: Methods for Non-Manufactured Gases Section 11.2: Determination of Potential Hydrocarbon Liquid Content Subsection 11.2.3: Volumetric Method. 11 pp.
ISO 6570 Pt 3-84. Natural Gas—Determination of Potential Hydrocarbon Liquid Content—Part 3: Volumetric Method First Edition. 9 pp.

—Natural Gas—Weight Measurement
BSI BS 3156: SUB SEC 11.2.2-86. 1986 Methods for the Analysis of Fuel Gases Part 11: Methods for Non-Manufactured Gases Section 11.2.2: Determination of Potential Hydrocarbon Liquid Content Subsection 11.2.2: Weighing Method. 11 pp.
ISO 6570 Pt 2-84. Natural Gas—Determination of Potential Hydrocarbon Liquid Content—Part 2: Weighing Method First Edition. 9 pp.

—Olefins—Gas Chromatography
ISO 6377-81. Light Olefins for Industrial Use—Determination of Hydrocarbon Impurities by Gas Chromatography—General Considerations First Edition. 6 pp.

—Petroleum Products—Bromine Number
CNS K6268-68. Method of Test for Unsaturated Hydrocarbons in Petroleum Distillates by Bromine Number (Oct)(2915)(R 1973).

—Petroleum Products—Fluorometric Analysis
BSI BS 2000: Part 156-93. 1993 Methods of Test for Petroleum and Its Products Part 156: Determination of Hydrocarbon Types—Fluorescent Indicator Adsorption Method (W). 6 pp.
BSI BS 2000: Part 156-85. 1985 Petroleum and Its Products Part 156: Hydrocarbon Types in Liquid Petroleum Products by Fluorescent Indicator Absorption. 8 pp.
CNS K6368-73. Method of Test for Hydrocarbon Types in Liquid Petroleum Products by Fluorescent Indicator Absorption (Nov)(3577).
JIS K 2536-80. Testing Method for Hydrocarbon Types in Petroleum Products by Fluorescent Indicator Adsorption (R 1991). 36 pp.

—Petroleum Products—Volumetric Analysis
BSI BS 7340-91. 1991 Schedule for Compressibility Factors for Petroleum Hydrocarbons in the Density Range 638kg/m3 to 1074 kg/m3. 14 pp.
ISO 9770-89. Crude Petroleum and Petroleum Products—Compressibility Factors for Hydrocarbons in the Range 638 kg/m3 to 1 074 kg/m3 First Edition; (Corrected and Reprinted -1989). 11 pp.

—Propane—Gas Chromatography
BSI BS 7276-90. 1990 Method for Gas Chromatographic Analysis of Liquified Petroleum Gas. 16 pp.
CEN PREN 242-85. Commercial Propane and Butane (LPG); Determination of Hydrocarbons Content by Gas Chromatographic Method. 24 pp.
ISO 7941-88. Commercial Propane and Butane—Analysis by Gas Chromatography First Edition. 15 pp.

—Toluene—Gas Chromatography
ISO 5279-80. Toluene—Determination of Hydrocarbon Impurities—Gas Chromatographic Method First Edition; (Erratum—Dec 1981). 7 pp.

Hydrocarbon Content Analysis (Cont.)

—Uranium Hexafluoride
CNS J2066-82. Method for Determining Hydrocarbons, Chlorocarbons and Partially Substituted Halohydrocarbons in UF6 (Sept)(9399).

—Waste Water—Infrared Spectrometry
DIN ENGL 38409 Pt 18-81. German Standard Methods for the Analysis of Water, Waste Water and Sludge; Summary Action and Material Characteristic Parameters (Group H); Determination of Hydrocarbons (H 18) (Feb). 6 pp.

—Water—Infrared Spectrometry
DIN ENGL 38409 Pt 18-81. German Standard Methods for the Analysis of Water, Waste Water and Sludge; Summary Action and Material Characteristic Parameters (Group H); Determination of Hydrocarbons (H 18) (Feb). 6 pp.

Hydrocarbons
See Also: Acetylene; Aliphatic Hydrocarbons; Anthracene; Aromatic Hydrocarbons; Benzene; Butadienes; Butanes; Butylene; Chlorofluorocarbons; Cyclohexane; Cyclohexane-995; Decanes; Dodecylbenzene; Ethylene; Fluorocarbons; Heptane; n-Heptane; n-Hexane; Hydrocarbon Analyzers; Hydrocarbon Content Analysis; Isooctane; Light Hydrocarbons; Liquid Hydrocarbons; Methane; Naphtha; Naphthalene; Nonane; Octane; Petroleum Products; Propane; Propylene; Styrene; Toluene; Viscous Hydrocarbons; Xylene

—Acidity
CNS K6518-80. Method of Test for Acidity of Distillation Residues or Hydrocarbon Liquid (Jul)(5835).

—Boiling Points
ISO 4626-80. Volatile Organic Liquids—Determination of Boiling Range of Organic Solvents Used as Raw Materials First Edition. 14 pp.

—Bromine Number—Electrometric Analysis
CNS K6522-80. Method of Test for Bromine Index of Petroleum Hydrocarbons by Electrometric Titration (Jul)(5839).

—Chemical Analysis
CNS K6364-73. Chemical Analysis of Rubber Hydrocarbons in Natural Rubber (May)(3573).

—Density
BSI BS 6665-85. 1985 Method for Determination of Density or Relative Density of Liquefied Petroleum Gases and Other Light Hydrocarbons by Pressure Hydrometer Method. 10 pp.
ISO 3993-84. Liquefied Petroleum Gas and Light Hydrocarbons—Determination of Density or Relative Density—Pressure Hydrometer Method First Edition. 8 pp.
JIS K 0519-78. Testing Method for Density of High-Purity Hydrocarbons by Bingham Pycnometer.

—Electrical Insulating Liquids—Oxidation Resistance
CENELEC HD 486 S1-88. Test Method for Evaluating the Oxidation Stability of Hydrocarbon Insulating Liquids. 3 pp.

—Flowmeters—Proving Systems
BSI BS 6864: Part 3-87. 1987 Laboratory Tests on Noise Emission from Appliances and Equipment Intended for Use in Water Supply Installations Part 3: Method for Mounting and Operating in Line Valves and Appliances. 8 pp.
BSI BS 6866: Part 1-90. 1990 Proving Systems for Meters Used in Dynamic Measurement of Liquid Hydrocarbons Part 1: Introduction. 12 pp.
BSI BS 6866: Part 2-90. 1990 Proving Systems for Meters Used in Dynamic Measurement of Liquid Hydrocarbons Part 2: Methods for Design, Installation and Calibration of Pipe Provers. 24 pp.
BSI BS 6866: Part 3-87. 1987 Proving Systems for Meters Used in Dynamic Measurement of Liquid Hydrocarbons Part 3: Methods of Pulse Interpolation. 12 pp.
ISO 7278 Pt 1-87. Liquid Hydrocarbons—Dynamic Measurement—Proving Systems for Volumetric Meters—Part 1: General Principles First Edition. 8 pp.
ISO 7278 Pt 2-88. Liquid Hydrocarbons—Dynamic Measurement—Proving Systems for Volumetric Meters—Part 2: Pipe Provers First Edition. 22 pp.
ISO 7278 Pt 3-86. Liquid Hydrocarbons—Dynamic Measurement—Proving Systems for Volumetric Meters—Part 3: Pulse Interpolation Techniques First Edition. 10 pp.

—Freezing Points
JIS K 0518-78. Testing Method for Freezing Points of High-Purity Hydrocarbons.

INDUSTRY STANDARDS

INTERNATIONAL AND NON-U.S. NATIONAL STANDARDS SUBJECT INDEX

Hydrocarbons

Hydrocarbons (Cont.)

—Heat of Combustion
BSI BS 2000: Part 12-82. 1982 Petroleum and Its Products Part 12: Heat of Combustion of Liquid Hydrocarbon Fuels. 16 pp.
CNS K6578-80. Method of Test for Heat of Combustion of Liquid Hydrocarbon Fuels by Bomb Colorimeter (Aug)(6359).
JIS K 2279-85. Testing Method for Heat of Combustion of Liquid Hydrocarbon Fuel by Bomb Calorimeter. 27 pp.

—Refractive Index
JIS K 0517-76. Testing Method for Refractive Index of High Purity Hydrocarbons.

—Sampling
BSI BS 3195: Part 1-89. 1989 Sampling Petroleum Products Part 1: Manual Sampling of Liquid Hydrocarbons. 29 pp.
BSI BS 3195: Part 2-89. 1989 Sampling Petroleum Products Part 2: Automatic Pipeline Sampling of Liquid Hydrocarbons. 60 pp.
ISO 3170-88. Petroleum Liquids—Manual Sampling Second Edition. 29 pp.
ISO 3171-88. Petroleum Liquids—Automatic Pipeline Sampling Second Edition. 60 pp.

—Storage—Underground Formations
CSA Z341-93. Storage of Hydrocarbons in Underground Formations; (Gen Instr 1). 86 pp.

—Storage—Underground Formations—Safety
CSA Z341-93. Storage of Hydrocarbons in Underground Formations; (Gen Instr 1). 86 pp.

—Tanks (Containers)—Ships—Calibration
BSI BS 7550: Part 1-92. 1992 Calibration of Spherical Tanks for Refrigerated Light-Hydrocarbon Fluids in Ships Part 1: Method of Stereo-Photogrammetry (ISO 9091-1: 1991). 19 pp.
BSI BS 7627-93. 1993 Method for Calibration by Physical Measurement of Membrane Tanks and Prismatic Tanks in Ships for Carriage of Refrigerated Light-Hydrocarbon Fluids (ISO 8311: 1989) (Q). 27 pp.
ISO 8311-89. Refrigerated Light Hydrocarbon Fluids—Calibration of Membrane Tanks and Independent Prismatic Tanks in Ships —Physical Measurement First Edition. 23 pp.
ISO 9091 Pt 1-91. Refrigerated Light-Hydrocarbon Fluids—Calibration of Spherical Tanks in Ships—Part 1: Stereo-Photogrammetry First Edition. 17 pp.

—Tanks (Containers)—Ships—Measurement—Stereophotogrammetry
BSI BS 7550: Part 1-92. 1992 Calibration of Spherical Tanks for Refrigerated Light-Hydrocarbon Fluids in Ships Part 1: Method of Stereo-Photogrammetry (ISO 9091-1: 1991). 19 pp.
ISO 9091 Pt 1-91. Refrigerated Light-Hydrocarbon Fluids—Calibration of Spherical Tanks in Ships—Part 1: Stereo-Photogrammetry First Edition. 17 pp.

—Thiol Content
BSI BS 2000: Part 104-83. (WITHDRAWN) 1983 Petroleum and Its Products Part 104: Mercaptan Sulphur Content of Light Hydrocarbon Products (Silver Nitrate Method). 8 pp.

—Viscosity
DIN ENGL 51550-78. Viscometry; Determination of Viscosity; General Principles (Dec). 4 pp.

—Volumetric Analysis—Displacement Meters
BSI BS 6169: Part 1-81. 1981 Volumetric Measurement of Liquid Hydrocarbons Part 1: Displacement Meter Systems (Other Than Dispensing Pumps). 12 pp.
ISO 2714-80. Liquid Hydrocarbons—Volumetric Measurement by Displacement Meter Systems Other Than Dispensing Pumps First Edition; (Erratum—July 1982). 12 pp.

—Volumetric Analysis—Turbine Meters
BSI BS 6169: Part 2-84. 1984 Volumetric Measurement of Liquid Hydrocarbons Part 2: Turbine Meter Systems. 19 pp.
ISO 2715-81. Liquid Hydrocarbons—Volumetric Measurement by Turbine Meter Systems First Edition. 18 pp.

Hydrocephalus Shunts
Use: Shunts (Medical)—Hydrocephalic

Hydrochloric Acid
Use For: Muriatic Acid *See Also:* Aqua Regia; Hazardous Materials; Hydrogen Chloride
BSI BS 3993-66. 1966 Hydrochloric Acid, Commercial, Types 1 and 2. 25 pp.
CGSB CAN/CGSB-15.33-92. Hydrochloric (Muriatic) Acid, Technical Grade. 6 pp.

Hydrochloric Acid (Cont.)
CNS K1002-81. Hydrochloric Acid for Industrial Use (Oct)(18). 1 p.
CNS K6083-63. Method of Test for Hydrochloric Acid (for Industrial Use) (Jun)(1002)(R 1973).
CNS K7246-67. Chemical Reagent (Hydrochloric Acid) (Oct)(1745).
JIS K 1310-59. Hydrochloric Acid, Synthetic. 5 pp.
JIS K 8180-88. Hydrochloric Acid.
MOD UK DSTAN 68-169-93. Acid, Hydrochloric, Pure Issue 1. 10 pp.
MOD UK CS 2532. Acid, Hydrochloric, Pure.

—Acid Resistance Testing—Enamels
BSI BS 1344: Part 10-87. 1987 Methods of Testing Vitreous Enamel Finishes Part 10: Determination of Resistance to Condensing Hydrochloric Acid Vapour. 6 pp.
ISO 2743-86. Vitreous and Porcelain Enamels—Determination of Resistance to Condensing Hydrochloric Acid Vapour Second Edition. 4 pp.

—Acid Resistance Testing—Glass
BSI BS 3473: Part 5-87. 1987 Chemical Resistance of Glass Used in the Production of Laboratory Glassware Part 5: Method for Determination of Resistance of Glass to Attack by 6 mol/L Hydrochloric Acid at 100 Degrees Celcius. 10 pp.
ISO 1776-85. Glass—Resistance to Attack by Hydrochloric Acid at 100 Degrees Celsius—Flame Emission or Flame Atomic Absorption Spectrometric Method First Edition. 7 pp.

—Acidity—Volumetric Analysis
ISO 904-76. Hydrochloric Acid for Industrial Use—Determination of Total Acidity—Titrimetric Method First Edition. 4 pp.

—Arsenic Content—Photometry
ISO 5785-78. Hydrochloric Acid for Industrial Use—Determination of Arsenic Content—Silver Diethyldithiocarbamate Photometric Method First Edition. 4 pp.

—Chlorine Content—Volumetric Analysis
ISO 908-80. Hydrochloric Acid for Industrial Use—Determination of Oxidizing or Reducing Substances Content—Titrimetric Method First Edition. 5 pp.

—Concentration—Density
ISO 905-76. Hydrochloric Acid for Industrial Use—Evaluation of Hydrochloric Acid Concentration by Measurement of Density First Edition. 4 pp.

—Density
BSI BS 976-87. 1987 Density-Composition Tables for Aqueous Solutions of Hydrochloric Acid. 34 pp.

—Highway Transportation—Emergency Procedures
SAA AS 1678.8.0. 001-83. Emergency Procedure Guides—Transport—Part 8.0.001: Hydrochloric Acid (Muriatic Acid).

—Iron Content—Spectrophotometry
ISO R909-68. Hydrochloric Acid for Industrial Use Determination of Iron Content 2,2'-Bipyridyl Spectrophotometric Method First Edition. 8 pp.

—Standard Solutions
CNS K0039-88. Arsenic Standard Solution (Nov)(12469).
JIS K 0026-83. Arsenic Standard Solution. 11 pp.

—Stripping—Textiles
DIN ENGL 54278 Pt 2-78. Testing of Textiles; Coatings and Attendant Materials; Determination of Condensation Products Which Can Be Stripped by Means of Hydrochloric Acid (Feb). 2 pp.

—Sulfate Content—Gravimetric Analysis
ISO 906-76. Hydrochloric Acid for Industrial Use—Determination of Sulphate Content—Barium Sulphate Gravimetric Method First Edition. 4 pp.

—Sulfate Content—Turbidimetry
ISO 2762-73. Hydrochloric Acid for Industrial Use—Determination of Soluble Sulphates—Turbidimetric Method First Edition. 5 pp.

—Sulfated Ash Content—Gravimetric Analysis
ISO 907-76. Hydrochloric Acid for Industrial Use—Determination of Sulphated Ash—Gravimetric Method First Edition. 4 pp.

—Sulfur Dioxide Content—Volumetric Analysis
ISO 908-80. Hydrochloric Acid for Industrial Use—Determination of Oxidizing or Reducing Substances Content—Titrimetric Method First Edition. 5 pp.

Hydrochloric Acid (Cont.)

—Water Treatment
DIN ENGL 19610-75. Hydrochloric Acid for Water Treatment for Supply Water; Technical Conditions of Delivery (Nov). 6 pp.

—Water Treatment Chemicals
CEN PREN 939-92. Hydrochloric Acid Used for Water Treatment Intended for Human Consumption. 18 pp.

Hydrochloric Acid Content Analysis

—Kaolin
MOD UK M 2002/59. Examination of Kaolin (Withdrawn).

Hydrocyanic Acid Content Analysis

—Legumes
BSI BS 4317: Part 11-76. 1976 Methods of Test for Cereals and Pulses Part 11: Determination of Glycosidic Hydrocyanic Acid in Pulses. 7 pp.
ISO 2164-75. Pulses—Determination of Glycosidic Hydrocyanic Acid First Edition; (Erratum—Dec 1981). 7 pp.

Hydroelectric Plants
Use: Hydroelectric Power Plants

Hydroelectric Power Plants
Scope Note: For additional listings, see the subheading Hydroelectric Power Plants under the specific equipment or products *See Also:* Dams; Hydraulic Structures; Power Plants
DIN ENGL 19752-86. Hydroelectric Power Plants; Rules for Planning and Operation (Apr) (Supersedes DIN 19753, DIN 19754, March 1956 Editions). 5 pp.

—Electromechanical Components
IEC 1116-92. Electromechanical Equipment Guide for Small Hydroelectric Installations First Edition. 112 pp.

—Hydraulic Turbines—Scale Effects
IEC 995-91. Determination of the Prototype Performance from Model Acceptance Tests of Hydraulic Machines with Consideration of Scale Effects First Edition. 68 pp.

Hydrofluoric Acid
See Also: Fluorspar
CNS K7247-67. Chemical Reagent (Hydrofluoric Acid) (Oct)(1746).
JIS K 1405-57. Hydrofluoric Acid HF = 20.01 (R 1985). 5 pp.
JIS K 8819-72. Hydrofluoric Acid.

—Acid Resistance Testing—Steels
CNS G2055-86. Method of Nitro-Hydrofluoric Acid Test for Stainless Steels (Jul)(4765).
JIS G 0574-80. Method of Nitric-Hydrofluoric Acid Test for Stainless Steels. 6 pp.

—Chromium Content—Emission Spectroscopy
DIN ENGL 50451 Pt 2-90. Determination of Cobalt, Chromium, Copper, Iron and Nickel as Impurities in Hydrofluoric Acid for Use in Semiconductor Technology by Plasma-Induced Emission Spectrometry (Oct). 2 pp.

—Cobalt Content—Emission Spectroscopy
DIN ENGL 50451 Pt 2-90. Determination of Cobalt, Chromium, Copper, Iron and Nickel as Impurities in Hydrofluoric Acid for Use in Semiconductor Technology by Plasma-Induced Emission Spectrometry (Oct). 2 pp.

—Copper Content—Emission Spectroscopy
DIN ENGL 50451 Pt 2-90. Determination of Cobalt, Chromium, Copper, Iron and Nickel as Impurities in Hydrofluoric Acid for Use in Semiconductor Technology by Plasma-Induced Emission Spectrometry (Oct). 2 pp.

—Highway Transportation—Emergency Procedures
SAA AS 1678.8.0. 008-87. Emergency Procedure Guides—Transport—Part 8.0.008: Hydrofluoric Acid.

—Iron Content—Emission Spectroscopy
DIN ENGL 50451 Pt 2-90. Determination of Cobalt, Chromium, Copper, Iron and Nickel as Impurities in Hydrofluoric Acid for Use in Semiconductor Technology by Plasma-Induced Emission Spectrometry (Oct). 2 pp.

Hydrofluoric Acid (Cont.)

—Nickel Content—Emission Spectroscopy
DIN ENGL 50451 Pt 2-90. Determination of Cobalt, Chromium, Copper, Iron and Nickel as Impurities in Hydrofluoric Acid for Use in Semiconductor Technology by Plasma-Induced Emission Spectrometry (Oct). 2 pp.

—Safety—Information Cards
SAA AS 2508.8.01 3-91. Safe Storage and Handling Information Cards for Hazardous Materials—Part 8.013: Hydrofluoric Acid (Aqueous) (in Professional Package 38).

—Sampling
BSI BS 5366-76. (WITHDRAWN) 1976 Amd 1 Methods of Sampling and Test for Aqueous Hydrofluoric Acid for Industrial Use (ISO 3139: 1976) (AMD 3784) October 30, 1981. 8 pp.

ISO 3139-76. Aqueous Hydrofluoric Acid for Industrial Use—Sampling and Methods of Test Second Edition; (Amendment 1-1980). 7 pp.

—Semiconductors
JIS K 1466-64. Testing Methods of Hydrofluoric Acid for Semi-Conductor. 12 pp.

—Volumetric Analysis
BSI BS 5366-76. (WITHDRAWN) 1976 Amd 1 Methods of Sampling and Test for Aqueous Hydrofluoric Acid for Industrial Use (ISO 3139: 1976) (AMD 3784) October 30, 1981. 8 pp.

ISO 3139-76. Aqueous Hydrofluoric Acid for Industrial Use—Sampling and Methods of Test Second Edition; (Amendment 1-1980). 7 pp.

Hydrofluosilicic Acid
Use: Fluosilicic Acid

Hydrogen
See Also: Hazardous Materials; Hydrogen Content Analysis; Tritium

CNS K1080-75. Compressed Hydrogen (Industrial Grade) (May)(2236).
CNS K6300-71. Method of Test for Compressed Hydrogen Gas (Apr)(3236).
JIS K 0512-74. Hydrogen.

—Coolants
IEC 842-88. Guide for Application and Operation of Turbine-Type Synchronous Machines Using Hydrogen as a Coolant First Edition. 23 pp.

—Impurities Content
CNS K6301-71. Method of Test for Impurities in Compressed Hydrogen (Jun)(3237).

Hydrogen Absorbing Alloys
Use: Hydrides

Hydrogen Chloride
See Also: Hydrochloric Acid

—Descaling Compounds—Hot Water Installations
MOD UK DSTAN 68-27-77. Scale Removing Compound (for Hot Water Installations) Issue 2. 6 pp.
MOD UK DSTAN 68-27-92. Scale Removing Compound (for Hot Water Installations) Issue 3. 10 pp.

Hydrogen Chloride Content Analysis

—Exhaust Gases
CNS K9063-81. Method of Test for Hydrogen Chloride in Exhaust Gas (Mar)(7055).
JIS K 0107-82. Methods for Determination of Hydrogen Chloride in Exhaust Gas. 15 pp.

Hydrogen Content Analysis
See Also: Hydrogen

—Coal
SAA AS 1038.6.1-86. Methods for the Analysis and Testing of Coal and Coke—Part 6.1: Ultimate Analysis of Higher Rank Coal—Determination of Carbon and Hydrogen (R 1992). 5 pp.
SAA AS 1038.7-81. Methods for the Analysis and Testing of Coal and Coke—Part 7: Ultimate Analysis of Coke. 21 pp.

—Coal—Combustion
CNS M3153-84. Method for Determination of Carbons and Hydrogen of Coal and Coke (Sheffield High Temperature Method) (Mar)(10834).
ISO 609-75. Coal and Coke—Determination of Carbon and Hydrogen—High Temperature Combustion Method First Edition. 9 pp.

—Coal—Liebig Method
CNS M3144-84. Method for Determination of Carbon and Hydrogen of Coal and Coke (Liebig Method) (Mar) (10825).

Hydrogen Content Analysis (Cont.)

—Coal—Liebig Method (Cont.)
ISO 625-75. Coal and Coke—Determination of Carbon and Hydrogen—Liebig Method First Edition. 9 pp.

—Coke—Combustion
CNS M3153-84. Method for Determination of Carbons and Hydrogen of Coal and Coke (Sheffield High Temperature Method) (Mar)(10834).
ISO 609-75. Coal and Coke—Determination of Carbon and Hydrogen—High Temperature Combustion Method First Edition. 9 pp.

—Coke—Liebig Method
CNS M3144-84. Method for Determination of Carbon and Hydrogen of Coal and Coke (Liebig Method) (Mar) (10825).
ISO 625-75. Coal and Coke—Determination of Carbon and Hydrogen—Liebig Method First Edition. 9 pp.

—Ethylene
CNS K6388-76. Method of Test for Ethylene, Other Olefins, Inerts and Hydrogen in High Purity Ethylene (Jun)(3947).

—Gases
CNS K6230-67. Method of Analysis for Carbon Dioxide, Oxygen, Carbon Monoxide Hydrogen and Methane in Gases (Mar)(2592)(R 1971).

—Low Alloy Steels
CSA W48.7-M1977. Diffusible Hydrogen in Mild Steel and Low-Alloy Steel Weld Metals: Test Method. 9 pp.

—Metals
JIS Z 2614-90. General Rules for Determination of Hydrogen in Metallic Materials. 45 pp.

—Mild Steels
CSA W48.7-M1977. Diffusible Hydrogen in Mild Steel and Low-Alloy Steel Weld Metals: Test Method. 9 pp.

—Natural Gas—Gas Chromatography
BSI BS 3156: SUB SEC 11.1.1-86. 1986 Methods for the Analysis of Fuel Gases Part 11: Methods for Non-Manufactured Gases Section 11.1: Hydro-carbons and Inert Gases Subsection 11.1.1: Determination of Hydro-carbons to $C<of12>8$ and Inert Gases by Gas Chromatography. 19 pp.
BSI BS 3156: SUB SEC 11.1.2-87. 1987 Methods for Analysis of Fuel Gases Part 11: Methods for Non-Manufactured Gases Sec 11.1: Hydrocarbons and Inert Gases Sub 11.1.2: Determination of Hydrocarbons from Butane ($C<of12>4$) to Hexadecane ($C<of12>16$) by Gas Chromatography. 13 pp.
ISO 6974-84. Natural Gas—Determination of Hydrogen, Inert Gases and Hydrocarbons up to C8—Gas Chromatographic Method First Edition. 16 pp.
ISO 6975-86. Natural Gas—Determination of Hydrocarbons from Butane (C4) to Hexadecane (C16)—Gas Chromatographic Method First Edition. 11 pp.

—Propylene—Gas Chromatography
ISO 6380-81. Propylene for Industrial Use—Determination of Hydrocarbon Impurities—Gas Chromatographic Method First Edition. 7 pp.

—Tantalum
JIS H 1696-76. Method for Determination of Hydrogen in Tantalum (R 1986). 5 pp.

—Titanium
JIS H 1619-88. Method for Determination of Hydrogen in Titanium and Titanium Alloys. 18 pp.

—Titanium Alloys
JIS H 1619-88. Method for Determination of Hydrogen in Titanium and Titanium Alloys. 18 pp.

—Weld Metal
BSI BS 6693: Part 1-86. 1986 Diffusable Hydrogen Part 1: Method for Determination of Hydrogen in Manual Metal-Arc Weld Metal Using 3 Day Collection. 8 pp.
BSI BS 6693: Part 2-86. 1986 Diffusable Hydrogen Part 2: Method for Determination of Hydrogen in Manual Metal-Arc Weld Metal. 9 pp.
BSI BS 6693: Part 3-88. 1988 Amd 1 Diffusible Hydrogen Part 3: Primary Method for the Determination of Diffusible Hydrogen in Manual Metal-Arc Ferritic Steel Weld Metal (AMD 6241) December 27, 1989. 16 pp.
BSI BS 6693: Part 4-88. 1988 Diffusable Hydrogen Part 4: Primary Method for the Determination of Diffusible Hydrogen in Submerged-Arc Steel Weld-Metal. 16 pp.

Hydrogen Content Analysis (Cont.)

—Weld Metal (Cont.)
BSI BS 6693: Part 5-88. 1988 Amd 1 Diffusable Hydrogen Part 5: Primary Method for the Determination of Diffusible Hydrogen in MIG, MAG, TIG or Cored Electrode Ferritic Steel Weld Metal (AMD 6242) January 31, 1990. 17 pp.
CSA W48.7-M1977. Diffusible Hydrogen in Mild Steel and Low-Alloy Steel Weld Metals: Test Method. 9 pp.
DIN ENGL 8572 Pt 1-81. Determination of Diffusible Hydrogen in Weld Metal; Manual Arc Welding (Mar). 6 pp.
DIN ENGL 8572 Pt 2-81. Determination of Diffusible Hydrogen in Weld Metal; Submerged Arc Welding (Mar). 6 pp.
ISO 3690-77. Welding—Determination of Hydrogen in Deposited Weld Metal Arising from the Use of Covered Electrodes for Welding Unalloyed and Low Alloy Steels First Edition; (Addendum 1-2:1983). 14 pp.

—Welded Joints
JIS Z 3118-92. Method of Measurement for Hydrogen Evolved from Steel Welds. 14 pp.

—Zirconium
JIS H 1664-88. Methods for Determination of Hydrogen in Zirconium and Zirconium Alloys. 14 pp.

—Zirconium Alloys
JIS H 1664-88. Methods for Determination of Hydrogen in Zirconium and Zirconium Alloys. 14 pp.

Hydrogen Cyanide Content Analysis

—Exhaust Gases
JIS K 0109-82. Methods for Determination of Hydrogen Cyanide in Exhaust Gas. 23 pp.

Hydrogen Embrittlement

—Copper
BSI BS 5899-80. 1980 Method for Hydrogen Embrittlement Test for Copper. 4 pp.
ISO 2626-73. Copper—Hydrogen Embrittlement Test First Edition. 4 pp.

—Steels
BSI BS EN 2831-93. 1993 Hydrogen Embrittlement of Steels—Test by Slow Bending (S). 10 pp.
BSI BS EN 2832-93. 1993 Hydrogen Embrittlement of Steels—Notched Specimen Test (S). 10 pp.
CEN PREN 2831-90. Hydrogen Embrittlement of Steels Test by Slow Bending. 4 pp.
CEN EN 2831-93. Aerospace Series—Hydrogen Embrittlement of Steels—Test by Slow Bending. 5 pp.
CEN PREN 2832-90. Hydrogen Embrittlement of Steels Notched Specimen Test. 3 pp.
CEN EN 2832-93. Aerospace Series—Hydrogen Embrittlement of Steels—Notched Specimen Test. 5 pp.

Hydrogen Embrittlement Testing
Use: Hydrogen Embrittlement

Hydrogen Fluoride
Scope Note: For additional listings, use a more specific term See Also: Anhydrous Hydrogen Fluoride; Hydrofluoric Acid

—Acidity—Volumetric Analysis
BSI BS 5365: Part 2-76. (WITHDRAWN) 1976 Methods of Sampling and Test for Anhydrous Hydrogen Fluoride for Industrial Use Part 2: Determination of Non-Volatile Acid Content: Titrimetric Method. 4 pp.
ISO 3138-74. Anhydrous Hydrogen Fluoride for Industrial Use—Determination of Non-Volatile Acid Content—Titrimetric Method First Edition. 4 pp.

—Fluosilicic Acid Content—Photometry
BSI BS 5365: Part 4-79. (WITHDRAWN) 1979 Methods of Sampling and Test for Anhydrous Hydrogen Fluoride for Industrial Use Part 4: Determination of Hexafluorosilicic Acid Content. 7 pp.
ISO 3701-76. Anhydrous Hydrogen Fluoride for Industrial Use—Determination of Hexafluorosilicic Acid Content—Reduced Molybdosilicate Photometric Method First Edition. 6 pp.

—Sampling
BSI BS 5365: Part 1-76. (WITHDRAWN) 1976 Methods of Sampling and Test for Anhydrous Hydrogen Fluoride for Industrial Use Part 1: Sampling. 9 pp.
ISO 3137-74. Anhydrous Hydrogen Fluoride for Industrial Use—Sampling First Edition. 8 pp.

Hydrogen

Hydrogen Fluoride (Cont.)

—Sulfur Dioxide Content—Iodometry
BSI BS 5365: Part 5-79. (WITHDRAWN) 1979 Methods of Sampling and Test for Anhydrous Hydrogen Fluoride for Industrial Use Part 5: Determination of Sulphur Dioxide Content. 4 pp.
ISO 3702-76. Anhydrous Hydrogen Fluoride for Industrial Use—Determination of Sulphur Dioxide Content—Iodometric Method First Edition. 4 pp.

—Water Content—Conductimetric Method
BSI BS 5365: Part 6-81. (WITHDRAWN) 1981 Methods of Sampling and Test for Anhydrous Fluoride for Industrial Use Part 6: Determination of Water Content (Conductimetric Method). 11 pp.
ISO 3700-80. Anhydrous Hydrogen Fluoride for Industrial Use—Determination of Water Content—Conductimetric Method First Edition. 10 pp.

—Water Content—Karl Fischer Method
BSI BS 5365: Part 3-79. (WITHDRAWN) 1979 Methods of Sampling and Test for Anhyrous Hydrogen Fluoride for Industrial Use Part 3: Determination of Water Content (Karl Fischer Method). 7 pp.
ISO 3699-76. Anhydrous Hydrogen Fluoride for Industrial Use—Determination of Water Content—Karl Fischer Method First Edition. 8 pp.

Hydrogen Fluoride Content Analysis

—Hydrofluoric Acid
BSI BS 5366-76. (WITHDRAWN) 1976 Amd 1 Methods of Sampling and Test for Aqueous Hydrofluoric Acid for Industrial Use (ISO 3139: 1976) (AMD 3784) October 30, 1981. 8 pp.

Hydrogen Generators

—Balloons
MOD UK TS 10188. Methanol/Water Hydrogen Generators AL-40 (Superseded by Def Stan 68-129).

—Balloons—Methanols
MOD UK DSTAN 68-129-90. Methanol: Water Mixture for Hydrogen Generators Joint Service Designation AL-40 Issue 1. 8 pp.

—Balloons—Methanols—Safety
NATO STANAG 4168 ED 1 AMD 1-85. Characteristics of Hydrogen Generating Equipment. 14 pp.

Hydrogen Hexachloroplatinate Hexahydrate
Use: Platinic Chloride Hexahydrate

Hydrogen Inorganic Compounds
Scope Note: Use a more specific term
See: Hydrobromic Acid; Hydrochloric Acid; Hydrofluoric Acid; Hydrogen Chloride; Hydrogen Fluoride; Hydrogen Peroxide; Hydrogen Sulfide

Hydrogen Ion Concentration
Use: pH

Hydrogen Ion Meters
Use: pH Meters

Hydrogen Peroxide
Use For: High Test Peroxide *See Also:* Hazardous Materials
CNS K1105-65. Hydrogen Peroxide for Industrial Grade (Sep)(2526)(R 1971). 1 p.
CNS K7249-68. Chemical Reagent (Hydrogen Peroxide, 30 Percent) (Nov)(1748).
JIS K 1463-71. Hydrogen Peroxide. 12 pp.
JIS K 8230-87. Hydrogen Peroxide.

—Acidity—Volumetric Analysis
BSI BS 7546: Part 3-92. 1992 Hydrogen Peroxide for Industrial Use Part 3: Method for Determination of Apparent Acidity (W). 6 pp.

—Aluminum
MOD UK DEF-60-58. Selection and Treatment of Aluminium Base Materials for Use with Concentrated Hydrogen Peroxide (H.T.P.) (Reprinted August, 1973) (Withdrawn). 6 pp.

—Aluminum Alloys
MOD UK DEF-60-58. Selection and Treatment of Aluminium Base Materials for Use with Concentrated Hydrogen Peroxide (H.T.P.) (Reprinted August, 1973) (Withdrawn). 6 pp.

—Bleaching Agents—Pulps
CPPA J.16P-92. Analysis of Peroxides. 3 pp.

Hydrogen Peroxide (Cont.)

—Corrosion Resistant Steels
MOD UK DEF-61-58. Selection and Treatment of Corrosion Resisting Steels for Use with Concentrated Hydrogen Peroxide (H.T.P.) (Withdrawn). 5 pp.

—Fabrics—Colorfastness Testing
CNS L3161-82. Method of Test for Colour Fastness to Bleaching with Peroxide (Mar)(8619).
JIS L 0857-75. Testing Method for Colour Fastness to Bleaching with Peroxide. 5 pp.

—Fuel Hose Fittings
MOD UK DEF-1403-61. Couplings, Pressure Fuelling, for High Test Peroxide. 12 pp.

—Highway Transportation—Emergency Procedures
SAA AS 1678.5.1. 005-93. Emergency Procedure Guides—Transport—Part 5.1.005: Hydrogen Peroxide (20% or More) (in Professional Package 37A).

—Hose Assemblies
MOD UK DEF-1214-60. PVC Hose and Hose Assemblies for Use with High Test Peroxide in Temperate and Tropical Conditions. 6 pp.

—Hose Fittings
MOD UK DEF-1215-60. Open-Ended Hose Couplings for High Test Peroxide. 16 pp.

—Hoses
MOD UK DEF-1214-60. PVC Hose and Hose Assemblies for Use with High Test Peroxide in Temperate and Tropical Conditions. 6 pp.

—Hydrogen Peroxide Content—Volumetric Analysis
BSI BS 7546: Part 1-92. 1992 Hydrogen Peroxide for Industrial Use Part 1: Method for Determination of Hydrogen Peroxide Content (W). 7 pp.

—Stability Testing
BSI BS 7546: Part 2-92. 1992 Hydrogen Peroxide for Industrial Use Part 2: Method for Determination of Stability at 100 Degrees C. 8 pp.

—Water Treatment
CEN PREN 902-92. Hydrogen Peroxide Used for Water Intended for Human Consumption. 19 pp.

Hydrogen Peroxide Content Analysis

—Fish—Dried
CNS N6152-81. Method of Test for Dehydrated Larval Fish, Chilled (Aug)(7839). 2 pp.

—Food
CNS N6189-88. Method of Test for Bactericides in Food-Test of Hydrogen Peroxide (Oct)(10893). 1 p.

—Hair Care Products
EC 82/434/EEC-82. Second Commission Directive on the Approximation of the Laws of the Member States Relating to Methods of Analysis Necessary for Checking the Composition of Cosmetic Products. 28 pp.

—Volumetric Analysis
BSI BS 7546: Part 1-92. 1992 Hydrogen Peroxide for Industrial Use Part 1: Method for Determination of Hydrogen Peroxide Content (W). 7 pp.

Hydrogen Sulfide
Use For: Hydrogen Sulphide *See Also:* Hydrogen Sulfide Content Analysis

—Connectors—Environmental Testing
BSI BS 2011: Part 2.1KD-77. 1977 Basic Environmental Testing Procedures Part 2.1: Tests Part 2.1KD: Test Kd. Hydrogen Sulfide Test for Contacts and Connections. 8 pp.
BSI BS 2011: Part 2.2KD-84. 1984 Basic Environmental Testing Procedures Part 2.2: Guidance Part 2.2KD: Test Kd. Guidance on Test Kd: Hydrogen Sulfide Test for Contacts and Connections. 12 pp.
CENELEC HD 323.2.46 S1-88. Basic Environmental Testing Procedures—Part 2: Tests. Guidance to Test Kd: Hydrogen Sulphide Test for Contacts and Connections. 3 pp.
IEC 68 Pt 2-43-76. Basic Environmental Testing Procedures Part 2: Tests Test Kd: Hydrogen Sulphide Test for Contacts and Connections First Edition. 16 pp.
IEC 68 Pt 2-46-82. Basic Environmental Testing Procedures Part 2: Tests Guidance to Test Kd: Hydrogen Sulphide Test for Contacts and Connections First Edition. 27 pp.
SAA AS 1099.2KD-81. Basic Environmental Testing Procedures for Electrotechnology—Part 2: Tests—Part 2Kd: Hydrogen Sulphide Test for Contacts and Connections. 6 pp.

Hydrogen Sulfide (Cont.)

—Connectors—Environmental Testing (Cont.)
SNZ IEC 68: Part 2-43-76. Basic Environmental Testing Procedures Part 2-43: Test Kd: Hydrogen Sulphide Test for Contacts and Connections. 14 pp.
SNZ IEC 68: Part 2-46-82. Basic Environmental Testing Procedures Part 2-46: Guidance to Test Kd: Hydrogen Sulphide Test for Contacts and Connections. 23 pp.

—Electric Contacts—Environmental Testing
BSI BS 2011: Part 2.1KD-77. 1977 Basic Environmental Testing Procedures Part 2.1: Tests Part 2.1KD: Test Kd. Hydrogen Sulfide Test for Contacts and Connections. 8 pp.
BSI BS 2011: Part 2.2KD-84. 1984 Basic Environmental Testing Procedures Part 2.2: Guidance Part 2.2KD: Test Kd. Guidance on Test Kd: Hydrogen Sulfide Test for Contacts and Connections. 12 pp.
CENELEC HD 323.2.46 S1-88. Basic Environmental Testing Procedures—Part 2: Tests. Guidance to Test Kd: Hydrogen Sulphide Test for Contacts and Connections. 3 pp.
IEC 68 Pt 2-43-76. Basic Environmental Testing Procedures Part 2: Tests Test Kd: Hydrogen Sulphide Test for Contacts and Connections First Edition. 16 pp.
IEC 68 Pt 2-46-82. Basic Environmental Testing Procedures Part 2: Tests Guidance to Test Kd: Hydrogen Sulphide Test for Contacts and Connections First Edition. 27 pp.
SAA AS 1099.2KD-81. Basic Environmental Testing Procedures for Electrotechnology—Part 2: Tests—Part 2Kd: Hydrogen Sulphide Test for Contacts and Connections. 6 pp.
SNZ IEC 68: Part 2-43-76. Basic Environmental Testing Procedures Part 2-43: Test Kd: Hydrogen Sulphide Test for Contacts and Connections. 14 pp.
SNZ IEC 68: Part 2-46-82. Basic Environmental Testing Procedures Part 2-46: Guidance to Test Kd: Hydrogen Sulphide Test for Contacts and Connections. 23 pp.

—Fabrics—Colorfastness Testing
DIN ENGL 53378-65. Testing of Plastic Films; Determination of Colour Fastness to Hydrogen Sulphide (June). 2 pp.

—Gas Detectors
JIS T 8204-86. Detector Tube Type Hydrogen Sulfide Measuring Instruments (Length-of-Stain Type). 13 pp.
JIS T 8205-86. Hydrogen Sulfide Indicator, Hydrogen Sulfide Alarm and Hydrogen Sulfide Indicator/Alarm. 15 pp.

—Plastic Sheets—Colorfastness Testing
DIN ENGL 53378-65. Testing of Plastic Films; Determination of Colour Fastness to Hydrogen Sulphide (June). 2 pp.

Hydrogen Sulfide Content Analysis
See Also: Hydrogen Sulfide

—Aliphatic Hydrocarbons—Potentiometric Analysis
CNS K6325-72. Method of Test for Hydrogen Sulfide and Mercaptan Sulfur in Petroleum Gas (Electric Potential Titration Method) (Oct)(3386).

—Aromatic Hydrocarbons—Qualitative Analysis
CNS K6258-67. Method of Test for Hydrogen Sulfide and Sulfur Dioxide Contents (Qualitative) of Aromatic Hydrocarbons (Mar)(2759)(R 1973).

—Coal Gas
CNS K6239-66. Method of Test for Hydrogen Sulfide in Coal Gas (Sep)(2633)(R 1971).

—Cresols
CNS K6219-65. Method of Test for Hydrogen Sulfide in Crude Methyl Phenol and M-Methyl Phenol (Sep)(2501)(R 1971).

—Cresylic Acids
ISO 1897 Pt VI-77. Phenol, o-Cresol, m-Cresol, p-Cresol, Cresylic Acid and Xylenols for Industrial Use—Methods of Test—Part VI: Test for Absence of Hydrogen Sulphide (Cresylic Acid and Xylenols Only) First Edition. 4 pp.

—Exhaust Gases
CNS K9070-81. Method of Test for Hydrogen Sulfide in Exhaust Gas (Aug)(7811).
JIS K 0108-83. Analytical Methods for Determination of Hydrogen Sulfide in Exhaust Gas. 23 pp.

—Liquefied Petroleum Gas
BSI BS 7026-89. 1989 Method for Detection of Hydrogen Sulphide in Liquefied Petroleum Gases Using Lead Acetate. 6 pp.

Hydrogen Sulfide Content Analysis (Cont.)

—Liquefied Petroleum Gas (Cont.)
ISO 8819-87. Liquefied Petroleum Gases—Detection of Hydrogen Sulfide—Lead Acetate Method First Edition. 6 pp.

—Liquefied Petroleum Gas—Electrometric Analysis
BSI BS 2000: Part 272-93. 1993 Methods of Test for Petroleum and Its Products Part 272: Determination of Mercaptan Sulphur and Hydrogen Sulphide Content of LPG—Electrometric Titration Method (W). 7 pp.
BSI BS 2000: Part 272-87. 1987 Petroleum and Its Products Part 272: Determination of Mercaptan Sulphur and Hydrogen Sulphide Content of Liquefied Petroleum Gases by Electometric Titration. 11 pp.

—Liquid Fuels
DIN ENGL 51766-74. Testing of Liquid Fuels; Testing for Hydrogen Sulphide (Sept). 1 p.

—Natural Gas
CNS K6327-72. Method of Test for Hydrogen Sulfide in Natural Gas (Tutwiler) (Oct)(3388).

—Natural Gas—Potentiometric Analysis
ISO 6326 Pt 3-89. Natural Gas—Determination of Sulfur Compounds—Part 3: Determination of Hydrogen Sulfide, Mercaptan Sulfur and Carbonyl Sulfide Sulfur by Potentiometry First Edition. 9 pp.

—Phosphoric Acid—Volumetric Analysis
ISO 7099-83. Phosphoric Acid for Industrial Use (Including Foodstuffs)—Determination of Hydrogen Sulfide Content —Titrimetric Method First Edition. 6 pp.

—Xylenols
ISO 1897 Pt VI-77. Phenol, o-Cresol, m-Cresol, p-Cresol, Cresylic Acid and Xylenols for Industrial Use—Methods of Test—Part VI: Test for Absence of Hydrogen Sulphide (Cresylic Acid and Xylenols Only) First Edition. 4 pp.

Hydrogen Sulphide
Use: Hydrogen Sulfide

Hydrography
See Also: Hydrology; Water

—Military Geographic Documentation—Inland
NATO STANAG 2256 ED 4 AMD 8-71. MGD Inland Hydrography. 29 pp.
NATO STANAG 2256 ED 4 AMD 9-71. MGD Inland Hydrography. 31 pp.

Hydroiodic Acid
JIS K 8917-75. Hydroiodic Acid.

Hydrology
See Also: Flood Protection; Ground Water; Hydrography; Precipitation (Meteorology); Surface Water; Water Supply Engineering

—Glossaries
DIN ENGL 4049 Pt 1-79. Hydrology; Concepts, Quantitative (Sept). 48 pp.

Hydrolysis

—Uranium Hexafluoride
CNS J2089-83. Method of Determining Technetium-99 in UF6 (Jan)(9888). 4 pp.

Hydrolytic Resistance

—Glass—Atomic Absorption Spectrometry
BSI BS 3473: Sec 4.2-89. 1989 Chemical Resistance of Glass Used in the Pro-duction of Laboratory Glassware: Part 4: Methods for determination of Hydrolytic Resistance of the Interior Surfaces of Glass Containers: Section 4.2:. 13 pp.
ISO 4802 Pt 2-88. Glassware—Hydrolytic Resistance of the Interior Surfaces of Glass Containers—Part 2: Determination by Flame Spectrometry and Classification First Edition. 11 pp.

—Glass—Atomic Emission Spectrometry
BSI BS 3473: Sec 4.2-89. 1989 Chemical Resistance of Glass Used in the Pro-duction of Laboratory Glassware: Part 4: Methods for determination of Hydrolytic Resistance of the Interior Surfaces of Glass Containers: Section 4.2:. 13 pp.
ISO 4802 Pt 2-88. Glassware—Hydrolytic Resistance of the Interior Surfaces of Glass Containers—Part 2: Determination by Flame Spectrometry and Classification First Edition. 11 pp.

Hydrolytic Resistance (Cont.)

—Glass—High Temperature Testing
BSI BS 3473: Part 2-87. 1987 Chemical Resistance of Glass Used in the Production of Laboratory Glassware Part 2: Method for Determination of Hydrolytic Resistance of Glass Grains at 98 Degrees Centigrade. 10 pp.
BSI BS 3473: Part 3-87. 1987 Chemical Resistance of Glass Used in the Production of Laboratory Glassware Part 3: Method for Determination of Hydrolytic Resistance of Glass Grains at 121 Degrees Centigrade. 10 pp.
ISO 719-85. Glass—Hydrolytic Resistance of Glass Grains at 98 Degrees C—Method of Test and Classification Second Edition. 7 pp.
ISO 720-85. Glass—Hydrolytic Resistance of Glass Grains at 121 Degrees C—Method of Test and Classification Second Edition. 7 pp.

—Glass—Volumetric Analysis
BSI BS 3473: Sec 4.1-89. 1989 Chemical Resistance of Glass Used in the Pro-duction of Laboratory Glassware: Part 4: Methods for Determination of hydrolytic Resistance of the Interior Surfaces of Glass Containers: Section 4.1:. 11 pp.
ISO 4802 Pt 1-88. Glassware—Hydrolytic Resistance of the Interior Surfaces of Glass Containers—Part 1: Determination by Titration Method and Classification First Edition. 11 pp.

Hydromassage Bathtub Units
Use: Hydrobaths

Hydrometer Method
Use: Hydrometry

Hydrometers
Use For: Density/Specific Gravity Meters
See Also: Chemical Analysis; Density; Pycnometers; Weight Measurement
CNS B6038-83. Specific Gravity Hydrometers (Jul)(4894).
ISO 387-77. Hydrometers—Principles of Construction and Adjustment First Edition; (Amendment Slip-1977). 8 pp.
ISO 650-77. Relative Density 60/60 Degrees F Hydrometers for General Purposes First Edition. 10 pp.
JIS B 7525-92. Specific Gravity Hydrometers. 9 pp.
SAA AS 2026-77. Density Hydrometers. 28 pp.

—Alcohol
EC 76/765/EEC-76. Council Directive on the Approximation of the Laws of the Member States Relating to Alcoholometers and Alcohol Hydrometers. 6 pp.
EC 82/624/EEC-82. Commission Directive Adapting to Technical Progress Council Directive 76/765/EEC on the Approximation of the Laws of the Member States Relating to Alcoholometers and Alcohol Hydrometers. 2 pp.
ISO 4801-79. Glass Alcoholometers and Alcohol Hydrometers Not Incorporating a Thermometer First Edition; (Erratum—April 1980). 10 pp.
ISO 4805-82. Laboratory Glassware—Thermo-Alcoholometers and Alcohol-Thermohydrometers First Edition. 9 pp.

—Alcohol—Ethanol Content
BSI BS 5470-77. 1977 Glass Alcohol Hydrometers Not Incorporating a Thermometer. 7 pp.

—Alcohol—Thermometers
BSI BS 5471-77. 1977 Amd 1 Thermometers for Use with Alcohol Hydrometers. 8 pp.
ISO 6152-82. Thermometers for Use with Alcoholometers and Alcohol Hydrometers First Edition. 6 pp.

—Battery Electrolytes
MOD UK DSTAN 66-3: Part 1-01. Testers, Specific Gravity Part 1: Tester, Battery Electrolyte Solution Issue 2; Amendment 1. 9 pp.

—Dairy Products
BSI BS 734: Part 1-73. 1973 Amd 1 Density Hydrometers for Use in Milk Part 1: Measurement of the Density of Milk Using a Hydrometer. 18 pp.
BSI BS 734: Part 2-59. 1959 Amd 1 Density Hydrometers for Use in Milk Part 2: Methods. 734C is an Extract From BS 734 Part 2. 32 pp.
ISO 2449-74. Milk and Liquid Milk Products—Density Hydrometers for Use in Products with a Surface Tension of Approximately 45 mN/m First Edition. 10 pp.
SAA AS N40 Part 2-62. Density Hydrometers for Use in Milk—Part 2: Methods. 32 pp.

—Glass
BSI BS 718-91. 1991 Density Hydrometers (H). 22 pp.
BSI BS 718-79. 1979 Amd 1 Density Hydrometers (AMD 5601) July 29, 1988. 21 pp.

Hydrometers (Cont.)

—Glass (Cont.)
ISO 649 Pt 1-81. Laboratory Glassware—Density Hydrometers for General Purposes—Part 1: Specification First Edition. 11 pp.
ISO 649 Pt 2-81. Laboratory Glassware—Density Hydrometers for General Purposes—Part 2: Test Methods and Use First Edition. 11 pp.

—Marine
MOD UK NES 2064-92. Specification for Hydrometers (Metal Type) NS Cat Nos 0253/2176, 220—5799 & 220—5800 Issue 1 (01.92). 10 pp.

—Thermal Expansion
ISO 1768-75. Glass Hydrometers—Conventional Value for the Thermal Cubic Expansion Coefficient (for Use in the Preparation of Measurement Tables for Liquids) First Edition. 4 pp.

Hydrometric Boats
Use: Boats (Marine)—Hydrometric

Hydrometric Telemetry Systems
See Also: Hydrometry; Telemetry
BSI BS 3680: Part 8G-92. 1992 Measurement of Liquid Flow in Open Channels Part 8: Measuring Instruments and Equipment Part 8G: Specification for System Requirements for Hydrometric Telemetry Systems (ISO 6419-2: 1992) (Q). 20 pp.
ISO 6419 Pt 2-92. Hydrometric Telemetry Systems—Part 2: Specification of System Requirements First Edition. 19 pp.

Hydrometry
Use For: Density Hydrometric Analysis; Specific Gravity Hydrometric Analysis
See Also: Hydrometric Telemetry Systems

—Coatings—Density
DIN ENGL 53217 Pt 4-91. Determination of Density of Paints, Varnishes and Similar Coating Materials by the Hydrometer Method (Mar). 3 pp.

—Crude Oils
BSI BS 4714-80. 1980 Laboratory Determination of Density or Relative Density of Crude Petroleum and Liquid Petroleum Products (Hydrometer Method). 9 pp.
CNS K6916-90. Method of Test for Density, Relative Density, or API Gravity of Crude Petroleum and Liquid Petroleum Products by Hydrometer Method (Aug)(12017).
ISO 3675-93. Crude Petroleum and Liquid Petroleum Products—Laboratory Determination of Density or Relative Density—Hydrometer Method Second Edition. 11 pp.

—Data Transmission Equipment
BSI BS 3680: Part 8F-86. 1986 Measurement of Liquid Flow in Open Channels: Measuring Instruments and Equipment: Hydrometric Data Transmission Systems: General. 21 pp.
ISO 6419 Pt 1-84. Hydrometric Data Transmission Systems—Part 1: General First Edition. 22 pp.

—Hydrocarbons
BSI BS 6665-85. 1985 Method for Determination of Density or Relative Density of Liquefied Petroleum Gases and Other Light Hydrocarbons by Pressure Hydrometer Method. 10 pp.

—Iron Ores
JIS M 8717-88. Test Methods for Determination of Density of Iron Ores. 9 pp.

—Liquefied Petroleum Gas—Density
BSI BS 6665-85. 1985 Method for Determination of Density or Relative Density of Liquefied Petroleum Gases and Other Light Hydrocarbons by Pressure Hydrometer Method. 10 pp.

—Milk
BSI BS 734: Part 2-59. 1959 Amd 1 Density Hydrometers for Use in Milk Part 2: Methods. 734C is an Extract From BS 734 Part 2. 32 pp.

—Petroleum Products
BSI BS 4714-80. 1980 Laboratory Determination of Density or Relative Density of Crude Petroleum and Liquid Petroleum Products (Hydrometer Method). 9 pp.
CNS K6916-90. Method of Test for Density, Relative Density, or API Gravity of Crude Petroleum and Liquid Petroleum Products by Hydrometer Method (Aug)(12017).
ISO 3675-93. Crude Petroleum and Liquid Petroleum Products—Laboratory Determination of Density or Relative Density—Hydrometer Method Second Edition. 11 pp.

Hydrometry (Cont.)
—Potassium Silicates
BSI BS 6092: Part 1-81. 1981 Sampling and Test for Sodium and Potassium Silicates for Industrial Use Part 1: Determination of Solution Density at 20 Degrees Celsius. 6 pp.
ISO 1687-76. Sodium and Potassium Silicates for Industrial Use—Determination of Density at 20 Degrees Celsius of Products in Solution—Method Using Density Hydrometer and Method Using Pyknometer First Edition. 5 pp.

—Sodium Silicates—Density
BSI BS 6092: Part 1-81. 1981 Sampling and Test for Sodium and Potassium Silicates for Industrial Use Part 1: Determination of Solution Density at 20 Degrees Celsius. 6 pp.
ISO 1687-76. Sodium and Potassium Silicates for Industrial Use—Determination of Density at 20 Degrees Celsius of Products in Solution—Method Using Density Hydrometer and Method Using Pyknometer First Edition. 5 pp.

Hydronic Heating Equipment
Use: Water Heaters

Hydronic Heating System Radiator Valves
Use: Radiator Valves

Hydronic Heating System Radiators
Use: Radiators (Heating)

Hydrophobia
Use: Rabies

Hydrophones
See Also: Acoustic Measuring Instruments; Electroacoustic Transducers; Microphones
IEC 500-74. IEC Standard Hydrophone First Edition. 27 pp.

—Calibration
BSI BS 5652-79. 1979 Amd 1 The Calibration of Hydrophones. 67 pp.
BSI BS 5653-78. 1978 Hydrophones for Calibration Purposes. 14 pp.
BSI BS 7041-89. 1989 Characteristics and Calibration of Hydrophones for Operation in the Frequency Range 0.5 MHz to 15 MHz. 34 pp.
IEC 565-77. Calibration of Hydrophones First Edition; (Supplement A-1980). 121 pp.
IEC 866-87. Characteristics and Calibration of Hydrophones for Operation in the Frequency Range 0.5 MHz to 15 MHz First Edition. 62 pp.
IEC 1101-91. Absolute Calibration of Hydrophones Using the Planar Scanning Technique in the Frequency Range 0,5 MHz to 15 MHz First Edition. 49 pp.

—Ultrasonic Frequencies
IEC 1220-93. Ultrasonics—Fields—Guidance for the Measurement and Characterization of Ultrasonic Fields Generated by Medical Ultrasonic Equipment Using Hydrophones in the Frequency Range 0,5 MHz to 15 MHz First Edition. 57 pp.

Hydropneumatic Accumulators
Use For: Gas Loaded Accumulators
See Also: Accumulators; Hydraulic Accumulators; Hydraulic Equipment; Pneumatic Equipment
ISO 5596-82. Hydraulic Fluid Power—Gas-Loaded Accumulators with Separators—Range of Pressures and Volumes, Characteristic Quantities and Identification First Edition. 6 pp.

—Steel
BSI BS 7201: Part 1-89. 1989 Hydro-Pneumatic Accumulators for Fluid Power Purposes Part 1: Seamless Steel Accumulator Bodies Above 0.5L Water Capacity. 25 pp.
BSI BS 7201: Part 1-01. 1989 Amd 1 Hydro-Pneumatic Accumulators for Fluid Power Purposes Part 1: Specification for Seamless Steel Accumulator Bodies Above 0.5 L Water Capacity (AMD 7186) July 15, 1992. 25 pp.

Hydroprobes
Use: Moisture Meters

Hydroquinone
Use For: p-Dihydroxybenzene
See Also: Hydroquinone Content Analysis; Phenol
CNS K7250-68. Chemical Reagent (Hydroquinone) (Nov)(1749).
JIS K 8738-79. Hydroquinone.

—Photographic Chemicals
BSI BS 3103-78. 1978 Photographic Grade Hydroquinone (P-Dihydroxybenzene). 9 pp.
CNS Z9009-80. Hydroquinone (Photographic Grade) (Jan)(5150).

Hydroquinone (Cont.)
—Photographic Chemicals (Cont.)
ISO 423-76. Photographic Grade Hydroquinone—Specification First Edition. 7 pp.
JIS K 7713-86. Photographic Grade Hydroquinone. 6 pp.

Hydroquinone Content Analysis
See Also: Hydroquinone; Phenol
—Photographic Wastes—Spectrophotometry
ISO 7760-86. Photographic Processing Waste—Determination of Hydroquinone Content—Spectrophotometric Method First Edition. 5 pp.

Hydrostatic Pressure Testing
Use: Hydrostatic Testing

Hydrostatic Testing
See Also: Compression Testing; Dynamic Testing; High Temperature Testing; Impact Testing; Static Testing; Tensile Testing
—Connectors
CNS C6192-88. Method of Test for Low Frequency (Below 3 MHz) Electrical Connectors (TP-39 Hydrostatic Test) (Apr)(9371).

—Hoses
BSI BS 5173: Sec 102.1-91. 1991 Rubber and Plastics Hoses and Hose Assemblies Part 102: Hydraulic Pressure Tests Section 102.1: Hydrostatic Tests. 10 pp.
ISO 1402-84. Rubber and Plastic Hoses and Hose Assemblies—Hydrostatic Testing Second Edition. 6 pp.
SAA AS 1180.5A-72. Methods of Test for Hose Made from Elastomeric Materials—Part 5A: Hydrostatic Pressure—Burst Test (Superseded by AS 1180.5—1993).
SAA AS 1180.5B-72. Methods of Test for Hose Made from Elastomeric Materials—Part 5B: Hydrostatic Pressure—Proof Test (Superseded by AS 1180—1993).
SAA AS 1180.5C-72. Methods of Test for Hose Made from Elastomeric Materials—Part 5C: Hydrostatic Pressure—Change-in-Length Test (Supresedo by AS 1180—1993).
SAA AS 1180.5D-72. Methods of Test for Hose Made from Elastomeric Materials—Part 5D: Hydrostatic Pressure—Leakage Test (Superseded by AS 1180—1993).
SAA AS 1180.5E-72. Methods of Test for Hose Made from Elastomeric Materials—Part 5E: Hydrostatic Pressure—Expansion and Distortion Test (Superseded by AS 1180—1993).

—Pipe Fittings—PVC—Unplasticized
SAA AS 1462.9-84. Methods of Test for Unplasticized PVC (UPVC) Pipes and Fittings—Part 9: Method of Hydrostatic Pressure Testing of UPVC Pressure Fittings. 3 pp.
SAA AS 1462.10-84. Methods of Test for Unplasticized PVC (UPVC) Pipes and Fittings—Part 10: Method for Hydrostatic Pressure Testing of UPVC Non-Pressure Fittings. 3 pp.

—Pipelines—Ductile Iron
ISO 10802-92. Ductile Iron Pipelines—Hydrostatic Testing After Installation First Edition. 7 pp.

—Plastic Pipes—Fiberglass Reinforced
DIN ENGL 53769 Pt 2-86. Testing of Glass Fibre Reinforced Plastics Pipes; Long-Term Hydrostatic Pressure Test (Dec). 4 pp.

—Plastic Pipes—PVC—Unplasticized
SAA AS 1462.6-89. Methods of Test for Unplasticized PVC (UPVC) Pipes and Fittings—Part 6: Method for Hydrostatic Pressure Testing of UPVC Pressure Pipes. 10 pp.

—Plastic Pipes—Thermoplastic
ISO TR9080-92. Thermoplastics Pipes for the Transport of Fluids—Methods of Extrapolation of Hydrostatic Stress Rupture Data to Determine the Log-Term Hydrostatic Strength of Thermoplastics Pipe Materials First Edition. 60 pp.

Hydrotherapy Pools
SAA AS 3979-91. Hydrotherapy Pools. 6 pp.

Hydroxide Ion Content Analysis
—Water
CNS K9041-80. Method of Test for Hydroxide Ion in Water (Jul)(5860).

Hydroxides
Scope Note: Use a more specific term
See: Aluminum Hydroxides; Barium Hydroxide; Basic Lead Carbonates; Calcium Hydroxides; Lime Water; Magnesium Carbonate, Basic; Magnesium Hydroxide; Potassium Hydroxides; Sodium Hydroxide;

Hydroxides (Cont.)
See: (Cont.)
Tetramethylammonium Hydroxide

2-Hydroxy-1-(2-hydroxy-4-sulfo-1-naphthylazo)-3-naphthoic Acid
JIS K 8776-72. 2-Hydroxyl-1-(2'-Hydroxy-4-Sulfo-1'-Naphthylazo)-3-Naphthoic Acid.

Hydroxy Methyl Furfurol Content Analysis
—Glucose
CNS N6055-72. Method of Test for Powdered Glucose (Jun)(3351). 3 pp.

1-Hydroxy-2-naphthoic Acid
JIS K 4131-85. Hydroxynaphthoic Acids (3-Hydroxy-2-Naphthoic Acid.1-Hydroxy-2-Naphthoic Acid).

3-Hydroxy-2-naphthoic Acid
CNS K2225-88. 3-Hydroxy-2-Naphthoic Acid (Jun)(12342).
JIS K 4131-85. Hydroxynaphthoic Acids (3-Hydroxy-2-Naphthoic Acid.1-Hydroxy-2-Naphthoic Acid).

2-Hydroxybenzaldehyde
Use: Salicylaldehyde

p-Hydroxybenzoic Acid Content Analysis
—Food—Gas Chromatography
CNS N6190-84. Method of Test for Preservatives in Food (Jun)(10949). 8 pp.

—Food—Thin Layer Chromatography
CNS N6190-84. Method of Test for Preservatives in Food (Jun)(10949). 8 pp.

o-Hydroxybenzoic Acid
Use: Salicylic Acid

Hydroxybenzoic Esters Content Analysis
—Food—Gas Chromatography
CNS N6190-84. Method of Test for Preservatives in Food (Jun)(10949). 8 pp.

—Food—Thin Layer Chromatography
CNS N6190-84. Method of Test for Preservatives in Food (Jun)(10949). 8 pp.

p-Hydroxybenzyl Isothiocyanate Content Analysis
—Mustard Seeds
CNS N4061-80. Determination of P-Hydroxybenzyl Isothiocyanate in Mustard Seed (Oct)(6605). 3 pp.

Hydroxyl Number
Use: Hydroxyl Value

Hydroxyl Value
—Binders (Materials)—Volumetric Analysis
BSI BS 6782: Part 4-87. 1987 Binders for Paints Part 4: Method for Determination of Hydroxyl Value (Titrimetric Method). 6 pp.
ISO 4629-78. Paint Media—Determination of Hydroxyl Value—Titrimetric Method First Edition; (Erratum—Nov 1979). 6 pp.

—Castor Oil
MOD UK M 9530/66. Examination of Oil, Caster, Hardened.

—Chemical Products
JIS K 0070-92. Test Methods for Acid Value, Saponification Value, Ester Value, Iodine Value, Hydroxyl Value and Unsaponifiable Matter of Chemical Products. 20 pp.

—Fats
BSI BS 684: Sec 2.9-77. 1977 Methods of Analysis of Fats and Fatty Oils Part 2: Other Methods Section 2.9: Determination of the Hydroxyl Value and the Acetyl Value. 4 pp.

—Fatty Acids
CNS K6648-81. Method of Test for Hydroxyl Value of Fatty Oils and Acid (Mar)(7167).

—Fatty Oils
CNS K6648-81. Method of Test for Hydroxyl Value of Fatty Oils and Acid (Mar)(7167).

—Polyester Resins
BSI BS 2782:Pt4: METH 432C-78. 1978 Methods of Testing Plastics Part 4: Chemical Properties Method 432C: Determination of Hydroxyl Value of Unsaturated Polyester Resins. 4 pp.

Hydroxyl Value (Cont.)

—Polyester Resins (Cont.)
ISO 2554-74. Plastics—Unsaturated Polyester Resins—Determination of Hydroxyl Value First Edition; (Erratum—Oct 1980). 5 pp.

—Polyoxyalkylene Resins—Esterification
ISO 6796-81. Polyglycols for Industrial Use—Determination of Hydroxyl Number—Phthalic Anhydride Esterification Method First Edition. 5 pp.

—Surfactants
BSI BS 6829: Sec 4.6-88. 1988 Analysis of Surface Active Agents (Raw Materials) Part 4: Ethylene Oxide Adducts Section 4.6: Method for Determination of Hydroxyl Value. 6 pp.
ISO 4326-80. Non-Ionic Surface Active Agents—Polyethoxylated Derivatives—Determination of Hydroxyl Value—Acetic Anhydride Method First Edition. 5 pp.
ISO 4327-79. Non-Ionic Surface Active Agents—Polyalkoxylated Derivatives-Determination of Hydroxyl Value—Phthalic Anhydride Method First Edition; (Erratum—May 1980). 11 pp.

—Vinyl Resins
CGSB 1-GP-71 METH 82.4-82. Methods of Testing Paints and Pigments Vinyl Resin Analysis Determination of Vinyl Alcohol. 2 pp.

Hydroxylamine Hydrochloride
CNS K7251-68. Chemical Reagent (Hydroxlamine Hydrochloride) (Nov)(1750).
JIS K 8201-79. Hydroxylamine Hydrochloride.

Hydroxylamine Sulfate
Use For: Hydroxylammonium Sulfate
JIS K 8993-80. Hydroxyl Ammonium Sulfate.

Hydroxylammonium Sulfate
Use: Hydroxylamine Sulfate

5-Hydroxymethylfurfural Content Analysis
Use For: 5-HMF Content Analysis

—Fruits
ISO 7466-86. Fruit and Vegetable Products—Determination of 5-Hydroxymethylfurfural (5-HMF) Content First Edition. 5 pp.

—Vegetables
ISO 7466-86. Fruit and Vegetable Products—Determination of 5-Hydroxymethylfurfural (5-HMF) Content First Edition. 5 pp.

Hydroxyproline
See Also: Hydroxyproline Content Analysis
JIS K 9102-76. L-Hydroxyproline.

Hydroxyproline Content Analysis
See Also: Hydroxyproline

—Meat
BSI BS 4401: Part 11-79. 1979 Amd 1 Methods of Test for Meat and Meat Products Part 11: Determination of L(-)-Hydroxyproline Content (Reference Method). 7 pp.
CNS N6147-80. Method of Test for Meat and Meat Products—Determination of L (-)—Hydroxyproline Content (Oct)(6608). 4 pp.
ISO 3496-78. Meat and Meat Products—Determination of L(-)-Hydroxyproline Content (Reference Method) First Edition. 5 pp.

8-Hydroxyquinoline Content Analysis
See Also: 8-Hydroxyquinoline

—Cosmetics
EC 83/514/EEC-83. Third Commission Directive on the Approximation of the Laws of the Member States Relating to Methods of Analysis Necessary for Checking the Composition of Cosmetic Products. 38 pp.

8-Hydroxyquinoline Sulfate Content Analysis

—Cosmetics
EC 83/514/EEC-83. Third Commission Directive on the Approximation of the Laws of the Member States Relating to Methods of Analysis Necessary for Checking the Composition of Cosmetic Products. 38 pp.

8-Hydroxyquinoline
Use For: 8-Quinolinol *See Also:* 8-Hydroxyquinoline Content Analysis
CNS K7252-68. Chemical Reagent (8-Hydroxyquinoline) (Nov)(1751).
JIS K 8775-80. 8-Quinolinol.

alpha-Hydroxytoluene
Use: Benzyl Alcohol

Hygiene
See Also: Sanitary Engineering

—Air Conditioning Equipment
SNZ NZS 4302-87. Control of Hygiene in Air and Water Systems in Buildings Amend: 1, 1991. 30 pp.

—Classification
BSI BS 1000: (613)-80. 1980 Universal Decimal Classification (UDC). English Full Edition (613): Hygiene Generally. Personal Health and Hygiene. 12 pp.
SNZ NZS/BS 1000 (613)-80. Universal Decimal Classification Hygiene Generally. Personal Health and Hygiene. 12 pp.

—Cooking Utensils
CNS Z6087-88. Method of Test for the Hygiene of Food Utensils, Containers and Packages Products (Jan)(12218).
CNS Z6088-88. Method of Test for the Hygiene of Food Utensils, Containers and Packages Metallic (Jan)(12219).
CNS Z6089-88. Method of Test for the Hygiene of Food Utensils, Containers and Packages Plastic Products (General Regulation) (Jan)(12220).
CNS Z6090-88. Method of Test for the Hygiene of Food Utensils, Containers and Packages Plastic Products (Classified Regulation) (Jan)(12221).

—Dairies—Sampling
ISO 8086-86. Dairy Plant—Hygiene Conditions—General Guidance on Inspection and Sampling Procedures First Edition. 8 pp.

—Dairies—Visual Inspection
ISO 8086-86. Dairy Plant—Hygiene Conditions—General Guidance on Inspection and Sampling Procedures First Edition. 8 pp.

—Food
EC 93/43/EEC-93. Council Directive on the Hygiene of Foodstuffs. 11 pp.

—Food Packaging
CNS Z6087-88. Method of Test for the Hygiene of Food Utensils, Containers and Packages Products (Jan)(12218).
CNS Z6088-88. Method of Test for the Hygiene of Food Utensils, Containers and Packages Metallic (Jan)(12219).
CNS Z6089-88. Method of Test for the Hygiene of Food Utensils, Containers and Packages Plastic Products (General Regulation) (Jan)(12220).
CNS Z6090-88. Method of Test for the Hygiene of Food Utensils, Containers and Packages Plastic Products (Classified Regulation) (Jan)(12221).

—Meat Processing Facilities
SNZ NZS 5102-81. Specification for Animal Offal and Meat Treatment Units for Hydatids Control. 11 pp.

—Medical Supplies—Rubber Products
JIS T 9010-82. Testing Methods of Hygienic Safety for Rubber Goods.

—Military—Training
NATO STANAG 2122 ED 2 AMD 4-75. Medical Training in First-Aid, Basic Hygiene and Emergency Care. 12 pp.
NATO STANAG 2122 ED 2 AMD 5-75. Medical Training in First-Aid, Basic Hygiene and Emergency Care. 12 pp.

—Nuclear, Biological and Chemical Warfare
NATO STANAG 2358 ED 2 AMD 1-89. First Aid and Hygiene Training in NBC Operations. 11 pp.

—Rubber
CNS T4017-85. Method of Test for Hygienic Safety for Rubber Goods: General Rules (Aug)(11350).
CNS T4017-1-85. Method of Test for Hygienic Safety for Rubber Goods: Material Test (Aug)(11350-1).

—Rubber—Leach Testing
CNS T4017-2-85. Method of Test for Hygienic Safety for Rubber Goods: Leaching Test (Aug)(11350-2).

—Water Supply Engineering
SNZ NZS 4302-87. Control of Hygiene in Air and Water Systems in Buildings Amend: 1, 1991. 30 pp.

Hygienic Valves
Use: Sanitary Valves

Hygrographs
See Also: Moisture Content Analysis; Recording Instruments
CNS B6070-82. Hygrographs (Jul)(9060).
JIS B 7306-89. Hygrographs. 7 pp.

Hygrometers
Scope Note: For additional listings, use a more specific term *See Also:* Dew Point Hygrometers; Hair Hygrometers; Measuring Instruments; Psychrometers; Whirling Hygrometers
BSI BS 3292-60. (WITHDRAWN) 1960 Direct Reading Hygrometers. 8 pp.
BSI BS 5248-90. 1990 Aspirated Hygrometers. 15 pp.
BSI BS 5248-01. 1990 Amd 1 Aspirated Hygrometer (AMD 7383) January 15, 1993. 17 pp.
DIN ENGL 50012 Pt 4-86. Climates and Their Technical Application; Methods of Measuring Humidity; Dew Point Hygrometer (Jan). 2 pp.
DIN ENGL 50012 Pt 5-88. Artificial Climates in Technical Applications; Methods of Measuring Humidity; LiCl Hygrometers (Nov). 3 pp.

Hygromycin B Content Analysis

—Animal Feed
CNS N4115-84. Method of Test for Feed Additives: Determination of Hygromycin B (May)(10884).

Hygroscopicity
Use: Water Absorption

Hylotrupes
Use: Coleoptera

Hyne (Brantly) Helicopters
See Also: Helicopters

—Foreign Airworthiness Directives
CAA. Hynes (Brantly) Series Helicopters (Foreign Airworthiness Directives). 1 p.

Hyperbaric Chambers
See Also: Health Care Facilities; Medical Electrical Equipment; Medical Equipment; Pressure Vessels
CSA Z275.1-M1982. Hyperbaric Facilities; (Amd 1-15 January 1987). 60 pp.
JIS T 7321-89. Hyperbaric Oxygen Chambers.

—Medical Equipment
NATO STANAG 1213 ED 1 AMD 3-79. Minimum Medical Equipment for Multiplace Multicompartment Hyperbaric Chambers. 8 pp.
NATO STANAG 1213 ED 1 AMD 4-79. Minimum Medical Equipment for Multiplace Multicompartment Hyperbaric Chambers. 8 pp.

Hyperbolic Navigation Systems
See Also: Air Navigation; Navigation; Omega Navigation Systems; Radio Navigation; Radio Navigation Equipment

—Aircraft—Radio Receivers
CAA Part 14 CAP 208. Hyperbolic Navigation Systems Receiving Apparatus (Civil Air Publications: Airborne Radio Apparatus). 8 pp.

—Maritime—MF/HF
CCIR RECMN 631-86. Use of Hyperbolic Maritime Radionavigation Systems in the Band 283.5-315 kHz—Section 8D—Radiodetermination, Global Maritime Distress and Safety System and Related Subjects. 1 p.
CCIR RECMN 631-1-92. Use of Hyperbolic Maritime Radionavigation Systems in the Band 283.5-315 kHz—Section 8D—Radiodetermination, Global Maritime Distress and Safety System and Related Subjects. 3 pp.

—Maritime—Radio Frequency Interference
CCIR Report 913-2-90. Technical Characteristics of Maritime Radiobeacons —Section 8D—Radiodetermination, Global Maritime Distress and Safety System and Related Subjects. 16 pp.

—Maritime—Radio Receivers
CCIR Report 913-2-90. Technical Characteristics of Maritime Radiobeacons —Section 8D—Radiodetermination, Global Maritime Distress and Safety System and Related Subjects. 16 pp.

Hypermedia/Time-Based Structuring Language
Use: HyTime

Hypo Number

—Pulps
CNS P3067-84. Method of Test for Hypo Number of Pulp (Apr)(10864). 5 pp.
CPPA G.32P-81. Hypo Number of Pulp. 3 pp.

Hypochlorites
See Also: Bleach Liquors; Calcium Hypochlorite; Hazardous Materials; Sodium Hypochlorite
JIS K 1425-59. Bleaching Powder and High Test Hypochlorite.

Hypochlorites (Cont.)
—Fabrics—Colorfastness Testing
CNS L3031-83. Method of Test for Colour Fastness to Bleaching with Hypochlorite (Mar)(1498).
JIS L 0856-83. Testing Methods for Colour Fastness to Bleaching with Hypochlorite. 7 pp.

Hypodermic Needles
See Also: Medical Equipment; Medical Instruments; Syringes
BSI BS 5081: Part 2-87. 1987 Amd 1 Sterile Hypodermic Syringes and Needles Part 2: Specification for Sterile Hypodermic Needles for Single Use (AMD 6123) November 30, 1989. 11 pp.
CGSB CAN/CGSB-47.5-M88. Syringes, Hypodermic, Single Use, Sterile, with or Without Needle. 26 pp.
CGSB CAN/CGSB-47.6-M87. Needles, Hypodermic, Single Use, Sterile. 20 pp.
CNS T1052-81. Injection Needles (Nov)(8165). 8 pp.
JIS T 3101-79. Injection Needles.
JIS T 3101-65. Injection Needles. 9 pp.
MOD UK DSTAN 65-5-77. Medical, Surgical and Dental Instruments, Equipment and Supplies Issue 2. 8 pp.

—Color Coding
BSI BS 7128-93. 1993 Colour Coding of Hypodermic Needles for Single Use (ISO 6009: 1982). 15 pp.
BSI BS 7128-89. 1989 Colour Coding of Hypodermic Needles for Single Use (ISO 6009: 1988). 7 pp.
ISO 6009-92. Hypodermic Needles for Single Use—Colour Coding for Identification Third Edition. 12 pp.

—Conical Fittings
BSI BS 3522-62. (OBSOLESCENT) 1962 Hypodermic Surgical Mounted Needles (Luer Fitting). 12 pp.
BSI BS 3930: Part 1-87. 1987 Conical Fittings with 6 Per Cent (Luer) Taper for Syringes, Needles and Other Medical Equipment Part 1: General Requirements. 11 pp.
BSI BS 3930: Part 2-91. 1991 Conical Fittings with a 6 Percent (Luer) Taper for Syringes, Needles and Other Medical Equipment Part 2: Specification for Lock Fittings (ISO 594-2: 1991). 15 pp.
ISO 594 Pt 1-86. Conical Fittings with a 6 % (Luer) Taper for Syringes, Needles and Certain Other Medical Equipment—Part 1: General Requirements First Edition; DoD Adopted. 9 pp.
ISO 594 Pt 2-91. Conical Fittings with a 6 % (Luer) Taper for Syringes, Needles and Certain Other Medical Equipment—Part 2: Lock Fittings First Edition; DoD Adopted. 12 pp.

—Dental
SAA AS 1264-85. Dental Equipment—Cartridge Hypodermic Needles (Sterile), Single-Use (This is a Joint Standard with SANZ NZS/AS 1264). 17 pp.
SNZ NZS/AS 1264-85. Dental Equipment—Cartridge Hypodermic Needles (Sterile), Single Use (This is a Joint Standard with SAA AS 1264). 17 pp.

—Luer Syringes
CGSB CAN/CGSB-47.2-M86. Needles, Hypodermic, for Luer Syringes,. 15 pp.

—Needle Tubing
BSI BS 7547-91. 1991 Stainless Steel Needle Tubing for the Manufacture of Medical Devices (ISO 9626: 1991). 15 pp.
ISO 9626-91. Stainless Steel Needle Tubing for Manufacture of Medical Devices First Edition. 13 pp.

—Single Use
ISO 7864-93. Sterile Hypodermic Needles for Single Use Third Edition. 16 pp.
SAA AS 1946-84. Hypodermic Equipment—Single-Use Needles (Sterile) for General Medical Use Amdt 1 May 1985. 25 pp.

Hypodermic Syringes
Use: Syringes

Hypoid Gear Pair
Use: Hypoid Gears

Hypoid Gears
Use For: Hypoid Gear Pair See Also: Gears; Helical Gears; Ring Gears; Worms (Mechanical Drives)

—Glossaries
CNS B1177-84. Glossary of Gears Geometrical Definitions (Bevel and Hypoid Gears and Gear Pairs) (Feb)(5719).
DIN ENGL 3971-80. Definitions and Parameters for Bevel Gears and Bevel Gear Pairs (July). 24 pp.
DIN ENGL 3998 Pt 3-76. Denominations on Gears and Gear Pairs; Bevel and Hypoid Gears and Gear Pairs (Sept). 14 pp.

Hypophosphorus Acid
JIS K 8440-61. Hypophosphorous Acid.

Hypothetical Reference Chains
See Also: Hypothetical Reference Circuits

—Analog—Satellite Links
CCIR QUESTION 29/CMTT-90. Analogue Television Reference Chains for Terrestrial and Fixed-Satellite Links—Questions Concerning the CMTT CCIR/CCITT Joint Study Group for Television and Sound Transmission. 1 p.

—Television Transmission—Long Distances
CCIR RECMN 603-82. Hypothetical Reference Chain for Television Transmissions over Very Long Distances—Section CMTT A—Television Transmission Standards and Performance Objectives. 1 p.

Hypothetical Reference Circuits
Use For: HRC; Hypothetical Reference Connections; Reference Circuits See Also: Circuits; Hypothetical Reference Chains; Hypothetical Reference Digital Paths; Telecommunication Circuits; Telegraph Circuits; Telephone Circuits
CCITT RECMN G.212-89. Hypothetical Reference Circuits for Analogue Systems—International Analogue Carrier Systems (Study Group XV) 3 pp. 3 pp.
CCITT RECMN G.215-89. Hypothetical Reference Circuit of 5000 km for Analogue Systems—International Analogue Carrier Systems (Study Group XV) 2 pp. 2 pp.

—Analog Transmission—Fixed Satellite Services
CCIR RECMN 352-4-82. Hypothetical Reference Circuit for Systems Using Analogue Transmission in the Fixed-Satellite Service—Section 4B2—Performance and Availability. 2 pp.

—Bit Error—Mobile Satellite Communications
CCIR QUESTION 86-1/8-90. Performance Objectives for Mobile-Satellite Services—Questions Concerning Study Group 8 —Mobile, Radiodetermination, Amateur and Related Satellite Services. 1 p.

—Carrier Systems
CCITT RECMN G.215-89. Hypothetical Reference Circuit of 5000 km for Analogue Systems—International Analogue Carrier Systems (Study Group XV) 2 pp. 2 pp.

—Circuit Noise
CCITT RECMN G.215-89. Hypothetical Reference Circuit of 5000 km for Analogue Systems—International Analogue Carrier Systems (Study Group XV) 2 pp. 2 pp.

—Fixed Satellite Services
CCIR Report 706-2-86. Availability of Circuits in the Fixed-Satellite Service—Section 4B2—Performance and Availability. 5 pp.

—Fixed Satellite Services—Interruption of Services
CCIR QUESTION 45/4-90. Interruptions to Traffic on Digital Paths or Circuits in the Fixed-Satellite Service—Questions of Study Group 4 Fixed-Satellite Service. 2 pp.

—Fixed Satellite Services—Noise Level
CCIR RECMN 354-2-74. Video Bandwidth and Permissible Noise Level in the Hypothetical Reference Circuit for the Fixed-Satellite Service—Section 4B2—Performance and Availability. 1 p.

—Fixed Satellite Services—Video Bandwidth
CCIR RECMN 354-2-74. Video Bandwidth and Permissible Noise Level in the Hypothetical Reference Circuit for the Fixed-Satellite Service—Section 4B2—Performance and Availability. 1 p.

—Frequency Division Multiplexers—Radio Relay Systems
CCITT RECMN G.431-89. Hypothetical Reference Circuits for Frequency-Division Multiplex Radio-Relay Systems—International Analogue Carrier Systems (Study Group XV) 2 pp. 2 pp.
CCITT RECMN G.433-89. Hypothetical Reference Circuit for Trans-Horizon Radio-Relay Systems for Telephony Using Frequency-Division Multiplex—International Analogue Carrier Systems (Study Group XV) 1 pp. 1 p.
CCITT RECMN G.441-89. Permissable Circuit Noise on Frequency-Division Multiplex Radio-Relay Systems—International Analogue Carrier Systems (Study Group XV) 1 pp. 1 p.

Hypothetical Reference Circuits (Cont.)
—Frequency Division Multiplexers—Radio Relay Systems (Cont.)
CCITT RECMN G.444-89. Allowable Noise Power in the Hypothetical Reference Circuit of Trans-Horizon Radio-Relay Systems for Telephony Using Frequency-Division Multiplex—International Analogue Carrier Systems (Study Group XV) 1 p.

—Frequency Division Multiplexers—Satellite Communications
CCITT RECMN G.445-89. Allowable Noise Power in the Hypothetical Reference Circuit for Frequency-Division Multiplex Telephony in the Fixed-Satellite Service—International Analogue Carrier Systems (Study Group XV) 1 pp. 1 p.

—Glossaries
CCITT RECMN G.212-89. Hypothetical Reference Circuits for Analogue Systems—International Analogue Carrier Systems (Study Group XV) 3 pp. 3 pp.

—Group Links—Intermodulation Noise
CCITT RECMN G.223-89. Assumptions for the Calculation of Noise on Hypothetical Reference Circuits for Telephony—International Analogue Carrier Systems (Study Group XV) 9 pp. 9 pp.

—Groups—Carrier Systems
CCITT RECMN G.325-89. General Characteristics Recommended for Systems Providing 12 Telephone Carrier Circuits on a Symmetric Cable Pair ((12 + 12) Systems)—International Analogue Carrier Systems (Study Group XV) 5 pp. 5 pp.

—Integrated Services Digital Networks—Fixed Satellite Services
CCIR RECMN 579-1-86. Availability Objectives for a Hypothetical Reference Circuit and a Hypothetical Reference Digital Path When Used for Telephony Using Pulse-Code Modulation, or Part of an Integrated Services Digital Network Hypothetical Reference Connection, in the. 2 pp.
CCIR RECMN 579-2-92. Availability Objectives for a Hypothetical Reference Circuit and a Hypothetical Reference Digital Path When Used for Telephony Using Pulse-Code Modulation, or Part of an Integrated Services Digital Network Hypothetical Reference Connection, in the. 6 pp.

—Mastergroup Links—Intermodulation Noise
CCITT RECMN G.223-89. Assumptions for the Calculation of Noise on Hypothetical Reference Circuits for Telephony—International Analogue Carrier Systems (Study Group XV) 9 pp. 9 pp.

—Mobile Satellite Communications
CCIR QUESTION 85-1/8-90. Availability of Circuits in Mobile-Satellite Services—Questions Concerning Study Group 8 —Mobile, Radiodetermination, Amateur and Related Satellite Services. 1 p.

—Noise—Frequency Division Multiplexing—Telephony—Fixed Satellites
CCIR RECMN 353-6-90. Allowable Noise Power in the Hypothetical Reference Circuit for Frequency-Division Multiplex Telephony in the Fixed-Satellite Service—Section 4B2—Performance and Availability. 2 pp.
CCIR RECMN 353-7-92. Allowable Noise Power in the Hypothetical Reference Circuit for Frequency-Division Multiplex Telephony in the Fixed-Satellite Service—Section 4B2—Performance and Availability. 7 pp.

—Noise Power—Mobile Satellite Communications
CCIR QUESTION 86-1/8-90. Performance Objectives for Mobile-Satellite Services—Questions Concerning Study Group 8 —Mobile, Radiodetermination, Amateur and Related Satellite Services. 1 p.

—Open Wire Lines—Carrier Systems
CCITT RECMN G.311-89. General Characteristics of Systems Providing 12 Carrier Telephone Circuits on an Open-Wire Pair—International Analogue Carrier Systems (Study Group XV) 6 pp. 6 pp.

—Radio Relay Systems
CCIR RECMN 557-2-90. Availability Objective for a Hypothetical Reference Circuit and a Hypothetical Reference Digital Path—Section 9A—Performance Objectives, Propagation and Interference Effects. 2 pp.
CCITT RECMN G.215-89. Hypothetical Reference Circuit of 5000 km for Analogue Systems—International Analogue Carrier Systems (Study Group XV) 2 pp. 2 pp.

INTERNATIONAL AND NON-U.S. NATIONAL STANDARDS
SUBJECT INDEX
Hypothetical

Hypothetical Reference Circuits (Cont.)

—Radio Relay Systems—Availability
CCIR RECMN 557-3-91. Availability Objective for Radio-Relay Systems over a Hypothetical Reference Circuit and a Hypothetical Reference Digital Path—Section 9A—Performance Objectives, Propagation and Interference Effects. 5 pp.

—Radio Relay Systems—Glossaries
CCIR RECMN 390-4-82. Definitions of Terms and References Concerning Hypothetical Reference Circuits and Hypothetical Reference Digital Paths for Radio-Relay Systems—Section 9A—Performance Objectives, Propagation and Interference Effects. 2 pp.

—Radiotelephony—Maritime Mobile Satellite Communications—Noise
CCIR RECMN 547-78. Noise Objectives in the Hypothetical Reference Circuit for Systems in the Maritime Mobile-Satellite Service—Section 8G—Availability, Performance Objectives and Interworking with Terrestrial Networks. 2 pp.

—Radiotelephony—Mobile Satellite Communications
CCIR RECMN 546-2-90. Hypothetical Telephone Reference Circuit in the Aeronautical, Land and Maritime Mobile-Satellite Services—Section 8F—Frequencies, Orbits and Systems. 3 pp.

—Satellite Communications
CCITT RECMN G.434-89. Hypothetical Reference Circuit for Systems Using Analogue Transmission in the Fixed-Satellite Service—International Analogue Carrier Systems (Study Group XV) 2 pp. 2 pp.

—Sound Transmission
CCIR RECMN 502-2-82. Hypothetical Reference Circuits for Sound-Programme Transmissions Terrestrial Systems and Systems in the Fixed-Satellite Service—Section CMTT C—Transmission Standards and Performance Objectives for Sound-Programme Channels. 3 pp.
CCIR QUESTION 60/CMTT-90. Definition of a Digital Hypothetical Reference Connection for Digital Sound-Programme Transmission—Questions Concerning the CMTT CCIR/CCITT Joint Study Group for Television and Sound Transmission. 1 p.
CCITT RECMN J.11-89. Hypothetical Reference Circuits for Sound-Programme Transmissions—Line Transmission of Non-Telephone Signals—Transmission of Sound-Programme and Television Signals (Study Group XV) 3 pp. 3 pp.

—Sound Transmission—Fixed Satellite Services
CCIR RECMN 502-2-82. Hypothetical Reference Circuits for Sound-Programme Transmissions Terrestrial Systems and Systems in the Fixed-Satellite Service—Section CMTT C—Transmission Standards and Performance Objectives for Sound-Programme Channels. 3 pp.

—Supergroup Links—Intermodulation Noise
CCITT RECMN G.223-89. Assumptions for the Calculation of Noise on Hypothetical Reference Circuits for Telephony—International Analogue Carrier Systems (Study Group XV) 9 pp. 9 pp.

—Telegraphy—Radio Relay Systems
CCITT RECMN G.442-89. Radio-Relay System Design Objectives for Noise at the Far End of a Hypothetical Reference Circuit with Reference to Telegraphy Transmission—International Analogue Carrier Systems (Study Group XV) 1 pp. 1 p.

—Telephone Cables
CCITT RECMN G.322-89. General Characteristics Recommended for Systems on Symmetric Pair Cables—International Analogue Carrier Systems (Study Group XV) 8 pp. 8 pp.

—Telephone Systems—HF
CCITT RECMN G.334-89. 18 MHz Systems on Standardized 2.6/9.5 mm Coaxial Cable Pairs—International Analogue Carrier Systems (Study Group XV) 9 pp. 9 pp.
CCITT RECMN G.343-89. 4 MHz Systems on Standardized 1.2/4.4 mm Coaxial Cable Pairs—International Analogue Carrier Systems (Study Group XV) 4 pp. 4 pp.
CCITT RECMN G.344-89. 6 MHz Systems on Standardized 1.2/4.4 mm Coaxial Cable Pairs—International Analogue Carrier Systems (Study Group XV) 4 pp. 4 pp.
CCITT RECMN G.345-89. 12 MHz Systems on Standardized 1.2/4.4 mm Coaxial Cable Pairs—International Analogue Carrier Systems (Study Group XV) 1 pp. 1 p.

Hypothetical Reference Circuits (Cont.)

—Telephone Systems—HF (Cont.)
CCITT RECMN G.346-89. 18 MHz Systems on Standardized 1.2/4.4 mm Coaxial Cable Pairs—International Analogue Carrier Systems (Study Group XV) 1 pp. 1 p.

—Telephone Systems—HF/VHF
CCITT RECMN G.333-89. 60 MHz Systems on Standardized 2.6/9.5 mm Coaxial Cable Pairs—International Analogue Carrier Systems (Study Group XV) 10 pp. 10 pp.

—Telephone Systems—MF
CCITT RECMN G.341-89. 1.3 MHz Systems on Standardized 1.2/4.4 mm Coaxial Cable Pairs—International Analogue Carrier Systems (Study Group XV) 5 pp. 5 pp.

—Telephone Systems—MF/HF
CCITT RECMN G.332-89. 12 MHz Systems on Standardized 2.6/9.5 mm Coaxial Cable Pairs—International Analogue Carrier Systems (Study Group XV) 11 pp. 11 pp.

—Telephony
CCITT RECMN G.212-89. Hypothetical Reference Circuits for Analogue Systems—International Analogue Carrier Systems (Study Group XV) 3 pp. 3 pp.

—Telephony—Fixed Satellite Services
CCIR RECMN 579-1-86. Availability Objectives for a Hypothetical Reference Circuit and a Hypothetical Reference Digital Path When Used for Telephony Using Pulse-Code Modulation, or as Part of an Integrated Services Digital Network Hypothetical Reference Connection, in the. 2 pp.
CCIR RECMN 579-2-92. Availability Objectives for a Hypothetical Reference Circuit and a Hypothetical Reference Digital Path When Used for Telephony Using Pulse-Code Modulation, or as Part of an Integrated Services Digital Network Hypothetical Reference Connection, in the. 6 pp.

—Telephony—Intermodulation Noise
CCITT RECMN G.223-89. Assumptions for the Calculation of Noise on Hypothetical Reference Circuits for Telephony—International Analogue Carrier Systems (Study Group XV) 9 pp. 9 pp.

—Telephony—Modulation—Frequency Translation Equipment
CCITT RECMN G.229-89. Unwanted Modulation and Phase Jitter—International Analogue Carrier Systems (Study Group XV) 2 pp. 2 pp.

—Telephony—Modulation—Power Supplies
CCITT RECMN G.229-89. Unwanted Modulation and Phase Jitter—International Analogue Carrier Systems (Study Group XV) 2 pp. 2 pp.

—Telephony—Noise
CCITT RECMN G.222-89. Noise Objectives for Design of Carrier-Transmission Systems of 2500 km—International Analogue Carrier Systems (Study Group XV) 4 pp. 4 pp.

—Telephony—Phase Jitter—Frequency Translation Equipment
CCITT RECMN G.229-89. Unwanted Modulation and Phase Jitter—International Analogue Carrier Systems (Study Group XV) 2 pp. 2 pp.

—Telephony—Power Levels
CCITT RECMN G.223-89. Assumptions for the Calculation of Noise on Hypothetical Reference Circuits for Telephony—International Analogue Carrier Systems (Study Group XV) 9 pp. 9 pp.

—Telephony—Psophometric Power
CCITT RECMN G.222-89. Noise Objectives for Design of Carrier-Transmission Systems of 2500 km—International Analogue Carrier Systems (Study Group XV) 4 pp. 4 pp.
CCITT RECMN G.223-89. Assumptions for the Calculation of Noise on Hypothetical Reference Circuits for Telephony—International Analogue Carrier Systems (Study Group XV) 9 pp. 9 pp.

—Telephony—Radio Relay Systems
CCIR RECMN 392-63. Hypothetical Reference Circuit for Radio-Relay Systems for Telephony Using Frequency-Division Multiplex with a Capacity of More Than 60 Telephone Channels—Section 9A—Performance Objectives, Propagation and Interference Effects. 2 pp.
CCIR RECMN 396-1-66. Hypothetical Reference Circuit for Trans-Horizon Radio-Relay Systems for Telephony Using Frequency-Division Multiplex—Section 9A—Performance Objectives, Propagation and Interference Effects. 1 p.

Hypothetical Reference Circuits (Cont.)

—Telephony—Radio Relay Systems (Cont.)
CCITT RECMN G.433-89. Hypothetical Reference Circuit for Trans-Horizon Radio-Relay Systems for Telephony Using Frequency-Division Multiplex—International Analogue Carrier Systems (Study Group XV) 1 pp. 1 p.
CCITT RECMN G.444-89. Allowable Noise Power in the Hypothetical Reference Circuit of Trans-Horizon Radio-Relay Systems for Telephony Using Frequency-Division Multiplex—International Analogue Carrier Systems (Study Group XV) 1 pp. 1 p.

—Telephony—Radio Relay Systems—Noise Power
CCIR RECMN 393-4-82. Allowable Noise Power in the Hypothetical Reference Circuit for Radio-Relay Systems for Telephony Using Frequency-Division Multiplex—Section 9A—Performance Objectives, Propagation and Interference Effects. 2 pp.
CCIR RECMN 397-3-78. Allowable Noise Power in the Hypothetical Reference Circuit of Trans-Horizon Radio-Relay Systems for Telephony Using Frequency-Division Multiplex—Section 9A—Performance Objectives, Propagation and Interference Effects. 1 p.

—Telephony—Satellite Communications
CCITT RECMN G.445-89. Allowable Noise Power in the Hypothetical Reference Circuit for Frequency-Division Multiplex Telephony in the Fixed-Satellite Service—International Analogue Carrier Systems (Study Group XV) 1 pp. 1 p.

—Television Systems—Radio Relay Systems—Noise Power
CCIR RECMN 555-78. Permissible Noise in the Hypothetical Reference Circuit of Radio-Relay Systems for Television—Section 9A—Performance Objectives, Propagation and Interference Effects. 2 pp.

—Transmission Systems
CCITT RECMN G.602-89. Reliability and Availability of Analogue Cable Transmission Systems and Associated Equipments—Transmission Characteristics—(Study Group XV) 4 pp. 4 pp.

Hypothetical Reference Connections
Use: Hypothetical Reference Circuits

Hypothetical Reference Digital Paths
Use For: HRDP *See Also:* Digital Paths; Hypothetical Reference Circuits; Hypothetical Reference Model

—Bit Error Rate—Fixed Satellite Services—Pulse Code Modulation
CCIR RECMN 522-3-90. Allowable Bit Error Ratios at the Output of the Hypothetical Reference Digital Path for Systems in the Fixed-Satellite Service Using Pulse-Code Modulation for Telephony —Section 4B2—Performance and Availability. 2 pp.
CCIR RECMN 522-4-92. Allowable Bit Error Ratios at the Output of the Hypothetical Reference Digital Path for Systems in the Fixed-Satellite Service Using Pulse-Code Modulation for Telephony —Section 4B2—Performance and Availability. 8 pp.

—Error Performance—Fixed Satellite Services—ISDN
CCIR RECMN 614-1-90. Allowable Error Performance for a Hypothetical Reference Digital Path in the Fixed-Satellite Service Operating Below 15 GHz When Forming Part of an International Connection in an Integrated Services Digital Network —Section 4B2—Perform-ance and Availability. 2 pp.
CCIR RECMN 614-2-92. Allowable Error Performance for a Hypothetical Reference Digital Path in the Fixed-Satellite Service Operating Below 15 GHz When Forming Part of an International Connection in an Integrated Services Digital Network —Section 4B2—Perform-ance and Availability. 42 pp.

—Fixed Satellite Services
CCIR Report 706-2-86. Availability of Circuits in the Fixed-Satellite Service—Section 4B2—Performance and Availability. 5 pp.

—Fixed Satellite Services—Digital Transmission
CCIR RECMN 521-2-86. Hypothetical Reference Digital Path for Systems Using Digital Transmission in the Fixed-Satellite Service—Section 4B2—Performance and Availability. 2 pp.

Hypothetical Reference Digital Paths *(Cont.)*

—Fixed Satellite Services—Geostationary—ISDN—Interference
CCIR RECMN 735-92. Maximum Permissible Levels of Interference in a Geostationary-Satellite Network for an HRDP When Forming Part of the ISDN in the Fixed-Satellite Service Caused by Other Networks of This Service Below 15 GHz—Section 4D1—Permissible Levels of Interference. 5 pp.

—Fixed Satellite Services—Interruption of Services
CCIR QUESTION 45/4-90. Interruptions to Traffic on Digital Paths or Circuits in the Fixed-Satellite Service—Questions of Study Group 4 Fixed-Satellite Service. 2 pp.

—Fixed Satellite Services—ISDN
CCIR RECMN 579-1-86. Availability Objectives for a Hypothetical Reference Circuit and a Hypothetical Reference Digital Path When Used for Telephony Using Pulse-Code Modulation, or as Part of an Integrated Services Digital Network Hypothetical Reference Connection, in the. 2 pp.

CCIR RECMN 579-2-92. Availability Objectives for a Hypothetical Reference Circuit and a Hypothetical Reference Digital Path When Used for Telephony Using Pulse-Code Modulation, or as Part of an Integrated Services Digital Network Hypothetical Reference Connection, in the. 6 pp.

CCIR Report 997-1-90. Characteristics of a Fixed-Satellite Service Hypothetical Reference Digital Path Forming Part of an Integrated Services Digital Network —Section 4B2—Performance and Availability. 33 pp.

—Mobile Satellite Communications—Feeder Links
CCIR RECMN 827-92. Hypothetical Reference Digital Path for Systems in the Mobile-Satellite Service Using Feeder Links. 2 pp.

—Radio Relay Systems
CCIR RECMN 557-2-90. Availability Objective for a Hypothetical Reference Circuit and a Hypothetical Reference Digital Path—Section 9A—Performance Objectives, Propagation and Interference Effects. 2 pp.

CCIR QUESTION 134/9-90. Hypothetical Reference Digital Paths and Performance Objectives for Digital Radio-Relay Systems—Questions of Group 9 Fixed Service. 1 p.

—Radio Relay Systems—Availability
CCIR RECMN 557-3-91. Availability Objective for Radio-Relay Systems over a Hypothetical Reference Circuit and a Hypothetical Reference Digital Path—Section 9A—Performance Objectives, Propagation and Interference Effects. 5 pp.

—Radio Relay Systems—Glossaries
CCIR RECMN 390-4-82. Definitions of Terms and References Concerning Hypothetical Reference Circuits and Hypothetical Reference Digital Paths for Radio-Relay Systems—Section 9A—Performance Objectives, Propagation and Interference Effects. 2 pp.

—Radio Relay Systems—ISDN
CCIR RECMN 556-1-86. Hypothetical Reference Digital Path for Radio-Relay Systems Which May Form Part of an Integrated Services Digital Network with a Capacity Above the Second Hierarchical Level—Section 9A—Performance Objectives, Propagation and Interference Effects. 2 pp.

—Radio Relay Systems—ISDN—Availability
CCIR RECMN 696-90. Error Performance and Availability Objectives for Hypothetical Reference Digital Sect. Utilizing Digital Radio-Relay Systems Forming Part or All of the Medium Grade Portion of an ISDN Connection—Section 9A—Performance Objectives, Propagation and Interference Effects. 3 pp.

CCIR RECMN 696-1-91. Error Performance and Availability Objectives for Hypothetical Reference Digital Sect. Utilizing Digital Radio-Relay Systems Forming Part or All of the Medium-Grade Portion of an ISDN Connection—Section 9A—Performance Objectives, Propagation and Interference Effects. 4 pp.

—Radio Relay Systems—ISDN—Bit Error Rate
CCIR RECMN 594-2-90. Allowable Bit Error Ratios at the Output of the Hypothetical Reference Digital Path for Radio-Relay Systems Which May Form Part of an Integrated Services Digital Network—Section 9A—Performance Objectives, Propagation and Interference Effects. 2 pp.

Hypothetical Reference Digital Paths *(Cont.)*

—Radio Relay Systems—ISDN—Bit Error Rate *(Cont.)*
CCIR RECMN 594-3-91. Allowable Bit Error Ratios at the Output of the Hypothetical Reference Digital Path for Radio-Relay Systems Which May Form Part of an Integrated Services Digital Network—Section 9A—Performance Objectives, Propagation and Interference Effects. 4 pp.

CCIR RECMN 696-90. Error Performance and Availability Objectives for Hypothetical Reference Digital Sect. Utilizing Digital Radio-Relay Systems Forming Part or All of the Medium Grade Portion of an ISDN Connection—Section 9A—Performance Objectives, Propagation and Interference Effects. 3 pp.

CCIR RECMN 696-1-91. Error Performance and Availability Objectives for Hypothetical Reference Digital Sect. Utilizing Digital Radio-Relay Systems Forming Part or All of the Medium-Grade Portion of an ISDN Connection—Section 9A—Performance Objectives, Propagation and Interference Effects. 4 pp.

—Telephony—Fixed Satellite Services
CCIR RECMN 579-1-86. Availability Objectives for a Hypothetical Reference Circuit and a Hypothetical Reference Digital Path When Used for Telephony Using Pulse-Code Modulation, or as Part of an Integrated Services Digital Network Hypothetical Reference Connection, in the. 2 pp.

CCIR RECMN 579-2-92. Availability Objectives for a Hypothetical Reference Circuit and a Hypothetical Reference Digital Path When Used for Telephony Using Pulse-Code Modulation, or as Part of an Integrated Services Digital Network Hypothetical Reference Connection, in the. 6 pp.

Hypothetical Reference Model
See Also: Hypothetical Reference Digital Paths; Network Capabilities (Telecommunications); Network Performance; Performance Parameters (Telecommunications); Telecommunication Networks; Telecommunication Systems

—Digital Transmission Systems
CCITT RECMN G.801-89. Digital Transmission Models—Digital Networks, Digital Sections and Digital Line Systems (Study Groups XV and XVIII) 6 pp. 6 pp.

—Earth Exploration Satellites
CCIR Report 1120-90. Method for Deriving Performance Criteria for the Earth Exploration-Satellite Service—Section 2F—Earth Exploration Satellites. 6 pp.

—Performance Parameters—Integrated Services Digital Networks
CCITT RECMN I.325-89. Reference Configurations for ISDN Connection Types—Integrated Services Digital Network (ISDN)—Overall Network Aspects and Functions, ISDN User-Network Interfaces (Study Group XVIII) 7 pp. 7 pp.

—Telephone Connections
CCITT RECMN G.103-89. Hypothetical Reference Connections—General Characteristics of International Telephone Connections and Circuits (Study Groups XII and XV) 10 pp. 10 pp.

—Telephone Networks
CCITT RECMN G.102-89. Transmission Performance Objectives and Recommendations—General Characteristics of International Telephone Connections and Circuits (Study Groups XII and XV) 4 pp. 4 pp.

CCITT RECMN G.103-89. Hypothetical Reference Connections—General Characteristics of International Telephone Connections and Circuits (Study Groups XII and XV) 10 pp. 10 pp.

—Telephone Networks—Crosstalk
CCITT RECMN G.105-89. Hypothetical Reference Connection for Crosstalk Studies—General Characteristics of International Telephone Connections and Circuits (Study Groups XII and XV) 6 pp. 6 pp.

—Telephone Networks—Signal Levels
CCITT RECMN G.103-89. Hypothetical Reference Connections—General Characteristics of International Telephone Connections and Circuits (Study Groups XII and XV) 10 pp. 10 pp.

Hypothetical Reference Model *(Cont.)*

—Videoconferencing Services—Data Transmission
CCITT RECMN H.110-89. Hypothetical Reference Connections for Videoconferencing Using Primary Digital Group Transmission—Line Transmission of Non-Telephone Signals—Transmission of Sound-Programme and Television Signals (Study Group XV) 6 pp. 6 pp.

Hypsometry

—Tables (Data)—Standard Atmospheres
ISO 2533-75. Standard Atmosphere (Identical with the ICAO and WMO Standard Atmospheres from—2 to 32 km) First Edition; (Erratum—April 1977) (Addendum 1-1985). 193 pp.

Hyssop Oil
See Also: Essential Oils

—Quality Assurance
ISO 9841-91. Oil of Hyssop (Hyssopus Officinalis Linnaeus) First Edition. 5 pp.

HyTime
Use For: Hypermedia/Time-Based Structuring Language *See Also:* Programming Languages; SGML

BSI BS ISO/IEC 10744-92. 1992 Information Technology—Hypermedia/Time-Based Structuring Language (HyTime). 144 pp.

IEC 10744-92. Information Technology—Hypermedia/Time-Based Structuring Language (HyTime) First Edition. 142 pp.

ISO 10744-92. Information Technology—Hypermedia/Time-Based Structuring Language (HyTime) First Edition. 142 pp.

JTC1 10744-92. Information Technology—Hypermedia/Time-Based Structuring Language (HyTime) First Edition. 145 pp.

I Beams
See Also: Structural Members; Trusses (Structural Members)

—Aluminum Alloys—Extruded
DIN ENGL 9712-69. Aluminium and Magnesium Beams; Extruded; Dimensions; Static Values (Aug). 3 pp.

—Aluminum—Extruded
DIN ENGL 9712-69. Aluminium and Magnesium Beams; Extruded; Dimensions; Static Values (Aug). 3 pp.

—Magnesium—Extruded
DIN ENGL 9712-69. Aluminium and Magnesium Beams; Extruded; Dimensions; Static Values (Aug). 3 pp.

—Square Washers
CNS B2017-78. Square Taper Washers for I-Section (Nov)(157).

CNS B2324-83. Square Taper Washers for High Prestressed I-Sections (Mar)(4646).

DIN ENGL 435-89. Square Taper Washers for Use with I Sections (Dec). 2 pp.

DIN ENGL 6917-89. Square Taper Washers for High-Strength Structural Bolting of Steel I Sections (Oct). 3 pp.

—Steel—Hot Rolled
CEN PREN 10024-92. Taper Flange I Sections—Tolerances on Shape and Dimensions. 7 pp.

DIN ENGL 997-70. Tracing Dimensions for Bars and Rolled Steel Sections (Oct). 6 pp.

—Steel—Pipe Clamps
DIN ENGL 3568-77. Clamping Plates for the Suspension of Pipes on I and IP Steel Beams (Sept). 3 pp.

IAFETs
Use: MOSFETs

IAI/North American Jet Commander 1121 Aircraft
See Also: Aircraft

—Accidents
CAA. Israeli Aircraft Industries/North American Rockwell Jet Commander 1121 and Series (World Airline Accident Summary). 2 pp.

IA2
Use: International Telegraph Alphabet No. 2

IA5
Use: International Reference Alphabet

INTERNATIONAL AND NON-U.S. NATIONAL STANDARDS
SUBJECT INDEX
Ice

IBIS
Use: Aircraft Safety—Bird Strikes

IC (Ion Chromatography)
Use: Ion Chromatography

IC Cards
Use: Integrated Circuit Cards

ICA Brasov Motor Glider Aircraft
See Also: Aircraft
—Foreign Airworthiness Directives
CAA. ICA Brasov Motor Gliders (Foreign Airworthiness Directives). 2 pp.

ICAO Committee on Aircraft Engine Emissions
Use For: CAEE (ICAO); International Civil Aviation Organization CAEE See Also: Aircraft Engines
ICAO 9259. Committee on Aircraft Engine Emissions First Meeting Montreal, 12-22 June 1978; (Supplement 1 12/15/78). 195 pp.
ICAO 9304. Committee on Aircraft Engine Emissions Second Meeting Montreal, 14-29 May 1980; (Supplement 1 06/27/80). 154 pp.

ICAO Committee on Aircraft Noise
Use For: CAN (ICAO); International Civil Aviation Organization CAN See Also: Aircraft Noise
ICAO 8993. Committee on Aircraft Noise Second Meeting Montreal 15-26 November 1971; (Supplement 02/28/72). 66 pp.
ICAO 9063. Committee on Aircraft Noise Third Meeting Montreal 5-23 March 1973; (Supplement 06/15/73). 112 pp.
ICAO 9133. Committee on Aircraft Noise Fourth Meeting Montreal, 27 January—14 February 1975; (Supplement 1 03/02/75) (Supplement 2 06/02/75). 175 pp.
ICAO 9197. Committee on Aircraft Noise Fifth Meeting Montreal, 15-30 November 1976; (Supplement 04/04/77). 115 pp.
ICAO 9286. Committee on Aircraft Noise Sixth Meeting Montreal, 23 May-7 June 1979; (Supplement 11/14/79). 214 pp.
ICAO 9419. Committee on Aircraft Noise Seventh Meeting Montreal, 2-13 May 1983; (Corrigendum) (Supplement 12/05/83). 152 pp.

ICAO Committee on Aviation Environmental Protection
Use For: CAEP (ICAO); International Civil Aviation Organization CAEP See Also: Aircraft Engines
ICAO 9499. Committee on Aviation Environmental Protection First Meeting Montreal, 9-20 June 1986. 230 pp.

ICAO Sonic Boom Committee
Use For: International Civil Aviation Organization SBC; SBC (ICAO) See Also: Aircraft Noise
ICAO 9011. Sonic Boom Committee Report of the First Meeting. Montreal, 9-19 May 1972; (Supplement 06/26/72). 40 pp.
ICAO 9064. Sonic Boom Committee Second Meeting Montreal 19-29 June 1973; (Supplement 12/11/73) (Supplement 2 01/23/74). 77 pp.

ICAO Special Committee on Future Air Navigation Systems
Use For: FANS (ICAO); International Civil Aviation Organization FANS See Also: Air Navigation
ICAO 9503. Special Committee on Future Air Navigation Systems Third Meeting Montreal, 3-21 November 1986 (Report); (Corrigendum/Supplement 1 07/31/87) (Corrigendum/Addendum 1 08/10/87). 190 pp.
ICAO 9524. Special Committee on Future Air Navigation Systems Fourth Meeting Montreal, 2-20 May 1988; (Supplement 1). 232 pp.

Ice Anchors
See Also: Anchors (Fasteners); Mountaineering Equipment; Pitons
—Safety
BSI BS EN 568-93. 1993 Mountaineering Equipment Ice Anchors Safety Requirements and Test Method (N). 10 pp.
CEN PREN 568-91. Mountaineering Equipment —Ice Anchors—Safety Requirements, Testing, Marking. 8 pp.
CEN EN 568-92. Mountaineering Equipment —Ice Anchors—Safety Requirements and Test Method. 5 pp.

Ice Control Agents
Use: Antifreezes

Ice Cream
Use For: Ice Milk See Also: Dairy Products; Sherbets
CGSB 32.163M-89. Ice Cream, Ice Milk and Sherbet (Supersedes 32-GP-176M October 1978). 8 pp.
CNS N5171-86. Ice Cream (Packaged) (Jan)(6508). 2 pp.
—Acidity
CNS N6143-86. Method of Test for Ice Cream (Jan)(6509). 2 pp.
—Artificial Sweeteners Content
CNS N6143-86. Method of Test for Ice Cream (Jan)(6509). 2 pp.
—Chemical Analysis
BSI BS 2472-66. 1966 Amd 1 Methods for the Chemical Analysis of Ice Cream (AMD 6191) July 31, 1990. 19 pp.
SAA AS N72-70. Methods for the Chemical Analysis of Ice Cream and Frozen Milk Products. 23 pp.
—Containers
BSI BS 4879-73. 1973 Waxed Board for Packaging Ice Cream and Frozen Confectionery. 9 pp.
—Fat Content
CNS N6143-86. Method of Test for Ice Cream (Jan)(6509). 2 pp.
—Fat Content—Gravimetric Analysis
BSI BS 2472: Part 3-89. 1989 Methods for Chemical Analysis of Ice Cream Part 3: Determination of Fat Content. 12 pp.
BSI BS 7142: Part 3-89. 1989 Analysis of Milk-Based Products Part 3: Determination of Fat Content of Milk-Based Edible Ices and Ice Mixes by the Weibull-Berntrop Gravimetric Method. 10 pp.
CNS N6202-86. Method of Test for Fat Content in Ice Cream Products-Rose-Gottlieb Gravimetric Method (Oct)(11736). 3 pp.
ISO 7328-84. Milk-Based Edible Ices and Ice-Mixes—Determination of Fat Content—Rose-Gottlieb Gravimetric Method (Reference Method) First Edition. 10 pp.
ISO 8262 Pt 2-87. Milk Products and Milk-Based Foods—Determination of Fat Content by the Weibull-Berntrop Gravimetric Method (Reference Method)—Part 2: Edible Ices and Ice-Mixes First Edition. 8 pp.
—Glucose Content—Enzymatic Activity
DIN ENGL 10326-86. Determination of Sucrose and Glucose Content in Milk Products and Ice Cream; Enzymatic Method (Feb). 5 pp.
—Mellorine
CNS N5179-86. Mellorine (Packaged) (Jun)(7523). 2 pp.
—Microbiological Analysis
BSI BS 4285: Sec 5.2-89. 1989 Microbiological Examination for Dairy Purposes Part 5: Ancillary Methods Section 5.2: Methylene Blue Reduction Test for Cream and Ice Cream (W). 4 pp.
CNS N6143-86. Method of Test for Ice Cream (Jan)(6509). 2 pp.
—Preservatives
CNS N6143-86. Method of Test for Ice Cream (Jan)(6509). 2 pp.
—Protein Content
CNS N6143-86. Method of Test for Ice Cream (Jan)(6509). 2 pp.
—Residue Content—Dehydration
ISO 3728-77. Ice-Cream and Milk Ice—Determination of Total Solids Content (Reference Method) First Edition. 5 pp.
—Sampling
CNS N6143-86. Method of Test for Ice Cream (Jan)(6509). 2 pp.
—Solids Content
CNS N6143-86. Method of Test for Ice Cream (Jan)(6509). 2 pp.
—Sucrose Content—Enzymatic Activity
DIN ENGL 10326-86. Determination of Sucrose and Glucose Content in Milk Products and Ice Cream; Enzymatic Method (Feb). 5 pp.
—Test Specimens—Microbiological Analysis
DIN ENGL 10191 Pt 5-85. Microbiological Analysis of Milk; Preparation of Samples; Method of Analysis of Edible Ices and Edible Ice Powder (July). 2 pp.
—Weight Measurement
CNS N6143-86. Method of Test for Ice Cream (Jan)(6509). 2 pp.

Ice Cream Makers
See Also: Appliances
—Household—Safety
BSI BS EN 60335-2-57-92. 1992 Safety of Household and Similar Electrical Appliances Part 2: Particular Requirements Section 2.57: Ice-Cream Appliances with Incorporated Motor-Compressors. 23 pp.
CENELEC EN 60335-2-57-92. Safety of Household and Similar Electrical Appliances Part 2: Particular Requirements for Ice-Cream Appliances with Incorporated Motor-Compressors. 8 pp.
IEC 335 Pt 2-57-89. Safety of Household and Similar Electrical Appliances Part 2: Particular Requirements for Ice-Cream Appliances with Incorporated Motor-Compressors First Edition. 36 pp.

Ice Formation
See Also: Antiicing Additives; Ice Makers; Ice Prevention
—Aerostats
CAA Chapter Q4-7 12.79. Flight in Precipitation Conditions (Non-Rigid Airships). 1 p.
—Aircraft
CAA Chapter K4-7 04.72. Flight in Ice Forming Conditions (Light Aeroplanes). 2 pp.
CAA Chapter P4-7. Flight in Ice Forming Conditions (Provisional Airworthiness Requirements for Civil Powered Lift Aircraft). 5 pp.
CAA CAP 512 Appendix. General Information Relating to Ground and Flight Operations in Conditions Conductive to Aircraft Icing (Civil Air Publications: Ground De-Icing of Aircraft). 6 pp.
—Aircraft Engines
CAA Chapter K5-5 04.74. Engine Air Intake and Ice Protection Systems (Light Aeroplanes). 3 pp.
CAA Chapter Q5-5 12.79. Engine Air Intake and Ice Protection Systems (Non-Rigid Airships). 2 pp.

Ice Loading
Use: Ice Loads

Ice Loads
See Also: Dynamic Loads; Loads (Forces)
—Buildings
DIN ENGL 1055 Pt 5-75. Design Loads for Buildings; Live Loads; Snow Load and Ice Load (June). 6 pp.

Ice Makers
See Also: Appliances; Freezers; Ice Formation; Refrigeration Equipment
—Household—Safety
DIN VDE 0700 Pt 240-83. Safety of Household and Similar Electrical Appliances; Particular Requirements for Refrigerators and Freezers for Special Purposes and Ice Makers (Feb) (Supersedes VDE 0730 Part 2 ZK/10.78). 61 pp.
IEC 335 Pt 2-24-92. Safety of Household and Similar Electrical Appliances Part 2: Particular Requirements for Refrigerators, Food-Freezers and Ice-Makers Third Edition: (CENELEC HD 269 S2:1986). 78 pp.

Ice Making Equipment
Use: Ice Makers

Ice Making Machines
Use: Ice Makers

Ice Milk
Use: Ice Cream

Ice Prevention
See Also: Antiicing Additives; Deicers; Deicing; Ice Formation
—Aircraft
CAA LEAFLET 13-6. Ground De-Icing of Aircraft. 18 pp.
—Heating Cables
IEC 800-92. Heating Cables with a Rated Voltage of 300/500 V for Comfort Heating and Prevention of Ice Formation Second Edition. 91 pp.
—Supersonic Transports
CAA STANDARD NO. 5-4 07.69. Ice Protection. 8 pp.

Ice Protection
Use: Ice Prevention

Ice Water Bags
See Also: Bags; Medical Equipment

INDUSTRY STANDARDS

Ice Water Bags (Cont.)

—Rubber
CNS T2003-80. Rubber Ice Water Bottles (Jul)(5944).
CNS T2004-80. Rubber Ice Water Bags (Jul)(5945).
JIS T 9103-75. Rubber Ice Water Bottles (R 1988). 6 pp.
JIS T 9104-75. Rubber Ice Water Bags.

Ice Water Bottles
Use: Ice Water Bags

Iceland Spar
See Also: Minerals
CNS K7253-68. Chemical Reagent (Iceland Spar) (Nov)(1752).

Icing Inhibitors
Use: Antiicing Additives

Identification, Friend or Foe Interrogators
Use: IFF Interrogators

Identification Cards
See Also: Credit Cards; Magnetic Cards
BSI BS 7298-90. 1990 Schedule of Requirements and Recommendations for Cards to Be Used. 8 pp.
BSI BS EN 27 810-91. 1991 Design, Construction and Use of Identification Cards (S). 10 pp.
IEC DIS 10373-92. Identification Cards—Test Methods ***CD-ROM ONLY***. 44 pp.
ISO 7810-85. Identification Cards—Physical Characteristics First Edition. 5 pp.
ISO DIS 10373-92. Identification Cards—Test Methods ***CD-ROM ONLY***. 44 pp.
JTC1 7810-85. Identification Cards—Physical Characteristics First Edition. 5 pp.
JTC1 7812-87. Identification Cards—Numbering System and Registration Procedure for Issuer Identifiers Second Edition; (Corrigendum 1-1988). 10 pp.

—Embossing
BSI BS 7106: Part 1-89. 1989 Guide to Recording Techniques for Information on Identification Cards Part 1: Embossing. 31 pp.
BSI BS 7106: Part 3-89. 1989 Guide to Recording Techniques for Information on Identification Cards Part 3: Location of Embossed Characters on ID-7 Cards. 9 pp.
CEN EN 27 811 (Part 1)-89. Identification Cards Recording Technique Part 1: Embossing. 28 pp.
CEN EN 27 811 (Part 3)-89. Identification Cards Recording Technique—Part 3: Location of Embossed Characters on ID-1 Cards. 6 pp.
CNS C5247-1-90. Identification Cards—Recording Technique—Part 1: Embossing (Dec)(12821-1).
CNS C5247-3-90. Identification Cards—Recording Technique—Part 3: Location of Embossed Characters (Dec)(12821-3).
IEC DIS 7811 Pt 1-92. Identification Cards—Recording Technique—Part 1: Embossing (Revision of First Edition (ISO 7811-1:1985)) ***CD-ROM ONLY***. 30 pp.
IEC DIS 7811 Pt 3-92. Identification Cards—Recording Technique—Part 3: Location of Embossed Characters on ID-1 Cards (Revision of First Edition (ISO 7811-3:1985)) ***CD-ROM ONLY***. 8 pp.
ISO 7811 Pt 1-85. Identification Cards—Recording Technique—Part 1: Embossing First Edition. 27 pp.
ISO DIS 7811 Pt 1-92. Identification Cards—Recording Technique—Part 1: Embossing (Revision of First Edition (ISO 7811-1:1985)) ***CD-ROM ONLY***. 30 pp.
ISO 7811 Pt 3-85. Identification Cards—Recording Technique—Part 3: Location of Embossed Characters on ID-1 Cards First Edition. 4 pp.
ISO DIS 7811 Pt 3-92. Identification Cards—Recording Technique—Part 3: Location of Embossed Characters on ID-1 Cards (Revision of First Edition (ISO 7811-3:1985)) ***CD-ROM ONLY***. 8 pp.
JTC1 7811 Pt 1-85. Identification Cards—Recording Technique—Part 1: Embossing First Edition. 27 pp.
JTC1 DIS7811 Pt 1-92. Identification Cards—Recording Technique—Part 1: Embossing (Revision of First Edition (ISO 7811-1:1985)) ***CD-ROM ONLY***. 30 pp.
JTC1 7811 Pt 3-85. Identification Cards—Recording Technique—Part 3: Location of Embossed Characters on ID-1 Cards First Edition. 4 pp.
JTC1 DIS7811 Pt 3-92. Identification Cards—Recording Technique—Part 3: Location of Embossed Characters on ID-1 Cards (Revision of First Edition (ISO 7811-3:1985)) ***CD-ROM ONLY***. 8 pp.
OSI ISO 7811-1-85. Identification Cards—Recording Technique—Part 1: Embossing. 28 pp.
OSI ISO 7811-3-85. Identification Cards—Recording Technique—Part 3: Location of Embossed Characters on ID-1 Cards. 6 pp.

Identification Cards (Cont.)

—Integrated Circuit
BSI BS EN 27 816 Part 2-91. 1991 Guide to Design and Use of Identification Cards Having Integrated Circuits with Contacts Part 2: Contact Locations and Minimum Size (S). 14 pp.
CEN PREN 742-92. Identification Cards—ID-1 Card Location of Contacts for Cards and Devices Used in Europe. 6 pp.
CEN EN 27 816 (Part 2)-89. Identification Cards Integrated Circuit(s) with Contacts Part 2: Dimensions and Location of the Contacts. 14 pp.
ISO 7816 Pt 2-88. Identification Cards—Integrated Circuit(s) Cards with Contacts—Part 2: Dimensions and Location of the Contacts First Edition. 12 pp.
JTC1 7816 Pt 2-88. Identification Cards—Integrated Circuit(s) Cards with Contacts—Part 2: Dimensions and Location of the Contacts First Edition. 12 pp.
OSI ISO 7816-2-88. Identification Cards—Integrated Circuits Cards with Contacts—Part 2: Dimensions and Location of the Contacts. 14 pp.

—Integrated Circuit—Numbering Systems
IEC DIS 7816 Pt 5-92. Identification Cards—Integrated Circuit(s) Cards with Contact—Part 5: Numbering System and Registration Procedure for Application Identifiers ***CD-ROM ONLY***. 10 pp.
ISO DIS 7816 Pt 5-92. Identification Cards—Integrated Circuit(s) Cards with Contact—Part 5: Numbering System and Registration Procedure for Application Identifiers ***CD-ROM ONLY***. 10 pp.
JTC1 DIS7816 Pt 5-92. Identification Cards—Integrated Circuit(s) Cards with Contact—Part 5: Numbering System and Registration Procedure for Application Identifiers ***CD-ROM ONLY***. 10 pp.

—Integrated Circuit—Physical Properties
BSI BS EN 27 816 Part 1-91. 1991 Guide to Design and Use of Identification Cards Having Integrated Circuits with Contacts Part 1: Physical Characteristics (S). 12 pp.
CEN EN 27 816 (Part 1)-89. Identification Cards Integrated Circuit(s) with Contacts Part 1: Physical Characteristics. 9 pp.
IEC 10536 Pt 1-92. Identification Cards—Contactless Integrated Circuit(s) Cards—Part 1: Physical Characteristics First Edition. 10 pp.
ISO 7816 Pt 1-87. Identification Cards—Integrated Circuit(s) Cards with Contacts—Part 1: Physical Characteristics First Edition. 7 pp.
ISO 10536 Pt 1-92. Identification Cards—Contactless Integrated Circuit(s) Cards—Part 1: Physical Characteristics First Edition. 10 pp.
JTC1 7816 Pt 1-87. Identification Cards—Integrated Circuit(s) Cards with Contacts—Part 1: Physical Characteristics First Edition. 7 pp.
JTC1 10536 Pt 1-92. Identification Cards—Contactless Integrated Circuit(s) Cards—Part 1: Physical Characteristics First Edition. 10 pp.
OSI ISO 7816-1-87. Indentification Cards—Intergrated Circuit(s) Cards with Contacts—Part 1: Physical Characteristics. 9 pp.
OSI ISO/IEC DIS 10536-1-90. Identification Cards—Contactless Integrated Circuit(s) Cards—Part 1: Physical Characteristics. 10 pp.

—Integrated Circuit—Protocols
BSI BS 7109: Part 3-90. 1990 Guide to Design and Use of Identification Cards Having Integrated Circuits with Contacts Part 3: Electronic Signals and Transmission Protocols. 21 pp.
BSI BS 7109: Part 3-01. 1990 Amd 1 Guide to Design and Use of Identification Cards Having Integrated Circuits with Contacts Part 3: Electronic Signals and Transmission Protocols (ISO/IEC 7816-3: 1989) (AMD 7325) November 15, 1992. 25 pp.
CEN EN 27816-3-92. Identification Cards—Integrated Circuit(s) Cards with Contacts—Part 3: Electronic Signals and Transmission Protocols. 21 pp.
IEC 7816 Pt 3-89. Identification Cards—Integrated Circuit(s) Cards with Contacts—Part 3: Electronic Signals and Transmission Protocols Amendment 1: Protocol Type T = 1, Asynchronous Half Duplex Block Transmission Protocol First Edition; (Amendment 1-1992). 35 pp.
ISO 7816 Pt 3-89. Identification Cards—Integrated Circuit(s) Cards with Contacts—Part 3: Electronic Signals and Transmission Protocols Amendment 1: Protocol Type T = 1, Asynchronous Half Duplex Block Transmission Protocol First Edition; (Amendment 1-1992). 35 pp.
JTC1 7816-89. Identification Cards—Integrated Circuit(s) Cards with Contacts—Part 3: Electronic Signals and Transmission Protocols Amendment 1: Protocol Type T = 1, Asynchronous Half Duplex Block Transmission Protocol First Edition; (Amendment 1-1992). 19 pp.
OSI ISO/IEC 7816-3 DAM 1-90. Identification Cards—Integrated Circuit(s) Cards with Contacts—Part 3: Electronic Signals and Transmission Protocols AMENDMENT 1. 27 pp.

Identification Cards (Cont.)

—Integrated Circuit—Protocols (Cont.)
OSI ISO IEC 7816-3-89. Identification Cards—Integrated Circuit(s) Cards with Contacts—Part 3: Electronic Signals and Transmission Protocols. 19 pp.
OSI ISO DIS 7816-3-87. Identification Cards—Integrated Circuit Cards with Contacts—Part 3: Electronic Signals and Exchange Protocols. 28 pp.
OSI ISO DIS 7816-3-87. Identification Cards—Integrated Circuit Cards with Contacts—Part 3: Electronic Signals and Exchange Protocols. 28 pp.
OSI ISO DIS 7816-3-89. Identification Cards—Integrated Circuit(s) Cards with Contacts—Part 3: Electronic Signals and Transmission Protocols. 19 pp.
OSI ISO IEC 7816-3-89. Identification Cards—Integrated Circuit(s) Cards with Contacts—Part 3: Electronic Signals and Transmission Protocols. 19 pp.

—Integrated Circuit—Registration
IEC DIS 7816 Pt 5-92. Identification Cards—Integrated Circuit(s) Cards with Contact—Part 5: Numbering System and Registration Procedure for Application Identifiers ***CD-ROM ONLY***. 10 pp.
ISO 7816 Pt 5-92. Identification Cards—Integrated Circuit(s) Cards with Contact—Part 5: Numbering System and Registration Procedure for Application Identifiers ***CD-ROM ONLY***. 10 pp.
JTC1 DIS7816 Pt 5-92. Identification Cards—Integrated Circuit(s) Cards with Contact—Part 5: Numbering System and Registration Procedure for Application Identifiers ***CD-ROM ONLY***. 10 pp.

—Magnetic
BSI BS 7106: Part 2-89. 1989 Guide to Recording Techniques for Information on Identification Cards Part 2: Magnetic Stripe. 16 pp.
BSI BS 7106: Part 4-89. 1989 Guide to Recording Techniques for Information on Identification Cards Part 4: Location of Read-Only Magnetic Tracks—Tracks 1 and 2. 9 pp.
BSI BS 7106: Part 5-89. 1989 Guide to Recording Techniques for Information on Identification Cards Part 5: Location of Read-Write Magnetic Track-Track 3. 9 pp.
CEN PREN 753-1-92. Identification Cards—Thin Flexible Cards—Part 1: Technical Specifications. 76 pp.
CEN EN 27 811 (Part 2)-89. Identification Cards Recording Technique Part 2: Magnetic Stripe. 2 pp.
CEN EN 27 811 (Part 4)-89. Identification Cards Recording Technique Part 4: Location of Read-Only Magnetic Tracks—Tracks 1 and 2. 6 pp.
CEN EN 27 811 (Part 5)-89. Identification Cards Recording Technique Part 5: Location of Read-White Magnetic Track Track 3. 2 pp.
CNS C5247-2-90. Identification Cards—Recording Technique—Part 2: Magnetic Stripe (Dec)(12821-2).
CNS C5247-4-90. Identification Cards—Recording Technique—Part 4: Location of Read-Only Magnetic Tracks—Track 1 and 2 (Dec)(12821-4).
CNS C5247-5-90. Identification Cards—Recording Technique—Part 5: Location of Read-Write Magnetic Track—Track 3 (Dec)(12821-5).
IEC DIS 7811 Pt 2-92. Identification Cards—Recording Technique—Part 2: Magnetic Stripe (Revision of First Edition (ISO 7811-2:1985)) ***CD-ROM ONLY***. 24 pp.
IEC DIS 7811 Pt 4-92. Identification Cards—Recording Technique—Part 4: Location of Read-Only Magnetic Tracks—Tracks 1 and 2 (Revision of First Edition (ISO 7811-4: 1985)) ***CD-ROM ONLY***. 7 pp.
IEC DIS 7811 Pt 5-92. Identification Cards—Recording Technique—Part 5: Location of Read-Write Magnetic Track—Track 3 (Revision of First Edition (ISO 7811-5:1985)) ***CD-ROM ONLY***. 7 pp.
ISO 7811 Pt 2-85. Identification Cards—Recording Technique—Part 2: Magnetic Stripe First Edition. 10 pp.
ISO DIS 7811 Pt 2-92. Identification Cards—Recording Technique—Part 2: Magnetic Stripe (Revision of First Edition (ISO 7811-2:1985)) ***CD-ROM ONLY***. 24 pp.
ISO 7811 Pt 4-85. Identification Cards—Recording Technique—Part 4: Location of Read-Only Magnetic Tracks—Tracks 1 and 2 First Edition. 4 pp.
ISO DIS 7811 Pt 4-92. Identification Cards—Recording Technique—Part 4: Location of Read-Only Magnetic Tracks—Tracks 1 and 2 (Revision of First Edition (ISO 7811-4: 1985)) ***CD-ROM ONLY***. 7 pp.
ISO 7811 Pt 5-85. Identification Cards—Recording Technique—Part 5: Location of Read-Write Magnetic Track—Track 3 First Edition. 4 pp.

INTERNATIONAL AND NON-U.S. NATIONAL STANDARDS
SUBJECT INDEX

Identification Cards *(Cont.)*
—Magnetic *(Cont.)*
ISO DIS 7811 Pt 5-92. Identification Cards—Recording Technique—Part 5: Location of Read-Write Magnetic Track—Track 3 (Revision of First Edition (ISO 7811-5:1985)) ***CD-ROM ONLY***. 7 pp.

JTC1 7811 Pt 2-85. Identification Cards—Recording Technique—Part 2: Magnetic Stripe First Edition. 10 pp.

JTC1 DIS7811 Pt 2-92. Identification Cards—Recording Technique—Part 2: Magnetic Stripe (Revision of First Edition (ISO 7811-2:1985)) ***CD-ROM ONLY***. 24 pp.

JTC1 7811 Pt 4-85. Identification Cards—Recording Technique—Part 4: Location of Read-Only Magnetic Tracks—Tracks 1 and 2 First Edition. 4 pp.

JTC1 DIS7811 Pt 4-92. Identification Cards—Recording Technique—Part 4: Location of Read-Only Magnetic Tracks—Tracks 1 and 2 (Revision of First Edition (ISO 7811-4: 1985)) ***CD-ROM ONLY***. 7 pp.

JTC1 7811 Pt 5-85. Identification Cards—Recording Technique—Part 5: Location of Read-Write Magnetic Track—Track 3 First Edition. 4 pp.

JTC1 DIS7811 Pt 5-92. Identification Cards—Recording Technique—Part 5: Location of Read-Write Magnetic Track—Track 3 (Revision of First Edition (ISO 7811-5:1985)) ***CD-ROM ONLY***. 7 pp.

OSI ISO 7811-2-85. Identification Cards—Recording Technique—Part 2: Magnetic Stripe. 12 pp.

OSI ISO 7811-4-85. Identification Cards—Recording Techniques—Part 4: Location of Read-Only Magnetic Tracks—Tracks 1 and 2. 6 pp.

OSI ISO 7811-5-85. Identification Cards—Recording Technique—Part 5: Location of Read-Write Magnetic Track—Track 3. 6 pp.

—Numbering Systems
BSI BS 7107-90. 1990 Allocating International Issuer Identification Numbers (IINs) for Use on Identification Cards. 12 pp.

BSI BS 7227-90. 1990 Procedures for the Application of ISO 7812 to Allocate UK Issuer Identification Numbers (UK IINs) for Use on Identification Cards. 14 pp.

CEN EN 27 812-89. Identification Cards Numbering System and Registration Procedure for Issuer Identifiers. 11 pp.

CNS C5245-90. Identification Cards—Numbering System (Dec)(12819).

IEC DIS 7812 Pt 1-92. Identification Cards—Identification of Issuers—Part 1: Numbering System ***CD-ROM ONLY***. 9 pp.

ISO 7812-87. Identification Cards—Numbering System and Registration Procedure for Issuer Identifiers Second Edition; (Corrigendum 1-1988). 10 pp.

ISO DIS 7812 Pt 1-92. Identification Cards—Identification of Issuers—Part 1: Numbering System ***CD-ROM ONLY***. 9 pp.

JTC1 DIS7812 Pt 1-92. Identification Cards—Identification of Issuers—Part 1: Numbering System ***CD-ROM ONLY***. 9 pp.

OSI ISO 7812-87. Identification Cards—Numbering System and Registration Procedure for Issuer Identifiers. 2 pp.

—Optical Memory
IEC DIS 11693-93. Identification Cards—Optical Memory Cards ***CD-ROM ONLY***. 9 pp.

ISO DIS 11693-93. Identification Cards—Optical Memory Cards ***CD-ROM ONLY***. 9 pp.

JTC1 DIS11693-93. Identification Cards—Optical Memory Cards ***CD-ROM ONLY***. 9 pp.

—Optical Memory—Layout
IEC DIS 11694 Pt 2-93. Identification Cards—Optical Memory Cards—Linear Recording Method—Part 2: Dimensions and Location of the Accessible Optical Area ***CD-ROM ONLY***. 10 pp.

ISO DIS 11694 Pt 2-93. Identification Cards—Optical Memory Cards—Linear Recording Method—Part 2: Dimensions and Location of the Accessible Optical Area ***CD-ROM ONLY***. 10 pp.

JTC1 DIS11694 Pt 2-93. Identification Cards—Optical Memory Cards—Linear Recording Method—Part 2: Dimensions and Location of the Accessible Optical Area ***CD-ROM ONLY***. 10 pp.

—Optical Memory—Optical Properties
IEC DIS 11694 Pt 3-93. Identification Cards—Optical Memory Cards—Linear Recording Method—Part 3: Optical Properties and Characteristics ***CD-ROM ONLY***. 8 pp.

ISO DIS 11694 Pt 3-93. Identification Cards—Optical Memory Cards—Linear Recording Method—Part 3: Optical Properties and Characteristics ***CD-ROM ONLY***. 8 pp.

JTC1 DIS11694 Pt 3-93. Identification Cards—Optical Memory Cards—Linear Recording Method—Part 3: Optical Properties and Characteristics ***CD-ROM ONLY***. 8 pp.

Identification Cards *(Cont.)*
—Optical Memory—Physical Properties
IEC DIS 11694 Pt 1-93. Identification Cards—Optical Memory Cards—Linear Recording Method—Part 1: Physical Characteristics ***CD-ROM ONLY***. 6 pp.

ISO DIS 11694 Pt 1-93. Identification Cards—Optical Memory Cards—Linear Recording Method—Part 1: Physical Characteristics ***CD-ROM ONLY***. 6 pp.

JTC1 DIS11694 Pt 1-93. Identification Cards—Optical Memory Cards—Linear Recording Method—Part 1: Physical Characteristics ***CD-ROM ONLY***. 6 pp.

—Photographs—Defense Contracts
MOD UK DEFCON 112BN-88. Conditions of Contract Supply of Identity Photographs (Army) 4/88. 3 pp.

—Physical Properties
CEN EN 27 810-89. Identification Cards Physical Characteristics. 7 pp.

CNS C5244-90. Identification Cards—Physical Characteristics (Dec)(12818).

IEC DIS 7810-92. Identification Cards—Physical Characteristics (Revision of First Edition (ISO 7810:1985)) ***CD-ROM ONLY***. 12 pp.

ISO DIS 7810-92. Identification Cards—Physical Characteristics (Revision of First Edition (ISO 7810:1985)) ***CD-ROM ONLY***. 12 pp.

JTC1 DIS7810-92. Identification Cards—Physical Characteristics (Revision of First Edition (ISO 7810:1985)) ***CD-ROM ONLY***. 12 pp.

OSI ISO 7810-85. Identification Cards—Physical Characteristics. 7 pp.

—Public Data Networks—Temporary
CEPT T/TG 19-87. Fourniture Temporaire De L'Identification De L'Usager De Reseau Pour L'Acces Aux Services Publics De Communication De Donnees. 3 pp.

—Registration
CEN EN 27 812-89. Identification Cards Numbering System and Registration Procedure for Issuer Identifiers. 11 pp.

IEC DIS 7812 Pt 2-92. Identification Cards—Identification of Issuers—Part 2: Application and Registration Procedures ***CD-ROM ONLY***. 20 pp.

ISO 7812-87. Identification Cards—Numbering System and Registration Procedure for Issuer Identifiers Second Edition; (Corrigendum 1-1988). 10 pp.

ISO DIS 7812 Pt 2-92. Identification Cards—Identification of Issuers—Part 2: Application and Registration Procedures ***CD-ROM ONLY***. 20 pp.

JTC1 DIS7812 Pt 1-92. Identification Cards—Identification of Issuers—Part 1: Numbering System ***CD-ROM ONLY***. 9 pp.

JTC1 DIS7812 Pt 2-92. Identification Cards—Identification of Issuers—Part 2: Application and Registration Procedures ***CD-ROM ONLY***. 20 pp.

OSI ISO 7812-87. Identification Cards—Numbering System and Registration Procedure for Issuer Identifiers. 2 pp.

—Transferred Account Services
CCITT RECMN. F.13-89. Operational Provisions for Participation in the Transferred Account Telegraph and Telematic Service—Telegraph and Mobile Services Operations and Quality of Service (Study Group I) 6 pp (Renumbered from F.41). 6 pp.

CCITT RECMN. F.41-91. Interworking Between the Telemessage Service and the International Public Telegram Service (Study Group I) 5 pp (See Recmn F.13) (Renumbered from Recmn F.51). 5 pp.

Identification Marking
Use: Identification Systems

Identification Marking Systems
Use: Identification Systems

Identification Numbers
Use For: Product Identification Numbers
See Also: Numbers; Personal Telecommunication Numbers; Ship Station Numbers

—Radio Equipment—Automotive
ISO 10486-92. Passenger Cars—Car Radio Identification Number (CRIN) First Edition. 5 pp.

Identification Plates
Use: Indicator Plates

Identification Signals
See Also: Identification Systems; Signaling Systems

Identification Signals *(Cont.)*
—Emergency Position Indicating Radio Beacons
CCIR Report 749-3-90. Future Use and Characteristics of Emergency Position-Indicating Radio Beacons in the Mobile and Mobile-Satellite Services—Section 8D—Radiodetermination, Global Maritime Distress and Safety System and Related Subjects. 7 pp.

—Sound-Program Connections
CCITT RECMN. N.16-89. Identification Signal—Maintenance of International Sound-Programme and Television Transmission Circuits (Study Group IV) 1 pp. 1 p.

Identification Systems
Scope Note: For identification marking of specific products or materials, see the subheading Identification Systems under the specific material or product. For additional references, consult the following list *Use For:* Branding; Identification Marking; Labeling; Marking; Marking Codes
See Also: Accounting Authority Identification Codes; Automatic Reporting and Identification System; Automatic Transmitter Identification Systems; Badges; Called Line Identification; Calling Line Identification; Calling Line Identification Presentation; Coast Station Identity; Codes; Color Coding; Data Network Identification Codes; Destination Indicators (Telegraphy); Identification Signals; IFF Interrogators; Indicator Plates; Inspection Stamps; Labels; Location Registers; Name Plates; Origin Indicators (Telegraphy); Originator/Recipient Addresses (Message Handling); Originator/Recipient Names (Message Handling); Packaging; Personal Identification Numbers; Personal Telecommunication Numbers; Reflective Materials; Ship Station Identity; Shipping Containers; Symbols; Telegram Identification Groups; Telex Message Identifiers; Telex Network Identification Codes; Terminal Identification; Traffic Markings; Vehicle Identification Numbers

BSI PP 666-91. 1991 Language of Labels. 65 pp.

—Aircraft
CAA LEAFLET 13-2 07.90. Nationality and Registration Marks. 10 pp.

Idler Wheels
Use: Idlers

Idlers
See Also: Bearings; Belt Conveyors; Gears; Pulleys

—Belt Conveyors
DIN ENGL 15207 Pt 1-88. Continuous Mechanical Handling Equipment; Idlers for Belt Conveyors Which Handle Loose Bulk Materials; Principal Dimensions (July). 4 pp.

DIN ENGL 22107-84. Continuous Mechanical Handling Equipment; Idler Sets for Belt Conveyors for Loose Bulk Materials; Principal Dimensions (Aug). 8 pp.

ISO 1537-75. Continuous Mechanical Handling Equipment for Loose Bulk Materials—Troughed Belt Conveyors (Other Than Portable Conveyors)—Idlers First Edition. 7 pp.

—Carrying
BSI BS 5934-80. 1980 Calculation of Operating Power and Tensile Forces in Belt Conveyors with Carrying Idlers on Continuous Mechanical Handling Equipment. 18 pp.

ISO 5048-89. Continuous Mechanical Handling Equipment—Belt Conveyors with Carrying Idlers—Calculation of Operating Power and Tensile Forces Second Edition. 17 pp.

—Carrying—Impact Rings
ISO 4123-79. Belt Conveyors—Impact Rings for Carrying Idlers and Discs for Return Idlers—Main Dimensions First Edition; (Erratum—Sept 1980). 6 pp.

—Cranes—Assembly
DIN ENGL 15090-82. Cranes; Driving Wheel Units and Idler Wheel Units; Assembly (July). 6 pp.

—Cranes—Bearing Rings
DIN ENGL 15094-82. Cranes; Driving Wheel Units and Idler Wheel Units; Bearing Cage Rings (July). 3 pp.

—Cranes—Traveling Wheels
DIN ENGL 15093-82. Cranes; Driving Wheel Units and Idler Wheel Units; Travelling Wheels (July). 5 pp.

—Cranes—Traveling Wheels—Shafts
DIN ENGL 15091-82. Cranes; Driving Wheel Units and Idler Wheel Units; Travelling Wheel Shafts (July). 5 pp.

Idlers (Cont.)

—Return—Discs
ISO 4123-79. Belt Conveyors—Impact Rings for Carrying Idlers and Discs for Return Idlers—Main Dimensions First Edition; (Erratum—Sept 1980). 6 pp.

IDN
Use: Integrated Digital Networks

IEC
Use For: CEI; Commission Electrotechnique Internationale; International Electrotechnical Commission *See Also:* Standards
SNZ NZMP 6001-85. New Zealand's Role in the International Electrotechnical Commission. 20 pp.

—CCIR—CCITT—Organization—Glossaries
CCIR RESOLUTION 113-90. Organisation of Vocabulary Work—Volume XIII—Vocabulary and Related Subjects. 3 pp.

—Certification Management Committee
IECQ CMC Sec 199-87. Certification Management Committee (CMC) Membership of CMC Working Groups. 6 pp.
IECQ CMC Sec 248-89. Certification Management Committee (CMC) Electronic Components Quality Assessment Procedure for IECQ Adoption of a CECC Specification. 10 pp.
IECQ CMC Sec 324A-92. Certification Management Committee (CMC) CMC Membership of CMC Working Groups. 12 pp.
IECQ CMC Sec 341-93. Certification Management Committee (CMC) CMC Officers and Permanent Delegates. 12 pp.
IECQ CMC Sec 349-93. Certification Management Committee (CMC) Membership of CMC Working Groups. 12 pp.

—Certification Systems
IEC Guide 16-78. Code of Principles on Third Party Certification Systems and Related Standards First Edition. 5 pp.
IEC Guide 44-85. General Rules for ISO or IEC International Third-Party Certification Schemes for Products First Edition. 16 pp.
ISO Guide 16-78. Code of Principles on Third Party Certification Systems and Related Standards First Edition. 5 pp.
ISO Guide 44-85. General Rules for ISO or IEC International Third-Party Certification Schemes for Products First Edition. 16 pp.
SAA HB18.16-91. Guidelines for Third-Party Certification and Accreditation—Guide 16: Code of Principles on Third Party Certification Systems and Related Standards (ISO/IEC Guide 16:1978) (SANZ HB18.16—1991). 2 pp.
SAA HB18.44-91. Guidelines for Third-Party Certification and Accreditation—Guide 44: General Rules for ISO or IEC International Third-Party Certification Schemes for Products (ISO/IEC Guide 44:1985) (SANZ HB18.44—1991). 13 pp.
SNZ SANZ/SAA HB 18.16-91. Guidelines for Third-Party Certification and Accreditation Guide 16 Code of Principles on Third-Party Certification Systems and Related Standards (ISO/IEC Guide 16-1978). 2 pp.
SNZ SANZ/SAA HB 18.44-85. Guidelines for Third-Party Certification and Accreditation Guide 44 General Rules for ISO or IEC International Third-Party Certification Schemes for Products (ISO/IEC Guide 44-1985). 13 pp.

—Cooperation—CCIR—Documentation
CCIR RESOLUTION 23-3-90. Collaboration with the International Electrotechnical Commission on Graphical Symbols and Documentation Used in Telecommunications—Volume XIII—Vocabulary and Related Subjects. 1 p.

—Cooperation—CCIR—Symbols
CCIR RESOLUTION 23-3-90. Collaboration with the International Electrotechnical Commission on Graphical Symbols and Documentation Used in Telecommunications—Volume XIII—Vocabulary and Related Subjects. 1 p.

—Cooperation—CCITT
CCITT RECMN A.20-89. Collaboration with Other International Organizations over Data Transmission—Data Communication over the Telephone Network (Study Group XVII) 2 pp. 2 pp.
CCITT RECMN A.22-89. Collaboration with Other International Organizations on Information Technology—Terminal Equipment and Protocols for Telematic Services (Study Group VIII) 2 pp. 2 pp.
CCITT FASCICLE I.2-88. Opinions and Resolutions Recommendations on the Organisation and Working Procedures of CCITT (Series A). 64 pp.

IEC (Cont.)

—Cooperation—CCITT—Diagrams
CCITT RECMN A.13-80. Collaboration with the International Electrotechnical Commission on Graphical Symbols and Diagrams Used in Telecommunications—Opinions and Resolutions—Recommendations on the Organization and Working Procedures of CCITT (Series A) 1 pp. 1 p.

—Cooperation—CCITT—Symbols
CCITT RECMN A.13-80. Collaboration with the International Electrotechnical Commission on Graphical Symbols and Diagrams Used in Telecommunications—Opinions and Resolutions—Recommendations on the Organization and Working Procedures of CCITT (Series A) 1 pp. 1 p.

—Coordination—CCIR—Glossaries
CCIR RESOLUTION 114-90. Coordination of Vocabulary and Related Subjects—Volume XIII—Vocabulary and Related Subjects. 2 pp.

—Coordination—CCIR—Radio Relay Systems
CCIR OPINION 50-74. Coordination of the Work of the CCIR and the IEC on Measurements for the Adjustment and Maintenance of Radio-Relay Systems—Volume IX-1—Fixed Service Using Radio-Relay Systems. 1 p.

—Coordination—CCIR—Symbols
CCIR RESOLUTION 114-90. Coordination of Vocabulary and Related Subjects—Volume XIII—Vocabulary and Related Subjects. 2 pp.

—Coordination—CCITT—Glossaries
CCITT RECMN A.12-88. Collaboration with the International Electrotechnical Commission on the Subject of Definitions for Telecommunications—Opinions and Resolutions—Recommendations on the Organization and Working Procedures of CCITT (Series A) 1 pp. 1 p.

—Directories
IEC. IEC Directory 1993. 320 pp.

—Electronic Components—Quality Assurance
BSI BS 9000: Part 3-91. 1991 General Requirements for a System for Electronic Components of Assessed Quality Part 3: Specification for the National Implementation of the IECQ System. 8 pp.
BSI BS 9000: Part 3-87. 1987 General Requirements for Electronic Components of Assessed Quality Part 3: National Implementation of IECQ Basic Rules and Rules of Procedure. 11 pp.
BSI BS QC 001002-91. 1991 Rules of Procedure of the IEC Quality Assessment System for Electronic Components (IECQ). 91 pp.
BSI BS QC 001002-01. 1991 Amd 1 Rules of Procedure of the IEC Quality Assessment System for Electronic Components (IECQ) (AMD 7711) June 15, 1993 (T). 94 pp.
IEC Guide 102-89. Electronic Components Specification Structures for Quality Assessment (Qualification Approval and Capability Approval) Third Edition (IECQ Guide 102). 64 pp.
IECQ QC 001001-86. Basic Rules of the IEC Quality Assessment System for Electronic Components (IECQ); (Amendment 1-1992). 50 pp.
IECQ QC 001002-86. Rules of Procedure of the IEC Quality Assessment System for Electronic Components (IECQ); (Amendment 1-1992). 184 pp.
IECQ QC 001003-88. Guidance Documents Documents—Guides; (Amendment 1-1990). 93 pp.
IECQ QC 001005-93. Register of Firms, Products and Services Approved Under the IECQ System, Including ISO 9000. 99 pp.
IECQ Guide 102-89. Electronic Components: Specification Structures for Quality Assessment (Qualification Approval and Capability Approval) (IEC GUIDE 102 ED 3). 64 pp.
QSS QC 001001-86. Basic Rules of the IEC Quality Assessment System for Electronic Components (IECQ); (Amendment 1-1992). 49 pp.
QSS Guide 102-89. Electronic Components Specification Structures for Quality Assessment (Qualification Approval and Capability Approval) Third Edition. 64 pp.

—Electronic Components—Quality Assurance—Specifications
IECQ QC 001004 ISSUE 1-93. Specifications List Information Current to 1992-12. 63 pp.

—Electronic Components—Quality Assurance—Specifications List
QSS QC 001004-92. IEC Quality Assessment System for Electronic Components (IECQ) Specifications List Information Current to 1992-06. 61 pp.

IEC (Cont.)

—Inspectorate Co-ordination Committee
IECQ ICC Sec 57-91. Inspectorate Co-Ordination Committee (ICC) Electronic Components Quality Assessment IECQ List of Primary Stages of Manufacture. 7 pp.
IECQ ICC Sec 78-93. Inspectorate Co-Ordination Committee (ICC) ICC Officers and Permanent Delegates. 7 pp.
IECQ ICC Sec 79-93. Inspectorate Co-Ordination Committee (ICC) Membership of Working Groups. 6 pp.

—Specifications Preparation—Electronic Components
IECQ Guide 102-89. Electronic Components Specification Structures for Quality Assessment (Qualification Approval and Capability Approval) Third Edition (IECQ Guide 102). 64 pp.
IECQ Guide 102-89. Electronic Components: Specification Structures for Quality Assessment (Qualification Approval and Capability Approval) (IEC GUIDE 102 ED 3). 64 pp.
QSS Guide 102-89. Electronic Components Specification Structures for Quality Assessment (Qualification Approval and Capability Approval) Third Edition. 64 pp.

—Standards Preparation
IEC Guide 15-77. ISO/IEC Code of Principles on "Reference to Standards" First Edition. 5 pp.
ISO Guide 15-77. ISO/IEC Code of Principles on "Reference to Standards" First Edition. 5 pp.

—Standards Preparation—Sound Recording
CCIR OPINION 16-3-86. Organizations Qualified to Set Standards on Sound and Television Recording—Volumes X and XI—Part 3—Sound and Television Recording. 1 p.

—Standards Preparation—Television Recording
CCIR OPINION 16-3-86. Organizations Qualified to Set Standards on Sound and Television Recording—Volumes X and XI—Part 3—Sound and Television Recording. 1 p.

—Standards Preparation—Testing
IEC Guide 36-82. Preparation of Standard Methods of Measuring Performance (SMMP) of Consumer Goods First Edition. 5 pp.
ISO Guide 36-82. Preparation of Standard Methods of Measuring Performance (SMMP) of Consumer Goods First Edition. 5 pp.

IEEE 802.3
Use: Ethernet (BTN)

IF Transformers
Use: Intermediate Frequency Transformers

IFF Interrogators
Use For: Identification, Friend or Foe Interrogators
See Also: Identification Systems; Radar Equipment
MOD UK DSTAN 07-107-81. Military Characteristics of IFF Systems Issue 1. 87 pp.

—Design
NATO STANAG 4193 Pt1 ED1 AMD0-85. Technical Characteristics of IFF Mk XA and Mk XII Interrogators and Transponders—Part I: General Description of the System (Withdrawn). 145 pp.
NATO STANAG 4193 Pt1 ED2 AMD0-90. Technical Characteristics of IFF Mk XA and Mk XII Interrogators and Transponders—Part I: General Description of the System. 152 pp.
NATO STANAG 4193 Pt3 ED1 AMD0-90. Technical Characteristics of IFF Mk XA and Mk XII Interrogators and Transponders—Part III: IFF Installed System Characteristics (Withdrawn 93-02). 43 pp.

IFF Transponders
See Also: Radar Equipment; Transponders

—Design
NATO STANAG 4193 Pt1 ED1 AMD0-85. Technical Characteristics of IFF Mk XA and Mk XII Interrogators and Transponders—Part I: General Description of the System (Withdrawn). 145 pp.
NATO STANAG 4193 Pt1 ED2 AMD0-90. Technical Characteristics of IFF Mk XA and Mk XII Interrogators and Transponders—Part I: General Description of the System. 152 pp.
NATO STANAG 4193 Pt3 ED1 AMD0-90. Technical Characteristics of IFF Mk XA and Mk XII Interrogators and Transponders—Part III: IFF Installed System Characteristics (Withdrawn 93-02). 43 pp.

IFRB
Use For: International Frequency Registration Board
See Also: CCIR; CCITT

INTERNATIONAL AND NON-U.S. NATIONAL STANDARDS
SUBJECT INDEX
Ignition

IFRB (Cont.)
—Cooperation—CCIR
CCIR RESOLUTION 83-2-90. Collaboration Between the CCIR and the IFRB—Volume XIV—Administrative Texts of the CCIR. 2 pp.

IFS
Use: Freephone Services

Ig
Use: Globulins

Ignitability Testing
Use: Flammability Testing

Igniters
See Also: Ignition Cables; Ignition Transformers; Initiators (Explosives); Spark Plugs

—Gas Burners
BSI BS 3929-65. 1965 Domestic Solid Fuel Ignition Pokers and Portable Undergrate Ignition Burners for Use with Commerical Butane. 9 pp.

—Oil Burners
DIN ENGL 4787 Pt 2-81. Atomizing Oil Burners; Flame Monitoring Devices; Flame Detectors and Automatic Firing Units; Safety Requirements, Testing, Marking (Sept). 11 pp.

—Solid State—Gas Burners
CSA CAN/CSA-C22. 2 NO 199-M89. Combustion Safety Controls and Solid-State Igniters for Gas-and Oil-Burning Equipment; (Gen Instr 1). 94 pp.

—Solid State—Oil Burners
CSA CAN/CSA-C22. 2 NO 199-M89. Combustion Safety Controls and Solid-State Igniters for Gas-and Oil-Burning Equipment; (Gen Instr 1). 94 pp.

Ignitibility Testing
Use: Flammability Testing

Ignition Cables
See Also: Cables (Electric); Igniters

—Aerospace—Physical Properties
BSI G 243-90. 1990 Igniter Cables for Engine Use (4kV d.c.). 10 pp.

—Automotive
ISO 3808 Pt 1-79. Road Vehicles—Unscreened High-Tension Ignition Cables—Part 1: Dimensions, General Requirements and Test Methods First Edition. 10 pp.
ISO 3808 Pt 2-80. Road Vehicles—Unscreened High-Tension Ignition Cables—Part 2: Cable Classes, Types, Applicable Tests and Special Requirements First Edition; (Amendment 1-1985). 6 pp.
ISO 6856-90. Road Vehicles—Unscreened High-Tension Ignition Cable Assemblies—General Requirements and Test Methods Second Edition. 8 pp.
MOD UK DSTAN 29-1-67. Cable Assemblies, Special Purpose, Electrical Ignition Cable Assemblies, Electrically Screened—5 mm Cable Issue 1. 7 pp.

—Automotive—Coils
BSI BS AU 167-78. (WITHDRAWN) 1978 Low Tension Cable Connections for Ignition Cells (Superseded by BS ISO 4024: 1992). 3 pp.

—Combat Vehicles
NATO STANAG 4006 Pt2 ED1 AMD1-65. Shielded Ignition Cables for Wheeled Tactical Vehicles. 11 pp.

—Gas Burners
CSA C22.2 NO 17-1973. Cable for Luminous-Tube Signs and for Oil-and Gas-Burner Ignition Equipment (R 1992); (Gen Instr 1) (Amd 1 February 1981) (Amd 2-8 March 1988). 28 pp.

—Oil Burners
CSA C22.2 NO 17-1973. Cable for Luminous-Tube Signs and for Oil-and Gas-Burner Ignition Equipment (R 1992); (Gen Instr 1) (Amd 1 February 1981) (Amd 2-8 March 1988). 28 pp.

—Photoflash Units
ISO 10330-92. Photography—Synchronizers, Ignition Circuits and Connectors for Cameras and Photoflash Units—Electrical Characteristics and Test Methods First Edition. 17 pp.

—Sleeves
SBAC AGS 1716 ISSUE 2. Insulating, Sleeve (For H.T. Ignition Cable Sparking Plug Terminal).

Ignition Coils
See Also: Electric Coils; Spark Plugs

Ignition Coils (Cont.)
—Automotive
CNS D2022-86. Ignition Coils for Automobiles (Nov)(5561). 10 pp.
ISO 10455-92. Road Vehicles—Dry Ignition Coils Using Rotating High-Voltage Distributor First Edition. 9 pp.
JIS D 5121-80. Ignition Coils for Automobiles (R 1986). 20 pp.

—Automotive—Bands
CNS D2022-1-86. Ignition Coils Fixing Band for Automobiles (Nov)(5561-1).

—Automotive—Connectors
BSI BS AU 167-78. (WITHDRAWN) 1978 Low Tension Cable Connections for Ignition Cells (Superseded by BS ISO 4024: 1992). 3 pp.
CNS D1066-87. Dimension of High Tension Connection for Ignition Coils and Distributors for Automobiles (Oct)(12123).
CNS D2022-2-86. High Tension Cable Connection for Ignition Coils and Distributors for Automobiles (Nov)(5561-2).
CNS D2022-3-86. Low Tension Cable Connections for Ignition Coils for Automobiles (Nov)(5561-3).
ISO 3553 Pt 1-87. Road Vehicles—High-Tension Connections for Ignition Coils and Distributors—Part 1: Socket-Type First Edition. 6 pp.
ISO 3553 Pt 2-87. Road Vehicles—High-Tension Connections for Ignition Coils and Distributors—Part 2: Plug-Type First Edition. 4 pp.
ISO 4024-92. Road Vehicles—Ignition Coils—Low-Tension Cable Connections Second Edition. 5 pp.

—Automotive—Mounting Brackets
ISO 3285-86. Road Vehicles—Ignition Coil Mounting Brackets Second Edition. 3 pp.

Ignition Loss
See Also: Combustion; Residue-on-Ignition Determination

—Calcium Carbonates
CNS K6721-82. Method of Test for Calcium Carbonate for Rubber (Jun)(9004). 4 pp.

—Chemical Products
CNS K0019-82. Method for Determining Loss and Residue of Chemical Products (May)(8838).
JIS K 0067-92. Test Methods for Loss and Residue of Chemical Products. 18 pp.

—Dolomite
CNS M3101-82. Method of Determining Ignition Loss in Dolomite (Sep)(9419).
ISO 10058-92. Magnesites and Dolomites —Chemical Analysis First Edition. 23 pp.

—Dolomite Refractories
ISO 10058-92. Magnesites and Dolomites —Chemical Analysis First Edition. 23 pp.

—Feldspars
CNS M3081-82. Method for Determination of Ignition Loss in Feldspar (Jan)(8436).

—Fiberglass Reinforced Plastics
BSI BS 2782:Pt10: METH 1002-77. 1977 Amd 1 Methods of Testing Plastics Part 10: Glass Reinforced Plastics Method 1002: Determination of Loss on Ignition. 8 pp.
CEN EN 60-77. Glass Reinforced Plastics: Determination of the Loss on Ignition. 6 pp.
ISO 1172-75. Textile Glass Reinforced Plastics—Determination of Loss on Ignition First Edition. 5 pp.

—Foot Powders
MOD UK M 9507/65. Determination of Foot Powder (No Information).

—Foundry Sands
CNS Z8032-82. Method for Determining Ignition Loss of Foundry Sand (May)(8899).
JIS Z 2606-63. Method for Determining Ignition Loss of Foundry Sands.

—Gypsum Plaster
MOD UK M 9519/66. Examination of Plaster of Paris (Withdrawn).

—Iron Ores
SAA AS 1673.5-82. Methods for the Analysis of Iron Ores—Part 5: Determination of Loss on Ignition at 1 000 degrees Celsius. 3 pp.

—Kaolin
MOD UK M 2002/59. Examination of Kaolin (Withdrawn).

—Limestone
CPPA J.4-92. Analysis of Limestone. 5 pp.

Ignition Loss (Cont.)
—Magnesite
ISO 10058-92. Magnesites and Dolomites —Chemical Analysis First Edition. 23 pp.

—Magnesite Refractories
ISO 10058-92. Magnesites and Dolomites —Chemical Analysis First Edition. 23 pp.

—Magnesium Oxides
MOD UK M 9541/69. Examination of Magnesium Oxide and Magnesium Oxide, Special, Lead Free (Withdrawn).

—Sands
CGSB 31-GP-0A METH 73.1-62. Methods of Testing Corrosion-Prevention Materials and Processes Ignition Loss. 1 p.

—Sludge (Sewage)
DIN ENGL 38414 Pt 3-85. German Standard Methods for the Examination of Water, Waste Water and Sludge; Sludge and Sediments (Group S); Determination of Loss on Ignition and Residue on Ignition of the Dry Matter of a Sludge (S 3) (Nov). 2 pp.

—Sodium Borates
BSI BS 5688: Part 3-79. 1979 Orthoboric Acid (Boric Acid), Diboron Trioxide (Boric Oxide), Disodium Tetraborates, Sodium Perborates and Crude Sodium Borates for Ind. Use Part 3: Detn. of Sodium Oxide and Boric Oxide Contents and Loss on Ingintion of Disodium Tetraborates. 4 pp.
ISO 1916-72. Disodium Tetraborates for Industrial Use—Determination of Sodium Oxide and Boric Oxide Contents and Loss on Ignition First Edition. 4 pp.

—Sodium Tripolyphosphate
BSI BS 4427: Part 4-69. 1969 Methods of Test for Sodium Tripolyphos-phate (Pentasodium Triphosphate) and Sodium Pyrophosphate (Tetrasodium Pyrophos-phate) for Industrial Use Part 4: Determination of Loss on Ignition. 7 pp.
CNS K6415-78. Method of Test for Sodium Tripolyphosphate (for Industrial Use) (Mar)(4285). 4 pp.
ISO 853-76. Sodium Tripolyphosphate and Sodium Pyrophosphate for Industrial Use—Determination of Loss on Ignition First Edition. 4 pp.

—Soil Analysis
BSI BS 1377: Part 3-90. 1990 Methods of Test for Soils for Civil Engineering Purposes Part 3: Chemical and Electro-Chemical Tests. 45 pp.

—Tetrasodium Pyrophosphate
BSI BS 4427: Part 4-69. 1969 Methods of Test for Sodium Tripolyphos-phate (Pentasodium Triphosphate) and Sodium Pyrophosphate (Tetrasodium Pyrophos-phate) for Industrial Use Part 4: Determination of Loss on Ignition. 7 pp.
ISO 853-76. Sodium Tripolyphosphate and Sodium Pyrophosphate for Industrial Use—Determination of Loss on Ignition First Edition. 4 pp.

—Zinc Oxides
MOD UK M 806/91. Examination of Zinc Oxide.

—Zinc Stearate
CNS K6296-71. Method of Test for Zinc Stearate Powder (Jun)(3215). 2 pp.

Ignition Pokers
Use For: Gas Pokers See Also: Ignition Systems

—Installation
BSI CP 335: Part 1-73. 1973 Selection and Installation of Miscellaneous Town Gas Appliances Part 1: Domestic Laundering and Miscellaneous Appliances. 17 pp.

—Solid Fuel Furnaces
BSI BS 3929-65. 1965 Domestic Solid Fuel Ignition Pokers and Portable Undergrate Ignition Burners for Use with Commerical Butane. 9 pp.

Ignition Switches
See Also: Ignition Systems

—Automotive—Steering Locks
CNS D2058-87. Steering Locks with Ignition Switches for Automobiles (Oct)(6810).
JIS D 5812-76. Steering Locks with Ignition Switches for Automobiles.

Ignition Systems
See Also: Automobiles; Breaker Points; Distributors (Electrical); Engines; Ignition Pokers; Ignition Switches; Spark Ignition Engines; Spark Plugs
ISO 6518 Pt 2-82. Ignition Systems—Part 2: Test Methods First Edition. 7 pp.

INTERNATIONAL AND NON-U.S. NATIONAL STANDARDS
SUBJECT INDEX

Ignition

Ignition Systems *(Cont.)*

—Automotive—Resistors—Noise Suppression
CNS D2021-86. External Resistor-Suppressors for Automobiles, Ignition Interference (Nov)(5560).
JIS D 5111-77. External Resistor-Suppressors for Automobiles, Ignition Interference (R 1985). 9 pp.

—Checkout Equipment
BSI BS AU 206: Part 3-86. 1986 Diagnostic Systems for Road Vehicles Part 3: General Requirements for Testing Ignition Systems. 8 pp.
ISO 7342-82. Road Vehicles—Diagnostic Systems—Equipment for Ignition Systems Testing First Edition. 8 pp.

—Glossaries
ISO 6518 Pt 1-80. Ignition Systems—Part 1: Vocabulary First Edition; (Erratum—Dec 1981). 9 pp.

—Internal Combustion Engines—Hazardous Locations
CSA C22.2 NO 30-M1986. Explosion-Proof Enclosures for Use in Class I Hazardous Locations (R 1992); (Gen Instr 1 Thru 2). 41 pp.
CSA CAN/CSA-C22. 2 NO 157-92. Intrinsically Safe and Non-Incendive Equipment for Use in Hazardous Locations; (Gen Instr 1). 61 pp.

—Internal Combustion Engines—Radio Frequency Interference
BSI BS 833-70. 1970 Radio Interference Limits and Measurements for the Electrical Ignition Systems of Internal Combustion Engines. 33 pp.

Ignition Temperature
See Also: Combustion; Residue-on-Ignition Determination; Temperature

—Combustible Materials
DIN ENGL 54836-84. Testing of Combustible Materials; Determination of Ignition Temperature (Feb). 4 pp.

—Explosive Atmospheres
IEC 79 Pt 4-75. Electrical Apparatus for Explosive Gas Atmospheres Part 4: Method of Test for Ignition Temperature Second Edition. 22 pp.

—Explosive Gases
DIN ENGL 51794-78. Testing of Mineral Oil Hydrocarbons; Determination of Ignition Temperature (Jan). 5 pp.

—Flammable Liquids
DIN ENGL 51794-78. Testing of Mineral Oil Hydrocarbons; Determination of Ignition Temperature (Jan). 5 pp.

—Gases
SAA AS 1896-76. Method of Test for Ignition Temperature of Gases and Vapours Corrig.. 20 pp.

—Vapors
SAA AS 1896-76. Method of Test for Ignition Temperature of Gases and Vapours Corrig.. 20 pp.

Ignition Temperature Testing
Use: Ignition Temperature

Ignition Transformers
See Also: Igniters; Specialty Transformers; Transformers
DIN VDE 0550 Pt 5 (S)-72. Regulations for Small Transformers; Special Regulations for Ignition Transformers (Sept). 10 pp.

—
CSA C22.2 NO 13-1962. Transformers for Luminous-Tube Signs, Oil-or Gas-Burner Ignition Equipment, Cold-Cathode Interior Lighting (R 1992). 46 pp.
CSA 1169 Bull. Electrical Bulletin 1169 June 27, 1978 to C22.2 NO 13. 2 pp.
CSA 1238 Bull. Electrical Bulletin 1238 September 17, 1979 to C22.2 NO 13. 2 pp.

—
CSA C22.2 NO 13-1962. Transformers for Luminous-Tube Signs, Oil-or Gas-Burner Ignition Equipment, Cold-Cathode Interior Lighting (R 1992). 46 pp.
CSA 1169 Bull. Electrical Bulletin 1169 June 27, 1978 to C22.2 NO 13. 2 pp.
CSA 1238 Bull. Electrical Bulletin 1238 September 17, 1979 to C22.2 NO 13. 2 pp.

—Gas Burners
JIS S 2142-85. Ignition Transformers for Burning Appliances.

—Oil Burners
JIS S 2142-85. Ignition Transformers for Burning Appliances.

Ignition Transformers *(Cont.)*

—5-15 kV—Gas Burners
CSA C22.2 NO 13-1962. Transformers for Luminous-Tube Signs, Oil-or Gas-Burner Ignition Equipment, Cold-Cathode Interior Lighting (R 1992). 46 pp.
CSA 1169 Bull. Electrical Bulletin 1169 June 27, 1978 to C22.2 NO 13. 2 pp.
CSA 1238 Bull. Electrical Bulletin 1238 September 17, 1979 to C22.2 NO 13. 2 pp.

—5-15 kV—Oil Burners
CSA C22.2 NO 13-1962. Transformers for Luminous-Tube Signs, Oil-or Gas-Burner Ignition Equipment, Cold-Cathode Interior Lighting (R 1992). 46 pp.
CSA 1169 Bull. Electrical Bulletin 1169 June 27, 1978 to C22.2 NO 13. 2 pp.
CSA 1238 Bull. Electrical Bulletin 1238 September 17, 1979 to C22.2 NO 13. 2 pp.

Ignitrons
See Also: Electron Tubes; Gas Discharge Tubes; Rectifiers

—Resistance Welding Equipment
CNS C4156-79. Ignition for Welding Machine (Nov)(5024). 10 pp.
IEC 237-67. Ignitrons to Be Used in Welding Machine Control First Edition. 27 pp.

Ilang-Ilang Oil
Use For: Cananga Oil; Ylang Ylang Oil; Ylang-Ylang Oil *See Also:* Oils
CNS K5090-80. Oil of Cananga (Sep)(6461).
ISO 3063-83. Oil of Ylang-Ylang (Cananga Odorata (Lamark) J.D. Hooker and Thomson) First Edition. 4 pp.
ISO 3523-76. Oil of Cananga First Edition. 4 pp.

ILL
Use: Interlibrary Loans

Illuminance
Use For: Illumination *See Also:* Luminance; Optical Density; Vignetting

—Clean Rooms
SAA AS 1807.15-89. Cleanrooms, Workstations and Safety Cabinets—Methods of Test—Part 15: Determination of Illuminance. 1 p.

—Lighting
BSI BS 5225: Part 1-75. 1975 Photometric Data for Luminaires Part 1: Photometric Measurements. 43 pp.
CNS C3069-88. Method of Illumination Measurements (Oct)(5065).
CNS Z1044-87. Recommended Levels of Illumination (Sep)(12112).
JIS C 7612-85. Illuminance Measurements for Lighting Installations.
JIS C 7612-68. Methods of Illumination Measurements (R 1977). 9 pp.
JIS Z 9110-79. Recommended Levels of Illumination. 21 pp.

—Lighting—Rolling Stock
JIS E 4016-92. Illuminance for Railway Rolling Stock—Recommended Levels and Measuring Methods.

—Marine
JIS F 8041-86. Recommended Levels of Illumination and Methods of Illumination Measurements for Marine Use.

—Photographic Prints
ISO 3664-75. Photography—Illumination Conditions for Viewing Colour Transparencies and Their Reproductions First Edition. 6 pp.

—Projectors
ISO 8341-89. Photography—Slide Projectors and Filmstrip Projectors—Illumination Test First Edition; (ANSI IT7.201-1991). 6 pp.

—Reproduction (Copying)
BSI BS 950: Part 2-67. 1967 Amd 1 Artificial Daylight for the Assessment of Colour Part 2: Viewing Conditions for Reproduction in the Graphic Arts. 16 pp.

Illuminance Meters
Use: Light Meters

Illuminants
Use: Light Sources

Illuminating Gas
Use: Coal Gas

Illumination
Use: Illuminance

Illuminators (Optics)
See Also: Light Sources

—Radiographic
BSI BS EN 25580-92. 1992 Minimum Requirements for Industrial Radiographic Illuminators for Non-Destructive Testing (ISO 5580: 1985). 10 pp.
CEN EN 25580-92. Non-Destructive Testing—Industrial Radiographic Illuminators—Minimum Requirements. 6 pp.
DIN ENGL EN 25580-92. Non-Destructive Testing; Minimum Requirements for Industrial Radiographic Illuminators (ISO 5580: 1985) (June). 6 pp.
ISO 5580-85. Non-Destructive Testing—Industrial Radiographic Illuminators—Minimum Requirements First Edition. 5 pp.
JIS Z 4561-92. Viewing Illuminators for Industrial Radiograph. 10 pp.

—Transparency Films
ISO 3664-75. Photography—Illumination Conditions for Viewing Colour Transparencies and Their Reproductions First Edition. 6 pp.

ILS
Use: Instrument Landing Systems

Image Converters
Use: Image Tubes

Image Document Processors
Use: Image Processing Systems

Image Intensifier Tubes
Use: Image Tubes

Image Intensifiers
Use For: Light Amplifiers (Intensifiers)
See Also: Cathode Ray Tubes; Image Tubes

—Quality Assurance
MOD UK DSTAN 59-60: 90/113-81. Valve Electronic, Image Intensifier Tube Issue 1. 43 pp.

—X-Ray
BSI BS 6251-82. 1982 Determining the Luminance Distribution of Electro-Optical X-Ray Image Intensifiers. 12 pp.
BSI BS 6251-01. 1982 Amd 1 Method for Determining the Luminance Distribution of Electro-Optical X-Ray Image Intensifiers (IEC 572: 1977) (AMD 7001) February 28, 1992. 13 pp.
BSI BS 6252-82. 1982 Measuring the Conversion Factor of Electro-Optical X-ray Image Intensifiers. 11 pp.
BSI BS 6252-01. 1982 Amd 1 Method for Measuring the Conversion Factor of Electro-Optical X-Ray Image Intensifiers (IEC 573: 1977) (AMD 7000) February 28, 1992. 12 pp.
BSI BS 6801-86. 1986 Determination of Image Distortion of Electro-Optical X-Ray Image Intensifiers (IEC 858: 1986). 9 pp.
BSI BS 6801-01. 1986 Amd 1 Method for Determination of Image Distortion of Electro-Optical X-Ray Image Intensifiers (IEC 858: 1986) (AMD 6996) February 28, 1992. 10 pp.
BSI BS 6983-88. 1988 Methods for Determination and Expression of Entrance Field Sizes of Electro-Optical X-Ray Image Intensifiers (N). 7 pp.
BSI BS 6983-01. 1988 Amd 1 Methods for Determination and Expression of Entrance Field Sizes of Electro-Optical X-Ray Image Intensifiers (IEC 520: 1975) (AMD 6999) February 28, 1992. 8 pp.
CENELEC HD 510 S1-88. Entrance Field Sizes of Electro-Optical X-Ray Image Intensifiers. 3 pp.
CENELEC HD 511 S1-88. Determination of the Luminance Distribution of Electro-Optical X-Ray Image Intensifiers. 3 pp.
CENELEC HD 512 S1-89. Measurement of the Conversion Factor of Electro-Optical X-Ray Intensifiers. 3 pp.
CENELEC HD 515 S1-89. Determination of the Image Distortion of Electro-Optical X-Ray Image Intensifiers. 3 pp.
IEC 520-75. Entrance Field Sizes of Electro-Optical X-Ray Image Intensifiers First Edition. 15 pp.
IEC 572-77. Determination of the Luminance Distribution of Electro-Optical X-Ray Image Intensifiers First Edition. 26 pp.
IEC 573-77. Measurement of the Conversion Factor of Electro-Optical X-Ray Image Intensifiers First Edition; (Corrigenda—June 1978). 26 pp.
IEC 858-86. Determination of the Image Distortion of Electro-Optical X-Ray Image Intensifiers First Edition. 21 pp.

—X-Ray—Medical
JIS Z 4721-87. Medical X-Ray Image Intensifiers. 20 pp.

Image Processing
Use For: Picture Processing *See Also:* Image Processing Systems; Optical Character Recognition

Image Processing (Cont.)
—Computer Graphics Software—Graphical Kernel System
IEC DIS 12087 Pt 2-92. Information Technology—Computer Graphics and Image Processing—Image Processing and Interchange (IPI)—Functional Specification —Part 2: Programmer's Imaging Kernel System Application Programme Interface ***CD-ROM ONLY***. 915 pp.

ISO DIS 12087 Pt 2-92. Information Technology—Computer Graphics and Image Processing—Image Processing and Interchange (IPI)—Functional Specification —Part 2: Programmer's Imaging Kernel System Application Programme Interface ***CD-ROM ONLY***. 915 pp.

JTC1 DIS12087 Pt 2-92. Information Technology—Computer Graphics and Image Processing—Image Processing and Interchange (IPI)—Functional Specification —Part 2: Programmer's Imaging Kernel System Application Programme Interface ***CD-ROM ONLY***. 915 pp.

—Digital Compression
IEC DIS 10918 Pt 1-92. Information Technology—Digital Compression and Coding of Continuous-Tone Still Images—Part 1: Requirements and Guidelines ***CD-ROM ONLY***. 203 pp.

IEC DIS 10918 Pt 2-93. Information Technology—Digital Compression and Coding of Continuous-Tone Still Images—Part 2: Compliance Testing ***CD-ROM ONLY***. 77 pp.

ISO DIS 10918 Pt 1-92. Information Technology—Digital Compression and Coding of Continuous-Tone Still Images—Part 1: Requirements and Guidelines ***CD-ROM ONLY***. 203 pp.

ISO DIS 10918 Pt 2-93. Information Technology—Digital Compression and Coding of Continuous-Tone Still Images—Part 2: Compliance Testing ***CD-ROM ONLY***. 77 pp.

JTC1 DIS10918 Pt 1-92. Information Technology—Digital Compression and Coding of Continuous-Tone Still Images—Part 1: Requirements and Guidelines ***CD-ROM ONLY***. 203 pp.

JTC1 DIS10918 Pt 2-93. Information Technology—Digital Compression and Coding of Continuous-Tone Still Images—Part 2: Compliance Testing ***CD-ROM ONLY***. 77 pp.

—Encoding
IEC DIS 10918 Pt 1-92. Information Technology—Digital Compression and Coding of Continuous-Tone Still Images—Part 1: Requirements and Guidelines ***CD-ROM ONLY***. 203 pp.

IEC DIS 10918 Pt 2-93. Information Technology—Digital Compression and Coding of Continuous-Tone Still Images—Part 2: Compliance Testing ***CD-ROM ONLY***. 77 pp.

ISO DIS 10918 Pt 1-92. Information Technology—Digital Compression and Coding of Continuous-Tone Still Images—Part 1: Requirements and Guidelines ***CD-ROM ONLY***. 203 pp.

ISO DIS 10918 Pt 2-93. Information Technology—Digital Compression and Coding of Continuous-Tone Still Images—Part 2: Compliance Testing ***CD-ROM ONLY***. 77 pp.

JTC1 DIS10918 Pt 1-92. Information Technology—Digital Compression and Coding of Continuous-Tone Still Images—Part 1: Requirements and Guidelines ***CD-ROM ONLY***. 203 pp.

JTC1 DIS10918 Pt 2-93. Information Technology—Digital Compression and Coding of Continuous-Tone Still Images—Part 2: Compliance Testing ***CD-ROM ONLY***. 77 pp.

—Glossaries
IEC DIS 2382 Pt 27-92. Information Technology—Vocabulary—Part 27: Office Automation ***CD-ROM ONLY***. 22 pp.

ISO DIS 2382 Pt 27-92. Information Technology—Vocabulary—Part 27: Office Automation ***CD-ROM ONLY***. 22 pp.

JTC1 DIS2382 Pt 27-92. Information Technology—Vocabulary—Part 27: Office Automation ***CD-ROM ONLY***. 22 pp.

—Image Interchange Facility
IEC DIS 12087 Pt 1-92. Information Technology—Computer Graphics and Image Processing—Image Processing and Interchange (IPI)—Functional Specification —Part 1: Common Architecture for Imaging ***CD-ROM ONLY***. 78 pp.

IEC DIS 12087 Pt 3-92. Information Technology—Computer Graphics and Image Processing—Image Processing and Interchange (IPI)—Functional Specification —Part 3: Image Interchange Facility (IIF) ***CD-ROM ONLY***. 173 pp.

ISO DIS 12087 Pt 1-92. Information Technology—Computer Graphics and Image Processing—Image Processing and Interchange (IPI)—Functional Specification —Part 1: Common Architecture for Imaging ***CD-ROM ONLY***. 78 pp.

ISO DIS 12087 Pt 3-92. Information Technology—Computer Graphics and Image Processing—Image Processing and Interchange (IPI)—Functional Specification —Part 3: Image Interchange Facility (IIF) ***CD-ROM ONLY***. 173 pp.

JTC1 DIS12087 Pt 1-92. Information Technology—Computer Graphics and Image Processing—Image Processing and Interchange (IPI)—Functional Specification —Part 1: Common Architecture for Imaging ***CD-ROM ONLY***. 78 pp.

JTC1 DIS12087 Pt 3-92. Information Technology—Computer Graphics and Image Processing—Image Processing and Interchange (IPI)—Functional Specification —Part 3: Image Interchange Facility (IIF) ***CD-ROM ONLY***. 173 pp.

—Information Interchange
NATO STANAG 3764 ED 2 AMD 0-89. Exchange of Imagery. 8 pp.
NATO STANAG 3764 ED 2 AMD 1-89. Exchange of Imagery. 9 pp.

—Information Interchange—Architecture
IEC DIS 12087 Pt 1-92. Information Technology—Computer Graphics and Image Processing—Image Processing and Interchange (IPI)—Functional Specification —Part 1: Common Architecture for Imaging ***CD-ROM ONLY***. 78 pp.

ISO DIS 12087 Pt 1-92. Information Technology—Computer Graphics and Image Processing—Image Processing and Interchange (IPI)—Functional Specification —Part 1: Common Architecture for Imaging ***CD-ROM ONLY***. 78 pp.

JTC1 DIS12087 Pt 1-92. Information Technology—Computer Graphics and Image Processing—Image Processing and Interchange (IPI)—Functional Specification —Part 1: Common Architecture for Imaging ***CD-ROM ONLY***. 78 pp.

Image Processing Systems
Use For: Optical Document Image Systems; Optical Image Processing Systems; Photo Imaging Systems
See Also: Computers; Data Processing Equipment; Image Processing; Laboratory Equipment; Optical Equipment; Optoelectronic Imaging Systems; Thermal Imaging Systems

—Optical Transfer Functions
NATO STANAG 4161 ED 1 AMD 1-82. The Optical Transfer Function of Imaging Systems. 58 pp.

Image Quality Indicators
Use: Radiographic Image Quality Indicators

Image Storage Tubes
Use: Direct View Storage Tubes

Image Tubes
Use For: Image Converters; Image Intensifier Tubes
See Also: Electron Tubes; Image Intensifiers
BSI BS CECC 12001-80. 1980 Image Converter and Image Intensifier Tubes: Blank Detail Specification. 11 pp.
CECC CECC 12 001 ISSUE 1-80. Blank Detail Specification: Image Converter and Image Intensifier Tubes (En, Fr). 20 pp.
MOD UK DSTAN 59-60: Part 15-01. Valves, Electronic (Electronic Tubes) (Listed on EPIC Database) Part 15: Image Converter and Image Intensifier Tubes Issue 1; Amendment 1. 17 pp.
MOD UK DSTAN 59-60: 90/077-77. Valve Electronic, Image Intersifier Tube Issue 1A. 26 pp.
MOD UK DSTAN 59-60: 90/086-77. Valve Electronic, Image Intensifier Tube Issue 1A. 23 pp.
MOD UK DSTAN 59-60: 90/089-01. Valve Electronic, Image Intensifier Tube Issue 2; Amendment 2. 41 pp.
MOD UK DSTAN 59-60: 90/098-01. Valve Electronic, Image Intensifier Tube Issue 1A; Amendment 1. 27 pp.
MOD UK DSTAN 59-60: 90/100-76. Valve Electronic, Image Intensifier Tube Issue 1. 26 pp.
MOD UK DSTAN 59-60: 90/107-01. Valve Electronic, Image Intensifier Tube Issue 1; Amendment 1. 32 pp.
MOD UK DSTAN 59-60: 90/115-01. Valve Electronic, Image Intensifier Tube Issue 1; Amendment 1. 45 pp.

—Night Vision Devices
CECC CECC 12 001-007 ISSUE 1-84. BS CECC 12 001-007; Image Intensifier: Inverting Proximity Focused Tube for Night Vision Equipment, Incorporating a Gallium Arsenide Photo-Cathode and a Filmed Micro-Channel Plate (En). 51 pp.

—Night Vision Goggles
NATO STANAG 4341 ED 1 AMD 0-89. High Performance Image Intensifier Tube (IIT) with Gallium Arsenide Photocathode for Aviation Night Vision Goggles. 11 pp.

—Quality Assurance
BSI BS CECC 12000-81. 1981 Image Converter and Image Intensifier Tubes. Generic Specification. 24 pp.
BSI BS CECC 12000-01. 1981 Amd 1 Generic Specification: Image Converter and Image Intensifier Tubes (AMD 6008) July 15, 1992. 35 pp.

—Three Stage—Night Vision Devices
CECC CECC 12 001-001 ISSUE 1-86. BS CECC 12 001-001; Image Intensifier: 3-Stage Cascade Tube for Night Vision Equipment with Enhanced MTF (En). 35 pp.

—Weapon Sights
NATO STANAG 4342 ED 1 AMD 0-90. (DRAFT) High Performance 18 mm Format Image Intensifier Tube (IIT) with Arsenide Photocathode for Weapon Sights. 12 pp.

Imaging Equipment (Medical)
See Also: Medical Equipment; Radiology; Ultrasonic Medical Equipment

—Symbols
CNS C1143-14-89. Graphical Symbols for Use on Electrical and Electronic Equipment (Image) (Feb)(12491-14).

2,2'-Iminodiethanol
Use: Diethanolamine

Imitation Leather
Use: Leathercloth

Immersion Coatings
Use For: Dipcoats; Dipped Coatings
See Also: Coatings
MOD UK CS 2486B. Protective PX 15.

—Conversion
CGSB 31-GP-101MA-89. Chemical Conversion Films for Aluminum and Aluminum Alloys. 9 pp.

—Phosphate
CGSB 31-GP-103MA-89. Heavy Phosphate Conversion Coatings for Iron and Steel (for Corrosion Resistance). 9 pp.
CGSB 31-GP-104MA-89. Heavy Phosphate Conversion Coatings for Iron and Steel (for Wear Resistance). 9 pp.

—Steel—Self Lifting Resistance
CGSB 1-GP-71 METH 132.3-79. Methods of Testing Paints and Pigments Self-Lifting Dipcoats. 1 p.

Immersion Corrosion Testing
Use: Immersion Testing

Immersion Heaters
DIN VDE 0720 Pt 2C-72. Specifications for Electrical Heating Appliances for Domestic Use and Similar Purposes; Part 2C: Particular Specifications for Immersion Heaters (July). 13 pp.
DIN VDE 0720 Pt 2 CA-74. Specifications for Electrical Heating Appliances for Domestic Use and Similar Purposes; Part 2: CA Particular Specifications for Immersion Heaters (Nov). 8 pp.
SNZ NZS 918-65. Immersion Heaters for Thermal Storage Electric Water Heaters. 11 pp.

—Portable
CENELEC HD 262-77. Particular Specification for Portable Immersion Heaters. 6 pp.
CENELEC HD 262 S1-75. Particular Specification for Portable Immersion Heaters. 4 pp.
CENELEC HD 262.2-78. Particular Specification for Portable Immersion Heaters. 4 pp.
CENELEC HD 262.3 S1-89. Particular Specification for Portable Immersion Heaters. 7 pp.
CENELEC HD 262.4 S1-90. Particular Specification for Portable Immersion Heaters. 6 pp.
SNZ NZS 952-52. Specification for Portable Electric Immersion Water-Heaters. 11 pp.
SNZ NZS 1880-64. Replaceable Sheathed Immersion Heating Elements for Portable Water Heating Appliances. 13 pp.

—Portable—Safety
BSI BS 3456: Sec 3.21-81. (OBSOLESCENT) 1981 Amd 2 1972-1981 Edition. Specification for Safety of Household Electrical Appliances Part 3: Complete Particular Specifications Section 3.21: Portable Immersion Heater (AMD 5895) August 31, 1988. 45 pp.

—Safety
BSI BS 3456: Sec 2.21-72. 1972 Amd 3 Safety of Household and Similar Electrical Appliances Part 2: Particular Requirements Section 2.21: Electric Immersion Heaters (AMD 6345) September 30, 1991. 21 pp.

INTERNATIONAL AND NON-U.S. NATIONAL STANDARDS
SUBJECT INDEX

Immersion

Immersion Hydrobaths
Use: Hydrobaths

Immersion Objectives
Use: Microscope Objectives—Immersion

Immersion Oils
See Also: Microscope Objectives

—**Microscopes**
BSI BS 7011: Sec 1.1-89. 1989 Consumable Accessories for Light Microscopes Part 1: Immersion Oils Section 1.1: Immersion Oils for General Use. 4 pp.
JIS K 2400-88. Immersion Oil for Microscope. 6 pp.

Immersion Suits
Use: Exposure Suits

Immersion Testing
Use For: Fluid Immersion Testing; Liquid Immersion Testing See Also: Accelerated Testing; Acid Resistance Testing; Atmospheric Corrosion Testing; Corrodkote Testing; Corrosion Testing; Environmental Testing; Heaters; Liquid Penetrant Testing; Materials Testing; Salt Spray Testing; Sealing; Water; Water Resistance Testing
MOD UK DSTAN 07-55: Pt 2:Sec 4-01. Environmental Testing of Service Materiel Part 2: Tests Section 4: Penetration and Immersion Issue 1; Amendment 3. 49 pp.

—**Acid—Anodic Coatings—Loss of Mass**
BSI BS 6161: Part 3-84. 1984 Methods of Test for Anodic Oxidation Coatings on Aluminium and Its Alloys Part 3: Assessment of Sealing Quality by Measurement of the Loss of Mass After Immersion in Phosphoric-Chromic Acid-Solution. 7 pp.
BSI BS 6161: Part 4-81. 1981 Methods of Test for Anodic Oxidaton Coatings on Aluminium and Its Alloys Part 4: Assessmentof Sealing Quality by Measurement of the Loss of Mass After Immersion in Acid Solution. 4 pp.
ISO 3210-83. Anodizing of Aluminium and Its Alloys—Assessment of Quality of Sealed Anodic Oxide Coatings by Measurement of the Loss of Mass After Immersion in Phosphoric-Chromic Acid Solution Second Edition. 5 pp.

—**Alloys**
BSI M 36-70. 1970 Amd 1 Ultrasonic Testing of Special Forgings by an Immersion Technique Using Flat-Bottomed Holes as a Reference Standard. 21 pp.

—**Connectors**
CNS C6177-88. Method of Test for Low Frequency (Below 3 MHz) Electrical Connector (TP-3 Altitude Immersion) (May)(9237).

—**Electronic Components**
JIS C 5032-75. Sealing (Immersion Cyclic) Testing Method for Electronic Components. 4 pp.

—**Electronic Equipment**
CNS C6246-85. Sealing (Immersion Cyclic) Testing Method for Electronic Components (Apr)(11237).

—**Fiber Optic Cables**
CNS C6251-85. Method of Test for Fiber Optic Devices (Fiber Optic Cable Fluid Immersion Test) (Aug)(11333).

—**Fiber Optic Equipment**
CNS C6225-84. Method of Test for Fiber Optic Devices (Fluid Immersion Test) (Jun)(10925).
CNS C6260-86. Method of Test for Fiber Optic Devices (FOTP-15 Altitude Immersion Test) (Sep)(11704).
CNS C6313-88. Method of Test for Fiber Optic Devices (FOTP-75 Fluid Immersion Test) (Jul)(12360).

—**Fuel—Sealants**
CGSB CAN2-19.0-M77METH 5.3-78. Methods of Testing Putty, Caulking and Sealing Compounds Change in Mass After Test Fuel Immersion. 1 p.

—**Metals**
DIN ENGL 50905 Pt 4-87. Corrosion of Metals; Corrosion Testing; Corrosion Testing in Liquids Under Laboratory Conditions Without Mechanical Stress (Jan). 4 pp.
JIS Z 2277-90. Tensile Testing Method for Metallic Materials in Liquid Helium. 15 pp.

—**Methylene Chloride—Pipes**
CEN PREN 27676-93. Unplasticized Poly(Vinyl Chloride) (PVC-U) Pipes—Dichloromethane Test (ISO 7676: 1990). 9 pp.
ISO 7676-90. Unplasticized Poly (Vinyl Chloride) (PVC-U) Pipes—Dichloromethane Test First Edition (CEN PREN 27676: 1993). 7 pp.

Immersion Testing (Cont.)
—**Mineral Oil—Cable Insulation**
BSI BS 6469: Sec 2.1-92. 1992 Insulation and Sheathing Materials of Electric Cables Part 2: Methods of Test Specific to Elastomeric Compounds Section 2.1: Ozone Resistance Test—Hot Set Test—Mineral Oil Immersion Test. 19 pp.
CENELEC HD 505.2.1 S1-88. Common Test Methods for Insulating and Sheathing Materials of Electric Cables Part 2: Methods Specific to Elastomeric Compounds Section One—Ozone Resistance Test—Hot Set Test—Mineral Oil Immersion Test. 3 pp.
IEC 811 Pt 2-1-86. Common Test Methods for Insulating and Sheathing Materials of Electric Cables Part 2: Methods Specific to Elastomeric Compounds Section 1: Ozone Resistance Test—Hot Set Test—Mineral Oil Immersion Test First Edition; (Amendment 1-1992) (Amendment 2-1993). 39 pp.

—**Oil—Electric Conductors**
DIN VDE 0472 Pt 803-86. Testing of Cables, Wires and Flexible Cords; Oil Immersion (Apr). 7 pp.

—**Plastic Pipes**
BSI BS 2782:Pt11: METH 1102A-81. 1981 Methods of Testing Plastics Part 11: Thermoplastic Pipes, Fittings and Valves Method 1102A: Longitudinal Reversion of Pipes: Immersion Bath Method. 2 pp.
DIN ENGL 16888 Pt 1-89. Determination of Chemical Resistance of Thermoplastics Pipes; Polyolefin Pipes (June). 7 pp.
ISO 4433-84. Polyolefin Pipes—Resistance to Chemical Fluids—Immersion Test Method—System for Preliminary Classification First Edition. 16 pp.

—**Plastics**
BSI BS 2782:Pt6: METH 620A-D-91. 1991 Methods of Testing Plastics Part 6: Dimensional Properties Method 620A to 620D: Determination of Density and Relative Density of Non-Cellular Plastics (ISO 1183: 1987. 13 pp.
BSI BS 2782:Pt8: METH 830A-86. 1986 Methods of Testing Plastics Part 8: Other Properties Method 830A: Determination of the Effects of Liquid Chemicals, Including Water. 15 pp.
BSI BS 4618: Sec 4.1-72. 1972 Recommendations for the Presentation of Plastics Design Data Part 4: Enviromental and Chemical Effects Section 4.1: Chemical Resistance to Liquids. 12 pp.
ISO 175-81. Plastics—Determination of the Effects of Liquid Chemicals, Including Water First Edition. 14 pp.

—**Sea Water—Paints**
CGSB 1-GP-71 METH 113.2-79. Methods of Testing Paints and Pigments Corrosion Resistance to Synthetic Sea Water. 1 p.

—**Sea Water—Pigments**
CGSB 1-GP-71 METH 113.2-79. Methods of Testing Paints and Pigments Corrosion Resistance to Synthetic Sea Water. 1 p.

—**Solvent**
CNS C6338-90. Basic Environmental Testing Procedures Part 2: Test, Test XA and Guidance: Immersion in Cleaning Solvents (Dec)(12817).

—**Solvent—Electrical Components**
BSI BS 2011: Part 2.1XA-81. 1981 Basic Environmental Testing Procedures Part 2.1: Tests Part 2.1XA: Tests. Test XA and Guidance. Immersion in Cleaning Solvents (Renumbered as BS EN 60068-2-45: 1993). 10 pp.
BSI BS EN 60068-2-45-93. 1993 Amd 1 Environmental Testing Part 2.1: Tests Test XA and Guidance. Immersion in Cleaning Solvents (IEC 68-2-45: 1980) (AMD 7553) April 15, 1993. 19 pp.
CENELEC HD 323.2.45 S1-88. Basic Environmental Testing Procedures-Part 2: Tests. Test XA and Guidance: Immersion in Cleaning Solvents. 3 pp.
CENELEC EN 60068-2-45-92. Basic Safety Publication—Environmental Testing Part 2: Test Methods Test XA and Guidance: Immersion in Cleaning Solvents. 5 pp.
CENELEC EN 60068-2-45-92. Basic Environmental Testing Procedures Part 2: Tests Test XA and Guidance: Immersion in Cleaning Solvents. 5 pp.
CENELEC EN 60068-2-45-92. Basic Safety Publication—Environmental Testing Part 2: Test Methods Test XA and Guidance: Immersion in Cleaning Solvents (IEC 68-2-45: 1980 + Corrigendum 1981). 12 pp.
IEC 68 Pt 2-45-80. Basic Environmental Testing Procedures Part 2: Tests Test XA and Guidance: Immersion in Cleaning Solvents First Edition; (Amendment 1-1993) (CENELEC EN 60068-2-45: 1992). 30 pp.
SNZ IEC 68: Part 2-45-80. Basic Environmental Testing Procedures Part 2-45: Test XA and Guidance: Immersion in Cleaning Solvents. 20 pp.

—**Solvent—Electronic Components**
JIS C 0052-86. Testing Method of Resistance to Solvents (Immersion in Cleaning Solvents). 12 pp.

Immersion Testing (Cont.)
—**Solvent—Plastics**
BSI BS 4443: Part 4-89. 1989 Methods of Test for Flexible Cellular Materials Part 4: Method 10. Determination of Solvent Swelling Method 11. Humidity Ageing at an Elevated Temperature Method 12. Heat Ageing. 12 pp.
BSI BS 4443: Part 4-76. 1976 Methods of Test for Flexible Cellular Materials Part 4: Method 10. Determination of Solvent Swelling Method 11. Humidity Ageing at an Elevated Temperature Method 12. Heat Ageing. 7 pp.

—**Uranium Dioxide Pellets**
ISO 9278-92. Uranium Dioxide Pellets—Determination of Density and Amount of Open and Closed Porosity—Boiling Water Method and Penetration Immersion Method First Edition. 11 pp.

—**Vulcanized Rubber**
BSI BS 903: Part A16-87. 1987 Methods of Testing Vulcanized Rubber Part A16: Determination of the Effect of Liquids. 16 pp.
CNS K6353-86. Method of Test for Immersion of Vulcanized Rubber (Jan)(3562).
ISO 1817-85. Rubber, Vulcanized—Determination of the Effect of Liquids Second Edition. 13 pp.
SAA AS 1683.23-83. Methods of Test for Elastomers—Part 23: Rubber—Vulcanized—Determination of Resistance to Liquids. 8 pp.

—**Water—Bitumens**
DIN ENGL 1996 Pt 10-77. Testing of Bituminous Materials for Road Building and Related Purposes; Testing of Behaviour of Mix in Respect of Immersion in Water (Dec). 3 pp.

—**Water—Bituminous Coatings**
DIN ENGL 52006 Pt 1-80. Testing of Bituminous Binders; Effect of Water on Binder Coatings; Bitumen Emulsion Binder Coating (Dec). 2 pp.
DIN ENGL 52006 Pt 2-80. Testing of Bituminous Binders; Effect of Water on Binder Coatings; Rapid Curing Cutback or Cold Tar Binder Coating (Dec). 2 pp.
DIN ENGL 52006 Pt 3-85. Bitumen and Coal Tar Pitch; Effect of Water on Binder Coatings; Testing of Fluxed Bitumen Binder Coatings (Sept). 2 pp.

—**Water—Connectors**
SBAC TS 344 ISSUE 1. Test for Electrical Connectors Immersion at Low Air Pressure.

—**Water—Electrical Components**
IEC 68 Pt 2-18-89. Environmental Testing Part 2: Tests—Test R and Guidance: Water First Edition; (Corrigendum—May 1991) (Amendment 1-1993). 96 pp.

—**Water—Fiber Board—Corrugated—Bonding Strength**
ISO 3038-75. Corrugated Fibreboard—Determination of the Water Resistance of the Glue Bond by Immersion First Edition. 6 pp.

—**Water—Fishing Nets—Dimensional Stability**
BSI BS 5171-74. 1974 Netting Yarns: Determination of Change in Length After Immersion in Water. 8 pp.
ISO 3090-74. Netting Yarns—Determination of Change in Length After Immersion in Water First Edition. 4 pp.

—**Water—Greases**
DIN ENGL 51807 Pt 1-79. Testing of Lubricants; Test of the Behaviour of Lubricating Greases in the Presence of Water; Static Test (Apr). 5 pp.

—**Water—Joint Sealants**
ISO 10590-91. Building Construction—Sealants—Determination of Adhesion/Cohesion Properties at Maintained Extension After Immersion in Water First Edition. 6 pp.
ISO 10591-91. Building Construction—Sealants—Determination of Adhesion/Cohesion Properties After Immersion in Water First Edition. 6 pp.

—**Water—Packaging**
BSI BS 4826: Part 13-87. 1987 Complete Filled Transport Packages Part 13: Method for Determination of Resistance to Immersion in Water (Renumbered as BS EN 28474: 1993). 4 pp.
BSI BS EN 28474-93. 1993 Amd 1 Complete, Filled Transport Packages. Method for Determination of Resistance to Immersion in Water (ISO 8474: 1986) (AMD 7464) March 15, 1993 (H). 11 pp.
CEN EN 28474-92. Packaging—Complete, Filled Transport Packages—Water Immersion Test (ISO 8474: 1986). 5 pp.
ISO 8474-86. Packaging—Complete, Filled Transport Packages—Water Immersion Test First Edition; (CEN EN 28474: 1992). 4 pp.

INTERNATIONAL AND NON-U.S. NATIONAL STANDARDS
SUBJECT INDEX
Impact

Immersion Testing *(Cont.)*
—**Water—Paints**
BSI BS 3900: Part G8-93. 1993 Methods of Test for Paints Part G8: Determination of Resistance to Liquids. Water Immersion Method (ISO 2812-2: 1993) (W). 8 pp.
ISO 1521-73. Paints and Varnishes—Determination of Resistance to Water—Water Immersion Method First Edition. 4 pp.
ISO 2812 Pt 2-93. Paints and Varnishes—Determination of Resistance to Liquids—Part 2: Water Immersion Method First Edition. 7 pp.

—**Water—Petroleum Products**
ISO 7120-87. Petroleum Products and Lubricants—Petroleum Oils and Other Fluids—Determination of Rust-Preventing Characteristics in the Presence of Water First Edition. 12 pp.

—**Water—Plastics**
BSI BS 2782:Pt6: METH 620A-D-91. 1991 Methods of Testing Plastics Part 6: Dimensional Properties Method 620A to 620D: Determination of Density and Relative Density of Non-Cellular Plastics (ISO 1183: 1987. 13 pp.

—**Water—Refractory Materials**
CNS R3014-83. Method of Test for True Specific Gravity of Refractory Material by Water Immersion (Apr) (620).

—**Water—Sealants**
CGSB CAN2-19.0-M77METH14.1-78. Methods of Testing Putty, Caulking and Sealing Compounds Tensile Extension After Heat Aging and Water Immersion. 6 pp.

—**Water—Varnishes**
BSI BS 3900: Part G8-93. 1993 Methods of Test for Paints Part G8: Determination of Resistance to Liquids. Water Immersion Method (ISO 2812-2: 1993) (W). 8 pp.
ISO 1521-73. Paints and Varnishes—Determination of Resistance to Water—Water Immersion Method First Edition. 4 pp.
ISO 2812 Pt 2-93. Paints and Varnishes—Determination of Resistance to Liquids—Part 2: Water Immersion Method First Edition. 7 pp.

Immersion Testing(Corrosion)
Use: Immersion Testing

Immunity to Interference (Electromagnetic Compatibility)
Use: Electromagnetic Compatibility

Immunization
Use For: Vaccination
—**Military Operations**
NATO STANAG 2037 ED 6 AMD 0-88. Vaccination of NATO Forces. 51 pp.

Immunoglobulins
Use: Globulins

Immunological Products
Use: Medicinal Products—Immunological

Impact Avalanche and Transit Time Diodes
Use: IMPATT Diodes

Impact Drills
See Also: Drills
—**Safety**
IEC 745 Pt 2-1-89. Safety of Hand-Held Motor-Operated Electric Tools Part 2: Particular Requirements for Drills First Edition; (Amendment 1-1992). 37 pp.

Impact Effects (Human Body)
Use: Biomechanical Effects (Human Body)

Impact Extruding
Scope Note: See the subheading Extruded under the specific metal or alloy

Impact Loads
Use: Impact Shock

Impact Printers
See Also: Line Printers; Printers
—**Continuous Forms**
BSI BS 4623-89. 1989 Folded Continuous Stationery for Impact Printers. 11 pp.

Impact Printers *(Cont.)*
—**Dot Matrix**
CNS C6290-87. Method of Test for Serical Impact Dot Matrix Printer (Jul)(12010).

Impact Resistance
Use: Impact Testing

Impact Shock
Use For: Impact Loads; Landing Impact
See Also: Impact Testing; Impact Velocity
—**Walls**
CEN PREN 596-91. Timber Structures—Timber Framed Walls—Soft Body Impact Tests. 9 pp.
—**Water—Ground Effect Machines**
CAA Chapter B4-2 App 01.91. Loading Cases. 3 pp.

Impact Strength Testing
Use: Impact Testing

Impact Testing
Use For: Impact Strength Testing; Tumbling Drum Testing *See Also:* Bend Testing; Brittleness Testing; Collision Research; Compression Testing; Deceleration Testing; Ductility; Dynamic Testing; Fatigue Testing; Hardness Testing; High Temperature Testing; Hydrostatic Testing; Impact Shock; Impact Testing Equipment; Impact Velocity; Low Temperature Testing; Mechanical Shock; Puncture Resistance; Shear Testing

—**Adhesive Bonding**
ISO 11343-93. Adhesives—Determination of Dynamic Resistance to Cleavage of High Strength Adhesive Bonds Under Impact Conditions—Wedge Impact Method First Edition. 9 pp.

—**Adhesives**
BSI BS 5350: Part C4-86. 1986 Methods of Test for Adhesives Group C: Adhesively Bonded Joints: Mechanical Tests Part C4: Determination of Impact Resistance of Adhesive Bonds. 6 pp.
CNS K6506-80. Methods of Test for Impact Shear Strength of Adhesives (Jul)(5810).
JIS K 6855-77. Testing Methods for Impact Shear Strength of Adhesives (R 1986). 8 pp.

—**Aggregates**
BSI BS 812: Part 112-90. 1990 Testing Aggregates Part 112: Methods for Determination of Aggregate Impact Value (AIV). 12 pp.

—**Automotive**
BSI BS AU 228: Part 1-89. 1989 Impact Tests on Road Vehicles Part 1: Measurement Techniques and General Instrumentation. 8 pp.
BSI BS AU 228: Part 2-89. 1989 Impact Tests on Road Vehicles Part 2: Measurement Techniques Using Optical Instrumentation. 6 pp.
ISO 6487-87. Road Vehicles—Measurement Techniques in Impact Tests—Instrumentation Second Edition. 8 pp.
ISO 8721-87. Road Vehicles—Measurement Techniques in Impact Tests—Optical Instrumentation First Edition. 6 pp.
JIS D 1050-86. Road Vehicles—Techniques of Measurement in Impact Tests—Instrumentation. 7 pp.

—**Automotive—Bumpers**
CNS D3158-83. Method of Impact Test for Bumper of Passenger Cars (Dec)(10703).

—**Automotive—Steering Gear**
EC 74/297/EEC-74. Council Directive on the Approximation of the Laws of the Member States Relating to the Interior Fittings of Motor Vehicles (the Behaviour of the Steering Mechanism in the Event of an Impact). 10 pp.
JIS D 1061-85. Steering Control System Laboratory Impact Test Procedure for Passenger Cars.

—**Bottles**
CNS Z6057-82. Method of Drop Testing of Glass Aerosol Bottles (Nov)(9650).

—**Bottles—Carbonated Beverages**
CNS S2044-83. Method of Impact Test for Carbonated Beverage Bottles (Jun)(6530).
JIS S 2303-78. Method of Impact Test for Carbonated Beverage Bottles.
JIS S 2303-74. Method of Impact Test for Carbonated Beverage Bottles. 5 pp.

—**Building Materials—Boards**
CNS A3178-83. Method of Impact Test for Boards of Buildings (Feb)(9961).
JIS A 1421-81. Method of Impact Test for Boards of Buildings. 9 pp.

Impact Testing *(Cont.)*
—**Butt Welded Joints**
CEN PREN 875-92. Welding—Welded Joints in Metallic Materials—Specimen Location and Notch Orientation for Impact Tests. 8 pp.
JIS Z 3128-86. Method of Impact Test for Welded Joint. 7 pp.

—**Cadmium Coatings**
CNS H2048-79. Dropping Test for Electroplated Coatings of Cadmium (Apr)(4830).

—**Cans**
CNS Z6070-83. Method of Test for Coating Materials of Metal Cans for Foods—Impact Bent Test of Lacquered Film (Sep)(10586). 2 pp.

—**Cathode Ray Tubes**
CSA CAN/CSA-C22. 2 NO 228-92. Cathode Ray Tubes (UL 1418); (Gen Instr 1 Thru 2). 48 pp.

—**Coated Fabrics**
ISO 4646-89. Rubber-or Plastics-Coated Fabrics—Low-Temperature Impact Test Second Edition. 9 pp.

—**Coatings**
CGSB 1-GP-71 METH 147.1-74. Methods of Testing Paints and Pigments Impact Resistance. 1 p.
CGSB 1-GP-71 METH 147.2-78. Methods of Testing Paints and Pigments Impact Resistance Falling Ball Method. 2 pp.

—**Coatings—Bicycles**
CNS B7070-75. Method of Test for Coating of Bicycle Parts (Dec)(3883). 1 p.

—**Connectors**
CNS C6182-88. Method of Test for Low Frequency (Below 3 MHz) Electrical Connectors (TP-42 Impact) (May)(9242).

—**Containers—Glass**
DIN ENGL 52295-83. Testing of Glass; Pendulum Impact Test on Containers; Inspection by Attributes and by Variables (Sept). 4 pp.

—**Control Cables (Electric)**
CSA CAN/CSA-C22. 2 NO 239-M91. Control and Instrumentation Cables; (Gen Instr 1 Thru 3). 50 pp.

—**Crushed Stone**
DIN ENGL 52115 Pt 2-88. Determination of Impact Resistance of Mineral Aggregates; Testing of Crushed Stone (Aug) (This Standard, Together with DIN 52115 Part 1, August 1988 Edition, Supersedes Parts of DIN 52109, October 1939 Edition, Withdrawn in 1985). 2 pp.

—**Doors**
CEN EN 85-80. Methods of Testing Doors: Hard Body Impact Test on Door Leaves. 4 pp.
CEN EN 162-85. Methods of Testing Doors: Soft and Heavy Body Impact Test on Door Leaves. 6 pp.
CEN PREN 949-2-92. Resistance to Soft and Heavy Body Impact—Test Method—Part 2: Hinged, Pivoted or Sliding Doors. 6 pp.
CEN PREN 950-2-92. Resistance to Hard Body Impact—Test Method—Part 2: Door Leaves. 6 pp.
ISO 8270-85. Doorsets—Soft Heavy Body Impact Test First Edition. 4 pp.
ISO 8271-85. Door Leaves—Hard Body Impact Test First Edition. 4 pp.
JIS A 1518-86. Soft Heavy Body Impact Test for Doors and Doorsets. 9 pp.

—**Elastomers**
DIN ENGL 53512-88. Determination of Rebound Resilience of Rubber (Dec). 4 pp.

—**Electric Wire**
DIN VDE 0472 Pt 611-85. Testing of Cables and Insulated Flexible Cords; Cold Impact Test (Jan). 7 pp.

—**Electrical Components**
BSI BS 2011: Part 2.1EC-77. 1977 Amd 1 Basic Environmental Testing Procedures Part 2.1: Tests Part 2.1EC: Test Ec. Drop and Topple, Primarily for Equipment-Type Specimens (Renumbered as BS EN 68-2-31: 1969). 7 pp.
BSI BS 2011: Part 2.1ED-92. 1992 Amd 2 Environmental Testing Part 2.1: Tests Test Ed: Free Fall (IEC 68-2-32: 1975) (G). 13 pp.
BSI BS 2011: Part 2.1ED-77. 1977 Amd 1 Basic Environmental Testing Procedures Part 2.1: Tests Part 2.1Ed: Test Ed. Free Fall (Renumbered as BS EN 60068-2-32: 1993). 11 pp.
BSI BS 2011: Part 2.1EF-92. 1992 Environmental Testing Part 2.1: Tests Test Ef: Impact, Pendulum Hammer (IEC 68-2-62: 1991) (G). 15 pp.

INDUSTRY STANDARDS

Impact Testing (Cont.)

—Electrical Components (Cont.)
BSI BS EN 60068-2-31-93. 1993 Amd 2 Environmental Testing Part 2.1: Tests Test Ec. Drop and Topple, Primarily for Equipment-Type Specimens (IEC 68-2-31: 1969) (AMD 7921) September 15, 1993 (G). 11 pp.
BSI BS EN 60068-2-32-93. 1993 Amd 1 Environmental Testing Part 2.1: Tests Test Ed. Free Fall (IEC 68-2-32: 1975) (AMD 7920) September 15, 1993 (G). 16 pp.
CENELEC HD 323.2.31 S1-88. Basic Environmental Testing Procedures—Part 2: Tests. Test Ec: Drop and Topple, Primarily for Equipment-Type Specimens (Superseded by EN 60068-2-31: 1993). 3 pp.
CENELEC HD 323.2.32 S1-88. Basic Environmental Testing Procedures—Part 2: Tests. Test Ed: Free Fall. 3 pp.
CENELEC EN 60068-2-31-93. Basic Environmental Testing Procedures Part 2: Tests Test Ec: Drop and Topple, Primarily for Equipment-Type Specimens (IEC 68-2-31: 1969 + A1: 1982) (Supersedes HD 323.2.31 S1: 1988). 8 pp.
CENELEC EN 60068-2-32-93. Basic Environmental Testing Procedures Part 2: Tests Test Ed: Free Fall (IEC 68-2-32: 1975 + A1: 1982 + A2: 1990) (Supersedes HD 323.2.32 S2: 1991). 12 pp.
IEC 68 Pt 2-31-69. Basic Environmental Testing Procedures Part 2: Tests—Test Ec: Drop and Topple, Primarily for Equipment-Type Specimens First Edition; (Amendment 1-1982) (CENELEC EN 60068-2-31: 1993). 13 pp.
IEC 68 Pt 2-32-75. Basic Environmental Testing Procedures Part 2: Tests Test Ed: Free Fall Second Edition; (Amendment 2-1990; Incorporating Amendment 1) (CENELEC EN 60068-2-32: 1993). 31 pp.
IEC 68 Pt 2-62-91. Environmental Testing Part 2: Test Methods Test Ef: Impact, Pendulum Hammer First Edition. 26 pp.
JIS C 0043-90. Basic Environmental Testing Procedures Part 2: Tests-Ec: Drop and Topple, Primarily for Equipment-Type Specimens. 10 pp.
JIS C 0044-90. Basic Environmental Testing Procedures Part 2: Tests-Test Ed: Free Fall. 10 pp.
SAA AS 1099.2ED-78. Basic Environmental Testing Procedures for Electrotechnology—Part 2: Tests—Part 2Ed: Free Fall Reconfirmed 1985. 3 pp.
SAA AS 1099.2.31-90. Basic Environmental Testing Procedures for Electrotechnology—Part 2: Tests—Part 2.31: Test Ec—Drop and Topple, Primarily for Equipment (IEC 68-2-31). 6 pp.
SNZ IEC 68: Part 2-31-69. Basic Environmental Testing Procedures Part 2-31: Test Ec: Drop and Topple, Primarily for Equipment-Type Specimens Amend: 1. 7 pp.
SNZ IEC 68: Part 2-32-75. Basic Environmental Testing Procedures Part 2-32: Test Ed: Free Fall Amend: 1. 12 pp.

—Electrical Insulation
CNS K6724-82. Method of Test for Impact Resistance of Plastics and Electrical Insulating Materials (Aug)(9284).

—Electromechanical Components
BSI BS 5772: Part 5-93. 1993 Electromechanical Components for Electronic Equipment: Testing Procedures and Measuring Methods Part 5: Impact Tests (Free Components), Static Load Tests (Fixed Components), Endurance Tests and Overload Tests (IEC 512-5: 1992) (S). 24 pp.
BSI BS 5772: Part 5-79. 1979 Basic Testing Procedures and Measuring Methods for Electromechanical Components for Electronic Equipment Part 5: Impact Tests (Free Components), Static Load Tests (Fixed Components), Endurance Tests and Overload Tests (Switches). 16 pp.
BSI BS 5772: Pt 5: Supp 1-81. 1981 Basic Testing Procedures and Measuring Methods for Electromechanical Components for Electronic Equip. Part 5: Impact Tests (Free Components), Static Load Tests (FixedComponents) Endurance Test and Over-Load Tests (Switches) Supp 1: Tests 7b and 10a. 6 pp.
BSI BS 5772: Pt 5: Supp 2-82. 1982 Basic Testing Procedures and Measuring Methods for Electromechanical Components for Electronic Equip. Part 5: Impact Tests (Free Components), Static Load Tests (FixedComp.), Enduranceand Overload Tests (Switches) Supp. 2:Test 10c. 6 pp.
IEC 512 Pt 5-92. Electromechanical Components for Electronic Equipment; Basic Testing Procedures and Measuring Methods Part 5: Impact Tests (Free Components), Static Load Tests (Fixed Components), Endurance Tests and Overload Tests Second Edition. 50 pp.

—Enamels
BSI BS 1344: Part 21-93. 1993 Methods of Testing Vitreous Enamel Finishes Part 21: Determination of the Resistance of Vitreous Enamelled Articles to Impact: Pistol Test (ISO 4532: 1991) (F). 11 pp.

Impact Testing (Cont.)

—Enamels (Cont.)
ISO 4532-91. Vitreous and Porcelain Enamels—Determination of the Resistance of Enamelled Articles to Impact—Pistol Test First Edition. 8 pp.

—Eye Protectors
CEN PREN 168-89. Personal Eye Protection: Non-Optical Test Methods. 24 pp.

—Ferritic Steels
SAA AS 1663-74. Method for Dropweight Test for Nil-Ductility Transition Temperature of Ferritic Steels Reconfirmed 1989. 24 pp.

—Fiber Optic Cables
CNS C6223-84. Method of Test for Fiber Optic Devices (Impact Testing of Cable Assemblies) (May)(10878).

—Fiber Optic Equipment
CNS C6217-87. Method of Test for Fiber Optic Devices (FOTP—2 Impact Test) (Jun)(10872).

—Fiberglass Reinforced Plastics
JIS K 7061-92. Testing Method for Charpy Impact Strength of Glass Fiber Reinforced Plastics. 15 pp.
JIS K 7062-92. Testing Method of Izod Impact Strength of Glass Fiber Reinforced Plastics. 14 pp.

—Fire Alarm and Extinguishing Equipment
CEN EN 54-5-76. Components of Automatic Fire Detection Systems Part 5 Heat Sensitive Detectors—Point Detectors Containing a Static Element. 20 pp.
CEN EN 54 (Part 5)-88. AMD 1 Components of Automatic Fire Detection Systems Part 5 Heat Sensitive Detectors—Point Detectors Containing a Static Element. 21 pp.
CEN EN 54-7-88. Components of Automatic Fire Detection Systems—Part 7: Point-Type Smoke Detectors; Detectors Using Scattered Light, Transmitted Light or Ionization. 30 pp.
CEN EN 54-8-88. Components of Automatic Fire Detection Systems—Part 8: Hig Temperature Heat Detectors. 23 pp.
CEN EN 54-9-82. Components of Automatic Fire Fetection Systems Part 9 Fire Sensitivity Test. 10 pp.

—Fire Extinguishers
CEN EN 3-5-84. Portable Fire Extinguishers—Part 5: Complementary Requirements and Tests. 17 pp.
CEN PREN 3-5-93. Fight Against Fire—Portable Fire Extinguishers—Part 5: Complementary Requirements and Tests. 23 pp.

—Footwear
BSI BS 5131: Sec 4.8-90. 1990 Methods of Test for Footwear and Footwear Materials Part 4: Other Components Section 4.8: Resistance of Heels of Ladies' Shoes to Lateral Impact. 7 pp.
BSI BS 5131: Sec 5.6-91. 1991 Methods of Test for Footwear and Footwear Materials Part 5: Testing of Complete Footwear Section 5.6: Impact Test for Rigid Units and Shoe Bottoms. 9 pp.

—Furniture
BSI BS 4875: Part 1-85. 1985 Strength and Stability of Furniture Part 1: Methods for Determination of Strength of Chairs and Stools. 20 pp.
BSI BS 4875: Part 3-85. 1985 Strength and Stability of Furniture Part 3: Methods for Determination of Strength of Settees. 20 pp.
BSI BS 4875: Part 5-85. 1985 Strength and Stability of Furniture Part 5: Methods for Determination of Strength of Tables and Trolleys. 12 pp.
BSI BS 4875: Part 7-85. 1985 Strength and Stability of Furniture Part 7: Methods for Determination of Strength of Storage Furniture. 20 pp.
BSI BS 5873: Part 3-85. 1985 Amd 1 Educational Furniture Part 3: Specification for Strength and Stability of Tables for Educational Institutions (AMD 6325) May 31, 1991. 16 pp.
ISO 4211 Pt 4-88. Furniture—Surfaces—Part 4: Assessment of Resistance to Impact First Edition. 7 pp.
ISO 7173-89. Furniture—Chairs and Stools—Determination of Strength and Durability First Edition. 23 pp.

—Gravel
DIN ENGL 52115 Pt 3-88. Determination of Impact Resistance of Mineral Aggregates; Testing of Gravel (Aug). 2 pp.

—Gray Iron
ISO 946-75. Grey Cast Iron—Beam Unnotched Impact Test First Edition. 5 pp.

—Instrumentation Cables
CSA CAN/CSA-C22. 2 NO 239-M91. Control and Instrumentation Cables; (Gen Instr 1 Thru 3). 50 pp.

Impact Testing (Cont.)

—Iron—Sintered
JIS M 8711-87. Test Method for Determination of Shatter Strength of Iron Ore Sinter. 7 pp.

—Lacquers
CNS K6031-87. Method of Test for Lacquer Enamel (May) (628). 6 pp.
SAA AS 1580.406. 1-93. Paints and Related Materials—Methods of Test—Part 406.1: Resistance to Impact—Falling Weight Test (Gardener-Type Tester) (in Professional Package 39A). 2 pp.
SNZ NZS/AS 1580. 406.1-86. Methods of Test for Paints and Related Materials Part 406.1: Resistance to Impact—Falling Weight Test (Gardner-Type Tester) (This is a Joint Standard with SAA AS 1580.406.1). 3 pp.

—Laminated Glass
DIN ENGL 52338-85. Methods of Testing Flat Glass for Use in Buildings; Ball Drop Test on Laminated Glass (Sept). 3 pp.

—Machine Guards—Saws
ISO 6534-92. Portable Chain-Saws—Hand-Guards—Mechanical Strength Second Edition. 6 pp.
ISO 8380-85. Forestry Machinery—Portable Brush-Saws—Circular Saw-Blade Guard—Strength First Edition. 4 pp.

—Metal Coatings (Made From Metal)
JIS Z 3111-86. Methods of Tension and Impact Test for Deposited Metal.
JIS Z 3111-70. Method of Tension Test for Deposited Metal. 4 pp.

—Metals
BSI BS 131: Part 1-61. 1961 Amd 2 Methods for Notched Bar Tests Part 1: The Izod Impact Test on Metals. 24 pp.
BSI BS 131: Part 2-72. (WITHDRAWN) 1972 Methods for Notched Bar Tests Part 2: The Charpy V-Notch Impact Test on Metals (Superseded by BS EN 10 045-1: 1990). 13 pp.
BSI BS 131: Part 3-72. (WITHDRAWN) 1972 Methods for Notched Bar Tests Part 3: The Charpy U-Notch Impact Test on Metals (Superseded by BS EN 10 045-1: 1990). 13 pp.
BSI BS 131: Part 6-89. 1989 Notched Bar Tests Part 6: Method for Precision Determinations of Charpy V-Notch Impact Energies for Metals. 10 pp.
BSI BS EN 10045-1-90. 1990 Charpy Impact Test on Metallic Materials Part 1: Test Method (V-and U-Notches) (V). 12 pp.
BSI 4A 4-66. 1966 Amd 1 Test Pieces and Test Methods for Metallic Materials for Aircraft. 13 pp.
CEN EN 10 045 (Part 1)-90. Metal Products, Mechanical Tests, Charpy Impact Tests, Test Specimens, Testing Conditions. 14 pp.
CNS G2022-84. Test Pieces for Metallic Materials (Aug)(3033).
CNS G2023-84. Method of Impact Test for Metallic Materials (Aug)(3034). 7 pp.
DIN ENGL 50115-91. Notched Bar Impact Testing of Metallic Materials Using Test Pieces Other Than ISO Test Pieces (Apr) (This Standard, Together with the April 1991 Edition of EN 10 045 Part 1, Supersedes February 1975 Edition). 3 pp.
DIN ENGL EN 10045 Pt 1-91. Charpy Impact Test on Metallic Materials; Test Method (Apr) (This Standard, Together with DIN 50115, April 1991 Edition, Supersedes DIN 50115, February 1975 Edition). 9 pp.
JIS Z 2202-80. Test Pieces for Impact Test for Metallic Materials. 6 pp.
JIS Z 2242-80. Method of Impact Test for Metallic Materials. 11 pp.
SAA AS 1544.1-77. Methods for Impact Tests on Metals—Part 1: IZOD (R 1993). 24 pp.
SAA AS 1544.2-89. Methods for Impact Tests on Metals—Part 2: Charpy V-Notch. 9 pp.
SAA AS 1544.3-75. Methods for Impact Tests on Metals—Part 3: Charpy U-Notch and Keyhole Notch (R 1993). 18 pp.

—Metals—Crystal Structure
BSI BS 131: Part 5-65. 1965 Methods for Notched Bar Tests Part 5: Determination of Crystallinity. 12 pp.

—Metals—Sintered
BSI BS EN 25754-93. 1993 Sintered Metal Materials, Excluding Hardmetals—Unnotched Impact Test Piece (ISO 5754: 1978) (V). 9 pp.
CEN EN 25754-93. Sintered Metal Materials, Excluding Hardmetals—Unnotched Impact Test Piece (ISO 5754: 1978). 4 pp.
ISO 5754-78. Sintered Metal Materials, Excluding Hardmetals—Unnotched Impact Test Piece First Edition (CEN EN 25754: 1993). 3 pp.

INTERNATIONAL AND NON-U.S. NATIONAL STANDARDS
SUBJECT INDEX
Impact

Impact Testing (Cont.)

—Mineral Aggregates
DIN ENGL 52115 Pt 1-88. Determination of Impact Resistance of Mineral Aggregates; Impact Testing Machine (Aug) (This Standard, Together with DIN 52115 Part 2, August 1988 Edition, Supersedes Parts of DIN 52109, October 1939 Edition, Withdrawn in 1985). 9 pp.

—Packaging
BSI BS 4826: Part 4-86. (WITHDRAWN) 1986 Complete, Filled Transport Packages Part 4: Method for Determination of Resistance to Vertical Impact by Dropping (Renumbered as BS EN 22248: 1993). 7 pp.

BSI BS 4826: Part 5-86. (WITHDRAWN) 1986 Complete, Filled Transport Packages Part 5: Method for Determination of Resistance to Horizontal Impact (Renumbered as BS EN 22244: 1993). 7 pp.

BSI BS 4826: Part 14-87. (WITHDRAWN) 1987 Complete Filled Transport Packages Part 14: Method for Determination of Resistance to Damage by Toppling (Renumbered as BS EN 28768: 1993). 8 pp.

BSI BS EN 22244-93. 1993 Amd 1 Complete, Filled Transport Packages. Method for Determination of Resistance to Horizontal Impact (ISO 2244: 1985) March 15, 1993 (H). 13 pp.

BSI BS EN 22248-93. 1993 Complete, Filled Transport Packages. Method for Determination of Resistance to Vertical Impact by Dropping (ISO 2248: 1985) (AMD 7456) March 15, 1993 (H). 13 pp.

BSI BS EN 28768-93. 1993 Amd 1 Complete, Filled Transport Packages. Method for Determination of Resistance to Damage by Toppling (ISO 8768: 1986) (AMD 7465) March 15, 1993 (H). 14 pp.

CEN EN 22244-92. Packaging—Complete, Filled Transport Packages—Horizontal Impact Tests (Horizontal or Inclined Plane Test—Pendulum Test) (ISO 2244: 1985). 5 pp.

CEN EN 22248-92. Packaging—Complete, Filled Transport Packages—Vertical Impact Test by Dropping (ISO 2248: 1985). 5 pp.

CEN EN 28768-92. Packaging—Complete, Filled Transport Packages—Toppling Test (ISO 8768: 1986). 6 pp.

CNS Z6012-76. Method of Drop Test for Packing and Shipping Containers (Dec)(2999).

ISO 2244-85. Packaging—Complete, Filled Transport Packages—Horizontal Impact Tests (Horizontal or Inclined Plane Test; Pendulum Test); (CEN EN 22244: 1992). 5 pp.

ISO 2248-85. Packaging—Complete, Filled Transport Packages—Vertical Impact Test by Dropping Second Edition; (CEN EN 22248: 1992). 5 pp.

ISO 8768-86. Packaging—Complete, Filled Transport Packages—Toppling Test First Edition; (CEN EN 28768: 1992). 6 pp.

JIS Z 0200-87. General Rules of Performance Testing for Packaged Freights. 12 pp.

JIS Z 0202-87. Method of Drop Test for Packages and Shipping Containers.

JIS Z 0205-61. Method of Incline Impact Test for Package and Shipping Container. 6 pp.

JIS Z 0209-76. Method of Revolving Hexagonal Drum Test for Package and Shipping Containers (R 1983). 11 pp.

—Paints
BSI BS 3900: Part E3-73. 1973 Methods of Test for Paints Part E3: Impact (Falling Weight) Resistance. 5 pp.

BSI BS 3900: Part E3-01. 1973 Amd 1 Methods of Test for Paints Part E3: Impact (Falling Weight) Resistance (AMD 6907) February 28, 1992 (W). 6 pp.

BSI BS 3900: Part E7-74. 1974 Amd 1 Methods of Test for Paints Part E7: Resistance to Impact (Falling Ball Test). 3 pp.

BSI BS 3900: Part E7-02. 1974 Amd 2 Methods of Test for Paints Part E7: Resistance to Impact (Falling Ball Test) (AMD 6909) February 28, 1992 (W). 4 pp.

BSI BS 3900: Part E8-74. 1974 Methods of Test for Paints Part E8: Resistance to Impact (Pendulum Test) (W). 4 pp.

BSI BS 3900: Part E8-01. 1974 Amd 1 Methods of Test for Paints Part E8: Resistance to Impact (Pendulum Test) (AMD 6908) February 28, 1992 (W). 5 pp.

BSI BS 3900: Part E13-93. 1993 Methods of Test for Paints Part E13: Falling Weight Resistance (Impact) (ISO 6272: 1993) (W). 11 pp.

DIN ENGL 55995 (P)-83. Testing of Paints, Varnishes and Related Products; Test for Impact Resistance to Grit and Gravel of Paints, Varnishes and Related Products Using Single-Impact Apparatus (Aug). 8 pp.

ISO 6272-93. Paints and Varnishes—Falling-Weight Test First Edition. 9 pp.

Impact Testing (Cont.)

—Paints (Cont.)
SAA AS 1580.406. 1-93. Paints and Related Materials—Methods of Test—Part 406.1: Resistance to Impact—Falling Weight Test (Gardner-Type Tester) (in Professional Package 39A). 2 pp.

SNZ NZS/AS 1580. 406.1-86. Methods of Test for Paints and Related Materials Part 406.1: Resistance to Impact—Falling Weight Test (Gardner-Type Tester) (This is a Joint Standard with SAA AS 1580.406.1). 3 pp.

—Pallets
CNS Z6041-81. Method of Test for Flat Pallet (Nov)(8170).

JIS Z 0602-88. Test Methods for Flat Pallets. 10 pp.

—Paper Bags
BSI BS 6838: Part 1-87. (WITHDRAWN) 1987 Testing of Sacks Part 1: Method for Determination of Resistance of Paper Sacks to Vertical Impact by Dropping (Superseded by BS EN 27965-1: 1993). 12 pp.

BSI BS EN 27965-1-93. 1993 Packaging—Sacks—Drop Test Part 1: Paper Sacks (ISO 7965/1: 1984) (Supersedes BS 6838: Part 1: 1987) (H). 15 pp.

CEN EN 27965-1-92. Packaging—Sacks—Drop Test—Part 1: Paper Sacks (ISO 7965-1: 1984). 10 pp.

CNS Z6077-83. Method of Test for Drop Kraft Paper Sack (Oct)(10635).

ISO 7965 Pt 1-84. Packaging—Sacks—Drop Test—Part 1: Paper Sacks First Edition; (CEN EN 27965-1:1992). 10 pp.

JIS Z 0217-63. Drop Testing Method for Kraft Paper Sacks.

—Particle Board
CEN PREN 1128-93. Cement-Bonded Particleboards—Determination of Impact Resistance. 8 pp.

—Phenol Formaldehyde
CNS K6273-74. Method of Test for Phenol Formaldehyde Resin Molding Compounds (Oct)(2988). 11 pp.

—Plastic Pipes
BSI BS 2782:Pt11: METH 1108A-89. 1989 Methods of Testing Plastics Part 11: Thermoplastics Pipes, Fittings and Valves Method 1108A: True Impact Rate (TIR) Boundaries of Pipes. 8 pp.

CNS K6641-81. Method of Test for Rigid Polyvinyl Chloride Corrugated Pipes (Mar)(7057). 6 pp.

ISO 3127-80. Unplasticized Polyvinyl Chloride (PVC) Pipes for the Transport of Fluids—Determination and Specification of Resistance to External Blows First Edition. 10 pp.

—Plastic Pipes—PVC—Unplasticized
SAA AS 1462.3-88. Methods of Test for Unplasticized PVC (UPVC) Pipes and Fittings—Part 3: Method for Determining the Impact Characteristics of UPVC Pipes. 6 pp.

—Plastic Sheets
BSI BS 2782:Pt3: METH 352D-79. 1979 Methods of Testing Plastics Part 3: Mechanical Properties Method 352D: Determination of Falling Weight Impact Resistance of Thin Flexible Sheet (Film). 8 pp.

CNS K6287-87. Method of Test for Polyvinyl Chloride Sheet (Jul)(3143). 7 pp.

CNS K6302-77. Method of Test for Polymethylmethacrylate Corrugated Sheet (Apr)(3273). 5 pp.

CNS K6306-71. Method of Test for Polymethylmethacrylate Corrugated Sheet (Impact Test) (Apr)(3277). 2 pp.

DIN ENGL 53373-70. Testing of Plastic Films; Impact Penetration Test with Electronic Data Recording (Sept). 4 pp.

ISO 7765 Pt 1-88. Plastics Film and Sheeting—Determination of Impact Resistance by the Free-Falling Dart Method Part 1: Staircase Methods First Edition. 8 pp.

JIS K 7124-87. Testing Method for Dart Impact Resistance of Plastic Film and Sheeting. 11 pp.

—Plastics
BSI BS 2782:Pt3: METH 350-93. 1993 Methods of Testing Plastics Part 3: Mechanical Properties Method 350: Determination of Izod Impact Strength (ISO 180: 1993) (V). 15 pp.

BSI BS 2782:Pt3: METH 350-84. 1984 Methods of Testing Plastics Part 3: Mechanical Properties Method 350: Determination of Izod Impact Strength of Rigid Materials. 12 pp.

BSI BS 2782:Pt3: METH 353A-91. 1991 Methods of Testing Plastics Part 3: Mechanical Properties Method 353A: Determination of Multi-Axial Impact Behaviour by the Falling Dart Method (ISO 6603-1: 1985) (V). 19 pp.

Impact Testing (Cont.)

—Plastics (Cont.)
BSI BS 2782:Pt3: METH 353B-91. 1991 Methods of Testing Plastics Part 3: Mechanical Properties Method 353B: Determination of Multi-Axial Impact Behaviour by the Instrumented Puncture Test (ISO 6603-2: 1989). 15 pp.

BSI BS 2782:Pt3: METH 354A-B-91. 1991 Methods of Testing Plastics Part 3: Mechanical Properties Methods 354A and 354B: Determination of Tensile-Impact Strength (ISO 8256: 1990) (V). 21 pp.

BSI BS 2782:Pt3: METH 354A-B-01. 1991 Amd 1 Methods of Testing Plastics Part 3: Mechanical Properties Methods 354A and 354B: Determination of Tensile-Impact Strength (ISO 8256: 1990) (AMD 6815) February 28, 1992 (V). 22 pp.

BSI BS 2782:Pt3: METH 359-93. 1993 Methods of Testing Plastics Part 3: Mechanical Properties Method 359: Determination of Charpy Impact Strength (ISO 179: 1993) (V). 17 pp.

BSI BS 2782:Pt3: METH 359-84. 1984 Methods of Testing Plastics Part 3: Mechanical Properties Method 359: Determination of Charpy Impact Strength of Rigid Materials (Charpy Impact Flexural Test). 12 pp.

BSI BS 4618: Sec 1.2-72. 1972 Recommendations for the Presentation of Plastics Design Data Part 1: Mechanical Properties Section 1.2: Impact Behaviour. 16 pp.

CEN PREN 987-92. Plastics Piping and Ducting Systems—Injection-Molded Fittings—Determination of the Charpy Impact Strength. 6 pp.

CNS K6527-80. Determination of the Charpy Impact Resistance of Rigid Plastics (Charpy Impact Flexural Test) (Jul)(5846).

CNS K6724-82. Method of Test for Impact Resistance of Plastics and Electrical Insulating Materials (Aug)(9284).

DIN ENGL 53435-83. Testing of Plastics; Bending Test and Impact Test on Dynstat Test Pieces (July). 4 pp.

DIN ENGL 53453-75. Testing of Plastics; Impact Flexural Test (May). 5 pp.

ISO 179-93. Plastics—Determination of Charpy Impact Strength Second Edition; (Corrected and Reprinted -1993). 16 pp.

ISO 180-93. Plastics—Determination of Izod Impact Strength Second Edition. 14 pp.

ISO 6603 Pt 1-85. Plastics—Determination of Multiaxial Impact Behaviour of Rigid Plastics—Part 1: Falling Dart Method First Edition. 16 pp.

ISO 6603 Pt 2-89. Plastics—Determination of Multiaxial Impact Behaviour of Rigid Plastics—Part 2: Instrumented Puncture Test First Edition. 15 pp.

ISO 8256-90. Plastics—Determination of Tensile-Impact Strength First Edition; (Corrigendum 1-1991). 21 pp.

JIS K 7110-84. Method of Izod Impact Test for Rigid Plastics. 31 pp.

JIS K 7111-84. Method of Charpy Impact Test for Rigid Plastics. 35 pp.

JIS K 7211-76. General Rules for Testing Impact Strength of Rigid Plastics by the Falling Weight Method (R 1979). 10 pp.

SAA AS 1146.1-90. Method for Impact Tests on Plastics—Part 1: Izod Impact Resistance of Rigid Plastics. 9 pp.

SAA AS 1146.2-90. Method for Impact Tests on Plastics—Part 2: Charpy Impact Resistance. 9 pp.

—Plastics—Safety
BSI BS 6206-81. 1981 Amd 2 Impact Performance Requirements for Flat Safety Glass and Safety Plastics for Use in Buildings. 14 pp.

BSI BS 6206-03. 1981 Amd 3 Impact Performance Requirements for Flat Safety Glass and Safety Plastics for Use in Buildings (AMD 7589) May 15, 1993 (R). 15 pp.

—Plywood
CNS O2044-82. Method of Test for Impact of Special Plywood (Mar)(8628). 1 p.

—Polystyrene
CNS K6395-76. Method of Test for Polystyrene Moulding Materials (Sep)(4013). 7 pp.

ISO 2897 Pt 2-81. Plastics—Impact-Resistant Polystyrenes—Part 2: Determination of Properties First Edition. 7 pp.

—Primers
CNS K6115-86. Method of Test for Oleoresinous Primer (Apr)(1249). 3 pp.

—PVC
CEN PREN 477-91. Unplasticized Polyvinylchloride (PVC-U) Profiles for the Construction of Windows—Determination of the Resistance to Impact by Falling Mass. 7 pp.

—Reinforced Plastics
JIS K 7077-91. Testing Method for Charpy Impact Strength of Carbon Fiber Reinforced Plastics. 11 pp.

INDUSTRY STANDARDS

INTERNATIONAL AND NON-U.S. NATIONAL STANDARDS
SUBJECT INDEX

Impact

Impact Testing *(Cont.)*

—Safety
IEC 68 Pt 2-63-91. Environmental Testing Part 2: Testing Methods Test Eg: Impact, Spring Hammer First Edition. 38 pp.

—Security Glazing
DIN ENGL 52290 Pt 4-88. Testing of Security Glazing for Impact Resistance (Nov). 3 pp.

—Sheet Glass
DIN ENGL 52337-85. Methods of Testing Flat Glass for Use in Buildings; Pendulum Impact Tests (Sept). 10 pp.

—Shipping Containers
CNS Z6007-76. Method of Inclined Impact Test for Packing and Shipping Containers (Dec)(2544).
CNS Z6012-76. Method of Drop Test for Packing and Shipping Containers (Dec)(2999).
JIS Z 0202-87. Method of Drop Test for Packages and Shipping Containers.
JIS Z 0205-61. Method of Incline Impact Test for Package and Shipping Container. 6 pp.
JIS Z 0209-76. Method of Revolving Hexagonal Drum Test for Package and Shipping Containers (R 1983). 11 pp.

—Ski Bindings
ISO 9465-91. Alpine Ski-Bindings—Lateral Release Under Impact Loading—Test Method First Edition. 9 pp.

—Steels
AECMA PREN2003-01-86. Test Methods for Steel Products—Part 1—Charpy Impact Test (U Notch).
CEN PREN 2003-01-91. Test Methods for Steel Products Part 1 Charpy Impact Test (U Notch). 1 p.
CEN PREN2003-02-75. Test Methods for Steel Products—Part 2—IZOD Impact Test (C5/24). 4 pp.
CEN PREN 2003-05-91. Test Methods for Steel Products Part 5 Charpy Impact Test (V Notch). 1 p.
ISO 83-76. Steel—Charpy Impact Test (U-Notch) First Edition; (DIN 50115). 5 pp.
ISO TR7705-91. Guidelines for Specifying Charpy V-Notch Impact Prescriptions in Steel Specifications Second Edition. 18 pp.

—Surfacing Materials—Playgrounds
BSI BS 7188-89. 1989 Amd 2 Methods of Test for Impact Absorbing Playground Surfaces (AMD 6855) September 30, 1991. 21 pp.

—Test Dummies
ISO TR9790 Pt 1-89. Road Vehicles—Anthropomorphic Side Impact Dummy—Part 1: Lateral Head Impact Response Requirements to Assess Biofidelity of Dummy First Edition. 10 pp.
ISO TR9790 Pt 2-89. Road Vehicles—Anthropomorphic Side Impact Dummy—Part 2—Lateral Neck Impact Response Requirements to Assess Biofidelity of Dummy First Edition. 21 pp.
ISO TR9790 Pt 3-89. Road Vehicles—Anthropomorphic Side Impact Dummy—Part 3: Lateral Thoracic Impact Response Requirements to Assess Biofidelity of Dummy First Edition. 30 pp.
ISO TR9790 Pt 4-89. Road Vehicles—Anthropomorphic Side Impact Dummy—Part 4: Lateral Shoulder Impact Response Requirements to Assess Biofidelity of Dummy First Edition. 15 pp.
ISO TR9790 Pt 5-89. Road Vehicles—Anthropomorphic Side Impact Dummy—Part 5: Lateral Abdominal Impact Response Requirements to Assess Biofidelity of Dummy First Edition. 24 pp.
ISO TR9790 Pt 6-89. Road Vehicles—Anthropomorphic Side Impact Dummy—Part 6: Lateral Pelvic Impact Response Requirements to Assess Biofidelity of Dummy First Edition. 33 pp.

—Toys
CNS Z8016-17-87. Method of Test for Toy Safety (Drop Test) (Dec)(4798-17). 2 pp.
CNS Z8016-24-87. Method of Test for Toy Safety (Impact Test for Projectiles Propelled by a Discharge Mechanism) (Dec) (4798-24). 1 p.
CNS Z8016-31-87. Method of Test for Toy Safety (Impact Test for Toys That Cover the Eyes) (Dec)(4798-31). 2 pp.

—Tractors
BSI BS 4063-73. (WITHDRAWN) 1973 Amd 1 Requirements and Testing of Protective Cabs and Frames for Agricultural Wheeled Tractors. 28 pp.

—Urea Resins
CNS K6272-74. Method of Test for Urea Resin Molding Compounds (Oct)(2986). 9 pp.

Impact Testing *(Cont.)*

—Varnishes
BSI BS 3900: Part E3-73. 1973 Methods of Test for Paints Part E3: Impact (Falling Weight) Resistance. 5 pp.
BSI BS 3900: Part E3-01. 1973 Amd 1 Methods of Test for Paints Part E3: Impact (Falling Weight) Resistance (AMD 6907) February 28, 1992 (W). 6 pp.
BSI BS 3900: Part E7-74. 1974 Amd 1 Methods of Test for Paints Part E7: Resistance to Impact (Falling Ball Test). 3 pp.
BSI BS 3900: Part E7-02. 1974 Amd 2 Methods of Test for Paints Part E7: Resistance to Impact (Falling Ball Test) (AMD 6909) February 28, 1992 (W). 4 pp.
BSI BS 3900: Part E8-74. 1974 Methods of Test for Paints Part E8: Resistance to Impact (Pendulum Test) (W). 4 pp.
BSI BS 3900: Part E8-01. 1974 Amd 1 Methods of Test for Paints Part E8: Resistance to Impact (Pendulum Test) (AMD 6908) February 28, 1992 (W). 5 pp.
BSI BS 3900: Part E13-93. 1993 Methods of Test for Paints Part E13: Falling Weight Resistance (Impact) (ISO 6272: 1993) (W). 11 pp.
CNS K6056-70. Method of Test for Baking Varnish (Jan)(771). 6 pp.
CNS K6057-70. Method of Test for Clear Varnish for Baking Varnish (Jan)(773). 3 pp.
CNS K6103-59. Method of Test for Low Temperature Baking Varnish (Apr)(1113). 3 pp.
DIN ENGL 55995 (P)-83. Testing of Paints, Varnishes and Related Products; Test for Impact Resistance to Grit and Gravel of Paints, Varnishes and Related Products Using Single-Impact Apparatus (Aug). 8 pp.
ISO 6272-93. Paints and Varnishes—Falling-Weight Test First Edition. 9 pp.
SAA AS 1580.406. 1-93. Paints and Related Materials—Methods of Test—Part 406.1: Resistance to Impact—Falling Weight Test (Gardener-Type Tester) (in Professional Package 39A). 2 pp.
SNZ NZS/AS 1580. 406.1-86. Methods of Test for Paints and Related Materials Part 406.1: Resistance to Impact—Falling Weight Test (Gardner-Type Tester) (This is a Joint Standard with SAA AS 1580.406.1). 3 pp.

—Vulcanized Rubber
BSI BS 903: Part A8-90. 1990 Amd 1 Methods of Testing Vulcanized Rubber Part A8: Determination of Rebound Resilience. 20 pp.
CNS K6352-85. Method of Test for Rebound of Vulcanized Rubber (Dec)(3561).
ISO 4662-86. Rubber—Determination of Rebound Resilience of Vulcanizates Second Edition. 13 pp.

—Walls
ISO 7892-88. Vertical Building Elements—Impact Resistance Tests—Impact Bodies and General Test Procedures First Edition. 8 pp.

—Welded Joints
DIN ENGL 50122-84. Testing of Metallic Materials; Impact Test on Welded Joints; Specimen Position, Notch Position (Aug). 4 pp.

—Wheelchairs
SAA AS 3696.8 (INT)-91. Wheelchairs—Part 8(Int): Static, Impact and Fatigue Strength Tests (Expires 9 December 1993). 72 pp.

—Wheels
ISO 7141-81. Road Vehicles—Wheels—Impact Test Procedure First Edition; (Corrigendum 1-1992). 7 pp.

—Wood
CNS O2008-81. Method of Test for Impact Bending of Wood (Mar)(457). 1 p.
ISO 3348-75. Wood—Determination of Impact Bending Strength First Edition. 4 pp.
ISO 3351-75. Wood—Determination of Resistance to Impact Indentation First Edition. 4 pp.
JIS Z 2116-63. Method of Impact Bending Test for Wood (R 1970). 3 pp.

—Zinc Coatings (Zinc)
CNS H2047-79. Dropping Test for Electroplated Coatings of Zinc (Apr)(4829).

Impact Testing Equipment
See Also: Impact Testing; Puncture Testers; Testing Equipment

—Calibration
BSI BS 7003-88. 1988 Spring-Operated Impact-Test Apparatus and Its Calibration. 12 pp.
CENELEC HD 495 S1-88. Spring-Operated Impact-Test Apparatus and its Calibration. 3 pp.

Impact Testing Equipment *(Cont.)*

—Calibration *(Cont.)*
DIN VDE 0700 Pt 1 A7 (D)-82. Safety of Household and Similar Electrical Appliances; Part 1: General Requirements—Calibration of Spring-Operated Impact Test Apparatus: Amendment 7 (VDE Specification) (Feb). 10 pp.
IEC 817-84. Spring-Operated Impact-Test Apparatus and Its Calibration First Edition. 21 pp.
SAA AS 1544.4-89. Methods for Impact Tests on Metals—Part 4: Calibration of the Testing Machine. 12 pp.
SNZ IEC 817-84. Spring-Operated Impact-Test Apparatus and Its Calibration. 17 pp.

—Charpy
BSI BS 131: Part 6-89. 1989 Notched Bar Tests Part 6: Method for Precision Determinations of Charpy V-Notch Impact Energies for Metals. 10 pp.
BSI BS 131: Part 7-90. 1990 Notched Bar Tests Part 7: Specification for Verification of the Test Machine Used for Precision Determination of Charpy V-Notch Impact Energies for Metals. 15 pp.
CNS B6082-83. Charpy Impact Testing Machines (Jul)(10424).
CNS B7255-83. Method of Test for Charpy Impact Testing (Jul)(10425).
JIS B 7722-90. Charpy Impact Testing Machines. 25 pp.
JIS B 7740-90. Standard Specimens for Charpy Impact Testing Machine. 10 pp.

—Charpy—Calibration
SAA AS 1146.3-90. Methods for Impact Tests on Plastics—Part 3: Calibration of the Testing Machine. 11 pp.

—Dynstat
DIN ENGL 51230-77. Material Testing Machines; Dynstat Apparatus for the Determination of Bending Strength and Impact Strength of Small Specimens (Dec). 3 pp.

—Izod
CNS B6068-82. Izod Impact Testing Machines (May)(8768).
JIS B 7723-76. Izod Impact Testing Machines (R 1979). 6 pp.

—Izod—Calibration
SAA AS 1146.3-90. Methods for Impact Tests on Plastics—Part 3: Calibration of the Testing Machine. 11 pp.

—Pendulum
BSI BS 131: Part 4-72. 1972 Methods for Notched Bar Tests Part 4: Calibration of Pendulum Impact Testing Machines for Metals. 17 pp.
BSI BS 131: Part 4-01. 1972 Amd 1 Methods for Notched Bar Tests Part 4: Calibration of Izod Pendulum Impact Testing Machines for Metals (AMD 7768) August 15, 1993 (V). 20 pp.
BSI BS EN 10045-2-93. 1993 Charpy Impact Test on Metallic Materials Part 2: Method for the Verification of Impact Testing Machines (V). 23 pp.
CEN PREN2003-03-75. Test Methods for Steel Products—Part 3—Calibration of Pendulum Impact Testing Machines (C5/24). 3 pp.
CEN PREN 10 045-2-90. Metallic Materials—Charpy Impact Test—Part 2: Verification of the Testing Machine (Pendulum Impact). 31 pp.
CEN EN 10045-2-92. Metallic Materials—Charpy Impact Test—Part 2: Verification of the Testing Machine (Pendulum Impact). 33 pp.
CEN EN 10045-2-92. Metallic Materials—Charpy Impact Test—Part 2: Verification of the Testing Machine (Pendulum Impact). 18 pp.
DIN ENGL 51222-85. Materials Testing Machines; Pendulum Impact Testing Machines (Jan). 9 pp.
DIN ENGL 51306-83. Material Testing Machines; Verification of Pendulum Impact Testing Machines (Sept). 10 pp.
ISO R442-65. Verification of Pendulum Impact Testing Machines for Testing Steels First Edition. 12 pp.
JIS B 7739-89. Pendulum-Type Impact Testing Machines for Non-Metallic Materials. 25 pp.

Impact Velocity
See Also: Impact Shock; Impact Testing

—Automotive—Collision Research
BSI BS AU 164-77. 1977 Measurement of Impact Velocity in Collision Tests on Road Vehicles. 4 pp.
ISO 3784-76. Road Vehicles—Measurement of Impact Velocity in Collision Tests First Edition. 5 pp.

Impact Wrenches
See Also: Hand Tools; Pipe Wrenches; Wrenches

Impact Wrenches (Cont.)
—Electric
BSI BS 2769: Sec 2.2-84. 1984 Hand-Held Electric-Motor-Operated Tools Part 2: Particular Requirements Section 2.2: Screwdrivers and Impact Wrenches. 7 pp.
CGSB 45-GP-6B-72. Wrench, Impact, Electric, Portable, Heavy Duty. 9 pp.
CSA CAN/CSA-C22. 2 NO71.1-M89. Portable Electric Tools; (Gen Instr 1 Thru 4). 89 pp.
—Pneumatic
CGSB 45-GP-5B-81. Wrenches, Impact, Pneumatic, Portable, Standard for. 15 pp.
—Safety
BSI BS 2769: Sec 2.2-84. 1984 Hand-Held Electric-Motor-Operated Tools Part 2: Particular Requirements Section 2.2: Screwdrivers and Impact Wrenches. 7 pp.
CEN PREN 792-6-92. Handheld Non-Electric Power Tools—Safety Requirements—Part 6: Assembly Power Tools for Threaded Fasteners. 9 pp.
IEC 745 Pt 2-2-82. Safety of Hand-Held Motor-Operated Electric Tools Part 2: Particular Requirements for Screwdrivers and Impact Wrenches First Edition; (Amendment 1-1991). 20 pp.

IMPATT Diodes
Use For: Impact Avalanche and Transit Time Diodes
See Also: Avalanche Diodes; Diodes
—Microwave—Preferred Products List
CECC CECC MUAHAG Vol 9 IS 3-90. Preferred Products List; Semiconductors (En, Fr, Ge) AMD 1 (En, Fr, Ge). 51 pp.

Impedance
Use: Electrical Impedance

Impedance Bonds
JIS E 3603-80. Impedance Bonds.
—Lead Wire
JIS E 3604-77. Lead Wire for Impedance Bonds.
—Railroad Equipment
JIS E 3018-78. Performance Test Methods for Impedance Bonds.

Impedance Measurement
See Also: Electrical Impedance; Electrical Measurement; Input Impedance; Transfer Impedance
—Busway Fittings
CEN PREN 2591 (Part G7)-92. Elements of Electrical and Optical Connection Test Methods Part G7—Measurement of Characteristic Impedance of a Bus or a Stub Terminator. 3 pp.
—Cable Couplers
CEN PREN 2591 (Part G1)-92. Elements of Electrical and Optical Connection Test Methods Part G1—Measurement of Open Circuit Impedance. 3 pp.

Impedance Measuring Relays
Use: Impedance Relays

Impedance Protected Motors
Use: Motor Protectors—Impedance

Impedance Relays
See Also: Relays
IEC 255 Pt 16-82. Electrical Relays Part 16: Impedance Measuring Relays First Edition. 30 pp.

Impeller Agitators
Use: Agitators

Implants (Medical)
Use: Surgical Implants

Implants (Surgical)
Use: Surgical Implants

Import Documents
—Declaration Forms
EC EEC 1900/85-85. Council Regulation Introducing Community Export and Import Declaration Forms. 2 pp.

Import Duties
Use: Duty Free Entry

Import Levies
Use: Import Tariffs

Import Tariffs
Use For: Customs Duties; Import Levies; Revenue Tariffs

Import Tariffs (Cont.)
—Dairy Products
EC EEC/735/79-79. Commission Regulation Fixing the Import Levies on Milk and Milk Products. 3 pp.
—Exemptions—Fuels—Automotive
EC 85/347/EEC-85. Council Directive Amending Directive 68/297/EEC on the Standardization of Provisions Regarding the Duty-Free Admission of Fuel Contained in the Fuel Tanks of Commercial Motor Vehicles. 2 pp.
—Exemptions—Postal Fees
EC EEC/1797/86-86. Council Regulation Abolishing Certain Postal Fees for Customs Presentation. 1 p.
—Milk
EC EEC/735/79-79. Commission Regulation Fixing the Import Levies on Milk and Milk Products. 3 pp.
—Reductions
EC 91/191/EEC-91. Council Directive Amending Directive 69/169/EEC on Tax-Paid Allowances in Intra-Cummunity Travel and as Regards a Derogation Granted to the Kingdom of Denmark and to Ireland Relating to the Rules Governing Travellers' Allowances on Imports. 3 pp.
—Reductions—Rum—France
EC 88/245/EEC-88. Council Decision Authorizing the French Republic to Apply in its Overseas Departments and in Metropolitan France, by Way of Derogation from Article 95 of the Treaty, a Reduced Rate of the Revenue Duty Imposed on the Consumption of 'Traditional' Rum. 2 pp.
—Suspensions—Industrial Products
EC EEC/1626/92-92. Council Regulation Temporarily Suspending the Autonomous Common Customs Tariff Duties on a Number of Industrial Products (in the Microelectronics and Related Sectors). 110 pp.

Importation
Scope Note: For importation of products from third countries, see the subheading Importation under the specific product

Impregnated Papers
Use For: Treated Papers *See Also:* Asphalt Papers; Laminated Papers; Packaging Papers; Papers; Test Papers; Waxed Papers
—Flammability Testing
CNS P3063-84. Method of Test for Flame Resistance of Treated Paper and Paperboard (Jan)(10760). 4 pp.
—Nitrogen Content
BSI BS 4497-69. 1969 Recommendations for the Detection and Estimation of Nitrogenous Treating Agents in Paper. 21 pp.
—Oil Based Varnishes
MOD UK CS 2957. Paper, Oil, Varnished (Synthetic Resin Type) and LF Quality.
—Polyethylene
CNS Z5101-82. Polyethylene Treated Paper (Feb)(8551).
CNS Z6043-82. Method of Test for Polyethylene Treated Paper (Feb)(8552).
JIS Z 1514-76. Polyethylene Treated Paper (R 1980). 9 pp.
—Silica Gel—Residual Vapor Detection
MOD UK DSTAN 68-101-88. Paper, Silica Gel (RVD) Issue 1. 19 pp.
—Vinylidene Chloride
JIS Z 1515-76. Vinylidene Chloride Treated Paper (R 1979). 10 pp.
—Waxes
MOD UK TS 10171. Wax Special No. 9.

Impulse Generators
Use: Pulse Generators

Impulse Noise
Use For: Impulsive Interference
See Also: Electromagnetic Noise; Impulse Noise Counters; Noise (Spurious Signals)
—Data Circuits—Pulse Code Modulation Systems
CCITT FASCICLE III.1. General Characteristics of International Telephone Connections and Circuits Recommendations G.101-G.181. 332 pp.

Impulse Noise (Cont.)
—Data Transmission
CCITT RECMN H.16-89. Characteristics of an Impulsive-Noise Measuring Instrument for Wideband Data Transmission—Line Transmission of Non-Telephone Signals—Transmission of Sound-Programme and Television Signals (Study Group XV) 1 pp (Same as Recmn O.72). 1 p.
CCITT RECMN O.72-89. Characteristics of an Impulsive Noise Measuring Instrument for Wideband Data Transmissions—Specifications for Measuring Equipment (Study Group IV) 1 pp (Same as Recmn H.16). 1 p.
—Radio Receivers
BSI BS 4054: Part 5-80. 1980 Measuring and Expressing the Performance of Radio Receivers for Sound Broadcasting Part 5: Measurement on Frequency-Modulated Receivers of the Response to Impulsive Interference. 12 pp.
IEC 315 Pt 5-71. Methods of Measurement on Radio Receivers for Various Classes of Emission Part 5: Specialized Radio-Frequency Measurements. Measurement on Frequency-Modulated Receivers of the Response to Impulsive Interference, Recm. First Edition. 19 pp.
—Telephone Circuits
CCITT FASCICLE III.1. General Characteristics of International Telephone Connections and Circuits Recommendations G.101-G.181. 332 pp.
CCITT RECMN H.13-89. Characteristics of an Impulsive Noise Measuring Instrument for Telephone-Type Circuits—Line Transmission of Non-Telephone Signals—Transmission of Sound-Programme and Television Signals (Study Group XV) 1 pp (Same as Recmn O.71). 1 p.
CCITT RECMN Q.45-89. Transmission Characteristics of an Analogue International Exchange—General Recommendations on Telephone Switching and Signalling—Functions and Information Flows for Services in the ISDN—Supplements. 1 p.
CCITT RECMN Q.45 BIS-89. Transmission Characteristics of an Analogue International Exchange—General Recommendations on Telephone Switching and Signalling—Functions and Information Flows for Services in the ISDN—Supplements. 9 pp.
CCITT RECMN V.55-89. Specification for an Impulsive Noise Measuring Instrument for Telephone-Type Circuits—Data Communication over the Telephone Network (Study Group XVII) 1 pp. 1 p.
—Voice Band Data Transmission
CCITT RECMN G.113-89. Transmission Impairments—General Characteristics of International Telephone Connections and Circuits (Study Groups XII and XV) 22 pp. 22 pp.

Impulse Noise Counters
Use For: Impulsive Noise Counters
See Also: Impulse Noise
—Telephone Circuits
CCITT RECMN O.71-89. Impulsive Noise Measuring Equipment for Telephone-Type Circuits—Specifications for Measuring Equipment (Study Group IV) 4 pp (Same as Recmn H.13). 4 pp.
—Telephone Transmission
CCITT RECMN P.54-89. Sound Level Meters (Apparatus for the Objective Measurement of Room Noise)—Telephone Transmission Quality (Study Group XII) 1 pp. 1 p.
CCITT RECMN P.55-89. Apparatus for the Measurement of Impulsive Noise—Telephone Transmission Quality (Study Group XII) 2 pp. 2 pp.

Impulse Sealing Equipment
See Also: Sealing
CNS C3116-81. Method of Test for Impulse Sealer (Apr)(7210).
CNS C4300-81. Hand Type Impulse Sealer (Apr)(7208).
CNS C4301-81. Foot Type Impulse Sealer (Apr)(7209).

Impulse Suppressors
Use: Surge Suppressors

Impulse Testing
Use: Impulse Voltage Testing

Impulse Testing (Hoses)
Use: Pressure Measurement

Impulse Testing (Tubes)
Use: Pressure Measurement

Impulse Voltage Testing
See Also: Electrical Properties; Voltage Fluctuations
CNS C6110-81. Techniques for Switching Impulse Testing (Apr)(7231).

Impulse Voltage Testing (Cont.)

—Cable Assemblies (Electric)
DIN VDE 0472 Pt 511-85. Testing of Cables, Wires and Flexible Cords; Impulse Test (Aug). 7 pp.

—Cables (Electric)
CENELEC HD 48 S1-88. Impulse Tests on Cables and Their Accessories. 3 pp.
DIN VDE 0472 Pt 511-85. Testing of Cables, Wires and Flexible Cords; Impulse Test (Aug). 7 pp.
IEC 230-66. Impulse Tests on Cables and Their Accessories First Edition. 11 pp.
IEC 840-88. Tests for Power Cables with Extruded Insulation for Rated Voltages Above 30 kV (Um = 36 kV) up to 150 kV (Um = 170 kV) First Edition; (Corrigendum—Nov 1988) (Amendment 2-1993, Including Amendment 1) (Corrigendum—July 1993). 71 pp.

—Digital Recorders
BSI BS 7528: Part 1-91. 1991 Digital Recorders for Measurements in High-Voltage Impulse Tests Part 1: Requirements for Digital Recorders (Renumbered as BS EN 61083-1: 1993). 32 pp.
BSI BS EN 61083-1-93. 1993 Amd 1 Digital Recorders for Measurements in High-Voltage Impulse Tests Part 1: Requirements for Digital Recorders (AMD 7925) September 15, 1993 (S). 41 pp.
CENELEC EN 61083-1-93. Digital Recorders for Measurements in High-Voltage Impulse Tests Part 1: Requirements for Digital Recorders (IEC 1083-1: 1991, Modified). 32 pp.
IEC 1083 Pt 1-91. Digital Recorders for Measurements in High-Voltage Impulse Tests Part 1: Requirements for Digital Recorders First Edition (CENELEC EN 61083-1: 1993). 58 pp.

—Electric Reactors
IEC 722-82. Guide to the Lightning Impulse and Switching Impulse Testing of Power Transformers and Reactors First Edition. 70 pp.

—Electrical Insulation
IEC 243 Pt 3-93. Methods of Test for Electric Strength of Solid Insulating Materials Part 3: Additional Requirements for Impulse Tests First Edition. 26 pp.

—Electrical Insulators—High Voltage
IEC 506-75. Standard Impulse Tests on High-Voltage Insulators First Edition. 23 pp.

—Low Voltage Installations
BSI BS 7640: Part 1-93. 1993 High-Voltage Test Techniques for Low-Voltage Equipment Part 1: Definitions, Test and Procedure Requirements (IEC 1180-1: 1992) (E). 31 pp.
IEC 1180 Pt 1-92. High-Voltage Test Techniques for Low-Voltage Equipment Part 1: Definitions, Test and Procedure Requirements First Edition. 60 pp.

—Oscilloscopes
BSI BS 6647-85. 1985 Guide to Oscilloscopes and Peak Voltmeters for Impulse Tests. 31 pp.
BSI BS EN 61083-1-93. 1993 Amd 1 Digital Recorders for Measurements in High-Voltage Impulse Tests Part 1: Requirements for Digital Recorders (AMD 7925) September 15, 1993 (S). 41 pp.
CENELEC HD 479-86. Oscilloscopes and Peak Voltmeters for Impulse Tests. 2 pp.
CENELEC EN 61083-1-93. Digital Recorders for Measurements in High-Voltage Impulse Tests Part 1: Requirements for Digital Recorders (IEC 1083-1: 1991, Modified). 32 pp.
IEC 790-84. Oscilloscopes and Peak Voltmeters for Impulse Tests First Edition. 59 pp.
IEC 1083 Pt 1-91. Digital Recorders for Measurements in High-Voltage Impulse Tests Part 1: Requirements for Digital Recorders First Edition (CENELEC EN 61083-1: 1993). 58 pp.

—Power Transformers
IEC 722-82. Guide to the Lightning Impulse and Switching Impulse Testing of Power Transformers and Reactors First Edition. 70 pp.

—Rotating Machines
BSI BS 4999: Part 115-92. 1992 Rotating Electrical Machines Part 115: Impulse Voltage Withstand Levels of Rotating a.c. Machines with Form-Wound Stator Coils (IEC 34-15: 1990). 13 pp.
CENELEC HD 53.15 S1-91. Rotating Electrical Machines Part 15: Impulse Voltage Withstand Levels of Rotating A.C. Machines with Form-Wound Stator Coils. 6 pp.
DIN VDE 0530 Pt 15 (D)-88. Rotating Electrical Machines: Impulse Voltage Withstand Levels of Rotating a.c. Machines with Form-Wound Stator Coils Identical to IEC 2(CO) 535 (Nov) (Supersedes Draft DIN VDE 0530 Part 1 A1/08. 86). 14 pp.
IEC 34 Pt 15-90. Rotating Electrical Machines Part 15: Impulse Voltage Withstand Levels of Rotating a.c. Machines with Form-Wound Stator Coils First Edition. 19 pp.

—Voltmeters
BSI BS 6647-85. 1985 Guide to Oscilloscopes and Peak Voltmeters for Impulse Tests. 31 pp.
CENELEC HD 479-86. Oscilloscopes and Peak Voltmeters for Impulse Tests. 2 pp.
IEC 790-84. Oscilloscopes and Peak Voltmeters for Impulse Tests First Edition. 59 pp.

Impulsive Interference
Use: Impulse Noise

Impulsive Noise Counters
Use: Impulse Noise Counters

Impurities Content Analysis
Use For: Extraneous Matter Content Analysis; Foreign Matter Content Analysis
See Also: Contamination; Metallic Impurities Content Analysis

—Aggregates
CNS A3028-83. Method of Test for Organic Impurities in Fine Aggregate (Jan)(1164).
JIS A 1105-76. Method of Test for Organic Impurities in Fine Aggregate. 5 pp.

—Beverages
CNS N6091-91. Method of Test for Fruit and Vegetable Juices and Drinks (General Rules) (Jun)(3736). 1 p.
CNS N6208-87. Method of Test for Sport Drinks (Oct)(12150). 3 pp.

—Canned Foods
CNS N6023-78. Method of Test for Canned Food (Inspection for Purity) (Jul)(979). 1 p.

—Cereals
CNS N4030-89. Methods of Test for Cereals (Nov)(3287). 10 pp.

—Chemical Pulps
ISO 5350 Pt 1-82. Pulps—Estimation of Dirt and Shives—Part 1: Unbleached Chemical Pulps First Edition. 7 pp.

—Chloromethanes—Gas Chromatography
BSI BS 5598: Part 10-83. 1983 Methods of Sampling and Test for Halogenated Hydrocarbons Part 10: Determination of Impurities in Methyl Chloride, Gas Chromatographic Method. 9 pp.
ISO 5916-82. Methyl Chloride for Industrial Use—Determination of Impurities Chromatographic Methods First Edition. 8 pp.

—Coffee
BSI BS 5752: Part 4-80. 1980 Coffee and Coffee Products Part 4: Green Coffee: Olfactory and Visual Examination and Determination of Foreign Matter and Defects. 5 pp.
ISO 4149-80. Green Coffee—Olfactory and Visual Examination and Determination of Foreign Matter and Defects First Edition. 4 pp.

—Condiments
CNS N6125-80. Spices and Condiments—Determination of Extraneous Matter (Jul)(5929). 2 pp.

—Cresols—Visual Testing
ISO 1897 Pt IV-77. Phenol, o-Cresol, m-Cresol, p-Cresol, Cresylic Acid and Xylenols for Industrial Use—Methods of Test—Part IV: Visual Test for Impurities Insoluble in Sodium Hydroxide Solution (Excluding Cresylic Acid and Xylenols) First Edition. 6 pp.

—Ethylene
JIS K 1800-78. Analytical Methods for Trace of Impurities in High Purity Ethylene.

—Fats—Solvent Extraction
BSI BS 684: Sec 2.3-93. 1993 Methods of Analysis of Fats and Fatty Oils Part 2: Other Methods Section 2.3: Determination of Insoluble Impurities Content (ISO 663: 1992). 6 pp.
BSI BS 684: Sec 2.3-83. 1983 Methods of Analysis of Fats and Fatty Oils Part 2: Other Methods Section 2.3: Determination of Insoluble Impurities. 4 pp.
ISO 663-92. Animal and Vegetable Fats and Oils—Determination of Insoluble Impurities Content Second Edition. 6 pp.

—Fish Liver Oil
CNS K6114-63. Method of Test for Fish-Liver Oil (Crude) (Jul)(1223) (R 1973). 2 pp.

—Fish Paste—Frozen
CNS N6107-82. Method of Test for Frozen Minced Fish Meat (Nov)(4641). 2 pp.

—Fruits—Dried
CNS N6049-85. Method of Test for Dehydrated Fruits and Vegetables (Mar)(2896). 2 pp.

—Hair (Animal)
CNS N4025-85. Method of Test for Animal Hairs (May)(2826). 7 pp.

—Honey
CNS N6027-74. Method of Test for Honey (Oct)(1344). 8 pp.

—Hydrogen
CNS K6301-71. Method of Test for Impurities in Compressed Hydrogen (Jun)(3237).

—Juices
CNS N6091-91. Method of Test for Fruit and Vegetable Juices and Drinks (General Rules) (Jun)(3736). 1 p.

—Legumes
BSI BS 4317: Part 4-91. 1991 Cereals and Pulses Part 4: Determination of Impurities, Size, Foreign Odours, Insects, and Species and Variety, in Pulses (ISO 605: 1991). 11 pp.

—Milk
BSI BS 4938: Part 1-82. 1982 Dirt Content of Milk Part 1: Method for the Determination of Dirt Content (Reference Method). 8 pp.
BSI BS 4938: Part 2-82. 1982 Dirt Content of Milk Part 2: Method for the Determination of Visible Dirt Content (Rapid Method). 12 pp.

—Musk
CNS N4029-70. Method of Test for Musk (Jan)(3083). 2 pp.

—Noodles
CNS N6112-79. Method of Test for Dry Noodles (Sep)(4992). 2 pp.

—Oilseeds
BSI BS 4289: Part 2-89. 1989 Methods for the Analysis of Oilseeds Part 2: Determination of Impurities Content. 8 pp.
ISO 658-88. Oilseeds—Determination of Impurities Content Second Edition. 7 pp.

—Paperboard
CNS P3058-83. Method of Test for Dirt in Paper, Paperboard and Pulp (Jul)(10471). 4 pp.
JIS P 8145-76. Testing Method for Dirt in Paper and Paperboard (R 1984). 5 pp.

—Papers
CNS P3058-83. Method of Test for Dirt in Paper, Paperboard and Pulp (Jul)(10471). 4 pp.
CPPA D.36P-80. Dirt in Paper. 2 pp.
CPPA B.1U-77. Identification of Spots and Specks in Paper. 4 pp.
JIS P 8145-76. Testing Method for Dirt in Paper and Paperboard (R 1984). 5 pp.
SNZ NZS/AS 1301. P204RP-93. Methods of Test for Pulp and Paper Estimation of Dirt and Shives (This is a Joint Standard with SAA AS 1301.P204RP). 9 pp.

—Papers—Visual Testing
BSI BS 5477-77. 1977 Method for Visual Assessment, by Grid Assay, of Dirt in Paper for Character Recognition. 2 pp.

—Phenols—Visual Testing
ISO 1897 Pt IV-77. Phenol, o-Cresol, m-Cresol, p-Cresol, Cresylic Acid and Xylenols for Industrial Use—Methods of Test—Part IV: Visual Test for Impurities Insoluble in Sodium Hydroxide Solution (Excluding Cresylic Acid and Xylenols) First Edition. 6 pp.
ISO 1897 Pt V-77. Phenol, o-Cresol, m-Cresol, p-Cresol, Cresylic Acid and Xylenols for Industrial Use—Methods of Test—Part V: Visual Test for Impurities Insoluble in Water (Phenol Only) First Edition. 6 pp.

—Phthalic Anhydride—Iodometry
ISO 1389 Pt IX-77. Phthalic Anhydride for Industrial Use—Methods of Test—Part IX: Determination of Impurities Oxidizable in the Cold by Potassium Permanganate—Iodometric Method First Edition. 4 pp.

—Pulps
BSI BS 7421-91. 1991 Method for Estimation of Dirt and Shives in Bleached Pulp. 8 pp.
CNS P3058-83. Method of Test for Dirt in Paper, Paperboard and Pulp (Jul)(10471). 4 pp.

INTERNATIONAL AND NON-U.S. NATIONAL STANDARDS
SUBJECT INDEX
Incandescent

Impurities Content Analysis *(Cont.)*
—**Pulps** *(Cont.)*
CNS P3059-83. Method of Test for Foreign Particulate Matter in Pulp (by Standard Dirt Estimation Chart) (Jul)(10472). 3 pp.
CPPA B.1U-77. Identification of Spots and Specks in Paper. 4 pp.
ISO 5350 Pt 2-90. Pulps—Estimation of Dirt and Shives—Part 2: Bleached Pulp First Edition. 6 pp.
JIS P 8208-76. Method of Testing Dirt, Shivers and Specks of Paper Pul
SNZ NZS/AS 1301. P204RP-93. Methods of Test for Pulp and Paper Estimation of Dirt and Shives (This is a Joint Standard with SAA AS 1301.P204RP). 9 pp.

—**Pulps—Visual Inspection**
CPPA D.22U-77. Dirt in Pulp—Dry Method. 2 pp.

—**PVC**
ISO 1265-79. Plastics—Polyvinyl Chloride Resins—Determination of Number of Impurities and Foreign Particles First Edition; (Corrected and Reprinted -1979). 5 pp.

—**Ramie**
CNS L3015-67. Method of Test for Ramie Raw Fiber (Aug)(1227) (R 1973). 2 pp.

—**Rice**
CNS N4043-82. Method of Test for Unmilled Rice (Jun) (3491). 4 pp.
CNS N4044-82. Method of Test for Milled Rice (Jun)(3492). 4 pp.
CNS N4045-82. Method of Test for Rough Rice (Jun)(3493). 2 pp.

—**Rubber**
BSI BS 7164: Part 8-90. 1990 Chemical Tests for Raw and Vulcanized Rubber Part 8: Method for Determination of Dirt Content of Raw Natural Rubber. 10 pp.
CNS K6357-73. Chemical Analysis of Dirt Matter in Natural Rubber (May)(3566).
ISO 249-87. Rubber, Raw Natural—Determination of Dirt Content Second Edition. 7 pp.

—**Seasonings**
BSI BS 4585: Part 1-83. 1983 Methods of Test for Spices and Condiments Part 1: Determination of Extraneous Matter. 3 pp.
BSI BS 4585: Part 14-83. 1983 Methods of Test for Spices and Condiments Part 14: Determination of Filth. 8 pp.
ISO 927-82. Spices and Condiments—Determination of Extraneous Matter Content Second Edition. 3 pp.
ISO 1208-82. Spices and Condiments—Determination of Filth First Edition. 7 pp.

—**Seeds**
CNS N4014-72. Method of Test for Seeds (Jun)(2166). 22 pp.

—**Semolina**
BSI BS 4317: Part 29-93. 1993 Methods of Test for Cereals and Pulses Part 29: Determination of Impurities of Animal Origin in Wheat Flour and Durum Wheat Semolina (ISO 11050: 1993) (W). 16 pp.
ISO 11050-93. Wheat Flour and Durum Wheat Semolina—Determination of Impurities of Animal Origin First Edition. 14 pp.

—**Shellfish—Frozen**
CNS N6043-86. Method of Test for Frozen Shrimp (Mar)(2301). 5 pp.

—**Shortening**
CNS N6111-86. Method of Test for Shortening (Nov)(4990). 2 pp.

—**Silver Indium Cadmium Alloys**
CNS J2051-82. Method for Determining the Trace Impurities in Silver-Indium-Cadmium Alloys (May)(8833).

—**Soy Sauce**
CNS N6008-88. Methods of Test for Soy Sauce (Nov)(955). 7 pp.

—**Soybeans**
CNS N6009-83. Method of Test for Soybean Paste (Dec)(956). 1 p.

—**Spices**
CNS N6125-80. Spices and Condiments—Determination of Extraneous Matter (Jul)(5929). 2 pp.

—**Uranium Hexafluoride—Mass Spectrometry**
CNS J2079-82. Method for Determining Impurities in UF6 Spark Source Mass Spectrographic Method (Dec)(9712).

Impurities Content Analysis *(Cont.)*
—**Vegetables—Dried**
CNS N6049-85. Method of Test for Dehydrated Fruits and Vegetables (Mar)(2896). 2 pp.

—**Vinegars**
CNS N6025-83. Method of Test for Vinegar (Jul)(1071). 1 p.

—**Wheat Flour**
BSI BS 4317: Part 29-93. 1993 Methods of Test for Cereals and Pulses Part 29: Determination of Impurities of Animal Origin in Wheat Flour and Durum Wheat Semolina (ISO 11050: 1993) (W). 16 pp.
ISO 11050-93. Wheat Flour and Durum Wheat Semolina—Determination of Impurities of Animal Origin First Edition. 14 pp.

—**Wood—Chips**
CPPA D.3U-77. Dirt in Chips. 1 p.

IN
Use: Intelligent Networks

In Vitro Analysis
See Also: Culture Media

—**Reagents—Diagnostic—Identification Systems**
BSI BS EN 375-92. 1992 Labelling of In Vitro Diagnostic Reagents for Professional Use. 12 pp.
BSI BS EN 376-92. 1992 Labelling of In Vitro Diagnostic Reagents for Self-Testing. 11 pp.
CEN PREN 375-90. In-Vitro Diagnostic Systems—Requirements for Labelling of In-Vitro Diagnostic Reagents for Professional Use. 11 pp.
CEN EN 375-92. In Vitro Diagnostic Systems—Requirements for Labelling of In Vitro Diagnostic Reagents for Professional Use. 11 pp.
CEN PREN 376-90. In-Vitro Diagnostic Systems—Requirements for Labelling of In-Vitro Diagnostic Reagents for Self Testing. 10 pp.
CEN EN 376-92. In Vitro Diagnostic Systems—Requirements for Labelling of In Vitro Diagnostic Reagents for Self-Testing. 10 pp.

Inband Signaling
CCITT RECMN Q.20-89. Maximum Permissible Value for the Absolute Power Level of a Signalling Pulse—General Recommendations on Telephone Switching and Signalling—Functions and Information Flows for Services in the ISDN—Supplements (Study Group XI) 2 pp. 2 pp.
CCITT RECMN Q.22-89. Frequencies to Be Used for In-Band Signalling—General Recommendations on Telephone Switching and Signalling—Functions and Information Flows for Services in the ISDN—Supplements (Study Group XI) 1 pp. 1 p.

—**Line—CCITT R2 Signaling Systems**
CCITT FASCICLE VI.4. Specifications of Signalling Systems R1 and R2 Recommendations Q.310—Q.490 (Study Group XI). 181 pp.

—**Splitting**
CCITT RECMN Q.25-89. Splitting Arrangements and Signal Recognition Times in "In-Band" Signalling Systems—General Recommendations on Telephone Switching and Signalling—Functions and Information Flows for Services in the ISDN—Supplements (Study Group XI) 3 pp. 3 pp.

Incandescent Lamps
Use: Incandescent Lighting

Incandescent Lighting
Use For: Filament Lamps *See Also:* Cap Lamps; Discharge Lamps; Dumet Wire; Gas Filled Lamps; Halogen Lamps; Infrared Lamps; Lamps; Lighting; Mercury Vapor Lamps; Resistance Lamps; Tungsten Halogen Lamps; Tungsten Lamps
BSI BS 4533: Sec 102.6-90. 1990 Luminaires Part 102: Particular Requirements Section 102.6: Specification for Luminaires with Built-in Transformers for Filament Lamps. 17 pp.
BSI BS 4533: Sec 102.6-01. 1990 Amd 1 Luminaires Part 102: Particular Requirements Section 102.6: Specification for Luminaires with Built-in Transformers for Filament Lamps (AMD 7712) May 15, 1993 (F). 23 pp.
BSI BS 6873-88. 1988 Amd 1 Filament Lamps for Cycles (AMD 6639) April 30, 1991. 28 pp.
BSI BS 6873-02. 1988 Amd 2 Filament Lamps for Cycles (AMD 7757) July 15, 1993 (E). 38 pp.
CEN PREN2756-89. Incandescent Lamps Technical Specification. 22 pp.
CENELEC HD 77-74. Lighting Fittings for Incandescent Lamps for Domestic and Similar Purposes. 2 pp.
CENELEC HD 77-87. Lighting Fittings for Incandescent Lamps for Domestic and Similar Purposes. 3 pp.

Incandescent Lighting *(Cont.)*
CENELEC EN 60 598-2-6-89. Luminaires Part 2: Particular Requirements Section Six—Luminaires with Built-in Transformers for Filament Lamps. 5 pp.
CENELEC EN 60598-2-6 /A2-91. AMD 2 Luminaires Part 2: Particular Requirements Section Six—Luminaires with Built-in Transformers for Filament Lamps. 5 pp.
CNS C3002-87. Incandescent Lamps Bulbs for General Lighting Service (Oct)(298).
CNS C3044-86. Method of Test for Incandescent Lamps for General Lighting Service (Jul)(3891).
CSA C22.2 NO 9-M1989. Luminaires; (Gen Instr 1) (Supercedes C22.2 No 97-1969). 83 pp.
CSA 601C Bull. Electrical Bulletin 601C December 22, 1966 to C22.2 NO 9. 2 pp.
CSA 705 Bull. Electrical Bulletin 705 January 30, 1968 to C22.2 NO 9. 1 p.
CSA 775A Bull. Electrical Bulletin 775A January 27, 1970 C22.2 NO 9. 2 pp.
CSA 1169 Bull. Electrical Bulletin 1169 June 27, 1978 to C22.2 NO 9. 2 pp.
CSA 1221 Bull. Electrical Bulletin 1221 March 26, 1979 to C22.2 NO 9. 2 pp.
CSA C22.2 NO 66-1988. Speciality Transformers; (Gen Instr 1). 53 pp.
CSA 1169 Bull. Electrical Bulletin 1169 June 27, 1978 to C22.2 NO 66. 2 pp.
CSA C22.2 NO 84-1974. Incandescent Lamps (R 1992); (Amd 1-2 May 1990). 19 pp.
CSA 976 Bull. Electrical Bulletin 976 December 9, 1974 to C22.2 NO 84. 2 pp.
CSA 613F Bull. Electrical Bulletin 613F November 22, 1978 to C22.2 NO 84. 5 pp.
IEC 598 Pt 2-6-79. Luminaires Part 2: Particular Requirements Section Six: Luminaires with Built-in Transformer for Filament Lamps First Edition; (Amendment 2-1990, Incorporating Amendment 1). 32 pp.
JIS C 7501-83. Incandescent Lamps for General Lighting Service. 17 pp.
JIS C 7523-83. Small Lamps for Household Use. 10 pp.
JIS C 8116-81. Table Study Lamps for Incandescent Lamps.
MOD UK DSTAN 62-6-01. Lamps, Electric Issue 2; Amendment 1. 109 pp.

—**Aerospace—Lampholders**
AECMA PREN2241-90. Lamp Base Dimensions. 16 pp.

—**Aircraft**
IEC 434-73. Aircraft Electrical Filament Lamps First Edition; (Amendment 1-1981) (Amendment 2-1984). 64 pp.

—**Aircraft—Indicating Instruments**
BSI 2G 191-88. 1988 Lighting for Aircraft Indicators Using Integral Filament Lamps. 4 pp.

—**Automotive**
BSI BS 941-70. (WITHDRAWN) 1970 Amd 2 Filament Lamps for Road Vehicles (Superseded by BS 6873: 1988). 55 pp.
BSI BS 6797-86. 1986 Performance Requirements for Lamps for Road Vehicles. 38 pp.
EC 76/761/EEC-76. Council Directive on the Approximation of the Laws of the Member States Relating to Motor-Vehicle Headlamps Which Function as Main-Beam and/or Dipped-Beam Headlamps and to Incandescent Electric Filament Lamps for Such Headlamps. 26 pp.
EC 89/517/EEC-89. Commission Directive Adapting to Technical Progress Council Directive 76/761/EEC on the Approximation of the Laws of the Mem. Sts. Rel. to Motor-Vehicle Headlamps Which Function as Main-Beam and/or Dipped-Beam Headlamps Incandescent Elec. Fil. Lamps for Such Headlamps. 2 pp.
IEC 810-86. Lamps for Road Vehicles Performance Requirements First Edition; (Amendment 1-1988) (Amendment 2-1992). 64 pp.
IEC 983-90. Road Vehicle Lamps for Supplementary Purposes First Edition; (Amendment 1-1991). 28 pp.

—**Automotive—Fog Lights**
EC 76/762/EEC-76. Council Directive on the Approximation of the Laws of the Member States Relating to Front Fog Lamps for Motor Vehicles and Filament Lamps for Such Lamps. 13 pp.

—**Automotive—Voltage Measurement**
ISO 3559-76. Road Vehicles—Working Voltages for Lights Fitted to Motor Vehicles and to Their Trailers First Edition. 9 pp.

—**Buses (Vehicles)**
CNS D2041-87. Incandescent Lamps for Buses (Aug)(5797).

—**Control Circuits**
CSA C22.2 NO 184-M1988. Solid-State Lighting Controls; (Gen Instr 1 Thru 3). 33 pp.

INDUSTRY STANDARDS

Incandescent Lighting (Cont.)

—Converters
BSI BS EN 61046-92. 1992 D.C. or a.c. Supplied Electronic Step-Down Convertors for Filament Lamps. General and Safety Requirements. 31 pp.
BSI BS EN 61046-01. 1992 Amd 1 D.C. or a.c. Supplied Electronic Step-Down Convertors for Filament Lamps. General and Safety Requirements (IEC 1046: 1991) (AMD 7608) March 15, 1993. 40 pp.
BSI BS EN 61047-93. 1993 D.C. or a.c. Supplied Electronic Step-Down Convertors for Filament Lamps Performance Requirements (IEC 1047: 1991) (E). 22 pp.
CENELEC EN 61046-92. D.C. or a.c. Supplied Electronic Step-Down Convertors for Filament Lamps General and Safety Requirements. 2 pp.
CENELEC EN 61046/A1-92. AMD 1 D.C. or a.c Supplied Electronic Step-Down Convertors for Filament Lamps—General and Safety Requirements; (IEC 1046; 1991/A1: 1991). 5 pp.
CENELEC EN 61047-92. D.C. or A.C. Supplied Electronic Step-Down Convertors for Filament Lamps—Performance Requirements; (IEC 1047: 1991). 5 pp.
IEC 1046-91. D.C. or a.c. Supplied Electronic Step-down Convertors for Filament Lamps General and Safety Requirements First Edition; (Amendment 1-1991) (CENELEC EN 61046/A1: 1992). 53 pp.
IEC 1047-91. D.C. or a.c. Supplied Electronic Step-Down Convertors for Filament Lamps Performance Requirements First Edition; (CENELEC EN 61047:1992). 36 pp.

—Converters—Safety
BSI BS EN 61046-92. 1992 D.C. or a.c. Supplied Electronic Step-Down Convertors for Filament Lamps. General and Safety Requirements. 31 pp.
BSI BS EN 61046-01. 1992 Amd 1 D.C. or a.c. Supplied Electronic Step-Down Convertors for Filament Lamps. General and Safety Requirements (IEC 1046: 1991) (AMD 7608) March 15, 1993. 40 pp.
CENELEC EN 61046-92. D.C. or a.c. Supplied Electronic Step-Down Convertors for Filament Lamps General and Safety Requirements. 2 pp.
CENELEC EN 61046/A1-92. AMD 1 D.C. or a.c Supplied Electronic Step-Down Convertors for Filament Lamps—General and Safety Requirements; (IEC 1046; 1991/A1: 1991). 5 pp.
IEC 1046-91. D.C. or a.c. Supplied Electronic Step-down Convertors for Filament Lamps General and Safety Requirements First Edition; (Amendment 1-1991) (CENELEC EN 61046/A1: 1992). 53 pp.

—Explosion Proof
CNS C1039-87. Explosion-Proof Construction of Incandescent Lamps for General Use (Jul)(3377). 7 pp.

—Fire Resistant
CNS C1039-88. Incandescent Lamp for General Use of Explosion-Proof Construction (Oct)(3377).

—Identification Systems
MOD UK DSTAN 62-10: Part 1-01. Lamps, Filament Issue 1; Amendment 3. 201 pp.

—Inspection
SAA AS 3118-86. Approval and Test Specification—Electric Inspection Handlamps Amdt 1 July 1989 (This is a Joint Standard with SANZ NZS 3118) (Superseded by AS/NZS 3118:1993).
SNZ NZS/AS 3118-93. Approval and Test Specification Electric Inspection Handlamps. 6 pp.

—Lampholders
CSA C22.2 NO 43-M1984. Lampholders (R 1992); (Gen Instr 1 Thru 6). 54 pp.
CSA 1233A Bull. Electrical Bulletin 1233A June 17, 1981 to C22.2 NO 43. 2 pp.
CSA 1416 Bull. Electrical Bulletin 1416 November 14, 1983 to C22.2 NO 84. 2 pp.

—Lampholders—Temperature Testing
CSA 601J Bull. Electrical Bulletin 601J November 29, 1978 to C22.2 NO 9. 4 pp.

—Lighting Chains
BSI BS EN 60598-2-20-92. 1992 Luminaires Part 2: Particular Requirements Section 2.20: Lighting Chains. 29 pp.
BSI BS EN 60598-2-20-01. 1992 Amd 1 Luminaires Part 2: Particular Requirements Section 2.20: Lighting Chains (AMD 7709) April 15, 1993 (F). 34 pp.
CENELEC EN 60598-2-20-91. Luminaires Part 2: Particular Requirements Section Twenty—Lighting Chains. 20 pp.
CENELEC EN 60598-2-20/A11-92. AMD 11 Luminaires Part 2: Particular Requirements Section Twenty—Lighting Chains. 4 pp.

—Lighting Chains—Portable
CSA C22.2 NO 12-1982. Portable Luminaires; (Gen Instr 1 Thru 4). 52 pp.
CSA 1169 Bull. Electrical Bulletin 1169 June 27, 1978 to C22.2 NO 12. 2 pp.

—Marine
CNS F5016-85. General Requirements for Construction of Marine Electric Lighting Fittings Incandesscent Lamp Type (Jul)(8268).
JIS F 8003-89. General Requirements for Construction of Marine Electric Lighting Fittings Incandescent Lamps Type.

—Marine—Bulkheads—Fire Resistant
CNS F5083-84. Explosion-Proof Bulkhead Lights for Marine Use (Feb)(10766).

—Marine—Searchlights
CNS F1005-83. Method of Test for Search Lights of Marine Use (Mar)(3913). 2 pp.
CNS F3013-76. Search Lights for Marine Use (Mar)(3912). 6 pp.

—Marine—Signal
CNS F5021-85. Marine Daylight Signalling Lamps (Nov)(8393).
CNS F5024-85. Marine Signalling Lights (Nov)(8396). 8 pp.
JIS F 8455-83. Marine Daylight Signalling Lamps.
JIS F 8455-64. Daylight Signalling Lamps for Marine Use. 9 pp.

—Marine—Signal Lights
NATO STANAG 1007 ED 4 AMD 2-86. Preferred Electric Lamps for Signalling Purposes on Naval Vessels of NATO Countries. 12 pp.

—Marine—Temperature Testing
CNS F5015-85. General Rules on the Temperature Test of Electric Lighting Fittings (Incandescent Lamps) for Marine Use (Jul)(8267).
JIS F 8002-89. General Rules on the Temperature Test of Electric Lighting Fittings (Incandescent Lamps) for Marine Use.

—Naval Ships
NATO STANAG 1006 ED 3 AMD 0-75. Preferred Electrical Filament Lamps for Use in Ships of NATO Navies. 11 pp.
NATO STANAG 1006 ED 4 AMD 3-86. Preferred Incandescent and Fluorescent Electric Lamps for Lighting Purposes on Naval Vessels of NATO Countries. 7 pp.

—Packaging
CSA 1187 Bull. Electrical Bulletin 1187 November 6, 1978 to C22.2 NO 9. 2 pp.
MOD UK DSTAN 62-10: Part 1-01. Lamps, Filament Issue 1; Amendment 3. 201 pp.

—Photometry
CNS C3070-80. Photometric Measurement on Working Standard Incandescent Lamps (Jan)(5118).
CNS C3071-80. Photometric Measurement on Standard Fluorescent Lamps (Feb)(5197).
JIS C 7613-85. Photometric Measurements on Incandescent Lamps Used for Photometric Standards. 15 pp.

—Portable
CSA C22.2 NO 12-1982. Portable Luminaires; (Gen Instr 1 Thru 4). 52 pp.
CSA 1169 Bull. Electrical Bulletin 1169 June 27, 1978 to C22.2 NO 12. 2 pp.
CSA C22.2 NO 84-1974. Incandescent Lamps (R 1992); (Amd 1-2 May 1990). 19 pp.
CSA 976 Bull. Electrical Bulletin 976 December 9, 1974 to C22.2 NO 84. 2 pp.

—Portable—Electric Plugs
CSA 1276 Bull. Electrical Bulletin 1276 July 16, 1980 to C22.2 NO 12. 2 pp.

—Shipping
CSA 1187 Bull. Electrical Bulletin 1187 November 6, 1978 to C22.2 NO 9. 2 pp.

—Signs
CSA CAN/CSA-C22. 2 NO 207-M89. Portable and Stationary Electric Signs and Displays; (Gen Instr 1). 50 pp.

—Stage Lighting
CSA C22.2 NO 166-M1983. Stage and Studio Luminaires; (Gen Instr 1) (Erratum May 1983). 26 pp.

—Studio Lighting
CSA C22.2 NO 166-M1983. Stage and Studio Luminaires; (Gen Instr 1) (Erratum May 1983). 26 pp.

—Submersible—Swimming Pools—Accessories
CSA C22.2 NO 89-1976. Swimming-Pool Luminaires, Submersible Luminaires and Accessories (R 1992). 23 pp.

—Temperature Measurement—Photometry
JIS Z 8725-87. Methods for Determining Distribution Temperature and Colour Temperature or Correlated Colour Temperature of Light Sources. 28 pp.

—Temperature Testing
CSA C22.2 NO 84-1974. Incandescent Lamps (R 1992); (Amd 1-2 May 1990). 19 pp.
CSA 613F Bull. Electrical Bulletin 613F November 22, 1978 to C22.2 NO 84. 5 pp.

—Trailers—Voltage Measurement
ISO 3559-76. Road Vehicles—Working Voltages for Lights Fitted to Motor Vehicles and to Their Trailers First Edition. 9 pp.

—Transformers
CSA C22.2 NO 66-1988. Speciality Transformers; (Gen Instr 1). 53 pp.
CSA 1169 Bull. Electrical Bulletin 1169 June 27, 1978 to C22.2 NO 66. 2 pp.

Incendiary Mixtures
See Also: Explosives

—Aluminum Magnesium Alloys
MOD UK DSTAN 68-157-93. Magnesium-Aluminium Alloy 50/50 Powdered, Sizes 120 and 200 Issue 1. 12 pp.

—Barium Nitrates
MOD UK DSTAN 68-74-01. Barium Nitrate Grades 1 and 1A Issue 1; Amendment 1. 8 pp.

—Iron Oxide
MOD UK DSTAN 13-121-91. Hammerscale (Magnetic Oxide of Iron) Sizes 25, 100 and 200 Issue 1. 8 pp.

Incentive Contracts
Use For: Target Cost Contracts See Also: Cost Reimbursement Contracts; Defense Contracts; Fixed Price Contracts
MOD UK DEFCON Guide no. 5-79. Incentive (Target Cost) Contracting 3/79. 48 pp.

Incident Investigations
See Also: Accident Investigations

—Aircraft
NATO STANAG 3531 ED 5 AMD 3-84. Investigation of Aircraft Missile Accidents/Incidents. 15 pp.
NATO STANAG 3531 ED 6 AMD 0-91. Safety Investigation and Reporting of Accidents/Incidents Involving Military Aircraft and/or Missiles. 11 pp.
NATO STANAG 3531 ED 6 AMD 1-91. Safety Investigation and Reporting of Accidents/Incidents Involving Military Aircraft and/or Missiles. 12 pp.
NATO STANAG 3531 ED 6 AMD 2-91. Safety Investigation and Reporting of Accidents/Incidents Involving Military Aircraft and/or Missiles. 13 pp.

—Aircraft—Human Factors
ICAO Circular 240. Human Factors Digest No. 7 Investigation of Human Factors in Accidents and Incidents—1993. 61 pp.

—Aircraft—Near Misses
NATO STANAG 3750 ED 4 AMD 0-91. Reporting and Investigation of Airmiss Incidents. 10 pp.
NATO STANAG 3750 ED 4 AMD 1-91. Reporting and Investigation of Airmiss Incidents. 11 pp.
NATO STANAG 3750 ED 4 AMD 2-91. Reporting and Investigation of Airmiss Incidents. 12 pp.

—Missiles
NATO STANAG 3531 ED 5 AMD 3-84. Investigation of Aircraft Missile Accidents/Incidents. 15 pp.
NATO STANAG 3531 ED 6 AMD 0-91. Safety Investigation and Reporting of Accidents/Incidents Involving Military Aircraft and/or Missiles. 11 pp.
NATO STANAG 3531 ED 6 AMD 1-91. Safety Investigation and Reporting of Accidents/Incidents Involving Military Aircraft and/or Missiles. 12 pp.
NATO STANAG 3531 ED 6 AMD 2-91. Safety Investigation and Reporting of Accidents/Incidents Involving Military Aircraft and/or Missiles. 13 pp.

—Naval Operations
NATO STANAG 1179 ED 2 AMD 5-77. Combined Investigation of Maritime Incidents. 27 pp.

Incinerators
Use For: Waste Combustion Systems
See Also: Furnaces; Solid Waste Disposal

Incinerators (Cont.)

—Medical Wastes
BSI BS 3316: Part 1-87. 1987 Incinerators Part 1: Standard Performance Requirements for Incineration Plant for the Destruction of Hospital Waste. 7 pp.
BSI BS 3316: Part 2-87. 1987 Incinerators Part 2: Methods of Test and Calculation for the Perfomance of Incineration Plant for the Destruction of Hospital Waste. 15 pp.
BSI BS 3316: Part 3-87. 1987 Incinerators Part 3: Method for Specifying Purchasers' Requirements for Incineration Plant for the Destruction of Hospital Waste. 12 pp.
BSI BS 3316: Part 4-87. 1987 Incinerators Part 4: Code of Practice for the Design Specification Installation and Commissioning of Incineration Plant for the Desctruction of Hospital Waste. 20 pp.

—Residential
BSI BS 3813: Part 1-64. 1964 Incinerators for Waste from Trade and Residental Premises Part 1: Capacities Between 50 Lb/H and 1000 Lb/H. 17 pp.
SNZ NZS 5202-79. Specification for Incinerators (Reconfirmed 1984). 20 pp.

—Residential—Safety
SAA AS 1875-76. Domestic Incinerators (Fire Safety). 8 pp.

—Ships
JIS F 7011-89. Application of Incinerator for Marine Use.
MOD UK NES 346-90. Requirements for Incinerators for R N Surface Ships Issue 1 (12.90). 16 pp.

Inclined Lifts
Use: Elevators (Lifts)

Inclined Seat Valves
Use: Y Valves

Inclusions
Scope Note: See the subheading Inclusions under the specific material See Also: Voids

Incoming Call Barring
Use: Call Barring

Incoming Lines
See Also: Internal Loss Probability; Telephone Circuits
CCITT RECMN E.146-89. Division of Circuits into Outgoing and Incoming Circuits—Telephone Network and ISDN—Operation, Numbering, Routing and Mobile Service (Study Group II) 1 pp. 1 p.

—Telecommunication Traffic—Measurement
CCITT RECMN E.502 (REV 1)-92. Traffic Measurement Requirements for Digital Telecommunication Exchanges (Study Group II) 23 pp. 23 pp.
CCITT RECMN E.502-89. Traffic Measurement Requirements for SPC (Especially Digital) Telecommunication Exchanges—Telephone Network and ISDN—Quality of Service, Network Management and Traffic Engineering (Study Group II) 15 pp. 15 pp.

Incoming Response Delay
See Also: Delay; Seizing (Telecommunications); Signaling Systems; Time

—Grade of Service—Digital Exchanges
CCITT RECMN E.543-89. Grades of Service in Digital International Telephone Exchanges—Telephone Network and ISDN—Quality of Service, Network Management and Traffic Engineering (Study Group II) 4 pp. 4 pp.

Incontinent Care Products
See Also: Diapers; Urinary Collection Bags

—Glossaries
BSI BS 7710: Part 1-93. 1993 Urine Absorbing Aids—Vocabulary—Part 1: Conditions of Urinary Incontinence (ISO 9949-1: 1993) (N). 6 pp.
BSI BS 7710: Part 2-93. 1993 Urine Absorbing Aids—Vocabulary—Part 2: Products (ISO 9949-2: 1993) (N). 6 pp.
BSI BS 7710: Part 3-93. 1993 Urine Absorbing Aids—Vocabulary—Part 3: Identification of Product Types (ISO 9949-3: 1993) (N). 7 pp.
ISO 9949 Pt 1-93. Urine Absorbing Aids—Vocabulary—Part 1: Conditions of Urinary Incontinence First Edition. 6 pp.
ISO 9949 Pt 2-93. Urine Absorbing Aids—Vocabulary—Part 2: Products First Edition. 6 pp.
ISO 9949 Pt 3-93. Urine Absorbing Aids—Vocabulary—Part 3: Identification of Product Types First Edition. 7 pp.

Incubators
See Also: Agricultural Equipment; Infant Incubators; Medical Electrical Equipment; Medical Equipment
JIS T 1702-90. Incubator. 7 pp.
JIS T 7303-68. Baby Incubators.

—Electric
CSA 1169 Bull. Electrical Bulletin 1169 June 27, 1978 to C22.2 NO 102. 2 pp.

—Electric—Construction
CSA C22.2 NO 102-1958. Construction and Test of Brooders and Incubators (R 1992). 41 pp.

—Electric—Dielectric Strength
CSA C22.2 NO 102-1958. Construction and Test of Brooders and Incubators (R 1992). 41 pp.

—Electric—Portable—Construction
CSA C22.2 NO 102-1958. Construction and Test of Brooders and Incubators (R 1992). 41 pp.

—Electric—Portable—Dielectric Strength
CSA C22.2 NO 102-1958. Construction and Test of Brooders and Incubators (R 1992). 41 pp.

—Electric—Portable—Temperature Testing
CSA C22.2 NO 102-1958. Construction and Test of Brooders and Incubators (R 1992). 41 pp.

—Electric—Temperature Testing
CSA C22.2 NO 102-1958. Construction and Test of Brooders and Incubators (R 1992). 41 pp.

Indentation Forming
Use: Pressing (Forming)

Indentation Hardness Testing
See Also: Brinell Hardness Testing; Compression Testing; Durometer Hardness Testing; Hardness Testing; Knoop Hardness Testing; Microhardness Testing; Rockwell Hardness Testing; Rockwell Superficial Hardness Testing

—Asphalts
DIN ENGL 1996 Pt 13-84. Testing of Asphalt; Indentation Testing Using a Flat-Ended Indentor Pin (July). 7 pp.

—Cable Insulation
DIN VDE 0472 Pt 619-83. Testing of Cables, Wires and Flexible Cords; Indentation Strength (Jan). 4 pp.

—Cellular Materials
BSI BS 4443: Part 2-88. 1988 Methods of Test for Flexible Cellular Materials Part 2: Method 7, Indentation Hardness Tests. 7 pp.
ISO 2439-80. Polymeric Materials, Cellular Flexible—Determination of Hardness (Indentation Technique) Second Edition; (Erratum—Nov 1981). 6 pp.

—Ebonite
BSI BS 2782:Pt3: METH 365D-91. 1991 Methods of Testing Plastics Part 3: Mechanical Properties Method 365D: Determination of Hardness by the Ball Indentation Method (ISO 2039-1: 1987). 6 pp.
ISO 2039 Pt 1-93. Plastics—Determination of Hardness—Part 1: Ball Indentation Method Second Edition. 8 pp.

—Floor Coverings
CEN PREN 433-90. Resilient Floorcoverings—Determination of Static Indentation. 5 pp.
CEN PREN 667-92. Rubber Floor Coverings—Determination of Indentation Hardness by Means of a Durometer (Shore A Hardness). 7 pp.

—Paints
BSI BS 3900: Part E9-76. 1976 Methods of Test for Paints Group E: Mechanical Tests on Paint Films Part E9: Buchholz Indentation Test. 4 pp.
BSI BS 3900: Part E12-86. 1986 Methods of Test for Paints Group E: Mechanical Tests on Paint Films Part E12: Indentation Test (Spherical or Pyramidal). 5 pp.
DIN ENGL 53153-77. Testing of Paints, Varnishes and Similar Coating Materials; Buchholz Indentation Test on Paint Coatings and Similar Coatings (Nov). 4 pp.
ISO 2815-73. Paints and Varnishes—Buchholz Indentation Test First Edition. 7 pp.
ISO 6441-84. Paints and Varnishes—Indentation Test (Spherical or Pyramidal) First Edition. 5 pp.

—Plastics
BSI BS 2782:Pt3: METH 365D-91. 1991 Methods of Testing Plastics Part 3: Mechanical Properties Method 365D: Determination of Hardness by the Ball Indentation Method (ISO 2039-1: 1987). 6 pp.
ISO 2039 Pt 1-93. Plastics—Determination of Hardness—Part 1: Ball Indentation Method Second Edition. 8 pp.

Indentation Hardness Testing (Cont.)

—Rubber
BSI BS 903: Part A57-89. 1989 Methods of Testing Vulcanized Rubber Part A57: Determination of Indentation Hardness by Means of Pocket Hardness Meters (ISO 7619: 1986). 8 pp.
DIN ENGL 53519 Pt 1-72. Testing of Elastomers; Determination of Indentation Hardness of Soft Rubber (IRHD); Hardness Testing on Standard Specimens (May). 4 pp.
ISO 7619-86. Rubber—Determination of Indentation Hardness by Means of Pocket Hardness Meters First Edition. 6 pp.

—Varnishes
BSI BS 3900: Part E9-76. 1976 Methods of Test for Paints Group E: Mechanical Tests on Paint Films Part E9: Buchholz Indentation Test. 4 pp.
BSI BS 3900: Part E12-86. 1986 Methods of Test for Paints Group E: Mechanical Tests on Paint Films Part E12: Indentation Test (Spherical or Pyramidal). 5 pp.
DIN ENGL 53153-77. Testing of Paints, Varnishes and Similar Coating Materials; Buchholz Indentation Test on Paint Coatings and Similar Coatings (Nov). 4 pp.
ISO 2815-73. Paints and Varnishes—Buchholz Indentation Test First Edition. 7 pp.
ISO 6441-84. Paints and Varnishes—Indentation Test (Spherical or Pyramidal) First Edition. 5 pp.

—Wood
ISO 3351-75. Wood—Determination of Resistance to Impact Indentation First Edition. 4 pp.

Independent Axiom of Inclusion
Use: Voids

Independent Chucks
Use: Chucks—Independent

Independent Sidebands
Use: Sidebands—Independent

Index Cards
CGSB CAN/CGSB-53.60-M89. Card, Index. 8 pp.

—Filing Cabinets
CGSB 44.16A-93. Metal Filing Cabinet for Index Cards. 9 pp.

Index of Cooperation
See Also: Diameters; FAX Communications; FAX Machines

—FAX Machines—CCITT Group 1
CCITT RECMN T.2-89. Standardization of Group 1 Facsimile Apparatus for Document Transmission—Terminal Equipment and Protocols for Telematic Services (Study Group VIII) 3 pp. 3 pp.

—Phototelegraph Equipment
CCITT RECMN T.1-89. Standardization of Phototelegraphic Apparatus—Terminal Equipment and Protocols for Telematic Services (Study Group VIII) 5 pp. 5 pp.

Index of Refraction
Use: Refractive Index

Indexable Inserts
See Also: Bushings; Fasteners; Washers (Fasteners)

—Boring Bars
BSI BS 4193: Part 14-84. 1984 Hardmetal Insert Tooling Part 14: Designation of Boring Bars (Tool Holders with Cylindrical Shank) for Indexable Inserts. 10 pp.
BSI BS 4193: Part 18-90. 1990 Hardmetal Insert Tooling Part 18: Dimensions of Boring Bars for Indexable Inserts. 10 pp.
ISO 5609-89. Boring Bars for Indexable Inserts—Dimensions AMENDMENT 1: Boring Bars Style Q and U, with Rhombic V-Shape Indexable Inserts Second Edition; (Amendment 1-1993). 11 pp.
ISO 6261-84. Boring Bars (Tool Holders with Cylindrical Shank) for Indexable Inserts—Designation First Edition. 8 pp.

—Carbide
JIS B 4121-85. Indexable Inserts Without Fixing Hole, with 0 Degree Normal Clearance. 17 pp.
JIS B 4122-85. Indexable Inserts Without Fixing Hole, with 11 Degree Normal Clearance. 18 pp.
JIS B 4123-85. Indexable Inserts with Cylindrical Fixing Hole, with 0 Degree Normal Clearance. 31 pp.

—Cartridges
BSI BS 4193: Part 8-90. 1990 Hardmetal Insert Tooling Part 8: Dimensions for Type A Cartridges for Indexable (Throwaway) Inserts. 12 pp.

INTERNATIONAL AND NON-U.S. NATIONAL STANDARDS
SUBJECT INDEX

Indexable

Indexable Inserts *(Cont.)*
—Cartridges *(Cont.)*
ISO 5611-89. Cartridges, Type A, for Indexable Inserts—Dimensions Second Edition. 10 pp.

—Cutting Tools—Carbide
ISO 6987 Pt 2-90. Indexable Inserts for Cutting Tools—Hardmetal (Carbide) Inserts with Rounded Corners, with Partly Cylindrical Fixing Hole—Part 2: Dimensions of Inserts with 11 Degree Normal Clearance First Edition. 12 pp.

ISO TR6987 Pt 3-90. Indexable Inserts for Cutting Tools—Hardmetal (Carbide) Inserts with Rounded Corners, with Partly Cylindrical Fixing Hole—Part 3: V-Shape Inserts First Edition. 9 pp.

—Cutting Tools—Designations
BSI BS 4193: Part 1-93. 1993 Hardmetal Insert Tooling Part 1: Specification for Designation of Indexable Inserts for Cutting Tools (ISO 1832: 1991). 20 pp.

BSI BS 4193: Part 1-86. 1986 Hardmetal Insert Tooling Part 1: Designation of Indexable Inserts for Cutting Tools. 16 pp.

ISO 1832-91. Indexable Inserts for Cutting Tools—Designation Third Edition. 18 pp.

JIS B 4120-85. Indexable Inserts for Cutting Tools-Designation. 21 pp.

—End Mills
BSI BS 4193: Part 10-82. 1982 Hardmetal Insert Tooling Part 10: Mills with Indexable Inserts and Morse Taper Shank. 4 pp.

ISO 6262 Pt 2-82. End Mills with Indexable Inserts—Part 2: End Mills with Morse Taper Shank First Edition. 3 pp.

—Milling Cutters
BSI BS 4193: Part 12-84. 1984 Hardmetal Insert Tooling Part 12: Specification for Dimensions of Side and Face Milling and Slotting Cutters with Indexable Inserts. 4 pp.

ISO 6986-83. Side and Face Milling (Slotting) Cutters with Indexable Inserts—Dimensions First Edition; (ANSI 6986-1983). 3 pp.

—Milling Cutters—Carbide
BSI BS 4193: Part 15-86. 1986 Hardmetal Insert Tooling Part 15: Dimensions for Indexable Hardmetal (Carbide) Inserts with Wiper Edgesbut Without Fixing Hole. 16 pp.

ISO 3365-85. Indexable Hardmetal (Carbide) Inserts with Wiper Edges, Without Fixing Hole—Dimensions Second Edition. 14 pp.

—Milling Cutters—Designations
BSI BS 4193: Part 16-86. 1986 Hardmetal Insert Tooling Part 16: Designation of Shank Type Milling Cutters for Indexable Inserts. 12 pp.

BSI BS 4193: Part 17-89. 1989 Hardmetal Insert Tooling Part 17: Designation of Bore Type Milling Cutters for Indexable Inserts. 10 pp.

ISO 7406-86. Bore Type Milling Cutters for Indexable Inserts—Designation First Edition; (ANSI 7406-1986). 8 pp.

ISO 7848-86. Shank Type Milling Cutters for Indexable Inserts—Designation First Edition; (ANSI 7848-1986). 10 pp.

—Tool Holders
JIS B 4126-90. Tool Holders for Indexable Inserts. 61 pp.

—Tool Holders—Copying Tools
BSI BS 4193: Part 7-90. 1990 Hardmetal Insert Tooling Part 7: Dimensions for Single Point Turning and Copying Tool Holders for Indexable Inserts. 12 pp.

ISO 5610-89. Single-Point Tool Holders for Turning and Copying, for Indexable Inserts—Dimensions AMENDMENT 1: Tool Holders Style H, J and V, with Rhombic V-Shape Indexable Inserts Third Edition; (Amendment 1-1993). 15 pp.

—Tool Holders—Designations
BSI BS 4193: Part 14-84. 1984 Hardmetal Insert Tooling Part 14: Designation of Boring Bars (Tool Holders with Cylindrical Shank) for Indexable Inserts. 10 pp.

ISO 6261-84. Boring Bars (Tool Holders with Cylindrical Shank) for Indexable Inserts—Designation First Edition. 8 pp.

JIS B 4125-90. Tool Holders for Indexable Inserts—Designation. 13 pp.

—Tool Holders—Turning Tools
BSI BS 4193: Part 7-90. 1990 Hardmetal Insert Tooling Part 7: Dimensions for Single Point Turning and Copying Tool Holders for Indexable Inserts. 12 pp.

ISO 5610-89. Single-Point Tool Holders for Turning and Copying, for Indexable Inserts—Dimensions AMENDMENT 1: Tool Holders Style H, J and V, with Rhombic V-Shape Indexable Inserts Third Edition; (Amendment 1-1993). 15 pp.

Indexable Inserts *(Cont.)*
—Turning/Boring Tools—Carbide
BSI BS 4193: Part 2-86. 1986 Hardmetal Insert Tooling Part 2: Dimensions for Indexable Hardmetal (Carbide) Inserts with Rounded Corners, Without Fixing Hole. 10 pp.

BSI BS 4193: Part 3-86. 1986 Hardmetal Insert Tooling Part 3: Dimensions of Indexable Hardmetal (Carbide) Inserts with Rounded Corners and with Cylindrical Fixing Hole. 12 pp.

BSI BS 4193: Part 13-84. 1984 Hardmetal Insert Tooling Part 13: Dimensions of Indexable Hardmetal Inserts Having No Normal Clearance Rounded Corners and Partly Cylindrical Fixing Hole. 14 pp.

ISO 883-85. Indexable Hardmetal (Carbide) Inserts with Rounded Corners, Without Fixing Hole—Dimensions Third Edition. 9 pp.

ISO 3364-85. Indexable Hardmetal (Carbide) Inserts with Rounded Corners, with Cylindrical Fixing Hole—Dimensions Second Edition. 11 pp.

ISO 6987 Pt 1-83. Indexable Hardmetal (Carbide) Inserts with Rounded Corners, with Partly Cylindrical Fixing Hole—Part 1: Dimensions of Inserts with 7 Degrees Normal Clearance First Edition. 12 pp.

—Turning/Boring Tools—Ceramic
BSI BS 4193: Part 19-93. 1993 Hardmetal Insert Tooling Part 19: Specification for Dimensions of Indexable Ceramic Inserts with Rounded Corners Without Fixing Hole (ISO 9361-1: 1991). 18 pp.

BSI BS 4193: Part 20-93. 1993 Hardmetal Insert Tooling Part 20: Specification for Dimensions of Ceramic Inserts with Rounded Corners with Fixing Hole (ISO 9361-2: 1991). 16 pp.

ISO 9361 Pt 1-91. Indexable Inserts for Cutting Tools—Ceramic Inserts with Rounded Corners—Part 1: Dimensions of Inserts Without Fixing Hole First Edition. 15 pp.

ISO 9361 Pt 2-91. Indexable Inserts for Cutting Tools—Ceramic Inserts with Rounded Corners—Part 2: Dimensions of Inserts with Cylindrical Fixing Hole First Edition. 15 pp.

Indexes (Documentation)
See Also: Directories; Subject Indexing
ISO 999-75. Documentation—Index of a Publication First Edition. 4 pp.

—Aeronautical Navigation Charts
NATO STANAG 3672 ED 1 AMD 5-73. Indexes to Series of Land Maps and Aeronautical Charts and Indexes to Military Geographic Information and Documentation (MGID). 10 pp.

NATO STANAG 3672 ED 1 AMD 6-73. Indexes to Series of Land Maps and Aeronautical Charts and Indexes to Military Geographic Information and Documentation (MGID). 12 pp.

NATO STANAG 3672 ED 1 AMD 7-73. Indexes to Series of Land Maps and Aeronautical Charts and Indexes to Military Geographic Information and Documentation (MGID). 13 pp.

—Aeronautics
BSI BS 185 Index-73. Obsolescent 1973 Glossary of Aeronautical and Astronautical Terms: Index. 38 pp.

—Agreed Firms Schedules
MOD UK. Sectional Index of Agreed Firms Schedules.

—Astronautics
BSI BS 185 Index-73. Obsolescent 1973 Glossary of Aeronautical and Astronautical Terms: Index. 38 pp.

—CCITT
CCITT FASCICLE I.4-88. Index of Blue Book. 680 pp.

—CNS
CNS. Alphabetical Subject Index.

—COM Documents—CEC (Commission of the European Communities)
EURO 1987. Index to 1987 COM Documents of the Commission of the European Communities. 370 pp.

—Connectors
CEN PREN 3660-002-93. Aerospace Series Cable Outlet Accessories for Circular and Rectangular Electrical and Optical Connectors Part 002—Index of Product Standards. 3 pp.

SBAC AS (V). AS Index for High Temperature Electrical Connectors. 32 pp.

SBAC AG (V). AGS Index for High Temperature Electrical Connectors. 13 pp.

SBAC TS (V). TS Index for High Temperature Electrical Connectors. 6 pp.

SBAC ESC (V). ESC Index for High Temperature Electrical Connectors. 1 p.

Indexes (Documentation) *(Cont.)*
—CS and TS Specifications
MOD UK. Alphabetical Index of Specifications for Non-Metallic Materials Issue 6.

—Electric Terminals
MOD UK DSTAN 59-40: Part 90-92. Inter-Circuit Terminations Part 90: Detail Specifications Cross-Reference Index Issue 1. 7 pp.

—Electron Tubes
MOD UK DSTAN 59-60: Part 90-01. Valves, Electronic (Electronic Tubes) (Listed on EPIC Database) Part 90: Detail Specifications Issue 2; Amendment 3 Corrigendum. 25 pp.

—Electronic Equipment
MOD UK DSTAN 59-20-01. Index to NATO Electronic Parts Recommendations (NEPRs) and NATO Electronic Technical Recommendations (NETRs) Issue 3; Amendment 1. 22 pp.

—European Communities
EURO 1989 13TH ED. Directory of Community Legislation in Force and Other Acts of the Community Institutions-Volume II. 192 pp.

EURO 1989 14TH ED VOLUME II. Directory of Community Legislation in Force and Other Acts of the Community Institutions—Volume II. 178 pp.

EURO 1990-2 Volume II. Directory of Community Legislation in Force and Other Acts of the Community Institutions-Volume II. 181 pp.

EURO 1991-2 Volume II. Directory of Community Legislation in Force and Other Acts of the Community Institutions Volume II. 198 pp.

EURO 1992-2 Volume II. Directory of Community Legislation in Force and Other Acts of the Community Institutions Volume II. 195 pp.

EURO 1992-2 Volume II. Directory of Community Legislation in Force and Other Acts of the Community Institutions Volume II. 197 pp.

EURO 1993-2 Volume II. Directory of Community Legislation in Force and Other Acts of the Community Institutions Volume II Chronological Index Alphabetical Index. 198 pp.

—Master Record Indexes—Naval Ships
MOD UK NES 1012-85. Requirements for Master Record Indexes (Warship Equipments) Issue 2 (10.85). 129 pp.

MOD UK NES 1012: Part 1-92. Master Record Indexes Part 1: Requirements for Preparation Issue 1 (10.92). 99 pp.

MOD UK NES 1012: Part 2-92. Master Record Indexes Part 2: Code of Practice —Specimen MRI Issue 1 (10.92). 101 pp.

—Military Geographic Documentation
NATO STANAG 3672 ED 1 AMD 5-73. Indexes to Series of Land Maps and Aeronautical Charts and Indexes to Military Geographic Information and Documentation (MGID). 10 pp.

NATO STANAG 3672 ED 1 AMD 6-73. Indexes to Series of Land Maps and Aeronautical Charts and Indexes to Military Geographic Information and Documentation (MGID). 12 pp.

NATO STANAG 3672 ED 1 AMD 7-73. Indexes to Series of Land Maps and Aeronautical Charts and Indexes to Military Geographic Information and Documentation (MGID). 13 pp.

—Military Geographic Information
NATO STANAG 3672 ED 1 AMD 5-73. Indexes to Series of Land Maps and Aeronautical Charts and Indexes to Military Geographic Information and Documentation (MGID). 10 pp.

NATO STANAG 3672 ED 1 AMD 6-73. Indexes to Series of Land Maps and Aeronautical Charts and Indexes to Military Geographic Information and Documentation (MGID). 12 pp.

NATO STANAG 3672 ED 1 AMD 7-73. Indexes to Series of Land Maps and Aeronautical Charts and Indexes to Military Geographic Information and Documentation (MGID). 13 pp.

—NES Related Documents
MOD UK NES 2: Part 8-91. Index of Standards for Ships and Weapons Part 8: Guide to Sources of Supply of NES Related Documents Issue 2 (09.91). 17 pp.

MOD UK NES 2: Part 11-91. Index of Standards for Ships and Weapons Part 11: Numerical List of Unclassified Naval Engineering Standards Issue 1 (03.91)., 39 pp.

MOD UK NES 2: Part 11-92. Index of Standards for Ships and Weapons Part 11: Numerical List of Unclassified Naval Engineering Standards Issue 2 (01.92). 41 pp.

MOD UK NES 2: Part 11-93. Index of Standards for Ships and Weapons Part 11: Numerical List of Unclassified Naval Engineering Standards Issue 3 (01.93). 42 pp.

INDEX and DIRECTORY of

INTERNATIONAL AND NON-U.S. NATIONAL STANDARDS SUBJECT INDEX

Indexes (Documentation) (Cont.)

—Newsletters
MOD UK Index TO ISSUES 1-72. Standards in Defence News. 2 pp.

—Procurement Standards—Aerospace
MOD UK DSTAN 00-00: Pt 3:Sec 1-90. Standards for Defence Part 3: Index of Standards for Defence Procurement Section 1: Preferred Standards for the Design and Procurement of Aerospace Vehicles and Equipments Issue 11. 185 pp.

MOD UK DSTAN 00-00: Pt 3:Sec 1-91. Standards for Defence Part 3: Index of Standards for Defence Procurement Section 1: Preferred Standards for the Design and Procurement of Aerospace Vehicles and Equipments Issue 12. 194 pp.

MOD UK DSTAN 00-00: Pt 3:Sec 1-92. Standards for Defence Part 3: Index of Standards for Defence Procurement Section 1: Preferred Standards for the Design and Procurement of Aerospace Vehicles and Equipments Issue 13. 200 pp.

—Procurement Standards—Land Service Equipment
MOD UK DSTAN 00-00: Pt 3:Sec 3-83. Standards for Defence Part 3: Index of Procurement Executive Standards Section 3: Preferred Standards for the Design and Procurement of Land Service Equipment Issue 1. 143 pp.

—Procurement Standards—MOD (UK)
MOD UK DSTAN 00-00: Pt 3:Sec 4-91. Standards for Defence Part 3: Index of Standards for Defence Procurement Section 4: Defence Standards Index Issue 7. 272 pp.

MOD UK DSTAN 00-00: Pt 3:Sec 4-92. Standards for Defence Part 3: Index of Standards for Defence Procurement Section 4: Defence Standards Index Issue 8. 226 pp.

MOD UK DSTAN 00-00: Pt 3:Sec 4-93. Standards for Defence Part 3: Index of Standards for Defence Procurement Section 4: Defence Standards Index Issue 9. 232 pp.

MOD UK DSTAN 00-00: Pt 3:Sec 5-85. Standards for Defence Part 3: Index of Standards for Defence Procurement Section 5: MoD Departmental Standards & Specifications Issue 1. 58 pp.

—Procurement Standards—NATO STANAGS
MOD UK DSTAN 00-00: Pt 3:Sec 6-01. Standards for Defence Part 3: Index of Standards for Defence Procurement Section 6: Index of NATO Standardization Agreements (STANAGs) and United Kingdom Implementing Documents Issue 2; Amendment 1. 393 pp.

MOD UK DSTAN 00-00: Pt 3:Sec 6-92. Standards for Defence Part 3: Index of Standards for Defence Procurement Section 6: Index of NATO STANAGs and United Kingdom Implementing Documents Issue 3. 220 pp.

MOD UK DSTAN 00-00: Pt 3:Sec 6-92. Standards for Defence Part 3: Index of Standards for Defence Procurement Section 6: Index of NATO STANAGs and United Kingdom Implementing Documents Issue 4. 220 pp.

MOD UK DSTAN 00-00: Pt 3:Sec 6-93. Standards for Defence Part 3: Index of Standards for Defence Procurement Section 6: Index of NATO STANAGs and United Kingdom Implementing Documents Issue 5. 226 pp.

—Procurement Standards—NES
MOD UK DSTAN 00-00: Pt 3:Sec 2-85. Standards for Defence Part 3: Index of Procurement Executive Standards Section 2: Controllerate of Navy List of Naval Engineering Standards Issue 4 (Superseded by NES 2: Part 1 (Classified)). 26 pp.

—Quality Assurance—Procedural Requirements
MOD UK QPR NUMERICA L Index. Numerical Index of Defence Quality Assurance Board Procedural Requirements (QPRs).
MOD UK NUMERICAL Index.

—Quality Assurance—Technical Memoranda
MOD UK NUMERICAL Index. QTM Numerical Index.

—Radio Component Standardization Committee Publications
MOD UK DSTAN 59-73-76. Index of Radio Component Standardization Committee Publications Issue 1. 11 pp.

—Radio Wave Propagation
CCIR Volume VI Index-90. Numerical Index of Texts. 1 p.

—Radio Wave Propagation—Ionized Media
CCIR Volume VI ANX Index-90. Numerical Index of Texts. 1 p.

Indexes (Documentation) (Cont.)

—Radio Wave Propagation—Ionized Media (Cont.)
CCIR Volume VI ANX DELETED-90. Index of Texts Deleted. 1 p.

—Radio Wave Propagation—Nonionized Media
CCIR Volume V NUM INDX TXT-90. Numerical Index of Texts. 1 p.
CCIR Volume V NUM IND TXTS-90. Numerical Index of Texts. 1 p.
CCIR Volume V IND TXTS DEL-90. Index of Texts Deleted. 1 p.

—SBAC
SBAC AS,AGS & RS Series. AS, AGS, & RS Series Alphabetical Index. 63 pp.
SBAC AS (V). AS Numerical Index. 23 pp.
SBAC AGS (V). AGS Numerical Index (V). 7 pp.
SBAC RS (V). RS Numerical Index. 4 pp.
SBAC TS (V). TS Numerical Index. 5 pp.
SBAC TR (V). TR Numerical Index. 1 p.
SBAC ESC (V). ESC Numerical Index. 1 p.
SBAC AG (V). AGS Index for High Temperature Electrical Connectors. 13 pp.
SBAC AS (ALPHA) (V).
SBAC AS (NUMERICA L) (V).
SBAC RS (NUMERICA L) (V).
SBAC TS (NUMERICA L) (V).

—Semiconductor Devices
MOD UK DSTAN 59-61: Part 100. Semiconductor Devices (Listed on EPIC Database) Part 99: Data Sheets for BS 9000. Detail Specifications Section 100: Cross Reference Index Issue 1.
MOD UK DSTAN 59-61: Part 100-81. Semiconductor Devices (Listed on EPIC Database) Part 100: Cross Reference Index Issue 1. 112 pp.

—Standards
ISO. KWIC INDEX of International Standards 1991 Fifth Edition. 712 pp.

—Telecommunication Networks
CCITT FASCICLE I.4-88. Index of Blue Book. 680 pp.

—Topographic Maps
NATO STANAG 3672 ED 1 AMD 5-73. Indexes to Series of Land Maps and Aeronautical Charts and Indexes to Military Geographic Information and Documentation (MGID). 10 pp.
NATO STANAG 3672 ED 1 AMD 6-73. Indexes to Series of Land Maps and Aeronautical Charts and Indexes to Military Geographic Information and Documentation (MGID). 12 pp.
NATO STANAG 3672 ED 1 AMD 7-73. Indexes to Series of Land Maps and Aeronautical Charts and Indexes to Military Geographic Information and Documentation (MGID). 13 pp.

Indexing Tables (Machine Tools)
See Also: Rotary Tables (Machine Tools); Tables (Stands)

—Rotary
CNS B2760-84. Rotary Indexing Tables (Horizontal Type) (Jan)(10738).
CNS B2761-84. Rotary Indexing Tables (Oblique Type) (Jan)(10739).
JIS B 6160-85. Rotary Indexing Tables. 22 pp.

India Ink
Scope Note: Black ink made of lampblack and blue binder (MGH4*) *See Also:* Carbon Black; Drawing Inks; Inks

—Water Based—Tracing Paper—Engineering Drawings
ISO 9957 Pt 1-92. Fluid Draughting Media—Part 1: Water-Based India Ink—Requirements and Test Conditions First Edition. 9 pp.

Indicating Devices
Use: Signal Devices

Indicating Instruments
See Also: Aircraft Instruments; Attitude Indicators; Azimuth Indicators; Compasses (Indicating Instruments); Dials; Digital Readouts; Display Devices; Gyrohorizons; Humidity Indicators; Indicator Lights; Level Indicators; Load Indicators; Maximum Demand Indicators; Optical Equipment; Peak Program Meters; Phase Sequence Indicators; Pointers; Process Control Systems; Recording Instruments; Revolution Counters; Smoke Density Controls; Temperature Indicators; Volume Unit Meters

BSI BS 5164-75. 1975 Indirect Acting Electrical Indicating and Recording Instruments and Their Accessories. 38 pp.
CNS B6034-82. Microindicators (Aug)(4757).

Indicating Instruments (Cont.)
CNS C1136-85. Tolerance of the Combined Indicator and the Driving Elements (Dec)(11443).
JIS B 7519-76. Microindicators (R 1984). 12 pp.

—Automotive
BSI BS AU 199-84. 1984 Location of Hand Controls, Indicators and Tell-Tales in Passenger Cars. 9 pp.
CNS D1034-82. General Rules for Automobiles Instruments (Part 1) (Nov)(9586).
CNS D1035-82. General Rules for Automobiles Instruments (Part 2) (Nov)(9587).
ISO 4040-83. Road Vehicles—Passenger Cars—Location of Hand Controls, Indicators and Tell-Tales Second Edition. 9 pp.
JIS D 0033-86. Location of Hand Controls, Indicators and Tell-Tales for Passenger Cars. 15 pp.

—Automotive—Symbols
BSI BS AU 143C-84. 1984 Symbols for Controls, Indicators and Tell-Tales for Road Vehicles. 8 pp.
EC 93/29/EEC-93. Council Directive on the Identification of Controls, Tell-Tales and the Indicators for Two-or Three-Wheel Motor Vehicles. 10 pp.
ISO 2575-82. Road Vehicles—Symbols for Controls, Indicators and Tell-Tales Fourth Edition. 9 pp.
ISO 4129-90. Mopeds—Symbols for Controls, Indicators and Tell-Tales Second Edition. 7 pp.
JIS D 0032-82. Symbols for Controls, Indicators and Tell-Tales for Automobiles. 17 pp.

—Dimmers—Marine
CNS F5049-83. Dimmers for Marine Instrument Illumination (Jan)(9858).
JIS F 8852-78. Dimmers for Marine Instrument Illumination.

—Doors—Ships
CNS F3047-80. Indicators for Watertight Sliding Doors (Aug)(6179).
JIS F 2315-68. Ships' Watertight Sliding Door Indicators. 4 pp.

—Electrical
CNS C3189-84. Method of Test for Electrical Indicating Instruments (Jun)(10908).
CNS C4196-80. Dimensions of Edgewise Electrical Indicating Instruments (Jul)(5743).
CNS C4406-84. Electrical Indicating Instruments (Jun)(10907).
JIS C 1102-81. Electrical Indicating Instruments. 71 pp.
JIS C 1103-84. Dimensions of Electrical Indicating Instruments for Switchboards. 11 pp.
MOD UK DSTAN 66-7: Part 1-01. Instruments, Electrical Indicating (Sealed) 50 mm to 100 mm Part 1: Specification Issue 3; Amendment 1. 53 pp.
MOD UK DSTAN 66-9: Part 1-01. Instruments, Electrical Indicating (Hermetically Sealed) 26 mm to 38 mm Part 1: Specification Issue 2; Amendment 2. 38 pp.
MOD UK DSTAN 66-9: Part 3-01. Instruments, Electrical Indicating (Hermetically Sealed) 16mm Part 3: Specification/List Issue 2; Amendment 1. 37 pp.

—Electrical—Shock
CNS C6035-79. Shocks Testing for Electrical Indicating Instruments (Jun)(4876).

—Electrical—Switchboards
CNS C4195-80. Dimensions of Electrical Indicating Instruments for Switchboards (Jul)(5742).
JIS C 1103-84. Dimensions of Electrical Indicating Instruments for Switchboards. 11 pp.

—Elevators (Lifts)
BSI BS 5655: Part 7-83. 1983 Amd 1 Lifts and Service Lifts Part 7: Manual Control Devices, Indicators and Additional Fittings. 10 pp.
ISO 4190 Pt 5-87. Lifts and Service Lifts (USA: Elevators and Dumbwaiters)—Part 5: Control Devices, Signals and Additional Fittings Second Edition. 10 pp.

—Ergonomics
DIN ENGL 33413 Pt 1-84. Ergonomic Aspects of Indicating Devices; Types, Observation Tasks, Suitability (June). 4 pp.

—Fire Alarm and Extinguishing Equipment
BSI BS 5839: Part 4-88. 1988 Amd 1 Fire Detection and Alarm Systems for Buildings Part 4: Specification for Control and Indicating Equipment (AMD 6654) December 21, 1990. 24 pp.
BSI BS 5839: Part 4-02. 1988 Amd 2 Fire Detection and Alarm Systems for Buildings Part 4: Specification for Control and Indicating Equipment (AMD 6875) May 1, 1992 (Supersedes BS 3116: Part 4: 1974). 26 pp.
BSI PD 6531-92. 1992 Queries and Interpretations on BS 5839: Parts 1 and 4 (as Amended). 13 pp.

INDUSTRY STANDARDS

Indicating Instruments (Cont.)

—Fire Alarm and Extinguishing Equipment (Cont.)
SAA AS 1603.4-87. Automatic Fire Detection and Alarm Systems—Part 4: Control and Indicating Equipment Amdt 1 June 1988 Amdt 2 October 1989. 32 pp.
SAA AS 4050(INT)-92. Fire Detection and Fire Alarm Systems—Firefighters' Control and Indicating Facilities (in Professional Package 44). 2 pp.

—Forklifts—Symbols
JIS D 6022-85. Control Symbols for Fork Lift Trucks. 20 pp.

—Gliders
CAA JAR-22 SUBPART F. Equipment (Joint Airworthiness Requirements). 4 pp.
CAA JAR-22 SUBPART G. Operating Limitations and Information (Joint Airworthiness Requirements). 13 pp.

—Identification Systems
DIN VDE 0199 A2 (D)-82. Marking of Mechanical Indicators Which Are Integral Parts of Electrical Equipment (IEC 16(SEC)275) (May). 7 pp.

—Marine
CNS F4012-85. Alarms and Indications for Marine Use (Jan)(11184).
JIS F 0412-90. Alarming and Indicating System of Equipment Used for Engine Department in Ships.
MOD UK NES 604-89. Electrical Indicating Instruments Issue 3 (02.89). 64 pp.

—Motorcycles
JIS D 0034-87. Location and Indication of Controls, Indicators and Tell-Tales for Motorcycles. 13 pp.

—Motorcycles—Symbols
ISO 6727-81. Road Vehicles—Motorcycles—Symbols for Controls, Indicators and Telltales First Edition. 5 pp.

—Packaging
MOD UK DEF-159-01. Packaging of Electrical Indicating Instruments; Amendment 1. 8 pp.

—Rudders—Angles
CNS F5093-84. Electric Rudder Angle Indicators (Jun)(10936).
JIS F 8522-90. Electric Rudder Angle Indicators.

—Safety
CENELEC HD 215-74. Safety Requirements for Indicating and Recording Electrical Instruments and Their Accessories. 4 pp.

—Shunt Assemblies (Electrical)
MOD UK DSTAN 66-13-78. Shunts, Instrument (External Type)-Metric for Use with Electrical Indicating Instruments Issue 1. 19 pp.
MOD UK DSTAN 81-35-90. Packaging of Shunts, Instrument (External Type) for Use with Electrical Indicating Instruments Issue 3. 16 pp.

—Volumetric Analysis
CPPA J.9-84. Indicators for Volumetric Analysis. 2 pp.

Indicator Lights
See Also: Annunciators; Brake Lights; Display Devices; Flashing Lights; Indicating Instruments; Lighting; Neon Glow Lamps; Parking Lights; Revolution Counters; Traffic Lights; Visual Signals
MOD UK DSTAN 62-9: Part 1-74. Lampholders; Lights, Indicator; and Lenses, Indicator Light; for Use in Equipments Part 1: Generic Specification Issue 1. 15 pp.
MOD UK DSTAN 62-9: Part 3-75. Lampholders; Lights, Indicator; and Lenses, Indicator Light; for Use in Equipments Part 3: Detail Specification, Standard Range, and Detail Specification Sheets for Lights, Indicator Issue 1. 38 pp.
SAA AS 1431.2-89. Low Voltages Switchgear and Controlgear—Control Circuit Devices and Switching Elements—Part 2: Additional Requirements for Push-Buttons and Indicator Lights (This is a Joint Standard with SNZ NZS 1431.2). 7 pp.
SNZ NZS/AS 1431. 2-89. Low Voltage Switchgear and Controlgear—Control Circuit Devices and Switching Elements Part 2: Additional Requirements for Push-Buttons and Indicator Lights (This is a Joint Standard with SAA AS 1431.2). 7 pp.

—Automotive—Oil Pressure
CNS D2051-87. Switches of Lubrication Oil Pressure Warning Lamps for Automobiles (Jan)(6173).
JIS D 5803-87. Warning Lamp Switches of Lubricating Oil Pressure for Automobiles. 11 pp.

—Automotive—Symbols
EC 78/316/EEC-77. Council Directive on the Approximation of the Laws of the Member States Relating to the Interior Fittings of Motor Vehicles (Identification of Controls, Tell-Tales and Indicators). 24 pp.

—Color Coding
BSI BS 4099: Part 1-86. (WITHDRAWN) 1986 Colours of Indicator Lights, Push Buttons, Annunciators and Digital Readouts Part 1: Colours of Indicator Lights and Push Buttons (Superseded by BS EN 60073: 1993). 12 pp.
BSI BS EN 60073-93. 1993 Coding of Indicating Devices and Actuators by Colours and Supplementary Means (IEC 73: 1991) (H). 29 pp.
CENELEC HD 354 S2-87. Colours of Indicator Lights and Push-Buttons. 4 pp.
CENELEC EN 60073-93. Coding of Indicating Devices and Actuators by Colours and Supplementary Means (IEC 73:1991) (Supersedes HD 354 S2:1987). 5 pp.
CNS C1067-80. Colors of Indicator Lights and Push Buttons (Jul)(5741).
CSA CAN/CSA-Z431-M89. Colours of Indicator Lights and Push-Buttons (IEC 73-1984). 23 pp.
DIN VDE 0199 (D)-88. Coding of Indicating Devices and Actuators by Colours and Supplementary Means; Requirements for Safety; Identical with IEC 16 (Secretariat) 302 (Dec). 36 pp.
IEC 73-91. Coding of Indicating Devices and Actuators by Colours and Supplementary Means Fourth Edition; (CENELEC EN 60073: 1993). 48 pp.
SNZ BS 4099: Part 1-86. Colours of Indicator Lights, Push-Buttons, Annunciators and Digital Readouts Part 1: Specification for Colours of Indicator Lights and Push-Buttons. 12 pp.

—Contactors
BSI BS 4794: Sec 2.5 & 2.6-78. (WITHDRAWN) 1978 Control Switches (Switching Devices, Including Contactor Relays, for Control and Auxiliary Circuits for Voltages up to and Including 1000 V a.c. and 1200 V d.c. Part 2: Special Requirements for Specific Types of. 12 pp.

—Control Circuits
BSI BS 4794: Sec 2.5 & 2.6-78. (WITHDRAWN) 1978 Control Switches (Switching Devices, Including Contactor Relays, for Control and Auxiliary Circuits for Voltages up to and Including 1000 V a.c. and 1200 V d.c. Part 2: Special Requirements for Specific Types of. 12 pp.

—Control Switches
DIN VDE 0660 Pt 205-82. Switchgear and Control Gear; Low-Voltage Switchgear and Control Gear; Control Switches; Additional Specification for Indicator Lights (Sept). 12 pp.

—Exit Signs
CNS Z1036-83. Exit Indicator Lights (Apr)(10207).

—Industrial
JIS C 8151-91. Indicator Lights for Industry Use. 25 pp.

—Lampholders
MOD UK DSTAN 62-9: Part 1-74. Lampholders; Lights, Indicator; and Lenses, Indicator Light; for Use in Equipments Part 1: Generic Specification Issue 1. 15 pp.
MOD UK DSTAN 62-9: Part 2-01. Lampholders; Lights, Indicator; and Lenses, Indicator Light; for Use in Equipments Part 2: Detail Specification, Standard Range, and Detail Specification Sheets for Lampholders Issue 1; Amendment 1. 36 pp.
MOD UK DSTAN 62-9: Part 4-75. Lampholders; Lights, Indicator; and Lenses, Indicator Light; for Use in Equipments Part 4: Detail Specification, Standard Range, and Detail Specification Sheets for Lenses, Indicator Light Issue 1. 33 pp.

—Lenses
MOD UK DSTAN 62-9: Part 1-74. Lampholders; Lights, Indicator; and Lenses, Indicator Light; for Use in Equipments Part 1: Generic Specification Issue 1. 15 pp.
MOD UK DSTAN 62-9: Part 4-75. Lampholders; Lights, Indicator; and Lenses, Indicator Light; for Use in Equipments Part 4: Detail Specification, Standard Range, and Detail Specification Sheets for Lenses, Indicator Light Issue 1. 33 pp.

—Low Voltage
CENELEC EN 50 007-81. Low Voltage Switchgear and Controlgear for Industrial Use: Single Hole Mounted Control Switches and Indicator Lights: Mounting Dimensions. 4 pp.

—Red—Medical Electrical Equipment
CSA CAN/CSA-C22. 2NO601.1-M90. Medical Electrical Equipment, Part 1: General Requirements for Safety; (Gen Instr 1). 240 pp.

—Ships—Color Coding
ISO 2412-82. Shipbuilding—Colours of Indicator Lights Second Edition. 3 pp.
NATO STANAG 4151 ED 1 AMD 3-86. Standard Colours for Indicating Lights. 8 pp.
NATO STANAG 4151 ED 1 AMD 4-86. Standard Colours for Indicating Lights. 5 pp.
NATO STANAG 4151 ED 1 AMD 5-86. Standard Colours for Indicating Lights. 6 pp.

—Ships—Navigation
CNS F5087-84. Navigation Light Indicators (Feb)(10770).
CNS F5088-84. Navigation Light Indicators (Simple Type) (Feb)(10771).
JIS F 8452-91. Navigation Light Indicators.

—Ships—Reflectors
CNS F5012-83. Glass Globes for Marine Indicator Lamps (Mar)(7991).
JIS F 8404-63. Glass Globes for Marine Indicator Lamps (R 1966). 7 pp.

—Switchboards
JIS C 7516-92. Indicator Lamps. 9 pp.

—Switchgear
BSI BS 6517-84. (WITHDRAWN) 1984 Low Voltage Switchgear and Controlgear for Industrial Use. Single Hole Mounted Control Switches and Indicator Lights. Mounting Dimensions (Superseded by BS EN 60947-5-1: 1992). 6 pp.
DIN VDE 0660 Pt 205-82. Switchgear and Control Gear; Low-Voltage Switchgear and Control Gear; Control Switches; Additional Specification for Indicator Lights (Sept). 12 pp.

—Telecommunication Equipment
BSI BS 1050-84. 1984 Indicator Lamps for Use in Telecommunication Apparatus and for Allied Purposes. 23 pp.

Indicator Papers
Use: Test Papers

Indicator Plates
Use For: Identification Plates; Rating Plates
See Also: Name Plates

—Bypass Valves—Meters
BSI BS 3251-76. 1976 Amd 1 Indicator Plates for Fire Hydrants and Emergency Water Supplies (AMD 6736) September 30, 1991. 14 pp.

—Emergency Water Supplies
BSI BS 3251-76. 1976 Amd 1 Indicator Plates for Fire Hydrants and Emergency Water Supplies (AMD 6736) September 30, 1991. 14 pp.

—Fire Hydrants
BSI BS 3251-76. 1976 Amd 1 Indicator Plates for Fire Hydrants and Emergency Water Supplies (AMD 6736) September 30, 1991. 14 pp.

Indicator Tubes
Use For: Selector Tubes *See Also:* Electron Tubes
MOD UK DSTAN 59-60: Part 3-77. Valves, Electronic (Electronic Tubes) (Listed on EPIC Database) Part 3: Counter, Selector, and Indicator Tubes Issue 2. 29 pp.

—Cold Cathode
CECC CECC 46 000 ISSUE 1-76. Generic Specification: Cold Cathode Indicator Tubes (En, Fr, Ge). 27 pp.
CECC CECC 46 001 ISSUE 1-76. Blank Detail Specification: Cold Cathode Indicator Tubes (En, Fr, Ge). 18 pp.

—Cold Cathode—Electrical Measurement
IEC 151 Pt 22-70. Measurements of the Electrical Properties of Electronic Tubes and Valves Part 22: Methods of Measurement for Cold Cathode Counting and Indicator Tubes First Edition. 28 pp.

—Quality Assurance
BSI BS 9015-70. (WITHDRAWN) 1970 Amd 1 Counter and Indicator Tubes of Assessed Quality: Generic Data and Methods of Test. 12 pp.
BSI BS 9016-70. (WITHDRAWN) 1970 Rules for the Preparation of Detail Specifications for Indicator Tubes of Assessed Quality. 10 pp.

Indicators (Instrumentation)
Use: Indicating Instruments

Indigo Carmine
Use For: Sodium Indigotin Disulfonate

Indigo Carmine (Cont.)
CNS K7254-68. Chemical Reagent (Indigo Carmine) (Nov)(1753).
JIS K 8092-72. Indigo Carmine (Sodium Indigotin Disulfonate).

Indirect Taxes
See Also: Taxes

—Securities
EC COM(87) 139-87. Amended Proposal for a Council Directive Relating to Indirect Taxes on Transactions in Securities. 9 pp.

Indium Content Analysis

—Gallium Arsenide
JIS H 1191-91. Methods for Chemical Analysis of Gallium Arsenide. 57 pp.

—Silver Indium Cadmium Alloys—Control Rods
CNS J2050-82. Method for Determining of Silver, Indium, and Cadmium in Silver-Indium-Cadmium Alloys Control Rod (May)(8832).

—Zinc Alloys—Polarographic Analysis
BSI BS 3630: Part 11-70. 1970 Methods for the Sampling and Analysis of Zinc and Zinc Alloys Part 11: Indium in Ingot Zinc and Zinc Alloys (Polarographic Method). 9 pp.

—Zinc—Polarographic Analysis
BSI BS 3630: Part 11-70. 1970 Methods for the Sampling and Analysis of Zinc and Zinc Alloys Part 11: Indium in Ingot Zinc and Zinc Alloys (Polarographic Method). 9 pp.

Individual Protective Equipment
Use: Personal Protective Equipment

Indole
CNS K7560-81. Chemical Reagent (Indole) (Jul)(7718).

Indole-2,3-Dione
Use: Isatin

Inducers
Use: Inductors

Induction, Electrostatic
Use: Electrostatic Induction

Induction Furnaces
See Also: Furnaces; Induction Heating Equipment

—Channel
CENELEC HD 610 S1-92. Test Methods for Induction Channel Furnaces; (IEC 396:1991). 5 pp.
IEC 396-91. Test Methods for Induction Channel Furnaces Second Edition; (CENELEC HD 610 S1:1992). 30 pp.

—Crucible
IEC 646-92. Test Methods for Crucible Induction Furnaces Second Edition. 28 pp.

Induction Hardening
Scope Note: See the subheading Hardening under the specific metal

Induction Heating Equipment
See Also: Dielectric Heating; Heating Equipment; Induction Furnaces

—Safety
BSI BS 7699: Part 3-93. 1993 Safety in Electroheat Installations Part 3: Particular Requirements for Induction and Conduction Heating and Induction Melting Installations (F). 26 pp.
CENELEC HD 491.3 S1-90. Safety in Electroheat Installations Part 3: Particular Requirements for Induction and Conduction Heating and Induction Melting Installations. 4 pp.
IEC 519 Pt 3-88. Safety in Electroheat Installations Part 3: Particular Requirements for Induction and Conduction Heating and Induction Melting Installations Second Edition. 46 pp.

Induction Loops
See Also: Armature Relays; Relays
BSI BS 7594-93. 1993 Code of Practice for Audio-Frequency Induction-Loop Systems (AFILS) (S). 88 pp.

—Hearing Aids—Magnetic Field Strength
BSI BS 6083: Part 4-81. 1981 Amd 1 Hearing Aids Part 4: Magnetic Field Strength in Audio-Frequency Induction-Loops for Hearing Aid Purposes (AMD 6323) May 31, 1991. 7 pp.
CENELEC HD 450.4-84. Hearing Aids Part 4: Magnetic Field Strength in Audio-Frequency Induction Loops for Hearing Aid Purposes. 2 pp.

—Hearing Aids—Magnetic Field Strength (Cont.)
IEC 118 Pt 4-81. Methods of Measurement of Electro-Acoustical Characteristics of Hearing Aids Part 4: Magnetic Field Strength in Audio-Frequency Induction Loops for Hearing Aid Purposes First Edition. 13 pp.
SAA AS 1088.4-87. Hearing Aids—Part 4: Magnetic Field Strength in Audio-Frequency Induction Loops for Hearing Aid Purposes. 3 pp.

Induction Melting
See Also: Fusion (Melting)

—Safety
BSI BS 7699: Part 3-93. 1993 Safety in Electroheat Installations Part 3: Particular Requirements for Induction and Conduction Heating and Induction Melting Installations (F). 26 pp.
CENELEC HD 491.3 S1-90. Safety in Electroheat Installations Part 3: Particular Requirements for Induction and Conduction Heating and Induction Melting Installations. 4 pp.
IEC 519 Pt 3-88. Safety in Electroheat Installations Part 3: Particular Requirements for Induction and Conduction Heating and Induction Melting Installations Second Edition. 46 pp.

Induction Motor/Generators
See Also: Induction Motors

—Radio Equipment—Marine
CNS C4303-81. Marine Inductor Type Motor-Generators for Wireless Sets (Apr)(7213).

Induction Motors
See Also: AC Motors; Induction Motor/Generators; Motors; Single Phase Motors; Squirrel Cage Motors; Three Phase Motors; Two Phase Motors
CNS C1009-88. Scheduled Information to Be Given on Enquiring and Ordering of Electric Machines (Medium and Small Size Induction Motors) (Jan)(106) (R 1971).

—Cage—Rotating Machines
IEC 34 Pt 17-92. Rotating Electrical Machines Part 17: Guide for the Application of Cage Induction Motors When Fed from Converters First Edition. 29 pp.

—Classification
CNS C1054-80. Classification of Starting KVA of Induction Motors (Apr)(5419).
CNS C1078-83. Classification of Starting KVA of Induction Motors (Jun)(7211).

—Hoists
CNS C1037-70. Standard Dimensions of Electric Motor for Hoist (Dec)(3202). 4 pp.

—Oil Burners
BSI BS 5000: Part 50-82. 1982 Rotating Electrical Machines of Particular Types or for Particular Applications Part 50: Dimensions for Oil Burner Motors. 13 pp.
IEC 681 Pt 1-80. Dimensions of Small Power Motors for Definite Purpose Application Part 1: Oil Burner Motors First Edition. 22 pp.

Induction Units
Use: Air Handling Equipment—Induction Units

Induction Valves
Use: Inlet Valves

Inductive Coordination
See Also: Communication Equipment

—Communication Equipment
CSA C22.3 NO 3-1954. Inductive Co-Ordination Definitions, Principles and Practices (R 1970). 22 pp.

—Communication Equipment—Glossaries
CSA C22.3 NO 3-1954. Inductive Co-Ordination Definitions, Principles and Practices (R 1970). 22 pp.

—Electric Wiring—Outside
CSA C22.3 NO 3-1954. Inductive Co-Ordination Definitions, Principles and Practices (R 1970). 22 pp.

—Electric Wiring—Outside—Glossaries
CSA C22.3 NO 3-1954. Inductive Co-Ordination Definitions, Principles and Practices (R 1970). 22 pp.

—Handbooks
CSA C22.3 NO 3.1-1974SP. Inductive Coordination Handbook for Use with CSA Standard C22.3 No. 3 of the Canadian Electrical Code, Part III. 54 pp.

Inductive Proximity Switches
Use: Proximity Switches—Inductive

Inductors
Use For: Inducers   See Also: Choke Coils; Electric Coils; Electronic Components; Filter Inductors; Fixed Inductors; Power Inductors; Transformers
BSI BS 6600: Part 10-85. 1985 Outline Dimensions of Transformers and Inductors for Use in Telecommunication Electronic Equipment Part 10: Outline Dimensions of Transformers and Inductors Using the Q Range of C-Cores. 19 pp.
BSI BS 7565-92. 1992 Traction Transformers and Inductors (IEC 310: 1991). 35 pp.
CNS C5190-85. General Rules of High Frequency Inductors and Intermediate Frequency Transformers for Electronic Equipment (Jan)(11182).
CNS C6241-85. Method of Test for High Frequency Inductors and Intermediate Frequency Transformers for Electronic Equipment (Jan)(11183).
IEC 310-91. Traction Transformers and Inductors. 68 pp.
IEC 852 Pt 1-86. Outline Dimensions of Transformers and Inductors for Use in Telecommunication and Electronic Equipment Part 1: Transformers and Inductors Using YEI-1 Laminations First Edition. 22 pp.
IEC 1007-90. Transformers and Inductors for Use in Electronic and Telecommunication Equipment—Measuring Methods and Test Procedures First Edition; (Amendment 1-1993). 113 pp.
JIS C 5320-87. General Rules of High Frequency Inductors and Intermediate Frequency Transformers for Electronic Equipment. 16 pp.
JIS C 5321-90. Methods of Test for High Frequency Inductors and Intermediate Frequency Transformers for Electronic Equipment. 44 pp.
JIS C 6435-89. Testing Methods for Low Frequency Transformers and Inductors. 50 pp.

—Custom
IECQ PQC 106-92. Generic Specification for: Custom-Built Transformers and Inductors. 45 pp.

—Ferrite Cores
IECQ QC 250101/JP 0001-88. Detail Specification for Electronic Components; Ferrite Cores; Assessment Level: A. 11 pp.
IECQ QC 250101/JP 0002-88. Detail Specification for Electronic Components; Ferrite Cores; Assessment Level: A. 11 pp.

—Ferrite Cores—RM Type
CECC CECC 25 100-019. NEN CECC 25 100-019 Edition 1; Magnetic Oxide Core Intended for Inductor Applications (En). 8 pp.

—Laminated Cores
BSI BS 7382: Part 1-90. 1990 Laminated Core Packages for Transformers and Inductors Used in Telecommunication and Electronic Equipment Part 1: Dimensions. 11 pp.
IEC 1021 Pt 1-90. Laminated Core Packages for Transformers and Inductors Used in Telecommunication and Electronic Equipment Part 1: Dimensions First Edition. 17 pp.

—Laminated Cores—Designations
IEC 1186 Pt 1-92. Transformers and Inductors for Use in Telecommunication and Electronic Equipment—Designations for Cores and Assemblies Part 1: Laminated Cores First Edition. 20 pp.

—Laminates
BSI BS 6554: Part 1-84. 1984 Laminations for Transformers and Inductors for Use in Telecommunications and Electronic Equipment Part 1: Characteristics and Electrical Testing. 24 pp.
BSI BS 6600: Part 1-87. 1987 Outline Dimensions of Transformers and Inductors for Use in Telecommunication Electronic Equipment Part 1: The Outline Dimensions of Transformers and Inductors Using YE1-1 Laminations. 14 pp.
IEC 740-82. Laminations for Transformers and Inductors for Use in Telecommunication and Electronic Equipment First Edition; (Amendment 1-1991). 57 pp.
IEC 852 Pt 1-86. Outline Dimensions of Transformers and Inductors for Use in Telecommunication and Electronic Equipment Part 1: Transformers and Inductors Using YEI-1 Laminations First Edition. 22 pp.
IEC 852 Pt 3-92. Outline Dimensions of Transformers and Inductors for Use in Telecommunication and Electronic Equipment Part 3: Transformers and Inductors Using YUI-1 Laminations First Edition. 21 pp.

—Magnetic Cores
BSI BS 5347-76. 1976 Amd 1 Silicon-Iron Strip-Wound Cores for Use in Transformers and Inductors for Telecommunication and Electronic Equipment. 16 pp.

INTERNATIONAL AND NON-U.S. NATIONAL STANDARDS
SUBJECT INDEX

Inductors

Inductors (Cont.)
—Magnetic Cores (Cont.)
CECC CECC 25 100-003. UTE C 83-311; Magnetic Oxide Core Intended for Inductor Applications; Class: 8 B; Pot Core: FP 14 X 8 (En, Fr). 10 pp.

CECC CECC 25 100-004. UTE C 83-311; Magnetic Oxide Core Intended for Inductor Applications; Class: 8 B; Pot Core: FP 18 X 11 (En, Fr). 10 pp.

CECC CECC 25 100-005. UTE C 83-311; Magnetic Oxide Core Intended for Inductor Applications; Class: 8 B; Pot Core: FP 22 X 13 (En, Fr). 10 pp.

CECC CECC 25 100-006. UTE C 83-311; Magnetic Oxide Core Intended for Inductor Applications; Class: 8 B; Pot Core: FP 26 X 16 (En, Fr). 10 pp.

CECC CECC 25 100-007. UTE C 83-311; Magnetic Oxide Core Intended for Inductor Applications; Class: 8 B; Pot Core: FP 30 X 19 (En, Fr). 10 pp.

CECC CECC 25 100-008. UTE C 83-311; Magnetic Oxide Core Intended for Inductor Applications; Class: 8 B; Pot Core: FP 36 X 27 (En, Fr). 10 pp.

CECC CECC 25 100-009. UTE C 83-311; Magnetic Oxide Core Intended for Inductor Applications; Class: 8 D; Pot Core: FP 11 X 7 (En, Fr). 10 pp.

CECC CECC 25 100-010. UTE C 83-311; Magnetic Oxide Core Intended for Inductor Applications; Class: 8 D; Pot Core: FP 14 X 8 (En, Fr). 10 pp.

CECC CECC 25 100-011. UTE C 83-311; Magnetic Oxide Core Intended for Inductor Applications; Class: 8 D; Pot Core: FP 18 X 11 (En, Fr). 10 pp.

CECC CECC 25 100-012. UTE C 83-311; Magnetic Oxide Core Intended for Inductor Applications; Class: 8 D; Pot Core: FP 22 Z 13 (En, Fr). 10 pp.

CECC CECC 25 100-013. UTE C 83-311; Magnetic Oxide Core Intended for Inductor Applications; Class: 8 D; Pot Core: FP 26 X 16 (En, Fr). 10 pp.

CECC CECC 25 100-014. UTE C 83-311; Magnetic Oxide Core Intended for Inductor Applications; Class: 8 D; Pot Core: FP 30 X 19 (En, Fr). 10 pp.

CECC CECC 25 100-015. UTE C 83-311; Magnetic Oxide Core Intended for Inductor Applications; Class: 8 D; Pot Core; FP 26 X 22 (En, Fr). 10 pp.

CECC CECC 25 100-016. UTE C 83-311; Magnetic Oxide Core Intended for Inductor Applications; Class: 8 D; Pot Core; RM-6 (En, Fr). 10 pp.

CECC CECC 25 100-017. UTE C 83-311; Magnetic Oxide Core Intended for Inductor Applications; Class: 8 D; Pot Core: RM 6 (En, Fr). 10 pp.

CECC CECC 25 100-027. NEN CECC 25 100-027 Issue 1; Magnetic Oxide Core Intended for Inductor Applications (En). 8 pp.

CECC CECC 25 100-049 ISSUE 1. NL CECC 25 100-049 Issue 1; Magnetic Oxide Core Intended for Inductor Applications Type Number: 4322 022 6726 (En). 1 p.

CECC EN 125 100-91. Sectional Specification: Magnetic Oxide Cores for Inductor Applications (Supersedes CECC 25 100: Issue 1: 1976). 16 pp.

CNS C3175-83. Method of Test for Ferrite Cores for Inductors and Transformers (Jun)(10319).

JIS C 2562-81. Measuring Methods for Ferrite Cores for Inductors and Transformers. 16 pp.

—Magnetic Cores—Adjusters
BSI BS CECC 25400-84. (WITHDRAWN) 1984 Adjusters Used with Magnetic Oxide Cores for Use in Inductors and Tuned Transformers: Sectional Specification (Renumbered as BS EN 125400: 1993). 9 pp.

BSI BS CECC 25401-84. 1984 Adjusters with Magnetic Oxide (Ferrite) Cores for Use in Inductors and Tuned Transformers: Blank Detail Specification. 9 pp.

BSI BS EN 125400-93. 1993 Amd 1 Sectional Specification: Adjusters Used with Magnetic Oxide Cores for Use in Inductors and Tuned Transformers (AMD 7579) February 15, 1993 (T). 13 pp.

CECC CECC 25 401 ISSUE 1-82. Blank Detail Specification: Adjusters Used with Magnetic Oxide (Ferrite) Cores for Use in Inductors and Tuned Transformers (En, Fr, Ge). 21 pp.

CECC EN 125 400-91. Sectional Specification: Ajusters Used with Magnetic Oxide Cores for Use in Inductors and Tuned Transformers (Supersedes CECC 25 400 Issue 1: 1982). 7 pp.

CENELEC EN 125400-91. Sectional Specification: Adjusters Used with Magnetic Oxide Cores for Use in Inductors and Tuned Transformers. 7 pp.

—Magnetic Cores—Adjusters—Telecommunications—Quality Assurance
IEC 723 Pt 5-93. Inductor and Transformer Cores for Telecommunications Part 5: Sectional Specification: Adjusters Used with Magnetic Oxide Cores for Use in Adjustable Inductors and Transformers First Edition. 24 pp.

IEC 723 Pt 5-1-93. Inductor and Transformer Cores for Telecommunications Part 5: Sectional Specification: Adjusters Used with Magnetic Oxide Cores for Use in Adjustable Inductors and Transformers Section 1: Blank Specifica-tion—Assessment Level A First Edition. 20 pp.

Inductors (Cont.)
—Magnetic Cores—Quality Assurance
BSI BS CECC 25100-77. (WITHDRAWN) 1977 Magnetic Oxide Cores for Inductor Applications: Sectional Specification (Renumbered as BS EN 125100: 1993). 18 pp.

BSI BS EN 125100-93. 1993 Amd 1 Sectional Specification: Magnetic Oxide Cores for Inductor Applications (AMD 7577) February 15, 1993 (T). 22 pp.

CENELEC EN 125100-91. Sectional Specification: Magnetic Oxide Cores for Inductor Applications. 16 pp.

—Magnetic Cores—RM Type
CECC CECC 25 100-020 ISSUE 1-79. BS CECC 25 100-020; Magnetic Oxide Cores in Material Class A13 Intenended for Inductors and Transformers in Tuned Circuits for Professional and Industrial Applications up to Approximately 300 KHz (En). 8 pp.

CECC CECC 25 100-021 ISSUE 1-79. BS CECC 25 100-021; Magnetic Oxide Cores in Material Class A13 Intended for Inductors and Transformers in Tuned Circuits for Professional and Industrial Applications up to Approximately 300 KHz (En). 8 pp.

CECC CECC 25 100-022 ISSUE 1-79. BS CECC 25 100-022; Magnetic Oxide Cores in Material Class A13 Intenended for Inductors and Transformers in Tuned Circuits for Professional and Industrial Applications up to Approximately 300 KHz (En). 8 pp.

CECC CECC 25 100-023 ISSUE 1-79. BS CECC 25 100-023; Magnetic Oxide Cores in Material Class A13 Intended for Inductors and Transformers in Tuned Circuits for Professional and Industrial Applications up to Approximately 300 KHz (En). 8 pp.

CECC CECC 25 100-024 ISSUE 1-79. BS CECC 25 100-024; Magnetic Oxide Cores in Material Class A13 Intended for Inductors and Transformers in Tuned Circuits for Professional and Industrial Applications up to Approximately 300 KHz (En). 8 pp.

—Magnetic Cores—Telecommunications
BSI BS 5938: Part 1-84. 1984 Cores for Inductors and Transformers for Telecommunications Part 1: Methods of Measurement. 69 pp.

BSI BS 5938: Part 2-80. 1980 Cores for Inductors and Transformers for Telecommunications Part 2: Guide to the Drafting of Performance Specifications. 19 pp.

BSI BS 7382: Part 1-90. 1990 Laminated Core Packages for Transformers and Inductors Used in Telecommunication and Electronic Equipment Part 1: Dimensions. 11 pp.

BSI BS 9925: Part 0-84. 1984 Inductor and Transformer Cores for Telecommunications Part 0: Generic Specification. 16 pp.

CECC CECC 25 000 ISSUE 1-76. Generic Specification: Inductor and Transformer Cores for Telecommunications (En, Fr, Ge) AMD 1 (En, Fr, Ge). 191 pp.

IEC 367 Pt 1-82. Cores for Inductors and Transformers for Telecommunications Part 1: Measuring Methods Second Edition; (Amendment 1-1984) (Amendment 2-1992). 159 pp.

IEC 367 Pt 2-74. Cores for Inductors and Transformers for Telecommunications Part 2: Guides for the Drafting of Performance Specifications First Edition; (Supplement A-1976) (Amendment 1-1983). 50 pp.

IEC 723 Pt 1-82. Inductor and Transformer Cores for Telecommunications Part 1: Generic Specification First Edition; (IECQ QC 250000). 33 pp.

IEC 723 Pt 2-83. Inductor and Transformer Cores for Telecommunications Part 2: Sectional Specification: Magnetic Oxide Cores for Inductor Applications First Edition; (Amendment 1-1989) (IECQ QC 250100). 27 pp.

IEC 723 Pt 2-1-83. Detail Specification for Electronic Components; Ferrite Cores; Assessment Level: A First Edition; (IECQ QC 250101). 11 pp.

IEC 723 Pt 4-87. Inductor and Transformer Cores for Telecommunications Part 4: Sectional Specification: Magnetic Oxide Cores for Transformers and Chokes for Power Applications First Edition; (IECQ QC 250300). 21 pp.

IEC 723 Pt 4-1-87. Inductor and Transformer Cores for Telecommunications Part 4: Blank Detail Specification: Magnetic Oxide Cores for Transformers and Chokes for Power Applications Assessment Level A First Edition; (IECQ QC 250301). 21 pp.

IEC 1021 Pt 1-90. Laminated Core Packages for Transformers and Inductors Used in Telecommunication and Electronic Equipment Part 1: Dimensions First Edition. 17 pp.

IECQ QC 250000-82. Inductor and Transformer Cores for Telecommunications Part 1: Generic Specifications (IEC 723-1 ED 1). 33 pp.

Inductors (Cont.)
—Magnetic Cores—Telecommunications (Cont.)
IECQ QC 250100-83. Inductor and Transformer Cores for Telecommunications Part 2: Sectional Specification: Magnetic Oxide Cores for Inductor Applications (Amendment 1-1989) (IEC 723-2 ED 1). 26 pp.

IECQ QC 250101-83. Inductor and Transformer Cores for Telecommunications Part 2: Blank Detail Specification: Magnetic Oxide Cores for Inductor Applications; Assessment Level A (IEC 723-2-1 ED 1). 23 pp.

—Magnetic Cores—Telecommunications—Quality Assurance
BSI BS 9925: Part 01.0-84. 1984 Inductor and Transformer Cores for Telecommunications Part 01.0: Magnetic Oxide Cores in Inductor Applications. 11 pp.

BSI BS 9925: Part 01.01-84. 1984 Inductors and Transformer Cores for Telecommunications Part 01.01: Magnetic Oxide Cores for Inductor Applications: Blank Detail Specification: Assessment Level A. 9 pp.

BSI BS CECC 25000-77. 1977 Amd 1 Inductor and Transformer Cores for Telecommunications: Generic Specification. 64 pp.

—Packaging
MOD UK DSTAN 81-80-89. Packaging of Transformers and Inductors Issue 1. 19 pp.

—Pot Cores
CECC CECC 25 100-018. NEN CECC 25 100-018 Edition 1; Magnetic Oxide Core Intended for Inductor Applications (En). 8 pp.

CECC CECC 25 100-044. DIN 45 970 Teil 1117; Magnetic Oxide Cores for Inductor Applications (Ge). 5 pp.

CECC CECC 25 100-045. DIN 45 970 Teil 1118; Magnetic Oxide Cores for Inductor Applications (Ge). 5 pp.

CECC CECC 25 100-046. DIN 45 970 Teil 1119; Magnetic Oxide Cores for Inductor Applications (Ge). 5 pp.

CECC CECC 25 100-048 ISSUE 1. NL CECC 25 100-048 Issue 1; Magnetic Oxide Core Intended for Inductor Applications Type Number: 4322 022 2526 (En). 1 p.

IECQ QC 250101/JP 0001-88. Detail Specification for Electronic Components; Ferrite Cores; Assessment Level: A. 11 pp.

IECQ QC 250101/JP 0002-88. Detail Specification for Electronic Components; Ferrite Cores; Assessment Level: A. 11 pp.

—Printed Circuit Board Hardware—Coil Forms
IEC 852 Pt 2-92. Outline Dimensions of Transformers and Inductors for Use in Telecommunication and Electronic Equipment Part 2: Transformers and Inductors Using YEx-2 Laminations for Printed Wiring Board Mounting First Edition. 23 pp.

—Quality Assurance
BSI BS 9720-89. 1989 Amd 1 Custom-Built Transformers and Inductors of Assessed Quality: Generic Data and Methods of Test (AMD 6348) September 30,. 50 pp.

BSI BS 9741-86. 1986 Sectional Specification for Inductors of Assessed Quality for Use in Electronic Equipment for Capability Approval. 13 pp.

—Reliability Assured—Procurement
MOD UK DSTAN 59-76-81. Transformers and Inductors General Requirements for the Procurement of Transformers and Inductors-Capability Approval Issue 1. 7 pp.

—Telecommunication Equipment
IEC 1007-90. Transformers and Inductors for Use in Electronic and Telecommunication Equipment—Measuring Methods and Test Procedures First Edition; (Amendment 1-1993). 113 pp.

—Wirewound—Magnetic Circuits—Preferred Products List
CECC CECC MUAHAG Vol 4 IS 3-90. Preferred Products List; Magnetic Components and Ferrite Materials (En, Fr, Ge). 46 pp.

—Wirewound—Magnetic Cores—Preferred Products List
CECC CECC MUAHAG Vol 4 IS 3-90. Preferred Products List; Magnetic Components and Ferrite Materials (En, Fr, Ge). 46 pp.

Industrial, Scientific and Medical Apparatus
Use: ISM Equipment

INTERNATIONAL AND NON-U.S. NATIONAL STANDARDS
SUBJECT INDEX
Industrial

Industrial Automation
Use: Automation

Industrial Buildings
Use: Buildings

Industrial Capacitors
Use: Power Capacitors

Industrial Clothing
Use: Protective Clothing

Industrial Control Equipment
Use: Process Control Equipment

Industrial Control Systems
Use: Process Control Systems

Industrial Controllers
Use: Process Controllers

Industrial Equipment
Scope Note: See the subheading Industrial Equipment under specific items. For additional references, consult the following list *See Also:* Agricultural Equipment; Caterpillars (Vehicles); Construction Equipment; Cranes; Dollies; Earthmoving Equipment; Eye Protectors; Forestry Equipment; Industrial Trucks; ISM Equipment; Lift Trucks; Machinery; Monorails; Pallet Trucks; Pallets; Platform Trucks; Process Control Equipment; Protective Clothing; Robots; Tractors

—**Brake Linings—Asbestos**
CNS R2179-85. Asbestos Brake Linings for Industrial Machines (May)(11268).
JIS R 3455-79. Asbestos Brake Linings for Industrial Machines. 15 pp.

—**Classification**
BSI BS 1000: (62/62.9)-86. 1986 Universal Decimal Classification (UDC). English Full Edition (62/62.9): Engineering, Technology in General. Characteristics and Details of Machines, Equipment, Processes and Products. 50 pp.
SNZ NZS/BS 1000 (62/62-9)-86. Universal Decimal Classification Engineering, Technology in General. Characteristics and Details of Machines, Equipment, Plant, Processes and Products. 52 pp.

—**Diesel Engines**
JIS B 8018-89. Test Method of Performance of Small Size Diesel Engines for Land Use. 25 pp.

—**Electrical Equipment**
CENELEC HD 93.2-74. Part 2: Electrical Equipment of Machines Used in Large Series Production Lines (Superseded by EN 60204-1-1985). 3 pp.
CENELEC EN 60 204-1/A1-88. AMD 1 Electrical Equipment of Industrial Machines—Part 1: General Requirements; (Supersedes HD 93.2: 1974). 4 pp.
DIN VDE 0113 Pt 1-86. Electrical Equipment of Industrial Machines; Part 1: General Requirements (Feb). 22 pp.
IEC 204 Pt 1-92. Electrical Equipment of Industrial Machines—Part 1: General Requirements Third Edition; (CENELEC EN 60204-1: 1992). 199 pp.
SAA AS 1543-85. Electrical Equipment of Industrial Machines. 26 pp.

—**Electrical Equipment—Diagrams**
IEC 204 Pt 2-84. Electrical Equipment of Industrial Machines Part 2: Item Designation and Examples of Drawings, Diagrams, Tables and Instructions (Appendices D and E of Publication 204-1) Second Edition. 67 pp.

—**Electronic Equipment**
BSI BS 2771: Part 1-86. 1986 Electronic Equipment for Industrial Machines Part 1: General Requirements. 94 pp.

—**Fuel Oils**
BSI BS 2869: Part 2-88. 1988 Amd 1 Fuel Oils for Non-Marine Use Part 2: Specification for Fuel Oil for Agricultural and Industrial Engines and Burners (Classes A2,C1, C2,D,E,F,G and H) (AMD 6505) February 28, 1991 (Supersedes BS 2869: 1993). 16 pp.

—**Harmonics**
SAA AS 2279.2-91. Disturbances in Mains Supply Networks—Part 2: Limitation of Harmonics Caused by Industrial Equipment. 12 pp.

—**Hydraulic Torque Converters**
JIS D 1007-76. Test Code of Hydraulic Torque Converters for Construction Machineries and Industrial Vehicles (R 1984). 22 pp.

—**Interfaces—Numerical Control**
BSI BS 5782-79. 1979 Interface Between Numerical Controls and Industrial Machines. 24 pp.

Industrial Equipment (Cont.)
—**Interfaces—Numerical Control** (Cont.)
CENELEC HD 373-78. Interface Between Numerical Controls and Industrial Machines. 1 p.
ISO 8867 Pt 1-88. Industrial Asynchronous Data Link and Physical Layer—Part 1: Physical Interconnection and Two-Way Alternate Communication First Edition. 17 pp.
JTC1 8867 Pt 1-88. Industrial Asynchronous Data Link and Physical Layer—Part 1: Physical Interconnection and Two-Way Alternate Communication First Edition. 17 pp.
OSI ISO 8867-1-88. Industrial Asynchronous Data Link and Physical Layer—Part 1: Physical Interconnection and Two-Way Alternate Communication. 17 pp.
OSI ISO DIS 8867-2-88. Industrial Asynchronous Data Link and Physical Layer—Part 2: Two-Way Simultaneous Communication. 16 pp.

—**Manual Controls—Symbols**
SAA AS 1064-87. Agricultural and Light Industrial Equipment—Operator Controls—Symbols. 7 pp.

—**Noise**
CSA Z107.51-M1980. Procedure for In-Situ Measurement of Noise from Industrial Equipment (R 1985); (Gen Instr 1). 19 pp.

—**Power**
SAA AS 2279.2-91. Disturbances in Mains Supply Networks—Part 2: Limitation of Harmonics Caused by Industrial Equipment. 12 pp.

—**Power System Disturbances**
SAA AS 2279.4-91. Disturbances in Mains Supply Networks—Part 4: Limitation of Voltage Fluctuations Caused by Industrial Equipment. 9 pp.

—**Roll Over Protective Structures**
CSA B352-M1980. Rollover Protective Structures (ROPS) for Agricultural, Construction, Earthmoving, Forestry, Industrial, and Mining Machines; (Gen Instr 1). 76 pp.

—**Safety**
CEN PREN 746-1-92. Industrial Thermoprocessing Equipment—Part 1: Common Safety Requirements for Industrial Thermoprocessing Equipment. 75 pp.
CEN PREN 746-2-92. Industrial Thermoprocessing Equipment—Part 2: Safety Requirements for Combustion and Fuel Handling Systems. 62 pp.
CEN PREN 746-3-92. Industrial Thermoprocessing Equipment—Part 3: Safety Requirements for Controlled Atmospheres. 25 pp.

—**Tires**
BSI BS AU 50: Pt 1: Sec 5D-77. 1977 Tyres and Wheels Part 1: Tyres Section 5D: Industrial Vehicle Tyres. 10 pp.
CNS K4030-80. Dimensions of Tires for Industrial Vehicles and off the Road Service (Feb)(3673). 33 pp.
CNS K4044-84. Dimensions of Solid Tires for Industrial Vehicles (Sep)(4878). 6 pp.
CNS K4047-89. Tires for Industrial Vehicles (Oct)(4881). 4 pp.
JIS D 6401-81. Dimensions and Load Ratings of Tires for Industrial Vehicles and off the Road Service. 45 pp.
JIS D 6403-87. Tires for Industrial Vehicles. 8 pp.
JIS D 6405-90. Dimensions of Solid Tires for Industrial Vehicles. 15 pp.

—**Tires—Load Ratings**
JIS D 6401-81. Dimensions and Load Ratings of Tires for Industrial Vehicles and off the Road Service. 45 pp.

—**Vibration—Vehicular**
BSI BS 6794-86. 1986 Reporting Measured Vibration Data for Land Vehicles. 14 pp.
ISO 8002-86. Mechanical Vibrations—Land Vehicles—Method for Reporting Measured Data First Edition. 12 pp.

—**Voltage Fluctuations**
SAA AS 2279.4-91. Disturbances in Mains Supply Networks—Part 4: Limitation of Voltage Fluctuations Caused by Industrial Equipment. 9 pp.

—**Wheels**
CNS D1060-83. Contours of Rims for Industrial Vehicles and off the Road Service (Oct)(10620).
CNS D2131-82. Fastening Dimensions of Two-Piece Divided Rims for Industrial Vehicles (May)(8816).
JIS D 6402-85. Contours of Rims for Industrial Vehicles and off the Road Service. 31 pp.
JIS D 6407-65. Fastening Dimensions of Two-Piece Divided Rims for Industrial Vehicles.

Industrial Jewel Bearings
Use: Jewel Bearings

Industrial Land Use
Use: Land Use Planning

Industrial Plants
Scope Note: For additional listings, see the subheading Industrial under specific types of equipment *Use For:* Factories; Manufacturing Plants; Processing Plants *See Also:* Asphalt Plants; Chemical Plants; Gas Processing Plants; Gasification Plants; Glass Plants; Nuclear Fuel Reprocessing Plants; Shops; Waste Water Treatment Plants

—**Asbestos Emissions**
ISO 10397-93. Stationary Source Emissions—Determination of Asbestos Plant Emissions—Method by Fibre Count Measurement First Edition. 26 pp.

—**Classification**
BSI BS 1000: (62/62.9)-86. 1986 Universal Decimal Classification (UDC). English Full Edition (62/62.9): Engineering, Technology in General. Characteristics and Details of Machines, Equipment, Processes and Products. 50 pp.
SNZ NZS/BS 1000 (62/62-9)-86. Universal Decimal Classification Engineering, Technology in General. Characteristics and Details of Machines, Equipment, Plant, Processes and Products. 52 pp.

—**Diagrams**
DIN ENGL 28004 Pt 1-88. Flow Sheets and Diagrams of Process Plants; Concepts, Types of Diagram, Information Content (May) (Supersedes DIN 28004 Part 10, August 1976 Edition). 16 pp.
DIN ENGL 28004 Pt 2-88. Flow Sheets and Diagrams of Process Plants; Drawing Instructions (May). 3 pp.

—**Ergonomics**
SAA AS 1837-76. Code of Practice for Application of Ergonomics to Factory and Office Work. 32 pp.

—**Flow Charts**
BSI BS 5070: Part 3-88. 1988 Engineering Diagram Drawing Practice Part 3: Recommendations for Mechanical/Fluid Flow Diagrams. 44 pp.
DIN ENGL 28004 Pt 1-88. Flow Sheets and Diagrams of Process Plants; Concepts, Types of Diagram, Information Content (May) (Supersedes DIN 28004 Part 10, August 1976 Edition). 16 pp.
DIN ENGL 28004 Pt 2-88. Flow Sheets and Diagrams of Process Plants; Drawing Instructions (May). 3 pp.
SNZ BS 5070: Part 3-88. Engineering Diagram Drawing Practice Part 2: Recommendations for Mechanical/Fluid Flow Diagrams. 4 pp.

—**Flow Charts—Symbols**
DIN ENGL 28004 Pt 3-88. Flow Sheets and Diagrams of Process Plants; Graphical Symbols (May). 27 pp.
DIN ENGL 28004 Pt 4-77. Flow Sheets and Diagrams of Process Plants; Symbols (May). 2 pp.

—**Noise**
BSI BS 4142-90. 1990 Method of Rating Industrial Noise Affecting Mixed Residential and Industrial Areas. 16 pp.
CSA CAN/CSA-Z107.55-M86. Recommended Practice for the Prediction of Sound Levels Received at a Distance from an Industrial Plant; (Gen Instr 1). 24 pp.
DIN ENGL 45635 Pt 49-86. Measurement of Noise Emitted by Machines; Airborne Noise Emission; Enveloping Surface Method; Surface Treatment Plants (May). 16 pp.

—**Power Supplies**
CSA CAN/CSA-C22. 2NO107.1-M91. Commercial and Industrial Power Supplies; (Gen Instr 1 Thru 2). 112 pp.

—**Reference Models**
BSI DD 203: Part 1-91. 1991 Industrial Automation: Shop Floor Production Part 1: Reference Model for Standardization and a Methodology for Indentification of Requirements. 31 pp.
BSI DD 203: Part 2-91. 1991 Industrial Automation: Shop Floor Production Part 2: Application of the Reference Model for Standardization and Methodology (G) (ISO/TR 10314-2: 1991). 61 pp.
ISO TR10314 Pt 1-90. Industrial Automation—Shop Floor Production—Part 1: Reference Model for Standardization and a Methodology for Identification of Requirements First Edition. 30 pp.
ISO TR10314 Pt 2-91. Industrial Automation—Shop Floor Production—Part 2: Application of the Reference Model for Standardization and Methodology First Edition. 60 pp.

—**Service Contracts**
MOD UK DEFCON 112BR-86. Conditions of Agreement Hire of Contractor's Plant 10/86. 7 pp.
MOD UK DEFCON 112J-83. Conditions of Contract—Plant 9/83. 1 p.

INDUSTRY STANDARDS

INTERNATIONAL AND NON-U.S. NATIONAL STANDARDS
SUBJECT INDEX

Industrial

Industrial Plants (Cont.)
—Tray Cables
 CSA C22.2 NO 230-M1988. Tray Cables (R 1993); (Gen Instr 1 Thru 2). 20 pp.

Industrial Process Control Systems
Use: Process Control Systems

Industrial Robots
Use: Robots

Industrial Safety
Use: Occupational Safety and Health

Industrial Standards
Use: Standards

Industrial Trucks
Scope Note: For additional listings, use a more specific term *See Also:* Automatic Guided Vehicles; Dollies; Electric Trucks; Electric Vehicles; Forklifts; Ground Vehicles; Hand Trucks; Industrial Equipment; Lift Trucks; Motor Vehicles; Pallet Trucks; Platform Trucks; Self Propelled Machinery; Stillage Trucks; Tank Trucks; Tow Tractors; Trucks
 DIN VDE 0525-83. Rotating Electrical Machines for Industrial Trucks Driven by Electric Motors (Mar). 14 pp.
 EC 86/663/EEC-86. Council Directive on the Approximation of the Laws of the Member States Relating to Self-Propelled Industrial Trucks. 40 pp.
 EC 89/240/EEC-88. Commission Directive Adapting to Technical Progress Council Directive 86/663/EEC on the Approximation of the Laws of the Member States Relating to Self-Propelled Industrial Trucks. 75 pp.
 SNZ NZS/ANSI/NFPA 505-92. Powered Industrial Trucks Including Type Designations, Areas of Use, Maintenance and Operation. 15 pp.
 SNZ NZS/ANSI/UL 558-84. Industrial Trucks, Internal Combustion Engine-Powered. 29 pp.

—Battery Operated
 DIN VDE 0117-86. Battery Powered Industrial Trucks (Apr). 22 pp.
 SNZ NZS/ANSI/UL 583-84. Electric-Battery-Powered Industrial Trucks. 31 pp.

—Brakes
 ISO 6292 Pt 1-81. Powered Industrial Trucks—Brake Performance—Part 1: High-Lift, Low-Lift and Non-Lifting First Edition. 5 pp.
 ISO 6500-80. Powered Industrial Trucks—Service Brakes—Component Strength-Performance Requirements First Edition; (Amendment Slip-1980). 4 pp.

—Designations
 SNZ NZS/ANSI/NFPA 505-92. Powered Industrial Trucks Including Type Designations, Areas of Use, Maintenance and Operation. 15 pp.

—Glossaries
 ISO 5053-87. Powered Industrial Trucks—Terminology First Edition. 40 pp.
 SAA AS 1763-85. Industrial Trucks—Glossary of Terms. 20 pp.

—Lead Acid Batteries
 BSI BS EN 60095-4-93. 1993 Lead-Acid Starter Batteries Part 4: Dimensions of Batteries for Heavy Commercial Vehicles (E). 14 pp.
 CENELEC EN 60095-4-93. Lead-Acid Starter Batteries Part 4: Dimensions of Batteries for Heavy Commercial Vehicles. 11 pp.
 CENELEC EN 60095-4-93. Lead-Acid Starter Batteries Part 4: Dimensions of Batteries for Heavy Commercial Vehicles (IEC 95-4: 1989, Modified). 9 pp.
 IEC 95 Pt 4-89. Lead-Acid Starter Batteries Part 4: Dimensions of Batteries for Heavy Trucks First Edition (CENELEC EN 60095-4: 1993). 16 pp.

—Maintenance
 SNZ NZS/ANSI/NFPA 505-92. Powered Industrial Trucks Including Type Designations, Areas of Use, Maintenance and Operation. 15 pp.

—Manual Controls—Symbols
 BSI BS 5829-79. 1979 Control Symbols for Powered Industrial Trucks. 12 pp.
 ISO 3287-78. Powered Industrial Trucks—Control Symbols First Edition. 11 pp.

—Operation
 SNZ NZS/ANSI/NFPA 505-92. Powered Industrial Trucks Including Type Designations, Areas of Use, Maintenance and Operation. 15 pp.

—Pedals
 BSI BS 7178-89. 1989 Construction and Layout of Pedals of Self-Propelled Industrial Trucks Sit-Down Rider-Controlled. 8 pp.

Industrial Trucks (Cont.)
—Pedals (Cont.)
 CEN PREN281-88. Self-Propelled Industrial Trucks Sit-Down Rider-Controlled; Rules for the Construction and Layout of Pedals. 8 pp.
 CEN EN 281-88. Self Propelled Industrial Trucks Sit-Down Rider-Controlled: Rules for the Construction and Layout of Pedals. 4 pp.

—Safety
 BSI BS 4430: Part 2-69. 1969 Safety of Powered Industrial Trucks Part 2: Operation and Maintenance. 16 pp.
 ISO 3691-80. Powered Industrial Trucks—Safety Code Second Edition; (Amendment 1-1983). 39 pp.

—Tires
 ISO 10499-91. Industrial Tyres and Rims—Solid Tyres (Metric Series) for Pneumatic Tyre Rims—Designation, Dimensions and Marking First Edition. 6 pp.
 ISO 10500-91. Industrial Tyres and Rims—Cylindrical and Conical Base Rubber Solid Tyres (Metric Series)—Designation, Dimensions and Marking First Edition. 8 pp.

Industrial Wastes
See Also: Air Pollution; Black Liquors; Effluents; Exhaust Gases; Solid Wastes; Spent Liquors; Waste Disposal; Waste Water Analysis

—Civil Engineering
 BSI BS 6543-85. 1985 Guide to Use of Industrial By-Products and Waste Materials in Building and Civil Engineering. 40 pp.

—Construction
 BSI BS 6543-85. 1985 Guide to Use of Industrial By-Products and Waste Materials in Building and Civil Engineering. 40 pp.

—Laboratory—Certification
 CSA Z201-M1990. Qualification Code for Ontario for Laboratories Analyzing Industrial Waste; (Gen Instr 1). 123 pp.

—Sampling
 BSI BS 6068: Sec 6.10-93. 1993 Water Quality Part 6: Sampling Section 6.10: Guidance on Sampling of Waste Waters (ISO 5667-10: 1992). 15 pp.
 ISO 5667 Pt 10-92. Water Quality—Sampling—Part 10: Guidance on Sampling of Waste Waters First Edition. 15 pp.
 JIS K 0060-92. Sampling Method of Industrial Wastes. 34 pp.

Industrial Water
Use For: Process Water, Industrial *See Also:* Boiler Water; Cooling Water; Feedwater; Waste Water Analysis; Water; Water Treatment
 JIS K 0101-91. Testing Methods for Industrial Water.

—Acidity
 CPPA H.4P(F)-67. Determination of Acidity in Process Waters. 2 pp.

—Alkalinity
 CPPA H.4P(D)-67. Determination of Alkalinity in Process Waters. 2 pp.

—Aluminum Content
 CPPA H.4P(J)-67. Analysis of Process Waters Aluminum. 2 pp.

—Calcium Content—Volumetric Analysis
 CPPA H.4P(K)-67. Analysis of Process Waters Calcium and Magnesium. 2 pp.

—Chemical Analysis
 CPPA H.4P(A)-67. Analysis of Process Waters. 1 p.

—Chloride Content
 CPPA H.4P(N)-67. Analysis of Process Waters Chloride. 2 pp.

—Chlorine Content
 CPPA H.4P(C)-67. Determination of Free Chlorine in Process Waters. 1 p.

—Corrosion Testing
 JIS K 0100-90. Testing Method for Corrosivity of Industrial Water. 32 pp.

—Halogenated Hydrocarbon Content
 JIS K 0125-90. Testing Methods for Determination of Low Molecular Weight Halogenated Hydrocarbons in Industrial Water and Waste Water. 21 pp.

—Iron Content
 CPPA H.4P(I)-67. Analysis of Process Waters Iron. 2 pp.

—Magnesium Content—Volumetric Analysis
 CPPA H.4P(K)-67. Analysis of Process Waters Calcium and Magnesium. 2 pp.

Industrial Water (Cont.)
—Manganese Content
 CPPA H.4P(L)-67. Analysis of Process Waters Manganese. 2 pp.

—Oil Content Analysis
 CPPA H.4P(G)-67. Determination of Oil in Process Waters. 2 pp.

—PH
 CPPA H.4P(E)-67. Hydrogen Ion Concentration (pH) of Process Waters. 2 pp.

—Sampling
 JIS K 0094-85. Sampling Methods for Industrial Water and Industrial Waste Water. 60 pp.

—Silica Content
 CPPA H.4P(H)-67. Analysis of Process Waters Silica. 3 pp.

—Sulfate Content
 CPPA H.4P(M)-67. Analysis of Process Waters Sulphate. 3 pp.

—Water Hardness
 CPPA H.4P(B)-67. Determination of Hardness in Process Waters. 2 pp.
 DIN ENGL 38409 Pt 6-86. German Standard Methods for the Examination of Water, Waste Water and Sludge; Summary Indices of Actions and Substances (Group H); Water Hardness (H 6) (Jan).

Industries
Scope Note: Use a more specific term
See: Chemical Industry; Food Industry; Gas Industry; Lumbering Industry; Metal Industry; Paper Industry; Petroleum Industry; Textile Industry

Ineffective Calls
Use: Unsuccessful Calls

Inert Gas Content Analysis
Use: Noble Gas Content Analysis

Inert Gas Shielded Arc Welding Equipment
Use: Inert Gas Welding Equipment

Inert Gas Welding Equipment
Use For: Inert Gas Shielded Arc Welding Equipment
See Also: Arc Welding Equipment; Gas Tungsten Arc Welding Equipment; Shielded Metal Arc Welding Equipment; Welding Equipment

—Electrodes
 JIS Z 3341-83. Copper and Copper Alloy Rods and Wires for Inert Gas Shielded Arc Welding. 10 pp.

—Welding Rods
 JIS Z 3341-83. Copper and Copper Alloy Rods and Wires for Inert Gas Shielded Arc Welding. 10 pp.

Inertial Navigation Systems
Use For: INS *See Also:* Air Navigation; Avionics; Navigation; Navigational Aids; Radio Navigation Equipment

—Aircraft
 NATO STANAG 4219 ED 1 AMD 1-84. System Test Procedure for an Aircraft Inertial Navigation System. 46 pp.

—Aircraft—Accelerometers
 NATO STANAG 4218 ED 1 AMD 2-84. Accelerometer Test Procedure for an Aircraft Inertial Navigation System. 31 pp.

—Positional Information—Aircraft Landing Areas
 NATO STANAG 7010 ED 1 AMD 1-89. Positions for INS Settings on Airfields. 9 pp.

Infant Foods
Use: Baby Foods

Infant Formulas
Use: Baby Foods—Formulas

Infant Incubators
Use For: Infant Warmers *See Also:* Incubators; Medical Equipment

—Electric
 SAA AS T47-70. Electrically Heated Incubators for Babies. 18 pp.

INTERNATIONAL AND NON-U.S. NATIONAL STANDARDS
SUBJECT INDEX

Infant Incubators *(Cont.)*
—Safety
BSI BS 5724: Sec 2.20-82. (WITHDRAWN) 1982 Medical Electrical Equipment Part 2: Particular Requirements for Safety Section 2.20: Nursing Incubators (Superseded by BS 5724: Section 2.119: 1991). 17 pp.
BSI BS 5724: Sec 2.21-83. (WITHDRAWN) 1983 Medical Electrical Equipment Part 2: Particular Requirements for Safety Section 2.21: Transport Incubators (Superseded by BS 5734: Section 2.120: 1991). 16 pp.
BSI BS 5724: Sec 2.119-91. 1991 Medical Electrical Equipment Part 2: Particular Requirements for Safety Section 2.119: Specification for Baby Incubators (N). 33 pp.
BSI BS 5724: Sec 2.119-01. 1991 Amd 1 Medical Electrical Equipment Part 2: Particular Requirements for Safety Section 2.119: Specification for Baby Incubators (IEC 601-2-19: 1990) (AMD 7319) December 15, 1992. 37 pp.
BSI BS 5724: Sec 2.120-91. 1991 Medical Electrical Equipment Part 2: Particular Requirements for Safety Section 2.120: Specification for Transport Incubators (IEC 601-2-20: 1990). 35 pp.
BSI BS 5724: Sec 2.120-01. 1991 Amd 1 Medical Electrical Equipment Part 2: Particular Requirements for Safety Section 2.120: Specification for Transport Incubators (IEC 601-2-20: 1990) (AMD 7320) December 15, 1992. 39 pp.
CENELEC HD 395.2.19 S1-92. Medical Electrical Equipment Part 2: Particular Requirements for Safety of Baby Incubators. 6 pp.
CENELEC HD 395.2.20 S1-92. Medical Electrical Equipment Part 2: Particular Requirements for Safety of Transport Incubators. 6 pp.
CSA CAN/CSA-C22. 2601.2.19-92. Medical Electrical Equipment Part 2: Particular Requirements for the Safety of Baby Incubators (IEC 601-2-19:1990); (Gen Instr 1). 47 pp.
CSA CAN/CSA-C22. 2601.2.20-92. Medical Electrical Equipment Part 2: Particular Requirements for the Safety of Transport Incubators (IEC 601-2-20:1990); (Gen Instr 1). 50 pp.
DIN VDE 0750 Pt 217-87. Medical Electrical Equipment; Transport Incubators; Special Safety Regulations (Feb). 39 pp.
DIN VDE 0750 Pt 227 (D)-88. Medical, Electrical Equipment; Infant Radiant Warmers; Special Regulations for Safety (Oct). 26 pp.
IEC 601 Pt 2-19-90. Medical Electrical Equipment Part 2: Particular Requirements for Safety of Baby Incubators First Edition. 61 pp.
IEC 601 Pt 2-20-90. Medical Electrical Equipment Part 2: Particular Requirements for Safety of Transport Incubators First Edition. 65 pp.

Infant Warmers
Use: Infant Incubators

Infection Hazards
Use: Biological Hazards

Infiltration Capacity
See Also: Permeability

—Plastic Pipes—PVC—Unplasticized
SAA AS 1462.8-84. Methods of Test for Unplasticized PVC (UPVC) Pipes and Fittings—Part 8: Method of Test for Liquid Infiltration. 2 pp.

Inflammability
Use: Flammability Testing

Inflatable Boats
Use: Boats (Marine)—Inflatable

Inflatable Life Rafts
Use: Life Rafts—Inflatable

Information Centers
—Explosive Ordnance Disposal
NATO STANAG 2834 ED 1 AMD 7-75. The Operation of the Explosive Ordnance Disposal Technical Information Centre (EODTIC). 10 pp.

Information Exchange
Use: Information Interchange

Information Interchange
Scope Note: For additional listings, use a more specific term *Use For:* Data Interchange
See Also: Alphanumeric Character Sets; Answerback; Bureaufax Communications; Call Holding; CD ROMs; CD WOMs; Coded Character Sets; Codes; Computer Tapes; Computers; Control Characters; Control Procedures; Data Blocks; Data Compression; Data Processing; Data Transmission; Disk Cartridges; Dot Matrix Printers; EDI Messaging Services; EDIFACT; FAX Communications; Flow Control Protocols; Information Security; Magnetic Tapes;

Information Interchange *(Cont.)*
See Also: (Cont.)
Message Handling Systems; Message Routing; Message Stores; Military Geographic Information; Mixed Mode; Open Systems Interconnection; Processable Mode Number One; Protocols; Punched Cards; Software; Telecommunication Connections; Teletex Communications; Videotex Communications
BSI PP 7315-87. (WITHDRAWN) 1987 Standardization for Information Technology. 50 pp.
CENELEC HD 40 001-85. Section 2: Requirements for Information Technology Equipment. 10 pp.
JIS X 4001-89. File Specification for Japanese Documents Interchange (Basic Type). 40 pp.
JIS X 4002-87. File Specification for Japanese Documents Interchange (Business Graph Type). 58 pp.
JIS X 4004-89. File Specification for Japanese Documents Interchange (Raster Graphics Type). 55 pp.

—Aerial Delivery
NATO STANAG 3428 ED 4 AMD 2-88. Exchange of Information on Aerial Delivery Systems. 10 pp.

—Air Cargo Handling Equipment—Transport Aircraft
NATO STANAG 3543 ED 3 AMD 3-86. Air Transport Cargo/Passenger Handling Systems—Request for Information. 14 pp.
NATO STANAG 3543 ED 3 AMD 4-86. Air Transport Cargo/Passenger Handling Systems—Request for Information. 17 pp.

—Airborne Operations
NATO STANAG 3572 ED 1 AMD 8-77. Exchange of Information on Tactical Air Transport Operations. 8 pp.
NATO STANAG 3982 ED 1 AMD 2-85. Messages Supplement to ATP-27—Offensive Air Support Operations (Withdrawn 93-02). 10 pp.

—Aircraft Accidents
NATO STANAG 3101 ED 8 AMD 1-89. Exchange of Accident/Incident Information Concerning Aircraft and Missiles. 35 pp.
NATO STANAG 3101 ED 8 AMD 2-89. Exchange of Accident/Incident Information Concerning Aircraft and Missiles. 36 pp.
NATO STANAG 3101 ED 9 AMD 0-93. Exchange of Accident/Incident Information Concerning Aircraft and Missiles. 36 pp.

—Architectural—Computer Graphics
SAA AS 3643.2-92. Computer Graphics—Initial Graphics Exchange Specification (IGES) for Digital Exchange of Product Definition Data—Part 2: Subset of AS 3643.1—Two-Dimensional Drawings for Architectural, Engineering and Construction (AEC) Industries. 37 pp.

—ATM Adaptation Layer (OSI)
CCITT RECMN I.363-91. B-ISDN ATM Adaptation Layer (AAL) Specification (Study Group XVIII) 21 pp. 21 pp.

—Bibliographic
OSI ISO DIS 6862.2-90. Documentation—Mathematical Coded Character Set for Bibliographic Information Interchange. 18 pp.

—Bibliographic Data Elements
BSI BS 7124: Part 1-89. 1989 Bibliographic Data Element Directory Part 1: Interloan Applications. 26 pp.
ISO 8459 Pt 1-88. Documentation—Bibliographic Data Element Directory—Part 1: Interloan Applications First Edition. 24 pp.

—Check Characters
BSI BS 6541-85. 1985 Check Character Systems for Use in Information Interchange, and Guidance on Choice and Methods of Application. 16 pp.
CNS C5200-85. Data Processing (Check Character Systems) (Nov)(11402).
ISO 7064-83. Data Processing—Check Character Systems First Edition. 15 pp.
JTC1 7064-83. Data Processing—Check Character Systems First Edition. 15 pp.
OSI ISO 7064-83. Data Processing Character Systems. 15 pp.

—Computer Graphics
SNZ NZS/AS 3643. 1-89. Computer Graphics—Initial Graphics Exchange Specification (IGES) for Digital Exchange of Product Definition Data Part 1: General (This is a Joint Standard with SAA AS 3643.1). 516 pp.

—Computer Programs—CCIR
CCIR RESOLUTION 104-90. Access to and Exchange of Computer Programs—Volume XIV—Administrative Texts of the CCIR. 2 pp.

Information Interchange *(Cont.)*
—Computer Programs—Radio Spectra
CCIR RECMN 668-90. Methods of Exchanging Computer Programs and Data for Spectrum Management Purposes—Section 1A—Spectrum Engineering and Computer-Aided Principles and Techniques. 1 p.
CCIR QUESTION 65/1-86. Improved Methods of Exchanging Computer Programs and Data for Spectrum Management Purposes—Questions Concerning Study Group 1 —Spectrum Management Techniques (Spectrum Engineering, Planning, Sharing, Monitoring and Utilization). 1 p.

—Container Equipment Data Exchange (CEDEX)—Codes
ISO 9897 Pt 1-90. Freight Containers—Container Equipment Data Exchange (CEDEX)—Part 1: General Communication Codes First Edition. 63 pp.
ISO 9897 Pt 3-90. Freight Containers—Container Equipment Data Exchange (CEDEX)—Part 3: Message Types for Electronic Data Interchange First Edition. 31 pp.
OSI ISO 9897-3-90. Freight Containers—Container Equipment Data Exchange (CEDEX)—Part 3: Message Types for Electronic Data Interchange. 31 pp.

—Corporate Body Names
JIS X 0802-89. Description of Name of Corporate Body for Information Interchange.

—Data Descriptive Files
BSI BS 6690-86. 1986 Data Descriptive File for Information Interchange. 30 pp.
IEC DIS 8211-93. Information Technology—Specification for a Data Descriptive File for Information Interchange (Revision of First Edition (ISO 8211:1985)) ***CD-ROM ONLY***. 77 pp.
ISO 8211-85. Information Processing—Specification for a Data Descriptive File for Information Interchange First Edition. 30 pp.
ISO DIS 8211-93. Information Technology—Specification for a Data Descriptive File for Information Interchange (Revision of First Edition (ISO 8211:1985)) ***CD-ROM ONLY***. 77 pp.
JIS X 0604-86. Specification for a Data Descriptive File for Information Interchange. 44 pp.
JTC1 8211-85. Information Processing—Specification for a Data Descriptive File for Information Interchange First Edition. 30 pp.
JTC1 DIS8211-93. Information Technology—Specification for a Data Descriptive File for Information Interchange (Revision of First Edition (ISO 8211:1985)) ***CD-ROM ONLY***. 77 pp.
OSI ISO 8211-85. Information Processing—Specification for a Data Descriptive File for Information Interchange.

—Data Elements
BSI BS 7151-89. (WITHDRAWN) 1989 Amd 1 Representation of Dates and Times in Information Interchange (G) (ISO 8601: 1988) (AMD 6816) September 30, 1991 (Renumbered as BS EN 28601: 1992). 20 pp.
BSI BS EN 28601-92. 1992 Data Elements and Interchange Formats—Information Interchange—Representation of Dates and Times (ISO 8601: 1988) (AMD 7825) June 15, 1993 (E). 27 pp.
BSI DD 75: Sec 2.1-82. (WITHDRAWN) 1982 Amd 1 Structure and Representation of Data for Interchange at the Application Level (DIAL) Part 2: Recommendations for the Basic Directory of Data Elements and Groups: Section 2.1: General Information About the Directory and Its Implementation. 9 pp.
BSI DD 75: Sec 2.2-82. (WITHDRAWN) 1982 Amd 1 Structure and Representation of Data for Interchange at the Application Level (DIAL) Part 2: Recommendations for the Basic Directory of Data Elements and Groups: Section 2.2: The DIAL Directory Addendum 1-1/31/84 (Sup. by BS 8104: 1992). 394 pp.
CEN EN 27 372-89. Trade Data Interchange—Trade Data Elements Directory. 5 pp.
CEN EN 28601-92. Data Elements and Interchange Formats—Information Interchange—Representation of Dates and Times (ISO 8601, 1st Edition 1988 and Technical Corrigendum 1: 1991). 16 pp.
CNS C5118-89. Representation for Calendar Date and Original Date for Information Interchange (Jul)(7648).
CSA CAN/CSA-Z234.5-89. Data Elements and Interchange Formats—Information Interchange—Representation of Dates and Times (ISO 8601:1988). 24 pp.
ISO 8601-88. Data Elements and Interchange Formats—Information Interchange—Representation of Dates and Times First Edition; (Supersedes 2014, 2015, 2711, 3307, and 4031) (Corrigendum 1-1991) (CEN EN 28601: 1992). 19 pp.

INDUSTRY STANDARDS

INTERNATIONAL AND NON-U.S. NATIONAL STANDARDS
SUBJECT INDEX
Information

Information Interchange *(Cont.)*

—**Data Elements** *(Cont.)*
  JTC1 8601-88. Data Elements and Interchange Formats—Information Interchange—Representation of Dates and Times First Edition; (Supersedes 2014, 2015, 2711, 3307, and 4031) (Corrigendum 1-1991). 19 pp.
  OSI ISO DP 7352-1.3-84. Guidelines for the Organization and Representation of Data Elements for Data Interchange—Part 1: Basis for the Interchange of Data.
  OSI ISO DP 7352-2.3-84. Guidelines for the Organization and Representation of Data Elements for Data Interchange—Part 2: Coding Methods and Principles.
  OSI ISO DIS 8601-86. Data Elements and Interchange Formats—Information Interchange-Representation of Dates and Times. 25 pp.
  SNZ NZS/ISO 8601-88. Data Elements and Interchange Formats (This is a Joint Standard with SAA AS 3802). 14 pp.

—**Data Elements—Directories**
  CGSB CAN/CGSB-200.9-91. Trade Data Elements Directory (UNTDED 1990). 326 pp.

—**Data File Formats—Magnetic Tapes**
  NATO STANAG 4191 ED 1 AMD 0-86. (DRAFT) Test Format for the Transferral of Multidimensional Data and Source Programmes. 20 pp.

—**Data Formats**
  BSI BS 7151-89. (WITHDRAWN) 1989 Amd 1 Representation of Dates and Times in Information Interchange (G) (ISO 8601: 1988) (AMD 6816) September 30, 1991 (Renumbered as BS EN 28601: 1992). 20 pp.
  BSI BS EN 28601-92. 1992 Data Elements and Interchange Formats—Information Interchange—Representation of Dates and Times (ISO 8601: 1988) (AMD 7825) June 15, 1993 (E). 27 pp.
  CEN EN 28601-92. Data Elements and Interchange Formats—Information Interchange—Representation of Dates and Times (ISO 8601, 1st Edition 1988 and Technical Corrigendum 1: 1991). 16 pp.
  CNS C5118-89. Representation for Calendar Date and Original Date for Information Interchange (Jul)(7648).
  CSA CAN/CSA-Z234.5-89. Data Elements and Interchange Formats—Information Interchange—Representation of Dates and Times (ISO 8601:1988). 24 pp.
  ISO 8601-88. Data Elements and Interchange Formats—Information Interchange—Representation of Dates and Times First Edition; (Supersedes 2014, 2015, 2711, 3307, and 4031) (Corrigendum 1-1991) (CEN EN 28601: 1992). 19 pp.
  JTC1 8601-88. Data Elements and Interchange Formats—Information Interchange—Representation of Dates and Times First Edition; (Supersedes 2014, 2015, 2711, 3307, and 4031) (Corrigendum 1-1991). 19 pp.
  OSI ISO DIS 8601-86. Data Elements and Interchange Formats—Information Interchange-Representation of Dates and Times. 25 pp.
  SNZ NZS/ISO 8601-88. Data Elements and Interchange Formats (This is a Joint Standard with SAA AS 3802). 14 pp.

—**Data Processing Equipment—Radio Links**
  NATO STANAG 4202 ED 2 AMD 0-88. Transmission Envelope Characteristics for High Reliability Data Exchange Between Land Tactical Data Processing Equipment over Single Channel Radio Links. 30 pp.

—**Electronic Data Interchange—Neutral Transfer Format**
  BSI BS 7567: Part 1-92. 1992 Electronic Transfer of Geographic Information (NTF) Part 1: Specification for NTF Stuctures. 76 pp.
  BSI BS 7567: Part 2-92. 1992 Electronic Transfer of Geographic Information (NTF) Part 2: Specification for Implementing Plain NTF. 52 pp.
  BSI BS 7567: Part 3-92. 1992 Electronic Transfer of Geographic Information (NTF) Part 3: Specification for Implementing NTF Using BS 6690. 29 pp.

—**Electronic Prepress Systems**
  BSI BS 7607-92. 1992 Graphic Technology—Prepress Digital Data Exchange—Colour Picture Data on Magnetic Tape (ISO 10755: 1992) (H). 15 pp.
  ISO 10755-92. Graphic Technology—Prepress Digital Data Exchange—Colour Picture Data on Magnetic Tape First Edition; (ANSI IT8.1-1988). 14 pp.
  JIS X 0651-91. User Exchange Format for the Exchange of Color Picture Data Between Electronic Prepress Systems Via Magnetic Tape. 24 pp.
  JTC1 10755-92. Graphic Technology—Prepress Digital Data Exchange—Colour Picture Data on Magnetic Tape First Edition; (ANSI IT8.1-1988). 14 pp.

Information Interchange *(Cont.)*

—**Electronic Prepress Systems** *(Cont.)*
  OSI ISO/DIS 10755-90. User Exchange Format (UEF00) for the Exchange of Colour Picture Data Between Electronic Prepress Systems Via Magnetic Tape (DDE500). 18 pp.
  OSI ISO/DIS 10756-90. User Exchange Format (UEF01) for the Exchange of Line of Data Between Electronic Prepress Systems Via Magnetic Tape (DDE500). 21 pp.
  OSI ISO/DIS 10757-90. User Exchange Format (UEF02) for the Exchange of Geometric Information Between Electronic Prepress Systems Via Magnetic Tape (DDE500). 48 pp.
  OSI ISO/DIS 10758-90. Device Exchange Format for the Online Transfer of Colour Proofs from Electronic Prepress Systems to Direct Digital Colour Proofing Systems. 45 pp.
  OSI ISO/DIS 10759-90. User Exchange Format (UEF03) for the Exchange of Monochrome Image Data Between Electronic Prepress Systems Via Magnetic Tape (DDE500). 21 pp.

—**EPROMs**
  ECMA ECMA 167-92. Volume and File Structure of Write-Once and Rewritable Media Using Non-Sequential Recording for Information Interchange. 120 pp.
  IEC DIS 13346-93. Information Technology—Volume and File Structure of Write-Once and Rewritable Media Using Non-Sequential Recording for Information Interchange ***CD-ROM ONLY***. 123 pp.
  ISO DIS 13346-93. Information Technology—Volume and File Structure of Write-Once and Rewritable Media Using Non-Sequential Recording for Information Interchange ***CD-ROM ONLY***. 123 pp.
  JTC1 DIS13346-93. Information Technology—Volume and File Structure of Write-Once and Rewritable Media Using Non-Sequential Recording for Information Interchange ***CD-ROM ONLY***. 123 pp.

—**Fonts**
  BSI BS ISO/IEC 9541-1-91. 1991 Information Technology—Font Information Interchange—Part 1: Architecture (S). 94 pp.
  BSI BS ISO/IEC 9541-1-01. 1991 Amd 1 Information Technology—Font Information Interchange—Part 1: Architecture (AMD 7571) February 15, 1993 (Technical Corr 1). 97 pp.
  BSI BS ISO/IEC 9541-2-91. 1991 Information Technology—Font Information Interchange—Part 2: Interchange Format. 35 pp.
  BSI BS ISO/IEC 9541-2-01. 1991 Amd 1 Information Technology—Font Information Interchange—Part 2: Interchange Format (AMD 7813) July 15, 1993 (Technical Corr 1) (S). 40 pp.
  IEC 9541 Pt 1-91. Information Technology—Font Information Interchange—Part 1: Architecture First Edition; (Corrigendum 1-1992). 94 pp.
  IEC 9541 Pt 2-91. Information Technology—Font Information Interchange—Part 2: Interchange Format First Edition; (Corrigendum 1-1993). 38 pp.
  IEC DIS 9541 Pt 3-91. Information Technology—Font Information Interchange—Part 3: Glyph Shape Representation ***CD-ROM ONLY***. 73 pp.
  IEC 10036-93. Information Technology—Font Information Interchange—Procedure for Registration of Glyph and Glyph Collection Identifiers First Edition. 18 pp.
  ISO 9541 Pt 1-91. Information Technology—Font Information Interchange—Part 1: Architecture First Edition; (Corrigendum 1-1992). 94 pp.
  ISO 9541 Pt 2-91. Information Technology—Font Information Interchange—Part 2: Interchange Format First Edition; (Corrigendum 1-1993). 38 pp.
  ISO DIS 9541 Pt 3-91. Information Technology—Font Information Interchange—Part 3: Glyph Shape Representation ***CD-ROM ONLY***. 73 pp.
  ISO 10036-93. Information Technology—Font Information Interchange—Procedure for Registration of Glyph and Glyph Collection Identifiers First Edition. 18 pp.
  JTC1 9541-91. Information Technology—Font Information Interchange—Part 1: Architecture First Edition; (Corrigendum 1-1992). 2 pp.
  JTC1 9541-91. Information Technology—Font Information Interchange—Part 2: Interchange Format First Edition; (Corrigendum 1-1993). 38 pp.
  JTC1 DIS9541 Pt 3-91. Information Technology—Font Information Interchange—Part 3: Glyph Shape Representation ***CD-ROM ONLY***. 73 pp.
  JTC1 10036-93. Information Technology—Font Information Interchange—Procedure for Registration of Glyph and Glyph Collection Identifiers First Edition. 18 pp.
  OSI ISO DIS 9541-1-87. Information Processing—Font and Character Information Interchange Part 1: Introduction. 13 pp.
  OSI ISO/IEC DIS 9541-1.2-90. Information Technology—Font Information Interchange—Part 1: Architecture. 91 pp.

Information Interchange *(Cont.)*

—**Fonts** *(Cont.)*
  OSI ISO/IEC DIS 9541-2.2-90. Information Technology—Font Information Interchange—Part 2: Interchange Format. 35 pp.
  OSI ISO/IEC DIS 10036-90. Information Technology—Procedure for Registration at Glyph and Glyph Collection Identifiers. 21 pp.

—**Fonts—Architecture**
  BSI BS ISO/IEC 9541-1-91. 1991 Information Technology—Font Information Interchange—Part 1: Architecture (S). 94 pp.
  BSI BS ISO/IEC 9541-1-01. 1991 Amd 1 Information Technology—Font Information Interchange—Part 1: Architecture (AMD 7571) February 15, 1993 (Technical Corr 1). 97 pp.
  IEC 9541 Pt 1-91. Information Technology—Font Information Interchange—Part 1: Architecture First Edition; (Corrigendum 1-1992). 94 pp.
  ISO 9541 Pt 1-91. Information Technology—Font Information Interchange—Part 1: Architecture First Edition; (Corrigendum 1-1992). 94 pp.
  JTC1 9541-91. Information Technology—Font Information Interchange—Part 1: Architecture First Edition; (Corrigendum 1-1992). 2 pp.
  JTC1 9541-91. Information Technology—Font Information Interchange—Part 2: Interchange Format First Edition; (Corrigendum 1-1993). 38 pp.
  OSI ISO/IEC DIS 9541-1.2-90. Information Technology—Font Information Interchange—Part 1: Architecture. 91 pp.

—**Geometry—Aerospace**
  EURO CTI/80/10832-87. Recommendation for Geometry Data Exchange. 48 pp.
  EURO CTI/85/15617-85. Report of Geometry Data Exchange Study Group. 88 pp.

—**Image Processing**
  IEC DIS 12087 Pt 1-92. Information Technology—Computer Graphics and Image Processing—Image Processing and Interchange (IPI) Functional Specification —Part 1: Common Architecture for Imaging ***CD-ROM ONLY***. 78 pp.
  ISO DIS 12087 Pt 1-92. Information Technology—Computer Graphics and Image Processing—Image Processing and Interchange (IPI) Functional Specification —Part 1: Common Architecture for Imaging ***CD-ROM ONLY***. 78 pp.
  JTC1 DIS12087 Pt 1-92. Information Technology—Computer Graphics and Image Processing—Image Processing and Interchange (IPI) Functional Specification —Part 1: Common Architecture for Imaging ***CD-ROM ONLY***. 78 pp.
  NATO STANAG 3764 ED 2 AMD 0-89. Exchange of Imagery. 8 pp.
  NATO STANAG 3764 ED 2 AMD 1-89. Exchange of Imagery. 9 pp.

—**Integrated Services Digital Networks**
  CCITT RECMN I.320-89. ISDN Protocol Reference Model—Integrated Services Digital Network (ISDN)—Overall Network Aspects and Functions, ISDN User-Network Interfaces (Study Group XVIII) 10 pp. 10 pp.
  CENELEC ETR 054-92. Integrated Services Digital Network (ISDN); the Protocol Reference Model and Its Relationship with the OSI Reference Model. 13 pp.
  ETSI ETR 054-92. Integrated Services Digital Network (ISDN); the Protocol Reference Model and Its Relationship with the OSI Reference Model. 13 pp.

—**International—Teletex Communications**
  CEPT T/TG 17 E-91. Exchange of Data on Telex/Teletex Subscribers Between Network Operators for Computerized Automatic Directory Enquiry Systems. 1 p.

—**International—Telex Communications**
  CEPT T/TG 17 E-91. Exchange of Data on Telex/Teletex Subscribers Between Network Operators for Computerized Automatic Directory Enquiry Systems. 1 p.

—**International—Videotex Communications**
  CCITT RECMN T.100-89. International Information Exchange for Interactive Videotex—Terminal Equipment and Protocols for Telematic Services (Study Group VIII) 40 pp. 40 pp.

—**International—Videotex Communications—Character Repertoire**
  CCITT RECMN T.100-89. International Information Exchange for Interactive Videotex—Terminal Equipment and Protocols for Telematic Services (Study Group VIII) 40 pp. 40 pp.

—**Interworking Units**
  BSI BS 7267-90. 1990 Open Systems Interconnection: Guide to Operation of an X.25 Interworking Unit. 12 pp.

INDEX and DIRECTORY of

INTERNATIONAL AND NON-U.S. NATIONAL STANDARDS
SUBJECT INDEX
Information

Information Interchange (Cont.)
—Interworking Units (Cont.)
IEC TR10029-89. Information Technology—Telecommunications and Information Exchange Between Systems—Operation of an X.25 Interworking Unit First Edition. 10 pp.

ISO TR10029-89. Information Technology—Telecommunications and Information Exchange Between Systems—Operation of an X.25 Interworking Unit First Edition. 10 pp.

JTC1 TR10029-89. Information Technology—Telecommunications and Information Exchange Between Systems—Operation of an X.25 Interworking Unit First Edition. 10 pp.

OSI ISO IEC TR 10029-89. Information Technology—Telecommunications and Information Exchange Between Systems—Operation of an X.25 Interworking Unit. 10 pp.

SAA AS 4015-92. Information Technology—Telecommunications and Information Exchange Between Systems—Operation of an X.25 Interworking Unit (ISO/IEC/TR 10029:1989) (in Professional Package 26A). 5 pp.

—Ionospheric Disturbances
CCIR RECMN 313-6-90. Exchange of Information for Short-Term Forecasts and Transmission of Ionospheric Disturbance Warnings—Section 6C—Ionospheric Propagation and Operational Forecasting. 5 pp.

CCIR RECMN 313-7-92. Exchange of Information for Short-Term Forecasts and Transmission of Ionospheric Disturbance Warnings—Section 6C—Ionospheric Propagation and Operational Forecasting. 6 pp.

—Japanese Characters
JIS X 4003-89. File Specification for Japanese Documents Interchange (Geometric Graphics). 111 pp.

—Liaison—Land Forces
NATO STANAG 2101 ED 7 AMD 0-91. Establishing Liaison. 13 pp.

—Magnetic Tapes
BSI BS EN 21864-91. 1991 Information Processing—Unrecorded 12,7 mm (0.5 in) Wide Magnetic Tape for Information Interchange—32 ftpmm (800 ftpi) NRZ1, 126 ftpmm (3 200 ftpi) Phase Encoded and 356 ftpmm (9 042 ftpi) NRZ1 (ISO 1864: 1985). 19 pp.

BSI BS ISO/IEC 8441-2-91. 1991 Information Technology—High Density Digital Recording (HDDR)—Part 2: Guide for Interchange Practice. 41 pp.

CEN EN 21864-91. Information Processing—Unrecorded 12,7 mm (0.5 in) Wide Magnetic Tape for Information Interchange—32 ftpmm (800 ftpi) NRZ1, 126 ftpmm (3 200 ftpi) Phase Encoded and 356 ftpmm (9 042 ftpi) NRZ1. 3 pp.

DIN ENGL EN 21864-92. Information Technology; Unrecorded 12,7 mm (0,5 in) Wide Magnetic Tape for Information Interchange, 32 ftpmm (800 ftpi), NRZ1, 126 ftpmm (3200 ftpi), Phase-Encoded, and 356 ftpmm (9042 ftpi), NRZ1 (ISO/IEC 1864: 1992) (Sept). 17 pp.

IEC 1864-92. Information Technology—Unrecorded 12,7 mm (0,5 in) Wide Magnetic Tape for Information Interchange—32 ftpmm (800 ftpi), NRZ1, 126 ftpmm (3 200 ftpi) Phase Encoded and 356 ftpmm (9 042 ftpi), NRZ1 Fourth Edition. 18 pp.

IEC 8441 Pt 2-91. Information Technology—High Density Digital Recording (HDDR)—Part 2: Guide for Interchange Practice First Edition. 38 pp.

ISO 1864-92. Information Technology—Unrecorded 12,7 mm (0,5 in) Wide Magnetic Tape for Information Interchange—32 ftpmm (800 ftpi), NRZ1, 126 ftpmm (3 200 ftpi) Phase Encoded and 356 ftpmm (9 042 ftpi), NRZ1 Fourth Edition. 18 pp.

ISO 8441 Pt 2-91. Information Technology—High Density Digital Recording (HDDR)—Part 2: Guide for Interchange Practice First Edition. 38 pp.

JIS X 6105-88. 9-Track, 12.7 mm Wide Magnetic Tape for Information Interchange—Format and Recording Using Group Coding at 246 Cpmm.

JTC1 1864-92. Information Technology—Unrecorded 12,7 mm (0,5 in) Wide Magnetic Tape for Information Interchange—32 ftpmm (800 ftpi), NRZ1, 126 ftpmm (3 200 ftpi) Phase Encoded and 356 ftpmm (9 042 ftpi), NRZ1 Fourth Edition. 18 pp.

JTC1 8441 Pt 2-91. Information Technology—High Density Digital Recording (HDDR)—Part 2: Guide for Interchange Practice First Edition. 38 pp.

OSI ISO/IEC DIS 1864-91. Information Processing—Unrecorded 12,7mm (0.5in) Wide Magnetic Tape for Information Interchange—32ftpmm (800 fti) NRZI, 126ftpmm (3200 ftpi) Phase Encoded and 356ftpmm (9 042 ftpi) NRZI. 18 pp.

SAA AS 3977.2-92. Information Processing Systems—High Density Digital Recording (HDDR) —Part 2: Guide for Interchange Practice (ISO/IEC 8441-2:1991) (in Professional Package 26A). 31 pp.

Information Interchange (Cont.)
—Magnetic Tapes (Cont.)
SNZ NZS/AS 1011-93. Information Processing Systems—Unrecorded 12.7 mm (0.5 in) Wide Magnetic Tape for Info. Interchange—32 ftpmm (800 ftpi), NRZ1, 126 ftpmm (3 200 ftpi) Phase Encoded and 356 ftpmm (9 042 ftpi), NRZ1, (ISO/IEC 1864:1992) 1992) (This is a Joint Stand. with SAA AS 1011). 15 pp.

—Magnetic Tapes—Military Geographic Information
NATO STANAG 3984 ED 1 AMD 0-86. (DRAFT) Header Record Format for Exchange of Digital Geographic Information. 20 pp.

NATO STANAG 3985 ED 1 AMD 3-87. Preferred Magnetic Tape Standards for the Exchange of Digital Geographic Information. 10 pp.

NATO STANAG 3985 ED 1 AMD 4-87. Preferred Magnetic Tape Standards for the Exchange of Digital Geographic Information. 11 pp.

NATO STANAG 3985 ED 1 AMD 5-87. Preferred Magnetic Tape Standards for the Exchange of Digital Geographic Information. 12 pp.

—Manufacturing—Aircraft
EURO PSC/Dec/83/3 984-83. Guide for Data Exchange Between Aviation Industry and Airlines (Second Edition). 34 pp.

EURO Dec/81/11953-81. Reliability Military Data Exchange Guide. 11 pp.

EURO Dec/81/11953-82. Reliability Military Data Exchange guide (2nd Edition). 20 pp.

EURO CTI/85/15918-88. Aerospace Business Communications—Guidelines for Aircraft Projects—Volume 1: Methods (Superseded by CTI/85/15918 Issue 6). 90 pp.

EURO CTI/85/15918-91. Aerospace Business Communications Guidelines Volume 1: Methods. 78 pp.

—Manufacturing—Aircraft—Glossaries
EURO CTI/86/17768-88. Aerospace Business Communications Guidelines for Aircraft Projects—Volume 2: Definitions (Superceded by CTI/86/17768 Issue 3). 277 pp.

EURO CTI/86/17768-91. Aerospace Business Communications Guidelines—Volume 2: Definitions. 278 pp.

—Mapping
NATO STANAG 2029 ED 7 AMD 1-89. Method of Describing Ground Locations, Areas and Boundaries. 8 pp.

NATO STANAG 2029 ED 7 AMD 2-89. Method of Describing Ground Locations, Areas and Boundaries. 8 pp.

—Materiel Management
NATO STANAG 4199 ED 1 AMD 2-83. Uniform System of Exchange of Materiel Management Data. 7 pp.

—Message Formats—Military Communications
NATO STANAG 5500 ED 1 AMD 0-80. NATO Message Text Formatting System (FORMETS). 5 pp.

NATO STANAG 5500 ED 3 AMD 0-87. NATO Message Text Formatting System (FORMETS)—ADatP-3. 5 pp.

—Microfilm
NATO STANAG 3836 ED 3 AMD 3-88. Microfilm. 26 pp.

—Military Geographic Information
NATO STANAG 3809 ED 2 AMD 2-87. Digital Terrain Elevation Data Exchange (Format). 30 pp.

NATO STANAG 3809 ED 2 AMD 3-87. Digital Terrain Elevation Data Exchange (Format). 30 pp.

—Military Geographic Information—Header Record
NATO STANAG 3984 ED 1 AMD 0-86. (DRAFT) Header Record Format for Exchange of Digital Geographic Information. 20 pp.

—Missile Accidents
NATO STANAG 3101 ED 8 AMD 1-89. Exchange of Accident/Incident Information Concerning Aircraft and Missiles. 35 pp.

NATO STANAG 3101 ED 8 AMD 2-89. Exchange of Accident/Incident Information Concerning Aircraft and Missiles. 36 pp.

NATO STANAG 3101 ED 9 AMD 0-93. Exchange of Accident/Incident Information Concerning Aircraft and Missiles. 36 pp.

—Oceanography
NATO STANAG 1317 ED 1 AMD 2-83. NATO Oceanographic Data Exchange Format (NODEF 1). 29 pp.

Information Interchange (Cont.)
—Oceanography (Cont.)
NATO STANAG 1317 ED 1 AMD 3-83. NATO Oceanographic Data Exchange Format (NODEF 1). 22 pp.

NATO STANAG 1317 ED 1 AMD 4-83. NATO Oceanographic Data Exchange Format (NODEF 1). 24 pp.

—Office Document Architecture
BSI BS 7299: Part 1-90. 1990 Office Document Architecture (ODA) and Interchange Format for Text and Office Systems Part 1: Introduction and General Principles. 44 pp.

BSI BS 7299: Part 2-90. 1990 Office Document Architecture (ODA) and Interchange Format for Text and Office Systems Part 2: Document Structures. 237 pp.

BSI BS 7299: Part 4-90. 1990 Office Document Architecture (ODA) and Interchange Format for Text and Office Systems Part 4: Document Profile. 25 pp.

BSI BS 7299: Part 6-90. 1990 Office Document Architecture (ODA) and Interchange Format for Text and Office Systems Part 6: Character Content Architectures. 96 pp.

BSI BS 7299: Part 7-90. 1990 Office Document Architecture (ODA) and Interchange Format for Text and Office Systems Part 7: Raster Graphics Content Architectures. 51 pp.

BSI BS 7299: Part 8-90. 1990 Office Document Architecture (ODA) and Interchange Format for Text and Office Systems Part 8: Geometric Graphics Content Architectures. 56 pp.

BSI BS ISO/IEC TR 10183-1-93. 1993 Information Technology—Text and Office Systems—Office Document Architecture (ODA) and Interchange Format—Technical Report on ISO 8613 Implementation Testing—Part 1: Testing Methodology (S). 21 pp.

BSI DD ENV 41509-91. 1991 Information Systems Interconnection—Office Document Architecture (ODA)—Document Application Profile—Processable and Formatted Documents—Basic Character Content. 46 pp.

BSI DD ENV 41510-91. 1991 Information Systems Interconnection—Office Document Architecture (ODA)—Document Application Profile—Processable and Formatted Documents—Extended Mixed Mode. 69 pp.

BSI DD ENV 41510-01. 1991 Amd 1 Information Systems Interconnection—Office Document Architecture (ODA)—Document Application Profile—Processable And Formatted Documents — Extended Mixed Mode (AMD 7225) August 15, 1992. 72 pp.

BSI DD ENV 41511-91. 1991 Information Systems Interconnection—Office Document Architecture (ODA)—Document Application Profile—Layout Independent Documents—Simple Messaging Profile. 25 pp.

CEN ENV 41 509-90. Information Systems Interconnection—Office Document Architecture (ODA)—Document Application Profile—Processable and Formatted Documents—Basic Character Content. 44 pp.

CEN ENV 41 510-90. Information Systems Interconnection—Office Document Architecture (ODA)—Document Application Profile—Processable and Formatted Documents—Extended Mixed Mode. 67 pp.

CEN ENV 41510/A1-91. AMD 1 Information Systems Interconnection—Office Document Architecture (ODA)—Document Application Profile—Processable and Formatted Documents—Extended Mixed Mode. 4 pp.

CEN ENV 41 511-90. Information Systems Interconnection—Office Document Architecture (ODA)—Document Application Profile—Layout Independent Documents—Simple Messaging Profile. 23 pp.

CSA CAN/CSA-Z243.221-90. Information Processing—Text and Office Systems—Office Document Architecture (ODA) and Interchange Format—Part 1: Introduction and General Principles (ISO 8613-1:1989). 52 pp.

CSA CAN/CSA-Z243.222-90. Information Processing—Text and Office Systems—Office Document Architecture (ODA) and Interchange Format—Part 2: Document Structures (ISO 8613-2:1989). 247 pp.

CSA CAN/CSA-Z243.224-90. Information Processing—Text and Office Systems—Office Document Architecture (ODA) and Interchange Format—Part 4: Document Profile (ISO 8613-4:1989). 33 pp.

CSA CAN/CSA-Z243.226-90. Information Processing—Text and Office Systems—Office Document Architecture (ODA) and Interchange Format—Part 6: Character Content Architectures (ISO 8613-6:1989). 107 pp.

INDUSTRY STANDARDS

Information
INTERNATIONAL AND NON-U.S. NATIONAL STANDARDS
SUBJECT INDEX

Information Interchange *(Cont.)*

—Office Document Architecture *(Cont.)*

CSA CAN/CSA-Z243.227-90. Information Processing—Text and Office Systems—Office Document Architecture (ODA) and Interchange Format—Part 7: Raster Graphics Content Architectures (ISO 8613-7:1989). 60 pp.

CSA CAN/CSA-Z243.228-90. Information Processing—Text and Office Systems—Office Document Architecture (ODA) and Interchange Format—Part 8: Geometric Graphics Content Architectures (ISO 8613-8:1989). 65 pp.

ECMA ECMA-TR 48-88. Study on the Translation of the ODA Formatted Form into Page Description Languages. 29 pp.

IEC 8613 Draft AMD 5. Information Processing—Text and Office Systems—Office Document Architecture (ODA) and Interchange Format—Amendment 5: Streams; (1991) ***CD-ROM ONLY***. 22 pp.

IEC 8613 Pt 10-91. Info. Processing—Text and Office Sys.—Office Document Architecture (ODA) and Interchange Format—Part 10: Formal Specs. Amd. 5: Formal Spec. of the Defaulting Mech. for Defaultable Attributes First Ed.; (Amendment 1-2:1991) (Amendment 3-4:1992) (Amendment 5-1993). 368 pp.

IEC TR10183 Pt 1-93. Information Technology—Text and Office Systems—Office Document Architecture (ODA) and Interchange Format—Technical Report on ISO 8613 Implementation Testing—Part 1: Testing Methodology First Edition. 18 pp.

ISO 8613 Draft AMD 5. Information Processing—Text and Office Systems—Office Document Architecture (ODA) and Interchange Format—Amendment 5: Streams; (1991) ***CD-ROM ONLY***. 22 pp.

ISO 8613 Pt 1-89. Information Processing—Text and Office Systems—Office Document Architecture (ODA) and Interchange Format—Part 1: Introduction and General Principles First Edition; (ANSI 8613-1989). 43 pp.

ISO 8613 Pt 2-89. Information Processing—Text and Office Systems—Office Document Architecture (ODA) and Interchange Format—Part 2: Document Structures First Edition; (ANSI 8613-1989). 239 pp.

ISO 8613 Pt 4-89. Information Processing—Text and Office Systems—Office Document Architecture (ODA) and Interchange Format—Part 4: Document Profile First Edition; (ANSI 8613-1989). 24 pp.

ISO 8613 Pt 6-89. Information Processing—Text and Office Systems—Office Document Architecture (ODA) and Interchange Format—Part 6: Character Content Architectures First Edition; (ANSI 8613-1989). 98 pp.

ISO 8613 Pt 8-89. Information Processing—Text and Office Systems—Office Document Architecture (ODA) and Interchange Format—Part 8: Geometric Graphics Content Architectures First Edition; (ANSI 8613-1989). 54 pp.

ISO TR10183 Pt 1-93. Information Technology—Text and Office Systems—Office Document Architecture (ODA) and Interchange Format—Technical Report on ISO 8613 Implementation Testing—Part 1: Testing Methodology First Edition. 18 pp.

JTC1 8613 Draft AMD 5. Information Processing—Text and Office Systems—Office Document Architecture (ODA) and Interchange Format—Amendment 5: Streams; (1991) ***CD-ROM ONLY***. 22 pp.

JTC1 8613 Pt 1-89. Information Processing—Text and Office Systems—Office Document Architecture (ODA) and Interchange Format—Part 1: Introduction and General Principles First Edition; (ANSI 8613-1989). 43 pp.

JTC1 8613 Pt 2-89. Information Processing—Text and Office Systems—Office Document Architecture (ODA) and Interchange Format—Part 2: Document Structures First Edition; (ANSI 8613-1989). 239 pp.

JTC1 8613 Pt 4-89. Information Processing—Text and Office Systems—Office Document Architecture (ODA) and Interchange Format—Part 4: Document Profile First Edition; (ANSI 8613-1989). 24 pp.

JTC1 8613 Pt 5-89. Information Processing—Text and Office Systems—Office Document Architecture (ODA) and Interchange Format—Part 5: Office Document Interchange Format (ODIF) First Edition; (ANSI 8613-1989). 93 pp.

JTC1 8613 Pt 6-89. Information Processing—Text and Office Systems—Office Document Architecture (ODA) and Interchange Format—Part 6: Character Content Architectures First Edition; (ANSI 8613-1989). 98 pp.

JTC1 8613 Pt 7-89. Information Processing—Text and Office Systems—Office Document Architecture (ODA) and Interchange Format—Part 7: Raster Graphics Content Architectures First Edition; (ANSI 8613-1989). 51 pp.

JTC1 8613 Pt 8-89. Information Processing—Text and Office Systems—Office Document Architecture (ODA) and Interchange Format—Part 8: Geometric Graphics Content Architectures First Edition; (ANSI 8613-1989). 54 pp.

Information Interchange *(Cont.)*

—Office Document Architecture *(Cont.)*

JTC1 8613-91. Info. Processing—Text and Office Sys.—Office Document Architecture (ODA) and Interchange Format—Part 10: Formal Specs. Amd. 5: Formal Spec. of the Defaulting Mech. for Defaultable Attributes First Ed.; (Amendment 1-2:1991) (Amendment 3-4:1992) (Amendment 5-1993). 368 pp.

JTC1 TR 10183 Pt 1-93. Information Technology—Text and Office Systems—Office Document Architecture (ODA) and Interchange Format—Technical Report on ISO 8613 Implementation Testing—Part 1: Testing Methodology First Edition. 18 pp.

OSI ISO 8613/DAM 2.2-90. Information Processing—Text and Office Systems—Office Document Architecture (ODA) and Interchange Format—Amendment 2: Colour. 80 pp.

OSI ISO 8613 DAD 6-90. Information Processing—Text and Office Systems—Office Document Architecture (ODA) and Interchange Format—Addendum 6: Styles Extension. 7 pp.

OSI ISO 8613 DAD 5-90. Information Processing—Text and Office Systems—Office Document Architecture (ODA) and Interchange Format—Addendum 5: Streams. 21 pp.

OSI ISO 8613 DAD 4-90. Information Processing—Text and Office Systems—Office Document Architecture (ODA) and Interchange Format—Addendum 4: Security. 48 pp.

OSI ISO 8613 DAD 3-90. Information Processing—Text and Office Systems—Office Document Architecture (ODA) and Interchange Format—Addendum 3: Alternate Representation. 13 pp.

OSI ISO 8613 DAD 2-90. Information Processing—Text and Office Systems—Office Document Architecture (ODA) and Interchange Format—Addendum 2: Colour. 84 pp.

OSI ISO 8613 DAD 1-90. Information Processing—Text and Office Systems—Office Document Architecture (ODA) and Interchange Format—Addendum 1: Tiled Raster Graphics. 23 pp.

OSI ISO IEC PDAD 8613-89. 2nd PDAD to ISO 8613—To Add Colour Capability. 107 pp.

OSI ISO IEC PDAD 8613-89. Proposed Draft Addendum to ISO 8613 on Styles Extension. 7 pp.

OSI ISO IEC PDAD 8613-89. Proposed Draft Addendum to ISO 8613 on Conformance Testing Methodology. 8 pp.

OSI ISO 8613-1 DAM 2-90. Information Processing—Text and Office Systems—Office Document Architecture (ODA) and Interchange Format—Part 1: Introduction and General Principles—Amendment 2: Conformance Testing Methodology. 8 pp.

OSI ISO 8613-1 DAM 1-90. Information Processing—Text and Office Systems—Office Document Architecture (ODA) and Interchange Format—Part 1: Introduction and General Principles—Amendment 1. 5 pp.

OSI ISO IEC 8613-1 DAD 1-90. Information Processing—Text and Office Systems—Office Document Architecture (ODA) and Interchange Format—Part 1: Introduction and General Principles—Addendum 1: Document Application Profile Proforma and Notation. 52 pp.

OSI ISO 8613-1-90. Information Processing—Text and Office Systems—Office Document Architecture (ODA) and Interchange Format—Part 1: Introduction and General Principles. 43 pp.

OSI ISO DIS 8613-1-86. Information Processing—Text and Office Systems—Office Document Architecture (ODA) and Interchange Format Part 1: General Introduction (Superseded by ISO 8613-1). 43 pp.

OSI ISO 8613-1-89. Information Processing—Text and Office Systems—Office Document Architecture (ODA) and Interchange Format—Part 1: Introduction and General Principles. 43 pp.

OSI ISO 8613-2-89. Information Processing—Text and Office Systems—Office Document Architecture (ODA) and Interchange Format—Part 2: Document Structures. 239 pp.

OSI ISO 8613-2/DAD 1-89. Information Processing—Text and Office Systems—Office Document Architecture (ODA) and Interchage Format—Part 2: Document Structures ADDENDUM 1: Formal Specification of ODA Document Structures. 143 pp.

OSI ISO DIS 8613-2-86. Information Processing—Text and Office Systems—Office Document Architecture (ODA) and Interchange Format—Part 2: Document Structures (Superseded by ISO 8613-2). 171 pp.

OSI ISO DIS 8613-3-87. Information Processing—Text and Office Systems—Office Document Architecture and Interchange Format Part 3: Document Processing Reference Model. 37 pp.

OSI ISO 8613-4-89. Information Processing-Text and Office Systems—Office Document Architecture (ODA) and Interchange Format—Part 4: Document Profile. 24 pp.

OSI ISO DIS 8613-4-85. Text and Office Systems—Office Document Architecture (ODA) and Interchange Format Part 4: Document Profile (Superseded by ISO 8613-4). 29 pp.

Information Interchange *(Cont.)*

—Office Document Architecture *(Cont.)*

OSI ISO 8613-6-89. Information Processing—Text and Office Systems—Office Document Architecture (ODA) and Interchange Format—Part 6: Character Content Architectures. 98 pp.

OSI ISO DIS 8613-6.2-87. Information Processing—Text and Office Systems—Office Document Architecture (ODA) and Interchange Format.

OSI ISO 8613-7-89. Information Processing—Text and Office Systems—Office Document Architecture (ODA) and Interchange Format—Part 7: Raster Graphics Content Architectures. 51 pp.

OSI ISO DIS 8613-7-87. Information Processing—Text and Office Systems—Office Document Architecture (ODA) and Interchange Format Part 7: Raster Graphics Content Architectures.

OSI ISO 8613-8-89. Information Processing—Text and Office Systems—Office Document Architecture (ODA) and Interchange Format Part 8: Geometric Graphics Content Architectures. 56 pp.

OSI ISO DIS 8613-8-88. Information Processing—Text and Office Systems—Office Document Architecture (ODA) and Interchange Format-Part 8: Geometric Graphics Content Architectures. 58 pp.

OSI ISO 8613-10/DAM 1-90. Information Processing—Text and Office Systems—Office Document Architecture (ODA) and Interchange Format-Part 10: Formal Specifications Amendment 1: Formal Specification of the Document Profile. 27 pp.

OSI ISO 8613-10/DAM 2-90. Information Processing—Text and Office Systems—Office Document Architecture (ODA) and Interchange Format-Part 10: Formal Specifications Amendment 2: Formal Specification of the Rasler Graphics Content Architectures. 16 pp.

OSI ISO IEC 8613-10-91. Information Processing—Text and Office Systems—Office Document Architecture (ODA) and Interchange Format—Part 10: Formal Specifications.

SAA AS 3951.1-91. Information Processing—Text and Office Systems—Office Document Architecture (ODA) and Interchange Format—Part 1: Introduction and General Principles (ISO 8613-1:1989) (in Professional Package 26A). 38 pp.

SAA AS 3951.2-91. Information Processing—Text and Office Systems—Office Document Architecture (ODA) and Interchange Format—Part 2: Document Structures (ISO 8613-2:1989) (in Professional Package 26A). 229 pp.

SAA AS 3951.4-91. Information Processing—Text and Office Systems—Office Document Architecture (ODA) and Interchange Format—Part 4: Document Profile (ISO 8613-4:1989) (in Professional Package 26A). 20 pp.

SAA AS 3951.6-91. Information Processing—Text and Office Systems—Office Document Architecture (ODA) and Interchange Format—Part 6: Character Content Architectures (ISO 8613-6:1989) (in Professional Package 26A). 92 pp.

SAA AS 3951.7-91. Information Processing—Text and Office Systems—Office Document Architecture (ODA) and Interchange Format—Part 7: Raster Graphics Content Architectures (ISO 8613-7:1989) (in Professional Package 26A). 46 pp.

SAA AS 3951.8-91. Information Processing—Text and Office Systems—Office Document Architecture (ODA) and Interchange Format—Part 8: Geometric Graphics Content Architectures (ISO 8613-8:1989) (in Professional Package 26A). 51 pp.

SNZ NZS/ISO 8613. 1-89. Information Processing—Text and Office Systems—Office Document Architecture (ODA) and Interchange Format. Introduction and General Principles Part 1: Introduction and General Principles. 38 pp.

SNZ NZS/ISO 8613. 2-89. Information Processing—Text and Office Systems—Office Document Architecture (ODA) and Interchange Format. Introduction and General Principles Part 2: Document Structures. 229 pp.

SNZ NZS/ISO 8613. 4-89. Information Processing—Text and Office Systems—Office Document Architecture (ODA) and Interchange Format. Introduction and General Principles Part 4: Document Profile. 20 pp.

SNZ NZS/ISO 8613. 6-89. Information Processing—Text and Office Systems—Office Document Architecture (ODA) and Interchange Format. Introduction and General Principles Part 6: Character Content Architectures. 92 pp.

SNZ NZS/ISO 8613. 8-89. Information Processing—Text and Office Systems—Office Document Architecture (ODA) and Interchange Format. Introduction and General Principles Part 8: Geometric Graphics Content Architectures. 51 pp.

Information Interchange *(Cont.)*

—Office Document Architecture—Interchangeable Storage Media

IEC 8613 Pt 1 Draft AMD 3. Information Processing—Text and Office Systems—Office Document Arch-itecture (ODA) and In-terchange Format—Pt 1: Introduction and General Principles Amd 3: Annex H: Recording of Docu-ments Conforming to ISO 8613 on Interchangeable Storage Media; (1992) ***CD-ROM ONLY***. 7 pp.

JTC1 8613 Pt1 Draft AMD 3. Information Processing—Text and Office Systems—Office Document Archi-tecture (ODA) and Inter-change Format—Pt 1: Introduction and General Principles Amd 3: Annex H: Recording of Docu-ments Conforming to ISO 8613 on Interchangeable Storage Media; (1992) ***CD-ROM ONLY***. 7 pp.

—Office Document Architecture—Telematic Services

CCITT FASCICLE VII.6. Terminal Equipment and Protocols for Telematic Services—Recommendations T.400—T.418. 482 pp.

—Office Document Interchange Format

BSI BS 7299: Part 1-90. 1990 Office Document Architecture (ODA) and Interchange Format for Text and Office Systems Part 1: Introduction and General Principles. 44 pp.

BSI BS 7299: Part 5-90. 1990 Office Document Architecture (ODA) and Interchange Format for Text and Office Systems Part 5: Office Document Interchange Format (ODIF). 92 pp.

BSI DD ENV 41509-91. 1991 Information Systems Interconnection—Office Document Architecture (ODA)—Document Application Profile—Processable and Formatted Documents—Basic Character Content. 46 pp.

BSI DD ENV 41510-91. 1991 Information Systems Interconnection—Office Document Architecture (ODA)—Document Application Profile—Processable and Formatted Documents—Extended Mixed Mode. 69 pp.

BSI DD ENV 41510-01. 1991 Amd 1 Information Systems Interconnection—Office Document Architecture (ODA)—Document Application Profile—Processable And Formatted Documents — Extended Mixed Mode (AMD 7225) August 15, 1992. 72 pp.

CSA CAN/CSA-Z243.221-90. Information Processing—Text and Office Systems—Office Document Architecture (ODA) and Interchange Format—Part 1: Introduction and General Principles (ISO 8613-1:1989). 52 pp.

CSA CAN/CSA-Z243.225-90. Information Processing—Text and Office Systems—Office Document Architecture (ODA) and Interchange Format—Part 5: Office Document Interchange Format (ODIF) (ISO 8613-5:1989). 101 pp.

ECMA ECMA-TR 48-88. Study on the Translation of the ODA Formatted Form into Page Description Languages. 29 pp.

ISO 8613 Pt 1-89. Information Processing—Text and Office Systems—Office Document Architecture (ODA) and Interchange Format—Part 1: Introduction and General Principles First Edition; (ANSI 8613-1989). 43 pp.

ISO 8613 Pt 5-89. Information Processing—Text and Office Systems—Office Document Architecture (ODA) and Interchange Format—Part 5: Office Document Interchange Format (ODIF) First Edition; (ANSI 8613-1989). 93 pp.

JTC1 8613 Pt 1-89. Information Processing—Text and Office Systems—Office Document Architecture (ODA) and Interchange Format—Part 1: Introduction and General Principles First Edition; (ANSI 8613-1989). 43 pp.

OSI ISO IEC PDAD 8613-89. 2nd PDAD to ISO 8613—To Add Colour Capability. 107 pp.

OSI ISO 8613-1 DAM 1-90. Information Processing—Text and Office Systems—Office Document Architecture (ODA) and Interchange Format—Part 1: Introduction and General Principles—Amendment 1. 5 pp.

OSI ISO IEC 8613-1 DAD 1-90. Information Processing—Text and Office Systems—Office Document Architecture (ODA) and Interchange Format—Part 1: Introduction and General Principles—Addendum 1: Document Application Profile Proforma and Notation. 52 pp.

OSI ISO 8613-1-90. Information Processing—Text and Office Systems—Office Document Architecture (ODA) and Interchange Format—Part 1: Introduction and General Principles. 43 pp.

OSI ISO DIS 8613-1-86. Information Processing—Text and Office Systems—Office Document Architecture (ODA) and Interchange Format Part 1: General Introduction (Superseded by ISO 8613-1). 43 pp.

OSI ISO 8613-1-89. Information Processing—Text and Office Systems—Office Document Architecture (ODA) and Interchange Format—Part 1: Introduction and General Principles. 43 pp.

Information Interchange *(Cont.)*

—Office Document Interchange Format *(Cont.)*

OSI ISO DIS 8613-3-87. Information Processing—Text and Office Systems—Office Document Architecture and Interchange Format Part 3: Document Processing Reference Model. 37 pp.

OSI ISO 8613-5-89. Information Processing—Text and Office Systems—Office Document Architecture (ODA) and Interchange Format—Part 5: Office Document Interchange Format (ODIF). 93 pp.

OSI ISO DIS 8613-5-86. Text and Office Systems—Office Document Architecture (ODA) and Interchange Format—Part 5: Office Document Interchange Format (Superseded by ISO 8613-5). 65 pp.

OSI ISO/IEC DIS 10033-89. Information Technology—Text and Office Systems—Recording of Documents Conforming to ISO 8613 on Flexible Disk Cartridges Conforming to ISO 9293. 6 pp.

SAA AS 3951.1-91. Information Processing—Text and Office Systems—Office Document Architecture (ODA) and Interchange Format—Part 1: Introduction and General Principles (ISO 8613-1:1989) (in Professional Package 26A). 38 pp.

SAA AS 3951.5-91. Information Processing—Text and Office Systems—Office Document Architecture (ODA) and Interchange Format—Part 5: Office Document Interchange Format (ODIF) (ISO 8613-5:1989) (in Professional Package 26A). 88 pp.

SNZ NZS/ISO 8613. 1-89. Information Processing—Text and Office Systems—Office Document Architecture (ODA) and Interchange Format. Introduction and General Principles Part 1: Introduction and General Principles. 38 pp.

SNZ NZS/ISO 8613. 5-89. Information Processing—Text and Office Systems—Office Document Architecture (ODA) and Interchange Format. Introduction and General Principles Part 5: Office Document Interchange Format (ODIF). 88 pp.

—Office Document Interchange Format—Character Repertoire

CCITT RECMN T.501-89. Document Application Profile MM for the Interchange of Formatted Mixed Mode Document—Terminal Equipment and Protocols for Telematic Services (Study Group VIII) 16 pp. 16 pp.

—Office Document Interchange Format—Characters

IEC DISP 10610 Pt 1-91. Information Technology—International Standardized Profile FOD11—Office Document Format—Simple Document Structure—Character Content Architecture Only—Part 1: Document Application Profile ***CD-ROM ONLY***. 64 pp.

ISO DISP 10610 Pt 1-91. Information Technology—International Standardized Profile FOD11—Office Document Format—Simple Document Structure—Character Content Architecture Only—Part 1: Document Application Profile ***CD-ROM ONLY***. 64 pp.

—Office Document Interchange Format—Characters Format

IEC DISP 11181 Pt 1-91. Information Technology—International Standard-ized Profile FOD26—Office Document Format—Enhanced Document Struc-ture—Character, Raster Graphics and Geometric Graphics Content Architectures—Part 1: Document Application Profile ***CD-ROM ONLY***. 134 pp.

IEC DISP 11182 Pt 1-91. Information Technology—International Standard-ized Profile FOD36—Office Document Format—Extended Document Struc-ture—Character, Raster Graphics and Geometric Graphics Content Architectures—Part 1: Document Application Profile ***CD-ROM ONLY***. 240 pp.

ISO DISP 11181 Pt 1-91. Information Technology—International Standard-ized Profile FOD26—Office Document Format—Enhanced Document Struc-ture—Character, Raster Graphics and Geometric Graphics Content Architectures—Part 1: Document Application Profile ***CD-ROM ONLY***. 134 pp.

ISO DISP 11182 Pt 1-91. Information Technology—International Standard-ized Profile FOD36—Office Document Format—Extended Document Struc-ture—Character, Raster Graphics and Geometric Graphics Content Architectures—Part 1: Document Application Profile ***CD-ROM ONLY***. 240 pp.

—Office Document Interchange Format—Geometric Graphics

IEC DISP 11181 Pt 1-91. Information Technology—International Standard-ized Profile FOD26—Office Document Format—Enhanced Document Struc-ture—Character, Raster Graphics and Geometric Graphics Content Architectures—Part 1: Document Application Profile ***CD-ROM ONLY***. 134 pp.

Information Interchange *(Cont.)*

—Office Document Interchange Format—Geometric Graphics *(Cont.)*

IEC DISP 11182 Pt 1-91. Information Technology—International Standard-ized Profile FOD36—Office Document Format—Extended Document Struc-ture—Character, Raster Graphics and Geometric Graphics Content Architectures—Part 1: Document Application Profile ***CD-ROM ONLY***. 240 pp.

ISO DISP 11181 Pt 1-91. Information Technology—International Standard-ized Profile FOD26—Office Document Format—Enhanced Document Struc-ture—Character, Raster Graphics and Geometric Graphics Content Architectures—Part 1: Document Application Profile ***CD-ROM ONLY***. 134 pp.

ISO DISP 11182 Pt 1-91. Information Technology—International Standard-ized Profile FOD36—Office Document Format—Extended Document Struc-ture—Character, Raster Graphics and Geometric Graphics Content Architectures—Part 1: Document Application Profile ***CD-ROM ONLY***. 240 pp.

—Office Document Interchange Format—Interchangeable Storage Media

IEC 8613 Pt 1 Draft AMD 3. Information Processing—Text and Office Systems—Office Document Arch-itecture (ODA) and In-terchange Format—Pt 1: Introduction and General Principles Amd 3: Annex H: Recording of Docu-ments Conforming to ISO 8613 on Interchangeable Storage Media; (1992) ***CD-ROM ONLY***. 7 pp.

JTC1 8613 Pt1 Draft AMD 3. Information Processing—Text and Office Systems—Office Document Archi-tecture (ODA) and Inter-change Format—Pt 1: Introduction and General Principles Amd 3: Annex H: Recording of Docu-ments Conforming to ISO 8613 on Interchangeable Storage Media; (1992) ***CD-ROM ONLY***. 7 pp.

—Office Document Interchange Format—Mixed Content

CCITT RECMN T.505-91. Document Application Profile PM-26 for the Interchange of Mixed Content Documents in Processable and Formatted Forms (Study Group VIII) 110 pp. 110 pp.

—Office Document Interchange Format—Mixed Mode

CCITT RECMN T.501-89. Document Application Profile MM for the Interchange of Formatted Mixed Mode Document—Terminal Equipment and Protocols for Telematic Services (Study Group VIII) 16 pp. 16 pp.

—Office Document Interchange Format—Raster Graphics

BSI BS ISO/IEC 8613-10-91. 1991 Information Processing—Text and Office Systems—Office Document Architecture (ODA) and Interchange Format—Part 10: Formal Specifications. 121 pp.

BSI BS ISO/IEC 8613-10-02. 1991 Amd 2 Information Processing—Text and Office Systems—Office Document Architecture (ODA) and Interchange Format—Part 10: Formal Specifications (AMD 7138) July 15, 1992 (S). 232 pp.

BSI BS ISO/IEC 8613-10-04. 1991 Amd 4 Information Processing—Text and Office Systems—Office Document Architecture (ODA) and Interchange Format—Part 10: Formal Specifications (AMD 7649) March 15, 1993. 406 pp.

CCITT RECMN T.501-89. Document Application Profile MM for the Interchange of Formatted Mixed Mode Document—Terminal Equipment and Protocols for Telematic Services (Study Group VIII) 16 pp. 16 pp.

IEC DISP 11181 Pt 1-91. Information Technology—International Standard-ized Profile FOD26—Office Document Format—Enhanced Document Struc-ture—Character, Raster Graphics and Geometric Graphics Content Architectures—Part 1: Document Application Profile ***CD-ROM ONLY***. 134 pp.

IEC DISP 11182 Pt 1-91. Information Technology—International Standard-ized Profile FOD36—Office Document Format—Extended Document Struc-ture—Character, Raster Graphics and Geometric Graphics Content Architectures—Part 1: Document Application Profile ***CD-ROM ONLY***. 240 pp.

ISO 8613 Pt 7-89. Information Processing—Text and Office Systems—Office Document Architecture (ODA) and Interchange Format—Part 7: Raster Graphics Content Architectures First Edition; (ANSI 8613-1989). 51 pp.

ISO 8613 Pt 10-91. Info. Processing—Text and Office Sys.—Office Document Architecture (ODA) and Interchange Format—Part 10: Formal Specs. Amd. 5: Formal Spec. of the Defaulting Mech. for Defaultable Attributes First Ed,; (Amendment 1-2:1991) (Amendment 3-4:1992) (Amendment 5-1993). 368 pp.

INDUSTRY STANDARDS
1383

INTERNATIONAL AND NON-U.S. NATIONAL STANDARDS SUBJECT INDEX

Information Interchange (Cont.)

—Office Document Interchange Format—Raster Graphics (Cont.)

ISO DISP 11181 Pt 1-91. Information Technology—International Standard-ized Profile FOD26—Office Document Format—Enhanced Document Structure—Character, Raster Graphics and Geometric Graphics Content Architectures—Part 1: Document Application Profile ***CD-ROM ONLY***. 134 pp.

ISO DISP 11182 Pt 1-91. Information Technology—International Standard-ized Profile FOD36—Office Document Format—Extended Document Structure—Character, Raster Graphics and Geometric Graphics Content Architectures—Part 1: Document Application Profile ***CD-ROM ONLY***. 240 pp.

SNZ NZS/ISO 8613. 7-89. Information Processing—Text and Office Systems—Office Document Architecture (ODA) and Interchange Format. Introduction and General Principles Part 7: Raster Graphics Content Architectures. 46 pp.

—Office Document Interchange Format—Raster Graphics—FAX

CCITT RECMN T.503-91. Document Application Profile for the Interchange of Group 4 Facsimile Documents (Study Group VIII) 11 pp. 11 pp.

CCITT RECMN T.503-89. Document Application Profile for the Interchange of Group 4 Facsimile Documents—Terminal Equipment and Protocols for Telematic Services (Study Group VIII) 9 pp. 9 pp.

—Office Document Interchange Format—Security

OSI ISO 8613 DAD 4-90. Information Processing—Text and Office Systems—Office Document Architecture (ODA) and Interchange Format—Addendum 4: Security. 48 pp.

—Office Document Interchange Format—Streams

IEC 8613 Draft AMD 5. Information Processing—Text and Office Systems—Office Document Architecture (ODA) and Interchange Format—Amendment 5: Streams; (1991) ***CD-ROM ONLY***. 22 pp.

ISO 8613 Draft AMD 5. Information Processing—Text and Office Systems—Office Document Architecture (ODA) and Interchange Format—Amendment 5: Streams; (1991) ***CD-ROM ONLY***. 22 pp.

JTC1 8613 Draft AMD 5. Information Processing—Text and Office Systems—Office Document Architecture (ODA) and Interchange Format—Amendment 5: Streams; (1991) ***CD-ROM ONLY***. 22 pp.

OSI ISO 8613 DAM 5.2-91. Information Processing—Text and Office Systems—Office Document Architecture (ODA) and Interchange Format—Amendment 5: Streams.

—Office Document Interchange Format—Telematic Services

CCITT FASCICLE VII.6. Terminal Equipment and Protocols for Telematic Services—Recommendations T.400—T.418. 482 pp.

—Open Document Architecture

CCITT RECMN T.410 Series-91. First Extension (January 1991) to the T.410 Series (1988) of Recommendations Contained in the CCITT Blue Book, Fascicle VII.6 (Study Group VIII) 58 pp. 58 pp.

ECMA ECMA 101-1-88. Open Document Architecture (ODA) and Interchange Format Volume 1. 277 pp.

ECMA ECMA 101-2-88. Open Document Architecture (ODA) and Interchange Format Volume 2. 244 pp.

—Open Document Architecture—Character Format

CCITT RECMN T.411-89. Open Document Architecture (ODA) and Interchange Format Introduction and General Principles—Terminal Equipment and Protocols for Telematic Services (Study Group VIII) 81 pp. 81 pp.

CCITT RECMN T.412-89. Open Document Architecture (ODA) and Interchange Format—Document Structures—Terminal Equipment and Protocols for Telematic Services (Study Group VIII) 198 pp. 198 pp.

CCITT RECMN T.414-89. Open Document Architecture (ODA) and Interchange Format—Document Profile—Terminal Equipment and Protocols for Telematic Services (Study Group VIII) 17 pp. 17 pp.

CCITT RECMN T.415-89. Open Document Architecture (ODA) and Interchange Format—Open Document Interchange Format (ODIF)—Terminal Equipment and Protocols for Telematic Services (Study Group VIII) 51 pp. 51 pp.

Information Interchange (Cont.)

—Open Document Architecture—Character Format (Cont.)

CCITT RECMN T.416-89. Open Document Architecture (ODA) and Interchange Format—Character Content Architectures—Terminal Equipment and Protocols for Telematic Services (Study Group VIII) 81 pp. 81 pp.

CCITT RECMN T.417-89. Open Document Architecture (ODA) and Interchange Format—Raster Graphics Content Architectures—Terminal Equipment and Protocols for Telematic Services (Study Group VIII) 37 pp. 37 pp.

CCITT RECMN T.418-89. Open Document Architecture (ODA) and Interchange Format—Geometric Graphics Content Architecture—Terminal Equipment and Protocols for Telematic Services (Study Group VIII) 46 pp. 46 pp.

—Open Document Architecture—Color

CCITT RECMN T.410 Series(REV2)-92. Revision (February, 1992) of the T-410 Series (1988) of Recommendations Contained in the CCITT Blue Book, Fascicle VII.6 on the Subject of "Colour" (Study Group VIII) 64 pp. 64 pp.

—Open Document Architecture—Conformance

CCITT RECMN T.411-89. Open Document Architecture (ODA) and Interchange Format Introduction and General Principles—Terminal Equipment and Protocols for Telematic Services (Study Group VIII) 81 pp. 81 pp.

—Open Document Architecture—Geometric Graphics

CCITT RECMN T.414-89. Open Document Architecture (ODA) and Interchange Format—Document Profile—Terminal Equipment and Protocols for Telematic Services (Study Group VIII) 17 pp. 17 pp.

CCITT RECMN T.415-89. Open Document Architecture (ODA) and Interchange Format—Open Document Interchange Format (ODIF)—Terminal Equipment and Protocols for Telematic Services (Study Group VIII) 51 pp. 51 pp.

CCITT RECMN T.416-89. Open Document Architecture (ODA) and Interchange Format—Character Content Architectures—Terminal Equipment and Protocols for Telematic Services (Study Group VIII) 81 pp. 81 pp.

CCITT RECMN T.417-89. Open Document Architecture (ODA) and Interchange Format—Raster Graphics Content Architectures—Terminal Equipment and Protocols for Telematic Services (Study Group VIII) 37 pp. 37 pp.

CCITT RECMN T.418-89. Open Document Architecture (ODA) and Interchange Format—Geometric Graphics Content Architecture—Terminal Equipment and Protocols for Telematic Services (Study Group VIII) 46 pp. 46 pp.

—Open Document Architecture—Raster Graphics

CCITT RECMN T.411-89. Open Document Architecture (ODA) and Interchange Format Introduction and General Principles—Terminal Equipment and Protocols for Telematic Services (Study Group VIII) 81 pp. 81 pp.

CCITT RECMN T.414-89. Open Document Architecture (ODA) and Interchange Format—Document Profile—Terminal Equipment and Protocols for Telematic Services (Study Group VIII) 17 pp. 17 pp.

CCITT RECMN T.415-89. Open Document Architecture (ODA) and Interchange Format—Open Document Interchange Format (ODIF)—Terminal Equipment and Protocols for Telematic Services (Study Group VIII) 51 pp. 51 pp.

CCITT RECMN T.416-89. Open Document Architecture (ODA) and Interchange Format—Character Content Architectures—Terminal Equipment and Protocols for Telematic Services (Study Group VIII) 81 pp. 81 pp.

CCITT RECMN T.417-89. Open Document Architecture (ODA) and Interchange Format—Raster Graphics Content Architectures—Terminal Equipment and Protocols for Telematic Services (Study Group VIII) 37 pp. 37 pp.

CCITT RECMN T.418-89. Open Document Architecture (ODA) and Interchange Format—Geometric Graphics Content Architecture—Terminal Equipment and Protocols for Telematic Services (Study Group VIII) 46 pp. 46 pp.

—Open Document Architecture—Telematic Services

CCITT RECMN T.411-89. Open Document Architecture (ODA) and Interchange Format Introduction and General Principles—Terminal Equipment and Protocols for Telematic Services (Study Group VIII) 81 pp. 81 pp.

Information Interchange (Cont.)

—Open Document Architecture—Telematic Services (Cont.)

CCITT RECMN T.412-89. Open Document Architecture (ODA) and Interchange Format—Document Structures—Terminal Equipment and Protocols for Telematic Services (Study Group VIII) 198 pp. 198 pp.

CCITT RECMN T.414-89. Open Document Architecture (ODA) and Interchange Format—Document Profile—Terminal Equipment and Protocols for Telematic Services (Study Group VIII) 17 pp. 17 pp.

CCITT RECMN T.415-89. Open Document Architecture (ODA) and Interchange Format—Open Document Interchange Format (ODIF)—Terminal Equipment and Protocols for Telematic Services (Study Group VIII) 51 pp. 51 pp.

CCITT RECMN T.416-89. Open Document Architecture (ODA) and Interchange Format—Character Content Architectures—Terminal Equipment and Protocols for Telematic Services (Study Group VIII) 81 pp. 81 pp.

CCITT RECMN T.417-89. Open Document Architecture (ODA) and Interchange Format—Raster Graphics Content Architectures—Terminal Equipment and Protocols for Telematic Services (Study Group VIII) 37 pp. 37 pp.

CCITT RECMN T.418-89. Open Document Architecture (ODA) and Interchange Format—Geometric Graphics Content Architecture—Terminal Equipment and Protocols for Telematic Services (Study Group VIII) 46 pp. 46 pp.

—Open Systems Interconnection

BSI BS 6568: Part 1-88. 1988 Amd 1 Reference Model of Open Systems Interconnection Part 1: Basic Reference Model (Incorporating Connectionless-Mode Transmission) (ISO 7498-1984) (AMD 6112) April 30, 1991. 62 pp.

—Packet Assembly/Disassembly Devices

CCITT RECMN X.39-92. Procedures for the Exchange of Control Information and User Data Between a Facsimile Packet Assembly/Disassembly (FPAD) Facility and a Packet Mode Data Terminal Equipment (DTE) or Another FPAD (Study Group VII) 20 pp. 20 pp.

—Packet Assembly/Disassembly Devices—Data Terminal Equipment

CCITT RECMN X.29-89. Procedures for the Exchange of Control Information and User Data Between a Packet Assembly/Disassembly (PAD) Facility and a Packet Mode DTE or Another Pad—Data Communication Networks: Services and Facilities, Interfaces (Study Group VII) 15 pp. 15 pp.

CCITT RECMN X.39-92. Procedures for the Exchange of Control Information and User Data Between a Facsimile Packet Assembly/Disassembly (FPAD) Facility and a Packet Mode Data Terminal Equipment (DTE) or Another FPAD (Study Group VII) 20 pp. 20 pp.

—Post Telephone and Telegraph Administrations

CEPT T/R 75-04-86. Exchange De Donnees Pour Les Systemes Spaciaux Entre Administrations Et Entre Administrations Et Organismes Internationaux. 6 pp.

CEPT T/SF 8-81. Services Et Facilities Que Les Administrations Pourraient Offrir a Leurs Clients Dans Le Domaine De La Communication de Donnees. 5 pp.

CEPT T/TG 17-86. Echange De Donnees Sur Les Abonnes Telex/Teletex Entre Les Administrations Pour Les Systemes De Renseignements Automatiques Informatises. 3 pp.

—Radio Spectra—Management

CCIR RECMN 668-90. Methods of Exchanging Computer Programs and Data for Spectrum Management Purposes—Section 1A—Spectrum Engineering and Computer-Aided Principles and Techniques. 1 p.

CCIR QUESTION 65/1-86. Improved Methods of Exchanging Computer Programs and Data for Spectrum Management Purposes—Questions Concerning Study Group 1—Spectrum Management Techniques (Spectrum Engineering, Planning, Sharing, Monitoring and Utilization). 1 p.

—ROMs

ECMA ECMA 167-92. Volume and File Structure of Write-Once and Rewritable Media Using Non-Sequential Recording for Information Interchange. 120 pp.

IEC DIS 13346-93. Information Technology—Volume and File Structure of Write-Once and Rewritable Media Using Non-Sequential Recording for Information Interchange ***CD-ROM ONLY***. 123 pp.

INTERNATIONAL AND NON-U.S. NATIONAL STANDARDS
SUBJECT INDEX

Information

Information Interchange *(Cont.)*
—ROMs *(Cont.)*
ISO DIS 13346-93. Information Technology—Volume and File Structure of Write-Once and Rewritable Media Using Non-Sequential Recording for Information Interchange ***CD-ROM ONLY***. 123 pp.
JTC1 DIS13346-93. Information Technology—Volume and File Structure of Write-Once and Rewritable Media Using Non-Sequential Recording for Information Interchange ***CD-ROM ONLY***. 123 pp.

—Shipping—Naval Operations
NATO STANAG 1212 ED 6 AMD 0-89. Naval Control of Shipping—Information on Ports, Authorities and NCS Publications—AAP-8(C). 5 pp.

—Symbols
JIS X 0124-81. Representation of Unit Symbols for Information Interchange.

—Syntax
BSI DD 75: Part 1-81. (WITHDRAWN) 1981 Amd 1 Structure and Representation of Data for Interchange at the Application Level (DIAL) Part 1: Recommendations for Syntax and Basic Principles (Superseded by BS EN 29735: 1992). 26 pp.

—Telecommunication Circuits—Teleprinters
CCITT RECMN U.4-89. Exchange of Information Regarding Signals Destined to be Used over International Circiuts Concerned with Switched Teleprinter Networks—Telegraph Switching (Study Group IX) 1 pp. 1 p.

—Teletex Communications
BSI DD ENV 41504-91. 1991 Information Systems Interconnection—Character Repertoire and Coding for Interworking with Telex Services (Incorporating Corrigendum June 1991). 14 pp.
CEN ENV 41504-90. Information Systems Interconnection—Character Repertoire and Coding for Interworking with Telex Services. 12 pp.
CEN PRENV 41506-91. Information Systems Interconnection—Data Stream Formats for Information Received from or Transmitted to the Teletex Service or Private Information Processing Systems Using Teletex Technology. 9 pp.
CEN ENV 41506-91. Information Systems Interconnection—Data Stream Formats for Information Received from or Transmitted to the Teletex Service or Private Information Processing Systems Using Teletex Technology. 9 pp.
CEPT T/TE 07-03 E-88. Interchange of Teletex Documents. 3 pp.

—Transport Aircraft—Bearing Capacity
NATO STANAG 3767 ED 1 AMD 10-75. Exchange of Data on Load Capabilities of Transport Aircraft. 20 pp.
NATO STANAG 3767 ED 1 AMD 11-75. Exchange of Data on Load Capabilities of Transport Aircraft. 20 pp.

Information Links
Scope Note: Use a more specific term *See:* Data Links; Group Links; Line Links; Mastergroup Links; Radio Links; Sound-Program Links; Supergroup Links; Supermastergroup Links; Television Links; 15 Supergroup Assembly Links

Information Networks
Use: Information Services

Information Processing
Use: Data Processing

Information Processing Equipment
Use: Data Processing Equipment

Information Resource Dictionary System
Use For: IRDS *See Also:* Open Systems Interconnection; Software
BSI BS ISO/IEC 10027-90. 1990 Information Technology—Information Resource Dictionary System (IRDS) Framework. 22 pp.
IEC 10027-90. Information Technology—Information Resource Dictionary System (IRDS) Framework First Edition. 20 pp.
ISO 10027-90. Information Technology—Information Resource Dictionary System (IRDS) Framework First Edition. 20 pp.
JTC1 10027-90. Information Technology—Information Resource Dictionary System (IRDS) Framework First Edition. 20 pp.
OSI ISO/IEC 10027-90. Information Technology—Information Resource Dictionary System (IRDS) Framework. 20 pp.
OSI ISO DP 10027-88. Information Processing Systems—Information Resource Dictionary System (IRDS) Framework. 21 pp.

Information Resource Dictionary System *(Cont.)*
OSI ISO DP 10027-88. Information Processing Systems-Information Resource Dictionary System (IRDS) Framework.
SNZ NZS/AS 4101-93. Information Technology—Information Resource Dictionary System Framework (This is a Joint Standard with SAA AS 4101). 15 pp.

—Interfaces
BSI BS ISO/IEC 10728-93. 1993 Information Technology—Information Resource Dictionary System (IRDS) Services Interface (S). 119 pp.
IEC 10728-93. Information Technology—Information Resource Dictionary System (IRDS) Services Interface First Edition. 117 pp.
ISO 10728-93. Information Technology—Information Resource Dictionary System (IRDS) Services Interface First Edition. 117 pp.
JTC1 10728-93. Information Technology—Information Resource Dictionary System (IRDS) Services Interface First Edition. 117 pp.

Information Retrieval Systems, Computer
Use: Computer Assisted Retrieval Systems

Information Security
Use For: Computer Security *See Also:* Hash Functions; Information Interchange; Passwords; Personal Identification Numbers; Security
CSA Z243.57-1991SP. Canadian Vocabulary for Information Technology Security; (Gen Instr 1). 113 pp.

—Authentication
BSI BS ISO/IEC 9798-1-91. 1991 Information Technology—Security Techniques—Entity Authentication Mechanisms—Part 1: General Model. 10 pp.
IEC 9798 Pt 1-91. Information Technology—Security Techniques—Entity Authentication Mechanisms—Part 1: General Model First Edition. 8 pp.
IEC DIS 9798 Pt 2-93. Information Technology—Security Techniques—Entity Authentication Mechanisms—Part 2: Entity Authentication Using Symmetric Techniques ***CD-ROM ONLY***. 12 pp.
IEC DIS 9798 Pt 3-92. Information Technology—Security Techniques—Entity Authentication Mechanisms—Part 3: Entity Authentication Using a Public Key Algorithm ***CD-ROM ONLY***. 13 pp.
ISO 9798 Pt 1-91. Information Technology—Security Techniques—Entity Authentication Mechanisms—Part 1: General Model First Edition. 8 pp.
ISO DIS 9798 Pt 2-93. Information Technology—Security Techniques—Entity Authentication Mechanisms—Part 2: Entity Authentication Using Symmetric Techniques ***CD-ROM ONLY***. 12 pp.
ISO DIS 9798 Pt 3-92. Information Technology—Security Techniques—Entity Authentication Mechanisms—Part 3: Entity Authentication Using a Public Key Algorithm ***CD-ROM ONLY***. 13 pp.
JTC1 9798-91. Information Technology—Security Techniques—Entity Authentication Mechanisms—Part 1: General Model First Edition. 8 pp.
JTC1 DIS 9798 Pt 2-93. Information Technology—Security Techniques—Entity Authentication Mechanisms—Part 2: Entity Authentication Using Symmetric Techniques ***CD-ROM ONLY***. 12 pp.
JTC1 DIS9798 Pt 3-92. Information Technology—Security Techniques—Entity Authentication Mechanisms—Part 3: Entity Authentication Using a Public Key Algorithm ***CD-ROM ONLY***. 13 pp.

—CAD
ISO 11442 Pt 1-93. Technical Product Documentation—Handling of Computer-Based Technical Information—Part 1: Security Requirements First Edition. 7 pp.

—Digital Signature
IEC 9796-91. Information Technology—Security Techniques—Digital Signature Scheme Giving Message Recovery First Edition. 17 pp.
ISO 9796-91. Information Technology—Security Techniques—Digital Signature Scheme Giving Message Recovery First Edition. 17 pp.
JTC1 9796-91. Information Technology—Security Techniques—Digital Signature Scheme Giving Message Recovery First Edition. 17 pp.

Information Security *(Cont.)*
—European Digital Cordless Telecommunications
CENELEC PRETS 300 175-7-91. Radio Equipment and Systems Digital European Cordless Telecommunications Common Interface Part 7: Security Features (DE/RES 3001-7). 99 pp.
CENELEC PRETS 300 175-7-92. Radio Equipment and Systems (RES); Digital European Cordless Telecommunications (DECT) Common Interface Part 7: Security Features. 102 pp.
CENELEC ETS 300 175-7-92. Radio Equipment and Systems (RES); Digital European Cordless Telecommunications (DECT) Common Interface Part 7: Security Features. 104 pp.
ETSI ETS 300 175-7-92. Radio Equipment and Systems (RES); Digital European Cordless Telecommunications (DECT) Common Interface Part 7: Security Features. 104 pp.
ETSI PRETS 300 175-7-92. Radio Equipment and Systems (RES); Digital European Cordless Telecommunications (DECT) Common Interface Part 7: Security Features. 102 pp.
ETSI PRETS 300 175-7-91. Radio Equipment and Systems Digital European Cordless Telecommunications Common Interface Part 7: Security Features (DE/RES 3001-7). 99 pp.

—Glossaries
CNS C5235-90. Information Processing Systems—Vocabulary (Part 8: Control, Integrity and Security) (May)(12717).
CSA Z243.57-1991SP. Canadian Vocabulary for Information Technology Security; (Gen Instr 1). 113 pp.
JIS X 0008-87. Glossary of Terms Used in Information Processing (Control, Integrity and Security).

—Interpersonal Messaging Services
CCITT RECMN F.420-92. Message Handling Services: Public Interpersonal Messaging Service (Study Group I) 16 pp. 16 pp.
CCITT RECMN F.420-89. Message Handling Services: the Public Interpersonal Messaging Service—Message Handling and Directory Services—Operations and Definition of Service (Study Group I) 15 pp. 15 pp.

—Message Handling Systems
CCITT RECMN F.400-92. Message Handling Services: Message Handling System and Service Overview (Study Group I) 82 pp (Same as Recmn X.400). 82 pp.
CCITT RECMN F.400-89. Message Handling System and Service Overview—Message Handling and Directory Services—Operations and Definition of Service (Study Group I) 73 pp. 73 pp.
CCITT RECMN X.400-93. Message Handling Services: Message Handling System and Service Overview (Study Group VII) 82 pp (Same as Recmn F.400). 82 pp.

—Office Document Interchange
OSI ISO 8613 DAD 4-90. Information Processing—Text and Office Systems—Office Document Architecture (ODA) and Interchange Format—Addendum 4: Security. 48 pp.

—Open Systems Interconnection
CCITT RECMN X.800-91. Security Architecture for Open Systems Interconnection for CCITT Applications (Study Group VII) 49 pp. 49 pp.
CSA CAN/CSA-Z243. 100.2-92. Information Processing Systems—Open Systems Interconnection—Basic Reference Model—Part 2: Security Architecture (ISO 7498-2:1989); (Gen Instr 1). 44 pp.
ECMA ECMA 138-89. Security on Open Systems—Data Elements and Service Definitions. 81 pp.
ECMA ECMA-TR 46-88. Security in Open Systems a Security Framework. 77 pp.
IEC 9594 Pt 8-90. Information Technology—Open Systems Interconnection—The Directory—Part 8: Authentication Framework First Edition; (Corrigendum 1-1991). 33 pp.
IEC DIS 10181 Pt 2-91. Information Technology—Open Systems Interconnection—Security Frameworks for Open Systems—Part 2: Authentication Framework ***CD-ROM ONLY***. 68 pp.
IEC DIS 10736-91. Information Technology—Telecommunications and Information Exchange Between Systems—Transport Layer Security Protocol ***CD-ROM ONLY***. 53 pp.
IEC DIS 10736 Draft AMD 1. Information Technology—Telecommunications and Information Exchange Between Systems—Transport Layer Security Protocol Amendment 1: Security Association Establishment; (1993) ***CD-ROM ONLY***. 31 pp.
IEC DIS 10745-92. Information Technology—Open Systems Interconnection—Upper Layers Security Model ***CD-ROM ONLY***. 34 pp.

INDUSTRY STANDARDS

INTERNATIONAL AND NON-U.S. NATIONAL STANDARDS
SUBJECT INDEX

Information Security (Cont.)
—Open Systems Interconnection (Cont.)
IEC DIS 11577-93. Information Technology—Telecommunications and Information Exchange Between Systems—Network Layer Security Protocol ***CD-ROM ONLY***. 73 pp.
ISO 7498 Pt 2-89. Information Processing Systems—Open Systems Interconnection—Basic Reference Model—Part 2: Security Architecture First Edition. 34 pp.
ISO DIS 10181 Pt 2-91. Information Technology—Open Systems Interconnection—Security Frameworks for Open Systems—Part 2: Authentication Framework ***CD-ROM ONLY***. 68 pp.
ISO DIS 10736-91. Information Technology—Telecommunications and Information Exchange Between Systems—Transport Layer Security Protocol ***CD-ROM ONLY***. 53 pp.
ISO DIS 10736 Draft AMD 1. Information Technology—Telecommunications and Information Exchange Between Systems—Transport Layer Security Protocol Amendment 1: Security Association Establishment; (1993) ***CD-ROM ONLY***. 31 pp.
ISO DIS 10745-92. Information Technology—Open Systems Interconnection—Upper Layers Security Model ***CD-ROM ONLY***. 34 pp.
ISO DIS 11577-93. Information Technology—Telecommunications and Information Exchange Between Systems—Network Layer Security Protocol ***CD-ROM ONLY***. 73 pp.
JIS X 5004-91. Information Processing Systems—Open Systems Interconnection—Basic Reference Model—Part 2: Security Architecture.
JTC1 7498 Pt 2-89. Information Processing Systems—Open Systems Interconnection—Basic Reference Model—Part 2: Security Architecture First Edition. 34 pp.
JTC1 DIS10736-91. Information Technology—Telecommunications and Information Exchange Between Systems—Transport Layer Security Protocol ***CD-ROM ONLY***. 53 pp.
JTC1 DIS10736 Draft AMD 1. Information Technology—Telecommunications and Information Exchange Between Systems—Transport Layer Security Protocol Amendment 1: Security Association Establishment; (1993) ***CD-ROM ONLY***. 31 pp.
JTC1 DIS10745-92. Information Technology—Open Systems Interconnection—Upper Layers Security Model ***CD-ROM ONLY***. 34 pp.
JTC1 DIS11577-93. Information Technology—Telecommunications and Information Exchange Between Systems—Network Layer Security Protocol ***CD-ROM ONLY***. 73 pp.
OSI ISO IEC DIS 7498-2-88. Final Text of DIS 7498-2, Information Processing Systems—OSI Reference Model Part 2: Security Architecture.
SNZ NZS/ISO 7498. 2-88. Information Processing Systems—Open Systems Interconnection—Basic Reference Model Part 2: Security Architecture. 5 pp.

—Open Systems Interconnection—Authentication
IEC DIS 10181 Pt 2-91. Information Technology—Open Systems Interconnection—Security Frameworks for Open Systems—Part 2: Authentication Framework ***CD-ROM ONLY***. 68 pp.
IEC DIS 10181 Pt 2.2-93. Information Technology—Open Systems Interconnection—Security Frameworks for Open Systems: Authentication Framework ***CD-ROM ONLY***. 74 pp.
ISO DIS 10181 Pt 2-91. Information Technology—Open Systems Interconnection—Security Frameworks for Open Systems—Part 2: Authentication Framework ***CD-ROM ONLY***. 68 pp.
ISO DIS 10181 Pt 2.2-93. Information Technology—Open Systems Interconnection—Security Frameworks for Open Systems: Authentication Framework ***CD-ROM ONLY***. 74 pp.
JTC1 DIS 10181 Pt 2.2-93. Information Technology—Open Systems Interconnection—Security Frameworks for Open Systems: Authentication Framework ***CD-ROM ONLY***. 74 pp.

—Open Systems Interconnection—Security Audit Trails
BSI BS ISO/IEC 10164-8-93. 1993 Information Technology—Open Systems Interconnection—Systems Management: Security Audit Trail Function (S). 34 pp.
IEC 10164 Pt 8-93. Information Technology—Open Systems Interconnection—Systems Management: Security Audit Trail Function First Edition; (CCITT RECMN X.740). 32 pp.
ISO 10164 Pt 8-93. Information Technology—Open Systems Interconnection—Systems Management: Security Audit Trail Function First Edition; (CCITT RECMN X.740). 32 pp.
JTC1 10164 Pt 8-93. Information Technology—Open Systems Interconnection—Systems Management: Security Audit Trail Function First Edition; (CCITT RECMN X.740). 32 pp.

Information Security (Cont.)
—Post Telephone and Telegraph Administrations
CEPT T/SF 13-81. Procedure D'Agrement Pour Les Materiels De Teleinformatique Ou Les Equipements De Transmission De Donnees Appartenant a Des Personnes Privees Et Susceptibles D'Etre Directement Connectes Aux Reseaux Publics Ou Aux Installations Des Administrations. 3 pp.

—Telecommunication Equipment
CEPT T/SF 13-81. Procedure D'Agrement Pour Les Materiels De Teleinformatique Ou Les Equipements De Transmission De Donnees Appartenant a Des Personnes Privees Et Susceptibles D'Etre Directement Connectes Aux Reseaux Publics Ou Aux Installations Des Administrations. 3 pp.

—Telecommunications
CCITT RECMN T.410 Series-91. First Extension (January 1991) to the T.410 Series (1988) of Recommendations Contained in the CCITT Blue Book, Fascicle VII.6 (Study Group VIII) 58 pp. 58 pp.

—Teleconference Services
CEPT T/SF 56 E-87. Services and Facilities for Information Security in Visual Telematic Services. 8 pp.

—Universal Personal Telecommunication Services
CENELEC ETR 055-4-92. Universal Personal Telecommunication (UPT); the Service Concept Part 4: Service Requirements on Security Mechanisms. 12 pp.
ETSI ETR 055-4-92. Universal Personal Telecommunication (UPT); the Service Concept Part 4: Service Requirements on Security Mechanisms. 12 pp.

Information Services
Use For: Information Networks; Technical Information Services *See Also:* Directory Assistance; Handbooks; Microfiche; Recorded Information Services (Telephone)
EC 88/524/EEC-88. Council Decision Concerning the Establishment of a Plan of Action for Setting up an Information Services Market. 5 pp.

—Aircraft Equipment
MOD UK DSTAN 05-123: Part 4-01. Supply of Technical Information Issue 1; Amendment 55. 175 pp.
MOD UK DSTAN 05-123: Part 4-02. Supply of Technical Information Issue 1; Amendment 57. 188 pp.
MOD UK DSTAN 05-123: Part 4-03. Supply of Technical Information Issue 1; Amendment 58. 179 pp.
MOD UK DSTAN 05-123: Part 4-04. Supply of Technical Information Issue 1; Amendment 59. 181 pp.

—Glossaries
ISO 2146-88. Documentation—Directories of Libraries, Archives, Information and Documentation Centres, and Their Data Bases Second Edition. 28 pp.

—Telecommunication Systems—Developing Countries
CCITT FASCICLE I.1-88. Minutes and Reports of the Plenary Assembly List of Study Groups and Questions Under Study. 249 pp.

Information Systems
Use For: Information Technology
See Also: Computer Equipment; Computer Systems; Computers; Consumer Information Systems; Data Processing Equipment; Documentation; Libraries; Systems Engineering; Telecommunication Equipment; Videotex Communications
EC SM 4. Information Technology. 1 p.

—Cooperation—CCITT
CCITT RECMN A.22-89. Collaboration with Other International Organizations on Information Technology—Terminal Equipment and Protocols for Telematic Services (Study Group VIII) 2 pp. 2 pp.

—European Communities
EURO 1989 Nov. Esprit: Key to the Technological Awakening of Europe. 11 pp.

—Flight Operations—Supersonic Transports
CAA STANDARD NO. 7-5&AP 03.76. Basic Flight Information Systems. 14 pp.

—Framework
BSI DD 210-92. 1992 Guide to a Framework for User Requirements for Information Technology. 28 pp.

Information Systems (Cont.)
—Glossaries
CSA CAN/CSA-Z243.58-92. Information Technology Vocabulary. 612 pp.

—Protocols
BSI DD ENV 41802-1-92. 1992 Information Systems Interconnection—X.25 Protocol Relaying Part 1: General Overview and Subnetwork Independent Requirements. 21 pp.

—Standards Preparation—Abbreviations
EURO STANDARDIZAT ION-REF ST 1-88. Glossary. 3 pp.

—Standards Preparation—Certification/Conformance
EURO 1987 Dec. IT Certification in Europe. 213 pp.
EURO STANDARDIZAT ION-FACTST 6. Conformance Testing and Certification Policy. 2 pp.
EURO STANDARDIZAT ION-FACTST 7-88. The CTS Programme. 3 pp.
EURO STANDARDIZAT ION-FACTST 8-88. Public Procurement. 2 pp.

—Standards Preparation—European Communities
EURO STANDARDIZAT ION-FACTST 0-88. Foreword. 2 pp.
EURO STANDARDIZAT ION-FACTST 1-88. The Challenge of Standardization. 3 pp.
EURO STANDARDIZAT ION-FACTST 2. Networks, Communication and Interworking. 3 pp.
EURO STANDARDIZAT ION-FACTST 3-88. Community Standardization Policy. 3 pp.
EURO STANDARDIZAT ION-FACTST 4-88. Standards Bodies: the Changing Scene. 3 pp.
EURO STANDARDIZAT ION-FACTST 5-88. European Standards. 3 pp.
EURO STANDARDIZAT ION-FACTST 9-88. International Cooperation. 1 p.
EURO STANDARDIZAT ION-REF ST 2-88. Milestones in Community Standardization policy. 1 p.
EURO STANDARDIZAT ION-REF ST 3-88. List of Standardization Organizations and Technical Bodies. 1 p.

Information Technology
Use: Information Systems

Information Technology Equipment
Use: Data Processing Equipment

Information Theory
See Also: Data Processing; Data Transmission; Systems Engineering

—Glossaries
CNS C5156-85. Data Processing Vocabulary (Part 16: Information Theory) (Nov)(10243).
JIS X 0016-87. Glossary of Terms Used in Information Processing (Information Theory).
SAA AS 1189.16-82. Data Processing—Vocabulary—Part 16: Information Theory. 12 pp.

Infrared Absorption Analysis
Use: Infrared Analysis

Infrared Absorption Photometry
Use: Infrared Spectrophotometry

Infrared Absorption Spectroscopy
Use: Infrared Spectroscopy

Infrared Analysis
Use For: Infrared Absorption Analysis
See Also: Gas Analysis; Gas Chromatography

—Admixtures
CEN PREN 480 (Part 6)-91. Admixtures for Concrete, Mortar and Grout—Test Methods—Part 6: Infrared Analysis. 5 pp.

—Coal
SAA AS 1038.6.3. 3-86. Methods for the Analysis and Testing of Coal and Coke—Part 6.3.3: Ultimate Analysis of Higher Rank Coal—Determination of Total Sulphur (Infrared Method) (R 1992). 3 pp.

—Ferronickel
BSI BS 6783: Part 5-86. 1986 Sampling and Analysis of Nickel, Ferronickel and Nickel Alloys Part 5: Method for Determining of Carbon in Nickel Ferronickel and Nickel Alloys by Infra-Red Absorption After Induction Furnace Combustion. 11 pp.

INTERNATIONAL AND NON-U.S. NATIONAL STANDARDS
SUBJECT INDEX
Infrared

Infrared Analysis (Cont.)
—Ferronickel (Cont.)
BSI BS 6783: Part 7-86. 1986 Sampling and Analysis of Nickel, Ferronickel and Nickel Alloys Part 7: Method for Determining of Sulphur in Nickel, Ferronickel and Nickel Alloy by Infra-Red Absorption After Induction Furnace Combustion. 11 pp.

BSI BS 6783: Part 7-01. 1986 Amd 1 Sampling and Analysis of Nickel, Ferronickel and Nickel Alloys Part 7: Method for Determination of Sulphur in Nickel, Ferronickel and Nickel Alloys by Infra-Red Absorption After Induction Furnace Combustion (ISO 7526: 1985). 14 pp.

CEN EN 27526-91. Nickel, Ferronickel and Nickel Alloys—Determination of Sulfur Content—Infra-Red Absorption Method After Induction Furnace Combustion. 3 pp.

DIN ENGL EN 27526-92. Determination of Sulfur Content of Nickel, Ferronickel and Nickel Alloys; Infrared Absorption Method After Induction Furnace Combustion (ISO 7526: 1985) (Feb). 10 pp.

ISO 7524-85. Nickel, Ferronickel and Nickel Alloys—Determination of Carbon Content—Infra-Red Absorption Method After Induction Furnace Combustion First Edition. 10 pp.

ISO 7526-85. Nickel, Ferronickel and Nickel Alloys—Determination of Sulfur Content—Infra-Red Absorption Method After Induction Furnace Combustion First Edition. 9 pp.

—Iron
BSI BS 6200: SUB SEC 3.28.2-90. 1990 Sampling and Analysis of Iron, Steel and Other Ferrous Metals Part 3: Methods of Analysis Sec. 3.28: Determination of Sulphur Subsec. 3.28.2: Steel and Cast Iron: Infra-Red Absorption Method After Combustion in and Induc-tion Furnace (V). 11 pp.

BSI BS 6200: SUB SEC 3.28.2-01. 1990 Amd 1 Sampling and Analysis of Iron, Steel and Other Ferrous Metals Part 3: Methods of Analysis Sec 3.28: Determin of Sulphur Subsec 3.28.2: Steel and Cast Iron: Infra-Red Absorption Method After Combustion in an Induction Furnace (ISO 4935: 1989). 16 pp.

ISO 9556-89. Steel and Iron—Determination of Total Carbon Content—Infrared Absorption Method After Combustion in an Induction Furnace First Edition; (Corrected and Reprinted -1989). 11 pp.

ISO 9686-92. Direct Reduced Iron—Determination of Carbon and/or Sulfur Content—High Frequency Combustion Method with Infrared Measurement First Edition. 15 pp.

SAA AS 1050.32-84. Methods for the Analysis of Iron and Steel—Part 32: Determination of Carbon Content (Infrared Method). 3 pp.

—Nickel
BSI BS 6783: Part 5-86. 1986 Sampling and Analysis of Nickel, Ferronickel and Nickel Alloys Part 5: Method for Determining of Carbon in Nickel Ferronickel and Nickel Alloys by Infra-Red Absorption After Induction Furnace Combustion. 11 pp.

BSI BS 6783: Part 7-86. 1986 Sampling and Analysis of Nickel, Ferronickel and Nickel Alloys Part 7: Method for Determining of Sulphur in Nickel, Ferronickel and Nickel Alloy by Infra-Red Absorption After Induction Furnace Combustion. 11 pp.

BSI BS 6783: Part 7-01. 1986 Amd 1 Sampling and Analysis of Nickel, Ferronickel and Nickel Alloys Part 7: Method for Determination of Sulphur in Nickel, Ferronickel and Nickel Alloys by Infra-Red Absorption After Induction Furnace Combustion (ISO 7526: 1985). 14 pp.

CEN EN 27526-91. Nickel, Ferronickel and Nickel Alloys—Determination of Sulfur Content—Infra-Red Absorption Method After Induction Furnace Combustion. 3 pp.

DIN ENGL EN 27526-92. Determination of Sulfur Content of Nickel, Ferronickel and Nickel Alloys; Infrared Absorption Method After Induction Furnace Combustion (ISO 7526: 1985) (Feb). 10 pp.

ISO 7524-85. Nickel, Ferronickel and Nickel Alloys—Determination of Carbon Content—Infra-Red Absorption Method After Induction Furnace Combustion First Edition. 10 pp.

ISO 7526-85. Nickel, Ferronickel and Nickel Alloys—Determination of Sulfur Content—Infra-Red Absorption Method After Induction Furnace Combustion First Edition. 9 pp.

—Nickel Alloys
BSI BS 6783: Part 5-86. 1986 Sampling and Analysis of Nickel, Ferronickel and Nickel Alloys Part 5: Method for Determining of Carbon in Nickel Ferronickel and Nickel Alloys by Infra-Red Absorption After Induction Furnace Combustion. 11 pp.

Infrared Analysis (Cont.)
—Nickel Alloys (Cont.)
BSI BS 6783: Part 7-86. 1986 Sampling and Analysis of Nickel, Ferronickel and Nickel Alloys Part 7: Method for Determining of Sulphur in Nickel, Ferronickel and Nickel Alloy by Infra-Red Absorption After Induction Furnace Combustion. 11 pp.

BSI BS 6783: Part 7-01. 1986 Amd 1 Sampling and Analysis of Nickel, Ferronickel and Nickel Alloys Part 7: Method for Determination of Sulphur in Nickel, Ferronickel and Nickel Alloys by Infra-Red Absorption After Induction Furnace Combustion (ISO 7526: 1985). 14 pp.

CEN EN 27526-91. Nickel, Ferronickel and Nickel Alloys—Determination of Sulfur Content—Infra-Red Absorption Method After Induction Furnace Combustion. 3 pp.

DIN ENGL EN 27526-92. Determination of Sulfur Content of Nickel, Ferronickel and Nickel Alloys; Infrared Absorption Method After Induction Furnace Combustion (ISO 7526: 1985) (Feb). 10 pp.

ISO 7524-85. Nickel, Ferronickel and Nickel Alloys—Determination of Carbon Content—Infra-Red Absorption Method After Induction Furnace Combustion First Edition. 10 pp.

ISO 7526-85. Nickel, Ferronickel and Nickel Alloys—Determination of Sulfur Content—Infra-Red Absorption Method After Induction Furnace Combustion First Edition. 9 pp.

—Nickel Castings
JIS H 1277-88. Methods for Determination of Sulfur in Nickel and Nickel Alloy Castings. 10 pp.

—Oilseeds
BSI BS 4289: Part 6-86. (WITHDRAWN) 1986 Methods for the Analysis of Oilseeds Part 6: Determination of Oil, Moisture and Volatile Matter, and Protein by Infrared Reflectance. 4 pp.

—Steels
BSI BS 6200: SUB SEC 3.28.2-90. 1990 Sampling and Analysis of Iron, Steel and Other Ferrous Metals Part 3: Methods of Analysis Sec. 3.28: Determination of Sulphur Subsec. 3.28.2: Steel and Cast Iron: Infra-Red Absorption Method After Combustion in and Induc-tion Furnace (V). 11 pp.

BSI BS 6200: SUB SEC 3.28.2-01. 1990 Amd 1 Sampling and Analysis of Iron, Steel and Other Ferrous Metals Part 3: Methods of Analysis Sec 3.28: Determin of Sulphur Subsec 3.28.2: Steel and Cast Iron: Infra-Red Absorption Method After Combustion in an Induction Furnace (ISO 4935: 1989). 16 pp.

ISO 9556-89. Steel and Iron—Determination of Total Carbon Content—Infrared Absorption Method After Combustion in an Induction Furnace First Edition; (Corrected and Reprinted -1989). 11 pp.

SAA AS 1050.32-84. Methods for the Analysis of Iron and Steel—Part 32: Determination of Carbon Content (Infrared Method). 3 pp.

Infrared Analyzers
Use For: Air Quality Infrared Analyzers
See Also: Gas Analyzers; Radiation Measuring Instruments

—Air Pollution Control Equipment
BSI BS 5849-80. 1980 Method of Expression of Performance of Air Quality Infra-Red Analysers. 20 pp.

IEC 528-75. Expression of Performance of Air Quality Infra-Red Analyzers First Edition. 37 pp.

—Nondispersive
CNS K0043-89. Non-Dispersed Infrared Gas Analyzer (Apr)(12510).

JIS K 0151-83. Non-Dispersive Infrared Gas Analyzer. 12 pp.

Infrared Beam Interruption Detectors
Use: Photoelectric Alarm Systems

Infrared Coatings
See Also: Coatings

—Metal—Military—Camouflage
NATO STANAG 2338 ED 1 AMD 1-74. NATO Infra Red Reflective (IRR) Green Colour for Painting Military Equipment. 15 pp.

Infrared Detectors
See Also: Passive Infrared Detectors; Photodetectors; Radiation Measuring Instruments; Radiometers
MOD UK DSTAN 59-61: 90/054-73. Semiconductor Device, Photocell Issue 1. 16 pp.

MOD UK DSTAN 59-99: Part 1-01. Coolers, Infrared Detector, Joule-Thomson Part 1: General Requirements Issue 2; Amendment 1. 29 pp.

—Air Cleaners
MOD UK DSTAN 58-96-92. Pure Air Systems for Detector Cooling Applications Issue 2. 17 pp.

Infrared Detectors (Cont.)
—Air Cleaners (Cont.)
MOD UK DSTAN 81-91-93. High Pressure Pure Air Equipment for Detector Cooling Applications Issue 1. 34 pp.

—Headphones—Sound Transmission
BSI BS 6418-83. 1983 Cordless Audio Transmission Devices Using Infra-Red Radiation. 4 pp.

CENELEC HD 455-85. Sound Transmission Using Infra-Red Radiation. 2 pp.

IEC 764-83. Sound Transmission Using Infra-Red Radiation First Edition. 11 pp.

Infrared Emitting Diode Arrays
See Also: Infrared Emitting Diodes

—Fiber Optic—Preferred Products List
CECC CECC MUAHAG Vol 10 IS 2-92. Preferred Products List; OPTO Electronic Devices (En, Fr, Ge). 48 pp.

—Preferred Products List
CECC CECC MUAHAG Vol 10 IS 2-92. Preferred Products List; OPTO Electronic Devices (En, Fr, Ge). 48 pp.

Infrared Emitting Diodes
Use For: IRED **See Also:** Diodes; Infrared Emitting Diode Arrays; Light Emitting Diodes; Optoelectronic Devices

BSI BS 6493: Sec 1.5-92. 1992 Semiconductor Devices Part 1: Discrete Devices Section 1.5: Recommendations for Optoelectronic Devices (IEC 747-5: 1992). 115 pp.

BSI BS 6493: Sec 1.5-85. 1985 Semiconductor Devices Part 1: Discrete Devices Section 1.5: Recommendations for Optoelectronic Devices. 34 pp.

CECC CECC 20 000 ISSUE 1-82. Generic Specification: Semiconductor Optoelectronic and Liquid Crystal Devices (En, Fr, Ge) AMD 3 (En, Fr, Ge) SUPP 1 (En, Fr, Ge) SUPP 2 (En, Fr, Ge). 401 pp.

IEC 747 Pt 5-92. Semiconductor Devices Discrete Devices and Integrated Circuits Part 5: Optoelectronic Devices Second Edition. 238 pp.

IECQ QC 720100-91. Semiconductor Devices Part 12: Sectional Specification for Optoelectronic Devices (IEC 747-12 ED 1). 44 pp.

MOD UK DSTAN 59-61: 80/021-01. Semiconductor Device, Infrared Emitting Diode Issue 1; Amendment 1. 13 pp.

MOD UK DSTAN 59-61: 90/168-77. Semiconductor Device—Light Emitting Diode Issue 1. 18 pp.

—Fiber Optic—Preferred Products List
CECC CECC MUAHAG Vol 10 IS 2-92. Preferred Products List; OPTO Electronic Devices (En, Fr, Ge). 48 pp.

—Preferred Products List
CECC CECC MUAHAG Vol 10 IS 2-92. Preferred Products List; OPTO Electronic Devices (En, Fr, Ge). 48 pp.

—Quality Assurance
BSI BS 9370-83. 1983 Light Emitting and Infra-Red Diode Arrays of Assessed Quality: Generic Data and Methods of Test. 43 pp.

BSI BS CECC 20002-88. 1988 Infrared Emitting Diodes, Infrared Emitting Diode Arrays: Blank Detail Specification. 14 pp.

IEC 747 Pt 12-91. Semiconductor Devices Part 12: Sectional Specification for Optoelectronic Devices First Edition (IECQ QC 720100). 44 pp.

Infrared Equipment
Scope Note: Use a more specific term **See:** Infrared Analyzers; Infrared Detectors; Infrared Emitting Diodes; Infrared Filters; Infrared Heating Equipment; Infrared Lamps; Infrared Lasers; Ultraviolet Spectrophotometers

Infrared Filters
See Also: Photographic Filters; Ultraviolet Filters

—Eye Protectors
BSI BS EN 171-92. 1992 Infra-Red Filters Used in Personal Eye-Protection Equipment. 9 pp.

CEN EN 171-86. Personal Eye-Protection: Infrared Filters; Transmittance Requirements and Recommended Use. 5 pp.

CEN EN 171-92. Personal Eye-Protection—Infra Red Filters—Transmittance Requirements and Recommended Use. 4 pp.

CEN PREN 172-89. Personal Eye Protection: Filters for General and Industrial Use. 25 pp.

DIN ENGL 4647 Pt 3-77. Glasses for Eye-Protection Equipment; Protective Filters Against Infra-Red Radiation (Protective IR Filters) (Feb). 4 pp.

ISO 4852-78. Personal Eye-Protectors—Infra-Red Filters—Utilisation and Transmittance Requirements First Edition. 6 pp.

INDUSTRY STANDARDS

Infrared Filters (Cont.)
—Eye Protectors (Cont.)
JIS T 8141-80. Eye Protectors for Radiations. 19 pp.
SAA AS 1338.3-81. Filters for Eye Protectors—Part 3: Filters for Protection Against Infrared Radiation (Superseded by AS/NZS 1338.3:1992).
SAA AS/NZS 1338. 3-92. Filters for Eye Protectors—Part 3: Filters for Protection Against Infra-Red Radiation (in Professional Packages 32, 36, 47) (Supersedes AS 1338.3—1981). 7 pp.
SNZ NZS/AS 1338. 3-92. Filters for Eye Protectors Part 3: Filters for Protection Against Infra-Red Radiation (This is a Joint Standard with SAA AS 1338.2). 7 pp.

Infrared Heating Equipment
See Also: Heating Equipment
IEC 240 Pt 1-92. Characteristics of Electric Infra-Red Emitters for Industrial Heating Part 1: Short Wave Infa-Red Emitters Second Edition. 20 pp.

—Industrial—Construction
CSA C22.2 NO 88-1958. Construction and Test of Industrial Heating Equipment (R 1992). 22 pp.
CSA 1169 Bull. Electrical Bulletin 1169 June 27, 1978 to C22.2 NO 88. 2 pp.

—Industrial—Testing
CSA C22.2 NO 88-1958. Construction and Test of Industrial Heating Equipment (R 1992). 22 pp.
CSA 1169 Bull. Electrical Bulletin 1169 June 27, 1978 to C22.2 NO 88. 2 pp.

Infrared Lamps
See Also: Incandescent Lighting; Lamps
CNS C4124-79. Infrared Lamps (Sep)(3737).
CNS C4216-80. Infrared Lamps (Aug)(6049).
JIS C 7514-85. Infrared Lamps. 18 pp.

—Cosmetics
CSA CAN/CSA-C22. 2 NO 224-M89. Radiant Heaters and Infrared and Ultraviolet Lamp Assemblies for Cosmetic or Hygienic Purposes in Nonmedical Applications; (Gen Instr 1 Thru 2). 70 pp.

—Food Service Equipment
CSA CAN/CSA-C22. 2 NO 224-M89. Radiant Heaters and Infrared and Ultraviolet Lamp Assemblies for Cosmetic or Hygienic Purposes in Nonmedical Applications; (Gen Instr 1 Thru 2). 70 pp.

—Hygiene
CSA CAN/CSA-C22. 2 NO 224-M89. Radiant Heaters and Infrared and Ultraviolet Lamp Assemblies for Cosmetic or Hygienic Purposes in Nonmedical Applications; (Gen Instr 1 Thru 2). 70 pp.

—Safety
BSI BS 3456: Sec 102.27-87. (WITHDRAWN) 1987 Safety of Household and Similar Electrical Appliances Part 102: Particular Requirements Section 102.27: Ultra-Violet and Infra-Red Radiation Skin Treatment Appliances for Household Use (Superseded by BS 3456: Sec. 202.27: 1991) 20 pp.
BSI BS 3456: Sec 202.27-91. 1991 Safety of Household Electrical Appliances Part 202: Particular Requirements Section 202.27: Ultra-Violet and Infra-Red Radiation Skin Treatment Appliances for Household and Similar Use. 34 pp.
BSI BS 3456: Sec 202.27-01. 1991 Amd 1 Safety of Household and Similar Electrical Appliances Part 202: Particular Requirements Section 202.27: Ultra-Violet and Infra-Red Radiation Skin Treatment Appliances for Household and Similar Use (AMD 7963) October 15, 1993 (L). 35 pp.
BSI BS EN 60335-2-27-92. 1992 Safety of Household and Similar Electrical Appliances Part 2: Particular Requirements Section 2.27: Ultra-Violet and Infra-Red Radiation Skin Treatment Appliances for Household and Similar Use. 28 pp.
BSI BS EN 60335-2-27-01. 1992 Amd 1 Safety of Household and Similar Electrical Appliances Part 2: Particular Requirements Section 2.27: Ultra-Violet and Infra-Red Radiation Skin Treatment Appliances for Household and Similar Use (AMD 7794) June 15, 1993 (L). 35 pp.
CENELEC HD 272 S3-87. Safety of Household and Similar Electrical Appliances—Part 2: Particular Requirements for Ultra-Violet and Infra-Red Radiation Skin Treatment Appliances for Household Use. 8 pp.
CENELEC EN 60 335-2-27-89. Safety of Household and Similar Electrical Appliance Part 2: Particular Requirements for Ultra-Violet and Infra-Red Radiation Skin Treatment Appliances for Household. 17 pp.
CENELEC EN 60335-2-27-92. Safety of Household and Similar Electrical Appliances Part 2: Particular Requirements for Ultra-Violet and Infra-Red Radiation Skin Treatment Appliances for Household Use; (IEC 335-2-27:1987 + A1:1989). 7 pp.

Infrared Lamps (Cont.)
—Safety (Cont.)
CENELEC EN 60335-2-27-92. Safety of Household and Similar Electrical Appliances Part 2: Particular Requirements for Ultra-Violet and Infra-Red Radiation Skin Treatment Appliances for Household and Similar Use; (IEC 335-2-27: 1987 + Amendment 1: 1989, Modified). 20 pp.
CENELEC EN 60335-2-27-92. Safety of Household and Similar Electrical Appliances Part 2: Particular Requirements for Ultra-Violet and Infra-Red Radiation Skin Treatment Appliances for Household Use; (IEC 335-2-27:1987 + A1:1989). 9 pp.
CENELEC EN 60335-2-27/A2-92. AMD 2 Safety of Household and Similar Electrical Appliances Part 2: Particular Requirements for Ultra-Violet and Infra-Red Radiation Skin Treatment Appliances for Household and Similar Use (IEC 335-2-27:1987/A2: 1991). 4 pp.
CENELEC EN 60335-2-37/A51-92. AMD 51 Safety of Household and Similar Electrical Appliances Part 2: Particular Requirements for Commercial Electric Deep Fat Fryers. 6 pp.
IEC 335 Pt 2-27-87. Safety of Household and Similar Electrical Appliances Part 2: Particular Requirements for Ultra-Violet and Infra-Red Radiation Skin Treatment Appliances for Household Use Second Edition; (Amendment 1-1989) (Amendment 2-1991); (CENELEC EN 60335-2-27:/A2:1992). 60 pp.

Infrared Lasers
See Also: Lasers
—Telecommunication
CCIR Report 680-2-90. Techniques for Telecommunication by Means of Electromagnetic Waves in the Infra-Red and Visible Regions of the Spectrum—Section 2B—Topics of General Interest. 16 pp.

Infrared Radiation
See Also: Electromagnetic Radiation; Light (Visible Radiation); Radiation; Solar Radiation; Wavelength
—Atmospheric Attenuation
CCIR Report 883-2-90. Attenuation of Visible and Infra-Red Radiation—Section 5C—Effects of the Atmosphere (Radiometeorology). 8 pp.

—Electromagnetic Interference—Sound Transmission
IEC 1147-93. Uses of Infra-Red Transmission and the Prevention or Control of Interference Between Systems First Edition. 43 pp.

—Medical Electrical Equipment—Safety
CSA CAN/CSA-C22. 2NO601.1-M90. Medical Electrical Equipment, Part 1: General Requirements for Safety; (Gen Instr 1). 240 pp.

Infrared Spectrography
Use: Infrared Spectroscopy

Infrared Spectrometry
See Also: Spectrometry
—Milk
SAA AS 2300.1.8. 2-91. Methods of Chemical and Physical Testing for the Dairying Industry—Part 1: General Methods and Principles—Part 1.8.2: Assessment of Instrumental Methods—Infared Spectrometric Analysis of Milk. 4 pp.

—Rubber
BSI BS 4181: Part 1-85. 1985 Identification of Rubbers by Infra-Red Spectrometry Part 1: Method for Identification of Hydrocarbon, Chloroprene, Nitrite and Chlorosulphonated Polyethylene Rubbers. 47 pp.
ISO 4650-84. Rubber—Identification—Infra-Red Spectrometric Method First Edition. 45 pp.

—Waste Water
DIN ENGL 38409 Pt 18-81. German Standard Methods for the Analysis of Water, Waste Water and Sludge; Summary Action and Material Characteristic Parameters (Group H); Determination of Hydrocarbons (H 18) (Feb). 6 pp.

—Water
DIN ENGL 38409 Pt 18-81. German Standard Methods for the Analysis of Water, Waste Water and Sludge; Summary Action and Material Characteristic Parameters (Group H); Determination of Hydrocarbons (H 18) (Feb). 6 pp.

Infrared Spectrophotometry
Use For: Infrared Absorption Photometry
See Also: Spectrophotometry
JIS K 0117-90. General Rules for Infrared Spectrophotometric Analysis. 19 pp.

Infrared Spectrophotometry (Cont.)
—Ammonia
BSI BS 4431-89. 1989 Methods of Sampling and Test for Liquefied Anhydrous Ammonia. 16 pp.
ISO 7106-85. Liquefied Anhydrous Ammonia for Industrial Use—Determination of Oil Content—Gravimetric and Infra-Red Spectrometric Methods First Edition. 7 pp.

—Aromatic Hydrocarbons
MOD UK DSTAN 05-50: Part 60-91. Methods for Testing Fuels, Lubricants and Associated Products Part 60: Aromatics Content in Hydrocarbon Solvents Issue 1. 7 pp.

Infrared Spectroscopy
Use For: Infrared Spectrography
—Gasoline
DIN ENGL 51414-85. Testing of Petroleum Products; Determination of the Benzene Content of Gasolines by Infrared Spectroscopy (June). 3 pp.

—Iron
DIN ENGL EN 24935-92. Determination of Sulfur Content of Steel and Iron by Infrared Absorption Spectroscopy After Combustion in an Induction Furnace; (ISO 4935: 1989) (July). 10 pp.
ISO 4935-89. Steel and Iron—Determination of Sulfur Content—Infrared Absorption Method After Combustion in an Induction Furnace First Edition; (Corrected and Reprinted -1990). 10 pp.

—Petroleum Products
CEN PREN 238-85. Liquid Petroleum Products Determination of the Benzene Content; Infrared Spectrometric Method. 7 pp.

—Polyvinyl Acetate
DIN ENGL 53742-71. Testing of Plastics; Determination of the Vinyl Acetate Content of Copolymers of Vinyl Chloride and Vinyl Acetate; Infrared Spectrographic Method (Sept). 2 pp.

—PVC
DIN ENGL 53742-71. Testing of Plastics; Determination of the Vinyl Acetate Content of Copolymers of Vinyl Chloride and Vinyl Acetate; Infrared Spectrographic Method (Sept). 2 pp.

—Silicon
DIN ENGL 50438 Pt 3-84. Testing of Materials for Use in Semiconductor Technology; Determination of Interstitial Atomic Boron and Phosphorus Content of Silicon by Infrared Absorption Spectroscopy (Feb). 4 pp.

—Steels
DIN ENGL EN 24935-92. Determination of Sulfur Content of Steel and Iron by Infrared Absorption Spectroscopy After Combustion in an Induction Furnace; (ISO 4935: 1989) (July). 10 pp.
ISO 4935-89. Steel and Iron—Determination of Sulfur Content—Infrared Absorption Method After Combustion in an Induction Furnace First Edition; (Corrected and Reprinted -1990). 10 pp.

—Toluene Diisocyanates
ISO TR9372-93. Plastics—Basic Materials for Polyurethanes—Determination of the Amounts of 2, 4-and 2,6-Isomers in Toluenediisocyanate by Infrared Spectroscopy First Edition. 7 pp.

Infrared Thermographs
Use: Thermographs—Infrared

Infusion Equipment (Medical)
Use: Intravenous Medical Equipment

Infusion Equipment (Parenteral)
Use: Parenteral Infusion Equipment

Infusion Medical Equipment
Use: Intravenous Medical Equipment

Infusion Pumps
See Also: Medical Electrical Equipment; Medical Equipment; Pumps
JIS T 1653-91. Infusion Pumps. 9 pp.

Ingots
Scope Note: See the subheading Ingots under the specific metal or alloy.

Inhalation Hazards
See Also: Safety
—Toys
CNS Z8016-14-87. Method of Test for Toy Safety (Testing for Ingestion or Inhalation Hazards) (Dec)(4798-14). 2 pp.

INTERNATIONAL AND NON-U.S. NATIONAL STANDARDS
SUBJECT INDEX
INMARSAT

Inhalation Hazards (Cont.)
—Toys (Cont.)
SNZ NZS 5822-92. Prevention of Ingestion and Inhalation Hazards in Toys Intended for Use by Children Under Three Years of Age. 48 pp.

Inhibitors
—Meat
CNS N6121-87. Method of Test for the Inhibitory Substances in Fresh Meat (May)(5916). 2 pp.

—Milk—Microbiological Analysis
BSI BS 4285: Sec 5.3-87. 1987 Microbiological Examination for Dairy Purposes Part 5: Ancillary Method Section 5.3: Test for Inhibitory Substances. 6 pp.

CNS N6069-87. Method of Test for the Inhibitory Substances in Raw Milk (May)(3453). 3 pp.

DIN ENGL 10182 Pt 1-81. Microbiological Analysis of Milk; Detection of Inhibitors in Milk; Reference Method (Oct). 4 pp.

DIN ENGL 10182 Pt 2-82. Microbiological Analysis of Milk; Detection of Inhibitors in Milk; Routine Method (Brilliant Black Reduction Test) (Oct). 4 pp.

DIN ENGL 10182 Pt 3-83. Microbiological Analysis of Milk; Detection of Inhibitors in Milk; Routine Method (Agar Diffusion Test) (Oct). 4 pp.

Initials
Use: Abbreviations

Initiators (Explosives)
Scope Note: For additional listings, use a more specific term *Use For:* Electroexplosive Devices; Electronic Explosive Devices *See Also:* Blasting Caps; Bombs (Ordnance); Detonators; Explosives; Fuses (Ordnance); Igniters; Ordnance; Pyrotechnics

—Barium Chloride Dihydrate
MOD UK DSTAN 68-34-91. Barium Chloride Dihydrate Issue 2. 8 pp.

—Carboxymethylcellulose Sodium
MOD UK DSTAN 68-47-91. Sodium Carboxymethylcellulose Issue 2. 8 pp.

—Electromagnetic Radiation
NATO STANAG 4234 ED 1 AMD 0-00. (DRAFT) Electromagnetic Radiation (Radio Frequency)—200 khz to 40 GHz Environment—Affecting the Design of Material for Use by NATO Forces. 9 pp.

NATO STANAG 4234 ED 1 AMD 0-92. Electromagnetic Radiation (Radio Frequency)—200 kHz to 40 GHz Environment—Affecting the Design of Materiel for Use by NATO Forces. 10 pp.

NATO STANAG 4324 ED 1 AMD 0-91. Electromagnetic Radiation (Radio Frequency) Test Information to Determine the Safety and Suitability for Service of Electro-Explosive Devices and Associated Electronic Systems in Munitions and Weapon Systems. 11 pp.

—Electromagnetic Radiation—Radiofrequency Interference
BSI BS 6657-91. 1991 Prevention of Inadvertent Initiation of Electro-Explosive Devices by Radio-Frequency Radiation. 33 pp.

BSI BS 6657-86. 1986 Guide to Prevention of Inadvertent Initiation of Electro-Explosive Devices by Radio-Frequency Radiation. 32 pp.

—Electrostatic Charge
NATO STANAG 4235 ED 1 AMD 0-00. (DRAFT) Electrostatic Environment Conditions Affecting the Design of Material for Use by NATO Forces. 5 pp.

NATO STANAG 4235 ED 1 AMD 0-93. Electrostatic Environmental Conditions Affecting the Design of Materiel for Use by NATO Forces. 9 pp.

—Lead Acetate
MOD UK DSTAN 68-112-85. Lead Acetate (Normal) Issue 1. 7 pp.

MOD UK DSTAN 68-112-91. Lead Acetate (Normal) Issue 2. 8 pp.

—Testing Conditions—Marine
MOD UK NES 1003-92. Requirement for Assessing Transient Coupled Energy into Stores Containing Electro-Explosive Devices (EED) Whilst Under Test Within Royal Naval Armament Depots Issue 1 (02.92). 13 pp.

MOD UK NES 1003-93. Requirement for Assessing Transient Coupled Energy into Stores Containing Electro-Explosive Devices (EED) Issue 2 (01.93). 16 pp.

—Voltage Detectors
MOD UK DSTAN 66-16-75. Test Set, Safety, Voltage Detection Issue 1. 26 pp.

Injection Equipment (Medical)
See Also: Medical Electrical Equipment; Medical Equipment

—Ampules
ISO 9187 Pt 1-91. Injection Equipment for Medical Use—Part 1: Ampoules for Injectables First Edition. 10 pp.

ISO 9187 Pt 2-93. Injection Equipment for Medical Use—Part 2: One-Point-Cut (OPC) Ampoules First Edition. 8 pp.

—Caps (Lids)
BSI BS 3263-60. (WITHDRAWN) 1960 Amd 1 Rubber Closures for Injectable Products (Superseded by BS EN 28362-2: 1993). 12 pp.

—Stoppers
BSI BS 3263-60. (WITHDRAWN) 1960 Amd 1 Rubber Closures for Injectable Products (Superseded by BS EN 28362-2: 1993). 12 pp.

—Vials—Caps (Lids)
ISO 8362 Pt 6-92. Injection Containers for Injectables and Accessories—Part 6: Caps Made of Aluminium-Plastics Combinations for Injection Vials First Edition. 8 pp.

ISO 10985-92. Caps Made of Aluminium-Plastics Combinations for Infusion Bottles and Injection Vials—Requirements and Test Methods First Edition. 8 pp.

Injection Molding Machines
See Also: Molding Machines

—Plastics
CNS B1297-86. Dimensions Relating to Molds for Plastics Injection Molding Machines (Feb)(9965).

CNS B1355-88. General Requirements for Construction of Plastics Injection Molding Machines (Apr)(12244).

CNS B4067-89. Method of Test for Plastics Injection Molding Machines (Dec)(11529).

DIN ENGL 16754-77. Tools and Machines for Injection Moulding of Plastic Moulding Materials; Connecting Dimensions (Nov). 6 pp.

JIS B 6701-92. Dimensions Relating to Molds for Plastics Injection Molding Machines. 13 pp.

—Plastics—Safety
BSI BS 6679-85. 1985 Technical Safety Requirements for the Design and Construction of Injection Moulding Machines for Plastics and Rubber. 18 pp.

CEN EN 201-85. Technical Safety Requirements for the Design and Construction of Injection Moulding Machines for Plastics and Rubber. 15 pp.

—Rubber—Safety
BSI BS 6679-85. 1985 Technical Safety Requirements for the Design and Construction of Injection Moulding Machines for Plastics and Rubber. 18 pp.

CEN EN 201-85. Technical Safety Requirements for the Design and Construction of Injection Moulding Machines for Plastics and Rubber. 15 pp.

Injuries
See Also: Cold Injuries; Electric Shock

—Statistical Classifications
NATO STANAG 2050 ED 5 AMD 2-89. Statistical Classification of Diseases, Injuries and Causes of Death. 28 pp.

Ink Absorption
See Also: Inks; Penetration Resistance; Printability

—Blotting Papers
CNS P3062-83. Method of Test for Ink Absorbency of Blotting Paper (Dec)(10727). 2 pp.

CPPA F.4-92. Absorption of Water and Ink by Bibulous and Blotting Paper. 2 pp.

Ink Removers
See Also: Inks; Solvents
CGSB CAN/CGSB-21.4-93. Lithographic Blanket and Roller Wash Solvent. 7 pp.

—Ethyl Acetate Content
MOD UK M 2003/59. Examination of Solvent, Cleaning.

—Karl Fischer Method
MOD UK M 2003/59. Examination of Solvent, Cleaning.

—Mineral Spirits Content
MOD UK M 2003/59. Examination of Solvent, Cleaning.

—Water Content
MOD UK M 2003/59. Examination of Solvent, Cleaning.

Inking Pads
Use: Stamp Pads

Inks
Scope Note: For additional listings, use a more specific term *See Also:* Drawing Inks; India Ink; Ink Absorption; Ink Removers; Marking Inks; Printing Inks

BSI BS 3484-62. 1962 Amd 1 Blue-Black Record Inks. 11 pp.

CNS S2002-59. Testing Standard for Blue Inks (Sep)(57)(R 1973).

—Archival
BSI BS 3484-91. 1991 Blue-Black Record Inks (G). 11 pp.

BSI BS 3484-62. 1962 Amd 1 Blue-Black Record Inks. 11 pp.

—Classification
BSI BS 1000: (667)-79. 1979 Universal Decimal Classification (UDC). English Full Edition (667): Colour Industries (Dyes, Inks, Paints Etc.). 20 pp.

SNZ NZS/BS 1000 (667)-79. Universal Decimal Classification Colour Industries (Dyes, Inks, Paints, Etc.). 20 pp.

—Magnetic
BSI BS 4810-80. 1980 Print for Magnetic Ink Character Recognition. 52 pp.

ISO 1004-77. Information Processing—Magnetic Ink Character Recognition—Print Specifications Second Edition; (Erratum—Aug 1978). 54 pp.

JTC1 1004-77. Information Processing—Magnetic Ink Character Recognition—Print Specifications Second Edition; (Erratum—Aug 1978). 54 pp.

OSI ISO 1004-77. Information Processing—Magnetic Ink Character Recognition—Print Specification. 53 pp.

—Red
CNS S1039-66. Red Ink (for Writing) (Mar)(2636).

—Seal
CNS S1162-83. Seal Ink (Apr)(10202).

CNS S2100-83. Method of Test for Seal Ink (Apr)(10203).

—Stamp Pads
CGSB 53-GP-12M-78. Ink, Stamp Pad, Regular, Standard for. 7 pp.

CNS S1021-64. Stamp Pad Ink (Tentative) (Mar)(2267)(R 1972).

—Stencils—Coatings
CGSB 1-GP-71 METH 130.9-79. Methods of Testing Paints and Pigments Behavior Towards Topcoats Resistance to Stencil Inks. 1 p.

Inland Containers
Use: Containers

Inlet Valves
Use For: Induction Valves; Intake Valves
See Also: Engine Cylinders; Exhaust Valves; Valves

—Motorcycles
CNS D2190-88. Intake and Exhaust Valves for Motorcycles (Jul)(12374).

—Oil Filters
BSI BS 7403: Part 8-91. 1991 Full-Flow Lubricating Oil Filters for Internal Combustion Engines Part 8: Method of Test for Inlet Anti-Drain Valves (ISO 4548-8: 1989). 8 pp.

ISO 4548 Pt 8-89. Methods of Test for Full-Flow Lubricating Oil Filters for Internal Combustion Engines—Part 8: Inlet Anti-Drain Valve Test First Edition; (Corrigendum 1-1990). 9 pp.

Inlets (Electrical)
Use: Electric Inlets

Inlets (Fluids)
Use: Fluid Inlets

INMARSAT
Use For: International Maritime Satellite System; MARISAT *See Also:* INMARSAT Mobile International Numbers; Mobile Satellite Communications; Satellite Communication Equipment; Satellite Communications; Satellites; Telecommunication Systems

CCITT FASCICLE II.2-88. Telephone Network and ISDN-Operation, Numbering, Routing and Mobile Service. Recommendations E.100—E.333. 362 pp.

—CCITT No. 5 Signaling Systems—Interworking
CCITT RECMN Q.1103-89. Interworking Between Signalling System No.5 and Inmarsat Standard A System—Interworking with Satellite Mobile Systems (Study Group XI) 8 pp. 8 pp.

INDUSTRY STANDARDS

INTERNATIONAL AND NON-U.S. NATIONAL STANDARDS
SUBJECT INDEX

INMARSAT

INMARSAT (Cont.)

—CCITT R2 Signaling Systems—Interworking
  CCITT Q.1102-89. Interworking Between Signalling System R2 and Inmarsat Standard A System—Interworking with Satellite Mobile Systems (Study Group XI) 10 pp. 10 pp.

—Earth Stations
  CCIR Report 921-2-90. System Aspects of Digital Ship Earth Stations—Section 8F—Frequencies, Orbits and Systems. 46 pp.
  CCIR Report 918-1-90. Availability of Communications Circuits in the Maritime Mobile-Satellite Service—Section 8G—Availability, Performance Objectives and Interworking with Terrestrial Networks. 12 pp.

—EPIRB
  CCIR RECMN 632-1-90. Transmission Characteristics of a Satellite Emergency Position-Indicating Radiobeacon (Satellite EPIRB) System Operating Through Geostationary Satellites in the 1.6 GHz Band—Section 8J—Technical and Operating Characteristics of Radiocommunications. 7 pp.
  CCIR Report 761-3-90. Technical and Operating Characteristics of Distress Systems in the Mobile-Satellite Service—Section 8J—Technical and Operating Characteristics of Radiocommunications Using Satellite Distress and Safety Operation and of Radio Determination Satellite Services. 19 pp.
  CCIR Report 1045-1-90. Satellite EPIRB Coordinated Trials Programme and Pre-Operational Demonstrations Using the Inmarsat Geostationary Space Segment Operating in the 1.6 GHz Band—Section 8J—Technical and Operating Characteristics of Radiocommunications. 12 pp.

—EPIRB—Duty Cycles
  CCIR RECMN 632-1-90. Transmission Characteristics of a Satellite Emergency Position-Indicating Radiobeacon (Satellite EPIRB) System Operating Through Geostationary Satellites in the 1.6 GHz Band—Section 8J—Technical and Operating Characteristics of Radiocommunications. 7 pp.

—EPIRB—Geostationary Satellites
  CCIR Report 1184-90. Pre-Operational Demonstrations of the 1.6 GHz Satellite EPIRB System Using the Inmarsat Geostationary Space Segment—Section 8J—Technical and Operating Characteristics of Radiocommunications Using Satellite Distress and Safety Operation and. 8 pp.

—ISDN—Communication Interfaces
  CCITT RECMN Q.1111-89. Interfaces Between the Inmarsat Standard B System and the International Public Switched Telephone Network/ISDN—Interworking with Satellite Mobile Systems (Study Group XI) 22 pp. 22 pp.
  CCITT RECMN Q.1151-89. Interfaces Between the Inmarsat Aeronautical Mobile-Satellite System and the International Public Switched Telephone Network/ISDN—Interworking with Satellite Mobile Systems (Study Group XI) 19 pp. 19 pp.

—ISDN—Signaling Systems—Interworking
  CCITT RECMN Q.1100-89. Structure of the Recommendations on the Inmarsat Mobile Satellite Systems—Interworking with Mobile Systems (Study Group XI) 3 pp. 3 pp.
  CCITT RECMN Q.1112-89. Procedures for Interworking Between Inmarsat Standard B System and the International Public Switched Telephone Network/ISDN—Interworking with Satellite Mobile Systems (Study Group XI) 51 pp. 51 pp.
  CCITT RECMN Q.1152-89. Procedures for Interworking Between Inmarsat Aeronautical Mobile-Satellite System and the International Public Switched Telephone Network/ISDN—Interworking with Satellite Mobile Systems (Study Group XI) 40 pp. 40 pp.

—Message Handling Systems
  CCIR Report 921-2-90. System Aspects of Digital Ship Earth Stations—Section 8F—Frequencies, Orbits and Systems. 46 pp.

—Mobile Satellite Communications
  CCIR Report 918-1-90. Availability of Communications Circuits in the Maritime Mobile-Satellite Service—Section 8G—Availability, Performance Objectives and Interworking with Terrestrial Networks. 12 pp.

—Mobile Satellite Communications—Country Codes
  CCITT RECMN E.215-89. Telephone/ISDN Numbering Plan for the Mobile-Satellite Services of INMARSAT—Telephone Network and ISDN—Operation, Numbering, Routing and Mobile Service (Study Group II)14 pp. 14 pp.

INMARSAT (Cont.)

—Mobile Satellite Communications—Data Transmission
  CCIR Report 1173-90. Technical and Operational Considerations for Aeronautical Mobile-Satellite Communications—Section 8F—Frequencies, Orbits and Systems. 40 pp.

—Mobile Satellite Communications—Data Transmission—PDN
  CCITT RECMN F.122-89. Operational Procedures for the Maritime Satellite Data Transmission Service—Telegraph and Mobile Services Operations and Quality of Service (Study Group I) 8 pp. 8 pp.

—Mobile Satellite Communications—Group Calls
  CCITT RECMN F.122-89. Operational Procedures for the Maritime Satellite Data Transmission Service—Telegraph and Mobile Services Operations and Quality of Service (Study Group I) 8 pp. 8 pp.

—Mobile Satellite Communications—Interworking—Data Networks
  CCIR Report 1176-90. Interworking Between the Mobile Satellite Systems and the Terrestrial Networks for Data Transmission Services—Section 8G—Availability, Performance Objectives and Interworking with Terrestrial Networks. 8 pp.

—Mobile Satellite Communications—Message Handling—Maintenance
  CCITT RECMN M.1150-92. Maritime Mobile Telecommunication Store-and-Forward Services (Packet Mode) Via Satellite (Study Group IV) 8 pp. 8 pp.

—Mobile Satellite Communications—Numbering Plans
  CCITT RECMN E.215-89. Telephone/ISDN Numbering Plan for the Mobile-Satellite Services of INMARSAT—Telephone Network and ISDN—Operation, Numbering, Routing and Mobile Service (Study Group II)14 pp. 14 pp.
  CCITT RECMN F.125-89. Telex Numbering Plan for the Mobile-Satellite Services of INMARSAT—Telegraph and Mobile Services Operations and Quality of Service (Study Group I) 9 pp. 9 pp.
  CCITT RECMN F.126-89. Selection Procedures for the INMARSAT Mobile-Satellite Telex Service—Telegraph and Mobile Services Operations and Quality of Service (Study Group I) 11 pp. 11 pp.

—Mobile Satellite Communications—Performance Measurement
  CCIR Report 920-2-90. Maritime Satellite System Performance at Low Elevation Angles—Section 8I—Technical and Operating Characteristics of Mobile Satellite Services. 14 pp.

—Mobile Satellite Communications—Radiotelephony
  CCIR Report 1173-90. Technical and Operational Considerations for Aeronautical Mobile-Satellite Communications—Section 8F—Frequencies, Orbits and Systems. 40 pp.

—Mobile Satellite Communications—SELCAL Services
  CCITT RECMN F.126-89. Selection Procedures for the INMARSAT Mobile-Satellite Telex Service—Telegraph and Mobile Services Operations and Quality of Service (Study Group I) 11 pp. 11 pp.
  CCITT RECMN F.127-89. Operational Procedures for Interworking Between the Telex Service and the Service Offered by INMARSAT Standard-C System—Telegraph and Mobile Services Operations and Quality of Service (Study Group I) 4 pp. 4 pp.

—Mobile Satellite Communications—SELCAL Services—ISDN
  CCITT RECMN E.215-89. Telephone/ISDN Numbering Plan for the Mobile-Satellite Services of INMARSAT—Telephone Network and ISDN—Operation, Numbering, Routing and Mobile Service (Study Group II)14 pp. 14 pp.
  CCITT RECMN E.216-89. Selection Procedures for the INMARSAT Mobile-Satellite Telephone and ISDN Services—Telephone Network and ISDN—Operation, Numbering, Routing and Mobile Service (Study Group II) 12 pp. 12 pp.

—Mobile Satellite Communications—SELCAL Services—PSTN
  CCITT RECMN E.215-89. Telephone/ISDN Numbering Plan for the Mobile-Satellite Services of INMARSAT—Telephone Network and ISDN—Operation, Numbering, Routing and Mobile Service (Study Group II)14 pp. 14 pp.

INMARSAT (Cont.)

—Mobile Satellite Communications—SELCAL Services—PSTN (Cont.)
  CCITT RECMN E.270-89. Monthly Telephone and Telex Accounts—Telephone Network and ISDN—Operation, Numbering, Routing and Mobile Service (Study Group II)1 pp (Same as Recmn D.170). 1 pp.

—Mobile Satellite Communications—Ship Motion—Statistical Analysis
  CCIR Report 921-2-90. System Aspects of Digital Ship Earth Stations—Section 8F—Frequencies, Orbits and Systems. 46 pp.

—Mobile Satellite Communications—Ship Station Identity
  CCITT RECMN E.215-89. Telephone/ISDN Numbering Plan for the Mobile-Satellite Services of INMARSAT—Telephone Network and ISDN—Operation, Numbering, Routing and Mobile Service (Study Group II)14 pp. 14 pp.
  CCITT RECMN F.125-89. Telex Numbering Plan for the Mobile-Satellite Services of INMARSAT—Telegraph and Mobile Services Operations and Quality of Service (Study Group I) 9 pp. 9 pp.

—Mobile Satellite Communications—Ship to Ship Calls
  CCITT RECMN E.216-89. Selection Procedures for the INMARSAT Mobile-Satellite Telephone and ISDN Services—Telephone Network and ISDN—Operation, Numbering, Routing and Mobile Service (Study Group II) 12 pp. 12 pp.
  CCITT RECMN F.126-89. Selection Procedures for the INMARSAT Mobile-Satellite Telex Service—Telegraph and Mobile Services Operations and Quality of Service (Study Group I) 11 pp. 11 pp.
  CCITT RECMN F.127-92. Operational Procedures for Interworking Between the International Telex Service and the Service Offered by the Inmarsat-C System (Study Group I) 6 pp. 6 pp.
  CCITT RECMN F.127-89. Operational Procedures for Interworking Between the Telex Service and the Service Offered by INMARSAT Standard-C System—Telegraph and Mobile Services Operations and Quality of Service (Study Group I) 4 pp. 4 pp.

—Mobile Satellite Communications—Ship to Shore Calls
  CCITT RECMN E.216-89. Selection Procedures for the INMARSAT Mobile-Satellite Telephone and ISDN Services—Telephone Network and ISDN—Operation, Numbering, Routing and Mobile Service (Study Group II) 12 pp. 12 pp.
  CCITT RECMN F.126-89. Selection Procedures for the INMARSAT Mobile-Satellite Telex Service—Telegraph and Mobile Services Operations and Quality of Service (Study Group I) 11 pp. 11 pp.
  CCITT RECMN F.127-92. Operational Procedures for Interworking Between the International Telex Service and the Service Offered by the Inmarsat-C System (Study Group I) 6 pp. 6 pp.
  CCITT RECMN F.127-89. Operational Procedures for Interworking Between the Telex Service and the Service Offered by INMARSAT Standard-C System—Telegraph and Mobile Services Operations and Quality of Service (Study Group I) 4 pp. 4 pp.

—Mobile Satellite Communications—Shore to Ship Calls
  CCITT RECMN E.216-89. Selection Procedures for the INMARSAT Mobile-Satellite Telephone and ISDN Services—Telephone Network and ISDN—Operation, Numbering, Routing and Mobile Service (Study Group II) 12 pp. 12 pp.
  CCITT RECMN F.122-89. Operational Procedures for the Maritime Satellite Data Transmission Service—Telegraph and Mobile Services Operations and Quality of Service (Study Group I) 8 pp. 8 pp.
  CCITT RECMN F.126-89. Selection Procedures for the INMARSAT Mobile-Satellite Telex Service—Telegraph and Mobile Services Operations and Quality of Service (Study Group I) 11 pp. 11 pp.
  CCITT RECMN F.127-92. Operational Procedures for Interworking Between the International Telex Service and the Service Offered by the Inmarsat-C System (Study Group I) 6 pp. 6 pp.
  CCITT RECMN F.127-89. Operational Procedures for Interworking Between the Telex Service and the Service Offered by INMARSAT Standard-C System—Telegraph and Mobile Services Operations and Quality of Service (Study Group I) 4 pp. 4 pp.

—Mobile Satellite Communications—Speech Encoding
  CCIR Report 509-5-90. Modulation and Coding Technique for Mobile Satellite Service—Section 8I—Technical and Operating Characteristics of Mobile Satellite Services. 23 pp.

INTERNATIONAL AND NON-U.S. NATIONAL STANDARDS
SUBJECT INDEX

INMARSAT (Cont.)
—Mobile Satellite Communications—Telex Communications
CCITT RECMN F.125-89. Telex Numbering Plan for the Mobile-Satellite Services of INMARSAT—Telegraph and Mobile Services Operations and Quality of Service (Study Group I) 9 pp. 9 pp.
CCITT RECMN F.126-89. Selection Procedures for the INMARSAT Mobile-Satellite Telex Service—Telegraph and Mobile Services Operations and Quality of Service (Study Group I) 11 pp. 11 pp.
CCITT RECMN F.127-89. Operational Procedures for Interworking Between the Telex Service and the Service Offered by INMARSAT Standard-C System—Telegraph and Mobile Services Operations and Quality of Service (Study Group I) 4 pp. 4 pp.
CCITT RECMN M.1150-92. Maritime Mobile Telecommunication Store-and-Forward Services (Packet Mode) Via Satellite (Study Group IV) 8 pp. 8 pp.

—Mobile Satellite Communications—Telex Communications—Interworking
CCITT RECMN F.127-92. Operational Procedures for Interworking Between the International Telex Service and the Service Offered by the Inmarsat-C System (Study Group I) 6 pp. 6 pp.
CCITT RECMN F.127-89. Operational Procedures for Interworking Between the Telex Service and the Service Offered by INMARSAT Standard-C System—Telegraph and Mobile Services Operations and Quality of Service (Study Group I) 4 pp. 4 pp.

—Mobile Satellite Communications—UHF
CCIR Report 1173-90. Technical and Operational Considerations for Aeronautical Mobile-Satellite Communications —Section 8F—Frequencies, Orbits and Systems. 40 pp.

—Numbering Plans
CCITT RECMN E.215-89. Telephone/ISDN Numbering Plan for the Mobile-Satellite Services of INMARSAT—Telephone Network and ISDN—Operation, Numbering, Routing and Mobile Service (Study Group II)14 pp. 14 pp.

—Numbering Plans—Group Calls
CCITT RECMN E.215-89. Telephone/ISDN Numbering Plan for the Mobile-Satellite Services of INMARSAT—Telephone Network and ISDN—Operation, Numbering, Routing and Mobile Service (Study Group II)14 pp. 14 pp.

—PSTN—Interfaces
CCITT RECMN Q.1111-89. Interfaces Between the Inmarsat Standard B System and the International Public Switched Telephone Network/ISDN—Interworking with Satellite Mobile Systems (Study Group XI) 22 pp. 22 pp.
CCITT RECMN Q.1151-89. Interfaces Between the Inmarsat Aeronautical Mobile-Satellite System and the International Public Switched Telephone Network/ISDN—Interworking with Satellite Mobile Systems (Study Group XI) 19 pp. 19 pp.

—PSTN—Signaling Systems—Interworking
CCITT RECMN Q.1100-89. Structure of the Recommendations on the Inmarsat Mobile Satellite Systems—Interworking with Mobile Systems (Study Group XI) 3 pp. 3 pp.
CCITT RECMN Q.1112-89. Procedures for Interworking Between Inmarsat Standard B System and the International Public Switched Telephone Network/ISDN—Interworking with Satellite Mobile Systems (Study Group XI) 51 pp. 51 pp.
CCITT RECMN Q.1152-89. Procedures for Interworking Between Inmarsat Aeronautical Mobile-Satellite System and the International Public Switched Telephone Network/ISDN—Interworking with Satellite Mobile Systems (Study Group XI) 40 pp. 40 pp.

—Signal Processing
CCIR Report 921-2-90. System Aspects of Digital Ship Earth Stations—Section 8F—Frequencies, Orbits and Systems. 46 pp.

—Telephone Networks—Signaling Systems—Interworking
CCITT RECMN Q.1101-89. General Requirements for the Interworking of the Terrestrial Telephone Network and Inmarsat Standard A System—Interworking with Satellite Mobile Systems (Study Group XI) 16 pp. 16 pp.

—Telex Communications—Interfaces
CCITT RECMN U.61-89. Detailed Requirements to be Met in Interfacing the International Telex Network with Maritime Satellite Systems—Telegraph Switching (Study Group IX) 10 pp. 10 pp.

—Telex Communications—Signaling Systems
CCITT FASCICLE VII.2. Telegraph Switching—Series U Recommendations. 282 pp.

INMARSAT (Cont.)
—Telex Communications—Signaling Systems—Norway
CCITT FASCICLE VII.2. Telegraph Switching—Series U Recommendations. 282 pp.

—Telex Destination Codes
CCITT RECMN F.125-89. Telex Numbering Plan for the Mobile-Satellite Services of INMARSAT—Telegraph and Mobile Services Operations and Quality of Service (Study Group I) 9 pp. 9 pp.

INMARSAT Mobile International Numbers
See Also: INMARSAT; INMARSAT Mobile Numbers
CCITT RECMN E.215-89. Telephone/ISDN Numbering Plan for the Mobile-Satellite Services of INMARSAT—Telephone Network and ISDN—Operation, Numbering, Routing and Mobile Service (Study Group II)14 pp. 14 pp.

—Telex Communications
CCITT RECMN F.125-89. Telex Numbering Plan for the Mobile-Satellite Services of INMARSAT—Telegraph and Mobile Services Operations and Quality of Service (Study Group I) 9 pp. 9 pp.

INMARSAT Mobile Numbers
See Also: INMARSAT Mobile International Numbers
CCITT RECMN E.215-89. Telephone/ISDN Numbering Plan for the Mobile-Satellite Services of INMARSAT—Telephone Network and ISDN—Operation, Numbering, Routing and Mobile Service (Study Group II)14 pp. 14 pp.

—Directories
CCITT RECMN E.215-89. Telephone/ISDN Numbering Plan for the Mobile-Satellite Services of INMARSAT—Telephone Network and ISDN—Operation, Numbering, Routing and Mobile Service (Study Group II)14 pp. 14 pp.
CCITT RECMN F.125-89. Telex Numbering Plan for the Mobile-Satellite Services of INMARSAT—Telegraph and Mobile Services Operations and Quality of Service (Study Group I) 9 pp. 9 pp.

—Public Data Networks
CCITT RECMN F.122-89. Operational Procedures for the Maritime Satellite Data Transmission Service—Telegraph and Mobile Services Operations and Quality of Service (Study Group I) 8 pp. 8 pp.

—Ship Station Identity
CCITT RECMN E.210-89. Ship Station Identification for VHF/UHF and Maritime Mobile-Satellite Services—Telephone Network and ISDN—Operation, Numbering, Routing and Mobile Service (Study Group II) 5 pp (Same as Recmn F.120). 5 pp.
CCITT RECMN E.215-89. Telephone/ISDN Numbering Plan for the Mobile-Satellite Services of INMARSAT—Telephone Network and ISDN—Operation, Numbering, Routing and Mobile Service (Study Group II)14 pp. 14 pp.

—Ship to Ship Calls
CCITT RECMN E.216-89. Selection Procedures for the INMARSAT Mobile-Satellite Telephone and ISDN Services—Telephone Network and ISDN—Operation, Numbering, Routing and Mobile Service (Study Group II) 12 pp. 12 pp.

—Shore to Ship Calls
CCITT RECMN E.216-89. Selection Procedures for the INMARSAT Mobile-Satellite Telephone and ISDN Services—Telephone Network and ISDN—Operation, Numbering, Routing and Mobile Service (Study Group II) 12 pp. 12 pp.

—Telex Communications
CCITT RECMN F.125-89. Telex Numbering Plan for the Mobile-Satellite Services of INMARSAT—Telegraph and Mobile Services Operations and Quality of Service (Study Group I) 9 pp. 9 pp.
CCITT RECMN F.126-89. Selection Procedures for the INMARSAT Mobile-Satellite Telex Service—Telegraph and Mobile Services Operations and Quality of Service (Study Group I) 11 pp. 11 pp.
CCITT RECMN F.127-89. Operational Procedures for Interworking Between the Telex Service and the Service Offered by INMARSAT Standard-C System—Telegraph and Mobile Services Operations and Quality of Service (Study Group I) 4 pp. 4 pp.

—Telex Communications—Group Calls
CCITT RECMN F.125-89. Telex Numbering Plan for the Mobile-Satellite Services of INMARSAT—Telegraph and Mobile Services Operations and Quality of Service (Study Group I) 9 pp. 9 pp.

Inner Tubes
Use For: Inner Tubes (Tires); Tire Tubes

Inner Tubes (Cont.)
See Also: Tires

—Aircraft
JIS W 2501-77. Rubber Tubes for Tires of Aircraft.

—Aircraft—Identification Systems
MOD UK DTD-1097-49. Marking of Aeroplane Tyres and Inner Tubes. 4 pp.

—Automotive
CGSB 20-GP-6E-81. Tubes, Inner, Vehicle and Mobile Ground Equipment, Standard for. 26 pp.
CNS D1003-82. Nomenclature of Tire, Tube, Rim, Rim Band and Flap for Automobiles (Nov)(4678).
CNS K4018-82. Tubes for Tires of Automobile (May)(1432). 2 pp.
JIS D 4201-84. Designation System of Tyre, Tube, Rim Band, and Flap for Automobiles. 21 pp.
JIS D 4231-87. Inner Tubes for Automobile Tires. 7 pp.

—Automotive—Designations
CNS D1003-82. Nomenclature of Tire, Tube, Rim, Rim Band and Flap for Automobiles (Nov)(4678).
JIS D 4201-84. Designation System of Tyre, Tube, Rim Band, and Flap for Automobiles. 21 pp.

—Bicycles
CNS K4002-83. Rubber Inner Tubes for Bicycles (Dec)(738). 2 pp.
CNS K4056-81. Rubber Valve Tubing (Jan)(6922). 3 pp.
CNS K6050-83. Method of Test for Rubber Inner Tubes of Bicycles (Dec)(739).
JIS K 6304-82. Rubber Inner Tubes for Bicycles.
JIS K 6304-76. Rubber Inner Tubes for Bicycles. 7 pp.
JIS K 6307-77. Valve Tubing (R 1980). 6 pp.

—Carts
CNS K4015-59. Wheel Tubes for Carts (Oct)(1155). 2 pp.
CNS K6105-59. Method of Test for Tubes for Carts (Oct)(1156). 2 pp.

—Glossaries
ISO 3877 Pt III-78. Tyres, Valves and Tubes—List of Equivalent Terms—Part III: Tubes First Edition. 6 pp.

—Ground Vehicles
CGSB 20-GP-6E-81. Tubes, Inner, Vehicle and Mobile Ground Equipment, Standard for. 26 pp.

—Motorcycles
CNS K4050-80. Rubber Inner Tubes for Motorcycles (Feb)(5216). 3 pp.
JIS K 6367-87. Inner Tubes for Motorcycle Tires. 7 pp.

—Storage
BSI BS AU 50: Pt 1: Sec 8-89. 1989 Tyres and Wheels Part 1: Tyres Section 8: Code of Practice for the Storage of Tyres, Inner Tubes and Flaps. 4 pp.

Inner Tubes (Tires)
Use: Inner Tubes

Inoculation
Use For: Seeding (Inoculation)
See Also: Crystallization; Therapy

—Water Analysis
BSI BS 6068: Sec 2.14-90. 1990 Water Quality Part 2: Physical, Chemical and Bio-Chemical Methods Section 2.14: Determination of Biochemical Oxygen Demand After 5 Days (BOD): Dilution and Seeding Method. 12 pp.

Inorganic Acid Content Analysis
—Rust Removers
CGSB 31-GP-0A METH 36.4-62. Methods of Testing Corrosion-Prevention Materials and Processes Orthophosphoric Acid. 1 p.

Inorganic Acids
Scope Note: Use a more specific term *See:* Boric Acid; Chlorosulfonic Acid; Chromium Trioxide; Fluoboric Acid; Fluosilicic Acid; Hydrobromic Acid; Hydrochloric Acid; Hydrofluoric Acid; Inorganic Acid Content Analysis; Nitric Acid; Perchloric Acid; Phosphoric Acid; Phosphoric Acid, Meta; Phosphorous Acid; Selenious Acid; Sulfuric Acid; Sulfurous Acid

Inorganic Compounds
—Spectrography
MOD UK M 9516/65. General Method of Qualitative Inorganic Analysis by Emission Spectrography (No Information).

Inorganic Content Analysis
—Coal
SAA AS 2434.9-91. Methods for the Analysis and Testing of Lower Rank Coal and its Chars—Part 9: Determination of Four Acid-Extractable Inorganic Ions in Lower Rank Coal (In Professional Package 32). 4 pp.

Inorganic Fiber Content Analysis
See Also: Fiber Content Analysis

—Air—Phase Contrast Microscopy
ISO 8672-93. Air Quality—Determination of the Number Concentration of Airborne Inorganic Fibres by Phase Contrast Optical Microscopy—Membrane Filter Method First Edition. 30 pp.

Inorganic Nitrates
Scope Note: Use a more specific term
See: Aluminum Nitrate; Ammonium Nitrate; Barium Nitrate; Calcium Ammonium Nitrate; Calcium Nitrate; Cobaltous Nitrate; Ferric Nitrate; Lithium Nitrate; Magnesium Nitrate; Mercuric Nitrate; Mercurous Nitrate; Nickel Nitrate; Potassium Nitrate; Sodium Nitrate; Strontium Nitrate; Thallium Nitrate; Thorium Nitrate; Uranyl Nitrate; Zirconyl Nitrate

Inorganic Peroxides
Scope Note: Use a more specific term *See:* Barium Peroxide; Hydrogen Peroxide; Lead Dioxide; Organic Peroxides; Sodium Peroxide; Strontium Peroxide

Inorganic Phosphates
Scope Note: Use a more specific term *See:* Calcium Phosphate, Dibasic; Calcium Phosphate, Monobasic; Calcium Phosphates; Phosphoric Acid; Sodium Phosphate, Monobasic; Sodium Phosphates; Sodium Tripolyphosphate; Superphosphates; Zinc Phosphates

Inorganic Silicates
Scope Note: Use a more specific term *See:* Calcium Silicate; Potassium Silicate; Sodium Silicate

Inorganic Sulfides
Scope Note: Use a more specific term
See: Ammonium Sulfide; Copper Sulfide; Ferrous Sulfides; Lead Sulfide; Molybdenum Disulfide; Zinc Sulfide

Inositol
See Also: Alcohols
CNS K7593-81. Chemical Reagent (Inositol) (Nov)(8138).
JIS K 8094-80. Inositol.

Input Impedance
See Also: Electrical Impedance; Impedance Measurement

—Cable Couplers
CEN PREN 2591 (Part G5)-92. Elements of Electrical and Optical Connection Test Methods Part G5—Measurement of Stub Input Impedance. 3 pp.

Input/Output Interfaces
Use: Communication Interfaces

Input Voltage
See Also: Voltage

—Medical Electrical Equipment
CSA CAN/CSA-C22. 2NO601.1-M90. Medical Electrical Equipment, Part 1: General Requirements for Safety; (Gen Instr 1). 240 pp.

Inquiry Characters
—Telex Communications—Data Terminal Equipment
CCITT RECMN F.73-90. Operational Principles for Communication Between Terminals of the International Telex Service and Data Terminal Equipment on Packet Switched Public Data Networks (Study Group I) 9 pp. 9 pp.
CCITT RECMN F.73-89. Operational Principles for Communication Between Terminals on Telex Networks and Data Terminal Equipment on Packet Switched Public Data Networks—Telegraph and Mobile Services Operations and Quality of Service (Study Group I) 5 pp. 5 pp.

—Telex Packet Interworking Function
CCITT RECMN F.73-90. Operational Principles for Communication Between Terminals of the International Telex Service and Data Terminal Equipment on Packet Switched Public Data Networks (Study Group I) 9 pp. 9 pp.

Inquiry Characters *(Cont.)*
—Telex Packet Interworking Function *(Cont.)*
CCITT RECMN F.73-89. Operational Principles for Communication Between Terminals on Telex Networks and Data Terminal Equipment on Packet Switched Public Data Networks—Telegraph and Mobile Services Operations and Quality of Service (Study Group I) 5 pp. 5 pp.

INS
Use: Inertial Navigation Systems

Insect Contamination
Use For: Bug Contamination (Insects)
See Also: Bees; Coleoptera; Larvae; Pest Control

—Damage—Coffee Beans
BSI BS 5752: Part 8-86. 1986 Coffee and Coffee Products Part 8: Green Coffee: Determination of Proportion of Insect-Damaged Beans. 25 pp.
ISO 6667-85. Green Coffee—Determination of Proportion of Insect-Damaged Beans First Edition. 14 pp.

—Food
BSI BS 4317: Part 18-88. 1988 Cereals and Pulses Part 18: Determination of Hidden Insect Infestation. 12 pp.
ISO 6639 Pt 1-86. Cereals and Pulses—Determination of Hidden Insect Infestation—Part 1: General Principles First Edition. 4 pp.
ISO 6639 Pt 3-86. Cereals and Pulses—Determination of Hidden Insect Infestation—Part 3: Reference Method First Edition. 6 pp.
ISO 6639 Pt 4-87. Cereals and Pulses—Determination of Hidden Insect Infestation—Part 4: Rapid Methods First Edition. (Supersedes 1162). 20 pp.

—Food—Sampling
ISO 6639 Pt 2-86. Cereals and Pulses—Determination of Hidden Insect Infestation—Part 2: Sampling First Edition. 6 pp.

Insect Control Equipment
See Also: Control Systems Equipment; Insect Electrocution Devices
NATO STANAG 2048 ED 2 AMD 2-82. Chemical Methods of Insect and Rodent Control (AMedP-3). 7 pp.

Insect Electrocution Devices
Use For: Bug Zappers; High Voltage Insect Killers
See Also: Insect Control Equipment; Insecticides; Pesticides
DIN VDE 0710 Pt 16 (D)-77. VDE Specification for Luminaires with Operating Voltages Below 1000 V; Luminaires with High Voltage Equipment for Killing Insects (Oct). 14 pp.
SAA AS 3150-89. Approval and Test Specification—Insect Electrocuters Amdt 1 May 1990 (This is a Joint Standard with SANZ NZS 3150). 5 pp.
SNZ NZS/AS 3150-89. Approval and Test Specification—Insect Electrocutors Amend: 1; 2, 1992 (This is a Joint Standard with SAA AS 3150). 5 pp.

—Indoor
CSA CAN/CSA-C22. 2 NO 189-M89. High-Voltage Insect Killers; (Gen Instr 1). 46 pp.
CSA 1304 Bull. Electrical Bulletin 1304 January 26, 1981 to C22.2 NO 189. 2 pp.

—Outdoor
CSA CAN/CSA-C22. 2 NO 189-M89. High-Voltage Insect Killers; (Gen Instr 1). 46 pp.
CSA 1304 Bull. Electrical Bulletin 1304 January 26, 1981 to C22.2 NO 189. 2 pp.

—Safety
IEC 335 Pt 2-59-90. Safety of Household and Similar Electrical Appliances Part 2: Particular Requirements for Insect Killers First Edition. 33 pp.

Insect Killers (High Voltage)
Use: Insect Electrocution Devices

Insect Repellents
—Diethyltoluamide
CGSB CAN/CGSB-15.19-92. Insect Repellent, Diethyltoluamide. 8 pp.
MOD UK TS 475A. Insect Repellant.

Insect Traps
Use For: Traps, Insect
CNS C4177-90. Electric Trap for Insect (Dec)(5425).

Insecticides
Scope Note: For additional listings, see also specific chemicals used for insect control *See Also:* Arsenic; Chlorobenzene; DDT; Insect Electrocution Devices; Pesticides

Insecticides *(Cont.)*
—Aerosol Containers
BSI BS 4172-67. (WITHDRAWN) 1967 Insecticidal Efficiency of Aerosols Against Flies (Superseded by BS 4172: Parts 1 & 2; 1993). 17 pp.
BSI BS 4172: Part 1-93. 1993 Hand-Held Pressurized Aerosol Dispensers Against Houseflies Part 1: Specification for Insecticidal Efficiency (Supersedes BS 4172: 1967). 6 pp.
BSI BS 4172: Part 2-93. 1993 Hand-Held Pressurized Aerosol Dispensers Against Houseflies Part 2: Method for Determination of Insecticidal Efficiency (Supersedes BS 4172: 1967). 20 pp.
MOD UK DSTAN 68-57-93. Insecticide Concentrate, Space Spray Issue 1. 12 pp.
MOD UK DSTAN 68-161-93. Dispensers, Insecticide Aerosol Flying Insect Killer Issue 1. 14 pp.
MOD UK CS 2833. Insecticide Concentrate Space Spray (Superseded by Def Stan 68-57).
MOD UK TS 10123. Dispensers, Insecticidal, Aerosol (Superseded by Def Stan 68-161).

—Particle Size Distribution
CNS Z6063-83. Method of Test for Particle-Size Distribution of Space-Insecticide (Feb)(10037).

—Piperonyl Butoxide Content—Absorptiometric Analysis
MOD UK M 5003/66. Examination of Insecticide Concentrate Space Spray and Insecticide Space Spray.

—Plywood
SNZ NZS 3611-70. Specification for Exterior Plywood Amend: 1, 1990 (Reconfirmed 1984). 36 pp.
SNZ NZS 3612-70. Specification for Interior Plywood Amend: 1, 1990 (Reconfirmed 1984). 33 pp.
SNZ NZS 3613-70. Specification for Plywood for Marine Craft (Reconfirmed 1984). 40 pp.
SNZ NZS 3614-71. Specification for the Manufacture of Construction Plywood Amend: 3, 1990. 40 pp.

—Pyrethrin Content—Distillation Methods
MOD UK M 5003/66. Examination of Insecticide Concentrate Space Spray and Insecticide Space Spray.

—Visual Inspection
MOD UK M 5003/66. Examination of Insecticide Concentrate Space Spray and Insecticide Space Spray.

—Wood Preservatives
BSI BS 7282-90. 1990 Field Test Method for Determining the Relative Protective Effectiveness of a Wood Preservative in Ground Contact. 15 pp.
BSI BS EN 20-1-92. 1992 Wood Preservatives—Determination of the Protective Effectiveness Against Lyctus Brunneus (Stephens)—Part 1: Application by Surface Treatment (Laboratory Method) (Supersedes BS 5217: 1975). 19 pp.
BSI BS EN 20-2-93. 1993 Wood Preservatives—Determination of the Protective Effectiveness Against Lyctus Brunneus (Stephens)—Part 2: Application by Impregnation (Laboratory Method) (W). 19 pp.
BSI BS EN 49-1-92. 1992 Wood Preservatives—Determination of the Protective Effectiveness Against Anobium Punctatum (De Geer) by Egg-Laying and Larval Survival Part 1: Application by Surface Treatment (Laboratory Method) (W). 16 pp.
BSI BS EN 49-2-92. 1992 Wood Preservatives—Determination of the Protective Effectiveness Against Anobium Punctatum (De Geer) by Egg-Laying and Larval Survival Part 2: Application by Impregnation (Laboratory Method) (Supersedes BS 5437: 1977). 16 pp.
BSI BS EN 273-92. 1992 Wood Preservatives—Determination of the Curative Action Against Lyctus Brunneus (Stephens) (Laboratory Method). 17 pp.
BSI BS EN 275-92. 1992 Wood Preservatives—Determination of the Protective Effectiveness Against Marine Borers. 22 pp.
BSI BS EN 370-93. 1993 Wood Preservatives—Determination of Eradicant Efficacy in Preventing Emergence of Anobium Punctatum (De Geer) (W). 18 pp.
CEN EN 20-74. Wood Preservatives: Determination of the Preventive Action Against Lyctus Brunneus: (Stephens) (Laboratory Method). 13 pp.
CEN EN 20-1-92. Wood Preservatives—Determination of the Protective Effectiveness Against Lyctus Brunneus (Stephens)—Part 1: Application by Surface Treatment (Laboratory Method). 23 pp.
CEN PREN 20 (Part 2)-90. Wood Preservatives—Determination of the Preventive Action Against Lyctus Brunneus (Stephens)—Part 2: Preservatives Application: Fully Impregnated Wood Treatment (Laboratory Method). 20 pp.
CEN EN 20-2-93. Wood Preservatives—Determination of the Protective Effectiveness Against Lyctus Brunneus (Stephens)—Part 2: Application by Impregnation (Laboratory Method). 14 pp.

Insecticides (Cont.)
—Wood Preservatives (Cont.)
CEN EN 21-74. Wood Preservatives Determination of the Toxic Values Against Anobium Punctatum (de Geer) by Larval Transfer (Laboratory Method). 13 pp.
CEN EN 21-88. Wood Preservatives; Determination of the Toxic Values Against Anobium Punctatum (De Geer) by Larval Transfer (Laboratory Method). 12 pp.
CEN EN 22-74. Wood Preservatives: Determination of Eradicant Action Against Hylotrupes Bajulus (Linnaeus) Larvae (Laboratory Method). 12 pp.
CEN EN 46-76. Wood Preservatives: Determination of the Preventive Action Against Recently Hatched Larvae of Hylotrupes Bajulus (Linnaeus) (Laboratory Method). 11 pp.
CEN EN 47-88. Wood Preservatives; Determination of the Toxic Values Against Larvae of Hylotrupes Bajulus (Linnaeus) (Laboratory Method). 10 pp.
CEN EN 48-76. Wood Preservatives; Determination of Eradicant Action Against Larvae of Anobium Punctatum (De Geer) (Laboratory Method). 12 pp.
CEN EN 48-88. Wood Preservatives; Determination of Eradicant Action Against Larvae of Anobium Punctatum (De Geer) (Laboratory Method). 13 pp.
CEN EN 49-76. Wood Preservatives: Determination of the Toxic Values Against Anobium Punctatum (De Geer) by Egg-Laying and Larval Survival (Laboratory Method). 11 pp.
CEN PREN 49 (Part 1)-89. Wood Preservatives: Part 1: Determination of the Toxic Valves Against Anobium Punctatum (De Geer) by Egg-Laying and Larval Survival (Laboratory Method). 12 pp.
CEN EN 49-1-92. Wood Preservatives—Determination of the Protective Effectiveness Against Anobium Punctatum (De Geer) by Egg-Laying and Larval Survival—Part 1: Application by Surface Treatment (Laboratory Method). 23 pp.
CEN PREN 49 (Part 2)-89. Wood Preservatives: Part 2: Determination of the Preventive Action Anobium Punctatum (De Geer) (Laboratory Method). 16 pp.
CEN EN 49-2-92. Wood Preservatives—Determination of the Protective Effectiveness Against Anobium Punctatum (De Geer) by Egg-Laying and Larval Survival—Part 2: Application by Impregnation (Laboratory Method). 24 pp.
CEN EN 117-81. Wood Preservatives Determination of Toxic Values Against Reticulitermes Santonensis de Feytaud (Laboratory Method). 11 pp.
CEN EN 117-89. Wood Preservatives: Determination of Toxic Values Against Reticulitermes Santonensis de Feytaud (Laboratory Method). 9 pp.
CEN EN 117-89. AMD 1 Wood Preservatives: Determination of Toxic Values Against Reticulitermes Santonensis de Feytaud (Laboratory Method). 2 pp.
CEN EN 118-86. AMD 1 Wood Preservatives; Determination of Preventive Action Against Reticulitermes Santonensis de Feytaud; (Laboratory Method). 14 pp.
CEN EN 118-90. Wood Preservatives: Determination of Preventive Action Against Reticulitermes Santonensis de Feytaud; (Laboratory Method). 9 pp.
CEN PREN 273-88. Wood Preservatives Determination of the Curative Action Against Lyctus Brunneus (Stephens) (Laboratory Method). 21 pp.
CEN EN 273-92. Wood Preservatives—Determination of the Curative Action Against Lyctus Brunneus (Stephens) (Laboratory Method). 22 pp.
CEN PREN 275-89. Test Method of Determining the Protective Effectiveness of a Preservative Against Marine Borers. 23 pp.
CEN EN 275-92. Wood Preservatives—Determination of the Protective Effectiveness Against Marine Borers. 17 pp.
CEN EN 275-92. Wood Preservatives—Determination of the Protective Effectiveness Against Marine Borers. 25 pp.
CEN PREN 370-90. Wood Preservatives: Determination of Eradicant Efficacy in Preventing Emergence of Anobium Punctatum. 24 pp.
CEN EN 370-93. Wood Preservatives—Determination of Eradicant Efficacy in Preventing Emergence of Anobium Punctatum (De Geer). 13 pp.
CENELEC PREN 20-89. Wood Preservatives; Determination of the Preventive Action Against Lyctus Brunneus (Stephens) (Laboratory Method). 21 pp.
DIN ENGL EN 49 Pt 1-92. Wood Preservatives; Determination of the Effectiveness Against Anobium Punctatum (de Geer) on the Basis of Egg-Laying and Larval Survival; Application by Surface Treatment (Laboratory Method) (Nov). 11 pp.
DIN ENGL EN 49 Pt 2-92. Wood Preservatives; Determination of the Effectiveness Against Anobium Punctatum (de Geer) on the Basis of Egg-Laying and Larval Survival; Application by Impregnation (Laboratory Method) (Nov). 11 pp.
DIN ENGL EN 117-92. Wood Preservatives; Determination of the Limit of Effectiveness Against Reticulitermes Santonensis (De Feytaud) (Laboratory Method) (Aug). 10 pp.

Insecticides (Cont.)
—Wood Preservatives (Cont.)
DIN ENGL EN 275-92. Wood Preservatives; Determination of the Protective Effectiveness Against Marine Borers (Nov). 14 pp.

Insert Bits
See Also: Bits (Tools)
—Hexagonal Drive Extensions
CNS B3336-89. Assembly Tools for Screws and Nuts Drive Extensions for Hexagon Insert Bits (Jul)(5302).
—Socket Wrenches
CNS B3306-79. Insert Bits,for Power Socket Wrenches (Nov)(5011).

Insert Nuts
See Also: Locking Nuts; Nuts (Fasteners)
—Closed—Plastic Moldings
CNS B2341-84. Closed Insert Nuts with Disk for Plastics Mouldings (Jul)(4663).
CNS B2342-84. Closed Insert Nuts with Dead Holes for Plastic Mouldings (Jul)(4664).
DIN ENGL 16903 Pt 2-74. Closed Insert Nuts with Disc for Plastics Mouldings (Dec). 3 pp.
DIN ENGL 16903 Pt 3-74. Closed Insert Nuts with Dead Hole for Plastics Mouldings (Dec). 3 pp.
—Heavy—Die Castings
CNS B2343-84. Insert Nuts Heavy Type for Plastics Moulding and Pressure Die Castings with Dead Hole (Jul)(4665).
—Heavy—Plastic Moldings
CNS B2343-84. Insert Nuts Heavy Type for Plastics Moulding and Pressure Die Castings with Dead Hole (Jul)(4665).
—Open—Plastic Moldings
CNS B2340-84. Open Insert Nuts for Plastic Mouldings (Jul)(4662).
DIN ENGL 16903 Pt 1-74. Open Insert Nuts for Plastic Mouldings (Dec). 3 pp.

Insert Retention
Use: Contact Force

Insertion Force
See Also: Contact Force
—Connectors
CEN PREN 2591 (Part D8)-92. Elements of Electrical and Optical Connection Test Methods Part D8—Mating and Unmating Forces. 4 pp.
CEN PREN 2591 (Part D12)-92. Elements of Electrical and Optical Connection Test Methods Part D12—Contact Insertion and Extraction Forces. 4 pp.
CNS C6185-88. Method of Test for Low Frequency (Below 3 MHz) Electrical Connectors (TP-5 Contact Insertion and Removal Forces) (Apr)(9364).
CNS C6187-88. Method of Test for Low Frequency (Below 3 MHz) Electrical Connectors (TP-13 Mating and Unmating Forces) (Apr)(9366).
—Electric Contacts—Gages
CEN PREN 2591 (Part D18)-92. Elements of Electrical and Optical Connection Test Methods Part D18—Gauge Insertion and Extraction Forces in a Female Contact. 4 pp.

Insertion Loss
See Also: Transmission Loss
—Connectors
CEN PREN 2591-F1-93. Aerospace Series Elements of Electrical and Optical Connection Test Methods Part F1—Optical Elements Insertion Loss. 7 pp.
—Echo Suppressors
CCITT RECMN G.164-89. Echo Suppressors—General Characteristics of International Telephone Connections and Circuits (Study Groups XII and XV) 36 pp. 36 pp.
—Hearing Protectors—Quality Assurance
BSI DD 192-90. 1990 Method for Measurement of Insertion Loss of Ear-Muff Type Hearing Protectors (Simplified Method for Quality Inspection Purposes). 11 pp.
ISO TR4869 Pt 3-89. Acoustics—Hearing Protectors—Part 3: Simplified Method for the Measurement of Insertion Loss of Ear-Muff Type Protectors for Quality Inspection Purposes First Edition. 11 pp.
—Silencers
ISO 7235-91. Acoustics—Measurement Procedures for Ducted Silencers—Insertion Loss, Flow Noise and Total Pressure Loss First Edition. 31 pp.

Insertion Loss (Cont.)
—Waveguide Connectors—Optical Waveguides
CNS C6119-81. Method of Test for Fiber Optic Devices (Test Procedure for Fiber Optics Bundle Connector Insertion Loss) (Jul)(7641).
CNS C6121-87. Method of Test for Fiber Optic Devices (FOTP—34 Interconnection Device Insertion Loss) (Jun)(7643).

Insertion Tools
See Also: Extraction Tools; Hand Tools; Hydraulic Tools; Pneumatic Tools; Tools
—Electric Contacts
CEN PREN 2591 (Part E6)-92. Elements of Electrical and Optical Connection Test Methods Part E6—Use of Tools. 4 pp.

Insertion Tubes
See Also: Tubes
SBAC AS 3152 ISSUE 2. Insertion Tube Switch Body Pilot's Control Wheel. 1 p.
—Aircraft—Control Wheels—Switches
SBAC AS 3153 ISSUE 2. Insertion Tube Switch Body Pilot's Control Wheel. 1 p.
SBAC AS 3154 ISSUE 2. Insertion Tube—Switch Body Pilot's Control Wheel. 1 p.
SBAC AS 3155 ISSUE 2. Insertion Tube—Switch Body Pilot's Control Wheel. 1 p.
SBAC AS 3156 ISSUE 2. Insertion Tube Switch Body Pilot's Control Wheel. 1 p.
SBAC AS 3157 ISSUE 2. Insertion Tube Switch Body Pilot's Control Wheel. 1 p.

Inserts (Fasteners)
See Also: Bushings; Fasteners; Fittings; Locking Thread Inserts; Spacers (Mechanical); Tapping Thread Inserts; Thread Inserts; Washers (Fasteners)
—Aircraft—Control Wheels—Pushbutton Switches
SBAC AS 3068 ISSUE 1. Insert for Press Button Pilots Control Wheel. 1 p.
—Aircraft—Filler Caps—Fittings
SBAC AS 2514 ISSUE 3. Insert for Seating (Filler Cap). 1 p.
SBAC AS 6359 ISSUE 1. Insert for Seating, (Filler Cap).
—Aircraft—Pushbutton Switches
SBAC AS 3143 ISSUE 1. Press Button Insert—Auto Pilot Switch Pilot's Control Switch. 1 p.

Inserts, Indexable
Use: Indexable Inserts

Insider Trading
Use: Securities—Insider Trading

Insoles (Footwear)
See Also: Footwear
—Foam Rubber
CNS S1135-82. Rubber Sponge Insole for Shoes (Feb)(8547). 1 p.
CNS S2069-82. Method of Test for Rubber Sponge Insole for Shoes (Feb)(8548). 2 pp.
—Leather
MOD UK DSTAN 83-3-82. Leather, Cattlehide, Shoulder, Vegetable Tanned, for Insoles and Welting; and Leather, Cattlehide, Belly, Vegetable Tanned, for Use in Other Footwear Components Issue 2. 10 pp.
—Shear Testing
BSI BS 5131: Sec 4.6-75. 1975 Methods of Test for Footwear and Footwear Materials Part 4: Other Components Sectiion 4.6: Shear Strength of Ribs Stuck to Insoles. 4 pp.
—Tensile Testing
BSI BS 5131: Sec 4.1-75. 1975 Methods of Test for Footwear and Footwear Materials Part 4: Other Components Section 4.1: Resistance to Peeling of Insole Board. 3 pp.
BSI BS 5131: Sec 4.7-75. 1975 Methods of Test for Footwear and Footwear Materials Part 4: Other Components Section 4.7: Resistance to Peeling of Ribs Stuck to Insoles. 3 pp.

Insoluble Matter Content Analysis
CGSB 2-GP-11M METH 10.4-83. Methods of Testing and Analysis of Soaps and Detergents Water Insoluble Matter in Chemicals. 1 p.
—Agar
CNS N6113-79. Method of Test for Agar-Agar (Oct)(5006). 2 pp.

Insoluble Matter Content Analysis (Cont.)

—Aluminum Sulfate
CNS K6162-79. Method of Test for Aluminium Sulfate (Industrial Use) (Jun)(2073). 3 pp.
CPPA J.6-86. Analysis of Alum Reprinted—1988. 5 pp.
MOD UK M 9508/67. Examination of Aluminium Sulphate.

—Ammonium Dihydrogen Orthophosphate
MOD UK M 9515/65. Examination of Ammonium Dihydrogen Orthophosphate (Withdrawn).

—Ammonium Nitrate
CNS K6081-62. Method of Test for Ammonium Nitrate (for Industrial Use) (Dec)(998)(R 1971). 1 p.

—Ammonium Nitrate—Gravimetric Analysis
BSI BS 4267: Part 5-87. 1987 Ammonium Nitrate Part 5: Method for Determination of Matter Insoluble in Water. 4 pp.
ISO 2995-74. Ammonium Nitrate for Industrial Use—Determination of Matter Insoluble in Water—Gravimetric Method First Edition. 4 pp.

—Ammonium Sulfate—Gravimetric Analysis
ISO 2994-74. Ammonium Sulphate for Industrial Use—Determination of Matter Insoluble in Water—Gravimetric Method First Edition. 4 pp.

—Asphalts
CNS K6206-65. Determination of Benzene-Insoluble Substance in Asphalt and Coal Tar (Sep)(2488)(R 1973).

—Barium Peroxide
MOD UK M 819/72. Examination of Barium Peroxide.

—Beeswax
MOD UK M 2012/71. Examination of Beeswax GS and Beeswax, Lead Free.
MOD UK M 9517/65. Examination of Beeswax and Beeswax, Lead Free (LAG 931A, No Information).

—Beverages
CNS N6091-91. Method of Test for Fruit and Vegetable Juices and Drinks (General Rules) (Jun)(3736). 1 p.
CNS N6173-82. Method of Test for Beverage: Determination of Water-Insoluble Solids (Sep)(9431). 1 p.

—Bronze Powder
BSI BS 5600: Sec 2.9-80. (WITHDRAWN) 1980 Powder Metallurgical Materials and Products Part 2: Methods of Sampling and Testing Metallic Powders Section 2.9: Determination of Acid-Insoluble Content in Iron, Copper, Tin and Bronze Powders. 4 pp.
BSI BS EN 24496-93. 1993 Metallic Powders—Determination of Acid Insoluble Content in Iron, Copper, Tin and Bronze Powders (ISO 4496: 1978) (V). 10 pp.
CEN EN 24496-93. Metallic Powders—Determination of Acid Insoluble Content in Iron, Copper, Tin and Bronze Powders (ISO 4496: 1978). 5 pp.
ISO 4496-78. Metallic Powders—Determination of Acid-Insoluble Content in Iron, Copper, Tin and Bronze Powders First Edition (CEN EN 24496: 1993). 5 pp.

—Calcium Carbonates
CNS K6721-82. Method of Test for Calcium Carbonate for Rubber (Jun)(9004). 4 pp.

—Carbon Disulfide
CGSB 1-GP-71 METH 85.1-82. Methods of Testing Paints and Pigments Matter Insoluble in Carbon Disulfide. 1 p.

—Cementitious Materials
SAA AS 3583.14-91. Methods of Test for Supplementary Cementitious Materials for Use with Portland Cement—Part 14: Determination of Insoluble Residue Content (In Professional Packages 30,58). 2 pp.

—Cereals
ISO 8129 Pt 1-84. Fruits, Vegetables and Derived Products—Determination of Alcohol-Insoluble Solids Content—Part 1: Method for Fresh or Quick-Frozen Maize First Edition. 4 pp.

—Chloroform
CNS K6657-81. Method of Test for Matter Insoluble in Chloroform in Oiticica Oils (Apr)(7258).

—Coal Tar
CNS K6206-65. Determination of Benzene-Insoluble Substance in Asphalt and Coal Tar (Sep)(2488)(R 1973).

Insoluble Matter Content Analysis (Cont.)

—Coffee
BSI BS 5752: Part 9-86. 1986 Coffee and Coffee Products Part 9: Instant Coffee: Determination of Insoluble Matter Content. 6 pp.
CNS N6180-83. Method of Test for Beverage: Determination of Methanol Content (May)(10291). 1 p.
CNS N6199-85. Instant Coffee—Determination of Insoluble Matter (Jun)(11289). 3 pp.
ISO 7534-85. Instant Coffee—Determination of Insoluble Matter Content First Edition. 5 pp.

—Copper Powder
BSI BS 5600: Sec 2.9-80. (WITHDRAWN) 1980 Powder Metallurgical Materials and Products Part 2: Methods of Sampling and Testing Metallic Powders Section 2.9: Determination of Acid-Insoluble Content in Iron, Copper, Tin and Bronze Powders. 4 pp.
BSI BS EN 24496-93. 1993 Metallic Powders—Determination of Acid Insoluble Content in Iron, Copper, Tin and Bronze Powders (ISO 4496: 1978) (V). 10 pp.
CEN EN 24496-93. Metallic Powders—Determination of Acid Insoluble Content in Iron, Copper, Tin and Bronze Powders (ISO 4496: 1978). 5 pp.
ISO 4496-78. Metallic Powders—Determination of Acid-Insoluble Content in Iron, Copper, Tin and Bronze Powders First Edition (CEN EN 24496: 1993). 5 pp.

—Creosote (Coal Tar)
CNS K6071-57. Method of Test for Benzene Insoluble Substances of Creosote Oil (Jul)(914)(R 1971).

—Cryolite
MOD UK M 874/68. Examination of Cryolite.

—Dammar
MOD UK M 9535/68. Examination of Gum, Dammar, and Gum, Dammar, Lead Free.

—Degreasers (Cleaning Agents)
MOD UK M 9522/66. Examination of Degreasing Compound, Alkaline, Type A (Superseded by Def Stan 68-160).

—Detergents
CGSB 2-GP-11M METH 9.2-83. Methods of Testing and Analysis of Soaps and Detergents Alcohol Insoluble Matter in Synthetic Detergents. 1 p.
CGSB 2-GP-11M METH 10.2-83. Methods of Testing and Analysis of Soaps and Detergents Water Insoluble Matter in Synthetic Detergents. 1 p.
CGSB 2-GP-11M METH 10.3-83. Methods of Testing and Analysis of Soaps and Detergents Water Insoluble Matter in Synthetic Detergents. 1 p.

—Dicyandiamide
MOD UK M 812/77. Examination of Dicyandiamide.

—Dyes
MOD UK M 818/63. Examination of Dyestuffs for Use in Pyrotechnic Compositions and HE Substitutes.

—Fats—Solvent Extraction
BSI BS 684: Sec 2.3-93. 1993 Methods of Analysis of Fats and Fatty Oils Part 2: Other Methods Section 2.3: Determination of Insoluble Impurities Content (ISO 663: 1992). 6 pp.
BSI BS 684: Sec 2.3-83. 1983 Methods of Analysis of Fats and Fatty Oils Part 2: Other Methods Section 2.3: Determination of Insoluble Impurities. 4 pp.
CNS N6078-82. Methods of Test for Edible Oils and Fats (Determination of Insoluble Impurities) (Jan)(3643). 2 pp.
ISO 663-92. Animal and Vegetable Fats and Oils—Determination of Insoluble Impurities Content Second Edition. 6 pp.

—Fish Meal
CNS N4015-78. Method of Test for Fish Meal and Paste (Mar)(2245). 2 pp.

—Fish Paste
CNS N4015-78. Method of Test for Fish Meal and Paste (Mar)(2245). 2 pp.

—Fruit Juices
CNS N6091-91. Method of Test for Fruit and Vegetable Juices and Drinks (General Rules) (Jun)(3736). 1 p.
CNS N6218-89. Method of Test for Fruit and Vegetable Juices and Drinks Determination of Water-Insoluble Solids (Nov) (12629).

—Fruits
CNS N6163-91. Method of Test for Fruit and Vegetable Products-Determination of Water-Insoluble Solids (Jun)(8622). 3 pp.
ISO 751-81. Fruit and Vegetable Products—Determination of Water-Insoluble Solids Content First Edition. 4 pp.

Insoluble Matter Content Analysis (Cont.)

—Hexachloroethane
BSI BS 577-66. (WITHDRAWN) 1966 Amd 1 Hexachloroethane. 17 pp.

—Iron Powder
BSI BS 5600: Sec 2.9-80. (WITHDRAWN) 1980 Powder Metallurgical Materials and Products Part 2: Methods of Sampling and Testing Metallic Powders Section 2.9: Determination of Acid-Insoluble Content in Iron, Copper, Tin and Bronze Powders. 4 pp.
BSI BS EN 24496-93. 1993 Metallic Powders—Determination of Acid Insoluble Content in Iron, Copper, Tin and Bronze Powders (ISO 4496: 1978) (V). 10 pp.
CEN EN 24496-93. Metallic Powders—Determination of Acid Insoluble Content in Iron, Copper, Tin and Bronze Powders (ISO 4496: 1978). 5 pp.
ISO 4496-78. Metallic Powders—Determination of Acid-Insoluble Content in Iron, Copper, Tin and Bronze Powders First Edition (CEN EN 24496: 1993). 5 pp.

—Lactose
MOD UK M 821/71. Examination of Lactose.

—Lead Monoxide
CNS K6095-58. Method of Test for Yellow-Lead Paint Powder (Jun)(1045) (R 1968). 1 p.

—Limestone
CPPA J.4-92. Analysis of Limestone. 5 pp.

—Magnesium Carbonates
MOD UK M 870/67. Examination of Magnesium Carbonate, Heavy.

—Magnesium Oxide
MOD UK M 9541/69. Examination of Magnesium Oxide and Magnesium Oxide, Special, Lead Free (Withdrawn).

—Malononitrile
MOD UK M 866/66. Examination of Malononitrile.

—Methanols
CNS K6263-80. Method of Test for Methyl Alcohol (Methanol) (Feb)(2790). 7 pp.

—Methenamine
MOD UK M 879/69. Examination of Hexamine.

—Milk
BSI BS 1743: Part 3-88. 1988 Analysis of Dried Milk and Dried Milk Products Part 3: Determination of Insolubility Index. 10 pp.
ISO 8156-87. Dried Milk and Dried Milk Products—Determination of Insolubility Index First Edition. 10 pp.

—Musk
CNS N4029-70. Method of Test for Musk (Jan)(3083). 2 pp.

—Nitrodiphenylamine
MOD UK M 824/92. Examination of 2-Nitrodiphenylamine.

—Organic Coatings
CNS K6804-4-84. Method of Test for Organic Coating Alcohol Insoluble Matter (Jun)(10880-4).
CNS K6804-8-84. Method of Test for Organic Coating Solvent Insolubles (Jun)(10880-8).
CNS K6804-17-84. Method of Test for Organic Coating (Chemical Analysis) — Sieving Residue of Solvent Insolubles (Jul)(10880-17).
CNS K6804-18-84. Method of Test for Organic Coating (Chemical Analysis) — Water Solubles in Solvent Insolubles (Jul)(10880-18).

—Organic Coatings—Acidity
CNS K6804-19-89. Method of Test for Organic Coating (Chemical Analysis) — Acidity Test of Water Extract of Solvent Insolubles (Jan)(10880-19).

—Organic Coatings—Quantitative Analysis
CNS K6804-20-85. Method of Test for Organic Coating (Chemical Analysis) — Quantative Test of Lead Oxide in Solvent Insolubles (Jan)(10880-20).
CNS K6804-21-85. Method of Test for Organic Coating (Chemical Analysis) — Quantative Test of Trilead Tetroxide in Solvent Insolubles (Jan)(10880-21).
CNS K6804-22-85. Method of Test for Organic Coating (Chemical Analysis) — Quantative Test of Lead in Solvent Insolubles (Jan)(10880-22).
CNS K6804-23-85. Method of Test for Organic Coating (Chemical Analysis) — Quantative Test of Zinc Oxide in Solvent Insolubles (Jan)(10880-23).
CNS K6804-24-89. Method of Test for Organic Coating (Chemical Analysis) — Quantative Test of Chromic Anhydride in Solvent Insolubles (Jan)(10880-24).

INTERNATIONAL AND NON-U.S. NATIONAL STANDARDS
SUBJECT INDEX

Insoluble Matter Content Analysis (Cont.)

—Organic Coatings—Quantitative Analysis (Cont.)
CNS K6804-25-89. Method of Test for Organic Coating (Chemical Analysis) — Quantative Test of Ferric Oxide in Solvent Insolubles (Jan)(10880-25).
CNS K6804-26-89. Method of Test for Organic Coating (Chemical Analysis) — Quantative Test of Titanium Oxide in Solvent Insolubles (Jan)(10880-26).

—Paints
CNS K6143-87. Method of Test for Traffic Paints (May)(1334). 19 pp.

—Paraformaldehyde
BSI BS 2941-57. (WITHDRAWN) 1957 Amd 1 Paraformaldehyde. 14 pp.
CNS K6551-80. Method of Test for Water-Insoluble Matter of Paraformaldehyde for Industrial Use (Aug)(6196).
ISO 1391 Pt IV-76. Paraformaldehyde for Industrial Use—Methods of Test—Part IV: Determination of Water-Insoluble Matter First Edition; (Corrected and Reprinted -1977). 3 pp.

—Phosphorus
CNS K6195-65. Method of Test for Red Phosphorus (Apr)(2452) (R 1971). 3 pp.

—Photographic Chemicals
ISO 10349 Pt 2-92. Photography—Photographic-Grade Chemicals—Test Methods —Part 2: Determination of Matter Insoluble in Water First Edition. 6 pp.
ISO 10349 Pt 3-92. Photography—Photographic-Grade Chemicals—Test Methods —Part 3: Determination of Matter Insoluble in Ammonium Hydroxide Solution First Edition. 6 pp.

—Pigments
CGSB 1-GP-71 METH 50.17-80. Methods of Testing Paints and Pigments Pigment Analysis Matter Insoluble in Hydrochloric Acid. 1 p.
CGSB 1-GP-71 METH 74.1-74. Methods of Testing Paints and Pigments Carbon and Insoluble Matter. 1 p.
CNS K6094-89. Method of Test for Red Lead (Pigment) (Jul)(1043). 2 pp.

—Potassium Silicates
BSI BS 6092: Part 6-81. 1981 Amd 1 Sampling and Test for Sodium and Potassium Silicates for Industrial Use Part 6: Preparation of Solutions of Not Readily Soluble Products and Determination of Matter Insoluble in Water. 6 pp.
BSI BS 6092: Part 10-81. 1981 Sampling and Test for Sodium and Potassium Silicates for Industrial Use Part 10: Determination of Matter Insoluble in Water. 2 pp.
ISO 2122-72. Sodium and Potassium Silicates for Industrial Use—Preparation of Solution of Products Not Easily Soluble in Boiling Water and Determination of Matter Insoluble in Water First Edition. 5 pp.

—Pulps—Gravimetric Analysis
ISO 699-82. Pulps—Determination of Alkali Resistance Second Edition. 5 pp.

—Quinoline—Gravimetric Analysis
BSI BS 6043: Sec 1.5-83. 1983 Methods of Sampling and Test for Carbonaceous Materials Used in Aluminium Manufacture Part 1: Electrode Pitch Section 1.5: Determination of Content of Quinoline-Insoluble Material. 7 pp.
ISO 6791-81. Carbonaceous Materials for the Production of Aluminium—Pitch for Electrodes—Determination of Content of Quinoline-Insoluble Material First Edition. 5 pp.

—Refrigeration Oils
DIN ENGL 51590 Pt 1-85. Testing of Lubricants; Determination of the R 12 Insolubles in Refrigerator Oils Using the-30 Degrees Celsius Precipitation Method (Sept). 3 pp.
DIN ENGL 51590 Pt 2-76. Testing of Lubricants; Determination of the Content of Material Insoluble in R 12 in Refrigerator Oils; Method at-40 Deg. C (233 K) (June). 4 pp.

—Selenious Acid
MOD UK M 9525/67. Examination of Selenious Acid Crystals.

—Soaps
BSI BS 1715: Sec 2.11-89. 1989 Analysis of Soaps Part 2: Quantitative Test Methods Section 2.11: Method for Determination of Ethanol-Insoluble Matter Content (ISO 673-1981). 4 pp.
CGSB 2-GP-11M METH 9.1-83. Methods of Testing and Analysis of Soaps and Detergents Alcohol Insoluble Matter in Soaps. 1 p.

—Soaps (Cont.)
CGSB 2-GP-11M METH 10.1-83. Methods of Testing and Analysis of Soaps and Detergents Water Insoluble Matter in Soaps. 1 p.
ISO 673-81. Soaps—Determination of Content of Ethanol-Insoluble Matter Second Edition. 4 pp.

—Sodium Azide
MOD UK M 807/91. Examination of Sodium Azide.

—Sodium Borates
BSI BS 5688: Part 10-79. 1979 Orthoboric Acid (Boric Acid), Diboron Trioxide (Boric Oxide), Disodium Tetraborates, Sodium Perborates and Crude Sodium Borates for Industrial Use Part 10: Determination of Matter Insoluble in Alkaline Med & Prep of Test Solu of Crude Sodium Borates. 4 pp.
ISO 2217-75. Crude Sodium Borates for Industrial Use—Determination of Matter Insoluble in Alkaline Medium and Preparation of Test Solutions First Edition. 4 pp.

—Sodium Carbonates
BSI BS 6070: Part 4-81. 1981 Methods of Sampling and Test for Sodium Carbonate for Industrial Use Part 4: Determination of Matter Insoluble in Water (ISO 746: 1976). 4 pp.
CPPA J.11-90. Analysis of Soda Ash. 3 pp.
ISO 746-76. Sodium Carbonate for Industrial Use—Determination of Matter Insoluble in Water at 50 Degrees Celsius First Edition. 4 pp.

—Sodium Chlorate
ISO 2461-73. Sodium Chlorate for Industrial Use—Determination of Matter Insoluble in Water First Edition. 4 pp.

—Sodium Chloride
BSI BS 7319: Part 3-90. 1990 Analysis of Sodium Chloride for Industrial Use Part 3: Method for Determination of Matter Insoluble in Water or in Acid. 6 pp.
ISO 2479-72. Sodium Chloride for Industrial Use—Determination of Matter Insoluble in Water or in Acid and Preparation of Principal Solutions for Other Determinations First Edition. 5 pp.

—Sodium Fluoride
BSI BS 5072: Part 1-74. 1974 Methods of Test for Sodium Fluoride for Industrial Use Part 1: Determination of Water-Insoluble Matter. 7 pp.
CNS K6555-80. Determination of Water-Insoluble Matter of Sodium Fluoride for Industrial Use (Aug)(6200).
ISO 2831-73. Sodium Fluoride for Industrial Use—Determination of Water-Insoluble Matter First Edition. 3 pp.

—Sodium Hydroxide
BSI BS 6075: Part 14-81. 1981 Sampling and Test for Sodium Hydroxide for Industrial Use Part 14: Determination of Matter Insoluble in Water. 2 pp.
MOD UK M 808/73. Examination of Sodium Hydroxide Pure (Withdrawn).

—Sodium Nitrate
MOD UK M 869/91. Examination of Sodium Nitrate, Grade 1.

—Sodium Silicates
BSI BS 6092: Part 6-81. 1981 Amd 1 Sampling and Test for Sodium and Potassium Silicates for Industrial Use Part 6: Preparation of Solutions of Not Readily Soluble Products and Determination of Matter Insoluble in Water. 6 pp.
BSI BS 6092: Part 10-81. 1981 Sampling and Test for Sodium and Potassium Silicates for Industrial Use Part 10: Determination of Matter Insoluble in Water. 2 pp.
ISO 2122-72. Sodium and Potassium Silicates for Industrial Use—Preparation of Solution of Products Not Easily Soluble in Boiling Water and Determination of Matter Insoluble in Water First Edition. 5 pp.

—Sodium Sulfate
CNS K6085-74. Methods of Test for Sodium Sulfate of Industrial Grade (Oct)(1004). 5 pp.
ISO 3235-75. Sodium Sulphate for Industrial Use—Determination of Acid-Insoluble Matter First Edition. 5 pp.

—Sodium Tripolyphosphate
BSI BS 4427: Part 1-69. 1969 Methods of Test for Sodium Tripolyphosphate (Pentasodium Triphosphate) and Sodium Pyrophosphate (Tetrasodium Pyrophosphate) for Indus Use: Part 1: Determ of Matter Insoluble in Water in Sodium Tripolyphosphate. 7 pp.

—Sodium Tripolyphosphate (Cont.)
CNS K6415-78. Method of Test for Sodium Tripolyphosphate (for Industrial Use) (Mar)(4285). 4 pp.
ISO 850-76. Sodium Tripolyphosphate for Industrial Use—Determination of Matter Insoluble in Water First Edition. 4 pp.

—Steam Cleaning Compounds
CGSB 31-GP-0A METH 42.1-57. Methods of Testing Corrosion-Prevention Materials and Processes Water Insoluble Matter. 1 p.

—Strontium Nitrate
MOD UK M 867/92. Examination of Strontium Nitrate (Anhydrous).

—Strontium Peroxide
MOD UK M 884/70. Examination of Strontium Peroxide (Anhydrous).

—Tin Powder
BSI BS 5600: Sec 2.9-80. (WITHDRAWN) 1980 Powder Metallurgical Materials and Products Part 2: Methods of Sampling and Testing Metallic Powders Section 2.9: Determination of Acid-Insoluble Content in Iron, Copper, Tin and Bronze Powders. 4 pp.
BSI BS EN 24496-93. 1993 Metallic Powders—Determination of Acid Insoluble Content in Iron, Copper, Tin and Bronze Powders (ISO 4496: 1978) (V). 10 pp.
CEN EN 24496-93. Metallic Powders—Determination of Acid Insoluble Content in Iron, Copper, Tin and Bronze Powders (ISO 4496: 1978). 5 pp.
ISO 4496-78. Metallic Powders—Determination of Acid-Insoluble Content in Iron, Copper, Tin and Bronze Powders First Edition (CEN EN 24496: 1993). 5 pp.

—Toluene—Gravimetric Analysis
BSI BS 6043: Sec 1.4-81. 1981 Methods of Sampling and Test for Carbonaceous Materials Used in Aluminium Manufacture Part 1: Electrode Pitch Section 1.4: Determination of Content of Toluene-Insoluble Material. 4 pp.
ISO 6376-80. Carbonaceous Materials for the Production of Aluminium—Pitch for Electrodes—Determination of Content of Toluene-Insoluble Material First Edition. 4 pp.

—Vegetable Juices
CNS N6091-91. Method of Test for Fruit and Vegetable Juices and Drinks (General Rules) (Jun)(3736). 1 p.
CNS N6218-89. Method of Test for Fruit and Vegetable Juices and Drinks Determination of Water-Insoluble Solids (Nov) (12629).

—Vegetable Oils—Solvent Extraction
CNS N6078-82. Methods of Test for Edible Oils and Fats (Determination of Insoluble Impurities) (Jan)(3643). 2 pp.

—Vegetables
CNS N6163-91. Method of Test for Fruit and Vegetable Products-Determination of Water-Insoluble Solids (Jun)(8622). 3 pp.
ISO 751-81. Fruit and Vegetable Products—Determination of Water-Insoluble Solids Content First Edition. 4 pp.
ISO 8129 Pt 2-84. Fruits, Vegetables and Derived Products—Determination of Alcohol-Insoluble Solids Content—Part 2: Method for Fresh or Quick-Frozen Peas First Edition. 4 pp.

—Zinc Oxides
MOD UK M 806/91. Examination of Zinc Oxide.

Inspection

Scope Note: For inspection of specific products or materials, see the material or product. For additional references, consult the following list
See Also: Analyzers; Chemical Analysis; Construction; Evaluation; Failure (Quality Control); Fracture Testing; Handholes; Inspection by Attributes; Inspection by Variables; Macroscopic Examination; Quality Assurance; Sampling; Screening Testing; Spark Testing; Tolerances; Visual Inspection

ISO Guide 39-88. General Requirements for the Acceptance of Inspection Bodies Second Edition. 12 pp.
SAA HB18.39-91. Guidelines for Third-Party Certification and Accreditation—Guide 39: General Requirements for the Acceptance of Inspection Bodies (ISO/IEC Guide 39:1988) (SANZ HB18.39—1991). 8 pp.
SNZ SANZ/SAA HB 18.39-91. Guidelines for Third-Party Certification and Accreditation Guide 39 General Requirements for the Acceptance of Inspection Bodies (ISO/IEC Guide 39-1988). 8 pp.

INTERNATIONAL AND NON-U.S. NATIONAL STANDARDS
SUBJECT INDEX

Inspection

Inspection (Cont.)

—Aircraft—Metal
CAA LEAFLET 6-3 07.90. Inspection of Metal Aircraft After Abnormal Occurrences. 8 pp.

—Aircraft—Wooden Structures
CAA LEAFLET 6-1 07.90. Inspection of Wooden Structures. 8 pp.

—Quality Assurance
CEN PREN 45004-92. General Criteria for the Operation of Bodies Performing Inspection. 15 pp.
ISO 9000 Pt 2-93. Quality Management and Quality Assurance Standards—Part 2: Generic Guidelines for the Application of ISO 9001, ISO 9002 and ISO 9003 First Edition. 22 pp.
ISO 9003-87. Quality Systems—Model for Quality Assurance in Final Inspection and Test First Edition; DoD Adopted. 6 pp.
SAA AS 3900.2 (INT)-92. Quality Management and Quality Assurance Standards—Part 2(Int): Generic Guidelines for the Application of AS 3901/NZS 9001/ISO 9001, AS 3902/NZS 9002/ISO 9002 and AS 3903/NZS 9003/ISO 9003 (Expires 15 June 1994) (in Professional. 15 pp.
SNZ AS 1823-85. Suppliers' Quality Inspection Systems. 17 pp.
SNZ AS 2000-87. Guide to AS 1821-23-Suppliers Quality Systems. 34 pp.
SNZ NZS 9000.2 (INT)-92. Quality Management and Quality Assurance Standards Part 2: Generic Guidelines for the Application of NZS 9001/AS 3901/ISO 9001, NZS 9002/AS 3902/ISO 9002 and NZS 9003/AS 3903/ISO 9003 (NZS 9000.2:1992 (Int)/AS 3900.2:1992 (Int)/. 35 pp.

—Quality Assurance—Contractors'—Military
MOD UK AQAP-119-93. NATO Guide to AQAPs-110, -120 and-130 Edn 1, 3/93. 39 pp.
MOD UK AQAP-130-93. NATO Quality Assurance Requirements for Inspection Edn 1, 3/93. 16 pp.
MOD UK AQAP-131-93. NATO Quality Assurance Requirements for Final Inspection Edn 1, 3/93. 9 pp.
NATO AQAP-4 ED 2 AMD 0-76. NATO Inspection System Requirements for Industry. 11 pp.
NATO AQAP-9 ED 2 AMD 0-76. NATO Basic Inspection Requirements for Industry (Amd 0). 8 pp.
NATO AQAP-119 ED 1 AMD 0-93. NATO Guide to AQAPs-110, -120 and-130. 39 pp.
NATO AQAP-130 ED 1 AMD 0-93. NATO Quality Assurance Requirements for Inspection. 16 pp.
NATO AQAP-131 ED 1 AMD 0-93. NATO Quality Assurance Requirements for Final Inspection. 9 pp.
QSS MIL-I-45208A-63. Inspection System Requirements (Includes Amendment 1). 5 pp.

—Quality Assurance—Contractors'—Military—Evaluation
NATO AQAP-5 ED 2 AMD 0-76. Guide for the Evaluation of a Contractor's Inspection System for Compliance with AQAP-4. 32 pp.

—Reports
ISO Guide 57-91. Guidelines for the Presentation of Inspection Results First Edition. 6 pp.
SAA HB18.57-92. Guidelines for Third-Party Certification and Accreditation—Guide 57: Guidelines for the Presentation of Inspection Results (ISO/IEC Guide 57:1991) (SANZ HB18.57—1992). 3 pp.
SNZ SANZ/SAA HB 18.57-92. Guidelines for Third-Party Certification and Accreditation Guide 57 Guidelines for the Presentation of Inspection Results (ISO/IEC Guide 57:1991). 3 pp.

Inspection by Attributes

Scope Note: For inspection by attributes of specific products or materials, see the material or product
See Also: Quality Assurance
CECC CECC 00 007 ISSUE 2-78. Basic Specification: Sampling Plans and Procedures for Inspection by Attributes (En, Fr, Ge). 15 pp.

—Sampling
BSI BS 6000-72. 1972 Guide to the Use of BS 6001. Sampling Procedures and Tables for Inspection by Attributes. 51 pp.
BSI BS 6001: Part 1-91. 1991 Sampling Procedures for Inspection by Attributes Part 1: Specification for Sampling Plans Indexed by Acceptable Quality Level (AQL) for Lot-by-Lot Inspection (ISO 2859-1: 1989). 77 pp.
BSI BS 6001: Part 2-93. 1993 Sampling Procedures for Inspection by Attributes Part 2: Specification for Sampling Plans Indexed by Limiting Quality (LQ) for Isolated Lot Inspection (ISO 2859-2: 1985) (G). 27 pp.
BSI BS 6001: Part 2-84. 1984 Amd 1 Sampling Procedures for Inspection by Attributes Part 2: Sampling Plans Indexed by Limiting Quality (LQ) for Isolated Lot Inspection. 26 pp.

Inspection by Attributes (Cont.)

—Sampling (Cont.)
BSI BS 6001: Part 3-93. 1993 Sampling Procedures for Inspection by Attributes Part 3: Specification for Skip-Lot Procedures (ISO 2859-3: 1991) (G). 23 pp.
BSI BS 6001: Part 3-86. 1986 Sampling Procedures for Inspection by Attributes Part 3: Skip-Lot Procedures. 20 pp.
BSI BS 6002-79. 1979 Sampling Procedures and Charts for Inspection by Variables for Percent Defective. 95 pp.
IEC 410-73. Sampling Plans and Procedures for Inspection by Attributes First Edition; (Corrigenda—Feb 1976). 85 pp.
ISO 2859 Pt 1-89. Sampling Procedures for Inspection by Attributes—Part 1: Sampling Plans Indexed by Acceptable Quality Level (AQL) for Lot-by-Lot Inspection First Edition; (Technical Corrigendum 1—1993). 74 pp.
ISO 2859 Pt 2-85. Sampling Procedures for Inspection by Attributes—Part 2: Sampling Plans Indexed by Limiting Quality (LQ) for Isolated Lot Inspection First Edition. 24 pp.
ISO 2859 Pt 3-91. Sampling Procedures for Inspection by Attributes —Part 3: Skip-Lot Sampling Procedures First Edition. 21 pp.
ISO 8422-91. Sequential Sampling Plans for Inspection by Attributes First Edition; (Corrigendum—Sept 1993). 50 pp.
JIS Z 9001-80. General Rules for Sampling Inspection Procedures. 87 pp.
JIS Z 9002-56. Single Sampling Inspection Plans Having Desired Operating Characteristics (Part 1. Sampling by Attributes).
JIS Z 9006-56. Single Sampling Inspection Plans with Screening by Attributes.
JIS Z 9008-57. Sampling Inspection Plans for Continuous Production by Attributes.
JIS Z 9009-62. Sequential Sampling Inspection Plans Having Desired Operating Characteristics by Attributes.
JIS Z 9011-63. Single Sampling Inspection Plans by Attributes with Adjustment. 70 pp.
JIS Z 9015-80. Sampling Inspection Procedures and Tables Attributes with Severity Adjustment (Receiving Inspection Where a Customer Can Select Suppliers).
JIS Z 9015-71. Sampling Inspection Plans by Attributes with Severity Adjustment. 31 pp.
MOD UK DSTAN 05-58-86. Sampling Procedures and Tables for Inspection by Attributes of Isolated Lots Issue 2. 146 pp.
MOD UK DSTAN 05-70-01. Guide to Defence Standard 05-58: Sampling Procedures and Tables for Inspection by Attributes of Isolated Lots Issue 1; Amendment 1. 37 pp.
QSS MIL-Std-105E-89. Sampling Procedures and Tables for Inspection by Attributes. 73 pp.
SAA AS 1199-88. Sampling Procedures and Tables for Inspection by Attributes. 104 pp.
SAA AS 1399-90. Guide to AS 1199—Sampling Procedures and Tables for Inspection by Attributes Amdt 1 March 1991 (In Professional Package 46). 57 pp.
SNZ AS 1199-88. Sampling Procedures and Tables for Inspection by Attributes. 104 pp.
SNZ AS 1399-73. Guide to AS 1199. Sampling Procedures and Tables for Inspection by Attributes. 51 pp.

—Tables (Data)
QSS MIL-Std-105E-89. Sampling Procedures and Tables for Inspection by Attributes. 73 pp.
SAA AS 1399-90. Guide to AS 1199—Sampling Procedures and Tables for Inspection by Attributes Amdt 1 March 1991 (In Professional Package 46). 57 pp.
SNZ AS 1399-73. Guide to AS 1199. Sampling Procedures and Tables for Inspection by Attributes. 51 pp.

Inspection by Variables

Scope Note: For inspection by variable of specific products or materials, see the material or product
See Also: Quality Assurance
CNS Z4023-87. Sampling Procedures and Tables for Inspection by Variables for Percent Nonconforming (May)(9445).
JIS Z 9001-80. General Rules for Sampling Inspection Procedures. 87 pp.
JIS Z 9003-79. Single Sampling Inspection Plans Having Desired Operating Characteristics by Variables (Standard Deviation Known).
JIS Z 9004-55. Normal Single Sampling Inspection Plans by Variables (Standard Deviation Unknown and Single Limit Specified) (R 1964). 19 pp.
JIS Z 9010-79. Sequential Sampling Inspection Plans for Percent Defectives Having Desired Operating Characteristics by Variables (Standard Deviation Known).

Inspection by Variables (Cont.)

—Sampling
ISO 3951-89. Sampling Procedures and Charts for Inspection by Variables for Percent Nonconforming Second Edition. 112 pp.
ISO 8423-91. Sequential Sampling Plans for Inspection by Variables for Percent Nonconforming (Known Standard Deviation) First Edition; (Corrigendum—Sept 1993). 45 pp.
JIS Z 9004-83. Single Sampling Inspection Plans Having Desired Operating Characteristics by Variables (Standard Deviation Unknown and Single Limit Specified).

Inspection Chambers
Use: Manholes

Inspection Gages
CNS B6011-55. Inspection Plug Gauge for Whitworth Thread Ring Gauge (Go End Limit for Fine, Medium and Coarse Fits) (Sep)(527)(R 1970).
CNS B6018-55. Inspection Plug Gauge for Metric Thread Ring Gauge (Go End Limit, Fine, Medium and Coarse Fits) (Sep)(538)(R 1970).
DIN ENGL 7162-65. Plain Workshop and Inspection Gauges; Manufacturing Tolerances and Permissible Wear (Dec). 6 pp.

—Paints—Adhesion Testing
SNZ NZS/AS 1580. 408.1-84. Methods of Test for Paints and Related Materials Part 408.1: Adhesion—Paint Inspection Gauge (This is a Joint Standard with SAA AS 1580.408.1). 4 pp.

Inspection Stamps
Use For: Test Marks **See Also:** Identification Systems

—Aircraft Equipment
SBAC TS 131 ISSUE 1. Guidelines for the Registration and Use of Final Inspection Stamps.

Installation
Scope Note: See the subheading Installation under specific types of equipment

Installation Boxes (Electrical Enclosures)
Use: Electric Conduit Boxes

Instant Coffee
Use: Coffee—Instant

Instant Noodles
Use For: Ramen Noodles **See Also:** Dry Noodles; Pasta
CNS N5198-82. Instant Noodles (Oct)(9537). 3 pp.

—Antioxidants Content
CNS N6094-85. Method of Test for Instant Noodles (Jan)(3962). 1 p.

—Iodine Number
CNS N6094-85. Method of Test for Instant Noodles (Jan)(3962). 1 p.

—Moisture Content
CNS N6094-85. Method of Test for Instant Noodles (Jan)(3962). 1 p.

—Preservatives Content
CNS N6094-85. Method of Test for Instant Noodles (Jan)(3962). 1 p.

—Saponification Number
CNS N6094-85. Method of Test for Instant Noodles (Jan)(3962). 1 p.

Instruction Manuals
Use: Technical Manuals

Instrument Approach Procedures
See Also: Aircraft Landing Approaches; Instrument Landing Systems; Standard Instrument Departure

—Technical Manuals
NATO STANAG 3759 ED 3 AMD 0-87. Criteria for the Preparation of Instrument Approach and Departure Procedures (APATC-1). 5 pp.
NATO STANAG 3759 ED 4 AMD 0-92. Criteria for the Preparation of Instrument Approach and Departure Procedures (APATC-1). 5 pp.
NATO STANAG 3759 ED 4 AMD 1-92. Criteria for the Preparation of Instrument Approach and Departure Procedures—APATC-1. 6 pp.
NATO STANAG 3970 ED 1 AMD 0-87. (DRAFT) Production Specifications for Flight Information Publications (Terminal High/Low Instrument Approach Procedures and Instrument Departure Procedures)—APATC-2. 3 pp.

INDEX and DIRECTORY of

INTERNATIONAL AND NON-U.S. NATIONAL STANDARDS
SUBJECT INDEX
Instrumentation

Instrument Approach Procedures (Cont.)
—Technical Manuals (Cont.)
NATO STANAG 3970 ED 1 AMD 0-92. Content and Format of Flight Information Publication (FLIP) Terminal High/Low Instrument Approach Procedures, Instrument Departure Procedures, and Aerodrome Diagrams/Layouts. 14 pp.
NATO STANAG 3970 ED 1 AMD 1-92. Content and Format of Flight Information Publication (FLIP) Terminal High/Low Instrument Approach Procedures, Instrument Departure Procedures, and Aerodrome Diagrams/Layouts. 15 pp.

Instrument Dials
Use: Dials

Instrument Flight
See Also: Air Navigation; Flight Control Systems
ICAO 7192 Part B-2. Training Manual Part B-2 Pilots—Helicopter Licenses, CVFR-Rating and Instrument Flight Rating First Edition—1977. 90 pp.
ICAO 8168 Vol II. Procedures for Air Navigation Services Aircraft Operations Volume II Construction of Visual and Instrument Flight Procedures Third Edition—1986; (Corrigendum 3 03/01/88) (Amendment 5 11/15/90). 513 pp.
ICAO 9368. Instrument Flight Procedures Construction Manual First Edition—1983. 251 pp.
ICAO 9371. Template Manual for Holding, Reversal and Racetrack Procedures Second Edition—1986. 195 pp.

Instrument Housings
Use For: Cases (Instrument) See Also: Electrical Enclosures; Housings
—Railroad Signaling Systems
JIS E 3101-80. Instrument Cases for Railway Signaling.

Instrument Landing Systems
See Also: Air Navigation; Air Traffic Control; Aircraft Landings; Glide Path Receiving Instruments; Instrument Approach Procedures; Localizers; Microwave Landing Systems; Standard Instrument Departure
CAA Chapter R4-6 App #4 04.74. Tests for VOR, ILS and Marker Systems (Radio). 2 pp.
CAA JAR-AWO SUBPART 1. Automatic Landing Systems (Joint Airworthiness Requirements). 11 pp.
ICAO 8071 Vol II. Manual on Testing of Radio Navigation Aids Volume II ILS (Instrument Landing System) Third Edition—1972. 146 pp.
—Aircraft—Radio Receivers
CAA Part 2 CAP 208. ILS Localiser Receiving Apparatus (Civil Air Publications: Airborne Radio Apparatus). 16 pp.
CAA Chapter R4-6 App #4 04.74. Tests for VOR, ILS and Marker Systems (Radio). 2 pp.
EUROCAE ED-46A 01.88. MPS for Airborne ILS Receiving Equipment (Localiser). 120 pp.
—Certificate of Airworthiness
CAA JAR-AWO SUBPART 3. Airworthiness Certification of Aeroplanes for Operations with Decision Heights Below 30M (100ft) or No Decision Height—Category 3 Operations (Joint Airworthiness Requirements). 10 pp.
—Collision Risk Model
ICAO 9274. Manual on the Use of the Collision Risk Model (CRM) for ILS Operations First Edition—1980. 307 pp.

Instrument Panels
See Also: Aircraft Instruments; Display Devices; Panel Boards (Electrical); Panels
—Aircraft
NATO STANAG 3319 ED 6 AMD 5-83. Dimensions for Cases of Control Boxes, Instruments and Panel Mounted Equipment. 29 pp.
NATO STANAG 3319 ED 6 AMD 6-83. Dimensions for Cases of Control Boxes, Instruments and Panel Mounted Equipment. 30 pp.
—Aircraft—Arrangement
NATO STANAG 3319 ED 6 AMD 5-83. Dimensions for Cases of Control Boxes, Instruments and Panel Mounted Equipment. 29 pp.
—Earthmoving Equipment
BSI BS 5768-79. 1979 Operating Instrumentation for Earth-Moving Machinery. 7 pp.
ISO 6011-87. Earth-Moving Machinery—Operating Instrumentation Second Edition. 6 pp.
JIS A 8110-88. Construction Machinery—Service Instruments. 7 pp.

Instrument Panels (Cont.)
—Flight Control Systems—Design
NATO STANAG 3221 ED 6 AMD 6-77. Automatic Flight Control System (AFCS) in Aircraft—Design Standards and Location of Controls. 7 pp.
NATO STANAG 3221 ED 6 AMD 7-77. Automatic Flight Control System (AFCS) in Aircraft—Design Standards and Location of Controls. 8 pp.

Instrument Scales (Indicators)
Use: Scales (Indicators)

Instrument Transformers
See Also: Capacitor Voltage Transformers; Current Transformers; Transformers; Voltage Transformers
BSI BS 7628-93. 1993 Combined Transformers (E). 19 pp.
CENELEC HD 548.3 S1-92. Instrument Transformers Part 3: Combined Transformers. 5 pp.
CNS C3201-85. Method of Test for Instrument Transformers for Metering Service (Dec)(11438).
CNS C4036-86. General Rules of Transformers for Instruments (Jul)(1328).
CNS C4435-85. Instrument Transformers for Metering Service (Dec)(11437).
CSA CAN3-C13-M83. Instrument Transformers (R 1992); (Gen Instr 1). 95 pp.
IEC 44 Pt 3-80. Instrument Transformers Part 3: Combined Transformers First Edition. 28 pp.
JIS C 1731-88. Instrument Transformers for Testing Purpose and Used with General Instrument. 50 pp.
JIS C 1736-88. Instrument Transformers for Metering Service. 66 pp.
—Burden
CNS C1135-85. Determination of the Range of the Burden for the Instrument Transformers (Dec)(11440).
—Electric Discharges
BSI BS 6184-81. 1981 Measuring Partial Discharges in Instrument Transformers. 7 pp.
IEC 44 Pt 4-80. Instrument Transformers Part 4: Measurement of Partial Discharges First Edition. 14 pp.
—Electric Measuring Instruments—Error Analysis
CNS C1134-85. Calculation Method for the Error of the Instrument Transformers for Metering Service (Dec)(11439).
—Error Analysis
CNS C1133-85. Tolerance of Allowable Error of Instrument Transformer (Dec)(11436).
—Glossaries
BSI BS 4727:Pt2: Group 16-91. 1991 Electrotechnical, Power, Telecommunication, Electronics, Lighting and Colour Terms Part 2: Terms Particular to Power Engineering Group 16: Instrument Transformers (IEC 50(321): 1986). 48 pp.
IEC 50 Chap 321-86. International Electrotechnical Vocabulary Chapter 321: Instrument Transformers. 51 pp.
SAA AS 1852.321-88. International Electrotechnical Vocabulary—Part 321: Instrument Transformers. 24 pp.
—Insulating Oils—Gas Analysis
BSI BS EN 61181-93. 1993 Impregnated Insulating Materials—Application of Dissolved Gas Analysis (DGA) to Factory Tests on Electrical Equipment (IEC 1181: 1993) (Q). 15 pp.
CENELEC PREN 61181-92. Impregnated Insulating Materials—Application of Dissolved Gas Analysis (DGA) to Factory Tests on Electrical Equipment. 13 pp.
IEC 1181-93. Impregnated Insulating Materials—Application of Dissolved Gas Analysis (DGA) to Factory Tests on Electrical Equipment First Edition. 21 pp.
—Symbols
CNS C1025-57. Symbols of Transformers for Instruments (Jan)(409) (R 1971).
DIN ENGL 40714 Pt 2-58. Heavy Current and Telecommunications Engineering; Graphical Symbols; Instrument Transformers (May). 3 pp.

Instrumentation
Use For: Instrumentation Systems; Instruments
—Classification
BSI BS 1000: (68/681.3)-82. 1982 Universal Decimal Classification (UDC). English Full Edition (68/681.3): Finished Articles in General. Precision Mechanisms. Horology. Instrumentation. Data Processing Equipment. 31 pp.
SNZ NZS/BS 1000 (68/681.3)-82. Universal Decimal Classification Finished Articles in General. Precision Mechanisms. Horology. Instrumentation. Data Processing Equipment. 32 pp.

Instrumentation (Cont.)
—Explosive Atmospheres
SAA AS 1076.13-77. Code of Practice for Selection, Installation and Maintenance of Electrical Apparatus and Associated Equipment for Use in Explosive Atmospheres (Other Than Mining Applications)—Part 13: Installation and Maintenance Requirements for Instrumentation. 36 pp.
—Glossaries
CNS Z7169-84. Glossary of Terms Used in Instrumentation (May)(10895).
JIS K 0215-89. Technical Terms for Analytical Chemistry (Analytical Instrument Part).
—Mounting Hardware—Indoor/Outdoor
CSA C22.2 NO 115-M1989. Meter-Mounting Devices; (Gen Instr 1 Thru 3). 55 pp.
CSA 1169 Bull. Electrical Bulletin 1169 June 27, 1978 to C22.2 NO 115. 2 pp.
—Nuclear Power Plants—Safety—Classification
IEC 1226-93. Nuclear Power Plants—Instrumentation and Control Systems Important for Safety—Classification First Edition. 47 pp.
—Pneumatic—Associated Gas—Safety
IEC 1081-91. Pneumatic Instruments Driven by Associated Process Gas—Safe Installation and Operating Procedures—Guidelines First Edition. 19 pp.
—Power Supplies—Nuclear Power Plants
BSI BS 7674-93. 1993 Nuclear Power Plants—Instrumentation and Control Systems Important for Safety—Requirements for Electrical Supplies (IEC 1225: 1993) (Q). 24 pp.
IEC 1225-93. Nuclear Power Plants—Instrumentation and Control Systems Important for Safety—Requirements for Electrical Supplies First Edition. 49 pp.
—Symbols
SAA AS 1101.6-89. Graphical Symbols for General Engineering—Part 6: Process Measurement Control Functions and Instrumentation. 49 pp.

Instrumentation Cables
See Also: Cables (Electric); Electric Conductors; Thermocouple Wire
—Control—Multiconductor—150 V
CSA CAN/CSA-C22. 2 NO 239-M91. Control and Instrumentation Cables; (Gen Instr 1 Thru 3). 50 pp.
—Control—Multiconductor—300 V
CSA CAN/CSA-C22. 2 NO 239-M91. Control and Instrumentation Cables; (Gen Instr 1 Thru 3). 50 pp.
—Control—Multiconductor—600 V
CSA CAN/CSA-C22. 2 NO 239-M91. Control and Instrumentation Cables; (Gen Instr 1 Thru 3). 50 pp.
—Dielectric Strength
CSA CAN/CSA-C22. 2 NO 239-M91. Control and Instrumentation Cables; (Gen Instr 1 Thru 3). 50 pp.
—Impact Testing
CSA CAN/CSA-C22. 2 NO 239-M91. Control and Instrumentation Cables; (Gen Instr 1 Thru 3). 50 pp.
—Insulation Resistance
CSA CAN/CSA-C22. 2 NO 239-M91. Control and Instrumentation Cables; (Gen Instr 1 Thru 3). 50 pp.
—Low Temperature Testing
CSA CAN/CSA-C22. 2 NO 239-M91. Control and Instrumentation Cables; (Gen Instr 1 Thru 3). 50 pp.
—Polyethylene Insulated—Lead Sheathed
BSI BS 5308: Part 1-86. 1986 Instrumentation Cables Part 1: Polyethylene Insulated Cables. 20 pp.
BSI BS 5308: Part 1-01. 1986 Amd 1 Instrumentation Cables Part 1: Specification for Polyethylene Insulated Cables (AMD 7617) July 15, 1993 (E). 22 pp.
—Polyethylene Insulated—PVC Sheathed
BSI BS 5308: Part 1-86. 1986 Instrumentation Cables Part 1: Polyethylene Insulated Cables. 20 pp.
BSI BS 5308: Part 1-01. 1986 Amd 1 Instrumentation Cables Part 1: Specification for Polyethylene Insulated Cables (AMD 7617) July 15, 1993 (E). 22 pp.

INDUSTRY STANDARDS

INTERNATIONAL AND NON-U.S. NATIONAL STANDARDS
SUBJECT INDEX

Instrumentation Cables *(Cont.)*
—**PVC Insulated—PVC Sheathed**
BSI BS 5308: Part 2-86. 1986 Instrumentation Cables Part 2: PVC Insulated Cables (Q). 16 pp.
BSI BS 5308: Part 2-01. 1986 Amd 1 Instrumentation Cables Part 2: Specification for PVC Insulated Cables (AMD 7618) July 15, 1993 (E). 18 pp.

—**Spark Testing**
CSA CAN/CSA-C22. 2 NO 239-M91. Control and Instrumentation Cables; (Gen Instr 1 Thru 3). 50 pp.

Instrumentation Systems
Use: Instrumentation

Instruments
Use: Instrumentation

Insulated Cables
Use: Cables (Electric)

Insulated Case Circuit Breakers
Use: Molded Case Circuit Breakers

Insulated Conductors
Use: Electric Conductors

Insulated Gate FETs
Use: MOSFETs

Insulated Pins, Electrical
Use: Electrical Insulator Pins

Insulated Wire
Use: Electric Wire

Insulating Blankets
Use: Blanket Insulation

Insulating Board
Use For: Insulating Boards; Insulating Panels; Insulating Slabs; Insulation Panels
See Also: Absorbents; Acoustical Insulation; Building Board; Noise Reduction; Particle Board; Pipe Insulation; Plasterboard; Pressboard; Sound Transmission; Thermal Insulation

—**Asbestos**
CNS R2178-85. Asbestos Mill Boards (May)(11267).
ISO TR1896-91. Products in Fibre-Reinforced Cement—Non-Combustible Fibre-Reinforced Boards of Calcium Silicate or Cement for Insulation and Fire Protection First Edition. 20 pp.
JIS R 3454-79. Asbestos Mill Boards. 6 pp.

—**Calcium Silicate**
JIS A 9510-89. Thermal Insulation Material Made of Calcium Silicate. 13 pp.

—**Fiberboard**
CNS O1027-83. Heat Insulation Fiberboards (Jan)(9911). 4 pp.
CNS O2054-83. Method of Test for Heat Insulation Fiberboards (Jan)(9912).
JIS A 5905-79. Insulation Fibreboards. 12 pp.

—**Foam Rubber**
CNS A2103-81. Rigid Foam Rubber Heat Insulating Material (May)(7331).

—**Glass Wool**
CNS R2059-85. Glass Fiber Heat Insulation Materials (May)(3065). 10 pp.
JIS A 9505-89. Thermal Insulation Material Made of Glass Wool. 13 pp.

—**Isocyanurate**
CGSB 51-GP-21M-78. Thermal Insulation, Urethane and Isocyanurate, Unfaced. 12 pp.
CGSB CAN/CGSB-51.26-M86. Thermal Insulation Urethane and Isocyanurate, Boards, Faced. 19 pp.

—**Magnesium Carbonates**
CNS R2040-76. Basic Magnesium Carbonate Heat Insulation (Mar) (1123).

—**Mineral Aggregate—Roofs**
CSA A284-1976. Mineral Aggregate Thermal Roof Insulation. 8 pp.

—**Mineral Wool—Boilers**
CGSB CAN/CGSB-51.10-92. Mineral Fibre Board Thermal Insulation. 8 pp.

—**Mineral Wool—Decks**
CGSB CAN/CGSB-51.31-M84. Thermal Insulation, Mineral Fibre Board for Above Roof Decks. 9 pp.

—**Mineral Wool—Ducts**
CGSB CAN/CGSB-51.10-92. Mineral Fibre Board Thermal Insulation. 8 pp.

Insulating Board *(Cont.)*
—**Mineral Wool—Machinery**
CGSB CAN/CGSB-51.10-92. Mineral Fibre Board Thermal Insulation. 8 pp.

—**Mineral Wool—Ships**
MOD UK NES 802: Part 1-01. Requirements for Acoustic and Thermal Insulation Material Part 1: Mineral Wool Products Mineral Wool Marine Board Issue 2 (04.89); Amendment 3. 19 pp.

—**Perlite**
JIS A 9512-89. Water Repellent Thermal Insulation Material Made of Perlite. 11 pp.

—**Phenolic**
CGSB CAN/CGSB-51.25-M87. Thermal Insulation, Phenolic, Faced. 12 pp.

—**Phenolic—Ducts**
CGSB 51-GP-29MA-88. Phenolic Thermal Insulation for Pipes and Ducts. 9 pp.

—**Polystyrene**
BSI BS 3837: Part 1-86. 1986 Expanded Polystyrene Boards Part 1: Boards Manufactured from Expandable Beads. 9 pp.
BSI BS 3837: Part 2-90. 1990 Expanded Polystyrene Boards Part 2: Specification for Extruded Boards. 15 pp.
BSI BS 3837: Part 2-01. 1990 Amd 1 Expanded Polystyrene Boards Part 2: Specification for Extruded Boards (AMD 6637) September 28, 1990. 16 pp.
CGSB CAN/CGSB-51.20-M87. Thermal Insulation, Polystyrene Boards and Pipe Coverings; (Amendment 1 Sept 1988). 12 pp.
CNS K3014-65. Foam Polystyrene Heat Insulation Material (Sep)(2535)(R 1971). 3 pp.
DIN ENGL 18184-91. Gypsum Plasterboard Composites with Polystyrene or Polyurethane Rigid Foam Insulation (June). 3 pp.
JIS A 9511-89. Thermal Insulation Material Made of Polystyrene Foam. 19 pp.

—**Polystyrene—Moisture Content**
CNS K6224-65. Method of Test for Foam Polystyrene Heat Insulating Material (Sep)(2536) (R 1971). 3 pp.

—**Polystyrene—Physical Properties**
CNS K6224-65. Method of Test for Foam Polystyrene Heat Insulating Material (Sep)(2536) (R 1971). 3 pp.

—**Polyurethane**
DIN ENGL 18184-91. Gypsum Plasterboard Composites with Polystyrene or Polyurethane Rigid Foam Insulation (June). 3 pp.

—**Polyurethane—Roofs**
BSI BS 4841: Part 3-87. 1987 Rigid Urethane Foam for Building Applications Part 3: Two Types of Laminated Board (Roof Boards) with Auto-Adhesively Bonded Reinforcing Facings for Use as Roof Board Thermal Insulation for Built-Up Roof. 16 pp.

—**Quality Assurance**
DIN ENGL 68750-58. Wood Fibre Building Boards; Porous and Hard Wood Fibre Building Boards; Quality Conditions (Apr). 2 pp.

—**Rock Wool**
CNS A2194-86. Rock Wool Sheathing Boards (Sep)(11701).
CNS A3243-86. Method of Test for Rock Wool Sheathing Boards (Sep)(11702).
CNS R2080-73. Rock Wool for Heat Insulation (Nov)(3657).
JIS A 5451-80. Rock Wool Sheathing Boards. 12 pp.
JIS A 9504-89. Thermal Insulation Material Made of Rock Wool. 14 pp.

—**Urethane**
CGSB 51-GP-21M-78. Thermal Insulation, Urethane and Isocyanurate, Unfaced. 12 pp.
CGSB CAN/CGSB-51.26-M86. Thermal Insulation Urethane and Isocyanurate, Boards, Faced. 19 pp.
CNS A2108-81. Rigid Foam Urethane Heat Insulating Materials (Aug)(7774).
JIS A 9514-89. Thermal Insulation Material Made of Rigid Urethane Foam. 15 pp.

—**Urethane Foam—Cavity Walls**
BSI BS 4841: Part 1-93. 1993 Rigid Polyurethane (PUR) and Polyisocyanurate (PIR) Foam for Building Applications Part 1: Specification for Laminated Board for General Purposes (R). 10 pp.
BSI BS 4841: Part 1-75. 1975 Rigid Urethane Foam for Building Applications Part 1: Laminated Board for General Purposes. 9 pp.
BSI BS 4841: Part 2-75. 1975 Rigid Urethane Foam for Building Applications Part 2: Laminated Board for Use as a Wall and Ceiling Insulation. 11 pp.

Insulating Board *(Cont.)*
—**Urethane Foam—Ceilings**
BSI BS 4841: Part 2-75. 1975 Rigid Urethane Foam for Building Applications Part 2: Laminated Board for Use as a Wall and Ceiling Insulation. 11 pp.

—**Wood Wool**
DIN ENGL 1101-89. Wood Wool Slabs and Sandwich Composite Panels for Use as Insulating Building Material; Requirements and Testing (Nov). 14 pp.

—**Wood Wool—Installation**
DIN ENGL 1102-89. Installation of DIN 1101; Wood Wool Slabs and Sandwich Composite Panels (Nov). 16 pp.

Insulating Boards
Use: Insulating Board

Insulating Bushings (Electrical)
Use: Electrical Insulating Bushings

Insulating Cements
See Also: Cements; Thermal Insulation
BSI BS 3958: Part 6-72. 1972 Amd 1 Thermal Insulating Materials Part 6: Finishing Materials; Hard Setting Composition, Self-Setting Cement and Gypsum Plaster. 19 pp.
CGSB CAN/CGSB-51.12-M86. Cement, Thermal Insulating and Finishing. 11 pp.

Insulating Films (Electrical)
Use: Electrical Insulating Films

Insulating Fluids
Use: Electrical Insulating Liquids

Insulating Glass
See Also: Coated Glass; Glass
CGSB CAN/CGSB-12.8-M90. Insulating Glass Units. 21 pp.
JIS R 3209-86. Sealed Insulating Glass. 14 pp.

—**Doors**
CNS A2217-87. Thermal Insulating Windows and Doors with Sealed Insulating Glasses (Sliding Windows and Doors) (Oct)(12115).

—**Gas Filled**
BSI BS 5713-79. 1979 Specification for Hermetically Sealed Flat Double Glazing Units. 15 pp.

—**Gas Filled—Tightness**
DIN ENGL 52293-87. Testing of Glass; Testing the Tightness of Gas-Filled Insulating Glazing Units (Dec). 3 pp.

—**Multilayer—Environmental Testing**
DIN ENGL 52344-84. Testing of Glass; Testing the Effect of Alternating Atmosphere on Multilayer Insulating Glass (May). 4 pp.

—**Windows**
CNS A2217-87. Thermal Insulating Windows and Doors with Sealed Insulating Glasses (Sliding Windows and Doors) (Oct)(12115).

Insulating Liquids
Use: Electrical Insulating Liquids

Insulating Liquids (Electrical)
Use: Electrical Insulating Liquids

Insulating Oils (Electrical)
Use: Electrical Insulating Oils

Insulating Pads (Thermal Insulation)
Use: Blanket Insulation

Insulating Panels
Use: Insulating Board

Insulating Papers (Electrical)
Use: Electrical Insulating Papers

Insulating Slabs
Use: Insulating Board

Insulating Sleeving
Use: Sleeving

Insulating Sticks
Use: Linemen's Sticks

Insulating Tapes (Electrical)
Use: Electrical Insulating Tapes

Insulating Tubing (Electrical)
Use: Electrical Flexible Tubing—Insulating

INTERNATIONAL AND NON-U.S. NATIONAL STANDARDS
SUBJECT INDEX
Insulation

Insulating Varnishes (Electrical)
Use: Electrical Insulating Varnishes

Insulation
Scope Note: Use a more specific term
See: Absorbents; Acoustical Insulation; Blanket Insulation; Cellular Materials; Electric Conductors; Electrical Insulating Fabrics; Electrical Insulating Films; Electrical Insulating Liquids; Electrical Insulating Oils; Electrical Insulating Papers; Electrical Insulating Tapes; Electrical Insulating Varnishes; Electrical Insulation; Gas Insulation; Insulating Board; Insulating Cements; Insulating Glass; Insulation Coordination; Insulation Resistance; Roof Insulation; Sleeving; Steam Pipes; Thermal Insulation

Insulation Board
Use: Insulating Board

Insulation Coordination
See Also: Electrical Insulation
  BSI BS 5622: Part 2-79. 1979 Insulation Co-Ordination Part 2: Application Guide. 76 pp.
  BSI BS 5622: Part 2-01. 1979 Amd 1 Insulation Co-Ordination Part 2: Application Guide (IEC 71-2: 1976) (AMD 7533) March 15, 1993. 87 pp.
  CENELEC HD 540.2 S1-91. Insulation Co-Ordination Part 2: Application Guide. 7 pp.
  CENELEC HD 540.3 S1-91. Insulation Co-Ordination Part 3: Phase-to-Phase Insulation Co-Ordination Principles, Rules and Application Guide. 8 pp.
  CSA CAN3-C308-M85. Principles and Practice of Insulation Coordination; (Gen Instr 1). 90 pp.
  IEC 71 Pt 2-76. Insulation Co-Ordination Part 2: Application Guide Second Edition. 129 pp.
  IEC 71 Pt 3-82. Insulation Co-Ordination Part 3: Phase-to-Phase Insulation Co-Ordination Principles, Rules and Application Guide First Edition. 38 pp.

—**Electrical Equipment**
  CSA C22.2 NO 0.2-93. Insulation Coordination; (Gen Instr 1). 43 pp.
  DIN VDE 0111 Pt 1-79. Insulation Co-Ordination to Equipment for Three-Phase a.c. Systems Above 1 kV Insulation Phase-To-Earth (VDE Specification) (Oct). 43 pp.
  DIN VDE 0111 Pt 1 A1-86. Insulation Co-Ordination to Equipment for Three-Phase a.c. Systems 1 kV Insulation Phase-To-Earth: Amendment 1 (Sept). 5 pp.
  DIN VDE 0111 Pt 2-83. Insulation Co-Ordination for Equipment in Three-Phase a.c. Systems Above 1 kV Phase-to-Insulation (VDE Specification) (Jan). 19 pp.
  DIN VDE 0111 Pt 3-82. Insulation Co-Ordination for Equipment in Three-Phase a.c. Systems Above 1 kV Application Guide (VDE Guide) (Nov). 76 pp.

—**Electrical Equipment—Glossaries**
  CSA C22.2 NO 0.2-93. Insulation Coordination; (Gen Instr 1). 43 pp.

—**Electrical Equipment—Spacing**
  CSA C22.2 NO 0.2-93. Insulation Coordination; (Gen Instr 1). 43 pp.

—**Glossaries**
  BSI BS 5622: Part 1-79. 1979 Insulation Co-Ordination Part 1: Terms, Definitions, Principles and Rules. 24 pp.
  IEC 71 Pt 1-76. Insulation Co-Ordination Part 1: Terms, Definitions, Principles and Rules Sixth Edition. 47 pp.

—**Low Voltage Installations**
  BSI PD 6499-81. 1981 Amd 1 Guide to Insulation Co-Ordination Within Low-Voltage Systems Including Clearances and Creepage Distances for Equipment. 51 pp.
  DIN VDE 0110 Pt 1-89. Insulation Co-Ordination for Equipment Within Low-Voltage Systems; Fundamental Requirements (Jan) (This Standard, Together with VDE 0110 Pt 2/01.89 Supersedes 0110/11.72 and 0110b/02.79). 27 pp.
  DIN VDE 0110 Pt 2-89. Insulation Co-Ordination for Equipment Within Low-Voltage Systems; Dimensioning of Clearances and Creepage Distances (Jan). 15 pp.
  IEC 664 Pt 1-92. Insulation Coordination for Equipment Within Low-Voltage Systems Part 1: Principles, Requirements and Tests First Edition. 130 pp.

—**Printed Circuit Boards—Coatings**
  IEC 664 Pt 3-92. Insulation Coordination for Equipment Within Low-Voltage Systems Part 3: Use of Coatings to Achieve Insulation Coordination of Printed Board Assemblies First Edition. 44 pp.

—**Railroad Traction Equipment**
  CENELEC PREN 50124-1-93. Insulation Co-Ordination Part 1: Basic Requirements—Clearances and Creepages. 21 pp.

Insulation Monitors
Use: Insulation Resistance Meters

Insulation Panels
Use: Insulating Board

Insulation Resistance
See Also: Dielectric Strength; Electrical Impedance; Electrical Insulation; Electrical Properties; Electrical Resistance; Electrical Resistivity; Insulation Resistance Meters
  BSI BS 5823-79. 1979 Method of Test for Electrical Resistance and Resistivity of Insulating Materials at Elevated Temperatures. 8 pp.
  BSI BS EN 60343-93. 1993 Recommended Test Methods for Determining the Relative Resistance of Insulating Materials to Breakdown by Surface Discharges (IEC 343: 1991) (F). 19 pp.
  CENELEC HD 438 S1-89. Method of Test for Electrical Resistance and Resistivity of Insulating Materials at Elevated Temperatures. 3 pp.
  CENELEC HD 568 S1-90. Methods of Test for the Determination of the Insulation Resistance of Solid Insulating Materials. 3 pp.
  CENELEC EN 60343-92. Recommended Test Methods for Determining the Relative Resistance of Insulating Materials to Breakdown by Surface Discharges. 5 pp.
  CENELEC EN 60343-92. Recommended Test Methods for Determining the Relative Resistance of Insulating Materials to Breakdown by Surface Discharges (IEC 343: 1991). 14 pp.
  CNS C3128-81. Method of Test for Surface Insulation Resistivity of Multi-Strip Specimens (Jul)(7631).
  CNS C6154-82. Method of Test for Surface Insulation Resistivity of Single Strip Specimens (Feb)(8477).
  IEC 167-64. Methods of Test for the Determination of the Insulation Resistance of Solid Insulating Materials First Edition. 21 pp.
  IEC 343-91. Recommended Test Methods for Determining the Relative Resistance of Insulating Materials to Breakdown by Surface Discharges Second Edition (CENELEC EN 60343: 1992). 30 pp.
  IEC 345-71. Method of Test for Electrical Resistance and Resistivity of Insulating Materials at Elevated Temperatures First Edition. 12 pp.

—**Cables (Electric)**
  BSI BS 6469: Sec 99.2-92. 1992 Insulating and Sheathing Materials of Electric Cables Part 99: Test Methods Used in the United Kingdom but Not Specified in Parts 1 to 5 Section 99: Electrical Tests. 10 pp.
  DIN VDE 0282 Pt 1000-83. Rubber Cables, Wires and Flexible Cords for Power Installation; Surface Resistance of Sheath (Feb). 4 pp.
  DIN VDE 0472 Pt 503-85. Testing of Cables and Insulated Flexible Cords; Surface Insulation Resistance of Sheath (May). 6 pp.

—**Cables (Electric)—Aircraft**
  CEN PREN 3475 (Part 303)-92. Cables, Electrical, Aircraft Use Test Methods Part 303—Insulation Resistance. 3 pp.
  CEN PREN 3475 (Part 501)-92. Cables, Electrical, Aircraft Use Test Methods Part 501—Dynamic Cut-Through. 3 pp.
  CEN PREN 3475 (Part 502)-92. Cables, Electrical, Aircraft Use Test Methods Part 502—Notch Propagation. 3 pp.

—**Cabling Systems—Aircraft**
  AECMA PREN2283-88. Testing of Aircraft Wiring. 9 pp.

—**Connectors**
  CEN PREN 2591 (Part B6)-92. Elements of Electrical and Optical Connection Test Methods Part B6—Measurement of Insulation Resistance. 3 pp.
  CNS C6130-86. Method of Test for Low Frequency (Below 3 MHz) Electrical Connectors (TP-21 Insulation Resistance) (Dec)(7660).
  SBAC TS 324 ISSUE 2. Test for Electrical Connectors Insulation Resistance.

—**Control Cables (Electric)**
  CSA CAN/CSA-C22. 2 NO 239-M91. Control and Instrumentation Cables; (Gen Instr 1 Thru 3). 50 pp.

—**Control Systems Equipment**
  CNS C3168-83. Method of Test for Control Gear Insulation (Jan)(9808).
  JIS C 0704-81. Insulation Test for Control Gear. 22 pp.

—**Dry Reed Switches**
  CNS C6152-87. Method of Test for Dry Reed Switches (Insulation Resistance) (Jul)(8230).

—**Electric Switches**
  CNS C6092-88. Method of Test for Electromechanical Switches (Actuator/Mounting Bushing Resistance) (Dec)(6155).

Insulation Resistance *(Cont.)*
—**Electrical Equipment—Aircraft**
  BSI 3G 100:Pt 4: SUBSEC 1.1-82. 1982 General Requirements for Equipment for Use in Aircraft Part 4: Electrical Equipment Section 1: Construction and General Subsection 1.1: Electrical Insulation Tests. 4 pp.
  ISO 2678-85. Environmental Tests for Aircraft Equipment—Insulation Resistance and High Voltage Tests for Electrical Equipment Second Edition. 5 pp.

—**Electromechanical Components**
  BSI BS 5772: Part 2-79. 1979 Basic Testing Procedures and Measuring Methods for Electromechanical Components for Electronic Equipment Part 2: General Examination, Electrical Continuity and Contact Resistance Tests, Insulation Tests and Voltage Stress Tests. 16 pp.
  BSI BS 5772: Pt 2: Supp 1-81. 1981 Electromechanical Components for Electronic Equipment: Basic Testing Procedures Part 2: General Examination, Electrical Continuity and Contract Resistance Tests, Insulation Tests and Voltage Stress Supp 1: 1981 Test 4c. 4 pp.
  IEC 512 Pt 2-85. Electromechanical Components for Electronic Equipment; Basic Testing Procedures and Measuring Methods Part 2: General Examination, Electrical Continuity and Contact Resistance Tests, Insulation Tests and Voltage Stress Tests Second Edition. 39 pp.

—**Electronic Equipment**
  CNS C6022-85. Insulation Resistance Testing Method for Electronic Components (Apr)(3635).

—**Instrumentation Cables**
  CSA CAN/CSA-C22. 2 NO 239-M91. Control and Instrumentation Cables; (Gen Instr 1 Thru 3). 50 pp.

—**Joints—Rails**
  JIS E 3023-85. Testing Methods for Insulating Materials of Insulated Rail Joint.

—**Magnetic Materials**
  BSI BS 6404: Part 11-91. 1991 Magnetic Materials Part 11: Method of Test for the Determination of Surface Insulation Resistance of Magnetic Sheet and Strip (IEC 404-11: 1991). 10 pp.
  IEC 404 Pt 11-91. Magnetic Materials Part 11: Method of Test for the Determination of Surface Insulation Resistance of Magnetic Sheet and Strip First Edition. 18 pp.

—**Phenolic Resins**
  CNS K6273-74. Method of Test for Phenol Formaldehyde Resin Molding Compounds (Oct)(2988). 11 pp.

—**Railroad Signaling Systems**
  JIS E 3021-81. Insulation Resistance and Withstand Voltage Testing Methods of Parts for Railway Signaling.

—**Relays**
  BSI BS 5992: Part 3-80. 1980 Electrical Relays Part 3: Insulation Testing of Electrical Relays. 20 pp.
  IEC 255 Pt 5-77. Electrical Relays Part 5: Insulation Tests for Electrical Relays First Edition. 38 pp.

—**Telephone Cables**
  CCITT RECMN G.611-89. Characteristics of Symmetric Cable Pairs for Analogue Transmission—Transmission Media Characteristics (Study Group XV) 5 pp. 5 pp.
  CCITT RECMN G.621-89. Characteristics of 0.7/2.9 mm Coaxial Cable Pairs—Transmission Media Characteristics (Study Group XV) 4 pp. 4 pp.
  CCITT RECMN G.622-89. Characteristics of 1.2/4.4 mm Coaxial Cable Pairs—Transmission Media Characteristics (Study Group XV) 8 pp. 8 pp.
  CCITT RECMN G.623-89. Characteristics of 2.6/9.5 mm Coaxial Cable Pairs—Transmission Media Characteristics (Study Group XV) 8 pp. 8 pp.

—**Telephone Repeaters**
  CCITT RECMN G.611-89. Characteristics of Symmetric Cable Pairs for Analogue Transmission—Transmission Media Characteristics (Study Group XV) 5 pp. 5 pp.

—**Urea Resins**
  CNS K6272-74. Method of Test for Urea Resin Molding Compounds (Oct)(2986). 9 pp.

—**Vulcanized Rubber**
  BSI BS 903: Part C5-92. 1992 Physical Testing of Rubber Part C5: Determination of Insulation Resistance (Including BS 2782: Part 2 Method 232: 1992) (IEC 167: 1964) (Supersedes BS 2782: Mehtod 232: 1992) (V). 16 pp.
  BSI BS 903: Part C5-83. 1983 Methods of Testing Vulcanized Rubber Part C5: Determination of Insulation Resistance (Incorporating BS 2782: Part 2: Method 232) (V). 10 pp.

INDUSTRY STANDARDS

Insulation Resistance (Cont.)
—Vulcanized Rubber (Cont.)
BSI BS 2782:Pt2: METH 232-92. 1992 Methods of Testing Plastics Part 2: Electrical Properties Method 232: Determination of Insulation Resistance (Including BS 903: Part C5: 1992) (IEC 167: 1992). 16 pp.
ISO 2951-74. Vulcanized Rubber—Determination of Insulation Resistance First Edition. 7 pp.

Insulation Resistance Meters
See Also: Ground Resistance Meters; Insulation Resistance
CNS C4167-90. High Insulation Resistance Meters (Jan)(5198).
CNS C4400-84. Insulation Resistance Testers (Magneto Generator Operated) (Mar)(10795).
DIN VDE 0413 Pt 1-80. Measurement and Control; Equipment for Testing the Protective Measures in Electrical Installations; Insulation Testers (Sept). 17 pp.
DIN VDE 0413 Pt 8-84. Measurement and Control; Instruments for Testing the Protective Measures in Electrical Installations; Insulation Monitors for ac Voltage Systems with Metallically Connected dc Circuits and for dc Voltage Systems (Feb). 14 pp.
JIS C 1303-72. High Insulation Resistance Meters. 11 pp.

—Battery Operated
CNS C4401-84. Insulation Resistance Testers (Battery Operated) (Mar)(10796).
JIS C 1302-82. Insulation Resistance Testers (Battery Operated). 18 pp.

Insulation Resistance Testers
Use: Insulation Resistance Meters

Insulation Resistance Testing
Use: Insulation Resistance

Insulator Bushings
Use: Electrical Insulating Bushings

Insulator Sets
Use: Insulator Strings

Insulator Strings
Use For: Insulator Sets; String Insulator Units
See Also: Cap and Pin Insulators; Electrical Insulators

—Ball and Socket Couplings
BSI BS 3288: Part 3-89. 1989 Insulator and Conductor Fittings for Overhead Power Lines Part 3: Dimensions of Ball and Socket Couplings of String Insulator Units. 30 pp.
BSI BS 3288: Part 4-89. 1989 Insulator and Conductor Fittings for Overhead Power Lines Part 4: Locking Devices for Ball and Socket Couplings of String Insulator Units: Dimensions and Tests. 18 pp.
CENELEC HD 474 S1-87. Dimensions of Ball and Socket Couplings of String Insulator Units. 3 pp.
IEC 120-84. Dimensions of Ball and Socket Couplings of String Insulator Units Third Edition; (Corrigendum—May 1989). 38 pp.
IEC 372-84. Locking Devices for Ball and Socket Couplings of String Insulator Units: Dimensions and Tests Third Edition; (Amendment 1-1991). 34 pp.

—Long Rod
IEC 433-80. Characteristics of String Insulator Units of the Long Rod Type Second Edition. 20 pp.
IEC 471-77. Dimensions of Clevis and Tongue Couplings of String Insulator Units Second Edition; (Amendment 1-1980). 14 pp.

—Mechanical Testing
IEC 575-77. Thermal-Mechanical Performance Test and Mechanical Performance Test on String Insulator Units First Edition. 53 pp.

—Overhead Contact Systems
IEC 383 Pt 2-93. Insulators for Overhead Lines with a Nominal Voltage Above 1 000 V Part 2: Insulator Strings and Insulator Sets for a.c. Systems—Definitions, Test Methods and Acceptance Criteria First Edition. 27 pp.

—Overhead Power Lines
IEC 383 Pt 2-93. Insulators for Overhead Lines with a Nominal Voltage Above 1 000 V Part 2: Insulator Strings and Insulator Sets for a.c. Systems—Definitions, Test Methods and Acceptance Criteria First Edition. 27 pp.

—Substations (Electric)
IEC 383 Pt 2-93. Insulators for Overhead Lines with a Nominal Voltage Above 1 000 V Part 2: Insulator Strings and Insulator Sets for a.c. Systems—Definitions, Test Methods and Acceptance Criteria First Edition. 27 pp.

Insulator Strings (Cont.)
—Thermal Testing
IEC 575-77. Thermal-Mechanical Performance Test and Mechanical Performance Test on String Insulator Units First Edition. 53 pp.

Insulators (Electrical)
Use: Electrical Insulators

Insurance
See Also: Credit Insurance; Health Insurance; Legal Expenses Insurance; Liability Insurance; Risk; Surety Insurance
EC COM(86) 768/2-87. Proposal for a Council Directive on the Coordination of Laws, Regulations and Administrative Provisions Relating to the Compulsory Winding-up of Direct Insurance Undertakings. 39 pp.
EC COM(88) 729-88. Proposal for a Second Council Directive on the Coordination of Laws, Regulations and Administrative Provisions Relating to Direct Life Assurance, Laying down Provisions to Facilitate the Effective Exercise of Freedom to Provide Services and Amending. 46 pp.
EC 88/357/EEC-88. Second Council Directive on the Coordination of Laws, Regulations and Administrative Provisions Relating to Direct Insurance Other Than Life Insurance and Laying Down Provisions to Facilitate the Effective Exercise of Freedom to Provide Services and Amending. 14 pp.
EC 90/619/EEC-90. Council Directive on the Coordination of Laws, Regulations and Administrative Provisions Relating to Direct Life Assurance, Laying Down, Provisions to Facilitate the Effective Exercise of Freedom to Provide Services and Amending Directive 79/267/EEC. 12 pp.

—Aircraft
CAA NOTICE #66 ISSUE 2. Aircraft Insurance (Airworthiness Notices). 1 p.

—Annual Accounts
EC COM(86) 764-87. Proposal for a Council Directive on the Annual Accounts and Consolidated Accounts of Insurance Undertakings. 129 pp.
EC 91/674/EEC-91. Council Directive on the Annual Accounts and Consolidated Accounts of Insurance Undertakings. 26 pp.

—Classification
BSI BS 1000: (368)-71. 1971 Universal Decimal Classification (UDC). English Full Edition (368): Insurance. 23 pp.
SNZ NZS/BS 1000 (368)-71. Universal Decimal Classification Insurance. 24 pp.

—Consolidated Accounts
EC COM(86) 764-87. Proposal for a Council Directive on the Annual Accounts and Consolidated Accounts of Insurance Undertakings. 129 pp.
EC 91/674/EEC-91. Council Directive on the Annual Accounts and Consolidated Accounts of Insurance Undertakings. 26 pp.

Intake Valves
Use: Inlet Valves

Integral Transmissions
Use: Hydraulic Transmission Systems

Integrated Circuit Cards
Use For: Function Cards *See Also:* Circuits; Integrated Circuits

—Bank
BSI BS EN 29992-1-93. 1993 Amd 1 Financial Transaction Cards—Messages Between the Integrated Circuit Card and the Card Accepting Device Part 1: Concepts and Structures (ISO 9992-1: 1990) (AMD 7943) September 15, 1993 (S). 19 pp.
CEN EN 29992-1-93. Financial Transaction Cards—Messages Between the Integrated Circuit Card and the Card Accepting Device—Part 1: Concepts and Structures (ISO 9992-1: 1990). 11 pp.
ISO 9992 Pt 1-90. Financial Transaction Cards—Messages Between the Integrated Circuit Card and the Card Accepting Device—Part 1: Concepts and Structures First Edition (CEN EN 29992-1: 1993). 9 pp.

—Identification
BSI BS EN 27 816 Part 2-91. 1991 Guide to Design and Use of Identification Cards Having Integrated Circuits with Contacts Part 2: Contact Locations and Minimum Size (S). 14 pp.
CEN PREN 742-92. Identification Cards—ID-1 Card Location of Contacts for Cards and Devices Used in Europe. 6 pp.
CEN EN 27 816 (Part 2)-89. Identification Cards Integrated Circuit(s) with Contacts Part 2: Dimensions and Location of the Contacts. 14 pp.

Integrated Circuit Cards (Cont.)
—Identification (Cont.)
ISO 7816 Pt 2-88. Identification Cards—Integrated Circuit(s) Cards with Contacts—Part 2: Dimensions and Location of the Contacts First Edition. 12 pp.
JTC1 7816 Pt 2-88. Identification Cards—Integrated Circuit(s) Cards with Contacts—Part 2: Dimensions and Location of the Contacts First Edition. 12 pp.
OSI ISO 7816-2-88. Identification Cards—Integrated Circuits Cards with Contacts—Part 2: Dimensions and Location of the Contacts. 14 pp.

—Identification—Numbering Systems
IEC DIS 7816 Pt 5-92. Identification Cards—Integrated Circuit(s) Cards with Contact—Part 5: Numbering System and Registration Procedure for Application Identifiers ***CD-ROM ONLY***. 10 pp.
ISO DIS 7816 Pt 5-92. Identification Cards—Integrated Circuit(s) Cards with Contact—Part 5: Numbering System and Registration Procedure for Application Identifiers ***CD-ROM ONLY***. 10 pp.

—Identification—Physical Properties
BSI BS EN 27 816 Part 1-91. 1991 Guide to Design and Use of Identification Cards Having Integrated Circuits with Contacts Part 1: Physical Characteristics (S). 12 pp.
CEN EN 27 816 (Part 1)-89. Identification Cards Integrated Circuit(s) with Contacts Part 1: Physical Characteristics. 9 pp.
IEC 10536 Pt 1-92. Identification Cards—Contactless Integrated Circuit(s) Cards—Part 1: Physical Characteristics First Edition. 10 pp.
ISO 7816 Pt 1-87. Identification Cards—Integrated Circuit(s) Cards with Contacts—Part 1: Physical Characteristics First Edition. 7 pp.
ISO 10536 Pt 1-92. Identification Cards—Contactless Integrated Circuit(s) Cards—Part 1: Physical Characteristics First Edition. 10 pp.
JTC1 7816 Pt 1-87. Identification Cards—Integrated Circuit(s) Cards with Contacts—Part 1: Physical Characteristics First Edition. 7 pp.
OSI ISO 7816-1-87. Indentification Cards—Intergrated Circuit(s) Cards with Contacts—Part 1: Physical Characteristics. 9 pp.
OSI ISO/IEC DIS 10536-1-90. Identification Cards—Contactless Integrated Circuit(s) Cards—Part 1: Physical Characteristics. 10 pp.

—Identification—Protocols
BSI BS 7109: Part 3-90. 1990 Guide to Design and Use of Identification Cards Having Integrated Circuits with Contacts Part 3: Electronic Signals and Transmission Protocols. 21 pp.
BSI BS 7109: Part 3-01. 1990 Amd 1 Guide to Design and Use of Identification Cards Having Integrated Circuits with Contacts Part 3: Electronic Signals and Transmission Protocols (ISO/IEC 7816-3: 1989) (AMD 7325) November 15, 1992. 25 pp.
CEN EN 27816-3-92. Identification Cards—Integrated Circuit(s) Cards with Contacts—Part 3: Electronic Signals and Transmission Protocols. 21 pp.
IEC 7816 Pt 3-89. Identification Cards—Integrated Circuit(s) Cards with Contacts—Part 3: Electronic Signals and Transmission Protocols Amendment 1: Protocol Type T = 1, Asynchronous Half Duplex Block Transmission Protocol First Edition; (Amendment 1-1992). 35 pp.
ISO 7816 Pt 3-89. Identification Cards—Integrated Circuit(s) Cards with Contacts—Part 3: Electronic Signals and Transmission Protocols Amendment 1: Protocol Type T = 1, Asynchronous Half Duplex Block Transmission Protocol First Edition; (Amendment 1-1992). 35 pp.
OSI ISO/IEC 7816-3 DAM 1-90. Identification Cards—Integrated Circuit(s) Cards with Contacts—Part 3: Electronic Signals and Transmission Protocols AMENDMENT 1. 27 pp.
OSI ISO IEC 7816-3-89. Identification Cards—Integrated Circuit(s) Cards with Contacts—Part 3: Electronic Signals and Transmission Protocols. 19 pp.
OSI ISO DIS 7816-3-87. Identification Cards—Integrated Circuit Cards with Contacts—Part 3: Electronic Signals and Exchange Protocols. 28 pp.
OSI ISO DIS 7816-3-87. Identification Cards—Integrated Circuit Cards with Contacts—Part 3: Electronic Signals and Exchange Protocols. 28 pp.
OSI ISO DIS 7816-3-89. Identification Cards—Integrated Circuit(s) Cards with Contacts—Part 3: Electronic Signals and Transmission Protocols. 19 pp.
OSI ISO IEC 7816-3-89. Identification Cards—Integrated Circuit(s) Cards with Contacts—Part 3: Electronic Signals and Transmission Protocols. 19 pp.

INTERNATIONAL AND NON-U.S. NATIONAL STANDARDS
SUBJECT INDEX
Integrated

Integrated Circuit Cards *(Cont.)*
—Identification—Registration
IEC DIS 7816 Pt 5-92. Identification Cards—Integrated Circuit(s) Cards with Contact—Part 5: Numbering System and Registration Procedure for Application Identifiers ***CD-ROM ONLY***. 10 pp.
ISO DIS 7816 Pt 5-92. Identification Cards—Integrated Circuit(s) Cards with Contact—Part 5: Numbering System and Registration Procedure for Application Identifiers ***CD-ROM ONLY***. 10 pp.

—Pay Telephones
CEN PREN 1038-93. IC Card Applications for Telecommunications—Part 1: IC Card Payphone. 15 pp.

—Physical Properties
JIS X 6303-88. Integrated Circuit Cards with Contacts-Physical Characteristics and Location of Contacts. 13 pp.

—Telecommunication Equipment
CEN PREN 726-1-92. Requirements for IC Cards and Terminals for Telecommunication Use—Part 1: System Overview. 13 pp.
CEN PREN 726-2-92. Requirements for IC Cards and Terminals for Telecommunication Use—Part 2: Application Independent Card Requirements. 108 pp.
CEN PREN 726-3-92. Requirements for IC Cards and Terminals for Telecommunication Use—Part 3: Application Independent Card Related Terminal Requirements. 20 pp.
CEPT T/SF 67 E-90. Identification and Payment Card Services. 6 pp.

—Telecommunication Equipment—Interfaces
CEN PREN 726-6-93. Requirements for IC Cards and Terminals for Telecommunications Use—Part 6: Telecommunication Features. 57 pp.

—Telecommunication Equipment—Security Framework
CEN PREN 726-2-93. Requirements for IC Cards and Terminals for Telecommunications Use—Part 2: Security Framework. 24 pp.

Integrated Circuit Memories
Use: Memory Circuits

Integrated Circuit Microprocessors
Use: Microprocessors

Integrated Circuit Packaging
Use: Semiconductor/Integrated Circuit Packaging

Integrated Circuit/Semiconductor Packaging
Use: Semiconductor/Integrated Circuit Packaging

Integrated Circuit Socket Adapters
See Also: Adapters (Fittings)
BSI BS 546-50. 1950 Amd 7 Two-Pole and Earthing Pin-Plugs, Socket-Outlets and Socket-Outlet Adaptors for Circuits up to 250 Volts. 49 pp.

Integrated Circuits
Scope Note: For additional listings, use a more specific term *Use For:* Digital Monolithic Integrated Circuits; Microcircuits; Microelectronic Circuits; Microelectronics; Microminiature Circuitry
See Also: Analog to Digital Converters; Broadband Amplifiers; Buffer/Line Drivers; Bus Drivers; Bus Registers; Circuits; Compandors; Delay Lines; Differential Amplifiers; Digital Circuits; Digital to Analog Converters; DIP Packages; Electrical Leads; Electronic Component Packaging; Electronic Components; Film Integrated Circuits; Gate Circuits; Hybrid Circuits; Integrated Circuit Cards; Interface Circuits; Line Buffers; Line Drivers; Line Receivers; Linear Circuits; Microwave Integrated Circuits; Monolithic Integrated Circuits; Multiplexers; NAND Gates; NOT Elements; PLAs; Printed Circuit Boards; Pulse Synchronizer/Drivers; Semiconductor Devices; Sense Amplifiers; Shift Registers; Surface Mount Technology; Voltage Reference Integrated Circuits; Voltage Regulator Integrated Circuits
BSI BS 6493: Sec 1.1-84. 1984 Semiconductor Devices Part 1: Discrete Devices Section 1.1: General. 58 pp.
BSI BS 6493: Sec 1.1-01. 1984 Amd 1 Semiconductor Devices: Discrete Devices and Integrated Circuits Part 1: Discrete Devices Section 1.1 General (IEC 747-1: 1983) (AMD 7213) November 15, 1992. 87 pp.
BSI BS 6493: Sec 2.1-85. 1985 Semiconductor Devices Part 2: Integrated Circuits Section 2.1: General (IEC 748-1: 1984). 21 pp.

Integrated Circuits *(Cont.)*
BSI BS 6493: Sec 2.1-01. 1985 Amd 1 Semiconductor Devices Part 2: Integrated Circuits Section 2.1: General (IEC 748-1: 1984) (AMD 6830) January 31, 1992. 41 pp.
BSI BS QC 700000-91. 1991 Semiconductor Devices. Generic Specification for Discrete Devices and Integrated Circuits (IEC 747-10: 1991). 48 pp.
CNS C5014-76. General Rules for Mechanical Dimension of Semiconductor Integrated Circuits (Sep)(3999). 3 pp.
CNS Z5079-82. Preparation for Delivery of Microcircuits (May)(6983).
IEC 747 Pt 1-83. Semiconductor Devices—Discrete Devices and Integrated Circuits Part 1: General First Edition; (Amendment 1-1991). 156 pp.
IEC 747 Pt 10-91. Semiconductor Devices Part 10: Generic Specification for Discrete Devices and Integrated Circuits Second Edition (IECQ QC 700000). 90 pp.
IEC 748 Pt 1-84. Semiconductor Devices Integrated Circuits Part 1: General First Edition; (Amendment 1-1991) (Amendment 2-1993). 94 pp.
IECQ QC 700000-91. Semiconductor Devices Part 10: Generic Specification for Discrete Devices and Integrated Circuits (IEC 747-10 ED 2). 90 pp.
IECQ QC 790100-90. Semiconductor Devices Integrated Circuits Part 11: Sectional Specification for Semiconductor Integrated Circuits Excluding Hybrid Circuits (IEC 748-11 ED 1). 68 pp.
IECQ PQC 82-89. Semiconductors for Use in Electronic Equipment: Sectional Specification: Semiconductor Integrated Circuits Excluding Hybrid Circuits. 26 pp.
MOD UK DSTAN 59-62: Part 1-82. Microcircuits Electronic (Integrated Circuits) (Listed on EPIC Database) Part 1: General Requirements Issue 2. 18 pp.
MOD UK DSTAN 59-62: Part 90-01. Microcircuits Electronic (Integrated Circuits) (Listed on EPIC Database) Part 90: Detail Specifications Issue 1; Amendment 1. 11 pp.

—Binary Counter—Quality Assurance
BSI BS 9442-73. (WITHDRAWN) 1973 Amd 1 Rules for the Preparation of Detail Specifications for Integrated Circuits of Assessed Quality: TTL Binary Counter Circuits. General Application Category. 13 pp.

—Binary Memory—Quality Assurance
BSI BS 9443-74. (WITHDRAWN) 1974 Amd 1 Rules for the Preparation of Detail Specifications for Integrated Circuits of Assessed Quality: TTL Binary Memory Circuits. General Application Category. 18 pp.

—Defense Contracts
MOD UK DEFCON 60-84. Special Microcircuits 8/84. 1 p.

—Differential Operational Amplifiers—Quality Assurance
BSI BS 9460-74. (OBSOLESCENT) 1974 Amd 2 Rules for the Preparation of Detail Specifications for Integrated Circuits of Assessed Quality: Differential Operational Amplifiers. General Application Category. 15 pp.

—Drafting—Package Outlines
SAA AS C379 Part 3-78. Mechanical Standardization of Semiconductor Devices—Part 3: General Rules for the Preparation of Outline Drawings of Integrated Circuits. 24 pp.

—Electrical Properties
IEC 147 Pt 4-76. Essential Ratings and Characteristics of Semiconductor Devices and General Principles of Measuring Methods Part 4: Acceptance and Reliability First Edition. 42 pp.

—Endurance Testing
JIS C 7022-79. Environmental Testing Methods and Endurance Testing Methods for Semiconductor Integrated Circuits (R 1984). 50 pp.

—Engineering Drawings
BSI BS 3934-65. (WITHDRAWN) 1965 (Includes Addenda 1,2,3, 4, and 5) Dimensions of Semiconductor Devices (Includes Addenda 1,2, and 3) (Superseded by BS 3934: Parts 1 to 6 and Part 10: 1992). 270 pp.

—Environmental Testing
JIS C 7022-79. Environmental Testing Methods and Endurance Testing Methods for Semiconductor Integrated Circuits (R 1984). 50 pp.

—Gages—Engineering Drawings
BSI BS 3934-65. (WITHDRAWN) 1965 (Includes Addenda 1,2,3, 4, and 5) Dimensions of Semiconductor Devices (Includes Addenda 1,2, and 3) (Superseded by BS 3934: Parts 1 to 6 and Part 10: 1992). 270 pp.

Integrated Circuits *(Cont.)*
—Glossaries
BSI BS 6493: Sec 2.2-86. 1986 Semiconductor Devices Part 2: Integrated Circuits Section 2.2: Recommendations for Digital Integrated Circuits. 143 pp.
BSI BS 6493: Sec 2.2-01. 1986 Amd 1 Semiconductor Devices Part 2: Integrated Circuits Section 2.2: Recommendations for Digital Integrated Circuits (IEC 748-2: 1985) (AMD 7003) July 15, 1992. 166 pp.
CNS C5174-83. Glossary of Integrated Circuit Introduction (Oct)(10617).
CNS C5175-83. Glossary of General Integrated Circuit (Oct)(10618).
CNS C5176-83. Glossary of Integrated Circuit Production Techniques (Oct)(10619).
IEC 50 Chap 521-84. International Electrotechnical Vocabulary Chapter 521: Semiconductor Devices and Integrated Circuits. 100 pp.
IEC 748 Pt 2-85. Semiconductor Devices Integrated Circuits Part 2: Digital Integrated Circuits First Edition; (Amendment 1-1991). 312 pp.
JIS C 5610-75. Glossary of Terms Used in Integrated Circuits.
SAA AS 1852.521-88. International Electrotechnical Vocabulary—Part 521: Semiconductor Devices and Integrated Circuits. 90 pp.
SNZ IEC 50: 50(521)-84. International Electrotechnical Vocabulary 50(521): Semiconductor Devices and Integrated Circuits. 90 pp.

—Handling
BSI BS 5783-87. (WITHDRAWN) 1987 Handling of Electrostatic Sensitive Devices (Superseded by BS CECC 00015: Part 1: 1991). 16 pp.

—Quality Assurance
BSI BS 9000: Part 5-89. (OBSOLESCENT) 1989 General Requirements for Electronic Components of Assessed Quality Part 5: Qualification Approval and Conformance Testing Procedures (Superseded by BS CECC 00114: Part 2: 1991). 17 pp.
BSI BS 9400-70. 1970 Amd 7 Integrated Electronic Circuits and Micro-Assemblies of Assessed Quality: (Qualification Approval Procedures): Generic Data and Methods of Test. 235 pp.
BSI BS 9400-08. (WITHDRAWN) 1970 Amd 8 Integrated Electronic Circuits and Micro-Assemblies of Assessed Quality: (Qualification Approval Procedures): Generic Data and Methods of Test (AMD 7224) November 15, 1992 (Superseded by BS CECC 90 000). 239 pp.
BSI BS 9450-75. 1975 Amd 5 Integrated Electronic Circuits and Micro-Assemblies of Assessed Quality (Capacity Approval Procedure): Generic Data and Methods of Test. 100 pp.
BSI BS QC 790100-91. 1991 Semiconductor Devices. Sectional Specification for Semiconductor Integrated Circuits Excluding Hybrid Circuits (IEC 748-11: 1990). 35 pp.
BSI BS QC 790101-92. 1992 Semiconductor Devices. Integrated Circuits. Sectional Specification for Semiconductor Integrated Circuits Excluding Hybrid Circuits Section 1: Internal Visual Examination for Semiconductor Integrated Circuits Excluding. 39 pp.
BSI BS CECC 90000-91. 1991 Harmonized System of Quality Assessment for Electronic Components Generic Specification: Monolithic Integrated Circuits. 181 pp.
BSI BS CECC 90000-01. 1991 Amd 1 Generic Specification: Monolithic Integrated Circuits (AMD 7386) January 15, 1993 (Supersedes BS 9400: 1970) (T). 178 pp.
BSI BS CECC 90000:Add 1-83. (OBSOLESCENT) 1983 Amd 1 Monolithic Integrated Circuits: Generic Specification Addendum 1: Internal Visual Inspection (AMD 5102) July 31, 1986. 48 pp.
BSI BS CECC 90000:Add 1-04. 1983 Amd 4 Generic Specification: Monolithic Integrated Circuits Addendum 1: Internal Visual Inspection (AMD 5923) February 28, 1990 (T). 54 pp.
BSI BS EN 190100-93. 1993 Amd 3 Harmonized System of Quality Assessment for Electronic Components: Sectional Specification: Digital Monolithic Integrated Circuits (AMD 7845) July 15, 1993 (T). 75 pp.
CENELEC EN 190100-93. Sectional Specification: Digital Monolithic Integrated Circuits. 23 pp.
IEC 748 Pt 11-90. Semiconductor Devices Integrated Circuits Part 11: Sectional Specification for Semiconductor Integrated Circuits Excluding Hybrid Circuits First Edition (IECQ QC 790100). 68 pp.
IEC 748 Pt 11-1-92. Semiconductor Devices Integrated Circuits Part 11: Section 1: Internal Visual Examination for Semiconductor Integrated Circuits Excluding Hybrid Circuits First Edition. 76 pp.

—Ratings—Quality Assurance
BSI BS CECC 90108-88. 1988 TTL Advanced SCHOTTKY Digital Integrated Circuits: Family Specification. 19 pp.

INDUSTRY STANDARDS

Integrated Circuits *(Cont.)*

—Symbols

BSI BS 3363-80. 1980 Letter Symbols for Semiconductor Devices and Integrated Microcircuits (Incorporating Supplements 1 and 2). 35 pp.

BSI BS 3363: Supp 1-81. 1981 Letter Symbols for Semiconductor Devices and Integrated Microcircuits Supplement No. 1:. 20 pp.

BSI BS 3363: Supp 2-81. 1981 Letter Symbols for Semiconductor Devices and Integrated Microcircuits Supplement No. 2:. 14 pp.

BSI BS 6493: Sec 2.2-86. 1986 Semiconductor Devices Part 2: Integrated Circuits Section 2.2: Recommendations for Digital Integrated Circuits. 143 pp.

BSI BS 6493: Sec 2.2-01. 1986 Amd 1 Semiconductor Devices Part 2: Integrated Circuits Section 2.2: Recommendations for Digital Integrated Circuits (IEC 748-2: 1985) (AMD 7003) July 15, 1992. 166 pp.

IEC 148-69. Letter Symbols for Semiconductor Devices and Integrated Microcircuits Second Edition; (Supplement A-1974) (Supplement B-1979). 130 pp.

IEC 748 Pt 2-85. Semiconductor Devices Integrated Circuits Part 2: Digital Integrated Circuits First Edition; (Amendment 1-1991). 312 pp.

—Tape Automated Bonding (TAB)

BSI BS 3934: Part 5-92. 1992 Mechanical Standardization of Semiconductor Devices Part 5: Recommendations for Tape Automated Bonding (TAB) of Integrated Circuits (IEC 191-5: 1987). 15 pp.

IEC 191 Pt 5-87. Mechanical Standardization of Semiconductor Devices Part 5: Recommendations Applying to Tape Automated Bonding (TAB) of Integrated Circuits First Edition. 29 pp.

Integrated Compilers

Use: Communication Terminal Equipment

Integrated Digital Networks

CEPT T/GSI 01-04-85. Application Des Recommendations Concernant Le Reseau Numerique Integre Au RNIS. 1 p.

CEPT T/GSI 01-04 E-85. Applicability of IDN Recommendations to ISDN. 1 p.

—Digital Exchanges

CCITT RECMN Q.500-89. Digital Local, Combined, Transit and International Exchanges Introduction and Field of Application—Digital Local, Transit, Combined and International Exchanges in Integrated Digital Networks and Mixed Analogue-Digital Networks—Supplements (Study Group XI) 4 pp. 4 pp.

CCITT RECMN Q.521-89. Exchange Functions—Digital Local, Transit, Combined and International Exchanges in Integrated Digital Networks and Mixed Analogue-Digital Networks—Supplements (Study Group XI) 3 pp. 3 pp.

—Digital Line Systems

CCITT FASCICLE III.5-89. Digital Networks, Digital Sections and Digital Line Systems—Recommendations G.801—G.961. 292 pp.

—Digital Sections

CCITT FASCICLE III.5-89. Digital Networks, Digital Sections and Digital Line Systems—Recommendations G.801—G.961. 292 pp.

—Jitter—Hierarchical Interfaces

CCITT FASCICLE III.5-89. Digital Networks, Digital Sections and Digital Line Systems—Recommendations G.801—G.961. 292 pp.

CCITT RECMN G.823-89. Control of Jitter and Wander Within Digital Networks Which Are Based on the 2048 kBit/s Hierarchy—Digital Networks, Digital Sections and Digital Line Systems (Study Groups XV and XVIII) 13 pp. 13 pp.

CCITT RECMN G.824-89. Control of Jitter and Wander Within Digital Networks Which Are Based on the 1544 kBit/s Hierarchy—Digital Networks, Digital Sections and Digital Line Systems (Study Groups XV and XVIII) 6 pp. 6 pp.

—Mixed Analog/Digital Networks—International Exchanges

CCITT FASCICLE VI.5. Digital Local, Transit, Combined and International Exchanges Integrated Digital Networks and Mixed Analogue—Digital Networks Supplements Recommendations Q.500—Q.554 (Study Group XI). 188 pp.

—Network Performance

CCITT RECMN I.350-89. General Aspects of Quality of Service and Network Performance in Digital Networks, Including ISDN—Integrated Services Digital Network (ISDN)—Overall Network Aspects and Functions, ISDN User-Network Interfaces (Study Group XVIII) 10 pp. 10 pp.

—Performance Parameters

CCITT RECMN I.350-89. General Aspects of Quality of Service and Network Performance in Digital Networks, Including ISDN—Integrated Services Digital Network (ISDN)—Overall Network Aspects and Functions, ISDN User-Network Interfaces (Study Group XVIII) 10 pp. 10 pp.

—Quality of Service

CCITT RECMN I.350-89. General Aspects of Quality of Service and Network Performance in Digital Networks, Including ISDN—Integrated Services Digital Network (ISDN)—Overall Network Aspects and Functions, ISDN User-Network Interfaces (Study Group XVIII) 10 pp. 10 pp.

—Teleservices

CEPT T/SF 62 E-88. Supplementary Service Digital Connectivity for Teleservices. 5 pp.

Integrated Service Digital Broadcasting

Use For: ISDB *See Also:* Broadcasting

—Broadcasting Satellite Services

CCIR Report 1227-90. Satellite Broadcasting Systems for ISDB (Integrated Service Digital Broadcasting)—Section 10/11B—Systems. 4 pp.

CCIR QUESTION 101/11-90. Integrated Service Digital Broadcasting in the Broadcasting-Satellite Service (Sound and Television)—Questions Concerning Study Group 11—Broadcasting Service (Television). 1 p.

Integrated Services Digital Networks

Use For: ISDN *See Also:* Bearer Services; Fixed Terminal Networks; Integrated Services Digital Networks User Part; Intelligent Networks; Private Integrated Services Networks; Teleservices; User to User Information Services; Voice/Data Systems

CCIR Report 1139-90. General System and Performance Aspects of Digital Transmission in the Fixed-Satellite Service—Section 4B1—Systems Aspects. 19 pp.

CCITT FASCICLE II.2-88. Telephone Network and ISDN-Operation, Numbering, Routing and Mobile Service. Recommendations E.100—E.333. 362 pp.

CCITT FASCICLE III.7-88. Integrated Services Digital Network (ISDN) General Structure and Service Capabilities Recommendations I.110-I.257 (Study Group XVIII). 353 pp.

CCITT FASCICLE III.8-88. Integrated Services Digital Network (ISDN) Overall Network Aspects and Functions, ISDN User-Network Interfaces Recommendations I.310—I.470 (Study Group XVIII). 280 pp.

CCITT RECMN I.310-89. ISDN—Network Functional Principles—Integrated Services Digital Network (ISDN)—Overall Network Aspects and Functions, ISDN User -Network Interfaces (Study Group XVIII) 22 pp. 22 pp.

CCITT RECMN I.340-89. ISDN Connection Types—Integrated Services Digital Network (ISDN)—Overall Network Aspects and Functions, ISDN User-Network Interfaces (Study Group XVIII) 17 pp. 17 pp.

CCITT FASCICLE VI.1. General Recommendations on Telephone Switching and Signalling--Functions and Information Flows for Services in the ISDN—Supplements Recommendations Q1—Q118 Bis. 502 pp.

CCITT RECMN Q.65-89. Stage 2 of the Method for the Characterization of Services Supported by an ISDN—General Recommendations on Telephone Switching and Signalling—Functions and Information Flows for Services in the ISDN —Supplements (Study Group XI) 12 pp. 12 pp.

CENELEC ETR 010-93. ISDN Standards Management (ISM); the ETSI Basic Guide on the European Integrated Services Digital Network. 41 pp.

CENELEC ETR 076-93. Integrated Services Digital Network (ISDN); Standards Guide. 88 pp.

CEPT ISDN USER HANDBOOK-92. European Conference of Postal and Telecommunications Administrations. 85 pp.

CEPT T/CAC S 10 E-92. Services and Facilities Aspects of an Integrated Services Digital Network (ISDN) (Revised in Odense 1986, Copenhagen 1987 and Vienna 1989(CAC) Athens 1992). 10 pp.

CEPT T/CAC S 10.1 E-92. General Aspects of an Integrated Services Digital Network (ISDN) (Revised in Odense 1986, Vienna 1989 (CAC) and Athens 1992). 14 pp.

CEPT T/GSI 01-01 E-85. Integrated Service Digital Networks. 3 pp.

CEPT T/GSI 01-04-85. Application Des Recommendations Concernant Le Reseau Numerique Integre Au RNIS. 1 p.

CEPT T/GSI 01-04 E-85. Applicability of IDN Recommendations to ISDN. 1 p.

CEPT T/GSI 01-05-85. Methodes Generales De Modelisation. 10 pp.

CEPT T/GSI 01-05 E-85. General Modelling Method. 10 pp.

CEPT T/GSI 01-06-87. Evolution Vers Le RNIS. 4 pp.

CEPT T/GSI 01-06 E-87. Evolution Towards ISDN. 4 pp.

CEPT T/GSI 03-01-87. Principles Fonctionnels Du Reseau RNIS. 20 pp.

CEPT T/GSI 03-01 E-87. Network Functional Principles. 20 pp.

CEPT T/GSI 03-08/1E-85. Network Performance in the ISDN (Revised by Correspondence, June 1987). 4 pp.

CEPT T/SF 31-83. Services Et Facilities D'un Reseau Numerique a Integration Des Services. 26 pp.

CEPT T/SF 31 E-88. Services and Facilities Aspects of an Integrated Services Digital Network (ISDN). 3 pp.

CEPT T/SF 31-01 E-88. General Aspects of an Integrated Services Digital Network (ISDN). 10 pp.

CEPT T/SF 31-04 E-88. Intercommunication Aspects of an Integrated Services Digital Network (ISDN). 4 pp.

CEPT T/SF 31-05 E-88. General Supplementary Service Aspects of an Integrated Services Digital Network (ISDN). 10 pp.

CEPT T/SF 31-06 E-87. Relevance of Supplementary Services to Be Provided in an Integrated Services Digital Network (ISDN). 5 pp.

CEPT T/SF 31-07 E-87. Operational Requirements of ISDN Supplementary Services. 89 pp.

ETSI ETR 010-93. ISDN Standards Management (ISM); the ETSI Basic Guide on the European Integrated Services Digital Network. 41 pp.

ETSI ETR 076-93. Integrated Services Digital Network (ISDN); Standards Guide. 88 pp.

—Accounting

CCITT RECMN D.250-91. General Charging and Accounting Principles for Non-Voice Services Provided by Interworking Between the ISDN and Existing Public Data Networks (Study Group III) 5 pp. 5 pp.

CCITT RECMN D.250-89. General Charging and Accounting Principles for Non-Voice Services Provided by Interworking Between the ISDN and Existing Public Data Networks—General Tariff Prin.—Charging and Accounting in Intl Telecommunications Services (Study Group III) 1 pp. 1 p.

CCITT RECMN D.260-91. Charging and Accounting Capabilities to Be Applied on the ISDN (Study Group III) 6 pp. 6 pp.

—Add On Conference

CENELEC PRETS 300 183-91. Integrated Services Digital Network (ISDN); Conference Call Add-on (CONF) Supplementary Service Service Description (T/NA1 (89)25). 17 pp.

CENELEC PRETS 300 183-92. Integrated Services Digital Network (ISDN); Conference Call, Add-On (CONF) Supplementary Service Service Description. 20 pp.

CENELEC ETS 300 183-92. Integrated Services Digital Network (ISDN); Conference Call, Add-on (CONF) Supplementary Service Service Description. 20 pp.

CENELEC PRETS 300 184-91. Integrated Services Digital Network (ISDN); Conference Call Add-on (CONF) Supplementary Service Functional Capabilities and Information Flows (T/S 22-12). 41 pp.

CENELEC PRETS 300 184-92. Integrated Services Digital Network (ISDN); Conference Call, Add-on (CONF) Supplementary Service Functional Capabilities and Information Flows. 40 pp.

CENELEC ETS 300 184-93. Integrated Services Digital Network (ISDN); Conference Call, Add-on (CONF) Supplementary Service Functional Capabilities and Information Flows. 40 pp.

ETSI ETS 300 183-92. Integrated Services Digital Network (ISDN); Conference Call, Add-on (CONF) Supplementary Service Service Description. 20 pp.

ETSI PRETS 300 183-91. Integrated Services Digital Network (ISDN); Conference Call Add-on (CONF) Supplementary Service Service Description (T/NA1 (89)25). 17 pp.

ETSI ETS 300 184-93. Integrated Services Digital Network (ISDN); Conference Call, Add-on (CONF) Supplementary Service Functional Capabilities and Information Flows. 40 pp.

ETSI PRETS 300 184-92. Integrated Services Digital Network (ISDN); Conference Call, Add-on (CONF) Supplementary Service Functional Capabilities and Information Flows. 40 pp.

ETSI PRETS 300 184-91. Integrated Services Digital Network (ISDN); Conference Call Add-on (CONF) Supplementary Service Functional Capabilities and Information Flows (T/S 22-12). 41 pp.

INTERNATIONAL AND NON-U.S. NATIONAL STANDARDS
SUBJECT INDEX
Integrated

Integrated Services Digital Networks *(Cont.)*

—Add On Conference—DSS1

CENELEC PRETS 300 185-91. Integrated Services Digital Network (ISDN); Conference Call Add-on (CONF) Supplementary Service Digital Subscriber Signalling System No. One (DSS1) Protocol (T/S 46-33J1). 38 pp.

CENELEC PRETS 300 185-92. Integrated Services Digital Network (ISDN); Conference Call, Add-on (CONF) Supplementary Service Digital Subscriber Signalling System No. One (DSS1) Protocol. 44 pp.

CENELEC ETS 300 185-93. Integrated Services Digital Network (ISDN); Conference Call, Add-on (CONF) Supplementary Service Digital Subscriber Signalling System No. One (DSS1) Protocol. 44 pp.

ETSI ETS 300 185-93. Integrated Services Digital Network (ISDN); Conference Call, Add-on (CONF) Supplementary Service Digital Subscriber Signalling System No. One (DSS1) Protocol. 44 pp.

ETSI PRETS 300 185-92. Integrated Services Digital Network (ISDN); Conference Call, Add-on (CONF) Supplementary Service Digital Subscriber Signalling System No. One (DSS1) Protocol. 44 pp.

ETSI PRETS 300 185-91. Integrated Services Digital Network (ISDN); Conference Call Add-on (CONF) Supplementary Service Digital Subscriber Signalling System No. One (DSS1) Protocol (T/S 46-33J1). 38 pp.

—Addressing Systems

CCITT RECMN I.330-89. ISDN Numbering and Addressing Principles—Integrated Services Digital Network (ISDN)—Overall Network Aspects and Functions, ISDN User-Network Interfaces (Study Group XVIII) 8 pp. 8 pp.

CENELEC ETR 006-90. Network Aspects; Numbering and Addressing for the Memorandum of Understanding (MOU) on Integrated Services Digital Network (ISDN) (Priorities 1 and 2). 15 pp.

CEPT T/GSI 03-07-85. Numerotation Et Adressage. 1 p.

CEPT T/GSI 03-07 E-85. Numbering and Addressing. 1 p.

CEPT T/N 21-02 E-88. Numbering and Addressing for Gap Phase II. 5 pp.

ETSI ETR 006-90. Network Aspects; Numbering and Addressing for the Memorandum of Understanding (MOU) on Integrated Services Digital Network (ISDN) (Priorities 1 and 2). 15 pp.

—Advice of Charge Services

CCITT RECMN I.256-89. Charging Supplementary Services—Integrated Services Digital Network (ISDN)—General Structure and Service Capabilities (Study Group XVIII) 16 pp. 16 pp.

CCITT RECMN I.256. 2-88. Advice of Charge—Integrated Services Digital Network (ISDN)—General Structure and Service Capabilities (Study Group XVIII) 15 pp. 15 pp.

CENELEC PRETS 300 178-91. Integrated Services Digital Network (ISDN); Advice of Charge: Charging Information at Call Set-up Time (AOC-S) Supplementary Service Service Description (T/NA1(89)13). 18 pp.

CENELEC PRETS 300 178-92. Integrated Services Digital Network (ISDN); Advice of Charge: Charging Information at Call Set-up Time (AOC-S) Supplementary Service Service Description. 18 pp.

CENELEC ETS 300 178-92. Integrated Services Digital Network (ISDN); Advice of Charge: Charging Information at Call Set-up Time (AOC-S) Supplementary Service Service Description. 18 pp.

CENELEC PRETS 300 179-91. Integrated Services Digital Network (ISDN); Advice of Charge: Charging Information During the Call (AOC-D) Supplementary Service Service Description (T/NA1(89)14). 16 pp.

CENELEC PRETS 300 179-92. Integrated Services Digital Network (ISDN); Advice of Charge: Charging Information During the Call (AOC-D) Supplementary Service Service Description. 16 pp.

CENELEC ETS 300 179-92. Integrated Services Digital Network (ISDN); Advice of Charge: Charging Information During the Call (AOC-D) Supplementary Service Service Description. 16 pp.

CENELEC PRETS 300 180-91. Integrated Services Digital Network (ISDN); Advice of Charge: Charging Information at the End of the Call (AOC-E) Supplementary Service Service Description (T/NA1(89)15). 16 pp.

CENELEC PRETS 300 180-92. Integrated Services Digital Network (ISDN); Advice of Charge: Charging Information at the End of the Call (AOC-E) Supplementary Service Service Description. 16 pp.

CENELEC ETS 300 180-92. Integrated Services Digital Network (ISDN); Advice of Charge: Charging Information at the End of the Call (AOC-E) Supplementary Service Service Description. 16 pp.

Integrated Services Digital Networks *(Cont.)*

—Advice of Charge Services *(Cont.)*

CENELEC PRETS 300 181-91. Integrated Services Digital Network (ISDN); Advice of Charge (AOC) Supplementary Service Functional Capabilities and Information Flows (T/S 22-04). 46 pp.

CENELEC PRETS 300 181-92. Integrated Services Digital Network (ISDN); Advice of Charge (AOC) Supplementary Service Functional Capabilities and Information Flows. 46 pp.

CENELEC ETS 300 181-93. Integrated Services Digital Network (ISDN); Advice of Charge (AOC) Supplementary Service Functional Capabilities and Information Flows. 46 pp.

CENELEC ETS 300 182-93. Integrated Services Digital Network (ISDN); Advice of Charge (AOC) Supplementary Service Digital Subscriber Signalling System No. One (DSS1) Protocol. 41 pp.

CEPT T/S 22-04 E-88. Advice of Charge Service. 24 pp.

ETSI ETS 300 178-92. Integrated Services Digital Network (ISDN); Advice of Charge: Charging Information at Call Set-up Time (AOC-S) Supplementary Service Service Description. 18 pp.

ETSI PRETS 300 178-91. Integrated Services Digital Network (ISDN); Advice of Charge: Charging Information at Call Set-up Time (AOC-S) Supplementary Service Service Description (T/NA1(89)13). 18 pp.

ETSI ETS 300 179-92. Integrated Services Digital Network (ISDN); Advice of Charge: Charging Information During the Call (AOC-D) Supplementary Service Service Description. 16 pp.

ETSI PRETS 300 179-91. Integrated Services Digital Network (ISDN); Advice of Charge: Charging Information During the Call (AOC-D) Supplementary Service Service Description (T/NA1(89)14). 16 pp.

ETSI ETS 300 180-92. Integrated Services Digital Network (ISDN); Advice of Charge: Charging Information at the End of the Call (AOC-E) Supplementary Service Service Description. 16 pp.

ETSI PRETS 300 180-91. Integrated Services Digital Network (ISDN); Advice of Charge: Charging Information at the End of the Call (AOC-E) Supplementary Service Service Description (T/NA1(89)15). 16 pp.

ETSI ETS 300 181-93. Integrated Services Digital Network (ISDN); Advice of Charge (AOC) Supplementary Service Functional Capabilities and Information Flows. 46 pp.

ETSI PRETS 300 181-92. Integrated Services Digital Network (ISDN); Advice of Charge (AOC) Supplementary Service Functional Capabilities and Information Flows. 46 pp.

ETSI PRETS 300 181-91. Integrated Services Digital Network (ISDN); Advice of Charge (AOC) Supplementary Service Functional Capabilities and Information Flows (T/S 22-04). 46 pp.

—Advice of Charge Services—DSS1

CENELEC PRETS 300 182-91. Integrated Services Digital Network (ISDN); Advice of Charge (AOC) Supplementary Service Digital Subscriber Signalling System No. One (DSS1) Protocol (T/S 46-33K). 37 pp.

CENELEC PRETS 300 182-92. Integrated Services Digital Network (ISDN); Advice of Charge (AOC) Supplementary Service Digital Subscriber Signalling System No. One (DSS1) Protocol. 41 pp.

CENELEC PRETS 300 182-92. Integrated Services Digital Network (ISDN); Advice of Charge (AOC) Supplementary Service Digital Subscriber Signalling System No. One (DSS1) Protocol. 41 pp.

ETSI ETS 300 182-93. Integrated Services Digital Network (ISDN); Advice of Charge (AOC) Supplementary Service Digital Subscriber Signalling System No. One (DSS1) Protocol. 41 pp.

ETSI PRETS 300 182-92. Integrated Services Digital Network (ISDN); Advice of Charge (AOC) Supplementary Service Digital Subscriber Signalling System No. One (DSS1) Protocol. 41 pp.

ETSI PRETS 300 182-91. Integrated Services Digital Network (ISDN); Advice of Charge (AOC) Supplementary Service Digital Subscriber Signalling System No. One (DSS1) Protocol (T/S 46-33K). 37 pp.

—Advice of Charge Services—Functions/ Information Flows

CCITT RECMN Q.86-89. Charging Supplementary Services—General Recommendations on Telephone Switching and Signalling—Functions and Information Flows for Services in the ISDN—Supplements (Study Group XI) 22 pp. 22 pp.

—Architectural Model

CCITT RECMN I.325-89. Reference Configurations for ISDN Connection Types—Integrated Services Digital Network (ISDN)—Overall Network Aspects and Functions, ISDN User-Network Interfaces (Study Group XVIII) 7 pp. 7 pp.

Integrated Services Digital Networks *(Cont.)*

—Audiographic Conferencing Services

CENELEC PRETS 300 101-90. Integrated Services Digital Network (ISDN) International Digital Audiographic Teleconference (T/N 33-01). 185 pp.

CENELEC PRI-ETS 300 101-92. Integrated Services Digital Network (ISDN); International Digital Audiographic Teleconference. 179 pp.

CENELEC I-ETS 300 101-93. Integrated Services Digital Network (ISDN); International Digital Audiographic Teleconference. 179 pp.

ETSI PRETS 300 101-90. Integrated Services Digital Network (ISDN) International Digital Audiographic Teleconference (T/N 33-01). 181 pp.

ETSI I-ETS 300 101-93. Integrated Services Digital Network (ISDN); International Digital Audiographic Teleconference. 179 pp.

ETSI PRI-ETS 300 101-92. Integrated Services Digital Network (ISDN); International Digital Audiographic Teleconference. 179 pp.

ETSI PRETS 300 101-90. Integrated Services Digital Network (ISDN) International Digital Audiographic Teleconference (T/N 33-01). 185 pp.

—Audiovisual Equipment

CENELEC PRETS 300 143-91. Integrated Services Digital Network (ISDN) and Other Digital Telecommunications Networks Audiovisual Teleservices System for Establishing Communication Between Audiovisual Terminals Using Digital Channels up to 2048 Kbit/s. 10 pp.

CENELEC PRETS 300 143-93. Integrated Services Digital Network (ISDN); Audiovisual Services Inband Signalling Procedures for Audiovisual Terminals Using Digital Channels up to 2 048 kbit/s. 27 pp.

CENELEC PRETS 300 145-91. Integrated Services Digital Network (ISDN) and Other Digital Telecommunications Networks Audiovisual Teleservices Narrowband Visual Telephone Systems (T/N 32-05). 28 pp.

CENELEC PRETS 300 145-93. Integrated Services Digital Network (ISDN); Audiovisual Services Videotelephone Systems and Terminal Equipment Operating on One or Two 64 kbit/s Channels. 24 pp.

ETSI PRETS 300 143-91. Integrated Services Digital Network (ISDN) and Other Digital Telecommunications Networks—Audiovisual Teleservices—System for Establishing Communication Between Audiovisual Terminals Using Digital Channels up to 2048 kbit/s (T/N 32-03).

ETSI PRETS 300 145-91. Integrated Services Digital Network (ISDN) and Other Digital Telecommunications Networks—Audiovisual Teleservices—Narrowband Visual Telephone Systems (T/N 32-05).

ETSI PRETS 300 143-91. Integrated Services Digital Network (ISDN) and Other Digital Telecommunications Networks Audiovisual Teleservices System for Establishing Communication Between Audiovisual Terminals Using Digital Channels up to 2048 Kbit/s. 10 pp.

ETSI PRETS 300 143-93.

ETSI PRETS 300 145-93.

—Audiovisual Equipment—Frame Structure

CENELEC PRETS 300 144-91. Integrated Services Digital Network (ISDN) and Other Digital Telecommunications Networks Audiovisual Teleservices Frame Structure for a 64 to 1920 Kbit/s Channel (T/N 32-04). 8 pp.

CENELEC PRETS 300 144-93. Integrated Services Digital Network (ISDN); Audiovisual Services Frame Structure for a 64 kbit/s to 1 920 kbit/s Channel and Associated Syntax for Inband Signalling. 51 pp.

CENELEC PRETS 300 146-91. Integrated Services Digital Network (ISDN) and Other Digital Telecommunications Networks Audiovisual Teleservices Frame Synchronous Control and Indication Signals for Audiovisual Systems (T/N 32-06). 9 pp.

ETSI PRETS 300 144-91. Integrated Services Digital Network (ISDN) and Other Digital Telecommunications Networks—Audiovisual Teleservices—Frame Structure for a 64 to 1920 kbit/s Channel (T/N 32-04).

ETSI PRETS 300 146-91. Integrated Services Digital Network (ISDN) and Other Digital Telecommunications Networks—Audiovisual Teleservices—Frame Synchronous Control and Indication Signals for Audiovisual Systems (T/N 32-06).

ETSI PRETS 300 144-91. Integrated Services Digital Network (ISDN) and Other Digital Telecommunications Networks Audiovisual Teleservices Frame Structure for a 64 to 1920 Kbit/s Channel (T/N 32-04). 8 pp.

ETSI PRETS 300 144-93. Integrated Services. 51 pp.

INDUSTRY STANDARDS
1403

Integrated
INTERNATIONAL AND NON-U.S. NATIONAL STANDARDS
SUBJECT INDEX

Integrated Services Digital Networks *(Cont.)*

—Audiovisual Equipment—Frame Structure *(Cont.)*

ETSI PRETS 300 145-91. Integrated Services Digital Network (ISDN) and Other Digital Telecommunications Networks Audiovisual Teleservices Narrowband Visual Telephone Systems (T/N 32-05). 28 pp.
ETSI PRETS 300 146-91. Integrated Services Digital Network (ISDN) and Other Digital Telecommunications Networks Audiovisual Teleservices Frame Synchronous Control and Indication Signals for Audiovisual Systems (T/N 32-06). 9 pp.

—Audiovisual Equipment—Video Coding

CENELEC PRETS 300 142-91. Integrated Services Digital Network (ISDN) and Other Digital Telecommunications Networks Audiovisual Teleservices Video Codec for Audiovisual Services at p* 64 Kbit/s (T/N 31-04). 6 pp.
ETSI PRETS 300 142-91. Integrated Services Digital Network (ISDN) and Other Digital Telecommunications Networks—Audiovisual Teleservices—Video Codec for Audiovisual Services at p* 64 kbit/s (T/N 31-04).
ETSI PRETS 300 142-91. Integrated Services Digital Network (ISDN) and Other Digital Telecommunications Networks Audiovisual Teleservices Video Codec for Audiovisual Services at p* 64 Kbit/s (T/N 31-04). 6 pp.

—Basic Access—Maintenance

CCITT RECMN I.603-89. Application of Maintenance Principles to ISDN Basic Accesses—Integrated Services Digital Network (ISDN)—Internetwork Interfaces and Maintenance Principles (Study Group XVIII) 6 pp. 6 pp.
CCITT RECMN I.605-89. Application of Maintenance Principles to Static Multiplexed ISDN Basic Accesses—Integrated Services Digital Network (ISDN)—Internetwork Interfaces and Maintenance Principles (Study Group XVIII) 4 pp. 4 pp.

—Basic Rate Access—NATO Reference Model

NATO STANAG 4460 ED 1 AMD 0-00. (Draft) NATO Reference Model for Open System Interconnection/Integrated Services Digital Network (ISDN) Layer 1 (Physical Layer) Specification for ISDN Basic Rate Access at the S/T Reference Point. 7 pp.
NATO STANAG 4462 ED 1 AMD 0-00. (Draft) NATO Reference Model for Open Systems Interconnection/Integrated Services Digital Network (ISDN) Layer 2 (Data Link Layer) Specification for ISDN Basic and Primary Rate Access at the S/T Reference Point. 17 pp.

—Bearer Services

CCITT RECMN D.210-89. General Charging and Accounting Principles for International Telecommunication Services Provided over the Integrated Services Digital Network (ISDN)—General Tariff Principles—Charging and Accounting in Intl Telecom. Serv. (Study Group III) 4 pp. 4 pp.
CCITT RECMN E.711-92. User Demand Modelling Study Group II) 11 pp. 11 pp.
CCITT RECMN I.230-89. Definition of Bearer Service Categories—Integrated Services Digital Network (ISDN)—General Structure and Service Capabilities (Study Group XVIII) 5 pp. 5 pp.
CCITT RECMN I.241-89. Teleservices Supported by an ISDN—Integrated Services Digital Network (ISDN)—General Structure and Service Capabilities (Study Group XVIII) 28 pp. 28 pp.
CCITT RECMN I.320-89. ISDN Protocol Reference Model—Integrated Services Digital Network (ISDN)—Overall Network Aspects and Functions, ISDN User-Network Interfaces (Study Group XVIII) 10 pp. 10 pp.
CENELEC PRETS 300 083-90. Integrated Services Digital Network (ISDN): Circuit Mode Structural Bearer Service Category Usable for Speech Information Transfer End-to-End Compatibility (T/TE 12-07). 7 pp.
CENELEC PRETS 300 083-92. Integrated Services Digital Network (ISDN); Circuit Mode Structured Bearer Service Category Usable for Speech Information Transfer Terminal Requirements Necessary for End-to-End Compatibility. 9 pp.
CENELEC ETS 300 083-93. Integrated Services Digital Network (ISDN); Circuit Mode Structured Bearer Service Category Usable for Speech Information Transfer Terminal Requirements Necessary for End-to-End Compatibility. 9 pp.
CENELEC PRETS 300 084-90. Integrated Services Digital Network (ISDN): Circuit Mode Structural Bearer Service Category Usable for 3.1 KHz Audio Information Transfer End-to-End Compatibility (T/TE 12-08). 7 pp.

Integrated Services Digital Networks *(Cont.)*

—Bearer Services *(Cont.)*

CENELEC PRETS 300 084-92. Integrated Services Digital Network (ISDN); Circuit Mode Structured Bearer Service Category Usable for 3,1 kHz Audio Information Transfer Terminal Requirements Necessary for End-to-End Compatibility. 9 pp.
CENELEC ETS 300 084-93. Integrated Services Digital Network (ISDN); Circuit Mode Structured Bearer Service Category Usable for 3,1 kHz Audio Information Transfer Terminal Requirements Necessary for End-to-End Compatibility. 9 pp.
CENELEC PRETS 300 108-90. Integrated Services Digital Network (ISDN) Circuit-Mode 64Kbit/s Unrestricted SKHz Structured Bearer Service Category (T/NA1 (89)35) Service Description. 6 pp.
CENELEC PRETS 300 108-92. Integrated Services Digital Network (ISDN); Circuit-Mode 64 kbit/s Unrestricted 8 kHz Structured Bearer Service Category Service Description. 15 pp.
CENELEC ETS 300 108-92. Integrated Services Digital Network (ISDN); Circuit-Mode 64 kbit/s Unrestricted 8 kHz Structured Bearer Service Category Service Description. 14 pp.
CENELEC PRETS 300 109-90. Integrated Services Digital Network (ISDN) Circuit-Mode 64Kbit/s 8KHz Structured Bearer Service Category (T/NA1 (89)36) Service Description. 6 pp.
CENELEC PRETS 300 109-92. Integrated Services Digital Network (ISDN); Circuit-Mode 64 kbit/s 8 kHz Structured Bearer Service Category Usable for Speech Information Transfer Service Description. 15 pp.
CENELEC ETS 300 109-92. Integrated Services Digital Network (ISDN); Circuit-Mode 64 kbit/s 8 kHz Structured Bearer Service Category Usable for Speech Information Transfer Service Description. 13 pp.
CENELEC PRETS 300 110-90. Integrated Services Digital Network (ISDN) Circuit-Mode 64 Kbit/s 8 KHz Structured Bearer Service Category (T/NA1 (89)37) Usable for 3.1 KHz Audio Information Transfer Service Description. 7 pp.
CENELEC PRETS 300 110-92. Integrated Services Digital Network (ISDN); Circuit-Mode 64 kbit/s 8 kHz Structured Bearer Service Category Usable for 3,1 kHz Audio Information Transfer Service Description. 15 pp.
CENELEC ETS 300 110-92. Integrated Services Digital Network (ISDN); Circuit-Mode 64 kbit/s 8 kHz Structured Bearer Service Category Usable for 3,1 kHz Audio Information Transfer Service Description. 14 pp.
CEPT T/CAC S 10.2 E-92. Bearer Services to Be Provided by an Integrated Services Digital Network (ISDN) (Revised in Odense 1986, Vienna 1989 (CAC) and Athens 1992). 70 pp.
CEPT T/CAC S 10.4 E-92. Intercommunication Aspects of an Integrated Services Digital Network (ISDN) (Revised in Odense 1986, Vienna 1989 (CAC) and Athens 1992). 11 pp.
CEPT T/GSI 02-02-87. Services Supports. 16 pp.
CEPT T/GSI 02-02 E-87. Bearer Services. 16 pp.
CEPT T/GSI 02-07-87. Services Supplementaires Associes Aux Services Supports. 5 pp.
CEPT T/GSI 02-07 E-87. Supplementary Services Associated with Bearer Services. 5 pp.
CEPT T/N 13-04 E-89. Supplementary Services Associated with Bearer Services. 5 pp.
CEPT T/SF 31-02 E-89. Bearer Services to Be Provided by Integrated Services Digital Network (ISDN) (Revised in Odense 1986 and Vienna 1989 (CAC)). 38 pp.
CEPT T/SF 69 E-91. International Switched 2 Mbit/s Bearer Service. 4 pp.
CSA CAN/CSA-T543-91. Integrated Services Digital Network (ISDN)—Minimal Set of Bearer Services for the Primary Rate Interface (ANSI T1.603-1990); (Gen Instr 1). 30 pp.
CSA CAN/CSA-T544-91. Integrated Services Digital Network (ISDN)—Minimal Set of Bearer Services for the Basic Rate Interface (ANSI T1.604-1990); (Gen Instr 1). 34 pp.
ETSI PRETS 300 083-90. Integrated Services Digital Network (ISDN); Circuit Mode Structured Bearer Service Category Usable for Speech Information Transfer—End-To-End Compatibility (T/TE 12-07). 7 pp.
ETSI PRETS 300 084-90. Integrated Services Digital Network (ISDN); Circuit Mode Structured Bearer Service Category Usable for 3.1 KHz Audio Information Transfer—End-To-End Compatibility (T/TE 12-08). 7 pp.
ETSI PRETS 300 108-90. Integrated Services Digital Network (ISDN); Circuit-Mode 64 Kbit/s Unrestricted 8kHz Structed Bearer Service Category (T/NA1 (89)35) Service Description. 10 pp.
ETSI PRETS 300 109-90. Integrated Services Digital Network (ISDN) Circuit-Mode 64 Kbit/s 8 KHz Structured Bearer Service Category (T/NA1 (89)36) Service Description. 10 pp.

Integrated Services Digital Networks *(Cont.)*

—Bearer Services *(Cont.)*

ETSI PRETS 300 110-90. Integrated Services Digital Network (ISDN) Circuit-Mode 64 Kbit/s 8 KHz Structured Bearer Service Category (T/NAI (89)37) Usable for 3.1 KHz Audio Information Transfer—Service Description. 10 pp.
ETSI ETS 300 083-93. Integrated Services Digital Network (ISDN); Circuit Mode Structured Bearer Service Category Usable for Speech Information Transfer Terminal Requirements Necessary for End-to-End Compatibility. 9 pp.
ETSI PRETS 300 083-92. Integrated Services Digital Network (ISDN); Circuit Mode Structured Bearer Service Category Usable for Speech Information Transfer Terminal Requirements Necessary for End-to-End Compatibility. 9 pp.
ETSI PRETS 300 083-90. Integrated Services Digital Network (ISDN): Circuit Mode Structural Bearer Service Category Usable for Speech Information Transfer End-to-End Compatibility (T/TE 12-07). 7 pp.
ETSI ETS 300 084-93. Integrated Services Digital Network (ISDN); Circuit Mode Structured Bearer Service Category Usable for 3,1 kHz Audio Information Transfer Terminal Requirements Necessary for End-to-End Compatibility. 9 pp.
ETSI PRETS 300 084-92. Integrated Services Digital Network (ISDN); Circuit Mode Structured Bearer Service Category Usable for 3,1 kHz Audio Information Transfer Terminal Requirements Necessary for End-to-End Compatibility. 9 pp.
ETSI PRETS 300 084-90. Integrated Services Digital Network (ISDN): Circuit Mode Structural Bearer Service Category Usable for 3.1 KHz Audio Information Transfer End-to-End Compatibility (T/TE 12-08). 7 pp.
ETSI PRETS 300 108-92. Integrated Services Digital Network (ISDN); Circuit-Mode 64 kbit/s Unrestricted 8 kHz Structured Bearer Service Category Service Description. 15 pp.
ETSI PRETS 300 108-90. Integrated Services Digital Network (ISDN) Circuit-Mode 64Kbit/s Unrestricted SKHz Structured Bearer Service Category (T/NA1 (89)35) Service Description. 6 pp.
ETSI ETS 300 109-92. Integrated Services Digital Network (ISDN); Circuit-Mode 64 kbit/s 8 kHz Structured Bearer Service Category Usable for Speech Information Transfer Service Description. 13 pp.
ETSI PRETS 300 109-92. Integrated Services Digital Network (ISDN); Circuit-Mode 64 kbit/s 8 kHz Structured Bearer Service Category Usable for Speech Information Transfer Service Description. 15 pp.
ETSI PRETS 300 109-90. Integrated Services Digital Network (ISDN) Circuit-Mode 64Kbit/s 8KHz Structured Bearer Service Category (T/NA1 (89)36) Service Description. 6 pp.
ETSI ETS 300 110-92. Integrated Services Digital Network (ISDN); Circuit-Mode 64 kbit/s 8 kHz Structured Bearer Service Category Usable for 3,1 kHz Audio Information Transfer Service Description. 14 pp.
ETSI PRETS 300 110-92. Integrated Services Digital Network (ISDN); Circuit-Mode 64 kbit/s 8 kHz Structured Bearer Service Category Usable for 3,1 kHz Audio Information Transfer Service Description. 15 pp.
ETSI PRETS 300 110-90. Integrated Services Digital Network (ISDN) Circuit-Mode 64 Kbit/s 8 KHz Structured Bearer Service Category (T/NA1 (89)37) Usable for 3.1 KHz Audio Information Transfer Service Description. 7 pp.

—Bearer Services—Classifications

CCITT RECMN I.210-89. Principles of Telecommunication Services Supported by an ISDN and the Means to Describe Them—Integrated Services Digital Network (ISDN)—General Structure and Service Capabilities (Study Group XVIII) 19 pp. 19 pp.

—Bearer Services—Congestion Management

CCITT RECMN I.370-91. Congestion Management for the ISDN Frame Relaying Bearer Service (Study Group XVIII) 13 pp. 13 pp.

—Bearer Services—Functions/Information Flows

CCITT RECMN Q.71-89. ISDN 64 kBit/s Circuit Mode Switched Bearer Services—General Recommendations on Telephone Switching and Signalling—Functions and Information Flows for Services in the ISDN—Supplements (Study Group XI) 44 pp. 44 pp.

—Bearer Services—NATO Reference Model

NATO STANAG 4459 ED 1 AMD 0-00. (Draft) NATO Reference Model for Open Systems Interconnection/Integrated Services Digital Networks—Bearer Services. 8 pp.

1404
INDEX and DIRECTORY of

INTERNATIONAL AND NON-U.S. NATIONAL STANDARDS
SUBJECT INDEX
Integrated

Integrated Services Digital Networks *(Cont.)*

—**Broadband Aspects**

CCIR Report 1139-90. General System and Performance Aspects of Digital Transmission in the Fixed-Satellite Service—Section 4B1—Systems Aspects. 19 pp.

CCITT RECMN I.121-91. Broadband Aspects of ISDN (Study Group XVIII) 6 pp. 6 pp.

CCITT RECMN I.121-89. Broadband Aspects of ISDN—Integrated Services Digital Network (ISDN)—General Structure and Service Capabilities (Study Group XVIII) 28 pp. 28 pp.

CCITT RECMN I.211-91. B-ISDN Service Aspects—(Study Group XVIII) 20 pp. 20 pp.

CCITT RECMN I.311-91. B-ISDN General Network Aspects (Study Group XVIII) 26 pp. 26 pp.

CCITT RECMN I.363-91. B-ISDN ATM Adaptation Layer (AAL) Specification (Study Group XVIII) 21 pp. 21 pp.

CENELEC ETR 073-93. Broadband Integrated Services Digital Network (B-ISDN); Evolution Towards B-ISDN. 43 pp.

ETSI ETR 073-93. Broadband Integrated Services Digital Network (B-ISDN); Evolution Towards B-ISDN. 43 pp.

—**Broadband Aspects—Asynchronous Transfer Mode**

CCITT RECMN I.150-91. B-ISDN Asynchronous Transfer Mode Functional Characteristics (Study Group XVIII) 12 pp. 12 pp.

CENELEC PRETS 300 298-1-93. Network Aspects (NA); Basic Characteristics and Functional Specification of Asynchronous Transfer Mode (ATM) Part 1: B-ISDN ATM Functional Specification. 15 pp.

CENELEC PRETS 300 298-2-93. Network Aspects (NA); Basic Characteristics and Functional Specification of Asynchronous Transfer Mode (ATM) Part 2: B-ISDN ATM Layer Specification. 16 pp.

ETSI PRETS 300 298-1-93. Network Aspects (NA); Basic Characteristics and Functional Specification of Asynchronous Transfer Mode (ATM) Part 1: B-ISDN ATM Functional Specification. 15 pp.

ETSI PRETS 300 298-2-93. Network Aspects (NA); Basic Characteristics and Functional Specification of Asynchronous Transfer Mode (ATM) Part 2: B-ISDN ATM Layer Specification. 16 pp.

—**Broadband Aspects—ATM Adaptation Layer**

CCITT RECMN I.362-91. B-ISDN ATM Adaptation Layer (AAL) Functional Description (Study Group XVIII) 7 pp. 7 pp.

CCITT RECMN I.363-91. B-ISDN ATM Adaptation Layer (AAL) Specification (Study Group XVIII) 21 pp. 21 pp.

—**Broadband Aspects—ATM Layer**

CCITT RECMN I.361-91. B-ISDN ATM Layer Specification (Study Group XVIII) 10 pp. 10 pp.

CENELEC PRETS 300 298-2-93. Network Aspects (NA); Basic Characteristics and Functional Specification of Asynchronous Transfer Mode (ATM) Part 2: B-ISDN ATM Layer Specification. 16 pp.

ETSI PRETS 300 298-2-93. Network Aspects (NA); Basic Characteristics and Functional Specification of Asynchronous Transfer Mode (ATM) Part 2: B-ISDN ATM Layer Specification. 16 pp.

—**Broadband Aspects—ATM Layer—Maintenance**

CCITT RECMN I.610-91. OAM Principles of the B-ISDN Access (Study Group XVIII) 18 pp. 18 pp.

—**Broadband Aspects—ATM Layer—Television Transmission**

CCIR Report 1240-90. Television Transmission in an ATM-Based Network—Section CMTT A—Television Transmission Standards and Performance Objectives. 5 pp.

CCIR QUESTION 38/CMTT-90. Service Requirements for Long-Distance Digital Television Connections on the Integrated Services Digital Network (ISDN)—Questions Concerning the CMTT CCIR/CCITT Joint Study Group for Television and Sound Transmission. 1 p.

—**Broadband Aspects—Classification**

CCITT RECMN I.211-91. B-ISDN Service Aspects—(Study Group XVIII) 20 pp. 20 pp.

—**Broadband Aspects—Congestion Management**

CENELEC PRETS 300 301-93. Network Aspects (NA); Traffic Control and Congestion Control in Broadband Integrated Services Digital Network (B-ISDN). 31 pp.

Integrated Services Digital Networks *(Cont.)*

—**Broadband Aspects—Congestion Management** *(Cont.)*

ETSI PRETS 300 301-93. Network Aspects (NA); Traffic Control and Congestion Control in Broadband Integrated Services Digital Network (B-ISDN). 31 pp.

—**Broadband Aspects—Connection Oriented Bearer Services**

CCITT RECMN F.811-92. Broadband Connection-Oriented Bearer Service (Study Group I) 12 pp. 12 pp.

—**Broadband Aspects—Connectionless Data Bearer Service**

CCITT RECMN F.812-92. Broadband Connectionless Data Bearer Service (Study Group I) 8 pp. 8 pp.

—**Broadband Aspects—Glossaries**

CCITT RECMN I.113-91. Vocabulary of Terms for Broadband Aspects of ISDN (Study Group XVIII) 22 pp. 22 pp.

CCITT RECMN I.113-89. Vocabulary of Terms for Broadband Aspects of ISDN—Integrated Services Digital Network (ISDN)—General Structure and Service Capabilities (Study Group XVIII) 10 pp. 10 pp.

—**Broadband Aspects—Maintenance**

CCITT RECMN I.610-91. OAM Principles of the B-ISDN Access (Study Group XVIII) 18 pp. 18 pp.

—**Broadband Aspects—Network Architecture**

CCITT RECMN I.327-91. B-ISDN Functional Architecture (Study Group XVIII) 15 pp. 15 pp.

—**Broadband Aspects—Network Layer**

CCITT RECMN I.311-91. B-ISDN General Network Aspects (Study Group XVIII) 26 pp. 26 pp.

—**Broadband Aspects—Network Performance**

CCITT RECMN I.211-91. B-ISDN Service Aspects—(Study Group XVIII) 20 pp. 20 pp.

—**Broadband Aspects—Operation**

CCITT RECMN I.610-91. OAM Principles of the B-ISDN Access (Study Group XVIII) 18 pp. 18 pp.

—**Broadband Aspects—Physical Layer**

CCITT RECMN I.413-91. B-ISDN User-Network Interface (Study Group XVIII) 13 pp. 13 pp.

—**Broadband Aspects—Physical Layer—Maintenance**

CCITT RECMN I.610-91. OAM Principles of the B-ISDN Access (Study Group XVIII) 18 pp. 18 pp.

—**Broadband Aspects—Protocol Reference Model**

CCITT RECMN I.321-91. B-ISDN Protocol Reference Model and Its Application (Study Group XVIII) 11 pp. 11 pp.

CENELEC ETR 054-92. Integrated Services Digital Network (ISDN); the Protocol Reference Model and Its Relationship with the OSI Reference Model. 13 pp.

ETSI ETR 054-92. Integrated Services Digital Network (ISDN); the Protocol Reference Model and Its Relationship with the OSI Reference Model. 13 pp.

—**Broadband Aspects—Quality of Service**

CCITT RECMN I.211-91. B-ISDN Service Aspects—(Study Group XVIII) 20 pp. 20 pp.

—**Broadband Aspects—Reference Configurations**

CENELEC ETR 072-93. Broadband Integrated Services Digital Network (B-ISDN); Connection Types and Their Reference Configurations. 18 pp.

ETSI ETR 072-93. Broadband Integrated Services Digital Network (B-ISDN); Connection Types and Their Reference Configurations. 18 pp.

—**Broadband Aspects—Resources Management**

CCITT RECMN I.311-91. B-ISDN General Network Aspects (Study Group XVIII) 26 pp. 26 pp.

—**Broadband Aspects—Signaling Principles**

CCITT RECMN I.311-91. B-ISDN General Network Aspects (Study Group XVIII) 26 pp. 26 pp.

—**Broadband Aspects—Sound-Program Signals**

CCIR Decision 68-2-89. Television and Sound-Programme Signals in the Broadband ISDN—Annex to Volume XII—Television and Sound Transmission (CMTT). 2 pp.

Integrated Services Digital Networks *(Cont.)*

—**Broadband Aspects—Telecommunication Traffic**

CCITT RECMN I.311-91. B-ISDN General Network Aspects (Study Group XVIII) 26 pp. 26 pp.

—**Broadband Aspects—Telecommunication Traffic—Management**

CENELEC PRETS 300 301-93. Network Aspects (NA); Traffic Control and Congestion Control in Broadband Integrated Services Digital Network (B-ISDN). 31 pp.

ETSI PRETS 300 301-93. Network Aspects (NA); Traffic Control and Congestion Control in Broadband Integrated Services Digital Network (B-ISDN). 31 pp.

—**Broadband Aspects—Television Signals**

CCIR Decision 68-2-89. Television and Sound-Programme Signals in the Broadband ISDN—Annex to Volume XII—Television and Sound Transmission (CMTT). 2 pp.

—**Broadband Aspects—Television Transmission**

CCIR Report 1241-90. Liaison Between the CMTT and Other Study Groups of the CCITT on Matters Related to Digital Transmission—Section CMTT A—Television Transmission Standards and Performance Objectives. 17 pp.

CCIR QUESTION 38/CMTT-90. Service Requirements for Long-Distance Digital Television Connections on the Integrated Services Digital Network (ISDN)—Questions Concerning the CMTT CCIR/CCITT Joint Study Group for Television and Sound Transmission. 1 p.

—**Broadband Aspects—Television Transmission—Bit Rates**

CCIR QUESTION 39/CMTT-90. Requirements for Channel Capacities Above the H4 Rate for High-Bit-Rate Television Services—Questions Concerning the CMTT CCIR/CCITT Joint Study Group for Television and Sound Transmission. 1 p.

—**Broadband Aspects—Television Transmission—Secondary Distribution**

CCIR Report 1239-90. Standards for Digital Secondary Distribution Systems—Section CMTT A—Television Transmission Standards and Performance Objectives. 4 pp.

—**Broadband Aspects—User Network Interfaces**

CCITT RECMN I.413-91. B-ISDN User-Network Interface (Study Group XVIII) 13 pp. 13 pp.

CCITT RECMN I.432-91. B-ISDN User-Network Interface—Physical Layer Specification (Study Group XVIII) 20 pp. 20 pp.

CENELEC PRETS 300 299-93. Network Aspects (NA); Cell Based User Network Access Physical Layer Interfaces for B-ISDN Applications. 50 pp.

CENELEC PRETS 300 300-93. Network Aspects (NA); SDH Based User Network Access Physical Layer Interfaces for B-ISDN Applications. 47 pp.

CENELEC ETR 072-93. Broadband Integrated Services Digital Network (B-ISDN); Connection Types and Their Reference Configurations. 18 pp.

ETSI PRETS 300 299-93. Network Aspects (NA); Cell Based User Network Access Physical Layer Interfaces for B-ISDN Applications. 50 pp.

ETSI PRETS 300 300-93. Network Aspects (NA); SDH Based User Network Access Physical Layer Interfaces for B-ISDN Applications. 47 pp.

ETSI ETR 072-93. Broadband Integrated Services Digital Network (B-ISDN); Connection Types and Their Reference Configurations. 18 pp.

—**Broadband Aspects—User Network Interfaces—Maintenance**

CCITT RECMN I.610-91. OAM Principles of the B-ISDN Access (Study Group XVIII) 18 pp. 18 pp.

—**Broadband Aspects—User Network Interfaces—Operation**

CCITT RECMN I.610-91. OAM Principles of the B-ISDN Access (Study Group XVIII) 18 pp. 18 pp.

—**Bureaufax Communications**

CCITT RECMN F.170-92. Operational Provisions for the International Public Facsimile Service Between Public Bureaux (Bureaufax) (Study Group I) 9 pp. 9 pp.

CCITT RECMN F.170-89. Operational Provisions for the International Public Facsimile Service Between Public Bureaux (Bureaufax)—Telematic, Data Transmission and Teleconference Services—Operations and Quality of Service—(Study Group I) 8 pp. 8 pp.

INDUSTRY STANDARDS

Integrated Services Digital Networks (Cont.)

—Busy Tones
CCITT RECMN I.221-89. Common Specific Characteristics of Services—Integrated Services Digital Network (ISDN)—General Structure and Service Capabilities (Study Group XVIII) 3 pp. 3 pp.

—Call Barring
CCITT RECMN I.255. 5-92. Outgoing Call Barring (Study Group I) 9 pp. 9 pp.

—Call Forwarding
CCITT RECMN I.252-89. Call Offering Supplementary Services—Integrated Services Digital Network (ISDN)—General Structure and Service Capabilities (Study Group XVIII) 37 pp. 37 pp.

CCITT RECMN I.252. 5-92. Call Deflection (Study Group I) 12 pp. 12 pp.

CENELEC PRETS 300 202-92. Integrated Services Digital Network (ISDN); Call Deflection (CD) Supplementary Service Service Description. 17 pp.

CENELEC PRETS 300 206-92. Integrated Services Digital Network (ISDN); Call Deflection (CD) Supplementary Service Functional Capabilities and Information Flows. 39 pp.

ETSI PRETS 300 202-92. Integrated Services Digital Network (ISDN); Call Deflection (CD) Supplementary Service Service Description. 17 pp.

ETSI PRETS 300 206-92. Integrated Services Digital Network (ISDN); Call Deflection (CD) Supplementary Service Functional Capabilities and Information Flows. 39 pp.

—Call Forwarding Busy
CCITT RECMN I.252. 2-92. Call Forwarding Busy (Study Group I) 17 pp. 17 pp.

CENELEC PRETS 300 199-92. Integrated Services Digital Network (ISDN); Call Forwarding Busy (CFB) Supplementary Service Service Description. 19 pp.

CENELEC PRETS 300 203-92. Integrated Services Digital Network (ISDN); Call Forwarding Busy (CFB) Supplementary Service Functional Capabilities and Information Flows. 45 pp.

ETSI PRETS 300 199-92. Integrated Services Digital Network (ISDN); Call Forwarding Busy (CFB) Supplementary Service Service Description. 19 pp.

ETSI PRETS 300 203-92. Integrated Services Digital Network (ISDN); Call Forwarding Busy (CFB) Supplementary Service Functional Capabilities and Information Flows. 45 pp.

—Call Forwarding—CCITT No. 7 Signaling
CCITT RECMN Q.730-89. ISDN Supplementary Services—Specifications of Signalling System No. 7 (Study Group XI) 59 pp. 59 pp.

—Call Forwarding—Functions/Information Flows
CCITT RECMN Q.82-89. Call Offering Supplementary Services—General Recommendations on Telephone Switching and Signalling—Functions and Information Flows for Services in the ISDN—Supplements (Study Group XI) 25 pp. 25 pp.

—Call Forwarding No Reply
CCITT RECMN I.252. 3-92. Call Forwarding No Reply (Study Group I) 12 pp. 12 pp.

CENELEC PRETS 300 201-92. Integrated Services Digital Network (ISDN); Call Forwarding No Reply (CFNR) Supplementary Service Service Description. 20 pp.

CENELEC PRETS 300 205-92. Integrated Services Digital Network (ISDN); Call Forwarding No Reply (CFNR) Supplementary Service Functional Capabilities and Information Flows. 51 pp.

ETSI PRETS 300 201-92. Integrated Services Digital Network (ISDN); Call Forwarding No Reply (CFNR) Supplementary Service Service Description. 20 pp.

ETSI PRETS 300 205-92. Integrated Services Digital Network (ISDN); Call Forwarding No Reply (CFNR) Supplementary Service Functional Capabilities and Information Flows. 51 pp.

—Call Forwarding Unconditional
CCITT RECMN I.252. 4-92. Call Forwarding Unconditional (Study Group I) 13 pp. 13 pp.

CENELEC PRETS 300 200-92. Integrated Services Digital Network (ISDN); Call Forwarding Unconditional (CFU) Supplementary Service Service Description. 19 pp.

CENELEC PRETS 300 204-92. Integrated Services Digital Network (ISDN); Call Forwarding Unconditional (CFU) Supplementary Service Functional Capabilities and Information Flows. 43 pp.

ETSI PRETS 300 200-92. Integrated Services Digital Network (ISDN); Call Forwarding Unconditional (CFU) Supplementary Service Service Description. 19 pp.

ETSI PRETS 300 204-92. Integrated Services Digital Network (ISDN); Call Forwarding Unconditional (CFU) Supplementary Service Functional Capabilities and Information Flows. 43 pp.

—Call Holding
CCITT RECMN I.253-89. Call Completion Supplementary Services—Integrated Services Digital Network (ISDN)—General Structure and Service Capabilities (Study Group XVIII) 17 pp. 17 pp.

CCITT RECMN I.253. 2-92. Call Hold (Study Group I) 10 pp. 10 pp.

CENELEC PRETS 300 139-90. Integrated Services Digital Network (ISDN) Call Hold (HOLD) Supplementary Service Service Description (T/NA1(89)27). 14 pp.

CENELEC PRETS 300 139-91. Integrated Services Digital Network (ISDN); Call Hold (HOLD) Supplementary Service Service Description. 16 pp.

CENELEC ETS 300 139-92. Integrated Services Digital Network (ISDN); Call Hold (HOLD) Supplementary Service Service Description. 15 pp.

CENELEC PRETS 300 140-90. Integrated Services Digital Network (ISDN); Call Hold (HOLD) Supplementary Service Functional Capabilities and Information Flows (T/S 22-19). 12 pp.

CENELEC PRETS 300 140-91. Integrated Services Digital Network (ISDN); Call Hold (HOLD) Supplementary Service Functional Capabilities and Information Flows. 27 pp.

CENELEC ETS 300 140-92. Integrated Services Digital Network (ISDN); Call Hold (HOLD) Supplementary Service Functional Capabilities and Information Flows. 26 pp.

ETSI PRETS 300 139-90. Integrated Services Digital Network (ISDN) Call Hold (HOLD) Supplementary Service Service Description (T/NA1(89)27). 13 pp.

ETSI PRETS 300 140-90. Integrated Services Digital Network (ISDN); Call Hold (HOLD) Supplementary Service Functional Capabilities and Information Flows (T/S 22-19). 12 pp.

ETSI ETS 300 139-92. Integrated Services Digital Network (ISDN); Call Hold (HOLD) Supplementary Service Service Description. 15 pp.

ETSI PRETS 300 139-91. Integrated Services Digital Network (ISDN); Call Hold (HOLD) Supplementary Service Service Description. 16 pp.

ETSI ETS 300 140-92. Integrated Services Digital Network (ISDN); Call Hold (HOLD) Supplementary Service Functional Capabilities and Information Flows. 26 pp.

ETSI PRETS 300 140-91. Integrated Services Digital Network (ISDN); Call Hold (HOLD) Supplementary Service Functional Capabilities and Information Flows. 27 pp.

—Call Holding—DSS1
CENELEC PRETS 300 141-90. Integrated Services Digital Network (ISDN): Call Hold (HOLD) Supplementary Service Digital Subscriber Signalling One (DSS1) Protocol (T/S 46-33S). 11 pp.

CENELEC PRETS 300 141-91. Integrated Services Digital Network (ISDN); Call Hold (HOLD) Supplementary Service Digital Subscriber Signalling System No. One (DSS1) Protocol. 30 pp.

CENELEC ETS 300 141-92. Integrated Services Digital Network (ISDN); Call Hold (HOLD) Supplementary Service Digital Subscriber Signalling System No. One (DSS1) Protocol. 19 pp.

ETSI PRETS 300 141-90. Integrated Services Digital Network (ISDN); Call Hold (HOLD) Supplementary Service Digital Subscriber Signalling One (DSS1) Protocol (T/S)46-33S). 11 pp.

ETSI ETS 300 141-92. Integrated Services Digital Network (ISDN); Call Hold (HOLD) Supplementary Service Digital Subscriber Signalling System No. One (DSS1) Protocol. 19 pp.

ETSI PRETS 300 141-91. Integrated Services Digital Network (ISDN); Call Hold (HOLD) Supplementary Service Digital Subscriber Signalling System No. One (DSS1) Protocol. 30 pp.

—Call Holding—Functions/Information Flows
CCITT RECMN Q.83-92. Stage 2 Description for Call Completion Supplementary Services Section 2—Call Hold (REV.1) (Study Group XI) 22 pp. 22 pp.

CCITT RECMN Q.83-91. Stage 2 Description for Call Completion Supplementary Services Section 1—Call Waiting (CW) Section 4—Terminal Portability (Study Group XI) 28 pp. 28 pp.

CCITT RECMN Q.83-89. Call Completion Supplementary Services—General Recommendations on Telephone Switching and Signalling—Functions and Information Flows for Services in the ISDN—Supplements (Study Group XI) 33 pp. 33 pp.

—Call Redirection
CCITT RECMN I.252-89. Call Offering Supplementary Services—Integrated Services Digital Network (ISDN)—General Structure and Service Capabilities (Study Group XVIII) 37 pp. 37 pp.

CCITT RECMN I.252. 1-88. Call Transfer—Integrated Services Digital Network (ISDN)—General Structure and Service Capabilities (Study Group XVIII) 9 pp. 9 pp.

—Call Release Time—Grade of Service
CCITT RECMN E.721-91. Network Grade of Service Parameters and Target Values for Circuit-Switched Services in the Evolving ISDN (Study Group II) 8 pp. 8 pp.

CCITT RECMN E.721-89. Network Grade of Service Parameters in ISDN—Telephone Network and ISDN—Quality of Service, Network Management and Traffic Engineering (Study Group II) 2 pp. 2 pp.

—Call Waiting
CCITT RECMN I.253-89. Call Completion Supplementary Services—Integrated Services Digital Network (ISDN)—General Structure and Service Capabilities (Study Group XVIII) 17 pp. 17 pp.

CCITT RECMN I.253. 1-90. Call Waiting (CW) Supplementary Service (Study Group I) 15 pp. 15 pp.

CENELEC PRETS 300 056-90. Integrated Services Digital Network (ISDN); Call Waiting (CW) Supplementary Service Service Description (T/NA(89)18). 14 pp.

CENELEC PRETS 300 056-91. Integrated Services Digital Network (ISDN); Call Waiting (CW) Supplementary Service Service Description. 16 pp.

CENELEC ETS 300 056-91. Integrated Services Digital Network (ISDN); Call Waiting (CW) Supplementary Service Service Description. 16 pp.

CENELEC PRETS 300 057-90. Integrated Services Digital Network (ISDN); Call Waiting (CW) Supplementary Service Functional Capabilities and Information Flows (T/S 22.02). 18 pp.

CENELEC PRETS 300 057-91. Integrated Services Digital Network (ISDN); Call Waiting (CW) Supplementary Service Functional Capabilities and Information Flows. 22 pp.

CENELEC ETS 300 057-92. Integrated Services Digital Network (ISDN); Call Waiting (CW) Supplementary Service Functional Capabilities and Information Flows. 21 pp.

ETSI PRETS 300 056-90. Integrated Services Digital Network (ISDN); Call Waiting (CW) Supplementary Service—Service Description (T/NA1 (89)18). 14 pp.

ETSI PRETS 300 057-90. Integrated Services Digital Network (ISDN); Call Waiting (CW) Supplementary Service Functional Capabilities and Information Flows (T/S 22-2). 18 pp.

ETSI ETS 300 056-91. Integrated Services Digital Network (ISDN); Call Waiting (CW) Supplementary Service Service Description. 16 pp.

ETSI PRETS 300 056-91. Integrated Services Digital Network (ISDN); Call Waiting (CW) Supplementary Service Service Description. 16 pp.

ETSI PRETS 300 056-90. Integrated Services Digital Network (ISDN); Call Waiting (CW) Supplementary Service Service Description (T/NAI(89)18). 14 pp.

ETSI ETS 300 057-92. Integrated Services Digital Network (ISDN); Call Waiting (CW) Supplementary Service Functional Capabilities and Information Flows. 21 pp.

ETSI PRETS 300 057-91. Integrated Services Digital Network (ISDN); Call Waiting (CW) Supplementary Service Functional Capabilities and Information Flows. 22 pp.

ETSI PRETS 300 057-90. Integrated Services Digital Network (ISDN); Call Waiting (CW) Supplementary Service Functional Capabilities and Information Flows (T/S 22.02). 18 pp.

—Call Waiting—CCITT No. 7 Signaling Systems
CCITT RECMN Q.733-92. Stage 3 Description for Call Completion Supplementary Services Using No. 7 Signalling System Section 1—Call Waiting (CW)—Specifications of Signalling System No. 7 (Study Group XI) 10 pp. 10 pp.

—Call Waiting—DSS1
CENELEC PRETS 300 058-90. Integrated Services Digital Network (ISDN); Call Waiting (CW) Supplementary Service Digital Subscriber Signalling One (DSS1) Protocol (T/S 46-33f). 14 pp.

CENELEC PRETS 300 058-91. Integrated Services Digital Network (ISDN); Call Waiting (CW) Supplementary Service Digital Subscriber Signalling System No. One (DSS1) Protocol. 18 pp.

CENELEC ETS 300 058-91. Integrated Services Digital Network (ISDN); Call Waiting (CW) Supplementary Service Digital Subscriber Signalling System No. One (DSS1) Protocol. 17 pp.

INTERNATIONAL AND NON-U.S. NATIONAL STANDARDS
SUBJECT INDEX
Integrated

Integrated Services Digital Networks *(Cont.)*

—Call Waiting—DSS1 *(Cont.)*

ETSI PRETS 300 058-90. Integrated Services Digital Network (ISDN); Call Waiting (CW) Supplementary Service Digital Subscriber Signalling One (DSS1) Protocol (T/S 46-33F). 14 pp.

ETSI ETS 300 058-91. Integrated Services Digital Network (ISDN); Call Waiting (CW) Supplementary Service Digital Subscriber Signalling System No. One (DSS1) Protocol. 17 pp.

ETSI PRETS 300 058-91. Integrated Services Digital Network (ISDN); Call Waiting (CW) Supplementary Service Digital Subscriber Signalling System No. One (DSS1) Protocol. 18 pp.

ETSI PRETS 300 058-90. Integrated Services Digital Network (ISDN); Call Waiting (CW) Supplementary Service Digital Subscriber Signalling One (DSS1) Protocol (T/S 46-33f). 14 pp.

—Call Waiting—Functions/Information Flows

CCITT RECMN Q.83-91. Stage 2 Description for Call Completion Supplementary Services Section 1—Call Waiting (CW) Section 4—Terminal Portability (Study Group XI) 28 pp. 28 pp.

CCITT RECMN Q.83-89. Call Completion Supplementary Services—General Recommendations on Telephone Switching and Signalling—Functions and Information Flows for Services in the ISDN—Supplements (Study Group XI) 33 pp. 33 pp.

—Called Line Identification

CCITT RECMN E.164-91. Numbering Plan for the ISDN Era (Study Group II) 19 pp (Same as Recmn I.331). 19 pp.

CCITT RECMN E.164-89. Numbering Plan for the ISDN Era—Telephone Network and ISDN—Operation, Numbering, Routing and Mobile Service (Study Group II) 6 pp (Same as Recmn I.331). 6 pp.

—Calling Line Identification

CCITT RECMN E.164-91. Numbering Plan for the ISDN Era (Study Group II) 19 pp (Same as Recmn I.331). 19 pp.

CCITT RECMN E.164-89. Numbering Plan for the ISDN Era—Telephone Network and ISDN—Operation, Numbering, Routing and Mobile Service (Study Group II) 6 pp (Same as Recmn I.331). 6 pp.

CEPT T/S 22-01-87. Service Supplementaire Identification. 9 pp.

—Calling Line Identification—Malicious Calls

CCITT RECMN I.251. 7 (REV 1)-92. Malicious Call Identification (Study Group I) 10 pp. 10 pp.

CENELEC PRETS 300 128-90. Integrated Services Digital Network (ISDN) Malicious Call Identification (MCID) Supplementary Service Service Description (T/NA1(89)03). 11 pp.

CENELEC PRETS 300 128-91. Integrated Services Digital Network (ISDN); Malicious Call Identification (MCID) Supplementary Service Service Description (T/NA1(89)03). 15 pp.

CENELEC ETS 300 128-92. Integrated Services Digital Network (ISDN); Malicious Call Identification (MCID) Supplementary Service Service Description. 14 pp.

CENELEC PRETS 300 129-90. Integrated Services Digital Network (ISDN): Malicious Call Identification (MCID) Supplementary Service Functional Capabilities and Information Flows (T/S 22-10). 24 pp.

CENELEC PRETS 300 129-91. Integrated Services Digital Network (ISDN); Malicious Call Identification (MCID) Supplementary Service Functional Capabilities and Information Flows. 24 pp.

CENELEC ETS 300 129-92. Integrated Services Digital Network (ISDN); Malicious Call Identification (MCID) Supplementary Service Functional Capabilities and Information Flows. 23 pp.

CEPT T/CS 21-05-83. Sequence De Traitement D'Appel Pour La Famille De Services Supplementaires: Identification D'Appels Malveillants. 13 pp.

CEPT T/CS 21-05 E-83. Call Handling Sequences for Supplementary Services Family (MCI) Malicious Call Identification. 13 pp.

ETSI PRETS 300 128-90. Intergrated Services Digital Network (ISDN) Malicious Call Identification (MCID) Supplementary Service Service Description (T/NA1(89)03). 11 pp.

ETSI PRETS 300 129-90. Intergrated Services Digital Network (ISDN); Malicious Call Identification (MCID) Supplementary Service Functional Capabilities and Information Flows (T/S 22-10). 24 pp.

Integrated Services Digital Networks *(Cont.)*

—Calling Line Identification—Malicious Calls *(Cont.)*

ETSI ETS 300 128-92. Integrated Services Digital Network (ISDN); Malicious Call Identification (MCID) Supplementary Service Service Description. 14 pp.

ETSI PRETS 300 128-91. Integrated Services Digital Network (ISDN); Malicious Call Identification (MCID) Supplementary Service Service Description (T/NA1(89)03). 15 pp.

ETSI ETS 300 129-92. Integrated Services Digital Network (ISDN); Malicious Call Identification (MCID) Supplementary Service Functional Capabilities and Information Flows. 23 pp.

ETSI PRETS 300 129-91. Integrated Services Digital Network (ISDN); Malicious Call Identification (MCID) Supplementary Service Functional Capabilities and Information Flows. 24 pp.

—Calling Line Identification—Malicious Calls—DSS1

CENELEC PRETS 300 130-90. Integrated Services Digital Network (ISDN); Malicious Call Identification (MCID) Supplementary Service Digital Subscriber Signalling One (DSSI) Protocol (T/S 46-33N). 14 pp.

CENELEC PRETS 300 130-91. Integrated Services Digital Network (ISDN); Malicious Call Identification (MCID) Supplementary Service Digital Subscriber Signalling System No. One (DSS1) Protocol. 16 pp.

CENELEC ETS 300 130-92. Integrated Services Digital Network (ISDN); Malicious Call Identification (MCID) Supplementary Service Digital Subscriber Signalling System No. One (DSS1) Protocol. 15 pp.

ETSI PRETS 300 130-90. Intergrated Services Digital Network (ISDN); Malicious Call Identification (MCID) Supplementary Service Digital Subscriber Signalling One (DSS1) Protocol (T/S 46-33N). 14 pp.

ETSI ETS 300 130-92. Integrated Services Digital Network (ISDN); Malicious Call Identification (MCID) Supplementary Service Digital Subscriber Signalling System No. One (DSS1) Protocol. 15 pp.

ETSI PRETS 300 130-91. Integrated Services Digital Network (ISDN); Malicious Call Identification (MCID) Supplementary Service Digital Subscriber Signalling System No. One (DSS1) Protocol. 16 pp.

—Calling Line Identification Presentation

CCITT RECMN I.251-89. Number Identification Supplementary Services—Integrated Services Digital Network (ISDN)—General Structure and Service Capabilities (Study Group XVIII) 28 pp. 28 pp.

CCITT RECMN I.251. 3 (REV 1)-92. Calling Line Identification Presentation (Study Group I) 12 pp. 12 pp.

CCITT RECMN Q.81-91. Stage 2 Description for Number Identification Supplementary Services (Study Group XI) 46 pp. 46 pp.

CCITT RECMN Q.81-89. Number Identification Supplementary Services—General Recommendations on Telephone Switching and Signalling—Functions and Information Flows for Services in the ISDN—Supplements (Study Group XI) 30 pp. 30 pp.

CENELEC PRETS 300 089-90. Integrated Services Digital Network (ISDN) Calling Line Identification Presentation (CLIP) Supplementary Service Service Description (T/NA1(89)07). 13 pp.

CENELEC PRETS 300 089-91. Integrated Services Digital Network (ISDN); Calling Line Identification Presentation (CLIP) Supplementary Service Service Description. 17 pp.

CENELEC ETS 300 089-92. Integrated Services Digital Network (ISDN); Calling Line Identification Presentation (CLIP) Supplementary Service Service Description. 17 pp.

CENELEC PRETS 300 091-90. Integrated Services Digital Network (ISDN) Calling Line Identification, Presentation and Restriction (CLIP and CLIR) Supplementary Services Functional Capabilities and Information Flows (T/S 22-01). 21 pp.

CENELEC PRETS 300 091-91. Integrated Services Digital Network (ISDN); Calling Line Identification Presentation (CLIP) and Calling Line Identification Restriction (CLIR) Supplementary Services Functional Capabilities and Information Flows. 39 pp.

CENELEC ETS 300 091-92. Integrated Services Digital Network (ISDN); Calling Line Identification Presentation (CLIP) and Calling Line Identification Restriction (CLIR) Supplementary Services Functional Capabilities and Information Flows. 43 pp.

CEPT T/S 22-01 E-87. Supplementary Service Calling Line Identification Presentation. 9 pp.

ETSI PRETS 300 089-90. Integrated Services Digital Network (ISDN) Calling Line Identification Presentation (CLIP) Supplementary Service—Service Description (T/NA1 (89)07). 13 pp.

Integrated Services Digital Networks *(Cont.)*

—Calling Line Identification Presentation *(Cont.)*

ETSI PRETS 300 091-90. Integrated Services Digital Network (ISDN) Calling Line Identification, Presentation and Restriction (CLIP and CLIR) Supplementary Services—Functional Capabilities and Information Flows (T/S 22-01). 21 pp.

ETSI ETS 300 089-92. Integrated Services Digital Network (ISDN); Calling Line Identification Presentation (CLIP) Supplementary Service Service Description. 17 pp.

ETSI ETS 300 091-92. Integrated Services Digital Network (ISDN); Calling Line Identification Presentation (CLIP) and Calling Line Identification Restriction (CLIR) Supplementary Services Functional Capabilities and Information Flows. 43 pp.

ETSI PRETS 300 091-91. Integrated Services Digital Network (ISDN); Calling Line Identification Presentation (CLIP) and Calling Line Identification Restriction (CLIR) Supplementary Services Functional Capabilities and Information Flows. 39 pp.

—Calling Line Identification Presentation—CCITT No. 7 Signaling

CCITT RECMN Q.730-89. ISDN Supplementary Services—Specifications of Signalling System No. 7 (Study Group XI) 59 pp. 59 pp.

—Calling Line Identification Presentation—DSS1

CENELEC PRETS 300 092-90. Integrated Services Digital Network (ISDN) Calling Line Identification, Presentation (CLIP) Supplementary Service Digital Subscriber Signalling One (DSS1) Protocol (T/S 46-33C). 18 pp.

CENELEC PRETS 300 092-91. Integrated Services Digital Network (ISDN); Calling Line Identification Presentation (CLIP) Supplementary Service Digital Subscriber Signalling System No. One (DSS1) Protocol. 4 pp.

CENELEC ETS 300 092-92. Integrated Services Digital Network (ISDN); Calling Line Identification Presentation (CLIP) Supplementary Service Digital Subscriber Signalling No. One (DSS1) Protocol. 22 pp.

CENELEC ETS 300 092 PRA1-92. AMD 1 Integrated Services Digital Network (ISDN); Calling Line Identification Presentation (CLIP) Supplementary Service Digital Subscriber Signalling System No. One (DSS1) Protocol. 4 pp.

CENELEC ETS 300 092/A1-93. AMD 1 Integrated Services Digital Network (ISDN); Calling Line Identification Presentation (CLIP) Supplementary Service Digital Subscriber Signalling System No. One (DSS1) Protocol. 4 pp.

ETSI PRETS 300 092-90. Integrated Services Digital Network (ISDN) Calling Line Identification Presentation (CLIP) Supplementary Service—Digital Subscriber Signalling One (DSS1) Protocol (T/S 46-33C). 18 pp.

ETSI ETS 300 092/A1-93. AMD 1 Integrated Services Digital Network (ISDN); Calling Line Identification Presentation (CLIP) Supplementary Service Digital Subscriber Signalling System No. One (DSS1) Protocol. 4 pp.

ETSI ETS 300 092 PRA1-92. AMD 1 Integrated Services Digital Network (ISDN); Calling Line Identification Presentation (CLIP) Supplementary Service Digital Subscriber Signalling System No. One (DSS1) Protocol. 4 pp.

ETSI ETS 300 092-92. Integrated Services Digital Network (ISDN); Calling Line Identification Presentation (CLIP) Supplementary Service Digital Subscriber Signalling No. One (DSS1) Protocol. 22 pp.

ETSI PRETS 300 092-91. Integrated Services Digital Network (ISDN); Calling Line Identification Presentation (CLIP) Supplementary Service Digital Subscriber Signalling System No. One (DSS1) Protocol. 4 pp.

—Calling Line Identification Restriction

CCITT RECMN I.251-89. Number Identification Supplementary Services—Integrated Services Digital Network (ISDN)—General Structure and Service Capabilities (Study Group XVIII) 28 pp. 28 pp.

CCITT RECMN I.251. 4 (REV 1)-92. Calling Line Identification Restriction (Study Group I) 8 pp. 8 pp.

CCITT RECMN Q.81-91. Stage 2 Description for Number Identification Supplementary Services (Study Group XI) 46 pp. 46 pp.

CCITT RECMN Q.81-89. Number Identification Supplementary Services—General Recommendations on Telephone Switching and Signalling—Functions and Information Flows for Services in the ISDN—Supplements (Study Group XI) 30 pp. 30 pp.

INDUSTRY STANDARDS
1407

INTERNATIONAL AND NON-U.S. NATIONAL STANDARDS SUBJECT INDEX

Integrated Services Digital Networks (Cont.)

—Calling Line Identification Restriction (Cont.)

CENELEC PRETS 300 090-90. Integrated Services Digital Network (ISDN) Calling Line Identification Restriction (CLIR) Supplementary Service Description (T/NA1(89) 08). 11 pp.

CENELEC PRETS 300 090-91. Integrated Services Digital Network (ISDN); Calling Line Identification Restriction (CLIR) Supplementary Service Service Description. 15 pp.

CENELEC ETS 300 090-92. Integrated Services Digital Network (ISDN); Calling Line Identification Restriction (CLIR) Supplementary Service Service Description. 13 pp.

CENELEC PRETS 300 091-90. Integrated Services Digital Network (ISDN) Calling Line Identification, Presentation and Restriction (CLIP and CLIR) Supplementary Services Functional Capabilities and Information Flows (T/S 22-01). 21 pp.

CENELEC PRETS 300 091-91. Integrated Services Digital Network (ISDN) Calling Line Identification Presentation (CLIP) and Calling Line Identification Restriction (CLIR) Supplementary Services Functional Capabilities and Information Flows. 39 pp.

CENELEC ETS 300 091-92. Integrated Services Digital Network (ISDN); Calling Line Identification Presentation (CLIP) and Calling Line Identification Restriction (CLIR) Supplementary Services Functional Capabilities and Information Flows. 43 pp.

ETSI PRETS 300 090-90. Integrated Services Digital Network (ISDN) Calling Line Identification Restriction (CLIR) Supplementary Service—Service Description (T/NA1 (89)08). 11 pp.

ETSI PRETS 300 091-90. Integrated Services Digital Network (ISDN) Calling Line Identification Presentation and Restriction (CLIP and CLIR) Supplementary Services—Functional Capabilities and Information Flows (T/S 22-01). 21 pp.

ETSI ETS 300 090-92. Integrated Services Digital Network (ISDN); Calling Line Identification Restriction (CLIR) Supplementary Service Service Description. 13 pp.

ETSI ETS 300 091-92. Integrated Services Digital Network (ISDN) Calling Line Identification Presentation (CLIP) and Calling Line Identification Restriction (CLIR) Supplementary Services Functional Capabilities and Information Flows. 43 pp.

ETSI PRETS 300 091-91. Integrated Services Digital Network (ISDN) Calling Line Identification Presentation (CLIP) and Calling Line Identification Restriction (CLIR) Supplementary Services Functional Capabilities and Information Flows. 39 pp.

—Calling Line Identification Restriction—CCITT No. 7 Signaling

CCITT RECMN Q.730-89. ISDN Supplementary Services—Specifications of Signalling System No. 7 (Study Group XI) 59 pp. 59 pp.

—Calling Line Identification Restriction—DSS1

CENELEC PRETS 300 093-90. Integrated Services Digital Network (ISDN) Calling Line Identification Restriction (CLIR) Supplementary Service Digital Subscriber Signalling One (DSS1) Protocol (T/S 46-33D). 9 pp.

CENELEC PRETS 300 093-91. Integrated Services Digital Network (ISDN); Calling Line Identification Restriction (CLIR) Supplementary Service Digital Subscriber Signalling System No. One (DSS1) Protocol. 14 pp.

CENELEC ETS 300 093-92. Integrated Services Digital Network (ISDN); Calling Line Identification Restriction (CLIR) Supplementary Service Digital Subscriber Signalling System No. One (DSS1) Protocol. 12 pp.

ETSI PRETS 300 093-90. Integrated Services Digital Network (ISDN) Calling Line Identification Restriction (CLIR) Supplementary Service—Digital Subscriber Signalling One (DSS1) Protocol (T/S 46-33D). 9 pp.

ETSI ETS 300 093-92. Integrated Services Digital Network (ISDN); Calling Line Identification Restriction (CLIR) Supplementary Service Digital Subscriber Signalling System No. One (DSS1) Protocol. 12 pp.

ETSI PRETS 300 093-91. Integrated Services Digital Network (ISDN); Calling Line Identification Restriction (CLIR) Supplementary Service Digital Subscriber Signalling System No. One (DSS1) Protocol. 14 pp.

ETSI ETS 300 094-92. Integrated Services Digital Network (ISDN); Connected Line Identification Presentation (COLP) Supplementary Service Service Description. 16 pp.

Integrated Services Digital Networks (Cont.)

—CCITT No. 7 Signaling

CCITT RECMN E.172-89. Call Routing in the ISDN Era—Telephone Network and ISDN—Operation, Numbering, Routing and Mobile Service (Study Group II) 14 pp. 14 pp.

—CCITT No. 7 Signaling Systems—Dimensioning

CCITT RECMN E.730-92. ISDN Dimensioning Methods Overview (Study Group II) 4 pp. 4 pp.

—CCITT No. 7 Signaling Systems—Grade of Service

CCITT RECMN E.723-92. Grade-of-Service Parameters for Signalling System No. 7 Networks (Study Group II) 8 pp. 8 pp.

—Circuit Mode Bearer Services

CCITT RECMN F.353-89. Provision of Telematic and Data Transmission Services on Integrated Services Digital Network (ISDN)—Telematic, Data Transmission and Teleconference Services—Operations and Quality of Service (Study Group I) 5 pp. 5 pp.

CCITT RECMN I.231-89. Circuit-Mode Bearer Service Categories—Integrated Services Digital Network (ISDN)—General Structure and Service Capabilities (Study Group XVIII) 25 pp. 25 pp.

CCITT RECMN I.231. 1-88. Circuit-Mode 64 kbit/s Unrestricted, 8 kHz Structured Bearer Service Category (Note 1)—Integrated Services Digital Network (ISDN)—General Structure and Service Capabilities (Study Group XVIII) 5 pp. 5 pp.

CCITT RECMN I.231. 5-88. Circuit-Mode 2 X 64 kbit/s Unrestricted, 8 kHz Structured Bearer Service Category—Integrated Services Digital Network (ISDN)—General Structure and Service Capabilities (Study Group XVIII) 2 pp. 2 pp.

CCITT RECMN I.231. 6-88. Circuit-Mode 384 kbit/s Unrestricted, 8 kHz Structured Bearer Service Category—Integrated Services Digital Network (ISDN)—General Structure and Service Capabilities (Study Group XVIII) 3 pp. 3 pp.

CCITT RECMN I.231. 7-88. Circuit-Mode 1536 kbit/s Unrestricted, 8 kHz Structured Bearer Service Category—Integrated Services Digital Network (ISDN)—General Structure and Service Capabilities (Study Group XVIII) 3 pp. 3 pp.

CCITT RECMN I.231. 8-88. Circuit-Mode 1920 kbit/s Unrestricted, 8 kHz Structured Bearer Service Category—Integrated Services Digital Network (ISDN)—General Structure and Service Capabilities (Study Group XVIII) 3 pp. 3 pp.

—Circuit Mode Bearer Services—Accounting

CCITT RECMN D.220-91. Charging and Accounting Principles to Be Applied to International Circuit-Mode Demand Bearer Services Provided over the Integrated Services Digital Network (ISDN) (Study Group III) 5 pp. 5 pp.

CCITT RECMN D.220-89. Charging and Accounting Principles to Be Applied to International Circuit Mode Demand Bearer Services Provided over the Integrated Services Digital Network (ISDN)—General Tariff Principles—Charging and Accounting of Intl Telecom. Serv. (Study Group III) 2 pp. 2 pp.

CEPT T/PGT 33 E-90. Charging and Accounting Principles Applicable to Circuit Mode on Demand Bearer Services over the International ISDN. 1 p.

—Circuit Mode Bearer Services—Audio

CCITT RECMN I.231. 3-88. Circuit-Mode 64 kbit/s, 8 kHz Structured Bearer Service Category Usable for 3.1 kHz Audio Information Transfer—Integrated Services Digital Network (ISDN)—General Structure and Service Capabilities (Study Group XVIII) 5 pp. 5 pp.

—Circuit Mode Bearer Services—Multiple Rate

CCITT RECMN I.231. 10-92. Circuit-Mode Multiple-Rate Unrestricted 8 kHz Structured Bearer Service Category (Study Group I) 8 pp. 8 pp.

—Circuit Mode Bearer Services—Speech Transmission

CCITT RECMN I.231. 2-88. Circuit-Mode 64 kbit/s, 8 kHz Structured Bearer Service Category Usable for Speech Information Transfer—Integrated Services Digital Network (ISDN)—General Structure and Service Capabilities (Study Group XVIII) 5 pp. 5 pp.

CCITT RECMN I.231. 4-88. Circuit-Mode, Alternate Speech /64 kbit/s Unrestricted, 8 kHz Structured Bearer Service Category—Integrated Services Digital Network (ISDN)—General Structure and Service Capabilities (Study Group XVIII) 5 pp. 5 pp.

Integrated Services Digital Networks (Cont.)

—Circuit Mode Bearer Services—Tariffs

CCITT RECMN D.220-91. Charging and Accounting Principles to Be Applied to International Circuit-Mode Demand Bearer Services Provided over the Integrated Services Digital Network (ISDN) (Study Group III) 5 pp. 5 pp.

CCITT RECMN D.220-89. Charging and Accounting Principles to Be Applied to International Circuit Mode Demand Bearer Services Provided over the Integrated Services Digital Network (ISDN)—General Tariff Principles—Charging and Accounting of Intl Telecom. Serv. (Study Group III) 2 pp. 2 pp.

—Circuit Switched Connections

CCITT RECMN I.120-89. Integrated Services Digital Networks (ISDNs) —Integrated Services Digital Network (ISDN)—General Structure and Service Capabilities (Study Group XVIII) 2 pp. 2 pp.

—Circuit Switched Connections—Call Processing Delays

CEPT T/GSI 03-08/2E-87. Call Processing Delays in ISDN Circuit-Switched Connections. 4 pp.

—Circuit Switched Connections—Controlled Slip Rate

CCITT RECMN G.822-89. Controlled Slip Rate Objectives on an International Digital Connection—Digital Networks, Digital Sections and Digital Line Systems (Study Groups XV and XVIII) 3 pp. 3 pp.

—Circuit Switched Connections—Error Performance

CCITT RECMN G.821-89. Error Performance of an International Digital Connection Forming Part of an Integrated Services Digital Network—Digital Networks, Digital Sections and Digital Line Systems (Study Groups XV and XVIII) 10 pp. 10 pp.

—Circuit Switched Connections—Processing Delays

CCITT RECMN I.352-89. Network Performance Objectives for Connection Processing Delays in an ISDN—Integrated Services Digital Network (ISDN)—Overall Network Aspects and Functions, ISDN User-Network Interfaces (Study Group XVIII) 14 pp. 14 pp.

—Circuit Switched Data Transmission Services—Grade of Service

CCITT RECMN E.701-92. Reference Connections for Traffic Engineering—Telephone Network and ISDN—Quality of Service, Network Management and Traffic Engineering (Study Group II) 4 pp. 4 pp.

CCITT RECMN E.721-91. Network Grade of Service Parameters and Target Values for Circuit-Switched Services in the Evolving ISDN (Study Group II) 8 pp. 8 pp.

CCITT RECMN E.721-89. Network Grade of Service Parameters in ISDN—Telephone Network and ISDN—Quality of Service, Network Management and Traffic Engineering (Study Group II) 2 pp. 2 pp.

CCITT RECMN E.723-92. Grade-of-Service Parameters for Signalling System No. 7 Networks (Study Group II) 8 pp. 8 pp.

—Circuit Switched Networks—Dimensioning

CCITT RECMN E.730-92. ISDN Dimensioning Methods Overview (Study Group II) 4 pp. 4 pp.

—Circuit Switched Public Data Networks—Interworking

CCITT RECMN E.166-89. Numbering Plan Interworking in the ISDN Era—Telephone Network and ISDN—Operation, Numbering, Routing and Mobile Service (Study Group II) 14 pp (Same as Recmn X.122). 14 pp.

CCITT RECMN I.540-89. General Arrangements For Interworking Between Circuit Switched Public Data Networks (CSPDNs) and Integrated Services Digital Networks (ISDNs) for the Provision of Data Transmission—Integrated Services Digital Network (ISDN)—Internetwork Interfaces and Maintenance. 1 p.

CCITT RECMN X.81-89. Interworking Between an ISDN Circuit-Switched and a Circuit-Switched Public Data Network (CSPDN)—Data Communication Networks—Transmission, Signalling and Switching, Network Aspects, Maintenance and Administrative Arrangements (Study Group VII) 18 pp. 18 pp.

INTERNATIONAL AND NON-U.S. NATIONAL STANDARDS
SUBJECT INDEX

Integrated

Integrated Services Digital Networks (Cont.)

—Circuit Switched Public Data Networks—Interworking (Cont.)

CCITT RECMN X.321-89. General Arrangements for Interworking Between Circuit Switched Public Data Networks (CSPDNs) and Integrated Service Digital Networks (ISDNs) for the Provision of Data Transmission Services—Data Communication Networks—Interworking Between Networks, Mobile Data. 7 pp.

—Closed User Groups

CCITT RECMN I.255. 1-92. Closed User Group (Study Group I) 17 pp. 17 pp.

CENELEC PRETS 300 136-90. Integrated Services Digital Network (ISDN) Closed User Group (CUG) Supplementary Service Service Description (T/NA1(89)21). 19 pp.

CENELEC PRETS 300 136-91. Integrated Services Digital Network (ISDN); Closed User Group (CUG) Supplementary Service Service Description. 18 pp.

CENELEC ETS 300 136-92. Integrated Services Digital Network (ISDN); Closed User Group (CUG) Supplementary Service Service Description. 17 pp.

CENELEC PRETS 300 137-90. Integrated Services Digital Network (ISDN); Closed User Group (CUG) Supplementary Service Functional Capabilities and Information Flows (T/S 22-03). 34 pp.

CENELEC PRETS 300 137-91. Integrated Services Digital Network (ISDN); Closed User Group (CUG) Supplementary Service Functional Capabilities and Information Flows. 33 pp.

CENELEC ETS 300 137-92. Integrated Services Digital Network (ISDN); Closed User Group (CUG) Supplementary Service Functional Capabilities and Information Flows. 32 pp.

ETSI PRETS 300 136-90. Intergrated Services Digital Network (ISDN) Closed User Group (CUG) Supplementary Service Service Description (T/NA1(89)21). 19 pp.

ETSI PRETS 300 137-90. Integrated Services Digital Network (ISDN); Closed User Group Supplementary Service Functional Capabilities and Information Flows (T/S 22-03). 34 pp.

ETSI ETS 300 136-92. Integrated Services Digital Network (ISDN); Closed User Group (CUG) Supplementary Service Service Description. 17 pp.

ETSI PRETS 300 136-91. Integrated Services Digital Network (ISDN); Closed User Group (CUG) Supplementary Service Service Description. 18 pp.

ETSI ETS 300 137-92. Integrated Services Digital Network (ISDN); Closed User Group (CUG) Supplementary Service Functional Capabilities and Information Flows. 32 pp.

ETSI PRETS 300 137-91. Integrated Services Digital Network (ISDN); Closed User Group (CUG) Supplementary Service Functional Capabilities and Information Flows. 33 pp.

—Closed User Groups—CCITT No. 7 Signaling

CCITT RECMN Q.730-89. ISDN Supplementary Services—Specifications of Signalling System No. 7 (Study Group XI) 59 pp. 59 pp.

—Closed User Groups—DSS1

CCITT RECMN I.255-89. Community of Interest Supplementary Services—Integrated Services Digital Network (ISDN)—General Structure and Service Capabilities (Study Group XVIII) 10 pp. 10 pp.

CENELEC PRETS 300 138-90. Integrated Services Digital Network (ISDN): Closed User Group (CUG) Supplementary Service Digital Subscriber Signalling One (DSS1) Protocol (T/S 46-33H). 29 pp.

CENELEC PRETS 300 138-91. Integrated Services Digital Network (ISDN); Closed User Group (CUG) Supplementary Service Digital Subscriber Signalling No. One (DSS1) Protocol. 30 pp.

CENELEC ETS 300 138-92. Integrated Services Digital Network (ISDN); Closed User Group (CUG) Supplementary Service Digital Subscriber Signalling System No. One (DSS1) Protocol. 29 pp.

CEPT T/S 22-03-87. Service Supplementaire "Groupe Ferme D'Usagers". 8 pp.

CEPT T/S 22-03 E-87. Supplementary Service "Closed User Group". 8 pp.

ETSI PRETS 300 138-90. Intergrated Services Digital Network (ISDN); Closed User Group (CUG) Supplementary Service Digital Subscriber Signalling One (DSS1) Protocol (T/S 46-33H). 29 pp.

ETSI ETS 300 138-92. Integrated Services Digital Network (ISDN); Closed User Group (CUG) Supplementary Service Digital Subscriber Signalling System No. One (DSS1) Protocol. 29 pp.

ETSI PRETS 300 138-91. Integrated Services Digital Network (ISDN); Closed User Group (CUG) Supplementary Service Digital Subscriber Signalling No. One (DSS1) Protocol. 30 pp.

Integrated Services Digital Networks (Cont.)

—Closed User Groups—Functions/Information Flows

CCITT RECMN Q.85-92. Stage 2 Description for Community of Interest Supplementary Services Section 1—Closed User Group (CUG) Section 3—Multi-Level Precedence and Preemption (MLPP) (Rev.1) (Study Group XI) 59 pp. 59 pp.

CCITT RECMN Q.85-89. Community of Interest Supplementary Services—General Recommendations on Telephone Switching and Signalling—Functions and Information Flows for Services in the ISDN—Supplements (Study Group XI) 14 pp. 14 pp.

—Communication Protocols

CENELEC PRETS 300 079-90. Integrated Services Digital Network (ISDN): Syntax-Based Videotex End-to-End Protocols (T/ TE 06-10). 81 pp.

CENELEC PRETS 300 079-91. Integrated Services Digital Network (ISDN); Syntax-Based Videotex End-to-End Protocols (T/TE 06-10). 82 pp.

CENELEC ETS 300 079-91. Integrated Services Digital Network (ISDN); Syntax-Based Videotex End-to-End Protocols Circuit Mode DTE—DTE. 78 pp.

CEPT T/GSI 04-02/1-87. Protocoles Et Interfaces Usager-Reseau (Couche 1). 1 p.

CEPT T/GSI 04-02/1E-87. User to Network Interfaces and Protocols (Layer 1). 1 p.

CEPT T/GSI 04-02/2-85. Protocoles Et Interfaces Usager-Reseau (Couche 1). 1 p.

CEPT T/GSI 04-02/2E-85. User to Network Interfaces and Protocols (Layer 2). 1 p.

CEPT T/GSI 04-02/3-85. Protocoles Et Interfaces Usager-Reseau (Couche 1). 1 p.

CEPT T/GSI 04-02/3E-85. User to Network Interfaces and Protocols (Layer 3). 1 p.

ETSI PRETS 300 079-90. Integrated Services Digital Network (ISDN); Syntax-Based Videotex—End-To-End Protocols (T/TE 06-10). 82 pp.

ETSI ETS 300 079-91. Integrated Services Digital Network (ISDN); Syntax-Based Videotex End-to-End Protocols Circuit Mode DTE—DTE. 78 pp.

—Communication Protocols—Reference Model

CEPT T/GSI 03-02-85. Modele De Reference Pour Un Protocole RNIS. 19 pp.

CEPT T/GSI 03-02 E-85. Protocol Reference Model. 19 pp.

—Communication Terminal Equipment

CCITT RECMN I.333-89. Terminal Selection in ISDN—Integrated Services Digital Network (ISDN)—Overall Network Aspects and Functions, ISDN User-Network Interfaces (Study Group XVIII) 18 pp. 18 pp.

CENELEC ETR 026-92. Network Aspects (NA); Terminal Selection Principles for Priority 1 and 2 Services of MoU—ISDN Applicable in Multi-Terminal Environments at Customer Premises. 21 pp.

ETSI ETR 026-92. Network Aspects (NA); Terminal Selection Principles for Priority 1 and 2 Services of MoU—ISDN Applicable in Multi-Terminal Environments at Customer Premises. 21 pp.

—Communication Terminal Equipment—Bearer Services

CCITT RECMN I.333-89. Terminal Selection in ISDN—Integrated Services Digital Network (ISDN)—Overall Network Aspects and Functions, ISDN User-Network Interfaces (Study Group XVIII) 18 pp. 18 pp.

CENELEC ETR 026-92. Network Aspects (NA); Terminal Selection Principles for Priority 1 and 2 Services of MoU—ISDN Applicable in Multi-Terminal Environments at Customer Premises. 21 pp.

ETSI ETR 026-92. Network Aspects (NA); Terminal Selection Principles for Priority 1 and 2 Services of MoU—ISDN Applicable in Multi-Terminal Environments at Customer Premises. 21 pp.

—Communication Terminal Equipment—Network Capabilities

CCITT RECMN I.333-89. Terminal Selection in ISDN—Integrated Services Digital Network (ISDN)—Overall Network Aspects and Functions, ISDN User-Network Interfaces (Study Group XVIII) 18 pp. 18 pp.

CENELEC ETR 026-92. Network Aspects (NA); Terminal Selection Principles for Priority 1 and 2 Services of MoU—ISDN Applicable in Multi-Terminal Environments at Customer Premises. 21 pp.

ETSI ETR 026-92. Network Aspects (NA); Terminal Selection Principles for Priority 1 and 2 Services of MoU—ISDN Applicable in Multi-Terminal Environments at Customer Premises. 21 pp.

Integrated Services Digital Networks (Cont.)

—Communication Terminal Equipment—Teleinformatic—Protocols

CCITT RECMN T.90-92. Characteristics and Protocols for Terminals for Telematic Services in ISDN (Study Group VIII) 61 pp. 61 pp.

—Communication Terminal Equipment—Teleinformatic Services

CCITT RECMN T.90-92. Characteristics and Protocols for Terminals for Telematic Services in ISDN (Study Group VIII) 61 pp. 61 pp.

—Communication Terminal Equipment—Teleinformatic Services—Protocol

CCITT RECMN T.90-89. Characteristics and Protocols for Terminals for Telematic Services in ISDN—Terminal Equipment and Protocols for Telematic Services (Study Group VIII) 32 pp. 32 pp.

CENELEC PRETS 300 080-90. Integrated Services Digital Network (ISDN): ISDN Lower Layer Protocols for Telematic Terminals (T/TE 12-04). 78 pp.

CENELEC PRETS 300 080-91. Intergrated Services Digital Network (ISDN); ISDN Lower Layer Protocols for Telematic Terminals (T/TE 12-04). 67 pp.

CENELEC PRETS 300 080-91. Proposed Revision to Subclause 7.1.2.1 on Page 19 of Final Draft prETS 300 080:1991. 2 pp.

CENELEC ETS 300 080-92. Integrated Services Digital Network (ISDN); ISDN Lower Layer Protocols for Telematic Terminals. 67 pp.

ETSI PRETS 300 080-90. Integrated Services Digital Network (ISDN); ISDN Lower Layer Protocols for Telematic Terminals (T/TE 12-04). 78 pp.

ETSI ETS 300 080-92. Integrated Services Digital Network (ISDN); ISDN Lower Layer Protocols for Telematic Terminals. 67 pp.

ETSI PRETS 300 080-91. Intergrated Services Digital Network (ISDN); ISDN Lower Layer Protocols for Telematic Terminals (T/TE 12-04). 67 pp.

—Communication Terminal Equipment—Telephony

CCITT RECMN I.330-89. ISDN Numbering and Addressing Principles—Integrated Services Digital Network (ISDN)—Overall Network Aspects and Functions, ISDN User-Network Interfaces (Study Group XVIII) 8 pp. 8 pp.

CENELEC PRETS 300 281-93. Integrated Services Digital Network (ISDN); Telephony 7 kHz Teleservice Terminal Requirements Necessary for End-to-End Compatibility. 53 pp.

ETSI PRETS 300 281-93. Integrated Services Digital Network (ISDN); Telephony 7 kHz Teleservice Terminal Requirements Necessary for End-to-End Compatibility. 53 pp.

—Communication Terminal Equipment—Teleservices

CCITT RECMN I.333-89. Terminal Selection in ISDN—Integrated Services Digital Network (ISDN)—Overall Network Aspects and Functions, ISDN User-Network Interfaces (Study Group XVIII) 18 pp. 18 pp.

CENELEC ETR 026-92. Network Aspects (NA); Terminal Selection Principles for Priority 1 and 2 Services of MoU—ISDN Applicable in Multi-Terminal Environments at Customer Premises. 21 pp.

ETSI ETR 026-92. Network Aspects (NA); Terminal Selection Principles for Priority 1 and 2 Services of MoU—ISDN Applicable in Multi-Terminal Environments at Customer Premises. 21 pp.

—Completion of Calls to Busy Subscribers

CCITT RECMN I.253-89. Call Completion Supplementary Services—Integrated Services Digital Network (ISDN)—General Structure and Service Capabilities (Study Group XVIII) 17 pp. 17 pp.

CCITT RECMN I.253. 3-88. Completion of Calls to Busy Subscribers—Integrated Services Digital Network (ISDN)—General Structure and Service Capabilities (Study Group XVIII) 3 pp. 3 pp.

CEPT T/S 22-08 E-88. Completion of Calls to Busy Subscriber Service. 29 pp.

—Conference Calling

CCITT RECMN I.254-89. Multiparty Supplementary Services—Integrated Services Digital Network (ISDN)—General Structure and Service Capabilities (Study Group XVIII) 30 pp. 30 pp.

CCITT RECMN I.254. 1-88. Conference Calling Service Description—Integrated Services Digital Network (ISDN)—General Structure and Service Capabilities (Study Group XVIII) 19 pp. 19 pp.

INDUSTRY STANDARDS

Integrated
INTERNATIONAL AND NON-U.S. NATIONAL STANDARDS
SUBJECT INDEX

Integrated Services Digital Networks *(Cont.)*

—Connected Line Identification Presentation

CCITT RECMN I.251. 5-88. Connected Line Identification Presentation (Study Group I) 5 pp. 5 pp.

CCITT RECMN Q.81-91. Stage 2 Description for Number Identification Supplementary Services (Study Group XI) 46 pp. 46 pp.

CCITT RECMN Q.81-89. Number Identification Supplementary Services—General Recommendations on Telephone Switching and Signalling—Functions and Information Flows for Services in the ISDN—Supplements (Study Group XI) 30 pp. 30 pp.

CENELEC PRETS 300 094-90. Integrated Services Digital Network Connected Line Identification Presentation (COLP) Supplementary Service Service Description (T/NA1 (89)09). 12 pp.

CENELEC PRETS 300 094-91. Integrated Services Digital Network (ISDN); Connected Line Identification Presentation (COLP) Supplementary Service Service Description. 17 pp.

CENELEC ETS 300 094-92. Integrated Services Digital Network (ISDN); Connected Line Identification Presentation (COLP) Supplementary Service Service Description. 16 pp.

CENELEC PRETS 300 096-90. Integrated Services Digital Network (ISDN) Connected Line Identification, Presentation and Restriction (COLP and COLR) Supplementary Services Functional Capabillities and Information Flows (T/S 22-05). 23 pp.

CENELEC PRETS 300 096-91. Integrated Services Digital Network (ISDN); Connected Line Identification Presentation (COLP) and Connected Line Identification Restriction (COLR) Supplementary Services Functional Capabilities and Information Flows. 34 pp.

CENELEC ETS 300 096-92. Integrated Services Digital Network (ISDN); Connected Line Identification Presentation (COLP) and Connected Line Identification Restriction (COLR) Supplementary Services Functional Capabilities and Information Flows. 39 pp.

CEPT T/S 22-05 E-89. Connected Line Identification Presentation Service. 11 pp.

ETSI PRETS 300 094-90. Integrated Services Digital Network Connected Line Identification Presentation (COLP) Supplementary Service—Service Description (T/NA1 (89)09). 12 pp.

ETSI PRETS 300 096-90. Integrated Services Digital Network (ISDN) Connected Line Identification, Presentation and Restriction (COLP and COLR) Supplementary Services—Functional Capabilities and Information Flows (T/S 22-05). 23 pp.

ETSI ETS 300 094-92. Integrated Services Digital Network (ISDN); Connected Line Identification Presentation (COLP) Supplementary Service Service Description. 16 pp.

ETSI PRETS 300 094-91. Integrated Services Digital Network (ISDN); Connected Line Identification Presentation (COLP) Supplementary Service Service Description. 17 pp.

ETSI ETS 300 096-92. Integrated Services Digital Network (ISDN); Connected Line Identification Presentation (COLP) and Connected Line Identification Restriction (COLR) Supplementary Services Functional Capabilities and Information Flows. 39 pp.

ETSI PRETS 300 096-91. Integrated Services Digital Network (ISDN); Connected Line Identification Presentation (COLP) and Connected Line Identification Restriction (COLR) Supplementary Services Functional Capabilities and Information Flows. 34 pp.

—Connected Line Identification Presentation—DSS1

CENELEC PRETS 300 097-90. Integrated Services Digital Network (ISDN) Connected Line Identification Presentation (COLP) Supplementary Service Digital Subscriber One (DSS1) Protocol (T/S 46-33L). 17 pp.

CENELEC PRETS 300 097-91. Integrated Services Digital Network (ISDN); Connected Line Identification Presentation (COLP) Supplementary Service Digital Subscriber Signalling System No. One (DSS1) Protocol. 24 pp.

ETSI PRETS 300 097-90. Integrated Services Digital Network (ISDN) Connected Line Identification, Presentation (COLP) Supplementary Service—Digital Subscriber Signalling One (DSSI) Protocol (T/S 46-33L). 17 pp.

ETSI PRETS 300 097-91. Integrated Services Digital Network (ISDN); Connected Line Identification Presentation (COLP) Supplementary Service Digital Subscriber Signalling System No. One (DSS1) Protocol. 24 pp.

Integrated Services Digital Networks *(Cont.)*

—Connected Line Identification Restriction

CCITT RECMN I.251. 6-88. Connected Line Identification Restriction (Study Group I) 7 pp. 7 pp.

CCITT RECMN Q.81-91. Stage 2 Description for Number Identification Supplementary Services (Study Group XI) 46 pp. 46 pp.

CCITT RECMN Q.81-89. Number Identification Supplementary Services—General Recommendations on Telephone Switching and Signalling—Functions and Information Flows for Services in the ISDN—Supplements (Study Group XI) 30 pp. 30 pp.

CENELEC PRETS 300 095-90. Integrated Services Digital Network (ISDN) Connected Line Identification Restriction (COLR) Supplementary Service Service Description (T/NA1(89)10). 12 pp.

CENELEC PRETS 300 095-91. Integrated Services Digital Network (ISDN); Connected Line Identification Restriction (COLR) Supplementary Service Service Description. 15 pp.

CENELEC ETS 300 095-92. Integrated Services Digital Network (ISDN); Connected Line Identification Restriction (COLR) Supplementary Service Service Description. 14 pp.

CENELEC PRETS 300 096-90. Integrated Services Digital Network (ISDN) Connected Line Identification, Presentation and Restriction (COLP and COLR) Supplementary Services Functional Capabillities and Information Flows (T/S 22-05). 23 pp.

CENELEC PRETS 300 096-91. Integrated Services Digital Network (ISDN); Connected Line Identification Presentation (COLP) and Connected Line Identification Restriction (COLR) Supplementary Services Functional Capabilities and Information Flows. 34 pp.

CENELEC ETS 300 096-92. Integrated Services Digital Network (ISDN); Connected Line Identification Presentation (COLP) and Connected Line Identification Restriction (COLR) Supplementary Services Functional Capabilities and Information Flows. 39 pp.

ETSI PRETS 300 095-90. Integrated Services Digital Network (ISDN) Connected Line Identification Restriction (COLR) Supplementary Service—Service Description (T/NA1 (89)10). 12 pp.

ETSI PRETS 300 096-90. Integrated Services Digital Network (ISDN) Connected Line Identification, Presentation and Restriction (COLP and COLR) Supplementary Services—Functional Capabilities and Information Flows (T/S 22-05). 23 pp.

ETSI ETS 300 095-92. Integrated Services Digital Network (ISDN); Connected Line Identification Restriction (COLR) Supplementary Service Service Description. 14 pp.

ETSI PRETS 300 095-91. Integrated Services Digital Network (ISDN); Connected Line Identification Restriction (COLR) Supplementary Service Service Description. 15 pp.

ETSI ETS 300 096-92. Integrated Services Digital Network (ISDN); Connected Line Identification Presentation (COLP) and Connected Line Identification Restriction (COLR) Supplementary Services Functional Capabilities and Information Flows. 39 pp.

ETSI PRETS 300 096-91. Integrated Services Digital Network (ISDN); Connected Line Identification Presentation (COLP) and Connected Line Identification Restriction (COLR) Supplementary Services Functional Capabilities and Information Flows. 34 pp.

—Connected Line Identification Restriction—DSS1

CENELEC ETS 300 097-92. Integrated Services Digital Network (ISDN); Connected Line Identification Presentation (COLP) Supplementary Service Digital Subscriber Signalling System No. One (DSS1) Protocol. 21 pp.

CENELEC PRETS 300 098-90. Integrated Services Digital Network (ISDN) Connected Line Identification Restriction (COLR) Supplementary Service Digital Subscriber Signalling One (DSS1) Protocol (T/S 46-33M). 9 pp.

CENELEC PRETS 300 098-91. Integrated Services Digital Network (ISDN); Connected Line Identification Restriction (COLR) Supplementary Service Digital Subscriber Signalling System No. One (DSS1) Protocol. 15 pp.

CENELEC ETS 300 098-92. Integrated Services Digital Network (ISDN); Connected Line Identification Restriction (COLR) Supplementary Service Digital Subscriber Signalling System No. One (DSS1) Protocol. 12 pp.

ETSI PRETS 300 098-90. Integrated Services Digital Network (ISDN)—Connected Line Identification Restriction (COLR) Supplementary Service—Digital Subscriber Signalling One (DSS1) (T/S 46-33M). 9 pp.

Integrated Services Digital Networks *(Cont.)*

—Connected Line Identification Restriction—DSS1 *(Cont.)*

ETSI ETS 300 097-92. Integrated Services Digital Network (ISDN); Connected Line Identification Presentation (COLP) Supplementary Service Digital Subscriber Signalling System No. One (DSS1) Protocol. 21 pp.

ETSI ETS 300 098-92. Integrated Services Digital Network (ISDN); Connected Line Identification Restriction (COLR) Supplementary Service Digital Subscriber Signalling System No. One (DSS1) Protocol. 12 pp.

ETSI PRETS 300 098-91. Integrated Services Digital Network (ISDN); Connected Line Identification Restriction (COLR) Supplementary Service Digital Subscriber Signalling System No. One (DSS1) Protocol. 15 pp.

—Connectors

BSI BS 7266: Part 1-90. 1990 Connectors for Access to the Integrated Services Digital Network (ISDN) Part 1: Basic Access Interface Connector and Its Contact Assignments. 16 pp.

CEN EN 28 877-90. Information Processing Systems—Interface Connector and Contact Assignments for ISDN Basic Access Interface Located at Reference Points S and T. 3 pp.

CENELEC ENV 41 001-87. ISDN Connector up to 8 Pins and up to 2,048 M bit/s. 59 pp.

IEC 8877-92. Information Technology—Telecommunications and Information Exchange Between Systems—Interface Connector and Contact Assignments for ISDN Basic Access Interface Located at Reference Points S and T Second Edition. 15 pp.

IEC 10173-91. Information Technology—Integrated Services Digital Network (ISDN) Primary Access Connector at Reference Points S and T First Edition. 15 pp.

ISO 8877-92. Information Technology—Telecommunications and Information Exchange Between Systems—Interface Connector and Contact Assignments for ISDN Basic Access Interface Located at Reference Points S and T Second Edition. 15 pp.

ISO 10173-91. Information Technology—Integrated Services Digital Network (ISDN) Primary Access Connector at Reference Points S and T First Edition. 15 pp.

JIS X 5110-90. Information Processing Systems—Interface Connector and Contact Assignments for ISDN Basic Access Interface Located at Reference Points S and T.

JTC1 8877-92. Information Technology—Telecommunications and Information Exchange Between Systems—Interface Connector and Contact Assignments for ISDN Basic Access Interface Located at Reference Points S and T Second Edition. 15 pp.

JTC1 10173-91. Information Technology—Integrated Services Digital Network (ISDN) Primary Access Connector at Reference Points S and T First Edition. 15 pp.

OSI ISO 8877 DAM 1.2-90. Information Processing Systems—Interface Connector and Contact Assignments for ISDN Basic Access Interface Located at Reference Points S and T Amendment 1: ISDN Basic Access TE Cord. 5 pp.

OSI ISO 8877-87. Information Processing Systems-Interface Connector and Contact Assignments for ISDN Basic Access Interface Located at Reference Points S and T. 11 pp.

OSI ISO/IEC DIS 10173-90. Information Technology—Integrated Services Digital Network (ISDN) Primary Access Connecter at Reference Points S and T. 17 pp.

SNZ NZS/AS 4102-93. Information Technology—Integrated Services Digital Network Primary Access Connector at Reference Points S and T (This is a Joint Standard with SAA AS 4102). 12 pp.

SNZ NZS/ISO 8877-87. Information Processing Systems—Interface Connector and Contact Assignments for ISDN Basic Access Interface Located at Reference Points S and T. 9 pp.

—Credit Card Calling

CCITT RECMN I.256-89. Charging Supplementary Services—Integrated Services Digital Network (ISDN)—General Structure and Service Capabilities (Study Group XVIII) 16 pp. 16 pp.

CCITT RECMN I.256. 1-88. Credit Card Calling—Integrated Services Digital Network (ISDN)—General Structure and Service Capabilities (Study Group XVIII) 1 pp. 1 p.

—Customer Administration

CENELEC PRI-ETS 300 291-93. Network Aspects (NA); Functional Specification of Customer Administration (CA) on the Operations System/Network Element (OS/NE) Interface. 131 pp.

INTERNATIONAL AND NON-U.S. NATIONAL STANDARDS
SUBJECT INDEX
Integrated

Integrated Services Digital Networks (Cont.)

—Customer Administration (Cont.)

ETSI PRI-ETS 300 291-93. Network Aspects (NA); Functional Specification of Customer Administration (CA) on the Operations System/Network Element (OS/NE) Interface. 131 pp.

—Customer Premises Equipment

CENELEC ETR 034-91. Business Telecommunications (BT); Approval Requirements for Complex Customer Premises Apparatus and Installations Connected to the Public ISDN (Including Principles for the Application of the Essential Requirements to Any Apparatus). 39 pp.

CSA CAN/CSA-T540-M92. Integrated Services Digital Network (ISDN) Primary Rate—Customer Installation Metallic Interfaces—Layer 1 Specification (ANSI T1.408-1990); (Gen Instr 1). 69 pp.

ETSI ETR 034-91. Business Telecommunications (BT); Approval Requirements for Complex Customer Premises Apparatus and Installations Connected to the Public ISDN (Including Principles for the Application of the Essential Requirements to Any Apparatus). 39 pp.

—Data Link Layer—Broadcast Procedures—DSS1

CCITT RECMN Q.921-89. ISDN User-Network Interface—Data Link Layer Specification—Digital Subscriber Signalling System No.1 (DSS 1), Data Link Layer (Study Group XI) 123 pp (Same as Recmn I.441). 123 pp.

—Data Link Layer—Frame Mode Bearer Services—DSS1

CCITT RECMN Q.922-92. ISDN Data Link Layer Specification for Frame Mode Bearer Services (Study Group XI) 112 pp. 112 pp.

—Data Link Layer—Interfaces

CCITT RECMN I.440-89. ISDN User-Network Interface Data Link Layer—General Aspects—Integrated Services Digital Network (ISDN)—Overall Network Aspects and Functions, ISDN User-Network Interfaces (Study Group XVIII) 1 pp (Same as Recmn Q.920). 1 p.

CCITT RECMN I.441-89. ISDN User-Network Interface, Data Link Layer Specification—Integrated Services Digital Network (ISDN)—Overall Network Aspects and Functions, ISDN User-Network Interfaces (Study Group XVIII) 1 pp (Same as Recmn Q.921). 1 p.

—Data Link Layer—Link Access Procedures—D Channel—DSS1

CCITT RECMN Q.920-89. ISDN User-Network Interface Data Link Layer—General Aspects—Digital Subscriber Signalling System No.1 (DSS 1), Data Link Layer (Study Group XI) 16 pp (Same as Recmn I.440). 16 pp.

CCITT RECMN Q.921-89. ISDN User-Network Interface—Data Link Layer Specification—Digital Subscriber Signalling System No.1 (DSS 1), Data Link Layer (Study Group XI) 123 pp (Same as Recmn I.441). 123 pp.

—Data Link Layer—PICS

CENELEC PRI-ETS 300 305-93. Integrated Services Digital Network (ISDN); Digital Subscriber Signalling System No. One (DSS1) Protocol Implementation Conformance Statement (PICS) Proforma for Basic-Access User for Data-Link-Layer Protocol for General Application. 19 pp.

CENELEC PRI-ETS 300 306-93. Integrated Services Digital Network (ISDN); Digital Subscriber Signalling System No. One (DSS1) Protocol Implementation Conformance Statement (PICS) Proforma for Primary-Rate-Access User for Data-Link-Layer Protocol for General Application. 20 pp.

CENELEC PRI-ETS 300 307-93. Integrated Services Digital Network (ISDN); Digital Subscriber Signalling System No. One (DSS1) Protocol Implementation Conformance Statement (PICS) Proforma for Basic-Access Network for Data-Link-Layer Protocol for General Application. 20 pp.

CENELEC PRI-ETS 300 308-93. Integrated Services Digital Network (ISDN); Digital Subscriber Signalling System No. One (DSS1) Protocol Implementation Conformance Statement (PICS) Proforma for Primary-Rate-Access Network for Data-Link-Layer Protocol for General Application. 19 pp.

ETSI PRI-ETS 300 305-93. Integrated Services Digital Network (ISDN); Digital Subscriber Signalling System No. One (DSS1) Protocol Implementation Conformance Statement (PICS) Proforma for Basic-Access User for Data-Link-Layer Protocol for General Application. 19 pp.

Integrated Services Digital Networks (Cont.)

—Data Link Layer—PICS (Cont.)

ETSI PRI-ETS 300 306-93. Integrated Services Digital Network (ISDN); Digital Subscriber Signalling System No. One (DSS1) Protocol Implementation Conformance Statement (PICS) Proforma for Primary-Rate-Access User for Data-Link-Layer Protocol for General Application. 20 pp.

ETSI PRI-ETS 300 307-93. Integrated Services Digital Network (ISDN); Digital Subscriber Signalling System No. One (DSS1) Protocol Implementation Conformance Statement (PICS) Proforma for Basic-Access Network for Data-Link-Layer Protocol for General Application. 20 pp.

ETSI PRI-ETS 300 308-93. Integrated Services Digital Network (ISDN); Digital Subscriber Signalling System No. One (DSS1) Protocol Implementation Conformance Statement (PICS) Proforma for Primary-Rate-Access Network for Data-Link-Layer Protocol for General Application. 19 pp.

—Data Link Layer—PIXIT

CENELEC PRI-ETS 300 309-93. Integrated Services Digital Network (ISDN); Digital Subscriber Signalling System No. One (DSS1) Protocol Implementation eXtra Information for Testing (PIXIT) Proforma for Basic-Access User for Data-Link-Layer Protocol for General Application. 11 pp.

CENELEC PRI-ETS 300 310-93. Integrated Services Digital Network (ISDN); Digital Subscriber Signalling System No. One (DSS1) Protocol Implementation eXtra Information for Testing (PIXIT) Proforma for Primary-Rate-Access User for Data-Link-Layer Protocol for General Application. 11 pp.

CENELEC PRI-ETS 300 311-93. Integrated Services Digital Network (ISDN); Digital Subscriber Signalling System No. One (DSS1) Protocol Implementation eXtra Information for Testing (PIXIT) Proforma for Basic-Access Network for Data-Link-Layer Protocol for General Application. 11 pp.

CENELEC PRI-ETS 300 312-93. Integrated Services Digital Network (ISDN); Digital Subscriber Signalling System No. One (DSS1) Protocol Implementation eXtra Information for Testing (PIXIT) Proforma for Primary-Rate-Access Network for Data-Link-Layer Protocol for General Application. 11 pp.

ETSI PRI-ETS 300 309-93. Integrated Services Digital Network (ISDN); Digital Subscriber Signalling System No. One (DSS1) Protocol Implementation eXtra Information for Testing (PIXIT) Proforma for Basic-Access User for Data-Link-Layer Protocol for General Application. 11 pp.

ETSI PRI-ETS 300 310-93. Integrated Services Digital Network (ISDN); Digital Subscriber Signalling System No. One (DSS1) Protocol Implementation eXtra Information for Testing (PIXIT) Proforma for Primary-Rate-Access User for Data-Link-Layer Protocol for General Application. 11 pp.

ETSI PRI-ETS 300 311-93. Integrated Services Digital Network (ISDN); Digital Subscriber Signalling System No. One (DSS1) Protocol Implementation eXtra Information for Testing (PIXIT) Proforma for Basic-Access Network for Data-Link-Layer Protocol for General Application. 11 pp.

ETSI PRI-ETS 300 312-93. Integrated Services Digital Network (ISDN); Digital Subscriber Signalling System No. One (DSS1) Protocol Implementation eXtra Information for Testing (PIXIT) Proforma for Primary-Rate-Access Network for Data-Link-Layer Protocol for General Application. 11 pp.

—Data Link Layer—Point to Point Procedures—DSS1

CCITT RECMN Q.921-89. ISDN User-Network Interface—Data Link Layer Specification—Digital Subscriber Signalling System No.1 (DSS 1), Data Link Layer (Study Group XI) 123 pp (Same as Recmn I.441). 123 pp.

—Data Link Layer—User Network Interface—DSS1

CCITT RECMN Q.920-89. ISDN User-Network Interface Data Link Layer—General Aspects—Digital Subscriber Signalling System No.1 (DSS 1), Data Link Layer (Study Group XI) 16 pp (Same as Recmn I.440). 16 pp.

CCITT RECMN Q.921-89. ISDN User-Network Interface—Data Link Layer Specification—Digital Subscriber Signalling System No.1 (DSS 1), Data Link Layer (Study Group XI) 123 pp (Same as Recmn I.441). 123 pp.

—Data Link Layer—User Network Interfaces

CENELEC PRETS 300 125-90. Integrated Services Digital Network (ISDN); User-Network Interface Data Link Layer Specifications Application of CCITT Recommendations Q.920/I.440 and Q.921/I.441. 162 pp.

Integrated Services Digital Networks (Cont.)

—Data Link Layer—User Network Interfaces (Cont.)

CENELEC ETS 300 125-91. Integrated Services Digital Network (ISDN); User-Network Interface Data Link Layer Specification Application of CCITT Recommendations Q.920/I.440 and Q.921/I.441. 161 pp.

ETSI PRETS 300 125-90. Integrated Services Digital Network (ISDN); User-Network Interface Data Link Layer Specifications—Application of CCITT Recommendations Q/920/I.440 and Q.921/I.441—((Formerly ETS(CC))). 162 pp.

ETSI ETS 300 125-91. Integrated Services Digital Network (ISDN); User-Network Interface Data Link Layer Specification Application of CCITT Recommendations Q.920/I.440 and Q.921/I.441. 161 pp.

—Data Network Identification Codes

CCITT RECMN E.167-89. ISDN Network Identification Codes—Telephone Network and ISDN—Operation, Numbering, Routing and Mobile Service (Study Group II) 2 pp. 2 pp.

—Data Network Identification Codes—Closed User Groups

CCITT RECMN E.167-89. ISDN Network Identification Codes—Telephone Network and ISDN—Operation, Numbering, Routing and Mobile Service (Study Group II) 2 pp. 2 pp.

—Data Terminal Equipment

CCITT RECMN I.470-89. Relationship of Terminal Functions to ISDN—Integrated Services Digital Network (ISDN)—Overall Network Aspects and Functions, ISDN User-Network Interfaces (Study Group XVIII) 4 pp. 4 pp.

CEPT NET 3 Part 1-88. Approval Requirements for Terminal Equipment to Connect to Integrated Services Digital Network (ISDN) Using ISDN Basic Access Part 1: Layers 1 and 2 Aspects. 93 pp.

CEPT NET 3 Part 2-89. Approval Requirements for Terminal Equipment to Connect to ISDN Using ISDN Basic Access, Layer 3 Aspect (ETS P (Rev.1)). 225 pp.

CEPT T/S 54-08-87. Maintenance Des Acces Et Installations Terminales D'Abonnes Du RNIS. 22 pp.

—Data Terminal Equipment—Control

CEPT T/SF 31-08 E-89. User Control Functions of ISDN Terminal Equipment. 6 pp.

—Data Terminal Equipment—Interfaces

CCITT RECMN E.331-91. Minimum User-Terminal Interface for a Human User Entering Address Information into an ISDN Terminal (Study Group I) 9 pp. 9 pp.

—Data Terminal Equipment—Interworking

CCITT RECMN I.320-89. ISDN Protocol Reference Model—Integrated Services Digital Network (ISDN)—Overall Network Aspects and Functions, ISDN User-Network Interfaces (Study Group XVIII) 10 pp. 10 pp.

—Data Terminal Equipment—Open Systems Interconnection

BSI BS ISO/IEC 9574-89. 1989 Information Technology—Telecommunications and Information Exchange Between Systems—Provision of the OSI Connection-Mode Network Service by Packet Mode Terminal Equipment Connected to an Inter-grated Services Digital Network (ISDN). 18 pp.

CCITT RECMN X.612-92. Information Technology—Provision of the OSI Connection-Mode Network Service by Packet-Mode Terminal Equipment Connected to an Integrated Services Digital Network (ISDN) 21 pp (ISO/CCIT Common Text). 21 pp.

CCITT RECMN X.612-90. Provision of the OSI Connection-Mode Network Service by Packet Mode Terminal Equipment Connected to an Integrated Services Digital Network (ISDN) for CCITT Applications (Study Group VII) 19 pp. 19 pp.

IEC 9797-89. Data Cryptographic Techniques—Data Integrity Mechanism Using a Cryptographic Check Function Employing a Block Cipher Algorithm First Edition. 7 pp.

ISO 9574-92. Information Technology—Provision of the OSI Connection-Mode Network Service by Packet Mode Terminal Equipment to an Integrated Services Digital Network (ISDN) Second Edition. 23 pp.

JTC1 9574-92. Information Technology—Provision of the OSI Connection-Mode Network Service by Packet Mode Terminal Equipment to an Integrated Services Digital Network (ISDN) Second Edition. 23 pp.

INDUSTRY STANDARDS

INTERNATIONAL AND NON-U.S. NATIONAL STANDARDS
SUBJECT INDEX

Integrated

Integrated Services Digital Networks (Cont.)

—**Data Terminal Equipment—Open Systems Interconnection** (Cont.)

OSI ISO 8880-2 DAM 1-91. Telecommunications and Information Exchange Between Systems—Protocol Combinations to Provide and Support and Support the OSI Network Service—Part 2: Provision and Support of the Connection-Mode Network Service-Amendment 1: Addition of the ISDN Environment.

OSI ISO IEC 9574-89. Information Technology—Telecommunications and Information Exchange Between Systems—Provision of the OSI Connection-Mode Network Service by Packet Mode Terminal Equipment Connected to an Integrated Services Digital Network (ISDN). 15 pp.

OSI ISO IEC DIS 9574-88. Information Processing Systems—Data Communications—Provision of the OSI Connection—Mode Network Service by Packet Mode Terminal Equipment Connected to an Integrated Services Digital Network. 16 pp.

—**Data Terminal Equipment—Terminal Adaptation Functions**

CENELEC PRETS 300 007-90. Integrated Services Digital Network (ISDN): Support of Packet Mode Terminal Equipment by an ISDN (T/S 46-50). 116 pp.

CENELEC PRETS 300 007-91. Integrated Services Digital Network (ISDN); Support of Packet-Mode Terminal Equipment by an ISDN. 104 pp.

CENELEC ETS 300 007-91. Integrated Services Digital Network (ISDN); Support of Packet-Mode Terminal Equipment by an ISDN. 103 pp.

CENELEC ETS 300 103-90. Integrated Services Digital Network (ISDN); Support of CCITT Recommendation X.21, X.21 bis and X.20 bis Based Data Terminal Equipments (DTEs) by an ISDN Synchronous and Asynchronous Terminal Adaption Functions. 105 pp.

CEPT T/S 46-40-87. Support Des Equipments Terminaux De Traitement De Donnees (ETTD) Dutype X.21, X.21 Bis X.20 Par Le Reseau Numerique a Integration De Services (RNIS) Fonction D'Adaptation Des Terminaux Synchrones Et Asynchrones. 59 pp.

CEPT T/S 46-40 E-87. Support of X.21, X.21 Bis and X.20 Bis Based Data Terminal Equipments (DTEs) by an Integrated Services Digital Network (ISDN) Synchronous and Asynchronous Terminal Adaptation Function. 59 pp.

CEPT T/S 46-50 E-87. Support of Packet Mode Terminal Equipment by an ISDN (X.31/I.462). 25 pp.

CSA CAN/CSA-T552-M91. Integrated Services Digital Network (ISDN) Terminal Adaptation Using Statistical Multiplexing (ANSI T1.612-1990); (Gen Instr 1). 49 pp.

ETSI PRETS 300 007-90. Integrated Services Digital Network (ISDN); Support Packet Mode Terminal Equipment by an ISDN Also Known as T/S 46-50. 116 pp.

ETSI ETS 300 007-91. Integrated Services Digital Network (ISDN); Support of Packet-Mode Terminal Equipment by an ISDN. 103 pp.

ETSI PRETS 300 007-91. Integrated Services Digital Network (ISDN); Support of Packet-Mode Terminal Equipment by an ISDN. 104 pp.

ETSI PRETS 300 007-90. Integrated Services Digital Network (ISDN); Support of Packet Mode Terminal Equipment by an ISDN (T/S 46-50). 116 pp.

ETSI ETS 300 103-90. Integrated Services Digital Network (ISDN); Support of CCITT Recommendation X.21, X.21 bis and X.20 bis Based Data Terminal Equipments (DTEs) by an ISDN Synchronous and Asynchronous Terminal Adaption Functions. 105 pp.

—**Data Terminal Equipment—Terminal Adapters**

CCITT RECMN I.461-89. Support of X.21, X.21 bis and X.20 bis Based Data Terminal Equipments (DTEs) by an Integrated Services Digital Network (ISDN)—Integrated Services Digital Network (ISDN)—Overall Network Aspects and Functions, ISDN User-Network Interfaces (Study Group XVIII) 1 pp. 1 p.

CCITT RECMN X.30-89. Support of X.21,X.21 bis and X.20 bis Based Data Terminal Equipments (DTEs) by an Integrated Services Digital Network (ISDN)—Data Communication Networks: Services and Facilities, Interfaces (Study Group VII) 43 pp (Same as Recmn I.461). 43 pp.

CENELEC PRETS 300 077-90. Integrated Services Digital Network (ISDN); Attachment Requirements for Terminal Adaptors to Connect to an ISDN at the S/T reference Point (T/TE 04-10) (Candidate NET 7). 187 pp.

CENELEC PRETS 300 104-90. Integrated Services Digital Network (ISDN); Attachment Requirements for Terminal Equipment to Connect to an ISDN Using ISDN Basic Access Layer 3 Aspects. 212 pp.

Integrated Services Digital Networks (Cont.)

—**Data Terminal Equipment—Terminal Adapters** (Cont.)

CENELEC ETS 300 104-91. Integrated Services Digital Network (ISDN); Attachment Requirements for Terminal Equipment to Connect to an ISDN Using ISDN Basic Access Layer 3 Aspects. 177 pp.

CENELEC PRETS 300 153-91. Integrated Services Digital Network (ISDN) Attachment Requirements for Terminal Equipment to Connect to an ISDN Using ISDN Basic Access (T/TE 04-08). 62 pp.

CENELEC PRETS 300 153-92. Integrated Services Digital Network (ISDN); Attachment Requirements for Terminal Equipment to Connect to an ISDN Using ISDN Basic Access. 69 pp.

CENELEC PRETS 300 153-92. Integrated Services Digital Network (ISDN); Attachment Requirements for Terminal Equipment to Connect to an ISDN Using ISDN Basic Access (Candidate NET 3 Part 1). 70 pp.

CENELEC PRETS 300 156-91. Integrated Services Digital Network (ISDN); Attachment Requirements for Terminal Equipment to Connect to an ISDN Using ISDN Primary Rate Access (T/TE 04-24). 38 pp.

CENELEC PRETS 300 156-92. Integrated Services Digital Network (ISDN); Attachment Requirements for Terminal Equipment to Connect to an ISDN Using ISDN Primary Rate Access. 41 pp.

CENELEC ETS 300 156-92. Integrated Services Digital Network (ISDN); Attachment Requirements for Terminal Equipment to Connect to an ISDN Using ISDN Primary Rate Access (Candidate NET 5). 41 pp.

ETSI PRETS 300 077-90. Integrated Services Digital Network (ISDN); Attachment Requirements for Terminal Adaptors to Connect to an ISDN at the S/T Reference Point (T/TE 04-10) (Candidate NET 7). 186 pp.

ETSI PRETS 300 104-90. Integrated Services Digital Network (ISDN); Attachment Requirements for Terminal Equipment to Connect to an ISDN Using ISDN Basic Access Layer 3 Aspects (The Text of This Draft ETS May Be Utilized, Wholly or in Part, for the Establishment of NET 3 Part 2). 212 pp.

ETSI PRETS 300 153-91. Integrated Services Digital Network (ISDN); Attachment Requirements for Terminal Equipment to Connect to an ISDN Using ISDN Basic Access (T/TE 04-08).

ETSI PRETS 300 156-91. Integrated Services Digital Network (ISDN); Attachment Requirements for Terminal Equipment to Connect to an ISDN Using ISDN Primary Rate Access (T/TE 04-24).

ETSI PRETS 300 077-90. Integrated Services Digital Network (ISDN); Attachment Requirements for Terminal Adaptors to Connect to an ISDN at the S/T reference Point (T/TE 04-10) (Candidate NET 7). 187 pp.

ETSI ETS 300 104-91. Integrated Services Digital Network (ISDN); Attachment Requirements for Terminal Equipment to Connect to an ISDN Using ISDN Basic Access Layer 3 Aspects. 177 pp.

ETSI PRETS 300 104-90. Integrated Services Digital Network (ISDN); Attachment Requirements for Terminal Equipment to Connect to an ISDN Using ISDN Basic Access Layer 3 Aspects. 212 pp.

ETSI ETS 300 153-92. Integrated Services Digital Network (ISDN); Attachment Requirements for Terminal Equipment to Connect to an ISDN Using ISDN Basic Access (Candidate NET 3 Part 1). 70 pp.

ETSI PRETS 300 153-92. Integrated Services Digital Network (ISDN); Attachment Requirements for Terminal Equipment to Connect to an ISDN Using ISDN Basic Access. 69 pp.

ETSI PRETS 300 153-91. Integrated Services Digital Network (ISDN) Attachment Requirements for Terminal Equipment to Connect to an ISDN Using ISDN Basic Access (T/TE 04-08). 62 pp.

ETSI ETS 300 156-92. Integrated Services Digital Network (ISDN); Attachment Requirements for Terminal Equipment to Connect to an ISDN Using ISDN Primary Rate Access (Candidate NET 5). 41 pp.

ETSI PRETS 300 156-92. Integrated Services Digital Network (ISDN); Attachment Requirements for Terminal Equipment to Connect to an ISDN Using ISDN Primary Rate Access. 41 pp.

ETSI PRETS 300 156-91. Integrated Services Digital Network (ISDN); Attachment Requirements for Terminal Equipment to Connect to an ISDN Using ISDN Primary Rate Access (T/TE 04-24). 38 pp.

—**Data Terminal Equipment—User Control Functions**

CEPT T/CAC S 10.8 E-92. User Control Functions of ISDN Terminal Equipment. 8 pp.

Integrated Services Digital Networks (Cont.)

—**Data Terminal Equipment—Voice/Data Systems**

CSA CAN/CSA-T531-M91. Acoustic-to-Digital and Digital-to-Acoustic Transmission Requirements for ISDN Terminals; (Gen Instr 1). 47 pp.

—**Data Transmission**

CCITT RECMN F.353-89. Provision of Telematic and Data Transmission Services on Integrated Services Digital Network (ISDN)—Telematic, Data Transmission and Teleconference Services—Operations and Quality of Service (Study Group I) 5 pp. 5 pp.

CCITT RECMN I.326-89. Reference Configuration for Relative Network Resource Requirements—Integrated Services Digital Network (ISDN)—Overall Network Aspects and Functions, ISDN User-Network Interfaces (Study Group XVIII) 3 pp. 3 pp.

—**Data Transmission Equipment**

CEPT T/L 05-03 E-88. Digital Transmission System on Metallic Local Lines for ISDN Basic Rate Access. 25 pp.

CEPT T/SF 58 E-87. Telematic and Data Transmission Services in the ISDN. 24 pp.

—**Data Transmission—Interworking**

CCITT RECMN X.320-89. General Arrangements for Interworking Between Integrated Services Digital Networks (ISDNs) for the Provision of Data Transmission Services—Data Communication Networks—Interworking Between Networks, Mobile Data Transmission Systems, Internetwork Management. 8 pp.

—**Dedicated Networks—Interworking**

CCITT RECMN I.500-89. General Structure of the ISDN Interworking Recommendations—Integrated Services Digital Network (ISDN)—Internetwork Interfaces and Maintenance Principles (Study Group XVIII) 4 pp. 4 pp.

CCITT RECMN I.510-89. Definitions and General Principles for ISDN Interworking—Integrated Services Digital Network (ISDN)—Internetwork Interfaces and Maintenance Principles (Study Group XVIII) 12 pp. 12 pp.

—**Dedicated—Numbering Plans—Interworking**

CCITT RECMN I.332-89. Numbering Principles for Interworking Between ISDNs and Dedicated Networks with Different Numbering Plans—Integrated Services Digital Network (ISDN)—Overall Network Aspects and Functions, ISDN User-Network Interfaces (Study Group XVIII) 4 pp. 4 pp.

—**Design—Grade of Service**

CCITT RECMN E.525 (REV 1)-92. Designing Networks to Control Grade of Service (Study Group II) 10 pp. 10 pp.

—**Digital Exchanges**

CCITT RECMN Q.500-89. Digital Local, Combined, Transit and International Exchanges Introduction and Field of Application—Digital Local, Transit, Combined and International Exchanges in Integrated Digital Networks and Mixed Analogue-Digital Networks—Supplements (Study Group XI) 4 pp. 4 pp.

CCITT RECMN Q.521-89. Exchange Functions—Digital Local, Transit, Combined and International Exchanges in Integrated Digital Networks and Mixed Analogue-Digital Networks—Supplements (Study Group XI) 3 pp. 3 pp.

CCITT RECMN Q.522-89. Digital Exchange Connections, Signalling and Ancillary Functions—Digital Local, Transit, Combined and International Exchanges in Integrated Digital Networks and Mixed Analogue-Digital Networks—Supplements (Study Group XI) 16 pp. 16 pp.

—**Digital Sections**

CCITT RECMN G.960-89. Digital Section for ISDN Basic Rate Access—Digital Networks, Digital Sections and Digital Line Systems (Study Groups XV and XVIII) 38 pp. 38 pp.

CENELEC PRETS 300 297-93. Integrated Services Digital Network (ISDN); Access Digital Section for ISDN Basic Rate. 45 pp.

ETSI PRETS 300 297-93. Integrated Services Digital Network (ISDN); Access Digital Section for ISDN Basic Rate. 45 pp.

—**Dimensioning**

CCITT RECMN E.730-92. ISDN Dimensioning Methods Overview (Study Group II) 4 pp. 4 pp.

INTERNATIONAL AND NON-U.S. NATIONAL STANDARDS
SUBJECT INDEX
Integrated

Integrated Services Digital Networks *(Cont.)*

—Direct Dialing In Services

CCITT RECMN E.164-91. Numbering Plan for the ISDN Era (Study Group II) 19 pp (Same as Recmn I.331). 19 pp.
CCITT RECMN E.164-89. Numbering Plan for the ISDN Era—Telephone Network and ISDN—Operation, Numbering, Routing and Mobile Service (Study Group II) 6 pp (Same as Recmn I.331). 6 pp.
CCITT RECMN I.251-89. Number Identification Supplementary Services—Integrated Services Digital Network (ISDN)—General Structure and Service Capabilities (Study Group XVIII) 28 pp. 28 pp.
CCITT RECMN I.251. 1 (REV 1)-92. Direct-Dialling-in (Study Group I) 7 pp. 7 pp.
CENELEC PRETS 300 062-90. Integrated Services Digital Network (ISDN); Direct Dialling in (DDI) Supplementary Service Description (T/NAI(89) 19). 10 pp.
CENELEC PRETS 300 062-91. Integrated Services Digital Network (ISDN); Direct Dialling In (DDI) Supplementary Service Service Description. 14 pp.
CENELEC ETS 300 062-91. Integrated Services Digital Network (ISDN); Direct Dialling in (DDI) Supplementary Service Service Description. 14 pp.
CENELEC PRETS 300 063-90. Integrated Services Digital Network (ISDN); Direct Dialling in (DDI) Supplementary Service Functional Capabilities and Information Flows (T/S 22-27). 11 pp.
CENELEC PRETS 300 063-91. Integrated Services Digital Network (ISDN); Direct Dialling In (DDI) Supplementary Service Functional Capabilities and Information Flows. 14 pp.
CENELEC ETS 300 063-91. Integrated Services Digital Network (ISDN); Direct Dialing in (DDI) Supplementary Service Functional Capabilities and Information Flows. 14 pp.
ETSI PRETS 300 062-90. Integrated Services Digital Network (ISDN); Direct Dialling in (DDT) Supplementary Service—Service Description (T/NA 1 (89) 19). 10 pp.
ETSI PRETS 300 063-90. Integrated Services Digital Network (ISDN); Direct Dialling in (DDI) Supplementary Service Functional Capabilities and Information Flows (T/S 22-27). 11 pp.
ETSI ETS 300 062-91. Integrated Services Digital Network (ISDN); Direct Dialling in (DDI) Supplementary Service Service Description. 14 pp.
ETSI PRETS 300 062-90. Integrated Services Digital Network (ISDN); Direct Dialling in (DDI) Supplementary Service Description (T/NAI(89) 19). 10 pp.
ETSI ETS 300 063-91. Integrated Services Digital Network (ISDN); Direct Dialing in (DDI) Supplementary Service Functional Capabilities and Information Flows. 14 pp.
ETSI PRETS 300 064-91. Integrated Services Digital Network (ISDN); Direct Dialling In (DDI) Supplementary Service Digital Subscriber Signalling System No. One (DSS1) Protocol. 10 pp.

—Direct Dialing In Services—CCITT No. 7 Signaling Systems

CCITT RECMN Q.730-89. ISDN Supplementary Services—Specifications of Signalling System No. 7 (Study Group XI) 59 pp. 59 pp.
CCITT RECMN Q.731-91. Stage 3 Description for Number Identification Supplementary Services Using Signalling System No. 7 Section 1—Direct Dialling in (DDI) Section 8—Sub-Addressing (SUB)—Specifications of Signalling System No. 7 (Study Group XI) 12 pp. 12 pp.

—Direct Dialing In Services—DSS1

CCITT RECMN Q.951-92. Stage 3 Description for Number Identification Supplementary Services Using DSS 1 Section 1—Direct-Dialling-in (DDI) Section 2—Multiple Subscriber Number (MSN) Section 8—Sub-Addressing (SUB) (Study Group XI) 17 pp. 17 pp.
CENELEC PRETS 300 064-90. Integrated Services Digital Network (ISDN); Direct Dialling in (DD1) Supplementary Service Digital Subscriber Signalling One (DSS1) Protocol (T/S 46-33A). 8 pp.
CENELEC PRETS 300 064-91. Integrated Services Digital Network (ISDN); Direct Dialling In (DDI) Supplementary Service Digital Subscriber Signalling System No. One (DSS1) Protocol. 10 pp.
CENELEC ETS 300 064-91. Integrated Services Digital Network (ISDN); Direct Dialling In (DDI) Supplementary Service Digital Subscriber Signalling System No. One (DSS1) Protocol. 10 pp.
ETSI PRETS 300 064-90. Integrated Services Digital Network (ISDN) Direct Dialling in (DDI) Supplementary Service Digital Subscriber Signalling One (DSS1) Protocol (T/S 46-33A). 8 pp.
ETSI ETS 300 064-91. Integrated Services Digital Network (ISDN); Direct Dialling In (DDI) Supplementary Service Digital Subscriber Signalling System No. One (DSS1) Protocol. 10 pp.

Integrated Services Digital Networks *(Cont.)*

—Direct Dialing In Services—End of Address

CCITT RECMN E.164-91. Numbering Plan for the ISDN Era (Study Group II) 19 pp (Same as Recmn I.331). 19 pp.
CCITT RECMN E.164-89. Numbering Plan for the ISDN Era—Telephone Network and ISDN—Operation, Numbering, Routing and Mobile Service (Study Group II) 6 pp (Same as Recmn I.331). 6 pp.

—Direct Dialing In Services—Functions/Information Flows

CCITT RECMN Q.81-91. Stage 2 Description for Number Identification Supplementary Services (Study Group XI) 46 pp. 46 pp.
CCITT RECMN Q.81-89. Number Identification Supplementary Services—General Recommendations on Telephone Switching and Signalling—Functions and Information Flows for Services in the ISDN—Supplements (Study Group XI) 30 pp. 30 pp.

—European Automatic Freephone Service

CENELEC PRETS 300 208-92. Integrated Services Digital Network (ISDN); Freephone (FPH) Supplementary Service Service Description. 16 pp.
CENELEC PRETS 300 209-92. Integrated Services Digital Network (ISDN); Freephone (FPH) Supplementary Service Functional Capabilities and Information Flows. 61 pp.
ETSI PRETS 300 208-92. Integrated Services Digital Network (ISDN); Freephone (FPH) Supplementary Service Service Description. 16 pp.
ETSI PRETS 300 209-92. Integrated Services Digital Network (ISDN); Freephone (FPH) Supplementary Service Functional Capabilities and Information Flows. 61 pp.

—European Automatic Freephone Service—DSS1

CENELEC PRETS 300 210-92. Integrated Services Digital Network (ISDN); Freephone (FPH) Supplementary Service Digital Subscriber Signalling System No. One (DSS1) Protocol. 22 pp.
ETSI PRETS 300 210-92. Integrated Services Digital Network (ISDN); Freephone (FPH) Supplementary Service Digital Subscriber Signalling System No. One (DSS1) Protocol. 22 pp.

—FAX Communications

CCITT RECMN I.241. 3-88. Telefax 4—Integrated Services Digital Network (ISDN)—General Structure and Service Capabilities (Study Group XVIII) 8 pp. 8 pp.

—FAX Communications—CCITT Group 4

CCITT RECMN F.184-89. Operational Provisions for the International Public Facsimile Service Between Subscriber Stations with Group 4 Facsimile Machines (Telefax 4)—Telematic, Data Transmission and Teleconference Services—Operations and Quality of Service (Study Group I) 9 pp. 9 pp.
CENELEC PRETS 300 280-92. Terminal Equipment (TE); Facsimile Group 4 Class 1 Equipment on the Integrated Services Digital Network (ISDN) Terminal Testing. 44 pp.
ETSI PRETS 300 280-92. Terminal Equipment (TE); Facsimile Group 4 Class 1 Equipment on the Integrated Services Digital Network (ISDN) Terminal Testing. 44 pp.

—FAX Machines

CCITT RECMN I.333-89. Terminal Selection in ISDN—Integrated Services Digital Network (ISDN)—Overall Network Aspects and Functions, ISDN User-Network Interfaces (Study Group XVIII) 18 pp. 18 pp.
CCITT RECMN T.563-91. Terminal Characteristics for Group 4 Facsimile Apparatus (Study Group VIII) 25 pp. 25 pp.
CCITT RECMN T.563-89. Terminal Characteristics for Group 4 Facsimile Apparatus—Terminal Equipment and Protocols for Telematic Services (Study Group VIII) 19 pp. 19 pp.
CENELEC PRETS 300 087-91. Integrated Services Digital Network (ISDN); Facsimile Group 4 Class 1 Equipment on the ISDN Functional Specification of the Equipment (T/TE 05.09). 12 pp.
CENELEC PRETS 300 088-91. Integrated Services Digital Network (ISDN); Facsimile Group 4 Class 1 Equipment on the ISDN General and Service Aspects (T/TE 05-06). 10 pp.
CENELEC PRETS 300 112-90. Integrated Services Digital Network (ISDN): Facsimile Group 4 Class 1 Equipment on the ISDN End-to-End Protocols (T/TE 05-07). 25 pp.
CENELEC PRETS 300 112-90. Integrated Services Digital Network (ISDN); Facsimile Group 4 Class 1 Equipment on the ISDN End-to-End Protocols (T/TE 05-07). 48 pp.

Integrated Services Digital Networks *(Cont.)*

—FAX Machines *(Cont.)*

CENELEC PRETS 300 120-90. Integrated Services Digital Network (ISDN). 15 pp.
CENELEC PRETS 300 120-92. Integrated Services Digital Network (ISDN); Service Requirements for Telefax Group 4. 16 pp.
CENELEC ETS 300 120-92. Integrated Services Digital Network (ISDN); Service Requirements for Telefax Group 4. 16 pp.
CENELEC PRETS 300 155-91. Integrated Services Digital Network (ISDN); Facsimile Group 4, Class 1 Equipment on the ISDN End-to-End Protocols Tests (T/TE 05-08). 201 pp.
ETSI PRETS 300 087-91. Integrated Services Digital Network (ISDN); Facsimile Group 4 Class 1 Equipment on the ISDN-Functional Specification of the Equipment (T/TE 05-09).
ETSI PRETS 300 088-91. Integrated Services Digital Network (ISDN); Facsimile Group 4 Class 1 Equipment on the ISDN—General and Service Aspects (T/TE 05-6).
ETSI PRETS 300 112-90. Integrated Services Digital Network (ISDN); Facsimile Group 4 Class 1 Equipment on the ISDN—End-To-End Protocols (T/TE 05-07). 48 pp.
ETSI PRETS 300 120-90. Integrated Services Digital Network (ISDN) Telefax G4 (T/NAI (90) 02). 14 pp.
ETSI PRETS 300 155-91. Integrated Services Digital Network (ISDN); Facsimile Group 4, Class 1 Equipment on the ISDN—End-to-End Protocols Test (T/TE 05-08). 201 pp.
ETSI PRETS 300 087-91. Integrated Services Digital Network (ISDN); Facsimile Group 4 Class 1 Equipment on the ISDN Functional Specification of the Equipment (T/TE 05.09). 12 pp.
ETSI PRETS 300 088-91. Integrated Services Digital Network (ISDN); Facsimile Group 4 Class 1 Equipment on the ISDN General and Service Aspects (T/TE 05-06). 10 pp.
ETSI PRETS 300 112-90. Integrated Services Digital Network (ISDN); Facsimile Group 4 Class 1 Equipment on the ISDN End-to-End Protocols (T/TE 05-07). 48 pp.
ETSI PRETS 300 112-90. Integrated Services Digital Network (ISDN): Facsimile Group 4 Class 1 Equipment on the ISDN End-to-End Protocols (T/TE 05-07). 25 pp.
ETSI ETS 300 120-92. Integrated Services Digital Network (ISDN); Service Requirements for Telefax Group 4. 16 pp.
ETSI PRETS 300 120-92. Integrated Services Digital Network (ISDN); Service Requirements for Telefax Group 4. 16 pp.
ETSI PRETS 300 120-90. Integrated Services Digital Network (ISDN). 15 pp.
ETSI PRETS 300 155-91. Integrated Services Digital Network (ISDN); Facsimile Group 4, Class 1 Equipment on the ISDN End-to-End Protocols Tests (T/TE 05-08). 201 pp.

—FAX Machines—CCITT Group 4—Interworking

CCITT RECMN F.353-89. Provision of Telematic and Data Transmission Services on Integrated Services Digital Network (ISDN)—Telematic, Data Transmission and Teleconference Services—Operations and Quality of Service (Study Group I) 5 pp. 5 pp.

—Fixed Satellite Services

CCIR RESOLUTION 103-90. Liasion and Joint Studies with the CCITT in Mobile and Fixed-Satellite Services and High-Definition Television in ISDNs—Volume XIV—Administrative Texts of the CCIR. 2 pp.

—Fixed Satellite Services—Hypothetical Reference Digital Path

CCIR Report 997-1-90. Characteristics of a Fixed-Satellite Service Hypothetical Reference Digital Path Forming Part of an Integrated Services Digital Network —Section 4B2—Performance and Availability. 33 pp.

—Fixed Satellites—Hypothetical Reference Digital Path—Error

CCIR RECMN 614-1-90. Allowable Error Performance for a Hypothetical Reference Digital Path in the Fixed-Satellite Service Operating Below 15 GHz When Forming Part of an International Connection in an Integrated Services Digital Network —Section 4B2—Perform-ance and Availability. 2 pp.
CCIR RECMN 614-2-92. Allowable Error Performance for a Hypothetical Reference Digital Path in the Fixed-Satellite Service Operating Below 15 GHz When Forming Part of an International Connection in an Integrated Services Digital Network —Section 4B2—Perform-ance and Availability. 42 pp.

INDUSTRY STANDARDS
1413

Integrated Services Digital Networks (Cont.)

—Flow Control Protocols
CCITT RECMN I.320-89. ISDN Protocol Reference Model—Integrated Services Digital Network (ISDN)—Overall Network Aspects and Functions, ISDN User-Network Interfaces (Study Group XVIII) 10 pp. 10 pp.
CENELEC ETR 054-92. Integrated Services Digital Network (ISDN); the Protocol Reference Model and Its Relationship with the OSI Reference Model. 13 pp.
ETSI ETR 054-92. Integrated Services Digital Network (ISDN); the Protocol Reference Model and Its Relationship with the OSI Reference Model. 13 pp.

—Frame Relaying Bearer Services
CCITT RECMN I.233-92. Frame Mode Bearer Services (Study Group XVIII) 49 pp. 49 pp.

—General Structure—Preamble
CCITT RECMN I.110-89. Preamble and General Structure of the I-Series Recommendations for the Integrated Services Digital Network (ISDN)—Integrated Services Degital Network (ISDN)—General Structure and Service Capabilities (Study Group XVIII) 7 pp. 7 pp.

—General Structure—Services
CCITT FASCICLE III.7-88. Integrated Services Digital Network (ISDN) General Structure and Service Capabilities Recommendations I.110-I.257 (Study Group XVIII). 353 pp.

—Glossaries
CCITT RECMN I.112-89. Vocabulary of Terms for ISDNs—Integrated Services Digital Network (ISDN)—General Structure and Service Capabilities (Study Group XVIII) 13 pp. 13 pp.
CEPT T/GSI 01-02 E-85. ISDN Definition. 1 p.
CEPT T/GSI 01-03-85. Vocabulaire. 1 p.
CEPT T/GSI 01-03 E-85. Vocabulary. 1 p.
CEPT T/GSI 02-01-87. Definition Des Services. 9 pp.
CEPT T/GSI 02-01 E-87. Service Definition. 9 pp.

—Grade of Service—Dimensioning
CCITT RECMN E.730-92. ISDN Dimensioning Methods Overview (Study Group II) 4 pp. 4 pp.

—High Definition Television Systems
CCIR RESOLUTION 103-90. Liasion and Joint Studies with the CCITT in Mobile and Fixed-Satellite Services and High-Definition Television in ISDNs—Volume XIV—Administrative Texts of the CCIR. 2 pp.

—Hypothetical Reference Circuits—Fixed Satellite Services
CCIR RECMN 579-1-86. Availability Objectives for a Hypothetical Reference Circuit and a Hypothetical Reference Digital Path When Used for Telephony Using Pulse-Code Modulation, or as Part of an Integrated Services Digital Network Hypothetical Reference Connection, in the. 2 pp.
CCIR RECMN 579-2-92. Availability Objectives for a Hypothetical Reference Circuit and a Hypothetical Reference Digital Path When Used for Telephony Using Pulse-Code Modulation, or as Part of an Integrated Services Digital Network Hypothetical Reference Connection, in the. 6 pp.

—Hypothetical Reference Digital Path—Fixed Satellite Services
CCIR RECMN 579-1-86. Availability Objectives for a Hypothetical Reference Circuit and a Hypothetical Reference Digital Path When Used for Telephony Using Pulse-Code Modulation, or as Part of an Integrated Services Digital Network Hypothetical Reference Connection, in the. 2 pp.
CCIR RECMN 579-2-92. Availability Objectives for a Hypothetical Reference Circuit and a Hypothetical Reference Digital Path When Used for Telephony Using Pulse-Code Modulation, or as Part of an Integrated Services Digital Network Hypothetical Reference Connection, in the. 6 pp.

—Hypothetical Reference Digital Paths—Radio Relay—Availability
CCIR RECMN 696-90. Error Performance and Availability Objectives for Hypothetical Reference Digital Sect. Utilizing Digital Radio-Relay Systems Forming Part or All of the Medium Grade Portion of an ISDN Connection—Section 9A—Performance Objectives, Propagation and Interference Effects. 3 pp.
CCIR RECMN 696-1-91. Error Performance and Availability Objectives for Hypothetical Reference Digital Sect. Utilizing Digital Radio-Relay Systems Forming Part or All of the Medium-Grade Portion of an ISDN Connection—Section 9A—Performance Objectives, Propagation and Interference Effects. 4 pp.

—Hypothetical Reference Digital Paths—Radio Relay—Bit Error Rate
CCIR RECMN 594-2-90. Allowable Bit Error Ratios at the Output of the Hypothetical Reference Digital Path for Radio-Relay Systems Which May Form Part of an Integrated Services Digital Network—Section 9A—Performance Objectives, Propagation and Interference Effects. 2 pp.
CCIR RECMN 594-3-91. Allowable Bit Error Ratios at the Output of the Hypothetical Reference Digital Path for Radio-Relay Systems Which May Form Part of an Integrated Services Digital Network—Section 9A—Performance Objectives, Propagation and Interference Effects. 4 pp.
CCIR RECMN 696-90. Error Performance and Availability Objectives for Hypothetical Reference Digital Sect. Utilizing Digital Radio-Relay Systems Forming Part or All of the Medium Grade Portion of an ISDN Connection—Section 9A—Performance Objectives, Propagation and Interference Effects. 3 pp.
CCIR RECMN 696-1-91. Error Performance and Availability Objectives for Hypothetical Reference Digital Sect. Utilizing Digital Radio-Relay Systems Forming Part or All of the Medium-Grade Portion of an ISDN Connection—Section 9A—Performance Objectives, Propagation and Interference Effects. 4 pp.

—Hypothetical Reference Digital Paths—Radio Relay Systems
CCIR RECMN 556-1-86. Hypothetical Reference Digital Path for Radio-Relay Systems Which May Form Part of an Integrated Services Digital Network with a Capacity Above the Second Hierarchical Level—Section 9A—Performance Objectives, Propagation and Interference Effects. 2 pp.

—Information Interchange
CCITT RECMN I.320-89. ISDN Protocol Reference Model—Integrated Services Digital Network (ISDN)—Overall Network Aspects and Functions, ISDN User-Network Interfaces (Study Group XVIII) 10 pp. 10 pp.
CENELEC ETR 054-92. Integrated Services Digital Network (ISDN); the Protocol Reference Model and Its Relationship with the OSI Reference Model. 13 pp.
ETSI ETR 054-92. Integrated Services Digital Network (ISDN); the Protocol Reference Model and Its Relationship with the OSI Reference Model. 13 pp.

—INMARSAT—Interfaces
CCITT RECMN Q.1111-89. Interfaces Between the Inmarsat Standard B System and the International Public Switched Telephone Network/ISDN—Interworking with Satellite Mobile Systems (Study Group XI) 22 pp. 22 pp.
CCITT RECMN Q.1151-89. Interfaces Between the Inmarsat Aeronautical Mobile-Satellite System and the International Public Switched Telephone Network/ISDN—Interworking with Satellite Mobile Systems (Study Group XI) 19 pp. 19 pp.

—INMARSAT—Interworking
CCITT RECMN Q.1100-89. Structure of the Recommendations on the Inmarsat Mobile Satellite Systems—Interworking with Mobile Systems (Study Group XI) 3 pp. 3 pp.
CCITT RECMN Q.1112-89. Procedures for Interworking Between Inmarsat Standard B System and the International Public Switched Telephone Network/ISDN—Interworking with Satellite Mobile Systems (Study Group XI) 51 pp. 51 pp.
CCITT RECMN Q.1152-89. Procedures for Interworking Between Inmarsat Aeronautical Mobile-Satellite System and the International Public Switched Telephone Network/ISDN—Interworking with Satellite Mobile Systems (Study Group XI) 40 pp. 40 pp.

—Installation Maintenance
CEPT T/S 54-08 E-87. ISDN Subscriber Access and Installation Maintenance. 22 pp.

—Interfaces
CCITT FASCICLE III.8-88. Integrated Services Digital Network (ISDN) Overall Network Aspects and Functions, ISDN User—Network Interfaces Recommendations I.310—I.470 (Study Group XVIII). 280 pp.
CCITT RECMN I.310-89. ISDN—Network Functional Principles—Integrated Services Digital Network (ISDN)—Overall Network Aspects and Functions, ISDN User -Network Interfaces (Study Group XVIII) 22 pp. 22 pp.
CCITT RECMN I.324-89. ISDN Network Architecture—Integrated Services Digital Network (ISDN)—Overall Network Aspects and Functions, ISDN User-Network Interfaces (Study Group XVIII) 12 pp. 12 pp.
CCITT RECMN I.325-89. Reference Configurations for ISDN Connection Types—Integrated Services Digital Network (ISDN)—Overall Network Aspects and Functions, ISDN User-Network Interfaces (Study Group XVIII) 7 pp. 7 pp.
CCITT RECMN I.410-89. General Aspects and Principles Relating to Recommendations on ISDN User-Network Interfaces—Integrated Services Digital Network (ISDN)—Overall Network Aspects and Functions, ISDN User -Network Interfaces (Study Group XVIII) 3 pp. 3 pp.
CCITT RECMN I.411-89. ISDN User-Network Interfaces—Reference Configurations—Integrated Services Digital Network (ISDN)—Overall Network Aspects and Functions, ISDN User-Network Interfaces (Study Group XVIII) 7 pp. 7 pp.
CCITT RECMN I.412-89. ISDN User-Network Interfaces Interface Structures and Access Capabilities—Integrated Services Digital Network (ISDN)—Overall Network Aspects and Functions, ISDN User -Network Interfaces (Study Group XVIII) 6 pp. 6 pp.
CCITT RECMN I.420-89. Basic User-Network Interface—Integrated Services Digital Network (ISDN)—Overall Network Aspects and Functions, ISDN User-Network Interfaces (Study Group XVIII) 1 pp. 1 p.
CCITT RECMN I.421-89. Primary Rate User-Network Interface—Integrated Services Digital Network (ISDN)—Overall Network Aspects and Functions, ISDN User-Network Interfaces (Study Group XVIII) 1 pp. 1 p.
CCITT FASCICLE III.9-88. Integrated Services Digital Network (ISDN) Internetwork Interfaces and Maintenance Principles Recommendations I.500—I.605 (Study Group XVIII). 103 pp.
CENELEC PRETS 300 011-90. Integrated Services Digital Network (ISDN): Primary rate User-Network Interface Layer 1 Specification and Test Principles (L 03-14). 35 pp.
CENELEC PRETS 300 011-91. Integrated Services Digital Network (ISDN); Primary Rate User-Network Interface Layer 1 Specification and Test Principles. 80 pp.
CENELEC ETS 300 011-92. Integrated Services Digital Network (ISDN); Primary Rate User-Network Interface Layer 1 Specification and Test Principles. 79 pp.
CENELEC PRETS 300 012-90. Integrated Services Digital Network (ISDN): Basic User-Network Interface Layer 1 Specification and Test Principles (L 03-07). 40 pp.
CENELEC PRETS 300 012-91. Integrated Services Digital Network (ISDN); Basic User-Network Interface Layer 1 Specification and Test Principles. 179 pp.
CENELEC ETS 300 012-92. Integrated Services Digital Network (ISDN); Basic User-Network Interface Layer 1 Specification and Test Principles. 177 pp.
CENELEC PRETS 300 102-2-90. Integrated Services Digital Network (ISDN): User-Network Interface Layer 3 Specification Description Language (SDL) Diagrams. 72 pp.
CENELEC ETS 300 102-2-90. Integrated Services Digital Network (ISDN); User-Network Interface Layer 3 Specifications for Basic Call Control Specification Description Language (SDL) Diagrams. 75 pp.
CEPT T/CAC S 10.4 E-92. Intercommunication Aspects of an Integrated Services Digital Network (ISDN) (Revised in Odense 1986, Vienna 1989 (CAC) and Athens 1992). 11 pp.
CEPT T/GSI 04-01-87. Interfaces Usager-Reseau RNIS Structures D'Interface Et Possibilites D'Acces. 6 pp.
CEPT T/L 03-13 E-88. Digital Section for ISDN Basic Access. 1 p.
CEPT T/L 03-14 E-88. Primary Rate User-Network Interface Layer 1 Specification. 7 pp.
CSA CAN/CSA-T540-M92. Integrated Services Digital Network (ISDN) Primary Rate—Customer Installation Metallic Interfaces—Layer 1 Specification (ANSI T1.408-1990); (Gen Instr 1). 69 pp.
ETSI PRETS 300 011-90. Integrated Service Digital Network (ISDN) Primary Rate User-Network Interface Layer 1 Specification and Test Principles. 35 pp.
ETSI PRETS 300 012-90. Integrated Services Digital Network (ISDN) Basic User-Network Interface Layer 1 Specification and Test Principles. 40 pp.
ETSI ETS 300 011-92. Integrated Services Digital Network (ISDN); Primary Rate User-Network Interface Layer 1 Specification and Test Principles. 79 pp.

INTERNATIONAL AND NON-U.S. NATIONAL STANDARDS
SUBJECT INDEX
Integrated

Integrated Services Digital Networks (Cont.)
—Interfaces (Cont.)

ETSI PRETS 300 011-91. Integrated Services Digital Network (ISDN); Primary Rate User-Network Interface Layer 1 Specification and Test Principles. 80 pp.

ETSI PRETS 300 011-90. Integrated Services Digital Network (ISDN): Primary rate User-Network Interface Layer 1 Specification and Test Principles (L 03-14). 35 pp.

ETSI ETS 300 012-92. Integrated Services Digital Network (ISDN); Basic User-Network Interface Layer 1 Specification and Test Principles. 177 pp.

ETSI PRETS 300 012-91. Integrated Services Digital Network (ISDN); Basic User-Network Interface Layer 1 Specification and Test Principles. 179 pp.

ETSI PRETS 300 012-90. Integrated Services Digital Network (ISDN): Basic User-Network Interface Layer 1 Specification and Test Principles (L 03-07). 40 pp.

ETSI PRETS 300 102-2-90. Integrated Services Digital Network (ISDN); User-Network Interface Layer 3—Specifications for Basic Call Control—Specification Description Language (SDL) Diagrams. 73 pp.

ETSI PRETS 300 121-90. Integrated Services Digital Network (ISDN); Application of the ISDN User Part of CCITT Signalling System No.7 for International ISDN Interconnections CCITT Recommendation Q.767 Draft Edition 3:1990—Modified (ISUP Version 1—T/S 43-14). 239 pp.

ETSI PRETS 300 126-90. Intergrated Services Digital Network (ISDN) Equipment with ISDN Interface at Basic and Primary Rate EMC Requirements (DE/EE-4001). 24 pp.

ETSI PRETS 300 102-1-90. Integrated Services Digital Network (ISDN); User-Network Interface Layer 3 Specifications for Basic Call Control. 233 pp.

ETSI ETS 300 102-2-90. Integrated Services Digital Network (ISDN); User-Network Interface Layer 3 Specifications for Basic Call Control Specification Description Language (SDL) Diagrams. 75 pp.

ETSI PRETS 300 102-2-90. Integrated Services Digital Network (ISDN): User-Network Interface Layer 3 Specification Description Language (SDL) Diagrams. 72 pp.

ETSI PRETS 300 126-90. Integrated Services Digital Network (ISDN) Equipment with ISDN Interface at Basic and Primary Rate EMC Requirements (DE/EE-4001). 24 pp.

—Interfaces—Communication Channels

CCITT RECMN I.412-89. ISDN User-Network Interfaces Interface Structures and Access Capabilities—Integrated Services Digital Network (ISDN)—Overall Network Aspects and Functions, ISDN User -Network Interfaces (Study Group XVIII) 6 pp. 6 pp.

CCITT RECMN I.431-89. Primary Rate User-Network Interface—Layer 1 Specification—Integrated Services Digital Network (ISDN)—Overall Network Aspects and Functions, ISDN User-Network Interfaces (Study Group XVIII) 29 pp. 29 pp.

—Interfaces—Communication Channels—Glossaries

CCITT RECMN I.431-89. Primary Rate User-Network Interface—Layer 1 Specification—Integrated Services Digital Network (ISDN)—Overall Network Aspects and Functions, ISDN User-Network Interfaces (Study Group XVIII) 29 pp. 29 pp.

—Interfaces—Communication Networks

CCITT FASCICLE VIII.2. Data Communications Networks: Services and Facilities, Interfaces Recommendations X.1–X.32 (Study Group VII). 563 pp.

—Interfaces—Communication Terminal Equipment

CCITT RECMN I.410-89. General Aspects and Principles Relating to Recommendations on ISDN User-Network Interfaces—Integrated Services Digital Network (ISDN)—Overall Network Aspects and Functions, ISDN User -Network Interfaces (Study Group XVIII) 3 pp. 3 pp.

—Interfaces—Data Link Layer—Private Branch Exchanges

CEPT T/S 49-20 E-87. ISDN ISPBX-ISPBX Interface Layer 2 Specification Application of CEPT Recommendation T/S 46-20. 9 pp.

—Interfaces—Data Links

CEPT T/S 46-20-87. Specification De La Couche De Liaisonde Donnees a L'Interface Usager-Reseau Dans Un RNIS; Application Des Recommandations CCITT Q.920/I.440 Et Q.921/I.441. 188 pp.

CEPT T/S 46-20 E-87. ISDN User-Network Interface Data Link Layer Specification Application of CCITT Recommendations Q.920/I.440 and Q.921/I.441. 168 pp.

Integrated Services Digital Networks (Cont.)
—Interfaces—Data Links (Cont.)

CSA CAN/CSA-T542-90. Integrated Services Digital Network (ISDN)—Data-Link Layer Signalling Specification for Application at the User-Network Interface (ANSI T1.602-1989). 199 pp.

—Interfaces—Data Terminal Equipment

CCITT RECMN I.463-89. Support of Data Terminal Equipments (DTEs) with V-Series Type Interfaces by an Integrated Services Digital Network (ISDN)—Integrated Services Network (ISDN)—Overall Network Aspects and Functions, ISDN User-Network Interfaces (Study Group XVIII) 1 pp. 1 p.

CCITT RECMN I.465-89. Support by an ISDN of Data Terminal Equipment with V-Series Type Interfaces with Provision for Statisical Multiplexing—Integrated Services Digital Network Aspects and Functions, ISDN User-Network Interfaces (Study Group XVIII) 1 pp (Same as Recmn V.120). 1 p.

CCITT RECMN V.110-92. Support of Data Terminal Equipments with V-Series Type Interfaces by an Integrated Services Digital Network (Study Group XVII) 60 pp. 60 pp.

CCITT RECMN V.110-89. Support of Data Terminal Equipments (DTEs) with V-Series Type Interfaces by an Integrated Services Digital Network (ISDN)—Data Communication Over the Telephone Network (Study Group XVII) 47 pp (Same as Recmn I.463). 47 pp.

CCITT RECMN V.120-92. Support by an ISDN of Data Terminal Equipment with V-Series Type Interfaces with Provision for Statistical Multiplexing (Study Group XVII) 38 pp. 38 pp.

CCITT RECMN V.120-89. Support by an ISDN of Data Terminal Equipment with V-Series Type Interfaces with Provision for Statistical Multiplexing—Data Communication over the Telephone Network (Study Group XVII) 26 pp (Same as Recmn I.465). 26 pp.

ECMA ECMA-TR 51-90. Requirements for Access to Integrated Voice and Data Local and Metropolitan Area Networks. 62 pp.

—Interfaces—Dedicated Services

CCITT RECMN I.410-89. General Aspects and Principles Relating to Recommendations on ISDN User-Network Interfaces—Integrated Services Digital Network (ISDN)—Overall Network Aspects and Functions, ISDN User -Network Interfaces (Study Group XVIII) 3 pp. 3 pp.

—Interfaces—Electromagnetic Compatibility

CENELEC PREN 50096-91. Integrated Services Digital Network (ISDN) Equipment with ISDN User-Network Interface at Basic and Primary Rate—EMC Requirements. 34 pp.

CENELEC PRETS 300 126-90. Integrated Services Digital Network (ISDN) Equipment with ISDN Interface at Basic and Primary Rate EMC Requirements (DE/EE-4001). 24 pp.

—Interfaces—Local Area Networks

ECMA ECMA-TR 51-90. Requirements for Access to Integrated Voice and Data Local and Metropolitan Area Networks. 62 pp.

—Interfaces—Metropolitan Area Networks

ECMA ECMA-TR 51-90. Requirements for Access to Integrated Voice and Data Local and Metropolitan Area Networks. 62 pp.

—Interfaces—Network Capabilities

CCITT RECMN I.412-89. ISDN User-Network Interfaces Interface Structures and Access Capabilities—Integrated Services Digital Network (ISDN)—Overall Network Aspects and Functions, ISDN User -Network Interfaces (Study Group XVIII) 6 pp. 6 pp.

—Interfaces—Network Services (OSI)

CCITT RECMN E.330-89. User Control of ISDN-Supported Services—Telephone Network and ISDN—Operation, Numbering, Routing and Mobile Service (Study Group II) 2 pp. 2 pp.

CCITT RECMN I.111-89. Relationship with Other Recommendations Relevant to ISDNs—Integrated Services Digital Network (ISDN)—General Structure and Service Capabilities (Study Group XVIII) 15 pp. 2 pp.

—Interfaces—Packet Switched Networks

CCITT RECMN X.32-89. Interface Between Data Terminal Equipment (DTE) and Data Circuit-Terminating Equipment (DCE) for Terminals Operating in the Packet Mode and Accessing a Packet Switched Public Data Network Through a Public Switched Telephone Network or an Integrated Services. 57 pp.

Integrated Services Digital Networks (Cont.)
—Interfaces—PDN

CCITT RECMN X.32-89. Interface Between Data Terminal Equipment (DTE) and Data Circuit-Terminating Equipment (DCE) for Terminals Operating in the Packet Mode and Accessing a Packet Switched Public Data Network Through a Public Switched Telephone Network or an Integrated Services. 57 pp.

—Interfaces—Protocols

CEPT T/GSI 04-02/1-87. Protocoles Et Interfaces Usager-Reseau (Couche 1). 1 p.

CEPT T/GSI 04-02/1E-87. User to Network Interfaces and Protocols (Layer 1). 1 p.

CEPT T/GSI 04-02/2-85. Protocoles Et Interfaces Usager-Reseau (Couche 1). 1 p.

CEPT T/GSI 04-02/2E-85. User to Network Interfaces and Protocols (Layer 2). 1 p.

CEPT T/GSI 04-02/3-85. Protocoles Et Interfaces Usager-Reseau (Couche 1). 1 p.

CEPT T/GSI 04-02/3E-85. User to Network Interfaces and Protocols (Layer 3). 1 p.

—Interfaces—Protocols—Data Links

ECMA ECMA 105-90. Data Link Layer Protocol for the D-Channel of the Interfaces at the Reference Point Between Terminal Equipment and Private Telecommunication Networks. 68 pp.

—Interfaces—Protocols—Videotex Communications

ETSI PRETS 300 079-90. Integrated Services Digital Network (ISDN); Syntax-Based Videotex—End-To-End Protocols (T/TE 06-10). 82 pp.

ETSI ETS 300 079-91. Integrated Services Digital Network (ISDN); Syntax-Based Videotex End-to-End Protocols Circuit Mode DTE—DTE. 78 pp.

—Interfaces—Public Land Mobile Networks

CCITT RECMN E.220-92. Interconnection of Public Land Mobile Networks (PLMN) (Study Group II) 8 pp. 8 pp.

—Interfaces—Public Land Mobile Networks—Routing

CCITT RECMN E.173-91. Routing Plan for Interconnection Between Public Land Mobile Networks and Fixed Terminal Networks (Study Group II) 15 pp. 15 pp.

—Interfaces—Public Switched Telephone Networks

CCITT RECMN D.251-89. General Charging and Accounting Principles for the Basic Telephone Service Provided over the ISDN or by Interconnection Between the ISDN and the Public Switched Telephone Network—General Tariff Principles—Charging and Accounting in International. 2 pp.

—Interfaces—Reference Configurations

CCITT RECMN I.411-89. ISDN User-Network Interfaces—Reference Configurations—Integrated Services Digital Network (ISDN)—Overall Network Aspects and Functions, ISDN User-Network Interfaces (Study Group XVIII) 7 pp. 7 pp.

—Interfaces—Safety

CENELEC PRETS 300 046-1-90. Integrated Services Digital Network (ISDN); Primary Rate Access—Safety and Protection Part 1: General (T/TE yyy-1). 19 pp.

CENELEC PRETS 300 046-1-92. Integrated Services Digital Network (ISDN); Primary Rate Access—Safety and Protection Part 1: General. 20 pp.

CENELEC ETS 300 046-1-92. Integrated Services Digital Network (ISDN); Primary Rate Access—Safety and Protection Part 1: General. 19 pp.

CENELEC PRETS 300 046-2-90. Integrated Services Digital Network (ISDN); Primary Rate Access—Safety and Protection Part 2: Interface 1a—Safety (T/TE yyy-2-1). 11 pp.

CENELEC PRETS 300 046-2-92. Integrated Services Digital Network (ISDN); Primary Rate Access—Safety and Protection Part 2: Interface Ia—Safety. 12 pp.

CENELEC ETS 300 046-2-92. Integrated Services Digital Network (ISDN); Primary Rate Access—Safety and Protection Part 2: Interface Ia—Safety. 11 pp.

CENELEC PRETS 300 046-3-90. Integrated Services Digital Network (ISDN); Primary Rate Access—Safety and Protection Part 3: Interface 1—Protection (T/TE yyy-2-2). 17 pp.

CENELEC PRETS 300 046-3-92. Integrated Services Digital Network (ISDN); Primary Rate Access—Safety and Protection Part 3: Interface Ia—Protection. 17 pp.

INDUSTRY STANDARDS

INTERNATIONAL AND NON-U.S. NATIONAL STANDARDS
SUBJECT INDEX

Integrated

Integrated Services Digital Networks (Cont.)

—**Interfaces—Safety** (Cont.)

CENELEC ETS 300 046-3-92. Integrated Services Digital Network (ISDN); Primary Rate Access—Safety and Protection Part 3: Interface Ia—Protection. 16 pp.

CENELEC PRETS 300 046-4-90. Integrated Services Digital Network (ISDN); Primary Rate Access—Safety and Protection Part 4: Interface 16—Safety (T/TE yyy-2-3). 11 pp.

CENELEC PRETS 300 046-4-92. Integrated Services Digital Network (ISDN); Primary Rate Access—Safety and Protection Part 4: Interface Ib—Safety. 11 pp.

CENELEC ETS 300 046-4-92. Integrated Services Digital Network (ISDN); Primary Rate Access—Safety and Protection Part 4: Interface Ib—Safety. 10 pp.

CENELEC PRETS 300 046-5-90. Integrated Services Digital Network (ISDN); Primary Rate Access—Safety and Protection Part 5: Interface 1b—Protection (T/TE yyy-2-4). 17 pp.

CENELEC PRETS 300 046-5-92. Integrated Services Digital Network (ISDN); Primary Rate Access—Safety and Protection Part 5: Interface Ib—Protection. 16 pp.

CENELEC ETS 300 046-5-92. Integrated Services Digital Network (ISDN); Primary Rate Access—Safety and Protection Part 5: Interface Ib—Protection. 16 pp.

CENELEC PRETS 300 047-1-90. Integrated Services Digital Network (ISDN); Basic Access—safety and Protection Part 1: General (T/TE zzz-1). 19 pp.

CENELEC PRETS 300 047-1-92. Integrated Services Digital Network (ISDN); Basic Access—Safety and Protection Part 1: General. 23 pp.

CENELEC ETS 300 047-1-92. Integrated Services Digital Network (ISDN); Basic Access—Safety and Protection Part 1: General. 21 pp.

CENELEC PRETS 300 047-2-90. Integrated Services Digital Network (ISDN); Basic Access—Safety and Protection Part 2: Interface Ia—Safety (T/TE zzz-2-1). 12 pp.

CENELEC PRETS 300 047-2-92. Integrated Services Digital Network (ISDN); Basic Access—Safety and Protection Part 2: Interface Ia—Safety. 12 pp.

CENELEC ETS 300 047-2-92. Integrated Services Digital Network (ISDN); Basic Access—Safety and Protection Part 2: Interface Ia—Safety. 11 pp.

CENELEC PRETS 300 047-3-90. Integrated Services Digital Network (ISDN); Basic Access—Safety and Protection Part 3: Interface Ia—Protection (T/TE zzz-2-2). 19 pp.

CENELEC PRETS 300 047-3-92. Integrated Services Digital Network (ISDN); Basic Access—Safety and Protection Part 3: Interface Ia—Protection. 20 pp.

CENELEC ETS 300 047-3-92. Integrated Services Digital Network (ISDN); Basic Access—Safety and Protection Part 3: Interface Ia—Protection. 19 pp.

CENELEC PRETS 300 047-4-90. Integrated Services Digital Network (ISDN) Basic Access—Safety and Protection Part 4: Interface 16—Safety (T/TE zzz-2-3). 11 pp.

CENELEC PRETS 300 047-4-92. Integrated Services Digital Network (ISDN); Basic Access—Safety and Protection Part 4: Interface Ib—Safety. 12 pp.

CENELEC ETS 300 047-4-92. Integrated Services Digital Network (ISDN); Basic Access—Safety and Protection Part 4: Interface Ib—Safety. 11 pp.

CENELEC PRETS 300 047-5-90. Integrated Services Digital Network (ISDN); Basic Access-Safety and Protection Part 5: Interface 16-Protection (T/TE zzz-2-4). 17 pp.

CENELEC PRETS 300 047-5-92. Integrated Services Digital Network (ISDN); Basic Access-Safety and Protection Part 5: Interface Ib—Protection. 18 pp.

CENELEC ETS 300 047-5-92. Integrated Services Digital Network (ISDN); Basic Access—Safety and Protection Part 5: Interface Ib—Protection. 17 pp.

CSA CAN/CSA-T540-M92. Integrated Services Digital Network (ISDN) Primary Rate—Customer Installation Metallic Interfaces—Layer 1 Specification (ANSI T1.408-1990); (Gen Instr 1). 69 pp.

ETSI PRETS 300 046-1-90. Integrated Services Digital Network (ISDN); Primary Rate Access—Safety and Protection—Part 1: General (T/TE yyy-1). 17 pp.

ETSI PRETS 300 046-2-90. Integrated Services Digital Network (ISDN); Primary Rate Access—Safety and Protection—Part 2: Interface Ia—Safety (T/TE yyy-2-1). 9 pp.

ETSI PRETS 300 046-3-90. Integrated Services Digital Network (ISDN); Primary Rate Access—Safety and Protection—Part 3: Interface Ia—Protection (T/TE yyy-2-2). 15 pp.

ETSI PRETS 300 046-4-90. Integrated Services Digital Network (ISDN); Primary Rate Access—Safety and Protection—Part 4: Interface Ib—Safety (T/TE yyy-2-3). 9 pp.

ETSI PRETS 300 046-5-90. Integrated Services Digital Network (ISDN); Primary Rate Access—Safety and Protection—Part 5: Interface Ib—Protection (T/TE yyy-2-4). 15 pp.

Integrated Services Digital Networks (Cont.)

—**Interfaces—Safety** (Cont.)

ETSI PRETS 300 047-1-90. Integrated Services Digital Network (ISDN); Basic Access—Safety and Protection—Part 1: General (T/TE ZZZ-1). 17 pp.

ETSI PRETS 300 047-2-90. Integrated Services Digital Network (ISDN); Basic Access—Safety and Protection—Part 2: Interface Ia-Safety (T/TE zzz-2-1). 10 pp.

ETSI PRETS 300 047-3-90. Integrated Services Digital Network (ISDN); Basic Access—Safety and Protection—Part 3: Interface Ia—Protection (T/TE zzz-2-2). 17 pp.

ETSI PRETS 300 047 004-90. Integrated Services Digital Network (ISDN); Basic Access—Safety and Protection—Part 4: Interface Ib—Safety (T/TE zzz-2-3). 9 pp.

ETSI PRETS 300 047-5-90. Integrated Services Digital Network (ISDN); Basic Access—Safety and Protection—Part 5: Interface Ib—Protection (T/TE zzz-2-4). 15 pp.

ETSI ETS 300 046-1-92. Integrated Services Digital Network (ISDN); Primary Rate Access—Safety and Protection Part 1: General. 19 pp.

ETSI PRETS 300 046-1-92. Integrated Services Digital Network (ISDN); Primary Rate Access—Safety and Protection Part 1: General. 20 pp.

ETSI PRETS 300 046-1-90. Integrated Services Digital Network (ISDN); Primary Rate Access—Safety and Protection Part 1: General (T/TE yyy-1). 19 pp.

ETSI ETS 300 046-2-92. Integrated Services Digital Network (ISDN); Primary Rate Access—Safety and Protection Part 2: Interface Ia—Safety. 11 pp.

ETSI PRETS 300 046-2-92. Integrated Services Digital Network (ISDN); Primary Rate Access—Safety and Protection Part 2: Interface Ia—Safety. 12 pp.

ETSI PRETS 300 046-2-90. Integrated Services Digital Network (ISDN); Primary Rate Access—Safety and Protection Part 2: Interface Ia—Safety (T/TE yyy-2-1). 11 pp.

ETSI ETS 300 046-3-92. Integrated Services Digital Network (ISDN); Primary Rate Access—Safety and Protection Part 3: Interface Ia—Protection. 16 pp.

ETSI PRETS 300 046-3-92. Integrated Services Digital Network (ISDN); Primary Rate Access—Safety and Protection Part 3: Interface Ia—Protection. 17 pp.

ETSI PRETS 300 046-3-90. Integrated Services Digital Network (ISDN); Primary Rate Access—Safety and Protection Part 3: Interface 1—Protection (T/TE yyy-2-2). 17 pp.

ETSI ETS 300 046-4-92. Integrated Services Digital Network (ISDN); Primary Rate Access—Safety and Protection Part 4: Interface Ib—Safety. 10 pp.

ETSI PRETS 300 046-4-92. Integrated Services Digital Network (ISDN); Primary Rate Access—Safety and Protection Part 4: Interface Ib—Safety. 11 pp.

ETSI PRETS 300 046-4-90. Integrated Services Digital Network (ISDN); Primary Rate Access—Safety and Protection Part 4: Interface 16—Safety (T/TE yyy-2-3). 11 pp.

ETSI ETS 300 046-5-92. Integrated Services Digital Network (ISDN); Primary Rate Access—Safety and Protection Part 5: Interface Ib—Protection. 16 pp.

ETSI PRETS 300 046-5-92. Integrated Services Digital Network (ISDN); Primary Rate Access—Safety and Protection Part 5: Interface Ib—Protection. 16 pp.

ETSI PRETS 300 046-5-90. Integrated Services Digital Network (ISDN); Primary Rate Access—Safety and Protection Part 5: Interface 1b—Protection (T/TE yyy-2-4). 17 pp.

ETSI ETS 300 047-1-92. Integrated Services Digital Network (ISDN); Basic Access—Safety and Protection Part 1: General. 21 pp.

ETSI PRETS 300 047-1-92. Integrated Services Digital Network (ISDN); Basic Access—Safety and Protection Part 1: General. 23 pp.

ETSI PRETS 300 047-1-90. Integrated Services Digital Network (ISDN); Basic Access—safety and Protection Part 1: General (T/TE zzz-1). 19 pp.

ETSI ETS 300 047-2-92. Integrated Services Digital Network (ISDN); Basic Access—Safety and Protection Part 2: Interface Ia—Safety. 11 pp.

ETSI PRETS 300 047-2-92. Integrated Services Digital Network (ISDN); Basic Access—Safety and Protection Part 2: Interface Ia—Safety. 12 pp.

ETSI PRETS 300 047-2-90. Integrated Services Digital Network (ISDN); Basic Access—Safety and Protection Part 2: Interface Ia—Safety (T/TE zzz-2-1). 12 pp.

ETSI ETS 300 047-3-92. Integrated Services Digital Network (ISDN); Basic Access—Safety and Protection Part 3: Interface Ia—Protection. 19 pp.

ETSI PRETS 300 047-3-92. Integrated Services Digital Network (ISDN); Basic Access—Safety and Protection Part 3: Interface Ia—Protection. 20 pp.

ETSI PRETS 300 047-3-90. Integrated Services Digital Network (ISDN); Basic Access—Safety and Protection Part 3: Interface Ia—Protection (T/TE zzz-2-2). 19 pp.

ETSI ETS 300 047-4-92. Integrated Services Digital Network (ISDN); Basic Access—Safety and Protection Part 4: Interface Ib—Safety. 11 pp.

Integrated Services Digital Networks (Cont.)

—**Interfaces—Safety** (Cont.)

ETSI PRETS 300 047-4-92. Integrated Services Digital Network (ISDN); Basic Access—Safety and Protection Part 4: Interface Ib—Safety. 12 pp.

ETSI PRETS 300 047-4-90. Integrated Services Digital Network (ISDN) Basic Access—Safety and Protection Part 4: Interface 16—Safety (T/TE zzz-2-3). 11 pp.

ETSI ETS 300 047-5-92. Integrated Services Digital Network (ISDN); Basic Access—Safety and Protection Part 5: Interface Ib—Protection. 17 pp.

ETSI PRETS 300 047-5-92. Integrated Services Digital Network (ISDN); Basic Access-Safety and Protection Part 5: Interface Ib—Protection. 18 pp.

ETSI PRETS 300 047-5-90. Integrated Services Digital Network (ISDN); Basic Access-Safety and Protection Part 5: Interface 16-Protection (T/TE zzz-2-4). 17 pp.

—**Interfaces—Structures**

CCITT RECMN I.412-89. ISDN User-Network Interfaces Interface Structures and Access Capabilities—Integrated Services Digital Network (ISDN)—Overall Network Aspects and Functions, ISDN User-Network Interfaces (Study Group XVIII) 6 pp. 6 pp.

—**Interfaces—Supplementary Services**

CCITT RECMN I.452-89. Generic Procedure for the Control of ISDN Supplementary Services—Integrated Services Digital Network (ISDN)—Overall Network Aspects and Functions, ISDN User-Network Interfaces (Study Group XVIII) 1 pp (Same as Recmn Q.932). 1 p.

—**Interfaces—Telecommunication Services**

CCITT RECMN I.340-89. ISDN Connection Types—Integrated Services Digital Network (ISDN)—Overall Network Aspects and Functions, ISDN User-Network Interfaces (Study Group XVIII) 17 pp. 17 pp.

—**Interfaces—Videotex Communications**

CENELEC PRETS 300 079-90. Integrated Services Digital Network (ISDN): Syntax-Based Videotex End-to-End Protocols (T/ TE 06-10). 81 pp.

CENELEC PRETS 300 079-91. Integrated Services Digital Network (ISDN); Syntax-Based Videotex End-to-End Protocols (T/TE 06-10). 82 pp.

CENELEC ETS 300 079-91. Integrated Services Digital Network (ISDN); Syntax-Based Videotex End-to-End Protocols Circuit Mode DTE—DTE. 78 pp.

—**International Numbers (Telephone)**

CCITT RECMN I.330-89. ISDN Numbering and Addressing Principles—Integrated Services Digital Network (ISDN)—Overall Network Aspects and Functions, ISDN User-Network Interfaces (Study Group XVIII) 8 pp. 8 pp.

—**Interworking**

CCITT RECMN I.310-89. ISDN—Network Functional Principles—Integrated Services Digital Network (ISDN)—Overall Network Aspects and Functions, ISDN User-Network Interfaces (Study Group XVIII) 22 pp. 22 pp.

CCITT RECMN I.320-89. ISDN Protocol Reference Model—Integrated Services Digital Network (ISDN)—Overall Network Aspects and Functions, ISDN User-Network Interfaces (Study Group XVIII) 10 pp. 10 pp.

CCITT RECMN I.326-89. Reference Configuration for Relative Network Resource Requirements—Integrated Services Digital Network (ISDN)—Overall Network Aspects and Functions, ISDN User-Network Interfaces (Study Group XVIII) 3 pp. 3 pp.

CCITT RECMN I.510-89. Definitions and General Principles for ISDN Interworking—Integrated Services Digital Network (ISDN)—Internetwork Interfaces and Maintenance Principles (Study Group XVIII) 12 pp. 12 pp.

CCITT RECMN I.520-89. General Arrangements for Network Interworking Between ISDNs—Integrated Services Digital Network (ISDN)—Internetwork Interfaces and Maintenance Principles (Study Group XVIII) 12 pp. 12 pp.

CENELEC ETR 031-92. Network Aspects (NA); Network Aspects of ISDN to ISDN and ISDN Internal Interworking. 14 pp.

CENELEC ETR 054-92. Integrated Services Digital Network (ISDN); the Protocol Reference Model and Its Relationship with the OSI Reference Model. 13 pp.

ETSI ETR 031-92. Network Aspects (NA); Network Aspects of ISDN to ISDN and ISDN Internal Interworking. 14 pp.

ETSI ETR 054-92. Integrated Services Digital Network (ISDN); the Protocol Reference Model and Its Relationship with the OSI Reference Model. 13 pp.

INDEX and DIRECTORY of

INTERNATIONAL AND NON-U.S. NATIONAL STANDARDS
SUBJECT INDEX

Integrated

Integrated Services Digital Networks (Cont.)

—**Interworking—Glossaries**
CCITT RECMN I.510-89. Definitions and General Principles for ISDN Interworking—Integrated Services Digital Network (ISDN)—Internetwork Interfaces and Maintenance Principles (Study Group XVIII) 12 pp. 12 pp.

—**Interworking—Mobile Satellite Communications**
CCIR Report 1176-90. Interworking Between the Mobile Satellite Systems and the Terrestrial Networks for Data Transmission Services—Section 8G—Availability, Performance Objectives and Interworking with Terrestrial Networks. 8 pp.
CCIR QUESTION 89-1/8-90. Compatibility for Interworking Between the Mobile-Satellite Systems and Terrestrial Networks Including ISDN—Questions Concerning Study Group 8—Mobile, Radiodetermination, Amateur and Related Satellite Services. 1 p.

—**Interworking—Private Switched Networks**
CENELEC ETR 045-92. Network Aspects (NA); General Configuration and Basic Functions for the Interconnection of Private Telecommunications Networks with the Public ISDN. 12 pp.
ETSI ETR 045-92. Network Aspects (NA); General Configuration and Basic Functions for the Interconnection of Private Telecommunications Networks with the Public ISDN. 12 pp.

—**Interworking—Public Data Networks—Accounting**
CEPT T/PGT 36 E-91. General Tariff and Accounting Principles for Interworking Between ISDN and Existing Public Data Networks. 2 pp.

—**Interworking—Public Data Networks—Tariffs (Telecommunications)**
CEPT T/PGT 36 E-91. General Tariff and Accounting Principles for Interworking Between ISDN and Existing Public Data Networks. 2 pp.

—**Land Mobile Services**
CCIR Report 1153-90. Future Public Land Mobile Telecommunication Systems—Section 8A—Land Mobile Service and Related Subjects. 62 pp.

—**Layer 1—Interworking**
CCITT RECMN I.511-89. ISDN-to-ISDN Layer 1 Internetwork Interface —Integrated Services Digital Network (ISDN)—Internetwork Interfaces and Maintenance Principles (Study Group XVIII) 5 pp. 5 pp.

—**Layer 1—Network Interfaces**
CCITT RECMN I.511-89. ISDN-to-ISDN Layer 1 Internetwork Interface —Integrated Services Digital Network (ISDN)—Internetwork Interfaces and Maintenance Principles (Study Group XVIII) 5 pp. 5 pp.

—**Line Hunting Services**
CCITT RECMN I.252-89. Call Offering Supplementary Services—Integrated Services Digital Network (ISDN)—General Structure and Service Capabilities (Study Group XVIII) 37 pp. 37 pp.
CCITT RECMN I.252. 6-88. Line Hunting—Integrated Services Digital Network (ISDN)—General Structure and Service Capabilities (Study Group XVIII) 5 pp. 5 pp.

—**Line Hunting Services—Functions/Information Flows**
CCITT RECMN Q.82-89. Call Offering Supplementary Services—General Recommendations on Telephone Switching and Signalling—Functions and Information Flows for Services in the ISDN—Supplements (Study Group XI) 25 pp. 25 pp.

—**Link Access Procedures**
CSA CAN/CSA-T542-90. Integrated Services Digital Network (ISDN)—Data-Link Layer Signalling Specification for Application at the User-Network Interface (ANSI T1.602-1989). 199 pp.

—**Local Area Networks**
CEPT T/SF 36-84. Relative Aux Services Et Facilities Pour Les Centraux Prives Aintegration De Services Et Pour Les Reseaux Locaux D'Entreprises (RLE) Relies Aun RNIS. 15 pp.
CEPT T/SF 45 E-86. Services and Facilities for Subscriber Complex Installations (Including ISPBX and LAN Installations) Connected to an Integrated Services Digital Network (ISDN). 6 pp.
CEPT T/SF 55 E-87. Supplementary Services Available to Subscriber Complex Installations Connected to an Integrated Services Digital Network (ISDN). 5 pp.

Integrated Services Digital Networks (Cont.)

—**Local Area Networks (Cont.)**
ECMA ECMA-TR 51-90. Requirements for Access to Integrated Voice and Data Local and Metropolitan Area Networks. 62 pp.

—**Maintenance**
CCITT RECMN M.36-89. Principles for the Maintenance of ISDN's—General Maintenance Principles—Maintenance of International Transmission Systems and Telephone Circuits (Study Group IV) 15 pp. 15 pp.

—**Meet Me Conference**
CENELEC PRETS 300 164-91. Integrated Services Digital Network (ISDN) Meet-Me Conferance (MMC) Supplementary Service Service Description (T/NA1(89)26). 15 pp.
CENELEC PRETS 300 164-92. Integrated Services Digital Network (ISDN); Meet-Me Conference (MMC) Supplementary Service Service Description. 14 pp.
CENELEC ETS 300 164-92. Integrated Services Digital Network (ISDN); Meet-Me Conference (MMC) Supplementary Service Service Description. 14 pp.
CENELEC PRETS 300 165-91. Integrated Services Digital Network (ISDN); Meet Me Conferance (MMC) Supplementary Service Functional Capabilities and Information Flows (T/S 22-11). 24 pp.
CENELEC PRETS 300 165-92. Integrated Services Digital Network (ISDN); Meet-Me Conference (MMC) Supplementary Service Functional Capabilities and Information Flows. 22 pp.
CENELEC ETS 300 165-93. Integrated Services Digital Network (ISDN); Meet-Me Conference (MMC) Supplementary Service Functional Capabilities and Information Flows. 22 pp.
ETSI PRETS 300 164-91. Integrated Services Digital Network (ISDN)—Meet Conference (MMC) Supplementary Service—Service Description (T/NA1(89)26).
ETSI PRETS 300 165-91. Integrated Services Digital Network (ISDN); Meet Me Conference (MMC) Supplementary Service-Functional Capabilities and Information Flows (T/S 22-11).
ETSI ETS 300 164-92. Integrated Services Digital Network (ISDN); Meet-Me Conference (MMC) Supplementary Service Service Description. 14 pp.
ETSI PRETS 300 164-92. Integrated Services Digital Network (ISDN); Meet-Me Conference (MMC) Supplementary Service Service Description. 14 pp.
ETSI PRETS 300 164-91. Integrated Services Digital Network (ISDN) Meet-Me Conferance (MMC) Supplementary Service Service Description (T/NA1(89)26). 15 pp.
ETSI ETS 300 165-93. Integrated Services Digital Network (ISDN); Meet-Me Conference (MMC) Supplementary Service Functional Capabilities and Information Flows. 22 pp.
ETSI PRETS 300 165-92. Integrated Services Digital Network (ISDN); Meet-Me Conference (MMC) Supplementary Service Functional Capabilities and Information Flows. 22 pp.
ETSI PRETS 300 165-91. Integrated Services Digital Network (ISDN); Meet Me Conferance (MMC) Supplementary Service Functional Capabilities and Information Flows (T/S 22-11). 24 pp.

—**Message Conversion Systems**
CCITT RECMN F.202-89. Interworking Between the Telex Service and the Teletex Service—General Procedures and Operational Requirements for the International Interconnection of Telex/Teletex Conversion Facilities—Telematic, Data Transmission and Teleconference Services—Operations and. 4 pp.

—**Metropolitan Area Networks**
ECMA ECMA-TR 51-90. Requirements for Access to Integrated Voice and Data Local and Metropolitan Area Networks. 62 pp.

—**Mixed Mode**
CCITT RECMN I.241. 4-88. Mixed Mode—Integrated Services Digital Network (ISDN)—General Structure and Service Capabilities (Study Group XVIII) 4 pp. 4 pp.

—**Mobile Satellite Communications**
CCIR RESOLUTION 103-90. Liasion and Joint Studies with the CCITT in Mobile and Fixed-Satellite Services and High-Definition Television in ISDNs—Volume XIV—Administrative Texts of the CCIR. 2 pp.

—**Mobile Satellite Communications—INMARSAT—SELCAL Services**
CCITT RECMN E.216-89. Selection Procedures for the INMARSAT Mobile-Satellite Telephone and ISDN Services—Telephone Network and ISDN—Operation, Numbering, Routing and Mobile Service (Study Group II) 12 pp. 12 pp.

Integrated Services Digital Networks (Cont.)

—**Multilevel Precedence and Preemption—Functions/Information Flows**
CCITT RECMN Q.85-92. Stage 2 Description for Community of Interest Supplementary Services Section 1—Closed User Group (CUG) Section 3—Multi-Level Precedence and Preemption (MLPP) (Rev.1) (Study Group XI) 59 pp. 59 pp.

—**Multiple Subscriber Number—Functions/Information Flows**
CCITT RECMN Q.81-92. Stage 2 Description for Number Identification Supplementary Services Section 2—Multiple Subscriber Number Section 8—Sub-Addressing (SUB)—(Addenda to Q.81) (Study Group XI) 19 pp. 19 pp.

—**Multiple Subscriber Numbers**
CCITT RECMN I.251-89. Number Identification Supplementary Services—Integrated Services Digital Network (ISDN)—General Structure and Service Capabilities (Study Group XVIII) 28 pp. 28 pp.
CCITT RECMN I.251. 2 (REV 1)-92. Multiple Subscriber Number (Study Group I) 8 pp. 8 pp.
CENELEC PRETS 300 050-90. Integrated Services Digital Network (ISDN): Multiple Subscriber Number (MSN) Supplementary Service Service Description (T/NAI(89)20). 11 pp.
CENELEC PRETS 300 050-91. Integrated Services Digital Network (ISDN); Multiple Subscriber Number (MSN) Supplementary Service Service Description. 16 pp.
CENELEC ETS 300 050-91. Integrated Services Digital Network (ISDN); Multiple Subscriber Number (MSN) Supplementary Service Service Description. 17 pp.
CENELEC PRETS 300 051-90. Integrated Services Digital Network (ISDN): Multiple Subscriber Number (MSN) Supplementary Service Functional Capabilities and Information Flows (T/S 22-15). 16 pp.
CENELEC PRETS 300 051-91. Integrated Services Digital Network (ISDN); Multiple Subscriber Number (MSN) Supplementary Service Functional Capabilities and Information Flows. 19 pp.
CENELEC ETS 300 051-91. Integrated Services Digital Network (ISDN); Multiple Subscriber Number (MSN) Supplementary Service Functional Capabilities and Information Flows. 19 pp.
ETSI PRETS 300 050-90. Integrated Services Digital Network (ISDN); Multiple Subscriber Number (MSN) Supplementary Service—Service Description (T/NA1 (89)20). 11 pp.
ETSI PRETS 300 051-90. Integrated Services Digital Network (ISDN); Multiple Subscriber Number (MSN) Supplementary Service Functional Capabilities and Information Flows (T/S 22-15). 16 pp.
ETSI ETS 300 050-91. Integrated Services Digital Network (ISDN); Multiple Subscriber Number (MSN) Supplementary Service Service Description. 17 pp.
ETSI PRETS 300 050-91. Integrated Services Digital Network (ISDN); Multiple Subscriber Number (MSN) Supplementary Service Service Description. 16 pp.
ETSI PRETS 300 050-90. Integrated Services Digital Network (ISDN): Multiple Subscriber Number (MSN) Supplementary Service Service Description (T/NAI(89)20). 11 pp.
ETSI ETS 300 051-91. Integrated Services Digital Network (ISDN); Multiple Subscriber Number (MSN) Supplementary Service Functional Capabilities and Information Flows. 19 pp.
ETSI PRETS 300 051-91. Integrated Services Digital Network (ISDN); Multiple Subscriber Number (MSN) Supplementary Service Functional Capabilities and Information Flows. 19 pp.
ETSI PRETS 300 051-90. Integrated Services Digital Network (ISDN): Multiple Subscriber Number (MSN) Supplementary Service Functional Capabilities and Information Flows (T/S 22-15). 16 pp.

—**Multiple Subscriber Numbers—DSS1**
CCITT RECMN Q.951-92. Stage 3 Description for Number Identification Supplementary Services Using DSS 1 Section 1—Direct-Dialling-in (DDI) Section 2—Multiple Subscriber Number (MSN) Section 8—Sub-Addressing (SUB) (Study Group XI) 17 pp. 17 pp.

—**Network Architecture**
CCITT RECMN I.324-91. ISDN Network Architecture (Study Group XVIII) 20 pp. 20 pp.
CCITT RECMN I.324-89. ISDN Network Architecture—Integrated Services Digital Network (ISDN)—Overall Network Aspects and Functions, ISDN User-Network Interfaces (Study Group XVIII) 12 pp. 12 pp.
CEPT T/GSI 03-03-87. Architecture Du Reseau. 13 pp.

INDUSTRY STANDARDS

INTERNATIONAL AND NON-U.S. NATIONAL STANDARDS
SUBJECT INDEX

Integrated

Integrated Services Digital Networks (Cont.)

—Network Architecture (Cont.)
CEPT T/GSI 03-03 E-87. Network Architecture. 13 pp.

—Network Capabilities
CCITT RECMN I.310-89. ISDN—Network Functional Principles—Integrated Services Digital Network (ISDN)—Overall Network Aspects and Functions, ISDN User-Network Interfaces (Study Group XVIII) 22 pp. 22 pp.

CCITT RECMN I.325-89. Reference Configurations for ISDN Connection Types—Integrated Services Digital Network (ISDN)—Overall Network Aspects and Functions, ISDN User-Network Interfaces (Study Group XVIII) 7 pp. 7 pp.

—Network Interfaces
CCITT RECMN I.500-89. General Structure of the ISDN Interworking Recommendations—Integrated Services Digital Network (ISDN)—Internetwork Interfaces and Maintenance Principles (Study Group XVIII) 4 pp. 4 pp.

CCITT RECMN I.520-89. General Arrangements for Network Interworking Between ISDNs—Integrated Services Digital Network (ISDN)—Internetwork Interfaces and Maintenance Principles (Study Group XVIII) 12 pp. 12 pp.

—Network Layer—Call Control Procedures—D Channel—DSS1
CCITT RECMN Q.931-89. ISDN User-Network Interface Layer 3 Specification for Basic Call Control—Digital Subscriber Signalling System No. 1 (DSS 1), Network Layer, User-Network Management (Study Group XI) 356 pp (Same as Recmn I.451). 356 pp.

—Network Layer—Call Control Procedures—DSS1
CENELEC PRI-ETS 300 314-93. Integrated Services Digital Network (ISDN); Digital Subscriber Signalling System No. One (DSS1) Protocol Implementation Conformance Statement (PICS) Proforma for Basic-Access User for Signalling-Network-Layer Protocol for Circuit-Mode Basic Call Control. 51 pp.

CENELEC PRI-ETS 300 315-93. Integrated Services Digital Network (ISDN); Digital Sub. Signalling Sys. No. One (DSS1) Protocol Implementation Conformance Statement (PICS) Proforma for Primary-Rate-Access User for Signalling-Network-Layer Protocol for Circuit-Mode Basic Call Control. 52 pp.

CENELEC PRI-ETS 300 316-93. Integrated Services Digital Network (ISDN); Digital Subscriber Signalling System No. One (DSS1) Protocol Implementation Conformance Statement (PICS) Proforma for Basic-Access Network for Signalling-Network-Layer Protocol for Circuit-Mode Basic Call Control. 47 pp.

CENELEC PRI-ETS 300 317-93. Integrated Services Digital Network (ISDN); Digital Subscriber Signalling System No. One (DSS1) Protocol Impl. Conformance Statement (PICS) Proforma for Primary-Rate-Access Network for Signalling-Network-Layer Protocol for Circuit-Mode Basic Call Control. 46 pp.

CENELEC PRI-ETS 300 318-93. Integrated Services Digital Network (ISDN); Digital Subscriber Signalling System No. One (DSS1) Protocol Implementation eXtra Information for Testing (PIXIT) Proforma for Basic-Access User for Signalling-Network-Layer Protocol for Circuit-Mode Basic Call Control. 13 pp.

CENELEC PRI-ETS 300 319-93. Integrated Services Digital Network (ISDN); Digital Subscriber Signalling System No. One (DSS1) Protocol Impl. eXtra Information for Testing (PIXIT) Proforma for Primary-Rate-Access User for Signalling-Network-Layer Protocol for Circuit-Mode Basic Call Control. 13 pp.

CENELEC PRI-ETS 300 320-93. Integrated Services Digital Network (ISDN); Digital Subscriber Signalling System No. One (DSS1) Protocol Implementation eXtra Information for Testing (PIXIT) Proforma for Basic-Access Network for Signalling-Network-Layer Protocol for Circuit-Mode Basic Call Control. 13 pp.

CENELEC PRI-ETS 300 321-93. Integrated Services Digital Network (ISDN); Digital Sub. Signalling Sys. No. One (DSS1) Protocol Implementation eXtra Information for Testing (PIXIT) Proforma for Primary-Rate-Access Network for Signalling-Network-Layer Protocol for Circuit-Mode Basic Call Control. 13 pp.

ETSI PRI-ETS 300 314-93. Integrated Services Digital Network (ISDN); Digital Subscriber Signalling System No. One (DSS1) Protocol Implementation Conformance Statement (PICS) Proforma for Basic-Access User for Signalling-Network-Layer Protocol for Circuit-Mode Basic Call Control. 51 pp.

Integrated Services Digital Networks (Cont.)

—Network Layer—Call Control Procedures—DSS1 (Cont.)
ETSI PRI-ETS 300 315-93. Integrated Services Digital Network (ISDN); Digital Sub. Signalling Sys. No. One (DSS1) Protocol Implementation Conformance Statement (PICS) Proforma for Primary-Rate-Access User for Signalling-Network-Layer Protocol for Circuit-Mode Basic Call Control. 52 pp.

ETSI PRI-ETS 300 316-93. Integrated Services Digital Network (ISDN); Digital Subscriber Signalling System No. One (DSS1) Protocol Implementation Conformance Statement (PICS) Proforma for Basic-Access Network for Signalling-Network-Layer Protocol for Circuit-Mode Basic Call Control. 47 pp.

ETSI PRI-ETS 300 317-93. Integrated Services Digital Network (ISDN); Digital Subscriber Signalling System No. One (DSS1) Protocol Impl. Conformance Statement (PICS) Proforma for Primary-Rate-Access Network for Signalling-Network-Layer Protocol for Circuit-Mode Basic Call Control. 46 pp.

ETSI PRI-ETS 300 318-93. Integrated Services Digital Network (ISDN); Digital Subscriber Signalling System No. One (DSS1) Protocol Implementation eXtra Information for Testing (PIXIT) Proforma for Basic-Access User for Signalling-Network-Layer Protocol for Circuit-Mode Basic Call Control. 13 pp.

ETSI PRI-ETS 300 319-93. Integrated Services Digital Network (ISDN); Digital Subscriber Signalling System No. One (DSS1) Protocol Impl. eXtra Information for Testing (PIXIT) Proforma for Primary-Rate-Access User for Signalling-Network-Layer Protocol for Circuit-Mode Basic Call Control. 13 pp.

ETSI PRI-ETS 300 320-93. Integrated Services Digital Network (ISDN); Digital Subscriber Signalling System No. One (DSS1) Protocol Implementation eXtra Information for Testing (PIXIT) Proforma for Basic-Access Network for Signalling-Network-Layer Protocol for Circuit-Mode Basic Call Control. 13 pp.

ETSI PRI-ETS 300 321-93. Integrated Services Digital Network (ISDN); Digital Sub. Signalling Sys. No. One (DSS1) Protocol Implementation eXtra Information for Testing (PIXIT) Proforma for Primary-Rate-Access Network for Signalling-Network-Layer Protocol for Circuit-Mode Basic Call Control. 13 pp.

—Network Layer—Interfaces
CCITT RECMN I.430-89. Basic User-Network Interface—Layer 1 Specification—Integrated Services Digital Network (ISDN)—Overall Network Aspects and Functions, ISDN User-Network Interfaces (Study Group XVIII) 70 pp. 70 pp.

CCITT RECMN I.450-89. ISDN User-Network Interface Layer 3—General Aspects—Integrated Services Digital Network (ISDN)—Overall Network Aspects and Functions, ISDN User-Network Interfaces (Study Group XVIII) 1 pp (Same as Recmn Q.930). 1 p.

—Network Layer—Interfaces—Call Control Procedures
CCITT RECMN I.451-89. ISDN User-Network Interface Layer 3 Specification for Basic Call Control—Integrated Services Digital Network (ISDN)—Overall Network Aspects and Functions, ISDN User-Network Interfaces (Study Group XVIII) 1 pp (Same as Recmn Q.931). 1 p.

CENELEC PRETS 300 102-1-90. Integrated Services Digital Network (ISDN); User-Network Interface Layer 3 Specifications for Basic Call Control. 233 pp.

CENELEC ETS 300 102-1-90. Integrated Services Digital Network (ISDN); User-Network Interface Layer 3 Specifications for Basic Call Control. 183 pp.

CENELEC ETS 300 102-1/A1-93. AMD 1 Integrated Services Digital Network (ISDN); User-Network Interface Layer 3 Specifications for Basic Call Control. 6 pp.

CENELEC ETS 300 102-1 PRA1-92. AMD 1 Integrated Services Digital Network (ISDN); User-Network Interface Layer 3 Specifications for Basic Call Control. 6 pp.

CENELEC ETS 300 102-1 PRA2-93. AMD prA2 Integrated Services Digital Network (ISDN); User-Network Interface Layer 3 Specifications for Basic Call Control. 4 pp.

ETSI PRETS 300 102-1-90. Integrated Services Digital Network (ISDN); User Network Interface Layer 3—Specifications for Basic Call Control. 233 pp.

ETSI ETS 300 102-1/A1-93. AMD 1 Integrated Services Digital Network (ISDN); User-Network Interface Layer 3 Specifications for Basic Call Control. 6 pp.

Integrated Services Digital Networks (Cont.)

—Network Layer—Interfaces—Call Control Procedures (Cont.)
ETSI ETS 300 102-1 PRA2-93. AMD prA2 Integrated Services Digital Network (ISDN); User-Network Interface Layer 3 Specifications for Basic Call Control. 4 pp.

ETSI ETS 300 102-1 PRA1-92. AMD 1 Integrated Services Digital Network (ISDN); User-Network Interface Layer 3 Specifications for Basic Call Control. 6 pp.

ETSI ETS 300 102-1-90. Integrated Services Digital Network (ISDN); User-Network Interface Layer 3 Specifications for Basic Call Control. 183 pp.

—Network Layer—Interfaces—Communication Channels
CCITT RECMN I.430-89. Basic User-Network Interface—Layer 1 Specification—Integrated Services Digital Network (ISDN)—Overall Network Aspects and Functions, ISDN User-Network Interfaces (Study Group XVIII) 70 pp. 70 pp.

—Network Layer—Interfaces—Glossaries
CCITT RECMN I.430-89. Basic User-Network Interface—Layer 1 Specification—Integrated Services Digital Network (ISDN)—Overall Network Aspects and Functions, ISDN User-Network Interfaces (Study Group XVIII) 70 pp. 70 pp.

—Network Layer—Interfaces—Reference Points
CCITT RECMN I.430-89. Basic User-Network Interface—Layer 1 Specification—Integrated Services Digital Network (ISDN)—Overall Network Aspects and Functions, ISDN User-Network Interfaces (Study Group XVIII) 70 pp. 70 pp.

CENELEC ETR 044-92. Network Aspects (NA); Reference Events for Network Performance Parameters in an ISDN. 13 pp.

ETSI ETR 044-92. Network Aspects (NA); Reference Events for Network Performance Parameters in an ISDN. 13 pp.

—Network Layer—Supplementary Services—DSS1
CCITT RECMN Q.932-89. Generic Procedures for the Control of ISDN Supplementary Services—Digital Subscriber Signalling System No. 1 (DSS 1), Network Layer, User-Network Management (Study Group XI) 61 pp (Same as Recmn I.452). 61 pp.

—Network Layer—User Network Interfaces—DSS1
CCITT RECMN Q.930-89. ISDN User-Network Interface Layer 3—General Aspects—Digital Subscriber Signalling System No.1 (DSS 1), Network Layer, User-Network Management (Study Group XI) 4 pp (Same as Recmn I.450). 4 pp.

CCITT RECMN Q.931-89. ISDN User-Network Interface Layer 3 Specification for Basic Call Control—Digital Subscriber Signalling System No. 1 (DSS 1), Network Layer, User-Network Management (Study Group XI) 356 pp (Same as Recmn I.451). 356 pp.

CCITT RECMN Q.932-89. Generic Procedures for the Control of ISDN Supplementary Services—Digital Subscriber Signalling System No. 1 (DSS 1), Network Layer, User-Network Management (Study Group XI) 61 pp (Same as Recmn I.452). 61 pp.

CCITT RECMN Q.940-89. ISDN User-Network Interface Protocol for Management—General Aspects Subscriber Signalling System No. 1 (DSS 1), Network Layer, User-Network Management (Study Group XI) 11 pp. 11 pp.

—Network Management Systems
CCITT RECMN E.410-92. International Network Management—General Information—Telephone Network and ISDN Quality of Service, Network Management and Traffic Engineering (Study Group II) 8 pp. 8 pp.

CCITT RECMN E.411-92. International Network Management—Operational Guidance—Telephone Network and ISDN Quality of Service, Network Management and Traffic Engineering (Study Group II) 16 pp. 16 pp.

—Network Performance
CCITT RECMN E.411-92. International Network Management—Operational Guidance—Telephone Network and ISDN Quality of Service, Network Management and Traffic Engineering (Study Group II) 16 pp. 16 pp.

CCITT RECMN I.350-89. General Aspects of Quality of Service and Network Performance in Digital Networks, Including ISDN—Integrated Services Digital Network (ISDN)—Overall Network Aspects and Functions, ISDN User-Network Interfaces (Study Group XVIII) 10 pp. 10 pp.

INDEX and DIRECTORY of

INTERNATIONAL AND NON-U.S. NATIONAL STANDARDS
SUBJECT INDEX
Integrated

Integrated Services Digital Networks (Cont.)

—**Network Performance** (Cont.)

CCITT RECMN I.351-89. Recommendations in Other Series Concerning Network Performance Objectives That Apply at Reference Point T of an ISDN—Integrated Services Digital Network (ISDN)—Overall Network Aspects and Functions, ISDN User-Network Interfaces (Study Group XVIII) 1 pp. 1 p.

CCITT RECMN I.352-89. Network Performance Objectives for Connection Processing Delays in an ISDN—Integrated Services Digital Network (ISDN)—Overall Network Aspects and Functions, ISDN User-Network Interfaces (Study Group XVIII) 14 pp. 14 pp.

—**Network Services (OSI)**

BSI BS ISO/IEC 9574-89. 1989 Information Technology—Telecommunications and Information Exchange Between Systems—Provision of the OSI Connection-Mode Network Service by Packet Mode Terminal Equipment Connected to an Inter-grated Services Digital Network (ISDN). 18 pp.

CCITT RECMN X.612-92. Information Technology—Provision of the OSI Connection-Mode Network Service by Packet-Mode Terminal Equipment Connected to an Integrated Services Digital Network (ISDN) 21 pp (ISO/CCIT Common Text). 21 pp.

IEC 9574-92. Information Technology—Provision of the OSI Connection-Mode Network Service by Packet Mode Terminal Equipment to an Integrated Services Digital Network (ISDN) Second Edition. 23 pp.

ISO 8880 Pt 2 Draft AMD 1. Information Technology—Telecommunications and Info. Exchange Between Systems—Protocol Com-binations to Provide and Support the OSI Network Service—Part 2: Provi-sion and Support of the Connection-Mode Network Service Amd 1: Addition of the ISDN Environment; (1991) ***CD-ROM ONLY*** 6 pp.

ISO 9574-92. Information Technology—Provision of the OSI Connection-Mode Network Service by Packet Mode Terminal Equipment to an Integrated Services Digital Network (ISDN) Second Edition. 23 pp.

JTC1 8880 Pt2 Draft AMD 1. Information Technology—Telecommunications and Info. Exchange Between Systems—Protocol Com-binations to Provide and Support the OSI Network Service—Part 2: Provi-sion and Support of the Connection-Mode Network Service Amd 1: Addition of the ISDN Environment; (1991) ***CD-ROM ONLY*** 6 pp.

JTC1 9574-92. Information Technology—Provision of the OSI Connection-Mode Network Service by Packet Mode Terminal Equipment to an Integrated Services Digital Network (ISDN) Second Edition. 23 pp.

OSI ISO IEC 9574-89. Information Technology—Telecommunications and Information Exchange Between Systems—Provision of the OSI Connection-Mode Network Service by Packet Mode Terminal Equipment Connected to an Integrated Services Digital Network (ISDN). 15 pp.

OSI ISO IEC DIS 9574-88. Information Processing Systems—Data Communications—Provision of the OSI Connection—Mode Network Service by Packet Mode Terminal Equipment Connected to an Integrated Services Digital Network. 16 pp.

SNZ NZS/AS 4099-93. Information Technology—Telecommunications and Information Exchange Between Systems—Provision of the OSI Connection-Mode Network Service by Packet Mode Terminal Equipment Connected to an Integrated Services Digital Network (This is a Joint. 11 pp.

—**Network Services (OSI)—Protocols**

CCITT RECMN I.130-89. Method for the Characterization of Telecommunication Services Supported by an ISDN and Network Capabilities of an ISDN—Integrated Services Digital Network (ISDN)—General Structure and Service Capabilities (Study Group XVIII) 5 pp. 5 pp.

—**Network Services (OSI)—Protocols—Circuit Switched Connections**

IEC 8473 Draft AMD 5. Information Processing Systems—Data Commun-ications—Protocol for Providing the Connec-tionless-Mode Network Service Amendment 5: Provision of the Underlying Service for Operation over ISDN Circuit-Switched B-Channel; (1992) ***CD-ROM ONLY*** 9 pp.

ISO 8473 Draft AMD 5. Information Processing Systems—Data Commun-ications—Protocol for Providing the Connec-tionless-Mode Network Service Amendment 5: Provision of the Underlying Service for Operation over ISDN Circuit-Switched B-Channel; (1992) ***CD-ROM ONLY*** 9 pp.

Integrated Services Digital Networks (Cont.)

—**Network Services (OSI)—Protocols—Circuit Switched Connections** (Cont.)

JTC1 8473 Draft AMD 5. Information Processing Systems—Data Commun-ications—Protocol for Providing the Connec-tionless-Mode Network Service Amendment 5: Provision of the Underlying Service for Operation over ISDN Circuit-Switched B-Channel; (1992) ***CD-ROM ONLY*** 9 pp.

—**Numbering Plans**

CCITT FASCICLE II.2-88. Telephone Network and ISDN-Operation, Numbering, Routing and Mobile Service. Recommendations E.100—E.333. 362 pp.

CCITT RECMN E.164-91. Numbering Plan for the ISDN Era (Study Group II) 19 pp (Same as Recmn I.331). 19 pp.

CCITT RECMN E.164-89. Numbering Plan for the ISDN Era—Telephone Network and ISDN—Operation, Numbering, Routing and Mobile Service (Study Group II) 6 pp (Same as Recmn I.331). 6 pp.

CCITT RECMN E.166-89. Numbering Plan Interworking in the ISDN Era—Telephone Network and ISDN—Operation, Numbering, Routing and Mobile Service (Study Group II) 14 pp (Same as Recmn X.122). 14 pp.

CCITT RECMN I.255-89. Community of Interest Supplementary Services—Integrated Services Digital Network (ISDN)—General Structure and Service Capabilities (Study Group XVIII) 10 pp. 10 pp.

CCITT RECMN I.255. 2-88. Private Numbering Plan—Integrated Services Digital Network (ISDN)—General Structure and Service Capabilities (Study Group XVIII) 4 pp. 4 pp.

CCITT RECMN I.330-89. ISDN Numbering and Addressing Principles—Integrated Services Digital Network (ISDN)—Overall Network Aspects and Functions, ISDN User-Network Interfaces (Study Group XVIII) 8 pp. 8 pp.

CCITT RECMN I.331-89. Numbering Plan for the ISDN Area—Integrated Services Digital Network (ISDN)—Overall Network Aspects and Functions, ISDN User-Network Interfaces (Study Group XVIII) 1 pp (Same as Recmn E.164). 1 p.

CCITT RECMN Q.11-89. Numbering Plan for the International Telephone Service—General Recommendations on Telephone Switching and Signalling—Functions and Information Flows for Services in the ISDN —Supplements (Study Group XI) 8 pp. 8 pp.

CCITT RECMN Q.11 BIS-89. Numbering Plan for the ISDN Era—General Recommendations on Telephone Switching and Signalling—Functions and Information Flows for Services in the ISDN —Supplements (Study Group XI) 6 pp. 6 pp.

CCITT RECMN Q.11 TER-89. Timetable for Coordinated Implementation of the Full Capability of the Numbering Plan for the ISDN Era (Recommendation E.164)—General Recommendations on Telephone Switching and Signalling—Functions and Information Flows for Services in the ISDN —Supplements. 2 pp.

CCITT RECMN X.122-89. Numbering Plan Interworking Between a Packet Switched Public Data Network (PSPDN) and an Integrated Services Digital Network (ISDN) or Public Switched Telephone Network (PSTN) in the Short-Term—Data Communication Networks—Transmission, Signalling and Switching, Network. 10 pp.

CENELEC ETR 006-90. Network Aspects; Numbering and Addressing for the Memorandum of Understanding (MOU) on Integrated Services Digital Network (ISDN) (Priorities 1 and 2). 15 pp.

ETSI ETR 006-90. Network Aspects; Numbering and Addressing for the Memorandum of Understanding (MOU) on Integrated Services Digital Network (ISDN) (Priorities 1 and 2). 15 pp.

—**Numbering Plans—Country Codes**

CCITT RECMN E.164-91. Numbering Plan for the ISDN Era (Study Group II) 19 pp (Same as Recmn I.331). 19 pp.

CCITT RECMN E.164-89. Numbering Plan for the ISDN Era—Telephone Network and ISDN—Operation, Numbering, Routing and Mobile Service (Study Group II) 6 pp (Same as Recmn I.331). 6 pp.

—**Numbering Plans—Destination Network Codes**

CCITT RECMN E.164-91. Numbering Plan for the ISDN Era (Study Group II) 19 pp (Same as Recmn I.331). 19 pp.

CCITT RECMN E.164-89. Numbering Plan for the ISDN Era—Telephone Network and ISDN—Operation, Numbering, Routing and Mobile Service (Study Group II) 6 pp (Same as Recmn I.331). 6 pp.

Integrated Services Digital Networks (Cont.)

—**Numbering Plans—Escape Characters**

CCITT RECMN E.166-89. Numbering Plan Interworking in the ISDN Era—Telephone Network and ISDN—Operation, Numbering, Routing and Mobile Service (Study Group II) 14 pp (Same as Recmn X.122). 14 pp.

—**Numbering Plans—Interworking**

CCITT RECMN E.166-92. Numbering Plan Interworking for the E.164 and X.121 Numbering Plans (Study Group II) 41 pp (Same as Recmn X.122). 41 pp.

CCITT RECMN I.330-89. ISDN Numbering and Addressing Principles—Integrated Services Digital Network (ISDN)—Overall Network Aspects and Functions, ISDN User-Network Interfaces (Study Group XVIII) 8 pp. 8 pp.

CCITT RECMN X.122-92. Numbering Plan Interworking for the E.164 and X.121 Numbering Plans (Study Group VII) 41 pp (Same as Recmn X.122). 41 pp.

—**Numbering Plans—Land Mobile Stations**

CCITT RECMN E.213-89. Telephone and ISDN Numbering Plan for Land Mobile Stations in Public Land Mobile Networks (PLMN)—Telephone Network and ISDN—Operation, Numbering, Routing and Mobile Service (Study Group II) 2 pp. 2 pp.

—**Numbering Plans—National Significant Numbers**

CCITT RECMN E.164-91. Numbering Plan for the ISDN Era (Study Group II) 19 pp (Same as Recmn I.331). 19 pp.

CCITT RECMN E.164-89. Numbering Plan for the ISDN Era—Telephone Network and ISDN—Operation, Numbering, Routing and Mobile Service (Study Group II) 6 pp (Same as Recmn I.331). 6 pp.

—**Numbering Plans—Network Layer—Addressing Systems**

CCITT RECMN I.334-89. Principles Relating ISDN Numbers/Subaddresses to the OSI Reference Model Network Layer Addresses—Integrated Services Digital Network (ISDN)—Overall Network Aspects and Functions, ISDN User-Network Interfaces (Study Group XVIII) 5 pp. 5 pp.

—**Numbering Plans—Scheduling**

CCITT RECMN E.165-89. Timetable for Coordinated Implementation of the Full Capability of the Numbering Plan for the ISDN Era (Recommendation E.164)—Telephone Network and ISDN—Operation, Numbering, Routing and Mobile Service (Study Group II) 2 pp. 2 pp.

—**Numbering Plans—Subscriber Numbers**

CCITT RECMN E.164-91. Numbering Plan for the ISDN Era (Study Group II) 19 pp (Same as Recmn I.331). 19 pp.

CCITT RECMN E.164-89. Numbering Plan for the ISDN Era—Telephone Network and ISDN—Operation, Numbering, Routing and Mobile Service (Study Group II) 6 pp (Same as Recmn I.331). 6 pp.

—**Numbering Plans—Terminal Selection**

CCITT RECMN I.333-89. Terminal Selection in ISDN—Integrated Services Digital Network (ISDN)—Overall Network Aspects and Functions, ISDN User-Network Interfaces (Study Group XVIII) 18 pp. 18 pp.

—**Numbering Plans—Trunk Codes**

CCITT RECMN E.164-91. Numbering Plan for the ISDN Era (Study Group II) 19 pp (Same as Recmn I.331). 19 pp.

CCITT RECMN E.164-89. Numbering Plan for the ISDN Era—Telephone Network and ISDN—Operation, Numbering, Routing and Mobile Service (Study Group II) 6 pp (Same as Recmn I.331). 6 pp.

—**Open Systems Interconnection—Interworking**

CCITT RECMN I.320-89. ISDN Protocol Reference Model—Integrated Services Digital Network (ISDN)—Overall Network Aspects and Functions, ISDN User-Network Interfaces (Study Group XVIII) 10 pp. 10 pp.

CENELEC ETR 054-92. Integrated Services Digital Network (ISDN); the Protocol Reference Model and Its Relationship with the OSI Reference Model. 13 pp.

ETSI ETR 054-92. Integrated Services Digital Network (ISDN); the Protocol Reference Model and Its Relationship with the OSI Reference Model. 13 pp.

INDUSTRY STANDARDS

Integrated Services Digital Networks (Cont.)

—Open Systems Interconnection—Reference Model

CCITT RECMN I.320-89. ISDN Protocol Reference Model—Integrated Services Digital Network (ISDN)—Overall Network Aspects and Functions, ISDN User-Network Interfaces (Study Group XVIII) 10 pp. 10 pp.

CENELEC ETR 054-92. Integrated Services Digital Network (ISDN); the Protocol Reference Model and Its Relationship with the OSI Reference Model. 13 pp.

ETSI ETR 054-92. Integrated Services Digital Network (ISDN); the Protocol Reference Model and Its Relationship with the OSI Reference Model. 13 pp.

—Overvoltage Protection—Data Terminal Equipment

CCITT RECMN K.22-89. Overvoltage Resistibility of Equipment Connected to an ISDN T/S Bus—Protection Against Interference (Study Group V) 3 pp. 3 pp.

—Packet Mode Bearer Services

CCITT RECMN F.353-89. Provision of Telematic and Data Transmission Services on Integrated Services Digital Network (ISDN)—Telematic, Data Transmission and Teleconference Services—Operations and Quality of Service (Study Group I) 5 pp. 5 pp.

CCITT RECMN I.122-89. Framework for Providing Additional Packet Mode Bearer Services—Integrated Services Digital Network (ISDN)—General Structure and Service Capabilities (Study Group XVIII) 17 pp. 17 pp.

CCITT RECMN I.232-89. Packet-Mode Bearer Services Categories—Integrated Services Digital Network (ISDN)—General Structure and Service Capabilities (Study Group XVIII) 5 pp. 5 pp.

CCITT RECMN I.232. 2-88. Connectionless Bearer Service Category—Integrated Services Digital Network (ISDN)—General Structure and Service Capabilities (Study Group XVIII) 1 pp. 1 p.

CCITT RECMN I.232. 3-88. User Signalling Bearer Service Category—Integrated Services Digital Network (ISDN)—General Structure and Service Capabilities (Study Group XVIII) 1 pp. 1 p.

CCITT RECMN I.462-89. Support of Packet Mode Terminal Equipment by an ISDN—Integrated Services Digital Network (ISDN)—Overall Network Aspects and Functions, ISDN User-Network Interfaces (Study Group XVIII) 1 pp (Same as Recmn X.31). 1 p.

CCITT RECMN X.31-89. Support of Packet Mode Terminal Equipment by an ISDN—Data Communication Networks: Services and Facilities, Interfaces (Study Group VII) 61 pp (Same as Recmn I.462). 61 pp.

CENELEC PRETS 300 048-90. Integrated Services Digital Network (ISDN): ISDN Packet Mode Bearer Services (PMBS) ISDN Virtual Call (VC) and Permanent Virtual Circuit (PVC) Bearer Services Provided by the B Channel of the User Access-Basic and Primary Rate (T/NAI(89)29). 13 pp.

CENELEC PRETS 300 048-91. Integrated Services Digital Network (ISDN); ISDN Packet Mode Bearer Services (PMBS) ISDN Virtual Call (VC) and Permanent Virtual Call (PVC) Bearer Services Provided by the B-Channel of the User Access—Basic and Primary Rate. 18 pp.

CENELEC ETS 300 048-92. Integrated Services Digital Network (ISDN); ISDN Packet Mode Bearer Sevices (PMBS) ISDN Virtual Call (VC) and Permanent Virtual Call (PVC) Bearer Services Provided by the B-Channel of the User Access—Basic and Primary Rate. 17 pp.

CENELEC PRETS 300 049-90. Integrated Services Digital Network (ISDN); ISDN Packet Mode Bearer Services (PMBS) ISDN Virtual Call (VC) and Permanent Virtual Circuit (PVC) Bearer Services Provided by the D Channel of the User Access-Basic and Primary Rate (T/NAI(89)30). 15 pp.

CENELEC PRETS 300 049-91. Integrated Services Digital Network (ISDN); ISDN Packet Mode Bearer Service (PMBS) ISDN Virtual Call (VC) and Permanent Virtual Call (PVC) Bearer Services Provided by the D-Channel of the User Access—Basic and Primary Rate. 20 pp.

CENELEC ETS 300 049-92. Integrated Services Digital Network (ISDN); ISDN Packet Mode Bearer Sevice (PMBS) ISDN Virtual Call (VC) and Permanent Virtual Call (PVC) Bearer Services Provided by the D-Channel of the User Access—Basic and Primary Rate. 18 pp.

ETSI PRETS 300 048-90. Integrated Services Digital Network (ISDN); ISDN Packet Mode Bearer Service (PMBS) ISDN Virtual Call (VC) and Permanent Virtual Circuit (PVC) Bearer Services Provided by the B Channel of the User Access—Basic and Primary Rate (T/N A1 (89)29). 11 pp.

ETSI PRETS 300 049-90. Integrated Services Digital Network (ISDN); ISDN Packet Mode Bearer Services (PMBS) and Permanent Virtual Circuit (PVC) Bearer Services Provided by the D Channel of the User Access—Basic and Primary Rate (T/NA1 (89) 30). 15 pp.

ETSI ETS 300 048-92. Integrated Services Digital Network (ISDN); ISDN Packet Mode Bearer Sevices (PMBS) ISDN Virtual Call (VC) and Permanent Virtual Call (PVC) Bearer Services Provided by the B-Channel of the User Access—Basic and Primary Rate. 17 pp.

ETSI PRETS 300 048-91. Integrated Services Digital Network (ISDN); ISDN Packet Mode Bearer Services (PMBS) ISDN Virtual Call (VC) and Permanent Virtual Call (PVC) Bearer Services Provided by the B-Channel of the User Access—Basic and Primary Rate. 18 pp.

ETSI PRETS 300 048-90. Integrated Services Digital Network (ISDN); ISDN Packet Mode Bearer Services (PMBS) ISDN Virtual Call (VC) and Permanent Virtual Circuit (PVC) Bearer Services Provided by the B Channel of the User Access-Basic and Primary Rate (T/NAI(89)29). 13 pp.

ETSI ETS 300 049-92. Integrated Services Digital Network (ISDN); ISDN Packet Mode Bearer Sevice (PMBS) ISDN Virtual Call (VC) and Permanent Virtual Call (PVC) Bearer Services Provided by the D-Channel of the User Access—Basic and Primary Rate. 18 pp.

ETSI PRETS 300 049-91. Integrated Services Digital Network (ISDN); ISDN Packet Mode Bearer Service (PMBS) ISDN Virtual Call (VC) and Permanent Virtual Call (PVC) Bearer Services Provided by the D-Channel of the User Access—Basic and Primary Rate. 20 pp.

ETSI PRETS 300 049-90. Integrated Services Digital Network (ISDN); ISDN Packet Mode Bearer Services (PMBS) ISDN Virtual Call (VC) and Permanent Virtual Circuit (PVC) Bearer Services Provided by the D Channel of the User Access-Basic and Primary Rate (T/NAI(89)30). 15 pp.

—Packet Mode Bearer Services—Virtual Call

CCITT RECMN I.232. 1-88. Virtual Call and Permanent Virtual Circuit Bearer Service Category—Integrated Services Digital Network (ISDN)—General Structure and Service Capabilities (Study Group XVIII) 5 pp. 5 pp.

—Packet Mode Services—Accounting

CEPT T/CAC E39 E-92. General Tariff and Accounting Principles for Packet Mode Services Provided over the ISDN. 2 pp.

—Packet Mode Services—Tariffs

CEPT T/CAC E39 E-92. General Tariff and Accounting Principles for Packet Mode Services Provided over the ISDN. 2 pp.

—Packet Switched Data Transmission Services

CENELEC PRETS 300 007-90. Integrated Services Digital Network (ISDN): Support of Packet Mode Terminal Equipment by an ISDN (T/S 46-50). 116 pp.

CENELEC PRETS 300 007-91. Integrated Services Digital Network (ISDN); Support of Packet-Mode Terminal Equipment by an ISDN. 104 pp.

CENELEC ETS 300 007-91. Integrated Services Digital Network (ISDN); Support of Packet-Mode Terminal Equipment by an ISDN. 103 pp.

CENELEC PRETS 300 099-90. Integrated Services Digital Network (ISDN) Specification of the Packet Handler Access Point Interface (PH1) for the Provision of prETS 300 007 (CCITT Recommendation X.31) Packet Mode Services. 135 pp.

CENELEC PRETS 300 099-91. Integrated Services Digital Network (ISDN) Specification of the Packet Handler Access Point Interface (PHI). 154 pp.

CENELEC ETS 300 099-92. Integrated Services Digital Network (ISDN); Specification of the Packet Handler Access Point Interface (PHI). 166 pp.

CEPT T/S 46-50-87. Support D'Equipement De Terminal En Mode-Paquet Par Un RNIS (X.31/I.462). 25 pp.

CEPT T/S 46-50 E-87. Support of Packet Mode Terminal Equipment by an ISDN (X.31/I.462). 25 pp.

ETSI PRETS 300 007-90. Integrated Services Digital Network (ISDN); Support Packet Mode Terminal Equipment by an ISDN Also Known as T/S 46-50. 116 pp.

ETSI ETS 300 007-91. Integrated Services Digital Network (ISDN); Support of Packet-Mode Terminal Equipment by an ISDN. 103 pp.

ETSI PRETS 300 007-91. Integrated Services Digital Network (ISDN); Support of Packet-Mode Terminal Equipment by an ISDN. 104 pp.

ETSI PRETS 300 007-90. Integrated Services Digital Network (ISDN); Support of Packet Mode Terminal Equipment by an ISDN (T/S 46-50). 116 pp.

ETSI PRETS 300 099-90. Integrated Services Digital Network (ISDN) Specification of the Packet Handler Access Point Interface (PHI) for the Provision of prETS 300 007 (CCITT Recommendation X.31) Packet Mode Services (T/NA2 (89)10). 135 pp.

ETSI ETS 300 099-92. Integrated Services Digital Network (ISDN); Specification of the Packet Handler Access Point Interface (PHI). 166 pp.

ETSI PRETS 300 099-91. Integrated Services Digital Network (ISDN) Specification of the Packet Handler Access Point Interface (PHI). 154 pp.

—Packet Switched Data Transmission Services—Grade of Service

CCITT RECMN E.701-92. Reference Connections for Traffic Engineering—Telephone Network and ISDN—Quality of Service, Network Management and Traffic Engineering (Study Group II) 4 pp. 4 pp.

CCITT RECMN E.721-91. Network Grade of Service Parameters and Target Values for Circuit-Switched Services in the Evolving ISDN (Study Group II) 8 pp. 8 pp.

CCITT RECMN E.721-89. Network Grade of Service Parameters in ISDN—Telephone Network and ISDN—Quality of Service, Network Management and Traffic Engineering (Study Group II) 2 pp. 2 pp.

—Packet Switched Data Transmission Services—Traffic

CCITT RECMN E.711-92. User Demand Modelling Study Group II) 11 pp. 11 pp.

—Packet Switched Networks

CCITT RECMN I.120-89. Integrated Services Digital Networks (ISDNs)—Integrated Services Digital Network (ISDN)—General Structure and Service Capabilities (Study Group XVIII) 2 pp. 2 pp.

CCITT RECMN I.462-89. Support of Packet Mode Terminal Equipment by an ISDN—Integrated Services Digital Network (ISDN)—Overall Network Aspects and Functions, ISDN User-Network Interfaces (Study Group XVIII) 1 pp (Same as Recmn X.31). 1 p.

CCITT RECMN X.31-89. Support of Packet Mode Terminal Equipment by an ISDN—Data Communication Networks: Services and Facilities, Interfaces (Study Group VII) 61 pp (Same as Recmn I.462). 61 pp.

—Packet Switched Networks—Addressing Systems

CENELEC ETR 020-91. Network Aspects Numbering and Addressing for X.31 Services. 17 pp.

ETSI ETR 020-91. Network Aspects Numbering and Addressing for X.31 Services. 17 pp.

—Packet Switched Networks—Dimensioning

CCITT RECMN E.730-92. ISDN Dimensioning Methods Overview (Study Group II) 4 pp. 4 pp.

—Packet Switched Networks—Numbering Plans

CENELEC ETR 020-91. Network Aspects Numbering and Addressing for X.31 Services. 17 pp.

ETSI ETR 020-91. Network Aspects Numbering and Addressing for X.31 Services. 17 pp.

—Packet Switched Public Data Networks—Interworking

CCITT RECMN E.166-92. Numbering Plan Interworking for the E.164 and X.121 Numbering Plans (Study Group II) 41 pp (Same as Recmn X.122). 41 pp.

CCITT RECMN E.166-89. Numbering Plan Interworking in the ISDN Era—Telephone Network and ISDN—Operation, Numbering, Routing and Mobile Service (Study Group II) 14 pp (Same as Recmn X.122). 14 pp.

CCITT RECMN I.550-89. General Arrangements for Interworking Between Packet Switched Public Data Networks (PSPDNs) and Integrated Services Digital Networks (ISDNs) for the Provision of Data Transmission—Integrated Services Digital Network (ISDN)—Internetwork Interfaces and. 1 p.

CCITT RECMN X.122-92. Numbering Plan Interworking for the E.164 and X.121 Numbering Plans (Study Group VII) 41 pp. 41 pp.

INTERNATIONAL AND NON-U.S. NATIONAL STANDARDS
SUBJECT INDEX

Integrated

Integrated Services Digital Networks (Cont.)

—Packet Switched Public Data Networks—Interworking (Cont.)

CCITT RECMN X.325-89. General Arrangements for Interworking Between Packet Switched Public Data Networks (PSPDNs) and Integrated Services Digital Networks (ISDNs) for the Provision of Data Transmission Services—Data Communication Networks—Interworking Between Networks, Mobile Data. 7 pp.

—Parameter Exchange

ECMA ECMA 123-90. In-Band Parameter Exchange in Private Pre-ISDN Networks Using Standard ECMA-102. 41 pp.

—Parameter Exchange—Interworking

CCITT RECMN I.515-89. Parameter Exchange for ISDN Interworking—Integrated Services Digital Network (ISDN)—Internetwork Interfaces and Maintenance Principles (Study Group XVIII) 17 pp. 17 pp.

—PDN—Interworking

CCITT RECMN D.250-91. General Charging and Accounting Principles for Non-Voice Services Provided by Interworking Between the ISDN and Existing Public Data Networks (Study Group III) 5 pp. 5 pp.

CCITT RECMN D.250-89. General Charging and Accounting Principles for Non-Voice Services Provided by Interworking Between the ISDN and Existing Public Data Networks—General Tariff Prin.—Charging and Accounting in Intl Telecommunications Services (Study Group III) 1 pp. 1 pp.

CCITT RECMN X.1-89. International User Classes of Service in Public Data Networks and Integrated Services Digital Networks (ISDNs)—Data Communication Networks: Services and Facilities, Interfaces (Study Group VII) 4 pp. 4 pp.

—Performance Parameters

CCITT RECMN E.411-92. International Network Management—Operational Guidance—Telephone Network and ISDN Quality of Service, Network Management and Traffic Engineering (Study Group II) 16 pp. 16 pp.

CCITT RECMN I.325-89. Reference Configurations for ISDN Connection Types—Integrated Services Digital Network (ISDN)—Overall Network Aspects and Functions, ISDN User-Network Interfaces (Study Group XVIII) 7 pp. 7 pp.

CCITT RECMN I.350-89. General Aspects of Quality of Service and Network Performance in Digital Networks, Including ISDN—Integrated Services Digital Network (ISDN)—Overall Network Aspects and Functions, ISDN User-Network Interfaces (Study Group XVIII) 10 pp. 10 pp.

—Performance Parameters—Hypothetical Reference Model

CCITT RECMN I.325-89. Reference Configurations for ISDN Connection Types—Integrated Services Digital Network (ISDN)—Overall Network Aspects and Functions, ISDN User-Network Interfaces (Study Group XVIII) 7 pp. 7 pp.

—Permanent Virtual Circuit Services

CENELEC PRETS 300 048-90. Integrated Services Digital Network (ISDN): ISDN Packet Mode Bearer Services (PMBS) ISDN Virtual Call (VC) and Permanent Virtual Circuit (PVC) Bearer Services Provided by the B Channel of the User Access-Basic and Primary Rate (T/NAI(89)29). 13 pp.

CENELEC PRETS 300 048-91. Integrated Services Digital Network (ISDN); ISDN Packet Mode Bearer Services (PMBS) ISDN Virtual Call (VC) and Permanent Virtual Call (PVC) Bearer Services Provided by the B-Channel of the User Access—Basic and Primary Rate. 18 pp.

CENELEC ETS 300 048-92. Integrated Services Digital Network (ISDN); ISDN Packet Mode Bearer Sevices (PMBS) ISDN Virtual Call (VC) and Permanent Virtual Call (PVC) Bearer Services Provided by the B-Channel of the User Access—Basic and Primary Rate. 17 pp.

CENELEC PRETS 300 049-90. Integrated Services Digital Network (ISDN): ISDN Packet Mode Bearer Services (PMBS) ISDN Virtual Call (VC) and Permanent Virtual Circuit (PVC) Bearer Services Provided by the D Channel of the User Access-Basic and Primary Rate (T/NAI(89)30). 15 pp.

CENELEC PRETS 300 049-91. Integrated Services Digital Network (ISDN); ISDN Packet Mode Bearer Service (PMBS) ISDN Virtual Call (VC) and Permanent Virtual Call (PVC) Bearer Services Provided by the D-Channel of the User Access—Basic and Primary Rate. 20 pp.

Integrated Services Digital Networks (Cont.)

—Permanent Virtual Circuit Services (Cont.)

CENELEC ETS 300 049-92. Integrated Services Digital Network (ISDN); ISDN Packet Mode Bearer Sevice (PMBS) ISDN Virtual Call (VC) and Permanent Virtual Call (PVC) Bearer Services Provided by the D-Channel of the User Access—Basic and Primary Rate. 18 pp.

ETSI PRETS 300 048-90. Integrated Services Digital Network (ISDN); ISDN Packet Mode Bearer Service (PMBS) ISDN Virtual Call (VC) and Permanent Virtual Circuit (PVC) Bearer Services Provided by the B Channel of the User Access—Basic and Primary Rate (T/N A1 (89)29). 11 pp.

ETSI PRETS 300 049-90. Integrated Services Digital Network (ISDN); ISDN Packet Mode Bearer Services (PMBS) ISDN Virtual Circuit (PVC) Bearer Services Provided by the D Channel of the User Access—Basic and Primary Rate (T/NA1 (89) 30). 15 pp.

ETSI ETS 300 048-92. Integrated Services Digital Network (ISDN); ISDN Packet Mode Bearer Sevices (PMBS) ISDN Virtual Call (VC) and Permanent Virtual Call (PVC) Bearer Services Provided by the B-Channel of the User Access—Basic and Primary Rate. 17 pp.

ETSI PRETS 300 048-91. Integrated Services Digital Network (ISDN); ISDN Packet Mode Bearer Services (PMBS) ISDN Virtual Call (VC) and Permanent Virtual Call (PVC) Bearer Services Provided by the B-Channel of the User Access—Basic and Primary Rate. 18 pp.

ETSI PRETS 300 048-90. Integrated Services Digital Network (ISDN): ISDN Packet Mode Bearer Services (PMBS) ISDN Virtual Call (VC) and Permanent Virtual Circuit (PVC) Bearer Services Provided by the B Channel of the User Access-Basic and Primary Rate (T/NAI(89)29). 13 pp.

ETSI ETS 300 049-92. Integrated Services Digital Network (ISDN); ISDN Packet Mode Bearer Sevice (PMBS) ISDN Virtual Call (VC) and Permanent Virtual Call (PVC) Bearer Services Provided by D-Channel of the User Access—Basic and Primary Rate. 18 pp.

ETSI PRETS 300 049-91. Integrated Services Digital Network (ISDN); ISDN Packet Mode Bearer Service (PMBS) ISDN Virtual Call (VC) and Permanent Virtual Call (PVC) Bearer Services Provided by the D-Channel of the User Access—Basic and Primary Rate. 20 pp.

ETSI PRETS 300 049-90. Integrated Services Digital Network (ISDN): ISDN Packet Mode Bearer Services (PMBS) ISDN Virtual Call (VC) and Permanent Virtual Circuit (PVC) Bearer Services Provided by the D Channel of the User Access-Basic and Primary Rate (T/NAI(89)30). 15 pp.

—Postselection Delay—Grade of Service

CCITT RECMN E.721-91. Network Grade of Service Parameters and Target Values for Circuit-Switched Services in the Evolving ISDN (Study Group II) 8 pp. 8 pp.

CCITT RECMN E.721-89. Network Grade of Service Parameters in ISDN—Telephone Network and ISDN—Quality of Service, Network Management and Traffic Engineering (Study Group II) 2 pp. 2 pp.

—Preselection Delay—Grade of Service

CCITT RECMN E.721-91. Network Grade of Service Parameters and Target Values for Circuit-Switched Services in the Evolving ISDN (Study Group II) 8 pp. 8 pp.

CCITT RECMN E.721-89. Network Grade of Service Parameters in ISDN—Telephone Network and ISDN—Quality of Service, Network Management and Traffic Engineering (Study Group II) 2 pp. 2 pp.

—Primary Rate Access

CENELEC PRETS 300 233-92. Transmission and Multiplexing (TM); Digital Section for ISDN Primary Rate Access. 72 pp.

ETSI PRETS 300 233-92. Transmission and Multiplexing (TM); Digital Section for ISDN Primary Rate Access. 72 pp.

—Primary Rate Access—Maintenance

CCITT RECMN I.604-89. Application of Maintenance Principles to ISDN Primary Rate Accesses—Integrated Services Digital Network (ISDN)—Internetwork Interfaces and Maintenance Principles (Study Group XVIII) 13 pp. 13 pp.

Integrated Services Digital Networks (Cont.)

—Primary Rate Access—NATO Reference Model

NATO STANAG 4461 ED 1 AMD 0-00. (Draft) NATO Reference Model for Open Systems Interconnection/Integrated Services Digital Network (ISDN) Layer 1 (Physical Layer) Specification for ISDN Primary Rate Access at the S/T Reference Point. 14 pp.

NATO STANAG 4462 ED 1 AMD 0-00. (Draft) NATO Reference Model for Open Systems Interconnection/Integrated Services Digital Network (ISDN) Layer 2 (Data Link Layer) Specification for ISDN Basic and Primary Rate Access at the S/T Reference Point. 17 pp.

—Priority Services

CCITT RECMN I.255. 4-90. Priority Service (Study Group I) 9 pp. 9 pp.

—Private Branch Exchanges

CEPT T/SF 36-84. Relative Aux Services Et Facilities Pour Les Centraux Prives Aintegration De Services Et Pour Les Reseaux Locaux D'Entreprises (RLE) Relies Aun RNIS. 15 pp.

CEPT T/SF 45 E-86. Services and Facilities for Subscriber Complex Installations (Including ISPBX and LAN Installations) Connected to an Integrated Services Digital Network (ISDN). 6 pp.

CEPT T/SF 55 E-87. Supplementary Services Available to Subscriber Complex Installations Connected to an Integrated Services Digital Network (ISDN). 5 pp.

—Protocols—Keyboards—DSS1

CENELEC PRETS 300 122-90. Integrated Services Digital Network (ISDN); Generic Keyboard Protocol for the Support of Supplementary Services Digital Subscriber Signaling One (DSSI) Protocol (T/S 46-32A). 8 pp.

CENELEC PRETS 300 122-91. Integrated Services Digital Network (ISDN); Generic Keypad Protocol for the Support of Supplementary Services Digital Subscriber Signalling No. One (DSS1) Protocol. 10 pp.

CENELEC ETS 300 122-92. Integrated Services Digital Network (ISDN); Generic Keypad Protocol for the Support of Supplementary Services Digital Subscriber Signalling System No. One (DSS1) Protocol. 10 pp.

ETSI ETS 300 122-92. Integrated Services Digital Network (ISDN); Generic Keypad Protocol for the Support of Supplementary Services Digital Subscriber Signalling System No. One (DSS1) Protocol. 10 pp.

—Protocols—Keypads—DSS1

ETSI PRETS 300 122-90. Integrated Services Digital Network (ISDN); Generic Keypad Protocol for the Support of Supplementary Services—Digital Subscriber Signalling One (DSS1) Protocol (T/S 46-32A). 8 pp.

—Protocols—Multiple Subscriber Numbers—DSS1

CENELEC PRETS 300 052-90. Integrated Services Digital Network (ISDN): Multiple Subscriber Number (MSN) Supplementary Service Digital Subscriber Signalling One (DSS1) Protocol (T/S 46-33B). 9 pp.

CENELEC PRETS 300 052-91. Integrated Services Digital Network (ISDN); Multiple Subscriber Number (MSN) Supplementary Service Digital Subscriber Signalling System No. One (DSS1) Protocol. 13 pp.

CENELEC ETS 300 052-91. Integrated Services Digital Network (ISDN); Multiple Subscriber Number (MSN) Supplementary Service Digital Subscriber Signalling System No. One (DSS1) Protocol. 14 pp.

ETSI PRETS 300 052-90. Integrated Services Digital Network (ISDN); Multiple Subscriber Number (MSN) Supplementary Service Digital Subscriber Signalling One (DSS1) Protocol (T/S 46-33B). 9 pp.

ETSI ETS 300 052-91. Integrated Services Digital Network (ISDN); Multiple Subscriber Number (MSN) Supplementary Service Digital Subscriber Signalling System No. One (DSS1) Protocol. 14 pp.

ETSI PRETS 300 052-91. Integrated Services Digital Network (ISDN); Multiple Subscriber Number (MSN) Supplementary Service Digital Subscriber Signalling System No. One (DSS1) Protocol. 13 pp.

ETSI PRETS 300 052-90. Integrated Services Digital Network (ISDN); Multiple Subscriber Number (MSN) Supplementary Service Digital Subscriber Signalling One (DSS1) Protocol (T/S 46-33B). 9 pp.

—Protocols—Reference Model

CCITT RECMN I.320-89. ISDN Protocol Reference Model—Integrated Services Digital Network (ISDN)—Overall Network Aspects and Functions, ISDN User-Network Interfaces (Study Group XVIII) 10 pp. 10 pp.

INDUSTRY STANDARDS

INTERNATIONAL AND NON-U.S. NATIONAL STANDARDS
SUBJECT INDEX

Integrated

Integrated Services Digital Networks (Cont.)
—Protocols—Reference Model (Cont.)
CENELEC ETR 054-92. Integrated Services Digital Network (ISDN); the Protocol Reference Model and Its Relationship with the OSI Reference Model. 13 pp.
ETSI ETR 054-92. Integrated Services Digital Network (ISDN); the Protocol Reference Model and Its Relationship with the OSI Reference Model. 13 pp.

—Protocols—Videotex Equipment
CENELEC PRETS 300 218-92. Integrated Services Digital Network (ISDN) Syntax-Based Videotex Lower Layers Protocols for ISDN Packet Mode (CCITT Recommendation X.31 Case A and Case B). 77 pp.
CENELEC PRETS 300 218-92. Integrated Services Digital Network (ISDN); Syntax-Based Videotex Lower Layers Protocols for ISDN Packet Mode (CCITT Recommendation X.31 Case A and Case B). 75 pp.
CENELEC ETS 300 218-93. Integrated Services Digital Network (ISDN); Syntax-Based Videotex Lower Layers Protocols for ISDN Packet Mode (CCITT Recommendation X.31 Case A and Case B). 75 pp.
ETSI ETS 300 218-93. Integrated Services Digital Network (ISDN); Syntax-Based Videotex Lower Layers Protocols for ISDN Packet Mode (CCITT Recommendation X.31 Case A and Case B). 75 pp.
ETSI PRETS 300 218-92. Integrated Services Digital Network (ISDN); Syntax-Based Videotex Lower Layers Protocols for ISDN Packet Mode (CCITT Recommendation X.31 Case A and Case B). 75 pp.
ETSI PRETS 300 218-92. Integrated Services Digital Network (ISDN) Syntax-Based Videotex Lower Layers Protocols for ISDN Packet Mode (CCITT Recommendation X.31 Case A and Case B). 77 pp.

—Public Data—Numbering Plans—Interworking
CCITT RECMN I.332-89. Numbering Principles for Interworking Between ISDNs and Dedicated Networks with Different Numbering Plans—Integrated Services Digital Network (ISDN)—Overall Network Aspects and Functions, ISDN User-Network Interfaces (Study Group XVIII) 4 pp. 4 pp.

—Public Land Mobile Networks—Interworking
CCITT RECMN I.510-89. Definitions and General Principles for ISDN Interworking—Integrated Services Digital Network (ISDN)—Internetwork Interfaces and Maintenance Principles (Study Group XVIII) 12 pp. 12 pp.
CENELEC GSM 09.07-92. General Requirements on Interworking Between the PLMN and the ISDN or PSTN. 54 pp.
ETSI GSM 09.07-92. General Requirements on Interworking Between the PLMN and the ISDN or PSTN. 54 pp.

—Public Land Mobile Networks—Public Switched Telephone Networks
CCITT FASCICLE VI.12. Public Land Mobile Network Interworking with ISDN and PSTN. Recommendations Q.1000—Q.1032 (Study Group XI). 95 pp.

—Public Land Mobile Networks—Signaling Systems—Interworking
CCITT RECMN Q.1031-89. General Signalling Requirements on Interworking Between the ISDN or PSTN and the PLMN—Public Land Mobile Network Interworking with ISDN and PSTN (Study Group XI) 3 pp. 3 pp.
CENELEC PRETS 300 303-93. Integrated Services Digital Network (ISDN); ISDN—Global System for Mobile Communications (GSM) Public Land Mobile Network (PLMN) Signalling Interface. 18 pp.
CENELEC GSM 09.03-92. European Digital Cellular Telecommunication System (Phase 1); Requirements on Interworking Between the ISDN or PSTN and the PLMN. 9 pp.
CENELEC GSM 09.09-92. European Digital Cellular Telecommunication System (Phase 1); Detailed Signalling Interworking Within the PLMN and with the PSTN/ISDN. 38 pp.
ETSI PRETS 300 303-93. Integrated Services Digital Network (ISDN); ISDN—Global System for Mobile Communications (GSM) Public Land Mobile Network (PLMN) Signalling Interface. 18 pp.
ETSI GSM 09.03-92. European Digital Cellular Telecommunication System (Phase 1); Requirements on Interworking Between the ISDN or PSTN and the PLMN. 9 pp.
ETSI GSM 09.09-92. European Digital Cellular Telecommunication System (Phase 1); Detailed Signalling Interworking Within the PLMN and with the PSTN/ISDN. 38 pp.

Integrated Services Digital Networks (Cont.)
—Public Switched Telephone Networks—Interworking
CCITT RECMN E.166-92. Numbering Plan Interworking for the E.164 and X.121 Numbering Plans (Study Group II) 41 pp (Same as Recmn X.122). 41 pp.
CCITT RECMN E.166-89. Numbering Plan Interworking in the ISDN Era—Telephone Network and ISDN—Operation, Numbering, Routing and Mobile Service (Study Group II) 14 pp (Same as Recmn X.122). 14 pp.
CCITT RECMN E.172-89. Call Routing in the ISDN Era—Telephone Network and ISDN—Operation, Numbering, Routing and Mobile Service (Study Group II) 14 pp. 14 pp.
CCITT RECMN I.510-89. Definitions and General Principles for ISDN Interworking—Integrated Services Digital Network (ISDN)—Internetwork Interfaces and Maintenance Principles (Study Group XVIII) 12 pp. 12 pp.
CCITT RECMN I.530-89. Network Interworking Between an ISDN and a Public Switched Telephone Network (PSTN) —Integrated Services Digital Network (ISDN)—Internetwork Interfaces and Maintenance Principles (Study Group XVIII) 8 pp. 8 pp.
CCITT RECMN X.122-92. Numbering Plan Interworking for the E.164 and X.121 Numbering Plans (Study Group VII) 41 pp. 41 pp.

—Quality Assurance
CEPT T/GSI 02-05-87. Qualite De Service Du Mis. 4 pp.
CEPT T/GSI 02-05 E-87. Quality of Service. 4 pp.

—Quality of Service
CCITT RECMN I.350-89. General Aspects of Quality of Service and Network Performance in Digital Networks, Including ISDN—Integrated Services Digital Network (ISDN)—Overall Network Aspects and Functions, ISDN User-Network Interfaces (Study Group XVIII) 10 pp. 10 pp.

—Radio Links—Radio Relay Systems—Availability
CCIR RECMN 695-90. Availability Objectives for Real Digital Radio-Relay Links Forming Part of a High-Grade Circuit Within an Integrated Services Digital Network—Section 9A—Performance Objectives, Propagation and Interference Effects. 2 pp.

—Radio Links—Radio Relay Systems—Bit Error Rate
CCIR RECMN 634-1-90. Error Performance Objectives for Real Digital Radio-Relay Links Forming Part of a High-Grade Circuit Within an Integrated Services Digital Network—Section 9A—Performance Objectives, Propagation and Interference Effects. 2 pp.
CCIR RECMN 634-2-91. Error Performance Objectives for Real Digital Radio-Relay Links Forming Part of a High-Grade Circuit Within an Integrated Services Digital Network—Section 9A—Performance Objectives, Propagation and Interference Effects. 12 pp.

—Radio Relay Systems—Availability
CCIR RECMN 697-1-91. Error Performance and Availability Objectives for the Local-Grade Portion at Each End of an ISDN Connection Utilizing Digital Radio-Relay Systems—Section 9A—Performance Objectives, Propagation and Interference Effects. 5 pp.

—Radio Relay Systems—Bit Error Rate
CCIR RECMN 697-90. Error Performance Objectives for the Local-Grade Portion at Each End of an ISDN Connection Utilizing Digital Radio-Relay Systems—Section 9A—Performance Objectives, Propagation and Interference Effects. 2 pp.
CCIR RECMN 697-1-91. Error Performance and Availability Objectives for the Local-Grade Portion at Each End of an ISDN Connection Utilizing Digital Radio-Relay Systems—Section 9A—Performance Objectives, Propagation and Interference Effects. 5 pp.

—Radio Relay Systems—Point to Multipoint Communications
CCIR Report 1193-90. Requirements for Point-to-Multipoint Systems Used in the Local Grade Portion of an ISDN Connection—Section 9E1—Line-of-Sight Radio-Relay Systems. 12 pp.
CCIR QUESTION 126/9-90. Requirements for Point-to-Multipoint Systems Used in the Local Grade Portion of an ISDN Connection—Questions of Study Group 9 Fixed Service. 1 p.

Integrated Services Digital Networks (Cont.)
—Reverse Charging
CCITT RECMN I.256-89. Charging Supplementary Services—Integrated Services Digital Network (ISDN)—General Structure and Service Capabilities (Study Group XVIII) 16 pp. 16 pp.
CCITT RECMN I.256. 3-92. Reverse Charging (Study Group I) 13 pp. 13 pp.

—Routing (Telecommunications)
CCITT FASCICLE II.2-88. Telephone Network and ISDN-Operation, Numbering, Routing and Mobile Service. Recommendations E.100—E.333. 362 pp.
CCITT RECMN E.172-89. Call Routing in the ISDN Era—Telephone Network and ISDN—Operation, Numbering, Routing and Mobile Service (Study Group II) 14 pp. 14 pp.
CCITT RECMN I.310-89. ISDN—Network Functional Principles—Integrated Services Digital Network (ISDN)—Overall Network Aspects and Functions, ISDN User -Network Interfaces (Study Group XVIII) 22 pp. 22 pp.
CCITT RECMN I.330-89. ISDN Numbering and Addressing Principles—Integrated Services Digital Network (ISDN)—Overall Network Aspects and Functions, ISDN User-Network Interfaces (Study Group XVIII) 8 pp. 8 pp.
CCITT RECMN I.335-89. ISDN Routing Principles—Integrated Services Digital Network (ISDN)—Overall Network Aspects and Functions, ISDN User-Network Interfaces (Study Group XVIII) 19 pp. 19 pp.
CENELEC PRETS 300 100-90. Integrated Services Digital Network (ISDN) Routing in support of the ISDN Memorandum of Understanding (T/N 23-03). 19 pp.
CENELEC PRETS 300 100-91. Integrated Services Digital Network (ISDN) Routing in Support of ISUP Version 1 Services. 22 pp.
CENELEC ETS 300 100-92. Integrated Services Digital Network (ISDN); Routing in Support of ISUP Version 1 Services. 18 pp.
CEPT T/GSI 03-06-85. Acheminement Des Communications Avec Le RNIS. 6 pp.
CEPT T/GSI 03-06 E-85. Routing. 6 pp.
CEPT T/N 23-02 E-89. Routing in the ISDN, for Gap Phase I Services. 17 pp.
CEPT T/S 26-01-87. Acheminement Dans Le RNIS Pour Les Services De La Phase 1 Du Gap. 17 pp.
CEPT T/S 26-01 E-87. Routing in the ISDN, for Gap Phase 1 Services. 17 pp.
ETSI PRETS 300 100-90. Integrated Services Digital Network (ISDN) Routing in Support of the ISDN Memorandum of Understanding (T/N 23-03). 19 pp.
ETSI ETS 300 100-92. Integrated Services Digital Network (ISDN); Routing in Support of ISUP Version 1 Services. 18 pp.
ETSI PRETS 300 100-91. Integrated Services Digital Network (ISDN) Routing in Support of ISUP Version 1 Services. 22 pp.

—Satellite Communications
CCIR QUESTION 69/4-90. Use of the Satellite Transmission Medium in the Framework of the ISDN—Questions of Study Group 4 Fixed-Satellite Service. 1 p.

—Signaling Systems
CCITT RECMN E.184-89. Indications to Users of ISDN Terminals—Telephone Network and ISDN—Operation, Numbering, Routing and Mobile Service (Study Group II) 2 pp. 2 pp.
CCITT RECMN I.326-89. Reference Configuration for Relative Network Resource Requirements—Integrated Services Digital Network (ISDN)—Overall Network Aspects and Functions, ISDN User-Network Interfaces (Study Group XVIII) 3 pp. 3 pp.
CCITT RECMN Q.65-89. Stage 2 of the Method for the Characterization of Services Supported by an ISDN—General Recommendations on Telephone Switching and Signalling—Functions and Information Flows for Services in the ISDN —Supplements (Study Group XI) 12 pp. 12 pp.
CENELEC PRETS 300 008-90. CCITT Signalling System Number 7: Message Transfer Part (MTP) to Support International Interconnection (T/S 43-01). 9 pp.
CENELEC PRETS 300 008-91. Integrated Services Digital Network (ISDN); CCITT Signalling System No. 7 Message Transfer Part (MTP) to Support International Interconnection. 20 pp.
CENELEC ETS 300 008-91. Integrated Services Digital Network (ISDN); CCITT Signalling System No. 7 Message Transfer Part (MTP) to Support International Interconnection. 20 pp.
CENELEC ETS 300 008 PRA1-93. AMD prA1 Integrated Services Digital Network (ISDN); CCITT Signalling System No. 7 Message Transfer Part (MTP) to Support International Interconnection. 11 pp.

INTERNATIONAL AND NON-U.S. NATIONAL STANDARDS
SUBJECT INDEX
Integrated

Integrated Services Digital Networks *(Cont.)*

—Signaling Systems *(Cont.)*

CENELEC PRETS 300 009-90. CCITT Signalling System Number 7: Signalling Connection Control Part (SCCP)-Connectionless Service-to Support International Interconnection (T/S 43-03). 6 pp.

CENELEC PRETS 300 009-91. Integrated Services Digital Network (ISDN); CCITT Signalling System No. 7 Signalling Connection Control Part (SCCP) (Connectionless Service) to Support International Interconnection. 9 pp.

CENELEC ETS 300 009-91. Integrated Services Digital Network (ISDN); CCITT Signalling System No. 7 Signalling Connection Control Part (SCCP) (Connectionless Service) to Support International Interconnection. 9 pp.

CENELEC PRETS 300 009-93. Integrated Services Digital Network (ISDN); CCITT Signalling System No. 7 Signalling Connection Control Part (SCCP) (Connectionless and Connection-Oriented) to Support International Interconnection. 34 pp.

CENELEC PRETS 300 134-90. Integrated Services Digital Network (ISDN): CCITT Signalling System No. 7 Transaction capabilities Application Part (TCAP) (T/S 43-05). 37 pp.

CENELEC PRETS 300 134-92. Integrated Services Digital Network (ISDN); CCITT Signalling System No. 7 Transaction Capabilities Application Part (TCAP). 39 pp.

CENELEC ETS 300 134-92. Integrated Services Digital Network (ISDN); CCITT Signalling System No. 7 Transaction Capabilities Application Part (TCAP). 38 pp.

CEPT T/N 21-02 E-88. Numbering and Addressing for Gap Phase II. 5 pp.

CEPT T/S 43-03 E-88. CCITT SS No. 7 Signalling Connection Control Part (SCCP). 2 pp.

CEPT T/S 43-11 E-88. Signalling Requirements on Interworking Between the ISDN or PSTN and the PLMN. 3 pp.

ETSI PRETS 300 008-90. CCITT Signalling System Number 7; Message Transfer Part (MTP) to Support International Interconnection Also Known as T/S 43-01. 9 pp.

ETSI PRETS 300 009-90. CCITT Signalling System Number 7; Signalling Connection Contgrol Part (SCCP) (Connectionless Service) to Support International Interconnection. Also Know as T/S 43-03. 6 pp.

ETSI ETS 300 008 PRA1-93. AMD prA1 Integrated Services Digital Network (ISDN); CCITT Signalling System No. 7 Message Transfer Part (MTP) to Support International Interconnection. 11 pp.

ETSI ETS 300 008-91. Integrated Services Digital Network (ISDN); CCITT Signalling System No. 7 Message Transfer Part (MTP) to Support International Interconnection. 20 pp.

ETSI PRETS 300 008-91. Integrated Services Digital Network (ISDN); CCITT Signalling System No. 7 Message Transfer Part (MTP) to Support International Interconnection. 20 pp.

ETSI PRETS 300 008-90. CCITT Signalling System Number 7: Message Transfer Part (MTP) to Support International Interconnection (T/S 43-01). 9 pp.

ETSI PRETS 300 009-93. Integrated Services Digital Network (ISDN); CCITT Signalling System No. 7 Signalling Connection Control Part (SCCP) (Connectionless and Connection-Oriented) to Support International Interconnection. 34 pp.

ETSI ETS 300 009-91. Integrated Services Digital Network (ISDN); CCITT Signalling System No. 7 Signalling Connection Control Part (SCCP) (Connectionless Service) to Support International Interconnection. 9 pp.

ETSI PRETS 300 009-91. Integrated Services Digital Network (ISDN); CCITT Signalling System No. 7 Signalling Connection Control Part (SCCP) (Connectionless Service) to Support International Interconnection. 9 pp.

ETSI PRETS 300 009-90. CCITT Signalling System Number 7: Signalling Connection Control Part (SCCP)-Connectionless Service-to Support International Interconnection (T/S 43-03). 6 pp.

ETSI PRETS 300 134-90. Integrated Services Digital Network (ISDN): CCITT Signalling System No. 7 Transaction Capabilities Application Part (TCAP)(T/S 43-05). 37 pp.

ETSI ETS 300 134-92. Integrated Services Digital Network (ISDN); CCITT Signalling System No. 7 Transaction Capabilities Application Part (TCAP). 38 pp.

ETSI PRETS 300 134-92. Integrated Services Digital Network (ISDN); CCITT Signalling System No. 7 Transaction Capabilities Application Part (TCAP). 39 pp.

ETSI PRETS 300 134-90. Integrated Services Digital Network (ISDN): CCITT Signalling System No. 7 Transaction capabilities Application Part (TCAP) (T/S 43-05). 37 pp.

Integrated Services Digital Networks *(Cont.)*

—Signaling Systems—Network Architecture

CCITT RECMN I.324-89. ISDN Network Architecture—Integrated Services Digital Network (ISDN)—Overall Network Aspects and Functions, ISDN User-Network Interfaces (Study Group XVIII) 12 pp. 12 pp.

—Signaling Systems—Private Networks

CEPT T/S 49-15-87. RNIS, Systeme De Signalisation Pour Reseaux Prives. 10 pp.

CEPT T/S 49-15 E-87. ISDN Signalling System in Private Network. 10 pp.

—Signaling Systems—Terminal to Terminal

CEPT T/SF 34 E-84. Customer Equipment Aspects of Terminal to Terminal Signalling. 2 pp.

—Sound-Program Signals

CCIR QUESTION 56/CMTT-90. Specific Requirements for the Integrated Services Digital Network (ISDN) Carrying Sound-Programme Services—Questions Concerning the CMTT CCIR/CCITT Joint Study Group for Television and Sound Transmission. 1 p.

—Sound Studios—Interfaces

CCIR QUESTION 61/CMTT-90. Standards for Digital Interfaces Between Studio and Dedicated or Integrated Services Digital Networks—Questions Concerning the CMTT CCIR/CCITT Joint Study Group for Television and Sound Transmission. 1 p.

—Spacecraft—Ionospheric Propagation

CCIR RECMN 531-1-90. Ionospheric Effects Influencing Radio Systems Involving Spacecraft—Section 6F—Ionospheric Propagation Prediction and Applications at Frequencies Above About 30 MHz. 1 p.

CCIR RECMN 531-2-92. Ionospheric Effects Influencing Radio Systems Involving Spacecraft—Section 6F—Ionospheric Propagation Prediction and Applications at Frequencies Above About 30 MHz. 14 pp.

—Specification and Description Language

CENELEC PRETS 300 102-2-90. Integrated Services Digital Network (ISDN): User-Network Interface Layer 3 Specification Description Language (SDL) Diagrams. 72 pp.

CENELEC ETS 300 102-2-90. Integrated Services Digital Network (ISDN); User-Network Interface Layer 3 Specifications for Basic Call Control Specification Description Language (SDL) Diagrams. 75 pp.

ETSI PRETS 300 102-2-90. Integrated Services Digital Network (ISDN); User-Network Interface Layer 3—Specifications for Basic Call Control—Specification Description Language (SDL) Diagrams. 73 pp.

ETSI ETS 300 102-2-90. Integrated Services Digital Network (ISDN); User-Network Interface Layer 3 Specifications for Basic Call Control Specification Description Language (SDL) Diagrams. 75 pp.

ETSI PRETS 300 102-2-90. Integrated Services Digital Network (ISDN): User-Network Interface Layer 3 Specification Description Language (SDL) Diagrams. 72 pp.

—Subaddressing

CCITT RECMN E.164-91. Numbering Plan for the ISDN Era (Study Group II) 19 pp (Same as Recmn I.331). 19 pp.

CCITT RECMN E.164-89. Numbering Plan for the ISDN Era—Telephone Network and ISDN—Operation, Numbering, Routing and Mobile Service (Study Group II) 6 pp (Same as Recmn I.331). 6 pp.

CCITT RECMN I.251. 8 (REV 1)-92. Sub-Addressing Supplementary Service (Study Group I) 9 pp. 9 pp.

CCITT RECMN I.330-89. ISDN Numbering and Addressing Principles—Integrated Services Digital Network (ISDN)—Overall Network Aspects and Functions, ISDN User-Network Interfaces (Study Group XVIII) 8 pp. 8 pp.

CENELEC PRETS 300 059-90. Integrated Services Digital Network (ISDN): Subaddressing (SUB) Supplementary Service Service Description (T/NA1(89)16). 12 pp.

CENELEC PRETS 300 059-91. Integrated Services Digital Network (ISDN); Subaddressing (SUB) Supplementary Service Service Description. 15 pp.

CENELEC ETS 300 059-91. Integrated Services Digital Network (ISDN); Subaddressing (SUB) Supplementary Service Service Description. 15 pp.

CENELEC PRETS 300 060-90. Integrated Services Digital Network (ISDN): Subaddressing (SUB) Supplementary Service Functional Capabilities and Information Flows (T/S 22-26). 17 pp.

CENELEC PRETS 300 060-91. Integrated Services Digital Network (ISDN); Subaddressing (SUB) Supplementary Service Functional Capabilities and Information Flows. 23 pp.

Integrated Services Digital Networks *(Cont.)*

—Subaddressing *(Cont.)*

CENELEC ETS 300 060-91. Integrated Services Digital Network (ISDN); Subaddressing (SUB) Supplementary Service Functional Capabilities and Information Flows. 23 pp.

ETSI PRETS 300 059-90. Integrated Services Digital Network (ISDN); Subaddressing (SUB) Supplementary Service—Service Description (T/NA1 (89)16). 12 pp.

ETSI PRETS 300 060-90. Integrated Services Digital Network (ISDN); Subaddressing (SUB) Supplementary Service Functional Capabilities and Information Flows (T/S 22-26). 17 pp.

ETSI ETS 300 059-91. Integrated Services Digital Network (ISDN); Subaddressing (SUB) Supplementary Service Service Description. 15 pp.

ETSI PRETS 300 059-91. Integrated Services Digital Network (ISDN); Subaddressing (SUB) Supplementary Service Service Description. 15 pp.

ETSI PRETS 300 059-90. Integrated Services Digital Network (ISDN): Subaddressing (SUB) Supplementary Service Service Description (T/NA1(89)16). 12 pp.

ETSI ETS 300 060-91. Integrated Services Digital Network (ISDN); Subaddressing (SUB) Supplementary Service Functional Capabilities and Information Flows. 23 pp.

ETSI PRETS 300 060-91. Integrated Services Digital Network (ISDN); Subaddressing (SUB) Supplementary Service Functional Capabilities and Information Flows. 23 pp.

ETSI PRETS 300 060-90. Integrated Services Digital Network (ISDN): Subaddressing (SUB) Supplementary Service Functional Capabilities and Information Flows (T/S 22-26). 17 pp.

—Subaddressing—CCITT No. 7 Signaling Systems

CCITT RECMN Q.731-91. Stage 3 Description for Number Identification Supplementary Services Using Signalling System No. 7 Section 1—Direct Dialling in (DDI) Section 8—Sub-Addressing (SUB)—Specifications of Signalling System No. 7 (Study Group XI) 12 pp. 12 pp.

—Subaddressing—DSS1

CCITT RECMN Q.951-92. Stage 3 Description for Number Identification Supplementary Services Using DSS 1 Section 1—Direct-Dialling-in (DDI) Section 2—Multiple Subscriber Number (MSN) Section 8—Sub-Addressing (SUB) (Study Group XI) 17 pp. 17 pp.

CENELEC PRETS 300 061-90. Integrated Services Digital Network (ISDN); Subaddressing (SUB) Supplementary Service Digital Subscriber Signalling One (DSS1) Protocol (T/S 46-331). 9 pp.

CENELEC PRETS 300 061-91. Integrated Services Digital Network (ISDN); Subaddressing (SUB) Supplementary Service Digital Subscriber Signalling System No. One (DSS1) Protocol. 11 pp.

CENELEC ETS 300 061-91. Integrated Services Digital Network (ISDN); Subaddressing (SUB) Supplementary Service Digital Subscriber Signalling System No. One (DSS1) Protocol. 11 pp.

ETSI PRETS 300 061-90. Integrated Services Digital Network (ISDN); Subaddressing (SUB) Supplementary Service Digital Subscriber Signalling One (DSS1) Protocol (T/S 46-331). 9 pp.

ETSI ETS 300 061-91. Integrated Services Digital Network (ISDN); Subaddressing (SUB) Supplementary Service Digital Subscriber Signalling System No. One (DSS1) Protocol. 11 pp.

ETSI PRETS 300 061-91. Integrated Services Digital Network (ISDN); Subaddressing (SUB) Supplementary Service Digital Subscriber Signalling System No. One (DSS1) Protocol. 11 pp.

ETSI PRETS 300 061-90. Integrated Services Digital Network (ISDN); Subaddressing (SUB) Supplementary Service Digital Subscriber Signalling One (DSS1) Protocol (T/S 46-331). 9 pp.

—Subaddressing—Functions/Information Flows

CCITT RECMN Q.81-92. Stage 2 Description for Number Identification Supplementary Services Section 2—Multiple Subscriber Number Section 8—Sub-Addressing (SUB)—(Addenda to Q.81) (Study Group XI) 19 pp. 19 pp.

—Subaddressing—Network Layer

CCITT RECMN I.334-89. Principles Relating ISDN Numbers/Subaddresses to the OSI Reference Model Network Layer Addresses—Integrated Services Digital Network (ISDN)—Overall Network Aspects and Functions, ISDN User-Network Interfaces (Study Group XVIII) 5 pp. 5 pp.

INDUSTRY STANDARDS
1423

Integrated Services Digital Networks (Cont.)

—Subscriber Access—Maintenance

CCITT FASCICLE III.9-88. Integrated Services Digital Network (ISDN) Internetwork Interfaces and Maintenance Principles Recommendations I.500—I.605 (Study Group XVIII). 103 pp.

CCITT RECMN I.601-89. General Maintenance Principles of ISDN Subscriber Access and Subscriber Installation—Integrated Services Digital Network (ISDN)—Internetwork Interfaces and Maintenance Principles (Study Group XVIII) 11 pp. 11 pp.

CCITT RECMN I.603-89. Application of Maintenance Principles to ISDN Basic Accesses—Integrated Services Digital Network (ISDN)—Internetwork Interfaces and Maintenance Principles (Study Group XVIII) 6 pp. 6 pp.

CCITT RECMN I.604-89. Application of Maintenance Principles to ISDN Primary Rate Accesses—Integrated Services Digital Network (ISDN)—Internetwork Interfaces and Maintenance Principles (Study Group XVIII) 13 pp. 13 pp.

CENELEC ETR 001-90. Integrated Services Network (ISDN); Customer Access Maintenance. 5 pp.

CEPT T/S 54-08 E-87. ISDN Subscriber Access and Installation Maintenance. 22 pp.

ETSI ETR 001-90. Integrated Services Network (ISDN); Customer Access Maintenance. 5 pp.

—Subscriber Dedicated Storage

CEPT T/CS 20-07-83. Memoire Specifique D'Abonne. 2 pp.

CEPT T/CS 20-07 E-83. Subscriber Dedicated Storage. 2 pp.

—Subscriber Installation—Maintenance

CCITT RECMN I.601-89. General Maintenance Principles of ISDN Subscriber Access and Subscriber Installation—Integrated Services Digital Network (ISDN)—Internetwork Interfaces and Maintenance Principles (Study Group XVIII) 11 pp. 11 pp.

CCITT RECMN I.602-89. Application of Maintenance Principles to ISDN Subscriber Installations—Integrated Services Digital Network (ISDN)—Internetwork Interfaces and Maintenance Principles (Study Group XVIII) 6 pp. 6 pp.

—Subscriber Numbers

CCITT RECMN I.330-89. ISDN Numbering and Addressing Principles—Integrated Services Digital Network (ISDN)—Overall Network Aspects and Functions, ISDN User-Network Interfaces (Study Group XVIII) 8 pp. 8 pp.

—Supplementary Services

CCITT RECMN I.241-89. Teleservices Supported by an ISDN—Integrated Services Digital Network (ISDN)—General Structure and Service Capabilities (Study Group XVIII) 28 pp. 28 pp.

CCITT RECMN I.250-89. Definition of Supplementary Services—Integrated Services Digital Network (ISDN)—General Structure and Service Capabilities (Study Group XVIII) 6 pp. 6 pp.

CCITT RECMN I.320-89. ISDN Protocol Reference Model—Integrated Services Digital Network (ISDN)—Overall Network Aspects and Functions, ISDN User-Network Interfaces (Study Group XVIII) 10 pp. 10 pp.

CCITT RECMN I.452-89. Generic Procedure for the Control of ISDN Supplementary Services—Integrated Services Digital Network (ISDN)—Overall Network Aspects and Functions, ISDN User-Network Interfaces (Study Group XVIII) 1 pp (Same as Recmn Q.932). 1 p.

CEPT T/CAC S 10.5 E-92. General Supplementary Service Aspects of an Integrated Services Digital Network (ISDN) (Revised in Odense 1986, Vienna 1989(CAC) and Athens 1992). 23 pp.

CEPT T/CAC S 10.6 E-92. Relevance of Supplementary Services to Be Provided in an Integrated Services Digital Network (ISDN) (Revised in Odense 1986, Vienna 1989(CAC) and Athens 1992). 13 pp.

CEPT T/CAC S 10.7 E-92. Operational Requirements of ISDN Supplementary Services (Revised in Vienna 1989 (CAC) and Athens 1992). 345 pp.

—Supplementary Services—Accounting

CCITT RECMN D.232 (REV 1)-92. Specific Tariff and Accounting Principles Applicable to ISDN Supplementary Services (Study Group III) 10 pp. 10 pp.

CCITT RECMN D.232-91. Specific Tariff and Accounting Principles Applicable to ISDN Supplementary Services (Study Group III) 9 pp. 9 pp.

CEPT T/PGT 34 E-91. Charging and Accounting Principles for Specific Supplementary Services Provided over the ISDN. 1 p.

Integrated Services Digital Networks (Cont.)

—Supplementary Services—Protocols—DSS1

CENELEC PRETS 300 195-92. Integrated Services Digital Network (ISDN); Supplementary Service Interactions Digital Subscriber Signalling System No. One (DSS1) Protocol. 41 pp.

CENELEC PRETS 300 196-92. Integrated Services Digital Network (ISDN); Generic Functional Protocol for the Support of Supplementary Services Digital Subscriber Signalling System No. One (DSS1) Protocol. 88 pp.

CENELEC PRETS 300 196-93. Integrated Services Digital Network (ISDN); Generic Functional Protocol for the Support of Supplementary Services Digital Subscriber Signalling System No. One (DSS1) Protocol. 108 pp.

CENELEC PRETS 300 207-92. Integrated Services Digital Network (ISDN); Diversion Supplementary Services Digital Subscriber Signalling No. One (DSS1) Protocol. 72 pp.

ETSI PRETS 300 195-92. Integrated Services Digital Network (ISDN); Supplementary Service Interactions Digital Subscriber Signalling System No. One (DSS1) Protocol. 41 pp.

ETSI PRETS 300 196-93. Integrated Services Digital Network (ISDN); Generic Functional Protocol for the Support of Supplementary Services Digital Subscriber Signalling System No. One (DSS1) Protocol. 108 pp.

ETSI PRETS 300 196-92. Integrated Services Digital Network (ISDN); Generic Functional Protocol for the Support of Supplementary Services Digital Subscriber Signalling System No. One (DSS1) Protocol. 88 pp.

ETSI PRETS 300 207-92. Integrated Services Digital Network (ISDN); Diversion Supplementary Services Digital Subscriber Signalling No. One (DSS1) Protocol. 72 pp.

—Supplementary Services—Tariffs (Telecommunications)

CCITT RECMN D.232 (REV 1)-92. Specific Tariff and Accounting Principles Applicable to ISDN Supplementary Services (Study Group III) 10 pp. 10 pp.

CCITT RECMN D.232-91. Specific Tariff and Accounting Principles Applicable to ISDN Supplementary Services (Study Group III) 9 pp. 9 pp.

CCITT RECMN I.140-89. Attribute Technique for the Characterization of Telecommunication Services Supported by an ISDN and Network Capabilities of an ISDN—Integrated Services Digital Network (ISDN)—General Structure and Service Capabilities (Study Group XVIII) 13 pp. 13 pp.

CEPT T/PGT 34 E-91. Charging and Accounting Principles for Specific Supplementary Services Provided over the ISDN. 1 p.

—Switching Systems

CCITT RECMN I.326-89. Reference Configuration for Relative Network Resource Requirements—Integrated Services Digital Network (ISDN)—Overall Network Aspects and Functions, ISDN User-Network Interfaces (Study Group XVIII) 3 pp. 3 pp.

CCITT RECMN Q.65-89. Stage 2 of the Method for the Characterization of Services Supported by an ISDN—General Recommendations on Telephone Switching and Signalling—Functions and Information Flows for Services in the ISDN —Supplements (Study Group XI) 12 pp. 12 pp.

—Switching Systems—Network Architecture

CCITT RECMN I.324-89. ISDN Network Architecture—Integrated Services Digital Network (ISDN)—Overall Network Aspects and Functions, ISDN User-Network Interfaces (Study Group XVIII) 12 pp. 12 pp.

—Synchronous Digital Hierarchy—Frame Structure

CCITT RECMN G.704-91. Synchronous Frame Structures Used at Primary and Secondary Hierarchical Levels (Study Group XVIII) 35 pp. 35 pp.

CCITT RECMN G.704-89. Synchronous Frame Structures Used at Primary and Secondary Hierarchical Levels—General Aspects of Digital Transmission Systems; Terminal Equipments (Study Groups XV and XVIII) 21 pp. 21 pp.

—Tariffs (Telecommunications)

CCITT RECMN D.250-91. General Charging and Accounting Principles for Non-Voice Services Provided by Interworking Between the ISDN and Existing Public Data Networks (Study Group III) 5 pp. 5 pp.

Integrated Services Digital Networks (Cont.)

—Tariffs (Telecommunications) (Cont.)

CCITT RECMN D.250-89. General Charging and Accounting Principles for Non-Voice Services Provided by Interworking Between the ISDN and Existing Public Data Networks—General Tariff Prin.—Charging and Accounting in Intl Telecommunications Services (Study Group III) 1 pp. 1 p.

CCITT RECMN D.260-91. Charging and Accounting Capabilities to Be Applied on the ISDN (Study Group III) 6 pp. 6 pp.

CCITT RECMN D.141-89. ISDN Network Charging Capabilities Attributes—Integrated Services Digital Network (ISDN)—General Structure and Service Capabilities (Study Group XVIII) 2 pp. 2 pp.

—Telecommunication Equipment

CENELEC PRETS 300 081-92. Integrated Services Digital Network (ISDN); Teletex End-to-End Protocol over the ISDN. 14 pp.

CEPT T/GSI 01-07-87. Methode De Caracterisation Des Services De Telecommunications Assures Par Le RNIS Et Possibilites Reseau D'un RNIS. 4 pp.

CEPT T/GSI 01-07 E-87. Method for the Characterization of Telecommunication Services Supported by an ISDN and Network Capabilities of an ISDN. 4 pp.

CEPT T/GSI 02-03-87. Teleservices. 13 pp.

CEPT T/GSI 02-03 E-87. Teleservices. 13 pp.

CEPT T/GSI 02-08-87. Services Supplementaires Associes Aux Teleservices. 5 pp.

CEPT T/GSI 02-08 E-87. Supplementary Services Associated with Teleservices. 5 pp.

CEPT T/N 13-05 E-87. Supplementary Services Associated with Teleservices. 5 pp.

CEPT T/SF 31-02 E-89. Bearer Services to Be Provided by Integrated Services Digital Network (ISDN) (Revised in Odense 1986 and Vienna 1989 (CAC)). 38 pp.

CEPT T/SF 58 E-87. Telematic and Data Transmission Services in the ISDN. 24 pp.

ETSI PRETS 300 082-90. Integrated Services Digital Network (ISDN); 3.1 KHz Telephony Teleservice—End-To-End Compatibility (T/TE 12-05). 10 pp.

ETSI PRETS 300 082-91. Integrated Services Digital Network (ISDN); 3,1 kHz Telephony Teleservice End-to-End Compatibility Requirements for Telephony Terminals. 35 pp.

—Telecommunication Services

CCITT RECMN I.200-89. Guidance to the I.200-Series of Recommendations—Integrated Services Digital Network (ISDN)—General Structure and Service Capabilities (Study Group XVIII) 3 pp. 3 pp.

CCITT RECMN I.240-89. Definition of Teleservices—Integrated Services Digital Network (ISDN)—General Structure and Service Capabilities (Study Group XVIII) 3 pp. 3 pp.

CCITT RECMN I.241-89. Teleservices Supported by an ISDN—Integrated Services Digital Network (ISDN)—General Structure and Service Capabilities (Study Group XVIII) 28 pp. 28 pp.

CCITT RECMN I.310-89. ISDN—Network Functional Principles—Integrated Services Digital Network (ISDN)—Overall Network Aspects and Functions, ISDN User -Network Interfaces (Study Group XVIII) 22 pp. 22 pp.

CCITT RECMN I.326-89. Reference Configuration for Relative Network Resource Requirements—Integrated Services Digital Network (ISDN)—Overall Network Aspects and Functions, ISDN User-Network Interfaces (Study Group XVIII) 3 pp. 3 pp.

CENELEC ETR 018-92. Integrated Services Digital Network (ISDN); Application of the BC-, HLC-, LLC-Information Elements by Terminals Supporting ISDN Services. 70 pp.

CEPT T/CAC S 10.3 E-92. Teleservices to Be Provided by an Integrated Services Digital Network (ISDN). 36 pp.

CEPT T/CAC S 10.4 E-92. Intercommunication Aspects of an Integrated Services Digital Network (ISDN) (Revised in Odense 1986, Vienna 1989 (CAC) and Athens 1992). 11 pp.

ETSI ETR 018-92. Integrated Services Digital Network (ISDN); Application of the BC-, HLC-, LLC-Information Elements by Terminals Supporting ISDN Services. 70 pp.

—Telecommunication Services—Accounting

CCITT RECMN D.210-89. General Charging and Accounting Principles for International Telecommunication Services Provided over the Integrated Services Digital Network (ISDN)—General Tariff Principles—Charging and Accounting in Intl Telecom. Serv. (Study Group III) 4 pp. 4 pp.

INTERNATIONAL AND NON-U.S. NATIONAL STANDARDS
SUBJECT INDEX

Integrated

Integrated Services Digital Networks (Cont.)

—Telecommunication Services—Accounting (Cont.)

CCITT RECMN D.230-89. General Charging and Accounting Principles for Supplementary Services Associated with International Telecommunication Services Provided over the Integrated Services Digital Network (ISDN)—General Tariff Principles—Charging and Accounting in. 2 pp.

—Telecommunication Services—Busy Tones

CCITT RECMN I.221-89. Common Specific Characteristics of Services—Integrated Services Digital Network (ISDN)—General Structure and Service Capabilities (Study Group XVIII) 3 pp. 3 pp.

—Telecommunication Services—Classifications

CCITT RECMN I.210-89. Principles of Telecommunication Services Supported by an ISDN and the Means to Describe Them—Integrated Services Digital Network (ISDN)—General Structure and Service Capabilities (Study Group XVIII) 19 pp. 19 pp.

—Telecommunication Services—Forecasting

CCITT RECMN E.508-89. Forecasting New International Services —Telephone Network and ISDN—Quality of Service, Network Management and Traffic Engineering (Study Group II) 9 pp. 9 pp.

—Telecommunication Services—Grade of Service

CCITT RECMN E.700-92. Framework of the E.700-Series Recommendations—Telephone Network and ISDN Service, Network Management and Traffic Engineering (Study Group II) 3 pp. 3 pp.

CCITT RECMN E.720-89. ISDN Grade of Service Concept—Telephone Network and ISDN—Quality of Service, Network Management and Traffic Engineering (Study Group II) 3 pp. 3 pp.

—Telecommunication Services—Interworking

CENELEC PRETS 300 345-93. Integrated Services Digital Network (ISDN); Interworking Between Public ISDNs and Private ISDNs for the Provision of Telecommunications Services General Aspects. 15 pp.

ETSI PRETS 300 345-93. Integrated Services Digital Network (ISDN); Interworking Between Public ISDNs and Private ISDNs for the Provision of Telecommunications Services General Aspects. 15 pp.

—Telecommunication Services—Protocols

CCITT RECMN I.130-89. Method for the Characterization of Telecommunication Services Supported by an ISDN and Network Capabilities of an ISDN—Integrated Services Digital Network (ISDN)—General Structure and Service Capabilities (Study Group XVIII) 5 pp. 5 pp.

—Telecommunication Services—Quality of Service

CCITT FASCICLE II.3-88. Telephone Network and ISDN Quality of Service, Network Management and Traffic Engineering. Recommendations E.401-E.880. 368 pp.

—Telecommunication Services—Routing

CCITT RECMN I.335-89. ISDN Routing Principles—Integrated Services Digital Network (ISDN)—Overall Network Aspects and Functions, ISDN User-Network Interfaces (Study Group XVIII) 19 pp. 19 pp.

—Telecommunication Services—Tariffs

CCITT RECMN D.210-89. General Charging and Accounting Principles for International Telecommunication Services Provided over the Integrated Services Digital Network (ISDN)—General Tariff Principles—Charging and Accounting in Intl Telecom. Serv. (Study Group III) 4 pp. 4 pp.

CCITT RECMN D.230-89. General Charging and Accounting Principles for Supplementary Services Associated with International Telecommunication Services Provided over the Integrated Services Digital Network (ISDN)—General Tariff Principles—Charging and Accounting in. 2 pp.

—Telecommunication Services—Tariffs—Europe

CCITT RECMN D.307R-91. Remuneration of Digital Systems and Channels Used in Telecommunication Relations Between the Countries of Europe and the Mediterranean Basin (Study Group III) 7 pp. 7 pp.

Integrated Services Digital Networks (Cont.)

—Telecommunication Services—Tariffs—Europe (Cont.)

CCITT RECMN D.307R-89. Remuneration of Digital Systems and Channels Used in Telecommunication Relations Between the Countries of Europe and the Mediterranean Basin—General Tariff Principles—Charging and Accounting in Intl Telecom. Serv. (Study Group III) 3 pp. 3 pp.

—Telecommunication Services—Tariffs—Mediterranean Basin

CCITT RECMN D.307R-91. Remuneration of Digital Systems and Channels Used in Telecommunication Relations Between the Countries of Europe and the Mediterranean Basin (Study Group III) 7 pp. 7 pp.

CCITT RECMN D.307R-89. Remuneration of Digital Systems and Channels Used in Telecommunication Relations Between the Countries of Europe and the Mediterranean Basin—General Tariff Principles—Charging and Accounting in Intl Telecom. Serv. (Study Group III) 3 pp. 3 pp.

—Telecommunication Systems—Reference Configurations

CENELEC ETR 061-93. Integrated Services Digital Network (ISDN); Rules for the Construction of Reference Configurations. 26 pp.

ETSI ETR 061-93. Integrated Services Digital Network (ISDN); Rules for the Construction of Reference Configurations. 26 pp.

—Telecommunication Traffic

CENELEC PRETS 300 251-92. Network Aspects (NA); Accessibility for 64 kbit/s Circuit Switched International End-to-End ISDN Traffic Relations. 11 pp.

CENELEC PRETS 300 251-93. Network Aspects (NA); Accessibility for 64 kbit/s Circuit Switched International End-to-End ISDN Traffic Relations. 11 pp.

ETSI PRETS 300 251-93. Network Aspects (NA); Accessibility for 64 kbit/s Circuit Switched International End-to-End ISDN Traffic Relations. 11 pp.

ETSI PRETS 300 251-92. Network Aspects (NA); Accessibility for 64 kbit/s Circuit Switched International End-to-End ISDN Traffic Relations. 11 pp.

—Teleinformatic Services

CCITT RECMN F.353-89. Provision of Telematic and Data Transmission Services on Integrated Services Digital Network (ISDN)—Telematic, Data Transmission and Teleconference Services—Operations and Quality of Service (Study Group I) 5 pp. 5 pp.

—Teleinformatic Services—Terminal Identification

CCITT RECMN F.351-89. General Principles on the Presentation of Terminal Identification to Users of the Telematic Services—Telematic, Data Transmission and Teleconference Services—Operations and Quality of Service (Study Group I) 2 pp. 2 pp.

—Telephone Cables—Digital Transmission Systems

CCITT RECMN G.961-89. Digital Transmission System on Metallic Local Lines for ISDN Basic Rate Access—Digital Networks, Digital Sections and Digital Line Systems (Study Groups XV and XVIII) 119 pp. 119 pp.

—Telephone Cables—Noise

CCITT RECMN K.23-89. Types of Induced Noise and Description of Noise Voltage Parameters for ISDN Basic User Networks Protection Against Interference (Study Group V) 4 pp. 4 pp.

—Telephone Connections

CCITT RECMN I.320-89. ISDN Protocol Reference Model—Integrated Services Digital Network (ISDN)—Overall Network Aspects and Functions, ISDN User-Network Interfaces (Study Group XVIII) 10 pp. 10 pp.

—Telephone Exchanges

CEPT T/GSI 03-09-87. Commutation. 1 p.

CEPT T/GSI 03-09 E-87. Switching. 1 p.

—Telephone Services

CCITT RECMN I.310-89. ISDN—Network Functional Principles—Integrated Services Digital Network (ISDN)—Overall Network Aspects and Functions, ISDN User -Network Interfaces (Study Group XVIII) 22 pp. 22 pp.

Integrated Services Digital Networks (Cont.)

—Telephone Services (Cont.)

CCITT RECMN Q.80-89. Introduction to Stage 2 Service Descriptions for Supplementary Services—General Recommendations on Telephone Switching and Signalling—Functions and Information Flows for Services in the ISDN —Supplements (Study Group XI) 4 pp. 4 pp.

—Telephone Services—Accounting

CCITT RECMN D.251-89. General Charging and Accounting Principles for the Basic Telephone Service Provided over the ISDN or by Interconnection Between the ISDN and the Public Switched Telephone Network—General Tariff Principles—Charging and Accounting in International. 2 pp.

—Telephone Services—Functions/Information Flows

CCITT RECMN Q.65-89. Stage 2 of the Method for the Characterization of Services Supported by an ISDN—General Recommendations on Telephone Switching and Signalling—Functions and Information Flows for Services in the ISDN —Supplements (Study Group XI) 12 pp. 12 pp.

—Telephone Services—Tariffs

CCITT RECMN D.251-89. General Charging and Accounting Principles for the Basic Telephone Service Provided over the ISDN or by Interconnection Between the ISDN and the Public Switched Telephone Network—General Tariff Principles—Charging and Accounting in International. 2 pp.

—Telephones

CENELEC PRETS 300 085-90. Integrated Services Digital Network (ISDN): 3,1 KHz Telephony Teleservice Attachment Requirements for Handset Terminals. 56 pp.

CENELEC ETS 300 085-90. Integrated Services Digital Network (ISDN); 3,1 kHz Telephony Teleservice Attachment Requirements for Handset Terminals. 54 pp.

ETSI PRETS 300 085-90. Integrated Services Digital Network (ISDN); 3,1 KHz Telephony Teleservices—Attachment Requirements for Handset Terminals (The Text of This Draft ETS May Be Utilized, Wholly or in Part, for the Establishment of NET 33). 56 pp.

ETSI ETS 300 085-90. Integrated Services Digital Network (ISDN); 3,1 kHz Telephony Teleservice Attachment Requirements for Handset Terminals. 54 pp.

ETSI PRETS 300 085-90. Integrated Services Digital Network (ISDN); 3,1 KHz Telephony Teleservice Attachment Requirements for Handset Terminals. 56 pp.

—Telephony

CCITT RECMN I.241. 1-88. Telephony—Integrated Services Digital Network (ISDN)—General Structure and Service Capabilities (Study Group XVIII) 4 pp. 4 pp.

CENELEC PRETS 300 082-90. Integrated Services Digital Network (ISDN): 3.1 KHz Telephony Teleservice End-to-End Compatibility (T/TE 12-05). 10 pp.

CENELEC PRETS 300 082-91. Integrated Services Digital Network (ISDN); 3,1 kHz Telephony Teleservice End-to-End Compatibility Requirements for Telephony Terminals. 35 pp.

CENELEC ETS 300 082-92. Integrated Services Digital Network (ISDN); 3,1 kHz Telephony Teleservice End-to-End Compatibility Requirements for Telephony Terminals. 35 pp.

CENELEC PRETS 300 111-90. Integrated Services Digital Network (ISDN) Telephony 3.1 KHz Teleservice Service Description. 6 pp.

CENELEC PRETS 300 111-92. Integrated Services Digital Network (ISDN); Telephony 3,1 kHz Teleservice Service Description. 14 pp.

CENELEC ETS 300 111-92. Integrated Services Digital Network (ISDN); Telephony 3,1 kHz Teleservice Service Description. 13 pp.

CENELEC PRETS 300 263-92. Integrated Services Digital Network (ISDN); Telephony 7 kHz Teleservice Service Description. 15 pp.

CENELEC PRETS 300 265-92. Integrated Services Digital Network (ISDN); Telephony 7 kHz Teleservice Functional Capabilities and Information Flows. 22 pp.

ETSI PRETS 300 111-90. Integrated Services Digital Network (ISDN)—Telephony 3.1 KHz Teleservice—Service Description. 9 pp.

ETSI ETS 300 082-92. Integrated Services Digital Network (ISDN); 3,1 kHz Telephony Teleservice End-to-End Compatibility Requirements for Telephony Terminals. 35 pp.

ETSI ETS 300 111-92. Integrated Services Digital Network (ISDN); Telephony 3,1 kHz Teleservice Service Description. 13 pp.

INTERNATIONAL AND NON-U.S. NATIONAL STANDARDS
SUBJECT INDEX

Integrated

Integrated Services Digital Networks (Cont.)

—Telephony (Cont.)

ETSI PRETS 300 111-92. Integrated Services Digital Network (ISDN); Telephony 3,1 kHz Teleservice Service Description. 14 pp.

ETSI PRETS 300 111-90. Integrated Services Digital Network (ISDN) Telephony 3.1 KHz Teleservice Service Description. 6 pp.

ETSI PRETS 300 263-92. Integrated Services Digital Network (ISDN); Telephony 7 kHz Teleservice Service Description. 15 pp.

ETSI PRETS 300 265-92. Integrated Services Digital Network (ISDN); Telephony 7 kHz Teleservice Functional Capabilities and Information Flows. 22 pp.

—Telephony—DSS1

CENELEC PRETS 300 267-92. Integrated Services Digital Network (ISDN); Telephony 1 kHz and Videotelephony Teleservices Digital Subscriber Signalling System No. One (DSS1). 36 pp.

ETSI PRETS 300 267-92. Integrated Services Digital Network (ISDN); Telephony 1 kHz and Videotelephony Teleservices Digital Subscriber Signalling System No. One (DSS1). 36 pp.

—Teleservices

CCITT RECMN D.210-89. General Charging and Accounting Principles for International Telecommunication Services Provided over the Integrated Services Digital Network (ISDN)—General Tariff Principles—Charging and Accounting in Intl Telecom. Serv. (Study Group III) 4 pp. 4 pp.

CCITT RECMN E.711-92. User Demand Modelling Study Group II) 11 pp. 11 pp.

CCITT RECMN F.353-89. Provision of Telematic and Data Transmission Services on Integrated Services Digital Network (ISDN)—Telematic, Data Transmission and Teleconference Services—Operations and Quality of Service (Study Group I) 5 pp. 5 pp.

CCITT RECMN I.241-89. Teleservices Supported by an ISDN—Integrated Services Digital Network (ISDN)—General Structure and Service Capabilities (Study Group XVIII) 28 pp. 28 pp.

CCITT RECMN I.320-89. ISDN Protocol Reference Model—Integrated Services Digital Network (ISDN)—Overall Network Aspects and Functions, ISDN User-Network Interfaces (Study Group XVIII) 10 pp. 10 pp.

CEPT T/SF 31-03 E-89. Teleservices to Be Provided by Integrated Services Digital Network (ISDN). 20 pp.

CEPT T/SF 62 E-88. Supplementary Service Digital Connectivity for Teleservices. 5 pp.

—Teleservices—Accounting

CCITT RECMN D.240-91. Charging and Accounting Principles for Teleservices Supported by the ISDN (Study Group III) 5 pp. 5 pp.

—Teleservices—Tariffs

CCITT RECMN D.240-91. Charging and Accounting Principles for Teleservices Supported by the ISDN (Study Group III) 5 pp. 5 pp.

—Teletex Communications

CCITT RECMN F.200-92. Teletex Service (Study Group I) 28 pp. 28 pp.

CCITT RECMN F.200-89. Teletex Service—Telematic, Data Transmission and Teleconference Services—Operations and Quality of Service (Study Group I) 20 pp. 20 pp.

CCITT RECMN I.241. 2-88. Teletex—Integrated Services Digital Network (ISDN)—General Structure and Service Capabilities (Study Group XVIII) 9 pp. 9 pp.

CENELEC PRETS 300 081-90. Integrated Services Digital Network (ISDN); Teletex End-to-End Protocol over the ISDN (T/TE 07-11). 10 pp.

CENELEC ETS 300 081-93. Integrated Services Digital Network (ISDN); Teletex End-to-End Protocol over the ISDN. 14 pp.

ETSI PRETS 300 081-90. Integrated Services Digital Network (ISDN); Teletex End-To-End Protocol over the ISDN (T/TE 07-11). 10 pp.

ETSI ETS 300 081-93. Integrated Services Digital Network (ISDN); Teletex End-to-End Protocol over the ISDN. 14 pp.

ETSI PRETS 300 081-92. Integrated Services Digital Network (ISDN); Teletex End-to-End Protocol over the ISDN. 14 pp.

ETSI PRETS 300 081-90. Integrated Services Digital Network (ISDN); Teletex End-to-End Protocol over the ISDN (T/TE 07-11). 10 pp.

Integrated Services Digital Networks (Cont.)

—Teletex Communications—Interworking

CCITT RECMN F.353-89. Provision of Telematic and Data Transmission Services on Integrated Services Digital Network (ISDN)—Telematic, Data Transmission and Teleconference Services—Operations and Quality of Service (Study Group I) 5 pp. 5 pp.

—Telex Communications

CCITT RECMN I.241. 6-88. Telex—Integrated Services Digital Network (ISDN)—General Structure and Service Capabilities (Study Group XVIII) 3 pp. 3 pp.

CCITT RECMN I.560-89. Requirements to be Met in Providing the Telex Service Within the ISDN—Integrated Services Digital Network (ISDN)—Internetwork Interfaces and Maintenance Principles (Study Group XVIII) 1 pp (Same as Recmn U.202). 1 p.

CCITT RECMN U.202-89. Requirements to be Met in Providing the Telex Service Within the ISDN—Telegraph Switching (Study Group IX) 9 pp (Same as Recmn I.560). 9 pp.

—Terminal Portability

CENELEC PRETS 300 053-90. Integrated Services Digital Network (ISDN); Terminal Portability (TP) Supplementary Service Service Description (T/NA1(89) 17). 11 pp.

CENELEC PRETS 300 053-91. Integrated Services Digital Network (ISDN); Terminal Portability (TP) Supplementary Service Service Description. 14 pp.

CENELEC ETS 300 053-91. Integrated Services Digital Network (ISDN); Terminal Portability (TP) Supplementary Service Service Description. 14 pp.

CENELEC PRETS 300 054-90. Integrated Services Digital Network (ISDN); Terminal Portability (TP) Supplementary Service Functional Capabilities and Information Flows (T/S 22-28). 16 pp.

CENELEC PRETS 300 054-91. Integrated Services Digital Network (ISDN); Terminal Portability (TP) Supplementary Service Functional Capabilities and Information Flows. 20 pp.

CENELEC ETS 300 054-91. Integrated Services Digital Network (ISDN); Terminal Portability (TP) Supplementary Service Functional Capabilities and Information Flows. 20 pp.

ETSI PRETS 300 053-90. Integrated Services Digital Network (ISDN); Terminal Portability (TP) Supplementary Service—Service Description (T/NA1 (89)17). 11 pp.

ETSI PRETS 300 054-90. Integrated Services Digital Network (ISDN); Terminal Portability (TP) Supplementary Service Functional Capabilities and Information Flows (T/S 22-28). 16 pp.

ETSI ETS 300 053-91. Integrated Services Digital Network (ISDN); Terminal Portability (TP) Supplementary Service Service Description. 14 pp.

ETSI PRETS 300 053-91. Integrated Services Digital Network (ISDN); Terminal Portability (TP) Supplementary Service Service Description. 14 pp.

ETSI PRETS 300 053-90. Integrated Services Digital Network (ISDN); Terminal Portability (TP) Supplementary Service Service Description (T/NA1(89) 17). 11 pp.

ETSI ETS 300 054-91. Integrated Services Digital Network (ISDN); Terminal Portability (TP) Supplementary Service Functional Capabilities and Information Flows. 20 pp.

ETSI PRETS 300 054-91. Integrated Services Digital Network (ISDN); Terminal Portability (TP) Supplementary Service Functional Capabilities and Information Flows. 20 pp.

ETSI PRETS 300 054-90. Integrated Services Digital Network (ISDN); Terminal Portability (TP) Supplementary Service Functional Capabilities and Information Flows (T/S 22-28). 16 pp.

—Terminal Portability—DSS1

CENELEC PRETS 300 055-90. Integrated Services Digital Network (ISDN); Terminal Portability (TP) Supplementary Service Digital Subscriber Signalling One (DSS1) Protocol (T/S 46-33E). 8 pp.

CENELEC PRETS 300 055-91. Integrated Services Digital Network (ISDN); Terminal Portability (TP) Supplementary Service Digital Subscriber Signalling System No. One (DSS1) Protocol. 11 pp.

CENELEC PRETS 300 055-91. Integrated Services Digital Network (ISDN); Terminal Portability (TP) Supplementary Service Digital Subscriber Signalling System No. One (DSS1) Protocol. 10 pp.

ETSI PRETS 300 055-90. Integrated Services Digital Network (ISDN); Terminal Portability (TP) Supplementary Service Digital Subscriber Signalling One (DSS1) Protocol (T/S 46-33E). 8 pp.

ETSI PRETS 300 055-91. Integrated Services Digital Network (ISDN); Terminal Portability (TP) Supplementary Service Digital Subscriber Signalling System No. One (DSS1) Protocol. 11 pp.

Integrated Services Digital Networks (Cont.)

—Terminal Portability—DSS1 (Cont.)

ETSI PRETS 300 055-91. Integrated Services Digital Network (ISDN); Terminal Portability (TP) Supplementary Service Digital Subscriber Signalling System No. One (DSS1) Protocol. 10 pp.

ETSI PRETS 300 055-90. Integrated Services Digital Network (ISDN); Terminal Portability (TP) Supplementary Service Digital Subscriber Signalling One (DSS1) Protocol (T/S 46-33E). 8 pp.

—Terminal Selection

CCITT RECMN I.333-89. Terminal Selection in ISDN—Integrated Services Digital Network (ISDN)—Overall Network Aspects and Functions, ISDN User-Network Interfaces (Study Group XVIII) 18 pp. 18 pp.

CENELEC ETR 026-92. Network Aspects (NA); Terminal Selection Principles for Priority 1 and 2 Services of MoU—ISDN Applicable in Multi-Terminal Environments at Customer Premises. 21 pp.

ETSI ETR 026-92. Network Aspects (NA); Terminal Selection Principles for Priority 1 and 2 Services of MoU—ISDN Applicable in Multi-Terminal Environments at Customer Premises. 21 pp.

—Terminating Exchange

CCITT RECMN I.333-89. Terminal Selection in ISDN—Integrated Services Digital Network (ISDN)—Overall Network Aspects and Functions, ISDN User-Network Interfaces (Study Group XVIII) 18 pp. 18 pp.

—Three Party Services

CCITT RECMN I.254-89. Multiparty Supplementary Services—Integrated Services Digital Network (ISDN)—General Structure and Service Capabilities (Study Group XVIII) 30 pp. 30 pp.

CCITT RECMN I.254. 2-92. Three-Party Supplementary Service (Study Group I) 13 pp. 13 pp.

CENELEC PRETS 300 186-92. Integrated Services Digital Network (ISDN); Three Party (3PTY) Supplementary Service Service Description. 16 pp.

CENELEC PRETS 300 186-93. Integrated Services Digital Network (ISDN); Three-Party (3PTY) Supplementary Service; Service Description. 15 pp.

CENELEC PRETS 300 187-92. Integrated Services Digital Network (ISDN); Three Party (3PTY) Supplementary Service Functional Capabilities and Information Flows. 26 pp.

CENELEC PRETS 300 187-93. Integrated Services Digital Network (ISDN); Three-Party (3PTY) Supplementary Service Functional Capabilities and Information Flows. 27 pp.

ETSI PRETS 300 186-93. Integrated Services Digital Network (ISDN); Three-Party (3PTY) Supplementary Service; Service Description. 15 pp.

ETSI PRETS 300 186-92. Integrated Services Digital Network (ISDN); Three Party (3PTY) Supplementary Service Service Description. 16 pp.

ETSI PRETS 300 187-93. Integrated Services Digital Network (ISDN); Three-Party (3PTY) Supplementary Service Functional Capabilities and Information Flows. 27 pp.

ETSI PRETS 300 187-92. Integrated Services Digital Network (ISDN); Three Party (3PTY) Supplementary Service Functional Capabilities and Information Flows. 26 pp.

ETSI PRETS 300 188-92. Integrated Services Digital Network (ISDN); Three Party (3PTY) Supplementary Service Digital Subscriber Signalling System No. One (DSS1) Protocol. 24 pp.

—Three Party Services—DSS1

CENELEC PRETS 300 188-92. Integrated Services Digital Network (ISDN); Three Party (3PTY) Supplementary Service Digital Subscriber Signalling System No. One (DSS1) Protocol. 24 pp.

CENELEC PRETS 300 188-93. Integrated Services Digital Network (ISDN); Three-Party (3PTY) Supplementary Service Digital Subscriber Signalling System No. One (DSS1) Protocol. 39 pp.

ETSI PRETS 300 188-93. Integrated Services Digital Network (ISDN); Three-Party (3PTY) Supplementary Service Digital Subscriber Signalling System No. One (DSS1) Protocol. 39 pp.

ETSI PRETS 300 188-92. Integrated Services Digital Network (ISDN); Three Party (3PTY) Supplementary Service Digital Subscriber Signalling System No. One (DSS1) Protocol. 24 pp.

—Three Party Services—Functions/Information Flows

CCITT RECMN Q.84-92. Stage 2 Description for Multiparty Supplementary Services Section 2—Three-Party Service (Study Group XI) 20 pp. 20 pp.

—Time Out—CCITT No. 7 Signaling

CCITT RECMN Q.730-89. ISDN Supplementary Services—Specifications of Signalling System No. 7 (Study Group XI) 59 pp. 59 pp.

INTERNATIONAL AND NON-U.S. NATIONAL STANDARDS
SUBJECT INDEX

Integrated

Integrated Services Digital Networks (Cont.)

—Tones (Telephone Services)

CCITT RECMN E.180-89. Technical Characteristics of Tones for the Telephone Service—Telephone Network and ISDN—Operation, Numbering, Routing and Mobile Service (Study Group II) 9 pp. 9 pp.

CCITT RECMN E.184-89. Indications to Users of ISDN Terminals—Telephone Network and ISDN—Operation, Numbering, Routing and Mobile Service (Study Group II) 2 pp. 2 pp.

—Traffic

CCITT RECMN E.700-92. Framework of the E.700-Series Recommendations—Telephone Network and ISDN Service, Network Management and Traffic Engineering (Study Group II) 3 pp. 3 pp.

CCITT RECMN E.710-92. ISDN Traffic Modelling Overview (STudy Group II) 4 pp. 4 pp.

CCITT RECMN E.711-92. User Demand Modelling Study Group II) 11 pp. 11 pp.

CCITT RECMN E.712-92. User Plane Traffic Modeling (Study Gruop II) 16 pp. 16 pp.

CCITT RECMN E.713-89. Control Plane Traffic Models—Telephone Network and ISDN—Quality of Service, Network Management and Traffic Engineering (Study Group II) 7 pp. 7 pp.

—Traffic—Call Attempts

CCITT RECMN E.711-92. User Demand Modelling Study Group II) 11 pp. 11 pp.

CCITT RECMN E.712-92. User Plane Traffic Modeling (Study Gruop II) 16 pp. 16 pp.

CCITT RECMN E.713-89. Control Plane Traffic Models—Telephone Network and ISDN—Quality of Service, Network Management and Traffic Engineering (Study Group II) 7 pp. 7 pp.

—Traffic—Dimensioning

CCITT RECMN E.730-92. ISDN Dimensioning Methods Overview (Study Group II) 4 pp. 4 pp.

—Traffic Engineering

CCITT FASCICLE II.3-88. Telephone Network and ISDN Quality of Service, Network Management and Traffic Engineering. Recommendations E.401-E.880. 368 pp.

CCITT RECMN E.700-92. Framework of the E.700-Series Recommendations—Telephone Network and ISDN Service, Network Management and Traffic Engineering (Study Group II) 3 pp. 3 pp.

—Traffic Engineering—Reference Connections

CCITT RECMN E.701-92. Reference Connections for Traffic Engineering—Telephone Network and ISDN—Quality of Service, Network Management and Traffic Engineering (Study Group II) 4 pp. 4 pp.

—Traffic—Holding Time

CCITT RECMN E.711-92. User Demand Modelling Study Group II) 11 pp. 11 pp.

—Traffic—Measurement

CCITT RECMN E.502 (REV 1)-92. Traffic Measurement Requirements for Digital Telecommunication Exchanges (Study Group II) 23 pp. 23 pp.

CCITT RECMN E.502-89. Traffic Measurement Requirements for SPC (Especially Digital) Telecommunication Exchanges—Telephone Network and ISDN—Quality of Service, Network Management and Traffic Engineering (Study Group II) 15 pp. 15 pp.

CCITT RECMN E.700-92. Framework of the E.700-Series Recommendations—Telephone Network and ISDN Service, Network Management and Traffic Engineering (Study Group II) 3 pp. 3 pp.

—Transmission Medium Requirements

CCITT RECMN E.172-89. Call Routing in the ISDN Era—Telephone Network and ISDN—Operation, Numbering, Routing and Mobile Service (Study Group II) 14 pp. 14 pp.

—Universal Synchronous Asynchronous Receiver/Transmitters

CENELEC PRETS 300 325-93. Integrated Services Digital Network (ISDN); Programming Communication Interface (PCI) for Euro-ISDN. 253 pp.

ETSI PRETS 300 325-93. Integrated Services Digital Network (ISDN); Programming Communication Interface (PCI) for Euro-ISDN. 253 pp.

—User Network Interfaces

CCITT RECMN I.363-91. B-ISDN ATM Adaptation Layer (AAL) Specification (Study Group XVIII) 21 pp. 21 pp.

Integrated Services Digital Networks (Cont.)

—User to User Information Services—Accounting

CCITT RECMN D.231-89. Charging and Accounting Principles Relating to the User-To-User Information (UUI) Supplementary Service—General Tariff Principles—Charging and Accounting in International Telecommunications Services (Study Group III) 2 pp. 2 pp.

—User to User Information Services—Tariffs

CCITT RECMN D.231-89. Charging and Accounting Principles Relating to the User-To-User Information (UUI) Supplementary Service—General Tariff Principles—Charging and Accounting in International Telecommunications Services (Study Group III) 2 pp. 2 pp.

—User to User Signaling

CCITT RECMN I.257-89. Additional Information Transfer—Integrated Services Digital Network (ISDN)—General Structure and Service Capabilities (Study Group XVIII) 9 pp. 9 pp.

CCITT RECMN I.257. 1-92. User-to-User Signalling (Study Group I) 18 pp. 18 pp.

CENELEC PRETS 300 284-93. Integrated Services Digital Network (ISDN); User-to-User Signalling (UUS) Supplementary Service; Service Description. 17 pp.

CENELEC PRETS 300 285-93. Integrated Services Digital Network (ISDN); User-to-User Signalling (UUS) Supplementary Service Functional Capabilities and Information Flows. 42 pp.

ETSI PRETS 300 284-93. Integrated Services Digital Network (ISDN); User-to-User Signalling (UUS) Supplementary Service; Service Description. 17 pp.

ETSI PRETS 300 285-93. Integrated Services Digital Network (ISDN); User-to-User Signalling (UUS) Supplementary Service Functional Capabilities and Information Flows. 42 pp.

—User to User Signaling—CCITT No. 7 Signaling

CCITT RECMN Q.730-89. ISDN Supplementary Services—Specifications of Signalling System No. 7 (Study Group XI) 59 pp. 59 pp.

—User to User Signaling—DSS1

CENELEC PRETS 300 286-93. Integrated Services Digital Network (ISDN); User-to-User Signalling (UUS) Supplementary Service Digital Subscriber Signalling System No. One (DSS1) Protocol. 79 pp.

ETSI PRETS 300 286-93. Integrated Services Digital Network (ISDN); User-to-User Signalling (UUS) Supplementary Service Digital Subscriber Signalling System No. One (DSS1) Protocol. 79 pp.

—User to User Signaling—Functions/ Information Flows

CCITT RECMN Q.87-89. Additional Information Transfer Supplementary Services—General Recommendations on Telephone Switching and Signalling—Functions and Information Flows for Services in the ISDN—Supplements (Study Group XI) 26 pp. 26 pp.

—Video Telephone Services

CCITT RECMN F.721-92. Videotelephony Teleservice for ISDN (Study Group I) 12 pp. 12 pp.

CCITT RECMN F.721-89. Basic Narrow Band Videophone Service in the ISDN—Telematic, Data Transmission and Teleconference Services—Operations and Quality of Service (Study Group I) 5 pp. 5 pp.

CENELEC PRETS 300 264-92. Integrated Services Digital Network (ISDN); Videotelephony Teleservice Service Description. 19 pp.

CENELEC PRETS 300 266-92. Integrated Services Digital Network (ISDN); Videotelephony Teleservice Functional Capabilities and Information Flows. 32 pp.

CEPT T/SF 65 E-89. Basic Narrow Band Videophone Service in the ISDN. 4 pp.

ETSI PRETS 300 264-92. Integrated Services Digital Network (ISDN); Videotelephony Teleservice Service Description. 19 pp.

ETSI PRETS 300 266-92. Integrated Services Digital Network (ISDN); Videotelephony Teleservice Functional Capabilities and Information Flows. 32 pp.

—Video Telephone Services—Call Set Up

CCITT RECMN F.721-92. Videotelephony Teleservice for ISDN (Study Group I) 12 pp. 12 pp.

CCITT RECMN F.721-89. Basic Narrow Band Videophone Service in the ISDN—Telematic, Data Transmission and Teleconference Services—Operations and Quality of Service (Study Group I) 5 pp. 5 pp.

Integrated Services Digital Networks (Cont.)

—Video Telephone Services—DSS1

CENELEC PRETS 300 267-92. Integrated Services Digital Network (ISDN); Telephony 1 kHz and Videotelephony Teleservices Digital Subscriber Signalling System No. One (DSS1). 36 pp.

ETSI PRETS 300 267-92. Integrated Services Digital Network (ISDN); Telephony 1 kHz and Videotelephony Teleservices Digital Subscriber Signalling System No. One (DSS1). 36 pp.

—Video Telephone Services—Electroacoustics

CENELEC PRI-ETS 300 302-1-93. Integrated Services Digital Network (ISDN); Videotelephony Teleservice Part 1: Electroacoustic Characteristics for Handset Terminals When Using PCM Encoding. 12 pp.

ETSI PRI-ETS 300 302-1-93. Integrated Services Digital Network (ISDN); Videotelephony Teleservice Part 1: Electroacoustic Characteristics for Handset Terminals When Using PCM Encoding. 12 pp.

—Videotex Communications

CCITT RECMN I.241. 5-88. Videotex—Integrated Services Digital Network (ISDN)—General Structure and Service Capabilities (Study Group XVIII) 4 pp. 4 pp.

CENELEC PRETS 300 262-92. Integrated Services Digital Network (ISDN); Syntax-Based Videotex Teleservice Service Description. 12 pp.

CENELEC PRETS 300 262-93. Integrated Services Digital Network (ISDN); Syntax-Based Videotex Teleservice Service Description. 12 pp.

ETSI PRETS 300 262-93. Integrated Services Digital Network (ISDN); Syntax-Based Videotex Teleservice Service Description. 12 pp.

ETSI PRETS 300 262-92. Integrated Services Digital Network (ISDN); Syntax-Based Videotex Teleservice Service Description. 12 pp.

—Virtual Call Services

CENELEC PRETS 300 048-90. Integrated Services Digital Network (ISDN); ISDN Packet Mode Bearer Services (PMBS) ISDN Virtual Call (VC) and Permanent Virtual Circuit (PVC) Bearer Services Provided by the B Channel of the User Access-Basic and Primary Rate (T/NAI(89)29). 13 pp.

CENELEC PRETS 300 048-91. Integrated Services Digital Network (ISDN); ISDN Packet Mode Bearer Services (PMBS) ISDN Virtual Call (VC) and Permanent Virtual Circuit (PVC) Bearer Services Provided by the B-Channel of the User Access—Basic and Primary Rate. 18 pp.

CENELEC ETS 300 048-92. Integrated Services Digital Network (ISDN); ISDN Packet Mode Bearer Sevices (PMBS) ISDN Virtual Call (VC) and Permanent Virtual Call (PVC) Bearer Services Provided by the B-Channel of the User Access—Basic and Primary Rate. 17 pp.

CENELEC PRETS 300 049-90. Integrated Services Digital Network (ISDN); ISDN Packet Mode Bearer Services (PMBS) ISDN Virtual Call (VC) and Permanent Virtual Circuit (PVC) Bearer Services Provided by the D Channel of the User Access—Basic and Primary Rate (T/NAI(89)30). 15 pp.

CENELEC PRETS 300 049-91. Integrated Services Digital Network (ISDN); ISDN Packet Mode Bearer Service (PMBS) ISDN Virtual Call (VC) and Permanent Virtual Call (PVC) Bearer Services Provided by the D-Channel of the User Access—Basic and Primary Rate. 20 pp.

CENELEC ETS 300 049-92. Integrated Services Digital Network (ISDN); ISDN Packet Mode Bearer Sevice (PMBS) ISDN Virtual Call (VC) and Permanent Virtual Call (PVC) Bearer Services Provided by the D-Channel of the User Access—Basic and Primary Rate. 18 pp.

ETSI PRETS 300 048-90. Integrated Services Digital Network (ISDN); ISDN Packet Mode Bearer Service (PMBS) ISDN Virtual Call (VC) and Permanent Virtual Circuit (PVC) Bearer Services Provided by the B Channel of the User Access—Basic and Primary Rate (T/N A1 (89)29). 11 pp.

ETSI PRETS 300 049-90. Integrated Services Digital Network (ISDN); ISDN Packet Mode Bearer Services (PMBS) ISDN Virtual Call (VC) and Permanent Virtual Circuit (PVC) Bearer Services Provided by the D Channel of the User Access—Basic and Primary Rate (T/NA1 (89) 30). 15 pp.

ETSI ETS 300 048-92. Integrated Services Digital Network (ISDN); ISDN Packet Mode Bearer Sevices (PMBS) ISDN Virtual Call (VC) and Permanent Virtual Call (PVC) Bearer Services Provided by the B-Channel of the User Access—Basic and Primary Rate. 17 pp.

ETSI PRETS 300 048-91. Integrated Services Digital Network (ISDN); ISDN Packet Mode Bearer Services (PMBS) ISDN Virtual Call (VC) and Permanent Virtual Call (PVC) Bearer Services Provided by the B-Channel of the User Access—Basic and Primary Rate. 18 pp.

INDUSTRY STANDARDS

Integrated Services Digital Networks (Cont.)

—Virtual Call Services (Cont.)

ETSI PRETS 300 048-90. Integrated Services Digital Network (ISDN): ISDN Packet Mode Bearer Services (PMBS) ISDN Virtual Call (VC) and Permanent Virtual Circuit (PVC) Bearer Services Provided by the B Channel of the User Access-Basic and Primary Rate (T/NAI(89)29). 13 pp.

ETSI ETS 300 049-92. Integrated Services Digital Network (ISDN); ISDN Packet Mode Bearer Service (PMBS) ISDN Virtual Call (VC) and Permanent Virtual Call (PVC) Bearer Services Provided by the D-Channel of the User Access—Basic and Primary Rate. 18 pp.

ETSI PRETS 300 049-91. Integrated Services Digital Network (ISDN); ISDN Packet Mode Bearer Service (PMBS) ISDN Virtual Call (VC) and Permanent Virtual Call (PVC) Bearer Services Provided by the D-Channel of the User Access—Basic and Primary Rate. 20 pp.

ETSI PRETS 300 049-90. Integrated Services Digital Network (ISDN): ISDN Packet Mode Bearer Services (PMBS) ISDN Virtual Call (VC) and Permanent Virtual Circuit (PVC) Bearer Services Provided by the D Channel of the User Access-Basic and Primary Rate (T/NAI(89)30). 15 pp.

—Virtual Private Networks—Numbering Plans

CENELEC ETR 033-92. Network Aspects (NA); Numbering Plans for the Interconnection of Virtual Private Networks and Public Networks. 27 pp.

ETSI ETR 033-92. Network Aspects (NA); Numbering Plans for the Interconnection of Virtual Private Networks and Public Networks. 27 pp.

Integrated Services Digital Networks User Part

Use For: Integrated Services User Part; ISDN-UP; ISDN User Part; ISUP *See Also:* CCITT No. 7 Signaling Systems; Common Channel Signaling Systems; Integrated Services Digital Networks; Transmission Medium Requirements

CCITT RECMN E.172-89. Call Routing in the ISDN Era—Telephone Network and ISDN—Operation, Numbering, Routing and Mobile Service (Study Group II) 14 pp. 14 pp.

CCITT RECMN Q.761-89. Functional Description of the ISDN User Part of Signalling System No. 7—Specifications of Signalling System No. 7 (Study Group XI) 4 pp. 4 pp.

CCITT RECMN Q.764-89. Signalling Procedures—Specifications of Signalling System No. 7 (Study Group XI) 178 pp. 178 pp.

CCITT RECMN Q.766-89. Performance Objectives in the Integrated Services Digital Network Application—Specifications of Signalling System No. 7 (Study Group XI) 4 pp. 4 pp.

—CCITT No. 7 Signaling Systems

CENELEC PRETS 300 121-90. Integrated Services Digital Network, (ISDN); Application of the ISDN User part of CCITT Signalling System No.7 for International ISDN Interconnections CCITT Recommendation Q.767 Draft edition 3: 1990-Modified. 239 pp.

CENELEC PRETS 300 121-92. Integrated Services Digital Network (ISDN); Application of the ISDN User Part (ISUP) of CCITT Signalling System No. 7 for International ISDN Interconnections (ISUP Version 1). 7 pp.

CENELEC ETS 300 121-92. Integrated Services Digital Network (ISDN); Application of the ISDN User Part (ISUP) of CCITT Signalling System No. 7 for International ISDN Interconnections (ISUP Version 1). 7 pp.

ETSI ETS 300 121-92. Integrated Services Digital Network (ISDN); Application of the ISDN User Part (ISUP) of CCITT Signalling System No. 7 for International ISDN Interconnections (ISUP Version 1). 7 pp.

ETSI PRETS 300 121-92. Integrated Services Digital Network (ISDN); Application of the ISDN User Part (ISUP) of CCITT Signalling System No. 7 for International ISDN Interconnections (ISUP Version 1). 7 pp.

—CCITT No. 7 Signaling Systems—Call Testing

CCITT RECMN Q.784-91. ISUP Basic Call Test Specification (Study Group XI) 83 pp. 83 pp.

CENELEC PRETS 300 335-93. Integrated Services Digital Network (ISDN); CCITT Signalling System No. 7 Integrated Services User Part (ISUP) Version 1 Test Specification. 120 pp.

ETSI PRETS 300 335-93. Integrated Services Digital Network (ISDN); CCITT Signalling System No. 7 Integrated Services User Part (ISUP) Version 1 Test Specification. 120 pp.

—CCITT No. 7 Signaling Systems—Routing (Telecommunications)

CENELEC PRETS 300 334-93. Integrated Services Digital Network (ISDN); Routeing in Support of ISDN User Part (ISUP) Version 2 Services. 18 pp.

ETSI PRETS 300 334-93. Integrated Services Digital Network (ISDN); Routeing in Support of ISDN User Part (ISUP) Version 2 Services. 18 pp.

—CCITT No. 7 Signaling Systems—Supplementary Services—Testing

CCITT RECMN Q.785-91. ISUP Protocol Test Specification for Supplementary Services (Study Group XI) 52 pp. 52 pp.

—CCITT No. 7 Signaling Systems—Telecommunication Connections

CCITT RECMN Q.767-91. Application of the ISDN User Part of CCITT Signalling System No. 7 for International ISDN Interconnections (Study Group XI) 274 pp. 274 pp.

—Codes

CCITT RECMN Q.763-89. Formats and Codes—Specifications of Signalling System No. 7 (Study Group XI) 50 pp. 50 pp.

—Message Formats

CCITT RECMN Q.763-89. Formats and Codes—Specifications of Signalling System No. 7 (Study Group XI) 50 pp. 50 pp.

—Preference Indicators

CCITT RECMN E.172-89. Call Routing in the ISDN Era—Telephone Network and ISDN—Operation, Numbering, Routing and Mobile Service (Study Group II) 14 pp. 14 pp.

—Preference Indicators—Bearer Services

CCITT RECMN E.172-89. Call Routing in the ISDN Era—Telephone Network and ISDN—Operation, Numbering, Routing and Mobile Service (Study Group II) 14 pp. 14 pp.

—Preference Indicators—Teleservices

CCITT RECMN E.172-89. Call Routing in the ISDN Era—Telephone Network and ISDN—Operation, Numbering, Routing and Mobile Service (Study Group II) 14 pp. 14 pp.

—Routing (Telecommunications)

CCITT RECMN E.172-89. Call Routing in the ISDN Era—Telephone Network and ISDN—Operation, Numbering, Routing and Mobile Service (Study Group II) 14 pp. 14 pp.

—Signaling Messages

CCITT RECMN Q.762-89. General Function of Messages and Signals—Specifications of Signalling System No. 7 (Study Group XI) 17 pp. 17 pp.

Integrated Services Private Branch Exchanges

Use For: ISPBX *See Also:* Private Branch Exchanges; Telephone Exchanges

BSI PD 0001-90. 1990 Outline ISPBX Specification. 58 pp.

CCITT RECMN I.320-89. ISDN Protocol Reference Model—Integrated Services Digital Network (ISDN)—Overall Network Aspects and Functions, ISDN User-Network Interfaces (Study Group XVIII) 10 pp. 10 pp.

Integrated Services User Part

Use: Integrated Services Digital Networks User Part

Intelligent Networks

Use For: IN *See Also:* Integrated Services Digital Networks

CENELEC ETR 023-91. Network Aspects Intelligent Networks: Framework. 122 pp.

ETSI ETR 023-91. Network Aspects Intelligent Networks: Framework. 122 pp.

—Global Functional Plane Architecture

CCITT RECMN I.329-92. Intelligent Network—Global Functional Plane Architecture (Study Group XVIII) 15 pp (Same as Recmn Q.1203). 15 pp.

CCITT RECMN Q.1203-92. Intelligent Network—Global Functional Plane Architecture (Study Group XVIII) 15 pp (Same as Recmn I.329). 15 pp.

—Network Architecture

CCITT RECMN I.312-92. Principles of Intelligent Network Architecture (Study Group XVIII) 36 pp (Same as Recmn Q.1201). 36 pp.

CCITT RECMN Q.1201-92. Principles of Intelligent Network Architecture (Study Group XVIII) 36 pp (Same as Recmn I.312). 36 pp.

—Service Plane Architecture

CCITT RECMN I.328-92. Intelligent Network—Service Plane Architecture (Study Group XVIII) 7 pp (Same as Recmn Q.1202). 7 pp.

CCITT RECMN Q.1202-92. Intelligent Network—Service Plane Architecture (Study Group XVIII) 7 pp (Same as Recmn I.328). 7 pp.

—Switching Systems

CENELEC ETR 024-91. Signalling Protocols & Switching (SPS); Intelligent Networks Switching Aspects. 92 pp.

ETSI ETR 024-91. Signalling Protocols & Switching (SPS); Intelligent Networks Switching Aspects. 92 pp.

—Telecommunications Management Networks

CENELEC ETR 062-93. Network Aspects (NA); Baseline Document on the Integration of Intelligent Network (IN) and Telecommunications Management Network (TMN). 56 pp.

ETSI ETR 062-93. Network Aspects (NA); Baseline Document on the Integration of Intelligent Network (IN) and Telecommunications Management Network (TMN). 56 pp.

Intelligent Peripheral Interfaces

Use: Peripheral Interfaces—Intelligent

Intensifying Screens

Use: Radiographic Intensifying Screens

Interband Telegraphy

See Also: Carrier Systems; Telecommunication Services; Telegraphy

—Open Wire Lines—Carrier Systems

CCITT RECMN G.361-89. Systems Providing Three Carrier Telephone Circuits on a Pair of Open-Wire Lines—International Analogue Carrier Systems (Study Group XV) 5 pp. 5 pp.

CCITT RECMN R.49-89. Interband Telegraphy over Open-Wire 3-Channel Carrier Systems—Telegraph Transmission (Study Group IX) 2 pp. 2 pp.

Interception (Aircraft)

Use: Aircraft Interception

Interchange Circuits

See Also: Circuits; Data Transmission

BSI BS ISO/IEC 9549-90. 1990 Information Technology—Galvanic Isolation of Balanced Interchange Circuits. 14 pp.

IEC 9549-90. Information Technology—Galvanic Isolation of Balanced Interchange Circuits First Edition. 11 pp.

ISO 9549-90. Information Technology—Galvanic Isolation of Balanced Interchange Circuits First Edition. 11 pp.

JTC1 9549-90. Information Technology—Galvanic Isolation of Balanced Interchange Circuits First Edition. 11 pp.

OSI ISO/IEC 9549-90. Information Technology—Galvanic Isolation of Balanced Interchange Circuits. 11 pp.

OSI ISO IEC DIS 9549-88. Information Processing Systems—Galvanic Isolation of Balanced Interchange Circuits. 11 pp.

—Data Circuit Terminating Equipment—Glossaries

CCITT RECMN X.24-89. List of Definitions for Interchange Circuits Between Data Terminal Equipment (DTE) and Data Circuit-Terminating Equipment (DCE) on Public Data Networks—Data Communication Networks: Services and Facilities, Interfaces (Study Group VII) 6 pp. 6 pp.

—Data Communication—Electrical Measurement

CCITT RECMN X.26-89. Electrical Characteristics for Unbalanced Double-Current Interchange Circuits for General Use with Integrated Circuit Equipment in the Field of Data Communications —Data Communication Networks: Services and Facilities, Interfaces (Study Group VII) 1 pp. 1 p.

CCITT RECMN X.27-89. Electrical Characteristics for Balanced Double-Current Interchange Circuits for General Use with Integrated Cir. Equipment in the Field of Data Communications—Data Comm. Networks:Services and Facilities, Inter. (Study Group VII) 1 pp. 1 p.

—Data Signaling Rates—Data Terminal Equipment

CCITT RECMN V.31-89. Electrical Characteristics for Single-Current Interchange Circuits Controlled by Contact Closure—Data Communication over the Telephone Network (Study Group XVII) 3 pp. 3 pp.

INTERNATIONAL AND NON-U.S. NATIONAL STANDARDS
SUBJECT INDEX

Interchange Circuits *(Cont.)*
—Data Signaling Rates—Data Terminal Equipment *(Cont.)*
CCITT RECMN V.31 BIS-89. Electrical Characteristics for Single-Current Interchange Circuits Using Optocouplers—Data Communication over the Telephone Network (Study Group XVII) 3 pp. 3 pp.

—Data Signaling Rates—Data Terminal Equipment—Optoisolators
CCITT RECMN V.31-89. Electrical Characteristics for Single-Current Interchange Circuits Controlled by Contact Closure—Data Communication over the Telephone Network (Study Group XVII) 3 pp. 3 pp.

CCITT RECMN V.31 BIS-89. Electrical Characteristics for Single-Current Interchange Circuits Using Optocouplers—Data Communication over the Telephone Network (Study Group XVII) 3 pp. 3 pp.

—Data Terminal Equipment—Glossaries
CCITT RECMN V.24-89. List of Definitions for Interchange Circuits Between Data Terminal Equipment (DTE) and Data Circuit-Terminating Equipment (DCE)—Data Communication over the Telephone Network (Study Group XVII) 18 pp. 18 pp.

CCITT RECMN X.24-89. List of Definitions for Interchange Circuits Between Data Terminal Equipment (DTE) and Data Circuit-Terminating Equipment (DCE) on Public Data Networks—Data Communication Networks: Services and Facilities, Interfaces (Study Group VII) 6 pp. 6 pp.

—Interfaces—Data Terminal Equipment—Electrical Measurement
CCITT RECMN V.10-89. Electrical Characteristics for Unbalanced Double-Current Interchange Circuits for General Use with Integrated Circuit Equipment in the Field of Data Communications—Data Communication Over the Telephone Network (Study Group XVII) 17 pp. 17 pp.

CCITT RECMN V.11-89. Electrical Characteristics for Balanced Double-Current Interchange Circuits for General Use with Integrated Circuit Equipment in the Field of Data Communications—Data Communication Over the Telephone Network (Study Group XVII) 12 pp. 12 pp.

—Interfaces—Signal Generators—Electrical Measurement
CCITT RECMN V.10-89. Electrical Characteristics for Unbalanced Double-Current Interchange Circuits for General Use with Integrated Circuit Equipment in the Field of Data Communications—Data Communication Over the Telephone Network (Study Group XVII) 17 pp. 17 pp.

CCITT RECMN V.11-89. Electrical Characteristics for Balanced Double-Current Interchange Circuits for General Use with Integrated Circuit Equipment in the Field of Data Communications—Data Communication Over the Telephone Network (Study Group XVII) 12 pp. 12 pp.

—Public Data Networks—Glossaries
CCITT RECMN X.24-89. List of Definitions for Interchange Circuits Between Data Terminal Equipment (DTE) and Data Circuit-Terminating Equipment (DCE) on Public Data Networks—Data Communication Networks: Services and Facilities, Interfaces (Study Group VII) 6 pp. 6 pp.

—Signal Generators—Electrical Measurement
CCITT RECMN V.28-89. Electrical Characteristics for Unbalanced Double-Current Interchange Circuits—Data Communication over the Telephone Network (Study Group XVII) 5 pp. 5 pp.

—Telephone Answering Equipment—Data Terminal Equipment
CCITT RECMN V.25-89. Automatic Answering Equipment and/or Parallel Automatic Calling Equipment on the General Switched Telephone Network Including Procedures for Disabling of Echo Control Devices for Both Manually and Automatically Established Calls—Data. 11 pp.

CCITT RECMN V.25 BIS-89. Automatic Calling and/or Answering Equipment on the General Switched Telephone Network (GSTN) Using the 100-Series Interchange Circuits—Data Communication over the Telephone Network (Study Group XVII) 23 pp. 23 pp.

Interchange Circuits *(Cont.)*
—Telephone Networks—Telephone Answering Equipment
CCITT RECMN V.25-89. Automatic Answering Equipment and/or Parallel Automatic Calling Equipment on the General Switched Telephone Network Including Procedures for Disabling of Echo Control Devices for Both Manually and Automatically Established Calls—Data. 11 pp.

CCITT RECMN V.25 BIS-89. Automatic Calling and/or Answering Equipment on the General Switched Telephone Network (GSTN) Using the 100-Series Interchange Circuits—Data Communication over the Telephone Network (Study Group XVII) 23 pp. 23 pp.

Interchange Signaling
Use: Signaling Systems

Intercom Equipment
Use For: Intercommunication Systems
See Also: Audio Equipment; Interphones; Microphones; Public Address Equipment; Radio Paging Systems; Telecommunication Equipment

—Aircraft
CAA Chapter R4-6 App #7 04.74. Tests for Intercommunication and Audio-Integration Systems (Radio). 2 pp.

—Ships
MOD UK NES 562-91. Guide to Internal Communication Equipment Non–VCS Units (RICE 1, CCU and SINBAD) Issue 3 (09.91). 122 pp.

—Submarines
MOD UK NES 562-91. Guide to Internal Communication Equipment Non–VCS Units (RICE 1, CCU and SINBAD) Issue 3 (09.91). 122 pp.

Intercommunication Equipment
Use: Intercom Equipment

Intercommunication Systems
Use: Intercom Equipment

Interconnection
Scope Note: See the subheading Interfaces under specific types of Communication Networks and equipment *Use:* Transit Traffic

Interconnection Diagrams
Use: Diagrams

Interconnection Systems
Use For: Interface Systems; Multiconnection Systems
See Also: Connectors; Electrical Equipment

—Aircraft Stores
NATO STANAG 3837 ED 2 AMD 2-86. Aircraft Stores Electrical Interconnection System. 7 pp.

—Electric—Aircraft
CEN PREN 3197-92. Installation of Aircraft Electrical and Optical Interconnection Systems. 39 pp.
CEN PREN 3197-93. Aerospace Series Installation of Aircraft Electrical and Optical Interconnection Systems. 31 pp.

—Fiber Optics—Aircraft
CEN PREN 3197-92. Installation of Aircraft Electrical and Optical Interconnection Systems. 39 pp.
CEN PREN 3197-93. Aerospace Series Installation of Aircraft Electrical and Optical Interconnection Systems. 31 pp.

—Printed Circuit Boards
BSI BS 7588-93. 1993 Hierarchical Interconnection Technology. 37 pp.

Intercontinental Automatic Transit Traffic
Use: Transit Traffic

Intercontinental Circuits
Use: Telecommunication Circuits

Interdepartmental Procurement
Use For: Cross Servicing (Military)
See Also: Aircraft Cross Servicing; Procurement

—Aerial Reconnaissance
NATO STANAG 3781 ED 1 AMD 1-85. Reconnaissance Cross-Servicing. 11 pp.

—Aircraft
NATO STANAG 3430 ED 8 AMD 0-91. Responsibilities for Aircraft Cross-Servicing. 24 pp.

Interdepartmental Procurement *(Cont.)*
—Aircraft—Air Hoses
NATO STANAG 3054 ED 4 AMD 4-73. Characteristics of Compressed Air for Technical Purposes, Supply Pressure and Hoses. 6 pp.
NATO STANAG 3054 ED 5 AMD 0-89. Characteristics of Compressed Air for Technical Purposes, Supply Pressure and Hoses. 6 pp.
NATO STANAG 3054 ED 5 AMD 1-89. Characteristics of Compressed Air for Technical Purposes, Supply Pressure and Hoses. 6 pp.

—Aircraft—Compressed Air
NATO STANAG 3054 ED 4 AMD 4-73. Characteristics of Compressed Air for Technical Purposes, Supply Pressure and Hoses. 6 pp.
NATO STANAG 3054 ED 5 AMD 0-89. Characteristics of Compressed Air for Technical Purposes, Supply Pressure and Hoses. 6 pp.
NATO STANAG 3054 ED 5 AMD 1-89. Characteristics of Compressed Air for Technical Purposes, Supply Pressure and Hoses. 6 pp.

—Aircraft—Grease Nipples
NATO STANAG 3766 ED 2 AMD 3-88. Grease Nipples. 8 pp.
NATO STANAG 3766 ED 2 AMD 4-88. Grease Nipples. 9 pp.

—Aircraft—Liquid Nitrogen
NATO STANAG 3546 ED 3 AMD 5-73. Liquid Nitrogen Characteristics. 6 pp.

—Aircraft—Liquid Oxygen
NATO STANAG 3545 ED 2 AMD 6-75. Characteristics of Breathable Liquid Oxygen. 6 pp.

—Aircraft—Oxygen—Hoses
NATO STANAG 3053 ED 4 AMD 2-70. Breathing Oxygen Characteristics, Supply Pressure and Hoses. 11 pp.
NATO STANAG 3053 ED 4 AMD 3-70. Breathing Oxygen Characteristics, Supply Pressure and Hoses. 7 pp.
NATO STANAG 3053 ED 5 AMD 0-92. Characteristics of Breathing Oxygen, Supply Pressure and Hoses. 6 pp.

—Blood Transfusion Equipment
NATO STANAG 2939 ED 2 AMD 1-87. Medical Requirements for Blood, Blood Donors and Associated Equipment. 8 pp.

—Cryogenic Containers
NATO STANAG 3056 ED 5 AMD 3-83. Marking of Airborne and Ground Gas and Cryogenic Fluid Containers. 7 pp.
NATO STANAG 3056 ED 5 AMD 4-83. Marking of Airborne and Ground Gas and Cryogenic Fluid Containers. 9 pp.

—Dental Equipment
NATO STANAG 2128 ED 3 AMD 3-82. Medical and Dental Supply Procedures. 6 pp.
NATO STANAG 2128 ED 4 AMD 0-91. Medical and Dental Supply Procedures. 5 pp.

—Emergency Medical Services
NATO STANAG 2127 ED 3 AMD 1-89. Medical, Surgical and Dental Instruments, Equipment and Supplies. 10 pp.

—Gas Cylinders
NATO STANAG 2121 ED 2 AMD 5-79. Cross Servicing of Medical Gas Cylinders. 13 pp.
NATO STANAG 3056 ED 5 AMD 3-83. Marking of Airborne and Ground Gas and Cryogenic Fluid Containers. 7 pp.
NATO STANAG 3056 ED 5 AMD 4-83. Marking of Airborne and Ground Gas and Cryogenic Fluid Containers. 9 pp.

—Helicopters
NATO STANAG 3430 ED 8 AMD 0-91. Responsibilities for Aircraft Cross-Servicing. 24 pp.

—Medical Equipment
NATO STANAG 2127 ED 3 AMD 1-89. Medical, Surgical and Dental Instruments, Equipment and Supplies. 10 pp.
NATO STANAG 2128 ED 3 AMD 3-82. Medical and Dental Supply Procedures. 6 pp.
NATO STANAG 2128 ED 4 AMD 0-91. Medical and Dental Supply Procedures. 5 pp.

—Radiographic Film
NATO STANAG 2357 ED 2 AMD 0-89. X-Ray Film Formats, Cassettes, Screens and Hangers. 6 pp.

Interexchange Signaling
Use: Signaling Systems

INTERNATIONAL AND NON-U.S. NATIONAL STANDARDS
SUBJECT INDEX

Interface

Interface Boards
See Also: Communication Interfaces; Printed Circuit Board Assemblies

—**Computer Numerical Control—Lathes**
CNS B3462-87. Interface Board for CNC Lathe (Aug)(12044).

Interface Circuits
See Also: Analog Switches; Analog to Digital Converters; Buffer/Line Drivers; Bus Controllers; Bus Drivers; Bus Transceivers; Display Drivers; Integrated Circuits; Level Translators; Line Drivers; Line Receivers; Memory Drivers; NAND Buffer Gates; Non-Inverting Buffer Gates; NOR Buffer Gates; Priority Interrupt Controllers; Pulse Synchronizer/Drivers; Schmitt Triggers; Subscriber Line Interface Circuits; Successive Approximation Registers; Universal Synchronous Asynchronous Receiver/Transmitters

MOD UK DSTAN 59-62: Part 6-83. Microcircuits Electronic (Integrated Circuits) (Listed on EPIC Database) Part 6: Interface Circuits Issue 1. 14 pp.
MOD UK DSTAN 59-62: 90/169-74. Integrated Circuit Issue 1. 17 pp.
MOD UK DSTAN 59-62: 90/170-74. Integrated Circuit Issue 1. 20 pp.

—**Quality Assurance**
BSI BS 9493-81. 1981 Rules for the Preparation of Detail Specifications for Integrated Circuits of Assessed Quality Which Perform Mixed Digital and/or Analogue Functions. 3 pp.
BSI BS CECC 90300-88. 1988 Interface Monolithic Integrated Circuits: Sectional Specification. 34 pp.

Interface Systems
Use: Interconnection Systems; Interfaces

Interfaces
Scope Note: For computer communication or computer systems interfaces, see the listings under Communication Interfaces. For additional listings see the subheading Interfaces under specific types of equipment *See Also:* CAMAC; Data Buses; Peripheral Interfaces; Protocols

Interfacial Tension
Use: Surface Tension

Interference
Scope Note: Use a more specific term *See:* Adjacent Channel Interference; Co-Channel Interference; Electromagnetic Interference; Intermodulation Interference; Physical Interference; Radio Frequency Interference; Signal to Interference Ratio; Single Tone Interference

Interference Cancellation Systems
Use For: Interference Cancellors
See Also: Crosstalk; Electromagnetic Interference; Electromagnetic Noise

CCIR Report 830-1-86. Use of Interference Cancellers in Order to Increase Spectrum Usage—Section 1D—Spectrum Utilization and Applications. 17 pp.
CCIR RECMN 856-92. Use of Interference Cancellers, Screens and Adaptive Antennas. 1 p.

—**Adaptive Antenna Arrays**
CCIR Report 1128-90. Adaptive Antenna Arrays for Interference Cancellation—Section 3Ab—Antennas Characteristics. 5 pp.

—**Earth Stations—Fixed Satellite Services**
CCIR RECMN 734-92. Application of Interference Cancellers in the Fixed-Satellite Service—Section 4C—Earth Station and Baseband Characteristics—Earth Station Antennas—Maintenance of Earth Stations. 10 pp.
CCIR QUESTION 58/4-90. Interference Reduction and Cancellation Techniques for the Earth Stations in the Fixed-Satellite Service—Questions of Study Group 4 Fixed-Satellite Service. 1 p.

—**Surveys—Fixed Satellite Services**
CCIR Report 875-1-90. Survey of Interference Cancellers for Application in the Fixed-Satellite Service—Section 4C—Earth Station and Baseband Characteristics—Earth Station Antennas—Maintenance of Earth Stations. 10 pp.

—**Telecommunication Systems**
CCIR Report 830-1-86. Use of Interference Cancellers in Order to Increase Spectrum Usage—Section 1D—Spectrum Utilization and Applications. 17 pp.

Interference Cancellors
Use: Interference Cancellation Systems

Interference Fits
See Also: Fits

Interference Fits *(Cont.)*
—**Oil Pressure Method**
DIN ENGL 15055-82. Steelworks, Rolling Mills and Cranes; Interference Fits Using Oil Under Pressure; Application, Dimensions, Design (July). 8 pp.

Interference Screens
See Also: Filters
CCIR RECMN 856-92. Use of Interference Cancellers, Screens and Adaptive Antennas. 1 p.

—**Antennas—Radio Links**
CCIR Report 831-1-90. Use of Special Screens to Improve Interference Immunity in Radio Links—Section 1D—Spectrum Utilization and Applications. 20 pp.

—**Radio Links**
CCIR Report 831-1-90. Use of Special Screens to Improve Interference Immunity in Radio Links—Section 1D—Spectrum Utilization and Applications. 20 pp.

Interference Suppressor Capacitors
Use: Filter Capacitors

Interferometry
—**Coatings—Thickness Measurement**
ISO 3868-76. Metallic and Other Non-Organic Coatings—Measurement of Coating Thicknesses—Fizeau Multiple-Beam Interferometry Method First Edition. 6 pp.

—**Thermal Expansion—Glass**
JIS R 3251-90. Measuring Method of the Linear Thermal Expansion Coefficient for Low Expansion Glass by Laser Interferometry. 14 pp.

Intergranular Corrosion Testing
See Also: Corrosion Testing

—**Austenitic Stainless Steels**
AECMA PREN2003-11-86. Test Methods for Austenitic Stainless Steels—Part 11—Determination of Resistance to Intergranular Corrosion by the Huey Method. 1 p.
AECMA PREN2003-12-86. Test Methods for Austenitic Stainless Steels—Part 12—Determination of Resistance to Intergranular Corrosion by the Monypenny-Strauss Method. 1 p.
BSI BS 5903-80. 1980 Determination of Resistance to Intergranular Corrosion of Austenitic Stainless Steels: Copper Sulphate-Sulphuric Acid Method (Moneypenny Strauss Test). 4 pp.
ISO 3651 Pt I-76. Austenitic Stainless Steels—Determination of Resistance to Intergranular Corrosion—Part I: Corrosion Test in Nitric Acid Medium by Measurement of Loss in Mass (Huey Test) First Edition. 4 pp.
ISO 3651 Pt II-76. Austenitic Stainless Steels—Determination of Resistance to Intergranular Corrosion—Part II: Corrosion Test in a Sulphuric Acid/Copper Sulphate Medium in the Presence of Copper Turnings (Monypenny Strauss Test) First Edition. 5 pp.
SAA AS 2038-77. Methods for Detecting the Susceptibility of Austenitic Stainless Steel to Intergranular Corrosion Reconfirmed 1986. 8 pp.

—**Nickel Alloys**
ISO 9400-90. Nickle-Based Alloys—Determination of Resistance to Intergranular Corrosion First Edition. 14 pp.

Interior Paints
Scope Note: See the subheading Interior under the specific kind of paint or item being painted

Interlibrary Loans
See Also: Libraries

—**Open Systems Interconnection**
BSI BS ISO 10160-93. 1993 Information and Documentation—Open Systems Interconnection—Interlibrary Loan Application Service Definition (S). 123 pp.
CSA Z243.45-88P. Interlibrary Loan (ILL) Service Definition. 67 pp.
ISO 10160-93. Information and Documentation—Open Systems Interconnection—Interlibrary Loan Application Service Definition First Edition. 121 pp.

—**Open Systems Interconnection—Protocols**
BSI BS ISO 10161-1-93. 1993 Information and Documentation—Open Systems Interconnection—Interlibrary Loan Application Protocol Specification—Part 1: Protocol Specification (S). 183 pp.
CSA Z243.46-88P. Interlibrary Loan (ILL) Protocol Specification. 93 pp.

Interlibrary Loans *(Cont.)*
—**Open Systems Interconnection—Protocols** *(Cont.)*
ISO 10161 Pt 1-93. Information and Documentation—Open Systems Interconnection—Interlibrary Loan Application Protocol Specification—Part 1: Protocol Specification First Edition. 181 pp.

Interlock Circuits
See Also: Circuits

—**Mining Equipment**
BSI BS 3101-86. 1986 Control and Interlock Circuits Primarily Associated with Flameproof Restrained Plugs and Sockets for Use in Coal Mines. 10 pp.

Interlocking Cutters
Use: Milling Cutters

Interlocking Joints
Use For: Finger Joints *See Also:* Joints

—**Softwoods**
BSI BS 5291-84. 1984 Finger Joints in Structural Softwood. 20 pp.

—**Structural Timber**
CEN PREN 385-90. Finger Jointed Structural Timber. 17 pp.
CEN PREN 387-90. Glued Laminated Timber-Production Requirements for Large Finger Joints. 12 pp.

—**Wood**
BSI BS 5291-84. 1984 Finger Joints in Structural Softwood. 20 pp.
DIN ENGL 68140-71. Wood Finger-Jointing (Oct). 3 pp.

Intermediate Frequencies
See Also: Frequencies; Frequency Bands; Radio Frequencies

—**Radio Relay Systems**
CCIR Report 788-2-90. Choice of Intermediate Frequencies for Digital Radio-Relay Systems—Section 9C—Interconnection Characteristics (Baseband and Intermediate Frequency). 3 pp.

—**Radio Relay Systems—Interfaces**
CCIR RECMN 403-3-78. Intermediate-Frequency Characteristics for the Interconnection of Analogue Radio-Relay Systems—Section 9C—Interconnection Characteristics (Baseband and Intermediate Frequency). 2 pp.
CCIR QUESTION 101/9-90. Analogue Radio-Relay Systems Using Amplitude or Frequency Modulation—Questions of Study Group 9 Fixed Service. 1 p.
CCIR QUESTION 137/9-90. Interconnection at Baseband and Intermediate Frequencies for Digital Radio-Relay Systems—Questions of Study Group 9 Fixed Service. 2 pp.

—**Receivers**
CNS C5040-80. Intermediate Frequencies for Entertainment Receivers (May)(5552).

—**Reference Receivers—Sound Broadcasting**
CCIR RECMN 703-90. Characteristics of AM Sound Broadcasting Reference Receivers for Planning Purposes—Section 10A-1—Amplitude-Modulation Sound-Broadcasting in Bands 5 (LF), 6 (MF) and 7 (HF). 6 pp.
CCIR RECMN 704-90. Characteristics of FM Sound Broadcasting Reference Receivers for Planning Purposes—Section 10B—Frequency-Modulation Sound Broadcasting in Bands 8 (VHF) and 9 (UHF). 4 pp.

Intermediate Frequency Transformers
Use For: IF Transformers *See Also:* Transformers

—**Electronic**
CNS C6241-85. Method of Test for High Frequency Inductors and Intermediate Frequency Transformers for Electronic Equipment (Jan)(11183).
JIS C 5320-87. General Rules of High Frequency Inductors and Intermediate Frequency Transformers for Electronic Equipment. 16 pp.

—**Radio Receivers**
CNS C5011-89. Dimensions and Winding for Medium Frequency Transformers and Oscillating Coils of Transistor Radios (Jun)(3770).
CNS C5013-89. General Rules of Intermediate Frequency Transformers for FM Radios (Tube Type) (Jun)(3773).
CNS C6001-90. Testing Standard of Intermediate Frequency Transformers for Broadcast Receivers (Oct)(1262).
CNS C6026-89. Method of Test for MF Transformers of Transistor Radios (Jun)(3771).

INTERNATIONAL AND NON-U.S. NATIONAL STANDARDS
SUBJECT INDEX
Internal

Intermediate Frequency Transformers (Cont.)
—Radio Receivers (Cont.)
CNS C6027-89. Method of Test for MF Transformers of Radios (for Vacuum Tube Frequency Modulation Use) (Jun)(3774).
JIS C 6421-88. Intermediate Frequency Transformers for Broadcast Receivers. 16 pp.

—Radio Receivers—Electrical Impedance
CNS C5012-89. Impedance and Ratios of Standard Q's for Medium Frequency Transformers of Transistor Radios (Jun)(3772).

—Radio Receivers—Quality Factor
CNS C5012-89. Impedance and Ratios of Standard Q's for Medium Frequency Transformers of Transistor Radios (Jun)(3772).

Intermediate Gears
See Also: Gears

—Joints—Automotive
DIN ENGL 75532 Pt 1-76. Transmission of Rotary Motions; Types of Connection to Gears, Intermediate Gears, Flexible Drive Shafts and Equipment (June). 8 pp.

—Trim Controls
SBAC AS 881 ISSUE 1. Intermediate Gear—Trim Control. 1 p.
SBAC AS 1232 ISSUE 4. Intermediate Gear—Trim Control. 1 p.

Intermodal Distortion
Use: Multimode Distortion

Intermodulation
See Also: Cross Modulation; Intermodulation Distortion; Intermodulation Noise

—Compandors—Telephone Circuits
CCITT RECMN G.162-89. Characteristics of Compandors for Telephony—General Characteristics of International Telephone Connections and Circuits (Study Groups XII and XV) 8 pp. 8 pp.
CCITT RECMN G.166-89. Characteristics of Syllabic Compandors for Telephony on High Capacity Long Distance Systems—General Characteristics of International Telephone Connections and Circuits (Study Groups XII and XV) 7 pp. 7 pp.

—Echo Suppressors
CCITT RECMN G.164-89. Echo Suppressors—General Characteristics of International Telephone Connections and Circuits (Study Groups XII and XV) 36 pp. 36 pp.

—Passive
CCIR Report 1049-1-90. Control of Passive Intermodulation Products —Section 8G—Availability, Performance Objectives and Interworking with Terrestrial Networks. 6 pp.

—Passive—Satellite Communications
CCIR Report 1049-1-90. Control of Passive Intermodulation Products —Section 8G—Availability, Performance Objectives and Interworking with Terrestrial Networks. 6 pp.

—Radio Receivers—Models
CCIR Report 522-2-90. Procedure for Modelling Receiver Intermodulation Characteristics—Section 1A—Spectrum Engineering and Computer-Aided Principles and Techniques. 12 pp.

—Radio Transmitters
CCIR Report 839-82. Procedure for Calculating the Amplitude of Transmitter Intermodulation (TIM) Products—Section 1A—Spectrum Engineering and Computer-Aided Principles and Techniques. 3 pp.

—Telephone Circuits
CCITT RECMN G.151-89. General Performance Objectives Applicable to All Modern International Circuits and National Extension Circuits—General Characteristics of International Telephone Connections and Circuits (Study Groups XII and XV) 6 pp. 6 pp.

Intermodulation Distortion
See Also: Distortion (Electrical); Intermodulation; Intermodulation Interference; Intermodulation Noise; Nonlinear Distortion

—Voice Band Data Transmission
CCITT RECMN G.113-89. Transmission Impairments—General Characteristics of International Telephone Connections and Circuits (Study Groups XII and XV) 22 pp. 22 pp.

Intermodulation Interference
See Also: Intermodulation Distortion; Intermodulation Noise; Radio Frequency Interference

—Cellular Mobile Radio Equipment
CCIR Report 740-2-86. General Aspects of Cellular Systems—Section 8A—Land Mobile Service and Related Subjects. 9 pp.

—Land Mobile Services
CCIR Report 739-1-86. Interference Due to Intermodulation Products in the Land Mobile Service Between 25 and 1000 MHz—Section 8A—Land Mobile Service and Related Subjects. 9 pp.
CCIR QUESTION 99/8-90. Interference Due to Intermodulation Products in the Land Mobile Services Between 25 and 3000 MHz—Questions Concerning Study Group 8—Mobile, Radiodetermination, Amateur and Related Satellite Services. 1 p.

—Mobile Radio Services
CCIR Report 655-2-90. Method of Calculating Intermodulation Interference in the Land Mobile Service—Section 1B—Spectrum Sharing and Planning Principles and Techniques. 9 pp.

—Radio Receivers
CCIR Report 739-1-86. Interference Due to Intermodulation Products in the Land Mobile Service Between 25 and 1000 MHz—Section 8A—Land Mobile Service and Related Subjects. 9 pp.
CCIR QUESTION 99/8-90. Interference Due to Intermodulation Products in the Land Mobile Services Between 25 and 3000 MHz—Questions Concerning Study Group 8—Mobile, Radiodetermination, Amateur and Related Satellite Services. 1 p.

Intermodulation Noise
See Also: Intermodulation; Intermodulation Distortion; Intermodulation Interference; Noise (Spurious Signals)

—Group Links—Hypothetical Reference Circuits
CCITT RECMN G.223-89. Assumptions for the Calculation of Noise on Hypothetical Reference Circuits for Telephony—International Analogue Carrier Systems (Study Group XV) 9 pp. 9 pp.

—Hypothetical Reference Circuits—Telephony
CCITT RECMN G.223-89. Assumptions for the Calculation of Noise on Hypothetical Reference Circuits for Telephony—International Analogue Carrier Systems (Study Group XV) 9 pp. 9 pp.

—Mastergroup Links—Hypothetical Reference Circuits
CCITT RECMN G.223-89. Assumptions for the Calculation of Noise on Hypothetical Reference Circuits for Telephony—International Analogue Carrier Systems (Study Group XV) 9 pp. 9 pp.

—Supergroup Links—Hypothetical Reference Circuits
CCITT RECMN G.223-89. Assumptions for the Calculation of Noise on Hypothetical Reference Circuits for Telephony—International Analogue Carrier Systems (Study Group XV) 9 pp. 9 pp.

Internal Combustion Engines
See Also: Aircraft Engines; Automotive Engines; Camshafts; Carburetors; Combustion Chambers; Crankcases; Crankshafts; Diesel Engines; Distributors (Electrical); Engine Cylinders; Engine Starters; Engine Valves; Exhaust Pipes; Gas Turbine Engines; Gasoline Engines; Internal Combustion Piston Engines; Marine Engines; Multifuel Engines; Oil Filters; Outboard Engines; Piston Pins; Piston Rings; Piston Rods; Pistons; Ramjet Engines; Spark Ignition Engines; Spark Plugs; Starter Batteries; Turbofan Engines; Turbojet Engines; Turboprop Engines; Turboshaft Engines

—Agricultural Equipment
CNS B7036-66. Testing Standard for Small Internal Combustion Engines of Land Use (Dec)(2696) (R 1973).
CNS B7037-66. Method of Test for Small Internal Combustion Engines of Land Use (Dec)(2697)(R 1973).

—Air Filters
BSI BS 1701-70. (WITHDRAWN) 1970 Air Filters for Air Supply to Internal Combustion Engines and Compressors Other Than for Aircraft (Superseded by BS 7226: 1989). 40 pp.
BSI BS 2806-56. (OBSOLESCENT) 1956 Limiting Dimensions of Air Filters for Internal Combustion Engines and Compressors, Other Than for Aircraft. 17 pp.

Internal Combustion Engines (Cont.)
—Air Filters (Cont.)
BSI BS 7226-89. 1989 Methods of Test for Performance of Inlet Air Cleaning Equipment for Internal Combustion Engines and Compressors. 28 pp.
DIN ENGL 24189-86. Testing of Air Cleaners for Internal Combustion Engines and Compressors; Test Methods (Jan). 10 pp.
ISO 5011-88. Inlet Air Cleaning Equipment for Internal Combustion Engines and Compressors—Performance Testing First Edition. 28 pp.

—Air Filters—Flammability Testing
BSI BS 7525-92. 1992 Flammability of Air Cleaner Elements for Internal Combustion Engines. 13 pp.

—Compressed Air Cylinders
DIN ENGL 6274-82. Internal Combustion Engines for General Purposes; Compressed Air Containers with Valve Block; 38 mm Bore; Assembly (Apr). 6 pp.
DIN ENGL 6275-82. Internal Combustion Engines for General Purposes; Compressed Air Containers for Permissible Working Overpressures up to 30 Bar (Apr). 4 pp.

—Cooling Systems—Hoses
BSI BS 2952-58. 1958 Amd 2 Rubber Hose for I.C. Engine Cooling Systems. 13 pp.

—Drive Shafts
BSI BS 5208-75. 1975 Dimensions of Power Take-Off Driving Shafts and Mounting Faces for Small Internal Combustion Engines. 13 pp.

—Earthmoving Equipment—Net Power
BSI BS 6911: Part 4-90. 1990 Testing Earth-Moving Machinery Part 4: Method for the Evaluation of Engine Power Output. 24 pp.
ISO 9249-89. Earth-Moving Machinery—Engine Test Code—Net Power First Edition. 20 pp.

—Flywheels—Housings
BSI BS 3529-62. 1962 Dimensions of Flywheel Housing Mounting Pads and Brackets for I.C. Engines. 12 pp.

—Friction Clutches
BSI BS 3092-73. 1973 Main Friction Clutches, Main Power-Take-Off Assemblies and Associated Attachment for Internal Combustion Engines. 37 pp.

—Fuel Lines
ISO 4639 Pt 1-87. Rubber Tubing and Hoses for Fuel Circuits for Internal Combustion Engines—Specification—Part 1: Conventional Liquid Fuels First Edition. 12 pp.

—Generators—Military Vehicles
MOD UK DSTAN 29-3-71. Generators, Engine Accessory for Tactical and Logistical Vehicles and IC Engine Powered Equipments Issue 1. 5 pp.
MOD UK DSTAN 29-3: Addendum. Generators, Engine Accessory, for Motor Transport Vehcules and IC Engine Powered Equipments Issue 1. 7 pp.

—Generators—Recreational Vehicles
CSA 946 Bull. Electrical Bulletin 946 April 19, 1974 to C22.2 NO 0. 8 pp.
CSA 946 Bull. Electrical Bulletin 946 April 19, 1974 to C22.2 NO 100. 8 pp.

—Glossaries
JIS B 0145-84. Glossary of Terms for Thermal Power Plant (Internal Combustion Engines and Auxiliary Equipments).

—Ignition Systems—Hazardous Locations
CSA C22.2 NO 30-M1986. Explosion-Proof Enclosures for Use in Class I Hazardous Locations (R 1992); (Gen Instr 1 Thru 2). 41 pp.
CSA CAN/CSA-C22. 2 NO 157-92. Intrinsically Safe and Non-Incendive Equipment for Use in Hazardous Locations; (Gen Instr 1). 61 pp.

—Ignition Systems—RFI
BSI BS 833-70. 1970 Radio Interference Limits and Measurements for the Electrical Ignition Systems of Internal Combustion Engines. 33 pp.

—Marine—Classification
CNS F3002-71. Classification of Small Internal Combustion Engines for Marine Use (Tentative) (Oct)(2145).

—Model
BSI BS 7328-90. 1990 Model Steam Engines and Internal Combustion Engines for Models. 8 pp.

—Motor Oils
CNS K6933-87. Method of Multicylinder Test Sequences for Evaluating Automotive Engine Oils-Sequence 2D Test (Dec)(12189).

INDUSTRY STANDARDS

Internal Combustion Engines (Cont.)

—Motor Oils (Cont.)
CNS K6934-87. Method of Multicylinder Test Sequences for Evaluating Automotive Engine Oils-Sequence 3D Test (Dec)(12190).

CNS K6935-87. Method of Multicylinder Test Sequences for Evaluating Automotive Engine Oils-Sequence VD Test (Dec)(12191).

JIS K 2215-91. Internal Combustion Engine Oils. 19 pp.

MOD UK DSTAN 91-43-01. Lubricating Oil, Engine: Moderate Duty, Diesel Engine Service: NATO Code No: O-176 (Grade 10) Joint Service Designation: OMD-40 NATO Code No: O-178 (Grade 20) Joint Service Designation: OMD-60 NATO Code No: O-180 (Grade 30) Joint Service Designation: OMD-110. 12 pp.

MOD UK TS 10033E. Lubricating Oil, Engine: OMD-30, OMD-80 and OMD-85.

MOD UK TS 10111A. Oils OMD-45 and OMD-175.

—Motor Oils—Sampling
DIN ENGL 51574-83. Testing of Lubricants; Sampling of Lubricating Oils from Internal Combustion Engines (Apr). 3 pp.

—Motorcycles
CNS D3012-66. Method of Test for Internal Combustion Engines of Motorcycles (Mar)(1129).

—Mounting Surfaces
BSI BS 5208-75. 1975 Dimensions of Power Take-Off Driving Shafts and Mounting Faces for Small Internal Combustion Engines. 13 pp.

—Noise
DIN ENGL 45635 Pt 11-87. Measurement of Noise Emitted by Machines; Airborne Noise Emission, Enveloping Surface Method; Internal Combustion Engines (Jan). 8 pp.

JIS B 8005-75. Measuring Method of Noise Emitted by Internal Combustion Engines (R 1984). 17 pp.

—Oil Filters
BSI BS 4836-72. 1972 Methods of Test for Full-Flow Lubricating Oil Filters for Internal Combustion Engines Within the Range of 180 l/h to 14000 l/h (40 gal/h to 3150 gal/h). 52 pp.

BSI BS 7403: Part 7-91. 1991 Full-Flow Lubricating Oil Filters for Internal Combustion Engines Part 7: Method of Test for Vibration Fatigue (ISO 4548-7: 1990). 10 pp.

ISO 4548 Pt 7-90. Methods of Test for Full-Flow Lubricating Oil Filters for Internal Combustion Engines—Part 7: Vibration Fatigue Test First Edition. 10 pp.

—Power Takeoffs
BSI BS 3092-73. 1973 Main Friction Clutches, Main Power-Take-Off Assemblies and Associated Attachment for Internal Combustion Engines. 37 pp.

—Spark Arresters
SAA AS 1019-85. Internal Combustion Engines—Spark Emission Control Devices. 7 pp.

—Valves
ISO 683 Pt 15-92. Heat-Treatable Steels, Alloy Steels and Free-Cutting Steels—Part 15: Valve Steels for Internal Combustion Engines Second Edition. 16 pp.

Internal Combustion Piston Engines
Use For: Reciprocating Internal Combustion Engines
See Also: Aircraft Engines; Distributors (Electrical); Exhaust Pipes; Internal Combustion Engines; Piston Engines; Piston Rings

BSI BS 5514: Part 1-87. 1987 Reciprocating Internal Combustion Engines: Performance Part 1: Standard Reference Conditions and Declarations of Power, Fuel Consumption and Lubricating Oil Consumption. 20 pp.

BSI BS 5514: Part 2-88. 1988 Reciprocating Internal Combustion Engines: Performance Part 2: Test Methods. 18 pp.

BSI BS 5514: Part 3-90. 1990 Reciprocating Internal Combustion Engines: Performance Part 3: Test Measurements. 10 pp.

ISO 3046 Pt 2-87. Reciprocal Internal Combustion Engines—Performance—Part 2: Test Methods Second Edition. 18 pp.

ISO 3046 Pt 3-89. Reciprocating Internal Combustion Engines—Performance—Part 3: Test Measurements Second Edition. 7 pp.

JIS B 8002-86. General Rules for Test Code of Reciprocating Internal Combustion Engines.

JIS B 8002-83. General Rules for Test Code of Reciprocating Internal Combustion Engines. 57 pp.

SNZ NZS/ISO 3046: Part 3-89. Reciprocating Internal Combustion Engines-Performance Part 3: Test Measurements. 4 pp.

—AC Generators
ISO 8528 Pt 3-93. Reciprocating Internal Combustion Engine Driven Alternating Current Generating Sets—Part 3: Alternating Current Generators for Generating Sets First Edition. 16 pp.

—Alternating Current Generators
BSI BS 5000: Part 3-80. 1980 Amd 1 Rotating Electrical Machines of Particular Types or for Particular Applications Part 3: Generators to Be Driven by Reciprocating Internal Combustion Engines. 9 pp.

DIN ENGL 6281-78. Generator Sets with Reciprocating Internal Combustion Engines; Connection Dimensions for Generators and Reciprocating Internal Combustion Engines (Apr). 3 pp.

—Alternators
BSI BS AU 250-93. 1993 Mounting Dimensions of Alternators, Types 1, 2 and 3 for Commercial Vehicles and Buses (ISO 7651: 1991) (E). 9 pp.

ISO 7651-91. Commercial Vehicles and Buses—Mounting Dimensions for Alternators of Types 1, 2 and 3 First Edition. 8 pp.

—Bell Housings
BSI BS AU 249-93. 1993 Clutch Housings for Reciprocating Internal Combustion Engines—Nominal Dimensions and Tolerances (ISO 7649: 1991) (E). 9 pp.

ISO 7649-91. Clutch Housings for Reciprocating Internal Combustion Engines—Nominal Dimensions and Tolerances First Edition. 7 pp.

—Cooling Systems—Glossaries
BSI BS 7016: Part 5-92. 1992 Components and Systems of Reciprocating Internal Combustion Engines Part 5: Glossary of Terms for Cooling Systems (ISO 7967-5: 1992). 19 pp.

ISO 7967 Pt 5-92. Reciprocating Internal Combustion Engines—Vocabulary of Components and Systems—Part 5: Cooling Systems First Edition. 20 pp.

—Direct Current Generators
BSI BS 5000: Part 3-80. 1980 Amd 1 Rotating Electrical Machines of Particular Types or for Particular Applications Part 3: Generators to Be Driven by Reciprocating Internal Combustion Engines. 9 pp.

—Direction of Rotation
BSI BS 5672-91. 1991 Designation of the Direction of Rotation and of Cylinders and Valves in Cylinder Heads, and Definition of Right-Hand and Left-Hand in-Line Engines and Locations on an Engine for Reciprocating Internal Combustion Engs. (ISO 1204: 1990). 16 pp.

ISO 1204-90. Reciprocating Internal Combustion Engines—Designation of the Direction of Rotation and of Cylinders and Valves in Cylinder Heads, and Definition of Right-Hand and Left-Hand in-Line Engines and Locations on an Engine Second Edition. 14 pp.

JIS B 8001-90. Vocabulary and Designation for Construction of Reciprocating Internal Combustion Engines. 19 pp.

—Drive Shafts—Torsional Vibration
BSI BS 5514: Part 5-79. 1979 Reciprocating Internal Combustion Engines: Performance Part 5: Torsional Vibrations. 4 pp.

ISO 3046 Pt V-78. Reciprocating Internal Combustion Engines: Performance—Part V: Torsional Vibrations First Edition; (Erratum—April 1979). 4 pp.

—Engine Cylinders
BSI BS 5673-79. (WITHDRAWN) 1979 Designation of the Cylinders of Reciprocating Internal Combustion Engines (Superseded by BS 5672: 1991). 7 pp.

—Exhaust Gases—Measurement
CEN PREN 28178-2-93. Reciprocating Internal Combustion Engines—Exhaust Emission Measurement Part 2: Measurement of Gaseous and Particulate Emissions at Site. 42 pp.

CEN PREN 28178-4-93. Reciprocating Internal Combustion Engines—Exhaust Emission Measurement Part 4: Test Cycles for Different Engine Applications. 26 pp.

—Fire Protection
BSI BS 6327-82. 1982 Fire Protection of Reciprocating Internal Combustion Engines. 4 pp.

ISO 6826-82. Reciprocating Internal Combustion Engines—Fire Protection First Edition. 5 pp.

—Flywheels—Housings
BSI BS AU 226-88. 1988 Nominal Dimensions and Tolerances of Flywheel Housings for Reciprocating Internal Combustion Engine. 11 pp.

ISO 7648-87. Flywheel Housings for Reciprocating Internal Combustion Engines—Nominal Dimensions and Tolerances First Edition. 11 pp.

—Fuel Consumption
ISO 3046 Pt 1-86. Reciprocating Internal Combustion Engines—Performance—Part 1: Standard Reference Conditions and Declarations of Power, Fuel Consumption and Lubricating Oil Consumption Third Edition; (Amendment 1-1987). 19 pp.

—Gages—Symbols
BSI BS 7712-93. 1993 Graphical Symbols for Reciprocating Internal Combustion Engines (ISO 8999: 1993) (G). 23 pp.

ISO 8999-93. Reciprocating Internal Combustion Engines—Graphical Symbols First Edition. 20 pp.

—Generator Sets
BSI BS 7698: Part 6-93. 1993 Reciprocating Internal Combustion Engine Driven Alternating Current Generating Sets Part 6: Test Methods (ISO 8528-6: 1993) (E). 12 pp.

ISO 8528 Pt 1-93. Reciprocating Internal Combustion Engine Driven Alternating Current Generating Sets—Part 1: Application, Ratings and Performance First Edition. 14 pp.

ISO 8528 Pt 2-93. Reciprocating Internal Combustion Engine Driven Alternating Current Generating Sets—Part 2: Engines First Edition. 14 pp.

ISO 8528 Pt 5-93. Reciprocating Internal Combustion Engine Driven Alternating Current Generating Sets—Part 5: Generating Sets First Edition. 32 pp.

ISO 8528 Pt 6-93. Reciprocating Internal Combustion Engine Driven Alternating Current Generating Sets—Part 6: Test Methods First Edition. 11 pp.

JIS B 8022-90. Standard Form of Specifications of Engine-Driven Generating Sets. 7 pp.

—Generator Sets—Controlgear
ISO 8528 Pt 4-93. Reciprocating Internal Combustion Engine Driven Alternating Current Generating Sets—Part 4: Controlgear and Switchgear First Edition. 16 pp.

—Generator Sets—Switchgear
ISO 8528 Pt 4-93. Reciprocating Internal Combustion Engine Driven Alternating Current Generating Sets—Part 4: Controlgear and Switchgear First Edition. 16 pp.

—Glossaries
BSI BS 5676-79. 1979 Reciprocating Internal Combustion Engines. 25 pp.

BSI BS 7016: Part 1-88. 1988 Components and Systems of Reciprocating Internal Combustion Engines Part 1: Glossary of Terms for Structure and External Covers. 16 pp.

BSI BS 7016: Part 2-88. 1988 Components and Systems of Reciprocating Internal Combustion Engines Part 2: Glossary of Terms for Main Running Gear. 23 pp.

BSI BS 7016: Part 3-88. 1988 Components and Systems of Reciprocating Internal Combustion Engines Part 3: Glossary of Terms of Valves, Camshaft Drive and Actuating Mechanisms. 19 pp.

BSI BS 7016: Part 4-88. 1988 Components and Systems of Reciprocating Internal Combustion Engines Part 4: Glossary of Terms for Pressure Charging and Air/Exhaust Gas Ducting Systems. 17 pp.

BSI BS 7016: Part 8-90. 1990 Components and Systems of Reciprocating Internal Combustion Engines Part 8: Glossary of Terms for Starting Systems. 12 pp.

ISO 2710-78. Reciprocating Internal Combustion Engines—Vocabulary Second Edition; (Corrected and Reprinted -1978) (Addendum 1-1982). 32 pp.

ISO 7967 Pt 1-87. Reciprocating Internal Combustion Engines—Vocabulary of Components and Systems—Part 1: Structure and External Covers First Edition. 15 pp.

ISO 7967 Pt 3-87. Reciprocating Internal Combustion Engines—Vocabulary of Components and Systems—Part 3: Valves, Camshaft Drive and Actuating Mechanisms First Edition. 19 pp.

ISO 7967 Pt 4-88. Reciprocating Internal Combustion Engines—Vocabulary of Components and Systems—Part 4: Pressure Charging and Air/Exhaust Ducting Systems First Edition. 18 pp.

JIS B 0108-84. Glossary of Terms Relating to Reciprocating Internal-Combustion Engines.

JIS B 0109-84. Glossary of Terms Relating to Reciprocating Internal-Combustion Engines (Main Parts).

JIS B 0110-84. Glossary of Terms Relating to Reciprocating Internal-Combustion Engines (Attachments).

INTERNATIONAL AND NON-U.S. NATIONAL STANDARDS
SUBJECT INDEX

Internal

Internal Combustion Piston Engines (Cont.)

—Lubricating Oils—Consumption Rate

ISO 3046 Pt 1-86. Reciprocating Internal Combustion Engines—Performance—Part 1: Standard Reference Conditions and Declarations of Power, Fuel Consumption and Lubricating Oil Consumption Third Edition; (Amendment 1-1987). 19 pp.

—Lubrication—Glossaries

BSI BS 7016: Part 6-92. 1992 Components and Systems of Reciprocating Internal Combustion Engines Part 6: Glossary of Terms for Lubricating Systems (ISO 7967-6: 1992). 19 pp.

ISO 7967 Pt 6-92. Reciprocating Internal Combustion Engines—Vocabulary of Components and Systems—Part 6: Lubricating Systems First Edition. 20 pp.

—Manual Controls

BSI BS 5674-79. (WITHDRAWN) 1979 Standard Direction of Motion of Hand Operated Control Devices of Reciprocating Internal Combustion Engines (Superseded by BS 5672: 1991). 7 pp.

—Manual Controls—Direction of Motion

CNS B1217-80. Reciprocating Internal Combustion Engines Hand Operated Control Devices Direction of Motion (Dec)(6733).

ISO 2261-72. Reciprocating Internal Combustion Engines—Hand Operated Control Devices—Standard Direction of Motion First Edition. 6 pp.

—Manual Controls—Symbols

BSI BS 7712-93. 1993 Graphical Symbols for Reciprocating Internal Combustion Engines (ISO 8999: 1993) (G). 23 pp.

ISO 8999-93. Reciprocating Internal Combustion Engines—Graphical Symbols First Edition. 20 pp.

—Marine—AC Generators

ISO 8528 Pt 3-93. Reciprocating Internal Combustion Engine Driven Alternating Current Generating Sets—Part 3: Alternating Current Generators for Generating Sets First Edition. 16 pp.

—Marine—Generator Sets

BSI BS 7698: Part 6-93. 1993 Reciprocating Internal Combustion Engine Driven Alternating Current Generating Sets Part 6: Test Methods (ISO 8528-6: 1993) (E). 12 pp.

ISO 8528 Pt 1-93. Reciprocating Internal Combustion Engine Driven Alternating Current Generating Sets—Part 1: Application, Ratings and Performance First Edition. 14 pp.

ISO 8528 Pt 2-93. Reciprocating Internal Combustion Engine Driven Alternating Current Generating Sets—Part 2: Engines First Edition. 14 pp.

ISO 8528 Pt 5-93. Reciprocating Internal Combustion Engine Driven Alternating Current Generating Sets—Part 5: Generating Sets First Edition. 32 pp.

ISO 8528 Pt 6-93. Reciprocating Internal Combustion Engine Driven Alternating Current Generating Sets—Part 6: Test Methods First Edition. 11 pp.

—Marine—Generator Sets—Controlgear

ISO 8528 Pt 4-93. Reciprocating Internal Combustion Engine Driven Alternating Current Generating Sets—Part 4: Controlgear and Switchgear First Edition. 16 pp.

—Marine—Generator Sets—Switchgear

ISO 8528 Pt 4-93. Reciprocating Internal Combustion Engine Driven Alternating Current Generating Sets—Part 4: Controlgear and Switchgear First Edition. 16 pp.

—Marine—Symbols

BSI BS 7712-93. 1993 Graphical Symbols for Reciprocating Internal Combustion Engines (ISO 8999: 1993) (G). 23 pp.

ISO 8999-93. Reciprocating Internal Combustion Engines—Graphical Symbols First Edition. 20 pp.

—Orientation

BSI BS 5675-79. (WITHDRAWN) 1979 Definition of Right-Hand and Left-Hand Single Bank Reciprocating Internal Combustion Engines (Superseded by BS 5672: 1991). 6 pp.

JIS B 8001-90. Vocabulary and Designation for Construction of Reciprocating Internal Combustion Engines. 19 pp.

—Overspeed Protection

BSI BS 5514: Part 6-90. 1990 Reciprocating Internal Combustion Engines: Performance Part 6: Overspeed Protection (E). 8 pp.

ISO 3046 Pt 6-90. Reciprocating Internal Combustion Engines—Performance—Part 6: Overspeed Protection Third Edition. 6 pp.

Internal Combustion Piston Engines (Cont.)

—Piston Rings

BSI BS 5341: Part 1-76. (WITHDRAWN) 1976 Piston Rings up to 200 mm Diameter for Reciprocating Internal Combustion Engines Part 1: Designs, Dimensions, Materials and Designations for Single Piece Rings (Superseded by BS 5341: Parts 6-8 & DD 208 & DD 209: 1992). 49 pp.

BSI BS 5341: Part 2-77. (WITHDRAWN) 1977 Piston Rings up to 200 mm Diameter for Reciprocating Internal Combustion Engines Part 2: Designs, Dimensions, Materials and Designations for Multi-Piece Oil Control Rings (Superseded by BS 5341: Parts 6-8 & DD 208 & DD 209: 1992). 10 pp.

BSI BS 5341: Part 3-76. (WITHDRAWN) 1976 Piston Rings up to 200 mm Diameter for Reciprocating Internal Combustion Engines Part 3: Quality Requirements (Superseded by BS 5341: Parts 6-8 & DD 208 & DD 209: 1992). 12 pp.

BSI BS 5341: Part 4-76. (WITHDRAWN) 1976 Piston Rings up to 200 mm Diameter for Reciprocating Internal Combustion Engines Part 4: Measuring and Testing Procedures (Superseded by BS 5341: Parts 6-8 & DD 208 & DD 209: 1992). 24 pp.

BSI BS 5341: Part 5-76. 1976 Piston Rings up to 200 mm Diameter for Reciprocating Internal Combustion Engines Part 5: Ring Grooves. 19 pp.

BSI BS 5341: Sec 6.2-92. 1992 Piston Rings up to 200 mm Diameter for Reciprocating Internal Combustion Engines Part 6: Vocabulary, Inspection Methods and Generic Specifications Section 6.2: Inspection Measuring Principles (ISO 6621-2: 1984). 27 pp.

BSI BS 5341: Sec 6.3-92. 1992 Piston Rings up to 200 mm Diameter for Reciprocating Internal Combustion Engines Part 6: Vocabulary, Inspection Methods and Generic Specifications Section 6.3: Material Specifications (ISO 6621-3: 1983). 7 pp.

BSI BS 5341: Sec 6.4-92. 1992 Piston Rings up to 200 mm Diameter for Reciprocating Internal Combustion Engines Part 6: Vocabulary, Inspection Methods and Generic Specifications Section 6.4: General Specifications (ISO 6621-4: 1988). 19 pp.

BSI BS 5341: Sec 6.5-92. 1992 Piston Rings up to 200 mm Diameter for Reciprocating Internal Combustion Engines Part 6: Vocabulary, Inspection Methods and Generic Specifications Section 6.5: Quality Requirements (ISO 6621-5: 1988). 15 pp.

BSI BS 5341: SUB SEC7.1.1-92. 1992 Piston Rings up to 200 mm Diameter for Reciprocating Internal Combustion Engines Part 7: Designs, Dimensions and Designations for Single Piece Rings Section 7.1: Rectangular Compression Rings Subsection 7.1.1: Specification for. 23 pp.

BSI BS 5341: Sec 7.2-92. 1992 Piston Rings up to 200 mm Diameter for Reciprocating Internal Combustion Engines Part 7: Designs, Dimensions and Designations for Single Piece Rings Section 7.2: Specification for Scraper Rings (ISO 6623: 1986). 19 pp.

BSI BS 5341: SUB SEC7.3.1-92. 1992 Piston Rings up to 200 mm Diameter for Reciprocating Internal Combustion Engines Part 7: Designs, Dimensions and Designations for Single Piece Rings Section 7.3: Keystone and Half Keystone Compression Rings. 22 pp.

BSI BS 5341: Sec 7.4-92. 1992 Piston Rings up to 200 mm Diameter for Reciprocating Internal Combustion Engines Part 7: Designs, Dimensions and Designations for Single Piece Rings Section 7.4: Specification for Oil Control Rings (ISO 6625: 1986). 25 pp.

BSI BS 5341: Sec 8.1-92. 1992 Piston Rings up to 200 mm Diameter for Reciprocating Internal Combustion Engines Part 8: Designs, Dimens. and Design. for Multi-Piece Oil Control Rings Sec 8.1: Specification for Coil-Spring-Loaded Oil Control Rings (ISO 6626: 1989). 59 pp.

BSI DD 208-92. 1992 Piston Rings up to 200 mm Diameter for Reciprocating Internal Combustion Engines: Specification for Rectangular Rings with Narrow Ring Width (ISO/TR 6622-2: 1988) (E). 19 pp.

BSI DD 209-92. 1992 Piston Rings up to 200 mm Diameter for Reciprocating Internal Combustion Engines: Specification for Half Keystone Rings (ISO/TR 6624-2: 1988). 15 pp.

BSI DD 212-92. 1992 Piston Rings up to 200 mm Diameter for Reciprocating Internal Combustion Engines: Specification for Expander/Segment Oil Control Rings (ISO/TR 6627: 1992). 12 pp.

CNS B4013-70. Piston Ring for Internal Combustion Engine (Tentative)) (Dec)(3176).

ISO 6621 Pt 2-84. Internal Combustion Engines—Piston Rings—Part 2: Inspection Measuring Principles First Edition. 23 pp.

ISO 6621 Pt 3-83. Internal Combustion Engines—Piston Rings—Part 3: Material Specifications First Edition. 4 pp.

Internal Combustion Piston Engines (Cont.)

—Piston Rings (Cont.)

ISO 6621 Pt 4-88. Internal Combustion Engines—Piston Rings—Part 4: General Specifications First Edition. 16 pp.

ISO 6621 Pt 5-88. Internal Combustion Engines—Piston Rings—Part 5: Quality Requirements First Edition; (SAE J 1996). 12 pp.

ISO 6622 Pt 1-86. Internal Combustion Engines—Piston Rings—Part 1: Rectangular Rings First Edition; (Corrected and Reprinted -1986). 19 pp.

ISO TR6622 Pt 2-88. Internal Combustion Engines—Piston Rings—Part 2: Rectangular Rings with Narrow Ring Width First Edition. 15 pp.

ISO 6623-86. Internal Combustion Engines—Piston Rings—Scraper Rings First Edition. 16 pp.

ISO 6624 Pt 1-86. Internal Combustion Engines—Piston Rings—Part 1: Keystone Rings First Edition. 18 pp.

ISO TR6624 Pt 2-88. Internal Combustion Engines—Piston Rings—Part 2: Half Keystone Rings First Edition. 11 pp.

ISO 6625-86. Internal Combustion Engines—Piston Rings—Oil Control Rings First Edition. 22 pp.

ISO 6626-89. Internal Combustion Engines—Piston Rings—Coil-Spring-Loaded Oil Control Rings First Edition. 55 pp.

ISO TR6627-92. Internal Combustion Engines—Piston Rings—Expander/Segment Oil Control Rings First Edition. 11 pp.

—Piston Rings—Glossaries

BSI BS 5341: Sec 6.1-92. 1992 Piston Rings up to 200 mm Diameter for Reciprocating Internal Combustion Engines Part 6: Vocabulary, Inspection Methods and Generic Specifications Section 6.1: Vocabulary (ISO 6621-1: 1986). 23 pp.

ISO 6621 Pt 1-86. Internal Combustion Engines—Piston Rings—Part 1: Vocabulary First Edition. 19 pp.

—Power Ratings

BSI BS 5514: Part 7-88. 1988 Reciprocating Internal Combustion Engines: Performance Part 7: Codes for Engine Power. 8 pp.

ISO 3046 Pt 1-86. Reciprocating Internal Combustion Engines—Performance—Part 1: Standard Reference Conditions and Declarations of Power, Fuel Consumption and Lubricating Oil Consumption Third Edition; (Amendment 1-1987). 19 pp.

ISO 3046 Pt 7-87. Reciprocating Internal Combustion Engines—Performance—Part 7: Codes for Engine Power First Edition. 7 pp.

—Pressure Measurement

JIS B 8006-86. Pressure Measurements of Performance for Reciprocating Internal Combustion Engines. 11 pp.

—Running Gear—Glossaries

ISO 7967 Pt 2-87. Reciprocating Internal Combustion Engines—Vocabulary of Components and Systems—Part 2: Main Running Gear First Edition. 23 pp.

—Signal Devices—Symbols

BSI BS 7712-93. 1993 Graphical Symbols for Reciprocating Internal Combustion Engines (ISO 8999: 1993) (G). 23 pp.

ISO 8999-93. Reciprocating Internal Combustion Engines—Graphical Symbols First Edition. 20 pp.

—Speed Control

BSI BS 5514: Part 4-79. 1979 Reciprocating Internal Combustion Engines: Performance Part 4: Speed Governing. 11 pp.

ISO 3046 Pt IV-78. Reciprocating Internal Combustion Engines: Performance—Part IV: Speed Governing First Edition; (Erratum—April 1979). 10 pp.

—Starters—Glossaries

ISO 7967 Pt 8-90. Reciprocating Internal Combustion Engines—Vocabulary of Components and Systems—Part 8: Starting Systems First Edition. 13 pp.

—Symbols

BSI BS 5676: Add 1-83. 1983 Reciprocating Internal Combustion Engines Addendum 1: Symbols. 5 pp.

BSI BS 7712-93. 1993 Graphical Symbols for Reciprocating Internal Combustion Engines (ISO 8999: 1993) (G). 23 pp.

ISO 8999-93. Reciprocating Internal Combustion Engines—Graphical Symbols First Edition. 20 pp.

—Temperature Measurement

JIS B 8007-87. Temperature Measurements of Performance for Reciprocating Internal Combustion Engines. 17 pp.

Internal Gear Pairs

Use: Internal Gears

Internal Gears
Use For: Internal Gear Pairs See Also: Gears; Helical Gears

—**Design**
DIN ENGL 3993 Pt 1-81. Geometrical Design of Cylindrical Internal Involute Gear Pairs; Basic Rules (Aug). 36 pp.
DIN ENGL 3993 Pt 2-81. Geometrical Design of Cylindrical Internal Involute Gear Pairs; Diagrams for Geometrical Limits of Internal Gear-Pinion Matings (Aug). 25 pp.

—**Design—Addendum Modification**
DIN ENGL 3993 Pt 3-81. Geometrical Design of Cylindrical Internal Involute Gear Pairs; Diagrams for the Determination of Addendum Modification Coefficients (Aug). 26 pp.

—**Design—Pinion Type Cutters**
DIN ENGL 3993 Pt 4-81. Geometrical Design of Cylindrical Internal Involute Gear Pairs; Diagrams for Limits of Internal Gear—Pinion Type Cutter Matings (Aug). 11 pp.

—**Hobbing Machines**
BSI BS 2007-75. 1975 Circular Gear Shaving Cutters, 1 to 8 Metric Module, Accuracy Requirements. 8 pp.

Internal Grinders
See Also: Grinders; Machine Tools
CNS B1121-79. Specified Items for Machine Tools (Internal Grinder) (Apr)(4670-24).

Internal Loss Probability
See Also: Call Attempts; Incoming Lines; Outgoing Lines; Telephone Networks; Telephone Services

—**Grade of Service—Digital Exchanges**
CCITT RECMN E.543-89. Grades of Service in Digital International Telephone Exchanges—Telephone Network and ISDN—Quality of Service, Network Management and Traffic Engineering (Study Group II) 4 pp. 4 pp.

—**Tables (Data)**
CCITT FASCICLE II.3-88. Telephone Network and ISDN Quality of Service, Network Management and Traffic Engineering. Recommendations E.401-E.880. 368 pp.

Internal Tab Washers
Use: Tab Washers—Internal

Internal Vibrators
Use: Concrete Vibrators—Internal

International Agreements
See Also: European Communities; Foreign Acquisition

—**Certification Systems**
IEC Guide 42-84. Guidelines for a Step-by-Step Approach to an International Certification System First Edition. 9 pp.
IEC Guide 44-85. General Rules for ISO or IEC International Third-Party Certification Schemes for Products First Edition. 16 pp.
ISO Guide 42-84. Guidelines for a Step-by-Step Approach to an International Certification System First Edition. 9 pp.
ISO Guide 44-85. General Rules for ISO or IEC International Third-Party Certification Schemes for Products First Edition. 16 pp.
SAA HB18.42-91. Guidelines for Third-Party Certification and Accreditation—Guide 42: Guidelines for a Step-by-Step Approach to an International Certification System (ISO/IEC Guide 42:1984) (SANZ HB18.42—1991). 6 pp.
SAA HB18.44-91. Guidelines for Third-Party Certification and Accreditation—Guide 44: General Rules for ISO or IEC International Third-Party Certification Schemes for Products (ISO/IEC Guide 44:1985) (SANZ HB18.44—1991). 13 pp.
SNZ SANZ/SAA HB 18.42-91. Guidelines for Third-Party Certification and Accreditation Guide 42 Guidelines for a Step-by-Step Approach to an International Certification System (ISO/IEC Guide 42-1984). 6 pp.
SNZ SANZ/SAA HB 18.44-85. Guidelines for Third-Party Certification and Accreditation Guide 44 General Rules for ISO or IEC International Third-Party Certification Schemes for Products (ISO/IEC Guide 44-1985). 13 pp.

—**Defense Contracts**
MOD UK DEFCON 126-73. International Collaboration Clause 6/73. 2 pp.

—**Quality Assurance**
MOD UK DSTAN 05-3-01. Mutual Acceptance of Government Quality Assurance Issue 4; Amendment 1. 30 pp.
MOD UK. Information on Quality Assurance in NATO.

International Air Navigation
Use: Air Navigation

International Alphabet No. 5
Use: International Reference Alphabet

International Atomic Time
Use For: International Atomic Time Scales; TAI
See Also: Atomic Time Scales; Time Scales
CCIR RECMN 460-4-86. Standard-Frequency and Time-Signal Emissions—Section 7B—Specifications for the Standard-Frequency and Time-Signal Services. 4 pp.
CCIR Report 439-5-90. Relativistic Effects in a Terrestrial Coordinate Time System—Section 7D—Characterization of Sources and Time Scales Formation. 5 pp.

—**Abbreviations**
CCIR RECMN 536-78. Time-Scale Notations—Section 7B—Specifications for the Standard-Frequency and Time-Signal Services. 2 pp.

—**Frequency Generators**
CCIR RECMN 486-1-78. Reference of Precisely Controlled Frequency Generators and Emissions to the International Atomic Time Scale—Section 7B—Specifications for the Standard-Frequency and Time-Signal Services. 1 p.

—**Standard Frequency Emissions**
CCIR RECMN 486-1-78. Reference of Precisely Controlled Frequency Generators and Emissions to the International Atomic Time Scale—Section 7B—Specifications for the Standard-Frequency and Time-Signal Services. 1 p.

—**Time Signal Emissions**
CCIR RECMN 486-1-78. Reference of Precisely Controlled Frequency Generators and Emissions to the International Atomic Time Scale—Section 7B—Specifications for the Standard-Frequency and Time-Signal Services. 1 p.

International Atomic Time Scales
Use: International Atomic Time

International Automatic Telephone Service
Use: Automatic Services (Telephone)

International Circuits
Use: Telecommunication Circuits

International Civil Aviation Organization CAEE
Use: ICAO Committee on Aircraft Engine Emissions

International Civil Aviation Organization CAEP
Use: ICAO Committee on Aviation Environmental Protection

International Civil Aviation Organization CAN
Use: ICAO Committee on Aircraft Noise

International Civil Aviation Organization FANS
Use: ICAO Special Committee on Future Air Navigation Systems

International Civil Aviation Organization SBC
Use: ICAO Sonic Boom Committee

International Commission on Illumination
Use: Commision International de l'Eclairage

International Data Transmission Services
Use: Data Transmission

International Data Transmission Systems
Use: Data Transmission Equipment

International Digital Connections
Use: Circuit Switched Connections

International Digital Leased Circuits
Use: International Leased Circuits—Digital

International Directory Databases
Use: Directory Databases

International Electrotechnical Commission
Use: IEC

International Exchanges
See Also: Demand Operating; Language Digits; Telephone Exchanges; Telephone Relations
CCITT RECMN G.142-89. Transmission Characteristics of Exchanges—General Characteristics of International Telephone Connections and Circuits (Study Groups XII and XV) 6 pp. 6 pp.

—**Answer Signals**
CCITT RECMN Q.109-89. Transmission of the Answer Signal in International Exchanges—General Recommendations on Telephone Switching and Signalling—Functions and Information Flows for Services in the ISDN—Supplements (Study Group XI) 1 pp. 1 p.

—**Design**
CCITT RECMN Q.541-89. Digital Exchange Design Objectives—General—Digital Local, Transit, Combined and International Exchanges in Integrated Digital Networks and Mixed Analogue-Digital Networks—Supplements (Study Group XI) 6 pp. 6 pp.
CCITT RECMN Q.542-89. Digital Exchange Design Objectives—Operations and Maintenance—Digital Local, Transit, Combined and International Exchanges in Integrated Digital Networks and Mixed Analogue-Digital Networks—Supplements (Study Group XI) 21 pp. 21 pp.
CCITT RECMN Q.543-89. Digital Exchange Performance Design Objectives—Digital Local, Transit, Combined and International Exchanges in Integrated Digital Networks and Mixed Analogue-Digital Networks—Supplements (Study Group XI) 40 pp. 40 pp.

—**Grade of Service—Failure (Quality Control)**
CCITT RECMN E.550-89. Grade-of-Service and New Performance Criteria Under Failure Conditions in International Telephone Exchanges—Telephone Network and ISDN—Quality of Service, Network Management and Traffic Engineering (Study Group II) 8 pp. 8 pp.

—**Integrated Digital Networks**
CCITT RECMN Q.500-89. Digital Local, Combined, Transit and International Exchanges Introduction and Field of Application—Digital Local, Transit, Combined and International Exchanges in Integrated Digital Networks and Mixed Analogue-Digital Networks—Supplements (Study Group XI) 4 pp. 4 pp.
CCITT RECMN Q.521-89. Exchange Functions—Digital Local, Transit, Combined and International Exchanges in Integrated Digital Networks and Mixed Analogue-Digital Networks—Supplements (Study Group XI) 3 pp. 3 pp.

—**Integrated Digital Networks—Mixed Analog/Digital Networks**
CCITT FASCICLE VI.5. Digital Local, Transit, Combined and International Exchanges Integrated Digital Networks and Mixed Analogue—Digital Networks Supplements Recommendations Q.500—Q.554 (Study Group XI). 188 pp.

—**Integrated Services Digital Networks**
CCITT RECMN Q.500-89. Digital Local, Combined, Transit and International Exchanges Introduction and Field of Application—Digital Local, Transit, Combined and International Exchanges in Integrated Digital Networks and Mixed Analogue-Digital Networks—Supplements (Study Group XI) 4 pp. 4 pp.
CCITT RECMN Q.521-89. Exchange Functions—Digital Local, Transit, Combined and International Exchanges in Integrated Digital Networks and Mixed Analogue-Digital Networks—Supplements (Study Group XI) 3 pp. 3 pp.

—**Interfaces**
CCITT RECMN Q.511-89. Exchange Interfaces Towards Other Exchanges—Digital Local, Transit, Combined and International Exchanges in Integrated Digital Networks and Mixed Analogue-Digital Networks—Supplements (Study Group XI) 4 pp. 4 pp.

INTERNATIONAL AND NON-U.S. NATIONAL STANDARDS
SUBJECT INDEX

International Exchanges *(Cont.)*

—**Interfaces—Operations, Administration and Maintenance**
CCITT RECMN Q.513-89. Exchange Interfaces for Operations, Administration and Maintenance—Digital Local, Transit, Combined and International Exchanges in Integrated Digital Networks and Mixed Analogue-Digital Networks—Supplements (Study Group XI) 4 pp. 4 pp.

—**Interfaces—Subscriber Access**
CCITT RECMN Q.512-89. Exchange Interfaces for Subscriber Access—Digital Local, Transit, Combined and International Exchanges in Integrated Digital Networks and Mixed Analogue-Digital Networks—Supplements (Study Group XI) 9 pp. 9 pp.

—**Interfaces—Transmission**
CCITT RECMN Q.552-89. Transmission Characteristics at 2-Wire Analogue Interfaces of Digital Exchange—Digital Local, Transit, Combined and International Exchanges in Integrated Digital Networks and Mixed Analogue-Digital Networks—Supplements (Study Group XI) 25 pp. 25 pp.
CCITT RECMN Q.553-89. Transmission Characteristics at 4-Wire Analogue Interfaces of a Digital Exchange—Digital Local, Transit, Combined and International Exchanges in Integrated Digital Networks and Mixed Analogue-Digital Networks—Supplements (Study Group XI) 12 pp. 12 pp.
CCITT RECMN Q.554-89. Transmission Characteristics at Digital Interfaces of a Digital Exchange—Digital Local, Transit, Combined and International Exchanges in Integrated Digital Networks and Mixed Analogue-Digital Networks—Supplements (Study Group XI) 3 pp. 3 pp.

—**Maintenance**
CCITT RECMN Q.542-89. Digital Exchange Design Objectives—Operations and Maintenance—Digital Local, Transit, Combined and International Exchanges in Integrated Digital Networks and Mixed Analogue-Digital Networks—Supplements (Study Group XI) 21 pp. 21 pp.

—**Measurement**
CCITT RECMN Q.544-89. Digital Exchange Measurements—Digital Local, Transit, Combined and International Exchanges in Integrated Digital Networks and Mixed Analogue-Digital Networks—Supplements (Study Group XI) 15 pp. 15 pp.

—**Mixed Analog/Digital Networks**
CCITT RECMN Q.500-89. Digital Local, Combined, Transit and International Exchanges Introduction and Field of Application—Digital Local, Transit, Combined and International Exchanges in Integrated Digital Networks and Mixed Analogue-Digital Networks—Supplements (Study Group XI) 4 pp. 4 pp.
CCITT RECMN Q.521-89. Exchange Functions—Digital Local, Transit, Combined and International Exchanges in Integrated Digital Networks and Mixed Analogue-Digital Networks—Supplements (Study Group XI) 3 pp. 3 pp.

—**Operations**
CCITT RECMN Q.542-89. Digital Exchange Design Objectives—Operations and Maintenance—Digital Local, Transit, Combined and International Exchanges in Integrated Digital Networks and Mixed Analogue-Digital Networks—Supplements (Study Group XI) 21 pp. 21 pp.

—**Registers (Switching)—CCITT R2 Signaling Systems**
CCITT RECMN Q.462-89. Signalling Procedures—Signalling Between the Outgoing International R2 Register and an Incoming R2 Register in an International Exchange—Specifications of Signalling Systems R1 and R2 (Study Group XI) 3 pp. 3 pp.

—**Signal Transfer—CCITT No. 6 Signaling Systems**
CCITT RECMN Q.265-89. 4.5 Speed of Switching and Signal Transfer in International Exchanges—Specifications of Signalling System No. 6 (Study Group XI) 1 pp. 1 p.

—**Switching Speed—CCITT No. 4 Signaling Systems**
CCITT RECMN Q.125-89. Speed of Switching in International Exchanges—Specifications of Signalling Systems Nos. 4 and 5 (Study Group XI) 1 pp. 1 p.

—**Switching Speed—CCITT No. 5 Signaling Systems**
CCITT RECMN Q.146-89. Speed of Switching in International Exchanges—Specifications of Signalling Systems Nos. 4 and 5 (Study Group XI) 1 pp. 1 p.

International Exchanges *(Cont.)*

—**Switching Speed—CCITT No. 6 Signaling Systems**
CCITT RECMN Q.265-89. 4.5 Speed of Switching and Signal Transfer in International Exchanges—Specifications of Signalling System No. 6 (Study Group XI) 1 pp. 1 p.

—**Switching Speed—CCITT R1 Signaling Systems**
CCITT RECMN Q.319-89. Speed of Switching in International Exchanges—Specifications of Signalling Systems R1 and R2 (Study Group XI) 1 pp. 1 p.

—**Telecommunication Traffic—Measurement**
CCITT RECMN E.500 (REV 1)-92. Traffic Intensity Measurement Principles (Study Group II) 17 pp. 17 pp.
CCITT RECMN E.500-89. Traffic Intensity Measurement Principles —Telephone Network and ISDN—Quality of Service, Network Management and Traffic Engineering (Study Group II) 12 pp. 12 pp.

—**Telex Communications**
CCITT RECMN U.10-89. Equipment of an International Telex Position—Telegraph Switching (Study Group IX) 1 pp. 1 p.

—**Telex Communications—Answering Time**
CCITT RECMN F.65-89. Time-to-Answer by Operators at International Telex Positions—Telegraph and Mobile Services Operations and Quality of Service (Study Group I) 1 pp. 1 p.

—**Traffic Units**
CCITT RECMN Q.544-89. Digital Exchange Measurements—Digital Local, Transit, Combined and International Exchanges in Integrated Digital Networks and Mixed Analogue-Digital Networks—Supplements (Study Group XI) 15 pp. 15 pp.

—**Transmission**
CCITT RECMN Q.551-89. Transmission Characteristics of Digital Exchanges—Digital Local, Transit, Combined and International Exchanges in Integrated Digital Networks and Mixed Analogue-Digital Networks—Supplements (Study Group XI) 14 pp. 14 pp.

International Freephone Services
Use: Freephone Services

International Frequency Registration Board
Use: IFRB

International Leased Circuits
See Also: Leased Circuits; Private Leased Circuits; Telecommunication Circuits; Telephone Circuits
CCITT RECMN M.1010-89. Constitution and Nomenclature of International Leased Circuits—Maintenance of International Telegraph, Phototelegraph and Leased Circuits—Maintenance of the International Public Telephone Network—Maintenance of Maritime. 4 pp.
CCITT RECMN M.1020-89. Characteristics of Special Quality International Leased Circuits with Special Bandwidth Conditioning—Maintenance of International Telegraph, Phototelegraph and Leased Circuits—Maintenance of the International Public Telephone Network-. 5 pp.
CCITT RECMN M.1025-89. Characteristics of Special Quality International Leased Circuits with Basic Bandwidth Conditioning—Maintenance of International Telegraph, Phototelegraph and Leased Circuits—Maintenance of the International Public Telephone Network-. 5 pp.
CCITT RECMN M.1040-89. Characteristics of Ordinary Quality International Leased Circuits—Maintenance of International Telegraph, Phototelegraph and Leased Circuits—Maintenance of the International Public Telephone Network—Maintenance of Maritime. 3 pp.

—**Analog—Private**
CCITT RECMN D.3-92. Principles for the Lease of Analogue International Circuits for Private Service (Study Group III) 5 pp (Replaces Recmn D2 and D3). 5 pp.

—**Analog—Private—Broadband Telecommunication Services**
CCITT RECMN D.3-92. Principles for the Lease of Analogue International Circuits for Private Service (Study Group III) 5 pp (Replaces Recmn D2 and D3). 5 pp.

International Leased Circuits *(Cont.)*

—**Analog—Private—Broadband Telecommunication Services—Tariffs**
CCITT RECMN D.3-92. Principles for the Lease of Analogue International Circuits for Private Service (Study Group III) 5 pp (Replaces Recmn D2 and D3). 5 pp.

—**Analog—Private—HF**
CCITT RECMN D.3-92. Principles for the Lease of Analogue International Circuits for Private Service (Study Group III) 5 pp (Replaces Recmn D2 and D3). 5 pp.

—**Analog—Private—HF—Tariffs**
CCITT RECMN D.3-92. Principles for the Lease of Analogue International Circuits for Private Service (Study Group III) 5 pp (Replaces Recmn D2 and D3). 5 pp.

—**Analog—Private—Special Grade**
CCITT RECMN D.3-92. Principles for the Lease of Analogue International Circuits for Private Service (Study Group III) 5 pp (Replaces Recmn D2 and D3). 5 pp.

—**Analog—Private—Special Grade—Tariffs**
CCITT RECMN D.3-92. Principles for the Lease of Analogue International Circuits for Private Service (Study Group III) 5 pp (Replaces Recmn D2 and D3). 5 pp.

—**Analog—Private—Tariffs**
CCITT RECMN D.3-92. Principles for the Lease of Analogue International Circuits for Private Service (Study Group III) 5 pp (Replaces Recmn D2 and D3). 5 pp.

—**Analog—Private—Telegraphy**
CCITT RECMN D.3-92. Principles for the Lease of Analogue International Circuits for Private Service (Study Group III) 5 pp (Replaces Recmn D2 and D3). 5 pp.

—**Analog—Private—Telegraphy—Tariffs**
CCITT RECMN D.3-92. Principles for the Lease of Analogue International Circuits for Private Service (Study Group III) 5 pp (Replaces Recmn D2 and D3). 5 pp.

—**Analog—Signaling Systems**
CCITT RECMN Q.8-89. Signalling Systems to Be Used for International Manual and Automatic Working on Analogue Leased Circuits—General Recommendations on Telephone Switching and Signalling—Functions and Infor. Flows for Services in the ISDN—Supplements (Study Group XI) 17 pp. 17 pp.

—**Data Transmission**
CCITT RECMN M.1015-89. Types of Transmission on Leased Circuits—Maintenance of International Telegraph, Phototelegraph and Leased Circuits—Maintenance of the International Public Telephone Network—Maintenance of Maritime Satellite and Data Transmission Systems. 3 pp.

—**Designations—Customer Circuits**
CCITT RECMN F.104-91. International Leased Circuit Services—Customer Circuit Designations (Study Group I) 5 pp. 5 pp.

—**Digital—Private**
CCITT RECMN D.8-89. Special Conditions for the Lease of International End-To-End Digital Circuits for Private Service (Study Group III) 2 pp. 2 pp.

—**Digital—Private—Tariffs**
CCITT RECMN D.8-89. Special Conditions for the Lease of International End-To-End Digital Circuits for Private Service (Study Group III) 2 pp. 2 pp.

—**Digital—Reliability**
CEPT T/TPH 44 E-90. Reliability of Digital Leased Circuits. 3 pp.

—**European Broadcasting Union**
CEPT T/TPH 6-86. Location Permanente Et Temporaire De Circuits Internationaux De Telecommunications Aux Membres De L'UER Et a Leur Utilisation Par Des Tiers Et Par Des Administrations Des Telecommunications. 4 pp.
CEPT T/TPH 6 E-90. Permanent and Occasional Lease of International Telecommunications Circuits to the EBU and Its Members and Their Use by Third Parties and Telecommunications Administrations. 3 pp.

—**FAX Communications**
CCITT RECMN M.1015-89. Types of Transmission on Leased Circuits—Maintenance of International Telegraph, Phototelegraph and Leased Circuits—Maintenance of the International Public Telephone Network—Maintenance of Maritime Satellite and Data Transmission Systems. 3 pp.

INDUSTRY STANDARDS

INTERNATIONAL AND NON-U.S. NATIONAL STANDARDS
SUBJECT INDEX

International

International Leased Circuits *(Cont.)*

—Glossaries
CCITT RECMN M.1010-89. Constitution and Nomenclature of International Leased Circuits—Maintenance of International Telegraph, Phototelegraph and Leased Circuits—Maintenance of the International Public Telephone Network—Maintenance of Maritime. 4 pp.

—Line Up—Exchange of Information
CCITT RECMN M.1045-89. Preliminary Exchange of Information for the Provision of International Leased Circuits—Maintenance of International Telegraph, Phototelegraph and Leased Circuits—Maintenance of the International Public Telephone Network-. 3 pp.

—Maintenance
CCITT RECMN M.1010-89. Constitution and Nomenclature of International Leased Circuits—Maintenance of International Telegraph, Phototelegraph and Leased Circuits—Maintenance of the International Public Telephone Network—Maintenance of Maritime. 4 pp.
CCITT RECMN M.1060-89. Maintenance of International Leased Circuits—Maintenance of International Telegraph, Phototelegraph and Leased Circuits—Maintenance of the International Public Telephone Network—Maintenance of Maritime Satellite and Data. 5 pp.

—Multiterminal—Line Up
CCITT RECMN M.1055-89. Lining up an International Multiterminal Leased Circuit—Maintenance of International Telegraph, Phototelegraph and Leased Circuits—Maintenance of the International Public Telephone Network—Maintenance of Maritime. 4 pp.

—Point to Point Transmission—Line Up
CCITT RECMN M.1050-89. Lining up an International Point-to-Point Leased Circuit—Maintenance of International Telegraph, Phototelegraph and Leased Circuits—Maintenance of the International Public Telephone Network—Maintenance of Maritime Satellite and Data. 10 pp.

—Post Telephone and Telegraph Administrations
CEPT T/TPH 6-86. Location Permanente Et Temporaire De Circuits Internationaux De Telecommunications Aux Membres De L'UER Et a Leur Utilisation Par Des Tiers Et Par Des Administrations Des Telecommunications. 4 pp.
CEPT T/TPH 6 E-90. Permanent and Occasional Lease of International Telecommunications Circuits to the EBU and Its Members and Their Use by Third Parties and Telecommunications Administrations. 3 pp.

—Private
CCITT RECMN D.1-91. General Principles for the Lease of International (Continental and Intercontinental) Private Telecommunication Circuits and Networks (Study Group III) 11 pp. 11 pp.

—Private—Accounting
CCITT RECMN D.1-91. General Principles for the Lease of International (Continental and Intercontinental) Private Telecommunication Circuits and Networks (Study Group III) 11 pp. 11 pp.
CCITT RECMN D.1-89. General Principles for the Lease of International (Continental and Intercontinental) Private Telecommunication Circuits—General Tariff Principles—Charging and Accounting in International Telecommunications. 8 pp.

—Private—Broadband Telecommunication Services
CCITT RECMN D.3-89. Special Conditions for the Lease of Intercontinental Telecommunication Circuits for Private Service—General Tariff Principles—Charging and Accounting in International Telecommunications Services (Study Group III) 4 pp. 4 pp.

—Private—Broadband Telecommunication Services—Tariffs
CCITT RECMN D.2-89. Special Conditions for the Lease of Continental Telecommunication Circuits for Private Service—General Tariff Principles—Charging and Accounting in International Telecommunications Services (Study Group III) 2 pp. 2 pp.
CCITT RECMN D.3-89. Special Conditions for the Lease of Intercontinental Telecommunication Circuits for Private Service—General Tariff Principles—Charging and Accounting in International Telecommunications Services (Study Group III) 4 pp. 4 pp.

International Leased Circuits *(Cont.)*

—Private—Data Processing
CCITT RECMN D.1-91. General Principles for the Lease of International (Continental and Intercontinental) Private Telecommunication Circuits and Networks (Study Group III) 11 pp. 11 pp.
CCITT RECMN D.1-89. General Principles for the Lease of International (Continental and Intercontinental) Private Telecommunication Circuits—General Tariff Principles—Charging and Accounting in International Telecommunications. 8 pp.

—Private—Diagrams
CCITT RECMN D.1-91. General Principles for the Lease of International (Continental and Intercontinental) Private Telecommunication Circuits and Networks (Study Group III) 11 pp. 11 pp.
CCITT RECMN D.1-89. General Principles for the Lease of International (Continental and Intercontinental) Private Telecommunication Circuits—General Tariff Principles—Charging and Accounting in International Telecommunications. 8 pp.

—Private—Digital Channels
CCITT RECMN D.3-92. Principles for the Lease of Analogue International Circuits for Private Service (Study Group III) 5 pp (Replaces Recmn D2 and D3). 5 pp.
CCITT RECMN D.3-89. Special Conditions for the Lease of Intercontinental Telecommunication Circuits for Private Service—General Tariff Principles—Charging and Accounting in International Telecommunications Services (Study Group III) 4 pp. 4 pp.

—Private—Digital Channels—Tariffs
CCITT RECMN D.3-92. Principles for the Lease of Analogue International Circuits for Private Service (Study Group III) 5 pp (Replaces Recmn D2 and D3). 5 pp.
CCITT RECMN D.3-89. Special Conditions for the Lease of Intercontinental Telecommunication Circuits for Private Service—General Tariff Principles—Charging and Accounting in International Telecommunications Services (Study Group III) 4 pp. 4 pp.

—Private—HF
CCITT RECMN D.3-89. Special Conditions for the Lease of Intercontinental Telecommunication Circuits for Private Service—General Tariff Principles—Charging and Accounting in International Telecommunications Services (Study Group III) 4 pp. 4 pp.

—Private—One Stop Shopping
CCITT RECMN D.7-92. Concept and Implementation of "One-Stop Shopping" for International Private Leased Telecommunication Circuits (Study Group III) 6 pp. 6 pp.
CEPT T/CAC 05 E-91. Service Specific Schedule for International Private Leased Circuits. 14 pp.

—Private—Public Networks
CCITT RECMN D.1-91. General Principles for the Lease of International (Continental and Intercontinental) Private Telecommunication Circuits and Networks (Study Group III) 11 pp. 11 pp.
CCITT RECMN D.1-89. General Principles for the Lease of International (Continental and Intercontinental) Private Telecommunication Circuits—General Tariff Principles—Charging and Accounting in International Telecommunications. 8 pp.

—Private—Public Telephone Networks
CCITT RECMN D.1-91. General Principles for the Lease of International (Continental and Intercontinental) Private Telecommunication Circuits and Networks (Study Group III) 11 pp. 11 pp.
CCITT RECMN D.1-89. General Principles for the Lease of International (Continental and Intercontinental) Private Telecommunication Circuits—General Tariff Principles—Charging and Accounting in International Telecommunications. 8 pp.

—Private—Sound-Programs
CCITT RECMN D.4-89. Special Conditions for the Lease of International (Continental and Intercontinental) Sound-and Television-Programme Circuits for Private Service—General Tariff Principles—Charging and Accounting in International. 5 pp.

—Private—Sound-Programs—Accounting
CCITT RECMN D.4 (REV 1)-92. Special Conditions for the Lease of International (Continental and Intercontinental) Sound-and Television-Programme Circuits for Private Service (Study Group III) 8 pp. 8 pp.

International Leased Circuits *(Cont.)*

—Private—Sound-Programs—Accounting (Cont.)
CCITT RECMN D.4-89. Special Conditions for the Lease of International (Continental and Intercontinental) Sound-and Television-Programme Circuits for Private Service—General Tariff Principles—Charging and Accounting in International. 5 pp.

—Private—Sound-Programs—Europe
CCITT RECMN D.310R-89. Determination of Rentals for the Lease of International Programme (Sound- and Television-) Circuits and Associated Control Circuits for Private Service in Relations Between Countries in Europe and the Mediterranean Basin—General Tariff Principles-. 2 pp.

—Private—Sound-Programs—Mediterranean Basin
CCITT RECMN D.310R-89. Determination of Rentals for the Lease of International Programme (Sound- and Television-) Circuits and Associated Control Circuits for Private Service in Relations Between Countries in Europe and the Mediterranean Basin—General Tariff Principles-. 2 pp.

—Private—Special Grade
CCITT RECMN D.3-89. Special Conditions for the Lease of Intercontinental Telecommunication Circuits for Private Service—General Tariff Principles—Charging and Accounting in International Telecommunications Services (Study Group III) 4 pp. 4 pp.

—Private Switched Networks
CCITT RECMN M.1030-89. Characteristics of Ordinary Quality International Leased Circuits Forming Part of Private Switched Telephone Networks—Maintenance of International Telegraph, Phototelegraph and Leased Circuits—Maintenance of the International Public. 4 pp.

—Private—Tariffs
CCITT RECMN D.1-91. General Principles for the Lease of International (Continental and Intercontinental) Private Telecommunication Circuits and Networks (Study Group III) 11 pp. 11 pp.
CCITT RECMN D.1-89. General Principles for the Lease of International (Continental and Intercontinental) Private Telecommunication Circuits—General Tariff Principles—Charging and Accounting in International Telecommunications. 8 pp.
CCITT RECMN D.2-89. Special Conditions for the Lease of Continental Telecommunication Circuits for Private Service—General Tariff Principles—Charging and Accounting in International Telecommunications Services (Study Group III) 2 pp. 2 pp.
CCITT RECMN D.3-89. Special Conditions for the Lease of Intercontinental Telecommunication Circuits for Private Service—General Tariff Principles—Charging and Accounting in International Telecommunications Services (Study Group III) 4 pp. 4 pp.

—Private—Telegraphy
CCITT RECMN D.3-89. Special Conditions for the Lease of Intercontinental Telecommunication Circuits for Private Service—General Tariff Principles—Charging and Accounting in International Telecommunications Services (Study Group III) 4 pp. 4 pp.

—Private—Telegraphy—Tariffs
CCITT RECMN D.2-89. Special Conditions for the Lease of Continental Telecommunication Circuits for Private Service—General Tariff Principles—Charging and Accounting in International Telecommunications Services (Study Group III) 2 pp. 2 pp.
CCITT RECMN D.3-89. Special Conditions for the Lease of Intercontinental Telecommunication Circuits for Private Service—General Tariff Principles—Charging and Accounting in International Telecommunications Services (Study Group III) 4 pp. 4 pp.

—Private—Television Transmission
CCITT RECMN D.4 (REV 1)-92. Special Conditions for the Lease of International (Continental and Intercontinental) Sound-and Television-Programme Circuits for Private Service (Study Group III) 8 pp. 8 pp.
CCITT RECMN D.4-89. Special Conditions for the Lease of International (Continental and Intercontinental) Sound-and Television-Programme Circuits for Private Service—General Tariff Principles—Charging and Accounting in International. 5 pp.

INTERNATIONAL AND NON-U.S. NATIONAL STANDARDS
SUBJECT INDEX
International

International Leased Circuits (Cont.)
—Private—Television Transmission—Accounting
CCITT RECMN D.4 (REV 1)-92. Special Conditions for the Lease of International (Continental and Intercontinental) Sound-and Television-Programme Circuits for Private Service (Study Group III) 8 pp. 8 pp.
CCITT RECMN D.4-89. Special Conditions for the Lease of International (Continental and Intercontinental) Sound-and Television-Programme Circuits for Private Service—General Tariff Principles—Charging and Accounting in International. 5 pp.

—Private—Television Transmission—Europe
CCITT RECMN D.310R-89. Determination of Rentals for the Lease of International Programme (Sound- and Television-) Circuits and Associated Control Circuits for Private Service in Relations Between Countries in Europe and the Mediterranean Basin—General Tariff Principles-. 2 pp.

—Private—Television Transmission—Mediterranean Basin
CCITT RECMN D.310R-89. Determination of Rentals for the Lease of International Programme (Sound- and Television-) Circuits and Associated Control Circuits for Private Service in Relations Between Countries in Europe and the Mediterranean Basin—General Tariff Principles-. 2 pp.

—Private—Telex Communications
CCITT RECMN D.1-91. General Principles for the Lease of International (Continental and Intercontinental) Private Telecommunication Circuits and Networks (Study Group III) 11 pp. 11 pp.
CCITT RECMN D.1-89. General Principles for the Lease of International (Continental and Intercontinental) Private Telecommunication Circuits—General Tariff Principles—Charging and Accounting in International Telecommunications. 8 pp.

—Reliability
CEPT T/TPH 41 E-88. Reliability of Analogue Leased Circuits. 3 pp.

—Service Avilability
CCITT RECMN M.1016-89. Assessment of the Service Availability Performance of International Leased Circuits—Maintenance of International Telegraph, Phototelegraph and Leased Circuits—Maintenance of the International Public Telephone Network—Maintenance of Maritime. 15 pp.

—Speech Transmission Systems
CCITT RECMN M.1015-89. Types of Transmission on Leased Circuits—Maintenance of International Telegraph, Phototelegraph and Leased Circuits—Maintenance of the International Public Telephone Network—Maintenance of Maritime Satellite and Data Transmission Systems. 3 pp.

—Subcontrol Stations
CCITT RECMN M.1013-89. Sub-Control Station for Leased and Special Circuits—Maintenance of International Telegraph, Phototelegraph and Leased Circuits—Maintenance of the International Public Telephone Network—Maintenance of Maritime Satellite and Data. 2 pp.

—Tariffs/Taxes
CEPT T/TPH 25 E-75. Determination of Charges Applicable to the Occasional Provision of Circuits for Sound-and Television-Programme Transmissions and Lease of Sound-Programme and Television Circuits and Associated Control Circuits for Private Service Between CEPT Member Countries. 1 p.

—Telecommunication Circuit Control Stations
CCITT RECMN M.1012-89. Circuit Control Station for Leased and Special Circuits—Maintenance of International Telegraph, Phototelegraph and Leased Circuits—Maintenance of the International Public Telephone Network—Maintenance of Maritime Satellite and Data. 2 pp.

—Telegraphy—Telephony
CCITT RECMN R.43-89. Simultaneous Communication by Telephone and Telegraph on a Telephone-Type Circuit—Telegraph Transmission (Study Group IX) 2 pp. 2 pp.

—Telephony
CCITT RECMN M.1040-89. Characteristics of Ordinary Quality International Leased Circuits—Maintenance of International Telegraph, Phototelegraph and Leased Circuits—Maintenance of the International Public Telephone Network—Maintenance of Maritime. 3 pp.

International Leased Circuits (Cont.)
—Television Transmission—Maintenance
CCITT RECMN N.73-89. Maintenance of Permanent International Television Circuits, Links and Connections—Maintenance of International Sound-Programme and Television Transmission Circuits (Study Group IV) 13 pp. 13 pp.

—Transmission Maintenance Point
CCITT RECMN M.1014-89. Transmission Maintenance Point (International Line) (TMP-IL)—Maintenance of International Telegraph, Phototelegraph and Leased Circuits—Maintenance of the International Public Telephone Network—Maintenance of Maritime Satellite and Data. 2 pp.

—Voice Frequency Telegraphy
CCITT RECMN M.1015-89. Types of Transmission on Leased Circuits—Maintenance of International Telegraph, Phototelegraph and Leased Circuits—Maintenance of the International Public Telephone Network—Maintenance of Maritime Satellite and Data Transmission Systems. 3 pp.

International Leased Group Links
Use: Group Links

International Leased Supergroup Links
Use: Supergroup Links

International Maritime Satellite System
Use: INMARSAT

International Numbers (Telephone)
See Also: Numbers

—FAX Communications—Notation
CCITT RECMN E.123-89. Notation for National and International Telephone Numbers—Telephone Network and ISDN—Operation, Numbering, Routing and Mobile Service (Study Group II) 5 pp. 5 pp.

—Integrated Services Digital Networks
CCITT RECMN I.330-89. ISDN Numbering and Addressing Principles—Integrated Services Digital Network (ISDN)—Overall Network Aspects and Functions, ISDN User-Network Interfaces (Study Group XVIII) 8 pp. 8 pp.

—Notation
CCITT RECMN E.123-89. Notation for National and International Telephone Numbers—Telephone Network and ISDN—Operation, Numbering, Routing and Mobile Service (Study Group II) 5 pp. 5 pp.

—Symbols
CCITT RECMN E.123-89. Notation for National and International Telephone Numbers—Telephone Network and ISDN—Operation, Numbering, Routing and Mobile Service (Study Group II) 5 pp. 5 pp.

International Organization for Standardization
Use: ISO

International Paging Systems
Use: Radio Paging Systems

International Phototelegraph Positions
Use For: IPP *See Also:* Phototelegraphy; Telecommunication Administrations
CCITT RECMN E.323-89. Rules for Phototelegraph Communications Set up Over Circuits Normally Used for Telephone Traffic—Telephone Network and ISDN—Operation, Numbering, Routing and Mobile Service (Study Group II) 1 pp (Same as Recmn F.82 and F.107). 1 p.
CCITT RECMN F.82-91. Operational Provisions to Permit Interworking Between the International Telex Service and the Intex Service (Study Group I) 5 pp. 5 pp.
CCITT RECMN F.85-89. Message Handling Service; Intercommunication Between the IPM Service and the Telex Service—Telegraph and Mobile Services Operations and Quality of Service (Study Group I) 4 pp (Same as Recmn F.421) (Renumbered from Recmn F.75). 4 pp.
CCITT RECMN F.107-89. Rules for Phototelegraph Calls Established over Circuits Normally Used for Telephone Traffic—Telegraph and Mobile Services Operations and Quality of Service (Study Group I) 4 pp (Renumbered from Recmn F.82) (Same as Recmn E.323). 4 pp.

International Prefixes
Use For: Prefixes, International

International Prefixes (Cont.)
See Also: Digits (Telecommunications); Prefixes (Communications)
CCITT RECMN Q.103-89. Numbering Used—General Recommendations on Telephone Switching and Signalling—Functions and Information Flows for Services in the ISDN—Supplements (Study Group XI) 1 pp. 1 p.

—Automatic Services
CCITT RECMN E.122-89. Measures to Reduce Customer Difficulties in the International Telephone Service—Telephone Network and ISDN—Operation, Numbering, Routing and Mobile Service (Study Group II) 2 pp. 2 pp.

—Numbering Plans—Telephone Services
CCITT RECMN E.163-89. Numbering Plan for the International Telephone Service—Telephone Network and ISDN—Operation, Numbering, Routing and Mobile Service (Study Group II) 8 pp. 8 pp.

—Recorded Information Services (Telephone)
CCITT RECMN E.122-89. Measures to Reduce Customer Difficulties in the International Telephone Service—Telephone Network and ISDN—Operation, Numbering, Routing and Mobile Service (Study Group II) 2 pp. 2 pp.

International Public Facsimile Services
Use: FAX Communications

International Public Mobile Satellite Systems
Use: Mobile Satellite Communications

International Public Telemessage Services
Use: Telemessage Services

International Public Telephone Networks
Use: Public Telephone Networks

International Radio Consultative Committee
Use: CCIR

International Reference Alphabet
Use For: IA5; International Alphabet No. 5; IRA
See Also: Character Sets; Data Communication; Data Transmission; International Telegraph Alphabet No. 5; Telegraphy

—Coded Character Sets
CCITT RECMN T.50-92. International Reference Alphabet (IRA) (Formerly International Alphabet No. 5 or IA5) Information Technology—7-Bit Coded Character Set for Information Interchange (Study Group VIII) 22 pp. 22 pp.
CCITT RECMN T.50-89. International Alphabet No. 5—Terminal Equipment and Protocols for Telematic Services (Study Group VIII) 16 pp. 16 pp.

—Data Transmission—Public Data Networks
CCITT RECMN X.4-89. General Structure of Signals of International Alphabet No. 5 Code for Character Oriented Data Transmission over Public Data Networks—Data Communication Networks: Services and Facilities, Interfaces (Study Group VII) 2 pp. 2 pp.

—Data Transmission—Public Telephone Networks
CCITT RECMN V.4-89. General Structure of Signals of International Alphabet No. 5 Code for Character Oriented Data Transmission over Public Telephone Networks—Data Communication over the Telephone Network (Study Group XVII) 2 pp. 2 pp.

—Intex Services
CCITT RECMN F.150-91. Service and Operational Provisions for the Intex Service (Study Group I) 7 pp. 7 pp.

International Rubber Hardness Degrees
See Also: Rubber

—Rollers
BSI BS 7442: Sec 3.1-91. 1991 Rubber or Plastics Covered Rollers Part 3: Methods of Test Section 3.1: Determination of Apparent Hardness (IRHD Method) (ISO 7267/1: 1986). 7 pp.

INDUSTRY STANDARDS

INTERNATIONAL AND NON-U.S. NATIONAL STANDARDS
SUBJECT INDEX

International

International Rubber Hardness Degrees *(Cont.)*
—Rollers *(Cont.)*
BSI BS 7442: Sec 4.1-92. 1992 Rubber or Plastics Covered Rollers Part 4: Requirements Section 4.1: Specification for Apparent Hardness. 8 pp.
ISO 7267 Pt 1-86. Rubber-Covered Rollers—Determination of Apparent Hardness—Part 1: IRHD Method First Edition. 4 pp.

—Vulcanized Rubber
DIN ENGL 53519 Pt 1-72. Testing of Elastomers; Determination of Indentation Hardness of Soft Rubber (IRHD); Hardness Testing on Standard Specimens (May). 4 pp.
DIN ENGL 53519 Pt 2-72. Testing of Elastomers; Determination of Indentation Hardness of Soft Rubber (IRHD); Hardness Testing on Specimens of Small Dimensions; Micro-Testing (May). 2 pp.

International Securities Identification Numbering System
Use For: ISIN *See Also:* Codes; Securities
BSI BS 6602-85. (WITHDRAWN) 1985 International Securities Identification Numbering System (ISIN) (Superseded by BS 7097: 1990). 12 pp.
BSI BS 7097-90. 1990 Procedure for Allocating an International Securities Identification Number (ISIN). 17 pp.
BSI BS 7103-89. 1989 Method for Numbering Securities Certificates. 4 pp.
ISO 6166-87. Securities—International Securities Identification Numbering System (ISIN) Fourth Edition. 14 pp.
ISO 9019-87. Securities—Numbering of Certificates First Edition. 4 pp.

—Data Transmission
BSI BS 6636: Part 2-87. (WITHDRAWN) 1987 Message Types for Securities Part 2: Messages for Orders to Buy/Sell (Superseded by BS ISO 7775: 1991). 19 pp.
BSI BS 6822-87. 1987 Format for Transmission of Certificate Numbers of Securities. 7 pp.
ISO 8532-86. Securities—Format for Transmission of Certificate Numbers First Edition. 6 pp.
JTC1 8532-86. Securities—Format for Transmission of Certificate Numbers First Edition. 6 pp.
OSI ISO 7775-2-86. Securities—Scheme for Message Types—Part 2: Order to Buy-Sell. 17 pp.
OSI ISO 8532-86. Securities—Format for Transmission of Certificate Numbers. 6 pp.

International Signaling Point Codes
See Also: Codes; Signaling Systems

—Numbering—CCITT No. 7 Signaling Systems
CCITT RECMN Q.708-89. Numbering of International Signalling Point Codes—Specifications of Signalling System No. 7 (Study Group XI) 7 pp. 7 pp.

International Sound-Program Centers
Use For: ISPC *See Also:* International Switching Centers; Sound-Program Circuits; Sound-Program Connections; Sound-Program Links

—Directories
CCITT RECMN D.180-89. Occasional Provision of Circuits for International Sound-and Television-Programme Transmissions—General Tariff Principles—Charging and Accounting in International Telecommunications Services (Study Group III) 13 pp. 13 pp.

—Monitoring—Sound Transmission
CCITT RECMN N.17-89. Monitoring the Transmission—Maintenance of International Sound-Programme and Television Transmission Circuits (Study Group IV) 1 pp. 1 p.

—Sound Transmission
CCITT RECMN N.11-89. Essential Transmission Performance Objectives for International Sound-Programme Centres (ISPC) —Maintenance of International Sound-Programme and Television Transmission Circuits (Study Group IV) 2 pp. 2 pp.

International Sound Program Exchange
Use: Sound-Program Exchange

International Sound Program Transmission
Use: Sound Transmission

International Special Committee on Radio Interference
Use: CISPR

International Standard Book Numbering
See Also: Books
ISO 2108-92. Information and Documentation—International Standard Book Numbering (ISBN) Third Edition. 6 pp.
JIS X 0305-88. International Standard Book Numbering (ISO 2108:1978).
SAA AS 1519-82. International Standard Book Numbering (ISBN). 8 pp.

International Standard Recording Code
Use For: ISRC *See Also:* Codes
ISO 3901-86. Documentation—International Standard Recording Code (ISRC) First Edition. 6 pp.

International Standard Serial Numbering
Use For: ISSN *See Also:* Periodicals
ISO 3297-86. Documentation—International Standard Serial Numbering (ISSN) Second Edition. 6 pp.
JIS X 0306-88. International Standard Serial Numbering (ISO 3207:1986).

International Standards
Use: Standards

International Switching Centers
Use For: ISC *See Also:* International Sound-Program Centers; International Television Program Centers; Telephone Exchanges

—Circuit Multiplication Equipment—Signaling Systems
CCITT RECMN Q.50-89. Signalling Between Circuit Multiplication Equipments (CME) and International Switching Centres (ISC)—General Recommendations on Telephone Switching and Signalling—Functions and Information Flows for Services in the ISDN—Supplements. 17 pp.

—Completion Ratio
CCITT RECMN E.426-92. General Guide to the Percentage of Effective Attempts Which Should Be Observed for International Telephone Calls—Telephone Network and ISDN Quality of Service, Network Management and Traffic Engineering (Study Group II) 3 pp. 3 pp.

—Echo Cancellors
CCITT RECMN Q.115-89. Control of Echo Suppressors and Echo Cancellers by International Switching Centres—General Recommendations on Telephone Switching and Signalling—Functions and Information Flows for Services in the ISDN —Supplements (Study Group XI) 7 pp. 7 pp.

—Echo Suppressors
CCITT RECMN Q.115-89. Control of Echo Suppressors and Echo Cancellers by International Switching Centres—General Recommendations on Telephone Switching and Signalling—Functions and Information Flows for Services in the ISDN —Supplements (Study Group XI) 7 pp. 7 pp.

—Incoming Test Facilities—Exchange of Information
CCITT RECMN M.734-89. Exchange of Information on Incoming Test Facilities at International Switching Centres—General Maintenance Principles—Maintenance of International Transmission Systems and Telephone Circuits (Study Group IV) 3 pp. 3 pp.
CCITT RECMN Q.327-89. Testing Arrangements-General Arrangements—Specifications of Signalling Systems R1 and R2 (Study Group XI) 1 pp. 1 p.

—Routing
CCITT RECMN E.171-89. International Telephone Routing Plan—Telephone Network and ISDN—Operation, Numbering, Routing and Mobile Service (Study Group II) 10 pp. 10 pp.

—Satellite Communications—Routing
CCITT RECMN E.171-89. International Telephone Routing Plan—Telephone Network and ISDN—Operation, Numbering, Routing and Mobile Service (Study Group II) 10 pp. 10 pp.

—Virtual Analog Switching Points
CCITT RECMN G.101-89. The Transmission Plan—General Characteristics of International Telephone Connections and Circuits (Study Groups XII and XV) 17 pp. 17 pp.

International System of Units
Use For: SI System *See Also:* Metric System; Units of Measurement
BSI BS 5555-93. 1993 SI Units and Recommendations for the Use of Their Multiples and of Certain Other Units (ISO 1000: 1992) (G). 28 pp.

International System of Units *(Cont.)*
BSI BS 5555-81. 1981 Amd 1 SI Units and Recommendations for the Use of Their Multiples and of Certain Other Units. 20 pp.
BSI BS 5775: Part 0-93. 1993 Specification for Quantities, Units and Symbols Part 0: General Principles (ISO 31-0: 1992) (G). 27 pp.
BSI BS 5775: Part 0-82. 1982 Amd 1 Specifications for Quantities, Units and Symbols Part 0: General Principles (AMD 5845) July 31, 1989. 17 pp.
CNS Z7170-84. International System of Units (SI) (Aug)(10987).
CNS Z7196-85. General Principles Concerning Quantities, Units and Symbols (Jun)(11296).
ISO 31 Pt 0-92. Quantities and Units—Part 0: General Principles Third Edition. 25 pp.
ISO 1000-92. SI Units and Recommendations for the Use of Their Multiples and of Certain Other Units Third Edition. 26 pp.
JIS Z 8203-85. SI Units and the Use of Their Multiples and of Certain Other Units. 29 pp.
SAA AS 1000-79. The International System of Units (SI) and Its Application. 24 pp.
SNZ NZS 6501-82. Units of Measurement. 258 pp.

—Calculations—Guides
SAA AS MP43-79. Guide for Calculations Using SI Units. 10 pp.

—Data Processing—Character Sets
BSI BS 6430-83. 1983 Representing SI and Other Units in Information Processing Systems with Limited Character Sets. 8 pp.
ISO 2955-83. Information Processing—Representation of SI and Other Units in Systems with Limited Character Sets Second Edition. 7 pp.
JTC1 2955-83. Information Processing—Representation of SI and Other Units in Systems with Limited Character Sets Second Edition. 7 pp.
OSI ISO 2955-83. Information Processing—Representation of SI and Other Unit in Systems with Limited Character Sets. 7 pp.
SAA AS 1340-75. Symbols for SI Units for Systems with Limited Character Set. 11 pp.

—Dosimetry—Military Operations
NATO STANAG 2957 ED 2 AMD 0-88. International System (SI) Units Used by the Armed Forces in the Nuclear Field. 7 pp.
NATO STANAG 2957 ED 2 AMD 1-88. International System (SI) Units Used by the Armed Forces in the Nuclear Field. 8 pp.

—Frequency Bands
CCITT RECMN B.15-89. Nomenclature of the Frequency and Wavelength Bands Used in Telecommunications—Terms and Definitions Abbreviations and Acronyms Recommendations on Means of Expression (Series B) General Telecommunications Statistics (Series C) 3 pp. 3 pp.

—Keyboards—Character Sets
BSI PD 6462-72. 1972 Modification of Keyboards to Include Symbols for SI Units. 4 pp.

—Marine
BSI PD 6430/1-70. 1970 Is an Extract from PD 6430 Units of the SI System, Their Application and Use in the Marine Industry. 13 pp.

—Nuclear, Biological, and Chemical Warfare
NATO STANAG 2957 ED 2 AMD 0-88. International System (SI) Units Used by the Armed Forces in the Nuclear Field. 7 pp.
NATO STANAG 2957 ED 2 AMD 1-88. International System (SI) Units Used by the Armed Forces in the Nuclear Field. 8 pp.

—Paper Products
BSI BS 5916-89. 1989 Units for Expressing Properties of Paper, Board and Pulps. 13 pp.

—Tables (Data)
BSI BS 350: Pt 2: Supp 1-67. 1967 Amd 2 Conversion Factors and Tables Part 2: Supplement 1: Additional Tables for SI Conversions (Supersedes BS 1359: 1947). 89 pp.

—Telecommunication Equipment
CCIR RECMN 430-3-90. Use of the International System of Units (SI)—Section C—Other Means of Expression. 1 p.

—Telecommunication Services
CCIR RECMN 430-3-90. Use of the International System of Units (SI)—Section C—Other Means of Expression. 1 p.

—Telecommunication Systems
CCIR RECMN 430-3-90. Use of the International System of Units (SI)—Section C—Other Means of Expression. 1 p.

INTERNATIONAL AND NON-U.S. NATIONAL STANDARDS
SUBJECT INDEX
Interpersonal

International System of Units *(Cont.)*
—Telecommunication Systems *(Cont.)*
CCITT RECMN B.3-89. Use of the International System of Units (SI)—Terms and Definitions Abbreviations and Acronyms Recommendations on Means of Expression (Series B) General Telecommunications Statistics (Series C) 2 pp. 2 pp.

—Wavelength
CCITT RECMN B.15-89. Nomenclature of the Frequency and Wavelength Bands Used in Telecommunications—Terms and Definitions Abbreviations and Acronyms Recommendations on Means of Expression (Series B) General Telecommunications Statistics (Series C) 3 pp. 3 pp.

International Telecommunication Charge Card Services
Use: Credit Card Calling Services

International Telecommunications Union
Use: ITU

International Telegraph Alphabet No. 2
Use For: IA2 *See Also:* Character Sets; Codes; Telegraphy

CCITT RECMN F.1-89. Operational Provisions for the International Public Telegram Service—Telegraph and Mobile Services Operations and Quality of Service (Study Group I) 54 pp. 54 pp.

—Conversion—No. 5
CCITT RECMN S.18-89. Conversion Between International Telegraph Alphabet No. 2 and International Alphabet No. 5—Telegraph Services Terminal Equipment (Study Group IX) 6 pp. 6 pp.

—Maritime Mobile Services
CCIR RECMN 490-74. Introduction of Direct-Printing Telegraph Equipment in the Maritime Mobile Service Equivalence of Terms—Section 8B—Maritime Mobile Service; Telegraphy and Related Subjects. 1 p.

—Regenerative Repeaters
CCITT RECMN R.60-89. Conditions to be Fulfilled by Regenerative Repeaters for Start-Stop Signals of International Telegraph Alphabet No. 2—Telegraph Transmission (Study Group IX) 1 pp. 1 p.

—Symbols
CCITT RECMN S.4-89. Special Use of Certain Characters of the International Telegraph Alphabet No. 2—Telegraph Services Terminal Equipment—(Study Group IX) 3 pp. 3 pp.

—Symbols—Coding
CCITT RECMN S.1-89. International Telegraph Alphabet No. 2—Telegraph Services Terminal Equipment (Study Group IX) 4 pp. 4 pp.

CCITT RECMN S.2-89. Coding Scheme Using International Telegraph Alphabet No. 2 (ITA2) to Allow the Transmission of Capital and Small Letters—Telegraph Services Terminal Equipment (Study Group IX) 4 pp. 4 pp.

—Symbols—Telegraph Channels
CCITT RECMN S.10-89. Transmission at Reduced Character Transfer Rate over a Standardized 50-Baud Telegraph Channel—Telegraph Services Terminal Equipment (Study Group IX) 1 pp. 1 p.

—Telegraph Equipment—Message Switching
CCITT RECMN F.35-89. Provisions Applying to the Operation of an International Public Automatic Message Switching Service for Equipments Utilizing the International Telegraph Alphabet No. 2—Telegraph and Mobile Services Operations and Quality of Service (Study Group I) 7 pp. 7 pp.

International Telegraph Alphabet No. 5
See Also: Character Sets; Codes; International Reference Alphabet; Telegraphy

—Communication Terminal Equipment—Interpersonal Messaging Services
CCITT RECMN F.420-92. Message Handling Services: Public Interpersonal Messaging Service (Study Group I) 16 pp. 16 pp.

CCITT RECMN F.420-89. Message Handling Services: the Public Interpersonal Messaging Service—Message Handling and Directory Services—Operations and Definition of Service (Study Group I) 15 pp. 15 pp.

International Telegraph Alphabet No. 5 *(Cont.)*
—Conversion—No. 2
CCITT RECMN S.18-89. Conversion Between International Telegraph Alphabet No. 2 and International Alphabet No. 5—Telegraph Services Terminal Equipment (Study Group IX) 6 pp. 6 pp.

—Data Terminal Equipment—Start-Stop
CCITT RECMN S.31-89. Transmission Characteristics for Start-Stop Data Terminal Equipment Using International Alphabet No. 5—Telegraph Services Terminal Equipment (Study Group IX) 2 pp. 2 pp.

—Page Printers
CCITT RECMN S.30-89. Standardization of Basic Model Page-Printing Machine Using International Alphabet No. 5—Telegraph Services Terminal Equipment (Study Group IX) 1 pp. 1 p.

International Telegraph and Telephone Consultative Committee
Use: CCITT

International Telegraph Circuits
Use: Telegraph Circuits

International Telephone Circuits
Use: Telephone Circuits

International Telephone Networks
Use: Telephone Networks

International Telephone Services
Use: Telephone Services

International Television Centers
Use For: ITC *See Also:* Television Links

—Television Transmission
CCITT RECMN N.55-89. Organization, Responsibilities and Functions of Control and Sub-Control ITCs and Control and Sub-Control Stations for International Television Connections, Links, Circuits and Circuit Sections—Maintenance of International Sound-. 6 pp.

International Television Connections
Use: Television Connections

International Television Links
Use: Television Links

International Television Program Centers
Use For: ITPC *See Also:* International Switching Centers

—Directories
CCITT RECMN D.180-89. Occasional Provision of Circuits for International Sound-and Television-Programme Transmissions—General Tariff Principles—Charging and Accounting in International Telecommunications Services (Study Group III) 13 pp. 13 pp.

International Television Program Exchange
Use: Television Program Exchange

International Telex Communications
Use: Telex Communications

International Telex Store and Forward
Use: Store and Forward Mode—Telex Communications

International Thermonuclear Experimental Reactor
See Also: Nuclear Reactors

—Engineering Drawings
EC 92/439/EURATOM-92. Commission Decision Concerning the Conclusion of an Agreement Between the European Atomic Energy Community, the Government of Japan, the Government of the Russian Federation, and the Government of the United States of America on Cooperation in the. 18 pp.

International Time Division Multiplexers
Use: Time Division Multiplexers

International Trade Documents
Use: Trade Documents

International Transmission Systems
Use: Transmission Systems

Internetwork Interfaces
Use: Network Interfaces

Internetwork Management Information
Use: Network Management Systems

Interpersonal Messaging Services
See Also: Interpersonal Messaging Systems; Message Handling Systems; Message Transfer Systems

CCITT RECMN F.400-92. Message Handling Services: Message Handling System and Service Overview (Study Group I) 82 pp (Same as Recmn X.400). 82 pp.

CCITT RECMN F.400-89. Message Handling System and Service Overview—Message Handling and Directory Services—Operations and Definition of Service (Study Group I) 73 pp. 73 pp.

CCITT RECMN F.420-92. Message Handling Services: Public Interpersonal Messaging Service (Study Group I) 16 pp. 16 pp.

CCITT RECMN F.420-89. Message Handling Services: the Public Interpersonal Messaging Service—Message Handling and Directory Services—Operations and Definition of Service (Study Group I) 15 pp. 15 pp.

CCITT RECMN X.400-93. Message Handling Services: Message Handling System and Service Overview (Study Group VII) 82 pp (Same as Recmn F.400). 82 pp.

—Abbreviations
CCITT RECMN F.420-92. Message Handling Services: Public Interpersonal Messaging Service (Study Group I) 16 pp. 16 pp.

CCITT RECMN F.420-89. Message Handling Services: the Public Interpersonal Messaging Service—Message Handling and Directory Services—Operations and Definition of Service (Study Group I) 15 pp. 15 pp.

CCITT RECMN F.421-89. Message Handling Services: Intercommunication Between the IPM Service and the Telex Service—Message Handling and Directory Services—Operations and Definition of Service (Study Group I) 12 pp (Same as Recmn F.85). 12 pp.

CCITT RECMN F.422-89. Message Handling Services: Intercommunication Between the IPM Service and the Teletex Service—Message Handling and Directory Services—Operations and Definition of Service (Study Group I) 6 pp. 6 pp.

—Accounting
CCITT RECMN D.35-89. General Charging and Accounting Principles in the International Public Interpersonal Messaging (IPM) Service—General Tariff Principles—Charging and Accounting in International Telecommunications Services (Study Group III) 1 pp. 1 p.

—Administration Management Domain
CCITT RECMN F.420-92. Message Handling Services: Public Interpersonal Messaging Service (Study Group I) 16 pp. 16 pp.

CCITT RECMN F.420-89. Message Handling Services: the Public Interpersonal Messaging Service—Message Handling and Directory Services—Operations and Definition of Service (Study Group I) 15 pp. 15 pp.

—Directory Services
CCITT RECMN F.420-92. Message Handling Services: Public Interpersonal Messaging Service (Study Group I) 16 pp. 16 pp.

CCITT RECMN F.420-89. Message Handling Services: the Public Interpersonal Messaging Service—Message Handling and Directory Services—Operations and Definition of Service (Study Group I) 15 pp. 15 pp.

—Distribution Lists
CCITT RECMN F.420-92. Message Handling Services: Public Interpersonal Messaging Service (Study Group I) 16 pp. 16 pp.

CCITT RECMN F.420-89. Message Handling Services: the Public Interpersonal Messaging Service—Message Handling and Directory Services—Operations and Definition of Service (Study Group I) 15 pp. 15 pp.

—Encoded Information Types
CCITT RECMN F.420-92. Message Handling Services: Public Interpersonal Messaging Service (Study Group I) 16 pp. 16 pp.

CCITT RECMN F.420-89. Message Handling Services: the Public Interpersonal Messaging Service—Message Handling and Directory Services—Operations and Definition of Service (Study Group I) 15 pp. 15 pp.

—FAX Communications—Intercommunication
CCITT RECMN F.423-92. Message Handling Services: Intercommunication Between the Interpersonal Messaging Service and the Telefax Service (Study Group I) 8 pp. 8 pp.

INTERNATIONAL AND NON-U.S. NATIONAL STANDARDS
SUBJECT INDEX

Interpersonal

Interpersonal Messaging Services (Cont.)

—**FAX Machines—CCITT Group 3**
CCITT RECMN F.420-92. Message Handling Services: Public Interpersonal Messaging Service (Study Group I) 16 pp. 16 pp.
CCITT RECMN F.420-89. Message Handling Services: the Public Interpersonal Messaging Service—Message Handling and Directory Services—Operations and Definition of Service (Study Group I) 15 pp. 15 pp.

—**FAX Machines—CCITT Group 4**
CCITT RECMN F.420-92. Message Handling Services: Public Interpersonal Messaging Service (Study Group I) 16 pp. 16 pp.
CCITT RECMN F.420-89. Message Handling Services: the Public Interpersonal Messaging Service—Message Handling and Directory Services—Operations and Definition of Service (Study Group I) 15 pp. 15 pp.

—**IA5—Communication Terminal Equipment**
CCITT RECMN F.420-92. Message Handling Services: Public Interpersonal Messaging Service (Study Group I) 16 pp. 16 pp.
CCITT RECMN F.420-89. Message Handling Services: the Public Interpersonal Messaging Service—Message Handling and Directory Services—Operations and Definition of Service (Study Group I) 15 pp. 15 pp.

—**Information Security**
CCITT RECMN F.420-92. Message Handling Services: Public Interpersonal Messaging Service (Study Group I) 16 pp. 16 pp.
CCITT RECMN F.420-89. Message Handling Services: the Public Interpersonal Messaging Service—Message Handling and Directory Services—Operations and Definition of Service (Study Group I) 15 pp. 15 pp.

—**Message Handling Systems**
CCITT RECMN F.420-92. Message Handling Services: Public Interpersonal Messaging Service (Study Group I) 16 pp. 16 pp.
CCITT RECMN F.420-89. Message Handling Services: the Public Interpersonal Messaging Service—Message Handling and Directory Services—Operations and Definition of Service (Study Group I) 15 pp. 15 pp.

—**Message Headers**
CCITT RECMN F.415-89. Message Handling Services: Intercommunication with Public Physical Delivery Services—Message Handling and Directory Services—Operations and Definition of Service (Study Group I) 15 pp (Erratum in Recmn F.410). 15 pp.

—**Message Stores**
CCITT RECMN F.420-92. Message Handling Services: Public Interpersonal Messaging Service (Study Group I) 16 pp. 16 pp.
CCITT RECMN F.420-89. Message Handling Services: the Public Interpersonal Messaging Service—Message Handling and Directory Services—Operations and Definition of Service (Study Group I) 15 pp. 15 pp.

—**Originator/Recipient Addresses**
CCITT RECMN F.420-92. Message Handling Services: Public Interpersonal Messaging Service (Study Group I) 16 pp. 16 pp.
CCITT RECMN F.420-89. Message Handling Services: the Public Interpersonal Messaging Service—Message Handling and Directory Services—Operations and Definition of Service (Study Group I) 15 pp. 15 pp.

—**Originator/Recipient Names**
CCITT RECMN F.420-92. Message Handling Services: Public Interpersonal Messaging Service (Study Group I) 16 pp. 16 pp.
CCITT RECMN F.420-89. Message Handling Services: the Public Interpersonal Messaging Service—Message Handling and Directory Services—Operations and Definition of Service (Study Group I) 15 pp. 15 pp.

—**Public Networks**
CCITT RECMN F.420-92. Message Handling Services: Public Interpersonal Messaging Service (Study Group I) 16 pp. 16 pp.
CCITT RECMN F.420-89. Message Handling Services: the Public Interpersonal Messaging Service—Message Handling and Directory Services—Operations and Definition of Service (Study Group I) 15 pp. 15 pp.

—**Quality of Service**
CCITT RECMN F.420-92. Message Handling Services: Public Interpersonal Messaging Service (Study Group I) 16 pp. 16 pp.
CCITT RECMN F.420-89. Message Handling Services: the Public Interpersonal Messaging Service—Message Handling and Directory Services—Operations and Definition of Service (Study Group I) 15 pp. 15 pp.
CCITT RECMN F.422-89. Message Handling Services: Intercommunication Between the IPM Service and the Teletex Service—Message Handling and Directory Services—Operations and Definition of Service (Study Group I) 6 pp. 6 pp.

Interpersonal Messaging Services (Cont.)

—**Tariffs**
CCITT RECMN D.35-89. General Charging and Accounting Principles in the International Public Interpersonal Messaging (IPM) Service—General Tariff Principles—Charging and Accounting in International Telecommunications Services (Study Group III) 1 pp. 1 pp.

—**Teletex Communications**
CCITT RECMN F.422-89. Message Handling Services: Intercommunication Between the IPM Service and the Teletex Service—Message Handling and Directory Services—Operations and Definition of Service (Study Group I) 6 pp. 6 pp.

—**Teletex Equipment**
CCITT RECMN F.420-92. Message Handling Services: Public Interpersonal Messaging Service (Study Group I) 16 pp. 16 pp.
CCITT RECMN F.420-89. Message Handling Services: the Public Interpersonal Messaging Service—Message Handling and Directory Services—Operations and Definition of Service (Study Group I) 15 pp. 15 pp.

—**Telex Communications**
CCITT RECMN F.421-89. Message Handling Services: Intercommunication Between the IPM Service and the Telex Service—Message Handling and Directory Services—Operations and Definition of Service (Study Group I) 12 pp (Same as Recmn F.85). 12 pp.
CCITT RECMN U.204-89. Interworking Between the Telex Service and the Public Interpersonal Messaging Service—Telegraph Switching (Study Group IX) 22 pp. 22 pp.

—**Telex Communications—Store and Forward Mode**
CCITT RECMN F.421-89. Message Handling Services: Intercommunication Between the IPM Service and the Telex Service—Message Handling and Directory Services—Operations and Definition of Service (Study Group I) 12 pp (Same as Recmn F.85). 12 pp.

—**Telex Equipment**
CCITT RECMN F.420-92. Message Handling Services: Public Interpersonal Messaging Service (Study Group I) 16 pp. 16 pp.
CCITT RECMN F.420-89. Message Handling Services: the Public Interpersonal Messaging Service—Message Handling and Directory Services—Operations and Definition of Service (Study Group I) 15 pp. 15 pp.

—**Videotex Equipment**
CCITT RECMN F.420-92. Message Handling Services: Public Interpersonal Messaging Service (Study Group I) 16 pp. 16 pp.
CCITT RECMN F.420-89. Message Handling Services: the Public Interpersonal Messaging Service—Message Handling and Directory Services—Operations and Definition of Service (Study Group I) 15 pp. 15 pp.

Interpersonal Messaging Systems
Use For: IPMS *See Also:* Encoded Information Types; Interpersonal Messaging Services; Message Oriented Text Interchange Systems

—**Abstract Service—Message Handling Systems**
CCITT RECMN X.420-89. Message Handling Systems: Interpersonal Messaging System—Data Communication Networks—Message Handling Systems (Study Group VII) 86 pp. 86 pp.

—**Open Systems Interconnection**
BSI BS ISO/IEC 10021-7-90. 1990 Amd 1 Information Technology—Text Communication—Message-Oriented Text Interchange Systems (MOTIS)—Part 7: Interpersonal Messaging System (AMD 6851) September 30, 1991. 122 pp.
BSI BS ISO/IEC 10021-7-02. 1990 Amd 2 Information Technology—Text Communication—Message-Oriented Text Interchange Systems (MOTIS)—Part 7: Interpersonal Messaging System (AMD 7144) July 15, 1992 (Technical Corr 2) (S). 130 pp.
BSI BS ISO/IEC 10021-7-03. 1990 Amd 3 Information Technology—Text Communication—Message-Oriented Text Interchange Systems (MOTIS)—Part 7: Interpersonal Messaging System (AMD 7277) August 15, 1992 (Technical Corr 3) (S). 142 pp.
BSI BS ISO/IEC 10021-7-04. 1990 Amd 4 Information Technology—Text Communication—Message-Oriented Text Interchange Systems (MOTIS)—Part 7: Interpersonal Messaging System (AMD 7545) December 15, 1992 (Technical Corr 4) (S). 148 pp.

Interpersonal Messaging Systems (Cont.)

—**Open Systems Interconnection** (Cont.)
BSI BS ISO/IEC 10021-7-05. 1990 Amd 5 Information Technology—Text Communication—Message-Oriented Text Interchange Systems (MOTIS)—Part 7: Interpersonal Messaging System (AMD 7686) March 15, 1993 (Technical Corr 5) (S). 152 pp.
IEC 10021 Pt 7 Draft AMD 1. Information Technology—Text Communication—Message-Oriented Text Interchange Systems (MOTIS)—Part 7: Interpersonal Messaging System Amendment 1: Minor Enhancements; (1992) ***CD-ROM ONLY***. 11 pp.
IEC 10021 Pt 7 Draft AMD1.2. Information Technology—Text Communication—Message-Oriented Text Interchange Systems (MOTIS)—Part 7: Interpersonal Messaging System Amendment 1: Minor Enhancements; (1993) ***CD-ROM ONLY***. 16 pp.
ISO 10021 Pt 7-90. Information Technology—Text Communication—Message-Oriented Text Interchange Systems (MOTIS)—Part 7: Interpersonal Messaging System First Edition; (Corrigendum 1-2:1991) (Corrigendum 3-5:1992) (Amendment 3-1993). 79 pp.
ISO 10021 Pt 7 Draft AMD 1. Information Technology—Text Communication—Message-Oriented Text Interchange Systems (MOTIS)—Part 7: Interpersonal Messaging System Amendment 1: Minor Enhancements; (1992) ***CD-ROM ONLY***. 11 pp.
ISO 10021 Pt 7 Draft AMD1.2. Information Technology—Text Communication—Message-Oriented Text Interchange Systems (MOTIS)—Part 7: Interpersonal Messaging System Amendment 1: Minor Enhancements; (1993) ***CD-ROM ONLY***. 16 pp.
JTC1 10021 Pt7 Draft AMD 1. Information Technology—Text Communication—Message-Oriented Text Interchange Systems (MOTIS)—Part 7: Interpersonal Messaging System Amendment 1: Minor Enhancements; (1992) ***CD-ROM ONLY***. 11 pp.
JTC1 10021 Pt7 Draft AMD1.2. Information Technology—Text Communication—Message-Oriented Text Interchange Systems (MOTIS)—Part 7: Interpersonal Messaging System Amendment 1: Minor Enhancements; (1993) ***CD-ROM ONLY***. 16 pp.
SAA AS 4033.7-92. Information Technology—Text Communication—Message-Oriented Text Interchange Systems—Part 7: Interpersonal Messaging System (ISO/IEC 10021-7:1990) (in Professional Package 26A). 121 pp.

—**Teleinformatic Services—Abstract Services**
CCITT RECMN T.330-89. Telematic Access to Interpersonal Message System—Terminal Equipment and Protocols for Telematic Services (Study Group VIII) 82 pp. 82 pp.

—**Teleinformatic Services—Interworking**
CCITT RECMN T.330-89. Telematic Access to Interpersonal Message System—Terminal Equipment and Protocols for Telematic Services (Study Group VIII) 82 pp. 82 pp.

Interphones
See Also: Intercom Equipment

—**Ships**
MOD UK NES 562-91. Guide to Internal Communication Equipment Non—VCS Units (RICE 1, CCU and SINBAD) Issue 3 (09.91). 122 pp.

—**Submarines**
MOD UK NES 562-91. Guide to Internal Communication Equipment Non—VCS Units (RICE 1, CCU and SINBAD) Issue 3 (09.91). 122 pp.

—**UHF—Ships**
CCIR RECMN 542-1-82. On-Board Communications by Means of Portable Radiotelephone Equipment—Section 8C—Maritime Mobile Service; Telephony and Related Subjects. 1 p.

Interpreting (Translating)
Use: Translating

Interregister Signaling
See Also: CCITT R1 Signaling Systems; CCITT R2 Signaling Systems; Signaling Systems

—**Backward Signals—Pulse Transmission—CCITT R2 Signaling Systems**
CCITT RECMN Q.442-89. Interregister Signalling—Pulse Transmission of Backward Signals A-3, A-4, A-6 or A-15. Multifrequency Signalling Equipment—Specifications of Signalling Systems R1 and R2 (Study Group XI) 2 pp. 2 pp.

INDEX and DIRECTORY of

INTERNATIONAL AND NON-U.S. NATIONAL STANDARDS SUBJECT INDEX

Interregister Signaling (Cont.)

—CCITT R2 Signaling Systems
CCITT RECMN Q.440-89. Interregister Signalling—General—Specifications of Signalling Systems R1 and R2 (Study Group XI) 4 pp. 4 pp.

—Range—Transmission Loss—Formulas—CCITT R2 Signaling Systems
CCITT FASCICLE VI.4. Specifications of Signalling Systems R1 and R2 Recommendations Q.310—Q.490 (Study Group XI). 181 pp.

CCITT RECMN Q.457-89. Interregister Signalling—Range of Interregister Sinalling—Specifications of Signalling Systems R1 and R2 (Study Group XI) 4 pp. 4 pp.

—Regeneration—Transit Exchanges—CCITT R2 Signaling Systems
CCITT RECMN Q.478-89. Signalling Procedures—Relay and Regeneration of R2 Interregister Signals by an Outgoing R2 Register in a Transit Exchange—Specifications of Signalling Systems R1 and R2 (Study Group XI) 2 pp. 2 pp.

—Relay—Transit Exchanges—CCITT R2 Signaling Systems
CCITT RECMN Q.478-89. Signalling Procedures—Relay and Regeneration of R2 Interregister Signals by an Outgoing R2 Register in a Transit Exchange—Specifications of Signalling Systems R1 and R2 (Study Group XI) 2 pp. 2 pp.

—Reliability—CCITT R2 Signaling Systems
CCITT RECMN Q.458-89. Interregister Signalling—Reliability of Interregister Signalling—Specifications of Signalling Systems R1 and R2 (Study Group XI) 5 pp. 5 pp.

—Signal Codes—CCITT R2 Signaling Systems
CCITT RECMN Q.441-89. Interregister Signalling—Signalling Code—Specifications of Signalling Systems R1 and R2 (Study Group XI) 10 pp. 10 pp.

—Speed—CCITT R2 Signaling Systems
CCITT RECMN Q.458-89. Interregister Signalling—Reliability of Interregister Signalling—Specifications of Signalling Systems R1 and R2 (Study Group XI) 5 pp. 5 pp.

—Termination—End of Pulsing Signals—CCITT R2 Signaling Systems
CCITT RECMN Q.473-89. Signalling Procedures—Use of End-of-Pulsing Signal I-15 in International Working—Specifications of Signalling Systems R1 and R2 (Study Group XI) 2 pp. 2 pp.

—Termination—Group B Signals—CCITT R2 Signaling Systems
CCITT RECMN Q.474-89. Signalling Procedures—Use of Group B Signals—Specifications of Signalling Systems R1 and R2 (Study Group XI) 3 pp. 3 pp.

—Termination—Telephone Exchanges—CCITT R2 Signaling Systems
CCITT RECMN Q.471-89. Signalling Procedures—at the Last Incoming R2 Register Situated in the Exchange to Which the Called Subscriber is Connected—Specifications of Signalling Systems R1 and R2 (Study Group XI) 2 pp. 2 pp.

—Termination—Transit Exchanges—CCITT R2 Signaling Systems
CCITT RECMN Q.470-89. Signalling Procedures—at an Incoming R2 Register Situated in a Transit Exchange—Specifications of Signalling Systems R1 and R2 (Study Group XI) 2 PP. 2 pp.

CCITT RECMN Q.472-89. Signalling Procedures—at the Last Incoming R2 Register Situated in a Transit Exchange—Specifications of Signalling Systems R1 and R2 (Study Group XI) 2 pp. 2 pp.

—Testing Points
CCITT RECMN M.719-89. Testing Point (Switching and Interregister Signalling)—General Maintenance Principles—Maintenance of International Transmission Systems and Telephone Circuits (Study Group IV) 3 pp. 3 pp.

Interrogation (Intelligence)
See Also: Military Intelligence; Military Operations

—Prisoners of War
NATO STANAG 2033 ED 5 AMD 0-86. Interrogation of Prisoners of War (PW). 15 pp.

Interrupt Control Units
Use: Interrupt Controllers

Interrupt Controllers
Use For: Interrupt Control Units

Interrupt Controllers (Cont.)
See Also: Controllers; Microprocessors; Priority Interrupt Controllers

—NMOS—Programmable—Microprocessors—Preferred Products List
CECC CECC MUAHAG Vol 7 IS 8-92. Preferred Products List; Active Microcircuits (En, Fe, Ge). 89 pp.

Interrupter Switches
Use For: Circuit Interrupters; Load Interrupter Switches; Load Interrupter Switchgear
See Also: Electric Switches; Switches; Switchgear

—41-100 kV
CSA CAN/CSA-C22. 2 NO 31-M89. Switchgear Assemblies; (Gen Instr 1 Thru 3). 54 pp.

CSA 1169 Bull. Electrical Bulletin 1169 June 27, 1978 to C22.2 NO 31. 2 pp.

—High Voltage
CSA C22.2 NO 193-M1983. High Voltage Full-Load Interrupter Switches (R 1992); (Gen Instr 1 Thru 2). 16 pp.

Interrupters
Scope Note: Use a more specific term *See:* Circuit Breakers; Control Systems Equipment; Electric Switches; Ground Fault Circuit Interrupters; Relays

Interruption Control
See Also: Interruption Counters; Interruption of Service; Line Signaling; Switching Systems

—Line Signaling—CCITT R2 Signaling Systems
CCITT RECMN Q.416-89. Line Signalling, Analogue Version—Interruption Control—Specifications of Signalling Systems R1 and R2 (Study Group XI) 5 pp. 5 pp.

—Receivers—Multifrequency Signaling—CCITT R2 Signaling Systems
CCITT FASCICLE VI.4. Specifications of Signalling Systems R1 and R2 Recommendations Q.310—Q.490 (Study Group XI). 181 pp.

Interruption Counters
See Also: Interruption Control; Interruption of Service; Transmission Measuring Equipment

—Telephone Circuits
CCITT RECMN O.61-89. Simple Equipment to Measure Interruptions on Telephone-Type Circuits—Specifications for Measuring Equipment (Study Group IV) 3 pp. 3 pp.

CCITT RECMN O.62-89. Sophisticated Equipment to Measure Interruptions on Telephone-Type Circuits—Specifications for Measuring Equipment (Study Group IV) 3 pp. 3 pp.

Interruption Monitors (Anesthesia)
Use: Anesthesia Breathing Systems—Interruption Monitors

Interruption Monitors, Anesthesia Breathing Gas
Use: Anesthesia Breathing Systems—Interruption Monitors

Interruption of Service
Use For: Disruption of Service; Disturbance of Service; Service Disruption; Service Disturbance; Service Interruption *See Also:* Interruption Control; Interruption Counters; Outages (Communications); Telecommunication Services; Telephone Services

CCITT FASCICLE I.2-88. Opinions and Resolutions Recommendations on the Organisation and Working Procedures of CCITT (Series A). 64 pp.

—Telecommunication Networks—Expenses
CCITT RECMN E.862 (REV 1)-92. Dependability Planning of Telecommunication Networks (Study Group II) 16 pp. 16 pp.

CCITT RECMN E.862-89. Dependability Planning of Telecommunication Networks—Telephone Network and ISDN—Quality of Service, Network Management and Traffic Engineering (Study Group II) 12 pp. 12 pp.

—Telegraphy
CCITT RECMN F.1-89. Operational Provisions for the International Public Telegram Service—Telegraph and Mobile Services Operations and Quality of Service (Study Group I) 54 pp. 54 pp.

Interruption of Service (Cont.)

—Telegraphy—Message Switching
CCITT RECMN F.35-89. Provisions Applying to the Operation of an International Public Automatic Message Switching Service for Equipments Utilizing the International Telegraph Alphabet No. 2—Telegraph and Mobile Services Operations and Quality of Service (Study Group I) 7 pp. 7 pp.

Interruption of Services

—Fixed Satellite Services
CCIR QUESTION 45/4-90. Interruptions to Traffic on Digital Paths or Circuits in the Fixed-Satellite Service—Questions of Study Group 4 Fixed-Satellite Service. 2 pp.

—Sound Transmission
CCIR Report 642-74. Interruption to Sound-Programme Services—Section CMTT D—Methods of Operation and Assessment of Performance of Sound-Programme Transmission Channels. 1 p.

Intersatellite Services
Use For: ISS *See Also:* Communications Satellites; Satellite Communications; Satellites; Space Communications

—Broadcasting Satellite Services—Frequency Band Sharing
CCIR Report 951-82. Sharing Between the Inter-Satellite Service and the Broadcasting-Satellite Service in the Vicinity of 23 GHz—Section 10/11E—Sharing. 7 pp.

—Fixed Satellite Services—Frequency Band Sharing
CCIR QUESTION 56/4-90. Frequency Sharing Between the Inter-Satellite Service When Used for Links of the Fixed-Satellite Service and Terrestrial Radiocommunication Services—Questions of Study Group 4 Fixed-Satellite Service. 1 p.

—Fixed Services (Radio Communications)—Frequency Band Sharing
CCIR QUESTION 62/4-90. Frequency Sharing of the Fixed-Satellite Service and the Inter-Satellite Service with the Fixed Service Under Provisions of RR Article 14—Questions of Study Group 4 Fixed-Satellite Service. 1 p.

—Radio Communications—Frequency Band Sharing
CCIR QUESTION 56/4-90. Frequency Sharing Between the Inter-Satellite Service When Used for Links of the Fixed-Satellite Service and Terrestrial Radiocommunication Services—Questions of Study Group 4 Fixed-Satellite Service. 1 p.

CCIR QUESTION 63/4-90. Frequency Sharing of the Fixed-Satellite Service and the Inter-Satellite Service with Terrestrial Radio Services Other Than the Fixed Service Under Provisions of RR Article 14—Questions of Study Group 4 Fixed-Satellite Service. 1 p.

—Satellite Communications—Frequency Band Sharing
CCIR QUESTION 68/4-90. Frequency Sharing of the Fixed-Satellite Service and the Inter-Satellite Service with Other Space Radio Services Under Provisions of RR Article 14—Questions of Study Group 4 Fixed-Satellite Service. 1 p.

Interval Time Switches
Use: Time Switches

Interval Timers
Use: Timers—Interval

Intervalometers
See Also: Arming Devices; Electrical Equipment; Timers

—Rocket Launchers—Connectors
NATO STANAG 3576 ED 3 AMD 0-91. Electrical Connector for Dispensers and Internal Intervalometer Type Rocket Launchers for Aircraft. 6 pp.

Interworking
Scope Note: See the subheading Interworking under specific subjects *See Also:* Interworking Units

Interworking Functional Units
Use: Interworking Units

Interworking Units
See Also: Open Systems Interconnection; Telecommunication Equipment

INTERNATIONAL AND NON-U.S. NATIONAL STANDARDS
SUBJECT INDEX

Interworking

Interworking Units *(Cont.)*
JTC1 TR10029-89. Information Technology—Telecommunications and Information Exchange Between Systems—Operation of an X.25 Interworking Unit First Edition. 10 pp.

—Local Area Networks
ECMA ECMA-TR 21-84. Local Area Networks Interworking Units for Distributed Systems. 13 pp.

—Network Services—Relay Function
BSI DD ENV 41801-1-92. 1992 Information Systems Interconnection—Relaying the Connectionless-Mode Network Service—Part 1: General Overview and Subnetwork Independent Requirements. 26 pp.
BSI DD ENV 41801-2-92. 1992 Information Systems Interconnection—Relaying the Connectionless-Mode Network Service—Part 2: LAN Subnetwork Dependent Media Independent Requirements. 35 pp.
BSI DD ENV 41802-2-92. 1992 Information Systems Interconnection—X.25 Protocol Relaying Part 2: LAN Subnetwork Dependent Media Independent Requirements. 60 pp.
CEN ENV 41801-1-92. Information Systems Interconnection—Relaying the Connectionless-Mode Network Service—Part 1: General Overview and Subnetwork Independent Requirements. 8 pp.
CEN ENV 41801-2-92. Information Systems Interconnection—Relaying the Connectionless-Mode Network Service—Part 2: LAN Subnetwork Dependent Media Independent Requirements. 43 pp.
CEN ENV 41802-1-92. Information Systems Interconnection—X.25 Protocol Relaying—Part 1: General Overview and Subnetwork Independent Requirements. 18 pp.
CEN ENV 41802-2-92. Information Systems Interconnection—X.25 Protocol Relaying—Part 2: LAN Subnetwork Dependent Media Independent Requirements. 56 pp.
IEC DISP 10613 Pt 1-92. Information Technology—International Standardized Profile RA—Relaying the Connectionless-Mode Network Service—Part 1: Relay Function General Overview and Subnetwork-Independent Requirements ***CD-ROM ONLY***. 29 pp.
IEC DISP 10613 Pt 2-92. Information Technology—International Standardized Profile RA—Relaying the Connectionless-Mode Network Service—Part 2: LAN Subnetwork Dependent Media Independent Requirements ***CD-ROM ONLY***. 20 pp.
IEC DISP 10613 Pt 3-92. Information Technology—International Standardized Profile RA—Relaying the Connectionless-Mode Network Service—Part 3: ISO 8802-3 CSMA/CD LAN Subnetwork Dependent Media Dependent Requirements ***CD-ROM ONLY***. 22 pp.
IEC DISP 10613 Pt 5-92. Information Technology—International Standardized Profile RA—Relaying the Connectionless-Mode Network Service—Part 5: Relaying the CLNS Between CSMA/CD LAN Subnetworks (RA51.51) ***CD-ROM ONLY***. 17 pp.
IEC DISP 10613 Pt 7-92. Information Technology—International Standardized Profile RA—Relaying the Connectionless-Mode Network Service—Part 7: PSDN Subnetwork Dependent Media Dependent Requirements for Virtual Calls over a Permanent Access ***CD-ROM ONLY***. 24 pp.
IEC DISP 10613 Pt 8-92. Information Technology—International Standardized Profile RA—Relaying the Connectionless-Mode Network Service—Part 8: Profile RA51.1111 ***CD-ROM ONLY***. 20 pp.
IEC DISP 10613 Pt 9-92. Information Technology—International Standardized Profile RA—Relaying the Connectionless-Mode Network Service—Part 9: Profile RA51.1121 ***CD-ROM ONLY***. 20 pp.
ISO DISP 10613 Pt 1-92. Information Technology—International Standardized Profile RA—Relaying the Connectionless-Mode Network Service—Part 1: Relay Function General Overview and Subnetwork-Independent Requirements ***CD-ROM ONLY***. 29 pp.
ISO DISP 10613 Pt 2-92. Information Technology—International Standardized Profile RA—Relaying the Connectionless-Mode Network Service—Part 2: LAN Subnetwork Dependent Media Independent Requirements ***CD-ROM ONLY***. 20 pp.
ISO DISP 10613 Pt 3-92. Information Technology—International Standardized Profile RA—Relaying the Connectionless-Mode Network Service—Part 3: ISO 8802-3 CSMA/CD LAN Subnetwork Dependent Media Dependent Requirements ***CD-ROM ONLY***. 22 pp.
ISO DISP 10613 Pt 5-92. Information Technology—International Standardized Profile RA—Relaying the Connectionless-Mode Network Service—Part 5: Relaying the CLNS Between CSMA/CD LAN Subnetworks (RA51.51) ***CD-ROM ONLY***. 17 pp.
ISO DISP 10613 Pt 7-92. Information Technology—International Standardized Profile RA—Relaying the Connectionless-Mode Network Service—Part 7: PSDN Subnetwork Dependent Media Dependent Requirements for Virtual Calls over a Permanent Access ***CD-ROM ONLY***. 24 pp.
ISO DISP 10613 Pt 8-92. Information Technology—International Standardized Profile RA—Relaying the Connectionless-Mode Network Service—Part 8: Profile RA51.1111 ***CD-ROM ONLY***. 20 pp.
ISO DISP 10613 Pt 9-92. Information Technology—International Standardized Profile RA—Relaying the Connectionless-Mode Network Service—Part 9: Profile RA51.1121 ***CD-ROM ONLY***. 20 pp.
JTC1 DISP10613 Pt 1-92. Information Technology—International Standardized Profile RA—Relaying the Connectionless-Mode Network Service—Part 1: Relay Function General Overview and Subnetwork-Independent Requirements ***CD-ROM ONLY***. 29 pp.
JTC1 DISP10613 Pt 2-92. Information Technology—International Standardized Profile RA—Relaying the Connectionless-Mode Network Service—Part 2: LAN Subnetwork Dependent Media Independent Requirements ***CD-ROM ONLY***. 20 pp.
JTC1 DISP10613 Pt 3-92. Information Technology—International Standardized Profile RA—Relaying the Connectionless-Mode Network Service—Part 3: ISO 8802-3 CSMA/CD LAN Subnetwork Dependent Media Dependent Requirements ***CD-ROM ONLY***. 22 pp.
JTC1 DISP10613 Pt 5-92. Information Technology—International Standardized Profile RA—Relaying the Connectionless-Mode Network Service—Part 5: Relaying the CLNS Between CSMA/CD LAN Subnetworks (RA51.51) ***CD-ROM ONLY***. 17 pp.
JTC1 DISP10613 Pt 7-92. Information Technology—International Standardized Profile RA—Relaying the Connectionless-Mode Network Service—Part 7: PSDN Subnetwork Dependent Media Dependent Requirements for Virtual Calls over a Permanent Access ***CD-ROM ONLY***. 24 pp.
JTC1 DISP10613 Pt 8-92. Information Technology—International Standardized Profile RA—Relaying the Connectionless-Mode Network Service—Part 8: Profile RA51.1111 ***CD-ROM ONLY***. 20 pp.
JTC1 DISP10613 Pt 9-92. Information Technology—International Standardized Profile RA—Relaying the Connectionless-Mode Network Service—Part 9: Profile RA51.1121 ***CD-ROM ONLY***. 20 pp.

—Network/Transport Protocols
BSI BS ISO/IEC TR 10172-91. 1991 Information Technology—Telecommunications and Information Exchange Between Systems—Network/Transport Protocol Interworking Specification. 32 pp.
ISO TR10172-91. Information Technology—Telecommunications and Information Exchange Between Systems—Network/Transport Protocol Interworking Specification First Edition. 30 pp.

—Packet Switched Networks
BSI BS 7267-90. 1990 Open Systems Interconnection: Guide to Operation of an X.25 Interworking Unit. 12 pp.
IEC TR10029-89. Information Technology—Telecommunications and Information Exchange Between Systems—Operation of an X.25 Interworking Unit First Edition. 10 pp.
ISO TR10029-89. Information Technology—Telecommunications and Information Exchange Between Systems—Operation of an X.25 Interworking Unit First Edition. 10 pp.
OSI ISO IEC TR 10029-89. Information Technology—Telecommunications and Information Exchange Between Systems—Operation of an X.25 Interworking Unit. 10 pp.
SAA AS 4015-92. Information Technology—Telecommunications and Information Exchange Between Systems—Operation of an X.25 Interworking Unit (ISO/IEC/TR 10029:1989) (in Professional Package 26A). 5 pp.

—Telex Communications—Status Enquiry Function
CCITT RECMN F.89-92. Status Enquiry Function in the International Telex Service (Study Group I) 6 pp. 6 pp.

Intex Services
See Also: Telecommunication Services; Telex Communications

—International Alphabet No. 5
CCITT RECMN F.150-91. Service and Operational Provisions for the Intex Service (Study Group I) 7 pp. 7 pp.

—Telex Communications—Interworking
CCITT RECMN F.82-91. Operational Provisions to Permit Interworking Between the International Telex Service and the Intex Service (Study Group I) 5 pp. 5 pp.

INTRASTAT System
See Also: Commerce; Statistical Analysis

INTRASTAT System *(Cont.)*
EC COM(88) 810-89. Proposal for a Council Regulation (EEC) on the Statistics Relating to the Trading of Goods Between Member States. 29 pp.

Intrauterine Devices
Use For: IUD
ISO TR7439-81. Mechanical Contraceptives—Intra-Uterine Devices First Edition. 4 pp.
SNZ NZS 7102-80. Specification for Intra-Uterine Contraceptive Devices. 8 pp.

—Breaking Load
ISO 7857 Pt 1-83. Intra-Uterine Devices—Part 1: Determination of Breaking Force First Edition. 3 pp.

—Identification Systems
ISO 7857 Pt 3-83. Intra-Uterine Devices—Part 3: Packaging and Labelling First Edition. 4 pp.

—Packaging
ISO 7857 Pt 3-83. Intra-Uterine Devices—Part 3: Packaging and Labelling First Edition. 4 pp.

Intravenous Medical Equipment
Use For: Infusion Equipment (Medical); Infusion Medical Equipment *See Also:* Medical Equipment

—Bottles—Caps (Lids)
ISO 8536 Pt 3-92. Infusion Equipment for Medical Use—Part 3: Aluminium Caps for Infusion Bottles First Edition. 10 pp.
ISO 8536 Pt 7-92. Infusion Equipment for Medical Use—Part 7: Caps Made of Aluminium-Plastics Combinations for Infusion Bottles First Edition. 8 pp.
ISO 10985-92. Caps Made of Aluminium-Plastics Combinations for Infusion Bottles and Injection Vials—Requirements and Test Methods First Edition. 8 pp.

—Bottles—Glass
ISO 8536 Pt 1-91. Infusion Equipment for Medical Use—Part 1: Infusion Glass Bottles First Edition. 9 pp.

—Bottles—Stoppers
ISO 8536 Pt 2-92. Infusion Equipment for Medical Use—Part 2: Closures for Infusion Bottles First Edition. 13 pp.

—Stands (Supports)
BSI BS 3619-76. 1976 Amd 1 Mobile Infusion Stands. 5 pp.

Intrinsic Viscosity
Use For: Limiting Viscosity Number
See Also: Dynamic Viscosity; Kinematic Viscosity; Rheological Properties; Viscosity

—Cellulose
BSI BS 6306: Part 1-82. 1982 Determination of Limiting Viscosity Number of Celulose in Dilute Solutions Part 1: Cupri-Ethylene-Diamine (CED) Method. 13 pp.
BSI BS 6306: Part 2-82. 1982 Determination of Limiting Viscosity Number of Cellulose in Dilate Solutions Part 2: Iron (III) Sodium Tartrate Complex (EWNN Model NaCI Method). 9 pp.
ISO 5351 Pt 1-81. Cellulose in Dilute Solutions—Determination of Limiting Viscosity Number—Part 1: Method in Cupri-Ethylene-Diamine (CED) Solution First Edition. 14 pp.
ISO 5351 Pt 2-81. Cellulose in Dilute Solutions—Determination of Limiting Viscosity Number—Part 2: Method in Iron(III) Sodium Tartrate Complex (EWNN mod NaCl) Solution First Edition. 8 pp.

—Polyalkylene Terephthalates
BSI BS 2782:Pt7: METH 732E-91. 1991 Plastics Part 7: Rheological Properties Method 732F: Determination of Viscosity Number of Poly(Alkylene Terephthalates) (ISO 1628-5: 1986). 8 pp.
ISO 1628 Pt 5-86. Plastics—Determination of Viscosity Number and Limiting Viscosity Number—Part 5: Poly(Alkylene Terephthalates) First Edition. 6 pp.

—Polycarbonate Resins
BSI BS 2782:Pt7: METH 732D-91. 1991 Methods of Testing Plastics Part 7: Rheological Properties Method 732D: Determination of Viscosity Number of Polycarbonate Molding and Extrusion Materials (ISO 1628-4: 1986). 8 pp.
ISO 1628 Pt 4-86. Plastics—Determination of Viscosity Number and Limiting Viscosity Number—Part 4: Polycarbonate (PC) Moulding and Extrusion Materials First Edition. 7 pp.

—Polymers
BSI BS 2782:Pt7: METH 732A-91. 1991 Methods of Testing Plastics Part 7: Rheological Properties Method 732A: Determination of Viscosity Number: General Conditions (ISO 1628-1: 1984) (Supersedes BS 2782: Method 730A: 1979). 13 pp.

Intrinsic Viscosity (Cont.)
—Polymers (Cont.)
ISO 1628 Pt 1-84. Guidelines for the Standardization of Methods for the Determination of Viscosity Number and Limiting Viscosity Number of Polymers in Dilute Solution—Part 1: General Conditions First Edition. 11 pp.

—Polymethyl Methacrylate
BSI BS 2782:Pt7: METH 732F-91. 1991 Methods of Testing Plastics Part 7: Rheological Properties Method 732F: Determination of Viscosity Number of Methyl Methacrylate Polymers (ISO 1628-6: 1990). 11 pp.
ISO 1628 Pt 6-90. Plastics—Determination of Viscosity Number and Limiting Viscosity Number—Part 6: Methyl Methacrylate Polymers First Edition; (Replaces 1233). 8 pp.

—PVC
BSI BS 2782:Pt7: METH 732B-91. 1991 Plastics Part 7: Rheological Properties Method 732B: Determination of Viscosity Number of Poly(Vinyl Chloride) Resins (ISO 1628-2: 1988). 6 pp.
ISO 1628 Pt 2-88. Plastics—Determination of Viscosity Number and Limiting Viscosity Number—Part 2: Poly (Vinyl Chloride) Resins First Edition. 4 pp.

Introducers (Medical Equipment)
See Also: Medical Electrical Equipment; Medical Equipment

—Needles—Epidural
BSI BS 6196-89. 1989 Sterile Epidural Catheters and Introducer Needles for Single Use. 16 pp.

Intruder Alarm Systems
Use: Burglar Detectors

Intruder Alarms
Use: Burglar Alarms

Intruder Detectors
Use: Burglar Detectors

Intrusion Tones
See Also: Tones (Telephone Services)
CCITT RECMN E.182-89. Application of Tones and Recorded Announcements in Telephone Services—Telephone Network and ISDN—Operation, Numbering, Routing and Mobile Service (Study Group II) 8 pp. 8 pp.

Intumescent Coatings
See Also: Coatings; Fire Retardant Paints

—Interior
CGSB CAN/CGSB-1.151-92. Fire-Retardant Intumescent Interior Coating. 9 pp.
CGSB 1-GP-152M-80. Standard for: Coating, Interior, Overcoat for Fire Retardant Coating; (Amendment 1 Aug 1982) (QPL Aug 1986). 13 pp.

—Metal
BSI BS 8202: Part 2-92. 1992 Coatings for Fire Protection of Building Elements Part 2: Code of Practice for the Use of Intumescent Coating Systems to Metallic Substrates for Providing Fire Resistance. 16 pp.

Inulin
CNS K7581-81. Chemical Reagent (Inulin) (Aug)(7816).

Inventory Control Systems
Use For: Material Control *See Also:* Accounting; Control Charts; Production Management

—Glossaries
CNS Z4031-84. Glossary of Terms for Computerized Production and Inventory Management (General) (Nov)(11143).
CNS Z4031-1-84. Glossary of Terms for Computerized Production and Inventory Management (Terms of Inventory Management) (Nov)(11143-1).
CNS Z4031-2-84. Glossary of Terms for Computerized Production and Inventory Management (Terms of Material Requirements Planning) (Nov)(11143-2).
CNS Z4031-3-84. Glossary of Terms for Computerized Production and Inventory Management (Terms of Production Management) (Nov)(11143-3).
CNS Z4031-4-84. Glossary of Terms for Computerized Production and Inventory Management (Terms of Capacity Planning) (Nov)(11143-4).
CNS Z4031-5-84. Glossary of Terms for Computerized Production and Inventory Management (Terms of Forecast) (Nov)(11143-5).
CNS Z4031-6-84. Glossary of Terms for Computerized Production and Inventory Management (Terms of Purchase) (Nov)(11143-6).

Invertebrates
Scope Note: Use a more specific term

Invertebrates (Cont.)
See: Aquatic Organisms; Coleoptera; Crustacea; Daphnia; Insect Contamination; Termites

Inverted Siphons
Use: Siphons

Inverter Circuits
Use: NOT Circuits

Inverters
Use For: Power Inverters *See Also:* Static Inverters
CECC CECC 90 104-108 ISSUE 1-81. BS CECC 90 104-108; Dual Complementary Pair Plus Inverter (En). 25 pp.
CECC CECC 90 106-043 ISSUE 1-85. UTE C 86-217; Digital Integrated Circuits in Accordance with FS 90 106; 54/74 ALS 05, 54/74 ALS 05U; Inverter with Open Collector Outputs (En, Fr) ADD 3 (En, Fr). 15 pp.
CECC CECC 90 109-706 ISSUE 1-87. Digital Integrated Circuits in Accordance with FS 90 109; 54/74 HCT 04; Inverters (En, Fr). 6 pp.
CSA CAN/CSA-C22. 2NO107.1-M91. Commercial and Industrial Power Supplies; (Gen Instr 1 Thru 2). 112 pp.

—Aircraft
BSI G 174-59. 1959 Invertors for Secondary Electrical Supplies for Aircraft. 16 pp.

—Hex
CECC CECC 90 102-003 ISSUE 1-81. BS CECC 90 102-003; Hex Inverters (En) AMD 1 (En). 17 pp.
CECC CECC 90 104-045 ISSUE 2. NL CECC 90 104-045 Issue 2; Digital Integrated Circuits in Accordance with FS 90 104; HEC/HEF 4069UB; Hex Inverter (En). 11 pp.
CECC CECC 90 107-003 ISSUE 2-89. UTE C 86-218 ADD 2/FA 3; Digital Integrated Circuits in Accordance with FS 90 107; 54/74 F 04, Inverter (En, Fr) ADD 3 (En, Fr). 9 pp.

—Radio Equipment—Marine
CNS C4302-81. Marine Converters and Inverters for Wireless Sets (Apr)(7212).

—Thyristors
CECC CECC 50 011-010 ISSUE 1-83. BS CECC 50 011-010; Type Numbers: BTW 63 Series (En). 18 pp.

Inverting Buffer Gates
See Also: Gate Circuits; Inverting Gates

—Hex
CECC CECC 90 104-037 ISSUE 2. NL CECC 90 104-037 Issue 2; Digital Integrated Circuits in Accordance with FS 90 104; HEC/HEF 4049B; Hex Inverting Buffers (En). 7 pp.
CECC CECC 90 104-091 ISSUE 2. NL CECC 90 104-091 Issue 2; Digital Integrated Circuits in Accordance with FS 90 104; HEC/HEF 40098B; 3-State Hex Inverting Buffer (En). 8 pp.
CECC CECC 90 104-133 ISSUE 1-81. BS CECC 90 104-133; Silicon Complementary MOS with (UB) Unbuffered Outputs and Cavity Packaging (En). 24 pp.
CECC CECC 90 104-133 ISSUE 1-86. CEI CECC 90 104-133; Silicon Complementary MOS with (UB) Unbuffered Outputs Cavity andNon Cavity Packaging (En). 2 pp.
CECC CECC 90 104-152 ISSUE 1-81. BS CECC 90 104-152; Strobed Hex Inverter/Buffer with Three State Outputs (En). 27 pp.
CECC CECC 90 104-153 ISSUE 1-86. CEI-CECC 90 104-153; Hex Non-Inverting 3-State Buffer (En). 2 pp.
CECC CECC 90 109-846 ISSUE 1-90. Digital Integrated Circuits; Silicon Monolithic C MOS, Cavity or Non-Cavity Packages; Type(s) 54/74 HCT 368 Hex Inverting 3-State Buffer Assessment Levels P,Y, L (En, Fr, Ge). 9 pp.
CECC CECC 90 109-858 ISSUE 1-90. Digital Integrated Circuits; Silicon Monolithic C MOS, Cavity or Non-Cavity Packages; Type(s) 54/74 HCT 366 Hex Inverting 3-State Buffer Assessment Levels P, Y, L (En, Fr, Ge). 9 pp.

—Octal
CECC CECC 90 104-227 ISSUE 1. NL CECC 90 104-227; Digital Integrated Circuits in Accordance with FS 90 104; HEC/HEF 40240B; Octal Inverting Buffers with 3-State Outputs (En). 8 pp.

Inverting Buffer/Line Drivers
See Also: Line Transceivers

—Octal
CECC CECC 90 109-616 ISSUE 1-86. Digital Integrated Circuits in Accordance with FS 90 109; 54 HC 240, 74 HC 240; Inverting Octal 3-State Buffer (En, Fr). 6 pp.

Inverting Buffers
—Hex
CECC CECC 90 104-061 ISSUE 2. NL CECC 90 104-061 Issue 2; Digital Integrated Circuits in Accordance with FS 90 104; HEC/HEF 4502B; Strobed Hex Inverter/Buffer (En). 8 pp.

Inverting Bus Buffer/Down Counters
—Integrated Circuit
CECC CECC 90 109-718 ISSUE 1-87. Digital Integrated Circuits in Accordance with FS 90 109; 54/74 HC 4049; Inverting Bus Buffer/Logic Level Down Converter (En, Fr). 6 pp.
CECC CECC 90 109-744 ISSUE 1-87. Digital Integrated Circuits in Accordance with FS 90 109; 54/74 HC 4049B; Inverting Bus Buffer/Logic Level Down Converter (En, Fr). 6 pp.
CECC CECC 90 109-745 ISSUE 1-87. Digital Integrated Circuits in Accordance with FS 90 109; 54/74 HC 4050B; Non Inverting Bus Buffer/Logic Level Down Converter (En, Fr). 6 pp.

Inverting Gates
Use For: Inverting NAND Gates *See Also:* Gate Circuits; Inverting Buffer Gates

—Hex
CECC CECC 90 109-633 ISSUE 1-86. Digital Integrated Circuits in Accordance with FS 90 109; 54/74 HC 04; Inverters (En, Fr). 5 pp.

—NAND
CECC CECC 90 106-001 ISSUE 1-85. UTE C 86-217; Digital Integrated Circuits in Accordance with FS 90 106; 54/74 ALS 00, 54/74 ALS 00A, 54/74 ALS 04, 54/74 ALS 04A, 54/74 ALS 20, 54/74 ALS 20A; NAND Gates and Inverter (En, Fr) ADD 3 (En, Fr). 16 pp.

Inverting NAND Gates
Use: Inverting Gates

Inverting NOR/NAND Gates
—Hex—2-Input
CECC CECC 90 104-202 ISSUE 1-81. BS CECC 90 104-202; Silicon Complementary MOS. with (UB) Unbuffered Outputs and Cavity Packaging (En). 24 pp.
CECC CECC 90 104-202 ISSUE 1-86. CEI CECC 90 104-202; Silicon Complementay MOS. with (UB) Unbuffered Outputs Cavity and Non Cavity Packaging (En). 2 pp.

Invertors
Use: Inverters

Investment Castings
Use For: Precision Castings *See Also:* Castings; Metal Products

—Aluminum Alloy
DIN ENGL 1725 Pt 2-86. Aluminium Alloys; Casting Alloys; Sand Casting; Gravity Die Casting; Pressure Die Casting; Investment Casting (Feb). 15 pp.

—Carbon Steel
BSI BS 3146: Part 1-74. 1974 Investment Castings in Metal Part 1: Carbon and Low Alloy Steels. 28 pp.

—Chromium Steel—Aerospace
MOD UK DTD-5259-67. Chromium-Nickel Corrosion-Resisting Steel Investment Castings (Not Stabilised) (Tensile Strength 47 kgf /mm2) (Not to Be Used for Applications at Temperatures Exceeding 350 Degrees C). 2 pp.
MOD UK DTD-5269-67. Chromium-Nickel Corrosion-Resisting Steel Investment Castings (Niobium Stabilized) (Tensile Strength 47 kgf /mm2). 2 pp.
MOD UK DTD-5279-68. Chromium-Nickel-2.5 Per Cent Molybdenum Heat-Resisting and Corrosion-Resisting Steel Investment Castings (50 Hbar) (High Temperature Properties Not Verified). 2 pp.
MOD UK DTD-5289-68. Chromium-Nickel-3.5 Per Cent Molybdenum Heat-Resisting and Corrosion-Resisting Steel Investment Castings (50 Hbar) (High Temperature Properties Not Verified). 2 pp.

—Cobalt Alloy
BSI BS 3146: Part 2-75. 1975 Amd 1 Investment Castings in Metal Part 2: Corrosion and Heat Resisting Steels, Nickel and Cobalt Base Alloys. 28 pp.

—Cobalt Alloy—Aerospace
BSI HC 100-72. 1972 Amd 3 Inspection and Testing Procedure for Iron, Nickel, Copper, Cobalt and Refractory Metal Base Alloy Castings (AMD 6574) May 28, 1991. 17 pp.

Investment Castings (Cont.)

—Copper Alloy—Aerospace
BSI HC 100-72. 1972 Amd 3 Inspection and Testing Procedure for Iron, Nickel, Copper, Cobalt and Refractory Metal Base Alloy Castings (AMD 6574) May 28, 1991. 17 pp.

—Ferroally—Aerospace
BSI HC 100-72. 1972 Amd 3 Inspection and Testing Procedure for Iron, Nickel, Copper, Cobalt and Refractory Metal Base Alloy Castings (AMD 6574) May 28, 1991. 17 pp.

—Low Alloy Steel
BSI BS 3146: Part 1-74. 1974 Investment Castings in Metal Part 1: Carbon and Low Alloy Steels. 28 pp.

—Magnesium Alloy—Aerospace—Mechanical Properties
AECMA PREN3349-88. Magnesium Alloy MG-C92 Solution Treated and Artificially Aged (T6) Casting. 5 pp.

—Maraging Steel—Aerospace
BSI HC 401-74. 1974 18% Nickel Maraging Steel Precision Castings (1600-1850 MPa). 3 pp.

—Nickel Alloy
BSI BS 3146: Part 2-75. 1975 Amd 1 Investment Castings in Metal Part 2: Corrosion and Heat Resisting Steels, Nickel and Cobalt Base Alloys. 28 pp.

—Nickel Alloy—Aerospace
AECMA PREN2192-78. Heat Resisting Nickel Base Alloy NI-C98-HT—as Cast—Precision Castings (C5/48). 3 pp.
AECMA PREN2198-79. Heat Resisting Nickel Base Alloy NI-C103-HT—Solution Treated and Precipitation Treated—Precision Castings (C5/48). 3 pp.
AECMA PREN2204-79. Heat Resisting Nickel Base Alloy NI-C105-HT—Solution Treated and Precipitation Treated—Precision Castings (C5/48). 3 pp.
AECMA PREN2233-79. Heat Resisting Nickel Base Alloy NI-C104-HT—as Cast—Precision Castings (C5/48). 3 pp.
AECMA PREN2403-80. Heat Resisting Nickel Base Alloy NI-C100-HT—Solution Treated and Precipitation Treated—Precision Castings (C5/48). 3 pp.
BSI HC 100-72. 1972 Amd 3 Inspection and Testing Procedure for Iron, Nickel, Copper, Cobalt and Refractory Metal Base Alloy Castings (AMD 6574) May 28, 1991. 17 pp.

—Nickel Alloy—Aerospace—Mechanical Properties
AECMA PREN2664-88. Heat Resisting Nickel Base Alloy Ni-C99HT Not Heat Treated Precision Castings. 5 pp.

—Nickel Alloy—Precipitation Hardening—Aerospace
BSI HC 206-75. 1975 Amd 1 Precipitation Hardening Nickel Base Chromium-Cobalt-Molybdenum-Titanium Alloy Precision Castings (Cr 20, Co 20, Mo 6, Ti 2). 3 pp.

—Nickel Steel—Aerospace
BSI HC 401-74. 1974 18% Nickel Maraging Steel Precision Castings (1600-1850 MPa). 3 pp.

—Refractory Metal Alloy—Aerospace
BSI HC 100-72. 1972 Amd 3 Inspection and Testing Procedure for Iron, Nickel, Copper, Cobalt and Refractory Metal Base Alloy Castings (AMD 6574) May 28, 1991. 17 pp.

—Steel
BSI BS 3146: Part 2-75. 1975 Amd 1 Investment Castings in Metal Part 2: Corrosion and Heat Resisting Steels, Nickel and Cobalt Base Alloys. 28 pp.

—Steel—Aerospace
AECMA PREN3363-88. Steel FE-CM68 Solution Treated Rm Greater Than or Equal to 485 Mpa Sand or Investment Casting. 4 pp.
AECMA PREN3483-88. Steel FE-CM61 Homogenised Solution Treated, Precipitation Hardened and Sub Zero Rm Greater Than or Equal to 1240 MPa High Strength Sand or Investment Casting. 4 pp.
AECMA PREN3485-88. Steel FE-CM61 Homogenised Solution Treated, Precipitation Hardened and Sub Zero Rm Greater Than or Equal to 830 MPa Medium Strength Sand or Investment Casting. 4 pp.

Investment Castings (Cont.)

—Titanium Alloy—Aerospace—Mechanical Properties
AECMA PREN3352-88. Titanium Alloy TI-C63 Hot Isostatic Pressed (Annealed) Rm Greater Than or Equal to 880 MPa High Strength Investment Casting a Less Than or Equal to 25 mm. 5 pp.

Invitation for Bids
See Also: Bids; Defense Contracts

—Cables (Electric)/Wire
MOD UK DEFCON 47B-91. Invitation to Tender Special Notices and Instructions—Wire, Cables Etc. 9/91. 1 p.

—Clothing
MOD UK DEFCON 47D-86. Invitation to Tender for Clothing and Textile Contracts 8/86. 2 pp.
MOD UK DEFCON 47D-91. Invitation to Tender for Clothing and Textile Contracts 9/91. 2 pp.

—Coal/Coke
MOD UK DEFCON 47C-89. Invitation to Tender for the Supply and Delivery of Coal and/or Coke 11/89. 2 pp.

—Dies
MOD UK DEFCON 47V-86. Invitation to Tender Supplementary Conditions—Jigs, Tools, Etc 10/86. 1 p.

—Equal Opportunity Certificate—Northern Ireland
MOD UK DEFCON 47AA-82. Fair Employment (Northern Ireland) Act 1976 Equal Opportunity Certificate 3/82. 1 p.

—European Communities
MOD UK DEFCON 47M-91. Invitation to Tender—Special Notices and Instructions—EC Works Directive 9/91. 1 p.

—Fabrics
MOD UK DEFCON 47D-86. Invitation to Tender for Clothing and Textile Contracts 8/86. 2 pp.
MOD UK DEFCON 47D-91. Invitation to Tender for Clothing and Textile Contracts 9/91. 2 pp.

—Food
MOD UK DEFCON 47G-89. Invitation to Tender for the Supply of Food 11/89. 2 pp.

—Gages
MOD UK DEFCON 47V-86. Invitation to Tender Supplementary Conditions—Jigs, Tools, Etc 10/86. 1 p.

—Highway Transportation
MOD UK DEFCON 47T-88. Invitation to Tender for the Hire of Motor Transport with or Without Drivers and Conditions of Contract 3/88. 2 pp.

—Jigs
MOD UK DEFCON 47V-86. Invitation to Tender Supplementary Conditions—Jigs, Tools, Etc 10/86. 1 p.

—Maintenance Services
MOD UK DEFCON 47-89. Invitation to Tender—Tenderer's Offer and Declaration 5/89. 2 pp.
MOD UK DEFCON 47A-89. Invitation to Tender Special Notices and Instructions to Tenderers, Including Consideration of Metric Equivalents 4/89. 1 p.
MOD UK DEFCON 47 LWS/A-89. Invitation to Tender for Local Works Services Routine Maintenance and Other Minor Jobs of a Recurring Nature 1/89. 3 pp.
MOD UK DEFCON 47 LWS/A-92. Invitation to Tender for Local Works Services Routine Maintenance and Other Minor Jobs of a Recurring Nature 1/92. 3 pp.
MOD UK DEFCON 47 LWS-89. Invitation to Tender for Local Works Services 5/89. 3 pp.
MOD UK DEFCON 47 LWS-91. Invitation to Tender for Local Works Services 11/91. 3 pp.

—Meat, Poultry, and Offals
MOD UK DEFCON 47K-84. Invitation to Tender for the Supply of Meat, Poultry and Offals 4/84. 2 pp.

—Molds (Casting)
MOD UK DEFCON 47V-86. Invitation to Tender Supplementary Conditions—Jigs, Tools, Etc 10/86. 1 p.

—Motor Vehicle Rental
MOD UK DEFCON 47T-88. Invitation to Tender for the Hire of Motor Transport with or Without Drivers and Conditions of Contract 3/88. 2 pp.

Invitation for Bids (Cont.)

—Oils/Greases
MOD UK DEFCON 47Q-87. Invitation to Tender for Supply of Oils and Greases 10/87. 2 pp.

—Personnel Employment
MOD UK DEFCON 47R-88. Invitation to Tender Special Notices and Instructions to Tenderers 7/88. 1 p.

—Procurement
MOD UK DEFCON 47L-90. Invitation to Tender 10/90. 3 pp.
MOD UK DEFCON 47N-90. Invitation to Tender Special Notices and Instructions EC Supplies Directives and GATT Agreement on Government Procurement 9/90. 1 p.

—Service Contracts
MOD UK DEFCON 47J-85. Invitation to Tender for Services 12/85. 2 pp.

—Software
MOD UK DEFCON 143-89. Software Development Questionnaire—for Inclusion with Invitations to Tender 8/89. 10 pp.

—Testing Equipment
MOD UK DEFCON 47V-86. Invitation to Tender Supplementary Conditions—Jigs, Tools, Etc 10/86. 1 p.

—Tools
MOD UK DEFCON 47V-86. Invitation to Tender Supplementary Conditions—Jigs, Tools, Etc 10/86. 1 p.

Invoices
Use For: Bills (Invoices); Statements (Accounts)
See Also: Forms (Paper); Record Layouts; Trade Documents

—Document Formats—Maritime Mobile Services
CCITT RECMN D.90-92. Charging, Accounting and Refunds in the Maritime Mobile Service (Study Group III) 23 pp. 23 pp.
CCITT RECMN D.90-89. Charging, Accounting and Refunds in the Maritime Mobile Service—General Tariff Principles—Charging and Accounting in International Telecommunications Services (Study Group III) 19 pp. 19 pp.

—Document Formats—Telephone Services
CCITT RECMN D.170-89. Monthly Telephone and Telex Accounts—General Tariff Principles—Charging and Accounting in International Telecommunications Services (Study Group III) 9 pp (Same as Recmn E.270). 9 pp.

—Document Formats—Telex Communications
CCITT RECMN D.170-89. Monthly Telephone and Telex Accounts—General Tariff Principles—Charging and Accounting in International Telecommunications Services (Study Group III) 9 pp (Same as Recmn E.270). 9 pp.

—Record Layouts—Reverse Charging
CCITT RECMN D.176 (REV 1)-92. Transmission in Encoded Form of Telephone Reversed Charge Billing and Accounting Information (Study Group III) 9 pp (Same as Recmn E.276). 9 pp.
CCITT RECMN D.176-89. Transmission in Encoded Form of Telephone Reversed Charge Billing and Accounting Information—General Tariff Principles—Charging and Accounting in International Telecommunications Services (Study Group III) 6 pp. 6 pp.

Involute Gear Teeth
See Also: Gear Teeth

—Cylindrical Gears
CNS B1249-84. Basic Rack Tooth Profile of Gear Tools for Cylindrical Gears with Involute Teeth for Fine Mechanics (Aug)(7572).
CNS B2023-84. Basic Rack of Cylindrical Gears with Involute Teeth for General and Heavy Engineering (Aug)(184).
CNS B2421-84. Basic Rack of Cylindrical Gears with Involute Teeth for Fine Mechanics (Aug)(5278).
DIN ENGL 867-86. Basic Rack Tooth Profiles for Involute Teeth of Cylindrical Gears for General Engineering and Heavy Engineering (Feb). 3 pp.
DIN ENGL 3977-81. Measuring Element Diameters for the Radial or Diametral Dimension for Testing Tooth Thickness of Cylindrical Gears (Feb). 7 pp.

—Cylindrical Gears—Engineering Drawings
DIN ENGL 3966 Pt 1-78. Information on Gear Teeth in Drawings; Information on Involute Teeth for Cylindrical Gears (Aug). 8 pp.

INTERNATIONAL AND NON-U.S. NATIONAL STANDARDS
SUBJECT INDEX

Involute Gear Teeth (Cont.)
—Cylindrical Gears—Inspection
ISO TR10064 Pt 1-92. Cylindrical Gears—Code of Inspection Practice—Part 1: Inspection of Corresponding Flanks of Gear Teeth First Edition. 63 pp.

—Helical Gears
BSI PD 6457-70. 1970 Guide to the Application of Addendum Modification to Involute Spur and Helical Gears. 20 pp.
JIS B 1701-73. Involute Gear Tooth Profile and Dimensions (R 1979). 7 pp.

—Hobbing Machines
BSI BS 5246-76. 1976 Amd 1 Pinion Type Cutters for Spur Gears; 1 to 8 Metric Module. 17 pp.
CNS B2425-84. Basic Rack of Gear Cutting Tools for Involute Tooth System (Aug)(5282).
JIS B 4232-85. Involute Gear Milling Cutters. 10 pp.

—Spur Gears
BSI PD 6457-70. 1970 Guide to the Application of Addendum Modification to Involute Spur and Helical Gears. 20 pp.
JIS B 1701-73. Involute Gear Tooth Profile and Dimensions (R 1979). 7 pp.

Involute Serrations
See Also: Involute Splines; Serrations
SAA AS B265-68. Involute Serrations. 75 pp.

Involute Splines
See Also: Involute Serrations; Splines
BSI BS 3550-63. 1963 Amd 2 Involute Splines. 81 pp.
CNS B1322-83. Involute Splines Summary of Data (Sep)(10524).
CNS B1323-83. Involute Splines Dimensions and Measurements (Module 0.6, 0.8 and 1) (Sep)(10525).
CNS B1324-83. Involute Splines Dimensions and Measurements (Module 1.25) (Sep)(10526).
CNS B1325-83. Involute Splines Dimensions and Measurements (Module 1.5) (Sep)(10527).
CNS B1326-83. Involute Splines Dimensions and Measurements (Module 2) (Sep)(10528).
CNS B1327-83. Involute Splines Dimensions and Measurements (Module 2.5) (Sep)(10529).
CNS B1328-83. Involute Splines Dimensions and Measurements (Module 3) (Sep)(10530).
CNS B1329-83. Involute Splines Dimensions and Measurements (Module 4) (Sep)(10531).
CNS B1330-83. Involute Splines Dimensions and Measurements (Module 5) (Sep)(10532).
CNS B1331-83. Involute Splines Dimensions and Measurements (Module 6) (Sep)(10533).
CNS B1332-83. Involute Splines Dimensions and Measurements (Module 8) (Sep)(10534).
CNS B1333-83. Involute Splines Dimensions and Measurements (Module 10) (Sep)(10535).
CNS B1334-83. Involute Splines Side Fits, Tolerances (Sep)(10536).
CNS B1335-83. Involute Splines Testing and Gauges for Side Fit (Sep)(10537).
DIN ENGL 5480 Pt 14-86. Involute Splines with 30 Degree Pressure Angle Side Fits; Tolerances (Mar). 7 pp.
DIN ENGL 5480 Pt 15-74. Involute Splines; Testing and Gauges for Side Fit (Sept). 12 pp.
SAA AS B213-66. Involute Splines Being BS 3550:1963, Endorsed without Amendment. 80 pp.

—Broaches
CNS B1336-83. Involute Splines Hobs, Pinion-Type Cutters, Broaches (Sep)(10538).
DIN ENGL 5480 Pt 16-86. Involute Splines with 30 Degree Pressure Angle; Hobs; Pinion Type Cutters; Broaches (Mar). 11 pp.
JIS B 4239-88. Involute Spline Broaches. 14 pp.

—Cylindrical
BSI BS 6186:Pt1: Add 1-93. 1993 Straight Cylindrical Involute Splines: Metric Module, Side Fit Part 1: Dimensions and Tolerances Addendum 1: Inspection (ISO 4156: 1981/Amendment 1: 1992). 153 pp.
BSI BS 6186: Part 1-81. 1981 Amd 1 Involute Splines: Metric Module Part 1: Dimensions and Tolerances. 145 pp.
ISO 4156-81. Straight Cylindrical Involute Splines—Metric Module, Side Fit—Generalities, Dimensions and Inspection Amendment 1: Section Three: Inspection First Edition; (Amendment 1-1992). 288 pp.

—Drive Ends—Socket Wrenches
ISO 4228-86. Spanners and Wrenches—Spline Drive Ends for Power Socket Wrenches First Edition. 5 pp.

—Drive Shafts—Aircraft
SBAC RS 623 (V). Accessory Drives and Mounting Pads (Taper Flanges—Involute Splines).

Involute Splines (Cont.)
—Drive Shafts—Automotive
JIS D 2001-59. Involute Spline for Automobiles (R 1974). 41 pp.

—Gages
CNS B1335-83. Involute Splines Testing and Gauges for Side Fit (Sep)(10537).
DIN ENGL 5480 Pt 15-74. Involute Splines; Testing and Gauges for Side Fit (Sept). 12 pp.

—Gear Hobs
CNS B1336-83. Involute Splines Hobs, Pinion-Type Cutters, Broaches (Sep)(10538).
DIN ENGL 5480 Pt 16-86. Involute Splines with 30 Degree Pressure Angle; Hobs; Pinion Type Cutters; Broaches (Mar). 11 pp.

—Glossaries
CNS B1314-83. Involute Splines (Definitions) (May)(10235).

—Joints
DIN ENGL 5480 Pt 1-91. Involute Spline Joints; Principles (Oct). 10 pp.
DIN ENGL 5480 Pt 2-91. Involute Spline Joints; 30 Degree Pressure Angle; Survey (Oct). 2 pp.
DIN ENGL 5480 Pt 3-91. Involute Spline Joints; 30 Degree Pressure Angle; Basic Dimensions and Test Dimensions for Modules 0,5, 0,6, 0,75, 0,8 and 1 (Oct). 11 pp.
DIN ENGL 5480 Pt 4-91. Involute Spline Joints; 30 Degree Pressure Angle; Basic Dimensions and Test Dimensions for Module 1,25 (Oct). 4 pp.
DIN ENGL 5480 Pt 5-91. Involute Spline Joints; 30 Degree Pressure Angle; Basic Dimensions and Test Dimensions for Modules 1,5 and 1,75 (Oct). 6 pp.
DIN ENGL 5480 Pt 6-91. Involute Spline Joints; 30 Degree Pressure Angle; Basic Dimensions and Test Dimensions for Module 2 (Oct). 4 pp.
DIN ENGL 5480 Pt 7-91. Involute Spline Joints; 30 Degree Pressure Angle; Basic Dimensions and Test Dimensions for Module 2,5 (Oct). 4 pp.
DIN ENGL 5480 Pt 8-91. Involute Spline Joints; 30 Degree Pressure Angle; Basic Dimensions and Test Dimensions for Module 3 (Oct). 4 pp.
DIN ENGL 5480 Pt 9-91. Involute Spline Joints; 30 Degree Pressure Angle; Basic Dimensions and Test Dimensions for Module 4 (Oct). 4 pp.
DIN ENGL 5480 Pt 10-91. Involute Spline Joints; 30 Degree Pressure Angle; Basic Dimensions and Test Dimensions for Module 5 (Oct). 4 pp.
DIN ENGL 5480 Pt 11-91. Involute Spline Joints; 30 Degree Pressure Angle; Basic Dimensions and Test Dimensions for Module 6 (Oct). 4 pp.
DIN ENGL 5480 Pt 12-91. Involute Spline Joints; 30 Degree Pressure Angle; Basic Dimensions and Test Dimensions for Module 8 (Oct). 4 pp.
DIN ENGL 5480 Pt 13-91. Involute Spline Joints; 30 Degree Pressure Angle; Basic Dimensions and Test Dimensions for Module 10 (Oct). 4 pp.

—Milling Cutters
CNS B1336-83. Involute Splines Hobs, Pinion-Type Cutters, Broaches (Sep)(10538).
DIN ENGL 5480 Pt 16-86. Involute Splines with 30 Degree Pressure Angle; Hobs; Pinion Type Cutters; Broaches (Mar). 11 pp.

—Taper Shafts—Aircraft
SBAC RS 623 (V). Accessory Drives and Mounting Pads (Taper Flanges—Involute Splines).

Iodate Content Analysis
—Sodium Nitrate
MOD UK M 869/91. Examination of Sodium Nitrate, Grade 1.

Iodic Acid
CNS K7255-68. Chemical Reagent (Iodic Acid) (Nov)(1754).

Iodic Acid Anhydride
Use: Iodine Pentoxide

Iodide Content Analysis
—Water—Colorimetric Analysis
CNS K9048-80. Method of Test for Iodide in Water (Colorimetric Method) (Aug)(6230).

Iodine
See Also: Iodine Content Analysis; Iodine Number; Nonmetals; Radioactive Iodine
CNS K7256-68. Chemical Reagent (Iodine) (Nov)(1755).
JIS K 8920-86. Iodine.

Iodine Content Analysis
See Also: Iodine; Radioactive Iodine Content Analysis

Iodine Content Analysis (Cont.)
—Milk
BSI BS 1741: Part 11-89. 1989 Methods for Chemical Analysis of Liquid Milk and Cream Part 11: Detection of Detergent/Disinfectant Residues. 4 pp.

Iodine Inorganic Compounds
Scope Note: Use a more specific term
See: Ammonium Iodide; Cadmium Iodide; Ethyl Iodide; Iodine Pentoxide; Iodine Trichloride; Mercuric Iodide; Methyl Iodide; Potassium Iodate; Potassium Iodides; Sodium Iodate; Sodium Iodide

Iodine Number
See Also: Iodine; Peroxide Number; Saponification Number; Unsaturation

—Amidoamines
CNS K6682-81. Method of Test for Iodine Value of Fatty Amines, Amidoamines, and Diamines (Jul)(7726).

—Boiled Oil
CNS K6804-7-84. Method of Test for Organic Coating Iodine Value of Boiled Oil (Jun)(10880-7).

—Butynediol
ISO 6793-81. But-2-Ene-1,4-Diol for Industrial Use—Determination of Iodine Value First Edition. 5 pp.

—Carbon Black
CNS K6514-80. Methods of Test for Iodine Absorption Number of Carbon Black (Jul)(5818).

—Carbon Black—Volumetric Analysis
BSI BS 5293: Part 10-90. 1990 Sampling and Testing Carbon Black for Use in the Rubber Industry Part 10: Method for Determination of Iodine Adsorption Number. 8 pp.
ISO 1304-85. Rubber Compounding Ingredients—Carbon Black—Determination of Iodine Adsorption Number—Titrimetric Method Second Edition. 6 pp.

—Castor Oil
MOD UK M 9530/66. Examination of Oil, Caster, Hardened.

—Chemical Products
JIS K 0070-92. Test Methods for Acid Value, Saponification Value, Ester Value, Iodine Value, Hydroxyl Value and Unsaponifiable Matter of Chemical Products. 20 pp.

—Coconut Oil
CNS K6180-74. Method of Test for Coconut Oil (Technical Grade) (Oct)(2241). 5 pp.

—Diamines
CNS K6682-81. Method of Test for Iodine Value of Fatty Amines, Amidoamines, and Diamines (Jul)(7726).

—Drying Oils
CNS K6646-81. Method of Test for Iodine Value of Drying Oils and Fatty Acid (Mar)(7165).
CNS K6671-81. Method of Test for Total Iodine Value of Drying Oils and Their Derivatives (May)(7435).

—Fats
BSI BS 684: Sec 2.13-90. 1990 Methods of Analysis of Fats and Fatty Oils Part 2: Other Methods Section 2.13: Determination of Iodine Value. 4 pp.
CNS N6081-86. Methods of Test for Edible Oils and Fats (Determination of Iodine Value) (Nov)(3646). 3 pp.
ISO 3961-89. Animal and Vegetable Fats and Oils—Determination of Iodine Value Second Edition. 4 pp.

—Fatty Acids
CNS K6646-81. Method of Test for Iodine Value of Drying Oils and Fatty Acid (Mar)(7165).

—Fatty Amines
CNS K6682-81. Method of Test for Iodine Value of Fatty Amines, Amidoamines, and Diamines (Jul)(7726).

—Instant Noodles
CNS N6094-85. Method of Test for Instant Noodles (Jan)(3962). 1 p.

—Linseed Oil
CNS K6055-75. Method of Test for Boiled Linseed Oil (Feb)(769). 4 pp.

—Oleic Acid—Volumetric Analysis
MOD UK M 9545/68. Examination of Oleic Acid, Special, Lead Free.

—Quaternary Ammonium Chloride
CNS K6681-81. Method of Test for Iodine Value of Fatty Quaternary Ammonium Chlorides (Jul)(7725).

INDUSTRY STANDARDS

Iodine Number (Cont.)

—Vegetable Oils
CNS N6081-86. Methods of Test for Edible Oils and Fats (Determination of Iodine Value) (Nov)(3646). 3 pp.

Iodine Pentoxide
Use For: Iodic Acid Anhydride
CNS K7257-68. Chemical Reagent (Iodine Pentoxide) (Nov)(1756).

Iodine Trichloride
CNS K7668-83. Chemical Reagent (Iodine Trichloride) (Mar)(10107).
JIS K 8403-61. Iodine Trichloride.

Iodine Value
Use: Iodine Number

Iodine 131
Use: Radioactive Iodine

Iodoethane
Use: Ethyl Iodide

Iodomethane
Use: Methyl Iodide

Iodometry
See Also: Volumetric Analysis
CNS K7500-78. Chemical Reagent (Starch Soluble (for Iodometry) (Nov)(1999).

—Acetic Acid
ISO 753 Pt 3-83. Acetic Acid for Industrial Use—Methods of Test—Part 3: Determination of Formic Acid Content—Iodometric Method First Edition. 6 pp.

—Benzene
DIN ENGL 51774 Pt 3-75. Testing of Liquid Fuels; Determination of Bromine Assimilation by the Iodometric Method (BC) (Aug). 2 pp.

—Fluorite
BSI BS 5659: Part 3-89. (WITHDRAWN) 1989 Acid-Grade Fluorspar Part 3: Determination of Sulphide Content. 8 pp.
ISO 4284-93. Acid-Grade and Ceramic-Grade Fluorspar—Determination of Sulfide Content—Iodometric Method Third Edition. 7 pp.
ISO 9501-91. Metallurgical-Grade Fluorspar—Determination of Total Sulfur Content—Iodometric Method After Combustion First Edition. 9 pp.

—Hydrogen Fluoride
BSI BS 5365: Part 5-79. (WITHDRAWN) 1979 Methods of Sampling and Test for Anhydrous Hydrogen Fluoride for Industrial Use Part 5: Determination of Sulphur Dioxide Content. 4 pp.
ISO 3702-76. Anhydrous Hydrogen Fluoride for Industrial Use—Determination of Sulphur Dioxide Content—Iodometric Method First Edition. 4 pp.

—Phenol Formaldehyde Resins
BSI BS 2782:Pt4: METH 451E-78. 1978 Methods of Testing Plastics Part 4: Chemical Properties Method 451E: Determination of Free Phenols in Phenol-ings Formaldehyde Mouldings (Iodometric Method). 4 pp.
ISO 119-77. Plastics—Phenol-Formaldehyde Mouldings—Determination of Free Phenols—Iodometric Method First Edition; (Erratum—Aug 1979). 5 pp.

—Phthalic Anhydride
ISO 1389 Pt IX-77. Phthalic Anhydride for Industrial Use—Methods of Test—Part IX: Determination of Impurities Oxidizable in the Cold by Potassium Permanganate—Iodometric Method First Edition. 4 pp.

—Rubber
BSI BS 903: Part B16-67. (WITHDRAWN) 1967 Methods of Testing Vulcanized Rubber Part B16: Determination of Antimony. 7 pp.
BSI BS 7164: Part 25-90. 1990 Chemical Tests for Raw and Vulcanized Rubber Part 25: Method for Determination of Sulphide Sulphur Content (ISO 8054: 1988). 11 pp.
ISO 8054-88. Rubber, Compounded or Vulcanized—Determination of Sulfide Sulfur Content—Iodometric Method First Edition. 9 pp.

—Sodium Hexafluorosilicate
BSI BS 5705: Part 7-81. (WITHDRAWN) 1981 Sodium Hexafluorosilicate for Industrial Use Part 7: Determination of Sulphate Content. 8 pp.

—Sulfuric Acid
ISO 3423-75. Sulphuric Acid and Oleums for Industrial Use—Determination of Sulphur Dioxide Content—Iodometric Method First Edition. 5 pp.

Iodometry (Cont.)

—Surfactants
BSI BS 6829: Sec 4.3-90. 1990 Analysis of Surface Active Agents (Raw Materials) Part 4: Ethylene Oxide Adducts Section 4.3: Methods for Determination of Oxyethylene Groups. 8 pp.
ISO 2270-89. Non-Ionic Surface Active Agents—Polyethoxylated Derivatives—Iodometric Determination of Oxyethylene Groups Second Edition. 8 pp.

—Water
BSI BS 2690: Part 101-84. (WITHDRAWN) 1984 Water Used in Industry Part 101: Dissolved Oxygen (Superseded by BS EN 25813). 10 pp.
BSI BS 6068: Sec 2.3-93. 1993 Amd 1 Water Quality—Determination of Dissolved Oxygen—Iodometric Method (ISO 5813: 1983) (AMD 7428) May 15, 1993 (N). 15 pp.
BSI BS 6068: Sec 2.3-84. (WITHDRAWN) 1984 Water Quality Part 2: Physical, Chemical and Bio-Chemical Methods Section 2.3: Determination of Dissolved Oxygen; Iodometric Method (Renumbered as BS EN 25813: 1993). 8 pp.
BSI BS 6068: Sec 2.27-90. 1990 Water Quality Part 2: Physical, Chemical and Bio-Chemical Methods Section 2.27: Method for Determination of Total Chlorine: Iodometric Titration Method. 12 pp.
BSI BS EN 25813-93. 1993 Amd 1 Water Quality—Determination of Dissolved Oxygen—Iodometric Method (ISO 5813: 1983) (AMD 7428) May 15, 1993 (N). 15 pp.
CEN EN 25813-92. Water Quality—Determination of Dissolved Oxygen—Iodometric Method (ISO 5813: 1983). 9 pp.
DIN ENGL EN 25813-93. Water Quality; Determination of Dissolved Oxygen by the Iodometric Method (ISO 5813: 1983) (Jan). 9 pp.
ISO 5813-83. Water Quality—Determination of Dissolved Oxygen—Iodometric Method First Edition (CEN EN 25813: 1992). 7 pp.
ISO 7393 Pt 3-90. Water Quality—Determination of Free Chlorine and Total Chlorine—Part 3: Iodometric Titration Method for the Determination of Total Chlorine Second Edition. 10 pp.

Iodophors

—Dairy Equipment
SNZ NZS 8341-72. Specification for Dairy Iodophors. 12 pp.

Iodoxyquinsulfonic Acid
CNS K7258-68. Chemical Reagent (Iodoxyquinsulfonic Acid) (Nov)(1757).

Ion Chambers
Use: Ionization Chambers

Ion Chromatography
See Also: Chromatography

—Water
JIS K 0127-92. General Rules for Ion Chromatographic Analysis. 19 pp.

Ion Exchange Capacity Measurement
Use: Exchange Capacity Measurement

Ion Exchange Chromatography

—Sediments
DIN ENGL 38414 Pt 17-89. German Standard Methods for the Examination of Water, Waste Water and Sludge; Sludge and Sediments (Group S); Determination of Strippable and Extractable Organically Bound Halogens (S 17) (Nov). 5 pp.
DIN ENGL 38414 Pt 18-89. German Standard Methods for the Examination of Water, Waste Water and Sludge; Sludge and Sediments (Group S); Determination of Adsorbed Organically Bound Halogens (AOX) (S18) (Nov). 4 pp.

—Sludge (Sewage)
DIN ENGL 38414 Pt 17-89. German Standard Methods for the Examination of Water, Waste Water and Sludge; Sludge and Sediments (Group S); Determination of Strippable and Extractable Organically Bound Halogens (S 17) (Nov). 5 pp.
DIN ENGL 38414 Pt 18-89. German Standard Methods for the Examination of Water, Waste Water and Sludge; Sludge and Sediments (Group S); Determination of Adsorbed Organically Bound Halogens (AOX) (S18) (Nov). 4 pp.

—Waste Water
DIN ENGL 38405 Pt 20-91. German Standard Methods for the Examination of Water, Waste Water and Sludge; Anions (Group D); Determination of Dissolved Bromide, Chloride, Nitrate, Nitrite, (Ortho) Phosphate and Sulfate Anions in Waste Water by Ion Chromatography (D 20) (Sept). 11 pp.

Ion Exchange Chromatography (Cont.)

—Water
JIS K 0556-90. Testing Methods for Determination of Anions in Highly Purified Water. 18 pp.

Ion Exchange Resins
See Also: Zeolites

—Water Treatment
DIN ENGL 19633-86. Ion Exchange Resins for Water Treatment; Technical Delivery Conditions (Jan). 4 pp.

Ion Exchangers
See Also: Anion Exchangers; Separators (Mechanical)

—Density
DIN ENGL 54410-85. Testing of Ion Exchangers; Determination of Density (Oct). 2 pp.

—Hydrochloric Acid—Water Treatment
DIN ENGL 19610-75. Hydrochloric Acid for Water Treatment for Supply Water; Technical Conditions of Delivery (Nov). 6 pp.

—Sodium Chloride—Water Treatment
DIN ENGL 19604-91. Sodium Chloride for Use in Water Treatment; Technical Delivery Conditions (Mar). 4 pp.

—Sodium Hydroxide—Water Treatment
DIN ENGL 19616 Pt 2-71. Caustic Soda Solution for Water Treatment for Regeneration of Ion Exchangers; Technical Conditions of Delivery (Dec). 5 pp.

—Water Treatment—Recycling Method
DIN ENGL 54411-87. Testing of Ion Exchangers; Recycling Method for Testing Ion Exchangers and Polymer Adsorption Resins (Apr). 4 pp.

Ion Exchanging
See Also: Water Treatment

—Uranium Hexafluoride
CNS J2084-83. Method for Determining Plutonium in UF6 Ion Exchange and Alpha Counting Method (Jan)(9884).

Ion Pumps
See Also: Pumps; Vacuum Pumps
DIN ENGL 28429-85. Vacuum Technology; Acceptance Specifications for Sputter Ion Pumps (Aug). 6 pp.

Ion Selective Electrodes
See Also: Electrodes; pH Electrodes
BSI BS 7310-90. 1990 Ion-Selective Electrodes, Reference Electrodes, Combination Electrodes and Ion-Selective Electrode Meters for Determination of Ions in Solution. 24 pp.
CNS K0041-89. General Rules for Ion-Selective Electrode Method (Apr)(12508).
JIS K 0122-81. General Rules for Ion-Selective Electrode Method (R 1986). 48 pp.

—Iron Ores
BSI BS 7020: Part 20-90. 1990 Analysis of Iron Ores Part 20: Method for the Determination of Water Soluble Chloride Content: Ion-Selective Electrode Method. 21 pp.
ISO 9517-89. Iron Ores—Determination of Water Soluble Chloride Content—Ion-Selective Electrode Method First Edition. 11 pp.

—Uranium
ISO 9892-92. Uranium Metal, Uranium Dioxide Powder and Pellets, and Uranyl Nitrate Solutions—Determination of Fluorine Content—Fluoride Ion Selective Electrode Method First Edition. 8 pp.

—Uranium Dioxide
ISO 9892-92. Uranium Metal, Uranium Dioxide Powder and Pellets, and Uranyl Nitrate Solutions—Determination of Fluorine Content—Fluoride Ion Selective Electrode Method First Edition. 8 pp.

—Uranyl Nitrate
ISO 9892-92. Uranium Metal, Uranium Dioxide Powder and Pellets, and Uranyl Nitrate Solutions—Determination of Fluorine Content—Fluoride Ion Selective Electrode Method First Edition. 8 pp.

Ionic Impurities

—Electrical Insulation
BSI BS 5591-78. 1978 Amd 1 Determination of Ionic Impurities in Electrical Insulating Materials by Extraction with Liquids. 9 pp.
CENELEC HD 381-78. Methods of Test for the Determination of Ionic Impurities in Electric Insulation Materials by Extraction with Liquids. 2 pp.

INTERNATIONAL AND NON-U.S. NATIONAL STANDARDS
SUBJECT INDEX

Ionosphere

Ionic Impurities *(Cont.)*
—Electrical Insulation *(Cont.)*
IEC 589-77. Methods of Test for the Determination of Ionic Purities in Electrical Insulating Materials by Extraction with Liquids First Edition; (Corrigendum—June 1978). 17 pp.

Ionization Chambers
See Also: Dosimeters; Neutron Counters; Proportional Counter Tubes; Proportional Counters; Radiation Measuring Instruments; Test Chambers
IEC 1145-92. Calibration and Usage of Ionization Chamber Systems for Assay of Radionuclides First Edition. 27 pp.

—Calibration
IEC 1145-92. Calibration and Usage of Ionization Chamber Systems for Assay of Radionuclides First Edition. 27 pp.

—Dosimeters
CENELEC HD 534 S1-89. Medical Electrical Equipment Dosimeters with Ionization Chambers as Used in Radiotherapy. 4 pp.
IEC 731-82. Medical Electrical Equipment Dosimeters with Ionization Chambers as Used in Radiotherapy First Edition; (Amendment 1-1987). 157 pp.

Ionization Gages
See Also: Gages; Vacuum Gages
JIS Z 8752-89. Measuring Methods of Low Pressures by Hot Cathode and Cold Cathode Ionization Gauges. 8 pp.

Ionized Media
—Radio Wave Propagation
CCIR Volume VI TOC-90. Table of Contents. 2 pp.
CCIR Volume VI TERMS-90. Terms of Reference of Study Group 6 and Introduction by the Chairman, Study Group 6. 6 pp.
CCIR Volume VI Annex TOC-90. Table of Contents. 4 pp.

—Radio Wave Propagation—Indexes
CCIR Volume VI ANX Index-90. Numerical Index of Texts. 1 p.
CCIR Volume VI ANX DELETED-90. Index of Texts Deleted. 1 p.

Ionizing Radiation
Use For: Delta Rays; Hard Radiation
See Also: Gamma Rays; Irradiation; Nonionizing Radiation; Radiation Measuring Instruments; Radiation Physics; Radioactive Materials; Solar Radiation; Sterilization; Ultraviolet Radiation; X-Rays

—Densitometers
CENELEC HD 435-83. Density Meters Utilizing Ionizing Radiation Definitions and Test Methods. 2 pp.
IEC 692-80. Density Meters Utilizing Ionizing Radiation Definitions and Test Methods First Edition. 29 pp.

—Electrical Insulation
IEC 544 Pt 1-77. Guide for Determining the Effects of Ionizing Radiation on Insulating Materials Part 1: Radiation Interaction First Edition. 45 pp.
IEC 544 Pt 2-91. Guide for Determining the Effects of Ionizing Radiation on Insulating Materials Part 2: Procedures for Irradiation and Test Second Edition. 45 pp.
IEC 544 Pt 3-79. Guide for Determining the Effects of Ionizing Radiation on Insulating Materials Part 3: Test Procedures for Permanent Effects First Edition. 19 pp.
IEC 544 Pt 4-85. Guide for Determining the Effects of Ionizing Radiation on Insulating Materials Part 4: Classification System for Service in Radiation Environments First Edition. 19 pp.

—Electron Tubes
IEC 562-76. Measurements of Incidental Ionizing Radiation from Electronic Tubes First Edition. 21 pp.

—Food
EC COM(87) 242-87. Amended proposal for a Council Directive Amending Directive 79/112/EEC on the Approximation of the Laws of the Member States Relating to the Labelling, Presentation and Advertising of Foodstuffs for Sale to the Ultimate Consumer (COM(86) 89 final of 22. 7 pp.
EC COM(88) 654-88. Proposal for a Council Directive on the Approximation of the Laws of the Member States Concerning Foods and Food Ingredients Treated with Ionizing Radiation. 19 pp.

—Glossaries
IEC 50 Chap 391-75. International Electrotechnical Vocabulary Chapter 391: Detection and Measurement of Ionizing Radiation by Electric Means. 126 pp.

Ionizing Radiation *(Cont.)*
—Glossaries *(Cont.)*
IEC 50 Chap 881-83. Advance Edition of the International Electrotechnical Vocabulary Chapter 881: Radiology and Radiological Physics. 228 pp.
SAA AS 1852.391-78. International Electrotechnical Vocabulary—Part 391: Detection and Measurement of Ionizing Radiation by Electric Means Being IEC 50(391). 124 pp.
SNZ IEC 50: 50(391)-75. International Electrotechnical Vocabulary 50(391): Detection and Measurement of Ionizing Radiation by Electric Means. 114 pp.
SNZ IEC 50: 50(881)-83. International Electrotechnical Vocabulary 50(881): Radiology and Radiological Physics. 218 pp.

—Lead Shields
ISO 7212-86. Enclosures for Protection Against Ionizing Radiation—Lead Shielding Units for 50 mm and 100 mm Thick Walls First Edition. 59 pp.
ISO 9404 Pt 1-91. Enclosures for Protection Against Ionizing Radiation—Lead Shielding Units for 150 mm, 200 mm and 250 mm Thick Walls—Part 1: Chevron Units of 150 mm and 200 mm Thickness First Edition. 49 pp.

—Measurement Systems
IEC 982-89. Level Measuring Systems Utilizing Ionizing Radiation with Continuous or Switching Output First Edition. 33 pp.

—Measurement Systems—Thickness Measurement
CENELEC HD 468-87. Ionizing Radiation Measurement Systems with Analogue or Digital Signal Processing for Thickness Measurements. 3 pp.
CENELEC HD 468 S1-89. Ionizing Radiation Measurement Systems with Analogue or Digital Signal Processing for Thickness Measurements. 3 pp.
IEC 769-83. Ionizing Radiation Measurement Systems with Analogue or Digital Signal Processing for Thickness Measurements First Edition. 97 pp.

—Medical Equipment—Symbols
BSI BS 7139-89. 1989 Guide to Graphical Symbols for Use on Medical Electrical Equipment. 51 pp.
IEC 878-88. Graphical Symbols for Electrical Equipment in Medical Practice First Edition. 63 pp.

—Radiation Measuring Instruments
IEC 181-64. Index of Electrical Measuring Apparatus Used in Connection with Ionizing Radiation First Edition; (Supplement A-B:1965) (Amendment 1-1967). 211 pp.
IEC 405-72. Nuclear Instruments: Constructional Requirements to Afford Personal Protection Against Ionizing Radiation First Edition. 21 pp.
IEC 476-93. Nuclear Instruments—Electrical Measuring Systems and Instruments Utilizing Ionizing Radiation Sources—General Aspects Second Edition. 47 pp.

—Radiation Measuring Instruments—Symbols
CNS C5100-81. Graphical Symbols of Detectors and Measuring Instrument Accessories for Ionizing Radiation (Mar)(7020).

—Symbols
BSI BS 3510-68. 1968 A Basic Symbol to Denote the Actual or Potential Presence of Ionizing Radiation. 8 pp.
BSI BS 7139-89. 1989 Guide to Graphical Symbols for Use on Medical Electrical Equipment. 51 pp.
IEC 878-88. Graphical Symbols for Electrical Equipment in Medical Practice First Edition. 63 pp.
ISO 361-75. Basic Ionizing Radiation Symbol First Edition. 3 pp.

—Units of Measurement
BSI BS 5775: Part 10-93. 1993 Quantities, Units and Symbols Part 10: Nuclear Reactions and Ionizing Radiations (ISO 31-10: 1992) (G). 34 pp.
BSI BS 5775: Part 10-82. 1982 Amd 1 Specification for Quantities, Units and Symbols Part 10: Nuclear Reactions and Ionizing Radiations. 22 pp.
CNS Z7196-10-85. Quantities and Units of Nuclear Reactions and Ionizing Radiations (Dec)(11296-10).
ISO 31 Pt 10-92. Quantities and Units—Part 10: Nuclear Reactions and Ionizing Radiations Third Edition. 32 pp.

Ionizing Radiation Shielding
Use: Radiation Shields (Protective)

Ionosondes
See Also: Measuring Instruments
—Ionosphere
CCIR Report 430-5-90. Improvement in the World-Wide Ionospheric Observing Programme for Numerical Mapping Purposes—Section 6G—Ionospheric Propagation Measurements and Data Banks. 8 pp.

Ionosphere
See Also: Ionospheric Channel Sounding Systems; Ionospheric Disturbances; Ionospheric Propagation; Ionospheric Sounding Systems; Magnetosphere
CCIR Report 725-3-90. Ionospheric Properties—Section 6A—Ionospheric Properties. 16 pp.
CCIR Report 340-6-91. CCIR Atlas of Ionospheric Characteristics—Section 6E—Ionospheric Propagation Prediction at Frequencies Between About 1.6 and 30 MHz. 116 pp.
CCIR Report 340-6-86. CCIR Atlas of Ionospheric Characteristics—Section 6E—Ionospheric Propagation Prediction at Frequencies Between About 1.6 and 30 MHz. 1 p.

—Above—Radio Astronomy—Frequencies
CCIR QUESTION 149/7-90. Frequency Utilization Above the Ionosphere and on the Far Side of the Moon—Questions Concerning Study Group 7—Science Services. 1 p.

—Cross Modulation—Sound Broadcasting
CCIR RECMN 498-2-90. Ionospheric Cross-Modulation in the LF and MF Broadcasting Bands—Section 10A-1—Amplitude-Modulation Sound Broadcasting in Bands 5 (LF), 6 (MF) and 7 (HF). 9 pp.

—Forecasting
CCIR RECMN 434-5-92. CCIR Reference Ionospheric Characteristics and Methods of Basic MUF, Operational MUF and Ray-Path Prediction Section 6E—Ionospheric Propagation Prediction at Frequencies Between About 1.6 and 30 MHz. 13 pp.
CCIR Report 888-2-90. Short-Term Forecasting of Critical Frequencies, Operational Maximum Usable Frequencies and Total Electron Content—Section 6C—Ionospheric Propagation and Operational Forecasting. 9 pp.
CCIR RECMN 846-92. Measurements of Ionospheric and Related Characteristics. 1 p.

—Forecasting—Computer Programs
CCIR RESOLUTION 63-3-90. Computer Programs for the Prediction of Ionospheric Characteristics, Sky-Wave Transmission Loss and Noise—Volume VI—Propagation in Ionized Media. 7 pp.

—High Latitudes—Radio Communications
CCIR Report 886-2-90. Special Properties of the High Latitude Ionosphere Affecting Radiocommunications—Section 6A—Ionospheric Properties. 25 pp.

—Indices—Forecasting
CCIR RECMN 371-6-90. Choice of Indices for Long-Term Ionospheric Predictions—Section 6E—Ionospheric Propagation Prediction at Frequencies Between About 1.6 and 30 MHz. 6 pp.
CCIR QUESTION 34/6-90. Long-Term Predictions of Solar and Ionospheric Indices—Questions Concerning Study Group 6—Radio Wave Propagation in Ionized Media. 1 p.

—Ionosondes
CCIR Report 430-5-90. Improvement in the World-Wide Ionospheric Observing Programme for Numerical Mapping Purposes—Section 6G—Ionospheric Propagation Measurements and Data Banks. 8 pp.

—Modification—Radio Transmission
CCIR RECMN 532-78. Ionospheric Modification by High Power Transmissions—Section 6A—Ionospheric Properties. 1 p.
CCIR RECMN 532-1-92. Ionospheric Effects and Operational Considerations Associated with Artificial Modification of the Ionosphere and the Radio-Wave Channel Section 6A—Ionospheric Properties. 12 pp.
CCIR Report 728-3-90. Ionospheric Modification by Ground-Based, High-Power Radio Transmissions—Section 6A—Ionospheric Properties. 10 pp.
CCIR QUESTION 39/6-90. Ionospheric Effects Caused by High-Power Transmissions—Questions Concerning Study Group 6—Radio Wave Propagation in Ionized Media. 1 p.

—Modification—Rocket Exhaust
CCIR Report 1011-1-90. Artificial Modification of the Ionosphere by Chemical Injections—Section 6A—Ionospheric Properties. 2 pp.

—Noise (Spurious Signals)
CCIR Report 342-6-90. Radio Noise Within and Above the Ionosphere—Section 6B—Radio Noise. 16 pp.

—Observation—Mapping
CCIR Report 430-5-90. Improvement in the World-Wide Ionospheric Observing Programme for Numerical Mapping Purposes—Section 6G—Ionospheric Propagation Measurements and Data Banks. 8 pp.

INDUSTRY STANDARDS

Ionosphere (Cont.)

—Observation Stations
CCIR Report 430-5-90. Improvement in the World-Wide Ionospheric Observing Programme for Numerical Mapping Purposes—Section 6G—Ionospheric Propagation Measurements and Data Banks. 8 pp.

—Radio Communications
CCIR QUESTION 25-2/6-90. Ionospheric Properties—Questions Concerning Study Group 6—Radio Wave Propagation in Ionized Media. 1 p.

—Radio Communications—Forecasting
CCIR QUESTION 27-1/6-90. Short-Term Forecasting of Operational Parameters for Ionospheric and Trans-Ionospheric Radiocommunications—Questions Concerning Study Group 6—Radio Wave Propagation in Ionized Media. 1 p.

—Solar Power Satellites
CCIR Report 893-1-90. Solar Power Satellites and the Ionosphere—Section 6A—Ionospheric Properties. 2 pp.

Ionospheric Channel Simulators

See Also: Ionospheric Channel Sounding Systems; Ionospheric Propagation

—Comparative Testing
CCIR RECMN 520-1-82. Use of High Frequency Ionospheric Channel Simulators—Section 3Ac—Influence of the Ionosphere. 2 pp.
CCIR RECMN 520-2-92. Use of High Frequency Ionospheric Channel Simulators—Section 3Ac—Influence of the Ionosphere. 4 pp.

—High Frequency
CCIR RECMN 520-1-82. Use of High Frequency Ionospheric Channel Simulators—Section 3Ac—Influence of the Ionosphere. 2 pp.
CCIR RECMN 520-2-92. Use of High Frequency Ionospheric Channel Simulators—Section 3Ac—Influence of the Ionosphere. 4 pp.
CCIR Report 549-3-90. HF Ionospheric Channel Simulators—Section 3Ac—Influence of the Ionosphere. 13 pp.

—Qualification Testing
CCIR RECMN 520-1-82. Use of High Frequency Ionospheric Channel Simulators—Section 3Ac—Influence of the Ionosphere. 2 pp.
CCIR RECMN 520-2-92. Use of High Frequency Ionospheric Channel Simulators—Section 3Ac—Influence of the Ionosphere. 4 pp.

Ionospheric Channel Sounding Systems

See Also: Ionosphere; Ionospheric Channel Simulators; Ionospheric Propagation; Ionospheric Sounding Systems

—Fixed Services (Radio Communications)—Frequencies
CCIR RECMN 613-86. Use of Ionospheric Channel Sounding Systems Operating in the Fixed Service at Frequencies Below About 30 MHz—Section 3Ac—Influence of the Ionosphere. 1 p.

Ionospheric Disturbances

See Also: Ionosphere; Magnetic Disturbances; Radio Frequency Interference

—Forecasting—Information Interchange
CCIR RECMN 313-6-90. Exchange of Information for Short-Term Forecasts and Transmission of Ionospheric Disturbance Warnings—Section 6C—Ionospheric Propagation and Operational Forecasting. 5 pp.
CCIR RECMN 313-7-92. Exchange of Information for Short-Term Forecasts and Transmission of Ionospheric Disturbance Warnings—Section 6C—Ionospheric Propagation and Operational Forecasting. 6 pp.

—Forecasting—Ionospheric Propagation
CCIR Report 763-3-90. Signal Level Variation Due to Multipath Effects and Blockage by Ship's Superstructure in Maritime Mobile-Satellite Service Links—Section 8I—Technical and Operating Characteristics of Mobile Satellite Services. 13 pp.

—Indices—Forecasting
CCIR QUESTION 34/6-90. Long-Term Predictions of Solar and Ionospheric Indices—Questions Concerning Study Group 6—Radio Wave Propagation in Ionized Media. 1 p.

Ionospheric Propagation

See Also: Ionosphere; Ionospheric Channel Simulators; Ionospheric Channel Sounding Systems; Ionospheric Sounding Systems; Maximum Usable Frequencies; Meteoric Scatter; Optimum Working Frequencies; Radio Transmission; Radio Wave Propagation; Scatter Propagation; Sporadic E Propagation; Wave Propagation

CCIR RECMN 434-4-82. CCIR Atlas of Ionospheric Characteristics—Section 6E—Ionospheric Propagation Prediction at Frequencies Between About 1.6 and 30 MHz. 1 p.
CCIR Report 340-6-91. CCIR Atlas of Ionospheric Characteristics—Section 6E—Ionospheric Propagation Prediction at Frequencies Between About 1.6 and 30 MHz. 116 pp.
CCIR Report 340-6-86. CCIR Atlas of Ionospheric Characteristics—Section 6E—Ionospheric Propagation Prediction at Frequencies Between About 1.6 and 30 MHz. 1 p.

—0-1.7 MHz—Field Strength
CCIR OPINION 69-82. Field-Strength Measurements for Frequencies Below About 1.7 MHz—Volume VI—Propagation in Ionized Media. 1 p.

—0-500 kHz—Field Strength
CCIR RECMN 684-90. Prediction of Field Strength at Frequencies Below About 500 khz—Section 6D—Ionospheric Propagation Prediction at Frequencies Below About 1.6 MHz. 1 p.

—0-500 kHz—Sky Waves—Field Strength
CCIR Decision 9-5-89. Radio Propagation and Circuit Performance at Frequencies Below About 500 khz—Annex to Volume VI—Propagation in Ionized Media. 2 pp.

—1.6 MHz+—Sky Waves—Signal Intensity
CCIR Report 253-5-90. Measurement of Sky-Wave Signal Intensities at Frequencies Above 1.6 MHz—Section 6G—Ionospheric Propagation Measurements and Data Banks. 21 pp.

—Antennas
CCIR QUESTION 40/6-90. Antenna Characteristics—Questions Concerning Study Group 6—Radio Wave Propagation in Ionized Media. 1 p.

—Atmospheric Sounding
CCIR Report 249-7-90. Use of Oblique Sounding for Propagation Analysis and Optimization—Section 6C—Ionospheric Propagation and Operational Forecasting. 6 pp.

—Backscattering
CCIR Report 726-2-90. Ground and Ionospheric Side- and Back-Scatter—Section 6C—Ionospheric Propagation and Operational Forecasting. 7 pp.
CCIR Report 890-2-90. Operational Use of Side-Scatter and Back-Scatter—Section 6C—Ionospheric Propagation and Operational Forecasting. 3 pp.
CCIR QUESTION 32/6-90. Progagation Via Side-and Back-Scatter—Questions Concerning Study Group 6—Radio Wave Propagation in Ionized Media. 1 p.

—Cross Modulation
CCIR Report 574-2-86. Ionospheric Cross-Modulation—Section 6A—Ionospheric Properties. 9 pp.

—Data Banks
CCIR QUESTION 42/6-90. Measurements and Data Banks—Questions Concerning Study Group 6—Radio Wave Propagation in Ionized Media. 1 p.

—ELF/VLF—Field Strength
CCIR Report 895-2-90. Radio Propagation and Circuit Performance at Frequencies Below About 30 khz—Section 6D—Ionospheric Propagation Prediction at Frequencies Below About 1.6 MHz. 29 pp.

—ELF/VLF/LF
CCIR Report 262-7-90. ELF, VLF and LF Propagation in and Through the Ionosphere—Section 6C—Ionospheric Propagation and Operational Forecasting. 5 pp.

—Forecasting
CCIR RECMN 434-5-92. CCIR Reference Ionospheric Characteristics and Methods of Basic MUF, Operational MUF and Ray-Path Prediction Section 6E—Ionospheric Propagation Prediction at Frequencies Between About 1.6 and 30 MHz. 13 pp.
CCIR Report 255-7-90. Long-Term Ionospheric Propagation Predictions—Section 6E—Ionospheric Propagation Prediction at Frequencies Between About 1.6 and 30 MHz. 9 pp.
CCIR QUESTION 33/6-90. Ionospheric Propagation Predictions—Questions Concerning Study Group 6—Radio Wave Propagation in Ionized Media. 1 p.
CCIR RECMN 846-92. Measurements of Ionospheric and Related Characteristics. 1 p.

—Forecasting—Observation Stations
CCIR OPINION 67-82. Geophysical and Solar Observations Needed for Short-Term Forecasting of Ionospheric Propagation—Volume VI—Propagation in Ionized Media. 1 p.

—Frequencies
CCIR RECMN 373-6-90. Definitions of Maximum and Minimum Transmission Frequencies—Section 6C—Ionospheric Propagation and Operational Forecasting. 1 p.

—HF—Data Banks
CCIR Decision 83-89. HF Measurements and Data Banks—Annex to Volume VI—Propagation in Ionized Media. 2 pp.

—HF—Field Strength
CCIR RESOLUTION 111-90. HF Field-Strength Measurement Campaign—Volume VI—Propagation in Ionized Media. 1 p.
CCIR Report 1149-90. HF Field Strength Measurements Specifications for a Field Strength Measurement Campaign Intended for Future Improvements in Prediction Methods, Particularly Those Used for HF Broadcasting—Section 6G—Ionospheric Propagation Measurements. 17 pp.
CCIR Decision 84-89. Potential for an HF Field-Strength Measurement Campaign to Provide a Significant Measurement Data Bank in Time for Preparations for a WARC HFBC in 1993—Annex to Volume VI—Propagation in Ionized Media. 1 p.

—HF—Forecasting
CCIR RESOLUTION 112-90. CCIR Study Group 6 Report to the WARC HFBC(93)—Volume VI—Propagation in Ionized Media. 1 p.
CCIR OPINION 45-3-90. Evaluation of the CCIR HF Propagation Prediction Methods—Volume VI—Propagation in Ionized Media. 1 p.
CCIR Report 894-2-90. CCIR HF Propagation Prediction Method Third CCIR Computer-Based Method for Estimation of MUF, Sky-Wave Field Strength, Signal-to-Noise Ratio, LUF and Basic Circuit Reliability—Section 6E—Ionospheric Propagation Prediction at Frequencies Between. 12 pp.
CCIR Decision 85-89. Studies of the Propagation Prediction Method for HF Broadcasting—Annex to Volume VI—Propagation in Ionized Media. 1 p.

—HF—High Latitudes
CCIR Report 1012-1-90. Operational Modelling of HF Radio Propagation Conditions at High Latitudes—Section 6E—Ionospheric Propagation Prediction at Frequencies Between About 1.6 and 30 MHz. 5 pp.

—HF/SHF—Space Communications
CCIR Report 263-7-90. Ionospheric Effects Upon Earth-Space Propagation—Section 6F—Ionospheric Propagation Prediction and Applications at Frequencies Above About 30 MHz. 31 pp.

—HF—Sky Waves—Signal Intensity
CCIR Report 1150-90. Standardized Procedure for Comparing Predicted and Observed HF Sky Wave Signal Intensities and Results of Such Comparisons—Section 6G—Ionospheric Propagation Measurements and Data Banks. 9 pp.

—HF—Sky Waves—Signal Intensity—Data Banks
CCIR OPINION 68-1-90. Data Bank of HF Sky-Wave Signal Intensity Measurements—Volume VI—Propagation in Ionized Media. 1 p.

—HF/VHF—Long Distance
CCIR Report 250-6-86. Long-Distance Ionospheric Propagation Without Intermediate Ground Reflection—Section 6C—Ionospheric Propagation and Operational Forecasting. 8 pp.

—Indices
CCIR RESOLUTION 4-4-90. Dissemination of Basic Indices for Ionospheric Propagation—Volume VI—Propagation in Ionized Media. 1 p.

—Indices—Observation Stations
CCIR OPINION 23-4-86. Observations Needed to Provide Basic Indices for Ionospheric Propagation—Volume VI—Propagation in Ionized Media. 1 p.

—Ionospheric Disturbances—Forecasting
CCIR Report 727-3-90. Short-Term Prediction of Solar-Induced Variations of Operational Parameters for Ionospheric Propagation—Section 6C—Ionospheric Propagation and Operational Forecasting. 10 pp.

Ionospheric Propagation (Cont.)

—ISDN—Spacecraft

CCIR RECMN 531-1-90. Ionospheric Effects Influencing Radio Systems Involving Spacecraft—Section 6F—Ionospheric Propagation Prediction and Applications at Frequencies Above About 30 MHz. 1 p.

CCIR RECMN 531-2-92. Ionospheric Effects Influencing Radio Systems Involving Spacecraft—Section 6F—Ionospheric Propagation Prediction and Applications at Frequencies Above About 30 MHz. 14 pp.

—LF/MF—Sky Waves—Field Strength

CCIR RECMN 435-6-90. Prediction of Sky-Wave Field Strength Between 150 and 1600 khz—Section 6D—Ionospheric Propagation Prediction at Frequencies Below About 1.6 MHz. 24 pp.

CCIR Report 265-7-90. Sky-Wave Propagation and Circuit Performance at Frequencies Between About 30 khz and 500 khz—Section 6D—Ionospheric Propagation Prediction at Frequencies Below About 1.6 MHz. 18 pp.

CCIR Report 431-5-90. Analysis of Sky-Wave Propagation Measurements for the Frequency Range 150 to 1600 khz—Section 6D—Ionospheric Propagation Prediction at Frequencies Below About 1.6 MHz. 21 pp.

CCIR Report 432-2-86. Accuracy of Predictions of Sky-Wave Field Strength in Bands 5 (LF) and 6 (MF)—Section 6D—Ionospheric Propagation Prediction at Frequencies Below About 1.6 MHz. 6 pp.

CCIR Report 575-4-90. Methods for Predicting Sky-Wave Field Strengths at Frequencies Between 150 khz and 1705 khz—Section 6D—Ionospheric Propagation Prediction at Frequencies Below About 1.6 MHz. 18 pp.

CCIR Decision 57-2-89. Sky-Wave Propagation at Frequencies Between 150 and 1700 khz—Annex to Volume VI—Propagation in Ionized Media. 2 pp.

—Magnetic Disturbances—Forecasting

CCIR Report 727-3-90. Short-Term Prediction of Solar-Induced Variations of Operational Parameters for Ionospheric Propagation—Section 6C—Ionospheric Propagation and Operational Forecasting. 10 pp.

—MF—Aircraft—Sky Waves—Field Strength

CCIR RECMN 683-90. Sky-Wave Field Strength Prediction Method for Propagation to Aircraft at About 500 khz—Section 6D—Ionospheric Propagation Prediction at Frequencies Below About 1.6 MHz. 7 pp.

—MF/HF—Sky Waves—Field Strength

CCIR RECMN 533-2-90. Estimating Sky-Wave Field Strength at Frequencies Between 2 and 30 MHz—Section 6E—Ionospheric Propagation Prediction at Frequencies Between About 1.6 and 30 MHz. 1 p.

CCIR RECMN 533-3-92. CCIR HF Propagation Prediction Method Section 6E—Ionospheric Propagation Prediction at Frequencies Between About 1.6 and 30 MHz. 15 pp.

CCIR Report 252-2-70. CCIR Interim Method for Estimating Sky-Wave Field Strength and Transmission Loss at Frequencies Between the Approximate Limits of 2 and 30 MHz—Section 6E—Ionospheric Propagation Prediction at Frequencies Between About 1.6 and 30 MHz. 1 p.

CCIR Report 252-2 Supplement-80. Second CCIR Computer-Based Interim Method for Estimating Sky-Wave Field Strength and Transmission Loss at Frequencies Between 2 and 30 MHz—Section 6E—Ionospheric Propagation Prediction at Frequencies Between About 1.6 and 30 MHz. 44 pp.

CCIR Report 252-2 Supp-82. Second CCIR Computer-Based Interim Method for Estimating Sky-Wave Field Strength and Transmission Loss Frequencies Between 2 and 30 MHz—Section 6E—Ionospheric Propagation Prediction at Frequencies Between About 1.6 and 30 MHz. 1 p.

CCIR Report 729-3-90. Developments in the Estimation of Sky-Wave Field Strength and Transmission Loss at Frequencies Above 1.5 MHz—Section 6E—Ionospheric Propagation Prediction at Frequencies Between About 1.6 and 30 MHz. 8 pp.

CCIR Decision 6-7-89. Sky-Wave Field Strength and Transmission Loss at Frequencies Above 1.6 MHz—Annex to Volume VI—Propagation in Ionized Media. 2 pp.

—MF/HF—Sky Waves—Signal Intensity

CCIR Report 571-4-90. Comparisons Between Observed and Predicted Sky-Wave Signal Intensities at Frequencies Between 2 and 30 MHz—Section 6G—Ionospheric Propagation Measurements and Data Banks. 9 pp.

Ionospheric Propagation (Cont.)

—MF/HF—Sky Waves—Transmission Loss

CCIR Report 252-2-70. CCIR Interim Method for Estimating Sky-Wave Field Strength and Transmission Loss at Frequencies Between the Approximate Limits of 2 and 30 MHz—Section 6E—Ionospheric Propagation Prediction at Frequencies Between About 1.6 and 30 MHz. 1 p.

CCIR Report 252-2 Supplement-80. Second CCIR Computer-Based Interim Method for Estimating Sky-Wave Field Strength and Transmission Loss at Frequencies Between 2 and 30 MHz—Section 6E—Ionospheric Propagation Prediction at Frequencies Between About 1.6 and 30 MHz. 44 pp.

CCIR Report 252-2 Supp-82. Second CCIR Computer-Based Interim Method for Estimating Sky-Wave Field Strength and Transmission Loss Frequencies Between 2 and 30 MHz—Section 6E—Ionospheric Propagation Prediction at Frequencies Between About 1.6 and 30 MHz. 1 p.

CCIR Report 729-3-90. Developments in the Estimation of Sky-Wave Field Strength and Transmission Loss at Frequencies Above 1.5 MHz—Section 6E—Ionospheric Propagation Prediction at Frequencies Between About 1.6 and 30 MHz. 8 pp.

CCIR Decision 6-7-89. Sky-Wave Field Strength and Transmission Loss at Frequencies Above 1.6 MHz—Annex to Volume VI—Propagation in Ionized Media. 2 pp.

—Radio Circuits—Real Time Channel Evaluation

CCIR Report 889-2-90. Real-Time Channel Evaluation of HF Ionospheric Radio Circuits—Section 6C—Ionospheric Propagation and Operational Forecasting. 17 pp.

—Radio Communications

CCIR QUESTION 35/6-90. Variations of Ionospheric Propagation Characteristics and Fading—Questions Concerning Study Group 6—Radio Wave Propagation in Ionized Media. 1 p.

—Radio Communications—Terrestrial—Frequency Band Sharing

CCIR QUESTION 38/6-90. Propagation Factors Affecting the Sharing of the Radio-Frequency Spectrum Between Terrestrial Systems Involving Ionospheric Propagation—Questions Concerning Study Group 6—Radio Wave Propagation in Ionized Media. 1 p.

—Radio Frequency Interference—Forecasting

CCIR Report 727-3-90. Short-Term Prediction of Solar-Induced Variations of Operational Parameters for Ionospheric Propagation—Section 6C—Ionospheric Propagation and Operational Forecasting. 10 pp.

—Radio Systems—Spacecraft

CCIR RECMN 531-1-90. Ionospheric Effects Influencing Radio Systems Involving Spacecraft—Section 6F—Ionospheric Propagation Prediction and Applications at Frequencies Above About 30 MHz. 1 p.

CCIR RECMN 531-2-92. Ionospheric Effects Influencing Radio Systems Involving Spacecraft—Section 6F—Ionospheric Propagation Prediction and Applications at Frequencies Above About 30 MHz. 14 pp.

—Sidescattering

CCIR Report 890-2-90. Operational Use of Side-Scatter and Back-Scatter—Section 6C—Ionospheric Propagation and Operational Forecasting. 3 pp.

CCIR Report 762-2-82. Effects of Multipath on Digital Transmission over Links in the Maritime Mobile-Satellite Service—Section 8I—Technical and Operating Characteristics of Mobile Satellite Services. 16 pp.

CCIR QUESTION 32/6-90. Propagation Via Side-and Back-Scatter—Questions Concerning Study Group 6—Radio Wave Propagation in Ionized Media. 1 p.

—Signal Intensity—Data Banks

CCIR QUESTION 42/6-90. Measurements and Data Banks—Questions Concerning Study Group 6—Radio Wave Propagation in Ionized Media. 1 p.

—Sky Waves—Antennas

CCIR Report 891-2-90. Antenna Characteristics Important for the Analysis and Prediction of Sky-Wave Propagation Paths—Section 6E—Ionospheric Propagation Prediction at Frequencies Between About 1.6 and 30 MHz. 6 pp.

—Sky Waves—Transmission Loss

CCIR RESOLUTION 63-3-90. Computer Programs for the Prediction of Ionospheric Characteristics, Sky-Wave Transmission Loss and Noise—Volume VI—Propagation in Ionized Media. 7 pp.

Ionospheric Propagation (Cont.)

—Space Communications

CCIR QUESTION 36/6-90. Ionospheric Influences on Space Communications at Frequencies Below About 1.6 MHz—Questions Concerning Study Group 6—Radio Wave Propagation in Ionized Media. 1 p.

CCIR QUESTION 37/6-90. Ionospheric Influences on Space Systems at Frequencies Above About 1.6 MHz—Questions Concerning Study Group 6—Radio Wave Propagation in Ionized Media. 1 p.

—VHF

CCIR Report 259-7-90. VHF Ionospheric Propagation—Section 6F—Ionospheric Propagation Prediction and Applications at Frequencies Above About 30 MHz. 20 pp.

—VHF—Frequency Band Sharing

CCIR RECMN 844-92. Ionospheric Factors Affecting Frequency Sharing in the VHF (30-300 MHz) Band. 6 pp.

—VHF/UHF

CCIR Decision 11-6-89. VHF and UHF Propagation by Sporadic E and Other Ionospheric Regions—Annex to Volume VI—Propagation in Ionized Media. 2 pp.

CCIR QUESTION 41/6-90. VHF and UHF Propagation by Way of Sporadic E and Other Ionization—Questions Concerning Study Group 6—Radio Wave Propagation in Ionized Media. 1 p.

Ionospheric Sounding Systems

See Also: Ionosphere; Ionospheric Channel Sounding Systems; Ionospheric Propagation

CCIR Report 357-2-86. Operational Ionospheric-Sounding Systems at Oblique Incidence—Section 3Ac—Influence of the Ionosphere. 3 pp.

CCIR OPINION 22-5-90. Routine Ionospheric Sounding—Volume VI—Propagation in Ionized Media. 1 p.

IPE

Use: Personal Protective Equipment

IPMS

Use: Interpersonal Messaging Systems

IPP

Use: International Phototelegraph Positions

IQI

Use: Radiographic Image Quality Indicators

IR (Isoprene Rubber)

Use: Isoprene Rubber

IRA

Use: International Reference Alphabet

IRDS

Use: Information Resource Dictionary System

IRED

Use: Infrared Emitting Diodes

IRHD

Use: International Rubber Hardness Degrees

Iron

Scope Note: For additional listings, see also specific products made from iro *See Also:* Cast Iron; Ferroalloys; Iron Chromium; Iron Coatings (On Iron); Iron Content Analysis; Iron Ores; Iron Powder; Metals; Pig Iron; Steels

EC SM 1. Iron and Steel. 1 p.

ISO TR9769-91. Steel and Iron—Review of Available Methods of Analysis Third Edition. 47 pp.

—Aircraft

MOD UK DTD-5092-01. Soft Iron for Dynamo-Electric Machines (Type A); Amendment 1. 4 pp.

MOD UK DTD-5102-01. Soft Iron for Dynamo-Electric Machines (Type B); Amendment 1. 4 pp.

—Aluminum Content

CNS G2242-85. Method of Determination for Aluminium in Iron and Steel (Apr)(11244).

JIS G 1224-81. Methods for Determination of Aluminium in Iron and Steel.

—Aluminum Content—Photometry

SAA AS K1.24-68. Methods for the Sampling and Analysis of Iron and Steel Method for the Determination of Aluminium in Iron and Steel (Photometric Method) Reconfirmed 1987. 7 pp.

—Annealing

CNS B1277-81. Normalizing and Annealing of Iron and Steel (Nov)(8103).

JIS B 6911-87. Normalizing and Annealing of Iron and Steel. 21 pp.

Iron (Cont.)

—Antimony Content
CNS G2257-85. Method of Determination for Antimony in Iron and Steel (Dec)(11449).
JIS G 1235-81. Methods for Determination of Antimony in Iron and Steel. 11 pp.

—Arsenic Content
CNS G2243-85. Method of Determination for Arsenic in Iron and Steel (Apr)(11245).
JIS G 1225-81. Methods for Determination of Arsenic in Iron and Steel.

—Arsenic Content—Spectrophotometry
CEN PREN 10212-91. Chemical Analysis of Ferrous Materials—Determination of Arsenic in Steel and Iron—Spectrophotometric Method. 12 pp.

—Atomic Absorption Spectrometry
BSI BS 6200: Sec 6.1-90. 1990 Sampling and Analysis of Iron, Steel and Other Ferrous Metals: Part 6: Guidelines on Atomic Absorption Spectrometric Techniques: Section 6.1: Recommendations for the Drafting of Standard Methods for the Chemical Anal. of Iron and Steel by Flame Atomic Absorp. 21 pp.
BSI BS 6200: Sec 6.2-90. 1990 Sampling and Analysis of Iron, Steel and Other Ferrous Metals: Part 6: Guidelines on Atomic Absorption Spectrometric Techniques: Section 6.2: Recom. for the Appl. of Flame Atomic Absorp. Spectrometry in Standard Methods for the Chemical Anal. of Iron and Steel. 17 pp.

—Bars—Magnetic
CNS C4191-88. Soft Magnetic Bars (Jul)(5524).
JIS C 2503-90. Soft Magnetic Iron Bars. 8 pp.

—Boron Content
CNS G2246-85. Method of Determination for Boron in Iron and Steel (Jul)(11303).
JIS G 1227-92. Methods for Determination of Boron in Iron and Steel. 19 pp.

—Carbon Content
CNS G2228-84. Method of Determination for Carbon in Iron and Steel (Oct)(11069).
JIS G 1211-81. Methods for Determination of Carbon in Iron and Steel. 14 pp.

—Carbon Content—Gravimetric Analysis
BSI BS EN 10 036-91. 1991 Chemical Analysis of Ferrous Materials Determination of Total Carbon in Steels and Irons. Gravimetric Method After Combustion in a Stream of Oxygen (V). 10 pp.
CEN EN 10 036-89. Chemical Analysis of Ferous Marterials Determination of Total Carbon in Steels and Irons Gravimetric After Combustion in a Stream of Oxygen. 6 pp.
DIN ENGL EN 10036-90. Chemical Analysis of Ferrous Materials; Determination of Total Carbon in Steel and Iron; Gravimetric Method After Combustion in a Stream of Oxygen (Apr). 7 pp.
SAA AS 1050.2-84. Methods for the Analysis of Iron and Steel—Part 2: Determination of Carbon Content (Gravimetric Method) See Also AS K1. 5 pp.

—Carbon Content—Infrared Analysis
BSI BS 6200: SUB SEC3.8.3-90. 1990 Sampling and Analysis of Iron, Steel and Other Ferrous Metals Part 3: Methods of Analysis Section 3.8: Determination of Carbon Subsection 3.8.3: Steel and Cast Iron: Infra-Red Absorption Method After Combustion in an Induction Furnace. 13 pp.
ISO 9556-89. Steel and Iron—Determination of Total Carbon Content—Infrared Absorption Method After Combustion in an Induction Furnace First Edition; (Corrected and Reprinted -1989). 11 pp.
ISO 9686-92. Direct Reduced Iron—Determination of Carbon and/or Sulfur Content—High Frequency Combustion Method with Infrared Measurement First Edition. 15 pp.
SAA AS 1050.32-84. Methods for the Analysis of Iron and Steel—Part 32: Determination of Carbon Content (Infrared Method). 3 pp.

—Chemical Analysis
BSI BS 4237-67. 1967 Report on Reproducibility of Methods of Chemical Analysis Used in the Iron and Steel Industry. 32 pp.
BSI BS 6200: Part 1-91. 1991 Sampling and Analysis of Iron, Steel and Other Ferrous Metals Part 1: Introduction and Contents. 12 pp.
BSI BS 6200: Sec 3.0-91. 1991 Sampling and Analysis of Iron, Steel and Other Ferrous Metals Part 3: Methods of Analysis Section 3.0: Summary of Methods. 10 pp.
CNS G2221-84. General Rules of Chemical Analysis for Iron and Steel (Sep)(11012).
JIS G 1201-92. General Rules for Chemical Analysis of Iron and Steel. 7 pp.

—Chromium Content
CNS G2245-85. Method of Determination for Chromium in Iron and Steel (Jul)(11302).
JIS G 1217-81. Methods for Determination of Chromium in Iron and Steel. 20 pp.

—Chromium Content—Atomic Absorption Spectrometry
BSI BS 6200: SUB SEC 3.10.2-89. 1989 Sampling and Analysis of Iron, Steel and Other Ferrous Metals Part 3: Methods of Analysis Section 3.10 Determination of Chromium Subsection 3.10.2: Steel and Cast Iron: Flame Atomic Absorption Spectrometric Method. 12 pp.
CEN EN 10 188-89. Chemical Analysis of Ferrous Materials Determination of Chromium in Steels and Irons Flame Atomic Absorption Spectrometic Method. 9 pp.
DIN ENGL EN 10188-90. Chemical Analysis of Ferrous Materials; Determination of Chromium in Steels and Iron; Flame Atomic Absorption Spectrometric Method (Apr). 10 pp.
ISO 10138-91. Steel and Iron—Determination of Chromium Content—Flame Atomic Absorption Spectrometric Method First Edition. 12 pp.

—Chromium Content—Potentiometric Analysis
JIS G 1238-92. Steel and Iron—Determination of Chromium Content—Potentiometric on Visual Titration Method.

—Cobalt Content
CNS G2237-84. Method of Determination for Cobalt in Iron and Steel (Dec)(11168).
JIS G 1222-81. Methods for Determination of Cobalt in Iron and Steel.

—Cobalt Content—Atomic Absorption Spectrometry
BSI BS 6200: Sec 6.1-90. 1990 Sampling and Analysis of Iron, Steel and Other Ferrous Metals: Part 6: Guidelines on Atomic Absorption Spectrometric Techniques: Section 6.1: Recommendations for the Drafting of Standard Methods for the Chemical Anal. of Iron and Steel by Flame Atomic Absorp. 21 pp.
SAA AS 1050.29-89. Methods for the Analysis of Iron and Steel—Part 29: Determination of Cobalt Content—Flame Atomic Absorption Spectrometric Method. 3 pp.

—Cobalt Content—Photometry
SAA AS K1.21-65. Methods for the Sampling and Analysis of Iron and Steel Cobalt in Iron and Steel (Photometric Method) Corrig. Reconfirmed 1987. 6 pp.

—Cobalt Content—Spectrophotometry
BSI BS 6200: SUB SEC 3.11.2-91. 1991 Sampling and Analysis of Iron, Steel and Other Ferrous Metals Part 3: Methods of Analysis Section 3.11: Determination of Cobalt Subsec 3.11.2: Steel, Irons and Steelmaking Materials: Spectrophotometric Method for Trace Amounts. 10 pp.

—Copper Content
CNS G2234-84. Method of Determination for Copper in Iron and Steel (Dec)(11165).
JIS G 1219-81. Methods for Determination of Copper in Iron and Steel.

—Corrosion Prevention
SAA AS 2312-85. Guide to the Protection of Iron and Steel Against Exterior Atmospheric Corrosion (This is a Joint Standard with SANZ NZS 2312). 59 pp.
SNZ NZS/AS 2312-85. Guide to the Protection of Iron and Steel Against Exterior Atmospheric Corrosion (This is a Joint Standard with SAA AS 2312). 59 pp.

—Emission Spectroscopy
CNS G1021-84. General Rules on Emission-Spectroscopic Analysis for Iron and Steel (Oct)(9705).
JIS G 1202-75. General Rules on Emission-Spectroscopic Analysis for Iron and Steel (R 1978). 23 pp.

—Fabrics—Colorfastness Testing
CNS L3190-82. Method of Test for Colour Fastness to Metals in the Dyebath: Iron and Copper (Aug)(9313).
JIS L 0871-75. Testing Method for Colour Fastness to Metals in the Dyebath: Iron and Copper.

—Fertilizers
CNS N3094-88. Chelated Iron Fertilizer (Jul)(12028).

—Fluorite—Chemical Analysis
JIS M 8514-89. Methods for Chemical Analysis of Metallurgical Grade Fluorspar. 34 pp.

—Glossaries
BSI BS 6562: Part 2-86. (WITHDRAWN) 1986 Terms Used in the Iron and Steel Industry Part 2: Glossary of Terms Used in Classifying and Defining Steel Industry Products by Shape and Dimensions (Superseded by BS EN 10079: 1993). 27 pp.
CNS G1025-87. Glossary of Terms Used in Iron and Steel (Products and Quality) (Feb) (11836).
JIS G 0202-87. Glossary of Terms Used in Iron and Steel (Testing). 161 pp.
JIS G 0203-84. Glossary of Terms Used in Iron and Steel (Products and Quality). 42 pp.

—Grit
BSI BS 2451-63. (OBSOLESCENT) 1963 Chilled Iron Shot and Grit. 8 pp.

—Hardening—Induction Heating
CNS B1350-85. Hardening by Induction Heating and Tempering of Iron and Steel (Jun)(11277).
JIS B 6912-85. Hardening by Induction Heating and Tempering of Iron and Steel. 19 pp.

—Hardening—Tempering
JIS B 6912-85. Hardening by Induction Heating and Tempering of Iron and Steel. 19 pp.

—Heat Treatment—Glossaries
BSI BS 6562: Part 1-85. 1985 Terms Used in the Iron and Steel Industry Part 1: Glossary of Heat Treatment Terms. 31 pp.
CEN PREN 10052-91. Vocabulary of Heat Treatment Terms for Ferrous Products. 31 pp.
CNS G1024-87. Glossary of Terms Used in Heat Treatment of Iron and Steel (Feb)(11835).
DIN ENGL 17014 Pt 1-88. Heat Treatment of Ferrous Materials; Terminology (Aug). 18 pp.
JIS G 0201-87. Glossary of Terms Used in Iron and Steel (Heat Treatment). 56 pp.

—Heat Treatment—Numbering Systems
DIN ENGL 17014 Pt 3-76. Heat Treatment of Ferrous Materials; Notation to Indicate Heat Treatment Processes (May). 2 pp.

—Inspection
BSI BS EN 10204-91. 1991 Metallic Products—Types of Inspection Documents. 11 pp.
CEN PREN 10 204-89. Steel and Iron and Steel Products: Inspection Documents. 11 pp.
CEN EN 10204-91. Metallic Products—Types of Inspection Documents. 9 pp.
DIN ENGL 50049-92. Inspection Documents for the Delivery of Metallic Products (Apr) (This Standard Incorporates the English Version of EN 10204.). 6 pp.

—Knoop Hardness Testing
JIS G 0563-93. Method of Measuring Surface Hardness for Nitrided Iron and Steel. 6 pp.

—Lead Content—Atomic Absorption Spectrometry
SAA AS 1050.25-86. Methods for the Analysis of Iron and Steel—Part 25: Determination of Lead Content (Flame Atomic Absorption Spectrometric Method). 3 pp.

—Magnesium Content—Atomic Absorption Spectrometry
SAA AS 1050.20-83. Methods for the Analysis of Iron and Steel—Part 20: Determination of Magnesium in Iron and Steel (Flame Atomic Absorption Spectrometric Method) Amdt 1 March 1983 Reconfirmed 1989. 6 pp.

—Magnetic Particle Testing
CNS G1017-73. Method of Flaw Inspection for Steel and Iron Materials by Means of Magnetic Powder (Dec)(3666).

—Manganese Content
CNS G2223-84. Method of Determination for Manganese in Iron and Steel (Sep)(11014).
JIS G 1213-81. Methods for Determination of Manganese in Iron and Steel. 16 pp.

—Manganese Content—Atomic Absorption Spectrometry
ISO TR10281-90. Steel and Iron—Determination of Manganese Content—Flame Atomic Absorption Spectrometric Method First Edition. 14 pp.

—Manganese Content—Electrometric Analysis
BSI BS EN 10 071-91. 1991 Chemical Analysis of Ferrous Materials Determination of Manganese in Steels and Irons Electrometric Titration Method. 9 pp.
CEN EN 10 071-89. Chemical Analysis of Ferrous Materials Determination of Manganese in Steels and Irons Electrometric Titration Method. 5 pp.

INTERNATIONAL AND NON-U.S. NATIONAL STANDARDS
SUBJECT INDEX

Iron

Iron (Cont.)

—**Manganese Content—Electrometric Analysis (Cont.)**
DIN ENGL EN 10071-90. Chemical Analysis of Ferrous Materials; Determination of Manganese in Steel and Iron; Electrometric Titration Method (Apr). 6 pp.

—**Manganese Content—Spectrophotometry**
SAA AS 1050.8-80. Methods for the Analysis of Iron and Steel—Part 8: Determination of Manganese Content in Iron and Steel (Spectrophotometric Method) Reconfirmed 1989. 10 pp.

—**Manganese Content—Volumetric Analysis**
SAA AS 1050.14-81. Methods for the Analysis of Iron and Steel—Part 14: Determination of Manganese in Iron and Steel (Titrimetric Method) Corrig. Reconfirmed 1989. 8 pp.

—**Mineral Supplements—Animal Feed**
CNS N4069-81. Method of Test for Purity of Ferrous Salts in Mineral Premix (For Feeding) (Jan)(6944).

—**Molybdenum Content**
CNS G2253-85. Method of Determination for Molybdenum in Iron and Steel (Oct)(11389).
JIS G 1218-80. Methods for Determination of Molybdenum in Iron and Steel.

—**Molybdenum Content—Photometry**
SAA AS K1 Part 23-67. Methods for the Sampling and Analysis of Iron and Steel Part 23: Molybdenum in Iron and Steel (Photometric Method). 5 pp.

—**Neutron Flux—Radioactivation Analysis**
CNS J2052-82. Method for Determining Fast-Neutron Flux by Radioactivation of Iron (Jun)(8978).

—**Nickel Content**
CNS G2252-85. Method of Determination for Nickel in Iron and Steel (Oct)(11388).
JIS G 1216-81. Methods for Determination of Nickel in Iron and Steel. 18 pp.

—**Nickel Content—Atomic Absorption Spectrometry**
BSI BS EN 10 136-91. 1991 Chemical Analysis of Ferrous Materials. Determination of Nickel in Steel and Irons. Flame Atomic Absorption Spectrometric Method (V). 11 pp.
CEN EN 10 136-89. Chemical Analysis of Ferrous Materials Determination of Nickel in Steels and Irons Flame Atomic Absorption Spectrometric Method. 7 pp.
DIN ENGL EN 10136-90. Chemical Analysis of Ferrous Materials; Determination of Nickel in Steel and Iron; Flame Atomic Absorption Spectrometric Method (Apr). 8 pp.

—**Nickel Content—Gravimetric Analysis**
CEN EN 24 938-90. Steel and Iron—Determination of Nickel Content—Gravimetric or Titrimetic Method. 2 pp.
DIN ENGL EN 24938-92. Determination of Nickel Content of Steel and Iron by Gravimetry or Volumetric Analysis; (ISO 4938: 1988) (Oct). 11 pp.
ISO 4938-88. Steel and Iron—Determination of Nickel Content—Gravimetric or Titrimetric Method First Edition. 11 pp.

—**Nickel Content—Volumetric Analysis**
CEN EN 24 938-90. Steel and Iron—Determination of Nickel Content—Gravimetric or Titrimetic Method. 2 pp.
DIN ENGL EN 24938-92. Determination of Nickel Content of Steel and Iron by Gravimetry or Volumetric Analysis; (ISO 4938: 1988) (Oct). 11 pp.
ISO 4938-88. Steel and Iron—Determination of Nickel Content—Gravimetric or Titrimetric Method First Edition. 11 pp.

—**Nitrided Case Depth**
JIS G 0562-93. Method of Measuring Nitrided Case Depth for Iron and Steel. 12 pp.

—**Nitrogen Content**
CNS G2247-85. Method of Determination for Nitrogen in Iron and Steel (Jul)(11304).
JIS G 1228-80. Methods for Determination of Nitrogen in Iron and Steel.

—**Nitrogen Content—Volumetric Analysis**
ISO 10702-93. Steer and Iron—Determination of Nitrogen Content—Titrimetric Method After Distillation First Edition. 12 pp.

—**Normalizing**
CNS B1277-81. Normalizing and Annealing of Iron and Steel (Nov)(8103).
JIS B 6911-87. Normalizing and Annealing of Iron and Steel. 21 pp.

Iron (Cont.)

—**Phosphorus Content**
CNS G2224-84. Method of Determination for Phosphorus in Iron and Steel (Sep)(11015).
JIS G 1214-80. Methods for Determination of Phosphorus in Iron and Steel. 22 pp.

—**Phosphorus Content—Spectrophotometry**
CEN EN 10 184-89. Chemical Analysis of Ferrous Materials Determination of Phosphorous in Steels and Irons Spectrophotometric Method. 9 pp.
DIN ENGL EN 10184-90. Chemical Analysis of Ferrous Materials; Determination of Phosphorus in Steels and Iron; Spectrophotometric Method (Including Corrigendum AC: 1991) (Apr). 10 pp.
ISO 10714-92. Steel and Iron—Determination of Phosphorus Content—Phosphovanadomolybdate Spectrophotometric Method First Edition. 11 pp.
SAA AS 1050.18-84. Methods for the Analysis of Iron and Steel—Part 18: Determination of Phosphorous (Spectrophotometric Method). 5 pp.

—**Plates—Magnetic**
CNS C4190-80. Soft Magnetic Iron Plates (May)(5523).
JIS C 2504-90. Soft Magnetic Iron Plates. 9 pp.

—**Rockwell Superficial Hardness Testing**
JIS G 0563-93. Method of Measuring Surface Hardness for Nitrided Iron and Steel. 6 pp.

—**Sampling**
BSI BS 1837-70. (WITHDRAWN) 1970 Methods for the Sampling of Iron, Steel, Permanent Magnet Alloys and Ferro-Alloys (Superseded by BS 6200: Section 2.2: 1993). 20 pp.
BSI BS 6200: Part 1-91. 1991 Sampling and Analysis of Iron, Steel and Other Ferrous Metals Part 1: Introduction and Contents. 12 pp.
CNS G2002-53. Sampling Method of Chemical Test for Steel and Iron (Feb)(268) (R 1973). 2 pp.
CNS K6361-73. Chemical Analysis of Iron in Natural Rubber (May)(3570).
SAA AS 1213-82. Iron and Steel—Methods of Sampling. 11 pp.

—**Scrap**
CNS G3071-71. Classification Standards for Iron and Steel Scraps (Tentative) (Jun)(3292).
JIS G 2401-79. Classification Standard for Iron and Steel Scraps.
JIS G 2401-55. Classification Standard for Iron and Steel Scraps (R 1971). 7 pp.

—**Shore Hardness Testing**
JIS G 0563-93. Method of Measuring Surface Hardness for Nitrided Iron and Steel. 6 pp.

—**Shot**
BSI BS 2451-63. (OBSOLESCENT) 1963 Chilled Iron Shot and Grit. 8 pp.

—**Silicon Content**
CNS G2222-84. Method of Determination for Silicon in Iron and Steel (Sep)(11013).
JIS G 1212-81. Methods for Determination of Silicon in Iron and Steel. 14 pp.

—**Silicon Content—Atomic Absorption Spectrometry**
CEN PREN 10 201-88. Chemical Analysis of Ferrous Materials Determination of Silicon in Steel and Iron Flame Atomic Absorbption Spectrometric Method. 13 pp.

—**Silicon Content—Gravimetric Analysis**
BSI BS 6200: SUB SEC 3.26.2-91. 1991 Sampling and Analysis of Iron, Steel and Other Ferrous Metals Part 3: Methods of Analysis Section 3.26: Determination of Silicon Subsection 3.26.2: Acid-Resisting High Silicon Iron: Gravimetric Method. 6 pp.

—**Silicon Content—Spectrophotometry**
SAA AS 1050.26-78. Methods for the Analysis of Iron and Steel—Part 26: Determination of Silicon in Iron and Steel (Spectrophotometric Method) Reconfirmed 1989. 10 pp.

—**Solubility**
CNS R3063-76. Method of Test for Iron Solubility in Acid of Chemical Stoneware (Jun)(3190).

—**Spectrochemical Analysis**
CNS G1022-84. General Rules for Photoelectric Emission Spectrochemical Analysis for Iron and Steel (Dec)(9706).
CNS G2167-84. Method of Photoelectric Emission Spectrochemical Analysis for Iron and Steel (Dec)(10006).
CNS G2238-85. Atomic Absorption Spectrochemical Analysis for Iron and Steel (Feb)(11206).
JIS G 1253-83. Method for Photoelectric Emission Spectrochemical Analysis for Iron and Steel. 14 pp.

Iron (Cont.)

—**Spectrochemical Analysis (Cont.)**
JIS G 1257-88. Methods for Atomic Absorption Spectrochemical Analysis of Iron and Steel. 71 pp.

—**Standard Solution**
JIS K 0016-83. Iron Standard Solution. 14 pp.

—**Sulfur Content**
CNS G2251-85. Method of Determination for Sulfur in Iron and Steel (Oct)(11387).
JIS G 1215-82. Methods for Determination of Sulfur in Iron and Steel. 27 pp.

—**Sulfur Content—Infrared Analysis**
BSI BS 6200: SUB SEC 3.28.2-90. 1990 Sampling and Analysis of Iron, Steel and Other Ferrous Metals Part 3: Methods of Analysis Sec. 3.28: Determination of Sulphur Subsec. 3.28.2: Steel and Cast Iron: Infra-Red Absorption Method After Combustion in and Induc-tion Furnace (V). 11 pp.
BSI BS 6200: SUB SEC 3.28.2-01. 1990 Amd 1 Sampling and Analysis of Iron, Steel and Other Ferrous Metals Part 3: Methods of Analysis Sec 3.28: Determin of Sulphur Subsec 3.28.2: Steel and Cast Iron: Infra-Red Absorption Method After Combustion in an Induction Furnace (ISO 4935: 1989). 16 pp.
ISO 9686-92. Direct Reduced Iron—Determination of Carbon and/or Sulfur Content—High Frequency Combustion Method with Infrared Measurement First Edition. 15 pp.

—**Sulfur Content—Infrared Spectroscopy**
DIN ENGL EN 24935-92. Determination of Sulfur Content of Steel and Iron by Infrared Absorption Spectroscopy After Combustion in an Induction Furnace; (ISO 4935: 1989) (July). 10 pp.
ISO 4935-89. Steel and Iron—Determination of Sulfur Content—Infrared Absorption Method After Combustion in an Induction Furnace First Edition; (Corrected and Reprinted -1990). 10 pp.

—**Symbols**
CNS G1001-47. Symbols of Steel and Iron (Mar)(109)(R 1972).

—**Tabular Layouts**
DIN ENGL 4000 Pt 23-88. Tabular Layouts of Article Characteristics for Iron and Steel (Dec). 4 pp.

—**Tin Content**
CNS G2244-85. Method of Determination for Tin in Iron and Steel (Apr)(11246).
JIS G 1226-80. Methods of Determination of Tin in Iron and Steel.

—**Titanium Content**
CNS G2241-85. Method of Determination for Titanium in Iron and Steel (Apr)(11243).
JIS G 1223-81. Methods for Determination of Titanium in Iron and Steel.

—**Titanium Content—Atomic Absorption Spectrometry**
CEN PREN 10211-91. Chemical Analysis of Ferrous Materials—Determination of Titanium in Steel and Iron—Flame Atomic Absorption Spectrometric Method. 16 pp.

—**Titanium Content—Spectrophotometry**
BSI BS 6200: SUB SEC 3.32.1-91. 1991 Sampling and Analysis of Iron, Steel and Other Ferrous Metals Part 3: Methods of Analysis Section 3.32: Determination of Titanium Subsection 3.32.1: Steel and Cast Iron: Spectrophotometric Method (ISO 10280: 1991). 13 pp.
BSI BS 6200: SUB SEC 3.32.1-86. 1986 Sampling and Analysis of Iron, Steel and Other Ferrous Metals Part 3: Methods of Analysis Section 3.32: Determination of Titanium Subsection 3.32.1: Steel Spectrophotometric Method. 6 pp.
ISO 10280-91. Steel and Iron—Determination of Titanium Content—Diantipyrylmethane Spectrophotometric Method First Edition. 11 pp.
SAA AS 1050.27-84. Methods for the Analysis of Iron and Steel—Part 27: Determination of Titanium in Iron and Steel (Spectrophotometric Method) Amdt 1 August 1985. 4 pp.

—**Tungsten Content**
CNS G2235-84. Method of Determination for Tungsten in Iron and Steel (Dec)(11166).
JIS G 1220-80. Methods for Determination of Tungsten in Iron and Steel.

—**Vanadium Content**
CNS G2236-84. Method of Determination for Vanadium in Iron and Steel (Dec)(11167).
JIS G 1221-81. Methods for Determination of Vanadium in Iron and Steel.

INDUSTRY STANDARDS

Iron (Cont.)

—Vanadium Content—Atomic Absorption Spectrometry
BSI BS 6200: SUB SEC 3.34.3-90. 1990 Sampling and Analysis of Iron, Steel and Other Ferrous Metals Part 3: Methods of Analysis Section 3.34: Determination of Vanadium Subsection 3. 34.3: Steel and Cast. Iron: Flame Atomic Absorption Spectrometric Method. 13 pp.
ISO 9647-89. Steel and Iron—Determination of Vanadium Content—Flame Atomic Absorption Spectrometric Method First Edition. 11 pp.

—Vanadium Content—Spectrophotometry
BSI BS 6200: SUB SEC 3.34.2-89. 1989 Sampling and Analysis of Iron, Steel and Other Ferrous Metals Part 3: Methods of Analysis Section 3.34: Determination of Vanadium Subsection 3.34.2: Steel and Cast Iron: Spectro-photometric Method (ISO 4942: 1988). 10 pp.
ISO 4942-88. Steel and Iron—Determination of Vanadium Content—N-BPHA Spectrophotometric Method First Edition. 8 pp.

—Vickers Hardness Testing
JIS G 0563-93. Method of Measuring Surface Hardness for Nitrided Iron and Steel. 6 pp.

—X-Ray Fluorescence Spectrometry
CNS G2180-84. General Rules for X-Ray Fluorescence Spectrometric Analysis of Iron and Steel (Oct)(10500).
CNS G2231-84. Method of X-Ray Fluorescence Spectrometric Analysis of Iron and Steel (Oct)(11072).
JIS G 1204-78. General Rules for Fluorescent X-Ray Analysis of Iron and Steel. 18 pp.
JIS G 1256-82. Method for X-Ray Fluorescence Spectrometric Analysis of Iron and Steel. 15 pp.

Iron (II) Chloride
Use: Ferrous Chloride

Iron (II) Sulfate
Use: Ferrous Sulfate

Iron (III) Chloric Sulfate
Use: Ferric Chloride Sulfate

Iron (III) Chloride
Use: Ferric Chloride

Iron (III) Sulfate
Use: Ferric Sulfate

Iron Alloys
Use: Ferroalloys

Iron Ammonium Citrate
Use: Ammonium Ferric Citrate

Iron Bars
Use: Iron—Bars

Iron Blue
See Also: Pigments

—Pigments
CNS K2167-87. Iron Blue (Pigment) (Nov)(12158). 1 p.
CNS K6929-87. Method of Test for Iron Blue (Pigment) (Nov)(12159). 2 pp.
ISO 2495-72. Iron Blue Pigments for Paints First Edition. 4 pp.
JIS K 5113-86. Iron Blue (Pigment). 6 pp.

Iron Castings
See Also: Austenitic Iron Castings; Cast Iron; Castings; Ductile Iron Castings; Ferroalloy Castings; Gray Iron Castings; Malleable Iron Castings; Metal Products; Steels; White Iron; White Iron Castings
BSI BS 1591-75. 1975 Corrosion Resisting High Silicon Iron Castings. 8 pp.
CNS B1040-87. Permissible Deviations in Dimensions Without Tolerance Indication (Iron Castings) (Jun)(4021).
DIN ENGL 28500-77. Cast Iron Pressure Pipes and Special Castings; Technical Conditions of Delivery (Aug). 4 pp.
JIS B 0407-78. Permissible Deviations in Dimensions Without Tolerance Indication for Iron Castings. 5 pp.

—Austenitic—Chemical Analysis
DIN ENGL 1694-81. Austenitic Cast Iron (Sept). 9 pp.

—Austenitic—Mechanical Properties
DIN ENGL 1694-81. Austenitic Cast Iron (Sept). 9 pp.

Iron Castings (Cont.)

—Austenitic—Mechanical Properties (Cont.)
DIN ENGL 1694 Suppl. 1-81. Austenitic Cast Iron; Reference Data on Mechanical and Physical Properties (Sept). 7 pp.
JIS G 5510-87. Austenitic Iron Castings; (Erratum). 24 pp.

—Austenitic—Physical Properties
DIN ENGL 1694-81. Austenitic Cast Iron (Sept). 9 pp.
DIN ENGL 1694 Suppl. 1-81. Austenitic Cast Iron; Reference Data on Mechanical and Physical Properties (Sept). 7 pp.
JIS G 5510-87. Austenitic Iron Castings; (Erratum). 24 pp.

—Austenitic—Symbols
JIS G 5510-87. Austenitic Iron Castings; (Erratum). 24 pp.

—Centrifugal—Aerospace
BSI HC 100-72. 1972 Amd 3 Inspection and Testing Procedure for Iron, Nickel, Copper, Cobalt and Refractory Metal Base Alloy Castings (AMD 6574) May 28, 1991. 17 pp.

—Graphite
JIS G 5504-92. Heavy-Walled Ferritic Spheroidal Graphite Iron Castings for Low Temperature Service. 8 pp.

—Investment—Aerospace
BSI HC 100-72. 1972 Amd 3 Inspection and Testing Procedure for Iron, Nickel, Copper, Cobalt and Refractory Metal Base Alloy Castings (AMD 6574) May 28, 1991. 17 pp.

—Radiography
DIN ENGL 54111 Pt 2-82. Non-Destructive Testing; Testing of Metallic Materials by X-Rays or Gamma Rays; Radiographic Techniques for Castings of Ferrous Materials (June). 16 pp.

—Sand—Aerospace
BSI HC 100-72. 1972 Amd 3 Inspection and Testing Procedure for Iron, Nickel, Copper, Cobalt and Refractory Metal Base Alloy Castings (AMD 6574) May 28, 1991. 17 pp.

Iron Chromium
See Also: Iron

—Chemical Analysis
JIS H 1411-76. Methods of Chemical Analysis for Iron Chromium Electric Heating Material.

Iron Citrate
Use: Ferric Citrate

Iron Coatings (On Iron)
See Also: Coatings; Iron; Metal Coatings (On Metal)

—Aluminum—Hot Dip
CNS H2058-82. Methods of Test for Aluminium Coating (Hot-Dipped) on Iron and Steel (Jan)(8296). 5 pp.
CNS H3099-82. Aluminum Coatings (Hot-Dipped) on Iron and Steel (Jan)(8295).
CNS H3100-82. Recommended Practice for Aluminum Coatings (Hot-Dipped) (Jan)(8297).
JIS H 8642-72. Aluminium Coating (Hot-Dipped) on Iron or Steel. 5 pp.
JIS H 8672-72. Methods of Test for Aluminium Coating (Hot-Dipped) on Iron or Steel. 12 pp.
JIS H 9126-68. Recommended Practice for Aluminium Coating (Hot-Dipped). 9 pp.

—Aluminum—Sprayed
BSI BS 2569: Part 1-64. 1964 Amd 1 Sprayed Metal Coatings Part 1: Protection of Iron and Steel by Aluminium and Zinc Against Atmospheric Corrosion. 14 pp.
CNS H2057-82. Methods of Test for Aluminium Spray Products (Jan)(8293). 6 pp.
CNS H3095-81. Aluminium Spray Coatings on Iron and Steel (Oct)(8009).
CNS H3098-82. Recommended Practice for Aluminum Spray Coatings on Iron and Steel (Jan)(8294).
CSA G189-1966. Sprayed Metal Coatings for Atmospheric Corrosion Protection (R 1992). 16 pp.
JIS H 8301-90. Aluminium Spraying on Iron and Steel. 6 pp.
JIS H 8663-61. Testing Method for Aluminium Spray Products. 9 pp.
JIS H 9301-90. Recommended Practice for Aluminium Spray Coatings on Iron or Steel. 9 pp.

—Bituminous
BSI BS 4147-80. 1980 Amd 1 Bitumen-Based Hot-Applied Coating Material for Protecting Iron and Steel, Including Suitable Primers Where Required. 11 pp.

Iron Coatings (On Iron) (Cont.)

—Black Oxide
CGSB CAN/CGSB-31.114-M91. Black Oxide Conversion Coatings for Ferrous Metals. 8 pp.
DIN ENGL 50938-87. Alkaline Blackening Treatment of Ferrous Components; Principles and Methods of Test (Nov). 3 pp.

—Cadmium—Electroplated
BSI BS 1706-90. 1990 Amd 1 Electroplated Coatings of Zinc and Cadmium on Iron and Steel (AMD 6731) May 31, 1991. 17 pp.
CNS H3080-87. Electroplated Coatings of Cadmium on Iron and Steel (Sep)(4828).
ISO 2082-86. Metallic Coatings—Electroplated Coatings of Cadmium on Iron or Steel Second Edition. 7 pp.
JIS H 8611-93. Electroplated Coatings of Cadmium on Steel. 8 pp.
SAA AS 1790-84. Electroplated Coatings—Cadmium on Iron or Steel (R 1993). 5 pp.

—Chromium—Electroplated
BSI BS 1224-70. 1970 Amd 1 Electroplated Coatings of Nickel and Chromium. 29 pp.

—Coal Tar
BSI BS 4164-87. 1987 Amd 2 Coal Tar Based Hot Applied Coating Materials for Protecting Iron and Steel, Including a Suitable Primer (AMD 6216) August 31, 1989. 26 pp.

—Conversion
CGSB CAN/CGSB-31.114-M91. Black Oxide Conversion Coatings for Ferrous Metals. 8 pp.

—Corrosion Inhibitive
BSI BS 5493-77. 1977 Amd 1 Code of Practice for Protective Coating or Iron and Steel Structures Against Corrosion. 113 pp.

—Epoxy
CGSB CAN/CGSB-1.183-92. Zinc-Rich Epoxy Coating. 10 pp.

—Lead—Electroplated
CNS H3146-87. Electroplated Coatings of Lead and Lead-Tia Alloys on Iron and Steel (Sep)(12089).

—Lead—Sprayed
CNS H2054-81. Methods of Test for Lead Spray Products (Oct)(8011). 3 pp.
JIS H 8662-61. Testing Methods for Lead Spray Products (R 1969). 6 pp.

—Metal—Sprayed
BSI BS 2569: Part 2-65. 1965 Sprayed Metal Coatings Part 2: Protection of Iron and Steel Against Corrosion and Oxidation at Elevated Temperatures. 14 pp.

—Nickel—Electroplated
BSI BS 1224-70. 1970 Amd 1 Electroplated Coatings of Nickel and Chromium. 29 pp.
DIN ENGL 50968-91. Electrodeposited Coatings of Nickel and Nickel Plus Copper (Jan). 5 pp.

—Paints (Enamel)—Interior/Exterior
JIS K 5980-86. Baking Enamel Film on Metal Substrate. 12 pp.

—Paints (Lead Chromate)—Corrosion Inhibitive
CNS K2157-87. Basic Lead Chromate Anticorrosive Paint (Oct)(12129). 2 pp.
CNS K6919-87. Method of Test for Basic Lead Chromate Anticorrosive Paint (Oct)(12130). 5 pp.
JIS K 5624-84. Basic Lead Chromate Anticorrosive Paint. 11 pp.

—Paints (Lead Cyanamide)—Corrosion Inhibitive
CNS K2158-87. Lead Cyanamide Anticorrosive Paint (Oct)(12131). 2 pp.
CNS K6920-87. Method of Test for Lead Cyanamide Anticorrosive Paint (Oct)(12132). 5 pp.
JIS K 5625-84. Lead Cyanamide Anticorrosive Paint. 12 pp.

—Paints (Lead Suboxide)—Corrosion Inhibitive
CNS K2060-86. Lead Suboxide Anticorrosive Paint (Apr)(4909). 3 pp.
CNS K6844-86. Method of Test for Lead Suboxide Anticorrosive Paint (Apr)(11564).
JIS K 5623-84. Lead Suboxide Anticorrosive Paint. 12 pp.

—Paints (Red Lead)—Corrosion Inhibitive
CNS K2026-89. Red Lead (Pigment) (Jul)(1042). 2 pp.
CNS K2058-86. Red-Lead Zinc Chromate Anticorrosive Paint (Apr)(4907). 2 pp.
CNS K6842-86. Method of Test for Red Lead Zinc Chromate Anticorrosive Paint (Apr)(11562).

INTERNATIONAL AND NON-U.S. NATIONAL STANDARDS
SUBJECT INDEX

Iron

Iron Coatings (On Iron) (Cont.)

—Paints (Red Lead)—Corrosion Inhibitive (Cont.)
JIS K 5622-84. Red-Lead Anticorrosive Paint. 11 pp.
JIS K 5628-84. Red-Lead Zinc Chromate Anticorrosive Paint. 11 pp.

—Paints (Zinc)—Corrosion Inhibitive
CNS K2159-87. Zinc Dust Anticorrosive Paint (Oct)(12133). 2 pp.
CNS K6921-87. Method of Test for Zinc Dust Anticorrosive Paint (Oct)(12134). 6 pp.

—Paints (Zinc Chromate)—Corrosion Inhibitive
CNS K2021-86. Zinc Chromate Anticorrosive Paint (Apr)(776). 2 pp.
CNS K2058-86. Red-Lead Zinc Chromate Anticorrosive Paint (Apr)(4907). 2 pp.
CNS K6842-86. Method of Test for Red Lead Zinc Chromate Anticorrosive Paint (Apr)(11562).
JIS K 5628-84. Red-Lead Zinc Chromate Anticorrosive Paint. 11 pp.

—Phosphate
BSI BS 3189-91. 1991 Phosphate Conversion Coatings for Metals (ISO 9717: 1990). 20 pp.
BSI BS 3189-73. 1973 Phosphate Treatment of Iron and Steel. 14 pp.
CGSB 31-GP-104MA-89. Heavy Phosphate Conversion Coatings for Iron and Steel (for Wear Resistance). 9 pp.
CGSB 31-GP-105MA-89. Zinc Phosphate Conversion Coatings for Paint Base. 10 pp.
CGSB 31-GP-106M-88. Coating, Conversion, Iron Phosphate, for Paint Base. 10 pp.
DIN ENGL 50942-87. Phosphating of Metals; Principles, Methods of Test (May). 12 pp.
JIS K 3151-78. Phosphatizing Compounds Under Painting.
JIS K 3151-68. Phosphatizing Compounds Under Painting. 8 pp.
MOD UK DSTAN 03-11-01. Phosphate Treatment of Iron and Steel Issue 2; Amendment 3. 29 pp.
MOD UK DSTAN 03-11-02. Phosphate Treatment of Iron and Steel Issue 2; Amendment 4. 30 pp.

—Phosphate—Dipped
CGSB 31-GP-103MA-89. Heavy Phosphate Conversion Coatings for Iron and Steel (for Corrosion Resistance). 9 pp.

—Phosphate—Surface Density
BSI BS 5411: Part 14-82. 1982 Method of Test for Metallic and Related Coatings Part 14: Gravimetric Method for Determination of Coating Mass per Unit Area of Conversion Coatings on Metallic Materials. 7 pp.
ISO 3892-80. Conversion Coatings on Metallic Materials—Determination of Coating Mass per Unit Area—Gravimetric Methods First Edition. 5 pp.

—Primers (Bituminous)
BSI BS 4147-80. 1980 Amd 1 Bitumen-Based Hot-Applied Coating Material for Protecting Iron and Steel, Including Suitable Primers Where Required. 11 pp.

—Primers (Coal Tar)
BSI BS 4164-87. 1987 Amd 2 Coal Tar Based Hot Applied Coating Materials for Protecting Iron and Steel, Including a Suitable Primer (AMD 6216) August 31, 1989. 26 pp.

—Primers (Red Lead)
BSI BS 2523-66. (OBSOLESCENT) 1966 Amd 2 Lead-Based Priming Paints. 11 pp.

—Primers (Zinc)
MOD UK DSTAN 80-4-81. Paint, Priming, Zinc Dust Type: Brushing Issue 2. 7 pp.

—Tin—Electrodeposited
MOD UK DTD-927A-56. Tin-Zinc Alloy Plating (Reprinted October 1962). 5 pp.

—Tin—Electroplated
JIS H 8624-90. Electroplated Coatings of Tin-Lead Alloys. 8 pp.

—Zinc
CNS H3118-83. Coatings of Zinc Mechanically Deposited on Iron and Steel (Mar)(10263).

—Zinc—Electroplated
BSI BS 1706-90. 1990 Amd 1 Electroplated Coatings of Zinc and Cadmium on Iron and Steel (AMD 6731) May 31, 1991. 17 pp.
CNS H3079-87. Electroplated Coatings of Zinc on Iron and Steel (Sep)(4827).

Iron Coatings (On Iron) (Cont.)

—Zinc—Electroplated (Cont.)
DIN ENGL 50961-87. Electroplated Coatings; Zinc and Cadmium Coatings on Iron and Steel; Chromate Treatment of Zinc and Cadmium Coatings (June) (Supersedes DIN 50941, May 1978 Edition, and DIN 50962, April 1976 Edition). 6 pp.
ISO 2081-86. Metallic Coatings—Electroplated Coatings of Zinc on Iron or Steel Second Edition. 7 pp.
JIS H 8610-91. Electroplated Coatings of Zinc on Iron or Steel. 8 pp.
SAA AS 1789-84. Electroplated Coatings—Zinc on Iron or Steel (R 1993). 4 pp.

—Zinc—Hot Dip
BSI BS 729-71. 1971 Hot Dip Galvanized Coatings on Iron and Steel Articles. 15 pp.
BSI BS 4479: Part 6-90. 1990 Design of Articles That Are to Be Coated Part 6: Recommendations for Hot-Dip Metal Coatings. 18 pp.
CEN PREN 1029-93. Hot Dip Galvanized Coatings on Fabricated Ferrous Products—Specification. 17 pp.
CNS H3102-82. Recommended Practice for Zinc Coatings (Hot-Dipped) (Feb)(8503).
CNS H3116-83. Zinc Coating (Hot Dipped) on Iron and Steel (Feb)(10007).
DIN ENGL 50976-89. Corrosion Protection; Hot-Dip Batch Galvanizing; Requirements and Testing (May). 7 pp.
ISO 1461-73. Metallic Coatings—Hot Dip Galvanized Coatings on Fabricated Ferrous Products—Requirements First Edition. 4 pp.
JIS H 8641-82. Zinc Hot Dip Galvanizings. 7 pp.
JIS H 9124-87. Recommended Practice for Zinc Coating (Hot-Dipped). 9 pp.
SAA AS 1650-89. Hot-Dipped Galvanized Coatings on Ferrous Articles (This is a Joint Standard with SANZ NZS 1650). 20 pp.

—Zinc—Hot Dip—Surface Density
ISO 1460-92. Metallic Coatings—Hot Dip Galvanized Coatings on Ferrous Materials—Gravimetric Determination of the Mass per Unit Area Second Edition. 5 pp.

—Zinc—Pipes
CSA CAN/CSA-B131.1-M88. External Zinc-Coating of Ductile Iron Pipe (ISO 8179-1985); (Gen Instr 1). 13 pp.
ISO 8179-85. Ductile Iron Pipes—External Zinc Coating First Edition. 4 pp.

—Zinc—Sherardized
BSI BS 4921-88. 1988 Sherardized Coatings on Iron and Steel. 8 pp.

—Zinc—Sprayed
BSI BS 2569: Part 1-64. 1964 Amd 1 Sprayed Metal Coatings Part 1: Protection of Iron and Steel by Aluminium and Zinc Against Atmospheric Corrosion. 14 pp.
CNS H2053-81. Methods of Test for Zinc Spray Products (Oct)(8010). 6 pp.
CNS H3103-82. Recommended Practice for Zinc Spray Coatings on Iron and Steel (Feb)(8504).
CSA G189-1966. Sprayed Metal Coatings for Atmospheric Corrosion Protection (R 1992). 16 pp.
ISO 2063-91. Metallic and Other Inorganic Coatings—Thermal Spraying—Zinc, Aluminium and Their Alloys Second Edition. 14 pp.
JIS H 8300-90. Zinc Spraying on Iron and Steel. 6 pp.
JIS H 8305-82. Sprayed Zinc-Aluminum Alloy Coatings on Iron or Steel. 9 pp.
JIS H 8661-61. Testing Methods for Zinc Spray Products. 10 pp.
JIS H 9300-91. Recommended Practice for Zinc Spraying on Iron and Steel. 11 pp.

Iron Content Analysis

See Also: Iron; Radioactive Iron Content Analysis

—Acetaldehyde—Photometry
ISO 2886-73. Acetaldehyde for Industrial Use—Determination of Iron Content—2,2'-Bipyridyl Photometric Method First Edition. 5 pp.

—Aluminum
JIS H 1353-72. Methods for Determination of Iron in Aluminium and Aluminium Alloy.

—Aluminum—Absorptiometric Analysis
BSI BS 1728: Part 8-57. 1957 Methods for the Analysis of Aluminium and Aluminium Alloys Part 8: Determination of Iron (Absorptiometric 1:10-Phenanthroline Method). 8 pp.

—Aluminum Alloys
JIS H 1353-72. Methods for Determination of Iron in Aluminium and Aluminium Alloy.

Iron Content Analysis (Cont.)

—Aluminum Alloys—Absorptiometric Analysis
BSI BS 1728: Part 8-57. 1957 Methods for the Analysis of Aluminium and Aluminium Alloys Part 8: Determination of Iron (Absorptiometric 1:10-Phenanthroline Method). 8 pp.

—Aluminum Alloys—Photometry
ISO 793-73. Aluminium and Aluminium Alloys—Determination of Iron—Orthophenanthroline Photometric Method First Edition. 6 pp.

—Aluminum Alloys—Volumetric Analysis
BSI BS 1728: Part 6-55. 1955 Methods for the Analysis of Aluminium and Aluminium Alloys Part 6: Determination of Iron (Volumetric: Titanous Chloride Method). 8 pp.

—Aluminum Fluoride—Photometry
BSI BS 4993: Part 2-74. 1974 Methods of Test for Aluminium Fluoride for Industrial Use Part 2: Determination of Iron Content. 12 pp.
ISO 2368-72. Aluminium Fluoride for Industrial Use—Determination of Iron Content—1,10-Phenanthroline Photometric Method First Edition. 6 pp.

—Aluminum Ores—Reduction (Chemistry)
ISO 10213-91. Aluminium Ores—Determination of Total Iron Content—Titanium Trichloride Reduction Method First Edition. 11 pp.

—Aluminum Ores—Volumetric Analysis
BSI BS 6870: Sec 2.5-87. 1987 Analysis of Aluminium Ores Part 2: Chemical Methods Section 2.5: Method for Determination of Iron Content: Titrimetric Method. 5 pp.
ISO 6609-85. Aluminium Ores—Determination of Iron Content—Titrimetric Method First Edition. 5 pp.

—Aluminum Ores—X-Ray Fluorescence
SAA AS 2564-82. Aluminium Ores—Determination of Aluminium, Silicon, Iron, Titanium and Phosphorus Contents—Wavelength Dispersive X-Ray Fluorescence Spectrometric Method. 48 pp.

—Aluminum Oxide—Photometry
BSI BS 4140: Part 7-86. 1986 Methods of Test for Aluminium Oxide Part 7: Determination of Iron Content. 6 pp.
ISO 805-76. Aluminium Oxide Primarily Used for the Production of Aluminium—Determination of Iron Content—1,10-Phenanthroline Photometric Method First Edition. 6 pp.

—Aluminum—Photometry
ISO 793-73. Aluminium and Aluminium Alloys—Determination of Iron—Orthophenanthroline Photometric Method First Edition. 6 pp.

—Aluminum Sulfate
CPPA J.6-86. Analysis of Alum Reprinted—1988. 5 pp.
MOD UK M 9508/67. Examination of Aluminium Sulphate.

—Aluminum—Volumetric Analysis
BSI BS 1728: Part 6-55. 1955 Methods for the Analysis of Aluminium and Aluminium Alloys Part 6: Determination of Iron (Volumetric: Titanous Chloride Method). 8 pp.

—Ammonium Hydroxide—Spectrometry
BSI BS 4651: Part 5-88. 1988 Ammonia Solution Part 5: Methods for Determination of Iron Content. 4 pp.

—Ammonium Sulfate—Spectrophotometry
ISO 2992-74. Ammonium Sulphate for Industrial Use—Determination of Iron Content—2,2'-Bipyridyl Photometric Method First Edition. 5 pp.

—Animal Fats—Colorimetry
BSI BS 684: Sec 2.17-76. 1976 Methods of Analysis of Fats and Fatty Oils Part 2: Other Methods Section 2.17: Determination of Iron. Colorimetric Method. 2 pp.

—Boric Acid—Photometry
BSI BS 5688: Part 17-79. 1979 Orthoboric Acid, Diboron Trioxide, Disodium Tetraborates, Sodium Perborates and Crude Sodium Borates for Industrial Use Part 17: Deter of Iron Content of Boric Acid, Boric Oxide, Disodium Tetraborates, Sodium Perborates and Crude Sodium Borates. 5 pp.
ISO 3122-76. Boric Acid, Boric Oxide, DiSodium Tetraborates, Sodium Perborates and Crude Sodium Borates for Industrial Use—Determination of Iron Content—2,2'-Bipyridyl Photometric Method First Edition. 6 pp.

INDUSTRY STANDARDS

Iron Content Analysis (Cont.)

—Boron Oxides—Photometry
BSI BS 5688: Part 17-79. 1979 Orthoboric Acid, Diboron Trioxide, Disodium Tetraborates, Sodium Perborates and Crude Sodium Borates for Industrial Use Part 17: Deter of Iron Content of Boric Acid, Boric Oxide, Disodium Tetraborates, Sodium Perborates and Crude Sodium Borates. 5 pp.

ISO 3122-76. Boric Acid, Boric Oxide, DiSodium Tetraborates, Sodium Perborates and Crude Sodium Borates for Industrial Use—Determination of Iron Content—2,2'-Bipyridyl Photometric Method First Edition. 6 pp.

—Calcium Silicide
MOD UK M 805/84. Examination of Calcium Silicide.

—Cementitious Materials
SAA AS 3583.10-91. Methods of Test for Supplementary Cementitous Materials for Use with Portland Cement—Part 10: Determination of Alumina and Total Iron Content (In Professional Packages 30,58). 4 pp.

—Chemical Products—Spectrophotometry
BSI BS 6337: Part 3-83. 1983 General Methods of Chemical Analysis Part 3: Method for Determination of Iron Content (1.10-Phenanthroline Spectrophotometric Method). 7 pp.

ISO 6685-82. Chemical Products for Industrial Use—General Method for Determination of Iron Content—1,10-Phenanthroline Spectrophotometric Method First Edition. 6 pp.

—Chromium Ores
CNS M3199-88. Method of Determination for Iron in Chrome Ores (May)(12316).

JIS M 8263-82. Methods for Determination of Iron in Chrome Ores.

—Chromium Ores—Volumetric Analysis
ISO 6130-85. Chromium Ores—Determination of Total Iron Content—Titrimetric Method After Reduction First Edition. 5 pp.

—Copper
CNS H2087-89. Methods of Determination for Iron in Copper and Copper Alloys (Jun)(12533).

JIS H 1054-84. Methods for Determination of Iron in Copper and Copper Alloys. 15 pp.

—Copper Alloys
CNS H2087-89. Methods of Determination for Iron in Copper and Copper Alloys (Jun)(12533).

JIS H 1054-84. Methods for Determination of Iron in Copper and Copper Alloys. 15 pp.

—Copper Alloys—Photometry
BSI BS 1748: Parts 1-5-61. 1961 Methods for the Analysis of Copper Alloys Parts 1-5: Determination of Copper, Lead, Iron, Aluminium and Nickel in Copper Alloys. 23 pp.

—Copper Alloys—Spectrophotometry
ISO 1812-76. Copper Alloys—Determination of Iron Content—1,10-Phenanthroline Spectrophotometric Method First Edition. 4 pp.

—Copper Alloys—Volumetric Analysis
ISO 4748-84. Copper Alloys—Determination of Iron Content—Na2EDTA Titrimetric Method First Edition. 4 pp.

—Cryolite
MOD UK M 874/68. Examination of Cryolite.

—Cryolite—Photometry
BSI BS 5050-74. 1974 Methods of Test for Cryolite. 20 pp.

CNS M3115-83. Method for Determining Iron Content in Natural and Artificial Cryolite (1.10-Phenanthroline Photometric Method) (Mar)(10114)).

ISO 1694-76. Cryolite, Natural and Artificial—Determination of Iron Content—1,10-Phenanthroline Photometric Method First Edition. 7 pp.

—Electrolytic Copper—Atomic Absorption Spectrophotometry
BSI BS 7317: Part 1-90. 1990 Methods for Analysis of High Purity Copper Cathode Cu-CATH-1 Part 1: Method for Determination of Cadmium Manganese and Silver (Screening Procedure for Chromium, Cobalt, Iron, Nickel and Zinc) by Atomic Absorption. 9 pp.

BSI BS 7317: Part 2-90. 1990 Methods for Analysis of High Purity Copper Cathode Cu-CATH-1 Part 2: Method for Determination of Chromium, Cobalt, Iron, Nickel and Zinc by Discrete Volume Nebulization Atomic Absorption Spectrophotometry. 9 pp.

Iron Content Analysis (Cont.)

—Electrolytic Copper—Atomic Absorption Spectrophotometry (Cont.)
BSI DD 95: Part 2-84. (WITHDRAWN) 1984 Amd 1 Analysis of Higher Purity Copper Cathode Cu-CATH-1 Part 2: Method for Determination of Chromium, Cobalt, Iron, Nickel, and Zinc by Discrete Volume Nebulization Atomic Absorption Spectrophotmetry. 10 pp.

—Fertilizers
EC COM(88) 562-88. Proposal for a Council Directive on the Approximation of the Laws of the Member States in Respect of the Trace Elements Boron, Cobalt, Copper, Iron, Manganese, Molybdenum and Zinc Contained in Fertilizers. 20 pp.

EC 89/530/EEC-89. Council Directive Supplementing and Amending Directive 76/116/EEC in Respect of the Trace Elements Boron, Cobalt, Copper Iron, Manganese, Molybdenum and Zinc Contained in Fertilizers. 9 pp.

—Fluorite—Spectrometry
BSI BS 5659: Part 9-89. (WITHDRAWN) 1989 Acid-Grade Fluorspar Part 9: Determination of Iron Content. 7 pp.

—Fluorspar—Spectrometry
ISO 9061-93. Acid-Grade and Ceramic-Grade Fluorspar—Determination of Iron Content—1,10-Phenanthroline Spectrometric Method Second Edition. 7 pp.

—Formaldehyde
BSI BS 2942-57. (WITHDRAWN) 1957 Amd 2 Formaldehyde Solution. 12 pp.

—Formaldehyde—Photometry
ISO 2226-72. Formaldehyde Solutions for Industrial Use—Determination of Iron Content—2,2'-Bipyridyl Photometric Method First Edition. 5 pp.

—Formic Acid—Photometry
BSI BS 4341: Add 1-71. 1971 Methods of Test for Formic Acid Addendum No.1:. 6 pp.

ISO 731 Pt VI-77. Formic Acid for Industrial Use—Methods of Test—Part VI: Determination of Iron Content—2,2'-Bipyridyl Photometric Method First Edition. 5 pp.

—Fruits—Atomic Absorption Spectrometry
ISO 9526-90. Fruits, Vegetables and Derived Products—Determination of Iron Content by Flame Atomic Absorption Spectrometry First Edition. 6 pp.

—Fruits—Photometry
ISO 5517-78. Fruits, Vegetables and Derived Products—Determination of Iron Content—1,10-Phenanthroline Photometric Method First Edition. 5 pp.

—Gallium Arsenide
JIS H 1191-91. Methods for Chemical Analysis of Gallium Arsenide. 57 pp.

—Glucose
CNS N6054-72. Method of Test for Solid Glucose (Jun)(3349). 5 pp.

—Glycerol—Photometry
BSI BS 5711: Part 16-79. 1979 Methods of Sampling and Test for Glycerol Part 16: Determination of Iron Content. 2 pp.

—Hard Metals—Atomic Absorption Spectrometry
BSI BS 5600: SUB SEC 4.17.3-86. (WITHDRAWN) 1986 Part 4: Methods of Testing and Chemical Analysis of Hardmetals Section 4.17: Chemical Analysis by Flame Atomic Absorption Spectrophotometry Subsection 4.17.3: De-termination of Cobalt, Iron, Manganese and Nickel in Contents from 0.01% to 0.05% (m/m). 4 pp.

BSI BS 5600: SUB SEC 4.17.5-86. (WITHDRAWN) 1986 Part 4: Methods of Testing and Chemical Analysis of Hardmetals Section 4.17: Chemical Analysis by Flame Atomic Absorption Spectrometry Sub 4.17.5:Determination of Cobalt, Iron, Manganese, Molybdenum, Nickel Titanium and Vanadium in Cont from 0.5% to 2% m/m. 4 pp.

BSI BS EN 27627-3-93. 1993 Hardmetals—Chemical Analysis by Flame Atomic Absorption Spectrometry Part 3: Determination of Cobalt, Iron, Manganese and Nickel in Contents from 0.01 to 0.5 % (m/m) (ISO 7627-3: 1983) (V). 10 pp.

BSI BS EN 27627-5-93. 1993 Hardmetals—Chemical Analysis by Flame Atomic Absorption Spectrometry Part 5: Determination of Cobalt, Iron, Manganese, Molybdenum, Nickel, Titanium and Vanadium in Contents from 0.5 to 2 % (m/m) (ISO 7627-5: 1985) (V). 10 pp.

Iron Content Analysis (Cont.)

—Hard Metals—Atomic Absorption Spectrometry (Cont.)
CEN EN 27627-3-93. Hardmetals—Chemical Analysis by Flame Atomic Absorption Spectrometry—Part 3: Determination of Cobalt, Iron, Manganese and Nickel in Contents from 0,01 to 0,5 % (m/m) (ISO 7627-3: 1983). 5 pp.

CEN EN 27627-5-93. Hardmetals—Chemical Analysis by Flame Atomic Absorption Spectrometry—Part 5: Determination of Cobalt, Iron, Manganese, Molybdenum, Nickel, Titanium and Vanadium in Contents from 0,5 to 2 % (m/m) (ISO 7627-5: 1983). 5 pp.

ISO 7627 Pt 3-83. Hardmetals—Chemical Analysis by Flame Atomic Absorption Spectrometry—Part 3: Determination of Cobalt, Iron, Manganese and Nickel in Contents from 0,01 to 0,5 % (m/m) First Edition (CEN EN 27627-3: 1993). 4 pp.

ISO 7627 Pt 5-83. Hardmetals—Chemical Analysis by Flame Atomic Absorption Spectrometry—Part 5: Determination of Cobalt, Iron, Manganese, Molybdenum, Nickel, Titanium and Vanadium in Contents from 0,5 to 2 % (m/m) First Edition (CEN EN 27627-5: 1993). 4 pp.

—Hydrochloric Acid—Spectrophotometry
ISO R909-68. Hydrochloric Acid for Industrial Use Determination of Iron Content 2,2'-Bipyridyl Spectrophotometric Method First Edition. 8 pp.

—Hydrofluoric Acid—Emission Spectroscopy
DIN ENGL 50451 Pt 2-90. Determination of Cobalt, Chromium, Copper, Iron and Nickel as Impurities in Hydrofluoric Acid for Use in Semiconductor Technology by Plasma-Induced Emission Spectrometry (Oct). 2 pp.

—Industrial Water
CPPA H.4P(I)-67. Analysis of Process Waters Iron. 2 pp.

—Iron Ores
JIS M 8212-83. Methods for Determination of Total Iron in Iron Ores. 19 pp.

—Iron Ores—Reduction (Chemistry)
BSI BS 7020: Sec 4.2-90. 1990 Analysis of Iron Ores Part 4: Methods for the Determination of Total Iron Content Section 4.2: Titanium (III) Chloride Reduction Methods. 15 pp.

BSI BS 7020: Sec 4.3-90. 1990 Analysis of Iron Ores Part 4: Methods for the Determination of Total Iron Content Section 4.3: Silver Reduction Titrimetric Method. 14 pp.

ISO 9507-90. Iron Ores—Determination of Total Iron Content—Titanium(III) Chloride Reduction Methods First Edition. 12 pp.

ISO 9508-90. Iron Ores—Determination of Total Iron Content—Silver Reduction Titrimetric Method First Edition. 11 pp.

—Iron Ores—Volumetric Analysis
BSI BS 7020: Sec 4.1-88. 1988 Analysis of Iron Ores Part 4: Methods for the Determination of Total Iron Content Section 4.1: Titrimetric Methods. 16 pp.

BSI BS 7020: Part 22-90. 1990 Analysis of Iron Ores Part 22: Method for the Determination of Acid-Soluble Iron (II) Content: Titrimetric Method. 11 pp.

ISO 2597-85. Iron Ores—Determination of Total Iron Content—Titrimetric Methods Second Edition. 14 pp.

ISO 5416-87. Direct Reduced Iron—Determination of Metallic Iron Content—Bromine-Methanol Titrimetric Method First Edition. 11 pp.

ISO 9035-89. Iron Ores—Determination of Acid-Soluble Iron(II) Content—Titrimetric Method First Edition. 10 pp.

—Kaolin
MOD UK M 2002/59. Examination of Kaolin (Withdrawn).

—Latex—Photometry
BSI BS 7164: Sec 27.2-90. 1990 Chemical Tests for Raw and Vulcanized Rubber Part 27: Methods for Determination of Iron Content Section 27.2: Photometric Method (ISO 1657: 1986). 8 pp.

ISO 1657-86. Rubber, Raw and Rubber Latex—Determination of Iron Content—1,10-Phenanthroline Photometric Method Second Edition. 5 pp.

—Lead Alloys—Photometry
BSI BS 3908: Part 15-72. 1972 Methods for the Sampling and Analysis of Lead and Lead Alloys Part 15: Iron in Lead and Lead Alloys (Photometric Method). 8 pp.

INTERNATIONAL AND NON-U.S. NATIONAL STANDARDS
SUBJECT INDEX
Iron

Iron Content Analysis (Cont.)

—Lead—Photometry

BSI BS 3908: Part 15-72. 1972 Methods for the Sampling and Analysis of Lead and Lead Alloys Part 15: Iron in Lead and Lead Alloys (Photometric Method). 8 pp.

—Limestone—Photometry

DIN ENGL 52240 Pt 5-85. Analysis of Raw Materials Used in Glass Production; Chemical Analysis of Limestone Containing Not Less Than 95% of Calcium Carbonate; Determination of Total Iron Content Expressed as Iron (III) Oxide (Sept). 2 pp.

—Magnesium Alloys

JIS H 1338-76. Methods for Determination of Iron in Magnesium Alloys.

—Magnesium Alloys—Photometry

BSI BS 3907: Part 2-66. 1966 Methods for the Analysis of Magnesium and Magnesium Alloys Part 2: Iron in Magnesium and Magnesium Alloys (Photometric-1, 10-Phenanthroline Method). 8 pp.

ISO 792-73. Magnesium and Magnesium Alloys—Determination of Iron—Orthophenanthroline Photometric Method First Edition. 5 pp.

—Magnesium Oxide

MOD UK M 9541/69. Examination of Magnesium Oxide and Magnesium Oxide, Special, Lead Free (Withdrawn).

—Magnesium—Photometry

BSI BS 3907: Part 2-66. 1966 Methods for the Analysis of Magnesium and Magnesium Alloys Part 2: Iron in Magnesium and Magnesium Alloys (Photometric-1, 10-Phenanthroline Method). 8 pp.

ISO 792-73. Magnesium and Magnesium Alloys—Determination of Iron—Orthophenanthroline Photometric Method First Edition. 5 pp.

—Maleic Anhydride—Photometry

CNS K6570-80. Method of Test for Iron Content of Maleic Anhydride for Industrial Use (2,2-Bipyridyl Photometric Method) (Aug)(6215).

ISO 1390 Pt VI-77. Maleic Anhydride for Industrial Use—Methods of Test—Part VI: Determination of Iron Content—2,2'-Bipyridyl Photometric Method First Edition. 5 pp.

—Manganese Ores

CNS M3066-81. Method for Determination of Iron in Manganese Ores (o-Phenanthroline Absorptiometric Method) (Aug)(7831).

JIS M 8234-92. Manganese Ores—Determination of Iron Content. 16 pp.

—Manganese Ores—Atomic Absorption Spectrometry

ISO 9681-90. Manganese Ores and Concentrates—Determination of Iron Content—Flame Atomic Absorption Spectrometric Method First Edition. 7 pp.

—Manganese Ores—Photometry

ISO 621-81. Manganese Ores—Determination of Metallic Iron Content (Metallic Iron Content Not Exceeding 2 %)—Sulphosalicylic Acid Photometric Method Second Edition. 4 pp.

—Manganese Ores—Spectrometry

ISO 9292-88. Manganese Ores and Concentrates—Determination of Total Iron Content—1,10-Phenanthroline Spectrometric Method First Edition. 6 pp.

—Manganese Ores—Spectrophotometry

ISO 7990-85. Manganese Ores and Concentrates—Determination of Total Iron Content—Titrimetric Method After Reduction and Sulfosalicylic Acid Spectrophotometric Method First Edition. 8 pp.

—Manganese Ores—Volumetric Analysis

BSI BS 3917: Part 3-65. (WITHDRAWN) 1965 Methods for the Analysis of Manganese Ores Part 3: Iron. 10 pp.

ISO 7990-85. Manganese Ores and Concentrates—Determination of Total Iron Content—Titrimetric Method After Reduction and Sulfosalicylic Acid Spectrophotometric Method First Edition. 8 pp.

—Milk—Spectrometry

BSI BS 6394: Part 2-85. 1985 Trace Elements in Milk and Milk Products Part 2: Method for the Determination of Iron Content (Reference Method). 9 pp.

ISO 6732-85. Milk and Milk Products—Determination of Iron Content—Spectrometric Method (Reference Method) First Edition. 8 pp.

—Nickel

JIS H 1425-79. Methods for Determination of Iron in Nickel Materials for Electron Tubes.

Iron Content Analysis (Cont.)

—Nickel Alloys—Atomic Absorption Spectrometry

BSI BS 7455: Part 5-91. 1991 Analysis of Nickel Alloys by Flame Atomic Absorption Spectroscopy Part 5: Method for the Determination of Iron (ISO 7530-5: 1990). 8 pp.

ISO 7530 Pt 5-90. Nickel Alloys—Flame Atomic Absorption Spectrometric Analysis—Part 5: Determination of Iron Content First Edition. 6 pp.

—Nickel Alloys—Volumetric Analysis

BSI BS 6783: Part 10-90. 1990 Sampling and Analysis of Nickel, Ferronickel and Nickel Alloys Part 10: Method for Determination of Iron in Nickel Alloys (Titrimetric Method Using Potassium Dichromate). 11 pp.

ISO 7528-89. Nickel Alloys—Determination of Iron Content—Titrimetric Method with Potassium Dichromate First Edition. 8 pp.

—Nickel—Atomic Absorption Spectrometry

BSI BS 6783: Part 1-86. 1986 Sampling and Analysis of Nickel, Ferronickel and Nickel Alloys Part 1: Method for Determination of Silver, Bismuth, Cadmium, Cobalt, Copper, Iron, Manganese, Lead and Zinc in Nickel by Flame Atomic Absorption Spectrometry. 15 pp.

ISO 6351-85. Nickel—Determination of Silver, Bismuth, Cadmium Cobalt, Copper, Iron, Manganese, Lead and Zinc Contents—Flame Atomic Absorption Spectrometric Method First Edition. 13 pp.

—Nickel Castings—Spectrochemical Analysis

JIS H 1273-88. Methods for Determination of Iron in Nickel and Nickel Alloy Castings. 12 pp.

—Nickel Castings—Volumetric Analysis

JIS H 1273-88. Methods for Determination of Iron in Nickel and Nickel Alloy Castings. 12 pp.

—Nickel—Photometry

BSI BS 3727: Part 7-64. 1964 Methods for the Analysis of Nickel for Use in Electronic Tubes and Valves Part 7: Determination of Iron (Photometric Method). 9 pp.

—Nitric Acid—Photometry

ISO R1982-71. Nitric Acid for Industrial Use Determination of Iron Content 2,2'-Bipyridyl Photometric Method First Edition. 6 pp.

—Papers—Atomic Absorption Spectrophotometry

CPPA G.34P-92. Determination of Sodium, Magnesium, Calcium, Manganese, Iron, Copper and Cadmium in Wood, Pulp or Paper by Atomic Absorption Spectrophotometry. 3 pp.

—Paraformaldehyde

BSI BS 2941-57. (WITHDRAWN) 1957 Amd 1 Paraformaldehyde. 14 pp.

—Paraformaldehyde—Photometry

CNS K6550-80. Method of Test for Iron Content of Paraformaldehyde for Industrial Use (2,2-Bipyridyl Photometric Method) (Aug)(6195).

ISO 1391 Pt III-76. Paraformaldehyde for Industrial Use—Methods of Test—Part III: Determination of Iron Content—2,2'-Bipyridyl Photometric Method First Edition; (Corrected and Reprinted -1977). 4 pp.

—Phosphoric Acids—Photometry

BSI BS 4258: Part 2-89. 1989 Methods of Test for Phosphoric Acid (Orthophosphoric Acid) for Industrial Use Part 2: Determination of Iron Content. 4 pp.

—Phosphoric Acids—Spectrophotometry

ISO R849-68. Phosphoric Acid for Industrial Use Determination of Iron Content 2,2'-Bipyridyl Spectrophotometric Method First Edition. 8 pp.

—Photographic Chemicals

ISO 10349 Pt 5-92. Photography—Photographic-Grade Chemicals—Test Methods —Part 5: Determination of Heavy Metals and Iron Content First Edition. 7 pp.

—Phthalic Anhydride—Photometry

ISO 1389 Pt XI-77. Phthalic Anhydride for Industrial Use—Methods of Test—Part XI: Determination of Iron Content—2,2'-Bipyridyl Photometric Method First Edition. 5 pp.

—Potassium Hydroxides—Photometry

ISO 994-73. Potassium Hydroxide for Industrial Use—Determination of Iron Content—1,10-Phenanthroline Photometric Method First Edition. 4 pp.

Iron Content Analysis (Cont.)

—Potassium Silicates—Photometry

BSI BS 6092: Part 9-81. 1981 Sampling and Test for Sodium and Potassium Silicates for Industrial Use Part 9: Determination of Iron Content. 4 pp.

ISO 3201-75. Sodium and Potassium Silicates for Industrial Use—Determination of Iron Content—1, 10-Phenanthroline Photometric Method First Edition. 4 pp.

—Pulps

CNS P3053-82. Method of Test for Iron in Pulp (Aug)(9321).

—Pulps—Atomic Absorption Spectrometry

BSI BS 4897: Part 3-83. 1983 Trace Metal Contents of Pulps Part 3: Method for Determination of Iron Content by 1,10-Phenanthroline Photometric and Flame Atomic Absorption Spectrometric Methods. 8 pp.

ISO 779-82. Pulps—Determination of Iron Content—1, 10-Phenanthroline Photometric and Flame Atomic Absorption Spectrometric Methods First Edition. 6 pp.

—Pulps—Atomic Absorption Spectrophotometry

CPPA G.34P-92. Determination of Sodium, Magnesium, Calcium, Manganese, Iron, Copper and Cadmium in Wood, Pulp or Paper by Atomic Absorption Spectrophotometry. 3 pp.

—Pulps—Photometry

BSI BS 4897: Part 3-83. 1983 Trace Metal Contents of Pulps Part 3: Method for Determination of Iron Content by 1,10-Phenanthroline Photometric and Flame Atomic Absorption Spectrometric Methods. 8 pp.

ISO 779-82. Pulps—Determination of Iron Content—1, 10-Phenanthroline Photometric and Flame Atomic Absorption Spectrometric Methods First Edition. 6 pp.

—Rosins

CNS K6463-80. Method of Test for Iron in Rosin (Mar)(5348).

—Rubber—Atomic Absorption Spectrometry

BSI BS 7164: Sec 27.1-91. 1991 Chemical Tests for Raw and Vulcanized Rubber Part 27: Methods for Determination of Iron Content Section 27.1: Atomic Absorption Spectrometry (ISO 6101-5: 1990). 9 pp.

ISO 6101 Pt 5-90. Rubber—Determination of Metal Content by Atomic Absorption Spectrometry—Part 5: Determination of Iron Content First Edition. 7 pp.

—Rubber—Photometry

BSI BS 7164: Sec 27.2-90. 1990 Chemical Tests for Raw and Vulcanized Rubber Part 27: Methods for Determination of Iron Content Section 27.2: Photometric Method (ISO 1657: 1986). 8 pp.

ISO 1657-86. Rubber, Raw and Rubber Latex—Determination of Iron Content—1,10-Phenanthroline Photometric Method Second Edition. 5 pp.

—Sodium Azide

MOD UK M 807/91. Examination of Sodium Azide.

—Sodium Bicarbonate—Photometry

ISO 2460-73. Sodium Hydrogen Carbonate for Industrial Use—Determination of Iron Content—1, 10-Phenanthroline Photometric Method First Edition. 4 pp.

—Sodium Borates—Photometry

BSI BS 5688: Part 17-79. 1979 Orthoboric Acid, Diboron Trioxide, Disodium Tetraborates, Sodium Perborates and Crude Sodium Borates for Industrial Use Part 17: Deter of Iron Content of Boric Acid, Boric Oxide, Disodium Tetraborates, Sodium Perborates and Crude Sodium Borates. 5 pp.

BSI BS 5688: Part 19-79. 1979 Orthoboric Acid (Boric Acid), Diboron Trioxide (Boric Oxide), Disodium Tetraborates, Sodium Perborates and Crude Sodium Borates for Industrial Use Part 19: Determination of Iron Soluble in Alkaline Medium in Crude Sodium Borates. 4 pp.

ISO 3122-76. Boric Acid, Boric Oxide, DiSodium Tetraborates, Sodium Perborates and Crude Sodium Borates for Industrial Use—Determination of Iron Content—2,2'-Bipyridyl Photometric Method First Edition. 6 pp.

ISO 3124-76. Crude Sodium Borates for Industrial Use—Determination of Iron Soluble in Alkaline Medium—2,2'-Bipyridyl Photometric Method First Edition. 5 pp.

—Sodium Carbonates—Photometry

ISO R744-68. Sodium Carbonate for Industrial Use—Determination of Iron Content-2,2'-Bipyridyl Photometric Method First Edition. 6 pp.

INDUSTRY STANDARDS

INTERNATIONAL AND NON-U.S. NATIONAL STANDARDS SUBJECT INDEX

Iron

Iron Content Analysis (Cont.)

—Sodium Chloride—Photometry
BSI BS 7319: Part 11-90. 1990 Analysis of Sodium Chloride for Industrial Use Part 11: Method for Determination of Iron Content. 6 pp.

—Sodium Fluoride—Photometry
BSI BS 5072: Part 5-80. 1980 Methods of Test for Sodium Fluoride for Industrial Use Part 5: Determination of Iron Content. 8 pp.
ISO 3429-76. Sodium Fluoride Primarily Used for the Production of Aluminium—Determination of Iron Content—1,10-Phenanthroline Photometric Method First Edition. 6 pp.

—Sodium Hexafluorosilicate—Spectrophotometry
BSI BS 5705: Part 3-79. (WITHDRAWN) 1979 Sodium Hexafluorosilicate for Industrial Use Part 3: Determination of Iron Content. 7 pp.
ISO 5443-78. Sodium Hexafluorosilicate for Industrial Use—Determination of Iron Content—1,10-Phenanthroline Spectrophotometric Method First Edition. 6 pp.

—Sodium Hexametaphosphate
MOD UK M 9520/66. Examination of Sodium Hexametaphosphate (Withdrawn).

—Sodium Hydroxide
MOD UK M 808/73. Examination of Sodium Hydroxide Pure (Withdrawn).

—Sodium Hydroxide—Photometry
BSI BS 6075: Part 3-81. 1981 Sampling and Test for Sodium Hydroxide for Industrial Use Part 3: Determination of Iron Content. 4 pp.
ISO 983-74. Sodium Hydroxide for Industrial Use—Determination of Iron Content—1,10-Phenanthroline Photometric Method First Edition. 4 pp.

—Sodium Nitrate
MOD UK M 869/91. Examination of Sodium Nitrate, Grade 1.

—Sodium Perborates—Photometry
BSI BS 5688: Part 17-79. 1979 Orthoboric Acid, Diboron Trioxide, Disodium Tetraborates, Sodium Perborates and Crude Sodium Borates for Industrial Use Part 17: Deter of Iron Content of Boric Acid, Boric Oxide, Disodium Tetraborates, Sodium Perborates and Crude Sodium Borates. 5 pp.
ISO 3122-76. Boric Acid, Boric Oxide, DiSodium Tetraborates, Sodium Perborates and Crude Sodium Borates for Industrial Use—Determination of Iron Content—2,2'-Bipyridyl Photometric Method First Edition. 6 pp.

—Sodium Silicates—Photometry
BSI BS 6092: Part 9-81. 1981 Sampling and Test for Sodium and Potassium Silicates for Industrial Use Part 9: Determination of Iron Content. 4 pp.
ISO 3201-75. Sodium and Potassium Silicates for Industrial Use—Determination of Iron Content—1,10-Phenanthroline Photometric Method First Edition. 4 pp.

—Sodium Sulfate
CNS K6085-74. Methods of Test for Sodium Sulfate of Industrial Grade (Oct)(1004). 5 pp.

—Sodium Sulfate—Photometry
ISO 3239-75. Sodium Sulphate for Industrial Use—Determination of Iron Content—1,10-Phenanthroline Photometric Method First Edition. 6 pp.

—Sodium Sulfite Anhydrous
CNS K6157-63. Method of Test for Anhydrous Sodium Sulfite of Industrial Grade (Nov)(1424) (R 1973). 2 pp.

—Sodium Thiosulfate
CNS K6156-62. Method of Test for Sodium Thiosulfate of Industrial Grade (May)(1422) (R 1973). 1 p.

—Sodium Tripolyphosphate
CNS K6415-78. Method of Test for Sodium Tripolyphosphate (for Industrial Use) (Mar)(4285). 4 pp.

—Sodium Tripolyphosphate—Spectrophotometry
BSI BS 4427: Part 3-89. 1989 Methods of Test for Sodium Tripolyphosphate (Pentasodium Triphosphate) and Sodium Pyrophosphate (Tetraso-dium Pyrophosphate) for Industrial Use Part 3: Condensed Phosphates for Indust Use (Including Foodstuffs): Determin of Iron Content. 2 pp.
ISO R852-68. Sodium Tripolyphosphate and Sodium Pyrophosphate for Industrial Use Determination of Iron Content 2,2'-Bipyridyl Spectrophotometric Method First Edition. 6 pp.

Iron Content Analysis (Cont.)

—Soil Analysis
DIN ENGL 19684 Pt 6-77. Methods of Soil Analysis for Water Management for Agricultural Purposes; Chemical Laboratory Tests; Determination of Oxalate Soluble Iron Content in a Soil (Feb). 2 pp.
DIN ENGL 19684 Pt 7-77. Methods of Soil Analysis for Water Management for Agricultural Purposes; Chemical Laboratory Tests; Determination of Easily Soluble Bivalent Iron in a Soil (Feb). 2 pp.

—Solders—Photometry
BSI BS 3338: Part 10-61. 1961 Methods for the Sampling and Analysis of Tin and Tin Alloys Part 10: Determination of Iron in Ingot Tin, Tin-Lead Solders and White Metal Bearing Alloys (Photometric Method). 8 pp.

—Spectrophotometry
CNS J2040-82. Iron by 1, 10-Phenanthroline Spectrophotometric Method (Feb)(8510).

—Strontium Peroxide
MOD UK M 884/70. Examination of Strontium Peroxide (Anhydrous).

—Sulfuric Acid—Spectrophotometry
ISO R915-68. Sulphuric Acid and Oleum for Industrial Use Determination of Iron Content 2,2'-Bipyridyl Spectrophotometric Method First Edition. 7 pp.

—Tantalum
JIS H 1683-76. Methods for Determination of Iron in Tantalum (R 1986). 9 pp.

—Tetrasodium Pyrophosphate—Spectrophotometry
BSI BS 4427: Part 3-89. 1989 Methods of Test for Sodium Tripolyphosphate (Pentasodium Triphosphate) and Sodium Pyrophosphate (Tetraso-dium Pyrophosphate) for Industrial Use Part 3: Condensed Phosphates for Indust Use (Including Foodstuffs): Determin of Iron Content. 2 pp.

—Tin Alloys—Photometry
BSI BS 3338: Part 10-61. 1961 Methods for the Sampling and Analysis of Tin and Tin Alloys Part 10: Determination of Iron in Ingot Tin, Tin-Lead Solders and White Metal Bearing Alloys (Photometric Method). 8 pp.

—Tin—Photometry
BSI BS 3338: Part 10-61. 1961 Methods for the Sampling and Analysis of Tin and Tin Alloys Part 10: Determination of Iron in Ingot Tin, Tin-Lead Solders and White Metal Bearing Alloys (Photometric Method). 8 pp.

—Titanium
JIS H 1614-76. Methods for Determination of Iron in Titanium.

—Titanium Ores
JIS M 8312-76. Determination of Iron in Titanium Ores.

—Titanium Tetrachloride
MOD UK M 888/72. Examination of Titanium Tetrachloride (Withdrawn).

—Urea
CNS K6158-75. Method of Test for Urea Technical Grade (Jan)(1426). 2 pp.

—Urea—Photometry
ISO R1595-70. Urea for Industrial Use Determination of Iron Content 2,2'-Bipyridyl Photometric Method First Edition. 5 pp.

—Vegetable Fats—Colorimetry
BSI BS 684: Sec 2.17-76. 1976 Methods of Analysis of Fats and Fatty Oils Part 2: Other Methods Section 2.17: Determination of Iron. Colorimetric Method. 2 pp.

—Vegetables—Atomic Absorption Spectrometry
ISO 9526-90. Fruits, Vegetables and Derived Products—Determination of Iron Content by Flame Atomic Absorption Spectrometry First Edition. 6 pp.

—Vegetables—Photometry
ISO 5517-78. Fruits, Vegetables and Derived Products—Determination of Iron Content—1,10-Phenanthroline Photometric Method First Edition. 5 pp.

—Water
BSI BS 2690: Part 1-64. 1964 Water Used in Industry Part 1: Copper and Iron. 16 pp.

Iron Content Analysis (Cont.)

—Water—Atomic Absorption Spectrometry
CNS K9032-80. Method of Test for Iron in Water (Atomic Absorption, Direct) (Apr)(5466).
JIS K 0553-90. Testing Methods for Determination of Metallic Elements in Highly Purified Water. 27 pp.

—Water—Photometry
BSI BS 6068: Sec 2.2-83. 1983 Water Quality Part 2: Physical, Chemical and Bio-Chemical Methods Section 2.2: Determination of Iron: 1,10-Phenanthroline Photometric Method. 8 pp.

—Water—Spectrometry
ISO 6332-88. Water Quality—Determination of Iron—Spectrometric Method Using 1,10-Phenanthroline Second Edition. 7 pp.

—White Metals—Photometry
BSI BS 3338: Part 10-61. 1961 Methods for the Sampling and Analysis of Tin and Tin Alloys Part 10: Determination of Iron in Ingot Tin, Tin-Lead Solders and White Metal Bearing Alloys (Photometric Method). 8 pp.

—Wood—Atomic Absorption Spectrophotometry
CPPA G.34P-92. Determination of Sodium, Magnesium, Calcium, Manganese, Iron, Copper and Cadmium in Wood, Pulp or Paper by Atomic Absorption Spectrophotometry. 3 pp.

—Zinc—Absorptiometric Analysis
JIS H 1109-89. Methods for Determination of Iron in Zinc Metal. 10 pp.

—Zinc Alloys—Photometry
BSI BS 3630: Part 5-63. 1963 Methods for the Sampling and Analysis of Zinc and Zinc Alloys Part 5: Iron in Ingot Zinc and Zinc Alloys (PhotometricMethod). 8 pp.

—Zinc Alloys—Spectrophotometry
ISO 1055-75. Zinc and Zinc Alloys—Determination of Iron Content—Spectrophotometric Method First Edition. 4 pp.

—Zinc—Photometry
BSI BS 3630: Part 5-63. 1963 Methods for the Sampling and Analysis of Zinc and Zinc Alloys Part 5: Iron in Ingot Zinc and Zinc Alloys (PhotometricMethod). 8 pp.
ISO 714-75. Zinc—Determination of Iron Content—Photometric Method First Edition. 4 pp.

—Zinc—Spectrochemical Analysis
JIS H 1109-89. Methods for Determination of Iron in Zinc Metal. 10 pp.

—Zinc—Spectrophotometry
ISO 1055-75. Zinc and Zinc Alloys—Determination of Iron Content—Spectrophotometric Method First Edition. 4 pp.

—Zirconium
JIS H 1654-89. Methods for Determination of Iron in Zirconium and Zirconium Alloys. 11 pp.

—Zirconium Alloys
JIS H 1654-89. Methods for Determination of Iron in Zirconium and Zirconium Alloys. 11 pp.

Iron Cores
Use: Magnetic Cores *Scope Note:* Use a more specific term

Iron Hydrogen
CNS K7259-68. Chemical Reagent (Iron Hydrogen) (Nov)(1758).

Iron Inorganic Compounds
Scope Note: Use a more specific term
See: Ammonium Ferrous Sulfate; Ferric Chloride; Ferric Nitrate; Ferric Oxide; Ferric Sulfate; Ferrous Chloride; Ferrous Sulfate

Iron Loss
Use: Core Loss

Iron Ores
See Also: Iron; Magnetite; Ores; Pyrite
ISO 9516-92. Iron Ores—Determination of Silicon, Calcium, Manganese, Aluminium, Titanium, Magnesium, Phosphorus, Sulfur and Potassium—Wavelength Dispersive X-Ray Fluorescence Spectrometric Method First Edition. 74 pp.

INTERNATIONAL AND NON-U.S. NATIONAL STANDARDS
SUBJECT INDEX

Iron Ores (Cont.)

—Aluminum Content—Atomic Absorption Spectrometry
BSI BS 7020: Sec 8.2-93. 1993 Analysis of Iron Ores Part 8: Methods for the Determination of Aluminium Content Section 8.2: Flame Atomic Absorption Spectrometric Method (ISO 4688-1: 1990) (Q). 15 pp.
ISO 4688 Pt 1-92. Iron Ores—Determination of Aluminium Content—Part 1: Flame Atomic Absorption Spectrometric Method First Edition. 12 pp.

—Aluminum Content—Volumetric Analysis
BSI BS 7020: Sec 8.1-88. 1988 Analysis of Iron Ores Part 8: Methods for the Determination of Aluminium Content Section 8.1: Titrimetric Method. 12 pp.
ISO 6830-86. Iron Ores—Determination of Aluminium Content—EDTA Titrimetric Method First Edition. 10 pp.

—Aluminum Content—X-Ray Fluorescence Spectrometry
ISO 9516-92. Iron Ores—Determination of Silicon, Calcium, Manganese, Aluminium, Titanium, Magnesium, Phosphorus, Sulfur and Potassium—Wavelength Dispersive X-Ray Fluorescence Spectrometric Method First Edition. 74 pp.

—Aluminum Oxide Content
JIS M 8220-83. Methods for Determination of Aluminium Oxide in Iron Ores. 20 pp.

—Aluminum Oxide Content—Volumetric Analysis
CNS M3128-83. Method for Determining Aluminium Oxide in Iron Ores (Zinc Titration Method) (Apr)(10183).
CNS M3129-83. Method for Determining Aluminium Oxide in Iron Ores (Oxine Method) (Apr) (10184).

—Arsenic Content
JIS M 8226-83. Methods for Determination of Arsenic in Iron Ores.
JIS M 8226-71. Methods for Determination of Arsenic in Iron Ores. 10 pp.

—Arsenic Content—Spectrophotometry
BSI BS 7020: Part 13-88. 1988 Analysis of Iron Ores Part 13: Method for the Determination of Arsenic Content: Spectrophotometric Method. 12 pp.
ISO 7834-87. Iron Ores—Determination of Arsenic Content—Molybdenum Blue Spectrophotometric Method First Edition. 12 pp.

—Bismuth Content
JIS M 8230-83. Methods for Determination of Bismuth in Iron Ores. 14 pp.

—Bulk Density
BSI BS 6147-89. 1989 Methods for Determination of Bulk Density of Iron Ores. 7 pp.
ISO 3852-88. Iron Ores—Determination of Bulk Density Second Edition; (Supersedes 5464). 6 pp.
JIS M 8716-90. Iron Ore Pellets—Determination of Apparent Density and Porosity. 5 pp.

—Calcium Content—Atomic Absorption Spectrometry
ISO 10203-92. Iron Ores—Determination of Calcium Content—Flame Atomic Absorption Spectrometric Method First Edition. 12 pp.

—Calcium Content—X-Ray Fluorescence Spectrometry
ISO 9516-92. Iron Ores—Determination of Silicon, Calcium, Manganese, Aluminium, Titanium, Magnesium, Phosphorus, Sulfur and Potassium—Wavelength Dispersive X-Ray Fluorescence Spectrometric Method First Edition. 74 pp.

—Calcium Oxide Content
JIS M 8221-83. Methods for Determination of Calcium Oxide in Iron Ores. 20 pp.

—Chemical Analysis
CNS M3014-80. General Rules for Chemical Analysis of Iron Ores (Mar)(5376).
JIS M 8202-83. General Rules for Chemical Analysis of Iron Ores. 9 pp.

—Chloride Content—Ion Selective Electrode Method
BSI BS 7020: Part 20-90. 1990 Analysis of Iron Ores Part 20: Method for the Determination of Water Soluble Chloride Content: Ion-Selective Electrode Method. 21 pp.
ISO 9517-89. Iron Ores—Determination of Water Soluble Chloride Content—Ion-Selective Electrode Method First Edition. 11 pp.

Iron Ores (Cont.)

—Chromium Content
JIS M 8224-83. Methods for Determination of Chromium in Iron Ores.
JIS M 8224-71. Methods for Determination of Chromium in Iron Ores. 11 pp.

—Chromium Content—Atomic Absorption Spectrometry
BSI BS 7020: Part 16-93. 1993 Analysis of Iron Ores Part 16: Method for the Determination of Nickel and/or Chromium Contents: Flame Atomic Absorption Spectrometric Method (ISO 9685: 1991) (Q). 16 pp.
ISO 9685-91. Iron Ores—Determination of Nickel and/or Chromium Contents—Flame Atomic Absorption Spectrometric Method First Edition. 14 pp.

—Cobalt Content
CNS M3016-80. Method for Determination of Cobalt in Iron Ores (Mar)(5378).
JIS M 8210-83. Methods for Determination of Cobalt in Iron Ores. 11 pp.

—Copper Content
JIS M 8218-83. Methods for Determination of Copper in Iron Ores. 14 pp.

—Copper Content—Absorptiometric Analysis
CNS M3079-81. Method for Determination of Copper in Iron Ores (BCOD Absorptiometric Method) (Oct)(8047).

—Copper Content—Atomic Absorption Spectrometry
BSI BS 7020: Sec 10.2-88. 1988 Analysis of Iron Ores Part 10: Methods for the Determination of Copper Content Section 10.2: Flame Atomic Absorption Spectrometric Method. 12 pp.
ISO 4693-86. Iron Ores—Determination of Copper Content—Flame Atomic Absorption Spectrometric Method First Edition. 10 pp.

—Copper Content—Spectrophotometry
BSI BS 7020: Sec 10.1-88. 1988 Analysis of Iron Ores Part 10: Methods for the Determination of Copper Content Section 10.1: Spectrophotometric Method. 12 pp.
ISO 5418-84. Iron Ores—Determination of Copper Content—2,2'-Biquinolyl Spectrophotometric Method First Edition. 10 pp.

—Copper Content—Volumetric Analysis
CNS M3078-81. Method for Determination of Copper in Iron Ores (Sodium Thiosulfate Titration Method) (Oct)(8046).

—Density
JIS M 8717-88. Test Methods for Determination of Density of Iron Ores. 9 pp.

—Ferrous Oxide Content
CNS M3018-80. Method for Determination of Ferrous Oxide in Iron Ores (Mar)(5380).
JIS M 8213-83. Method for Determination of Ferrous Oxide in Iron Ores.
JIS M 8213-71. Method for Determination of Ferrous Oxide in Iron Ores. 6 pp.

—Fluorine Content—Electrolytic Analysis
BSI BS 7020: Part 19-88. 1988 Analysis of Iron Ores Part 19: Method for the Determination of Fluorine Content: Ion-Selective Electrode Method. 12 pp.
ISO 4694-87. Iron Ores—Determination of Fluorine Content—Ion-Selective Electrode Method First Edition. 11 pp.

—Ignition Loss
SAA AS 1673.5-82. Methods for the Analysis of Iron Ores—Part 5: Determination of Loss on Ignition at 1 000 degrees Celsius. 3 pp.

—Impact Testing
JIS M 8711-87. Test Method for Determination of Shatter Strength of Iron Ore Sinter. 7 pp.

—Iron Content
JIS M 8212-83. Methods for Determination of Total Iron in Iron Ores. 19 pp.

—Iron Content—Reduction (Chemistry)
BSI BS 7020: Sec 4.2-90. 1990 Analysis of Iron Ores Part 4: Methods for the Determination of Total Iron Content Section 4.2: Titanium (III) Chloride Reduction Methods. 15 pp.
BSI BS 7020: Sec 4.3-90. 1990 Analysis of Iron Ores Part 4: Methods for the Determination of Total Iron Content Section 4.3: Silver Reduction Titrimetric Method. 14 pp.
ISO 9507-90. Iron Ores—Determination of Total Iron Content—Titanium(III) Chloride Reduction Methods First Edition. 12 pp.

Iron Ores (Cont.)

—Iron Content—Reduction (Chemistry) (Cont.)
ISO 9508-90. Iron Ores—Determination of Total Iron Content—Silver Reduction Titrimetric Method First Edition. 11 pp.

—Iron Content—Volumetric Analysis
BSI BS 7020: Sec 4.1-88. 1988 Analysis of Iron Ores Part 4: Methods for the Determination of Total Iron Content Section 4.1: Titrimetric Methods. 16 pp.
BSI BS 7020: Part 22-90. 1990 Analysis of Iron Ores Part 22: Method for the Determination of Acid-Soluble Iron (II) Content: Titrimetric Method. 11 pp.
ISO 2597-85. Iron Ores—Determination of Total Iron Content—Titrimetric Methods Second Edition. 14 pp.
ISO 5416-87. Direct Reduced Iron—Determination of Metallic Iron Content—Bromine-Methanol Titrimetric Method First Edition. 11 pp.
ISO 9035-89. Iron Ores—Determination of Acid-Soluble Iron(II) Content—Titrimetric Method First Edition. 10 pp.

—Karl Fischer Method
BSI BS 7020: Part 2-88. 1988 Analysis of Iron Ores Part 2: Method for the Determination of Hygroscopic Moisture in Analytical Samples. 15 pp.
ISO 2596-84. Iron Ores—Determination of Hygroscopic Moisture in Analytical Samples—Gravimetric and Karl Fischer Methods Third Edition. 13 pp.

—Lead Content
JIS M 8229-83. Methods for Determination of Lead in Iron Ores. 15 pp.

—Lead Content—Absorptiometric Analysis
CNS M3136-83. Method for Determining Lead in Iron Ores (Dithizone Absorptiometric Method) (Apr)(10191).

—Lead Content—Atomic Absorption Spectrometry
BSI BS 7020: Part 17-88. 1988 Analysis of Iron Ores Part 17: Method for the Determination of Lead and/or Zinc Contents: Flame Atomic Absorption Spectrometric Method. 14 pp.
ISO 8753-87. Iron Ores—Determination of Lead and/or Zinc Content—Flame Atomic Absorption Spectrometric Method First Edition. 12 pp.

—Low Temperature Testing
ISO 4696-84. Iron Ores—Low-Temperature Disintegration Test—Method Using Cold Tumbling After Static Reduction First Edition. 9 pp.

—Magnesium Content—Atomic Absorption Spectrometry
ISO 10204-92. Iron Ores—Determination of Magnesium Content—Flame Atomic Absorption Spectrometric Method First Edition. 12 pp.

—Magnesium Content—X-Ray Fluorescence Spectrometry
ISO 9516-92. Iron Ores—Determination of Silicon, Calcium, Manganese, Aluminium, Titanium, Magnesium, Phosphorus, Sulfur and Potassium—Wavelength Dispersive X-Ray Fluorescence Spectrometric Method First Edition. 74 pp.

—Magnesium Oxide Content
JIS M 8222-83. Methods for Determination of Magnesium Oxide in Iron Ores. 14 pp.

—Manganese Content
CNS M3020-80. Method for Determination of Manganese in Iron Ores (Mar)(5382).
JIS M 8215-83. Methods for Determination of Manganese in Iron Ores.
JIS M 8215-71. Methods for Determination of Manganese in Iron Ores. 7 pp.

—Manganese Content—Atomic Absorption Spectrometry
BSI BS 7020: Sec 9.2-93. 1993 Analysis of Iron Ores Part 9: Methods for the Determination of Manganese Content Section 9.2: Flame Atomic Absorption Spectrometric Method (ISO 9682-1: 1991) (Q). 16 pp.
ISO 9682 Pt 1-91. Iron Ores—Determination of Manganese Content—Part 1: Flame Atomic Absorption Spectrometric Method First Edition. 14 pp.

—Manganese Content—Spectrophotometry
BSI BS 7020: Sec 9.1-88. 1988 Analysis of Iron Ores Part 9: Methods for the Determination of Manganese Content Section 9.1: Spectrophotometric Method. 12 pp.

Iron Ores (Cont.)

—Manganese Content—Spectrophotometry (Cont.)
ISO 3886-86. Iron Ores—Determination of Manganese Content—Periodate Spectrophotometric Method Second Edition. 11 pp.

—Manganese Content—X-Ray Fluorescence Spectrometry
ISO 9516-92. Iron Ores—Determination of Silicon, Calcium, Manganese, Aluminium, Titanium, Magnesium, Phosphorus, Sulfur and Potassium—Wavelength Dispersive X-Ray Fluorescence Spectrometric Method First Edition. 74 pp.

—Moisture Content
BSI BS 5663: Part 1-87. 1987 Methods of Testing Iron Ores Part 1: Determination of Moisture Content. 15 pp.
ISO 3087-87. Iron Ores—Determination of Moisture Content of a Consignment Second Edition. 15 pp.
JIS M 8705-85. Iron Ores—Determination of Moisture Content of Consignment. 18 pp.

—Moisture Content—Gravimetric Analysis
BSI BS 7020: Part 2-88. 1988 Analysis of Iron Ores Part 2: Method for the Determination of Hygroscopic Moisture in Analytical Samples. 15 pp.
ISO 2596-84. Iron Ores—Determination of Hygroscopic Moisture in Analytical Samples—Gravimetric and Karl Fischer Methods Third Edition. 13 pp.

—Nickel Content
JIS M 8223-83. Methods for Determination of Nickel in Iron Ores. 16 pp.

—Nickel Content—Atomic Absorption Spectrometry
BSI BS 7020: Part 16-93. 1993 Analysis of Iron Ores Part 16: Method for the Determination of Nickel and/or Chromium Contents: Flame Atomic Absorption Spectrometric Method (ISO 9685: 1991) (Q). 16 pp.
ISO 9685-91. Iron Ores—Determination of Nickel and/or Chromium Contents—Flame Atomic Absorption Spectrometric Method First Edition. 14 pp.

—Pellets—Compression Testing
BSI BS 6599-85. 1985 Method for Determination of Crushing Strength of Iron Ore Pellets. 4 pp.
ISO 4700-83. Iron Ore Pellets—Determination of Crushing Strength First Edition. 4 pp.

—Pellets—Volumetric Analysis
JIS M 8719-90. Iron Ore Pellets—Determination of Volume. 11 pp.

—Phosphorus Content
JIS M 8216-83. Methods for Determination of Phosphorous in Iron Ores.
JIS M 8216-71. Methods for Determination of Phosphorus in Iron Ores (R 1974). 11 pp.

—Phosphorus Content—Photometry
BSI BS 4158: Part 2-70. 1970 Methods for the Analysis of Iron Ores Part 2: Method for the Determination of Phosphorus in Iron Ores. 8 pp.

—Phosphorus Content—Spectrophotometry
BSI BS 7020: Sec 6.2-93. 1993 Analysis of Iron Ores Part 6: Methods for the Determination of Phosphorus Content Section 6.2: Molybdenum Blue Spectrophotometric Method (ISO 4687-1: 1992) (Q). 15 pp.
ISO 4687 Pt 1-92. Iron Ores—Determination of Phosphorus Content—Part 1: Molybdenum Blue Spectrophotometric Method First Edition. 12 pp.
SAA AS 1673.4-82. Methods for the Analysis of Iron Ores—Part 4: Determination of Phosphorus Content (Spectrophotometric Method). 4 pp.

—Phosphorus Content—Volumetric Analysis
BSI BS 7020: Sec 6.1-88. 1988 Analysis of Iron Ores Part 6: Methods for the Determination of Phosphorus Content Section 6.1: Titrimetric Method. 12 pp.
ISO 2599-83. Iron Ores—Determination of Phosphorus Content—Titrimetric Method Second Edition. 10 pp.

—Phosphorus Content—X-Ray Fluorescence Spectrometry
ISO 9516-92. Iron Ores—Determination of Silicon, Calcium, Manganese, Aluminium, Titanium, Magnesium, Phosphorus, Sulfur and Potassium—Wavelength Dispersive X-Ray Fluorescence Spectrometric Method First Edition. 74 pp.

—Porosity
JIS M 8716-90. Iron Ore Pellets—Determination of Apparent Density and Porosity. 5 pp.

Iron Ores (Cont.)

—Potassium Content—Atomic Absorption Spectrometry
BSI BS 7020: Part 14-88. 1988 Analysis of Iron Ores Part 14: Method for the Determination of Sodium and/or Potassium Contents: Flame Atomic Absorption Spectrometric Method. 12 pp.
ISO 6831-86. Iron Ores—Determination of Sodium and/or Potassium Contents—Flame Atomic Absorption Spectrometric Method First Edition. 11 pp.

—Potassium Content—X-Ray Fluorescence Spectrometry
ISO 9516-92. Iron Ores—Determination of Silicon, Calcium, Manganese, Aluminium, Titanium, Magnesium, Phosphorus, Sulfur and Potassium—Wavelength Dispersive X-Ray Fluorescence Spectrometric Method First Edition. 74 pp.

—Potassium Oxide Content
CNS M3015-80. Methods for Determination of Sodium Oxide and Potassium Oxide in Iron Ores (Mar)(5377).
JIS M 8208-83. Methods for Determination of Potassium Oxide in Iron Ores.

—Reduction (Chemistry)
BSI BS 6598-85. 1985 Determination of Reducibility of Iron Ores. 10 pp.
ISO 4695-84. Iron Ores—Determination of Reducibility First Edition. 9 pp.
ISO 7215-85. Iron Ores—Determination of Relative Reducibility First Edition. 10 pp.
ISO 7992-92. Iron Ores—Determination of Reduction Properties Under Load First Edition. 13 pp.
JIS M 8713-87. Test Methods for Determination of Reducibility of Iron Ores. 21 pp.

—Sampling
BSI BS 5660: Part 1-87. 1987 Methods of Sampling Iron Ores Part 1: Manual Method of Increment Sampling. 15 pp.
BSI BS 5660: Part 2-87. 1987 Methods for Sampling Iron Ores Part 2: Mechanical Method of Increment Sampling and Sample Preparation. 43 pp.
BSI BS 5661-87. 1987 Method for Preparation of Samples of Iron Ores by Manual Means. 19 pp.
BSI BS 5662: Part 2-87. 1987 Methods of Sampling Iron Ores Part 2: Experimental Methods for Checking the Precision of Sampling. 16 pp.
BSI BS 7020: Part 1-88. 1988 Analysis of Iron Ores Part 1: Method for the Preparation of Pre-Dried Test Samples for Chemical Analysis. 4 pp.
ISO 3081-86. Iron Ores—Increment Sampling—Manual Method Second Edition. 14 pp.
ISO 3082-87. Iron Ores—Increment Sampling and Sample Preparation—Mechanical Method First Edition. 44 pp.
ISO 3083-86. Iron Ores—Preparation of Samples—Manual Method Second Edition. 19 pp.
ISO 3085-86. Iron Ores—Experimental Methods for Checking the Precision of Sampling Second Edition. 15 pp.
ISO 7764-85. Iron Ores—Preparation of Predried Test Samples for Chemical Analysis First Edition. 4 pp.
JIS M 8701-85. Iron Ores—Increment Sampling—Manual Method.
JIS M 8702-85. Iron Ores—Increment Sampling and Sample Preparation—Mechanical Method.
JIS M 8703-85. Iron Ores—Preparation of Samples—Manual Method. 30 pp.

—Sampling—Quality Control
BSI BS 5662: Part 1-87. 1987 Methods of Sampling Iron Ores Part 1: Experimental Methods for Evaluation of Quality Variation. 14 pp.
BSI BS 5662: Part 3-87. 1987 Methods of Sampling Iron Ores Part 3: Experimental Methods for Checking the Bias of Sampling. 9 pp.
ISO 3084-86. Iron Ores—Experimental Methods for Evaluation of Quality Variation Second Edition. 13 pp.
ISO 3086-86. Iron Ores—Experimental Methods for Checking the Bias of Sampling Second Edition. 8 pp.

—Sieve Analysis
BSI BS 5663: Part 2-87. 1987 Methods of Testing Iron Ores Part 2: Determination of Size Distribution by Sieving. 30 pp.
ISO 4701-85. Iron Ores—Determination of Size Distribution by Sieving First Edition. 29 pp.
JIS M 8706-85. Iron Ores—Determination of Size Distribution by Sieving. 25 pp.

—Silicon Content—Gravimetric Analysis
BSI BS 4158: Part 3-72. 1972 Methods for the Analysis of Iron Ores Part 3: Silicon (Gravimetric Method). 8 pp.
ISO 2598 Pt 1-92. Iron Ores—Determination of Silicon Content—Part 1: Gravimetric Methods First Edition. 12 pp.

Iron Ores (Cont.)

—Silicon Content—Spectrophotometry
BSI BS 7020: Sec 5.2-93. 1993 Analysis of Iron Ores Part 5: Methods for the Determination of Silicon Content Section 5.2: Reduced Molybdosilicate Spectrophotometric Method (ISO 2598-2: 1992) (Q). 14 pp.
ISO 2598 Pt 2-92. Iron Ores—Determination of Silicon Content—Part 2: Reduced Molybdosilicate Spectrophotometric Method First Edition. 11 pp.
ISO 4686-80. Iron Ores—Determination of Silicon Content—Reduced Molybdosilicate Spectrophotometric Method First Edition. 8 pp.

—Silicon Content—X-Ray Fluorescence Spectrometry
ISO 9516-92. Iron Ores—Determination of Silicon, Calcium, Manganese, Aluminium, Titanium, Magnesium, Phosphorus, Sulfur and Potassium—Wavelength Dispersive X-Ray Fluorescence Spectrometric Method First Edition. 74 pp.

—Silicon Dioxide Content
CNS M3019-80. Methods for Determination of Silicon Dioxide in Iron Ores (Mar)(5381).
JIS M 8214-83. Methods for Determination of Silicon Dioxide in Iron Ores. 10 pp.

—Sintering
BSI BS 5663: Part 3-93. 1993 Methods of Testing Iron Ores Part 3: Iron Ore Fines—Presentation of the Results of Sintering Tests (ISO 8263: 1992) (W). 21 pp.
ISO 8263-92. Iron Ore Fines—Method for Presentation of the Results of Sintering Tests First Edition; (Corrected and Reprinted -1993). 18 pp.

—Sodium Content—Atomic Absorption Spectrometry
BSI BS 7020: Part 14-88. 1988 Analysis of Iron Ores Part 14: Method for the Determination of Sodium and/or Potassium Contents: Flame Atomic Absorption Spectrometric Method. 12 pp.
ISO 6831-86. Iron Ores—Determination of Sodium and/or Potassium Contents—Flame Atomic Absorption Spectrometric Method First Edition. 11 pp.

—Sodium Oxide Content
CNS M3015-80. Methods for Determination of Sodium Oxide and Potassium Oxide in Iron Ores (Mar)(5377).
JIS M 8207-83. Methods for Determination of Sodium Oxide in Iron Ores.

—Sulfur Content
CNS M3021-80. Methods for Determination of Sulphur in Iron Ores (Mar)(5383).
JIS M 8217-83. Methods for Determination of Sulfur in Iron Ores. 16 pp.

—Sulfur Content—Combustion Analysis
BSI BS 7020: Sec 7.2-88. 1988 Analysis of Iron Ores Part 7: Methods for the Determination of Sulphur Content Section 7.2: Combustion Method. 12 pp.
ISO 4690-86. Iron Ores—Determination of Sulfur Content—Combustion Method First Edition. 10 pp.

—Sulfur Content—Gravimetric Analysis
BSI BS 7020: Sec 7.1-88. 1988 Analysis of Iron Ores Part 7: Methods for the Determination of Sulphur Content Section 7.1: Gravimetric Method. 12 pp.
ISO 4689-86. Iron Ores—Determination of Sulfur Content—Barium Sulfate Gravimetric Method First Edition. 10 pp.

—Sulfur Content—X-Ray Fluorescence Spectrometry
ISO 9516-92. Iron Ores—Determination of Silicon, Calcium, Manganese, Aluminium, Titanium, Magnesium, Phosphorus, Sulfur and Potassium—Wavelength Dispersive X-Ray Fluorescence Spectrometric Method First Edition. 74 pp.

—Swelling Index
JIS M 8715-90. Iron Ore Pellets—Determination of Swelling Index. 7 pp.

—Tin Content
JIS M 8227-83. Methods for Determination of Tin in Iron Ores. 15 pp.

—Tin Content—Absorptiometric Analysis
CNS M3131-83. Method for Determining Tin in Iron Ores (Phenylfluorone Absorptiometric Method) (Apr)(10186).

—Tin Content—Polarographic Analysis
CNS M3132-83. Method for Determining Tin in Iron Ores (Alternating Current Polarographic Method) (Apr)(10187).

Iron Ores (Cont.)

—Tin Content—Volumetric Analysis
CNS M3130-83. Method for Determining Tin in Iron Ores (Iodine Titration Method) (Apr)(10185).

—Titanium Content—Spectrophotometry
BSI BS 7020: Part 12-88. 1988 Analysis of Iron Ores Part 12: Method for the Determination of Titanium Content: Spectrophotometric Method. 14 pp.
ISO 4691-85. Iron Ores—Determination of Titanium Content—Diantipyrylmethane Spectrophotometric Method First Edition. 12 pp.

—Titanium Content—X-Ray Fluorescence Spectrometry
ISO 9516-92. Iron Ores—Determination of Silicon, Calcium, Manganese, Aluminium, Titanium, Magnesium, Phosphorus, Sulfur and Potassium—Wavelength Dispersive X-Ray Fluorescence Spectrometric Method First Edition. 74 pp.

—Titanium Dioxide Content
JIS M 8219-83. Methods for Determination of Titanium Dioxide in Iron Ores.
JIS M 8219-71. Methods for Determination of Titanium Dioxide in Iron Ores. 9 pp.

—Titanium Dioxide Content—Absorptiometric Analysis
CNS M3126-83. Method for Determining Titanium Dioxide in Iron Ores (Diantipyrylmethane Absorptiometric Method) (Apr)(10181).

—Titanium Dioxide Content—Volumetric Analysis
CNS M3127-83. Method for Determining Titanium Dioxide in Iron Ores (Ferric Sulfate Titration Method) (Apr) (10182).

—Tumbling
BSI BS 6212-87. 1987 Method for Determination of Tumbler Strength of Iron Ores. 9 pp.
ISO 3271-85. Iron Ores—Determination of Tumbler Strength Second Edition. 8 pp.
ISO 4696-84. Iron Ores—Low-Temperature Disintegration Test—Method Using Cold Tumbling After Static Reduction First Edition. 9 pp.
JIS M 8712-71. Method for Measuring the Tumbler Strength of Iron Ore Pellets and Sinter. 5 pp.

—Vanadium Content—Atomic Absorption Spectrometry
ISO 9684-91. Iron Ores—Determination of Vanadium Content—Flame Atomic Absorption Spectrometric Methods First Edition. 14 pp.

—Vanadium Content—Spectrophotometry
BSI BS 7020: Sec 11.1-93. 1993 Analysis of Iron Ores Part 11: Methods for the Determination of Vanadium Content Section 11.1: BPHA Spectrophotometric Method (ISO 9683: 1991) (Q). 15 pp.
ISO 9683-91. Iron Ores—Determination of Vanadium Content—BPHA Spectrophotometric Method First Edition. 12 pp.
JIS M 8225-83. Methods for Determination of Vanadium in Iron Ores.
JIS M 8225-71. Methods for Determination of Vanadium in Iron Ores. 9 pp.

—Water Content—Karl Fischer Method
BSI BS 7020: Part 3-88. 1988 Analysis of Iron Ores Part 3: Method for the Determination of Combined Water Content. 12 pp.
CNS M3017-80. Methods for Determination of Combined Water in Iron Ores (Mar)(5379).
ISO 7335-87. Iron Ores—Determination of Combined Water Content—Karl Fischer Titrimetric Method First Edition. 12 pp.
JIS M 8211-83. Method for Determination of Combined Water in Iron Ores. 16 pp.

—X-Ray Fluorescence Spectrometry
CNS M3196-88. X-Ray Fluorescence Spectrometric Analysis for Iron Ores (May)(12313).
JIS M 8205-83. X-Ray Fluorescence Spectrometric Analysis for Iron Ores.

—Zinc Content
JIS M 8228-83. Methods for Determination of Zinc in Iron Ores.
JIS M 8228-71. Methods for Determination of Zinc in Iron Ores. 11 pp.

—Zinc Content—Absorptiometric Analysis
CNS M3134-83. Method for Determining Zinc in Iron Ores (Zincon Absorptiometric Method) (Apr)(10189).

—Zinc Content—Atomic Absorption Spectrometry
BSI BS 7020: Part 17-88. 1988 Analysis of Iron Ores Part 17: Method for the Determination of Lead and/or Zinc Contents: Flame Atomic Absorption Spectrometric Method. 14 pp.

Iron Ores (Cont.)

—Zinc Content—Atomic Absorption Spectrometry (Cont.)
ISO 8753-87. Iron Ores—Determination of Lead and/or Zinc Content—Flame Atomic Absorption Spectrometric Method First Edition. 12 pp.

—Zinc Content—Polarographic Analysis
CNS M3135-83. Method for Determining Zinc in Iron Ores (Polarographic Method) (Apr)(10190).

—Zinc Content—Volumetric Analysis
CNS M3133-83. Method for Determining Zinc in Iron Ores (EDTA Titration Method) (Apr)(10188).

Iron Organic Compounds
Scope Note: Use a more specific term
See: Ammonium Ferric Citrate; Ferric Citrate; Organic Compounds

Iron Oxide
See Also: Ferric Oxide; Magnetic Oxides; Micaceous Iron Oxide Pigments; Red Oxide Primers; Sienna
CNS K6381-75. Method of Test for Iron Oxide (Aug)(3836).
MOD UK DSTAN 68-90-84. Iron Oxide, Special Purpose Issue 1. 6 pp.
MOD UK DSTAN 68-117-89. Iron Oxide, Synthetic Issue 1. 8 pp.

—Incendiary Mixtures
MOD UK DSTAN 13-121-91. Hammerscale (Magnetic Oxide of Iron) Sizes 25, 100 and 200 Issue 1. 8 pp.

—Pigments
BSI BS 3981-76. 1976 Iron Oxide Pigments for Paints. 10 pp.
CNS K2053-75. Iron Oxide (Pigment) (Aug)(3835). 1 p.
ISO 1248-74. Iron Oxide Pigments for Paints First Edition. 8 pp.

—Pigments—Calcium Content
CGSB 1-GP-71 METH 50.15-81. Methods of Testing Paints and Pigments Pigment Analysis Soluble Calcium Compounds. 1 p.

—Pigments—Lead Tetroxide Content
CGSB 1-GP-71 METH 50.3-80. Methods of Testing Paints and Pigments Pigment Analysis Red Lead with Graphite or Iron Oxide. 1 p.
CGSB 1-GP-71 METH 50.13-80. Methods of Testing Paints and Pigments Pigment Analysis Red Lead with Zinc Chromate Yellow and Iron Oxide. 2 pp.

—Pigments—Zinc Chromate Content
CGSB 1-GP-71 METH 50.12-81. Methods of Testing Paints and Pigments Pigment Analysis Zinc Chromate Yellow. 1 p.

—Pigments—Zinc Oxide Content
CGSB 1-GP-71 METH 50.16-80. Methods of Testing Paints and Pigments Pigment Analysis Zinc Oxide. 1 p.

Iron Oxide Brown

—Pigments
CNS K2114-80. Iron Oxide Pigments (Red and Brown) (Jul)(5879). 2 pp.

Iron Oxide Content Analysis

—Dolomite
ISO 10058-92. Magnesites and Dolomites—Chemical Analysis First Edition. 23 pp.

—Dolomite—Absorptiometric Analysis
CNS M3105-82. Method for Determining Iron Oxide in Dolomite (O-Phenanthroline Absorptiometric Method) (Sep)(9423).
CNS M3106-82. Method for Determining Calcium Oxide in Dolomite (EDTA Titration Method) (Sep)(9424).

—Feldspars—Absorptiometric Analysis
CNS M3086-82. Method for Determination of Iron Oxide in Feldspar (O-Phenanthroline Absorptiometric Method) (Jan)(8441).

—Glass—Atomic Absorption Spectrometry
BSI BS 7709: Part 5-93. 1993 Analysis of Extract Solutions of Glass Part 5: Method for Determination of Iron(III) Oxide by Molecular Absorption Spectrometry and Flame Atomic Absorption Spectrometry (ISO 10136-5: 1993) (V). 13 pp.
ISO 10136 Pt 5-93. Glass and Glassware—Analysis of Extract Solutions—Part 5: Determination of Iron(III) Oxide by Molecular Absorption Spectrometry and Flame Atomic Absorption Spectrometry First Edition. 10 pp.

Iron Oxide Content Analysis (Cont.)

—Glass—Molecular Absorption Spectrophotometry
BSI BS 7709: Part 5-93. 1993 Analysis of Extract Solutions of Glass Part 5: Method for Determination of Iron(III) Oxide by Molecular Absorption Spectrometry and Flame Atomic Absorption Spectrometry (ISO 10136-5: 1993) (V). 13 pp.
ISO 10136 Pt 5-93. Glass and Glassware—Analysis of Extract Solutions—Part 5: Determination of Iron(III) Oxide by Molecular Absorption Spectrometry and Flame Atomic Absorption Spectrometry First Edition. 10 pp.

—Glassware—Atomic Absorption Spectrometry
BSI BS 7709: Part 5-93. 1993 Analysis of Extract Solutions of Glass Part 5: Method for Determination of Iron(III) Oxide by Molecular Absorption Spectrometry and Flame Atomic Absorption Spectrometry (ISO 10136-5: 1993) (V). 13 pp.
ISO 10136 Pt 5-93. Glass and Glassware—Analysis of Extract Solutions—Part 5: Determination of Iron(III) Oxide by Molecular Absorption Spectrometry and Flame Atomic Absorption Spectrometry First Edition. 10 pp.

—Glassware—Molecular Absorption Spectrophotometry
BSI BS 7709: Part 5-93. 1993 Analysis of Extract Solutions of Glass Part 5: Method for Determination of Iron(III) Oxide by Molecular Absorption Spectrometry and Flame Atomic Absorption Spectrometry (ISO 10136-5: 1993) (V). 13 pp.
ISO 10136 Pt 5-93. Glass and Glassware—Analysis of Extract Solutions—Part 5: Determination of Iron(III) Oxide by Molecular Absorption Spectrometry and Flame Atomic Absorption Spectrometry First Edition. 10 pp.

—Limestone
CPPA J.4-92. Analysis of Limestone. 5 pp.

—Magnesite
ISO 10058-92. Magnesites and Dolomites—Chemical Analysis First Edition. 23 pp.

—Pigments
CGSB 1-GP-71 METH 50.11-81. Methods of Testing Paints and Pigments Pigment Analysis Iron Oxide and Siliceous Matter. 1 p.

Iron Oxide Paints
Use: Red Oxide Primers

Iron Oxide Red
See Also: Pigments
MOD UK CS 2753B. Red Oxide of Iron, Manufactured LF Quality (Withdrawn).

—Fillers
MOD UK TS 566B. Iron Oxide, Synthetic (Superseded by Def Stan 68-117).

—Pigments
CNS K2114-80. Iron Oxide Pigments (Red and Brown) (Jul)(5879). 2 pp.
JIS K 5109-72. Red Iron Oxide (Pigment). 6 pp.

Iron Oxide Yellow

—Pigments
CNS K2105-80. Yellow Iron Oxide Hydrated (Jul)(5870). 2 pp.

—Propellants
MOD UK TS 10254. Iron Oxide, Synthetic, Yellow.

Iron Phosphate Coatings (Made From Iron Phosphate)
See Also: Coatings; Conversion Coatings

—Iron
CGSB 31-GP-106M-88. Coating, Conversion, Iron Phosphate, for Paint Base. 10 pp.

—Steel
CGSB 31-GP-106M-88. Coating, Conversion, Iron Phosphate, for Paint Base. 10 pp.

Iron Pipes
See Also: Cast Iron Pipes; Ductile Iron Pipes; Gray Iron Pipes; Pipes

—Pressure—Design
BSI BS 806-93. 1993 Design and Construction of Ferrous Piping Installations for and in Connection with Land Boilers (Q). 138 pp.

Iron Plates
Scope Note: See the subheading Plates under Iron

Iron Powder
See Also: Iron; Metal Powders
CNS G3234-89. Iron Powder (Jun)(12529).
JIS H 2601-83. Iron Powder. 8 pp.

—Insoluble Matter Content
BSI BS 5600: Sec 2.9-80. (WITHDRAWN) 1980 Powder Metallurgical Materials and Products Part 2: Methods of Sampling and Testing Metallic Powders Section 2.9: Determination of Acid-Insoluble Content in Iron, Copper, Tin and Bronze Powders. 4 pp.
BSI BS EN 24496-93. 1993 Metallic Powders—Determination of Acid Insoluble Content in Iron, Copper, Tin and Bronze Powders (ISO 4496: 1978) (V). 10 pp.
CEN EN 24496-93. Metallic Powders—Determination of Acid Insoluble Content in Iron, Copper, Tin and Bronze Powders (ISO 4496: 1978). 5 pp.
ISO 4496-78. Metallic Powders—Determination of Acid-Insoluble Content in Iron, Copper, Tin and Bronze Powders First Edition (CEN EN 24496: 1993). 5 pp.

Iron Pyrite
Use: Pyrite

Iron Sulfates
Scope Note: Use a more specific term
See: Ammonium Ferrous Sulfate; Ferric Sulfate; Ferrous Sulfate

Iron Sulfides
Use: Ferrous Sulfides

Iron Wire
See Also: Electric Conductors; Electric Wire; Wire
CNS K7260-68. Chemical Reagent (Iron Wire) (Nov)(1759).

—Electric—Compressed Connectors
BSI BS 4579: Part 2-73. 1973 Amd 1 Performance of Mechanical and Compression Joints in Electric Cable and Wire Connectors Part 2: Compression Joints in Nickel, Iron and Platted Copper Conductors. 11 pp.

Iron-59 Content Analysis
Use: Radioactive Iron Content Analysis

Ironing
Use For: Hot Pressing (Ironing); Pressing (Ironing)
See Also: Irons (Electric); Steaming

—Fabrics
BSI BS 7305-90. 1990 Method for Determination of the Maximum Safe Ironing Temperature of Fabrics. 4 pp.
CGSB CAN/CGSB-4.2 NO.33-M86. Textile Test Methods Pressing—Selection of Methods. 6 pp.
CGSB CAN/CGSB-4.2 NO.33.1-M86. Textile Test Methods Method of Pressing Ironing. 8 pp.
CGSB CAN/CGSB-4.2 NO.57-M90. Textile Test Methods Determination of Maximum Safe Ironing Temperature. 9 pp.

—Fabrics—Colorfastness Testing
CGSB CAN/CGSB-4.2 NO.31-M89. Textile Test Methods Textiles—Tests for Colourfastness—Part X11: Colourfastness to Hot Pressing (ISO 105/X11-1987). 11 pp.
CNS L3156-83. Method of Test for Colour Fastness to Hot Pressing (Jun)(8532).
ISO 105 Pt X11-87. Textiles—Tests for Colour Fastness—Part X11: Colour Fastness to Hot Pressing Third Edition. 4 pp.
ISO 105 Pt X13-87. Textiles—Tests for Colour Fastness—Part X13: Colour Fastness of Wool Dyes to Processes Using Chemical Means for Creasing, Pleating and Setting Third Edition. 6 pp.
JIS L 0850-75. Testing Method for Colour Fastness to Hot Pressing (R 1978). 9 pp.

—Fabrics—Shrinkage
CGSB CAN/CGSB-4.2 NO.33.4-M86. Textile Test Methods Method of Pressing Tension Pressing. 11 pp.
JIS L 1057-92. Testing Methods for Shrinkage Percentage by Ironing of Woven and Knitted Fabrics. 7 pp.

Ironing Machines
Use: Irons (Electric)

Irons (Electric)
See Also: Appliances; Ironing; Steam Irons
BSI BS 3999: Part 7-92. 1992 Methods of Measuring the Performance of Household Electrical Appliances Part 7: Electric Irons. 26 pp.
BSI BS 3999: Part 7-69. 1969 Amd 1 Methods of Measuring the Performance of Household Electrical Appliances Part 7: Electric Irons. 23 pp.
CNS C3063-87. Method of Test for Electric Irons (Apr)(4779). 7 pp.
CNS C4145-87. Electric Irons (Sep)(4778). 10 pp.
IEC 508-75. Methods for Measuring the Performance of Electric Ironing Machines for Household and Similar Purposes First Edition. 29 pp.
JIS C 9203-92. Electric Irons. 40 pp.
JIS C 9204-76. Japanese Style Electric Irons.
SAA AS 1805-75. Electric Irons for Household Use Corrig. Amdt 1 February. 24 pp.

—Acceptance Testing
SAA AS 3307-92. Approval and Test Specification—Particular Requirements for Electric Irons (NZS 6303:1992) Amdt 1 December 1992 (in Professional Package 28). 6 pp.
SNZ NZS 6303-92. Approval and Test Specification—Particular Requirements for Electric Irons Amend: 1, 1992 (NZS 6303:1992/AS 3307-1992). 5 pp.

—Commercial
CSA CAN/CSA-C22. 2 NO 81-M90. Electric Irons; (Gen Instr 1). 42 pp.
CSA 1169 Bull. Electrical Bulletin 1169 June 27, 1978 to C22.2 NO 81. 2 pp.

—Flatwork
BSI BS 6246: Part 3-85. 1985 Industrial Laundry Machinery Part 3: Methods for the Assessment of the Effect of Flatwork Ironing Machines on Textiles. 18 pp.

—Flatwork—Capacity Measurement
ISO 9398 Pt 1-93. Specifications for Industrial Laundry Machines—Definitions and Testing of Capacity and Consumption Characteristics—Part 1: Flatwork Ironing Machines First Edition. 8 pp.

—Flatwork—Energy Consumption
ISO 9398 Pt 1-93. Specifications for Industrial Laundry Machines—Definitions and Testing of Capacity and Consumption Characteristics—Part 1: Flatwork Ironing Machines First Edition. 8 pp.

—Household
CSA CAN/CSA-C22. 2 NO 81-M90. Electric Irons; (Gen Instr 1). 42 pp.
CSA 1169 Bull. Electrical Bulletin 1169 June 27, 1978 to C22.2 NO 81. 2 pp.
CSA C22.2 NO 99-1954. Construction and Test of Domestic Electric Ironing Machines (R 1992). 22 pp.
CSA 1169 Bull. Electrical Bulletin 1169 June 27, 1978 to C22.2 NO 99. 2 pp.
IEC 311-88. Methods of Measurement of Performance of Electric Irons for Household or Similar Use Second Edition; (Amendment 1-1991) (Amendment 2-1993). 66 pp.

—Household—Construction
CSA C22.2 NO 99-1954. Construction and Test of Domestic Electric Ironing Machines (R 1992). 22 pp.
CSA 1169 Bull. Electrical Bulletin 1169 June 27, 1978 to C22.2 NO 99. 2 pp.

—Household—Portable
CSA C22.2 NO 99-1954. Construction and Test of Domestic Electric Ironing Machines (R 1992). 22 pp.
CSA 1169 Bull. Electrical Bulletin 1169 June 27, 1978 to C22.2 NO 99. 2 pp.

—Household—Portable—Construction
CSA C22.2 NO 99-1954. Construction and Test of Domestic Electric Ironing Machines (R 1992). 22 pp.
CSA 1169 Bull. Electrical Bulletin 1169 June 27, 1978 to C22.2 NO 99. 2 pp.

—Household—Safety
IEC 335 Pt 2-3-93. Safety of Household and Similar Electrical Appliances Part 2: Particular Requirements for Electric Irons Fourth Edition. 38 pp.

—Safety
BSI BS 3456: Sec 102.3-90. (WITHDRAWN) 1990 Safety of Household Electrical Appliances Part 102: Particular Requirements Section 102.3: Electric Irons (Superseded by BS EN 60335-2-3: 1991). 20 pp.
BSI BS EN 60335-2-3-91. 1991 Safety of Household and Similar Electrical Appliances Part 2: Particular Requirements Section 2.3: Electric Irons (L). 25 pp.
BSI BS EN 60335-2-3-01. 1991 Amd 1 Safety of Household and Similar Electrical Appliances Part 2: Particular Requirements Section 2.3: Electric Irons (AMD 7280) October 15, 1992. 32 pp.
BSI BS EN 60335-2-3-02. 1991 Amd 2 Safety of Household and Similar Electrical Appliances Part 2: Particular Requirements Section 2.3: Electric Irons (AMD 7694) June 15, 1993 (L). 38 pp.
BSI BS EN 60335-2-44-91. 1991 Safety of Household and Similar Electrical Appliances Part 2: Particular Requirements Section 2.44: Electric Ironers. 22 pp.
CENELEC HD 253-74. Particular Specification for Electric Irons, Ironess and Pressing Machines. 6 pp.
CENELEC HD 253 S3-86. Safety of Household and Similar Electrical Appliances Part 2: Particular Requirements for Electric Irons. 3 pp.
CENELEC HD 253 /A1-87. AMD 1 Safety of Household and Similar Electrical Appliances Part 2: Particular Requirements for Electric Irons. 5 pp.
CENELEC EN 60 335-2-3-90. Safety of Household and Similar Electrical Appliances Part 2: Particular Requirements for Electric Irons. 18 pp.
CENELEC EN 60335-2-3 /PRAB-91. AMD prAB Safety of Household and Similar Electrical Appliances Part 2: Particular Requirements for Electric Irons. 1 p.
CENELEC EN 60335-2-3 /A1-92. AMD 1 Safety of Household and Similar Electrical Appliances Part 2: Particular Requirements for Electric Irons. 6 pp.
CENELEC EN 60335-2-3 /A52-92. AMD 52 Safety of Household and Similar Electrical Appliances Part 2: Particular Requirements for Electric Irons. 4 pp.
CENELEC EN 60335-2-3-92. CORRIGENDUM Safety of Household and Similar Electrical Appliances Part 2: Particular Requirements for Electric Irons. 1 p.
CENELEC EN 60 335-2-44-91. Safety of Household and Similar Electrical Appliances Part 2 Particular Requirements for Electric Ironers. 9 pp.
DIN VDE 0700 Pt 212-81. Safety of Household and Similar Electrical Appliances; Particular Requirements for Rotary Ironers and Flat Bed Ironers (July). 16 pp.
IEC 335 Pt 2-44-87. Safety of Household and Similar Electrical Appliances Part 2: Particular Requirements for Electric Ironers First Edition. 27 pp.

Irradiance
Use: Optical Density

Irradiation
See Also: Alpha Irradiation; Electron Irradiation; Gamma Irradiation; Ionizing Radiation; Preservation; Radiography

—Electrical Insulation
IEC 544 Pt 2-91. Guide for Determining the Effects of Ionizing Radiation on Insulating Materials Part 2: Procedures for Irradiation and Test Second Edition. 45 pp.

Irrigation
See Also: Drainage; Drains; Surface Water; Water Pipelines; Water Supply
DIN ENGL 19655-82. Irrigation; Objectives, Basic Principles and Methods (Oct). 9 pp.

Irrigation Equipment (Agricultural)
Use For: Irrigation Machines (Agricultural)
See Also: Agricultural Equipment; Irrigation Pipe Fittings; Irrigation Pipes; Sprinklers
BSI BS 7562: Part 2-92. 1992 Planning, Design and Installation of Irrigation Schemes Part 2: Guide for Acquisition of Site Data. 18 pp.
BSI BS 7562: Part 4-92. 1992 Planning, Design and Installation of Irrigation Schemes Part 4: Guide to Water Resources. 25 pp.
BSI BS 7562: Part 6-92. 1992 Planning, Design and Installation of Irrigation Schemes Part 6: Guide for Feasibility and Implementation Procedures. 14 pp.

—Athletic Fields
DIN ENGL 18035 Pt 2-79. Sports Grounds; Watering of Turfed Areas and Tamped Areas (Jan). 4 pp.

—Automatic—Hydraulic Control
ISO TR8059-86. Irrigation Equipment—Automatic Irrigation Systems—Hydraulic Control First Edition. 7 pp.

—Emitters
BSI BS 7605: Part 1-92. 1992 Agricultural Irrigation Equipment Part 1: Specification for Emitters (ISO 9260: 1991). 12 pp.
ISO 9260-91. Agricultural Irrigation Equipment—Emitters—Specification and Test Methods First Edition. 9 pp.

—Filter/Strainers
ISO 9912 Pt 2-92. Agricultural Irrigation Equipment—Filters—Part 2: Strainer-Type Filters First Edition. 9 pp.
ISO 9912 Pt 3-92. Agricultural Irrigation Equipment—Filters—Part 3: Automatic Self-Cleaning Strainer-Type Filters First Edition. 9 pp.

INTERNATIONAL AND NON-U.S. NATIONAL STANDARDS
SUBJECT INDEX
ISO

Irrigation Equipment (Agricultural) (Cont.)

—Front Boom—Safety
CEN PREN 909-92. Safety Requirements for Agricultural and Forestry Machinery—Irrigating Machines—Pivot Type and Front Boom Linear Move Type. 9 pp.

—Glossaries
BSI BS 7562: Part 1-92. 1992 Planning, Design and Installation of Irrigation Schemes Part 1: Glossary of Terms. 37 pp.

—Hose Reel—Safety
CEN PREN 908-92. Safety Requirements for Agricultural and Forestry Machinery—Hose Reel Type Irrigating Machines. 10 pp.

—Hoses
DIN ENGL 19658 Pt 1-88. Polyethylene (PE) Coiling Pipes and Hoses for Use in Irrigation Systems; Coiling Pipes; Dimensions and Technical Delivery Conditions (Oct). 3 pp.
DIN ENGL 19658 Pt 3-88. Polyethylene (PE) Coiling Pipes and Hoses for Use in Irrigation Systems; Hoses of Nonrigid Material with Woven Fabric Inner Layer; Dimensions and Technical Delivery Conditions (Oct). 2 pp.

—Hydraulic Valves
ISO 9635-90. Irrigation Equipment—Hydraulically Operated Irrigation Valves First Edition. 8 pp.

—Overhead—Safety
DIN ENGL 11001 Pt 10-80. Agricultural Machines and Tractors; Overhead Irrigation Machines; Special Technical Safety Requirements and Testing (Jan). 3 pp.

—Pivot—Safety
CEN PREN 909-92. Safety Requirements for Agricultural and Forestry Machinery—Irrigating Machines—Pivot Type and Front Boom Linear Move Type. 9 pp.

—Sprayers
ISO 8026-85. Irrigation Equipment—Irrigation Sprayers—General Requirements and Test Methods First Edition. 9 pp.

—Traveller
ISO 8224 Pt 1-85. Traveller Irrigation Machines—Part 1: Laboratory and Field Test Methods First Edition. 12 pp.

—Traveller—Hose Couplings—Physical Properties
ISO 8224 Pt 2-91. Traveller Irrigation Machines—Part 2: Softwall Hose and Couplings—Test Methods First Edition. 7 pp.

—Traveller—Hoses—Physical Properties
ISO 8224 Pt 2-91. Traveller Irrigation Machines—Part 2: Softwall Hose and Couplings—Test Methods First Edition. 7 pp.

—Volumetric Valves
ISO 7714-85. Irrigation Equipment—Volumetric Valves—General Requirements and Test Methods First Edition. 6 pp.

—Water Valves
ISO 9911-93. Agricultural Irrigation Equipment—Manually Operated Small Plastics Valves First Edition. 11 pp.

—Water Valves—Check
ISO 9952-93. Agricultural Irrigation Equipment—Check Valves First Edition. 11 pp.

—Water Valves—Pressure Loss
ISO 9644-93. Agricultural Irrigation Equipment—Pressure Losses in Irrigation Valves—Test Method First Edition. 9 pp.

Irrigation Machines (Agricultural)
Use: Irrigation Equipment (Agricultural)

Irrigation Pipe Fittings
See Also: Irrigation Equipment (Agricultural); Irrigation Pipes; Pipe Fittings
BSI BS 7605: Part 2-92. 1992 Agricultural Irrigation Equipment Part 2: Specification for Emitting Pipe Systems (ISO 9261: 1991). 12 pp.
ISO 9261-91. Agricultural Irrigation Equipment—Emitting-Pipe Systems—Specification and Test Methods First Edition. 10 pp.

—Socket—Concrete
JIS A 5322-90. Reinforced Spun-Concrete Pipes with Socket. 9 pp.
JIS A 5330-90. Concrete Pipes with Socket. 9 pp.

Irrigation Pipe Joints
See Also: Pipe Joints

—Polyethylene
ISO 9625-93. Mechanical Joint Fittings for Use with Polyethylene Pressure Pipes for Irrigation Purposes First Edition. 8 pp.

Irrigation Pipes
See Also: Irrigation Equipment (Agricultural); Irrigation Pipe Fittings; Pipes; Water Pipes
BSI BS 7605: Part 2-92. 1992 Agricultural Irrigation Equipment Part 2: Specification for Emitting Pipe Systems (ISO 9261: 1991). 12 pp.
ISO 9261-91. Agricultural Irrigation Equipment—Emitting-Pipe Systems—Specification and Test Methods First Edition. 10 pp.

—Aluminum Alloy
SAA AS 3975-91. Aluminium Alloys—Irrigation Tube (in Professional Package 61). 10 pp.

—Concrete
JIS A 5322-90. Reinforced Spun-Concrete Pipes with Socket. 9 pp.
JIS A 5330-90. Concrete Pipes with Socket. 9 pp.

—Polyethylene
DIN ENGL 19658 Pt 1-88. Polyethylene (PE) Coiling Pipes and Hoses for Use in Irrigation Systems; Coiling Pipes; Dimensions and Technical Delivery Conditions (Oct). 3 pp.
DIN ENGL 19658 Pt 3-88. Polyethylene (PE) Coiling Pipes and Hoses for Use in Irrigation Systems; Hoses of Nonrigid Material with Woven Fabric Inner Layer; Dimensions and Technical Delivery Conditions (Oct). 2 pp.
ISO 8779-92. Polyethylene (PE) Pipes for Irrigation Laterals—Specifications First Edition. 10 pp.

—Polyethylene—Environmental Stress Cracking
ISO 8796-89. Polyethylene (PE) 25 Pipes for Irrigation Laterals—Susceptibility to Environmental Stress-Cracking Induced by Insert-Type Fittings—Test Method and Specification First Edition. 6 pp.

Irrigation Systems
Use: Irrigation Equipment (Agricultural)

Irrigators, Oral
Use: Oral Irrigators

Irritants Testing

—Toys
CNS Z8016-6-86. Method of Test for Toy Safety (Testing for Irritant Substances Content) (Jun)(4798-6). 4 pp.

Isatin
Use For: Indole-2,3-Dione
CNS K7261-68. Chemical Reagent (Isatin) (Nov)(1760).
JIS K 8089-78. Isatin, Indole-2,3-Dione.

ISB
Use: Sidebands—Independent

ISBN
Use: International Standard Book Numbering

ISC
Use: International Switching Centers

ISDB
Use: Integrated Service Digital Broadcasting

ISDN
Use: Integrated Services Digital Networks

ISDN-UP
Use: Integrated Services Digital Networks User Part

ISDN User Part
Use: Integrated Services Digital Networks User Part

ISIN
Use: International Securities Identification Numbering System

ISM Equipment
Use For: Industrial, Scientific and Medical Apparatus
See Also: Industrial Equipment; Medical Electrical Equipment; Medical Equipment

—Electromagnetic Interference
SAA AS 2064-90. Electromagnetic Interference—Industrial, Scientific and Medical (ISM) Radiofrequency Equipment—Limits and Methods of Measurement (Superseded (in Part) by AS/NZS 2064.1—1992 and AS/NZS 2064.2—1992).

ISM Equipment (Cont.)

—Radio Frequency Interference
BSI BS EN 55011-91. 1991 Limits and Methods of Measurement of Radio Disturbance Characteristics of Industrial, Scientific and Medical (ISM) Radio-Frequency Equipment. 38 pp.
CCIR Report 1104-90. Limitation of Radiation from Industrial, Scientific and Medical (ISM) Equipment—Section 1D—Spectrum Utilization and Applications. 28 pp.
CCIR Decision 54-2-89. Radiation from ISM Equipment—Annex to Volume I—Spectrum Utilization and Monitoring. 2 pp.
CENELEC EN 55 011-91. Limits and Methods of Measurement of Radio Disturbance Characteristics of Industrial, Scientific and Medical (ISM) Radio-Frequency Equipment. 8 pp.
DIN VDE 0871 Pt 11 (D)-87. Limits and Measurements of Radio Interference Characteristics of Industrial, Scientific and Medical (ISM) Radio Frequency Equipment: German Version Pr EN 55011: 1987 (Sept). 53 pp.
IEC CISPR 11-90. Limits and Methods of Measurement of Electromagnetic Disturbance Characteristics of Industrial, Scientific and Medical (ISM) Radio-Frequency Equipment Second Edition. 60 pp.
IEC CISPR 23-87. Determination of Limits for Industrial, Scientific and Medical Equipment First Edition. 29 pp.
SAA AS/NZS 2064. 1-92. Limits and Methods of Measurement of Electro-magnetic Disturbance Characteristics of Industrial, Scientific and Medical (ISM) Radio-Frequency Equipment—Part 1: General Require-ments (IEC/CISPR 11:1990) Bound with AS/NZS 2064. 2:1992) (Supersedes AS 2064—1990 (in Part)). 25 pp.
SAA AS/NZS 2064. 2-92. Limits and Methods of Measurement of Electro-magnetic Disturbance Characteristics of Industrial, Scientific and Medical (ISM) Radio-Frequency Equipment—Part 2: Additional Re-quirements for Australia Bound with AS/NZS 2064.1: 1992 (Supersedes AS 2064—1990 (in Part)). 1 p.
SNZ NZS/AS 2064. 1-92. Limits and Methods of Measurement of Electromagnetic Disturbance Characteristics of Industrial, Scientific and Medical (ISM) Radio-Frequency Equipment Part 1: General Requirements (This is a Joint Standard with SAA AS 1053.1). 25 pp.
SNZ NZS/AS 2064. 2-92. Limits and Methods of Measurement of Electromagnetic Disturbance Characteristics of Industrial, Scientific and Medical (ISM) Radio-Frequency Equipment Part 2: Additional Requirements for Australia and New Zealand. 1 p.

—Radio Frequency Interference—Suppression
CCIR QUESTION 70/1-90. Limitation of Radiation from Industrial, Scientific and Medical (ISM) Equipment—Questions Concerning Study Group 1—Spectrum Management Techniques (Spectrum Engineering, Planning, Sharing, Monitoring and Utilization). 1 p.
DIN VDE 0871-78. Radio Interference Suppression of Radio Frequency for Equipment; for Industrial, Scientific and Medical (ISM) and Similar Purposes (June). 44 pp.
DIN VDE 0871 Pt 1 (D)-85. Radio Interference Suppression of Radio Frequency Equipment; for Industrial, Scientific, and Medical and Similar Purposes; ISM Equipment (Aug). 55 pp.

ISO
Use For: International Organization for Standardization *See Also:* Standards

—Catalogs
ISO. ISO Catalogue 1993; (Supplement 1-1993). 1004 pp.
OSI ISO TP 1989-89. Technical Programme. 217 pp.

—Certification Systems
IEC Guide 16-78. Code of Principles on Third Party Certification Systems and Related Standards First Edition. 5 pp.
IEC Guide 44-85. General Rules for ISO or IEC International Third-Party Certification Schemes for Products First Edition. 16 pp.
ISO Guide 16-78. Code of Principles on Third Party Certification Systems and Related Standards First Edition. 5 pp.
ISO Guide 44-85. General Rules for ISO or IEC International Third-Party Certification Schemes for Products First Edition. 16 pp.
SAA HB18.16-91. Guidelines for Third-Party Certification and Accreditation—Guide 16: Code of Principles on Third Party Certification Systems and Related Standards (ISO/IEC Guide 16:1978) (SANZ HB18.16—1991). 2 pp.

INTERNATIONAL AND NON-U.S. NATIONAL STANDARDS SUBJECT INDEX

ISO (Cont.)

—Certification Systems (Cont.)
SAA HB18.44-91. Guidelines for Third-Party Certification and Accreditation—Guide 44: General Rules for ISO or IEC International Third-Party Certification Schemes for Products (ISO/IEC Guide 44:1985) (SANZ HB18.44—1991). 13 pp.

SNZ SANZ/SAA HB 18.16-91. Guidelines for Third-Party Certification and Accreditation Guide 16 Code of Principles on Third-Party Certification Systems and Related Standards (ISO/IEC Guide 16-1978). 2 pp.

SNZ SANZ/SAA HB 18.44-85. Guidelines for Third-Party Certification and Accreditation Guide 44 General Rules for ISO or IEC International Third-Party Certification Schemes for Products (ISO/IEC Guide 44-1985). 13 pp.

—Cooperation—CCITT
CCITT FASCICLE I.1-88. Minutes and Reports of the Plenary Assembly List of Study Groups and Questions Under Study. 249 pp.

CCITT RECMN A.20-89. Collaboration with Other International Organizations over Data Transmission—Data Communication over the Telephone Network (Study Group XVII) 2 pp. 2 pp.

CCITT RECMN A.21-89. Collaboration with Other International Organizations on CCITT-Defined Telematic Services—Terminal Equipment and Protocols for Telematic Services (Study Group VIII) 2 pp. 2 pp.

CCITT RECMN A.22-89. Collaboration with Other International Organizations on Information Technology—Terminal Equipment and Protocols for Telematic Services (Study Group VIII) 2 pp. 2 pp.

CCITT FASCICLE I.2-88. Opinions and Resolutions Recommendations on the Organisation and Working Procedures of CCITT (Series A). 64 pp.

—Standards Preparation
IEC Guide 15-77. ISO/IEC Code of Principles on "Reference to Standards" First Edition. 5 pp.

ISO Guide 15-77. ISO/IEC Code of Principles on "Reference to Standards" First Edition. 5 pp.

—Standards Preparation—SGML
BSI BS ISO/IEC TR 9573-11-92. 1992 Information Processing—SGML Support Facilities—Techniques for Using SGML—Part 11: Application at ISO Central Secretariat for International Standards and Technical Reports. 84 pp.

IEC TR9573 Pt 11-92. Information Processing—SGML Support Facilities—Techniques for Using SGML—Part 11: Application at ISO Central Secretariat for International Standards and Technical Reports First Edition. 82 pp.

ISO TR9573 Pt 11-92. Information Processing—SGML Support Facilities—Techniques for Using SGML—Part 11: Application at ISO Central Secretariat for International Standards and Technical Reports First Edition. 82 pp.

JTC1 TR 9573 Pt 11-92. Information Processing—SGML Support Facilities—Techniques for Using SGML—Part 11: Application at ISO Central Secretariat for International Standards and Technical Reports First Edition. 82 pp.

—Standards Preparation—Sound Recording
CCIR OPINION 16-3-86. Organizations Qualified to Set Standards on Sound and Television Recording—Volumes X and XI—Part 3—Sound and Television Recording. 1 p.

—Standards Preparation—Television Recording
CCIR OPINION 16-3-86. Organizations Qualified to Set Standards on Sound and Television Recording—Volumes X and XI—Part 3—Sound and Television Recording. 1 p.

—Standards Preparation—Testing
IEC Guide 36-82. Preparation of Standard Methods of Measuring Performance (SMMP) of Consumer Goods First Edition. 5 pp.

ISO Guide 36-82. Preparation of Standard Methods of Measuring Performance (SMMP) of Consumer Goods First Edition. 5 pp.

—Technical Reports—SGML
BSI BS ISO/IEC TR 9573-11-92. 1992 Information Processing—SGML Support Facilities—Techniques for Using SGML—Part 11: Application at ISO Central Secretariat for International Standards and Technical Reports. 84 pp.

IEC TR9573 Pt 11-92. Information Processing—SGML Support Facilities—Techniques for Using SGML—Part 11: Application at ISO Central Secretariat for International Standards and Technical Reports First Edition. 82 pp.

ISO TR9573 Pt 11-92. Information Processing—SGML Support Facilities—Techniques for Using SGML—Part 11: Application at ISO Central Secretariat for International Standards and Technical Reports First Edition. 82 pp.

JTC1 TR 9573 Pt 11-92. Information Processing—SGML Support Facilities—Techniques for Using SGML—Part 11: Application at ISO Central Secretariat for International Standards and Technical Reports First Edition. 82 pp.

—Translations
ISO Guide 47-86. Presentation of Translations of ISO Publications First Edition. 8 pp.

ISO/CCI
Use: Photographic Lenses—Color Contribution Index

Isoamyl Alcohol
Use: Isopentyl Alcohol

Isoamyl Ethyl Ketone
Use: Ethyl Isoamyl Ketone

Isoamyl Nitrite
Use: Amyl Nitrite

Isobutene Isoprene Rubber
Use: Butyl Rubber

Isobutyl Acetate
See Also: Acetates; n-Butyl Acetate; sec-Butyl Acetate; tert-Butyl Acetate; Hazardous Materials
CNS K1214-79. Isobutyl Acetate (95% Grade) (Jun)(4883).

—Storage and Handling—Information Cards
SAA AS 2508.3.00 2-82. Safe Storage and Handling Information Cards for Hazardous Materials—Part 3.002: Acetates (Ethyl Acetate, Propyl Acetates, Butyl Acetates, Amyl Acetate) Double Sided Card.

Isobutyl Alcohol
Use For: 2-Methyl-1-propanol *See Also:* Alcohols
CNS K1224-80. Isobutyl Alcohol (Feb)(5230).
CNS K7263-68. Chemical Reagent (Isobutyl Alcohol) (Nov)(1762).
CNS K7650-82. Chemical Reagent (2-Methyl-1-Propanol) (Oct)(9507).
JIS K 8811-79. 2-Methyl-1-Propanol.

Isobutyl Carbinol
Use: Isopentyl Alcohol

Isobutyl Methyl Ketone
Use: Methyl Isobutyl Ketone

Isobutylene Isoprene Rubber
Use: Butyl Rubber

Isochronous Distortion
See Also: Distortion (Electrical); Distortion Analyzers; Modulation; Telegraph Distortion

—Telegraph Circuits— 75 BPS
CCITT RECMN R.120-89. Tolerable Limits for the Degree of Isochronous Distortion of Code-Independent Telegraph Circuits Operating at Modulation Rates of 75, 100 and 200 Bauds—Telegraph Transmission (Study Group IX) 1 pp. 1 p.

—Telegraph Circuits—100 BPS
CCITT RECMN R.120-89. Tolerable Limits for the Degree of Isochronous Distortion of Code-Independent Telegraph Circuits Operating at Modulation Rates of 75, 100 and 200 Bauds—Telegraph Transmission (Study Group IX) 1 pp. 1 p.

—Telegraph Circuits—200 BPS
CCITT RECMN R.120-89. Tolerable Limits for the Degree of Isochronous Distortion of Code-Independent Telegraph Circuits Operating at Modulation Rates of 75, 100 and 200 Bauds—Telegraph Transmission (Study Group IX) 1 pp. 1 p.

Isocitric Acid Content Analysis
See Also: Citric Acid Content Analysis; Organic Acid Content Analysis

—Fruit Juices—Enzymatic Method
CNS N6226-91. Method of Test for Fruit and Vegetable Juices and Drinks-Determination of Isocitric Acid by Enzymatic Analysis (Jun)(12637). 3 pp.

—Vegetable Juices—Enzymatic Method
CNS N6226-91. Method of Test for Fruit and Vegetable Juices and Drinks-Determination of Isocitric Acid by Enzymatic Analysis (Jun)(12637). 3 pp.

Isocyanate Adhesives
Use: Polyurethane Adhesives

Isocyanate Resins
Use: Polyurethane Resins

Isocyanurate
Use For: Polyisocyanurate *See Also:* Isocyanurate Foams

—Insulating Board
CGSB CAN/CGSB-51.26-M86. Thermal Insulation Urethane and Isocyanurate, Boards, Faced. 19 pp.

—Thermal Insulation
CGSB 51-GP-21M-78. Thermal Insulation, Urethane and Isocyanurate, Unfaced. 12 pp.

SAA AS 1366.2-92. Rigid Cellular Plastics Sheets for Thermal Insulation—Part 2: Rigid Cellular Polyisocyanurate (RC/PIR). 5 pp.

—Thermal Insulation—Physical Properties
BSI BS 5241: Part 2-91. 1991 Rigid Polyurethane (PUR) and Polyisocyanurate (PIR) Foam When Dispensed or Sprayed on a Construction Site Part 2: Specification for Dispensed Foam for Thermal Insulation or Buoyancy Applications. 10 pp.

Isocyanurate Foams
Use For: PIR Foams; Polyisocyanurate Foams
See Also: Isocyanurate

—Insulation—Construction Sites
BSI BS 5241: Part 1-89. 1989 Rigid Polyurethane (PUR) and Polyisocyanurate (PIR) Foam When Dispensed or Sprayed on a Construction Site Part 1: Sprayed Foam Thermal Insulation Applied Externally. 8 pp.

—Pipe Insulation
BSI BS 5608-86. 1986 Preformed Rigid Polyurethane (PUR) and Polyisocyanurate (PIR) Foams for Thermal Insulation of Pipework and Equipment. 8 pp.

—Roof Insulation
BSI BS 7021-89. 1989 Code of Practice for Thermal Insulation of Roofs Externally by Means of Sprayed Rigid Polyurethane (PUR) or Polyisocyanurate (PIR) Foam. 18 pp.

Isodecyl Pelargonate
See Also: Lubricants

—Propellants
MOD UK TS 10200. Isodecyl Pelargonate.

Isolated Phase Busways
Use: Busways—Isolated Phase

Isolating Switches
Use For: Isolation Switches *See Also:* Bypass Switches; Disconnecting Switches; Electric Switches; Switches

—Automatic Transfer—Bypass—120-240 V AC
CSA C22.2 NO 178-1978. Automatic Transfer Switches (R 1992); (Amd 1-2 July 1982) (Amd 3 June 1987). 51 pp.

—Automatic Transfer—Bypass—120-240 V DC
CSA C22.2 NO 178-1978. Automatic Transfer Switches (R 1992); (Amd 1-2 July 1982) (Amd 3 June 1987). 51 pp.

—Automatic Transfer—Bypass—250-500 V AC
CSA C22.2 NO 178-1978. Automatic Transfer Switches (R 1992); (Amd 1-2 July 1982) (Amd 3 June 1987). 51 pp.

—Automatic Transfer—Bypass—250-500 V DC
CSA C22.2 NO 178-1978. Automatic Transfer Switches (R 1992); (Amd 1-2 July 1982) (Amd 3 June 1987). 51 pp.

INTERNATIONAL AND NON-U.S. NATIONAL STANDARDS
SUBJECT INDEX
Isopropylamine

Isolating Switches (Cont.)
—Automatic Transfer—Bypass—550-1000 V AC
CSA C22.2 NO 178-1978. Automatic Transfer Switches (R 1992); (Amd 1-2 July 1982) (Amd 3 June 1987). 51 pp.

—Automatic Transfer—Bypass—550-1000 V DC
CSA C22.2 NO 178-1978. Automatic Transfer Switches (R 1992); (Amd 1-2 July 1982) (Amd 3 June 1987). 51 pp.

—Electrical Installations
DIN VDE 0100 Pt 460-88. Erection of Power Installations with Nominal Voltages up to 1000 V; Protective Measures; Isolation and Switching (Oct) (Supersedes VDE 0100/05.73 Sec 31 b)2, VDE 0100g/07. 76 Sec 31 a)1.2 and DIN 57100 Part 410/VDE 0100 Part 410/11.83, Sub-clause 9. 12 pp.
DIN VDE 0100 Pt 537-88. Erection of Power Installations with Nominal Voltages up to 1000 V; Selection and Erection of Electrical Equipment; Devices for Isolation and Switching (Oct) (Supersedes VDE 0100/05.73 Sec 31 b)1, VDE 0100g/07.76 Sec 31 a)1.1.1.3. With DIN VDE 0100 Part 460/10.88. 12 pp.

—High Voltage
CSA C22.2 NO 58-M1989. High-Voltage Isolating Switches; (Gen Instr 1). 23 pp.
CSA 1169 Bull. Electrical Bulletin 1169 June 27, 1978 to C22.2 NO 58. 2 pp.

Isolating Transformers
Use: Isolation Transformers

Isolating Valves
Use: Isolation Valves

Isolation Gowns
Use: Gowns (Protective)—Isolation

Isolation Switches
Use: Isolating Switches

Isolation Transformers
Use For: Isolating Transformers See Also: Safety Isolating Transformers; Separating Transformers; Transformers
BSI BS 3535-62. (WITHDRAWN) 1962 Amd 1 Safety Isolating Transformers for Industrial and Domestic Purposes (Superseded by BS 3535: Part 1: 1990). 58 pp.
BSI BS 3535: Part 1-90. 1990 Isolating Transformers and Safety Isolating Transformers Part 1: General Requirements. 126 pp.
BSI BS 3535: Part 2-90. 1990 Isolating Transformers and Safety Isolating Transformers Part 2: Transformers for Reduced System Voltage. 8 pp.
BSI BS 3535: Part 2-01. 1990 Amd 1 Isolating Transformers and Safety Isolating Transformers Part 2: Specification for Transformers for Reduced System Voltage (AMD 6881) February 28, 1992. 9 pp.
CENELEC EN 60 742-89. Isolating Transformers and Safety Isolating Transformers—Requirements. 29 pp.
DIN VDE 0550 Pt 3-69. Regulations for Small Transformers; Particular Regulations for Isolating and Control Transformers with Network Connection and Isolating Transformers Above 1000 V (Dec) (Partially Superseded by Draft 0550 Part 1/11.87). 10 pp.
IEC 742-83. Isolating Transformers and Safety Isolating Transformers—Requirements First Edition; (Amendment 1-1992). 273 pp.
SAA AS 3108-90. Approval and Test Specification—Particular Requirements for Isolating Transformers and Safety Isolating Transformers (This is a Joint Standard with SANZ NZS 3108) Amdt 1 March 1992 Amdt 2 October 1992 Amdt 3 Sept. 1993 (in Professional Package 28). 1 p.
SNZ NZS/AS 3108-90. Approval and Test Specification—Particular Requirements for Isolating Transformers and Safety Isolating Transformers Amend: 1; 2, 1992 (This is a Joint Standard with SAA AS 3108). 89 pp.

—Power Supplies
CSA CAN/CSA-C22. 2 NO 223-M91. Power Supplies with Extra-Low-Voltage Class 2 Outputs; (Gen Instr 1). 53 pp.

—Series—Airport Lighting
CSA C22.2 NO 180-M1983. Series Isolating Transformers for Airport Lighting (R 1993); (Gen Instr 1 Thru 2). 14 pp.

Isolation Valves
Use For: Block Valves; Isolating Valves
See Also: Valves

Isolation Valves (Cont.)
CEN PREN 1074-2-93. Valves for Water Supply—Specification for Use and Appropriate Verification Tests—Part 2: Isolating Valves. 12 pp.

Isolator Switches
Use: Disconnecting Switches

Isolator Valves
Use: Isolation Valves

Isolators
Scope Note: Use a more specific term
See: Anechoic Chambers; Attenuators; Coaxial Isolators; Ferrite Isolators; Microwave Isolators; Noise Reduction; Optoisolators; Shock Absorbers; Sound Transmission; Vibration Isolators

L-Isoleucine
CNS K7657-82. Chemical Reagent (L-Isoleucine) (Oct)(9514).
JIS K 9055-72. L-Isoleucine.

Isometric Projections
Use: Axonometric Projections

Isooctane
Use For: 2,2,4-Trimethyl Pentane
See Also: Hydrocarbons
CNS K7264-68. Chemical Reagent (Isooctane) (Nov)(1763).
JIS K 9703-83. 2,2,4-Trimethyl Pentane (Isooctane).

Isopentyl Alcohol
Use For: Isoamyl Alcohol; Isobutyl Carbinol; 3-Methylbutyl Alcohol
CNS K7262-68. Chemical Reagent (Isoamyl Alcohol) (Nov)(1761).
JIS K 8051-78. Isoamyl Alcohol, 3-Methybutyl Alcohol.

Isophorone
CNS K1238-80. Isophorone (Apr)(5457).

Isoprene Rubber
Use For: IR (Isoprene Rubber); Polyisoprene Rubber
See Also: Rubber
MOD UK DSTAN 83-17-88. Isoprene Rubber Compound for Moulded-On Soling Issue 4. 10 pp.
MOD UK DSTAN 83-18-88. Isoprene Rubber Compound for Pre-Moulded Soles and Heels Issue 3. 10 pp.
MOD UK DSTAN 93-44-86. Natural and Synthetic Isoprene Rubber Compounds Issue 1. 11 pp.

—Polyisoprene Content
DIN ENGL 53621 Pt 1-75. Testing of Rubber and Elastomers; Quantitative Determination of Polymers; Determination of the Polyisoprene Content (Jan). 4 pp.

—Thermogravimetric Analysis
ISO 9924 Pt 1-93. Rubber and Rubber Products—Determination of the Composition of Vulcanizates and Uncured Compounds by Thermogravimetry—Part 1: Butadiene, Ethylene-Propylene Copolymer and Terpolymer, Isobutene-Isoprene, Isoprene and Styrene-Butadiene. 8 pp.

—Tubing (Flexible)—Air—Aircraft
BSI 2F 65-81. (WITHDRAWN) 1981 Amd 1 Flexible Corrugated Oxygen and Air Tubes for Aircraft (Natural or Synthetic Polyisoprene) (AMD 4871) July 31, 1985. 6 pp.

—Tubing (Flexible)—Oxygen—Aircraft
BSI 2F 65-81. (WITHDRAWN) 1981 Amd 1 Flexible Corrugated Oxygen and Air Tubes for Aircraft (Natural or Synthetic Polyisoprene) (AMD 4871) July 31, 1985. 6 pp.

—Vulcanization
BSI BS 5474-91. 1991 Evaluation of Non-Oil-Extended Solution-Polymerized Isoprene Rubber (IR). 8 pp.
ISO 2303-83. Rubber, Polyisoprene (IR)—Non Oil-Extended, Solution-Polymerized Types—Test Recipe and Evaluation of Vulcanization Characteristics Second Edition. 4 pp.
MOD UK TS 490B. Vulcanized Polyisoprene Rubber, Non Dermatitic Hardness 40 Degrees, Black (Superseded by TS 10087 Grade MD 4).

Isopropanol
Use: Isopropyl Alcohol

Isopropanol Content Analysis
Use: Isopropyl Alcohol Content Analysis

Isopropyl Acetate
See Also: Acetates; Hazardous Materials; n-Propyl Acetate
BSI BS 1834-68. 1968 Isopropyl Acetate. 11 pp.

Isopropyl Acetate (Cont.)
BSI BS 1834-01. 1968 Amd 1 Isopropyl Acetate (AMD 7219) September 15, 1992. 13 pp.
CNS K1204-78. Isopropyl Acetate (98% Grade) (May)(4388).
CNS K1213-79. Isopropyl Acetate (95% Grade) (Jun)(4882).

—Storage and Handling—Information Cards
SAA AS 2508.3.00 2-82. Safe Storage and Handling Information Cards for Hazardous Materials—Part 3.002: Acetates (Ethyl Acetates, Propyl Acetates, Butyl Acetates, Amyl Acetate) Double Sided Card.

Isopropyl Alcohol
Use For: Isopropanol; Propan-2-ol; 2-Propanol
See Also: Alcohols
BSI BS 1595: Part 1-86. 1986 Amd 2 Propan-2-Ol (Isopropyl Alcohol) for Industrial Use Part 1: Propan-2-01 (Isopropyl Alcohol). 4 pp.
BSI BS 1595: Part 2-84. 1984 Propan-2-Ol (Isopropyl Alcohol) for Industrial Use Part 2: Methods of Test. 8 pp.
CNS K1209-79. Isopropyl Alcohol (99% Grade) (May)(4862).
CNS K7265-68. Chemical Reagent (Isopropyl Alcohol) (Nov)(1764).
ISO 756 Pt 1-81. Propan-2-ol for Industrial Use—Methods of Test—Part 1: General First Edition. 5 pp.
JIS K 1522-78. Isopropyl Alcohol (Isopropanol).
JIS K 1522-63. Isopropyl Alcohol (Isopropanol). 9 pp.
JIS K 8839-81. 2-Propanol.

—Acidity—Volumetric Analysis
ISO 756 Pt 2-81. Propan-2-ol for Industrial Use—Methods of Test—Part 2: Determination of Acidity—Titrimetric Method First Edition. 4 pp.

—Antiicing Additives
CGSB 3-GP-525MA-88. Isopropanol. 8 pp.

—Deicers
CGSB 3-GP-525MA-88. Isopropanol. 8 pp.

—Miscibility
ISO 756 Pt 3-81. Propan-2-ol for Industrial Use—Methods of Test—Part 3: Test for Miscibility with Water First Edition. 4 pp.

—Safety
SAA AS 2508.3.00 6-91. Safe Storage and Handling Information Cards for Hazardous Materials—Part 3.006: Butanols, Propanols, Amyl Alcohols (In Professional Package 38) Double Sided Card.

Isopropyl Alcohol Content Analysis
Use For: Isopropanol Content Analysis

—Ethylene—Gas Chromatography
ISO 8174-86. Ethylene and Propylene for Industrial Use—Determination of Acetone, Acetonitrile, Propan-2-ol and Methanol—Gas Chromatographic Method First Edition. 8 pp.

—Methanols
CNS K6263-80. Method of Test for Methyl Alcohol (Methanol) (Feb)(2790). 7 pp.

—Propylene—Gas Chromatography
ISO 8174-86. Ethylene and Propylene for Industrial Use—Determination of Acetone, Acetonitrile, Propan-2-ol and Methanol—Gas Chromatographic Method First Edition. 8 pp.

Isopropyl Ether
See Also: Ethers
CNS K7659-82. Chemical Reagent (Isopropyl Ether) (Oct)(9516).
JIS K 9528-76. Isopropyl Ether.

N-Isopropyl-N'-Phenyl-p-Phenylenediamine
See Also: Diamines

—Antioxidants—Rubber
CNS K4083-87. Rubber Antioxidant IPPD (N—Isopropyl—N'-Phenyl—P Phenylenediamine) (Mar)(11881).
CNS K6901-87. Method of Test for Rubber Antioxidant IPPD (N-Isopropyl-N'-Phenyl P-Phenylenediamine) (Mar)(11882).

Isopropylamine
See Also: Hazardous Materials

—Storage and Handling—Information Cards
SAA AS 2508.3.01 0-82. Safe Storage and Handling Information Cards for Hazardous Materials—Part 3.010: Amines (Isopropylamine, Diethylamine, Triethylamine) Double Sided Card.

Isosafrole Content Analysis
—Nutmeg Oil—Gas Chromatography
ISO 7355-85. Oils of Sassafras and Nutmeg—Determination of Safrole and cis-and trans-isosafrole Content—Gas Chromatographic Method on Packed Columns First Edition. 6 pp.

—Sassafras Oil—Gas Chromatography
ISO 7355-85. Oils of Sassafras and Nutmeg—Determination of Safrole and cis-and trans-isosafrole Content—Gas Chromatographic Method on Packed Columns First Edition. 6 pp.

Isothermal Atmospheres
Use For: Exponential Atmospheres
—Ground Wave Propagation
CCIR Report 714-2-90. Ground-Wave Propagation in an Exponential Atmosphere—Section 5B—Effects of the Ground (Including Ground-Wave Propagation). 3 pp.

Isothiocyanates Content Analysis
—Oilseeds
BSI BS 4325: Part 8-93. 1993 Methods for Analysis of Oilseed Residues Part 8: Determination of Total Isothiocyanate Content and Vinylthiooxazolidone Content (ISO 5504: 1992). 16 pp.

BSI BS 4325: Part 8-84. 1984 Analysis of Oilseed Residues Part 8: Determination of Isothiocyanates and Vinyl Thiooxazolidone. 10 pp.

ISO 5504-92. Oilseed Residues—Determination of Total Isothiocyanate Content and Vinylthiooxazolidone Content Second Edition. 13 pp.

Isotopes
Scope Note: Use a more specific term *See:* Isotopic Content Analysis; Radioactive Iodine; Radioisotopes; Uranium Isotopes

Isotopic Content Analysis
See Also: Radioisotopes
—Nuclear Fuels—Mass Spectrometry
ISO 8299-93. Determination of Isotopic Content and Concentration of Uranium and Plutonium in Nitric Acid Solution—Mass Spectrometric Method First Edition. 17 pp.

Isotropic Radiated Power
See Also: Effective Isotropic Radiated Power
—Radio Relay Systems—Fixed Satellites—Frequency Band Sharing
CCIR RECMN 406-6-90. Maximum Equivalent Isotropically Radiated Power of Line-of-Sight Radio-Relay System Transmitters Operating in the Frequency Bands Shared with the Fixed-Satellite Service—Section 4/9A—Sharing Conditions. 3 pp.

CCIR RECMN 406-7-92. Maximum Equivalent Isotropically Radiated Power of Line-of-Sight Radio-Relay System Transmitters Operating in the Frequency Bands Shared with the Fixed-Satellite Service—Section 4/9A—Sharing Conditions. 2 pp.

ISPBX
Use: Integrated Services Private Branch Exchanges

ISPC
Use: International Sound-Program Centers

Israeli Aircraft Industries IAI 101 Arava Aircraft
See Also: Aircraft
CAA. 1AI, 101, and 201 Arava (World Airline Accident Summary). 1 p.

Israeli Aircraft Industries IAI 102 Arava Aircraft
See Also: Aircraft
CAA. 1AI, 101, and 201 Arava (World Airline Accident Summary). 1 p.

Israeli Aircraft Industries IAI 201 Arava Aircraft
See Also: Aircraft
CAA. 1AI, 101, and 201 Arava (World Airline Accident Summary). 1 p.

ISRC
Use: International Standard Recording Code

ISS
Use: Intersatellite Services

ISSN
Use: International Standard Serial Numbering

ISUP
Use: Integrated Services Digital Networks User Part

ITC
Use: International Television Centers

ITPC
Use: International Television Program Centers

ITU
Use For: International Telecommunications Union
See Also: CCIR; CCITT

—CCIR—CCITT—Organization—Glossaries
CCIR RESOLUTION 113-90. Organisation of Vocabulary Work—Volume XIII—Vocabulary and Related Subjects. 3 pp.

—CCIR—Physical Quantities—Glossaries
CCIR RECMN 663-1-90. Use of Certain Terms Linked with Physical Quantities—Section A—Terminology. 5 pp.

—Cooperation—CCITT
CCITT FASCICLE I.2-88. Opinions and Resolutions Recommendations on the Organisation and Working Procedures of CCITT (Series A). 64 pp.

—Documentation—Gentex Networks—Routing
CCITT RECMN F.20-89. International Gentex Service—Telegraph and Mobile Services Operations and Quality of Service (Study Group I) 5 pp. 5 pp.

CCITT RECMN F.93-89. Routing Table for Offices Connected to the Gentex Service—Telegraph and Mobile Services Operations and Quality of Service (Study Group I) 1 pp. 1 p.

IUD
Use: Intrauterine Devices

Ivory Board
See Also: Paperboard
CNS P2043-79. Ivory Board (Aug)(2920). 1 p.

Izod Impact Testing
Use: Impact Testing

Izod Impact Testing Equipment
Use: Impact Testing Equipment—Izod

J Acid
Use: 2-Amino-5-naphthol-7-sulfonic Acid

J-K Bistables
Use: J-K Flip Flops

J-K Flip Flops
Use For: J-K Bistables *See Also:* Digital Circuits; J-K Master/Slave Flip Flops

CECC CECC 90 106-018 ISSUE 1-85. UTE C 86-217; Digital Integrated Circuits in Accordance with FS 90 106;54/74 ALS 112, 54/74 ALS 112A; J K Negative Edge Triggered Bistable (En, Fr) ADD 3 (En, Fr). 16 pp.

CECC CECC 90 106-063 ISSUE 1-85. UTE C 86-217; Digital Integrated Circuits in Accordance with FS 90 106; 54/74 ALS 114, 54/74 ALS 114A; J K Negative Edge Triggered Bistable with Preset and Clear (En, Fr) ADD 3 (En, Fr). 16 pp.

CECC CECC 90 109-627 ISSUE 1-86. Digital Integrated Circuits in Accordance with FS 90 109: 54/74 HC 109; J K Bistable with Preset and Clear (En, Fr). 6 pp.

CECC CECC 90 109-638 ISSUE 1-86. Digital Integrated Circuits in Accordance with FS 90 109; 54/74 HC 112; J-K Negative Edge Triggered Bistable (En, Fr). 6 pp.

CECC CECC 90 109-771 ISSUE 1-87. Digital Integrated Circuits in Accordance with FS 90 109; 54/74 HCT 109; JK Bistable with Preset and Clear (En, Fr). 6 pp.

CECC CECC 90 109-772 ISSUE 1-87. Digital Integrated Circuits in Accordance with FS 90 109; 54/74 HCT 112; J-K Negative Edge Triggered Bistable (En, Fr). 6 pp.

MOD UK DSTAN 59-62: 90/010-68. Integrated Circuit Issue 1. 15 pp.
MOD UK DSTAN 59-62: 90/011-68. Integrated Circuit Issue 1. 3 pp.
MOD UK DSTAN 59-62: 90/012-68. Integrated Circuit Issue 1. 2 pp.
MOD UK DSTAN 59-62: 90/013-68. Integrated Circuit Issue 1. 2 pp.
MOD UK DSTAN 59-62: 90/014-68. Integrated Circuit Issue 1. 3 pp.
MOD UK DSTAN 59-62: 90/016-68. Integrated Circuit Issue 1. 13 pp.
MOD UK DSTAN 59-62: 90/017-68. Integrated Circuit Issue 1. 3 pp.
MOD UK DSTAN 59-62: 90/018-68. Integrated Circuit Issue 1. 2 pp.

J-K Flip Flops *(Cont.)*
MOD UK DSTAN 59-62: 90/035-69. Integrated Circuit Issue 1. 2 pp.
MOD UK DSTAN 59-62: 90/036-69. Integrated Circuit Issue 1. 3 pp.
MOD UK DSTAN 59-62: 90/037-69. Integrated Circuit Issue 1. 2 pp.
MOD UK DSTAN 59-62: 90/039-69. Integrated Circuit Issue 1. 2 pp.
MOD UK DSTAN 59-62: 90/040-69. Integrated Circuit Issue 1. 3 pp.
MOD UK DSTAN 59-62: 90/041-69. Integrated Circuit Issue 1. 2 pp.
MOD UK DSTAN 59-62: 90/044-69. Integrated Circuit Issue 1. 2 pp.
MOD UK DSTAN 59-62: 90/045-69. Integrated Circuit Issue 1. 3 pp.
MOD UK DSTAN 59-62: 90/046-69. Integrated Circuit Issue 1. 2 pp.
MOD UK DSTAN 59-62: 90/048-69. Integrated Circuit Issue 1. 2 pp.
MOD UK DSTAN 59-62: 90/049-69. Integrated Circuit Issue 1. 3 pp.
MOD UK DSTAN 59-62: 90/050-69. Integrated Circuit Issue 1. 2 pp.
MOD UK DSTAN 59-62: 90/060-69. Integrated Circuit Issue 1. 2 pp.
MOD UK DSTAN 59-62: 90/061-69. Integrated Circuit Issue 1. 2 pp.
MOD UK DSTAN 59-62: 90/062-69. Integrated Circuit Issue 1. 2 pp.
MOD UK DSTAN 59-62: 90/087-68. Integrated Circuit Issue 1. 13 pp.
MOD UK DSTAN 59-62: 90/088-68. Integrated Circuit Issue 1. 1 pp.
MOD UK DSTAN 59-62: 90/091-68. Integrated Circuit Issue 1. 3 pp.
MOD UK DSTAN 59-62: 90/092-68. Integrated Circuit Issue 1. 3 pp.
MOD UK DSTAN 59-62: 90/094-69. Integrated Circuit Issue 1. 2 pp.
MOD UK DSTAN 59-62: 90/095-69. Integrated Circuit Issue 1. 2 pp.
MOD UK DSTAN 59-62: 90/096-69. Integrated Circuit Issue 1. 2 pp.
MOD UK DSTAN 59-62: 90/098-69. Integrated Circuit Issue 1. 2 pp.
MOD UK DSTAN 59-62: 90/099-69. Integrated Circuit Issue 1. 2 pp.
MOD UK DSTAN 59-62: 90/100-69. Integrated Circuit Issue 1. 2 pp.
MOD UK DSTAN 59-62: 90/105-70. Integrated Circuit Issue 1. 20 pp.
MOD UK DSTAN 59-62: 90/106-70. Integrated Circuit Issue 1. 2 pp.
MOD UK DSTAN 59-62: 90/107-70. Integrated Circuit Issue 1. 3 pp.
MOD UK DSTAN 59-62: 90/108-70. Integrated Circuit Issue 1. 2 pp.
MOD UK DSTAN 59-62: 90/161-71. Integrated Circuit Issue 1. 10 pp.

—Dual
CECC CECC 90 101-024 ISSUE 1-87. UTE C 86-213/B 44; Digital Integrated Circuits in Accordance with FS 90 101; 54/64/74 73; Dual J-K Bistable with Clear (En). 16 pp.

CECC CECC 90 101-027 ISSUE 1-87. UTE C 86-213/B 47; Digital Integrated Circuits in Accordance with FS 90 101; 54/64/74 76; Dual J-K Bistable with Preset and Clear (En). 16 pp.

CECC CECC 90 102-023 ISSUE 1-81. BS CECC 90 102-023; Dual J-K Bistables with Preset Common Clear and Common Clock (En) AMD 1 (En). 19 pp.

CECC CECC 90 102-047 ISSUE 1-82. BS CECC 90 102-047; Dual J-K Negative-Edge Triggered Bistables with Preset and Clear (En). 18 pp.

CECC CECC 90 102-048 ISSUE 1-82. BS CECC 90 102-048; Dual J-K Negative-Edge-Triggered Bistable Circuit with Preset (En). 19 pp.

CECC CECC 90 103-036 ISSUE 1-81. BS CECC 90 103-036; Dual J-K Bistable with Preset and Clear (En) AMD 1 (En). 21 pp.

CECC CECC 90 103-037 ISSUE 1-81. BS CECC 90 103-037; Dual J-K Bistables with Preset, Common Clear and Common Clock (En). 19 pp.

CECC CECC 90 103-047 ISSUE 1-81. BS CECC 90 103-047; Dual J-K Positive Edge Triggered Bistable with Preset, Clear (En) AMD 1 (En). 19 pp.

CECC CECC 90 103-140 ISSUE 1-81. BS CECC 90 103-140; Dual J-K Bistables with Clear (En). 18 pp.

CECC CECC 90 104-024 ISSUE 2. NL CECC 90 104-024 Issue 2; Digital Integrated Circuits in Accordance with FS 90 104; HEC/HEF 4027B; Dual JK Flip-Flop (En). 9 pp.

CECC CECC 90 104-122 ISSUE 1-81. BS CECC 90 104-122; Silicon Complementary MOS with (B) Buffered Outputs and Cavity Packaging (En). 25 pp.

CECC CECC 90 104-122 ISSUE 1-86. CEI CECC 90 104-122; Silicon Complementary MOS with (B) Buffered Outputs Cavity and Non Cavity Packaging (En). 2 pp.

J-K Flip Flops (Cont.)
—Dual (Cont.)
CECC CECC 90 106-061 ISSUE 1-85. UTE C 86-217; Digital Integrated Circuits in Accordance with FS 90 106; 54/74 ALS 109, 54/74 ALS 109A; J K Positive Edge Triggered Bistable (En, Fr) ADD 3 (En, Fr). 16 pp.

CECC CECC 90 107-012 ISSUE 2-89. UTE C 86-218 ADD 2/FA 12 Digital Integrated Circuits in Accordance with FS 90 107; 54/74 F 109; J K Positive Edge-Triggered Bistable (En, Fr) ADD 3 (En, Fr). 9 pp.

CECC CECC 90 109-635 ISSUE 1-86. Digital Integrated Circuits in Accordance with FS 90 109; 54/74 HC 73; Dual J-K Bistable with Clear (En, Fr). 6 pp.

CECC CECC 90 109-636 ISSUE 1-86. Digital Integrated Circuits in Accordance with FS 90 109; 54/74 HC 76; Dual J-K Bistable with Preset and Clear (En, Fr). 6 pp.

CECC CECC 90 109-637 ISSUE 1-86. Digital Integrated Circuits in Accordance with FS 90 109; 54/74 HC 107; Dual J-K Bistable with Clear (En, Fr). 6 pp.

CECC CECC 90 109-639 ISSUE 1-86. Digital Integrated Circuits in Accordance with FS 90 109; 54/74 HC 113; Dual J-K Bistable with Preset (En, Fr). 6 pp.

CECC CECC 90 109-679 ISSUE 1-87. Digital Integrated Circuits in Accordance with FS 90 109; 54/74 HC 114; Dual J-K Bistable with Clear (En, Fr). 6 pp.

CECC CECC 90 109-741 ISSUE 1-87. Digital Integrated Circuits in Accordance with FS 90 109; 54 HCT 73, 74 HCT 73; Dual J-K Bistable with Clear (En, Fr). 6 pp.

CECC CECC 90 109-770 ISSUE 1-87. Digital Integrated Circuits in Accordance with FS 90 109; 54/74 HCT 107; Dual J-K Bistable with Clear (En, Fr). 6 pp.

—Dual—Preferred Products List
CECC CECC MUAHAG Vol 7 IS 8-92. Preferred Products List; Active Microcircuits (En, Fe, Ge). 89 pp.

J-K Master/Slave Flip Flops
See Also: Digital Circuits; J-K Flip Flops
—Dual
CECC CECC 90 103-046 ISSUE 1-81. BS CECC 90 103-046; Dual J-K Master-Slave Bistable with Clear (En) AMD 1 (En). 18 pp.

—Dual—Preferred Products List
CECC CECC MUAHAG Vol 7 IS 8-92. Preferred Products List; Active Microcircuits (En, Fe, Ge). 89 pp.

Jack Pads
See Also: Jacks (Lifts)
—Aircraft
BSI 2C 12-80. 1980 Jacking Pads for Aircraft. 6 pp.
ISO 43-76. Aircraft—Jacking Pads First Edition. 6 pp.
JIS W 3901-91. Aircraft—Jacking Pads (ISO 43:1976).
MOD UK DSTAN 17-3-01. Adaptors, Aircraft Jacking Point (Detachable Jacking Pads) Issue 1; Amendment 1. 8 pp.
NATO STANAG 3098 ED 9 AMD 0-88. Aircraft Jacking. 7 pp.
NATO STANAG 3098 ED 9 AMD 1-88. Aircraft Jacking. 8 pp.
NATO STANAG 3098 ED 10 AMD 0-00. Aircraft Jacking. 6 pp.

Jackets
See Also: Blazers; Clothing; Coats; Outerwear
—Dress Uniforms
CGSB 38-GP-43M-78. Construction Details for Uniform Coats, Jackets, and Tunics (Fusible Interlining), Standard for. 29 pp.

—Dress Uniforms—Mens'
CGSB 38-GP-110MA-83. Tunics and Jackets, Men's Uniform Dress Single-Breasted—Dimensions, Standard for. 10 pp.

—Firefighters'
MOD UK DEF-1103-54. Jackets & Trousers, Dungaree, Blue, Fire Brigade (Pattern No. A.125 and A.126). 7 pp.

—Lightweight—Boys'
CGSB CAN/CGSB-49.30-M91. Canada Standard Children's Sizes 2 to 6X, Girls' Sizes 7 to 16 and Boys' Sizes 7 to 20, Unlined or Lightweight-Lined Outerwear Coats and Jackets, Regular Range—Dimensions. 19 pp.

Jackets (Cont.)
—Lightweight—Children's
CGSB CAN/CGSB-49.30-M91. Canada Standard Children's Sizes 2 to 6X, Girls' Sizes 7 to 16 and Boys' Sizes 7 to 20, Unlined or Lightweight-Lined Outerwear Coats and Jackets, Regular Range—Dimensions. 19 pp.

—Lightweight—Girls'
CGSB CAN/CGSB-49.30-M91. Canada Standard Children's Sizes 2 to 6X, Girls' Sizes 7 to 16 and Boys' Sizes 7 to 20, Unlined or Lightweight-Lined Outerwear Coats and Jackets, Regular Range—Dimensions. 19 pp.

—Weather Resistant—Men's—Safety
DIN ENGL 61536-88. Mens' Coated Winterproof Outfits; Safety Requirements and Testing (Feb) (Together with DIN 61537, February 1988 Edition, Supersedes DIN 61536, February 1983 Edition). 7 pp.
DIN ENGL 61539-88. Weatherproof Outfits; Weatherproof Jackets and Trousers; Safety Requirements and Testing (Nov). 8 pp.

—Winter—Boys'
CGSB CAN/CGSB-49.59-M80. Jackets, Winter, Children's and Boys', Regular Range—Dimensions; (Amendment 1 Nov 1983). 10 pp.

—Winter—Children's
CGSB CAN/CGSB-49.59-M80. Jackets, Winter, Children's and Boys', Regular Range—Dimensions; (Amendment 1 Nov 1983). 10 pp.

—Winter—Girls'
CGSB CAN/CGSB-49.56-M80. Jackets, Winter, Girls', Regular Range—Dimensions; (Amendment 1 Nov 1983). 13 pp.

Jacking Pads
Use: Jack Pads

Jacks (Electric)
See Also: Electric Outlets; Phone Jacks
—Plugs—Telecommunication Equipment
BSI BS 6312-85. 1985 Plugs for Use with British Telecommunications Line Jack Units. 14 pp.

Jacks (Lifts)
Use For: Screw Jacks See Also: Hydraulic Jacks; Jack Pads; Scissor Lifts; Track Jacks
MOD UK DSTAN 51-1-67. Jacks, Screws, Hand (Ratchet Operated) Issue 1. 8 pp.
MOD UK DSTAN 51-2-67. Jacks, Rack Bar Issue 1. 8 pp.
MOD UK DSTAN 51-3-67. Jacks, Screw, Hand (Triple Lift) Issue 1. 7 pp.
MOD UK DSTAN 51-4-01. Jacks, Hydraulic, Hand Issue 1; Amendment 2. 11 pp.

—Aircraft
BSI 2M 50-86. 1986 Clearance Dimensions for Tripod Jacks for Aircraft. 6 pp.
ISO 1464-85. Aerospace—Tripod Jacks—Clearance Dimensions Second Edition. 6 pp.
JIS W 3902-91. Aerospace—Tripod Jacks—Clearance Dimensions (ISO 1464:1985).
NATO STANAG 3098 ED 9 AMD 0-88. Aircraft Jacking. 7 pp.
NATO STANAG 3098 ED 9 AMD 1-88. Aircraft Jacking. 8 pp.
NATO STANAG 3098 ED 10 AMD 0-00. Aircraft Jacking. 6 pp.

—Aircraft—Piston Rods
SBAC RS 200 ISSUE 1. Standard, Diameters of Jack Piston Rods.

—Aircraft—Shock Absorbers
SBAC RS 185 ISSUE 1. Standard Jack and Shock Absorber Data.

—Aircraft—Standard Diameters
SBAC RS 199 ISSUE 1. Standard, Diameters of Shock Absorbers and Jack Cylinders.

—Automotive
BSI BS AU 154A-89. 1989 Hydraulic Trolley Jacks. 8 pp.
BSI BS AU 165-77. (WITHDRAWN) 1977 Original Equipment Passenger Car Jacks (Superseded by BS AU 165A: 1990). 3 pp.
BSI BS AU 165A-90. 1990 Performance Requirements for Original Equipment Jacks for Passenger Cars and Light Vans. 6 pp.
BSI BS AU 172A-79. (WITHDRAWN) 1979 Amd 1 Performance Requirements for Accessory Jacks for Passenger Cars (Superseded by BS AU 172B: 1990). 4 pp.
BSI BS AU 172B-90. 1990 Performance Requirements for Accessory Jacks for Vehicles. 4 pp.

Jacks (Lifts) (Cont.)
—Automotive (Cont.)
BSI BS AU 223-88. 1988 Vehicle Support Stands for Cars and Light Vans. 8 pp.
CNS D4003-85. Portable Screw Jacks for Automobiles (Jul)(4076). 8 pp.
CNS D4006-85. Hydraulic Trolley Jacks (Jun)(5432). 3 pp.
JIS D 8103-77. Portable Screw Jacks for Automobiles (R 1983). 12 pp.

—Automotive—Safety
ISO 8720-91. Passenger Cars—Specifications for Mechanical Jacks First Edition. 5 pp.

—Drawbars—Agricultural Equipment
BSI BS 6792-86. 1986 Agricultural Trailer and Trailed Machinery Drawbar Jacks. 4 pp.

—Poles (Supports)
MOD UK DEF-1132-01. Jacks, Pole Lifting; Amendment 1. 9 pp.

—Poles (Supports)—Bars
MOD UK DEF-1134-57. Bars, Operating, for Use with Jacks, Pole Lifting. 2 pp.

—Poles (Supports)—Bases
MOD UK DEF-1135-57. Bases, Joist for Use with Jacks, Pole Lifting. 2 pp.

—Ships
MOD UK NES 113: Part 9-89. Requirements for Mechanical Handling Part 9: Jacks Issue 2 (11.89). 11 pp.

—Streetcars
SNZ NZS/AS 2615-87. Trolley Jacks (This is a Joint Standard with SAA AS 2615). 6 pp.

—Submarines
MOD UK NES 113: Part 9-89. Requirements for Mechanical Handling Part 9: Jacks Issue 2 (11.89). 11 pp.

Jacob's Ladders
Use: Ladders

Jacobs Tapers
See Also: Drill Chucks
BSI BS 1660: Part 10-88. 1988 Machine Tapers Part 10: Drill Chuck Tapers (Jacob and Short Morse). 6 pp.
ISO 239-74. Drill Chuck Tapers First Edition. 5 pp.

Jacquard Weaving Machinery
See Also: Textile Machinery; Weaving Machines
—Cards
JIS L 6311-59. Card for Jacquard.

—Harnesses
BSI BS 6876-87. (WITHDRAWN) 1987 Numbering of Machine Parts for Drawing-in on Jacquard Machines (Superseded by ISO 7506). 12 pp.
ISO 7506-84. Textile Machinery and Accessories—Numbering of Harnesses for Drawing-in on Jacquard Machines First Edition; (Corrected and Reprinted -1992). 11 pp.

—Heddles
BSI BS 3182: Part 2-83. 1983 Textile Machinery and Accessories: Healds, Heald Frames and Reeds Part 2: Twin Wire Healds with Inset Mail for Jacquard Weaving. 4 pp.
ISO 365-82. Textile Machinery and Accessories—Twin Wire Healds with Inset Mail for Jacquard Weaving Second Edition. 5 pp.
JIS L 6312-83. Wire Healds for Jacquard Weaving.

—Lingoes
BSI BS 5198-75. 1975 Lingoes for Jacquard Weaving. 6 pp.
ISO 2748-83. Textile Machinery and Accessories—Lingoes for Jacquard Weaving Second Edition. 4 pp.
JIS L 6516-83. Lingoes for Jacquard Weaving.

Jacuzzis (BTN)
Use: Whirlpool Baths

Jacuzzis (TM)
Use: Whirlpool Baths

Jam Nuts
Use: Locknuts

Jamming
See Also: Electromagnetic Interference; Electronic Warfare; Radio Frequency Interference; Warfare; White Noise

Jamming (Cont.)
—Reporting—Warning Systems
NATO STANAG 6004 ED 3 AMD 0-92. Meaconing, Intrusion Jamming and Interference Report. 5 pp.

Jams (Food)
See Also: Food; Jellies (Food)
CGSB 32.236M-89. Jams, Marmalades, Jellies and Cranberry Sauce. 9 pp.
EC 79/693/EEC-79. Council Directive on the Approximation of the Laws of the Member States Relating to Fruit Jams, Jellies and Marmalades and Chestnut Puree. 12 pp.
EC 88/593/EEC-88. Council Directive Amending Directive 79/693/EEC on the Approximation of the Laws of the Member States Relating to Fruit Jams, Jellies and Marmalades and Chestnut Puree. 4 pp.

—Canned
CNS N5154-79. Jams, Packaged (Jun)(4892). 3 pp.

Janitorial Supplies
Use: Cleaning Equipment and Supplies

Japanese Characters
Use For: HIRAGANA Characters
See Also: Graphic Characters

—Character Sets—Optical Character Recognition
JIS X 9003-80. KATAKANA Character Set for Optical Recognition (R 1985). 45 pp.
JIS X 9005-79. Handprinted KATAKANA Characters for Optical Character Recognition.
JIS X 9009-91. Handprinted HIRAGANA Characters for Optical Character Recognition. 26 pp.

—Coded Character Sets
JIS X 0207-79. Code of the Control Character Set for Japanese Graphic Characters for Information Interchange.
JIS X 0208-90. Code of the Japanese Graphic Character Set for Information Interchange. 83 pp.
JIS X 0212-90. Code of the Supplementary Japanese Graphic Character Set for Information Interchange. 86 pp.

—Documents Interchange
JIS X 4003-89. File Specification for Japanese Documents Interchange (Geometric Graphics). 111 pp.

—Engraving
JIS Z 8904-76. Standard Type of Letters Used in Mechanical Engraving (Katakana Characters) (R 1989). 8 pp.
JIS Z 8906-77. Standard Type of Letters Used in Mechanical Engraving (Hiragana Characters) (R 1988). 8 pp.

—Romanization
BSI BS 4812-72. 1972 The Romanization of Japanese. 16 pp.
ISO 3602-89. Documentation—Romanization of Japanese (Kana Script) First Edition. 11 pp.

—Typewriters
JIS B 9501-79. Japanese Character Typewriter.
JIS B 9502-77. Type of Japanese Character Typewriter.
JIS B 9509-64. KATAKANA-Alphabet Keyboards Arrangement for Typewriters.

Japanese Industrial Standards
Use: JIS

Japanese Lacquers
Use For: Rhus Lacquers *See Also:* Lacquers
JIS K 5950-79. Refined Rhus Lacquer.

Japanese Plums
Use: Plums

Jaquet Chronometric Tachometers
Use: Chronometric Tachometers

JAR Administration
Use: Joint Airworthiness Requirements Administration

Jars
See Also: Containers; Lead Acid Batteries

—Canning—Caps (Lids)
CGSB CAN/CGSB-143.1-M85. Lids, Home Canning Jar. 8 pp.

—Food
CNS S1147-82. Food Jar (Oct)(9545).

Jars (Cont.)
—Food—Vacuum Insulated
BSI BS 6672: Part 1-86. 1986 Insulated Domestic Food Containers Part 1: Vacuum Ware and Insulated Flasks, Jars and Jugs. 12 pp.
BSI BS 6672: Part 1-01. 1986 Amd 1 Insulated Domestic Food Containers Part 1: Specification for Vacuum Ware and Insulated Flasks, Jars and Jugs (AMD 7199) June 15, 1992. 14 pp.
SNZ NZS 5847: Part 1-88. Insulated Domestic Food Containers Part 1: Specification for Vacuum Ware and Insulated Flasks, Jars and Jugs Amend: A, 1988. 12 pp.

—Ice
CNS S1148-82. Ice Jar (Oct)(9546).

Jellies (Food)
See Also: Food; Jams (Food)
CGSB 32.236M-89. Jams, Marmalades, Jellies and Cranberry Sauce. 9 pp.
EC 79/693/EEC-79. Council Directive on the Approximation of the Laws of the Member States Relating to Fruit Jams, Jellies and Marmalades and Chestnut Puree. 12 pp.
EC 88/593/EEC-88. Council Directive Amending Directive 79/693/EEC on the Approximation of the Laws of the Member States Relating to Fruit Jams, Jellies and Marmalades and Chestnut Puree. 4 pp.

—Grass
CNS N5130-89. Canned Grass Jelly (Dec)(4147). 5 pp.

Jerricans
Use: Jerry Cans

Jerry Cans
Use For: Jerricans *See Also:* Containers

—Gasoline—Baking Paints
MOD UK DSTAN 80-65-76. Paint, Finishing, Stoving, Jerricans Paint, Finishing, Stoving, Interior, Jerricans Types: Spraying Flowcoating Paint, Finishing, Stoving, Exterior, Jerricans Types: Spraying Flowcoating Dipping Issue 1. 11 pp.

Jet Aircraft Noise
Use: Aircraft Noise

Jet Engine Fuels
Use For: Aviation Turbine Fuels *See Also:* Aircraft Fuels; Aircraft Gasoline; Gas Turbine Fuels; Gasoline; Kerosene; Liquid Fuels
CGSB 3-GP-24MB-90. Turbine Fuel, Aviation, High-Flash Type; (Amendment 1 May 1991). 12 pp.
CNS K5040-83. Aviation Turbine Fuel (Oct)(2558).
EUROCAE ED-42 10.83. MPS for a Fuel Flowmeter to Aircraft Standards. 24 pp.
NATO STANAG 4270 ED 1 AMD 1-84. Aviation Fuel Design Parameter for Future NATO Land Based Turbine Powered Military Aircraft. 5 pp.

—Anticing Additives
CGSB CAN/CGSB-3.526-M87. Icing Inhibitor for Aviation Fuels. 9 pp.

—Distillation Methods
BSI BS 7392-90. 1990 Method for Determination of Distillation Characteristics of Petroleum Products. 26 pp.
ISO 3405-88. Petroleum Products—Determination of Distillation Characteristics Second Edition. 24 pp.

—Gum Testing—Jet Evaporation Method
BSI BS 4348-76. 1976 Determination of Existent Gum in Fuels by Jet Evaporation. 12 pp.
CEN EN 5-74. Determination of Existent Gum in Fuels by Jet Evaporation. 10 pp.
CNS K6321-72. Method of Test for Existing Gum Content in Jet and Motor Fuels (Oct)(3382).
ISO 6246-81. Petroleum Products—Motor Gasoline and Aviation Fuels—Determination of Existent Gum—Jet Evaporation Method First Edition. 10 pp.

—High Flash
CGSB 3-GP-24MB-90. Turbine Fuel, Aviation, High-Flash Type; (Amendment 1 May 1991). 12 pp.
MOD UK DERD 2452-01. Turbine Fuel, Aviation: High Flash Type Containing Fuel System Icing Inhibitor NATO Code No: F-44 Joint Service Designation: AVCAT/FSII Issue 2; Amendment 1. 14 pp.
MOD UK DERD 2498-01. Turbine Fuel, Aviation: High Flash Type NATO Code No: F-43 Issue 7; Amendment 1. 12 pp.

—Kerosene
CGSB CAN/CGSB-3.23-93. Aviation Turbine Fuel, Kerosine Type. 12 pp.
MOD UK DERD 2453-90. Aviation Turbine Fuel AVTUR/FS11 Issue 6. 20 pp.

Jet Engine Fuels (Cont.)
—Kerosene—Thermal Stability
MOD UK DERD 2494-88. Aviation Turbine Fuel Kerosene Type AVTUR F-35 Issue 10. 19 pp.
MOD UK DERD 2494-01. Aviation Turbine Fuel Kerosene Type AVTUR F-35 Issue 10; Amendment 1. 23 pp.

—Moisture Content—Solubility
CNS K6544-80. Method of Test for Water Reaction of Aviation Fuels (Aug)(6189).

—Oxidation Resistance
CNS K6966-89. Method of Test for Thermal Oxidation Stability of Aviation Turbine Fuels (JETOT Procedure) (Oct)(12617).

—Smoke Point
JIS K 2537-92. Petroleum Products—Aviation Turbine Fuels and Kerosine—Determination of Smoke Point. 13 pp.

—Thermal Stability
ISO 6249-84. Petroleum Products—Gas Turbine Fuels—Determination of Thermal Oxidation Stability—JFTOT Method First Edition. 25 pp.

—Water
DIN ENGL 51415-81. Testing of Liquid Fuels; Testing of the Behaviour of Aviation Fuels in the Presence of Water (July). 3 pp.

—Water Content—Solubility
ISO 6250-82. Aviation Fuels—Determination of Water Reaction First Edition. 4 pp.

—Wide Cut
CGSB CAN/CGSB-3.22-93. Aviation Turbine Fuel, Wide Cut Type. 10 pp.
MOD UK DERD 2454-01. Aviation Turbine Fuel AVTAG/FSII F-40 Issue 4; Amendment 1. 14 pp.
MOD UK DERD 2486-01. Turbine Fuel, Aviation: Wide Cut Type Joint Service Designation: AVTAG Issue 9; Amendment 1. 12 pp.

Jet Fans
See Also: Fans

—Regulators
IEC 535-77. Jet Fans and Regulators First Edition. 27 pp.

—Regulators—Safety
CENELEC HD 280.3 S1-90. Safety Requirements for Electric Fans and Regulators Part 3: Jet Fans. 3 pp.
IEC 342 Pt 3-82. Safety Requirements for Electric Fans and Regulators Part 3: Jet Fans Second Edition. 34 pp.

Jet Fuels
Use: Jet Engine Fuels

Jet Propulsion
See Also: Aircraft Engines

—Glossaries
BSI BS 185: Sec 8-70. 1970 Glossary of Aeronautical and Astronautical Terms Section 8: Power Plant (Piston Engines, Gas Turbines and Jet Propulsion). 20 pp.

Jet Regulators
Use: Flow Regulators

Jet Seaplanes
Use: Seaplanes

Jetties
See Also: Quays

—Design
BSI BS 6349: Part 2-88. 1988 Code of Practice for Maritime Structures Part 2: Design of Quay Walls, Jetties and Dolphins. 110 pp.

Jewel Balls
Use: Jewel Bearings

Jewel Bearings
Use For: Industrial Jewel Bearings
See Also: Bearings; Time Measuring Instruments
ISO 1112-74. Horology—Functional and Non-Functional Jewels First Edition. 4 pp.

Jewellery
Use: Jewelry

INTERNATIONAL AND NON-U.S. NATIONAL STANDARDS
SUBJECT INDEX

JIS

Jewelry

—Gold Alloys
ISO 11426-93. Determination of Gold in Gold Jewellery Alloys—Cupellation Method (Fire Assay) First Edition. 7 pp.

—Gold Castings
ISO 10713-92. Jewellery—Gold Alloy Coatings First Edition. 6 pp.

—Gold Coatings—Composition
SAA AS 2335-80. Rolled Gold, Gold-Plated and Silver-Plated Jewellery—Composition and Marking (This is a Joint Standard with SANZ NZS 2335). 8 pp.
SNZ NZS/AS 2335-80. Rolled Gold, Gold-Plated and Silver-Plated Jewellery—Composition and Marking (This is a Joint Standard with SAA AS 2335). 8 pp.

—Gold Coatings—Identification Systems
SAA AS 2335-80. Rolled Gold, Gold-Plated and Silver-Plated Jewellery—Composition and Marking (This is a Joint Standard with SANZ NZS 2335). 8 pp.
SNZ NZS/AS 2335-80. Rolled Gold, Gold-Plated and Silver-Plated Jewellery—Composition and Marking (This is a Joint Standard with SAA AS 2335). 8 pp.

—Metals—Composition
CNS H3038-90. Composition and Marking Requirements for Metal of Jewelry (Sep)(2968).

—Metals—Identification Systems
CNS H3038-90. Composition and Marking Requirements for Metal of Jewelry (Sep)(2968).

—Nickel—Restricted Use
EC COM(93) 134-93. Proposal for a Council Directive Amending for the Fourteenth Time Directive 76/769/EEC on the Approximation of the Laws, Regulations and Administrative Provisions of the Member States Relating to Restrictions on the Marketing and Use of Certain Dangerous. 19 pp.

—Precious Metals—Fineness
BSI BS EN 29202-93. 1993 Jewellery—Fineness of Precious Metal Alloys (ISO 9202: 1991). 7 pp.
CEN EN 29202-92. Jewellery—Fineness of Precious Metal Alloys (ISO 9202: 1991). 6 pp.
ISO 9202-91. Jewelery—Fineness of Precious Metal Alloys First Edition; (CEN EN 29202:1992). 4 pp.

—Rings
BSI BS 6820-87. (WITHDRAWN) 1987 Amd 1 Measuring and Designating the Size of Rings Used in Jewellery (AMD 5672) September 30, 1987 (Superseded by BS EN 28653: 1993). 4 pp.
BSI BS EN 28653-93. 1993 Jewellery—Ring-Sizes—Definition, Measurement and Designation (Supersedes BS 6820: 1987) (L). 7 pp.
CEN EN 28653-92. Jewellery—Ring-Sizes—Definition, Measurement and Designation (ISO 8653: 1986). 2 pp.
ISO 8653-86. Jewellery—Ring-Sizes—Definition, Measurement and Designation First Edition; (CEN EN 28653:1992). 3 pp.

—Silver Alloys
ISO 11427-93. Determination of Silver in Silver Jewellery Alloys—Volumetric (Potentiometric) Method Using Potassium Bromide First Edition. 6 pp.

—Silver Coatings—Composition
SAA AS 2335-80. Rolled Gold, Gold-Plated and Silver-Plated Jewellery—Composition and Marking (This is a Joint Standard with SANZ NZS 2335). 8 pp.
SNZ NZS/AS 2335-80. Rolled Gold, Gold-Plated and Silver-Plated Jewellery—Composition and Marking (This is a Joint Standard with SAA AS 2335). 8 pp.

—Silver Coatings—Identification Systems
SAA AS 2335-80. Rolled Gold, Gold-Plated and Silver-Plated Jewellery—Composition and Marking (This is a Joint Standard with SANZ NZS 2335). 8 pp.
SNZ NZS/AS 2335-80. Rolled Gold, Gold-Plated and Silver-Plated Jewellery—Composition and Marking (This is a Joint Standard with SAA AS 2335). 8 pp.

JFETs
Use For: Junction FETs; Junction Field Effect Transistors *See Also:* FETs; Transistors
MOD UK DSTAN 59-61: 90/095-01. Field Effect Transistor Issue 2; Amendment 1. 23 pp.
MOD UK DSTAN 59-61: 90/215-82. Transistor Issue 1. 13 pp.

JFETs *(Cont.)*

—N Channel
MOD UK DSTAN 59-61: 80/023-78. Semiconductor Device Transistor Issue 1. 17 pp.
MOD UK DSTAN 59-61: 80/024-78. Semiconductor Device Transistor Issue 1. 14 pp.
MOD UK DSTAN 59-61: 90/159-74. Field Effect Transistor Issue 1. 15 pp.
MOD UK DSTAN 59-61: 90/163-74. Field Effect Transistor Issue 1. 13 pp.

—N-Channel—HF
CECC CECC 50 012-049 ISSUE 1-88. NL CECC 50 012-049; Single Gate Field-Effect-Transistor (En). 8 pp.

—N-Channel—LF
CECC CECC 50 012-025 ISSUE 1-88. BS CECC 50 012-025; Dual Single Gate Junction (En). 14 pp.
CECC CECC 50 012-068 ISSUE 1-89. BS CECC 50 012-068; Single Gate Junction FET (En). 8 pp.

—N-Channel—Switching
CECC CECC 50 012-026 ISSUE 1-84. BS CECC 50 012-026; Single Gate Junction F.E.T. (En). 14 pp.

Jib Cranes
See Also: Cranes
BSI BS 357-58. 1958 Amd 3 Power-Driven Travelling Jib Cranes (Rail-Mounted Low Carriage Type). 52 pp.
BSI BS 2452-54. 1954 Amd 2 High Pedestal or Portal Jib Cranes. 45 pp.
BSI BS 7333-90. 1990 Slewing Jib Cranes. 25 pp.
ISO 4301 Pt 4-89. Cranes and Related Equipment—Classification—Part 4: Jib Cranes First Edition. 5 pp.
ISO 9374 Pt 4-89. Cranes—Information to Be Provided—Part 4: Jib Cranes First Edition. 8 pp.

—Control Systems
ISO 7752 Pt 4-89. Cranes—Controls—Layout and Characteristics—Part 4: Jib Cranes First Edition. 7 pp.

—Ships—Materials Handling
BSI BS MA 79-78. 1978 Jib Cranes: Ship Mounted Type. 7 pp.
ISO 8431-88. Shipbuilding—Fixed Jib Cranes—Ship-Mounted Type for General Cargo Handling First Edition. 8 pp.

Jig Bushings
Use For: Drill Bushes; Drill Bushings
See Also: Bushings; Machine Tools
CNS B2058-54. Insert Type Drill Bushing (Metric System) (Jul)(390)(R 1970).
CNS B2059-54. Insert Type Drill Bushing (Inch System) (Jul)(391)(R 1970).
CNS B3375-80. Bushings for Jigs (Sep)(6401).
JIS B 5201-89. Jig Bushings and Accessories. 18 pp.

—Adaptable
CNS B2056-54. Adaptable Drill Bushing with Set Screw (Metric System) (Jul)(388)(R 1970).
CNS B2057-54. Adaptable Drill Bushing with Set Screw (Inch System) (Jul)(389)(R 1970).

—Glossaries
ISO 4248-78. Jig Bushes—Definitions and Nomenclature First Edition. 14 pp.

—Reamers
BSI BS 1098: Part 1-67. 1967 Amd 1 Jig Bushes Part 1: Inch Units. 20 pp.
BSI BS 1098: Part 2-77. 1977 Jig Bushes Part 2: Metric Units. 12 pp.

—Set Screws
CNS B3376-80. Set Screw for Jigs Bushings (Sep)(6402).

—Twist Drills
BSI BS 1098: Part 1-67. 1967 Amd 1 Jig Bushes Part 1: Inch Units. 20 pp.
BSI BS 1098: Part 2-77. 1977 Jig Bushes Part 2: Metric Units. 12 pp.
CNS B3407-81. Parallel Shank Twist Drills for Use in Jig Bushes (Mar)(7004).
DIN ENGL 339-78. Parallel Shank Twist Drills for Use in Jig Bushes (Mar). 3 pp.
DIN ENGL 341-78. Morse Taper Shank Twist Drills, Long Series for Use in Jig Bushes (Mar). 4 pp.
ISO 4247-77. Jig Bushes and Accessories for Drilling Purposes—Dimensions First Edition. 11 pp.

Jig Clamps
Use: Jigs (Positioners)—Clamps

Jig Feet
Use: Jigs (Positioners)—Feet

Jig Nuts
Use: Jigs (Positioners)—Nuts (Fasteners)

Jig Pins
Use: Jigs (Positioners)—Pins

Jig Washers
Use: Jigs (Positioners)—Washers

Jigs (Positioners)
See Also: Chucks; Drill Jigs; Holders; Jig Bushings; Locating Pins; Mechanical Guides; Separators (Mechanical); Tools
JIS B 5216-89. Locating Pins for Jigs and Fixtures. 8 pp.

—Abbreviations
DIN ENGL 6300-70. Jigs and Fixtures for Form-Modifying Production Processes; Denominations and Their Abbreviations (June). 2 pp.

—Clamps
BSI BS 5078-74. 1974 Jig and Fixture Components. 58 pp.
CNS B3379-80. Clamps for Jigs (Sep)(6405).
JIS B 5227-89. Clamps for Jigs and Fixtures. 12 pp.

—Components
BSI BS 5078-74. 1974 Jig and Fixture Components. 58 pp.

—Defense Contracts
MOD UK DEFCON 23-76. Special Jigs, Tools, Etc 6/76. 1 p.

—Feet
CNS B3315-83. Feet with Thread for Jig and Fixtures (Mar)(5057).
DIN ENGL 6320-71. Feet with Threaded Shank for Jigs and Fixtures (Feb). 1 p.

—Hexagonal Nuts
BSI BS 5078-74. 1974 Jig and Fixture Components. 58 pp.
JIS B 5226-89. Hexagon Nuts for Jigs and Fixtures. 6 pp.

—Invitation for Bids
MOD UK DEFCON 47V-86. Invitation to Tender Supplementary Conditions—Jigs, Tools, Etc 10/86. 1 p.

—Master Reference—Aircraft
SBAC RS 211 ISSUE 1. Master Reference Jig.

—Nuts (Fasteners)
BSI BS 5078-74. 1974 Jig and Fixture Components. 58 pp.

—Pins
BSI BS 5078-74. 1974 Jig and Fixture Components. 58 pp.

—Plates (Supports)
ISO 6753-82. Machined Plates for Press Tools, Moulds, Jigs and Fixtures—Nominal Dimensions First Edition. 3 pp.

—Washers
BSI BS 5078-74. 1974 Jig and Fixture Components. 58 pp.
CNS B3377-80. C Washers for Jigs (Sep)(6403).
CNS B3378-80. Swing C Washers Jigs (Sep)(6404).
JIS B 5211-89. Washers for Jigs and Fixtures. 10 pp.

Jigsaws
See Also: Saws; Scroll Saws; Tools
BSI BS 2769: Sec 2.10-84. 1984 Hand-Held Electric-Motor-Operated Tools Part 2: Particular Requirements Section 2.10: Jig Saws. 5 pp.

—Electric—Portable
MOD UK DSTAN 51-19: Part 2-92. Hand Tools, Powered Part 2: Electric Issue 1. 31 pp.

—Safety
BSI BS 2769: Sec 2.10-84. 1984 Hand-Held Electric-Motor-Operated Tools Part 2: Particular Requirements Section 2.10: Jig Saws. 5 pp.
CENELEC HD 400.3-81. Hand-Held Motor Operated Tools—Part 2: Particular Specifications. 28 pp.
IEC 745 Pt 2-11-84. Safety of Hand-Held Motor-Operated Electric Tools Part 2: Particular Requirements for Reciprocating Saws (Jig and Sabre Saws) First Edition. 19 pp.

JIS
Use For: Japanese Industrial Standards
See Also: Standards

—Standards Preparation
JIS Z 8301-90. Rules for the Drafting Presentation of Japanese Industrial Standards.

—Yearbooks
JIS. JIS Yearbook 1992. 304 pp.

INDUSTRY STANDARDS

Jitter

See Also: Distortion (Electrical); Judder; Phase Jitter; Slip; Telecommunication Equipment; Timing Jitter; Vibration

—**Glossaries**
CNS C5169-83. Glossary of Pulse Distortion and Jitter (Sep)(10550).

—**Integrated Digital Networks—Hierarchical Interfaces**
CCITT FASCICLE III.5-89. Digital Networks, Digital Sections and Digital Line Systems—Recommendations G.801—G.961. 292 pp.
CCITT RECMN G.823-89. Control of Jitter and Wander Within Digital Networks Which Are Based on the 2048 kBit/s Hierarchy—Digital Networks, Digital Sections and Digital Line Systems (Study Groups XV and XVIII) 13 pp. 13 pp.
CCITT RECMN G.824-89. Control of Jitter and Wander Within Digital Networks Which Are Based on the 1544 kBit/s Hierarchy—Digital Networks, Digital Sections and Digital Line Systems (Study Groups XV and XVIII) 6 pp. 6 pp.

—**Reference Clocks—Measurement**
CCITT FASCICLE III.5-89. Digital Networks, Digital Sections and Digital Line Systems—Recommendations G.801—G.961. 292 pp.

—**Restoration Switching Equipment**
CCITT RECMN G.180-89. Characteristics of N + M Type Direct Transmission Restoration Systems for Use on Digital Sections, Links or Equipment—General Characteristics of International Telephone Connections and Circuits (Study Groups XII and XV) 11 pp. 11 pp.

—**Slave Clocks—Measurement**
CCITT FASCICLE III.5-89. Digital Networks, Digital Sections and Digital Line Systems—Recommendations G.801—G.961. 292 pp.

—**Television Signals—Digital**
CCIR QUESTION 62/11-90. Protection Against Errors, Jitter and Slip for Digital Television Signals—Questions Concerning Study Group 11—Broadcasting Service (Television). 1 p.

Jo-Bolts
See Also: Fasteners

—**Aircraft—Countersunk Head**
SBAC AGS 3818 (V). 'Jo-Bolt' Fastners 100 Degree Countersunk Head.

—**Aircraft—Inspection**
SBAC TS 50 ISSUE 4. Alloy Steel "JO—BOLT" Fasteners.

—**Aircraft—Manufacture**
SBAC TS 50 ISSUE 4. Alloy Steel "JO—BOLT" Fasteners.

Jodel (CEA) 253 Aircraft
See Also: Aircraft

—**Antenna Positions**
CAA. Jodel (CEA) 253, 315, 340, 360, 380 (Approved Aerial Positions). 1 p.

Jodel (CEA) 315 Aircraft
See Also: Aircraft

—**Antenna Positions**
CAA. Jodel (CEA) 253, 315, 340, 360, 380 (Approved Aerial Positions). 1 p.

Jodel (CEA) 340 Aircraft
See Also: Aircraft

—**Antenna Positions**
CAA. Jodel (CEA) 253, 315, 340, 360, 380 (Approved Aerial Positions). 1 p.

Jodel (CEA) 360 Aircraft
See Also: Aircraft

—**Antenna Positions**
CAA. Jodel (CEA) 253, 315, 340, 360, 380 (Approved Aerial Positions). 1 p.

Jodel (CEA) 380 Aircraft
See Also: Aircraft

—**Antenna Positions**
CAA. Jodel (CEA) 253, 315, 340, 360, 380 (Approved Aerial Positions). 1 p.

Jodel D 117 Aircraft
See Also: Aircraft

—**Antenna Positions**
CAA. Jodel DR 100A, D 117, D 120, D 150, DR 200, DR 221, DR 250/160, DR 1050, DR 1051 (Approved Aerial Positions). 1 p.

Jodel D 120 Aircraft
See Also: Aircraft

—**Antenna Positions**
CAA. Jodel DR 100A, D 117, D 120, D 150, DR 200, DR 221, DR 250/160, DR 1050, DR 1051 (Approved Aerial Positions). 1 p.

Jodel D 140 Aircraft
See Also: Aircraft

—**Antenna Positions**
CAA. Jodel D.140 (Approved Aerial Positions). 1 p.

Jodel D 150 Aircraft
See Also: Aircraft

—**Antenna Positions**
CAA. Jodel DR 100A, D 117, D 120, D 150, DR 200, DR 221, DR 250/160, DR 1050, DR 1051 (Approved Aerial Positions). 1 p.

Jodel DR 100A Aircraft
See Also: Aircraft

—**Antenna Positions**
CAA. Jodel DR 100A, D 117, D 120, D 150, DR 200, DR 221, DR 250/160, DR 1050, DR 1051 (Approved Aerial Positions). 1 p.

Jodel DR 1050 Aircraft
See Also: Aircraft

—**Antenna Positions**
CAA. Jodel DR 100A, D 117, D 120, D 150, DR 200, DR 221, DR 250/160, DR 1050, DR 1051 (Approved Aerial Positions). 1 p.

Jodel DR 1051 Aircraft
See Also: Aircraft

—**Antenna Positions**
CAA. Jodel DR 100A, D 117, D 120, D 150, DR 200, DR 221, DR 250/160, DR 1050, DR 1051 (Approved Aerial Positions). 1 p.

Jodel DR 200 Aircraft
See Also: Aircraft

—**Antenna Positions**
CAA. Jodel DR 100A, D 117, D 120, D 150, DR 200, DR 221, DR 250/160, DR 1050, DR 1051 (Approved Aerial Positions). 1 p.

Jodel DR 221 Aircraft
See Also: Aircraft

—**Antenna Positions**
CAA. Jodel DR 100A, D 117, D 120, D 150, DR 200, DR 221, DR 250/160, DR 1050, DR 1051 (Approved Aerial Positions). 1 p.

Jodel DR 250/160 Aircraft
See Also: Aircraft

—**Antenna Positions**
CAA. Jodel DR 100A, D 117, D 120, D 150, DR 200, DR 221, DR 250/160, DR 1050, DR 1051 (Approved Aerial Positions). 1 p.

Jodel Series Aircraft
See Also: Aircraft

—**Foreign Airworthiness Directives**
CAA. Jodel Series (Foreign Airworthiness Directives). 11 pp.

Jogging Suits
Use: Sweat Suits

Johnson Noise
Use: Thermal Noise

Joiners' Squares
Use: Squares (Instruments)

Joinery
See Also: Woodworking

—**Construction Contracts**
DIN ENGL 18355-88. Tendering and Performance Stipulations in Contracts for Construction Works (VOB); Part C: General Technical Specifications in Contracts for Construction Works (ATV); Joinery (Sept) (This Standard, Together with DIN 18299, September 1988 Edition,. 8 pp.

—**Glossaries**
BSI BS 6100: Sec 4.4-92. 1992 Glossary of Building and Civil Engineering Terms Part 4: Forest Products Section 4.4: Carpentry and Joinery. 20 pp.
BSI BS 6100: Sec 4.4-85. 1985 Amd 1 Glossary of Building and Civil Engineering Terms Part 4: Forest Products Section 4.4: Carpentry and Joinery (AMD 5552) July 29, 1988. 33 pp.

—**Pine Wood**
SAA AS 1497-73. Milled Radiata Pine for Joinery Amdt 1 April 1977.

—**Quality Assurance**
BSI BS 8000: Part 5-90. 1990 Workmanship on Building Sites Part 5: Code of Practice for Carpentry, Joinery and General Fixings. 33 pp.

—**Softwoods**
SAA AS 1786-75. Joinery Timber Milled from Australian-Grown Conifers (Softwoods) (Excluding Radiata Pine and Cypress Pine).

—**Timber—Quality Assurance**
BSI BS 1186: Part 1-91. 1991 Timber for and Workmanship in Joinery Part 1: Specification for Timber. 30 pp.
BSI BS 1186: Part 2-88. 1988 Amd 1 Quality of Timber and Workmanship in Joinery Part 2: Quality of Workmanship. 15 pp.
BSI BS 1186: Part 3-90. 1990 Quality of Timber and Workmanship in Joinery Part 3: Wood Trim and Its Fixing (V). 35 pp.
CEN PREN 942-92. Timber Joinery—Classification of Timber Quality. 18 pp.
DIN ENGL 68360 Pt 1-81. Timber in Joinery; Quality Requirements for Exterior Use (May). 3 pp.
DIN ENGL 68360 Pt 2-81. Timber in Joinery; Quality Requirements for Interior Use (May). 3 pp.

—**Wood Preservation**
SAA AS 1606-74. Code of Practice for Water-Repellent Treatment of Timber, Joinery and Other Timber Products Bound with AS 1607. 18 pp.

—**Wood Preservatives**
SAA AS 1607-74. Water-Repellent Solutions for the Treatment of Timber, Joinery and Other Timber Products Bound with AS 1606. 18 pp.

Joining
Scope Note: Use a more specific term *See:* Adhesive Bonding; Brazing; Sealing; Soldering; Thermal Cutting; Welding

Joint Airworthiness Requirements Administration
Scope Note: For standards on Joint Airworthiness Requirements, see the specific aircraft type
Use For: JAR Administration
See Also: Airworthiness Leaflets; Airworthiness Notices

CAA 23B-21. Proposals to Amend the Handling Requirements of BCAR 23 to Incorporate Commuter Class Aeroplanes. 7 pp.
CAA 23CDE-22. Proposals to Amend the Structural Requirements of BCAR 23, Subpart C, D and E to Tare Account of Commuter Class Aeroplanes. 6 pp.
CAA 23B-23. Proposal to Amend the Performance Requirements of BCAR 23, Subpart B to Incorporate Commuter Class Aeroplanes. 8 pp.
CAA 23C,D,E,F-24. Miscellaneous Amendments. 10 pp.
CAA 23,D,E,F-25. Proposals to Amend the Systems Requirements of BCAR 23, Subparts D, E & F to Incorporate Commuter Aeroplanes. 31 pp.
CAA NOTICE #14 ISSUE 3. Approved Maintenance Organisations (JAR 145)—Implementation Procedure. 6 pp.
CAA. Check List of Pages. 2 pp.
CAA. Contents. 3 pp.
CAA. Foreword. 1 p.
CAA. Check List of Pages. 2 pp.
CAA. Preambles. 1 p.
CAA JAR-145 Section 1. Requirements. 8 pp.
CAA JAR-145 Section 2. Acceptable Means of Compliance and Interpretations (ACJ). 17 pp.
CAA JAR-145 Sec 2: App 1. Organisations Approval Rating System. 1 p.
CAA JAR-145 Sec 2: App 2. Example Maintenance Organisation Exposition Only. 3 pp.
CAA JAR-145 Sec 2: App 3. Guidance Material for the Authorised Release Certificate/Airworthiness Approval Tag. 6 pp.
CAA JAR-145 Sec 2: App 4. Some Outline Examples of Organisational Structures Possible Under JAR-145. 2 pp.
CAA JAR-145 Sec 2: App 5. JAR-FAR Comparison Information. 3 pp.
CAA. Contents (Joint Airworthiness Requirements). 1 p.

INTERNATIONAL AND NON-U.S. NATIONAL STANDARDS
SUBJECT INDEX

Joint Airworthiness Requirements Administration *(Cont.)*
CAA. Foreword (Joint Airworthiness Requirements). 2 pp.
CAA. Check List of Pages. 1 p.
CAA. Preambles (Joint Airworthiness Requirements). 1 p.
CAA JAR-TSO SUBPART A. General (Joint Airworthiness Requirements). 5 pp.
CAA JAR-TSO SUBPART B. List of JTSOs (Index 1 and Index 2). 6 pp.
CAA JAR-TSO SUBPART C. Joint Technical Standard Orders (JTSO). 1 p.
CAA JAR-TSO Index 1. JTSOs. 206 pp.
CAA JAR-TSO Index 2. JTSOs. 39 pp.
CAA JAR-VLA Section 2. Acceptable Means of Compliance and Interpretations (ACJ). 14 pp.
CAA. Check List of Pages (Administrative Procedures). 2 pp.
CAA. Contents (Administrative Procedures). 2 pp.
CAA. Preambles (Administrative Procedures). 2 pp.
CAA AP Section 1. Historical and General Organization (Administrative Procedures). 2 pp.
CAA AP Section 2. Formation and Terms of Reference of Groups (Administrative Procedures). 4 pp.
CAA AP Section 3. Amendment Procedures (Administrative Procedures). 13 pp.
CAA AP Section 4. Preparation of Reports by Technical Study Groups (Administrative Procedures). 3 pp.
CAA AP Section 5. Printing and Publication (Administrative Procedures). 5 pp.
CAA AP Section 6. Retention of Documents (Administrative Procedures). 3 pp.

—**Type Certificates**
CAA Chapter B4-2 06.90. Type Certification or Validation of Engines.
CAA Chapter B4-4 07.89. Type Certification or Validation of Propellers. 3 pp.
CAA. Check List of Pages (Guidance on Administrative Procedures. 1 p.
CAA. Foreword (Guidance on Administrative Procedures. 1 p.
CAA GAP SUBPART A. General (Guidance on Administrative Procedures. 3 pp.
CAA GAP SUBPART B. Type Certificate Procedures (Guidance on Administrative Procedures. 3 pp.
CAA GAP SUBPART C. Change to Type Certificates (Guidance on Administrative Procedures. 1 p.

Joint Boxes
Use: Junction Boxes

Joint Fillers
See Also: Fillers; Joints; Sealants

—**Concretes**
CNS A3212-84. Method of Testing Preformed Expansion Joint Fillers for Concrete (Nonextruding and Resilient Types) (Sep)(10993).

—**Cork**
ISO 3867-82. Agglomerated Cork Material of Expansion Joints for Construction and Building—Test Methods First Edition. 7 pp.
ISO 3869-81. Agglomerated Cork—Filler Material of Expansion Joints for Construction and Buildings—Characteristics, Sampling and Packing First Edition. 3 pp.

Joint Prostheses
See Also: Ankle Prostheses; Arm Prostheses; Foot Prostheses; Hip Prostheses; Knee Prostheses; Leg Prostheses; Prosthetic Devices; Wrist Prostheses
BSI BS 7251: Part 1-90. 1990 Orthopaedic Joint Prostheses Part 1: General Requirements. 8 pp.
ISO 5839-85. Implants for Surgery—Orthopaedic Joint Prostheses—Basic Requirements First Edition. 7 pp.

Joint Rings
Use: Sealing Rings

Joint Sealants
Use For: Building Joint Sealants; Jointing Compounds; Jointing Materials; Water Stops
See Also: Construction Materials; Joints; Lutes (Material); Sealants
BSI BS 6956: Part 5-92. 1992 Jointing Materials and Compounds Part 5: Specification for Jointing Compounds for Use with Water, Low Pressure Saturated Steam, 1st Family Gases (Excluding Coal Gas) and 2nd Family Gases. 10 pp.
BSI BS 6956: Part 6-92. 1992 Jointing Materials and Compounds Part 6: Specification for Jointing Compounds for 3rd Family Gases. 11 pp.
BSI BS EN 26927-91. 1991 Building Construction—Jointing Products—Sealants—Vocabulary (ISO 6927: 1981). 10 pp.
BSI BS EN 28394-91. 1991 Building Construction—Jointing Products—Determination of Extrudability of One-Component Sealants (ISO 8394: 1988). 9 pp.
CEN EN 28 394-90. Building Construction—Jointing Products—Determination of Extrudability of One—Component Sealants. 6 pp.
CGSB CAN/CGSB-19.24-M90. Multi-Component, Chemical-Curing Sealing Compound. 9 pp.
CNS A2136-82. Sealing Compounds for Sealing and Glazing in Building (Jun)(8903).
CNS A3154-82. Method of Test for Sealing Compounds for Sealing and Glazing in Building (Jun)(8904).
DIN ENGL 30660-82. Jointing Material for Gas and Water Supplies and for Hot Water Installations; Non-Hardening Jointing Material for Metallic Threaded Joints in Interior Installations (June). 8 pp.
ISO 8394-88. Building Construction—Jointing Products—Determination of Extrudability of One-Component Sealants First Edition. 4 pp.
JIS A 5757-75. Sealing Compounds for Sealing, Glazing and Caulking in Buildings.
JIS A 5758-92. Sealing Compounds for Sealing and Glazing in Building. 36 pp.
JIS K 6820-77. Fluid Sealants. 21 pp.

—**Acrylic**
CGSB 19-GP-5M-76. Sealing Compound, One Component, Acrylic Base, Solvent Curing, Standard for (R 1984). 8 pp.
CGSB CAN/CGSB-19.17-M90. One-Component, Acrylic Emulsion Base Sealing Compound. 9 pp.
CNS A2209-87. Joint Treatment Materials for Gypsum Boards (Jun)(11990).
JIS A 6914-85. Joint Treatment Materials for Gypsum Boards. 9 pp.

—**Adhesive Strength**
BSI BS 3712: Part 4-85. 1985 Building and Construction Sealants Part 4: Methods of Test for Adhesion in Peel, Tensile Extension and Recovery and Loss of Mass After Heat Ageing. 13 pp.
BSI BS EN 29046-91. 1991 Building Construction—Sealants—Determination of Adhesion/Cohesion Properties at Constant Temperature (ISO 9046: 1987). 10 pp.
CEN EN 29 046-90. Building Construction—Jointing Products—Determination of Adhesion Properties at Constant Temperatures. 9 pp.
DIN ENGL 52455 Pt 1-87. Testing of Building Construction Sealants; Determination of Adhesion/Cohesion After Conditioning in Standard Atmosphere, Water, or at Elevated Temperature (Apr). 4 pp.
DIN ENGL 52455 Pt 4-87. Testing of Building Construction Sealants; Determination of Adhesion/Cohesion When Subjected to an Extension/Compression Cycle at Alternating Temperature (Apr). 3 pp.
DIN ENGL EN 29046-91. Sealants in Building Construction; Determination of Adhesion/Cohesion Properties at Constant Temperature (ISO 9046: 1987) (May). 6 pp.
ISO 9046-87. Building Construction—Sealants—Determination of Adhesion/Cohesion Properties at Constant Temperature First Edition. 6 pp.
ISO 9047-89. Building Construction—Sealants—Determination of Adhesion/Cohesion Properties at Variable Temperatures First Edition. 6 pp.
ISO 10590-91. Building Construction—Sealants—Determination of Adhesion/Cohesion Properties at Maintained Extension After Immersion in Water First Edition. 6 pp.
ISO 10591-91. Building Construction—Sealants—Determination of Adhesion/Cohesion Properties After Immersion in Water First Edition. 6 pp.

—**Aircraft**
MOD UK DTD-5604-69. Pigmented Jointing Compound, Non-Hardening Type. 5 pp.
SBAC TS 110 ISSUE 1. British Metric Swaged Pipe Coupling: Swaging Process Specification for AS 43022 and AS 43160 (Nipples) and AS 43024 and AS 43161 (Unions) Incorporating Use of a Jointing Compound.

—**Aramid Fiber—Aerospace**
BSI F 130-87. 1987 Rubber Bonded Compressed Aramid Fibre Jointing Material for Aerospace Purposes. 12 pp.

—**Asbestos**
BSI BS 1832-91. 1991 Compressed Asbestos Fibre Jointing (Supersedes BS 2815: 1973). 20 pp.
BSI BS 1832-01. 1991 Amd 1 Compressed Asbestos Fibre Jointing (AMD 7558) March 15, 1993 (Supersedes BS 2815: 1973). 21 pp.
BSI BS 2815-73. (WITHDRAWN) 1973 Compressed Asbestos Fiber Jointing (Superseded by BS 1832: 1991). 19 pp.

—**Asbestos—Oil Resistant**
BSI BS 1832-72. 1972 Oil Resistant Compressed Asbestos Fibre Jointing. 20 pp.

—**Asbestos—Rubber Bonded—Aerospace**
BSI F 125-73. 1973 Amd 2 Rubber Bonded Compressed Asbestos Fibre Jointing (AMD 5816) January 31, 1989. 13 pp.

—**Backfill Material—Water Absorption**
DIN ENGL 52459-81. Testing of Building Sealants; Determination of the Water Absorption of Backfill Material; Retentive Capacity (June). 2 pp.

—**Bathtubs—Mildew Resistant**
CGSB CAN/CGSB-19.22-M89. Mildew-Resistant Sealing Compound for Tubs and Tiles. 9 pp.

—**Belt Conveyors**
BSI BS 6593-85. 1985 Code of Practice for on-Site Non-Mechanical Jointing of Plied Textile and Steel Reinforced Conveyor Belting. 26 pp.

—**Bituminous**
MOD UK CS 2699A. Triethylamine (Withdrawn).

—**Bituminous—Adhesion Testing**
DIN ENGL 1996 Pt 19-84. Testing of Asphalt; Determination of Extensibility and Adhesion Using a Rabe Joint Model (May). 6 pp.

—**Bituminous—Extensibility**
DIN ENGL 1996 Pt 19-84. Testing of Asphalt; Determination of Extensibility and Adhesion Using a Rabe Joint Model (May). 6 pp.

—**Bituminous—Segregation**
DIN ENGL 1996 Pt 16-75. Testing of Bituminous Materials for Road Building and Related Purposes; Determination of Segregation Tendency (Dec). 4 pp.

—**Bituminous—Softening Points**
DIN ENGL 1996 Pt 15-75. Testing of Bituminous Materials for Road Building and Related Purposes; Determination of Softening Point in Accordance with Wilhelmi (Dec). 4 pp.

—**Classification**
BSI BS 6213-82. 1982 Amd 1 Selection of Constructional Sealants. 32 pp.
EC 83/265/EEC-83. Council Directive Amending Directive 77/728/EEC on the Approximation of the Laws, Regulations and Administrative Provisions of the Member States Relating to the Classification, Packaging and Labelling of Paints, Varnishes, Printing Inks, Adhesives. 7 pp.

—**Cohesion**
BSI BS EN 29046-91. 1991 Building Construction—Sealants—Determination of Adhesion/Cohesion Properties at Constant Temperature (ISO 9046: 1987). 10 pp.
CEN EN 29 046-90. Building Construction—Jointing Products—Determination of Adhesion Properties at Constant Temperatures. 9 pp.
DIN ENGL 52455 Pt 1-87. Testing of Building Construction Sealants; Determination of Adhesion/Cohesion After Conditioning in Standard Atmosphere, Water, or at Elevated Temperature (Apr). 4 pp.
DIN ENGL 52455 Pt 4-87. Testing of Building Construction Sealants; Determination of Adhesion/Cohesion When Subjected to an Extension/Compression Cycle at Alternating Temperature (Apr). 3 pp.
DIN ENGL EN 29046-91. Sealants in Building Construction; Determination of Adhesion/Cohesion Properties at Constant Temperature (ISO 9046: 1987) (May). 6 pp.
ISO 9046-87. Building Construction—Sealants—Determination of Adhesion/Cohesion Properties at Constant Temperature First Edition. 6 pp.
ISO 9047-89. Building Construction—Sealants—Determination of Adhesion/Cohesion Properties at Variable Temperatures First Edition. 6 pp.
ISO 10590-91. Building Construction—Sealants—Determination of Adhesion/Cohesion Properties at Maintained Extension After Immersion in Water First Edition. 6 pp.
ISO 10591-91. Building Construction—Sealants—Determination of Adhesion/Cohesion Properties After Immersion in Water First Edition. 6 pp.

—**Comprehensive Testing**
BSI BS 3712: Part 2-73. 1973 Amd 2 Building and Construction Sealants Part 2: Methods of Test for Seepage, Staining, Shrinkage, Shelf Life and Paintability (AMD 6818) September 30, 1991. 14 pp.
BSI BS 3712: Part 3-74. 1974 Amd 2 Building and Construction Sealants Part 3: Methods of Test for Application Life, Skinning Properties and Tack-Free Time (AMD 6819) September 30, 1991. 12 pp.

—**Concrete Pavements**
BSI BS 2499: Part 1-93. 1993 Hot-Applied Joint Sealant Systems for Concrete Pavements Part 1: Specification for Joint Sealants (R). 10 pp.

INDUSTRY STANDARDS

Joint Sealants (Cont.)

—Concrete Pavements (Cont.)
BSI BS 2499: Part 2-92. 1992 Hot-Applied Joint Sealant Systems for Concrete Pavements Part 2: Code of Practice for the Application and Use of Joint Sealants (Superseded BS 2499: 1973). 14 pp.
BSI BS 2499: Part 3-93. 1993 Hot-Applied Joint Sealant Systems for Concrete Pavements Part 3: Methods of Test (Superseded BS 2499: 1973) (R). 24 pp.
BSI BS 2499-73. (WITHDRAWN) 1973 Hot Applied Joint Sealants for Concrete Pavements (Superseded by BS 2499: Parts 1 & 3: 1992 & Part 2: 1992). 28 pp.
BSI BS 5212-75. (WITHDRAWN) 1975 Cold Poured Joint Sealants for Concrete Pavements (Superseded by BS 5212: Parts 1, 2, & 3: 1990). 19 pp.
BSI BS 5212: Part 1-90. 1990 Cold Applied Joint Sealant Systems for Concrete Pavements Part 1: Joint Sealants. 11 pp.
BSI BS 5212: Part 2-90. 1990 Cold Applied Joint Sealant Systems for Concrete Pavements Part 2: Code of Practice for the Application and Use of Joint Sealants. 14 pp.

—Concrete Pavements—Comprehensive Testing
BSI BS 5212: Part 3-90. 1990 Cold Applied Joint Sealant Systems for Concrete Pavements Part 3: Methods of Test. 22 pp.

—Concrete—Strips
DIN ENGL 7865 Pt 1-82. Elastomeric Joint Sealing Strip for Sealing Joints in Concrete; Shape and Dimensions (Feb). 12 pp.
DIN ENGL 7865 Pt 2-82. Elastomeric Joint Sealing Strip for Sealing Joints in Concrete; Material Requirements and Testing (Feb). 3 pp.

—Cork—Gaskets
BSI BS 4332-89. 1989 Phenol-Formaldehyde Resin-Bonded Cork Jointing. 12 pp.

—Cork/Paper—Gaskets
BSI BS 4243-67. (WITHDRAWN) 1967 Amd 1 Cork/Paper Jointing (Superseded by BS 4249: 1989). 24 pp.

—Elastic Properties
BSI BS EN 27389-91. 1991 Building Construction—Jointing Products—Determination of Elastic Recovery (ISO 7389: 1987). 10 pp.
CEN EN 27 389-90. Building Construction—Jointing Products—Determination of Elastic Recovery. 9 pp.
DIN ENGL EN 27389-91. Jointing Products in Building Construction; Determination of Elastic Recovery (ISO 7389:1987) (May) (Supersedes DIN 52458, April 1987 Edition). 6 pp.
ISO 7389-87. Building Construction—Jointing Products—Determination of Elastic Recovery Second Edition. 6 pp.

—Elastomers
BSI BS 7531-92. 1992 Compressed Non-Asbestos Fibre Jointing. 19 pp.

—Extrudability
BSI BS EN 29048-91. 1991 Building Construction—Jointing Products—Determination of Extrudability of Sealants Using Standardized Apparatus (ISO 9048: 1987). 11 pp.
CEN EN 29 048-90. Building Construction—Jointing Products—Determination of Extrudability of Sealants Using Standardized Apparatus. 9 pp.
DIN ENGL EN 28394-91. Jointing Products in Building Construction; Determination of Extrudability of One-Component Sealants (ISO 8394:1988) (May). 5 pp.
DIN ENGL EN 29048-91. Jointing Products in Building Construction; Determination of Extrudability of Sealants Using Standardized Apparatus (ISO 9048:1987) (May) (Supersedes DIN 52456, May 1976 Edition). 7 pp.
ISO 9048-87. Building Construction—Jointing Products—Determination of Extrudability of Sealants Using Standardized Apparatus First Edition. 7 pp.

—Gas Pipes
BSI BS 5292-80. 1980 Amd 2 Jointing Materials and Compounds for Installations Using Water, Low-Pressure Steam or 1st, 2nd and 3rd Family Gases (AMD 5960) November 30, 1988. 22 pp.
BSI BS 5292-03. 1980 Amd 3 Jointing Materials and Compounds for Installations Using Water, Low-Pressure Steam or 1st, 2nd and 3rd Family Gases (AMD 6918) January 31, 1992 (Q). 24 pp.
BSI BS 5292-04. 1980 Amd 4 Jointing Materials and Compounds for Installations Using Water, Low-Pressure Steam or 1st, 2nd and 3rd Family Gases (AMD 7047) April 1, 1992. 26 pp.

—Glazing
DIN ENGL 18545 Pt 2-92. Glazing with Sealants; Sealants (Feb). 3 pp.

Joint Sealants (Cont.)

—Glazing (Cont.)
DIN ENGL 18545 Pt 3-92. Glazing with Sealants; Glazing Systems (Feb). 3 pp.

—Glazing—Adhesion Testing
DIN ENGL 52455 Pt 3-74. Testing of Materials for Joint and Glazing Seals in Building Construction; Adhesion and Extension Test; Exposure to Light (Sept). 2 pp.

—Glossaries
BSI BS 6100: SUBSEC 1.5.2-87. 1987 Glossary of Building and Civil Engineering Terms Part 1: General and Miscellaneous Section 1.5: Operations; Associated Plant and Equipment Subsection 1.5.2: Jointing. 13 pp.
BSI BS 6100: SUBSEC 1.5.2-01. 1987 Amd 1 Glossary of Building and Civil Engineering Terms Part 1: General and Miscellaneous Section 1.5: Operations; Associated Plant and Equipment Subsection 1.5.2: Jointing (AMD 7241) August 15, 1992. 15 pp.
CEN EN 26927-90. Building Construction—Jointing Products—Sealants—Vocabulary. 6 pp.
DIN ENGL EN 26927-91. Jointing Products in Building Construction; Sealants; Vocabulary (ISO 6927:1981) (May) (Supersedes DIN 52460, August 1979 Edition). 6 pp.
ISO 6927-81. Building Construction—Jointing Products—Sealants—Vocabulary First Edition. 6 pp.

—Graphite
MOD UK DSTAN 68-145-92. Jointing Compound, Graphited Issue 1. 13 pp.
MOD UK DEF-174-69. Jointing Compound, Graphited (Superseded by Def Stan 68-145). 5 pp.

—Heat Shrink
BSI BS 6910: Part 1-88. 1988 Cold Pour Resin Compound and Heat-Shrink Cable Joints in the Voltage Range up to 1000V a.c. and 1500V d.c. Part 1: Materials. 27 pp.
BSI BS 6910: Part 2-89. 1989 Cold Pour Resin Compound and Heat-Shrink Cable Joints in the Voltage Range up to 1000 V a.c. and 1500 V d.c.: Part 2: Code of Practice for on-Site Installation. 16 pp.

—Identification Systems
EC 83/265/EEC-83. Council Directive Amending Directive 77/728/EEC on the Approximation of the Laws, Regulations and Administrative Provisions of the Member States Relating to the Classification, Packaging and Labelling of Paints, Varnishes, Printing Inks, Adhesives. 7 pp.

—Immersion Testing
ISO 10590-91. Building Construction—Sealants—Determination of Adhesion/Cohesion Properties at Maintained Extension After Immersion in Water First Edition. 6 pp.
ISO 10591-91. Building Construction—Sealants—Determination of Adhesion/Cohesion Properties After Immersion in Water First Edition. 6 pp.

—Mass
ISO 10563-91. Building Construction—Sealants for Joints—Determination of Change in Mass and Volume First Edition. 5 pp.

—Oil Base
CGSB CAN/CGSB-19.6-M87. Caulking Compound, Oil Base. 8 pp.
CNS A2135-82. Oil Based Caulking Compounds for Building (Jun)(8901).
CNS A3153-82. Method of Test for Oil Based Caulking Compounds for Building (Jun)(8902).
JIS A 5751-75. Oil Based Caulking Compounds for Buildings (R 1983). 15 pp.

—Packaging
EC 83/265/EEC-83. Council Directive Amending Directive 77/728/EEC on the Approximation of the Laws, Regulations and Administrative Provisions of the Member States Relating to the Classification, Packaging and Labelling of Paints, Varnishes, Printing Inks, Adhesives. 7 pp.

—Paper
BSI BS 4249-89. 1989 Paper and Cork/Paper Jointing. 14 pp.

—Physical Properties
BSI BS 3712: Part 1-91. 1991 Building and Construction Sealants Part 1: Methods of Test for Homogeneity, Relative Density and Penetration. 14 pp.
BSI BS 3712: Part 1-85. (WITHDRAWN) 1985 Building and Construction Sealants Part 1: Methods of Test for Homogeneity, Relative Density, Extrudability, Penetration and Slump. 14 pp.
BSI BS 3712: Part 4-91. 1991 Building and Construction Sealants Part 4: Method of Test for Adhesion in Peel. 10 pp.

Joint Sealants (Cont.)

—Physical Properties (Cont.)
BSI BS 6956: Part 5-92. 1992 Jointing Materials and Compounds Part 5: Specification for Jointing Compounds for Use with Water, Low Pressure Saturated Steam, 1st Family Gases (Excluding Coal Gas) and 2nd Family Gases. 10 pp.
BSI BS 6956: Part 6-92. 1992 Jointing Materials and Compounds Part 6: Specification for Jointing Compounds for 3rd Family Gases. 11 pp.

—Polyisobutylene
CGSB 19-GP-14M-76. Sealing Compound, One Component, Butyl-Polyisobutylene Polymer Base, Solvent Curing, Standard for (R 1984). 9 pp.

—Polysulfide
BSI BS 4254-83. 1983 Amd 1 Two-Part Polysulphide-Based Sealants. 12 pp.
BSI BS 5215-86. 1986 Amd 1 One-Part Gun-Grade Polysulphide-Based Sealants. 9 pp.
SAA AS 1526-74. One-Part Polysulphide-Based Sealing Compounds for the Building Industry Corrig.. 39 pp.
SAA AS 1527-74. Two-Part Polysulphide-Based Sealing Compounds for the Building Industry Corrig.. 36 pp.

—Polyurethane
CNS A2090-85. Polyurethane for Building Joint Sealant (Feb)(6985). 2 pp.

—PVC
CGSB 41-GP-35M-83. Polyvinyl Chloride Waterstop, Standard for. 15 pp.
CNS K6384-90. Method of Test for Flexible PVC Water-Stops (Jun)(3896).

—PVC—Construction Joints
CNS K3031-90. Flexible Polyvinylchloride Water-Stops (Jun)(3895).
CNS K6384-90. Method of Test for Flexible PVC Water-Stops (Jun)(3896).
JIS K 6773-77. Flexible Polyvinylchloride Water-Stops. 16 pp.

—Red Lead
BSI BS 217-61. 1961 Red Lead for Paints and Jointing Compounds. 17 pp.

—Resin
BSI BS 6910: Part 1-88. 1988 Cold Pour Resin Compound and Heat-Shrink Cable Joints in the Voltage Range up to 1000V a.c. and 1500V d.c. Part 1: Materials. 27 pp.
BSI BS 6910: Part 2-89. 1989 Cold Pour Resin Compound and Heat-Shrink Cable Joints in the Voltage Range up to 1000 V a.c. and 1500 V d.c.: Part 2: Code of Practice for on-Site Installation. 16 pp.

—Rubber Asphalt
CGSB CAN/CGSB-37.29-M89. Rubber-Asphalt Sealing Compound. 9 pp.

—Silicone
CGSB CAN/CGSB-19.18-M87. Sealing Compound, One-Component, Silicone Base, Solvent Curing. 9 pp.

—Tapes—Chromate
MOD UK CS 3080. Tape, Jointing, Chromated (Withdrawn).

—Tensile Testing
BSI BS EN 28339-91. 1991 Building Construction—Jointing Products—Sealants—Determination of Tensile Properties (ISO 8339: 1984). 10 pp.
BSI BS EN 28340-91. 1991 Building Construction—Jointing Products—Sealants—Determination of Tensile Properties at Maintained Extension (ISO 8340: 1984). 10 pp.
CEN EN 28 339-90. Building Construction—Jointing Products—Sealants—Determination of Tensile Properties. 8 pp.
CEN EN 28 340-90. Building Construction—Jointing Products—Sealants—Determination of Tensile Properties at Maintained Extension. 7 pp.
DIN ENGL EN 28339-91. Jointing Products in Building Construction; Sealants; Determination of Resistance to Flow (ISO 8339:1984) (May). 6 pp.
DIN ENGL EN 28340-91. Jointing Products in Building Construction; Sealants; Determination of Tensile Properties at Maintained Extension (ISO 8340:1984) (May) (Supersedes DIN 52455 Part 2, July 1987 Edition). 6 pp.
ISO 8339-84. Building Construction—Jointing Products—Sealants—Determination of Tensile Properties First Edition. 5 pp.
ISO 8340-84. Building Construction—Jointing Products—Sealants—Determination of Tensile Properties at Maintained Extension First Edition. 5 pp.

INTERNATIONAL AND NON-U.S. NATIONAL STANDARDS
SUBJECT INDEX

Joints

Joint Sealants (Cont.)

—Tiles—Mildew Resistant
CGSB CAN/CGSB-19.22-M89. Mildew-Resistant Sealing Compound for Tubs and Tiles. 9 pp.

—Varnishes—Aircraft
MOD UK DTD-369B-01. Pigmented Varnish Jointing Compound; Amendment 1. 6 pp.

—Viscosity
BSI BS EN 27390-91. 1991 Building Construction—Jointing Products—Determination of Resistance to Flow (ISO 7390: 1987). 9 pp.
CEN EN 27 390-90. Building Construction—Jointing Products—Determination of Resistance to Flow. 7 pp.
DIN ENGL EN 27390-91. Jointing Products in Building Construction; Determination of Resistance to Flow (ISO 7390:1987) (May) (Supersedes DIN 52454, September 1987 Edition). 5 pp.
ISO 7390-87. Building Construction—Jointing Products—Determination of Resistance to Flow Second Edition. 4 pp.

—Volume
ISO 10563-91. Building Construction—Sealants for Joints—Determination of Change in Mass and Volume First Edition. 5 pp.

—Volume—Thermal Stresses
DIN ENGL 52451-83. Testing of Building Sealants; Determination of the Change in Volume After Thermal Stress; Dipping and Weighing Method (Feb). 2 pp.

—Wall Joints
DIN ENGL 18540-88. Sealing of Joints in External Walls Using Building Sealants (Oct) (Supersedes DIN 18540 Parts 1 to 3, January 1980 Edition). 5 pp.

—Water Pipes
BSI BS 5292-80. 1980 Amd 2 Jointing Materials and Compounds for Installations Using Water, Low-Pressure Steam or 1st, 2nd and 3rd Family Gases (AMD 5960) November 30, 1988. 22 pp.
BSI BS 5292-03. 1980 Amd 3 Jointing Materials and Compounds for Installations Using Water, Low-Pressure Steam or 1st, 2nd and 3rd Family Gases (AMD 6918) January 31, 1992 (Q). 24 pp.
BSI BS 5292-04. 1980 Amd 4 Jointing Materials and Compounds for Installations Using Water, Low-Pressure Steam or 1st, 2nd and 3rd Family Gases (AMD 7047) April 1, 1992. 26 pp.

—Wax Polyisobutylene
MOD UK TS 10052A. Sealing Composition Medium—LF Quality (Composition: RD 1108) (Withdrawn).

Jointers
Use For: Veneer Jointers *See Also:* Tools; Woodworking Equipment
CNS B1310-89. Nominal Dimension of Veneer Jointers (Dec)(10062).
CNS B7249-89. Test Code for Accuracy of Veneer Jointers (Jun)(10229).
CNS B7267-89. Test Code for Performance of Veneer Jointers (Jun)(10790).

—Accuracy Testing
CNS B7112-79. Accuracy Inspection Method of Jointer (Apr)(4810). 2 pp.

Jointing
Use: Joint Sealants

Jointing Compounds
Use: Joint Sealants

Jointing Materials
Use: Joint Sealants

Joints
Scope Note: For standards on joints between concrete structures, see Construction Joints
See Also: Adhesives; Angle Joints; Ball Joints; Bed Joints; Bolted Joints; Bonded Joints; Brazed Joints; Butt Joints; Butt Strap Joints; Butt Welded Joints; Cable Joints; Compression Joints; Connectors; Construction Joints; Couplings; Duct Joints; Expansion Joints; Fasteners; Fillet Welds; Fork Joints; Glass Joints; Grommets; Interlocking Joints; Joint Fillers; Joint Sealants; Lap Joints; Pipe Joints; Projection Welds; Resistance Spot Welds; Rotary Joints (Electrical); Seam Welds; Sleeves (Fittings); Slip Joints; Soldered Joints; Spot Welds; Structural Members; Swivels; T Joints; Truss Clips; Tube Joints; Universal Joints; Wall Joints; Welded Joints

Joints (Cont.)

—Adhesive Strength—Cleavage Strength
BSI BS 5350: Part C1-86. 1986 Methods of Test for Adhesives Group C: Adhesively Bonded Joints: Mechanical Tests Part C1: Determination of Cleavage Strength of Adhesive Bonds. 6 pp.

—Adhesive Strength—Impact Testing
BSI BS 5350: Part C4-86. 1986 Methods of Test for Adhesives Group C: Adhesively Bonded Joints: Mechanical Tests Part C4: Determination of Impact Resistance of Adhesive Bonds. 6 pp.

—Adhesive Strength—Peel Strength
BSI BS 5350: Part C9-90. 1990 Methods of Test for Adhesives Group C: Adhesively Bonded Joints: Mechanical Tests Part C9: Floating Roller Peel Test. 4 pp.
BSI BS 5350: Part C10-91. (WITHDRAWN) 1991 Methods of Test for Adhesives Group C: Adhesively Bonded Joints: Mechanical Tests Part C10: 90 Degree Peel Test for a Flexible-to-Rigid Assembly (ISO 8510-1: 1990) (Renumbered as BS EN 28510-1: 1993). 9 pp.
BSI BS 5350: Part C11-91. (WITHDRAWN) 1991 Methods of Test for Adhesives Group C: Adhesively Bonded Joints: Mechanical Tests Part C11: 180 Degree Peel Test for a Flexible-to-Rigid Assembly (ISO 8510-2: 1990) (Renumbered as BS EN 28510-2: 1993). 9 pp.
BSI BS 5350: Part C12-79. 1979 Methods of Test for Adhesives Group C: Adhesively Bonded Joints: Mechanical Tests Part C12: 'T' Peel Test for a Flexible Assembly. 4 pp.
BSI BS 5350: Part C13-90. 1990 Methods of Test for Adhesives Group C: Adhesively Bonded Joints: Mechanical Tests Part 13: Climbing Drum Peel Test. 6 pp.
BSI BS 5350: Part C14-79. 1979 Methods of Test for Adhesives Group C: Adhesively Bonded Joints: Mechanical Tests Part C14: 90 Degree Peel Test for a Rigid-To-Rigid Assembly. 4 pp.
BSI BS EN 28510-1-93. 1993 Amd 1 Adhesives—Peel Test for a Flexible-Bonded-to-Rigid Test Specimen Assembly Part 1: 90 Degree Peel (ISO 8510-1: 1990) (AMD 7559) August 15, 1993 (W). 12 pp.
BSI BS EN 28510-2-93. 1993 Amd 1 Adhesives—Peel Test for a Flexible-Bonded-to-Rigid Test Specimen Assembly Part 2: 180 Degree Peel (ISO 8510-2: 1990) (AMD 7560) August 15, 1993 (W). 17 pp.
CEN EN 28510-1-93. Adhesives—Peel Test for a Flexible-Bonded-to-Rigid Test Specimen Assembly—Part 1: 90 Degree Peel (ISO 8510-1: 1990). 7 pp.
CEN EN 28510-2-93. Adhesives—Peel Test for a Flexible-Bonded-to-Rigid Test Specimen Assembly—Part 2: 180 Degree Peel (ISO 8510-2: 1990). 7 pp.
ISO 8510 Pt 1-90. Adhesives—Peel Test for a Flexible-Bonded-to-Rigid Test Specimen Assembly—Part 1: 90 Degree Peel First Edition (CEN EN 28510-1: 1993). 8 pp.
ISO 8510 Pt 2-90. Adhesives—Peel Test for a Flexible-Bonded-to-Rigid Test Specimen Assembly—Part 2: 180 Degree Peel First Edition (CEN EN 28510-2: 1993). 8 pp.

—Adhesives—Creep Testing
BSI BS 5350: Part C7-90. 1990 Methods of Test for Adhesives Group C: Adhesively Bonded Joints: Mechanical Tests Part C7: Determination of Creep and Resistance to Sustained Application of Force. 4 pp.

—Adhesives—Shear Strength
BSI BS 5350: Part C5-90. 1990 Methods of Test for Adhesives Group C: Adhesively Bonded Joints: Mechanical Tests Part C5: Determination of Bond Strength in Longitudinal Shear. 8 pp.
BSI BS 5350: Part C15-90. 1990 Methods of Test for Adhesives Group C: Adhesively Bonded Joints: Mechanical Tests Part C15: Determination of Bond Strength in Compressive Shear. 5 pp.

—Aircraft
SBAC AS 8662 ISSUE 1. Pitot and Static, Manifold, Connection Assembly.
SBAC AS 8663 ISSUE 3. Connection, Pitot.
SBAC AS 8664 ISSUE 3. Connection, Static.

—Aircraft—Adapters
SBAC AS 6290 ISSUE 1. Adaptor—1 1/4 Inch. 1 p.
SBAC AS 6314 ISSUE 1. Adaptor. 1 p.

—Aircraft—Bolts
CAA LEAFLET 6-7 07.90. Assembly and Maintenance of Critical Bolted Joints. 7 pp.

—Aircraft—Brackets
SBAC AS 8665 ISSUE 1. Bracket.

—Aircraft—Drain Valves
SBAC AS 6289 ISSUE 1. Tank Drainage Connection Drain Valve AS 6270, 1 1/4 Inch. 1 p.
SBAC AS 6313 ISSUE 1. Tank Drainage Connection Drain Valve AS 6295, 2 Inches. 1 p.

Joints (Cont.)

—Aircraft—Marine
SBAC AS 2805 ISSUE 1. De-Oiling Connection Naval Aircraft. 1 p.

—Automotive
DIN ENGL 75532 Pt 1-76. Transmission of Rotary Motions; Types of Connection to Gears, Intermediate Gears, Flexible Drive Shafts and Equipment (June). 8 pp.

—Automotive—Flexible Shafts
DIN ENGL 75532 Pt 1-76. Transmission of Rotary Motions; Types of Connection to Gears, Intermediate Gears, Flexible Drive Shafts and Equipment (June). 8 pp.

—Automotive—Gears
DIN ENGL 75532 Pt 1-76. Transmission of Rotary Motions; Types of Connection to Gears, Intermediate Gears, Flexible Drive Shafts and Equipment (June). 8 pp.

—Automotive—Gears—Intermediate
DIN ENGL 75532 Pt 1-76. Transmission of Rotary Motions; Types of Connection to Gears, Intermediate Gears, Flexible Drive Shafts and Equipment (June). 8 pp.

—Construction (Excludes Concrete Structures)
DIN ENGL 1052 Pt 2-88. Structural Use of Timber; Mechanically Fastened Joints (Apr) (This Standard, Together with DIN 1052 Part 1, April 1988 Edition, Supersedes DIN 1052 Part 1, October 1969 Edition). 27 pp.

—Construction (Excludes Concrete Structures)—Air Permeability
BSI BS 6181-81. 1981 Air Permeability of Joints in Building. 8 pp.
CNS B2768-85. Sets of High Strength Hexagon Bolt, Hexagon Nut and Plain Washers for Friction Grip Joints (Aug)(11328).
CNS B7272-85. Method of Test for the Sets of High Strength Hexagon Bolt, Hexagon Nut and Plain Washers for Friction Grip Joints (Aug)(11329).
ISO 2445-72. Joints in Building—Fundamental Principles for Design First Edition. 3 pp.
ISO 3447-75. Joints in Building—General Check-List of Joint Functions First Edition. 4 pp.
ISO 6589-83. Joints in Building—Laboratory Method of Test for Air Permeability of Joints Second Edition. 5 pp.
ISO 7727-84. Joints in Building—Principles for Jointing of Building Components—Accommodation of Dimensional Deviations During Construction First Edition. 7 pp.
JIS B 1186-79. Sets of High Strength Hexagon Bolt, Hexagon Nut and Plain Washers for Friction Grip Joints. 21 pp.

—Construction (Excludes Concrete Structures)—Compression Testing
ISO 8969-90. Timber Structures—Testing of Unilateral Punched Metal Plate Fasteners and Joints First Edition. 12 pp.

—Construction (Excludes Concrete Structures)—Design
BSI BS 6093-93. 1993 Code of Practice for Design of Joints and Jointing in Building Construction. 47 pp.
BSI BS 6093-81. 1981 Design of Joints and Jointing in Building Construction. 48 pp.

—Construction (Excludes Concrete Structures)—Glossaries
ISO 2444-88. Joints in Building—Vocabulary Second Edition. 6 pp.

—Construction (Excludes Concrete Structures)—Shear Strength
ISO 8969-90. Timber Structures—Testing of Unilateral Punched Metal Plate Fasteners and Joints First Edition. 12 pp.

—Construction (Excludes Concrete Structures)—Stiffness Testing
ISO 8969-90. Timber Structures—Testing of Unilateral Punched Metal Plate Fasteners and Joints First Edition. 12 pp.

—Construction (Excludes Concrete Structures)—Tensile Testing
ISO 8969-90. Timber Structures—Testing of Unilateral Punched Metal Plate Fasteners and Joints First Edition. 12 pp.

INDUSTRY STANDARDS

INTERNATIONAL AND NON-U.S. NATIONAL STANDARDS
SUBJECT INDEX

Joints (Cont.)

—Construction (Excludes Concrete Structures)—Waterproof Coatings
DIN ENGL 18195 Pt 8-83. Waterproofing of Buildings and Structures; Waterproofing over Movement Joints (Aug). 4 pp.

—Construction (Excludes Concrete Structures)—Wood Density
BSI BS EN 28970-91. 1991 Timber Structures—Testing of Joints Made with Mechanical Fasteners—Requirements for Wood Density (ISO 8970: 1989). 9 pp.
CEN EN 28970-91. Timber Structures—Testing of Joints Made with Mechanical Fasteners—Requirements for Wood Density. 5 pp.
DIN ENGL EN 28970-91. Timber Structures; Testing of Joints Made with Mechanical Fasteners; Requirements for Wood Density (ISO 8970: 1989) (July). 5 pp.
ISO 8970-89. Timber Structures—Testing of Joints Made with Mechanical Fasteners—Requirements for Wood Density First Edition. 4 pp.

—Drill Rods—Rock Drills
CNS M2052-80. Shape and Dimension of Joint in Extension Rod for Rock Drill (May) (5625).
JIS M 3914-76. Shape and Dimension of Joint in Extension Rod for Rock Drill.

—Flanged
DIN ENGL 2695-72. Diaphragm Packings and Diaphragm Weld Seals for Flanged Connections; Nominal Pressures 64 to 400 (Jan). 2 pp.
DIN ENGL 2696-72. Lenticular Gaskets and Lenticular Seals for ND 64 to ND 400 Flanged Connections (Apr). 2 pp.
DIN ENGL 2697-72. Grooved O-Rings and Seals for Flanged Joints; Nominal Pressures 64 to 400 (Jan). 2 pp.
DIN ENGL 4922 Pt 3-75. Steel Filter Pipes for Drilled Wells; Flanged Connection, NW 500 to NW 1000 (Nominal Diameter 500 to 1000) (Dec). 2 pp.
DIN ENGL 28030 Pt 1-89. Flanged Joints for Chemical Apparatus; Design and Construction (Feb). 7 pp.

—Flanged—Gaskets
DIN ENGL 28040-89. Gaskets for Use with Flanged Joints (Feb). 3 pp.

—Gaskets
CNS A3279-88. Method of Test for Sponge Gaskets for Windows, Doors and Joints of Panel in Building (Jul)(12352).

—Gaskets—Panels
CNS A2154-88. Gaskets for Windows, Doors and Joints of Panel in Buildings (Dec)(10209).
CNS A2226-88. Sponge Gaskets for Windows, Doors and Joints of Panel in Buildings (Jul)(12351).
CNS A3186-88. Method of Test for Gaskets for Windows, Doors and Joints of Panel in Buildings (Dec)(10210).
JIS A 5750-87. Sponge Gaskets for Windows, Doors and Joints of Panel in Buildings. 21 pp.
JIS A 5756-89. Gaskets for Windows, Doors and Joints of Panel in Buildings. 26 pp.

—Involute Splines
DIN ENGL 5480 Pt 1-91. Involute Spline Joints; Principles (Oct). 10 pp.
DIN ENGL 5480 Pt 2-91. Involute Spline Joints; 30 Degree Pressure Angle; Survey (Oct). 2 pp.
DIN ENGL 5480 Pt 3-91. Involute Spline Joints; 30 Degree Pressure Angle; Basic Dimensions and Test Dimensions for Modules 0,5, 0,6, 0,75, 0,8 and 1 (Oct). 11 pp.
DIN ENGL 5480 Pt 4-91. Involute Spline Joints; 30 Degree Pressure Angle; Basic Dimensions and Test Dimensions for Module 1,25 (Oct). 4 pp.
DIN ENGL 5480 Pt 5-91. Involute Spline Joints; 30 Degree Pressure Angle; Basic Dimensions and Test Dimensions for Modules 1,5 and 1,75 (Oct). 6 pp.
DIN ENGL 5480 Pt 6-91. Involute Spline Joints; 30 Degree Pressure Angle; Basic Dimensions and Test Dimensions for Module 2 (Oct). 4 pp.
DIN ENGL 5480 Pt 7-91. Involute Spline Joints; 30 Degree Pressure Angle; Basic Dimensions and Test Dimensions for Module 2,5 (Oct). 4 pp.
DIN ENGL 5480 Pt 8-91. Involute Spline Joints; 30 Degree Pressure Angle; Basic Dimensions and Test Dimensions for Module 3 (Oct). 4 pp.
DIN ENGL 5480 Pt 9-91. Involute Spline Joints; 30 Degree Pressure Angle; Basic Dimensions and Test Dimensions for Module 4 (Oct). 4 pp.
DIN ENGL 5480 Pt 10-91. Involute Spline Joints; 30 Degree Pressure Angle; Basic Dimensions and Test Dimensions for Module 5 (Oct). 4 pp.
DIN ENGL 5480 Pt 11-91. Involute Spline Joints; 30 Degree Pressure Angle; Basic Dimensions and Test Dimensions for Module 6 (Oct). 4 pp.

Joints (Cont.)

—Involute Splines (Cont.)
DIN ENGL 5480 Pt 12-91. Involute Spline Joints; 30 Degree Pressure Angle; Basic Dimensions and Test Dimensions for Module 8 (Oct). 4 pp.
DIN ENGL 5480 Pt 13-91. Involute Spline Joints; 30 Degree Pressure Angle; Basic Dimensions and Test Dimensions for Module 10 (Oct). 4 pp.

—Plates—PVC
CNS K3045-87. Polyvinyl Chloride Joint (for Plate) (Sep)(6480). 4 pp.
CNS K6590-87. Method of Test for PVC Joint (for Plate) (Sep)(6481).

—Pressure—Concrete
CEN PREN 639-92. Common Requirements for Concrete Pressure Pipes Including Joints and Fittings. 15 pp.
CEN PREN 640-92. Reinforced Concrete Pressure Pipes Non-Cylinder Type, Including Joints and Fittings. 13 pp.
CEN PREN 641-92. Reinforced Concrete Pressure Pipes, Cylinder Type, Including Joints and Fittings. 14 pp.

—Pressure—Fiber Cement
CEN PREN 512-91. Fibre-Cement Products—Pressure Pipes and Joints. 24 pp.

—Timber Connectors
CEN PREN 912-92. Timber Fasteners—Specifications for Connectors for Timber. 37 pp.

—Wood
BSI BS 6948-89. 1989 Methods of Test for Mechanically Fastened Joints in Timber and Wood-Based Materials. 39 pp.
BSI BS 6948-01. 1989 Amd 1 Methods of Test for Mechanically Fastened Joints in Timber and Wood-Based Materials (AMD 6916) February 28, 1992. 40 pp.
SNZ NZS 3616-78. Specification for Finger-Jointed Timber. 21 pp.

—Wooden Structures—Fasteners—Design Testing
BSI BS EN 26891-91. 1991 Timber Structures—Joints Made with Mechanical Fasteners—General Principles for the Determination of Strength and Deformation Characteristics. 11 pp.
CEN EN 26891-91. Timber Structures—Joints Made with Mechanical Fasteners—General Principles for the Determination of Strength and Deformation Characteristics. 7 pp.
DIN ENGL EN 26891-91. Timber Structures; Joints Made with Mechanical Fasteners; General Principles for the Determination of Strength and Deformation Characteristics (ISO 6891:1983) (July). 6 pp.
ISO 6891-83. Timber Structures—Joints Made with Mechanical Fasteners—General Principles for the Determination of Strength and Deformation Characteristics First Edition. 6 pp.

—Wooden Structures—Fasteners—Stiffness Testing
CEN PREN 1075-93. Timber Structures—Test Methods—Joints Made of Punched Metal Plate Fasteners. 14 pp.

—Wooden Structures—Fasteners—Strength Testing
BSI BS EN 26891-91. 1991 Timber Structures—Joints Made with Mechanical Fasteners—General Principles for the Determination of Strength and Deformation Characteristics. 11 pp.
CEN PREN 1075-93. Timber Structures—Test Methods—Joints Made of Punched Metal Plate Fasteners. 14 pp.
CEN EN 26891-91. Timber Structures—Joints Made with Mechanical Fasteners—General Principles for the Determination of Strength and Deformation Characteristics. 7 pp.
DIN ENGL EN 26891-91. Timber Structures; Joints Made with Mechanical Fasteners; General Principles for the Determination of Strength and Deformation Characteristics (ISO 6891:1983) (July). 6 pp.
ISO 6891-83. Timber Structures—Joints Made with Mechanical Fasteners—General Principles for the Determination of Strength and Deformation Characteristics First Edition. 6 pp.

Joist Hangers

—Masonry—Stiffness Testing
CEN PREN 846-8-92. Methods of Test for Ancillary Components for Masonry—Part 8: Determination of Strength and Stiffness of Joist Hangers. 10 pp.

—Masonry—Strength
CEN PREN 846-8-92. Methods of Test for Ancillary Components for Masonry—Part 8: Determination of Strength and Stiffness of Joist Hangers. 10 pp.

Joist Hangers (Cont.)

—Residential Buildings
BSI BS 6178: Part 1-90. 1990 Joist Hangers Part 1: Joist Hangers for Building into Masonry Walls of Domestic Dwellings. 16 pp.

Joists

Use For: Stringers *See Also:* Beams (Supports); Structural Members

—Filler—Concrete—Floors
DIN ENGL 4158-78. Filler Concrete Joists for Reinforced and Prestressed Concrete Floors (May). 7 pp.

—Pine Wood
SAA AS 1491-73. Laminated and/or Finger-Jointed Radiata Pine Scantlings (Metric Units) (Incorporating Amdt 1). 21 pp.

—Steel—Hot Rolled
BSI BS 4: Part 1-80. 1980 Amd 5 Structural Steel Sections Part 1: Hot-Rolled Sections. 30 pp.
SNZ NZS/BS 4: Part 1-80. Structural Steel Sections Part 1: Specification for Hot-Rolled Sections. 27 pp.

—Timber
SAA AS O46-60. Round Section Stringers from Western Australian Timbers. 4 pp.

—Timber—Ceilings—Tables (Data)
BSI BS 5268: Sec 7.3-89. 1989 Structural Use of Timber Part 7: Recommendations for the Calculation Basis for Span Tables Section 7.3: Ceiling Joists. 16 pp.

—Timber—Floors—Tables (Data)
BSI BS 5268: Sec 7.1-89. 1989 Structural Use of Timber Part 7: Recommendations for the Calculation Basis for Span Tables Section 7.1: Domestic Floor Joists. 15 pp.

—Timber—Roofing—Tables (Data)
BSI BS 5268: Sec 7.2-89. 1989 Structural Use of Timber Part 7: Recommendations for the Calculation Basis for Span Tables Section 7.2: Joists for Flat Roofs. 16 pp.

Jominy Testing

Use For: End Quenching Testing
See Also: Ferroalloys; Hardenability; Hardness Testing; Microstructures; Physical Testing; Static Testing

—Steels
AECMA PREN2003-14-86. Test Methods for Steel Products—Part 14—Hardenability Test by End Quenching (Jominy Test). 1 p.
BSI BS 4437-87. 1987 Method for Determining Hardenability of Steel by End Quenching (Jominy Test). 16 pp.
ISO 642-79. Steel—Hardenability Test by End Quenching (Jominy Test) First Edition. 14 pp.
JIS G 0561-83. Method of Hardenability Test for Steel (End Quenching Method). 10 pp.
SAA AS 1770-75. Method for the End-Quench Test for Hardenability of Steel (Jominy Test) Corrig. Reconfirmed 1986. 16 pp.

Journal Bearings

See Also: Antifriction Bearings; Bearings; Magnetic Bearings; Plain Bearings; Radial Bearings; Sleeve Bearings

—Ball
CNS B2472-85. Self-Aligning Ball Bearing; Wide Inner Race; Inner Race with Clamping Sleeve (Mar)(5501).
DIN ENGL 628 Pt 1-73. Angular Contact Ball Bearings; Single-Row and Double-Row (Mar). 5 pp.
DIN ENGL 630 Pt 1-60. Self-Aligning Ball (Journal) Bearings; Parallel and Tapered Bore (May). 2 pp.

—Bushings
JIS B 1582-76. Bushes for Journal Bearings. 11 pp.

—Circular Cylindrical—Design
DIN ENGL 31652 Pt 1-83. Plain Bearings; Hydrodynamic Plain Journal Bearings Designed for Operation Under Steady-State Conditions; Design of Circular Cylindrical Bearings (Apr). 16 pp.
DIN ENGL 31652 Pt 2-83. Plain Bearings; Hydrodynamic Plain Journal Bearings Designed for Operation Under Steady-State Conditions; Functions Necessary When Designing Circular Cylindrical Bearings (Feb). 19 pp.
DIN ENGL 31652 Pt 3-83. Plain Bearings; Hydrodynamic Plain Journal Bearings Designed for Operation Under Steady-State Conditions; Operational Parameters Necessary When Designing Circular Cylindrical Bearings (Apr). 7 pp.

INTERNATIONAL AND NON-U.S. NATIONAL STANDARDS
SUBJECT INDEX

Journal Bearings (Cont.)
—Roller
DIN ENGL 635 Pt 1-87. Rolling Bearings; Single-Row Self-Aligning Roller Bearings (Aug). 3 pp.

Journals
Use: Periodicals

Journals (Machine Elements)
See Also: Shafts (Machine Elements)
—Greases
CNS K5041-73. Lubricating Grease for Journal (Tentative) (May)(2590).
CNS K6229-73. Method of Test for Lubricating Grease for Journal (Aug)(2591).

Joystick Switches
See Also: Electric Switches; Switches
—Preferred Products List
CECC CECC MUAHAG Vol 8 IS 4-91. Preferred Products List; Switches (En, Ge, Fr). 29 pp.

Judder
See Also: FAX Communications; FAX Machines; Jitter; Reproduction (Copying)
—Phototelegraph Equipment
CCITT RECMN T.1-89. Standardization of Phototelegraphic Apparatus—Terminal Equipment and Protocols for Telematic Services (Study Group VIII) 5 pp. 5 pp.

Jugs
See Also: Appliances; Containers
—Electric
BSI BS 3999: Part 1-93. 1993 Methods of Measuring the Performance of Household Electrical Appliances Part 1: Electric Kettles and Jugs (IEC 530: 1975). 8 pp.
BSI BS 3999: Part 1-66. 1966 Amd 3 Methods of Measuring the Performance of Household Electrical Appliances Part 1: Electric Kettles. 14 pp.
—Food—Vacuum Insulated
BSI BS 6672: Part 1-86. 1986 Insulated Domestic Food Containers Part 1: Vacuum Ware and Insulated Flasks, Jars and Jugs. 12 pp.
BSI BS 6672: Part 1-01. 1986 Amd 1 Insulated Domestic Food Containers Part 1: Specification for Vacuum Ware and Insulated Flasks, Jars and Jugs (AMD 7199) June 15, 1992. 14 pp.
SNZ NZS 5847: Part 1-88. Insulated Domestic Food Containers Part 1: Specification for Vacuum Ware and Insulated Flasks, Jars and Jugs Amend: A, 1988. 12 pp.
—Plastic—Oils—Export
CGSB 43-GP-163M-91. Plastic Jugs for Export of Oil as Food Aid. 9 pp.

Jugs (Water Boilers)
Use: Teakettles

Juice Extractors
Use: Juicers

Juicers
See Also: Appliances; Food Processing Equipment
CNS C4438-86. Electric Citrus Juicer (Jan)(11474).
—Household
JIS C 9609-90. Electric Blenders and Electric Juicers for Household Use. 29 pp.
—Rice
CNS B4061-83. Electric Been (Rice) Juice Grinding Machine (Oct)(10596).

Juices
Scope Note: Use a more specific term See: Fruit Juices; Juicers; Vegetable Juices

Julian Date
See Also: Calendars
—Modified
CCIR RECMN 808-92. Broadcasting of Time and Date Information in Coded Form. 5 pp.
—Modified—Standard Frequency Satellite Services
CCIR RECMN 457-1-74. Use of the Modified Julian Date by the Standard-Frequency and Time-Signal Services—Section 7B—Specifications for the Standard-Frequency and Time-Signal Services. 1 p.

Julian Date (Cont.)
—Modified—Time Signal Services
CCIR RECMN 457-1-74. Use of the Modified Julian Date by the Standard-Frequency and Time-Signal Services—Section 7B—Specifications for the Standard-Frequency and Time-Signal Services. 1 p.

Jump Suits
Use: Coveralls

Jumper Cables
See Also: Automotive Equipment; Batteries; Cables (Electric); Motor Vehicles
—Automotive
BSI BS AU 237: Part 1-90. 1990 Jumper Lead Sets for Automotive Starting Part 1: Specification for Jumper Lead Sets for Light Domestic Duty. 10 pp.
BSI BS AU 237A-93. 1993 Light Domestic Duty Jumper Lead Sets for Automotive Starting (E). 15 pp.
—Couplers
CNS E1017-82. Jumper Couplers (Dec)(9698).
JIS E 4202-91. Jumper Couplers.

Jumper Lead Sets
Use: Jumper Cables

Jumpers
Use: Dresses

Jumpers (Electric)
See Also: Connectors
—Communication Systems and Equipment
DIN VDE 0881-86. Jumper Wires and Stranded Hook-Up Wires with Extended Temperature Range for Telecommunications Systems and Data Processing Systems (Mar). 34 pp.
—Data Processing Equipment
DIN VDE 0881-86. Jumper Wires and Stranded Hook-Up Wires with Extended Temperature Range for Telecommunications Systems and Data Processing Systems (Mar). 34 pp.
—Telecommunication Systems and Equipment
DIN VDE 0881-86. Jumper Wires and Stranded Hook-Up Wires with Extended Temperature Range for Telecommunications Systems and Data Processing Systems (Mar). 34 pp.

Jumpsuits
Use: Coveralls

Junction Boxes
Use For: Joint Boxes See Also: Circuit Boxes; Electric Conduit Boxes; Electric Outlet Boxes; Electrical Enclosures
CSA 1208 Bull. Electrical Bulletin 1208 January 3, 1979 to C22.2 NO 85. 1 p.
—Cable Assemblies—Controlgear
DIN VDE 0660 Pt 505-90. Switchgear and Controlgear; Low-Voltage Switchgear and Controlgear Assemblies; Specification for Domestic Connection Boxes and Fuseboxes (July). 38 pp.
—Cable Assemblies—Switchgear
DIN VDE 0660 Pt 505-90. Switchgear and Controlgear; Low-Voltage Switchgear and Controlgear Assemblies; Specification for Domestic Connection Boxes and Fuseboxes (July). 38 pp.
—Cables (Electric)
BSI BS 2562-79. 1979 Cable Boxes for Transformers and Reactors. 63 pp.
CNS C4422-84. Junction Box for Indoor Wiring (for Polyvinyl-Chloride Insulated and Sheathed Cables: VVF) (Nov)(11093).
DIN ENGL 47600 Pt 3-74. Cast Metal Joint Boxes for Power Cables up to 10 kV; Correlation of Joint Boxes with Paper-Insulated Cables; Correlation of Bare Stranded Copper Bonding Wire (Oct). 2 pp.
DIN ENGL 47600 Pt 4-74. Cast Metal Joint Boxes for Power Cables up to 10 kV; Internal Design for Paper-Insulated Cables (Oct). 3 pp.
DIN ENGL 47600 Pt 5-74. Cast Metal Joint Boxes for Power Cables up to 10 kV; Installation Instructions for Paper-Insulated Cables (Oct). 4 pp.
DIN ENGL 47600 Pt 6-74. Cast Metal Joint Boxes for Power Cables up to 10 kV; Correlation of Joint Boxes with Plastic Insulated Cables 0.6/1 kV (Oct). 2 pp.
DIN ENGL 47600 Pt 7-74. Cast Metal Joint Boxes for Power Cables up to 10 kV; Internal Design for Plastic Insulated Cables 0.6/1 kV (Oct). 2 pp.
JIS C 8365-88. Junction Box for Indoor Wiring (for 600 V Grade Polyvinyl Chloride Insulated and Sheathed Cables; VVF). 12 pp.

Junction Boxes (Cont.)
—Cables (Electric)—Plastic Insulated
BSI BS 6220-83. 1983 Junction Boxes for Use in Electrical Installations with Rated Voltages Not Exceeding 250V. 14 pp.
—Cables (Electric)—Ships
MOD UK NES 514-81. Guide to Cable Entry, Termination and Junction Components for Equipment Issue 2 (01.81). 101 pp.
MOD UK NES 514-92. Guide to Cable Entry, Termination and Junction Components for Equipment Issue 3 (04.92). 79 pp.
—Ducts
CNS C4243-87. Junction Boxes for Underfloor Duct (Jan)(6099).
—Electric Conduit Boxes—Fittings
CSA CAN/CSA-C22. 2 NO 85-M89. Rigid PVC Boxes and Fittings; (Gen Instr 1 Thru 2). 37 pp.
—Electrical Enclosures—Fittings
CSA CAN/CSA-C22. 2 NO 85-M89. Rigid PVC Boxes and Fittings; (Gen Instr 1 Thru 2). 37 pp.
—Hinged Covers
CSA C22.2 NO 40-M1989. Cutout, Junction, and Pull Boxes; (Gen Instr 1 Thru 2). 46 pp.
CSA 896 Bull. Electrical Bulletin 896 September 18, 1972 to C22.2 NO 40. 1 p.
—Internal
DIN ENGL 47600 Pt 2-74. Cast Metal Joint Boxes for Power Cables up to 10 kV; Internal Joint Boxes (Oct). 2 pp.
—Mines
DIN VDE 0279-82. Accessories for Underground Mining Cables; Joints Uo/U = 0.6/1 kV (VDE Specification) (Oct). 11 pp.
—Protective
DIN ENGL 47600 Pt 1-74. Cast Metal Joint Boxes for Power Cables up to 10 kV; Protective Joint Boxes (Oct). 3 pp.
—Ships
CNS F5056-87. Marine Joint Boxes (Watertight Type) (Jul)(10164).
CNS F5057-87. Marine Joint Boxes (Non-Watertight Type) (Jul)(10165).
JIS F 8821-83. Marine Watertight Type Joint Boxes.
JIS F 8822-90. Non-Watertight Type Joint Boxes for Marine Use.
MOD UK NES 514-81. Guide to Cable Entry, Termination and Junction Components for Equipment Issue 2 (01.81). 101 pp.
MOD UK NES 514-92. Guide to Cable Entry, Termination and Junction Components for Equipment Issue 3 (04.92). 79 pp.
—Transformers
BSI BS 2562-79. 1979 Cable Boxes for Transformers and Reactors. 63 pp.
BSI BS 6436-84. 1984 Ground Mounted Distribution Transformers for Cable Box or Unit Substation Connection. 11 pp.

Junction FETs
Use: JFETs

Junction Field Effect Transistors
Use: JFETs

Junction Transistors
Scope Note: Use a more specific term See: NPN Transistors; Phototransistors; Planar Transistors; PNP Transistors; Semiconductor Devices; Thyristors; Transistors

Junctions (Cables Electric)
Use: Cable Junctions

Juniper Berries
ISO 7377-84. Juniper Berries (Juniperus Communis Linnaeus)—Specification First Edition. 5 pp.

Juniper Berry Oil
See Also: Essential Oils
—Quality Assurance
ISO 8897-91. Oil of Juniper Berry (Juniperus Communis Linnaeus) First Edition. 8 pp.

Juniper Oil
Use: Juniper Berry Oil

Junkers JV52/3M
See Also: Aircraft

INDUSTRY STANDARDS

Junkers JV52/3M *(Cont.)*
—Accidents
CAA. Junkers JU52/3M (World Airline Accident Summary). 1 p.

Jute
Scope Note: For products made from jute, see the specific products *See Also:* Fabrics; Hemp; Sisal
CEN PREN 766-92. Sacks for the Transport of Food Aid—Sacks Made of Jute Fabric. 8 pp.
CNS L3020-62. Method of Test for Jute and Ambria Hemp Fibers (Jul)(1407)(R 1973).
CNS L4005-88. Bast Fiber Bag (Sep)(1388).
CNS L4008-62. Jute Fiber (Tentative) (Jul)(1405)(R 1973).
CNS L4018-62. Long Fiber of Jute (Tentative) (Jul)(2049) (R 1973).
MOD UK DSTAN 83-25-01. Cloths, Jute, and Mixtures Containing Jute Issue 2; Amendment 1. 13 pp.

—Oil Content
BSI BS 3845-90. 1990 Determination of Added Oil Content of Jute Yarn, Rove and Fabric. 8 pp.

Kanamycin Content Analysis
—Animal Feed
CNS N4130-84. Method of Test for Feed Additives: Determination of Kanamycin (Nov)(11136).

Kaolin
Use For: China Clay *See Also:* Refractory Materials
JIS K 8746-61. Kaolin.
MOD UK DSTAN 96-2-90. Kaolin Types 1 and 2 Issue 2. 14 pp.
MOD UK DEF-118. Kaolin, Sizes 120, 200, 240 and 300 (Superseded by Def Stan 96-2).
MOD UK M 2002/59. Examination of Kaolin (Withdrawn).

—Content Analysis
MOD UK M 2002/59. Examination of Kaolin (Withdrawn).

—Ignition Loss
MOD UK M 2002/59. Examination of Kaolin (Withdrawn).

—Rubber—Compounding Ingredients
ISO 5795 Pt 1-88. Rubber Compounding Ingredients—Kaolin Clay—Part 1: Methods of Test (Excluding Tests in Rubber) First Edition. 15 pp.

—Sampling
CNS R3049-69. Method of Sampling for China Clays (Jan)(2885). 1 p.

—Solubility
MOD UK M 2002/59. Examination of Kaolin (Withdrawn).

Kapok Oil
See Also: Oils; Vegetable Oils
CNS K5109-81. Crude Kapok Oil (Sep)(7420).
CNS N5185-88. Kapok Oil (Edible) (May)(7923). 2 pp.

Kappa Number
—Chemical Pulps
CPPA G.8U-77. Rapid Determination of Kappa Number. 2 pp.

—Pulps
BSI BS 4498-82. 1982 Determination of the Kappa Number of Pulp (Degree of Delignification). 8 pp.
CNS P3038-91. Method of Test for Kappa Number of Pulp (Jun)(5470). 4 pp.
CPPA G.18-84. Kappa Number of Pulp (Supersedes G.17H). 2 pp.
DIN ENGL 54357-78. Testing of Pulp; Determination of the Kappa Number (Aug). 4 pp.
ISO 302-81. Pulps—Determination of Kappa Number First Edition. 6 pp.
JIS P 8211-76. Testing Method for Kappa Number of Pulp (R 1984). 10 pp.
SAA AS 1301.P201 M-86. Methods of Test for Pulp and Paper (Metric Units)—Part P201m: Kappa Number of Pulp (This is a Joint Standard with SANZ NZS 1301).
SNZ NZS/AS 1301. P201M-86. Methods of Test for Pulp and Paper Kappa Number of Pulp (This is a Joint Standard with SAA AS 1301.P201M). 4 pp.

Karl Fischer Method
See Also: Conductimetric Method; Volumetric Analysis; Water Content Analysis
BSI BS 2511-70. 1970 Methods for the Determination of Water (Karl Fischer Method). 36 pp.
CNS K6424-78. Determination of Water by the Karl Fischer Method (Jul)(4446).
ISO 760-78. Determination of Water—Karl Fischer Method (General Method) First Edition. 15 pp.

Karl Fischer Method *(Cont.)*
—Aluminum Fluoride
BSI BS 4993: Part 5-80. 1980 Methods of Test for Aluminium Fluoride for Industrial Use Part 5: Determination of Moisture Content (Karl Fischer Method). 9 pp.

—Ammonia
BSI BS 4431-89. 1989 Methods of Sampling and Test for Liquefied Anhydrous Ammonia. 16 pp.
ISO 7105-85. Liquefied Anhydrous Ammonia for Industrial Use—Determination of Water Content—Karl Fischer Method First Edition. 6 pp.

—Ammonium Nitrate
BSI BS 4267: Part 6-87. 1987 Ammonium Nitrate Part 6: Method for Determination of Water Content. 4 pp.
ISO 5791-78. Ammonium Nitrate for Industrial Use—Determination of Water Content—Karl Fischer Method First Edition. 4 pp.

—Anhydrous Hydrogen Fluoride
BSI BS 5365: Part 3-79. (WITHDRAWN) 1979 Methods of Sampling and Test for Anhyrous Hydrogen Fluoride for Industrial Use Part 3: Determination of Water Content (Karl Fischer Method). 7 pp.
ISO 3699-76. Anhydrous Hydrogen Fluoride for Industrial Use—Determination of Water Content—Karl Fischer Method First Edition. 8 pp.

—Crude Oils
JIS K 2275-89. Testing Methods for Water Content of Crude Oil and Petroleum Products. 36 pp.

—Cryolite
BSI BS 5050: Part 9-80. 1980 Methods of Test for Cryolite Part 9: Determination of Moisture Content (Karl Fischer Method). 10 pp.

—Dielectrics
BSI BS 6725-86. 1986 Determination of Water in Liquid Dielectrics by Automatic Coulometric Karl Fischer Titration. 16 pp.

—Electrical Insulating Liquids
CENELEC HD 487 S1-88. Determination of Water in Insulating Liquids by Automatic Coulometric Karl Fischer Titration. 3 pp.
IEC 814-85. Determination of Water in Insulating Liquids by Automatic Coulometric Karl Fischer Titration First Edition. 27 pp.

—Fats
BSI BS 684: Sec 2.1-76. 1976 Methods of Analysis of Fats and Fatty Oils Part 2: Other Methods Section 2.1: Determination of Water by the (Karl Fischer). 6 pp.

—Glycerol
BSI BS 5711: Part 8-79. 1979 Methods of Sampling and Test for Glycerol Part 8: Determination of Water Content: Karl Fischer Method. 2 pp.
ISO 2097-72. Glycerols for Industrial Use—Determination of Water Content—Karl Fischer Method First Edition. 4 pp.

—Ink Removers
MOD UK M 2003/59. Examination of Solvent, Cleaning.

—Iron Ores
BSI BS 7020: Part 2-88. 1988 Analysis of Iron Ores Part 2: Method for the Determination of Hygroscopic Moisture in Analytical Samples. 15 pp.
BSI BS 7020: Part 3-88. 1988 Analysis of Iron Ores Part 3: Method for the Determination of Combined Water Content. 12 pp.
CNS M3017-80. Methods for Determination of Combined Water in Iron Ores (Mar)(5379).
ISO 2596-84. Iron Ores—Determination of Hygroscopic Moisture in Analytical Samples—Gravimetric and Karl Fischer Methods Third Edition. 9 pp.
ISO 7335-87. Iron Ores—Determination of Combined Water Content—Karl Fischer Titrimetric Method First Edition. 12 pp.
JIS M 8211-83. Method for Determination of Combined Water in Iron Ores. 16 pp.

—Manganese Ores
JIS M 8231-82. Method for Determination of Combined Water in Manganese Ores. 12 pp.

—Mineral Oils
DIN ENGL 51777 Pt 2-74. Testing of Mineral Oil Hydrocarbons and Solvents; Determination of Water Content According to Karl Fischer; Indirect Method (Sept). 5 pp.

—Olefins
ISO 6191-81. Light Olefins for Industrial Use—Determination of Traces of Water—Karl Fischer Method First Edition. 10 pp.

Karl Fischer Method *(Cont.)*
—Petroleum Products
CGSB 31-GP-0A METH 48.5-62. Methods of Testing Corrosion-Prevention Materials and Processes Water by Electrometric Titration with Karl Fischer Reagent. 3 pp.
JIS K 2275-89. Testing Methods for Water Content of Crude Oil and Petroleum Products. 36 pp.

—Phenols
ISO 1897 Pt I-77. Phenol, o-Cresol, m-Cresol, p-Cresol, Cresylic Acid and Xylenols for Industrial Use—Methods of Test—Part I: General First Edition. 4 pp.

—Solvents
DIN ENGL 51777 Pt 2-74. Testing of Mineral Oil Hydrocarbons and Solvents; Determination of Water Content According to Karl Fischer; Indirect Method (Sept). 5 pp.

—Starches
ISO 5381-83. Starch Hydrolysis Products—Determination of Water Content—Modified Karl Fischer Method First Edition. 8 pp.

—Surfactants
BSI BS 6829: Sec 1.5-90. 1990 Analysis of Surface Active Agents (Raw Materials) Part 1: General Methods Section 1.5: Methods for Determination of Water Content. 9 pp.
ISO 4317-91. Surface-Active Agents and Detergents—Determination of Water Content—Karl Fischer Method Second Edition. 9 pp.

—Urea
CNS K6564-80. Determination of Water Content of Urea for Industrial Use by Karl Fischer Method (Aug)(6209).
ISO 2753-73. Urea for Industrial Use—Determination of Water Content—Karl Fischer Method First Edition. 3 pp.

KATAKANA Character Sets
Use: Optical Character Recognition—Japanese Character Sets—KATAKANA

Kelp
Use: Algae

Kerbs
Use: Curbs

Kerma Rate Meters
See Also: Counting Rate Meters; Exposure Rate Meters; Radiation Detection and Measurement

—Gamma Rays
IEC 1017 Pt 1-91. Portable, Transportable or Installed X or Gamma Radiation Ratemeters for Environmental Monitoring Part 1: Ratemeters First Edition. 63 pp.

—Nuclear Reactors
IEC 1017 Pt 1-91. Portable, Transportable or Installed X or Gamma Radiation Ratemeters for Environmental Monitoring Part 1: Ratemeters First Edition. 63 pp.

—X-Rays
IEC 1017 Pt 1-91. Portable, Transportable or Installed X or Gamma Radiation Ratemeters for Environmental Monitoring Part 1: Ratemeters First Edition. 63 pp.

Kerosene
Use For: Lamp Oils *See Also:* Automotive Fuels; Fuel Oils; Gasoline; Jet Engine Fuels; Liquid Fuels; Petroleum Products
CGSB CAN/CGSB-3.3-M89. Kerosine. 9 pp.
CNS K5023-83. Kerosene (Dec)(1470). 2 pp.
JIS K 2203-92. Kerosine. 7 pp.
MOD UK DERD 2453-90. Aviation Turbine Fuel AVTUR/FS11 Issue 6. 20 pp.

—Burning Quality
BSI BS 2000: Part 10-93. 1993 Methods of Test for Petroleum and Its Products Part 10: Determination of Kerosine Burning Characteristics—24 Hour Method. 5 pp.
BSI BS 2000: Part 10-82. 1982 Petroleum and Its Products Part 10: Burning Test (24-Hour) for Kerosine. 7 pp.
CNS K6957-88. Method of Test for Burning Quality of Kerosine (Jul)(12377).

—Hoses
CNS K4064-82. Oil Discharge Rubber Hose (Jun)(9011).
JIS K 6343-82. Rubber Hoses for Oil Discharge. 7 pp.

INTERNATIONAL AND NON-U.S. NATIONAL STANDARDS
SUBJECT INDEX

Kerosene (Cont.)

—Jet Engine Fuels
CGSB CAN/CGSB-3.23-93. Aviation Turbine Fuel, Kerosine Type. 12 pp.

—Plastic Containers
JIS Z 1710-77. Blow Moulded Polyethylene Containers for Kerosine (R 1980). 11 pp.

—Rocket Fuels
MOD UK DERD 2495-01. Fuel: Rocket Engines: Kerosine Type Issue 1; Amendment 1. 5 pp.

—Smoke Point
BSI BS 2000: Part 57-93. 1993 Methods of Test for Petroleum and Its Products Part 57: Determination of Smoke Point. 5 pp.

BSI BS 2000: Part 57-82. 1982 Petroleum and Its Products Part 57: Smoke Point of Kerosine. 7 pp.

—Solubility—Cleaning Agents
CGSB 31-GP-0A METH 54.1-57. Methods of Testing Corrosion-Prevention Materials and Processes Solubility (in Kerosine). 1 p.

—Stoves
CNS B4002-81. Kerosene Stove (Apr)(707).

—Thiol Content—Potentiometric Analysis
CNS K6547-80. Method of Test for Mercaptan Sulfur in Gasoline, Kerosine, Aviation Turbine and Distillate Fuels (Potentiometric Method) (Aug)(6192).

Kerosine
Use: Kerosene

Ketchup
Use: Catsup

Ketohexamethylene
Use: Cyclohexanone

Ketones
See Also: Acetone; Acetophenones; Amyl Ethyl Ketone; Anthraquinones; Benzophenones; Butyl Methyl Ketone; Camphor; Cyclohexanone; 5,5-Dimethyl-1,3-Cyclohexanedione; 5-(4-Dimethylaminobenzilidene)-2-Thioxo-4-Thiazolidinone; Diphenylbenzidine; Ethyl Isoamyl Ketone; Methyl Ethyl Ketone; Methyl Isoamyl Ketone; Methyl Isobutyl Ketone; Solvents

—Boiling Points
ISO 4626-80. Volatile Organic Liquids—Determination of Boiling Range of Organic Solvents Used as Raw Materials First Edition. 14 pp.

—Purity
CNS K6476-80. Method of Test for Purity of Aldehydes and Ketones (Apr)(5459).

Kettles
See Also: Appliances; Autoclaves; Cooking Appliances; Food Service Equipment; Steam Jacketed Kettles

BSI BS 6557-85. 1985 Kettles. 8 pp.

—Aluminum
CNS H3011-86. Aluminium Kettles, Drum Shape (Sep)(753).

CNS H3012-86. Aluminium Kettles, Flat Bottom, Shallow Shape (Jul)(754).

CNS H3035-86. Aluminium Kettles, Flat Bottom, Deep Shape (Jul)(2702).

—Electric
BSI BS 3999: Part 1-93. 1993 Methods of Measuring the Performance of Household Electrical Appliances Part 1: Electric Kettles and Jugs (IEC 530: 1975). 8 pp.

BSI BS 3999: Part 1-66. 1966 Amd 3 Methods of Measuring the Performance of Household Electrical Appliances Part 1: Electric Kettles. 14 pp.

CSA CAN/CSA-C22. 2 NO 64-M91. Household Cooking and Liquid-Heating Appliances; (Gen Instr 1 Thru 2). 91 pp.

—Enamel
CNS R2098-79. Enamel Kettle (Nov)(5042). 1 p.

—Stainless Steel
CNS S1196-86. Stainless Steel Kettle (Feb)(11513).

Key Cocks
Use: Plug Valves

Keyboard Switches
Use: Keyswitches

Keyboards
See Also: Consoles; Control Characters; Data Processing Equipment; Function Keys; Graphic Characters; Keypads; Keyswitches; Office Machines; Typewriters

BSI BS 5231-75. 1975 Principles Governing the Positioning of Control Keys on Keyboards of Office Machines and Data Processing Equipment. 9 pp.

BSI BS 5448-77. 1977 Keyboard Layouts for Numeric Applications on Office Machines and Data Processing Equipment. 6 pp.

BSI BS 5959-80. 1980 Key Numbering System and Layout Charts for Keyboards on Office Machines. 12 pp.

CNS C5151-83. Office Machines Basic Arrangement for the Alphanumeric Section of Keyboards Operated with Both Hands (Apr)(10154).

CNS C5191-85. Office Machines and Data Processing Equipment—Principles Governing the Positioning of Control Keys on Keyboards (Mar)(11215).

CNS C5192-85. Office Machines and Data Processing Equipment—Keyboard Layouts for Numeric Applications (Mar)(11216).

CNS C6291-87. Method of Test for Keyboard (Aug)(12045).

IEC DIS 9995 Pt 1-91. Information Technology—Keyboard Layouts for Text and Office Systems—Part 1: General Principles Governing Keyboard Layouts ***CD-ROM ONLY***. 18 pp.

ISO 2126-75. Office Machines—Basic Arrangement for the Alphanumeric Section of Keyboards Operated with Both Hands First Edition. 5 pp.

ISO 3243-75. Keyboards for Countries Whose Languages Have Alphabetic Extenders—Guidelines for Harmonization First Edition. 6 pp.

ISO 3244-84. Office Machines and Data Processing Equipment—Principles Governing the Positioning of Control Keys on Keyboards Second Edition. 6 pp.

ISO 4169-79. Office Machines—Keyboards—Key Numbering System and Layout Charts First Edition. 11 pp.

ISO DIS 9995 Pt 1-91. Information Technology—Keyboard Layouts for Text and Office Systems—Part 1: General Principles Governing Keyboard Layouts ***CD-ROM ONLY***. 18 pp.

JTC1 2126-75. Office Machines—Basic Arrangement for the Alphanumeric Section of Keyboards Operated with Both Hands First Edition. 5 pp.

JTC1 3243-75. Keyboards for Countries Whose Languages Have Alphabetic Extenders—Guidelines for Harmonization First Edition. 6 pp.

JTC1 3244-84. Office Machines and Data Processing Equipment—Principles Governing the Positioning of Control Keys on Keyboards Second Edition. 6 pp.

JTC1 3791-76. Office Machines and Data Processing Equipment—Keyboard Layouts for Numeric Applications First Edition. 5 pp.

JTC1 4169-79. Office Machines—Keyboards—Key Numbering System and Layout Charts First Edition. 11 pp.

JTC1 DIS9995 Pt 8-93. Information Technology—Keyboard Layouts for Text and Office Systems—Part 8: Allocation of Letters to the Keys of a Numeric Keyboard ***CD-ROM ONLY***. 6 pp.

OSI ISO 2126-75. Office Machines—Basic Arrangements for the Alphanumeric Section of Keyboards Operated with Both Hands. 5 pp.

OSI ISO 3243-75. Keyboards for Countries Whose Languages Have Alphabetic Extenders—Guidelines for Harmonization. 6 pp.

OSI ISO 3244-84. Office Machines and Data Processing Equipment—Principles Governing the Positioning of Control Keys on Keyboards. 6 pp.

OSI ISO 4169-79. Office Machines—Keyboards—Key Numbering System and Layout Charts. 10 pp.

SAA AS 1659-74. Alphanumeric Section of Keyboards Operated with Both Hands Amdt 1 December 1976. 4 pp.

—Adding Machines
BSI BS 5478: Part 1-77. 1977 Calculators and Adding Mahchines Part 1:Numeric Section of Ten-Key Keyboards for Calculators and Adding Machines. 3 pp.

BSI BS 5478: Part 2-77. 1977 Calculators and Adding Machines Part 2: Layout of Function Keys on Adding Machines. 4 pp.

CNS C5119-81. 10-Key Keyboard for Adding and Calculating Machines (Jul)(7649).

ISO 1092-74. Adding Machines and Calculating Machines—Numeric Section of Ten-Key Keyboards First Edition. 3 pp.

ISO 3792-76. Adding Machines—Layout of Function Keyboard First Edition. 4 pp.

JIS B 9517-90. Keyboards Arrangement for Calculating Machines with Ten-Key. 6 pp.

JTC1 1092-74. Adding Machines and Calculating Machines—Numeric Section of Ten-Key Keyboards First Edition. 3 pp.

JTC1 3792-76. Adding Machines—Layout of Function Keyboard First Edition. 4 pp.

OSI ISO 1092-74. Adding Machines and Calculating Machines—Numeric Sectionb of Ten-Key Keyboards. 3 pp.

Keyboards (Cont.)

—Adding Machines (Cont.)
OSI ISO 3792-76. Adding Machines—Layout of Function Keyboard. 4 pp.

—Alphanumeric Character Sets
IEC DIS 9995 Pt 2-91. Information Technology—Keyboard Layouts for Text and Office Systems—Part 2: Alphanumeric Section ***CD-ROM ONLY***. 20 pp.

IEC DIS 9995 Pt 3-91. Information Technology—Keyboard Layouts for Text and Office Systems—Part 3: Common Secondary Layout of the Alphanumeric Zone of the Alphanumeric Section ***CD-ROM ONLY***. 9 pp.

ISO DIS 9995 Pt 2-91. Information Technology—Keyboard Layouts for Text and Office Systems—Part 2: Alphanumeric Section ***CD-ROM ONLY***. 20 pp.

ISO DIS 9995 Pt 3-91. Information Technology—Keyboard Layouts for Text and Office Systems—Part 3: Common Secondary Layout of the Alphanumeric Zone of the Alphanumeric Section ***CD-ROM ONLY***. 9 pp.

—Automatic Control Equipment—Household
BSI BS 7245-89. 1989 Numeric Keyboard for Home Electronic Systems (HES). 4 pp.

CENELEC EN 60 948-90. Numeric Keyboard for Home Electronic Systems (HES). 5 pp.

CNS C7209-90. Numeric Keyboard for Home Electronic Systems (May)(12710).

IEC 948-88. Numeric Keyboard for Home Electronic Systems (HES) First Edition. 12 pp.

JTC1 948-88. Numeric Keyboard for Home Electronic Systems (HES) First Edition. 12 pp.

—Calculators
BSI BS 5478: Part 1-77. 1977 Calculators and Adding Mahchines Part 1:Numeric Section of Ten-Key Keyboards for Calculators and Adding Machines. 3 pp.

CNS C5119-81. 10-Key Keyboard for Adding and Calculating Machines (Jul)(7649).

CNS C5195-85. Calculating Machines—Numeric Section of Ten-Key Keyboards (Mar)(11219).

CNS C5196-85. Calculating Machines—Keytop and Printed or Displayed Symbols (Mar)(11220).

ISO 1092-74. Adding Machines and Calculating Machines—Numeric Section of Ten-Key Keyboards First Edition. 3 pp.

JIS B 9517-90. Keyboards Arrangement for Calculating Machines with Ten-Key. 6 pp.

JTC1 1092-74. Adding Machines and Calculating Machines—Numeric Section of Ten-Key Keyboards First Edition. 3 pp.

OSI ISO 1092-74. Adding Machines and Calculating Machines—Numeric Sectionb of Ten-Key Keyboards. 3 pp.

—Cash Registers
JIS B 9511-77. Registration Symbols for Keyboard Arrangements for Cash Register.

—Character Sets—International System of Units
BSI PD 6462-72. 1972 Modification of Keyboards to Include Symbols for SI Units. 4 pp.

—Data Processing
BSI BS 4822-80. 1980 Keyboard Arrangements of the Graphic Characters of the United Kingdom 7-Bit Data Code, for Data Processing. 8 pp.

CNS C5150-83. Keyboard for International Information Processing Interchange Using 7-Bit Coded Character Set (Alphanumeric Area) (Apr)(10153).

CSA CAN/CSA-Z243.200-92. Canadian Keyboard Standard for the English and French Languages; (Gen Instr 1). 43 pp.

ECMA ECMA 115-86. Common Secondary Keyboard Layout for Languages Using a Latin Alphabet. 15 pp.

ISO 2530-75. Keyboard for International Information Processing Interchange Using the ISO 7-Bit Coded Character Set—Alphanumeric Area First Edition. 9 pp.

JIS X 6002-80. Keyboard Layout for Information Processing Using the JIS 7 Bit Coded Character Set.

JTC1 2530-75. Keyboard for International Information Processing Interchange Using the ISO 7-Bit Coded Character Set—Alphanumeric Area First Edition. 9 pp.

OSI ISO 2530-75. Keyboard for International Information Processing Interchange Using the ISO 7-Bit Coded Character Set-Alphanumeric Area. 8 pp.

—Display Terminals
JIS X 6041-87. CRT Display and Key Boards Units for Business Use.

INDUSTRY STANDARDS

INTERNATIONAL AND NON-U.S. NATIONAL STANDARDS
SUBJECT INDEX

Keyboards (Cont.)
—Display Tubes—Ergonomics
BSI BS 7179: Part 4-90. 1990 Ergonomics of Design and Use of Visual Display Terminals (VDTs) in Offices Part 4: Keyboards. 16 pp.

—Numeric Character Sets
IEC DIS 9995 Pt 4-91. Information Technology—Keyboard Layouts for Text and Office Systems—Part 4: Numeric Section ***CD-ROM ONLY***. 11 pp.
ISO 3791-76. Office Machines and Data Processing Equipment—Keyboard Layouts for Numeric Applications First Edition. 5 pp.
ISO DIS 9995 Pt 4-91. Information Technology—Keyboard Layouts for Text and Office Systems—Part 4: Numeric Section ***CD-ROM ONLY***. 11 pp.
OSI ISO 3791-76. Office Machines and Data Processing Equipment—Keyboard Layouts for Numeric Applications. 5 pp.

—Numeric Character Sets—Allocation of Letters
IEC DIS 9995 Pt 8-93. Information Technology—Keyboard Layouts for Text and Office Systems—Part 8: Allocation of Letters to the Keys of a Numeric Keyboard ***CD-ROM ONLY***. 6 pp.
ISO DIS 9995 Pt 8-93. Information Technology—Keyboard Layouts for Text and Office Systems—Part 8: Allocation of Letters to the Keys of a Numeric Keyboard ***CD-ROM ONLY***. 6 pp.

—Protocols—Integrated Services Digital Networks
CENELEC PRETS 300 122-90. Integrated Services Digital Network (ISDN); Generic Keyboard Protocol for the Support of Supplementary Services Digital Subscriber Signaling One (DSSI) Protocol (T/S 46-32A). 8 pp.
CENELEC PRETS 300 122-91. Integrated Services Digital Network (ISDN); Generic Keypad Protocol for the Support of Supplementary Services Digital Subscriber Signalling No. One (DSS1) Protocol. 10 pp.
CENELEC ETS 300 122-92. Integrated Services Digital Network (ISDN); Generic Keypad Protocol for the Support of Supplementary Services Digital Subscriber Signalling System No. One (DSS1) Protocol. 10 pp.
ETSI ETS 300 122-92. Integrated Services Digital Network (ISDN); Generic Keypad Protocol for the Support of Supplementary Services Digital Subscriber Signalling System No. One (DSS1) Protocol. 10 pp.

—Ships—Format
JIS F 9011-92. Ships Keyboard—Format.

—Typewriters
BSI BS 2481: Part 1-82. 1982 Typewriters Part 1: Keyboard Arrangements. 4 pp.
BSI BS 2481: Part 2-82. 1982 Typewriters Part 2: Function Key Symbols. 4 pp.
BSI BS 2481: Part 3-82. 1982 Typewriters Part 3: Layout of Printing and Function Keys. 4 pp.
JIS B 9509-64. KATAKANA-Alphabet Keyboards Arrangement for Typewriters.

—Word Processing
BSI BS 7292-90. 1990 Layout and Operation of Keyboards for Multiple Latin-Alphabet Languages. 18 pp.
CSA CAN/CSA-Z243.200-92. Canadian Keyboard Standard for the English and French Languages; (Gen Instr 1). 43 pp.
IEC DIS 9995 Pt 5-91. Information Technology—Keyboard Layouts for Text and Office Systems—Part 5: Editing Section ***CD-ROM ONLY***. 11 pp.
ISO 8884-89. Information Processing—Text and Office Systems—Keyboards for Multiple Latin-Alphabet Languages—Layout and Operation First Edition. 14 pp.
ISO DIS 9995 Pt 5-91. Information Technology—Keyboard Layouts for Text and Office Systems—Part 5: Editing Section ***CD-ROM ONLY***. 11 pp.
JIS X 6003-84. Keyboard Layout for Japanese Text Processing.
JIS X 6004-86. Basic Keyboard Layout for Japanese Test Processing Using Kana-Kanji Translation Method.
JTC1 8884-89. Information Processing—Text and Office Systems—Keyboards for Multiple Latin-Alphabet Languages—Layout and Operation First Edition. 14 pp.
OSI ISO 8884-89. Information Processing-Text and Office Systems-Keyboards for Multiple Latin-Alphabet Languages—Layout and Operation. 14 pp.
OSI ISO DIS 8884-87. Information Processing—Text and Office Systems—Keyboards for Multiple Latin-Alphabet Languages—Layout and Operation (Superseded by ISO 8884). 18 pp.
OSI ISO DIS 8884-86. Information Processing—Text and Office Systems—Keyboards for Multiple Latin-Alphabet Languages Using Four Levels (Superseded by ISO 8884 1987). 17 pp.
SAA AS 3970-91. Information Processing—Text and Office Systems—Keyboards for Multiple Latin-Alphabet Languages —Layout and Operation (ISO 8884:1989) (in Professional Package 26A). 10 pp.

Keyhole Saws
See Also: Compass Saws; Hand Tools; Saws
MOD UK DSTAN 51-11: Part 13-76. Hand Tools, General Purpose Part 13: Saw, Keyhole and Blade, Keyhole Saw Issue 1. 10 pp.

—Blades
MOD UK DSTAN 51-11: Part 13-76. Hand Tools, General Purpose Part 13: Saw, Keyhole and Blade, Keyhole Saw Issue 1. 10 pp.

—Woodworking
CNS B3141-59. Keyhole Saw for Carpenter (Apr)(1106)(R 1970).

Keypads
Use For: Switchpads See Also: Keyboards

—Signaling Protocols—Supplementary Services—ISDN—DSS1
ETSI PRETS 300 122-90. Integrated Services Digital Network (ISDN); Generic Keypad Protocol for the Support of Supplementary Services—Digital Subscriber Signalling One (DSS1) Protocol (T/S 46-32A). 8 pp.

—Signaling Protocols—Supplementary Services—PTN
CENELEC PRETS 300 190-91. Generic Stimulus Procedure for the Control of Supplementary Services Using the Keypad Protocol at the S Reference Point. 24 pp.
CENELEC PRETS 300 190-92. Private Telecommunications Network (PTN); Signalling at the S-Reference Point Generic Keypad Protocol for the Support of Supplementary Services. 23 pp.
CENELEC ETS 300 190-92. Private Telecommunication Network (PTN); Signalling at the S-Reference Point Generic Keypad Protocol for the Support of Supplementary Servics. 22 pp.
ECMA ECMA 156-91. Generic Stimulus Procedure for the Control of Supplementary Services Using the Keypad Protocol at the S Reference Point. 21 pp.
ETSI ETS 300 190-92. Private Telecommunication Network (PTN); Signalling at the S-Reference Point Generic Keypad Protocol for the Support of Supplementary Servics. 22 pp.
ETSI PRETS 300 190-92. Private Telecommunications Network (PTN); Signalling at the S-Reference Point Generic Keypad Protocol for the Support of Supplementary Services. 23 pp.
ETSI PRETS 300 190-91. Generic Stimulus Procedure for the Control of Supplementary Services Using the Keypad Protocol at the S Reference Point. 24 pp.

Keys (Engineering)
Use: Machine Keys

Keys (Locks)
See Also: Keyways; Locks (Security)

—Glossaries
BSI BS 3827: Part 1-64. (WITHDRAWN) 1964 Glossary of Terms Relating to Builders' Hardware Part 1: Locks (Including Locks and Latches in One Case) (Superseded by BS 6100: Subsection 1.3.6: 1991). 34 pp.

Keys (Machinery)
Use: Machine Keys

Keys (Switches)
Use: Keyswitches

Keyswitches
Use For: Keyboard Switches; Keys (Switches); Keytops See Also: Electric Switches; Keyboards; Pushbutton Switches; Switches

—Preferred Products List
CECC CECC MUAHAG Vol 8 IS 4-91. Preferred Products List; Switches (En, Ge, Fr). 29 pp.

Keytops
Use: Keyswitches

Keyways
See Also: Axles; Keys (Locks); Machine Keys; Pulleys; Shaft Ends; Shafts (Machine Elements); Splines; Woodruff Keyways

CSA CAN3-B232-75. Keys, Keyseats and Keyways (R 1989). 22 pp.

—Broaches
JIS B 4238-88. Keyway Broaches. 11 pp.

—Machine Tools
CNS B2631-82. Parallel Keys for Machine Tools, Dimension and Application (Jan)(8193).

—Milling Cutters
CNS B3411-81. Tools for Metal-Working; Bores, Keys and Driving Features for Tools with Parallel Bore and with 1 in 30 Taper Bore (Apr)(7195).

—Parallel
BSI BS 46: Part 1-58. (OBSOLESCENT) 1958 Amd 3 Keys and Keyways and Taper Pins Part 1: Keys and Keyways. 36 pp.
BSI BS 4235: Part 1-72. 1972 Amd 2 Metric Keys and Keyways Part 1: Paralell and Taper Keys. 27 pp.
ISO R773-69. Rectangular or Square Parallel Keys and Their Corresponding Keyways (Dimensions in Millimetres) First Edition. 6 pp.
ISO 2491-74. Thin Parallel Keys and Their Corresponding Keyways (Dimensions in Millimetres) First Edition. 5 pp.
SNZ NZS/BS 4235: Part 1-72. Specification for Metric Keys and Keyways Part 1: Parallel and Taper Keys. 28 pp.

—Shaft Ends
BSI BS 46: Part 1-58. (OBSOLESCENT) 1958 Amd 3 Keys and Keyways and Taper Pins Part 1: Keys and Keyways. 36 pp.

—Ships—Tailshafts
BSI BS 46: Part 1-58. (OBSOLESCENT) 1958 Amd 3 Keys and Keyways and Taper Pins Part 1: Keys and Keyways. 36 pp.

—Sliding
CNS B3034-83. Screw and Hole for Sliding Key (Aug)(171).
JIS B 1303-76. Sliding Keys and Their Corresponding Keyways. 8 pp.

—Sunk
JIS B 1301-76. Sunk Keys and Their Corresponding Keyways. 10 pp.

—Tangential
BSI BS 46: Part 1-58. (OBSOLESCENT) 1958 Amd 3 Keys and Keyways and Taper Pins Part 1: Keys and Keyways. 36 pp.
DIN ENGL 268-74. Tangential Keys and Tangential Keyways for Alternating Shock Loads (Sept). 3 pp.
DIN ENGL 271-74. Tangential Keys and Tangential Keyways for Constant Loads (Sept). 3 pp.
ISO 3117-77. Tangential Keys and Keyways First Edition. 6 pp.

—Taper
BSI BS 46: Part 1-58. (OBSOLESCENT) 1958 Amd 3 Keys and Keyways and Taper Pins Part 1: Keys and Keyways. 36 pp.
BSI BS 4235: Part 1-72. 1972 Amd 2 Metric Keys and Keyways Part 1: Paralell and Taper Keys. 27 pp.
DIN ENGL 138-55. Metal-Cutting Tools; Bores, Keyways and Driving Features for Tools with Parallel Bore and with 1 in 30 Taper Bore (Sept). 4 pp.
DIN ENGL 6886-67. Stressed-Type Fastenings with Taper Action; Taper Keys; Keyways; Dimensions and Application (Dec). 2 pp.
DIN ENGL 6887-68. Stressed-Type Fastenings with Taper Action; Taper Keys with Gib-Head; Keyways; Dimensions and Application (Apr). 2 pp.
ISO R774-69. Taper Keys with or Without Gib Head and Their Corresponding Keyways (Dimensions in Millimetres) First Edition. 7 pp.
ISO 2492-74. Thin Taper Keys with or Without Gib Head and Their Corresponding Keyways (Dimensions in Millimetres) First Edition. 5 pp.
SNZ NZS/BS 4235: Part 1-72. Specification for Metric Keys and Keyways Part 1: Parallel and Taper Keys. 28 pp.

Keyword Indexing
Use: Subject Indexing

Kickstands
—Motorcycles
SNZ NZS/AS 2538-85. Vehicle Support Stands (This is a Joint Standard with SAA AS 2538). 10 pp.

Kidney Beans
See Also: Beans; Green Beans; Seeds

INTERNATIONAL AND NON-U.S. NATIONAL STANDARDS
SUBJECT INDEX

Kidney Beans (Cont.)
—Grading
CNS N1047-63. Grades of Kidney Beans (for Export) (Mar) (2132)(R 1973). 2 pp.

Killed Steels
See Also: Ferroalloys; Steels

—Carbon Steels—Cold Worked
DIN ENGL 1654 Pt 2-89. Cold Heading and Cold Extruding Steels; Technical Delivery Conditions for Killed Unalloyed Steels Not Intended for Heat Treatment (Oct). 4 pp.

Killion Harmonics
Use: Frequencies

Kilns
See Also: Furnaces; Refractory Materials

—Calcium Oxide—Heat Balance
CNS R3124-85. Method for Calculating Heat Balance of Kiln and Furnace for Lime (May) (11274).
JIS R 0305-91. Heat Balancing of Kiln and Furnace for Lime. 25 pp.

—Electric—Safety
CSA C22.2 NO 221-M1986. Electrically Heated Hobby and Educational Type Kilns (R 1992); (Gen Instr 1). 30 pp.

—Firebricks
BSI BS 3056: Part 3-86. 1986 Amd 1 Sizes of Refractory Bricks Part 3: Bricks for Rotary Cement Kilns. 17 pp.
CNS R1019-86. Shape and Dimension of Refractory Bricks for Rotary Kiln (Jul)(11636).
DIN ENGL 1082 Pt 4-89. Refractory Materials; Refractory Arch Bricks for Use in Rotary Kilns; Dimensions (Feb). 2 pp.
ISO 5417-86. Refractory Bricks for Use in Rotary Kilns—Dimensions First Edition. 4 pp.
JIS R 2103-83. Shape and Dimension of Refractory Bricks for Rotary Kiln. 9 pp.

—Firebricks—Identification Systems
ISO 9205-88. Refractory Bricks for Use in Rotary Kilns—Hot-Face Identification Marking First Edition. 7 pp.

—Oil Burners
BSI BS 5410: Part 3-76. 1976 Amd 1 Code of Practice for Oil Firing Part 3: Installations for Furnaces, Kilns, Ovens and Other Industrial Purposes. 40 pp.

—Periodic—Heat Balance
CNS R3120-85. Method for Calculating Heat Balance of Firing Periodic Kiln for Pottery and Refractory (Apr)(11249).
JIS R 0302-91. Heat Balancing of Pottery and Refractory Firing Periodic Kiln. 44 pp.

—Rotary
BSI BS 3056: Part 3-86. 1986 Amd 1 Sizes of Refractory Bricks Part 3: Bricks for Rotary Cement Kilns. 17 pp.
CNS R1019-86. Shape and Dimension of Refractory Bricks for Rotary Kiln (Jul)(11636).
ISO 5417-86. Refractory Bricks for Use in Rotary Kilns—Dimensions First Edition. 4 pp.
ISO 9205-88. Refractory Bricks for Use in Rotary Kilns—Hot-Face Identification Marking First Edition. 7 pp.
JIS R 2103-83. Shape and Dimension of Refractory Bricks for Rotary Kiln. 9 pp.

—Rotary—Firebricks
DIN ENGL 1082 Pt 4-89. Refractory Materials; Refractory Arch Bricks for Use in Rotary Kilns; Dimensions (Feb). 2 pp.

—Rotary—Heat Balance
CNS R3118-84. Method for Calculating Heat Balancing of Cement Rotary Kiln (Oct)(11081).
JIS R 0303-91. Heat Balancing of Cement Rotary Kiln. 46 pp.

—Tunnel—Heat Balance
CNS R3119-85. Method for Calculating Heat Balancing of Pottery and Refractory Firing Tunnel Kiln (Jan)(11191).
JIS R 0301-91. Heat Balancing of Pottery and Refractory Firing Tunnel Kiln. 47 pp.

Kinematic Diagrams
See Also: Diagrams

—Symbols
ISO 3952 Pt 1-81. Kinematic Diagrams—Graphical Symbols—Part 1 First Edition. 21 pp.
ISO 3952 Pt 2-81. Kinematic Diagrams—Graphical Symbols—Part 2 First Edition. 15 pp.
ISO 3952 Pt 3-79. Kinematic Diagrams—Graphical Symbols—Part 3 First Edition. 12 pp.
ISO 3952 Pt 4-84. Kinematic Diagrams—Graphical Symbols—Part 4 First Edition. 9 pp.

Kinematic Quantities, Cutting Tools
Use: Cutting Tools—Geometric and Kinematic Quantities

Kinematic Viscosity
See Also: Dynamic Viscosity; Intrinsic Viscosity; Rheological Properties; Saybolt Viscosity; Viscosity
CNS K6328-87. Method of Test for Calculation of Viscosity Index from Kinematic Viscosity at 40 and 100 Degree C (Jul)(3389).
DIN ENGL 51562 Pt 1-83. Viscometry; Determination of Kinematic Viscosity Using the Standard Design Ubbelohde Viscometer (Jan). 4 pp.
DIN ENGL 51562 Pt 2-88. Viscometry; Determination of Kinematic Viscosity Using the Ubbelohde Microviscometer (Dec). 3 pp.
DIN ENGL 51562 Pt 3-85. Viscometry; Determination of Kinematic Viscosity Using the Ubbelohde Viscometer; Viscosity Relative Increment at Short Flow Times (May). 5 pp.

—Crude Oil
JIS K 2283-83. Testing Methods for Kinematic Viscosity and Calculating Method for Viscosity Index of Crude Oil and Petroleum Products. 75 pp.

—Liquid Fuels
DIN ENGL 51561-78. Testing of Mineral Oils, Liquid Fuels and Related Liquids; Measurement of Viscosity Using the Vogel-Ossag Viscometer; Temperature Range: Approximately 10 to 150 Deg. C (Dec). 4 pp.
DIN ENGL 51569-78. Testing of Mineral Oils, Liquid Fuels and Related Liquids; Measurement of Viscosity Using the Vogel-Ossag Viscometer; Temperature Range:-55 to Approximately +10 Deg. C (Dec). 5 pp.

—Liquids
BSI BS 188-77. 1977 Determination of the Viscosity of Liquids. 18 pp.
CNS K6329-72. Method of Test for Viscosities of Transparent and Opaque Liquids (Kinematic and Dynamic Viscosities) (Oct)(3390).

—Lubricants
BSI BS 4231-92. 1992 Viscosity Grades of Industrial Liquid Lubricants (ISO 3448: 1992). 12 pp.
BSI BS 4231-82. 1982 Viscosity Grades of Industrial Liquid Lubricants. 6 pp.
ISO 3448-92. Industrial Liquid Lubricants—ISO Viscosity Classification Second Edition. 9 pp.
JIS K 2001-83. Viscosity Classification for Industrial Liquid Lubricants. 6 pp.

—Mineral Oils
DIN ENGL 51561-78. Testing of Mineral Oils, Liquid Fuels and Related Liquids; Measurement of Viscosity Using the Vogel-Ossag Viscometer; Temperature Range: Approximately 10 to 150 Deg. C (Dec). 4 pp.
DIN ENGL 51569-78. Testing of Mineral Oils, Liquid Fuels and Related Liquids; Measurement of Viscosity Using the Vogel-Ossag Viscometer; Temperature Range:-55 to Approximately +10 Deg. C (Dec). 5 pp.

—Petroleum Products
BSI BS 2000: Part 71-90. 1990 Petroleum and Its Products Part 71: Kinematic Viscosity of Transparent and Opaque Liquids and Calculation of Dynamic Viscosity. 16 pp.
BSI BS 4459-78. 1978 Method for Calculating Viscosity Index of Petroleum Products and Lubricants from Kinematic Viscosity. 10 pp.
ISO 2909-81. Petroleum Products—Calculation of Viscosity Index from Kinematic Viscosity Second Edition. 8 pp.
ISO 3104-76. Petroleum Products—Transparent and Opaque Liquids—Determination of Kinematic Viscosity and Calculation of Dynamic Viscosity First Edition; (Erratum—Aug 1976) (Erratum—July 1978). 9 pp.
JIS K 2283-83. Testing Methods for Kinematic Viscosity and Calculating Method for Viscosity Index of Crude Oil and Petroleum Products. 75 pp.

Kinematics
—Gears—Glossaries
CNS B1182-84. Definitions and Denominations for Glossary of Gear Terms — Kinematics (May)(5724).

Kingpins
Use For: Hinge Pins; Pivot Pins; Swivel Pins
See Also: Fasteners; Hinges; Pins
BSI BS AU 1B-77. (WITHDRAWN) 1977 Dimensions of 50 mm Fifth Wheel King Pin of Semi-Trailers (Superseded by BS AU 1C: 1989). 4 pp.
BSI BS AU 1C-89. 1989 Dimensions of '50' Fifth Wheel King Pin for Semi-Trailers. 4 pp.
BSI BS AU 2A-70. 1970 Dimensions of 3 1/2-4 1/2 Inch Diameter Fifth Wheel King Pin for Use with Extra Heavy Duty Semi-Trailers. 2 pp.
MOD UK DSTAN 25-4-78. Kingpins, Fifth Wheel Issue 2. 12 pp.

—Aerospace
DIN ENGL LN 9373-84. Aerospace; Pins, Hook Form, for Hinges (Feb). 2 pp.

—Aircraft—Parachutes
SBAC AS 479 ISSUE 2(I). Pivot Pin Anti-Spin Parachute Jettison Slip. 1 p.

—Ball Thrust Bearings—Automotive
CNS D2018-80. Kingpin Thrust Ball Bearing for Automobiles (Mar)(5324).

—Fifth Wheels
NATO STANAG 4009 Pt1 ED1 AMD4-88. Fifth Wheel Attachments (Part I—NATO Kingpin). 6 pp.
SNZ NZS 5451-89. Coupling Devices for Articulated Vehicles—Fifth Wheel Kingpins. 12 pp.

—Semitrailers—Fifth Wheels
NATO STANAG 4009 Pt2 ED1 AMD3-58. Fifth Wheel Attachments (Part II) Location and Height of Fifth Wheel Attachments on Tractors and Semi-Trailers, Class I. 7 pp.
NATO STANAG 4009 Pt3 ED1 AMD7-59. Fifth Wheel Attachments (Part III—Location and Height of Fifth Wheel Attachments on Tractors and Semi-Trailers, Class II). 18 pp.
NATO STANAG 4009 Pt4 ED1 AMD5-60. Fifth Wheel Kingpin (for Semi-Trailers of a Gross Weight from 22 to 72 Tonnes). 12 pp.

—Tractor Trucks—Fifth Wheels
NATO STANAG 4009 Pt2 ED1 AMD3-58. Fifth Wheel Attachments (Part II) Location and Height of Fifth Wheel Attachments on Tractors and Semi-Trailers, Class I. 7 pp.
NATO STANAG 4009 Pt3 ED1 AMD7-59. Fifth Wheel Attachments (Part III—Location and Height of Fifth Wheel Attachments on Tractors and Semi-Trailers, Class II). 18 pp.

Kipp Gas Generators
Use: Gas Generators

Kissing Gates
Use: Gates (Barriers)—Pedestrian

Kitasamycin Content Analysis
Use: Leucomycins Content Analysis

Kitchen Cabinets
See Also: Cabinets (Furniture); Kitchens
BSI BS 1195: Part 1-73. (WITHDRAWN) 1973 Amd 1 Kitchen Fitments and Equipment Part 1: Imperial Units with Metric Equivalents (Superseded by BS 6222: Part 1: 1982, Part 2: 1992 & Part 3: 1988). 41 pp.
BSI BS 1195: Part 2-72. (WITHDRAWN) 1972 Amd 1 Kitchen Fitments and Equipment Part 2: Metric Units. 36 pp.
CSA CAN3-A278-M82. Kitchen Cabinets and Bathroom Vanities; (Gen Instr 1 Thru 3). 23 pp.

—Safety
CEN PREN 1153-93. Kitchen Furniture—Safety Requirements and Test Methods for Built-in and Free Standing Kitchen Cabinets and Work Tops. 16 pp.

Kitchen Equipment
Use: Food Preparation Equipment; Food Processing Equipment

Kitchen Fitments
Use: Food Processing Equipment

Kitchen Trays
See Also: Trays

—Fiberboard—Tomatoes
BSI BS 3789-75. 1975 Amd 1 Non-Returnable Fibreboard Tomato Trays. 8 pp.

—Wooden—Tomatoes
BSI BS 2892-74. 1974 Non-Returnable Wooden Trays for Tomatoes. 8 pp.

Kitchens
See Also: Kitchen Cabinets; Sinks; Sinks (Plumbing Fixtures)

Kitchens (Cont.)
JIS A 4420-92. Kitchen Equipment—System Components. 21 pp.

—Design
SAA AS 1351.1-74. Spaces in Dwellings (Metric Units)—Part 1: Kitchens. 4 pp.

—Ergonomics
ISO 3055-85. Kitchen Equipment—Coordinating Sizes Second Edition. 18 pp.

—Modular Construction
BSI BS 5957-80. 1980 Recommendations for Modules for the Coordinating Dimensions of Catering Equipment Using Containers to BS 4874. 11 pp.
CNS A1022-78. Modular Coordinating Sizes of Kitchen Unit for Dwellings (Jul) (4440).
JIS A 0017-92. Kitchen Equipment—Coordinating Sizes. 11 pp.

—Panels—Decorative—Sizes
CEN PREN 1116-93. Kitchen Furniture—Co-Ordinating Sizes for Kitchen Furniture and Kitchen Appliances. 8 pp.

—Units—Fitted
BSI BS 6222: Part 4-88. 1988 Domestic Kitchen Equipment Part 4: Code of Practice for Protection, Storage and Installation of Fitted Kitchen Units. 24 pp.

—Units—Safety
CEN PREN 1153-93. Kitchen Furniture—Safety Requirements and Test Methods for Built-in and Free Standing Kitchen Cabinets and Work Tops. 16 pp.

—Units—Wall
JIS A 4411-84. Wall Type Kitchen Unit for Dwellings.

—Work Surfaces—Durability
BSI BS 6222: Part 3-93. 1993 Domestic Kitchen Equipment Part 3: Specification for Performance Requirements for Durability of Surface Finish and Adhesion of Surfacing and Edging Materials (SUPERSEDES BS 1195: PT 1: 1973 & PT 2: 1972). 15 pp.
BSI BS 6222: Part 3-88. 1988 Domestic Kitchen Equipment Part 3: Performance Requirements for Durability of Surface Finish. 12 pp.

—Work Surfaces—Safety
CEN PREN 1153-93. Kitchen Furniture—Safety Requirements and Test Methods for Built-in and Free Standing Kitchen Cabinets and Work Tops. 16 pp.

Kiwifruit
See Also: Fruits
EC EEC/305/92-92. Commission Regulation Amending Regulation (EEC) No 410/90 Laying down Quality Standards for Kiwi Fruit. 2 pp.

—Canned
SNZ NZS 8407-85. Canned Sliced Kiwifruit. 9 pp.
SNZ NZS 8407C-85. Canned Sliced Kiwifruit. 1 p.

Kjeldahis Method
Use: Kjeldahl Method

Kjeldahl Method
See Also: Amido Black Method

—Agricultural Products
ISO 1871-75. Agricultural Food Products—General Directions for the Determination of Nitrogen by the Kjeldahl Method First Edition. 7 pp.

—Animal Feed
BSI BS 5766: Part 4-81. 1981 Amd 1 Methods for Analysis of Animal Feeding Stuffs Part 4: Determination of Nitrogen Content and Calculation of Crude Protein Content. 9 pp.
BSI BS 5766: Part 4-02. 1981 Amd 2 Methods for Analysis of Animal Feeding Stuffs Part 4: Determination of Nitrogen Content and Calculation of Crude Protein Content (ISO 5983: 1979) (AMD 7725) June 15, 1993 (W). 10 pp.
ISO 5983-79. Animal Feeding Stuffs—Determination of Nitrogen Content and Calculation of Crude Protein Content First Edition. 6 pp.

—Coal
ISO 333-83. Coal—Determination of Nitrogen—Semi-Micro Kjeldahl Method Third Edition. 6 pp.

—Coke
CNS M3147-84. Method for Determination of Nitrogen of Coal and Coke (Kjeldahl Method) (Mar)(10828).

—Crude Oil
JIS K 2609-90. Crude Petroleum and Petroleum Products—Determination of Nitrogen Content. 50 pp.

—Dairy Products
SAA AS 2300.1.2. 1-91. Methods of Chemical and Physical Testing for the Dairying Industry—Part 1: General Methods and Principles—Part 1.2.1: Determination of Nitrogen—Reference Kjeldahl Method. 7 pp.

—Elastomers
DIN ENGL 53621 Pt 3-76. Testing of Rubber and Elastomers; Quantitative Determination of Polymers; Determination of the Acrylonitrile Content and Calculation of the Acrylonitrile-Butadiene Copolymer Content (July). 3 pp.

—Fertilizers
CNS N4027-70. Method of Test for Compound Fertilizer (May)(3077). 6 pp.

—Petroleum Products
JIS K 2609-90. Crude Petroleum and Petroleum Products—Determination of Nitrogen Content. 50 pp.

—Phenolic Resins
ISO 8988-89. Plastics—Phenolic Resins—Determination of Hexamethylenetetramine Content First Edition. 6 pp.

—Photographic Processing Wastes
ISO 6851-87. Photography—Processing Waste—Determination of Total Amino Nitrogen—Microdiffusion Kjeldahl Method First Edition; (ANSI PH4.47-1991). 7 pp.
ISO 6853-87. Photography—Processing Waste—Determination of Ammoniacal Nitrogen Content—Microdiffusion Method First Edition; (ANSI PH4.48-1990). 6 pp.

—Reagents
CNS K7009-82. Chemical Reagent (Kjeldahl Nitrogen Determination Method) (Jun)(1509).

—Rubber
DIN ENGL 53621 Pt 3-76. Testing of Rubber and Elastomers; Quantitative Determination of Polymers; Determination of the Acrylonitrile Content and Calculation of the Acrylonitrile-Butadiene Copolymer Content (July). 3 pp.

—Soy Sauce
CNS N6008-88. Methods of Test for Soy Sauce (Nov)(955). 7 pp.

—Starches
ISO 3188-78. Starches and Derived Products—Determination of Nitrogen Content by the Kjeldahl Method—Titrimetric Method First Edition. 5 pp.
ISO 5378-78. Starches and Derived Products—Determination of Nitrogen Content by the Kjeldahl Method—Spectrophotometric Method First Edition. 5 pp.

—Testing Equipment
BSI BS 1428: Part B1-64. 1964 Microchemical Apparatus: Group B: Apparatus for the Determination of Elements by Other Than Combustion Methods Part B1: Nitrogen Determination Apparatus (Micro-Kjeldahl). 16 pp.
BSI BS 1428: Part B2-60. 1960 Microchemical Apparatus: Group B: Apparatus for the Determination of Elements by Other Than Combustion Methods Part B2: Ammonia Distillation Apparatus (Markham). 8 pp.
BSI BS 1428: Part B3-64. 1964 Microchemical Apparatus: Group B: Apparatus for the Determination of Elements by Other Than Combustion Methods Part B3: Nitrogen Determination Apparatus (Non-Transference Micro-Kjeldahl). 16 pp.

—Water
ISO 5663-84. Water Quality—Determination of Kjeldahl Nitrogen—Method After Mineralization with Selenium First Edition. 6 pp.

Klystron Amplifiers
Use For: Amplifier Klystrons See Also: Amplifiers; Microwave Amplifiers; Power Amplifiers

—Preferred Products List
CECC CECC MUAHAG Vol 12 IS 1-90. Preferred Products List; Microwave Components (En, Fr, Ge). 76 pp.

Klystron Oscillators
Use: Velocity Modulated Oscillators

Klystrons
See Also: Electron Tubes; Microwave Tubes; Reflex Klystrons
MOD UK DSTAN 59-60: Part 7-81. Valves, Electronic (Electronic Tubes) (Listed on EPIC Database) Part 7: Klystrons Issue 1. 22 pp.
MOD UK DSTAN 59-60: 07/004-83. Valve, Electronic, Klystron Issue 1. 19 pp.
MOD UK DSTAN 59-60: 90/011-70. Electronic Valve, Klystron Issue 1. 6 pp.
MOD UK DSTAN 59-60: 90/057-01. Valve, Electronic, Klystron Issue 1; Amendment 1. 14 pp.
MOD UK DSTAN 59-60: 90/059-72. Valve Electronic, Klystron Issue 1. 7 pp.
MOD UK DSTAN 59-60: 90/073-73. Valve Electronic, Klystron Issue 1. 13 pp.

—Electrical Properties
IEC 235 Pt 5-72. Measurement of the Electrical Properties of Microwave Tubes Part 5: Low-Power Oscillator Klystrons First Edition. 15 pp.
IEC 235 Pt 6-72. Measurement of the Electrical Properties of Microwave Tubes Part 6: High-Power Klystrons First Edition. 21 pp.

—Quality Assurance
BSI BS 9035-79. (WITHDRAWN) 1979 Klystrons of Assessed Quality: Generic Data and Methods of Test. 53 pp.
BSI BS 9036-79. (WITHDRAWN) 1979 Detail Specification for Amplifier Klystrons of Assessed Quality. Full or Basic Assessment Level. 16 pp.
BSI BS 9037-79. (WITHDRAWN) 1979 Oscillator Klystrons of Assessed Quality. Full or Basic Assessment Level. 15 pp.

Knapsack Sprayers
See Also: Sprayers
BSI BS 4115-93. 1993 Compression Knapsack Sprayers (L). 12 pp.
BSI BS 4115: Part 1-67. (OBSOLESCENT) 1967 Compression Knapsack Sprayers: Non-Pressure Retaining Type. 29 pp.

—Pesticides
BSI BS 7411-01. 1991 Amd 1 Lever Operated Knapsack Sprayers (AMD 7431) December 15, 1992. 10 pp.
BSI BS 7411-91. 1991 Lever Operated Knapsack Sprayers (L). 9 pp.

Knebworth Corvus Chaff System
See Also: Ammunition; Weapons Systems

—Interchangeable—Naval
NATO STANAG 1350 ED 1 AMD 0-86. Knebworth Corvus Chaff System Interchangeable Within NATO Naval Forces. 15 pp.

Knee Prostheses
See Also: Joint Prostheses; Leg Prostheses
BSI BS 2574: Part 2-85. 1985 Amd 1 Knee-Ankle-Foot Orthoses Part 2: Metal Knee Joints (AMD 6511) July 31, 1990. 9 pp.
BSI DD 157-87. (WITHDRAWN) 1987 Methods for the Determination of Mech-anical Properties of Metal Knee Joint and Side Member Assemblies for Use in Lower Limb Orthoses (Superseded by BS 2574: Part 3: 1990). 14 pp.
JIS T 9213-85. Artificial Knees.
JIS T 9216-91. Metallic Knee Joints for Lower Extremity Orthoses. 11 pp.

—Classification
BSI BS 7251: Part 2-90. 1990 Orthopaedic Joint Prostheses Part 2: Classification and Designation of Dimensions for Knee Joint Prostheses. 11 pp.
ISO 7207 Pt 1-85. Implants for Surgery—Partial and Total Knee Joint Prostheses—Part 1: Classification, Definitions and Designation of Dimensions First Edition. 10 pp.

—Designations
BSI BS 7251: Part 2-90. 1990 Orthopaedic Joint Prostheses Part 2: Classification and Designation of Dimensions for Knee Joint Prostheses. 11 pp.
ISO 7207 Pt 1-85. Implants for Surgery—Partial and Total Knee Joint Prostheses—Part 1: Classification, Definitions and Designation of Dimensions First Edition. 10 pp.

—Glossaries
ISO 7207 Pt 1-85. Implants for Surgery—Partial and Total Knee Joint Prostheses—Part 1: Classification, Definitions and Designation of Dimensions First Edition. 10 pp.

—Mechanical Properties
BSI BS 2574: Part 3-90. 1990 Metal Knee Joints Part 3: Method for the Determination of Mechanical Properties of Metal Knee Joint and Side Member Assemblies. 21 pp.

Knife Files
See Also: Files (Tools)
CNS B3173-78. Knife File for Smith (Mar)(1201). 1 p.

Knife Grinders
See Also: Grinders; Knives; Machine Tools
CNS B1301-83. Nominal Dimension of Veneer Knife Grinders (Mar)(10053).
CNS B1302-83. Nominal Dimension of Scraper Knife Grinders (Mar)(10054).
CNS B7244-86. Test Code for Accuracy of Knife Grinders (Oct)(10224).
CNS B7261-86. Test Code for Performance of Knife Grinders (Oct)(10544).
JIS B 6543-91. Veneer Knife Grinders—Test and Inspection Methods. 13 pp.

—Inspection
JIS B 6516-89. Test Methods for Performance and Accuracy of Planer Knife Grinding Machines. 13 pp.

Knife Switches
See Also: Air Switches; Electric Switches
CNS C4016-84. Open Type Knife Switch (May)(696). 6 pp.
CNS C4044-84. Knife Switches (with Cover) (May)(1488).
JIS C 8308-88. Knife Switches with Cover. 17 pp.

—Rolling Stock
JIS E 4902-91. Knife Switches for Railway Rolling Stock.

Knife Testing
See Also: Brittleness Testing; Toughness

—Coatings
CGSB 1-GP-71 METH 135.1-74. Methods of Testing Paints and Pigments Adhesion Knife Test. 1 p.

Knitted Fabrics
Use: Fabrics

Knitting
See Also: Knitting Machinery; Knitting Needles; Weaving

—Glossaries
JIS L 0211-86. Glossary of Terms Used in Textile Industry (Knitting Section).

—Stitches—Symbols
JIS L 0200-76. Method of Representation for Knitting Stitch (R 1984). 14 pp.
JIS L 0201-78. Letter Symbols for Knitting Stitch (R 1983). 32 pp.

Knitting Beams
See Also: Beams (Textile Machinery); Knitting Machinery; Textile Machinery

—Ribbon
ISO 8116 Pt 6-90. Textile Machinery and Accessories—Beams for Winding—Part 6: Beams for Ribbon Weaving and Ribbon Knitting—Terminology and Main Dimensions First Edition. 8 pp.

—Ribbon—Glossaries
ISO 8116 Pt 6-90. Textile Machinery and Accessories—Beams for Winding—Part 6: Beams for Ribbon Weaving and Ribbon Knitting—Terminology and Main Dimensions First Edition. 8 pp.

Knitting Machinery
Use For: Knitting Machines *See Also:* Knitting; Knitting Beams; Knitting Needles; Textile Machinery; Warp Knitting Machines
ISO 8117-86. Textile Machinery—Knitting Machines—Nominal Diameters of Circular Machines First Edition. 3 pp.
JIS B 9091-78. Hand Knitting Machines (R 1988). 7 pp.
JIS L 6602-72. Circular Interlock Knitting Machines.

—Glossaries
BSI BS 6875-87. 1987 Terms for and Classification of Knitting Machines. 18 pp.
ISO 7839-84. Textile Machinery and Accessories—Knitting Machines—Classification and Vocabulary First Edition. 18 pp.
JIS L 0202-78. Nomenclature of Hand Knitting Machine.
JIS L 0307-85. Glossary of Terms Used in Knitting, Braiding and Related Machinery.

—Name Plates
BSI BS ISO 8121-86. 1986 Textile Machinery—Knitting Machines—Nameplate Information (L). 7 pp.
ISO 8121-86. Textile Machinery—Knitting Machines—Nameplate Information First Edition; (Corrigendum 1-1992). 5 pp.

Knitting Machinery (Cont.)
—Points
JIS L 1802-57. Knitting Needle and Point. 17 pp.

Knitting Machines
Use: Knitting Machinery

Knitting Needles
See Also: Knitting; Knitting Machinery
ISO 8122-88. Textile Machinery—Knitting Machines—Number of Needles for Circular Knitting Machines of Large Nominal Diameter First Edition. 5 pp.
ISO 8188-86. Textile Machinery and Accessories—Pitches of Knitting Machine Needles First Edition. 6 pp.
JIS B 9092-78. Latch Needles for Hand Knitting Machines (R 1983). 6 pp.
JIS B 9093-78. Crochet Needles for Hand Knitting Machines (R 1983). 5 pp.
JIS L 1802-57. Knitting Needle and Point. 17 pp.

—Knitting Machinery—Glossaries
ISO 8119 Pt 1-89. Textile Machinery and Accessories—Needles for Knitting Machines—Terminology—Part 1: Latch-Type Needles First Edition; (Corrected and Reprinted -1989). 25 pp.
ISO 8119 Pt 2-89. Textile Machinery and Accessories—Needles for Knitting Machines—Terminology—Part 2: Bearded Needles First Edition. 15 pp.
ISO 8119 Pt 3-92. Textile Machinery and Accessories—Needles for Knitting Machines—Terminology—Part 3: Compound Needles First Edition. 23 pp.

Knives
Scope Note: Use a more specific term
See Also: Circular Knives; Cutting Tools; Hand Tools; Knife Grinders; Paper Cutters; Putty Knives; Surgical Knives; Tools

—Kitchen
CNS S2037-80. Method of Test for Chopper & Knife for Kitchen (Aug)(6267).

—Lathes
JIS B 4708-72. Veneer Lathe Knives (R 1987). 7 pp.

—Meat Slicers
CSA C22.2 NO 195-M1987. Motor Operated Food Processing Appliances (Household and Commercial); (Gen Instr 1 Thru 2). 55 pp.

—Oilstones
CNS S2038-80. Method of Test for Oil Stone for Sharping Knives (Aug)(6268).

Knobs
Use For: Control Knobs *See Also:* Doorknobs; Handles; Levers
CNS Z7071-81. Knobs Form B (May)(7460).
CNS Z7072-81. Knobs Form C and D (May)(7461).
CNS Z7073-81. Knobs Form H (May)(7462).
CNS Z7074-81. Knobs Form J (May)(7463).
CNS Z7075-81. Knobs Form Riveting on A (May)(7464).
CNS Z7076-81. Knobs Form Screwing in B (May)(7465).
CNS Z7077-81. Blech Knob with Fastened Plate (May)(7466).
JIS B 2602-82. Handles. 11 pp.
MOD UK DEF-5221-61. Knobs, Finger. 42 pp.

—Aircraft
SBAC AS 1909 ISSUE 6. Knob (General Purpose). 1 p.
SBAC AS 2311 ISSUE 3. Rotating Knob. 1 p.
SBAC AS 2313 ISSUE 3. Rotating Knob. 1 p.
SBAC AS 6387 ISSUE 3. Knob (General Purpose).
SBAC AS 6389 ISSUE 2. Knob (Flap Control).
SBAC AS 6396 ISSUE 2. Knob (Mixture Control).
SBAC AS 6536 ISSUE 1. Outer Handle Assembly—Single Handle Lock.

—Aircraft—Air Brakes
SBAC AS 6384 ISSUE 2. Knob (Airbrake Control).

—Aircraft—Air Shutters
SBAC AS 6388 ISSUE 2. Knob (Air Shutter Control).

—Aircraft—Air Valves
SBAC AS 1910 ISSUE 6. Knob (Air Shutter Control). 1 p.

—Aircraft—Arrester Hooks
SBAC AS 3195 ISSUE 4. Knob (Arrestor Hook Control). 1 p.
SBAC AS 6394 ISSUE 2. Knob (Arrestor Hook Control).

—Aircraft—Canopies—Winding Gear
SBAC AS 3197 ISSUE 3. Knob (Canopy Winding Gear). 1 p.

Knobs (Cont.)
—Aircraft—Carburetors
SBAC AS 6386 ISSUE 2. Knob (Carburetter Hot/Cold Air Control).

—Aircraft—Cockpits
SBAC AS 6397 ISSUE 2. Knob (For Cockpit Heating Control).

—Aircraft—Emergency Oxygen
SBAC AS 4559 ISSUE 1. Control, Knob for Emergency Oxygen. 1 p.

—Aircraft—Filler Caps
SBAC AS 2409 ISSUE 7. Knob (Filler Cap). 1 p.

—Aircraft—Flaps
SBAC AS 1911 ISSUE 7. Knob (Flap Control). 1 p.
SBAC AS 1913 ISSUE 6. Knob (Undercarriage Control). 1 p.

—Aircraft—Fuel Cocks
SBAC AS 3196 ISSUE 1. Fuel Cock, Knob (Low Pressure). 1 p.
SBAC AS 6395 ISSUE 1. Fuel Cock, Knob (Low Pressure).

—Aircraft Fuel Systems
SBAC RS 434 ISSUE 2. Knob for High Pressure Fuel Control, (Incorporating Re-Light Switch).

—Aircraft—Gliders
SBAC AS 6393 ISSUE 2. Knob (Glider Release Control).

—Aircraft—Parachute Brakes
SBAC RS 437 ISSUE 3. Knob Braking Parachute Control.

—Aircraft—Propellers
SBAC AS 1914 ISSUE 6. Knob (Propeller Control). 1 p.
SBAC AS 6391 ISSUE 2. Knob (Propeller Control).

—Aircraft—Reciprocating Engines—Fuel Air Mixture
SBAC AS 3199 ISSUE 3. Knob (Mixture Control). 1 p.

—Aircraft—Screws
SBAC AS 2314 ISSUE 2. Screw for Rotating Knob. 1 p.

—Aircraft—Superchargers
SBAC AS 1915 ISSUE 6. Knob (Supercharger Control). 1 p.
SBAC AS 6392 ISSUE 2. Knob (Supercharger Control).

—Aircraft—Throttles
SBAC AS 1907 ISSUE 6. Knob (Carburetter Hot/Cold Air Control). 1 p.
SBAC AS 1908 ISSUE 6. Knob (Throttle Control). 1 p.
SBAC AS 6385 ISSUE 2. Knob (Throttle Control).

—Aircraft—Undercarriage
SBAC AS 6390 ISSUE 3. Knob (Undercarriage Control).

—Ball
CNS B2395-80. Ball Knobs (Jan)(5109).
DIN ENGL 319-78. Control Elements; Ball Knobs (Dec). 3 pp.

—Furniture Hardware—Glossaries
ISO 8555 Pt 7-87. Hardware for Furniture—Terms for Furniture Fittings—Part 7: Handles, Knobs, Escutcheons, Escutcheon Insets First Edition. 8 pp.

—Gliders—Release
SBAC AS 3194 ISSUE 3. Knob (Glider Release Control). 1 p.

—Spring
CNS B2657-82. Double Wing Type Spring Knob (May)(8774).
CNS B2658-82. Single Wing Type Spring Knob (May)(8775).

Knobs (Handles)
Use: Handles

Knock Rating
See Also: Cetane Number; Octane Number

—Fuels—Automotive
CNS K6910-87. Method of Test for Knock Characteristics of Motor Fuels by the Research Method (Jul)(12011).

Knockouts
See Also: Press Dies

INTERNATIONAL AND NON-U.S. NATIONAL STANDARDS
SUBJECT INDEX

Knockouts (Cont.)
—Press Dies
CNS B3366-87. Knockout Equipment for Press Dies (Apr)(5992).
JIS B 5011-55. Knock-out Equipment for Press Dies.

Knoop Hardness
Use: Knoop Hardness Testing

Knoop Hardness Testers
See Also: Hardness Testers; Knoop Hardness Testing
—Metals
ISO 4546-93. Metallic Materials—Hardness Test—Verification of Knoop Hardness Testing Machines First Edition. 7 pp.
—Metals—Blocks—Calibration
ISO 4547-93. Metallic Materials—Hardness Test—Calibration of Standardized Blocks to Be Used for Knoop Hardness Testing Machines First Edition. 6 pp.

Knoop Hardness Testing
See Also: Indentation Hardness Testing; Knoop Hardness Testers
CNS B6065-82. Micro Hardness Testing Machine for Vickers and Knoop Hardness (May)(8762).
—Glass
CNS Z9068-86. Method of Test for Knoop Hardness of Optical Glasses (Dec)(9939).
ISO 9385-90. Glass and Glass-Ceramics —Knoop Hardness Test First Ediiton. 7 pp.
—Glass Ceramics
ISO 9385-90. Glass and Glass-Ceramics —Knoop Hardness Test First Ediiton. 7 pp.
—Iron
JIS G 0563-93. Method of Measuring Surface Hardness for Nitrided Iron and Steel. 6 pp.
—Metal Coatings (Made From Metal)
BSI BS 5411: Part 6-81. 1981 Methods of Test for Metallic and Related Coatings Part 6: Vickers and Knoop Microhardness Tests. 11 pp.
ISO 4516-80. Metallic and Related Coatings—Vickers and Knoop Microhardness Tests First Edition. 9 pp.
—Metals
CNS B7189-83. Method of Test for Micro Hardness Testing Machines for Vickers and Knoop Hardness (Jan)(8763).
CNS Z8017-83. Method of Micro Hardness Test for Vickers and Knoop Hardness (Jul)(7094).
ISO 4545-93. Metallic Materials—Hardness Test—Knoop Test First Edition. 7 pp.
JIS Z 2251-92. Method of Knoop Hardness Test. 8 pp.
—Steels
JIS G 0563-93. Method of Measuring Surface Hardness for Nitrided Iron and Steel. 6 pp.
—Testers
JIS B 7734-91. Micro Hardness Testing Machines for Vickers and Knoop Hardness. 10 pp.

Knoop Indentation Testing
Use: Knoop Hardness Testing

KNOR
Use: Exclusive NOR Gates

Knotters
See Also: Paper Machines; Papermaking Equipment
—Sieve Analysis
CPPA D.4U-77. Coarse Screen Rejects from Sulphite Knotter Screens. 2 pp.

Knotting
See Also: Varnishes; Wood Coatings
BSI BS 1336-71. 1971 Knotting. 9 pp.

Knuckle Threads
See Also: Screw Threads
CNS B2087-78. Knuckle Thread (Profiles, Nominal Sizes) (May)(508).
DIN ENGL 76 Pt 3-77. Runouts; Undercuts for Trapezoidal Threads, Buttress Threads and Knuckle Threads and Other Threads of Coarse Pitch (Jan). 3 pp.
DIN ENGL 405 Pt 1-75. Knuckle Threads; Thread Profiles, Nominal Dimensions, Thread Series (Nov). 4 pp.
DIN ENGL 405 Pt 2-81. Knuckle Screw Threads; Deviations and Tolerances (Oct). 5 pp.
DIN ENGL 20400-51. Knuckle Threads with Large Load-Bearing Depth (Mar). 2 pp.

Knuckle Threads (Cont.)
—Lifting Hooks
CNS B2213-78. Lifting Hooks, Knuckle Threads (May)(4368).
—Profiles
CNS B2087-78. Knuckle Thread (Profiles, Nominal Sizes) (May)(508).
DIN ENGL 405 Pt 1-75. Knuckle Threads; Thread Profiles, Nominal Dimensions, Thread Series (Nov). 4 pp.
—Protective Masks
CNS B2088-55. Knuckle Thread for Poison Gas Mask (Sep)(509)(R 1970).
—Runouts
CNS B1066-78. Runout and Undercut for Metric Trapezoidal Screw Threads, Kunkle Screw Threads, Bultress Screw Threads and Other Screw Threads of Coarse Pitches (Mar)(4325).
—Sheet Steel
CNS B2212-78. Knuckle Thread for Steel Sheet Pieces (up to 0.5mm Thickness and Appropriated Coupling Dimensions Tolerances (May)(4367).
DIN ENGL 7273 Pt 1-70. Knuckle Threads for Steel Sheet Pieces up to 0.5 mm Thickness and Appropriated Couplings; Dimensions; Tolerances (July). 2 pp.
—Undercuts
CNS B1066-78. Runout and Undercut for Metric Trapezoidal Screw Threads, Kunkle Screw Threads, Bultress Screw Threads and Other Screw Threads of Coarse Pitches (Mar)(4325).

Knurled Head Screws
See Also: Screws
—Thumb
CNS B2172-80. Knurled Thumb Screws (May)(4246).
DIN ENGL 464-86. Knurled Thumb Screws (Sept). 4 pp.
—Thumb—Flat
CNS B2173-80. Flat Knurled Thumb Screws (May)(4247).
—Thumb—Slotted
CNS B2174-80. Slotted Knurled Thumb Screws (May)(4248).
—Thumb—Thin
DIN ENGL 653-86. Knurled Thin Thumb Screws (Sept). 4 pp.

Knurled Nuts
See Also: Nuts (Fasteners)
CNS B2267-78. Knurled Nuts (Aug)(4474).
DIN ENGL 467-86. Knurled Nuts (Sept). 3 pp.
DIN ENGL 6303-86. Knurled Nuts (Nov). 3 pp.
DIN ENGL 6324-71. Operating Elements for Clamping Devices; Survey (Jan). 2 pp.
—Collar Type
DIN ENGL 466-86. Knurled Nuts with Collar (Sept). 3 pp.
—Drawing Equipment
CNS B2364-84. Knurled Nuts for Drawing Instruments (Jul)(4731).
—Rings
CNS B2365-84. Knurled Nuts with Ring (Jul)(4732).
—Thin
CNS B2255-78. Knurled Nuts (Thin Type) (Aug)(4462).

Knurling
JIS B 0951-62. Knurling (R 1968). 4 pp.

Knurling Tools
Use For: Knurling Wheels See Also: Tools
BSI BS 1759-69. 1969 Knurling Wheels. 16 pp.
BSI BS 1759-01. 1969 Amd 1 Knurling Wheels (AMD 7812) October 15, 1993 (F). 18 pp.
CNS B3005-82. Knurling Wheels (Mar)(76).

Knurling Wheels
Use: Knurling Tools

Knurls
See Also: Handles
CNS B1005-82. Knurls (Mar)(75).
DIN ENGL 82-73. Knurls (Jan). 4 pp.

KOH Number
Use: Saponification Number

Kohlrausch Flasks
Use: Flasks—Volumetric

Korean Characters
—Romanization
BSI PD 6505-82. 1982 Romanization of Korean. 49 pp.

Kraft Liquors
Use For: Sulfate Liquors See Also: Black Liquors; Green Liquors; White Liquors
—Green—Chemical Analysis
CPPA J.12-84. Analysis of Sulphate Green and White Liquors. 6 pp.
—White—Chemical Analysis
CPPA J.12-84. Analysis of Sulphate Green and White Liquors. 6 pp.

Kraft Paper
See Also: Packaging Papers; Papers; Tissue Papers
CNS P2013-84. Kraft Paper (Oct)(1458). 2 pp.
JIS P 3401-92. Kraft Papers. 9 pp.
—Adhesive Tapes
CNS Z5021-72. Adhesive Tapes of Kraft Paper (Tentative) (Jun)(2389).
—Adhesive Tapes—Water Soluble
CNS Z7020-72. Water Soluble Kraft Paper Adhesive Tapes for Plywood Use (Tentative) (Oct)(2647).
—Ammunition
MOD UK DEF-99-A-64. Paper Kraft, Polythene Coated (Heat Sealable) (Reprinted January 1974, Incorporating Amendment No. 1). 6 pp.
MOD UK DEF-137-01. Paper, Laminated (Kraft-Polythene-Kraft) (Reprinted January 1963, Incorporating Correction); Amendment 1. 7 pp.
MOD UK TS 10084. Paper, Kraft Bleached (Withdrawn).
—Bags
CNS P2008-84. Cement Bag Paper (Jan)(1394). 2 pp.
CNS P2065-84. Sack Kraft Paper (Oct)(10759). 2 pp.
CNS Z6077-83. Method of Test for Drop Kraft Paper Sack (Oct)(10635).
JIS Z 0217-63. Drop Testing Method for Kraft Paper Sacks.
JIS Z 1505-75. Sewn Kraft Paper Sacks for Cement.
JIS Z 1509-90. Kraft Paper Sacks for Potato Starch. 12 pp.
JIS Z 1530-82. Sewn Kraft Paper Sacks for Sweet Potato Starch. 8 pp.
JIS Z 1531-76. General Rules for Sewn Kraft Paper Sacks (R 1985). 9 pp.
JIS Z 1532-90. General Rules for Pasted Kraft Paper Sacks. 18 pp.
—Bags—Glossaries
CNS Z5063-78. Glossary of Terms Used in Kraft Paper Sack Industry (Mar)(4340).
JIS Z 0102-90. Glossary of Terms Used in Kraft Paper Sack Industry.
—Covers
CGSB 9-GP-31MA-87. Paper, Golden Kraft, for Envelopes. 8 pp.
CGSB 9-GP-32MA-87. Paper, Kraft, for File Folders. 8 pp.
—Crepe
MOD UK DEF-1238-57. Paper, Kraft, Creped, Waxed (Reprinted June 1971, Incorporating Amendment No. 1). 3 pp.
MOD UK DEF-1252-A-64. Paper, Kraft, Creped (Anti-Bleed) (Reprinted May, 1970 Incorporating Amendments Nos. 1 and 2). 5 pp.
—Crepe—Waterproof
MOD UK DEF-1239-A-66. Paper, Creped, Kraft Union, Reinforced. 6 pp.
—Crepe—Wet Strength
MOD UK DEF-1247-01. Paper, Kraft, Creped, High Wet-Strength; Amendment 1. 5 pp.
—Cushioning
MOD UK DSTAN 81-110-93. Paper, Kraft, Embossed, Four Ply (for Cushioning or Space Filling) Issue 1. 13 pp.
MOD UK TS 10077. Paper Kraft Embossed, Four Ply (for Cushioning).
—Envelopes
CGSB 9-GP-31MA-87. Paper, Golden Kraft, for Envelopes. 8 pp.
—Extensible
JIS P 3412-76. Kraft Extensible Paper.
—File Folders
CGSB 9-GP-32MA-87. Paper, Kraft, for File Folders. 8 pp.

INTERNATIONAL AND NON-U.S. NATIONAL STANDARDS
SUBJECT INDEX
Laboratories

Kraft Paper *(Cont.)*
—Laminated
MOD UK DEF-137-01. Paper, Laminated (Kraft-Polythene-Kraft) (Reprinted January 1963, Incorporating Correction); Amendment 1. 7 pp.

—Packaging
MOD UK DSTAN 81-48-01. Paper, Kraft Union Types A and B Issue 1; Amendment 1. 13 pp.

—Tapes—Gummed
MOD UK DEF-1299-58. Tape, Kraft Paper, Gummed (Reprinted December 1965). 3 pp.
MOD UK TS 447. Tape, Kraft Paper, Gummed, Non-Corrosive (Withdrawn).

—Waterproof
MOD UK DSTAN 81-48-01. Paper, Kraft Union Types A and B Issue 1; Amendment 1. 13 pp.
MOD UK DEF-1240-A-66. Paper, Kraft Union (Superseded by Def Stan 81-48). 4 pp.
MOD UK TS 10154. Paper, Kraft, Waterproof.

—Waxed
MOD UK DEF-135-01. Paper, Kraft, Dry Paraffin-Waxed; Amendment 1. 5 pp.
MOD UK DEF-1238-57. Paper, Kraft, Creped, Waxed (Reprinted June 1971, Incorporating Amendment No. 1). 3 pp.

—Wrapping
CGSB 9-GP-5MA-87. Paper, Unbleached Kraft, Wrapping. 7 pp.
MOD UK DEF-1243-A-67. Paper, Wrapping, Kraft. 3 pp.

Kraft Pulps
Use For: Sulfate Pulps *See Also:* Chemical Pulps; Dissolving Pulps; Pulps
CNS P2040-85. Unbleached Sulfate Softwood Pulp (NUKP) (Dec)(2774). 1 p.
CNS P2041-85. Semi-Bleached Sulfate Softwood Pulp (NSKP) (Dec)(2775). 1 p.
CNS P2042-85. Bleached Sulfate Softwood Pulp (NBKP) (Dec)(2776). 1 p.
CNS P2051-85. Bleached Sulfate Hardwood Pulp (LBKP) (Dec)(3501). 1 p.

—Brightness
CPPA E.1U-77. Brightness Reversion of Bleached Kraft Pulp. 1 p.

—Cartridge Cases
MOD UK TS 10002B. Pulp Kraft, Bleached for Combustible Charge Components. Type 1 and Type 2.
MOD UK TS 10259. Pulp, Kraft, Unbleached for Combustible Case Components.

—Papermaking
JIS P 2102-62. Kraft Pulp for Paper (R 1972). 4 pp.

Krypton
See Also: Hazardous Materials; Noble Gases

—Storage and Handling—Information Cards
SAA AS 2508.2.01 2-85. Safe Storage and Handling Information Cards for Hazardous Materials—Part 2.012: Non-Flammable, Inert, Compressed Gases (Includes: Argon, Helium, Krypton, Neon, Nitrogen and Xenon) Double Sided Card.

Kubelka Munk Equation
—Paints
BSI BS 3900: Part D7-83. 1983 Methods of Test for Paints Group D: Optical Tests on Paint Films Part D7: Determination of Hiding Power of White and Light-Coloured Paints by the Kubelka-Munk Method. 33 pp.
ISO 6504 Pt 1-83. Paints and Varnishes—Determination of Hiding Power—Part 1: Kubelka-Munk Method for White and Light Coloured Paints First Edition. 33 pp.

L 40 Meta Sokol Aircraft
See Also: Aircraft

—Foreign Airworthiness Directives
CAA. L.40 Meta Sokol (Foreign Airworthiness Directives). 1 p.

Label Tapes
Use: Labels

Labeling
Use: Identification Systems

Labels
See Also: Decals; Identification Systems; Name Plates
MOD UK DEF-1300-01. Paper, Label (Gummed and Ungummed) (Reprinted November 1967); Amendment 1. 6 pp.

Labels *(Cont.)*
MOD UK DEF-1300-58. Paper, Label (Gummed and Ungummed) (Reprinted June, 1970 Incorporating Amendment No. 1). 4 pp.

—Adhesive
BSI BS 4781: Part 2-79. (WITHDRAWN) 1979 Self-Adhesive Plastics Labels for Permanent Use Part 2: Requirements for Stringent Conditions (Superseded by BS 4781: 1990). 8 pp.
BSI BS 4781-90. 1990 Pressure Sensitive Adhesive Plastics Labels for Permanent Use. 10 pp.
CGSB CAN/CGSB-43.31-M86. Labels and Label Tapes, Adhesive, Pressure-Sensitive. 20 pp.
CGSB 53.205M-88. Labels, File Tray, Desk. 8 pp.
JIS Z 1529-91. Pressure Sensitive Adhesive Films for Printing. 11 pp.
MOD UK DSTAN 80-87-83. Adhesive for Affixing Paper Labels to Tinplate Issue 1 (Withdrawn). 7 pp.
MOD UK DSTAN 80-115-85. Adhesive for Affixing Paper Labels to Aluminium Issue 1. 8 pp.
MOD UK DSTAN 80-132-90. Adhesive, Polyvinyl Acetate Emulsion, Type I and Type II Issue 1. 14 pp.
MOD UK DSTAN 80-132-01. Adhesive, Polyvinyl Acetate Emulsion, Type I and Type II Issue 1; Amendment 1. 15 pp.
MOD UK DSTAN 80-132-02. Adhesive, Polyvinyl Acetate Emulsion, Type I and Type II Issue 1; Amendment 2. 16 pp.
MOD UK DEF-156-01. Adhesive, Emulsion, Polyvinyl Acetate Types 'A' and 'B'; Amendment 1. 6 pp.
MOD UK DEF-1300-01. Paper, Label (Gummed and Ungummed) (Reprinted November 1967); Amendment 1. 6 pp.
MOD UK DEF-1300-58. Paper, Label (Gummed and Ungummed) (Reprinted June, 1970 Incorporating Amendment No. 1). 4 pp.
MOD UK TS 10098A. Labels, Paper, Self-Adhesive.
MOD UK TS 10196. Labels, Plastic, Self Adhesive.

—Adhesive—Aerospace
BSI 2J 12-75. (WITHDRAWN) 1975 Pressure-Sensitive Adhesive Identification Tape (Superseded by 3J 12: 1991). 8 pp.

—Adhesive—Electrical Equipment
CSA CAN/CSA-C22. 2 NO0.15-M90. Adhesive Labels; (Gen Instr 1). 35 pp.

—Adhesive—Marine
BSI BS 5609-86. 1986 Printed Pressure-Sensitive, Adhesive-Coated Labels for Marine Use, Including Requirements for Label Base Material. 20 pp.

—Aircraft
SBAC AS 549 ISSUE 1. Labels. 1 p.
SBAC AS 600-649 ISSUE 4. Labels. 1 p.
SBAC AS 650 ISSUE 1. Labels. 1 p.
SBAC AS 750-799 ISSUE 1. Labels. 1 p.

—Aircraft—Paint Containers
SBAC RS 721 ISSUE 1. Paint Container Label.

—Aircraft—Tanks (Containers)
SBAC AS 2484-2489 ISSUE 2. Label for De-Icing Fluid Tanks. 1 p.
SBAC AS 2490-2495 ISSUE 2. Label for Drinking Water Tanks. 1 p.

—Bedding
SAA AS 1957-87. Care Labelling of Clothing, Household Textiles, Furnishings, Upholstered Furniture, Bedding, Piece Goods and Yarns. 7 pp.

—Clothing
SAA AS 1957-87. Care Labelling of Clothing, Household Textiles, Furnishings, Upholstered Furniture, Bedding, Piece Goods and Yarns. 7 pp.

—Eyelets
CNS Z7095-81. Eyelets for Cardboard Labels (Jul)(7772).

—Fabrics—Household
SAA AS 1957-87. Care Labelling of Clothing, Household Textiles, Furnishings, Upholstered Furniture, Bedding, Piece Goods and Yarns. 7 pp.

—Furniture
SAA AS 1957-87. Care Labelling of Clothing, Household Textiles, Furnishings, Upholstered Furniture, Bedding, Piece Goods and Yarns. 7 pp.

—Hazardous Materials—Transportation
SNZ NZS 5417-86. Specification for Transportation Labels for Hazardous Substances Amend: 1, 1988. 36 pp.

—Holders
MOD UK DSTAN 99-2-81. Holders, Card Label, Tinplated Issue 2. 6 pp.

—Lacquers
MOD UK DSTAN 80-64-76. Lacquer, Finishing, Paper Labels Type: Brushing Issue 1. 6 pp.

Labels *(Cont.)*
—Paper—Milk Bottles
CNS Z5027-66. Paper Slips for Milk Bottles (Mar)(2651) (R 1972).

—Tie
MOD UK DEF-1301-58. Labels, Tie-On (Stringed). 3 pp.
MOD UK DEF-1302-58. Paper, Label (for Stringed Tie-On Labels) (Reprinted February 1968). 3 pp.

—Yarns
SAA AS 1957-87. Care Labelling of Clothing, Household Textiles, Furnishings, Upholstered Furniture, Bedding, Piece Goods and Yarns. 7 pp.

Labor Unions
Use For: Trade Unions

—Standards Preparation
EURO MEMORANDUM 5-89. Trade Unions and the Preparation of European Standards. 8 pp.

Laboratories
Use For: Testing Laboratories *See Also:* Aircraft; Experimental Design; Laboratory Accreditation; Laboratory Animals; Laboratory Equipment; Language Laboratories; Microbiology Laboratories

—Calibration—Operation/Management
ISO Guide 25-90. General Requirements for the Competence of Calibration and Testing Laboratories Third Edition. 11 pp.
ISO Guide 58-93. Calibration and Testing Laboratory Accreditation Systems—General Requirements for Operation and Recognition First Edition. 9 pp.
MOD UK DSTAN 05-55-88. Ministry of Defence Calibration Laboratories Operation and Management Issue 3. 31 pp.

—Certification
IEC Guide 38-83. General Requirements for the Acceptance of Testing Laboratories First Edition. 11 pp.
ISO Guide 38-83. General Requirements for the Acceptance of Testing Laboratories First Edition. 11 pp.

—Chemical Analysis—Quality Assurance
CSA Q641-1992SP. Guide for Quality Assurance Programs for Physical and Chemical Testing Laboratories; (Gen Instr 1). 35 pp.

—Concrete Testing—Certification
CSA A283-1980. Qualification Code for Concrete Testing Laboratories; (Gen Instr 2 Thru 6) (Supplement 1 1987). 27 pp.

—Design
BSI BS 3202-59. (WITHDRAWN) 1959 Amd 1 Recommendations on Laboratory Furniture and Fittings (Superseded by BS 3202: Parts 1,2,3 and 4 1991). 113 pp.

—Glossaries
ISO Guide 2-91. General Terms and Their Definitions Concerning Standardization and Related Activities Sixth Edition. 65 pp.
SAA HB18.2-92. Guidelines for Third-Party Certification and Accreditation—Guide 2: General Terms and Their Definitions Concerning Standardization and Related Activities (ISO/IEC Guide 2:1991) (SANZ HB18.2—1992). 60 pp.
SNZ SANZ/SAA HB 18.2-92. Guidelines for Third-Party Certification and Accreditation Guide 2 General Terms and Their Definitions Concerning Standardization and Related Activities (ISO/IEC Guide 2:1991). 60 pp.

—Industrial Waste Analysis—Certification
CSA Z201-M1990. Qualification Code for Ontario for Laboratories Analyzing Industrial Waste; (Gen Instr 1). 123 pp.

—Nuclear Physics—Design—Submarine
MOD UK NES 795-90. Requirements for Health Physics Laboratories in Nuclear Submarines Issue 1 (08.90). 44 pp.

—Open Systems Interconnection—Conformance
BSI BS ISO/IEC 9646-5-91. 1991 Information Technology—Open Systems Interconnection—Conformance Testing Methodology and Framework—Part 5: Requirements on Test Laboratories and Clients for the Conformance Assessment Process. 35 pp.
BSI BS ISO/IEC 9646-5-00. 1991 Amd 0 Information Technology—Open Systems Interconnection—Conformance Testing Methodology and Framework—Part 5: Reqmnts on Test Laboratories and Clients for the Conformance Assessment Process (AMD 7304) November 15, 1992. 40 pp.

INTERNATIONAL AND NON-U.S. NATIONAL STANDARDS
SUBJECT INDEX

Laboratories

Laboratories (Cont.)
—**Open Systems Interconnection—Conformance** (Cont.)

CCITT RECMN X.294-92. OSI Conformance Testing Methodology and Framework for Protocol Recommendations for CCITT Applications—Requirements on Test Laboratories and Clients for the Conformance Assessment Process (Study Group VII) 34 pp. 34 pp.

CEN EN 29646-5-92. Information Technology—Open Systems Interconnection—Conformance Testing Methodology and Framework—Part 5: Requirements on Test Laboratories and Clients for the Conformance Assessment Process. 34 pp.

IEC 9646 Pt 5-91. Information Technology—Open Systems Interconnection—Conformance Testing Methodology and Framework—Part 5: Requirements on Test Laboratories and Clients for the Conformance Assessment Process First Edition. 32 pp.

IEC 9646 Pt 5 Draft AMD 1. Information Technology—Open Systems Intercon-nection—Conformance Testing Methodology and Framework—Part 5: Requirements on Test Laboratories and Clients for the Conformance Assessment Process Amend-ment 1: Protocol Profile Testing Methodology; (1992) ***CD-ROM ONLY***. 17 pp.

IEC 9646 Pt 5 Draft AMD 2. Information Technology—Open Systems Intercon-nection—Conformance Testing Methodology and Framework—Part 5: Requirements on Test Laboratories and Clients for the Conformance Assessment Process Amend-ment 2: Multi-Party Testing Methodology; (1992) ***CD-ROM ONLY***. 5 pp.

ISO 9646 Pt 5-91. Information Technology—Open Systems Interconnection—Conformance Testing Methodology and Framework—Part 5: Requirements on Test Laboratories and Clients for the Conformance Assessment Process First Edition. 32 pp.

ISO 9646 Pt 5 Draft AMD 1. Information Technology—Open Systems Intercon-nection—Conformance Testing Methodology and Framework—Part 5: Requirements on Test Laboratories and Clients for the Conformance Assessment Process Amend-ment 1: Protocol Profile Testing Methodology; (1992) ***CD-ROM ONLY***. 17 pp.

ISO 9646 Pt 5 Draft AMD 2. Information Technology—Open Systems Intercon-nection—Conformance Testing Methodology and Framework—Part 5: Requirements on Test Laboratories and Clients for the Conformance Assessment Process Amend-ment 2: Multi-Party Testing Methodology; (1992) ***CD-ROM ONLY***. 5 pp.

OSI ISO/TC97/SC21-4-87. Working Draft of OSI Conformance Testing Methodology and Framework Part 4: Requirements on Suppliers and Clients of Test Laboratories. 7 pp.

OSI ISO/TC97/SC21-5-87. Working Draft of OSI Conformance Testing Methodology and Framework Part 5: Test Laboratory Operations. 27 pp.

OSI ISO IEC DP 9646-5-88. Information Processing Systems—OSI Conformance Testing Methodology and Framework—Part 5: Requirements on Test Laboratories and Clients for the Conformance Assessment Process. 33 pp.

SNZ NZS/AS 4103. 5-93. Information Technology—Open Systems Interconnection—Conformance Testing Methodology and Framework Part 5: Requirements on Test Laboratorires and Clients for the Conformance Assessment Process (This is a Joint. 25 pp.

—**Paper Products**

ISO 4094-91. Paper, Board and Pulps—International Calibration of Testing Apparatus—Nomination and Acceptance of Standardizing and Authorized Laboratories First Edition. 10 pp.

—**Physical Testing—Quality Assurance**

CSA Q641-1992SP. Guide for Quality Assurance Programs for Physical and Chemical Testing Laboratories; (Gen Instr 1). 35 pp.

—**Proficiency Testing**

IEC Guide 43-84. Development and Operation of Laboratory Proficiency Testing First Edition. 9 pp.

ISO Guide 43-84. Development and Operation of Laboratory Proficiency Testing First Edition. 9 pp.

SAA HB18.43-91. Guidelines for Third-Party Certification and Accreditation—Guide 43: Development and Operation of Laboratory Proficiency Testing (ISO/IEC Guide 43:1984) (SANZ HB18.43—1991). 6 pp.

SNZ SANZ/SAA HB 18.43-91. Guidelines for Third-Party Certification and Accreditation Guide 43 Development and Operation of Laboratory Proficiency Testing (ISO/IEC Guide 43-1984). 6 pp.

Laboratories (Cont.)
—**Pulps**

ISO 4094-91. Paper, Board and Pulps—International Calibration of Testing Apparatus—Nomination and Acceptance of Standardizing and Authorized Laboratories First Edition. 10 pp.

—**Quality Assurance**

BSI BS 7501-89. 1989 General Criteria for the Operation of Testing Laboratories. 14 pp.

BSI BS 7502-89. 1989 General Criteria for the Assessment of Testing Laboratories. 13 pp.

BSI BS 7503-89. 1989 General Criteria for Laboratory Accreditation Bodies. 12 pp.

CEN EN 45001-89. General Criteria for the Operation of Testing Laboratories. 10 pp.

CEN EN 45002-89. General Criteria for the Assessment of Testing Laboratories. 9 pp.

CEN EN 45003-89. General Criteria for Laboratory Accreditation Bodies. 8 pp.

SAA HB18.25-91. Guidelines for Third-Party Certification and Accreditation—Guide 25: General Requirements for the Competence of Calibration and Testing Laboratories (ISO/IEC Guide 25:1990) (SANZ HB18.25—1991). 7 pp.

SNZ SANZ/SAA HB 18.25-91. Guidelines for Third-Party Certification and Accreditation Guide 25 General Requirements for the Competence of Calibration and Testing Laboratories (ISO/IEC Guide 25-1990). 7 pp.

—**Safety**

SAA AS 2243.7-91. Safety in Laboratories—Part 7: Electrical Aspects (in Professional Packages 34C, 47). 22 pp.

—**Water**

CNS K1165-74. Purified Water for Chemical Analysis (Mar)(3699).

ISO 3696-87. Water for Analytical Laboratory Use—Specification and Test Methods First Edition. 8 pp.

Laboratory Accreditation

Use For: Accreditation (Laboratories)
See Also: Laboratories

BSI BS 6460: Part 1-83. (OBSOLESCENT) 1983 Accreditation of Testing Laboratories Part 1: General Requirements for the Technical Competence of Testing Laboratories. 7 pp.

BSI BS 7501-89. 1989 General Criteria for the Operation of Testing Laboratories. 14 pp.

BSI BS 7502-89. 1989 General Criteria for the Assessment of Testing Laboratories. 13 pp.

BSI BS 7503-89. 1989 General Criteria for Laboratory Accreditation Bodies. 12 pp.

CEN EN 45001-89. General Criteria for the Operation of Testing Laboratories. 10 pp.

CEN EN 45002-89. General Criteria for the Assessment of Testing Laboratories. 9 pp.

CEN EN 45003-89. General Criteria for Laboratory Accreditation Bodies. 8 pp.

ISO Guide 55-88. Testing Laboratory Accreditation Systems—General Recommendations for the Acceptance of Operation First Edition. 7 pp.

ISO Guide 58-93. Calibration and Testing Laboratory Accreditation Systems—General Requirements for Operation and Recognition First Edition. 9 pp.

SAA HB18.55-91. Guidelines for Third-Party Certification and Accreditation—Guide 55: Testing Laboratory Accreditation Systems—General Recommendations for Operation (ISO/IEC Guide 55:1988) (SANZ HB18.55—1991). 4 pp.

SNZ SANZ/SAA HB 18.55-91. Guidelines for Third-Party Certification and Accreditation Guide 55 Testing Laboratory Accreditation Systems—General Recommendations for Operation (ISO/IEC Guide 55-1988). 4 pp.

—**Accreditation Bodies—Acceptance**

SAA HB18.54-91. Guidelines for Third-Party Certification and Accreditation—Guide 54: Testing Laboratory Accreditation Systems—General Requirements for the Acceptance of Accreditation Bodies (ISO/IEC Guide 54:1988) (SANZ HB18.54—1991). 3 pp.

SNZ SANZ/SAA HB 18.54-88. Guidelines for Third-Party Certification and Accreditation Guide 54 Testing Laboratory Accreditation Systems—General Recommendations for the Acceptance of Accreditation Bodies (ISO/IEC Guide 54-1988). 3 pp.

—**Data Terminal Equipment**

CEPT T/G 01-01-87. Criteres D'Accreditation Des Laboratoires D'Essais D'Equipements Terminaux De Telecommunication. 4 pp.

CEPT T/G 01-01 E-87. Criteria for Accreditation of Testing Laboratories for Telecommunications Terminal Equipment. 4 pp.

—**NAMAS**

CAA NOTICE #69 ISSUE 1. CAA Approval of Test Houses Holding—Names Accreditation (Airworthiness Notices). 3 pp.

Laboratory Animals

See Also: Laboratories; Rabbits

—**Humane Treatment**

ISO 10993 Pt 2-92. Biological Evaluation of Medical Devices—Part 2: Animal Welfare Requirements First Edition. 10 pp.

Laboratory Apparatus

Use: Laboratory Equipment

Laboratory Buckets

Use For: Laboratory Pails *See Also:* Buckets (Pails)

—**Enameled**

CNS R2155-85. Enamel Bucket for Laboratory (Jun)(9030).

Laboratory Burners

Use: Burners—Laboratory

Laboratory Chemicals

Scope Note: Use a more specific term *See:* Benzoyl Peroxide; 2,3-Butanedione Monoxime; tert-Dodecyl Mercaptan; Ferrous Chloride; Ferrous Sulfate; Lead Tetroxide; Lithium Chloride; Lithium Nitrate; Lithium Sulfate; L-Lysine Monohydrochloride; Manganese Nitrate; Molecular Sieves; alpha-Naphtholbenzene; alpha-Naphtholphthalein; N-1-Naphthyl-N'-diethylenediamine Oxalate; N-1-Naphthylethylenediamine Dihydrochloride; Neutral Red; Nitromethane; Pentaerythritol Tetranitrate; Quinaldic Acid; Quinalizarin; Quinhydrone; Quinoline; Reagents; Selenium Oxide; Tetrasodium Pyrophosphate; Xylenol Orange

Laboratory Clamps

See Also: Clamps; Laboratory Equipment

CNS T1024-80. Medical Forceps (Aug)(6283).
CNS Z7099-81. Metal Laboratory Ware; Clamps (Nov)(8176).
JIS T 2302-53. Medical Forceps.

Laboratory Containers

See Also: Containers; Laboratory Dishes; Laboratory Equipment

—**Enamel**

CNS R2156-85. Enamel Round Vessel for Laboratory (Jun)(9031).
CNS R2157-85. Enamel Reaction Vessel with Three Necks for Laboratory (Jun)(9032).

Laboratory Dishes

See Also: Beakers; Laboratory Containers; Laboratory Equipment; Laboratory Ware; Petri Dishes

—**Enamel**

CNS R2154-85. Enamel Dish for Laboratory (Jun)(9029).

—**Evaporating—Coatings**

CGSB 1-GP-71 METH 20.1-78. Methods of Testing Paints and Pigments Residue on Evaporation. 1 p.

—**Glass**

CNS R2126-86. Glass Dishes for Chemical Analysis (Mar)(7302).

—**Platinum**

JIS H 6202-86. Platinum Dishes for Chemical Analysis. 6 pp.

Laboratory Equipment

Use For: Laboratory Apparatus
See Also: Autoclaves; Casseroles; Circulators (Laboratory); Combustion Boats; Combustion Tubes; Condensers (Laboratory); Containment Cabinets; Crucibles; Diagnostic Equipment; Electrochemical Analyzers; Filter Paper; Funnels; Glove Boxes; Image Processing Systems; Laboratories; Laboratory Clamps; Laboratory Containers; Laboratory Dishes; Laboratory Filters; Laboratory Furniture; Laboratory Ovens; Laboratory Sinks; Laboratory Ware; Measuring Cylinders; Measuring Instruments; Microscope Slides; Nessler Cylinders; Petri Dishes; Photographic Processing Equipment; Photometers; Pipettes; Polarimeters; Sample Handling Equipment; Test Tubes; Water Baths

CSA C22.2 NO 151-M1986. Laboratory Equipment (R 1992); (Gen Instr 1 Thru 2). 61 pp.

—**Drainage Systems**

CSA CAN/CSA-B181.3-M86. Polyolefin Laboratory Drainage Systems (R 1992) (Gen Instr 1). 19 pp.

—**Electrical Grounding—Safety**

CSA C22.2 NO 151-M1986. Laboratory Equipment (R 1992); (Gen Instr 1 Thru 2). 61 pp.

INDEX and DIRECTORY of

INTERNATIONAL AND NON-U.S. NATIONAL STANDARDS
SUBJECT INDEX
Laboratory

Laboratory Equipment (Cont.)
—Electrical Safety
BSI BS EN 61010-1-93. 1993 Safety Requirements for Electrical Equipment for Measurement, Control, and Laboratory Use Part 1: General Requirements. 123 pp.

CENELEC EN 61010-1-93. Safety Requirements for Electrical Equipment for Measurement, Control and Laboratory Use Part 1: General Requirements (IEC 1010-1: 1990 + A1: 1992, Modified). 117 pp.

CSA CAN/CSA-C22. 2NO1010.1-92. Safety Requirements for Electrical Equipment for Measurement, Control, and Laboratory Use, Part 1: General Requirements (IEC 1010-1:1990); (Gen Instr 1). 160 pp.

DIN VDE 0411 Pt 100-86. Measurement and Control; Safety Requirements for Electrically Operated Measuring, Control and Laboratory Equipment; General Requirements; Identical with IEC 66E (Sec) 22 (Aug). 137 pp.

IEC 1010 Pt 1-90. Safety Requirements for Electrical Equipment for Measurement, Control and Laboratory Use Part 1: General Requirements First Edition (CENELEC EN 61010-1: 1993). 190 pp.

—Enamel
JIS R 4203-78. Enamel Apparatus for Laboratory.

—Glossaries
BSI BS 6711: Part 1-86. 1986 Vocabulary Relating to Laboratory Apparatus Made Essentially from Glass, Porcelain or Vitreous Silica Part 1: Names for Items of Apparatus. 296 pp.

ISO 4791 Pt 1-85. Laboratory Apparatus—Vocabulary Relating to Apparatus Made Essentially from Glass, Porcelain or Vitreous Silica—Part 1: Names for Items of Apparatus First Edition. 297 pp.

—Heaters—Electrical Safety
BSI BS 7687: Sec 2.10-93. 1993 Safety Requirements for Electrical Equipment for Measurement, Control and Laboratory Use Part 2: Particular Requirements Section 2.10: Specification for Laboratory Equipment for the Heating of Materials (IEC 1010-2-010: 1992) (S). 21 pp.

IEC 1010 Pt 2-010-92. Safety Requirements for Electrical Equipment for Measurement, Control, and Laboratory Use Part 2-010: Particular Requirements for Laboratory Equipment for the Heating of Materials First Edition. 38 pp.

—Hoses
BSI BS 2775-87. 1987 Rubber Stoppers and Tubing for General Laboratory Use. 11 pp.

—Polytetrafluoroethylene Materials
CNS K3069-83. Polytetrafluoroethylene Sheets (Aug)(10511). 7 pp.

JIS K 6887-77. Polytetrafluoroethylene Tapes (R 1980). 12 pp.

JIS K 6888-77. Polytetrafluoroethylene Sheets (R 1980). 10 pp.

JIS K 6889-77. Polytetrafluoroethylene Rods (R 1980). 11 pp.

JIS K 6890-77. Polytetrafluoroethylene Tubes (R 1980). 11 pp.

—Porcelain
CNS R4001-80. Porcelain Laboratory Apparatus (Apr)(5473).

ISO 1775-75. Porcelain Laboratory Apparatus—Requirements and Methods of Test First Edition. 7 pp.

ISO 4791 Pt 1-85. Laboratory Apparatus—Vocabulary Relating to Apparatus Made Essentially from Glass, Porcelain or Vitreous Silica—Part 1: Names for Items of Apparatus First Edition. 297 pp.

—Porcelain—Glazes
CNS R3085-80. Method of Test for Porosity of Body and Imperfections of Glaze for Porcelain Laboratory Apparatus (Apr)(5474).

—Porcelain—Glazes—Acid Resistance Testing
CNS R3089-80. Method of Test for Resistance of Glaze to Acid and Alkali for Porcelain Laboratory Apparatus (Apr)(5478).

—Porcelain—Glazes—Alkali Resistance Testing
CNS R3089-80. Method of Test for Resistance of Glaze to Acid and Alkali for Porcelain Laboratory Apparatus (Apr)(5478).

—Porcelain—Glazes—High Temperature Testing
CNS R3087-80. Method of Test for Resistance of Glaze to High Temperature for Porcelain Laboratory Apparatus (Apr)(5476).

Laboratory Equipment (Cont.)
—Porcelain—Heat Resistance
CNS R3086-80. Method of Test for Resistance to Heat and to Sudden Change in Temperature for Porcelain Laboratory Apparatus (Apr)(5475).

—Porcelain—Loss of Mass—High Temperature Testing
CNS R3088-80. Method of Test for Constancy of Mass on Ignition for Porcelain Laboratory Apparatus (Apr)(5477).

—Porcelain—Porosity
CNS R3085-80. Method of Test for Porosity of Body and Imperfections of Glaze for Porcelain Laboratory Apparatus (Apr)(5474).

—Porcelain—Temperature Change Testing
CNS R3086-80. Method of Test for Resistance to Heat and to Sudden Change in Temperature for Porcelain Laboratory Apparatus (Apr)(5475).

—Stoppers
BSI BS 2775-87. 1987 Rubber Stoppers and Tubing for General Laboratory Use. 11 pp.

—Vitreous Silica
ISO 4791 Pt 1-85. Laboratory Apparatus—Vocabulary Relating to Apparatus Made Essentially from Glass, Porcelain or Vitreous Silica—Part 1: Names for Items of Apparatus First Edition. 297 pp.

—Volumetric—Piston Operated
BSI BS 7653: Part 2-93. 1993 Piston and/or Plunger Operated Volumetric Apparatus Part 2: Methods of Operation (H). 11 pp.

BSI BS 7653: Part 3-93. 1993 Piston and/or Plunger Operated Volumetric Apparatus Part 3: Methods of Test (H). 11 pp.

BSI BS 7653: Part 4-93. 1993 Piston and/or Plunger Operated Volumetric Apparatus Part 4: Specification for Conditions of Test, Safety and Supply (H). 7 pp.

—Volumetric—Piston Operated—Glossaries
BSI BS 7653: Part 1-93. 1993 Piston and/or Plunger Operated Volumetric Apparatus Part 1: Glossary of Terms (H). 12 pp.

—Volumetric—Piston Operated—Reliability
DIN ENGL 12650 Pt 6-83. Mechanical, Physical and Electrical Laboratory Apparatus; Piston Operated Volumetric Apparatus; Gravimetric Assessment of Metrological Reliability (Apr). 4 pp.

—Volumetric—Plunger Operated
BSI BS 7653: Part 2-93. 1993 Piston and/or Plunger Operated Volumetric Apparatus Part 2: Methods of Operation (H). 11 pp.

BSI BS 7653: Part 3-93. 1993 Piston and/or Plunger Operated Volumetric Apparatus Part 3: Methods of Test (H). 11 pp.

BSI BS 7653: Part 4-93. 1993 Piston and/or Plunger Operated Volumetric Apparatus Part 4: Specification for Conditions of Test, Safety and Supply (H). 7 pp.

—Volumetric—Plunger Operated—Glossaries
BSI BS 7653: Part 1-93. 1993 Piston and/or Plunger Operated Volumetric Apparatus Part 1: Glossary of Terms (H). 12 pp.

Laboratory Filters
See Also: Laboratory Equipment

—Glass
CNS R2135-86. Glass Filters for Chemical Analysis (Mar) (7311).

—Sintered
BSI BS 1752-83. 1983 Laboratory Sintered or Fritted Filters Including Porosity Grading. 11 pp.

ISO 4793-80. Laboratory Sintered (Fritted) Filters—Porosity Grading, Classification and Designation First Edition. 9 pp.

Laboratory Furniture
See Also: Furniture; Laboratory Equipment

—Design
BSI BS 3202: Part 3-91. 1991 Laboratory Furniture and Fittings Part 3: Recommendations for Design Supersedes BS 3202: 1959 (H). 16 pp.

—Installation
BSI BS 3202: Part 4-91. 1991 Laboratory Furniture and Fittings Part 4: Recommendations for Installation (Supersedes BS 3202: 1959) (H). 14 pp.

—Safety
BSI BS 3202: Part 1-91. 1991 Laboratory Furniture and Fittings Part 1: Introduction (H) (Supersedes BS 3202: 1959). 18 pp.

Laboratory Furniture (Cont.)
—Stability Testing
BSI BS 3202: Part 2-91. 1991 Laboratory Furniture and Fittings Part 2: Specifiication for Performance. 40 pp.

Laboratory Glassware
Use: Laboratory Ware

Laboratory Medicine
Use: Clinical Pathology

Laboratory Ovens
See Also: Laboratory Equipment; Laboratory Ovens; Ovens

—Drying
BSI BS 2648-55. 1955 Amd 1 Performance Requirements for Electrically-Heated Laboratory Drying Ovens. 11 pp.

Laboratory Pails
Use: Laboratory Buckets

Laboratory Resistors
Use: Resistors—Laboratory

Laboratory Sinks
See Also: Laboratory Equipment; Sinks

—Ceramic
CNS R2061-5-82. Sanitary Ceramic Ware Laboratory Sink (Feb) (3220-5).

Laboratory Thermometers
See Also: Thermometers
BSI BS 593-89. 1989 Laboratory Thermometers. 16 pp.

—Liquid In Glass
ISO 386-77. Liquid-in-Glass Laboratory Thermometers—Principles of Design, Construction and Use First Edition; (Erratum—Aug 1978). 16 pp.

—Liquid In Glass—Contact
DIN ENGL 12878-80. Electrical Laboratory Equipment; Adjustable Liquid-In-Glass Contact Thermometers and Control Devices; General and Safety Requirements, Testing (Dec). 6 pp.

—Liquid In Glass—Contact—Control Systems Equipment
DIN ENGL 12878-80. Electrical Laboratory Equipment; Adjustable Liquid-In-Glass Contact Thermometers and Control Devices; General and Safety Requirements, Testing (Dec). 6 pp.

—Liquid In Glass—Contact—Control Systems Equipment—Safety
DIN ENGL 12878-80. Electrical Laboratory Equipment; Adjustable Liquid-In-Glass Contact Thermometers and Control Devices; General and Safety Requirements, Testing (Dec). 6 pp.

—Liquid In Glass—Contact—Safety
DIN ENGL 12878-80. Electrical Laboratory Equipment; Adjustable Liquid-In-Glass Contact Thermometers and Control Devices; General and Safety Requirements, Testing (Dec). 6 pp.

Laboratory Ware
Use For: Laboratory Glassware See Also: Beakers; Bottles; Burets; Centrifuge Tubes; Extractors (Glassware); Flasks; Funnels; Glass Joints; Laboratory Dishes; Laboratory Equipment; Measuring Cylinders; Micropipettes; Pipettes; Reagent Bottles; Stoppers; Suction Bottles; Test Tubes; Wash Bottles; Weighing Bottles

JIS R 3502-87. Testing Method of Glass Apparatus for Chemical Analysis.

JIS R 3503-87. Glass Apparatus for Chemical Analysis. 93 pp.

JIS R 3504-76. Glass Material for Chemical Analysis.

—Adapters
CNS R2143-86. Glass Adapters for Chemical Analysis (Mar)(7319).

—Alkali Resistance Testing
CNS R3104-82. Standard Test Method for Glass Apparatus Chemical Analysis (May)(8861). 4 pp.

—Connectors
CNS R2134-86. Glass Connectors for Chemical Analysis (Mar) (7310).

—Enamel Paints—Color Coding—Chemical Resistance
BSI BS 3996-78. 1978 Colour Coding for One-Mark and Graduated Pipettes (Including Requirements for the Service Performance of the Colour Coding Enamels). 7 pp.

INDUSTRY STANDARDS

INTERNATIONAL AND NON-U.S. NATIONAL STANDARDS
SUBJECT INDEX

Laboratory

Laboratory Ware *(Cont.)*
—Enamel Paints—Color Coding—Chemical Resistance *(Cont.)*
ISO 4794-82. Laboratory Glassware—Methods for Assessing Chemical Resistance of Enamels Used for Colour Coding and Colour Marking First Edition. 4 pp.

—Engineering Drawings
BSI BS 2774-83. 1983 Drawing Conventions for Laboratory Glass Apparatus. 10 pp.

—Glossaries
ISO 4791 Pt 1-85. Laboratory Apparatus—Vocabulary Relating to Apparatus Made Essentially from Glass, Porcelain or Vitreous Silica—Part 1: Names for Items of Apparatus First Edition. 297 pp.

—Pressure Measurement—Inspection by Attributes
DIN ENGL 52320 Pt 1-78. Testing of Glass; Internal Pressure Test on Hollow Glassware, Especially Glass Containers; Inspection by Attributes (Nov). 2 pp.

—Pressure Measurement—Inspection by Variables
DIN ENGL 52320 Pt 2-81. Testing of Glass; Internal Pressure Test on Hollow Glassware, Especially Glass Containers; Inspection by Variables (Oct). 2 pp.

—Thermal Shock
BSI BS 3517-91. 1991 Thermal Shock Tests on Laboratory Glassware (ISO 718: 1990). 8 pp.
CNS R3104-82. Standard Test Method for Glass Apparatus Chemical Analysis (May)(8861). 4 pp.
ISO 718-90. Laboratory Glassware—Thermal Shock and Thermal Shock Endurance—Test Methods Second Edition. 6 pp.

—Towers—Drying
CNS R2137-86. Glass Gas-Drying Towers for Chemical Analysis (Mar)(7313).

—Vacuum—Safety
BSI BS 3423-86. 1986 Recommendations for Design of Glass Vacuum Vessels (Including Desiccators) for Laboratory Use. 18 pp.

—Volumetric
BSI BS 5898-80. 1980 Principles of Design and Construction of Volumetric Glassware for Laboratory Use. 16 pp.
BSI BS 6696-86. 1986 Methods for Use and Testing of Capacity of Volumetric Glassware. 18 pp.
ISO 384-78. Laboratory Glassware—Principles of Design and Construction of Volumetric Glassware First Edition. 16 pp.
ISO 4787-84. Laboratory Glassware—Volumetric Glassware—Methods for Use and Testing of Capacity First Edition. 15 pp.

—Volumetric—Calibration
BSI BS 1797-87. 1987 Schedule for Tables for Use in the Calibration of Volumetric Glassware. 82 pp.

Laboratory Water
Use: High Purity Water

Lac
See Also: Resins; Seed Lac
CNS K2127-82. Orange Shellac and Other Lacs (Jan)(8423). 1 p.

—Bleached
BSI BS 1284-60. 1960 Amd 1 Bleached Lac (AMD 6368) June 28, 1991. 33 pp.
ISO 57-75. Bleached Lac—Specification First Edition. 28 pp.

Laces (Fabrics)
See Also: Fabrics

—Glossaries
JIS L 0214-83. Glossary of Terms Used in Lace.

Laces (Footwear)
See Also: Footwear
BSI BS 7141: Part 6-89. 1989 Narrow Fabrics Part 6: Laces for Footwear and Other Purposes. 8 pp.

—Abrasion Testing
BSI BS 5131: Sec 3.6-91. 1991 Methods of Test for Footwear and Footwear Materials Part 3: Uppers, Textiles and Threads Section 3.6: Abrasion Resistance of Shoe Laces. 6 pp.

—Boots—Length Measurement
CGSB CAN/CGSB-4.2 NO.4.3-M87. Textile Test Methods Bootlaces and Shoelaces—Measurement of Length. 8 pp.

Laces (Footwear) *(Cont.)*
—Breaking Load
BSI BS 5131: Sec 3.7-91. 1991 Methods of Test for Footwear and Footwear Materials Part 3: Uppers, Textiles and Threads Section 3.7: Breaking Strength of Shoe Laces. 4 pp.

Lacing Cords
See Also: Electric Wiring
MOD UK DSTAN 40-8-83. Cord, Lacing Issue 2. 16 pp.

Lacmoid
CNS K7266-68. Chemical Reagent (Lacmoid) (Resorcinol Blue) (Nov)(1765).

Lacquer Enamels
See Also: Enamels; Lacquers

—Chemical Resistance
CNS K6031-87. Method of Test for Lacquer Enamel (May) (628). 6 pp.

—Colorimetric Analysis
CNS K6031-87. Method of Test for Lacquer Enamel (May) (628). 6 pp.

—Environmental Testing
CNS K6031-87. Method of Test for Lacquer Enamel (May) (628). 6 pp.

—Nonvolatile Matter Content
CNS K6031-87. Method of Test for Lacquer Enamel (May) (628). 6 pp.

—Physical Properties
CNS K6031-87. Method of Test for Lacquer Enamel (May) (628). 6 pp.

—Soluble Matter Content—Qualitative Analysis
CNS K6031-87. Method of Test for Lacquer Enamel (May) (628). 6 pp.

—Water Resistance Testing
CNS K6031-87. Method of Test for Lacquer Enamel (May) (628). 6 pp.

Lacquer Primers
See Also: Primers (Coatings)
CNS K2049-86. Lacquer Primer (Jan)(2948). 2 pp.
CNS K6819-86. Method of Test for Lacquer Primer (Jan)(11477).

Lacquer Sealants
CNS K2063-86. Lacquer Sanding Sealer (Apr)(4912). 2 pp.
CNS K6831-86. Method of Test for Lacquer Sanding Sealer (Apr)(11544).

—Wood
CNS K2062-86. Lacquer Wood Sealer (Apr)(4911). 2 pp.
CNS K6830-86. Method of Test for Lacquer Wood Sealer (Apr)(11543).

Lacquers
See Also: Acrylic Lacquers; Baking Lacquers; Chinese Lacquers; Clear Coatings; Coatings; Dopes; Enamel Paints; Ethyl Cellulose Lacquers; Japanese Lacquers; Lacquer Enamels; Organic Coatings; Paints; Phenolic Resin Lacquers; Polyvinyl Alcohol Lacquers; Shellacs; Solvents; Varnishes
CNS K2064-86. Clear Lacquer (May)(4913). 2 pp.
CNS K6333-86. Method of Test for Lustless Lacquer (Jan)(3399).
CNS K6874-86. Method of Test for Clear Lacquer (May)(11604).

—Abrasion Testing
CNS K6729-82. Method of Test for Abrasion Resistance of Coating of Paint, Varnish, Lacquer and Related Products with the Air Blast Abrasion Tester (Sep)(9405).

—Adhesion Testing
SAA AS 1580.408. 2-93. Paints and Related Materials—Methods of Test—Part 408.2: Adhesion—Knife Test (in Professional Package 39). 3 pp.
SAA AS 1580.408. 4-93. Paints and Related Materials—Methods of Test—Part 408.4: Adhesion (Cross-Cut) (in Professional Package 39A). 5 pp.
SNZ NZS/AS 1580. 408.2-84. Methods of Test for Paints and Related Materials Part 408.2: Adhesion—Knife Test (This is a Joint Standard with SAA AS 1580.408.2). 4 pp.
SNZ NZS/AS 1580. 408.4-80. Methods of Test for Paints and Related Materials Part 408.4: Adhesion (Cross-Cut) (This is a Joint Standard with SAA AS 1580.408.4). 4 pp.

—Aerosols
CNS K2240-89. Canned Aerosol Lacquer (Jan)(12484).

Lacquers *(Cont.)*
—Alkali Resistance Testing
SAA AS 1580.460. 2-82. Paints and Related Materials—Methods of Test—Part 460.2: Resistance to Alkaline Conditions (Immersion Test) (R 1992). 1 p.
SNZ NZS/AS 1580. 460.2-82. Methods of Test for Paints and Related Materials Part 460.2: Resistance to Alkaline Conditions (Immersion Test) (This is a Joint Standard with SAA AS 1580.460.2). 1 p.

—Aluminum—Oil Resistance Testing
CGSB 1-GP-71 METH 109.2-79. Methods of Testing Paints and Pigments Oil Resistance Lacquers on Primed Aluminum; (Amended September 1980) (Re-Edited September 1982). 1 p.

—Aluminum—Self Lifting Resistance
CGSB 1-GP-71 METH 132.2-79. Methods of Testing Paints and Pigments Self-Lifting Lacquers. 1 p.
CGSB 1-GP-71 METH 133.1-79. Methods of Testing Paints and Pigments Behavior Towards Undercoats and Self-Lifting Finishes on Primed Aluminum; (Amended September 1980) (Re-Edited September 1982). 1 p.

—Aluminum—Solvent Resistance Testing—Deicers
CGSB 1-GP-71 METH 108.3-79. Methods of Testing Paints and Pigments Solvent Resistance Resistance of Lacquers to De-Icing Fluid; (Amended September 1980) (Re-Edited September 1982). 1 p.

—Aluminum—Sprayed
CGSB 1-GP-71 METH 131.1-79. Methods of Testing Paints and Pigments Behavior Towards Applied Films Lacquer on Zinc Chromate Primer. 1 p.

—Aluminum—Undercoatings
CGSB 1-GP-71 METH 133.1-79. Methods of Testing Paints and Pigments Behavior Towards Undercoats and Self-Lifting Finishes on Primed Aluminum; (Amended September 1980) (Re-Edited September 1982). 1 p.

—Ammunition
MOD UK DSTAN 80-45-85. Lacquer, Shellac/Castor Oil Type QX Types: Brushing Dipping Issue 2. 11 pp.

—Application
SAA AS 1580.107. 8-79. Paints and Related Materials—Methods of Test—Part 107.8: Application of Paint at a Nominated Spreading Rate by Determination of Wet Film Mass (Brush Application) (R 1991). 1 p.
SNZ NZS/AS 1580. 107.8-79. Methods of Test for Paints and Related Materials Part 107.8: Application of Paint at a Nominated Spreading Rate by Determination of Wet Film Mass (Brush Application) (This is a Joint Standard with SAA AS 1580.107.8). 2 pp.

—Application—Air Spraying
SAA AS 1580.205. 2-81. Paints and Related Materials—Methods of Test—Part 205.2: Application Properties—Conventional Air Spraying (R 1992). 1 p.
SNZ NZS/AS 1580. 205.2-81. Methods of Test for Paints and Related Materials Part 205.2: Application Properties. Conventional Air Spraying (This is a Joint Standard with SAA AS 1580.205.2). 1 p.

—Application—Airless Spraying
SAA AS 1580.205. 4-92. Paints and Related Materials—Methods of Test—Part 205.4: Application Properties—Airless Spraying (in Professional Packages 30, 39). 2 pp.
SNZ NZS/AS 1580. 205.4-92. Methods of Test for Paints and Related Materials Part 205.4: Application Properties. Airless Spraying (This is a Joint Standard with SAA AS 1580.205.4). 2 pp.

—Application—Brushing
SAA AS 1580.205. 1-81. Paints and Related Materials—Methods of Test—Part 205.1: Application Properties—Brushing (R 1992). 1 p.
SNZ NZS/AS 1580. 205.1-81. Methods of Test for Paints and Related Materials Part 205.1: Application Properties. Brushing (This is a Joint Standard with SAA AS 1580.205.1). 2 pp.

—Application—Rolling
SAA AS 1580.205. 3-81. Paints and Related Materials—Methods of Test—Part 205.3: Application Properties—Roller Coating (R 1992). 1 p.
SNZ NZS/AS 1580. 205.3-81. Methods of Test for Paints and Related Materials Part 205.3: Application Properties. Roller Coating (This is a Joint Standard with SAA AS 1580.205.3). 1 p.

—Bend Testing
SAA AS 1580.402. 1-92. Paints and Related Materials—Methods of Test—Part 402.1: Bend Test (in Professional Packages 30, 39). 3 pp.

INDEX and DIRECTORY of

INTERNATIONAL AND NON-U.S. NATIONAL STANDARDS
SUBJECT INDEX

Lacquers

Lacquers (Cont.)

—Bend Testing (Cont.)

SNZ NZS/AS 1580. 402.1-92. Methods of Test for Paints and Related Materials Part 402.1: Bend Test (This is a Joint Standard with SAA AS 1580.402.1). 3 pp.

—Blistering

SNZ NZS/AS 1580. 481.2-75. Methods of Test for Paints and Related Materials Part 481.2: Assessment of Blistering of Paint Films (This is a Joint Standard with SAA AS 1581.481.2). 4 pp.

—Blushing

CGSB 1-GP-71 METH 38.2-80. Methods of Testing Paints and Pigments Blushing Resistance Lacquers at Normal Humidity. 1 p.

CGSB 1-GP-71 METH 38.3-80. Methods of Testing Paints and Pigments Blushing Resistance Lacquers at High Temperature and Humidity. 1 p.

—Cellulose Nitrate

MOD UK DSTAN 80-127-89. Lacquer, Cellulose Nitrate, Type QX Grades 1.0 and 1.1: Brushing/Sealing Grade 2: Spraying Grades 3.0 and 3.1: Dipping Issue 1. 15 pp.

—Chemical Resistance—Sodium Carbonate

MOD UK DEF-1053: METH NO. 66. Standard Methods of Testing Paint, Varnish, Lacquer and Related Products. Indices Method 66: Resistance to Sodium Carbonate Solution (Withdrawn).

—Colorimetric Analysis

CNS K6386-76. Method of Test for Lacquer in Surface of Leather Shoes (Mar) (3922). 2 pp.

—Comprehensive Testing

MOD UK DEF-1053. Standard Methods of Testing Paint, Varnish, Lacquer and Related Products. Indices (Withdrawn).

SAA AS K41. Methods of Test for Paints, Varnishes, Laquers, and Related Materials See Also AS 1050.

SAA AS 1580. Paints and Related Materials—Methods of Test Complete Set in Binder (This is a Joint Standard with SANZ NZS 1580).

—Consistency

SAA AS 1580.214. 1-90. Paints and Related Materials—Methods of Test—Part 214.1: Consistency—Stormer Viscometer Amdt 1 January/February 1992 Amdt 2 January/February 1993 (in Professional Packages 30, 39). 1 p.

SAA AS 1580.214. 2-90. Paints and Related Materials—Methods of Test—Part 214.2: Consistency—Flow Cup. 6 pp.

SAA AS 1580.214. 4-90. Paints and Related Materials—Methods of Test—Part 214.4: Consistency—Rotothinner. 4 pp.

SAA AS 1580.214. 5-90. Paints and Related Materials—Methods of Test—Part 214.5: Consistency—Ratational Viscometer. 10 pp.

SNZ NZS/AS 1580. 214.1-90. Methods of Test for Paints and Related Materials Part 214.1: Consistency—Stormer Viscometer Amend: 1, 1992 (This is a Joint Standard with SAA AS 1580.214.1). 6 pp.

SNZ NZS/AS 1580. 214.2-90. Methods of Test for Paints and Related Materials Part 214.2: Consistency—Flow Cup (This is a Joint Standard with SAA AS 1580.214.2). 6 pp.

SNZ NZS/AS 1580. 214.4-90. Methods of Test for Paints and Related Materials Part 214.4: Consistency—Rotothinner (This is a Joint Standard with SAA AS 1580.214.4). 4 pp.

SNZ NZS/AS 1580. 214.5-90. Methods of Test for Paints and Related Materials Part 214.5: Consistency—Rotational Viscometer (This is a Joint Standard with SAA AS 1580.214.5). 10 pp.

—Corrosion Inhibitive

MOD UK DSTAN 03-30: Part 3-90. Treatments for the Protection of Metal Parts of Service Stores and Equipment Against Corrosion Part 3: Schedule of Paint, Lacquer and Varnish Finishes Issue 1. 42 pp.

MOD UK DG-8: Pt 1: Sec III. Defence Guide: Treatments for the Protection of Metal Parts of Service Stores and Equipments Against Corrosion Part 1: Schedule of Protective Coatings Section 3: Index of Specifications Issue 5 (Superseded by Def Stan 03-30: Part 3).

MOD UK DG-8: Pt 1: Sec III. Defence Guide: Treatments for the Protection of Metal Parts of Service Stores and Equipments Against Corrosion Part 1: Schedule of Protective Coatings Section 3: Paint, Lacquer and Varnish Finishes Issue 2 (Superseded by Def Stan 03-30: Part 3).

—Corrosion Testing

SAA AS 1580.452. 2-92. Paints and Related Materials—Methods of Test—Part 452.2: Resistance to Corrosion-Salt Droplet Test (in Professional Packages 30, 39). 5 pp.

Lacquers (Cont.)

—Corrosion Testing (Cont.)

SNZ NZS/AS 1580. 452.2-92. Methods of Test for Paints and Related Materials Part 452.2: Resistance to Corrosion—Salt Droplet Test (This is a Joint Standard with SAA AS 1580.452.2). 5 pp.

—Density

CNS K6665-81. Method of Test for Density of Paint Varnish, Lacquer and Related Products (May)(7405).

SAA AS 1580.202. 1-80. Paints and Related Materials—Methods of Test—Part 202.1: Density (R 1991). 2 pp.

SNZ NZS/AS 1580. 202.1-80. Methods of Test for Paints and Related Materials Part 202.1: Density (This is a Joint Standard with SAA AS 1580.202.1). 2 pp.

—Drying

CGSB 1-GP-71 METH 5.1-78. Methods of Testing Paints and Pigments Drying Time General Method. 2 pp.

SAA AS 1580.101. 1-86. Paints and Related Materials—Methods of Test—Part 101.1: Air Drying Conditions (R 1992). 4 pp.

SAA AS 1580.401. 1-81. Paints and Related Materials—Methods of Test—Part 401.1: Surface Dry Condition (R 1992). 2 pp.

SAA AS 1580.401. 3-90. Paints and Related Materials—Methods of Test—Part 401.3: Drying Times Using a BK-Type Recorder. 4 pp.

SNZ NZS/AS 1580. 101.1-86. Methods of Test for Paints and Related Materials Part 101.1: Air Drying Conditions (This is a Joint Standard with SAA AS 1580.101.1). 4 pp.

SNZ NZS/AS 1580. 401.1-81. Methods of Test for Paints and Related Materials Part 401.1: Surface Dry Condition (This is a Joint Standard with SAA AS 1580.401.1). 2 pp.

SNZ NZS/AS 1580. 401.3-90. Methods of Test for Paints and Related Materials Part 401.3: Drying Times Using a BK-Type Recorder (This ia a Joint Standard with SAA AS 1580.401.3). 4 pp.

—Dyes

MOD UK DEF-1430-64. Dyestuffs for R.D. Varnishes and Lacquers. 4 pp.

—Enamels

CNS K2014-87. Lacquer Enamel (May)(609). 3 pp.

—Environmental Testing

SAA AS 1580.457. 1-88. Paints and Related Materials—Methods of Test—Part 457.1: Resistance to Natural Weathering. 6 pp.

SNZ NZS/AS 1580. 457.1-88. Methods of Test for Paints and Related Materials Part 457.1: Resistance to Outdoor Weathering (This is a Joint Standard with SAA AS 1580.457.1). 6 pp.

—Environmental Testing—Artificial

SAA AS 1580.483. 1-92. Paints and Related Materials—Methods of Test—Part 483.1: Resistance to Artificial Weathering (Carbon-Arc Type Instruments) (in Professional Packages 30, 39). 8 pp.

SNZ NZS/AS 1580. 483.1-92. Methods of Test for Paints and Related Materials Part 483.1: Resistance to Artificial Weathering (Carbon-Arc Type Instruments) (This is a Joint Standard with SAA AS 1580.483.1). 8 pp.

—Fineness

SAA AS 1580.204. 1-90. Paints and Related Materials—Methods of Test—Part 204.1: Fineness of Grind. 8 pp.

SNZ NZS/AS 1580. 204.1-90. Methods of Test for Paints and Related Materias Part 204.1: Fineness of Grind (This is a Joint Standard with SAA AS 1580.204.1). 8 pp.

—Finish

SAA AS 1580.603. 1-92. Paints and Related Materials—Methods of Test—Part 603.1: Finish (in Professional Packages 30, 39). 1 p.

SNZ NZS/AS 1580. 603.1-75. Methods of Test for Paints and Related Materials Part 603.1: Finish (Reconfirmed 1988) (This is a Joint Standard with SAA AS 1580.603.1). 1 p.

—Furniture—Print Resistance

CGSB 1-GP-71 METH 142.2-78. Methods of Testing Paints and Pigments Print Resistance Furniture Lacquers. 1 p.

—Gloss

SAA AS 1580.602. 1-75. Paints and Related Materials—Methods of Test—Part 602.1: Visual Assessment of Gloss Reconfirmed 1988. 2 pp.

SAA AS 1580.602. 2-74. Paints and Related Materials—Methods of Test—Part 602.2: Specular Gloss Corrig. Reconfirmed 1988. 4 pp.

SNZ NZS/AS 1580. 602.1-75. Methods of Test for Paints and Related Materials Part 602.1: Visual Assessment of Gloss (Reconfirmed 1988) (This is a Joint Standard with SAA AS 1580.602.1). 2 pp.

Lacquers (Cont.)

—Gloss (Cont.)

SNZ NZS/AS 1580. 602.2-74. Methods of Test for Paints and Related Materials Part 602.2: Specular Gloss (Reconfirmed 1988) (This is a Joint Standard with SAA AS 1580.602.2). 4 pp.

—Hardness Testing

SAA AS 1580.401. 5-93. Paints and Related Materials—Methods of Test—Part 401.5: Hard Dry Condition—Sanding Test (in Professional Package 39A). 2 pp.

SAA AS 1580.401. 6-92. Paints and Related Materials—Methods of Test—Part 401.6: Hard Dry Condition (Mechanical Thumb Test) (in Professional Packages 30, 39). 6 pp.

SNZ NZS/AS 1580. 401.5-80. Methods of Test for Paints and Related Materials Part 401.5: Hard Dry Condition—Sanding Test (This is a Joint Standard with SAA AS 1580.401.5). 2 pp.

SNZ NZS/AS 1580. 401.6-92. Methods of Test for Paints and Related Materials Part 401.6: Hard Dry Condition (Mechanical Thumb Test) (This is a Joint Standard with SAA AS 1580.401.6). 6 pp.

—Heat Resistance

SAA AS 1580.407. 1-93. Paints and Related Materials—Methods of Test—Part 407.1: Heat Resistance—Slow Cooling (in Professional Package 39A). 2 pp.

SAA AS 1580.407. 2-93. Paints and Related Materials—Methods of Test—Part 407.2: Heat Resistance—Thermal Shock (in Professional Package 39A). 2 pp.

SNZ NZS/AS 1580. 407.1-81. Methods of Test for Paints and Related Materials Part 407.1: Heat Resistance—Slow Cooling (This is a Joint Standard with SAA AS 1580.407.1). 2 pp.

SNZ NZS/AS 1580. 407.2-81. Methods of Test for Paints and Related Materials Part 407.2: Heat Resistance—Thermal Shock (This is a Joint Standard with SAA AS 1580.407.2). 2 pp.

—Humidity

SAA AS 1580.101. 5-86. Paints and Related Materials—Methods of Test—Part 101.5: Conditions of Test, Temperature and Humidity Controlled (R 1992). 1 p.

SAA AS 1580.452. 1-92. Paints and Related Materials—Methods of Test—Part 452.1: Resistance to Humidity Under Condensation Conditions (in Professional Packages 30, 39). 3 pp.

SNZ NZS/AS 1580. 101.5-86. Methods of Test for Paints and Related Materials Part 101.5: Conditions of Test, Temperature and Humidity Controlled (This is a Joint Standard with SAA AS 1580.101.5). 1 p.

SNZ NZS/AS 1580. 452.1-92. Methods of Test for Paints and Related Materials Part 452.1: Resistance to Humidity (This is a Joint Standard with SAA AS 1580.452.1). 3 pp.

—Impact Testing

SAA AS 1580.406. 1-93. Paints and Related Materials—Methods of Test—Part 406.1: Resistance to Impact—Falling Weight Test (Gardener-Type Tester) (in Professional Package 39A). 2 pp.

SNZ NZS/AS 1580. 406.1-86. Methods of Test for Paints and Related Materials Part 406.1: Resistance to Impact—Falling Weight Test (Gardner-Type Tester) (This is a Joint Standard with SAA AS 1580.406.1). 3 pp.

—Knife Testing

SAA AS 1580.408. 2-93. Paints and Related Materials—Methods of Test—Part 408.2: Adhesion—Knife Test (in Professional Package 39). 3 pp.

SNZ NZS/AS 1580. 408.2-84. Methods of Test for Paints and Related Materials Part 408.2: Adhesion—Knife Test (This is a Joint Standard with SAA AS 1580.408.2). 4 pp.

—Labels

MOD UK DSTAN 80-64-76. Lacquer, Finishing, Paper Labels Type: Brushing Issue 1. 6 pp.

—Lead Content—Colorimetric Analysis

MOD UK DEF-1053: METH NO. 81. Standard Methods of Testing Paint, Varnish, Lacquer and Related Products. Indices Method 81: Determination of Lead in 'Lead Free' Paints, Varnishes and Allied Products and Their Containers. (Withdrawn).

—Leather Shoes

CNS K6386-76. Method of Test for Lacquer in Surface of Leather Shoes (Mar) (3922). 2 pp.

—Light Testing

SAA AS 1580.482. 1-75. Paints and Related Materials—Methods of Test—Part 482.1: Fastness to Light Reconfirmed 1988. 2 pp.

SNZ NZS/AS 1580. 482.1-75. Methods of Test for Paints and Related Materials Part 482.1: Fastness to Light (Reconfirmed 1988) (This is a Joint Standard with SAA AS 1580.482.1). 2 pp.

INDUSTRY STANDARDS

INTERNATIONAL AND NON-U.S. NATIONAL STANDARDS
SUBJECT INDEX

Lacquers

Lacquers (Cont.)

—Metal
MOD UK DSTAN 80-92-84. Lacquer, Phenol-Formaldehyde, Stoving, Type QX Issue 1. 12 pp.
MOD UK DSTAN 80-92-92. Lacquer, Phenol-Formaldehyde, Stoving, Type QX, Grades 22%, 30%, 39% Types: Dipping Spraying Issue 2. 18 pp.

—Metal—Corrosion Testing
SAA AS 1580.481. 4-75. Paints and Related Materials—Methods of Test—Part 481.4: Assessment of Corrosion of an Underlying Iron or Steel Surface (Superseded by AS 1580.481.3—1992).
SNZ NZS/AS 1580. 481.4-75. Methods of Test for Paints and Related Materials Part 481.4: Assessment of Corrosion of an Underlying Iron or Steel Surface (This is a Joint Standard with SAA AS 1580.481.4). 2 pp.

—Mineral Oils—Resistance
SAA AS 1580.454. 1-92. Paints and Related Materials—Methods of Test—Part 454.1: Resistance to Mineral Oil and Other Organic Liquids (in Professional Package 39). 2 pp.
SNZ NZS/AS 1580. 454.1-82. Methods of Test for Paints and Related Materials Part 454.1: Resistance to Mineral Oil (This is a Joint Standard with SAA AS 1580.454.1). 2 pp.

—Nonvolatile Matter Content
CGSB 1-GP-71 METH 17.1-80. Methods of Testing Paints and Pigments Volatile Matter Paints, Enamels and Lacquers. 1 p.
SAA AS 1580.301. 1-92. Paints and Related Materials—Methods of Test—Part 301.1: Non-Volatile Content by Mass (in Professional Packages 30, 39). 3 pp.
SAA AS 1580.301. 2-90. Paints and Related Materials—Methods of Test—Part 301.2: Non-Volatile Content by Volume (Volume Solids). 12 pp.
SNZ NZS/AS 1580. 301.1-92. Methods of Test for Paints and Related Materials Part 301.1: Non-Volatile Content by Mass (This is a Joint Standard with SAA AS 1580.301.1). 3 pp.
SNZ NZS/AS 1580. 301.2-90. Methods of Test for Paints and Related Materials Part 301.2: Non-Volatile Content by Volume (Volume Solids) (This is a Joint Standard with SAA AS 1580.301.2). 12 pp.

—Opacity
MOD UK DEF-1053: METH NO. 12-01. Standard Methods of Testing Paint, Varnish, Lacquer and Related Products Method 12: Opacity (Contrast Ratio) (Reprinted November 1960, Incorporating Amendments Nos. 1, 2 & 3); Amendment 5 (Withdrawn). 7 pp.

—Petroleum Spirits—Resistance
SAA AS 1580.453. 1-92. Paints and Related Materials—Methods of Test—Part 453.1: Resistance to Petroleum Spirit (in Professional Packages 30, 39). 2 pp.
SNZ NZS/AS 1580. 453.1-92. Methods of Test for Paints and Related Materials Part 453.1: Resistance to Petroleum Spirit (This is a Joint Standard with SAA AS 1580.453.1). 2 pp.

—Physical Properties
CNS K6386-76. Method of Test for Lacquer in Surface of Leather Shoes (Mar) (3922). 2 pp.

—Pigment Content
SAA AS 1580.302. 1-93. Paints and Related Materials—Methods of Test—Part 302.1: Pigment Content (in Professional Package 39). 3 pp.
SNZ NZS/AS 1580. 302.1-85. Methods of Test for Paints and Related Materials Part 302.1: Pigment Content (This is a Joint Standard with SAA AS 1580.302.1). 2 pp.

—Print Resistance
CGSB 1-GP-71 METH 142.1-78. Methods of Testing Paints and Pigments Print Resistance Lacquers. 1 p.

—Putty
CNS K2151-86. Lacquer Putty (Dec)(11803). 2 pp.
CNS K6897-86. Method of Test for Lacquer Putty (Dec)(11804). 3 pp.

—Pyrotechnics
MOD UK DSTAN 80-43-84. Lacquer, Shellac, Kaolin, Stearated Type QX Issue 2. 10 pp.
MOD UK DSTAN 80-108-85. Lacquer, Natural Shellac 40% Type QX Issue 1. 8 pp.

—Recoatability
SAA AS 1580.404. 1-81. Paints and Related Materials—Methods of Test—Part 404.1: Recoating Properties (R 1992). 2 pp.
SNZ NZS/AS 1580. 404.1-81. Methods of Test for Paints and Related Materials Part 404.1: Recoating Properties (This is a Joint Standard with SAA AS 1580.404.1). 2 pp.

Lacquers (Cont.)

—Reincorporation
SAA AS 1580.211. 2-80. Paints and Related Materials—Methods of Test—Part 211.2: Ease of Manual Re-Incorporation (R 1991). 2 pp.
SNZ NZS/AS 1580. 211.2-80. Methods of Test for Paints and Related Materials Part 211.2: Ease of Manual Re-Incorporation (This is a Joint Standard with SAA AS 1580.211.2). 2 pp.

—Retardants
CNS K2065-86. Retarder for Lacquer (Oct)(4914). 2 pp.
CNS K6888-86. Method of Test for Retarder for Lacquer (Oct)(11721).

—Rheological Properties
SAA AS 1580.214. 0-90. Paints and Related Materials—Methods of Test—Part 214.0: Rheological Properties—Guide to Test Methods. 5 pp.
SNZ NZS/AS 1580. 214.0-90. Methods of Test for Paints and Related Materials Part 214.0: Rheological Properties—Guide to Test Methods (This is a Joint Standard with SAA AS 1580.214.0). 5 pp.

—Rusting
SAA AS 1580.481. 3-92. Paints and Related Materials—Methods of Test—Part 481.3: Coatings—Exposed to Weathering—Degree of Corrosion of Coated Metal Substrates (in Professional Packages 30, 39) (Supersedes AS 1580.481.4—1975). 6 pp.
SNZ NZS/AS 1580. 481.3-92. Methods of Test for Paints and Related Materials Part 481.3: Coatings—Exposed to Weathering—Degree of Corrosion of Coated Metal Substrates (This is a Joint Standard with SAA AS 1580.481.3). 6 pp.

—Sampling
SAA AS 1580.102. 1-92. Paints and Related Materials—Methods of Test—Part 102.1: Sampling Procedure (in Professional Packages 30, 39). 2 pp.
SNZ NZS/AS 1580. 102.1-92. Methods of Test for Paints and Related Materials Part 102.1: Sampling Procedure (This is a Joint Standard with SAA AS 1580.102.1). 2 pp.

—Scratch Hardness Testing
SAA AS 1580.403. 1-81. Paints and Related Materials—Methods of Test—Part 403.1: Scratch Resistance (R 1992). 2 pp.
SNZ NZS/AS 1580. 403.1-81. Methods of Test for Paints and Related Materials Part 403.1: Scratch Resistance (This is a Joint Standard with SAA AS 1580.403.1). 2 pp.

—Settling
SAA AS 1580.211. 1-80. Paints and Related Materials—Methods of Test—Part 211.1: Degree of Settling (R 1991). 2 pp.
SNZ NZS/AS 1580. 211.1-80. Methods of Test for Paints and Related Materials Part 211.1: Degree of Settling (This is a Joint Standard with SAA AS 1580.211.1). 2 pp.

—Skinning Properties
SAA AS 1580.203. 1-93. Paints and Related Materials—Methods of Test—Part 203.1: Skin Formation (in Professional Package 39). 2 pp.
SNZ NZS/AS 1580. 203.1-80. Methods of Test for Paints and Related Materials Part 203.1: Skin Formation (This is a Joint Standard with SAA AS 1580.203.1). 2 pp.

—Solvents
CNS K6543-80. Method of Test for Acidity in Volatile Solvents and Chemical Intermediates Used in Paint, Varnish, Lacquer and Related Products (Aug)(6188).

—Solvents—Miscibility
CNS K6656-81. Method of Test for Heptane Mescibility of Lacquer Solvents (Apr)(7257).

—Surfacers
CNS K2148-86. Lacquer Surfacer (Oct)(11731). 2 pp.
CNS K6892-86. Method of Test for Lacquer Surfacer (Oct)(11732). 4 pp.

—Temperature Testing
SAA AS 1580.101. 4-86. Paints and Related Materials—Methods of Test—Part 101.4: Conditions of Test, Temperature Controlled (R 1992). 1 p.
SAA AS 1580.101. 5-86. Paints and Related Materials—Methods of Test—Part 101.5: Conditions of Test, Temperature and Humidity Controlled (R 1992). 1 p.
SNZ NZS/AS 1580. 101.4-86. Methods of Test for Paints and Related Materials Part 101.4: Conditions of Test, Temperature Controlled (This is a Joint Standard with SAA AS 1580.101.4). 1 p.
SNZ NZS/AS 1580. 101.5-86. Methods of Test for Paints and Related Materials Part 101.5: Conditions of Test, Temperature and Humidity Controlled (This is a Joint Standard with SAA AS 1580.101.5). 1 p.

Lacquers (Cont.)

—Test Panels
SAA AS 1580.104. 1-92. Paints and Related Materials—Methods of Test—Part 104.1: Recommended Materials for Test Panels Amdt 1 July 1993 (in Professional Packages 30, 39). 2 pp.
SNZ NZS/AS 1580. 104.1-92. Methods of Test for Paints and Related Materials Part 104.1: Recommended Materials for Test Panels (This is a Joint Standerd with SAA AS 1580.104.1). 2 pp.

—Test Panels—Metals
SAA AS 1580.105. 1-92. Paints and Related Materials—Methods of Test—Part 105.1: Pretreatment of Metal Test Panels—Solvent Cleaning (in Professional Packages 30, 39). 1 p.
SAA AS 1580.105. 2-93. Paints and Related Materials—Methods of Test—Part 105.2: Pretreatment of Metal Test Panels—Sanding (in Professional Package 39). 2 pp.
SNZ NZS/AS 1580. 105.1-92. Methods of Test for Paints and Related Materials Part 105.1: Pretreatment of Metal Test Panels. Solvent Cleaning (This is a Joint Standard with SAA AS 1580.105.1). 1 p.
SNZ NZS/AS 1580. 105.2-80. Methods of Test for Paints and Related Materials Part 105.2: Pretreatment of Metal Test Panels. Sanding (This is a Joint Standard with SAA AS 1580.105.2). 2 pp.
SNZ NZS/AS 1580. 105.3-80. Methods of Test for Paints and Related Materials Part 105.3: Pretreatment of Metal Test Panels. Chromic Acid Dipping (This is a Joint Standard with SAA AS 1580.105.3). 2 pp.

—Test Panels—Timber
SAA AS 1580.106. 1-79. Paints and Related Materials—Methods of Test—Part 106.1: Preparation of Timber Test Panels for Outdoor Weathering Test. 1 p.
SNZ NZS/AS 1580. 106.1-79. Methods of Test for Paints and Related Materials Part 106.1: Preparation of Timber Test Panels for Outdoor Weathering Test (This is a Joint Standard with SAA AS 1580.106.1). 2 pp.

—Test Preparation
SAA AS 1580.103. 1-80. Paints and Related Materials—Methods of Test—Part 103.1: Preliminary Examination and Preparation for Testing (R 1991). 2 pp.
SNZ NZS/AS 1580. 103.1-80. Methods of Test for Paints and Related Materials Part 103.1: Preliminary Examination and Preparation for Testing (This is a Joint Standard with SAA AS 1580.103.1). 2 pp.

—Test Specimens—Formulas
CGSB 1-GP-71 METH 150.4-74. Methods of Testing Paints and Pigments Standard Formulas for Test Materials Formula for Standard Test Lacquer; (Re-Edited September 1982). 1 p.

—Thickness Measurement
SAA AS 1580.107. 1-92. Paints and Related Materials—Methods of Test—Part 107.1: Determination of Wet Film Thickness from Dry Film Mass (in Professional Packages 30, 39). 2 pp.
SAA AS 1580.107. 2-80. Paints and Related Materials—Methods of Test—Part 107.2: Determination of Wet Film Thickness from Wet Film Mass (R 1991). 2 pp.
SAA AS 1580.107. 3-92. Paints and Related Materials—Methods of Test—Part 107.3: Determination of Wet Film Thickness by Wheel Gauge (in Professional Packages 30, 39). 2 pp.
SAA AS 1580.108. 1-80. Paints and Related Materials—Methods of Test—Part 108.1: Determination of Dry Film Thickness on Iron and Steel Substrates (Permanent Magnet Instruments). 2 pp.
SAA AS 1580.108. 2-93. Paints and Related Materials—Methods of Test—Part 108.2: Dry Film Thickness—Paint Inspection Guage (in Professional Package 39A). 4 pp.
SNZ NZS/AS 1580. 107.1-92. Methods of Test for Paints and Related Materials Part 107.1: Determination of Wet Film Thickness from Dry Film Mass (This is a Joint Standard with SAA AS 1580.107.1). 2 pp.
SNZ NZS/AS 1580. 107.2-80. Methods of Test for Paints and Related Materials Part 107.2: Determination of Wet Film Thickness from Wet Film Mass (This is a Joint Standard with SAA AS 1580.107.2). 2 pp.
SNZ NZS/AS 1580. 107.3-92. Methods of Test for Paints and Related Materials Part 107.3: Determination of Wet Film Thickness by Wheel Gauge (This is a Joint Standard with SAA AS 1580.107.3). 2 pp.

INDEX and DIRECTORY of

INTERNATIONAL AND NON-U.S. NATIONAL STANDARDS
SUBJECT INDEX

Lacquers (Cont.)
—**Thickness Measurement** (Cont.)
  SNZ NZS/AS 1580. 108.1-80. Methods of Test for Paints and Related Materials Part 108.1: Determination of Dry Film Thickness on Iron and Steel Substrates (Permanent Magnet Instruments) (This is a Joint Standard with SAA AS 1580.108.1). 2 pp.
  SNZ NZS/AS 1580. 108.2-85. Methods of Test for Paints and Related Materials Part 108.2: Dry Film Thickness. Paint Inspection Gauge (This is a Joint Standard with SAA AS 1580.108.2). 3 pp.

—**Thinner Compatibility**
  SAA AS 1580.208. 1-81. Paints and Related Materials—Methods of Test—Part 208.1: Thinner Compatibility (R 1992). 1 p.
  SNZ NZS/AS 1580. 208.1-81. Methods of Test for Paints and Related Materials Part 208.1: Thinning or Mixing Properties (This is a Joint Standard with SAA AS 1580.208.1). 1 p.

—**Thinners**
  CGSB CAN/CGSB-1.110-M91. General Purpose Thinner for Lacquers; (Corrigendum—May 1991). 8 pp.
  CNS K2005-86. Lacquer Thinner (Oct)(554). 2 pp.
  CNS K6065-86. Method of Test for Lacquer Thinner (Oct)(886).
  MOD UK DSTAN 80-38-85. Thinners for: Paint Epoxy Two-Pack, Cellulose Nitrate Paints, Dopes and Lacquers Issue 2. 8 pp.
  MOD UK DSTAN 80-99-01. Thinners for Paint Finishing Acrylic, Spraying Quality Issue 1; Amendment 1. 7 pp.

—**Thinners—Blushing**
  CGSB 1-GP-71 METH 38.1-80. Methods of Testing Paints and Pigments Blushing Resistance Lacquer Thinner. 1 p.

—**Thinners—Hydrocarbon Content**
  CGSB 1-GP-71 METH 66.2-82. Methods of Testing Paints and Pigments Hydrocarbons and Esters Hydrocarbons in Lacquer Thinners. 1 p.

—**Thinners—Print Resistance**
  CGSB 1-GP-71 METH 142.1-78. Methods of Testing Paints and Pigments Print Resistance Lacquers. 1 p.

—**Viscosity**
  SAA AS 1580.214. 3-93. Paints and Related Materials—Methods of Test—Part 214.3: Viscosity—Cone-and-Plate (in Professional Package 39A). 7 pp.
  SNZ NZS/AS 1580. 214.3-90. Methods of Test for Paints and Related Materials Part 214.3: Viscosity—Cone and Plate (This is a Joint Standard with SAA AS 1580.214.3). 6 pp.

—**Volatile Matter Content**
  CGSB 1-GP-71 METH 17.1-80. Methods of Testing Paints and Pigments Volatile Matter Paints, Enamels and Lacquers. 1 p.
  CNS K6386-76. Method of Test for Lacquer in Surface of Leather Shoes (Mar) (3922). 2 pp.

—**Washing Testing**
  SAA AS 1580.459. 1-83. Paints and Related Materials—Methods of Test—Part 459.1: Resistance to Washing (R 1992). 3 pp.
  SNZ NZS/AS 1580. 459.1-83. Methods of Test for Paints and Related Materials Part 459.1: Resistance to Washing (This is a Joint Standard with SAA AS 1580.459.1). 3 pp.

—**Water Resistance Testing**
  SAA AS 1580.455. 1-82. Paints and Related Materials—Methods of Test—Part 455.1: Resistance to Water at Room Temperature (R 1992). 2 pp.
  SAA AS 1580.456. 1-93. Paints and Related Materials—Methods of Test—Part 456.1: Resistance to Boiling Water (in Professional Package 39). 2 pp.
  SNZ NZS/AS 1580. 455.1-82. Methods of Test for Paints and Related Materials Part 455.1: Resistance to Water at Room Temperature (This is a Joint Standard with SAA AS 1580.455.1). 2 pp.
  SNZ NZS/AS 1580. 456.1-82. Methods of Test for Paints and Related Materials Part 456.1: Resistance to Boiling Water (This is a Joint Standard with SAA AS 1580.456.1). 2 pp.

—**Wood—Checking—Temperature Change Testing**
  CGSB 1-GP-71 METH 121.2-79. Methods of Testing Paints and Pigments Resistance to Temperature Changes Cold-Check Resistance. 1 p.

—**Wood—Cracking—Temperature Change Testing**
  CGSB 1-GP-71 METH 121.2-79. Methods of Testing Paints and Pigments Resistance to Temperature Changes Cold-Check Resistance. 1 p.

Lacquers (Cont.)
—**Yellowing**
  SAA AS 1580.461. 1-77. Paints and Related Materials—Methods of Test—Part 461.1: Determination of Resistance to Yellowing (Dark Chamber) Reconfirmed 1988. 1 p.
  SNZ NZS/AS 1580. 461.1-77. Methods of Test for Paints and Related Materials Part 461.1: Determination of Resistance to Yellowing (Dark Chamber) (This is a Joint Standard with SAA AS 1580.461.1). 1 p.

Lactate Analyzers
See Also: Analyzers

—**Blood**
  JIS K 3601-89. Lactate Analyzers. 12 pp.

—**Food**
  JIS K 3601-89. Lactate Analyzers. 12 pp.

Lactates Content Analysis
—**Milk**
  EC 79/1067/EEC-79. First Commission Directive Laying Down Community Methods of Analysis for Testing Certain Partly or Wholly Dehydrated Preserved Milk for Human Consumption. 24 pp.

—**Milk—Spectrometry**
  BSI BS 1743: Part 8-87. 1987 Analysis of Dried Milk and Dried Milk Products Part 8: of the Lactic Acid and Lactates Content of Dried Milk. 6 pp.
  ISO 8069-86. Dried Milk—Determination of Lactic Acid and Lactates Content—Enzymatic Method First Edition; (Corrected and Reprinted -1988). 6 pp.

Lactic Acid
See Also: Lactic Acid Content Analysis
  CNS K7267-68. Chemical Reagent (Lactic Acid) (Nov)(1766).
  JIS K 1353-86. Lactic Acid. 9 pp.
  JIS K 8726-81. Lactic Acid.
  MOD UK TS 10261. Lactic Acid (Superseded by Def Stan 68-144).

—**Propellants**
  MOD UK DSTAN 68-144-93. Lactic Acid Issue 1. 15 pp.

Lactic Acid Content Analysis
See Also: Lactic Acid

—**Blood—Electrolytic Analysis**
  JIS K 3601-89. Lactate Analyzers. 12 pp.

—**Food—Electrolytic Analysis**
  JIS K 3601-89. Lactate Analyzers. 12 pp.

—**Milk**
  EC 79/1067/EEC-79. First Commission Directive Laying Down Community Methods of Analysis for Testing Certain Partly or Wholly Dehydrated Preserved Milk for Human Consumption. 24 pp.

—**Milk—Spectrometry**
  BSI BS 1743: Part 8-87. 1987 Analysis of Dried Milk and Dried Milk Products Part 8: of the Lactic Acid and Lactates Content of Dried Milk. 6 pp.
  ISO 8069-86. Dried Milk—Determination of Lactic Acid and Lactates Content—Enzymatic Method First Edition; (Corrected and Reprinted -1988). 6 pp.

Lactose
See Also: Fructose
  CNS K7268-68. Chemical Reagent (Lactose) (Nov)(1767).
  CNS N5216-85. Lactose (Sep)(11371). 1 p.

—**Test Specimens—Microbiological Analysis**
  DIN ENGL 10191 Pt 2-86. Microbiological Examination of Milk; Preparation of Samples of Dried Milk, Dried Whey, and Lactose (Oct). 2 pp.

Lactose Content Analysis
—**Caseins—Photometry**
  BSI BS 6248: Part 6-82. 1982 Caseins and Caseinates Part 6: Method for Determination of Lactose Content (Photometric Method). 6 pp.
  ISO 5548-80. Caseins and Caseinates—Determination of Lactose Content—Photometric Method First Edition; (Erratum—Jan 1982). 6 pp.
  SAA AS 2300.10.7-91. Methods of Chemical and Physical Testing for the Dairying Industry—Part 10: Caseins, Caseinates and Coprecipitates—Part 10.7: Determination of Lactose—Photometric Method (Supersedes AS N60—1970 (in Part)). 4 pp.

Lactose Content Analysis (Cont.)
—**Cream**
  BSI BS 1741: Sec 7.1-89. 1989 Methods for Chemical Analysis of Liquid Milk and Cream Part 7: Determination of Lactose Content Section 7.1: Reference Method. 4 pp.

—**Dairy Products**
  CNS N6061-88. Method of Test for Milk and Milk Products (Determination of Lactose) (Sep)(3445). 4 pp.

—**Lactose Monohydrate**
  MOD UK M 821/71. Examination of Lactose.

—**Milk**
  BSI BS 1741: Sec 7.1-89. 1989 Methods for Chemical Analysis of Liquid Milk and Cream Part 7: Determination of Lactose Content Section 7.1: Reference Method. 4 pp.
  BSI BS 1741: Sec 7.2-89. 1989 Methods for Chemical Analysis of Liquid Milk and Cream Part 7: Determination of Lactose Content Section 7.2: Routine Method. 4 pp.
  CNS N6061-88. Method of Test for Milk and Milk Products (Determination of Lactose) (Sep)(3445). 4 pp.

Lactose Monohydrate
  JIS K 8728-82. Lactose Monohydrate.

—**Comprehensive Testing**
  MOD UK M 821/71. Examination of Lactose.

Ladders
Use For: Accommodation Ladders; Embarkation Ladders; Jacob's Ladders; Pilot Ladders
See Also: Construction Equipment; Dog Step Ladders; Extension Ladders; Shelf Ladders; Step Irons; Stepladders; Trestles (Supports)
  BSI BS MA 92-81. (OBSOLESCENT) 1981 Pilot Ladders. 7 pp.
  BSI BS EN 131-1-93. 1993 Ladders Part 1: Specification for Terms, Types, Functional Sizes (R). 25 pp.
  BSI BS EN 131-1-01. 1993 Amd 1 Ladders Part 1: Specification for Terms, Types, Functional Sizes (AMD 7873) May 15, 1993 (Corr 1) (AMD 7873) May 15, 1993 (R). 26 pp.
  BSI BS EN 131: Part 2-93. 1993 Ladders Part 2: Specification for Requirements, Testing, Marking (R). 15 pp.
  BSI BS EN 131-2-01. 1993 Amd 1 Ladders Part 2: Specification for Requirements, Testing, Marking (AMD 7874) May 15, 1993 (Corr 1) (R). 16 pp.
  CEN PREN 131 (Part 1)-79. Ladders Terms, Types, Functional Sizes. 23 pp.
  CEN EN 131-1-93. Ladders—Terms, Types, Functional Sizes. 17 pp.
  CEN EN 131-2-93. Ladders—Requirements, Testing, Marking. 14 pp.
  CSA CAN3-Z11-M81. Portable Ladders; (Gen Instr 1 Thru 3). 46 pp.
  ISO 799-86. Shipbuilding—Pilot Ladders Second Edition. 7 pp.
  JIS F 2615-82. Pilot Ladders.
  JIS F 2615-69. Pilot Ladders. 6 pp.
  JIS F 2622-85. Pilot Accommodation Ladders.
  SAA AS 1657-92. Fixed Platforms, Walkways, Stairways and Ladders—Design, Construction and Installation (in Professional Packages 20, 21, 30, 41, 47, 55, 62, 63, 64, 65, 66, 67, 68, 69) (This is a Joint Standard with SNZ NZS 1657). 27 pp.
  SAA AS 1892.1-86. Portable Ladders—Part 1: Metal Amdt 1 November 1986. 27 pp.
  SNZ NZS/AS 1657-85. Fixed Platforms, Walkways, Stairways and Ladders. Design, Construction and Installation (Known as the SAA Code for Fixed Platforms, Walkways, Stairways and Ladders) (This is a Joint Standard with SAA AS 1657). 17 pp.
  SNZ NZS 5233-86. Specification for Portable Ladders (Other Than Timber Ladders) (Reconfirmed 1991). 25 pp.

—**Agricultural Buildings**
  BSI BS 4211-87. 1987 Ladders for Permanent Access to Chimneys, Other High Structures, Silos and Bins. 16 pp.
  BSI BS 4211-01. 1987 Amd 1 Ladders for Permanent Access to Chimneys, Other High Structures, Silos and Bins (AMD 7064) June 15, 1992 (R). 18 pp.

—**Aluminum**
  BSI BS 2037-90. 1990 Amd 1 Portable Aluminium Ladders, Steps, Trestles and Lightweight Stagings (AMD 6394) September 30, 1991. 21 pp.
  BSI BS 2037-02. 1990 Amd 2 Portable Aluminium Ladders, Steps, Trestles and Lightweight Stagings (AMD 7895) September 15, 1993 (R). 22 pp.
  BSI BS 2037-84. (WITHDRAWN) 1984 Amd 1 Portable Aluminium Ladders, Steps, Trestles and Lightweight Stagings. 22 pp.

INDUSTRY STANDARDS

Ladders (Cont.)

—Aluminum Alloy—Ships
CNS F3229-83. Aluminium Alloy Accommodation Ladders (Jun)(10344).
MOD UK NES 2007-89. Specification for Extending Ladders and Folding Platform Stepladders Aluminium Alloy Issue 2 (04.89). 11 pp.

—Aluminum Alloy—Wharves
CNS F3224-83. Aluminum Alloy Wharf Ladders (Jun)(10339).
JIS F 2613-90. Aluminium Alloy Wharf Ladders.
JIS F 2613-67. Aluminium Alloy Wharf Ladders (R 1976). 10 pp.

—Chimneys
BSI BS 4211-87. 1987 Ladders for Permanent Access to Chimneys, Other High Structures, Silos and Bins. 16 pp.
BSI BS 4211-01. 1987 Amd 1 Ladders for Permanent Access to Chimneys, Other High Structures, Silos and Bins (AMD 7064) June 15, 1992 (R). 18 pp.

—Glossaries
BSI BS EN 131-1-93. 1993 Ladders Part 1: Specification for Terms, Types, Functional Sizes (R). 25 pp.
BSI BS EN 131-1-01. 1993 Amd 1 Ladders Part 1: Specification for Terms, Types, Functional Sizes (AMD 7873) May 15, 1993 (Corr 1) (AMD 7873) May 15, 1993 (R). 26 pp.
CEN PREN 131 (Part 1)-79. Ladders Terms, Types, Functional Sizes. 23 pp.
CEN EN 131-1-93. Ladders—Terms, Types, Functional Sizes. 17 pp.

—Identification Systems
BSI BS EN 131: Part 2-93. 1993 Ladders Part 2: Specification for Requirements, Testing, Marking (R). 15 pp.
BSI BS EN 131-2-01. 1993 Amd 1 Ladders Part 2: Specification for Requirements, Testing, Marking (AMD 7874) May 15, 1993 (Corr 1) (R). 16 pp.
CEN EN 131-2-93. Ladders—Requirements, Testing, Marking. 14 pp.

—Industrial Buildings—Design
BSI BS 5395: Part 3-85. 1985 Stairs, Ladders and Walkways Part 3: Code of Practice for the Design of Industrial Type Stairs, Permanent Ladders and Walkways. 17 pp.

—Insulated—Linemen's Equipment
CNS Z2070-88. Insulated Ladder for Electric Work (Oct)(12452).

—Lofts
BSI BS 7553-92. 1992 Loft Ladders. 16 pp.
BSI BS 7553-01. 1992 Amd 1 Loft Ladders (AMD 7417) December 15, 1992. 19 pp.

—Maintenance
SAA AS 1892.4 (INT)-92. Portable Ladders—Part 4(Int): Selection, Safe Use and Care (in Professional Packages 47, 61) (Supersedes AS 1689—1974). 4 pp.

—Residential Buildings
CNS Z2071-88. Aluminum Alloy Ladder for Home Use (Oct)(12453).
CNS Z3030-88. Testing Method of Aluminum Alloy Ladder for Home Use (Oct)(12454).

—Rope—Ships
CNS F3227-83. Pilot Ladder (Jun)(10342).
JIS F 3612-90. Rope Ladders.
JIS F 3612-74. Rope Ladders. 4 pp.

—Safety
SAA AS 1892.4 (INT)-92. Portable Ladders—Part 4(Int): Selection, Safe Use and Care (in Professional Packages 47, 61) (Supersedes AS 1689—1974). 4 pp.

—Selection
SAA AS 1892.4 (INT)-92. Portable Ladders—Part 4(Int): Selection, Safe Use and Care (in Professional Packages 47, 61) (Supersedes AS 1689—1974). 4 pp.

—Ships
BSI BS MA 89-80. (OBSOLESCENT) 1980 Accommodation Ladders. 6 pp.
BSI BS MA 90-80. 1980 Embarkation Ladders. 6 pp.
CNS F3086-83. Jacobs' Ladder (Mar)(6916).
CNS F3228-83. Embarkation Ladders (Jun)(10343).
ISO 5488-79. Shipbuilding—Accommodation Ladders First Edition. 6 pp.
ISO 5489-86. Shipbuilding—Embarkation Ladders Second Edition. 6 pp.
JIS F 2617-74. Embarkation Ladders. 7 pp.
JIS F 2621-85. Accommodation Ladders.
JIS F 2623-90. Wharf Ladders for Small Ships.

—Ships—Inclined
SAA AS 1036-83. Steel Ladders for Ships—Inclined Ladders for Machinery Spaces. 11 pp.
SAA AS 1037-83. Steel Ladders for Ships—Inclined Ladders for Use in Other Than Machinery Spaces. 13 pp.

—Ships—Rungs
ISO 9519-90. Shipbuilding and Marine Structures—Rungs for Dog-Step Ladders First Edition. 6 pp.

—Ships—Vertical
SAA AS 1035-83. Steel Ladders for Ships—Vertical Ladders. 10 pp.

—Ships—Winches
ISO 7364-83. Shipbuilding and Marine Structures—Deck Machinery—Accommodation Ladder Winches First Edition. 5 pp.

—Silos
BSI BS 4211-87. 1987 Ladders for Permanent Access to Chimneys, Other High Structures, Silos and Bins. 16 pp.
BSI BS 4211-01. 1987 Amd 1 Ladders for Permanent Access to Chimneys, Other High Structures, Silos and Bins (AMD 7064) June 15, 1992 (R). 18 pp.

—Steel—Decks
JIS F 2603-70. Steel Deck Ladders. 12 pp.

—Steel—Ships
BSI BS MA 39: Part 1-73. 1973 Ships' Ladders Part 1: Ladders, Steel Vertical. 5 pp.
BSI BS MA 39: Part 2-73. 1973 Ships' Ladders Part 2: Ladders, Steel Sloping. 5 pp.
CNS F3036-80. Ships' Steel Vertical Ladder (May)(5569). 2 pp.
CNS F3037-83. Steel Deck Ladders (Mar)(5570). 7 pp.
CNS F3230-83. Steel Accommodation Ladders (Jun)(10345).
ISO 3797-76. Shipbuilding—Vertical Steel Ladders First Edition. 4 pp.
JIS F 2602-87. Ships' Steel Vertical Ladders.
JIS F 2602-75. Ships' Steel Vertical Ladders. 4 pp.
JIS F 2605-90. Small Size Steel Accommodation Ladders.
JIS F 2605-75. Steel Accommodation Ladders for Small Ships. 23 pp.
JIS F 7502-89. Steel Ladders and Steel Handrails for Marine Use.

—Steel—Wharves
CNS F3223-83. Steel Wharf Ladders (Jun)(10338).
JIS F 2612-90. Steel Wharf Ladders.
JIS F 2612-67. Steel Wharf Ladders. 17 pp.

—Wharves
JIS F 2617-89. Embarkation Ladders.
JIS F 2617-74. Embarkation Ladders. 7 pp.
JIS F 2623-90. Wharf Ladders for Small Ships.

—Wood
BSI BS 1129-90. 1990 Portable Timber Ladders, Steps, Trestles and Lightweight Stagings. 36 pp.
BSI BS 1129-01. 1990 Amd 1 Portable Timber Ladders, Steps, Trestles and Lightweight Stagings (AMD 7414) January 15, 1993. 38 pp.
BSI BS 1129-02. 1990 Amd 2 Portable Timber Ladders, Steps, Trestles and Lightweight Stagings (AMD 7896) September 15, 1993 (R). 39 pp.
SAA AS 1892.2-92. Portable Ladders—Part 2: Timber (in Professional Packages 47, 61) (Supersedes AS 1688—1974). 28 pp.
SNZ NZS 3609-78. Specification for Timber Ladders Amend: 1, 1984. 59 pp.

Ladles (Materials Handling)

—Firebricks—Steel Making
BSI BS 3056: Part 8-87. 1987 Size of Refractory Bricks Part 8: Bricks for Ladles. 10 pp.

—Lifting Hooks
DIN ENGL 15407 Pt 2-89. Laminated Point Hooks for Spigot Ladles; Components (Aug). 6 pp.

—Molten Metal
CNS B3240-73. Small Ladles for Molten Metal (Nov)(3617).

Ladles (Utensils)
See Also: Cooking Utensils
CNS S2035-80. Method of Test for Tuning Shovel (Soup Ladle) for Kitchen (Aug)(6265). 2 pp.

Lag Bolts
Use For: Coach Screws; Lag Screws
See Also: Screws; Tapping Screws

—Hexagonal Head
SAA AS 1393-73. Coach Screws (Metric Series) (with ISO Hexagon Heads). 15 pp.

Lag Screws
Use: Lag Bolts

Lagrangian Points

—L2—Radio Astronomy
CCIR QUESTION 147/7-90. Radioastronomy in the Vicinity of the L2 Sun-Earth Lagrangian Point—Questions Concerning Study Group 7—Science Services. 1 p.

Lags

—Thermal Insulation
BSI BS 3958: Part 1-82. 1982 Thermal Insulating Materials Part 1: Magnesia Preformed Insulation. 7 pp.
BSI BS 3958: Part 2-82. 1982 Thermal Insulating Materials Part 2: Calcium Silicate Preformed Insulation. 8 pp.
SNZ BS 3958: Part 1-82. Thermal insulating Materials Part 1: Magnesia Preformed Insulation. 8 pp.
SNZ BS 3958: Part 2-82. Thermal insulating Materials Part 2: Calcium Silicate Preformed Insulation. 8 pp.

—Wood—Beaters
JIS L 5163-60. Spiked Wood Lag for Kirschner Beater.

—Wool—Lappers
JIS L 5162-60. Wool Lag for Cylinder of Lapper.

Lake LA 4 Aircraft
See Also: Aircraft

—Antenna Positions
CAA. Lake LA4 (Approved Aerial Positions). 1 p.

Lake Red C
See Also: Pigments

—Pigments
CNS K2136-83. Lake Red C (Jun)(10357). 3 pp.
JIS K 5207-71. Lake Red C. 5 pp.

Lake Red D
See Also: Pigments

—Pigments
CNS K2137-83. Lake Red D (Jun)(10358). 3 pp.
JIS K 5208-71. Lake Red D. 5 pp.

Lakes
See Also: Ponds; Reservoirs (Lakes); Sedimentation

—Sampling
BSI BS 6068: Sec 6.4-87. 1987 Water Quality Part 6: Sampling Section 6.4: Guidance on Sampling from Lakes, Natural and Man-Made. 8 pp.
ISO 5667 Pt 4-87. Water Quality—Sampling—Part 4: Guidance on Sampling from Lakes, Natural and Man-Made First Edition. 8 pp.

Lamb
See Also: Food
CGSB 32.48-92. Lamb Cuts. 12 pp.

—Organ Meat
CGSB 32.56M-90. Fresh or Frozen Organs. 9 pp.

—Organ Meat—Frozen
CGSB 32.56M-90. Fresh or Frozen Organs. 9 pp.

Laminar Flow
Use For: Poiseuille Flow; Streamline Flow

—Clean Rooms
SAA AS 1386.2-89. Cleanrooms and Clean Workstations—Part 2: Laminar Flow Cleanrooms. 6 pp.

Laminated Cores
See Also: Magnetic Cores

—Electric Coils
BSI BS 7382: Part 1-90. 1990 Laminated Core Packages for Transformers and Inductors Used in Telecommunication and Electronic Equipment Part 1: Dimensions. 11 pp.
IEC 1021 Pt 1-90. Laminated Core Packages for Transformers and Inductors Used in Telecommunication and Electronic Equipment Part 1: Dimensions First Edition. 17 pp.

—Electric Coils—Designations
IEC 1186 Pt 1-92. Transformers and Inductors for Use in Telecommunication and Electronic Equipment—Designations for Cores and Assemblies Part 1: Laminated Cores First Edition. 20 pp.

—Transformers
BSI BS 7382: Part 1-90. 1990 Laminated Core Packages for Transformers and Inductors Used in Telecommunication and Electronic Equipment Part 1: Dimensions. 11 pp.

INTERNATIONAL AND NON-U.S. NATIONAL STANDARDS
SUBJECT INDEX

Laminated

Laminated Cores *(Cont.)*
—Transformers *(Cont.)*
IEC 1021 Pt 1-90. Laminated Core Packages for Transformers and Inductors Used in Telecommunication and Electronic Equipment Part 1: Dimensions First Edition. 17 pp.

—Transformers—Designations
IEC 1186 Pt 1-92. Transformers and Inductors for Use in Telecommunication and Electronic Equipment—Designations for Cores and Assemblies Part 1: Laminated Cores First Edition. 20 pp.

Laminated Fabrics
Use For: Fabric Laminates *See Also:* Bonded Fabrics; Coated Fabrics; Fabrics; Foam Laminated Fabrics; Fused Fabrics; Laminates

DIN VDE 0318 Pt 1-75. VDE Specification for Laminated Products: Paper Base Laminates, Fabric Base Laminates and Glass-Mat Base Laminates; Test Methods (Sept). 19 pp.

DIN VDE 0318 Pt 2-75. VDE Specification for Laminated Products: Paper Base Laminates, Fabric Base Laminates and Glass-Mat Base Laminates (DIN 7735) Requirements, Types (Sept). 21 pp.

JIS L 1089-70. Testing Method for Laminated Fabrics (R 1988). 13 pp.

—Adhesion Testing
DIN ENGL 53530-81. Testing of Organic Materials; Separation Test on Fabric Plies Bonded Together (Feb). 4 pp.

—Bars
DIN ENGL 40625-67. Laminated Products; Solid Bars of Paper-Base Laminate or Fabric-Base Laminate (July). 2 pp.

—Bonding Strength
CGSB CAN/CGSB-4.2 NO.65-M91. Textile Test Methods Determination of Strength of Bonds of Bonded, Laminated and Fused Fabrics. 9 pp.

—Dimensional Stability
CGSB CAN/CGSB-4.2 NO.67-M90. Textile Test Methods Dimensional Change and Appearance After Laundering of Coated, Bonded, Laminated and Fused Fabrics. 11 pp.

—Dry Cleaning—Dimensional Stability
CGSB CAN/CGSB-4.2 NO.66-M91. Textile Test Methods Dimensional Change and Appearance After Dry Cleaning of Coated, Bonded, Laminated and Fused Fabrics. 9 pp.

—Phenolic Resins—Moldings
MOD UK DSTAN 93-18-89. Phenolic Resin Bonded Fabric Mouldings, Natural, Type QX Issue 2. 13 pp.

—Phenolic Resins—Rods
MOD UK DSTAN 93-17-89. Phenolic Resin Bonded Fabric Round Rod, Natural, Type QX Issue 2. 10 pp.

—Phenolic Resins—Sheets
MOD UK DSTAN 93-15-89. Phenolic Resin Bonded Fabric Sheet, Natural, Type QX Issue 2. 10 pp.

—Phenolic Resins—Tubes
MOD UK DSTAN 93-16-89. Phenolic Resin Bonded Fabric Round Tube, Natural, Type QX Issue 2. 10 pp.

—Polyimide Resins—Copper Clad—Sheets
BSI BS 4584: Sec 102.17-92. 1992 Metal-Clad Base Materials for Printed Wiring Boards Part 102: Copper-Clad Base Materials Section 102.17: Specification for Thin Polyimide Woven Glass Fabric Copper-Clad Laminated Sheet of Defined Flammability for Use in. 16 pp.

IEC 249 Pt 2-17-92. Base Materials for Printed Circuits Part 2: Specifications Spec. No. 17: Thin Polyimide Woven Glass Fabric Copper-Clad Laminated Sheet of Defined Flammability for Use in the Fabrication of Multilayer Printed Board First Ed.; (Cor—June 1992) (Amd 1-1993). 40 pp.

—Rods
DIN ENGL 40624-67. Laminated Products; Solid Rods of Paper-Base Laminate or Fabric-Base Laminate (July). 2 pp.

—Roofing Membranes
CNS A2153-83. Synthetic Polymer Roofing Sheets Laminated with Cloth and Others (Apr)(10145). 5 pp.

CNS A3183-83. Method of Test for Synthetic Polymer Roofing Sheets Laminated with Cloth and Others (Apr)(10146).

—Sheets
DIN ENGL 40606-69. Laminated Products; Laminated Sheets and Strips of Fabric-Base Laminate or Glass-Mat-Base Laminate (Jan). 3 pp.

Laminated Fabrics *(Cont.)*
—Strips
DIN ENGL 40606-69. Laminated Products; Laminated Sheets and Strips of Fabric-Base Laminate or Glass-Mat-Base Laminate (Jan). 3 pp.

—Thermosetting Resins—Rods
BSI BS 6128: Part 1-81. 1981 Industrial Laminated Rods and Tubes Based on Thermosetting Resins Part 1: Classification and Methods of Test. 8 pp.

JIS K 6913-77. Laminated Thermosetting Rods (R 1980). 14 pp.

—Thermosetting Resins—Rods—Molded
BSI BS 6128: Part 2-81. 1981 Industrial Laminated Rods and Tubes Based on Thermosetting Resins Part 2: Round Moulded Rods. 4 pp.

BSI BS 6128: Part 4-81. 1981 Industrial Laminated Rods and Tubes Based on Thermosetting Resins Part 4: Rectangular Moulded Rods. 4 pp.

BSI BS 6128: Part 6-81. 1981 Industrial Laminated Rods and Tubes Based on Thermosetting Resins Part 6: Hexagonal Moulded Rods. 4 pp.

—Thermosetting Resins—Rods—Pultruded
BSI BS 6128: Part 3-81. (WITHDRAWN) 1981 Amd 1 Industrial Laminated Rods and Tubes Based on Thermosetting Resins Part 3: Round Pultruded Rods. 5 pp.

BSI BS 6128: Part 5-81. (WITHDRAWN) 1981 Amd 1 Industrial Laminated Rods and Tubes Based on Thermosetting Resins Part 5: Rectangular Pultruded Rods. 5 pp.

BSI BS 6128: Part 7-81. (WITHDRAWN) 1981 Amd 1 Industrial Laminated Rods and Tubes Based on Thermosetting Resins Part 7: Hexagonal Pultruded Rods. 5 pp.

—Thermosetting Resins—Tubes
BSI BS 6128: Part 1-81. 1981 Industrial Laminated Rods and Tubes Based on Thermosetting Resins Part 1: Classification and Methods of Test. 8 pp.

JIS K 6914-77. Laminated Thermosetting Tubes (R 1980). 22 pp.

—Thermosetting Resins—Tubes—Molded
BSI BS 6128: Part 9-82. 1982 Industrial Laminated Rods and Tubes Based on Thermosetting Resins Part 9: Round and Moulded Tubes. 8 pp.

BSI BS 6128: Part 13-83. 1983 Industrial Laminated Rods and Tubes Based on Thermosetting Resins Part 13: Specification for Rectangular Moulded Tubes. 8 pp.

—Thermosetting Resins—Tubes—Rolled
BSI BS 6128: Part 8-82. 1982 Industrial Laminated Rods and Tubes Based on Thermosetting Resins Part 8: Round Rolled Tubes. 8 pp.

—Tubes—Molded
DIN ENGL 40615-67. Laminated Products; Compression Moulded Circular Tubes of Paper-Base Laminate or Fabric-Base Laminate (July). 2 pp.

DIN ENGL 40616-67. Laminated Products; Compression Moulded Square Tubes of Paper-Base Laminate or Fabric-Base Laminate (July). 2 pp.

DIN ENGL 40617-67. Laminated Products; Compression Moulded Hexagon Tubes of Paper-Base Laminate or Fabric-Base Laminate (July). 2 pp.

DIN ENGL 40618-71. Laminated Products; Laminated Moulded Tubes of Rectangular Cross-Section; Paper-Base Laminate or Fabric-Base Laminate (June). 2 pp.

—Tubes—Rolled
DIN ENGL 40607-67. Laminated Products; Tubes, Rolled, Not Moulded of Laminated Paper or Laminated Fabric (July). 4 pp.

—Urethane Foams
SAA AS K180-69. Urethane Foam Laminates for Garment Applications. 7 pp.

Laminated Glass
Use For: Glass Laminates *See Also:* Glass; Laminates; Safety Glass; Sheet Glass

CGSB CAN/CGSB-12.1-M90. Tempered or Laminated Safety Glass. 15 pp.

CNS R2042-86. Laminated Glasses (Oct) (1183).

CNS R3043-86. Method of Test for Laminated Glass (Oct) (1184). 6 pp.

DIN VDE 0318 Pt 1-75. VDE Specification for Laminated Products: Paper Base Laminates, Fabric Base Laminates and Glass-Mat Base Laminates; Test Methods (Sept). 19 pp.

DIN VDE 0318 Pt 2-75. VDE Specification for Laminated Products: Paper Base Laminates, Fabric Base Laminates and Glass-Mat Base Laminates (DIN 7735) Requirements, Types (Sept). 21 pp.

JIS R 3205-89. Laminated Glasses. 16 pp.

—Aircraft
MOD UK DTD-218D-66. Laminated Safety Glass. 4 pp.

Laminated Glass *(Cont.)*
—Aircraft *(Cont.)*
MOD UK DTD-869A-57. Laminated Safety Glass, High Light Transmission. 4 pp.

—Aircraft—Electrically Heated
MOD UK DTD-5576A-66. Laminated Safety Glass Heated by Electrically Conducting Film(s). 6 pp.

—Automotive
SNZ NZS 5443-87. Safety Glass for Land Vehicles. 32 pp.

—Daylight Testing
JIS R 3106-85. Testing Method on Transmittance and Reflectance for Daylight and Solar Radiation and Solar Heat Gain Coefficient of Flat Glass. 19 pp.

—Electrical Insulation—Polyester Resin
BSI BS 6673-86. 1986 Polyester Glass Mat Sheets for Electrical Purposes. 10 pp.

—Electrical Insulation—Silicone Resin
CNS C4353-82. Silicone Resin Glass Laminated Sheets for Electric Insulation (Jul)(9087).

JIS C 2241-78. Silicone Resin Glass Laminated Sheets for Electrical Insulation. 13 pp.

—Impact Testing
DIN ENGL 52338-85. Methods of Testing Flat Glass for Use in Buildings; Ball Drop Test on Laminated Glass (Sept). 3 pp.

—Sheets
DIN ENGL 40606-69. Laminated Products; Laminated Sheets and Strips of Fabric-Base Laminate or Glass-Mat-Base Laminate (Jan). 3 pp.

—Solar Radiation
JIS R 3106-85. Testing Method on Transmittance and Reflectance for Daylight and Solar Radiation and Solar Heat Gain Coefficient of Flat Glass. 19 pp.

—Strips
DIN ENGL 40606-69. Laminated Products; Laminated Sheets and Strips of Fabric-Base Laminate or Glass-Mat-Base Laminate (Jan). 3 pp.

Laminated Metals
See Also: Clad Metals

—Aluminum Foil—Packaging
MOD UK TS 429. Foil, Metal, Laminated Sheet (Heat Sealable) (Withdrawn).

—PVC
CNS K3092-87. Polyvinyl Chloride-Metal Laminated Sheets (Jun)(12005). 4 pp.

CNS K6909-87. Method of Test for Polyvinyl Chloride-Metal Laminated Sheets (Jun)(12006). 5 pp.

JIS K 6744-92. Polyvinyl Chloride Coated and Laminated Metal Sheets. 14 pp.

Laminated Papers
See Also: Foil Papers; Impregnated Papers; Laminates; Papers

DIN VDE 0318 Pt 1-75. VDE Specification for Laminated Products: Paper Base Laminates, Fabric Base Laminates and Glass-Mat Base Laminates; Test Methods (Sept). 19 pp.

DIN VDE 0318 Pt 2-75. VDE Specification for Laminated Products: Paper Base Laminates, Fabric Base Laminates and Glass-Mat Base Laminates (DIN 7735) Requirements, Types (Sept). 21 pp.

—Asphalt
CNS P2025-84. Asphalt Laminating Paper (Jan)(2305). 2 pp.

—Bars
DIN ENGL 40625-67. Laminated Products; Solid Bars of Paper-Base Laminate or Fabric-Base Laminate (July). 2 pp.

—Kraft—Ammunition
MOD UK DEF-137-01. Paper, Laminated (Kraft-Polythene-Kraft) (Reprinted January 1963, Incorporating Correction); Amendment 1. 7 pp.

—Kraft—Tubes
MOD UK DEF-137-01. Paper, Laminated (Kraft-Polythene-Kraft) (Reprinted January 1963, Incorporating Correction); Amendment 1. 7 pp.

—Phenolic Resins—Sheets
BSI BS 5102-74. (WITHDRAWN) 1974 Amd 2 Phenolic Resin Bonded Paper Laminated Sheets for Electrical Applications (Superseded by BS 2572: 1990). 38 pp.

MOD UK CS 3117B. Phenolic Resin Bonded Paper Laminated Sheet (Withdrawn).

INDUSTRY STANDARDS

INTERNATIONAL AND NON-U.S. NATIONAL STANDARDS SUBJECT INDEX

Laminated

Laminated Papers *(Cont.)*

—Rods

DIN ENGL 40624-67. Laminated Products; Solid Rods of Paper-Base Laminate or Fabric-Base Laminate (July). 2 pp.

—Sheets

DIN ENGL 40605-67. Laminated Products; Sheet and Strip of Paper-Base Laminate (July). 2 pp.

—Strips

DIN ENGL 40605-67. Laminated Products; Sheet and Strip of Paper-Base Laminate (July). 2 pp.

—Thermosetting Resins—Rods

BSI BS 6128: Part 1-81. 1981 Industrial Laminated Rods and Tubes Based on Thermosetting Resins Part 1: Classification and Methods of Test. 8 pp.

JIS K 6913-77. Laminated Thermosetting Rods (R 1980). 14 pp.

—Thermosetting Resins—Rods—Molded

BSI BS 6128: Part 2-81. 1981 Industrial Laminated Rods and Tubes Based on Thermosetting Resins Part 2: Round Moulded Rods. 4 pp.

BSI BS 6128: Part 4-81. 1981 Industrial Laminated Rods and Tubes Based on Thermosetting Resins Part 4: Rectangular Moulded Rods. 4 pp.

BSI BS 6128: Part 6-81. 1981 Industrial Laminated Rods and Tubes Based on Thermosetting Resins Part 6: Hexagonal Moulded Rods. 4 pp.

—Thermosetting Resins—Rods—Pultruded

BSI BS 6128: Part 3-81. (WITHDRAWN) 1981 Amd 1 Industrial Laminated Rods and Tubes Based on Thermosetting Resins Part 3: Round Pultruded Rods. 5 pp.

BSI BS 6128: Part 5-81. (WITHDRAWN) 1981 Amd 1 Industrial Laminated Rods and Tubes Based on Thermosetting Resins Part 5: Rectangular Pultruded Rods. 5 pp.

BSI BS 6128: Part 7-81. (WITHDRAWN) 1981 Amd 1 Industrial Laminated Rods and Tubes Based on Thermosetting Resins Part 7: Hexagonal Pultruded Rods. 5 pp.

—Thermosetting Resins—Tubes

BSI BS 6128: Part 1-81. 1981 Industrial Laminated Rods and Tubes Based on Thermosetting Resins Part 1: Classification and Methods of Test. 8 pp.

JIS K 6914-77. Laminated Thermosetting Tubes (R 1980). 22 pp.

—Thermosetting Resins—Tubes—Molded

BSI BS 6128: Part 9-82. 1982 Industrial Laminated Rods and Tubes Based on Thermosetting Resins Part 9: Round and Moulded Tubes. 8 pp.

BSI BS 6128: Part 13-83. 1983 Industrial Laminated Rods and Tubes Based on Thermosetting Resins Part 13: Specification for Rectangular Moulded Tubes. 8 pp.

—Thermosetting Resins—Tubes—Rolled

BSI BS 6128: Part 8-82. 1982 Industrial Laminated Rods and Tubes Based on Thermosetting Resins Part 8: Round Rolled Tubes. 8 pp.

—Tubes—Molded

DIN ENGL 40615-67. Laminated Products; Compression Moulded Circular Tubes of Paper-Base Laminate or Fabric-Base Laminate (July). 2 pp.

DIN ENGL 40616-67. Laminated Products; Compression Moulded Square Tubes of Paper-Base Laminate or Fabric-Base Laminate (July). 2 pp.

DIN ENGL 40617-67. Laminated Products; Compression Moulded Hexagon Tubes of Paper-Base Laminate or Fabric-Base Laminate (July). 2 pp.

DIN ENGL 40618-71. Laminated Products; Laminated Moulded Tubes of Rectangular Cross-Section; Paper-Base Laminate or Fabric-Base Laminate (June). 2 pp.

—Tubes—Rolled

DIN ENGL 40607-67. Laminated Products; Tubes, Rolled, Not Moulded of Laminated Paper or Laminated Fabric (July). 4 pp.

Laminated Plastics

See Also: Fiberglass Reinforced Plastics; Laminates; Plastics

BSI BS 4965-91. 1991 Decorative Laminated Plastics Sheet Veneered Boards and Panels. 15 pp.

BSI BS 4965-83. 1983 Amd 2 Decorative Laminated Plastics Sheet Veneered Boards and Panels. 14 pp.

—Aerospace—Compression Testing

CEN PREN 2850-90. Unidirectional Thermosetting Carbon Laminate Compression Test Parallel to the Fibre Direction. 14 pp.

CEN PREN 2850-91. Carbon Thermosetting Resin Unidirectional Laminates Comression Test Parallel to the Fibre Direction. 12 pp.

Laminated Plastics *(Cont.)*

—Aerospace—Flexural Strength

AECMA PREN2562-89. Unidirectional Laminates Carbon-Thermosetting Resin Flexural Test. 9 pp.

—Aerospace—Shear Strength

AECMA PREN2563-89. Unidirectional Laminates Carbon-Thermosetting Resin Test Method—Determination of Apparent Interlaminar Shear Strength. 10 pp.

—Aerospace—Tensile Testing

AECMA PREN2561-89. Aerospace Series Unidirectional Laminates Carbon-Thermosetting Resin Tensile Test Parallel to the Fibre Direction. 12 pp.

—Fiberglass Reinforced

BSI BS 2782:Pt9: METH 920A-C-77. 1977 Methods of Testing Plastics Part 9: Sampling and Test Specimen Preparation Methods 920A to 920C: Preparation of Glass Fibre Reinforced, Resin Bonded, Low-Pressure Laminated Plates or Panels for Test Purposes. 8 pp.

ISO 1268-74. Plastics—Preparation of Glass Fibre Reinforced, Resin Bonded, Low-Pressure Laminated Plates or Panels for Test Purposes First Edition. 8 pp.

—Fiberglass Reinforced—Aircraft

MOD UK DTD-933A-64. Glass Fabric Reinforced Polyester Laminates for Aircraft Structures and Airborne Radomes. 6 pp.

—Fiberglass Reinforced—Quality Assurance

MOD UK DSTAN 93-51-88. Quality Assurance Requirements for the Manufacture of Fibre Reinforced Plastics Laminated Components Issue 1. 12 pp.

—Fiberglass Reinforced—Quality Assurance—Technical Memoranda

MOD UK QTM 32. Manufacture of Fibre Reinforced Laminated Components Issue 1 (Superseded by Def Stan 93-51).

—Fiberglass Reinforced—Shear Strength

CNS K6981-90. Method of Test for Apparent Interlaminar Shear Strength of Glass Fiber Reinforced Plastics (Sep)(12782).

—Glowing Cigarette Test

DIN ENGL 51961-84. Testing of Plastics Surfaces; Behaviour on Exposure to Glowing Cigarettes (Aug). 2 pp.

—Paints (Oleoresinous)—Exterior

MOD UK DSTAN 80-29-73. Paint, Finishing, General Service, Gloss, for Marine Use Type: Brushing Issue 1. 8 pp.

—Sachets

MOD UK TS 10263. Metal-Foil/Plastics Laminate for Detergent Sachets.

—Shear Strength

BSI BS 2782:Pt3: METH 341A-77. 1977 Methods of Testing Plastics Part 3: Mechanical Properties Method 341A: Determination of Apparent Interlaminar Shear Strength of Reinforced Plastics. 4 pp.

—Tubes—Cohesion

BSI BS 2782:Pt3: METH 346A-84. 1984 Methods of Testing Plastics Part 3: Mechanical Properties Method 346A: Determination of Cohesion Between Layers of Laminated Tube. 4 pp.

Laminated Pressboard

Use: Pressboard—Laminated

Laminated Springs

See Also: Laminates; Springs (Elastic)

—Bolts

CNS B2599-81. Center Bolts for Laminated (Leaf) Springs (Sep)(7860). 3 pp.

—Clips—Alignment

CNS B2600-81. Alignment Clips for Laminated (Leaf) Spring (Sep)(7861). 3 pp.

—Rolling Stock—Ends

CNS B2601-81. Ends of Laminated Springs for Rail Vehicles (Sep)(7862). 3 pp.

DIN ENGL 5542-75. Ends of Laminated Springs for Rail Vehicles (June). 2 pp.

—Steel—Dimples

CNS B2602-81. Dimple for Laminated Spring (Sep)(7863).

Laminated Wood

See Also: Composite Materials; Laminates; Lumber; Plywood; Veneers; Wood Products

Laminated Wood *(Cont.)*

BSI BS 3444-72. 1972 Blockboard and Laminboard. 18 pp.

CEN PREN 390-90. Glued Laminated Timber-Dimensions-Permissable Deviations. 5 pp.

CNS O1032-84. Decorative Glued-Laminated Timber (Nonstructure Members) (Sep)(11030).

CNS O1043-87. Laminated Veneer Lumber (Jan)(11818). 3 pp.

CNS O2059-87. Method of Test for Laminated Veneer Lumber (Jan)(11819). 2 pp.

SNZ NZS 3606-87. Specification for the Manufacture of Glue Laminated Timber. 20 pp.

—Delamination—Quality Assurance

CEN PREN 391-90. Glued Laminated Timber-Delamination Test of Glue Lines. 8 pp.

—Electrical Equipment

IEC 1061 Pt 1-91. Specification for Non-Impregnated, Densified Laminated Wood for Electrical Purposes Part 1: Definitions, Designation and General Requirements First Edition. 19 pp.

IEC 1061 Pt 2-92. Specification for Non-Impregnated, Densified Laminated Wood for Electrical Purposes Part 2: Methods of Test First Edition. 33 pp.

—Electrical Equipment—Designations

IEC 1061 Pt 1-91. Specification for Non-Impregnated, Densified Laminated Wood for Electrical Purposes Part 1: Definitions, Designation and General Requirements First Edition. 19 pp.

—Electrical Equipment—Glossaries

IEC 1061 Pt 1-91. Specification for Non-Impregnated, Densified Laminated Wood for Electrical Purposes Part 1: Definitions, Designation and General Requirements First Edition. 19 pp.

—Plastic

JIS A 5703-78. Plastic Laminated or Printed Boards for Inside Use. 16 pp.

—Shear Strength

CEN PREN 392-90. Glued Laminated Timber-Glue Line Shear Test. 10 pp.

—Structural Timber

BSI BS 4169-88. 1988 Manufacture of Glued-Laminated Timber Structural Members. 16 pp.

CEN PREN 386-90. Glued Laminated Timber—Production Requirements. 15 pp.

SAA AS 1328-87. Glued-Laminated Structural Timber. 17 pp.

SAA AS 1491-73. Laminated and/or Finger-Jointed Radiata Pine Scantlings (Metric Units) (Incorporating Amdt 1). 21 pp.

—Structural Timber—Interlocking Joints

CEN PREN 387-90. Glued Laminated Timber-Production Requirements for Large Finger Joints. 12 pp.

—Structural Timber—Mechanical Properties

CEN PREN 408-90. Timber Structures—Solid Timber and Glued Laminated Timber—Determination of Same Physical and Mechanical Properties for Structural Purposes. 27 pp.

—Structural Timber—Physical Properties

CEN PREN 408-90. Timber Structures—Solid Timber and Glued Laminated Timber—Determination of Same Physical and Mechanical Properties for Structural Purposes. 27 pp.

Laminates

Scope Note: For additional listings, use a more specific term *See Also:* Clad Metals; Coatings; Copper Clad Laminates; Corrugated Board; Epoxy Laminates; Fabrics; Foil Papers; Laminated Fabrics; Laminated Glass; Laminated Metals; Laminated Papers; Laminated Plastics; Laminated Springs; Laminated Wood; Papers; Phenolic Laminates; Plywood; Polyester Laminates; Sheet Laminates; Thermosetting Resins

—Electric Coils

BSI BS 2857-76. 1976 Nickel-Iron Transformer and Choke Laminations. 30 pp.

BSI BS 6554: Part 1-84. 1984 Laminations for Transformers and Inductors for Use in Telecommunications and Electronic Equipment Part 1: Characteristics and Electrical Testing. 24 pp.

BSI BS 6600: Part 1-87. 1987 Outline Dimensions of Transformers and Inductors for Use in Telecommunication Electronic Equipment Part 1: The Outline Dimensions of Transformers and Inductors Using YE1-1 Laminations. 14 pp.

IEC 740-82. Laminations for Transformers and Inductors for Use in Telecommunication and Electronic Equipment First Edition; (Amendment 1-1991). 57 pp.

INTERNATIONAL AND NON-U.S. NATIONAL STANDARDS
SUBJECT INDEX

Laminates (Cont.)
—Electric Coils (Cont.)
IEC 852 Pt 3-92. Outline Dimensions of Transformers and Inductors for Use in Telecommunication and Electronic Equipment Part 3: Transformers and Inductors Using YUI-1 Laminations First Edition. 21 pp.
MOD UK DSTAN 59-55-01. Laminations, Magnetic Issue 1; Amendment 1. 68 pp.

—Electronic Transformers
BSI BS 2857-76. 1976 Nickel-Iron Transformer and Choke Laminations. 30 pp.
BSI BS 6554: Part 1-84. 1984 Laminations for Transformers and Inductors for Use in Telecommunications and Electronic Equipment Part 1: Characteristics and Electrical Testing. 24 pp.
BSI BS 6600: Part 1-87. 1987 Outline Dimensions of Transformers and Inductors for Use in Telecommunication Electronic Equipment Part 1: The Outline Dimensions of Transformers and Inductors Using YE1-1 Laminations. 14 pp.
BSI BS 7382: Part 1-90. 1990 Laminated Core Packages for Transformers and Inductors Used in Telecommunication and Electronic Equipment Part 1: Dimensions. 11 pp.
IEC 740-82. Laminations for Transformers and Inductors for Use in Telecommunication and Electronic Equipment First Edition; (Amendment 1-1991). 57 pp.
IEC 852 Pt 1-86. Outline Dimensions of Transformers and Inductors for Use in Telecommunication and Electronic Equipment Part 1: Transformers and Inductors Using YEI-1 Laminations First Edition. 22 pp.
IEC 852 Pt 3-92. Outline Dimensions of Transformers and Inductors for Use in Telecommunication and Electronic Equipment Part 3: Transformers and Inductors Using YUI-1 Laminations First Edition. 21 pp.
IEC 1021 Pt 1-90. Laminated Core Packages for Transformers and Inductors Used in Telecommunication and Electronic Equipment Part 1: Dimensions First Edition. 17 pp.

—Magnetic Materials
CNS C3127-81. Method of Test for Lamination Factor of Magnetic Materials (Jul)(7630).

—Magnetically Soft Materials—Permeability
IEC 740 Pt 2-93. Laminations for Transformers and Inductors for Use in Telecommunication and Electronic Equipment Part 2: Specification for the Minimum Permeabilities of Laminations Made of Soft Magnetic Metallic Materials First Edition. 43 pp.

—Packaging
BSI BS 1133: Sec 21-91. 1991 Packaging Code Section 21: Regenerated Cellulose Film, Films Made of Plastics, Aluminium Foil, Flexible Multilayer Structures and Metallized Materials (H). 36 pp.

—Printed Circuit Base Materials
BSI BS 6221: Part 12-92. 1992 Printed Wiring Boards Part 12: Specification for Mass Lamination Panels (Semi-Manufactured Multilayer Printed Boards) (IEC 326-12: 1992). 20 pp.
CNS C5173-85. General Rules of Copper-Clad Laminates for Printed Circuits (Dec)(10554).
CNS C6206-85. Method of Test for Copper-Clad Laminates for Printed Circuits (Dec)(10555).
CNS C7157-85. Copper-Clad Laminates for Printed Circuits (Paper Base, Epoxy Resin) (Dec)(10556).
CNS C7158-85. Copper-Clad Laminates for Printed Circuits (Synthetic Fiber Fabric Base, Epoxy Resin) (Dec)(10557).
CNS C7159-85. Copper-Clad Laminates for Printed Circuits (Glass Fabric Base, Epoxy Resin) (Dec)(10558).
CNS C7160-85. Copper-Clad Laminates for Printed Circuits (Paper Base, Phenolic Resin) (Dec)(10559).
CNS C7162-85. Epoxy Resin-Impregnated Glass Cloth (Pre-Preg) for Multilayer Printed Circuits (Dec)(10561).
IEC 326 Pt 12-92. Printed Boards Part 12: Specification for Mass Lamination Panels (Semi-Manufactured Multilayer Printed Boards) First Edition. 37 pp.
JIS C 6471-90. Test Methods of Copper-Clad Laminates for Flexible Printed Wiring Boards. 25 pp.
JIS C 6472-90. Copper-Clad Laminates for Flexible Printed Wiring Boards (Polyester Film, Polyimide Film). 11 pp.
JIS C 6480-89. General Rules of Copper-Clad Laminates for Printed Wiring Boards. 9 pp.
JIS C 6481-90. Test Methods of Copper-Clad Laminates for Printed Wiring Boards. 33 pp.
JIS C 6482-91. Copper-Clad Laminates for Printed Wiring Boards—Paper Base, Epoxy Resin. 14 pp.

Laminates (Cont.)
—Printed Circuit Base Materials (Cont.)
JIS C 6483-91. Copper-Clad Laminates for Printed Wiring Boards—Synthetic Fiber Fabric Base, Epoxy Resin. 13 pp.
JIS C 6484-91. Copper-Clad Laminates for Printed Wiring Boards—Glass Fabric Base, Epoxy Resin. 13 pp.
JIS C 6485-91. Copper-Clad Laminates for Printed Wiring Boards—Paper Base, Phenolic Resin. 14 pp.
JIS C 6486-90. Thin Copper-Clad Laminates for Multilayer Printed Wiring Boards—Glass Fabric Base, Epoxy Resin. 14 pp.
JIS C 6488-89. Copper-Clad Laminates for Printed Wiring Boards (Glass Cloth Surfaces, Cellulose Paper Core, Epoxy Resin). 13 pp.
JIS C 6489-89. Copper-Clad Laminates for Printed Wiring Boards (Glass Cloth Surfaces, Nonwoven Glass Core, Epoxy Resin). 13 pp.

—Resins—Gel Time
BSI BS 2782:Pt8: METH 835E-80. 1980 Methods of Testing Plastics Part 8: Other Properties Method 835E: Determination of Gelation Time of Resin in Impregnated Materials Used for Laminates. 2 pp.

—Rods—Electrical
CENELEC PREN 61212-1-92. Specification for Industrial Rigid Round Laminated Tubes and Rods Based on Thermosetting Resins for Electrical Purposes Part 1: Definitions, Designations and General Requirements. 11 pp.

—Saturable Reactors
MOD UK DSTAN 59-55-01. Laminations, Magnetic Issue 1; Amendment 1. 68 pp.

—Tensile Testing—Aerospace
CEN PREN 2597-90. Carbon Thermosetting Resin Undirectional Laminates Tensile Test Perpendicular to the Fibre Direction. 11 pp.

—Thermal Insulation—Thermal Resistance
DIN ENGL 52612 Pt 3-79. Testing of Thermal Insulating Materials; Determination of Thermal Conductivity by the Guarded Hot Plate Apparatus; Thermal Resistance of Laminated Materials for Use in Building Practice (Sept). 3 pp.

—Thickness Measurement—Radiation Meters
BSI BS 5868-80. 1980 Guide to Ionizing Radiation Thickness Meters for Materials in the Form of Sheets, Coatings or Laminates. 27 pp.

—Transformers
BSI BS 2857-76. 1976 Nickel-Iron Transformer and Choke Laminations. 30 pp.
BSI BS 6554: Part 1-84. 1984 Laminations for Transformers and Inductors for Use in Telecommunications and Electronic Equipment Part 1: Characteristics and Electrical Testing. 24 pp.
BSI BS 6600: Part 1-87. 1987 Outline Dimensions of Transformers and Inductors for Use in Telecommunication Electronic Equipment Part 1: The Outline Dimensions of Transformers and Inductors Using YE1-1 Laminations. 14 pp.
IEC 740-82. Laminations for Transformers and Inductors for Use in Telecommunication and Electronic Equipment First Edition; (Amendment 1-1991). 57 pp.
MOD UK DSTAN 59-55-01. Laminations, Magnetic Issue 1; Amendment 1. 68 pp.

Laminations
Use: Laminates

Lamp Bases
Use: Lampholders

Lamp Caps
Use: Lampholders

Lamp Holders
Use: Lampholders

Lamp Oils
Use: Kerosene

Lamp Poles
Use: Lighting Poles

Lamp Posts
Use: Lighting Poles

Lampblack
Use: Carbon Black

Lampcaps
Use: Lampholders

Lampholders
Use For: Lamp Bases; Lamp Caps; Lamp Holders
See Also: Lamps
CNS C1129-84. Types and Dimensions of Bases and Sockets for Lamps (Jun)(10902).
CSA C22.2 NO 12-1982. Portable Luminaires; (Gen Instr 1 Thru 4). 52 pp.
CSA 1169 Bull. Electrical Bulletin 1169 June 27, 1978 to C22.2 NO 12. 2 pp.
CSA CAN/CSA-C22. 2 NO 18-92. Outlet Boxes, Conduit Boxes, and Fittings; (Gen Instr 1 Thru 2). 118 pp.
IEC 838 Pt 1-93. Miscellaneous Lampholders Part 1: General Requirements and Tests Second Edition; (Corrigendum—Aug 1993). 62 pp.
JIS C 7709-89. Lamp Caps and Holders. 126 pp.
SNZ NZS 212-63. Specification for Lamp Caps and Lampholders for Double-Capped Tubular Lamps Amend: 1, 1963; 2, 1963. 8 pp.

—Automotive
CNS D2159-82. Kinds and Dimensions of Bases and Sockets of Electric Lamps for Automobiles (Nov)(9588). 6 pp.

—Barrel Thread
CENELEC HD 222-74. Standard Sheets for Barrel Thread for E14 and E27 Lampholder with Shade Holder Ring. 2 pp.
IEC 399-72. Standard Sheets for Barrel Thread for E14 and E27 Lampholders with Shade Holder Ring First Edition. 7 pp.

—Bayonet
BSI BS 5042-87. 1987 Bayonet Lampholder. 28 pp.
IEC 1184-93. Bayonet Lampholders First Edition. 107 pp.
SAA AS 3117-88. Approval and Test Specification—Bayonet Lampholders Amdt 1 1990 (This is a Joint Standard with SANZ NZS 3117). 12 pp.
SNZ NZS/AS 3117-88. Approval and Test Specification—Bayonet Lampholders Amend: 1, 1990 (This is a Joint Standard with SAA AS 3117). 12 pp.

—Bayonet—Aircraft
BSI G 193-66. 1966 Aircraft Lampholders. 5 pp.

—Discharge Lamps
CSA CAN/CSA-C22. 2 NO 74-92. Equipment for Use with Electric Discharge Lamps (Incorporating Electrical Bulletin Nos. 523F, 753, 846, 1124A, 1125A, 1325, and 1326); (Gen Instr 1). 82 pp.

—Electric Terminals
CSA 943B Bull. Electrical Bulletin 943B June 25, 1976 to C22.2 NO 55. 2 pp.

—Fans
CSA C22.2 NO 113-M1984. Fans and Ventilators (R 1993); (Gen Instr 1 Thru 6). 51 pp.
CSA 1169 Bull. Electrical Bulletin 1169 June 27, 1978 to C22.2 NO 113. 2 pp.

—Fluorescent Lighting
BSI BS 6702-91. (WITHDRAWN) 1991 Lampholders for Tubular Fluorescent Lamps and Starterholders (Superseded by BS EN 60400: 1992). 59 pp.
BSI BS EN 60400-92. 1992 Lampholders for Tubular Fluorescent Lamps and Starterholders. 82 pp.
CENELEC EN 60 400-85. Lampholders for Tubular Fluorescent Lamps and Starterholders. 42 pp.
CENELEC EN 60 400-89. Lampholders for Tubular Fluorescent Lamps and Starterholders. 8 pp.
CENELEC EN 60400-92. Lampholders for Tubular Fluorescent Lamps and Starterholders. 7 pp.
CNS C3094-80. Method of Test for Lampholders and Starterholders for Fluorescent Lamps (Aug)(6055).
CNS C4220-80. Lampholders and Starterholders for Fluorescent Lamps (Aug)(6054).
CSA C22.2 NO 43-M1984. Lampholders (R 1992); (Gen Instr 1 Thru 6). 54 pp.
CSA 1233A Bull. Electrical Bulletin 1233A June 17, 1981 to C22.2 NO 43. 2 pp.
IEC 400-91. Lampholders for Tubular Fluorescent Lamps and Starterholders Fourth Edition; (Amendment 1-1993, Including the Corrigendum of June 1992) (Corrigendum—Aug 1993). 130 pp.
JIS C 8324-79. Lampholders and Starterholders for Fluorescent Lamps. 21 pp.
SNZ IEC 400-87. Lampholders for Tubular Fluorescent Lamps and Startholders. 80 pp.
SNZ NZS 1123-53. Specification for Bi-Pin Lamp Caps and Lampholders for Tubular Fluorescent Lamps for Use in Circuits, the Declared Voltage of Which Does Not Exceed 250 Volts Amend: 1, 1963; 2, 1963; 3, 1981; 3A, 1981. 16 pp.

—Gages
BSI BS 5101: Part 1-75. (WITHDRAWN) 1975 Amd 5 Lamp Caps and Holders Together with Gauges for the Control of Interchangeability and Safety Part 1: Lamp Caps (IEC 61-1) (AMD 6519) January 31, 1991 (Superseded by BS EN 60061-2: 1993). 184 pp.

INDUSTRY STANDARDS

INTERNATIONAL AND NON-U.S. NATIONAL STANDARDS
SUBJECT INDEX

Lampholders

Lampholders (Cont.)
—Gages (Cont.)

BSI BS 5101: Part 2-75. (WITHDRAWN) 1975 Amd 5 Lamp Caps and Holders Together with Gauges for the Control of Interchangeability and Safety Part 2: Lamp Holders (IEC 61-2) (AMD 6521) January 31, 1991 (Superseded by BS EN 60061-2: 1993). 137 pp.

BSI BS 5101: Part 3-75. (WITHDRAWN) 1975 Amd 5 Lamp Caps and Holders Together with Gauges for the Control of Interchangeability and Safety Part 3: Gauges (AMD 6520) January 31, 1991 (Superseded by BS EN 60061-3: 1993). 338 pp.

BSI BS 5101: Part 4-80. 1980 Amd 1 Lamp Caps and Holders Together with Gauges for the Control of Interchangeability and Safety Part 4: Lamp Caps, Lampholders and Gauges Used in the United Kingdom but Not Specified in Parts 1, 2 or 3. 18 pp.

BSI BS 5101: Part 5-90. (WITHDRAWN) 1990 Lamp Caps and Holders Together with Gauges for the Control of Interchangeability and Safety Part 5: Guidelines and General Information (Renumbered as BS EN 60061-4: 1992). 22 pp.

BSI BS EN 60061-3-93. 1993 Lamp Caps and Holders Together with Gauges for the Control of Interchangeability and Safety Part 3: Gauges (E). 425 pp.

BSI BS EN 60061-4-92. 1992 Amd 1 Lamp Caps and Holders Together with Gauges for the Control of Interchangeability and Safety Part 4: Guidelines and General Information (AMD 7556) December 15, 1992. 24 pp.

CENELEC HD 65-82. Lamp Caps and Holders Together with Gauges for the Control of Interchangeability and Safety. 4 pp.

CENELEC HD 65.1 S1-88. Lamp Caps and Holders Together with Gauges for the Control of Interchangeability and Safety Part 1: Lamp Caps. 3 pp.

CENELEC HD 65.2-87. Lamp Caps and Holders Together with Gauges for the Control of Interchangeability and Safety Part 2: Lampholders. 3 pp.

CENELEC EN 60061-3-93. Lamp Caps and Holders Together with Gauges for the Control of Interchangeability and Safety Part 3: Gauges (IEC 61-3: 1969 + Supplements A: 1970 to M: 1992, Modified). 420 pp.

CENELEC EN 60061-4-92. Lamp Caps and Holders Together with Gauges for the Control of Interchangeability and Safety Part 4: Guidelines and General Information; (IEC 61-4: 1990, Mod). 4 pp.

CENELEC EN 60061-4-92. Lamp Caps and Holders Together with Gauges for the Control of Interchangeability and Safety Part 4: Guidelines and General Information. 13 pp.

IEC 61 Pt 1-69. Lamp Caps and Holders Together with Gauges for the Control of Interchangeability and Safety Part 1: Lamp Caps Third Edition; (Supplements A-J-1970-1980)(Supplement K-1983) (Supplement L-1987) (Supplement M-1989) (Supplement N-1992); (CENELEC EN60061-1:1993). 279 pp.

IEC 61 Pt 2-69. Lamp Caps and Holders Together with Gauges for the Control of Inter-changeability and Safety Part 2: Lampholders Third Edition; (Supplements A-F-1970-1980)(Supplement G-1983) (Supplement H-1987) (Supplement J-1989) (Supplement K-1992) (CENE EN 60061-2: 1993). 202 pp.

IEC 61 Pt 3-69. Lamp Caps and Holders Together with Gauges for the Control of Interchangeability and Safety Part 3: Gauges Third Edition; (Errata—Dec 1984) (Supplement K-1987) (Supplement L-1989) (Supplement M-1992) (CENE EN 60061-3: 1993). 506 pp.

IEC 61 Pt 4-90. Lamp Caps and Holders Together with Gauges for the Control of Interchangeability and Safety Part 4: Guidelines and General Information First Edition; (Supplement A-1992) (CENELEC EN 60061-4: 1992). 85 pp.

—Glassware

CNS C1104-82. Glass Ware for Lighting Fittings (Dimensions for Fitted Portions) (Jul)(9115).

—Incandescent Lighting

CSA C22.2 NO 43-M1984. Lampholders (R 1992); (Gen Instr 1 Thru 6). 54 pp.

CSA 1233A Bull. Electrical Bulletin 1233A June 17, 1981 to C22.2 NO 43. 2 pp.

CSA 1416 Bull. Electrical Bulletin 1416 November 14, 1983 to C22.2 NO 84. 2 pp.

—Incandescent Lighting—Aerospace

AECMA PREN2241-90. Lamp Base Dimensions. 16 pp.

—Incandescent Lighting—Temperature Testing

CSA 601J Bull. Electrical Bulletin 601J November 29, 1978 to C22.2 NO 9. 4 pp.

Lampholders (Cont.)
—Indicator Lights

MOD UK DSTAN 62-9: Part 1-74. Lampholders; Lights, Indicator; and Lenses, Indicator Light; for Use in Equipments Part 1: Generic Specification Issue 1. 15 pp.

MOD UK DSTAN 62-9: Part 2-01. Lampholders; Lights, Indicator; and Lenses, Indicator Light; for Use in Equipments Part 2: Detail Specification, Standard Range, and Detail Specification Sheets for Lampholders Issue 1; Amendment 1. 36 pp.

MOD UK DSTAN 62-9: Part 4-75. Lampholders; Lights, Indicator; and Lenses, Indicator Light; for Use in Equipments Part 4: Detail Specification, Standard Range, and Detail Specification Sheets for Lenses, Indicator Light Issue 1. 33 pp.

—Interchangeability

BSI BS 7001-88. 1988 Interchangeability and Safety of a Standardized Luminaire Supporting Coupler. 14 pp.

BSI BS 7001-01. 1988 Amd 1 Interchangeability and Safety of a Standardized Luminaire Supporting Coupler (AMD 7625) June 15, 1993 (E). 23 pp.

—Laryngoscopes

ISO 7376 Pt 2-84. Laryngoscopic Fittings—Part 2: Miniature Electric Lamps—Screw Threads and Sockets First Edition. 5 pp.

—Lighting Chains

CSA C22.2 NO 37-M1989. Christmas Tree and Other Decorative Lighting Outfits; (Gen Instr 1 Thru 2). 37 pp.

—Miniature—Lighting Chains

CSA C22.2 NO 37-M1989. Christmas Tree and Other Decorative Lighting Outfits; (Gen Instr 1 Thru 2). 37 pp.

—Portable

SAA AS/NZS 3128-93. Approval and Test Specification—Portable Lamp Standards and Brackets (in Professional Package 28). 7 pp.

SNZ NZS/AS 3128-93. Approval and Test Specification—Portable Lamp Standards and Brackets Amend: 4, 1992 (This is a Joint Standard with SAA AS 3128). 6 pp.

—Safety

BSI BS 5101: Part 1-75. (WITHDRAWN) 1975 Amd 5 Lamp Caps and Holders Together with Gauges for the Control of Interchangeability and Safety Part 1: Lamp Caps (IEC 61-1) (AMD 6519) January 31, 1991 (Superseded by BS EN 60061-2: 1993). 184 pp.

BSI BS 5101: Part 2-75. (WITHDRAWN) 1975 Amd 5 Lamp Caps and Holders Together with Gauges for the Control of Interchangeability and Safety Part 2: Lamp Holders (IEC 61-2) (AMD 6521) January 31, 1991 (Superseded by BS EN 60061-2: 1993). 137 pp.

BSI BS 5101: Part 3-75. (WITHDRAWN) 1975 Amd 5 Lamp Caps and Holders Together with Gauges for the Control of Interchangeability and Safety Part 3: Gauges (AMD 6520) January 31, 1991 (Superseded by BS EN 60061-3: 1993). 338 pp.

BSI BS 5101: Part 4-80. 1980 Amd 1 Lamp Caps and Holders Together with Gauges for the Control of Interchangeability and Safety Part 4: Lamp Caps, Lampholders and Gauges Used in the United Kingdom but Not Specified in Parts 1, 2 or 3. 18 pp.

BSI BS 5101: Part 5-90. (WITHDRAWN) 1990 Lamp Caps and Holders Together with Gauges for the Control of Interchangeability and Safety Part 5: Guidelines and General Information (Renumbered as BS EN 60061-4: 1992). 22 pp.

BSI BS 6972-88. 1988 General Requirements for Luminaire Supporting Couplers for Domestic, Light Industrial and Commercial Use. 22 pp.

BSI BS 7001-88. 1988 Interchangeability and Safety of a Standardized Luminaire Supporting Coupler. 14 pp.

BSI BS 7001-01. 1988 Amd 1 Interchangeability and Safety of a Standardized Luminaire Supporting Coupler (AMD 7625) June 15, 1993 (E). 23 pp.

BSI BS EN 60061-1-93. 1993 Lamp Caps and Holders Together with Gauges for the Control of Interchangeability and Safety Part 1: Lamp Caps (Supersedes BS 5101: part 1: 1975) (E). 202 pp.

BSI BS EN 60061-2-93. 1993 Lamp Caps and Holders Together with Gauges for the Control of Interchangebility and Safety Part 2: Lampholders (Supersedes BS 5101: part 2: 1975) (E). 158 pp.

BSI BS EN 60061-3-93. 1993 Lamp Caps and Holders Together with Gauges for the Control of Interchangeability and Safety Part 3: Gauges (E). 425 pp.

BSI BS EN 60061-4-92. 1992 Amd 1 Lamp Caps and Holders Together with Gauges for the Control of Interchangeability and Safety Part 4: Guidelines and General Information (AMD 7556) December 15, 1992. 24 pp.

Lampholders (Cont.)
—Safety (Cont.)

CENELEC HD 65-82. Lamp Caps and Holders Together with Gauges for the Control of Interchangeability and Safety. 4 pp.

CENELEC HD 65.1 S1-88. Lamp Caps and Holders Together with Gauges for the Control of Interchangeability and Safety Part 1: Lamp Caps. 3 pp.

CENELEC HD 65.2-87. Lamp Caps and Holders Together with Gauges for the Control of Interchangeability and Safety Part 2: Lampholders. 3 pp.

CENELEC EN 60061-1-93. Lamp Caps and Holders Together with Gauges for the Control of Interchangeability and Safety Part 1: Lamp Caps (IEC 61-1: 1969 + Supplements A: 1970 to N: 1992, Modified) (Supersedes HD 65.1 S1: 1978). 197 pp.

CENELEC EN 60061-2-93. Lamp Caps and Holders Together with Gauges for the Control of Interchangeability and Safety Part 2: Lampholders (IEC 61-2: 1969 + Supplements A: 1970 to K: 1992, Modified). 152 pp.

CENELEC EN 60061-3-93. Lamp Caps and Holders Together with Gauges for the Control of Interchangeability and Safety Part 3: Gauges (IEC 61-3: 1969 + Supplements A: 1970 to M: 1992, Modified). 420 pp.

CENELEC EN 60061-4-92. Lamp Caps and Holders Together with Gauges for the Control of Interchangeability and Safety Part 4: Guidelines and General Information; (IEC 61-4: 1990, Mod). 4 pp.

CENELEC EN 60061-4-92. Lamp Caps and Holders Together with Gauges for the Control of Interchangeability and Safety Part 4: Guidelines and General Information. 13 pp.

IEC 61 Pt 1-69. Lamp Caps and Holders Together with Gauges for the Control of Interchangeability and Safety Part 1: Lamp Caps Third Edition; (Supplements A-J-1970-1980)(Supplement K-1983) (Supplement L-1987) (Supplement M-1989) (Supplement N-1992); (CENELEC EN60061-1:1993). 279 pp.

IEC 61 Pt 2-69. Lamp Caps and Holders Together with Gauges for the Control of Inter-changeability and Safety Part 2: Lampholders Third Edition; (Supplements A-F-1970-1980)(Supplement G-1983) (Supplement H-1987) (Supplement J-1989) (Supplement K-1992) (CENE EN 60061-2: 1993). 202 pp.

IEC 61 Pt 3-69. Lamp Caps and Holders Together with Gauges for the Control of Interchangeability and Safety Part 3: Gauges Third Edition; (Errata—Dec 1984) (Supplement K-1987) (Supplement L-1989) (Supplement M-1992) (CENE EN 60061-3: 1993). 506 pp.

IEC 61 Pt 4-90. Lamp Caps and Holders Together with Gauges for the Control of Interchangeability and Safety Part 4: Guidelines and General Information First Edition; (Supplement A-1992) (CENELEC EN 60061-4: 1992). 85 pp.

IEC 838 Pt 1-93. Miscellaneous Lampholders Part 1: General Requirements and Tests Second Edition; (Corrigendum—Aug 1993). 62 pp.

—Screw

BSI BS 6776-90. (WITHDRAWN) 1990 Amd 1 Edison Screw Lampholders (AMD 6304) September 28, 1990 (Superseded by BS EN 60238: 1992). 67 pp.

BSI BS EN 60238-92. 1992 Edison Screw Lampholders (E). 73 pp.

BSI BS EN 60238-01. 1992 Amd 1 Edison Screw Lampholders (AMD 7445) November 15, 1992 (Supersedes BS 6776: 1990). 74 pp.

CENELEC EN 60 238-86. Edison Screw Lampholders. 58 pp.

CENELEC EN 60 238-89. Edison Screw Lampholders. 7 pp.

CENELEC EN 60 238/A1-89. AMD 1 Edison Screw Lampholders. 7 pp.

CENELEC EN 60238-92. Edison Screw Lampholders. 8 pp.

CENELEC EN 60238-92. CORRIGENDUM Edison Screw Lampholders. 1 p.

CNS C4013-83. Screw Type Lamp Holders for Indoor Use (Jun)(692).

IEC 238-91. Edison Screw Lampholders Fifth Edition; (Corrigendum—June 1992). 123 pp.

JIS C 8302-88. Edison Screw Lampholders. 23 pp.

SAA AS 3140-88. Approval and Test Specification—Edison-Type Screw Lampholders Amdt 1 June 1990 (This is a Joint Standard with SANZ NZS 3140). 5 pp.

SNZ IEC 238-87. Edison Screw Lampholders. 95 pp.

SNZ NZS/AS 3140-88. Approval and Test Specification—Edison-Type Screw Lampholders Amend: 1, 1990 (This is a Joint Standard with SAA AS 3140). 5 pp.

—Screw Threads

CNS B2284-78. Electric Lighting Fitting (Screw Thread for Cover Glasses and Caps) (Oct)(4550).

INTERNATIONAL AND NON-U.S. NATIONAL STANDARDS
SUBJECT INDEX

Lampholders *(Cont.)*
—Screw Threads *(Cont.)*
DIN ENGL 49689-66. External Screw Threads for Lamp Holders and Internal Screw Threads for Lamp Shade Carrier Rings Made of Metal or Plastic Moulded Material; Limiting Dimensions (Mar). 1 p.

—Screw—Waterproof
CNS C4014-81. Water-Proof Screw Type Lamp Holders (Jun)(693).

—Ships
CNS F5009-81. Lamp Holders for Marine Use (Oct)(7988).
JIS F 8401-90. Marine Lamp Holders.

—Temperature Measurement
BSI BS 5371-90. 1990 Standard Method of Measurement of Lamp Cap Temperature Rise. 23 pp.
BSI BS 5371-76. (WITHDRAWN) 1976 Amd 1 Standard Method of Measurement of Lamp Cap Temperature Rise. 13 pp.
CENELEC HD 403-80. Standard Method of Measurement of Lamp Cap Temperature Rise. 3 pp.
CENELEC EN 60 360-89. Standard Method of Measurement of Lamp Cap Temperature Rise. 5 pp.
CNS C3187-84. Method of Test for Temperature Rise of Lamp Socket (Jun)(10905).
IEC 360-87. Standard Method of Measurement of Lamp Cap Temperature Rise Second Edition; (Amendment 1-1993). 38 pp.

—Xenon Arc Lamps
IEC 1127-92. High Pressure Xenon Short Arc Lamps—Dimensional, Electrical and Photometric Data and Cap Types First Edition. 52 pp.

Lampposts
Use: Lighting Poles

Lamps
Scope Note: For additional listings, use a more specific term *Use For:* Electric Lamps
See Also: Ballasts (Electric); Bayonet Lamps; Cap Lamps; Dimmers; Discharge Lamps; Exciter Lamps; Flashlights; Flashtubes; Fluorescent Lighting; Gaseous Tritium Lighting; Glow Lamps; Halogen Lamps; Heat Test Source Lamps; Incandescent Lighting; Infrared Lamps; Lampholders; Lampshades; Lanterns; Light (Visible Radiation); Light Bulbs; Lighting; Mercury Vapor Lamps; Metal Halide Lamps; Motor Vehicle Lighting; Neon Glow Lamps; Neon Tubes; Pilot Lamps; Projection Lamps; Quartz Tungsten Halogen Lamps; Resistance Lamps; Safety Lamps; Signal Lights; Sodium Vapor Lamps; Timing Lamps; Trouble Lights; Tungsten Halogen Lamps; Tungsten Lamps; Ultraviolet Lamps; Xenon Arc Lamps; Xenon Lamps

CNS C1130-84. General Rules of Testing Methods for Lamps (Jun)(10904).
CNS C4349-82. Working Lamps (May)(8803).
IEC 259-68. Miscellaneous Lamps and Ballasts First Edition; (Supplement A-1972). 37 pp.
JIS C 7801-88. General Rules of Testing Method for Lamps. 20 pp.

—Aircraft—Miniature
CNS C4175-84. Miniature Lamps for Aircraft (Jun)(5420). 17 pp.
JIS C 7522-74. Miniature Lamps for Aircraft.

—Airport
JIS W 8302-76. Lamps for Airport and Airway Lighting.

—Bicycles
BSI BS 3648-63. 1963 Amd 2 Cycle Rear Lamps. 21 pp.
CNS C3014-71. Testing Standard for Illuminating Equipments of Bicycles (Apr)(804). 3 pp.
CNS C4089-84. Electrical Bulbs for Dynamo Lamps of Bicycles (Jun)(2944). 4 pp.
JIS C 7510-92. Incandescent Lamps for Bicycle Dynamo Lamps. 11 pp.
JIS C 9502-81. Dynamo Lamps for Bicycles.
SNZ NZS 5441: Part 1-87. Lighting and Retroreflectors for Pedal Cycles Part 1: Specification for Lamp Units and Retroreflectors Suitable for Fitting to Pedal Cycles. 13 pp.

—Bicycles—Installation
SNZ NZS 5441: Part 2-87. Lighting and Retroreflectors for Pedal Cycles Part 2: Code of Practice for the Fitting of Lamp Units and Retroreflectors to Pedal Cycles. 7 pp.

—Bicycles—Miniature
CNS C3013-84. Method of Test for Miniature Lamps (Jun)(720). 1 p.

Lamps *(Cont.)*
—Caps (Lids)
BSI BS EN 60309-1-92. 1992 Plugs, Socket-Outlets and Couplers for Industrial Purposes Part 1: General Requirements. 67 pp.
BSI BS EN 60309-1-01. 1992 Amd 1 Plugs, Socket-Outlets and Couplers for Industrial Purposes Part 1: General Requirements (AMD 7889) August 15, 1993 (E). 68 pp.
CENELEC EN 60309-1-92. Plugs, Socket-Outlets and Couplers for Industrial Purposes Part 1: General Requirements. 8 pp.
JIS C 7709-89. Lamp Caps and Holders. 126 pp.

—Children's
BSI BS 4533: Sec 102.10-90. 1990 Luminaires Part 102: Particular Requirements Section 102.10: Specification for Portable Child-Appealing Luminaires. 11 pp.
BSI BS 4533: Sec 102.10-01. 1990 Amd 1 Luminaires Part 102: Particular Requirements Section 102.10: Specification for Portable Child-Appealing Luminaires (AMD 7713) May 15, 1993 (F). 19 pp.
CENELEC EN 60 598-2-10-89. Luminaires Part 2: Particular Requirements Section Ten—Portable Child-Appealing Luminaires. 6 pp.
CENELEC EN 60598-2-10/A1-91. AMD 1 Luminaires Part 2: Particular Requirements Section Ten—Portable Child-Appealing Luminaires. 5 pp.
IEC 598 Pt 2-10-87. Luminaires Part 2: Particular Requirements Section Ten: Portable Child-Appealing Luminaires First Edition; (Amendment 1-1990). 16 pp.

—Christmas Trees
CNS C4045-77. Small Lamp Bulbs Used for Decoration (Aug)(2059). 4 pp.

—Classification
BSI BS 1000: (682/683)-73. 1973 Universal Decimal Classification (UDC). English Full Edition (682/683): Smithery. Farriery. Hand-Forged Ironwork. Ironmongery. Hardware. Locksmithing. Gunsmithing. Bottling. Lamps. Stoves. 27 pp.
SNZ NZS/BS 1000 (682/683)-73. Universal Decimal Classification Smithery. Farriery. Hand-Forged Ironwork. Ironmongery. Hardware. Locksmithing. Gunsmithing. Bottling. Lamps. Stoves. 28 pp.

—Defense Contracts
MOD UK DEFCON 112H-87. Basic Set of Conditions of Contract—Electric Lamps 3/87. 1 p.

—Designations
CNS C1073-84. Method for the Designation of Miniature Lamps (Jun)(6432).

—Diagnostic Instruments (Medical)—Screw Threads
BSI BS 1929-53. (OBSOLESCENT) 1953 Screw Threads and External Dimensions of Lamps for Endoscopic and Diagnostic Instruments. 8 pp.

—Engineering Drawings
IEC 1126-92. Procedure for Use in the Preparation of Maximum Lamp Outlines First Edition. 84 pp.

—Globes
CNS C4405-84. Ball Lamps (Jun)(10903).
JIS C 7530-87. Ball Lamps. 21 pp.

—Hand Held
CENELEC EN 60 598-2-8-89. Luminaires Part 2: Particular Requirements Section Eight-Handlamps. 8 pp.
DIN VDE 0710 Pt 2-59. Specifications for Luminaires with Operating Voltages Below 1000 V; Special Rules for Hand-Lamps and Cavity Lamps (Inspection Lamps for Hollow Spaces) (Oct). 6 pp.
IEC 598 Pt 2-8-81. Luminaires Part 2: Particular Requirements Section Eight: Handlamps First Edition; (Amendment 2-1990, Incorporating Amendment 1). 41 pp.

—Hand Held—Explosion Proof
JIS C 8004-85. Hand Lanterns for Explosive Atmospheres. 10 pp.

—Hand Held—Fireproof
BSI BS 889-65. (WITHDRAWN) 1965 Amd 3 Flameproof Electric Lighting Fittings (AMD 5050) May 30, 1986. 65 pp.

—Hand Held—Marine
CNS F5018-85. Marine Watertight Type Hand Lamps (Sep)(8390).
CNS F5019-82. Hand Lamps for Marine Use (Non-Watertight Type) (Jan)(8391).

—Identification Systems
IEC 1231-93. International Lamp Coding System (ILCOS) First Edition. 59 pp.
MOD UK DSTAN 62-10: Part 2-74. Marking of Lamps, Electric and Their Packages Issue 1. 5 pp.

Lamps *(Cont.)*
—Kerosene—Highway Lighting—Warning Lights
BSI BS 3143: Part 1-85. 1985 Amd 1 Road Danger Lamps Part 1: Kerosine Burning Lamps (AMD 4846) February 28, 1985. 11 pp.

—Luminance
CNS C4168-84. Standard Lamps of Luminous Intensity (Jun)(5200).

—Marine
CNS F5018-85. Marine Watertight Type Hand Lamps (Sep)(8390).
CNS F5019-82. Hand Lamps for Marine Use (Non-Watertight Type) (Jan)(8391).
CNS F5066-83. Explosion-Proof Ceiling Light for Marine Use (Jun)(10336).
JIS F 8411-83. Marine Ceiling Lights (Non-Watertight Type).

—Miniature
JIS C 7712-65. Method for the Designation of Miniature Lamps. 7 pp.

—Miniature—Laryngoscopes
BSI BS 6578: Part 2-85. 1985 Laryngoscopic Fittings Part 2: Screw Threads for Miniature Electric Lamps and Sockets. 6 pp.
ISO 7376 Pt 2-84. Laryngoscopic Fittings—Part 2: Miniature Electric Lamps—Screw Threads and Sockets First Edition. 5 pp.

—Packaging—Identification Systems
MOD UK DSTAN 62-10: Part 2-74. Marking of Lamps, Electric and Their Packages Issue 1. 5 pp.

—Photosensitivity Measurement
CNS C4261-84. Electric Lamps for Measurement on Photo-Sensitivity of Photographic Sensitive Materials (Jun)(6427).

—Portable
CSA B140.9.4-M1979. Portable, Pressurized-Type, Liquid Petroleum Fuelled Lamps and Lanterns (R 1991); (Amd 1-2 August 1982) (Errata August 1982). 25 pp.
CSA C22.2 NO 12-1982. Portable Luminaires; (Gen Instr 1 Thru 4). 52 pp.
SAA AS/NZS 3128-93. Approval and Test Specification—Portable Lamp Standards and Brackets (in Professional Package 28). 7 pp.

—Portable—Brackets
SAA AS/NZS 3128-93. Approval and Test Specification—Portable Lamp Standards and Brackets (in Professional Package 28). 7 pp.
SNZ NZS/AS 3128-90. Approval and Test Specification—Portable Lamp Standards and Brackets Amend: 4, 1992 (This is a Joint Standard with SAA AS 3128). 6 pp.

—Portable—Lampholders
SAA AS/NZS 3128-93. Approval and Test Specification—Portable Lamp Standards and Brackets (in Professional Package 28). 7 pp.
SNZ NZS/AS 3128-90. Approval and Test Specification—Portable Lamp Standards and Brackets Amend: 4, 1992 (This is a Joint Standard with SAA AS 3128). 6 pp.

—Portable—Marine
CNS F5071-85. Portable Lamps (Simple Type) for Marine Use (Sep)(10499).

—Radio Equipment
CNS C4263-84. Radio Panel Lamps (Jun)(6429).

—Reflector
CNS C3159-82. Method of Test for Reflectors for Lighting Fittings (Jul)(9121).
CNS C4102-84. Reflection Lamps (Jun)(3261).
JIS C 7525-92. Reflector Lamps. 11 pp.

—Sewing Machines—Household
CNS C3101-84. Method of Test for Household Sewing Machine Lamps (Jun)(6431).
CNS C4264-84. Household Sewing Machine Lamps (Jun)(6430).

—Swimming Pools
BSI BS 4533: Sec 102.18-90. 1990 Luminaires Part 102: Particular Requirements Section 102.18: Specification for Luminaires for Swimming Pools and Similar Applications. 14 pp.
BSI BS 4533: Sec 102.18-01. 1990 Amd 1 Luminaires Part 102: Particular Requirements Section 102.18: Specification for Luminaires for Swimming Pools and Similar Applications (AMD 7718) May 15, 1993 (F). 20 pp.
CENELEC EN 60 598-2-18-89. Luminaires Part 2: Particular Requirements Section Eighteen—Luminaires for Swimming Pools and Similar Applications. 6 pp.

INDUSTRY STANDARDS

Lamps (Cont.)

—Swimming Pools (Cont.)
CENELEC PREN 60598-2-18-93. Luminaires Part 2: Particular Requirements Section 18: Luminaires for Swimming Pools and Similar Applications (IEC 598-2-18: 1993, Modified). 7 pp.
CENELEC EN 60598-2-18/A2-91. AMD 2 Luminaires Part 2: Particular Requirements Section Eighteen—Luminaires for Swimming Pools and Similar Aplications. 5 pp.
IEC 598 Pt 2-18-93. Luminaires Part 2: Particular Requirements Section 18: Luminaires for Swimming Pools and Similar Applications Second Edition (CENELEC PREN 60598-2-18: 1993). 20 pp.

—Symbols—Architectural Drawings
CNS C1100-87. Symbols of Lamps and Socket of Interior Wiring Diagram for Architectural Plans (Dec)(9111).

—Symbols—Diagrams
BSI BS 3939: Part 8-85. 1985 Graphical Symbols for Electrical Power, Telecommunications and Electronics Diagrams Part 8: Measuring Instruments, Lamps and Signalling Devices (G). 15 pp.
IEC 617 Pt 8-83. Graphical Symbols for Diagrams Part 8: Measuring Instruments, Lamps and Signalling Devices First Edition. 26 pp.
SNZ IEC 617: Part 8-83. Graphical Symbols for Diagrams Part 8: Measuring Instruments, Lamps and Signalling Devices. 23 pp.

—Telephone Exchanges
CNS C4262-84. Telephone Exchange Lamps (Jun)(6428).
JIS C 7521-83. Telephone Exchange Lamps. 9 pp.

Lampshades
See Also: Lamps

—Carrier Rings—Screw Threads
DIN ENGL 49689-66. External Screw Threads for Lamp Holders and Internal Screw Threads for Lamp Shade Carrier Rings Made of Metal or Plastic Moulded Material; Limiting Dimensions (Mar). 1 p.

—Enameled
CNS R2024-58. Enamel Lamp Shape (Nov) (901)(R 1973).

LAN
Use: Local Area Networks

LAN Cables
Use: Local Area Network Cables

Lancets
See Also: Surgical Equipment; Surgical Instruments; Surgical Knives

—Blood
CNS T1026-80. Blood Lancet (Aug)(6285).
JIS T 2601-79. Blood Lancet.

Land Contamination
Use: Contamination—Land

Land Distress and Safety Systems
Use: Global Land Distress and Safety Systems

Land Maps
Use: Maps

Land Minefields
Use: Minefields

Land Mobile Stations
Use: Mobile Stations (Communications)—Land

Land Pollution
Use: Environmental Protection

Land Service Equipment
Scope Note: See specific types of equipment used in land service

Land Use Planning
Use For: Commercial Land Use; Industrial Land Use; Residential Land Use See Also: Civil Engineering; Urban Areas
BSI BS 7370: Part 1-91. 1991 Grounds Maintenance Part 1: Recommendations for Establishing and Managing Grounds Maintenance Organizations and for Design Considerations Related to Maintenance. 57 pp.
BSI BS 7370: Part 3-91. 1991 Grounds Maintenance Part 3: Recommendations for Maintenance of Amenity and Functional Turf (Other Than Sports Turf). 74 pp.
BSI BS 7370: Part 4-93. 1993 Grounds Maintenance Part 4: Recommendations for Maintenance of Soft Landscape (Other Than Amenity Turf). 94 pp.
SNZ NZS 4404-81. Code of Practice for Urban Land Subdivision. 82 pp.

—Acoustics
SAA AS 1055.3-89. Acoustics—Description and Measurement of Environment Noise—Part 3: Acquisition of Data Pertinent to Land Use. 5 pp.

—Airports
SNZ NZS 6805-92. Airport Noise Management and Land Use Planning. 28 pp.

—Classification
BSI BS 1000: (71)-76. 1976 Universal Decimal Classification (UDC). English Full Edition (71): Physical Planning. 18 pp.
SNZ NZS/BS 1000 (71)-76. Universal Decimal Classification Physical Planning. 20 pp.

—Glossaries
BSI BS 6100: Sec 1.8-90. 1990 Glossary of Building and Civil Engineering Terms Part 1: General and Miscellaneous Section 1.8: Environment and Physical Planning. 11 pp.
BSI BS 6100: Sec 1.8-01. 1990 Amd 1 Glossary of Building and Civil Engineering Terms Part 1: General and Miscellaneous Section 1.8: Environment and Physical Planning (AMD 7249) August 15, 1992. 13 pp.

Landfills
See Also: Solid Wastes; Waste Disposal

—Geomembranes
CNS K3093-89. High Density Polyethylene Geomembrane for Sanitary Landfill (Feb)(12493). 3 pp.
CNS K6961-89. Method of Test for High Density Polyethylene Geomembrane for Sanitary Landfill (Feb)(12494). 5 pp.

Landing Approaches
Use: Aircraft Landing Approaches

Landing Beaches
Use: Beaches

Landing Fields
Use: Aircraft Landing Areas

Landing Gear
See Also: Aircraft Arresting Systems; Aircraft Equipment; Aircraft Landings; Airframes; Brakes; Floats; Nose Wheels; Tires; Wheels

—Aerostats
CAA Chapter Q4-5 12.79. Landing Gear Design (Non-Rigid Airships). 1 p.

—Aerostats—Loads (Forces)
CAA Chapter Q3-5 12.79. Ground Loads (Non-Rigid Airships). 3 pp.

—Control Systems Equipment
CAA Chapter K4-8 04.72. Control System Loads and Design (Light Aeroplanes). 7 pp.

—Design
CAA Chapter K4-5 04.74. Landing Gear Design (Light Aeroplanes). 4 pp.
CAA Chapter K4-5 App 03.67. Landing Gear Design Position Indicators (Light Aeroplanes). 1 p.
CAA Chapter P4-5. Landing Gear Design (Provisional Airworthiness Requirements for Civil Powered Lift Aircraft). 19 pp.
CAA CAP 482 SUB-Part C 03.83. Structure (Small Light Aeroplanes). 12 pp.

—Loads (Forces)
CAA Chapter K3-5 04.72. Ground Loads (Light Aeroplanes). 8 pp.
CAA Chapter P3-5. Ground Loads (Provisional Airworthiness Requirements for Civil Powered Lift Aircraft). 18 pp.

—Rotary Wing Aircraft—Design
CAA Chapter G4-5 12.80. Landing Gear Design (Rotocraft). 3 pp.
CAA Chapter G4-5 App 12.80. Landing Gear Design (Rotocraft). 2 pp.
CAA CAP 524 SUB-Part D 12.86. Design and Construction (Rotorcraft). 23 pp.

—Rotary Wing Aircraft—Loads (Forces)
CAA Chapter G3-5 12.80. Ground and Water Loads (Rotocraft). 4 pp.

—Supersonic Transports—Reliability
CAA STANDARD NO. 5-6. Landing Gear. 6 pp.

Landing Impact
Use: Impact Shock

Landing Lamps
Use: Landing Lights

Landing Lights
See Also: Aircraft Lighting

—Glide Slope Guidance Systems
NATO STANAG 1236 ED 2 AMD 2-88. Glide Slope Indicators for Helicopter Operations from NATO Ships. 10 pp.
NATO STANAG 1236 ED 2 AMD 3-92. Glide Slope Indicators for Helicopter Operations from NATO Ships. 11 pp.

—Sealed Beam
ISO 1198-72. Sealed-Beam Landing Lamps for Aircraft—Dimensions First Edition. 6 pp.

Landing Zones
Use: Aircraft Landing Areas

Landscape Lighting
Use For: Lawn and Garden Lighting
See Also: Decorative Lighting; Lawn and Garden Equipment; Lighting
BSI BS 4533: Sec 102.18-90. 1990 Luminaires Part 102: Particular Requirements Section 102.18: Specification for Luminaires for Swimming Pools and Similar Applications. 14 pp.
BSI BS 4533: Sec 102.18-01. 1990 Amd 1 Luminaires Part 102: Particular Requirements Section 102.18: Specification for Luminaires for Swimming Pools and Similar Applications (AMD 7718) May 15, 1993 (F). 20 pp.

—Portable
BSI BS 4533: Sec 102.7-90. 1990 Luminaires Part 102: Particular Requirements Section 102.7: Specification for Portable Luminaires for Garden Use. 15 pp.
CENELEC EN 60 598-2-7-89. Luminaires Part 2: Particular Requirements Section Seven-Portable Luminaires for Garden Use. 8 pp.
IEC 598 Pt 2-7-82. Luminaires Part 2: Particular Requirements Section Seven—Portable Luminaires for Garden Use First Edition; (Amendment 1-1987). 19 pp.

—Wet Locations
CENELEC EN 60 598-2-18-89. Luminaires Part 2: Particular Requirements Section Eighteen—Luminaires for Swimming Pools and Similar Applications. 6 pp.
CENELEC PREN 60598-2-18-93. Luminaires Part 2: Particular Requirements Section 18: Luminaires for Swimming Pools and Similar Applications (IEC 598-2-18: 1993, Modified). 7 pp.
CENELEC EN 60598-2-18/A2-91. AMD 2 Luminaires Part 2: Particular Requirements Section Eighteen—Luminaires for Swimming Pools and Similar Aplications. 5 pp.
IEC 598 Pt 2-18-93. Luminaires Part 2: Particular Requirements Section 18: Luminaires for Swimming Pools and Similar Applications Second Edition (CENELEC PREN 60598-2-18: 1993). 20 pp.

Landscaping
See Also: Highways; Playgrounds; Recreational Facilities
BSI BS 4428-89. 1989 Amd 1 Code of Practice for General Landscape Operations (Excluding Hard Surfaces) (AMD 6784) September 30, 1991. 35 pp.

—Construction Contracts
DIN ENGL 18320-88. Tendering and Performance Stipulations in Contracts for Construction Works (VOB); Part C: General Technical Specifications in Contracts for Construction Works (ATV); Landscape Works (Sept) (This Standard, Together with DIN 18299, September 1988 Edition,. 5 pp.

—Drawings
BSI BS 1192: Part 4-84. 1984 Construction Drawing Practice Part 4: Recommendations for Landscape Drawings. 40 pp.

—Maintenance
BSI BS 7370: Part 1-91. 1991 Grounds Maintenance Part 1: Recommendations for Establishing and Managing Grounds Maintenance Organizations and for Design Considerations Related to Maintenance. 57 pp.
BSI BS 7370: Part 3-91. 1991 Grounds Maintenance Part 3: Recommendations for Maintenance of Amenity and Functional Turf (Other Than Sports Turf). 74 pp.
BSI BS 7370: Part 4-93. 1993 Grounds Maintenance Part 4: Recommendations for Maintenance of Soft Landscape (Other Than Amenity Turf). 94 pp.

Landscaping (Cont.)
—Scheduling
BSI BS 1192: Part 4-84. 1984 Construction Drawing Practice Part 4: Recommendations for Landscape Drawings. 40 pp.
—Topsoil
BSI BS 3882-65. 1965 Amd 1 Recommendations and Classification for Top Soil. 9 pp.
—Trees (Plants)
BSI BS 5837-91. 1991 Trees in Relation to Construction. 37 pp.
BSI BS 5837-80. 1980 Code of Practice for Trees in Relation to Construction. 28 pp.
—Turf
BSI BS 3969-90. 1990 Recommendations for Turf for General Purposes. 13 pp.
BSI BS 3969-65. (WITHDRAWN) 1965 Recommendations for Turf for General Landscape Purposes. 9 pp.

Language Digits
See Also: Discriminating Digits; International Exchanges; Telephone Systems
CCITT RECMN Q.104-89. Language Digit or Discriminating Digit—General Recommendations on Telephone Switching and Signalling—Functions and Information Flows for Services in the ISDN—Supplements (Study Group XI) 2 pp. 2 pp.

Language Laboratories
See Also: Laboratories
—Tape Recorders
BSI BS 5817: Part 17-90. 1990 Audio-Visual, Video and Television Equipment and Systems Part 17: Methods for Specifying and Measuring the Performance Characteristics of Audio Learning Systems. 22 pp.
IEC 574 Pt 17-89. Audiovisual, Video and Television Equipment and Systems Part 17: Audio-Learning Systems First Edition. 41 pp.

Languages
—Classification
BSI BS 1000: (8)-71. 1971 Universal Decimal Classification (UDC). English Full Edition (8): Language. Linguistics. Literature. 35 pp.
SNZ NZS/BS 1000 (8)-71. Universal Decimal Classification Language. Linguistics. Literature. 36 pp.
—Proficiency Levels—Military Personnel
NATO STANAG 6001 ED 1 AMD 0-76. Language Proficiency Levels. 10 pp.

Lanolin
BSI BS 3488-87. 1987 Amd 1 Technical Anhydrous Lanolin (AMD 6162) February 28, 1990. 5 pp.

Lantern Slides
Use: Transparencies

Lanterns
See Also: Lamps; Lighting
CSA B140.9.4-M1979. Portable, Pressurized-Type, Liquid Petroleum Fuelled Lamps and Lanterns (R 1991); (Amd 1-2 August 1982) (Errata August 1982). 25 pp.
JIS C 8104-91. Hand Lanterns. 29 pp.
—Battery Operated
MOD UK DSTAN 62-2: Part 3-74. Battery Operated Lights, Torches, and Lanterns, Electric Part 3: Lanterns, Electric Issue 1. 10 pp.
—Highway Lighting—Light Distribution
BSI BS 4533: Sec 103.1-81. 1981 Amd 1 Luminaires Part 103: Performance Requirements Section 103.1: Light Distribution from Road-Lighting Lanterns. 10 pp.
—Safety Signs
CNS Z1035-87. Lantern-Type Safety Signs (Mar)(9648).
JIS Z 9109-87. Lantern-Type Safety Signs. 14 pp.

Lanthanum Chloride
JIS K 9060-61. Lanthanum Chloride.

Lanyards
See Also: Cables (Mechanical); Ropes; Safety Belts
BSI BS 1397-79. (WITHDRAWN) 1979 Amd 3 Industrial Safety Belts, Harnesses and Safety Lanyards (AMD 4905) October 31,. 1985 (Superseded by BS EN 354, 355, 358, 361, 362, 363, 364, 365: 1993). 21 pp.
BSI BS EN 354-93. 1993 Personal Protective Equipment Against Falls from a Height—Lanyards (Supersedes BS 1397: 1979) (N). 10 pp.

Lanyards (Cont.)
CEN PREN 354-90. Personal Fall Arresting Systems: Lanyards. 5 pp.
CEN EN 354-92. Personal Protective Equipment Against Falls from a Height—Lanyards. 4 pp.
DIN ENGL 7471-85. Protective Equipment Against Fall; Lanyards; Safety Requirements, Testing (July). 4 pp.
—Aerospace—Connectors—Contact Force
CEN PREN 2591 (Part D14)-92. Elements of Electrical and Optical Connection Test Methods Part D14—Unmating of Lanyard Release Connectors. 4 pp.
—Aircraft—Handhole Covers—Lugs
SBAC AS 3177 ISSUE 1. Lug for Lanyard. 1 p.
—Anchors (Fasteners)—Self Locking
DIN ENGL 32769-86. Protective Equipment Against Fall; Self-Locking Safety Anchorages; Safety Requirements, Testing (Feb). 3 pp.
—Construction Equipment
CSA Z259.1-1976. Fall Arresting Safety Belts and Lanyards for the Construction and Mining Industries; (Amd 1-9 May 1979) (Amd 10-11 July 1993) (Gen Instr 2). 29 pp.
—Energy Absorbers
DIN ENGL 32766-81. Protective Equipment Against Fall; Energy Absorbers; Safety Requirements, Testing (Dec). 2 pp.
—Mining Equipment
CSA Z259.1-1976. Fall Arresting Safety Belts and Lanyards for the Construction and Mining Industries; (Amd 1-9 May 1979) (Amd 10-11 July 1993) (Gen Instr 2). 29 pp.

LAP
Use: Link Access Procedures

Lap Cements
Use For: Lap Sealants See Also: Asphalt Cements; Lap Joints; Roofing Cements
CGSB CAN/CGSB-37.4-M89. Fibrated, Cutback Asphalt, Lap Cement for Asphalt Roofing. 9 pp.
CGSB 37-GP-10MA-84. Application of Asphalt Lap Cement, Standard for. 6 pp.
CGSB CAN/CGSB-37.29-M89. Rubber-Asphalt Sealing Compound. 9 pp.

Lap Joints
See Also: Brazed Joints; Butt Joints; Butt Welded Joints; Joints; Lap Cements; Pipe Joints; Tube Joints; Welded Joints
—Aluminum
ISO 3777-76. Radiographic Inspection of Resistance Spot Welds for Aluminium and Its Alloys—Recommended Practice First Edition. 7 pp.

Lap Sealants
Use: Lap Cements

Lappers
Use: Lapping Machines

Lappets
See Also: Looms; Textile Machinery
—Fine Spinning
CNS L2066-82. Fine Spinning Lappet (Dec)(9728).
JIS L 5122-92. Lappets for Spinning Frames. 7 pp.
—Twisting Frames
CNS L2068-82. Twisting Lappet (Dec)(9730).
CNS L2069-82. Twisting Lappet Porcelain (Dec)(9731).
JIS L 5124-57. Lappet for Twisting Frame (R 1983). 6 pp.
JIS L 5125-57. Lappet Porcelain for Twisting Frame (R 1983). 5 pp.

Lapping Machines
Use For: Lappers See Also: Finishing Machinery; Honing Machinery
—Electric—Portable
CSA CAN/CSA-C22. 2 NO71.1-M89. Portable Electric Tools; (Gen Instr 1 Thru 4). 89 pp.

Laps (Textiles)
See Also: Textile Machinery
—Rods
JIS L 5164-92. Lap Rod.

Lard
See Also: Animal Fats
CGSB 32.77M-89. Lard. 7 pp.
CNS N5069-88. Lard (Edible) (May)(2421). 2 pp.

Lard (Cont.)
CNS N5192-88. Rendered Pork Fat (Edible) (May)(8155). 2 pp.
CNS N6074-83. Methods of Test for Edible Oils and Fats (General Rules) (Mar)(3639). 2 pp.

Larvae
See Also: Insect Contamination
—Wood Preservatives
BSI BS EN 49-1-92. 1992 Wood Preservatives—Determination of the Protective Effectiveness Against Anobium Punctatum (De Geer) by Egg-Laying and Larval Survival Part 1: Application by Surface Treatment (Laboratory Method) (W). 16 pp.
BSI BS EN 49-2-92. 1992 Wood Preservatives—Determination of the Protective Effectiveness Against Anobium Punctatum (De Geer) by Egg-Laying and Larval Survival Part 2: Application by Impregnation (Laboratory Method) (Supersedes BS 5437: 1977). 16 pp.
CEN EN 49-76. Wood Preservatives: Determination of the Toxic Values Against Anobium Punctatum (De Geer) by Egg-Laying and Larval Survival (Laboratory Method). 11 pp.
CEN PREN 49 (Part 1)-89. Wood Preservatives: Part 1: Determination of the Toxic Valves Against Anobium Punctatum (De Geer) by Egg-Laying and Larval Survival (Laboratory Method). 12 pp.
CEN EN 49-1-92. Wood Preservatives—Determination of the Protective Effectiveness Against Anobium Punctatum (De Geer) by Egg-Laying and Larval Survival—Part 1: Application by Surface Treatment (Laboratory Method). 23 pp.
CEN PREN 49 (Part 2)-89. Wood Preservatives: Part 2: Determination of the Preventive Action Against Anobium Punctatum (De Geer) (Laboratory Method). 16 pp.
CEN EN 49-2-92. Wood Preservatives—Determination of the Protective Effectiveness Against Anobium Punctatum (De Geer) by Egg-Laying and Larval Survival—Part 2: Application by Impregnation (Laboratory Method). 24 pp.
DIN ENGL EN 20 Pt 1-92. Wood Preservatives; Determination of the Effectiveness Against Lyctus Brunneus (Stephens); Application by Surface Treatment (Laboratory Method) (Sept). 11 pp.
DIN ENGL EN 46-90. Wood Preservatives; Determination of the Preventive Action Against Recently Hatched Larvae of Hylotrupes Bajulus (Linnaeus) (Laboratory Method) (Apr). 11 pp.
DIN ENGL EN 48-90. Wood Preservatives; Determination of Eradicant Action Against Larvae of Anobium Punctatum (De Geer) (Laboratory Method) (Apr). 13 pp.
DIN ENGL EN 49 Pt 1-92. Wood Preservatives; Determination of the Effectiveness Against Anobium Punctatum (de Geer) on the Basis of Egg-Laying and Larval Survival; Application by Surface Treatment (Laboratory Method) (Nov). 11 pp.
DIN ENGL EN 49 Pt 2-92. Wood Preservatives; Determination of the Effectiveness Against Anobium Punctatum (de Geer) on the Basis of Egg-Laying and Larval Survival; Application by Impregnation (Laboratory Method) (Nov). 11 pp.

Laryngoscopes
See Also: Medical Electrical Equipment; Medical Equipment
—Fittings
BSI BS 6578: Part 1-85. 1985 Laryngoscopic Fittings Part 1: Specification for Hook-on Type Handle-Blade Fittings (ISO 7376/1: 1984). 7 pp.
ISO 7376 Pt 1-84. Laryngoscopic Fittings—Part 1: Hook-on Type Handle-Blade Fittings First Edition. 5 pp.
—Screw Threads—Lampholders
BSI BS 6578: Part 2-85. 1985 Laryngoscopic Fittings Part 2: Screw Threads for Miniature Electric Lamps and Sockets. 6 pp.
ISO 7376 Pt 2-84. Laryngoscopic Fittings—Part 2: Miniature Electric Lamps—Screw Threads and Sockets First Edition. 5 pp.
—Screw Threads—Lamps
BSI BS 6578: Part 2-85. 1985 Laryngoscopic Fittings Part 2: Screw Threads for Miniature Electric Lamps and Sockets. 6 pp.
ISO 7376 Pt 2-84. Laryngoscopic Fittings—Part 2: Miniature Electric Lamps—Screw Threads and Sockets First Edition. 5 pp.

Laser Audio Disk Players
Use: Compact Disk Players

Laser Beams
See Also: Laser Gyros
—Optical Power Meters
JIS C 6182-91. Test Methods of Optical Power Meters for Laser Beam. 14 pp.

INTERNATIONAL AND NON-U.S. NATIONAL STANDARDS SUBJECT INDEX

Laser

Laser Cutting
See Also: Cutting; Thermal Cutting

—Metals
DIN ENGL 2310 Pt 5-90. Thermal Cutting; Laser Cutting of Metallic Materials; Principles of Process, Quality and Dimensional Tolerances (Dec). 4 pp.

Laser Diodes
See Also: Diode Lasers; Diodes; Laser Equipment; Lasers; Light Emitting Diodes; Light Sources
CECC CECC 20 000 ISSUE 1-82. Generic Specification: Semiconductor Optoelectronic and Liquid Crystal Devices (En, Fr, Ge) AMD 3 (En, Fr, Ge) SUPP 1 (En, Fr, Ge) SUPP 2 (En, Fr, Ge). 401 pp.

—Fiber Optics
JIS C 5940-89. General Rules of Laser Diodes for Fiber Optic Transmission. 14 pp.
JIS C 5941-89. Test Methods of Laser Diodes for Fiber Optic Transmission. 18 pp.

—Modules—Optical Fiber Pigtails—Preferred Products List
CECC CECC MUAHAG Vol 10 IS 2-92. Preferred Products List; OPTO Electronic Devices (En, Fr, Ge). 48 pp.

—Modules—Preferred Products List
CECC CECC MUAHAG Vol 10 IS 2-92. Preferred Products List; OPTO Electronic Devices (En, Fr, Ge). 48 pp.

—Quality Assurance
IEC 747 Pt 12-91. Semiconductor Devices Part 12: Sectional Specification for Optoelectronic Devices First Edition (IECQ QC 720100). 44 pp.

—Video Disk Recording
JIS C 5942-90. General Rules of Laser Diodes Used for Recording and Playback. 14 pp.
JIS C 5943-90. Test Methods of Laser Diodes Used for Recording and Playback. 19 pp.

Laser Equipment
See Also: Compact Disk Players; Diode Lasers; Laser Diodes; Laser Gyros; Laser Instruments (Distance Measurement); Laser Printers; Laser Radiation; Laser Target Designators; Lasers; Medical Electrical Equipment

—Documentation
ISO 11252-93. Lasers and Laser-Related Equipment—Laser Device —Minimum Requirements for Documentation First Edition. 6 pp.

—Identification Systems
ISO 11252-93. Lasers and Laser-Related Equipment—Laser Device —Minimum Requirements for Documentation First Edition. 6 pp.

—Laser Radiation—Safety
CEN PREN 31553-93. Optics and Optical Instruments—Lasers and Laser Related Equipment—Safety of Machines Using Radiation to Process Materials. 24 pp.

—Mechanical Interfaces
ISO 11253-93. Lasers and Laser-Related Equipment—Laser Device —Mechanical Interfaces First Edition. 5 pp.

—Medical—Diagnostic—Safety
BSI BS 5724: Sec 2.122-93. 1993 Medical Electrical Equipment Part 2: Particular Requirements for Safety Section 2.122: Specification for Diagnostic and Therapeutic Laser Equipment (ALSO KNOWN AS 60601-2-22: 1993). 27 pp.
BSI BS EN 60601-2-22-93. 1993 Medical Electrical Equipment Part 2: Particular Requirements for Safety Section 2.122: Specification for Diagnostic and Therapeutic Laser Equipment (IEC 601-2-22: 1992). 27 pp.
CENELEC EN 60601-2-22-92. Medical Electrical Equipment Part 2: Particular Requirements for the Safety of Diagnostic and Therapeutic Laser Equipment. 5 pp.
CENELEC EN 60601-2-22-92. Medical Electrical Equipment Part 2: Particular Requirements for the Safety of Diagnostic and Therapeutic Laser Equipment; (IEC 601-2-22: 1992). 22 pp.
IEC 601 Pt 2-22-92. Medical Electrical Equipment Part 2: Particular Requirements for the Safety of Diagnostic and Therapeutic Laser Equipment First Edition; (CENELEC EN 60601-2-22: 1992). 46 pp.
SNZ NZS/AS 3200. 2.22-93. Approval and Test Specification—Medical Electrical Equipment Part 2.22: Diagnostic and Therapeutic Laser Equipment (This is a Joint Standard with SAA AS 3200.2.22). 19 pp.

—Medical—Safety
CSA CAN/CSA-Z386-92. Laser Safety in Health Care Facilities; (Gen Instr 1). 70 pp.

—Medical—Therapeutic—Safety
BSI BS 5724: Sec 2.122-93. 1993 Medical Electrical Equipment Part 2: Particular Requirements for Safety Section 2.122: Specification for Diagnostic and Therapeutic Laser Equipment (ALSO KNOWN AS 60601-2-22: 1993). 27 pp.
BSI BS EN 60601-2-22-93. 1993 Medical Electrical Equipment Part 2: Particular Requirements for Safety Section 2.122: Specification for Diagnostic and Therapeutic Laser Equipment (IEC 601-2-22: 1992). 27 pp.
CENELEC EN 60601-2-22-92. Medical Electrical Equipment Part 2: Particular Requirements for the Safety of Diagnostic and Therapeutic Laser Equipment. 5 pp.
CENELEC EN 60601-2-22-92. Medical Electrical Equipment Part 2: Particular Requirements for the Safety of Diagnostic and Therapeutic Laser Equipment; (IEC 601-2-22: 1992). 22 pp.
IEC 601 Pt 2-22-92. Medical Electrical Equipment Part 2: Particular Requirements for the Safety of Diagnostic and Therapeutic Laser Equipment First Edition; (CENELEC EN 60601-2-22: 1992). 46 pp.
SNZ NZS/AS 3200. 2.22-93. Approval and Test Specification—Medical Electrical Equipment Part 2.22: Diagnostic and Therapeutic Laser Equipment (This is a Joint Standard with SAA AS 3200.2.22). 19 pp.

—Optical Materials
CEN PREN 31151-1-93. Optics and Optical Instruments—Lasers and Laser Related Equipment—Standard Optical Components—Part 1: Components for the UV, Visible and Near-Infrared Spectral Range. 12 pp.

—Safety
BSI BS 7192-89. (WITHDRAWN) 1989 Radiation Safety of Laser Products (Superseded by BS EN 60825: 1992). 79 pp.
BSI BS EN 60825-92. 1992 Radiation Safety of Laser Products, Equipment Classification, Requirements and User's Guide. 86 pp.
CENELEC HD 194-74. Requirements Concerning the Electrical Safety of Laser—Apparatus and Installations. 55 pp.
CENELEC HD 482 S1-88. Radiation Safety of Laser Products, Equipment, Classification, Requirements, and Users Guide. 8 pp.
CENELEC EN 60825-91. Radiation Safety of Laser Products, Equipment Classification, Requirements and User's Guide. 83 pp.
DIN VDE 0411 Pt 100-86. Measurement and Control; Safety Requirements for Electrically Operated Measuring, Control and Laboratory Equipment; General Requirements; Identical with IEC 66E (Sec) 22 (Aug). 137 pp.
DIN VDE 0836-77. VDE Specification for the Electrical Safety of Laser Equipment and Installations (Feb). 65 pp.
IEC 820-86. Electrical Safety of Laser Equipment and Installations First Edition. 81 pp.
IEC 825-84. Radiation Safety of Laser Products, Equipment Classification, Requirements and User's Guide First Edition; (Amendment 1-1990). 192 pp.
JIS C 6802-91. Radiation Safety Standards for Laser Products. 61 pp.

Laser Gyros
See Also: Laser Beams; Laser Equipment

—Single Axis
NATO STANAG 4220 ED 1 AMD 1-88. Specification Format Guide and Test Procedure for Single-Axis Laser Gyros. 9 pp.

Laser Instruments (Distance Measurement)
See Also: Distance Measuring Equipment; Laser Equipment; Surveying Instruments

—Construction—Accuracy Testing
BSI BS 7334: Part 6-92. 1992 Measuring Instruments for Building Construction Part 6: Methods for Determining Accuracy in Use of Laser Instruments (ISO 8322-6: 1991). 19 pp.
ISO 8322 Pt 6-91. Building Construction—Measuring Instruments—Procedures for Determining Accuracy in Use—Part 6: Laser Instruments First Edition. 17 pp.

Laser Photocoagulators
Use: Photocoagulators—Laser

Laser Printers
See Also: Laser Equipment; Page Printers; Printers

—Toners—Recycling
CGSB CAN/CGSB-53.148-M90. Rejuvenation of Laser Printer Cartridges. 11 pp.

Laser Radiation
See Also: Laser Equipment; Lasers

—Laser Equipment—Safety
CEN PREN 31553-93. Optics and Optical Instruments—Lasers and Laser Related Equipment—Safety of Machines Using Radiation to Process Materials. 24 pp.

—Measuring Instruments
BSI BS EN 61040-93. 1993 Power and Energy Measuring Detectors, Instruments and Equipment for Laser Radiation (IEC 1040: 1990) (E). 22 pp.
CENELEC EN 61040-92. Power and Energy Measuring Detectors, Instruments and Equipment for Laser Radiation; (IEC 1040: 1990). 5 pp.
DIN VDE 0835-81. Power and Energy Measuring Equipment for Laser Radiation (July). 16 pp.
IEC 1040-90. Power and Energy Measuring Detectors, Instruments and Equipment for Laser Radiation First Edition; (CENELEC EN 61040:1992). 34 pp.

—Medical Surveillance—Over Exposure
NATO STANAG 2900 ED 2 AMD 0-88. Laser Radiation—Medical Surveillance and Evaluation of Over-Exposure. 6 pp.
NATO STANAG 2900 ED 2 AMD 1-88. Laser Radiation—Medical Surveillance and Evaluation of Over-Exposure. 6 pp.

—Output Power—Measurement
JIS C 6180-91. Measuring Methods for Laser Output Power. 23 pp.

Laser Target Designators
See Also: Laser Equipment; Weapons Systems
NATO STANAG 3875 ED 1 AMD 0-82. Criteria for Categorization of Laser Designator Systems. 6 pp.

—Flight Crews—Safety
NATO STANAG 3828 ED 1 AMD 5-81. Minimum Requirements for Aircrew Protection Against the Hazards of Laser Target Designators. 12 pp.

Laser Video Disks
Use: Video Disks

Laser Vision Video Disks
Use: Video Disks

Lasers
See Also: Diode Lasers; Infrared Lasers; Laser Diodes; Laser Equipment; Laser Radiation; Light Transmission; Masers; Medical Equipment; Optical Glass; Space Communications

—Documentation
ISO 11252-93. Lasers and Laser-Related Equipment—Laser Device —Minimum Requirements for Documentation First Edition. 6 pp.

—Eye Protectors
ISO 6161-81. Personal Eye-Protectors—Filters and Eye-Protectors Against Laser Radiation First Edition; (Amendment Slip-1982). 7 pp.

—Glossaries
JIS C 6801-88. Glossary of Terms Used in Laser Safety. 18 pp.

—Hazards—Control and Evaluation—Safety
NATO STANAG 3606 ED 4 AMD 1-83. Evaluation and Control of Laser Hazards. 38 pp.
NATO STANAG 3606 ED 5 AMD 0-91. Evaluation and Control of Laser Hazards on Military Ranges. 41 pp.

—Identification Systems
ISO 11252-93. Lasers and Laser-Related Equipment—Laser Device —Minimum Requirements for Documentation First Edition. 6 pp.

—Safety
CNS Z1043-88. Safe Use of Lasers (Jun)(11640).
JIS C 6801-88. Glossary of Terms Used in Laser Safety. 18 pp.
MOD UK DSTAN 05-40-84. Evaluation and Control of Laser Hazards Issue 4. 39 pp.
SAA AS 2211-91. Laser Safety (in Professional Packages 34B, 34C). 64 pp.
SNZ NZS/AS 2211-91. Laser Safety (This is a Joint Standard with SAA AS 2211). 64 pp.
SNZ NZS 5821: Part 3-81. Laser Safety Part 3: Plain Language Code of Practice for the Safe Use of Lasers in Teaching (Reconfirmed 1990). 7 pp.

INTERNATIONAL AND NON-U.S. NATIONAL STANDARDS
SUBJECT INDEX

Lasers (Cont.)
—Surveying—Safety
SNZ NZS 5821: Part 2-81. Laser Safety Part 2: Plain Language Code of Practice for the Safe Use of Lasers in Surveying, Levelling and Alignment (Reconfirmed 1990). 8 pp.

—Symbols
CNS C5133-82. Graphic Symbols for Microwave Technology Masers and Lasers (Apr)(8665).

Lashing Ties
Use: Cable Ties

Lasts (Shoes)
Use: Shoe Machinery

Latch/Decoder/Drivers
See Also: Decoders; Display Drivers; Latch/Decoders; Latches (Circuits)

—BCD to 7-Segment
CECC CECC 90 104-083 ISSUE 2. NL CECC 90 104-083 Issue 2; Digital Integrated Circuits in Accordance with FS 90 104; HEC/HEF 4543B; BCD to 7-Segment Latch/Decoder/Driver (En). 11 pp.
CECC CECC 90 104-157 ISSUE 1-81. BS CECC 90 104-157; BCD-to-Seven Segment Latch/Decoder/Driver (En). 26 pp.
CECC CECC 90 104-157 ISSUE 1-86. CEI CECC 90 104-157; BCD-to-Seven Segment Latch /Decoder /Driver (En). 2 pp.
CECC CECC 90 104-220 ISSUE 1-85. BS CECC 90 104-220; BCD-To-Seven Segment Latch/Decoder/Driver for Liquid Crystals (En). 31 pp.
CECC CECC 90 104-220 ISSUE 1-86. CEI-CECC 90 104-220; Silicon with (B) Buffered Outputs BCD-to-Seven Segment Latch/Decoder/Driver for Liquid Crystals (En). 2 pp.

—Integrated Circuit—BCD to 7-Segment
CECC CECC 90 104-065 ISSUE 2. NL CECC 90 104-065 Issue 2; Digital Integrated Circuits in Accordance with FS 90 104; HEC/HEF 4511B; BCD to 7-Segment Latch/Decoder/Driver (En). 16 pp.
CECC CECC 90 109-755 ISSUE 1-87. Digital Integrated Circuits in Accordance with FS 90 109; 54/74 HC 4511; BCD-to-7 Segment Latch/Decoder/Driver (En, Fr). 6 pp.
CECC CECC 90 109-895 ISSUE 1-90. Digital Integrated Circuits; Silicon Monolithic C MOS, Cavity or Non-Cavity Packages; Type(s): 54/74 HCT 4511 BCD-to-7 Segment Latch/Decoder/Driver; Assessment Levels P, Y, L (En, Fr). 9 pp.

Latch/Decoders
See Also: Decoders; Latch/Decoder/Drivers; Latches (Circuits)

—Integrated Circuit—4 to 16 Lines—Preferred Products List
CECC CECC MUAHAG Vol 7 IS 8-92. Preferred Products List; Active Microcircuits (En, Fe, Ge). 89 pp.

—3 to 8 Lines
CECC CECC 90 109-731 ISSUE 1-87. Digital Integrated Circuits in Accordance with FS 90 109; 54 HC 131, 74 HC 131; 3 to 8 Line Decoder Latch (En, Fr). 5 pp.
CECC CECC 90 109-826 ISSUE 1-89. Digital Integrated Circuits; Silicon Monolithic C MOS, Cavity or Non-Cavity Packages; Type(s): 54/74 HCT 259 8-Bit Addressable Latch and 3 to 8 Line Decoder (En, Fr). 10 pp.

—8-Bit—Addressable
CECC CECC 90 109-826 ISSUE 1-89. Digital Integrated Circuits; Silicon Monolithic C MOS, Cavity or Non-Cavity Packages; Type(s); 54/74 HCT 259 8-Bit Addressable Latch and 3 to 8 Line Decoder (En, Fr). 10 pp.

Latch Needles
Use: Knitting Needles

Latches (Circuits)
See Also: Binary Counter/Latches; Bistable Latches; D-Type Latches; Decade/Biquinary Counter/Latches; Digital Circuits; Flip Flops; Latch/Decoder/Drivers; Latch/Decoders; R-S Latches; S-R Latches

—Octal
CECC CECC 90 104-230 ISSUE 1. NL CECC 90 104-230 Issue 1; Digital Integrated Circuits in Accordance with FS 90 104; HEC/HEF 40373B; Octal Transparent Latch with 3-State Outputs (En). 10 pp.

Latches (Circuits) (Cont.)
—Octal (Cont.)
CECC CECC 90 107-040 ISSUE 2-89. UTE C 86-218 ADD 3/FA 40; Digital Integrated Circuits in Accordance with FS 90 107; 54/74 F 373; Octal Transparent Latch with 3-State Non Inverting Outputs (En, Fr) ADD 3 (En, Fr). 10 pp.

—Quad
CECC CECC 90 101-026 ISSUE 1-87. UTE C 86-213/B 46; Digital Integrated Circuits in Accordance with FS 90 101; 54/64/74 75; Quad Latch (En). 16 pp.
CECC CECC 90 104-131 ISSUE 1-86. CEI CECC 90 104-131; Silicon Complementary MOS with (B) Buffered Outputs Cavity and Non Cavity Packaging (En). 2 pp.
CECC CECC 90 104-185 ISSUE 1-81. BS CECC 90 104-185; Quad Latch (En). 23 pp.
CECC CECC 90 104-185 ISSUE 1-86. CEI CECC 90 104-185; Quad Latch (En). 2 pp.

—4-Bit—Dual
CECC CECC 90 104-155 ISSUE 1-81. BS CECC 90 104-155; Dual 4-Bit Latch with Three State Outputs (En). 27 pp.
CECC CECC 90 104-155 ISSUE 1-86. CEI CECC 90 104-155; Dual 4-Bit Latch with Three State Outputs (En). 2 pp.

—4-Bit—Octal
CECC CECC 90 104-063 ISSUE 2. NL CECC 90 104-063 Issue 2; Digital Integrated Circuits in Accordance with FS 90 104; HEC/HEF 4508B; Dual 4-Bit Latch (En). 12 pp.

—8-Bit
CECC CECC 90 104-089 ISSUE 2. NL CECC 90 104-089 Issue 2; Digital Integrated Circuits in Accordance with FS 90 104; HEC/HEF 4724B; 8-Bit Addressable Latch (En). 10 pp.
CECC CECC 90 104-150 ISSUE 1-81. BS CECC 90 104-150; 8 Bit Addressable Latch Serial Input/Parallel Output (En). 26 pp.
CECC CECC 90 104-150 ISSUE 1-86. CEI CECC 90 104-150; 8 Bit Addressable Latch Serial Input/Parallel Output (En). 2 pp.

—8-Bit—Addressable
CECC CECC 90 103-100 ISSUE 1-81. BS CECC 90 103-100; 8-Bit Addressable Latches (En) AMD 1 (En). 19 pp.
CECC CECC 90 103-289 ISSUE A-85. BS CECC 90 103-289; 8-Bit Addressable Latches (En). 17 pp.
CECC CECC 90 109-684 ISSUE 1-87. Digital Integrated Circuits in Accordance with FS 90 109; 54/74 HC 259; 8-Bit Addressable Latch and 3 to 8 Line Decoder (En, Fr). 7 pp.

—8-Bit—Addressable—Preferred Products List
CECC CECC MUAHAG Vol 7 IS 8-92. Preferred Products List; Active Microcircuits (En, Fe, Ge). 89 pp.

Latches (Fasteners)
See Also: Door Bolts; Door Latches; Fasteners; Holders; Locks (Hardware); Pins; Quick Release Fasteners; Slide Latches; Spring Catches
BSI BS 3827: Part 3-67. (WITHDRAWN) 1967 Glossary of Terms Relating to Builders' Hardware Part 3: Catches (Superseded by BS 6100: Subsection 1.3.6:1991). 12 pp.
CGSB CAN/CGSB-69.21-M90. Auxiliary Locks and Associated Products (ANSI/BHMA A156.5-1984). 61 pp.
CGSB CAN/CGSB-69.28-M90. Interconnected Locks and Latches (ANSI/BHMA A156.12-1986). 23 pp.

—Aircraft—Harness Release Gears
SBAC AS 2829 ISSUE 4. Catch for Harness Release Gear. 1 p.

—Automotive
CNS Z7104-81. Wooden Catch for Automobile Vehicle Transportation (Nov)(8181).

—Automotive—Hoods
CNS D3161-83. Method of Test for Hood Latch System of Automobiles (Dec)(10707).

—Comprehensive Testing
CGSB CAN/CGSB-69.17-M86. Bored and Preassembled Locks and Latches. 65 pp.

—Doors
BSI BS 5872-80. 1980 Locks and Latches for Doors in Buildings. 22 pp.
CGSB CAN/CGSB-69.17-M86. Bored and Preassembled Locks and Latches. 65 pp.
CGSB CAN/CGSB-69.21-M90. Auxiliary Locks and Associated Products (ANSI/BHMA A156.5-1984). 61 pp.

Latches (Fasteners) (Cont.)
—Doors (Cont.)
CNS A2060-91. Locks and Latches for Doors in Buildings (Feb)(4349). 20 pp.
SNZ NZS/BS 5872-80. Specification for Locks and Latches for Doors in Buildings. 24 pp.

—Furniture
BSI BS 4951-73. 1973 Builders' Hardware: Lock and Latch Furniture (Doors). 17 pp.

—Glossaries
BSI BS 3827: Part 1-64. (WITHDRAWN) 1964 Glossary of Terms Relating to Builders' Hardware Part 1: Locks (Including Locks and Latches in One Case) (Superseded by BS 6100: Subsection 1.3.6: 1991). 34 pp.
BSI BS 3827: Part 2-67. (WITHDRAWN) 1967 Glossary of Terms Relating to Builders' Hardware Part 2: Latches (Superseded by BS 6100: Subsection 1.3.6.: 1991). 20 pp.

—Mortise
CGSB CAN/CGSB-69.29-M90. Mortise Locks and Latches (ANSI/BHMA A156.13-1980). 21 pp.

—Ships—Hatch Covers
CNS F3170-82. Fittings for Ships' Small Size Steel Hatch Covers (Apr)(8684).
JIS F 2322-87. Fittings of Ships' Steel Small Hatch Covers.
JIS F 2322-61. Fittings of Ships' Steel Small Hatch Covers (R 1970). 9 pp.

Latching Relays
See Also: Bistable Relays; Electric Switches; Relays

—Packaging
MOD UK DSTAN 81-33-01. Packaging of Relays, Electrical Issue 2; Amendment 1. 16 pp.

Latecoere 631 Aircraft
See Also: Aircraft

—Accidents
CAA. Latecoere 631 (World Airline Accident Summary). 1 p.

Lateral Protection (Motor Vehicles)
Use: Sideguards

Latex
Scope Note: For products made from latex, see the specific product *Use For:* Latices; Natural Rubber Latex; Rubber Latices *See Also:* Butyl Rubber; Cements; Chloroprene Rubber Latex; Elastomers; Nitrile Rubber Latex; Paints; Rubber
BSI BS 1672-03. 1972 Amd 3 Methods of Testing Natural Rubber Latices (AMD 7326) December 15, 1992. 26 pp.
BSI BS 1672-72. 1972 Amd 2 Methods of Testing Natural Rubber Latices (AMD 6590) September 28, 1990. 26 pp.
BSI BS 3397-76. 1976 Amd 4 Methods of Test for Synthetic Rubber Latices (AMD 6589) January 31, 1991. 25 pp.
BSI BS 6057: Part 0-81. 1981 Rubber Latices Part 0: General Introduction. 4 pp.
CNS K4021-74. Natural Rubber Latex (Jun)(2533). 1 p.
CNS K6223-74. Method of Test for Latex (Jun)(2534)(R 1974).
JIS K 6381-82. Natural Rubber Latex. 18 pp.

—Alkalinity
BSI BS 6057: Sec 3.3-90. 1990 Rubber Latices Part 3: Methods of Test Section 3.3: Determination of Alkalinity of Natural Rubber Latex Concentrate. 6 pp.
ISO 125-90. Natural Rubber Latex Concentrate—Determination of Alkalinity Fourth Edition. 4 pp.

—Binders
MOD UK TS 10255. Acrylonitrile/Butadiene/ Styrene Latex, Type QX.

—Boric Acid Content
BSI BS 6057: Sec 3.12-92. 1992 Rubber Latices Part 3: Methods of Test Section 3.12: Determination of Boric Acid Content of Natural Rubber Latex Concentrate (ISO 1802: 1992). 6 pp.
BSI BS 6057: Sec 3.12-88. 1988 Rubber Latices Part 3: Methods of Test Section 3.12: Determination of Boric Acid Content of Natural Rubber Latex Concentrate. 4 pp.
DIN ENGL 53605-77. Testing of Latex; Determination of Boric Acid Content in Natural Rubber Latex (Feb). 3 pp.
ISO 1802-92. Natural Rubber Latex Concentrate—Determination of Boric Acid Content Third Edition. 5 pp.

Latex

Latex (Cont.)

—Centrifuged
BSI BS 6057: Sec 1.1-89. 1989 Rubber Latices Part 1: Specifications and Coding Section 1.1: Specification for Centrifuged or Creamed, Ammonia-Preserved Natural Rubber Latices (ISO 2004: 1988). 6 pp.

ISO 2004-88. Natural Rubber Latex Concentrate—Centrifuged or Creamed, Ammonia-Preserved Types—Specification Third Edition. 4 pp.

—Coagulant Content
SAA AS 1683.5-74. Methods of Test for Elastomers—Part 5: Coagulum Content of Rubber Latices Reconfirmed 1989.

—Coagulant Content—Sieve Analysis
BSI BS 6057: Sec 3.8-87. 1987 Rubber Latices Part 3: Methods of Test Section 3.8: Determination of Coagulum Content (Sieve Residue). 4 pp.

ISO 706-85. Rubber Latex—Determination of Coagulum Content (Sieve Residue) Third Edition. 4 pp.

—Coding
BSI BS 6057: Sec 1.3-82. 1982 Rubber Latices Part 1: Specifications and Coding Section 1.3: Coding for Synthetic Rubber Latices. 4 pp.

—Copper Content—Photometry
BSI BS 7164: Sec 28.2-90. 1990 Chemical Tests for Raw and Vulcanized Rubber Part 28: Methods for Determination of Copper Content Section 28.2: Photometric Method (ISO 8053: 1986). 10 pp.

ISO 8053-86. Rubber and Latex—Determination of Copper Content—Photometric Method First Edition. 6 pp.

—Creamed
BSI BS 6057: Sec 1.1-89. 1989 Rubber Latices Part 1: Specifications and Coding Section 1.1: Specification for Centrifuged or Creamed, Ammonia-Preserved Natural Rubber Latices (ISO 2004: 1988). 6 pp.

ISO 2004-88. Natural Rubber Latex Concentrate—Centrifuged or Creamed, Ammonia-Preserved Types—Specification Third Edition. 4 pp.

—Density
BSI BS 6057: Sec 3.7-82. 1982 Rubber Latices Part 3: Methods of Test Section 3.7: Determination of Density of Natural Rubber Latices. 4 pp.

DIN ENGL 53597-72. Latex; Determination of Density (Sept). 2 pp.

ISO 705-74. Natural Rubber Latex—Determination of Density First Edition. 5 pp.

—Dry Films
BSI BS 6057: Sec 3.24-92. 1992 Rubber Latices Part 3: Methods of Test Section 3.24: Preparation of Dry Films from Natural Rubber Latex Concentrate (ISO 498: 1992). 6 pp.

ISO 498-92. Natural Rubber Latex Concentrate—Preparation of Dry Films Second Edition. 5 pp.

—Dry Polymers
BSI BS 6057: Sec 3.16-90. 1990 Rubber Latices Part 3: Methods of Test Section 3.16: Preparation of Dry Polymer from Synthetic Latices. 6 pp.

ISO 2028-89. Rubber—Synthetic Latices—Preparation of Dry Polymer Third Edition. 4 pp.

—Evaporated
BSI BS 6057: Sec 1.2-90. 1990 Rubber Latices Part 1: Specifications and Coding Section 1.2: Specification for Evaporated, Preserved Natural Rubber Latices. 6 pp.

ISO 2027-90. Natural Rubber Latex Concentrate, Evaporated, Preserved—Specification Third Edition. 4 pp.

—Fatty Acid Content
BSI BS 6057: Sec 3.6-93. 1993 Rubber Latices Part 3: Methods of Test Section 3.6: Determination of Volatile Fatty Acid Number of Natural Rubber Latices (ISO 506: 1992). 8 pp.

BSI BS 6057: Sec 3.6-87. 1987 Rubber Latices Part 3: Methods of Test Section 3.6: Determination of Volatile Fatty Acid Number. 6 pp.

ISO 506-92. Rubber Latex, Natural, Concentrate—Determination of Volatile Fatty Acid Number Third Edition. 7 pp.

SAA AS 1683.6-74. Methods of Test for Elastomers—Part 6: Volatile Fatty Acid Number of Natural Rubber Latex Reconfirmed 1989.

—Glossaries
BSI BS 3502: Part 3-78. (WITHDRAWN) 1978 Schedule of Common Names and Abbreviations for Plastics and Rubbers Part 3: Rubber and Latices (Supersedes by BS 3502: Part 2: 1991). 4 pp.

ISO 1629-87. Rubber and Latices—Nomenclature Second Edition. 5 pp.

Latex (Cont.)

—Iron Content—Photometry
BSI BS 7164: Sec 27.2-90. 1990 Chemical Tests for Raw and Vulcanized Rubber Part 27: Methods for Determination of Iron Content Section 27.2: Photometric Method (ISO 1657: 1986). 8 pp.

ISO 1657-86. Rubber, Raw and Rubber Latex—Determination of Iron Content—1,10-Phenanthroline Photometric Method Second Edition. 5 pp.

—Manganese Content—Photometry
BSI BS 7164: Sec 26.2-90. 1990 Chemical Tests for Raw and Vulcanized Rubber Part 26: Methods for Determination of Manganese Content Section 26.2: Sodium Periodate Photometric Methods (ISO 7780: 1987). 12 pp.

ISO 7780-87. Rubbers and Rubber Latices—Determination of Manganese Content—Sodium Periodate Photometric Methods First Edition; (Supersedes 1397 and 1655). 8 pp.

—Microbiological Analysis
BSI BS 6057: Sec 3.23-90. 1990 Rubber Latices Part 3: Methods of Test Section 3.23: Microbiological Examination of Synthetic Rubber Latex. 8 pp.

ISO 9252-89. Synthetic Rubber Latex—Microbiological Examination First Edition. 9 pp.

—Nitrogen Content
BSI BS 7164: Part 21-90. 1990 Chemical Tests for Raw and Vulcanized Rubber Part 21: Determination of Nitrogen Content. 16 pp.

ISO 1656-88. Rubber, Raw Natural, and Rubber Latex, Natural—Determination of Nitrogen Content Second Edition. 13 pp.

—PH
BSI BS 6057: Sec 3.9-90. 1990 Rubber Latices Part 3: Methods of Test Section 3.9: Determination of pH. 6 pp.

ISO 976-86. Rubber Latices—Determination of pH Second Edition. 4 pp.

—Residue Content—Dehydration
DIN ENGL 53563-79. Testing of Latex; Determination of Total Solids Content (May). 2 pp.

—Rubber Content
BSI BS 6057: Sec 3.4-90. 1990 Rubber Latices Part 3: Methods of Test Section 3.4: Determination of Dry Rubber Content of Natural Rubber Latices. 6 pp.

BSI BS 6057: Sec 3.4-01. 1990 Amd 1 Rubber Latices Part 3: Methods of Test Section 3.4: Determination of Dry Rubber Content of Natural Rubber Latices (ISO 126: 1989) (AMD 7321) October 15, 1992. 7 pp.

ISO 126-89. Natural Rubber Latex Concentrate—Determination of Dry Rubber Content Third Edition; (Corrigendum 1-1992). 6 pp.

SAA AS 1683.3-74. Methods of Test for Elastomers—Part 3: Dry Rubber Content of Natural Rubber Latex Reconfirmed 1989.

—Sampling
BSI BS 6057: Part 2-87. 1987 Rubber Latices Part 2: Sampling. 6 pp.

ISO 123-85. Rubber Latex—Sampling Second Edition. 6 pp.

SAA AS 1683.1-74. Methods of Test for Elastomers—Part 1: Sampling of Latex Bound Together with AS 1683.2-AS 1683.9 Reconfirmed 1989.

—Saponification Number
BSI BS 6057: Sec 3.5-85. 1985 Rubber Latices Part 3: Methods of Test Section 3.5: Determination of KOH Number of Natural Rubber Latices. 4 pp.

ISO 127-84. Rubber Latex, Natural—Determination of KOH Number Second Edition. 4 pp.

—Sludge Content
BSI BS 6057: Sec 3.13-93. 1993 Rubber Latices Part 3: Methods of Test Section 3.13: Determination of Sludge Content (ISO 2005: 1992). 6 pp.

BSI BS 6057: Sec 3.13-87. 1987 Rubber Latices Part 3: Methods of Test Section 3.13: Determination of Sludge Content. 4 pp.

ISO 2005-92. Rubber Latex, Natural, Concentrate—Determination of Sludge Content Third Edition. 5 pp.

—Solids Content
BSI BS 6057: Sec 3.2-92. 1992 Rubber Latices Part 3: Methods of Test Section 3.2: Determination of Total Solids Content of Rubber Lactices (ISO 124: 1992). 6 pp.

BSI BS 6057: Sec 3.2-87. 1987 Rubber Latices Part 3: Methods of Test Section 3.2: Determination of Total Solids Content. 4 pp.

ISO 124-92. Rubber Latices—Determination of Total Solids Content Third Edition. 5 pp.

SAA AS 1683.2-74. Methods of Test for Elastomer—Part 2: Solids Content of Latex Reconfirmed 1989.

Latex (Cont.)

—Stability Testing
BSI BS 6057: Sec 3.1-90. 1990 Rubber Latices Part 3: Methods of Test Section 3.1: Determination of Mechanical Stability of Natural Rubber Latex Concentrate (ISO 35: 1989). 6 pp.

BSI BS 6057: Sec 3.14-87. 1987 Rubber Latices Part 3: Methods of Test Section 3.14: Determination of High-Speed Mechanical Stability of Synthetic Rubber Latices. 4 pp.

ISO 35-89. Natural Rubber Latex Concentrate—Determination of Mechanical Stability Third Edition. 6 pp.

ISO 2006-85. Rubber Latex, Synthetic—Determination of High-Speed Mechanical Stability Second Edition. 4 pp.

—Styrene Content
BSI BS 6057: Sec 3.17-90. 1990 Rubber Latices Part 3: Methods of Test Section 3.17: Determination of Bound Styrene Content of Non-Reinforced Styrene Content of Non-Reinforced Styrene Butadiene Rubber Latices. 4 pp.

BSI BS 6057: Sec 3.19-86. 1986 Rubber Latices Part 3: Methods of Test Section 3.19: Determination of Total Bound Styrene Content of Reinforced Styrene-Butadiene Rubber Latices. 12 pp.

ISO 3136-83. Rubber Latex—Styrene-Butadiene—Determination of Bound Styrene Content Second Edition. 3 pp.

ISO 4655-85. Rubber—Reinforced Styrene-Butadiene Latex—Determination of Total Bound Styrene Content Second Edition. 12 pp.

—Surface Tension
BSI BS 6057: Sec 3.10-87. 1987 Rubber Latices Part 3: Methods of Test Section 3.10: Determination of Surface Tension. 4 pp.

ISO 1409-83. Rubber Latex—Determination of Surface Tension Third Edition. 4 pp.

—Synthetic Rubber—Codes
DIN ENGL 53549-77. Testing of Latex; Synthetic Rubber Latices; Codification (Feb). 2 pp.

—Viscosity
BSI BS 6057: Sec 3.11-87. 1987 Rubber Latices Part 3: Methods of Test Section 3.11: Determination of Viscosity of Rubber Latices. 6 pp.

ISO 1652-85. Rubber Latex—Determination of Viscosity Second Edition. 5 pp.

—Zinc Content—Complexometric Titrations
DIN ENGL 53581-74. Testing of Rubber and Elastomers; Determination of the Zinc Content (Mar). 4 pp.

—Zinc Content—Polarographic Analysis
DIN ENGL 53581-74. Testing of Rubber and Elastomers; Determination of the Zinc Content (Mar). 4 pp.

Latex Foams
See Also: Cellular Materials; Cellular Plastics

—Aging Testing
DIN ENGL 53578-88. Testing the Ageing Behaviour of Latex and Polyurethane Foams (Dec). 3 pp.

Latex Paints
See Also: Latex Stains; Paints

—Blocking Resistance Testing
SAA AS 1580.409. 3-92. Paints and Related Materials—Methods of Test—Part 409.3: Blocking Resistance of Latex Paint Finishes (in Professional Packages 30, 39). 2 pp.

SNZ NZS/AS 1580. 409.3-92. Methods of Test for Paints and Related Materials Part 409.3: Blocking Resistance to Latex Paint Finished (This is a Joint Standard with SAA AS 1580.409.3). 2 pp.

—Exterior—Flat
SAA AS 3730.7-92. Guide to the Properties of Paints for Buildings—Part 7: Latex—Exterior—Flat (in Professional Package 30) (Supersedes SAA TR1.7—1982 (AS 2313—1982)). 5 pp.

—Exterior—Gloss
SAA AS 3730.10-92. Guide to the Properties of Paints for Buildings—Part 10: Latex—Exterior—Gloss (in Professional Package 30) (Supersedes SAA TR1.10—1982 (AS 2316—1982)). 6 pp.

—Exterior—Low Gloss
SAA AS 3730.8-92. Guide to the Properties of Paints for Buildings—Part 8: Latex—Exterior—Low-Gloss (in Professional Package 30) (Supersedes SAA TR1.8—1982 (AS 2314—1982)). 5 pp.

—Exterior—Semigloss
SAA AS 3730.9-92. Guide to the Properties of Paints for Buildings—Part 9: Latex—Exterior—Semi-Gloss (in Professional Package 30) (Supersedes SAA TR1.9—1982 (AS 2315—1982)). 5 pp.

INTERNATIONAL AND NON-U.S. NATIONAL STANDARDS
SUBJECT INDEX

Latex Paints (Cont.)

—Floors—Concrete
CGSB CAN/CGSB-1.154-M89. Latex Type Paint for Concrete Floors. 10 pp.

—Interior
CGSB CAN/CGSB-1.100-M89. Interior Latex Type, Flat Paint. 10 pp.

CGSB CAN/CGSB-1.195-M90. Interior Semigloss Latex Paint. 11 pp.

—Interior—Flat
SAA AS 3730.1-92. Guide to the Properties of Paints for Buildings—Part 1: Latex—Interior—Flat (in Professional Package 30) (Supersedes SAA TR1.1—1982 (AS 2304—1982)). 4 pp.

—Interior—Gloss
SAA AS 3730.12-91. Guide to the Properties of Paints for Buildings—Part 12: Latex—Interior—Gloss. 6 pp.

—Interior—Low Gloss
SAA AS 3730.3-92. Guide to the Properties of Paints for Buildings—Part 3: Latex—Interior—Low-Gloss (in Professional Package 30) (Supersedes SAA TR1.3—1982 (AS 2306—1982)). 4 pp.

—Interior—Scrub Resistance
CNS K6624-89. Method of Test for Scrub Resistance of Interior Latex Flat Wall Paints (Jul)(6928).

—Interior—Semigloss
SAA AS 3730.2-92. Guide to the Properties of Paints for Buildings—Part 2: Latex—Interior—Semi-Gloss (in Professional Package 30) (Supersedes SAA TR1.2—1982 (AS 2305—1982)). 4 pp.

—Masonry—Exterior
CGSB CAN/CGSB-1.138-93. Exterior Latex Type, Flat Paint. 9 pp.

—Masonry—Flat
CGSB 1-GP-71 METH 134.18-78. Methods of Testing Paints and Pigments Applicability and Appearance One Coat over Simulated Masonry Surface. 1 p.

—Showers—Water Resistance Testing
CGSB 1-GP-71 METH 110.11-79. Methods of Testing Paints and Pigments Water Resistance Shower Resistance. 1 p.

—Steel Structures
SAA AS 3885-91. Paints for Steel Structures—Galvanized and Zinc Primed—Latex. 12 pp.

—Wallboard—Brushed
CGSB 1-GP-71 METH 134.7-78. Methods of Testing Paints and Pigments Applicability and Appearance Paints and Enamels on Sealed Boards, Brush and Roller Application. 1 p.

—Wallboard—Brushed—Hiding Power
CGSB 1-GP-71 METH 134.7-78. Methods of Testing Paints and Pigments Applicability and Appearance Paints and Enamels on Sealed Boards, Brush and Roller Application. 1 p.

—Wallboard—Roller Applied
CGSB 1-GP-71 METH 134.7-78. Methods of Testing Paints and Pigments Applicability and Appearance Paints and Enamels on Sealed Boards, Brush and Roller Application. 1 p.

—Wallboard—Roller Applied—Hiding Power
CGSB 1-GP-71 METH 134.7-78. Methods of Testing Paints and Pigments Applicability and Appearance Paints and Enamels on Sealed Boards, Brush and Roller Application. 1 p.

—Wood—Exterior
CGSB CAN/CGSB-1.138-93. Exterior Latex Type, Flat Paint. 9 pp.

—Wood—Primers—Exterior
CGSB 1-GP-71 METH 134.11-78. Methods of Testing Paints and Pigments Applicability and Appearance Solvent Based Primers. 1 p.

Latex Primers
See Also: Primers (Coatings)

—Interior
CGSB CAN/CGSB-1.119-M89. Primer-Sealer, Wall, Interior Latex Type; (Corrigendum—Oct 1989). 11 pp.

—Wood—Exterior
CGSB CAN/CGSB-1.203-M91. Exterior Latex Wood Primer. 8 pp.

Latex Primers (Cont.)

—Zinc
SAA AS 3730.15-91. Guide to the Properties of Paints for Buildings—Part 15: Primer—Latex—for Metallic Zinc Surfaces Amdt 1 January/February 1993. 6 pp.

Latex Stains
See Also: Latex Paints

—Wood—Exterior
CGSB 1-GP-204M-78. Stain, Pigmented, Exterior, Latex Type, Standard for. 9 pp.

Lathe Centers
Use: Lathes—Centers

Lathe Chucks
See Also: Scroll Chucks

BSI BS 1983: Part 5-89. 1989 Amd 1 Chucks for Machine Tools and Portable Power Tools Part 5: Code of Practice for the Safe Operation of Workholding Chucks Used on Lathes (Amd 6664) November 30, 1990. 57 pp.

—Backplates
DIN ENGL 6352 Pt 1-68. Chuck Backplates with 1: 4 Locating Taper; Basic Flanges (Aug). 3 pp.

—Hand Operated
DIN ENGL 6350 Pt 1-68. Lathe Chucks, Hand-Operated; Jaws Not Individually Adjustable with Cylindrical Mounting Recess (Aug). 2 pp.

DIN ENGL 6350 Pt 2-69. Lathe Chucks, Hand-Operated; Jaws Not Individually Adjustable; Mounting Recess with 1: 4 Taper (Dec). 4 pp.

DIN ENGL 6351 Pt 1-68. Lathe Chucks, Hand-Operated; Jaws Also Individually Adjustable with Cylindrical Mounting Recess (Aug). 2 pp.

DIN ENGL 6351 Pt 2-69. Lathe Chucks, Hand-Operated; Jaws Also Individually Adjustable Mounting Recess with 1: 4 Taper (Dec). 4 pp.

—Power Operated
DIN ENGL 6353-77. Lathe Chucks Power-Operated Without Through-Hole (Apr). 4 pp.

Lathes
See Also: Bench Lathes; Cutting Tools; Engine Lathes; Grinders; Horizontal Lathes; Machine Tools; Planishing Lathes; Tool Posts; Tools; Turning Tools; Turret Lathes; Wood Lathes; Woodworking Equipment

BSI BS 4656: Part 1-81. 1981 Accuracy of Machine Tools and Methods of Test Part 1: Lathes, General Purpose Type. 15 pp.

BSI BS 4656: Part 1-01. 1981 Amd 1 Accuracy of Machine Tools and Methods of Test Part 1: Specification for Lathes, General Purpose Type (AMD 7743) August 15, 1993 (F). 16 pp.

BSI BS 4656: Part 29-81. 1981 Accuracy of Machine Tools and Methods of Test Part 29: Automatic Lathes Multi-Spindle, (Indexing Drum) Type. 18 pp.

CNS B1101-79. Specified Items of Machine Tools Single Spindle Automatic Lathe (Stationary Head Type) (Jan)(4670-4). 1 p.

CNS B1102-79. Specified Items for Machine Tools (Single Spindle Automatic Lathe) (Sliding Head Type) (Jan)(4670-5).

CNS B1103-79. Specified Items for Machine Tools (Multiple Spindle Automatic Lathe) (Jan)(4670-6).

CNS B3458-87. General System of the Automatic Compensation Tools for CNC Lathe (Aug)(12040).

CNS B7001-82. Accuracy Inspection of Lathes (Dec)(94).

CNS B7151-80. Test Code for Performance of Lathes (Aug)(5975).

CNS B7161-81. Test Code for Performance of Numerically Controlled Lathes (Mar)(6999).

CNS B7162-86. Test Code for Accuracy of Numerically Controlled Lathes (Feb)(7000).

CNS B7217-82. Test Code for Performance of Single Spindle Automatic Lathes (Fixed Type) (Nov)(9563).

CNS B7218-82. Test Code for Accuracy of Single Spindle Automatic Lathes (Fixed Type) (Nov)(9564).

CNS B7219-86. Test Code for Accuracy of Single Spindle Automatic Lathes (Sliding Type) (Aug)(9565).

CNS B7220-82. Test Code for Accuracy of Multi Spindle Automatic Lathes (Nov)(9566).

CNS B7221-82. Test Code for Performance of Multi Spindle Automatic Lathes (Dec)(9663).

CNS B7222-86. Test Code for Performance of Single Spindle Automatic Lathes (Aug)(9664).

DIN ENGL 8605-76. Machine Tools; Lathes of High Accuracy; Swing up to 500 mm, Turning Length up to 1500 mm; Acceptance Conditions (June). 7 pp.

DIN ENGL 8606-76. Machine Tools; Lathes of Normal Accuracy; Swing up to 800 mm; Acceptance Conditions (June). 8 pp.

DIN ENGL 8607-76. Machine Tools; Lathes of Normal Accuracy; Swing Above 800 up to 1600 mm; Acceptance Conditions (June). 8 pp.

Lathes (Cont.)

ISO 1708-89. Acceptance Conditions for General Purpose Parallel Lathes—Testing of the Accuracy Fourth Edition. 16 pp.

JIS B 6202-86. Test Code for Performance and Accuracy of Lathes. 27 pp.

JIS B 6217-90. Test Code for Performance and Accuracy of Single Spindle Automatic Lathes (Fixed Headstock Type). 26 pp.

JIS B 6218-90. Test Code for Performance and Accuracy of Single Spindle Automatic Lathes (Sliding Headstock Type). 17 pp.

JIS B 6219-90. Test Code for Performance and Accuracy of Multi Spindle Automatic Lathes. 17 pp.

JIS B 6331-86. Test Code for Performance and Accuracy of Numerically Controlled Lathes. 34 pp.

—Bars
CNS B7216-82. Test Bar for Automatic Lathes (Nov)(9562).

—Benchtop
CNS B7164-81. Accuracy Inspection of Bench Lathes (Aug)(7777).

—Centers
DIN ENGL 806-71. 60 Degree Lathe Centres Without Extractor Nut (Feb). 2 pp.

DIN ENGL 807-66. Lathe Centres 60 Degree with Extractor Nut (Nov). 2 pp.

DIN ENGL 8012-72. Carbide Inserts for Centre Points (May). 1 p.

ISO 298-73. Machine Tools—Lathe Centres—Sizes for Interchangeability First Edition. 5 pp.

—Centers—Spigots
DIN ENGL 6785-82. Manufacturing Spigots on Parts Turned on Lathes; Indications on Drawings (May). 3 pp.

—Change Gears—Gear Teeth
CNS B2025-84. Number of Teeth of Change Gears for Geared Lathe, Milling Machine and Gear Generating Equipment (Aug)(186).

—Computer Numerical Control—Interface Boards
CNS B3462-87. Interface Board for CNC Lathe (Aug)(12044).

—Copying Attachments
BSI BS 4656: Part 2-88. 1988 Accuracy of Machine Tools and Methods of Test Part 2: Copying Lathes and Copying Attachments. 11 pp.

ISO 8956-86. Acceptance Conditions for Copying Attachments, Integral or Otherwise, for Lathes—Testing of the Accuracy First Edition. 13 pp.

—Faceplates (Machine Tool Components)
BSI BS 4442: Part 1-69. 1969 Lathe Spindle Noses and Faceplates Part 1: Types A and Camlock. 19 pp.

BSI BS 4442: Part 2-69. (WITHDRAWN) 1969 Amd 1 Lathe Spindle Noses and Faceplates Part 2:. 13 pp.

CNS B3398-83. Spindle Noses and Face Plates of Latches (May)(6876).

ISO 702 Pt I-75. Machine Tools—Spindle Noses and Face Plates—Sizes for Interchangeability—Part I: Type A First Edition. 7 pp.

ISO 702 Pt II-75. Machine Tools—Spindle Noses and Face Plates—Sizes for Interchangeability—Part II: Camlock Type First Edition. 16 pp.

ISO 702 Pt III-75. Machine Tools—Spindle Noses and Face Plates—Sizes for Interchangeability—Part III: Bayonet Type First Edition. 10 pp.

JIS B 6109-82. Spindle Noses and Face Plates of Lathes. 19 pp.

—Glossaries
BSI BS 4361: Part 21-89. 1989 Woodworking Machines Part 21: Turning Lathes. 14 pp.

BSI BS 4361: Part 35-90. 1990 Woodworking Machines Part 35: Nomenclature for Machines for Production of Core Stock from Laths. 11 pp.

CNS B1220-80. Glossary of Terms Relating to Parts of Lathes (Dec)(6744).

ISO 7987-85. Woodworking Machines—Turning Lathes—Nomenclature and Acceptance Conditions First Edition. 12 pp.

ISO 9535-89. Woodworking Machines—Machines for Production of Core Stock from Laths—Nomenclature First Edition. 9 pp.

—Heads
CNS B3461-87. Receive Head for CNC Lathe (Aug)(12043).

—High Speed Steel Tools
DIN ENGL 4951-62. Straight Cutting Tools for Lathe Work with High Speed Steel Cutting Edges (Sept). 2 pp.

INDUSTRY STANDARDS

Lathes (Cont.)

—Interface Boards
CNS B3462-87. Interface Board for CNC Lathe (Aug)(12044).

—Noise
DIN ENGL 45635 Pt 1601-78. Measurement of Airborne Noise Emitted by Machines; Enveloping Surface Method; Metal Processing Machine Tools; Special Stipulations for Lathes (July). 2 pp.

—Probes
CNS B3459-87. Touch Probes for Measuring Outside Used in CNC Lathe (Aug)(12041).
CNS B3460-87. Boring Bar Probes Used in CNC Lathe (Aug)(12042).

—Shanks
DIN ENGL 770 Pt 1-62. Shank Sections for Lathe and Planer Tools; Rolled and Forged Shanks (Aug). 2 pp.
DIN ENGL 770 Pt 2-62. Shank Sections for Lathe and Planer Tools; Machined All over; Not Required to Meet Special Standards of Accuracy (Aug). 2 pp.

—Spindle Noses
BSI BS 4442: Part 1-69. 1969 Lathe Spindle Noses and Faceplates Part 1: Types A and Camlock. 19 pp.
BSI BS 4442: Part 2-69. (WITHDRAWN) 1969 Amd 1 Lathe Spindle Noses and Faceplates Part 2:. 13 pp.
CNS B3398-83. Spindle Noses and Face Plates of Latches (May)(6876).
ISO 702 Pt I-75. Machine Tools—Spindle Noses and Face Plates—Sizes for Interchangeability—Part I: Type A First Edition. 7 pp.
ISO 702 Pt II-75. Machine Tools—Spindle Noses and Face Plates—Sizes for Interchangeability—Part II: Camlock Type First Edition. 16 pp.
ISO 702 Pt III-75. Machine Tools—Spindle Noses and Face Plates—Sizes for Interchangeability—Part III: Bayonet Type First Edition. 10 pp.
JIS B 6109-82. Spindle Noses and Face Plates of Lathes. 19 pp.

—Spinning—Acceptance Testing
DIN ENGL 55803-79. Machine Tools; Spinning and Planishing Lathes; Acceptance Conditions (Dec). 6 pp.

—Tips
DIN ENGL 771-62. High Speed Steel Tips for Lathe and Planer Tools (Sept). 2 pp.

—Vertical Turning and Boring
BSI BS 4656: Part 22-88. 1988 Accuracy of Machine Tools and Methods of Test Part 22: Vertical Boring and Turning Lathes, Single and Double Column Types. 20 pp.
ISO 3655-86. Acceptance Conditions for Vertical Turning and Boring Lathes with One or Two Columns and a Single Fixed or Movable Table—General Introduction and Testing of the Accuracy Second Edition. 19 pp.
JIS B 6223-82. Test Code for Performance and Accuracy of Vertical Turning and Boring Lathes. 19 pp.

Lathing

See Also: Construction Materials; Gypsum Laths; Plastering
BSI BS 1369: Part 1-87. 1987 Steel Lathing for Internal Plastering and External Rending Part 1: Expanded Metal and Ribbed Lathing. 7 pp.
BSI BS 1369-47. 1947 Metal lathing (Steel) for Plastering. 8 pp.
CNS A2141-82. Metal Lath (Oct)(9455).
CSA A82.30-M1980. Interior Furring, Lathing, and Gypsum Plastering (R 1992); (Gen Instr 1). 24 pp.
JIS A 5504-75. Wire Laths. 6 pp.
JIS A 5505-78. Metal Lath. 8 pp.
JIS A 5524-77. Lath Sheets.
SAA AS 1292-80. Laths, Folding Rods and Multi-Folding Rods. 8 pp.

Latices
Use: Latex

Latin Characters
Use: Roman Characters

Latitude

See Also: Datum (Geodetic); Grids (Coordinates); Position (Location)

—Geographic Representation
BSI BS 5249: Part 3-83. 1983 Representation of Elements of Data in Interchanges Using Data Processing Systems Part 3: Representation of Latitude, Longitude and Altitude for Geographical Point Locations. 4 pp.

Latitude (Cont.)

—Geographic Representation (Cont.)
ISO 6709-83. Standard Representation of Latitude, Longitude and Altitude for Geographic Point Locations First Edition. 5 pp.
JTC1 6709-83. Standard Representation of Latitude, Longitude and Altitude for Geographic Point Locations First Edition. 5 pp.
OSI ISO 6709-83. Standard Representation of Latitude, Longitude and Altitude for Geographic Part Locations. 5 pp.

Lattice Masts
Use: Lattice Towers

Lattice Parameters
Use For: Lattice Spacing

—Nuclear Graphite
CNS J2028-81. Method for Measurement of Lattice Spacing of Nuclear Graphite (Nov)(8128).

Lattice Spacing
Use: Lattice Parameters

Lattice Towers

—Loads (Forces)
BSI BS 8100: Part 1-86. 1986 Lattice Towers and Masts Part 1: Code of Practice for Loading. 68 pp.
BSI BS 8100: Part 2-86. 1986 Lattice Towers for Masts Part 2: Guide to the Background and Use of Part 1 'Code of Practice for Loading'. 163 pp.

—Precast Concrete—Reinforced
DIN ENGL 4228-89. Precast Concrete Lattice Towers, Masts and Columns (Feb). 10 pp.

—Steel—Design
SAA AS 3995(INT)-91. Design of Steel Lattice Towers and Masts (Expires 16 December 1993). 27 pp.

—Structural Members—Strength
BSI DD 133-86. 1986 Amd 1 Code of Practice for Strength Assessment of Members of Lattice Towers and Masts. 30 pp.

Laundering

Use For: Washing *See Also:* Cleaning; Laundering Testing; Laundry Facilities

—Fabrics—Commercial
CGSB CAN/CGSB-4.2 NO.34-M89. Textile Test Methods Standard Commercial Laundering Procedures. 13 pp.

—Service Contracts
MOD UK DEFCON 112AH-91. Conditions of Contract for Laundry/Dry Cleaning Services 5/91. 4 pp.

Laundering Effects
Use: Laundering Testing

Laundering Testing

Use For: Wash Testing (Laundry)
See Also: Detergent Resistance Testing; Laundering; Soap Resistance Testing

—Creasing—Fabrics
CGSB CAN/CGSB-4.2 NO.59.3-M88. Textile Test Methods Appearance After Repeated Domestic Launderings—Pressed-in Creases. 11 pp.
CNS L3149-82. Method of Test for Crease Evaluation of Woven and Knitted Fabrics After Laundering (Jan)(8313).

—Durable Press Fabrics
CGSB CAN/CGSB-4.2 NO.59.1-M88. Textile Test Methods Appearance After Repeated Domestic Laundering—Smoothness of Fabric. 11 pp.
ISO 7768-85. Textiles—Method for Assessing the Appearance of Durable Press Fabrics After Domestic Washing and Drying First Edition. 6 pp.
ISO 7769-92. Textiles—Method for Assessing the Appearance of Creases in Durable-Press Products After Domestic Washing and Drying Second Edition. 9 pp.
ISO 7770-85. Textiles—Method for Assessing the Appearance of Seams in Durable Press Products After Domestic Washing and Drying First Edition. 6 pp.

—Fabrics
BSI BS 4923-91. 1991 Methods for Individual Domestic Washing and Drying for Use in Textile Testing. 17 pp.
BSI BS 5742-89. 1989 Textile Labels Requiring to Be Washed and/or Dry Cleaned. 8 pp.
BSI BS 6246: Part 1-87. 1987 Industrial Laundry Machinery Part 1: Methods for the Assessment of the Effect of Rotary Washing Machines on Textiles. 26 pp.

Laundering Testing (Cont.)

—Fabrics (Cont.)
BSI BS 6246: Part 2-83. 1983 Industrial Laundry Machinery Part 2: Methods for the Assessment of the Effect of Extracting Machines on Textiles. 20 pp.
BSI BS 6246: Part 3-85. 1985 Industrial Laundry Machinery Part 3: Methods for the Assessment of the Effect of Flatwork Ironing Machines on Textiles. 18 pp.
BSI BS 6246: Part 4-83. 1983 Industrial Laundry Machinery Part 4: Methods for the Assessment of the Effect of Batch Drying Tumblers on Textiles. 11 pp.
BSI BS 6246: Part 5-91. 1991 Industrial Laundry Machinery Part 5: Methods for the Assessment of the Effect of Shaped Garment Finishing Machinery on Textiles. 32 pp.
CGSB CAN/CGSB-4.2 NO.24-M91. Textile Test Methods Colourfastness and Dimensional Change in Commercial Laundering. 7 pp.
CGSB CAN/CGSB-4.2 NO.24.2-M91. Textile Test Methods Dimensional Change in Commercial Type Laundering of Textiles (Washwheel). 13 pp.
CNS L3218-85. Method of Test for Soil Release of Fabric (Jul)(11309).
ISO 6330-84. Textiles—Domestic Washing and Drying Procedures for Textile Testing First Edition. 8 pp.

—Fabrics—Colorfastness
BSI BS EN 20105-C01-93. 1993 Textiles—Tests for Colour Fastness Part C01: Colour Fastness to Washing: Test 1 (ISO 105-C01: 1989) (L). 10 pp.
BSI BS EN 20105-C02-93. 1993 Textiles—Tests for Colour Fastness Part C02: Colour Fastness to Washing: Test 2 (ISO 105-C02: 1989) (L). 10 pp.
BSI BS EN 20105-C03-93. 1993 Textiles—Tests for Colour Fastness Part C03: Colour Fastness to Washing: Test 3 (ISO 105-C03: 1989) (L). 10 pp.
BSI BS EN 20105-C04-93. 1993 Textiles—Tests for Colour Fastness Part C04: Colour Fastness to Washing: Test 4 (ISO 105-C04: 1989) (L). 10 pp.
BSI BS EN 20105-C05-93. 1993 Textiles—Tests for Colour Fastness Part C05: Colour Fastness to Washing: Test 5 (ISO 105-C05: 1989) (L). 10 pp.
CEN EN 20105-C01-92. Textiles—Tests for Colour Fastness Part C01: Colour Fastness to Washing: Test 1 (ISO 105-C01: 1989). 7 pp.
CEN EN 20105-C03-92. Textiles—Tests for Colour Fastness Part C03: Colour Fastness to Washing: Test 3 (ISO 105-C03: 1989). 7 pp.
CEN EN 20105-C04-92. Textiles—Tests for Colour Fastness—Part C04: Colour Fastness to Washing: Test 4 (ISO 105-C04: 1989). 7 pp.
CEN EN 20105-C05-92. Textiles—Tests for Colour Fastness—Part C05: Colour Fastness to Washing: Test 5 (ISO 105-C05: 1989). 5 pp.
CGSB CAN/CGSB-4.2 NO.58-M90. Textile Test Methods Colourfastness and Dimensional Change in Domestic Laundering of Textiles. 18 pp.
CNS L2027-81. Apparatus for Testing of Color Fastness to Washing (Nov)(8149).
CNS L3027-83. Method of Test for Colour Fastness to Washing (Mar)(1494). 6 pp.
DIN ENGL EN 20105 Pt C01-93. Tests for Colour Fastness of Textiles; Colour Fastness to Washing: Test 1 (ISO 105-C01: 1989) (Mar). 6 pp.
DIN ENGL EN 20105 Pt C02-93. Tests for Colour Fastness of Textiles; Colour Fastness to Washing: Test 2 (ISO 105-C02: 1989) (Mar). 6 pp.
DIN ENGL EN 20105 Pt C03-93. Tests for Colour Fastness of Textiles; Colour Fastness to Washing: Test 3 (ISO 105-C03: 1989) (Mar). 6 pp.
DIN ENGL EN 20105 Pt C04-93. Tests for Colour Fastness of Textiles; Colour Fastness to Washing: Test 4 (ISO 105-C04: 1989) (Mar). 6 pp.
DIN ENGL EN 20105 Pt C05-93. Tests for Colour Fastness of Textiles; Colour Fastness to Washing: Test 5 (ISO 105-C05: 1989) (Mar). 6 pp.
ISO 105 Pt C02-89. Textiles—Tests for Colour Fastness—Part C02: Colour Fastness to Washing: Test 2 Fourth Edition. 6 pp.
ISO 105 Pt C03-89. Textiles—Tests for Colour Fastness—Part C03: Colour Fastness to Washing: Test 3 Fourth Edition; (CEN EN 20105-C03:1992). 6 pp.
ISO 105 Pt C04-89. Textiles—Tests for Colour Fastness—Part C04: Colour Fastness to Washing: Test 4 Fourth Edition; (CEN EN 20105-C04:1992). 6 pp.
ISO 105 Pt C05-89. Textiles—Tests for Colour Fastness—Part C05: Colour Fastness to Washing: Test 5 Fourth Edition; (CEN EN 20105-C05:1992). 6 pp.
JIS L 0821-83. Apparatus for Testing of Colour Fastness to Washing. 5 pp.
JIS L 0844-86. Testing Methods for Colour Fastness to Washing and Laundering. 14 pp.

—Fabrics—Dimensional Stability
BSI BS 3424: Part 36-93. 1993 Testing Coated Fabrics Part 36: Method 39. Method for Determination of the Dimensional Stability of Coated Fabrics to Domestic Washing. 7 pp.

INTERNATIONAL AND NON-U.S. NATIONAL STANDARDS
SUBJECT INDEX

Laundering Testing (Cont.)
—Fabrics—Dimensional Stability (Cont.)
BSI BS 5807-87. 1987 Determination of Dimensional Change of Textiles in Domestic Washing and Drying. 4 pp.
CGSB CAN/CGSB-4.2 NO.24.1-M88. Textile Test Methods Dimensional Change in Washing of Woven Fabrics—Accelerated Method. 9 pp.
CGSB CAN/CGSB-4.2 NO.58-M90. Textile Test Methods Colourfastness and Dimensional Change in Domestic Laundering of Textiles. 18 pp.
ISO 675-79. Textiles—Woven Fabrics—Determination of Dimensional Change on Commercial Laundering Near the Boiling Point First Edition; (Corrigendum 1-1990). 7 pp.
ISO 5077-84. Textiles—Determination of Dimensional Change in Washing and Drying First Edition. 5 pp.

—Fabrics—Enamel Finishes
BSI BS 1344: Part 5-84. 1984 Methods of Testing Vitreous Enamel Finishes Part 5: Determination of Resistance to Hot Detergent Solutions Used for Washing Textiles. 7 pp.
BSI BS 1344: Part 19-84. 1984 Methods of Testing Vitreous Enamel Finishes Part 19: Apparatus for Determination of Resistance to Hot Detergent Solutions Used for Washing Textiles. 8 pp.
ISO 4535-83. Vitreous and Porcelain Enamels—Apparatus for Determination of Resistance to Hot Detergent Solutions Used for Washing Textiles First Edition. 7 pp.

—Surfactants—Cotton Fabrics
BSI BS 5377: Part 1-88. 1988 Assessment of Laundering Effects by Means of Cotton Control Cloth Part 1: Preparation and Use of the Cotton Control Cloth. 10 pp.
BSI BS 5377: Part 2-90. 1990 Assessment of Laundering Effects by Means of a Cotton Control Cloth Part 2: Method of Analysis and Test for the Unsoiled Cotton Control Cloth. 36 pp.
ISO 2267-86. Surface Active Agents—Evaluation of Certain Effects of Laundering—Methods of Preparation and Use of Unsoiled Cotton Control Cloth Third Edition. 9 pp.
ISO 4312-89. Surface Active Agents—Evaluation of Certain Effects of Laundering—Methods of Analysis and Test for Unsoiled Cotton Control Cloth Second Edition. 37 pp.

—Toys
CNS Z8016-16-87. Method of Test for Toy Safety (Testing for Toys Which Are Labelled as Being Washable) (Dec)(4798-16). 2 pp.

—Washing Machines—Seams
CGSB CAN/CGSB-4.2 NO.59.2-M88. Textile Test Methods Appearance After Repeated Domestic Launderings—Seams. 11 pp.

—Wool Fabrics—Dimensional Stability
BSI BS 1955-81. 1981 Determination of Dimensional Changes of Wool-Containing Fabrics During Washing. 10 pp.

Laundry Carts
Use For: Linen Trolleys See Also: Carts; Laundry Equipment; Medical Equipment
—Hospitals
BSI BS 2854-79. (WITHDRAWN) 1979 Soiled Linen Trolleys. 8 pp.

Laundry Detergents
See Also: Detergents
BSI BS 5610-78. 1978 Surface Active Agents. Detergents for Washing Fabrics. Guide for Comparative Testing for Performance. 18 pp.
CNS S1031-83. Synthetic Detergents for Laundry (Jun)(2477). 3 pp.
ISO 4319-77. Surface Active Agents—Detergents for Washing Fabrics—Guide for Comparative Testing of Performance First Edition. 16 pp.
JIS K 3371-76. Synthetic Detergents for Home Laundering.
SAA AS 1658-74. Household Synthetic Laundry Detergent Powder (Composition Basis) Amdt 1 October 1978. 6 pp.

—Granular
CGSB CAN/CGSB-2.115-M89. Built Powder Laundry Detergent. 10 pp.

—Optical Brighteners
CGSB 2-GP-11M METH 40-88. Methods of Testing and Analysis of Soaps and Detergents Optical Brighteners Substantivity Test. 1 p.

—Powder
CGSB CAN/CGSB-2.115-M89. Built Powder Laundry Detergent. 10 pp.

Laundry Detergents (Cont.)
—Radiation Decontamination
ISO 9271-92. Decontamination of Radioactively Contaminated Surfaces—Testing of Decontamination Agents for Textiles First Edition. 21 pp.

Laundry Equipment
Scope Note: For additional listings, use a more specific term See Also: Agitators; Clothes Dryers; Laundry Carts; Laundry Facilities; Sinks; Sinks (Plumbing Fixtures); Steam Cleaning Equipment; Towel Rails (Heated); Washing Machines; Washing Tunnels; Water Extractors

—Naval Ships
MOD UK NES 123: Part 2-88. Laundries and Laundry Equipment Part 2: Laundry Machinery, Equipment and Fittings Issue 3 (11.88). 31 pp.
MOD UK NES 123: Part 2-01. Requirements for Laundries and Associated Compartments Part 2: Laundry Machinery, Equipment and Fittings Issue 3 (11.88); Amendment 1. 33 pp.

—Submarines
MOD UK NES 123: Part 2-88. Laundries and Laundry Equipment Part 2: Laundry Machinery, Equipment and Fittings Issue 3 (11.88). 31 pp.
MOD UK NES 123: Part 2-01. Requirements for Laundries and Associated Compartments Part 2: Laundry Machinery, Equipment and Fittings Issue 3 (11.88); Amendment 1. 33 pp.

—Symbols
CNS C1143-11-89. Graphical Symbols for Use on Electrical and Electronic Equipment (Laundry Machine, Dish Washer) (Feb)(12491-11).

—Troughs
SAA AS 1229-91. Laundry Troughs. 15 pp.

Laundry Facilities
See Also: Laundering; Laundry Equipment
—Ships
MOD UK NES 123: Part 1-88. Requirements for Laundries and Associated Compartments Part 1: Laundries and Associated Compartments Issue 2 (01.88). 35 pp.

—Submarines
MOD UK NES 123: Part 1-88. Requirements for Laundries and Associated Compartments Part 1: Laundries and Associated Compartments Issue 2 (01.88). 35 pp.

Laundry Soaps
See Also: Soaps
—Bar
JIS K 3302-85. Laundry Bar Soaps. 5 pp.
SAA AS 1878-76. Laundry Tablet or Bar Soap Amdt 1 July 1978. 16 pp.

—Chip
CGSB CAN/CGSB-2.3-92. Unbuilt Laundry Soap. 6 pp.

—Granular
CGSB CAN/CGSB-2.3-92. Unbuilt Laundry Soap. 6 pp.

—Powder
CGSB CAN/CGSB-2.3-92. Unbuilt Laundry Soap. 6 pp.
CGSB CAN/CGSB-2.33-92. Built Laundry Soap. 6 pp.
JIS K 3303-84. Laundry Powdered Soaps. 13 pp.

—Tablet
SAA AS 1878-76. Laundry Tablet or Bar Soap Amdt 1 July 1978. 16 pp.

Laundry Sour
CGSB CAN/CGSB-2.45-M87. Laundry Sour, Fluoride Type. 10 pp.

Laurel
Use For: Laurel Leaf; Laurel Leaves
See Also: Spices
ISO 6576-84. Laurel (Laurus Nobilis Linnaeus)—Whole and Pounded Leaves—Specification First Edition. 6 pp.

Laurel Leaf
Use: Laurel

Laurel Leaves
Use: Laurel

Lavandin Oil
Use: Lavender Oil

Lavatories
Use: Sinks

Lavender Oil
See Also: Essential Oils
CNS K5087-80. Oil of French Lavender (Sep)(6458).
ISO 3054-87. Oil of Lavandin Abrialis (Lavandula Angustifolia P. Miller X Lavandula Latifolia (Linnaeus F.) Medikus), France Second Edition. 7 pp.
ISO 3515-87. Oil of French Lavender (Lavandula angustifolia P. Miller) Second Edition. 7 pp.
ISO 8902-87. Oil of Lavandin Grosso (Lavandula Angustifolia P. Miller X Lavandula Latifolia (Linnaeus F.) Medikus) First Edition. 7 pp.

Lavender Spike Oil
Use: Spike Oil

Lawn and Garden Equipment
See Also: Landscape Lighting; Lawn and Garden Hoses; Lawn Mowers; Outdoor Furniture; Shredders; Spreaders; Trimmers
DIN VDE 0701 Pt 2-82. Repair, Modification and Testing of Household and Similar Electrical Appliances; Particular Requirements for Lawnmowers and Gardening Appliances (Apr). 7 pp.
DIN VDE 0730 Pt 2 ZP-80. Electric Motor-Operated Appliances for Domestic and Similar Purposes; Particular Requirements for Gardening Appliances (July). 37 pp.
ISO 9191-91. Lawn and Garden Ride-on (Riding) Tractors—Three-Point Hitch First Edition. 8 pp.
ISO 9192-91. Lawn and Garden Ride-on (Riding) Tractors—One-Point Tubular Sleeve Hitch First Edition. 8 pp.

—Electric
CSA CAN/CSA-C22. 2 NO 147-M90. Motor-Operated Gardening Appliances; (Gen Instr 1 Thru 3). 85 pp.

—Electrical Codes
CSA 1376 Bull. Electrical Bulletin 1376 June 30, 1982 to C22.2 NO 147. 1 p.

—Manual Controls
ISO 3789 Pt 1-82. Tractors, Machinery for Agriculture and Forestry, Powered Lawn and Garden Equipment—Location and Method of Operation of Operator Controls—Part 1: Common Controls First Edition. 6 pp.
ISO 3789 Pt 3-89. Tractors, Machinery for Agriculture and Forestry, Powered Lawn and Garden Equipment—Location and Method of Operation of Operator Controls—Part 3: Controls for Powered Lawn and Garden Equipment Second Edition. 8 pp.

—Manual Controls—Symbols
BSI BS 4964: Part 1-93. 1993 Symbols for Control Markings and Displays on Tractors and Machinery for Agricultural and Forestry, and on Powered Lawn and Garden Equipment Part 1: Specification for Common Symbols (ISO 3767-1: 1991) (E). 37 pp.
BSI BS 4964: Part 1-81. 1981 Symbols for Control Markings and Displays on Tractors and Machinery for Agricultural and Forestry, and on Powered Lawn and Garden Equipment Part 1: Common Symbols. 24 pp.
ISO 3767 Pt 1-91. Tractors, Machinery for Agriculture and Forestry, Powered Lawn and Garden Equipment—Symbols for Operator Controls and Other Displays—Part 1: Common Symbols Second Edition; (Incorporating Addendum 1). 33 pp.
ISO 3767 Pt 3-88. Tractors, Machinery for Agriculture and Forestry, Powered Lawn and Garden Equipment—Symbols for Operator Controls and Other Displays—Part 3: Symbols for Powered Lawn and Garden Equipment First Edition. 4 pp.
SNZ NZS 5104: Part 1-86. Tractors, Machinery for Agriculture and Forestry, Powered Lawn and Garden Equipment Part 1: Common Symbols Addendum 1; Amend: A, 1986; 1A, 1986. 5 pp.

—Repair/Maintenance
DIN VDE 0701 Pt 2-82. Repair, Modification and Testing of Household and Similar Electrical Appliances; Particular Requirements for Lawnmowers and Gardening Appliances (Apr). 7 pp.

—Safety
CEN PREN 836-92. Safety Requirements for Agricultural and Forestry Machinery—Powered Lawn Mowers, Lawn Tractors, Lawn and Garden Tractors, Professional Mowers, and Lawn and Garden Tractors with Mowing Attachments. 82 pp.

INTERNATIONAL AND NON-U.S. NATIONAL STANDARDS
SUBJECT INDEX

Lawn

Lawn and Garden Equipment *(Cont.)*
—Safety—Glossaries
ISO 5395-90. Power Lawn-Mowers, Lawn Tractors, Lawn and Garden Tractors, Profes-sional Mowers, and Lawn and Garden Tractors with Mowing Attachments—Definitions, Safety Re-quirements and Test Pro-cedures First Edition; (Cancels and Replaces ISO 5395 Parts 1 Thru 4) (Amendment 1-1992). 55 pp.

Lawn and Garden Hoses
See Also: Hoses; Lawn and Garden Equipment
BSI BS 3746-90. 1990 PVC Garden Hose. 9 pp.
JIS K 6337-77. Garden Hose.

Lawn and Garden Lighting
Use: Landscape Lighting

Lawn Furniture
Use: Outdoor Furniture

Lawn Mowers
See Also: Lawn and Garden Equipment; Mowers
DIN VDE 0701 Pt 2-82. Repair, Modification and Testing of Household and Similar Electrical Appliances; Particular Requirements for Lawnmowers and Gardening Appliances (Apr). 7 pp.
DIN VDE 0730 Pt 2 U-78. Special Regulations for Lawn Mowers (July). 14 pp.
DIN VDE 0730 Pt 2 ZJ-78. Regulations Concerning Appliances Driven by Electrical Motors for Domestic and Similar Purposes; Part 2 ZJ: Special Regulations for Manually Guided Battery-Operated Lawn Mowers (June). 8 pp.
SNZ NZS 5234-90. Powered Rotary Lawnmowers Amend: A, 1990. 20 pp.

—Electric
CSA CAN/CSA-C22. 2 NO 147-M90. Motor-Operated Gardening Appliances; (Gen Instr 1 Thru 3). 85 pp.
SAA AS 3156-92. Approval and Test Specification—Electric Lawnmowers Amdt 1 June 1992 Amdt 2 June 1993 (in Professional Package 28). 7 pp.
SNZ NZS/AS 3156-92. Approval and Test Specification—Electric Lawnmowers Amend: 1, 1992 (This is a Joint Standard with SAA AS 3156). 6 pp.

—Electric—Walk Behind—Safety
DIN VDE 0700 Pt 206 (D)-88. Safety of Household and Similar Electrical Appliances Particular Requirements for Pedestrian Controlled Mains Operated Electrical Lawn-Mowers and Similar Gardening Appliances; Identical to IEC 61F(Central Office) 58 (Jan). 37 pp.

—Repair/Maintenance
DIN VDE 0701 Pt 2-82. Repair, Modification and Testing of Household and Similar Electrical Appliances; Particular Requirements for Lawnmowers and Gardening Appliances (Apr). 7 pp.

—Safety
BSI BS 3456: Sec 2.32-74. 1974 Amd 3 1972-1981 Edition. Specification for Safety of Household Electrical Appliances Part 2: Section 2.32: Mains-Operated Electric Lawnmowers. 23 pp.
BSI BS 3456: Sec 2.42-77. 1977 1972-1981 Edition. Specification for Safety of Household Electrical Appliances Part 2: Section 2.42: Battery-Operated Lawnmowers. 10 pp.
BSI BS 5107-74. 1974 Amd 5 Powered Lawnmowers. 21 pp.
CEN PREN 836-92. Safety Requirements for Agricultural and Forestry Machinery—Powered Lawn Mowers, Lawn Tractors, Lawn and Garden Tractors, Professional Mowers, and Lawn and Garden Tractors with Mowing Attachments. 82 pp.

—Safety—Glossaries
ISO 5395-90. Power Lawn-Mowers, Lawn Tractors, Lawn and Garden Tractors, Profes-sional Mowers, and Lawn and Garden Tractors with Mowing Attachments—Definitions, Safety Re-quirements and Test Pro-cedures First Edition; (Cancels and Replaces ISO 5395 Parts 1 Thru 4) (Amendment 1-1992). 55 pp.

—Sound Power
EC 84/538/EEC-84. Council Directive on the Approximation of the Laws of the Member States Relating to the Permissible Sound Power Level of Lawnmowers. 8 pp.
EC 88/180/EEC-88. Council Directive Amending Directive 84/538/EEC on the Approximation of the Laws of the Member States Relating to the Permissible Sound Power Level of Lawnmowers. 2 pp.

Lawn Mowers *(Cont.)*
—Sound Power *(Cont.)*
EC 88/181/EEC-88. Council Directive Amending Directive 84/538/EEC on the Approximation of the Laws of the Member States Relating to the Permissible Sound Power Level of Lawnmowers. 4 pp.

—Sound Pressure
EC 88/180/EEC-88. Council Directive Amending Directive 84/538/EEC on the Approximation of the Laws of the Member States Relating to the Permissible Sound Power Level of Lawnmowers. 2 pp.
EC 88/181/EEC-88. Council Directive Amending Directive 84/538/EEC on the Approximation of the Laws of the Member States Relating to the Permissible Sound Power Level of Lawnmowers. 4 pp.
ISO 11094-91. Acoustics—Test Code for the Measurement of Airborne Noise Emitted by Power Lawn Mowers, Lawn Tractors, Lawn and Garden Tractors, Professional Mowers, and Lawn and Garden Tractors with Mowing Attachments First Edition. 15 pp.

Lawn Trimmers
Use: Trimmers—Grass

Laws
Use: Legislation

Layouts
Scope Note: For tabular layouts of specific products, see also the subheading Tabular Layouts under the product *Use For:* Lofting (Layouts)

—Aircraft
SBAC RS 193 ISSUE 1. Standard Layout Table.
SBAC RS 432 (V). Lofting Procedure.

—Aircraft—Abbreviations
SBAC RS 196 (V). Full Scale, Layout Reproduction, Recommended Terms and Abbreviations.

—Aircraft—Bend Allowances
SBAC RS 208 ISSUE 1. Bend Allowances for Full Scale, Layout.
SBAC RS 209 ISSUE 1. Bend Allowances for Full Scale Layout.
SBAC RS 210 ISSUE 1. Bend Allowances for Full Scale Layout.

—Aircraft—Blanks
SBAC RS 194 ISSUE 1. Standard Layout Blank.
SBAC RS 195 ISSUE 1. Preparation of Surface of, Layout Blanks for Photographic Reproduction.
SBAC RS 202 ISSUE 1. Preparation of Surface of Layout Blank for Lithographic Reproduction.

—Aircraft—Fuselage Lines—Identification Systems
SBAC RS 198 (V). Full Scale Layout, Reproduction Fuselage Lines, Identification and Layout.

—Aircraft—Glossaries
SBAC RS 196 (V). Full Scale, Layout Reproduction, Recommended Terms and Abbreviations.

—Aircraft—Negative Sizes
SBAC RS 197 ISSUE 1. Full Scale, Layout Reproduction, Negative Sizes.

—Aircraft—Symbols
SBAC RS 409 (V). Full Scale Layout Reproduction Recommended Standard Symbols.

—Tabular
DIN ENGL 4000 Pt 1 Suppl. 1-75. Tabular Lay-Outs of Article Characteristics; Principles; Application in Article Characteristics Lists (Feb). 3 pp.

—Tabular—Glossaries
DIN ENGL 4000 Pt 1-81. Tabular Layouts of Article Characteristics; Definitions and Principles (Apr). 7 pp.

LCD
Use: Liquid Crystal Displays

LCS
Use: Call Systems (Medical)

Leach Testing
See Also: Leachate Extraction

—Metals—Drinking Water
BSI BS 6920: Sec 2.6-90. 1990 Suitability of Non-Metallic Products for Use in Contact with Water Intended for Human Consumption with Regard to Their Effect on the Quality of the Water Part 2: Methods of Test Section 2.6: The Extraction of Metals. 8 pp.

Leach Testing *(Cont.)*
—Metals—Drinking Water *(Cont.)*
BSI BS 6920: Part 3-90. 1990 Suitability of Non-Metallic Products for Use in Contact with Water Intended for Human Consumption with Regard to Their Effect on the Quality of the Water Part 3: High Temperature Tests. 6 pp.
SNZ NZS/BS 6920: Pt 2:Sec 2.6-90. Suitability of Non-Metallic Products for Use in Contact with Water Intended for Human Consumption with Regard to Their Effect on the Quality of Water Part 2: Methods of Test Section 2: Taste of Water Section 2.6: The Extraction of Metals. 8 pp.
SNZ NZS/BS 6920: Part 3-90. Suitability of Non-Metallic Products for Use in Contact with Water Intended for Human Consumption with Regard to Their Effect on the Quality of Water Part 3: High Temperature Tests. 8 pp.

—Radioactive Wastes
ISO 6961-82. Long-Term Leach Testing of Solidified Radioactive Waste Forms First Edition. 8 pp.

—Rubber
CNS T4017-2-85. Method of Test for Hygienic Safety for Rubber Goods: Leaching Test (Aug)(11350-2).

—Sediments
DIN ENGL 38414 Pt 4-84. German Standard Methods for the Examination of Water, Waste Water and Sludge; Sludge and Sediments (Group S); Determination of Leachability by Water (S 4) (Oct). 6 pp.

—Sludge (Sewage)
DIN ENGL 38414 Pt 4-84. German Standard Methods for the Examination of Water, Waste Water and Sludge; Sludge and Sediments (Group S); Determination of Leachability by Water (S 4) (Oct). 6 pp.

—Wood Preservatives
BSI BS 5761: Part 2-90. 1990 Wood Preservatives. Accelerated Ageing of Treated Wood Prior to Biological Testing Part 2: Leaching Procedure. 8 pp.
CEN EN 84-79. Wood Preservatives: Accelerated Ageing of Treated Wood Prior to Biological Testing; Leaching Procedure. 8 pp.
CEN EN 84-89. Wood Preservatives: Accelerated Ageing of Treated Wood Prior to Biological Testing; Leaching Procedure. 7 pp.

Leachate Extraction
See Also: Leach Testing

—Solid Wastes
CGSB 164-GP-1MP-87. Leachate Extraction Procedure. 12 pp.

Lead
Scope Note: For additional listings, see also specific products made from lead *See Also:* Heavy Metals; Lead Coatings (Made From Lead); Lead Content Analysis; Lead Ores; Metals; Nonferrous Metals; White Metals
CNS K7269-68. Chemical Reagent (Lead) (Nov)(1768).
JIS K 8701-61. Lead.

—Antimony Content—Atomic Absorption Spectrometry
SAA AS 1671.1-87. Lead and Lead Alloys—Part 1: Determination of Antimony Content—Flame Atomic Absorption Spectrometric Method. 8 pp.

—Antimony Content—Photometry
BSI BS 3908: Part 13-70. 1970 Methods for the Sampling and Analysis of Lead and Lead Alloys Part 13: Antimony in Lead and Lead Alloys (Low Contents) (Photometric Method). 8 pp.

—Antimony Content—Volumetric Analysis
BSI BS 3908: Part 10-89. 1989 Sampling and Analysis of Lead and Lead Alloys Part 10: Antimony in Lead Alloys (Titrimetric Method). 6 pp.

—Arsenic Content—Photometry
BSI BS 3908: Part 2-67. 1967 Methods for the Sampling and Analysis of Lead and Lead Alloys Part 2: Arsenic inLead and Lead Alloys (Photometric Method). 8 pp.

—Bismuth Content—Photometry
BSI BS 3908: Part 3-67. 1967 Methods for the Sampling and Analysis of Lead and Lead Alloys Part 3: Bismuth in Lead and Lead Alloys (Photometric Method). 8 pp.

—Bismuth Content—Polarographic Analysis
MOD UK M 9331/63. Determination of Bismuth, Copper, Cadmium, and Zinc in Pure Lead (Polarographic Method) (No Information) (Withdrawn).

INTERNATIONAL AND NON-U.S. NATIONAL STANDARDS
SUBJECT INDEX
Lead

Lead (Cont.)

—Cadmium Content—Polarographic Analysis
MOD UK M 9331/63. Determination of Bismuth, Copper, Cadmium, and Zinc in Pure Lead (Polarographic Method) (No Information) (Withdrawn).

—Chemical Analysis
BSI BS 334-82. 1982 Compositional Requirements of Chemical Lead. 7 pp.
CNS H2014-53. Chemical Analysis of Lead (Feb)(277). 3 pp.
DIN ENGL 1719-86. Lead; Composition (Jan). 2 pp.
JIS H 1121-74. Methods for Chemical Analysis of Lead Metal.

—Copper Content—Photometry
BSI BS 3908: Part 4-67. 1967 Methods for the Sampling and Analysis of Lead and Lead Alloys Part 4: Copper in Lead and Lead Alloys (Photometric Method). 8 pp.

—Copper Content—Polarographic Analysis
MOD UK M 9331/63. Determination of Bismuth, Copper, Cadmium, and Zinc in Pure Lead (Polarographic Method) (No Information) (Withdrawn).

—Cosmetics—Hygiene
CNS S2104-84. Method of Hygienic Test for Cosmetics Lead (Jan)(10382).

—Extraction Analysis—Plastic Pipes
SAA AS 1462.7-84. Methods of Test for Unplasticized PVC (UPVC) Pipes and Fittings—Part 7: Method for Determining Extractability of Lead and Tin from UPVC Pipes and Fittings. 2 pp.

—Ingots
CNS H3002-84. Lead Ingot (Nov)(7). 2 pp.
SAA AS 1812-75. Lead Ingot (R 1993). 12 pp.

—Ingots—Chemical Analysis
CNS H2071-84. Method of Chemical Analysis for Lead Ingot (Nov)(11112).

—Ingots—Radiation Shielding
BSI BS 3909-65. 1965 Ingot Lead for Radiation Shielding. 12 pp.

—Ingots—Sampling
BSI BS 3908: Part 1-65. 1965 Amd 1 Methods for the Sampling and Analysis of Lead and Lead Alloys Part 1: Sampling of Ingot Lead, Lead Alloy Ingots, Sheet Pipe, and Cable Sheathing Alloys. 8 pp.

—Iron Content—Photometry
BSI BS 3908: Part 15-72. 1972 Methods for the Sampling and Analysis of Lead and Lead Alloys Part 15: Iron in Lead and Lead Alloys (Photometric Method). 8 pp.

—Nickel Content—Photometry
BSI BS 3908: Part 5-68. 1968 Methods for the Sampling and Analysis of Lead and Lead Alloys Part 5: Nickel in Lead Alloys (Photometric Method). 8 pp.

—Pigs (Castings)
CNS H3140-86. Hand Lead Plates (Dec)(11795).
JIS H 2105-55. Pig Lead. 4 pp.

—Plates
CNS H3049-85. Lead Plates (Feb)(4006).

—Sheets
JIS H 4301-55. Lead Sheets (R 1965). 5 pp.
SAA AS 1804-76. Soft Lead Sheet and Strip (R 1993). 12 pp.

—Sheets—Rolled
DIN ENGL 59610-67. Lead Sheet; Dimensions (Jan). 2 pp.

—Spectrochemical Analysis
JIS H 1123-76. Methods for Emission Spectrochemical Analysis of Lead Metal.

—Standard Solutions
JIS K 0015-83. Lead Standard Solution. 13 pp.

—Strips
SAA AS 1804-76. Soft Lead Sheet and Strip (R 1993). 12 pp.

—Sulfur Content—Volumetric Analysis
BSI BS 3908: Part 9-91. 1991 Sampling and Analysis of Lead and Lead Alloys Part 9: Sulphur in Lead and Lead Alloys. 8 pp.

Lead (Cont.)

—Tellurium Content—Photometry
BSI BS 3908: Part 6-71. 1971 Methods for the Sampling and Analysis of Lead and Lead Alloys Part 6: Tellurium in Lead and Lead Alloys (Photometric Method). 9 pp.

—Tin Content—Atomic Absorption Spectrometry
SAA AS 1671.4-88. Lead and Lead Alloys—Part 4: Determination of Tin in Antimonial Lead—Flame Atomic Absorption Spectrometric Method. 3 pp.

—Tin Content—Photometry
BSI BS 3908: Part 11-68. 1968 Methods for the Sampling and Analysis of Lead and Lead Alloys Part 11: Tin in Lead and Lead Alloys (Volumetric Method). 8 pp.

—Toys—Migration Testing
BSI BS 5665: Part 3-89. 1989 Safety of Toys Part 3: Migration of Certain Elements (Supersedes BS 3443: 1968). 11 pp.
CEN EN 71-3-88. Safety of Toys Part 3 Migration of Certain Elements. 14 pp.
CEN PREN 71-3-92. Safety of Toys—Part 3: Migration of Certain Elements. 24 pp.

—Zinc Content—Polarographic Analysis
MOD UK M 9331/63. Determination of Bismuth, Copper, Cadmium, and Zinc in Pure Lead (Polarographic Method) (No Information) (Withdrawn).

Lead Acetate
See Also: Hazardous Materials
CNS K7270-68. Chemical Reagent (Lead Acetate) (Nov)(1769).

—Initiators (Explosives)
MOD UK DSTAN 68-112-85. Lead Acetate (Normal) Issue 1. 7 pp.
MOD UK DSTAN 68-112-91. Lead Acetate (Normal) Issue 2. 8 pp.

Lead Acetate Trihydrate
See Also: Acetates
JIS K 8374-79. Lead Acetate Trihydrate.

Lead Acid Batteries
See Also: Batteries; Jars; Lead Acid Cells; Storage Batteries
BSI BS 6290: Part 1-83. (OBSOLESCENT) 1983 Lead-Acid Stationery Cells and Batteries Part 1: General Requirements. 11 pp.
BSI BS 6290: Part 2-84. 1984 Lead-Acid Stationery Cells and Batteries Part 2: Lead-Acid High Performance. Plantea Positive Type. 12 pp.
BSI BS 6290: Part 3-86. 1986 Lead-Acid Stationery Cells and Batteries Part 3: Lead-Acid Pasted Positive Plate Type. 10 pp.
BSI BS EN 60896-1-92. 1992 Stationary Lead-Acid Batteries General Requirements and Methods of Test Part 1: Vented Types (IEC 896-1: 1987). 20 pp.
BSI BS EN 60896-1-01. 1992 Amd 1 Stationary Lead-Acid Batteries General Requirements and Methods of Test Part 1: Vented Types (IEC 896-1: 1987) (AMD 7863) July 15, 1993 (E). 33 pp.
CENELEC EN 60896-1-91. Stationary Lead-Acid Batteries—General Requirements and Methods of Test Part 1: Vented Types. 6 pp.
CENELEC EN 60896-1/A2-92. AMD 2 Stationary Lead-Acid Batteries—General Requirements and Methods of Test Part 1: Vented Types. 4 pp.
CNS C3090-80. Method of Test for Stationary Lead-Acid Batteries (Aug)(6039).
CNS C4041-82. Micro-Porous Rubber Separator for Lead Storage Batteries (Dec)(1409).
CNS C4042-82. Strengthened Fiber Separator for Lead Storage Batteries (Dec)(1410).
CNS C4206-80. Lead Acid Batteries for General Service (Aug)(6034).
CNS C4208-80. Stationary Lead-Acid Batteries (Aug)(6038).
IEC 896 Pt 1-87. Stationary Lead-Acid Batteries General Requirements and Methods of Test Part 1: Vented Types First Edition; (Amendment 2-1990). 46 pp.
JIS C 2313-90. Separators for Lead-Acid Batteries. 11 pp.
JIS C 8701-75. Lead Acid Batteries for General Service.
JIS C 8701-69. Lead-Acid Batteries for General Service (R 1972) (Revised Nov 1975). 6 pp.
JIS C 8704-89. Stationary Lead-Acid Batteries. 23 pp.
SAA AS 1981-81. Stationary Batteries of the Lead-Acid Pasted Plate Type Corrig. (Superseded by AS 4029.3—1993).

—Aircraft
BSI 4G 205: Part 1-87. (WITHDRAWN) 1987 Secondary Batteries for Aircraft Part 1: Lead-Acid Batteries (S) (Superseded by BS 5G 205: Part 1: 1990). 20 pp.

Lead Acid Batteries (Cont.)

—Aircraft (Cont.)
BSI 5G 205: Part 1-90. 1990 Secondary Batteries for Aircraft Part 1: Lead-Acid Batteries. 21 pp.
CEN PREN3436-88. Lead Acid Batteries for Aircraft Particular Specification of Format Types: 3436. 10 pp.
CEN PREN3437-88. Lead Acid Batteries for Aircraft Particular Specification of Format Types: 3437. 10 pp.
CEN PREN3438-88. Lead Acid Batteries for Aircraft Particular Specification of Format Types: 3438. 9 pp.
CEN PREN3439-88. Lead Acid Batteries—for Aircarft Particular Specification of Format Types: 3439. 10 pp.

—Aircraft—Bolts
SBAC AS 415 ISSUE 5(I). Bolt Assembly—Attachment of Lead-Acid Accumulators. 1 p.

—Aircraft—Vented
CENELEC EN 60952-2-93. Aircraft Batteries Part 2: Design and Construction Requirements (IEC 952-2:1991). 5 pp.
IEC 952 Pt 1-88. Aircraft Batteries Part 1: General Test Requirements and Performance Levels First Edition. 30 pp.
IEC 952 Pt 2-91. Aircraft Batteries Part 2: Design and Construction Requirements First Edition; (CENELEC EN 60952-2: 1993). 43 pp.

—Automotive
BSI BS EN 60095-2-93. 1993 Lead-Acid Starter Batteries Part 2: Dimensions of Batteries and Dimensions and Marking of Terminals (E). 19 pp.
CENELEC EN 60095-2-93. Lead-Acid Starter Batteries Part 2: Dimensions of Batteries and Dimensions and Marking of Terminals. 16 pp.
CENELEC EN 60095-2-93. Lead-Acid Starter Batteries Part 2: Dimensions of Batteries and Dimensions and Marking of Terminals (IEC 95-2: 1984, Modified). 14 pp.
CNS D1004-87. Configuration and Mounting Dimension of Distributor Contact Breakers for Automobiles (Oct)(5559). 1 p.
CNS D2001-87. Lead Storage Batteries for Automobiles (Mar)(422). 13 pp.
IEC 95 Pt 1-88. Lead-Acid Starter Batteries Part 1: General Requirements and Methods of Test Fifth Edition; (Corrigendum—April 1990) (Amendment 1-1993). 39 pp.
IEC 95 Pt 2-84. Lead-Acid Starter Batteries Part 2: Dimensions of Batteries and Dimensions and Marking of Terminals Third Edition; (Amendment 1-1991) (Amendment 2-1993) (CENELEC EN 60095-2: 1993). 49 pp.
JIS D 5301-91. Lead-Acid Batteries for Automobiles. 16 pp.
MOD UK DSTAN 25-3-01. Starter Battery Spaces for Wheeled Vehicles Issue 1; Amendment 1 (Withdrawn). 7 pp.
SNZ NZS 6102-76. Code of Practice for the Storage, Handling and Maintenance of Automotive Lead-Acid Starter Batteries (Reconfirmed 1991). 12 pp.

—Automotive—Cable Clamps
CNS C4350-82. Cables Clamps for Automobiles Lead Storage Batteries (Jul)(9077).

—Automotive—Containers
SAA AS D23-70. Containers and Cell Covers for Lead-Acid Batteries of the Automobile Type. 20 pp.

—Automotive—Covers
SAA AS D23-70. Containers and Cell Covers for Lead-Acid Batteries of the Automobile Type. 20 pp.

—Battery Chargers
NATO STANAG 4247 ED 1 AMD 0-00. (DRAFT) Battery Chargers, Non-Rotating, for Lead/Acid and Nickel/Cadmium Batteries. 7 pp.

—Containers
CNS C4043-82. Container for Lead Acid Batteries (Dec)(1411).
JIS C 2335-91. Containers for Lead-Acid Batteries. 10 pp.

—Electric Vehicles
BSI BS 2550-83. 1983 Amd 2 Lead-Acid Traction Batteries (AMD 4911) September 30, 1985. 14 pp.
BSI BS 7483-91. 1991 Lead-Acid Batteries for the Propulsion of Light Electric Vehicles. 8 pp.
CENELEC HD 465.1-87. Lead—Acid Traction Batteries Part 1: General Requirements and Methods of Test. 3 pp.
CNS D2180-84. Lead-Acid Traction Batteries (Aug)(10980).
IEC 254 Pt 1-83. Lead-Acid Traction Batteries Part 1: General Requirements and Methods of Test Second Edition; (Amendment 1-1990). 29 pp.
JIS D 5303-86. Lead-Acid Traction Batteries. 14 pp.

INDUSTRY STANDARDS

Lead Acid Batteries *(Cont.)*

—Glass Mats
JIS C 2202-78. Glass Mats for Lead Acid Storage Battery (R 1983). 7 pp.

—Industrial Trucks
BSI BS EN 60095-4-93. 1993 Lead-Acid Starter Batteries Part 4: Dimensions of Batteries for Heavy Commercial Vehicles (E). 14 pp.

CENELEC EN 60095-4-93. Lead-Acid Starter Batteries Part 4: Dimensions of Batteries for Heavy Commercial Vehicles. 11 pp.

CENELEC EN 60095-4-93. Lead-Acid Starter Batteries Part 4: Dimensions of Batteries for Heavy Commercial Vehicles (IEC 95-4: 1989, Modified). 9 pp.

IEC 95 Pt 4-89. Lead-Acid Starter Batteries Part 4: Dimensions of Batteries for Heavy Trucks First Edition (CENELEC EN 60095-4: 1993). 16 pp.

—Internal Combustion Engines
BSI BS 3911: Part 1-82. (WITHDRAWN) 1982 Amd 1 Lead-Acid Starter Batteries for Internal Combustion Engines Part 1: Batteries Requiring Regular Maintenance (AMD 5943) April 28, 1989 (Superseded by BS EN 60095-2: 1993). 19 pp.

BSI BS 3911: Part 2-87. (WITHDRAWN) 1987 Lead-Acid Starter Batteries for Internal Combustion Engines Part 2: Maintenance-Free and Low-Maintenance Batteries (Superseded by BS EN 60095-2: 1993) (E). 12 pp.

—Motorcycles
CNS C4107-84. Small Type Lead Storage Batteries for Motorcycles (Aug)(3330).

JIS D 5302-91. Lead-Acid Batteries for Motorcycle. 12 pp.

—Opportunity Charging
BSI BS EN 61044-93. 1993 Opportunity-Charging of Lead-Acid Traction Batteries (IEC 1044: 1990) (E). 12 pp.

CENELEC EN 61044-92. Opportunity-Charging of Lead-Acid Traction Batteries. 5 pp.

CENELEC EN 61044-92. Opportunity-Charging of Lead-Acid Traction Batteries (IEC 1044: 1990). 7 pp.

IEC 1044-90. Opportunity-Charging of Lead-Acid Traction Batteries First Edition. 15 pp.

—Portable
BSI BS 6745: Part 1-86. (WITHDRAWN) 1986 Portable Lead-Acid Cells and Batteries Part 1: Performance, Design and Construction of Valve Regulated Sealed Type (Superseded by BS EN 61056-1: 1993). 8 pp.

BSI BS EN 61056-1-93. 1993 Portable Lead-Acid Cells and Batteries (Valve-Regulated Types) Part 1: General Requirements, Functional Characteristics—Methods of Test (IEC 1056-1: 1993) (E). 17 pp.

CENELEC EN 61056-1-93. Portable Lead-Acid Cells and Batteries (Valve-Regulated Types) Part 1: General Requirements, Functional Characteristics—Methods of Test (IEC 1056-1:1991). 5 pp.

IEC 1056 Pt 1-91. Portable Lead-Acid Cells and Batteries (Valve-Regulated Types) Part 1: General Requirements, Functional Characteristics—Methods of Test First Edition; (CENELEC EN 61056-1: 1993). 24 pp.

IEC 1056 Pt 3-91. Portable Lead-Acid Cells and Batteries (Valve-Regulated Types) Part 3: Safety Recommendations for Use in Electric Appliances First Edition. 12 pp.

MOD UK DSTAN 61-9: Part 1-01. Batteries, Secondary Part 1: Batteries Secondary—Portable Lead-Acid Type Issue 3; Amendment 8. 53 pp.

MOD UK DSTAN 61-9: Pt 1: Supp 1-73. Battery, Secondary—Portable, Lead-Acid Type 6V, 40 Ah (Fully Dry Charged) Issue 2. 4 pp.

MOD UK DSTAN 61-9: Pt 1: Supp 2-73. Battery, Secondary—Portable, Lead-Acid Type 6V, 170 Ah (Fully Dry Charged) Issue 2. 4 pp.

MOD UK DSTAN 61-9: Pt 1: Supp 3-73. Battery, Secondary—Portable, Lead-Acid Type 12V, 75 Ah (Fully Dry Charged) Issue 2. 4 pp.

MOD UK DSTAN 61-9: Pt 1: Supp 4-73. Battery, Secondary—Portable, Lead-Acid Type 12V, 22 Ah (Fully Dry Charged) Issue 2. 4 pp.

MOD UK DSTAN 61-9: Pt 1: Supp 5-73. Battery, Secondary—Portable, Lead-Acid Type 6V, 85 Ah (Fully Dry Charged) Issue 2. 4 pp.

MOD UK DSTAN 61-9: Pt 1: Supp 6-73. Battery, Secondary—Portable, Lead-Acid Type 6V, 100 Ah (Fully Dry Charged) No 4 MK 3 (UK/6TN) Issue 2. 4 pp.

MOD UK DSTAN 61-9: Pt 1: Supp 7-01. Battery, Secondary—Portable, Lead-Acid Type 12V, 100 Ah (Fully Dry Charged) No 4, MK 3 (UN/6TN) Issue 2; Amendment 9. 10 pp.

MOD UK DSTAN 61-9: Pt 1: Supp 8-01. Battery, Secondary—Portable, Lead-Acid Type 12V, 50 Ah (Fully Dry Charged) No 5, MK 1 (UK/2HN) Issue 1; Amendment 7. 8 pp.

Lead Acid Batteries *(Cont.)*

—Portable *(Cont.)*
MOD UK DSTAN 61-9: Pt 1: Supp 9-76. Battery, Secondary—Portable, Lead-Acid Type 12V, 9Ah (Filled Uncharged) No 6, Mk 1 Issue 1. 5 pp.

MOD UK DSTAN 61-9: Part 2-01. Batteries Secondary Part 2: Batteries Secondary—Portable Lead Acid and Alkaline Types Issue 2; Amendment 4 (Superseded by Def Stan 61-9: Part 1). 51 pp.

—Portable—Safety
IEC 1056 Pt 3-91. Portable Lead-Acid Cells and Batteries (Valve-Regulated Types) Part 3: Safety Recommendations for Use in Electric Appliances First Edition. 12 pp.

—Rechargeable
MOD UK DSTAN 61-9: Part 1-93. Generic Specification for Batteries, Rechargeable, Secondary Part 1: General Requirements Issue 4. 41 pp.

—Rechargeable—Preferred Product List
CECC CECC MUAHAG Vol 15 IS 1-91. Preferred Products List; Batteries (En). 20 pp.

—Rechargeable—Sealed—Preferred Product List
CECC CECC MUAHAG Vol 15 IS 1-91. Preferred Products List; Batteries (En). 20 pp.

—Safety
BSI BS 6133-85. 1985 Amd 1 Code of Practice for Safe Operation of Lead-Acid Stationary Cells and Batteries. 19 pp.

—Sealed
BSI BS 6290: Part 4-87. 1987 Lead-Acid Stationary Cells and Batteries Part 4: Lead-Acid Valve Regulated Sealed Type. 11 pp.

BSI BS 6745: Part 1-86. (WITHDRAWN) 1986 Portable Lead-Acid Cells and Batteries Part 1: Performance, Design and Construction of Valve Regulated Sealed Type (Superseded by BS EN 61056-1: 1993). 8 pp.

JIS C 8702-88. Small-Sized Sealed Lead-Acid Batteries. 18 pp.

MOD UK DSTAN 61-9: Pt 1: Supp 6-73. Battery, Secondary—Portable, Lead-Acid Type 6V, 100 Ah (Fully Dry Charged) No 4 MK 3 (UK/6TN) Issue 2. 4 pp.

—Ships
CNS F3043-80. Lead-Acid Marine Batteries (Aug)(6035). 3 pp.

JIS F 8101-76. Lead-Acid Marine Batteries.

—Submarines—Comprehensive Testing
NATO STANAG 4390 ED 1 AMD 0-90. Tests and Requirements for Submarine Main Lead-Acid Batteries. 15 pp.

—Submarines—Cooling Water
NATO STANAG 4391 ED 1 AMD 0-91. Cooling Water for Submarine Main Lead Acid Batteries. 4 pp.

—Submarines—Electrolytes
NATO STANAG 4287 ED 1 AMD 2-88. Electrolyte for Submarine Main Lead Acid Batteries. 21 pp.

—Submarines—Glossaries
NATO STANAG 4389 ED 1 AMD 0-90. Terms and Definitions Covering Submarine Main Lead Acid Batteries. 20 pp.

—Submarines—Purified Water
NATO STANAG 4248 ED 1 AMD 2-88. Purified Water for Submarine Main Lead Acid Batteries. 18 pp.

—Valve Regulated
CENELEC PREN 50105-92. Stationary Lead-Acid Batteries—General Requirements and Test Methods—Valve Regulated Types. 23 pp.

—Valve Regulated—Sealed
JIS C 8707-92. Stationary Sealed Load-Acid Batteries (Valve Regulated Type). 15 pp.

SAA AS 4029.2-92. Stationary Batteries—Lead-Acid—Part 2: Valve-Regulated Sealed Type (in Professional Package 50). 13 pp.

—Water
BSI BS 4974-75. 1975 Amd 1 Water for Lead-Acid Batteries. 17 pp.

Lead Acid Cells
See Also: Lead Acid Batteries

BSI BS 6290: Part 1-83. (OBSOLESCENT) 1983 Lead-Acid Stationery Cells and Batteries Part 1: General Requirements. 11 pp.

BSI BS 6290: Part 2-84. 1984 Lead-Acid Stationery Cells and Batteries Part 2: Lead-Acid High Performance. Plantea Positive Type. 12 pp.

Lead Acid Cells *(Cont.)*

BSI BS 6290: Part 3-86. 1986 Lead-Acid Stationery Cells and Batteries Part 3: Lead-Acid Pasted Positive Plate Type. 10 pp.

BSI BS EN 60896-1-92. 1992 Stationary Lead-Acid Batteries General Requirements and Methods of Test Part 1: Vented Types (IEC 896-1: 1987). 20 pp.

BSI BS EN 60896-1-01. 1992 Amd 1 Stationary Lead-Acid Batteries General Requirements and Methods of Test Part 1: Vented Types (IEC 896-1: 1987) (AMD 7863) July 15, 1993 (E). 33 pp.

BSI BS EN 61056-1-93. 1993 Portable Lead-Acid Cells and Batteries (Valve-Regulated Types) Part 1: General Requirements, Functional Characteristics—Methods of Test (IEC 1056-1: 1993) (E). 17 pp.

CENELEC HD 465.2 S1-89. Lead-Acid Traction Batteries-Part 2: Dimensions of Cells and Terminals and Marking of Polarity on Cells. 3 pp.

CENELEC EN 60896-1-91. Stationary Lead-Acid Batteries—General Requirements and Methods of Test Part 1: Vented Types. 6 pp.

CENELEC EN 60896-1/A2-92. AMD 2 Stationary Lead-Acid Batteries—General Requirements and Methods of Test Part 1: Vented Types. 4 pp.

CENELEC EN 61056-1-93. Portable Lead-Acid Cells and Batteries (Valve-Regulated Types) Part 1: General Requirements, Functional Characteristics—Methods of Test (IEC 1056-1:1991). 5 pp.

IEC 254 Pt 2-85. Lead-Acid Traction Batteries Part 2: Dimensions of Cells and Terminals and Marking of Polarity on Cells Second Edition. 12 pp.

IEC 896 Pt 1-87. Stationary Lead-Acid Batteries General Requirements and Methods of Test Part 1: Vented Types First Edition; (Amendment 2-1990). 46 pp.

IEC 1056 Pt 1-91. Portable Lead-Acid Cells and Batteries (Valve-Regulated Types) Part 1: General Requirements, Functional Characteristics—Methods of Test First Edition; (CENELEC EN 61056-1: 1993). 24 pp.

—Portable
BSI BS 6745: Part 1-86. (WITHDRAWN) 1986 Portable Lead-Acid Cells and Batteries Part 1: Performance, Design and Construction of Valve Regulated Sealed Type (Superseded by BS EN 61056-1: 1993). 8 pp.

IEC 1056 Pt 3-91. Portable Lead-Acid Cells and Batteries (Valve-Regulated Types) Part 3: Safety Recommendations for Use in Electric Appliances First Edition. 12 pp.

—Portable—Safety
IEC 1056 Pt 3-91. Portable Lead-Acid Cells and Batteries (Valve-Regulated Types) Part 3: Safety Recommendations for Use in Electric Appliances First Edition. 12 pp.

—Safety
BSI BS 6133-85. 1985 Amd 1 Code of Practice for Safe Operation of Lead-Acid Stationary Cells and Batteries. 19 pp.

—Sealed
BSI BS 6290: Part 4-87. 1987 Lead-Acid Stationary Cells and Batteries Part 4: Lead-Acid Valve Regulated Sealed Type. 11 pp.

BSI BS 6745: Part 1-86. (WITHDRAWN) 1986 Portable Lead-Acid Cells and Batteries Part 1: Performance, Design and Construction of Valve Regulated Sealed Type (Superseded by BS EN 61056-1: 1993). 8 pp.

Lead Alkyls
See Also: Hazardous Materials

—Highway Transportation—Emergency Procedures
SAA AS 1678.6.0. 017-93. Emergency Procedure Guides—Transport—Part 6.0.017: Motor Fuel Anti-Knock Mixtures (Lead Aklyls, N.O.S.) (in Professional Package 37A).

Lead Alloys
See Also: Hard Lead; Solders

—Antimony Content—Atomic Absorption Spectrometry
SAA AS 1671.1-87. Lead and Lead Alloys—Part 1: Determination of Antimony Content—Flame Atomic Absorption Spectrometric Method. 3 pp.

—Antimony Content—Photometry
BSI BS 3908: Part 13-70. 1970 Methods for the Sampling and Analysis of Lead and Lead Alloys Part 13: Antimony in Lead and Lead Alloys (Low Contents) (Photometric Method). 8 pp.

—Antimony Content—Volumetric Analysis
BSI BS 3908: Part 10-89. 1989 Sampling and Analysis of Lead and Lead Alloys Part 10: Antimony in Lead Alloys (Titrimetric Method). 6 pp.

INTERNATIONAL AND NON-U.S. NATIONAL STANDARDS
SUBJECT INDEX
Lead

Lead Alloys (Cont.)

—Antimony Content—Volumetric Analysis (Cont.)
SAA AS 1671.2-87. Lead Alloys—Part 2: Determination of Low Concentrations of Antimony in Lead and Lead Alloys Containing not More than 2.5 Percent Arsenic and 0.10 Percent Copper—Titrimetric Method. 4 pp.
SAA AS 1671.3-87. Lead and Lead Alloys—Part 3: Determination of High Concentrations of Antimony in Lead Alloys Containing not More than 2.5 Percent Arsenic and 1.0 Percent Copper-Titrimetric Method. 5 pp.

—Arsenic Content—Photometry
BSI BS 3908: Part 2-67. 1967 Methods for the Sampling and Analysis of Lead and Lead Alloys Part 2: Arsenic inLead and Lead Alloys (Photometric Method). 8 pp.

—Bismuth Content—Photometry
BSI BS 3908: Part 3-67. 1967 Methods for the Sampling and Analysis of Lead and Lead Alloys Part 3: Bismuth in Lead and Lead Alloys (Photometric Method). 8 pp.

—Copper Content—Photometry
BSI BS 3908: Part 4-67. 1967 Methods for the Sampling and Analysis of Lead and Lead Alloys Part 4: Copper in Lead and Lead Alloys (Photometric Method). 8 pp.

—Ingots—Chemical Analysis
DIN ENGL 17640 Pt 1-86. Lead Alloys for General Purposes (Jan). 3 pp.

—Ingots—Sampling
BSI BS 3908: Part 1-65. 1965 Amd 1 Methods for the Sampling and Analysis of Lead and Lead Alloys Part 1: Sampling of Ingot Lead, Lead Alloy Ingots, Sheet Pipe, and Cable Sheathing Alloys. 8 pp.

—Ingots—Storage Batteries—Chemical Analysis
DIN ENGL 17640 Pt 3-86. Lead Alloys for Storage Batteries (Jan). 3 pp.

—Iron Content—Photometry
BSI BS 3908: Part 15-72. 1972 Methods for the Sampling and Analysis of Lead and Lead Alloys Part 15: Iron in Lead and Lead Alloys (Photometric Method). 8 pp.

—Nickel Content—Photometry
BSI BS 3908: Part 5-68. 1968 Methods for the Sampling and Analysis of Lead and Lead Alloys Part 5: Nickel in Lead Alloys (Photometric Method). 8 pp.

—Printing Types
CNS H3022-85. Lead Alloys for Printing (Feb)(2154). 3 pp.

—Sheets—Sampling
BSI BS 3908: Part 1-65. 1965 Amd 1 Methods for the Sampling and Analysis of Lead and Lead Alloys Part 1: Sampling of Ingot Lead, Lead Alloy Ingots, Sheet Pipe, and Cable Sheathing Alloys. 8 pp.

—Sulfur Content—Volumetric Analysis
BSI BS 3908: Part 9-91. 1991 Sampling and Analysis of Lead and Lead Alloys Part 9: Sulphur in Lead and Lead Alloys. 8 pp.

—Tellurium Content—Photometry
BSI BS 3908: Part 6-71. 1971 Methods for the Sampling and Analysis of Lead and Lead Alloys Part 6: Tellurium in Lead and Lead Alloys (Photometric Method). 9 pp.

—Tin Content—Volumetric Analysis
BSI BS 3908: Part 11-68. 1968 Methods for the Sampling and Analysis of Lead and Lead Alloys Part 11: Tin in Lead and Lead Alloys (Volumetric Method). 8 pp.

Lead Arsenate
See Also: Hazardous Materials

—Highway Transportation—Emergency Procedures
SAA AS 1678.6.0. 014-84. Emergency Procedure Guides—Transport—Part 6.0.014: Lead Arsenate and Lead Arsenite.

Lead Arsenite
See Also: Hazardous Materials

—Highway Transportation—Emergency Procedures
SAA AS 1678.6.0. 014-84. Emergency Procedure Guides—Transport—Part 6.0.014: Lead Arsenate and Lead Arsenite.

Lead Assemblies
Use: Electrical Leads

Lead Bearing Alloys
Use: Bearing Alloys—Lead

Lead Carbonates
Use: Basic Lead Carbonates

Lead Castings
See Also: Castings; Metal Products
CNS H3141-86. Hard Lead Castings (Dec)(11796).
JIS H 2105-55. Pig Lead. 4 pp.
JIS H 5601-90. Hard Lead Castings. 9 pp.

Lead Chloride
CNS K7272-68. Chemical Reagent (Lead Chloride) (Nov)(1771).

Lead Chromate
See Also: Lead Chromate Content Analysis
BSI BS 282, 389-63. 1963 Lead Chromes and Zinc Chromes for Paints. 36 pp.
CNS K7273-68. Chemical Reagent (Lead Chromate) (Nov)(1772).
JIS K 8314-61. Lead Chromate, Precipitated.

—Pigments
BSI BS 7446-91. 1991 Lead Chromate Pigments and Lead Chromate-Molybdate Pigments for Paints (ISO 3711: 1990). 15 pp.
ISO 3711-90. Lead Chromate Pigments and Lead Chromate-Molybdate Pigments—Specifications and Methods of Test Second Edition; (Corrected and Reprinted -1991). 11 pp.

—Pyrotechnics
MOD UK DSTAN 68-105-88. Lead Chromate Issue 1. 8 pp.

Lead Chromate Content Analysis
See Also: Lead Chromate

—Pigments
CGSB 1-GP-71 METH 50.19-81. Methods of Testing Paints and Pigments Pigment Analysis Lead Chromate. 2 pp.

Lead Chromate Paints
See Also: Paints

—Metal—Corrosion Inhibitive
CNS K2157-87. Basic Lead Chromate Anticorrosive Paint (Oct)(12129). 2 pp.
CNS K6919-87. Method of Test for Basic Lead Chromate Anticorrosive Paint (Oct)(12130). 5 pp.
JIS K 5624-84. Basic Lead Chromate Anticorrosive Paint. 11 pp.

Lead Chrome Green

—Pigments
BSI BS 303-91. 1991 Lead Chrome Green Pigments for Paints (ISO 3710: 1990). 10 pp.
ISO 3710-90. Lead Chrome Green Pigments—Specifications and Methods of Test Second Edition. 6 pp.

Lead Coatings (Made From Lead)
Scope Note: Includes coatings made from lead alloys
Use For: Terne *See Also:* Coatings; Lead; Metal Coatings (Made From Metal)

—Carbon Steels—Hot Dip
BSI BS 6582-85. 1985 Continuously Hot-Dip Alloy (Terne) Coated Cold Reduced Carbon Steel Flat Rolled Products. 12 pp.
ISO 4999-91. Continuous Hot-Dip Terne (Lead Alloy) Coated Cold-Reduced Carbon Steel Sheet of Commercial and Drawing Qualities Second Edition. 17 pp.

—Iron—Electroplated
CNS H3146-87. Electroplated Coatings of Lead and Lead-Tia Alloys on Iron and Steel (Sep)(12089).

—Iron—Sprayed
CNS H2054-81. Methods of Test for Lead Spray Products (Oct)(8011). 3 pp.
JIS H 8662-61. Testing Methods for Lead Spray Products (R 1969). 6 pp.

—Steel—Electroplated
CNS H3146-87. Electroplated Coatings of Lead and Lead-Tia Alloys on Iron and Steel (Sep)(12089).

—Steel—Sprayed
CNS H2054-81. Methods of Test for Lead Spray Products (Oct)(8011). 3 pp.
JIS H 8662-61. Testing Methods for Lead Spray Products (R 1969). 6 pp.

Lead Content Analysis
See Also: Lead; Tetraethyl Lead Content Analysis; Tetramethyl Lead Content Analysis
MOD UK M 9506/65. Determination of Lead in Non-Pigmented 'Lead Free' Materials (No Information).

—Air—Atomic Absorption Spectrometry
BSI BS 6069: Sec 3.2-91. 1991 Characterization of Air Quality Part 3: Workplace Atmospheres Section 3.2: Method for the Determination of Particulate Lead and Lead Compounds by Flame Atomic Absorption Spectrometry (ISO 8518: 1990). 15 pp.
ISO 8518-90. Workplace Air—Determination of Particulate Lead and Lead Compounds—Flame Atomic Absorption Spectrometric Method First Edition. 13 pp.

—Algae
CNS N6106-78. Method of Test for Edible Chlorella (Oct)(4597). 9 pp.

—Aluminum Alloys
JIS H 1364-71. Methods for Determination of Bismuth and Lead in Aluminium Alloy.

—Aluminum Alloys—Atomic Absorption Spectrometry
ISO 4192-81. Aluminium and Aluminium Alloys—Determination of Lead Content—Flame Atomic Absorption Spectrometric Method First Edition. 5 pp.

—Aluminum Alloys—Atomic Absorption Spectrophotometry
BSI BS 1728: Part 20-71. 1971 Methods for the Analysis of Aluminium and Aluminium Alloys: Part 20: Lead (Atomic Absorption Method). 9 pp.

—Aluminum—Atomic Absorption Spectrometry
ISO 4192-81. Aluminium and Aluminium Alloys—Determination of Lead Content—Flame Atomic Absorption Spectrometric Method First Edition. 5 pp.

—Aluminum—Atomic Absorption Spectrophotometry
BSI BS 1728: Part 20-71. 1971 Methods for the Analysis of Aluminium and Aluminium Alloys: Part 20: Lead (Atomic Absorption Method). 9 pp.

—Ammonium Bicarbonate—Atomic Absorption Spectrometry
ISO 7110-85. Ammonium Bicarbonate (Ammonium Hydrogencarbonate) for Industrial Use (Including Foodstuffs)—Determination of Lead Content—Flame Atomic Absorption Method First Edition. 5 pp.

—Animal Feed
CNS N4024-20-86. Method of Test for Feeds: Determination of Lead (Aug)(2770-20).

—Basic Lead Silicochromate
CNS K6535-80. Method of Test for Total Lead in Basic Lead Silica-Chromate (Jul)(5864).

—Beeswax
MOD UK M 9517/65. Examination of Beeswax and Beeswax, Lead Free (LAG 931A, No Information).

—Carbon Steels—Gravimetric Analysis
BSI BS 6200: SUB SEC 3.16.1-86. 1986 Sampling and Analysis of Iron, Steel and Other Ferrous Metals Part 3: Methods of Analysis Section 3.16: Determination of Lead Subsection 3.16.1: Carbon Steels and Low Alloy Steels: Gravimetric Method. 4 pp.

—Carbonated Beverage Bottles
CNS S2111-83. Method of Test for Lead, Arsenic and Alkali Content Extracted from Carbonated Beverage Bottles (Oct)(10633).

—Ceramic Ware—Food Contamination
BSI BS 6748-86. 1986 Limits of Metal Release from Ceramic Ware, Glassware, Glass Ceramic Ware and Vitreous Enamel Ware. 8 pp.
CNS R2081-74. Lead and Cadmium Content from Glazed Ceramic Surface for Table Wares (Jun)(3725). 1 p.
ISO 6486 Pt 1-81. Ceramic Ware in Contact with Food—Release of Lead and Cadmium—Part 1: Method of Test First Edition. 6 pp.
ISO 6486 Pt 2-81. Ceramic Ware in Contact with Food—Release of Lead and Cadmium—Part 2: Permissible Limits First Edition. 4 pp.
ISO 7086 Pt 1-82. Glassware and Glass Ceramic Ware in Contact with Food—Release of Lead and Cadmium—Part 1: Method of Test First Edition. 7 pp.

INDUSTRY STANDARDS

Lead

Lead Content Analysis (Cont.)

—Ceramic Ware—Food Contamination (Cont.)
ISO 7086 Pt 2-82. Glassware and Glass Ceramic Ware in Contact with Food—Release of Lead and Cadmium—Part 2: Permissible Limits First Edition. 3 pp.
ISO 8391 Pt 1-86. Ceramic Cookware in Contact with Food—Release of Lead and Cadmium—Part 1: Method of Test First Edition. 6 pp.
ISO 8391 Pt 2-86. Ceramic Cookware in Contact with Food—Release of Lead and Cadmium—Part 2: Permissible Limits First Edition. 4 pp.

—Ceramics—Atomic Absorption Spectrometry
CNS R3073-86. Method of Test for Lead and Cadmium Extracted from Glazed Ceramic Surfaces (Apr)(3503). 4 pp.

—Coatings
CGSB 1-GP-71 METH 52.3-81. Methods of Testing Paints and Pigments Metal Content. 2 pp.
CGSB CAN2-1.500-75 METH 1-73. Methods of Test for Toxic Trace Elements in Protective Coatings Determination of Lead (Pb) in Low Concentration. 2 pp.

—Copper
CNS H2086-89. Methods of Determination for Lead in Copper and Copper Alloys (Jun)(12532).
JIS H 1053-84. Methods for Determination of Lead in Copper and Copper Alloys. 17 pp.

—Copper Alloys
CNS H2086-89. Methods of Determination for Lead in Copper and Copper Alloys (Jun)(12532).
JIS H 1053-84. Methods for Determination of Lead in Copper and Copper Alloys. 17 pp.

—Copper Alloys—Atomic Absorption Spectrometry
ISO 4749-84. Copper Alloys—Determination of Lead Content—Flame Atomic Absorption Spectrometric Method First Edition. 6 pp.
SAA AS K209.1-70. Methods for the Analysis of Copper Alloys—Part 1: Lead in Copper Alloys (Atomic Absorption Spectrometric Method). 5 pp.

—Copper Alloys—Electrolytic Analysis
BSI BS 1748: Parts 1-5-61. 1961 Methods for the Analysis of Copper Alloys Parts 1-5: Determination of Copper, Lead, Iron, Aluminium and Nickel in Copper Alloys. 23 pp.
BSI BS 1748: Parts 11-12-64. 1964 Methods for the Analysis of Copper Alloys Parts 11-12: The Determination of Copper and Lead in Leaded Bronze Alloys. 12 pp.

—Copper Alloys—Volumetric Analysis
ISO 3112-75. Copper and Copper Alloys—Determination of Lead—Extracting Titration Method First Edition. 6 pp.

—Copper—Volumetric Analysis
ISO 3112-75. Copper and Copper Alloys—Determination of Lead—Extracting Titration Method First Edition. 6 pp.

—Dammar
MOD UK M 9535/68. Examination of Gum, Dammar, and Gum, Dammar, Lead Free.

—Drinking Water—Voltametry
DIN ENGL 38406 Pt 16-90. German Standard Methods for the Examination of Water, Waste Water and Sludge; Cations (Group E); Determination of Zinc, Cadmium, Lead, Copper, Thallium, Nickel, Cobalt by Voltammetry (E16) (Mar). 8 pp.

—Ebonite—Atomic Absorption Spectroscopy
DIN ENGL 53599 Pt 1-78. Testing of Rubber and Elastomers; Determination of the Lead Content; Determination by Atomic Absorption Spectroscopy for Lead Contents up to 1000 mg/kg (0.1%) (May). 4 pp.

—Ebonite—Electrogravimetric Analysis
DIN ENGL 53599 Pt 2-78. Testing of Rubber and Elastomers; Determination of the Lead Content; Electrogravimetric and Gravimetric Method for Lead Contents over 1000 mg/kg (0.1%) (May). 3 pp.

—Ebonite—Gravimetric Analysis
DIN ENGL 53599 Pt 2-78. Testing of Rubber and Elastomers; Determination of the Lead Content; Electrogravimetric and Gravimetric Method for Lead Contents over 1000 mg/kg (0.1%) (May). 3 pp.

Lead Content Analysis (Cont.)

—Elastomers—Atomic Absorption Spectroscopy
DIN ENGL 53599 Pt 1-78. Testing of Rubber and Elastomers; Determination of the Lead Content; Determination by Atomic Absorption Spectroscopy for Lead Contents up to 1000 mg/kg (0.1%) (May). 4 pp.

—Elastomers—Electrogravimetric Analysis
DIN ENGL 53599 Pt 2-78. Testing of Rubber and Elastomers; Determination of the Lead Content; Electrogravimetric and Gravimetric Method for Lead Contents over 1000 mg/kg (0.1%) (May). 3 pp.

—Elastomers—Gravimetric Analysis
DIN ENGL 53599 Pt 2-78. Testing of Rubber and Elastomers; Determination of the Lead Content; Electrogravimetric and Gravimetric Method for Lead Contents over 1000 mg/kg (0.1%) (May). 3 pp.

—Elastomers—Photometry
DIN ENGL 53599 Pt 3-78. Testing of Rubber and Elastomers; Determination of the Lead Content; Photometric Method for Lead Contents up to 1000 mg/kg (0.1%) (May). 4 pp.

—Electrolytic Copper—Atomic Absorption Spectrophotometry
BSI BS 7317: Part 7-90. 1990 Methods for Analysis of High Purity Copper Cathode Cu-CATH-1 Part 7: Method for Determination of Lead by Lanthanum Hydroxide Separation and Atomic Absorption Spectrophotometry. 9 pp.
BSI DD 95: Part 7-86. (WITHDRAWN) 1986 Analysis of Higher Purity Copper Cathode Cu-CATH-1 Part 7: Method for Determination of Lead by Lanthanum Hydroxide Separation and Atomic Absorption Spectrophotometry (Superseded by BS 7317: Part 7: 1990). 8 pp.

—Electrolytic Copper—Lanthanum Hydroxide Separation
BSI BS 7317: Part 7-90. 1990 Methods for Analysis of High Purity Copper Cathode Cu-CATH-1 Part 7: Method for Determination of Lead by Lanthanum Hydroxide Separation and Atomic Absorption Spectrophotometry. 9 pp.
BSI DD 95: Part 7-86. (WITHDRAWN) 1986 Analysis of Higher Purity Copper Cathode Cu-CATH-1 Part 7: Method for Determination of Lead by Lanthanum Hydroxide Separation and Atomic Absorption Spectrophotometry (Superseded by BS 7317: Part 7: 1990). 8 pp.

—Enameled Surfaces
CNS R3137-86. Method of Test for Lead and Cadmium Release from Porcelain Enamel Surface (Apr)(11554).

—Enameled Surfaces—Food Contamination
BSI BS 6748-86. 1986 Limits of Metal Release from Ceramic Ware, Glassware, Glass Ceramic Ware and Vitreous Enamel Ware. 8 pp.

—Flue Gases
JIS K 0097-79. Method for Determination of Cadmium and Lead in Stack Gas (R 1984). 22 pp.

—Fluorite—Extraction Analysis—Atomic Absorption Spectrometry
ISO 9779-93. Metallurgical-Grade Fluorspar—Determination of Lead Content—Solvent Extraction Atomic Absorption Spectrometric Method Second Edition. 9 pp.

—Food Contamination—Permissible Limits
DIN ENGL 51031-86. Testing of Articles Intended for Use in Contact with Foodstuffs; Determination of Release of Lead and Cadmium from Silicate Surfaced Articles Intended for Use in Contact with Foodstuffs (Feb). 5 pp.
DIN ENGL 51032-86. Ceramics, Glass, Glass Ceramics, Vitreous Enamels; Permissable Limits for the Release of Lead and Cadmium from Articles Intended for Use in Contact with Foodstuffs (Feb). 4 pp.

—Fruits—Atomic Absorption Spectrometry
ISO 6633-84. Fruits, Vegetables and Derived Products—Determination of Lead Content—Flameless Atomic Absorption Spectrometric Method First Edition. 5 pp.

—Fuels—Complexometric Titrations
DIN ENGL 51769 Part 5-84. Testing of Petroleum Products; Determination of Lead Content (Total Lead) of Gasolines; Complexometric Method (June). 7 pp.

Lead Content Analysis (Cont.)

—Gasoline—Iodine Monochloride Method
BSI BS 5657-88. 1988 Amd 1 Method for Determination of Lead Content of Gasoline by the Iodine Monochloride Method (AMD 6512) February 28, 1991. 11 pp.
CEN EN 23 830-89. Petroleum Products, Gasoline, Chemical Analysis, Determination of Content, Lead, Analysis Method. 2 pp.
CNS K6912-87. Method of Test for Lead in Gasoline by Atomic Absorption Spectrometry (Jul)(12013).
ISO 3830-81. Petroleum Products—Gasoline—Determination of Lead Content—Iodine Monochloride Method Second Edition. 5 pp.
JIS K 2255-87. Testing Methods for Lead in Gasoline. 22 pp.

—Gasoline—Volumetric Analysis
BSI BS 5290-76. 1976 Determination of Lead Content of Gasoline Volumetric Chromate Method (Supersedes BS 2878: 1976). 11 pp.
CEN EN 13-74. Determination of Lead Content of Gasoline: Volumetric Chromate Method. 9 pp.

—Gasoline—X-Ray Spectrometry
CNS K6973-90. Method of Test for Lead in Gasoline by X-Ray Spectrometry (Aug)(12762).
DIN ENGL 51769 Pt 6-90. Determination of Total Lead Content of Petrol with a Lead Content Exceeding 25mg/l by Wavelength-Dispersive X-Ray Spectrometry (XRS) (Nov). 3 pp.

—Glassware—Food Contamination
BSI BS 6748-86. 1986 Limits of Metal Release from Ceramic Ware, Glassware, Glass Ceramic Ware and Vitreous Enamel Ware. 8 pp.

—Glucose
CNS N6053-73. Method of Test for Dried Glucose and Glucose Syrup (Jan)(3218). 9 pp.

—Glucose Syrups
CNS N6053-73. Method of Test for Dried Glucose and Glucose Syrup (Jan)(3218). 9 pp.

—Glycerol
BSI BS 5711: Part 17-83. 1983 Methods of Sampling and test for Glycerol Part 17: Limit Test for Lead. 2 pp.

—Greases
CNS K6759-83. Method of Test for Lead in New and Used Greases (Mar)(10094).

—Greases—Atomic Absorption Spectrometry
DIN ENGL 51827-89. Determination of Lead Content of Lubricating Greases by Atomic Absorption Spectrometry After Incineration (Jan). 3 pp.

—Ground Water—Voltametry
DIN ENGL 38406 Pt 16-90. German Standard Methods for the Examination of Water, Waste Water and Sludge; Cations (Group E); Determination of Zinc, Cadmium, Lead, Copper, Thallium, Nickel, Cobalt by Voltammetry (E16) (Mar). 8 pp.

—Iron Ores—Absorptiometric Analysis
CNS M3136-83. Method for Determining Lead in Iron Ores (Dithizone Absorptiometric Method) (Apr)(10191).
JIS M 8229-83. Methods for Determination of Lead in Iron Ores. 15 pp.

—Iron Ores—Atomic Absorption Spectrometry
BSI BS 7020: Part 17-88. 1988 Analysis of Iron Ores Part 17: Method for the Determination of Lead and/or Zinc Contents: Flame Atomic Absorption Spectrometric Method. 14 pp.
ISO 8753-87. Iron Ores—Determination of Lead and/or Zinc Content—Flame Atomic Absorption Spectrometric Method First Edition. 12 pp.
SAA AS 1050.25-86. Methods for the Analysis of Iron and Steel—Part 25: Determination of Lead Content (Flame Atomic Absorption Spectrometric Method). 3 pp.

—Iron Ores—Atomic Absorption Spectrophotometry
JIS M 8229-83. Methods for Determination of Lead in Iron Ores. 15 pp.

—Kaolin—Volumetric Analysis
MOD UK M 2002/59. Examination of Kaolin (Withdrawn).

—Lead Sulfide—Solvent Extraction
SAA AS 4030.1-92. Methods for the Analysis of Lead Sulfide Concentrates—Part 1: Determination of Lead Content—Acid Dissolution Solvent Extraction EDTA Titration Method. 7 pp.

Lead Content Analysis (Cont.)

—Low Alloy Steels—Gravimetric Analysis
BSI BS 6200: SUB SEC 3.16.1-86. 1986 Sampling and Analysis of Iron, Steel and Other Ferrous Metals Part 3: Methods of Analysis Section 3.16: Determination of Lead Subsection 3.16.1: Carbon Steels and Low Alloy Steels: Gravimetric Method. 4 pp.

—Magnesium Alloys—Atomic Absorption Spectrophotometry
BSI BS 3907: Part 15-76. 1976 Methods for the Analysis of Magnesium and Magnesium Alloys Part 15: Lead in Magnesium and Magnesium Alloys (Atomic Absorption Method). 7 pp.

—Magnesium—Atomic Absorption Spectrophotometry
BSI BS 3907: Part 15-76. 1976 Methods for the Analysis of Magnesium and Magnesium Alloys Part 15: Lead in Magnesium and Magnesium Alloys (Atomic Absorption Method). 7 pp.

—Magnesium Oxides—Volumetric Analysis
MOD UK M 9541/69. Examination of Magnesium Oxide and Magnesium Oxide, Special, Lead Free (Withdrawn).

—Manganese Ores—Atomic Absorption Spectrometry
ISO 5889-83. Manganese Ores and Concentrates—Determination of Aluminium, Copper, Lead and Zinc Contents—Flame Atomic Absorption Spectrometric Method First Edition. 6 pp.

—Nickel—Atomic Absorption Spectrometry
BSI BS 6783: Part 1-86. 1986 Sampling and Analysis of Nickel, Ferronickel and Nickel Alloys Part 1: Method for Determination of Silver, Bismuth, Cadmium, Cobalt, Copper, Iron, Manganese, Lead and Zinc in Nickel by Flame Atomic Absorption Spectrometry. 15 pp.

BSI BS 6783: Part 4-86. 1986 Sampling and Analysis of Nickel, Ferronickel and Nickel Alloys: Part 4: Method for Determination of Silver, Arsenic, Bismuth, Cadmium, Lead, Antimony, Selenium, Tin, Tellurium and Thallium in Nickel by Electrothermal Atomic Absorption Spectrometry. 11 pp.

ISO 6351-85. Nickel—Determination of Silver, Bismuth, Cadmium Cobalt, Copper, Iron, Manganese, Lead and Zinc Contents—Flame Atomic Absorption Spectrometric Method First Edition. 13 pp.

ISO 7523-85. Nickel—Determination of Silver, Arsenic, Bismuth, Cadmium, Lead, Antimony, Selenium, Tin, Tellurium and Thallium Contents—Electrothermal Atomic Absorption Spectrometric Method First Edition. 10 pp.

—Oleic Acid
MOD UK M 9545/68. Examination of Oleic Acid, Special, Lead Free.

—Ores
JIS M 8123-77. Methods for Determination of Lead in Ores (R 1983). 16 pp.

—Organic Coatings—Insoluble Matter—Quantitative Analysis
CNS K6804-22-85. Method of Test for Organic Coating (Chemical Analysis) — Quantative Test of Lead in Solvent Insolubles (Jan)(10880-22).

—Paint Driers
CNS K6627-81. Method of Test for Lead in Paint Drier by EDTA Method (Jan)(6931).

—Painted Surfaces—Food Contamination—Permissible Limits
BSI BS 7557-92. 1992 Limits of Metal Release from Painted Surfaces of Articles, Liable to Come into Contact with Foodstuffs (W). 8 pp.

—Paints
BSI BS 3900: Part B15-87. 1987 Methods of Test for Paints Group B: Tests Involving Chemcial Examination of Liquid Paints and Dried Paint Films Part B15: Rapid Method for Estimation of Lead in Liquid Paints. 6 pp.

BSI BS 4310-68. 1968 Amd 1 Permissible Limit in Low-Lead Paints and Similar Materials. 14 pp.

CGSB CAN2-1.500-75 METH 1-73. Methods of Test for Toxic Trace Elements in Protective Coatings Determination of Lead (Pb) in Low Concentration. 2 pp.

CNS K6490-80. Method of Test for Detection of Lead in Paint and Dried Paint Films (May)(5594).

CNS K6492-80. Method of Test for Determination of Low Concentrations of Lead in Paints (May)(5596).

CNS K6710-82. Method of Test for Lead in Lead-Free Paints (Feb)(8522).

Lead Content Analysis (Cont.)

—Paints (Cont.)
SAA AS 1580.501. 1-80. Paints and Related Materials—Methods of Test—Part 501.1: Soluble Lead Content (Gravimetric Method). 2 pp.

SNZ NZS/AS 1580. 501.1-80. Methods of Test for Paints and Related Materials Part 501.1: Soluble Lead Content (Gravimetric Method) (This is a Joint Standard with SAA AS 1580.501.1). 2 pp.

—Paints—Atomic Absorption Spectrometry
BSI BS 3900: Part B3-83. 1983 Group B: Tests Involving Chemical Examination of Liquid Paints and Dried Paint Films: Part B3: Determination of 'Soluble' Lead in Solid Matter in Liquid Paints: Meth for Use in Conj with the Control of Lead at Work Regul, 1980 (S.I. 1980 No. 1248). 8 pp.

BSI BS 3900: Part B4-86. 1986 Methods of Test for Paints Group B: Tests Involving Chemical Examination of Liquid Paints and Dried Paint Films Part B4: Determination of Total Lead in Paints and Similar Materials. 8 pp.

BSI BS 3900: Part B6-86. 1986 Methods of Test for Paints Group B: Tests Involving Chemical Examination of Liquid Paints and Dried Paint Films Part B6: Determination of Content 'Soluble' Lead Content. 9 pp.

ISO 3856 Pt 1-84. Paints and Varnishes—Determination of "Soluble" Metal Content—Determination of Lead Content—Flame Atomic Absorption Spectrometric Method and Dithizone Spectrophotometric Method Third Edition. 8 pp.

ISO 6503-84. Paints and Varnishes—Determination of Total Lead—Flame Atomic Absorption Spectrometric Method First Edition. 9 pp.

—Paints—Atomic Absorption Spectrophotometry
ISO 3856 Pt 1-84. Paints and Varnishes—Determination of "Soluble" Metal Content—Determination of Lead Content—Flame Atomic Absorption Spectrometric Method and Dithizone Spectrophotometric Method Third Edition. 8 pp.

—Paints—Identification Systems
EC 86/508/EEC-86. Commn. Directive Adapting to Tech. Progress for the Second Time Council Directive 77/728/EEC on the Approximation of the Laws, Regs. and Adm. Provisions of the Member States Rel. to the CL., Packaging and Labeling of Paints, Varnishes, Printing Inks, Adhesives and Similar Products. 1 p.

EC 89/451/EEC-89. Commission Directive Adapting to Tech. Prog. for the Third Time Coun. Directive 77/728/EEC on the Approximation of the Laws, Regs. and Ad. Provs. of the Member Sts. Rel. to the Class., Packaging and Labelling of Paints, Varnishes, Ptg. Inks, Adhesives and Similar Products. 2 pp.

—Paints—Spectrography
MOD UK M 9512/65. Spectrographic Determination of Lead in Paint Ashes and Similar Materials (No Information).

—Papers—Solvent Extraction
MOD UK DSTAN 13-10-68. Paper, Wrapping, Unglazed, and Paper, Wrapping, Unglazed, Lead Free Issue 1. 9 pp.

MOD UK M 6512/61. Examination of Paper Wrapping, Unglazed and Paper, Wrapping, Unglazed, LF Quality (Superseded by Def Stan 13-10).

—Paraffin Wax
MOD UK M 2009/67. Examination of Paraffin Wax.

—Petroleum Products—Atomic Absorption Spectrometry
CEN PREN 237-85. Liquid Petroleum Products, Determination of Low Lead Concentrations; Atomic Absorption Spectronmetric Method. 8 pp.

—Phosphoric Acids—Atomic Absorption Spectrometry
BSI BS 4258: Part 11-82. 1982 Methods of Test for Phosphoric Acid (Orthophosphoric Acid) for Industrial Use Part 11: Determination of Lead Content (Atomic Absorption Spectrometric Method). 4 pp.

ISO 6678-81. Phosphoric Acid for Industrial Use—Determination of Lead Content—Atomic Absorption Spectrometric Method First Edition. 4 pp.

—Pigments
CGSB 1-GP-71 METH 50.9-74. Methods of Testing Paints and Pigments Pigment Analysis Soluble Lead. 1 p.

CNS K6094-89. Method of Test for Red Lead (Pigment) (Jul)(1043). 2 pp.

Lead Content Analysis (Cont.)

—Precipitation—Voltametry
DIN ENGL 38406 Pt 16-90. German Standard Methods for the Examination of Water, Waste Water and Sludge; Cations (Group E); Determination of Zinc, Cadmium, Lead, Copper, Thallium, Nickel, Cobalt by Voltammetry (E16) (Mar). 8 pp.

—Rubber—Atomic Absorption Spectrometry
BSI BS 7164: Sec 30.1-90. 1990 Chemical Tests for Raw and Vulcanized Rubber Part 30: Methods for Determination of Lead Content Section 30.1: Atomic Absorption Spectrometry. 12 pp.

BSI BS 7165-91. 1991 Recommendations for Achievement of Quality in Software. 30 pp.

ISO 6101 Pt 2-86. Rubber—Determination of Metal Content—Flame Atomic Absorption Spectrometric Method—Part 2: Determination of Lead Content First Edition. 8 pp.

—Rubber—Atomic Absorption Spectroscopy
DIN ENGL 53599 Pt 1-78. Testing of Rubber and Elastomers; Determination of the Lead Content; Determination by Atomic Absorption Spectroscopy for Lead Contents up to 1000 mg/kg (0.1%) (May). 4 pp.

—Rubber—Electrogravimetric Analysis
DIN ENGL 53599 Pt 2-78. Testing of Rubber and Elastomers; Determination of the Lead Content; Electrogravimetric and Gravimetric Method for Lead Contents over 1000 mg/kg (0.1%) (May). 3 pp.

—Rubber—Gravimetric Analysis
DIN ENGL 53599 Pt 2-78. Testing of Rubber and Elastomers; Determination of the Lead Content; Electrogravimetric and Gravimetric Method for Lead Contents over 1000 mg/kg (0.1%) (May). 3 pp.

—Rubber—Photometry
DIN ENGL 53599 Pt 3-78. Testing of Rubber and Elastomers; Determination of the Lead Content; Photometric Method for Lead Contents up to 1000 mg/kg (0.1%) (May). 4 pp.

—Sodium Chloride—Atomic Absorption Spectrometry
BSI BS 7319: Part 8-90. 1990 Analysis of Sodium Chloride for Industrial Use Part 8: Method for Determination of Lead Content. 7 pp.

—Sodium Hexametaphosphate
MOD UK M 9520/66. Examination of Sodium Hexametaphosphate (Withdrawn).

—Solders—Photometry
BSI BS 3338: Part 5-61. 1961 Methods for the Sampling and Analysis of Tin and Tin Alloys Part 5: Determination of Lead in Ingot Tin and Tin-Antimony Solders (Photometric Method). 8 pp.

—Steels
CNS G2248-85. Method of Determination for Lead in Steel (Jul)(11305).

JIS G 1229-80. Methods for Determination of Lead in Steel.

—Steels—Atomic Absorption Spectrometry
BSI BS 6200: SUB SEC 3.16.4-87. 1987 Sampling and Analysis of Iron, Steel and Other Ferrous Metals Part 3: Methods of Analysis Section 3.16: Determination of Lead Subsection 3.16.4: Steel: Flame Atomic Absorption Spectrometric Method. 8 pp.

BSI BS 6200: SUB SEC 3.16.4-01. 1987 Amd 1 Sampling and Analysis of Iron, Steel and Other Ferrous Metals Part 3: Methods of Analysis Sec 3.16: Determination of Lead Subsec 3.16.4: Steel: Flame Atomic Absorption Spectrometric Method (AMD 7072) February 28, 1992 (V). 13 pp.

CEN EN 10 181-89. Chemical Analysis of Ferrous Materials Determination of Lead in Steels Flame Atomic Absorption Spectrometric Method. 6 pp.

DIN ENGL EN 10181-90. Chemical Analysis of Ferrous Materials; Determination of Lead in Steels; Flame Atomic Absorption Spectrometric Method (Apr). 7 pp.

SAA AS 1050.25-86. Methods for the Analysis of Iron and Steel—Part 25: Determination of Lead Content (Flame Atomic Absorption Spectrometric Method). 3 pp.

—Steels—Spectrophotometry
BSI BS 6200: SUB SEC 3.16.2-86. 1986 Sampling and Analysis of Iron, Steel and Other Ferrous Metals Part 3: Methods of Analysis Section 3.16: Determination of Lead Subsection 3.16.2: Steel: Spectrophotometric Method. 4 pp.

BSI BS 6200: SUB SEC 3.16.3-91. 1991 Sampling and Analysis of Iron, Steel and Other Ferrous Metals Part 3: Methods of Analysis Section 3.16: Determination of Lead Subsection 3.16.3: Steel: Spectrophotometric Method for Trace Amounts. 9 pp.

INDUSTRY STANDARDS

Lead

Lead Content Analysis (Cont.)

—Sulfuric Acid—Photometry
ISO 2717-73. Sulphuric Acid and Oleum for Industrial Use—Determination of Lead Content—Dithizone Photometric Method First Edition. 5 pp.

—Surface Waters—Voltametry
DIN ENGL 38406 Pt 16-90. German Standard Methods for the Examination of Water, Waste Water and Sludge; Cations (Group E); Determination of Zinc, Cadmium, Lead, Copper, Thallium, Nickel, Cobalt by Voltammetry (E16) (Mar). 8 pp.

—Tin Alloys—Photometry
BSI BS 3338: Part 5-61. 1961 Methods for the Sampling and Analysis of Tin and Tin Alloys Part 5: Determination of Lead in Ingot Tin and Tin-Antimony Solders (Photometric Method). 8 pp.

—Tin Coatings—Quantitative Analysis
BSI BS 6534-84. 1984 Quantitative Determination of Lead in Tin Coatings. 8 pp.

—Tin—Photometry
BSI BS 3338: Part 5-61. 1961 Methods for the Sampling and Analysis of Tin and Tin Alloys Part 5: Determination of Lead in Ingot Tin and Tin-Antimony Solders (Photometric Method). 8 pp.

—Toothpastes
BSI BS 5136-81. 1981 Toothpastes. 15 pp.

—Toys
CNS Z8016-10-86. Method of Test for Toy Safety (Testing for Lead and Other Hazardous Heavy Metal Contents) (Nov)(4798-10). 4 pp.

—Tribasic Lead Phosphosilicate
CNS K6572-80. Method of Test for Lead in Tribasic Lead Phosphosilicate (Aug)(6238).

—Varnishes—Colorimetric Analysis
MOD UK DEF-1053: METH NO. 81. Standard Methods of Testing Paint, Varnish, Lacquer and Related Products. Indices Method 81: Determination of Lead in 'Lead Free' Paints, Varnishes and Allied Products and Their Containers. (Withdrawn).

—Varnishes—Identification Systems
EC 86/508/EEC-86. Commn. Directive Adapting to Tech. Progress for the Second Time Council Directive 77/728/EEC on the Approximation of the Laws, Regs. and Adm. Provisions of the Member States Rel. to the CL., Packaging and Labeling of Paints, Varnishes, Printing Inks, Adhesives and Similar Products. 1 p.
EC 89/451/EEC-89. Commission Directive Adapting to Tech. Prog. for the Third Time Coun. Directive 77/728/EEC on the Approximation of the Laws, Regs. and Ad. Provs. of the Member Sts. Rel. to the Class., Packaging and Labelling of Paints, Varnishes, Ptg. Inks, Adhesives and Similar Products. 2 pp.

—Vegetable Oils
CNS N6105-78. Method of Test for Edible Vegetable Oils (Determination of Lead and Copper) (Aug)(4529). 3 pp.

—Vegetables—Atomic Absorption Spectrometry
ISO 6633-84. Fruits, Vegetables and Derived Products—Determination of Lead Content—Flameless Atomic Absorption Spectrometric Method First Edition. 5 pp.

—Waste Water—Atomic Absorption Spectrometry
DIN ENGL 38406 Pt 6-81. German Standard Methods for the Analysis of Water, Waste Water and Sludge; Cations (Group E); Determination of Lead (E 6) (May). 7 pp.

—Water—Atomic Absorption Spectrometry
BSI BS 6068: Sec 2.29-87. 1987 Water Quality Part 2: Physical, Chemical and Bio-Chemical Methods Section 2.29: Determination of Cobalt, Nickel, Copper, Zinc, Cadmium and Lead: Flame Atomic Absorption Spectrometric Methods. 13 pp.
CNS K9028-80. Method of Test for Lead in Water (Atomic Absorption, Direct) (Apr)(5462).
CNS K9034-80. Method of Test for Lead in Water (Atomic Absorption, Chelation-Extraction) (Apr)(5468).
DIN ENGL 38406 Pt 6-81. German Standard Methods for the Analysis of Water, Waste Water and Sludge; Cations (Group E); Determination of Lead (E 6) (May). 7 pp.
ISO 8288-86. Water Quality—Determination of Cobalt, Nickel, Copper, Zinc, Cadmium and Lead—Flame Atomic Absorption Spectrometric Methods First Edition. 13 pp.

Lead Content Analysis (Cont.)

—Water—Atomic Absorption Spectrometry (Cont.)
JIS K 0553-90. Testing Methods for Determination of Metallic Elements in Highly Purified Water. 27 pp.

—Water—Extraction Analysis
CNS K9034-80. Method of Test for Lead in Water (Atomic Absorption, Chelation-Extraction) (Apr)(5468).

—White Metals—Electrolytic Analysis
BSI BS 3338: Part 15-65. 1965 Methods for the Sampling and Analysis of Tin and Tin Alloys Part 15: Determination of Copper and Lead in White Metal Bearing Alloys (Electrodeposition Method). 8 pp.

—Zinc Alloys—Atomic Absorption Spectrometry
DIN ENGL 50551-90. Determination of the Lead, Cadmium and Copper Content of Zinc and Zinc Alloys by Atomic Absorption Spectrometry (Oct). 5 pp.
SAA AS 1329.7-80. Methods for the Analysis of Zinc and Zinc Alloys—Part 7: Determination of Lead Content—Flame Atomic Absorption Spectrometric Method. 10 pp.

—Zinc Alloys—Polarographic Analysis
BSI BS 3630: Part 8-71. 1971 Methods for the Sampling and Analysis of Zinc and Zinc Alloys: Lead and Cadmium in Zinc (Grades Zn1 and Zn2) and Zinc Alloys (Polarographic Method). 11 pp.
ISO 2576-72. Chemical Analysis of Zinc Alloys—Polarographic Determination of Lead and Cadmium in Zinc Alloys Containing Copper First Edition. 5 pp.

—Zinc—Atomic Absorption Spectrometry
DIN ENGL 50551-90. Determination of the Lead, Cadmium and Copper Content of Zinc and Zinc Alloys by Atomic Absorption Spectrometry (Oct). 5 pp.

—Zinc Oxides
CNS K6054-74. Method of Test for Zinc Oxide of Paints (Oct)(767). 3 pp.

—Zinc—Polarographic Analysis
BSI BS 3630: Part 7-67. 1967 Methods for the Sampling and Analysis of Zinc and Zinc Alloys Part 7: Lead in Zinc (Grades Zn3 and Zn4) (Polarographic Method). 8 pp.
BSI BS 3630: Part 8-71. 1971 Methods for the Sampling and Analysis of Zinc and Zinc Alloys: Lead and Cadmium in Zinc (Grades Zn1 and Zn2) and Zinc Alloys (Polarographic Method). 11 pp.
ISO 713-75. Zinc—Determination of Lead and Cadmium Contents—Polarographic Method First Edition. 4 pp.
ISO 715-75. Zinc—Determination of Lead Content—Polarographic Method First Edition. 4 pp.

—Zinc—Spectrochemical Analysis
JIS H 1108-89. Methods for Determination of Lead in Zinc Metal. 11 pp.

—Zirconium
JIS H 1673-85. Methods for Determination of Lead in Zirconium and Zirconium Alloys. 8 pp.

—Zirconium Alloys
JIS H 1673-85. Methods for Determination of Lead in Zirconium and Zirconium Alloys. 8 pp.

Lead Crystal
Use: Lead Glass

Lead Cyanamide Paints
See Also: Paints

—Metal—Corrosion Resistant
CNS K2158-87. Lead Cyanamide Anticorrosive Paint (Oct)(12131). 2 pp.
CNS K6920-87. Method of Test for Lead Cyanamide Anticorrosive Paint (Oct)(12132). 5 pp.
JIS K 5625-84. Lead Cyanamide Anticorrosive Paint. 12 pp.

Lead-2,4-Dihydroxybenzoate
See Also: Alcohols
MOD UK DSTAN 68-124-89. Lead 2, 4—Dihydroxybenzoate Issue 1. 11 pp.

Lead Dioxide
Use For: Lead Peroxide
CNS K7274-68. Chemical Reagent (Lead Dioxide) (Nov)(1773).
JIS K 8704-76. Lead Dioxide (Lead Peroxide).

—Explosives
MOD UK DSTAN 68-78-84. Lead Dioxide Issue 1. 6 pp.

Lead 2-Ethylhexanoate
See Also: Curing Agents; Esters
MOD UK DSTAN 68-130-90. Lead 2-Ethylhexanoate Issue 1. 8 pp.
MOD UK TS 524B. Lead 2-Ethylhexanoate (Superseded by Def Stan 68-130).

Lead 2-Ethylhexoate
CNS K1268-81. Lead 2-Ethylhexoate (Oct)(8018).
CNS K6697-81. Method of Test for Lead 2-Ethylhexoate (Oct)(8019).

Lead Glass
Use For: Crystal Glass; Glass Crystal; Lead Crystal
See Also: Glass; Glassware
BSI BS 3828-73. 1973 Amd 1 Crystal Glass. 15 pp.

—Chemical Analysis
BSI BS 2649: Part 3-88. 1988 Methods for the Analysis of Glass Part 3: Glasses of the Potassium Oxide-Lead Oxide-Silica Type. 16 pp.

—X-Ray Protection
JIS R 3701-90. Lead Glass for X-Ray Protection. 8 pp.

Lead II Chloride
Use: Plumbous Chloride

Lead Inorganic Compounds
Scope Note: Use a more specific term See: Lead Dioxide; Lead Monoxide; Lead Nitrates; Lead Sulfide

Lead Methacrylic Resins
Scope Note: For products made from lead methacrylic resins, see specific products

Lead Monoxide
Use For: Lead Oxide Yellow; Litharge; Yellow Lead
CNS K7275-66. Chemical Reagent (Lead Monoxide) (Dec)(1774).
JIS K 1456-70. Litharge (Lead Monoxide). 11 pp.
JIS K 8090-61. Lead Monoxide.
MOD UK TS 560A. Lead Monoxide, Yellow, Special (Superseded by Def Stan 13-124).

—Cements
MOD UK DEF-111. Lead Oxide, Yellow (Withdrawn).

—Insoluble Matter Content
CNS K6095-58. Method of Test for Yellow-Lead Paint Powder (Jun)(1045) (R 1968). 1 p.

—Lead Oxide Content
CNS K6095-58. Method of Test for Yellow-Lead Paint Powder (Jun)(1045) (R 1968). 1 p.

—Powder—Paints
CNS K2027-58. Yellow Lead Powder (Jun)(1044)(R 1968). 1 p.
CNS K6095-58. Method of Test for Yellow-Lead Paint Powder (Jun)(1045) (R 1968). 1 p.

—Propellants
MOD UK DSTAN 13-124-91. Lead Monoxide, Yellow, Special Issue 1. 8 pp.

—Volatile Matter Content
CNS K6095-58. Method of Test for Yellow-Lead Paint Powder (Jun)(1045) (R 1968). 1 p.

Lead Nitrates
CNS K7276-66. Chemical Reagent (Lead Nitrate) (Dec)(1775).
JIS K 8563-79. Lead Nitrate.
MOD UK DSTAN 68-71-84. Lead Nitrate Issue 1. 13 pp.

Lead Ores
See Also: Lead; Spiess; Zinc Ores

—Dry Matter Content
JIS M 8101-88. Methods for Sampling, Preparation and Determination of Moisture Content of Non-Ferrous Metal Bearing Ores. 25 pp.

—Moisture Content
JIS M 8101-88. Methods for Sampling, Preparation and Determination of Moisture Content of Non-Ferrous Metal Bearing Ores. 25 pp.

—Sampling
CNS M3032-81. Methods for Sampling of Metal Bearing Ores of Copper, Lead, Zinc, Tin, Gold, Silver and Others (Apr)(7280).
JIS M 8101-88. Methods for Sampling, Preparation and Determination of Moisture Content of Non-Ferrous Metal Bearing Ores. 25 pp.

INTERNATIONAL AND NON-U.S. NATIONAL STANDARDS
SUBJECT INDEX
Lead

Lead Oxide Content Analysis
—Lead Monoxide
CNS K6095-58. Method of Test for Yellow-Lead Paint Powder (Jun)(1045) (R 1968). 1 p.

—Organic Coatings—Insoluble Matter—Quantitative Analysis
CNS K6804-20-85. Method of Test for Organic Coating (Chemical Analysis) — Quantative Test of Lead Oxide in Solvent Insolubles (Jan)(10880-20).

Lead Oxide Red
Use: Lead Tetroxide

Lead Oxide Yellow
Use: Lead Monoxide

Lead Peroxide
Use: Lead Dioxide

Lead Pipes
See Also: Pipes
BSI BS 602 & 1085-70. 1970 Lead and Lead Alloy Pipes for Other Than Chemical Purposes. 26 pp.
CNS H3032-84. Lead Pipe (for General Purpose) (Dec)(2674). 3 pp.
JIS H 4313-56. Hard Lead Pipes. 5 pp.

—Drainpipes
CSA B67-1972. Lead Service Pipe, Waste Pipe, Traps, Bends and Accessories; (Rev 1-2 October 1977). 14 pp.
DIN ENGL 1263-66. Lead Waste Pipes and Bends for Drainage Systems (Sept). 1 p.

—Explosives
MOD UK DSTAN 47-24-87. Piping, Lead (for Explosive Filled Cord) Issue 2. 8 pp.

—Pressure—Water
DIN ENGL 1262-77. Pressure Pipes of Lead for Non-Potable Water Pipelines (Mar). 3 pp.

—Sampling
BSI BS 3908: Part 1-65. 1965 Amd 1 Methods for the Sampling and Analysis of Lead and Lead Alloys Part 1: Sampling of Ingot Lead, Lead Alloy Ingots, Sheet Pipe, and Cable Sheathing Alloys. 8 pp.

—Sewer
CSA B67-1972. Lead Service Pipe, Waste Pipe, Traps, Bends and Accessories; (Rev 1-2 October 1977). 14 pp.
DIN ENGL 1263-66. Lead Waste Pipes and Bends for Drainage Systems (Sept). 1 p.

—Water
CNS H3033-84. Lead Pipe (for Water Supply) (Dec)(2675). 4 pp.
CSA B67-1972. Lead Service Pipe, Waste Pipe, Traps, Bends and Accessories; (Rev 1-2 October 1977). 14 pp.

Lead Powder
See Also: Metal Powders
CNS K7674-83. Chemical Reagent (Lead Powder) (May)(10278).
MOD UK DSTAN 68-107-88. Lead Powder Issue 1. 8 pp.

Lead Salicylate
See Also: Esters
CNS K1271-81. Lead Salicylate (Oct)(8024).
CNS K6700-81. Method of Test for Lead Salicylate (Oct)(8025).

Lead Shields
See Also: Radiation Shields (Protective); Shields (Protective)
CNS J2091-83. Method of Test for Lead Equivalent for X-Ray Protectors (Jun)(10349). 2 pp.
JIS Z 4501-88. Testing Method of Lead Equivalent for X-Ray Protective Devices. 6 pp.

—Bricks
BSI BS 4513-69. 1969 Lead Bricks for Radiation Shielding. 24 pp.
JIS Z 4817-74. Radiation Shielding Lead Bricks.

—Enclosures
ISO 9404 Pt 1-91. Enclosures for Protection Against Ionizing Radiation—Lead Shielding Units for 150 mm, 200 mm and 250 mm Thick Walls—Part 1: Chevron Units of 150 mm and 200 mm Thickness First Edition. 49 pp.

—Filler Rods—Ships
MOD UK NES 124: Part 2-90. Requirements for Shielding Materials for Nuclear Purposes Part 2: Lead Slabs, Sheet, Filler Rods and Wool Issue 2 (06.90). 12 pp.

Lead Shields (Cont.)
—Ingots
BSI BS 3909-65. 1965 Ingot Lead for Radiation Shielding. 12 pp.

—Plates—Ships
MOD UK NES 124: Part 1-90. Requirements for Shielding Materials for Nuclear Purposes Part 1: Cast Lead and Steelplate Assemblies Issue 2 (06.90). 12 pp.

—Sheets
JIS K 6736-88. Lead-Containing Methacrylate Sheets for Protection Against X-Rays and Gamma-Rays. 12 pp.
JIS Z 4801-91. Lead Rubber Sheets and Lead Polyvinyl Chloride Sheets for X-Ray Shield. 8 pp.

—Sheets—Ships
MOD UK NES 124: Part 2-90. Requirements for Shielding Materials for Nuclear Purposes Part 2: Lead Slabs, Sheet, Filler Rods and Wool Issue 2 (06.90). 12 pp.

—Slabs—Ships
MOD UK NES 124: Part 2-90. Requirements for Shielding Materials for Nuclear Purposes Part 2: Lead Slabs, Sheet, Filler Rods and Wool Issue 2 (06.90). 12 pp.

—Wool—Ships
MOD UK NES 124: Part 2-90. Requirements for Shielding Materials for Nuclear Purposes Part 2: Lead Slabs, Sheet, Filler Rods and Wool Issue 2 (06.90). 12 pp.

Lead Silicochromate
—Chromium Trioxide Content
CNS K6536-80. Method of Test for Chromium Trioxide in Basic Lead Silica-Chromate (Jul)(5865).

—Lead Content
CNS K6535-80. Method of Test for Total Lead in Basic Lead Silica-Chromate (Jul)(5864).

—Pigments
CNS K2109-80. Basic Lead Silicochromate (Jul)(5874). 1 p.

—Silica Content
CNS K6534-80. Method of Test for Silica in Basic Lead Silica-Chromate (Jul)(5863).

Lead Silicochromate Alkyd Enamel Paints
See Also: Enamel Paints; Paints
—Metal—Weather Resistant
CGSB CAN/CGSB-1.167-M90. Exterior Basic Lead Silicochromate Enamel, Alkyd Type. 14 pp.

Lead Silicochromate Primers
CGSB CAN/CGSB-1.166-M90. Basic Lead Silicochromate Primer, Oil Alkyd Type. 17 pp.

Lead Stearate
MOD UK DSTAN 68-134-90. Lead Stearate Issue 1. 11 pp.

Lead Subacetate
See Also: Acetates
JIS K 8167-61. Lead Subacetate.

—Sugar Analysis
CNS K7278-66. Chemical Reagent (Lead Subacetate)(for Sugar Analysis) (Dec)(1777).

Lead Suboxide Paints
See Also: Paints
—Metal—Corrosion Inhibitive
CNS K2060-86. Lead Suboxide Anticorrosive Paint (Apr)(4909). 3 pp.
CNS K6844-86. Method of Test for Lead Suboxide Anticorrosive Paint (Apr)(11564).
JIS K 5623-84. Lead Suboxide Anticorrosive Paint. 12 pp.

Lead Sulfate
See Also: Sulfates
CNS K7279-66. Chemical Reagent (Lead Sulfate) (Dec)(1778).

—Pigments
CNS K2002-84. Lead Sulfate for Paint (Feb)(14). 2 pp.
CNS K6091-84. Method of Test for Lead Sulfate for Paint (Mar)(1039).

Lead Sulfate, Blue Basic
Use For: Blue Lead, Basic Sulfate See Also: Sulfates

Lead Sulfate, Blue Basic (Cont.)
—Pigments
CNS K2102-80. Blue Lead, Basic Sulfate (Jul)(5867). 1 p.

Lead Sulfide
—Lead Content—Solvent Extraction
SAA AS 4030.1-92. Methods for the Analysis of Lead Sulfide Concentrates—Part 1: Determination of Lead Content—Acid Dissolution Solvent Extraction EDTA Titration Method. 7 pp.

—Moisture Content—Gravimetric Analysis
ISO 9599-91. Copper, Lead and Zinc Sulfide Concentrates—Determination of Hygroscopic Moisture in the Analysis Sample—Gravimetric Method First Edition. 8 pp.

Lead Tetroxide
Use For: Lead Oxide Red; Red Lead
See Also: Lead Tetroxide Content Analysis; Pigments; Red Lead Primers
CNS K7277-66. Chemical Reagent (Lead Oxide, Red) (Red Lead Minium) (Dec)(1776).
JIS K 1457-70. Red Lead. 12 pp.
MOD UK DSTAN 68-48-77. Lead Oxide, Red Issue 1 (Withdrawn). 6 pp.

—Joint Sealants
BSI BS 217-61. 1961 Red Lead for Paints and Jointing Compounds. 17 pp.

—Pigments
BSI BS 217-61. 1961 Red Lead for Paints and Jointing Compounds. 17 pp.
ISO 510-77. Red Lead for Paints First Edition. 6 pp.
JIS K 5108-65. Red Lead (Pigment). 7 pp.

—Pigments—Colorimetric Analysis
CNS K6094-89. Method of Test for Red Lead (Pigment) (Jul)(1043). 2 pp.

—Pigments—Insoluble Matter Content
CNS K6094-89. Method of Test for Red Lead (Pigment) (Jul)(1043). 2 pp.

—Pigments—Lead Content
CNS K6094-89. Method of Test for Red Lead (Pigment) (Jul)(1043). 2 pp.

—Pigments—Particle Content
CNS K6094-89. Method of Test for Red Lead (Pigment) (Jul)(1043). 2 pp.

—Pigments—Red Lead Content
CNS K6094-89. Method of Test for Red Lead (Pigment) (Jul)(1043). 2 pp.

—Pigments—Soluble Matter Content
CNS K6094-89. Method of Test for Red Lead (Pigment) (Jul)(1043). 2 pp.

—Pigments—Volatile Matter Content
CNS K6094-89. Method of Test for Red Lead (Pigment) (Jul)(1043). 2 pp.

Lead Tetroxide Content Analysis
Use For: Red Lead Content Analysis
See Also: Lead Tetroxide
—Pigments
CGSB 1-GP-71 METH 50.3-80. Methods of Testing Paints and Pigments Pigment Analysis Red Lead with Graphite or Iron Oxide. 1 p.
CGSB 1-GP-71 METH 50.13-80. Methods of Testing Paints and Pigments Pigment Analysis Red Lead with Zinc Chromate Yellow and Iron Oxide. 2 pp.
CNS K6094-89. Method of Test for Red Lead (Pigment) (Jul)(1043). 2 pp.

Lead Wire
See Also: Motor Lead Wire; Wire
CNS C2087-80. Cambric Insulated Wires (Dec)(6645). 12 pp.
CNS C3105-80. Method of Test for Compensating Lead Wires (Dec)(6652).
CNS C4265-82. Compensating Lead Wires (Mar)(6651).
JIS C 1610-81. Compensating Lead Wires. 13 pp.
JIS C 3403-78. Cambric Insulated Wires.
JIS C 3403-71. Cambric Insulated Wires. 14 pp.

—Ballasts (Electric)—300 V
CSA C22.2 NO 116-1980. Coil-Lead Wires (R 1992); (Gen Instr 1 Thru 8). 52 pp.
CSA 1409 Bull. Electrical Bulletin 1409 July 28, 1983 to C22.2 NO 116. 9 pp.

—Ballasts (Electric)—600 V
CSA C22.2 NO 116-1980. Coil-Lead Wires (R 1992); (Gen Instr 1 Thru 8). 52 pp.
CSA 1227 Bull. Electrical Bulletin 1227 May 16, 1979 to C22.2 NO 116. 9 pp.

INDUSTRY STANDARDS

Lead Wire *(Cont.)*

—Ballasts (Electric)—600 V *(Cont.)*
CSA 1236 Bull. Electrical Bulletin 1236 September 12, 1979 to C22.2 NO 116. 9 pp.
CSA 1333 Bull. Electrical Bulletin 1333 August 17, 1981 to C22.2 NO 116. 9 pp.
CSA 1383 Bull. Electrical Bulletin 1383 November 29, 1982 to C22.2 NO 116. 9 pp.

—Cambric Insulated
JIS C 3403-78. Cambric Insulated Wires.
JIS C 3403-71. Cambric Insulated Wires. 14 pp.

—Dielectric Strength
CSA 1277 Bull. Electrical Bulletin 1277 July 17, 1980 to C22.2 NO 116. 2 pp.

—Electric Coils—300 V
CSA C22.2 NO 116-1980. Coil-Lead Wires (R 1992); (Gen Instr 1 Thru 8). 52 pp.
CSA 1409 Bull. Electrical Bulletin 1409 July 28, 1983 to C22.2 NO 116. 9 pp.

—Electric Coils—600 V
CSA C22.2 NO 116-1980. Coil-Lead Wires (R 1992); (Gen Instr 1 Thru 8). 52 pp.
CSA 1227 Bull. Electrical Bulletin 1227 May 16, 1979 to C22.2 NO 116. 9 pp.
CSA 1236 Bull. Electrical Bulletin 1236 September 12, 1979 to C22.2 NO 116. 9 pp.
CSA 1333 Bull. Electrical Bulletin 1333 August 17, 1981 to C22.2 NO 116. 9 pp.
CSA 1383 Bull. Electrical Bulletin 1383 November 29, 1982 to C22.2 NO 116. 9 pp.

—Electrical Codes
CSA 1334 Bull. Electrical Bulletin 1334 August 28, 1981 to C22.2 NO 16. 1 p.
CSA 851 Bull. Electrical Bulletin 851 November 17, 1971 to C22.2 NO 38. 2 pp.
CSA 1213A Bull. Electrical Bulletin 1213A October 25, 1979 to C22.2 NO 38. 3 pp.
CSA 851 Bull. Electrical Bulletin 851 November 17, 1971 to C22.2 NO 116. 2 pp.
CSA 1213A Bull. Electrical Bulletin 1213A October 25, 1979 to C22.2 NO 116. 3 pp.
CSA 1334 Bull. Electrical Bulletin 1334 August 28, 1981 to C22.2 NO 116. 1 p.

—Electrical Equipment—Automotive
CNS D2153-86. Lead Wires for Electrical Equipments for Automobiles (Nov)(9380).

—Ethylene Insulated
CSA C22.2 NO 116-1980. Coil-Lead Wires (R 1992); (Gen Instr 1 Thru 8). 52 pp.

—Ethylene Propylene Rubber Insulated—600 V
CSA C22.2 NO 116-1980. Coil-Lead Wires (R 1992); (Gen Instr 1 Thru 8). 52 pp.
CSA 1236 Bull. Electrical Bulletin 1236 September 12, 1979 to C22.2 NO 116. 9 pp.
CSA 1333 Bull. Electrical Bulletin 1333 August 17, 1981 to C22.2 NO 116. 9 pp.
CSA 1383 Bull. Electrical Bulletin 1383 November 29, 1982 to C22.2 NO 116. 9 pp.

—Impedance Bonds
JIS E 3604-77. Lead Wire for Impedance Bonds.

—Polychloroprene Insulated
CSA C22.2 NO 116-1980. Coil-Lead Wires (R 1992); (Gen Instr 1 Thru 8). 52 pp.

—Polyethylene Insulated
CSA C22.2 NO 116-1980. Coil-Lead Wires (R 1992); (Gen Instr 1 Thru 8). 52 pp.

—PVC Insulated
CSA C22.2 NO 116-1980. Coil-Lead Wires (R 1992); (Gen Instr 1 Thru 8). 52 pp.

—Rubber Insulated
CSA C22.2 NO 116-1980. Coil-Lead Wires (R 1992); (Gen Instr 1 Thru 8). 52 pp.

—Rubber Insulated—Electrical Equipment
CNS C2066-85. Rubber Insulated Lead Wires for Electric Machinery and Apparatus (May)(5747). 15 pp.
JIS C 3315-87. Rubber Insulated Lead Wires for Electric Machinery and Apparatus. 20 pp.

—Silicone Rubber Insulated
CSA C22.2 NO 116-1980. Coil-Lead Wires (R 1992); (Gen Instr 1 Thru 8). 52 pp.

—Solenoids—300 V
CSA C22.2 NO 116-1980. Coil-Lead Wires (R 1992); (Gen Instr 1 Thru 8). 52 pp.
CSA 1409 Bull. Electrical Bulletin 1409 July 28, 1983 to C22.2 NO 116. 9 pp.

—Solenoids—600 V
CSA C22.2 NO 116-1980. Coil-Lead Wires (R 1992); (Gen Instr 1 Thru 8). 52 pp.
CSA 1227 Bull. Electrical Bulletin 1227 May 16, 1979 to C22.2 NO 116. 9 pp.
CSA 1236 Bull. Electrical Bulletin 1236 September 12, 1979 to C22.2 NO 116. 9 pp.
CSA 1333 Bull. Electrical Bulletin 1333 August 17, 1981 to C22.2 NO 116. 9 pp.
CSA 1383 Bull. Electrical Bulletin 1383 November 29, 1982 to C22.2 NO 116. 9 pp.

—Teflon/Fiberglass Insulated—600 V
CSA C22.2 NO 116-1980. Coil-Lead Wires (R 1992); (Gen Instr 1 Thru 8). 52 pp.

—Thermocouples
JIS C 2533-83. Conductors for Compensating Lead Wires of Thermocouples. 18 pp.

—Thermoplastic Insulated
CSA C22.2 NO 116-1980. Coil-Lead Wires (R 1992); (Gen Instr 1 Thru 8). 52 pp.

—Thermoplastic Insulated—300 V
CSA C22.2 NO 116-1980. Coil-Lead Wires (R 1992); (Gen Instr 1 Thru 8). 52 pp.
CSA 1409 Bull. Electrical Bulletin 1409 July 28, 1983 to C22.2 NO 116. 9 pp.

—Thermoplastic Insulated—600 V
CSA C22.2 NO 116-1980. Coil-Lead Wires (R 1992); (Gen Instr 1 Thru 8). 52 pp.
CSA 1227 Bull. Electrical Bulletin 1227 May 16, 1979 to C22.2 NO 116. 9 pp.

—Transformers
CSA C22.2 NO 116-1980. Coil-Lead Wires (R 1992); (Gen Instr 1 Thru 8). 52 pp.

—Transformers—300 V
CSA C22.2 NO 116-1980. Coil-Lead Wires (R 1992); (Gen Instr 1 Thru 8). 52 pp.
CSA 1409 Bull. Electrical Bulletin 1409 July 28, 1983 to C22.2 NO 116. 9 pp.

—Transformers—600 V
CSA C22.2 NO 116-1980. Coil-Lead Wires (R 1992); (Gen Instr 1 Thru 8). 52 pp.
CSA 1227 Bull. Electrical Bulletin 1227 May 16, 1979 to C22.2 NO 116. 9 pp.
CSA 1236 Bull. Electrical Bulletin 1236 September 12, 1979 to C22.2 NO 116. 9 pp.
CSA 1333 Bull. Electrical Bulletin 1333 August 17, 1981 to C22.2 NO 116. 9 pp.
CSA 1383 Bull. Electrical Bulletin 1383 November 29, 1982 to C22.2 NO 116. 9 pp.

—Windings—300 V
CSA C22.2 NO 116-1980. Coil-Lead Wires (R 1992); (Gen Instr 1 Thru 8). 52 pp.
CSA 1409 Bull. Electrical Bulletin 1409 July 28, 1983 to C22.2 NO 116. 9 pp.

—Windings—600 V
CSA C22.2 NO 116-1980. Coil-Lead Wires (R 1992); (Gen Instr 1 Thru 8). 52 pp.
CSA 1227 Bull. Electrical Bulletin 1227 May 16, 1979 to C22.2 NO 116. 9 pp.
CSA 1236 Bull. Electrical Bulletin 1236 September 12, 1979 to C22.2 NO 116. 9 pp.
CSA 1333 Bull. Electrical Bulletin 1333 August 17, 1981 to C22.2 NO 116. 9 pp.
CSA 1383 Bull. Electrical Bulletin 1383 November 29, 1982 to C22.2 NO 116. 9 pp.

Leaded Brasses
Use For: Leaded Red Brass *See Also:* Brasses; Copper Alloys

—Plates
CNS H3066-84. Leaded Brass Sheets and Plates, Strips and Coiled Sheets (Oct)(4384).

—Sheets
CNS H3066-84. Leaded Brass Sheets and Plates, Strips and Coiled Sheets (Oct)(4384).

—Strips
CNS H3066-84. Leaded Brass Sheets and Plates, Strips and Coiled Sheets (Oct)(4384).

Leaded Bronze Castings
See Also: Bronze Castings; Castings; Leaded Tin Bronze Castings
CNS H3059-89. Leaded Bronze Castings (Jan)(4127).

—Ingots
CNS H3056-89. Leaded Bronze Ingots for Castings (Jan)(4083).
JIS H 2207-85. Leaded Bronze Ingots for Castings. 6 pp.

Leaded Bronzes
See Also: Copper Alloys

—Copper Content
BSI BS 1748: Parts 11-12-64. 1964 Methods for the Analysis of Copper Alloys Parts 11-12: The Determination of Copper and Lead in Leaded Bronze Alloys. 12 pp.

—Lead Content
BSI BS 1748: Parts 11-12-64. 1964 Methods for the Analysis of Copper Alloys Parts 11-12: The Determination of Copper and Lead in Leaded Bronze Alloys. 12 pp.

Leaded Red Brass
Use: Leaded Brasses

Leaded Tin Bronze Castings
See Also: Bronze Castings; Castings; Leaded Bronze Castings
DIN ENGL 1716-81. Copper-Lead-Tin Casting Alloys; (Cast Tin-Lead Bronze); Castings (Nov). 6 pp.
JIS H 5115-88. Leaded Tin Bronze Castings. 9 pp.

—Chemical Analysis
DIN ENGL 1716-81. Copper-Lead-Tin Casting Alloys; (Cast Tin-Lead Bronze); Castings (Nov). 6 pp.

—Mechanical Properties
DIN ENGL 1716-81. Copper-Lead-Tin Casting Alloys; (Cast Tin-Lead Bronze); Castings (Nov). 6 pp.
DIN ENGL 1716 Suppl. 1-81. Copper-Lead-Tin Casting Alloys; (Cast Tin-Lead Bronze); Castings; Reference Data on Mechanical and Physical Properties (Nov). 3 pp.

—Physical Properties
DIN ENGL 1716 Suppl. 1-81. Copper-Lead-Tin Casting Alloys; (Cast Tin-Lead Bronze); Castings; Reference Data on Mechanical and Physical Properties (Nov). 3 pp.

Leaded Zinc Oxide
See Also: Zinc Oxides
CNS K2121-80. Leaded Zinc Oxide (Aug)(6236). 1 p.

Leading Reins (Babies)
Use For: Walking Reins (Babies) *See Also:* Child Restraining Devices; Safety Belts

—Safety Belts
BSI BS 3785-64. 1964 Amd 3 Webbing Safety Harness for Baby Carriages and Chairs and Walking Reins. 13 pp.
BSI BS 6684-89. 1989 Amd 1 Safety Harnesses (Including Detachable Walking Reins) for Restraining Children When in Perambulators (Baby Carriages), Pushchairs and High Chairs and When Walking (AMD 6531) June 29, 1990. 16 pp.
SAA AS 3747-89. Harnesses for Use in Prams, Strollers, and High Chairs (Including a Detachable Walking Rein) (This is a Joint Standard with SANZ NZS 3747). 8 pp.
SNZ NZS/AS 3747-89. Harnesses for Use in Prams, Strollers and High Chairs, (Including a Detachable Walking Rein) (This is a Joint Standard with SAA AS 3747). 8 pp.

Leads (Electrical)
Use: Electrical Leads

Leadthrough Terminals
Use: Feedthrough Insulators

Leaf Chains
Use: Lifting Chains

Leaf Springs
Use For: Flat Springs *See Also:* Helical Springs; Spiral Springs; Springs (Elastic); Torsion Springs
CNS B1271-81. Design of Leaf Spring (Sep)(7859).
JIS B 2701-86. Leaf Springs.
JIS B 2701-58. Leaf Springs (R 1970). 8 pp.
JIS B 2710-86. Design of Leaf Springs. 24 pp.

—Automobiles
CNS D2160-82. Leaf Springs for Automobiles (Nov)(9595).

—Bolts
CNS B2599-81. Center Bolts for Laminated (Leaf) Springs (Sep)(7860). 3 pp.

—Clips—Alignment
CNS B2600-81. Alignment Clips for Laminated (Leaf) Spring (Sep)(7861). 3 pp.

—Rolling Stock
CNS B2530-82. Laminated Springs; Leaf Springs for Railway Rolling Stock (Nov)(6644).

INTERNATIONAL AND NON-U.S. NATIONAL STANDARDS
SUBJECT INDEX

Leaf Springs (Cont.)
—Steel
DIN ENGL 4620-54. Spring Steel; Hot Rolled for the Production of Laminated Springs (Apr). 2 pp.
DIN ENGL 59145-85. Hot Rolled Steel Strip with Semicircular Edges for Leaf Springs; Dimensions, Masses, Permissible Deviations, Moments of Inertia (June). 5 pp.

—Steel—Central Dimples
DIN ENGL 1571-79. Central Dimples for Spring Leaves (Jan). 1 p.

—Trailers—Agricultural
CNS B2674-82. Agricultural Machinery; Laminated Leaf Spring for Trailers (Jun)(8931).

Leak Detectors
See Also: Flow Measurement; Gas Detectors

—Calibration
BSI BS 5914-80. 1980 Methods of Calibrating Leak-Detectors of the Mass-Spectrometer Type Used in the Field of Vacuum Technology. 13 pp.
ISO 3530-79. Vacuum Technology—Mass Spectrometer-Type Leak-Detector Calibration First Edition. 12 pp.
JIS Z 8754-88. Mass-Spectrometer-Type Leak-Detector Calibration. 9 pp.

—Mass Spectrometers—Partial Pressure
DIN ENGL 28410-68. Vacuum Technology; Mass Spectrometer Partial Pressure Gauges; Definitions, Characteristics, Operating Conditions (Nov). 3 pp.

—Storage Tanks
DIN ENGL 6618 Pt 3-89. Vertical Double-Wall Steel Tanks with Liquid-Based Leak Detection System, for the Above Ground Storage of Flammable and Non-Flammable Water Polluting Liquids (Sept). 4 pp.

Leak Testing
Use: Leakage

Leakage
Use For: Leak Testing; Leaks See Also: Air Leakage; Fuel Leakage; Gas Permeability; Hermetic Seals; Porosity; Seepage

—Air Filters
SAA AS 1132.9-73. Methods of Test for Air Filters for Use in Air Conditioning and General Ventilation—Part 9: Determination of Local Area Flaws and Pinhole Leaks.

—Bubble Emission
JIS Z 2329-91. Methods for Bubble Leak Testing. 10 pp.

—Cables (Electric)
DIN VDE 0472 Pt 505-83. Testing of Cables and Insulated Flexible Cords; Loss Factor, Dielectric Loss Coefficient and Leakage (Apr). 9 pp.

—Cans—Food
CNS N6205-87. Method of Test for Canned Food: Leaking Test (Feb)(11867). 2 pp.

—Condoms
CEN PREN 600-5-91. Latex Rubber Condoms—Part 5: Testing for Perforations. 10 pp.
ISO 4074 Pt 5-84. Rubber Condoms—Part 5: Testing for Holes Second Edition. 3 pp.

—Connectors
CNS C6176-88. Method of Test for Low Frequency (Below 3 MHz) Electrical Connector (TP-2 Air Leakage) (May)(9236).

—Cords (Electric)
DIN VDE 0472 Pt 505-83. Testing of Cables and Insulated Flexible Cords; Loss Factor, Dielectric Loss Coefficient and Leakage (Apr). 9 pp.

—Electrical Components
SAA AS 1099.2Q-80. Basic Environmental Testing Procedures for Electrotechnology—Part 2: Tests—Part 2Q: Sealing. 20 pp.
SAA AS 1099.3.6-80. Basic Environmental Testing Procedures for Electrotechnology—Part 3: Background Information—Part 3.6: Section 6 Test Q—Sealing. 4 pp.

—Electrical Components—Environmental Testing
BSI BS 2011: Part 2.1Q-81. 1981 Amd 1 Basic Environmental Testing Procedures Part 2.1: Tests Part 2.1Q: Test Q. Sealing (Amd 6093) November 30, 1990. 41 pp.
CENELEC HD 323.2.17 S2-87. Basic Environmental Testing Procedures—Part 2: Tests. Test Q: Sealing. 3 pp.

Leakage (Cont.)
—Electrical Components—Environmental Testing (Cont.)
CENELEC HD 323.2.17 S3-88. Basic Environmental Testing Procedures-Part 2: Tests. Test Q: Sealing. 3 pp.
CENELEC HD 323.2.17 S4-90. Basic Environmental Testing Procedures Part 2: Tests Test Q: Sealing. 3 pp.
IEC 68 Pt 2-17-78. Basic Environmental Testing Procedures Part 2: Tests—Test Q: Sealing Third Edition; (Amendment 4-1991 Incorporating Amendments 1 Thru 3). 118 pp.
JIS C 0026-89. Basic Environmental Testing Procedures Part 2: Tests, Test Q: Sealing. 34 pp.
SNZ IEC 68: Part 2-17-78. Basic Environmental Testing Procedures Part 2-17: Test Q: Sealing Amend: 1. 71 pp.

—Electromechanical Components
IEC 512 Pt 7-93. Electromechanical Components for Electronic Equipment; Basic Testing Procedures and Measuring Methods Part 7: Mechanical Operating Tests and Sealing Tests Third Edition. 40 pp.

—Electronic Equipment
CNS C6018-85. Sealing Test Methods for Electronic Components (Apr)(3631). 2 pp.

—Gas Cutting Equipment
BSI BS 7278-90. (WITHDRAWN) 1990 Gas Tightness of Equipment for Gas Welding and Allied Processes (Superseded by BS EN 29090: 1992). 8 pp.
BSI BS EN 29090-92. 1992 Gas Tightness of Equipment for Gas Welding and Allied Processes (ISO 9090: 1989) (F). 11 pp.
CEN EN 29090-92. Gas Tightness of Equipment for Gas Welding and Allied Processes. 6 pp.
ISO 9090-89. Gas Tightness of Equipment for Gas Welding and Allied Processes First Edition. 7 pp.

—Gas—Fiber Optic Cables
CNS C6329-89. Method of Test for Fiber Optic Devices (FOTP-100 Gas Leakage Test for Gas—Blocked Fiber Optic Cables) (Jul)(12560).

—Gas Welding Equipment
BSI BS 7278-90. (WITHDRAWN) 1990 Gas Tightness of Equipment for Gas Welding and Allied Processes (Superseded by BS EN 29090: 1992). 8 pp.
BSI BS EN 29090-92. 1992 Gas Tightness of Equipment for Gas Welding and Allied Processes (ISO 9090: 1989) (F). 11 pp.
CEN EN 29090-92. Gas Tightness of Equipment for Gas Welding and Allied Processes. 6 pp.
ISO 9090-89. Gas Tightness of Equipment for Gas Welding and Allied Processes First Edition. 7 pp.

—Glass Joints
CNS R3091-80. Leakage Test for Conical Joints (Apr) (5481).

—Helium
JIS Z 2330-92. Standard Recommended Guide for the Selection of Helium Leak Testing. 14 pp.
JIS Z 2331-92. Method for Helium Leak Testing. 53 pp.

—Hoses
BSI BS 5173: Sec 103.12-88. 1988 Rubber and Plastics Hoses and Hose Assemblies Part 103: Physical Tests Section 103.12: Determination of Vapour Transmission of Liquids Through Walls. 6 pp.
ISO 8308-93. Rubber and Plastics Hoses and Tubing—Determination of Transmission of Liquids Through Hose and Tubing Walls Second Edition. 8 pp.

—Liquefied Petroleum Gas
CNS Z3021-81. Leakage Test for Liquefied Petroleum Gas Alarm (Jun)(7536).

—Nuclear Containment Structures
CSA CAN3-N287.6-M80. Pre-Operational Proof and Leakage Rate Testing Requirements for Concrete Containment Structures for CANDU Nuclear Power Plants; (Gen Instr 1 Thru 2). 15 pp.
CSA CAN3-N287.7-M80. In-Service Examination and Testing Requirements for Concrete Containment Structures for CANDU Nuclear Power Plants; (Gen Instr 1). 16 pp.

—Ostomy Collection Bags
ISO 8670 Pt 2-91. Ostomy Collection Bags—Part 2: Determination of Freedom from Leakage First Edition. 5 pp.

—Pipe Fittings
BSI BS 2782:Pt11: METH 1112A-89. 1989 Methods of Testing Plastics Part 11: Thermoplastic Pipes, Fittings and Valves Method 1112A: Leaktight-ness of Thermoplastics Pipes and Fittings for Non-Pressure Applications. 4 pp.

Leakage (Cont.)
—Pipes
BSI BS 2782:Pt11: METH 1112A-89. 1989 Methods of Testing Plastics Part 11: Thermoplastic Pipes, Fittings and Valves Method 1112A: Leaktight-ness of Thermoplastics Pipes and Fittings for Non-Pressure Applications. 4 pp.

—Pressure Pipe Fittings
ISO 3603-77. Fittings for Unplasticized Polyvinyl Chloride (PVC) Pressure Pipes with Elastic Sealing Ring Type Joints—Pressure Test for Leakproofness First Edition. 4 pp.
ISO 3604-76. Fitting for Unplasticized Polyvinyl Chloride (PVC) Pressure Pipes with Elastic Sealing Ring Type Joints—Pressure Test for Leakproofness Under Conditions of External Hydraulic Pressure First Edition. 4 pp.

—Pressure Pipe Joints
CEN PREN 911-92. Plastics Piping Systems—Elastomeric Sealing Ring Type Joints and Mechanical Joints for Thermoplastics Pressure Piping—Test Method for Leaktightness Under External Hydrostatic Pressure. 6 pp.
CEN PREN 1016-93. Plastics Piping Systems—Unplasticized Poly(Vinyl Chloride) (PVC-U) End-Load Bearing Double Socket Joints—Test Method for Leaktightness and Strength While Subjected to Bending and Internal Pressure. 8 pp.
CEN PREN 1119-93. Plastics Piping Systems Joints for Glass-Reinforced Thermosetting Plastics (GRP) Pipes and Fittings—Test Method for Leaktightness and Resistance to Damage of Flexible Joints. 11 pp.
ISO 3458-76. Assembled Joints Between Fittings and Polyethylene (PE) Pressure Pipes—Test of Leakproofness Under Internal Pressure First Edition. 4 pp.
ISO 3503-76. Assembled Joints Between Fittings and Polyethylene (PE) Pressure Pipes—Test of Leakproofness Under Internal Pressure When Subjected to Bending First Edition. 4 pp.

—Radioactive Materials—Containers
ISO 2855-76. Radioactive Materials—Packagings—Tests for Contents Leakage and Radiation Leakage First Edition; (Amendment Slip-1976). 7 pp.
ISO TR4826-79. Sealed Radioactive Sources—Leak Test Methods First Edition. 4 pp.
ISO 9978-92. Radiation Protection—Sealed Radioactive Sources—Leakage Test Methods First Edition. 15 pp.

—Residential Buildings
JIS A 1704-76. Method of Test for Leakage of Equipment Units for Dwellings.

—Respirators
JIS T 8159-83. Leakage Rate Testing Method for Facepieces of Respirators.

—Seals—Fiber Optics
CNS C6120-87. Method of Test for Fiber Optic Devices (FOTP-23 Air Leakage Testing of Fiber Optic Component Seals) (Apr)(7642).

—Smoke—Fire Dampers
JIS A 1314-92. Smoke-Proof Test Method for Fire Damper. 9 pp.

—Smoke—Fire Doors
DIN ENGL 18095 Pt 2-91. Smoke Control Doors; Type Testing for Durability and Leakage (Mar). 8 pp.

—Urinary Collection Bags
ISO 8669 Pt 4-90. Urine Collection Bags—Part 4: Determination of Freedom from Leakage First Edition. 4 pp.

—Valves
CEN PREN 917-92. Plastics Piping Systems—Thermoplastics Valves—Test Methods for Resistance to Internal Pressure. 9 pp.

—Watch Batteries
ISO TR10220-89. Batteries for Watches—Leakage Tests First Edition. 6 pp.

—Waveguides
CENELEC HD 138 S2-90. Sealing Test for Professional Waveguide Tubing and Assemblies. 4 pp.
IEC 261-89. Sealing Test for Pressurized Waveguide Tubing and Assemblies Second Edition. 22 pp.

Leakage Currents
See Also: Leaky Cables

—Medical Electrical Equipment
CSA CAN/CSA-C22. 2NO601.1-M90. Medical Electrical Equipment, Part 1: General Requirements for Safety; (Gen Instr 1). 240 pp.

Leakage Currents (Cont.)

—Medical Electrical Equipment—Testing
CSA CAN/CSA-C22. 2NO601.1-M90. Medical Electrical Equipment, Part 1: General Requirements for Safety; (Gen Instr 1). 240 pp.

—Patient—Medical Electrical Equipment
CSA CAN/CSA-C22. 2NO601.1-M90. Medical Electrical Equipment, Part 1: General Requirements for Safety; (Gen Instr 1). 240 pp.

Leakage Flux

—Ferromagnetic Materials
DIN ENGL 54136 Pt 1-88. Non-Destructive Testing; Magnetic Leakage Flux Testing by Scanning with Probes; Principles (Oct). 4 pp.

—Steel Bars
JIS Z 2319-91. Methods for Magnetic Leakage Flux Testing. 9 pp.

—Steel Tubes
JIS Z 2319-91. Methods for Magnetic Leakage Flux Testing. 9 pp.

Leakage Paths
Use For: Creepage Distances

—Controlgear
CNS C1046-82. Clearances and Creepage Distances for Control Gear (May)(4380).

—Electrical Equipment
DIN VDE 0110 Pt 2-89. Insulation Co-Ordination for Equipment Within Low-Voltage Systems; Dimensioning of Clearances and Creepage Distances (Jan). 15 pp.

Leaks
Use: Leakage

Leaky Cables
See Also: Coaxial Cables; Leakage Currents; Radiating Cables; Telephone Cables; Telephone Lines; Transmission Lines

—Bifilar—Land Mobile Services
CCIR Report 902-1-90. Leaky-Feeder Systems in the Land Mobile Service—Section 8A—Land Mobile Service and Related Subjects. 10 pp.

—Coaxial—Land Mobile Services
CCIR Report 902-1-90. Leaky-Feeder Systems in the Land Mobile Service—Section 8A—Land Mobile Service and Related Subjects. 10 pp.

—Land Mobile Services
CCIR Report 902-1-90. Leaky-Feeder Systems in the Land Mobile Service—Section 8A—Land Mobile Service and Related Subjects. 10 pp.

Leaky Feeder Cables
Use: Leaky Cables

Learjet Series Aircraft
See Also: Aircraft

—Accidents
CAA. Gates Learjet (World Airline Accident Summary). 1 p.
CAA. Gates Learjet. 2 pp.

—Foreign Airworthiness Directives
CAA. Learjet Series Aircraft (Foreign Airworthiness Directives). 1 p.

Learjet 25B Aircraft
See Also: Aircraft

—Antenna Positions
CAA. Learjet 25B (Approved Aerial Positions). 1 p.

Learjet 35A Aircraft
See Also: Aircraft

—Antenna Positions
CAA. Learjet 35A and 36A (Approved Aerial Positions). 1 p.

—Certification
CAA. Gates Learjet Models 35A and 36A. 16 pp.

—Data Sheets
CAA FA23 ISSUE 1. Gates Learjet 35A and 36A. 3 pp.

Learjet 36A Aircraft
See Also: Aircraft

—Antenna Positions
CAA. Learjet 35A and 36A (Approved Aerial Positions). 1 p.

—Certification
CAA. Gates Learjet Models 35A and 36A. 16 pp.

—Data Sheets
CAA FA23 ISSUE 1. Gates Learjet 35A and 36A. 3 pp.

Learjet 55 Aircraft
See Also: Aircraft

—Antenna Positions
CAA. Learjet 55 (Approved Aerial Positions). 1 p.

Leased Circuits
Use For: Leased Lines See Also: International Leased Circuits; Private Leased Circuits; Telecommunication Circuits; Telephone Circuits

—Data Transmission—Circuit Noise
CCITT RECMN G.143-89. Circuit Noise and the Use of Compandors—General Characteristics of International Telephone Connections and Circuits (Study Groups XII and XV) 5 pp. 5 pp.

—FAX Communications
CCITT RECMN H.43-89. Document Facsimile Transmissions on Leased Telephone-Type Circuits—Line Transmission of Non-Telephone Signals—Transmission of Sound-Programme and Television Signals (Study Group XV) 1 pp. 1 p.
CCITT RECMN T.10-89. Document Facsimile Transmissions on Leased Telephone-Type Circuits—Terminal Equipment and Protocols for Telematic Services (Study Group VIII) 2 pp. 2 pp.

—Maintenance
CCITT FASCICLE IV.2. Maintenance of International Telegraph, Phototelegraph and Leased Circuits Maintenance of Mantime Satellite and Data Transmission Systems Recommendations M.800—M.1375 (Study Group IV). 130 pp.

—Modems—Public Data Networks—Data Signaling Rates
CCITT RECMN V.36-89. Modems for Synchronous Data Transmission Using 60-108 kHz Group Band Circuits—Data Communication over the Telephone Network (Study Group XVII) 10 pp. 10 pp.

—Open Network Provision
CENELEC ETR 038-92. Business Telecommunications (BT); Open Network Provision (ONP) Technical Requirements Standardisation Requirements for ONP Leased Lines. 158 pp.
CENELEC ETR 087-93. Business Telecommunications (BT); Open Network Provision (ONP) Technical Requirements; Standardisation Requirements for ONP Leased Lines Higher Order Leased Lines. 48 pp.
ETSI ETR 038-92. Business Telecommunications (BT); Open Network Provision (ONP) Technical Requirements Standardisation Requirements for ONP Leased Lines. 158 pp.
ETSI ETR 087-93. Business Telecommunications (BT); Open Network Provision (ONP) Technical Requirements; Standardisation Requirements for ONP Leased Lines Higher Order Leased Lines. 48 pp.

—Outband Signaling
CCITT FASCICLE III.6-89. Line Transmission of Non-Telephone Signals Transmission of Sound-Programme and Television Signals—Series H and J Recommendations. 237 pp.

—Signaling Systems—Private Automatic Branch Exchanges
CCITT RECMN Q.8-89. Signalling Systems to Be Used for International Manual and Automatic Working on Analogue Leased Circuits—General Recommendations on Telephone Switching and Signalling—Functions and Infor. Flows for Services in the ISDN—Supplements (Study Group XI) 17 pp. 17 pp.

—Signaling Systems—Telephone Exchanges
CCITT RECMN Q.8-89. Signalling Systems to Be Used for International Manual and Automatic Working on Analogue Leased Circuits—General Recommendations on Telephone Switching and Signalling—Functions and Infor. Flows for Services in the ISDN—Supplements (Study Group XI) 17 pp. 17 pp.

—Speech—Phototelegraphy
CCITT RECMN F.85-89. Message Handling Service; Intercommunication Between the IPM Service and the Telex Service—Telegraph and Mobile Services Operations and Quality of Service (Study Group I) 4 pp (Same as Recmn F.421) (Renumbered from Recmn F.75). 4 pp.

—Telephone Signals
CCITT RECMN H.12-89. Characteristics of Telephone-Type Leased Circuits—Line Transmission of Non-Telephone Signals—Transmission of Sound-Programme and Television Signals (Study Group XV) 1 pp. 1 p.

Leased Lines
Use: Leased Circuits

Leasing
Scope Note: See the subheading Leasing under specific types of equipment or services

Leather
Scope Note: For additional listings, see also specific products made from leather See Also: Animal Products; Chamois Leather; Hides; Leather Working Equipment; Leathercloth; Pelts; Tannery Equipment; Tanning Materials

CNS K6128-81. Method of Test for Hide Substance of Leather (Aug)(1284).
CNS K6131-60. Method of Test for Leather (Combined Tannin Test) (May)(1287)(R 1973).
CNS K6137-60. Method of Test for Leather (P-Nitrophenol Test) (May)(1293)(R 1973).

—Abrasion Testing
BSI BS 3144: Add 1-81. 1981 Methods of Sampling and Physical Testing of Leather Addendum 1: Determination of Resistance to Bending and Abrasion of Heavy Leather. 12 pp.

—Adhesives
CEN PREN 522-91. Adhesives for Leather and Footwear Materials—Bond Strength—Minimum Requirements and Adhesive Classification. 17 pp.

—Ash Content
CNS K6130-81. Method of Test for Soluble Ash of Leather (Aug)(1286).
CNS K6134-81. Method of Test for Total Ash of Leather (Aug)(1290).

—Belts (Clothing)
MOD UK DSTAN 83-14-88. Leather, Cattlehide, Hydraulic Packing Leather, Cattlehide, Sole Leather, Cattlehide, for Hoses, Bellows, Straps, Harness and Saddlery, Leather Belting, Leather Butts, Special Issue 3. 19 pp.

—Bend Testing
BSI BS 3144: Add 1-81. 1981 Methods of Sampling and Physical Testing of Leather Addendum 1: Determination of Resistance to Bending and Abrasion of Heavy Leather. 12 pp.
CNS K6120-60. Method of Test for Leather (Creasing Test) (May)(1276)(R 1973).

—Bookbinding—Archival Storage
BSI BS 7451-91. 1991 Archival Quality Bookbinding Leather. 11 pp.

—Buffings—Ammunition
MOD UK DSTAN 83-78-87. Leather Buffings, Vegetable-Tanned Issue 1. 8 pp.

—Bulk Density
ISO 2420-72. Leather—Determination of Apparent Density First Edition. 4 pp.

—Bursting Strength
ISO 3379-76. Leather—Determination of Distension and Strength of Grain—Ball Burst Test First Edition. 5 pp.

—Chemical Analysis
BSI BS 1309-74. 1974 Amd 2 Methods for Sampling and Chemical Testing of Leather. 61 pp.
DIN ENGL 53300 Pt 2-68. Testing of Leather; Summary of Test Methods and Presentation of Results; Chemical Test Methods (Oct). 2 pp.

—Chrome/Tanned
CNS K8012-73. Dressed Chromed Inner Leather (Tentative) (Aug)(3548).
MOD UK DSTAN 83-1-82. Leather, Horsehide, Soft Dressed, Chrome Tanned, and Leather, Cattlehide, Soft Dressed, Chrome Tannned Issue 2. 7 pp.
MOD UK DSTAN 83-5-81. Leather, Sheepskin, Cape, Chrome Tanned, and Leather, Sheepskin, Cape, Chrome/Glutaraldehyde Combination Tanned Issue 2. 8 pp.
MOD UK DSTAN 83-10-82. Leather, Sheep, Basil Issue 2. 7 pp.

—Chromium Oxide Content
CNS K6135-81. Method of Test for Chromium Oxide Content of Leather (Aug)(1291).

Leather (Cont.)

—Classification
- BSI BS 1000: (675)-83. 1983 Universal Decimal Classification (UDC). English Full Edition (675): Leather Industry. 30 pp.
- SNZ NZS/BS 1000 (675)-83. Universal Decimal Classification Leather Industry. 32 pp.

—Clothing
- BSI BS 6453-84. 1984 Performance of Leathers for Garments. 12 pp.
- JIS K 6552-77. Testing Method for Clothing Leather.
- JIS K 6553-77. Clothing Leathers.

—Coatings—Thickness Measurement
- BSI BS 3144: Part 28-87. 1987 Methods of Sampling and Physical Testing of Leather Part 28: Method for Measurement of Thickness of Surface Coatings on Leather. 4 pp.

—Colorfastness Testing
- BSI BS 1006-90. 1990 Methods of Test for Colour Fastness of Textiles and Leather (L). 234 pp.
- BSI BS 1006-01. 1990 Amd 1 Methods of Test for Colour Fastness of Textiles and Leather (AMD 7284) October 15, 1992 (L). 247 pp.
- BSI BS 1006-02. 1990 Amd 2 Methods of Test for Colour Fastness of Textiles and Leather (AMD 7201) January 15, 1993 (L). 248 pp.
- CNS K6117-60. Method of Test for Leather (Decoloration Test) (May)(1273).
- SNZ NZS/BS 1006-78. Methods of Test for Colour Fastness of Textiles and Leather Amend: 7. 212 pp.

—Colorfastness Testing—Rubbing
- JIS K 6547-76. Testing Method for Colour Fastness to Rubbing of Leathers.
- JIS K 6547-67. Testing Method for Color Fastness to Rubbing of Leather. 6 pp.

—Composite Materials—Preservation
- MOD UK DEF-1234-A: LEAF C24-C30. Preservation of Leather and Leather Parts of Composite Materiel with 4-Nitrophenol Solution, Def Stan 68-6 Issue 2 (Withdrawn).

—Comprehensive Testing
- JIS K 6550-76. Testing Methods for Leather (R 1984). 16 pp.

—Compression Testing
- CNS K6121-60. Method of Test for Leather (Compression Test) (May)(1277). (R 1973).

—Conditioning
- ISO 2419-72. Leather—Conditioning of Test Pieces for Physical Tests First Edition. 3 pp.

—Crack Propagation
- CNS K6675-81. Method of Test for Grain Crack of Leather (Jul)(7701).
- ISO 3378-75. Leather—Determination of Resistance to Grain Cracking, and of Crack Index First Edition. 6 pp.

—Dry Cleaning
- BSI BS 7269: Part 1-91. 1991 Drycleanability of Leather Garments Part 1: Specification for Drycleanability and for Appropriate Care Labels. 11 pp.
- BSI BS 7269: Part 2-91. 1991 Drycleanability of Leather Garments Part 2: Method for Assessing Drycleanability. 12 pp.

—Elongation
- CNS K6125-81. Method of Test for Elongation Rate of Leather (Aug)(1281).
- ISO 3376-76. Leather—Determination of Tensile Strength and Elongation First Edition. 5 pp.

—Fertilizers
- CNS N3050-88. Steamed Leather Waste Fertilizer (Jun)(11916).

—Finishing Compounds—Adhesion Testing
- CNS K6818-85. Method of Test for Adhesion of Finish to Leather (Dec)(11455).
- JIS K 6555-79. Testing Method for Adhesion of Finish to Leathers.

—Flexural Strength
- CNS K6679-85. Method of Test for Flexing Endurance of Leather (Dec)(7705).
- JIS K 6545-70. Testing Method for Flexing Endurance of Light Leathers and Their Surface Finishes.

—Folding Endurance
- CNS K6119-60. Method of Test for Leather (Folding Test) (May)(1275).

—Footwear Linings
- MOD UK DSTAN 83-9-01. Leathers, Lining, for Footwear Issue 2; Amendment 2. 14 pp.

Leather (Cont.)

—Fungus Resistance Testing
- SAA AS 1157.6-71. Methods of Testing Materials for Resistance to Fungal Growth—Part 6: Resistance of Leather to Surface Fungal Growth. 3 pp.

—Furniture
- BSI BS 6608-85. 1985 Cattle Hide Leathers for Upholstered Furniture. 8 pp.

—Glossaries
- BSI BS 2780-83. 1983 Leather Terms. 24 pp.

—Grain Crack Testing
- JIS K 6548-76. Testing Method for Grain Crack of Leather (by Ball-Bursting Tester).

—Hats
- MOD UK DSTAN 83-4-92. Leather, Sheep, Head Leathers Issue 2. 12 pp.

—Identification Systems
- ISO 2418-72. Leather—Laboratory Samples—Location and Identification First Edition. 5 pp.
- SNZ NZS 8721-88. Care Labelling of Clothing, Household Textiles, Furnishings, Upholstered Furniture, Bedding, Piece Goods and Yarns Amend: A, 1988. 7 pp.
- SNZ NZS 8722-88. Care Labelling—Guide to the Selection of Correct Care Labelling Instructions from NZS 8721 Amend: A, 1988. 15 pp.

—Insoles (Footwear)
- MOD UK DSTAN 83-3-82. Leather, Cattlehide, Shoulder, Vegetable Tanned, for Insoles and Welting; and Leather, Cattlehide, Belly, Vegetable Tanned, for Use in Other Footwear Components Issue 2. 10 pp.

—Low Temperature Testing
- CNS K6676-81. Method of Test for Cold Resistance of Leather (Jul)(7702).
- JIS K 6542-74. Testing Methods for Cold Resistance of Leather.

—Magnesium Sulfate Content—Volumetric Analysis
- ISO 5399-84. Leather—Determination of Water-Soluble Magnesium Salts—EDTA Titrimetric Method First Edition. 5 pp.

—Methylene Chloride Extract Content
- CNS K6127-81. Method of Test for Dichloromethane Extract of Leather (Aug)(1283).

—Moisture Content
- CNS K6132-81. Method of Test for Moisture of Leather (Aug)(1288).

—Nitrogen Content
- CNS K6133-60. Method of Test for Leather (Nitrogen Content Test) (May)(1289).

—Nitrogen Content—Volumetric Analysis
- ISO 5397-84. Leather—Determination of Nitrogen Content and "Hide Substance"—Titrimetric Method First Edition. 5 pp.

—PH
- CNS K6138-81. Method of Test for pH Value of Leather (Aug)(1294).
- ISO 4045-77. Leather—Determination of pH First Edition. 4 pp.

—Physical Testing
- BSI BS 3144-68. 1968 Methods of Sampling and Physical Testing of Leather. 84 pp.

—Polyurethane
- CNS K5129-83. Polyurethane Leather (May)(10269).

—Polyurethane—Physical Properties
- CNS K6771-83. Method of Test for Polyurethane Leather (May)(10270). 3 pp.

—Preservation
- MOD UK DEF-1234-A: LEAF C24-C30. Preservation of Leather and Leather Parts of Composite Materiel with 4-Nitrophenol Solution, Def Stan 68-6 Issue 2 (Withdrawn).

—Sampling
- BSI BS 1309-74. 1974 Amd 2 Methods for Sampling and Chemical Testing of Leather. 61 pp.
- BSI BS 3144-68. 1968 Methods of Sampling and Physical Testing of Leather. 84 pp.
- CNS K6116-81. Method of Test for Leather (General Rule and Sampling Method) (Aug)(1272).
- ISO 2418-72. Leather—Laboratory Samples—Location and Identification First Edition. 5 pp.
- ISO 2588-85. Leather—Sampling—Number of Items for a Gross Sample Second Edition. 3 pp.

Leather (Cont.)

—Shrinkage
- CNS K6124-81. Method of Test for Shrinkage Temperature in Liquid of Leather (Aug)(1280).
- ISO 3380-75. Leather—Determination of Shrinkage Temperature First Edition. 5 pp.

—Silicon Content—Spectrometry
- ISO 5400-84. Leather—Determination of Total Silicon Content—Reduced Molybdosilicate Spectrometric Method First Edition. 4 pp.

—Soles (Footwear)
- MOD UK DSTAN 83-14-88. Leather, Cattlehide, Hydraulic Packing Leather, Cattlehide, Sole Leather, Cattlehide, for Hoses, Bellows, Straps, Harness and Saddlery, Leather Belting, Leather Butts, Special Issue 3. 19 pp.

—Solubility
- CNS K6129-81. Method of Test for Water Soluble Substance of Leather (Aug)(1285).

—Sulfated Ash Content
- ISO 4047-77. Leather—Determination Sulphated Total Ash and Sulphated Water-Insoluble Ash First Edition. 4 pp.
- ISO 4048-77. Leather—Determination of Matter Soluble in Dichloromethane First Edition. 4 pp.

—Tanning
- CNS K6136-81. Method of Test for Tanning Degree of Leather (Aug)(1292).

—Tear Strength
- CNS K6123-81. Method of Test for Tearing Strength of Leather (Aug)(1279).
- ISO 3377-75. Leather—Determination of Tearing Load First Edition. 4 pp.

—Tensile Testing
- CNS K6122-81. Method of Test for Tensile Strength of Leather (Aug)(1278).
- ISO 3376-76. Leather—Determination of Tensile Strength and Elongation First Edition. 5 pp.

—Test Specimens
- ISO 4044-77. Leather—Preparation of Chemical Test Samples First Edition. 3 pp.

—Thickness Measurement
- CNS K6118-81. Method of Test for Thickness of Leather (Aug)(1274).
- ISO 2589-72. Leather—Physical Testing—Measurement of Thickness First Edition. 4 pp.

—Trimming
- CNS K8015-80. Raw Hides of Cattle-Method of Trim (Jul)(5822).
- ISO 2820-74. Leather—Raw Hides of Cattle and Horses—Method of Trim First Edition. 4 pp.

—Uppers (Footwear)
- MOD UK DSTAN 83-2-82. Leather, Cattlehide, Side Chrome Tanned: and Leather, Calf, Chrome Tanned, for Footwear Uppers Issue 2. 11 pp.
- MOD UK DSTAN 83-2-01. Leather, Cattlehide, Side Chrome Tanned: and Leather, Calf, Chrome Tanned, for Footwear Uppers Issue 2 (Reprinted 23 March 1990 —Amendments 1 & 2); Amendment 3. 12 pp.

—Water Absorption
- BSI BS 3144: Add 2-81. 1981 Methods of Sampling and Physical Testing of Leather Addendum 2: Measurement of Water Vapor Absorption. 2 pp.
- CNS K6677-81. Method of Test for Water Vapor Absorption of Leather (Jul)(7703).
- CNS K6678-81. Method of Test for Water Vapor Permeability of Leather (Jul)(7704).
- CNS K6688-81. Method of Test for Water Absorption of Leather (Aug)(7819).
- ISO 2417-72. Leather—Determination of Absorption of Water First Edition. 4 pp.
- JIS K 6544-77. Testing Method for Water Vapour Absorption of Leather.
- JIS K 6549-77. Testing Method for Water Vapour Permeability of Leather.

—Water Resistance Testing
- CNS K6126-81. Method of Test for Water Resistance Leather (Aug)(1282).

Leather Working Equipment
See Also: Leather; Shoe Machinery

—Safety
- CEN PREN 930-92. Footwear, Leather and Imitation Leather Goods Manufacturing Machines—Roughing, Scouring, Polishing and Trimming Machines—Safety Requirements. 33 pp.

Leatherboards
See Also: Chipboard; Millboards

Leatherboards

Leatherboards (Cont.)
—**Ammunition**
MOD UK DEF-41-A-67. Imitation Leatherboard. 4 pp.

Leathercloth
Use For: Artificial Leather; Imitation Leather; Leatherette; Synthetic Leather *See Also:* Coated Fabrics; Leather
MOD UK DSTAN 83-14-88. Leather, Cattlehide, Hydraulic Packing Leather, Cattlehide, Sole Leather, Cattlehide, for Hoses, Bellows, Straps, Harness and Saddlery, Leather Belting, Leather Butts, Special Issue 3. 19 pp.

—**Folding Endurance**
DIN ENGL 53359-57. Testing of Artificial Leather; Repeated Flexure Test (Nov). 2 pp.

—**Mass**
DIN ENGL 53358-71. Testing of Artificial Leather and Similar Sheet Materials; Determination of Mass per Unit Area of the Surface Layer (June). 2 pp.

—**Thickness Measurement**
DIN ENGL 53353-71. Testing of Artificial Leather and Similar Sheet Materials; Determination of Thickness with Mechanical Feelers (June). 2 pp.

Leatherette
Use: Leathercloth

Leatheroid (R)
Scope Note: Leatheroid is a registered trademark
Use: Vulcanized Fibers

Lecithin
—**Cosmetics—Hygiene**
CNS S2096-83. Methods of Hygienic Test for Cosmetics Lecithin.

Leclanche Batteries
Use For: Carbon Zinc Batteries; Dry Cell Batteries; Dry Cells *See Also:* Alkaline Batteries; Batteries; Primary Batteries
CNS C3086-85. Method of Test for Manganese Dry Cells and Batteries (Jun)(6029).
CNS C3087-80. Method of Test for Air Dry Batteries (Aug)(6031).
CNS C4203-85. Manganese Dry Cells and Batteries (Jun)(6028).
CNS C4204-80. Air Dry Batteries (Aug)(6030).
JIS C 8501-84. Manganese Dry Cells and Batteries. 14 pp.
MOD UK DSTAN 61-3: Pt 1: Supp 1-01. Battery, Dry (Leclanche) 1.5 V, No 1 Issue 2; Amendment 4. 5 pp.
MOD UK DSTAN 61-3: Pt 1: Supp 3-73. Battery, Dry (Leclanche), 1.5 V, No 3 Issue 2. 4 pp.
MOD UK DSTAN 61-3: Pt 1: Supp 5-73. Battery, Dry (Leclanche), 1.5 V, No 5 Issue 2. 4 pp.
MOD UK DSTAN 61-3: Pt 1: Supp 7-01. Battery, Dry (Leclanche), 1.5 V, No 7 Issue 2; Amendment 2. 4 pp.
MOD UK DSTAN 61-3: Pt 1: Supp 8-73. Battery, Dry (Leclanche), 3 V, No 1 Issue 2. 4 pp.
MOD UK DSTAN 61-3: Pt 1: Supp 9-73. Battery, Dry (Leclanche), 3 V, No 2 Issue 2. 4 pp.
MOD UK DSTAN 61-3: Pt1: Supp 11-73. Battery, Dry (Leclanche), 6 V, No 1 Issue 2. 4 pp.
MOD UK DSTAN 61-3: Pt1: Supp 19-73. Battery, Dry (Leclanche), 45 V, No 1 Issue 2. 4 pp.
MOD UK DSTAN 61-3: Pt1: Supp 21-01. Battery, Dry (Leclanche), 45 V, No 3 Issue 2; Corrigendum. 4 pp.
MOD UK DSTAN 61-3: Pt1: Supp 22-73. Battery, Dry (Leclanche), 63 V, No 1 Issue 2. 4 pp.
MOD UK DSTAN 61-3: Pt1: Supp 23-73. Battery, Dry (Leclanche), 67.5 V, No 1 Issue 2. 4 pp.
MOD UK DSTAN 61-3: Pt1: Supp 24-73. Battery, Dry (Leclanche), 90 V, No 1 Issue 2. 4 pp.
MOD UK DSTAN 61-3: Pt1: Supp 27-73. Battery, Dry (Leclanche), 126 V, No 1 Issue 2. 4 pp.
MOD UK DSTAN 61-3: Pt1: Supp 28-73. Battery, Dry (Leclanche), 135 V, No 1 Issue 2. 4 pp.
MOD UK DSTAN 61-3: Pt1: Supp 31-01. Battery, Dry (Leclanche), 1.5 V, No 9 Issue 2; Amendment 4. 4 pp.
MOD UK DSTAN 61-3: Pt1: Supp 33-73. Battery, Dry (Leclanche), 4 1/2 V, No 3 Issue 2. 4 pp.
MOD UK DSTAN 61-3: Pt1: Supp 34-73. Battery, Dry (Leclanche), 6 V, No 6 Issue 2. 4 pp.
MOD UK DSTAN 61-3: Pt1: Supp 35-73. Battery, Dry (Leclanche), 9 V, No 2 Issue 2. 4 pp.
MOD UK DSTAN 61-3: Pt1: Supp 36-73. Battery, Dry (Leclanche), 22.5 V, No 4 Issue 2. 4 pp.
MOD UK DSTAN 61-3: Pt1: Supp 39-73. Battery, Dry (Leclanche), 90/1.5 V, No 1 Issue 2(R 1982).
MOD UK DSTAN 61-3: Pt1: Supp 40-73. Battery, Dry (Leclanche), 150/3 V, No 1 Issue 2. 4 pp.
MOD UK DSTAN 61-3: Pt1: Supp 42-73. Battery, Dry (Leclanche), 162/3 V, No 1 Issue 2. 4 pp.

Leclanche Batteries (Cont.)
MOD UK DSTAN 61-3: Pt1: Supp 45-01. Battery, Dry (Leclanche), 1.5 V, No 16 Issue 2; Amendment 2. 4 pp.
MOD UK DSTAN 61-3: Pt1: Supp 46-73. Battery, Dry (Leclanche), 22.5 V, No 5 Issue 2. 4 pp.
MOD UK DSTAN 61-3: Pt1: Supp 47-73. Battery, Dry (Leclanche), 15 V, No 8 Issue 2. 4 pp.
MOD UK DSTAN 61-3: Pt1: Supp 50-73. Battery, Dry (Leclanche), 4.5 V, No 4 Issue 2. 4 pp.
MOD UK DSTAN 61-3: Pt1: Supp 51-01. Battery, Dry (Leclanche), 1.5 V, No 12 Issue 2; Corrigendum. 4 pp.
MOD UK DSTAN 61-3: Pt1: Supp 54-73. Battery, Dry (Leclanche), 12 V, No 1 Issue 2. 4 pp.
MOD UK DSTAN 61-3: Pt1: Supp 55-73. Battery, Dry (Leclanche), 9 V, No 1 Issue 2. 4 pp.
MOD UK DSTAN 61-3: Pt1: Supp 56-73. Battery, Dry (Leclanche), 30 V, No 1 Issue 2. 4 pp.
MOD UK DSTAN 61-3: Pt1: Supp 57-73. Battery, Dry (Leclanche), 30 V, No 2 Issue 2. 4 pp.
MOD UK DSTAN 61-3: Pt1: Supp 58-01. Battery, Dry (Leclanche), 150 V, No 1 Issue 2; Amendment 4. 5 pp.
MOD UK DSTAN 61-3: Pt1: Supp 60-01. Battery, Dry (Leclanche), 15 V, No 1 Issue 2; Corrigendum. 4 pp.
MOD UK DSTAN 61-3: Pt1: Supp 62-73. Battery, Dry (Leclanche), 4 1/2 V, No 5 Issue 2. 4 pp.
MOD UK DSTAN 61-3: Pt1: Supp 64-73. Battery, Dry (Leclanche), 4 1/2 V, No 6 Issue 2. 4 pp.
MOD UK DSTAN 61-3: Pt1: Supp 66-73. Battery, Dry (Leclanche), 75 V, No 1 Issue 2. 4 pp.
MOD UK DSTAN 61-3: Pt1: Supp 70-01. Battery, Dry (Leclanche), 9 V Issue 2; Amendment 5. 4 pp.
MOD UK DSTAN 61-3: Pt1: Supp 74-73. Battery, Dry (Leclanche), 12 V Issue 2. 4 pp.
MOD UK DSTAN 61-3: Pt1: Supp 75-01. Battery, Dry (Leclanche), 90/45/1.5-3 V, No 1 Issue 2; Amendment 5. 7 pp.
MOD UK DSTAN 61-3: Pt1: Supp 76-01. Battery, Dry (Leclanche), 90/60/4.5 V, No 2 Issue 2; Amendment 5. 5 pp.
MOD UK DSTAN 61-3: Pt1: Supp 78-73. Battery, Dry (Leclanche), 9 V Issue 2. 4 pp.
MOD UK DSTAN 61-3: Pt1: Supp 79-73. Battery, Dry (Leclanche), 12 V Issue 2. 4 pp.
MOD UK DSTAN 61-3: Pt1: Supp 80-01. Battery, Dry (Leclanche), 135/67.5/1.5/-6 V, No 1 Issue 2; Amendment 5. 5 pp.
MOD UK DSTAN 61-3: Pt1: Supp 81-73. Battery, Dry (Leclanche), 1.5 V Issue 2. 4 pp.
MOD UK DSTAN 61-3: Part 2-01. Batteries Primary Part 2: Batteries Dry and Batteries Water Activated Issue 2; Amendment 4. 41 pp.

—**Acetylene Black**
CNS K1143-72. Acetylene Black (Jan)(3139). 1 p.
CNS K6286-70. Method of Test for Acetylene Black (Sep)(3140).
JIS K 1469-84. Acetylene Black.
JIS K 1469-66. Acetylene Black. 9 pp.

—**Ammonium Chloride**
CNS K1152-70. Ammonium Chloride for Dry Cell (Dec)(3179).

—**Blotting Papers**
CNS C3034-85. Method of Test for Blotting Paper for Dry Batteries (Jun)(3257).
CNS C4100-85. Blotting Paper for Dry Batteries (Jun)(3256). 1 p.

—**Carbon Electrodes**
CNS C4062-85. Carbon Electrode for Dry Batteries (Aug)(2654).

—**Copper Wire**
CNS C2045-80. Connecting Wire for Dry Battery (Tentative) (Jul)(2548). 1 p.

—**Manganese Dioxide**
JIS K 1467-84. Electrolytic Manganese Dioxide.
JIS K 1467-65. Electrolytic Manganese Dioxide. 15 pp.

—**Mercuric Chloride**
CNS K1151-70. Mercury Chloride for Dry Cell (Dec)(3177).
CNS K6291-70. Method of Test for Mercury Chloride (for Dry Battery) (Dec)(3178).

—**Plastic Pipes—Heat Shrink**
CNS C4058-82. Heat Shrinking PVC Pipes (for Dry Battery Manufacture) (Tentative) (Dec)(2547).

—**Sealants**
CNS C4082-68. Sealing Compound for Dry Cell (Nov)(2903) (R 1982).
CNS C4101-82. Sealing Reagent for Dry Batteries (Dec)(3258). 2 pp.

—**Varnishes**
CNS C3037-82. Method of Test for Varnish for Dry Battery (Dec)(3431).

Leclanche Batteries (Cont.)
—**Varnishes** (Cont.)
CNS C4117-82. Varnish for Dry Batteries (Dec)(3430).

LED
Use: Light Emitting Diodes

LED Displays
Use: Light Emitting Diode Displays

Ledger Cards
Use: Ledger Papers

Ledger Papers
Use For: Ledger Cards *See Also:* Papers
CGSB 9-GP-41M-79. Paper, Ledger, Standard for. 10 pp.

Leeks
See Also: Food; Scallions; Vegetables

—**Grading**
CNS N1110-83. Grades of Green Garlic (Apr)(10194). 2 pp.

—**Refrigerated Transportation**
ISO 7922-85. Leeks—Guide to Cold Storage and Refrigerated Transport First Edition. 4 pp.

—**Storage**
ISO 7922-85. Leeks—Guide to Cold Storage and Refrigerated Transport First Edition. 4 pp.

Left Hand Screw Threads
See Also: Screw Threads

—**Aerospace—Bolts—Identification Systems**
DIN ENGL LN 29962-76. Bolts and Accessories; Additional Procurement Data for Special Designs and Arrangement of Code Letters in Characteristic Blocks (Apr). 3 pp.

—**Aerospace—Nuts (Fasteners)— Identification Systems**
DIN ENGL LN 29962-76. Bolts and Accessories; Additional Procurement Data for Special Designs and Arrangement of Code Letters in Characteristic Blocks (Apr). 3 pp.

Leg Prostheses
See Also: Ankle Prostheses; Foot Prostheses; Hip Prostheses; Joint Prostheses; Knee Prostheses
BSI BS 2574: Part 1-91. 1991 Lower Limb Orthoses Part 1: Guide to the Design and Manufacture of Lower Limb Orthoses, Excluding Foot Orthoses (Supersedes BS 3330: 1961). 22 pp.

—**Pipes**
JIS T 9211-90. Pipes for Endoskeletal Type of Lower Limb Prostheses. 12 pp.

—**Structural Analysis**
JIS T 0111-92. General Rules for Structural Testing of Lower Limb Prostheses.

—**Stumps**
ISO 8548 Pt 2-93. Prosthetics and Orthotics—Limb Deficiencies—Part 2: Method of Describing Lower Limb Amputation Stumps First Edition. 20 pp.

Legal Expenses Insurance
See Also: Insurance
EC 87/344/EEC-87. Council Directive on the Coordination of Laws, Regulations and Administrative Provisions Relating to Legal Expenses Insurance. 4 pp.

Leggings
Use: Gaiters

Legionella Pneumophilla
See Also: Bacteria

—**Water**
SAA AS 3896-91. Waters—Examination for Legionellae (in Professional Packages 17F, 50B). 7 pp.

—**Water—Sampling**
BSI BS 7592-92. 1992 Methods for Sampling for Legionella Organisms in Water and Related Materials. 8 pp.
BSI DD 211-92. 1992 Method for Detection and Enumeration of Legionella Organisms in Water and Related Materials. 16 pp.

Legislation
Use For: Laws *See Also:* Radio Regulations

INTERNATIONAL AND NON-U.S. NATIONAL STANDARDS
SUBJECT INDEX

Legislation (Cont.)
—Agricultural Buildings
BSI BS 5502: Sec 3.1-86. (WITHDRAWN) 1986 Design of Buildings and Structures for Agriculture Part 3: Appendices: Legislation, Technical Data and References Section 3.1: Legislation (Superseded by BS 5502: Part 11: 1990). 8 pp.

—Distortion Analyzers—Units of Measurement
CCITT RECMN R.9-89. How the Laws Governing Distribution of Distortion Should Be Arrived at—Telegraph Transmission (Study Group IX) 2 pp. 2 pp.

—Noise Reduction—Construction/Demolition
BSI BS 5228: Part 2-84. 1984 Code of Practice for Noise Control on Construction and Demolition Sites Part 2: Guide to Noise Control Legislation for Construction and Demolition, Including Road Construction and Maintenance. 8 pp.

Legumes
Use For: Pulses (Food) *See Also:* Food; Soybeans
EC EEC/452/92-92. Commission Regulation Adding a Temporary Provision to the Detailed Rules for the Application of the Special Measures for Peas, Field Beans and Sweet Lupins. 1 p.
ISO 605-91. Pulses—Determination of Impurities, Size, Foreign Odours, Insects, and Species and Variety—Test Methods Second Edition. 8 pp.

—Ash Content
BSI BS 4317: Part 10-93. 1993 Methods of Test for Cereals and Pulses Part 10: Determination of Ash of Cereals and Milled Cereal Products (ISO 2171: 1993) (W). 11 pp.
BSI BS 4317: Part 10-81. 1981 Methods of Test for Cereals and Pulses Part 10: Determination of Ash in Cereals, Pulses and Their Derived Products. 7 pp.
ISO 2171-93. Cereals and Milled Cereal Products—Determination of Total Ash Third Edition. 8 pp.

—Bacteria Count Methods
ISO 7698-90. Cereals, Pulses and Derived Products—Enumeration of Bacteria, Yeasts and Moulds First Edition. 10 pp.

—Bagged—Sampling
BSI BS 4511-80. 1980 Methods for Sampling Pulses (In Bags). 9 pp.
ISO 951-79. Pulses in Bags—Sampling First Edition. 8 pp.

—Fungi—Count Methods
ISO 7698-90. Cereals, Pulses and Derived Products—Enumeration of Bacteria, Yeasts and Moulds First Edition. 10 pp.

—Glossaries
BSI BS 6860-87. 1987 Nomenclature for Cereals, Pulses and Other Food Grains. 15 pp.
ISO 5526-86. Cereals, Pulses and Other Food Grains—Nomenclature First Edition. 19 pp.

—Hydrocyanic Acid Content
BSI BS 4317: Part 11-76. 1976 Methods of Test for Cereals and Pulses Part 11: Determination of Glycosidic Hydrocyanic Acid in Pulses. 7 pp.
ISO 2164-75. Pulses—Determination of Glycosidic Hydrocyanic Acid First Edition; (Erratum—Dec 1981). 7 pp.

—Impurities Content
BSI BS 4317: Part 4-91. 1991 Cereals and Pulses Part 4: Determination of Impurities, Size, Foreign Odours, Insects, and Species and Variety, in Pulses (ISO 605: 1991). 11 pp.

—Insect Contamination
BSI BS 4317: Part 18-88. 1988 Cereals and Pulses Part 18: Determination of Hidden Insect Infestation. 12 pp.
ISO 6639 Pt 1-86. Cereals and Pulses—Determination of Hidden Insect Infestation—Part 1: General Principles First Edition. 4 pp.
ISO 6639 Pt 3-86. Cereals and Pulses—Determination of Hidden Insect Infestation—Part 3: Reference Method First Edition. 6 pp.
ISO 6639 Pt 4-87. Cereals and Pulses—Determination of Hidden Insect Infestation—Part 4: Rapid Methods First Edition; (Supersedes 1162). 20 pp.

—Insect Contamination—Sampling
ISO 6639 Pt 2-86. Cereals and Pulses—Determination of Hidden Insect Infestation—Part 2: Sampling First Edition. 6 pp.

—Mass
BSI BS 4317: Part 1-80. 1980 Methods of Test for Cereals and Pulses Part 1: Determination of the Mass of 1000 Grains. 4 pp.

Legumes (Cont.)
—Mass (Cont.)
ISO 520-77. Cereals and Pulses—Determination of the Mass of 1 000 Grains First Edition. 4 pp.

—Milled—Sampling
BSI BS 5333-81. 1981 Sampling Cereals and Pulses (As Milled Products). 11 pp.
ISO 2170-80. Cereals and Pulses—Sampling of Milled Products Second Edition. 10 pp.

—Storage
BSI BS 6279: Part 1-82. 1982 Storage of Cereals and Pulses Part 1: Guide to Particular Problems Encountered in the Storage of Cereals and Pulses. 11 pp.
BSI BS 6279: Part 2-82. 1982 Storage of Cereals and Pulses Part 2: Code of Practice for the Storage of Cereals and Pulses. 9 pp.
BSI BS 6279: Part 2-01. 1982 Amd 1 Storage of Cereals and Pulses Part 2: Code of Practice for the Storage of Cereals and Pulses (ISO 6322/2-1981) (AMD October 15, 1993 (W). 11 pp.
ISO 6322 Pt 1-81. Storage of Cereals and Pulses—Part 1: General Considerations in Keeping Cereals First Edition. 10 pp.

—Storage—Pest Control
BSI BS 6279: Part 3-90. 1990 Storage of Cereals and Pulses Part 3: Guide to Control of Attack by Pests. 10 pp.
ISO 6322 Pt 3-89. Storage of Cereals and Pulses—Part 3: Control of Attack by Pests Second Edition. 8 pp.

—Yeasts—Count Methods
ISO 7698-90. Cereals, Pulses and Derived Products—Enumeration of Bacteria, Yeasts and Moulds First Edition. 10 pp.

Leisure Vehicles
Use: Recreational Vehicles

Lemon Oil
See Also: Essential Oils
CNS K5069-80. Oil of Lemon, Expressed, Italy (Aug)(6348).
ISO 855-81. Oil of Lemon, Italy, Obtained by Expression First Edition. 4 pp.

—Citral Content—Gas Chromatography
ISO 7611-85. Oils of Lemon and Petitgrain Citronnier, and Oil of Lime Obtained by a Mechanical Process—Determination of Citral (Neral + Geranial) Content—Gas Chromatographic Method on Capillary Columns First Edition. 7 pp.

Lemon Petitgrain Oil
Use: Petitgrain Oil

Lemongrass Oil
See Also: Essential Oils
CNS K5085-81. Oil of Lemongrass (Cymbopogon Citratus) (Nov)(6456).
CNS K5120-81. Oil of Lemograss (Cymbopogon Flexuous) (Nov)(8132).
ISO 3217-74. Oil of Lemongrass (Cymbopogon citratus) First Edition. 4 pp.
ISO 4718-81. Oil of Lemongrass (Cymbopogon Flexuosus) First Edition. 4 pp.

Lemons
See Also: Citrus Fruits; Fruits

—Grading
CNS N1087-70. Grade and Packing of Lemon (for Export) (Feb)(3053). 4 pp.

—Packaging
CNS N1087-70. Grade and Packing of Lemon (for Export) (Feb)(3053). 4 pp.

—Storage
ISO 3631-78. Citrus Fruits—Guide to Storage First Edition. 11 pp.

Lempseed Oil
See Also: Oils
CNS K5126-82. Lempseed Oil for Industrial Use (Feb)(8530).

Length Measurement
See Also: Measurement; Pneumatic Length Measurement; Width Measurement
BSI PP 7309-85. 1985 Introduction to the Tolerancing of Functional Length Dimensions. 42 pp.

—Adhesive Tapes
SAA AS 1635.8.1-74. Methods of Testing Pressure Sensitive Adhesive Tape—Part 8.1: Length. 1 p.
SAA AS 1635.8.2-74. Methods of Testing Pressure Sensitive Adhesive Tape—Part 8.2: Length of Double Faced Tape. 1 p.

Length Measurement (Cont.)
—Articulated Vehicles
EC 89/461/EEC-89. Council Directive Amending, with a View to Fixing Certain Maximum Authorized Dimensions for Articulated Vehicles, Directive 85/3/EEC on the Weights, Dimensions and Certain Other Technical Characteristics of Certain Road Vehicles. 1 p.

—Belt Conveyors
SAA AS 1334.1-82. Methods of Testing Conveyor and Elevator Belting—Part 1: Determination of Length of Endless Belting. 1 p.

—Cables (Electric)
DIN VDE 0100 Suppl. 5 (D)-88. Erection of Power Installations with Rated Voltages up to 1000 V; Maximum Permissible Lengths of Cables and Cords Taking into Consideration Protection Against Electric Shock in the Event of Fault, Short Circuit of Voltage Drop (Mar). 54 pp.

—Carpet Tiles
CEN PREN 994-93. Textile Floorcoverings—Determination of the Length of the Edges and Squareness of Tiles. 5 pp.

—Casings (Meat)
CNS N6031-81. Method of Test for Sausage Casings (Oct)(1487). 2 pp.

—Chains
BSI BS 7615-92. 1992 Motor Cycle Chains (ISO 10190: 1992) (E). 14 pp.
ISO 10190-92. Motor Cycle Chains—Characteristics and Test Methods First Edition. 11 pp.

—Condoms
CEN PREN 600-2-91. Latex Rubber Condoms—Part 2: Determination of Length. 5 pp.
ISO 4074 Pt 2-80. Rubber Condoms—Part 2: Determination of Length First Edition. 4 pp.

—Electrical
DIN ENGL 32876 Pt 1-86. Electrical Length Measurement with Analogue Data Acquisition; Concepts, Requirements, Testing (Apr). 6 pp.
DIN ENGL 32876 Pt 1 Suppl. 1-86. Electrical Length Measurement; General Information and Examples of Application (Apr). 4 pp.
JIS B 7450-89. Digital Position Readout. 18 pp.

—Error Analysis—Glossaries
DIN ENGL 2257 Pt 2-74. Definitions of Length Verification Practice; Measurement Errors and Uncertainties (Aug). 7 pp.

—Fabrics
CGSB CAN/CGSB-4.2 NO.4.2-M87. Textile Test Methods Textiles Fabrics—Measurement of Length of Pieces. 11 pp.
SAA AS 1587-73. Methods for Measurements of Textile Fabrics—Length, Width, Thickness, Mass Per Unit Length and Mass Per Unit Area Corrig. Amdt 1 June 1987. 12 pp.

—Fiberboard
BSI BS EN 324-1-93. 1993 Wood-Based Panels—Determination of Dimensions of Boards Part 1: Determination of Thickness, Width and Length (R). 10 pp.
CEN PREN 324 (Part 1)-89. Wood-Based Panels—Determination of Dimensions of Boards—Part 1: Determination of Thickness, Width and Length. 5 pp.
CEN EN 324-1-93. Wood-Based Panels—Determination of Dimensions of Boards Part 1: Determination of Thickness, Width and Length. 5 pp.
ISO 9426 Pt 1-89. Wood-Based Panels—Determination of Dimensions—Part 1: Determination of Thickness, Width and Length First Edition. 4 pp.

—Fibers
BSI BS 5182-75. 1975 Measurement of the Length of Wool Fibres Processed on the Worsted System, Using a Fibre Diagram Machine. 10 pp.

—Fibers—Comb Sorter Method
BSI BS 4044-89. 1989 Determination of Fibre Length by Comb Sorter Diagram. 12 pp.

—Floor Coverings
CEN PREN 426-90. Resilient Floorcoverings—Determination of Width, Length, Flatness and Straightness. 5 pp.

—Floor Tiles
CEN PREN 427-90. Resilient Floor Coverings—Determination of the Side Length and Squareness of Tiles. 5 pp.

—Graduations
DIN ENGL 2268-75. Length Measurements with Graduations; Parameters; Tolerancing (Oct). 8 pp.

Length Measurement (Cont.)

—Laces (Footwear)
CGSB CAN/CGSB-4.2 NO.4.3-M87. Textile Test Methods Bootlaces and Shoelaces—Measurement of Length. 8 pp.

—Motor Vehicles
EC 85/3/EEC-84. Council Directive on the Weights, Dimensions and Certain Other Technical Characteristics of Certain Road Vehicles. 5 pp.

—Particle Board
BSI BS EN 324-1-93. 1993 Wood-Based Panels—Determination of Dimensions of Boards Part 1: Determination of Thickness, Width and Length (R). 10 pp.
CEN PREN 324 (Part 1)-89. Wood-Based Panels—Determination of Dimensions of Boards—Part 1: Determination of Thickness, Width and Length. 5 pp.
CEN EN 324-1-93. Wood-Based Panels—Determination of Dimensions of Boards Part 1: Determination of Thickness, Width and Length. 5 pp.
ISO 9426 Pt 1-89. Wood-Based Panels—Determination of Dimensions—Part 1: Determination of Thickness, Width and Length First Edition. 4 pp.

—Plastic Sheets
ISO 4592-92. Plastics—Film and Sheeting—Determination of Length and Width Second Edition. 6 pp.

—Plywood
BSI BS EN 324-1-93. 1993 Wood-Based Panels—Determination of Dimensions of Boards Part 1: Determination of Thickness, Width and Length (R). 10 pp.
CEN PREN 324 (Part 1)-89. Wood-Based Panels—Determination of Dimensions of Boards—Part 1: Determination of Thickness, Width and Length. 5 pp.
CEN EN 324-1-93. Wood-Based Panels—Determination of Dimensions of Boards Part 1: Determination of Thickness, Width and Length. 5 pp.
ISO 9426 Pt 1-89. Wood-Based Panels—Determination of Dimensions—Part 1: Determination of Thickness, Width and Length First Edition. 4 pp.

—Temperature
ISO 1-75. Standard Reference Temperature for Industrial Length Measurements First Edition. 3 pp.

—Thermal Insulation
CEN PREN 822-92. Thermal Insulating Products for Building Applications—Determination of Length and Width. 6 pp.

—Thread (Textiles)
CGSB CAN/CGSB-4.2 NO.4.4-M87. Textile Test Methods Determination of Length of Yarns and Threads. 12 pp.

—Wool Fibers
BSI BS 5182-75. 1975 Measurement of the Length of Wool Fibres Processed on the Worsted System, Using a Fibre Diagram Machine. 10 pp.
CGSB CAN/CGSB-4.2 NO.73.2-M91. Textile Test Methods Wool—Determination of Fibre Length Distribution Parameters—Electronic Method (ISO 2648:1974). 24 pp.
ISO 2648-74. Wool—Determination of Fibre Length Distribution Parameters—Electronic Method First Edition. 21 pp.
SNZ NZS 8719-92. Method for the Measurement of the Fibre Length After Carding of Scoured Wool. 44 pp.

—Yarns
CGSB CAN/CGSB-4.2 NO.4.4-M87. Textile Test Methods Determination of Length of Yarns and Threads. 12 pp.

Length Measuring Instruments
Use For: Measures of Length *See Also:* Dimensional Measuring Instruments; Measuring Instruments; Measuring Tapes; Pneumatic Length Measuring Instruments; Rulers
EC 73/362/EEC-73. Council Directive on the Approximation of the Laws of the Member States Relating to Material Measures of Length. 8 pp.
EC 78/629/EEC-78. Council Directive Adapting to Technical Progress Directive 73/362/EEC on the Approximation of the Laws of the Member States Relating to Material Measures of Length. 4 pp.
EC 85/146/EEC-85. Commission Directive Adapting to Technical Progress Council Directive 73/362/EEC on the Approximation of the Laws of the Member States Relating to Material Measures of Length. 5 pp.

Length Measuring Instruments (Cont.)

—Rods
BSI BS 5317-76. 1976 Amd 1 Metric Length Bars and Their Accessories. 12 pp.
EC 73/362/EEC-73. Council Directive on the Approximation of the Laws of the Member States Relating to Material Measures of Length. 8 pp.

Lens Meters
Use: Light Meters

Lens Mounts

—Motion Picture Cameras
CNS Z9058-81. Threaded Lens Mounted and Flange Focal Distance for 8mm and 16mm Motion Picture Cameras (Mar)(7098).
JIS B 7127-68. Threaded Lens Mounts and Flange Focal Distances for 8 mm and 16 mm Motion Picture Cameras. 4 pp.

Lensatic Compasses
Use: Compasses (Indicating Instruments)

Lenses
Scope Note: For additional listings, use a more specific term *See Also:* Contact Lenses; Focal Length; Magnification; Optical Equipment; Optical Lenses; Photographic Lenses; Trial Lenses
CNS T1042-81. Settled Lens Set (Jan)(6973).
CNS T1043-81. Test Lens Set (Jan)(6974).

LET L 410AG Turbolet Aircraft
See Also: Aircraft

—Data Sheets
CAA FA8 ISSUE 11. LET Model L-410AG Turbolet. 3 pp.

Let Z 37 Aircraft
See Also: Aircraft

—Antenna Positions
CAA. Let Z.37 (Approved Aerial Positions). 1 p.

Letter Boxes
Use: Mailboxes

Letter Contracts
Scope Note: Written preliminary contractual instruments that authorize the contractor to begin immediately manufacturing supplies or performing services. For letter contracts pertaining to specific products or materials, see the material or product

—Small Value Orders
MOD UK DEFCON 53N (EA)-85. Small Value Order—Covering Letter—D of C/RE Branches at Ensleigh 11/85. 2 pp.
MOD UK DEFCON 53Q-86. Small Value Covering Letter for RNSPDC Eaglescliffe 2/86. 2 pp.
MOD UK DEFCON 53R-84. Small Value Order—Covering Letter for Leeds 7/84. 1 p.
MOD UK DEFCON 53V-85. Small Value Order—Covering Letter for CB/RE 7/85. 2 pp.

Letter Folding Machines
See Also: Office Machines

—Glossaries
BSI BS 6191: Part 2-81. 1981 Mail Processing Machines Part 2: Glossary of Terms for Letter Folding Machines. 8 pp.
ISO 5138 Sec 05-81. Office Machines—Vocabulary—Section 05: Letter Folding Machines First Edition. 18 pp.
JTC1 5138 Sec 05-81. Office Machines—Vocabulary—Section 05: Letter Folding Machines First Edition. 18 pp.
OSI ISO 5138-5-81. Office Machines—Vocabulary—Section 05: Letter Folding Machines. 18 pp.

Letter Opening Machines
See Also: Office Machines

—Glossaries
BSI BS 6191: Part 1-81. 1981 Mail Processing Machines Part 1: Glossary of Terms for Letter Opening Machines. 8 pp.
ISO 5138 Sec 04-81. Office Machines—Vocabulary—Section 04: Letter Opening Machines First Edition. 14 pp.
JTC1 5138 Sec 04-81. Office Machines—Vocabulary—Section 04: Letter Opening Machines First Edition. 14 pp.
OSI ISO 5138-4-81. Office Machines—Vocabulary—Section 04: Letter Opening Machines. 14 pp.

Letter Plates
Use: Mail Drops

Letterheads
BSI BS 1808-85. 1985 Amd 1 Cut Business Forms and Letterheads (AMD 5612) July 31, 1987. 10 pp.

Letterpress Printing
Use For: Relief Printing *See Also:* Printing Presses

—Glossaries
BSI BS 3814-64. (OBSOLESCENT) 1964 Glossary of Letterpress Rotary Printing Terms Used in Connection with Newspaper, Magazine and Similar Machines. 36 pp.

—Inks
BSI BS 3020-59. 1959 Improved Inks for Letterpress Four-Colour Printing. 18 pp.
BSI BS 4160-67. 1967 Amd 1 Inks for Letterpress Three—or Four—Colour Printing. 10 pp.
CNS K2130-83. Press Ink (Mar)(10086).
CNS K6753-83. Method of Test for Press Ink (Mar)(10087).
ISO 2845-75. Set of Printing Inks for Letterpress Printing—Colorimetric Characteristics First Edition. 7 pp.
JIS K 5701-80. Testing Methods for Lithographic and Letterpress Inks.
JIS K 5702-60. Testing Method for Press Ink.
JIS K 5703-58. Press Ink.

—Inks—Viscosity
CGSB 21-GP-7M-78. Measuring Viscosity of Letterpress and Lithographic Inks with the Falling Rod Viscometer (R 1989). 78 pp.

—Papers
DIN ENGL 19306-77. Uncoated Printing Papers; Letterpress Paper, Offset Paper, Rotogravure Paper; Technical Conditions of Delivery (Sept). 2 pp.

Letterpress Printing Inks
Use: Printing Inks—Letterpress

Letters (Symbols)
Use: Symbols

Lettuce

—Head—Grading
CNS N1112-83. Grades of Head Lettuce (Apr)(10196). 2 pp.

—Leaf—Grading
CNS N1114-83. Grades of Non-Head Lettuce (Leaf Lettuce) (Apr)(10198). 2 pp.

—Refrigerated Transportation
ISO 8683-88. Lettuce—Guide to Precooling and Refrigerated Transport First Edition. 4 pp.

—Stem—Grading
CNS N1109-83. Grades of Stem Lettuce (Asparagus Lettuce) (Apr)(10193). 2 pp.

—Storage
ISO 8683-88. Lettuce—Guide to Precooling and Refrigerated Transport First Edition. 4 pp.

Leucaena Meal
See Also: Animal Feed; Bone Meal; Cornmeal; Oil Meal

—Animal Feed
CNS N2013-80. Leucaena Whole Plant Meal (Feed Grade) (Apr) (2293). 1 p.
CNS N2044-80. Leucaena Seed Meal (Feed Grade) (Apr) (4185). 2 pp.

L-Leucine
CNS K7656-82. Chemical Reagent (L-Leucine) (Oct)(9513).
JIS K 9054-72. L-Leucine.

Leucomycins Content Analysis
Use For: Kitasamycin Content Analysis

—Animal Feed
CNS N4129-84. Method of Test for Feed Additives: Determination of Kitasamycin (Nov)(11135).

Leucosis
Use: Leukosis

Leukosis

—Cattle—Control Measures
EC 87/58/EEC-86. Council Decision Introducing a Supplementary Community Measure for the Eradication of Brucellosis, Tuberculosis and Leucosis in Cattle. 3 pp.

Level Gages
Use: Level Indicators

INTERNATIONAL AND NON-U.S. NATIONAL STANDARDS
SUBJECT INDEX

Level Indicators
Use For: Level Gages; Liquid Level Gages; Liquid Level Indicators *See Also:* Capacitance Level Indicators; Dipsticks; Float Gages; Fuel Gages; Gages; Indicating Instruments
CNS B7136-80. Method of Test for Level Gauge (Jul)(5653).

—**Agricultural Spraying Equipment**
ISO 9357-90. Equipment for Crop Protection—Agricultural Sprayers—Tank Nominal Volume and Filling Hole Diameter First Edition. 4 pp.

—**Oil—Ships**
JIS F 7212-87. Marine Oil Level Gauges with Self Closing Valves.
JIS F 7215-89. Marine Flat Glass Oil Level Gauges.
JIS F 7216-89. Marine Self Closing Valves for Oil Level Gauges.

—**Oil—Transformers**
DIN ENGL 42552-69. Oil Level Indicator Type B for Transformers (Dec). 4 pp.

—**Refrigerants—Refrigeration Systems—Plants—Safety**
DIN ENGL 8975 Pt 8-79. Refrigeration Plants; Safety Requirements for Construction, Equipment, Installation and Operation; Level Indicating Devices for Refrigerant Vessels, Liquid Level Indicators (Apr). 6 pp.

—**Refrigeration Equipment**
CSA C22.2 NO 140.3-M1987. Refrigerant-Containing Components for Use in Electrical Equipment (R 1993). 19 pp.

—**Ships**
JIS F 7211-87. Marine 5k Level Gauges with Valves.
MOD UK NES 605: Part 1-89. Guide to the Selection of Sensors for the Measurement of System Parameters Part 1: Selection of Liquid Level Sensors Issue 3 (04.89). 79 pp.

—**Submarines**
MOD UK NES 605: Part 1-89. Guide to the Selection of Sensors for the Measurement of System Parameters Part 1: Selection of Liquid Level Sensors Issue 3 (04.89). 79 pp.

Level Rods
See Also: Levels (Instruments); Surveying Instruments
SAA AS 1298-80. Levelling Staffs. 4 pp.

—**Graduations/Markings**
CGSB CAN/CGSB-88.21-M90. Graduations and Markings on Linear Measuring Instruments for Construction Surveying and Special Applications. 15 pp.

Level Shifters
—**Hes—CMOS to CMOS**
CECC CECC 90 104-219 ISSUE 1-86. CEI-CECC 90 104-219; Dual 4-Channel Data Selector/Multiplexer (En). 2 pp.

—**Hes—TTL to CMOS**
CECC CECC 90 104-219 ISSUE 1-86. CEI-CECC 90 104-219; Dual 4-Channel Data Selector/Multiplexer (En). 2 pp.

Level Translators
See Also: Converters; Interface Circuits; Pulse Generators

—**TTL/CMOS**
CECC CECC 90 104-060 ISSUE 2. NL CECC 90 104-060 Issue 2; Digital Integrated Circuits in Accordance with FS 90 104; HEC/HEF 4104B; Quadruple Low-to-High Voltage Translator with 3-State Outputs (En). 8 pp.

Levels (Instruments)
Use For: Spirit Levels *See Also:* Geodetic Instruments; Level Rods; Measuring Instruments; Surveying Instruments; Tools
BSI BS 958-68. 1968 Spirit Levels for Use in Precision Engineering. 12 pp.
CGSB 39-GP-28B-75. Levels and Plumbs, and Levels, Standard for. 23 pp.
CNS B6039-81. Levels (Apr)(7199).
CNS B6044-81. Hand Levels (May)(7342).
CNS B6062-82. Precision Square Levels (Jan)(8190).
CNS B6063-82. Precision Levels (Jan)(8191).
DIN ENGL 877-86. Inclination Measuring Instruments (Levels) (June). 4 pp.
DIN ENGL 2277-61. Circular Levels; Terms; Construction (Nov). 3 pp.
JIS B 7510-72. Precision Square Levels (R 1983). 8 pp.
JIS B 7901-93. Levels. 10 pp.
SAA AS 2054-77. Spirit Levels for Use in Precision Engineering. 12 pp.

Levels (Instruments) *(Cont.)*
—**Electronic Measuring Instruments**
DIN ENGL 2276 Pt 2-86. Inclination Measuring Systems; Electronic Inclination Measuring Systems; Types and Requirements (June). 2 pp.

—**Optical**
BSI BS 7334: Part 3-90. 1990 Measuring Instruments for Building Construction Part 3: Methods for Determining Accuracy in Use: Optical Levelling Instruments. 15 pp.
ISO 8322 Pt 3-89. Building Construction—Measuring Instruments—Procedures for Determining Accuracy in Use—Part 3: Optical Levelling Instruments First Edition. 12 pp.

—**Vials**
BSI BS 3509-62. 1962 Spirit Level Vials. 24 pp.
DIN ENGL 2276 Pt 1-86. Inclination Measuring Systems; Cylindrical Spirit Level Vials; Dimensions and Requirements (June). 2 pp.

Levels and Plumbs
Use: Levels (Instruments)

Lever Gauges
Use: Dial Gages

Lever Switches
Use For: Actuating Levers; Signal Keys; Switch Levers *See Also:* Electric Switches; Toggle Switches
CECC CECC 96 000 ISSUE 2-92. Generic Specification: Electromechanical Switches (En, Fr, Ge). 137 pp.
CECC CECC 96 200 ISSUE 1-88. Sectional Specification: Including BDS CECC 96 201 and CECC 96 202; Lever Switches (En, Fr) AMD 1 (En, Fr, Ge). 93 pp.
CECC CECC 96 201-001 ISSUE 1-88. BS CECC 96 201-001; Lever Switch Slow Make Slow Break (S.M.S.B.); Environmentally Sealed Body; Panel Sealed (Optional); Lever Lock (Optional); Screw Terminations (En). 22 pp.
CECC CECC 96 201-002 ISSUE 1-89. BS CECC 96 201-002; Lever Switch (Q.M.Q.B.); Lever and Panel Sealed or Non-Sealed; Solder Lug Terminations (En). 19 pp.
CECC CECC 96 201-003 ISSUE 1-87. BS CECC 96 201-003; Lever Switch (S.M.S.B.); Lever and Panel Sealed or Non-Sealed; Solder Lug, Screw or Push-on Terminations (En). 19 pp.

—**Aircraft**
ISO 44-75. Aerospace—Lever-Operated, Two-Position, ON/OFF Switches—Directions of Operation First Edition. 3 pp.
ISO 493-75. Aircraft—Dimensions for Single-Hole Mounting (Class 1 and Class 2) Lever-Operated Manual Switches First Edition. 3 pp.
ISO 1466-73. Lever-Operated Manual Switches for Aircraft—Performance Requirements First Edition. 10 pp.
ISO 3282-76. Aircraft—Dimensions for Single-Hole and Triple-Hole Mounting (Class 3) Lever-Operated Manual Switches First Edition. 4 pp.
ISO 3456-75. Aircraft—Lever-Operated Manual Switches (Class 3)—Performance Requirements First Edition. 8 pp.

—**Aircraft—Control Wheels**
SBAC AS 3146 ISSUE 1. Switch Lever Auto Pilot's Switch Pilot's Control Wheel. 1 p.
SBAC AS 3147 ISSUE 1. Switch Lever Auto Pilot Switch Pilot's Control Wheel. 1 p.
SBAC AS 6591 ISSUE 1. Switch, Lever: Auto-Pilot Switch, Pilot's Control Wheel.
SBAC AS 6597 ISSUE 1. Switch, Lever Auto Pilot Switch, Pilot's Control Wheel.

—**Aircraft—Lockplates**
SBAC AS 3132 ISSUE 2. Locking Plate—Switch Lever. 1 p.
SBAC AS 3306 ISSUE 2. Locking, Plate—Switch Lever. 1 p.

—**Automotive—Turn Signals**
CNS D3065-86. Performance Inspection of Turn Signal Switches for Automobiles (Nov)(6665).
JIS D 1605-88. Performance Requirements of Turn Signal Switches for Automobiles. 10 pp.

—**Miniature**
CECC CECC 96 201-012 ISSUE 1-89. BS CECC 96 201-012; Lever Switch-Miniature; 1/4in or 5/16in Mounting Bush; Lever and Panel Sealed; Solder Lug or Printed Circuit Terminations (En). 18 pp.

—**Pivots—Control Wheels—Aircraft**
SBAC AS 6594 ISSUE 1. Switch Lever Pivot Pilot's Control Wheel.

—**Preferred Products List**
CECC CECC MUAHAG Vol 8 IS 4-91. Preferred Products List; Switches (En, Ge, Fr). 29 pp.

Lever Switches *(Cont.)*
—**Quality Assurance**
BSI BS 9561-79. (OBSOLESCENT) 1979 Amd 1 Lever Operated Switches of Assessed Quality: Generic Data and Methods of Test: General Rules for the Preparation of Detail Specifications (AMD 4711) April 28, 1989 (Superseded by BS CECC 96200). 30 pp.
BSI BS 9571 N/F000-73. (WITHDRAWN) 1973 Detail Specification for Lever Switches of Assessed Quality: Quick-Make Quick-Break. 13 pp.
BSI BS CECC 96200-88. 1988 Sectional Specification Including Blank Detail Specification: Lever Switches. 57 pp.

—**Reliability Assured**
MOD UK DSTAN 59-75: Part 2-80. Switches, of Assessed Quality (Listed on EPIC Database) Part 2: Switches, Lever Operated List of Items Conforming to BS 9561 Issue 2. 21 pp.

—**Screw/Solder—Preferred Products List**
CECC CECC MUAHAG Vol 8 IS 4-91. Preferred Products List; Switches (En, Ge, Fr). 29 pp.

—**Screw/Solder—Sealed—Preferred Products List**
CECC CECC MUAHAG Vol 8 IS 4-91. Preferred Products List; Switches (En, Ge, Fr). 29 pp.

—**Sealed—Preferred Products List**
CECC CECC MUAHAG Vol 8 IS 4-91. Preferred Products List; Switches (En, Ge, Fr). 29 pp.

—**Sealed—Printed Circuit Mount—Preferred Products List**
CECC CECC MUAHAG Vol 8 IS 4-91. Preferred Products List; Switches (En, Ge, Fr). 29 pp.

—**Ships—Signal Lights—Morse Code**
JIS F 8451-83. Keys for Marine Morse Signal Lights.
JIS F 8451-62. Keys for Morse Signal Lamps for Marine Use (R 1968). 5 pp.

—**Soldering Lugs—Preferred Products List**
CECC CECC MUAHAG Vol 8 IS 4-91. Preferred Products List; Switches (En, Ge, Fr). 29 pp.

—**Soldering Lugs—Printed Circuit Mount—Preferred Products List**
CECC CECC MUAHAG Vol 8 IS 4-91. Preferred Products List; Switches (En, Ge, Fr). 29 pp.

—**Soldering Lugs—Printed Circuit Mount—Sealed—PPL**
CECC CECC MUAHAG Vol 8 IS 4-91. Preferred Products List; Switches (En, Ge, Fr). 29 pp.

—**Soldering Lugs—Sealed—Preferred Products List**
CECC CECC MUAHAG Vol 8 IS 4-91. Preferred Products List; Switches (En, Ge, Fr). 29 pp.

—**Telecommunication Equipment**
CECC CECC 96 000 ISSUE 2-92. Generic Specification: Electromechanical Switches (En, Fr, Ge). 137 pp.

—**Telephones**
MOD UK DSTAN 59-11: Part 2-01. Switches, Electric Part 2: Switches, Telephone Key (Including Plates, Mounting, and Handles, Switch) Issue 1; Corrigendum to Amendment 1. 25 pp.

Lever Tumbler Locks
See Also: Locks (Security)

—**Mortise**
CNS A2241-90. Lever Tumbler Mortise Locks (Apr)(12696). 5 pp.
CNS A3315-90. Method of Test for Lever Tumbler Mortise Locks (Apr)(12697). 3 pp.
JIS A 5515-79. Lever Tumbler Mortise Locks.

Levers
See Also: Handles; Knobs; Manual Controls; Marline Spikes

—**Aircraft**
SBAC AS 149 ISSUE 5. Lever—Stream. 1 p.
SBAC AS 153 ISSUE 3. Lever—Hand. 1 p.
SBAC AS 2269 ISSUE 3. Control Lever. 1 p.
SBAC AS 2272 ISSUE 2. Lever. 1 p.
SBAC AS 2273-2274 ISSUE 3. Body—Control Lever. 1 p.

—**Aircraft—Brake—Control Wheels**
SBAC AS 3123 ISSUE 3. Brake Lever for Pilot's Control Wheel. 1 p.
SBAC AS 3124 ISSUE 3. Brake Lever for Pilot's Control Wheel. 1 p.
SBAC AS 3125 ISSUE 2. Brake Lever for Pilot's Control Wheel. 1 p.

INDUSTRY STANDARDS

Levers (Cont.)

—Aircraft—Brake Lock—Control Wheels
SBAC AS 3052 ISSUE 2. Brake Lock Lever for Pilot's Control Wheel. 1 p.

—Aircraft—Filler Caps
SBAC AS 2594 ISSUE 7. Filler Cap Locking Lever. 1 p.
SBAC AS 6748 ISSUE 2. Filler Cap Lever.

—Aircraft—Hand Assemblies
SBAC AS 134 ISSUE 4. Lever—Hand-Assembly. 1 p.

—Aircraft—Jettison
SBAC AS 136 ISSUE 5. Lever—Jettison. 1 p.

—Aircraft—Parachutes
SBAC AS 3341 ISSUE 1. Control—Single Lever—for Anti-Spin Parachutes. 1 p.
SBAC AS 3342 ISSUE 2. Box Single Lever Parachute Control. 1 p.
SBAC AS 6380 ISSUE 1. Control—Single Lever—for Anti Spin Parachutes.

—Aircraft—Pistons
SBAC AS 2271 ISSUE 3. Piston—Control Lever. 1 p.

—Aircraft—Sockets
SBAC AS 2275 ISSUE 3. Stop Socket—Control Lever. 1 p.

—Aircraft—Springs
SBAC AS 2270 ISSUE 3. Spring—Control Lever. 1 p.

—Aircraft—Wings
SBAC RS 438 ISSUE 2. Wing Spreading and Folding Control, Lever.

—Clamping
CNS B3103-79. Clamping Levers (Nov) (321).

—Pickup—Magnetic
CNS B3331-80. Magnetic Pick-Up Lever (Mar)(5272).

—Pickup—Mechanical
CNS B3332-80. Mechanical Pick-Up Leaver (Mar)(5273).

—Sleeve Bearings
CNS B2062-54. Bearing Sleeve with Hard Lever (Jul)(394)(R 1970).

—Wrenches
CNS B3108-54. Wrench Lever with Elliptical Handle (Jul)(326)(R 1973).
CNS B3109-54. Two-Way Wrench Lever with Elliptical Handle (Jul)(327)(R 1973).
CNS B3110-54. Straight Wrench Lever (Jul)(328)(R 1973).

Lexicography
See Also: Subject Indexing; Thesauri
CSA CAN/CSA-Z780-92. Principles and Methods of Terminology (ISO 704-1987); (Gen Instr 1 Thru 2). 31 p.
CSA CAN/CSA-Z781-92. Terminology—Vocabulary (ISO 1087:1990); (Gen Instr 1 Thru 2). 36 p.
ISO 704-87. Principles and Methods of Terminology First Edition; (CAN/CSA-Z780-92). 20 pp.
ISO R860-68. International Unification of Concepts and Terms First Edition. 14 pp.
ISO 1087-90. Terminology—Vocabulary First Edition. 23 pp.

—Symbols
ISO 1951-73. Lexicographical Symbols Particularly for Use in Classified Defining Vocabularies First Edition. 36 pp.

Liability, Contractor
Use: Contractor Liability

Liability Insurance
See Also: Insurance

—Aircraft—Software
EURO CE/CTI/89/19 487-89. Recommendation on Property and Liability in Software Matters. 109 pp.

—Automotive
EC COM(88) 644-88. Proposal for a Third Council Directive on the Approximation of the Laws of the Member States Relating to Insurance Against Civil Liability in Respect of the Use of Motor Vehicles. 18 pp.
EC COM(88) 791-88. Proposal for a Council Directive Amending First Council Directive 73/239/EEC, Second council Directive 88/357/EEC and Amending Directive 73/239/EEC. 20 pp.
EC 90/232/EEC-90. Council Directive on the Approximation of the Laws of the Member States Relating to Insurance Against Civil Liability in Respect of the Use of Motor Vehicles. 4 pp.

Liability Insurance (Cont.)

—Automotive (Cont.)
EC 90/618/EEC-90. Council Directive Amending Particularly as Regards Motor Vehicle Liability Insurance, Directive 73/239/EEC and Directive 88/357/EEC Which Concern the Coordination of Laws, Regulations and Administrative Provisions Relating to Direct Insurance. 6 pp.

Libraries
See Also: Archives; Document Storage and Retrieval Systems; Documents; Information Systems; Interlibrary Loans

—Books—Price Indexes
BSI BS ISO 9230-91. 1991 Information and Documentation—Determination of Price Indexes for Books and Serials Purchased by Libraries. 15 pp.
ISO 9230-91. Information and Documentation—Determination of Price Indexes for Books and Serials Purchased by Libraries First Edition. 13 pp.

—Directories
ISO 2146-88. Documentation—Directories of Libraries, Archives, Information and Documentation Centres, and Their Data Bases Second Edition. 28 pp.

—Model Bylaws
SNZ NZS 9201: Chapter 15-72. Model General Bylaws Chapter 15: Public Libraries (Reconfirmed 1980). 11 pp.

—Periodicals—Price Indexes
BSI BS ISO 9230-91. 1991 Information and Documentation—Determination of Price Indexes for Books and Serials Purchased by Libraries. 15 pp.
ISO 9230-91. Information and Documentation—Determination of Price Indexes for Books and Serials Purchased by Libraries First Edition. 13 pp.

—Statistical Analysis
BSI BS ISO 2789-91. 1991 Information and Documentation—International Library Statistics. 15 pp.
ISO 2789-91. Information and Documentation—International Library Statistics Second Edition. 13 pp.

License Plates
Use For: Marker Plates; Number Plates; Registration Plates (Vehicles); Statuatory Plates *See Also:* Motor Vehicles; Trailers
BSI BS AU 145A-72. 1972 Reflex-Reflecting Number Plates. 15 pp.
JIS D 4902-66. Standard Form of Licence Number Plates for Motorcycles (Under 125 cc) (R 1983). 5 pp.
MOD UK DSTAN 23-4-74. Marking of Service Vehicles Issue 3 (Withdrawn). 18 pp.

—Automotive
BSI BS AU 145B-89. 1989 Retroreflecting Number Plates (E). 36 pp.
EC 76/114/EEC-75. Council Directive on the Approximation of the Laws of the Member States Relating to Statutory Plates and Inscriptions for Motor Vehicles and Their Trailers, and Their Location and Method of Attachment. 5 pp.
EC 78/507/EEC-78. Commission Directive Adapting to Technical Progress Council Directive 76/114/EEC on the Approximation of the Laws of the Member States Relating to Statutory Plates and Inscriptions for Motor Vehicles and Their Trailers, and Their Loc. and Method of Attachment. 3 pp.
ISO 7591-82. Road Vehicles—Retro-Reflective Registration Plates for Motor Vehicles and Trailers—Specification First Edition. 6 pp.
SAA AS 4001.1-92. Motor Vehicles—Rear Marker Plates—Part 1: Manufacturing Requirements Amdt 1 April 1993. 24 pp.
SAA AS 4001.2-92. Motor Vehicles—Rear Marker Plates—Part 2: Fitting Requirements. 9 pp.

—Automotive—Lighting
EC 76/760/EEC-76. Council Directive on the Approximation of the Laws of the Member States Relating to the Rear Registration Plate Lamps for Motor Vehicles and Their Trailers. 11 pp.

—Trailers
EC 76/114/EEC-75. Council Directive on the Approximation of the Laws of the Member States Relating to Statutory Plates and Inscriptions for Motor Vehicles and Their Trailers, and Their Location and Method of Attachment. 5 pp.
EC 78/507/EEC-78. Commission Directive Adapting to Technical Progress Council Directive 76/114/EEC on the Approximation of the Laws of the Member States Relating to Statutory Plates and Inscriptions for Motor Vehicles and Their Trailers, and Their Loc. and Method of Attachment. 3 pp.

License Plates (Cont.)

—Trailers—Lighting
EC 76/760/EEC-76. Council Directive on the Approximation of the Laws of the Member States Relating to the Rear Registration Plate Lamps for Motor Vehicles and Their Trailers. 11 pp.

Licenses
Scope Note: See specific types of licenses

Lids
See Also: Caps (Lids)

—Aircraft
SBAC AS 2525 ISSUE 4. Top Cover. 1 p.

—Containers
BSI BS 1133: Sec 18-91. 1991 Packaging Code Section 18: Packaging in Glass (H). 14 pp.

—Containers—Apertures
BSI BS 5567-78. 1978 Apertures in Tinplate Containers to Receive Plug-In Plastics Closures (Supersedes BS 2878: 1976). 2 pp.

—Food Processing Equipment—Aluminum
BSI BS 5313-76. 1976 Aluminium Catering Containers and Lids. 8 pp.

—Food Processing Equipment—Plastic
BSI BS 5496-77. 1977 Plastics Catering Containers and Lids. 8 pp.

—Food Processing Equipment—Stainless Steel
BSI BS 5312-76. 1976 Stainless Steel Catering Containers and Lids. 8 pp.

Lie Detectors
Use For: Polygraphs
JIS T 1309-86. Polygraphs for Clinical Use.

Life (Durability)
Use For: Pot Life; Service Life *See Also:* Cyclic Testing; Dynamic Testing; High Temperature Testing; Low Temperature Testing; Reliability; Structural Design; Wear Testing

—Adhesives
BSI BS 5350: Part B4-93. 1993 Methods of Test for Adhesives Part B4: Determination of Pot Life (ISO 10364: 1993) (W). 6 pp.
BSI BS 5350: Part B4-76. 1976 Amd 1 Methods of Test for Adhesives Part B4: Determination of Pot Life (AMD 5358) February 27, 1987. 3 pp.
CNS K6444-80. Method of Test for Working Life of Adhesives (Jan)(5134).
ISO 10354-92. Adhesives—Characterization of Durability of Structural -Adhesive-Bonded Assemblies—Wedge Rupture Test First Edition. 7 pp.
ISO 10364-93. Adhesives—Determination of Working Life (Pot Life) of Multi-Component Adhesives First Edition. 6 pp.

—Aircraft Engines
CAA NOTICE #44 ISSUE 5. Gas Turbine Engine Parts Subject to Retirement or Ultimate (Scrap) Lives (Airworthiness Notices). 3 pp.

—Ball Bearings
BSI PD 6487-79. 1979 Calculating the Life of Ball Bearings When Used in Contact with Fire Resistant Hydraulic Fluids. 7 pp.

—Ball Screws
BSI BS 6101: Part 1-81. 1981 Machine Tool Ball Screws Part 1: Methods of Calculating Dynamic Load and Life Ratings. 6 pp.

—Bidirectional Triode Thyristors
CNS C6204-88. Method of Test for Reliability Assured Bi-Directional Triode Thyristors Durability (Dec)(9844).

—Concretes—Polyester Resin
JIS A 1186-78. Measuring Methods for Working Life of Polyester Resin Concrete. 8 pp.

—Connectors
CNS C6131-87. Method of Test for Low Frequency (Below 3 MHz) Electrical Connectors (TP-22 Life Test) (Oct)(7661).
CNS C6178-88. Method of Test for Low Frequency (Below 3 MHz) Electrical Connectors (TP-9 Durability) (May)(9238).
CNS C6179-82. Method of Test for Low Frequency (Below 3MHz) Electrical Connectors (TP-17 Temperature Life) (Aug)(9239).

—Dry Reed Switches
CNS C6170-87. Method of Test for Dry Reed Switches (Contact Life Testing) (Jul)(9230).

INTERNATIONAL AND NON-U.S. NATIONAL STANDARDS
SUBJECT INDEX
Lift

Life (Durability) (Cont.)
—Fiber Optics
CNS C6220-87. Method of Test for Fiber Optic Devices (FOTP—4 Temperature Life Test) (Jun)(10875).

—Fire Doors
DIN ENGL 18095 Pt 2-91. Smoke Control Doors; Type Testing for Durability and Leakage (Mar). 8 pp.

—Geotextiles
BSI PD 6533-93. 1993 Methods for Assessing the Durability of Geotextiles—an Interim Document (L). 18 pp.

—Glossaries
CNS B8006-5-85. Glossary of Terms for Reliability (Terms of Time) (Oct)(11381-5).

—Grout
BSI DD 88-83. 1983 Method for the Assessment of Pot Life of Non-Flowing Resin Compositions for Use in Civil Engineering. 6 pp.

—Heating Elements
CNS C3030-75. Method of Life for Electric Heating Wires and Bands (Feb)(2963).
JIS C 2524-79. Accelerated Life Test of Electric Heating Wires and Rolled Wires.
JIS C 2524-65. Methods of Life Test for Electric Heating Wires and Bands (R 1971). 6 pp.

—Mortars
BSI DD 88-83. 1983 Method for the Assessment of Pot Life of Non-Flowing Resin Compositions for Use in Civil Engineering. 6 pp.
CEN PREN 1015-9-93. Methods of Test for Mortar for Masonry—Part 9: Determination of Service Life of Fresh Mortar. 10 pp.

—Rotary Wing Aircraft
CAA CAP524 SUB-Part C 12.86. (Rotorcraft). 13 pp.

—Structural Design
ISO 4356-77. Bases for the Design of Structures—Deformations of Buildings at the Serviceability Limit States First Edition. 21 pp.

—Surgical Knives
JIS T 0201-76. Cut and Durability Test Methods for Surgical Knives.

—Surgical Scissors
JIS T 0202-79. Cut and Durability Test Methods for Dressing Scissors.

—Tools
CNS B7089-83. Method of Life Test for Single Point Carbide Tools (Jan)(4262).
CNS B7092-78. Method of Life Test for High Speed Steel Single Point Tools (Mar)(4265).
JIS B 4011-71. Method of Life Test for Single Point Carbide Tools. 10 pp.

—Tools—High Speed Steels
ISO 8688 Pt 1-89. Tool Life Testing in Milling—Part 1: Face Milling First Edition. 32 pp.
ISO 8688 Pt 2-89. Tool Life Testing in Milling—Part 2: End Milling First Edition. 31 pp.

—Turbine Engines—Ground Effect Machines
CAA Chapter B7-2 App 04.79. Engine Design and Construction. 3 pp.

Life Jackets
Use: Life Preservers

Life Preservers
Use For: Life Jackets; Life Vests See Also: Marine Safety Equipment; Personal Flotation Devices; Swimming Aids
BSI BS 3595-81. 1981 Amd 1 Lifejackets. 18 pp.
CEN PREN 340-90. General Requirements for Protective Clothing Including Hand and Arm Protection and Lifejackets. 20 pp.
CEN PREN 393-90. Life Jackets and Personal Buoyancy Aids—Buoyancy Aids 50N. 27 pp.
CEN PREN 394-90. Life Jackets and Personal Buoyancy Aids—Additional Items. 8 pp.
CEN PREN 395-90. Life Jackets and Personal Buoyancy Aids-Life Jackets, 100N. 34 pp.
CEN PREN 396-90. Life Jackets and Personal Buoyancy Aids-Life Jackets, 150N. 36 pp.
CEN PREN 399-90. Life Jackets and Personal Buoyancy Aids-Life Jackets, 275N. 31 pp.
CGSB CAN/CGSB-65.7-M88. Lifejackets, Inherently Buoyant Type; (Amendment 1 Nov 1990) (Amendment 2 Feb 1993). 24 pp.
CGSB 65-GP-14M-78. Life Jackets, Inherently Buoyant, Standard Type. 17 pp.

Life Preservers (Cont.)
CNS F1008-83. Methods for Testing of Inflatable Lifejackets for Marine Use (for Adults) (Oct)(10630).
CNS F1013-87. Method of Test for Non-Inflatable Lifejackets for Marine Use (Feb)(11519).
CNS F4011-83. Inflatable Lifejacket for Marine Use (for Adult) (Oct)(10629).
CNS F4017-86. Non-Inflatable Lifejackets for Marine Use (Mar)(11518).
MOD UK DSTAN 42-10-79. Life Preservers, Yoke, and Life Preservers, Waistcoat (Life Jackets) Issue 2. 6 pp.
SNZ NZS 5823-89. Specification for Bouyancy Aids and Marine Safety Harnesses and Lines Amend: 1, 1991. 56 pp.

—Aircraft
CAA LEAFLET 5-2 07.90. Lifejackets. 7 pp.
CAA Spec. NO. 5 ISSUE 2.

—Aircraft—Lighting
BSI G 247: Part 1-92. 1992 Locator Lights for Airborne Applications Part 1: Specification for Lifejacket Lights. 15 pp.

—Cellular Plastics
CGSB CAN/CGSB-65.18-M86. Closed-Cell Foamed Polymeric Materials. 17 pp.

—Lighting
CNS F1012-86. Method of Test for Lifejacket Light (Feb)(11506).
CNS F4016-86. Lifejacket Light (Feb)(11505).

—Military Personnel—Interchangeability
NATO STANAG 2997 ED 1 AMD 0-93. (DRAFT) Life Jackets and Personal Floatation Devices. 6 pp.
NATO STANAG 2997 ED 1 AMD 0-00. Life Jackets and Personal Floatation Devices. 4 pp.

—Naval Ships
MOD UK NES 148-92. Requirements for Lifesaving Equipment Issue 1 (06.92). 68 pp.

—Polymeric Films
CGSB 41-GP-10M-79. Film, Plastic, Flexible, Vinyl, Standard for. 10 pp.

Life Rafts
See Also: Floats; Lifeboats; Marine Safety Equipment; Rafts; Survival Craft

—Aerostats
CAA Chapter Q6-6 12.79. Life Rafts (Non-Rigid Airships). 1 p.

—Aircraft
CAA Chapter K6-6 10.69. Life Rafts (Light Aeroplanes). 1 p.
CAA Chapter P6-6. Life Rafts and Escape Chutes (Provisional Airworthiness Requirements for Civil Powered Lift Aircraft). 4 pp.

—Emergency Rations
CNS N5124-76. Emergency Rations for Life Boats and Life Rafts (Tentative) (Sep)(4015). 2 pp.

—First Aid Kits
NATO STANAG 1185 ED 1 AMD 8-77. Minimum Essential Medical and Survival Equipment for Ship Life Rafts Including Guidelines for Survival at Sea. 10 pp.
NATO STANAG 1185 ED 1 AMD 9-77. Minimum Essential Medical and Survival Equipment for Ship Life Rafts Including Guidelines for Survival at Sea. 12 pp.

—Inflatable—Aircraft
CAA Spec. NO. 2 ISSUE 2.

—Inflatable—Aircraft—Coated Fabrics
MOD UK DTD-537E-61. Proofed Fabric and Tape for Inflatable Liferaft Equipment (Reprinted January 1966, Incorporating Amendment No. 1). 7 pp.

—Inflatable—Aircraft—Cotton Fabrics
BSI 3F 57-75. (WITHDRAWN) 1975 Scoured Cotton Fabric for Inflatable Equipments (Superseded by BS 4F 57: 1990). 3 pp.
BSI 4F 57-90. 1990 Scoured Cotton Fabric for Inflatable Equipment for Aerospace Purposes. 4 pp.

—Inflatable—Ships—Fabrics
ISO TR6065-91. Shipbuilding and Marine Structures—Inflatable Liferafts—Materials First Edition. 15 pp.

—Naval Ships
MOD UK NES 148-92. Requirements for Lifesaving Equipment Issue 1 (06.92). 68 pp.

—Rotary Wing Aircraft
CAA 785 (G). Section G Rotorcraft Revision of Chapters G6-6 and G4-3 Launching and Boarding Liferafts (Blue Papers). 2 pp.

Life Rafts (Cont.)
—Rotary Wing Aircraft (Cont.)
CAA Chapter G6-6 12.80. Life Rafts and Escape Chutes/Slides (Rotorcraft). 2 pp.
CAA Chapter G6-6 App 12.80. Liferafts (Rotorcraft). 1 p.
CAA CAP 524 SUB-Part F 12.86. Equipment (Rotorcraft). 17 pp.

—Ships
ISO 4001-77. Shipbuilding—Inland Navigation—Raft-Type Life-Saving Apparatus First Edition. 4 pp.

—Survival Kits
NATO STANAG 1185 ED 1 AMD 8-77. Minimum Essential Medical and Survival Equipment for Ship Life Rafts Including Guidelines for Survival at Sea. 10 pp.
NATO STANAG 1185 ED 1 AMD 9-77. Minimum Essential Medical and Survival Equipment for Ship Life Rafts Including Guidelines for Survival at Sea. 12 pp.

Life Saving Cushions
Use: Personal Flotation Devices—Cushions

Life Testing
Use: Life (Durability)

Life Vests
Use: Life Preservers

Lifeboats
See Also: Boats (Marine); Life Rafts; Marine Safety Equipment; Rafts; Survival Craft
ISO 4143-81. Shipbuilding—Inland Vessels—Open Rowing Lifeboats First Edition. 6 pp.

—Emergency Rations
CNS N5124-76. Emergency Rations for Life Boats and Life Rafts (Tentative) (Sep)(4015). 2 pp.

—Ladders
CNS F3228-83. Embarkation Ladders (Jun)(10343).
JIS F 2617-74. Embarkation Ladders. 7 pp.

—Pulley Blocks
CNS F3028-80. Lifeboats' Steel Blocks (Apr)(5438). 10 pp.

—Winches
ISO 6067-85. Shipbuilding and Marine Structures—Winches for Lifeboats First Edition. 7 pp.

Lifebuoys
See Also: Personal Flotation Devices
CNS F1010-86. Method of Test for Lifebuoys (Feb)(11502).
CNS F4014-86. Lifebuoys (Feb)(11501).
SNZ NZS 5823-89. Specification for Bouyancy Aids and Marine Safety Harnesses and Lines Amend: 1, 1991. 56 pp.

—Lighting
CNS F1011-86. Method of Test for Lifebuoys Self-Igniting Lights (Feb)(11504).
CNS F4015-86. Lifebuoys Self-Igniting Lights (Feb)(11503).

—Naval Ships
MOD UK NES 148-92. Requirements for Lifesaving Equipment Issue 1 (06.92). 68 pp.

—Smoke Signal Devices—Self-Activating
CNS F1016-90. Method of Test for Lifebuoy Self-Activating Smoke Signal for Marine Use (Aug)(12768).
CNS F4020-90. Lifebuoy Self-Activating Smoke Signals for Marine Use (Aug)(12767).

Lift Check Valves
See Also: Check Valves; Valves; Y Valves

—Bronze
CNS B2507-85. Bronze Screwed Lift Check Valves (10kgf/cm2) (Mar)(5968).

—Bronze—Ships
CNS F3166-82. Marine Bronze 5 kgf/cm2 Lift Check Valves (Feb)(8496).
JIS F 7356-87. Marine Bronze 5 K Lift Check Valves.

—Cast Iron—Ships
CNS F3173-88. Cast Iron Lift Check Globe Valves for Marine Use (5 kgf/cm2) (Apr)(8689).
JIS F 7358-88. Cast Iron 5 K Lift Check Globe Valves for Marine Use.

—Copper Alloy
BSI BS 5154-91. 1991 Copper Alloy Globe, Globe Stop and Check, Check and Gate Valves. 22 pp.

INDUSTRY STANDARDS

INTERNATIONAL AND NON-U.S. NATIONAL STANDARDS
SUBJECT INDEX

Lift Check Valves (Cont.)
—Copper Alloy (Cont.)
SNZ NZS/BS 5154-89. Specification for Copper Alloy Globe, Globe Stop and Check, Check and Gate Valves. 20 pp.
SNZ NZS/BS 5154-91. Specification for Copper Alloy Globe, Globe Stop and Check, Check and Gate Valves. 24 pp.

—Steel
BSI BS 5160-89. 1989 Steel Globe Valves, Globe Stop and Check Valves and Lift Type Check Valves. 24 pp.

Lift-Off Tapes, Typewriter
Use: Typewriters—Correction Tapes

Lift Tables
Scope Note: For additional listings, use a more specific term See Also: Machine Tool Tables; Rotary Tables (Machine Tools); Woodworking Equipment

—Glossaries
BSI BS 4361: Part 40-90. 1990 Woodworking Machines Part 40: Nomenclature for Lifting Tables and Stages. 10 pp.
ISO 9617-89. Woodworking Machines—Lifting Tables and Stages—Nomenclature First Edition. 7 pp.

Lift Trucks
Use For: High-Lift Trucks See Also: Bulk Transporters; Dollies; Electric Trucks; Electric Vehicles; Forklifts; Ground Vehicles; Industrial Equipment; Industrial Trucks; Lifting Equipment; Materials Handling Equipment; Motor Vehicles; Stillage Trucks; Trucks
CNS B4055-82. Hand Lift Trucks (Sep)(9347). 5 pp.
JIS B 8924-90. Hand Lift Trucks. 9 pp.

—Falling Object Protective Structures
BSI BS 5933-80. 1980 Overhead Guards for High-Lift Rider Trucks. 4 pp.
ISO 6055-79. High-Lift Rider Trucks—Overhead Guards—Specification and Testing First Edition. 4 pp.

—Glossaries
BSI BS 3810: Part 1-64. 1964 Glossary of Terms Used in Materials Handling Part 1: Terms Used in Connection with Pallets, Stillages, Hand and Powered Trucks. 37 pp.
SNZ NZS 2000: Part 1-65. Glossary of Terms Used in Materials Handling Part 1: Terms Used in Connection with Pallets, Stillages, Hand and Powered Trucks Amend: A, 1970 (Reconfirmed 1978). 36 pp.

—Safety
SNZ NZSR 27-66. Recommendation for the Handling of Standard Pallets (Including a Recommended Safety Code for Industrial Lift Truck Operators) (Reconfirmed 1980). 11 pp.
SNZ NZS/ASME/ANSI B56.1-88. Safety Standard for Low Lift and High Lift Trucks. 49 pp.

—Stability Testing
BSI BS 3726-78. 1978 Counterbalanced Lift Trucks. Stability, Basic Tests. 7 pp.
BSI BS 5777-79. 1979 Methods of Test for Verification of Stability of Pallet Stackers and High Lift Platform Trucks. 11 pp.
ISO 1074-91. Counterbalanced Fork-Lift Trucks—Stability Tests Second Edition. 8 pp.
ISO 5766-90. Pallet Stackers and High-Lift Platform Trucks—Stability Tests Second Edition. 11 pp.
SNZ NZS/ISO 1074-75. Counterbalanced Lift Trucks. Stability. Basic Tests. 3 pp.

Liftcranes
Use: Cranes

Lifting (Coatings)
Scope Note: See the subheading Lifting under specific coating

Lifting Chains
Use For: Leaf Chains See Also: Chains; Lifting Equipment
BSI BS 4942: Part 1-81. 1981 Short Link Chain for Lifting Purposes Part 1: General Conditions of Acceptance. 11 pp.
BSI BS 5594-93. 1993 Leaf Chains, Clevises and Sheaves (ISO 4347: 1992). 22 pp.
BSI BS 5594-86. 1986 Leaf Chains, Clevises and Sheaves. 16 pp.
DIN ENGL 8152 Pt 3-89. Leaf Chains; Heavy Series LH (Feb). 4 pp.
DIN ENGL 8152 Pt 4-89. Leaf Chains; Clevises and Sheaves for Heavy Series LH Chains; Connecting Dimensions (Feb). 3 pp.

Lifting Chains (Cont.)
ISO 1834-80. Short Link Chain for Lifting Purposes—General Conditions of Acceptance First Edition. 10 pp.
ISO 4347-92. Leaf Chains, Clevises and Sheaves Third Edition. 19 pp.
JIS B 1804-83. Leaf Chains. 24 pp.
SNZ NZS/ISO 1834-80. Short Link Chain for Lifting Purposes—General Conditions of Acceptance. 8 pp.

—Alloy Steel
BSI BS 3113-59. 1959 Alloy Steel Chain, Grade 60. Short Link, for Lifting Purposes. 20 pp.

—Alloy Steel—Pulley Blocks
BSI BS 3114-59. 1959 Amd 2 Alloy Steel Chain, Grade 80. Polished Short Link Calibrated Load Chain for Pulley Blocks. 25 pp.

—Hoisting Slings
BSI BS 4942: Part 2-81. 1981 Short Link Chain for Lifting Purposes Part 2: Grade M (4) Non-Calibrated Chain. 11 pp.
BSI BS 4942: Part 3-81. 1981 Short Link Chain for Lifting Purposes Part 3: Grade M (4) Calibrated Chain. 12 pp.
BSI BS 4942: Part 4-81. 1981 Short Link Chain for Lifting Purposes Part 4: Grade S (6) Non-Calibrated Chain. 11 pp.
BSI BS 4942: Part 5-81. 1981 Amd 1 Short Link Chain for Lifting Purposes Part 5: Grade T (8) Non-Calibrated Chain (AMD 4934) March 31, 1986. 13 pp.
BSI BS 4942: Part 6-81. 1981 Amd 1 Short Link Chain for Lifting Purposes Part 6: Grade T (8) Calibrated Chain (AMD 4953) March 31, 1986. 13 pp.
CNS B2713-82. Round-Steel Link Chains for General Purposes and for Haisting Equipments, Chain in Fixed Length, Short Linked (Nov)(9557).
CNS B2714-82. Round-Steel Link Chains Haisting Equipments Calibrated, Tested, Short Linked (Nov)(9558).
CNS F3068-80. Chain Slings (Oct)(6565).
CNS F3070-80. Small Size Chain Slings (Oct)(6567).
DIN ENGL 5687 Pt 3-80. Round Steel Chains; Quality Grade 8; Not-True-to-Gauge-Size; Tested (June). 6 pp.
DIN ENGL 5688 Pt 1-86. Grade 5 Chain Slings with Hook or Ring Type Terminal Fittings (July). 8 pp.
ISO 1835-80. Short Link Chain for Lifting Purposes—Grade M (4), Non-Calibrated, for Chain Slings, Etc. First Edition. 10 pp.
ISO 1836-80. Short Link Chain for Lifting Purposes—Grade M (4), Calibrated, for Chain Hoists and Other Lifting Appliances First Edition. 11 pp.
ISO 3075-80. Short Link Chain for Lifting Purposes—Grade S (6), Non-Calibrated, for Chain Slings, Etc. First Edition. 10 pp.
ISO 3076-84. Short Link Chain for Lifting Purposes—Grade T (8), Non-Calibrated, for Chain Slings, Etc. Second Edition. 11 pp.
ISO 3077-84. Short Link Chain for Lifting Purposes—Grade T (8), Calibrated, for Chain Hoists and Other Lifting Appliances Second Edition. 11 pp.
ISO 8539-86. Forged Steel Lifting Components for Use with Grade T(8) Chain First Edition. 8 pp.
SNZ NZS/ISO 1835-80. Short Link Chain for Lifting Purposes—Grade M (4), Non-Calibrated, for Chain Slings Etc.. 8 pp.
SNZ NZS/ISO 1836-80. Short Link Chain for Lifting Purposes—Grade M (4), Calibrated, for Chain Hoists and Other Lifting Appliances. 9 pp.
SNZ NZS/ISO 3075-80. Short Link Chain for Lifting Purposes-Grade S(6) Non-Calibrated, for Chain Slings Etc.. 9 pp.
SNZ NZS/ISO 3076-84. Short Link Chain for Lifting Purposes-Grade T(8), Non-Calibrated, for Chain Slings Etc.. 9 pp.
SNZ NZS/ISO 3077-84. Short Link Chain for Lifting Purposes-Grade T(8), Calibrated, for Chain Hoists and Other Lifting Appliances. 9 pp.

—Hoisting Slings—Quality Assurance
CEN PREN 818-1-92. Short Link Chain for Lifting Purposes—Safety—Part 1: General Conditions of Acceptance. 20 pp.

—Hoisting Slings—Repair—Boats
MOD UK NES 744-90. Requirements for the Manufacture and Repair of Rings and Links Issue 2 (10.90). 11 pp.

—Hoisting Slings—Safety
CEN PREN 818-2-92. Short Link Chain for Lifting Purposes—Safety—Part 2: Medium Tolerance Chain for Chain Slings—Grade 8. 18 pp.

—Steel
BSI BS 1663-50. 1950 Amd 4 Higher Tensile Steel Chain Grade 40 (Short Link and Pitched or Calibrated) for Lifting Purposes. 22 pp.
BSI BS 2837-88. 1988 Steel Link and Strap Assemblies for Lifting Attachments for Packing Cases. 14 pp.

Lifting Chains (Cont.)
—Steel (Cont.)
BSI BS 6521-84. 1984 Proper Use and Maintenance of Calibrated Round Steel Link Lifting Chains. 7 pp.
BSI BS 6968-88. 1988 Guide for Use and Maintenance of Non-Calibrated Round Steel Lifting Chain and Chain Slings. 13 pp.
CNS B2713-82. Round-Steel Link Chains for General Purposes and for Haisting Equipments, Chain in Fixed Length, Short Linked (Nov)(9557).
CNS B2714-82. Round-Steel Link Chains Haisting Equipments Calibrated, Tested, Short Linked (Nov)(9558).
CNS B2715-82. Round-Steel Link Chains for Lifting, Grade 5, Calibrated, Tested, (Nov)(9559).
CNS B2716-82. Round-Steel Link Chains for Lifting, Grade 6, Calibrated, Tested, (Nov)(9560).
CNS B2717-82. Round-Steel Link Chains for Lifting, Grade 8, Calibrated, Tested, (Nov)(9561).
DIN ENGL 5684 Pt 1-84. Calibrated and Tested Round Steel Link Chains for Lifting Purposes; Grade 5 (May). 5 pp.
DIN ENGL 5684 Pt 2-84. Calibrated and Tested Round Steel Link Chains for Lifting Purposes; Grade 6 (May). 5 pp.
DIN ENGL 5684 Pt 3-84. Calibrated and Tested Round Steel Link Chains for Lifting Purposes; Grade 8 (May). 5 pp.
DIN ENGL 5687 Pt 3-80. Round Steel Chains; Quality Grade 8; Not-True-to-Gauge-Size; Tested (June). 6 pp.
ISO 3056-86. Non-Calibrated Round Steel Link Lifting Chain and Chain Slings—Use and Maintenance Second Edition. 12 pp.
ISO 7592-83. Calibrated Round Steel Link Lifting Chains—Guidelines to Proper Use and Maintenance First Edition. 6 pp.

Lifting Equipment
See Also: Cap Lifters; Corner Fittings; Cranes; Eyeplates; Gas Lifts; Hoisting Drums; Hoisting Slings; Hoists; Lift Trucks; Lifting Chains; Lifting Hooks; Lifting Tackle; Materials Handling Equipment; Patient Lifting Equipment; Rope Drives; Tongs
EC 84/528/EEC-84. Council Directive on the Approximation of the Laws of the Member States Relating to Common Provisions for Lifting and Mechanical Handling Appliances. 14 pp.
ISO 2374-83. Lifting Appliances—Range of Maximum Capacities for Basic Models First Edition. 3 pp.
ISO 4301 Pt 1-86. Cranes and Lifting Appliances—Classification—Part 1: General Second Edition. 6 pp.
ISO 7363-86. Cranes and Lifting Appliances—Technical Characteristics and Acceptance Documents First Edition. 14 pp.
SAA AS 1418.1-86. Cranes (Including Hoists and Winches) (Known as the SAA Crane Code)—Part 1: General Requirements Amdt 1 November 1987. 147 pp.

—Accessories
DIN ENGL 82003 Pt 1-72. Cargo Lifting Gear; Accessories and Fittings; Summary (Feb). 12 pp.
DIN ENGL 82003 Pt 2-72. Cargo Lifting Gear; Accessories and Fittings; Technical Conditions of Delivery (Feb). 4 pp.

—Automotive
JIS D 8108-87. Lifts Above Ground for Automobiles. 20 pp.
SNZ NZS 5462-88. Specification for Vehicle Hoists Amend: A, 1988. 22 pp.

—Barges
BSI BS MA 87: Part 3-87. 1987 Main Dimensions of Shipborne Barges Part 3: The Arrangement, Dimensions and Method of Test for Lifting Post Castings. 7 pp.
ISO 6764-85. Shipbuilding—Shipborne Barges, Series 1—Lifting Post Casting—Arrangement, Dimensions and Method of Testing First Edition. 7 pp.

—Color Coding—Safety
DIN ENGL 15026-78. Lifting Appliances; Marking of Points of Hazard (Jan). 2 pp.

—Control Devices
ISO 7752 Pt 1-83. Lifting Appliances—Controls—Layout and Characteristics—Part 1: General Principles First Edition. 4 pp.

—Electrical Installations
DIN VDE 0100 Pt 726-90. Erection of Power Installations with Nominal Voltages up to 1000 V; Lifting and Hoisting Devices (Mar). 35 pp.

—Glossaries
DIN ENGL 15002-80. Lifting Appliances; Load Carrying Devices; Vocabulary (Apr). 20 pp.
DIN ENGL 15003-70. Lifting Appliances; Load Suspending Devices, Loads and Forces; Definitions (Feb). 3 pp.

INTERNATIONAL AND NON-U.S. NATIONAL STANDARDS
SUBJECT INDEX

Lifting Equipment (Cont.)
—Glossaries (Cont.)
ISO 4306 Pt 1-90. Cranes—Vocabulary—Part 1: General Third Edition. 61 pp.

—Hand Signals
SNZ NZS 5818-82. Specification for Hand Signals for the Direction of Cranes and Similar Lifting Devices Including Helicopters (Other Than for Cargo Handling in Wharf Areas) (Reconfirmed 1989). 12 pp.

—Rope Drives
DIN ENGL 15020 Pt 1-74. Lifting Appliances; Principles Relating to Rope Drives; Calculation and Construction (Feb). 12 pp.
DIN ENGL 15020 Pt 2-74. Lifting Appliances; Principles Relating to Rope Drives; Supervision During Operation (Apr). 9 pp.

—Safety
CNS A1009-75. Safety Codes for Construction Contractors (Dec)(2857). 14 pp.

—Shackles
BSI BS 6994-88. 1988 Steel Shackles for Lifting and General Engineering Purposes: Grade M(4). 12 pp.

—Sheave Blocks—Cross Pieces
DIN ENGL 15412 Pt 1-83. Bottom Blocks for Lifting Appliances; Cross Pieces; Unmachined Parts (Aug). 3 pp.
DIN ENGL 15412 Pt 2-83. Bottom Blocks for Lifting Appliances; Cross Pieces; Finished Parts (Aug). 4 pp.

—Shipping Containers
BSI BS 3951: Sec 1.5-89. 1989 Freight Containers Part 1: General Section 1.5: Guide to the Handling and Securing of Series 1 Freight Containers. 32 pp.
BSI BS 3951: Sec 1.5-01. 1989 Amd 1 Freight Containers Part 1: General Section 1.5: Guide to the Handling and Securing of Series 1 Freight Containers (ISO 3874: 1988) (AMD 7350) December 15, 1992 (Q). 33 pp.
ISO 3874-88. Series 1 Freight Containers—Handling and Securing Amendment 2: 1AAA and 1BBB Containers Fourth Edition; (Amendment 1-1990) (Amendment 2-1993). 39 pp.
JIS Z 1617-79. Lifting and Securing Devices for Freight Containers for International Trade (R 1984). 9 pp.
SNZ NZS/ISO 3874-88. Series 1 Freight Containers. Handling and Securing. 28 pp.

—Wire Rope
ISO 4308 Pt 1-86. Cranes and Lifting Appliances—Selection of Wire Ropes—Part 1: General Second Edition. 7 pp.
ISO 4309-90. Cranes—Wire Ropes—Code of Practice for Examination and Discard Second Edition. 30 pp.

Lifting Hooks
See Also: Hooks; Lifting Equipment
ISO 1837-73. Lifting Hooks—Nomenclature First Edition. 3 pp.
ISO 2308-72. Hooks for Lifting Freight Containers of up to 30 Tonnes Capacity—Basic Requirements First Edition. 4 pp.

—Alloy Steel
BSI BS 3458-62. 1962 Alloy Steel Chain Slings. 49 pp.

—Forged
DIN ENGL 15404 Pt 1-89. Lifting Hooks; Technical Delivery Conditions for Forged Hooks (Dec). 4 pp.

—Knuckle Threads
CNS B2213-78. Lifting Hooks, Knuckle Threads (May)(4368).
DIN ENGL 15403-69. Lifting Hooks for Hoists; Knuckle Threads (Dec). 1 p.

—Laminated
DIN ENGL 15404 Pt 2-88. Lifting Hooks; Technical Delivery Conditions for Laminated Hooks (Nov). 4 pp.

—Laminated—Inspection
DIN ENGL 15405 Pt 2-88. Lifting Hooks; In-Service Inspection of Laminated Hooks (Nov). 2 pp.

—Laminated—Spigot Ladles
DIN ENGL 15407 Pt 2-89. Laminated Point Hooks for Spigot Ladles; Components (Aug). 6 pp.

—Nuts (Fasteners)
DIN ENGL 15413-83. Bottom Blocks for Lifting Appliances; Lifting Hook Nuts (Aug). 4 pp.

—Point
DIN ENGL 15105-85. Lifting Hooks for Lifting Appliances; Point Hooks With Collar (Aug). 3 pp.

Lifting Hooks (Cont.)
—Ramshorn
DIN ENGL 15402 Pt 1-82. Lifting Hooks for Lifting Appliances; Ramshorn Hooks; Unmachined Parts (Nov). 4 pp.
DIN ENGL 15402 Pt 2-83. Lifting Hooks for Lifting Appliances; Ramshorn Hooks; Finished Parts with Threaded Shank (Sept). 5 pp.

—Sheave Blocks
DIN ENGL 15411-83. Lifting Appliances; Lifting Hook Suspensions for Bottom Blocks (Aug). 6 pp.

—Sheave Blocks—Cross Pieces
DIN ENGL 15412 Pt 2-83. Bottom Blocks for Lifting Appliances; Cross Pieces; Finished Parts (Aug). 4 pp.

—Sheave Blocks—Securing Plates
DIN ENGL 15414-83. Bottom Blocks for Lifting Appliances; Securing Plates (Aug). 2 pp.

—Ships—Cargo
JIS F 2105-79. Ships' Cargo Hooks (R 1984). 12 pp.

—Single
DIN ENGL 15401 Pt 1-82. Lifting Hooks for Lifting Appliances; Single Hooks; Unmachined Parts (Nov). 4 pp.
DIN ENGL 15401 Pt 2-83. Lifting Hooks for Lifting Appliances; Single Hooks; Finished Parts with Threaded Shank (Sept). 5 pp.

—Steel
BSI BS 2903-80. 1980 Amd 1 Higher Tensile Steel Hooks for Chains, Slings, Blocks and General Engineering Purposes. 28 pp.
SNZ NZS/BS 2903-80. Specification for Higher Tensile Steel Hooks for Chains, Slings, Blocks and General Engineering Purposes Amend: 1. 28 pp.

—Steel—Forged
ISO 4779-86. Forged Steel Lifting Hooks with Point and Eye for Use with Steel Chains of Grade M(4) First Edition. 7 pp.
ISO 7597-87. Forged Steel Lifting Hooks with Point and Eye for Use with Steel Chains of Grade T(8) First Edition. 8 pp.

—Steel—Forged—Point—Lifting Capacities
DIN ENGL 15400-90. Lifting Hooks; Materials, Mechanical Properties, Lifting Capacity and Stresses (June). 8 pp.

—Steel—Forged—Point—Mechanical Properties
DIN ENGL 15400-90. Lifting Hooks; Materials, Mechanical Properties, Lifting Capacity and Stresses (June). 8 pp.

—Steel—Forged—Point—Stresses
DIN ENGL 15400-90. Lifting Hooks; Materials, Mechanical Properties, Lifting Capacity and Stresses (June). 8 pp.

—Steel—Forged—Ramshorn—Lifting Capacities
DIN ENGL 15400-90. Lifting Hooks; Materials, Mechanical Properties, Lifting Capacity and Stresses (June). 8 pp.

—Steel—Forged—Ramshorn—Mechanical Properties
DIN ENGL 15400-90. Lifting Hooks; Materials, Mechanical Properties, Lifting Capacity and Stresses (June). 8 pp.

—Steel—Forged—Ramshorn—Stresses
DIN ENGL 15400-90. Lifting Hooks; Materials, Mechanical Properties, Lifting Capacity and Stresses (June). 8 pp.

—Steel—Laminated—Lifting Capacities
DIN ENGL 15400-90. Lifting Hooks; Materials, Mechanical Properties, Lifting Capacity and Stresses (June). 8 pp.

—Steel—Laminated—Mechanical Properties
DIN ENGL 15400-90. Lifting Hooks; Materials, Mechanical Properties, Lifting Capacity and Stresses (June). 8 pp.

—Steel—Laminated—Stresses
DIN ENGL 15400-90. Lifting Hooks; Materials, Mechanical Properties, Lifting Capacity and Stresses (June). 8 pp.

—Steel—Shipping Containers
BSI BS 4654-70. 1970 Hooks for Lifting Freight Containers of up to 30 Tonnes. 11 pp.

Lifting Machinery
Use: Lifting Equipment

Lifting Platforms
Use: Elevators (Lifts)

Lifting Slings
Use: Hoisting Slings

Lifting Tackle
See Also: Eyebolts; Hoisting Slings; Lifting Equipment
MOD UK DSTAN 40-14-92. Lifting Tackle Issue 1. 16 pp.

—Glossaries
BSI BS 3810: Part 5-71. 1971 Glossary of Terms Used in Materials Handling Part 5: Terms Used in Connection with Lifting Tackle. 37 pp.

Lifts
Scope Note: Use a more specific term
See: Automotive Lifts; Elevators (Lifts); Gas Lifts; Manlifts; Shiplifts

Lifts (Elevators)
Use: Elevators (Lifts)

Ligatures
See Also: Sutures

—Carrier
CNS T1038-81. Ligature Carrier (Jan)(6969).
JIS T 2613-90. Ligature Carrier. 5 pp.

Light (Visible Radiation)
Use For: Visible Radiation See Also: Attenuation; Color; Electromagnetic Radiation; Infrared Radiation; Lamps; Lighting; Luminance; Opacity; Optical Measurement; Optical Properties; Photometry; Photosensitivity; Reflection; Refraction; Solar Radiation; Transmittance; Tristimulus Values; Ultraviolet Radiation; White Light

—Atmospheric Attenuation
CCIR Report 883-2-90. Attenuation of Visible and Infra-Red Radiation—Section 5C—Effects of the Atmosphere (Radiometeorology). 8 pp.

Light Absorption
—Cellulose Acetate
BSI BS 2782:Pt5: METH 553A-91. 1991 Methods of Testing Plastics: Part 5: Optical and Colour Properties, Weathering Method 553A: Determination of Light Absorption of Cellulose Acetate, Before and After Further Heating (ISO 1600: 1990). 9 pp.
ISO 1600-90. Plastics—Cellulose Acetate—Determination of Light Absorption on Moulded Specimens Produced Using Different Periods of Heating Second Edition. 8 pp.

Light Activated Semiconductor Controlled Rectifiers
Use: Silicon Controlled Rectifiers—Light Activated

Light Alloys
Use: Light Metal Alloys

Light Amplifiers (Intensifiers)
Use: Image Intensifiers

Light Bulbs
See Also: Lamps
BSI BS 535-87. 1987 Amd 1 Light Sources for Miners' Portable Electric Lamps (AMD 5863) March 31, 1989. 23 pp.
BSI BS 7132-89. 1989 Nomenclature for Glass Bulb Designation System for Lamps. 11 pp.
CNS C1052-80. Designation Methods for Glass Bulbs of Electric Lamps (Mar)(5312).
CNS C3013-84. Method of Test for Miniature Lamps (Jun)(720). 1 p.
CNS C4089-84. Electrical Bulbs for Dynamo Lamps of Bicycles (Jun)(2944). 4 pp.
CNS C4108-83. Small Lamps Bulb (for Coal Miner Helmet Use) (Feb)(3375).
CNS C4416-84. Small Lamp Bulbs for Household Use (Sep)(11006).
CNS D2004-86. Automotive Lamp Bulbs (Feb)(2945).
IEC 887-88. Glass Bulb Designation System for Lamps First Edition. 20 pp.
JIS C 7502-83. Bulbs for Miners' Electric Cap Lamps. 10 pp.
JIS C 7510-92. Incandescent Lamps for Bicycle Dynamo Lamps. 11 pp.
JIS C 7710-88. Designation Method for Glass Bulbs of Lamps. 19 pp.

—Marine
CNS F5017-85. Marine Lamp Bulbs (Jul)(8269).

—Packaging—Export
CNS Z5089-81. Packaging and Packing of Electric Bulb for Export (Jun)(7541).

INDUSTRY STANDARDS

Light

Light Bulbs (Cont.)
—Radio Panels
CNS C7006-62. Electric Bulbs for Radio Panel (Dec)(2058) (R 1973). 3 pp.

—Rolling Stock
CNS C4418-84. Railway Car Lamp Bulbs (Sep)(11008).

Light Dimmers
Use: Dimmers

Light Emitting Diode Arrays
See Also: Diodes; Light Emitting Diodes
CECC CECC 20 001 ISSUE 2-91. Blank Detail Specification: Light Emitting Diodes, Light Emitting Diode Arrays, Light Emitting Diode Displays Without Internal Logic and Resistor (En, Fr, Ge). 36 pp.
IECQ PQC 105-92. Blank Detail Specification for Light Emitting Diodes, Light Emitting Diode Arrays. 18 pp.

—Bar Graph—Preferred Products List
CECC CECC MUAHAG Vol 10 IS 2-92. Preferred Products List; OPTO Electronic Devices (En, Fr, Ge). 48 pp.

—Bars—Preferred Products List
CECC CECC MUAHAG Vol 10 IS 2-92. Preferred Products List; OPTO Electronic Devices (En, Fr, Ge). 48 pp.

—Preferred Products List
CECC CECC MUAHAG Vol 10 IS 2-92. Preferred Products List; OPTO Electronic Devices (En, Fr, Ge). 48 pp.

Light Emitting Diode Displays
See Also: Display Devices
CECC CECC 20 001 ISSUE 2-91. Blank Detail Specification: Light Emitting Diodes, Light Emitting Diode Arrays, Light Emitting Diode Displays Without Internal Logic and Resistor (En, Fr, Ge). 36 pp.

—Dot Matrix—Preferred Products List
CECC CECC MUAHAG Vol 10 IS 2-92. Preferred Products List; OPTO Electronic Devices (En, Fr, Ge). 48 pp.

—7-Segment—Preferred Products List
CECC CECC MUAHAG Vol 10 IS 2-92. Preferred Products List; OPTO Electronic Devices (En, Fr, Ge). 48 pp.

Light Emitting Diodes
Use For: LED *See Also:* Infrared Emitting Diodes; Laser Diodes; Light Emitting Diode Arrays; Light Sources; Optoelectronic Devices
BSI BS 6493: Sec 1.5-92. 1992 Semiconductor Devices Part 1: Discrete Devices Section 1.5: Recommendations for Optoelectronic Devices (IEC 747-5: 1992). 115 pp.
BSI BS 6493: Sec 1.5-85. 1985 Semiconductor Devices Part 1: Discrete Devices Section 1.5: Recommendations for Optoelectronic Devices. 34 pp.
CECC CECC 20 000 ISSUE 1-82. Generic Specification: Semiconductor Optoelectronic and Liquid Crystal Devices (En, Fr, Ge) AMD 3 (En, Fr, Ge) SUPP 1 (En, Fr, Ge) SUPP 2 (En, Fr, Ge). 401 pp.
CECC CECC 20 001 ISSUE 2-91. Blank Detail Specification: Light Emitting Diodes, Light Emitting Diode Arrays, Light Emitting Diode Displays Without Internal Logic and Resistor (En, Fr, Ge). 36 pp.
CNS C6278-87. Measuring Method for Light Emitting Diodes (for Indication) (Feb)(11830).
CNS C7185-87. Light Emitting Diodes (for Indication) (Feb)(11829).
IEC 747 Pt 5-92. Semiconductor Devices Discrete Devices and Integrated Circuits Part 5: Optoelectronic Devices Second Edition. 238 pp.
IECQ QC 720100-91. Semiconductor Devices Part 12: Sectional Specification for Optoelectronic Devices (IEC 747-12 ED 1). 44 pp.
IECQ PQC 105-92. Blank Detail Specification for Light Emitting Diodes, Light Emitting Diode Arrays. 18 pp.
JIS C 7035-85. Light Emitting Diodes (for Indication). 19 pp.
JIS C 7036-85. Measuring Methods for Light Emitting Diodes (for Indication); (Erratum). 23 pp.
MOD UK DSTAN 59-58: 90/002-82. Display Device—Light Emitting Diode Display Issue 1. 23 pp.
MOD UK DSTAN 59-58: 90/003-84. Semiconductor Device—Light Emitting Diode Display Issue 1. 25 pp.
MOD UK DSTAN 59-61: Part 7-01. Semiconductor Devices (Listed on EPIC Database) Part 7: Optoelectronic Devices Issue 1; Amendment 1. 17 pp.

Light Emitting Diodes (Cont.)
MOD UK DSTAN 59-61: 80/018-01. Semiconductor Device—Light Emitting Diode Issue 1; Amendment 1. 15 pp.
MOD UK DSTAN 59-61: 80/028-01. Semiconductor Device—Light Emitting Diode Issue 1; Amendment 1. 15 pp.
MOD UK DSTAN 59-61: 80/033-84. Semiconductor Device—Light Emitting Diode Modules Issue 1. 12 pp.
MOD UK DSTAN 59-61: 80/042-86. Semiconductor Device—Light Emitting Diode Issue 1. 13 pp.
MOD UK DSTAN 59-61: 90/167-76. Semiconductor Device, Light Emitting Diode Array Issue 1. 14 pp.
MOD UK DSTAN 59-61: 90/185-01. Semiconductor Device—Light Emitting Diode Issue 1; Amendment 1. 15 pp.
MOD UK DSTAN 59-61: 90/187-01. Semiconductor Device—Light Emitting Diode Display Issue 1; Amendment 1. 27 pp.
MOD UK DSTAN 59-61: 90/195-01. Semiconductor Device—Light Emitting Diode Issue 1; Amendment 1. 17 pp.

—Fiber Optic Equipment
JIS C 5950-89. General Rules of Light Emitting Diodes for Fiber Optic Transmission. 12 pp.
JIS C 5951-89. Test Methods of Light Emitting Diodes for Fiber Optic Transmission. 16 pp.

—Light Strips—Preferred Products List
CECC CECC MUAHAG Vol 10 IS 2-92. Preferred Products List; OPTO Electronic Devices (En, Fr, Ge). 48 pp.

—Preferred Products List
CECC CECC MUAHAG Vol 10 IS 2-92. Preferred Products List; OPTO Electronic Devices (En, Fr, Ge). 48 pp.

—Quality Assurance
BSI BS 9370-83. 1983 Light Emitting and Infra-Red Diode Arrays of Assessed Quality: Generic Data and Methods of Test. 43 pp.
BSI BS CECC 20001-83. 1983 Light Emitting Diodes, Light Emitting Diode Arrays: Blank Detail Specification. 14 pp.
BSI BS CECC 20001-91. 1991 Light Emitting Diodes, Light Emitting Diode Arrays, Light Emitting Diode Displays Without Internal Logic and Resistor (T). 12 pp.
IEC 747 Pt 12-91. Semiconductor Devices Part 12: Sectional Specification for Optoelectronic Devices First Edition (IECQ QC 720100). 44 pp.

—Seven Segment
MOD UK DSTAN 59-61: 90/199-81. Semiconductor Device—Light Emitting Diode Display Issue 1. 21 pp.

Light Exposure Apparatus
Use: Environmental Testing Equipment—Light Exposure

Light Fastness Testing
Use: Light Testing

Light Filters
Use: Optical Filters

Light Hydrocarbons
See Also: Hydrocarbons; Petroleum Products

—Tanks (Containers)—Refrigerated—Capacitance Level Indicators
ISO 8309-91. Refrigerated Light Hydrocarbon Fluids—Measurement of Liquid Levels in Tanks Containing Liquefied Gases—Electrical Capacitance Gauges First Edition. 15 pp.

—Tanks (Containers)—Refrigerated—Float Gages
ISO 10574-93. Refrigerated Light-Hydrocarbon Fluids—Measurement of Liquid Levels in Tanks Containing Liquefied Gases—Float-Type Level Gauges First Edition. 10 pp.

—Tanks (Containers)—Refrigerated—Resistance Thermometers
BSI BS 7575-92. 1992 Resistance Thermometers and Thermocouples for the Measurement of Temperature in Tanks Containing Refrigerated Light Hydrocarbon Fluids (ISO 8310: 1991). 15 pp.
ISO 8310-91. Refrigerated Light Hydrocarbon Fluids—Measurement of Temperature in Tanks Containing Liquefied Gases—Resistance Thermometers and Thermocouples First. 13 pp.

Light Hydrocarbons (Cont.)
—Tanks (Containers)—Refrigerated—Thermocouples
BSI BS 7575-92. 1992 Resistance Thermometers and Thermocouples for the Measurement of Temperature in Tanks Containing Refrigerated Light Hydrocarbon Fluids (ISO 8310: 1991). 15 pp.
ISO 8310-91. Refrigerated Light Hydrocarbon Fluids—Measurement of Temperature in Tanks Containing Liquefied Gases—Resistance Thermometers and Thermocouples First. 13 pp.

—Tanks (Containers)—Refrigerated—Triangulation Measurement
BSI BS 7550: Part 2-93. 1993 Calibration of Spherical Tanks for Refrigerated Light-Hydrocarbon Fluids in Ships Part 2: Triangulation Method (ISO 9091-2: 1992) (Q). 27 pp.
ISO 9091 Pt 2-92. Refrigerated Light Hydrocarbon Fluids—Calibration of Spherical Tanks in Ships—Part 2: Triangulation Measurement First Edition. 26 pp.

Light Metal Alloy Coatings (On Light Metal Alloys)
See Also: Coatings; Metal Coatings (On Metal)
MOD UK DEF-2331-A-61. Protective PX-1, Dyed Protective PX-1, Undyed (Consolidated Edition August 1976, Incorporating Amendments Nos. 1 and 2). 3 pp.
MOD UK DEF-2331-A-01. Protective PX-1, Dyed Protective PX-1, Undyed (Consolidated Edition August 1976 Incorporating Amendments Nos. 1 and 2); Amendment 3. 5 pp.

Light Metal Alloys
Scope Note: For additional listings, see also specific products made from light metal alloys
Use For: Light Alloys *See Also:* Alloys; Light Metals

—Chemical Analysis—Statistical Interpretation
ISO TR7242-81. Chemical Analysis of Light Metals and Their Alloys—Statistical Interpretation of Inter-Laboratory Trials First Edition. 15 pp.

—Forgings—Aircraft—Surface Properties
SBAC RS 724 (V). Forgings, Tolerances and Surface Condition.

—Glossaries
BSI BS EN 23134-1-91. 1991 Light Metals and Their Alloys—Terms and Definitions Part 1: Materials (ISO 3134-1: 1985). 13 pp.
BSI BS EN 23134-2-91. 1991 Light Metals and Their Alloys—Terms and Definitions Part 2: Unwrought Products (ISO 3134-2: 1985). 9 pp.
BSI BS EN 23134-3-91. 1991 Light Metals and Their Alloys—Terms and Definitions Part 3: Wrought Products (ISO 3134-3: 1985). 17 pp.
BSI BS EN 23134-4-91. 1991 Light Metals and Their Alloys—Terms and Definitions Part 4: Castings (ISO 3134-4: 1985). 9 pp.
CEN EN 23134-1-91. Light Metals and Their Alloys—Terms and Definitions—Part 1: Materials. 3 pp.
CEN EN 23134-2-91. Light Metals and Their Alloys—Terms and Definitions—Part 2: Unwrought Products. 3 pp.
CEN EN 23134-3-91. Light Metals and Their Alloys—Terms and Definitions—Part 3: Wrought Products. 3 pp.
CEN EN 23134-4-91. Light Metals and Their Alloys—Terms and Definitions—Part 4: Castings. 3 pp.
DIN ENGL EN 23134 Pt 1-91. Light Metals and Their Alloys; Terms and Definitions; Materials (ISO 3134-1:1985) (Nov) (Supersedes Parts of DIN 17600 Part 1, December 1969 Edition). 9 pp.
DIN ENGL EN 23134 Pt 2-91. Light Metals and Their Alloys; Terms and Definitions; Unwrought Products (ISO 3134-2: 1985) (Nov). 5 pp.
DIN ENGL EN 23134 Pt 3-91. Light Metals and Their Alloys; Terms and Definitions; Wrought Products (ISO 3134-3: 1985) (Nov). 13 pp.
DIN ENGL EN 23134 Pt 4-91. Light Metals and Their Alloys; Terms and Definitions; Castings (ISO 3134-4: 1985) (Nov). 5 pp.
ISO 3134 Pt 1-85. Light Metals and Their Alloys—Terms and Definitions—Part 1: Materials First Edition; (Corrected and Reprinted -1989). 8 pp.
ISO 3134 Pt 2-85. Light Metals and Their Alloys—Terms and Definitions—Part 2: Unwrought Products First Edition; (Corrected and Reprinted -1989). 4 pp.
ISO 3134 Pt 3-85. Light Metals and Their Alloys—Terms and Definitions—Part 3: Wrought Products First Edition; (Corrected and Reprinted -1989). 12 pp.

INTERNATIONAL AND NON-U.S. NATIONAL STANDARDS
SUBJECT INDEX

Light

Light Metal Alloys *(Cont.)*
—Glossaries *(Cont.)*
ISO 3134 Pt 4-85. Light Metals and Their Alloys—Terms and Definitions—Part 4: Castings First Edition; (Corrected and Reprinted -1989). 4 pp.

ISO 3134 Pt 5-81. Light Metals and Their Alloys—Terms and Definitions—Part 5: Methods of Processing and Treatment First Edition. 3 pp.

—Symbols
ISO 2092-81. Light Metals and Their Alloys—Code of Designation Based on Chemical Symbols First Edition. 4 pp.

Light Metals
Scope Note: For additional listings, see also specific products made from light metals *See Also:* Light Metal Alloys; Metals

—Chemical Analysis—Statistical Interpretation
ISO TR7242-81. Chemical Analysis of Light Metals and Their Alloys—Statistical Interpretation of Inter-Laboratory Trials First Edition. 15 pp.

—Glossaries
BSI BS EN 23134-1-91. 1991 Light Metals and Their Alloys—Terms and Definitions Part 1: Materials (ISO 3134-1: 1985). 13 pp.

BSI BS EN 23134-2-91. 1991 Light Metals and Their Alloys—Terms and Definitions Part 2: Unwrought Products (ISO 3134-2: 1985). 9 pp.

BSI BS EN 23134-3-91. 1991 Light Metals and Their Alloys—Terms and Definitions Part 3: Wrought Products (ISO 3134-3: 1985). 17 pp.

BSI BS EN 23134-4-91. 1991 Light Metals and Their Alloys—Terms and Definitions Part 4: Castings (ISO 3134-4: 1985). 9 pp.

CEN EN 23134-1-91. Light Metals and Their Alloys—Terms and Definitions—Part 1: Materials. 3 pp.

CEN EN 23134-2-91. Light Metals and Their Alloys—Terms and Definitions—Part 2: Unwrought Products. 3 pp.

CEN EN 23134-3-91. Light Metals and Their Alloys—Terms and Definitions—Part 3: Wrought Products. 3 pp.

CEN EN 23134-4-91. Light Metals and Their Alloys—Terms and Definitions—Part 4: Castings. 3 pp.

DIN ENGL EN 23134 Pt 1-91. Light Metals and Their Alloys; Terms and Definitions; Materials (ISO 3134-1:1985) (Nov) (Supersedes Parts of DIN 17600 Part 1, December 1969 Edition). 9 pp.

DIN ENGL EN 23134 Pt 2-91. Light Metals and Their Alloys; Terms and Definitions; Unwrought Products (ISO 3134-2: 1985) (Nov). 5 pp.

DIN ENGL EN 23134 Pt 3-91. Light Metals and Their Alloys; Terms and Definitions; Wrought Products (ISO 3134-3: 1985) (Nov). 13 pp.

DIN ENGL EN 23134 Pt 4-91. Light Metals and Their Alloys; Terms and Definitions; Castings (ISO 3134-4: 1985) (Nov). 5 pp.

ISO 3134 Pt 1-85. Light Metals and Their Alloys—Terms and Definitions—Part 1: Materials First Edition; (Corrected and Reprinted -1989). 8 pp.

ISO 3134 Pt 2-85. Light Metals and Their Alloys—Terms and Definitions—Part 2: Unwrought Products First Edition; (Corrected and Reprinted -1989). 4 pp.

ISO 3134 Pt 3-85. Light Metals and Their Alloys—Terms and Definitions—Part 3: Wrought Products First Edition; (Corrected and Reprinted -1989). 12 pp.

ISO 3134 Pt 4-85. Light Metals and Their Alloys—Terms and Definitions—Part 4: Castings First Edition; (Corrected and Reprinted -1989). 4 pp.

ISO 3134 Pt 5-81. Light Metals and Their Alloys—Terms and Definitions—Part 5: Methods of Processing and Treatment First Edition. 3 pp.

—Symbols
ISO 2092-81. Light Metals and Their Alloys—Code of Designation Based on Chemical Symbols First Edition. 4 pp.

Light Meters
Use For: Illuminance Meters; Lens Meters
See Also: Exposure Meters; Luminance; Photometers
CNS C4165-88. Illuminance Meters (Oct)(5119).
JIS B 7183-75. Lens-Meters.
JIS C 1609-83. Illuminance Meters. 17 pp.

—Packaging—Export
CNS Z5076-81. Packaging and Packing of Lens Meter for Export (Jan)(6980).

Light Microscopes
Use: Optical Microscopes

Light Poles
Use: Lighting Poles

Light Scattering Automatic Particle Counters
See Also: Particulate Monitors
JIS B 9921-89. Light Scattering Automatic Particle Counter. 22 pp.

JIS B 9925-91. Light Scattering Automatic Particle Counter for Liquid. 14 pp.

Light Scattering Equipment
Scope Note: Use a more specific term *See:* Light Scattering Automatic Particle Counters

Light Sensitive Detectors
Use: Photodetectors

Light Sensitive Resistors
Use: Photoconductive Cells

Light Signalling Devices
Use: Signal Lights

Light Sources
Use For: Illuminants *See Also:* Fiber Optic Light Sources; Flashtubes; Illuminators (Optics); Laser Diodes; Light Emitting Diodes; Lighting; Optical Equipment

—Colorimetry
BSI BS 950: Part 1-67. 1967 Artificial Daylight for the Assessment of Colour Part 1: Illuminant for Colour Matching and Colour Appraisal. 13 pp.

CGSB CAN/CGSB-4.2 NO.41-M91. Textile Test Methods Standard Light Sources for Colour Matching of Textiles. 6 pp.

CNS Z7194-85. Standard Illuminants and Sources for Colorimetry (Jun)(11294).

CNS Z7199-85. Method of Measurement for Light Source Colour (Aug)(11353).

ISO 4911-80. Textiles—Cotton Fibres—Equipment and Artificial Lighting for Cotton Classing Rooms First Edition. 9 pp.

ISO CIE10526-91. CIE Standard Colorimetric Illuminants First Edition. 15 pp.

JIS Z 8110-91. Names of Light-Source Colours. 8 pp.

JIS Z 8716-91. Fluorescent Lamp as a Simulator of CIE Standard Illuminant D65 for a Visual Comparison of Surface Colours—Type and Characteristics. 8 pp.

JIS Z 8720-83. Standard Illuminants and Sources for Colorimetry. 25 pp.

JIS Z 8724-83. Methods of Measurement for Light Source Colour. 25 pp.

JIS Z 8726-90. Method of Specifying Colour Rendering Properties of Light Sources. 21 pp.

Light Strips
See Also: Lighting

—Light Emitting Diodes—Preferred Products List
CECC CECC MUAHAG Vol 10 IS 2-92. Preferred Products List; OPTO Electronic Devices (En, Fr, Ge). 48 pp.

Light Susceptability
Use: Light Testing

Light Testing
Use For: Light Susceptability *See Also:* Daylight Testing

—Artificial
CNS B6087-85. Glass-Enclosed Carbon-Arc Type Apparatus for Artificial Light Exposure Tests (Apr)(11229).

JIS B 7751-90. Light-Exposure and Light-and-Water-Exposure Apparatus (Enclosed Carbon-Arc Type). 33 pp.

—Artificial—Adhesive Tapes
SAA AS 1635.21.2-85. Methods of Testing Pressure Sensitive Adhesive Tape—Part 21.2: Determination of Resistance to Accelerated Ageing by Artificial Light. 2 pp.

—Artificial—Aluminum Coatings (On Aluminum)
ISO TR11728-93. Anodized Aluminium and Aluminium Alloys—Accelerated Test of Weather Fastness of Coloured Anodic Oxide Coatings Using Cyclic Artificial Light and Pollution Gas First Edition. 9 pp.

—Artificial—Coatings—Bicycles
CNS B7072-75. Method of Test for Surface Treatment of Bicycle Parts (Dec)(3885). 2 pp.

Light Testing *(Cont.)*
—Artificial—Elastomers
DIN ENGL 53387-89. Artificial Weathering and Ageing of Plastics and Elastomers by Exposure to Filtered Xenon Arc Radiation (Apr) (Supersedes June 1982 Edition and the April 1974 Edition of DIN 53389, Withdrawn in 1986). 8 pp.

—Artificial—Plastics
BSI BS 2782:Pt5: METH 540B-82. 1982 Methods of Testing Plastics: Part 5: Optical and Colour Properties, Weathering: Method 540B: Methods of Exposure to Laboratory Light Sources, (Xenon Arc Lamp, Enclosed Carbon Arc Lamp, Open-Flame Carbon Arc Lamp, Flourescent Tube Lamps). 20 pp.

BSI BS 2782:Pt5: METH 540C-88. 1988 Methods of Testing Plastics Part 5: Optical and Colour Properties Method 540C: Determination of Ultraviolet Radiation Intensity Using Polysulphane Film. 8 pp.

BSI BS 2782:Pt5: METH 552A-81. 1981 Methods of Testing Plastics: Part 5: Optical and Colour Properties, Weathering: Method 552A: Determination of Changes in Colour Variations in Prop After Exposure to Daylight Under Glass, Natural Weathering or Artificial Light. 10 pp.

BSI BS 4618: Sec 4.3-74. 1974 Recommendations for the Presentation of Plastics Design Data Part 4: Enviromental and Chemical Effects Section 4.3: Resistance to Colour Change Produced by Exposure to Light. 8 pp.

DIN ENGL 53384-89. Artificial Weathering and Ageing of Plastics by Exposure to Laboratory UV Radiation Sources (Apr). 5 pp.

DIN ENGL 53387-89. Artificial Weathering and Ageing of Plastics and Elastomers by Exposure to Filtered Xenon Arc Radiation (Apr) (Supersedes June 1982 Edition and the April 1974 Edition of DIN 53389, Withdrawn in 1986). 8 pp.

ISO 4582-80. Plastics—Determination of Changes in Colour and Variations in Properties After Exposure to Daylight Under Glass, Natural Weathering or Artificial Light First Edition. 8 pp.

ISO 4892-81. Plastics—Methods of Exposure to Laboratory Light Sources First Edition; (Amendment Slip-1982). 21 pp.

—Artificial—Sealants
ISO 11431-93. Building Construction—Sealants—Determination of Adhesion/Cohesion Properties After Exposure to Artificial Light Through Glass First Edition. 7 pp.

—Artificial—Varnishes
CNS K6056-70. Method of Test for Baking Varnish (Jan)(771). 6 pp.

—Artificial—Vulcanized Rubber
BSI BS 903: Part A54-89. 1989 Methods of Testing Vulcanized Rubber Part A54: Methods for Exposure to Artificial Light, for Use in Determining Resistance to Weathering (ISO 4665-3: 1987) (V). 16 pp.

BSI BS 903: Part A55-89. 1989 Methods of Testing Vulcanized Rubber Part A55: Methods for Assessment of Changes in Properties After Exposure to Natural Weathering or Artificial Light, for Use in Determining Resistance to Weathering (ISO 4665/1: 1985) (V). 8 pp.

ISO 4665 Pt 1-85. Rubber, Vulcanized—Resistance to Weathering—Part 1: Assessment of Changes in Properties After Exposure to Natural Weathering or Artificial Light First Edition. 7 pp.

ISO 4665 Pt 3-87. Rubber, Vulcanized—Resistance to Weathering—Part 3: Methods of Exposure to Artificial Light First Edition. 16 pp.

—Clear Coatings
CNS K6708-82. Method of Test for Light Stability of Clear Coatings (Feb)(8515).

—Draperies
JIS L 1055-87. Testing Methods for Light Blocking Effect of Curtain Materials. 8 pp.

—Fabrics
BSI BS EN 20105-B02-93. 1993 Textiles—Tests for Colour Fastness Part B02: Colour Fastness to Artificial Light (Xenon Arc Fading Lamp Test) (ISO 105-B02: 1988) (L). 18 pp.

CEN EN 20105-B02-92. Textiles—Tests for Colour Fastness Part B02. Colour Fastness to Artificial Light: Xenon Arc Fading Lamp Test (ISO 105-B02: 1988). 13 pp.

CGSB CAN/CGSB-4.2 NO.18.1-M90. Textile Test Methods Colourfastness to Artificial Light: Carbon-Arc Radiation. 20 pp.

CGSB CAN/CGSB-4.2 NO.18.2-M90. Textile Test Methods Textiles—Tests for Colourfastness—Part B01: Colourfastness to Light: Daylight (ISO 105-B01:1988). 20 pp.

INDUSTRY STANDARDS

INTERNATIONAL AND NON-U.S. NATIONAL STANDARDS
SUBJECT INDEX — Light

Light Testing (Cont.)
—Fabrics (Cont.)
- CGSB CAN/CGSB-4.2 NO.18.3-M90. Textile Test Methods—Tests for Colourfastness—Part B02: Colourfastness to Artificial Light: Xenon Arc Fading Lamp Test (ISO 105-B02:1988). 28 pp.
- CNS L3026-82. Method of Test for Colour Fastness to Sunlight and Daylight (Jun)(1493).
- CNS L3074-82. Method of Test for Colour Fastness to Carbon Arc Lamp Light (Jun)(3845). 4 pp.
- CNS L3075-82. Method of Test for Colour Fastness to Xenon Arc Lamp Light (Jun)(3846).
- CNS L3192-83. Method of Test for Colour Fastness to Light of Fluorescent Whitening Agents and Fluorescent Whitened Textiles (Jan)(9901).
- CNS L3193-83. Method of Test for Colour Fastness to Light and Perspiration (Jan)(9902).
- CNS L3194-83. Method of Test for Detection and Colour Fastness to Photochromism (Jan) (9903).
- ISO 105 Pt B01-89. Textiles—Tests for Colour Fastness—Part B01: Colour Fastness to Light: Daylight Fourth Edition. 10 pp.
- ISO 105 Pt B02-88. Textiles—Tests for Colour Fastness—Part B02: Colour Fastness to Artifical Light: Xenon Arc Fading Lamp Test Third Edition; (CEN EN 20105-B02:1992). 14 pp.
- ISO 105 Pt B03-88. Textiles—Tests for Colour Fastness—Part B03: Colour Fastness to Weathering: Outdoor Exposure Third Edition. 8 pp.
- ISO 105 Pt B04-88. Textiles—Tests for Colour Fastness—Part B04: Colour Fastness to Weathering: Xenon Arc Third Edition. 8 pp.
- ISO 105 Pt B05-88. Textiles—Tests for Colour Fastness—Part B05: Detection and Assessment of Photochromism Third Edition. 7 pp.
- ISO 105 Pt B06-92. Textiles—Tests for Colour Fastness—Part B06: Colour Fastness to Artificial Light at High Temperatures: Xenon Arc Fading Lamp Test First Edition. 15 pp.
- JIS L 0841-92. Testing Methods for Colour Fastness to Daylight. 12 pp.
- JIS L 0842-88. Testing Methods for Colour Fastness to Carbon Arc Lamp Light. 6 pp.
- JIS L 0843-88. Testing Methods for Colour Fastness to Xenon Arc Lamp Light. 9 pp.
- JIS L 0886-92. Testing Methods for the Detection and Assessment of Photochromism. 7 pp.
- JIS L 0887-75. Testing Method for Colour Fastness to Light of Fluorescent Whitening Agents and Fluorescent Whitened Textiles.
- JIS L 0888-88. Testing Method for Colour Fastness to Light and Perspiration.

—Fiber Optics
- CNS C6250-85. Method of Test for Fiber Optic Devices (Ambient Light Susceptibility) (Aug)(11332).
- CNS C6307-87. Method of Test for Fiber Optic Devices (FOTP-22 Ambient Light Susceptibility) (Dec)(12171).

—Fluorescent Whitening Agents
- CNS L3192-83. Method of Test for Colour Fastness to Light of Fluorescent Whitening Agents and Fluorescent Whitened Textiles (Jan)(9901).
- JIS L 0887-75. Testing Method for Colour Fastness to Light of Fluorescent Whitening Agents and Fluorescent Whitened Textiles.

—Lacquers
- SAA AS 1580.482. 1-75. Paints and Related Materials—Methods of Test—Part 482.1: Fastness to Light Reconfirmed 1988. 2 pp.
- SNZ NZS/AS 1580. 482.1-75. Methods of Test for Paints and Related Materials Part 482.1: Fastness to Light (Reconfirmed 1988) (This is a Joint Standard with SAA AS 1580.482.1). 2 pp.

—Paints
- BSI BS 3900: Part F5-72. 1972 Methods of Test for Paints Group F: Durability Tests on Paint Films Part F5: Determination of Light Fastness of Paints for Interior Use (Exposed to Artificial Light Sources). 5 pp.
- ISO 2809-76. Paints and Varnishes—Determination of Light Fastness of Paints for Interior Use First Edition. 5 pp.
- SAA AS 1580.482. 1-75. Paints and Related Materials—Methods of Test—Part 482.1: Fastness to Light Reconfirmed 1988. 2 pp.
- SNZ NZS/AS 1580. 482.1-75. Methods of Test for Paints and Related Materials Part 482.1: Fastness to Light (Reconfirmed 1988) (This is a Joint Standard with SAA AS 1580.482.1). 2 pp.

—Plastic Sheets
- BSI BS 2782:Pt5: METH 521A-92. 1992 Methods of Testing Plastics Part 5: Optical and Colour Properties, Weathering Method 521A: Determination of Haze of Film and Sheet (L). 8 pp.
- CNS K6302-77. Method of Test for Polymethylmethacrylate Corrugated Sheet (Apr)(3273). 5 pp.

Light Testing (Cont.)
—Plastic Sheets (Cont.)
- CNS K6304-71. Method of Test for Polymethylmethacrylate Corrugated Sheet (Determination of Light Penetration) (Apr)(3275). 3 pp.

—Plastics
- BSI BS 2782:Pt5: METH 552A-81. 1981 Methods of Testing Plastics: Part 5: Optical and Colour Properties, Weathering: Method 552A: Determination of Changes in Colour Variations in Prop After Exposure to Daylight Under Glass, Natural Weathering or Artificial Light. 10 pp.
- BSI BS 4618: Sec 4.3-74. 1974 Recommendations for the Presentation of Plastics Design Data Part 4: Enviromental and Chemical Effects Section 4.3: Resistance to Colour Change Produced by Exposure to Light. 8 pp.
- ISO 4582-80. Plastics—Determination of Changes in Colour and Variations in Properties After Exposure to Daylight Under Glass, Natural Weathering or Artificial Light First Edition. 8 pp.

—Printing Inks
- ISO 2835-74. Prints and Printing Inks—Assessment of Light Fastness First Edition. 5 pp.

—Varnishes
- CNS K6056-70. Method of Test for Baking Varnish (Jan)(771). 6 pp.
- CNS K6103-59. Method of Test for Low Temperature Baking Varnish (Apr)(1113). 3 pp.
- SAA AS 1580.482. 1-75. Paints and Related Materials—Methods of Test—Part 482.1: Fastness to Light Reconfirmed 1988. 2 pp.
- SNZ NZS/AS 1580. 482.1-75. Methods of Test for Paints and Related Materials Part 482.1: Fastness to Light (Reconfirmed 1988) (This is a Joint Standard with SAA AS 1580.482.1). 2 pp.

Light Transmission
See Also: Lasers; Wave Propagation

—Carbon Black
- BSI BS 5293: Part 14-91. 1991 Sampling and Testing Carbon Black for Use in the Rubber Industry Part 14: Method for Determination of Light Transmittance of Toluene Extract (Rapid Method) (ISO 3858-1: 1990). 7 pp.
- BSI BS 5293: Part 15-91. 1991 Sampling and Testing Carbon Black for Use in the Rubber Industry Part 15: Method for Determination of Light Transmittance of Toluene Extract for Product Evaluation (ISO 3858-2: 1990). 7 pp.
- ISO 3858 Pt 1-90. Carbon Black for Use in the Rubber Industry—Determination of Light Transmittance of Toluene Extract—Part 1: Rapid Method Third Edition. 6 pp.
- ISO 3858 Pt 2-90. Carbon Black for Use in the Rubber Industry—Determination of Light Transmittance of Toluene Extract—Part 2: Method for Product Evaluation Third Edition. 6 pp.

—Glazes—Aircraft
- AECMA PREN2155-05-84. Test Methods for Transparent Materials for Aircraft Glazing Part 5—Determination of Visible Light Transmission. 7 pp.
- BSI BS EN 2155-5-89. 1989 Test Methods for Transparent Materials for Aircraft Glazing Part 5: Determination of Visible Light Transmission. 11 pp.
- CEN EN 2155 (Part 5)-89. Test Methods for Transparent Materials for Aircraft Glazing Part 5: Determination of Visible Light Transmission. 8 pp.

—Passive Devices
- JIS C 5860-90. General Rules of Passive Device for Light Beam Transmission. 10 pp.

—Plastic Pipes—PVC—Unplasticized
- SAA AS 1462.14-86. Methods of Test for Unplasticized PVC (UPVC) Pipes and Fittings—Part 14: Method for Determination of the Light Transmission of UPVC Pipe. 5 pp.

Light Water Reactors
See Also: Nuclear Reactors

—Accidents—Radiation Monitors
- IEC 951 Pt 4-91. Radiation Monitoring Equipment for Accident and Post-Accident Conditions in Nuclear Power Plants Part 4: Process Stream in Light Water Nuclear Power Plants First Edition. 31 pp.

—Decay Heat—Nuclear Fuels
- ISO 10645-92. Nuclear Energy—Light Water Reactors—Calculation of the Decay Heat Power in Nuclear Fuels First Edition. 16 pp.

—Radiation Measurement
- CENELEC HD 461-87. Process Stream Radiation Monitoring Equipment in Light Water Nuclear Reactors for Normal Operating and Incident Conditions. 3 pp.

Light Water Reactors (Cont.)
—Radiation Measurement (Cont.)
- CENELEC HD 462-87. Process Stream Radiation Monitoring Equipment in Light Water Nuclear Reactors for Normal Operating and Incident Conditions. 4 pp.
- IEC 768-83. Process Stream Radiation Monitoring Equipment in Light Water Nuclear Reactors for Normal Operating and Incident Conditions First Edition. 23 pp.
- IEC 910-88. Containment Monitoring Instrumentation for Early Detection of Developing Deviations from Normal Operation in Light Water Reactors First Edition. 15 pp.
- IEC 911-87. Measurements for Monitoring Adequate Cooling Within the Core of Pressurized Light Water Reactors First Edition. 37 pp.

—Sound Level Measurement
- IEC 988-90. Acoustic Monitoring Systems for Loose Parts Detection—Characteristics, Design Criteria and Operational Procedures First Edition. 48 pp.

Lightening Holes
See Also: Hole Size

—Aircraft—Flanged
- SBAC RS 445 (V). Flanged Lightening Holes.

Lightening Power
Use: Reducing Power

Lighter Fluid
- CNS K5033-64. Lighter Fluid (Mar)(2264). 1 p.

Lighter-Than-Air Aircraft
Use: Aerostats

Lighters
Use For: Cigarette Lighters; Firelighters; Smokers Lighters
- CNS B7135-80. Method of Test for Handy Lighter (Jul)(5650).
- CNS S1103-80. Cigarette Lighter (Dec)(6853).

—Automotive
- CNS D2027-86. Cigarette Lighters for Automobiles (Jun)(5782).
- JIS D 5807-91. Cigar Lighters for Automobiles. 12 pp.

—Disposable
- CNS S1174-83. Disposable Gas-Lighter (Nov)(10666).

—Electric—Safety
- BSI BS 3456: Sec 2.19-70. (WITHDRAWN) 1970 Amd 1 1972-1981 Edition. Specification for Safety of Household Electrical Appliances Part 2: Section 2.19: Electric Firelighters (Superseded by BS EN 60335-2-45: 1991). 13 pp.

—Safety
- BSI BS 6908-90. 1990 Safety Requirements for Gas Fuelled Smokers' Lighters. 19 pp.
- CEN EN 123-80. Gas Fuelled Smokers' Lighters: Safety Requirements. 10 pp.
- CEN EN 29 994-90. Lighters—Safety Specification. 16 pp.
- DIN ENGL EN 29994-91. Lighters; Safety Requirements (ISO 9994: 1989) (Feb). 15 pp.
- ISO 9994-89. Lighters—Safety Specification First Edition. 16 pp.

Lighthouses
See Also: Beacon Lights; Marine Lighting; Navigation Lights; Navigational Aids; Signal Lights; Warning Lights

—Monitors
- CCIR Report 1168-90. Beacon and Buoy Remote-Monitoring System—Section 8D—Radiodetermination, Global Maritime Distress and Safety System and Related Subjects. 4 pp.

—Telephone Answering Equipment
- CCIR Report 1168-90. Beacon and Buoy Remote-Monitoring System—Section 8D—Radiodetermination, Global Maritime Distress and Safety System and Related Subjects. 4 pp.

Lighting
Scope Note: For additional listings, use a more specific term. Lighting is indoor unless otherwise specified *Use For:* Lighting Equipment; Lighting Fittings; Lighting Fixtures; Luminaires
See Also: Aircraft Lighting; Airport Lighting; Ballasts (Electric); Bayonet Lamps; Beacon Lights; Daylighting; Decorative Lighting; Dimmers; Discharge Lamps; Display Lighting; Emergency Lighting; Flasher Units; Flashlights; Floodlights; Fluorescent Lighting; Fog Lights; Gaseous Tritium Lighting; Heat Test Source Lamps; High Intensity Discharge Lamps;

INTERNATIONAL AND NON-U.S. NATIONAL STANDARDS
SUBJECT INDEX

Lighting

Lighting *(Cont.)*
See Also: (Cont.)
Highway Lighting; Incandescent Lighting; Indicator Lights; Lamps; Landscape Lighting; Lanterns; Light (Visible Radiation); Light Sources; Light Strips; Lighting Chains; Lighting Circuits; Lighting Poles; Lighting Tracks; Luminance; Luminous Tube Signs; Marine Lighting; Motion Picture Lighting; Motor Vehicle Lighting; Navigation Lights; Neon Glow Lamps; Night Lights; Photographic Lighting Equipment; Power Reducers; Projection Lamps; Public Lighting; Quartz Tungsten Halogen Lamps; Reflectors; Running Lights; Runway Lighting; Searchlights; Signal Lights; Spotlights; Stage Lighting; Studio Lighting; Taxiway Lighting; Timing Lamps; Trouble Lights; Tungsten Lamps; Warning Lights

BSI BS 4533: Sec 102.1-90. 1990 Luminaires Part 102: Particular Requirements Section 102.1: Specification for Fixed General Purpose Luminaires. 13 pp.
BSI BS EN 60598-1-93. 1993 Luminaires Part 1: General Requirements and Tests (F). 161 pp.
CENELEC EN 60 598-1-89. Luminaires Part 1: General Requirements and Tests. 49 pp.
CENELEC EN 60 598-1-91. Luminaires Part 1: General Requirements and Tests. 53 pp.
CENELEC EN 60598-1-93. Luminaires Part 1: General Requirements and Tests (IEC 598-1: 1992). 14 pp.
CENELEC EN 60598-1-93. Luminaires Part 1: General Requirements and Tests (IEC 598-1: 1992, Modified). 154 pp.
CENELEC EN 60 598-2-1-89. Luminaires Part 2: Particular Requirements Section One—Fixed General Purpose Luminaires. 4 pp.
CSA C22.2 NO 9-M1989. Luminaires; (Gen Instr 1) (Supercedes C22.2 No 97-1969). 83 pp.
CSA 1169 Bull. Electrical Bulletin 1169 June 27, 1978 to C22.2 NO 9. 2 pp.
DIN VDE 0710 Pt 1-69. Specification for Lighting Fittings with Service Voltages Below 1000 V; Part 1: General Specification Incorporating Amendments: A/9.60; B/12.64; C/3.69 and Draft of D/...71 (Mar). 97 pp.
DIN VDE 0710 Pt 1 A5 (D)-80. Luminaires with Operating Voltages Below 1000V; Part 1: General Requirements and Tests; 5th Amendment (July). 6 pp.
DIN VDE 0710 Pt 5-83. Luminaires with Service Voltages Below 1000 V; Luminaires with Limited Surface Temperatures (Feb). 13 pp.
DIN VDE 0710 Pt 8-79. Luminaires with Service Voltages Below 1000 V; Luminaires with Igniters (June). 10 pp.
DIN VDE 0711 Pt 500 (D)-89. Luminaires; Special Requirements; Low-Voltage Lighting Systems (Oct). 15 pp.
IEC 598 Pt 1-92. Luminaires Part 1: General Requirements and Tests Third Edition; (CENELEC EN 60598-1: 1993). 280 pp.
IEC 598 Pt 2-1-79. Luminaires Part 2: Particular Requirements Section One—Fixed General Purpose Luminaires First Edition; (Amendment 1-1987). 17 pp.
JIS C 8105-87. Luminaires. 50 pp.
MOD UK DSTAN 62-5-69. Lights, Extension. 16 pp.
SAA AS 1680.1-90. Interior Lighting—Part 1: Principles and Recommendations Amdt 1 June 1993 (in Professional Package 62). 88 pp.
SAA AS 1680.2.0-90. Interior Lighting—Part 2.0: Recommendations for Specific Tasks and Interiors Amdt 1 December 1992 (Superseded by AS 1680.2.1—1993 (in Part) but Will Remain Current) (in Professional Packages 10, 15, 17A, 17F, 26,.
SAA AS 3137-92. Approval and Test Specification—Luminaires (Light Fittings) Amdt 1 October 1992 (in Professional Package 28). 8 pp.
SNZ NZS/AS 3137-92. Approval and Test Specification—Luminaires (Lighting Fittings) AMENDMENT No. 1, 1992. 7 pp.
SNZ NZS 6703-84. Code of Practice for Interior Lighting Design. 82 pp.
SNZ NZS 6705: Part 1-86. Luminaires Part 1: General Requirements and Tests Amend: A, 1986; 1, 1986; 1A, 1986; 2, 1986; 2A 1986. 193 pp.

—Accident Prevention
CNS Z1027-87. General Rules of Coloured Light for Safety (Mar)(9331).
JIS Z 9104-87. General Rules of Coloured Light for Safety. 9 pp.
JIS Z 9106-90. Fluorescent Safety Colours—General Rules for Application.

—Air Handling Equipment—Safety
BSI BS 4533: Sec 102.19-90. 1990 Luminaires Part 102: Particular Requirements Section 102.19: Specification for Air-Handling Luminaires (Safety Requirements). 19 pp.

—Battery Operated
MOD UK DSTAN 62-2: Part 1-74. Battery Operated Lights, Torches, and Lanterns, Electric Part 1: Lights, Electric Issue 1. 11 pp.

Lighting *(Cont.)*
—Buildings—Design
BSI BS 8206: Part 1-85. 1985 Lighting for Buildings Part 1: Code of Practice for Artifical Lighting. 40 pp.

—Cables (Electric)
BSI BS 6726-91. 1991 Festoon and Temporary Lighting Cables and Cords. 27 pp.
BSI BS 6726-01. 1991 Amd 1 Festoon and Temporary Lighting Cables and Cords (AMD 6939) May 1, 1992. 31 pp.
BSI BS 6726-02. 1991 Amd 2 Festoon and Temporary Lighting Cables and Cords (AMD 7888) September 15, 1993 (F). 37 pp.
DIN VDE 0250 Pt 211-86. Cables, Wires and Flexible Cords for Power Installation; PVC—Lighting Cable with Sheath (June) (Partially Supersedes VDE 0250/03.69). 10 pp.
DIN VDE 0282 Pt 604-90. Rubber Cables, Wires and Flexible Cords for Power Installation; Flexible Cable for Illumination Chains (Jan) (Supersedes DIN 57250 Part 604/VDE 0250 Part 604/10.83). 8 pp.
MOD UK DSTAN 61-12: Part 2-01. Wires, Cords, and Cables, Electrical—Metric Units Part 2: Cables, Electrical (for Power and Lighting) Issue 4; Amendment 1. 19 pp.

—Caps (Lids)—Screw Threads
DIN ENGL 40450-70. Electric Light Fittings; Glass Screw Threads for Glass Guards and Caps (Dec). 1 p.

—Ceiling—Compatibility
SAA AS 2946-91. Suspended Ceilings, Recessed Luminaires and Air Diffusers—Interface Requirements for Physical Compatability (In Professional Package 50). 22 pp.

—Classification
BSI BS 1000: (628)-82. 1982 Universal Decimal Classification (UDC). English Full Edition (628): Public Health Engineering. Water. Illuminating Engineering. 36 pp.
BSI BS 7216-89. 1989 Amd 1 Classification and Interpretation of New Lighting Products (IEC 972: 1989) (AMD 6767) September 30, 1991. 11 pp.
IEC 972-89. Classification and Interpretation of New Lighting Products First Edition; (Amendment 1-1991). 22 pp.
SNZ NZS/BS 1000 (628)-82. Universal Decimal Classification Public Health Engineering. Water. Illuminating Engineering. 36 pp.

—Cold Cathode—Transformers
CSA C22.2 NO 13-1962. Transformers for Luminous-Tube Signs, Oil-or Gas-Burner Ignition Equipment, Cold-Cathode Interior Lighting (R 1992). 46 pp.
CSA 1169 Bull. Electrical Bulletin 1169 June 27, 1978 to C22.2 NO 13. 2 pp.
SAA AS 3143-82. Approval and Test Specification—for Transformers for Cold-Cathode Electric Discharge Lamps and Lighting Systems (This is a Joint Standard with SANZ NZS 3143). 4 pp.
SNZ NZS/AS 3143-82. Approval and Test Specification for Transformers for Cold-Cathode Electric Discharge Lamps and Lighting Systems (This is a Joint Standard with SAA AS 3143). 4 pp.

—Color Temperature
JIS Z 8725-87. Methods for Determining Distribution Temperature and Colour Temperature or Correlated Colour Temperature of Light Sources. 28 pp.

—Colorimetry
BSI BS 950: Part 1-67. 1967 Artificial Daylight for the Assessment of Colour Part 1: Illuminant for Colour Matching and Colour Appraisal. 13 pp.
CNS Z7194-85. Standard Illuminants and Sources for Colorimetry (Jun)(11294).
ISO 4911-80. Textiles—Cotton Fibres—Equipment and Artificial Lighting for Cotton Classing Rooms First Edition. 9 pp.
JIS Z 8724-83. Methods of Measurement for Light Source Colour. 25 pp.
JIS Z 8726-90. Method of Specifying Colour Rendering Properties of Light Sources. 21 pp.

—Comprehensive Testing
BSI BS 4533: Part 101-90. 1990 Luminaires Part 101: General Requirements and Tests. 139 pp.

—Cords (Electric)
BSI BS 6726-91. 1991 Festoon and Temporary Lighting Cables and Cords. 27 pp.
BSI BS 6726-01. 1991 Amd 1 Festoon and Temporary Lighting Cables and Cords (AMD 6939) May 1, 1992. 31 pp.
BSI BS 6726-02. 1991 Amd 2 Festoon and Temporary Lighting Cables and Cords (AMD 7888) September 15, 1993 (F). 37 pp.

Lighting *(Cont.)*
—Covers—Screw Threads
DIN ENGL 40450-70. Electric Light Fittings; Glass Screw Threads for Glass Guards and Caps (Dec). 1 p.

—Darkroom
CSA C22.2 NO 118-1959. Construction and Test of Picture Machines and Appliances (R 1992); (Erratum April 1959) (Rev 1-5 March 1969). 21 pp.
CSA 649J Bull. Electrical Bulletin 649J November 17, 1975 to C22.2 NO 118. 1 p.
CSA 1169 Bull. Electrical Bulletin 1169 June 27, 1978 to C22.2 NO 118. 2 pp.

—Desks
MOD UK DSTAN 62-8-73. Light, Desk Issue 1. 13 pp.
MOD UK DSTAN 62-8: Addendum. Light, Desk. 3 pp.

—Dusk-to-Dawn
CSA CAN/CSA-C239-M89. Dusk-To-Dawn Luminaires; (Gen Instr 1). 26 pp.

—Electrical Codes
CSA 711 Bull. Electrical Bulletin 711 February 9, 1968 to C22.2 NO 9. 1 p.
CSA 923 Bull. Electrical Bulletin 923 September 12, 1973 to C22.2 NO 9. 5 pp.
CSA 979 Bull. Electrical Bulletin 979 December 9, 1974 to C22.2 NO 9. 4 pp.
CSA 979A Bull. Electrical Bulletin 979A December 24, 1975 to C22.2 NO 9. 1 p.
CSA 1217A Bull. Electrical Bulletin 1217A November 5, 1980 to C22.2 NO 9. 2 pp.
CSA 1427 Bull. Electrical Bulletin 1427 November 28, 1984 to C22.2 NO 9. 2 pp.
CSA 1444 Bull. Electrical Bulletin 1444 December 22, 1988 to C22.2 NO 9. 4 pp.
CSA 1444 Bull. Electrical Bulletin 1444 December 22, 1988 to C22.2 NO 12. 4 pp.

—Electrical Installations
DIN VDE 0100 Pt 559-83. Installation of Power Plant with Rated Voltages up to 1000 V; Luminaires and Lighting Equipment (Mar). 15 pp.

—Ergonomics
ISO 8995-89. Principles of Visual Ergonomics—the Lighting of Indoor Work Systems First Edition. 31 pp.

—Explosion Proof
JIS C 8001-91. Luminaires for Explosive Atmospheres. 28 pp.

—Field Hospitals—Surgical Equipment
NATO STANAG 2978 ED 1 AMD 1-83. Essential Performance Characteristics of Field Surgical Lights. 5 pp.

—Fireproof
BSI BS 889-65. (WITHDRAWN) 1965 Amd 3 Flameproof Electric Lighting Fittings (AMD 5050) May 30, 1986. 65 pp.
SNZ NZS 380-68. Specification for Flameproof Electric Lighting Fittings (Reconfirmed 1978). 64 pp.

—Furniture
DIN VDE 0710 Pt 14-82. Luminaires with Operating Voltages Below 1000 V; Luminaires to Be Built into Furniture (Apr). 17 pp.

—Gas Discharge Tubes—Cables (Electric)
CSA C22.2 NO 17-1973. Cable for Luminous-Tube Signs and for Oil-and Gas-Burner Ignition Equipment (R 1992); (Gen Instr 1) (Amd 1 February 1981) (Amd 2-8 March 1988). 28 pp.

—Glossaries
BSI BS 4727:Pt4: Group 03-72. (WITHDRAWN) 1972 Amd 1 Electrotechnical, Power Telecommunications, Electronics, Lighting and Colour Terms Part 4: Terms Particular to Lighting and Colour Group 03: Lighting Technology Terminology. 35 pp.
BSI BS 6100: Sec 3.4-85. 1985 Glossary of Building and Civil Engineering Terms Part 3: Services Section 3.4: Lighting. 14 pp.
BSI BS 6100: Sec 3.4-01. 1985 Amd 1 Glossary of Building and Civil Engineering Terms Part 3: Services Section 3.4: Lighting (AMD 7259) August 15, 1992. 15 pp.
DIN ENGL 5035 Pt 1-90. Artificial Lighting; Terminology and General Requirements (June). 12 pp.
IEC 50 Chap 845-87. International Electrotechnical Vocabulary Chapter 845: Lighting. 380 pp.
JIS Z 8113-88. Glossary of Lighting Terms.
SAA AS 1852.845-89. International Electrotechnical Vocabulary—Part 845: Lighting. 305 pp.

INDUSTRY STANDARDS

INTERNATIONAL AND NON-U.S. NATIONAL STANDARDS
SUBJECT INDEX

Lighting

Lighting (Cont.)

—Hazardous Locations
CSA C22.2 NO 137-M1981. Electric Luminaires for Use in Hazardous Locations (R 1993); (Gen Instr 1 Thru 3). 26 pp.

—Hazardous Locations—Flammability Testing
CSA C22.2 NO 137-M1981. Electric Luminaires for Use in Hazardous Locations (R 1993); (Gen Instr 1 Thru 3). 26 pp.

—Hazardous Locations—Temperature Testing
CSA C22.2 NO 137-M1981. Electric Luminaires for Use in Hazardous Locations (R 1993); (Gen Instr 1 Thru 3). 26 pp.

—Health Care Facilities
CSA CAN/CSA-Z317.5-M89. Illumination Systems in Health Care Facilities; (Gen Instr 1). 39 pp.
SAA AS 1765-75. Artificial Lighting for Clinical Observation Amdt 1 September 1979. 24 pp.

—Helicopters—In Flight Refueling
NATO STANAG 1251 ED 2 AMD 3-83. Shipboard Helicopter in-Flight Refuelling (HIFR) Operating Area Marking, Clearances and Lighting. 19 pp.

—Heliports
NATO STANAG 1162 ED 4 AMD 0-91. Vertical Replenishment (Vertrep) Operating Area Marking, Clearances, and Lighting. 20 pp.
NATO STANAG 1275 ED 2 AMD 0-91. Minimum Standard Requirements for the Marking and Lighting of Helicopter Deck Landing Areas. 13 pp.
NATO STANAG 1275 ED 2 AMD 1-92. Minimum Standard Requirements for the Marking and Lighting of Helicopter Deck Landing Areas. 14 pp.
NATO STANAG 3652 ED 1 AMD 6-74. Helipad Lighting (VMC). 15 pp.
NATO STANAG 3652 ED 2 AMD 0-92. Helipad Lighting (VMC). 14 pp.
NATO STANAG 3652 ED 2 AMD 1-92. Helipad Lighting (VMC). 15 pp.

—Heliports—Identification Systems
NATO STANAG 3535 ED 1 AMD 3-81. Heliport Marking and Lighting. 18 pp.

—Hospitals
DIN ENGL 5035 Pt 3-88. Artificial Lighting of Interiors; Hospital Lighting (Sept). 7 pp.
SAA AS 1765-75. Artificial Lighting for Clinical Observation Amdt 1 September 1979. 24 pp.

—Hot Tubs
CSA CAN/CSA-C22. 2NO218.1-M89. Spas, Hot Tubs, and Associated Equipment; (Gen Instr 1). 74 pp.

—Illuminance
BSI BS 5225: Part 1-75. 1975 Photometric Data for Luminaires Part 1: Photometric Measurements. 43 pp.
CNS C3069-88. Method of Illumination Measurements (Oct)(5065).
CNS Z1044-87. Recommended Levels of Illumination (Sep)(12112).
JIS C 7612-85. Illuminance Measurements for Lighting Installations.
JIS C 7612-68. Methods of Illumination Measurements (R 1977). 9 pp.
JIS F 8041-86. Recommended Levels of Illumination and Methods of Illumination Measurements for Marine Use.
JIS Z 9110-79. Recommended Levels of Illumination. 21 pp.

—Lighting Poles
CSA C22.2 NO 206-M1987. Lighting Poles (R 1993); (Gen Instr 1). 17 pp.

—Luminance
BSI BS 5225: Part 1-75. 1975 Photometric Data for Luminaires Part 1: Photometric Measurements. 43 pp.

—Mining Equipment—Glossaries
BSI BS 3618: Sec 7-73. 1973 Glossary of Mining Terms Section 7: Electrical Engineering and Lighting. 11 pp.

—Molybdenum Materials
CNS H2051-85. General Rules for Test of Tungsten and Molybdenum Materials for Lighting and Electronic Equipments (Nov)(6337).
CNS H3090-85. Molybdenum Wires for Lighting and Electronic Equipments (Nov)(6343).
CNS H3091-85. Molybdenum Rods for Lighting and Electronic Equipments (Nov)(6344).
CNS H3092-85. Molybdenum Sheets for Lighting and Electronic Equipments (Nov)(6345).

Lighting (Cont.)

—Molybdenum Materials (Cont.)
JIS H 4460-84. General Rules for Test of Tungsten and Molybdenum Materials for Lighting and Electronic Equipments. 9 pp.
JIS H 4471-89. Tungsten-Molybdenum Alloy Wires for Lighting and Electronic Equipments. 8 pp.
JIS H 4481-89. Molybdenum Wires for Lighting and Electronic Equipments. 10 pp.
JIS H 4482-89. Molybdenum Rods for Lighting and Electronic Equipments. 7 pp.
JIS H 4483-84. Molybdenum Sheets for Lighting and Electronic Equipments. 7 pp.

—Outdoor—Hazardous Locations
CSA C22.2 NO 137-M1981. Electric Luminaires for Use in Hazardous Locations (R 1993); (Gen Instr 1 Thru 3). 26 pp.

—Photometric Properties
DIN ENGL 5035 Pt 6-90. Artificial Lighting; Measurement and Evaluation (Dec). 12 pp.

—Photometry
BSI BS 5225: Part 1-75. 1975 Photometric Data for Luminaires Part 1: Photometric Measurements. 43 pp.
SAA AS 1680.3-91. Interior Lighting—Part 3: Measurement, Calculation and Presentation of Photometric Data (Supersedes AS 1190—1972 (Withdrawn April 1988)). 48 pp.

—Portable
BSI BS 3879-81. 1981 Amd 3 Portable Appliances Operating at Vapour Pressure from Liquefied Petroleum Gas Containers. 14 pp.
BSI BS 4533: Sec 102.4-90. 1990 Luminaires Part 102: Particular Requirements Section 102.4: Specification for Portable General Purpose Luminaires. 15 pp.
BSI BS 4533: Sec 102.4-01. 1990 Amd 1 Luminaires Part 102: Particular Requirements Section 102.4: Specification for Portable General Purpose Luminaires (AMD 7847) July 15, 1993 (F). 16 pp.
BSI BS 4533: Sec 102.7-90. 1990 Luminaires Part 102: Particular Requirements Section 102.7: Specification for Portable Luminaires for Garden Use. 15 pp.
CENELEC EN 60 598-2-4-89. Luminaires Part 2: Particular Requirements Section Four-Portable General Purpose Luminaires. 6 pp.
CENELEC EN 60598-2-4 /A3-93. AMD 3 Luminaires Part 2: Particular Requirements Section Four—Portable General Purpose Luminaires (IEC 598-2-4: 1979/A3: 1990). 5 pp.
CENELEC EN 60 598-2-7-89. Luminaires Part 2: Particular Requirements Section Seven-Portable Luminaires for Garden Use. 8 pp.
CSA C22.2 NO 9-M1989. Luminaires; (Gen Instr 1) (Supercedes C22.2 No 97-1969). 83 pp.
CSA 1169 Bull. Electrical Bulletin 1169 June 27, 1978 to C22.2 NO 9. 2 pp.
IEC 598 Pt 2-4-79. Luminaires Part 2: Particular Requirements Section Four—Portable General Purpose Luminaires First Edition; (Amendment 3-1990, Incorporating Amendment 1 and 2) (CENELEC EN 60598-2-4/A3: 1993). 26 pp.
IEC 598 Pt 2-7-82. Luminaires Part 2: Particular Requirements Section Seven—Portable Luminaires for Garden Use First Edition; (Amendment 1-1987). 19 pp.

—Radio Frequency Interference
BSI BS 6345-83. 1983 Radio Inteference Terminal Voltage of Lighting Equipment. 10 pp.
BSI BS EN 55015-93. 1993 Limits and Methods of Measurement of Radio Disturbance Characteristics of Electrical Lighting and Similar Equipment (S). 48 pp.
BSI BS EN 55015-01. 1993 Amd 1 Limits and Methods of Measurement of Radio Disturbance Characteristics of Electrical Lighting and Similar Equipment (AMD 7878) July 15, 1993 (S). 49 pp.
CENELEC EN 55 015-87. Limits and Methods of Measurement of Radio Interference Characteristics of Fluorescent Lamps and Luminaires. 11 pp.
CENELEC EN 55015-93. Limits and Methods of Measurement of Radio Disturbance Characteristics of Electrical Lighting and Similar Equipment (IEC CISPR 15: 1992). 5 pp.
CENELEC EN 55 015/A1-90. AMD 1 Limits and Methods of Measurement of Radio Interference Characteristics of Fluorescent Lamps and Luminaires. 4 pp.

—Recessed
BSI BS 4533: Sec 102.2-90. 1990 Luminaires Part 102: Particular Requirements Section 102.2: Recessed Luminaires. 16 pp.
CENELEC EN 60 598-2-2-89. Luminaires Part 2: Particular Requirements Section Two-Recessed Luminaires. 5 pp.
IEC 598 Pt 2-2-79. Luminaires Part 2: Particular Requirements Section Two—Recessed Luminaires First Edition; (Amendment 2-1991, Incorporating Amendment 1). 28 pp.

Lighting (Cont.)

—Recessed (Cont.)
SAA AS 2946-91. Suspended Ceilings, Recessed Luminaires and Air Diffusers—Interface Requirements for Physical Compatability (In Professional Package 50). 22 pp.

—Reflectors
CNS C4367-82. Reflectors for Lighting Fittings (Jul)(9120).

—Replenishment at Sea
NATO STANAG 1328 ED 1 AMD 4-87. Standard Distance Line Lighting. 11 pp.

—Residential
SNZ NZS 6703-84. Code of Practice for Interior Lighting Design. 82 pp.

—Revision Changes
CSA 1408 Bull. Electrical Bulletin 1408 July 29, 1983 to C22.2 NO 9. 6 pp.

—Rolling Stock
JIS E 4016-92. Illuminance for Railway Rolling Stock—Recommended Levels and Measuring Methods.

—Safety
BSI BS 4533: Sec 102.51-86. 1986 Amd 1 Luminaires Part 102: Particular Requirements Section 102.51: Luminaires with Type of Protection N (AMD 6427) April 30, 1990. 20 pp.
CNS C1027-87. General Rules of Coloured Light for Safety (Mar)(9331).
JIS Z 9104-87. General Rules of Coloured Light for Safety. 9 pp.

—Schools
DIN ENGL 5035 Pt 4-83. Artificial Lighting of Interiors; Special Recommendations for Lighting Educational Establishments (Feb). 6 pp.

—Ships
MOD UK NES 592-89. Requirements for Electric Lighting Fittings Issue 2 (02.89). 34 pp.

—Spas
CSA CAN/CSA-C22. 2NO218.1-M89. Spas, Hot Tubs, and Associated Equipment; (Gen Instr 1). 74 pp.

—Sports Facilities
DIN VDE 0710 Pt 13-81. Lighting Fittings with Service Voltage Below 1000 V; Lighting Fittings Resistant to Ball Throwing (May). 9 pp.
JIS Z 9122-90. Lighting for Sports Halls. 12 pp.

—Submarines
MOD UK NES 592-89. Requirements for Electric Lighting Fittings Issue 2 (02.89). 34 pp.

—Submersible—Swimming Pools—Accessories
CSA C22.2 NO 89-1976. Swimming-Pool Luminaires, Submersible Luminaires and Accessories (R 1992). 23 pp.
CSA 1060 Bull. Electrical Bulletin 1060 June 4, 1976 to C22.2 NO 89. 1 p.

—Swimming Pools
JIS Z 9123-91. Lighting for Swimming Pools. 16 pp.

—Swivels—Certification
CSA 1329 Bull. Electrical Bulletin 1329 June 24, 1981 to C22.2 NO 9. 2 pp.

—Theaters
BSI CP 1007-55. 1955 Maintained Lighting for Cinemas. 28 pp.

—Tungsten Materials
CNS H2051-85. General Rules for Test of Tungsten and Molybdenum Materials for Lighting and Electronic Equipments (Nov)(6337).
CNS H3085-85. Tungsten Wires for Lighting and Electronic Equipments (Nov)(6338).
CNS H3086-85. Tungsten Rods for Lighting and Electronic Equipments (Nov)(6339).
CNS H3088-85. Thoriated Tungsten Wires and Rods for Lighting and Electronic Equipments (Nov)(6341).
CNS H3089-85. Tungsten-Molybdenum Alloy Wires for Lighting and Electronic Equipments (Nov)(6342).
JIS H 4460-84. General Rules for Test of Tungsten and Molybdenum Materials for Lighting and Electronic Equipments. 9 pp.
JIS H 4461-89. Tungsten Wires for Lighting and Electronic Equipments. 11 pp.
JIS H 4462-89. Tungsten Rods for Lighting and Electronic Equipments. 8 pp.
JIS H 4463-84. Thoriated Tungsten Wires and Rods for Lighting and Electronic Equipments. 8 pp.
JIS H 4471-89. Tungsten-Molybdenum Alloy Wires for Lighting and Electronic Equipments. 8 pp.

INTERNATIONAL AND NON-U.S. NATIONAL STANDARDS
SUBJECT INDEX
Lighting

Lighting (Cont.)
—Underground—Hazardous Locations
CSA C22.2 NO 137-M1981. Electric Luminaires for Use in Hazardous Locations (R 1993); (Gen Instr 1 Thru 3). 26 pp.

—Video Display Terminals
DIN ENGL 5035 Pt 7-88. Artificial Lighting of Interiors; Lighting of Rooms with VDU Work Stations or VDU Assisted Workplaces (Sept). 8 pp.

—Work Areas
DIN ENGL 5035 Pt 2-90. Artificial Lighting; Recommended Values for Lighting Parameters for Indoor and Outdoor Workspaces (Sept). 13 pp.
SNZ NZS 6703-84. Code of Practice for Interior Lighting Design. 82 pp.

Lighting Busways
See Also: Busways
JIS C 8366-88. Lighting Busways. 18 pp.

Lighting Chains
Use For: Christmas Tree Lights; Lighting Strings; Ski Track Lights; String Lighting *See Also:* Cords (Electric); Lighting
CSA 1439 Bull. Electrical Bulletin 1439 May 30, 1986 to C22.2 NO 37. 2 pp.
IEC 598 Pt 2-20-82. Luminaires Part 2: Particular Requirements Section Twenty—Lighting Chains First Edition; (Amendment 2-1992, Incorporating Amendment 1). 48 pp.

—Flashing
CSA C22.2 NO 37-M1989. Christmas Tree and Other Decorative Lighting Outfits; (Gen Instr 1 Thru 2). 37 pp.

—Incandescent
BSI BS EN 60598-2-20-92. 1992 Luminaires Part 2: Particular Requirements Section 2.20: Lighting Chains. 29 pp.
BSI BS EN 60598-2-20-01. 1992 Amd 1 Luminaires Part 2: Particular Requirements Section 2.20: Lighting Chains (AMD 7709) April 15, 1993 (F). 34 pp.
CENELEC EN 60598-2-20-91. Luminaires Part 2: Particular Requirements Section Twenty—Lighting Chains. 20 pp.
CENELEC EN 60598-2-20/A11-92. AMD 11 Luminaires Part 2: Particular Requirements Section Twenty—Lighting Chains. 4 pp.

—Incandescent—Portable
CSA C22.2 NO 12-1982. Portable Luminaires; (Gen Instr 1 Thru 4). 52 pp.
CSA 1169 Bull. Electrical Bulletin 1169 June 27, 1978 to C22.2 NO 12. 2 pp.

—Lampholders
CSA C22.2 NO 37-M1989. Christmas Tree and Other Decorative Lighting Outfits; (Gen Instr 1 Thru 2). 37 pp.

—Lampholders—Miniature
CSA C22.2 NO 37-M1989. Christmas Tree and Other Decorative Lighting Outfits; (Gen Instr 1 Thru 2). 37 pp.

—Polychloroprene Coated
CENELEC HD 22.8 S1-89. Polychloroprene or Equivalent Synthetics Elastomer Sheathed Cables for Use as a Decorative Chain. 10 pp.
CENELEC HD 22.8 S1-92. Rubber Insulated Cables of Rated Voltages up to and Including 450/750 V—Part 8: Polychloroprene or Equivalent Synthetic Elastomer Sheathed Cables for Use as Decorative Chains (Reprint Incorporating A1). 10 pp.
CENELEC HD 22.8 S1/A2-92. AMD 2 Rubber Insulated Cables of Rated Voltages up to and Including 450/750 V—Part 8: Polychloroprene or Equivalent Synthetic Elastomer Sheathed Cables for Use as Decorative Chains. 4 pp.

—PVC Insulated
CENELEC HD 21.8 S1-89. Polyvinyl Chloride Insulated Cables of Rated Voltages up to and Including 450/750V Part 8: Non Sheathed Cable for Decorative Chain. 8 pp.
CENELEC HD 21.8 S1/A1-90. AMD 1 Polyvinyl Chloride Insulated Cables of Rated Voltages up to and Including 450/750 V Part 8: Non Sheathed Cable for Decorative Chain. 6 pp.
CENELEC HD 21.8 S1-90. Polyvinyl Chloride Insulated Cables of Rated Voltages up to and Including 450/750 V Single Core Non-Sheathed Cable for Decorative Chains. 10 pp.

—Safety
BSI BS 4647-70. (WITHDRAWN) 1970 Amd 5 Lighting Sets for Christmas Trees and Decorative Purposes for Indoor Use (AMD 6397) May 31, 1991 (Superseded by BS EN 60598-2-20: 1992). 22 pp.

Lighting Circuits
See Also: Branch Circuits; Lighting

—Dry Type Transformers
CSA CAN/CSA-C22. 2 NO 47-M90. Air-Cooled Transformers (Dry Type); (Gen Instr 1 Thru 2). 32 pp.
CSA 1169 Bull. Electrical Bulletin 1169 June 27, 1978 to C22.2 NO 47. 2 pp.

—Panel Boards (Electrical)
CSA 655 Bull. Electrical Bulletin 655 October 12, 1966 to C22.2 NO 4. 4 pp.
CSA C22.2 NO 29-M1989. Panelboards and Enclosed Panelboards; (Gen Instr 1 Thru 6). 100 pp.
CSA 397 Bull. Electrical Bulletin 397 May 27, 1957 to C22.2 NO 29. 2 pp.
CSA 655 Bull. Electrical Bulletin 655 October 12, 1966 to C22.2 NO 29. 4 pp.
CSA 1120 Bull. Electrical Bulletin 1120 May 26, 1977 to C22.2 NO 29. 8 pp.
CSA 1169 Bull. Electrical Bulletin 1169 June 27, 1978 to C22.2 NO 29. 2 pp.

—Switchgear
CSA CAN/CSA-C22. 2 NO 31-M89. Switchgear Assemblies; (Gen Instr 1 Thru 3). 54 pp.
CSA 1169 Bull. Electrical Bulletin 1169 June 27, 1978 to C22.2 NO 31. 2 pp.

Lighting Columns
Use: Lighting Poles

Lighting Equipment
Use: Lighting

Lighting Fittings
Use: Lighting

Lighting Fixtures
Use: Lighting

Lighting Poles
Use For: Lamp Posts; Lampposts; Light Poles; Lighting Columns; Lighting Posts
See Also: Lighting; Poles (Supports); Power Poles; Sign Posts
BSI BS 5649: Part 2-78. 1978 Amd 1 Lighting Columns Part 2: Dimensions and Tolerances. 13 pp.
BSI BS 5649: Part 3-82. 1982 Lighting Columns Part 3: Materials and Welding Requirements. 4 pp.
BSI BS 5649: Part 5-82. 1982 Lighting Columns Part 5: Base Compartments and Cableways. 7 pp.
CEN EN 40-2-76. Lighting Columns—Part 2: Dimensions and Tolerances. 11 pp.
CEN EN 40-3-82. Lighting columns—Part 3: Materials. 3 pp.
CEN EN 40-5-82. Lighting Columns—Part 5: Base Compartments and Cableways. 5 pp.
SAA AS 1798-92. Lighting Poles and Bracket Arms—Preferred Dimensions. 18 pp.

—Aluminum
CSA C22.2 NO 206-M1987. Lighting Poles (R 1993); (Gen Instr 1). 17 pp.

—Aluminum—Accessories
CSA C22.2 NO 206-M1987. Lighting Poles (R 1993); (Gen Instr 1). 17 pp.

—Brackets
SAA AS 1798-92. Lighting Poles and Bracket Arms—Preferred Dimensions. 18 pp.

—Concrete
BSI BS 5649: Part 9-82. 1982 Lighting Columns Part 9: Special Requirements for Reinforced and Prestressed Concrete Lighting Columns. 10 pp.
CEN EN 40-9-82. Lighting Columns—Part 9: Special Requirements for Reinforced and Prestressed Concrete Lighting Columns. 8 pp.
CSA C22.2 NO 206-M1987. Lighting Poles (R 1993); (Gen Instr 1). 17 pp.

—Concrete—Accessories
CSA C22.2 NO 206-M1987. Lighting Poles (R 1993); (Gen Instr 1). 17 pp.

—Concrete—Highway Lighting
SAA AS 4065-92. Concrete Poles for Overhead Lines and Street Lighting. 20 pp.

—Design Loads
BSI BS 5649: Part 6-82. 1982 Lighting Columns Part 6: Design Loads. 14 pp.
BSI BS 5649: Part 8-82. 1982 Lighting Columns Part 8: Method for Verification of Structural Design by Testing. 11 pp.
CEN EN 40-6-82. Lighting Columns—Part 6: Design Loads. 8 pp.
CEN EN 40-8-82. Lighting Columns—Part 8: Verification of Structural Design by Testing. 9 pp.

Lighting Poles (Cont.)
—Electric Raceways
BSI BS 5649: Part 5-82. 1982 Lighting Columns Part 5: Base Compartments and Cableways. 7 pp.
CEN EN 40-5-82. Lighting Columns—Part 5: Base Compartments and Cableways. 5 pp.

—Ferrous Metal
CSA C22.2 NO 206-M1987. Lighting Poles (R 1993); (Gen Instr 1). 17 pp.

—Ferrous Metal—Accessories
CSA C22.2 NO 206-M1987. Lighting Poles (R 1993); (Gen Instr 1). 17 pp.

—Fiberglass Reinforced Plastics
CNS K3083-86. Glassfiber Reinforced Plastic Lamp Pole (Aug)(11652). 7 pp.
CNS K6882-86. Method of Test for Glassfiber Reinforced Plastic Lamp Pole (Aug)(11653). 8 pp.

—Glossaries
BSI BS 5649: Part 1-78. (WITHDRAWN) 1978 Lighting Columns Part 1: Definitions and Terms (Superseded by BS EN 40-1: 1992). 11 pp.
BSI BS EN 40-1-92. 1992 Lighting Columns Part 1: Definitions and Terms. 13 pp.
CEN EN 40 (Part 1)-76. Lighting Columns Part 1 Definitions and Terms. 9 pp.
CEN EN 40-1-91. Lighting Columns—Part 1: Definitions and Terms. 8 pp.
DIN ENGL EN 40 Pt 1-92. Lighting Columns; Definitions and Terms (Feb). 6 pp.

—Grounding
BSI BS 5649: Part 5-82. 1982 Lighting Columns Part 5: Base Compartments and Cableways. 7 pp.
CEN EN 40-5-82. Lighting Columns—Part 5: Base Compartments and Cableways. 5 pp.

—Highway Lighting
CNS C4134-87. Lamp Pole for Illumination of Road (Jul)(4117).
CNS C4134-77. Lamp Pole for Illumination of Road (Jul)(4117). 6 pp.

—Metal Coatings
BSI BS 5649: Part 4-82. 1982 Lighting Columns Part 4: Surface Protection of Metal Lighting Columns. 8 pp.
CEN EN 40-4-82. Lighting Columns—Part 4: Surface Protection of Metal Lighting Columns. 8 pp.

—Oil Immersed
CNS C4019-83. Oil-Immersed Lamp Poles (for Taiwan Area) (Feb)(923).

—Signs
CSA C22.2 NO 206-M1987. Lighting Poles (R 1993); (Gen Instr 1). 17 pp.

—Structural Design
BSI BS 5649: Part 7-85. 1985 Lighting Columns Part 7: Method for Verification of Structural Design by Calculation. 20 pp.

—Traffic Lights
CSA C22.2 NO 206-M1987. Lighting Poles (R 1993); (Gen Instr 1). 17 pp.

—Welding
BSI BS 5649: Part 3-82. 1982 Lighting Columns Part 3: Materials and Welding Requirements. 4 pp.

—Wood
CSA C22.2 NO 206-M1987. Lighting Poles (R 1993); (Gen Instr 1). 17 pp.

—Wood—Accessories
CSA C22.2 NO 206-M1987. Lighting Poles (R 1993); (Gen Instr 1). 17 pp.

Lighting Posts
Use: Lighting Poles

Lighting Strings
Use: Lighting Chains

Lighting Tracks
See Also: Connectors; Lighting
BSI BS 4533: Sec 2.6-79. (WITHDRAWN) 1979 Amd 2 Luminaires Part 2: Detail Requirements Section 2.6: Electrical Supply Track Systems for Luminaires (F) (Superseded by BS 4533: Section 102.57: 1990). 36 pp.
BSI BS 4533: Sec 102.57-90. 1990 Luminaires Part 102: Particular Requirements Section 102.57: Electrical Supply Track Systems for Luminaires. 26 pp.
CENELEC EN 60 570-89. Electrical Supply Track Systems for Luminaires. 7 pp.
DIN VDE 0711 Pt 300 A2 (D)-80. Electrical Supply Systems for Luminaires; Amendment A2 to DIN IEC 570/VDE 0711 Part 300 (Sept). 4 pp.

INTERNATIONAL AND NON-U.S. NATIONAL STANDARDS
SUBJECT INDEX — Lighting

Lighting Tracks (Cont.)
DIN VDE 0711 Pt 300 A4 (D)-81. Electrical Supply Systems for Luminaires; Amendment 4 to DIN IEC 570/VDE 0711 Part 300 (Oct). 5 pp.
IEC 570-85. Electrical Supply Track Systems for Luminaires Second Edition; (Amendment 2-1993, Including Amendment 1). 46 pp.

Lightness
See Also: Reflectance
—Paints
DIN ENGL 53778 Pt 3-83. Emulsion Paints; Determination of Contrast Ratio and Lightness of Coatings (Aug). 3 pp.

Lightning Arresters
Use: Surge Arresters

Lightning Protection
See Also: Electromagnetic Protection; Lightning Protection Equipment; Overload Protection; Overvoltage Protection
IEC 1024 Pt 1-1-93. Protection of Structures Against Lightning Part 1: General Principles Section 1: Guide A—Selection of Protection Levels for Lightning Protection Systems First Edition. 47 pp.
SAA AS 1768-91. Lightning Protection (NZS 1768—1991) (in Professional Packages 21, 30, 57, 66). 91 pp.
SNZ NZS/AS 1768-91. Lightning Protection (This is a Joint Standard with SAA AS 1768). 91 pp.

—Aerospace
JIS W 2009-78. Bonding, Electrical and Lightning Protection, for Aerospace Systems.

—Aerostats
CAA Chapter Q4-6 12.79. Electrical Bonding and Lightning Discharge Protection (Non-Rigid Airships). 5 pp.
CAA Chapter Q4-6 App 12.79. Primary Conductors (Non-Rigid Airships). 1 p.

—Aircraft
CAA Chapter K4-6 03.67. Electrical Bonding and Lightning Discharge Protection (Light Aeroplanes). 5 pp.
CAA Chapter K4-6 App 03.67. Electrical Bonding and Lightning Discharge Protection (Light Aeroplanes). 2 pp.
CAA Chapter P4-6. Electrical Bonding and Lightning Discharge Protection (Provisional Airworthiness Requirements for Civil Powered Lift Aircraft). 18 pp.
NATO STANAG 3659 ED 1 AMD 6-75. Bonding and in-Flight Lightning Protection for Aircraft. 17 pp.
NATO STANAG 3659 ED 1 AMD 7-75. Bonding and In-Flight Lightning Protection for Aircraft. 19 pp.

—Aircraft—Radio Equipment
CAA Chapter R4-5 04.74. Bonding and Lightning Discharge Protection (Radio). 1 p.

—Broadcast Stations—Sound Broadcasting
CCIR Report 943-1-86. Protection of Sound-Broadcasting Stations Against Atmospheric Electricity—Section 10A-1—Amplitude-Modulation Sound Braodcasting in Bands 5 (LF), 6 (MF) and 7 (HF). 4 pp.
CCIR QUESTION 48-1/10-86. Protection of Sound-Broadcasting Stations Against Atmospheric Electricity—Questions Concerning Study Group 10—Broadcasting Service (Sound). 1 p.

—Broadcast Stations—Television Broadcasting
CCIR QUESTION 38/11-82. Protection of Television Broadcasting Stations Against Lightning—Questions Concerning Study Group 11—Broadcasting Service (Television). 1 p.

—Communication Cables
CCITT RECMN K.29-92. Coordinated Protection Schemes for Telecommunication Cables Below Ground (Study Group V) 7 pp. 7 pp.

—Fiber Optic Cables
CCITT RECMN K.25-89. Lightning Protection of Optical Fibre Cables—Protection Against Interference (Study Group V) 4 pp. 4 pp.

—Fixed Services—Transmitting Stations—Netherlands
CCIR Report 861-82. Protection of Radio Stations Against Lightning and Other Electromagnetic Disturbances—Section 3Ad—Operational Questions. 17 pp.

—Munitions
NATO STANAG 4236 ED 1 AMD 0-90. (DRAFT) Lighting Environmental Conditions Affecting the Design of Material, for Use by NATO Forces. 17 pp.
NATO STANAG 4236 ED 1 AMD 0-93. Lightning Environmental Conditions, Affecting the Design of Materiel, for Use by the NATO Forces. 16 pp.

—Power Lines
CCITT RECMN K.11-91. Principles of Protection Against Overvoltages and Overcurrents (Study Group V) 14 pp. 14 pp.

—Radio Equipment
CCIR QUESTION 62/1-82. Lightning Protection of Radio Equipment—Questions Concerning Study Group 1—Spectrum Management Techniques (Spectrum Engineering, Planning, Sharing, Monitoring and Utilization). 1 p.

—Radio Receivers—Italy
CCIR Report 861-82. Protection of Radio Stations Against Lightning and Other Electromagnetic Disturbances—Section 3Ad—Operational Questions. 17 pp.

—Radio Relay Systems
CCIR Report 932-82. Protection of Radio-Relay Stations Against Lightning Discharges—Section 9A—Performance Objectives, Propagation and Interference Effects. 10 pp.

—Radio Stations
CCIR Report 861-82. Protection of Radio Stations Against Lightning and Other Electromagnetic Disturbances—Section 3Ad—Operational Questions. 17 pp.
CCIR QUESTION 153/9-90. Protection of Radio Stations Against Lightning and Other Electromagnetic Disturbances—Questions of Study Group 9 Fixed Service. 1 p.

—Receiving Stations—Japan
CCIR Report 861-82. Protection of Radio Stations Against Lightning and Other Electromagnetic Disturbances—Section 3Ad—Operational Questions. 17 pp.

—Rotary Wing Aircraft
CAA Chapter G4-6 11.75. Electrical Bonding and Lightning Discharge Protection (Rotorcraft). 4 pp.
CAA Chapter G4-6 App #1 11.75. Electrical Bonding and Lightning Discharge Protection (Rotorcraft). 5 pp.
CAA CAP 524 SUB-Part D 12.86. Design and Construction (Rotorcraft). 23 pp.
CAA CAP524 SUB-Part D 12.86. (Rotorcraft). 31 pp.

—Ships
MOD UK NES 516-88. Guide to Lightning Protection in Surface Ships Issue 2 (10.88). 46 pp.
MOD UK NES 516-92. Guide to Lightning Protection in Surface Ships Issue 3 (01.92). 50 pp.

—Structures
BSI BS 6651-92. 1992 Code of Practice for Protection of Structures Against Lightning. 115 pp.
BSI BS 6651-90. 1990 Code of Practice for Protection of Structures Against Lightning. 82 pp.
BSI BS 6651-85. (WITHDRAWN) 1985 Amd 1 Code of Practice for Protection of Structures Against Lightning (R). 86 pp.
DIN VDE 0185 Pt 100 (D)-87. Specifications for the Lightning Protection of Structures; General Principles; Indentical to IEC 81(CO)6 (Oct). 41 pp.
IEC 1024 Pt 1-90. Protection of Structures Against Lightning Part 1: General Principles First Edition. 50 pp.
JIS A 4201-92. Protection of Structures Against Lightning. 20 pp.

—Supersonic Transports
CAA STANDARD NO. 8-6&AP 07.69. Electrical Bonding and Lightning Discharge Protection. 13 pp.

—Telecommunication Systems
CCITT RECMN K.11-91. Principles of Protection Against Overvoltages and Overcurrents (Study Group V) 14 pp. 14 pp.
DIN VDE 0845 Pt 1-87. Protection of Telecommunication Installations Against Lightning, Electrostatic Discharges and Overvoltages from Power Installations; Measures Against Overvoltages (Oct) (Partially Supersedes DIN 57845/VDE 0845/04.76, DIN 57 845 A1/VDE 0845A1/04.81). 79 pp.

—Telephone Cables
CCITT RECMN K.13-89. Induced Voltages in Cables with Plastic-Insulated Conductors—Protection Against Interference (Study Group V) 2 pp. 2 pp.

—Telephone Cables—Repeaters
CCITT RECMN K.15-89. Protection of Remote-Feeding Systems and Line Repeaters Against Lightning and Interference from Neighbouring Electricity Lines—Protection Against Interference (Study Group V) 4 pp. 4 pp.

—Telephones
CCITT RECMN K.7-89. Protection Against Acoustic Shock—Protection Against Interference (Study Group V) 1 pp. 1 p.

—Transmitters—Japan
CCIR Report 861-82. Protection of Radio Stations Against Lightning and Other Electromagnetic Disturbances—Section 3Ad—Operational Questions. 17 pp.

Lightning Protection Equipment
Use For: Lightning Protection Systems and Equipment See Also: Electrical Protection Equipment; Lightning Protection; Surge Arresters

—Boats (Marine)
ISO 10134-93. Small Craft—Electrical Devices—Lightning Protection First Edition. 9 pp.

—Installation
CSA CAN/CSA-B72-M87. Installation Code for Lightning Protection Systems; (Gen Instr 1 Thru 2). 51 pp.
DIN VDE 0185 Pt 1-82. Lightning Protection System; General with Regard to Installation (Nov). 69 pp.
DIN VDE 0185 Pt 2-82. Lightning Protection System; Installation of Special Structures (Nov). 52 pp.

—Installation—Construction Contracts
DIN ENGL 18384-88. Tendering and Performance Stipulations in Contracts for Construction Works (VOB); Part C: General Technical Specifications in Contracts for Construction Works (ATV); Installation of Lightning Protection Systems (Sept) (This. 3 pp.

Lightning Protection Systems and Equipment
Use: Lightning Protection Equipment

Lightweight Aggregates
Use: Aggregates

Lightweight Concrete
See Also: Concretes
DIN ENGL 4219 Pt 1-79. Lightweight Concrete and Reinforced Lightweight Concrete of Dense Structure; Properties, Manufacture and Inspection (Dec). 5 pp.
DIN ENGL 4219 Pt 2-79. Lightweight Concrete and Reinforced Lightweight Concrete of Dense Structure; Design and Construction (Dec). 7 pp.
SNZ NZS 3152-74. Specification for the Manufacture and Use of Structural and Insulating Lightweight Concrete Amend: 1, 1985 (Reconfirmed 1980). 23 pp.

Lignin Content Analysis
See Also: Chlorine Number

—Pulps
CNS P3021-87. Method of Test for Acid-Insoluble Lignin in Wood and Pulp (Sep)(2721). 4 pp.
CNS P3089-87. Method of Test for Acid-Soluble Lignin in Wood and Pulp (Sep)(12108).
CPPA G.9-84. Acid-Insoluble Lignin in Wood Pulp. 1 p.
SAA AS 1301.P6RP-78. Methods of Test for Pulp and Paper (Metric Units)—Part P6rp: Halse Lignin in Wood and Pulp (This is a Joint Standard with SANZ NZS 1301).
SAA AS 1301.P11S-78. Methods of Test for Pulp and Paper (Metric Units)—Part P11s: Klason Lignin in Wood and Pulp (This is a Joint Standard with SANZ NZS 1301).
SNZ NZS/AS 1301. P6RP-78. Methods of Test for Pulp and Paper Halse Lignin in Wood and Pulp (This is a Joint Standard with SAA AS 1301.P6RP). 2 pp.
SNZ NZS/AS 1301. P11S-78. Methods of Test for Pulp and Paper Klason Lignin in Wood and Pulp (This is a Joint Standard with SAA AS 1301.P11S). 3 pp.

—Wood
CNS P3021-87. Method of Test for Acid-Insoluble Lignin in Wood and Pulp (Sep)(2721). 4 pp.
CNS P3089-87. Method of Test for Acid-Soluble Lignin in Wood and Pulp (Sep)(12108).
CPPA G.8-84. Acid-Insoluble Lignin in Wood Reprinted—1988. 1 p.
SAA AS 1301.P6RP-78. Methods of Test for Pulp and Paper (Metric Units)—Part P6rp: Halse Lignin in Wood and Pulp (This is a Joint Standard with SANZ NZS 1301).

Lignin Content Analysis (Cont.)
—Wood (Cont.)
SAA AS 1301.P11S-78. Methods of Test for Pulp and Paper (Metric Units)—Part P11s: Klason Lignin in Wood and Pulp (This is a Joint Standard with SANZ NZS 1301).

SNZ NZS/AS 1301. P6RP-78. Methods of Test for Pulp and Paper Halse Lignin in Wood and Pulp (This is a Joint Standard with SAA AS 1301.P6RP). 2 pp.

SNZ NZS/AS 1301. P11S-78. Methods of Test for Pulp and Paper Klason Lignin in Wood and Pulp (This is a Joint Standard with SAA AS 1301.P11S). 3 pp.

Lignite
Use For: Brown Coal *See Also:* Coal; Fossil Fuels; Fuels; Rocks; Solid Fuels; Solid Mineral Fuels; Sub-Bituminous Coal

CGSB 18-GP-7-69. Coal, Bituminous, Subbituminous and Lignite; (Amendment 1 Feb 1980). 11 pp.

—Coal Gas Content—Distillation Methods
ISO 647-74. Brown Coals and Lignites—Determination of the Yields of Tar, Water, Gas and Coke Residue by Low Temperature Distillation First Edition. 8 pp.

—Coke Content—Distillation Methods
ISO 647-74. Brown Coals and Lignites—Determination of the Yields of Tar, Water, Gas and Coke Residue by Low Temperature Distillation First Edition. 8 pp.

—Extraction Analysis
ISO 1952-76. Brown Coals and Lignites—Method of Extraction for the Determination of Sodium and Potassium Soluble in Dilute Hydrochloric Acid First Edition. 4 pp.

—Humic Acids Content
ISO 5073-85. Brown Coals and Lignites—Determination of Humic Acids First Edition. 5 pp.

—Moisture Content
ISO 2950-74. Brown Coals and Lignites—Classification by Types on the Basis of Total Moisture Content and Tar Yield First Edition. 4 pp.

—Moisture Content—Gravimetric Analysis
ISO 5068-83. Brown Coals and Lignites—Determination of Moisture Content—Indirect Gravimetric Method First Edition. 7 pp.

—Moisture Content—Volumetric Analysis
ISO 1015-92. Brown Coals and Lignites—Determination of Moisture Content—Direct Volumetric Method Second Edition. 6 pp.

—Sampling
ISO 5069 Pt 1-83. Brown Coals and Lignites—Principles of Sampling—Part 1: Sampling for Determination of Moisture Content and for General Analysis First Edition. 6 pp.

ISO 5069 Pt 2-83. Brown Coals and Lignites—Principles of Sampling—Part 2: Sample Preparation for Determination of Moisture Content and for General Analysis First Edition. 10 pp.

—Solid Fuel Burning Equipment
SAA AS 4014.3-92. Domestic Solid Fuel Burning Appliances—Test Fuels—Part 3: Lignite Briquettes (NZS 7404.3:1992). 1 p.

SNZ NZS 7404.3-92. Domestic Solid Fuel Burning Appliances Part 3: Lignite Briquettes (NZS 7404.3:1992/AS 4014.3-1992). 1 p.

—Soluble Matter Content
ISO 975-85. Brown Coals and Lignites—Determination of Yield of Toluene-Soluble Extract Second Edition. 5 pp.

ISO 1017-85. Brown Coals and Lignites—Determination of Acetone-Soluble Material ("Resinous Substances") in the Toluene-Soluble Extract Second Edition. 4 pp.

—Tar Content
ISO 2950-74. Brown Coals and Lignites—Classification by Types on the Basis of Total Moisture Content and Tar Yield First Edition. 4 pp.

—Tar Content—Distillation Methods
ISO 647-74. Brown Coals and Lignites—Determination of the Yields of Tar, Water, Gas and Coke Residue by Low Temperature Distillation First Edition. 8 pp.

—Water Content—Distillation Methods
ISO 647-74. Brown Coals and Lignites—Determination of the Yields of Tar, Water, Gas and Coke Residue by Low Temperature Distillation First Edition. 8 pp.

Lignite Content Analysis
—Sand
CNS A3036-59. Method of Test for Coal and Lignite in Sand (Sep)(1172)(R 1970).

Ligroin
Use For: Ligroine; Petroleum Ether
See Also: Mineral Spirits; Naphtha; Petroleum Products

CNS K7350-69. Chemical Reagent (Petroleum Ether) (Jul)(1849).

JIS K 8593-80. Petroleum Ether.

JIS K 8937-61. Ligroin.

MOD UK CS 5438A. Petroleum Ether (Dormant).

Ligroine
Use: Ligroin

Lime
Use: Calcium Oxide

Lime Oil
See Also: Essential Oils; Oils

CNS K5089-80. Oil of Lime, Obtained by Distillation (Sep)(6460).

CNS K5097-80. Oil of Lime (Obtained by Expression of the Whole Fruit) (Sep)(6468).

ISO 3519-76. Oil of Lime, Obtained by Distillation First Edition. 4 pp.

ISO 3809-87. Oil of Lime, Mexico (Citrus Aurantiifolia (Christmann) Swingle) Obtained by Mechanical Means Second Edition. 8 pp.

—Citral Content—Gas Chromatography
ISO 7611-85. Oils of Lemon and Petitgrain Citronnier, and Oil of Lime Obtained by a Mechanical Process—Determination of Citral (Neral + Geranial) Content—Gas Chromatographic Method on Capillary Columns First Edition. 7 pp.

Lime Plaster
Use For: Plastering Lime *See Also:* Plaster

JIS A 6902-76. Plastering Lime.

Lime Putty
Use: Calcium Oxide

Lime Water
See Also: Calcium Hydroxides

—Chemical Analysis
CNS K6019-53. Method of Chemical Test for Lime Water (Feb)(281)(R 1973).

Limes
See Also: Citrus Fruits; Fruits

—Chemical Analysis
CNS K6024-53. Method of Chemical Test for Limes (Feb)(287)(R 1973).

—Storage
ISO 3631-78. Citrus Fruits—Guide to Storage First Edition. 11 pp.

Limestone
See Also: Aggregates; Dolomite; Rocks; Siliceous Limestone

CNS M1009-70. Limestone (Feb)(2207).

—Aluminum Oxide Content
CPPA J.4-92. Analysis of Limestone. 5 pp.

—Aluminum Oxide Content—Photometry
DIN ENGL 52240 Pt 4-85. Analysis of Raw Materials Used in Glass Production; Chemical Analysis of Limestone Containing Not Less Than 95% of Calcium Carbonate; Determination of Aluminium Oxide (Sept). 2 pp.

—Bituminous Pavements
CNS A2094-85. Limestone Filler for Bituminous Paving Mixtures (Oct)(6990). 3 pp.

JIS A 5008-76. Limestone Filler for Bituminous Paving Mixtures (R 1984). 7 pp.

—Calcium Oxide Content
CPPA J.4-92. Analysis of Limestone. 5 pp.

—Calcium Oxide Content—Volumetric Analysis
DIN ENGL 52240 Pt 8-85. Analysis of Raw Materials Used in Glass Production; Chemical Analysis of Limestone Containing Not Less Than 95% of Calcium Carbonate; Determination of Calcium Oxide and Magnesium Oxide (Sept). 4 pp.

—Carbon Dioxide Content
CGSB 1-GP-71 METH 50.10-80. Methods of Testing Paints and Pigments Pigment Analysis Carbon Dioxide in Carbonates. 1 p.

Limestone (Cont.)
—Chemical Analysis
CNS M3184-85. Method of Chemical Analysis for Limestone (Oct)(11393).

CPPA J.1U-77. Limestone Analysis. 1 p.

DIN ENGL 52240 Pt 1-85. Analysis of Raw Materials Used in Glass Production; Chemical Analysis of Limestone Containing Not Less Than 95% of Calcium Carbonate; General Information and Test Report (Sept). 2 pp.

DIN ENGL 52240 Pt 2-85. Analysis of Raw Materials Used in Glass Production; Chemical Analysis of Limestone Containing Not Less Than 95% of Calcium Carbonate; Digestion of Limestone (Sept). 2 pp.

JIS M 8850-82. Methods for Chemical Analysis of Limestone. 32 pp.

—Glass
BSI BS 3108-80. 1980 Amd 1 Limestone for Making Colourless Glasses. 10 pp.

—Ignition Loss
CPPA J.4-92. Analysis of Limestone. 5 pp.

—Insoluble Matter Content
CPPA J.4-92. Analysis of Limestone. 5 pp.

—Iron Content—Photometry
DIN ENGL 52240 Pt 5-85. Analysis of Raw Materials Used in Glass Production; Chemical Analysis of Limestone Containing Not Less Than 95% of Calcium Carbonate; Determination of Total Iron Content Expressed as Iron (III) Oxide (Sept). 2 pp.

—Iron Oxide Content
CPPA J.4-92. Analysis of Limestone. 5 pp.

—Magnesium Oxide Content
CPPA J.4-92. Analysis of Limestone. 5 pp.

—Magnesium Oxide Content—Atomic Absorption Spectrometry
DIN ENGL 52240 Pt 8-85. Analysis of Raw Materials Used in Glass Production; Chemical Analysis of Limestone Containing Not Less Than 95% of Calcium Carbonate; Determination of Calcium Oxide and Magnesium Oxide (Sept). 4 pp.

—Magnesium Oxide Content—Volumetric Analysis
DIN ENGL 52240 Pt 8-85. Analysis of Raw Materials Used in Glass Production; Chemical Analysis of Limestone Containing Not Less Than 95% of Calcium Carbonate; Determination of Calcium Oxide and Magnesium Oxide (Sept). 4 pp.

—Manganese Content—Atomic Absorption Spectrometry
DIN ENGL 52240 Pt 7-85. Analysis of Raw Materials Used in Glass Production; Chemical Analysis of Limestone Containing Not Less Than 95% of Calcium Carbonate; Determination of Manganese Content Expressed as MnO (Sept). 2 pp.

—Manganese Content—Photometry
DIN ENGL 52240 Pt 7-85. Analysis of Raw Materials Used in Glass Production; Chemical Analysis of Limestone Containing Not Less Than 95% of Calcium Carbonate; Determination of Manganese Content Expressed as MnO (Sept). 2 pp.

—Reserves
JIS M 1003-57. Calculation of Limestone Reserves.

—Silicon Oxide Content
DIN ENGL 52240 Pt 3-85. Analysis of Raw Materials Used in Glass Production; Chemical Analysis of Limestone Containing Not Less Than 95% of Calcium Carbonate; Determination of Silicon (IV) Oxide (Sept). 2 pp.

—Titanium Oxide Content—Photometry
DIN ENGL 52240 Pt 6-85. Analysis of Raw Materials Used in Glass Production; Chemical Analysis of Limestone Containing Not Less Than 95% of Calcium Carbonate; Determination of Titanium (IV) Oxide (Sept). 2 pp.

Limestone Content Analysis
—Portland Cements
CNS A3301-88. Method for Determination of Granulated (Water Quenched) Blast-Furnace Slag, Silica Material, Fly Ash and Limestone Content in Portland Cement (Nov)(12459).

SNZ NZS 3125-91. Specification for Portland-Limestone Filler Cement. 21 pp.

Limit Design Method
Use For: Limit States Design

CSA S408-1981SP. Guidelines for the Development of Limit States Design. 34 pp.

INTERNATIONAL AND NON-U.S. NATIONAL STANDARDS
SUBJECT INDEX

Limit Design Method (Cont.)

—Building Codes—Australian
SAA AS MP38-79. Proposed Timetable for the Coordinated Revision of Australian Structural Codes into Limit State Terms. 4 pp.

—Offshore Platforms
CSA CAN/CSA-S473-92. Steel Structures; (Gen Instr 1). 115 pp.
CSA S473.1-1992 SP. Commentary to CSA Standard CAN/CSA-S473-92, Steel Structures; (Gen Instr 1). 70 pp.

—Steel Structures
CSA CAN/CSA-S16.1-M89. Limit States Design of Steel Structures; (Gen Instr 1). 154 pp.
CSA S16.1.1-1992. Commentary on CSA Standard CAN/CSA-S16.1-M89, Limit States Design of Steel Structures; (Gen Instr 1). 97 pp.

—Structural Members
CSA CAN/CSA-S136-M89. Cold Formed Steel Structural Members; (Gen Instr 1 Thru 2). 78 pp.
CSA S136.1-M1991SP. Commentary on CSA Standard CAN/CSA-S136-M89, Cold Formed Steel Structural Members. 88 pp.

—Wooden Structures
CSA CAN/CSA-O86.1-M89. Engineering Design in Wood (Limit States Design); (Gen Instr 1 Thru 2) (Supplement 1 1993). 262 pp.
CSA O86.1.1-1992. Commentary on CSA Standard CAN/CSA-O86.1-M89, Engineering Design in Wood (Limit States Design); (Gen Instr 1). 89 pp.

Limit Gages
Scope Note: For additional listings, use a more specific term *Use For:* Limit Gauges *See Also:* Go Gages; No Go Gages; Plug Gages; Ring Gages; Thread Gages
BSI BS 969-82. 1982 Limits and Tolerances on Plain Limit Gauges. 12 pp.
DIN ENGL 7150 Pt 2-77. ISO Systems of Limits and Fits; Testing of Workpiece Elements with Cylindrical and Parallel Mating Surfaces (Aug). 10 pp.
ISO R1938-71. ISO System of Limits and Fits Part II: Inspection of Plain Workpieces First Edition. 28 pp.
JIS B 7420-80. Limit Gauges. 27 pp.
JIS B 7421-71. Tolerances, Allowable Deviations and Permissible Wears of Limit Gauges (R 1986). 14 pp.
SAA AS B129-68. Designs for Geometric Limit Gauges (Plain and Screwed in Inch Units) Reconfirmed 1987. 63 pp.
SAA AS 1997-77. Plain Limit Gauges (Metric Series). 32 pp.
SNZ NZS/BS 969-82. Specification for Limits and Tolerances on Plain Limit Gauges. 12 pp.
SNZ NZS/ISO R1938-71. ISO System of Limits and Fits. Part 2 Inspection of Plain Workpieces. 28 pp.

Limit Gauges
Use: Limit Gages

Limit States Design
Use: Limit Design Method

Limit Switches
See Also: Air Conditioners; Control Switches; Electric Switches
DIN VDE 0660 Pt 206-86. Switchgear and Control Gear; Control Switches; Additional Requirements for Position Switches with Positive Opening Operation (Limit Switch) (Oct). 8 pp.

Limited Liability Companies

—Annual Accounts
EC COM(86) 238-86. Proposal for a Council Directive Amending Directive 78/660/EEC on Annual Accounts and Directive 83/349/EEC on Consolidated Accounts as Regards the Scope of Those Directives. 8 pp.
EC 90/605/EEC-90. Council Directive Amending Directive 78/660/EEC on Annual Accounts and Directive 83/349/EEC on Consolidated Accounts as Regards the Scopes of Those Directives. 3 pp.

—Consolidated Accounts
EC COM(86) 238-86. Proposal for a Council Directive Amending Directive 78/660/EEC on Annual Accounts and Directive 83/349/EEC on Consolidated Accounts as Regards the Scope of Those Directives. 8 pp.
EC 90/605/EEC-90. Council Directive Amending Directive 78/660/EEC on Annual Accounts and Directive 83/349/EEC on Consolidated Accounts as Regards the Scopes of Those Directives. 3 pp.

Limiters
See Also: Microwave Limiters

Limiters (Cont.)

—Sound Transmission—FM
CCIR RECMN 642-1-90. Limiters for High-Quality Sound-Programme Signals—Section 10B—Frequency-Modulation Sound Broadcasting in Bands 8 (VHF) and 9 (UHF). 3 pp.

Limiting Viscosity Number
Use: Intrinsic Viscosity

Limits
Use: Tolerances

Linch Pins
Use: Linchpins

Linchpins
Use For: Lynch Pins; Socket Pins
See Also: Couplings; Drawbars; Pins

—Hitches—Tractors
ISO 7072-93. Tractors and Machinery for Agriculture and Forestry—Linch Pins and Spring Pins—Dimensions and Requirements Second Edition. 6 pp.

Lincompex Systems
Use: Compandors

Lincomycin Content Analysis

—Animal Feed
CNS N4104-82. Method of Test for Feed Additives (Determination of Lincomycin) (Oct)(9534).

Lindane
Use: Benzene Hexachloride; Chlorobenzenes

Lindane Content Analysis
Use: Benzene Hexachloride Content Analysis

Line Amplifiers
See Also: Amplifiers; Electronic Equipment

—Repeater Stations
CCITT RECMN G.323-89. Typical Transistorized System on Symmetric Cable Pairs—International Analogue Carrier Systems (Study Group XV) 3 pp. 3 pp.

—Telephone Cables
CCITT RECMN G.323-89. Typical Transistorized System on Symmetric Cable Pairs—International Analogue Carrier Systems (Study Group XV) 3 pp. 3 pp.
CCITT RECMN G.324-89. General Characteristics for Valve-Type Systems on Symmetric Cable Pairs —International Analogue Carrier Systems (Study Group XV) 1 pp. 1 p.

—Telephone Cables—Frequencies
CCITT RECMN G.323-89. Typical Transistorized System on Symmetric Cable Pairs—International Analogue Carrier Systems (Study Group XV) 3 pp. 3 pp.

—Telephone Cables—Pilot Channels
CCITT RECMN G.323-89. Typical Transistorized System on Symmetric Cable Pairs—International Analogue Carrier Systems (Study Group XV) 3 pp. 3 pp.

—Telephone Cables—Signal Levels
CCITT RECMN G.323-89. Typical Transistorized System on Symmetric Cable Pairs—International Analogue Carrier Systems (Study Group XV) 3 pp. 3 pp.

Line Buffers
See Also: Buffers (Data); Integrated Circuits

—Noninverting—Hex
CECC CECC 90 101-060 ISSUE 1-87. UTE C 86-213/B 80; Digital Integrated Circuits in Accordance with FS 90 101 54/64/74 367A; Non-Inverting Hex Buffers 2-Line and 4-Line Enable Inputs 3-State Outputs (En). 21 pp.

Line Drawings
See Also: Drawings

—Transistors
CNS C5081-89. General Rules for Transistors (Oct)(6808).
CNS C5082-89. Outlines Drawing Methods for Transistors (Oct)(6809).

Line Driver/Receivers
Use: Line Transceivers

Line Drivers
Use For: Line Transmitters *See Also:* Buffer/Line Drivers; Integrated Circuits; Interface Circuits

Line Drivers (Cont.)
BSI BS 6493: Sec 2.4-89. 1989 Semiconductor Devices Part 2: Integrated Circuits Section 2.4: Recommendations for Interface Integrated Circuits. 64 pp.
BSI BS 6493: Sec 2.4-01. 1989 Amd 1 Semiconductor Devices Part 2: Integrated Circuits Section 2.4: Recommendations for Interface Integrated Circuits (IEC 748-4: 1987) (AMD 7036) August 15, 1992. 122 pp.
CECC CECC 90 101-038 ISSUE 1-87. UTE C 86-213/B 58; Digital Integrated Circuits in Accordance with FS 90 101; 54/64/74 128; 75 Ohm Line Driver (En). 18 pp.
IEC 748 Pt 4-87. Semiconductor Devices Integrated Circuits Part 4: Interface Integrated Circuits First Edition; (Amendment 1-1991). 194 pp.

—Integrated Circuit
CECC CECC 90 102-025 ISSUE 1-81. BS CECC 90 102-025; Dual 4-Input NAND 50-OHM Line Driver (En) AMD 1 (En). 18 pp.

—Preferred Products List
CECC CECC MUAHAG Vol 7 IS 8-92. Preferred Products List; Active Microcircuits (En, Fe, Ge). 89 pp.

—Quad
CECC CECC 90 301-002 ISSUE 1-91. UTE C 86-901-001; Analogue Integrated Circuits in Accordance with FS 90 301 (En, Fr) ERRATUM (En, Fr). 38 pp.
CECC CECC MUAHAG Vol 7 IS 8-92. Preferred Products List; Active Microcircuits (En, Fe, Ge). 89 pp.

Line Filters (Electric)
Use For: Power Line Filters *See Also:* Electric Filters; Electromagnetic Interference Filters

—Passive—Electromagnetic Interference—Military—PPL
CECC CECC MUAHAG Vol 11 IS 2-92. Preferred Products List; Filters (En, Fr, Ge). 88 pp.

—Passive—Radio Frequency Interference—Military—PPL
CECC CECC MUAHAG Vol 11 IS 2-92. Preferred Products List; Filters (En, Fr, Ge). 88 pp.

Line Hunting Services
See Also: Telephone Services

—Integrated Services Digital Networks
CCITT RECMN I.252-89. Call Offering Supplementary Services—Integrated Services Digital Network (ISDN)—General Structure and Service Capabilities (Study Group XVIII) 37 pp. 37 pp.
CCITT RECMN I.252. 6-88. Line Hunting—Integrated Services Digital Network (ISDN)—General Structure and Service Capabilities (Study Group XVIII) 5 pp. 5 pp.

—Integrated Services Digital Networks—Functions/Information Flows
CCITT RECMN Q.82-89. Call Offering Supplementary Services—General Recommendations on Telephone Switching and Signalling—Functions and Information Flows for Services in the ISDN—Supplements (Study Group XI) 25 pp. 25 pp.

—Subscriber Lines
CEPT T/CS 20-11-81. Caracteristique Recherche De Ligne De Type Analogique. 1 p.
CEPT T/CS 20-11 E-81. Analogue Subscriber's Line Hunting. 1 p.

Line Impedance Stabilization Networks
Use For: LISN *See Also:* Electromagnetic Compatibility; Radio Frequency Interference
CSA CAN3-C108 .1.5-M85. Line Impedance Stabilization Network (LISN). 17 pp.

Line Isolation Monitors
See Also: Medical Electrical Equipment; Medical Equipment; Monitors
CSA C22.2 NO 204-M1984. Line Isolation Monitors (R 1992); (Gen Instr 1). 38 pp.

—Housings
CSA C22.2 NO 204-M1984. Line Isolation Monitors (R 1992); (Gen Instr 1). 38 pp.

—Power Supplies
CSA C22.2 NO 204-M1984. Line Isolation Monitors (R 1992); (Gen Instr 1). 38 pp.

Line Links
See Also: Communication Transmission Lines; Group Links; Main Repeater Stations; Mastergroup Links; Supergroup Links; Supermastergroup Links; Through

Line

Line Links (Cont.)

See Also: (Cont.)
Connection Points; 15 Supergroup Assembly Links

—Carrier Systems—Telephone Networks
CCITT RECMN G.211-89. Make-up of a Carrier Link—International Analogue Carrier Systems (Study Group XV) 8 pp. 8 pp.

—Designations
CCITT RECMN M.140-89. Designations of International Circuits, Groups, Group and Line Links, Digital Blocks, Digital Paths, Data Transmission Systems and Related Information—General Maintenance Principles—Maintenance of International Transmission Systems and Telephone Circuits. 53 pp.

—Interfaces—Main Repeater Stations
CCITT RECMN G.213-89. Interconnection of Systems in a Main Repeater Station—International Analogue Carrier Systems (Study Group XV) 4 pp. 4 pp.

Line of Sight Propagation

Use For: LOS Propagation *See Also:* Radio Wave Propagation; Wave Propagation

—Radio Relay Systems—Antennas—Radiation Patterns
CCIR RECMN 699-90. Reference Radiation Patterns for Line-of-Sight Radio-Relay System Antennas for Use in Coordination Studies and Interference Assessment in the Frequency Range from 1 to About 40 GHz—Section 9B2—System General Characteristics. 2 pp.
CCIR RECMN 699-1-92. Reference Radiation Patterns for Line-of-Sight Radio-Relay System Antennas for Use in Coordination Studies and Interference Assessment in the Frequency Range from 1 to About 40 GHz—Section 9B2—System General Characteristics. 2 pp.

—Radio Relay Systems—Design
CCIR RECMN 530-3-90. Propagation Data and Prediction Methods Required for the Design of Terrestrial Line-of-Sight Systems—Section 5E—Aspects Relative to the Terrestrial Fixed Service. 1 p.
CCIR Report 338-6-90. Propagation Data and Prediction Methods Required for Terrestrial Line-of-Sight Systems—Section 5E—Aspects Relative to the Terrestrial Fixed Service. 66 pp.
CCIR Report 784-3-90. Effects of Propagation on the Design and Operation of Line-of-Sight Radio-Relay Systems—Section 9A—Performance Objectives, Propagation and Interference Effects. 62 pp.
CCIR RECMN 530-4-92. Propagation Data and Prediction Methods Required for the Design of Terrestrial Line-of-Sight Systems. 23 pp.

—Radio Relay Systems—Operation
CCIR Report 784-3-90. Effects of Propagation on the Design and Operation of Line-of-Sight Radio-Relay Systems—Section 9A—Performance Objectives, Propagation and Interference Effects. 62 pp.

Line Outs (Ships)
Use: Ship Models

Line Pipes
See Also: Pipelines

—Aluminum—Gas—Coils
CSA CAN/CSA-Z245.6-92. Coiled Aluminum Line Pipe and Accessories; (Gen Instr 1). 28 pp.

—Aluminum—Oil—Coils
CSA CAN/CSA-Z245.6-92. Coiled Aluminum Line Pipe and Accessories; (Gen Instr 1). 28 pp.

—Steel
CSA Z245.1-93. Steel Line Pipe; (Gen Instr 1). 140 pp.
ISO 3183-80. Oil and Natural Gas Industries—Steel Line Pipe First Edition; (Erratum—Jan 1981). 44 pp.

Line Poles
Use: Poles (Supports)

Line Printers
See Also: Impact Printers; Printers; Printing Presses

—Ribbons—Fabric
BSI BS 5519: Part 2-82. 1982 Ribbons and Spools Used on Office Machines and Data Processing Equipment Part 2: Specification for Widths of Fabric Printing Ribbons on Spools up to 19 mm Wide (ISO 2257: 1980). 4 pp.

Line Printers (Cont.)
—Ribbons—Fabric (Cont.)
BSI BS 5519: Part 5-77. 1977 Ribbons and Spools Used on Office Machines and Data Processing Equipment Part 5: Recommendations for Widths of Fabric Printing Ribbons on Spools Exceeding 19 mm Wide. 3 pp.
CNS C5148-83. Office Machines and Printing Machines Used for Information Processing of Fabric Printing Ribbons on Spools (Feb)(9981).
CNS C5210-86. Office Machines and Printing Machines Used for Information Processing—Widths of Fabric Printing Ribbons on Spools Exceeding 19mm (Oct)(11719).
ISO 2257-80. Office Machines and Printing Machines Used for Information Processing—Widths of Fabric Printing Ribbons on Spools Second Edition; (Corrected and Reprinted -1981). 3 pp.
ISO 3866-77. Office Machines and Printing Machines Used for Information Processing—Widths of Fabric Printing Ribbons on Spools Exceeding 19 mm First Edition. 3 pp.
JIS B 9520-85. Fabric—Inked Ribbons for Line Printer.
JTC1 2257-80. Office Machines and Printing Machines Used for Information Processing—Widths of Fabric Printing Ribbons on Spools Second Edition; (Corrected and Reprinted -1981). 3 pp.
JTC1 3866-77. Office Machines and Printing Machines Used for Information Processing—Widths of Fabric Printing Ribbons on Spools Exceeding 19 mm First Edition. 3 pp.
OSI ISO 2257-80. Office Machines and Printing Machines Used for Information Processing—Widths of Fabric Printing Ribbons on Spools. 3 pp.
OSI ISO 3866-77. Office Machines and Printing Machines Used for Information Processing—Widths of Fabric Printing Ribbons on Spools exceeding 19mm. 3 pp.

—Ribbons—Paper
BSI BS 5519: Part 3-77. 1977 Ribbons and Spools Used on Office Machines and Data Processing Equipment Part 3: Widths of One-Time Paper or Plastic Printing Ribbons up to 19 mm Wide and Marking to Indicate the End of the Ribbons. 3 pp.
CNS C5193-85. Office Machines and Printing Machines Used for Information Processing—Widths of One-Time Paper or Plastic Printing Ribbons and Marking to Indicate the End of the Ribbons (Mar)(11217).
ISO 2775-77. Office Machines and Printing Machines Used for Information Processing—Widths of One-Time Paper or Plastic Printing Ribbons and Marking to Indicate the End of the Ribbons Second Edition. 3 pp.
JTC1 2775-77. Office Machines and Printing Machines Used for Information Processing—Widths of One-Time Paper or Plastic Printing Ribbons and Marking to Indicate the End of the Ribbons Second Edition. 3 pp.
OSI ISO 2775-77. Office Machines and Printing Machines Used for Information Processing—Widths of One-Time Paper or Plastic Printing Ribbons and Marking to Indicate the End of the Ribbons. 3 pp.

—Ribbons—Plastic
BSI BS 5519: Part 3-77. 1977 Ribbons and Spools Used on Office Machines and Data Processing Equipment Part 3: Widths of One-Time Paper or Plastic Printing Ribbons up to 19 mm Wide and Marking to Indicate the End of the Ribbons. 3 pp.
CNS C5193-85. Office Machines and Printing Machines Used for Information Processing—Widths of One-Time Paper or Plastic Printing Ribbons and Marking to Indicate the End of the Ribbons (Mar)(11217).
ISO 2775-77. Office Machines and Printing Machines Used for Information Processing—Widths of One-Time Paper or Plastic Printing Ribbons and Marking to Indicate the End of the Ribbons Second Edition. 3 pp.
JTC1 2775-77. Office Machines and Printing Machines Used for Information Processing—Widths of One-Time Paper or Plastic Printing Ribbons and Marking to Indicate the End of the Ribbons Second Edition. 3 pp.
OSI ISO 2775-77. Office Machines and Printing Machines Used for Information Processing—Widths of One-Time Paper or Plastic Printing Ribbons and Marking to Indicate the End of the Ribbons. 3 pp.

Line Receivers
See Also: Integrated Circuits; Interface Circuits
BSI BS 6493: Sec 2.4-89. 1989 Semiconductor Devices Part 2: Integrated Circuits Section 2.4: Recommendations for Interface Integrated Circuits. 64 pp.
BSI BS 6493: Sec 2.4-01. 1989 Amd 1 Semiconductor Devices Part 2: Integrated Circuits Section 2.4: Recommendations for Interface Integrated Circuits (IEC 748-4: 1987) (AMD 7036) August 15, 1992. 122 pp.

Line Receivers (Cont.)
IEC 748 Pt 4-87. Semiconductor Devices Integrated Circuits Part 4: Interface Integrated Circuits First Edition; (Amendment 1-1991). 194 pp.

—Preferred Products List
CECC CECC MUAHAG Vol 7 IS 8-92. Preferred Products List; Active Microcircuits (En, Fe, Ge). 89 pp.

—Quad
CECC CECC 90 301-003 ISSUE 1-91. UTE C 86-901-002; Analogue Integrated Circuits in Accordance with FS 90 301 (En, Fr) ERRATUM (En, Fr). 38 pp.
CECC CECC MUAHAG Vol 7 IS 8-92. Preferred Products List; Active Microcircuits (En, Fe, Ge). 89 pp.

—Quality Assurance
BSI BS CECC 90301-85. 1985 Integrated Line Transmitters and Receivers. 23 pp.

Line Regulation Systems (Telecommunications)

See Also: Frequency Response; Regulators; Telecommunication Equipment; Telecommunication Systems; Telephone Cables; Telephone Lines
CCITT RECMN G.214-89. Line Stability of Cable Systems—International Analogue Carrier Systems (Study Group XV) 1 pp. 1 p.

—Frequency Division Multiplexers—Radio Relay Systems
CCITT RECMN G.423-89. Interconnection at the Baseband Frequencies of Frequency-Division Multiplex Radio-Relay Systems—International Analogue Carrier Systems (Study Group XV) 8 pp. 8 pp.

—Group Links—Maintenance
CCITT RECMN M.530-89. Readjustment to the Nominal Value of an International Group, Supergroup, Etc., Link—General Maintenance Principles—Maintenance of International Transmission Systems and Telephone Circuits (Study Group IV) 3 pp. 3 pp.

—Supergroup Links—Maintenance
CCITT RECMN M.530-89. Readjustment to the Nominal Value of an International Group, Supergroup, Etc., Link—General Maintenance Principles—Maintenance of International Transmission Systems and Telephone Circuits (Study Group IV) 3 pp. 3 pp.

Line Relays
See Also: Relays; Track Relays

—Railroad Signals
BSI BS 1659-50. 1950 Amd 3 Tractive Armature Direct-Current Neutral Track and Line Relays for Railway Signalling. 16 pp.
BSI BS 1745-51. 1951 Amd 2 Alternating-Current Relays for Railway Signalling: Track Relays (Double-Element, 2-Position), Line Relays (Single-Element, 2-Position). 15 pp.

Line Service Marking
Use: Subscriber Lines—Identification Systems

Line Shears
Use: Shearing Machines

Line Signaling
See Also: Almost Differential Quasi-Ternary Code; Interruption Control; Signaling Systems

—Analog
CEPT T/CS 42-02-85. Signalisation De Ligne Du Systeme De Signalisation R2, Version Analogique. 1 p.
CEPT T/CS 42-02 E-86. Systeme R2 Line Signalling, Analogue Version. 1 p.

—Analog—Both Way Operation—CCITT R2 Signaling Systems
CCITT FASCICLE VI.4. Specifications of Signalling Systems R1 and R2 Recommendations Q.310—Q.490 (Study Group XI). 181 pp.

—Analog—Conversion Equipment—CCITT R2 Signaling Systems
CCITT RECMN Q.430-89. Line Signalling, Digital Version—Conversion Between Analogue and Digital Versions of System R2 Line Signalling—Specifications of Signalling Systems R1 and R2 (Study Group XI) 21 pp. 21 pp.

—Analog/Digital Conversion
CEPT T/CS 42-04-84. Conversion Entre La Version Analogique Et La Version Numerique De La Signalisation De Ligne Du Systeme R2. 1 p.

Line Signaling (Cont.)

—Analog/Digital Conversion (Cont.)
CEPT T/CS 42-04 E-84. System R2 Line Signalling Conversion Between Analogue Version and Digital Version. 1 p.

—Analog—Exchange Lines—CCITT R2 Signaling Systems
CCITT RECMN Q.412-89. Line Signalling, Analogue Version—Clauses for Exchange Line Signalling Equipment. Clauses for Transmission Line Signalling Equipment—Specifications of Signalling Systems R1 and R2 (Study Group XI) 7 pp. 7 pp.

—Analog—Interruption Control—CCITT R2 Signaling Systems
CCITT RECMN Q.416-89. Line Signalling, Analogue Version—Interruption Control—Specifications of Signalling Systems R1 and R2 (Study Group XI) 5 pp. 5 pp.

—Analog—Metering—CCITT R2 Signaling Systems
CCITT FASCICLE VI.4. Specifications of Signalling Systems R1 and R2 Recommendations Q.310–Q.490 (Study Group XI). 181 pp.

—Analog—Pulse Code Modulation—CCITT R1 Signaling Systems
CCITT FASCICLE VI.4. Specifications of Signalling Systems R1 and R2 Recommendations Q.310–Q.490 (Study Group XI). 181 pp.

—Analog—Pulse Code Modulation—CCITT R2 Signaling Systems
CCITT FASCICLE VI.4. Specifications of Signalling Systems R1 and R2 Recommendations Q.310–Q.490 (Study Group XI). 181 pp.

—Analog—Receivers—CCITT R2 Signaling Systems
CCITT RECMN Q.415-89. Line Signalling, Analogue Version—Signal Receiver—Specifications of Signalling Systems R1 and R2 (Study Group XI) 2 pp. 2 pp.

—Analog—Signal Codes—CCITT R2 Signaling Systems
CCITT RECMN Q.411-89. Line Signalling, Analogue Version-Line Signalling Code—Specifications of Signalling Systems R1 and R2 (Study Group XI) 2 pp. 2 pp.

—Analog—Signal Senders—CCITT R2 Signaling Systems
CCITT RECMN Q.414-89. Line Signalling, Analogue Version—Signal Sender—Specifications of Signalling Systems R1 and R2 (Study Group XI) 3 PP. 3 pp.

—CCITT R1 Signaling Systems
CCITT RECMN Q.311-89. Line Signalling-2600 Hz Line Signalling—Specifications of Signalling Systems R1 and R2 (Study Group XI) 2 pp. 2 pp.
CCITT RECMN Q.317-89. Further Specification Clauses Relative to Line Signalling—Specifications of Signalling Systems R1 and R2 (Study Group XI) 2 pp. 2 pp.

—Digital
CEPT T/CS 42-03-85. Signalisation De Ligne Du Systeme De Signalisation R2, Version Numerique. 1 p.
CEPT T/CS 42-03 E-86. System R2 Line Signalling, Digital Version. 1 p.

—Digital—CCITT R2 Signaling Systems
CCITT FASCICLE VI.4. Specifications of Signalling Systems R1 and R2 Recommendations Q.310–Q.490 (Study Group XI). 181 pp.

—Digital—Conversion Equipment—CCITT R2 Signaling Systems
CCITT RECMN Q.430-89. Line Signalling, Digital Version—Conversion Between Analogue and Digital Versions of System R2 Line Signalling—Specifications of Signalling Systems R1 and R2 (Study Group XI) 21 pp. 21 pp.

—Digital—Exchange Lines—CCITT R2 Signaling Systems
CCITT RECMN Q.422-89. Line Signalling, Digital Version—Clauses for Exchange Line Signalling Equipment—Specifications of Signalling Systems R1 and R2 (Study Group XI) 6 pp. 6 pp.

—Digital—Faulty Transmission—CCITT R2 Line Signaling
CCITT RECMN Q.424-89. Line Signalling, Digital Version—Protection Against the Effects of Faulty Transmission—Specifications of Signalling Systems R1 and R2 (Study Group XI) 2 pp. 2 pp.

Line Signaling (Cont.)

—Digital—Metering—CCITT R2 Signaling Systems
CCITT FASCICLE VI.4. Specifications of Signalling Systems R1 and R2 Recommendations Q.310–Q.490 (Study Group XI). 181 pp.

—Digital—Signal Codes—CCITT R2 Signaling Systems
CCITT RECMN Q.421-89. Line Signalling, Digital Version—Digital Line Signalling Code—Specifications of Signalling Systems R1 and R2 (Study Group XI) 2 pp. 2 pp.

—Inband—CCITT R2 Signaling Systems
CCITT FASCICLE VI.4. Specifications of Signalling Systems R1 and R2 Recommendations Q.310–Q.490 (Study Group XI). 181 pp.

—Pulse Code Modulation—CCITT R1 Signaling Systems
CCITT RECMN Q.314-89. PCM Line Signalling Specifications of Signalling Systems R1 and R2 (Study Group XI) 1 pp. 1 p.

—Pulse Code Modulation—Receivers—CCITT R1 Signaling Systems
CCITT RECMN Q.316-89. PCM Line Signal Receiver—Specifications of Signalling Systems R1 and R2 (Study Group XI) 1 pp. 1 p.

—Pulse Code Modulation—Signal Senders—CCITT R1 Signaling Systems
CCITT RECMN Q.315-89. PCM Line Signal Sender (Transmitter)—Specifications of Signalling Systems R1 and R2 (Study Group XI) 2 pp. 2 pp.

—Receivers—CCITT No. 5 Signaling Systems
CCITT RECMN Q.144-89. Line Signal Receiver—Specifications of Signalling Systems Nos. 4 and 5 (Study Group XI) 3 pp. 3 pp.

—Receivers—CCITT R1 Signaling Systems
CCITT RECMN Q.313-89. 2600 Hz Line Signal Receiving Equipment—Specifications of Signalling Systems R1 and R2 (Study Group XI). 3 pp.

—Signal Codes—CCITT No. 5 Signaling Systems
CCITT RECMN Q.141-89. Signal Code for Line Signalling—Specifications of Signalling Systems Nos. 4 and 5 (Study Group XI) 5 pp. 5 pp.

—Signal Senders—CCITT No. 5 Signaling Systems
CCITT RECMN Q.143-89. Line Signal Sender—Specifications of Signalling Systems Nos. 4 and 5 (Study Group XI) 1 pp. 1 p.

—Signal Senders—CCITT R1 Signaling Systems
CCITT RECMN Q.312-89. Line Signalling-2600 Hz Line Signal Sender (Transmitter)—Specifications of Signalling Systems R1 and R2 (Study Group XI) 1 pp. 1 p.

—Testing Points
CCITT RECMN M.718-89. Test Point (Line Signalling)—General Maintenance Principles—Maintenance of International Transmission Systems and Telephone Circuits (Study Group IV) 2 pp. 2 pp.

Line Transceivers
Use For: Line Driver/Receivers; Transmission Line Driver/Receivers *See Also:* Inverting Buffer/Line Drivers; Transceivers
CECC CECC 90 301 ISSUE 1-85. Blank Detail Specification: Integrated Line Transmitters and/or Receivers (En, Fr, Ge) ERRATUM (En, Fr, Ge). 70 pp.

—Octal
CECC CECC 90 109-611 ISSUE 1-86. Digital Integrated Circuits in Accordance with FS 90 109; 54/74 HC 245 54/74 HC 645; Octal 3-State Transceiver (En, Fr). 6 pp.

Line Transmitters
Use: Line Drivers

Line Traps
Use For: Called Party Hold Plugs; Nuisance Caller Traps *See Also:* Filters; Transmission Lines
IEC 353-89. Line Traps for A.C. Power Systems Second Edition. 82 pp.

—Power Line Carriers
BSI BS 4996-73. 1973 Line Traps for Power Line Carrier Systems. 18 pp.

Line Up (Telecommunications)
Use For: Lining Up *See Also:* Telecommunication Circuits; Transmission Systems

—Analog Channels
CCITT RECMN M.470-89. Setting up and Lining up Analogue Channels for International Telecommunication Services—General Maintenance Principles—Maintenance of International Transmission Systems and Telephone Circuits (Study Group IV) 1 pp. 1 p.
CCITT RECMN M.475-89. Setting up and Lining up Mixed Analogue/Digital Channels for International Telecommunication Services—General Maintenance Principles—Maintenance of International Transmission Systems and Telephone Circuits (Study Group IV) 3 pp. 3 pp.

—Communication Transmission Lines
CCITT RECMN M.35-89. Principles Concerning Line-up and Maintenance Limits—General Maintenance Principles—Maintenance of International Transmission Systems and Telephone Circuits (Study Group IV) 1 pp. 1 p.

—Data Transmission Equipment
CCITT RECMN M.1350-89. Setting up, Lining up and Characteristics of International Data Transmission Systems Operating in the Range 2.4 kBit/s to 14.4 kBit/s—Maintenance of International Telegraph, Phototelegraph and Leased Circuits—Maintenance of the International Public. 3 pp.
CCITT RECMN M.1370-89. Setting up and Lining up of International Data Transmission Systems Operating at 48 kBit/s and Above—Maintenance of International Telegraph, Phototelegraph and Leased Circuits—Maintenance of the International Public Telephone Network-. 6 pp.

—Demand Assignment Circuits
CCITT RECMN M.675-89. Lining up and Maintaining International Demand Assignment Circuits (Spade)—General Maintenance Principles—Maintenance of International Transmission Systems and Telephone Circuits (Study Group IV) 6 pp. 6 pp.

—Digital Channels
CCITT RECMN M.475-89. Setting up and Lining up Mixed Analogue/Digital Channels for International Telecommunication Services—General Maintenance Principles—Maintenance of International Transmission Systems and Telephone Circuits (Study Group IV) 3 pp. 3 pp.

—Group Links—Data Transmission
CCITT RECMN M.910-89. Setting up and Lining up an International Leased Group Link for Wide-Spectrum Signal Transmission—Maintenance of International Telegraph, Phototelegraph and Leased Circuits—Maintenance of the International Public Telephone Network-. 4 pp.

—International Leased Circuits
CCITT RECMN M.1050-89. Lining up an International Point-to-Point Leased Circuit—Maintenance of International Telegraph, Phototelegraph and Leased Circuits—Maintenance of the International Public Telephone Network—Maintenance of Maritime Satellite and Data. 10 pp.
CCITT RECMN M.1055-89. Lining up an International Multiterminal Leased Circuit—Maintenance of International Telegraph, Phototelegraph and Leased Circuits—Maintenance of the International Public Telephone Network—Maintenance of Maritime. 4 pp.

—International Leased Circuits—Exchange of Information
CCITT RECMN M.1045-89. Preliminary Exchange of Information for the Provision of International Leased Circuits—Maintenance of International Telegraph, Phototelegraph and Leased Circuits—Maintenance of the International Public Telephone Network-. 3 pp.

—Sound-Program Circuits
CCITT RECMN N.21-89. Limits and Procedures for the Lining-up of a Sound-Programme Circuit—Maintenance of International Sound-Programme and Television Transmission Circuits (Study Group IV) 11 pp. 11 pp.

—Sound-Program Connections
CCITT RECMN N.10-89. Limits for the Lining-up of International Sound-Programme Links and Connections—Maintenance of International Sound-Programme and Television Transmission Circuits (Study Group IV) 8 pp. 8 pp.

INTERNATIONAL AND NON-U.S. NATIONAL STANDARDS
SUBJECT INDEX

Line Up (Telecommunications) (Cont.)
—Sound-Program Links
CCITT RECMN N.4-89. Definition and Duration of the Line-Up Period and the Preparatory Period—Maintenance of International Sound-Programme and Television Transmission Circuits (Study Group IV) 1 pp. 1 p.
CCITT RECMN N.10-89. Limits for the Lining-up of International Sound-Programme Links and Connections—Maintenance of International Sound-Programme and Television Transmission Circuits (Study Group IV) 8 pp. 8 pp.

—Sound-Program Links—Frequency Measurement
CCITT RECMN N.12-89. Measurements to Be Made During the Line-up Period That Precedes a Sound-Programme Transmission—Maintenance of International Sound-Programme and Television Transmission Circuits (Study Group IV) 1 pp. 1 p.

—Telephone Circuits
CCITT RECMN M.35-89. Principles Concerning Line-up and Maintenance Limits—General Maintenance Principles—Maintenance of International Transmission Systems and Telephone Circuits (Study Group IV) 1 pp. 1 p.
CCITT RECMN M.580-89. Setting up and Lining up an International Circuit for Public Telephony—General Maintenance Principles—Maintenance of International Transmission Systems and Telephone Circuits (Study Group IV) 15 pp. 15 pp.

—Telephone Circuits—Compandors
CCITT RECMN M.590-89. Setting up and Lining up a Circuit Fitted with a Compandor—General Maintenance Principles—Maintenance of International Transmission Systems and Telephone Circuits (Study Group IV) 3 pp. 3 pp.

—Television Links
CCITT RECMN N.54-89. Definition and Duration of the Line-up Period and the Preparatory Period—Maintenance of International Sound-Programme and Television Transmission Circuits (Study Group IV) 2 pp. 2 pp.

—Transfer Links—CCITT No. 6 Signaling Systems
CCITT RECMN M.761-89. Setting up and Lining up a Transfer Link for Common Channel Signalling System No. 6 (Analogue Version)—General Maintenance Principles—Maintenance of International Transmission Systems and Telephone Circuits (Study Group IV) 4 pp. 4 pp.

—Transmission Systems
CCITT RECMN M.450-89. Bringing a New International Transmission System into Service—General Maintenance Principles—Maintenance of International Transmission Systems and Telephone Circuits (Study Group IV) 8 pp. 8 pp.

—Videoconferencing Systems
CCITT RECMN N.86-89. Line-up and Service Commissioning of International Videoconference Systems Operating at Transmission Bit Rates of 1544 and 2048 kBit/s—Maintenance of International Sound-Programme and Television Transmission Circuits (Study Group IV) 4 pp. 4 pp.

—Voice Frequency Telegraphy—Data Links—Telegraph Circuits
CCITT RECMN M.810-89. Setting up and Lining up an International Voice-Frequency Telegraph Link for Public Telegraph Circuits (for 50, 100 and 200 Baud Modulation Rates)—Maintenance of International Telegraph, Phototelegraph and Leased Circuits—Maintenance of Maritime Satellite and Data. 11 pp.

Linear Actuators
See Also: Actuators

—Aircraft
BSI 2G 143-65. 1965 Amd 1 A.C. and D.C. Rotary and Linear Actuators for Aircraft. 10 pp.

Linear Amplifiers
See Also: Amplifiers

—Preferred Products List
CECC CECC MUAHAG Vol 7 IS 8-92. Preferred Products List; Active Microcircuits (En, Fe, Ge). 89 pp.

—Quality Assurance
BSI BS CECC 90200-88. 1988 Analogue Monolithic Integrated Circuits: Sectional Specification. 63 pp.

Linear Circuits
See Also: Crosspoint Switches; Decade Counter/Dividers; Display Drivers; Frequency Synthesizers; Integrated Circuits; Octal Counter/Dividers; Phase Locked Loops; Phototransistor Arrays; Radio Circuits; Ripple Counters; Transistor Arrays; Voltage Comparators; Voltage Controlled Oscillators; Voltage to Frequency Converters
MOD UK DSTAN 59-62: Part 4-83. Microcircuits Electronic (Integrated Circuits) (Listed on EPIC Database) Part 4: Linear Integrated Circuits Issue 1. 15 pp.

—Glossaries
BSI BS 4727:Pt3: Group 14-92. 1992 Electrotechnical, Power, Telecommunication, Electronics, Lighting and Colour Terms Part 3: Terms Particular to Telecommunications and Electronics Group 14: Oscillations, Signals and Related Devices (IEC 50(702): 1992) (G). 235 pp.
IEC 50 Chap 702-92. International Electrotechnical Vocabulary Chapter 702: Oscillations, Signals and Related Devices First Edition. 241 pp.

—Voltage Regulators—Adjustable Positive—Preferred Products List
CECC CECC MUAHAG Vol 7 IS 8-92. Preferred Products List; Active Microcircuits (En, Fe, Ge). 89 pp.

—Voltage Regulators—Negative—Preferred Products List
CECC CECC MUAHAG Vol 7 IS 8-92. Preferred Products List; Active Microcircuits (En, Fe, Ge). 89 pp.

—Voltage Regulators—Positive—Preferred Products List
CECC CECC MUAHAG Vol 7 IS 8-92. Preferred Products List; Active Microcircuits (En, Fe, Ge). 89 pp.

Linear Coefficient of Thermal Expansion
Use: Thermal Expansion

Linear Crosstalk
Use: Crosstalk

Linear Density
See Also: Density; Tex System

—Carbon Fibers
BSI BS 7658: Part 2-93. 1993 Carbon Fibre Part 2: Method for Determination of Linear Density (ISO 10120: 1991) (V). 8 pp.
ISO 10120-91. Carbon Fibre—Determination of Linear Density First Edition. 5 pp.

—Cordage
CGSB 40-GP-1M METH 4-78. Methods of Sampling and Testing Cordage Linear Density. 1 p.

—Cotton Yarns
SAA AS 1506.1-78. Preferred Linear Density of Yarn in Tex Units—Part 1: Cotton. 8 pp.

—Fibers
BSI BS 947-70. 1970 Amd 1 Universal System for Designating Linear Density of Textiles (Tex System). 16 pp.
CEN PREN 21973-2-93. Textile Fibres—Determination of Linear Density—Gravimetric Method and Vibroscope Method. 15 pp.
ISO 1144-73. Textiles—Universal System for Designating Linear Density (Tex System) First Edition. 9 pp.
ISO 1973-76. Textile Fibres—Determination of Linear Density—Gravimetric Method First Edition. 4 pp.
JIS L 0101-78. Tex System to Designate Linear Density of Fibres, Yarn Intermediates, Yarns and Other Textile Materials.

—Fibers—Gravimetric Analysis
BSI BS 2016-73. 1973 Determination of the Linear Density of Textile Fibres: Gravimetric Methods. 10 pp.

—Rovings
ISO 1889-87. Textile Glass—Continuous Filament Yarns, Staple Fibre Yarns, Textured Yarns and Rovings (Packages)—Determination of Linear Density Second Edition. 6 pp.

—Wool Fibers
CGSB CAN/CGSB-4.2 NO.73.3-M91. Textile Test Methods Wool—Determination of Short-Term Irregularity of Linear Density of Slivers, Rovings and Yarns, by Means of an Electronic Evenness Tester (ISO 2649:1974). 19 pp.

Linear Density (Cont.)
—Wool Fibers (Cont.)
ISO 2649-74. Wool—Determination of Short-Term Irregularity of Linear Density of Slivers, Rovings and Yarns, by Means of an Electronic Evenness Tester First Edition. 16 pp.

—Yarns
BSI BS 2010-63. 1963 Amd 1 Determination of the Linear Density of Yarns from Packages. 15 pp.
BSI BS 2862-84. 1984 Methods for Determination of Number of Threads per Unit Length. 10 pp.
BSI BS 2865-84. 1984 Determination of Linear Density of Yarn Removed from Fabric. 8 pp.
CGSB CAN/CGSB-4.2 NO.5.2-M87. Textile Test Methods Linear Density of Yarn in SI Units. 9 pp.
ISO 1889-87. Textile Glass—Continuous Filament Yarns, Staple Fibre Yarns, Textured Yarns and Rovings (Packages)—Determination of Linear Density Second Edition. 6 pp.
ISO 2060-72. Textiles—Yarn from Packages—Determination of Linear Density (Mass per Unit Length)—Skein Method First Edition. 14 pp.
ISO 7211 Pt 5-84. Textiles—Woven Fabrics-Construction—Methods of Analysis—Part 5: Determination of Linear Density of Yarn Removed from Fabric First Edition. 6 pp.
JIS L 0101-78. Tex System to Designate Linear Density of Fibres, Yarn Intermediates, Yarns and Other Textile Materials.
SAA AS 1505 Supp 1-79. Designation of Yarns—Supplement 1: Yarn Notation Based on the Resultant linear Density (Folded-to-Single notation) (Supplement to AS 1505—1978). 6 pp.
SAA AS 1506.2-79. Preferred Linear Density of Yarn in Tex Units—Part 2: Worsted Yarn Counts for Weaving and Knitting Corrig.. 8 pp.
SAA AS 2001.2.23-90. Methods of Test for Textiles—Part 2: Physical Tests—Part 2.23: Determination of Linear Density of Textile Yarn from Packages. 5 pp.

Linear Dimensions
Use: Dimensions

Linear Expansion
Use: Thermal Expansion

Linear Receivers
See Also: Radio Receivers; Receivers

—Noise (Spurious Signals)
CCIR RECMN 331-4-78. Noise and Sensitivity of Receivers—Section 1A—Spectrum Engineering and Computer-Aided Principles and Techniques. 9 pp.

—Sensitivity (Electrical)
CCIR RECMN 331-4-78. Noise and Sensitivity of Receivers—Section 1A—Spectrum Engineering and Computer-Aided Principles and Techniques. 9 pp.

Linear Thermal Expansion Coefficient
Use: Thermal Expansion

Linear Transformers
See Also: Transformers

—Magnetic Cores
CECC CECC 25 200 ISSUE 1-77. Sectional Specification: Magnetic Oxide Cores for Linear Transformers (En, Fr, Ge). 39 pp.

—Magnetic Cores—Quality Assurance
BSI BS CECC 25200-78. (WITHDRAWN) 1978 Magnetic Oxide Cores for Linear Transformers: Sectional Specification (Renumbered as BS EN 1125200: 1993). 16 pp.
BSI BS EN 125200-93. 1993 Amd 1 Sectional Specification: Magnetic Oxide Cores for Linear Transformers (AMD 7578) February 15, 1993 (T). 20 pp.
CENELEC EN 125200-91. Sectional Specification: Magnetic Oxide Cores for Linear Transformers. 14 pp.

Linearity
See Also: Compensation; Electrical Properties; Electronic Equipment

—Communication Terminal Equipment—Groups
CCITT RECMN G.232-89. 12-Channel Terminal Equipments—International Analogue Carrier Systems (Study Group XV) 13 pp. 13 pp.
CCITT RECMN G.235-89. 16-Channel Terminal Equipments—International Analogue Carrier Systems (Study Group XV) 4 pp. 4 pp.

Linemen's Equipment
Use For: Live Wire Working Equipment

INTERNATIONAL AND NON-U.S. NATIONAL STANDARDS
SUBJECT INDEX

Linemen's

Linemen's Equipment (Cont.)
See Also: Electrical Protection Equipment; Linemen's Sticks; Linemen's Tools

DIN VDE 0680 Pt 1-83. Body Protection, Protective Devices and Apparatus for Working on Live Equipment up to 1000 V; Insulating Equipment and Insulating Protective Devices (Jan). 34 pp.

—**Glossaries**

IEC 743-83. Terminology for Tools and Equipment to Be Used in Live Working First Edition. 72 pp.

—**Ladders**

CNS Z2070-88. Insulated Ladder for Electric Work (Oct)(12452).

—**Platforms**

IEC 1057-91. Aerial Devices with Insulating Boom Used for Live Working First Edition. 119 pp.

—**Protective Clothing**

CENELEC HD 547 S1-90. Conductive Clothing for Live Working at a Nominal Voltage up to 800 KV a.c.. 3 pp.

DIN VDE 0680 Pt 1-83. Body Protection, Protective Devices and Apparatus for Working on Live Equipment up to 1000 V; Insulating Equipment and Insulating Protective Devices (Jan). 34 pp.

IEC 895-87. Conductive Clothing for Live Working at a Nominal Voltage up to 800 kV A.C. First Edition. 56 pp.

—**Protective Clothing—Gloves**

BSI BS 697-86. 1986 Rubber Gloves for Electrical Purposes. 11 pp.

BSI BS EN 60903-93. 1993 Gloves and Mitts of Insulating Material for Live Working (N). 54 pp.

CENELEC PREN 60903-92. Specification for Gloves and Mitts of Insulating Material for Live Working. 40 pp.

CENELEC EN 60903-92. Specification for Gloves and Mitts of Insulating Material for Live Working; (IEC 903: 1988). 12 pp.

IEC 903-88. Specification for Gloves and Mitts of Insulating Material for Live Working First Edition; (CENELEC EN 60903:1992). 73 pp.

—**Voltage Testers**

DIN VDE 0682 Pt 411 (D)-89. Apparatus & Equipment for Live Working; Voltage Detector, Capacitive Type to Be Used for Voltages Exceeding 1 kV AC (Apr). 69 pp.

Linemen's Poles
Use: Linemen's Sticks

Linemen's Sticks
Use For: Hook Sticks; Hotsticks; Insulating Sticks; Linemen's Poles; Switch Sticks *See Also:* Linemen's Equipment; Linemen's Tools; Power Line Hardware; Tools

CENELEC HD 542 S1-90. Insulating Poles (Insulating Sticks) and Universal Tool Attachments (Fittings) for Live Working. 3 pp.

IEC 832-88. Insulating Poles (Insulating Sticks) and Universal Tool Attachments (Fittings) for Live Working First Edition. 104 pp.

—**Fittings**

CENELEC HD 542 S1-90. Insulating Poles (Insulating Sticks) and Universal Tool Attachments (Fittings) for Live Working. 3 pp.

IEC 832-88. Insulating Poles (Insulating Sticks) and Universal Tool Attachments (Fittings) for Live Working First Edition. 104 pp.

Linemen's Tools
See Also: Linemen's Equipment; Linemen's Sticks; Tools

DIN VDE 0680 Pt 201 (D)-83. Hand Tools for Live Working up to 1000 V a.c. and 1500 V d.c. (July). 42 pp.

IEC 900-87. Hand Tools for Live Working up to 1000 V A.C. and 1500 V D.C. First Edition. 54 pp.

—**Insulating Rods**

CENELEC HD 496 S1-88. Insulating Foam-Filled Tubes and Solid Rods for Live Working. 4 pp.

IEC 855-85. Insulating Foam-Filled Tubes and Solid Rods for Live Working First Edition. 39 pp.

—**Insulating Tubes**

CENELEC HD 496 S1-88. Insulating Foam-Filled Tubes and Solid Rods for Live Working. 4 pp.

IEC 855-85. Insulating Foam-Filled Tubes and Solid Rods for Live Working First Edition. 39 pp.

IEC 1235-93. Live Working—Insulating Hollow Tubes for Electrical Purposes First Edition. 55 pp.

Linen
See Also: Fabrics; Fibers; Flax

BSI BS 1781-81. 1981 Linen and Linen Union Textiles. 8 pp.

Linen (Cont.)

CNS L3037-82. Method of Test for Linen Fabrics (Sep)(2272).

—**Aerospace**

BSI 8F 1-75. (WITHDRAWN) 1975 Amd 1 140 g Linen Fabric and Serrated Edge Strip (Superseded by BS 9f 1: 1992). 4 pp.

BSI 9F 1-92. 1992 140 g/m2 Linen (Flax) Fabric and Serrated Edge Strip for Aerospace Purposes. 9 pp.

—**Canvas**

CNS L3128-81. Method of Test for Linen Canvas (Jun)(7504).

CNS L4131-81. Linen Canvas (Jun)(7503).

JIS L 3403-61. Linen Canvas (R 1970). 6 pp.

—**Canvas—Automotive**

CNS L3133-81. Method of Test for Linen or Jute Canvas for Waterproof Roofing of Car (Jul)(7741).

CNS L4136-81. Linen or Tute Canvas for Waterproof Roofing of Car (Jul)(7740).

—**Cuprammonium Viscosity**

BSI BS 3090-78. 1978 Amd 1 Method of Test for the Determination of the Cuprammonium Fluidity of Linen Materials. 9 pp.

—**Linings**

CNS L3131-81. Method of Test for Linen Interlining Cloth (Unfinished) (Jul)(7737).

CNS L3132-83. Method of Test for Flax or Ramie Interlining Cloth (Jun)(7739).

CNS L4134-81. Linen Interlining Cloth, Unfinished (Jul)(7736).

—**Threads**

BSI BS 7318-90. 1990 Industrial Sewing Threads Made from Linen (Flax) or Cotton. 13 pp.

—**Yarns**

CNS L4153-86. Linen and Ramie Yarns, Cotton System, Unfinished (Feb) (11508).

Linen Trolleys
Use: Laundry Carts

Liner Board
See Also: Corrugated Board; Packaging Materials; Shipping Containers

CNS P2010-87. Liner Board (Sep)(1455). 3 pp.

JIS P 3902-85. Linerboards. 7 pp.

—**Burst Testers**

CPPA M-18-84. Perkins Model 'AH' Mullen Tester. 1 p.

—**Bursting Strength**

CPPA D.19P-90. Bursting Strength of Board. 2 pp.

JIS P 8131-77. Testing Method for Bursting Strength of Paper and Paperboard by Mullen High-Pressure Tester. 7 pp.

—**Softwoods**

SAA AS 1783-75. Lining Boards Milled from Australian-Grown Conifers (Softwoods) (Excluding Radiata Pine and Cypress Pine).

Lines (Ropes)
See Also: Ropes; Safety Lines

—**Cotton**

BSI BS 6125-81. 1981 Natural Fibre Cords, Lines and Twines. 10 pp.

—**Flax**

BSI BS 6125-81. 1981 Natural Fibre Cords, Lines and Twines. 10 pp.

—**Hemp**

BSI BS 6125-81. 1981 Natural Fibre Cords, Lines and Twines. 10 pp.

—**Remote Release—Yachts**

SAA AS 3989-92. Yachting Harnesses and Lines—Remote-Release Lines. 12 pp.

Lingoes
See Also: Looms

BSI BS 5198-75. 1975 Lingoes for Jacquard Weaving. 6 pp.

ISO 2748-83. Textile Machinery and Accessories—Lingoes for Jacquard Weaving Second Edition. 4 pp.

JIS L 6516-83. Lingoes for Jacquard Weaving.

Lining Up
Use: Line Up (Telecommunications)

Linings
Scope Note: For additional listings, use a more specific term *See Also:* Brake Linings; Bushings; Coatings; Footwear Linings; Rennet; Sheathing; Shields (Protective); Well Casings

Linings (Cont.)

—**Cypress Wood**

SAA AS O94-64. Unseasoned Cypress Pine Milled Lining.

—**Dentures**

BSI BS 7589: Part 1-92. 1992 Resilient Lining Materials for Removable Dentures Part 1: Specification for Short-Term Materials (ISO 10139-1: 1991). 11 pp.

ISO 10139 Pt 1-91. Dentistry—Resilient Lining Materials for Removable Dentures—Part 1: Short-Term Materials First Edition. 9 pp.

—**Organic—Chemical Process Equipment**

DIN ENGL 28055 Pt 1-90. Organic Linings for Application to Metallic Components of Chemical Apparatus; Requirements (Sept) (Supersedes Parts of VDI-Richtlinie (VDI Code of Practice) 2534 Suppl. 1, Aug. 1966 Ed., Suppl. 2, June 1972 Ed., Suppl. 3, Jan. 1967 Ed., and VDI-Richtlinie 2537, Jan. 1976 Ed.). 4 pp.

DIN ENGL 28055 Pt 2-91. Organic Linings for Application to Metallic Components of Chemical Apparatus; Testing (Feb) (Supersedes VDI-Richtlinie (VDI Code of Practice) 2539, August 1967 Edition). 4 pp.

Link Access Procedures
See Also: Data Links; Protocols

—**D Channel—Data Link Layer—ISDN—DSS1**

CCITT RECMN Q.920-89. ISDN User-Network Interface Data Link Layer—General Aspects—Digital Subscriber Signalling System No.1 (DSS 1), Data Link Layer (Study Group XI) 16 pp (Same as Recmn I.440). 16 pp.

CCITT RECMN Q.921-89. ISDN User-Network Interface—Data Link Layer Specification—Digital Subscriber Signalling System No.1 (DSS 1), Data Link Layer (Study Group XI) 123 pp (Same as Recmn I.441). 123 pp.

—**Half Duplex Transmission—Public Switched Telephone Networks**

CCITT RECMN T.71-89. Link Access Protocol Balanced (LAPB) Extended for Half-Duplex Physical Level Facility—Terminal Equipment and Protocols for Telematic Services (Study Group VIII) 7 pp. 7 pp.

—**Half Duplex Transmission—Teletex Equipment**

CCITT RECMN T.71-89. Link Access Protocol Balanced (LAPB) Extended for Half-Duplex Physical Level Facility—Terminal Equipment and Protocols for Telematic Services (Study Group VIII) 7 pp. 7 pp.

—**Integrated Services Digital Networks**

CEPT T/S 46-20-87. Specification De La Couche De Liaisonde Donnees a L'Interface Usager-Reseau Dans Un RNIS; Application Des Recommendations CCITT Q.920/I.440 Et Q.921/I.441. 188 pp.

CEPT T/S 46-20 E-87. ISDN User-Network Interface Data Link Layer Specification Application of CCITT Recommendations Q.920/I.440 and Q.921/I.441. 168 pp.

CSA CAN/CSA-T542-90. Integrated Services Digital Network (ISDN)—Data-Link Layer Signalling Specification for Application at the User-Network Interface (ANSI T1.602-1989). 199 pp.

Link Chains
See Also: Chains; Cranked Link Chains

CNS B2137-72. 2mm Chain (Double Link) (Tentative) (Jun)(3370).

—**Aircraft—Blanking Caps**

SBAC AS 2944 ISSUE 2. Chain Assembly. 1 p.

SBAC AS 2949 ISSUE 2. Chain Assembly. 1 p.

—**Alloy Steel**

BSI BS 3113-59. 1959 Alloy Steel Chain, Grade 60. Short Link, for Lifting Purposes. 20 pp.

—**Alloy Steel—Hoisting Slings**

BSI BS 3458-62. 1962 Alloy Steel Chain Slings. 49 pp.

—**Alloy Steel—Pulley Blocks**

BSI BS 3114-59. 1959 Amd 2 Alloy Steel Chain, Grade 80. Polished Short Link Calibrated Load Chain for Pulley Blocks. 25 pp.

—**Anchors (Marine)**

BSI BS MA 70: Part 1-75. 1975 Amd 1 Dimensions of Anchor Chain Cables Part 1: Stud Link Anchor Chain Cables. 12 pp.

CEN EN 24 565-89. Shipbuilding, Yachting, Ships, Ship Anchors, Chains, Specifications, Mechanical Properties, Breaking Load, Dimensions, Marking. 2 pp.

INTERNATIONAL AND NON-U.S. NATIONAL STANDARDS
SUBJECT INDEX
Link

Link Chains (Cont.)

—Anchors (Marine) (Cont.)
ISO 1704-91. Shipbuilding—Stud-Link Anchor Chains Second Edition. 16 pp.
JIS F 3303-82. Flash Butt Welded Anchor Chain Cables (R 1987). 16 pp.
MOD UK NES 171-01. Requirements for Copper—Based Alloy Stud Link Anchor Cables and Associated Items of Equipment Issue 1 (04.91); Amendment 1. 17 pp.
MOD UK NES 171-02. Requirements for Copper—Based Alloy Stud Link Anchor Cables and Associated Items of Equipment Issue 1 (04.91); Amendment 2. 18 pp.
MOD UK NES 172-01. Requirements for Forged Steel Stud Link Chain Cable Grades 1 and 2 Issue 1 (04.91); Amendment 1. 18 pp.
MOD UK NES 172-02. Requirements for Forged Steel Stud Link Chain Cable Grades 1, 2 and 3 Issue 2 (02.92); Amendment 1. 26 pp.
MOD UK NES 172-92. Requirements for Forged Steel Stud Link Chain Cable Grades 1, 2 and 3 Issue 2 (02.92). 19 pp.
MOD UK NES 175-91. Requirements for Equipment Associated with Forged Steel Chain Cable Issue 1 (06.91). 18 pp.
MOD UK NES 175-01. Requirements for Equipment Associated with Forged Steel Chain Cable Issue 1 (06.91); Amendment 1. 21 pp.
MOD UK NES 176-01. Requirements for Mooring Chain, Open—Link Chain Cable, Cast and Forged Steel and Associated Equipment Issue 1 (06.91); Amendment 1. 18 pp.

—Bucket Elevators
BSI BS 6318: Part 2-82. 1982 Bucket Elevators Part 2: Dimensions of Vertical Bucket Eelevators with Calibrated Round Steel Link Chains. 4 pp.

—Bucket Elevators—Chain Ends
CNS B2608-81. Round Steel Link Chains for Conveyor, Chain-Ends for Bucket Conveyers, Half Long Link (Oct)(7941).

—Conveyors
CNS B2607-81. Round Steel Link Chains for Conveyer, Long Link (Oct)(7940).
CNS B2608-81. Round Steel Link Chains for Conveyer, Chain-Ends for Bucket Conveyers, Half Long Link (Oct)(7941).

—Hoisting Slings
BSI BS 4942: Part 2-81. 1981 Short Link Chain for Lifting Purposes Part 2: Grade M (4) Non-Calibrated Chain. 11 pp.
BSI BS 4942: Part 4-81. 1981 Short Link Chain for Lifting Purposes Part 4: Grade S (6) Non-Calibrated Chain. 11 pp.
BSI BS 4942: Part 5-81. 1981 Amd 1 Short Link Chain for Lifting Purposes Part 5: Grade T (8) Non-Calibrated Chain (AMD 4934) March 31, 1986. 13 pp.
BSI BS 6968-88. 1988 Guide for Use and Maintenance of Non-Calibrated Round Steel Lifting Chain and Chain Slings. 13 pp.
ISO 1835-80. Short Link Chain for Lifting Purposes—Grade M (4), Non-Calibrated, for Chain Slings, Etc. First Edition. 10 pp.
ISO 3056-86. Non-Calibrated Round Steel Link Lifting Chain and Chain Slings—Use and Maintenance Second Edition. 12 pp.
ISO 3075-80. Short Link Chain for Lifting Purposes—Grade S (6), Non-Calibrated, for Chain Slings, Etc. First Edition. 10 pp.
ISO 3076-84. Short Link Chain for Lifting Purposes—Grade T (8), Non-Calibrated, for Chain Slings, Etc. Second Edition. 11 pp.
SNZ NZS/ISO 1835-80. Short Link Chain for Lifting Purposes—Grade M (4), Non-Calibrated, for Chain Slings Etc.. 8 pp.
SNZ NZS/ISO 3075-80. Short Link Chain for Lifting Purposes-Grade S(6) Non-Calibrated, for Chain Slings Etc.. 8 pp.
SNZ NZS/ISO 3076-84. Short Link Chain for Lifting Purposes-Grade T(8), Non-Calibrated, for Chain Slings Etc.. 9 pp.

—Hoisting Slings—Quality Assurance
CEN PREN 818-1-92. Short Link Chain for Lifting Purposes—Safety—Part 1: General Conditions of Acceptance. 20 pp.

—Hoisting Slings—Repair—Boats
MOD UK NES 744-90. Requirements for the Manufacture and Repair of Rings and Links Issue 2 (10.90). 11 pp.

—Hoisting Slings—Safety
CEN PREN 818-2-92. Short Link Chain for Lifting Purposes—Safety—Part 2: Medium Tolerance Chain for Chain Slings—Grade 8. 18 pp.

—Hoists
BSI BS 4942: Part 1-81. 1981 Short Link Chain for Lifting Purposes Part 1: General Conditions of Acceptance. 11 pp.
BSI BS 4942: Part 3-81. 1981 Short Link Chain for Lifting Purposes Part 3: Grade M (4) Calibrated Chain. 12 pp.
BSI BS 4942: Part 6-81. 1981 Amd 1 Short Link Chain for Lifting Purposes Part 6: Grade T (8) Calibrated Chain (AMD 4953) March 31, 1986. 13 pp.
ISO 1834-80. Short Link Chain for Lifting Purposes—General Conditions of Acceptance First Edition. 10 pp.
ISO 1836-80. Short Link Chain for Lifting Purposes—Grade M (4), Calibrated, for Chain Hoists and Other Lifting Appliances First Edition. 11 pp.
ISO 3077-84. Short Link Chain for Lifting Purposes—Grade T (8), Calibrated, for Chain Hoists and Other Lifting Appliances Second Edition. 11 pp.
JIS B 8812-89. Link Chains for Manually Operated Chain Hoists. 17 pp.
SNZ NZS/ISO 1834-80. Short Link Chain for Lifting Purposes—General Conditions of Acceptance. 8 pp.
SNZ NZS/ISO 1836-80. Short Link Chain for Lifting Purposes—Grade M (4), Calibrated, for Chain Hoists and Other Lifting Appliances. 9 pp.
SNZ NZS/ISO 3077-84. Short Link Chain for Lifting Purposes-Grade T(8), Calibrated, for Chain Hoists and Other Lifting Appliances. 9 pp.

—Hoists—Ships
MOD UK NES 113: Part 7-89. Requirements for Mechanical Handling Part 7: Lever Hoists Issue 2 (11.89). 13 pp.

—Hoists—Submarines
MOD UK NES 113: Part 7-89. Requirements for Mechanical Handling Part 7: Lever Hoists Issue 2 (11.89). 13 pp.

—Knotted
CNS B2756-83. Knotted Link Chains Without Quality Requirements (Dec) (10684).

—Mining Equipment
CNS M2011-86. Welded Three-Link Chain (Nov)(3524).
CNS M2039-80. Forged Three-Link Chain (Mar)(5369).
CNS M2043-80. Round Link Chain for Mine Double Chain Conveyor (Mar)(5373).
CNS M2084-80. Round Link Chain for Coal Plough (Dec)(6845).
CNS M2085-80. Shackle of Round Link Chain for Coal Plough (Dec)(6846).
DIN ENGL 762 Pt 1-82. Calibrated and Tested Round Steel Link Chains for Continuous Conveyors; Grade 2, Pitch 5d (Jan). 4 pp.
DIN ENGL 762 Pt 2-82. Calibrated and Tested Round Steel Link Chains for Continuous Conveyors; Grade 3, Pitch 5d (Jan). 4 pp.
DIN ENGL 764 Pt 1-82. Calibrated and Tested Round Steel Link Chains for Continuous Conveyors; Grade 2, Pitch 3,5d (Jan). 3 pp.
DIN ENGL 764 Pt 2-82. Calibrated and Tested Round Steel Link Chains for Continuous Conveyors; Grade 3, Pitch 3,5d (Jan). 3 pp.
JIS M 6504-76. Round Link Chain for Mine Double Chain Conveyors.
JIS M 6509-76. Round Link Chain for Coal Ploughs.
JIS M 6510-75. Shackle of Round Link Chain for Coal Plough.

—Steel
BSI BS 6405-84. 1984 Non-Calibrated Short Link Steel Chain (Grade 30) for General Engineering Purposes: Class 1 and 2. 12 pp.
BSI BS 6521-84. 1984 Proper Use and Maintenance of Calibrated Round Steel Link Lifting Chains. 7 pp.
BSI DD 173-87. 1987 Non-Calibrated Long Link Steel Chain for General Engineering Purposes. 16 pp.
DIN ENGL 763-90. Tested, Non-Calibrated, Long-Link Round Steel Chains (Dec). 3 pp.
DIN ENGL 766-86. Calibrated and Tested Grade 3 Round Steel Link Chains (Jan). 4 pp.
DIN ENGL 5685-91. Untested, Round Steel Chains with Long or Medium-Length Links (Feb). 2 pp.
DIN ENGL 17115-87. Steels for Welded Round Link Chains; Technical Delivery Conditions (Feb). 16 pp.

—Steel—Hoisting Slings
BSI BS 2902-57. 1957 Amd 2 Higher Tensile Steel Chain Slings and Rings, Links Alternative to Rings, Egg Links and Intermediate Links. 52 pp.

—Steel—Mining Equipment
BSI BS 2969-80. 1980 High-Tensile Steel Chains (Round Link) for Chain Conveyors and Coal Ploughs. 22 pp.
CNS M2040-82. Connector Shackle for Double Chains Conveyor and High-Tensile Round-Link Steel Chains for Mining (Sep)(5370).
CNS M2088-82. High-Tensile Round-Link Steel Chains for Mining (Dimension and Requirements) (Sep) (9415).
CNS M3099-82. Method of Test for High-Tensile Round-Link Steel Chains for Mining (Sep) (9416).
CNS M3100-82. Method of Test for Connector Shackle for Double Chains Conveyor and High-Tensile Round-Link Steel Chains for Mining (Sep)(9418).
DIN ENGL 22252-83. Calibrated and Tested Round Steel Link Chains for Conveyors and Machines Used in Mining (Sept) (Supersedes DIN 22252 Parts 1 and 2, December 1973 Editions). 12 pp.
ISO 610-90. High-Tensile Steel Chains (Round Link) for Chain Conveyors and Coal Ploughs Second Edition. 21 pp.

—Steel—Safety—Mining Equipment
CNS M2089-82. High-Tensile Round-Link Steel Chains for Mining (Safety Chain) (Sep) (9417).

—Steel—Span
DIN ENGL 82056-78. Round Steel Chains for Span; Tested Uncalibrated Long Link (Nov). 2 pp.

Link Control Procedures
See Also: Basic Mode Control Procedures; Control Procedures; Data Links; Logical Link Control; Protocols
JIS X 5107-85. Multilink Procedures.

—Cellular Mobile Radio Equipment
CENELEC PRI-ETS 300 034-90. European Digital Cellular Telecommunications System (Phase 1); Radio Sub-System Link Control. 50 pp.
CENELEC PRI-ETS 300 034-91. European Digital Cellular Telecommunications System (Phase 1); Radio Sub-System Link Control. 43 pp.
CENELEC I-ETS 300 034/A1-91. AMD 1 European Digital Cellular Telecommunications System (Phase 1); Radio Sub-System Link Control (GSM 05.08DCS). 25 pp.
CENELEC I-ETS 300 034-92. European Digital Cellular Telecommunications System (Phase 1); Radio Sub-System Link Control (GSM 05.08). 34 pp.
CENELEC GSM 05.08-92. See PRI-ETS 300 034. 39 pp.
CENELEC GSM 05.08-DCS-92. See I-ETS 300 034 A1. 22 pp.
ETSI PRI-ETS 300 034-90. European Digital Cellular Telecommunications System (Phase 1); Radio Sub-System Link Control (GSM 05.08). 50 pp.
ETSI I-ETS 300 034-92. European Digital Cellular Telecommunications System (Phase 1); Radio Sub-System Link Control (GSM 05.08). 34 pp.
ETSI I-ETS 300 034/A1-91. AMD 1 European Digital Cellular Telecommunications System (Phase 1); Radio Sub-System Link Control (GSM 05.08DCS). 25 pp.
ETSI PRI-ETS 300 034-91. European Digital Cellular Telecommunications System (Phase 1); Radio Sub-System Link Control. 43 pp.
ETSI GSM 05.08-92. See PRI-ETS 300 034. 39 pp.
ETSI GSM 05.08-DCS-92. See I-ETS 300 034 A1. 22 pp.

—Open Systems Interconnection
BSI BS 5397: Part 6-90. 1990 High-Level Data Link Control Procedures for Data Communication Part 6: Multilink Procedures. 15 pp.
BSI BS 7235-90. 1990 Open Systems Interconnection: Specification for Protocol to Provide the Connectionless-Mode Network Service. 55 pp.
BSI BS 7235-90. 1990 AD3 Open Systems Interconnection: Specification for Protocol to Provide the Connectionless-Mode Network Service (AMD 7207) September 15, 1992 (S). 62 pp.
CSA CAN/CSA-Z243.52-88. Information Processing Systems—Data Communication—Multilink Procedures (ISO 7478-1987). 23 pp.
ECMA ECMA 105-90. Data Link Layer Protocol for the D-Channel of the Interfaces at the Reference Point Between Terminal Equipment and Private Telecommunication Networks. 68 pp.
ECMA ECMA 106-91. Layer 3 Protocol for Signalling over the D-Channel of Interfaces at the S Reference Point Between Terminal Equipment and Private Telecommunication Networks for the Control of Circuit-Switched Calls. 39 pp.
ECMA ECMA 141-90. Data Link Layer Protocol at the Q Reference Point for the Signalling Channel Between Two Private Telecommunication Network Exchanges. 24 pp.
IEC 7478-89. Corrigendum 1 Information Processing Systems—Data Communication—Multilink Procedures. 2 pp.
IEC 8473 COR 1. Corrigendum 1-Information Processing Systems—Data Communications—Protocol for Providing the Connectionless-Mode Network Service; (1992). 9 pp.

INDUSTRY STANDARDS

INTERNATIONAL AND NON-U.S. NATIONAL STANDARDS SUBJECT INDEX

Link

Link Control Procedures *(Cont.)*
—Open Systems Interconnection *(Cont.)*
ISO 7478-87. Information Processing Systems—Data Communication—Multilink Procedures First Edition; (Corrigendum 1-1989). 16 pp.

ISO 8473-88. Information Processing Systems—Data Communications—Protocol for Providing the Connectionless-Mode Network Service First Edition; (Addendum 3-1989) (Corrigendum 1-1992). 64 pp.

JTC1 7478-87. Information Processing Systems—Data Communication—Multilink Procedures First Edition; (Corrigendum 1-1989). 16 pp.

JTC1 8473-88. Information Processing Systems—Data Communications—Protocol for Providing the Connectionless-Mode Network Service First Edition; (Addendum 3-1989) (Corrigendum 1-1992). 61 pp.

OSI ISO 7478. Information Processing Systems—Data Communication—Multilink Procedures.

OSI ISO 8473-88. Information Processing Systems—Data Communications—Protocol for Providing the Connectionless-Mode Network Service. 55 pp.

OSI ISO 8473 AD 3-89. Information Processing Systems—Data Communications—Protocol for Providing the Connectionless—Mode Network Service—Addendum 3: Provision of the Underlying Service Assumed be ISO 8473 over Subnetworks. 6 pp.

SNZ NZS/ISO 7478-87. Information Processing Systems—Data Communication—Multilink Procedures. 11 pp.

Link Fuses
Use: Fuse Links

Link Protocols
Scope Note: Use a more specific term *See:* Data Links; High Level Data Link Control; Link Access Procedures; Link Control Procedures; Protocols; Radio Link Protocols

Linkages
See: Anchors (Fasteners); Clevis; Couplings; Fasteners; Joints; Latches (Fasteners); Mechanical Components; Quick Release Fasteners; Turnbuckles; Yokes

Linked Compressor and Expandor Systems
Use: Compandors

Links (Communication)
Scope Note: Use a more specific term *See:* Data Links; Group Links; Line Links; Mastergroup Links; Radio Links; Sound-Program Links; Supergroup Links; Supermastergroup Links; Television Links; 15 Supergroup Assembly Links

Linnet Aircraft
See Also: Aircraft

—Antenna Positions
CAA. Scintex, Linnet and Piel CP.301A (Approved Aerial Positions). 1 p.

Linoleum
See Also: Carpets; Floor Coverings

BSI BS 6263: Part 2-91. 1991 Care and Maintenance of Floor Surfaces Part 2: Code of Practice for Resilient Sheet and Tile Flooring. 14 pp.

BSI BS 6826-87. 1987 Linoleum and Cork Carpet Sheet and Tiles. 12 pp.

BSI BS 8203-87. 1987 Code of Practice for Installation of Sheet and Tile Flooring. 31 pp.

CEN PREN 548-91. Linoleum Floorcoverings—Specification for Plain and Decorative Linoleum. 11 pp.

CEN PREN 686-92. Linoleum Floorcoverings—Specification for Linoleum on a Foam Backing. 11 pp.

CEN PREN 687-92. Linoleum Floorcoverings—Specification for Linoleum with Corkment Backing. 12 pp.

CEN PREN 688-92. Linoleum Floorcoverings—Specification for Cork Linoleum. 11 pp.

CSA A126.10-M1984. Test Methods for Resilient Flooring; (Gen Instr 1 Thru 2). 26 pp.

CSA A146-1965. Linoleum Products (R 1971); (Rev 1 May 1972). 11 pp.

—Ash Content
CEN PREN 670-92. Identification and Composition of Linoleum—Determination of Cement Content and Ash Residue. 6 pp.

—Cement Content
CEN PREN 670-92. Identification and Composition of Linoleum—Determination of Cement Content and Ash Residue. 6 pp.

Linoleum *(Cont.)*
—Ships
MOD UK NES 153-91. Requirements for the Production Decorative Marbled Flexible Linoleum Sheet and Tiles Issue 4 (03.91). 15 pp.

—Submarines
MOD UK NES 153-91. Requirements for the Production Decorative Marbled Flexible Linoleum Sheet and Tiles Issue 4 (03.91). 15 pp.

—Tiles—Dimensional Stability
CEN PREN 669-92. Linoleum Floorcoverings—Determination of the Dimensional Changes of Tiles Caused by Atmospheric Humidity Changes. 5 pp.

Linseed
See Also: Flax; Linseed Oil; Seeds

—Fatty Acids
CNS K1265-81. Distilled Linseed Fatty Acids (May)(7433).

Linseed Meal
Use: Oil Meal

Linseed Oil
See Also: Boiled Oil; Linseed; Oils; Tung Oil

CGSB CAN/CGSB-1.1-M89. Raw Linseed Oil. 8 pp.
CGSB CAN/CGSB-1.2-M89. Boiled Linseed Oil. 8 pp.
CNS K5013-79. Boiled Linseed Oil (Nov)(768). 1 p.
CNS K5100-81. Crude Linseed Oil (Sep)(7411).
CNS K5125-82. Linseed Oil for Industrial Use (Feb)(8529).
ISO 5513-82. Linseed for the Manufacture of Oil—Specification First Edition. 3 pp.
JIS K 5421-83. Boiled Oil and Boiled Linseed Oil. 7 pp.

—Colorimetric Analysis
CNS K6055-75. Method of Test for Boiled Linseed Oil (Feb)(769). 4 pp.

—Iodine Number
CNS K6055-75. Method of Test for Boiled Linseed Oil (Feb)(769). 4 pp.

—Lead Free
MOD UK DEF-1428-64. Oil, Linseed, Boiled, Lead-Free. 3 pp.

—Paints
BSI BS 4725-87. 1987 Linseed Stand Oil for Paints and Varnishes. 6 pp.
BSI BS 6900-87. 1987 Raw, Refined and Boiled Linseed Oils for Paints and Varnishes. 15 pp.
CNS K5004-89. Raw Linseed Oil (for Paint) (Jul)(92). 2 pp.
CNS K6008-89. Method of Test for Raw Linseed Oil (Jul)(93). 2 pp.
ISO 150-80. Raw, Refined and Boiled Linseed Oil for Paints and Varnishes—Specifications and Methods of Test First Edition. 13 pp.
ISO 276-81. Linseed Stand Oil for Paints and Varnishes—Specifications and Methods of Test First Edition. 5 pp.

—Physical Properties
CNS K6055-75. Method of Test for Boiled Linseed Oil (Feb)(769). 4 pp.

—Putty
BSI BS 544-69. 1969 Amd 3 Linseed Oil Putty for Use in Wooden Frames. 27 pp.
CGSB CAN/CGSB-19.1-M87. Putty, Linseed Oil Type. 9 pp.

—Pyrotechnics
MOD UK TS 10224. Polymerized Linseed Oil.

—Residue Content—Gravimetric Analysis
CNS K6673-81. Method of Test for Foots in Raw Linseed Oil (Gravimetric Method) (May)(7437).

—Residue Content—Volumetric Analysis
CNS K6672-81. Method of Test for Foots in Raw Linseed Oil (Volumetric Method) (May)(7436).

—Saponification Number
CNS K6055-75. Method of Test for Boiled Linseed Oil (Feb)(769). 4 pp.

—Unsaponifiable Matter Content
CNS K6055-75. Method of Test for Boiled Linseed Oil (Feb)(769). 4 pp.

—Varnishes
BSI BS 4725-87. 1987 Linseed Stand Oil for Paints and Varnishes. 6 pp.
BSI BS 6900-87. 1987 Raw, Refined and Boiled Linseed Oils for Paints and Varnishes. 15 pp.
ISO 150-80. Raw, Refined and Boiled Linseed Oil for Paints and Varnishes—Specifications and Methods of Test First Edition. 13 pp.

Linseed Oil *(Cont.)*
—Varnishes *(Cont.)*
ISO 276-81. Linseed Stand Oil for Paints and Varnishes—Specifications and Methods of Test First Edition. 5 pp.

Lintels
See Also: Beams (Supports); Walls

—Loads (Forces)
BSI BS 5977: Part 1-81. 1981 Amd 1 Lintels Part 1: Method of Assessment of Load. 13 pp.

—Masonry—Flexural Strength
CEN PREN 846-9-92. Methods of Test for Ancillary Components for Masonry—Part 9: Determination of Flexural Resistance, Shear Load Resistance and Stiffness of Lintels. 14 pp.

—Masonry—Shear Strength
CEN PREN 846-9-92. Methods of Test for Ancillary Components for Masonry—Part 9: Determination of Flexural Resistance, Shear Load Resistance and Stiffness of Lintels. 14 pp.

—Masonry—Stiffness Testing
CEN PREN 846-9-92. Methods of Test for Ancillary Components for Masonry—Part 9: Determination of Flexural Resistance, Shear Load Resistance and Stiffness of Lintels. 14 pp.

—Masonry—Straightness
CEN PREN 846-11-93. Methods of Test for Ancillary Components for Masonry—Part 11: Determination of Dimensions and Straightness or Bow of Lintels. 6 pp.

—Prefabricated
BSI BS 5977: Part 2-83. 1983 Amd 1 Lintels Part 2: Prefabricated Lintels. 20 pp.

—Prefabricated—Masonry
CEN PREN 845-2-92. Specification for Ancillary Components for Masonry—Part 2: Lintels. 26 pp.

Linters
See Also: Cellulose; Fibers; Textile Machinery

JIS P 9001-76. Testing Methods for Refined Cotton Linter (R 1985). 25 pp.

MOD UK TS 10034A. Cotton Linters, Bleached Types 1, 2 and 3.

Linting
Use For: Dusting (Paper) *See Also:* Defects; Surface Properties; Surfaces

—Newsprint
CPPA L.2U-77. Lint and Dust Test for Newsprint. 1 p.

CPPA L.3U-77. Indopol Lint Test. 2 pp.

Lip Seals
See Also: Sealing Rings; Seals

—Shafts (Machine Elements)
BSI BS 1399: Part 1-70. 1970 Rotary Shaft Lip Seals Part 1: Dimensions of Shafts and Housings. 20 pp.

BSI BS 1399: Part 2-70. 1970 Rotary Shaft Lip Seals Part 2: Methods of Performance Rating. 25 pp.

ISO 6194 Pt 1-82. Rotary Shaft Lip Type Seals—Part 1: Nominal Dimensions and Tolerances First Edition. 9 pp.

ISO 6194 Pt 3-88. Rotary Shaft Lip Type Seals—Part 3: Storage, Handling and Installation First Edition. 10 pp.

ISO 6194 Pt 4-88. Rotary Shaft Lip Type Seals—Part 4: Performance Test Procedures First Edition. 6 pp.

ISO 6194 Pt 5-90. Rotary Shaft Lip Type Seals—Part 5: Identification of Visual Imperfections First Edition. 11 pp.

—Shafts (Machine Elements)—Glossaries
BSI BS 1399: Part 3-70. 1970 Rotary Shaft Lip Seals Part 3: Terminology and Recommendations for Handling. 17 pp.

ISO 6194 Pt 2-91. Rotary Shaft Lip Type Seals—Part 2: Vocabulary First Edition. 22 pp.

—Shafts (Machine Elements)—Handling
BSI BS 1399: Part 3-70. 1970 Rotary Shaft Lip Seals Part 3: Terminology and Recommendations for Handling. 17 pp.

Lipase
—Lipolytic Activity
JIS K 0601-88. Determination of Lipolytic Activity of Lipase for Industrial Use. 12 pp.

INTERNATIONAL AND NON-U.S. NATIONAL STANDARDS
SUBJECT INDEX

Lipophilic Substance Content Analysis

—Waste Water—Gravimetric Analysis

DIN ENGL 38409 Pt 17-81. German Standard Methods for the Analysis of Water, Waste Water and Sludge; Summary Action and Material Characteristic Parameters (Group H); Detn. of Not Easily Volatile and Lipophilic Substances (Boiling Points Greater Than 250 Deg. C) (H 17) (May). 3 pp.

—Waste Water—Separation

DIN ENGL 38409 Pt 19-86. German Standard Methods for the Examination of Water, Waste Water and Sludge; Summary Indices of Actions and Substances (Group H); Determination of Directly Separable Lipophilic Substances (H 19) (Feb). 4 pp.

—Water—Gravimetric Analysis

DIN ENGL 38409 Pt 17-81. German Standard Methods for the Analysis of Water, Waste Water and Sludge; Summary Action and Material Characteristic Parameters (Group H); Detn. of Not Easily Volatile and Lipophilic Substances (Boiling Points Greater Than 250 Deg. C) (H 17) (May). 3 pp.

—Water—Separation

DIN ENGL 38409 Pt 19-86. German Standard Methods for the Examination of Water, Waste Water and Sludge; Summary Indices of Actions and Substances (Group H); Determination of Directly Separable Lipophilic Substances (H 19) (Feb). 4 pp.

Liquefied Ammonia

Use: Ammonia

Liquefied Anhydrous Ammonia

Use: Ammonia

Liquefied Gas Content Analysis

—Cosmetics

CNS S2074-82. Method of Hygienic Test for Cosmetics Liquefied Gases Used as Aerosol Propellents (Jun)(9037).

Liquefied Gases

Use For: Liquid Gases *See Also:* Fuels; Gases; Liquefied Natural Gas; Liquefied Petroleum Gas; Liquid Nitrogen; Liquid Oxygen; Liquids

BSI BS 5355-76. 1976 Filling Ratios and Developed Pressures for Liquefiable and Permanent Gases. 40 pp.

—Aircraft—Sampling—Technical Manuals

NATO STANAG 3977 ED 1 AMD 0-91. Manual of Techniques of Sampling and Analysis of Gases and Liquefied Gases for Aircraft Servicing—AEP-6. 5 pp.

NATO STANAG 3977 ED 1 AMD 1-91. Manual of Techniques of Sampling and Analysis of Gases and Liquefied Gases for Aircraft Servicing—AEP-6. 6 pp.

—Halogenated Hydrocarbons

BSI BS 5598: Part 2-79. 1979 Methods of Sampling and Test for Halogenated Hydrocarbons Part 2: Sampling of Liquefied Gases. 7 pp.

ISO 3427-76. Gaseous Halogenated Hydrocarbons (Liquefied Gases)—Taking of a Sample First Edition. 6 pp.

—Highway Transportation—Emergency Procedures

SAA AS 1678. Emergency Procedure Guide—Transport Group Text EPGs for Class 2 Substances—Compressed and Liquefied Gases Pads of 20 Forms.

SAA AS 1678.2A2-87. Emergency Procedure Guide—Transport Group Test EPGs for Class 2 Substances—Compressed and Liquefied Gases—Part 2A2: Flammable, Liquefied Gas (Supersedes AS 1678.2.1.002—1984 and AS 1678.2.1.006—1984).

SAA AS 1678.2A3-87. Emergency Procedure Guide—Transport Group Text EPGs for Class 2 Substances—Compressed and Liquefied Gases—Part 2A3: Falmmable, Low -Temperature, Liquefied Gas.

SAA AS 1678.2A4-87. Emergency Precedure Guide—Transport Group Test EPGs for Class 2 Substances—Compressed and Liquefied Gases—Part 2A4: Flammable, Poisonous Gas.

SAA AS 1678.2A5-87. Emergency Procedure Guide—Transport Group Text EPGs for Class 2 Substances—Compressed and Liquefied Gases—Part 2A5: Flammable, Poisonous—Special Low-Temperature Control Gas.

SAA AS 1678.2A6-87. Emergency Precedure Guide—Transport Group Text EPGs for Class 2

Liquefied Gases *(Cont.)*
—Highway Transportation—Emergency Procedures *(Cont.)*

SAA AS 1678.2B1-87. Emergency Precedure Guide—Transport Group Text EPGs for Class 2 Substances—Compressed and Liquefied Gases—Part 2B1: Poisonous (Non-Flammable) Gas.

SAA AS 1678.2B2-87. Emergency Procedure Guide—Transport Group Text EPGs for Class 2 Substances—Compressed and Liquefied Gases—Part 2B2: Poisonous, Flammable Gas.

SAA AS 1678.2B3-87. Emergency Procedure Guide—Transport Group Text EPGs for Class 2 Substances—Compressed and Liquefied Gases—Part 2B3: Poisonous Gas (Will Burn, Corrode) (Supersedes AS 1678.8.0.010—1984 (in Part)).

SAA AS 1678.2B4-87. Emergency Procedure Guide—Transport Group Text EPGs for Class 2 Substances—Compressed and Liquefied Gases—Part 2B4: Poisonous, Flammable (Spontaneously Combustible) Gas.

SAA AS 1678.2B5-87. Emergency Procedure Guide—Transport Group Text EPGs for Class 2 Substances—Compressed and Liquefied Gases—Part 2B5: Poisonous, Oxidizing Gas.

SAA AS 1678.2B6-87. Emergency Procedure Guide—Transport Group Text EPGs for Class 2 Substances—Compressed and Liquefied Gases—Part 2B6: Poisonous, Oxidizing Gas (Flourine-Type).

SAA AS 1678.2B7-87. Emergency Procedure Guide—Transport Group Text EPGs for Class 2 Substances—Compressed and Liquefied Gases—Part 2B7: Poisonous, Oxiding Gas, Corrosive Gas.

SAA AS 1678.2B8-87. Emergency Procedure Guide—Transport Group Text EPGs for Class 2 Substances—Compressed and Liquefied Gases—Part 2B8: Poisonous and Corrosive Gas.

SAA AS 1678.2C2-87. Emergency Procedure Guide—Transport Group Text EPGs for Class 2 Substances—Compressed and Liquefied Gases—Part 2C2: Non-Flammable, Liquefied Gas (Supersedes AS 1678.2.2.003—1984).

SAA AS 1678.2C3-87. Emergency Precedure Guide—Transport Group Text EPGs for Class 2 Substances—Compressed and Liquefied Gases—Part 2C3: Non-Flammable, Low-Temperature, Liquefied Gas.

SAA AS 1678.2C4-87. Emergency Procedure Guide—Transport Group Text EPGs for Class 2 Substances—Compressed and Liquefied Gases—Part 2C4: Non-Flammable, Corrosive, Liquefied Gas (Fluorine-Type).

SAA AS 1678.2C7-87. Emergency Procedure Guide—Transport Group Text EPGs for Class 2 Substances—Compressed and Liquefied Gases—Part 2C7: Non-Flammable, Oxidizing, Low-Temperature, Liquefied Gas—Breathable.

SAA AS 1678.2C9-87. Emergency Procedure Guide—Transport Group Text EPGs for Class 2 Substances—Compressed and Liquefied Gases—Part 2C9: Non-Flammable, Oxidizing, Low-Temperature, Liquefied Gas—Asphyxiating.

SAA AS 1678.2D1-87. Emergency Procedure Guide—Transport Group Text EPGs for Class 2 Substances—Compressed and Liquefied Gases—Part 2D1: Aerosol Dispensers.

—Shipping Containers

BSI BS 3951: Sec 2.3-92. 1992 Freight Containers Part 2: Specification and Testing of Series 1 Freight Containers Section 2.3: Tank Containers for Liquids, Gases and Pressurized Dry Bulk (ISO 1496-3: 1991) (Q). 33 pp.

BSI BS 3951: Sec 2.3-83. 1983 Freight Containers Part 2: Specification and Testing of Series 1 Freight Containers Section 2.3: Tank Containers for Liquids and Gases (Q). 26 pp.

ISO 1496 Pt 3-91. Series 1 Freight Containers—Specification and Testing—Part 3: Tank Containers for Liquids, Gases and Pressurized Dry Bulk Third Edition. 32 pp.

—Storage Tanks

BSI BS 5387-76. (WITHDRAWN) 1976 Vertical Cylindrical Welded Storage Tanks for Low-Temperature Services: Double-Wall Tanks for Temperature Down to-196 Degrees Celsius (Superseded by BS 7777: Parts 1, 2, 3 and 4: 1993). 57 pp.

BSI BS 5429-76. 1976 Code of Practice for Safe Operation of Small-Scale Storage Facilities for Cryogenic Liquids. 9 pp.

BSI BS 7777: Part 1-93. 1993 Flat-Bottomed, Vertical, Cylindrical Storage Tanks for Low Temperature Service Part 1: Guide to the General Povisions Applying for Design, Construction, Installation and Operation (Q). 36 pp.

BSI BS 7777: Part 2-93. 1993 Flat-Bottomed, Vertical, Cylindrical Storage Tanks for Low Temperature Service Part 2: Specification for the Design and Construction of Single, Double and Full Containment Metal Tanks for the Storage of Liquefied Gas at. 66 pp.

Liquefied Gases *(Cont.)*
—Storage Tanks *(Cont.)*

BSI BS 7777: Part 3-93. 1993 Flat-Bottomed, Vertical, Cylindrical Storage Tanks for Low Temperature Service Part 3: Recommendations for the Design and Construction of Prestressed and Reinforced Concrete Tanks and Tank Foundations, and for the. 35 pp.

BSI BS 7777: Part 4-93. 1993 Flat-Bottomed, Vertical, Cylindrical Storage Tanks for Low Temperature Service Part 4: Specification for the Design and Construction of Single Containment Tanks for the Storage of Liquid Oxygen, Liguid Nitrogen or Liquid Argon (Q). 16 pp.

DIN ENGL 4119 Pt 1-79. Above-Ground Cylindrical Flat-Bottomed Tank Installations of Metallic Materials; Fundamentals, Design, Tests (June). 14 pp.

DIN ENGL 4119 Pt 2-80. Above-Ground Cylindrical Flat-Bottom Tank Structures of Metallic Materials; Calculation (Feb). 13 pp.

—Sulfur Content

DIN ENGL 51400 Pt 1-78. Testing of Mineral Oils and Fuels; Determination of the Sulfur Content (Total Sulfur); General Working Principles (Feb). 6 pp.

DIN ENGL 51400 Pt 3-78. Testing of Mineral Oils and Fuels; Determination of the Sulfur Content (Total Sulfur); Combustion According to Schoniger; Thorin-Sulfonazo-III Titration (Feb). 3 pp.

DIN ENGL 51400 Pt 4-90. Determination of Total Sulfur Content of Gaseous Petroleum Products by the Lingener Combustion Method (Oct). 8 pp.

DIN ENGL 51400 Pt 8-78. Testing of Mineral Oils and Fuels; Determination of Sulfur Content (Total Sulfur); Nickel Reduction Method; Dithizone Titration (Feb). 4 pp.

—Tanks (Containers)

BSI BS 7122-89. 1989 Welded Steel Tanks for the Road Transport of Liquefiable Gases. 17 pp.

Liquefied Methane

Use: Liquefied Natural Gas

Liquefied Natural Gas

Use For: Liquefied Methane; LNG *See Also:* Fossil Fuels; Liquefied Gases; Liquefied Petroleum Gas; Liquid Hydrocarbons; Natural Gas

CEN PREN 1160-93. Installations and Equipment for Liquefied Natural Gas—General Characteristics of Liquefied Natural Gas. 50 pp.

CSA CAN/CSA-Z276-M89. Liquefied Natural Gas (LNG)—Production, Storage, and Handling; (Gen Instr 1). 69 pp.

—Automotive

SNZ NZS 5422: Part 2-87. Code of Practice for the Use of LPG and CNG Fuels in Internal Combustion Engines Part 2: CNG Fuel. 32 pp.

SNZ NZS 5422: Part 3-91. Code of Practice for the Use of LPG and CNG Fuels in Internal Combustion Engines Part 3: In Heavy Vehicles. 48 pp.

—Cylinders—Automotive

SNZ NZS 5454-89. Standard Requirements for Lightweight Steel Automotive Compressed Natural Gas Cylinders for Use in New Zealand. 20 pp.

—Cylinders—Refueling

SNZ NZS 5425: Part 1-80. Code of Practice for CNG Compressor and Refuelling Stations Part 1: On Site Storage and Location of Equipment Amend: 1, 1986. 27 pp.

—Sampling

BSI BS 7576-92. 1992 Method for Continuous Sampling of Liquified Natural Gas (LNG) in Transfer Pipelines (ISO 8943: 1991). 18 pp.

ISO 8943-91. Refrigerated Light Hydrocarbon Fluids—Sampling of Liquefied Natural Gas—Continuous Method First Edition. 17 pp.

—Storage and Handling

SAA AS 3961-91. Liquefied Natural Gas—Storage and Handling (in Professional Package 47). 43 pp.

—Storage Tanks

CNS Z2066-88. Construction of Welded Aluminium Alloy Liquefied Natural Gas Storage Tank (Aug)(12405).

JIS B 8251-81. Construction of Welded Aluminium Alloy Liquefied Natural Gas Storage Tanks (R 1986). 41 pp.

Liquefied Petroleum Gas

Use For: LPG *See Also:* Butanes; Butylene; Fossil Fuels; Fuels; Gaseous Fuels; Hazardous Materials; Liquefied Gases; Liquefied Natural Gas; Liquefied Petroleum Gas Cylinders; Liquid Hydrocarbons; Natural Gas; Petroleum Products; Propane; Propylene

INDUSTRY STANDARDS

Liquefied Petroleum Gas (Cont.)

CNS S1032-72. Liquified Petroleum Gas (for Household) (Jun)(2532).
ISO 8216 Pt 3-87. Petroleum Products—Fuels (Class F)—Classification—Part 3: Family L (Liquefied Petroleum Gases) First Edition. 4 pp.
ISO 9162-89. Petroleum Products—Fuels (Class F)—Liquefied Petroleum Gases—Specifications First Edition. 6 pp.
JIS K 2240-91. Liquefied Petroleum Gases. 112 pp.
SNZ NZS 5435-84. Specification for Liquefied Petroleum Gas (LPG). 11 pp.

—**Appliances**
BSI BS 3879-81. 1981 Amd 3 Portable Appliances Operating at Vapour Pressure from Liquefied Petroleum Gas Containers. 14 pp.
JIS S 2092-91. General Constructions of Gas Burning Appliances for Domestic Use. 34 pp.

—**Automotive**
BSI BS 4250: Part 2-87. (WITHDRAWN) 1987 Amd 1 Liquified Petroleum Gas Part 2: Automotive LPG (AMD 6230) February 28, 1990 (Superseded by BS EN 589: 1993). 9 pp.
BSI BS EN 589-93. 1993 Automotive Liquefied Petroleum Gas (W). 14 pp.
CEN PREN 589-91. Automotive Fuels—LPG—Requirements and Methods of Test. 14 pp.
CEN EN 589-93. Automotive Fuels—LPG—Requirements and Methods of Test. 7 pp.
SNZ NZS 5422: Part 1-87. Code of Practice for the Use of LPG and CNG Fuels in Internal Combustion Engines Part 1: LPG Fuel. 36 pp.
SNZ NZS 5422: Part 3-91. Code of Practice for the Use of LPG and CNG Fuels in Internal Combustion Engines Part 3: In Heavy Vehicles. 48 pp.

—**Automotive—Fuel Systems**
SAA AS 1425-89. LP Gas Fuel Systems for Vehicle Engines (Known as the SAA Automotive LP Gas Code) Amdt 1 August 1990 Amdt 2 March 1991. 17 pp.

—**Automotive—Refueling**
SNZ NZS 5434-86. Code of Practice for LPG Vehicle Refuelling Stations. 25 pp.

—**Barbecues**
CEN PREN 498-91. Barbecues Burning Liquefied Petroleum Gases for Outdoor Use. 44 pp.

—**Burners**
CEN PREN 497-91. Boiling Burners Burning Liquefied Petroleum Gases for Outdoor Use. 41 pp.
CNS B7050-81. Method of Test for Liquescent Petroleum Gas Burner (May)(3662).

—**Classification**
BSI BS 6843: Part 3-88. 1988 Classification of Petroleum Fuels Part 3: Liquefied Petroleum Gases. 3 pp.

—**Compressors**
CNS Z2067-88. Compressor for Liquefied Petroleum Gas Use (Aug)(12406).

—**Cooking Appliances**
CNS S1083-81. Liquefied Petroleum Gas Burner Ovens (Apr)(3660).
JIS S 2103-91. Gas Burning Cooking Appliances for Domestic Use. 93 pp.
JIS S 2108-80. Rice Cookers Used with Liquefied Petroleum Gas.

—**Corrosion Testing—Copper**
BSI BS 6924-88. 1988 Method for Determination of Corrosiveness of Liquefied Petroleum Gases to Copper. 7 pp.
CNS K6249-74. Method of Test for Corrosion of Liquefied Petroleum Gases (Copper Strip) (Jun)(2750).
ISO 6251-82. Liquefied Petroleum Gases—Corrosiveness to Copper—Copper Strip Test First Edition. 6 pp.

—**Cylinders—Boats**
SNZ NZS 5428-83. Code of Practice for the Use of LPG for Domestic Purposes in Caravans and Boats (Reconfirmed 1988). 27 pp.

—**Cylinders—Recreational Vehicles**
SNZ NZS 5428-83. Code of Practice for the Use of LPG for Domestic Purposes in Caravans and Boats (Reconfirmed 1988). 27 pp.

—**Density**
BSI BS 6665-85. 1985 Method for Determination of Density or Relative Density of Liquefied Petroleum Gases and Other Light Hydrocarbons by Pressure Hydrometer Method. 10 pp.
ISO 3993-84. Liquefied Petroleum Gas and Light Hydrocarbons—Determination of Density or Relative Density—Pressure Hydrometer Method First Edition. 8 pp.

—**Density—Hydrometry**
BSI BS 6665-85. 1985 Method for Determination of Density or Relative Density of Liquefied Petroleum Gases and Other Light Hydrocarbons by Pressure Hydrometer Method. 10 pp.

—**Fuel Dispensing Equipment—Electrical Equipment**
SNZ NZS 6109: Part 2-88. Electrical Systems of Dispensing Equipment for Explosive Atmospheres Part 2: Liquefied Petroleum Gas Dispensing Equipment. 16 pp.

—**Gas Alarms**
CNS Z2039-81. Alarm Equipments of Liquefied Petroleum Gas Leakage (Jul)(7760).
CNS Z3021-81. Leakage Test for Liquefied Petroleum Gas Alarm (Jun)(7536).

—**Highway Transportation—Emergency Procedures**
SAA AS 1678.2.1. 001-86. Emergency Procedure Guide—Transport—Part 2.1.001: Liquefied Petroleum Gas (LPG).

—**Hose Assemblies**
BSI BS 4089-89. 1989 Hoses and Hose Assemblies for Liquefied Petroleum Gas. 21 pp.
ISO 2928-86. Rubber Hoses and Hose Assemblies for Liquefied Petroleum Gases (LPG)—Bulk Transfer Applications—Specification Second Edition. 4 pp.
JIS B 8261-80. Hose Assemblies for Liquefied Petroleum Gas (R 1985). 33 pp.

—**Hose Assemblies—Safety**
BSI BS 3212-91. 1991 Flexible Rubber Tubing, Rubber Hose and Rubber Hose Assemblies for Use in LPG Vapour Phase and LPG/Air Installations. 13 pp.

—**Hoses**
BSI BS 4089-89. 1989 Hoses and Hose Assemblies for Liquefied Petroleum Gas. 21 pp.
CNS K4072-82. Liquefied Petroleum Gas Hoses (Nov)(9621).
ISO 2928-86. Rubber Hoses and Hose Assemblies for Liquefied Petroleum Gases (LPG)—Bulk Transfer Applications—Specification Second Edition. 4 pp.

—**Hoses—Safety**
BSI BS 3212-91. 1991 Flexible Rubber Tubing, Rubber Hose and Rubber Hose Assemblies for Use in LPG Vapour Phase and LPG/Air Installations. 13 pp.

—**Hydrogen Sulfide Content**
BSI BS 7026-89. 1989 Method for Detection of Hydrogen Sulphide in Liquefied Petroleum Gases Using Lead Acetate. 6 pp.
ISO 8819-87. Liquefied Petroleum Gases—Detection of Hydrogen Sulfide—Lead Acetate Method First Edition. 6 pp.

—**Hydrogen Sulfide Content—Electrometric Analysis**
BSI BS 2000: Part 272-93. 1993 Methods of Test for Petroleum and Its Products Part 272: Determination of Mercaptan Sulphur and Hydrogen Sulphide Content of LPG—Electrometric Titration Method (W). 7 pp.
BSI BS 2000: Part 272-87. 1987 Petroleum and Its Products Part 272: Determination of Mercaptan Sulphur and Hydrogen Sulphide Content of Liquefied Petroleum Gases by Electometric Titration. 11 pp.

—**Industrial Plants—Safety**
CNS Z1023-86. Safety Code for Liquefied Petroleum Gas Filled Plant (Sep)(8069).

—**Leakage**
CNS Z3021-81. Leakage Test for Liquefied Petroleum Gas Alarm (Jun)(7536).

—**Plastic Pipes**
JIS K 6774-89. Polyethylene Pipes for the Supply of Gaseous Fuels. 33 pp.

—**Pressure Regulators**
CNS Z2030-84. Regulator for Liquefied Petroleum Gas (Sep)(7088).
CNS Z3017-84. Method of Test for Regulator of Liquefied Petroleum Gas (Sep)(7089).

—**Pumps**
CNS Z2068-88. Pump for Liquefied Petroleum Gas Use (Aug)(12407).

—**Recreational Vehicles—Safety**
ISO 7421-91. Leisure Accommodation Vehicles—Liquefied Petroleum Gas Systems First Edition. 7 pp.

—**Residue Content**
CNS K6294-70. Method of Test for Volatile Residue of Liquefied Petroleum Gases (Dec)(3183).

—**Safety**
CNS Z1004-86. Safety Code for Liquefied Petroleum Gas (Sep)(1332). 9 pp.
SAA AS 1596-89. LP Gas—Storage and Handling (Known as the SAA LP Gas Code) Amdt 1 November 1990 Amdt 2 July 1991.

—**Sampling**
ISO 4257-88. Liquefied Petroleum Gases—Method of Sampling First Edition. 7 pp.

—**Storage Tanks**
CEN PREN 976-1-92. Underground Tanks of Glass-Reinforced Plastics (GRP)—Horizontal Cylindrical Tanks for the Non-Pressure Storage of Liquid Petroleum Base Fuels—Part 1: Requirements and Test Methods. 17 pp.
CEN PREN 976-2-92. Underground Tanks of Glass-Reinforced Plastics (GRP)—Horizontal Cylindrical Tanks for the Non-Pressure Storage of Liquid Petroleum Base Fuels—Part 2: Transport, Handling, Storage and Installation. 22 pp.
DIN ENGL 4680 Pt 1-92. Stationary, Above-Ground, Steel Pressure Vessels for the Storage of Liquefied Petroleum Gas; Dimensions and Mountings (May). 7 pp.
DIN ENGL 4680 Pt 2-92. Stationary, Partially Sunk, Steel Pressure Vessels for the Storage of Liquefied Petroleum Gas; Dimensions and Mountings (May). 5 pp.
JIS B 8242-91. Horizontal Type Cylindrical Storage Tanks Used for Liquefied Petroleum Gas—Construction. 52 pp.

—**Storage Tanks—Refueling**
SNZ NZS 5434-86. Code of Practice for LPG Vehicle Refuelling Stations. 25 pp.

—**Sulfur Content**
CNS K6248-74. Method of Test for Sulfur in Liquefied Petroleum Gas (Lamp Method) (Jun)(2749).

—**Tank Trucks—Trailers**
CNS Z3018-88. Liquefied Petroleum Gas Tank on Truck of Trailer (Jul)(7248).

—**Thiol Content—Electrometric Analysis**
BSI BS 2000: Part 272-93. 1993 Methods of Test for Petroleum and Its Products Part 272: Determination of Mercaptan Sulphur and Hydrogen Sulphide Content of LPG—Electrometric Titration Method (W). 7 pp.
BSI BS 2000: Part 272-87. 1987 Petroleum and Its Products Part 272: Determination of Mercaptan Sulphur and Hydrogen Sulphide Content of Liquefied Petroleum Gases by Electrometric Titration. 11 pp.

—**Vapor Pressure**
BSI BS 3324-80. 1980 Determination of Vapour Pressure of Liquefied Petroleum Gases (LPG Method). 11 pp.
CNS K6247-74. Method of Test for Vapor Pressures of Liquefied Petroleum Gas (Jun)(2748).
ISO 4256-78. Liquefied Petroleum Gases—Determination of Vapour Pressure—LPG Method First Edition. 8 pp.

—**Volatile Matter Content**
CNS K6250-74. Method of Test for Volatility of Butane Propane Gases in Liquefied Petroleum Gases (Jun)(2751).
CNS K6294-70. Method of Test for Volatile Residue of Liquefied Petroleum Gases (Dec)(3183).

—**Water Heaters**
JIS S 2109-91. Gas Burning Water Heaters for Domestic Use. 84 pp.
JIS S 2111-82. Bath Boilers Used with Liquefied Petroleum Gas.
JIS S 2116-91. Water Heaters Containing Water Under Atmospheric Pressure. 30 pp.
JIS S 2117-81. Water Heaters Containing Water Under Pressure Used with Liquefied Petroleum Gas.

—**Wood Preservatives**
CNS K1193-81. Volatile Petroleum Solvent (Liquefied Petroleum Gas) for Wood Preservatives (May)(4164).

Liquefied Petroleum Gas Cylinders

See Also: Gas Cylinders; Liquefied Petroleum Gas
BSI BS 5329-88. (WITHDRAWN) 1988 Non-Refillable Metallic Containers up to 1.4 Liters Capacity Filled with Liquefied Petroleum Gases (Superseded by BS EN 417: 1992). 14 pp.
CNS B5047-90. Structure for Liquefied Petroleum Gas Cylinder (Cylinder Body) (Apr)(2448).

INTERNATIONAL AND NON-U.S. NATIONAL STANDARDS
SUBJECT INDEX

Liquefied Petroleum Gas Cylinders (Cont.)
- CNS B7029-90. Method of Test for Liquefied Petroleum Gas Cylinder (Cylinder Body) (Mar)(1323).
- CNS Z2065-88. Sphere Holder and It's Accessories for Liquefied Petroleum Gas (Aug)(12403).
- JIS B 8233-86. Refillable Welded Steel Gas Cylinders for Liquefied Petroleum Gas. 34 pp.

—Appliances—Portable
- BSI BS EN 417-92. 1992 Non-Refillable Metallic Gas Cartridges for Liquefied Petroleum Gases, with or Without a Valve, for Use with Portable Appliances Construction, Inspection, Testing and Marking (E). 19 pp.
- CEN PREN 417-90. Non-Refillable Metallic Gas Cartridges for Liquefied Petroleum Gases, with or Without a Valve, for Use with Portable Appliances—Construction, Inspection, Testing and Marking. 27 pp.
- CEN EN 417-92. Non-Refillable Metallic Gas Cartridges for Liquefied Petroleum Gases, with or Without a Valve, for Use with Portable Appliances—Construction, Inspection, Testing and Marking. 27 pp.
- CEN PREN 521-91. Dedicated Liquefied Petroleum Gas Appliances—Portable Appliances Operating at Vapour Pressure from Liquefied Petroleum Gas Containers. 54 pp.

—Ranges—Portable
- JIS S 2147-91. Portable Cookers Attached to Liquefied Petroleum Gas Cylinder. 41 pp.

—Steel
- JIS B 8233-86. Refillable Welded Steel Gas Cylinders for Liquefied Petroleum Gas. 34 pp.
- MOD UK DEF-132. Cylinder Liquefied Gas, Steel (Suitable for Containing Butane or Propane) (Withdrawn).

—Valves
- CNS B7030-89. Inspection Standard of Valves for Liquefied Petroleum Gas Cylinder (Dec)(1324).
- JIS B 8245-89. Valves for Liquefied Petroleum Gas Cylinder. 10 pp.

Liquid Ammonia
Use: Ammonia

Liquid Chillers
Use: Chillers

Liquid Chromatography
Use For: High Performance Liquid Chromatography; HPLC
- CNS K6963-89. General Rules for Analytical Methods in High Performance Liquid Chromatography (Apr)(12512).
- JIS K 0124-83. General Rules for Analytical Methods in High Performance Liquid Chromatography. 20 pp.

—Apple Juice
- ISO 8128 Pt 1-93. Apple Juice, Apple Juice Concentrates and Drinks Containing Apple Juice—Determination of Patulin Content—Part 1: Method Using High-Performance Liquid Chromatography First Edition. 7 pp.

—Butadienes
- ISO 8176-86. Butadiene for Industrial Use—Determination of Active Tert-Butyl-Catechol (TBC) (4-(1,1-Dimethylethyl)-1,2-Benzenediol)—High Performance Liquid Chromatographic Method First Edition. 6 pp.

—Cheeses
- ISO 9233-91. Cheese and Cheese Rind—Determination of Natamycin Content—Method by Molecular Absorption Spectrometry and by High-Performance Liquid Chromatography First Edition. 13 pp.

—Chilies
- ISO 7543 Pt 2-93. Chillies and Chilli Oleoresins—Determination of Total Capsaicinoid Content—Part 2: Method Using High-Performance Liquid Chromatography First Edition. 8 pp.

—Coffee
- BSI BS 5752: Part 12-92. 1992 Methods of Test for Coffee and Coffee Products Part 12: Coffee: Determination of Caffeine Content (Routine Method by HPLC) (ISO 10095: 1992). 12 pp.
- ISO 10095-92. Coffee—Determination of Caffeine Content—Method Using High-Performance Liquid Chromatography First Edition. 9 pp.

—Electrical Insulating Oils
- CENELEC PREN 61198-92. Mineral Insulating Oils—Methods for the Determination of 2-Furfural and Related Compounds. 17 pp.

Liquid Chromatography (Cont.)
—Electrical Insulating Oils (Cont.)
- IEC 1198-93. Mineral Insulating Oils—Methods for the Determination of 2-Furfural and Related Compounds First Edition. 32 pp.

—Essential Oils
- ISO 8432-87. Essential Oils—Analysis by High Performance Liquid Chromatography—General Method First Edition. 7 pp.

—Food
- CNS N6191-86. Method of Test for Artificial Sweeteners in Food (Sep)(10950). 4 pp.

—Fruit Juices
- CNS N6223-91. Method of Test for Fruit and Vegetable Juices and Drinks-Sugar Analysis by HPLC Method (Jun)(12634). 2 pp.

—Nuclear Fuel Dissolver Solutions
- ISO 10981-93. Determination of Uranium in Reprocessing Plant Dissolver Solution—Liquid Chromatography Method First Edition. 9 pp.

—Pepper
- ISO 11027-93. Pepper and Pepper Oleoresins—Determination of Pipeline Content—Method Using High-Performance Liquid Chromatography First Edition. 8 pp.

—Petroleum Products
- DIN ENGL 51527 Pt 1-87. Testing of Petroleum Products; Determination of Polychlorinated Biphenyls (PCB); Preseparation by Liquid Chromatography and Determination of Six Selected PCB Compounds by Gas Chromatography Using an Electron Capture Detector (May). 6 pp.

—Rapeseeds
- BSI BS 4289: Part 9-93. 1993 Methods for the Analysis of Oilseeds Part 9: Determination of Glucosinolates Content of Rapeseed by High Performance Liquid Chromatography (ISO 9167-1: 1992) (W). 15 pp.
- ISO 9167 Pt 1-92. Rapeseed—Determination of Glucosinolates Content—Part 1: Method Using High-Performance Liquid Chromatography First Edition. 12 pp.

—Vegetable Juices
- CNS N6223-91. Method of Test for Fruit and Vegetable Juices and Drinks-Sugar Analysis by HPLC Method (Jun)(12634). 2 pp.

—Water
- DIN ENGL 38405 Pt 19-88. German Standard Methods for the Examination of Water, Waste Water and Sludge; Anions (Group D) Determination of Fluo-ride, Chloride, Nitrite, (Ortho)Phosphate, Brom-ide, Nitrate and Sulfate Anions in Water with a Low Pollution Level by Ion Exchange Chromatography (D 19) (Feb). 11 pp.
- ISO 10304 Pt 1-92. Water Quality—Determination of Dissolved Fluoride, Chloride, Nitrite, Orthophosphate, Bromide, Nitrate and Sulfate Ions, Using Liquid Chromatography of Ions—Part 1: Method for Water with Low Contamination First Edition. 16 pp.

Liquid Coolers
Use: Fluid Coolers

Liquid Crystal Display Modules
Use: Liquid Crystal Displays

Liquid Crystal Displays
Use For: LCD *See Also:* Display Devices; Optoelectronic Devices
- BSI BS CECC 20007-91. 1991 Liquid Crystal Displays Monochrome LCDs Without Electronic Circuit (T). 12 pp.
- CECC CECC 20 000 ISSUE 1-82. Generic Specification: Semiconductor Optoelectronic and Liquid Crystal Devices (En, Fr, Ge) AMD 3 (En, Fr, Ge) SUPP 1 (En, Fr, Ge) SUPP 2 (En, Fr, Ge). 401 pp.
- CECC CECC 20 007 ISSUE 1-91. Blank Detail Specification: Liquid Crystal Displays; Monochrome LCDs Without Electronic Circuit (En, Fr, Ge). 36 pp.
- CNS C5226-89. General Rule of Liquid Crystal Display Panel (Apr)(12506).
- JIS C 7071-88. General Rule of Liquid Crystal Display Panel. 9 pp.
- MOD UK DSTAN 59-58: Part 90-85. Optoelectronic Displays Part 90: Detail Specifications Issue 1. 7 pp.
- MOD UK DSTAN 59-58: 90/001-82. Opto-Electronic Display Issue 2. 17 pp.
- MOD UK DSTAN 59-58: 90/005-82. Opto-Electronic Display Issue 1. 42 pp.
- MOD UK DSTAN 59-58: 90/006-82. Opto-Electronic Display Issue 1. 39 pp.

Liquid Crystal Displays (Cont.)
—Dot Matrix—Preferred Products List
- CECC CECC MUAHAG Vol 10 IS 2-92. Preferred Products List; OPTO Electronic Devices (En, Fr, Ge). 48 pp.

—Measurement
- CNS C6327-89. Measuring Methods for Liquid Crystal Display Panels (Apr)(12507).
- JIS C 7072-88. Measuring Methods for Liquid Crystal Display Panels. 15 pp.

—Semiconductor—Quality Assurance
- BSI BS CECC 20000-83. 1983 Amd 1 Semiconductor Optoelectronic and Liquid Crystal Devices: Generic Specification. 86 pp.

—7-Segment—Preferred Products List
- CECC CECC MUAHAG Vol 10 IS 2-92. Preferred Products List; OPTO Electronic Devices (En, Fr, Ge). 48 pp.

Liquid Dielectrics
Use: Dielectrics

Liquid Driers
Use: Driers

Liquid Filled Transformers
Use For: Fluid Filled Transformers; Liquid Immersed Transformers *See Also:* Oil Filled Transformers; Transformers

—Bushings
- BSI BS 7616-93. 1993 Bushings for Liquid Filled Transformers Above 1 kV and up to 36 kV (E). 18 pp.
- CENELEC HD 506 S1-89. Bushings for Liquid Transformers Above 1kv up to 33kv. 12 pp.
- CENELEC HD 506 S1/A1-92. AMD 1 Bushings for Liquid Transformers Above 1kv up to 33kv. 5 pp.

Liquid Flow Measurement
Use: Flow Measurement

Liquid Flow Measurement Equipment
Use: Flow Measurement Equipment

Liquid Fuels
See Also: Aircraft Gasoline; Alcohol Fuels; Bunker Fuel Oil; Diesel Fuels; Fuel Oils; Gas Turbine Fuels; Gasohol; Gasoline; Jet Engine Fuels; Kerosene; Liquid Hydrocarbons; Liquid Rocket Fuels; Petroleum Products

—Calorific Value
- BSI BS 7420-91. 1991 Determination of Calorific Values of Solid, Liquid and Gaseous Fuels (Including Definitions). 11 pp.
- DIN ENGL 51900 Pt 1-89. Determination of Gross Calorific Value of Solid and Liquid Fuels by the Bomb Calorimeter and Calculation of Net Calorific Value; General Information (Nov). 11 pp.
- DIN ENGL 51900 Pt 2-77. Testing of Solid and Liquid Fuels; Determination of the Gross Calorific Value by the Bomb Calorimeter and Calculation of the Net Calorific Value; Method Using Isothermal Water Jacket (Aug). 8 pp.
- DIN ENGL 51900 Pt 3-77. Testing of Solid and Liquid Fuels; Determination of the Gross Calorific Value by the Bomb Calorimeter and Calculation of the Net Calorific Value; Method Using Adiabatic Jacket (Aug). 5 pp.

—Cargo Transportation
- MOD UK DSTAN 47-18: Part 1-01. Equipment for Receipt and Delivery of Liquid Fuels Part 1: Bulk Transportation and Transfer Issue 1; Amendment 1. 19 pp.

—Dynamic Viscosity—Viscometers
- DIN ENGL 51561-78. Testing of Mineral Oils, Liquid Fuels and Related Liquids; Measurement of Viscosity Using the Vogel-Ossag Viscometer; Temperature Range: Approximately 10 to 150 Deg. C (Dec). 4 pp.
- DIN ENGL 51569-78. Testing of Mineral Oils, Liquid Fuels and Related Liquids; Measurement of Viscosity Using the Vogel-Ossag Viscometer; Temperature Range:-55 to Approximately +10 Deg. C (Dec). 5 pp.

—Electrostatic Safety
- NATO STANAG 3682 ED 4 AMD 0-90. Electrostatic Safety Connection Procedures for Liquid Fuel Loading/ Unloading Operations During Ground Transfer. 7 pp.
- NATO STANAG 3682 ED 4 AMD 1-90. Electrostatic Safety Connection Procedures for Liquid Fuel Loading/ Unloading Operations During Ground Transfer. 7 pp.

INTERNATIONAL AND NON-U.S. NATIONAL STANDARDS
SUBJECT INDEX

Liquid

Liquid Fuels (Cont.)

—**Electrostatic Safety** (Cont.)
NATO STANAG 3682 ED 4 AMD 2-90. Electrostatic Safety Connection Procedures for Liquid Fuel Loading/ Unloading Operations During Ground Transfer. 8 pp.

—**Fire Hazards—Munitions**
NATO STANAG 4240 ED 1 AMD 0-91. Liquid Fuel Fire Tests for Munitions. 20 pp.

—**Fire Hazards—Weapons Systems**
NATO STANAG 4240 ED 1 AMD 0-91. Liquid Fuel Fire Tests for Munitions. 20 pp.

—**Flash Point—Pensky-Martens Apparatus**
DIN ENGL 51755-74. Testing of Mineral Oils and Other Combustible Liquids; Determination of Flash Point by the Closed Tester According to Abel-Pensky (Mar). 5 pp.

—**Heat of Combustion**
BSI BS 2000: Part 12-93. 1993 Methods of Test for Petroleum and Its Products Part 12: Determination of Specific Energy. 11 pp.

—**Hydrogen Sulfide Content**
DIN ENGL 51766-74. Testing of Liquid Fuels; Testing for Hydrogen Sulphide (Sept). 1 p.

—**Installation—Fuel Systems**
NATO STANAG 3756 ED 3 AMD 0-90. Facilities and Equipment for Receipt and Delivery of Liquid Fuels. 13 pp.
NATO STANAG 3756 ED 3 AMD 1-90. Facilities and Equipment for Receipt and Delivery of Liquid Fuels. 22 pp.
NATO STANAG 3756 ED 3 AMD 2-90. Facilities and Equipment for Receipt and Delivery of Liquid Fuels. 15 pp.

—**Kinematic Viscosity—Viscometers**
DIN ENGL 51561-78. Testing of Mineral Oils, Liquid Fuels and Related Liquids; Measurement of Viscosity Using the Vogel-Ossag Viscometer; Temperature Range: Approximately 10 to 150 Deg. C (Dec). 4 pp.
DIN ENGL 51569-78. Testing of Mineral Oils, Liquid Fuels and Related Liquids; Measurement of Viscosity Using the Vogel-Ossag Viscometer; Temperature Range:-55 to Approximately +10 Deg. C (Dec). 5 pp.

—**Oxidation Resistance**
DIN ENGL 51780-63. Testing of Liquid Fuels; Determination of Oxidation Stability; (Induction Period) (July). 4 pp.

—**Pressure Switches**
DIN ENGL 3398 Pt 4-86. Pressure Cut-Off Switches for Liquid Fuels and Heat Transfer Oils (Oct). 8 pp.

—**Sulfur Content**
EC 75/716/EEC-75. Council Directive on the Approximation of the Laws of the Member States Relating to the Sulphur Content of Certain Liquid Fuels. 3 pp.

—**Vanadium Content—Atomic Absorption Spectroscopy**
DIN ENGL 51790 Pt 3-78. Testing of Liquid Fuels; Determination of Vanadium Content; Vanadium Content in the Range of 0.4 to 4.0 mg/kg; Determination by Flameless Atomic Absorption Spectroscopy After Incineration (Apr). 6 pp.

Liquid Gases
Use: Liquefied Gases

Liquid Hydrocarbons
See Also: Hydrocarbons; Liquefied Natural Gas; Liquefied Petroleum Gas; Liquid Fuels; Petroleum Products; Viscous Hydrocarbons

—**Refractive Index**
ISO 5661-83. Petroleum Products—Hydrocarbon Liquids—Determination of Refractive Index First Edition. 4 pp.

—**Refrigerated—Calorific Value—Calculation Methods**
BSI BS 7577-92. 1992 Calculation Procedures for Static Measurement of Refrigerated Light Hydrocarbon Fluids (ISO 6578: 1991). 28 pp.
ISO 6578-91. Refrigerated Hydrocarbon Liquids—Static Measurement—Calculation Procedure First Edition. 26 pp.

—**Refrigerated—Volume—Calculation Methods**
BSI BS 7577-92. 1992 Calculation Procedures for Static Measurement of Refrigerated Light Hydrocarbon Fluids (ISO 6578: 1991). 28 pp.

Liquid Hydrocarbons (Cont.)

—**Refrigerated—Volume—Calculation Methods** (Cont.)
ISO 6578-91. Refrigerated Hydrocarbon Liquids—Static Measurement—Calculation Procedure First Edition. 26 pp.

Liquid Immersed Transformers
Use: Liquid Filled Transformers

Liquid Immersion Testing
Use: Immersion Testing

Liquid-In-Glass Thermometers
Use: Thermometers

Liquid Level Gages
Use: Level Indicators

Liquid Level Indicators
Use: Level Indicators

Liquid Limit
See Also: Moisture Content Analysis

—**Soil Analysis**
CNS A3086-86. Method of Test for Liquid Limit of Soils (Dec)(5087).
JIS A 1205-90. Test Method for Liquid Limit and Plastic Limit of Soils. 8 pp.
SAA AS 1289.C1.1-77. Methods of Testing Soil for Engineering Purposes—Part C1.1: Soil Classification Tests—Determination of the Liquid Limit of a Soil—Oven Drying Method (Standard Method) Corrig.. 4 pp.
SAA AS 1289.C1.2-77. Methods of Testing Soil for Engineering Purposes—Part C1.2: Soil Classification Tests—Determination of the Liquid Limit of a Soil—One Point Method (Subsidiary Method) Corrig.. 2 pp.

—**Subsoil**
DIN ENGL 18122 Pt 1-76. Subsoil; Testing of Soil Samples; Consistency Limits; Determination of Liquid Limit and Plastic Limit (Apr). 6 pp.

Liquid Metals
See Also: Liquids; Lubricants; Metals; Shape Memory Alloys

—**Aluminum**
CEN PREN 577-91. Aluminium and Aluminium Alloys—Liquid Metal. 8 pp.

—**Aluminum Alloys**
CEN PREN 577-91. Aluminium and Aluminium Alloys—Liquid Metal. 8 pp.

—**Aluminum Alloys—Alloying**
DIN ENGL 1725 Pt 5 Suppl. 1-86. Aluminium Alloys; Casting Alloys; Ingots (Pigs); Liquid Metal; Composition; Information on Alloying Processes (Feb). 2 pp.

—**Aluminum Alloys—Chemical Analysis**
DIN ENGL 1725 Pt 5-86. Aluminium Alloys; Casting Alloys; Ingots (Pigs); Liquid Metal; Composition (Feb). 6 pp.

Liquid Mixers
See Also: Food Processing Equipment

—**Commercial**
CSA C22.2 NO 195-M1987. Motor Operated Food Processing Appliances (Household and Commercial); (Gen Instr 1 Thru 2). 55 pp.

—**Household**
CSA C22.2 NO 195-M1987. Motor Operated Food Processing Appliances (Household and Commercial); (Gen Instr 1 Thru 2). 55 pp.

Liquid Nitrogen
See Also: Liquefied Gases; Nitrogen

—**Aircraft**
MOD UK DSTAN 68-94-90. Nitrogen, Liquid: for Aircraft, Ship Systems and General Purpose Useage Issue 2. 7 pp.
MOD UK DSTAN 68-94-01. Nitrogen, Liquid: for Aircraft, Ship Systems and General Purpose Useage Issue 2; Amendment 1. 8 pp.
MOD UK DSTAN 68-94-02. Nitrogen, Liquid: for Aircraft, Ship Systems and General Purpose Useage Issue 2; Amendment 2. 9 pp.

—**Aircraft—Interdepartmental Procurement—Characteristics**
NATO STANAG 3546 ED 3 AMD 5-73. Liquid Nitrogen Characteristics. 6 pp.

Liquid Nitrogen (Cont.)

—**Aircraft—Replenishment Equipment**
NATO STANAG 3547 ED 3 AMD 3-83. Characteristics of Replenishment Equipment for Liquid Nitrogen. 10 pp.

—**Electronic Equipment**
MOD UK DSTAN 68-94-90. Nitrogen, Liquid: for Aircraft, Ship Systems and General Purpose Useage Issue 2. 7 pp.
MOD UK DSTAN 68-94-01. Nitrogen, Liquid: for Aircraft, Ship Systems and General Purpose Useage Issue 2; Amendment 1. 8 pp.
MOD UK DSTAN 68-94-02. Nitrogen, Liquid: for Aircraft, Ship Systems and General Purpose Useage Issue 2; Amendment 2. 9 pp.

—**Embarkation—Marine**
MOD UK NES 792: Part 1-88. Bulk Embarkation of Compressed and Liquefied Gases Part 1: Nitrogen Issue 2 (07.88). 15 pp.

—**Fertilizers**
CNS N3075-88. By-Product Nitrogen Fertilizer, Fluid (Jul)(11962).

—**Missiles—Replenishment Equipment**
NATO STANAG 3547 ED 3 AMD 3-83. Characteristics of Replenishment Equipment for Liquid Nitrogen. 10 pp.

—**Weapons**
MOD UK DSTAN 68-94-90. Nitrogen, Liquid: for Aircraft, Ship Systems and General Purpose Useage Issue 2. 7 pp.
MOD UK DSTAN 68-94-01. Nitrogen, Liquid: for Aircraft, Ship Systems and General Purpose Useage Issue 2; Amendment 1. 8 pp.
MOD UK DSTAN 68-94-02. Nitrogen, Liquid: for Aircraft, Ship Systems and General Purpose Useage Issue 2; Amendment 2. 9 pp.

Liquid Oxygen
Use For: LOX *See Also:* Hazardous Materials; Liquefied Gases; Oxygen
CNS K6086-65. Method of Test for Liquid Oxygen (Aug)(1006)(R 1971).
MOD UK DSTAN 68-22-01. Oxygen, Liquid (for Breathing) Issue 1; Amendment 1 (Withdrawn). 8 pp.

—**Aircraft Filling Device Couplings**
BSI 2C 5-76. 1976 Mating Dimensions of Liquid Oxygen Replenishment Couplings. 5 pp.
ISO 1465-89. Aircraft—Liquid Oxygen Replenishment Couplings—Mating Dimensions Second Edition. 5 pp.

—**Aircraft—Interdepartmental Procurement—Characteristics**
NATO STANAG 3545 ED 2 AMD 6-75. Characteristics of Breathable Liquid Oxygen. 6 pp.

—**Aircraft—Oxygen Supply Equipment**
NATO STANAG 3499 ED 2 AMD 0-89. Characteristics of Supply Equipment for Liquid Oxygen. 11 pp.
NATO STANAG 3499 ED 2 AMD 1-89. Characteristics of Supply Equipment for Liquid Oxygen. 11 pp.

—**Cleanliness—Naval Ships**
MOD UK NES 372: Part 3-88. Oxygen Clean Standards (OX) Part 3: Liquid Oxygen (LOX) Plants Issue 2 (10.88). 46 pp.

—**Embarkation—Marine**
MOD UK NES 792-82. Embarkation of Nitrogen for HP Electrolyser Use Issue 1 (11.82) (Superseded by NES 792: Parts 1 & 2).
MOD UK NES 792: Part 2-88. Bulk Embarkation of Compressed and Liquefied Gases Part 2: Oxygen Issue 2 (07.88). 25 pp.

—**Highway Transportation—Emergency Procedures**
SAA AS 1678.2.2. 000-86. Emergency Procedure Guides—Transport—Part 2.2.000: Oxygen (Refrigerated Liquid).

Liquid Paints
Use: Paints

Liquid Paraffin
Use: Paraffin Oils

Liquid Penetrant Testing
Use For: Penetrant Inspection; Penetrant Testing
See Also: Fluorescent Penetration Testing; Immersion Testing; Liquid Penetrant Testing Equipment; Penetration Resistance
BSI BS 6443-84. 1984 Amd 1 Penetrant Flaw Detection. 15 pp.

INTERNATIONAL AND NON-U.S. NATIONAL STANDARDS
SUBJECT INDEX

Liquids

Liquid Penetrant Testing *(Cont.)*
CEN PREN 571-1-91. Non Destructive Testing—Penetrant Testing—Part 1: General Principles for the Examination. 19 pp.
CEN PREN 956-92. Non-Destructive Testing—Penetrant Testing—Equipment. 8 pp.
CGSB 48-GP-12M-77. Manual on Liquid Penetrant Inspection. 60 pp.
CNS Z8048-84. General Rules for Liquid Penetrant Testing (Sep)(11047).
DIN ENGL 54152 Pt 1-89. Non-Destructive Testing; Penetrant Inspection; Procedure (July). 5 pp.
DIN ENGL 54152 Pt 2-89. Non-Destructive Testing; Penetrant Inspection; Verification of Penetrant Inspection Materials (July). 8 pp.
DIN ENGL 54152 Pt 3-89. Non-Destructive Testing; Penetrant Inspection; Reference Blocks for Determination of the Sensitivity of Penetrant Systems (July). 6 pp.
ISO 3452-84. Non-Destructive Testing—Penetrant Inspection—General Principles First Edition. 13 pp.
ISO 3453-84. Non-Destructive Testing—Liquid Penetrant Inspection—Means of Verification First Edition. 5 pp.
SAA AS 2062-77. Methods for Non-Destructive Penetrant Testing of Products and Components. 16 pp.

—Aerospace
BSI M 39-72. 1972 Method for Penetrant Inspection of Aerospace Materials and Components. 12 pp.
JIS W 0904-90. Inspection Process of Liquid Penetrant for Aerospace Use.

—Aircraft
MOD UK DTD-929-57. Penetrant Methods of Flaw Detection (Reprinted December 1965). 1 p.

—Aircraft Equipment
CAA LEAFLET 4-4 07.90. Performance Testing of Penetrant Testing Materials. 3 pp.

—Aluminum Castings
ISO 9916-91. Aluminium Alloy and Magnesium Alloy Castings —Liquid Penetrant Inspection First Edition. 16 pp.

—Castings
CNS Z8054-85. Liquid Penetrant Test for Castings (Mar)(11225).

—Ceramics
CEN PREN 623-1-92. Advanced Technical Ceramics—General and Textural Properties of Monolithic Ceramics—Part 1: Determination of the Presence of Defects by Dye Penetration Tests. 8 pp.

—Fiber Optic Cables
CNS C6301-87. Method of Test for Fiber Optic Devices (FOTP-82 Fluid Penetration Test for Filled Fiber Optic Cable) (Sep)(12087).

—Forgings
CNS Z8056-85. Liquid Penetrant Testing for Forgings (Sep)(11376).

—Glossaries
BSI BS 3683: Part 1-85. 1985 Glossary of Terms Used in Non-Destructive Testing: Penetrant Flaw Detection. 8 pp.
CNS Z8065-86. Glossary of Terms Related to Nondestructive Testing (Liquid Penetrant Testing Terms) (Oct)(11749).

—Magnesium Castings
ISO 9916-91. Aluminium Alloy and Magnesium Alloy Castings —Liquid Penetrant Inspection First Edition. 16 pp.

—Ships
MOD UK NES 729: Part 4-84. Requirements for Non-Destructive Examination Methods Part 4: Liquid Penetrant Issue 1 (11.84). 18 pp.
MOD UK NES 729: Part 4-91. Requirements for Non-Destructive Examination Methods Part 4: Liquid Penetrant Issue 2 (12.91). 22 pp.

—Steel Castings
BSI BS 4080: Part 2-89. 1989 Severity Levels for Discontinuities in Steel Castings Part 2: Surface Discontinuities Revealed by Penetrant Flaw Detection. 8 pp.
ISO 4987-92. Steel Castings—Penetrant Inspection First Edition. 14 pp.

—Submarines
MOD UK NES 729: Part 4-84. Requirements for Non-Destructive Examination Methods Part 4: Liquid Penetrant Issue 1 (11.84). 18 pp.
MOD UK NES 729: Part 4-91. Requirements for Non-Destructive Examination Methods Part 4: Liquid Penetrant Issue 2 (12.91). 22 pp.

Liquid Penetrant Testing *(Cont.)*
—Testing Personnel—Certification
CGSB 48-GP-9M-79. Certification of Nondestructive Testing Personnel (Liquid Penetrant Method) (Incorporates Amendment 1); (Amendment 5 November 1990 Incorporates Amendments 2, 3 and 4) (Amendment 8 November 1992 Incorporates Amendment 7). 154 pp.

—Welded Joints
BSI M 42-72. 1972 Methods for Non-Destructive Testing of Fusion Welds in Thin Gauge Materials. 10 pp.
CNS Z8060-85. Method of Liquid Penetrant Test for Welds (Oct)(11398).

Liquid Penetrant Testing Equipment
Use For: Penetrant Flaw Detectors; Penetrant Testing Equipment *See Also:* Liquid Penetrant Testing
ISO 9935-92. Non-Destructive Testing—Penetrant Flaw Detectors —General Technical Requirements First Edition. 7 pp.

—Electrotechnical Equipment
BSI BS 2011: Part 2.1R-90. 1990 Basic Environmental Testing Procedures Part 2.1: Tests Part 2.1R: Test R and Guidance. Water. 43 pp.

Liquid Permeability
—Clothing
BSI BS 4724: Part 1-86. 1986 Resistance of Clothing Materials to Permeation by Liquids Part 1: Method for Assessment of Breakthrough Time. 8 pp.
BSI BS 4724: Part 2-88. (WITHDRAWN) 1988 Resistance of Clothing Material to Permeation by Liquids Part 2: Method for Determination of Liquid Permeating After Breakthrough (Superseded by BS EN 369: 1993). 11 pp.

—Protective Clothing
BSI BS 7182-89. 1989 Air-Impermeable Chemical Protective Clothing. 15 pp.
BSI BS 7184-89. 1989 Selection, Use and Maintenance of Chemical Protective Clothing. 25 pp.
BSI BS EN 368-93. 1993 Protective Clothing—Protection Against Liquid Chemicals—Test Method: Resistance of Materials to Penetration by Liquids (N). 11 pp.
BSI BS EN 369-93. 1993 Protective Clothing—Protection Against Liquid Chemicals—Test Method: Resistance of Materials to Permeation by Liquids (N). 14 pp.
CEN PREN 368-90. Protective Clothing: Protection Against Liquid Chemicals: Resistance of Materials to Penetration by Liquids. 10 pp.
CEN PREN 369-90. Protective Clothing: Protection Against Liquid Chemicals: Resistance of Air-Impermeable Materials to Permeation by Liquids. 15 pp.
CEN EN 369-93. Protective Clothing—Test Method: Resistance of Materials to Permeation by Liquids. 12 pp.
CEN PREN 464-91. Chemical Protective Clothing—Protective Against Gases and Vapours—Method of Test—Determination of Leak—Tightness (Internal Pressure Test). 6 pp.
CEN PREN 465-91. Protective Clothing—Protection Against Liquid Chemicals—Performance Requirements —Type 4 Equipment—Protective Suits with Spray—Tight Connections Between Different Parts of the Protective Suit. 12 pp.
CEN PREN 466-91. Chemical Protection Clothing—Protection Against Liquid Chemicals (Including Liquid Aerosols)—Performance Requirements —Type 3 Equipment—Chemical Protective Clothing with Liquid Tight—Connections Between Different Parts of the Clothing. 12 pp.
CEN PREN 467-91. Protective Clothing—Protective Against Liquid Chemicals—Performance Requirements —Type 5 Equipment—Garments Providing Chemical Protection to Parts of the Body. 11 pp.
CEN PREN 945-92. Protective Clothing for Use Against Liquid and Gaseous Chemicals, Including Liquid Aerosols and Solid Particles—Performance Requirements for Air Fed Protective Clothing with Non Gas Tight Connections (Type 2 Equipment). 9 pp.
DIN ENGL EN 368-93. Protective Clothing for Use Against Liquid Chemicals; Method of Determining the Resistance of Materials to Penetration by Liquids (Jan) (Supersedes Parts of DIN 32763, September 1986 Edition). 7 pp.
ISO 6529-90. Protective Clothing—Protection Against Liquid Chemicals—Determination of Resistance of Air-Impermeable Materials to Permeation by Liquids First Edition. 11 pp.

Liquid Permeability *(Cont.)*
—Sintered Materials
BSI BS 5600: Sec 3.6-88. 1988 Powder Metallurgical Materials and Products Part 3: Methods of Testing Sintered Metal Materials and Products, Excluding Hardmetals Section 3.6: Determination of Fluid Permeability. 12 pp.

Liquid Pumps
Use: Pumps

Liquid Resistance Testing Equipment
See Also: Testing Equipment
—Enamels
BSI BS 1344: Part 14-84. 1984 Methods of Testing Vitreous Enamel Finishes Part 14: Apparatus for Testing with Acid and Neutral Liquids and Their Vapours. 9 pp.
ISO 2733-83. Vitreous and Porcelain Enamels—Apparatus for Testing with Acid and Neutral Liquids and Their Vapours Second Edition. 7 pp.

Liquid Rocket Fuels
See Also: Liquid Fuels
—Water Reaction
DIN ENGL 51415-81. Testing of Liquid Fuels; Testing of the Behaviour of Aviation Fuels in the Presence of Water (July). 3 pp.

Liquids
See Also: Canning Liquids; Flammable Liquids; Fluids; Liquefied Gases; Liquid Metals; Solids; Solids Content Analysis

—Clear—Colorimetry
CNS K6644-81. Method of Test for Color of Transparent Liquids (Gardner Color Scale) (Mar)(7163).
CNS K6663-81. Method of Test for Color of Clear Liquids (Platinum Scale) (May)(7403).
ISO 6271-81. Clear Liquids—Estimation of Colour by the Platinum-Cobalt Scale First Edition. 5 pp.

—Density
BSI BS 4522-88. 1988 Method for Determination of Absolute Density at 20 Degrees of Liquid Chemical Products for Industrial Use. 4 pp.
CNS Z8070-88. Method for Measuring Specific Gravity of Liquid (Oct)(12450).
JIS Z 8804-76. Methods of Measuring Specific Gravity of Liquid (R 1989). 12 pp.

—Density—Glossaries
CNS Z7183-84. Definitions of Terms Relating to Density and Specific Gravity of Solids Liquids and Gases (Oct)(11027).

—Opaque—Dynamic Viscosity
CNS K6329-72. Method of Test for Viscosities of Transparent and Opaque Liquids (Kinematic and Dynamic Viscosities) (Oct)(3390).

—Opaque—Kinematic Viscosity
CNS K6329-72. Method of Test for Viscosities of Transparent and Opaque Liquids (Kinematic and Dynamic Viscosities) (Oct)(3390).

—Sampling
BSI BS 5309: Part 3-76. 1976 Methods for Sampling Chemical Products Part 3: Sampling of Liquids. 41 pp.

—Storage Tanks
BSI BS 1564-75. 1975 Amd 1 Pressed Steel Sectional Rectangular Tanks. 16 pp.
BSI BS 2594-75. 1975 Amd 1 Carbon Steel Welded Horizontal Cylindrical Storage Tanks. 17 pp.
BSI BS 5387-76. (WITHDRAWN) 1976 Vertical Cylindrical Welded Storage Tanks for Low-Temperature Services: Double-Wall Tanks for Temperature Down to-196 Degrees Celsius (Superseded by BS 7777: Parts 1, 2, 3 and 4: 1993). 57 pp.
BSI BS 5429-76. 1976 Code of Practice for Safe Operation of Small-Scale Storage Facilities for Cryogenic Liquids. 9 pp.
DIN ENGL 4119 Pt 1-79. Above-Ground Cylindrical Flat-Bottomed Tank Installations of Metallic Materials; Fundamentals, Design, Tests (June). 14 pp.
DIN ENGL 4119 Pt 2-80. Above-Ground Cylindrical Flat-Bottom Tank Structures of Metallic Materials; Calculation (Feb). 13 pp.
DIN ENGL 6608 Pt 1-89. Horizontal Single-Wall Steel Tanks for the Underground Storage of Flammable and Non-Flammable Water Polluting Liquids (Sept). 6 pp.
DIN ENGL 6608 Pt 2-89. Horizontal Double-Wall Steel Tanks for the Underground Storage of Flammable and Non-Flammable Water Polluting Liquids (Sept). 3 pp.

INDUSTRY STANDARDS

INTERNATIONAL AND NON-U.S. NATIONAL STANDARDS
SUBJECT INDEX

Liquids

Liquids (Cont.)
— Storage Tanks (Cont.)
DIN ENGL 6616-89. Horizontal Single-Wall and Double-Wall Steel Tanks for the Above Ground Storage of Flammable and Non-Flammable Water Polluting Liquids (Sept). 10 pp.
DIN ENGL 6618 Pt 1-89. Vertical Single-Wall Steel Tanks for the Above Ground Storage of Flammable and Non-Flammable Water Polluting Liquids (Sept). 10 pp.
DIN ENGL 6618 Pt 2-89. Vertical Double-Wall Steel Tanks with Vacuum-Based Leak Detection System, for the Above Ground Storage of Flammable and Non-Flammable Water Polluting Liquids (Sept). 6 pp.
DIN ENGL 6619 Pt 1-89. Vertical Single-Wall Steel Tanks for the Underground Storage of Flammable and Non-Flammable Water Polluting Liquids (Sept). 4 pp.
DIN ENGL 6619 Pt 2-89. Vertical Double-Wall Steel Tanks for the Underground Storage of Flammable and Non-Flammable Water Polluting Liquids (Sept). 4 pp.
DIN ENGL 6623 Pt 1-89. Vertical Single-Wall Steel Tanks with Less Than 1000 Litre Capacity, for the Above Ground Storage of Flammable and Non-Flammable Water Polluting Liquids (Sept). 5 pp.
DIN ENGL 6623 Pt 2-89. Vertical Double-Wall Steel Tanks with Less Than 1000 Litre Capacity, for the Above Ground Storage of Flammable and Non-Flammable Water Polluting Liquids (Sept). 4 pp.
DIN ENGL 6624 Pt 1-89. Horizontal Single-Wall Steel Tanks with Capacities Between 1000 and 5000 Litres, for the Above Ground Storage of Flammable and Non-Flammable Water Polluting Liquids (Sept). 5 pp.
DIN ENGL 6624 Pt 2-89. Horizontal Double-Wall Steel Tanks with Capacities Between 1000 and 5000 Litres, for the Above Ground Storage of Flammable and Non-Flammable Water Polluting Liquids (Sept). 3 pp.
DIN ENGL 6625 Pt 1-89. Steel Tanks Erected on Site, for the Above Ground Storage of Hazardous Flammable Water Polluting Liquids of Class A III and of Non-Flammable Water Polluting Liquids; Requirements and Testing (Sept). 3 pp.

— Storage Tanks—Design
DIN ENGL 6625 Pt 2-89. Steel Tanks Erected on Site, for the Above Ground Storage of Hazardous Flammable Water Polluting Liquids of Class A III and of Non-Flammable Water Polluting Liquids; Design (Sept). 4 pp.

— Storage Tanks—Inspection
DIN ENGL 6600-89. Steel Tanks for the Storage of Flammable and Non-Flammable Water Polluting Liquids; Concepts and Inspection (Sept). 4 pp.

— Storage Tanks—Leak Detectors
DIN ENGL 6618 Pt 3-89. Vertical Double-Wall Steel Tanks with Liquid-Based Leak Detection System, for the Above Ground Storage of Flammable and Non-Flammable Water Polluting Liquids (Sept). 4 pp.

— Storage Tanks—Manholes
DIN ENGL 6626-89. Steel Manhole Shafts for Underground Tanks Designed for the Storage of Flammable and Non-Flammable Water Polluting Liquids (Sept). 9 pp.

— Storage Tanks—Manholes—Collars
DIN ENGL 6627-89. Collars for Masonry Manhole Shafts for Underground Tanks Designed for the Storage of Flammable and Non-Flammable Water Polluting Liquids (Sept). 2 pp.

— Tank Trucks—Measuring Systems
EC 82/625/EEC-82. Commission Directive Adapting to Technical Progress Council Directive 77/313/EEC on the Approximation of the Laws of the Member States Relating to Measuring Systems for Liquids Other Than Water. 20 pp.

— Transparent—Dynamic Viscosity
CNS K6329-72. Method of Test for Viscosities of Transparent and Opaque Liquids (Kinematic and Dynamic Viscosities) (Oct)(3390).

— Transparent—Kinematic Viscosity
CNS K6329-72. Method of Test for Viscosities of Transparent and Opaque Liquids (Kinematic and Dynamic Viscosities) (Oct)(3390).

— Viscosity
BSI BS 188-77. 1977 Determination of the Viscosity of Liquids. 18 pp.

Liquidtight Flexible Conduit Fittings
See Also: Electric Conduit Fittings; Liquidtight Flexible Conduits

— Nonmetallic—Safety
CSA C22.2 NO 227.2-93. Flexible Liquid-Tight Nonmetallic Conduit; (Gen Instr 1). 29 pp.

Liquidtight Flexible Conduits
See Also: Electric Conduits; Electric Raceways; Flexible Metal Conduits; Flexible Nonmetallic Conduits; Liquidtight Flexible Conduit Fittings

— Electrical Codes
CSA 851 Bull. Electrical Bulletin 851 November 17, 1971 to C22.2 NO 38. 2 pp.
CSA 851 Bull. Electrical Bulletin 851 November 17, 1971 to C22.2 NO 56. 2 pp.

— Metal—Thermoplastic Covering
CSA 983B Bull. Electrical Bulletin 983B April 25, 1977 to C22.2 NO 38. 39 pp.
CSA C22.2 NO 56-1977. Flexible Metal Conduit and Liquid-Tight Flexible Metal Conduit (R 1992); (Amd 1-2 February 1990) (Amd 3 August 1990) (Amd 4-6 January 1993). 28 pp.
CSA 983B Bull. Electrical Bulletin 983B April 25, 1977 to C22.2 NO 56. 39 pp.

— Nonmetallic—Safety
CSA C22.2 NO 227.2-93. Flexible Liquid-Tight Nonmetallic Conduit; (Gen Instr 1). 29 pp.

Liquors (Papermaking)
Scope Note: Use a more specific term *See:* Bisulfite Liquors; Black Liquors; Bleach Liquors; Green Liquors; Kraft Liquors; Pulps; Spent Liquors; White Liquors

LISN
Use: Line Impedance Stabilization Networks

Listening Tests
Use: Acoustic Testing

Listeria Monocytogenes
See Also: Bacteria

— Food
SAA AS 1766.2.15 (INT)-91. Method for the Microbiological Examination of Food—Part 2: Examination for Specific Organisms—Part 2.15(Int): Listeria Monocytogenes in Dairy Products. 11 pp.

— Milk
ISO 10560-93. Milk and Milk Products—Detection of Listeria Monocytogenes First Edition. 15 pp.

Litchi Nuts
Use: Litchis

Litchis
Use For: Lychee Nuts *See Also:* Fruits

— Grading
CNS N1095-82. Grading and Packaging of Litchi (Jun)(6848). 2 pp.

— Packaging
CNS N1095-82. Grading and Packaging of Litchi (Jun)(6848). 2 pp.

Literature (Fine Arts)
See Also: Arts

— Classification
BSI BS 1000: (8)-71. 1971 Universal Decimal Classification (UDC). English Full Edition (8): Language. Linguistics. Literature. 35 pp.
SNZ NZS/BS 1000 (8)-71. Universal Decimal Classification Language. Linguistics. Literature. 36 pp.

Literature Surveys
Use: Reviews

Litharge
Use: Lead Monoxide

Lithia Content Analysis
Use: Lithium Oxide Content Analysis

Lithium Batteries
See Also: Batteries; Lithium Copper Oxide Batteries; Lithium Manganese Dioxide Batteries; Lithium Thionyl Chloride Batteries; Primary Batteries

— Aircraft
BSI G 239-87. (WITHDRAWN) 1987 Primary Active Lithium Batteries for Use in Aircraft (Superseded by BS 2G 239: 1992). 16 pp.
BSI 2G 239-92. Primary Active Lithium Batteries for Use in Aircraft. 22 pp.

— Packaging
MOD UK DSTAN 61-19-89. Storage, Handling and Disposal of Lithium Batteries Issue 1. 15 pp.
MOD UK DSTAN 81-69-01. Packaging of Primary Batteries Containing Lithium Issue 2; Amendment 1. 13 pp.

Lithium Batteries (Cont.)
— Safety
MOD UK DSTAN 61-19-89. Storage, Handling and Disposal of Lithium Batteries Issue 1. 15 pp.

— Shipping
MOD UK DSTAN 61-19-89. Storage, Handling and Disposal of Lithium Batteries Issue 1. 15 pp.

— Storage
MOD UK DSTAN 61-19-89. Storage, Handling and Disposal of Lithium Batteries Issue 1. 15 pp.

— Waste Disposal
MOD UK DSTAN 61-19-89. Storage, Handling and Disposal of Lithium Batteries Issue 1. 15 pp.

Lithium Chloride
CNS K7281-66. Chemical Reagent (Lithium Chloride) (Dec)(1780).
JIS K 8162-75. Lithium Chloride, Anhydrous.

Lithium Content Analysis
— Water—Photometry
BSI BS 2690: Part 125-81. 1981 Water Used in Industry Part 125: Lithium: Flame Photometric Method. 2 pp.
CNS K9047-80. Method of Test for Lithium in Water (Flame Photometric Method) (Aug)(6229).

Lithium Copper Oxide Batteries
See Also: Lithium Batteries

— Preferred Product List
CECC CECC MUAHAG Vol 15 IS 1-91. Preferred Products List; Batteries (En). 20 pp.

Lithium Inorganic Compounds
Scope Note: Use a more specific term *See:* Lithium Chloride; Lithium Nitrate; Lithium Sulfate

Lithium Manganese Dioxide Batteries
Use For: Lithium Manganese Dioxide Cells; Manganese Dioxide Lithium Batteries
See Also: Batteries; Lithium Batteries; Primary Batteries
JIS C 8512-89. Manganese Dioxide Lithium Primary Cells and Batteries. 11 pp.

— Preferred Product List
CECC CECC MUAHAG Vol 15 IS 1-91. Preferred Products List; Batteries (En). 20 pp.

Lithium Manganese Dioxide Cells
Use: Lithium Manganese Dioxide Batteries

Lithium Nitrate
CNS K7282-66. Chemical Reagent (Lithium Nitrate) (Dec)(1781).

Lithium Oxide Content Analysis
— Dolomite
ISO 10058-92. Magnesites and Dolomites —Chemical Analysis First Edition. 23 pp.

— Magnesite
ISO 10058-92. Magnesites and Dolomites —Chemical Analysis First Edition. 23 pp.

Lithium Sulfate
CNS K7283-66. Chemical Reagent (Lithium Sulfate) (Dec)(1782).
JIS K 8994-75. Lithium Sulfate.

Lithium Thionyl Chloride Batteries
See Also: Batteries; Lithium Batteries

— Preferred Product List
CECC CECC MUAHAG Vol 15 IS 1-91. Preferred Products List; Batteries (En). 20 pp.

Lithographic Varnishes
See Also: Varnishes
MOD UK DSTAN 80-94-01. Varnish, Lithographic, Thin, Type QX Issue 1; Amendment 1. 12 pp.

Lithography
See Also: Offset Printing Equipment

— Papers
CGSB 9-GP-47M-79. Paper, Bond, for Magnetic Ink Character Recognition Cheque Printing, Standard for; (Amendment 1 Mar 1980). 9 pp.

— Solvents
CGSB CAN/CGSB-21.4-93. Lithographic Blanket and Roller Wash Solvent. 7 pp.

Lithol Red B
See Also: Pigments

Lithol Red B (Cont.)
—Pigments
CNS K2132-83. Lithol Red B (Jun)(10353). 3 pp.
JIS K 5202-71. Lithol Red B. 5 pp.

Lithopone
See Also: Pigments
CNS K2003-84. Lithopone for Paint (Feb)(15). 2 pp.
CNS K6092-84. Method of Test for Lithopone for Lacquer Paint (Mar)(1040).
ISO 473-82. Lithopone Pigments for Paints—Specifications and Methods of Test Second Edition. 6 pp.
JIS K 5105-65. Lithopone (Pigment). 9 pp.

Litmus Papers
See Also: Papers; Test Papers
CNS K7284-66. Chemical Reagent (Litmus Paper) (Dec)(1783).
CNS K7285-66. Chemical Reagent (Litmus Paper, Blue) (Dec)(1784).
CNS K7286-66. Chemical Reagent (Litmus Paper, Red) (Dec)(1785).
JIS K 9071-61. Litmus Paper.

Live Wire Working
Use For: Hot Line Maintenance *See Also:* Electric Wiring; Electrical Safety; Occupational Safety and Health
—Covers
IEC 1229-93. Rigid Protective Covers for Live Working on a.c. Installations First Edition. 66 pp.
—Grounding Devices—Portable
IEC 1230-93. Live Working—Portable Equipment for Earthing or Earthing and Short-Circuiting First Edition. 77 pp.
—Hand Tools
CENELEC PREN 60900-92. Hand Tools for Live Working up to 1000 V A.C. and 1500 V D.C.. 39 pp.
—Pole Clamps
IEC 1236-93. Saddles, Pole Clamps (Stick Clamps) and Accessories for Live Working First Edition. 71 pp.
—Protective Clothing
BSI BS 7696-93. 1993 Conductive Clothing for Live Working at a Nominal Voltage up to 800 kV a.c. (IEC 895: 1987) (N). 37 pp.
—Saddles
IEC 1236-93. Saddles, Pole Clamps (Stick Clamps) and Accessories for Live Working First Edition. 71 pp.
—Short Circuiting/Grounding Devices—Portable
IEC 1230-93. Live Working—Portable Equipment for Earthing or Earthing and Short-Circuiting First Edition. 77 pp.
—Sleeving
BSI BS EN 60984-93. 1993 Sleeves of Insulating Material for Live Working (N). 50 pp.
CENELEC EN 60984-92. Sleeves of Insulating Material for Live Working; (IEC 984: 1990). 8 pp.

Live Wire Working Equipment
Use: Linemen's Equipment

Liver
See Also: Food; Meat; Organ Meat
—Dried
CNS N6041-87. Method of Test for Cured Meat and Dried Smoked Liver (Jul)(2214). 6 pp.

Livestock
Scope Note: For additional listings, use a more specific term *See Also:* Beef Cattle; Dairy Cattle; Equines; Goats; Horses; Poultry; Sheep; Swine
—Contaminants
EC 86/469/EEC-86. Council Directive Concerning the Examination of Animals and Fresh Meat for the Presence of Residues. 10 pp.
—Growth Hormones—Control Measures
EC 85/358/EEC-85. Council Directive Supplementing Directive 81/602/EEC Concerning the Prohibition of Certain Substances Having a Hormonal Action and of any Substances Having a Thyrostatic Action. 4 pp.
EC 88/146/EEC-88. Council Directive Prohibiting the Use in Livestock Farming of Certain Substances Having a Hormonal Action. 3 pp.

Livestock (Cont.)
—Inspection—Animal Health
EC COM(88) 383-88. Proposals for Council Regulations (EEC)—Concerning Veterinary Checks in Intro-Community Trade with a View to the Completion of the Internal Market—on Intensifying Controls on the Application of the Veterinary Rules—Amending Regulation (EEC) No 1468/81. 40 pp.
—Inspection—Public Health
EC COM(88) 383-88. Proposals for Council Regulations (EEC)—Concerning Veterinary Checks in Intro-Community Trade with a View to the Completion of the Internal Market—on Intensifying Controls on the Application of the Veterinary Rules—Amending Regulation (EEC) No 1468/81. 40 pp.
—Medicinal Products—Residues
EC 86/469/EEC-86. Council Directive Concerning the Examination of Animals and Fresh Meat for the Presence of Residues. 10 pp.
—Model Bylaws
SNZ NZS 9201: Chapter 13-72. Model General Bylaws Chapter 13: The Keeping of Animals, Poultry and Bees. 12 pp.
—Shipping Containers—Air Cargo
CAA NOTICE #92 ISSUE 1. (This Notice Gives Details of a Mandatory Action) Cargo Containment (Airworthiness Notices). 3 pp.

Livestock Crates
See Also: Containers; Crates (Shipping Containers); Livestock Equipment
—Highway Transportation
SNZ NZS 5413-77. Specification for Stock Crates Amend: 1 (Reconfirmed 1984). 12 pp.

Livestock Equipment
See Also: Agricultural Equipment; Cattle Grids; Farms; Livestock Crates; Materials Handling Equipment; Watering Troughs
—Stalls—Scraper Conveyors
ISO 5710-80. Equipment for Internal Farm Work and Husbandry—Continuous Manure Scraper Conveyors for Stalls First Edition. 4 pp.

LLC
Use: Logical Link Control

LNA
Use: Low Noise Amplifiers

LNG
Use: Liquefied Natural Gas

Load Break Switches
Use For: Load Interrupter Switches; Puffer Switches
See Also: Disconnecting Switches; Electric Switches; Electrical Protection Equipment; Switchgear
CNS C3114-81. Method of Test for A.C. Load Break Switches with Tripping Device (Mar)(7124).
CNS C4299-81. A.C. Load Break Switches with Tripping Device (Mar)(7123).
—Teleswitches—Radio
BSI BS 7647-93. 1993 Radio Teleswitches for Tariff and Load Control (E). 35 pp.
—3.3-6.6 kV
JIS C 4605-87. AC Load Break Switches for 3.3 kV or 6.6 kV. 64 pp.
JIS C 4607-91. AC Load Break Switches with Tripping Device for 3.3 kV or 6.6 kV. 32 pp.

Load Break Switchgear
Use: Load Break Switches

Load Break Tools
Use: Load Break Switches

Load Capacity
Use: Bearing Capacity

Load Cells
Use For: Loadcells; Pressure Cells *See Also:* Load Indicators; Measuring Instruments; Strain Gages; Weight Indicators
JIS B 7602-83. Test Method for Load Cell.

Load Indicators
Use For: Load Rings *See Also:* Indicating Instruments; Load Cells; Weight Indicators
—Cranes
BSI BS 7262-90. 1990 Automatic Safe Load Indicators. 24 pp.

Load Interrupter Switches
Use: Interrupter Switches; Load Break Switches

Load Interrupter Switchgear
Use: Interrupter Switches

Load Rings
Use: Load Indicators

Load Testers (Force)
Use: Load Indicators

Loadbearing Capacity
Use: Bearing Capacity

Loadbearing Walls
Use: Walls

Loadbreak Switches
Use: Load Break Switches

Loadcells
Use: Load Cells

Loaded Ebonite
Use: Ebonite

Loaders
See Also: Backhoes; Bulk Transporters; Crawler Loaders; Earthmoving Equipment; Front End Loaders; Materials Handling Equipment; Pallet Loaders; Palletizers; Unloaders
JIS A 8421-90. Terminology and Commercial Specifications on Loaders. 29 pp.
JIS D 0005-83. Standard Form of Specifications of Loaders. 27 pp.
JIS D 6505-82. Testing Method of Loaders. 40 pp.
—Bulk Handling
ISO 5699-79. Agricultural Machines, Implements and Equipment—Dimensions for Mechanical Loading with Bulk Goods First Edition. 3 pp.
—Glossaries
BSI BS 6685-86. 1986 Terms for Loaders Used for Earth-Moving. 28 pp.
ISO 7131-84. Earth-Moving Machinery—Loaders—Terminology and Commercial Specifications First Edition. 27 pp.
JIS A 8421-90. Terminology and Commercial Specifications on Loaders. 29 pp.
—Lift Capacity
BSI BS 6911: Part 6-92. 1992 Testing Earth-Moving Machinery Part 6: Method for the Evaluation of Lift Capacity of Pipelayers and Wheel Tractors or Loaders Equipped with Side Boom (ISO 8813: 1992). 22 pp.
ISO 8813-92. Earth-Moving Machinery—Lift Capacity of Pipelayers and Wheeled Tractors or Loaders Equipped with Side Boom First Edition. 19 pp.
—Loads (Forces)
BSI BS 6819-87. (WITHDRAWN) 1987 Measurement of Tool Forces and Tipping Loads of Loaders Used for Earth-Moving (Superseded by BS 6911: Part 2: 1990). 11 pp.
BSI BS 6911: Part 2-90. 1990 Testing Earth-Moving Machinery Part 2: Measurement of Tool Forces and Tipping Loads of Loaders. 12 pp.
BSI BS 6912: Part 2-89. 1989 Safety of Earth Moving Machinery Part 2: Rated Operating Load for Crawler and Wheel Loaders. 4 pp.
ISO 5998-86. Earth-Moving Machinery—Rated Operating Load for Crawler and Wheel Loaders Second Edition. 3 pp.
ISO 8313-89. Earth-Moving Machinery—Loaders—Methods of Measuring Tool Forces and Tipping Loads Second Edition. 10 pp.
JIS A 8917-92. Earth-Moving Machinery—Rated Operating Load for Crawler and Wheel Loaders.
—Manual Controls
CSA CAN/CSA-M3789.4-91. Controls for Forestry Log Loaders—System A (ISO 3789-4-1989). 16 pp.
ISO 3789 Pt 4-89. Tractors, Machinery for Agriculture and Forestry, Powered Lawn and Garden Equipment—Location and Method of Operation of Operator Controls—Part 4: Controls for Forestry Log Loaders First Edition. 6 pp.
JIS A 8918-92. Earth-Moving Machinery—Crawler Tractors and Crawler Loaders—Operators Controls.
—Roll Over Protective Structures
BSI BS 5527: Part 1-87. 1987 Roll-Over Protective Structures on Earth-Moving Machinery: Laboratory Tests and Preformance Requirements Part 1: Crawler, Wheel Loaders and Tractors, Backhoe Loaders, Graders, Tractor Scrapers and Articulated Steel Dumpers. 19 pp.
CNS A3126-81. Earth-Moving Machinery-Roll-Over Protective Structures for Driver Laboratory Tests and Performance Requirements (Sep)(7849).

Loaders (Cont.)

—Roll Over Protective Structures (Cont.)
ISO 3471 Pt 1-86. Earth-Moving Machinery—Roll-over Protective Structures—Laboratory Tests and Performance Requirements—Part 1: Crawler, Wheel Loaders and Tractors, Backhoe Loaders, Graders, Tractor Scrapers, Articulated Steer Dumpers First Edition. 18 pp.
JIS A 8910-89. Earth-Moving Machinery—Roll-Over Protective Structures—Laboratory Tests and Performance Requirements. 25 pp.

—Safety
CEN PREN 474-3-93. Earth-Moving Machinery—Safety—Part 3: Requirements for Loaders. 14 pp.

—Tool Forces
BSI BS 6819-87. (WITHDRAWN) 1987 Measurement of Tool Forces and Tipping Loads of Loaders Used for Earth-Moving (Superseded by BS 6911: Part 2: 1990). 11 pp.
BSI BS 6911: Part 2-90. 1990 Testing Earth-Moving Machinery Part 2: Measurement of Tool Forces and Tipping Loads of Loaders. 12 pp.
ISO 8313-89. Earth-Moving Machinery—Loaders—Methods of Measuring Tool Forces and Tipping Loads Second Edition. 10 pp.
JIS A 8916-91. Earth-Moving Machinery—Loaders—Methods of Measuring Tool Forces and Tipping Loads.

Loaders, Front End
Use: Front End Loaders

Loading Ramps
Use: Ramps (Loading)

Loads (Electric)
Use For: Loads (Electronic) *See Also:* Electronic Equipment; Overload Testing; Testing Equipment
CNS Z1031-84. Power Load (Nov)(9643).

—Distribution Transformers
CSA C9.1-M1981. Guide for Loading Dry-Type Distribution and Power Transformers; (Gen Instr 1). 18 pp.

—Dry Type Transformers
CSA C9.1-M1981. Guide for Loading Dry-Type Distribution and Power Transformers; (Gen Instr 1). 18 pp.

—Electrical Installations—Aircraft
CAA 706 (K). Section K Light Aeroplanes Electrical Supply, Systems and Equipment (Blue Papers). 54 pp.
CAA 707 (G). Section G Rotorcraft Electrical Supply, Systems and Equipment (Blue Papers). 47 pp.
CAA Chapter J2-1 App #1 09.66. Electrical Load Analyses. 13 pp.
CAA Chapter G6-13 11.85. Electrical Generation, Supply and Distribution (Rotocraft). 11 pp.
CAA Chap G6-13 App #3 11.85. Electrical Load Analyses (Rotocraft). 3 pp.
CAA Chapter K6-1 2 App3 10.92. Electrical Load Analyses. 3 pp.

—Oil Filled Transformers
BSI CP 1010-75. 1975 Loading Guide for Oil-Immersed Transformers. 50 pp.
IEC 354-91. Loading Guide for Oil-Immersed Power Transformers Second Edition; (Corrigendum—March 1992). 162 pp.
SAA AS 1078-84. Guide to Loading of Oil-Immersed Transformers. 50 pp.

—Power Lines—Overhead
IEC 826-91. Loading and Strength of Overhead Transmission Lines Second Edition; (Corrigendum—Sept 1991). 236 pp.

—Power Transformers
CSA C9.1-M1981. Guide for Loading Dry-Type Distribution and Power Transformers; (Gen Instr 1). 18 pp.

—Ripple Control Receivers
BSI BS 7401-90. (WITHDRAWN) 1990 Electronic Ripple Control Receivers for Tariff and Load Control (IEC 1037: 1990) (Renumbered as BS EN 61037: 1992). 44 pp.
BSI BS EN 61037-01. 1992 Amd 1 Electronic Ripple Control Receivers for Tarrif and Load Control (AMD 7673) April 15, 1993 (S). 56 pp.
CENELEC EN 61037-92. Electronic Ripple Control Receivers for Tariff and Load Control (IEC 1037:1990). 7 pp.
CENELEC EN 61037-92. Electronic Ripple Control Receivers for Tariff and Load Control (IEC 1037: 1990, Modified). 42 pp.
IEC 1037-90. Electronic Ripple Control Receivers for Tariff and Load Control First Edition; (CENELEC EN 61037:1992). 86 pp.

Loads (Electric) (Cont.)

—Ripple Control Receivers (Cont.)
SAA AS 1284.6-92. Electricity Metering—Part 6: Ripple Control Receivers for Tariff and Load Control. 43 pp.

—Time Switches
BSI BS 7415-91. (WITHDRAWN) 1991 Time Switches for Tariff and Load Control (Renumbered as BS EN 61038: 1993). 35 pp.
BSI BS EN 61038-93. 1993 Amd 1 Time Switches for Tariff and Load Control (AMD 7674) May 15, 1993 (E). 46 pp.
CENELEC EN 61038-92. Time Switches for Tariff and Load Control (IEC 1038:1990). 7 pp.
CENELEC EN 61038-92. Time Switches for Tariff and Load Control (IEC 1038: 1990, Modified). 34 pp.
IEC 1038-90. Time Switches for Tariff and Load Control First Edition; (CENELEC EN 61038:1992). 66 pp.

—X-Ray Tubes
CENELEC EN 60613-90. Electrical, Thermal and Loading Characteristics of Rotating Anode X-Ray Tubes for Medical Diagnosis. 5 pp.
CENELEC EN 60613-91. Electrical, Thermal and Loading Characteristics of Rotating Anode X-Ray Tubes for Medical Diagnosis. 3 pp.

Loads (Electronic)
Use: Loads (Electric)

Loads (Forces)
Scope Note: For load ratings of specific products or materials, see the product or material. For additional references, consult the following list *Use For:* Cone Proof Load Testing *See Also:* Aerodynamic Loads; Axial Loads; Bearing Capacity; Bearing Strength; Breaking Load; Civil Engineering; Dynamic Loads; Earth Pressure; Force; Ground Loads; Gust Loads; Ice Loads; Pressure; Snow Loads; Static Loads; Stresses; Wind Loads
SAA AS 1170.1-81. Minimum Design Loads on Structures (Known as the SAA Loading Code)—Part 1: Dead and Live Loads Obsolescent 1989 See AS 1170.1—1989. 19 pp.
SAA AS 1170.1-89. Minimum Design Loads on Structures (Known as the SAA Loading Code)—Part 1: Dead and Live Loads and Load Combinations Amdt 1 January/February 1993. 28 pp.

—Glossaries
DIN ENGL 15003-70. Lifting Appliances; Load Suspending Devices, Loads and Forces; Definitions (Feb). 3 pp.

Lobsters
Use For: Rock Lobsters *See Also:* Shellfish
SNZ NZS 8406-75. Code of Practice for the Handling, Processing, Storage and Distribution of Fresh and Frozen Rock Lobsters and Frozen Rock Lobster Tails. 20 pp.

—Frozen
SNZ NZS 8406-75. Code of Practice for the Handling, Processing, Storage and Distribution of Fresh and Frozen Rock Lobsters and Frozen Rock Lobster Tails. 20 pp.

LOC
Use: Localizers

Local Area Network Cables
Use For: LAN Cables *See Also:* Cables (Electric); Communication Cables; Local Area Networks

—Coaxial—Carrier Sense Multiple Access/CD
ECMA ECMA 80-84. Local Area Networks (CSMA/CD Baseband) Coaxial Cable System. 18 pp.
ECMA ECMA-TR 26-90. Planning and Installation Guide for CSMA/CD 10 Mbit/s Baseband LAN Coaxial Cable Systems. 64 pp.

Local Area Networks
Use For: LAN *See Also:* Data Highways; Ethernet (BTN); Fiber Distributed Data Interface; Hybrid Rings; Local Area Network Cables; Metropolitan Area Networks; Open Systems Interconnection; Radio Local Area Networks; Token Rings
BSI BS 7247-90. 1990 Guide for Characteristics of Local Area Networks (LAN). 13 pp.
CSA CAN/CSA-C22. 2 NO 220-M91. Information Processing and Business Equipment; (Gen Instr 1). 119 pp.
IEC 847-88. Characteristics of Local Area Networks (LAN) First Edition. 17 pp.
JTC1 847-88. Characteristics of Local Area Networks (LAN) First Edition. 17 pp.

Local Area Networks (Cont.)
SAA AS 4802.1-92. Information Processing Systems—Local Area Networks—Part 1: Overview and Architecture (IEEE 802—1990) (in Professional Package 26C). 20 pp.

—Carrier Sense Multiple Access/CD
BSI BS 6531: Part 1-84. (WITHDRAWN) 1984 10 Mbps Slotted Ring Local Area Network Part 1: Specification for the Coding of Bits and Structure of Slots and Mini-Packets (Superseded by BS ISO 8802-7: 1991). 9 pp.
BSI BS 7246-90. 1990 Guide for Local Area Networks CSMA/DC 10 Mbits/s Baseband Planning and Installation. 44 pp.
BSI BS ISO 8802/3-89. (WITHDRAWN) 1989 Carrier Sense Multiple Access with Collision Detection (CSMA/CD) Access Method and Physical Layer Specifications (Superseded by BS ISO/IEC 8802-3: 1992). 180 pp.
BSI BS ISO/IEC 8802-3-92. 1992 Information Technology—Local and Metropolitan Area Networks—Part 3: Carrier Sense Multiple Access with Collision Detection (CSMA/CD) Access Method and Physical Layer Specifications. 314 pp.
BSI DD 98: Part 1-84. (WITHDRAWN) 1984 CSMA/CD Local Area Network Part 1: Technical Specification (Superseded by BS ISO 8802-3: 1989). 194 pp.
BSI DD 98: Part 2-84. (WITHDRAWN) 1984 CSMA/CD Local Area Network Part 2: Guidance for Implementors (superseded by BS ISO 8802-3: 1989). 102 pp.
BSI DD ENV 41801-3-92. 1992 Information Systems Interconnection—Relaying the Connectionless-Mode Network Service—Part 3: ISO 8802-3 CSMA/CD Subnetwork Dependent Media Dependent Requirements. 18 pp.
BSI DD ENV 41802-3-92. 1992 Information Systems Interconnection—X.25 Protocol Relaying Part 3: ISO 8802-3 CSMA/CS Subnetwork Dependent Media Dependent Requirements. 13 pp.
CEN ENV 41801-3-92. Information Systems Interconnection—Relaying the Connectionless-Mode Network Service—Part 3: ISO 8802-3 CSMA/CD Subnetwork Dependent Media Dependent Requirements. 26 pp.
CEN ENV 41802-3-92. Information Systems Interconnection—X.25 Protocol Relaying—Part 3: ISO 8802-3 CSMA/CD Subnetwork Dependent Media Dependent Requirements. 10 pp.
ECMA ECMA 81-84. Local Area Networks (CSMA/CD Baseband) Physical Layer. 17 pp.
ECMA ECMA 82-84. Local Area Networks (CSMA/CD Baseband) Link Layer. 40 pp.
IEC 907-89. Local Area Networks CSMA/CD 10 Mbit/s Baseband Planning and Installation Guide First Edition. 86 pp.
IEC 8802 Pt 3 Draft AMD 6. Information Processing Systems—Local Area Networks—Part 3: Carrier Sense Multiple Access with Collision Detection (CSMA/CD) Access Method and Physical Layer Specifications Amendment 6; (1991) ***CD-ROM ONLY***. 5 pp.
IEC 8802 Pt 3 Draft AMD 7. Information Processing Systems—Local Area Networks—Part 3: Carrier Sense Multiple Access with Collision Detection (CSMA/CD) Access Method and Physi-cal Layer Specifications Amendment 7: Addition of Layer Management (Section 5); (1991) ***CD-ROM ONLY***. 40 pp.
IEC 8802 Pt 3 Draft AMD 9. Information Processing Systems—Local Area Networks—Part 3: Carrier Sense Multiple Access with Collision Detection (CSMA/CD) Access Method and Physical Layer Specifications Amendment 9: 10Base-T; (1991) ***CD-ROM ONLY***. 54 pp.
ISO 8802 Pt 3 Draft AMD 6. Information Processing Systems—Local Area Networks—Part 3: Carrier Sense Multiple Access with Collision Detection (CSMA/CD) Access Method and Physical Layer Specifications Amendment 6; (1991) ***CD-ROM ONLY***. 5 pp.
ISO 8802 Pt 3 Draft AMD 7. Information Processing Systems—Local Area Networks—Part 3: Carrier Sense Multiple Access with Collision Detection (CSMA/CD) Access Method and Physical Layer Specifications Amendment 7: Addition of Layer Management (Section 5); (1991)***CD-ROM ONLY***. 40 pp.
ISO 8802 Pt 3 Draft AMD 9. Information Processing Systems—Local Area Networks—Part 3: Carrier Sense Multiple Access with Collision Detection (CSMA/CD) Access Method and Physical Layer Specifications Amendment 9: 10Base-T; (1991) ***CD-ROM ONLY***. 54 pp.
JIS X 5252-90. Local Area Networks—Carrier Sense Multiple Access with Collision Detection (CSMA/CD) Access Method and Physical Layer Specifications (ISO 8802/3:1989).
JTC1 907-89. Local Area Networks CSMA/CD 10 Mbit/s Baseband Planning and Installation Guide First Edition. 86 pp.

INTERNATIONAL AND NON-U.S. NATIONAL STANDARDS
SUBJECT INDEX
Local

Local Area Networks *(Cont.)*

—Carrier Sense Multiple Access/CD *(Cont.)*

JTC1 8802 Pt3 Draft AMD 6. Information Processing Systems—Local Area Networks—Part 3: Carrier Sense Multiple Access with Collision Detection (CSMA/CD) Access Method and Physical Layer Specifications Amendment 6; (1991) ***CD-ROM ONLY***. 5 pp.

JTC1 8802 Pt3 Draft AMD 7. Information Processing Systems—Local Area Networks—Part 3: Carrier Sense Multiple Access with Collision Detection (CSMA/CD) Access Method and Physical Layer Specifications Amendment 7: Addition of Layer Management (Section 5); (1991)***CD-ROM ONLY***. 40 pp.

JTC1 8802 Pt3 Draft AMD 9. Information Processing Systems—Local Area Networks—Part 3: Carrier Sense Multiple Access with Collision Detection (CSMA/CD) Access Method and Physical Layer Specifications Amendment 9: 10Base-T; (1991) ***CD-ROM ONLY***. 54 pp.

OSI ISO 8802-3 DAM 4-90. Part 3: Carrier Sense Multiple Access with Collision Detection (CSMA/CD) Access Method and Physical Layer Specifications Amendment 4: Physical Signalling, Medium Attachment and Baseband Medium Specifications, StarLAN, Type 1 Bases. 62 pp.

OSI ISO 8802-3 DAM 3-90. Information Processing Systems—Local Area Networks—Part 3: Carrier Sense Multiple Access with Collision Detection. Amendment 3: Broadband Medium Attachment Unit and Broadband Medium Specification, Type10 BROAD 36. 43 pp.

OSI ISO DIS 8802/3 PDAD-87. Local Area Networks—Part 3: CSMA/CD—Addendum Broadband Medium Attachment Unit and Broadband Medium Specifications, Type IOBROAD36.

OSI ISO 8802-3 DAD 1-87. Information Processing—Local Area Networks PT3: Carrier Sense Multiple Access with Collision Detection Addendum 1: Medium Attachment Unit and Baseband Medium Specifications for Type 10 Base 2. 36 pp.

OSI ISO DIS 8802-3 PDAD-87. Local Area Networks—Part 3: CSMA/CD—Addendum Broadband Medium Attachment Unit and Broadband Medium Specifications, Type IOBROAD36. 230 pp.

OSI ISO/IEC 8802-3-90. Information Processing Systems—Local Area Networks—Part 3: Carrier Sense Multiple Access with Collision Detection (CSMA/CD) Access Method and Physical Layer Specifications (with Correction Sheet 09-90). 217 pp.

OSI ISO DIS 8802-3-87. Information Processing Systems—Local Area Networks Part 3: Carrier Sense Multiple Access with Collision Detection—Access Method and Physical Layer Specifications. 5 pp.

OSI ISO 8802-3-89. Information Processing Systems—Local Area Networks Part 3: Carrier Sense Multiple Access with Collision Detection (CSMA/CD) Access Method and Physical Layer Specifications. 178 pp.

OSI ISO 8802 3 DAD 1-87. Information Processing-Local Area Networks PT3: Carrier Sense Multiple Access with Collision Detection Addendum 1: Medium Attachment Unit and Baseband Medium Specifications for Type 10 Base 2.

OSI ISO 8802 3 DAD 2-87. Information Processing Systems-Local Area Networks Part 3: Carrier Sense Multiple Access with Collision Detection Addendum 2: Repeater Set and Repeater Unit Specification Foruse with 10 Base 5 and 10 Base 2 Networks.

OSI ISO/IEC DISP 10608-2-90. Information Technology—International Standardized Profile TAnnnn—Connection-Mode Transport Service over Connectionless-Mode Network Service—Part 2: TA51 Profile Including Subnetwork-Dependent Requirements for CSMS/CD Local Area Networks (LANs). 37 pp.

SAA AS 4802.3-91. Information Processing Systems—Local Area Networks—Part 3: Carrier Sense Multiple Access with Collision Detection (CSMA/CD) Access Method and Physical Layer Specifications (ISO/IEC 8802-3:1990/IEEE 802.3-1990) (in Prof. Packages 26A, 26C). 191 pp.

SAA AS 4802.3 Supp 1-91. Information Processing Systems—Local Area Networks—Part 3: Carrier Sense Multiple Access with Collision Detection (CSMA/CD) Access Method and Physical Layer Specifications—Supplement 1: Multisegment 10 Mb/s Baseband Networks:. 50 pp.

SNZ NZS/ISO 8802. 3-90. Information Processing Systems—Local Area Networks Part 3: Carrier Sense Multiple Access with Collision Detection (CSMA/CD) Access Method and Physical Layer Specifications. 191 pp.

Local Area Networks *(Cont.)*

—Carrier Sense Multiple Access/CD—Repeaters

IEC 8802 Pt 3 Draft AMD 11. Information Technology—Local and Metropolitan Area Networks—Part 3: Carrier Sense Multiple Access with Collision Detection (CSMA/CD) Access Meth. and Phys. Layer Specs. Amd 11: Layer Mgmt. for 10 Mb/s Baseband Repeaters (Sec. 19); (1993) (IEEE Std. 802.3k) ***CD-ROM ONLY**. 31 pp.

ISO 8802 Pt 3 Draft AMD 11. Information Technology—Local and Metropolitan Area Networks—Part 3: Carrier Sense Multiple Access with Collision Detection (CSMA/CD) Access Meth. and Phys. Layer Specs. Amd 11: Layer Mgmt. for 10 Mb/s Baseband Repeaters (Sec. 19); (1993) (IEEE Std. 802.3k) ***CD-ROM ONLY**. 31 pp.

JTC1 8802 Pt3 Draft AMD 11. Information Technology—Local and Metropolitan Area Networks—Part 3: Carrier Sense Multiple Access with Collision Detection (CSMA/CD) Access Meth. and Phys. Layer Specs. Amd 11: Layer Mgmt. for 10 Mb/s Baseband Repeaters (Sec. 19); (1993) (IEEE Std. 802.3k) ***CD-ROM ONLY**. 31 pp.

OSI ISO IEC 8802-3 DAD 5-89. Information Processing Systems—Local Area Networks Part 3: Carrier Sense Multiple Access with Collision Detection Addendum 5: Medium Attachment Unit and Baseband Medium Specifi-cation for a Vendor Independent Fibre Optic Inter Repeater Link (FOIRL-Standard). 24 pp.

OSI ISO 8802-3 DAD 2-87. Information Processing Systems—Local Area Networks Part 3: Carrier Sense Multiple Access with Collision Detection Addendum 2: Repeater Set and Repeater Unit Specification for Use with 10 Base 5 and 10 Base 2 Networks. 19 pp.

—Carrier Sense Multiple Access/CD—Subnetworks—PSDN

IEC DISP 10614 Pt 3-92. Information Technology—International Standardized Profile RC—X.25 Protocol Relaying —Part 3: ISO 8802-3 CSMA/CD Subnetwork-Dependent, Media-Dependent Requirements ***CD-ROM ONLY***. 37 pp.

ISO DISP 10614 Pt 3-92. Information Technology—International Standardized Profile RC—X.25 Protocol Relaying —Part 3: ISO 8802-3 CSMA/CD Subnetwork-Dependent, Media-Dependent Requirements ***CD-ROM ONLY***. 37 pp.

JTC1 DISP10614 Pt 3-92. Information Technology—International Standardized Profile RC—X.25 Protocol Relaying —Part 3: ISO 8802-3 CSMA/CD Subnetwork-Dependent, Media-Dependent Requirements ***CD-ROM ONLY***. 37 pp.

—Carrier Sense Multiple Access/CD—Subnetworks—Relay Function

IEC DISP 10612 Pt 2-92. Information Technology—International Standardized Profile RD5p.5q—Relaying the MAC Service Using Transparent Bridging—Part 2: CSMA/CD LAN Subnetwork-Dependent, Media-Dependent Requirements ***CD-ROM ONLY***. 32 pp.

IEC DISP 10612 Pt 4-92. Information Technology—International Standardized Profile RD5p.5q—Relaying the MAC Service Using Transparent Bridging—Part 4: Profile RD51.51 (CSMA/CD-CSMA/CD) ***CD-ROM ONLY***. 28 pp.

IEC DISP 10613 Pt 3-92. Information Technology—International Standardized Profile RA—Relaying the Connectionless-Mode Network Service—Part 3: ISO 8802-3 CSMA/CD LAN Subnetwork Dependent Media Dependent Requirements ***CD-ROM ONLY***. 22 pp.

IEC DISP 10613 Pt 5-92. Information Technology—International Standardized Profile RA—Relaying the Connectionless-Mode Network Service—Part 5: Relaying the CLNS Between CSMA/CD LAN Subnetworks (RA51.51) ***CD-ROM ONLY***. 17 pp.

IEC DISP 10613 Pt 8-92. Information Technology—International Standardized Profile RA—Relaying the Connectionless-Mode Network Service—Part 8: Profile RA51.1111 ***CD-ROM ONLY***. 20 pp.

IEC DISP 10613 Pt 9-92. Information Technology—International Standardized Profile RA—Relaying the Connectionless-Mode Network Service—Part 9: Profile RA51.1121 ***CD-ROM ONLY***. 20 pp.

ISO DISP 10612 Pt 2-92. Information Technology—International Standardized Profile RD5p.5q—Relaying the MAC Service Using Transparent Bridging—Part 2: CSMA/CD LAN Subnetwork-Dependent, Media-Dependent Requirements ***CD-ROM ONLY***. 32 pp.

ISO DISP 10612 Pt 4-92. Information Technology—International Standardized Profile RD5p.5q—Relaying the MAC Service Using Transparent Bridging—Part 4: Profile RD51.51 (CSMA/CD-CSMA/CD) ***CD-ROM ONLY***. 28 pp.

Local Area Networks *(Cont.)*

—Carrier Sense Multiple Access/CD—Subnetworks—Relay Function *(Cont.)*

ISO DISP 10613 Pt 3-92. Information Technology—International Standardized Profile RA—Relaying the Connectionless-Mode Network Service—Part 3: ISO 8802-3 CSMA/CD LAN Subnetwork Dependent Media Dependent Requirements ***CD-ROM ONLY***. 22 pp.

ISO DISP 10613 Pt 5-92. Information Technology—International Standardized Profile RA—Relaying the Connectionless-Mode Network Service—Part 5: Relaying the CLNS Between CSMA/CD LAN Subnetworks (RA51.51) ***CD-ROM ONLY***. 17 pp.

ISO DISP 10613 Pt 8-92. Information Technology—International Standardized Profile RA—Relaying the Connectionless-Mode Network Service—Part 8: Profile RA51.1111 ***CD-ROM ONLY***. 20 pp.

ISO DISP 10613 Pt 9-92. Information Technology—International Standardized Profile RA—Relaying the Connectionless-Mode Network Service—Part 9: Profile RA51.1121 ***CD-ROM ONLY***. 20 pp.

JTC1 DISP10612 Pt 2-92. Information Technology—International Standardized Profile RD5p.5q—Relaying the MAC Service Using Transparent Bridging—Part 2: CSMA/CD LAN Subnetwork-Dependent, Media-Dependent Requirements ***CD-ROM ONLY***. 32 pp.

JTC1 DISP10612 Pt 4-92. Information Technology—International Standardized Profile RD5p.5q—Relaying the MAC Service Using Transparent Bridging—Part 4: Profile RD51.51 (CSMA/CD-CSMA/CD) ***CD-ROM ONLY***. 28 pp.

JTC1 DISP10613 Pt 3-92. Information Technology—International Standardized Profile RA—Relaying the Connectionless-Mode Network Service—Part 3: ISO 8802-3 CSMA/CD LAN Subnetwork Dependent Media Dependent Requirements ***CD-ROM ONLY***. 22 pp.

JTC1 DISP10613 Pt 5-92. Information Technology—International Standardized Profile RA—Relaying the Connectionless-Mode Network Service—Part 5: Relaying the CLNS Between CSMA/CD LAN Subnetworks (RA51.51) ***CD-ROM ONLY***. 17 pp.

JTC1 DISP10613 Pt 8-92. Information Technology—International Standardized Profile RA—Relaying the Connectionless-Mode Network Service—Part 8: Profile RA51.1111 ***CD-ROM ONLY***. 20 pp.

JTC1 DISP10613 Pt 9-92. Information Technology—International Standardized Profile RA—Relaying the Connectionless-Mode Network Service—Part 9: Profile RA51.1121 ***CD-ROM ONLY***. 20 pp.

—Carrier Sense Multiple Access/CD—Subnetworks—Transport Services

BSI DD ENV 41103-2-93. 1993 Information Systems Interconnection—Provision of the OSI Connection-Mode Transport Service Using the OSI Connection-Mode Network Service in an End System Attached to a LAN Part 2: ISO 8802-3 CSMA/CD Subnetwork Dependent Media Reqmts.. 12 pp.

CEN ENV 41 101-86. Information Systems Interconnection: Local Area Networks; Provision of the OSI Connection-Mode Transport Service Using Connectionless-Mode Network Service on a CSMA/CD Single LAN. 16 pp.

CEN ENV 41 102-90. Information Systems Interconnection—Local Area Networks—Provision of the OSI Connection-Mode Transport Service Using the OSI Connectionless-Mode Network Service in an End System Attached to a CSMA/CD LAN. 61 pp.

CEN ENV 41 103-87. Information Systems Interconnection: Local Area Networks; Provision of the OSI Connection-Mode Transport Service and the OSI Connection-Mode Network Service in an End Systems on a CSMA/CD LAN. 23 pp.

CEN ENV 41103-2-92. Information Systems Interconnection—Provision of the OSI Connection-Mode Transport Service Using the OSI Connection-Mode Network Service in an End System Attached to a LAN—Part 2: ISO 8802-3 CSMA/CD Subnetwork Dependent Media Requirements. 10 pp.

CEN ENV 41103-2-92. Information Systems Interconnection—Provision of the OSI Connection-Mode Transport Service Using the OSI Connection-Mode Network Service in an End System Attached to a LAN Part 2: ISO 8802-3 CSMA/CD Subnetwork Dependent Media Requirements. 8 pp.

IEC ISP10608 Pt 2-92. Information Technology—International Standard-ized Profile TAnnnn—Connection-Mode Trans-port Service over Connectionless-Mode Network Service—Part 2: TA51 Profile Including Sub-network-Dependent Requirements for CSMA/CD Local Area Networks (LANs) First Edition. 15 pp.

ISO ISP10608 Pt 2-92. Information Technology—International Standard-ized Profile TAnnnn—Connection-Mode Trans-port Service over Connectionless-Mode Network Service—Part 2: TA51 Profile Including Sub-network-Dependent Requirements for CSMA/CD Local Area Networks (LANs) First Edition. 15 pp.

INDUSTRY STANDARDS
1545

INTERNATIONAL AND NON-U.S. NATIONAL STANDARDS SUBJECT INDEX

Local Area Networks (Cont.)

—Carrier Sense Multiple Access/CD—Subnetworks—Transport Services (Cont.)

JTC1 ISP 10608 Pt 2-92. Information Technology—International Standard-ized Profile TAnnnn—Connection-Mode Trans-port Service over Connectionless-Mode Network Service—Part 2: TA51 Profile Including Sub-network-Dependent Requirements for CSMA/CD Local Area Networks (LANs) First Edition. 15 pp.

—Class Nodes

BSI BS 6531: Part 4-84. (WITHDRAWN) 1984 10 Mbps Slotted Ring Local Area Network Part 4: Specification for Basic and Enhanced Class Nodes with Type 1 Node/DTE Interface (Superseded by BS ISO 8802-7: 1991). 31 pp.

—Connectors

BSI BS ISO/IEC TR 9578-90. 1990 Information Technology—Communication Interface Connectors Used in Local Area Networks. 49 pp.

IEC 807 Pt 8-92. Rectangular Connectors for Frequencies Below 3 MHz Part 8: Detail Specification for Connectors, Four-Signal Contacts and Earthing Contacts for Cable Screen First Edition. 98 pp.

IEC TR9578-90. Information Technology—Communication Interface Connectors Used in Local Area Networks First Edition. 47 pp.

ISO TR9578-90. Information Technology—Communication Interface Connectors Used in Local Area Networks First Edition. 47 pp.

JTC1 TR9578-90. Information Technology—Communication Interface Connectors Used in Local Area Networks First Edition. 47 pp.

OSI ISO/IEC TR 9578-90. Information Technology—Communication Interface Connectors Used in Local Area Networks. 47 pp.

SAA AS 4028-92. Information Technology--Communication Interface Connectors Used in Local Area Networks (ISO/IEC/TR 9578:1990) (in Professional Package 26A). 42 pp.

—Data Link Layer

ECMA ECMA 82-84. Local Area Networks (CSMA/CD Baseband) Link Layer. 40 pp.

ECMA ECMA-TR 14-82. Local Area Networks Layers 1 to 4—Architecture and Protocols. 18 pp.

—Data Processing Equipment

CSA CAN/CSA-C22. 2 NO 220-M91. Information Processing and Business Equipment; (Gen Instr 1). 119 pp.

—Data Terminal Equipment

BSI BS 6531: Part 4-84. (WITHDRAWN) 1984 10 Mbps Slotted Ring Local Area Network Part 4: Specification for Basic and Enhanced Class Nodes with Type 1 Node/DTE Interface (Superseded by BS ISO 8802-7: 1991). 31 pp.

BSI BS 6532: Part 1-84. (WITHDRAWN) 1984 Data Terminal Equipment for Attachment to 10 Mbps Slotted Ring Local Area Network Part 1: Specification for Media Access Control Procedures for Data Terminal Equipment (Superseded by BS ISO 8802-7: 1991). 30 pp.

BSI BS 6532: Part 2-84. (WITHDRAWN) 1984 Data Terminal Equipment for Attachment to 10 Mbps Slotted Ring Local Area Network Part 2: Specification for Implementation Requirements for Media Access Control in General Purpose Data Terminal Equipment. 16 pp.

—Distributed Processing Systems

ECMA ECMA-TR 21-84. Local Area Networks Interworking Units for Distributed Systems. 13 pp.

—Glossaries

BSI BS ISO/IEC 2382-25-92. 1992 Information Technology—Vocabulary—Part 25: Local Area Networks. 35 pp.

IEC 2382 Pt 25-92. Information Technology—Vocabulary—Part 25: Local Area Networks First Edition. 29 pp.

ISO 2382 Pt 25-92. Information Technology—Vocabulary—Part 25: Local Area Networks First Edition. 29 pp.

JTC1 2382-92. Information Technology—Vocabulary—Part 25: Local Area Networks First Edition. 29 pp.

OSI ISO/IEC DIS 2382-25.2-90. Information Technology—Vocabulary—Part 25: Local Area Networks. 53 pp.

SAA AS 1189.25-92. Data Processing—Vocabulary—Part 25: Local Area Networks (ISO/IEC 2382-25:1992) (in Professional Package 26C). 11 pp.

—Implementors

BSI DD 98: Part 2-84. (WITHDRAWN) 1984 CSMA/CD Local Area Network Part 2: Guidance for Implementors (superseded by BS ISO 8802-3: 1989). 102 pp.

—Implementors (Cont.)

BSI DD 99: Part 2-86. (WITHDRAWN) 1986 Logical Link Control for Local Area Networks Part 2: Guidance for Implementors (Superseded by BS ISO 8802-2: 1989). 33 pp.

BSI DD 136: Part 2-86. 1986 Token Ring Local Area Network Part 2: Guidance for Implementors. 46 pp.

—Integrated Services Digital Networks

CEPT T/SF 36-84. Relative Aux Services Et Facilities Pour Les Centraux Prives Aintegration De Services Et Pour Les Reseaux Locaux D'Entreprises (RLE) Relies Aun RNIS. 15 pp.

CEPT T/SF 45 E-86. Services and Facilities for Subscriber Complex Installations (Including ISPBX and LAN Installations) Connected to an Integrated Services Digital Network (ISDN). 6 pp.

CEPT T/SF 55 E-87. Supplementary Services Available to Subscriber Complex Installations Connected to an Integrated Services Digital Network (ISDN). 5 pp.

ECMA ECMA-TR 51-90. Requirements for Access to Integrated Voice and Data Local and Metropolitan Area Networks. 62 pp.

—Interfaces—Data Terminal Equipment—Voice/Data Systems

ECMA ECMA-TR 51-90. Requirements for Access to Integrated Voice and Data Local and Metropolitan Area Networks. 62 pp.

—Interfaces—Integrated Services Digital Networks

ECMA ECMA-TR 51-90. Requirements for Access to Integrated Voice and Data Local and Metropolitan Area Networks. 62 pp.

—Layer Management

ISO 8802 Pt 3 Draft AMD 7. Information Processing Systems—Local Area Networks—Part 3: Carrier Sense Multiple Access with Collision Detection (CSMA/CD) Access Method and Physical Layer Specifications Amendment 7: Addition of Layer Management (Section 5); (1991)***CD-ROM ONLY***. 40 pp.

JTC1 8802 Pt3 Draft AMD 7. Information Processing Systems—Local Area Networks—Part 3: Carrier Sense Multiple Access with Collision Detection (CSMA/CD) Access Method and Physical Layer Specifications Amendment 7: Addition of Layer Management (Section 5); (1991)***CD-ROM ONLY***. 40 pp.

—Logging Stations

BSI BS 6531: Part 6-84. (WITHDRAWN) 1984 10 Mbps Slotted Ring Local Area Network Part 6: Specification for Logging Station (Superseded by BS ISO 8802-7: 1991). 12 pp.

—Logical Link Control

BSI BS ISO 8802/2-89. 1989 Information Processing Systems—Local Area Networks Part 2: Logical Link Control. 118 pp.

BSI BS ISO/IEC TR 10178-92. 1992 Information Technology—Telecommunications and Information Exchange Between Systems—the Structure and Coding of Logical Link Control Addresses in Local Area Networks. 19 pp.

BSI DD 99: Part 1-85. (WITHDRAWN) 1985 Logical Link Control for Local Area Networks Part 1: Technical Specification (Superseded by BS ISO 8802-2: 1989). 104 pp.

IEC 8802 Pt 2-92. Corrigendum 1 Information Processing Systems—Local Area Networks—Part 2: Logical Link Control. 2 pp.

IEC 8802 Pt 2 Draft AMD 5. Information Processing Systems—Local Area Networks—Part 2: Logical Link Control Amendment 5: Bridged LAN Source Routeing Operation by End Systems; (1993) (IEEE 802.5p) ***CD-ROM ONLY***. 51 pp.

IEC TR10178-92. Information Technology—Telecommunications and Information Exchange Between Systems—the Structure and Coding of Logical Link Control Addresses in Local Area Networks First Edition. 17 pp.

ISO 8802 Pt 2-92. Corrigendum 1 Information Processing Systems—Local Area Networks—Part 2: Logical Link Control. 2 pp.

ISO 8802 Pt 2 Draft AMD 5. Information Processing Systems—Local Area Networks—Part 2: Logical Link Control Amendment 5: Bridged LAN Source Routeing Operation by End Systems; (1993) (IEEE 802.5p) ***CD-ROM ONLY***. 51 pp.

ISO TR10178-92. Information Technology—Telecommunications and Information Exchange Between Systems—the Structure and Coding of Logical Link Control Addresses in Local Area Networks First Edition. 17 pp.

JIS X 5251-89. Local Area Networks—Logical Link Control.

JTC1 8802-92. Corrigendum 1 Information Processing Systems—Local Area Networks—Part 2: Logical Link Control. 2 pp.

—Logical Link Control (Cont.)

JTC1 8802 Pt2 Draft AMD 5. Information Processing Systems—Local Area Networks—Part 2: Logical Link Control Amendment 5: Bridged LAN Source Routeing Operation by End Systems; (1993) (IEEE 802.5p) ***CD-ROM ONLY***. 51 pp.

OSI ISO 8802-2 CS28-0656. Information Processing Systems—Local Area Networks—Part 2: Logical Link Control. 16 pp.

OSI ISO 8802-2 DAM 4-90. Information Processing Systems—Local Area Networks—Part 2: Logical Link Control Amendment 4: Editorial Changes and Technical Corrections.

OSI ISO 8802-2 DAD 2-89. Information Processing Systems—Local Area Networks—Part 2: Logical Link Control Addendum 2: Acknowledged Connectionless-Mode Service and Protocol Type 3 Operation. 39 pp.

OSI ISO IEC 8802-2 DAD 1-88. Information Processing Systems—Local Area Networks—Part 2: Logical Link Control—Addendum 1: Flow Control Techniques for Bridged Local Area Networks. 5 pp.

OSI ISO DIS 8802-2.2-87. Information Processing Systems—Local Area Networks—Part 2: Logical Link Control. 94 pp.

OSI ISO DIS 8881-2.2-87. Information Processing Systems—Data Communications—Use of the X.2S Packet Level Protocol in Local Area Networks—Part 2: Use with LLC Type 2 Procedures. 10 pp.

OSI ISO IEC DIS 8881-3-88. Information Processing Systems—Data Communications—Use of the X.25 Packet Level Protocol in Local Area Networks. 19 pp.

SAA AS 4802.2-91. Information Processing Systems—Local Area Networks—Part 2: Logical Link Control (ISO 8802-2:1989) (in Professional Packages 26A, 26C). 101 pp.

SNZ NZS/ISO 8802. 2-90. Information Processing Systems—Local Area Networks Part 2: Logical Link Control. 101 pp.

—Media Access Control

BSI BS ISO/IEC 10039-91. 1991 Information Technology—Open Systems Interconnection—Local Area Networks—Medium Access Control (MAC) Service Definition. 22 pp.

IEC 10039-91. Information Technology—Open Systems Interconnection—Local Area Networks—Medium Access Control (MAC) Service Definition First Edition. 19 pp.

ISO 10039-91. Information Technology—Open Systems Interconnection—Local Area Networks—Medium Access Control (MAC) Service Definition First Edition. 19 pp.

JTC1 10039-91. Information Technology—Open Systems Interconnection—Local Area Networks—Medium Access Control (MAC) Service Definition First Edition. 19 pp.

SAA AS 4802.1 Supp 1-92. Information Processing Systems—Local Area Networks—Part 1: Overview and Architecture—Supplement 1: Media Access Control Bridges (Supplement to AS 4802.1—1992) (IEEE 802.1d—1990) (in Professional Package 26C). 162 pp.

SNZ NZS/AS 4094-93. Information Technology—Open Systems Interconnection—Local Area Networks—Medium Access Control Service Definition (This is a Joint Standard with SAA AS 4094). 13 pp.

—Media Access Control—Bridges (Electrical)

IEC DIS 10038-92. Information Technology—Telecommunications and Information Exchange Between Systems—Local Area Networks—Media Access Control (MAC) Bridges ***CD-ROM ONLY***. 163 pp.

IEC DIS 10038 Draft AMD 2. Information Technology—Telecommunications and Information Exchange Between Systems—Local Area Networks—Media Access Control (MAC) Bridges Amendment 2: Source Routeing Supplement; (1992) ***CD-ROM ONLY***. 44 pp.

IEC DISP 10612 Pt 1-92. Information Technology—International Standardized Profile RD5p.5q—Relaying the MAC Service Using Transparent Bridging—Part 1: Subnetwork-Independent Requirements ***CD-ROM ONLY***. 31 pp.

ISO DIS 10038-92. Information Technology—Telecommunications and Information Exchange Between Systems—Local Area Networks—Media Access Control (MAC) Bridges ***CD-ROM ONLY***. 163 pp.

ISO DIS 10038 Draft AMD 2. Information Technology—Telecommunications and Information Exchange Between Systems—Local Area Networks—Media Access Control (MAC) Bridges Amendment 2: Source Routeing Supplement; (1992) ***CD-ROM ONLY***. 44 pp.

ISO DISP 10612 Pt 1-92. Information Technology—International Standardized Profile RD5p.5q—Relaying the MAC Service Using Transparent Bridging—Part 1: Subnetwork-Independent Requirements ***CD-ROM ONLY***. 31 pp.

Local Area Networks *(Cont.)*

—Media Access Control—Bridges (Electrical) *(Cont.)*

JTC1 DISP10612 Pt 1-92. Information Technology—International Standardized Profile RD5p.5q—Relaying the MAC Service Using Transparent Bridging—Part 1: Subnetwork-Independent Requirements ***CD-ROM ONLY***. 31 pp.

—Monitors

BSI BS 6531: Part 5-84. (WITHDRAWN) 1984 10 Mbps Slotted Ring Local Area Network Part 5: Specification for Monitor (Superseded by BS ISO 8802-7: 1991). 12 pp.

—Network Layer

ECMA ECMA-TR 14-82. Local Area Networks Layers 1 to 4—Architecture and Protocols. 18 pp.

—Network Services

IEC DISP 10608 Pt 6-93. Information Technology—International Standardized Profile TAnnnn—Connection-Mode Transport Service over Connectionless-Mode Network Service—Part 6: Definition of Profile TA54 for Operation over an FDDI LAN Subnetwork ***CD-ROM ONLY***. 20 pp.

IEC DISP 10609 Pt 10-92. Information Technology—International Standardized Profiles TB, TC, TD and TE—Connection-Mode Transport Service over Connection-Mode Network Service—Part 10: LAN Subnetwork-Dependent, Media-Independent Requirements ***CD-ROM ONLY***. 71 pp.

IEC DISP 10609 Pt 11-92. Information Technology—International Standardized Profiles TB, TC, TD and TE—Connection-Mode Transport Service over Connection-Mode Network Service—Part 11: CSMA/CD Subnetwork-Dependent, Media-Dependent Requirements ***CD-ROM ONLY***. 37 pp.

ISO DISP 10608 Pt 6-93. Information Technology—International Standardized Profile TAnnnn—Connection-Mode Transport Service over Connectionless-Mode Network Service—Part 6: Definition of Profile TA54 for Operation over an FDDI LAN Subnetwork ***CD-ROM ONLY***. 20 pp.

ISO DISP 10609 Pt 10-92. Information Technology—International Standardized Profiles TB, TC, TD and TE—Connection-Mode Transport Service over Connection-Mode Network Service—Part 10: LAN Subnetwork-Dependent, Media-Independent Requirements ***CD-ROM ONLY***. 71 pp.

ISO DISP 10609 Pt 11-92. Information Technology—International Standardized Profiles TB, TC, TD and TE—Connection-Mode Transport Service over Connection-Mode Network Service—Part 11: CSMA/CD Subnetwork-Dependent, Media-Dependent Requirements ***CD-ROM ONLY***. 37 pp.

JTC1 DISP10608 Pt 6-93. Information Technology—International Standardized Profile TAnnnn—Connection-Mode Transport Service over Connectionless-Mode Network Service—Part 6: Definition of Profile TA54 for Operation over an FDDI LAN Subnetwork ***CD-ROM ONLY***. 20 pp.

JTC1 DISP10609 Pt 10-92. Information Technology—International Standardized Profiles TB, TC, TD and TE—Connection-Mode Transport Service over Connection-Mode Network Service—Part 10: LAN Subnetwork-Dependent, Media-Independent Requirements ***CD-ROM ONLY***. 71 pp.

JTC1 DISP10609 Pt 11-92. Information Technology—International Standardized Profiles TB, TC, TD and TE—Connection-Mode Transport Service over Connection-Mode Network Service—Part 11: CSMA/CD Subnetwork-Dependent, Media-Dependent Requirements ***CD-ROM ONLY***. 37 pp.

—Office Machines

CSA CAN/CSA-C22. 2 NO 220-M91. Information Processing and Business Equipment; (Gen Instr 1). 119 pp.

—Open Systems Interconnection

BSI BS ISO 8802/3-89. (WITHDRAWN) 1989 Carrier Sense Multiple Access with Collision Detection (CSMA/CD) Access Method and Physical Layer Specifications (Superseded by BS ISO/IEC 8802-3: 1992). 180 pp.

BSI BS ISO/IEC 8802-3-92. 1992 Information Technology—Local and Metropolitan Area Networks—Part 3: Carrier Sense Multiple Access with Collision Detection (CSMA/CD) Access Method and Physical Layer Specifications. 314 pp.

BSI BS ISO/IEC 8802-4-90. 1990 Information Processing Systems—Local Area Networks—Part 4: Token-Passing Bus Access Method and Physical Layer Specifications. 283 pp.

BSI BS ISO/IEC 10039-91. 1991 Information Technology—Open Systems Interconnection—Local Area Networks—Medium Access Control (MAC) Service Definition. 22 pp.

Local Area Networks *(Cont.)*

—Open Systems Interconnection *(Cont.)*

BSI DD ENV 41114-93. 1993 Information Technology—Functional Standard for Profile T/A52—Local Area Network—Token Bus (COTS + CLNS) (S). 66 pp.

BSI DD ENV 41801-3-92. 1992 Information Systems Interconnection—Relaying the Connectionless-Mode Network Service—Part 3: ISO 8802-3 CSMA/CD Subnetwork Dependent Media Dependent Requirements. 18 pp.

CEN ENV 41801-3-92. Information Systems Interconnection—Relaying the Connectionless-Mode Network Service—Part 3: ISO 8802-3 CSMA/CD Subnetwork Dependent Media Dependent Requirements. 26 pp.

IEC 8802 Pt 3 Draft AMD 6. Information Processing Systems—Local Area Networks—Part 3: Carrier Sense Multiple Access with Collision Detection (CSMA/CD) Access Method and Physical Layer Specifications Amendment 6; (1991) ***CD-ROM ONLY***. 5 pp.

IEC 8802 Pt 3 Draft AMD 7. Information Processing Systems—Local Area Networks—Part 3: Carrier Sense Multiple Access with Collision Detection (CSMA/CD) Access Method and Physi-cal Layer Specifications Amendment 7: Addition of Layer Management (Section 5); (1991) ***CD-ROM ONLY***. 40 pp.

IEC 8802 Pt 3 Draft AMD 9. Information Processing Systems—Local Area Networks—Part 3: Carrier Sense Multiple Access with Collision Detection (CSMA/CD) Access Method and Physical Layer Specifications Amendment 9: 10Base-T; (1991) ***CD-ROM ONLY***. 54 pp.

IEC 10039-91. Information Technology—Open Systems Interconnection—Local Area Networks—Medium Access Control (MAC) Service Definition First Edition. 19 pp.

ISO 10039-91. Information Technology—Open Systems Interconnection—Local Area Networks—Medium Access Control (MAC) Service Definition First Edition. 19 pp.

JIS X 5252-90. Local Area Networks—Carrier Sense Multiple Access with Collision Detection (CSMA/CD) Access Method and Physical Layer Specifications (ISO 8802/3:1989).

JTC1 10039-91. Information Technology—Open Systems Interconnection—Local Area Networks—Medium Access Control (MAC) Service Definition First Edition. 19 pp.

OSI ISO 8802-3 DAM 4-90. Part 3: Carrier Sense Multiple Access with Collision Detection (CSMA/CD) Access Method and Physical Layer Specifications Amendment 4: Physical Signalling, Medium Attachment and Baseband Medium Specifications, StarLAN, Type 1 Bases. 62 pp.

OSI ISO 8802-3 DAM 3-90. Information Processing Systems—Local Area Networks—Part 3: Carrier Sense Multiple Access with Collision Detection. Amendment 3: Broadband Medium Attachment Unit and Broadband Medium Specification, Type10 BROAD 36. 43 pp.

OSI ISO DIS 8802/3 PDAD-87. Local Area Networks—Part 3: CSMA/CD—Addendum Broadband Medium Attachment Unit and Broadband Medium Specifications, Type IOBROAD36.

OSI ISO IEC 8802-3 DAD 5-89. Information Processing Systems—Local Area Networks Part 3: Carrier Sense Multiple Access with Collision Detection Addendum 5: Medium Attachment Unit and Baseband Medium Specifi-cation for a Vendor Independent Fibre Optic Inter Repeater Link (FOIRL-Standard). 24 pp.

OSI ISO 8802-3 DAD 2-87. Information Processing Systems—Local Area Networks Part 3: Carrier Sense Multiple Access with Collision Detection Addendum 2: Repeater Set and Repeater Unit Specification for Use with 10 Base 5 and 10 Base 2 Networks. 19 pp.

OSI ISO 8802-3 DAD 1-87. Information Processing—Local Area Networks PT3: Carrier Sense Multiple Access with Collision Detection Addendum 1: Medium Attachment Unit and Baseband Medium Specifications for Type 10 Base 2. 36 pp.

OSI ISO DIS 8802-3 PDAD-87. Local Area Networks—Part 3: CSMA/CD—Addendum Broadband Medium Attachment Unit and Broadband Medium Specifications, Type IOBROAD36. 230 pp.

OSI ISO/IEC 8802-3-90. Information Processing Systems—Local Area Networks—Part 3: Carrier Sense Multiple Access with Collision Detection (CSMA/CD) Access Method and Physical Layer Specifications (with Correction Sheet 09-90). 217 pp.

OSI ISO 8802-3-89. Information Processing Systems—Local Area Networks Part 3: Carrier Sense Multiple Access with Collision Detection (CSMA/CD) Access Method and Physical Layer Specifications. 178 pp.

Local Area Networks *(Cont.)*

—Open Systems Interconnection *(Cont.)*

OSI ISO/IEC 8802-4-90. Information Processing Systems—Local Area Networks Part 4: Token-Passing Bus Access Method and Physical Layer Specifications (with Correction Sheet August 1990). 282 pp.

OSI ISO DIS 8802-5-87. Local Area Networks—Part 5: Token Ring Access Method and Physical Layer Specification. 63 pp.

OSI ISO/IEC DIS 8802 5.2-90. Information Processing Systems—Local Area Networks—Part 5: Token Ring Access Method and Physical Layer Specifications. 75 pp.

SAA AS 4802.3-91. Information Processing Systems—Local Area Networks—Part 3: Carrier Sense Multiple Access with Collision Detection (CSMA/CD) Access Method and Physical Layer Specifications (ISO/IEC 8802-3:1990/IEEE 802.3-1990) (in Prof. Packages 26A, 26C). 191 pp.

SAA AS 4802.3 Supp 1-91. Information Processing Systems—Local Area Networks—Part 3: Carrier Sense Multiple Access with Collision Detection (CSMA/CD) Access Method and Physical Layer Specifications—Supplement 1: Multisegment 10 Mb/s Baseband Networks:. 50 pp.

SNZ NZS/AS 4094-93. Information Technology—Open Systems Interconnection—Local Area Networks—Medium Access Control Service Definition (This is a Joint Standard with SAA AS 4094). 13 pp.

—Open Systems Interconnection—Bridges (Electrical)

IEC DIS 10038-92. Information Technology—Telecommunications and Information Exchange Between Systems—Local Area Networks—Media Access Control (MAC) Bridges ***CD-ROM ONLY***. 163 pp.

ISO DIS 10038-92. Information Technology—Telecommunications and Information Exchange Between Systems—Local Area Networks—Media Access Control (MAC) Bridges ***CD-ROM ONLY***. 163 pp.

ISO DIS 10038 Draft AMD 2. Information Technology—Telecommunications and Information Exchange Between Systems—Local Area Networks—Media Access Control (MAC) Bridges Amendment 2: Source Routeing Supplement; (1992) ***CD-ROM ONLY***. 44 pp.

—Packet Level Protocol

BSI BS ISO/IEC 8881-89. 1989 Information Processing Systems—Data Communications—Use of the X.25 Packet Level Protocol in Local Area Networks. 14 pp.

BSI BS ISO/IEC 8881-01. 1989 Amd 1 Information Processing Systems—Data Communications—Use of the X.25 Packet Level Protocol in Local Area Networks (AMD 7188) July 15, 1992. 15 pp.

CSA CAN/CSA-Z243. 136.3-90. Information Processing Systems—Data Communications—Use of the X.25 Packet Level Protocol in Local Area Networks (ISO/IEC 8881-1989 (E)). 20 pp.

IEC 8881-89. Information Processing Systems—Data Communications—Use of the X.25 Packet Level Protocol in Local Area Networks First Edition; (Corrigendum 1-1991). 12 pp.

ISO 8881-89. Information Processing Systems—Data Communications—Use of the X.25 Packet Level Protocol in Local Area Networks First Edition; (Corrigendum 1-1991). 12 pp.

JTC1 8881-89. Information Processing Systems—Data Communications—Use of the X.25 Packet Level Protocol in Local Area Networks First Edition; (Corrigendum 1-1991). 12 pp.

OSI ISO/TC/97/SC6/N 4661-87. Working Draft of Final Editor's Report on DIS 8881-1.2 and 881-2.2 (Use of X.25 PLP on LANs). 5 pp.

OSI ISO IEC 8881-89. Information Processing Systems—Date Communications—Use of the X.25 Packet Level Protocol in Local Area Networks. 10 pp.

OSI ISO DIS 8881-1.2-87. Information Processing Systems—Data Communications—Use of the X.2S Packet Level Protocol in Local Area Networks—Part 1: Use with LLC Type 1 Procedures. 14 pp.

OSI ISO DIS 8881-2.2-87. Information Processing Systems—Data Communications—Use of the X.2S Packet Level Protocol in Local Area Networks—Part 2: Use with LLC Type 2 Procedures. 10 pp.

OSI ISO IEC DIS 8881-3-88. Information Processing Systems—Data Communications—Use of the X.25 Packet Level Protocol in Local Area Networks. 19 pp.

—Physical Layer

BSI BS ISO/IEC 8802-3-92. 1992 Information Technology—Local and Metropolitan Area Networks—Part 3: Carrier Sense Multiple Access with Collision Detection (CSMA/CD) Access Method and Physical Layer Specifications. 314 pp.

ECMA ECMA 81-84. Local Area Networks (CSMA/CD Baseband) Physical Layer. 17 pp.

INDUSTRY STANDARDS
1547

Local
INTERNATIONAL AND NON-U.S. NATIONAL STANDARDS
SUBJECT INDEX

Local Area Networks *(Cont.)*

—Physical Layer *(Cont.)*

ECMA ECMA-TR 14-82. Local Area Networks Layers 1 to 4—Architecture and Protocols. 18 pp.

IEC 8802 Pt 3 Draft AMD 11. Information Technology—Local and Metropolitan Area Networks—Part 3: Carrier Sense Multiple Access with Collision Detection (CSMA/CD) Access Meth. and Phys. Layer Specs. Amd 11: Layer Mgmt. for 10 Mb/s Baseband Repeaters (Sec. 19); (1993) (IEEE Std. 802.3k) ***CD-ROM ONLY**. 31 pp.

ISO 8802 Pt 3 Draft AMD 6. Information Processing Systems—Local Area Networks—Part 3: Carrier Sense Multiple Access with Collision Detection (CSMA/CD) Access Method and Physical Layer Specifications Amendment 6; (1991) ***CD-ROM ONLY***. 5 pp.

ISO 8802 Pt 3 Draft AMD 7. Information Processing Systems--Local Area Networks—Part 3: Carrier Sense Multiple Access with Collision Detection (CSMA/CD) Access Method and Physical Layer Specifications Amendment 7: Addition of Layer Management (Section 5); (1991)***CD-ROM ONLY***. 40 pp.

ISO 8802 Pt 3 Draft AMD 9. Information Processing Systems—Local Area Networks—Part 3: Carrier Sense Multiple Access with Collision Detection (CSMA/CD) Access Method and Physical Layer Specifications Amendment 9: 10Base-T; (1991) ***CD-ROM ONLY***. 54 pp.

ISO 8802 Pt 3 Draft AMD 11. Information Technology—Local and Metropolitan Area Networks—Part 3: Carrier Sense Multiple Access with Collision Detection (CSMA/CD) Access Meth. and Phys. Layer Specs. Amd 11: Layer Mgmt. for 10 Mb/s Baseband Repeaters (Sec. 19); (1993) (IEEE Std. 802.3k) ***CD-ROM ONLY**. 31 pp.

JTC1 8802 Pt3 Draft AMD 6. Information Processing Systems—Local Area Networks—Part 3: Carrier Sense Multiple Access with Collision Detection (CSMA/CD) Access Method and Physical Layer Specifications Amendment 6; (1991) ***CD-ROM ONLY***. 5 pp.

JTC1 8802 Pt3 Draft AMD 7. Information Processing Systems—Local Area Networks—Part 3: Carrier Sense Multiple Access with Collision Detection (CSMA/CD) Access Method and Physical Layer Specifications Amendment 7: Addition of Layer Management (Section 5); (1991)***CD-ROM ONLY***. 40 pp.

JTC1 8802 Pt3 Draft AMD 9. Information Processing Systems—Local Area Networks—Part 3: Carrier Sense Multiple Access with Collision Detection (CSMA/CD) Access Method and Physical Layer Specifications Amendment 9: 10Base-T; (1991) ***CD-ROM ONLY***. 54 pp.

JTC1 8802 Pt3 Draft AMD 11. Information Technology—Local and Metropolitan Area Networks—Part 3: Carrier Sense Multiple Access with Collision Detection (CSMA/CD) Access Meth. and Phys. Layer Specs. Amd 11: Layer Mgmt. for 10 Mb/s Baseband Repeaters (Sec. 19); (1993) (IEEE Std. 802.3k) ***CD-ROM ONLY**. 31 pp.

OSI ISO 8802-3 DAM 3-90. Information Processing Systems—Local Area Networks—Part 3: Carrier Sense Multiple Access with Collision Detection. Amendment 3: Broadband Medium Attachment Unit and Broadband Medium Specification, Type10 BROAD 36. 43 pp.

OSI ISO DIS 8802-3-87. Information Processing Systems—Local Area Networks Part 3: Carrier Sense Multiple Access with Collision Detection—Access Method and Physical Layer Specifications. 5 pp.

OSI ISO 8802 3 DAD 1-87. Information Processing-Local Area Networks PT3: Carrier Sense Multiple Access with Collision Detection Addendum 1: Medium Attachment Unit and Baseband Medium Specifications for Type 10 Base 2.

OSI ISO 8802 3 DAD 2-87. Information Processing Systems-Local Area Networks Part 3: Carrier Sense Multiple Access with Collision Detection Addendum 2: Repeater Set and Repeater Unit Specification Foruse with 10 Base 5 and 10 Base 2 Networks.

OSI ISO DIS 8802 5 D-87. Local Area Networks-Part 5: Token Ring Access Method and Physical Layer Specifications.

OSI ISO DIS 8802-7-86. Local Area Networks: Part 7: Slotted Ring Access Method and Physical Layer Specification. 203 pp.

SNZ NZS/ISO 8802. 3-90. Information Processing Systems—Local Area Networks Part 3: Carrier Sense Multiple Access with Collision Detection (CSMA/CD) Access Method and Physical Layer Specifications. 191 pp.

—Power Supplies

BSI BS 6531: Part 7-84. (WITHDRAWN) 1984 10 Mbps Slotted Ring Local Area Network Part 7: Specification for Slave Power Supplies (Superseded by BS ISO 8802-7: 1991). 7 pp.

Local Area Networks *(Cont.)*

—Relay Function—Subnetworks

BSI DD ENV 41801-2-92. 1992 Information Systems Interconnection—Relaying the Connectionless-Mode Network Service—Part 2: LAN Subnetwork Dependent Media Independent Requirements. 35 pp.

BSI DD ENV 41802-2-92. 1992 Information Systems Interconnection—X.25 Protocol Relaying Part 2: LAN Subnetwork Dependent Media Independent Requirements. 60 pp.

CEN ENV 41801-2-92. Information Systems Interconnection—Relaying the Connectionless-Mode Network Service—Part 2: LAN Subnetwork Dependent Media Independent Requirements. 43 pp.

CEN ENV 41802-2-92. Information Systems Interconnection—X.25 Protocol Relaying—Part 2: LAN Subnetwork Dependent Media Independent Requirements. 56 pp.

IEC DISP 10613 Pt 2-92. Information Technology—International Standardized Profile RA—Relaying the Connectionless-Mode Network Service—Part 2: LAN Subnetwork Dependent Media Independent Requirements ***CD-ROM ONLY***. 20 pp.

ISO DISP 10613 Pt 2-92. Information Technology—International Standardized Profile RA—Relaying the Connectionless-Mode Network Service—Part 2: LAN Subnetwork Dependent Media Independent Requirements ***CD-ROM ONLY***. 20 pp.

JTC1 DISP10613 Pt 2-92. Information Technology—International Standardized Profile RA—Relaying the Connectionless-Mode Network Service—Part 2: LAN Subnetwork Dependent Media Independent Requirements ***CD-ROM ONLY***. 20 pp.

—Repeaters

BSI BS 6531: Part 3-84. (WITHDRAWN) 1984 10 Mbps Slotted Ring Local Area Network Part 3: Specification for Free-Standing Repeaters (Superseded by BS ISO 8802-7: 1991). 16 pp.

—Safety

ECMA ECMA 97-85. Local Area Networks—Safety Requirements. 21 pp.

ECMA ECMA 97-92. Local Area Networks—Safety Requirements. 18 pp.

—Slotted Rings

BSI BS 6531: Part 1-84. (WITHDRAWN) 1984 10 Mbps Slotted Ring Local Area Network Part 1: Specification for the Coding of Bits and Structure of Slots and Mini-Packets (Superseded by BS ISO 8802-7: 1991). 9 pp.

BSI BS 6531: Part 2-84. (WITHDRAWN) 1984 10 Mbps Slotted Ring Local Area Network Part 2: Specification for Configuration (Superseded by BS ISO 8802-7: 1991). 14 pp.

BSI BS 6531: Part 3-84. (WITHDRAWN) 1984 10 Mbps Slotted Ring Local Area Network Part 3: Specification for Free-Standing Repeaters (Superseded by BS ISO 8802-7: 1991). 16 pp.

BSI BS 6531: Part 4-84. (WITHDRAWN) 1984 10 Mbps Slotted Ring Local Area Network Part 4: Specification for Basic and Enhanced Class Nodes with Type 1 Node/DTE Interface (Superseded by BS ISO 8802-7: 1991). 31 pp.

BSI BS 6531: Part 5-84. (WITHDRAWN) 1984 10 Mbps Slotted Ring Local Area Network Part 5: Specification for Monitor (Superseded by BS ISO 8802-7: 1991). 12 pp.

BSI BS 6531: Part 6-84. (WITHDRAWN) 1984 10 Mbps Slotted Ring Local Area Network Part 6: Specification for Logging Station (Superseded by BS ISO 8802-7: 1991). 12 pp.

BSI BS 6531: Part 7-84. (WITHDRAWN) 1984 10 Mbps Slotted Ring Local Area Network Part 7: Specification for Slave Power Supplies (Superseded by BS ISO 8802-7: 1991). 7 pp.

BSI BS 6532: Part 1-84. (WITHDRAWN) 1984 Data Terminal Equipment for Attachment to 10 Mbps Slotted Ring Local Area Network Part 1: Specification for Media Access Control Procedures for Data Terminal Equipment (Superseded by BS ISO 8802-7: 1991). 30 pp.

BSI BS 6532: Part 2-84. (WITHDRAWN) 1984 Data Terminal Equipment for Attachment to 10 Mbps Slotted Ring Local Area Network Part 2: Specification for Implementation Requirements for Media Access Control in General Purpose Data Terminal Equipment. 16 pp.

OSI ISO DIS 8802-7-86. Local Area Networks: Part 7: Slotted Ring Access Method and Physical Layer Specification. 203 pp.

—Slotted Rings—Physical Layer

BSI BS ISO 8802-7-91. 1991 Information Technology—Local Area Networks—Part 7: Slotted Ring Access Method and Physical Layer Specification. 115 pp.

ISO 8802 Pt 7-91. Information Technology—Local Area Networks—Part 7: Slotted Ring Access Method and Physical Layer Specification First Edition. 112 pp.

Local Area Networks *(Cont.)*

—Slotted Rings—Physical Layer *(Cont.)*

JTC1 8802 Pt 7-91. Information Technology—Local Area Networks—Part 7: Slotted Ring Access Method and Physical Layer Specification First Edition. 112 pp.

OSI ISO 8802-7-91. Information Technology—Local Area Networks—Part 7: Slotted Ring Access Method and Physical Layer Specification.

OSI ISO DIS 8802-7-86. Local Area Networks: Part 7: Slotted Ring Access Method and Physical Layer Specification. 203 pp.

SNZ NZS 4407: Pt 4.2.1 CONT-91. Direct Transmission Mode Amend: 1, 1991.

—System Load Protocol

IEC DIS 15802 Pt 3-93. Information Technology—Local and Metropolitan Area Networks—Overview —Part 3: System Load Protocol ***CD-ROM ONLY***. 78 pp.

ISO DIS 15802 Pt 3-93. Information Technology—Local and Metropolitan Area Networks—Overview —Part 3: System Load Protocol ***CD-ROM ONLY***. 78 pp.

JTC1 DIS15802 Pt 3-93. Information Technology—Local and Metropolitan Area Networks—Overview —Part 3: System Load Protocol ***CD-ROM ONLY***. 78 pp.

SAA AS 4802.1 Supp 2-92. Information Processing Systems—Local Area Networks—Part 1: Overview and Architecture—Supplement 2: Systems Load Protocol (Supplement to AS 4802.1—1992) (IEEE 802.1e—1990) (in Professional Package 26C). 50 pp.

—Token Buses

BSI DD 100: Part 1-85. (WITHDRAWN) 1985 Token Bus Local Area Networks Part 1: Technical Specification (Superseded by BS ISO/IEC 8802-4: 1990). 287 pp.

BSI DD ENV 41114-93. 1993 Information Technology—Functional Standard for Profile T/A52—Local Area Network—Token Bus (COTS + CLNS) (S). 66 pp.

—Token Rings

BSI DD 99: Part 2-86. (WITHDRAWN) 1986 Logical Link Control for Local Area Networks Part 2: Guidance for Implementors (Superseded by BS ISO 8802-2: 1989). 33 pp.

BSI DD 136: Part 1-86. 1986 Token Ring Local Area Network Part 1: Technical Specification. 67 pp.

JIS X 5254-89. Local Area Networks—Token Ring Access Method and Physical Layer Specification.

—Token Rings—Physical Layer

JIS X 5254-89. Local Area Networks—Token Ring Access Method and Physical Layer Specification.

OSI ISO DIS 8802-5-87. Local Area Networks—Part 5: Token Ring Access Method and Physical Layer Specification. 63 pp.

OSI ISO/IEC DIS 8802 5.2-90. Information Processing Systems—Local Area Networks—Part 5: Token Ring Access Method and Physical Layer Specifications. 75 pp.

SNZ NZS/AS 4802. 5-92. Information Processing Systems—Local Area Networks Part 5: Token Ring Access Method and Physical Layer Specifications (This is a Joint Standard with SAA AS 4802.5). 94 pp.

—Token Rings—Transport Services

BSI DD ENV 41103-4-93. 1993 Information Systems Interconnection—Provision of the OSI Connection-Mode Transport Service Using the OSI Connection-Mode Network Service in an End System Attached to a LAN Part 4: Local Area Subnetwork-Type Dependent Requirements. 14 pp.

BSI DD ENV 41109-91. 1991 Information Systems Interconnection; Local Area Networks; Provision of the OSI Connection-Mode Transport Service Using Connectionless-Mode Network Service in an End System on a Token Ring Single LAN. 36 pp.

BSI DD ENV 41110-93. 1993 Information Technology—Functional Standard for Profile T/A53—Local Area Network—Token Ring (COTS + CLNS) (S). 17 pp.

CEN ENV 41103-4-92. Information Systems Interconnection—Prov. of the OSI Connection-Mode Transport Service Using the OSI Connection -Mode Network Service in an End System Att. to a LAN—Part 4: Local Area Subnetwork-Type Dependent Requirements for the Token Ring Med. Access Control Sublayer. 12 pp.

CEN ENV 41103-4-92. Information Systems Interconnection—Prov. of the OSI Conn.-Mode Trans. Serv. Using the OSI Conn.-Mode Network Serv. in an End System Att. to a LAN Part 4: Local Area Subnetwork-Type Dependent Reqmts. for the Token Ring Med. Access Contr. Sublayer and Physical Layer. 11 pp.

1548
INDEX and DIRECTORY of

Local Area Networks (Cont.)
—Token Rings—Transport Services (Cont.)
CEN ENV 41 108-88. Information Systems Interconnection: Local Area Networks; Provision of the OSI Connection-Mode Transport Service and the OSI Connection-Mode Network Service in an End System on a Token Ring LAN. 40 pp.

CEN ENV 41 109-88. Information Systems Interconnection; Local Area Networks; Provision of the OSI Connection-Mode. Transport Service using Connectionless-Mode Network Service in an End System on a Token Ring Single LAN. 33 pp.

CEN ENV 41 110-88. Information Systems Interconnection: Local Area Networks; Provision of the OSI Connection—Mode Transport Service and the OSI Connection Less—Mode Network Service in and End System on a Token Ring LAN in a Single or Multiple Configuration. 34 pp.

CEN ENV 41110-92. Information Technology—Functional Standard for Profile T/A53—Local Area Network—Token Ring (COTS + CLNS). 14 pp.

CEN ENV 41114-92. Information Technology—Functional Standard for Profile T/A52—Local Area Network—Token Bus (COTS + CLNS). 63 pp.

IEC DISP 10608 Pt 4-92. Information Technology—International Standardized Profile TAnnnn—Connection-Mode Transport Service over Connectionless-Mode Network Service—Part 4: Definition of Profile TA53 for Operation over a Token Ring LAN Subnetwork ***CD-ROM ONLY***. 19 pp.

IEC DISP 10608 Pt 13-92. Information Technology—International Standardized Profile TAnnnn—Connection-Mode Transport Service over Connectionless-Mode Network Service—Part 13: MAC Sublayer and for Physical Layer Dependent Requirements for Token Ring Local Area Network ***CD-ROM ONLY***. 20 pp.

ISO DISP 10608 Pt 4-92. Information Technology—International Standardized Profile TAnnnn—Connection-Mode Transport Service over Connectionless-Mode Network Service—Part 4: Definition of Profile TA53 for Operation over a Token Ring LAN Subnetwork ***CD-ROM ONLY***. 19 pp.

ISO DISP 10608 Pt 13-92. Information Technology—International Standardized Profile TAnnnn—Connection-Mode Transport Service over Connectionless-Mode Network Service—Part 13: MAC Sublayer and for Physical Layer Dependent Requirements for Token Ring Local Area Network ***CD-ROM ONLY***. 20 pp.

JTC1 DISP10608 Pt 4-92. Information Technology—International Standardized Profile TAnnnn—Connection-Mode Transport Service over Connectionless-Mode Network Service—Part 4: Definition of Profile TA53 for Operation over a Token Ring LAN Subnetwork ***CD-ROM ONLY***. 19 pp.

JTC1 DISP10608 Pt 13-92. Information Technology—International Standardized Profile TAnnnn—Connection-Mode Transport Service over Connectionless-Mode Network Service—Part 13: MAC Sublayer and for Physical Layer Dependent Requirements for Token Ring Local Area Network ***CD-ROM ONLY***. 20 pp.

—Transport Layer
ECMA ECMA-TR 14-82. Local Area Networks Layers 1 to 4—Architecture and Protocols. 18 pp.

—Transport Services
BSI DD ENV 41103-1-93. 1993 Information Systems Interconnection—Provision of the OSI Connection-Mode Transport Service Using the OSI Connection-Mode Network Service in an End System Attached to a LAN Part 1: LAN Subnetwork Dependent Media Independent Reqmts. 65 pp.

BSI DD ENV 41103-3-93. 1993 Information Systems Interconnection—Provision of the OSI Connection-Mode Transport Service Using the OSI Connection-Mode Network Service in an End System Attached to a LAN Part 3: Profile TC51 (M-IT-02 Profile T/611) (S). 9 pp.

BSI DD ENV 41103-5-93. 1993 Information Systems Interconnection—Provision of the OSI Connection-Mode Transport Service Using the OSI Connection-Mode Network Service in an End System Attached to a LAN Part 5: Profile TC53 (M-IT-02 Profile T/613) (S). 11 pp.

BSI DD ENV 41109-91. 1991 Information Systems Interconnection; Local Area Networks; Provision of the OSI Connection-Mode Transport Service Using Connectionless-Mode Network Service in an End System on a Token Ring Single LAN. 36 pp.

BSI DD ENV 41110-93. 1993 Information Technology—Functional Standard for Profile T/A53—Local Area Network—Token Ring (COTS + CLNS) (S). 17 pp.

CEN ENV 41103-1-92. Information Systems Interconnection—Provision of the OSI Connection-Mode Transport Service Using the OSI Connection-Mode Network Service in an End System Attached to a LAN-Part 1: LAN Subnetwork Dependent Media Independent Requirements. 62 pp.

CEN ENV 41103-3-92. Information Systems Interconnection—Provision of the OSI Connection-Mode Transport Service Using the OSI Connection-Mode Network Service in an End System Attached to a LAN—Part 3: Profile TC51 (M-IT-02 Profile T/ 611). 7 pp.

CEN ENV 41103-3-92. Information Systems Interconnection—Provision of the OSI Connection-Mode Transport Service Using the OSI Connection-Mode Network Service in an End System Attached to a LAN—Part 3: Profile TC51 (M-IT-02 Profile T/ 611). 6 pp.

CEN ENV 41103-5-92. Information Systems Interconnection—Provision of the OSI Connection-Mode Transport Service Using the OSI Connection-Mode Network Service in an End System Attached to a LAN—Part 5: Profile TC53 (M-IT-02 Profile T/613). 9 pp.

CEN ENV 41103-5-92. Information Systems Interconnection—Provision of the OSI Connection-Mode Transport Service Using the OSI Connection-Mode Network Service in an End System Attached to a LAN Part 5: Profile TC53 (M-IT-02 Profile T/ 613). 8 pp.

CEN ENV 41 108-88. Information Systems Interconnection: Local Area Networks; Provision of the OSI Connection-Mode Transport Service and the OSI Connection-Mode Network Service in an End System on a Token Ring LAN. 40 pp.

IEC DISP 10608 Pt 6-93. Information Technology—International Standardized Profile TAnnnn—Connection-Mode Transport Service over Connectionless-Mode Network Service—Part 6: Definition of Profile TA54 for Operation over an FDDI LAN Subnetwork ***CD-ROM ONLY***. 20 pp.

IEC DISP 10608 Pt 14-93. Information Technology—International Standardized Profile TAnnnn—Connection-Mode Transport Service over Connectionless-Mode Network Service—Part 14: MAC, PHY and PMD Sublayer Dependent and Stn. Mgmt. Reqmts. over an FDDI LAN Subnetwork ***CD-ROM ONLY***. 38 pp.

IEC DISP 10609 Pt 10-92. Information Technology—International Standardized Profiles TB, TC, TD and TE—Connection-Mode Transport Service over Connection-Mode Network Service—Part 10: LAN Subnetwork-Dependent, Media-Independent Requirements ***CD-ROM ONLY***. 71 pp.

IEC DISP 10609 Pt 11-92. Information Technology—International Standardized Profiles TB, TC, TD and TE—Connection-Mode Transport Service over Connection-Mode Network Service—Part 11: CSMA/CD Subnetwork-Dependent, Media-Dependent Requirements ***CD-ROM ONLY***. 37 pp.

ISO DISP 10608 Pt 6-93. Information Technology—International Standardized Profile TAnnnn—Connection-Mode Transport Service over Connectionless-Mode Network Service—Part 6: Definition of Profile TA54 for Operation over an FDDI LAN Subnetwork ***CD-ROM ONLY***. 20 pp.

ISO DISP 10608 Pt 14-93. Information Technology—International Standardized Profile TAnnnn—Connection-Mode Transport Service over Connectionless-Mode Network Service—Part 14: MAC, PHY and PMD Sublayer Dependent and Stn. Mgmt. Reqmts. over an FDDI LAN Subnetwork ***CD-ROM ONLY***. 38 pp.

ISO DISP 10609 Pt 10-92. Information Technology—International Standardized Profiles TB, TC, TD and TE—Connection-Mode Transport Service over Connection-Mode Network Service—Part 10: LAN Subnetwork-Dependent, Media-Independent Requirements ***CD-ROM ONLY***. 71 pp.

ISO DISP 10609 Pt 11-92. Information Technology—International Standardized Profiles TB, TC, TD and TE—Connection-Mode Transport Service over Connection-Mode Network Service—Part 11: CSMA/CD Subnetwork-Dependent, Media-Dependent Requirements ***CD-ROM ONLY***. 37 pp.

JTC1 DISP10608 Pt 6-93. Information Technology—International Standardized Profile TAnnnn—Connection-Mode Transport Service over Connectionless-Mode Network Service—Part 6: Definition of Profile TA54 for Operation over an FDDI LAN Subnetwork ***CD-ROM ONLY***. 20 pp.

JTC1 DISP10608 Pt 14-93. Information Technology—International Standardized Profile TAnnnn—Connection-Mode Transport Service over Connectionless-Mode Network Service—Part 14: MAC, PHY and PMD Sublayer Dependent and Stn. Mgmt. Reqmts. over an FDDI LAN Subnetwork ***CD-RON ONLY***. 38 pp.

JTC1 DISP10609 Pt 10-92. Information Technology—International Standardized Profiles TB, TC, TD and TE—Connection-Mode Transport Service over Connection-Mode Network Service—Part 10: LAN Subnetwork-Dependent, Media-Independent Requirements ***CD-ROM ONLY***. 71 pp.

JTC1 DISP10609 Pt 11-92. Information Technology—International Standardized Profiles TB, TC, TD and TE—Connection-Mode Transport Service over Connection-Mode Network Service—Part 11: CSMA/CD Subnetwork-Dependent, Media-Dependent Requirements ***CD-ROM ONLY***. 37 pp.

OSI ISO/IEC DISP 10608-2-90. Information Technology—International Standardized Profile TAnnnn—Connection-Mode Transport Service over Connectionless-Mode Network Service—Part 2: TA51 Profile Including Subnetwork-Dependent Requirements for CSMS/CD Local Area Networks (LANs). 37 pp.

Local Cables
See Also: Cables (Electric); Communication Cables

—Copper Wire
IEC 488-74. Dimensions of Copper Conductors in Local Cables First Edition. 4 pp.

Local Exchanges
See Also: Telephone Exchanges

CEPT T/CS 62-01-86. Introduction, Domaine D'Application Et Fonctions De Base Dans Les Commutateurs Numeriques Principaux D'Abonne Ou Mixtes. 4 pp.

CEPT T/CS 62-01 E-86. Introduction, Field of Application and Basic Functions for Digital Local and Combined Exchanges. 4 pp.

CEPT T/S 60-00 E-88. Introduction and Field of Application for Digital Local, Transit and Combined Exchanges. 1 p.

—Call Handling
CEPT T/CS 62-03-86. Connexions, Signalisation, Commande, Traitement Des Appels Et Fonctions Auxiliaires Pour Commutateurs Numeriques Principaux D'Abonne Ou Mixtes. 2 pp.

CEPT T/CS 62-03 E-86. Connections, Signalling, Control, Call Handling and Ancillary Functions for Digital Local and Combined Exchanges. 2 pp.

—Design
CCITT RECMN Q.541-89. Digital Exchange Design Objectives—General—Digital Local, Transit, Combined and International Exchanges in Integrated Digital Networks and Mixed Analogue-Digital Networks—Supplements (Study Group XI) 6 pp. 6 pp.

CCITT RECMN Q.542-89. Digital Exchange Design Objectives—Operations and Maintenance—Digital Local, Transit, Combined and International Exchanges in Integrated Digital Networks and Mixed Analogue-Digital Networks—Supplements (Study Group XI) 21 pp. 21 pp.

CCITT RECMN Q.543-89. Digital Exchange Performance Design Objectives—Digital Local, Transit, Combined and International Exchanges in Integrated Digital Networks and Mixed Analogue-Digital Networks—Supplements (Study Group XI) 40 pp. 40 pp.

CEPT T/CS 62-04-86. Objectifs Nominaux De Qualite Et De Disponibilite Pour Commutateurs Numeriques Principaux D'Abonne Ou Mixtes. 5 pp.

CEPT T/CS 62-04 E-86. Performance and Availability Design Ojectives for Digital Local and Combined Exchanges. 5 pp.

CEPT T/S 64-30 E-88. Digital Exchange Performance Design Objectives. 1 p.

—Integrated Digital Networks
CCITT RECMN Q.500-89. Digital Local, Combined, Transit and International Exchanges Introduction and Field of Application—Digital Local, Transit, Combined and International Exchanges in Integrated Digital Networks and Mixed Analogue-Digital Networks—Supplements (Study Group XI) 4 pp. 4 pp.

CCITT RECMN Q.521-89. Exchange Functions—Digital Local, Transit, Combined and International Exchanges in Integrated Digital Networks and Mixed Analogue-Digital Networks—Supplements (Study Group XI) 3 pp. 3 pp.

—Integrated Services Digital Networks
CCITT RECMN Q.500-89. Digital Local, Combined, Transit and International Exchanges Introduction and Field of Application—Digital Local, Transit, Combined and International Exchanges in Integrated Digital Networks and Mixed Analogue-Digital Networks—Supplements (Study Group XI) 4 pp. 4 pp.

Local Exchanges (Cont.)
—Integrated Services Digital Networks (Cont.)
CCITT RECMN Q.521-89. Exchange Functions—Digital Local, Transit, Combined and International Exchanges in Integrated Digital Networks and Mixed Analogue-Digital Networks—Supplements (Study Group XI) 3 pp. 3 pp.

CCITT RECMN Q.522-89. Digital Exchange Connections, Signalling and Ancillary Functions—Digital Local, Transit, Combined and International Exchanges in Integrated Digital Networks and Mixed Analogue-Digital Networks—Supplements (Study Group XI) 16 pp. 16 pp.

CEPT T/GSI 03-09-87. Commutation. 1 p.

CEPT T/GSI 03-09 E-87. Switching. 1 p.

—Interfaces
CCITT RECMN Q.511-89. Exchange Interfaces Towards Other Exchanges—Digital Local, Transit, Combined and International Exchanges in Integrated Digital Networks and Mixed Analogue-Digital Networks—Supplements (Study Group XI) 4 pp. 4 pp.

CEPT T/CS 62-02-86. Interfaces Pour Commutateurs Numeriques Principaux D'Abonne Ou Mixtes. 2 pp.

CEPT T/CS 62-02 E-86. Interfaces for Digital Local and Combined Exchanges. 2 pp.

CEPT T/S 61-10 E-88. Exchange Interfaces Towards Other Exchanges. 1 p.

CEPT T/S 61-20 E-88. Exchange Interfaces to Subscribers. 9 pp.

—Interfaces—Operations, Administration and Maintenance
CCITT RECMN Q.513-89. Exchange Interfaces for Operations, Administration and Maintenance—Digital Local, Transit, Combined and International Exchanges in Integrated Digital Networks and Mixed Analogue-Digital Networks—Supplements (Study Group XI) 4 pp. 4 pp.

—Interfaces—Subscriber Access
CCITT RECMN Q.512-89. Exchange Interfaces for Subscriber Access—Digital Local, Transit, Combined and International Exchanges in Integrated Digital Networks and Mixed Analogue-Digital Networks—Supplements (Study Group XI) 9 pp. 9 pp.

—Interfaces—Transmission
CCITT RECMN Q.552-89. Transmission Characteristics at 2-Wire Analogue Interfaces of Digital Exchange—Digital Local, Transit, Combined and International Exchanges in Integrated Digital Networks and Mixed Analogue-Digital Networks—Supplements (Study Group XI) 25 pp. 25 pp.

CCITT RECMN Q.553-89. Transmission Characteristics at 4-Wire Analogue Interfaces of a Digital Exchange—Digital Local, Transit, Combined and International Exchanges in Integrated Digital Networks and Mixed Analogue-Digital Networks—Supplements (Study Group XI) 12 pp. 12 pp.

CCITT RECMN Q.554-89. Transmission Characteristics at Digital Interfaces of a Digital Exchange—Digital Local, Transit, Combined and International Exchanges in Integrated Digital Networks and Mixed Analogue-Digital Networks—Supplements (Study Group XI) 3 pp. 3 pp.

—Interfaces—V5.1—Access Networks
CENELEC PRETS 300 324-1-93. Signalling Protocols and Switching (SPS); V5.1 Interface Specification for the Support of Access Network. 270 pp.

ETSI PRETS 300 324-1-93. Signalling Protocols and Switching (SPS); V5.1 Interface Specification for the Support of Access Network. 270 pp.

—Maintenance
CCITT RECMN Q.542-89. Digital Exchange Design Objectives—Operations and Maintenance—Digital Local, Transit, Combined and International Exchanges in Integrated Digital Networks and Mixed Analogue-Digital Networks—Supplements (Study Group XI) 21 pp. 21 pp.

CEPT T/CS 62-06-86. Fonctions D'Exploitation Et De Maintenance Pour Commutateurs Numeriques Principaux D'Abonne Ou Mixtes. 2 pp.

CEPT T/CS 62-06 E-86. Digital Local and Combined Exchange Operation and Maintenance Functions. 2 pp.

—Measurement
CCITT RECMN Q.544-89. Digital Exchange Measurements—Digital Local, Transit, Combined and International Exchanges in Integrated Digital Networks and Mixed Analogue-Digital Networks—Supplements (Study Group XI) 15 pp. 15 pp.

Local Exchanges (Cont.)
—Mixed Analog/Digital Networks
CCITT RECMN Q.500-89. Digital Local, Combined, Transit and International Exchanges Introduction and Field of Application—Digital Local, Transit, Combined and International Exchanges in Integrated Digital Networks and Mixed Analogue-Digital Networks—Supplements (Study Group XI) 4 pp. 4 pp.

CCITT RECMN Q.521-89. Exchange Functions—Digital Local, Transit, Combined and International Exchanges in Integrated Digital Networks and Mixed Analogue-Digital Networks—Supplements (Study Group XI) 3 pp. 3 pp.

—Operation
CCITT RECMN Q.542-89. Digital Exchange Design Objectives—Operations and Maintenance—Digital Local, Transit, Combined and International Exchanges in Integrated Digital Networks and Mixed Analogue-Digital Networks—Supplements (Study Group XI) 21 pp. 21 pp.

CEPT T/CS 62-06-86. Fonctions D'Exploitation Et De Maintenance Pour Commutateurs Numeriques Principaux D'Abonne Ou Mixtes. 2 pp.

CEPT T/CS 62-06 E-86. Digital Local and Combined Exchange Operation and Maintenance Functions. 2 pp.

—Overload Protection
CCITT RECMN K.11-91. Principles of Protection Against Overvoltages and Overcurrents (Study Group V) 14 pp. 14 pp.

—Overvoltage Protection
CCITT RECMN K.11-91. Principles of Protection Against Overvoltages and Overcurrents (Study Group V) 14 pp. 14 pp.

—Signaling Systems
CEPT T/CS 62-03-86. Connexions, Signalisation, Commande, Traitement Des Appels Et Fonctions Auxiliaires Pour Commutateurs Numeriques Principaux D'Abonne Ou Mixtes. 2 pp.

CEPT T/CS 62-03 E-86. Connections, Signalling, Control, Call Handling and Ancillary Functions for Digital Local and Combined Exchanges. 2 pp.

CEPT T/S 62-20 E-88. Digital Exchange Connections, Signalling and Ancillary Functions. 5 pp.

—Signaling Systems—Interworking
CCITT RECMN Q.522-89. Digital Exchange Connections, Signalling and Ancillary Functions—Digital Local, Transit, Combined and International Exchanges in Integrated Digital Networks and Mixed Analogue-Digital Networks—Supplements (Study Group XI) 16 pp. 16 pp.

—Telephone Circuits
CCITT RECMN G.101-89. The Transmission Plan—General Characteristics of International Telephone Connections and Circuits (Study Groups XII and XV) 17 pp. 17 pp.

—Traffic—Measurement
CEPT T/CS 54-12-84. Mesures De Trafic. 23 pp.

CEPT T/CS 54-12 E-84. Traffic Measurements. 22 pp.

—Traffic Units
CCITT RECMN Q.544-89. Digital Exchange Measurements—Digital Local, Transit, Combined and International Exchanges in Integrated Digital Networks and Mixed Analogue-Digital Networks—Supplements (Study Group XI) 15 pp. 15 pp.

—Transmission
CCITT RECMN Q.551-89. Transmission Characteristics of Digital Exchanges—Digital Local, Transit, Combined and International Exchanges in Integrated Digital Networks and Mixed Analogue-Digital Networks—Supplements (Study Group XI) 14 pp. 14 pp.

Local Loop Plants
See Also: Loop Transmission; Loopback Testing Equipment

—Overload Protection
CCITT RECMN K.11-91. Principles of Protection Against Overvoltages and Overcurrents (Study Group V) 14 pp. 14 pp.

—Overvoltage Protection
CCITT RECMN K.11-91. Principles of Protection Against Overvoltages and Overcurrents (Study Group V) 14 pp. 14 pp.

Local Node Clocks
Use: Slave Clocks

Localisers
Use: Localizers

Localizers
Use For: LOC *See Also:* Air Navigation Equipment; Aircraft Equipment; Instrument Landing Systems
CAA Chapter R4-6 App #4 04.74. Tests for VOR, ILS and Marker Systems (Radio). 2 pp.

—Design
EUROCAE ED-51 10.83. MPS for Airborne Automatic Direction Finding Equipment. 31 pp.

—Electrical Properties
EUROCAE ED-51 10.83. MPS for Airborne Automatic Direction Finding Equipment. 31 pp.

—Environmental Testing
EUROCAE ED-51 10.83. MPS for Airborne Automatic Direction Finding Equipment. 31 pp.

—Radio Receivers
EUROCAE ED-51 10.83. MPS for Airborne Automatic Direction Finding Equipment. 31 pp.

Locating Pins
See Also: Dowels; Jigs (Positioners); Pins
CNS B3317-83. Supporting Pin and Locating Pin (Mar)(5059).

DIN ENGL 6321-73. Support and Location Pins (Dec). 2 pp.

—Aerospace—Hole Size
AECMA PREN3368-90. Aerospace Design Standard Holes for Locating Pins. 8 pp.

—Aircraft
SBAC RS 686 (V). Application of Locating Pins.

—Dies
CNS B2785-87. Locating Pin of Die Casting Molds (Sep)(12072).

—Jigs (Positioners)
JIS B 5216-89. Locating Pins for Jigs and Fixtures. 8 pp.

Location
Use: Position (Location)

Location Indicators (Aeronautical)
Use: Aeronautical Information Services—Location Indicators

Location Registers
See Also: Address Codes; Identification Systems; Location Registration; Mobile Radio Services; Mobile Stations (Communications)

—Data Restoration
CENELEC GSM 03.07-92. European Digital Cellular Telecommunication System (Phase 1); Restoration Procedures. 11 pp.

ETSI GSM 03.07-92. European Digital Cellular Telecommunication System (Phase 1); Restoration Procedures. 11 pp.

—Home—Cellular Mobile Radio Equipment
CENELEC GSM 11.31-92. European Digital Cellular Telecommunication System (Phase 1); Home Location Register Specification. 11 pp.

ETSI GSM 11.31-92. European Digital Cellular Telecommunication System (Phase 1); Home Location Register Specification. 11 pp.

—Public Land Mobile Networks
CCITT RECMN Q.1003-89. Location Registration Procedures—Public Land Mobile Network Interworking with ISDN and PSTN (Study Group XI) 18 pp. 18 pp.

CCITT RECMN Q.1004-89. Location Register Restoration Procedures—Public Land Mobile Network Interworking with ISDN and PSTN (Study Group XI) 4 pp. 4 pp.

—Subscriber Data
CENELEC GSM 03.08-92. European Digital Cellular Telecommunication System (Phase 1); Organization of Subscriber Data. 20 pp.

ETSI GSM 03.08-92. European Digital Cellular Telecommunication System (Phase 1); Organization of Subscriber Data. 20 pp.

—Visitor—Cellular Mobile Radio Equipment
CENELEC GSM 11.32-92. European Digital Cellular Telecommunication System (Phase 1); Visitor Location Register Specification. 15 pp.

ETSI GSM 11.32-92. European Digital Cellular Telecommunication System (Phase 1); Visitor Location Register Specification. 15 pp.

Location Registration
See Also: Location Registers; Mobile Stations (Communications)

INTERNATIONAL AND NON-U.S. NATIONAL STANDARDS
SUBJECT INDEX
Lockheed

Location Registration *(Cont.)*
—Maritime Mobile Services— Radiotelephony
CCIR RECMN 586-1-86. Automated VHF/UHF Maritime Mobile Telephone System—Section 8C—Maritime Mobile Service; Telephony and Related Subjects. 44 pp.
CCIR RECMN 587-1-86. Coast Station Identities and Initiation of Location Registration in an Automated VHF/UHF Maritime Mobile Telephone System—Section 8C—Maritime Mobile Service; Telephony and Related Subjects. 1 p.

—Public Land Mobile Networks
CCIR RECMN 624-86. Public Land Mobile Communication Systems Location Registration—Section 8A—Land Mobile Service and Related Subjects. 5 pp.
CCITT RECMN Q.1003-89. Location Registration Procedures—Public Land Mobile Network Interworking with ISDN and PSTN (Study Group XI) 18 pp. 18 pp.
CENELEC GSM 03.03-92. European Digital Cellular Telecommunication System (Phase 1); Numbering, Addressing and Identification. 16 pp.
CENELEC GSM 03.12-92. European Digital Cellular Telecommunication System (Phase 1); Location Registration Procedures. 11 pp.
CENELEC GSM 03.12-DCS-92. European Digital Cellular Telecommunication System (Phase 1); Location Registration Procedures. 8 pp.
ETSI GSM 03.03-92. European Digital Cellular Telecommunication System (Phase 1); Numbering, Addressing and Identification. 23 pp.
ETSI GSM 03.12-92. European Digital Cellular Telecommunication System (Phase 1); Location Registration Procedures. 11 pp.
ETSI GSM 03.12-DCS-92. European Digital Cellular Telecommunication System (Phase 1); Location Registration Procedures. 8 pp.

Lock Caps
Use: Lock Covers

Lock Cases
See Also: Locks (Security)
—Doors
CNS A2012-89. Lock Cases for Doors (Jul)(862). 2 pp.

Lock Covers
Use For: Lock Caps *See Also:* Locks (Security)
—Doors
CNS A2011-89. Coverplate for Door Locks (Jun)(861). 2 pp.
CNS A2013-89. Lock Mouth Covers for Door Locks (Jun)(863). 2 pp.

Lock Gates
See Also: Gate Valves; Gates (Barriers); Locks (Waterways)
—Design
BSI BS 6349: Part 3-88. 1988 Code of Practice for Maritime Structures; Part 3: Design of Dry Docks, Locks, Slipways and Shipbuilding Berths, Shiplifts and Dock and Lock Gates. 75 pp.

Lock Nuts
Use: Locknuts

Lock Plates
Use: Lockplates

Lock Tooth Washers
Use For: Star Washers; Tooth Lock Washers
See Also: Fasteners; Lock Washers; Washers (Fasteners)
CNS B2223-78. Toothed Lock Washers (Jun)(4402).
DIN ENGL 6797-88. Toothed Lock Washers (July). 4 pp.
DIN ENGL 6906-72. Lock Washers for Screw Assemblies (Dec). 2 pp.
JIS B 1255-77. Toothed Lock Washers. 10 pp.
SAA AS 1969-76. Tooth Lock Washers (Metric Series). 12 pp.

—Aerospace—Serrated—Control Rods
AECMA PREN2327-87. Washers, Lock with Radial Serrations in Alloy Steel Dimensions.
AECMA PREN2546-81. Washers, Lock, with Radial Serrations, in Corrosion Resisting Steel—Dimensions. 5 pp.
AECMA PREN2596-84. Washers, Lock, with Radial Serrations, in Corrosion Resisting Steel, Cadmium Plated Dimensions. 5 pp.
BSI BS EN 2596-90. 1990 Washers, Lock, with Radial Serrations, in Corrosion Resisting Steel, Cadmium Plated. Dimensions. 8 pp.

Lock Tooth Washers *(Cont.)*
—Aerospace—Serrated—Control Rods *(Cont.)*
CEN PREN 2546-87. Washers, Lock with Radial Serrations in Corrosion Resisting Still Dimensions. 5 pp.
CEN EN 2596-88. Washers, Lock, with Radial Serrations in Corrosion Resisting Steel, Cadmium Plated Dimensions. 7 pp.

—External—Aircraft—Countersunk—Steel
SBAC AGS 2036 ISSUE 8. Lock/Washers Shakeproof Countersunk, External Teeth, Steel.

—External—Aircraft—Steel
SBAC AGS 2034 ISSUE 9. Lockwashers Shakeproof Flat, External Teeth, Steel.

—Internal—Aircraft—Bronze
SBAC AGS 2037 ISSUE 6(I). Lockwashers Shakeproof Flat, Internal Teeth, Phos. Bronze.

—Internal—Aircraft—Steel
SBAC AGS 2035 ISSUE 9. Lockwasher Shakeproof Flat, Internal Teeth, Steel.

—Screw Assemblies
CNS B2224-78. Toothed Lock Washers for Screw Assemblies (Jun)(4403).
CNS B2228-78. Curved Spring Washers for Screw Assemblies (Jun)(4407).
DIN ENGL 6900 Pt 4-90. Screw and Washer Assemblies; Coarse Threaded Screws with Captive Serrated Lock Washer (Dec). 3 pp.
DIN ENGL 6907-90. Serrated Lock Washers with External Teeth for Screw and Washer Assemblies (Dec). 3 pp.

—Serrated
CNS B2225-78. Serrated Lock Washers (Jun)(4404).
DIN ENGL 6798-88. Serrated Lock Washers (July). 4 pp.

—Serrated—Screw Assemblies
CNS B2226-78. Serrated Lock Washers for Screw Assemblies (Jun)(4405).

Lock Washers
See Also: Fasteners; Lock Tooth Washers; Spring Lock Washers; Tab Washers; Washers (Fasteners)
MOD UK DSTAN 53-31: Part 2-01. Washers for Locking Purposes Part 2: Washer, Lock, Inch (Single Coil, Double Coil, and Shakeproof) Issue 1; Amendment 2. 50 pp.

—Aerospace—Alloy Steel—Control Rods
BSI BS EN 2327-89. 1989 Washers, Lock with Radial Serrations in Alloy Steel. Dimensions. 8 pp.
CEN EN 2327-87. Washers, Lock with Radial Serrations in Alloy Steel Dimensions. 5 pp.

—Aerospace—Aluminum Alloys
CEN PREN 3904-92. Washers, Bent Tab, in Aluminium Alloy 2024, Clad, Anodized. 5 pp.

—Aerospace—Control Rods
AECMA PREN2586-86. Washers, Lock for Flight Control Rods Dimensions. 3 pp.
BSI BS EN 2586-91. 1991 Washers, Lock for Flight Control Rods—Dimensions. 9 pp.
CEN EN 2586-91. Washers, Lock for Flight Control Rods—Dimensions. 5 pp.

—Aerospace—Corrosion Resistant Steel—Control Rods
BSI BS EN 2546-90. 1990 Washers, Lock with Radial Serrations in Corrosion Resisting Steel. Dimensions. 8 pp.
CEN EN 2546-88. Washers, Lock with Radial Serrations in Corrosions Resisting Steel Dimensions. 5 pp.

—Aerospace—Steel
AECMA PREN3432-89. Lockwashers Steel, Cadmium Plated. 6 pp.
AECMA PREN3433-89. Lockwashers Heat Resisting Steel, Passivated. 6 pp.

—Aircraft—Brake Lock
SBAC AS 2951 ISSUE 2. Lock Washer Brake Lock Adjustment. 1 p.

—Aircraft—Cup
SBAC AS 8690-8699 ISSUE 3(I). Washer, Cup, Lock.

—Aircraft—Terminals
BSI G 203-67. 1967 Unified Screws with Captive Facing and Locking Washers. 4 pp.

—Antifriction Bearings
BSI BS 5646: Part 1-78. 1978 Rolling Bearings—Accessories Part 1: Locknuts, Narrow Series, and Lockwashers with Straight Inner Tab. 4 pp.

Lock Washers *(Cont.)*
—Antifriction Bearings *(Cont.)*
BSI BS 5646: Part 2-78. 1978 Rolling Bearings—Accessories Part 2: Locknuts, Wide Series, and Lockwashers with Bent Inner Tab. 4 pp.
CNS B2720-82. Rolling Bearings Accessories (Lockwashers) (Nov)(9569).
DIN ENGL 5406-77. Rolling Bearing Accessories; Lockwashers (Apr). 2 pp.
ISO 2982-72. Rolling Bearings—Locknuts, Narrow Series, and Lockwashers with Straight Inner Tab First Edition. 4 pp.
ISO 2983-75. Rolling Bearings—Locknuts, Wide Series, and Lockwashers with Bent Inner Tab First Edition. 4 pp.
JIS B 1554-93. Rolling Bearings—Locknuts, Lockwashers and Lockplates. 20 pp.
JIS B 1555-79. Lock Washers and Lock Plates for Rolling Bearings. 10 pp.

—Rectangular—Bolts
CNS B2018-47. Rectangular Lock Washers for Metric Bolts and Nuts (M3-M52) (Mar)(158)(R 1970).

—Rectangular—Nuts (Fasteners)
CNS B2018-47. Rectangular Lock Washers for Metric Bolts and Nuts (M3-M52) (Mar)(158)(R 1970).

—Shafts (Machine Elements)
DIN ENGL 6799-81. Lock Washers (Retaining Washers) for Shafts (Sept). 7 pp.

Lockers
Use For: Clothes Lockers *See Also:* Office Furniture; Recreational Facilities
BSI BS 4680-71. 1971 Clothes Lockers. 18 pp.
SNZ NZS 1187-69. Specification for Clothes Lockers Amend: 1, 1976 (Reconfirmed 1982). 10 pp.

—Hospital Bedside
BSI BS 1765: Part 1-90. 1990 Hospital Bedside Lockers Part 1: General Purpose Bedside Lockers for Patients. 13 pp.
BSI BS 1765: Part 2-76. 1976 Hospital Bedside Lockers Part 2: General Purpose Lockers of Wooden Construction with Facilities for Hanging Day Clothes. 8 pp.

—Office—Steel
CNS S1068-83. Office Steel Lockers (Jun)(2997).

—Schools
JIS S 1084-81. School Furnitures (Lockers for Pupil).

—Schools—Stability
BSI BS 5873: Part 4-91. 1991 Educational Furniture Part 4: Specification for Strength and Stability of Storage Furniture for Educational Institutions. 24 pp.

—Steel
CGSB CAN/CGSB-44.40-92. Steel Clothing Locker. 10 pp.

Lockheed Constellation Aircraft
See Also: Aircraft
—Accidents
CAA. Lockheed Constellation and Super Constellation (World Airline Accident Summary). 3 pp.
CAA. Lockheed Constellation and Super Constellation. 1 p.

Lockheed Hudson Aircraft
See Also: Aircraft
—Accidents
CAA. Lockheed Hudson (World Airline Accident Summary). 1 p.

Lockheed L 18 Lodestar Aircraft
See Also: Aircraft
—Accidents
CAA. Lockheed L18 Lodestar (World Airline Accident Summary). 1 p.
CAA. Lockheed L18 Lodestar. 1 p.
CAA. Lockheed L18 Lodestar. 1 p.

Lockheed L 100 Hercules Aircraft
See Also: Aircraft
—Accidents
CAA. Lockheed L100 and L382-Hercules. 1 p.

Lockheed L 188 Electra Aircraft
See Also: Aircraft
—Accidents
CAA. Lockheed L188 Electra (World Airline Accident Summary). 1 p.
CAA. Lockheed L188 Electra. 1 p.

INDUSTRY STANDARDS

Lockheed L 382 Hercules Aircraft
See Also: Aircraft
—Accidents
CAA. Lockheed L100 and L382-Hercules. 1 p.

Lockheed L 382B Hercules Aircraft
See Also: Aircraft
—Accidents
CAA. Lockheed L100 and L382B Hercules (World Airline Accident Summary). 1 p.

Lockheed L 1011 Aircraft
See Also: Aircraft
—Antenna Positions
CAA. Lockheed L1011 (Approved Aerial Positions). 1 p.
—Data Sheets
CAA FA9 ISSUE 4. Lockheed Model L1011-385-1 Configurations 193A, 193K, 193N, Model L1011-385-1-15 Configurations 193N, 193T, 193U, 293C and Model L1011-385-3 Configuration 193V. 6 pp.

Lockheed L 1011-385-1 Aircraft
See Also: Aircraft
—Certification
CAA. Lockheed L-1011-385-1 and-3. 10 pp.

Lockheed L 1011-385-3 Aircraft
See Also: Aircraft
—Certification
CAA. Lockheed L-1011-385-1 and-3. 10 pp.

Lockheed Super Constellation Aircraft
See Also: Aircraft
—Accidents
CAA. Lockheed Constellation and Super Constellation (World Airline Accident Summary). 3 pp.
CAA. Lockheed Constellation and Super Constellation. 1 p.

Lockheed Tri-Star L1011 Aircraft
See Also: Aircraft
—Accidents
CAA. Lockheed L-1011 Tristar (World Airline Accident Summary). 1 p.
CAA. Lockheed L-1011 Tristar. 1 p.
—Foreign Airworthiness Directives
CAA. Lockheed L-1011 Tristar (Foreign Airworthiness Directives). 26 pp.

Locking Collars
Use: Collars (Mechanical Components)—Locking

Locking Devices
Use: Locks (Hardware)

Locking Fasteners
Use For: Self Locking Fasteners
See Also: Fasteners; Locking Screws
CNS A2020-89. Fastenings for Locks (with Rotary Eye Pin) (Jun)(870). 2 pp.
CNS A2021-89. Fastenings for Locks (Serrated Edge Form) (Jun)(871). 2 pp.
CNS A2022-89. Fastenings for Locks (Plain Edge Form) (Jun)(872). 2 pp.
—Aircraft—Brake—Spindles
SBAC AS 2950 ISSUE 2. Spindle for Brake Lock. 1 p.
SBAC AS 6602 ISSUE 2. Spindle for Brake Lock. 1 p.
—Aircraft—Wire
SBAC AS 4561 ISSUE 1. Wire Locking Tabs. 1 p.

Locking Flanges
See Also: Flanges
—Aircraft
SBAC AS 6545 ISSUE 1. Locking Flange.
SBAC AS 6554 ISSUE 1. Locking Flange.

Locking Nuts
Use For: Self Locking Nuts; Stiffnuts
See Also: Anchor Nuts; Insert Nuts; Locknuts; Nuts (Fasteners)
CNS B2444-80. Self-Locking Nuts (Apr)(5399).
DIN ENGL 7967-70. Self Locking Counter Nuts (Nov). 2 pp.
—Aerospace
AECMA PREN3240-89. Nuts, Self-Locking, Clip in Heat Resisting Steel FE-PA92HT (A286), Uncoated Classification: 1100 MPa/425 Degrees Celsius. 4 pp.
AECMA PREN3241-89. Nuts, Self-Locking, Clip in Heat Resisting Steel FE-PA92HT (A286), Silvercoated, Classification: 1100 MPa/425 Degrees Celsius. 4 pp.
BSI A 293-83. 1983 Amd 1 Procurement of Self-Locking Nuts with Non-Metallic Locking Element. Metric Series (AMD 5767) April 30, 1990. 20 pp.
BSI A 295-85. 1985 Methods of Test for Self-Locking Nuts with Maximum Operating Temperature Less than or Equal to 425 Degrees C. 13 pp.
BSI A 298-88. 1988 Methods of Test for Self-Locking Nuts with Maximum Operating Temprature Greater than 425 Degrees C. 14 pp.
BSI A 302-92. 1992 Self-Locking Nuts with Maximum Operating Temperature Less Than or Equal to 425 Degrees C—Procurement Specification (ISO 5858: 1991). 23 pp.
CEN PREN 3726-91. Nuts, Self-Locking, Clip, in Heat Resisting Steel FE-PA92HT (A286), MoS2 Coated Classification: 1100 MPa (at Ambient Temperature)/425 Degrees C. 6 pp.
CEN PREN 3726-93. Aerospace Series Nuts, Self-Locking, Clip, in Heat Resisting Steel FE-PA92HT (A286), MoS2 Coated Classification: 1 100 MPa (at Ambient Temperature) /425 Degrees C. 4 pp.
CEN PREN 3752-91. Nuts, Self-Locking, in Heat Resisting Steel FE-PA92HT (A286), MoS2 Coated Classification: 1100 MPa/425 Degrees C Technical Specification. 24 pp.
DIN ENGL LN 65016-70. Self-Locking Nuts for Temperatures up to 425 Degrees C; Procurement Specification (Dec). 10 pp.
DIN ENGL LN 65100-75. Self-Locking Nuts for Temperature Class 650 Degrees C; Procurement Specification (Jan). 10 pp.
ISO 5858-91. Aerospace—Self-Locking Nuts with Maximum Operating Temperature Less Than or Equal to 425 Degrees Celsius—Procurement Specification First Edition. 23 pp.
ISO 7481-84. Aerospace—Fasteners—Self-Locking Nuts with Maximum Operating Temperature Less Than or Equal to 425 Degrees Celsius—Test Methods First Edition. 14 pp.
ISO 8641-87. Aerospace—Self-Locking Nuts with Maximum Operating Temperature Greater Than 425 Degrees Celsius—Procurement Specification First Edition. 19 pp.
ISO 8642-86. Aerospace—Self-Locking Nuts with Maximum Operating Temperature Greater Than 425 Degrees Celsius—Test Methods First Edition. 14 pp.
MOD UK DSTAN 53-89-01. Requirements for the Substitutability of Unified Self-Locking Nuts (Stiffnuts) Issue 1; Amendment 2. 89 pp.
—Aerospace—Acceptance Forms
DIN ENGL LN 65018 Pt 2-68. Manufacturer's Acceptance Certificate for Nuts and Self-Locking Nuts (Sept). 1 p.
—Aerospace—Anchor
CEN PREN 3653-92. Nuts, Anchor, Self-Locking, Floating, Self-Aligning, One Lug, in Steel, Cadmium Plated, MoS2 Lubricated Classificaton: 900 MPa (at Ambient Temperature) /235 Degrees C. 5 pp.
—Aerospace—Anchor—Counterbore
CEN PREN 2862-90. Nuts Anchor, Self-Locking, Fixed, 90 Degrees Corner, with Counterbore, in Alloy Steel, Cadmium Plated, MO2 Lubricated Classification: 1100 MPa (at Ambient Temperature) /235 Degrees Celsius. 6 pp.
CEN PREN 2862-93. Aerospace Series Nuts, Anchor, Self-Locking, Fixed, 90 Degree Corner, with Counterbore, in Alloy Steel, Cadmium Plated, MoS2 Lubricated Classification: 1 100 MPa (at Ambient Temperature)/235 Degrees C. 5 pp.
CEN PREN 2862-93. CORRIGENDUM Aerospace Series Nuts, Anchor, Self-Locking, Fixed, 90 Degree Corner, with Counterbore, in Alloy Steel, Cadmium Plated, MoS2 Lubricated Classification: 1 100 MPa (at Ambient Temperature) /235 Degrees C. 1 p.
CEN PREN 2863-90. Nuts, Anchor Self-Locking, Fixed, 90 Degrees Corner, with Counterbore, in Heat Resisting Steel, Passivated, MoS2 Lubricated Classification: 1100 MPa (at Ambient Temperature) /315 Degrees Celsius. 6 pp.
CEN PREN 2863-93. Aerospace Series Nuts, Anchor, Self-Locking, Fixed, 90 Degree Corner, with Counterbore, in Heat Resisting Steel, MoS2 Lubricated Classification: 1 100 MPa (at Ambient Temperature)/315 Degrees C. 5 pp.
CEN PREN 2863-93. CORRIGENDUM Aerospace Series Nuts, Anchor, Self-Locking, Fixed, 90 Degree Corner, with Counterbore, in Heat Resisting Steel, MoS2 Lubricated Classification: 1 100 MPa (at Ambient Temperature) /315 Degrees C. 1 p.
CEN PREN 2865-93. Aerospace Series Nuts, Anchor, Self-Locking, Floating, Two Lug, with Counterbore, in Heat Resisting Steel, MoS2 Lubricated Classification: 1 100 MPa (at Ambient Temperature)/315 Degrees C. 5 pp.
CEN PREN 2865-93. CORRIGENDUM Aerospace Series Nuts, Anchor, Self-Locking, Floating, Two Lug, with Counterbore, in Heat Resisting Steel, MoS2 Lubricated Classification: 1 100 MPa (at Ambient Temperature). 1 p.
CEN PREN 2866-92. Nuts, Anchor, Self-Locking, Floating, One Lug, with Counterbore, in Steel, Cadmium Plated, MoS2 Lubricated Classification: 1 110 MPa (at Ambient Temperature) /235 Degrees C. 5 pp.
CEN PREN 2867-92. Nuts, Anchor, Self-Locking, Floating, One Lug, with Counterbore, in Heat Resisting Steel, MoS2 Lubricated Classification: 1 110 MPa (at Ambient Temperature) /315 Degrees C. 5 pp.
CEN PREN 3435-93. Aerospace Series Nuts, Anchor, Self-Locking, Floating, Two Lug, Reduced Series, with Counterbore, in Heat Resisting Steel, MoS2 Lubricated Classification: 1 100 MPa (at Ambient Temperature)/315 Degrees C. 5 pp.
CEN PREN 3435-93. CORRIGENDUM Aerospace Series Nuts, Anchor, Self-Locking, Floating, Two Lug, Reduced Series, with Counterbore, in Heat Resisting Steel, MoS2 Lubricated Classification: 1 100 MPa (at Ambient Temperature) /315 Degrees C. 1 p.
CEN PREN 3537-90. Nuts, Anchor, Self-Locking, Fixed, Two Lug, with Counterbore, in Heat Resisting Steel, Passivated, MoS2 Lubricated Classification:1100 MPa (at Ambient Temperature) /315 Degrees Celsius. 6 pp.
CEN PREN 3537-93. Aerospace Series Nuts, Anchor, Self-Locking, Fixed, Two Lug, with Counterbore, in Heat Resisting Steel, MoS2 Lubricated Classification: 1 100 MPa (at Ambient Temperature)/315 Degrees C. 5 pp.
CEN PREN 3537-93. CORRIGENDUM Aerospace Series Nuts, Anchor, Self-Locking, Fixed, Two Lug, with Counterbore, in Heat Resisting Steel, MoS2 Lubricated Classification: 1 100 MPa (at Ambient Temperature) /315 Degrees C. 1 p.
CEN PREN 3538-90. Nuts, Anchor, Self-Locking, Fixed, Two Lug, Reduced Series with Counterbore, in Heat Resisting Steel Passivated, MoS 2 Lubricated Classification: 100 MPa (at Ambient Temperature) /315 Degrees Celsius. 6 pp.
CEN PREN 3538-93. Aerospace Series Nuts, Anchor, Self-Locking, Fixed, Two Lug, Reduced Series, with Counterbore, in Heat Resisting Steel, MoS2 Lubricated Classification: 1 100 MPa (at Ambient Temperature)/315 Degrees C. 5 pp.
CEN PREN 3538-93. CORRIGENDUM Aerospace Series Nuts, Anchor, Self-Locking, Fixed, Two Lug, Reduced Series, with Counterbore, in Heat Resisting Steel, MoS2 Lubricated Classification: 1 100 MPa (at Ambient Temperature) /315 Degrees C. 1 p.
CEN PREN 3539-90. Nuts Anchor, Self-Locking, One Lug, Fixed, with Counterbore, in Heat Resisting Steel, MoS2 Lubricated Classification: 1100 MPa (at Ambient Temperature) /315 Degrees Celsius. 5 pp.
CEN PREN 3539-93. Aerospace Series Nuts, Anchor, Self-Locking, One Lug, Fixed, with Counterbore, in Heat Resisting Steel, MoS2 Lubricated Classification: 1 100 MPa (at Ambient Temperature) /315 Degrees C. 5 pp.
CEN PREN 3712-91. Nuts, Anchor, Self-Locking, One Lug Fixed, Reduced Series, with Counterbore, in Steel, Cadmium Plated, MoS2 Lubricated Classification: 1100 MPa (at Ambient Temperature)/235 Degrees C. 6 pp.
CEN PREN 3714-93. Nuts, Anchor, Self-Locking, Floating, Two Lug, with Counterbore, in Heat Resisting Steel, Silver Plated Classification: 1 100 MPa (at Ambient Temperature)/425 Degrees C. 5 pp.
CEN PREN 3714-93. CORRIGENDUM Aerospace Series Nuts, Anchor, Self-Locking, Floating, Two Lug, with Counterbore, in Heat Resisting Steel, Silver Plated Classification: 1 100 MPa (at Ambient Temperature) /425 Degrees C. 1 p.
CEN PREN 3750-90. Nuts, Anchor, Self-Locking, Fixed 90 Degrees Corner, Reduced Series with Counterbore, in Heat Resisting Steel, Passivated, MoS2 Lubricated/Classification:100 MPa (at Ambient Temperature) /315 Degrees Celsius. 6 pp.
CEN PREN 3750-93. Aerospace Series Nuts, Anchor, Self-Locking, Fixed, 90 Degree Corner, Reduced Series, with Counterbore, in Heat Resisting Steel, MoS2 Lubricated Classification: 1 100 MPa (at Ambient Temperature)/315 Degrees C. 5 pp.
CEN PREN 3750-93. CORRIGENDUM Aerospace Series Nuts, Anchor, Self-Locking, Fixed, 90 Degree Corner, Reduced Series, with Counterbore, in Heat Resisting Steel, MoS2 Lubricated Classification: 1 100 MPa (at Ambient Temperature) /315 Degrees C. 1 p.
CEN PREN 3751-90. Nuts, Anchor, Self-Locking, Fixed, Closed Corner, Reduced Series, with Counterbore, in Heat Resisting Steel, Passivated, MoS2 Lubricated Classification: 1100 MPa (at Ambient Temperature)/315 Degrees Celsius. 6 pp.

INTERNATIONAL AND NON-U.S. NATIONAL STANDARDS
SUBJECT INDEX
Locking

Locking Nuts *(Cont.)*

—Aerospace—Anchor—Counterbore *(Cont.)*

CEN PREN 3751-93. Aerospace Series Nuts, Anchor, Self-Locking, Fixed, Closed Corner, Reduced Series, with Counterbore, in Heat Resisting Steel, MoS2 Lubricated Classification: 1 100 MPa (at Ambient Temperature)/315 Degrees C. 5 pp.

CEN PREN 3751-93. CORRIGENDUM Aerospace Series Nuts, Anchor, Self-Locking, Fixed, Closed Corner, Reduced Series, with Counterbore, in Heat Resisting Steel, MoS2 Lubricated Classification: 1 100 MPa (at Ambient Temperature) /315 Degrees C. 1 p.

CEN PREN 3753-90. Nuts, Anchor, Self-Locking, Fixed, Closed Corner, with Counterbore, in Alloy Steel, Cadmium Plated, MoS2 Lubricated Classification:1100 MPa (at Ambient Temperature) /235 Degrees Celsius. 6 pp.

CEN PREN 3753-93. Aerospace Series Nuts, Anchor, Self-Locking, Fixed, 60 Degree Corner, with Counterbore, in Alloy Steel, Cadmium Plated, MoS2 Lubricated Classification: 1 100 MPa (at Ambient Temperature)/235 Degrees C. 5 pp.

CEN PREN 3753-93. CORRIGENDUM Aerospace Series Nuts, Anchor, Self-Locking, Fixed, 60 Degree Corner, with Counterbore, in Alloy Steel, Cadmium Plated, MoS2 Lubricated Classification: 1 100 MPa (at Ambient Temperature) /235 Degrees C. 1 p.

CEN PREN 3754-90. Nuts, Anchor, Self-Locking, Fixed, Closed Corner, with Counterbore, in Heat Resisting Steel, Passivated, MoS2 Lubricated Classification: 100 MPa (at Ambient Temperature) /315 Degrees Celsius. 6 pp.

CEN PREN 3754-93. Aerospace Series Nuts, Anchor, Self-Locking, Fixed, 60 Degree Corner, with Counterbore, in Heat Resisting Steel, MoS2 Lubricated Classification: 1 100 MPa (at Ambient Temperature)/315 Degrees C. 5 pp.

CEN PREN 3754-93. CORRIGENDUM Aerospace Series Nuts, Anchor, Self-Locking, Fixed, 60 Degree Corner, with Counterbore, in Heat Resisting Steel, MoS2 Lubricated Classification: 1 100 MPa (at Ambient Temperature) /315 Degrees C. 1 p.

CEN PREN 3757-93. Aerospace Series Nuts, Anchor, Self-Locking, Floating, Self-Aligning, Two Lug, in Heat Resisting Steel, MoS2 Lubricated Classification: 900 MPa (at Ambient Temperature) /315 Degrees C. 5 pp.

CEN PREN 3757-93. CORRIGENDUM Aerospace Series Nuts, Anchor, Self-Locking, Floating, Self-Aligning, Two Lug, in Heat Resisting Steel, MoS2 Lubricated Classification: 900 MPa (at Ambient Temperature) /315 Degrees C. 1 p.

CEN PREN 3768-90. Nuts, Anchor, Self-Locking, One Lug, Fixed, Reduced Series, with Counterbore, in Heat Resisting Steel, MoS2 Lubricated Classification: 1100 MPa (at Ambient Temperature) /315 Degrees Celsius. 5 pp.

CEN PREN 3768-93. Aerospace Series Nuts, Anchor, Self-Locking, One Lug, Fixed, Reduced Series, with Counterbore, in Heat Resisting Steel, MoS2 Lubricated Classification: 1 100 MPa (at Ambient Temperature) /315 Degrees C. 5 pp.

CEN PREN 3834-91. Nuts, Anchor, Self-Locking, Floating, Two Lug, Incremental Counterbore, in Heat Resisting Steel, MoS2 Lubricated Classification: 900 MPa (at Ambient Temperature)/315 Degrees C. 7 pp.

CEN PREN 3834-93. Aerospace Series Nuts, Anchor, Self-Locking, Floating, Two Lug, Incremental Counterbore, in Heat Resisting Steel, MoS2 Lubricated Classification: 900 MPa (at Ambient Temperature) /315 Degrees C. 5 pp.

CEN PREN 4084-93. Aerospace Series Nuts, Anchor, Self-Locking, Fixed, Two Lug, with Counterbore, in Alloy Steel, Cadmium Plated, MoS2 Lubricated Classification: 1 100 MPa (at Ambient Temperature) /235 Degrees C. 5 pp.

DIN ENGL LN 29982-78. Aerospace; Nuts Anchor; Self-Locking; Deep Counterbore; Double Lug for Temperatures up to 235 Degrees C (Mar). 3 pp.

DIN ENGL LN 29983-77. Nuts Anchor; Self-Locking; Deep Counterbore; Double Lug Reduced for Temperatures up to 235 Degrees C (July). 3 pp.

DIN ENGL LN 29984-77. Nuts Anchor; Self-Locking; Deep Counterbore; Single Lug for Temperatures up to 235 Degrees C (July). 3 pp.

DIN ENGL LN 29985-77. Nuts Anchor; Self-Locking; Floating; Deep Counterbore; Double Lug for Temperatures up to 235 Degrees C (Dec). 3 pp.

DIN ENGL LN 29986-77. Anchor Nuts; Self-Locking, Floating, Deep Counterbore, Double Lug, Miniature, for Temperatures up to 235 Degrees C (May). 3 pp.

DIN ENGL LN 29987-77. Nuts Anchor; Self-Locking; Floating; Deep Counterbore; Single Lug for Temperatures up to 235 Degrees C (Dec). 3 pp.

DIN ENGL LN 29988-77. Nuts Anchor; Self-Locking; Deep Counterbore; Corner Lug for Temperatures up to 235 Degrees C (July). 3 pp.

DIN ENGL LN 29989-77. Nuts Anchor; Self-Locking; Deep Counterbore; Corner Lug Reduced for Temperatures up to 235 Degrees C (July). 3 pp.

Locking Nuts *(Cont.)*

—Aerospace—Anchor—Counterbore *(Cont.)*

DIN ENGL LN 29990-77. Anchor Nuts; Self-Locking, Deep Counterbore, Double Lug, for Temperatures up to 315 Degrees C and up to 425 Degrees C (July). 3 pp.

DIN ENGL LN 29991-77. Anchor Nuts; Self-Locking, Deep Counterbore, Double Lug, Miniature for Temperatures up to 315 Degrees C and up to 425 Degrees C (July). 3 pp.

DIN ENGL LN 29992-77. Nuts, Anchor; Self-Locking, Deep Counterbore, Single Lug for Temperatures up to 315 Degrees C and up to 425 Degrees C (July). 3 pp.

DIN ENGL LN 29996-77. Nuts Anchor; Self-Locking; Deep Counterbore; Single Lug Reduced for Temperatures up to 315 Degrees C and up to 425 Degrees C (May). 3 pp.

—Aerospace—Anchor—Counterbore—Miniature

DIN ENGL LN 29986-77. Anchor Nuts; Self-Locking, Floating, Deep Counterbore, Double Lug, Miniature, for Temperatures up to 235 Degrees C (May). 3 pp.

DIN ENGL LN 29991-77. Anchor Nuts; Self-Locking, Deep Counterbore, Double Lug, Miniature for Temperatures up to 315 Degrees C and up to 425 Degrees C (July). 3 pp.

—Aerospace—Castle

CEN PREN 3434-91. Nuts, Hexagon, Slotted, Self-Locking, in Steel, Cadmium Plated, MoS2 Lubricated Classification: 900 MPa (at Ambient Temperature) /235 Degrees C. 6 pp.

—Aerospace—Channel—Counterbore

DIN ENGL LN 29993-74. Gang Channel Nuts; Deep Counterbore, Self-Locking, Floating for Temperatures up to 120 Degrees C (Oct). 2 pp.

—Aerospace—Clip

CEN PREN 3240-92. Nuts, Self-Locking, Clip, in Heat Resisting Steel FE-PA92HT (A286), Uncoated Classification: 1 100 MPa (at Ambient Temperature) /425 Degrees C. 4 pp.

CEN PREN 3241-92. Nuts, Self-Locking, Clip, in Heat Resisting Steel FE-PA92HT (A286), Silver Coated Classification: 1 100 MPa (at Ambient Temperature) /425 Degrees C. 4 pp.

—Aerospace—Double Hexagonal

AECMA PREN2906-86. Nuts, Self Locking, Bi Hexagonal, in Heat Resisting Steel FE-PA92HT (A286) Unplated Classification 1100 MPa/650 Degrees. 4 pp.

AECMA PREN2907-86. Nuts, Self Locking, Bi Hexagonal, in Heat Resisting Steel FE-PA92HT (A286) Silver Plated Classification 1100 MPa/650 Degrees Celsius. 4 pp.

AECMA PREN3012-88. Nuts, Self-Locking, Bihexagonal in Heat Resisting Nickel Base Alloy NI-P101HT (Waspaloy) Unplated Classification: 1210 MPa/730 Degrees Celsius. 5 pp.

AECMA PREN3013-88. Nuts, Self-Locking, Bihexagonal in Heat Resisting Nickel Base Alloy NI-P101HT (Waspaloy) Completely Silver Plated: Classification: 1210 MPa/730 Degrees Celsius. 5 pp.

AECMA PREN3239-88. Nuts, Self-Locking, Bihexagonal in Heat Resisting Nickel Base Alloy NI-P101HT (Waspaloy) Silver Plated Thread Classification: 1210 MPa/730 Degrees Celsius. 4 pp.

AECMA PREN3637-90. Nuts, Self-Locking, Bi-Hexagonal (Double Reduced), in Heat Resisting Nickel Base Alloy N1-P101HT (Waspaloy), Silver Plated Classification: 1210 MPa/730 Degrees Celsius. 5 pp.

CEN PREN 3713-91. Nuts, Anchor, Self-Locking, Bihexagonal, in Steel, Cadmium Plated, MoS2 Lubricated Classification: 1550 MPa (at Ambient Temperature)/235 Degrees C. 6 pp.

CEN PREN 3720-91. Nuts, Bihexagonal, Self-Locking, in Heat Resisting Steel FE-PA92HT (A286), MoS2 Coated Classification:1100 (at Ambient Temperature) /425 Degrees C. 5 pp.

CEN PREN 4011-93. Aerospace Series Nuts, Bihexagonal, Self-Locking, in Heat Resisting Nickel Base Alloy NI-P100HT (Inconel 718), Silver Plated Classification: 1 550 MPa (at Ambient Temperature) /600 Degrees C. 4 pp.

CEN PREN 4012-93. Aerospace Series Nuts, Bihexagonal, Self-Locking, in Heat Resisting Nickel Base Alloy NI-P100HT (Inconel 718), MoS2 Coated Classification: 1 550 MPa (at Ambient Temperature) /425 Degrees C. 4 pp.

DIN ENGL LN 29528-80. Aerospace; Nut Bihexagonal with Flange; Selflocking for Temperatures up to 235 Degrees C for Bolts and Screws with Nominal Tensile Strength up to 1800 N/mm2 (Nov). 2 pp.

DIN ENGL LN 29942-85. Aerospace; Nuts, Bi-Hexagon Flange, Deep Counterbore, Self-Locking for Temperatures up to 235 Degrees C for Bolts and Screws with Nominal Tensile Strength of 1550 and 1800 N/mm2 (Feb). 3 pp.

Locking Nuts *(Cont.)*

—Aerospace—Double Hexagonal *(Cont.)*

ISO 9199-87. Aerospace—Self-Locking Bihexagonal Nuts, Classificatons 1 100 MPa/650 Degrees Celsius, 1 250 MPa/760 Degrees Celsius, 1 550 MPa/235 Degrees Celsius and 1 550 MPa/650 Degrees Celsius—Dimensions First Edition. 4 pp.

—Aerospace—Double Hexagonal—Counterbore

AECMA PREN2908-86. Nuts, Self Locking, Bi Hexagonal, Deep Counterbored, in Heat Resisting Steel FE-PA92HT (A286) Unplated Classification 1100 MPa/650 Degrees Celsius. 4 pp.

AECMA PREN2909-86. Nuts, Self Locking, Bi Hexagonal, Deep Counterbored, in Heat Resisting Steel FE-PA92HT (A286) Silver Plated Classification 1100 MPa/650 Degrees Celsius. 4 pp.

CEN PREN 3721-91. Nuts, Bihexagonal, Self-Locking, Deep Counterbored, in Heat Resisting Steel FE-PA92HT (A286), MoS Coated Classification:1100 MpA (at Ambient Temperature)/Degrees C. 5 pp.

CEN PREN 3772-90. Nuts, Bi-Hexagonal, with Flange, Deep Counterbore, Self Locking, in Heat Resisting Steel, Passivated Classification: 1100 MPa (at Ambient Temperature) /425 Degrees Celsius. 5 pp.

CEN PREN 3843-93. Aerospace Series Nuts, Bi-Hexagonal, Self-Locking, with Counterbore, in Heat Reisisting Steel, Passivated Classification: 1 100 MPa (at Ambient Temperature) /650 Degrees C. 5 pp.

—Aerospace—Hexagonal

AECMA PREN3034-89. Nuts, Self-Locking, Hexagonal with Captive Washer, in Heat Resisting Steel FE-PA92HT (A286), Silver Coated Classification: 1100 MPA/425 Degrees Celsius. 5 pp.

AECMA PREN3196-87. Nuts, Self-Locking, Hexagonal, in Heat Resisting Steel FE-PA92HT (A286), Silver Coated Classification 1100 MPa/425 Degrees Celsius. 4 pp.

AECMA PREN3377-89. Nuts, Self-Locking, Hexagonal in Heat Resisting Steel FE-PA92HT (A286), Uncoated Classification: 1100 MPa/425 Degrees Celsius. 5 pp.

BSI A 275-A 280-81. 1981 Amd 1 Hexagon Self-Locking Nuts with Non-Metallic Locking Inserts. Metric Series. 10 pp.

CEN PREN 3723-91. Nuts, Hexagonal,Self-Locking, in Heat Resisting Steel FE-PA92HT (A286), MoS2 Coated Classification: 1100 MPa (at Ambient Temperature)/425 Degrees C. 5 pp.

CEN PREN 3763-90. Nuts, Hexagonal, Self-Locking, Ball Seat, in Heat Resisting Steel, Passivated, MOS2 Lubricated Classification; 900MPa (at Ambient Temperature) /315 Degrees Celsius. 7 pp.

CEN PREN 3763-93. Aerospace Series Nuts, Hexagonal, Self-Locking, Ball Seat, in Heat Resisting Steel, MoS2 Lubricated Classification: 900 MPa (at Ambient Temperature) /315 Degrees C. 6 pp.

CEN PREN 3763-93. CORRIGENDUM Aerospace Series Nuts, Hexagonal, Self-Locking, Ball Seat, in Heat Resisting Steel, MoS2 Lubricated Classification: 900 MPa (at Ambient Temperature) /315 Degrees C. 1 p.

—Aerospace—Hexagonal—Counterbore

SBAC TS 135 ISSUE 1. All Metal Self-Locking Nuts (MJ Threads) Reduced Hexagon—Deep Counterbored Tensile Strength:—1100 MPa at Ambient Temperature.

—Aerospace—Hexagonal—Counterbore—Washers

CEN PREN2882-90. Nuts, Hexagonal, Self-Locking, with Counterbore and Captive Washer, in Steel, Cadmium Plated, MoS2 Lubricated Classification: 1100MPa (at Ambient Temperature) /235 Degrees Celsius. 9 pp.

—Aerospace—Hexagonal—Flanged

CEN PREN 3536-91. Nuts, Hexagon, Self-Locking, in Heat Resisting Steel, MoS2 Lubricated Classification: 1100MPa (at Ambient Temperature)/315 Degrees C. 5 pp.

CEN PREN 3626-91. Nuts, Hexagon, Self-Locking, in Steel, Cadmium Plated, MoS2 Lubricated Classification: 1100 MPa (at Ambient Temperature)/ 235 Degrees C. 5 pp.

DIN ENGL LN 9161-77. Nuts Hexagon Flanged; Self-Locking for Temperatures up to 315 Degrees Celsius and up to 425 Degrees Celsius (May). 3 pp.

DIN ENGL LN 9338-85. Aerospace; Nuts Hexagon Flanged; Self-Locking for Temperatures up to 235 Degrees C (Feb). 3 pp.

—Aerospace—Hexagonal—Self Aligning

DIN ENGL LN 29795-80. Aerospace; Hexagon Nuts; Self-Aligning; Self-Locking for Temperatures up to 235 Degrees C (Sept). 3 pp.

INDUSTRY STANDARDS
1553

INTERNATIONAL AND NON-U.S. NATIONAL STANDARDS
SUBJECT INDEX

Locking

Locking Nuts *(Cont.)*

—Aerospace—Hexagonal—Thin
BSI A 281-A 286-81. 1981 Hexagon (Thin) Self-Locking Nuts with Non-Metallic Locking Inserts. Metric Series. 8 pp.

—Aerospace—Hexagonal—Washers
AECMA PREN3033-89. Nuts, Self-Locking, Hexagonal with Captive Washer, in Heat Resisting Steel FE-PA92HT (A286), Uncoated Classification: 1100 MPa/425 Degrees Celsius. 5 pp.
DIN ENGL LN 29790-76. Nuts Hexagon with Captive Washer; Self-Locking for Temperatures up to 235 Degrees C (Oct). 3 pp.

—Aerospace—Plate
CEN PREN 3019-92. Self-Locking Plate Nuts, Floating, Two-Lug, in Heat Resisting Steel FE-PA92HT (A286) Classification: 1 100 MPa (at Ambient Temperature) /650 Degrees C. 4 pp.
CEN PREN 3020-92. Self-Locking Plate Nuts, Floating, Two-Lug, in Heat Resisting Steel FE-PA92HT (A286), Silver Plated Classification: 1 100 MPa (at Ambient Temperature) /650 Degrees C. 4 pp.

—Aerospace—Quality Assurance
AECMA PREN3004-89. Nuts, Self Locking, in Heat Resisting Steel FE-FE-PA92HT (A286) Classification: 1100 MPa/650 Degrees Celsius Technical Specification. 22 pp.
AECMA PREN3005-89. Nuts, Self-Locking, in Heat Resisting Nickel Base Alloy N1-P101HT (Waspaby) Classification: 1210 MPa/730 Degrees Celsius Technical Specification. 22 pp.
AECMA PREN3152-89. Propulsion Standard Parts Nuts, Self-Locking, in Heat Resisting Steel FE-PA92HT (A286) Classification: 1100 MPa/425 Degrees Celsius Technical Specification. 22 pp.

—Aerospace—Shaft
CEN PREN 3295-92. Shaft-Nuts, Self-Locking, in FE-PA92HT (A286), Silver Plated. 6 pp.

—Aerospace—Shank
AECMA PREN2910-86. Shank Nuts, Self Locking, Flange Restrained, in Heat Resisting Steel FE-PA92HT (A286) Unplated Classification 1100 MPa/650 Degrees Celsius. 4 pp.
AECMA PREN2911-86. Shank Nuts, Self Locking, Flange Restrained, in Heat Resisting Steel FE-PA92HT (A286) Silver Plated Classification 1100 MPa/650 Degrees Celsius. 4 pp.
AECMA PREN3014-87. Self-Locking Serrated Shank Nuts in Heat Resisting Steel FE-PA92HT Classification 1100 MPa/650 Degrees Celsius. 4 pp.
AECMA PREN3015-87. Self-Locking Serrated Shank Nuts in Heat Resisting Steel FE-PA92HT Silver Plated Classification 1100 MPa/650 Degrees Celsius. 4 pp.
CEN PREN 3722-91. Shank Nuts, Self-Locking, in Heat Resisting Steel FE-PA92HT (A286), MoS2 Coated Classification: 1100 MPa (at Ambient Temperature)/425 Degrees C. 5 pp.
CEN PREN 4013-93. Aerospace Series Shank Nuts, Self-Locking, in Heat Resisting Nickel Base Alloy NI-P100HT (Inconel 718), Silver Plated Classification: 1 550 MPa (at Ambient Temperature) /600 Degrees C. 4 pp.

—Aerospace—Shank—Assembly
AECMA PREN3064-88. Self-Locking Serrated Shank Nuts Assembly Procedure. 4 pp.

—Aerospace—Shank—Installation
AECMA PREN3148-89. Shank Nuts, Self-Locking, Flange Restrained Installation Procedure. 4 pp.

—Aerospace—Shank—Installation Holes
AECMA PREN3065-88. Installation Holes for Self-Locking Serrated Shank Nuts Design Specifications. 5 pp.
AECMA PREN3149-87. Shank Nuts, Metric Installation Holes. 5 pp.

—Aerospace—Spline
ISO 9157-88. Aerospace—Nuts, Spline-Drive, Self-Locking, with MJ Threads, Coated or Uncoated, Classification 1 100 MPa/650 Degrees Celsius, 1 250 MPa/760 Degrees Celsius, 1 550 MPa/235 Degrees Celsius or 1 550 MPa/650 Degrees Celsius—Dimensions First Edition. 5 pp.

—Aircraft
SBAC AGS 2000 ISSUE 14. AGS, Stiffnuts Application.
SBAC TS 2 ISSUE 4. All Metal Self-Locking Nuts (Unified Threads) 125,000 lbf/in2 Min. Tensile (56 Ton f/in2), 250 Degrees Celsius.
SBAC TS 3 ISSUE 4. All Metal Self-Locking Nuts (Unified Threads) 125,000 lbf/in2 Min. Tensile (56 Ton f/in2), 450 Degrees Celsius.
SBAC TS 4 ISSUE 4. All Metal Self-Locking Nuts (Unified Threads) 160,000 lbf/in2 Min. Tensile (71 Ton f/in2), 250 Degrees Celsius.

Locking Nuts *(Cont.)*

—Aircraft *(Cont.)*
SBAC TS 8 ISSUE 6. All Metal Self-Locking Nuts (UNJ Threads) 1100 MPa (160,000 lbf/in2) Minimum Tensile Strength at Room Temperature.
SBAC TS 26 ISSUE 2. All Metal Self-Locking Nuts (UNJ Threads) 1210 MPa (175,000 Lbf/in2) Minimum Tensile Strength at Ambient Temperature.
SBAC TS 120 ISSUE 2. Technical Specifications for AGS 2000 Series Stiffnuts (Including Clinch Nuts) (BA and BSF Threads) for Aircraft.
SBAC TS 128 ISSUE 3. All Metal Self-Locking Nuts (UNJ Threads) 1100 MPa (160,000 lbf/in2) Minimum Tensile Strength at Ambient Temperature.
SBAC TS 149 ISSUE 1. All Metal Self-Locking Nuts (UNJ Threads) 1550 MPa (225,000 lbf/in2) Minimum Tensile Strength at Ambient Temperature.

—Aircraft—Anchor
SBAC AS 8602 ISSUE 4. Self-Locking Nuts, Double Lug Anchor, Steel, Cadmium Plated, 125,000 lbf/sq.in. 250 Degrees C.
SBAC AS 8603 ISSUE 4. Self-Locking Nuts, Miniature Double Lug Anchor Steel, Cadmium Plated, 125,000 lbf/sq. in. 250 Degrees C.
SBAC AS 8604 ISSUE 4. Self-Locking Nuts, Single Long Lug Anchor, Steel, Cadmium Plated, 125,000 lbf/sq.in. 250 Degrees C.
SBAC AS 8605 ISSUE 4. Self-Locking Nuts, Miniature, Single Long Lug Anchor, Steel, Cadmium Plated, 125,000 lbf/sq.in. 250 Degrees C.
SBAC AS 8606 ISSUE 4. Self-Locking Nuts, Miniature, Single Short Lug Anchor Steel, Cadmium Plated, 125,000 lbf/sq.in. 250 Degrees C.
SBAC AS 8607 ISSUE 4. Self-Locking Nuts, Corner Anchor Steel, Cadmium Plated, 125,000 lbf/sq. in. 250 Degrees C.
SBAC AS 8608 ISSUE 4. Self-Locking Nuts, Miniature Corner Anchor, Steel, Cadmium Plated, 125,000 lbf/sq.in. 250 Degrees C.
SBAC AS 8609 ISSUE 4. Self-Locking Nuts, Double Lug Floating Anchor, Steel, Cadmium Plated, 125, 000 lbf/sq.in. 250 Degrees C.
SBAC AS 8610 ISSUE 5(I). Self-Locking Nuts, Miniature, Double Lug Floating Anchor, Steel, Cadmium Plated, 125,000 lbf/sq.in. 250 Degrees C.
SBAC AS 8611 ISSUE 4. Self-Locking Nuts, Single Long Lug Floating Anchor Steel, Cadmium Plated, 125,000 lbf/sq.in. 250 Degrees C.
SBAC AS 8612 ISSUE 5. Self-Locking Nuts, Strip Steel, Cadmium Plated, 125,000 lbf/sq.in. 250 Degrees C.
SBAC AS 8625 ISSUE 2. Self-Locking Nuts, Double Lug Anchor, CR Steel, Silver Plated, lbf/in.2 450 Degrees C.
SBAC AS 8626 ISSUE 2. Self-Locking Nuts, Miniature, Double Lug Anchor, CR Steel, Silver Plated, 125,000 lbf/in.2 450 Degrees C.
SBAC AS 8627 ISSUE 2. Self-Locking Nuts, Single Long Lug Anchor, CR Steel, Silver Plated, 125,000 lbf/in.2 450 Degrees C.
SBAC AS 8628 ISSUE 2. Self-Locking Nuts, Miniature, Single, Long Lug Anchor, CR Steel, Silver Plated, 125,000 lbf/in.2 450 Degrees C.
SBAC AS 8629 ISSUE 2. Self-Locking Nuts, Miniature, Single, Short Lug Anchor, CR Steel, Silver Plated, 125,000 lbf/in.2 450 Degrees C.
SBAC AS 8630 ISSUE 2. Self-Locking Nuts, Corner Anchor, CR Steel, Silver Plated 125,000 lbf/in.2 450 Degrees C.
SBAC AS 8631 ISSUE 2. Self-Locking Nuts, Miniature, Corner Anchor, CR Steel, Silver Plated, 125,000 lbf/in.2 450 Degrees C.
SBAC AS 8632 ISSUE 2. Self-Locking Nuts, Double Lug Anchor, CR Steel, Silver Plated, 125,000 lbf/in.2.
SBAC AS 8633 ISSUE 3(I). Self-Locking Nuts, Miniature, Double Lug Floating Anchor C.R. Steel, Silver-Plated, 125,000 lbf/in2 450 Degrees C.
SBAC AS 8634 ISSUE 2. Self-Locking Nuts, Single Long Lug Floating Anchor, CR Steel, Silver Plated, 125,000 lbf/in.2 450 Degrees C.
SBAC AS 8652 ISSUE 4. Self-Locking Nuts, Double Lug Anchor, CR Steel, 125,000 lbf/in.2 250 Degrees C.
SBAC AS 8653 ISSUE 4. Self-Locking Nuts, Miniature, Double Lug Anchor, CR Steel, 125,000 lbf/in.2 250 Degrees C.
SBAC AS 8654 ISSUE 4. Self-Locking Nuts, Single, Long Lug Anchor, CR Steel, 125,000 lbf/in.2 250 Degrees C.
SBAC AS 8655 ISSUE 4. Self-Locking Nuts, Miniature, Single Long Lug Anchor, CR Steel, 125, 000 lbf/in.2 250 Degrees C.
SBAC AS 8656 ISSUE 4. Self-Locking Nuts, Miniature, Single Short Lug Anchor, CR Steel, 125, 000 lbf/in.2 250 Degrees C.
SBAC AS 8657 ISSUE 4. Self-Locking Nuts, Corner Anchor, CR Steel, 125,000 lbf/in.2 250 Degrees C.
SBAC AS 8658 ISSUE 4. Self-Locking Nuts, Miniature, Corner Anchor, CR Steel, 125,000 lbf/in.2 250 Degrees C.

Locking Nuts *(Cont.)*

—Aircraft—Anchor *(Cont.)*
SBAC AS 8659 ISSUE 4. Self-Locking Nuts, Double Lug Floating Anchor, CR Steel, 125,000 lbf/in.2 250 Degrees C.
SBAC AS 8660 ISSUE 5(I). Self-Locking Nuts, Miniature, Double Lug Floating Anchor CR Steel, 125,000 lbf/in.2 250 Degrees C.
SBAC AS 8661 ISSUE 4. Self-Locking Nuts, Single, Long Lug Floating Anchor, CR Steel, 125,000 lbf/in.2 250 Degrees C.
SBAC AS 28003 ISSUE 2. Nut, Self-Locking, Non C.R. Steel, Long Lug Anchor, Cadmium Coated.
SBAC AS 28004 ISSUE 2. Nut—Self-Locking, Double Lug Anchor, Cadmium Coated.
SBAC AS 28005 ISSUE 2. Nut, Self-Locking, Non C.R. Steel, Corner Anchor, Cadmium Coated.
SBAC AS 28006 ISSUE 2. Nut, Self-Locking, Non C.R. Steel, Floating, Long Lug Anchor, Cadmium Coated.
SBAC AS 28007 ISSUE 4. Nut, Self-Locking, Non C.R. Steel, Floating, Double Lug Anchor, Cadmium Coated.
SBAC AS 28009 ISSUE 2. Nut, Self-Locking, Non C.R. Steel, Miniature, Single Long Lug Anchor, Cadmium Coated.
SBAC AS 28010 ISSUE 2. Nut, Self-Locking, Miniature, Double Lug Anchor, Cadmium Coated.
SBAC AS 28019 ISSUE 2. Nut, Self-Locking, Non C.R. Steel, Corner Anchor, Miniature, Cadmium Coated.
SBAC AS 46767 (V). Nut, Self Locking, Miniature, Double Lug Floating Anchor, Carbon Steel, Cadmium Coated (12500LBF/IN2) 235 Degrees Celsius.
SBAC AS 46768 (V). Nut, Self Locking, Miniature, Double Lug Floating Anchor, Cres. Steel, Silver Plated (125000 lbf/in2) 425 Degrees C.
SBAC AS 46769 (V). Nut, Self Locking, Miniature, Double Lug Floating Anchor, Cres. Steel, (125000 lbf/in2) 235 Degrees C.
SBAC AGS 2011 ISSUE 18(I). Stiffnuts—Clinch.
SBAC AGS 2023 ISSUE 11(I). Stiffnuts Double Anchor, with Translucent Ca

—Aircraft—Anchor—Countersunk
SBAC AGS 2009 ISSUE 14(I). Stiffnuts—Double Anchor—Countersunk.
SBAC AGS 2014 ISSUE 14(I). Stiffnuts—Floating Anchor—Countersunk.
SBAC AGS 2020 ISSUE 13(I). Stiffnuts—Single Anchor—Countersunk.

—Aircraft—Anchor—Thick
SBAC AGS 2007 ISSUE 14(I). Stiffnuts—Double Anchor—Thick.
SBAC AGS 2008 ISSUE 14(I). Stiffnuts—Double Anchor—Thin.
SBAC AGS 2012 ISSUE 14(I). Stiffnuts—Floating Anchor—Thick.
SBAC AGS 2018 ISSUE 12(I). Stiffnuts—Single Anchor—Thick.

—Aircraft—Anchor—Thin
SBAC AGS 2013 ISSUE 15(I). Stiffnuts—Floating Anchor—Thin.
SBAC AGS 2019 ISSUE 12(I). Stiffnuts—Single Anchor—Thin.

—Aircraft—Caps (Lids)
SBAC AGS 2021 ISSUE 11(I). Stiffnuts—Translucent Ca

—Aircraft—Clip Assemblies
SBAC AS 41100-109 (V). Nut, Self-Locking, Clip, Assembly of.

—Aircraft—Conversion Tables
SBAC AGS 2000 ISSUE 14. AGS, Stiffnuts Application.

—Aircraft—Countersunk
SBAC AGS 2017 ISSUE 14(I). Stiffnuts—Stripnut—Countersunk.

—Aircraft—Double Hexagonal
SBAC AS 42910-915 ISSUE 2. Nut, Self-Locking—Bi-Hexagon Waspaloy.

—Aircraft—Flange Restrained
SBAC AS 27850-899 (V). Nut, Self Locking, Shank, Flange Restrained.

—Aircraft—Hexagonal
ISO 7995-88. Aerospace—Nuts, Hexagonal, Self-Locking, with MJ Threads, Coated or Uncoated, Classification 1 100 MPa/235 Degrees Celsius, 1 100 MPa/315 Degrees Celsius or 1 100 MPa/425 Degrees Celsius—Dimensions Second Edition. 5 pp.
SBAC AS 8624 ISSUE 2(I). Stiffnuts—Hexagon, (Standard Hexagon).
SBAC AS 8651 ISSUE 2(I). Stiffnuts—Hexagon, (Standard Hexagon).

INTERNATIONAL AND NON-U.S. NATIONAL STANDARDS
SUBJECT INDEX
Locknuts

Locking Nuts (Cont.)

—Aircraft—Hexagonal—Captive Washers
SBAC AS 28000 ISSUE 2. Nut, Self-Locking, Non CR Steel, Captive Washer, Cadmium Coated.

—Aircraft—Hexagonal—Counterbore
ISO 8538-86. Aerospace—Self-Locking Hexagon Nuts with Counterbore and Captive Washer, Classification 1 100 MPa/235 Degrees Celsius First Edition. 5 pp.
SBAC AS 52600-699 ISSUE 1. Nut—Reduced—Hexagon—Self Locking—Deep Counterbored—Metric Series.
SBAC AS 54361-54460 (V). Nut—Reduced Hexagon—Self Locking—Deep Counterbored—Inch Series.

—Aircraft—Hexagonal—Countersunk
SBAC AGS 2003 ISSUE 14(I). Stiffnuts—Hexagon—Countersunk.

—Aircraft—Hexagonal—Flanged
SBAC AS 8600 ISSUE 4. Self-Locking Nuts, Hexagon, Flange Type, Steel, Cadmium Plated, 160,000LBS/SQ.IN. 250 Degrees Celsius.
SBAC AS 8623 ISSUE 2. Self-Locking Nuts, Hexagon, Flange Type, C.R. Steel, Silver Plated, 125,000 lbf/in2. 450 Degrees C.
SBAC AS 8650 ISSUE 4. Self-Locking Nuts, Hexagon, Flange Type, CR Steel, 125,000 lbf/in.2 250 Degrees C.

—Aircraft—Hexagonal—Thick
SBAC AGS 2001 ISSUE 14(I). Stiffnuts—Hexagon—Thick.

—Aircraft—Hexagonal—Thin
SBAC AGS 2002 ISSUE 14(I). Stiffnuts—Hexagon—Thin.

—Aircraft—Holders
SBAC RS 698 ISSUE 1. Application of Self Locking Nut Retainers.

—Aircraft—Plate
SBAC AS 46720-722 (V). Nut, Self Locking, Plate.

—Aircraft—Shafts (Machine Elements)
SBAC AS 27300-349 (V). Nut, Self-Locking, Shaft.
SBAC AS 42440-458 (V). Nut, Self-Locking, Shaft.
SBAC TS 9 ISSUE 3. All Metal Self Locking Shaft Nuts (Unified Threads) Non-Corrosion Resistant Steel Rated Temperature 235 Degrees C.
SBAC TS 129 ISSUE 2. SBAC Performance Specification for All Metal Self Locking Shaft Nuts (Unified Threads) Non-Corrosion Resistant Steel Rated Temperature 350 Degrees C.

—Aircraft—Shank
SBAC AS 44708-714 (V). Nut, Self Locking, Shank, Flanged Restrained.
SBAC AS 44750-753 (V). Nut, Self Locking, Shank, Flange Restrained.
SBAC TS 8 ISSUE 6. All Metal Self-Locking Nuts (UNJ Threads) 1100 MPa (160,000 lbf/in2) Minimum Tensile Strength at Room Temperature.
SBAC TS 26 ISSUE 2. All Metal Self-Locking Nuts (UNJ Threads) 1210 MPa (175,000 Lbf/in2) Minimum Tensile Strength at Ambient Temperature.
SBAC TS 128 ISSUE 3. All Metal Self-Locking Nuts (UNJ Threads) 1100 MPa (160,000 lbf/in2) Minimum Tensile Strength at Ambient Temperature.
SBAC TS 149 ISSUE 1. All Metal Self-Locking Nuts (UNJ Threads) 1550 MPa (225,000 lbf/in2) Minimum Tensile Strength at Ambient Temperature.

—Aircraft—Shank—Flange Restrained
SBAC AS 54470 ISSUE 1. Nut, Self Locking, Shank, Flange Restrained 0.1900—32 UNJF—3B BS 4084, Based on AS44750 with 1g, 2g, 3g and 6g of Additional Mass.
SBAC AS 54471 ISSUE 2. Nut, Self Locking, Shank, Flange Restrained, 0.2500—28 UNJF—3B BS4084, Based on AS44751 with 2g, 5g and 12g of Additional Mass.
SBAC AS 54472 ISSUE 2. Nut, Self Locking, Shank, Flange Restrained, 0.3125—24 UNJF—3B BS4084, Based on AS44752 with 2g, 5g and 14g of Additional Mass.
SBAC AS 54473 ISSUE 2. Nut, Self Locking, Shank, Flange Restrained, 0.3750—24 UNJF—3B BS4084, Based on AS44753 with 3g, 5g and 15g of Additional Mass.
SBAC AS 55200 ISSUE 1. Nut, Self Locking, Shank, Flange Restrained, 0.1900-32 UNJF-3B BS4084. Based on AS27852 with 1g, 2g and 3g of Additional Mass.
SBAC AS 55201 ISSUE 1. Nut, Self Locking, Shank, Flange Restrained, 0.2500-28 UNJF-3B BS4084. Based on AS27857 with 2g, 4g and 6g of Additional Mass.
SBAC AS 55203 ISSUE 1. Nut, Self Locking, Shank, Flange Restrained, 0.3750-24 UNJF-3B BS4084. Based on AS27883 with 3g, 6g and 10g of Additional Mass.
SBAC AS 62300-349 ISSUE 1. Nut, Self Locking, Shank, Flange Restrained, Waspaloy.

Locking Nuts (Cont.)

—Aircraft—Spline
SBAC AS 44860-869 ISSUE 1. Nut, Self Locking, Spline Drive—BS HR 650 (AMS 5737 or AMS 5732).

—Aircraft—Thick
SBAC AGS 2015 ISSUE 14(I). Stiffnuts—Stripnut—Thick.

—Aircraft—Thin
SBAC AGS 2016 ISSUE 15(I). Stiffnuts—Stripnut—Thin.

—Cap
CNS B2194-83. Self-Locking Domed Cap Nuts (Feb)(4315).

—Hexagonal—Heavy
CNS B2192-83. Self-Locking Hexagon Nuts Heavy Type (Feb)(4313).

—Hexagonal—Thin
CNS B2193-83. Self-Locking Hexagon Nuts Thin Type (Feb)(4314).

Locking Plates
Use: Lockplates

Locking Rings
Use: Lockrings

Locking Screws
Use For: Self Locking Screws *See Also:* Fasteners; Locking Fasteners; Screws

—Aircraft
SBAC AS 6553 ISSUE 1. Locking, Screw.
SBAC AS 52790-799 ISSUE 1. Insert—Inch, Thinwall—Long Screw Locking (INCO 718).

Locking Thread Inserts
Use For: Self Locking Thread Inserts
See Also: Inserts (Fasteners); Thread Inserts

—Aerospace
CEN PREN 2942-90. Insert, Screw Thread, Helical Coil, Self-Locking, in Heat Resisting Alloy, Nickel Base Alloy (INCO X750-EN 3018) Silver Plated. 7 pp.
CEN PREN 2943-90. Wire Thread Inserts, Prevailing Torque (Self-Locking) Technical Specification. 18 pp.
CEN PREN 2944-90. Insert, Screw Thread, Helical Coil, Self-Locking in Corrosion Resisting Alloy (Z10CN18-09-EN2947) Unplated. 7 pp.
CEN PREN 2945-90. Assembly with Self-Locking Thread Inserts. 7 pp.
CEN PREN 3044-90. Installation of Self-Locking Thread Inserts Design Dimensions. 6 pp.
CEN PREN 3542-90. Insert, Screw Thread, Helical Coil, Self-Locking, in Heat Resisting Alloy, Nickel Base Alloy (INCO X750-EN 3018), Unplated. 7 pp.
DIN ENGL LN 9499 Pt 1-86. Aerospace; Inserts, Helical Coil Threads; Screw-Locking; Class: 1100 MPa/235 Degrees C/425 Degrees C (Dec). 8 pp.
DIN ENGL LN 9499 Pt 2-86. Aerospace; Inserts, Helical Coil Threads, Screw-Locking, Assembly and Special Tools (Dec). 3 pp.
DIN ENGL LN 29534 Pt 1-70. Threaded Inserts; Locked and Self-Locking for Temperatures up to 800 Degrees Celsius (Nov). 2 pp.
DIN ENGL LN 29538-67. Thread Inserts for Locking with Serrated Ring; Internal Thread Locked for Temperatures up to 260 Degrees C (Oct). 2 pp.
DIN ENGL LN 29539-75. Thread Inserts for Locking with Serrated Ring; Internal Thread Locked; Corrosion Resistant for Temperatures up to 425 Degrees C (Nov). 2 pp.
DIN ENGL LN 29540 Pt 1-75. Threaded Inserts; Locked and Self-Locking for Temperatures up to 235 Degrees C, Corrosion Resistant (May). 2 pp.
DIN ENGL LN 29540 Pt 2-67. Threaded Inserts; Locked and Self-Locking; Directions for Design and Installation (Oct). 2 pp.
DIN ENGL LN 29908-80. Thread Inserts; Locked and Self-Locking for Temperatures up to 235 Degrees C (Aug). 3 pp.

—Aerospace—Thick Wall
CEN PREN 3831-91. Inserts, Thick Wall, Self-Locking, in Heat Resisting Steel FE-PM61 (17-4PH), MoS2 Lubricated Classification: 1100 MPa (at Ambient Temperature)/350 Degrees C. 5 pp.
CEN PREN 3899-91. Inserts, Thick Wall, Self-Locking, in Heat Resisting Steel FE-PM61 (17-4PH), MoS2 Lubricated Classification: 1100 MPa/350 Degrees C Technical Specification. 28 pp.

—Aerospace—Thin Wall
AECMA PREN3236-89. Inserts, Thin Wall, Short Length in Heat Resisting Nickel Base Alloy N1-P100HT (Inconel 718). 5 pp.

Locking Thread Inserts (Cont.)

—Aerospace—Thin Wall (Cont.)
AECMA PREN3297. Inserts, Thin Wall, Self-Locking in Heat Resisting Nickel Base Alloy N1-P100HT (Inconel 718) Classification: 1275 MPa/550 Degrees Celsius. 21 pp.
CEN PREN 3236-93. Aerospace Series Inserts, Thin Wall, Short, in Heat Resisting Nickel Base Alloy NI-P100HT (Inconel 718). 4 pp.
CEN PREN 3237-91. Inserts, Thin Wall, Long, in Heat Resisting Nickel Base Alloy NI-P100HT (Inconel 718). 7 pp.
CEN PREN 3237-93. Aerospace Series Inserts, Thin Wall, Long, in Heat Resisting Nickel Base Alloy NI-P100HT (Inconel 718). 4 pp.
SBAC AS 62130-139 ISSUE 1. Insert Salvage—Metric, Thinwall—Short, Screw Locking (INCO 718).
SBAC AS 62140-149 ISSUE 1. Insert Salvage—Inch Thinwall—Short, Screw Locking (INCO 718).
SBAC AS 62150-159 ISSUE 1. Insert Salvage—Metric, Thinwall—Long Screw Locking (INCO 718).
SBAC AS 62160-169 ISSUE 1. Insert Salvage—Inch Thinwall—Long Screw Locking (INCO 718).
SBAC TS 146 ISSUE 1. Performance of Insert—Metric Thinwall, Screw Locking, EN 2404 (Inconel 718) Silver Coated.

—Aerospace—Thin Wall—Design
CEN PREN 3676-93. Aerospace Series Inserts, Thin Wall Design Standard. 12 pp.

—Aerospace—Thin Wall—Installation
CEN PREN 3298-93. Aerospace Series Inserts, Thin Wall Installation and Removal Procedure. 10 pp.
SBAC TS 152 ISSUE 1. Installation and Removal of Inserts—Inch Threaded, Thinwall, Short and Long Series.

—Aerospace—Thin Wall—Quality Assurance
SBAC TS 153 ISSUE 1. Insert Salvage, Imperial, Thinwall Screw Locking, Inconel 718 Silver Coated.
SBAC TS 156 ISSUE 1. Insert Salvage Metric, Thinwall, Screw Locking, Inconel 718 Silver Coated.

—Aerospace—Thin Wall—Removal
CEN PREN 3298-93. Aerospace Series Inserts, Thin Wall Installation and Removal Procedure. 10 pp.
SBAC TS 152 ISSUE 1. Installation and Removal of Inserts—Inch Threaded, Thinwall, Short and Long Series.

—Aircraft
SBAC AS 52760-769 ISSUE 2. Insert-Inch, Thinwall—Short, Screw Locking (INCO 718).
SBAC AS 52770-780 ISSUE 2. Insert-Metric, Thinwall—Long, Screw Locking (INCO 718).
SBAC AGS 4678 (V). Thread Insert—Metric—Solid Wall S/Locking—Alloy Steel, Cad. Plated.
SBAC TS 84 ISSUE 1. Insert, Threaded, Metric, Solid Wall, Self Locking, Alloy Steel, Cadmium Plated.
SBAC TS 138 ISSUE 3. Insert, Imperial, Thinwall, Screw Locking, DTD 5638 (Inconel 718) Silver Coated.

Locknuts
Use For: Jam Nuts; Lock Nuts; Palnuts; Stiffnuts
See Also: Fasteners; Hexagonal Nuts; Locking Nuts; Nuts (Fasteners); Square Nuts
CSA CAN/CSA-C22. 2 NO 18-92. Outlet Boxes, Conduit Boxes, and Fittings; (Gen Instr 1 Thru 2). 118 pp.

—Aircraft
BSI 3A 125-132-81. 1981 3A 136-143, 3A 147-168, 3A 180-181, 3A 186-187, 3A 192-193, 3A 200-201, 2A 213-216 (Unified Threads) for Aircraft. 30 pp.
SBAC AGS 1710 ISSUE 5. Locknut L.T. Union.

—Aircraft—Union—Fuel Filters
SBAC AS 3272 ISSUE 2. Union, Locknut—Filter Fuel Type 'B'. 1 p.

—Antifriction Bearings
BSI BS 5646: Part 1-78. 1978 Rolling Bearings—Accessories Part 1: Locknuts, Narrow Series, and Lockwashers with Straight Inner Tab. 4 pp.
BSI BS 5646: Part 2-78. 1978 Rolling Bearings—Accessories Part 2: Locknuts, Wide Series, and Lockwashers with Bent Inner Tab. 4 pp.
CNS B2257-78. Rolling Bearing Accessories (Lock Nuts) (Aug)(4464).
DIN ENGL 981-83. Rolling Bearing Accessories; Locknuts (Jan). 2 pp.
ISO 2982-72. Rolling Bearings—Locknuts, Narrow Series, and Lockwashers with Straight Inner Tab First Edition. 4 pp.
ISO 2983-75. Rolling Bearings—Locknuts, Wide Series, and Lockwashers with Bent Inner Tab First Edition. 4 pp.
JIS B 1554-93. Rolling Bearings—Locknuts, Lockwashers and Lockplates. 20 pp.

INDUSTRY STANDARDS

Locknuts (Cont.)

—Electric Conduit Fittings
CNS C4227-86. Lock Nuts for Rigid Steel Conduit (Oct)(6083).

—Hexagonal—Prevailing Torque
CNS B2166-83. Torque-Type Steel Hexagon Locknuts (Feb)(4230). 3 pp.

—Hexagonal—Prevailing Torque—Steel—Mechanical Properties
CNS B2168-78. Torque Type Steel Hexagon Locknuts and Performance Properties (Mar)(4232).
CNS B7084-78. Method of Test for Mechanical and Performance Properties of Torque Type Steel Hexagon Locknuts (Mar)(4233).

—Machine Tools
BSI BS 4185: Part 1-67. 1967 Amd 2 Machine Tool Components Part 1: Locknuts ('C' Type, Socket Set Screw Locking). 14 pp.
BSI BS 4185: Part 9-76. 1976 Amd 1 Machine Tool Components Part 9: Levelling Screws, Nuts and Seating Plates. 8 pp.

Lockplates

Use For: Locking Plates *See Also:* Fasteners; Washers (Fasteners)

—Aircraft
SBAC AGS 3421 ISSUE 3. Locking Plate.

—Aircraft—Lever Switches
SBAC AS 3132 ISSUE 2. Locking Plate—Switch Lever. 1 p.
SBAC AS 3306 ISSUE 2. Locking, Plate—Switch Lever. 1 p.

—Antifriction Bearings
JIS B 1554-93. Rolling Bearings—Locknuts, Lockwashers and Lockplates. 20 pp.
JIS B 1555-79. Lock Washers and Lock Plates for Rolling Bearings. 10 pp.

—Groove Nuts
CNS B2260-78. Groove Nuts (Securing by Locking Plates) (Aug)(4467).

Lockrings

Use For: Locking Rings; Ring Locks
See Also: Rings

—Aerospace—Serrated
DIN ENGL LN 29558-74. Serrated Lockrings for Temperatures up to 260 Degrees C (June). 2 pp.

—Aerospace—Serrated—Corrosion Resistant
DIN ENGL LN 29559-67. Serrated Lockrings for Temperatures up to 700 Degrees C; Corrosion Resistant (Oct). 2 pp.

—Aerospace—Serrated—Corrosion Resistant—Studs (Fasteners)
DIN ENGL LN 29519-67. Studs for Locking with Serrated Ring with Minimum Tensile Strength of 125kg/mm2 for Temperatures up to 700 Degrees C, Corrosion Resistant (Oct). 3 pp.

—Aerospace—Serrated—Corrosion Resistant—Thread Inserts
DIN ENGL LN 29539-75. Thread Inserts for Locking with Serrated Ring; Internal Thread Locked; Corrosion Resistant for Temperatures up to 425 Degrees C (Nov). 2 pp.

—Aerospace—Serrated—Studs (Fasteners)
DIN ENGL LN 29518-67. Studs for Locking with Serrated Ring with Minimum Tensile Strength of 90kg/mm2 for Temperatures up to 260 Degrees C (Oct). 3 pp.

—Aerospace—Serrated—Thread Inserts
DIN ENGL LN 29538-67. Thread Inserts for Locking with Serrated Ring; Internal Thread Locked for Temperatures up to 260 Degrees C (Oct). 2 pp.

—Aerospace—Threaded
AECMA PREN3296-88. Rings, Threaded, Self-Locking, Silver-Coated in FE-PA92HT(A286). 5 pp.
CEN PREN 3296-92. Rings, Threaded, Self-Locking, in FE-PA92HT (A286), Silver Plated. 6 pp.

—Control Wheels—Stops—Aircraft
SBAC AS 3131 ISSUE 1. Ring Locking for Brake Stop Pilot's Control Wheel. 1 p.

—Retaining—Aircraft
SBAC AS 42951 (V). Ring, Retaining, (Friction Locking).

—Slotted—Face Wrenches
CNS B3325-80. Assembly Tools for Screws and Nuts for Slotted Lock Rings (Jan)(5104).

Locks (Hardware)

Use For: Locking Devices *See Also:* Hardware; Latches (Fasteners); Locks (Security); Slide Latches; Spring Catches

—Earthmoving Equipment—Frame
BSI BS 6912: Part 8-92. 1992 Safety of Earth-Moving Machinery Part 8: Specification for Performance Requirements of an Articulated Frame Locking Device (ISO 10570: 1992). 6 pp.
ISO 10570-92. Earth-Moving Machinery—Articulated Frame Lock—Performance Requirements First Edition. 4 pp.

—Flap Valves—Ships
DIN ENGL 87106-75. Locking Devices for Non-Return Flaps in Accordance with DIN 87 101 (June). 3 pp.

—Furniture
BSI BS 4951-73. 1973 Builders' Hardware: Lock and Latch Furniture (Doors). 17 pp.

—Furniture—Glossaries
ISO 8554 Pt 1-87. Hardware for Furniture—Terms for Furniture Locks—Part 1: Latch Lock, Dead Lock, Rod-Operating Lock, Central Locking System, Cylinder Lock, Combination Lock First Edition. 16 pp.
ISO 8554 Pt 2-87. Hardware for Furniture—Terms for Furniture Locks—Part 2: Surface-Mounted Lock, Inset-Type Lock, Mortise Lock First Edition. 6 pp.
ISO 8554 Pt 3-87. Hardware for Furniture—Terms for Furniture Locks—Part 3: Left Hand Lock, Right Hand Lock, down Lock, up Lock First Edition. 6 pp.
ISO 8554 Pt 4-87. Hardware for Furniture—Terms for Furniture Locks—Part 4: Key, Rotary Lock-Handle, Cylinder First Edition. 10 pp.

—Glossaries
BSI BS 3827: Part 1-64. (WITHDRAWN) 1964 Glossary of Terms Relating to Builders' Hardware Part 1: Locks (Including Locks and Latches in One Case) (Superseded by BS 6100: Subsection 1.3.6: 1991). 34 pp.

—Sewing Machine Tables—Household
CNS B2514-89. Hinges and Locks for Household Sewing Machine Table (Sep)(6000).
JIS B 9037-78. Hinges and Locks for Household Sewing Machine Table (R 1983). 13 pp.

—Taper—Molds (Casting)
JIS B 5119-89. Taper Lock of Moulds for Plastics. 8 pp.

Locks (Security)

See Also: Automotive Locks; Cylinder Locks; Door Chains; Fasteners; Keys (Locks); Lever Tumbler Locks; Lock Cases; Lock Covers; Locks (Hardware); Locksmithing; Padlocks; Pin Tumbler Locks; Tubular Locks

BSI BS 7480-92. 1992 Security Seals (H). 15 pp.
CGSB CAN/CGSB-69.21-M90. Auxiliary Locks and Associated Products (ANSI/BHMA A156.5-1984). 61 pp.
CGSB CAN/CGSB-69.28-M90. Interconnected Locks and Latches (ANSI/BHMA A156.12-1986). 23 pp.

—Aircraft Door
SBAC AS 2542 ISSUE 5. Main Body Left Hand. 1 p.
SBAC AS 2543 ISSUE 5. Main Body Right Hand. 1 p.

—Aircraft Door—Assembly
SBAC AS 6520 ISSUE 1. Main Body—Left Hand.
SBAC AS 6521 ISSUE 1. Main Body—Right Hand.

—Aircraft Door—Bolt Rods
SBAC AS 6529 ISSUE 1. Short Bolt Rod.
SBAC AS 6530 ISSUE 1. Long Bolt Rod.

—Aircraft Door—Pivot Bolts
SBAC AS 6531 ISSUE 1. Pivot Bolt.

—Aircraft Door—Plates (Supports)
SBAC AS 6522 ISSUE 1. Cover Plate Assembly.

—Aircraft Door—Pushbuttons
SBAC AS 6534 ISSUE 2. Push Button Outer, Door Lock.

—Aircraft Door—Screws
SBAC AS 6535 ISSUE 1. Special Mushroom Head, Screw.

—Aircraft Door—Spiders
SBAC AS 6526 ISSUE 1. Spider Assembly R.H..
SBAC AS 6527 ISSUE 1. Spider.
SBAC AS 6528 ISSUE 1. Spider Assembly L.H..

—Aircraft Door—Stop Bolts
SBAC AS 6532 ISSUE 1. Stop, Bolt.

Locks (Security) (Cont.)

—Aircraft Door—Stops
SBAC AS 6533 ISSUE 1. Sto

—Aircraft—Handles
SBAC AS 6523 ISSUE 1. Outer Handle Assembly—Double Handle Lock.
SBAC AS 6536 ISSUE 1. Outer Handle Assembly—Single Handle Lock.

—Aircraft—Spindles
SBAC AS 6524 ISSUE 1. Spindle Assembly—Double Handle Lock.
SBAC AS 6525 ISSUE 1. Spindle—Double Handle Lock.
SBAC AS 6537 ISSUE 1. Spindle Assembly Single Handle Lock.
SBAC AS 6538 ISSUE 1. Spindle—Single Handle Lock.

—Bicycles
JIS D 9456-61. Locks for Bicycles (R 1973). 5 pp.

—Cabinet
CGSB CAN/CGSB-69.27-M90. Cabinet Locks (ANSI/BHMA A156.11-1985). 33 pp.

—Cocks—Marine
CNS F3192-82. Marine Cocks with Lock (Aug)(9254).
JIS F 7390-89. Marine Cocks with Locks.

—Comprehensive Testing
CGSB CAN/CGSB-69.17-M86. Bored and Preassembled Locks and Latches. 65 pp.

—Doors
BSI BS 3621-80. 1980 Amd 2 Thief Resistant Locks. 12 pp.
BSI BS 3621-03. 1980 Amd 3 Thief Resistant Locks (AMD 7282) August 15, 1992. 17 pp.
BSI BS 5872-80. 1980 Locks and Latches for Doors in Buildings. 22 pp.
CGSB CAN/CGSB-69.17-M86. Bored and Preassembled Locks and Latches. 65 pp.
CGSB CAN/CGSB-69.21-M90. Auxiliary Locks and Associated Products (ANSI/BHMA A156.5-1984). 61 pp.
SNZ NZS/BS 5872-80. Specification for Locks and Latches for Doors in Buildings. 24 pp.

—Doors—Automotive
BSI BS AU 209: Part 1-86. (WITHDRAWN) 1986 Vehicle Security Part 1: Mechanical Locking System for Passenger Cars and Car Derived Vehicles (Superseded by BS AU 209: Part 1a: 1992). 7 pp.
BSI BS AU 209: Part 1A-92. 1992 Vehicle Security Part 1a: Specification for Locking Systems for Passenger Cars and Car Derived Vehicles (E). 9 pp.

—Doors—Buses (Vehicles)
CNS D2107-87. Door Locks for Buses (Jul)(7969).

—Doors—Electric
IEC 730 Pt 2-12-93. Automatic Electrical Controls for Household and Similar Use Part 2: Particular Requirements for Electrically Operated Door Locks First Edition. 31 pp.

—Doors—Exit Devices
CGSB CAN/CGSB-69.19-M89. Exit Devices (ANSI/BHMA A156.3-1984). 27 pp.

—Luggage Compartments—Automotive
BSI BS AU 209: Part 1A-92. 1992 Vehicle Security Part 1a: Specification for Locking Systems for Passenger Cars and Car Derived Vehicles (E). 9 pp.

—Mortise
CGSB CAN/CGSB-69.29-M90. Mortise Locks and Latches (ANSI/BHMA A156.13-1980). 21 pp.

—Mortise—Doors
CGSB CAN/CGSB-69.21-M90. Auxiliary Locks and Associated Products (ANSI/BHMA A156.5-1984). 61 pp.

—Mortise—Lever Tumbler
CNS A2241-90. Lever Tumbler Mortise Locks (Apr)(12696). 5 pp.
CNS A3315-90. Method of Test for Lever Tumbler Mortise Locks (Apr)(12697). 3 pp.
JIS A 5515-79. Lever Tumbler Mortise Locks.

—Windows—Buses (Vehicles)
CNS D2095-87. Sash Locks for Buses (Jul)(7678).
JIS D 4704-76. Sash Locks for Buses (R 1984). 7 pp.

—Windows—Rolling Stock
JIS E 7201-80. Sash-Locks for Passenger-Carrying Cars.

Locks (Waterways)

See Also: Concrete Structures; Hydraulic Structures; Lock Gates; Waterways

INTERNATIONAL AND NON-U.S. NATIONAL STANDARDS
SUBJECT INDEX

Locks (Waterways) *(Cont.)*
—Design
BSI BS 6349: Part 3-88. 1988 Code of Practice for Maritime Structures; Part 3: Design of Dry Docks, Locks, Slipways and Shipbuilding Berths, Shiplifts and Dock and Lock Gates. 75 pp.

Locksmithing
See Also: Locks (Security)
—Classification
BSI BS 1000: (682/683)-73. 1973 Universal Decimal Classification (UDC). English Full Edition (682/683): Smithery. Farriery. Hand-Forged Ironwork. Ironmongery. Hardware. Locksmithing. Gunsmithing. Bottling. Lamps. Stoves. 27 pp.
SNZ NZS/BS 1000 (682/683)-73. Universal Decimal Classification Smithery. Farriery. Hand-Forged Ironwork. Ironmongery. Hardware. Locksmithing. Gunsmithing. Bottling. Lamps. Stoves. 28 pp.
—Construction Contracts
DIN ENGL 18360-88. Tendering and Performance Stipulations in Contracts for Construction Works (VOB); Part C: General Technical Specifications in Contracts for Construction Works (ATV); Metalwork (Sept) (This Standard, Together with DIN 18299, September 1988 Edition,. 10 pp.

Lockwashers
Use: Lock Washers

Lockwire
See Also: Fasteners; Lockwire Holes; Wire
—Aerospace—Corrosion Resistant
CEN PREN 3628-91. Lockwire Drawn Corrosion Resisting Steel. 4 pp.
DIN ENGL LN 9424-87. Aerospace; Lockwire, Drawn, Corrosion-Resisting; Dimensions, Masses (Aug). 4 pp.
MOD UK DSTAN 95-7-01. Wire for Locking Purposes and Wire Locking Practice for Aerospace Use Issue 1; Amendment 2. 7 pp.
—Aircraft
ISO 245-86. Aircraft—Lockwire First Edition. 3 pp.
NATO STANAG 3625 ED 3 AMD 2-89. Aircraft Lockwire. 7 pp.
NATO STANAG 3625 ED 3 AMD 3-89. Aircraft Lockwire. 8 pp.
—Aircraft—Control Systems Equipment
NATO STANAG 3752 ED 2 AMD 3-89. Witness (Breaking) Wire for Aircraft Emergency Controls and Equipment. 8 pp.
NATO STANAG 3752 ED 2 AMD 4-89. Witness (Breaking) Wire for Aircraft Emergency Controls and Equipment. 9 pp.
—Aircraft—Corrosion Resistant
SBAC AS 44725 ISSUE 1. Wire, Locking Corrosion Resisting Steel for Temperatures up to 600 Degrees C.

Lockwire Holes
See Also: Fasteners; Lockwire
—Aerospace—Bolts—Identification Systems
DIN ENGL LN 29962-76. Bolts and Accessories; Additional Procurement Data for Special Designs and Arrangement of Code Letters in Characteristic Blocks (Apr). 3 pp.
—Aircraft—Bolts
SBAC RS 704 (V). Recommended Size and Location of Locking Wire Holes in Bolts Screws and Nuts.
—Aircraft—Electric Connectors
SBAC TS 375 ISSUE 1. Locking Wire Hole Strength.
—Aircraft—Fasteners
SBAC RS 697 (V). Wire Locking of Threaded Items.
—Aircraft—Nuts (Fasteners)
SBAC RS 704 (V). Recommended Size and Location of Locking Wire Holes in Bolts Screws and Nuts.
—Aircraft—Screws
SBAC RS 704 (V). Recommended Size and Location of Locking Wire Holes in Bolts Screws and Nuts.

Locomotives
Use For: Mine Locomotives; Railroad Engines; Trolley Locomotives See Also: Battery Locomotives; Mine Cars; Mining Equipment; Railroad Equipment; Rolling Stock
—Diesel—Inspection
JIS E 4044-90. General Rules for the Inspection of Diesel Locomotives on Completion of Construction.
JIS E 4044-67. General Rules for the Inspection of Diesel Locomotives on Completion of Construction (R 1976). 10 pp.

Locomotives *(Cont.)*
—Electric—Inspection
JIS E 4042-90. General Rules for the Inspection of Electric Locomotives on Completion of Construction.
—Grease Nipples
BSI BS 1486: Part 2-61. (OBSOLESCENT) 1961 Lubricating Nipples Part 2: Heavy Duty Lubricating Nipples. 19 pp.
—Lubricants
MOD UK DSTAN 91-42-78. Lubricating Oil, Petroleum: Compressor, Light Joint Service Designation: OM-58 Lubricating Oil, Petroleum: Compressor, Medium Joint Service Designation: OM-160 Issue 1. 6 pp.

Lofting (Layouts)
Use: Layouts

Log Splitters
—Safety
CEN PREN 609-91. Safety Requirements for Agricultural and Forestry Machinery—Log Splitters. 16 pp.

Logarithms
—Symbols
CENELEC HD 245.3-76. Letter Symbols to be Used in Electrical Technology Part 3: Logarithmic Quantities and Units. 2 pp.
CENELEC HD 245.3 S2-91. Letter Symbols to Be Used in Electrical Technology Part 3: Logarithmic Quantities and Units. 5 pp.
IEC 27 Pt 3-89. Letter Symbols to Be Used in Electrical Technology Part 3: Logarithmic Quantities and Units Second Edition. 26 pp.
SAA AS 1046.3-91. Letter Symbols for Use in Electrotechnology—Part 3: Logarithmic Quantities and Units (IEC 27-3:1989) (in Professional Package 56). 9 pp.

Logging Equipment
Scope Note: Use a more specific term See: Log Splitters; Tree Harvesters

Logging Stations (Data Transmission)
See Also: Data Logging
—Local Area Networks
BSI BS 6531: Part 6-84. (WITHDRAWN) 1984 10 Mbps Slotted Ring Local Area Network Part 6: Specification for Logging Station (Superseded by BS ISO 8802-7: 1991). 12 pp.

Logic Analyzers
Use For: Performance Monitor Systems
BSI BS 6653-85. 1985 Method for Expression of the Properties of Logic Analyzers. 18 pp.
CENELEC HD 469-87. Expression of the Properties of Logic Analysers. 3 pp.
IEC 776-83. Expression of the Properties of Logic Analyzers First Edition. 36 pp.

Logic Circuits
See Also: Digital Circuits; Gate Circuits; Logic Elements; Majority Gates; NOT Circuits
—Bipolar
IECQ QC 790132/SU 0005-90. Detail Specification for Electronic Components Digital Integrated Circuits KII02AP4—KII02AP14. 66 pp.

Logic Diagrams
See Also: Diagrams
BSI BS 5070: Part 4-90. 1990 Engineering Diagram Drawing Practice Part 4: Recommendations for Logic Diagrams. 42 pp.
CENELEC HD 246.7-84. Diagrams, Charts Tables Preparation of Logic Diagrams. 2 pp.
IEC 113 Pt 7-81. Diagrams, Charts, Tables Part 7: Preparation of Logic Diagrams First Edition; (Corrigenda—Aug 1982). 57 pp.
—Electrical Components
SAA AS 1103.8-86. Diagrams, Charts and Tables for Electrotechnology—Part 8: Guiding Principles for the Preparation of Logic Diagrams. 26 pp.
SNZ NZS/AS 1103. 8-86. Diagrams, Charts and Tables for Electrotechnology Part 8: Guiding Principles for the Preparation of Logic Diagrams (This is a Joint Standard with SAA AS 1103.8). 26 pp.
—Signaling Systems
CCITT RECMN Q.611-89. Logic Procedures for Incoming Signalling System No. 4—Interworking of Signalling Systems (Study Group XI) 4 pp. 4 pp.
CCITT RECMN Q.612-89. Logic Procedures for Incoming Signalling System No. 5—Interworking of Signalling Systems (Study Group XI) 5 pp. 5 pp.

Logic Diagrams *(Cont.)*
—Signaling Systems *(Cont.)*
CCITT RECMN Q.613-89. Logic Procedures for Incoming Signalling System No. 6—Interworking of Signalling Systems (Study Group XI) 10 pp. 10 pp.
CCITT RECMN Q.614-89. Logic Procedures for Incoming Signalling System No. 7 (TUP)—Interworking of Signalling Systems (Study Group XI) 14 pp. 14 pp.
CCITT RECMN Q.615-89. Logic Procedures for Incoming Signalling System R1—Interworking of Signalling Systems (Study Group XI) 3 pp. 3 pp.
CCITT RECMN Q.616-89. Logic Procedures for Incoming Signalling System R2—Interworking of Signalling Systems (Study Group XI) 5 pp. 5 pp.
CCITT RECMN Q.621-89. Logic Procedures for Outgoing Signalling System No. 4—Interworking of Signalling Systems (Study Group XI) 5 pp. 5 pp.
CCITT RECMN Q.622-89. Logic Procedures for Outgoing Signalling System No. 5—Interworking of Signalling Systems (Study Group XI) 4 pp. 4 pp.
CCITT RECMN Q.623-89. Logic Procedures for Outgoing Signalling System No. 6—Interworking of Signalling Systems (Study Group XI) 6 pp. 6 pp.
CCITT RECMN Q.624-89. Logic Procedures for Outgoing Signalling System No. 7 (TUP)—Interworking of Signalling Systems (Study Group XI) 11 pp. 11 pp.
CCITT RECMN Q.625-89. Logic Procedures for Outgoing Signalling System R1—Interworking of Signalling Systems (Study Group XI) 3 pp. 3 pp.
CCITT RECMN Q.626-89. Logic Procedures for Outgoing Signalling System R2—Interworking of Signalling Systems (Study Group XI) 5 pp. 5 pp.
CCITT RECMN Q.653-89. Logic Procedures for Interworking of Signalling System No. 6 to No. 7 (TUP)—Interworking of Signalling Systems (Study Group XI) 4 pp. 4 pp.
CCITT RECMN Q.654-89. Logic Procedures for Interworking of Signalling System No. 6 to R1—Interworking of Signalling Systems (Study Group XI) 2 pp. 2 pp.
CCITT RECMN Q.655-89. Logic Procedures for Interworking of Signalling System No. 6 to R2—Interworking of Signalling Systems (Study Group XI) 3 pp. 3 pp.
CCITT RECMN Q.662-89. Logic Procedures for Interworking of Signalling System No. 7 (TUP) to No. 5—Interworking of Signalling Systems (Study Group XI) 3 pp. 3 pp.
CCITT RECMN Q.663-89. Logic Procedures for Interworking of Signalling System No. 7 (TUP) to No. 6—Interworking of Signalling Systems (Study Group XI) 3 pp. 3 pp.
CCITT RECMN Q.664-89. Logic Procedures for Interworking of Signalling System No. 7 (TUP) to No. 7 (TUP)—Interworking of Signalling Systems (Study Group XI) 5 pp. 5 pp.
CCITT RECMN Q.665-89. Logic Procedures for Interworking of Signalling System No. 7 (TUP) to R1—Interworking of Signalling Systems (Study Group XI) 3 pp. 3 pp.
CCITT RECMN Q.666-89. Logic Procedures for Interworking of Signalling System No. 7 (TUP) to R2—Interworking of Signalling Systems (Study Group XI) 3 pp. 3 pp.
CCITT RECMN Q.671-89. Logic Procedures for Interworking of Signalling System R1 to No. 5—Interworking of Signalling Systems (Study Group XI) 3 pp. 3 pp.
CCITT RECMN Q.672-89. Logic Procedures for Interworking of Signalling System R1 to No. 6—Interworking of Signalling Systems (Study Group XI) 3 pp. 3 pp.
CCITT RECMN Q.673-89. Logic Procedures for Interworking of Signalling System R1 to No. 7 (TUP)—Interworking of Signalling Systems (Study Group XI) 3 pp. 3 pp.
CCITT RECMN Q.674-89. Logic Procedures for Interworking of Signalling System R1 to R2—Interworking of Signalling Systems (Study Group XI) 3 pp. 3 pp.
CCITT RECMN Q.681-89. Logic Procedures for Interworking of Signalling System R2 to No. 4—Interworking of Signalling Systems (Study Group XI) 2 pp. 2 pp.
CCITT RECMN Q.682-89. Logic Procedures for Interworking of Signalling System R2 to No. 5 (TUP)—Interworking of Signalling Systems (Study Group XI) 3 pp. 3 pp.
CCITT RECMN Q.683-89. Logic Procedures for Interworking of Signalling System R2 to No. 6—Interworking of Signalling Systems (Study Group XI) 3 pp. 3 pp.
CCITT RECMN Q.684-89. Logic Procedures for Interworking of Signalling System R2 to No. 7 (TUP)—Interworking of Signalling Systems (Study Group XI) 4 pp. 4 pp.
CCITT RECMN Q.685-89. Logic Procedures for Interworking of Signalling System R2 to R1—Interworking of Signalling Systems (Study Group XI) 2 pp. 2 pp.

INDUSTRY STANDARDS

Logic Diagrams (Cont.)
—Signaling Systems—Interworking
CCITT RECMN Q.634-89. Logic Procedures for Interworking of Signalling System No. 4 to R2—Interworking of Signalling Systems (Study Group XI) 3 pp. 3 pp.
CCITT RECMN Q.642-89. Logic Procedures for Interworking of Signalling System No. 6—Interworking of Signalling Systems (Study Group XI) 4 pp. 4 pp.
CCITT RECMN Q.643-89. Logic Procedures for Interworking of Signalling System No. 5 to No. 7 (TUP)—Interworking of Signalling Systems (Study Group XI) 4 pp. 4 pp.
CCITT RECMN Q.644-89. Logic Procedures for Interworking of Signalling System No. 5 to R1—Interworking of Signalling Systems (Study Group XI) 2 pp. 2 pp.
CCITT RECMN Q.645-89. Logic Procedures for Interworking of Signalling System No. 5 to R2—Interworking of Signalling Systems (Study Group XI) 3 pp. 3 pp.
CCITT RECMN Q.652-89. Logic Procedures for Interworking of Signalling System No. 6 to No. 5—Interworking of Signalling Systems (Study Group XI) 3 pp. 3 pp.

Logic Elements
Use For: Binary Logic Elements *See Also:* Logic Circuits; NOT Elements
—Quality Assurance
BSI BS CECC 90114-90. 1990 Blank Detail Specification: Programmable Logic Arrays (PLA). 27 pp.
—Symbols
JIS X 0122-86. Graphical Symbols for Binary Logic Elements. 108 pp.
SAA AS 1102.9-86. Graphical Symbols for Electrotechnology—Part 9: Binary Logic Elements. 81 pp.
—Symbols—Diagrams
BSI BS 3939: Part 12-91. 1991 Graphical Symbols for Electrical Power, Telecommunications and Electronics Diagrams Part 12: Guide for Binary Logic Elements (G) (IEC 617-12: 1991). 207 pp.
BSI BS 3939: Part 12-01. 1991 Amd 1 Graphical Symbols for Electrical Power, Telecommunications and Electronics Diagrams Part 12: Guide for Binary Logic Elements (AMD 7432) Februray 15, 1993 (IEC 617-12: 1991). 212 pp.
IEC 617 Pt 12-91. Graphical Symbols for Diagrams Part 12: Binary Logic Elements Second Edition; (Amendment 1-1992). 224 pp.
SNZ IEC 617: Part 12-83. Graphical Symbols for Diagrams Part 12: Binary Logic Elements. 148 pp.

Logic Inverters
Use: NOT Circuits

Logic Level Down Converters
Use: Down Converters

Logic Levels
See Also: Emitter Coupled Logic
—Nuclear Instruments
BSI BS 5251-75. 1975 Analogue Voltage Ranges and Logic Levels for Mains Operated Nuclear Instruments. 7 pp.
IEC 323-70. Analogue Voltage Ranges and Logic Levels for Mains Operated Nuclear Instruments First Edition; (Amendment 1-1974). 10 pp.

Logic Procedures
Scope Note: See the subheading Logic Procedures under the specific application. For additional references, consult the following list *See:* Logic Diagrams

Logical Link Control
Use For: LLC *See Also:* Link Control Procedures; Protocols; Virtual Call Services
—Local Area Networks
BSI BS ISO 8802/2-89. 1989 Information Processing Systems—Local Area Networks Part 2: Logical Link Control. 118 pp.
BSI BS ISO/IEC TR 10178-92. 1992 Information Technology—Telecommunications and Information Exchange Between Systems—the Structure and Coding of Logical Link Control Addresses in Local Area Networks. 19 pp.
BSI DD 99: Part 1-85. (WITHDRAWN) 1985 Logical Link Control for Local Area Networks Part 1: Technical Specification (Superseded by BS ISO 8802-2: 1989). 104 pp.
IEC 8802 Pt 2-92. Corrigendum 1 Information Processing Systems—Local Area Networks—Part 2: Logical Link Control. 2 pp.
IEC 8802 Pt 2 Draft AMD 5. Information Processing Systems—Local Area Networks—Part 2: Logical Link Control Amendment 5: Bridged LAN Source Routeing Operation by End Systems; (1993) (IEEE 802.5p) ***CD-ROM ONLY***. 51 pp.
IEC TR10178-92. Information Technology—Telecommunications and Information Exchange Between Systems—the Structure and Coding of Logical Link Control Addresses in Local Area Networks First Edition. 17 pp.
ISO 1151 Pt 8-92. Flight Dynamics—Concepts, Quantities and Symbols—Part 8: Concepts and Quantities Used in the Study of the Dynamic Behaviour of the Aircraft First Edition. 12 pp.
ISO 8802 Pt 2-92. Corrigendum 1 Information Processing Systems—Local Area Networks—Part 2: Logical Link Control. 2 pp.
ISO 8802 Pt 2 Draft AMD 5. Information Processing Systems—Local Area Networks—Part 2: Logical Link Control Amendment 5: Bridged LAN Source Routeing Operation by End Systems; (1993) (IEEE 802.5p) ***CD-ROM ONLY***. 51 pp.
ISO TR10178-92. Information Technology—Telecommunications and Information Exchange Between Systems—the Structure and Coding of Logical Link Control Addresses in Local Area Networks First Edition. 17 pp.
JIS X 5251-89. Local Area Networks—Logical Link Control.
JTC1 8802-92. Corrigendum 1 Information Processing Systems—Local Area Networks—Part 2: Logical Link Control. 2 pp.
JTC1 8802 Pt2 Draft AMD 5. Information Processing Systems—Local Area Networks—Part 2: Logical Link Control Amendment 5: Bridged LAN Source Routeing Operation by End Systems; (1993) (IEEE 802.5p) ***CD-ROM ONLY***. 51 pp.
OSI ISO 8802-2 CS28-0656. Information Processing Systems—Local Area Networks—Part 2: Logical Link Control. 16 pp.
OSI ISO 8802-2 DAM 4-90. Information Processing Systems—Local Area Networks—Part 2: Logical Link Control Amendment 4: Editorial Changes and Technical Corrections.
OSI ISO 8802-2 DAD 2-89. Information Processing Systems—Local Area Networks—Part 2: Logical Link Control Addendum 2: Acknowledged Connectionless-Mode Service and Protocol Type 3 Operation. 39 pp.
OSI ISO IEC 8802-2 DAD 1-88. Information Processing Systems—Local Area Networks—Part 2: Logical Link Control—Addendum 1: Flow Control Techniques for Bridged Local Area Networks. 5 pp.
OSI ISO DIS 8802-2.2-87. Information Processing Systems—Local Area Networks—Part 2: Logical Link Control. 94 pp.
OSI ISO DIS 8881-2.2-87. Information Processing Systems—Data Communications—Use of the X.2S Packet Level Protocol in Local Area Networks—Part 2: Use with LLC Type 2 Procedures. 10 pp.
OSI ISO IEC DIS 8881-3-88. Information Processing Systems—Data Communications—Use of the X.25 Packet Level Protocol in Local Area Networks. 19 pp.
SAA AS 4802.2-91. Information Processing Systems—Local Area Networks—Part 2: Logical Link Control (ISO 8802-2:1989) (in Professional Packages 26A, 26C). 101 pp.
SNZ NZS/ISO 8802. 2-90. Information Processing Systems—Local Area Networks Part 2: Logical Link Control. 101 pp.

Logistics Operations
See Also: Formations (Military); Medical Services; Military Operations
—Land Forces
NATO STANAG 2406 ED 1 AMD 0-91. Land Forces Logistic Doctrine—ALP-9. 4 pp.
—Marine Terminals—Naval
NATO STANAG 1200 ED 4 AMD 2-86. Procedures for Logistic Support Between NATO Navies and Naval Port Information—ALP-1(D). 6 pp.
NATO STANAG 1200 ED 5 AMD 0-92. Procedures for Logistic Support Between NATO Navies and Naval Port Information—ALP-1(D). 6 pp.
—Warfare—Assistance
NATO STANAG 2135 ED 3 AMD 4-82. Procedures for Emergency Logistic Assistance. 13 pp.

Logs, Borehole
Use: Borehole Logs

Logs (Reports)
Use For: Modification Record Books
See Also: Documents
—Aircraft
CAA Chapter A7-8 07.89. Technical Logs. 2 pp.
CAA Chapter A7-9 07.89. Modification Record Book. 2 pp.
—Aircraft—Balancing
CAA Chapter A7-10 07.89. Weight and Balance Report. 5 pp.
—Aircraft Engines
CAA LEAFLET 1-5 07.90. Aircraft, Engine and Propeller Log Books. 8 pp.
—Aircraft Propellers
CAA LEAFLET 1-5 07.90. Aircraft, Engine and Propeller Log Books. 8 pp.
—Aircraft—Weight Measurement
CAA Chapter A7-10 07.89. Weight and Balance Report. 5 pp.

Logs (Wood)
See Also: Lumber; Pulpwood; Wood
ISO 4476-83. Coniferous and Broadleaved Sawlogs—Sizes—Vocabulary First Edition. 5 pp.
ISO 4480-83. Coniferous Sawlogs—Measurement of Sizes and Determination of Volume First Edition. 3 pp.
—Defects—Classification
ISO 4473-88. Coniferous and Broadleaved Tree Sawlogs—Visible Defects—Classification First Edition. 4 pp.
—Defects—Glossaries
ISO 4474-89. Coniferous and Broadleaved Tree Sawlogs—Visible Defects—Terms and Definitions First Edition. 19 pp.
—Defects—Measurement
ISO 4475-89. Coniferous and Broadleaved Tree Sawlogs—Visible Defects—Measurement First Edition. 16 pp.
—Glossaries
ISO 4476-83. Coniferous and Broadleaved Sawlogs—Sizes—Vocabulary First Edition. 5 pp.
—Grading
CNS O1020-82. Grading Rules of Logs (for Taiwan Area) (Oct)(4748). 3 pp.
—Hardwoods—Glossaries
ISO 4476-83. Coniferous and Broadleaved Sawlogs—Sizes—Vocabulary First Edition. 5 pp.
—Hardwoods—Grading
CNS O1020-82. Grading Rules of Logs (for Taiwan Area) (Oct)(4748). 3 pp.
—Length Measurement
CNS O1004-79. Commercial Lengths of Logs (Jan)(445). 2 pp.
—Sampling
ISO 4471-82. Wood—Sampling Sample Trees and Logs for Determination of Physical and Mechanical Properties of Wood in Homogeneous Stands First Edition. 9 pp.
—Softwoods
DIN ENGL 4074 Pt 2-58. Building Timber for Wood Building Components; Quality Conditions for Building Logs (Softwood) (Dec). 3 pp.
ISO 4480-83. Coniferous Sawlogs—Measurement of Sizes and Determination of Volume First Edition. 3 pp.
—Softwoods—Glossaries
ISO 4476-83. Coniferous and Broadleaved Sawlogs—Sizes—Vocabulary First Edition. 5 pp.
—Softwoods—Grading
CNS O1020-82. Grading Rules of Logs (for Taiwan Area) (Oct)(4748). 3 pp.
—Structural
DIN ENGL 4074 Pt 2-58. Building Timber for Wood Building Components; Quality Conditions for Building Logs (Softwood) (Dec). 3 pp.

Long Taps
See Also: Taps (Threading Tools)
ISO 2283-72. Long Shank Machine Taps with Nominal Diameters from 3 to 24 mm and 1/8 to 1 in First Edition; (Amendment 1-1977) (Erratum-Aug 1976). 9 pp.
ISO 8051-89. Long Shank Taps with Nominal Diameters from M3 to M10—Taps with Reinforced Shank and Recess First Edition. 5 pp.
—Machine
CNS B3299-79. Long Shank Machine Taps with Nominal Diameters from 3 to 24 mm (Oct)(4998). 3 pp.

Longitude

See Also: Datum (Geodetic); Grids (Coordinates); Position (Location)

—Geographic Representation

BSI BS 5249: Part 3-83. 1983 Representation of Elements of Data in Interchanges Using Data Processing Systems Part 3: Representation of Latitude, Longitude and Altitude for Geographical Point Locations. 4 pp.

ISO 6709-83. Standard Representation of Latitude, Longitude and Altitude for Geographic Point Locations First Edition. 5 pp.

JTC1 6709-83. Standard Representation of Latitude, Longitude and Altitude for Geographic Point Locations First Edition. 5 pp.

OSI ISO 6709-83. Standard Representation of Latitude, Longitude and Altitude for Geographic Part Locations. 5 pp.

Longitudinal Reversion

Use: Dimensional Stability

Longwall Mining Equipment

Use: Mining Equipment

Look Ahead Carry Generators

See Also: Arithmetic Logic Unit/Function Generators; Digital Circuits; Generators

CECC CECC 90 102-056 ISSUE 1-82. BS CECC 90 102-056; Look-Ahead Carry Generators (En). 19 pp.

CECC CECC 90 109-833 ISSUE 1-89. Digital Integrated Circuits; Silicon Monolithic C MOS, Cavity or Non-Cavity Packages; Type(s) 54/74 HC 182 Look-Ahead Carry Generator Assessment Levels P, Y, L (En, Fr, Ge). 11 pp.

—Preferred Products List

CECC CECC MUAHAG Vol 7 IS 8-92. Preferred Products List; Active Microcircuits (En, Fe, Ge). 89 pp.

Looms

See Also: Harnesses (Weaving); Heddles; Lappets; Lingoes; Reeds (Textiles); Shuttles; Textile Machinery; Weaving; Weaving Machines

BSI BS 4392: Part 4-78. 1978 Textile Machinery and Accessories. Definition of Left and Right Side Part 4: Weaving Looms. 3 pp.

CNS L1021-86. Definition of Side (Left or Right) of Weaving Preparatory Machines and Looms (Apr)(11553).

CNS L2028-82. Standard Working Widths of Looms (Mar)(8604).

CNS L2049-82. Weight for Filament Looms (Apr)(8724).

ISO 108-76. Textile Machinery and Accessories—Weaving Looms—Definition of Left and Right Sides First Edition. 3 pp.

JIS L 0303-63. Definition of Side (Left or Right) of Weaving Preparatory Machines and Looms.

JIS L 6113-83. HATAGUSA for Silk Loom.

JIS L 6113-61. HATAKUSA for Looms. 4 pp.

JIS L 6205-77. Standard Working Widths of Looms.

JIS L 6512-75. Weight for Filament Looms.

—Beams (Textile Machinery)

CNS L2043-82. Yarn Beams for Woollen and Worsted Loom (Apr)(8718).

—Bobbins (Thread)

JIS L 6416-92. Leno Selvage Bobbins for Looms. 6 pp.

—Bolts

CNS B2245-81. Cup Square Bolts with Enlarged Head for Looms (Apr)(4425).

—Cards—Shuttle Changing

JIS L 6517-61. Cards for Shuttle Change Apparatus (for Silk Loom). 4 pp.

—Change Gears

JIS L 6207-88. Take-Up Change Gears for Looms. 6 pp.

—Comber Boards

JIS L 6114-61. Comber Board for Silk Loom. 4 pp.

—Drop Wires

CNS L2030-82. Drop Wires for Looms (Mar)(8606).
JIS L 6212-86. Drop Wires for Looms.

—Flanges

JIS L 6412-88. Warp Beams and Beam Flanges for Looms. 12 pp.

—Forks

CNS L2048-82. Weft Fork for Filament Loom (Apr)(8723).
JIS L 6505-73. Weft Fork for Filament Loom.

Looms (Cont.)

—Handles

CNS L2032-82. Handles for Filament Loom (Mar)(8608).
CNS L2033-82. Reel (Mar)(8609).
JIS L 6211-72. Handles for Filament Loom.

—Heddle Hooks

CNS L2044-82. Heald Hook (for Woollen and Worsted Loom) (Apr)(8719).

—Heddles

BSI BS 4802: Part 3-78. 1978 Specification for Terminology and Classification of Weaving Machinery Part 3: Textile Machinery and Accessories. Numbering of Heald Frames in a Loom. 3 pp.

CNS L2047-82. Flat Heads for Looms (Apr)(8722).
CNS L2061-84. Wire Heads for Looms (May)(9528).

ISO 5243-77. Textile Machinery and Accessories—Numbering of Heald Frames in a Loom First Edition. 3 pp.

JIS L 6405-78. Wire Heads for Frame Weaving of Looms.
JIS L 6405-61. Wire Heald. 6 pp.
JIS L 6507-79. Flat Heads for Looms.

—Pickers

CNS L2040-82. Picking Sticks for Cotton Looms (Apr)(8715).
CNS L2041-82. Picking Sticks for Woollen and Worsted Looms (Apr)(8716).
CNS L2042-82. Picking Sticks for Looms (Apr)(8717).
CNS L2050-82. Picker for Cotton Loom (Apr)(8725).
CNS L2051-82. Pickers for Woollen and Worsted Loom (Apr)(8726).
CNS L2052-82. Pickers for Looms (Apr)(8727).

ISO 227-78. Textile Machinery and Accessories—Single Box Pickers for Centre Tip Shuttles for Automatic Looms and Related Picking Stick Dimensions First Edition. 3 pp.

JIS L 6503-81. Picking Sticks for Looms.
JIS L 6509-81. Pickers for Looms.

—Quills

BSI BS 2098-78. (WITHDRAWN) 1978 Textile Machinery and Accessories: Weft Pirns for Automatic Looms. 4 pp.

BSI BS 5753: Part 1-79. 1979 Textile Machinery and Accessories: Pirns for Winding at the Loom Part 1: Specification for Weft Pirns Without Rings. 3 pp.

ISO 143-77. Textile Machinery and Accessories—Weft Pirns for Automatic Looms First Edition. 4 pp.

ISO 1131-76. Textile Machinery and Accessories—Weft Pirns for Box-Loaders for Automatic Looms—Dimensions of Prin Tip First Edition. 3 pp.

ISO 5245-77. Textile Machinery and Accessories—Weft Pirns with Rings (27 mm and 30 mm) for Automatic Winding at the Loom First Edition. 4 pp.

ISO 5246-77. Textile Machinery and Accessories—Ringless Weft Pirns (24 mm and 27 mm) for Automatic Winding at the Loom First Edition. 4 pp.

—Reeds

CNS L2046-84. Reeds for Looms (May)(8721).
JIS L 6506-86. Reeds for Looms.

—Rollers

CNS L2029-82. Cloth Rollers for Looms (Mar)(8605).
CNS L2031-82. Letting-off Rollers for Filament Loom (Mar)(8607).
JIS L 6208-79. Cloth Rollers for Looms.

—Shuttles

BSI BS 3240-77. 1977 Textile Machinery and Accessories. Shuttles for Pirn Changing Automatic Looms. Dimensions. 4 pp.

CNS L2001-81. Shuttle for Cotton Looms (Apr)(576).
CNS L2004-81. Shuttle for Filament Looms (Apr)(579).

ISO 227-78. Textile Machinery and Accessories—Single Box Pickers for Centre Tip Shuttles for Automatic Looms and Related Picking Stick Dimensions First Edition. 3 pp.

ISO 572-76. Textile Machinery and Accessories—Shuttles for Pirn Changing Automatic Looms—Dimensions First Edition; (Erratum—Oct 1976). 5 pp.

JIS L 6401-76. Shuttles for Filament Looms.

—Side Levers

CNS L2045-82. Side Lever for Cotton Loom (Apr)(8720).
JIS L 6410-82. Side Levers for Looms.

—Take Up Strips

JIS L 6418-92. Take-up Strips for Looms. 8 pp.

—Warp Beams

CNS L2062-82. Warp Beams for Cotton Looms (Oct)(9529).
JIS L 6412-88. Warp Beams and Beam Flanges for Looms. 12 pp.

Looms (Cont.)

—Worm Wheels and Worms

JIS L 6310-79. Pick Change Worm Wheels and Worms for Silk Loom.
JIS L 6310-62. Pick Change Worm Wheels and Worms for Silk Loom. 6 pp.

Loop Clamps

See Also: Clamps; Fasteners

CNS B2751-83. Clamp, Loop-Steel, Plain and Cushioned (Jul)(10421).

DIN VDE 0220 Pt 3-77. VDE Regulations for Single and Multiple Cable Clamps with Insulated Parts in Electrical Power Cable Installations up to 1000 V (Oct). 19 pp.

MOD UK DSTAN 53-15-01. Clamps, Loop (Rigid) Issue 2; Amendment 1. 11 pp.

—Aerospace

AECMA PREN2901-89. Clamps, Loop (P Type) with Rubber Cushioning in Corrosion Resisting Steel Dimensions-Masses. 6 pp.

AECMA PREN2902-89. Clamps, Loop (P type) with Rubber Cushioning in Aluminium Alloy Dimensions-Masses. 6 pp.

—Aerospace—Fluid Systems

AECMA PREN2901-89. Clamps, Loop (P Type) with Rubber Cushioning in Corrosion Resisting Steel Dimensions-Masses. 6 pp.

AECMA PREN2902-89. Clamps, Loop (P type) with Rubber Cushioning in Aluminium Alloy Dimensions-Masses. 6 pp.

—Aircraft—Cushions

SBAC AS 46760 (V). Clamp, Loop Corrosion Resisting Steel Fluorosilicone Rubber Cushion.

SBAC AS 46761 (V). Clamp, Loop, Aluminium Alloy Fluorosilicone Rubber Cushion.

SBAC AS 46782 (V). Clamp Loop, Corrosion Resistant Steel Fluorosilicone Rubber Cushion (Imperial Dimensions).

SBAC AS 46783 (V). Clamp Loop, Aluminium Alloy Fluorosilicone Rubber Cushion (Imperial Dimensions).

SBAC AS 61900-934 (V) (I). Clamp, Loop Style, Cushion Reinforced (Electrical Harness) (for Use with 5mm or 0.190 Bolts).

SBAC AS 61935-969 (V) (I). Clamp, Loop Style, Cushion Reinforced (Electrical Harness) (for Use with 6 mm or 0.250 Bolts).

SBAC AS 62200-234 (V). Clamp, Loop Style (Snap Lock) Cushion Reinforced (Electrical Harness) (for 5.00mm & 0.1900 Inch Fasteners).

SBAC AS 62250-284 ISSUE 2. Clamp, Loop Style (Snap Lock) Cushion Reinforced (Electrical Harness) (for 6.00mm & 0.2500 Inch Fasteners).

SBAC AS 62403-432 (V). Clamp, Loop Style, Box Cushion, Assembly of.

SBAC AS 62500-534 (V). Clamp, Loop Style, Cushion Reinforced (Electrical Harness) (for Use with 5mm or 0.190 Bolts).

SBAC AS 62600-634 (V). Clamp, Loop Style, Cushion Reinforced (Electrical Harness) (for Use with 6 mm or 0.250 Bolts).

SBAC AS 62700-742 (V). Clamp Loop Style (Grounding Strip) Cushion Reinforced (Electrical Harness) (for Use with 5mm or 0.190 Bolts).

SBAC AS 62800-842 ISSUE 2. Clamp Loop Style (Grounding Strip) Cushion Reinforced (Electrical Harness) (for Use with 6mm or 0.250 Bolts).

—Cable

DIN VDE 0220 Pt 3-77. VDE Regulations for Single and Multiple Cable Clamps with Insulated Parts in Electrical Power Cable Installations up to 1000 V (Oct). 19 pp.

MOD UK DSTAN 53-15-01. Clamps, Loop (Rigid) Issue 2; Amendment 1. 11 pp.

—Cable—Aerospace

AECMA PREN2901-89. Clamps, Loop (P Type) with Rubber Cushioning in Corrosion Resisting Steel Dimensions-Masses. 6 pp.

AECMA PREN2902-89. Clamps, Loop (P type) with Rubber Cushioning in Aluminium Alloy Dimensions-Masses. 6 pp.

—Cushioned

CNS B2751-83. Clamp, Loop-Steel, Plain and Cushioned (Jul)(10421).

Loop Transmission

See Also: Local Loop Plants; Loopback Testing Equipment; Telephone Transmission

—Telephone Circuits—Testing Equipment—CCITT No. 4 Signaling

CCITT RECMN Q.136-89. Loop Transmission Measurements—Specifications of Signalling Systems Nos. 4 and 5 (Study Group XI) 1 pp. 1 p.

Loopback Testing Equipment
See Also: Local Loop Plants; Loop Transmission; Testing Equipment

—Interfaces
CCITT RECMN I.430-89. Basic User-Network Interface—Layer 1 Specification—Integrated Services Digital Network (ISDN)—Overall Network Aspects and Functions, ISDN User-Network Interfaces (Study Group XVIII) 70 pp. 70 pp.

—Modems
CCITT RECMN V.54-89. Loop Test Devices for Modems—Data Communication over the Telephone Network (Study Group XVII) 17 pp. 17 pp.

Loops
—Wire
CNS Z7089-81. Loops Round Wire (Jul)(7766).
CNS Z7090-81. Loops Half-Round Wire (Jul)(7767).

Loose Leaf Binders
Use: Binders (Files)

Loose Leaf Papers
Use: Writing Papers—Loose Leaf

Loquats
See Also: Food; Fruits; Plums

—Grading
CNS N1115-83. Grades of Loquat (Sep)(10571). 3 pp.

Loran C
See Also: Navigational Aids; Radio Navigation; Radio Navigation Equipment; Radio Receivers; Receivers

—Aeronautical—Power Line Carriers—Radio Frequency Interference
CCIR Report 1174-90. Potential for Interference from Power-Line-Carrier Systems to Loran-C Aeronautical Receivers—Secton 8K—Aeronautical Mobile Service (Terrestrial). 6 pp.

—Radio Frequency Interference
CCIR Report 915-2-90. Interference Between Fixed, Maritime Mobile and Radionavigation Services in the Bands Between 70 kHz and 130 kHz—Section 8D—Radiodetermination, Global Maritime Distress and Safety System and Related Subjects. 17 pp.

—Receivers—Merchant Ships
BSI BS 7538-91. (WITHDRAWN) 1991 Loran-C Receivers for Ships (IEC 1075: 1991) (Renumbered as BS EN 61075: 1993). 24 pp.
BSI BS EN 61075-93. 1993 Amd 1 Loran-C Receivers for Ships—Minimum Performance Standards—Methods of Testing and Required Test Results (AMD 7886) August 15, 1993 (Q). 30 pp.
CENELEC EN 61075-93. Loran-C Receivers for Ships—Minimum Performance Standards—Methods of Testing and Required Test Results (IEC 1075: 1991). 22 pp.
IEC 1075-91. Loran-C Receivers for Ships Minimum Performance Standards—Methods of Testing and Required Test Results First Edition (CENELEC EN 61075: 1993). 48 pp.

Lorries
Use: Trucks

LOS Propagation
Use: Line of Sight Propagation

Loss Factor
See Also: Power Transmission

—Cables (Electric)
DIN VDE 0472 Pt 505-83. Testing of Cables and Insulated Flexible Cords; Loss Factor, Dielectric Loss Coefficient and Leakage (Apr). 9 pp.

—Cords (Electric)
DIN VDE 0472 Pt 505-83. Testing of Cables and Insulated Flexible Cords; Loss Factor, Dielectric Loss Coefficient and Leakage (Apr). 9 pp.

Loss on Ignition
Use: Ignition Loss

Loss on Ignition Testing
Use: Volatile Matter Content Analysis

Loss Tangent
Use: Dielectric Dissipation Factor

Lotus Roots
—Grading
CNS N1042-63. Grades of Lotus Root (for Export) (Mar) (2127)(R 1973). 1 p.

Loudness Ratings
Use For: LR See Also: Acoustics; Sound Power; Sound Pressure; Sound Pressure Levels; Telephone Connections

—Carbon Microphones—Telephone Transmission
CCITT RECMN P.75-89. Standard Conditioning Method for Handsets with Carbon Microphones—Telephone Transmission Quality (Study Group XII) 1 pp. 1 p.

—Conference Calling
CCITT RECMN G.172-89. Transmission Plan Aspects of International Conference Calls—General Characteristics of International Telephone Connections and Circuits (Study Groups XII and XV) 3 pp. 3 pp.

—Echo (Telecommunications)
CCITT RECMN G.111-89. Loudness Ratings (LRs) in an International Connection—General Characteristics of International Telephone Connections and Circuits (Study Groups XII and XV) 22 pp. 22 pp.

—Intermediate Reference System—Telephone Transmission
CCITT Volume V. Telephone Transmission Quality Series P Recommendations (Study Group XII). 438 pp.

—Sidetone
CCITT RECMN G.111-89. Loudness Ratings (LRs) in an International Connection—General Characteristics of International Telephone Connections and Circuits (Study Groups XII and XV) 22 pp. 22 pp.
CCITT RECMN G.121-89. Loudness Ratings (LRs) of National Systems—General Characteristics of International Telephone Connections and Circuits (Study Groups XII and XV) 17 pp. 17 pp.

—Speech Circuits
CCITT RECMN G.111-89. Loudness Ratings (LRs) in an International Connection—General Characteristics of International Telephone Connections and Circuits (Study Groups XII and XV) 22 pp. 22 pp.

—Speech Circuits—Attenuation Distortion
CCITT RECMN G.111-89. Loudness Ratings (LRs) in an International Connection—General Characteristics of International Telephone Connections and Circuits (Study Groups XII and XV) 22 pp. 22 pp.

—Telephone Circuits
CCITT RECMN G.121-89. Loudness Ratings (LRs) of National Systems—General Characteristics of International Telephone Connections and Circuits (Study Groups XII and XV) 17 pp. 17 pp.
CCITT RECMN G.122-89. Influence of National Systems on Stability, Talker Echo, and Listener Echo in International Connections—General Characteristics of International Telephone Connections and Circuits (Study Groups XII and XV) 13 pp. 13 pp.

—Telephone Circuits—Private Switched Networks
CCITT RECMN G.171-89. Transmission Plan Aspects of Privately Operated Networks—General Characteristics of International Telephone Connections and Circuits (Study Groups XII and XV) 19 pp. 19 pp.
CENELEC PRETS 300 283-92. Business Telecommunications (BT); Planning of Loudness Rating and Echo Values for Private Networks Digitally Connected to the Public Network. 13 pp.
ETSI PRETS 300 283-92. Business Telecommunications (BT); Planning of Loudness Rating and Echo Values for Private Networks Digitally Connected to the Public Network. 13 pp.

—Telephone Circuits—Virtual Analog Switching Points
CCITT RECMN G.111-89. Loudness Ratings (LRs) in an International Connection—General Characteristics of International Telephone Connections and Circuits (Study Groups XII and XV) 22 pp. 22 pp.

—Telephone Connections
CCITT RECMN G.111-89. Loudness Ratings (LRs) in an International Connection—General Characteristics of International Telephone Connections and Circuits (Study Groups XII and XV) 22 pp. 22 pp.
CCITT RECMN G.121-89. Loudness Ratings (LRs) of National Systems—General Characteristics of International Telephone Connections and Circuits (Study Groups XII and XV) 17 pp. 17 pp.

—Telephone Networks
CCITT RECMN G.111-89. Loudness Ratings (LRs) in an International Connection—General Characteristics of International Telephone Connections and Circuits (Study Groups XII and XV) 22 pp. 22 pp.

Loudness Ratings (Cont.)
—Telephone Networks—Crosstalk
CCITT RECMN G.111-89. Loudness Ratings (LRs) in an International Connection—General Characteristics of International Telephone Connections and Circuits (Study Groups XII and XV) 22 pp. 22 pp.

—Telephone Systems
CCITT RECMN P.64-89. Determination of Sensitivity/Frequency Characteristics of Local Telephone Systems to Permit Calculation of Their Loudness Ratings—Telephone Transmission Quality (Study Group XII) 10 pp. 10 pp.
CCITT RECMN P.79-89. Calculation of Loudness Ratings—Telephone Transmission Quality (Study Group XII) 21 pp. 21 pp.

—Telephone Transmission
CCITT Volume V. Telephone Transmission Quality Series P Recommendations (Study Group XII). 438 pp.
CCITT RECMN P.65-89. Objective Instrumentation for the Determination of Loudness Ratings—Telephone Transmission Quality (Study Group XII) 5 pp. 5 pp.
CCITT RECMN P.76-89. Determination of Loudness Ratings; Fundamental Principles—Telephone Transmission Quality (Study Group XII) 12 pp. 12 pp.
CCITT RECMN P.78-89. Subjective Testing Method for Determination of Loudness Ratings in Accordance with Recommendation P.76—Telephone Transmission Quality (Study Group XII) 17 pp. 17 pp.

—Telephones
CCITT RECMN G.111-89. Loudness Ratings (LRs) in an International Connection—General Characteristics of International Telephone Connections and Circuits (Study Groups XII and XV) 22 pp. 22 pp.
CCITT RECMN G.121-89. Loudness Ratings (LRs) of National Systems—General Characteristics of International Telephone Connections and Circuits (Study Groups XII and XV) 17 pp. 17 pp.
CCITT RECMN P.62-89. Measurements on Subscribers' Telephone Equipment—Telephone Transmission Quality (Study Group XII) 2 pp. 2 pp.

Loudspeakers
Use: Speakers

Louvers
See Also: Air Diffusion; Shields (Protective); Shutters (Blinds); Ventilation Equipment

—Power Operated
CSA CAN/CSA-C22. 2 NO 247-92. Operators and Systems of Doors, Gates, Draperies, and Louvres; (Gen Instr 1). 98 pp.

—Punkah—Ships
JIS F 2902-81. Ships' Punkah-Louvre.
JIS F 2902-60. Ships' Punkah-Louvers (R 1975). 7 pp.

Low Air Pressure Testing
Use: Low Pressure Testing

Low Alloy Steel Castings
See Also: Castings; Low Alloy Steels; Steel Castings
CNS G2075-86. Method of Test for High Tensile Strength Carbon Steel Castings and Low Alloy Steel Castings for Structural Purpose (Jul)(7146).
CNS G3136-86. High Tensile Strength Carbon Steel Castings and Low Alloy Steel Castings for Structural Purpose (Jul)(7145).
JIS G 5111-91. High Tensile Strength Carbon Steel Castings and Low Alloy Steel Castings for Structural Purposes. 9 pp.

—Aerospace
BSI HC 3-73. 1973 1% Chromium-Molybdenum Low Alloy Steel Castings (700. N/Square mm). 2 pp.
BSI HC 4-73. 1973 3% Chromium-Molybdenum Steel Castings (620-770 N/Square mm). 3 pp.

—Investment
BSI BS 3146: Part 1-74. 1974 Investment Castings in Metal Part 1: Carbon and Low Alloy Steels. 28 pp.

Low Alloy Steel Coatings
See Also: Coatings; Low Alloy Steels; Metal Coatings (On Metal); Steel Coatings

—Cadmium—Electroplated
MOD UK DTD-904C-63. Cadmium Plating (Reprinted December 1966, Incorporating Amendment No. 1) (Superseded by Def Stan 03-19). 5 pp.

Low Alloy Steels
Scope Note: For additional listings, see also specific products made from low alloy steels

Low Alloy Steels (Cont.)

See Also: Alloy Steels; Aluminum Chromium Molybdenum Steels; Carbon Manganese Steels; Chromium Molybdenum Steels; Chromium Molybdenum Vanadium Steels; Chromium Steels; Corrosion Resistant Steels; Electrical Steels; Ferroalloys; Free Machining Steels; High Alloy Steels; High Strength Steels; Low Alloy Steel Castings; Low Alloy Steel Coatings; Manganese Steels; Nickel Chromium Molybdenum Steels; Nickel Chromium Steels; Nickel Steels; Spring Steels; Steels; Structural Steels; Tungsten Steels

—Bars—Bolts
BSI BS 1506-90. 1990 Carbon, Low Alloy and Stainless Steel Bars and Billets for Bolting Material to be Used in Pressure Retaining Applications. 26 pp.

—Billets—Semi-Finished Products
BSI BS 1506-90. 1990 Carbon, Low Alloy and Stainless Steel Bars and Billets for Bolting Material to be Used in Pressure Retaining Applications. 26 pp.

—Crystal Structure—Impact Testing
BSI BS 131: Part 5-65. 1965 Methods for Notched Bar Tests Part 5: Determination of Crystallinity. 12 pp.

—Decarburizing
ISO 3887-76. Steel, Non-Alloy and Low-Alloy—Determination of Depth of Decarburization First Edition. 5 pp.
SAA AS 2003-91. Carbon and Low Alloy Steel—Measurement of Decarburization. 8 pp.

—Emission Spectroscopy
CNS G2230-84. Emission-Spectroscopic Analysis for Carbon Steel and Low Alloy Steel (Oct)(11071).
JIS G 1252-75. Emission-Spectroscopic Analysis for Carbon Steel and Low Alloy Steel. 8 pp.

—Forgings
BSI BS 1503-89. 1989 Amd 1 Steel Forgings for Pressure Purposes (AMD 6739) September 30, 1991. 35 pp.
BSI BS 1503-02. 1989 Amd 2 Steel Forgings for Pressure Purposes (AMD 7744) July 15, 1993 (Q). 36 pp.
SNZ NZS/BS 1503-89. Steel Forgings for Pressure Purposes Amend: 1, 1991. 32 pp.

—Forgings—Marine
MOD UK NES 848: Part 2-91. Requirements for Carbon Manganese and Low Alloy Steel Forgings Part 2: Low Alloy Steel Forgings Issue 1 (08.91). 19 pp.

—Forgings—Ultrasonic Testing
JIS G 0587-87. Methods for Ultrasonic Examination for Carbon and Low Alloy Steel Forgings. 24 pp.
SAA AS 1065-88. Non-Destructive Testing—Ultrasonic Testing of Carbon and Low Alloy Steel Forgings. 29 pp.

—Gage Lengths
ISO 2566 Pt 1-84. Steel—Conversion of Elongation Values—Part 1: Carbon and Low Alloy Steels Second Edition. 30 pp.

—Lead Content—Gravimetric Analysis
BSI BS 6200: SUB SEC 3.16.1-86. 1986 Sampling and Analysis of Iron, Steel and Other Ferrous Metals Part 3: Methods of Analysis Section 3.16: Determination of Lead Subsection 3.16.1: Carbon Steels and Low Alloy Steels: Gravimetric Method. 4 pp.

—Plates
ISO 9328 Pt 2-91. Steel Plates and Strips for Pressure Purposes—Technical Delivery Conditions—Part 2: Unalloyed and Low-Alloyed Steels with Specified Room Temperature and Elevated Temperature Properties First Edition. 16 pp.

—Plates—Heat Resistant
ISO 9328 Pt 2-91. Steel Plates and Strips for Pressure Purposes—Technical Delivery Conditions—Part 2: Unalloyed and Low-Alloyed Steels with Specified Room Temperature and Elevated Temperature Properties First Edition. 16 pp.

—Plates—High Strength
ISO 9328 Pt 1-91. Steel Plates and Strips for Pressure Purposes—Technical Delivery Conditions—Part 1: General Requirements First Edition. 24 pp.

—Plates—Ultrasonic Testing
BSI BS 5996-80. 1980 Amd 2 Ultrasonic Testing and Specifying Quality Grades of Ferritic Steel Plate. 10 pp.
SAA AS 1710-86. Non-Destructive Testing of Carbon and Low Alloy Steel Plate—Test Methods and Quality Classification. 8 pp.

Low Alloy Steels (Cont.)

—Semifinished Products
ISO 683 Pt 1-87. Heat-Treatable Steels, Alloy Steels and Free-Cutting Steels—Part 1: Direct-Hardening Unalloyed and Low-Alloyed Wrought Steel in Form of Different Black Products First Edition. 36 pp.

—Spectrography
MOD UK M 9071/63. Spectrographic Analysis of Low Alloy Steels (No Information) (Withdrawn).

—Strips
ISO 9328 Pt 2-91. Steel Plates and Strips for Pressure Purposes—Technical Delivery Conditions—Part 2: Unalloyed and Low-Alloyed Steels with Specified Room Temperature and Elevated Temperature Properties First Edition. 16 pp.

—Strips—Heat Resistant
ISO 9328 Pt 2-91. Steel Plates and Strips for Pressure Purposes—Technical Delivery Conditions—Part 2: Unalloyed and Low-Alloyed Steels with Specified Room Temperature and Elevated Temperature Properties First Edition. 16 pp.

—Strips—High Strength
ISO 9328 Pt 1-91. Steel Plates and Strips for Pressure Purposes—Technical Delivery Conditions—Part 1: General Requirements First Edition. 24 pp.

—Tubes—Seamless—Cold Drawn—Aerospace
DIN ENGL LN 9369-86. Aerospace; Close Tolerance Structural Tubes in Low Alloy Steels, Seamless Cold Drawn; Dimensions, Masses (Dec). 5 pp.
DIN ENGL LN 9397-87. Aerospace; Structural Tubes in Low Alloy Steels, Seamless, Cold Drawn; Dimensions, Masses (Aug). 4 pp.

—Weld Metal
JIS Z 3183-88. Quality Classification and Test Methods of Submerged Arc Deposited Metal for Carbon Steel and Low Alloy Steel. 18 pp.

—Weld Metal—Hydrogen Content
BSI BS 6693: Part 1-86. 1986 Diffusable Hydrogen Part 1: Method for Determination of Hydrogen in Manual Metal-Arc Weld Metal Using 3 Day Collection. 8 pp.
BSI BS 6693: Part 2-86. 1986 Diffusable Hydrogen Part 2: Method for Determination of Hydrogen in Manual Metal-Arc Weld Metal. 9 pp.
BSI BS 6693: Part 3-88. 1988 Amd 1 Diffusable Hydrogen Part 3: Primary Method for the Determination of Diffusible Hydrogen in Manual Metal-Arc Ferritic Steel Weld Metal (AMD 6241) December 22, 1989. 16 pp.
BSI BS 6693: Part 4-88. 1988 Diffusable Hydrogen Part 4: Primary Method for the Determination of Diffusible Hydrogen in Submerged-Arc Steel Weld-Metal. 16 pp.
BSI BS 6693: Part 5-88. 1988 Amd 1 Diffusable Hydrogen Part 5: Primary Method for the Determination of Diffusible Hydrogen in MIG, MAG, TIG or Cored Electrode Ferritic Steel Weld Metal (AMD 6242) January 31, 1990. 17 pp.
CSA W48.7-M1977. Diffusible Hydrogen in Mild Steel and Low-Alloy Steel Weld Metals: Test Method. 9 pp.
ISO 3690-77. Welding—Determination of Hydrogen in Deposited Weld Metal Arising from the Use of Covered Electrodes for Welding Unalloyed and Low Alloy Steels First Edition; (Addendum 1-2:1983). 14 pp.

—X-Ray Fluorescence Spectrometry
CNS G2182-84. Method for Fluorescent X-Ray Fluorescence Spectrochemical Analysis of Pig Iron, Cast Iron, Carbon Steel and Low Alloy Steel (Oct)(10502).

—Yield Strength
SAA AS 2069-77. Method for Verifying the Minimum Elevated Temperature Lower Yield or Proof Stress Properties of Carbon and Low Alloy Steel Products Reconfirmed 1983. 8 pp.

Low Carbon Steel Coatings

See Also: Carbon Steel Coatings; Coatings; Low Carbon Steels; Metal Coatings (On Metal)

—Chromium—Electrodeposited
CEN PREN 10 170-87. Single Cold Reduced Electrolytic Chromium/Chromium Oxide Coated Steel: Sheet. 27 pp.

—Chromium Oxide—Electrodeposited
CEN PREN 10 170-87. Single Cold Reduced Electrolytic Chromium/Chromium Oxide Coated Steel: Sheet. 27 pp.

—Tin
CNS G3097-87. Tinplate and Blackplate (Dec)(4155).

Low Carbon Steel Coatings (Cont.)

—Tin (Cont.)
DIN ENGL 1616 (S)-84. Tinplate and Blackplate Sheet; Grades, Dimensions and Permissible Deviations (Oct) (Superseded DIN EN 10203 August 1991 and DIN EN 10205 January 1992). 12 pp.
JIS G 3303-87. Tinplate and Blackplate. 40 pp.

—Tin—Electrodeposited
ISO 1111 Pt 1-83. Single Cold-Reduced Tinplate and Single Cold-Reduced Blackplate—Part 1: Electrolytic and Hot-Dipped Tinplate Sheet and Blackplate Sheet First Edition. 16 pp.
ISO 4977 Pt 2-88. Double Cold-Reduced Electrolytic Tinplate—Part 2: Coil for Subsequent Cutting into Sheets First Edition. 19 pp.
ISO 5950-91. Continuous Electrolytic Tin-Coated Cold-Reduced Carbon Steel Sheet of Commercial and Drawing Qualities Second Edition; DoD Adopted. 14 pp.

—Tin—Hot Dip
ISO 1111 Pt 1-83. Single Cold-Reduced Tinplate and Single Cold-Reduced Blackplate—Part 1: Electrolytic and Hot-Dipped Tinplate Sheet and Blackplate Sheet First Edition. 16 pp.

Low Carbon Steels

Scope Note: For additional listings, see also specific products made from lowcarbon steels
See Also: Carbon Steels; High Strength Steels; Low Carbon Steel Coatings; Steels

—Case Hardening—Bars—Forged
CEN PREN 10152-91. Electrolytically Zinc Coated Cold Rolled Steel Flat Products—Technical Delivery Conditions. 29 pp.
DIN ENGL 17210-86. Case Hardening Steels; Technical Delivery Conditions (Sept). 20 pp.

—Case Hardening—Bars—Hot Rolled
DIN ENGL 17210-86. Case Hardening Steels; Technical Delivery Conditions (Sept). 20 pp.

—Case Hardening—Flats—Hot Rolled
DIN ENGL 17210-86. Case Hardening Steels; Technical Delivery Conditions (Sept). 20 pp.

—Case Hardening—Forgings
DIN ENGL 17210-86. Case Hardening Steels; Technical Delivery Conditions (Sept). 20 pp.

—Case Hardening—Plates—Rolled
DIN ENGL 17210-86. Case Hardening Steels; Technical Delivery Conditions (Sept). 20 pp.

—Case Hardening—Sheets—Rolled
DIN ENGL 17210-86. Case Hardening Steels; Technical Delivery Conditions (Sept). 20 pp.

—Case Hardening—Strips—Rolled
DIN ENGL 17210-86. Case Hardening Steels; Technical Delivery Conditions (Sept). 20 pp.

—Coils—Cold Worked
CEN PREN 10 172-87. Single Cold Reduced Electrolytic Chromium/Chromium Oxide Coated Steel: Coil for Subsequent Cutting into Sheets. 29 pp.
CEN PREN 10 173-87. Double Cold Reduced Electrolytic Chromium/Chromium Oxide Coated Steel: Coil for Subsequent Cutting into Sheets. 29 pp.

—Flats—Cold Rolled
BSI BS EN 10130-91. 1991 Cold Rolled Low Carbon Steel Flat Products for Cold Forming: Technical Delivery Conditions (V). 18 pp.
BSI BS EN 10130-01. 1991 Amd 1 Cold Rolled Low Carbon Steel Flat Products for Cold Forming: Technical Delivery Conditions (AMD 7638) May 15, 1993 (V). 19 pp.
CEN PREN 10 130-89. Cold Rolled Low Carbon Steel Flat Products for Cold Forming Technical Delivery Conditions. 23 pp.
CEN EN 10130-91. Cold Rolled Low Carbon Steel Flat Products for Cold Forming—Technical Delivery Conditions. 25 pp.
CEN PREN 10209-92. Cold Rolled Low Carbon Steel Flat Products for Vitreous Enamelling—Technical Delivery Conditions. 30 pp.
DIN ENGL EN 10130-91. Cold Rolled Flat Steel Products for Cold Forming; Technical Delivery Conditions (Oct) (Supersedes DIN 1623 Part 1. February 1983 Edition). 11 pp.

—Projection Welding
BSI BS 2630-82. 1982 Amd 1 Resistance Projection Welding of Uncoated Low Carbon Steel Sheet and Strip Using Embossed Projections. 12 pp.
BSI BS 2630-02. 1982 Amd 2 Resistance Projection Welding of Uncoated Low Carbon Steel Sheet and Strip Using Embossed Projections (AMD 7331) August 15, 1992. 14 pp.

INTERNATIONAL AND NON-U.S. NATIONAL STANDARDS
SUBJECT INDEX

Low

Low Carbon Steels (Cont.)
—**Resistance Welding**
  BSI BS 1140-93. 1993 Resistance Spot Welding of Uncoated and Coated Low Carbon Steel. 23 pp.
  BSI BS 1140-80. 1980 Amd 2 Resistance Spot Welding of Uncoated and Coated Low Carbon Steel. 14 pp.
  BSI BS 2630-82. 1982 Amd 1 Resistance Projection Welding of Uncoated Low Carbon Steel Sheet and Strip Using Embossed Projections. 12 pp.
  BSI BS 2630-02. 1982 Amd 2 Resistance Projection Welding of Uncoated Low Carbon Steel Sheet and Strip Using Embossed Projections (AMD 7331) August 15, 1992. 14 pp.
  BSI BS 6265-82. 1982 Resistance Seam Welding of Uncoated and Coated Low Carbon Steel (Supersedes BS 2937: 1957). 12 pp.

—**Rods**
  CNS G3166-88. Low Carbon Steel Wire Rods (May)(8693). 3 pp.
  JIS G 3505-80. Low Carbon Steel Wire Rods. 6 pp.

—**Rods—Cold Rolled**
  CEN PREN 10016-3-92. Non-Alloy Steel Rod for Drawing and/or Cold Rolling—Part 3: Specific Requirements for Rimmed and Rimmed Substitute Low Carbon Steel Rod. 6 pp.

—**Rods—Drawn**
  CEN PREN 10016-3-92. Non-Alloy Steel Rod for Drawing and/or Cold Rolling—Part 3: Specific Requirements for Rimmed and Rimmed Substitute Low Carbon Steel Rod. 6 pp.

—**Sheets**
  DIN ENGL 1616 (S)-84. Tinplate and Blackplate Sheet; Grades, Dimensions and Permissible Deviations (Oct) (Superseded DIN EN 10203 August 1991 and DIN EN 10205 January 1992). 12 pp.

—**Sheets—Aerospace**
  BSI S 536-70. 1970 Low Carbon 18/10 Chromium-Nickel Corrosion-Resisting Steel Sheet and Strip (50 Hbar). 3 pp.
  BSI S 537-70. 1970 Low Carbon 17/12 Chromium-Nickel-Molybdenum Corrosion-Resisting Steel Sheet and Strip (50 Hbar). 4 pp.

—**Sheets—Cold Rolled**
  BSI BS EN 10131-91. 1991 Cold-Rolled Uncoated Low Carbon and High Yield Strength Steel Flat Products for Cold Forming—Tolerances on Dimensions and Shape (V). 7 pp.
  BSI BS EN 10131-01. 1991 Amd 1 Cold-Rolled Uncoated Low Carbon and High Yield Strength Steel Flat Products for Cold Forming—Tolerances on Dimensions and Shape (AMD 7327) February 15, 1993 (Supersedes BS 1449: Section 1.7: 1991). 8 pp.
  CEN PREN 10111-93. Continuously Hot-Rolled Low Carbon Steel Sheet and Strip for Cold Bending—Technical Delivery Conditions. 11 pp.
  CEN PREN 10 131-89. Cold Rolled Uncoated Low Carbon and High Yield Strength Steel Flat Products for Cold Forming: Tolerances on Dimensions and Shape. 13 pp.
  CEN EN 10131-91. Cold Rolled Uncoated Low Carbon and High Yield Strength Steel Flat Products for Cold Forming—Tolerances on Dimensions and Shape. 15 pp.
  DIN ENGL EN 10131-92. Cold Rolled Uncoated Low Carbon and High Yield Strength Steel Flats for Cold Forming; Tolerances on Size and Geometrical Tolerances; English Version of DIN EN 10 131 (Jan) (Supersedes DIN 1541, August 1975 Edition). 8 pp.
  SAA AS 1595-81. Cold-Rolled Unalloyed Low Carbon Steel Sheet and Strip Amdt 1 December 1982. 12 pp.

—**Sheets—Cold Worked**
  CEN PREN 10 170-87. Single Cold Reduced Electrolytic Chromium/Chromium Oxide Coated Steel: Sheet. 27 pp.

—**Sheets—Hot Dip**
  CEN PREN 10 214-92. Continuously Hot-Dip Zinc-Aluminium (ZA) Coated Steel Sheet and Strip; Technical Delivery Conditions. 20 pp.
  CEN PREN 10 215-92. Continuously Hot-Dip Aluminium-Zinc (AZ) Coated Steel Sheet and Strip; Technical Delivery Conditions. 19 pp.

—**Strips—Aerospace**
  BSI S 536-70. 1970 Low Carbon 18/10 Chromium-Nickel Corrosion-Resisting Steel Sheet and Strip (50 Hbar). 3 pp.
  BSI S 537-70. 1970 Low Carbon 17/12 Chromium-Nickel-Molybdenum Corrosion-Resisting Steel Sheet and Strip (50 Hbar). 4 pp.

Low Carbon Steels (Cont.)
—**Strips—Cold Rolled**
  CEN PREN 10111-93. Continuously Hot-Rolled Low Carbon Steel Sheet and Strip for Cold Bending—Technical Delivery Conditions. 11 pp.
  SAA AS 1595-81. Cold-Rolled Unalloyed Low Carbon Steel Sheet and Strip Amdt 1 December 1982. 12 pp.

—**Strips—Hot Dip**
  CEN PREN 10 214-92. Continuously Hot-Dip Zinc-Aluminium (ZA) Coated Steel Sheet and Strip; Technical Delivery Conditions. 20 pp.
  CEN PREN 10 215-92. Continuously Hot-Dip Aluminium-Zinc (AZ) Coated Steel Sheet and Strip; Technical Delivery Conditions. 19 pp.

—**Tubes—Seamless—Cold Drawn**
  BSI BS 7416-91. 1991 Precision Finished Seamless Cold-Drawn Low Carbon Steel Tubes for Use in Hydraulic Fluid Power Systems. 10 pp.

Low Frequency Transformers
See Also: Transformers
  JIS C 6435-89. Testing Methods for Low Frequency Transformers and Inductors. 50 pp.

—**Electronic**
  CNS C7147-81. General Rules of Low Frequency Transformer for Electronic Equipment (Jul)(7635).

Low Melting Alloys
Use: Alloys—Low Melting

Low Noise Amplifiers
Use For: LNA *See Also:* Amplifiers

—**Earth Stations**
  BSI BS 7573: Sec 3.4-93. 1993 Methods of Measurement for Equipment Used in Digital Microwave Radio Transmission Systems Part 3: Measurements on Satellite Earth Stations Section 3.4: Low Noise Amplifier (IEC 835-3-4: 1993) (S). 8 pp.
  CENELEC HD 467.2.3 S1-90. Methods of Measurement for Radio Equipment Used in Satellite Earth Stations Part 2: Measurements for Sub-Systems Section Three—Low-Noise Amplifier. 3 pp.
  IEC 510 Pt 2-3-89. Methods of Measurement for Radio Equipment Used in Satellite Earth Stations Part 2: Measurements for Sub-Systems Section Three—Low-Noise Amplifier First Edition. 15 pp.
  IEC 835 Pt 3-4-93. Methods of Measurement of Equipment Used in Digital Microwave Radio Transmission Systems Part 3: Measurements on Satellite Earth Stations Section 4: Low Noise Amplifier First Edition. 20 pp.

—**Tubes—Quality Assurance**
  BSI BS 9026-70. (WITHDRAWN) 1970 Amd 1 Rules for the Preparation of Detail Specifications for Low Noise Signal Amplifier Tubes of Assessed Quality with Integral Permanent Magnet Focusing. 21 pp.

Low Pass Filters
Use For: Lowpass Filters *See Also:* Electric Filters; Radio Frequency Filters

—**Sound Level Meters—Ultrasonic Frequencies**
  BSI BS 7422-91. 1991 Filters for the Measurement of Audible Sound in the Presence of Ultrasound. 12 pp.
  IEC 1012-90. Filters for the Measurement of Audible Sound in the Presence of Ultrasound First Edition. 17 pp.

Low Pressure Testing
See Also: Environmental Testing; Pressure Measurement

—**Connectors**
  CEN PREN 2591 (Part C3)-92. Elements of Electrical and Optical Connection Test Methods Part C3—Cold/Low Pressure and Damp Heat. 6 pp.
  CEN PREN 2591 (Part C11)-92. Elements of Electrical and Optical Connection Test Methods Part C11—Low Air Pressure. 4 pp.
  CEN PREN 2591 (Part C14)-92. Elements of Electrical and Optical Connection Test Methods Part C14—Immersion at Low Air Pressure. 5 pp.
  CEN PREN 2591-FC3-93. Aerospace Series Elements of Electrical and Optical Connection Test Methods Part FC3—Optical Elements Cold/Low Pressure and Damp Heat. 3 pp.
  CEN PREN 2591-FC14-93. Aerospace Series Elements of Electrical and Optical Connection Test Methods Part FC14—Optical Elements Immersion at Low Air Pressure. 3 pp.

Low Pressure Testing (Cont.)
—**Electrical Components**
  BSI BS 2011: Part 2.1M-84. 1984 Basic Environmental Testing Procedures Part 2.1 Tests Part 2.1M: Low Air Pressure. 7 pp.
  BSI BS 2011: Part 2.1Z/AM-77. 1977 Amd 1 Basic Environmental Testing Procedures Part 2.1: Tests Part 2.1Z/AM: Test Z/AM. Combined Cold/Low Air Pressure Tests. 10 pp.
  BSI BS 2011: Pt 2.1Z/AMD-77. 1977 Basic Environmental Testing Procedures Part 2.1: Tests Part 2.1Z/AMD: Test Z/AMD. Combined Sequential Cold, Low Air Pressure and Damp Heat Test. 4 pp.
  BSI BS 2011: Part 2.1Z/BM-77. 1977 Amd 1 Basic Environmental Testing Procedures Part 2.1: Tests Part 2.1Z/BM: Test Z/BM. Combined Dry Heat/Low Air Pressure Tests. 12 pp.
  BSI BS 2011:Pt 3 Z/AM & Z/BM-77. 1977 Basic Environmental Testing Procedures Part 3: Background Information Part 3Z/AM and Z/BM: Test Z/AM and Z/BM. Combined Temperature/Low Air Pressure Tests. 6 pp.
  CENELEC HD 323.2.13-87. Basic Environmental Testing Procedures Part 2: Tests Test M: Low air Pressure. 3 pp.
  CENELEC HD 323.2.39 S1-76. Basic Environmental Testing Procedures-Part 2: Tests. Test Z/Amd: Combined Sequential Cold, Low Air Pressure, and Damp Heat Test. 2 pp.
  CENELEC HD 323.2.40 S1-88. Basic Environmental Testing Procedures—Part 2: Tests Test Z/AM: Combined Cold/Low Air Pressure Tests. 3 pp.
  CENELEC HD 323.2.41 S1-88. Basic Environmental Testing Procedures—Part 2: Tests. Test Z/BM: Combined Dry Heat/Low Air Pressure Tests. 3 pp.
  CENELEC HD 323.3.2 S1-88. Basic Environmental Testing Procedures-Part 3: Background Information. Section Two: Combined Temperature/Low Air Pressure Tests. 3 pp.
  IEC 68 Pt 2-13-83. Basic Environmental Testing Procedures Part 2: Tests Test M: Low Air Pressure Fourth Edition. 17 pp.
  IEC 68 Pt 2-40-76. Basic Environmental Testing Procedures Part 2: Tests—Test Z/AM: Combined Cold/Low Air Pressure Tests First Edition; (Amendment 1-1983). 22 pp.
  IEC 68 Pt 2-41-76. Basic Environmental Testing Procedures Part 2: Tests—Test Z/BM: Combined Dry Heat/ Low Air Pressure Tests First Edition; (Amendment 1-1983). 24 pp.
  IEC 68 Pt 3-2-76. Basic Environmental Testing Procedures Part 3: Background Information Section Two-Combined Temperature/Low Air Pressure Tests First Edition. 14 pp.
  JIS C 0029-89. Basic Environmental Testing Procedures Part 2: Tests, Test M: Low Air Pressure. 7 pp.
  JIS C 0030-90. Basic Environmental Testing Procedures Part 2: Tests-Test Z/AM Combined Cold/Low Air Pressure Tests. 9 pp.
  JIS C 0031-90. Basic Environmental Testing Procedures Part 2: Tests-Test Z/BM Combined Dry Heat/Low Air Pressure Tests. 10 pp.
  SAA AS 1099.2Z/A M-80. Basic Environmental Testing Procedures for Electrotechnology—Part 2: Tests—Part 2Z/AM: Combined Cold/Low Air Pressure Tests Reconfirmed 1985. 6 pp.
  SAA AS 1099.2Z/A MD-80. Basic Environmental Testing Procedures for Electrotechnology—Part 2: Tests—Part 2Z/AMD: Combined Sequential Cold, Low Air Pressure and Damp Heat Test Reconfirmed 1985. 2 pp.
  SAA AS 1099.2Z/B M-80. Basic Environmental Testing Procedures for Electrotechnology—Part 2: Tests—Part 2Z/BM: Combined Dry Heat/Low Air Pressure Tests Reconfirmed 1985. 6 pp.
  SAA AS 1099.2.13-90. Basic Environmental Testing Procedures for Electrotechnology—Part 2: Tests—Part 2.13: Test M—Low Air Pressure (IEC 68-2-13). 4 pp.
  SAA AS 1099.3.2-80. Basic Environmental Testing Procedures for Electrotechnology—Part 3: Background Information—Part 3.2: Section 2—Combined Temperature/Low Air Pressure Tests. 4 pp.
  SNZ IEC 68: Part 2-13-83. Basic Environmental Testing Procedures Part 2-13: Test M: Low Air Pressure. 13 pp.
  SNZ IEC 68: Part 2-40-76. Basic Environmental Testing Procedures Part 2-40: Test 2/Am: Combined Cold/Low Air Pressure Tests Amend: 1. 17 pp.
  SNZ IEC 68: Part 2-41-76. Basic Environmental Testing Procedures Part 2-41: Test Z/Bm: Combined Dry Heat/Low Air Pressure Tests Amend: 1. 19 pp.
  SNZ IEC 68: Part 3-2-76. Basic Environmental Testing Procedures Part 3-2: Background Information. Section Two —Combined Temperature/Low Air Pressure Tests. 12 pp.

—**Electronic Equipment**
  CNS C6245-85. Low Air Pressure Testing Method for Electronic Components (Apr)(11236).

INDEX and DIRECTORY of

INTERNATIONAL AND NON-U.S. NATIONAL STANDARDS
SUBJECT INDEX

Low Temperature Steels
See Also: Steels

—Bars
DIN ENGL 17280-85. Steels with Low Temperature Toughness; Technical Delivery Conditions for Plate, Sheet, Strip, Wide Flats, Sections, Bars and Forgings (July) (Partially Superseded by DIN EN 10028 Part 1). 18 pp.

—Flats
DIN ENGL 17280-85. Steels with Low Temperature Toughness; Technical Delivery Conditions for Plate, Sheet, Strip, Wide Flats, Sections, Bars and Forgings (July) (Partially Superseded by DIN EN 10028 Part 1). 18 pp.

—Forgings
DIN ENGL 17280-85. Steels with Low Temperature Toughness; Technical Delivery Conditions for Plate, Sheet, Strip, Wide Flats, Sections, Bars and Forgings (July) (Partially Superseded by DIN EN 10028 Part 1). 18 pp.

—Plates
DIN ENGL 17280-85. Steels with Low Temperature Toughness; Technical Delivery Conditions for Plate, Sheet, Strip, Wide Flats, Sections, Bars and Forgings (July) (Partially Superseded by DIN EN 10028 Part 1). 18 pp.

—Sections
DIN ENGL 17280-85. Steels with Low Temperature Toughness; Technical Delivery Conditions for Plate, Sheet, Strip, Wide Flats, Sections, Bars and Forgings (July) (Partially Superseded by DIN EN 10028 Part 1). 18 pp.

—Sheets
DIN ENGL 17280-85. Steels with Low Temperature Toughness; Technical Delivery Conditions for Plate, Sheet, Strip, Wide Flats, Sections, Bars and Forgings (July) (Partially Superseded by DIN EN 10028 Part 1). 18 pp.

—Strips
DIN ENGL 17280-85. Steels with Low Temperature Toughness; Technical Delivery Conditions for Plate, Sheet, Strip, Wide Flats, Sections, Bars and Forgings (July) (Partially Superseded by DIN EN 10028 Part 1). 18 pp.

Low Temperature Testing
Use For: Cold Testing *See Also:* Bend Testing; Brittleness Testing; Compression Testing; Cryostats; Dynamic Testing; Environmental Testing; Fatigue Testing; Freezing Points; Heat of Fusion; High Temperature Testing; Impact Testing; Life (Durability); Materials Testing; Melting Points; Physical Testing; Shear Testing; Static Testing; Stiffness Testing; Stress (Testing); Temperature; Temperature Controllers; Temperature Testing; Thermal Expansion; Thermal Shock; Thermal Stability; Thermal Stresses; Thermodynamic Properties; Wear Testing

—Adhesives
BSI BS 5350: Part H2-82. 1982 Methods of Test for Adhesives Group H: Physical Tests on Hot Melt Adhesives Part H2: Determination of Low Temperature Flexibility or Cold Crack Temperature. 4 pp.

—Automotive
CNS D3047-89. Method of High and Low Temperature Test for Automobile Parts (Jul)(5434).
JIS D 0204-67. Method of High and Low Temperature Test for Automobile Parts (R 1979). 6 pp.

—Cable Insulation
BSI BS 6469: Sec 1.4-92. 1992 Insulation and Sheathing Materials of Electric Cables Part 1: Methods of Test for General Application Section 1.4: Test at Low Temperature. 22 pp.
CENELEC HD 505.1.4 S1-88. Common Test Methods for Insulating and Sheathing Materials of Electric Cables Part 1 Methods for General Application Section Four—Tests at Low Temperature. 3 pp.
IEC 811 Pt 1-4-85. Common Test Methods for Insulating and Sheathing Materials of Electric Cables Part 1: Methods for General Application Section Four—Tests at Low Temperature First Edition; (Corrigendum—May 1986) (Amendment 1-1993). 40 pp.

—Carburetors
CNS D3090-88. Method of Low Temperature Test of Carburetors for Automobiles (Dec)(8247).

Low Temperature Testing (Cont.)
—Cellular Plastics
ISO 4897-85. Cellular Plastics—Determination of the Coefficient of Linear Thermal Expansion of Rigid Materials at Sub-Ambient Temperatures First Edition. 11 pp.

—Coated Fabrics
BSI BS 3424: Part 8-83. 1983 Testing Coated Fabrics Part 8: Methods 10A, 10B and 10C. Methods for Determination of Low Temperature Performance. 10 pp.
ISO 4646-89. Rubber-or Plastics-Coated Fabrics—Low-Temperature Impact Test Second Edition. 9 pp.
ISO 4675-90. Rubber-or Plastics-Coated Fabrics—Low-Temperature Bend Test Second Edition. 8 pp.

—Connectors
CEN PREN 2591 (Part C10)-92. Elements of Electrical and Optical Connection Test Methods Part C10—Cold. 4 pp.
SBAC TS 326 ISSUE 1. Test for Electrical Connectors Low Temperature Handling.

—Control Cables (Electric)
CSA CAN/CSA-C22. 2 NO 239-M91. Control and Instrumentation Cables; (Gen Instr 1 Thru 3). 50 pp.

—Diesel Fuels
BSI BS 6380-83. (WITHDRAWN) 1983 Amd 1 Low Temperature Properties and Cold Weather Use of Diesel Fuels and Gas Oils (Classes A1, A2 and D of BS 2869) (AMD 5452) September 30, 1986. 16 pp.

—Diesel Fuels—Cold Filter Plugging Point
CEN EN 116-81. Determination of Cold Filter Plugging Point of Diesel and Domestic Heating Fuels. 13 pp.

—Elastomeric Sheets
CEN PREN 495-5-91. Thermoplastic and Elastomeric Roofing and Sealing Sheets—Low Temperature Folding Test. 5 pp.

—Electric Switches
CNS C6091-88. Method of Test for Electromechanical Switches (High/Low Temperature Operation) (Dec)(6154).

—Electrical Components
BSI BS 2011: Part 2.1A-90. (WITHDRAWN) 1990 Amd 2 Basic Environmental Testing Procedures Part 2.1: Tests Part 2.1A: Test A. Cold (Renumbered as BS EN 60068-2-1: 1993). 26 pp.
BSI BS EN 60068-2-1-93. 1993 Amd 1 Environmental Testing Part 2.1: Tests Tests A. Cold (AMD 7783) July 15, 1993 (G). 39 pp.
BSI BS 2011: Part 2.1Z/AD-77. 1977 Basic Environmental Testing Procedures Part 2.1: Tests Part 2.1Z/AD: Test Z/AD. Composite Temperature/Humidity Cyclic Test. 12 pp.
BSI BS 2011: Pt 2.1Z/AFC-84. 1984 Basic Environmental Testing Procedures Part 2.1: Tests: Part 2.1Z/AFc: Test Combined Cold/Vibration (Sinusoidal) Tests for Both Heat-Dissipating and Non-Heat-Dissipating Specimens. 12 pp.
BSI BS 2011: Part 2.1Z/AM-77. 1977 Amd 1 Basic Environmental Testing Procedures Part 2.1: Tests Part 2.1Z/AM: Test Z/AM. Combined Cold/Low Air Pressure Tests. 10 pp.
BSI BS 2011: Pt 2.1Z/AMD-77. 1977 Basic Environmental Testing Procedures Part 2.1: Tests Part 2.1Z/AMD: Test Z/AMD. Combined Sequential Cold, Low Air Pressure and Damp Heat Test. 4 pp.
BSI BS 2011:Pt2. 2Z/AFC/BFC-86. 1986 Basic Environmental Testing Procedures Part 2.2: Guidance Part 2.2Z/AFc and Z/BFc: Test Z/AFc and Z/BFc. Guidance on Combined Temperature (Cold and Dry Heat) and Vibration (Sinusoidal) Tests. 8 pp.
BSI BS 2011: Part 3A & B-77. 1977 Amd 1 Basic Environmental Testing Procedures Part 3: Background Information Part 3A and B: Tests A (Cold) and Tests B (Dry Heat). 37 pp.
BSI BS 2011:Pt 3 A & B:Supp 1-80. 1980 Basic Environmental Testing Procedures Part 3: Background Information Part 3A and B: Tests A (Cold) and Tests B (Dry Heat): Supplement No. 1:. 5 pp.
CENELEC HD 323.2.1 S2-87. Basic Environmental Testing Procedures Part 2: Tests Test A: Cold. 3 pp.
CENELEC HD 323.2.38 S1-76. Basic Environmental Testing Procedures—Part 2: Tests. Test Z/AD: Composite Temperature/Humidity Cyclic Test. 2 pp.
CENELEC HD 323.2.39 S1-76. Basic Environmental Testing Procedures-Part 2: Tests. Test Z/Amd: Combined Sequential Cold, Low Air Pressure, and Damp Heat Test. 2 pp.
CENELEC HD 323.2.40 S1-88. Basic Environmental Testing Procedures—Part 2: Tests Test Z/AM: Combined Cold/Low Air Pressure Tests. 3 pp.

Low Temperature Testing (Cont.)
—Electrical Components (Cont.)
CENELEC HD 323.2.50 S1-87. Basic Environmental Testing Procedures Part 2: Tests Tests Z/AFc: Combined Cold/Vibration (Sinusoidal) Tests for Both Heat-Dissipating and Non-Heat-Dissipating Specimens. 4 pp.
CENELEC HD 323.3.1 S1-85. Basic Environmental Testing Procedures—Part 3: Background Information. Section One:-Cold and Dry Heat Tests. 2 pp.
CENELEC HD 323.3.1 S1-88. Basic Environmental Testing Procedures Part J: Background Information Section One—Cold and Dry Heat Tests. 2 pp.
CENELEC EN 60068-2-1-93. Environmental Testing Part 2: Tests Tests A: Cold (IEC 68-2-1:1990) (Supersedes HD 323.2.1 S2:1987). 5 pp.
CENELEC EN 60068-2-1-93. Environmental Testing Part 2: Tests Tests A: Cold (IEC 68-2-1: 1990). 26 pp.
IEC 68 Pt 2-1-90. Environmental Testing Part 2: Tests—Test A: Cold Fifth Edition; (Amendment 1-1993) (CENELEC EN 60068-2-1: 1993). 53 pp.
IEC 68 Pt 2-38-74. Basic Environmental Testing Procedures Part 2: Tests—Test Z/AD: Composite Temperature/Humidity Cyclic Test First Edition. 23 pp.
IEC 68 Pt 2-39-76. Basic Environmental Testing Procedures Part 2: Tests—Test Z/AMD: Combined Sequential Cold, Low Air Pressure, and Damp Heat Test First Edition. 12 pp.
IEC 68 Pt 2-40-76. Basic Environmental Testing Procedures Part 2: Tests—Test Z/AM: Combined Cold/Low Air Pressure Tests First Edition; (Amendment 1-1983). 22 pp.
IEC 68 Pt 2-50-83. Basic Environmental Testing Procedures Part 2: Tests Tests Z/AFc: Combined Cold/Vibration (Sinusoidal) Tests for Both Heat-Dissipating and Non-Heat Dissipating Specimens First Edition. 25 pp.
IEC 68 Pt 2-53-84. Basic Environmental Testing Procedures Part 2: Tests Guidance to Tests Z/AFc and Z/BFc: Combined Temperature (Cold and Dry Heat) and Vibration (Sinusoidal) Tests First Edition. 14 pp.
IEC 68 Pt 3-1-74. Basic Environmental Testing Procedures Part 3: Background Information Section One—Cold and Dry Heat Tests First Edition; (Supplement A-1978) (Corrigenda—March 1980). 73 pp.
JIS C 0020-87. Basic Environmental Testing Procedures Part 2: Tests, Tests A: Cold. 22 pp.
JIS C 0030-90. Basic Environmental Testing Procedures Part 2: Tests-Test Z/AM Combined Cold/Low Air Pressure Tests. 9 pp.
SAA AS 1099.2AA-80. Basic Environmental Testing Procedures for Electrotechnology—Part 2: Tests—Part 2Aa: Cold Test for Non-Heat-Dissipating Specimens with Gradual Change of Temperature.
SAA AS 1099.2AB-80. Basic Environmental Testing Procedures for Electrotechnology—Part 2: Tests—Part 2Ab: Cold Test for Non-Heat-Dissipating Specimens with Gradual Change of Temperature.
SAA AS 1099.2AD-80. Basic Environmental Testing Procedures for Electrotechnology—Part 2: Tests—Part 2Ad: Cold Test for Heat-Dissipating Specimens with Gradual Change of Temperature (Bound Together). 20 pp.
SAA AS 1099.2Z/A D-80. Basic Environmental Testing Procedures for Electrotechnology—Part 2: Tests—Part 2Z/AD: Composite Temperature/Humidity Cyclic Test Reconfirmed 1985. 8 pp.
SAA AS 1099.2Z/A M-80. Basic Environmental Testing Procedures for Electrotechnology—Part 2: Tests—Part 2Z/AM: Combined Cold/Low Air Pressure Tests Reconfirmed 1985. 6 pp.
SAA AS 1099.2Z/A MD-80. Basic Environmental Testing Procedures for Electrotechnology—Part 2: Tests—Part 2Z/AMD: Combined Sequential Cold, Low Air Pressure and Damp Heat Test Reconfirmed 1985. 2 pp.
SAA AS 1099.3.1-80. Basic Environmental Testing Procedures for Electrotechnology—Part 3: Background Information—Part 3.1: Section 1 Tests A and B-Cold and Dry Heat Tests. 32 pp.
SNZ IEC 68: Part 2-1-74. Basic Environmental Testing Procedures Part 2-1: Tests A: Cold Amend: 1. 43 pp.
SNZ IEC 68: Part 2-1A-76. Basic Environmental Testing Procedures Part 2-1A: Tests A: Cold. 2 pp.
SNZ IEC 68: Part 2-38-74. Basic Environmental Testing Procedures Part 2-38: Test Z/AD: Composite Temperature/Humidity Cyclic Test. 20 pp.
SNZ IEC 68: Part 2-39-76. Basic Environmental Testing Procedures Part 2-39: Test 2/AMD: Combined Sequential Cold, Low Air Pressure, and Damp Heat Test. 8 pp.
SNZ IEC 68: Part 2-40-76. Basic Environmental Testing Procedures Part 2-40: Test 2/Am: Combined Cold/Low Air Pressure Tests Amend: 1. 17 pp.

INDUSTRY STANDARDS

INTERNATIONAL AND NON-U.S. NATIONAL STANDARDS
SUBJECT INDEX
Low

Low Temperature Testing *(Cont.)*
—Electrical Components *(Cont.)*
SNZ IEC 68: Part 2-50-83. Basic Environmental Testing Procedures Part 2-50: Test 2/AFc: Combined Cold-Vibration (Sinusoidal) Tests for Both Heat Dissipating and Non-Heat-Dissipating Specimens. 21 pp.

SNZ IEC 68: Part 2-53-84. Basic Environmental Testing Procedures Part 2-53: Guidance to Tests Z/AFc and Z/BFc: Combined Temperature (Cold and Dry Heat) and Vibration (Sinusoidal) Tests. 10 pp.

SNZ IEC 68: Part 3-1-74. Basic Environmental Testing Procedures Part 3-1: Background Information. Section One —Cold and Dry Heat Tests. 59 pp.

SNZ IEC 68: Part 3-1A-78. Basic Environmental Testing Procedures Part 3-1A: Background Information. Section One —Cold and Dry Heat Tests. 7 pp.

—Electrical Insulation—Heat Shrinkable Tubing
CNS C3144-88. Method of Test for Low-Temperature Characteristics of Heat Shrinkable Tubing for Electrical Insulation (Sep)(8791).

—Electronic Equipment
CNS C6242-89. Cold Testing Procedures for Electronic Components (Jul)(11233).

—Electronic Equipment—Storage
CNS C6244-85. Storage (Low Temperature) Testing Method for Electronic Components (Apr)(11235).

—Fiber Optic Cables
CNS C6317-88. Method of Test for Fiber Optic Devices (FOTP-98 Fiber Optic Cable External Freezing Test) (Jul)(12364).

—Fiber Optic Cables—Bend Testing
CNS C6276-86. Method of Test for Fiber Optic Devices (FOTP-37 Cable Bend Test Low and High Temperature) (Dec)(11788).

—Fire Alarms
CEN EN 54-5-76. Components of Automatic Fire Detection Systems Part 5 Heat Sensitive Detectors—Point Detectors Containing a Static Element. 20 pp.

CEN EN 54 (Part 5)-88. AMD 1 Components of Automatic Fire Detection Systems Part 5 Heat Sensitive Detectors—Point Detectors Containing a Static Element. 21 pp.

CEN EN 54-7-88. Components of Automatic Fire Detection Systems—Part 7: Point-Type Smoke Detectors; Detectors Using Scattered Light, Transmitted Light or Ionization. 30 pp.

CEN EN 54-8-88. Components of Automatic Fire Detection Systems—Part 8: Hig Temperature Heat Detectors. 23 pp.

CEN EN 54-9-82. Components of Automatic Fire Fetection Systems Part 9 Fire Sensitivity Test. 10 pp.

—Gas Oils
BSI BS 6380-83. (WITHDRAWN) 1983 Amd 1 Low Temperature Properties and Cold Weather Use of Diesel Fuels and Gas Oils (Classes A1, A2 and D of BS 2869) (AMD 5452) September 30, 1986. 16 pp.

—Heating Oils—Cold Filter Plugging Point
CEN EN 116-81. Determination of Cold Filter Plugging Point of Diesel and Domestic Heating Fuels. 13 pp.

—Hoses
BSI BS 5173: Sec 106.1-89. (WITHDRAWN) 1989 Rubber and Plastics Hoses and Hose Assemblies Part 106: Environmental Tests Section 106.1: Determination of Low Temperature Flexibility (Renumbered as BS EN 24672: 1993). 6 pp.

BSI BS EN 24672-93. 1993 Amd 1 Rubber and Plastics Hoses—Sub-Ambient Temperature Flexibility Tests (ISO 4672: 1988) (AMD 7524) April 15, 1993. 13 pp.

CEN EN 24672-93. Rubber and Plastics Hoses—Sub-Ambient Temperature Flexibility Tests (ISO 4672: 1988). 6 pp.

ISO 4672-88. Rubber and Plastics Hoses—Sub-Ambient Temperature Flexibility Tests Second Edition (CEN EN 24672: 1993). 6 pp.

—Instantaneous Water Heaters
BSI DD 84-82. 1982 Method of Test for the Resistance to Freezing of Gas-Fired Instantaneous Water Heaters. 5 pp.

—Instrumentation Cables
CSA CAN/CSA-C22. 2 NO 239-M91. Control and Instrumentation Cables; (Gen Instr 1 Thru 3). 50 pp.

Low Temperature Testing *(Cont.)*
—Iron Ores
ISO 4696-84. Iron Ores—Low-Temperature Disintegration Test—Method Using Cold Tumbling After Static Reduction First Edition. 9 pp.

—Leather
CNS K6676-81. Method of Test for Cold Resistance of Leather (Jul)(7702).

JIS K 6542-74. Testing Methods for Cold Resistance of Leather.

—Plastic Sheets
BSI BS 2782:Pt1: METH 150B-76. (WITHDRAWN) 1976 Amd 1 Methods of Testing Plastics Part 1: Thermal Properties Method 150B: Determination of Cold Flex Temperature of Flexible Polyvinyl Compound (Superseded by BS 2782: Method 153B: 1991). 5 pp.

BSI BS 2782:Pt1: METH 150C-83. 1983 Methods of Testing Plastics Part 1: Thermal Properties Method 150C: Determination of Low Temperature Extensibility of Flexible Polyvinyl Chloride Sheet. 4 pp.

DIN ENGL 53372-70. Testing of Plastic Films; Determination of Low-Temperature Breakage of Non-Rigid Polyvinyl Chloride (PVC) Films (Dec). 2 pp.

—Railroad Signals
JIS E 3019-79. High and Low Temperature Testing Methods for Parts of Railway Signaling.

—Refrigeration Oils
DIN ENGL 51590 Pt 2-76. Testing of Lubricants; Determination of the Content of Material Insoluble in R 12 in Refrigerator Oils; Method at-40 Deg. C (233 K) (June). 4 pp.

—Semiconductor Devices—Storage
CNS C6073-88. Environmental Testing Methods and Endurance Testing Methods for Discrete Semiconductor Devices (Test of Storage Under Low Temperature) (Nov)(6118).

—Soaps
CGSB 2-GP-11M METH 43.1-83. Methods of Testing and Analysis of Soaps and Detergents Cold Test. 1 p.

—Thermoplastic Sheets
CEN PREN 495-5-91. Thermoplastic and Elastomeric Roofing and Sealing Sheets—Low Temperature Folding Test. 5 pp.

—Tinplate Panels
CGSB 1-GP-71 METH 119.4-79. Methods of Testing Paints and Pigments Flexibility Low-Temperature. 1 p.

—Vulcanized Rubber
BSI BS 903: Part A6-92. 1992 Physical Testing of Rubber Part A6: Method for Determination of Compression Set at Ambient, Elevated or Low Temperatures (ISO 815: 1991) (V). 16 pp.

BSI BS 903: Part A25-92. 1992 Physical Testing of Rubber Part A25: Determination of Low-Temperature Brittleness (ISO 812: 1991) (V). 11 pp.

BSI BS 903: Part A25-68. 1968 Amd 1 Methods of Testing Vulcanized Rubber Part A25: Determination of Impact Brittleness Temperature. 11 pp.

BSI BS 903: Part A29-84. 1984 Methods of Testing Vulcanized Rubber Part A29: Determination of Low Temperature Characteristics by Temperature-Retraction Procedure (TR Test). 8 pp.

BSI BS 903: Part A39-80. (WITHDRAWN) 1980 Methods of Testing Vulcanized Rubber Part A39: Determination of Compression Set Under Constant Deflection at Low Temperatures (Superseded by BS 903: Part A6: 1992). 6 pp.

CNS K6355-85. Method of Test for Brittleness Temperature of Vulcanized Rubber by Impact (Dec)(3564).

CNS K6749-83. Method of Test for Torsion Under Low Temperature of Vulcanized Rubber (Feb)(10021).

CNS K6750-83. Method of Test for Permanent Compression Set Under Low Temperature of Vulcanized Rubber (Feb)(10022).

ISO 812-91. Rubber, Vulcanized—Determination of Low-Temperature Brittleness First Edition. 10 pp.

ISO 815-91. Rubber, Vulcanized or Thermoplastic—Determination of Compression Set at Ambient, Elevated or Low Temperatures Second Edition; (Corrigendum 1-1993). 15 pp.

ISO 2921-91. Rubber, Vulcanized—Determination of Low Temperature Characteristics—Temperature-Retraction Procedure (TR Test) Second Edition. 6 pp.

Low Voltage Avalanche Zener Diodes
Use: Voltage Reference Diodes

Low Voltage Installations
Use For: Low Voltage Systems **See Also:** Electrical Installations

—Dielectric Testing
BSI BS 7640: Part 1-93. 1993 High-Voltage Test Techniques for Low-Voltage Equipment Part 1: Definitions, Test and Procedure Requirements (IEC 1180-1: 1992) (E). 31 pp.

IEC 1180 Pt 1-92. High-Voltage Test Techniques for Low-Voltage Equipment Part 1: Definitions, Test and Procedure Requirements First Edition. 60 pp.

—Electrical Equipment
BSI BS EN 50065-1-92. 1992 Signalling on Low-Voltage Electrical Installations in the Frequency Range 3 kHz to 148.5 kHz Part 1: General Requirements, Frequency Bands and Electromagnetic Disturbances. 15 pp.

BSI BS EN 50065-1-01. 1992 Amd 1 Signalling on Low-Voltage Electrical Installations in the Frequency Range 3 kHz to 148.5 kHz Part 1: General Requirements, Frequency Bands and Electromagnetic Disturbances (AMD 7950) September 15, 1993 (S). 24 pp.

CENELEC EN 50 065-1-91. Signalling on Low—Voltage Electrical Installations in the Frequency Range 3/kHz to 148.5 kHz Part 1: General Requirements, Frequency Bands and Electromagnetic Disturbances. 23 pp.

CENELEC EN 50065-1/A1-92. AMD 1 Signalling on Low-Voltage Electrical Installations in the Frequency Range 3 khz to 148.5 khz Part 1: General Requirements, Frequency Bands and Electromagnetic Disturbances. 11 pp.

IEC 1204-93. Low-Voltage Power Supply Devices, d.c. Output—Performance Characteristics and Safety Requirements First Edition. 59 pp.

—Electrical Equipment—Safety
IEC 1204-93. Low-Voltage Power Supply Devices, d.c. Output—Performance Characteristics and Safety Requirements First Edition. 59 pp.

—Electrical Faults—High Voltage Installations
IEC 364 Pt 4-442-93. Electrical Installations of Buildings Part 4: Protection for Safety Chapter 44: Protection Against Overvoltages Section 442—Protection of Low-Voltage Installations Against Faults Between High-Voltage Systems and Earth First Edition. 45 pp.

—High Voltage Testing
BSI BS 7640: Part 1-93. 1993 High-Voltage Test Techniques for Low-Voltage Equipment Part 1: Definitions, Test and Procedure Requirements (IEC 1180-1: 1992) (E). 31 pp.

IEC 1180 Pt 1-92. High-Voltage Test Techniques for Low-Voltage Equipment Part 1: Definitions, Test and Procedure Requirements First Edition. 60 pp.

—Impulse Voltage Testing
BSI BS 7640: Part 1-93. 1993 High-Voltage Test Techniques for Low-Voltage Equipment Part 1: Definitions, Test and Procedure Requirements (IEC 1180-1: 1992) (E). 31 pp.

IEC 1180 Pt 1-92. High-Voltage Test Techniques for Low-Voltage Equipment Part 1: Definitions, Test and Procedure Requirements First Edition. 60 pp.

—Insulation Coordination
BSI PD 6499-81. 1981 Amd 1 Guide to Insulation Co-Ordination Within Low-Voltage Systems Including Clearances and Creepage Distances for Equipment. 51 pp.

DIN VDE 0110 Pt 1-89. Insulation Co-Ordination for Equipment Within Low-Voltage Systems; Fundamental Requirements (Jan) (This Standard, Together with VDE 0110 Pt 2/01.89 Supersedes 0110/11.72 and 0110b/02.79). 27 pp.

DIN VDE 0110 Pt 2-89. Insulation Co-Ordination for Equipment Within Low-Voltage Systems; Dimensioning of Clearances and Creepage Distances (Jan). 15 pp.

IEC 664 Pt 1-92. Insulation Coordination for Equipment Within Low-Voltage Systems Part 1: Principles, Requirements and Tests First Edition. 130 pp.

—Recreational Vehicles
BSI BS 6765: Part 3-89. 1989 Caravans Part 3: 12 V Direct Current Extra Low Voltage Electrical Installations. 12 pp.

ISO 8818-88. Leisure Accommodation Vehicles—Caravans—12 V Direct Current Extra Low Voltage Electrical Installations First Edition. 12 pp.

Low Voltage Systems
Use: Low Voltage Installations

Lowpass Filters
Use: Low Pass Filters

INTERNATIONAL AND NON-U.S. NATIONAL STANDARDS
SUBJECT INDEX
Lubricants

LOX
Use: Liquid Oxygen

LPG
Use: Liquefied Petroleum Gas

LR
Use: Loudness Ratings

Lubricant Additives
Use: Lubricating Oil Additives

Lubricant Content Analysis
—Metal Powders—Extraction Analysis
BSI BS 5600: Sec 2.11-80. 1980 Powder Metallurgical Materials and Products Part 2: Methods of Sampling and Testing Metallic Powders Section 2.11: Determination of the Lubricant Content of Lubricated Metallic Powders: Soxhlet Extraction Method. 4 pp.
ISO 4495-78. Lubricated Metallic Powders—Determination of Lubricant Content—Soxhlet Extraction Method First Edition. 4 pp.

Lubricants
See Also: Additives; Bearing Lubricants; Compressor Lubricants; Crankcase Oil; Cutting Fluids; Extreme Pressure Lubricants; Fabric Sizings; Four Ball Testers; Gear Lubricants; Greases; Isodecyl Pelargonate; Liquid Metals; Lubricating Oils; Lubrication; Lubricators; Mineral Oils; Synthetic Oils; Tool Lubricants; Turbine Oils; Vegetable Oils
CNS K6955-88. Method of Test for Extreme-Properties of Lubricating Fluids (Timken Method) (Jul)(12375).
MOD UK DSTAN 01-5-89. Fuels, Lubricants and Associated Products Issue 8. 328 pp.
MOD UK DSTAN 01-5-91. Fuels, Lubricants and Associated Products Issue 9. 319 pp.

—Aircraft
SBAC TS 68 ISSUE 7. Tables of Aerospace Fluid-Based Lubricants.

—Aircraft Engines—Spectrography
NATO STANAG 7017 ED 1 AMD 0-88. (DRAFT) Spectrographic Analysis of Aircraft Engine Lubricants. 3 pp.

—Appliances
BSI BS EN 377-93. 1993 Lubricants for Applications in Appliances and Associated Controls Using Combustible Gases Except Those Designed for Use in Industrial Processes (W). 15 pp.
CEN PREN 377-90. Lubricants for Applications in Domestic Appliances Using Combustible Gases. 20 pp.
CEN EN 377-93. Lubricants for Applications in Appliances and Associated Controls Using Combustible Gases Except Those Designed for Use in Industrial Processes. 10 pp.

—Ash Content
CNS K6002-47. Method of Test for Lubricants (Mar)(41). 4 pp.

—Asphalt Content
CNS K6002-47. Method of Test for Lubricants (Mar)(41). 4 pp.

—Channeling
CNS K6942-88. Method of Test for Channeling Characteristics of Lubricants (Jan)(12215).

—Chlorine Content
CNS K6760-83. Method of Test for Chlorine in New and Used Lubricants (Sodium Alcoholate Method) (Mar)(10095).

—Classification
BSI BS 6413: Part 0-83. 1983 Lubricants, Industrial Oils and Related Products (Class L) Part 0: Classification (General). 4 pp.
BSI BS 6413: Part 1-83. 1983 Lubricants, Industrial Oils and Related Products (Class L) Part 1: Classification for Family A (Total Loss Systems). 4 pp.
ISO 6743 Pt 0-81. Lubricants, Industrial Oils and Related Products (Class L)—Classification—Part 0: General First Edition. 4 pp.
ISO 6743 Pt 1-81. Lubricants, Industrial Oils and Related Products (Class L)—Classification—Part 1: Family A (Total Loss Systems) First Edition. 3 pp.
ISO 6743 Pt 10-89. Lubricants, Industrial Oils and Related Products (Class L)—Classification—Part 10: Family Y (Miscellaneous) First Edition. 4 pp.
ISO 8681-86. Petroleum Products and Lubricants—Method of Classification—Definition of Classes First Edition. 5 pp.

Lubricants *(Cont.)*
—Clutches
BSI BS 6413: Part 2-83. 1983 Lubricants, Industrial Oils and Related Products (Class L) Part 2: Classification for Family F (Spindle Bearings, Bearings and Associated Clutches). 3 pp.
ISO 6743 Pt 2-81. Lubricants, Industrial Oils and Related Products (Class L)—Classification—Part 2: Family F (Spindle Bearings, Bearings and Associated Clutches) First Edition. 3 pp.

—Control Equipment—Gas Appliances
DIN ENGL 3536-82. Lubricants for Gas Valves and Controls; Requirements, Testing (Nov). 5 pp.

—Cooling—Corrosion Prevention
DIN ENGL 51360 Pt 1-85. Testing of Cooling Lubricants; Determination of Corrosion Preventing Characteristics of Cooling Lubricants Mixed with Water; Herbert Corrosion Test (Aug). 5 pp.
DIN ENGL 51360 Pt 2-81. Testing of Cooling Lubricants; Determination of Corrosion Preventing Characteristics of Cooling Lubricants Mixed with Water; Chip/Filter Paper Method (July). 4 pp.

—Cooling—pH
DIN ENGL 51369-81. Testing of Cooling Lubricants; Determination of the pH Value of Water-Mixed Cooling Lubricants (July). 2 pp.

—Cooling—Stability
DIN ENGL 51367-78. Testing of Cooling Lubricants; Determination of the Stability of Emulsifiable Cooling Lubricants in Mixture with Hard Water (May). 4 pp.

—Cooling—Water Content—Acidification
DIN ENGL 51368-90. Determination of Fraction Separated by Hydrochloric Acid from Water Mix Metal Working Fluids (Nov). 3 pp.

—Corrosion Prevention
BSI BS 6413: Part 8-88. 1988 Lubricants, Industrial Oils and Related Products (Class L) Part 8: Classification for Family R (Temporary Protection Against Corrosion). 6 pp.
ISO 6743 Pt 8-87. Lubricants, Industrial Oils and Related Products (Class L)—Classification—Part 8: Family R (Temporary Protection Against Corrosion) First Edition. 6 pp.

—Dry Film—Aerospace—Corrosion Testing
CEN PREN 3026-92. Aerospace Series Test Method for Dry Film Lubricants Corrosion Test on Steels Specimens. 2 pp.

—Dry Film—Aerospace—Salt Spray Testing
CEN PREN 3027-92. Aerospace Series Test Method for Dry Film Lubricants Salt Spray Test. 3 pp.

—Dry Film—Aerospace—Solids Content Analysis
CEN PREN 3030-92. Aerospace Series Test Method for Dry Film Lubricants Solids Content. 2 pp.

—Dry Film—Aerospace—Thickness Measurement
CEN PREN 3032-92. Aerospace Series Test Method for Dry Film Lubricants Thickness Measurement. 3 pp.

—Electric Contacts—Telecommunication Equipment
MOD UK DSTAN 68-7-01. Cleaning and Lubricating Product, Electrical Contact, ZX-33; Amendment 1. 8 pp.

—Gas Valves
DIN ENGL 3536-82. Lubricants for Gas Valves and Controls; Requirements, Testing (Nov). 5 pp.

—Heat Transfer Fluids
ISO 6743 Pt 12-89. Lubricants, Industrial Oils and Related Products (Class L)—Classification—Part 12: Family Q (Heat Transfer Fluids) First Edition. 4 pp.

—Hydraulic Equipment
BSI BS 6413: Part 4-83. 1983 Lubricants, Industrial Oils and Related Products (Class L) Part 4: Classification for Family H (Hydraulic Systems). 4 pp.
ISO 6743 Pt 4-82. Lubricants, Industrial Oils and Related Products (Class L)—Classification—Part 4: Family H (Hydraulic Systems) First Edition. 4 pp.

—Identification Systems
DIN ENGL 51502-90. Designation of Lubricants and Marking of Lubricant Containers, Equipment and Lubricating Points (Aug). 7 pp.

—Kinematic Viscosity
BSI BS 4231-92. 1992 Viscosity Grades of Industrial Liquid Lubricants (ISO 3448: 1992). 12 pp.

Lubricants *(Cont.)*
—Kinematic Viscosity *(Cont.)*
BSI BS 4231-82. 1982 Viscosity Grades of Industrial Liquid Lubricants. 6 pp.
ISO 3448-92. Industrial Liquid Lubricants—ISO Viscosity Classification Second Edition. 9 pp.

—Metal Working
BSI BS 6413: Part 7-88. 1988 Lubricants, Industrial Oils and Related Products (Class L) Part 7: Classification for Family M (Metal-Working). 8 pp.
ISO 6743 Pt 7-86. Lubricants, Industrial Oils and Related Products (Class L)—Classification—Part 7: Family M (Metalworking) First Edition. 7 pp.

—Military Operations
NATO STANAG 2845 ED 2 AMD 3-81. Guide Specifications for NATO Army Fuels, Lubricants and Associated Products. 27 pp.
NATO STANAG 2845 ED 3 AMD 1-90. Guide Specifications for NATO Army Fuels, Lubricants and Associated Products. 42 pp.

—Molybdenum Disulfide—Powder
MOD UK DSTAN 68-62-79. Molybdenum Disulphide Powder, Lubricating NATO Code No: S-740 Joint Service Designation: ZX-35 Issue 1. 24 pp.
MOD UK DSTAN 68-62-93. Molybdenum Disulphide Powder, Lubricating NATO Code: S-740 Joint Service Designation: ZX-35 Issue 2. 21 pp.

—Neutralization Number
BSI BS 2000: Part 1-93. 1993 Methods of Test for Petroleum and Its Products Part 1: Determination of Acidity (W). 4 pp.
BSI BS 2000: Part 1-82. 1982 Amd 1 Methods of Test for Petroleum and Its Products Part 1: Acidity of Petroleum Products and Lubricants (Neutralization Value) (AMD 6372) June 28, 1991. 7 pp.
CNS K6002-47. Method of Test for Lubricants (Mar)(41). 4 pp.
JIS K 2501-92. Petroleum Products and Lubricants—Determination of Neutralization Number. 57 pp.

—Neutralization Number—Potentiometric Analysis
BSI BS 2000: Part 177-93. 1993 Methods of Test for Petroleum and Its Products Part 177: Determination of Neutralization Number—Potentiometric Titration Method (W). 10 pp.
BSI BS 2000: Part 177-84. 1984 Petroleum and Its Products Part 177: Neutralization Number of Potentiometric Titration. 15 pp.

—Neutralization Number—Volumetric Analysis
BSI BS 2000: Part 139-93. 1993 Methods of Test for Petroleum Products and Lubricants—Neutralization Number—Colour-Indicator Titration Method (ISO 6618: 1987) (W). 9 pp.
BSI BS 2000: Part 139-83. 1983 Petroleum and Its Products Part 139: Neutralization Number of Petroleum Products and Lubricants (Colour Indicator Titration Method) (Supersedes BS 2834: 1969). 9 pp.
ISO 6618-87. Petroleum Products and Lubricants—Neutralization Number—Colour-Indicator Titration Method First Edition. 10 pp.
ISO 6619-88. Petroleum Products and Lubricants—Neutralization Number—Potentiometric Titration Method First Edition. 11 pp.

—Oxygen Supply Equipment
NATO STANAG 3976 ED 1 AMD 0-87. Lubricants for Use in Oxygen Systems with Oxygen Rich Environments—AEP-15. 7 pp.

—Procurement
NATO STANAG 2504 ED 1 AMD 0-90. NATO Standardization of Petroleum Specifications. 9 pp.

—Saponification Number
CNS K6002-47. Method of Test for Lubricants (Mar)(41). 4 pp.

—Saponification Number—Volumetric Analysis
BSI BS 2000: Part 136-93. 1993 Methods of Test for Petroleum and Its Products Part 136: Determination of Saponification Number—Titration Method (W). 5 pp.
BSI BS 2000: Part 136-91. 1991 Methods of Test for Petroleum and Its Products Part 136: Saponification Number of Petroleum Products. 10 pp.
BSI BS 7393-90. 1990 Method for Determination of Total Acid Number in Petroleum Products by Semi-Micro Colour Indicator Titration. 11 pp.
ISO 7537-89. Petroleum Products—Determination of Total Acid Number—Semi-Micro Colour-Indicator Titration Method First Edition. 8 pp.

INDUSTRY STANDARDS

INTERNATIONAL AND NON-U.S. NATIONAL STANDARDS SUBJECT INDEX

Lubricants (Cont.)

—Slideways
ISO 6743 Pt 13-89. Lubricants, Industrial Oils and Related Products (Class L)—Classification—Part 13: Family G (Slideways) First Edition. 4 pp.

—Test Panels—Preparation
MOD UK DSTAN 05-50: Part 25-87. Methods for Testing Fuels, Lubricants and Associated Products Part 25: Preparation of Steel Panels for Test Purposes Issue 1. 8 pp.

—Turbines
BSI BS 6413: Part 5-88. 1988 Lubricants, Industrial Oils and Related Products (Class L) Part 5: Classification for Family T (Turbines). 6 pp.
ISO 6743 Pt 5-88. Lubricants, Industrial Oils and Related Products (Class L)—Classification—Part 5: Family T (Turbines) First Edition. 6 pp.

—Wear Testing
DIN ENGL 51350 Pt 5-84. Testing of Lubricants; Testing by the Shell Four-Ball Tester; Determination of Wear Characteristics for Consistent Lubricants (Jan). 3 pp.

—Wire Ropes
CNS M3025-80. Stranded Wire Ropes for Mine Hoisting Impregnating Compounds Lubricants and Service Dressing Characteristics and Tests (Oct)(6596).
ISO 3156-76. Stranded Wire Ropes for Mine Hoisting—Impregnating Compounds, Lubricants and Service Dressings—Characteristics and Tests First Edition. 6 pp.

Lubricating Cloths
Use: Cleaning Cloths

Lubricating Greases
Use: Greases

Lubricating Nipples
Use: Grease Nipples

Lubricating Oil Additives
See Also: Additives; Lubricating Oils; Pour Point Depressants

—Aging Testing
DIN ENGL 51587-74. Testing of Lubricants; Determination of the Ageing Behaviour of Steam Turbine Oils and Hydraulic Oils Containing Additives (Aug). 6 pp.

—Chlorine Content—X-Ray Fluorescence Spectrometry
DIN ENGL 51577 Pt 2-85. Testing of Petroleum Products; Determination of the Chlorine Content of Lubricating Oils and Lubricating Oil Additives by X-Ray Fluorescence Analysis (Sept). 3 pp.

—Phosphorus Content
BSI BS 2000: Part 149-93. 1993 Methods of Test for Petroleum and Its Products Part 149: Petroleum Products—Lubricating Oils and Additives—Determination of Phosphorus Content—Quinoline Phosphomolybdate Method (ISO 4265: 1986) (W). 6 pp.
BSI BS 7032-88. (WITHDRAWN) 1988 Method for Determination of Phosphorus Content of Lubricating Oils and Additives by a Quinoline Phosphomolybdate Procedure (Superseded by BS 2000: Part 149: 1993). 8 pp.
ISO 4265-86. Petroleum Products—Lubricating Oils and Additives—Determination of Phosphorus Content—Quinoline Phosphomolybdate Method First Edition. 6 pp.

—Phosphorus Content—X-Ray Spectrometry
DIN ENGL 51363 Pt 2-87. Testing of Petroleum Products; Determination of Phosphorus Content of Lubricating Oils and Additives by X-Ray Spectrometry (Sept). 3 pp.

—Sulfated Ash Content
BSI BS 2000: Part 163-93. 1993 Methods of Test for Petroleum and Its Products Part 163: Determination of Sulphated Ash of Turbine Lubricating Oils and Additives (W). 5 pp.
BSI BS 2000: Part 163-82. 1982 Petroleum and Its Products Part 163: Sulfated Ash from Lubrication Oils and Additives. 6 pp.
ISO 3987-80. Petroleum Products—Lubricating Oils and Additives—Determination of Sulphated Ash First Edition. 5 pp.

—Zinc Content—Polarographic Analysis
CNS K6761-83. Method of Test for Zinc in Lubricating Oils and Additives (Polarographic Method) (Mar)(10096).

Lubricating Oils
Scope Note: For additional listings, use a more specific term *Use For:* Machine Oils
See Also: Bearing Oils; Detergents; Diesel Motor Oil; Extreme Pressure Lubricants; Gear Oils; Graphite Oils; Hydraulic Oils; Lubricants; Lubricating Oil Additives; Mineral Oils; Motor Oils; Oil Filters; Oils; Penetrating Oils; Petroleum Products; Polyalkylene Glycol Oil; Refrigeration Oils; Synthetic Oils; Turbine Oils; Vegetable Oils

CNS K5056-88. Machine Oil (Jan)(2981). 2 pp.
CNS K5138-83. Lubricating Base Oil (Dec)(10723).
DIN ENGL 51501-79. Lubricants; Lubricating Oils L-AN; Minimum Requirements (Nov). 4 pp.
DIN ENGL 51513-86. Lubricants; B Lubricating Oils; Minimum Requirements (Feb). 2 pp.
DIN ENGL 51517 Pt 1-89. C Lubricating Oils; Minimum Requirements (Sept). 3 pp.
DIN ENGL 51517 Pt 2-89. CL Lubricating Oils; Minimum Requirements (Sept). 4 pp.
DIN ENGL 51517 Pt 3-89. CLP Lubricating Oils; Minimum Requirements (Sept). 4 pp.
JIS K 2238-83. Machine Oils. 7 pp.
MOD UK DSTAN 91-44-01. Lubricating Oil, General Purpose: Petroleum, Light NATO Code No: O-134 Joint Service Designation: OM-13 Issue 1; Amendment 1. 16 pp.
MOD UK DSTAN 91-47-80. Lubricating Oil, General Purpose: Low Temperature NATO Code No: O-142 Joint Service Designation: OM-12 Issue 1. 11 pp.
MOD UK DSTAN 91-49-80. Lubricating Oil, Instrument: Synthetic NATO Code No: O-147 Joint Service Designation: OX-14 Issue 1. 11 pp.
MOD UK DSTAN 91-59-91. Lubricating Oil, Extreme Pressure Grade 75W NATO Code 0—186 Joint Service Designation OEP—38 Lubricating Oil, Extreme Pressure Grade 80W/90 NATO Code 0—226 Joint Service Designation OEP—220 Issue 1. 21 pp.
MOD UK DEF-2182-57. Oil OX-10. 4 pp.
MOD UK CS 3118. Oil, Lubricating and Protective OX-18 (Dormant).

—Aging Testing
DIN ENGL 51352 Pt 1-85. Testing of Lubricants; Determination of Ageing Characteristics of Lubricating Oils; Increase in Conradson Carbon Residue After Ageing by Passing Air Through the Lubricating Oil (Aug). 4 pp.
DIN ENGL 51352 Pt 2-85. Testing of Lubricants; Determination of Ageing Characteristics of Lubricating Oils; Conradson Carbon Residue After Ageing by Passing Air Through the Lubricating Oil in the Presence of Iron (III) Oxide (Aug). 3 pp.
DIN ENGL 51586-77. Testing of Lubricants; Determination of the Ageing Properties of Lubricating Oils for High Pressure Load (Aug). 4 pp.

—Air Compressors
CNS K5049-88. Compressor Oil (Jan)(2974). 2 pp.
DIN ENGL 51506-85. Lubricants; VB and VC Lubricating Oils with and Without Additives and VDL Lubricating Oils; Classification and Requirements (Sept). 3 pp.
MOD UK DSTAN 91-42-78. Lubricating Oil, Petroleum: Compressor, Light Joint Service Designation: OM-58 Lubricating Oil, Petroleum: Compressor, Medium Joint Service Designation: OM-160 Issue 1. 6 pp.

—Air Release Value
BSI BS 2000: Part 313-93. 1993 Methods of Test for Petroleum and Its Products Part 313: Determination of Air Release Value of Hydraulic, Turbine and Lubricating Oils (W). 5 pp.
BSI BS 2000: Part 313-82. 1982 Petroleum and Its Products Part 313: Air Relase Value of Industrial Oils. 6 pp.

—Aircraft
JIS K 2503-80. Testing Methods of Lubricating Oil for Aircraft (R 1985). 21 pp.
MOD UK DTD-581C-72. Lubricating Oil, Gear: Aircraft Grade: Light/Medium NATO Code Number: 0-153/0-155 Joint Service Designation: OEP-30/OEP-70. 3 pp.
MOD UK DERD 2490-01. Lubricating Oil, Aircraft Turbine Engine, Petroleum Oil OM-11 NATO Code No. 0-135 Issue 2; Amendment 1. 23 pp.

—Aircraft Engines
MOD UK DERD 2450-01. Lubricating Oil—Aircraft Piston Engines: Ashless Dispersant Type Issue 2; Amendment 2. 11 pp.
MOD UK DERD 2487. Directorate of Engine Research and Development Issue 4.
MOD UK DERD 2487: APL-82. Lubricating Oil, Aircraft Turbine Engine, Synthetic Joint Service Designation OX-38 NATO Code No 0-149 Issue 5. 2 pp.

Lubricating Oils (Cont.)

—Ash Content
CEN EN 7-74. Determination of Ash from Petroleum Products. 5 pp.

—Asphaltene Content
BSI BS 2000: Part 143-93. 1993 Methods of Test for Petroleum and Its Products Part 143: Determination of Asphaltenes (Heptane Insolubles) (W). 6 pp.
BSI BS 2000: Part 143-85. 1985 Petroleum and Its Products Part 143: Asphaltenes in Petroleum Products (Precipitation with Normal Heptane). 8 pp.

—Atmospheric Corrosion Testing
DIN ENGL 51386 Pt 1-86. Testing of Corrosion Preventive Oils in a Condensation Water Alternating Atmosphere (Mar). 5 pp.

—Barium Content—Atomic Absorption Spectrometry
DIN ENGL 51391 Pt 1-85. Testing of Lubricants; Determination of the Barium, Calcium and Zinc Content of Lubricating Oils; Direct Determination by Atomic Absorption Spectrometry (Dec). 4 pp.

—Calcium Content—Atomic Absorption Spectrometry
DIN ENGL 51391 Pt 1-85. Testing of Lubricants; Determination of the Barium, Calcium and Zinc Content of Lubricating Oils; Direct Determination by Atomic Absorption Spectrometry (Dec). 4 pp.

—Carbon Residue Testing
DIN ENGL 51551-86. Testing of Lubricants and Liquid Fuels; Determination of Conradson Carbon Residue (Mar). 4 pp.

—Chip Detectors—Housings
MOD UK DSTAN 47-3-82. Metal Chip Detectors (Housings) Issue 4. 6 pp.

—Chlorine Content—X-Ray Fluorescence Spectrometry
DIN ENGL 51577 Pt 2-85. Testing of Petroleum Products; Determination of the Chlorine Content of Lubricating Oils and Lubricating Oil Additives by X-Ray Fluorescence Analysis (Sept). 3 pp.

—Classification
BSI BS 4475-75. 1975 Straight Mineral Lubricating Oils. 11 pp.

—Compressors
CNS K5049-88. Compressor Oil (Jan)(2974). 2 pp.

—Copper Content—Atomic Absorption Spectrometry
DIN ENGL 51404 Pt 2-85. Testing of Lubricants and Fuels; Determination of Copper Content of Lubricating Oils and Liquid Fuels; Direct Determination by Atomic Absorption Spectrometry (May). 4 pp.

—Corrosion Prevention
MOD UK DSTAN 80-34-01. Corrosion Preventive Compound, Oil Film Type Joint Service Designation: PX-4 Issue 2; Amendment 1. 38 pp.
MOD UK DSTAN 91-40-81. Corrosion Preventive Oil, Aircraft Engine: Piston, Metallic NATO Code No: C-615 Joint Service Designation: PX-27 Issue 2. 10 pp.

—Cylinders
CNS K5054-88. Cylinder Oil (Jan)(2979). 2 pp.

—Demulsification
BSI BS 2000: Part 19-93. 1993 Methods of Test for Petroleum and Its Products Part 19: Determination of Demulsibility Characteristics of Lubricating Oil. 4 pp.
BSI BS 2000: Part 19-90. 1990 Petroleum and Its Products Part 19: Demulsification Number of Lubricating Oil. 6 pp.
DIN ENGL 51599-75. Testing of Lubricating Oils; Determination of Demulsification Capacity According to the Stirring Method (Oct). 3 pp.
JIS K 2520-91. Petroleum Products—Lubricating Oils—Determination of Demulsibility Characteristics. 17 pp.

—Distillation Methods
DIN ENGL 51356-78. Testing of Lubricating Oils and Liquid Fuels; Determination of the Distillation Range Under Reduced Pressure According to Grosse-Oetringhaus (Feb). 7 pp.

—Emulsification
BSI BS 2000: Part 19-93. 1993 Methods of Test for Petroleum and Its Products Part 19: Determination of Demulsibility Characteristics of Lubricating Oil. 4 pp.
BSI BS 2000: Part 19-90. 1990 Petroleum and Its Products Part 19: Demulsification Number of Lubricating Oil. 6 pp.

INTERNATIONAL AND NON-U.S. NATIONAL STANDARDS
SUBJECT INDEX
Lubrication

Lubricating Oils (Cont.)

—Emulsification (Cont.)
DIN ENGL 51599-75. Testing of Lubricating Oils; Determination of Demulsification Capacity According to the Stirring Method (Oct). 3 pp.
JIS K 2520-91. Petroleum Products—Lubricating Oils—Determination of Demulsibility Characteristics. 17 pp.

—Evaporation Loss
CNS K6769-83. Method of Test for Evaporation Loss of Lubricating Greases and Oils (May)(10267).
DIN ENGL 51581-83. Testing of Lubricants; Determination of Evaporation Loss of Lubricating Oils; (Noack Test) (Sept). 4 pp.

—Flash Point
BSI BS 6664: Part 5-90. 1990 Flashpoint of Petroleum and Related Products Part 5: Method for Determination of Flashpoint by Pensky-Martens Closed Tester. 16 pp.
ISO 2719-88. Petroleum Products and Lubricants—Determination of Flash Point—Pensky-Martens Closed Cup Method Second Edition. 14 pp.

—Flow Measurement
DIN ENGL 51568-74. Testing of Lubricating Oils; Determination of Ability to Flow; U-Tube Method (June). 3 pp.

—Foaming Power
BSI BS 2000: Part 146-93. 1993 Methods of Test for Petroleum and Its Products Part 146: Determination of Foaming Characteristics of Lubricating Oils (W). 7 pp.
BSI BS 2000: Part 146-90. 1990 Petroleum and Its Products Part 146: Foaming Characteristics of Lubricating Oils. 11 pp.
CNS K6323-72. Method of Test for Foaming Characteristics of Lubricating Oils (Oct)(3384).
JIS K 2518-80. Petroleum Products—Lubricating Oils—Determination of Foaming Characteristics.

—Food Preparation Equipment
MOD UK DSTAN 91-36-77. Lubricating Oil: White Joint Service Designation: OM-17 Issue 1. 7 pp.

—Gun Components
MOD UK DSTAN 91-42-78. Lubricating Oil, Petroleum: Compressor, Light Joint Service Designation: OM-58 Lubricating Oil, Petroleum: Compressor, Medium Joint Service Designation: OM-160 Issue 1. 6 pp.

—Hoses—Ships
MOD UK NES 2016-88. Specification for the Manufacture of 64mm (2 1/2 ins) Bore Hose for Replenishment at Sea Issue 1 (06.88). 21 pp.
MOD UK NES 2017-88. Specification for the Manufacture of 76mm (3 ins) Bore Hose for Replenishment at Sea Issue 1 (06.88). 21 pp.

—Hoses—Submarines
MOD UK NES 2016-88. Specification for the Manufacture of 64mm (2 1/2 ins) Bore Hose for Replenishment at Sea Issue 1 (06.88). 21 pp.
MOD UK NES 2017-88. Specification for the Manufacture of 76mm (3 ins) Bore Hose for Replenishment at Sea Issue 1 (06.88). 21 pp.

—Interchangeability
NATO STANAG 1135 ED 3 AMD 6-83. Interchangeability of Fuels, Lubricants and Associated Products Used by the Armed Forces of the North Atlantic Treaty Nations. 82 pp.

—Loads (Forces)
CNS K6958-88. Method of Test for Load-Carrying Capacity for Lubricating Oil (FZG Gear Machine Method) (Jul)(12378).
JIS K 2519-87. Testing Methods for Load Carrying Capacity of Lubricating Oil. 35 pp.

—Locomotives
MOD UK DSTAN 91-42-78. Lubricating Oil, Petroleum: Compressor, Light Joint Service Designation: OM-58 Lubricating Oil, Petroleum: Compressor, Medium Joint Service Designation: OM-160 Issue 1. 6 pp.

—Machine Tools
ISO TR3498-86. Lubricants, Industrial Oils and Related Products (Class L)—Recommendations for the Choice of Lubricants for Machine Tools Second Edition. 6 pp.

—Magnesium Content—Atomic Absorption Spectrometry
DIN ENGL 51431-86. Testing of Lubricants; Determination of the Magnesium Content of Lubricating Oils; Direct Determination by Atomic Absorption Spectrometry (AAS) (Feb). 4 pp.

Lubricating Oils (Cont.)

—Marine—Gears
MOD UK DSTAN 91-74-93. Lubricating Oil, Steam Turbine and Gear, Extreme Pressure Joint Service Designation: OEP-80 Issue 1. 16 pp.

—Marine—Steam Turbines
MOD UK DSTAN 91-74-93. Lubricating Oil, Steam Turbine and Gear, Extreme Pressure Joint Service Designation: OEP-80 Issue 1. 16 pp.

—Military—Interchangeable
NATO STANAG 2410 ED 1 AMD 0-88. (DRAFT) List of NATO Civil/Military Ground Engine Lubricating Oil Interchangeability (Withdrawn 93-06). 7 pp.

—Molybdenum Content—Atomic Absorption Spectrometry
DIN ENGL 51379 Pt 3-85. Testing of Lubricants; Determination of the Molybdenum Content of Lubricating Oils by Atomic Absorption Spectrometry (AAS) (July). 3 pp.

—Molybdenum Content—X-Ray Spectrometry
DIN ENGL 51379 Pt 2-90. Determination of Molybdenum Content of Lubricating Oils by Wavelength-Dispersive X-Ray Spectrometry (Nov). 3 pp.

—Oxidation Resistance
BSI BS 2000: Part 48-93. 1993 Methods of Test for Petroleum and Its Products Part 48: Determination of Oxidation Characteristics of Lubricating Oil. 5 pp.
BSI BS 2000: Part 48-82. 1982 Petroleum and Its Products Part 48: Oxidation Test for Lubricating Oil. 4 pp.
CNS K6938-87. Method of Test for Oxidation Characteristic of Crankcase Lubricating Oils-L38 Test (Dec)(12194).
CNS K6947-88. Method of Test for Oxidation Characterisitics of Extreme-Pressure Lubricating Oils (Apr)(12260).

—Phosphorus Content
CNS K6767-83. Method of Test for Phosphorus in Lubricating Oils and Additives (May)(10265).

—Phosphorus Content—Quinoline Phosphomolybdate Method
BSI BS 2000: Part 149-93. 1993 Methods of Test for Petroleum and Its Products Part 149: Petroleum Products—Lubricating Oils and Additives—Determination of Phosphorus Content—Quinoline Phosphomolybdate Method (ISO 4265: 1986) (W). 6 pp.
BSI BS 7032-88. (WITHDRAWN) 1988 Method for Determination of Phosphorus Content of Lubricating Oils and Additives by a Quinoline Phosphomolybdate Procedure (Superseded by BS 2000: Part 149: 1993). 8 pp.
ISO 4265-86. Petroleum Products—Lubricating Oils and Additives—Determination of Phosphorus Content—Quinoline Phosphomolybdate Method First Edition. 6 pp.

—Phosphorus Content—X-Ray Spectrometry
DIN ENGL 51363 Pt 2-87. Testing of Petroleum Products; Determination of Phosphorus Content of Lubricating Oils and Additives by X-Ray Spectrometry (Sept). 3 pp.

—Precipitation Number
CNS K6948-88. Method of Test for Precipitation Number of Lubricatings Oils (Apr)(12261).

—Radar Equipment
MOD UK DSTAN 91-42-78. Lubricating Oil, Petroleum: Compressor, Light Joint Service Designation: OM-58 Lubricating Oil, Petroleum: Compressor, Medium Joint Service Designation: OM-160 Issue 1. 6 pp.

—Removers
MOD UK DSTAN 91-33-01. Flushing Oil Joint Service Designation: OM-24 Issue 2; Amendment 1. 10 pp.

—Rust Prevention
JIS K 2510-87. Testing Method for Rust-Preventing Characteristics of Lubricating Oil. 16 pp.
MOD UK DSTAN 05-50: Part 43-89. Methods for Testing Fuels, Lubricants and Associated Products Part 43: Rust Inhibiting Properties of Oils Issue 1. 9 pp.
MOD UK DSTAN 91-21-89. Lubricating Oil, Compounded NATO Code: 0-254 Joint Service Designation: OC-160 Issue 3. 12 pp.
MOD UK CS 3118. Oil, Lubricating and Protective OX-18 (Dormant).

Lubricating Oils (Cont.)

—Silicon Content—X-Ray Fluorescence
DIN ENGL 51390 Pt 2-87. Testing of Petroleum Products; Determination of Silicon Content by X-Ray Fluorescence Analysis (Aug). 3 pp.

—Spindles
CNS K5051-88. Spindle Oil (Jan)(2976). 2 pp.

—Steam Emulsion Testing
DIN ENGL 51589 Pt 1-91. Determination of Water Separation Ability of Lubricating Oils and Low-Flammability Fluids After Contact With Steam (Mar). 6 pp.

—Steel Cables
CNS K5057-88. Steel Cable Oil (Jan)(2982)(R 1973). 2 pp.

—Steel Wire Rope
CNS K5057-88. Steel Cable Oil (Jan)(2982)(R 1973). 2 pp.

—Sulfated Ash Content
BSI BS 2000: Part 163-93. 1993 Methods of Test for Petroleum and Its Products Part 163: Determination of Sulphated Ash of Lubricating Oils and Additives (W). 5 pp.
BSI BS 2000: Part 163-82. 1982 Petroleum and Its Products Part 163: Sulfated Ash from Lubrication Oils and Additives. 6 pp.
ISO 3987-80. Petroleum Products—Lubricating Oils and Additives—Determination of Sulphated Ash First Edition. 5 pp.

—Sulfur Content
DIN ENGL 51400 Pt 1-78. Testing of Mineral Oils and Fuels; Determination of the Sulfur Content (Total Sulfur); General Working Principles (Feb). 6 pp.
DIN ENGL 51400 Pt 3-78. Testing of Mineral Oils and Fuels; Determination of the Sulfur Content (Total Sulfur); Combustion According to Schoniger; Thorin-Sulfonazo-III Titration (Feb). 3 pp.
DIN ENGL 51400 Pt 4-90. Determination of Total Sulfur Content of Gaseous Petroleum Products by the Lingener Combustion Method (Oct). 8 pp.
DIN ENGL 51400 Pt 8-78. Testing of Mineral Oils and Fuels; Determination of Sulfur Content (Total Sulfur); Nickel Reduction Method; Dithizone Titration (Feb). 4 pp.

—Thermal Stability
JIS K 2540-89. Testing Method for Thermal Stability of Lubricating Oils. 9 pp.

—Vacuum Pumps
CNS K5049-88. Compressor Oil (Jan)(2974). 2 pp.
DIN ENGL 51506-85. Lubricants; VB and VC Lubricating Oils with and Without Additives and VDL Lubricating Oils; Classification and Requirements (Sept). 3 pp.

—Viscosity
JIS K 2001-83. Viscosity Classification for Industrial Liquid Lubricants. 6 pp.

—Zinc Content—Atomic Absorption Spectrometry
DIN ENGL 51391 Pt 1-85. Testing of Lubricants; Determination of the Barium, Calcium and Zinc Content of Lubricating Oils; Direct Determination by Atomic Absorption Spectrometry (Dec). 4 pp.

—Zinc Content—Polarographic Analysis
CNS K6761-83. Method of Test for Zinc in Lubricating Oils and Additives (Polarographic Method) (Mar)(10096).

Lubrication
Use For: Greasing; Oil Systems (Lubrication); Oiling (Lubrication) *See Also:* Bearings; Grease Guns; Grease Nipples; Lubricants; Lubricators; Oil Coolers; Oil Cups; Self Lubricating Materials; Tribology
CNS B2767-84. Lubrication Holes, Grooves and Bore Reliefs (Dec)(11157).

—Aircraft Engines
CAA Chapter G5-3 06.75. Oil Systems (Rotorcraft). 5 pp.
CAA Chapter K5-3 04.74. Oil Systems (Light Aeroplanes). 3 pp.
CAA Chapter K5-3 App 03.67. Oil Systems (Light Aeroplanes). 1 p.
CAA Chapter P5-3. Oil Systems (Provisional Airworthiness Requirements for Civil Powered Lift Aircraft). 5 pp.
CAA Chapter Q5-3 12.79. Oil Systems (Non-Rigid Airships). 5 pp.
CAA Chapter Q5-3 App 12.79. Oil Systems (Non-Rigid Airships). 1 p.
CAA CAP 482 SUB-Part E 03.83. Power-Plant (Small Light Aeroplanes). 5 pp.

INDUSTRY STANDARDS

INTERNATIONAL AND NON-U.S. NATIONAL STANDARDS
SUBJECT INDEX
Lubrication

Lubrication (Cont.)
— **Aircraft Engines—Couplings**
ISO 451-76. Aircraft—Pressure Re-Oiling Connection First Edition; (Erratum—Feb 1980). 5 pp.

— **Aircraft Engines—Rotary Wing**
CAA CAP 524 SUB-Part E 12.86. Power-Plant (Rotorcraft). 27 pp.

— **Automotive**
JIS D 8003-77. Test Code of Centralized Lubricating Systems for Automobiles.

— **Classification**
BSI BS 1000: (621.8)-76. 1976 Amd 1 Universal Decimal Classification (UDC). English Full Edition (621.8): Mechanical Power Transmission. Machine Elements. Gearing. Materials Handling. Mechanical Attachments, Fixing, Fasteners. Lubrication. 30 pp.
DIN ENGL 24271 Pt 1-82. Centralized Lubrication Systems; Terminology; Classification (Apr). 12 pp.
SNZ NZS/BS 1000 (621.8)-76. Universal Decimal Classification Mechanical Power Transmission. Machine Elements. Gearing. Materials Handling. Mechanical Attachments, Fixing, Fasteners. Lubrication Amend: 1. 28 pp.

— **Glossaries**
DIN ENGL 24271 Pt 1-82. Centralized Lubrication Systems; Terminology; Classification (Apr). 12 pp.

— **Internal Combustion Piston Engines—Glossaries**
BSI BS 7016: Part 6-92. 1992 Components and Systems of Reciprocating Internal Combustion Engines Part 6: Glossary of Terms for Lubricating Systems (ISO 7967-6: 1992). 19 pp.
ISO 7967 Pt 6-92. Reciprocating Internal Combustion Engines—Vocabulary of Components and Systems—Part 6: Lubricating Systems First Edition. 20 pp.

— **Machine Tools**
BSI BS 5739-79. 1979 Presentation of Lubrication Instructions for Machine Tools. 7 pp.
ISO 5169-77. Machine Tools—Presentation of Lubrication Instructions First Edition. 5 pp.
ISO 5170-77. Machine Tools—Lubrication Systems First Edition. 10 pp.
JIS B 6016-90. Machine Tools—General Rule for Lubrication. 29 pp.

— **Machinery**
BSI BS 4807-91. 1991 Centralized Lubrication Systems. 30 pp.
BSI BS 4807-72. 1972 Recommendations for Centralized Lubrication as Applied to Plant and Machinery. 33 pp.

— **Metal**
CNS Z7021-66. Plastic Sponges for Metal Oiling (Tentative) (Dec)(2722)(R 1972).

— **Photographic Films**
ISO 5769-84. Photography—Processed Films—Method for Determining Lubrication First Edition. 9 pp.

— **Plain Bearings**
ISO TR6281-90. Plain Bearings—Testing Under Conditions of Hydrodynamic and Mixed Lubrication in Test Rigs —Guidelines First Edition. 16 pp.

— **Ships**
MOD UK NES 303-01. Lubricating Oil Systems for Propulsion and Generation in Surface Ships and Submarines Issue 3 (09.89); Amendment 2. 79 pp.

— **Ships—Automatic Control Equipment**
JIS F 0804-79. Methods of Onboard Test on Automatic Control of Lubricating Oil System for Smaller Ships.

— **Submarines**
MOD UK NES 303-01. Lubricating Oil Systems for Propulsion and Generation in Surface Ships and Submarines Issue 3 (09.89); Amendment 2. 79 pp.

— **Symbols**
DIN ENGL 24271 Pt 2-82. Centralized Lubrication Systems; Graphic Symbols for Technical Drawings (Apr). 8 pp.

— **Units of Measurement**
DIN ENGL 24271 Pt 3-82. Centralized Lubrication Systems; Technical Quantities and Units (Apr). 5 pp.

Lubrication Dispersal Systems
Use: Lubricators

Lubricators
Use For: Oilers; Pneumatic Lubricators
See Also: Grease Guns; Grease Nipples; Lubricants; Lubrication; Oil Cups; Oil Rings

Lubricators (Cont.)
ISO 6301 Pt 1-89. Pneumatic Fluid Power—Compressed Air Lubricators—Part 1: Main Characteristics to Be Included in Commercial Literature and Specific Requirements First Edition. 9 pp.
JIS B 8378-81. Lubricators for Pneumatic Use. 17 pp.

— **Guns**
MOD UK DSTAN 49-2-01. Lubricating Guns, Hand, and Guns, Fluid, Direct Delivery Issue 2; Amendment 3. 21 pp.

— **Ships**
MOD UK NES 1021-01. List of Lubricators Issue 1 (07.84); Amendment 1. 80 pp.

— **Submarines**
MOD UK NES 1021-01. List of Lubricators Issue 1 (07.84); Amendment 1. 80 pp.

Lug Nuts
See Also: Nuts (Fasteners)

— **Wheels**
MOD UK DSTAN 53-6-68. Nuts, Cone Seat for Wheel Fastening Issue 1. 9 pp.

Luggage
Use For: Hand Luggage *See Also:* Bags; Briefcases; Containers; Luggage Compartments; Luggage Racks; Suitcases

— **Rail Transportation—Customs Regulations**
EC 89/339/EEC-89. Council Decisions Accepting on Behalf of the Community the Recommendation of 5 June 1962 of the Customs Cooperation Council Concerning the Customs Treatment of Registered Baggage Carried by Rail as Amended of 21 June. 1 p.

Luggage Carriers
Use: Luggage Racks

Luggage Compartments
See Also: Containers; Luggage

— **Automotive—Capacity Measurement**
ISO 3832-91. Passenger Cars—Luggage Compartments—Method of Measuring Reference Volume Second Edition; (Corrected and Reprinted -1991). 5 pp.
JIS D 0303-82. Method of Measuring the Reference Volume for the Luggage Compartments of Passenger Cars.

— **Locks (Security)—Automotive**
BSI BS AU 209: Part 1A-92. 1992 Vehicle Security Part 1a: Specification for Locking Systems for Passenger Cars and Car Derived Vehicles (E). 9 pp.

Luggage Racks
Use For: Car Top Carriers; Cartop Carriers; Luggage Carriers *See Also:* Bicycle Racks; Luggage; Racks (Storage)

— **Automotive**
SAA AS 1235-91. Roof Racks and Roof Bars for Passenger Vehicles Amdt 1—1993. 17 pp.

— **Bars**
SAA AS 1235-91. Roof Racks and Roof Bars for Passenger Vehicles Amdt 1—1993. 17 pp.

Lugs (Fasteners)
See Also: Fasteners; Holders; Studs (Fasteners)

— **Aircraft—Handhole Covers—Lanyards**
SBAC AS 3177 ISSUE 1. Lug for Lanyard. 1 p.

— **Aircraft—Harnesses**
SBAC AS 2111 ISSUE 1. Attachment Lug for 'Z' Type Harness. 1 p.

— **Aircraft—Suspension Systems**
MOD UK DSTAN 13-8-78. Dimensional Requirements for Airborne Stores, Associated Suspension and Release Systems, and Electrical Control Connections Issue 4 (Withdrawn). 12 pp.
MOD UK DSTAN 13-45-01. Lug, Suspension, Airborne Store for Aircraft Twin Lug Suspension and Release Systems Issue 3; Amendment 2 (Withdrawn). 15 pp.
MOD UK DSTAN 13-46-69. Lug, Suspension, Airborne Store, for Aircraft 'Single Lug' Suspension and Release Systems Issue 1 (Withdrawn). 6 pp.
NATO STANAG 3726 ED 3 AMD 4-83. Bail (Portal) Lugs for the Suspension of Aircraft Stores. 8 pp.
NATO STANAG 3726 ED 4 AMD 0-92. Bail (Portal) Lugs for the Suspension of Aircraft Stores. 9 pp.

Lugs (Fasteners) (Cont.)
— **Automotive—Drawbars**
BSI BS AU 24A-89. 1989 Towing Connections for Trailers of up to 5000 kg Gross Mass. 8 pp.

— **Military Vehicles—Towing**
NATO STANAG 4019 ED 1 AMD 3-57. Emergency Towing Facilities. 5 pp.
NATO STANAG 4019 ED 2 AMD 0-92. Emergency Towing Facilities. 9 pp.

— **Roofing Tiles**
SAA AS 4046.2-92. Methods of Testing Roof Tiles—Part 2: Determination of Batten Lugs and Squareness (in Professional Packages 20, 21, 40, 41, 58, 62, 63, 64, 65, 66, 67, 68, 69) (Supersedes AS 1757—1989 (in Part) and AS 2049—1989 (in Part)). 2 pp.

— **Swivels**
DIN ENGL 82008-71. Double Lugs with Screwed Shank for Swivels and Turnbuckles (Apr). 2 pp.

— **Turnbuckles**
DIN ENGL 82008-71. Double Lugs with Screwed Shank for Swivels and Turnbuckles (Apr). 2 pp.

Lumber
Use For: Sawn Lumber *See Also:* Construction Materials; Hardwoods; Laminated Wood; Logs (Wood); Lumbering Industry; Siding; Structural Timber; Wood
CNS O1006-66. Dimensions of Sawn Lumber of Hardwoods (for Taiwan Area) (Sep)(447). 2 pp.
CSA S442.1-M1985. Method of In-Grade Bending Tests of Lumber (S442 Series-M1985); (Gen Instr 1). 13 pp.
ISO 4472-83. Coniferous and Broadleaved Sawn Timber—Transportation Packages First Edition. 4 pp.
SAA AS O70-58. Sawn House Stumps, Sole Plates, Fence Posts and Struts from South-Eastern Australian Hardwoods. 7 pp.
SAA AS O82-65. Sawn Eastern Australian Hardwoods (Incorporating Amdt 1). 13 pp.
SAA AS O84-67. Sawn Australian Rainforest Timber (Incorporating Amdt 1). 20 pp.

— **Chains**
JIS F 2102-90. Chains for Lumber Lashing.

— **Compression Testing**
ISO 8906-88. Sawn Timber—Test Methods—Determination of Resistance to Local Transverse Compression First Edition. 6 pp.

— **Defects—Classification**
ISO 2299-73. Sawn Timber of Broadleaved Species—Defects—Classification First Edition. 9 pp.

— **Defects—Glossaries**
ISO 1031-74. Coniferous Sawn Timber—Defects—Terms and Definitions First Edition. 32 pp.
ISO 2300-73. Sawn Timber of Broadleaved Species—Defects—Terms and Definitions First Edition. 34 pp.
ISO 2301-73. Sawn Timber of Broadleaved Species—Defects—Measurement First Edition. 13 pp.

— **Glossaries**
CEN PREN 844-1-92. Terminology—Round and Sawn Timber—Part 1: General Terms Common to Round Timber and Sawn Timber. 6 pp.
CEN PREN 844-3-92. Terminology—Round and Sawn Timber—Part 3: General Terms Relating to Sawn Timber. 19 pp.

— **Grading**
CNS O1003-82. Grading Rules of Lumber (for Taiwan Area) (Oct)(444).
SAA AS 1748-78. Mechanically Stress-Graded Timber Bound with AS 1749. 5 pp.
SAA AS 1749-78. Rules for Mechanical Stress Grading of Timber Bound with AS 1748 Corrig.. 11 pp.
SNZ NZS 3631-88. New Zealand National Timber Grading Rules. 108 pp.

— **Marine Craft**
SAA AS 1738-75. Timber for Marine Craft. 22 pp.

— **Measurement**
ISO 8904-90. Broadleaved Sawn Timber—Sizes—Methods of Measurement First Edition. 4 pp.

— **Moisture Content**
ISO 4470-81. Sawn Timber—Determination of the Average Moisture Content of a Lot First Edition. 4 pp.

— **Packaging**
ISO 4472-83. Coniferous and Broadleaved Sawn Timber—Transportation Packages First Edition. 4 pp.

INTERNATIONAL AND NON-U.S. NATIONAL STANDARDS
SUBJECT INDEX

Lumber (Cont.)

—Preserved—Identification Systems
CSA O322-1976. Procedure for Certification of Pressure-Treated Wood Materials for Use in Preserved Wood Foundations; (Amd 1-4 July 1986) (Erratum July 1986). 17 pp.

—Scaffolds
BSI BS 2482-81. 1981 Amd 2 Timber Scaffold Boards (AMD 6258) August 31, 1990. 14 pp.
SAA AS 1577-74. Solid Timber Scaffold Planks. 15 pp.

—Shear Strength
ISO 8905-88. Sawn Timber—Test Methods—Determination of Ultimate Strength in Shearing Parallel to Grain First Edition. 4 pp.

—Softwoods
ISO 3179-74. Coniferous Sawn Timber—Nominal Dimensions First Edition. 4 pp.
SAA AS 1781-75. Sawn Boards from Australian-Grown Conifers (Softwoods) (Excluding Radiata Pine and Cypress Pine). 16 pp.

—Softwoods—Cypress
SAA AS O91-64. Unseasoned Cypress Pine Sawn Boards Bound with AS O93 and AS O94.

—Softwoods—Defects
ISO 1029-74. Coniferous Sawn Timber—Defects—Classification First Edition. 9 pp.
ISO 1030-75. Coniferous Sawn Timber—Defects—Measurement First Edition. 12 pp.
ISO 1031-74. Coniferous Sawn Timber—Defects—Terms and Definitions First Edition. 32 pp.

—Softwoods—Defects—Classification
ISO 2299-73. Sawn Timber of Broadleaved Species—Defects—Classification First Edition. 9 pp.

—Softwoods—Dimensional Stability
ISO 738-81. Coniferous Sawn Timber—Sizes—Permissible Deviations and Shrinkage First Edition. 5 pp.

—Softwoods—Eucalyptus
SAA AS O83-63. Sawn South-Eastern Australian Eucalypt Hardwoods. 8 pp.

—Softwoods—Glossaries
ISO 1032-74. Coniferous Sawn Timber—Sizes—Terms and Definitions First Edition. 9 pp.
ISO 2300-73. Sawn Timber of Broadleaved Species—Defects—Terms and Definitions First Edition. 34 pp.
ISO 2301-73. Sawn Timber of Broadleaved Species—Defects—Measurement First Edition. 13 pp.
ISO 4472-83. Coniferous and Broadleaved Sawn Timber—Transportation Packages First Edition. 4 pp.

—Softwoods—Measurement
ISO 737-75. Coniferous Sawn Timber—Sizes—Methods of Measurement First Edition. 3 pp.
ISO 1030-75. Coniferous Sawn Timber—Defects—Measurement First Edition. 12 pp.
ISO 8904-90. Broadleaved Sawn Timber—Sizes—Methods of Measurement First Edition. 4 pp.

—Softwoods—Packaging
ISO 4472-83. Coniferous and Broadleaved Sawn Timber—Transportation Packages First Edition. 4 pp.

—Stress Graded—Color Coding
SAA AS 1613-74. Colours for Marking Stress Graded Timber. 3 pp.

—Surfaces
SAA AS 1728-75. Types of Timber Surfaces. 23 pp.

—Tensile Testing
CSA S442.2-M1985. Method of In-Grade Tension Tests of Lumber (S442 Series-M1985); (Gen Instr 1). 14 pp.

—Truss Clips
CSA S347-M1980. Method of Test for Evaluation of Truss Plates Used in Lumber Joints. 19 pp.

—Turnbuckles
JIS F 2016-87. Cast Steel Pawl Type Chain Cable Stoppers for Grade 2 Chain Cable.
JIS F 2101-90. Turnbuckles for Lumber Lashing.

—Veneers
CNS O1043-87. Laminated Veneer Lumber (Jan)(11818). 3 pp.
CNS O2059-87. Method of Test for Laminated Veneer Lumber (Jan)(11819). 2 pp.

—Wood Preservation
SAA AS 1604-80. Preservative Treatment of Sawn Timber, Veneer and Plywood. 22 pp.

Lumber (Cont.)

—Wood Preservatives
CEN PREN 351 (Part 1)-90. Durability of Wood and Wood Based Products—Preservatives-Treated Solid Wood—Part 1 "Requirements for Preservative-Treated Wood According to Hazard Classes". 15 pp.
JIS A 9112-89. Treated Lumbers by Diffusion Process. 7 pp.

—Wood Preservatives—Identification Systems
CEN PREN 351 (Part 3)-90. Durability of Wood and Wood Based Products—Preservative-Treated Wood—Part 3: "Identification of Preservative-Treated Wood". 7 pp.

—Wood Preservatives—Sampling
CEN PREN 351 (Part 2)-90. Durability of Wood and Wood Based Products—Preservative-Treated Solid Wood—Part 2 "Sampling and Analysis of Preservative-Treated Wood". 11 pp.
SAA AS 1605-74. Methods for the Sampling and Analysis of Wood Preservatives and Preservative-Treated Wood. 39 pp.

Lumber Crayons
Use: Crayons—Lumber

Lumbering Industry
See Also: Forestry Equipment; Lumber; Trees (Plants)

—Glossaries
ISO 8965-87. Logging Industry—Technology—Terms and Definitions First Edition. 11 pp.
ISO 8966-87. Logging Industry—Products—Terms and Definitions First Edition. 10 pp.

Luminaire Track Systems
Use: Lighting Tracks

Luminaires
Use: Lighting

Luminance
Use For: Luminance Measurement; Luminance Ratio; Luminous Flux See Also: Illuminance; Light (Visible Radiation); Light Meters; Lighting; Optical Properties; Photometric Properties; Photometry
CNS C3068-79. Method of Luminance Measurements (Dec)(5064).
CNS C3188-84. Method of Test for Luminance Ratio (Jun)(10906).
JIS C 7614-70. Methods of Luminance Measurements. 6 pp.

—Electric Switches
CNS C6238-86. Method of Test for Electromechanical Switches—TP-3 Transmittancy (Luminance) (Dec)(11101).

—Flashtubes
ISO 10503-91. Photography—Expendable Reflectored Photoflash Lamp Arrays—Definitions and Requirements for Luminous Flux/Time Characteristics First Edition. 6 pp.

—Lamps
CNS C4168-84. Standard Lamps of Luminous Intensity (Jun)(5200).

—Screens (Projection)
BSI BS 5550: Sec 6.5-84. 1984 Cinematography Part 6: Television Usage Section 6.5: Colours, Luminances and Dimensions for Viewing Conditions for the Evaluation of Films and Slides for Television. 7 pp.
BSI BS 5550: SUBSEC 7.2.1-78. (WITHDRAWN) 1978 Cinematography Part 7: Production and Presentation Sec. 7.2: Screens and Screen Luminance Subsec. 7.2.1: Screen Luminance in Cinematograph Laboratory and Studio Review Rooms (Superseded by BS 5550: Subsec: 7.2.6: 1991). 2 pp.
BSI BS 5550: SUBSEC 7.2.2-78. (WITHDRAWN) 1978 Cinematography Part 7: Production and Presentation Sec. 7.2: Screens and Screen Luminance Subsec. 7.2.2: Screen Luminance for the Projection of 16 mm Film (Superseded by BS 5550: Subsec. 7.2.6: 1991). 4 pp.
BSI BS 5550: SUBSEC 7.2.3-78. (WITHDRAWN) 1978 Cinematography Part 7: Production and Presentation Sec. 7.2: Screens and Screen Luminance Subsec. 7.2.3: Screen Luminance for the Projection of 35 mm Film on Matt and Directional Screens (Superseded by BS 5550: Subsec 7..2.6: 1991). 13 pp.
BSI BS 5550: SUBSEC 7.2.4-78. (WITHDRAWN) 1978 Cinematography Part 7: Production and Presentation Section 7.2: Screens and Screen Luminance Subsection 7.2.4: Screen Luminance for the Projection of 70 mm Film on Directional Screens (Spsd by BS 5550:Subsec7.2.6: 1991). 8 pp.

Luminance (Cont.)

—Screens (Projection) (Cont.)
BSI BS 5550: SUBSEC. 7.2.5-80. 1980 Amd 1 Cinematography Part 7: Production and Presentation Section 7.2: Screens and Screen Luminance Subsection 7.2.5: Cinematograph Screens. 9 pp.
BSI BS 5550: SUBSEC 7.2.6-91. 1991 Cinematography Part 7: Production and Presenta-tion Section 7.2: Screens and Screen Lum-inance Subsection 7.2.6: Specification for Screen Luminance and Colour for the Projection of Motion -Picture Prints in Indoor Theatres and Review Rooms (H). 7 pp.
BSI BS 6354-83. 1983 Measuring the Screen Luminance, Contrast and Reflectance of Microform Readers. 9 pp.
ISO 2910-74. Cinematography—Screen Luminance for the Projection of Motion-Picture Films in Indoor Theatres First Edition; (Replaces 2895). 3 pp.
ISO 6035-83. Cinematography—Viewing Conditions for the Evaluation of Films and Slides for Television—Colours, Luminances and Dimensions First Edition. 5 pp.

Luminance Measurement
Use: Luminance

Luminance Ratio
Use: Luminance

Luminance Signal
See Also: Television Broadcasting
CNS C6099-80. Measurement of Luminance Signal Levels (Dec)(6799).

Luminous Call Systems
Use: Call Systems (Medical)

Luminous Discharge Tube Signs
Use: Luminous Tube Signs

Luminous Flux
Use: Luminance

Luminous Paints
Use For: Fluorescent Paints; Radioluminescent Deposits; Radioluminous Paints See Also: Paints
CNS K2039-86. Fluorescent Paint (Jan)(2364). 3 pp.
CNS K2168-87. Luminous Paints (Nov)(12160). 3 pp.
CNS K6187-86. Method of Test for Fluorescent Paint (Jan)(2365).
CNS K6930-87. Method of Test for Luminous Paints (Nov)(12161). 3 pp.
JIS K 5673-83. Fluorescent Paint for Safety Colour. 14 pp.
MOD UK CS 3140. Paint System, Luminous, Promethium Based (Withdrawn).
MOD UK TS 10282. Paint System, Fluorescent (Superseded by Def Stan 80-164).

—Aerospace
MOD UK DTD-5587-01. Paint System, Luminous, Tritium Activated 1. Paint, Undercoat for Luminous Paint: (A) Air Drying; (B) Stoving 2. Paint Medium for Luminous Paint: (A) Air Drying; (B) Stoving 3. Compound, Luminous, Tritium Acitvated; Amendment 1. 8 pp.

—Time Measuring Instruments
ISO 3157-91. Radioluminescence for Time Measurement Instruments—Specifications Second Edition. 9 pp.
ISO 4168-79. Time Measurement Instruments—Conditions for Carrying out Checks on Radioluminescent Deposits First Edition. 4 pp.

Luminous Tube Sign Transformers
Use: Luminous Tube Transformers

Luminous Tube Signs
Use For: High Voltage Luminous Tube Signs
See Also: Discharge Lamps; Display Devices; Gas Discharge Tubes; Lighting; Neon Signs; Signs
BSI BS 559-91. 1991 Electric Signs and High-Voltage Luminous-Discharge-Tube Installations. 22 pp.
BSI BS 559-01. 1991 Amd 1 Electric Signs and High-Voltage Luminous-Discharge-Tube Installations (AMD 7392) December 15, 1992. 23 pp.

—Cables (Electric)
CSA C22.2 NO 17-1973. Cable for Luminous-Tube Signs and for Electric Oil-and Gas-Burner Ignition Equipment (R 1992); (Gen Instr 1) (Amd 1 February 1981) (Amd 2-8 March 1988). 28 pp.

—Electrical Components
CENELEC PREN 50107-92. Signs and Luminous-Discharge-Tube Installations Operating from a No-Load Output Voltage Exceeding 1,000 V. 35 pp.

INDUSTRY STANDARDS

Luminous Tube Signs *(Cont.)*

—Electrode Fittings—Thermal Shock
CSA C22.2 NO 34-M1987. Electrode Receptacles, Fittings, and Connectors for Gas Tubes (R 1993); (Gen Instr 1 Thru 2). 23 pp.

—Portable—Indoor/Outdoor
CSA CAN/CSA-C22. 2 NO 207-M89. Portable and Stationary Electric Signs and Displays; (Gen Instr 1). 50 pp.

—Stationary—Indoor/Outdoor
CSA CAN/CSA-C22. 2 NO 207-M89. Portable and Stationary Electric Signs and Displays; (Gen Instr 1). 50 pp.

—Wiring Methods
CENELEC PREN 50107-92. Signs and Luminous-Discharge-Tube Installations Operating from a No-Load Output Voltage Exceeding 1,000 V. 35 pp.

Luminous Tube Transformers

See Also: Specialty Transformers; Transformers
JIS C 8109-91. Luminous-Tube Transformers. 13 pp.

—
CSA C22.2 NO 13-1962. Transformers for Luminous-Tube Signs, Oil-or Gas-Burner Ignition Equipment, Cold-Cathode Interior Lighting (R 1992). 46 pp.
CSA 1169 Bull. Electrical Bulletin 1169 June 27, 1978 to C22.2 NO 13. 2 pp.

—5-15 kV
CSA C22.2 NO 13-1962. Transformers for Luminous-Tube Signs, Oil-or Gas-Burner Ignition Equipment, Cold-Cathode Interior Lighting (R 1992). 46 pp.
CSA 1169 Bull. Electrical Bulletin 1169 June 27, 1978 to C22.2 NO 13. 2 pp.

Lunch Boxes

See Also: Boxes (Containers); Containers

Lunch Boxes *(Cont.)*

—Metal
CNS S1212-88. Metal Lunch Box (May)(12324).

Lung Ventilators

See Also: Medical Electrical Equipment; Medical Equipment; Respirators
BSI BS 5724: Sec 3.12-91. 1991 Medical Electrical Equipment Part 3: Particular Requirements for Performance Section 3.12: Method of Declaring Parameters for Lung Ventilators. 88 pp.
JIS T 7204-89. Lung Ventilators for Medical Use. 31 pp.

—Critical Care
CSA Z168.5.2-M1991P. Critical Care Ventilators; (Gen Instr 1). 57 pp.

—Safety
BSI BS 5724: Sec 2.12-90. 1990 Medical Electrical Equipment Part 2: Particular Requirements for Safety Section 2.12: Lung Ventilators. 20 pp.
BSI BS 5724: Sec 2.12-01. 1990 Amd 1 Medical Electrical Equipment Part 2: Particular Requirements for Safety Section 2.12: Specification for Lung Ventilators (IEC 601-2-12: 1988) (AMD 7446) January 15, 1993. 20 pp.
CEN PREN 794-1-92. Medical Electrical Equipment—Particular Requirements for Lung Ventilators—Part 1: Lung Ventilators for Medical Use. 55 pp.
CENELEC HD 395.2.12 S1-89. Medical Electrical Equipment Part 2: Particular Requirements for the Safety of Lung Ventilators for Medical Use. 3 pp.
ISO 10651 Pt 1-93. Lung Ventilators for Medical Use—Part 1: Requirements First Edition; (Cancels and Replaces ISO 5369, Published in 1988, and IEC 601-2-12:1988). 31 pp.

Lutes (Material)

See Also: Joint Sealants

Lutes (Material) *(Cont.)*

MOD UK DSTAN 68-173-93. Luting (Thick) MK 4 Luting (Thin) MK 5 Issue 1. 11 pp.

—Ammunition
MOD UK DSTAN 68-173-93. Luting (Thick) MK 4 Luting (Thin) MK 5 Issue 1. 11 pp.
MOD UK CS 2677G. Luting Mark 8, LF Quality (Composition RD 1284) (Withdrawn).

—Carnauba Wax
MOD UK DSTAN 68-126-90. Wax, Carnauba Issue 1. 10 pp.

—Polyisobutylene
MOD UK DSTAN 68-23-81. Polyisobutene (Low Molecular Weight) Polyisobutene (Low Molecular Weight, LF Quality) and Polyisobutene (Medium Molecular Weight, LF Quality, Types 1, 2 and 3) Issue 2. 10 pp.

Luting

Use: Lutes (Material)

Lychee Nuts

Use: Litchis

Lyctus

Use: Coleoptera

Lynch Pins

Use: Linchpins

Lysine Content Analysis

—Animal Feed
ISO 5510-84. Animal Feeding Stuffs—Determination of Available Lysine First Edition. 7 pp.

L-Lysine Monohydrochloride

JIS K 9053-72. L-Lysine Monohydrochloride.

NOTES

NOTES

NOTES

NOTES

NOTES

NOTES